Atlas of Woody Plants in China

Distribution and Climate

Volume I

Edited by

Jingyun Fang

Zhiheng Wang

Zhiyao Tang

高等教育出版社·北京

HIGHER EDUCATION PRESS BEIJING

内容提要

准确翔实的物种分布资料是宏观生态学、生物地理学以及保护生物学等学科的研究基础，也是农业、林业生产实践的重要基础。北美、欧洲和日本在20世纪七八十年代就出版了详细的动、植物物种分布图，为这些区域的生物地理学、宏观生态学研究以及生物多样性保护实践提供了不可缺少的基础资料。中国是全球12个“巨大生物多样性国家”之一，拥有超过30 000种维管植物，其中包括约11 000种木本植物、3200种树木，物种数量远高于同处中高纬度的北美洲和欧洲。但中国目前还没有较为完善的植物分布图集。

本图集在忠实于标本记录的基础上，收集了我国目前已出版的全国、各省区和地方植物志、树木志，以及区域科学考察报告和学术论文中有关的木本植物分布资料，并在此基础上邀请全国各地20多位专家对现有资料进行修订和补充，最终编制了我国已知的全部11 405种木本植物的详尽分布图。图集中的物种名录采用了当前最新的分类系统，与国际标准一致。同时，图集还提供了物种的生活型以及物种分布区的13个气候和初级生产力指标，为生态学、生物地理学和保护生物学的基础研究和实践提供了重要的基础资料。

本图集是目前中国第一部具有完整木本植物名录、包含翔实分布信息和气候特征的专著，可作为生态学、植物学、地理学等相关学科的广大师生以及农林科技工作者的参考书。

图书在版编目（CIP）数据

中国木本植物分布图集 = Atlas of Woody Plants in China: Distribution and Climate: 中、英文 / 方精云，王志恒，唐志尧编著 . — 北京：高等教育出版社，2009.9

ISBN 978-7-04-026859-1

I. 中⋯ II. ①方⋯ ②王⋯ ③唐⋯ III. 木本植物—中国—图集 Ⅳ . S717. 2–64

中国版本图书馆 CIP 数据核字（2009）第 170374 号

策划编辑 林金安 王 莉 **责任编辑** 王 莉 刘思涵 **封面设计** 张志奇
版式设计 张志奇 **责任排版** 高 瑾 张志奇 **责任印制** 陈伟光

出版发行	高等教育出版社	购书热线	010-58581118
社　　址	北京市西城区德外大街4号	咨询电话	400-810-0598
邮政编码	100120	网　　址	http://www.hep.edu.cn http://www.hep.com.cn
经　　销	蓝色畅想图书发行有限公司	网上订购	http://www.landraco.com http://www.landraco.com.cn
印　　刷	涿州市星河印刷有限公司		
开　　本	839×1194 1/16	版　　次	2009年10月第1版
印　　张	41.75	印　　次	2009年10月第1次印刷
字　　数	1 300 000	定　　价	580.00元（全4册）

本书如有缺页、倒页、脱页等质量问题，请到所购图书销售部门联系调换。

物 料 号 26859-00

Atlas of Woody Plants in China

Distribution and Climate

Edited by

Jingyun Fang

Zhiheng Wang

Zhiyao Tang

Other Contributors

Xiulian Chi, Zhaodi Guo, Yue Jiang, Lai Jiang, Jing Li, Liping Li, Xin Lin, Chang Liu, Jinglei Liu, Yining Liu, Xiujuan Qiao, Yanlin Ren, Zehao Shen, Kun Tan, Luying Tang, Jie Wang, Shaopeng Wang, Yetao Yan, Xinping Zhang, Yanming Zhang, Rui Zhou, Jiangling Zhu

Review Experts (Focal Areas)

Sizhao Fang (*Guizhou*), Ying Hai (*Xinjiang*), Shushen Lai (*Jiangxi*), Fazeng Li (*Shandong*), Guangzhao Li (*Guangxi*), Zhenyu Li (*Species checklist*), Cunzhu Liang (*Inner Mongolia and Ningxia*), Chudian Liu (*Fujian*), Anwar Mohammat (*Xinjiang*), Yi Ren (*Shaanxi*), Kun Sun (*Gansu*), Shizhou Sun (*North China and Tibet*), Zemin Wu (*Anhui*), Chengdong Xu (*Yunnan*), Qinzhou Yang (*Sichuan and Chongqing*), Huagu Ye (*Guangdong*), Yongzhong Ye (*Henan*), Xunlin Yu (*Hunan*), Zien Zhao (*Hubei*), Xingmin Zhou (*Qinghai*), Zhanjiang Zong (*Northeast China*)

中国木本植物分布图集

方精云　王志恒　唐志尧 编著

参与数据收集和数据库建立的人员（按姓氏拼音排序）

池秀莲、郭兆迪、江玥、姜来、李晶、李利平、林鑫、刘畅、刘晶磊、刘怿宁、乔秀娟、任艳林、沈泽昊、谭琨、汤璐瑛、汪洁、王少鹏、闫业涛、张新平、张延明、周睿、朱江玲

审校专家及负责区域（按姓氏拼音排序）

方嗣昭(贵州)、海鹰(新疆)、赖书绅(江西)、李法曾(山东)、李光照(广西)、李振宇(植物名录)、梁存柱(内蒙古、宁夏)、刘初钿(福建)、安尼瓦尔·买买提(新疆)、任毅(陕西)、孙坤(甘肃)、孙世洲(华北地区和西藏)、吴泽民(安徽)、徐成东(云南)、杨钦周(四川、重庆)、叶华谷(广东)、叶永忠(河南)、喻勋林(湖南)、赵子恩(湖北)、周兴民(青海)、宗占江(东北地区)

Biosketches of Editors | 作者简介

Jingyun Fang | 方精云

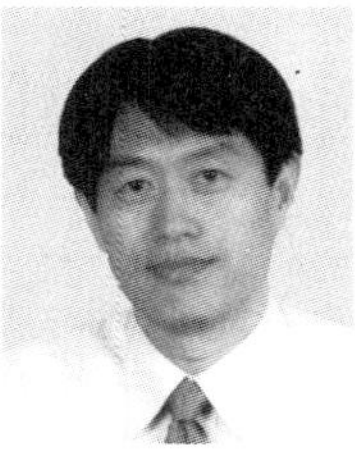

Dr. Jingyun Fang is a Cheung Kong Professor at Department of Ecology, Peking University. He obtained his PhD from Osaka City University (Japan) in 1989. His research interests cover vegetation ecology, biogeography of plants, carbon cycle and the application of remote sensing. He is an author of more than 230 publications including five books. He has received several awards including He-Liang-He-Li Science and Technology Progress Award, Cheung Kong Scholars Achievement Award, and National Natural Science Award of the State Council. He is a member of the Chinese Academy of Sciences and the Academy of Science for the Developing World (TWAS), and serves as the editorial board member for several journals including *Frontiers in Ecology and the Environment, Ecosystems, Global Environmental Change, Science China, Ecological Research, BMC Ecology* and others.

北京大学长江特聘教授、中科院院士，第三世界科学院院士。1989 年获日本大阪市立大学生物学博士学位。主要从事植被生态学、生物多样性、生态系统碳循环等方面的科研和教学工作。共发表论文 230 余篇，出版专著 5 部。曾获首届国家杰出青年科学基金 (1994)、教育部自然科学一等奖 (2003)，国家自然科学二等奖 (2004)，长江学者成就奖 (2006)，何梁何利科学技术进步奖 (2007) 等奖项以及全国野外科技工作先进个人称号 (2009)。是国内外多个专业杂志的编委或副主编，包括 *Frontiers in Ecology and the Environment, Ecosystems, Global Environmental Change, Science China, Ecological Research, BMC Ecology*。

Zhiheng Wang | 王志恒

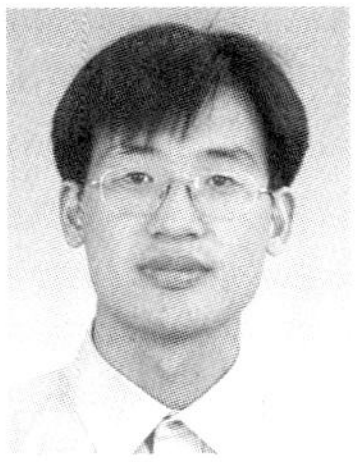

Dr. Zhiheng Wang is a postdoc researcher in Institute of Environmental Sciences, University of Zurich, Switzerland. He got his PhD at Peking University in 2008, and has research interests in all issues of macroecology. He has been working on the Database of China's Woody Plants for the last six years. His recent focus is on macro-scale patterns of species richness, species range size, species body size, patterns in community structure, and the mechanistic causes of such patterns. His current research involves testing the influence of contemporary climate, evolutionary history, human activities, and spatial scale on species diversity patterns using the synthesis of climatic, phylogenetic, and anthropogenic data. He has published 20 papers, including a recent paper in the Proceedings of the National Academy of Sciences, USA.

目前在瑞士苏黎世大学环境科学研究所从事博士后研究。2008 年毕业于北京大学，获理学博士学位。过去 6 年以来一直致力于"中国木本植物分布数据库"的建设。主要从事宏观生态学和全球变化生态学研究，内容包括物种多样性、物种分布区、个体大小和群落结构的大尺度格局及其进化和气候的形成机制，以及气候变化对物种多样性的影响。发表论文 20 余篇，包括一篇美国科学院院刊论文。

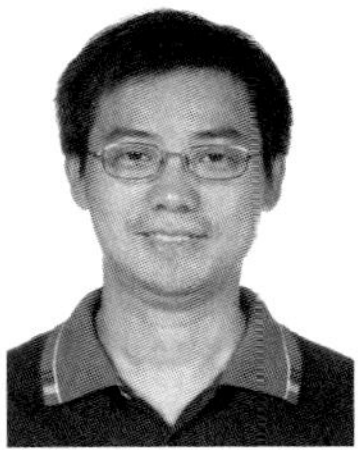

Zhiyao Tang | 唐志尧

Dr. Zhiyao Tang is an associate professor at the Department of Ecology, Peking University. He got his PhD at Peking University in 2003. His research interests include plant community ecology, biodiversity patterns, climate-vegetation interaction, and application of remote sensing in ecology. He has authored more than 40 peer-reviewed articles and two books.

北京大学生态学系副教授。2003 年毕业于北京大学，获理学博士学位。主要从事植被植物群落生态学、生物多样性、植被气候关系以及生态遥感等方面的科研与教学工作。发表论文 40 余篇，专著 2 部。

Preface | 前言

China is one of the earth's 12 mega-biodiversity countries, and the most diverse country in the northern hemisphere. With more than 30,000 species of vascular plants, more than 1/3 of which are woody plants, the number of plant species in China is higher than in Europe and North America combined. However, a comprehensive atlas of the distribution of woody plants in China has not been yet available though atlases of the species distribution in many western countries were compiled several decades ago. For example, an atlas of the United States trees was published in the 1970s. The atlas of vascular plants in Japan with a vertical resolution of 50 m was compiled in the 1970s. A project for mapping distribution of vascular plants in Europe was launched in 1965, and 13 volumes of the atlas have been published to date. These atlases of plants are of both scientific and practical importance, and provide an indispensable database for understanding the patterns and mechanisms of species distribution, utilizing the biological resources, protecting biological diversity, implementing regional development planning, and predicting ecological responses of plants to global environmental changes.

Why has China lacked of such atlas? This might be primarily due to there being too many species and limited information on the distributions of those species. These make some people think that it is not right time to compile the national atlas for China's plants. However, I don't agree with this opinion. Nothing is perfect, especially for the information on species distribution. For example, in 1817, Alexander von Humboldt drew the first isothermal map based on temperature data from only 58 climatic stations globally, which became the basis for climatic division and physiogeographical study for decades. Japan has completed the atlas for all its vascular plants with a vertical resolution of 50 m in the 1970s with limited information on distribution of plants. Compared with these, China has accumulated copious data on plant distribution through the efforts of the several generations of botanists and ecologists over the past half century. These hard-earned data provide an essential base to systematically compile the atlas of China's plants. Therefore, in my opinion, now it is the time to compile the atlas of China's plants, which is an urgent need for the conservation and sustainable use of biological resources.

The idea of compiling the atlas of China's plants initially came into my mind in the mid-1980s when I was a PhD student in Japan. My dissertation was to examine the relationships between climate and vegetation in China and I strongly felt that China was lacking of the species distribution data. I was very impressed when I saw the detailed spatial explicit distribution maps of vascular plants in Japan, and made a copy of the maps of major species and brought them back to China. They are still kept on my bookshelf now. In 1996, I was touched again when I saw the atlas of the United States trees and European species in McGill University, Canada, where I was a visiting scholar. These two shocks made me long hope that we should have an atlas for China's plants in the near future. In 2003 as the first step of compiling an atlas for all the plants of China, we initiated a project to compile the *Atlas for All the China's Woody Plants*, although without specific funds, because distribution information is relatively rich and the checklist is well-documented for the woody plants in China. As the basic data of the atlas, we first started to construct "*Database of China's Woody Plants*". We documented the distribution sites for all the species based on specimen records and all the validated records documented in publications. After six years' efforts, this atlas which documents more than 11,000 woody species was eventually published.

In this atlas, we provide distribution maps for more than 11,000 species at the level of counties. We also documented information on climate, primary production and life forms for each species. This atlas contains information on both species and environmental variables, compared with the previous atlases published in other countries or regions.

As above mentioned, we launched this project without specific financial support. However, we have been lucky to have creative human resources and excellent students, who have been dedicated to the collection and compilation of these basic data, which is difficult to yield articles but is critically significant in scientific research and practical use. Here, I would like especially to mention two major contributors of this atlas, Dr. Zhiheng Wang and Dr. Zhiyao Tang, who took major responsibility to complete the atlas. Zhiheng was my PhD student whose PhD project was to examine patterns of China's woody plants and the mechanisms that shape those patterns. He joined my group in his senior year, and I asked him to postpone his defense for his PhD twice to complete the construction of the database of China's woody plants. For this reason, he stayed in my group as a PhD student for eight years. During this period he led a group of students to overcome many problems that data compilations posed and spent six years and finally completed the database, which makes publishing this atlas possible. Zhiyao donated uncounted hours of his time to collect data, check distribution records, collate species name, design maps, and so on. In addition, many staff and students contributed their time in data collection, data input and species name collation. To all of them, I am grateful.

The atlas was based on many sources of information, including a Chinese version (*Flora Reipublicae Popularis Sinicae*) and English version of "*Flora of China*", and "*Seed Plants of China*" edited by Zhengyi Wu, "*Higher Plants of China*" edited by Liguo Fu, and nu-

merous local flora and species surveying reports. The compilation of this atlas would be impossible without these basic, most essential works. Here, I am greatly indebted to all the authors for their efforts. I would also like to thank the botanists and field biologists who reviewed and corrected the distribution records of species. I want to acknowledge many students in the Department of Ecology, Peking Universiy, who assisted this project by documenting distribution data. Mr. Jin'an Lin and Ms. Li Wang of Higher Education Press (China) spent lots of time editing, collating and publishing this atlas. I will like to express my many thanks for all these people. The compilation of this atlas was partly supported by the National Natural Science Foundation of China (Nos: 40638039, 90711002, 30721140306, 40871030, 40501025, 49971002, 39970044, 39425003).

In spite of our best efforts, errors still remain in this atlas due to our limited knowledge and limited information on species distribution. Please contact us (Email: zytang@urban.pku.edu.cn)and send us your comments so that we can update the atlas in the near future.

Jingyun Fang
Professor of Plant Ecology
September, 2009

中国是世界上物种多样性特别丰富的国家之一，拥有维管植物约30 000余种，其中1/3以上是木本植物，比北美和欧洲的总和还要多。但中国至今尚没有全面记述这些物种地理分布的图集。这应该说是件十分缺憾的事。早在几十年前，西方国家就已经完成了物种分布图的编制工作。比如，美国于20世纪70年代完成了全美木本植物分布图的编制；日本在20世纪70年代出版了所有维管植物的分布图，物种垂直分布范围的精度达50 m；欧洲也于1965年开始了维管植物分布图的编制工作。这些分布图为揭示物种的分布规律、开发和利用生物资源、开展生物多样性保护、进行区域发展规划，以及预测气候变化的生态影响等提供了不可或缺的基本资料，其科学价值和实践意义是不言而喻的。

为什么中国缺乏这类分布图？物种多、物种分布资料不足又散布于各类文献中，造成了编制工作难度大，这可能是最主要的原因。此外，国内对编制物种分布图的保守认识也是一个重要的限制因素。

据我所知，国内一些人认为，我国物种分布资料不足，目前尚不具备编制植物分布图集的条件。但我不同意这种说法。科学技术的进步是无止境的，大自然的很多现象和规律是逐渐被认知的，否则人类文明就不会进步到今天。1817年洪堡依据全球仅有的58个气象站的气温资料，编绘了世界首张全球等温线分布图，是气候分区和自然地理学研究的一个重要里程碑，成为相当长时期内开展相关研究的基石。又如，日本在30多年前就能出版垂直分布范围高达50 m精度的维管植物分布图，不能说他们当时已经掌握了所有物种准确的分布信息。因此，如果等物种分布资料齐全了才编制其分布图，那我们所待何时？事实上，经过广大植物学家、生态学家半个多世纪的努力，我国的物种分布资料已经有了相当深厚的积累，系统编制物种分布图的时机已经成熟，这也是时代发展的迫切要求。当然，随着知识的积累和认识水平的提高，这类分布图的信息需要进一步补充和完善。

编制中国植物分布图是我在学生时代就萌生的愿望。早在20世纪80年代的中后期，我在日本读书，博士期间的研究课题就是关于中国植物分布与气候关系的研究，实感中国物种分布数据之缺乏。当我看到日本详尽的物种分布图时，那种欣喜若狂至今还记忆深刻，当时我就不加思索地把主要物种的分布图全都复印了下来，带回国内，至今还保存着。1996年，我去加拿大McGill大学做访问学者，看到美国和欧洲也有详尽的物种分布图，又一次受到触动。于是，就产生了将来我们也要编制中国植物分布图的想法。2003年，在没有专项经费支持的情况下，我们决定先编制分布资料较多、种名相对稳定的木本植物分布图，于是启动了“中国木本植物分布图的编制”自设课题。次年，作为编制物种分布图的基础工作，课题组开始了“中国木本植物分布数据库”的建设。历经6年的艰苦工作，这本多达2 000余页的图集终于与读者见面了。

本图集的特点是为11 405个物种提供县级分布信息；每个物种的分布信息在忠实于标本记录的基础上，最大可能地吸收了所有物种分布研究的成果；提供物种分布区的气候资料和分布区初级生产力的平均状况。此外，我们也提供了绝大多数物种的主要生活型信息。本图集的这些特点，既克服了西方国家仅提供物种分布信息的局限性，也满足了人们对分布区气候和物种生活型等基础信息的需求，使其更具实用性。

前面讲过，我们是在没有专项经费的条件下开始此项工作的。我们虽然没有专项经费，但有丰厚的人力资源，有非常优秀的同学，他们愿意献身于这类发表成果难、但有重要科学和应用价值的基础性数据的收集和整理工作。这里，我要特别提到本图集的两位主要作者：王志恒和唐志尧，他们是本图集的主要贡献者。王志恒当时是我的直博生，我给他的博士论文题目就是中国木本植物多样性的分布格局及其机制的研究。他自大三就进入我的研究组，直到博士毕业，一共呆了8年。为了完成数据库的建设，我两度让他推延博士毕业。他带领一批同学，克服物种多、

物种名称杂乱、物种分布信息零散不全等困难，历时6年，终于建成了“中国木本植物分布数据库”，使本图集的编制成为可能。另一名主要作者唐志尧，在物种分布数据收集、分布区和种名校核、分布图制作与设计、编辑和出版等方面，做了主要的工作。可以说，他们二位是本图集编制出版的主要贡献者。此外，很多老师和同学在数据收集和录入、种名核校等方面也做了大量繁杂的工作。

本图集凝聚了几代人的心血和汗水，它是在中英文版《中国植物志》、吴征镒先生主编的《中国植物分布数据库》所记载的物种名录、傅立国主编的《中国高等植物》以及众多地方植物志和调查报告中所记载的物种分布信息的基础上编制出版的。因此，我要特别感谢这些作者们，没有他们的工作，就没有本图集的出版。北京大学生态学系的多位同学参与了数据的收集、整理和校对，以及分布图的制作等。数据库建成后邀请各地的有关专家对种名和分布区进行了校验和补充。高等教育出版社的林金安先生、王莉女士为本图集的制作、编辑和出版做了大量艰苦而有成效的工作。在此，对他们表示衷心的感谢。本图集在数据收集、整理和出版过程中，得到国家自然科学基金委的部分资助（编号：40638039, 90711002, 30721140306, 40871030, 40501025, 49971002, 39970044, 39425003）。

由于我们水平有限，以及现有资料的不足，本图集肯定存在一些纰漏，请读者们批评指正（请联系zytang@urban.pku.edu.cn），以便在修订时改进。

方精云
2009年9月
于北京大学

How to Use This Atlas | 本图集使用说明

1 Arrangement of families, genera and species

The order of families, genera, and species were arranged as follows: the families are organized according to the phylogenetic system of *Flora of China*; within each family, the genera and species are arranged alphabetically. The atlas is published in three volumes, plus a separate volume of species checklist.

2 Arrangement of the climatic variables

The atlas lists mean and range of 12 climatic variables and vegetation net primary productivity. They are: mean annual temperature (MAT, °C), mean temperature of the coldest month (MTCM, °C), mean temperature of the warmest month (MTWM, °C), annual biotemperature (ABT, °C), warmth index (WI, °C), coldness index (CI, °C), potential evapotranspiration (PET, mm), annual precipitation (AP, mm), precipitation of warmest quarter (PWQ, mm), precipitation of coldest quarter (PCQ, mm), moisture index (Im), annual actual evaportranspiration (AET, mm), vegetation net primary productivity (NPP, $g \cdot a^{-1} \cdot m^{-2}$). For those species distributed in less than five counties, only the mean values of the climatic variables are listed. To save space, names of these variables and their units are not showed in the illustration of each species; these variables are listed as the following order, and three categories of variables (thermal index, humid/arid index, and integrative index) are separated by mark " ▌ ".

MAT (°C); MTCM (°C); MTWM (°C); ABT (°C); WI (°C); CI (°C); PET (mm) ▌ AP(mm); PWQ (mm); PCQ (mm) ▌ Im; AET (mm); NPP ($g \cdot a^{-1} \cdot m^{-2}$)

The following illustration explains the content of each map, using *Amentotaxus argotaenia* (Hance) Pilg. as an example. Following the illustration, MAT is 18.0 °C (Range is 9.3~22.9 °C); MTCM 8.0 °C (−1.7~15.2 °C); MTWM 26.8 °C (19.6 ~29.8 °C); ABT 18.0 °C (9.5~22.9 °C); WI 157.1 °C (69.8 ~214.8 °C); CI 0.7 °C (0.0 ~18.0 °C); PET 952 mm (603 ~1232 mm) ▌ AP 1530 mm (711 ~2435 mm); PWQ 605 mm (357 ~1428 mm); PCQ 150 mm (15 ~250mm) ▌ Im 61.9 (21.1 ~145.8); AET 947 mm (603 ~1232 mm); NPP 228.0 $g \cdot a^{-1} \cdot m^{-2}$ (134.6~447.4 $g \cdot a^{-1} \cdot m^{-2}$).

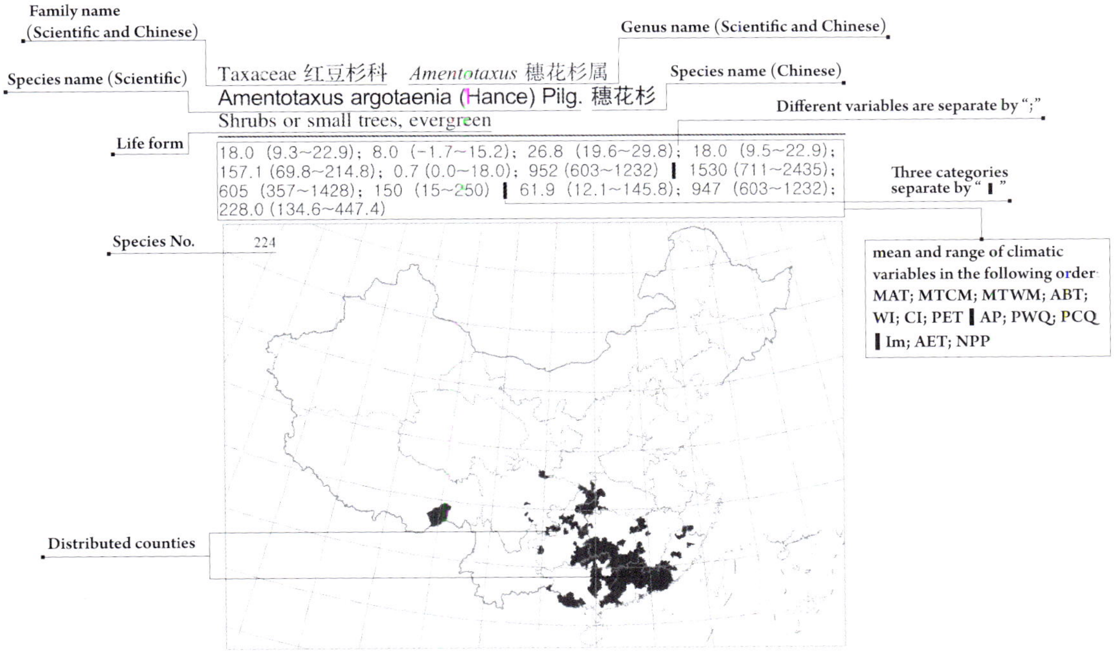

1 科、属、种排列说明

本图集科、属、种排列顺序如下：科序按*Flora of China*系统排列；科内（属、种）按字母顺序排列。本图集分三册出版，并附种名索引一册。

2 气候指标表示方法及使用说明

本图集给出每个种分布区的12个气候变量和植被净生产力的均值和范围。这些指标的意义和计算见第xxvi~xxvii页说明。对于分布区小于5个县域的物种，我们仅列出各气候指标的平均值。为节省空间，这些指标按以下顺序给出，并在各类气候指标（温度指标、水分指标和水热综合指标）之间隔以“▌”符号。另外，各指标都省去单位。顺序（含单位）如下：

年均温（℃）；最冷月均温（℃）；最热月均温（℃）；年生物温度（℃）；温暖指数（℃）；寒冷指数（℃）；潜在蒸散量（mm）▌年降水（mm）；最暖季降水（mm）；最冷季降水（mm）▌湿润指数；年实际蒸散量（mm）；植被净第一性生产力（$g \cdot a^{-1} \cdot m^{-2}$）

以穗花杉［*Amentotaxus argotaenia*（Hance）Pilg, 红豆杉科 Taxaceae, 穗花杉属 *Amentotaxus*］为例，用下面的示意图说明分布图及气候指标的含义。穗花杉在其分布区的气候指标（平均值和范围）如下：年均温为18.0℃（范围为9.3~22.9℃），最冷月均温8.0℃（−1.7~15.2℃），最热月均温26.8℃（19.6~29.8℃），年生物温度18.0℃（9.5~22.9℃），温暖指数157.1℃（69.8~214.8℃），寒冷指数0.7℃（0.0~18.0℃）、潜在蒸散量952 mm（603~1232 mm）▌年降水1530 mm（711~2435 mm），最暖季降水605 mm（357~1428 mm），最冷季降水150 mm（15~250 mm）▌湿润指数61.9（21.1~145.8），年实际蒸散量947 mm（603~1232 mm），植被净第一性生产力228.0 $g \cdot a^{-1} \cdot m^{-2}$（134.6~447.4 $g \cdot a^{-1} \cdot m^{-2}$）。

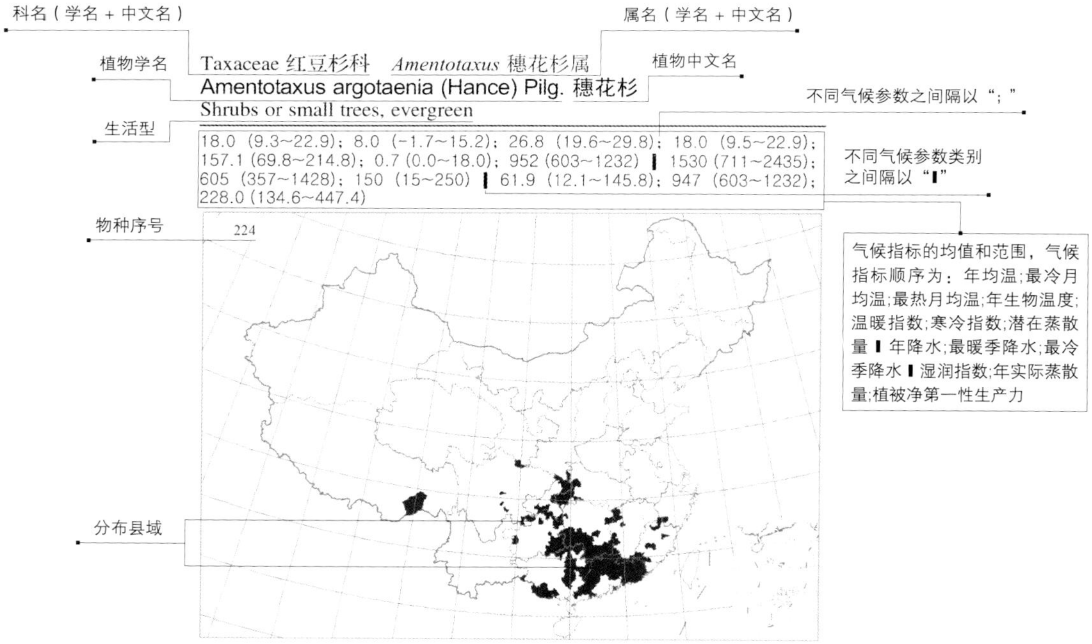

Contents | 目录

1 Introduction to the Compilation of This Atlas

中国木本植物分布图编制说明

1.1 Construction of "Database for Distribution of Woody Plants in China"

Accurate data for distribution ranges of species are fundamental in the study of macroecology and biogeography. In the past several decades, detailed range maps of vascular plants and terrestrial vertebrate animals (birds, mammals, amphibians and reptiles) have been published for North America, Europe and Japan, and these data have been widely used for the studies on patterns of species diversity and spatial turnover, and other large-scale ecological patterns (Brown, 1995; Brown & Lomolino, 1998).

China is one of the world's "mega-biodiversity countries" (McNeely *et al.*, 1990), harboring more than 30, 000 vascular plants, of which c.a. 11,000 species are woody plants, including c.a. 3200 trees. These numbers follow Brazil and Columbia (Chen *et al.*, 1998), and are substantially higher than North America and Europe (Li, 1952; Latham & Ricklefs, 1993). Nonetheless, no comprehensive and accurate databases are available for the plant distributions in China yet.

In the past century, a great deal of field surveys, such as key comprehensive projects of the Tibetan Plateau, Hengduan Mountains, and the Three-Gorge Region, have been conducted in China. These field surveys have accumulated enormous amounts of data on plant distributions. Numerous national floras, including "*Flora Reipublicae Popularis Sinicae*" (volume: 1-80), *Flora of China* (volume: 1-25), "*Sylva Sinica* (volume: 1-4)", "*Higher Plants of China* (volume: 1-14)", and "*Vegetation Atlas of China: 1:1,000,000*", have been published, together with dozens of provincial and local floras and checklists. These have provided essential data for the construction of databases about distributions of plants in China.

As woody plants have been relatively intensively investigated, we initiated a project to compile maps for "*China's woody plants*" in 2003, and dozens of local experts had been invited to check and correct the information in the database. The database was completed in the late 2008. It contains the distribution of all 13,570 woody plants (including variations and subspecies), which are defined as vascular plants with xylem in roots and stems, including trees (height > 4m), shrubs (height < 4m) and woody vines (Wu, 1980). However, the number of species was changed to 11,405 following newly-published "*Flora of China*" (English version) and "*Catalogue of Life, China: 2008 Annual Checklist*". Therefore, this database documents all these 11,405 species. The database is the most comprehensive and up-to-date document for the distribution of woody plants in China. It provides Chinese and scientific names, a distribution map for each species, as well as the climatic conditions for the distribution area of each species.

1.2 Processes to construct the database

The database was constructed by the following four steps:

First, a checklist for China's woody plants was documented based on species names and the phylogenetic system of "*Flora Reipublicae Popularis Sinicae*" (Volume 1-80) (Flora Reipublicae Popularis Sinicae Agendae Academiae Sinicae Edita, 1959-2006) and "*Seed Plants of China*" (Wu and Ding, 1999). All cultivated species were excluded from this checklist. The "*Flora Reipublicae Popularis Sinicae*" is the product made by more than 450 well-qualified Chinese taxonomists over the past 50 years, and is used national wide and worldwide. The "*Seed Plants of China*" is completed by a number of Chinese leading taxonomical experts over 10 years, led by Prof. Zhengyi Wu (a world-class taxonomist and the co-chief-in-editor of the "*Flora of China*"). Based on these two major sources, we documented a checklist of China's woody species. In order to improve the quality of the checklist, several well-known taxonomists were invited to further check the checklist. During the establishment of the database, the *Flora of China* (Volume 1-25) (Editorial Committee of *Flora of China*, 1994-2008), which is the revised and English version of "*Flora Reipublicae Popularis Sinicae*" was published, and the "*Catalogue of Life, China: 2008 Annual Checklist*" (Species 2000 China's Node, 2008, available at http://www.sp2000.cn/) was released. We followed these two to update the checklist. In total, 11,405 woody plants (including 1,355 variations and subspecies) from 1,175 genera and 170 families are documented in our database.

Second, we documented information on county-level distribution for each species using all the available literature, such as national, provincial, and local floras and checklists, monographs of field campaign for plants, and articles, as well as specimen distribution records (http://www.cvh.org.cn). We also provided life forms for most species according to published literatures.

Third, we invited 21 local experts in the area of botany and ecology to improve and correct information on species distribution and Chinese and scientific names of species, genera and families for each species from all the regions.

Fourth, using Visual Basic language and MapObject (ESRI Int.), we developed a software for the management of the database. As illustrated in Figure 1, the interface of the software provides basic functions for the database management, for example, data edition/modification, indication of distribution ranges of species, data input/output, searching and printing.

The database provides the information on the county-level distribution for the woody plants (Figure 2). China has 2,408 (based on the data from 1998) counties; the county area is greatly right skewed (skewness = 9.93, median area = 2,081 km^2), suggesting that a few counties are large but most counties are very small. Large counties are all located in Tibetan Plateau and Taklimakan Desert, where species richness is very low due to extremely arid and/or cold environment. In order to improve the precision of species distribution, we divided these large counties into several parts to document the distribution range for each species .

1.3 Special notes on the atlas

(1) The information on species distribution in this atlas was documented primarily based on specimen records, as well as valid provincial and local floras, investigation reports and research articles, and knowledge of local experts. Thus, the distribution ranges in this atlas are somewhat conservative, as we used only the evident records of the distribution.

(2) According to the species distribution-range theory and niche theory, a species can occur in all the counties with the similar habitats within its distribution range. However, we did not add the counties which lack evident records of distribution to the distribution range of the species. Thus, some species are distributed sporadically and disjunctly in China, such as *Fagus hayatae*.

(3) Some species occur in a large area, but in some cases that species might not have been documented in a few counties in the area. In this case, we determined if these counties were in their distribution area by checking if the topographic and climatic conditions of these counties are similar to their surrounding counties (with evident records of distribution). If yes, we took these counties as the distribution sites.

(4) Basically we document natural distribution range for all the species in the atlas. However, for some important and widely cultivated economic plants (such as *Ginkgo biloba* and most fruit trees), the natural and cultivated distribution range are not clear because of the long history of cultivation, and therefore the distribution range of the species in the atlas may not be natural. For example, according to *Flora of China*, *Ginkgo biloba* is only naturally distributed in the Mt. Tianmu, Zhejiang Province. However, it has been now planted all over China, and in most countries in the Northern Hemisphere. In this case, the readers should be careful to take the information on the distribution of this species.

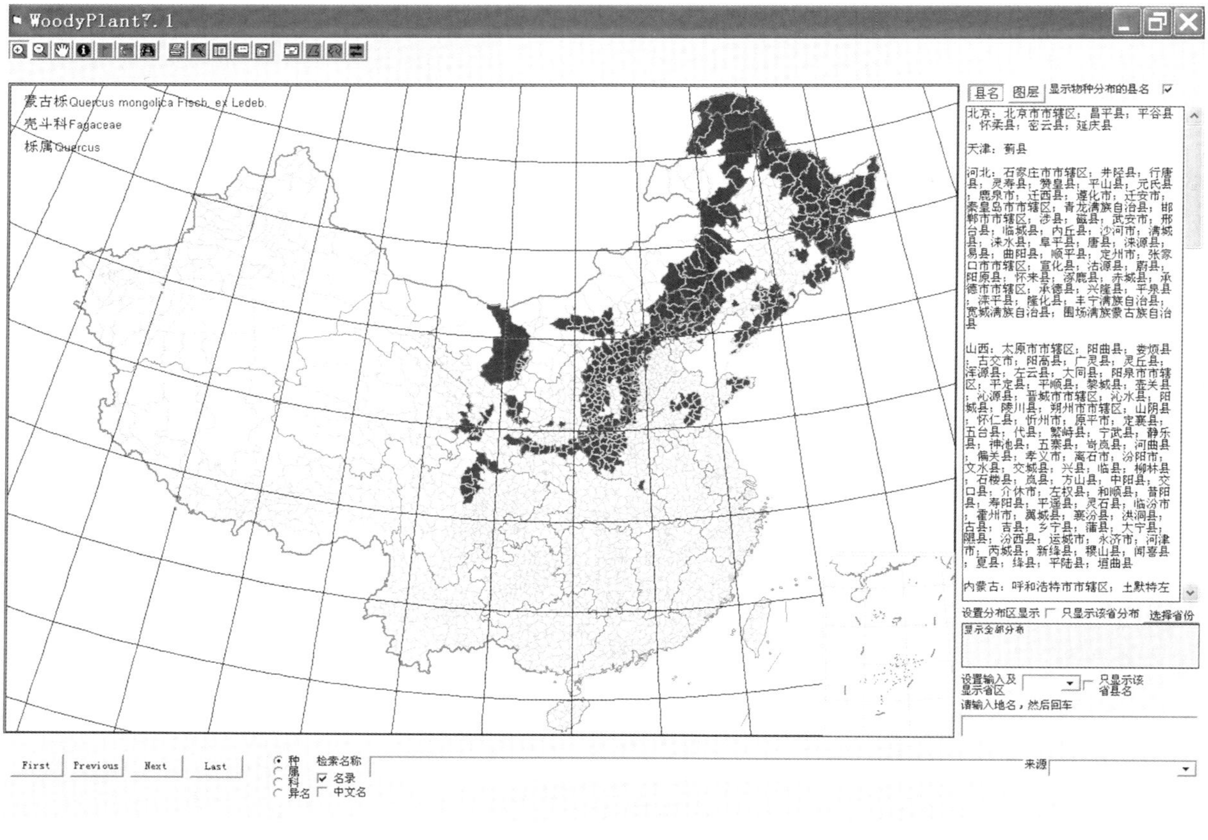

Figure 1 The interface of the software for "*Database of China's Woody Plants*". Using the interface, we can conveniently search and illustrate the distribution ranges of species, print range maps, input new data and edit the data in the database.

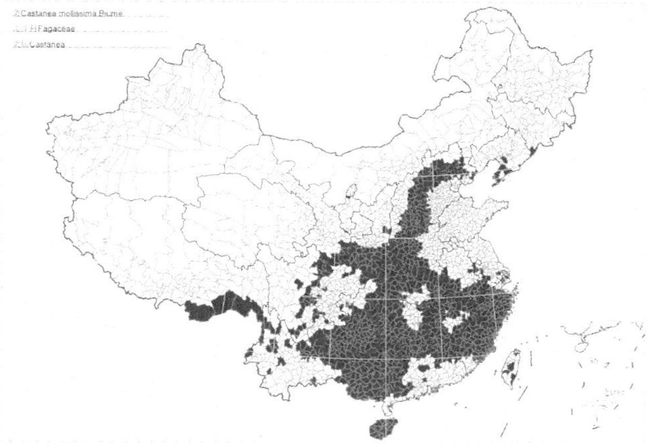

Figure 2 Distribution of *Castanea mollissima* in China.

(5) We provided climate information in this atlas for all the species, which is useful to understand the distribution and conservation of the species.

(6) We emphasize here again that there may be some incorrect information in this atlas because of a huge number of species and the limited information for some species. However, this will not prevent it from ecological and practical applications for the most cases.

1.1 “中国木本植物分布数据库”的建设

准确的物种分布信息是宏观生态学 (macroecology) 和生物地理学 (biogeography) 研究的重要基础。在过去的几十年中，随着物种分布资料的积累以及大量物种分布图的出版，基于物种分布区资料研究物种丰富度、物种组成更替以及其他生态格局已经成为宏观生态学和生物地理学中的通用方法 (Brown, 1995; Brown & Lomolino, 1998)。北美、欧洲和日本在几十年前就出版了详细的动、植物物种分布数据集或分布图，为这些区域的生物地理学和宏观生态学研究提供了不可缺少的基础资料。

中国是全球 12 个“巨大生物多样性国家”之一 (McNeely *et al.*, 1990)，拥有超过 30 000 种维管植物，其中包括约 11 000 种木本植物、3 200 种树木，物种数量仅次于位于南美洲的巴西和哥伦比亚，位居全球第三 (陈昌笃等 , 1998)，而远高于同处北温带的北美洲和欧洲地区 (Li, 1952; Latham & Ricklefs, 1993)。但中国目前还没有一个较为完善的植物分布数据库。

在过去的一个世纪中，中国的植物学家和生态学家对植物的分布进行了广泛调查，开展过多次大型科学考察，比如青藏高原考察、横断山区考察、三峡地区植物资源考察等，积累了大量的第一手资料。最近二十年来，《中国植物志》、《中国树木志》、《中国植被图集：1∶100 万》以及多个省区的植物志和树木志相继出版；一大批地方植物志和植物资源考察报告也纷纷出版发行。这为建立中国植物分布数据库提供了不可或缺的基础资料。考虑到木本植物分布资料较多、种名相对稳定，我们自 2003 年开始，先期组织实施了“中国木本植物分布数据库”建设项目，并邀请全国各地熟知当地植物分布的专家数十人参与物种分布区数据的校验。该数据库于 2008 年底建成，它收录了我国已知的全部木本植物，共包括 13 570 种 (包括变种、亚种和变型)。后按英文版《中国植物志》和“中国生物物种名录 2008 版”，对物种进行了归并，最终收录 11 405 种木本植物，是目前最为完善和全面的中国木本植物分布数据库。数据库中，木本植物是指根和茎含有木质部的植物，包括乔木 (高度 > 4 m)、灌木 (高度 < 4 m) 和木质藤本植物 (吴征镒 , 1980)。数据库包含了每种植物的中文名称、学名及其在中国境内的县级分布。

1.2 数据库编制过程及数据形式

数据库的建设主要包括以下四个步骤：

首先，以《中国植物志》和“中国种子植物数据库”为基础，建立中国木本植物名录，其分类系统参考《中国植物志》(共 80 卷 125 册，中国植物志编委会，1959–2006)，所有栽培种排除在外。《中国植物志》是超过 450 名中国植物学家近 50 年工作的积累，是国内外公认的有关中国植物最为权威的图志。“中国种子植物数据库”是由吴征镒院士领导、大量植物分类学家共同经过 10 多年工作的成果。在这两个数据源的基础上，我们建立了中国木本植物名录，并邀请国内有关分类学家进行校订。在数据库构建期间，英文版《中国植物志》“*Flora of China*” (1-25 卷) (Editorial Committee of *Flora of China*, 1994–2008) 陆续出版，以及全球“物种 2000”组织 (http://www.sp2000.org/) 在中国节点发布了“中国生物物种名录 2008 版”(Catalogue of Life, China: 2008

Annual Checklist）。为了体现最新成果，我们采用了“*Flora of China*”中的物种名录及分类系统，对异名按照“中国生物物种名录 2008 版”对应的物种进行合并。本名录共包含了 11 405 种木本植物（含 1 355 个种下单位），隶属于 170 科的 1 175 属。

第二，广泛收集物种分布信息，建立每个物种的分布区。物种分布信息参考了目前中国已出版的全部植物志，包括《中国植物志》各卷册以及其他全国性植物志（如《中国高等植物》，傅立国主编）、各省植物志和区域植物志，同时参考了已出版的《中国树木志》和各省树木志、木本植物名录、各个区域的植物资源考察和调查报告以及部分期刊上发表的文章。

第三，为了确保数据库的准确性和权威性，将全国划分为 20 个区域，在不同地区邀请 1~2 名熟知植物分布的植物学专家（共 21 位专家），对数据库中每个物种的分类信息（包括物种中文名、学名及所属科、属信息）以及地理分布信息进行校验、补遗和审定，最终完成数据库。

第四，为便于数据库的管理，使用 Visual basic 以及 MapObject 等工具开发了一款使用便捷的工具软件，其界面如 Figure 1 所示。该界面方便地实现了对物种分布数据库的编辑、修改、浏览显示、输入输出以及搜索等功能。

该数据库提供以县为分布单元的物种分布区 (Figure 2)。中国共有 2 408 个县 (1998 年资料)，各县的面积呈现极度的右偏分布 (skewness=9.93)，绝大多数县的面积较小，全部县的面积中值为 2 031 km^2，而青藏高原和塔克拉玛干沙漠以东地区 2 230 个县面积的中值仅 1 864 km^2；面积较大的县集中在新疆南部塔克拉玛干沙漠周边地区以及青藏高原北部高原腹地。在确定物种分布区时，为了减小这些面积较大的县对于物种分布区精度的影响，我们根据这些县的地形特征以及植被的分布，将每个县细分为几部分。由于特殊的环境条件（极干荒漠以及 4 000 m 以上的高原腹地），这些区域内的物种数量很少，因而这种处理对数据库整体的分布精度影响很小。

1.3 图集编制的几点说明

“中国木本植物分布数据库”建设及分布图编制时，以下几点需要特别说明：

(1) 本图集给出的物种分布区是在忠实于标本记录的基础上，参考地方植物志、植物资源调查报告和相关研究论文等资料，结合各地专家的实践经验确定的。所以，本图集给出的分布区信息应该是偏保守的。

(2) 按照物种分布区理论和生态位理论，一个物种在其分布范围，具有相似生境的区县都有可能是其分布区。但在确定该物种的分布地点时，如果没有充分的证据，本图集并没有将那些缺乏分布记录但生境相近的区县划成分布区。同样，不少物种在其分布区内，存在明显的零星或隔离或间断分布（如台湾水青冈是间断分布的典型例子）。本图集忠实于文献记录，对其分布区未作增减。

(3) 一些物种的分布图会出现这种情况：周围地区有该物种分布的记录，但中间少数区县没有分布记录。对这类情况，本图集对照地形图和气候信息，决定是否增补这些空白区县为分布区。这种处理带有一定的主观性。

(4) 本图集给出的物种分布基本反映自然分布，但某些重要经济和药用植物（如银杏和大多数有重要经济价值的果树等），由于其栽培历史悠久，自然分布区和栽培分布区难以分辨。例如，据记载，银杏仅在我国浙江天目山有野生分布，但它现已在我国大部分地区和北半球大多数国家都有栽培。在这种情况下，分布区的概念已经失去其原有的意义，不仅难以界定其精确的分布范围，其意义也不大。因此，我们所给的分布区仅供读者参考。

(5) 本图集提供的气候信息很重要，也基本准确，即使对那些分布区不连续的物种来说，本图集给出的气候指标也能够基本反映该物种的气候特征。

(6) 再次强调，由于中国木本植物种类多，分布信息不足，本图集一定会有这样或那样的问题，但编制其分布图是时代发展的迫切需要，不能再等了。我们已经落后于其他主要国家几十年了。

2 Characteristics of Distribution of Woody Plants in China

中国木本植物分布的基本特征

2.1 Composition of woody plants

China has 11,405 woody species, belonging to 1,175 genera and 170 families, of which, 6,735 (59.5%) are shrubs, belonging to 754 genera and 142 families; 3,165 species (27.8%) are trees, falling under 498 genera and 107 families; and 1,035 (9.1%) are woody vines, falling under 176 genera and 49 families; and 470 are bamboos, falling under 33 genera and 1 family(Table 1).

The numbers of woody species of gymnosperms, dicotyledons, and monocotyledons are 244 (2.1%), 10,480 (91.9%) and 664 (5.8%), respectively (Table 1). The woody monocotyledon species are mainly composed of bamboos (Gramineae) (470 species), liliaceous shrubs (Liliaceae) (93 species) and palms (Palmae) (89 species).

According to the definition of biogeographic types for plant families in China proposed by Wu *et al.* (2003), we divided all woody species into three categories according to their families: cosmopolitan, tropical, and temperate species. Nearly half (46.1%) of woody plants are tropical species, whereas the proportions of temperate and cosmopolitan species are 19.8% and 33.9% respectively (Table 1). China has only two endemic woody plant families, Eucommiaceae and Ginkgoaceae.

The frequency distributions of species/genus numbers per family, and species numbers per genus are all significantly right skewed (Figure 3), suggesting that most families/genera have a few species (small families and genera), whereas only a few families/genera have many species (large families and genera). More than one third of families have fewer than 10 species, and 24 families are represented by only a single species. Thirteen families (7.6%) have more than 200 species. The top five largest families are Rosaceae (947 species), Ericaceae (845 species), Fabaceae (682 species), Lauraceae (477) and Poaceae (470) (Table 2). The skewness of frequency distribution for species numbers per genus is much larger than that of species/genus numbers per family (skewness 11.1 *vs.* 3.5 and 3.7) (Figure 3). Four hundred and fifty genera (38.3%)

Table 1 Composition of woody plants in China

Group	No. of Species / %	No. of Genera	No. of Families
Woody plants	**11,405 / 100**	**1,175**	**170**
Trees	3,165 / 27.8	498	107
Shrubs	6,735 / 59.0	754	142
Woody vines	1,035 / 9.1	176	49
Bamboos	470 / 4.1	33	1
Ferns	17 / 0.1	5	3
Gymnosperms	244 / 2.1	35	10
Dicotyledons	10,480 / 91.9	1,076	150
Monocotyledons	664 / 5.8	59	7
Cosmopolitan	3,867 / 33.9	371	25
Temperate origin	2,259 / 19.8	134	44
Tropical origin	5,258 / 46.1	663	96

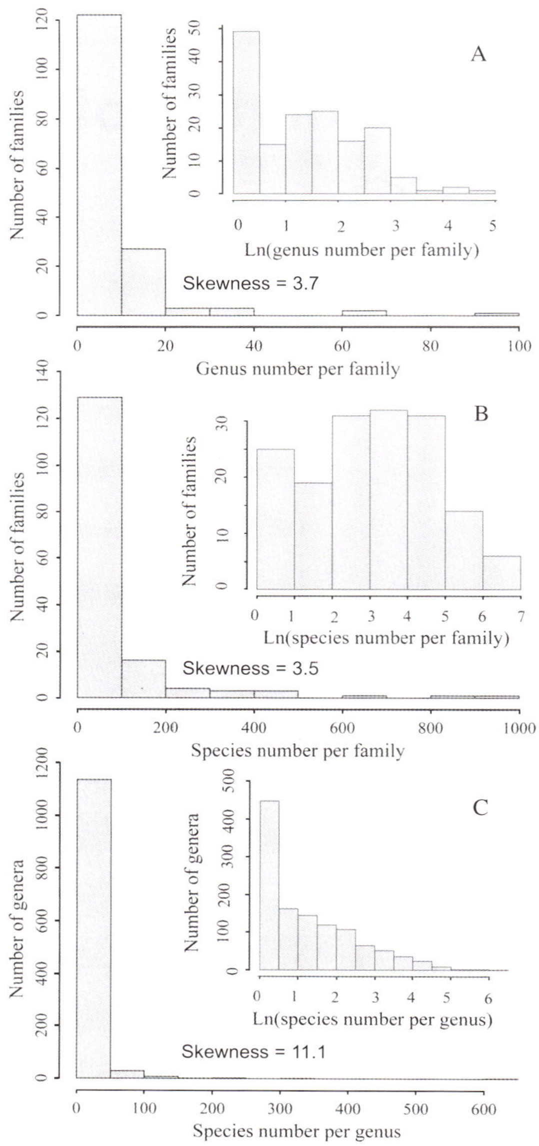

Figure 3 The frequency distribution of genus number per family (A), species number per family (B) and species number per genus (C). The subsets are log-transformed numbers of lower taxa per higher taxon, respectively.

have only one species, whereas only 12 genera (1.0%) have more than 100 species. The top five most species-rich genera are *Rhododendron* (624 species), *Salix* (331), *Rubus* (260), *Ilex* (224), and *Berberis* (175) (Table 2).

2.2 Richness pattern of woody plants

Species richness of woody plants exhibits a significant latitudinal trend in the east half part of the country, decreasing from south to north. Meanwhile, richness pattern is highly affected by the topography of China, with higher species richness in mountainous regions (Figure 4). Several regions with high species richness occur in mountainous areas, such as the Hengduan Mountains, South-Southeast Yunnan, mountains at the borders of Guangxi and Yunnan, Nanling Mountains, Wuyi Mountains, and Wuling Mountains. In contrast, species richness is very low in the plain and desert areas, such as the Northeast Plain, North China Plain, Tibetan Plateau, the Taklimakan Desert, and the Sichuan Basin (Figure 4).

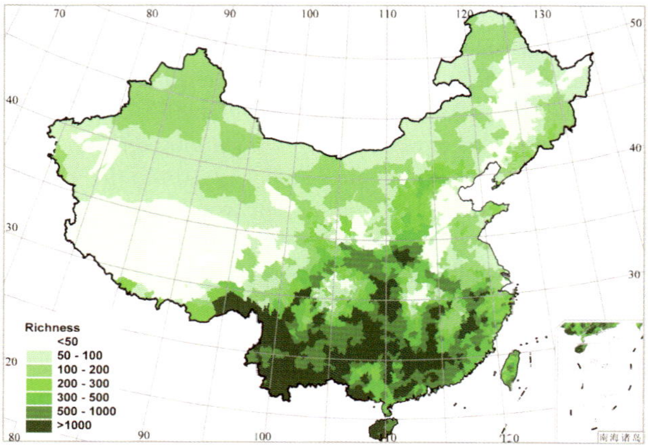

Figure 4 Species richness of woody plants in China.

Table 2 The top 10 families and genera in China's woody plants

Order	Family	No. of Species / Genera	Order	Genus	No. of Species
1	Rosaceae	947/40	1	*Rhododendron*	624
2	Ericaceae	845/16	2	*Salix*	331
3	Fabaceae	682/93	3	*Rubus*	260
4	Lauraceae	477/21	4	*Ilex*	224
5	Poaceae	470/33	5	*Berberis*	175
6	Salicaceae	420/3	6	*Lithocarpus*	131
7	Rubiaceae	374/67	7	*Clematis*	130
8	Theaceae	312/12	8	*Camellia*	127
9	Fagaceae	303/7	9	*Ficus*	118
10	Euphorbiaceae	279/63	10	*Acer*	114

2.1 物种组成的简要分析

中国木本植物共有 11 405 种，隶属 1 175 属、170 科。其中多数物种为灌木，共 6 735 种，占全部木本植物的 59.5%，隶属于 754 属、142 科；乔木共 3 165 种，占木本植物总数的 27.8%，隶属于 498 属、107 科；木质藤本植物的种类少，为 1 035 种，占木本植物总数的 9.1%，隶属 176 属、49 科；竹类 470 种，占木本植物总数的 4.1%，隶属 33 属、1 科 (表 1)。

中国木本植物中，裸子植物有 244 种，占总数的 2.1%；被子植物中，双子叶植物 10 480 种，占总数的 91.9%，单子叶植物较少，为 664 种，占 5.8% (表 1)；在单子叶植物中，以禾本科的竹类 (共 470 种)、百合类小灌木 (93 种) 和棕榈植物 (89 种) 为主。

根据吴征镒对中国植物科分布型的定义，本文将全部木本植物科分为四大类：全球广布成分、温带成分、热带成分以及中国特有成分 (表 1)。其中，热带成分种类最多，约占木本植物总数的一半 (46.1%)，而温带成分和全球广布成分则分别占 19.8% 和 33.9%。中国仅有两个特有科，分别为杜仲科 (Eucommiaceae) 和银杏科 (Ginkgoaceae)。

在中国木本植物中，每科的物种数、属数以及每属的物种数，均呈显著的右偏分布 (Figure 3)。从图中可以看到，绝大部分科拥有很少的物种或属，而拥有较多的物种数或属数的科很少 (Figure 3 A, B)；其中，拥有 10 种以下的木本科共 62 科，占总数的 1/3 强，有 24 科仅拥有一种木本植物；而拥有 200 种以上的仅 13 科，占总数的 7.6%；物种数最大的 5 个科分别是蔷薇科 (Rosaceae，947 种)、杜鹃花科 (Ericaceae，845 种)、豆科 (Fabaceae, 682 种)、樟科 (Lauraceae, 477 种) 和禾本科 (Poaceae, 470 种) (表 2)。每属中物和数的频度分布与科的情况相似，但其右偏程度则更高 (偏度系数 11.1 *vs.* 3.5 和 3.7) (Figure 3)；其中，仅拥有一种木本植物的属有 450 属，占总属数的 38.3%；而拥有 100 种以上的仅 12 属，占总数的 1.0%；最大的五属分别为杜鹃花属 (*Rhododendron*, 624 种)、柳属 (*Salix*, 331 种)、悬钩子属 (*Rubus*, 260 种)、冬青属 (*Ilex*, 224) 和小蘖属 (*Berberis*, 175 种)(表 2)。

2.2 我国物种丰富度的地理分布

中国木本植物的物种丰富度呈现明显的纬度梯度 (Figure 4)。总体上，南部地区物种丰富度较高，向北逐渐降低；同时，物种丰富度的分布格局与中国的地形特征具有良好的对应关系，即物种丰富度较高的地区多集中在山区。物种丰富度在云南南部和东南部、广西与贵州交界山地、南岭山地、横断山区、武夷山、武陵山区以及伏牛山区等山地最高；这些地区是中国木本植物多样性的热点 (hotspot) 地区。与此形成鲜明对照，物种多样性在东北平原、华北平原、青藏高原、塔克拉玛干沙漠及四川盆地内部等地区为最低。

表1 中国木本植物物种组成

类群	种数 / 比例(%)	属数	科数
木本植物	**11 405/100**	**1175**	**170**
乔木	3 165/27.8	498	107
灌木	6 735/59.0	754	142
木质藤本	1 035/9.1	176	49
竹类	470/4.1	33	1
蕨类植物	17/0.1	5	3
裸子植物	244/2.1	35	10
双子叶植物	10 480/91.9	1 076	150
单子叶植物	664/5.8	59	7
全球广布成分	3 867/33.9	371	25
温带成分	2 259/19.8	134	44
热带成分	5 258/46.1	663	96

表2 前10个最大的木本植物科、属

顺序	科名(拉丁名)	种数 / 属数	顺序	属名(拉丁名)	种数
1	蔷薇科(Rosaceae)	947/40	1	杜鹃花属(*Rhododendron*)	624
2	杜鹃花科(Ericaceae)	845/16	2	柳属(*Salix*)	331
3	豆科(Fabaceae)	682/93	3	悬钩子属(*Rubus*)	260
4	樟科(Lauraceae)	477/21	4	冬青属(*Ilex*)	224
5	禾本科(Poaceae)	470/33	5	小檗属(*Berberis*)	175
6	杨柳科(Salicaceae)	420/3	6	石栎属(*Lithocarpus*)	131
7	茜草科(Rubiaceae)	374/67	7	铁线莲属(*Clematis*)	130
8	山茶科(Theaceae)	312/12	8	山茶属(*Camellia*)	127
9	壳斗科(Fagaceae)	303/7	9	榕属(*Ficus*)	118
10	大戟科(Euphorbiaceae)	279/63	10	槭属(*Acer*)	114

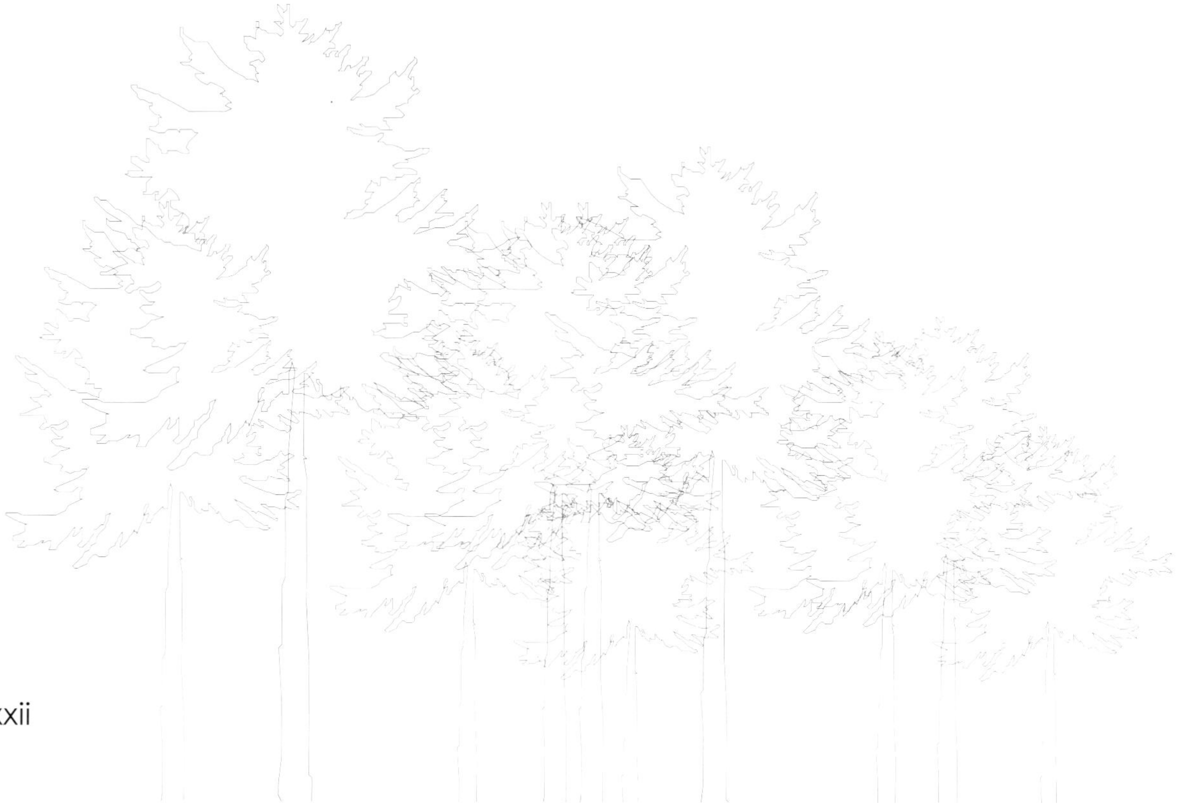

3 Climatic Variables and Their Distribution in China

气候指标的计算和分布图

3.1 Climate data

The climate data used in this atlas are based on the gridded spatial database available from the world climate website (http://www.worldclim.org/). The climate database, with 30 arc-second resolution (ca. 1 km at the equator), was developed using monthly mean temperature and monthly precipitation at climatic stations (ca. 7000 stations for temperature and 20,000 stations for precipitation) during 1950–2000 (Hijmans *et al.*, 2005). This database is more accurate and precise than previous databases, as it includes altitude as one of parameters when interpolating the climatic variables, compared to the 2- dimensional interpolation used in previous estimations (Hijmans *et al.*, 2005). We calculated mean values of all the variables for each county by using ArcGIS (ESRI Inc.), and then calculated the mean and range of these climatic variables based on the county-level data.

3.2 Definition and calculation of climatic variables

In this atlas, we provided following 12 climatic variables and one productivity variable for the distribution area of each species: mean annual temperature (MAT, °C), mean temperature of the coldest month (MTCM, °C), mean temperature of the warmest month (MTWM, °C), annual biotemperature (ABT, °C), warmth index (WI, °C), coldness index (CI, °C), potential evapotranspiration (PET, mm), annual precipitation (AP, mm), precipitation of the warmest quarter (PWQ, mm), precipitation of the coldest quarter (PCQ, mm), moisture index (Im), annual actual evapotranspiration (AET, mm), and vegetation net primary productivity ($g \cdot a^{-1} \cdot m^{-2}$). These 13 variables this atlas provides are their average over the year of 1950–2000. The definition and calculation of these variables are as follows:

Thermal index

(1) **Mean annual temperature (MAT, °C)** The average of monthly mean temperature over the year.

(2) **Mean temperature of the coldest month (MTCM, °C)** Normally, it is the same as January mean temperature.

(3) **Mean temperature of the warmest month (MTWM, °C)** Normally, it is the same as July mean temperature.

(4) **Annual biotemperature (ABT, °C)** An index for accumulated effective temperature. It is calculated from the months with mean temperature of 0–30 °C (Holdridge, 1947).

$$\text{ABT}=\frac{1}{12}\Sigma T \quad \text{(for months in which } 0<T<30\ ^{\circ}\text{C)}$$

(5) **Warmth index (WI, °C)** Similar to ABT, it is an index for accumulated effective temperature, and is a summation of monthly mean temperature above 5 °C (Kira, 1945).

$$\text{WI}=\Sigma(T-5) \quad \text{(for months in which } T>5\ ^{\circ}\text{C)}$$

(6) **Coldness index (CI, °C)** Contrary to WI, it represents the summation of winter coldness. It is a summation of monthly mean temperature below 5 °C (Kira, 1991).

$$\text{CI}=\Sigma(5-T) \quad \text{(for months in which } T<5\ ^{\circ}\text{C)}$$

(7) **Potential evapotranspiration (PET, mm)** An index in the Thornthwaite climate system. It is the potential water evapotranspirated from the earth surface (vegetation, soil and waterbody surface) to the atmosphere, and represents ambient energy in an area. PET is calculated using the method proposed by Thornthwaite (1955). For details, see Fang (2000) and Fang & Yoda (1990).

Humid/arid index

(8) **Annual precipitation (AP, mm)** Summation of monthly mean precipitation over the year.

(9) **Precipitation of the warmest quarter (PWQ, mm)** The summation of precipitation for the warmest quarter of the year. It represents summer precipitation in China.

(10) **Precipitation of the coldest quarter (PCQ, mm)** The summation of precipitation for the coldest quarter of the year. It represents winter precipitation in China.

Integrative index

(11) **Moisture index (Im)** Im and AET are also important indices proposed by Thornthwaite (1948). Im represents the moisture condition in an area and is calculated based on the water equivalence proposed by Thornthwaite (1948). For details, see Fang & Yoda (1990).

(12) **Annual actual evaportranspiration (AET,**

mm) Different from PET, AET represents the actual water evapotranpiration from the earth surface (vegetation, soil and waterbody surface) to the atmosphere. AET is a water-energy interaction index; it is determined by the energy where water is sufficient, whereas it is determined by the precipitation in water-deficient areas. In China, AET is more related to water than energy.

(13) **Vegetation net primary productivity (NPP, $g \cdot a^{-1} \cdot m^{-2}$)** NPP is calculated using the Carnegie-Ames-Stanford Approach (CASA) (Fang *et al.* 2003; Piao *et al.* 2005). CASA is a carbon process based model, using climate variables, such as temperature, precipitation and solar radiation, and vegetation and soil texture as inputs, with one month-step (Potter *et al.* 1993; Field *et al.* 1995). The NPP map illustrated in this atlas is a mean value between 1982 and 1999.

These 13 variables can be categorized into three groups: 1–7 are temperature-related variables, representing heat or energy conditions in the study area; 8–10 are water-related variables, representing moisture conditions; and 11–13 are water-energy interaction, which also represents vegetation primary productivity.

3.3 Distribution of climatic variables and some related factors

We present maps of topography, vegetation, vegetation zonation, and aforementioned 13 climatic and productivity variables.

3.1 气候数据来源

本图集使用的气候数据来自世界气候数据网站 (http://www.worldclim.org/)，其分辨率为 0.0083° * 0.0083° (在赤道附近约为 1 km*1 km) (Hijmans *et al.*, 2005)。该数据使用 1950–2000 年全球已有的气候站点数据进行插值，其中温度站点约有 7 000 个，降水站点约有 20 000 个。以往的数据插值通常是在二维地理空间，也即经度和纬度空间中进行的，忽略了海拔对气候的影响。与这些插值方法不同，该数据库采用了三维地理空间插值，即经度、纬度和海拔高度。因此，与以往的气候数据相比，本数据库更准确 (Hijmans *et al.*, 2005)。本图集利用此数据库，应用 ArcGIS (ESRI Inc.) 软件，计算了每个县域内主要气候要素的平均值，并以此计算分布区内气候的平均值与范围。

3.2 主要气候指标及计算

本图集为每种植物提供了 12 种气候指标和 1 个生产力指标，它们是：年均温、最冷月均温、最热月均温、年生物温度、温暖指数、寒冷指数、潜在蒸散量、年降水、最暖季降水、最冷季降水、湿润指数、年实际蒸散量、植被净第一性生产力。本图集给出的这些指标均为 1950–2000 年的均值。它们的定义和计算公式如下：

温度指标

(1) **年均温** (Mean annual temperature, MAT)： 各年平均温度的均值。

(2) **最冷月均温**(Mean temperature of the coldest month, MTCM)：各年1月份的均温值。

(3) **最热月均温**(Mean temperature of the warmest month, MTWM)：各年7月份的均温值。

(4) **年生物温度** (Annual biotemperature, ABT)： 有效积温的简易替代指标，由月均温 (T) 在 0~30 °C 之间的月份计算得到 (Holdridge, 1947)，即：

$$ABT=\frac{1}{12}\sum T \quad (T \text{ 为 } 0<T<30\,^\circ C \text{ 的月均温})$$

(5) **温暖指数** (Warmth index, WI)： 由月均温 (T) 在 5 °C 以上的月份计算得到 (Kira, 1945)，即：

$$WI=\sum (T-5) \quad (T \text{ 为 } >5\,^\circ C \text{ 的月均温})$$

(6) **寒冷指数** (Coldness index, CI)： 与温暖指数相反，指示冬季累计温度的大小，由月均温 (T) 在 5 °C 以下的月份计算得到 (Kira, 1991)，即：

$$CI=\sum (5-T) \quad (T \text{ 为 } <5\,^\circ C \text{ 的月均温})$$

(7) **潜在蒸散量** (Potential evaportranspiration, PET)：是 Thornthwaite 气候分类系统中的重要指标，表示在理想状态下，由地表(植被、土壤和水表面)向大气蒸发和蒸腾的最大可能水量，反映一个地区环境能量的多少。PET的计算采用 Thornthwaite(1955)提出的方法，具体计算方法也可参见方精云(2000) 和Fang & Yoda (1990)。

水分指标

(8) **年降水** (Annual precipitation, AP)： 各年降水量的均值。

(9) **最暖季降水** (Precipitation of the warmest quarter, PWQ)：一年中最暖季的各月降水量之和。在我国，实际上反映一个地区夏季降水量的多少。

(10) **最冷季降水** (Precipitation of the coldest quarter, PCQ)：一年中最冷季的各月降水量之和。在我国，实际上反映一个地区冬季降水量的多少。

水热综合指标

(11) **湿润指数** (Moisture index, Im)： 湿润指数 (Im) 和年实际蒸散量 (AET) 也是 Thornthwaite 气候分类系统中两个重要指标。Im 反映了一个地区的水分气候状况，它的计算是基于 Thornthwaite 的水分平衡法 (Thornthwaite, 1948)。具体算法见 Fang & Yoda (1990)。

(12) **年实际蒸散量** (Annual actual evaportranspiration, AET)：与潜在蒸散量 (PET) 不同，AET 反映一个地区在一年内，由

地表（植被、土壤和水表面）向大气移动的实际水量，其大小受能量和水分的双重控制。当水分充足时，AET 受能量控制；而当水分不足时，则主要受水分控制。因此，一般认为，AET 反映了一个地区的水热相互作用。在中国，AET 与水分的相关性更强。

（13）**植被净第一性生产力**(Vegetation net primary productivity, NPP, g · a^{-1} · m^{-2})：由CASA模型(Carnegie-Ames-Stanford Approach)计算得到(Fang *et al.* 2003; Piao *et al.* 2005)。该模型是一种过程模型，由归一化植被指数(NDVI)，以及温度、降水和太阳辐射等气象数据，结合植被和土壤质地等资料作为输入变量，时间步长为月 (Potter *et al.* 1993; Field *et al.* 1995)。本图集所显示的NPP值为1982–1999年的平均值。

可以将上述 13 项指标归为 3 类：1~7 项指标为温度指标，反映热量或能量状况；8~10 项为水分指标，反映水分状况；11~13 项为水热综合指标，同时也常常反映植被生产力的高低。

3.3 主要气候指标分布图和相关分布图

在这节里，我们给出中国的地形图、植被图、植被区划图以及上述 13 个气候和生产力指标的分布图。

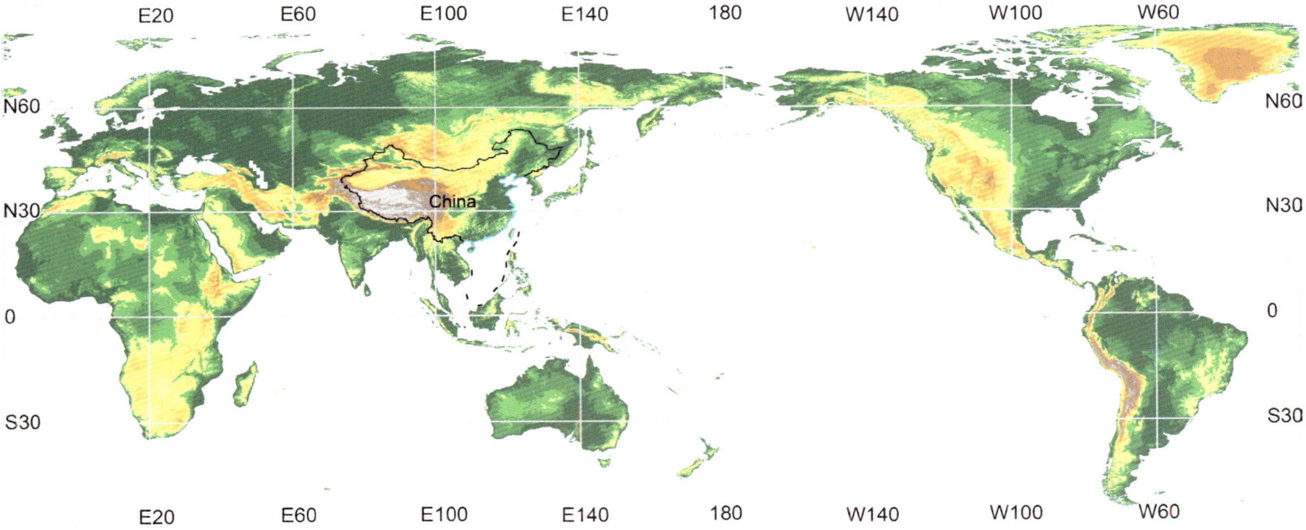

Location of China 中国的地理位置

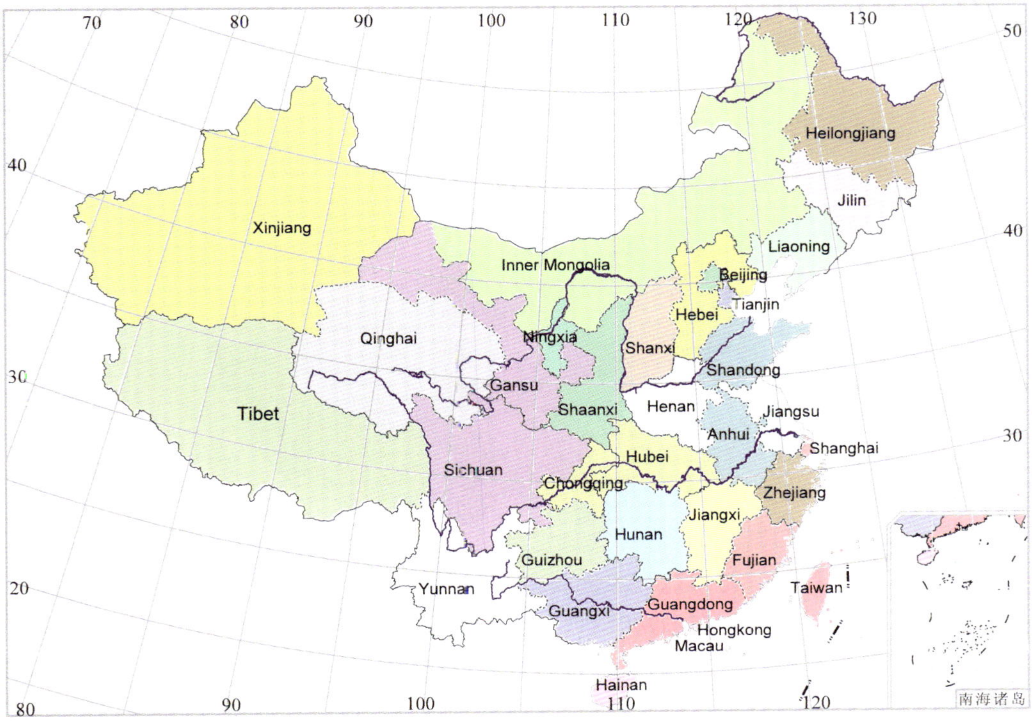

Provinces of China 中国政区

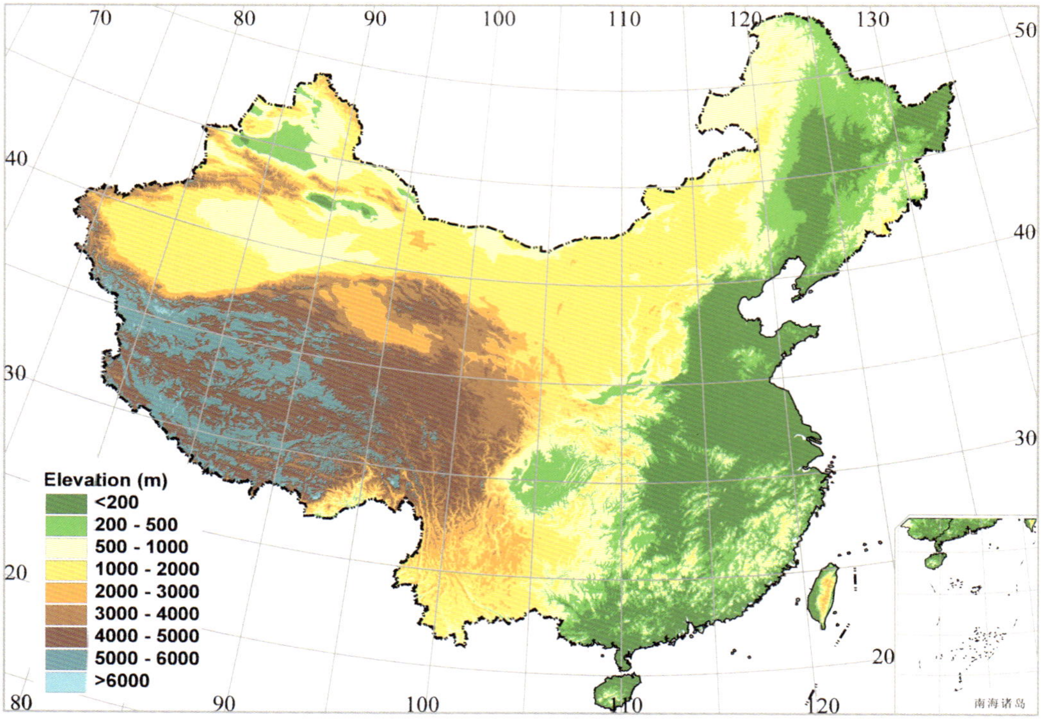

Topographic map of China 中国地形

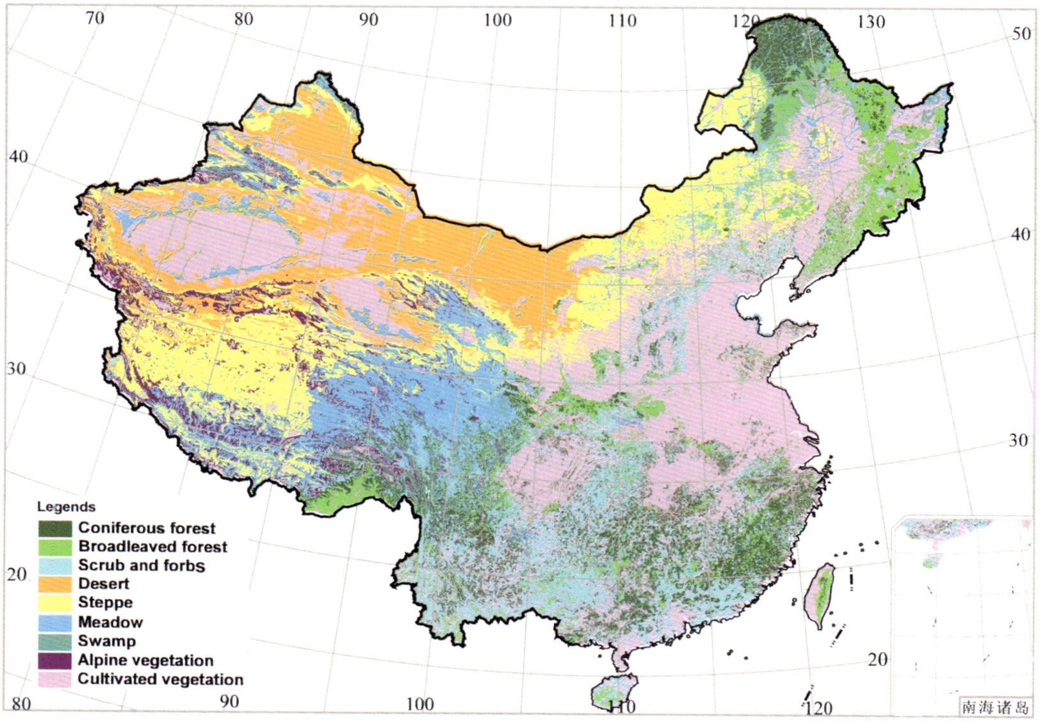

Vegetation map of China (Vegetation map of the People's Republic of China 1 : 1 000 000)
中国植被（基于中华人民共和国植被图 1∶1 000 000）

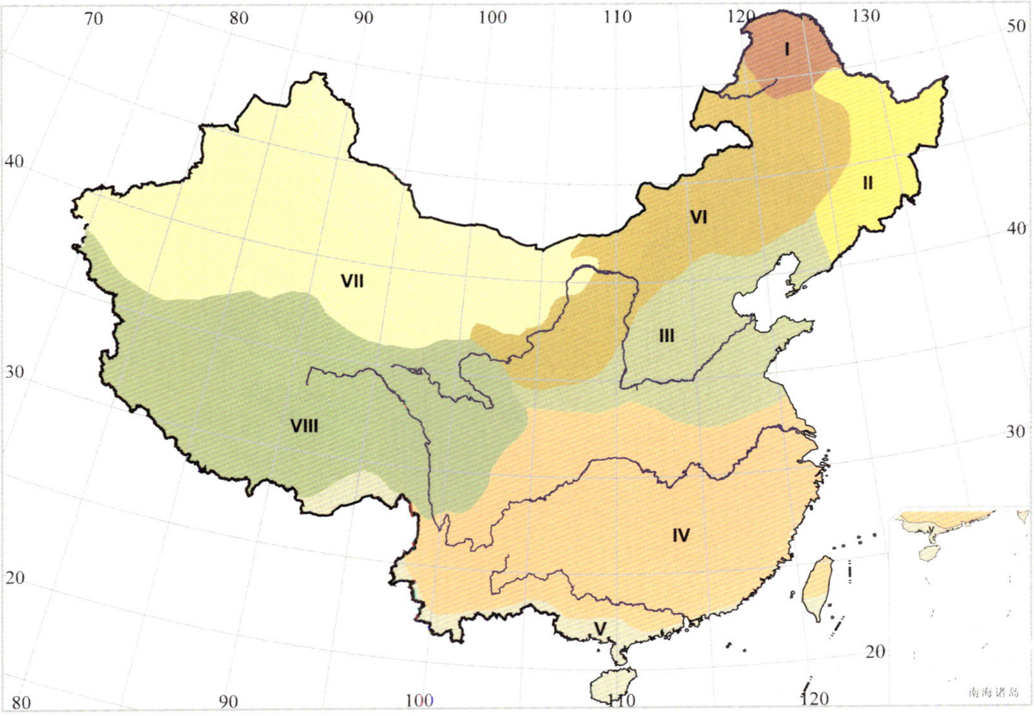

Vegetation regionalization map of China. I. Cold-temperate needleleaf deciduous forest region; II. Temperate mixed needleleaf and deciduous broadleaf forest region; III. Warm temperate deciduous broadleaf forest region; IV. Subtropical broadleaf evergreen forestion region; V. Tropical monsoon rain forest and rain forest region; VI. Temperate steppe region; VII. Temperate desert region; VIII. Qinghai-Xizang Plateau alping vegetation region. (Vegetation regionalization map of China 1∶6 000 000)

中国植被区划：I. 寒温带针叶林区域；II. 温带针叶、落叶阔叶混交林区域；III. 暖温带落叶阔叶林区域；IV. 亚热带常绿阔叶林区域；V. 热带季雨林、雨林区域；VI. 温带草原区域；VII. 温带荒漠区域；VIII. 青藏高原高寒植被区域（中国植被区划图 1∶6 000 000）

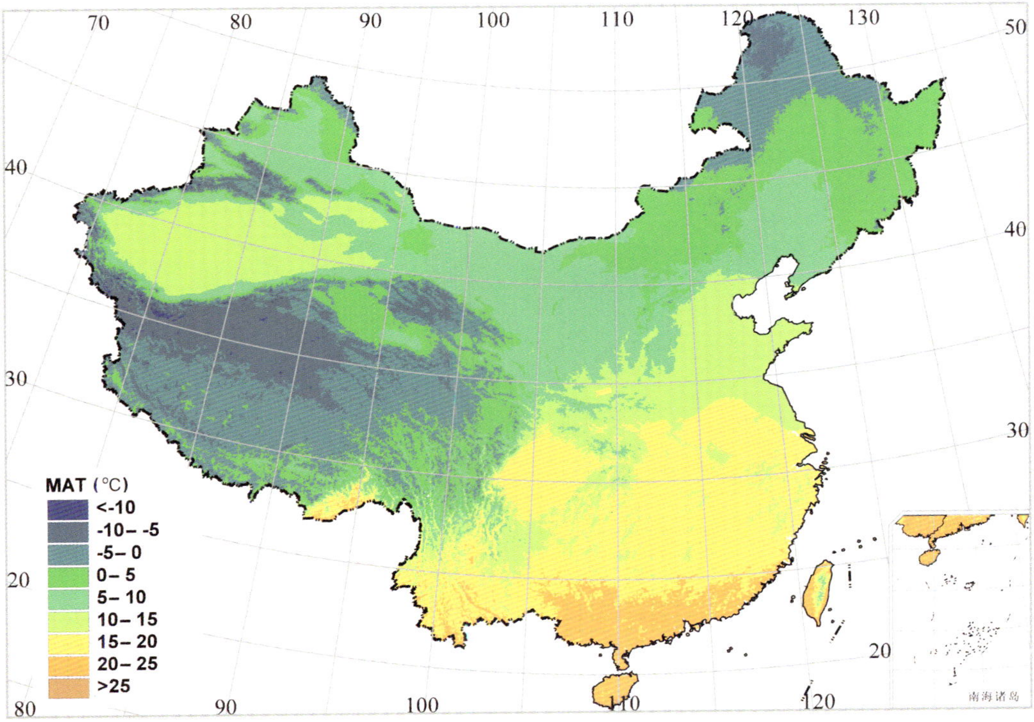

Mean annual temperature (MAT, °C) 年均温

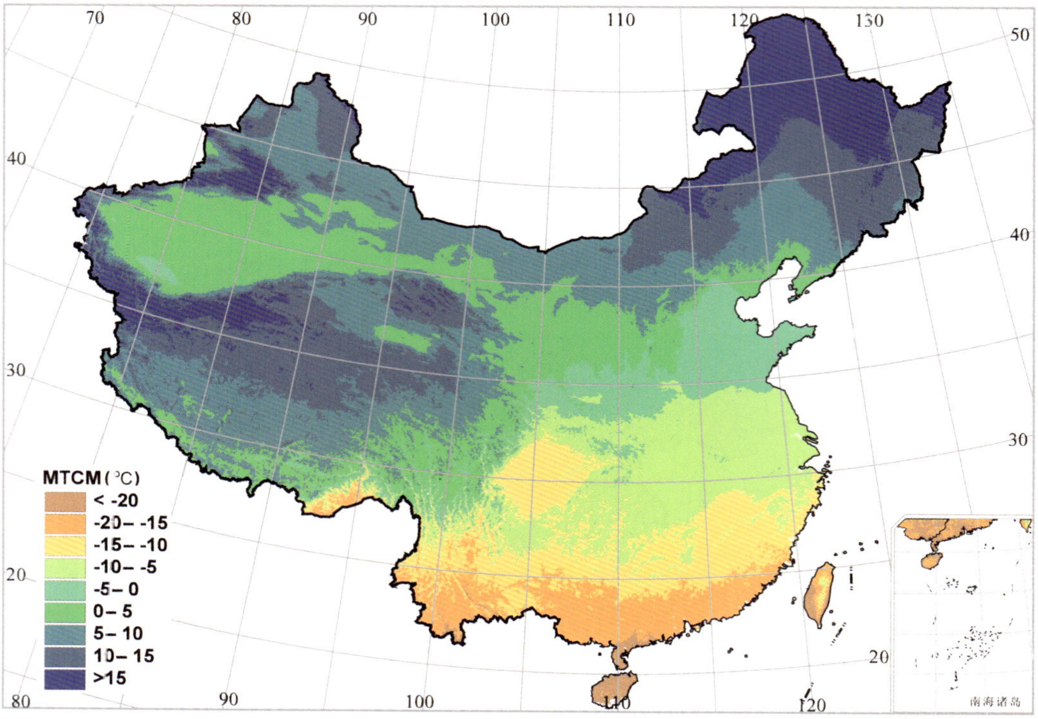

Mean temperature of the coldest month (MTCM, °C) 最冷月均温

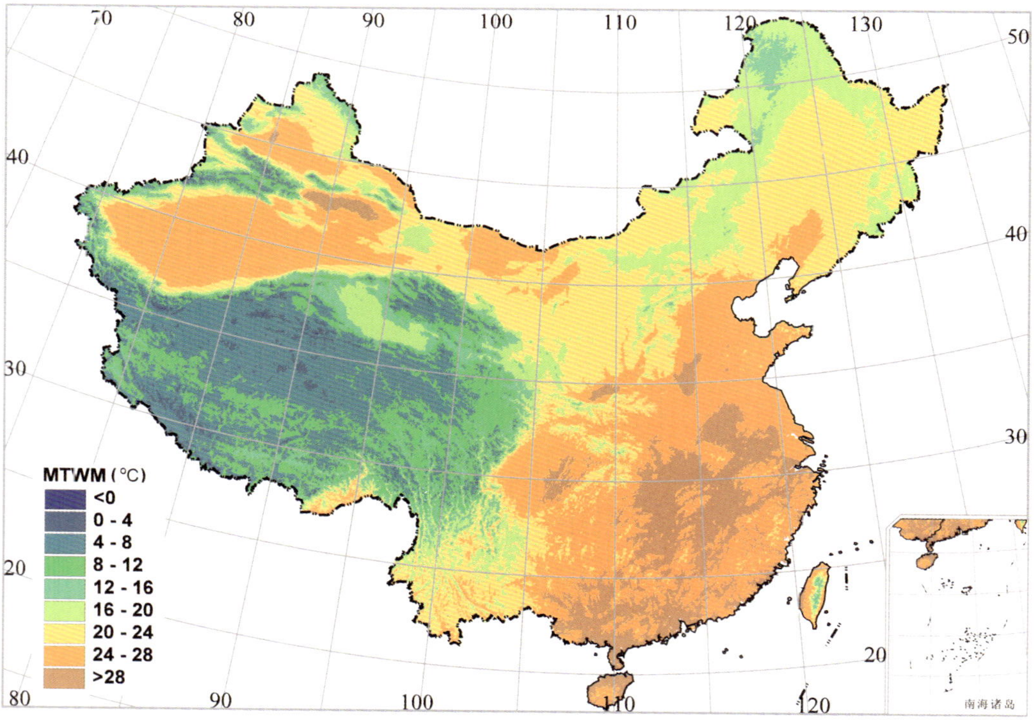

Mean temperature of the warmest month (MTWM, °C) 最暖月均温

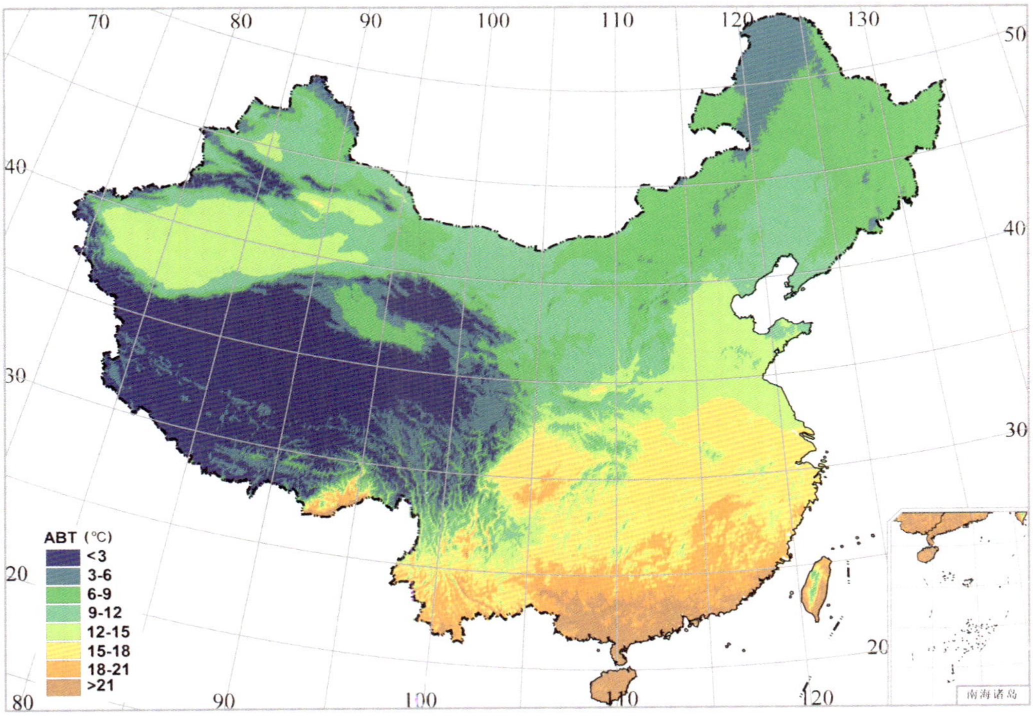

Annual biotemperature (ABT, °C) 年生物温度

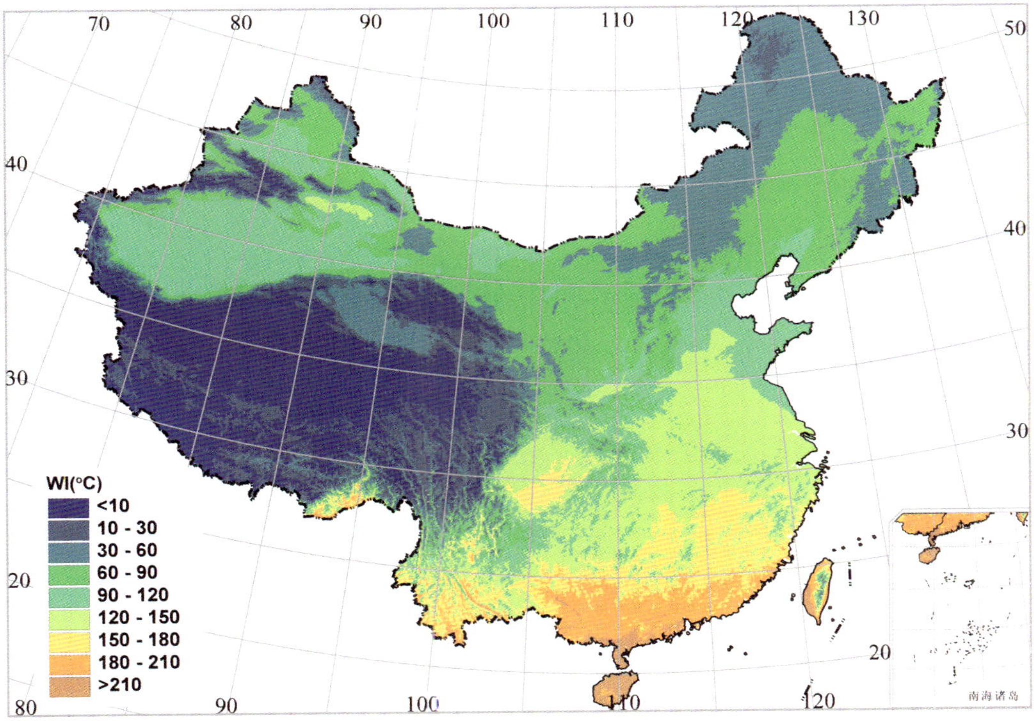

Warmth index (WI, °C) 温暖指数

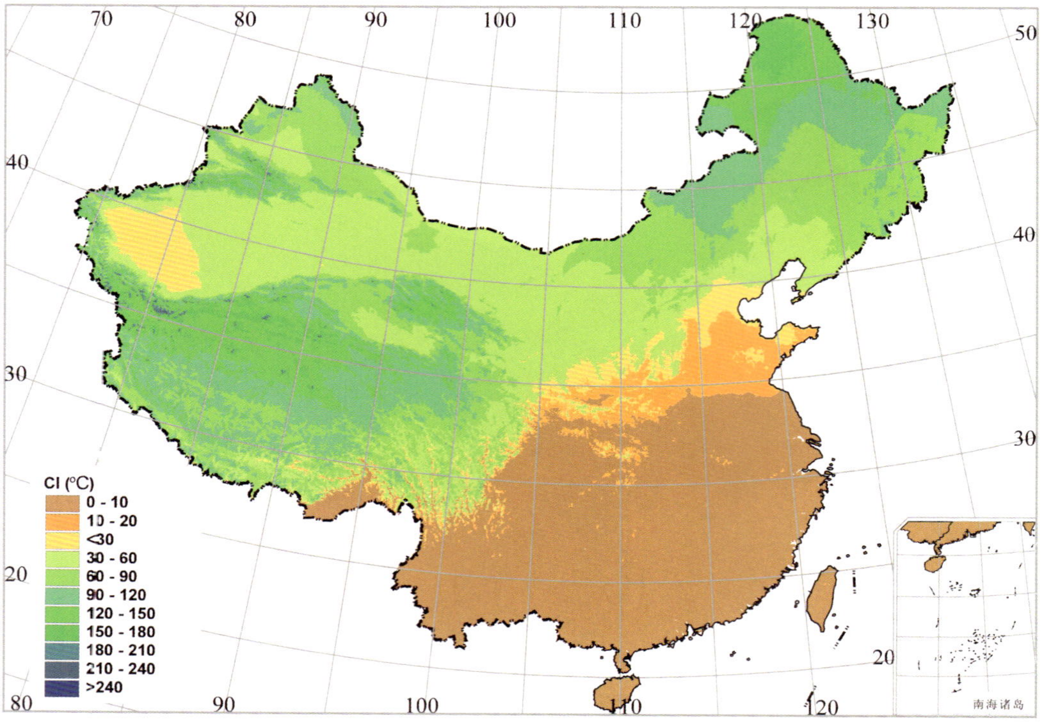

Coldness index (CI, °C) 寒冷指数

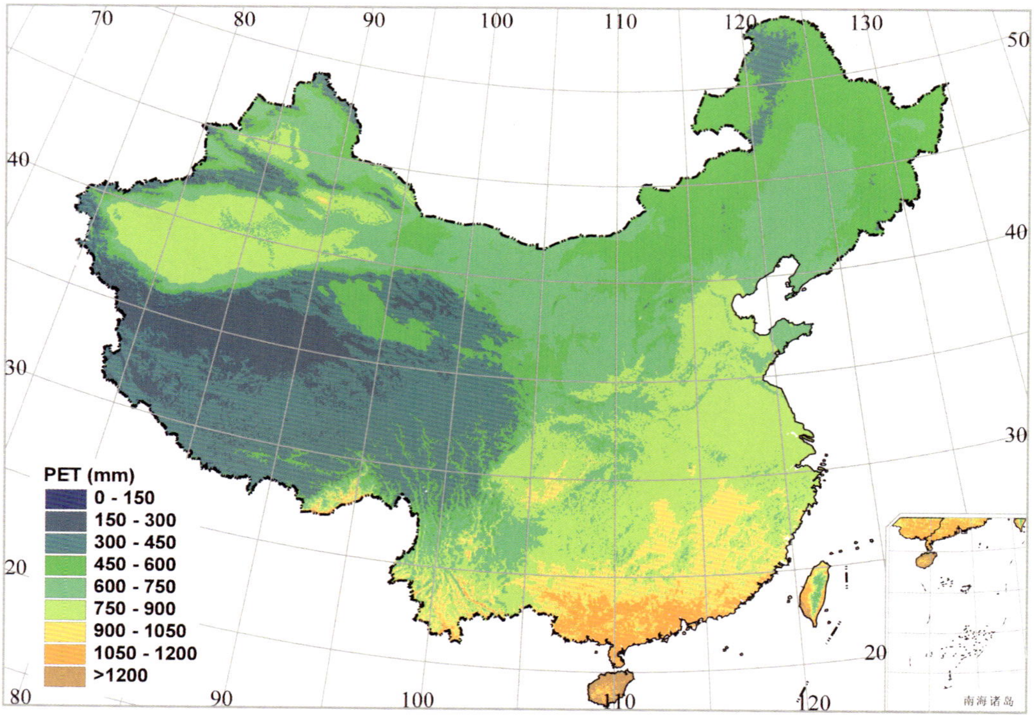

Potential evapotranspiration (PET, mm) 潜在蒸散量

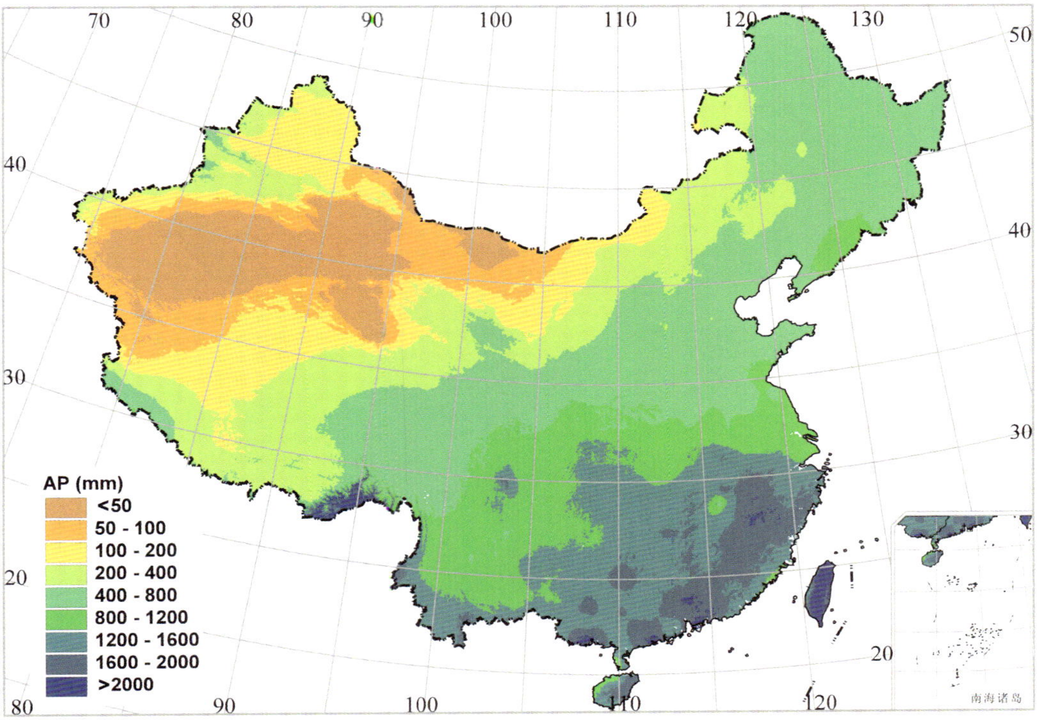

Annul precipitation (AP, mm) 年降水

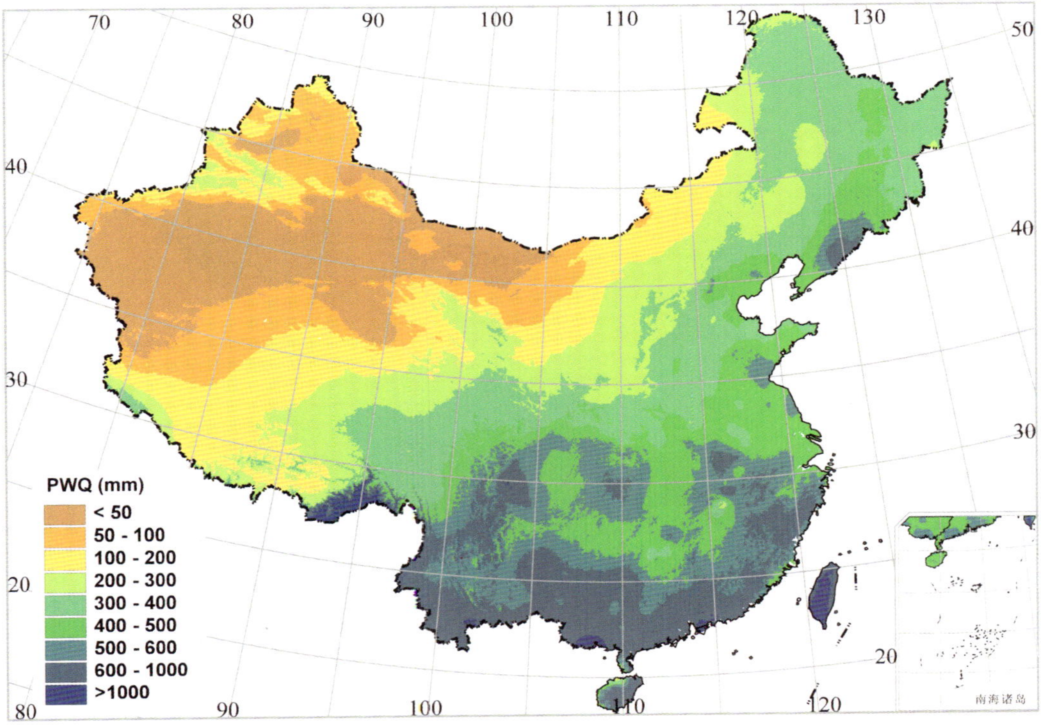

Precipitation of the warmest quarter (PWQ, mm) 最暖季降水

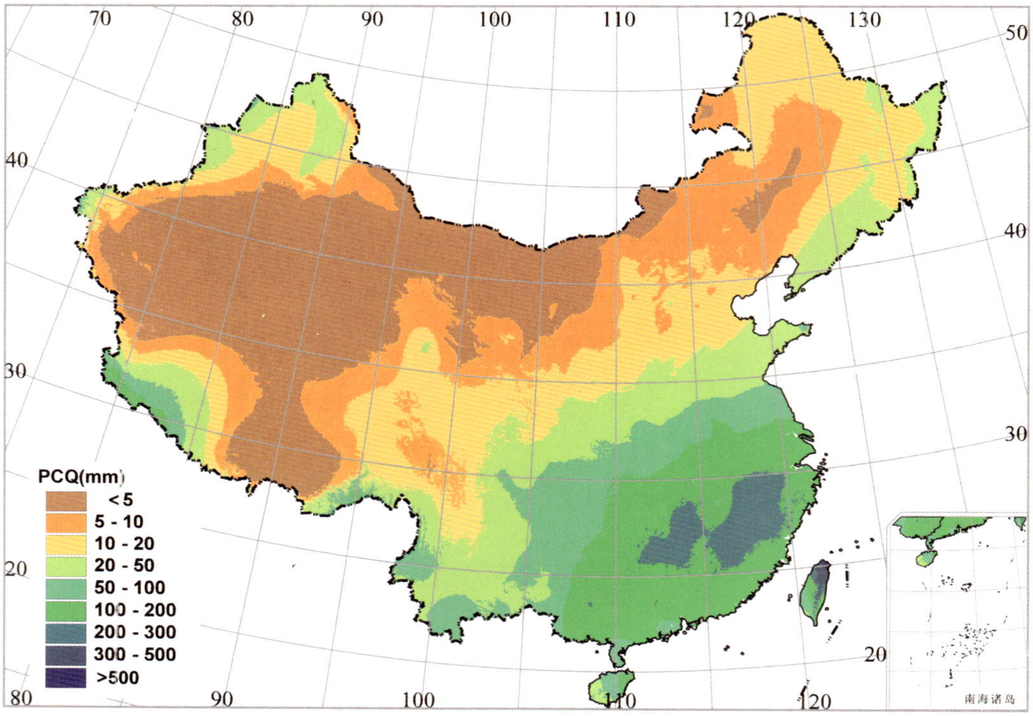

Precipitation of the coldest quarter (PCQ, mm) 最冷季降水

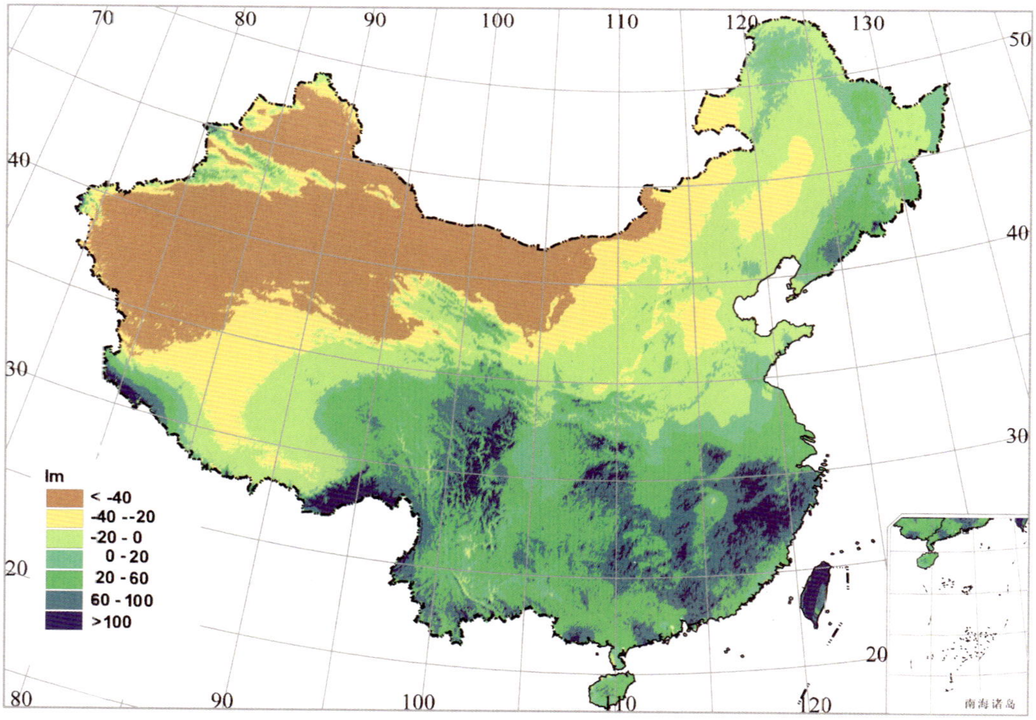

Moisture index (Im) 湿润指数

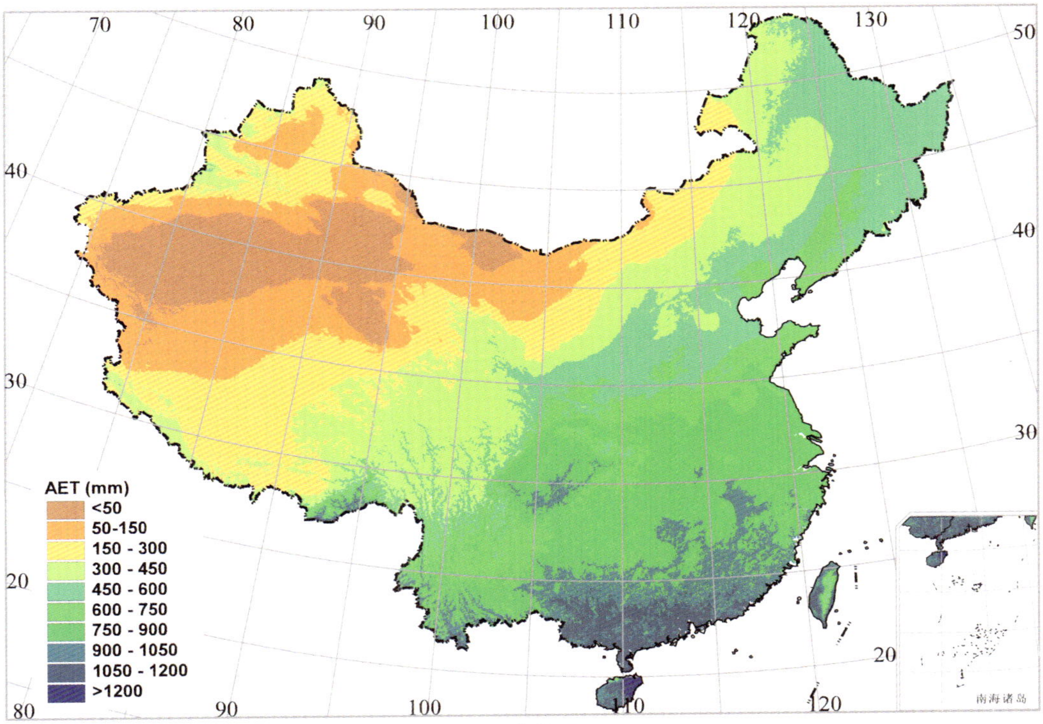

Annual actual evapotranspiration (AET, mm) 年实际蒸散量

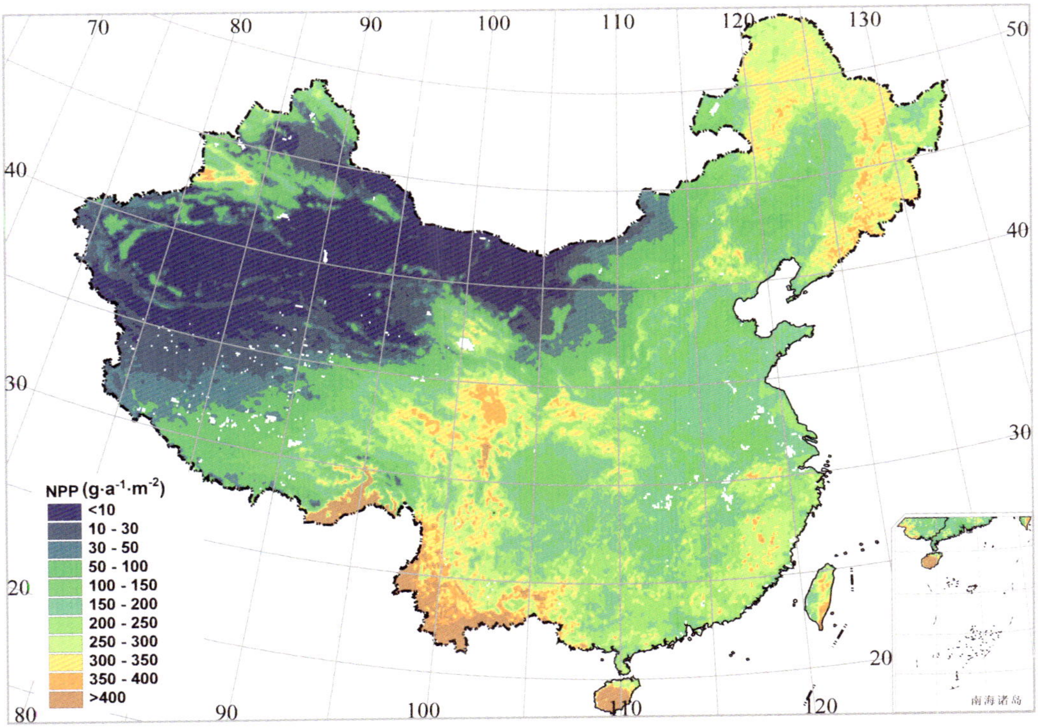

Vegetation net primary production (NPP, $g \cdot a^{-1} \cdot m^{-2}$) 植被净第一性生产力

Species Distribution and Climates

Cyatheaceae 桫椤科 *Alsophila* 桫椤属
Alsophila andersoni Scott ex Bedd. 毛叶桫椤
Trees

20.0 (13.3~24.7); 13.2 (5.4~19.8); 24.8 (19.1~29.0); 20.0 (13.4~24.7); 180.4 (100.4~236.5); 0.0 (0.0~0.0); 1018 (706~1376) | 1510 (1161~1777); 723 (444~1007); 77 (41~182) | 52.4 (-2.2~145.8); 990 (706~1326); 429.0 (194.5~659.0)

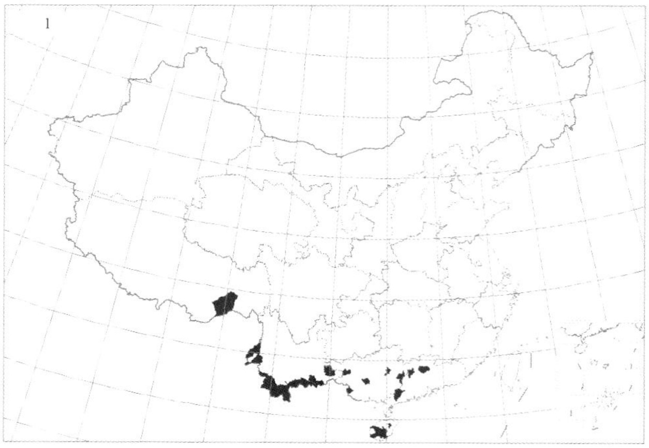

Cyatheaceae 桫椤科 *Alsophila* 桫椤属
Alsophila austro-yunnanensis S. G. Lu 滇南桫椤
Trees

17.7; 10.4; 22.9; 17.7; 152.9; 0.0; 857 | 1395; 785; 56 | 62.8; 857; 425.6

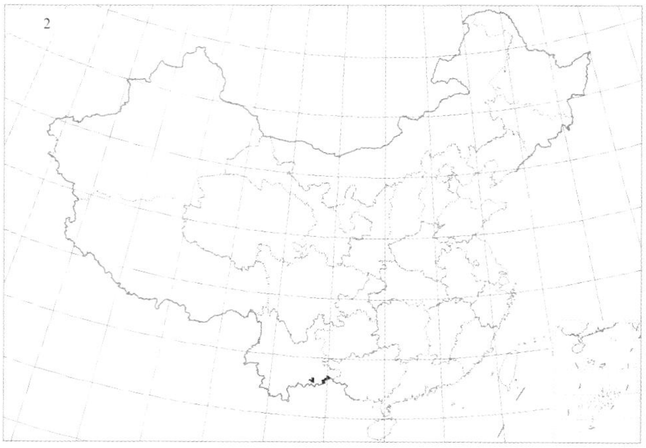

Cyatheaceae 桫椤科 *Alsophila* 桫椤属
Alsophila costularis Baker 中华桫椤
Trees

17.4 (8.6~21.0); 10.7 (1.3~14.9); 22.0 (15.0~25.6); 17.4 (8.7~21.0); 149.4 (52.6~191.8); 0.3 (0.0~8.8); 852 (574~1053) | 1338 (1007~1736); 732 (539~1007); 53 (41~68) | 58.7 (12.5~145.8); 839 (574~1053); 430.0 (277.0~627.0)

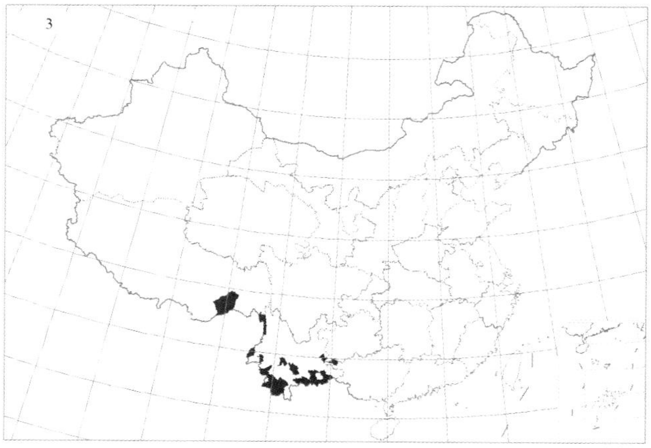

Cyatheaceae 桫椤科 *Alsophila* 桫椤属
Alsophila denticulata Baker 粗齿桫椤
Shrubs

18.3 (15.9~21.1); 10.0 (5.9~14.6); 25.9 (22.4~28.6); 18.4 (16.0~21.1); 160.6 (131.9~193.7); 0.0 (0.0~0.0); 930 (831~1104) | 1862 (1254~3495); 695 (425~960); 240 (48~897) | 98.9 (14.0~224.4); 927 (831~1077); 287.4 (206.0~491.5)

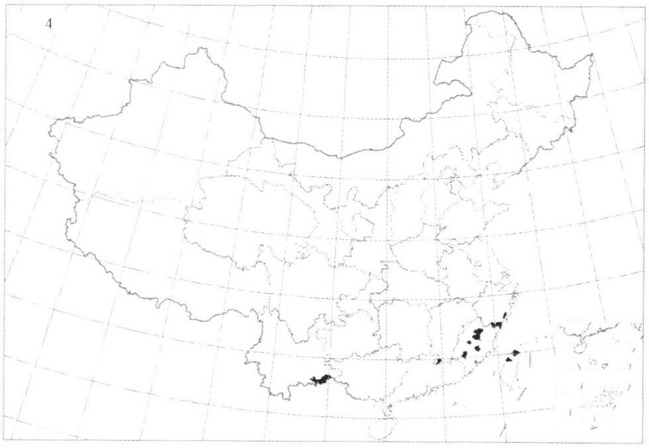

Cyatheaceae 桫椤科 *Alsophila* 桫椤属
Alsophila fenici (Copel) C. Chr. 兰屿桫椤
Shrubs

20.3; 15.9; 23.6; 20.4; 184.4; 0.0; 993 | 2853; 1618; 147 | 190.4; 993; 330.3

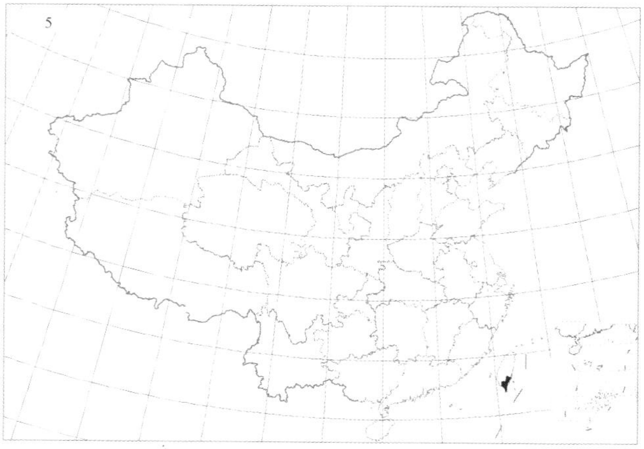

Cyatheaceae 桫椤科 *Alsophila* 桫椤属
Alsophila gigantea Wall. ex Hook. 大叶黑桫椤
Small trees

20.7 (17.2~24.7); 13.8 (8.5~19.8); 25.6 (21.6~29.0); 20.7 (17.3~24.7); 188.4 (147.0~236.5); 0.0 (0.0~0.0); 1059 (833~1376) | 1501 (1161~1777); 698 (444~996); 83 (41~182) | 44.5 (-2.2~88.8); 1028 (833~1326); 410.7 (194.5~659.0)

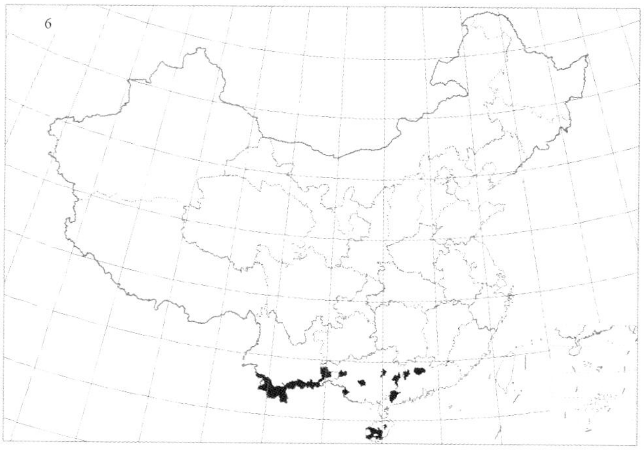

Cyatheaceae 桫椤科 *Alsophila* 桫椤属
Alsophila khasyana T. Moore ex Kuhn 西亚桫椤
Small trees

15.1 (8.6~19.1); 8.2 (1.3~13.0); 20.2 (15.0~23.3); 15.1 (8.7~19.1); 122.9 (52.6~169.7); 1.5 (0.0~8.8); 768 (574~917) ‖ 1346 (1041~1736); 762 (585~1007); 54 (44~70) ‖ 78.2 (26.1~145.8); 760 (574~917); 441.6 (352.3~506.2)

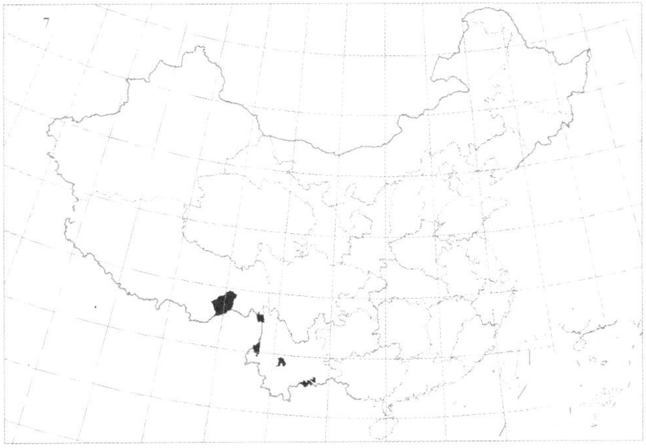

Cyatheaceae 桫椤科 *Alsophila* 桫椤属
Alsophila latebrosa Wall. ex Hook. 阴生桫椤
Small trees

23.2 (17.7~24.7); 17.4 (10.9~19.8); 27.3 (22.5~28.7); 23.2 (17.8~24.7); 218.6 (153.4~236.5); 0.0 (0.0~0.0); 1247 (854~1376) ‖ 1446 (868~1854); 557 (312~818); 69 (44~115) ‖ 17.8 (-34.3~60.1); 1132 (854~1360); 472.0 (288.3~620.0)

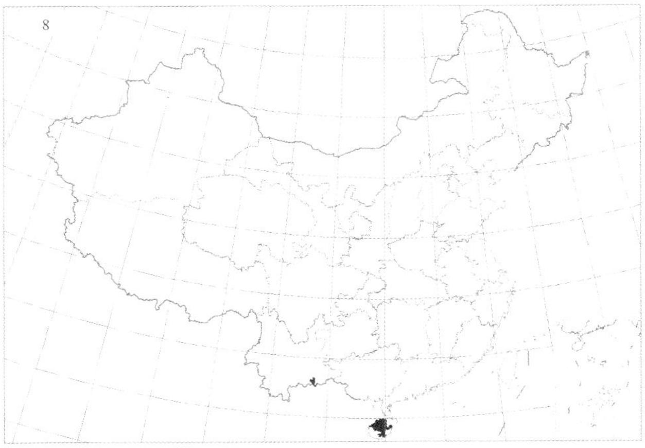

Cyatheaceae 桫椤科 *Alsophila* 桫椤属
Alsophila loheri Christ 南洋桫椤
Trees

20.3; 15.9; 23.6; 20.4; 184.4; 0.0; 993 ‖ 2853; 1618; 147 ‖ 190.4; 993; 330.3

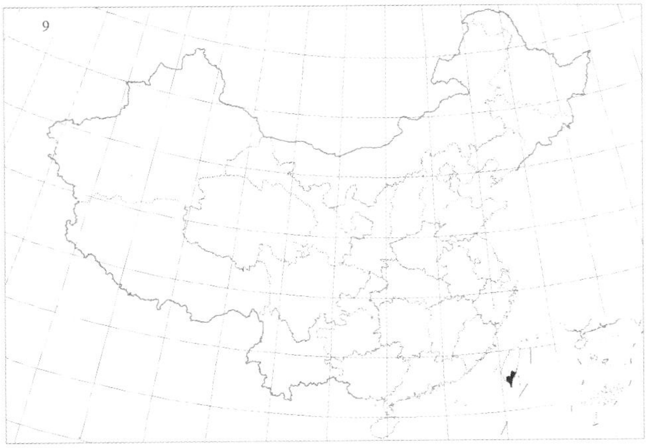

Cyatheaceae 桫椤科 *Alsophila* 桫椤属
Alsophila metteniana Hance 小黑桫椤
Small trees

17.5 (14.0~22.2); 8.6 (4.1~17.7); 25.3 (20.1~28.3); 17.5 (14.1~22.2); 150.2 (109.8~206.7); 0.1 (0.0~0.9); 896 (747~1123) ‖ 1695 (988~3495); 716 (458~1847); 174 (46~897) ‖ 86.8 (18.0~246.5); 896 (747~1123); 236.3 (126.1~413.5)

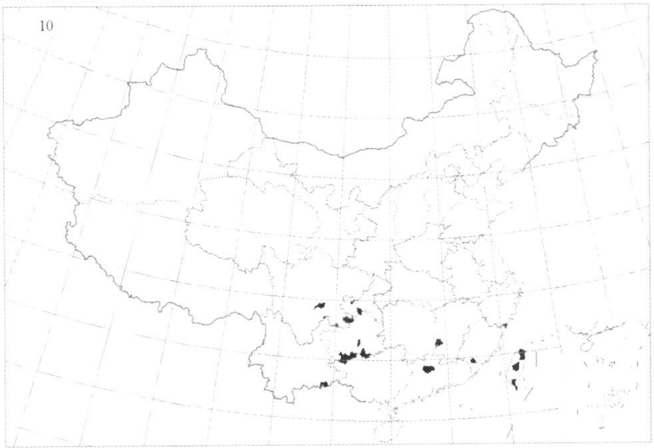

Cyatheaceae 桫椤科 *Alsophila* 桫椤属
Alsophila podophylla Hook. 黑桫椤
Small trees

20.5 (16.2~24.0); 13.3 (7.4~19.1); 26.3 (20.1~28.9); 20.6 (16.2~24.0); 187.2 (134.2~228.1); 0.0 (0.0~0.0); 1059 (773~1302) ‖ 1823 (1178~3495); 768 (455~1847); 175 (42~897) ‖ 76.1 (6.7~246.5); 1044 (773~1276); 331.4 (201.4~659.0)

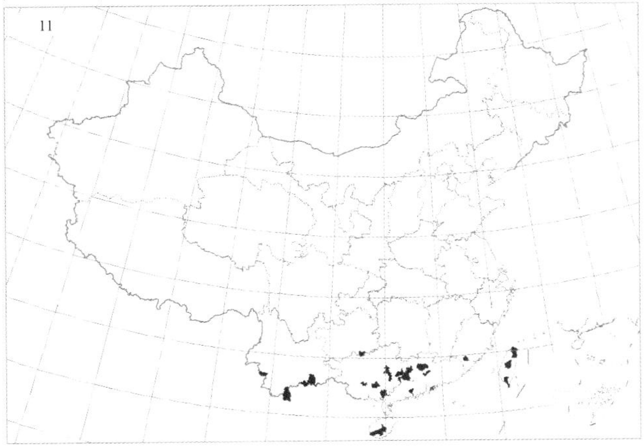

Cyatheaceae 桫椤科 *Alsophila* 桫椤属
Alsophila spinulosa Wall. ex Hook. 桫椤
Trees

18.5 (8.6~24.1); 10.2 (1.3~18.9); 25.6 (15.0~29.0); 18.6 (8.7~24.1); 162.9 (52.6~229.7); 0.2 (0.0~8.8); 956 (574~1322) ‖ 1592 (976~3495); 717 (448~1847); 138 (35~897) ‖ 68.3 (5.8~246.5); 949 (574~1216); 263.3 (135.3~627.0)

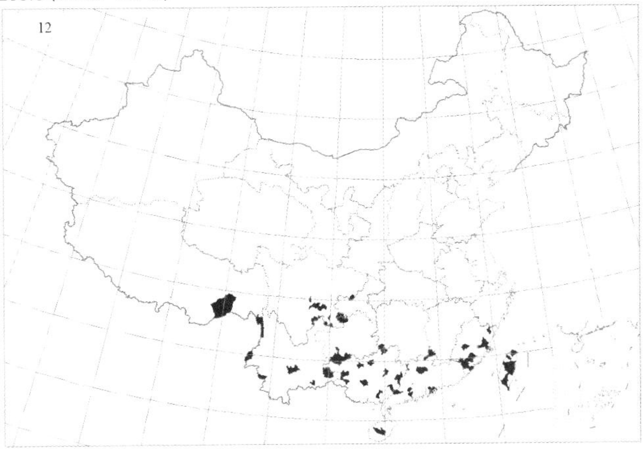

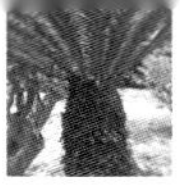

Cyatheaceae 桫椤科 *Sphaeropteris* 白桫椤属

Sphaeropteris brunoniana (Hook.) R.M. 白桫椤

Trees, evergreen

20.4 (13.3~24.8); 14.4 (5.4~20.1); 24.3 (19.6~28.0); 20.4 (13.4~24.8); 185.1 (100.4~237.6); 0.0 (0.0~0.0); 1040 (706~1385) ❙ 1512 (1242~1736); 723 (444~1007); 56 (41~77) ❙ 52.9 (-2.2~145.8); 977 (706~1212); 523.2 (439.7~659.0)

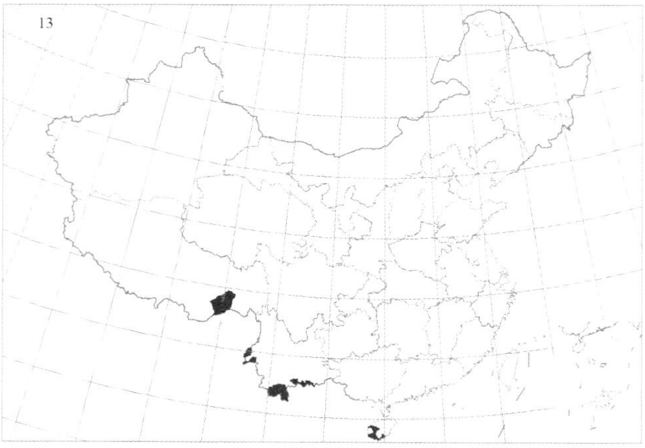

Cyatheaceae 桫椤科 *Sphaeropteris* 白桫椤属

Sphaeropteris lepifera (Hook.) R.M. 笔筒树

Trees, evergreen

18.5 (16.2~22.2); 13.3 (11.0~17.7); 22.7 (20.1~25.7); 18.5 (16.2~22.2); 162.1 (134.2~206.7); 0.0 (0.0~0.0); 895 (773~1123) ❙ 2659 (2208~2992); 1180 (750~1847); 264 (107~402) ❙ 201.3 (125.7~246.5); 895 (773~1123); 296.6 (260.5~363.7)

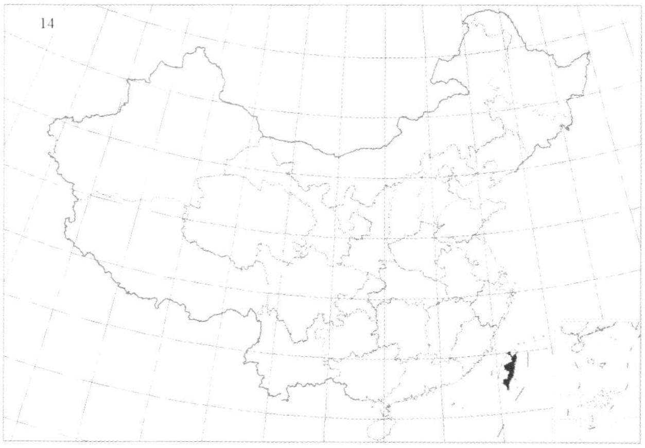

Dicksoniaceae 蚌壳蕨科 *Cibotium* 金毛狗属

Cibotium barometz (L.) J. Sm. 金毛狗

Shrubs, evergreen

18.8 (8.6~24.3); 10.0 (1.3~18.7); 26.2 (15.0~29.3); 18.9 (8.7~24.3); 166.7 (52.6~231.9); 0.1 (0.0~8.8); 979 (574~1338) ❙ 1503 (868~2680); 670 (312~1428); 118 (37~254) ❙ 55.4 (-34.3~246.5); 970 (574~1326); 271.9 (112.1~659.0)

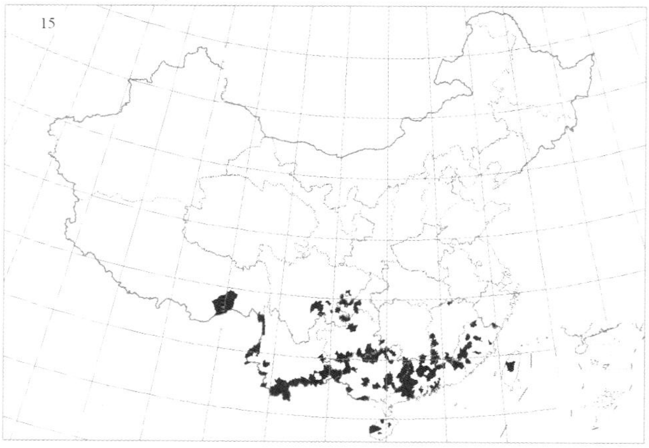

Blechnaceae 乌毛蕨科 *Brainea* 苏铁蕨属

Brainea insignis (Hook.) J. Sm. 苏铁蕨

Shrubs

20.0 (16.2~24.3); 12.7 (7.7~18.7); 25.8 (20.1~28.8); 20.2 (16.2~24.3); 181.8 (134.2~231.9); 0.0 (0.0~0.0); 1025 (773~1338) ❙ 1589 (1179~2680); 750 (392~1186); 103 (41~245) ❙ 57.4 (-11.9~246.5); 1013 (773~1216); 339.9 (139.6~659.0)

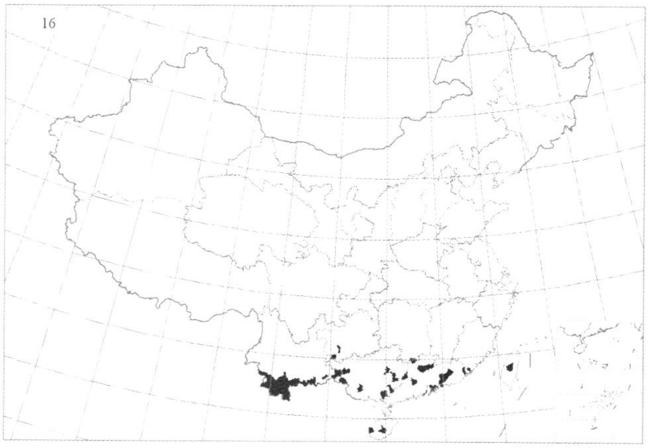

Blechnaceae 乌毛蕨科 *Diploblechnum* 扫把蕨属

Diploblechnum fraseri (A. Cunn.) De Vol 扫把蕨

Shrubs

20.3; 15.9; 23.6; 20.4; 184.4; 0.0; 993 ❙ 2853; 1618; 147 ❙ 190.4; 993; 330.3

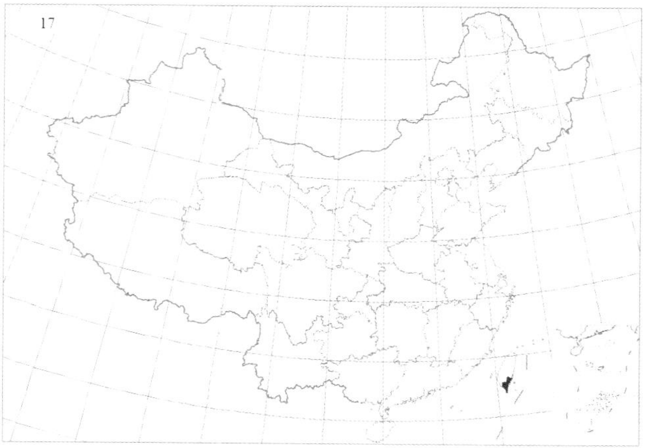

Cycadaceae 苏铁科 *Cycas* 苏铁属

Cycas balansae Warb. 宽叶苏铁

Trees, evergreen

18.3 (5.4~22.7); 11.4 (-2.1~15.1); 23.2 (12.3~28.6); 18.4 (5.8~22.8); 161.2 (31.5~213.1); 0.7 (0.0~25.6); 918 (481~1221) ❙ 1378 (741~2435); 752 (426~1428); 65 (28~173) ❙ 51.1 (5.8~99.5); 909 (481~1221); 391.6 (237.8~659.0)

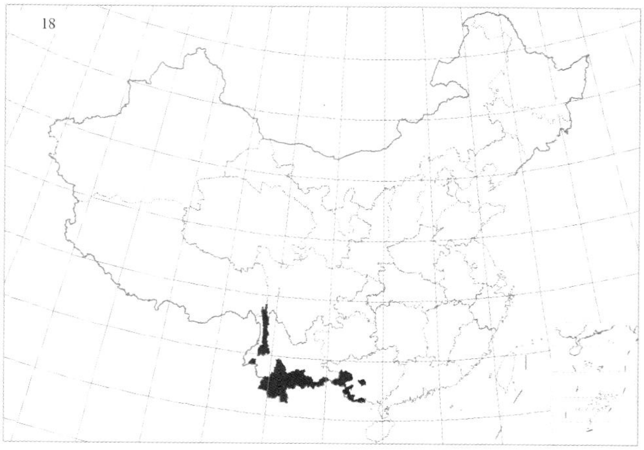

Cycadaceae 苏铁科 *Cycas* 苏铁属
Cycas changjiangensis N. Liu 葫芦苏铁
Trees, evergreen

24.0；18.1；28.4；24.1；229.0；0.0；1316 ▎1116；399；54 ▎−15.1；1039；398.8

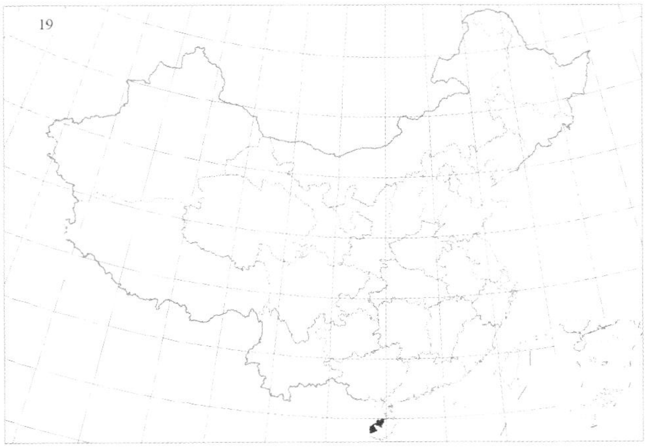

Cycadaceae 苏铁科 *Cycas* 苏铁属
Cycas debaoensis Y. C. Zhong et C. J. Chen 德保苏铁
Trees, evergreen

20.3；11.8；26.6；20.3；183.9；0.0；1042 ▎1326；726；65 ▎28.7；1041；315.1

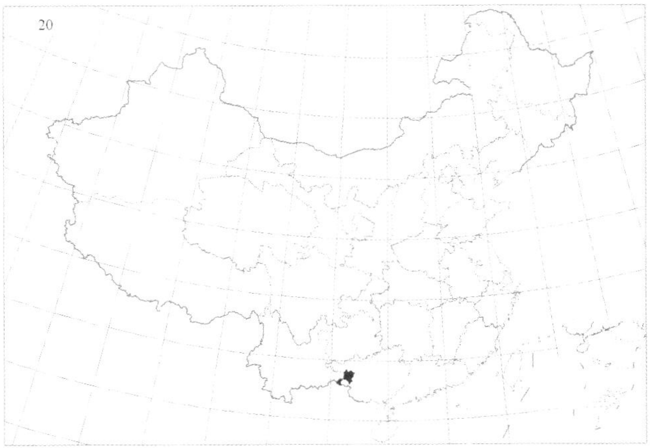

Cycadaceae 苏铁科 *Cycas* 苏铁属
Cycas ferruginea F. N. Wei 锈毛苏铁
Trees, evergreen

21.9 (21.5~22.2)；13.5 (12.9~13.9)；28.1 (27.4~28.4)；21.9 (21.6~22.3)；202.8 (199.1~207.5)；0.0 (0.0~0.0)；1162 (1131~1190) ▎1461 (1367~1581)；767 (708~850)；102 (87~116) ▎25.8 (15.4~39.9)；1162 (1131~1190)；298.2 (273.1~335.1)

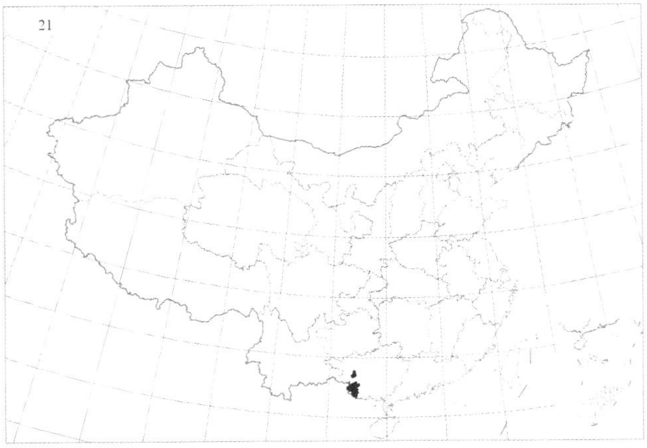

Cycadaceae 苏铁科 *Cycas* 苏铁属
Cycas hainanensis C. J. Chen 海南苏铁
Trees, evergreen

24.1 (23.0~24.8)；18.7 (17.6~20.1)；28.0 (26.7~28.8)；24.2 (23.1~24.8)；230.3 (217.0~237.6)；0.0 (0.0~0.0)；1326 (1217~1385) ▎1599 (1242~1854)；556 (444~622)；77 (41~115) ▎20.7 (−2.2~41.8)；1236 (1091~1360)；474.3 (284.3~620.0)

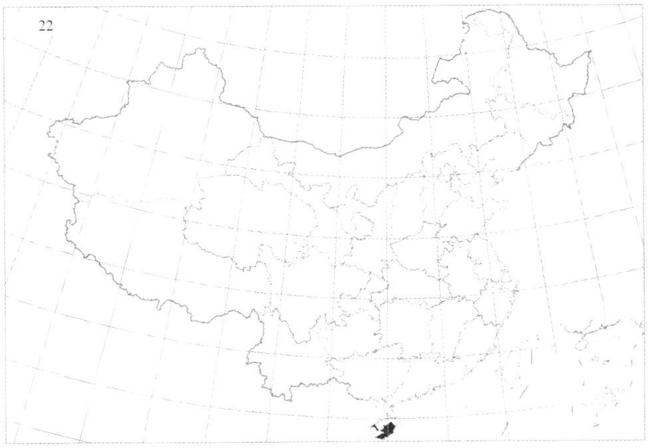

Cycadaceae 苏铁科 *Cycas* 苏铁属
Cycas hongheensis S. Y. Yang et S. L. Yang ex D. Y. Wang 灰干苏铁
Trees, evergreen

18.1；11.7；22.3；18.0；156.5；0.0；863 ▎1163；641；48 ▎34.8；863；406.7

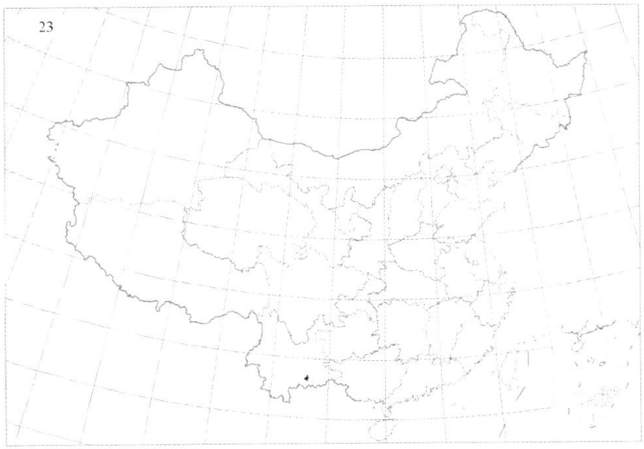

Cycadaceae 苏铁科 *Cycas* 苏铁属
Cycas micholitzii Dyer 叉叶苏铁
Trees, evergreen

21.3 (16.9~24.6)；13.9 (10.3~19.5)；26.6 (21.4~28.5)；21.3 (16.9~24.7)；195.6 (142.7~235.9)；0.0 (0.0~0.0)；1115 (813~1372) ▎1442 (1163~1854)；709 (567~850)；83 (48~116) ▎30.1 (5.8~44.9)；1108 (813~1360)；370.8 (273.1~551.6)

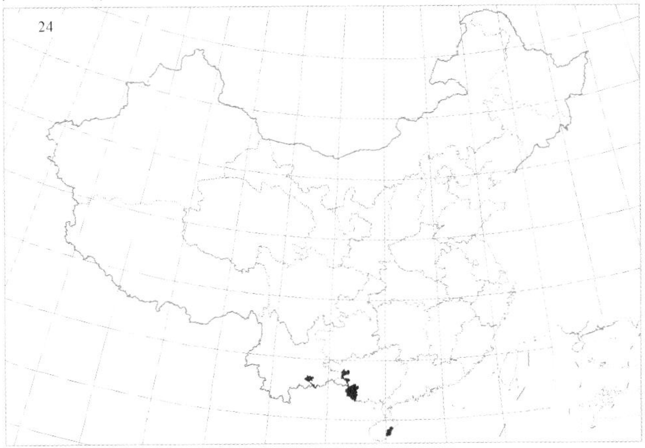

Cycadaceae 苏铁科 *Cycas* 苏铁属
Cycas miquelii Warb. 石山苏铁
Trees, evergreen

21.1 (18.3~22.2); 12.5 (9.4~13.9); 27.7 (25.5~28.6); 21.2 (18.4~22.3); 193.8 (160.9~207.5); 0.0 (0.0~0.0); 1112 (924~1190) ❙ 1435 (1250~1625); 747 (670~850); 104 (58~173) ❙ 29.3 (12.9~40.9); 1112 (924~1190); 282.6 (237.8~335.1)

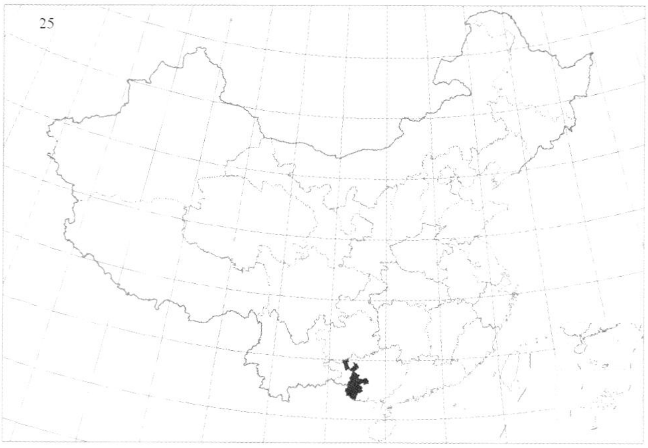

Cycadaceae 苏铁科 *Cycas* 苏铁属
Cycas multipinnata C. J. Chen et S. Y. Yang 多歧苏铁
Trees, evergreen

18.5 (16.5~21.3); 11.8 (9.0~14.2); 23.1 (21.4~27.9); 18.6 (16.6~21.3); 162.7 (139.2~196.0); 0.0 (0.0~0.0); 903 (810~1123) ❙ 1183 (1007~1522); 657 (539~818); 51 (44~68) ❙ 31.5 (5.8~60.1); 890 (810~1121); 381.7 (294.5~551.6)

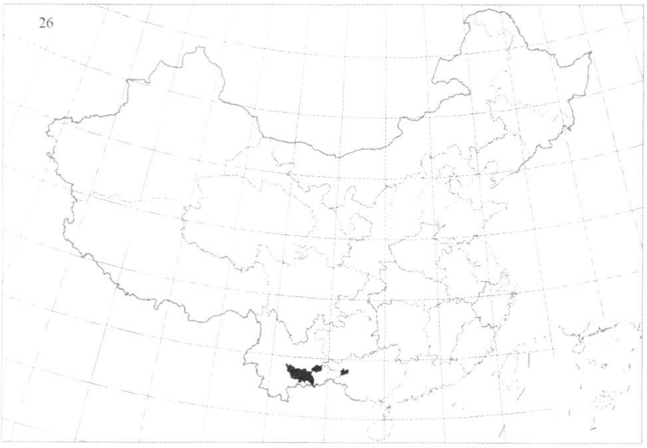

Cycadaceae 苏铁科 *Cycas* 苏铁属
Cycas panzhihuaensis L. Zhou et S. Y. Yang 攀枝花苏铁
Trees, evergreen

15.6 (11.0~18.9); 8.3 (3.7~12.0); 21.1 (17.0~24.3); 15.6 (11.1~19.0); 127.4 (74.6~167.8); 0.1 (0.0~1.8); 783 (640~929) ❙ 971 (797~1190); 581 (411~656); 25 (15~50) ❙ 25.4 (-14.2~54.7); 720 (638~797); 287.6 (218.0~347.3)

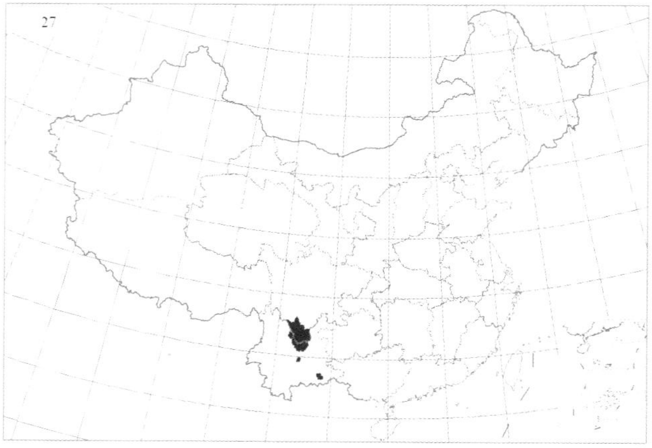

Cycadaceae 苏铁科 *Cycas* 苏铁属
Cycas pectinata Buch.-Ham. 篦齿苏铁
Trees, evergreen

20.1 (15.4~24.8); 14.1 (8.1~20.1); 24.1 (20.6~28.5); 20.2 (15.5~24.8); 182.2 (125.7~237.6); 0.0 (0.0~0.0); 1022 (767~1385) ❙ 1465 (1007~1725); 692 (392~959); 58 (41~109) ❙ 48.2 (-11.9~112.2); 970 (767~1321); 482.6 (274.9~659.0)

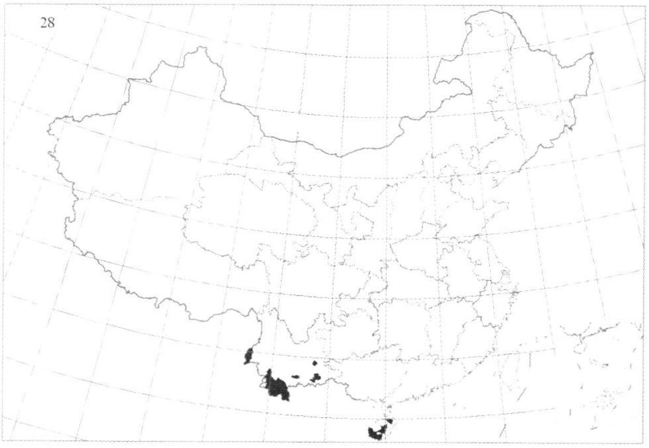

Cycadaceae 苏铁科 *Cycas* 苏铁属
Cycas revoluta Thunb. 苏铁
Trees, evergreen

18.8 (14.6~24.8); 9.2 (2.3~20.1); 27.6 (23.6~29.7); 18.9 (14.7~24.8); 167.5 (117.7~237.6); 0.5 (0.0~4.1); 1009 (779~1385) ❙ 1489 (939~1965); 583 (363~994); 153 (41~255) ❙ 50.9 (-12.0~137.5); 988 (779~1331); 256.2 (110.1~659.0)

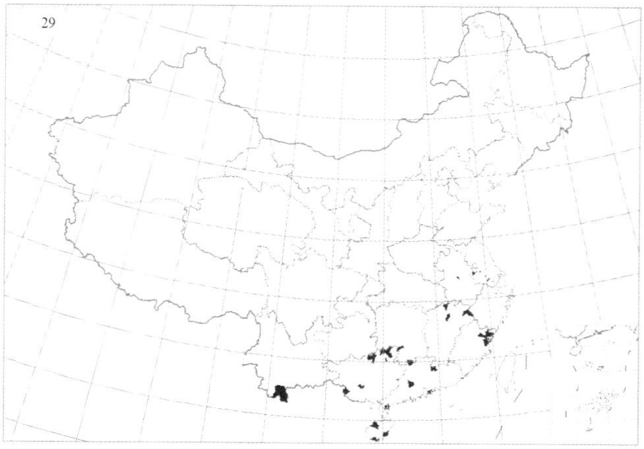

Cycadaceae 苏铁科 *Cycas* 苏铁属
Cycas segmentifida D. Y. Wang et C. Y. Deng 叉孢苏铁
Trees, evergreen

19.1 (16.8~21.3); 10.4 (7.8~12.6); 25.8 (23.6~27.9); 19.2 (16.8~21.3); 169.9 (141.5~196.0); 0.0 (0.0~0.0); 971 (836~1123) ❙ 1251 (1172~1334); 684 (634~729); 56 (43~73) ❙ 29.9 (5.8~59.6); 971 (836~1121); 309.0 (226.6~411.1)

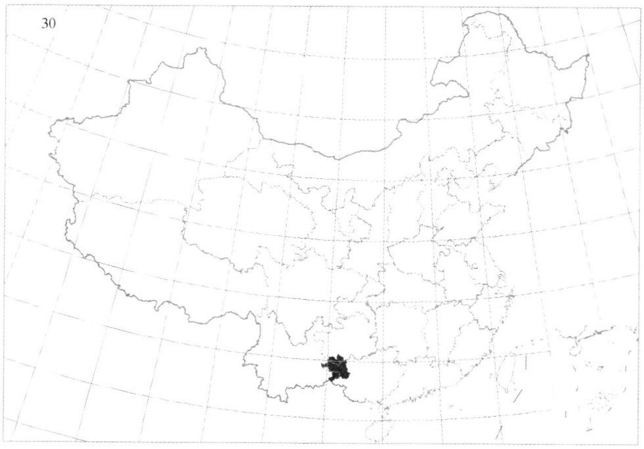

Cycadaceae 苏铁科 *Cycas* 苏铁属

Cycas szechuanensis W. C. Cheng et L. K. Fu 南盘江苏铁

Trees, evergreen

17.5 (13.3~19.8); 8.7 (3.2~11.7); 24.7 (20.7~27.7); 17.5 (13.2~19.8); 150.2 (100.9~177.9); 0.1 (0.0~2.5); 886 (723~1031) ❙ 1303 (950~1726); 631 (531~816); 89 (27~204) ❙ 46.9 (12.1~86.9); 885 (723~1031); 280.6 (133.8~411.1)

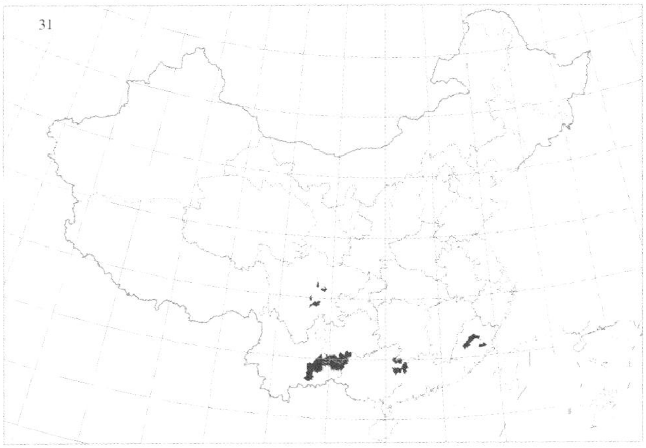

Cycadaceae 苏铁科 *Cycas* 苏铁属

Cycas taitungensis C. F. Shen et al. 台东苏铁

Trees, evergreen

18.1; 13.4; 21.6; 18.1; 157.3; 0.0; 853 ❙ 2657; 1310; 206 ❙ 212.0; 853; 306.5

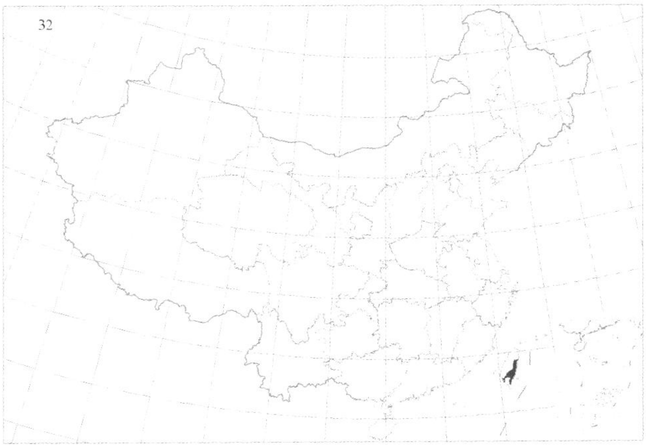

Cycadaceae 苏铁科 *Cycas* 苏铁属

Cycas taiwaniana Carruth. 广东苏铁

Trees, evergreen

21.2 (17.1~24.8); 13.8 (6.6~20.1); 27.1 (21.7~29.0); 21.3 (17.1~24.8); 195.3 (145.0~237.6); 0.0 (0.0~0.0); 1122 (816~1385) ❙ 1611 (1059~2713); 655 (392~1388); 120 (41~244) ❙ 48.2 (-12.0~214.4); 1074 (816~1329); 335.4 (180.3~620.0)

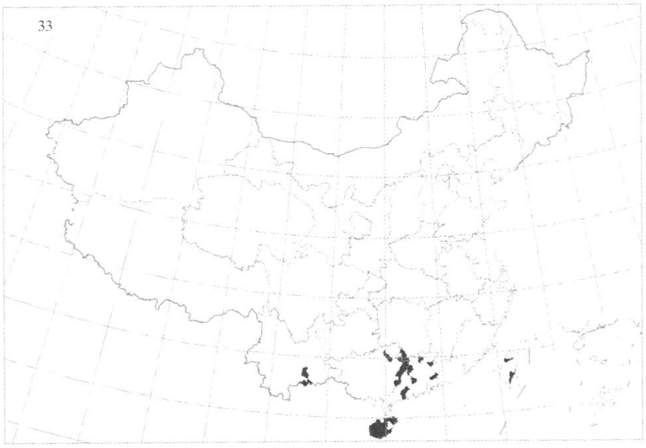

Ginkgoaceae 银杏科 *Ginkgo* 银杏属

Ginkgo biloba L. 银杏

Trees, deciduous

15.6 (6.2~21.8); 4.3 (-11.9~12.7); 25.9 (15.8~29.9); 15.7 (7.0~21.8); 130.3 (45.2~202.0); 2.8 (0.0~53.4); 844 (522~1168) ❙ 1295 (428~1997); 527 (264~792); 123 (12~259) ❙ 52.6 (-37.2~130.7); 831 (428~1134); 216.0 (110.1~487.2)

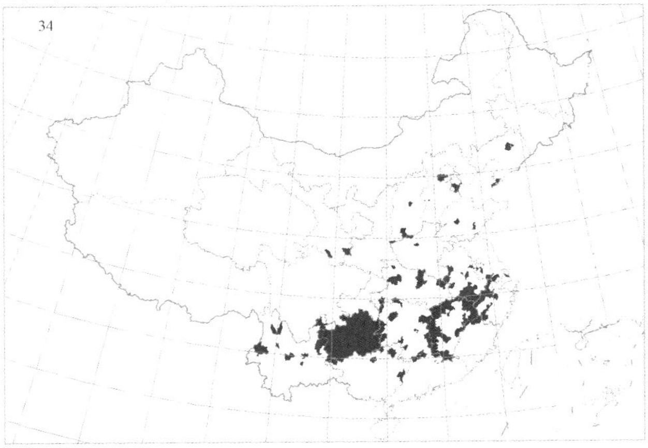

Pinaceae 松科 *Abies* 冷杉属

Abies beshanzuensis M. H. Wu 百山祖冷杉

Trees, evergreen

15.3; 4.9; 25.4; 15.4; 124.4; 0.1; 808 ❙ 1888; 687; 218 ❙ 133.9; 808; 275.4

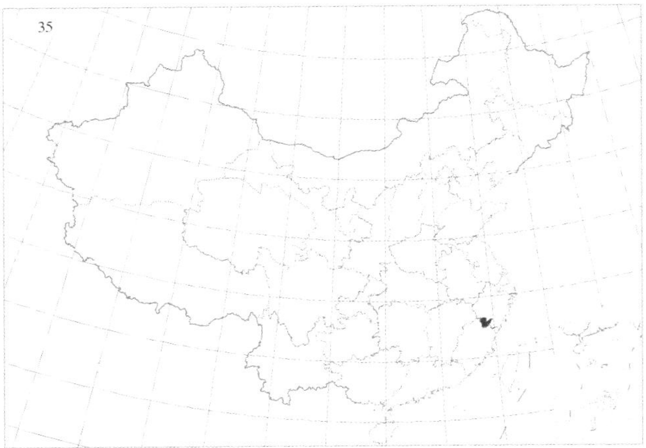

Pinaceae 松科 *Abies* 冷杉属

Abies beshanzuensis var. ziyuanensis (L. K. Fu et S. L. Mo) L. K. Fu et Nan Li 资源冷杉

Trees, evergreen

15.8 (14.3~17.6); 4.7 (3.3~6.6); 26.2 (24.3~28.1); 15.8 (14.3~17.6); 130.8 (114.1~151.3); 0.7 (0.0~2.3); 840 (765~928) ❙ 1549 (1415~1671); 512 (419~591); 197 (177~221) ❙ 85.6 (55.1~106.4); 834 (765~922); 225.0 (177.4~258.0)

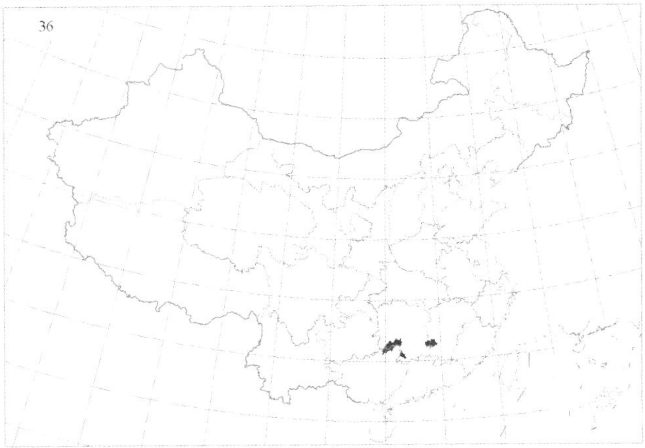

G

P

Pinaceae 松科 *Abies* 冷杉属

Abies chayuensis C. Y. Cheng et L. K. Fu 察隅冷杉

Trees, evergreen

5.3; -2.4; 12.2; 5.7; 30.5; 26.2; 479 ❙ 961; 558; 31 ❙ 100.7; 479; 234.2

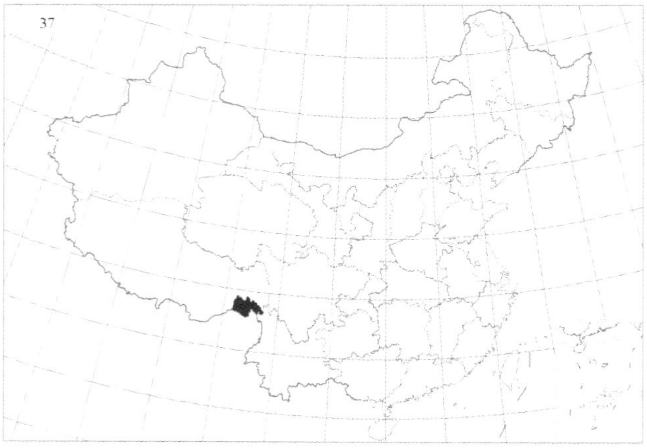

Pinaceae 松科 *Abies* 冷杉属

Abies chensiensis Tiegh. 秦岭冷杉

Trees, evergreen

11.6 (1.9~14.7); -0.7 (-9.2~2.3); 23.2 (11.7~27.4); 11.8 (4.3~14.8); 93.5 (21.1~124.9); 14.1 (4.9~57.8); 703 (427~852) ❙ 822 (590~1246); 379 (298~525); 32 (12~83) ❙ 18.2 (-14.2~101.4); 690 (427~817); 245.2 (140.4~334.4)

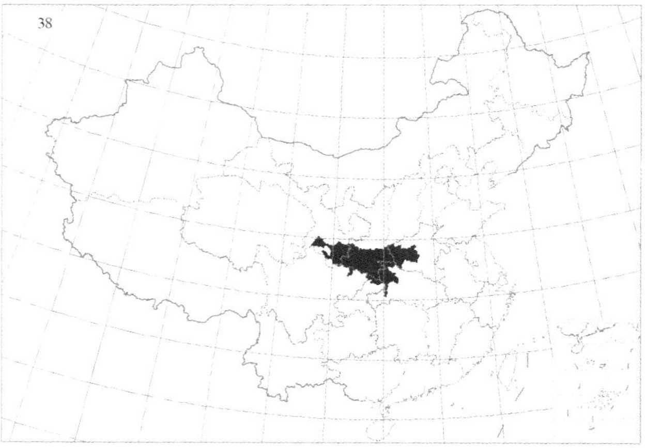

Pinaceae 松科 *Abies* 冷杉属

Abies delavayi Franch. 苍山冷杉

Trees, evergreen

11.3 (5.3~15.3); 4.3 (-2.4~8.7); 17.0 (12.2~20.5); 11.4 (5.7~15.4); 80.6 (30.5~124.3); 4.9 (0.0~26.2); 644 (479~761) ❙ 1089 (741~1736); 626 (426~1007); 42 (15~70) ❙ 69.6 (28.3~145.8); 640 (479~730); 347.9 (234.2~487.2)

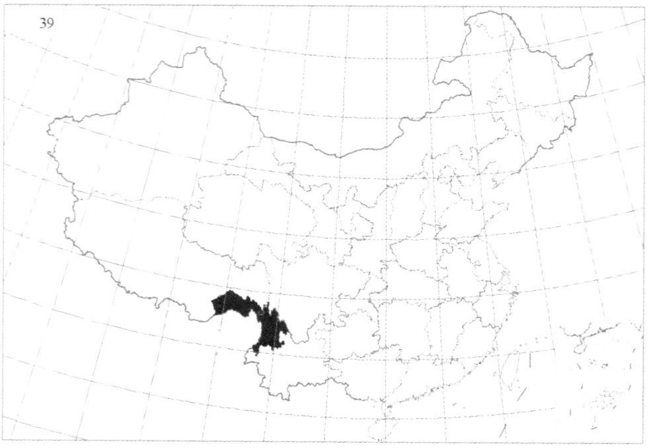

Pinaceae 松科 *Abies* 冷杉属

Abies delavayi var. motuoensis C. Y. Cheng et L. K. Fu 墨脱冷杉

Trees, evergreen

10.3; 2.4; 16.7; 10.4; 72.8; 9.0; 619 ❙ 1376; 794; 38 ❙ 119.8; 619; 348.4

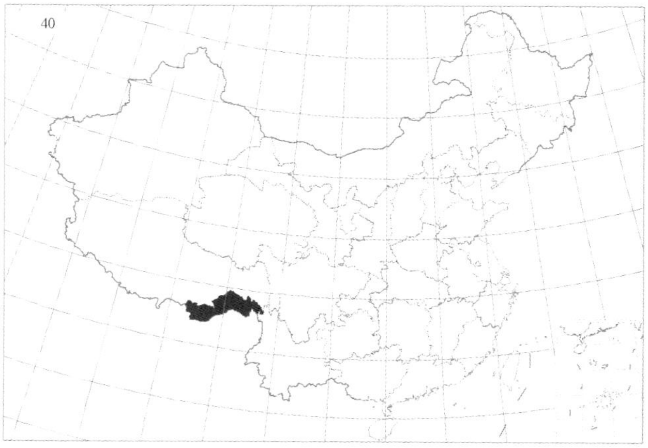

Pinaceae 松科 *Abies* 冷杉属

Abies densa Griff. 锡金冷杉

Trees, evergreen

0.3; -8.6; 8.3; 2.9; 9.6; 65.4; 368 ❙ 389; 246; 11 ❙ 0.0; 368; 142.5

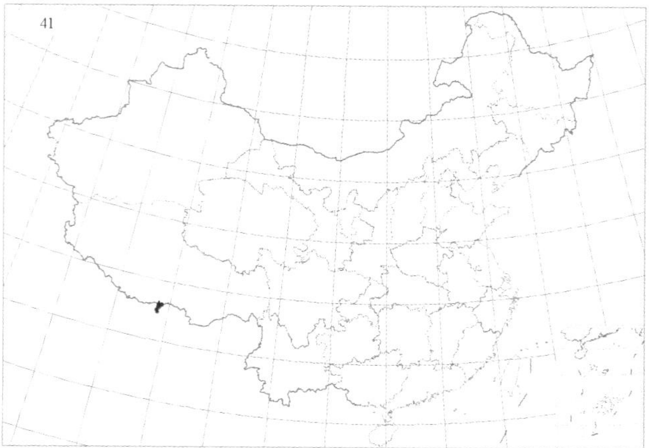

Pinaceae 松科 *Abies* 冷杉属

Abies ernestii Rehder 黄果冷杉

Trees, evergreen

5.3 (-1.0~18.5); -3.6 (-11.8~10.5); 13.1 (8.5~24.3); 6.5 (2.7~18.3); 39.0 (8.3~160.0); 34.5 (0.0~79.5); 497 (363~906) ❙ 845 (535~1337); 477 (317~727); 23 (9~71) ❙ 71.6 (5.5~128.6); 494 (363~786); 292.1 (200.8~407.1)

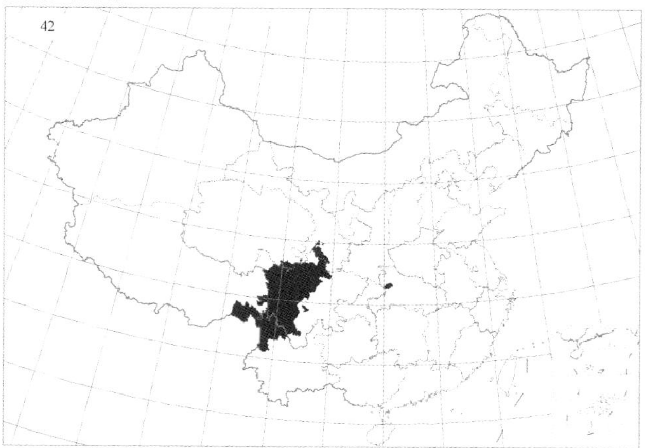

Pinaceae 松科　*Abies* 冷杉属

Abies ernestii var. salouenensis (Bordères et Gaussen) W. C. Cheng et L. K. Fu 大黄果冷杉　Trees, evergreen

9.7 (2.2~18.5); 2.3 (-6.3~10.5); 15.9 (9.5~24.3); 9.9 (3.9~18.3); 67.6 (15.8~160.0); 11.0 (0.0~48.6); 605 (411~906) | 1004 (729~1337); 593 (426~727); 31 (11~69) | 69.1 (5.5~121.7); 596 (411~786); 312.5 (208.8~407.1)

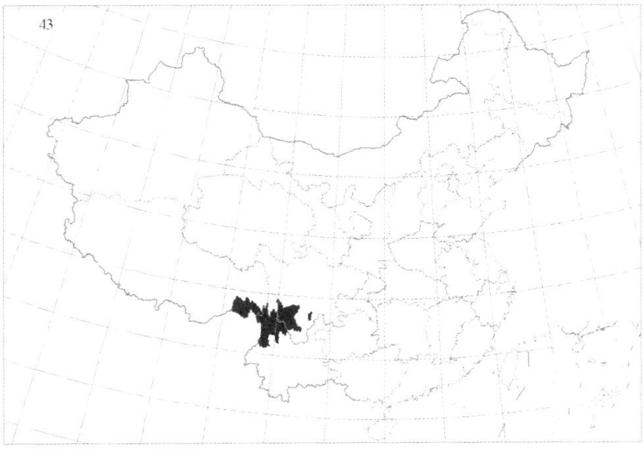

Pinaceae 松科　*Abies* 冷杉属

Abies fabri (Mast.) Craib 冷杉　Trees, evergreen

10.3 (2.0~17.4); 1.1 (-7.2~7.5); 18.6 (9.9~27.1); 10.6 (3.8~17.4); 77.5 (15.4~149.1); 13.3 (0.0~51.4); 641 (413~915) | 1071 (812~1414); 597 (420~817); 31 (12~53) | 72.3 (8.3~128.6); 641 (413~915); 243.8 (127.1~337.3)

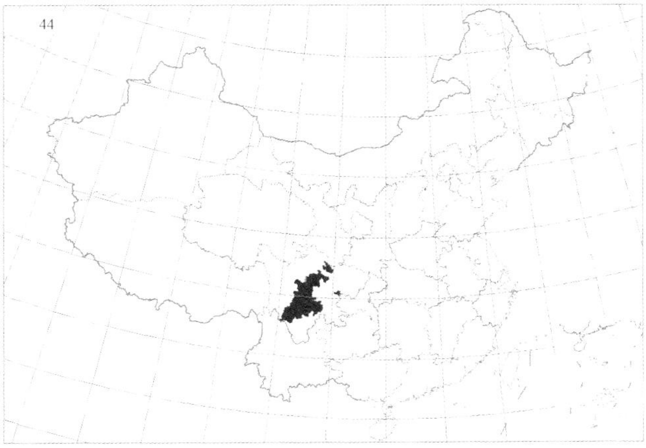

Pinaceae 松科　*Abies* 冷杉属

Abies fanjingshanensis W. L. Huang, Y. L. Tu et S. T. Fang 梵净山冷杉　Trees, evergreen

15.0; 4.0; 25.6; 15.0; 121.1; 1.2; 801 | 1231; 479; 86 | 53.6; 801; 191.1

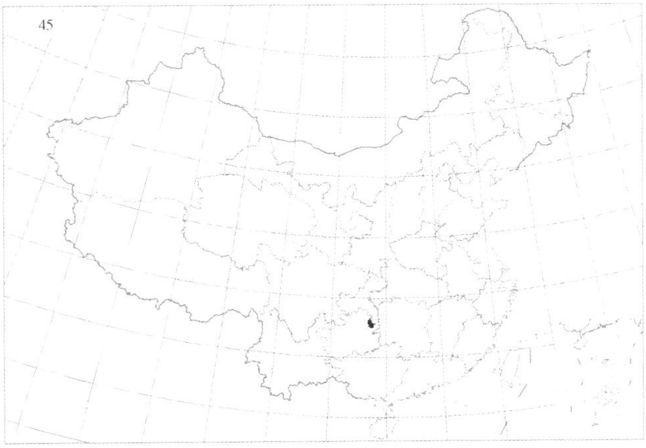

Pinaceae 松科　*Abies* 冷杉属

Abies fargesii Franch. 巴山冷杉　Trees, evergreen

11.2 (0.8~15.1); -1.2 (-10.6~2.3); 22.8 (10.7~27.5); 11.5 (3.8~15.2); 91.7 (16.2~128.0); 16.5 (4.9~66.0); 697 (411~865) | 793 (375~1246); 370 (203~525); 31 (8~83) | 15.9 (-35.3~101.4); 667 (375~817); 236.1 (95.7~344.4)

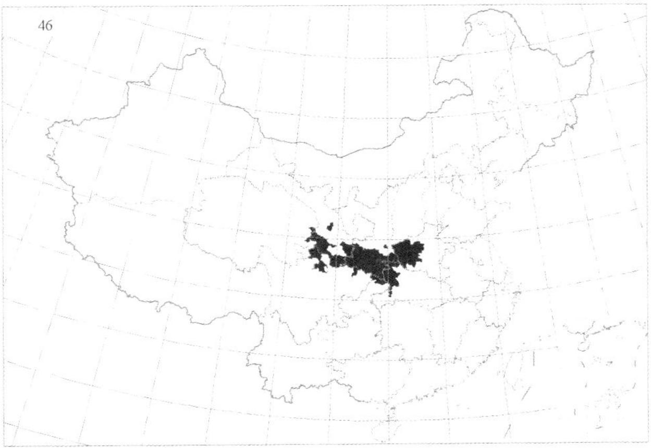

Pinaceae 松科　*Abies* 冷杉属

Abies fargesii var. faxoniana (Rehder et E. H. Wilson) Liu 岷江冷杉　Trees, evergreen

6.0 (-1.0~14.8); -4.6 (-11.8~4.5); 15.4 (8.5~25.3); 7.3 (2.7~14.8); 48.2 (8.3~119.2); 35.5 (0.5~79.5); 529 (363~795) | 787 (429~1358); 419 (228~692); 20 (8~88) | 52.1 (-22.9~128.6); 522 (363~795); 245.6 (85.5~357.7)

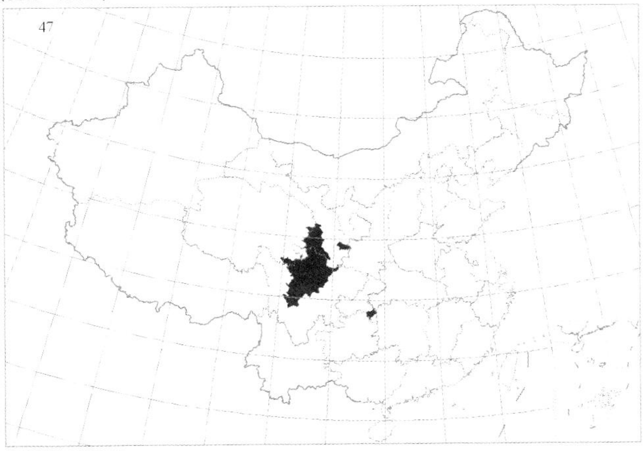

Pinaceae 松科　*Abies* 冷杉属

Abies ferreana Bordères et Gaussen 中甸冷杉　Trees, evergreen

10.7 (3.0~18.5); 3.5 (-5.6~10.5); 16.8 (10.5~24.3); 10.9 (4.3~18.3); 77.3 (20.1~160.0); 8.0 (0.0~43.9); 635 (423~906) | 992 (628~1337); 586 (410~727); 33 (11~69) | 58.3 (5.5~93.0); 620 (423~786); 327.6 (236.3~407.1)

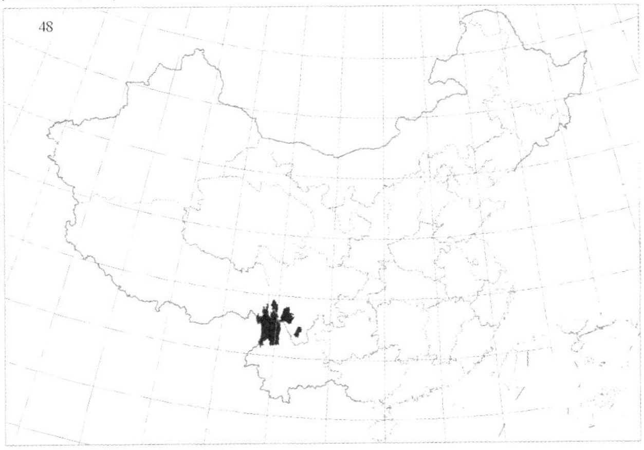

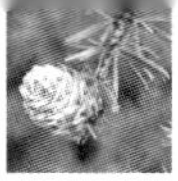

Pinaceae 松科　*Abies* 冷杉属

Abies ferreana var. longibracteata L. K. Fu et Nan 长苞中甸冷杉

Trees, evergreen

6.1；-1.5；12.7；6.3；35.2；21.8；496 | 801；496；21 | 61.4；496；282.9

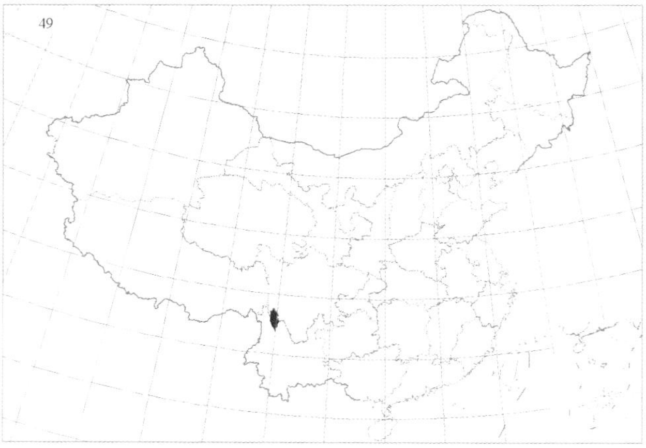

Pinaceae 松科　*Abies* 冷杉属

Abies forrestii Coltm.-Rog. 川滇冷杉

Trees, evergreen

8.9 (-1.0~15.4)；1.0 (-11.8~8.2)；15.7 (8.3~21.1)；9.4 (2.7~15.4)；62.9 (8.3~125.4)；15.5 (0.0~79.5)；586 (363~770) | 939 (389~1243)；557 (246~696)；24 (9~58) | 61.0 (0.0~121.7)；580 (363~717)；287.3 (142.5~384.7)

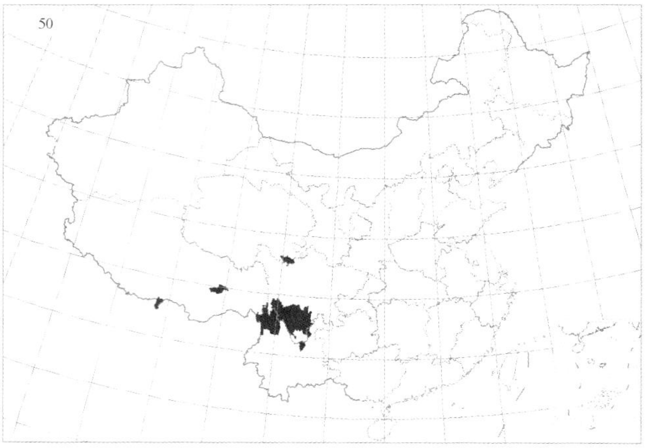

Pinaceae 松科　*Abies* 冷杉属

Abies georgei Orr 长苞冷杉

Trees, evergreen

7.6 (1.4~20.5)；-0.1 (-7.6~14.9)；14.2 (9.2~23.7)；8.3 (3.5~20.5)；54.3 (13.4~185.9)；22.3 (0.0~55.6)；552 (402~992) | 890 (473~1464)；522 (287~727)；27 (7~69) | 61.7 (5.0~125.9)；546 (402~959)；292.4 (169.5~570.1)

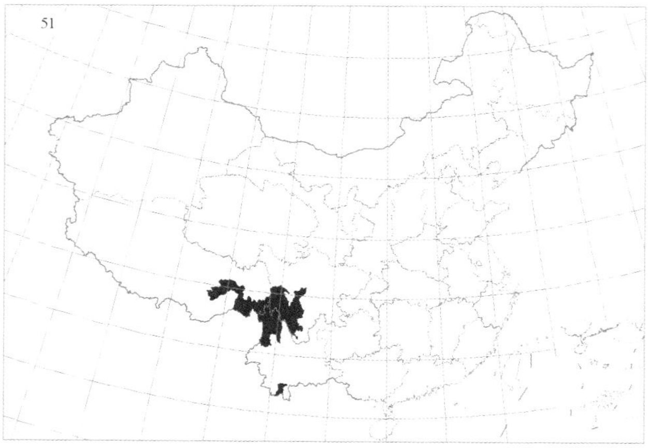

Pinaceae 松科　*Abies* 冷杉属

Abies georgei var. smithii (Viguié et Gaussen) W. C. Cheng et L. K. Fu 急尖长苞冷杉

Trees, evergreen

7.5 (1.4~15.4)；-0.2 (-7.6~8.2)；14.1 (9.2~21.1)；8.1 (3.5~15.4)；52.2 (13.4~125.4)；21.6 (0.0~55.6)；546 (402~770) | 890 (473~1337)；532 (287~727)；25 (7~69) | 63.9 (5.0~125.9)；541 (402~707)；283.5 (169.5~407.1)

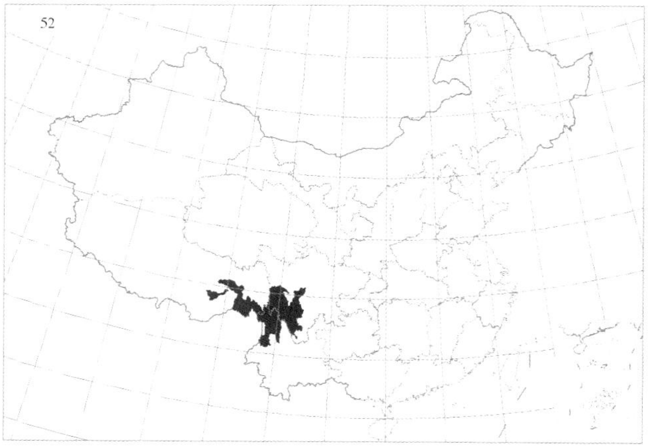

Pinaceae 松科　*Abies* 冷杉属

Abies holophylla Maxim. 杉松

Trees, evergreen

5.2 (1.4~9.5)；-14.2 (-19.9~-7.2)；21.9 (18.4~25.0)；8.7 (6.4~10.8)；69.4 (45.8~91.5)；65.9 (35.9~93.6)；605 (501~701) | 748 (384~1073)；465 (276~691)；27 (5~40) | 24.1 (-37.4~74.1)；594 (384~698)；275.2 (104.4~346.5)

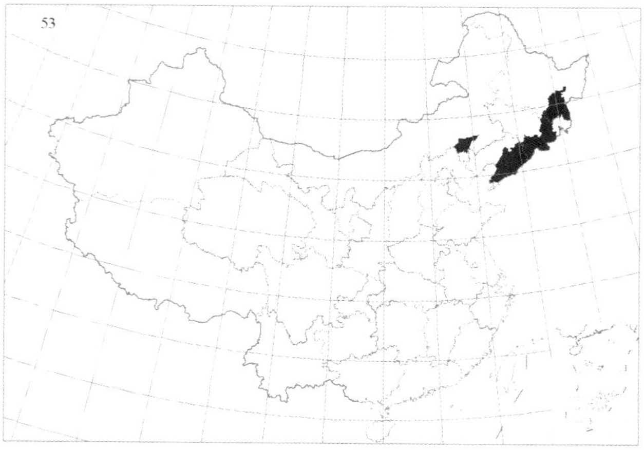

Pinaceae 松科　*Abies* 冷杉属

Abies kawakamii (Hayata) T. Ito 台湾冷杉

Trees, evergreen

18.3 (16.2~21.7)；12.7 (11.0~15.3)；22.9 (20.1~27.0)；18.3 (16.2~21.7)；160.1 (134.2~200.2)；0.0 (0.0~0.0)；894 (773~1118) | 2450 (1836~2859)；1097 (947~1412)；237 (110~402) | 182.9 (64.2~246.5)；886 (773~1075)；258.4 (202.8~303.6)

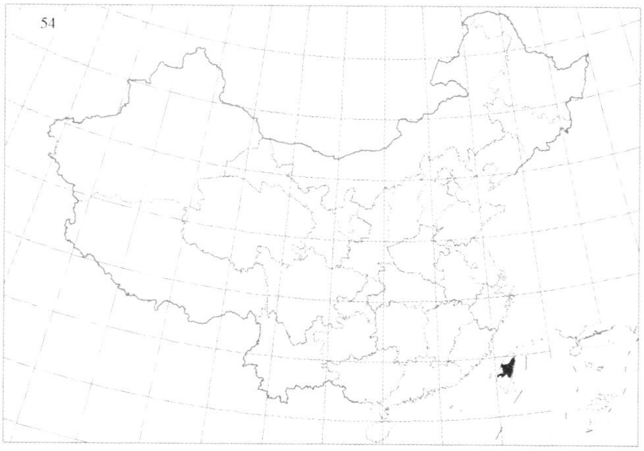

Pinaceae 松科 *Abies* 冷杉属
Abies nephrolepis (Trautv. ex Maxim.) Maxim. 臭冷杉
Trees, evergreen

5.2 (−0.2~10.2); −14.1 (−23.8~−6.2); 21.7 (18.4~24.8); 8.7 (6.4~11.4); 68.6 (45.8~95.3); 66.3 (32.8~112.6); 603 (501~722) ❙ 683 (397~1073); 430 (254~691); 24 (9~40) ❙ 13.8 (−36.6~74.1); 574 (397~690); 266.8 (111.2~346.5)

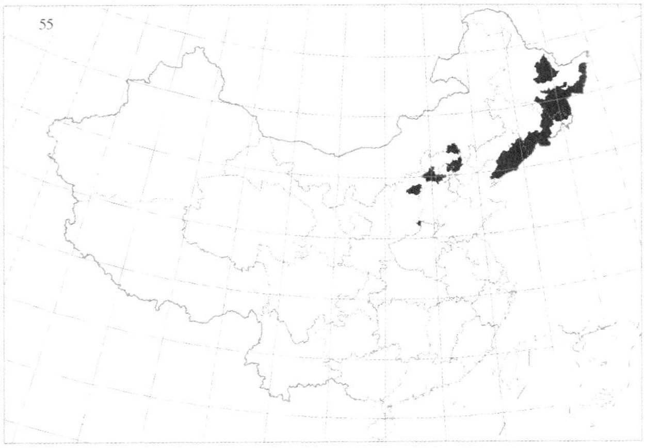

Pinaceae 松科 *Abies* 冷杉属
Abies nukiangensis C. Y. Cheng et L. K. Fu 怒江冷杉
Trees, evergreen

10.7; 3.6; 16.5; 10.7; 72.3; 3.7; 628 ❙ 1157; 647; 53 ❙ 83.7; 628; 374.0

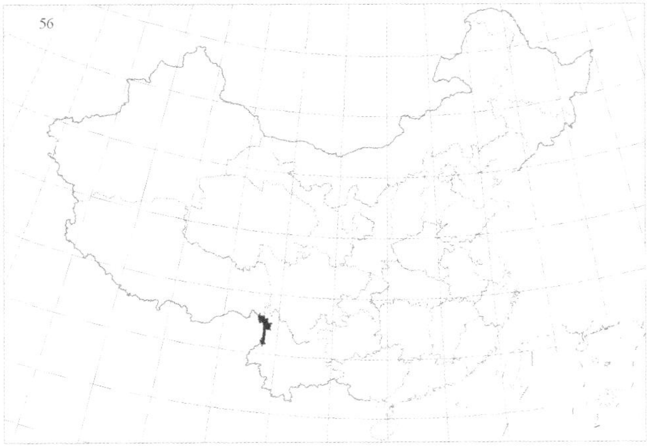

Pinaceae 松科 *Abies* 冷杉属
Abies recurvata Mast. 紫果冷杉
Trees, evergreen

4.7 (−0.3~14.8); −5.2 (−10.9~3.8); 13.6 (9.0~25.3); 6.0 (3.0~14.8); 36.2 (10.4~119.2); 39.0 (1.4~73.4); 487 (378~795) ❙ 810 (637~1358); 423 (306~546); 21 (11~88) ❙ 71.2 (−0.2~128.6); 487 (378~795); 275.2 (187.9~357.7)

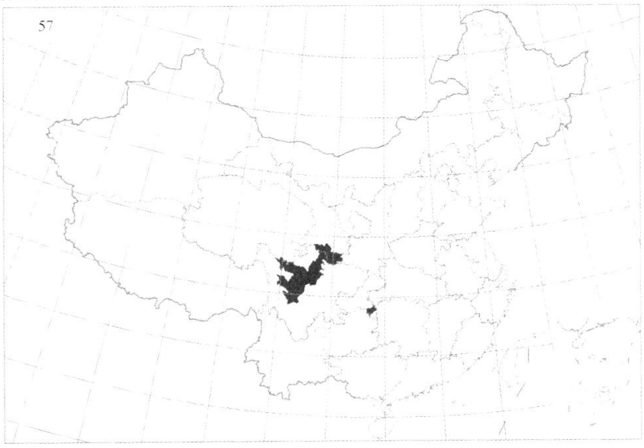

Pinaceae 松科 *Abies* 冷杉属
Abies sibirica Ledeb. 西伯利亚冷杉
Trees, evergreen

1.9; −16.2; 18.3; 6.8; 49.3; 86.5; 536 ❙ 216; 91; 27 ❙ −59.4; 216; 112.8

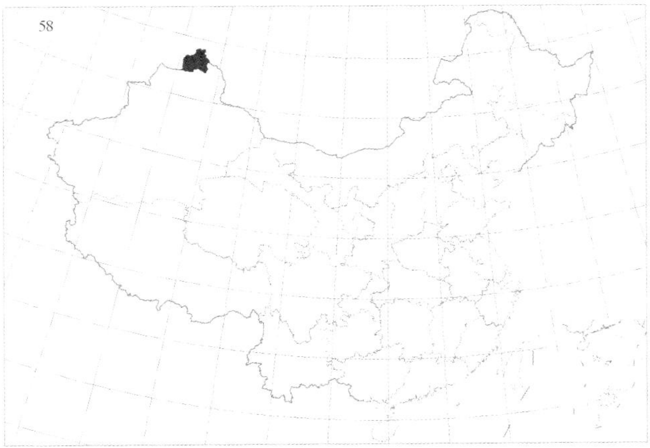

Pinaceae 松科 *Abies* 冷杉属
Abies spectabilis (D. Don) Spach 西藏冷杉
Trees, evergreen

−0.4 (−1.8~3.3); −9.0 (−10.5~−5.4); 7.6 (6.1~11.1); 2.6 (1.8~4.6); 7.6 (1.9~22.3); 71.3 (41.8~82.8); 352 (329~427) ❙ 359 (311~473); 237 (193~287); 16 (7~33) ❙ −0.7 (−6.3~5.0); 343 (311~427); 101.3 (61.6~230.7)

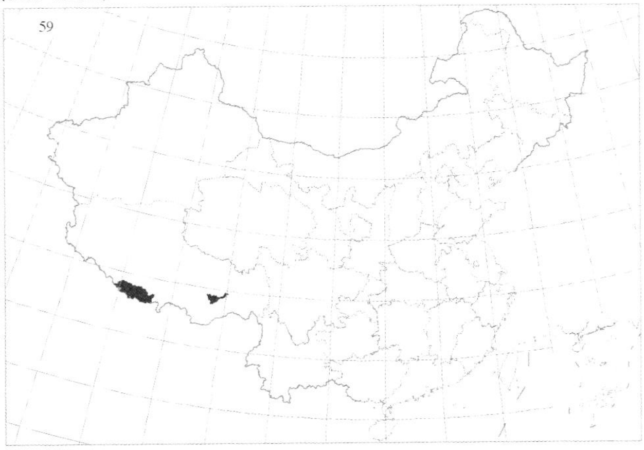

Pinaceae 松科 *Abies* 冷杉属
Abies squamata Mast. 鳞皮冷杉
Trees, evergreen

2.2 (−1.2~6.6); −7.0 (−11.8~−1.3); 10.4 (8.5~13.3); 4.1 (2.6~6.8); 18.4 (8.1~39.6); 51.1 (19.7~82.3); 418 (347~514) ❙ 724 (491~984); 426 (301~601); 15 (9~28) ❙ 72.8 (27.1~128.6); 418 (347~514); 275.6 (130.1~357.7)

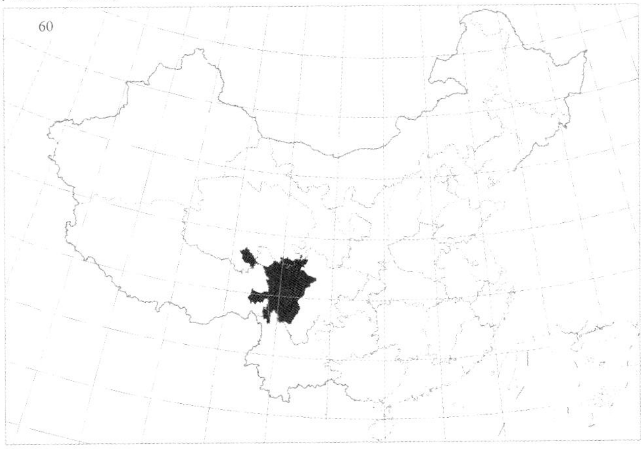

Pinaceae 松科 *Abies* 冷杉属
Abies yuanbaoshanensis Y. J. Lu et L. K. Fu 元宝山冷杉
Trees, evergreen

17.6；7.3；26.3；17.6；151.6；0.0；909 ❙ 1496；639；142 ❙ 64.7；909；225.9

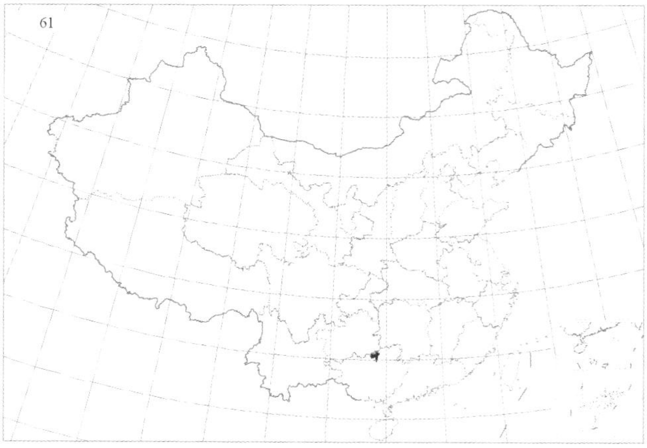

Pinaceae 松科 *Cathaya* 银杉属
Cathaya argyrophylla Chun et Kuang 银杉
Trees, evergreen

15.3 (13.4~18.0)；4.6 (2.7~8.5)；25.3 (23.5~27.2)；15.3 (13.4~18.0)；124.6 (104.5~156.4)；1.0 (0.0~3.7)；811 (735~913) ❙ 1431 (1132~1742)；531 (465~672)；139 (65~221) ❙ 76.9 (32.5~128.9)；811 (735~913)；213.6 (176.4~257.1)

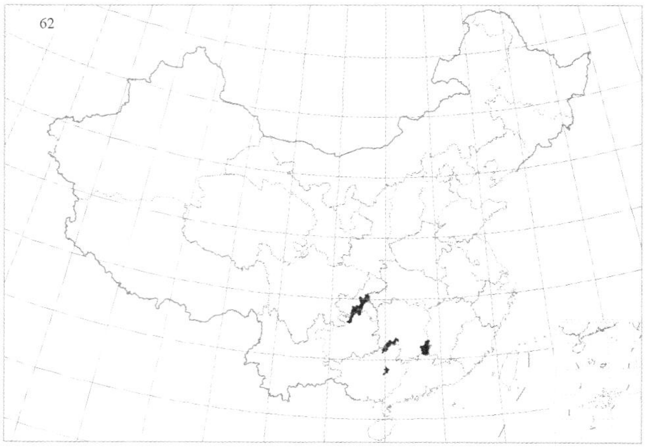

Pinaceae 松科 *Cedrus* 雪松属
Cedrus deodara (Roxburgh) G. Don 雪松
Trees, evergreen

9.3；1.5；15.9；9.6；65.5；13.1；592 ❙ 1348；783；41 ❙ 123.3；592；340.8

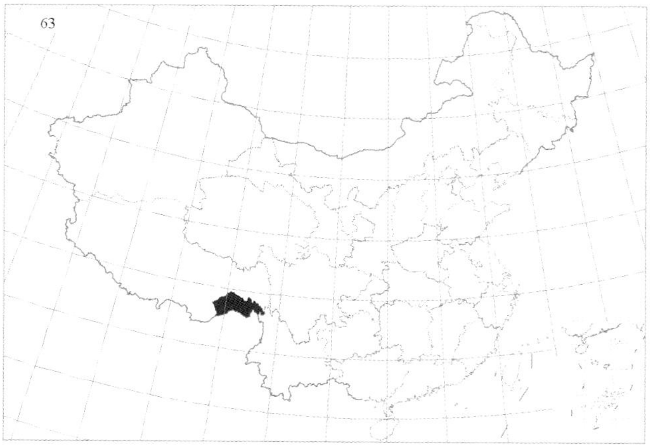

Pinaceae 松科 *Keteleeria* 油杉属
Keteleeria davidiana (Bertrand) Beissn. 铁坚杉
Trees, evergreen

15.1 (9.1~22.2)；4.3 (-2.5~17.7)；25.1 (17.7~29.9)；15.2 (9.2~22.2)；125.0 (64.1~206.7)；3.1 (0.0~20.6)；819 (595~1130) ❙ 1252 (677~2993)；530 (336~1847)；94 (14~547) ❙ 52.2 (-0.2~203.4)；816 (595~1130)；230.3 (141.7~506.2)

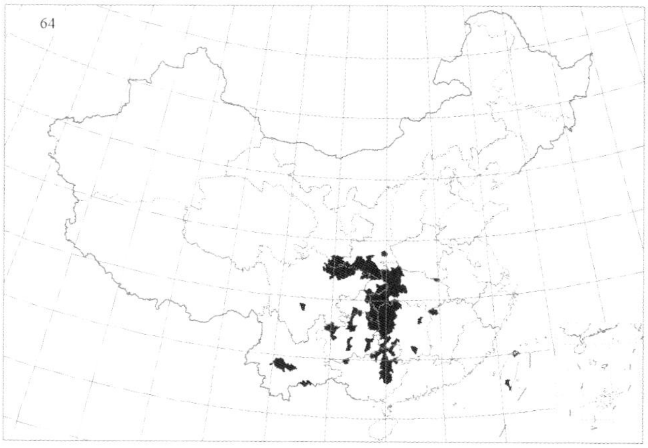

Pinaceae 松科 *Keteleeria* 油杉属
Keteleeria davidiana var. calcarea (C. Y. Cheng et L. K. Fu) Silba 黄枝油杉
Trees, evergreen

17.6 (14.6~20.3)；7.2 (3.8~10.2)；26.6 (24.4~28.7)；17.6 (14.7~20.3)；151.8 (117.7~183.8)；0.1 (0.0~1.5)；919 (779~1078) ❙ 1521 (1283~1776)；607 (476~718)；158 (73~197) ❙ 66.2 (30.9~101.8)；919 (779~1078)；235.0 (180.3~322.1)

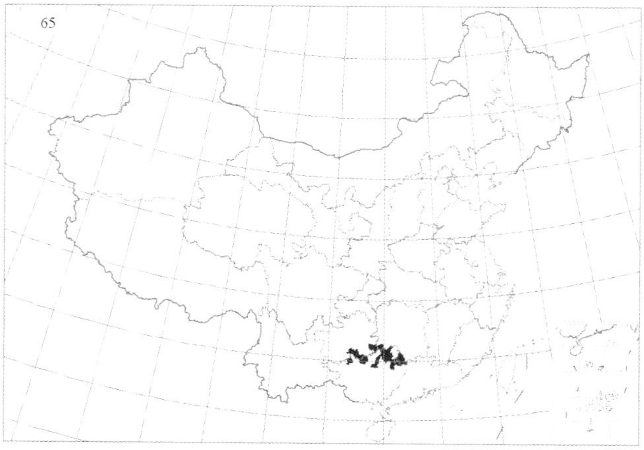

Pinaceae 松科 *Keteleeria* 油杉属
Keteleeria evelyniana Mast. 云南油杉
Trees, evergreen

15.2 (5.4~21.3)；7.7 (-2.1~13.3)；21.1 (12.3~27.9)；15.3 (5.8~21.3)；124.1 (31.5~196.0)；0.9 (0.0~25.6)；777 (481~1123) ❙ 1075 (741~1495)；600 (411~836)；44 (12~107) ❙ 39.3 (-14.2~78.3)；760 (481~1121)；294.9 (142.9~516.7)

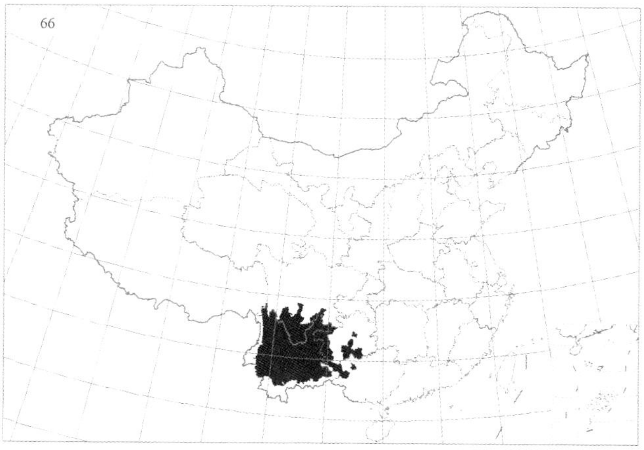

Pinaceae 松科 *Keteleeria* 油杉属
Keteleeria fortunei (A. Murray bis) Carrière 油杉
Trees, evergreen

19.9 (14.9~23.0); 11.2 (3.6~15.2); 27.4 (23.1~29.0); 20.0 (15.0~23.0); 179.7 (120.8~216.2); 0.0 (0.0~1.6); 1041 (789~1247) ❙ 1507 (1070~2435); 655 (356~1428); 144 (43~257) ❙ 46.6 (-5.0~147.0); 1033 (789~1247); 247.5 (123.9~354.5)

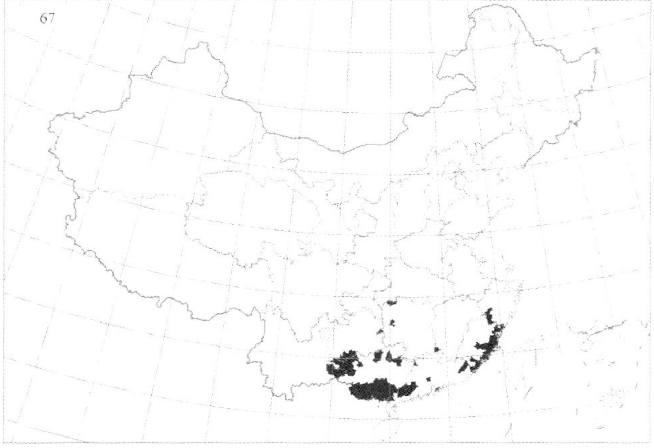

Pinaceae 松科 *Keteleeria* 油杉属
Keteleeria fortunei var. cyclolepis (Flous) Silba 江南油杉
Trees, evergreen

18.0 (14.3~21.9); 7.9 (3.3~13.5); 26.8 (23.4~29.4); 18.0 (14.3~22.0); 156.1 (113.2~203.8); 0.1 (0.0~2.3); 938 (761~1176) ❙ 1575 (1172~1997); 619 (403~848); 166 (43~254) ❙ 70.4 (5.3~147.0); 934 (761~1176); 251.0 (151.0~411.1)

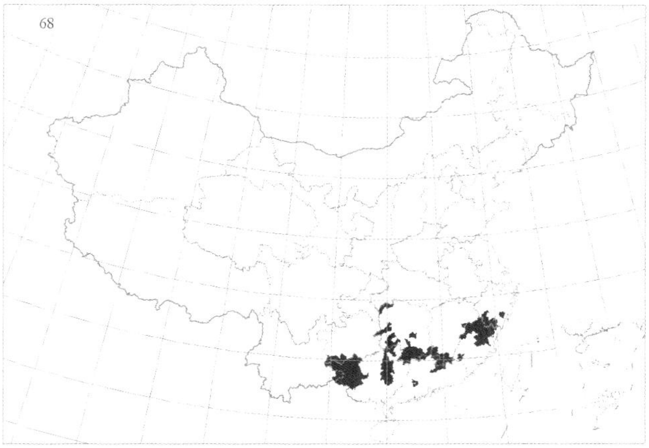

Pinaceae 松科 *Keteleeria* 油杉属
Keteleeria fortunei var. oblonga (W. C. Cheng et L. K. Fu) L. K. Fu et Nan Li 矩鳞油杉
Trees, evergreen

21.4; 12.7; 28.1; 21.5; 197.9; 0.0; 1138 ❙ 1323; 692; 85 ❙ 16.2; 1138; 276.1

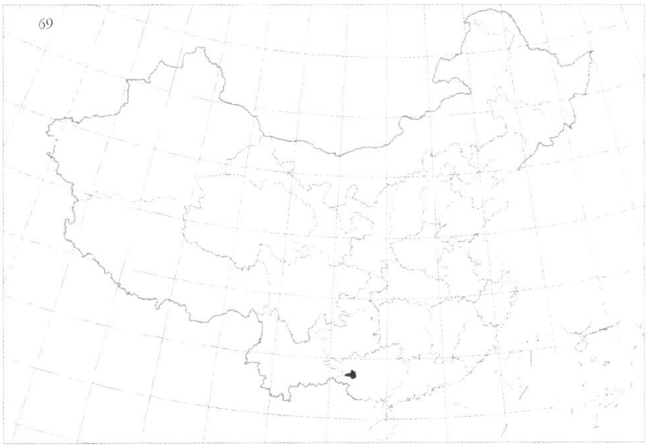

Pinaceae 松科 *Keteleeria* 油杉属
Keteleeria hainanensis Chun et Tsiang 海南油杉
Trees, evergreen

23.9 (23.1~24.3); 18.0 (17.5~18.7); 28.1 (27.2~28.7); 23.9 (23.2~24.3); 226.8 (218.4~231.9); 0.0 (0.0~0.0); 1299 (1232~1338) ❙ 1191 (868~1487); 428 (312~544); 54 (42~71) ❙ -7.9 (-34.3~20.7); 1059 (868~1142); 426.6 (288.3~538.2)

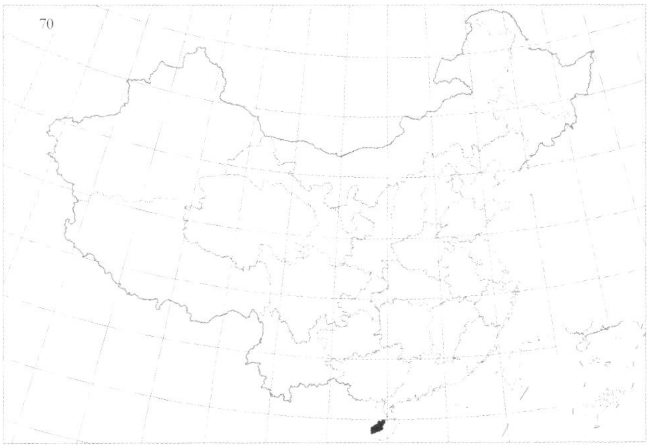

Pinaceae 松科 *Keteleeria* 油杉属
Keteleeria pubescens W. C. Cheng et L. K. Fu 柔毛油杉
Trees, evergreen

16.5 (14.3~20.5); 6.1 (3.3~10.7); 25.8 (23.5~28.6); 16.5 (14.3~20.5); 138.7 (112.6~186.1); 0.3 (0.0~2.3); 863 (756~1083) ❙ 1307 (1159~1625); 547 (431~724); 106 (51~211) ❙ 52.1 (27.5~101.8); 862 (756~1083); 228.9 (146.0~322.1)

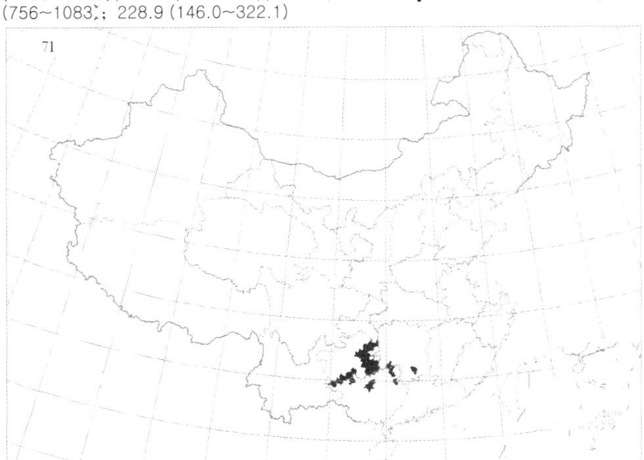

Pinaceae 松科 *Larix* 落叶松属
Larix gmelinii (Rupr.) Kuzen. 落叶松
Trees, deciduous

0.8 (-5.8~8.3); -20.8 (-29.8~-8.6); 19.8 (15.3~23.1); 7.0 (4.4~10.0); 52.7 (28.0~83.0); 102.4 (42.9~157.4); 537 (426~659) ❙ 494 (137~694); 329 (92~499); 13 (3~21) ❙ -7.8 (-77.5~25.8); 469 (137~659); 254.4 (25.4~333.3)

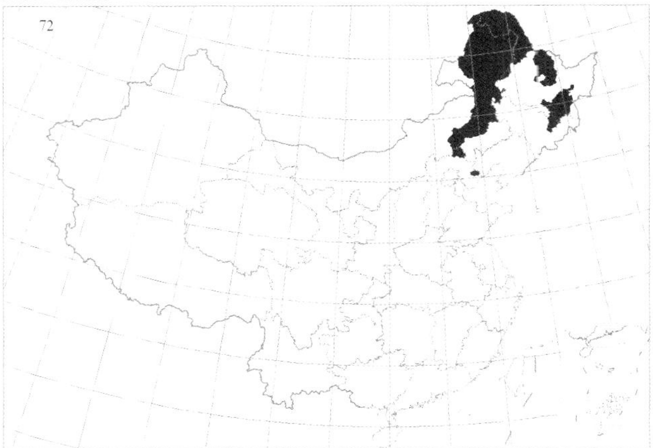

Pinaceae 松科 *Larix* 落叶松属
Larix gmelinii var. principis-rupprechtii (Mayr) Pilg. 华北落叶松
Trees, deciduous

6.4 (0.8~13.5); -10.6 (-20.7~1.4); 21.4 (11.7~27.1); 8.9 (4.3~13.7); 70.3 (21.1~117.5); 53.5 (8.0~101.4); 610 (427~820) ❙ 486 (137~1135); 303 (92~644); 13 (3~37) ❙ -19.8 (-77.5~101.2); 471 (137~711); 169.2 (25.4~311.0)

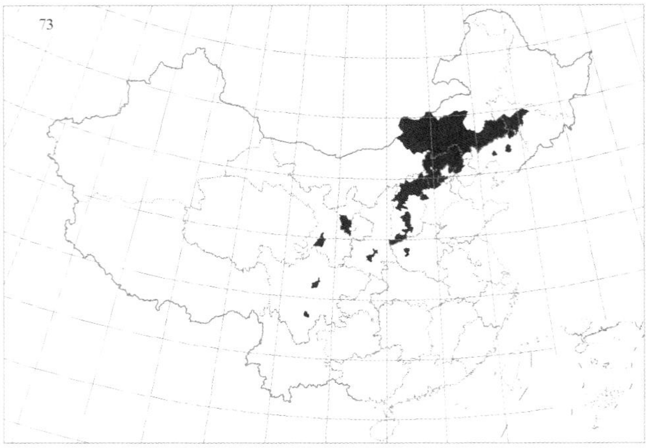

Pinaceae 松科 *Larix* 落叶松属
Larix griffithii Hook. f. 藏红杉
Trees, deciduous

0.5 (-2.5~4.4); -8.5 (-12.7~-4.3); 8.8 (6.1~12.1); 3.2 (1.8~5.3); 12.1 (1.9~28.5); 65.3 (34.6~93.0); 374 (329~455) ❙ 414 (290~583); 268 (212~348); 10 (5~20) ❙ 7.5 (-14.7~44.2); 367 (290~455); 147.8 (61.6~280.0)

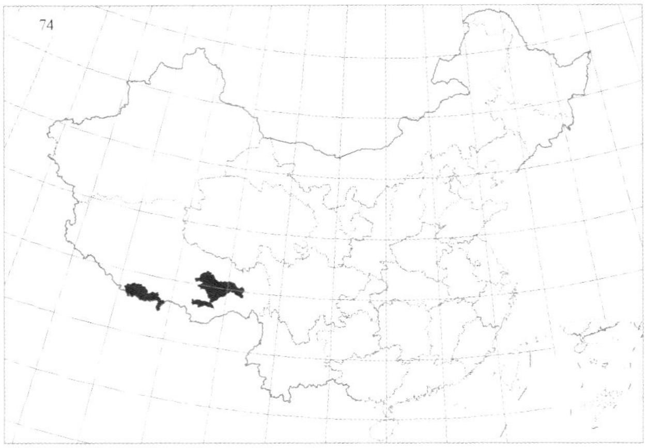

Pinaceae 松科 *Larix* 落叶松属
Larix himalaica W. C. Cheng et L. K. Fu 喜马拉雅红杉
Trees, deciduous

-1.3; -9.9; 6.8; 2.1; 4.0; 78.7; 333 ❙ 330; 225; 18 ❙ -2.1; 322; 68.9

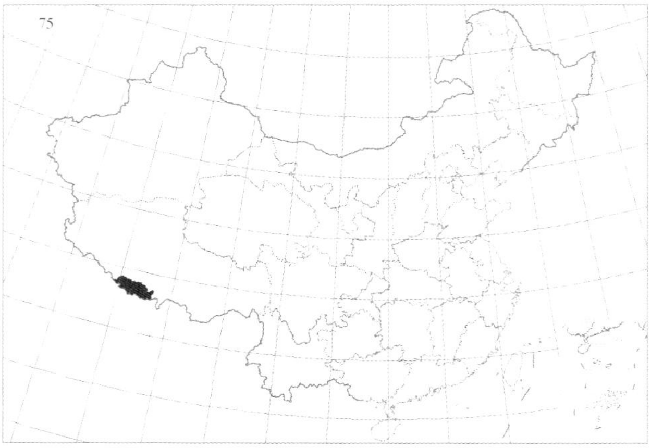

Pinaceae 松科 *Larix* 落叶松属
Larix kongboensis R. R. Mill 贡布红杉
Trees, deciduous

-0.4; -10.0; 8.4; 2.8; 9.4; 73.3; 360 ❙ 369; 243; 5 ❙ 0.0; 360; 205.9

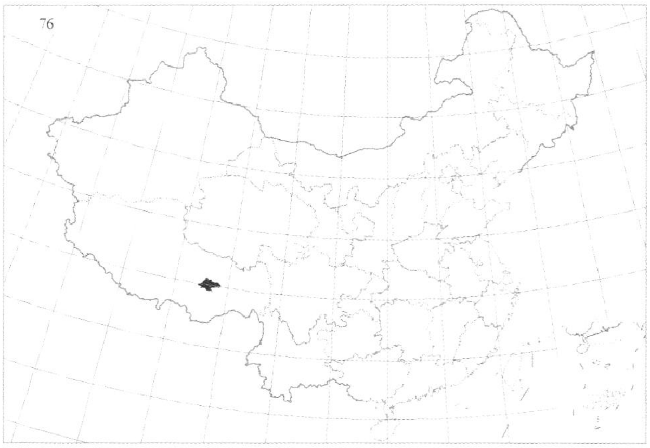

Pinaceae 松科 *Larix* 落叶松属
Larix mastersiana Rehder et E. H. Wilson 四川红杉
Trees, deciduous

6.6 (2.0~13.3); -2.8 (-7.2~3.2); 15.1 (10.1~22.1); 7.3 (3.8~13.2); 46.7 (15.4~100.9); 26.8 (2.5~51.4); 530 (413~723) ❙ 964 (822~1081); 507 (420~606); 26 (19~32) ❙ 87.1 (38.2~128.6); 530 (413~723); 245.8 (171.0~275.4)

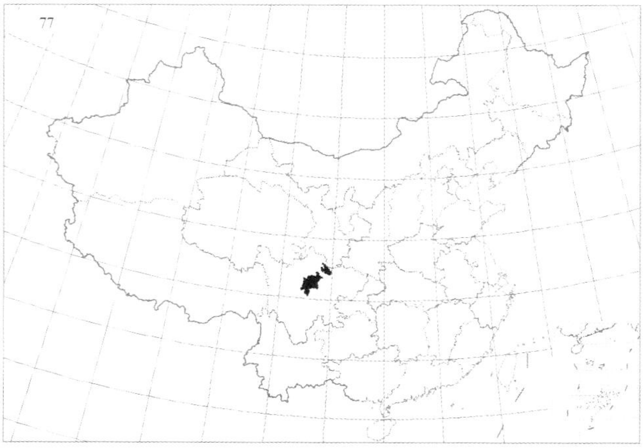

Pinaceae 松科 *Larix* 落叶松属
Larix olgensis A. Henry 黄花落叶松
Trees, deciduous

6.2 (1.4~10.5); -12.9 (-20.6~-4.9); 22.7 (18.4~25.1); 9.2 (6.4~11.3); 75.3 (45.8~93.2); 60.6 (26.8~93.8); 631 (501~710) ❙ 679 (457~1073); 429 (309~691); 22 (7~40) ❙ 8.8 (-28.0~74.1); 593 (457~698); 244.8 (136.6~346.5)

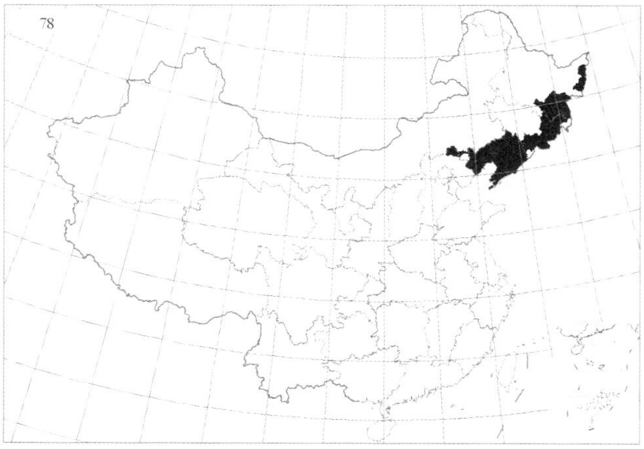

Pinaceae 松科 *Larix* 落叶松属

Larix potaninii Batalin 红杉

Trees, deciduous

4.9 (−1.0~18.5); −4.3 (−11.8~10.5); 13.0 (8.5~24.3); 6.2 (2.7~18.3); 36.1 (8.3~160.0); 36.7 (0.0~79.5); 488 (363~906) | 801 (516~1081); 450 (283~619); 19 (9~45) | 67.1 (5.5~128.6); 486 (363~786); 278.7 (171.0~357.7)

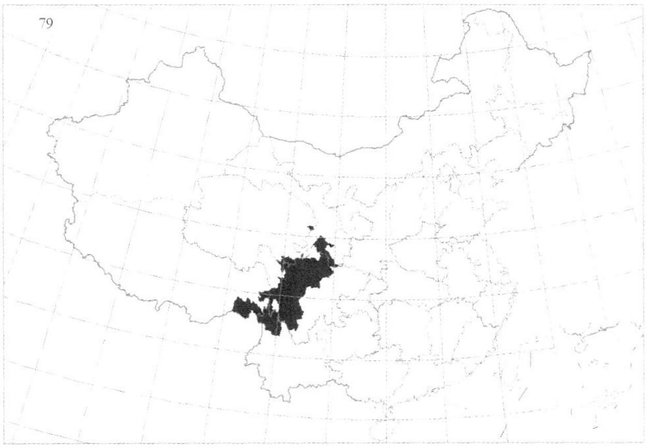

Pinaceae 松科 *Larix* 落叶松属

Larix potaninii var. australis A. Henry ex Hand.-Mazz. 大果红杉

Trees, deciduous

5.8 (1.4~18.5); −2.3 (−7.6~10.5); 12.9 (9.2~24.3); 6.6 (3.5~18.3); 39.5 (13.4~160.0); 29.3 (0.0~55.6); 503 (402~906) | 805 (535~1036); 488 (326~619); 18 (10~40) | 62.6 (5.5~121.7); 497 (402~786); 264.0 (201.9~351.7)

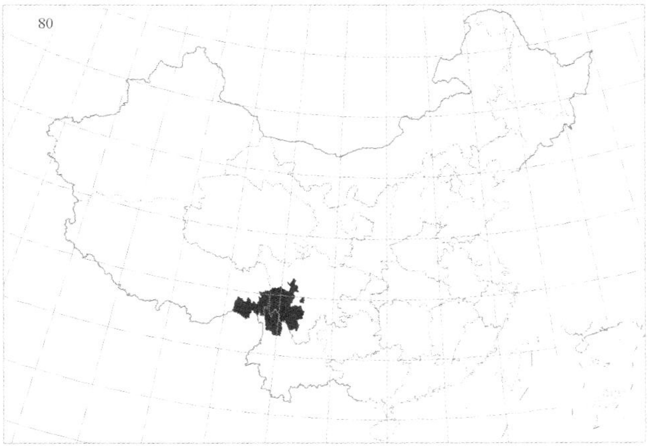

Pinaceae 松科 *Larix* 落叶松属

Larix potaninii var. chinensis (Beissn.) L. K. Fu et Nan Li 秦岭红杉

Trees, deciduous

10.0 (5.6~12.2); −2.3 (−6.4~0.2); 21.9 (16.5~24.2); 10.4 (6.9~12.3); 80.6 (46.9~97.9); 19.4 (10.8~39.0); 649 (517~717) | 771 (528~892); 351 (270~391); 25 (11~35) | 19.2 (0.0~52.6); 649 (517~717); 273.2 (192.9~334.4)

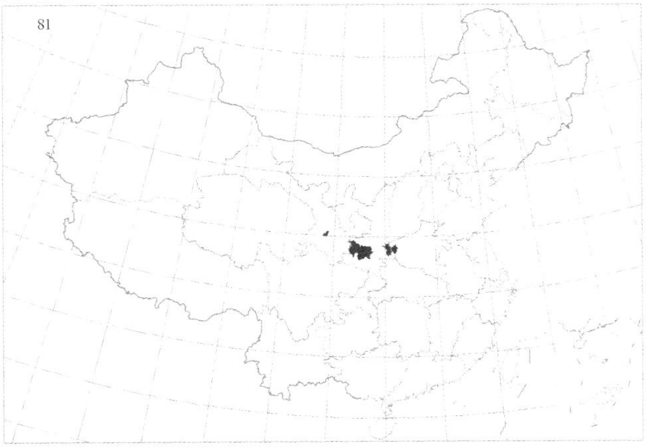

Pinaceae 松科 *Larix* 落叶松属

Larix sibirica Ledeb. 西伯利亚落叶松

Trees, deciduous

3.1 (−0.6~6.5); −15.7 (−19.5~−12.0); 19.6 (15.9~23.0); 7.5 (5.4~9.4); 57.6 (37.5~77.9); 80.3 (57.7~104.2); 567 (475~655) | 179 (69~315); 78 (44~115); 20 (4~51) | −67.8 (−89.2~−46.4); 179 (69~312); 83.3 (14.2~173.1)

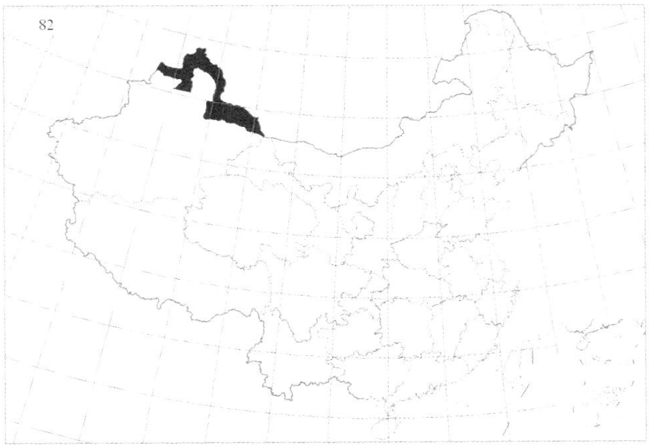

Pinaceae 松科 *Larix* 落叶松属

Larix speciosa W. C. Cheng et Y. W. Law 怒江红杉

Trees, deciduous

9.3 (1.3~13.3); 2.0 (−7.1~6.4); 15.5 (9.9~19.6); 9.6 (3.8~13.4); 63.2 (15.7~100.4); 11.1 (0.0~53.3); 591 (404~706) | 1059 (583~1736); 610 (348~1007); 40 (13~69) | 77.3 (44.2~145.8); 591 (404~706); 329.3 (169.5~447.4)

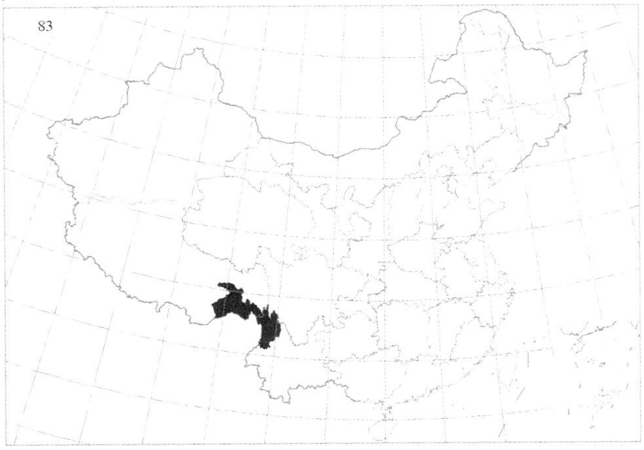

Pinaceae 松科 *Picea* 云杉属

Picea asperata Mast. 云杉

Trees, evergreen

4.3 (−1.0~12.3); −6.2 (−11.8~1.1); 13.6 (8.5~22.7); 5.8 (2.7~12.3); 34.7 (8.3~96.2); 42.6 (8.1~79.5); 479 (363~709) | 726 (471~1023); 385 (247~601); 16 (9~28) | 56.1 (−13.0~128.6); 475 (363~707); 276.6 (101.3~357.7)

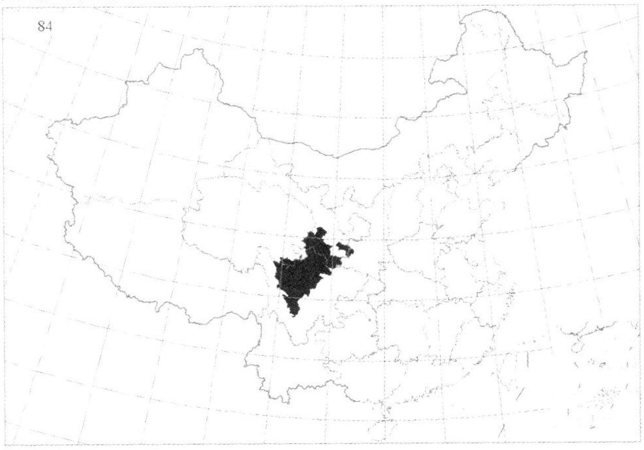

Pinaceae 松科 *Picea* 云杉属
Picea asperata var. aurantiaca (Masters) Boom 白皮云杉
Trees, evergreen

2.8 (1.6~5.7); -6.5 (-8.0~-3.5); 11.0 (9.8~14.2); 4.4 (3.7~6.4); 20.8 (15.2~38.1); 46.1 (28.1~56.2); 431 (408~504) ❙ 838 (687~950); 462 (400~546); 19 (11~25) ❙ 95.0 (68.4~128.6); 431 (408~504); 287.3 (248.3~357.7)

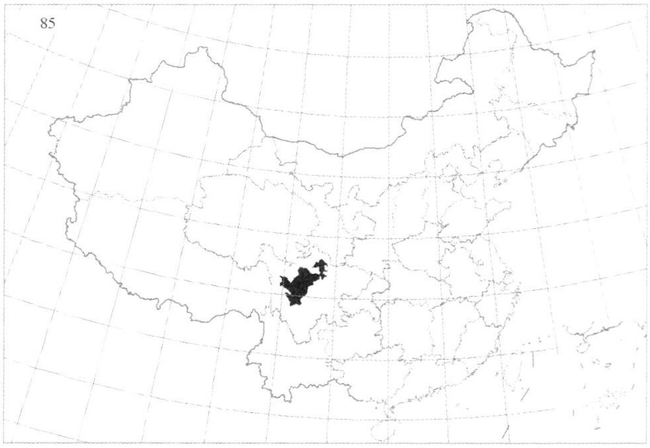

Pinaceae 松科 *Picea* 云杉属
Picea brachytyla (Franch.) E. Pritz. 麦吊杉
Trees, evergreen

10.1 (-1.0~18.5); -0.4 (-12.5~10.5); 19.7 (8.5~27.5); 10.5 (2.7~18.3); 78.6 (8.3~160.0); 17.1 (0.0~79.5); 647 (363~906) ❙ 895 (373~1435); 467 (201~817); 33 (7~88) ❙ 43.1 (-37.6~125.9); 627 (363~817); 245.0 (88.1~407.1)

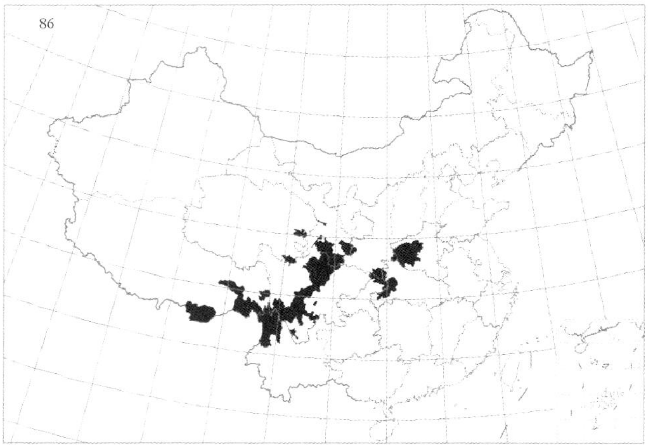

Pinaceae 松科 *Picea* 云杉属
Picea brachytyla var. complanata (Mast.) W. C. Cheng ex Rehder 油麦吊杉
Trees, evergreen

9.4 (1.8~18.5); 1.2 (-7.1~10.5); 16.6 (9.4~24.6); 9.8 (3.7~18.3); 67.8 (14.2~160.0); 14.6 (0.0~53.3); 602 (402~906) ❙ 1010 (535~1431); 579 (326~817); 30 (10~69) ❙ 69.2 (5.5~125.9); 597 (402~822); 278.3 (169.5~407.1)

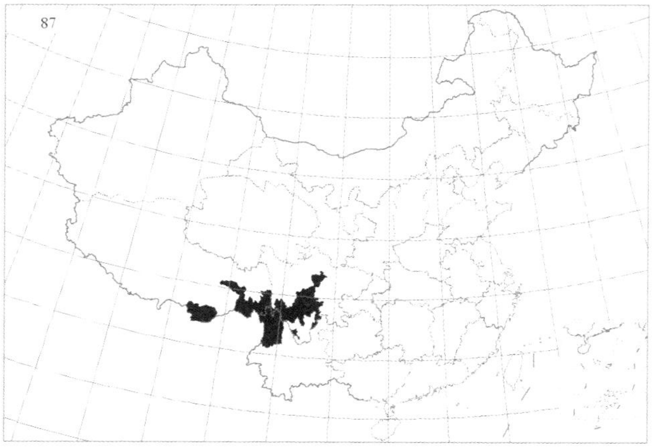

Pinaceae 松科 *Picea* 云杉属
Picea crassifolia Kom. 青海云杉
Trees, evergreen

4.8 (-4.1~12.3); -10.0 (-17.3~1.1); 17.9 (7.2~24.2); 7.3 (2.0~12.3); 53.1 (3.7~96.2); 54.9 (8.1~114.2); 540 (339~709) ❙ 351 (72~707); 201 (46~362); 8 (1~15) ❙ -29.9 (-89.1~55.3); 325 (72~707); 135.4 (7.5~364.1)

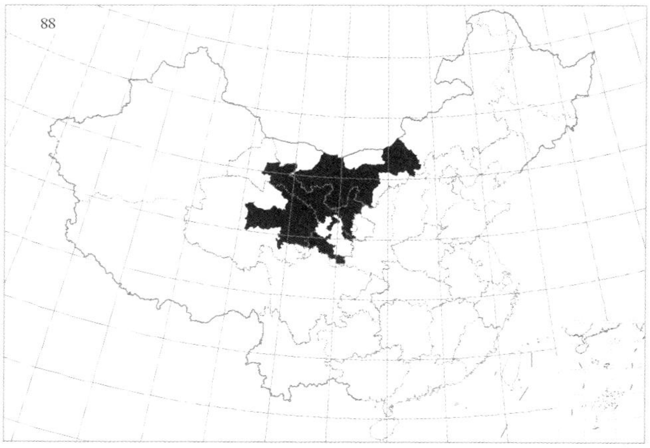

Pinaceae 松科 *Picea* 云杉属
Picea farreri C. N. Page et Rushforth 缅甸云杉
Trees, evergreen

14.4; 7.5; 19.1; 14.4; 113.1; 0.0; 727 ❙ 1454; 792; 70 ❙ 99.9; 727; 487.2

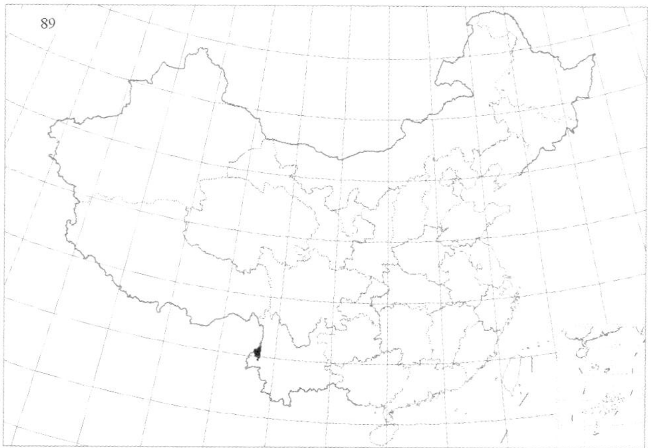

Pinaceae 松科 *Picea* 云杉属
Picea jezoensis var. komarovii (V. N. Vassil.) W. C. Cheng et L. K. Fu 长白鱼鳞云杉
Trees, evergreen

5.9 (1.4~9.5); -12.9 (-18.8~-7.2); 22.1 (18.4~24.8); 8.9 (6.4~10.8); 71.7 (45.8~91.5); 60.5 (35.9~88.8); 613 (501~701) ❙ 789 (561~1073); 490 (333~691); 29 (16~40) ❙ 29.9 (-5.7~74.1); 610 (501~690); 282.5 (164.3~346.5)

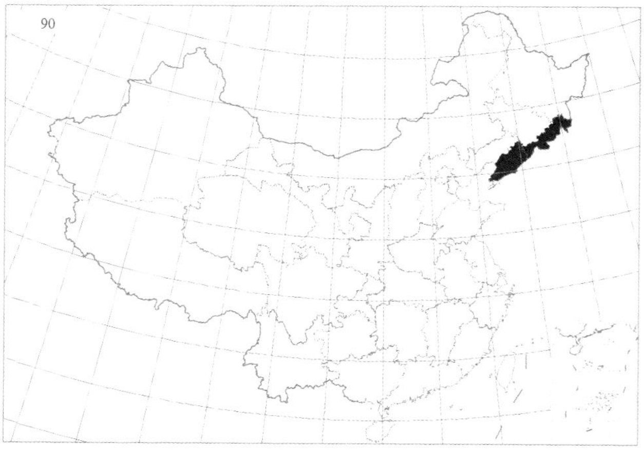

Pinaceae 松科 *Picea* 云杉属

Picea jezoensis var. microsperma (Lindl.) W. C. Cheng et L. K. Fu 兴安鱼鳞云杉 Trees, evergreen

2.0 (−5.8~10.5); −19.2 (−29.8~−4.9); 20.3 (15.3~25.0); 7.4 (4.4~11.3); 56.7 (28.0~93.2); 92.4 (26.8~157.4); 553 (426~705) ❙ 640 (401~1073); 400 (275~691); 22 (7~40) ❙ 15.3 (−30.9~74.1); 539 (401~698); 287.7 (177.8~346.5)

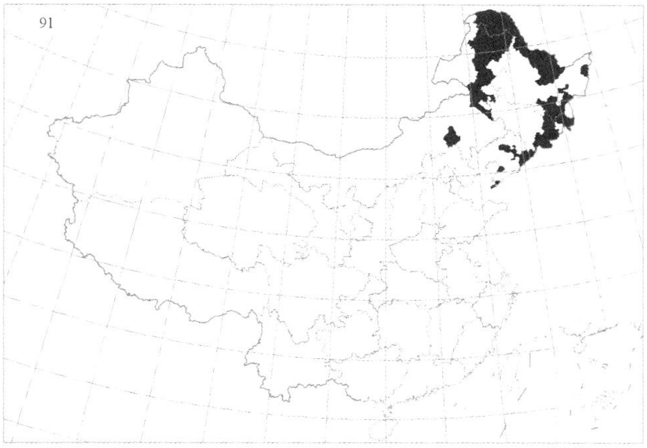

Pinaceae 松科 *Picea* 云杉属

Picea koraiensis Nakai 红皮云杉
Trees, evergreen

3.1 (−5.8~10.5); −16.5 (−29.8~−4.9); 20.2 (15.3~25.0); 7.7 (4.4~11.3); 58.7 (28.0~93.2); 80.9 (26.8~157.4); 560 (426~705) ❙ 566 (297~1073); 359 (203~691); 17 (7~40) ❙ 1.1 (−44.3~74.1); 516 (297~698); 250.5 (115.9~346.5)

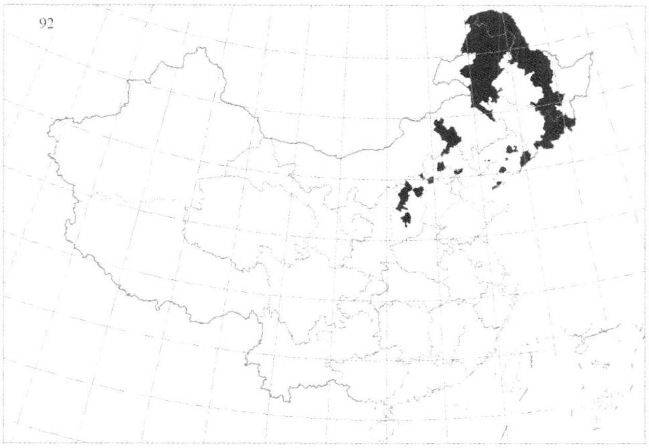

Pinaceae 松科 *Picea* 云杉属

Picea likiangensis (Franch.) E. Pritz. 丽江云杉
Trees, evergreen

3.5 (−2.2~15.4); −5.4 (−12.8~8.2); 11.3 (7.6~21.1); 5.2 (2.2~15.4); 27.5 (5.5~125.4); 45.6 (0.0~91.3); 449 (336~770) ❙ 699 (320~1431); 420 (225~817); 15 (3~45) ❙ 54.7 (−10.6~128.6); 446 (320~696); 261.0 (111.0~363.7)

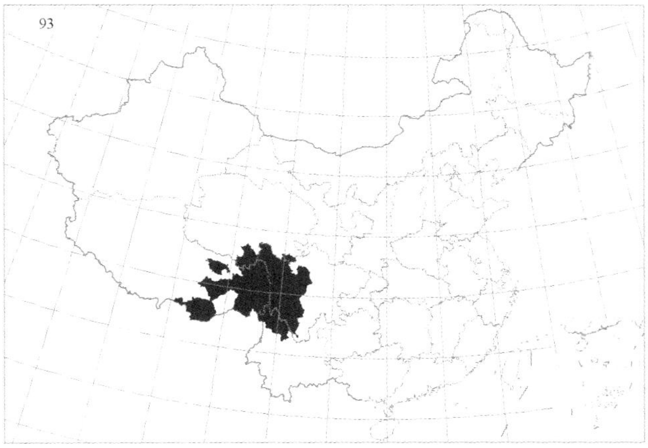

Pinaceae 松科 *Picea* 云杉属

Picea likiangensis var. hirtella (Rehder et E. H. Wilson) W. C. Cheng 黄果云杉 Trees, evergreen

5.0; −4.2; 13.1; 6.3; 37.9; 38.0; 492 ❙ 883; 520; 20 ❙ 83.8; 492; 230.2

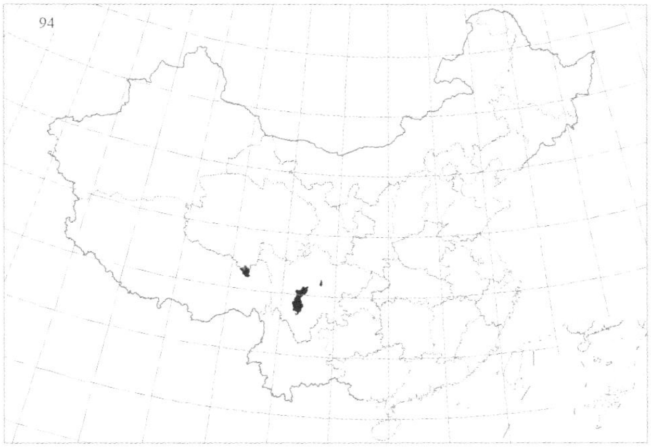

Pinaceae 松科 *Picea* 云杉属

Picea likiangensis var. linzhiensis W. C. Cheng et L. K. Fu 林芝云杉 Trees, evergreen

2.1 (−0.4~4.4); −6.9 (−10.0~−4.3); 10.2 (8.1~12.1); 4.0 (2.8~5.3); 18.0 (8.9~28.5); 52.1 (34.6~73.3); 406 (358~455) ❙ 468 (320~583); 293 (225~348); 8 (3~13) ❙ 13.2 (−10.6~44.2); 399 (320~455); 195.5 (118.8~230.0)

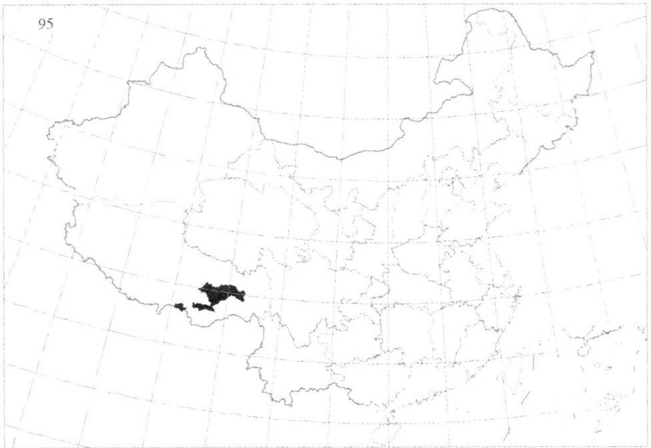

Pinaceae 松科 *Picea* 云杉属

Picea likiangensis var. montigena (Mast.) W. C. Cheng 康定云杉
Trees, evergreen

2.2; −6.6; 9.9; 3.8; 15.5; 49.1; 413 ❙ 932; 546; 19 ❙ 125.9; 413; 248.3

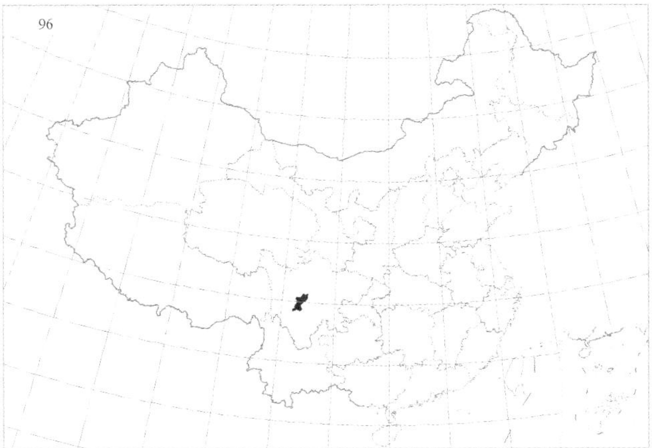

Pinaceae 松科 *Picea* 云杉属

Picea likiangensis var. rubescens Rehder et E. H. Wilson 川西云杉

Trees, evergreen

1.7 (-2.2~6.6); -7.5 (-12.8~-1.3); 10.0 (7.6~13.3); 3.9 (2.2~6.8); 16.8 (5.5~39.6); 55.3 (19.7~91.3); 408 (336~514) ❙ 653 (462~984); 392 (299~601); 14 (8~31) ❙ 59.5 (20.7~128.6); 408 (336~514); 256.1 (111.0~357.7)

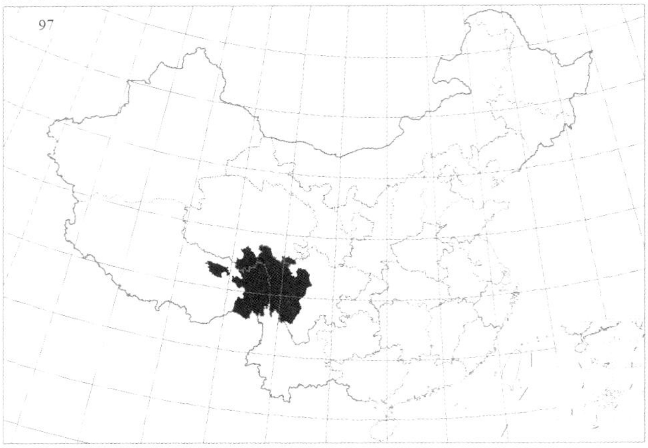

Pinaceae 松科 *Picea* 云杉属

Picea meyeri Rehder et E. H. Wilson 白扦

Trees, evergreen

5.8 (0.1~11.6); -11.9 (-21.5~-4.8); 21.5 (17.5~26.1); 8.8 (6.4~12.5); 69.4 (45.4~105.6); 59.2 (25.4~108.6); 605 (506~782) ❙ 440 (111~718); 301 (74~499); 10 (2~29) ❙ -27.4 (-82.3~9.1); 439 (111~698); 168.7 (13.9~287.6)

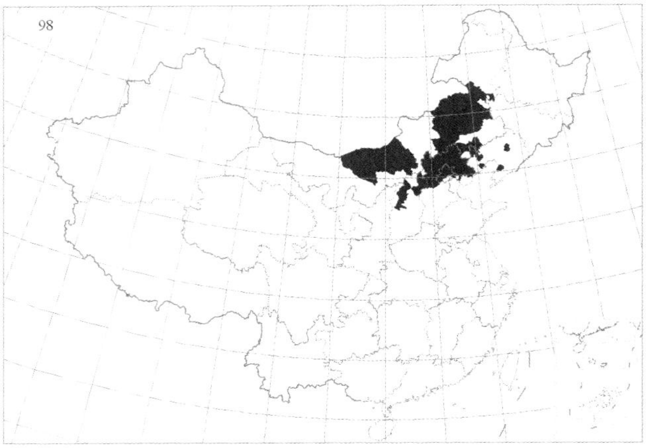

Pinaceae 松科 *Picea* 云杉属

Picea morrisonicola Hayata 台湾云杉

Trees, evergreen

18.6 (16.2~22.8); 13.3 (11.0~16.7); 22.8 (20.1~27.8); 18.6 (16.2~22.8); 163.5 (134.2~213.7); 0.0 (0.0~0.0); 903 (773~1209) ❙ 2495 (1625~2859); 1185 (947~1594); 214 (93~402) ❙ 185.5 (34.4~246.5); 894 (773~1131); 267.9 (179.2~363.7)

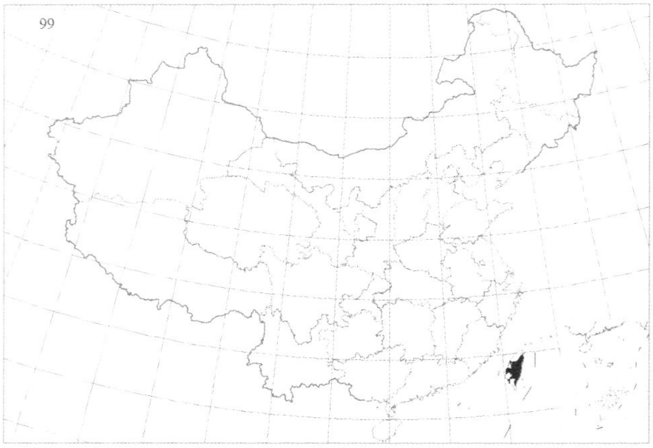

Pinaceae 松科 *Picea* 云杉属

Picea neoveitchii Mast. 大果青扦

Trees, evergreen

10.9 (1.9~14.6); -1.5 (-9.2~2.6); 22.6 (11.7~26.8); 11.3 (4.3~14.7); 89.0 (21.1~122.7); 17.1 (4.0~57.8); 685 (427~834) ❙ 784 (101~1246); 365 (61~525); 34 (5~83) ❙ 16.2 (-83.8~101.4); 638 (101~817); 236.2 (58.4~334.4)

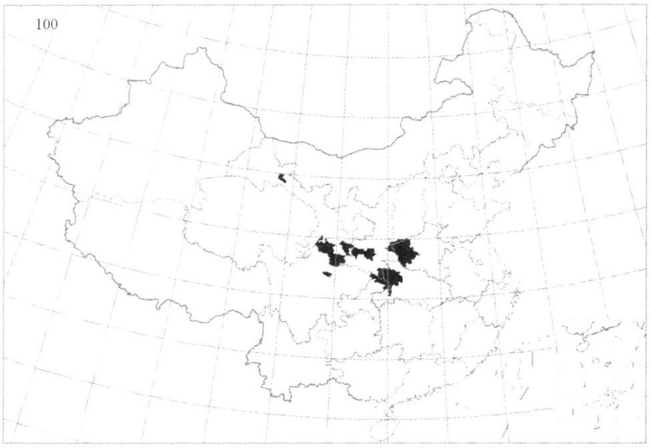

Pinaceae 松科 *Picea* 云杉属

Picea obovata Ledeb. 西伯利亚云杉

Trees, evergreen

0.8 (-0.6~2.1); -17.8 (-19.5~-16.0); 17.2 (15.9~18.7); 6.2 (5.4~6.9); 44.1 (37.5~50.8); 94.7 (85.3~104.2); 509 (475~544) ❙ 194 (161~251); 85 (74~106); 23 (15~33) ❙ -61.9 (-68.1~-52.0); 194 (161~251); 118.5 (74.6~139.1)

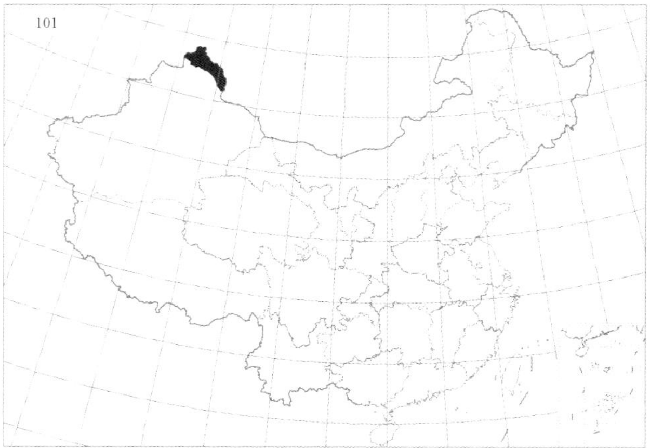

Pinaceae 松科 *Picea* 云杉属

Picea purpurea Mast. 紫果云杉

Trees, evergreen

3.4 (-1.3~12.3); -7.6 (-12.5~-1.1); 13.1 (8.3~22.7); 5.4 (2.5~12.3); 30.9 (7.0~96.2); 49.5 (8.1~82.3); 462 (347~709) ❙ 652 (375~1023); 344 (203~535); 14 (6~28) ❙ 44.4 (-32.2~128.6); 451 (347~707); 264.9 (95.7~364.1)

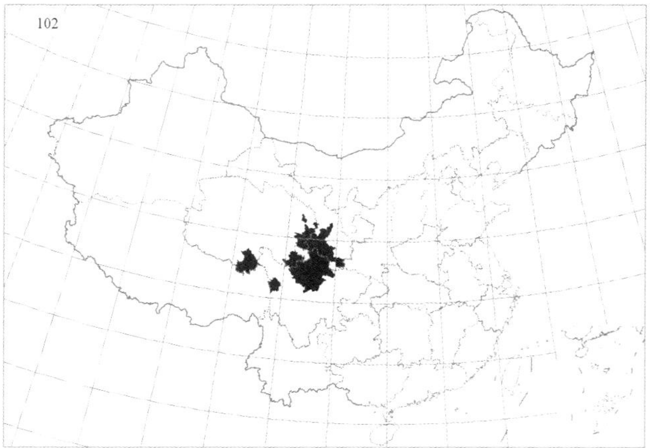

Pinaceae 松科 *Picea* 云杉属

Picea schrenkiana Fisch. et C. A. Mey. 雪岭云杉

Trees, evergreen

5.4 (-4.3~12.5); -13.2 (-20.9~-5.1); 20.6 (9.6~27.5); 8.7 (2.5~13.2); 69.3 (9.7~114.0); 63.5 (23.1~121.1); 615 (329~817) | 172 (42~450); 81 (17~211); 14 (3~48) | -68.9 (-94.7~3.9); 171 (42~406); 75.5 (7.6~297.3)

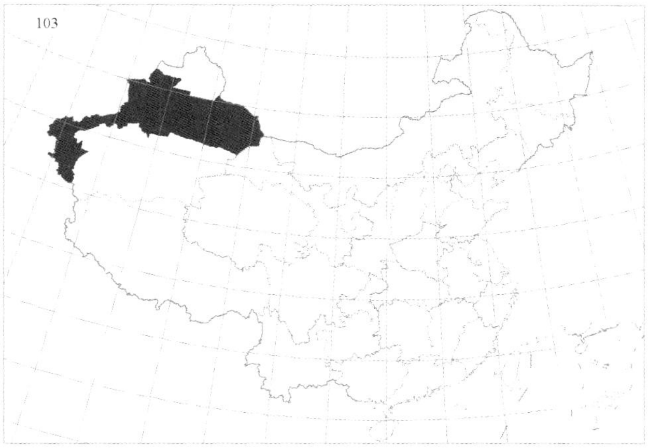

Pinaceae 松科 *Picea* 云杉属

Picea smithiana (Wall.) Boiss. 长叶云杉

Trees, evergreen

-1.1; -9.1; 6.2; 2.0; 2.5; 75.6; 334 | 321; 193; 33 | -4.0; 321; 79.3

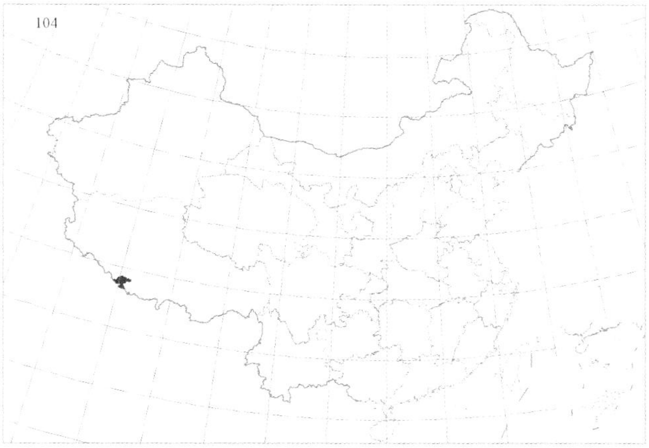

Pinaceae 松科 *Picea* 云杉属

Picea spinulosa (Griff.) A. Henry 喜马拉雅云杉

Trees, evergreen

6.0 (-1.1~13.3); -2.1 (-9.1~5.4); 12.9 (6.2~19.6); 7.2 (2.0~13.4); 46.1 (2.5~100.4); 33.6 (0.0~75.6); 512 (334~706) | 968 (321~1736); 564 (193~1007); 31 (11~51) | 71.1 (-4.0~145.8); 509 (321~706); 253.4 (79.3~447.4)

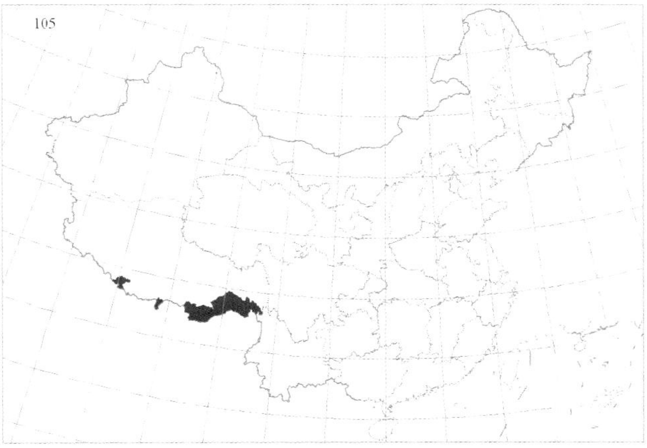

Pinaceae 松科 *Picea* 云杉属

Picea wilsonii Mast. 青扦

Trees, evergreen

7.6 (-1.3~15.9); -6.8 (-21.5~4.8); 20.6 (9.7~27.1); 9.3 (2.9~16.0); 71.4 (11.3~131.6); 39.5 (0.2~108.6); 613 (359~855) | 602 (132~1358); 315 (81~543); 19 (2~88) | -1.5 (-80.1~128.6); 512 (132~855); 197.4 (13.6~344.4)

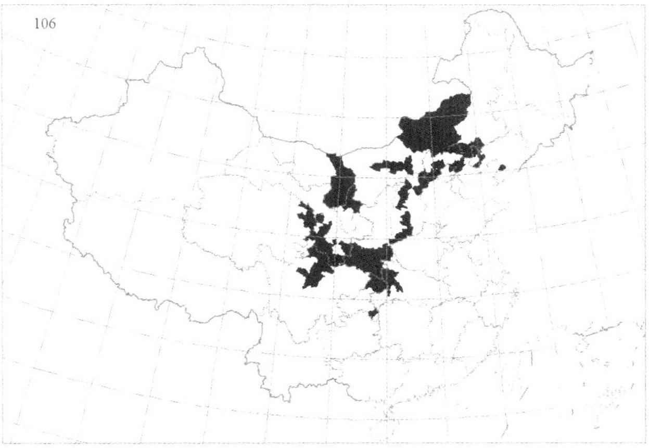

Pinaceae 松科 *Pinus* 松属

Pinus armandii Franch. 华山松

Trees, evergreen

12.3 (0.8~21.7); 1.0 (-13.7~15.3); 22.3 (9.2~29.3); 12.8 (3.5~21.7); 102.6 (13.4~200.2); 14.6 (0.0~71.7); 723 (402~1118) | 902 (341~2993); 463 (190~1412); 43 (8~547) | 24.1 (-40.9~214.4); 671 (341~1075); 227.6 (61.2~516.7)

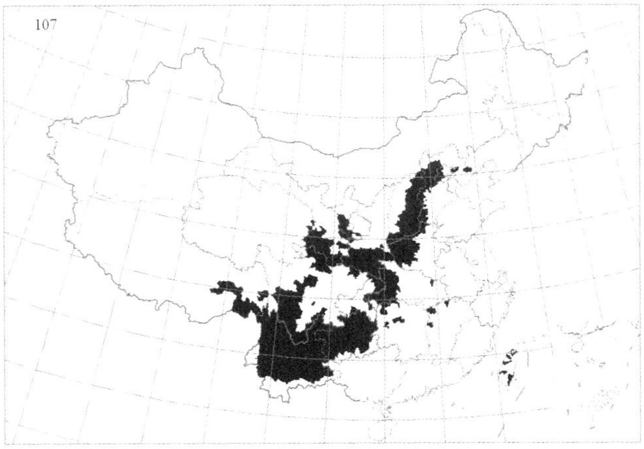

Pinaceae 松科 *Pinus* 松属

Pinus armandii var. mastersiana (Hayata) Hayata 台湾果松

Trees, evergreen

18.5; 13.2; 22.5; 18.5; 161.7; 0.0; 896 | 2563; 1299; 178 | 193.3; 896; 248.4

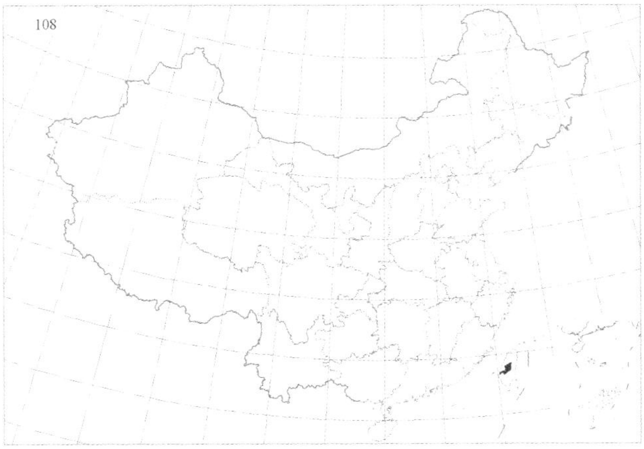

Pinaceae 松科　*Pinus* 松属

Pinus bhutanica Grierson 不丹松

Trees, evergreen

4.3 (-1.1~13.3); -4.0 (-9.6~5.4); 11.7 (6.2~19.6); 5.8 (2.0~13.4); 33.3 (2.5~100.4); 40.8 (0.0~75.6); 465 (334~706) ❙ 747 (320~1736); 445 (193~1007); 22 (3~51) ❙ 46.9 (-10.6~145.8); 461 (320~706); 219.8 (61.6~447.4)

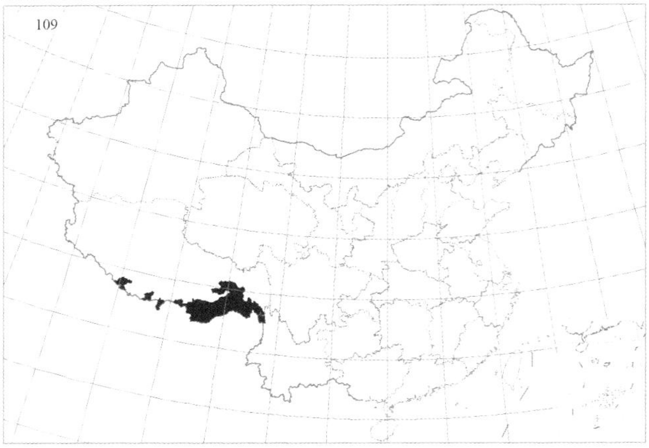

Pinaceae 松科　*Pinus* 松属

Pinus bungeana Zucc. ex Endl. 白皮松

Trees, evergreen

10.2 (3.3~16.1); -4.7 (-13.7~5.7); 23.6 (17.5~28.0); 11.2 (6.6~16.1); 91.2 (45.4~133.5); 28.0 (0.0~71.7); 697 (507~880) ❙ 591 (367~1337); 331 (235~573); 21 (8~113) ❙ -15.1 (-40.9~101.4); 572 (367~845); 170.0 (70.3~334.4)

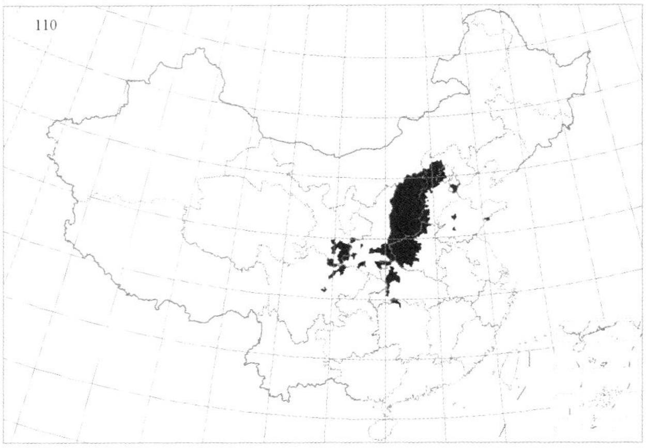

Pinaceae 松科　*Pinus* 松属

Pinus densata Mast. 高山松

Trees, evergreen

5.0 (-3.6~18.5); -3.9 (-15.6~10.7); 12.8 (7.0~24.3); 6.6 (1.9~18.3); 41.3 (3.6~160.0); 40.4 (0.0~107.2); 497 (328~906) ❙ 790 (369~1736); 459 (223~1007); 20 (5~59) ❙ 60.9 (0.0~145.8); 488 (328~786); 276.3 (135.0~447.4)

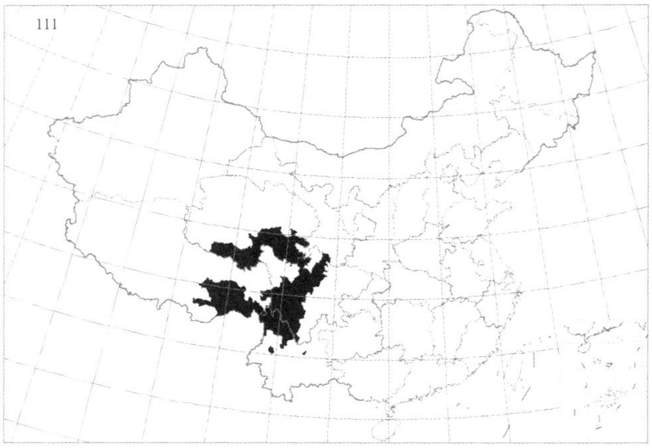

Pinaceae 松科　*Pinus* 松属

Pinus densiflora Siebold et Zucc. 赤松

Trees, evergreen

10.5 (1.4~16.5); -5.3 (-18.8~4.4); 24.6 (18.4~28.9); 11.8 (6.4~16.6); 97.8 (45.8~139.8); 31.0 (0.6~88.8); 728 (501~910) ❙ 756 (401~1377); 457 (262~691); 38 (9~197) ❙ 5.5 (-34.2~74.1); 681 (401~910); 200.0 (101.9~346.5)

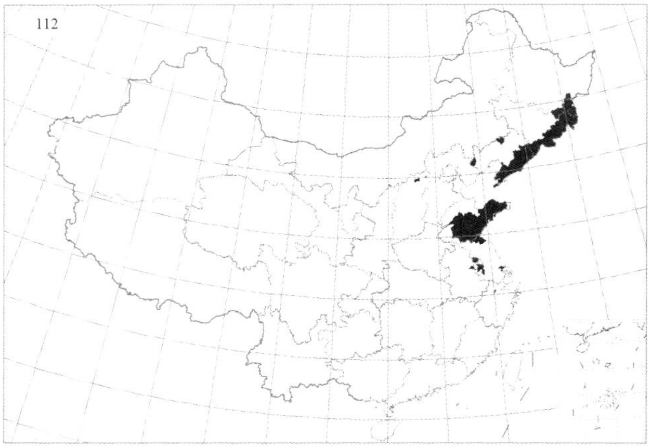

Pinaceae 松科　*Pinus* 松属

Pinus densiflora var. ussuriensis Liou et Z. Wang 兴凯赤松

Trees, evergreen

3.2; -17.3; 20.9; 7.9; 60.3; 80.8; 576 ❙ 568; 333; 22 ❙ -2.1; 560; 271.4

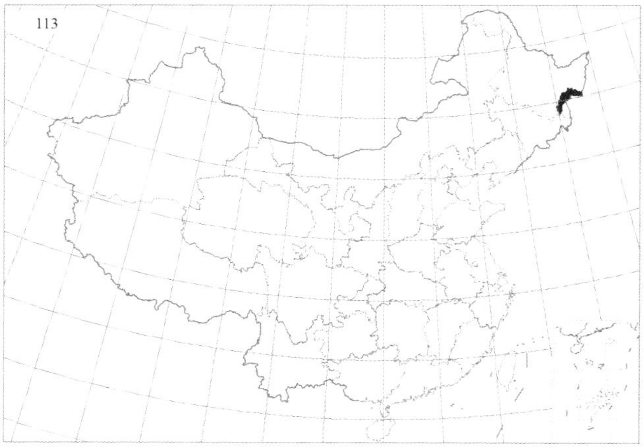

Pinaceae 松科　*Pinus* 松属

Pinus fenzeliana Hand.-Mazz. 海南五针松

Trees, evergreen

18.8 (14.2~24.8); 9.4 (3.6~20.1); 26.8 (22.8~28.9); 18.8 (14.2~24.8); 165.8 (112.3~237.6); 0.2 (0.0~1.7); 995 (751~1385) ❙ 1405 (868~1758); 589 (312~839); 118 (41~198) ❙ 44.3 (-34.3~101.8); 965 (751~1202); 255.0 (142.5~620.0)

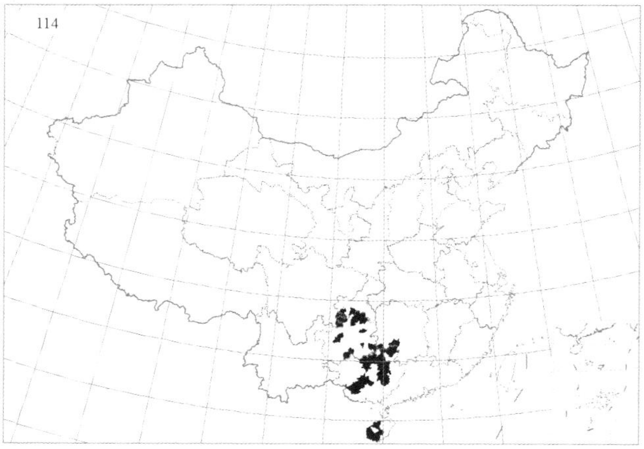

Pinaceae 松科 *Pinus* 松属

Pinus gerardiana Wall. ex D. Don 喜马拉雅白皮松

Trees, evergreen

-1.8; -12.0; 8.6; 2.4; 7.9; 88.8; 333 ❙ 509; 233; 98 ❙ 52.9; 333; 32.2

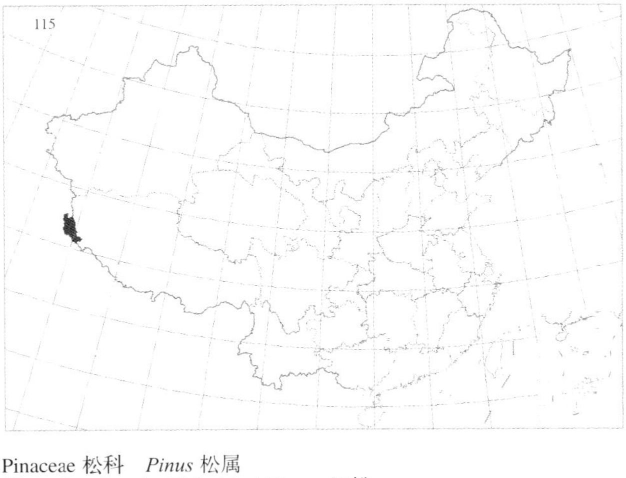

Pinaceae 松科 *Pinus* 松属

Pinus kesiya Royle ex Gordon 思茅松

Trees, evergreen

17.8 (13.1~24.1); 11.3 (5.4~18.3); 22.1 (18.4~28.2); 17.8 (13.2~24.1); 153.5 (98.1~229.4); 0.0 (0.0~0.0); 867 (692~1321) ❙ 1346 (917~1751); 734 (444~1007); 53 (18~92) ❙ 56.8 (-2.2~145.8); 850 (692~1294); 437.1 (242.3~659.0)

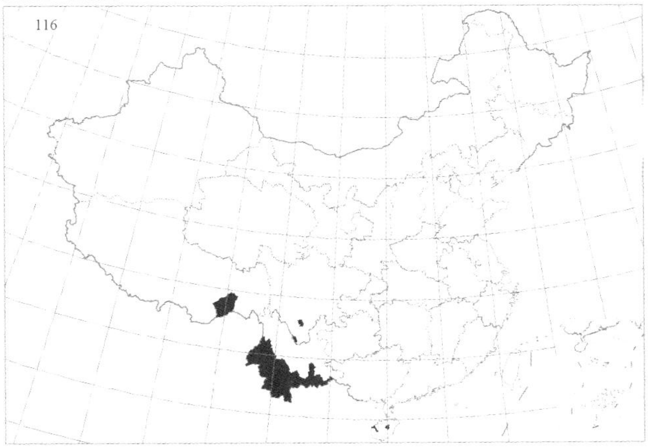

Pinaceae 松科 *Pinus* 松属

Pinus koraiensis Siebold et Zucc. 红松

Trees, evergreen

3.3 (-1.5~10.5); -17.5 (-25.0~-4.9); 21.0 (18.4~25.0); 8.0 (6.2~11.3); 61.4 (45.8~93.2); 81.7 (26.8~124.5); 576 (501~705) ❙ 689 (543~1073); 423 (309~691); 25 (12~40) ❙ 19.4 (-10.4~74.1); 570 (501~698); 292.2 (180.8~346.5)

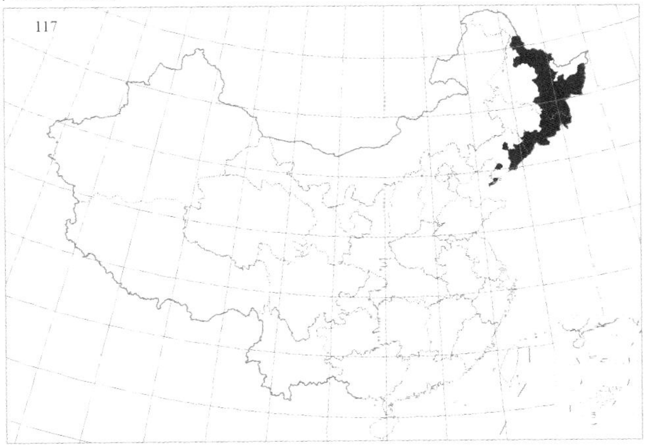

Pinaceae 松科 *Pinus* 松属

Pinus kwangtungensis Chun et Tsiang 华南五针松

Trees, evergreen

19.7 (14.3~24.1); 10.4 (3.3~19.1); 27.3 (24.3~29.8); 19.7 (14.3~24.2); 176.7 (114.1~230.1); 0.0 (0.0~2.3); 1034 (765~1326) ❙ 1528 (1172~2435); 667 (419~1428); 141 (43~203) ❙ 49.5 (5.8~101.8); 1028 (765~1326); 255.9 (160.7~620.0)

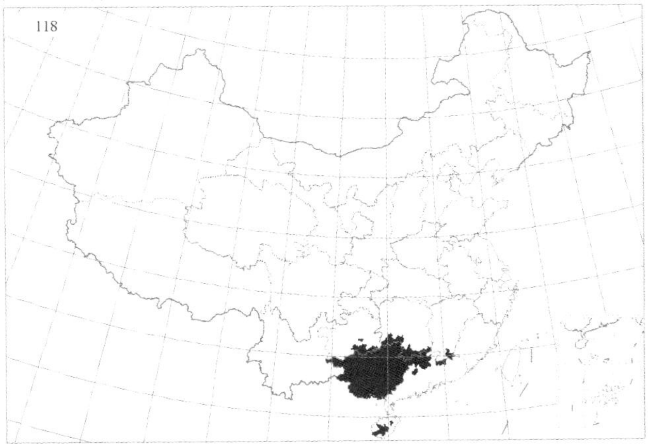

Pinaceae 松科 *Pinus* 松属

Pinus kwangtungensis var. varifolia Nan Li et Y. C. Zhong 变叶华南五针松

Trees, evergreen

19.4; 10.9; 25.9; 19.5; 174.1; 0.0; 991 ❙ 1372; 750; 79 ❙ 38.9; 991; 296.0

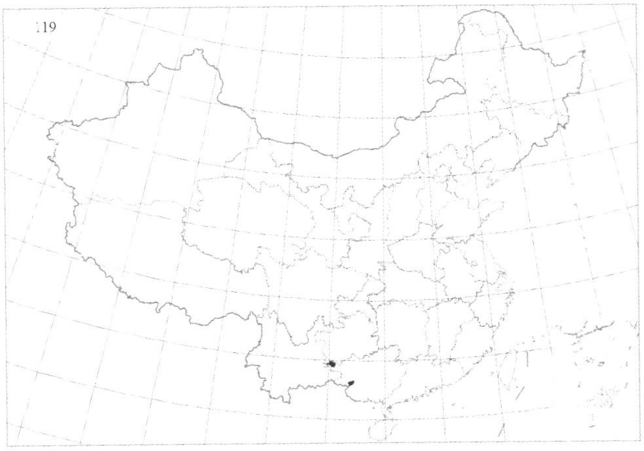

Pinaceae 松科 *Pinus* 松属

Pinus latteri Mason 南亚松

Trees, evergreen

23.4 (21.8~24.3); 16.8 (14.0~19.1); 28.2 (26.7~29.2); 23.4 (21.9~24.3); 221.0 (202.2~231.9); 0.0 (0.0~0.0); 1266 (1150~1338) ❙ 1593 (868~2435); 717 (312~1428); 84 (42~130) ❙ 27.0 (-34.3~99.5); 1170 (868~1326); 374.4 (178.7~620.0)

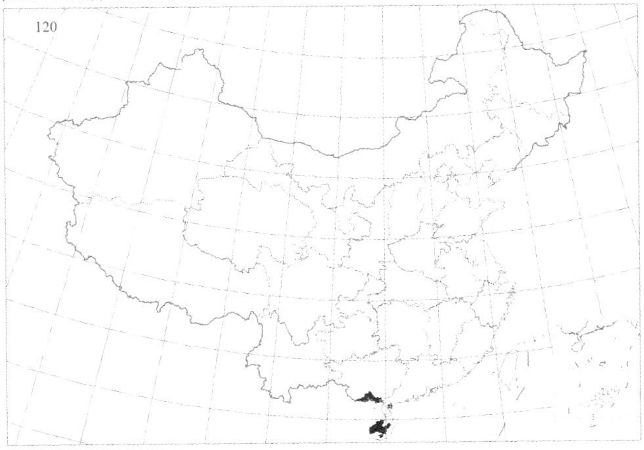

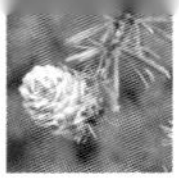

Pinaceae 松科 *Pinus* 松属

Pinus massoniana Lambert 马尾松

Trees, evergreen

17.1 (2.6~24.1); 6.4 (-7.0~18.3); 27.0 (11.2~30.1); 17.2 (4.3~24.1); 147.5 (20.0~229.4); 1.6 (0.0~48.2); 920 (427~1321) ❙ 1399 (512~3495); 560 (258~1428); 139 (15~897) ❙ 52.5 (-35.3~246.5); 908 (427~1294); 215.3 (95.3~474.7)

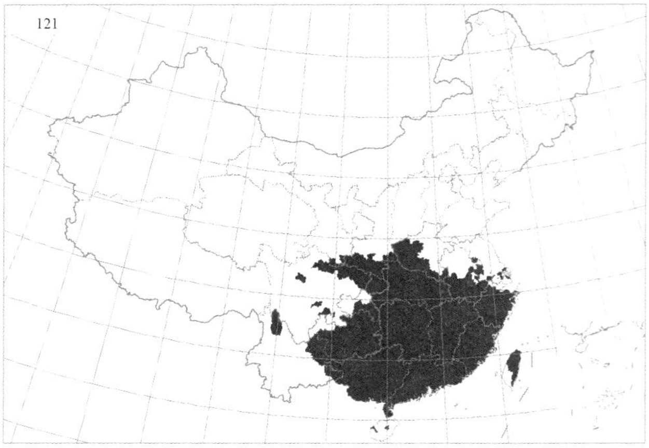

Pinaceae 松科 *Pinus* 松属

Pinus massoniana var. hainanensis W. C. Cheng et L. K. Fu 雅加松

Trees, evergreen

23.3; 17.7; 27.5; 23.4; 220.8; 0.0; 1252 ❙ 1365; 494; 48 ❙ 9.2; 1117; 489.0

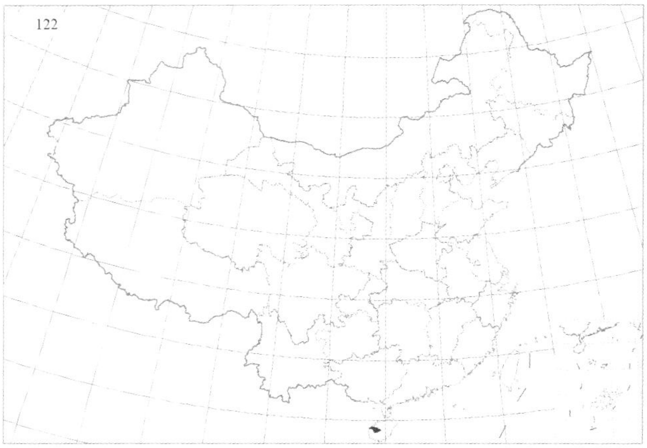

Pinaceae 松科 *Pinus* 松属

Pinus massoniana var. shaxianensis D. X. Zhou 沙黄松

Trees, evergreen

18.4; 9.0; 27.2; 18.4; 161.2; 0.0; 950 ❙ 1726; 587; 204 ❙ 81.7; 950; 279.7

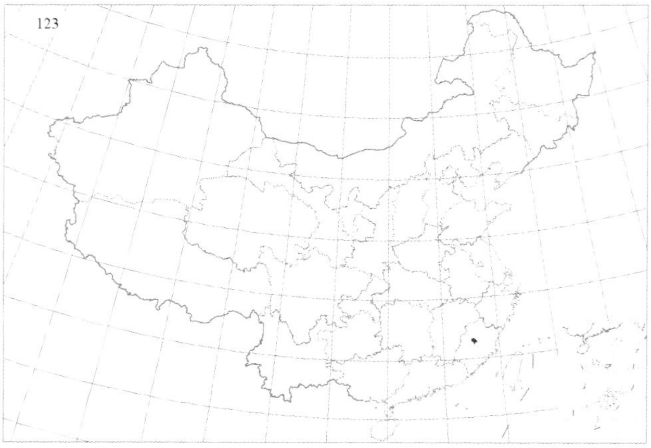

Pinaceae 松科 *Pinus* 松属

Pinus morrisonicola Hayata 台湾五针松

Trees, evergreen

18.8 (16.2~21.1); 12.9 (11.0~15.3); 23.9 (20.1~27.4); 18.8 (16.2~21.1); 165.9 (134.2~193.1); 0.0 (0.0~0.0); 925 (773~1077) ❙ 2706 (2355~3495); 1010 (821~1412); 401 (110~897) ❙ 194.7 (140.0~246.5); 925 (773~1077); 261.6 (207.7~303.6)

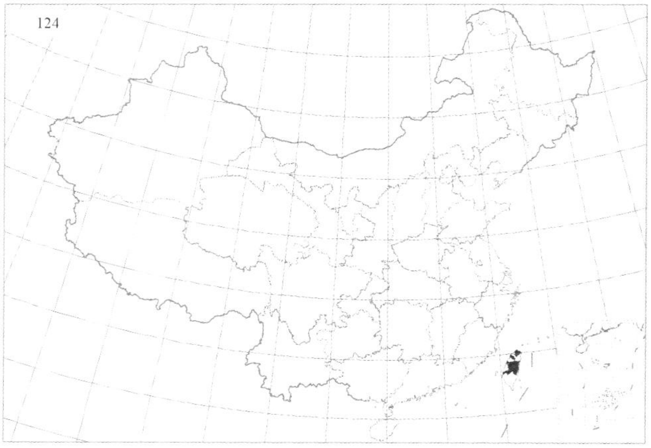

Pinaceae 松科 *Pinus* 松属

Pinus pumila (Pall.) Regel 偃松

Trees, evergreen

0.2 (-5.8~3.7); -21.5 (-29.8~-14.1); 19.2 (15.3~21.6); 6.6 (4.4~8.0); 49.0 (28.0~61.6); 105.7 (70.4~157.4); 519 (426~587) ❙ 577 (404~853); 367 (276~527); 18 (8~35) ❙ 10.6 (-26.8~57.4); 504 (404~573); 300.3 (239.3~346.5)

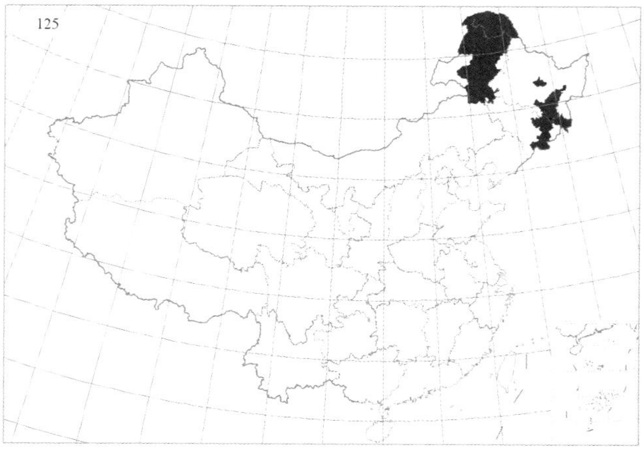

Pinaceae 松科 *Pinus* 松属

Pinus roxburghii Sarg. 喜马拉雅长叶松

Trees, evergreen

-1.1; -9.1; 6.2; 2.0; 2.5; 75.6; 334 ❙ 321; 193; 33 ❙ -4.0; 321; 79.3

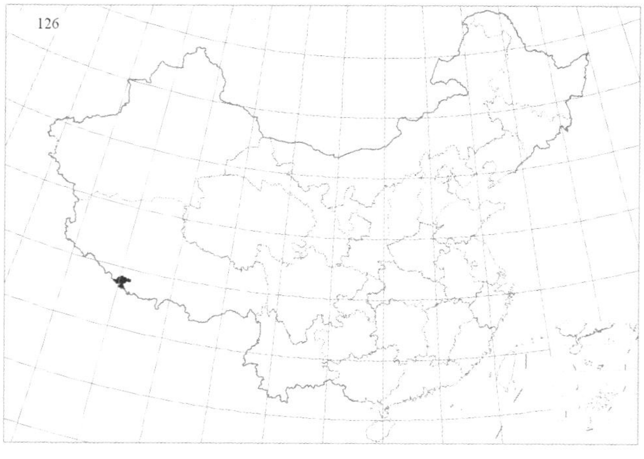

Pinaceae 松科 *Pinus* 松属

Pinus sibirica Du Tour 西伯利亚五针松

Trees, evergreen

-1.4 (-5.8~8.9); -23.1 (-29.8~-8.8); 18.0 (15.3~24.2); 6.0 (4.4~10.7); 44.3 (28.0~90.2); 120.1 (42.6~157.4); 497 (426~694) | 428 (187~569); 270 (79~390); 17 (9~33) | -13.3 (-65.1~15.9); 409 (187~569); 243.5 (120.8~308.8)

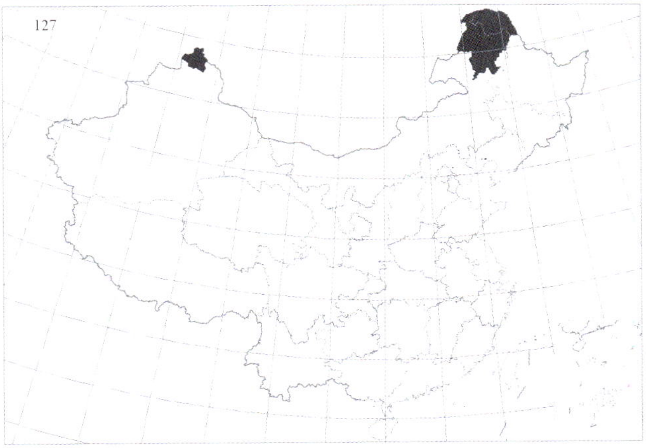

Pinaceae 松科 *Pinus* 松属

Pinus squamata X. W. Li 巧家五针松

Trees, evergreen

12.5; 4.3; 19.3; 12.5; 91.2; 0.7; 682 | 893; 521; 25 | 30.3; 682; 211.5

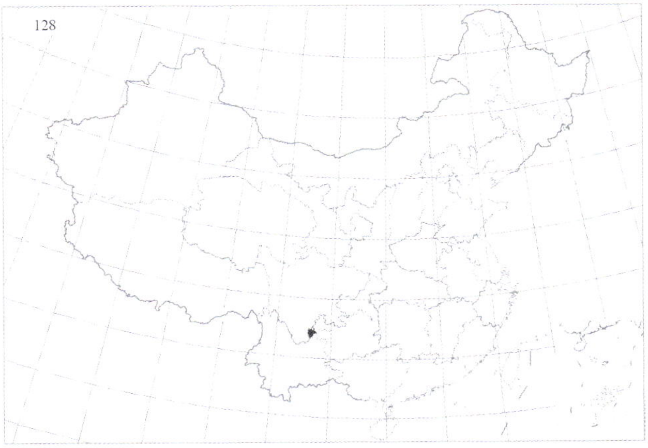

Pinaceae 松科 *Pinus* 松属

Pinus sylvestris L. 欧洲赤松

Trees, evergreen

-0.6 (-5.8~4.3); -23.3 (-29.8~-16.8); 19.2 (15.3~23.0); 6.5 (4.4~8.9); 49.1 (28.0~71.3); 116.1 (78.6~157.4); 521 (426~620) | 479 (372~629); 327 (258~417); 13 (6~16) | -7.7 (-28.7~15.9); 461 (372~620); 252.1 (185.5~308.8)

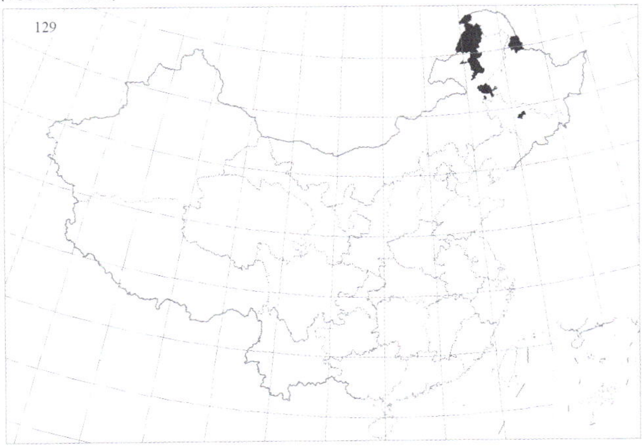

Pinaceae 松科 *Pinus* 松属

Pinus sylvestris var. mongolica Litv. 樟子松

Trees, evergreen

-1.1 (-5.8~3.9); -24.1 (-29.8~-16.1); 19.1 (15.3~21.9); 6.4 (4.4~8.3); 47.6 (28.0~64.9); 119.8 (77.4~157.4); 513 (426~597) | 501 (318~657); 328 (222~416); 15 (7~21) | -2.6 (-38.7~23.8); 469 (318~583); 274.0 (167.3~333.3)

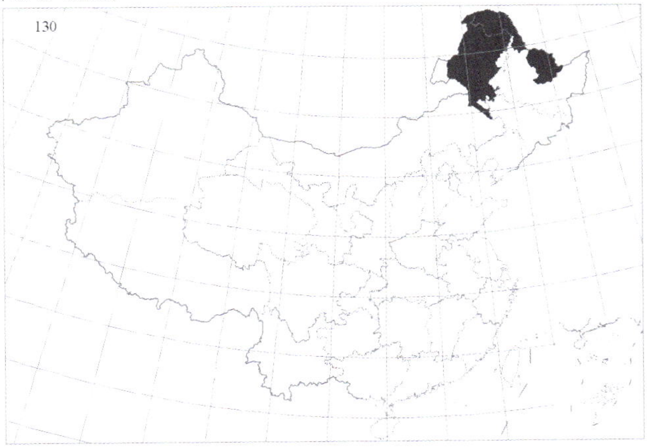

Pinaceae 松科 *Pinus* 松属

Pinus sylvestris var. sylvestriformis (Taken.) W. C. Cheng et C. D. Chu 长白松

Trees, evergreen

2.2; -17.2; 19.2; 6.9; 50.3; 83.8; 525 | 668; 415; 24 | 27.4; 525; 322.3

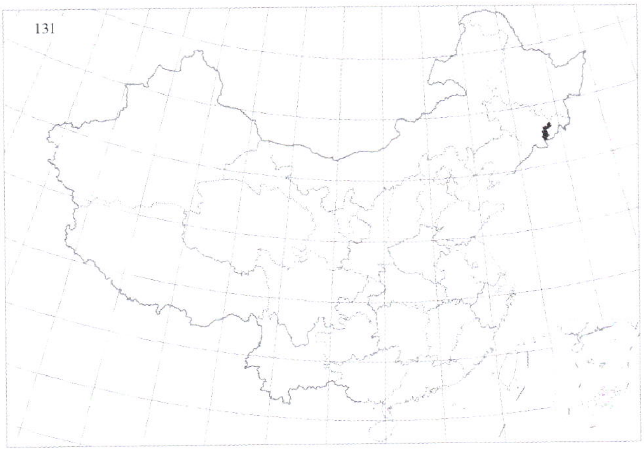

Pinaceae 松科 *Pinus* 松属

Pinus tabuliformis Carrière 油松

Trees, evergreen

9.1 (-4.1~16.4); -6.0 (-18.4~5.6); 22.5 (7.8~28.7); 10.4 (2.1~16.5); 83.7 (5.2~139.4); 34.3 (0.0~114.2); 668 (340~921) | 597 (127~1502); 340 (81~661); 21 (2~142) | -9.7 (-81.2~128.6); 548 (127~921); 178.1 (13.6~357.7)

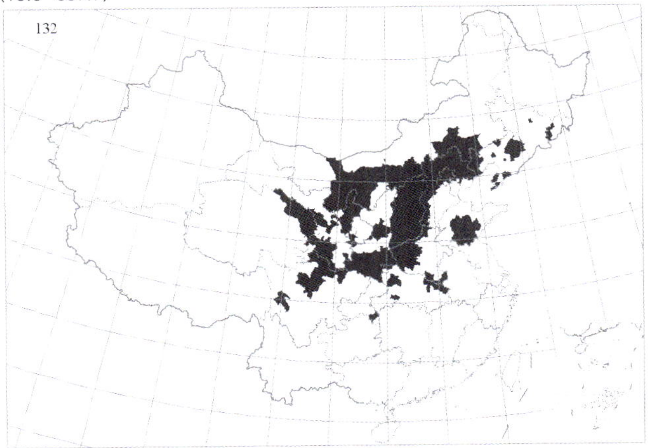

Pinaceae 松科 *Pinus* 松属

Pinus tabuliformis var. henryi (Mast.) C. T. Kuan 巴山松

Trees, evergreen

13.8 (9.3~17.1); 2.3 (-1.7~5.1); 24.8 (20.1~29.3); 13.8 (9.5~17.1); 111.0 (69.8~145.7); 5.2 (0.0~18.0); 771 (603~941) ❙ 1173 (760~1590); 498 (352~685); 72 (17~184) ❙ 52.4 (5.0~101.4); 771 (603~941); 240.3 (173.7~305.6)

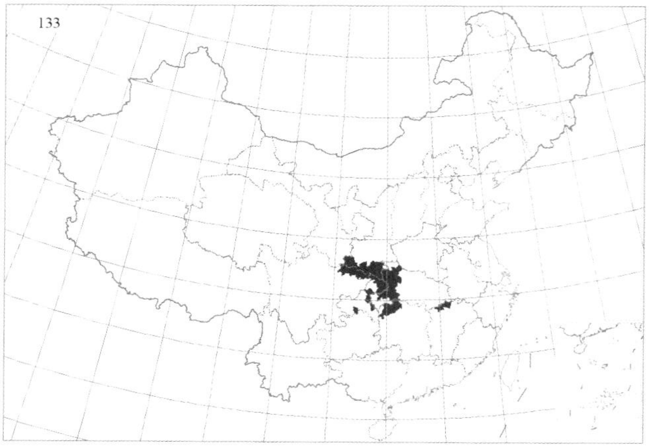

Pinaceae 松科 *Pinus* 松属

Pinus tabuliformis var. mukdensis (Uyeki ex Nakai) Uyeki 黑皮油松

Trees, evergreen

10.0 (2.5~13.5); -5.9 (-15.6~-1.6); 24.1 (18.6~27.1); 11.2 (6.9~13.7); 92.4 (48.9~117.5); 31.7 (14.6~78.3); 704 (523~820) ❙ 580 (457~735); 332 (258~457); 19 (9~26) ❙ -16.8 (-35.9~8.4); 577 (457~698); 190.0 (128.6~269.8)

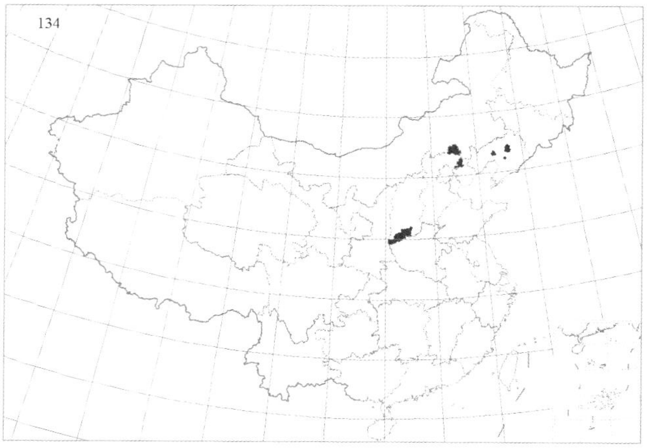

Pinaceae 松科 *Pinus* 松属

Pinus tabuliformis var. umbraculifera Liou et Z. Wang 扫帚油松

Trees, evergreen

8.2; -10.3; 24.3; 10.3; 87.0; 47.7; 678 ❙ 735; 457; 25 ❙ 8.4; 678; 219.8

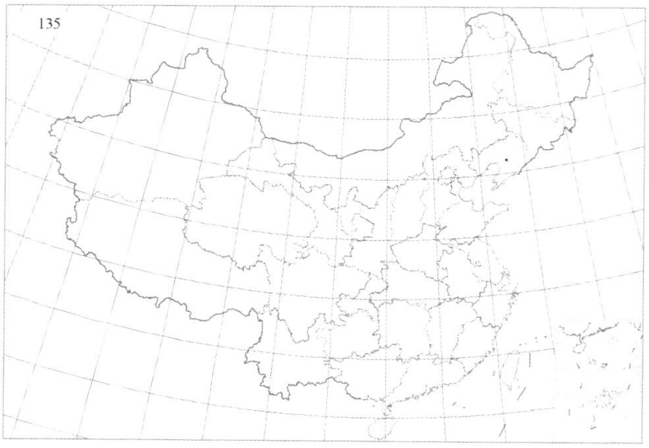

Pinaceae 松科 *Pinus* 松属

Pinus taiwanensis Hayata 黄山松

Trees, evergreen

16.9 (13.4~22.3); 5.8 (1.5~15.3); 27.3 (20.1~29.7); 16.9 (13.4~22.3); 143.5 (108.3~208.1); 0.7 (0.0~7.3); 901 (755~1202) ❙ 1569 (1076~2713); 569 (392~1412); 183 (55~303) ❙ 75.3 (20.3~246.5); 893 (755~1202); 229.2 (134.6~413.5)

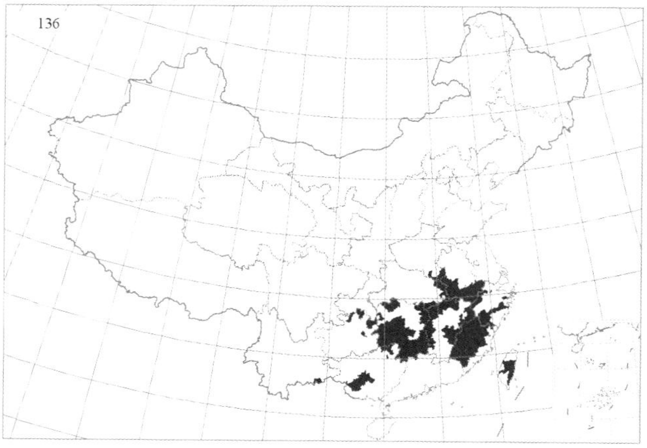

Pinaceae 松科 *Pinus* 松属

Pinus wallichiana A. B. Jacks. 乔松

Trees, evergreen

5.2 (-1.1~13.3); -2.9 (-9.6~5.4); 12.4 (6.2~19.6); 6.4 (2.0~13.4); 37.5 (2.5~100.4); 34.3 (0.0~75.6); 487 (334~706) ❙ 804 (321~1736); 473 (193~1007); 26 (9~51) ❙ 54.0 (-4.0~145.8); 486 (321~706); 243.6 (61.6~447.4)

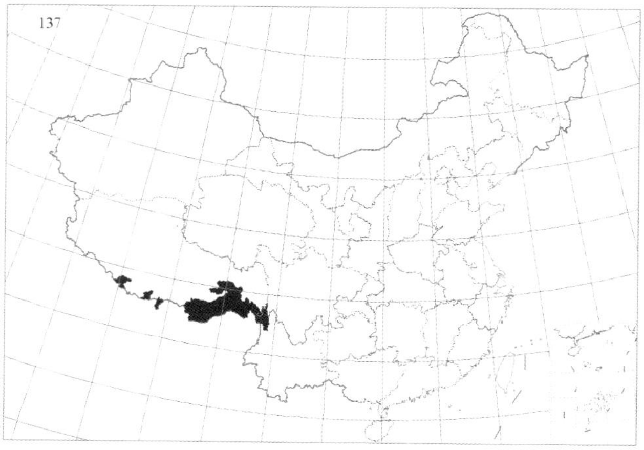

Pinaceae 松科 *Pinus* 松属

Pinus wangii Hu et W. C. Cheng 毛枝五针松

Trees, evergreen

17.4; 9.7; 22.9; 17.5; 149.7; 0.0; 848 ❙ 1338; 756; 50 ❙ 57.6; 848; 348.0

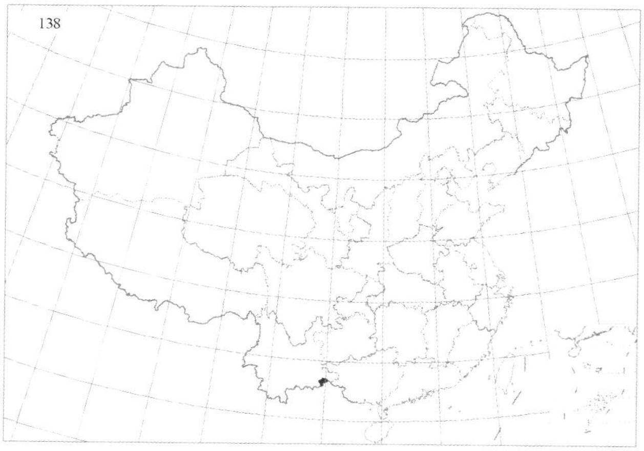

Pinaceae 松科 *Pinus* 松属

Pinus yunnanensis Franch. 云南松

Trees, evergreen

14.4 (0.0~21.6); 6.5 (-8.7~14.2); 20.8 (8.1~28.8); 14.5 (2.7~21.7); 117.6 (8.0~199.8); 4.5 (0.0~67.9); 763 (363~1153) ❙ 1094 (520~1900); 607 (317~1056); 47 (8~174) ❙ 45.3 (-14.2~125.9); 751 (363~1153); 282.9 (111.0~551.6)

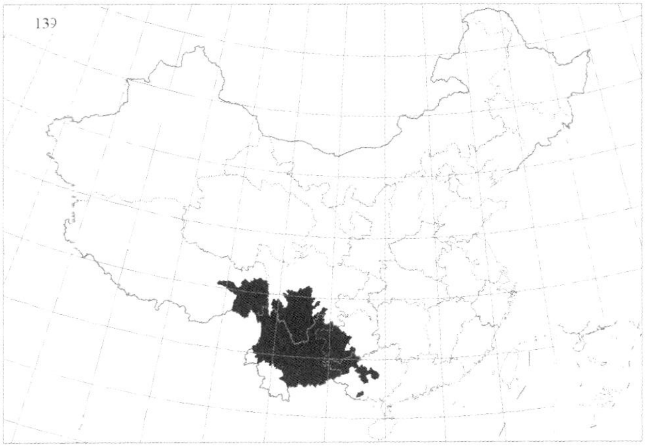

Pinaceae 松科 *Pinus* 松属

Pinus yunnanensis var. pygmaea (J. R. Xue) J. R. Xue 地盘松

Trees, evergreen

12.5 (6.6~17.5); 5.0 (-1.3~10.9); 18.4 (13.3~22.3); 12.5 (6.8~17.6); 94.0 (39.6~151.2); 4.1 (0.0~19.7); 686 (514~846) ❙ 1002 (862~1093); 588 (531~630); 25 (12~44) ❙ 49.4 (13.3~78.3); 674 (514~836); 306.6 (279.1~340.7)

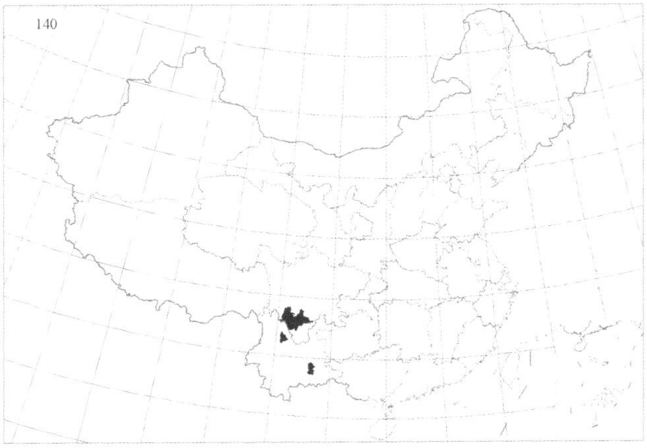

Pinaceae 松科 *Pinus* 松属

Pinus yunnanensis var. tenuifolia W. C. Cheng et Y. W. Law 细叶云南松

Trees, evergreen

17.7 (12.5~21.3); 8.7 (3.6~12.6); 24.8 (20.1~27.9); 17.7 (12.5~21.3); 152.7 (92.0~196.0); 0.1 (0.0~1.4); 901 (685~1123) ❙ 1272 (1071~1567); 669 (546~775); 65 (43~141) ❙ 42.7 (5.8~70.6); 901 (685~1121); 274.1 (189.4~491.5)

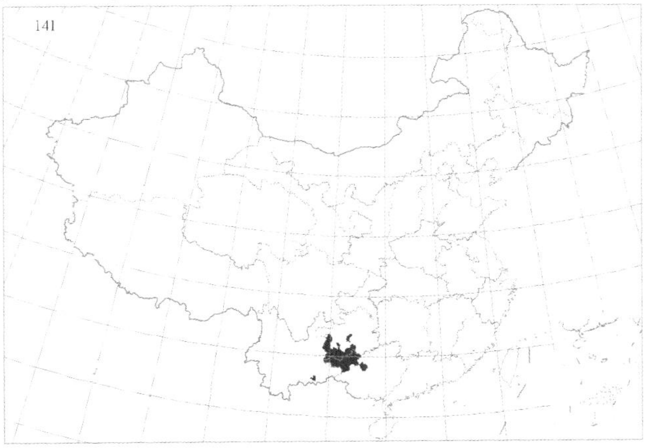

Pinaceae 松科 *Pseudolarix* 金钱松属

Pseudolarix amabilis (J. Nelson) Rehder 金钱松

Trees, evergreen

15.9 (10.6~21.1); 4.3 (-0.2~13.2); 27.1 (21.1~29.3); 16.0 (10.6~21.1); 133.6 (79.4~193.7); 1.8 (0.0~12.1); 869 (643~1104) ❙ 1363 (807~1997); 532 (349~753); 136 (27~254) ❙ 58.1 (5.6~130.7); 867 (643~1047); 207.2 (108.2~319.9)

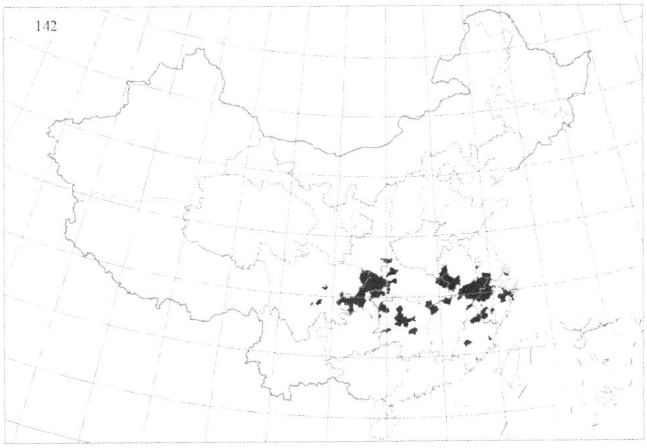

Pinaceae 松科 *Pseudotsuga* 黄杉属

Pseudotsuga brevifolia W. C. Cheng et L. K. Fu 短叶黄杉

Trees, evergreen

18.8 (16.0~22.2); 9.6 (5.7~13.9); 26.0 (23.6~28.3); 18.8 (16.0~22.3); 165.5 (132.2~207.5); 0.0 (0.0~0.0); 962 (829~1190) ❙ 1388 (1250~1495); 698 (528~848); 91 (57~142) ❙ 45.6 (15.4~64.7); 962 (829~1190); 281.5 (225.9~371.7)

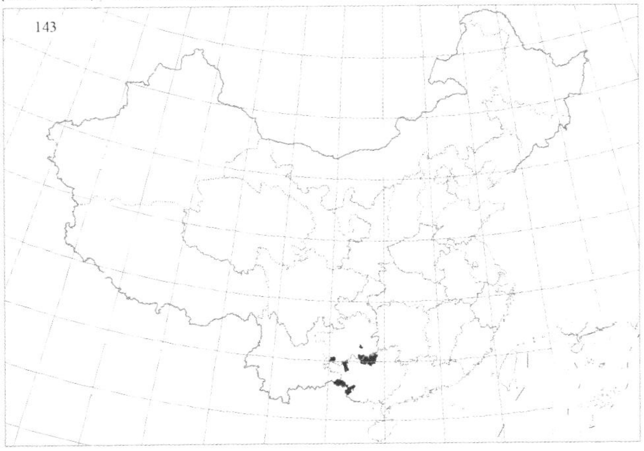

Pinaceae 松科 *Pseudotsuga* 黄杉属

Pseudotsuga forrestii Craib 澜沧黄杉

Trees, evergreen

13.3 (5.3~19.6); 6.4 (-2.4~13.9); 18.4 (12.2~22.7); 13.4 (5.7~19.6); 105.3 (30.5~175.1); 5.7 (0.0~26.2); 714 (479~933) ❙ 1184 (741~1500); 661 (426~816); 38 (18~52) ❙ 67.2 (40.1~100.7); 703 (479~882); 379.6 (221.3~530.1)

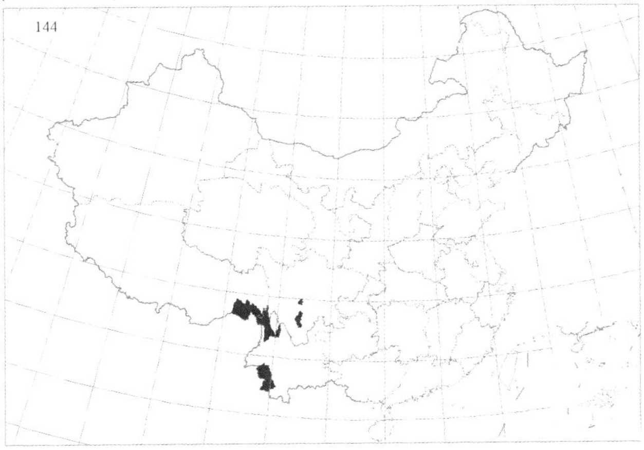

Pinaceae 松科 *Pseudotsuga* 黄杉属
Pseudotsuga sinensis Dode 黄杉
Trees, evergreen

15.1 (6.2~22.2); 4.8 (−4.1~13.9); 24.6 (14.1~29.3); 15.2 (6.7~22.3); 123.6 (39.4~207.5); 1.5 (0.0~29.9); 811 (509~1190) ❙ 1313 (669~1959); 551 (317~820); 109 (15~258) ❙ 61.9 (15.4~147.0); 809 (509~1190); 230.1 (152.4~371.7)

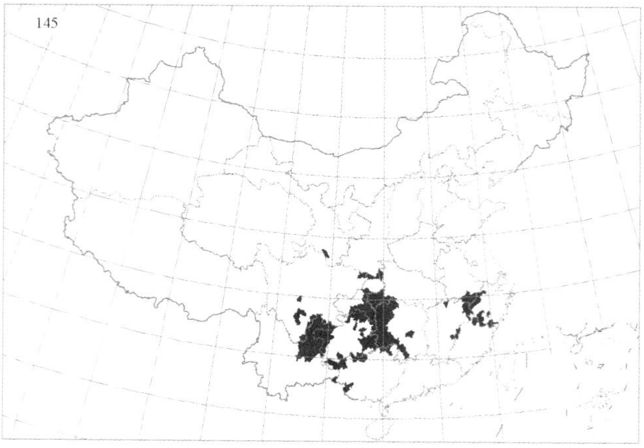

Pinaceae 松科 *Pseudotsuga* 黄杉属
Pseudotsuga sinensis var. wilsoniana (Hayata) L. K. Fu et Nan Li 台湾黄杉
Trees, evergreen

18.2; 12.3; 23.1; 18.2; 158.8; 0.0; 893 ❙ 2313; 1041; 229 ❙ 168.7; 882; 264.4

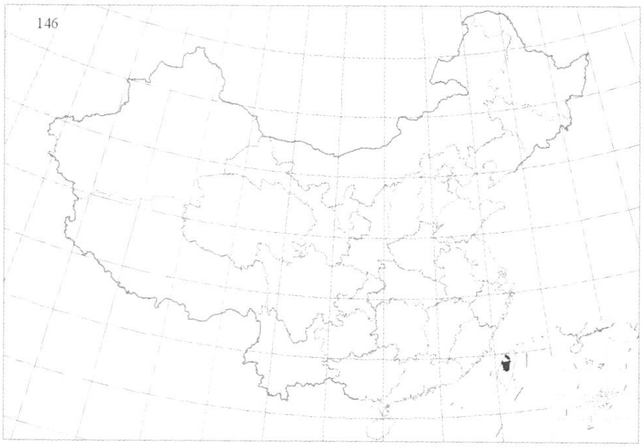

Pinaceae 松科 *Tsuga* 铁杉属
Tsuga chinensis (Franch.) Pritz. 铁杉
Trees, evergreen

14.0 (0.9~24.1); 3.0 (−9.5~18.9); 24.3 (9.8~29.8); 14.2 (3.6~24.1); 116.1 (14.0~229.7); 7.3 (0.0~63.1); 790 (403~1322) ❙ 1232 (512~2992); 520 (258~1847); 103 (14~402) ❙ 55.3 (−35.3~246.5); 776 (403~1130); 233.6 (121.6~413.5)

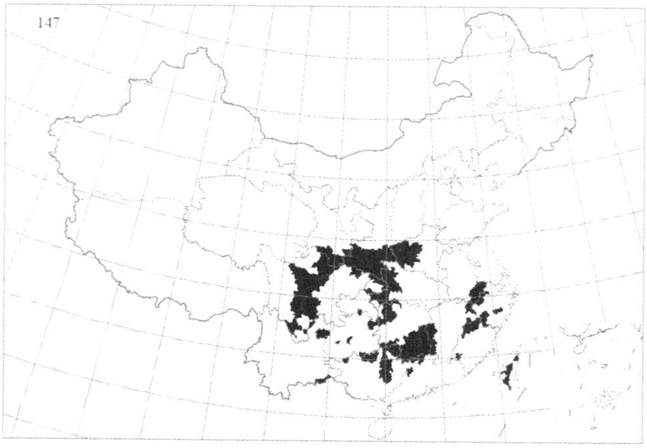

Pinaceae 松科 *Tsuga* 铁杉属
Tsuga chinensis var. forrestii (Downie) Silba 丽江铁杉
Trees, evergreen

9.6 (3.8~14.3); 2.1 (−4.3~7.6); 15.8 (10.8~19.7); 9.7 (4.7~14.3); 65.3 (21.6~111.9); 10.1 (0.0~35.4); 599 (444~728) ❙ 965 (801~1042); 585 (496~641); 22 (12~40) ❙ 64.0 (36.1~121.7); 594 (444~707); 306.0 (265.2~351.7)

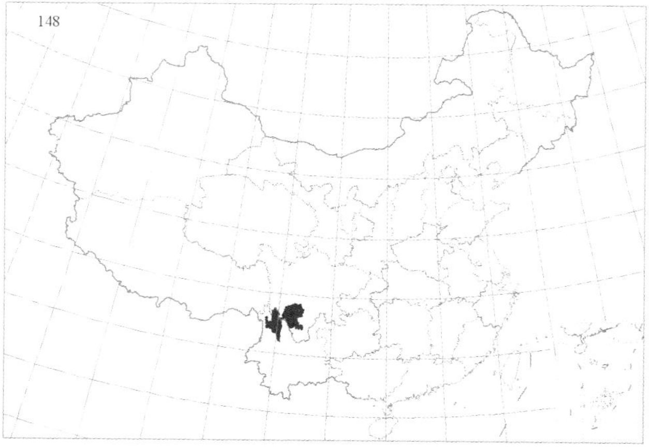

Pinaceae 松科 *Tsuga* 铁杉属
Tsuga chinensis var. robusta C. Y. Cheng et L. K. Fu 大果铁杉
Trees, evergreen

13.0; 1.9; 23.8; 13.1; 102.9; 6.1; 730 ❙ 1202; 500; 83 ❙ 64.6; 730; 262.4

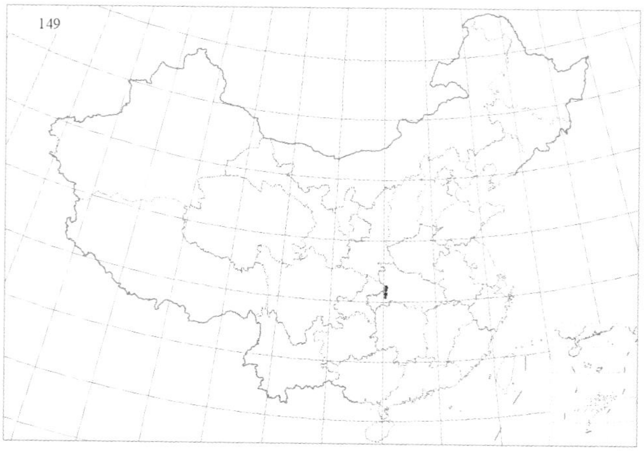

Pinaceae 松科 *Tsuga* 铁杉属
Tsuga dumosa (D. Don) Eichler 云南铁杉
Trees, evergreen

11.7 (−1.8~18.9); 3.8 (−10.5~12.5); 18.4 (6.1~26.6); 12.0 (1.8~19.0); 91.0 (1.9~167.8); 10.0 (0.0~82.8); 672 (329~929) ❙ 998 (311~1736); 574 (193~1007); 33 (9~70) ❙ 50.8 (−14.2~145.8); 653 (311~912); 276.9 (61.6~487.2)

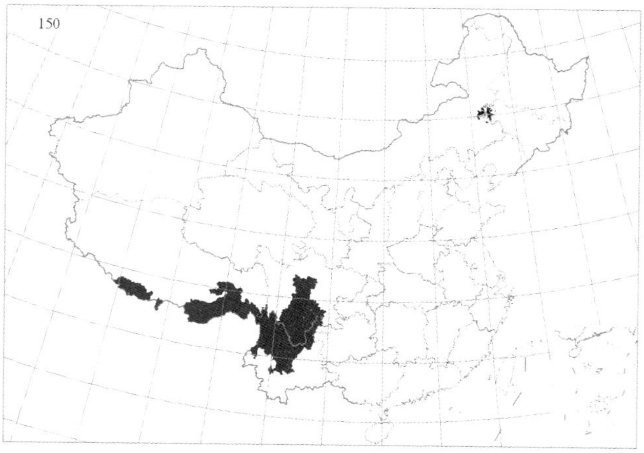

Pinaceae 松科 *Tsuga* 铁杉属

Tsuga long bracteata W. C. Cheng 长苞铁杉

Trees, evergreen

17.8 (14.3~22.3); 7.4 (3.3~13.8); 27.1 (23.5~29.0); 17.8 (14.3~22.3); 154.4 (112.6~208.1); 0.2 (0.0~2.3); 939 (756~1202) ❙ 1535 (1211~1865); 581 (403~839); 172 (76~258) ❙ 64.9 (20.3~104.6); 936 (756~1202); 224.2 (151.0~335.1)

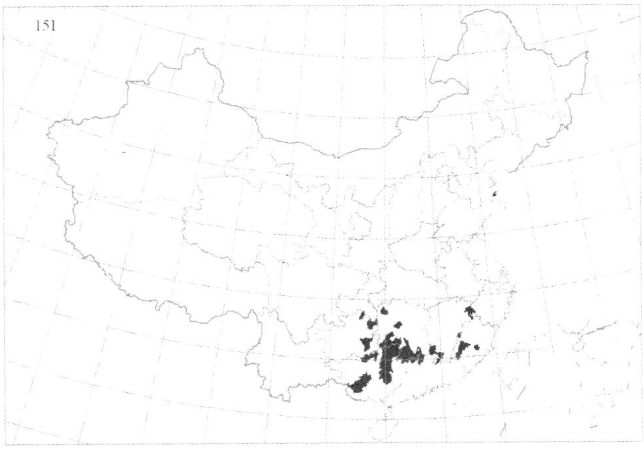

Pinaceae 松科 *Tsuga* 铁杉属

Tsuga oblongisquamata (W. C. Cheng et L. K. Fu) L. K. Fu et Nan Li 矩鳞铁杉

Trees, evergreen

6.9 (2.2~13.0); -3.2 (-7.7~1.9); 16.0 (9.9~23.8); 7.6 (3.8~13.1); 51.0 (15.5~102.9); 28.0 (6.1~49.3); 542 (413~730) ❙ 869 (656~1246); 434 (317~546); 26 (14~83) ❙ 65.4 (-0.2~125.9); 542 (413~730); 273.5 (200.8~357.7)

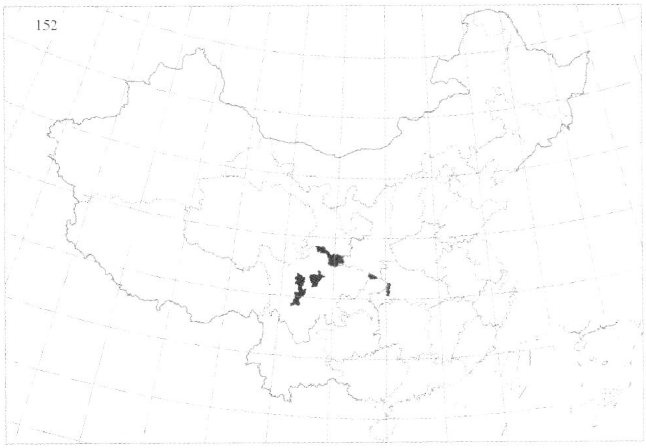

Taxodiaceae 杉科 *Cryptomeria* 柳杉属

Cryptomeria japonica (Thunb. ex L. f.) D. Don 日本柳杉

Trees, evergreen

15.9 (5.4~21.0); 6.6 (-3.5~14.2); 24.1 (12.3~29.9); 15.9 (5.8~21.0); 132.0 (31.5~191.8); 0.8 (0.0~28.1); 827 (481~1108) ❙ 1305 (741~1997); 576 (323~996); 106 (15~259) ❙ 57.3 (-14.2~147.0); 818 (481~1108); 257.3 (139.5~551.6)

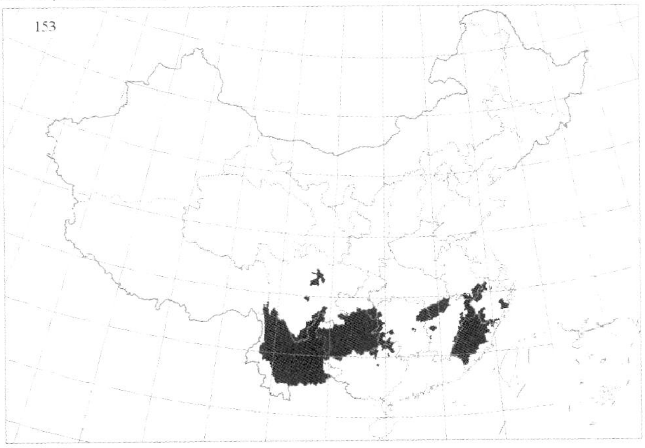

Taxodiaceae 杉科 *Cunninghamia* 杉属

Cunninghamia lanceolata (Lamb.) Hook. 杉木

Trees, evergreen

16.5 (2.0~24.7); 6.1 (-7.2~19.8); 25.9 (9.9~30.1); 16.6 (3.8~24.7); 140.8 (15.4~236.5); 2.0 (0.0~51.4); 885 (413~1376) ❙ 1330 (626~2435); 557 (299~1428); 123 (15~259) ❙ 50.7 (-22.2~147.0); 871 (413~1257); 235.2 (110.1~659.0)

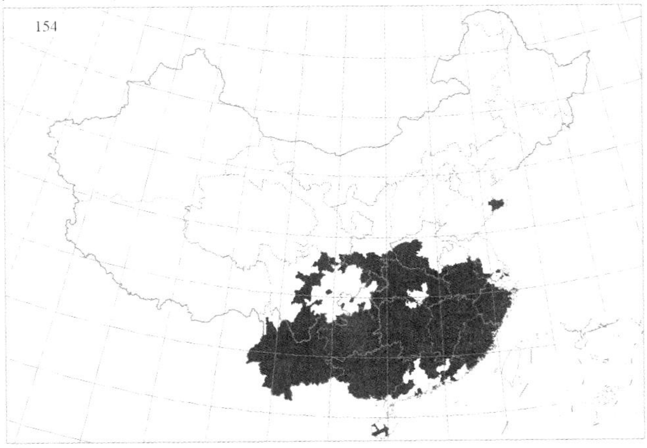

Taxodiaceae 杉科 *Glyptostrobus* 水松属

Glyptostrobus pensilis (Staunton ex D. Don) K. Koch 水松

Trees, semievergreen

18.8 (10.6~23.0); 9.0 (-0.2~15.2); 27.6 (21.1~29.7); 18.8 (10.6~23.0); 166.2 (79.4~216.2); 0.3 (0.0~12.1); 996 (643~1247) ❙ 1526 (950~2148); 629 (411~1234); 141 (27~259) ❙ 54.6 (8.0~137.5); 993 (643~1247); 208.2 (95.3~551.6)

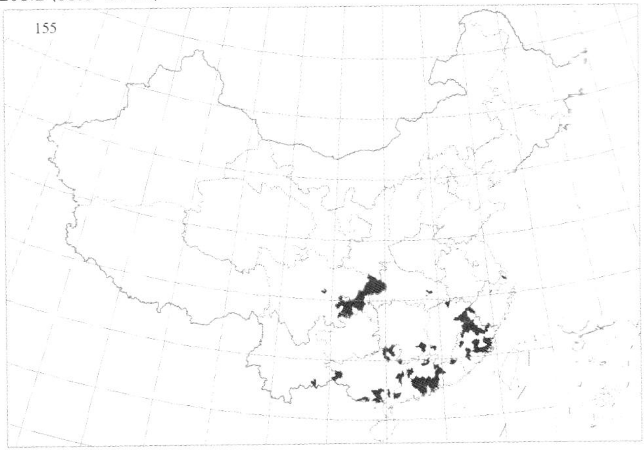

Taxodiaceae 杉科 *Metasequoia* 水杉属

Metasequoia glyptostroboides Hu et W. C. Cheng 水杉

Trees, deciduous

18.1 (12.7~22.1); 7.9 (2.1~13.7); 27.2 (23.0~29.8); 18.1 (12.8~22.1); 157.9 (99.0~205.6); 0.5 (0.0~5.6); 954 (713~1182) ❙ 1569 (1106~1852); 579 (439~779); 166 (64~216) ❙ 65.6 (22.1~107.2); 950 (713~1182); 224.9 (135.3~290.3)

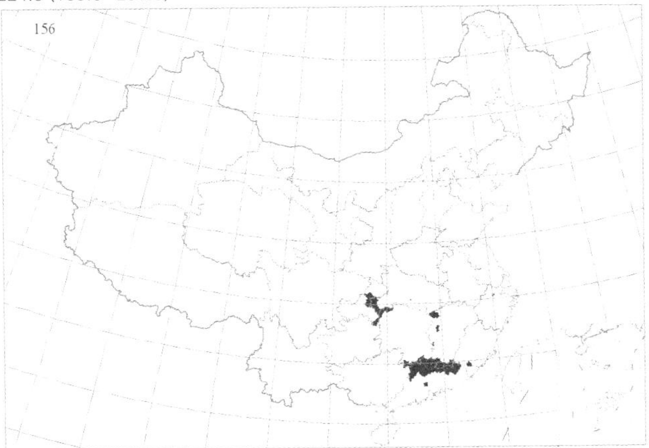

T

Taxodiaceae 杉科 *Taiwania* 台湾杉属

Taiwania cryptomerioides Hayata 台湾杉

Trees, evergreen

14.8 (5.3~20.8); 6.9 (−2.4~15.3); 21.6 (12.2~25.7); 14.9 (5.7~20.8); 120.2 (30.5~189.2); 1.8 (0.0~26.2); 771 (479~1019) ❙ 1625 (950~2859); 770 (476~1594); 122 (27~402) ❙ 108.2 (12.1~246.5); 771 (479~1019); 279.1 (133.8~487.2)

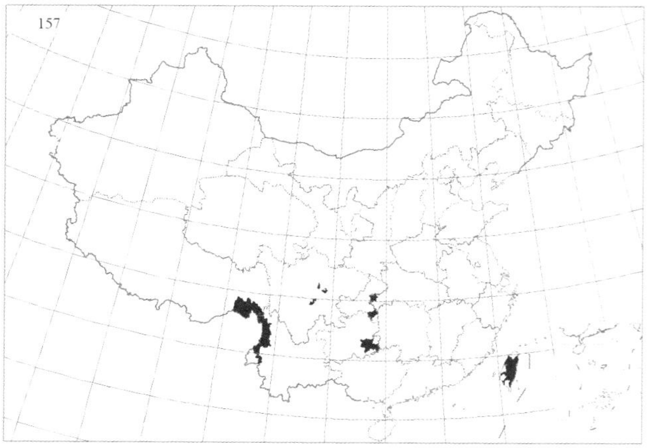

Cupressaceae 柏科 *Calocedrus* 翠柏属

Calocedrus macrolepis Kurz 翠柏

Trees, evergreen

17.9 (14.3~24.3); 10.2 (4.0~18.7); 24.1 (19.6~28.7); 18.0 (14.3~24.3); 155.9 (112.6~231.9); 0.1 (0.0~1.1); 911 (736~1338) ❙ 1362 (868~2993); 638 (312~1186); 88 (28~547) ❙ 53.0 (−34.3~246.5); 871 (690~1163); 310.3 (148.9~572.9)

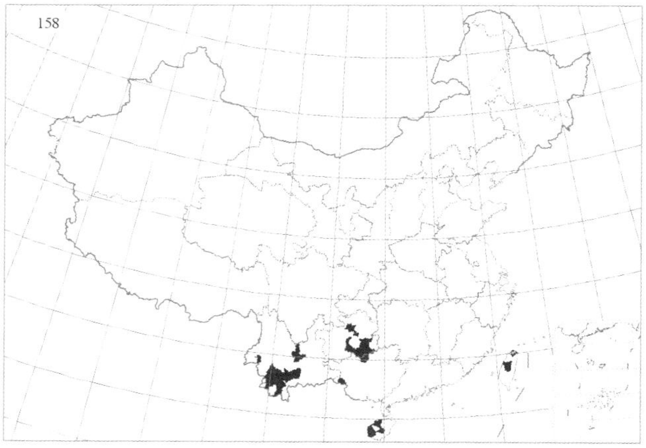

Cupressaceae 柏科 *Calocedrus* 翠柏属

Calocedrus macrolepis var. formosana (Florin) W. C. Cheng et L. K. Fu 台湾翠柏

Trees, evergreen

18.8 (16.2~22.1); 12.7 (11.0~15.3); 24.1 (20.1~28.8); 18.9 (16.2~22.1); 166.4 (134.2~205.6); 0.0 (0.0~0.0); 928 (773~1182) ❙ 2481 (1784~2993); 973 (750~1412); 313 (110~551) ❙ 173.3 (50.9~246.5); 928 (773~1182); 259.0 (157.7~303.6)

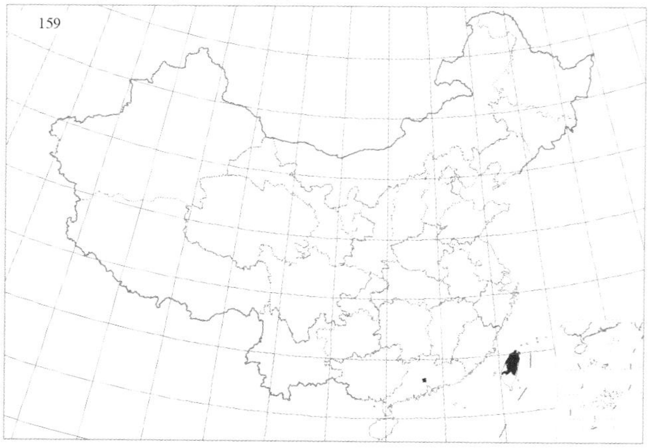

Cupressaceae 柏科 *Chamaecyparis* 扁柏属

Chamaecyparis formosensis Matsum. 红桧

Trees, evergreen

18.9 (16.2~21.1); 13.3 (11.0~15.3); 23.8 (20.1~27.4); 19.0 (16.2~21.1); 167.9 (134.2~193.1); 0.0 (0.0~0.0); 929 (773~1077) ❙ 2810 (2445~3495); 1060 (821~1412); 405 (110~897) ❙ 205.4 (140.0~246.5); 929 (773~1077); 269.6 (207.7~363.7)

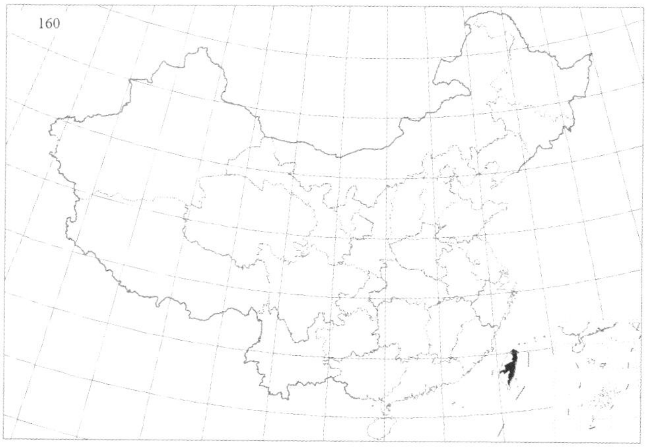

Cupressaceae 柏科 *Chamaecyparis* 扁柏属

Chamaecyparis obtusa var. formosana (Hayata) Rehder 台湾扁柏

Trees, evergreen

19.4 (16.2~21.7); 13.4 (11.0~15.3); 24.7 (20.1~27.4); 19.5 (16.2~21.7); 173.5 (134.2~200.2); 0.0 (0.0~0.0); 965 (773~1118) ❙ 2686 (1836~3495); 1028 (821~1412); 389 (110~897) ❙ 183.2 (64.2~246.5); 960 (773~1077); 246.7 (202.8~289.1)

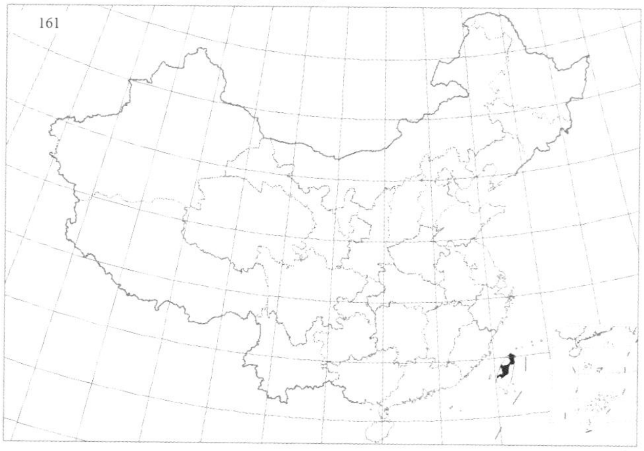

Cupressaceae 柏科 *Cupressus* 柏属

Cupressus chengiana S. Y. Hu 岷江柏木

Trees, evergreen

6.2 (2.0~16.2); −3.4 (−7.2~5.6); 14.9 (9.9~26.2); 7.1 (3.8~16.3); 45.6 (15.4~135.4); 30.2 (0.0~51.4); 525 (413~865) ❙ 808 (626~1023); 431 (305~578); 20 (11~40) ❙ 61.1 (−2.2~128.6); 524 (413~865); 267.4 (164.4~357.7)

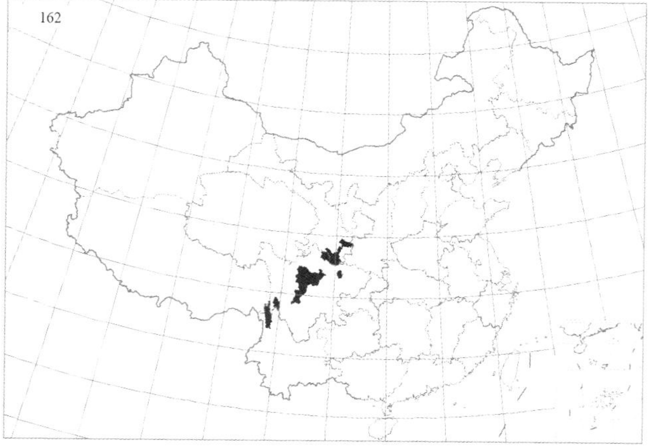

Cupressaceae 柏科 *Cupressus* 柏属

Cupressus chengiana var. jiangeensis (N. Zhao) Silba 剑阁柏木

Trees, evergreen

16.2；5.6；26.2；16.3；135.4；0.0；865 ▎945；533；24 ▎9.2；865；164.4

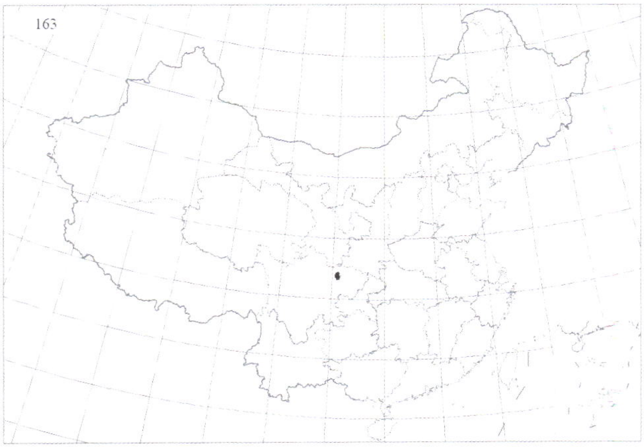

Cupressaceae 柏科 *Cupressus* 柏属

Cupressus duclouxiana B. Hickel 干香柏

Trees, evergreen

13.8 (2.2~20.7)；6.0 (-6.3~12.0)；20.1 (9.5~29.0)；13.9 (3.9~20.8)；109.2 (15.8~189.3)；3.2 (0.0~48.6)；731 (411~1108) ▎1028 (628~1688)；580 (317~959)；39 (11~135) ▎41.8 (-14.2~112.2)；714 (411~1108)；285.0 (142.9~627.0)

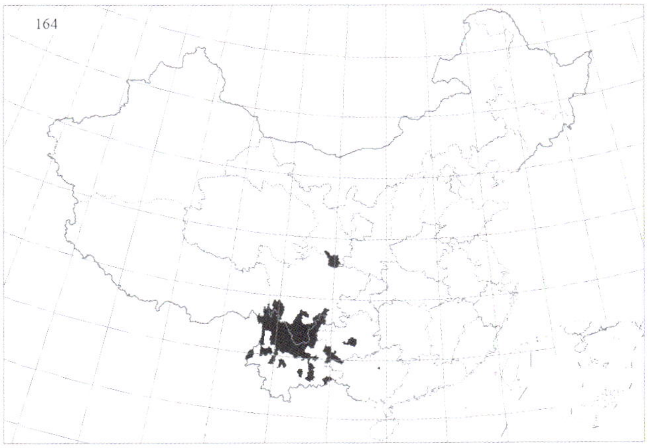

Cupressaceae 柏科 *Cupressus* 柏属

Cupressus funebris Endl. 柏木

Trees, evergreen

16.3 (6.2~24.1)；5.5 (-2.5~18.3)；26.2 (14.1~29.9)；16.3 (6.7~24.1)；137.3 (39.4~229.4)；1.4 (0.0~24.1)；872 (509~1321) ▎1362 (677~1997)；535 (336~821)；129 (14~259) ▎56.3 (-5.3~130.7)；866 (509~1294)；217.5 (108.2~474.7)

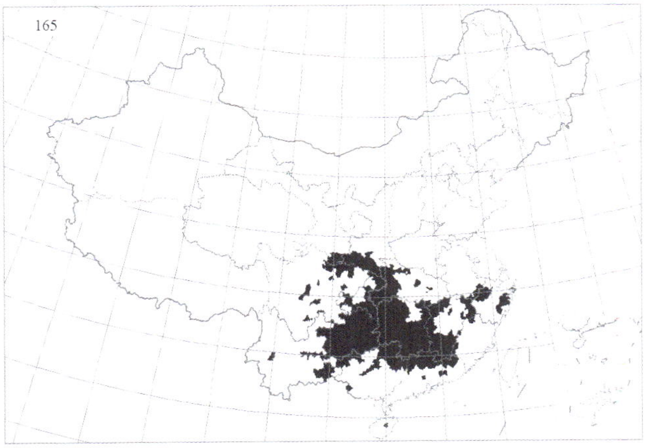

Cupressaceae 柏科 *Cupressus* 柏属

Cupressus gigantea W. C. Cheng et L. K. Fu 巨柏

Trees, evergreen

2.6；-6.3；10.6；4.2；19.5；47.6；416 ▎481；295；8 ▎14.1；401；224.4

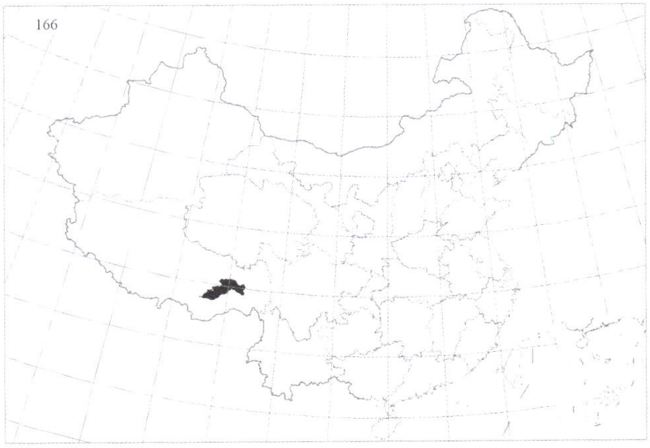

Cupressaceae 柏科 *Cupressus* 柏属

Cupressus torulosa D. Don ex Lamb. 西藏柏木

Trees, evergreen

6.9；-1.3；13.9；7.7；49.2；26.3；530 ▎1020；594；31 ▎81.4；530；296.8

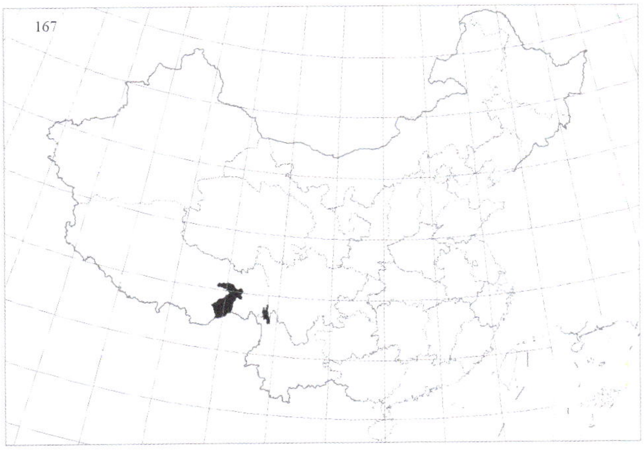

Cupressaceae 柏科 *Fokienia* 福建柏属

Fokienia hodginsii (Dunn) A. Henry et H. H. Thomas 福建柏

Trees, evergreen

17.7 (11.7~22.7)；7.9 (1.8~15.0)；26.3 (19.3~29.3)；17.7 (11.7~22.8)；152.9 (84.6~213.1)；0.3 (0.0~5.0)；924 (661~1221) ▎1502 (942~2435)；622 (419~1428)；146 (27~259) ▎63.2 (12.1~147.0)；923 (661~1221)；245.6 (126.1~551.6)

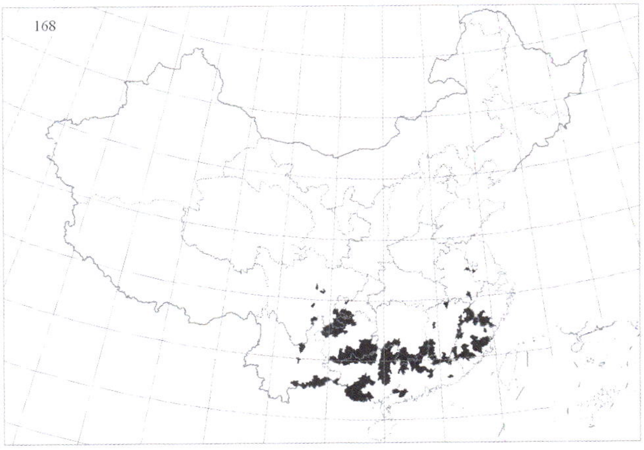

Cupressaceae 柏科 *Juniperus* 刺柏属
Juniperus baimashanensis Y. F. Yu et L. K. Fu 德钦柏
Trees, evergreen

5.4；-2.1；12.3；5.8；31.5；25.6；481 ❙ 741；426；28 ❙ 54.0；481；273.5

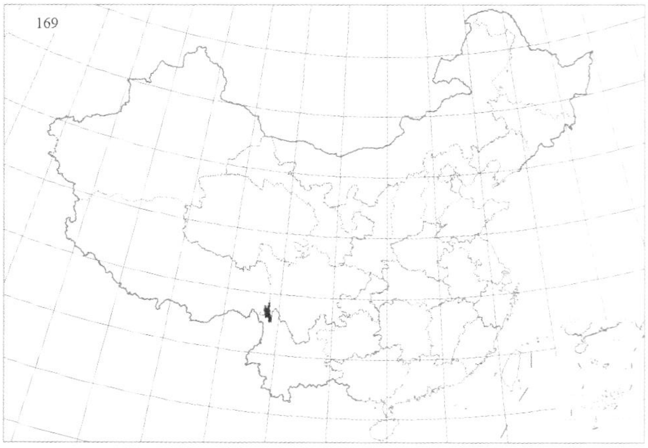

Cupressaceae 柏科 *Juniperus* 刺柏属
Juniperus centrasiatica Kom. 昆仑方枝柏
Trees, evergreen

5.6 (-4.3~12.4)；-11.2 (-20.1~-5.2)；19.5 (9.6~26.1)；8.5 (2.5~13.2)；66.9 (9.7~114.1)；58.8 (24.3~121.1)；597 (329~811) ❙ 74 (38~215)；29 (17~80)；9 (3~26) ❙ -84.5 (-95.3~-51.1)；74 (38~215)；25.2 (2.8~65.1)

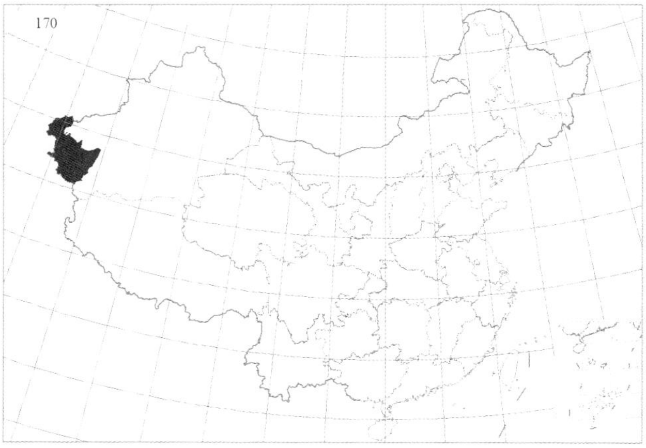

Cupressaceae 柏科 *Juniperus* 刺柏属
Juniperus chengii L. K. Fu et Y. F. Yu 万钧柏
Trees, evergreen

6.1；-1.5；12.7；6.3；35.2；21.8；496 ❙ 801；496；21 ❙ 61.4；496；282.9

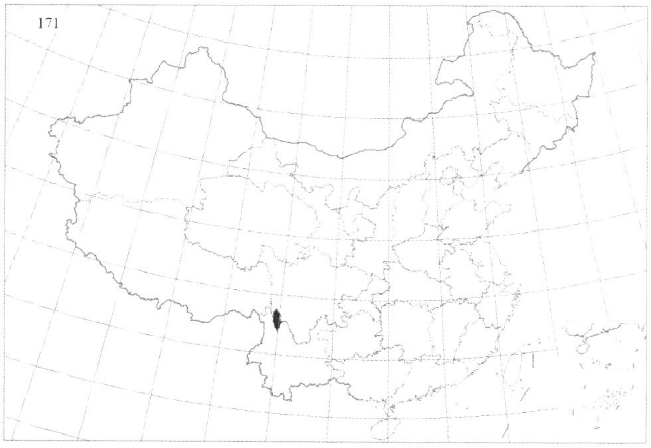

Cupressaceae 柏科 *Juniperus* 刺柏属
Juniperus chinensis L. 圆柏
Trees, evergreen

14.0 (0.8~24.1)；2.1 (-16.1~18.3)；25.0 (9.9~30.1)；14.4 (3.7~24.1)；119.3 (15.1~229.4)；10.4 (0.0~81.1)；803 (409~1321) ❙ 1069 (70~2499)；468 (23~947)；88 (2~303) ❙ 30.6 (-91.1~219.3)；759 (70~1294)；200.4 (13.9~474.7)

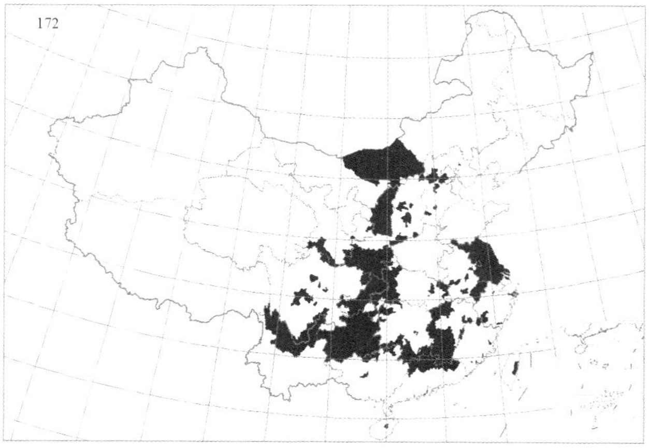

Cupressaceae 柏科 *Juniperus* 刺柏属
Juniperus chinensis var. tsukusiensis (Masam.) Masam. 清水圆柏
Shrubs or trees, evergreen

16.4；11.5；20.3；16.5；137.8；0.0；783 ❙ 2499；947；303 ❙ 219.3；783；303.6

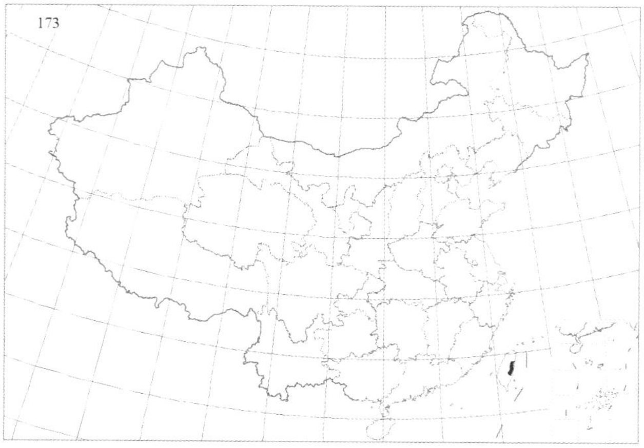

Cupressaceae 柏科 *Juniperus* 刺柏属
Juniperus convallium Rehder et E. H. Wilson 密枝圆柏
Trees, evergreen

1.9 (-3.2~7.3)；-7.7 (-14.4~-2.0)；10.5 (7.3~15.8)；4.0 (2.0~7.5)；18.3 (4.0~48.2)；55.0 (20.4~101.5)；414 (335~537) ❙ 664 (425~1005)；379 (259~546)；15 (8~27) ❙ 59.8 (14.7~128.6)；414 (335~537)；263.5 (111.0~357.7)

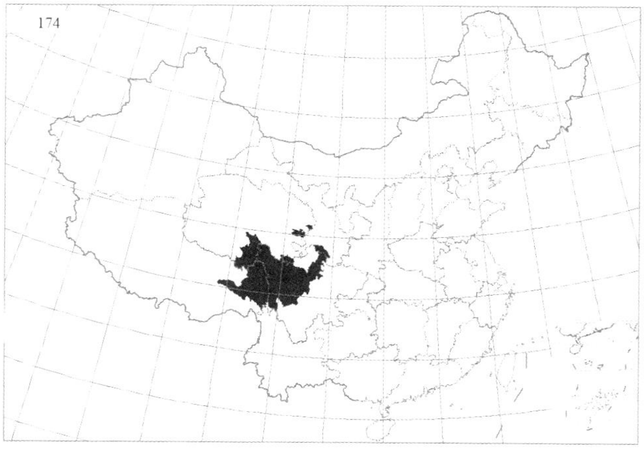

Cupressaceae 柏科 *Juniperus* 刺柏属

Juniperus convallium var. microsperma (W. C. Cheng et L. K. Fu) Silba 小子圆柏

Small trees, evergreen

1.3; -8.1 9.9; 3.7; 15.7; 59.1; 394 ❙ 534; 322; 11 ❙ 36.2; 394; 246.5

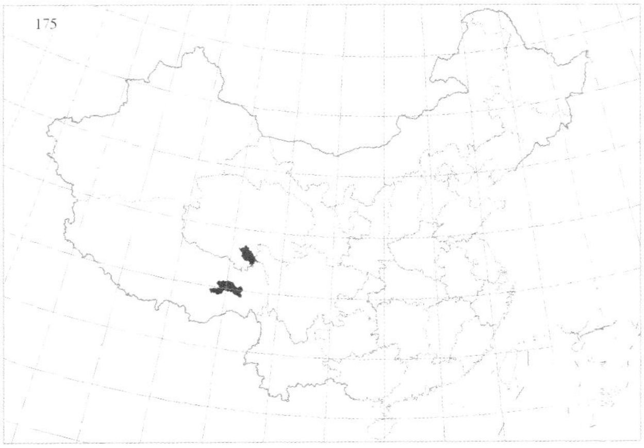

Cupressaceae 柏科 *Juniperus* 刺柏属

Juniperus davurica Pall. 兴安圆柏

Shrubs, evergreen

2.3 (-5.8~10.5); -18.1 (-29.8~-4.9); 20.2 (15.3~25.0); 7.5 (4.4~11.3); 57.4 (28.0~93.2); 88.9 (26.8~157.4); 554 (426~705) ❙ 581 (297~1018); 381 (203~648); 18 (5~40) ❙ 4.5 (-44.3~68.8); 493 (297~698); 253.1 (127.8~346.5)

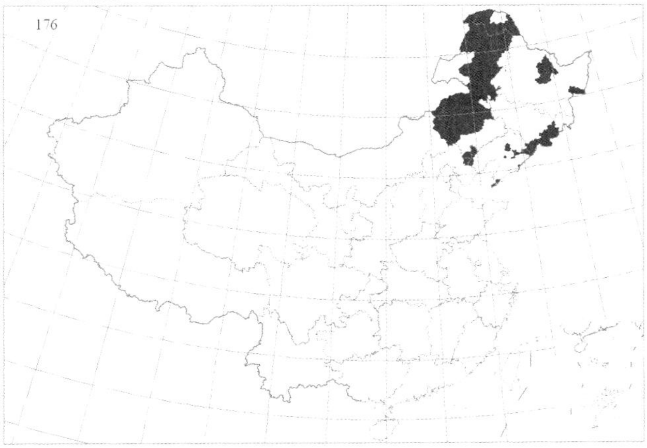

Cupressaceae 柏科 *Juniperus* 刺柏属

Juniperus formosana Hayata 刺柏

Trees, evergreen

14.7 (-2.3~21.7); 3.5 (-13.5~15.3); 25.1 (5.4~29.9); 14.9 (1.6~2 .7); 124.3 (0.4~200.2); 7.8 (0.0~89.7); 826 (324~1128) ❙ 1264 (252~2859); 500 (144~1412); 129 (1~402) ❙ 51.3 (-59.3~246.5); 808 (252~11 4); 218.7 (46.9~487.2)

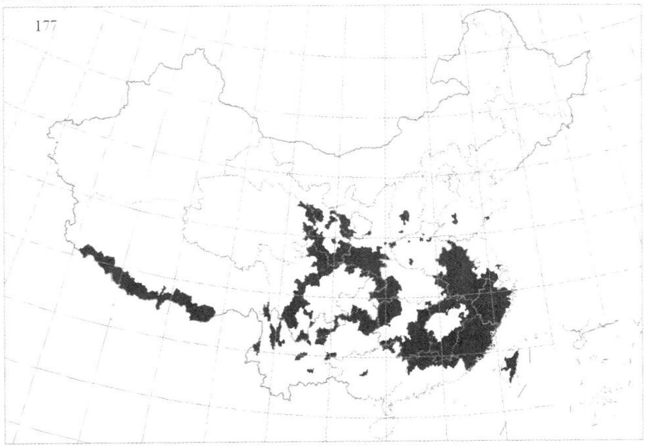

Cupressaceae 柏科 *Juniperus* 刺柏属

Juniperus gaussenii W. C. Cheng 昆明柏

Shrubs or small trees, evergreen

13.8 (5.4~18.9); 6.2 (-2.1~12.0); 20.1 (12.3~26.6); 13.9 (5.8~19.0); 107.8 (31.5~167.8); 1.8 (0.0~25.6); 727 (481~929) ❙ 967 (741~1254); 563 (411~699); 31 (15~57) ❙ 34.5 (-14.2~70.8); 698 (481~912); 272.5 (142.9~351.7)

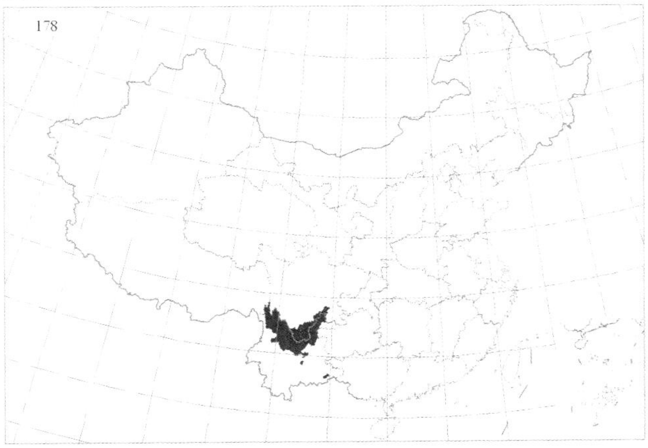

Cupressaceae 柏科 *Juniperus* 刺柏属

Juniperus indica Bertol. 滇藏方枝柏

Shrubs, evergreen

4.7 (-1.8~13 3); -3.5 (-10.5~6.4); 11.9 (6.1~19.6); 6.1 (1.8~13.4); 35.0 (1.9~100.4); 38.5 (0.0~82.8); 477 (329~706) ❙ 795 (311~1736); 472 (193~1007); 26 (10~69) ❙ 57.4 (-6.3~145.8); 474 (311~706); 250.0 (61.6~447.4)

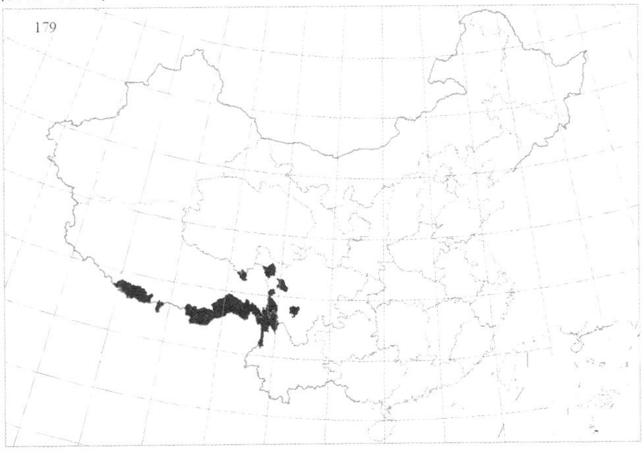

Cupressaceae 柏科 *Juniperus* 刺柏属

Juniperus komarovii Florin 塔枝圆柏

Small trees, evergreen

5.8 (-1.7~18.1); -3.9 (-12.3~8.1); 14.3 (8.0~27.3); 7.1 (2.4~18.2); 46.4 (6.4~158.1); 36.3 (0.0~86.1); 524 (344~949) ❙ 844 (502~1212); 477 (301~692); 21 (9~65) ❙ 67.0 (12.1~128.6); 523 (344~949); 250 6 (133.8~357.7)

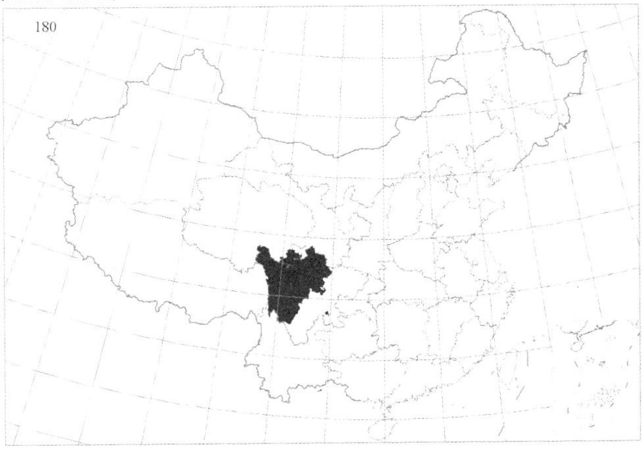

Cupressaceae 柏科 *Juniperus* 刺柏属
Juniperus pingii W. C. Cheng ex Ferré 垂枝香柏
Trees, evergreen

7.9 (−2.2~18.9); −0.6 (−11.9~12.0); 15.4 (5.7~26.6); 9.0 (1.7~19.0); 62.9 (1.0~167.8); 27.1 (0.0~91.0); 570 (327~929) ❙ 807 (212~1337); 466 (156~727); 25 (0~83) ❙ 41.0 (−35.9~128.6); 545 (212~912); 244.5 (61.6~407.1)

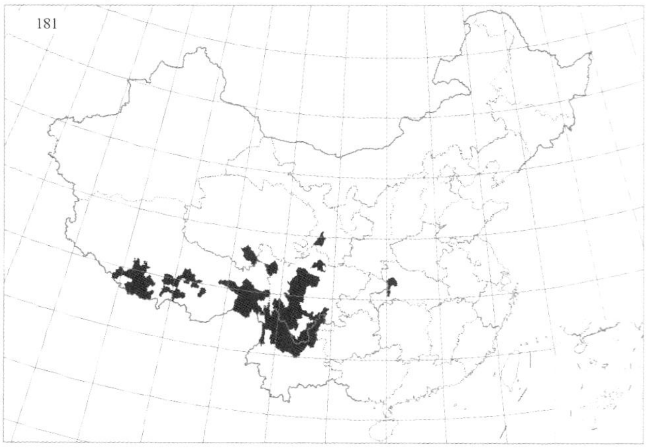

Cupressaceae 柏科 *Juniperus* 刺柏属
Juniperus pingii var. wilsonii (Rehder) Silba 香柏
Shrubs or trees, evergreen

5.8 (−2.2~16.2); −3.5 (−11.9~7.6); 14.2 (5.7~26.7); 7.2 (1.7~16.3); 46.7 (1.0~135.2); 36.6 (0.0~91.0); 519 (327~869) ❙ 770 (212~1265); 423 (156~619); 22 (0~68) ❙ 45.0 (−35.9~128.6); 501 (212~869); 224.8 (61.6~372.3)

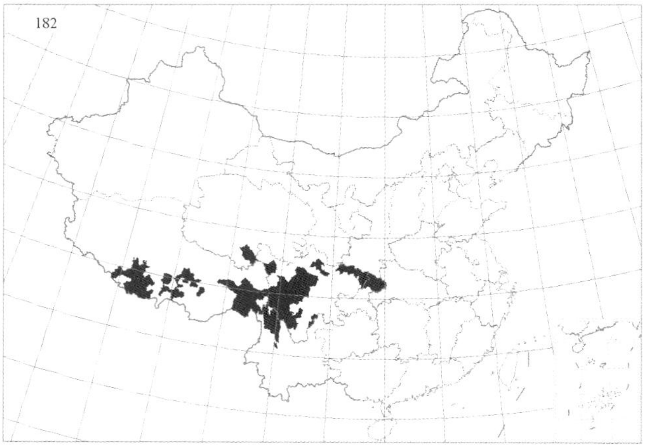

Cupressaceae 柏科 *Juniperus* 刺柏属
Juniperus procumbens (Siebold ex Endl.) Miq. 铺地柏
Shrubs, evergreen

16.9; 6.1; 26.9; 16.8; 143.3; 1.1; 892 ❙ 1232; 525; 119 ❙ 39.4; 862; 230.4

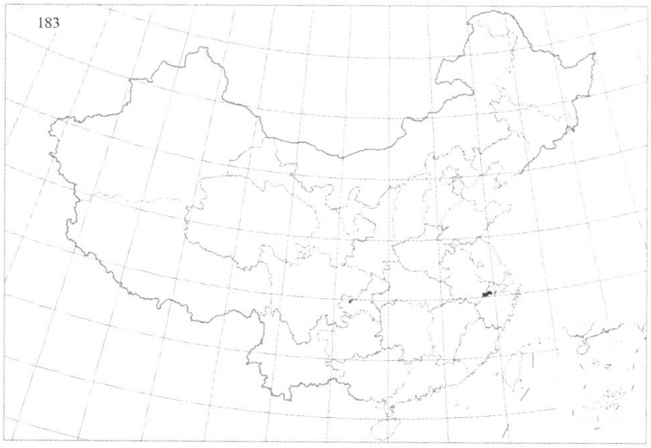

Cupressaceae 柏科 *Juniperus* 刺柏属
Juniperus przewalskii Kom. 祁连圆柏
Trees, evergreen

1.4 (−4.8~7.5); −11.3 (−18.0~−4.1); 12.5 (7.2~23.5); 4.6 (1.9~9.8); 25.7 (3.7~81.3); 68.7 (29.9~121.6); 434 (332~654) ❙ 431 (63~829); 242 (45~400); 9 (1~23) ❙ 2.0 (−87.1~94.4); 354 (63~522); 195.0 (7.5~364.1)

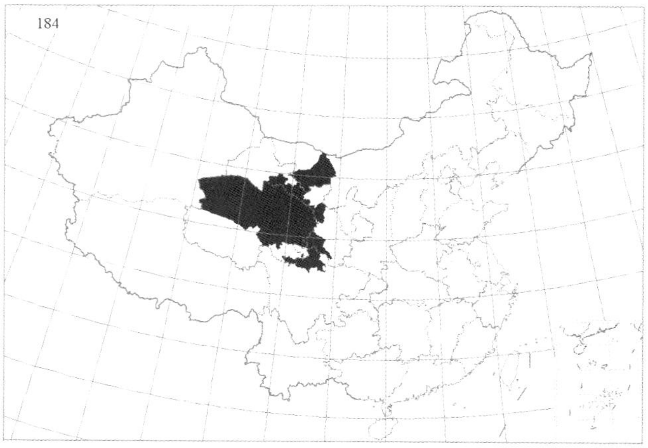

Cupressaceae 柏科 *Juniperus* 刺柏属
Juniperus pseudosabina Fisch. et C. A. Mey. 新疆方枝柏
Shrubs, evergreen

4.8 (−4.3~12.3); −14.0 (−20.9~−5.2); 20.3 (9.6~27.5); 8.4 (2.5~13.0); 66.1 (9.7~112.3); 68.0 (24.3~121.1); 602 (329~813) ❙ 178 (42~450); 82 (17~211); 16 (3~51) ❙ −67.7 (−94.7~3.9); 178 (42~406); 78.7 (7.6~297.3)

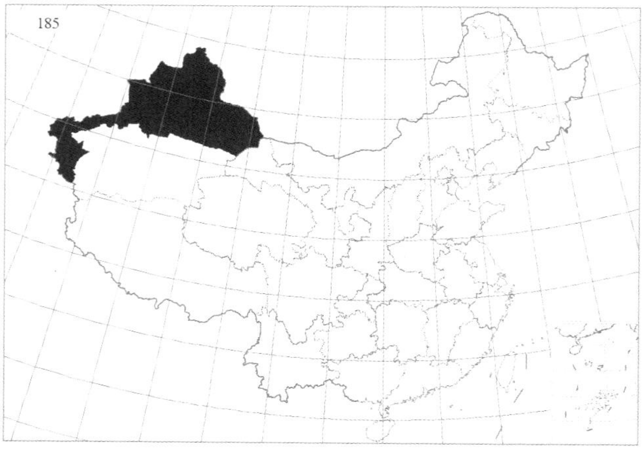

Cupressaceae 柏科 *Juniperus* 刺柏属
Juniperus pseudosabina var. turkestanica (Kom.) Silba 喀什方枝柏
Trees or shrubs, evergreen

3.5 (−4.3~11.9); −14.9 (−20.9~−5.6); 18.7 (9.6~25.5); 7.5 (2.5~12.7); 56.9 (9.7~107.9); 73.9 (25.2~121.1); 562 (329~783) ❙ 203 (44~406); 93 (19~211); 19 (3~51) ❙ −61.6 (−94.0~−7.6); 203 (44~406); 104.1 (15.7~297.3)

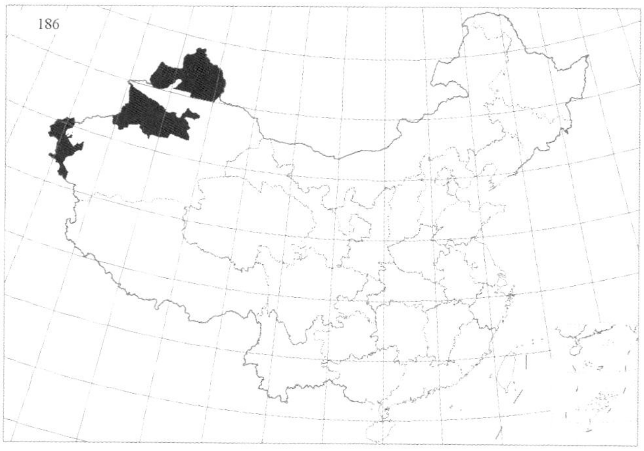

Cupressaceae 柏科 *Juniperus* 刺柏属

Juniperus recurva Buch.-Ham. ex D. Don 垂枝柏

Small trees, evergreen

7.3 (−1.8~14.5); −0.3 (−10.5~7.8); 13.9 (6.1~19.7); 8.3 (1.8~14.5); 55.3 (1.9~113.5); 27.3 (0.0~82.8); 544 (329~730) ❙ 875 (290~1736); 511 (193~1007); 34 (5~69) ❙ 51.7 (−14.7~145.8); 538 (290~730); 266.0 (61.6~447.4)

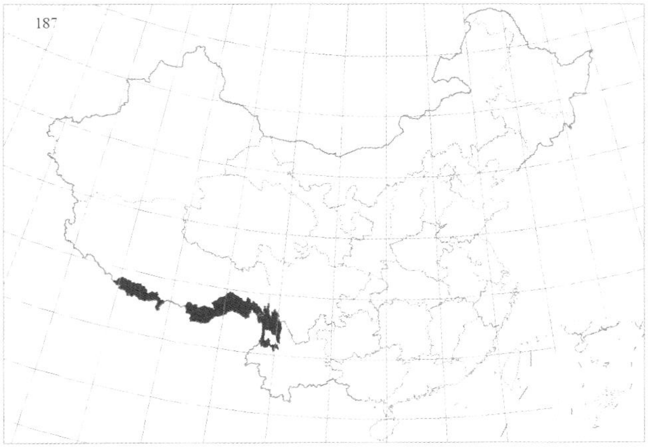

Cupressaceae 柏科 *Juniperus* 刺柏属

Juniperus recurva var. coxii (A. B. Jacks.) Melville 小果垂枝柏

Shrubs, evergreen

10.6 (5.3~14.5); 3.6 (−2.4~7.8); 16.5 (12.2~19.7); 10.7 (5.7~14.5); 74.7 (30.5~113.5); 6.8 (0.0~26.2); 626 (479~730) ❙ 1050 (741~1337); 595 (426~727); 44 (24~69) ❙ 69.0 (36.1~100.7); 624 (479~730); 336.9 (234.2~407.1)

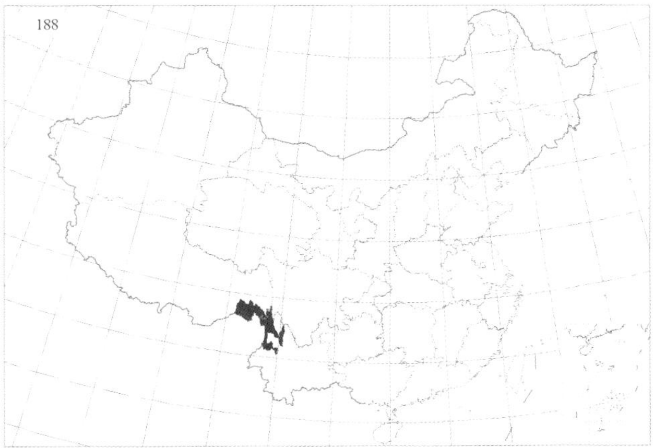

Cupressaceae 柏科 *Juniperus* 刺柏属

Juniperus rigida Siebold et Zucc. 杜松

Shrubs or small trees, evergreen

6.4 (−5.8~17.1); −10.8 (−29.8~4.0); 21.7 (15.3~29.4); 9.1 (4.4~17.1); 71.7 (28.0~146.1); 54.5 (1.0~157.4); 612 (426~952) ❙ 488 (111~1235); 287 (74~691); 15 (2~123) ❙ −20.5 (−82.3~74.1); 449 (111~921); 160.2 (13.6~346.5)

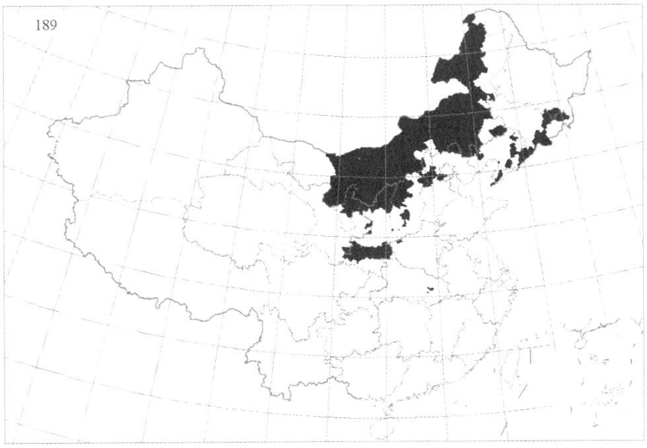

Cupressaceae 柏科 *Juniperus* 刺柏属

Juniperus sabina L. 叉子圆柏

Shrubs, evergreen

3.9 (−5.9~10.2); −13.9 (−29.8~−6.2); 19.3 (5.4~27.5); 7.6 (1.3~12.1); 58.3 (0.4~107.6); 71.4 (32.8~157.4); 561 (294~813) ❙ 266 (45~829); 156 (27~453); 12 (1~98) ❙ −49.1 (−94.0~94.4); 257 (45~590); 100.6 (7.5~346.5)

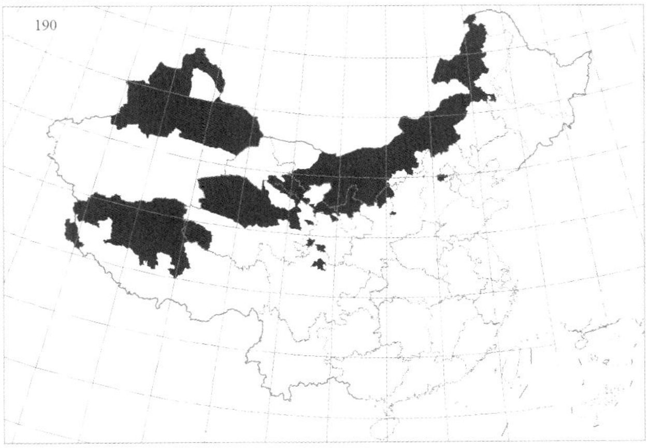

Cupressaceae 柏科 *Juniperus* 刺柏属

Juniperus sabina var. erectopatens (W. C. Cheng et L. K. Fu) Y. F. Yu et L. K. Fu 松潘圆柏

Small trees, evergreen

2.6; −7.0; 11.2; 4.3; 20.0; 48.2; 427 ❙ 829; 400; 23 ❙ 94.4; 427; 252.2

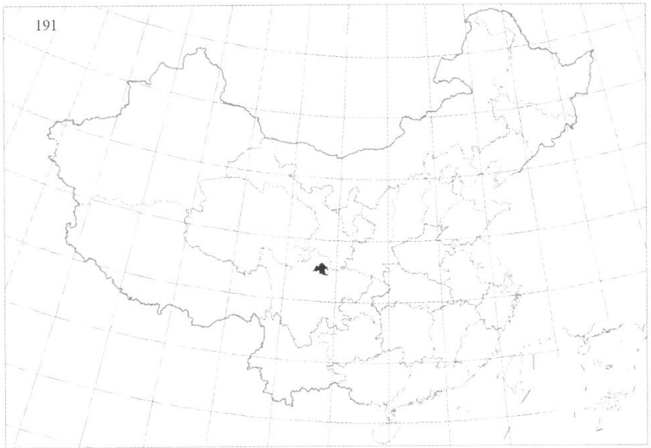

Cupressaceae 柏科 *Juniperus* 刺柏属

Juniperus saltuaria Rehder et E. H. Wilson 方枝柏

Trees, evergreen

4.6 (−1.3~16.9); −5.1 (−12.3~7.0); 13.1 (8.1~25.9); 6.0 (2.5~17.0); 35.7 (7.0~143.7); 40.4 (0.0~82.2); 483 (354~881) ❙ 777 (476~1577); 437 (249~932); 19 (9~60) ❙ 61.2 (−11.1~128.6); 481 (354~881); 262.6 (111.0~357.7)

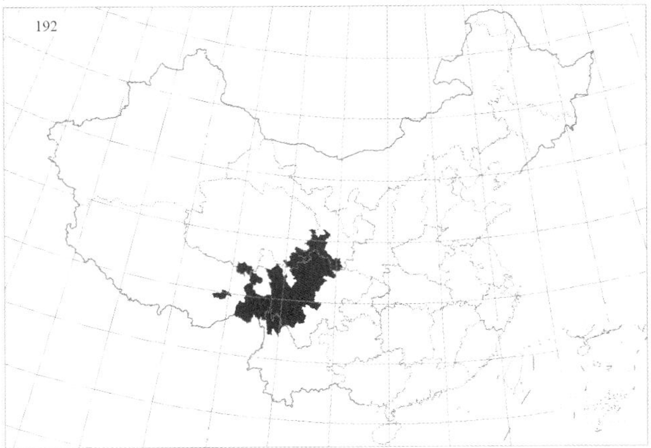

Cupressaceae 柏科 *Juniperus* 刺柏属

Juniperus semiglobosa Regel 昆仑多子柏

Trees, evergreen

2.4 (-4.3~12.3); -13.4 (-20.1~-5.3); 15.7 (8.6~26.0); 6.3 (2.4~13.0); 44.8 (7.9~112.3); 75.8 (24.7~121.1); 499 (329~801) ❙ 156 (44~509); 64 (19~233); 25 (4~98) ❙ -59.3 (-94.5~52.9); 131 (44~333); 35.8 (9.3~120.2)

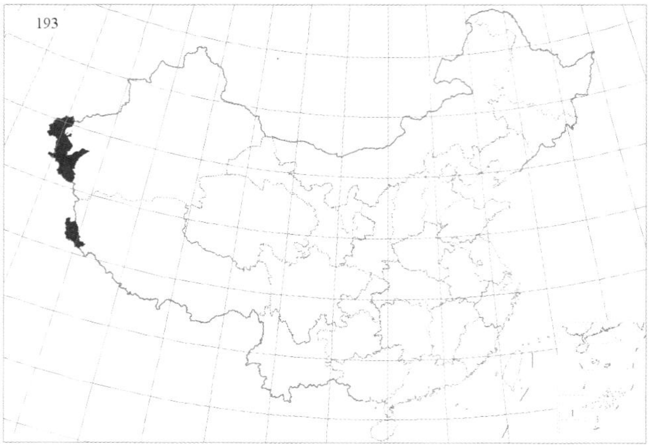

Cupressaceae 柏科 *Juniperus* 刺柏属

Juniperus sibirica Burgsd. 西伯利亚刺柏

Shrubs, evergreen

2.9 (-5.8~10.5); -17.0 (-29.8~-4.9); 19.8 (10.9~27.5); 7.7 (3.5~12.1); 59.1 (16.3~107.6); 83.2 (26.8~157.4); 570 (390~813) ❙ 337 (70~777); 190 (44~473); 19 (3~51) ❙ -37.1 (-90.6~55.1); 313 (70~632); 166.0 (10.9~333.3)

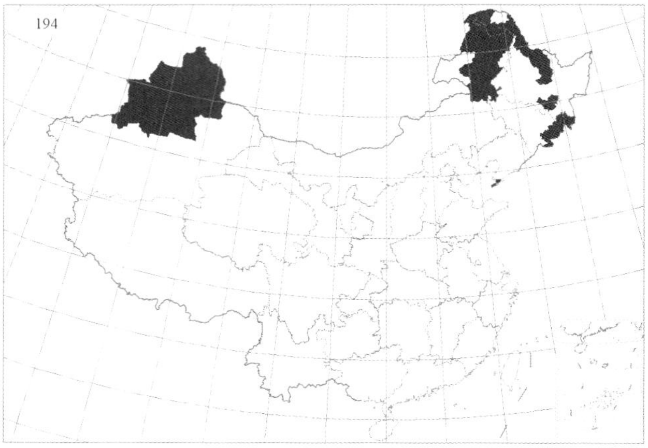

Cupressaceae 柏科 *Juniperus* 刺柏属

Juniperus squamata Buch.-Ham. ex D. Don 高山柏

Shrubs, evergreen

10.4 (-1.8~24.2); 0.5 (-13.7~18.1); 19.4 (6.1~28.7); 11.0 (1.8~24.3); 83.5 (1.9~231.2); 18.1 (0.0~86.1); 655 (329~1335) ❙ 955 (63~2859); 469 (45~1186); 48 (2~402) ❙ 46.4 (-84.9~246.5); 640 (63~1139); 264.3 (8.4~411.8)

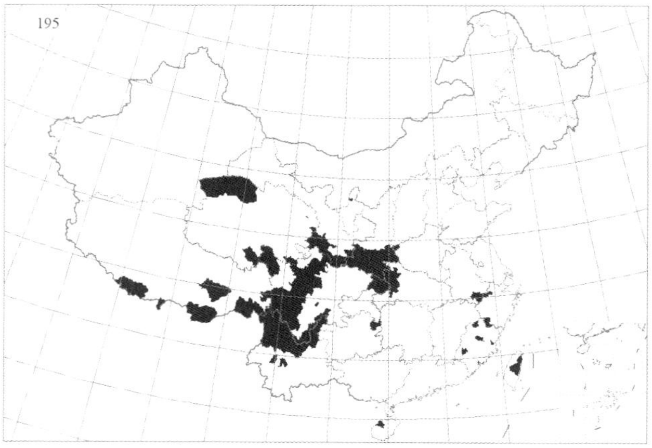

Cupressaceae 柏科 *Juniperus* 刺柏属

Juniperus squamata var. fargesii Rehder et E. H. Wilson 长叶高山柏

Shrubs or trees, evergreen

13.8; 2.8; 24.3; 14.0; 114.7; 8.4; 774 ❙ 1369; 521; 128 ❙ 74.3; 774; 275.7

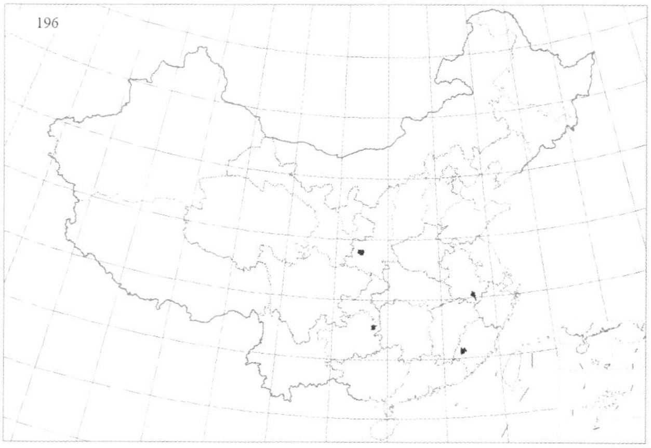

Cupressaceae 柏科 *Juniperus* 刺柏属

Juniperus squamata var. hongxiensis Y. F. Yu et L. K. Fu 洪溪高山柏

Shrubs, evergreen

9.0; 0.6; 16.5; 9.0; 58.7; 10.3; 584 ❙ 1178; 663; 34 ❙ 101.6; 584; 293.1

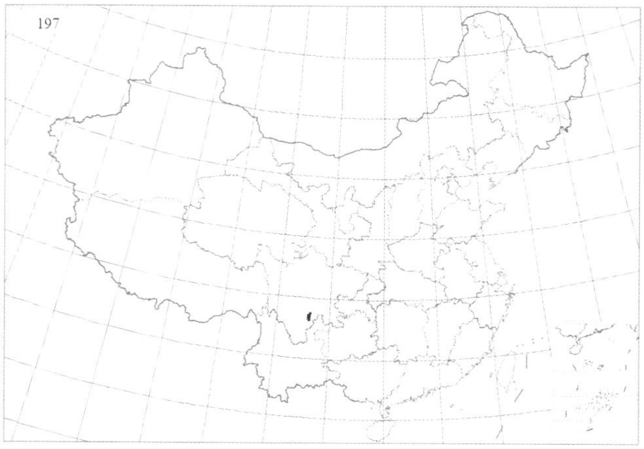

Cupressaceae 柏科 *Juniperus* 刺柏属

Juniperus tibetica Kom. 大果圆柏

Trees, evergreen

1.6 (-5.5~14.8); -8.2 (-17.1~3.8); 10.4 (5.8~25.3); 4.0 (1.3~14.8); 18.9 (1.2~119.2); 58.7 (1.4~127.1); 412 (299~795) ❙ 620 (246~1358); 358 (183~606); 14 (0~88) ❙ 47.9 (-28.5~128.6); 404 (246~795); 243.0 (62.8~357.7)

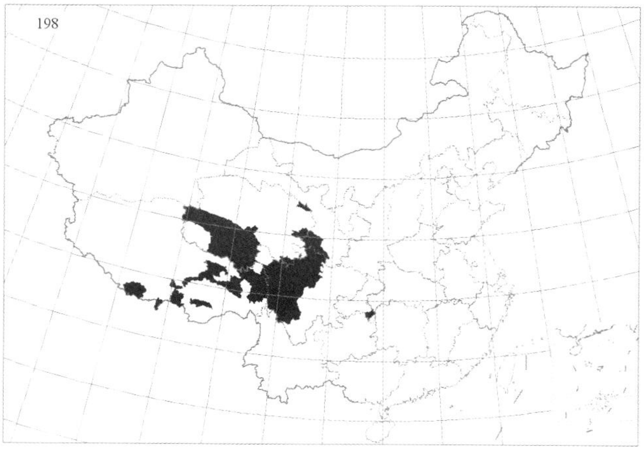

Cupressaceae 柏科 *Platycladus* 侧柏属

Platycladus orientalis (L.) Franco 侧柏

Trees, evergreen

12.7 (0.5~24.6); -0.3 (-17.2~19.5); 24.5 (9.4~30.1); 13.3 (3.3~24.7); 110.1 (13.6~235.9); 16.8 (0.0~83.8); 770 (376~1372) ❙ 876 (111~1997); 430 (67~911); 58 (1~259) ❙ 11.8 (-82.3~130.7); 685 (111~1331); 198.2 (13.9~659.0)

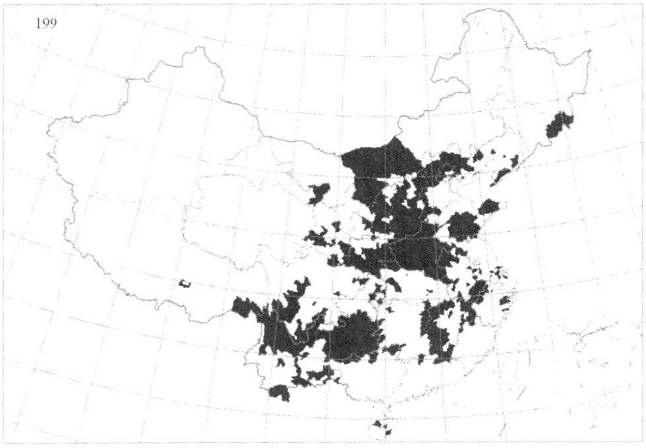

Cupressaceae 柏科 *Thuja* 崖柏属

Thuja koraiensis Nakai 朝鲜崖柏

Trees, evergreen

3.8 (1.4~10.5); -15.6 (-19.0~-4.9); 20.6 (18.4~25.0); 8.0 (6.4~11.3); 60.9 (45.8~93.2); 74.3 (26.8~88.8); 572 (501~705) ❙ 668 (558~976); 408 (330~608); 25 (16~40) ❙ 17.5 (-10.4~66.0); 565 (501~698); 289.7 (180.8~346.5)

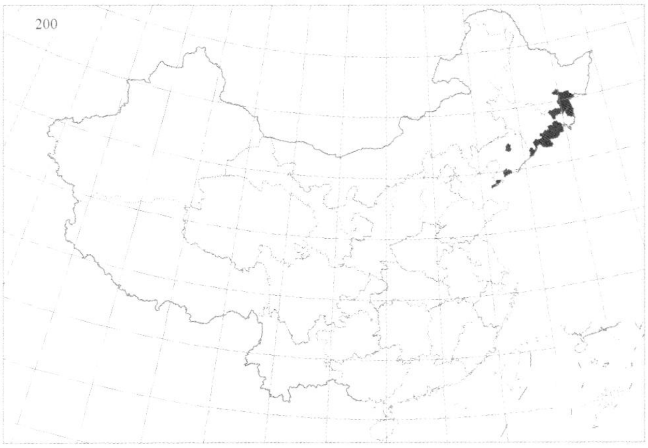

Cupressaceae 柏科 *Thuja* 崖柏属

Thuja sutchuenensis Franch. 崖柏

Shrubs or trees, evergreen

10.6; -0.2; 21.1; 10.6; 79.4; 12.1; 643 ❙ 1246; 525; 46 ❙ 93.8; 643; 284.3

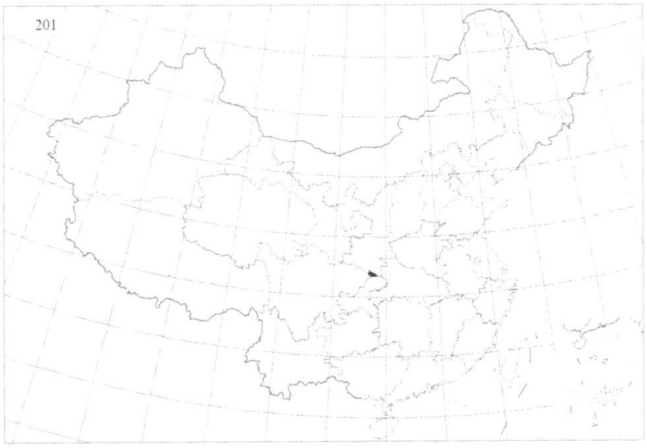

Podocarpaceae 罗汉松科 *Dacrycarpus* 鸡毛松属

Dacrycarpus imbricatus var. patulus de Laubenfels 鸡毛松

Trees, evergreen

21.1 (16.3~24.8); 13.5 (7.3~20.1); 26.8 (21.4~29.0); 21.1 (16.4~24.8); 193.2 (136.4~237.6); 0.0 (0.0~0.0); 1104 (797~1385) ❙ 1536 (868~2435); 686 (312~1428); 104 (41~189) ❙ 41.9 (-34.3~99.5); 1067 (797~1360); 351.7 (157.7~659.0)

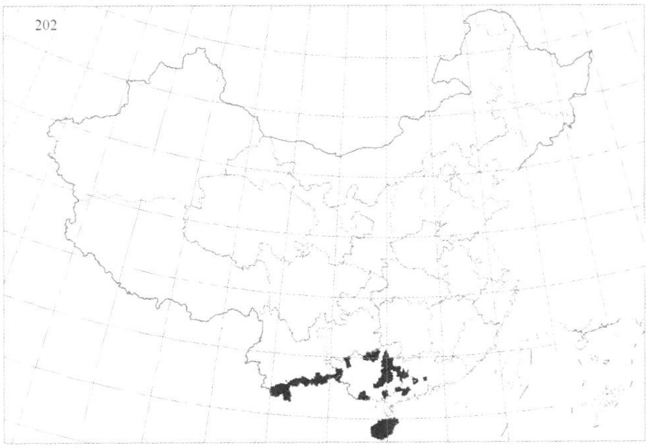

Podocarpaceae 罗汉松科 *Dacrydium* 陆均松属

Dacrydium pectinatum de Laub. 陆均松

Trees, evergreen

23.9 (22.2~24.8); 18.5 (17.2~20.1); 27.7 (25.7~28.7); 23.9 (22.3~24.8); 227.0 (207.2~237.6); 0.0 (0.0~0.0); 1297 (1127~1385) ❙ 1420 (868~1767); 493 (312~610); 56 (41~88) ❙ 10.1 (-34.3~47.8); 1119 (868~1299); 495.2 (288.3~620.0)

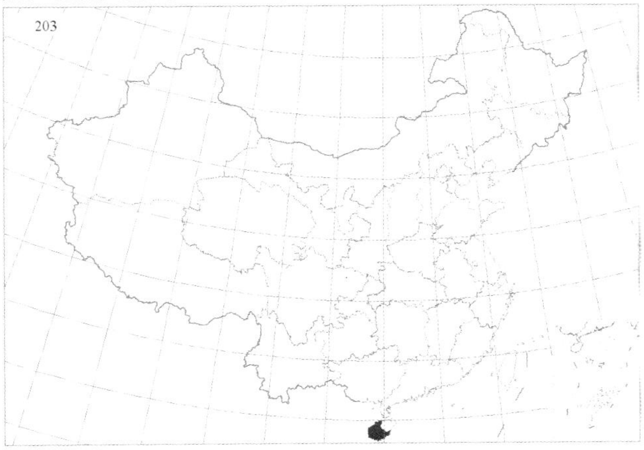

Podocarpaceae 罗汉松科 *Nageia* 竹柏属

Nageia fleuryi (Hickel) de Laub. 长叶竹柏

Trees, evergreen

22.0 (16.9~24.8); 15.1 (10.3~20.1); 27.2 (21.4~29.0); 22.0 (16.9~24.8); 204.2 (142.7~237.6); 0.0 (0.0~0.0); 1160 (813~1385) ❙ 1713 (868~3495); 704 (312~997); 144 (41~897) ❙ 50.9 (-34.3~224.4); 1102 (813~1299); 365.9 (157.7~659.0)

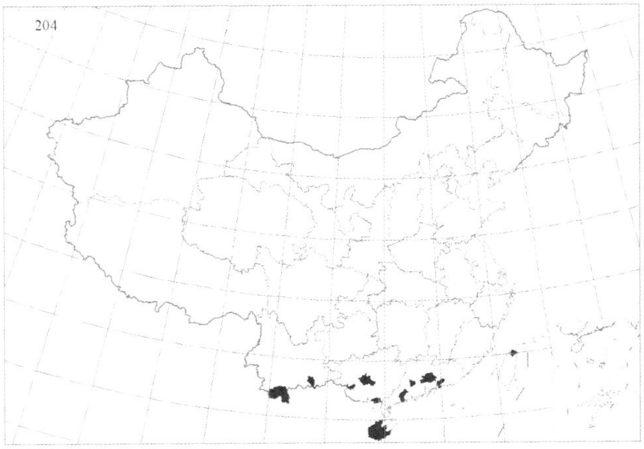

P

Podocarpaceae 罗汉松科 *Nageia* 竹柏属

Nageia nagi (Thunb.) Kuntze 竹柏

Trees, evergreen

19.7 (14.3~24.8); 10.7 (3.3~20.1); 27.6 (20.3~29.9); 19.8 (14.3~24.8); 177.2 (113.2~237.6); 0.0 (0.0~2.3); 1039 (761~1385) ❙ 1624 (868~3495); 636 (312~1428); 168 (20~897) ❙ 59.5 (-34.3~224.4); 1026 (761~1360); 244.4 (95.3~620.0)

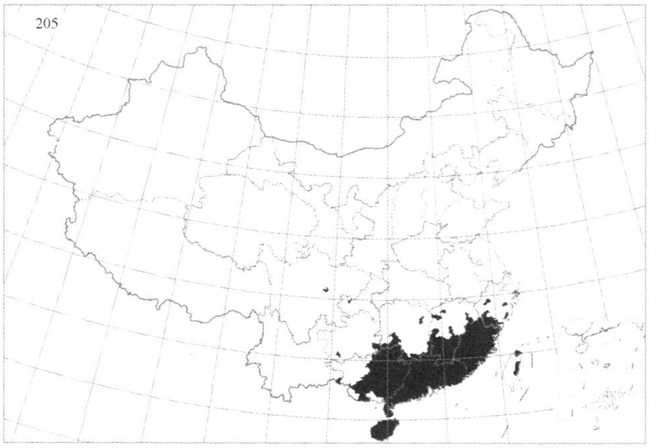

Podocarpaceae 罗汉松科 *Nageia* 竹柏属

Nageia wallichiana (C. Presl) Kuntze 肉托竹柏

Trees, evergreen

20.0; 14.6; 23.2; 20.1; 180.8; 0.0; 963 ❙ 1523; 681; 67 ❙ 58.4; 942; 581.0

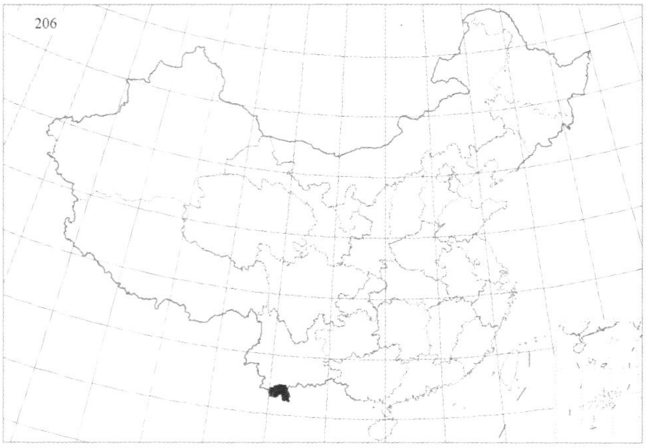

Podocarpaceae 罗汉松科 *Podocarpus* 罗汉松属

Podocarpus annamiensis N. E. Gray 海南罗汉松

Trees, evergreen

23.8 (22.2~24.8); 18.4 (17.2~20.1); 27.6 (25.7~28.7); 23.8 (22.3~24.8); 226.1 (207.2~237.6); 0.0 (0.0~0.0); 1291 (1127~1385) ❙ 1388 (868~1725); 486 (312~610); 53 (41~71) ❙ 8.4 (-34.3~47.8); 1102 (868~1212); 490.4 (288.3~620.0)

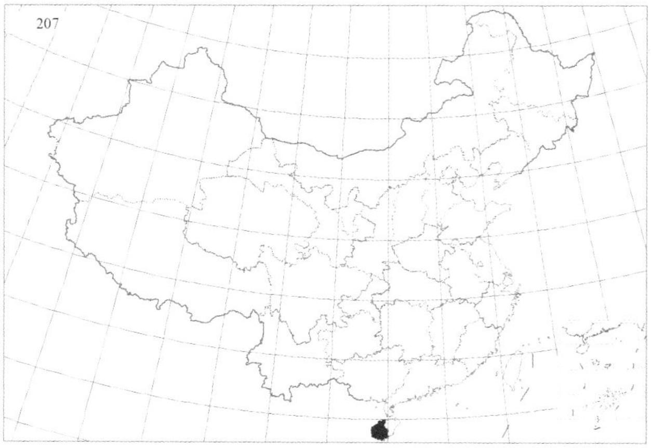

Podocarpaceae 罗汉松科 *Podocarpus* 罗汉松属

Podocarpus costalis C. Presl 兰屿罗汉松

Shrubs or small trees, evergreen

18.5; 14.1; 21.7; 18.5; 162.1; 0.0; 863 ❙ 2713; 1388; 188 ❙ 214.4; 863; 363.7

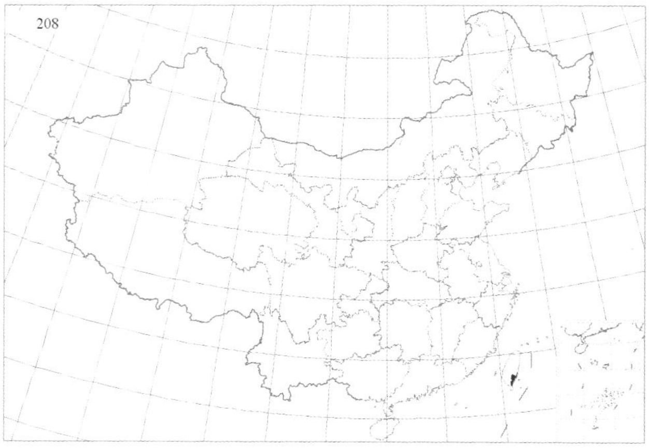

Podocarpaceae 罗汉松科 *Podocarpus* 罗汉松属

Podocarpus forrestii Craib et W. W. Sm. 大理罗汉松

Shrubs, evergreen

14.0; 7.5; 19.2; 14.1; 108.8; 0.0; 718 ❙ 1043; 581; 50 ❙ 45.3; 718; 377.4

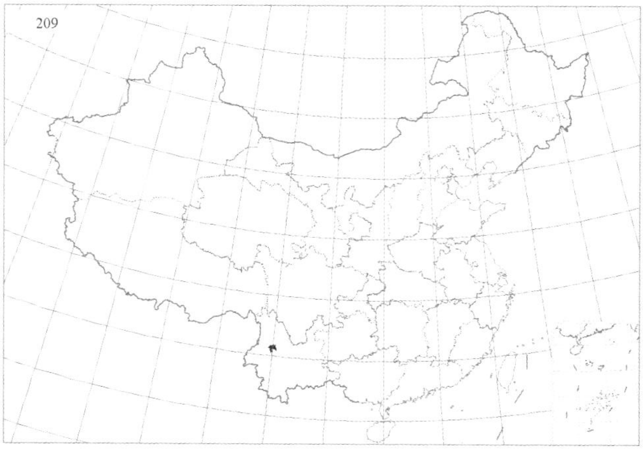

Podocarpaceae 罗汉松科 *Podocarpus* 罗汉松属

Podocarpus macrophyllus D. Don 罗汉松

Trees, evergreen

17.6 (11.1~23.0); 7.4 (1.7~17.7); 26.8 (19.4~29.9); 17.7 (11.1~23.1); 152.4 (79.8~217.0); 0.6 (0.0~6.9); 932 (648~1221) ❙ 1537 (862~2992); 634 (299~1847); 153 (27~253) ❙ 66.3 (-6.7~246.5); 926 (648~1221); 238.5 (111.9~572.9)

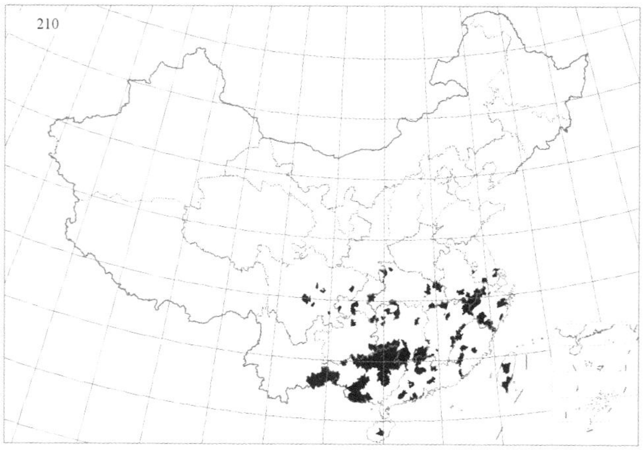

Podocarpaceae 罗汉松科 *Podocarpus* 罗汉松属
Podocarpus macrophyllus var. angustifolius Blume 狭叶罗汉松
Trees, evergreen

16.2；6.0；25.6；16.3；135.1；0.0；855 ❙ 1155；616；46 ❙ 36.8；854；158.1

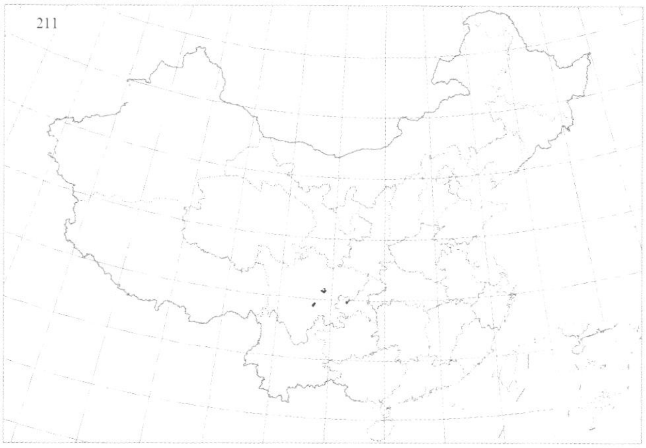

Podocarpaceae 罗汉松科 *Podocarpus* 罗汉松属
Podocarpus macrophyllus var. chingii N. E. Gray 柱冠罗汉松
Trees, evergreen

18.7 (15.0~22.1)；9.1 (5.0~13.7)；27.1 (24.7~28.8)；18.7 (15.1~22.1)；165.0 (120.8~205.6)；0.0 (0.0~0.0)；982 (789~1182) ❙ 1713 (1512~1950)；645 (507~779)；181 (144~223) ❙ 78.2 (50.9~147.0)；982 (789~1182)；235.7 (157.7~283.9)

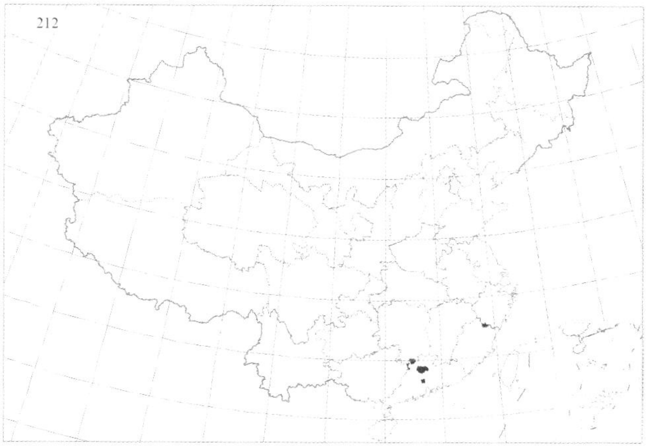

Podocarpaceae 罗汉松科 *Podocarpus* 罗汉松属
Podocarpus macrophyllus var. maki Siebold et Zucc. 短叶罗汉松
Trees, evergreen

14.7 (11.1~17.7)；4.7 (1.8~6.9)；23.8 (19.4~27.8)；14.7 (11.1~17.7)；118.7 (79.8~152.2)；1.9 (0.0~6.9)；793 (648~937) ❙ 1122 (950~1398)；607 (458~812)；40 (27~63) ❙ 43.5 (12.1~72.0)；793 (648~937)；150.3 (126.1~246.3)

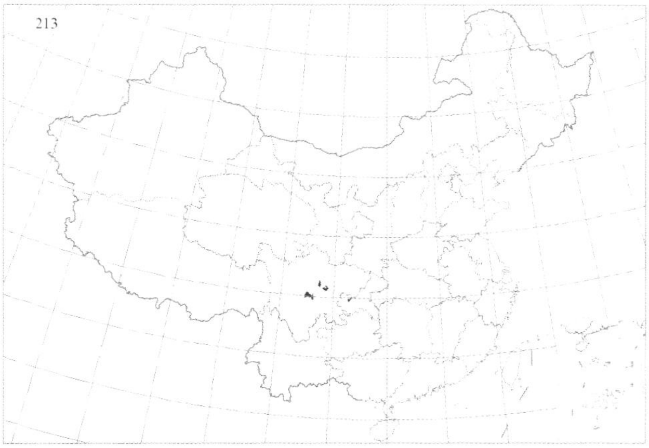

Podocarpaceae 罗汉松科 *Podocarpus* 罗汉松属
Podocarpus nakaii Hayata 台湾罗汉松
Trees, evergreen

16.2；11.0；20.1；16.2；134.2；0.0；773 ❙ 2680；1186；245 ❙ 246.5；773；289.1

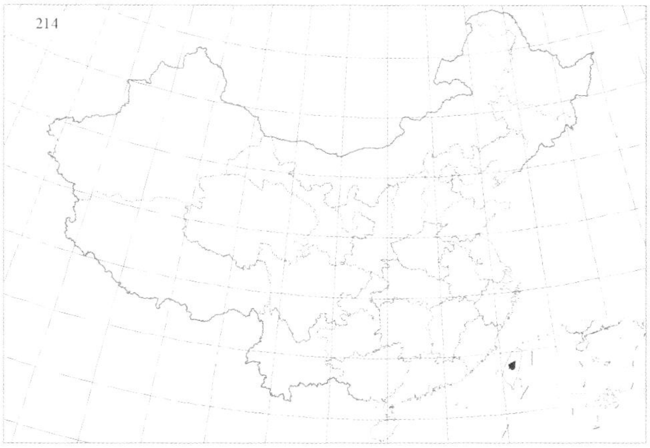

Podocarpaceae 罗汉松科 *Podocarpus* 罗汉松属
Podocarpus neriifolius D. Don 百日青
Trees, evergreen

18.5 (7.3~24.8)；9.7 (-2.0~20.1)；25.9 (15.6~28.9)；18.6 (7.5~24.3)；163.1 (48.2~237.6)；0.3 (0.0~20.4)；965 (537~1385) ❙ 1523 (868~2992)；679 (312~1847)；118 (27~259) ❙ 60.6 (-34.3~246.5)；949 (537~1221)；283.9 (139.6~659.0)

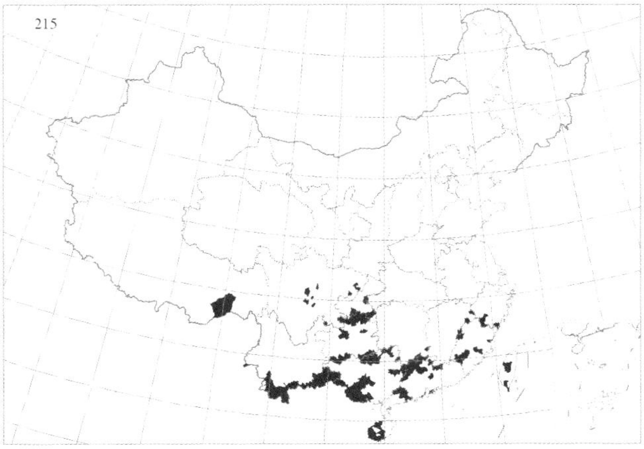

Podocarpaceae 罗汉松科 *Podocarpus* 罗汉松属
Podocarpus wangii C. C. Chang 小叶罗汉松
Trees, evergreen

21.4 (16.3~24.7)；13.8 (7.3~19.8)；27.5 (21.4~29.1)；21.5 (16.4~24.7)；197.7 (136.4~236.5)；0.0 (0.0~0.0)；1128 (797~1376) ❙ 1673 (1190~2037)；767 (455~1033)；112 (48~188) ❙ 49.2 (6.7~83.4)；1116 (797~1299)；263.7 (95.3~620.0)

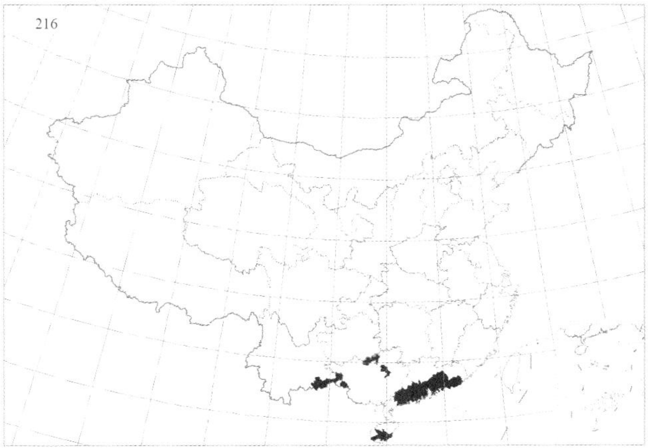

Cephalotaxaceae 三尖杉科 *Cephalotaxus* 粗榧属

Cephalotaxus fortunei Hook. 三尖杉

Trees, evergreen

16.1 (2.2~23.0); 5.6 (-7.0~17.7); 25.7 (9.9~30.1); 16.2 (3.8~23.1); 136.5 (15.5~217.0); 2.7 (0.0~49.1); 868 (413~1221) | 1327 (512~2435); 541 (258~1428); 125 (14~259) | 52.9 (-35.3~147.0); 854 (413~1221); 231.1 (121.6~572.9)

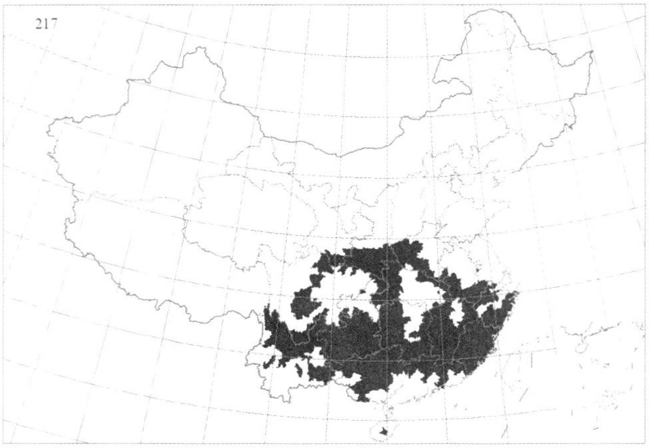

Cephalotaxaceae 三尖杉科 *Cephalotaxus* 粗榧属

Cephalotaxus fortunei var. alpina H. L. Li 高山三尖杉

Trees, evergreen

9.9 (1.9~16.7); 1.6 (-9.2~9.8); 17.1 (10.8~22.0); 10.3 (4.3~16.8); 72.5 (20.0~141.1); 13.0 (0.0~57.8); 615 (427~818) | 926 (620~1454); 522 (301~792); 27 (12~70) | 52.9 (0.0~121.7); 606 (427~727); 276.0 (190.3~487.2)

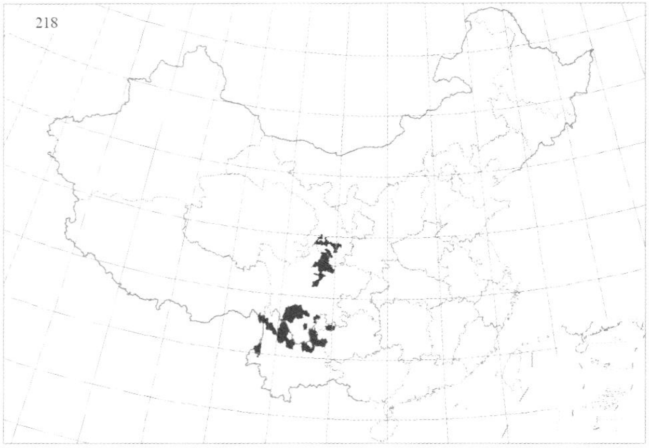

Cephalotaxaceae 三尖杉科 *Cephalotaxus* 粗榧属

Cephalotaxus lanceolata K. M. Feng 贡山三尖杉

Trees, evergreen

9.7 (2.2~18.5); 1.8 (-7.0~10.5); 16.3 (9.9~24.3); 10.2 (3.8~18.3); 72.0 (15.5~160.0); 15.0 (0.0~49.1); 613 (413~906) | 987 (731~1243); 564 (346~696); 29 (16~58) | 67.7 (5.5~125.9); 594 (413~786); 309.7 (248.3~384.7)

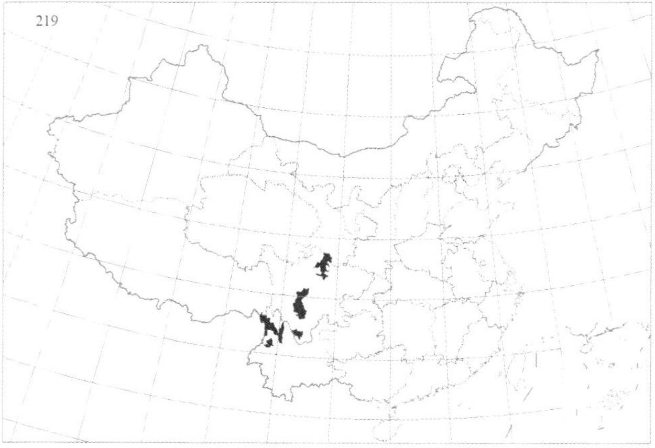

Cephalotaxaceae 三尖杉科 *Cephalotaxus* 粗榧属

Cephalotaxus latifolia L. K. Fu et R. R. Mill. 宽叶粗榧

Shrubs or small trees, evergreen

16.1 (9.0~21.2); 5.7 (0.6~12.8); 25.5 (16.5~28.5); 16.1 (9.0~21.3); 134.9 (58.7~195.4); 1.6 (0.0~10.3); 854 (584~1115) | 1464 (1132~1997); 587 (475~848); 135 (34~254) | 73.0 (16.9~130.7); 853 (584~1115); 251.2 (176.4~371.7)

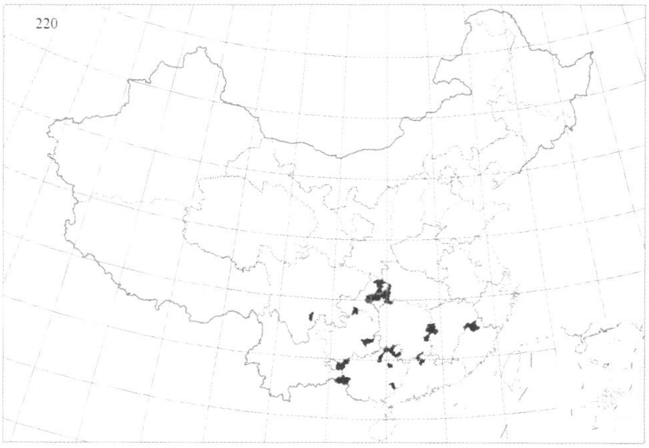

Cephalotaxaceae 三尖杉科 *Cephalotaxus* 粗榧属

Cephalotaxus mannii Hook. f. 海南粗榧

Trees, evergreen

20.4 (13.1~24.8); 12.9 (5.4~20.1); 26.1 (18.4~28.7); 20.4 (13.2~24.8); 185.1 (98.1~237.6); 0.0 (0.0~0.0); 1066 (692~1385) | 1538 (868~2435); 688 (312~1428); 100 (41~189) | 48.8 (-34.3~145.8); 1016 (692~1326); 375.9 (185.5~659.0)

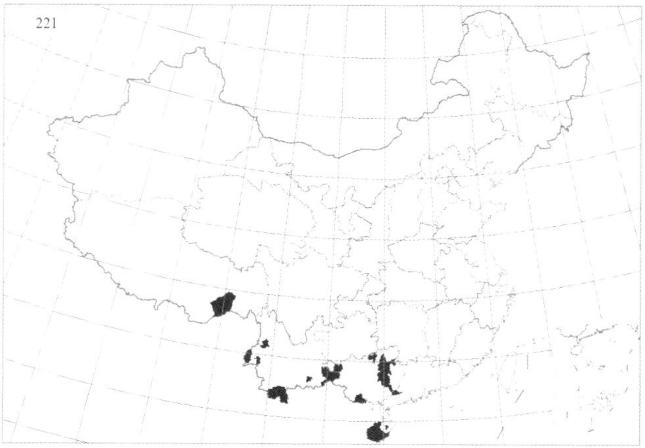

Cephalotaxaceae 三尖杉科 *Cephalotaxus* 粗榧属

Cephalotaxus oliveri Mast. 篦子三尖杉

Shrubs, evergreen

16.7 (9.3~21.4); 6.0 (-1.7~14.2); 26.6 (18.4~29.8); 16.8 (9.5~21.4); 142.0 (69.8~196.9); 0.7 (0.0~18.0); 888 (603~1130) | 1418 (955~1922); 537 (392~860); 145 (26~259) | 60.1 (12.5~128.9); 880 (603~1130); 222.1 (134.6~551.6)

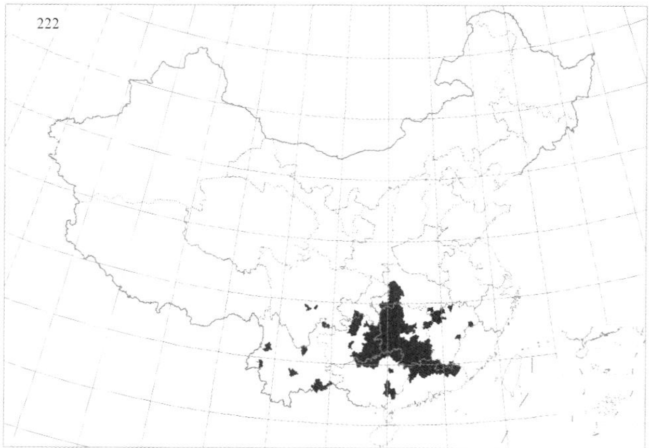

Cephalotaxaceae 三尖杉科 *Cephalotaxus* 粗榧属

Cephalotaxus sinensis (Rehder et E. H. Wilson) H. L. Li 粗榧

Shrubs or small trees, evergreen

15.4 (2.6~24.7); 4.3 (-7.0~19.8); 25.7 (11.2~29.8); 15.5 (4.3~24.7); 129.6 (20.0~236.5); 4.2 (0.0~48.2); 845 (427~1376) ▎1295 (581~2859); 527 (257~1412); 119 (12~402) ▎52.8 (-32.0~246.5); 833 (427~1225); 233.2 (130.9~494.7)

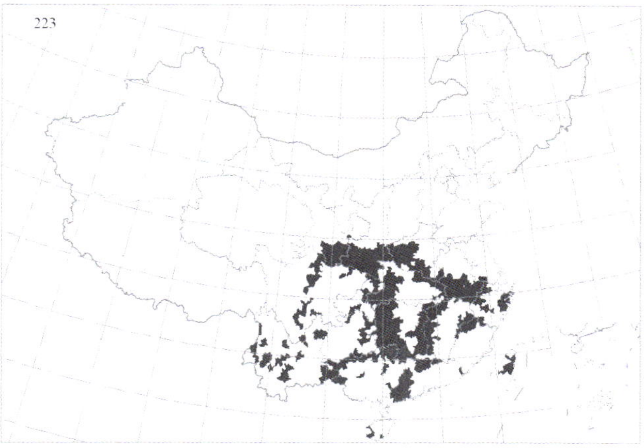

Taxaceae 红豆杉科 *Amentotaxus* 穗花杉属

Amentotaxus argotaenia (Hance) Pilg. 穗花杉

Shrubs or small trees, evergreen

18.0 (9.3~22.9); 8.0 (-1.7~15.2); 26.8 (19.6~29.8); 18.0 (9.5~22.9); 157.1 (69.8~214.8); 0.7 (0.0~18.0); 952 (603~1232) ▎1530 (711~2435); 605 (357~1428); 150 (15~250) ▎61.9 (12.1~145.8); 947 (603~1232); 228.0 (134.6~447.4)

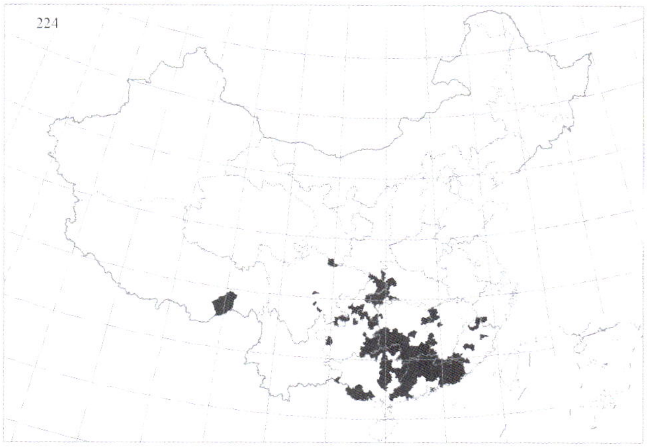

Taxaceae 红豆杉科 *Amentotaxus* 穗花杉属

Amentotaxus formosana H. L. Li 台湾穗花杉

Small trees, evergreen

18.5; 14.1; 21.7; 18.5; 162.1; 0.0; 863 ▎2713; 1388; 188 ▎214.4; 863; 363.7

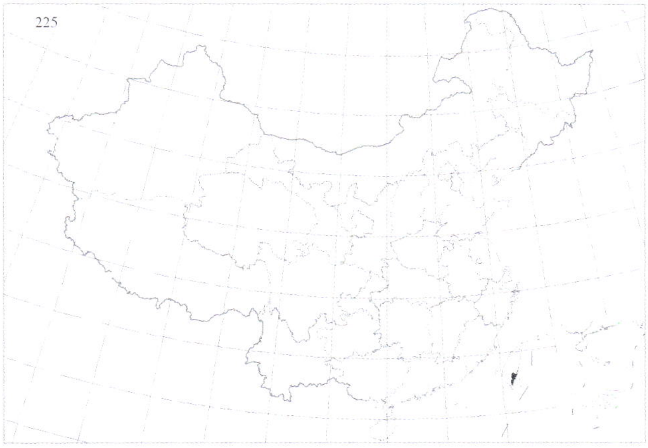

Taxaceae 红豆杉科 *Amentotaxus* 穗花杉属

Amentotaxus yunnanensis H. L. Li 云南穗花杉

Trees, evergreen

16.9 (12.5~21.0); 9.2 (3.6~14.2); 22.5 (20.1~25.6); 16.9 (12.5~21.0); 143.4 (92.0~191.8); 0.1 (0.0~1.4); 835 (685~1053) ▎1226 (1007~1522); 672 (539~860); 52 (44~68) ▎47.1 (19.0~80.6); 834 (685~1053); 338.2 (186.7~551.6)

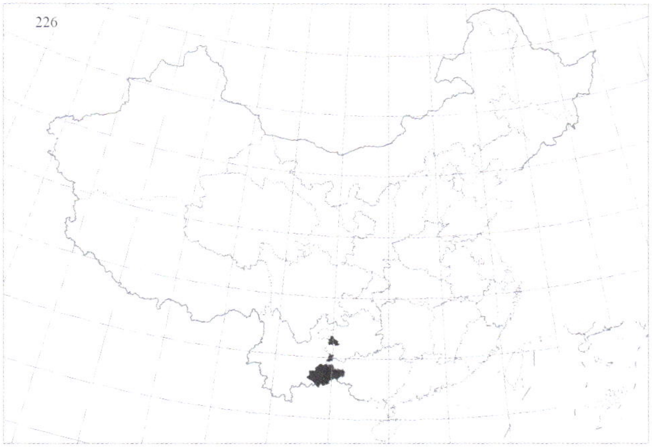

Taxaceae 红豆杉科 *Pseudotaxus* 白豆杉属

Pseudotaxus chienii (W. C. Cheng) W. C. Cheng 白豆杉

Shrubs, evergreen

18.6 (14.5~22.3); 8.5 (3.6~13.8); 27.4 (24.6~29.8); 18.7 (14.5~22.3); 164.0 (115.9~208.1); 0.1 (0.0~1.6); 981 (779~1202) ▎1614 (1337~1997); 623 (476~839); 179 (86~254) ▎66.6 (20.3~130.7); 979 (779~1202); 234.4 (174.9~335.1)

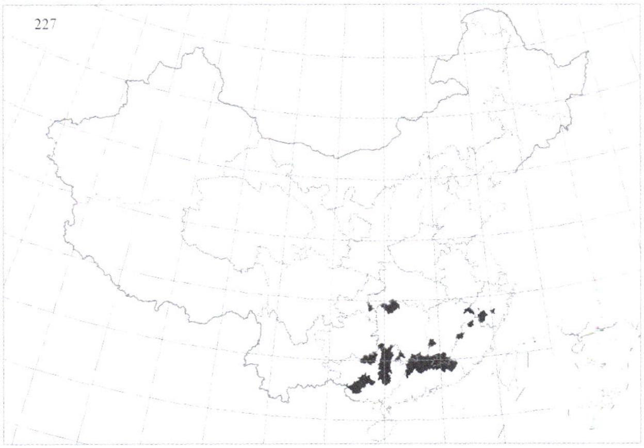

Taxaceae 红豆杉科 *Taxus* 红豆杉属

Taxus cuspidata Siebold et Zucc. 东北红豆杉

Trees, evergreen

3.1 (-0.2~6.5); -17.0 (-23.8~-12.0); 20.4 (18.4~22.8); 7.6 (6.4~9.2); 57.6 (45.8~75.2); 80.4 (55.1~112.6); 557 (501~624) ▎700 (561~1073); 431 (333~691); 24 (16~40) ▎25.2 (-2.6~74.1); 556 (501~624); 310.2 (265.9~346.5)

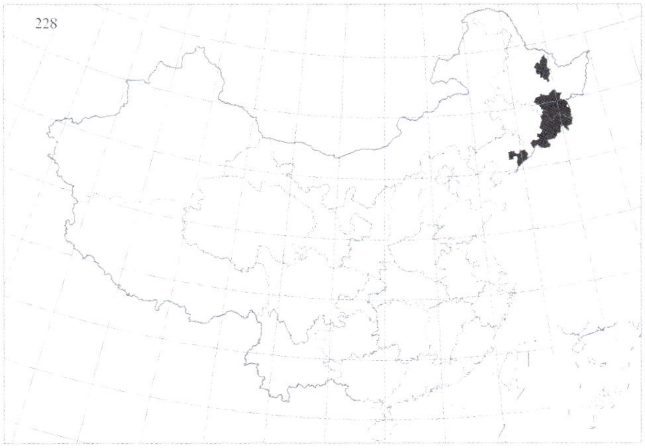

Taxaceae 红豆杉科 *Taxus* 红豆杉属

Taxus fuana Nan Li et R. R. Mill 密叶红豆杉

Trees, evergreen

-1.1; -9.1; 6.2; 2.0; 2.5; 75.6; 334 ❙ 321; 193; 33 ❙ -4.0; 321; 79.3

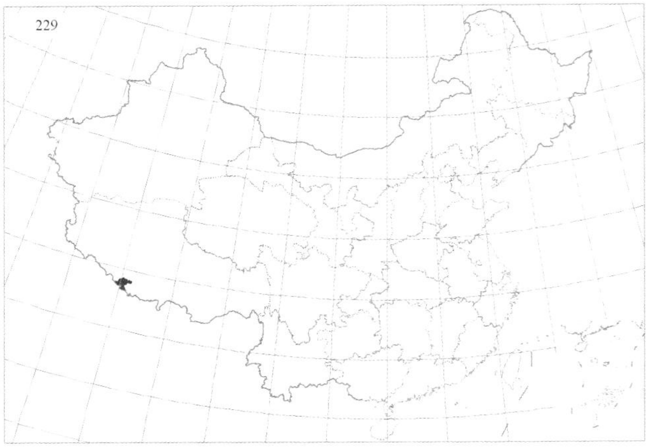

Taxaceae 红豆杉科 *Taxus* 红豆杉属

Taxus wallichiana Zucc. 喜马拉雅红豆杉

Trees, evergreen

16.2 (0.3~21.0); 6.4 (-8.6~12.5); 24.9 (8.3~30.1); 16.3 (2.9~21.0); 136.3 (9.6~192.6); 1.5 (0.0~65.4); 864 (368~1113) ❙ 1373 (389~1950); 553 (246~1007); 138 (11~259) ❙ 58.2 (-14.2~145.8); 841 (368~1107); 250.3 (142.5~487.2)

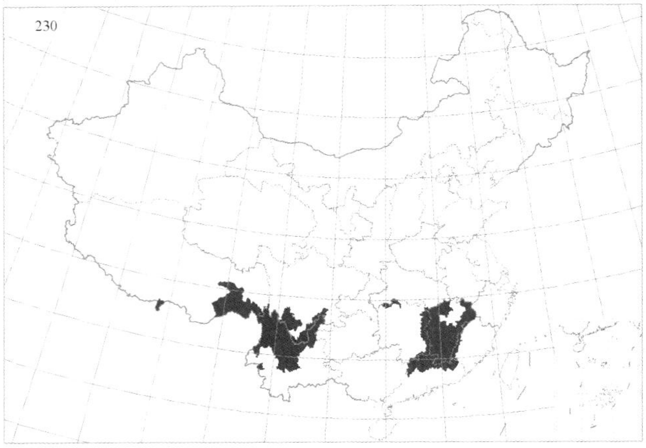

Taxaceae 红豆杉科 *Taxus* 红豆杉属

Taxus wallichiana var. chinensis (Pilg.) Florin 红豆杉

Trees, evergreen

14.4 (2.0~21.7); 4.0 (-7.2~15.3); 23.9 (9.9~29.8); 14.5 (3.8~21.7); 118.0 (15.4~200.2); 4.9 (0.0~51.4); 788 (413~1118) ❙ 1174 (587~2445); 525 (290~1412); 88 (12~254) ❙ 48.9 (-24.6~191.8); 773 (413~1075); 241.9 (131.2~394.7)

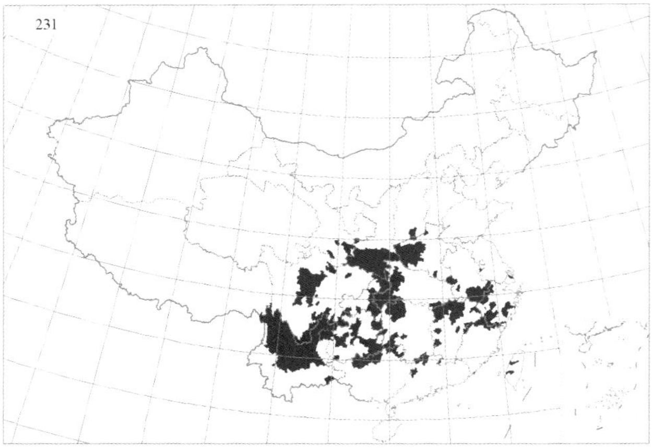

Taxaceae 红豆杉科 *Taxus* 红豆杉属

Taxus wallichiana var. mairei (Lemée et H. Lév.) L. K. Fu et Nan Li 南方红豆杉

Trees, evergreen

15.2 (2.0~19.4); 4.3 (-7.2~10.1); 25.4 (9.9~29.8); 15.3 (3.8~19.4); 126.4 (15.4~173.4); 3.0 (0.0~51.4); 829 (413~1029) ❙ 1339 (583~1950); 532 (298~816); 127 (17~259) ❙ 61.8 (-24.6~147.0); 820 (413~1029); 221.0 (142.5~366.3)

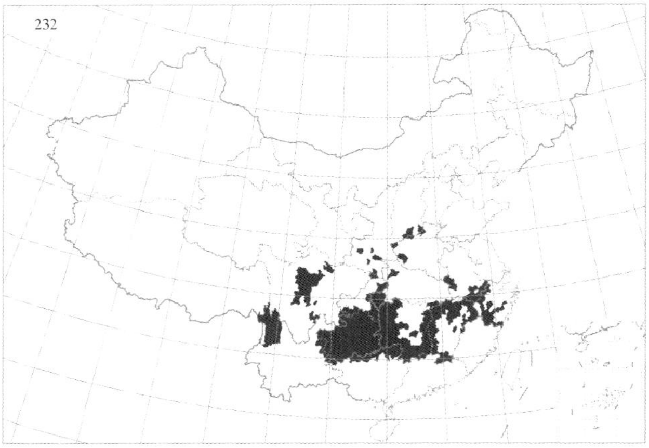

Taxaceae 红豆杉科 *Torreya* 榧树属

Torreya fargesii Franch. 巴山榧树

Trees, evergreen

13.1 (7.3~16.2); 1.7 (-2.5~6.2); 24.0 (15.6~27.9); 13.2 (7.5~16.2); 105.1 (48.2~134.8); 7.0 (0.0~20.6); 746 (537~883) ❙ 1095 (644~1571); 484 (321~816); 60 (14~173) ❙ 46.9 (-8.1~108.0); 744 (537~883); 246.1 (176.4~320.6)

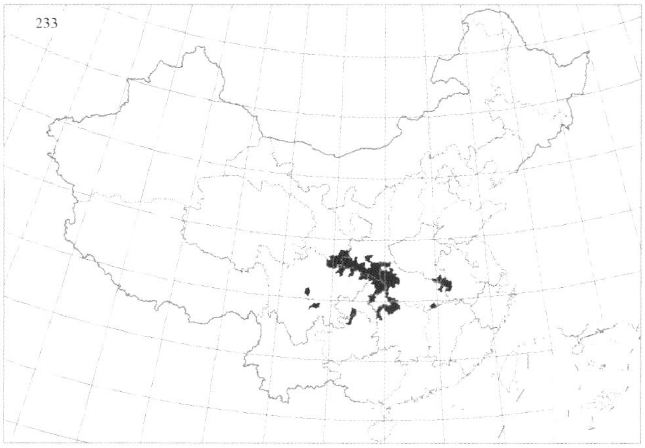

Taxaceae 红豆杉科 *Torreya* 榧树属

Torreya fargesii var. yunnanensis (C. Y. Cheng et L. K. Fu) N. Kang 云南榧

Trees, evergreen

11.1 (6.1~14.3); 4.2 (-1.5~7.6); 16.8 (12.7~19.7); 11.1 (6.3~14.3); 77.0 (35.2~111.9); 3.6 (0.0~21.8); 638 (496~728) ❙ 1072 (801~1337); 617 (496~727); 43 (21~69) ❙ 68.3 (36.1~93.0); 636 (496~707); 345.7 (282.9~407.1)

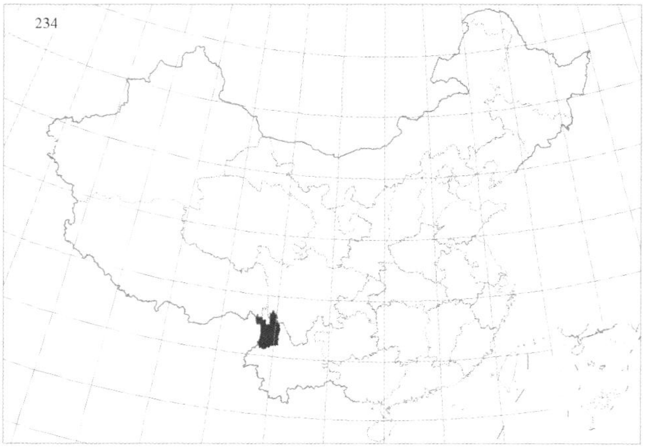

Taxaceae 红豆杉科 *Torreya* 榧树属

Torreya grandis Fortune ex Lindl. 榧树

Trees, evergreen

16.0 (12.7~18.2); 4.3 (0.5~9.0); 27.4 (23.0~29.3); 16.1 (12.8~18.2); 134.3 (99.0~158.8); 1.4 (0.0~9.2); 871 (713~984) ❙ 1510 (730~1997); 544 (415~677); 181 (27~259) ❙ 73.6 (-16.9~130.7); 865 (713~961); 227.5 (131.2~319.9)

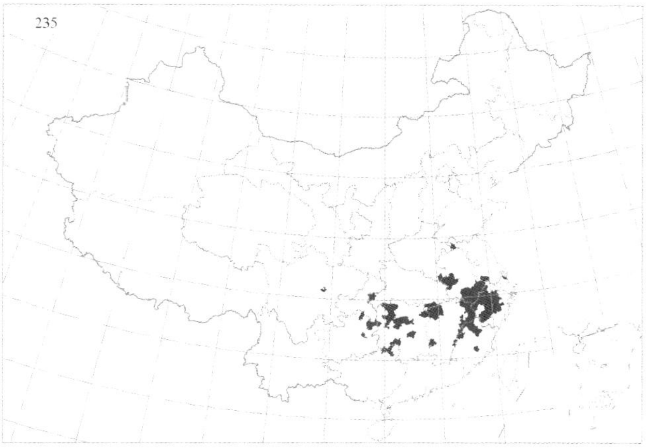

Taxaceae 红豆杉科 *Torreya* 榧树属

Torreya grandis var. jiulongshanensis Z. Y. Li et al. 九龙山榧树

Trees, evergreen

15.1; 4.0; 25.9; 15.2; 122.9; 1.0; 811 ❙ 1813; 632; 226 ❙ 123.6; 811; 252.7

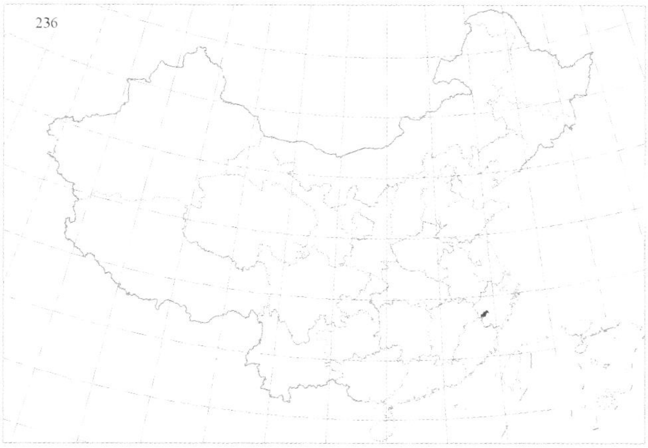

Taxaceae 红豆杉科 *Torreya* 榧树属

Torreya jackii Chun 长叶榧树

Trees, evergreen

16.2 (15.0~17.5); 5.1 (3.5~7.2); 26.9 (24.7~28.1); 16.2 (15.1~17.3); 135.0 (120.8~150.9); 0.3 (0.0~1.8); 862 (789~916) ❙ 1736 (1490~1959); 611 (513~713); 213 (185~250) ❙ 101.8 (72.8~147.0); 862 (789~916); 253.6 (217.4~283.9)

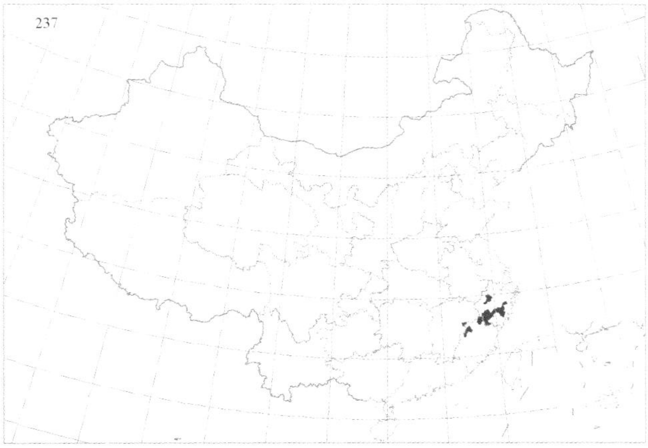

Ephedraceae 麻黄科 *Ephedra* 麻黄属

Ephedra distachya L. 双穗麻黄

Shrubs, evergreen

4.2 (-1.1~9.3); -15.5 (-19.5~-12.0); 20.9 (13.7~27.5); 8.4 (4.5~12.1); 66.3 (27.2~107.6); 75.4 (51.6~104.2); 606 (433~813) ❙ 200 (107~450); 85 (51~185); 21 (6~51) ❙ -65.4 (-85.3~3.9); 198 (107~404); 75.1 (16.6~173.1)

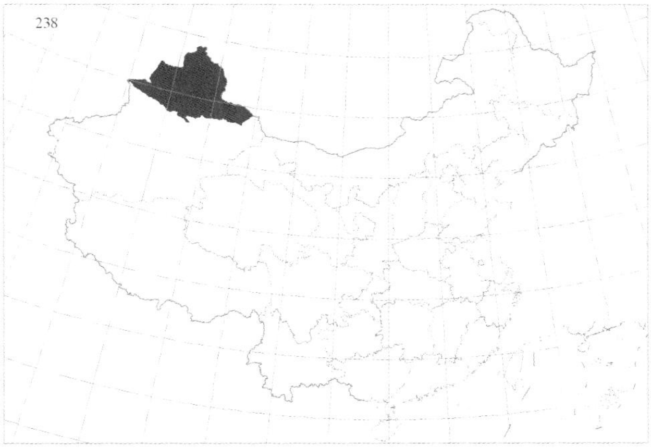

Ephedraceae 麻黄科 *Ephedra* 麻黄属

Ephedra equisetina Bunge 木贼麻黄

Small shrubs, evergreen

5.0 (-5.8~13.3); -12.6 (-29.8~-1.9); 20.3 (9.6~27.5); 8.2 (2.5~[illegible]3.5); 63.9 (9.7~1[illegible]5.5); 63.4 (15.7~157.4); 585 (329~813) ❙ 334 (45~694); 203 (24~499); 12 (1~51) ❙ -41.4 (-94.0~63.7); 329 (45~659); 122.9 (7.5~364.1)

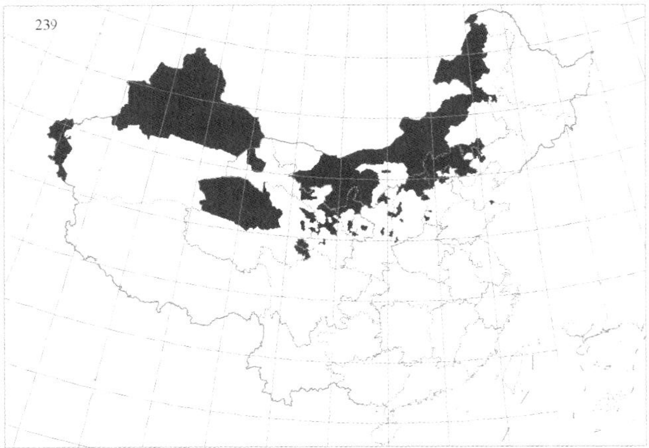

Ephedraceae 麻黄科 *Ephedra* 麻黄属

Ephedra fedtschenkoae Paulsen 雌雄麻黄

Small shrubs, evergreen

1.9 (-5.9~12.4); -14.5 (-20.9~-5.3); 15.9 (5.4~26.1); 6.2 (1.3~13.2); 44.6 (0.4~114.1); 81.1 (24.7~133.0); 499 (294~811) ❙ 131 (38~450); 72 (19~190); 9 (1~46) ❙ -69.4 (-95.3~3.9); 129 (38~404); 46.3 (2.8~297.3)

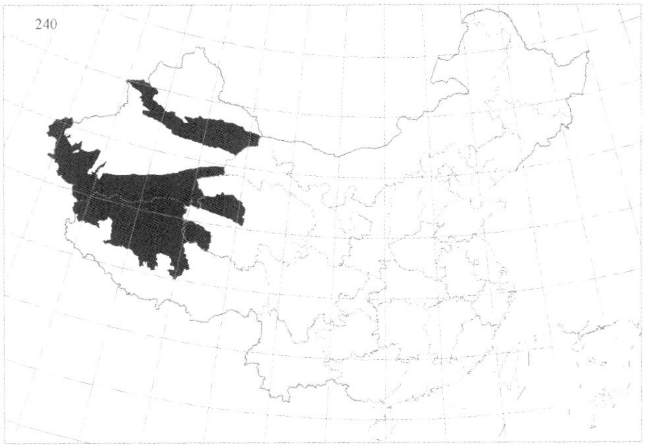

Ephedraceae 麻黄科 *Ephedra* 麻黄属
Ephedra gerardiana Wall. ex C. A. Mey. 山岭麻黄
Small shrubs, evergreen

0.0 (-5.9~11.0); -10.6 (-20.1~3.9); 9.8 (5.4~24.5); 3.6 (1.3~11.8); 17.2 (0.4~98.0); 76.5 (1.3~133.0); 388 (294~730) ❙ 382 (44~990); 239 (19~619); 17 (0~103) ❙ -3.0 (-94.0~91.4); 305 (44~638); 122.6 (8.4~319.8)

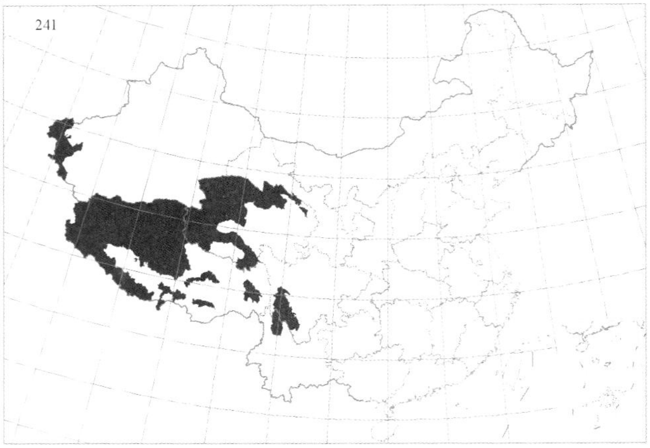

Ephedraceae 麻黄科 *Ephedra* 麻黄属
Ephedra intermedia Schrenk ex C. A. Mey. 中麻黄
Shrubs, evergreen

4.4 (-5.9~13.3); -12.2 (-29.8~5.4); 18.9 (5.4~27.5); 7.7 (1.3~13.4); 58.0 (0.4~114.0); 64.3 (0.0~157.4); 561 (294~822) ❙ 322 (28~1736); 189 (16~1007); 12 (0~103) ❙ -38.8 (-96.6~145.8); 302 (28~706); 120.6 (1.6~447.4)

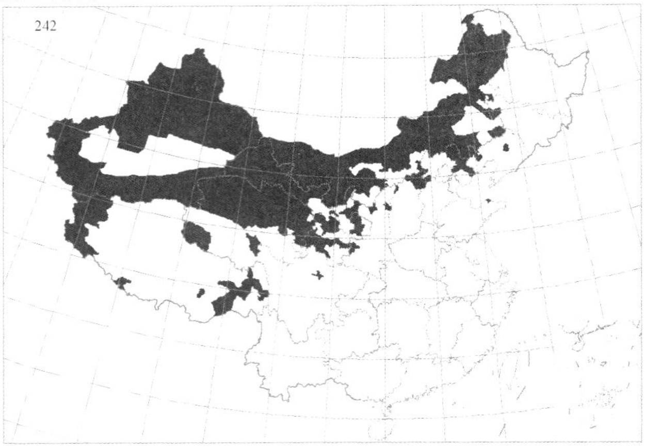

Ephedraceae 麻黄科 *Ephedra* 麻黄属
Ephedra likiangensis Florin 丽江麻黄
Shrubs, evergreen

10.3 (-1.8~18.9); 2.1 (-10.2~12.0); 17.2 (6.1~26.6); 10.8 (1.8~19.0); 79.4 (1.9~167.8); 15.8 (0.0~82.8); 632 (329~929) ❙ 904 (320~1252); 520 (217~686); 28 (10~59) ❙ 48.1 (-14.2~128.6); 613 (320~912); 259.2 (72.2~357.7)

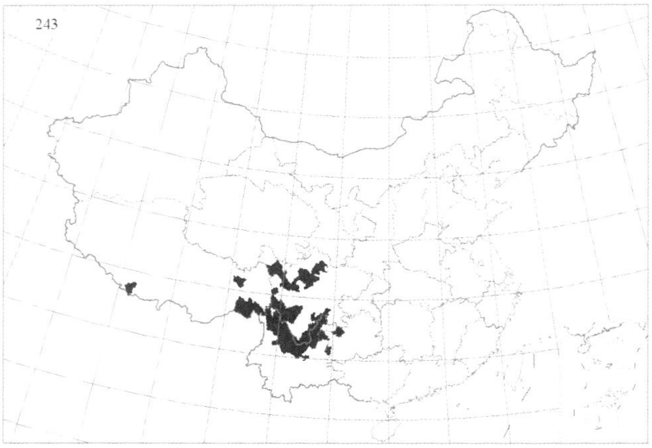

Ephedraceae 麻黄科 *Ephedra* 麻黄属
Ephedra lomatolepis Schrenk 窄膜麻黄
Shrubs, evergreen

3.9 (-1.3~9.3); -15.8 (-20.9~-11.6); 20.2 (13.4~27.5); 8.1 (4.7~12.1); 63.1 (28.1~107.6); 76.2 (51.6~104.2); 593 (443~813) ❙ 214 (120~357); 94 (51~171); 22 (12~45) ❙ -62.3 (-85.3~-29.7); 214 (120~357); 101.4 (16.6~297.3)

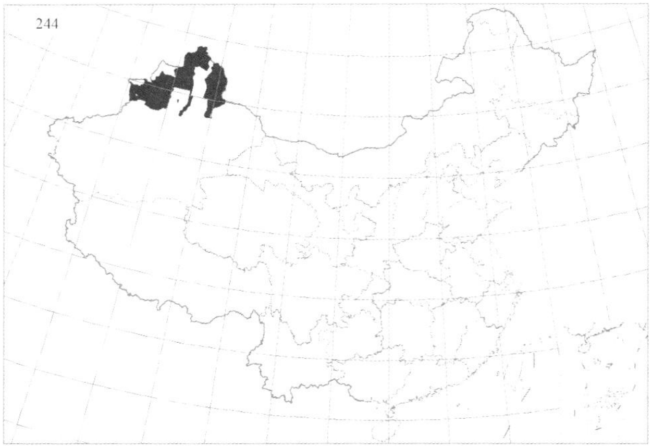

Ephedraceae 麻黄科 *Ephedra* 麻黄属
Ephedra minuta Florin 矮麻黄
Small shrubs, evergreen

1.7 (-4.8~10.2); -9.0 (-18.0~2.4); 11.1 (7.2~18.1); 4.3 (1.9~10.3); 21.3 (3.7~68.2); 60.4 (5.1~121.6); 423 (332~620) ❙ 659 (213~1036); 372 (135~606); 14 (5~28) ❙ 54.1 (-40.5~128.6); 414 (213~620); 256.0 (32.7~364.1)

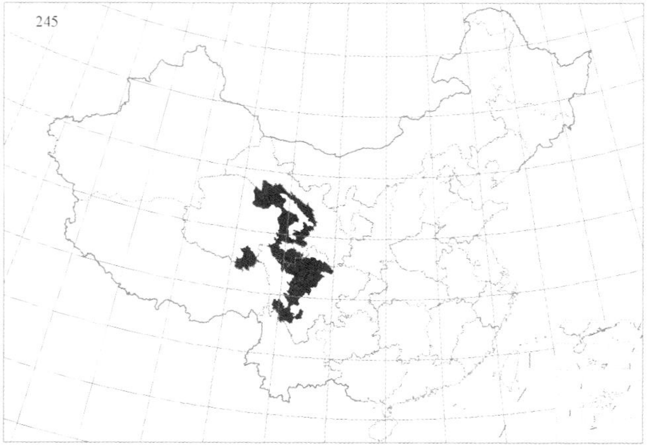

Ephedraceae 麻黄科 *Ephedra* 麻黄属
Ephedra monosperma Gemlin ex C. A. Mey. 单子麻黄
Small shrubs, evergreen

2.4 (-5.9~13.6); -14.0 (-29.8~-2.0); 16.7 (5.4~27.5); 6.4 (1.3~13.8); 45.6 (0.4~119.7); 76.1 (15.6~157.4); 511 (299~841) ❙ 369 (21~932); 224 (16~546); 12 (0~51) ❙ -21.6 (-97.3~125.9); 329 (21~632); 162.8 (3.0~364.1)

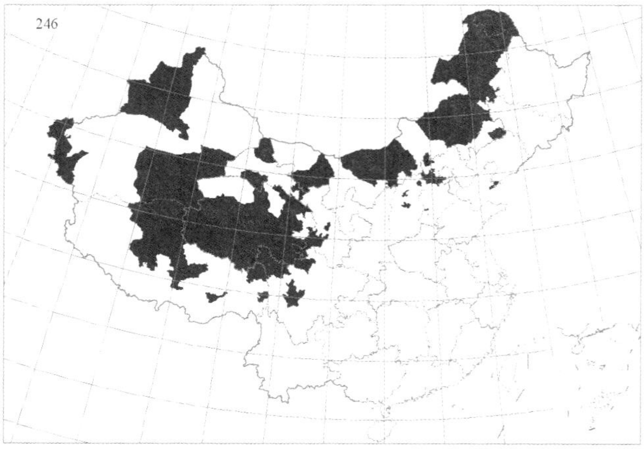

Ephedraceae 麻黄科 *Ephedra* 麻黄属
Ephedra przewalskii Stapf 膜果麻黄
Shrubs, evergreen

6.1 (-5.7~12.5); -11.6 (-20.3~-5.1); 21.0 (7.4~27.5); 9.0 (1.7~13.2); 72.2 (4.3~114.1); 58.3 (23.1~133.0); 624 (294~822) ❙ 133 (21~450); 71 (15~250); 8 (0~51) ❙ -76.0 (-97.3~3.9); 133 (21~407); 45.3 (1.6~247.5)

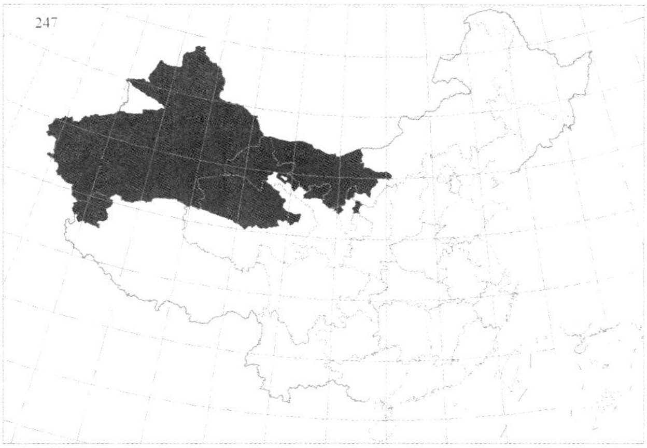

Ephedraceae 麻黄科 *Ephedra* 麻黄属
Ephedra regeliana Florin 细子麻黄
Small shrubs, evergreen

5.5 (-4.3~12.4); -13.0 (-20.9~-5.2); 20.5 (9.6~27.5); 8.7 (2.5~13.2); 69.7 (9.7~114.1); 63.6 (24.3~121.1); 616 (329~813) ❙ 143 (21~450); 68 (15~211); 12 (0~48) ❙ -74.0 (-97.3~3.9); 142 (21~406); 60.7 (1.6~297.3)

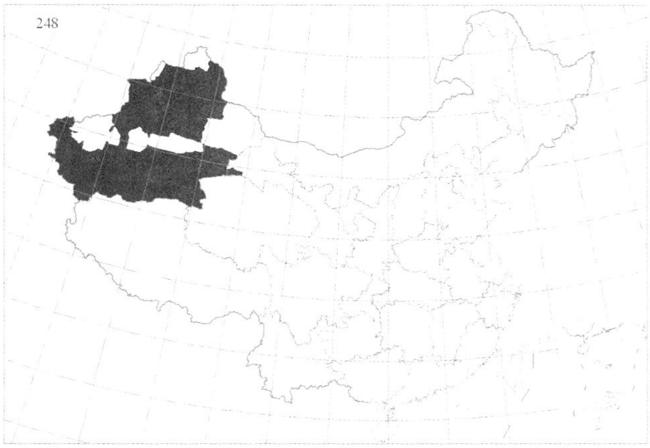

Ephedraceae 麻黄科 *Ephedra* 麻黄属
Ephedra rhytidosperma Pachom. 斑子麻黄
Small shrubs, evergreen

7.9 (7.0~8.8); -8.9 (-10.7~-7.0); 22.9 (21.4~24.3); 9.8 (9.2~10.3); 79.7 (72.8~85.9); 43.9 (35.3~52.0); 644 (612~679) ❙ 201 (49~300); 119 (34~168); 4 (1~7) ❙ -68.6 (-92.8~-51.0); 201 (49~300); 53.3 (5.1~116.9)

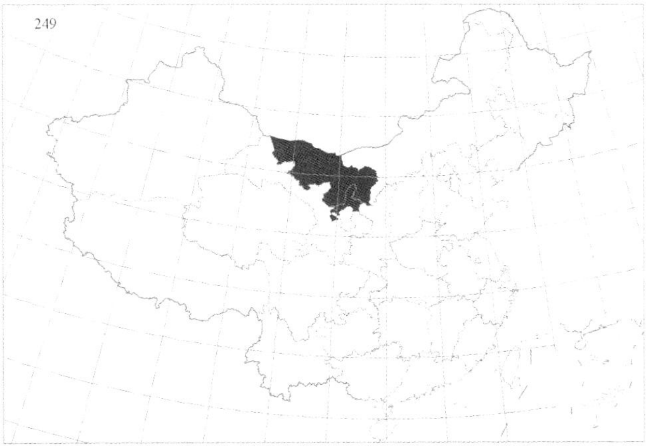

Ephedraceae 麻黄科 *Ephedra* 麻黄属
Ephedra rituensis Y. Yang, D. Z. Fu et G. Zhu 日土麻黄
Shrubs, evergreen

-1.9; -16.0; 11.0; 3.4; 17.1; 99.3; 374 ❙ 147; 88; 6 ❙ -62.1; 147; 58.5

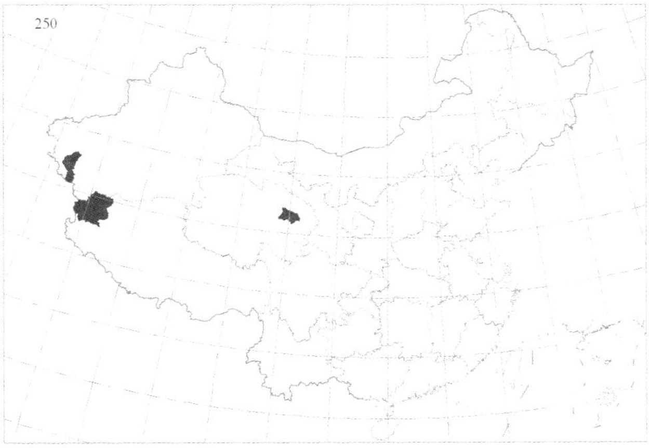

Ephedraceae 麻黄科 *Ephedra* 麻黄属
Ephedra saxatilis (Stapf) Royle ex Florin 藏麻黄
Small shrubs, evergreen

-0.9 (-5.7~3.7); -10.4 (-18.7~-6.1); 8.0 (6.1~12.1); 2.6 (1.7~5.1); 8.8 (1.9~27.8); 79.4 (43.5~133.0); 350 (294~441) ❙ 323 (74~620); 216 (46~306); 19 (1~103) ❙ -7.7 (-75.0~91.4); 290 (74~338); 103.9 (16.7~217.3)

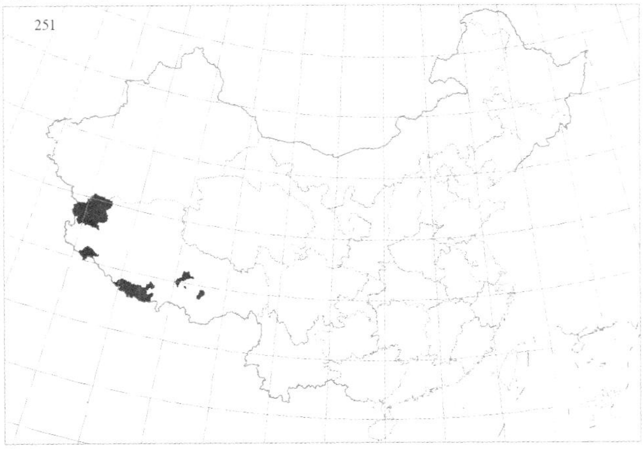

Ephedraceae 麻黄科 *Ephedra* 麻黄属
Ephedra sinica Stapf 草麻黄
Shrubs, evergreen

7.9 (-5.8~15.5); -8.9 (-29.8~0.5); 22.8 (9.2~29.5); 10.0 (2.9~15.5); 81.2 (10.1~136.0); 45.6 (9.3~157.4); 658 (374~932) ❙ 491 (111~795); 299 (65~499); 14 (2~48) ❙ -24.8 (-82.3~54.6); 485 (111~736); 156.8 (13.9~364.1)

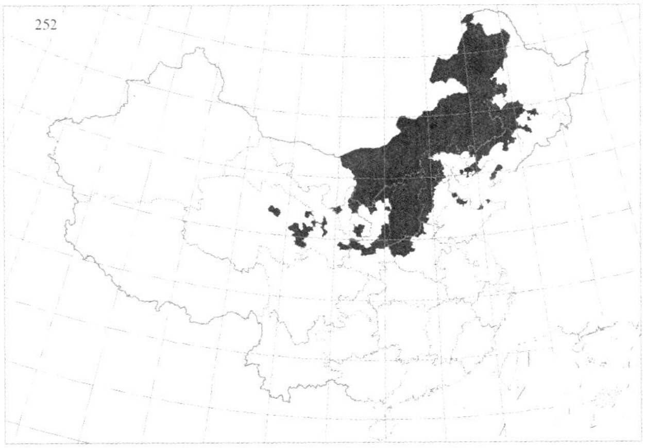

Gnetaceae 买麻藤科 *Gnetum* 买麻藤属
Gnetum catasphaericum H. Shao 球子买麻藤
Vines, evergreen

21.3; 13.2; 27.3; 21.3; 195.9; 0.0; 1111 ❙ 1900; 1056; 125 ❙ 71.0; 1111; 297.2

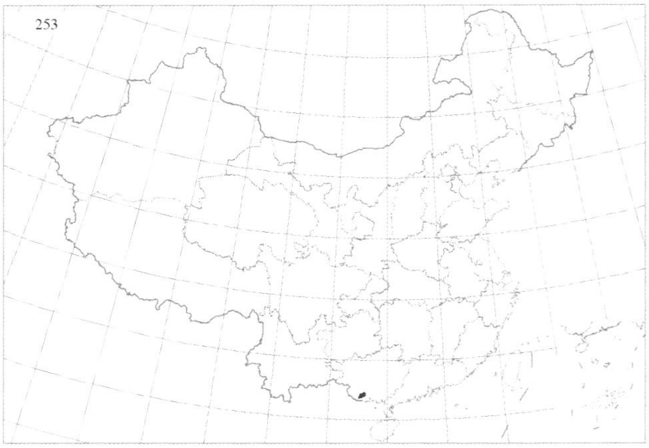

Gnetaceae 买麻藤科 *Gnetum* 买麻藤属
Gnetum giganteum H. Shao 巨子买麻藤
Vines, evergreen

21.8; 13.6; 27.8; 21.8; 201.8; 0.0; 1152 ❙ 1525; 813; 99 ❙ 33.1; 1152; 309.6

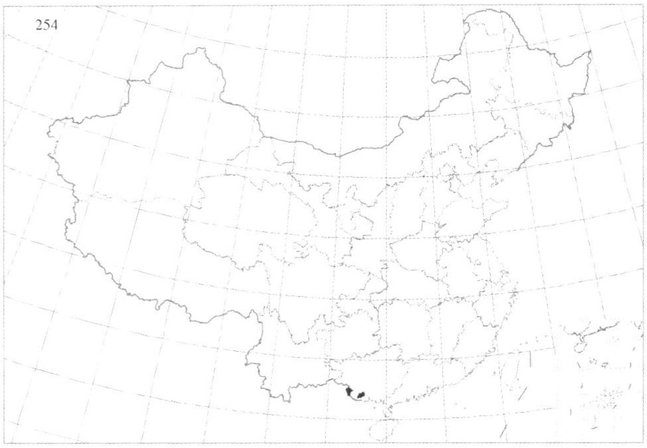

Gnetaceae 买麻藤科 *Gnetum* 买麻藤属
Gnetum gnemon L. 灌状买麻藤
Shrubs or small trees, evergreen

16.4; 9.6; 20.6; 16.4; 137.3; 0.0; 795 ❙ 1688; 959; 62 ❙ 112.2; 795; 627.0

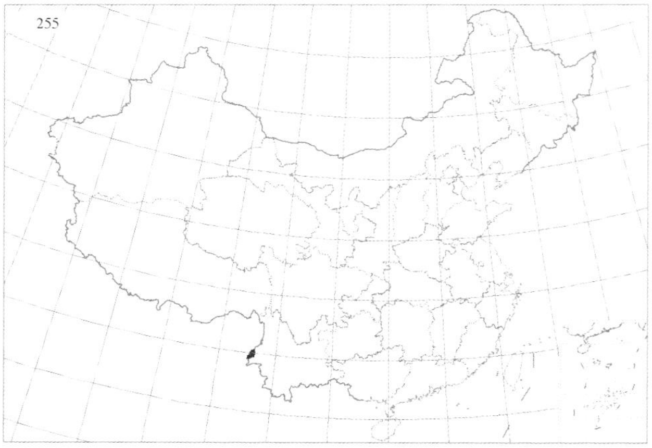

Gnetaceae 买麻藤科 *Gnetum* 买麻藤属
Gnetum gracilipes C. Y. Cheng 细柄买麻藤
Vines, evergreen

21.5 (18.9~22.7); 13.3 (10.6~15.0); 27.5 (25.0~28.6); 21.5 (19.0~22.8); 198.1 (167.5~213.1); 0.0 (0.0~0.0); 1129 (939~1221) ❙ 1681 (1292~2435); 926 (670~1428); 103 (49~130) ❙ 48.7 (12.9~99.5); 1129 (939~1221); 300.8 (235.1~371.7)

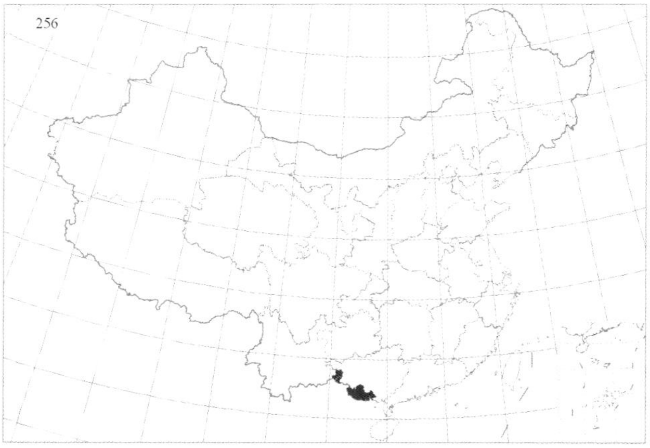

Gnetaceae 买麻藤科 *Gnetum* 买麻藤属
Gnetum hainanense C. Y. Cheng ex L. K. Fu, Y. F. Yu et M. G. Gilbert 海南买麻藤
Vines, evergreen

20.9 (17.3~24.7); 12.4 (6.4~19.8); 27.9 (25.9~28.9); 21.0 (17.3~24.7); 191.4 (147.9~236.5); 0.0 (0.0~0.0); 1107 (907~1376) ❙ 1679 (1176~2014); 686 (447~954); 146 (53~199) ❙ 53.9 (-12.0~89.4); 1089 (907~1257); 271.9 (187.2~572.9)

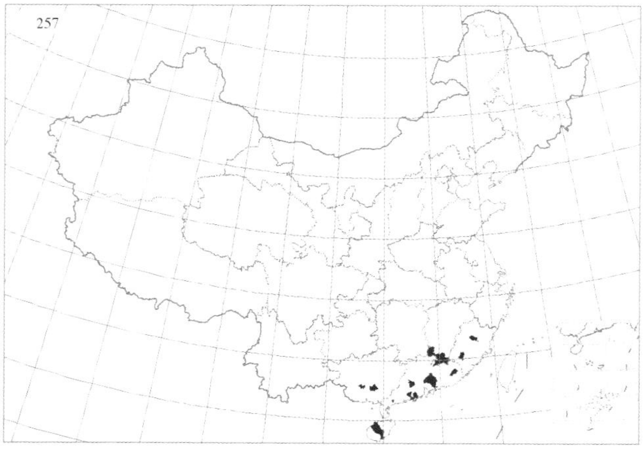

Gnetaceae 买麻藤科 *Gnetum* 买麻藤属
Gnetum luofuense C. Y. Cheng 罗浮买麻藤
Vines, evergreen

20.5 (18.3~22.0); 12.1 (9.2~15.2); 27.9 (25.9~28.8); 20.8 (18.4~22.7); 189.1 (160.5~212.7); 0.0 (0.0~0.0); 1088 (921~1216) ❙ 1803 (1636~1990); 754 (591~994); 163 (104~199) ❙ 66.8 (40.1~89.4); 1088 (921~1216); 241.6 (203.4~300.0)

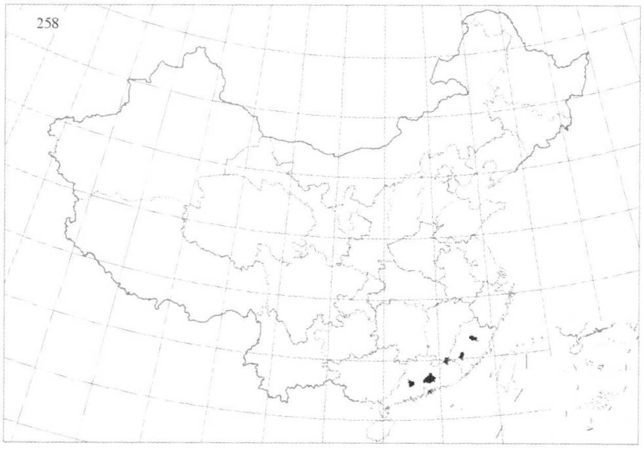

Gnetaceae 买麻藤科 *Gnetum* 买麻藤属

Gnetum montanum Markgr. 买麻藤

Large climbers, evergreen

19.6 (14.6~24.8); 11.4 (3.8~20.1); 26.0 (20.3~29.0); 19.7 (14.7~24.3); 175.9 (117.7~237.6); 0.0 (0.0~1.5); 1009 (767~1385) ❙ 1483 (939~2435); 708 (448~1428); 113 (39~255) ❙ 47.8 (-1.0~112.2); 1002 (751~1360); 298.3 (139.6~659.0)

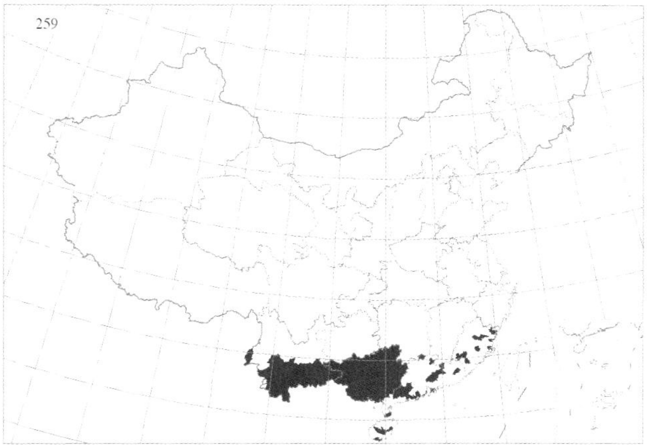

Gnetaceae 买麻藤科 *Gnetum* 买麻藤属

Gnetum parvifolium (Warb.) Chun 小叶买麻藤

Vines, evergreen

20.0 (14.3~24.8); 11.3 (3.3~20.1); 27.4 (21.4~29.8); 20.1 (14.3~24.8); 181.1 (113.2~237.6); 0.0 (0.0~2.3); 1050 (761~1385) ❙ 1567 (868~2435); 656 (312~1428); 144 (41~255) ❙ 51.2 (-34.3~128.9); 1038 (761~1360); 260.2 (95.3~659.0)

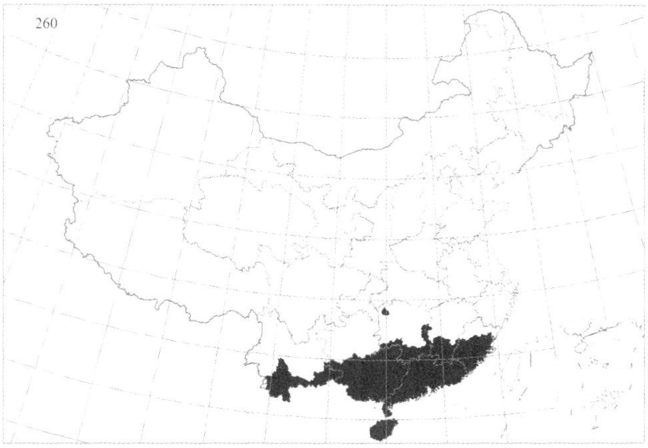

Gnetaceae 买麻藤科 *Gnetum* 买麻藤属

Gnetum pendulum C. Y. Cheng 垂子买麻藤

Large climbers, evergreen

17.6 (2.9~22.2); 10.4 (-5.7~15.1); 22.8 (10.6~28.5); 17.7 (4.3~22.3); 153.0 (19.9~207.5); 0.9 (0.0~45.2); 874 (421~1190) ❙ 1362 (535~1751); 745 (326~1007); 60 (10~129) ❙ 57.1 (15.4~145.8); 869 (421~1190); 403.5 (222.1~659.0)

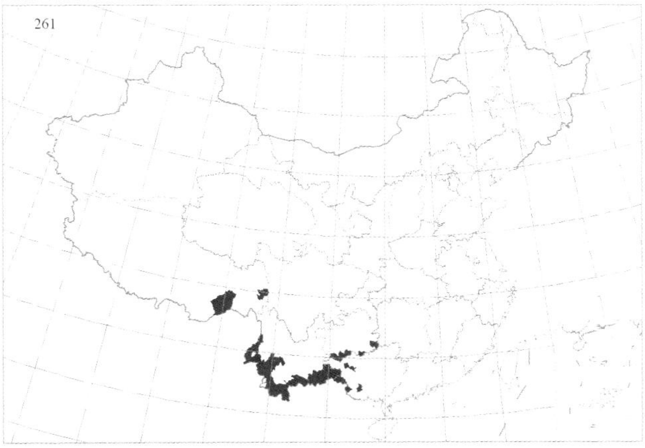

Piperaceae 胡椒科 *Piper* 胡椒属

Piper austrosinense Y. Q. Tseng 华南胡椒

Vines, evergreen

20.8 (14.1~24.7); 12.4 (3.5~19.8); 27.7 (24.2~28.8); 20.9 (14.2~24.7); 190.5 (111.5~236.5); 0.0 (0.0~1.7); 1101 (762~1376) ❙ 1607 (868~2435); 704 (312~1428); 132 (42~194) ❙ 47.7 (-34.3~99.5); 1076 (762~1221); 266.1 (150.2~620.0)

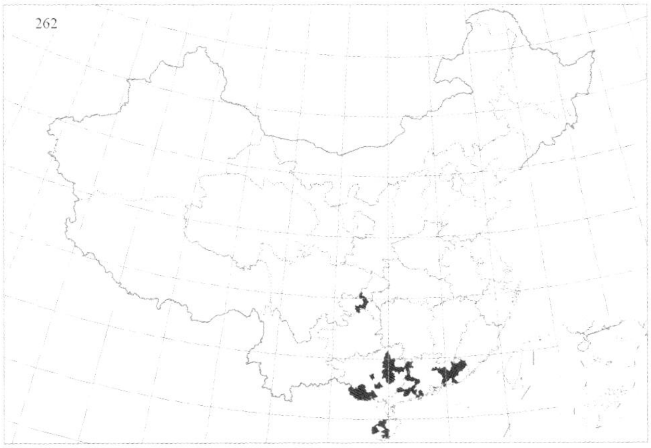

Piperaceae 胡椒科 *Piper* 胡椒属

Piper boehmeriifolium (Miq.) Wall. ex C. DC. 苎叶蒟

Subshrubs, evergreen

19.8 (8.6~24.8); 11.7 (1.3~20.1); 26.2 (15.0~29.3); 19.8 (8.7~24.8); 178.1 (52.6~237.6); 0.1 (0.0~8.8); 1029 (574~1385) ❙ 1520 (939~2148); 728 (444~1234); 101 (37~194) ❙ 49.0 (-2.2~112.2); 1019 (574~1326); 288.8 (95.3~659.0)

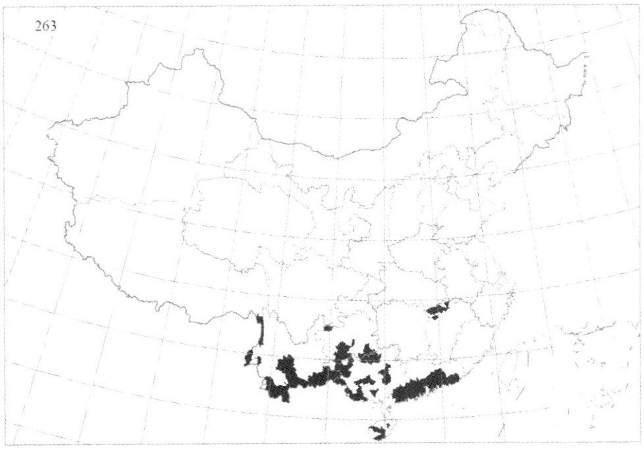

Piperaceae 胡椒科 *Piper* 胡椒属

Piper chinense Miq. 中华胡椒

Vines, evergreen

21.8 (21.0~22.2); 13.8 (13.2~14.7); 28.1 (26.9~28.8); 21.8 (21.0~22.2); 201.6 (192.3~206.9); 0.0 (0.0~0.0); 1149 (1081~1177) ❙ 1703 (1618~1876); 789 (722~891); 122 (107~132) ❙ 48.5 (40.1~64.3); 1149 (1081~1177); 229.4 (203.4~273.3)

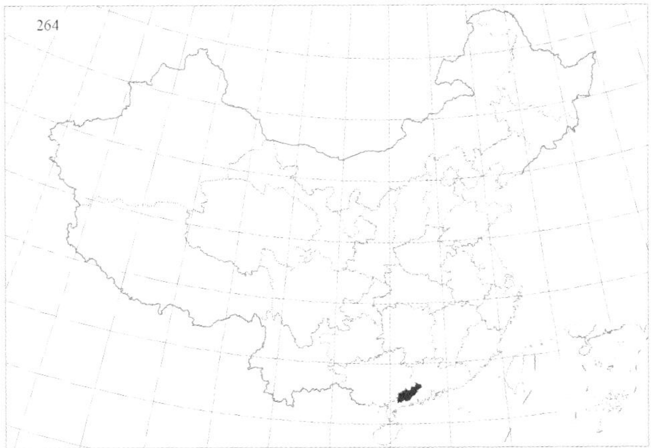

Piperaceae 胡椒科 *Piper* 胡椒属
Piper hainanense Hemsl. 海南蒟
Vines, evergreen

22.9 (15.5~24.8); 16.0 (4.8~20.1); 28.0 (25.1~29.0); 22.9 (15.6~24.8); 215.0 (127.2~237.6); 0.0 (0.0~0.2); 1236 (815~1385) ❙ 1561 (868~2435); 664 (312~1428); 95 (41~177) ❙ 28.1 (-34.3~99.5); 1171 (815~1360); 353.4 (194.5~620.0)

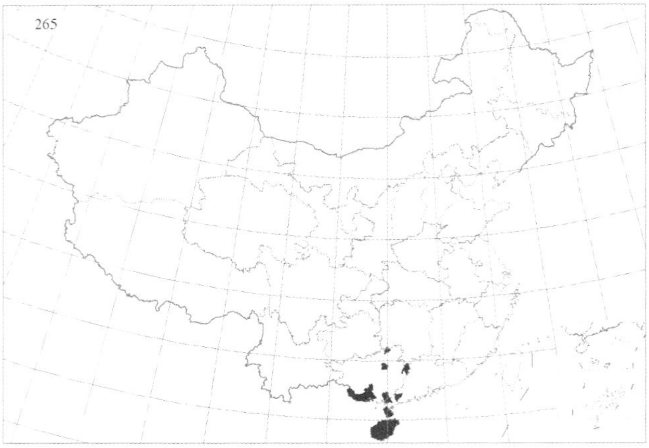

Piperaceae 胡椒科 *Piper* 胡椒属
Piper kadsura (Choisy) Ohwi 风藤
Vines, evergreen

18.1 (14.1~23.8); 8.4 (3.3~18.1); 27.2 (20.1~29.7); 18.2 (14.2~23.8); 157.9 (111.5~225.8); 0.1 (0.0~1.9); 948 (761~1297) ❙ 1603 (1055~2993); 585 (356~1847); 194 (52~547) ❙ 71.9 (-5.0~246.5); 930 (761~1190); 221.4 (119.4~363.7)

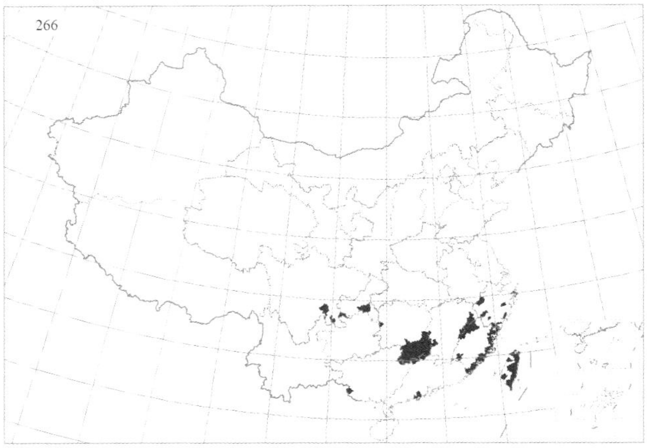

Piperaceae 胡椒科 *Piper* 胡椒属
Piper laetispicum C. DC. 大叶蒟
Vines, evergreen

23.7 (21.8~24.8); 17.6 (13.3~20.1); 28.1 (25.7~28.9); 23.7 (21.9~24.8); 224.7 (202.2~237.6); 0.0 (0.0~0.0); 1289 (1127~1385) ❙ 1522 (868~2148); 593 (312~1234); 80 (41~131) ❙ 18.8 (-34.3~86.7); 1195 (868~1360); 402.2 (194.5~620.0)

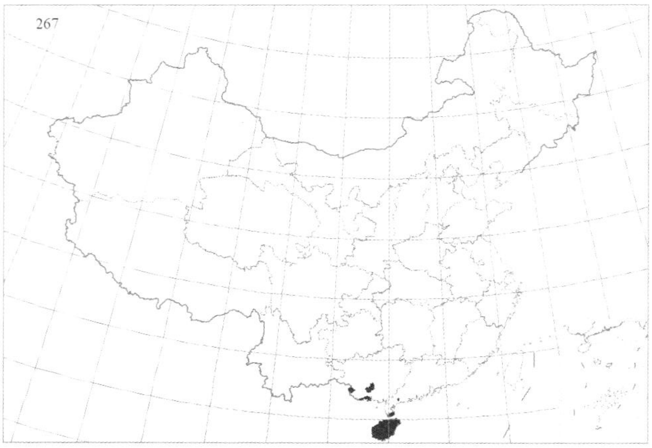

Piperaceae 胡椒科 *Piper* 胡椒属
Piper lingshuiense Y. Q. Tseng 陵水胡椒
Vines, evergreen

24.4; 19.5; 27.8; 24.5; 233.7; 0.0; 1351 ❙ 1458; 487; 49 ❙ 7.9; 1154; 492.5

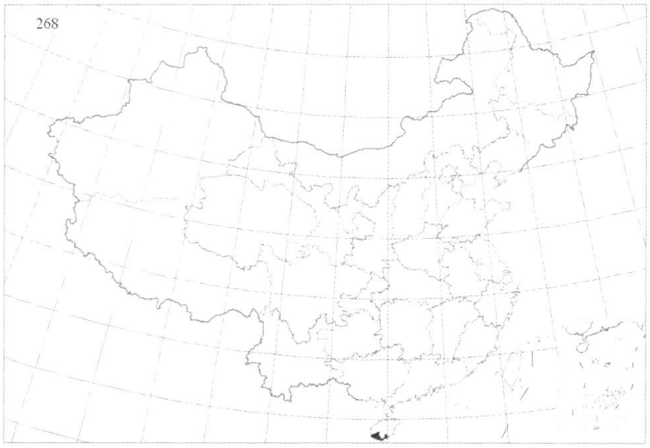

Piperaceae 胡椒科 *Piper* 胡椒属
Piper mullesua Buch.-Ham. ex D. Don 短蒟
Vines, evergreen

17.2 (6.6~23.0); 10.6 (-1.3~17.7); 21.7 (13.3~26.7); 17.3 (6.8~23.1); 147.9 (39.6~217.0); 0.6 (0.0~19.7); 848 (514~1217) ❙ 1337 (862~1751); 740 (524~1007); 51 (12~77) ❙ 58.9 (5.5~145.8); 831 (514~1163); 433.0 (296.0~659.0)

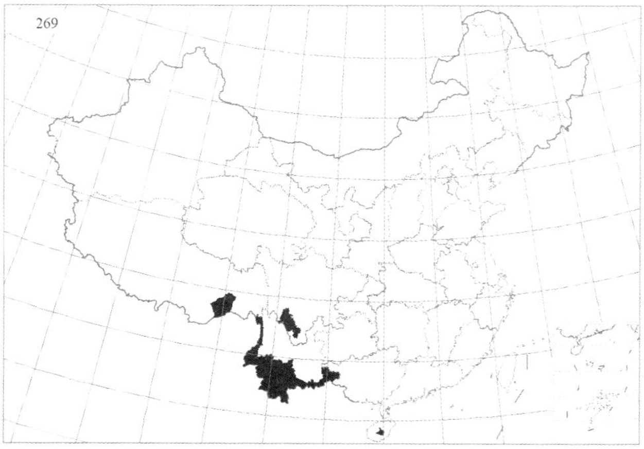

Piperaceae 胡椒科 *Piper* 胡椒属
Piper pedicellatum C. DC. 角果胡椒
Vines, evergreen

18.5 (16.4~21.0); 12.2 (9.5~15.1); 22.6 (20.6~25.6); 18.6 (16.4~21.0); 162.8 (137.3~191.8); 0.0 (0.0~0.0); 892 (795~1053) ❙ 1447 (1049~1751); 794 (598~996); 56 (41~77) ❙ 62.6 (26.1~112.2); 879 (783~1053); 472.4 (336.3~659.0)

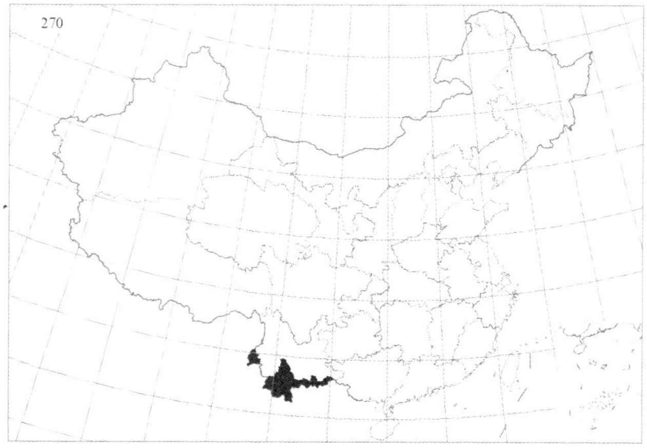

Piperaceae 胡椒科 *Piper* 胡椒属

Piper pleiocarpum C. C. Chang ex Y. Q. Tseng 线梗胡椒

Vines, evergreen

17.1 (14.4~19.6); 10.7 (7.5~13.9); 21.2 (19.1~22.7); 17.1 (14.4~19.6); 145.4 (113.1~175.1); 0.0 (0.0~0.0); 824 (727~933) ❙ 1327 (1083~1530); 722 (594~816); 51 (41~70) ❙ 61.2 (34.7~99.9); 812 (727~882); 438.0 (335.1~530.1)

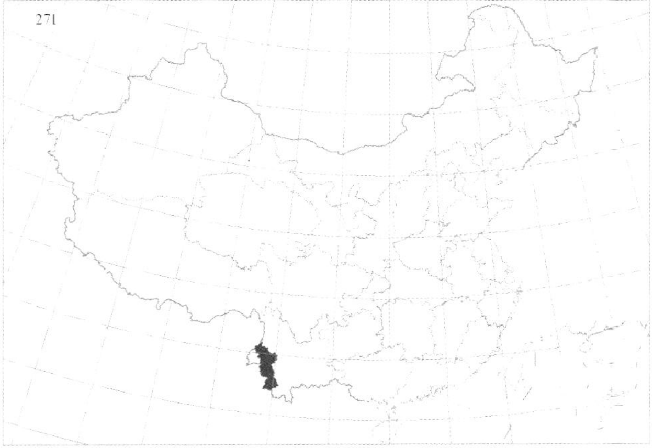

Piperaceae 胡椒科 *Piper* 胡椒属

Piper polysyphonum C. DC. 樟叶胡椒

Subshrubs, evergreen

18.3 (16.1~21.0); 11.6 (7.8~15.1); 22.9 (20.3~26.6); 18.3 (16.2~21.0); 160.2 (133.9~191.8); 0.0 (0.0~0.0); 890 (785~1053) ❙ 1356 (1049~1751); 728 (594~996); 55 (41~77) ❙ 52.5 (26.1~88.8); 877 (779~1053); 426.6 (189.5~659.0)

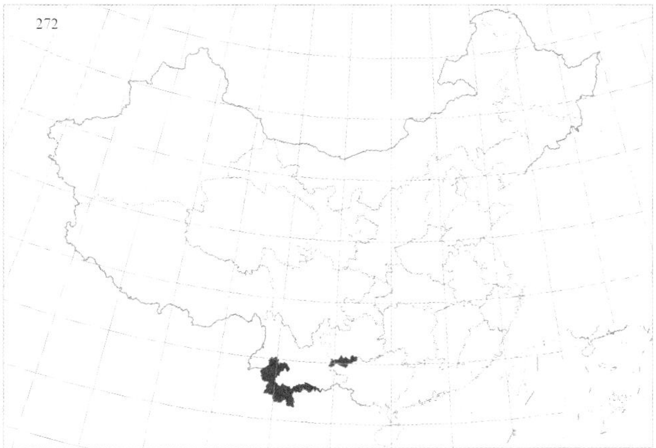

Piperaceae 胡椒科 *Piper* 胡椒属

Piper senporeiense Yamam. 斜叶蒟

Vines, evergreen

23.8 (22.2~24.8); 18.6 (17.2~20.1); 27.4 (25.7~28.2); 23.8 (22.3~24.3); 225.7 (207.2~237.6); 0.0 (0.0~0.0); 1286 (1127~1385) ❙ 1535 (1242~1757); 529 (444~610); 56 (41~88) ❙ 19.9 (-2.2~47.8); 1151 (1055~1299); 531.9 (439.7~620.0)

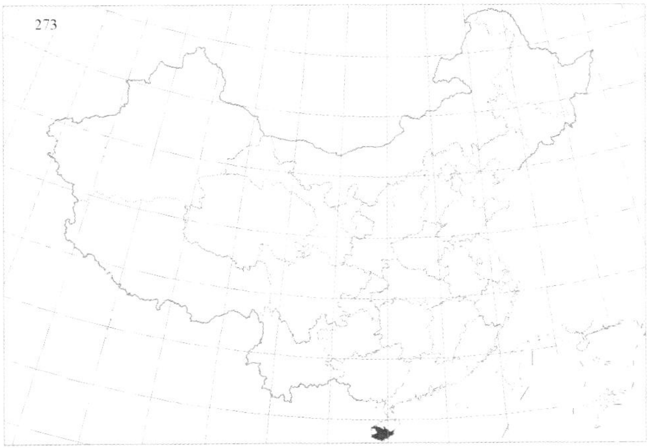

Piperaceae 胡椒科 *Piper* 胡椒属

Piper stipitiforme C. C. Chang ex Y. Q. Tseng 短柄胡椒

Vines, evergreen

19.2 (17.9~20.5); 13.6 (11.9~15.1); 22.5 (21.6~23.7); 19.3 (18.0~20.5); 171.3 (155.8~185.9); 0.0 (0.0~0.0); 920 (854~992) ❙ 1486 (1434~1607); 719 (655~803); 54 (41~77) ❙ 61.9 (47.6~74.2); 892 (833~986); 527.4 (439.5~659.0)

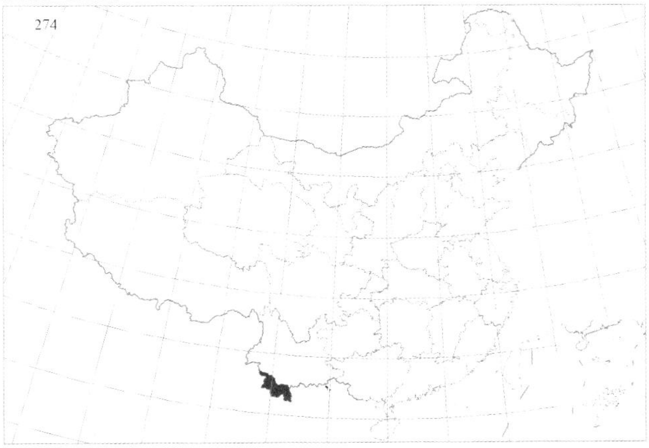

Piperaceae 胡椒科 *Piper* 胡椒属

Piper umbellatum L. 大胡椒

Subshrubs

19.6 (16.2~24.1); 14.7 (11.0~18.9); 23.3 (20.1~28.0); 19.7 (16.2~24.7); 176.0 (134.2~229.7); 0.0 (0.0~0.0); 971 (773~1322) ❙ 2349 (1751~2992); 1363 (947~1847); 161 (47~303) ❙ 174.5 (32.5~246.5); 938 (773~1123); 266.8 (154.2~363.7)

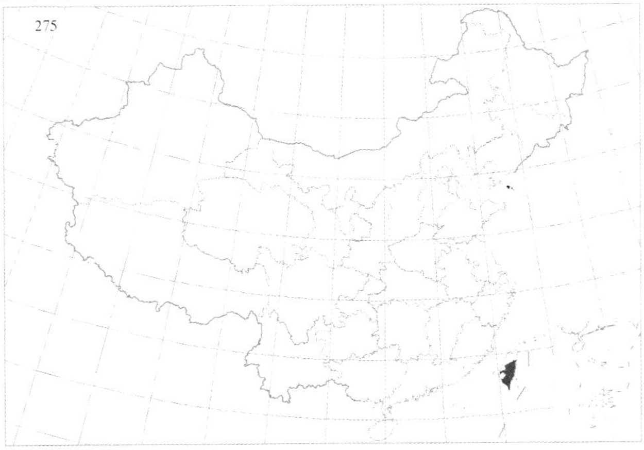

Chloranthaceae 金粟兰科 *Chloranthus* 金粟兰属

Chloranthus erectus (Buch.-Ham.) Verdc. 鱼子兰

Subshrubs

15.5 (5.3~22.2); 7.8 (-2.4~14.9); 21.6 (12.2~28.7); 15.6 (5.7~22.3); 128.2 (30.5~207.5); 1.5 (0.0~26.2); 802 (479~1190) ❙ 1124 (741~1776); 628 (411~1007); 45 (15~188) ❙ 41.4 (-14.2~145.8); 782 (479~1190); 298.9 (142.9~570.1)

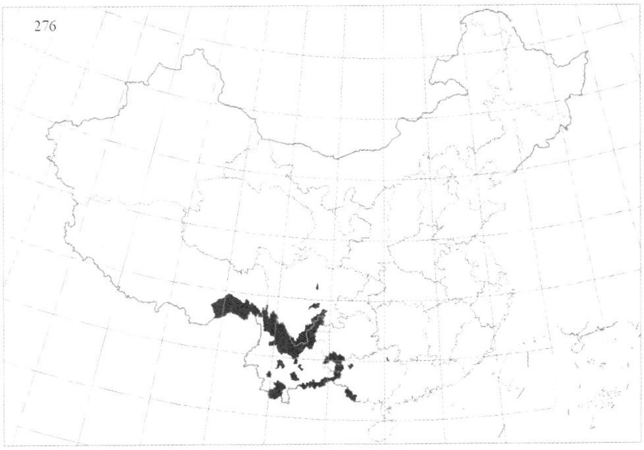

Chloranthaceae 金粟兰科 *Chloranthus* 金粟兰属
Chloranthus spicatus (Thunb.) Makino 金粟兰
Subshrubs, evergreen

19.2 (14.3~24.6); 10.3 (2.0~19.5); 27.0 (21.5~29.8); 19.2 (14.3~24.7); 170.9 (117.5~235.9); 0.1 (0.0~5.4); 995 (779~1372) ❙ 1487 (946~1997); 598 (356~858); 154 (38~257) ❙ 51.9 (-5.0~132.1); 981 (763~1299); 264.2 (123.9~659.0)

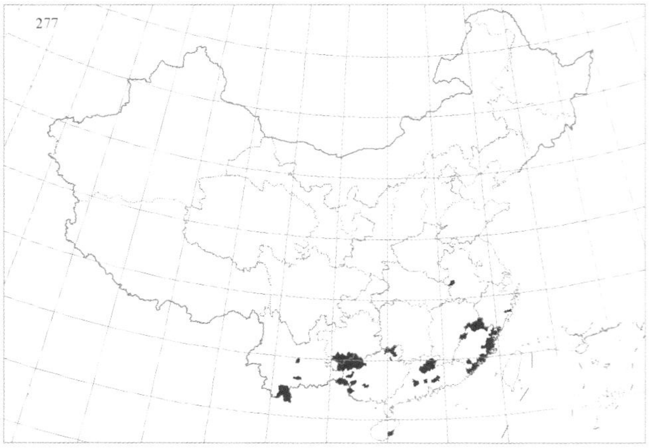

Chloranthaceae 金粟兰科 *Hedyosmum* 雪香兰属
Hedyosmum orientale Merr. et Chun 雪香兰
Herbs or subshrubs

22.6 (21.3~24.7); 15.4 (13.2~19.8); 27.9 (26.7~28.6); 22.7 (21.3~24.7); 211.8 (195.9~236.5); 0.0 (0.0~0.0); 1212 (1111~1376) ❙ 1719 (1302~2435); 841 (534~1428); 101 (48~130) ❙ 42.5 (12.9~99.5); 1183 (1111~1326); 356.6 (235.1~620.0)

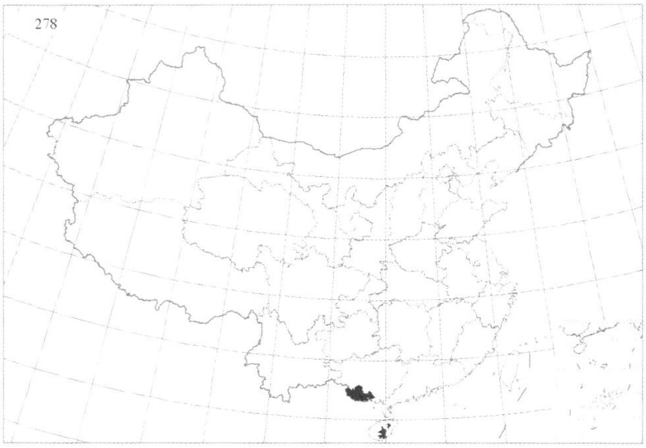

Chloranthaceae 金粟兰科 *Sarcandra* 草珊瑚属
Sarcandra glabra (Thunb.) Nakai 草珊瑚
Subshrubs, evergreen

17.9 (11.1~23.6); 8.0 (0.6~17.7); 26.8 (18.6~30.1); 18.0 (11.1~23.6); 155.8 (79.8~223.5); 0.3 (0.0~9.7); 945 (648~1281) ❙ 1471 (863~2993); 598 (299~1847); 142 (25~547) ❙ 56.8 (-6.9~246.5); 936 (648~1273); 226.4 (95.3~572.9)

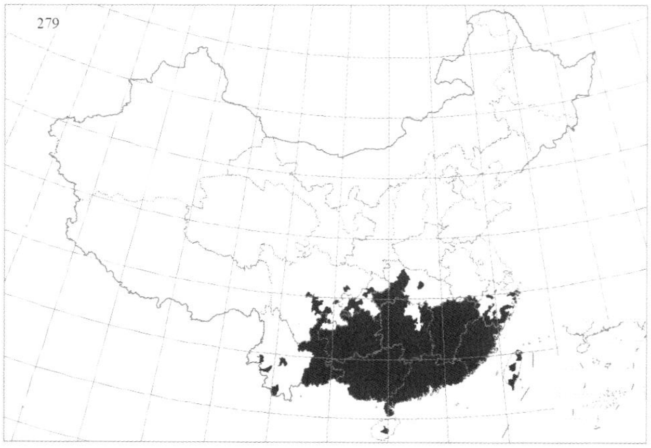

Chloranthaceae 金粟兰科 *Sarcandra* 草珊瑚属
Sarcandra glabra subsp. brachystachys (Blume) Verdc. 海南草珊瑚
Subshrubs, evergreen

20.1 (14.9~24.8); 12.5 (3.6~20.1); 26.1 (20.8~29.4); 20.2 (15.0~24.8); 181.9 (119.6~237.6); 0.0 (0.0~1.6); 1045 (782~1385) ❙ 1501 (868~1991); 647 (312~996); 103 (41~221) ❙ 47.5 (-34.3~106.4); 1007 (782~1360); 358.0 (180.3~659.0)

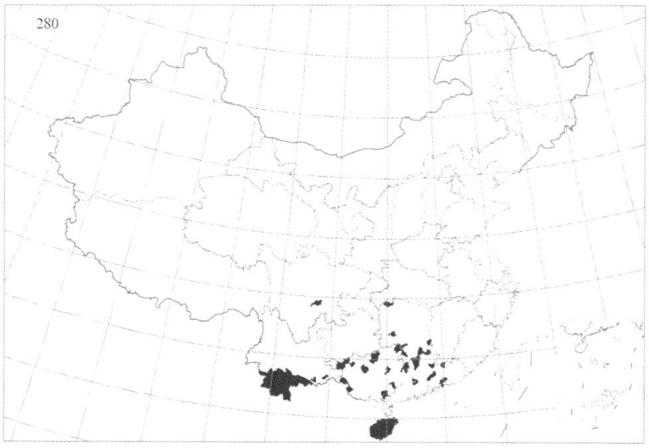

Salicaceae 杨柳科 *Chosenia* 钻天柳属
Chosenia arbutifolia (Pall.) A. K. Skvortsov 钻天柳
Trees, deciduous

2.5 (-5.8~12.7); -18.7 (-29.8~-4.2); 20.9 (15.3~27.4); 7.7 (4.4~13.3); 59.8 (28.0~114.5); 89.4 (22.0~157.4); 570 (426~829) ❙ 551 (311~1073); 354 (214~691); 16 (5~40) ❙ -3.1 (-42.0~74.1); 506 (311~661); 251.4 (97.2~335.3)

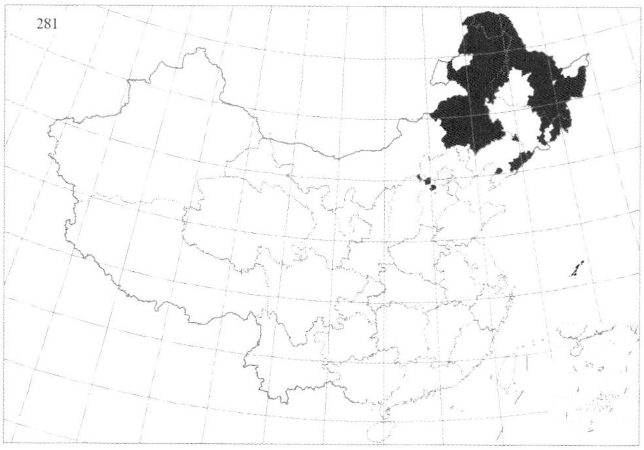

Salicaceae 杨柳科 *Populus* 杨属
Populus adenopoda Maxim. 响叶杨
Trees, deciduous

15.3 (0.0~22.0); 4.0 (-11.4~13.4); 25.9 (9.7~29.9); 15.4 (3.3~22.0); 128.5 (12.2~204.3); 3.9 (0.0~72.2); 841 (395~1176) ❙ 1184 (127~1997); 499 (82~832); 100 (2~259) ❙ 40.5 (-81.2~130.7); 822 (127~1176); 199.4 (41.6~364.1)

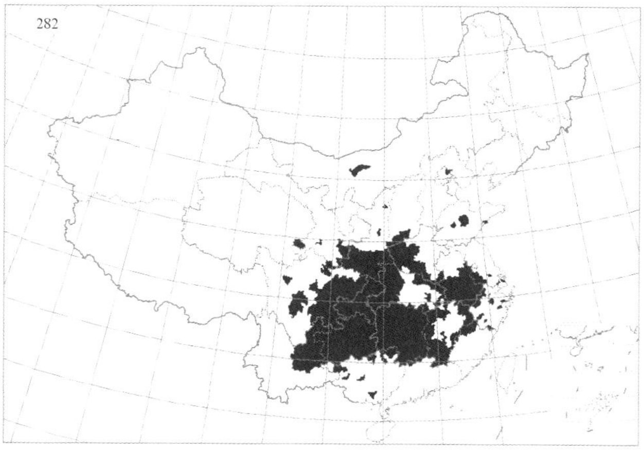

Salicaceae 杨柳科 *Populus* 杨属

Populus afghanica (Aiton et Hemsl.) C. K. Schneid. 阿富汗杨

Trees, deciduous

5.8 (-4.3~12.4); -11.0 (-20.1~-5.2); 19.6 (9.6~26.1); 8.7 (2.5~13.2); 69.0 (9.7~114.1); 58.5 (24.3~121.1); 606 (329~811) ❙ 67 (30~215); 27 (16~80); 8 (1~26) ❙ -86.1 (-96.2~-51.1); 67 (30~215); 28.9 (2.6~120.2)

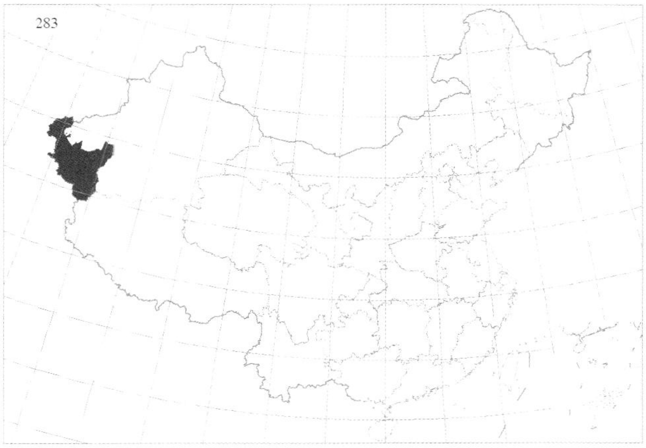

Salicaceae 杨柳科 *Populus* 杨属

Populus afghanica var. tajikistanica (Kom.) C. Wang et C. Y. Yang 喀什阿富汗杨

Trees, deciduous

8.6; -8.8; 22.4; 10.5; 85.7; 42.3; 686 ❙ 110; 41; 14 -80.4; 110; 62.4

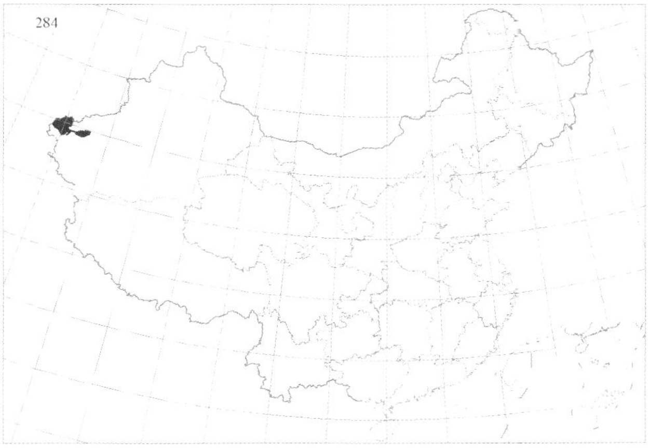

Salicaceae 杨柳科 *Populus* 杨属

Populus alaschanica Kom. 阿拉善杨

Trees, deciduous

7.7; -10.0; 23.6; 9.9; 82.5; 49.0; 658 ❙ 108; 67; 2 ❙ -83.6; 108; 10.6

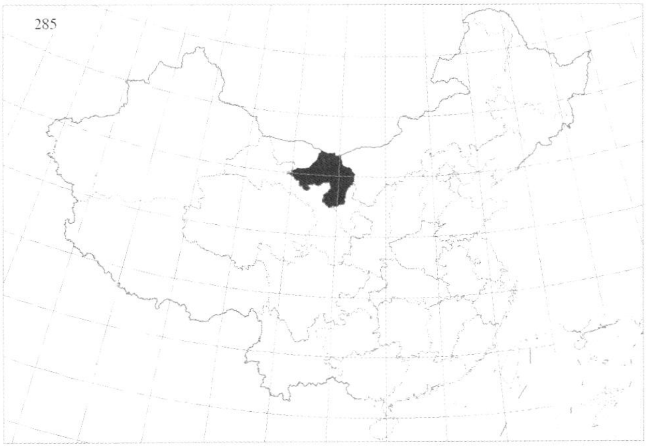

Salicaceae 杨柳科 *Populus* 杨属

Populus alba L. 银白杨

Trees, deciduous

7.6 (-4.3~16.5); -8.7 (-20.1~4.4); 21.8 (9.6~28.9); 9.7 (2.5~16.6); 77.8 (9.7~139.8); 45.7 (0.6~121.1); 646 (329~910) ❙ 384 (21~1377); 228 (16~499); 15 (0~197) ❙ -39.4 (-97.3~51.2); 373 (21~910); 134.5 (1.6~289.4)

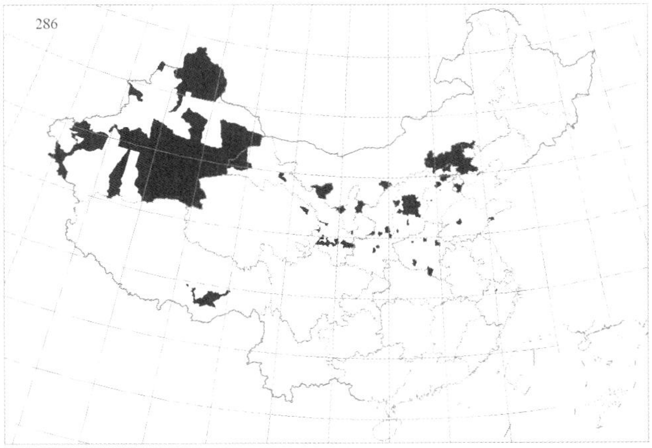

Salicaceae 杨柳科 *Populus* 杨属

Populus amurensis Kom. 黑龙江杨

Trees, deciduous

-0.7 (-4.0~0.8); -24.1 (-28.2~-20.9); 19.5 (17.6~20.8); 6.6 (5.3~7.3); 49.5 (38.6~56.0); 117.6 (105.6~146.2); 524 (463~557) ❙ 560 (470~657); 362 (324~415); 15 (11~20) ❙ 6.2 (-8.4~23.8); 518 (463~557); 291.6 (276.7~316.0)

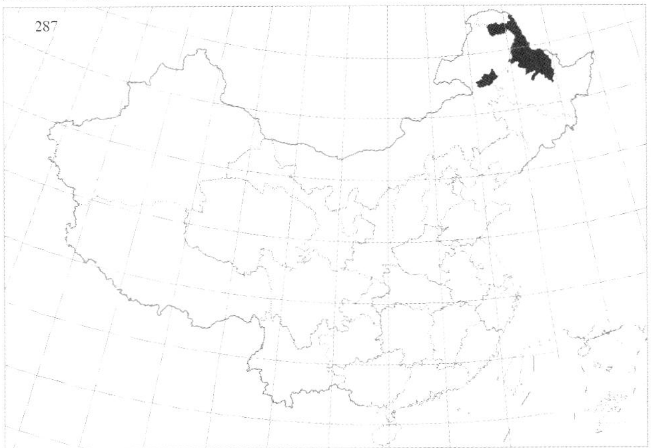

Salicaceae 杨柳科 *Populus* 杨属

Populus candicans Aiton 欧洲大叶杨

Trees, deciduous

5.1 (-1.3~14.5); -13.6 (-20.9~-0.2); 20.4 (13.2~28.2); 8.5 (4.6~14.6); 67.1 (26.7~125.3); 65.8 (10.4~104.2); 607 (440~864) ❙ 239 (53~551); 110 (33~245); 19 (2~48) ❙ -58.5 (-92.4~-7.6); 239 (53~551); 121.1 (10.3~297.3)

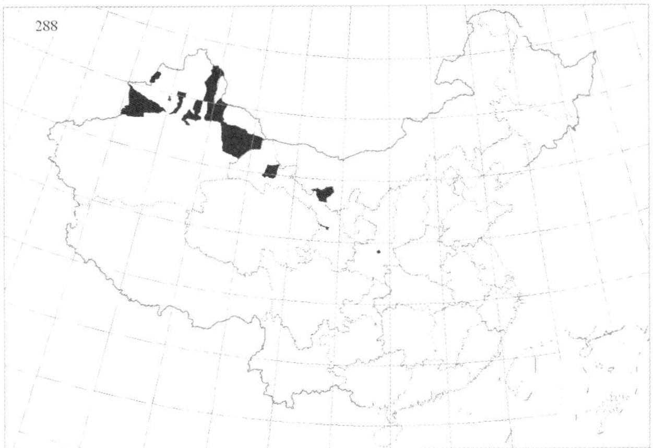

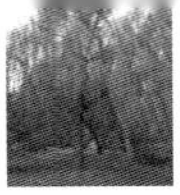

Salicaceae 杨柳科 *Populus* 杨属
Populus canescens (Aiton) Sm. 银灰杨
Trees, deciduous

2.6 (−0.6~8.0); −16.2 (−19.5~−9.0); 19.1 (15.9~23.3); 7.3 (5.4~9.8); 55.3 (37.5~80.9); 83.5 (43.6~104.2); 556 (475~668) ❙ 205 (156~410); 93 (56~247); 21 (8~33) ❙ −63.2 (−76.7~−36.4); 205 (156~410); 79.7 (16.6~139.1)

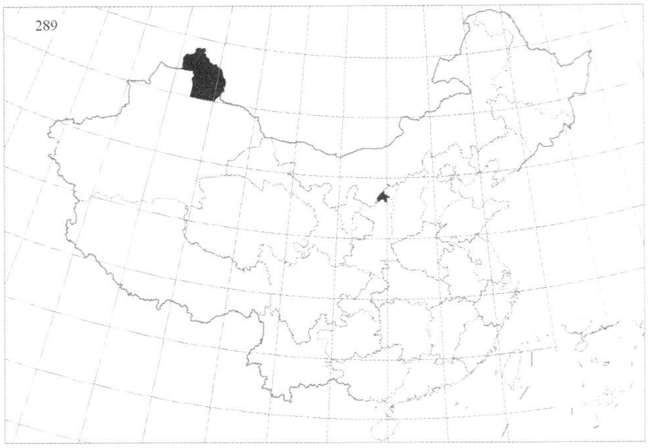

Salicaceae 杨柳科 *Populus* 杨属
Populus cathayana Rehder 青杨
Trees, deciduous

7.8 (−4.8~17.1); −6.6 (−20.9~7.1); 20.5 (7.3~28.5); 9.4 (1.9~17.2); 72.6 (4.0~146.0); 38.9 (0.0~121.6); 621 (332~918) ❙ 556 (21~1865); 301 (17~816); 18 (0~258) ❙ −8.6 (−97.3~128.6); 487 (21~918); 178.4 (1.6~364.1)

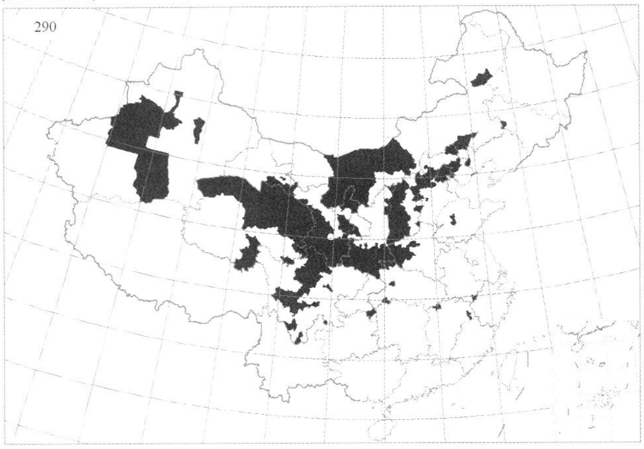

Salicaceae 杨柳科 *Populus* 杨属
Populus cathayana var. latifolia (C. Wang et C. Y. Yu) C. Wang et S. L. Tung 宽叶青杨
Trees, deciduous

4.9 (−1.4~11.3); −7.4 (−15.4~−0.1); 16.0 (8.5~23.2); 6.8 (2.6~11.4); 46.1 (8.1~89.6); 46.6 (11.7~92.1); 513 (347~683) ❙ 538 (282~785); 285 (182~381); 13 (5~39) ❙ 5.9 (−29.3~48.8); 482 (282~679); 219.7 (81.2~306.1)

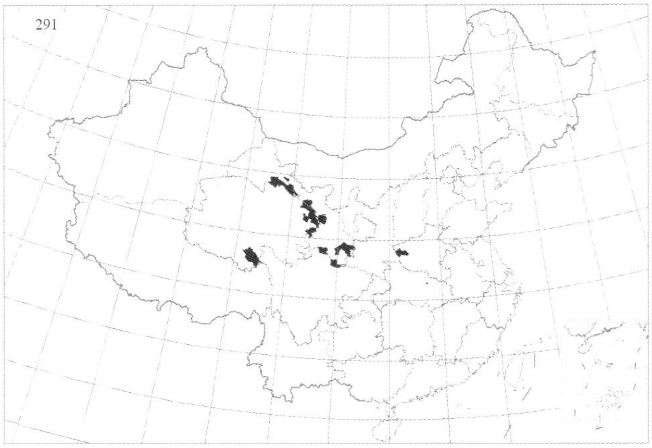

Salicaceae 杨柳科 *Populus* 杨属
Populus charbinensis C. Wang et Skvortsov 哈青杨
Trees, deciduous

6.7 (3.1~10.5); −10.1 (−19.6~−2.5); 21.3 (15.8~23.7); 9.1 (7.0~11.3); 70.7 (45.2~93.2); 50.0 (20.9~88.6); 618 (522~705) ❙ 606 (442~718); 356 (301~446); 18 (8~29) ❙ −1.7 (−27.6~28.2); 568 (442~658); 205.2 (166.0~271.9)

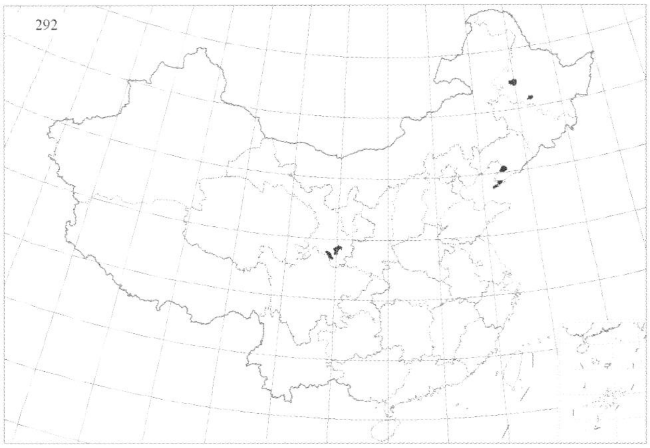

Salicaceae 杨柳科 *Populus* 杨属
Populus charbinensis var. pachydermis C. Wang et S. L. Tung 厚皮哈青杨
Trees, deciduous

3.1; −19.6; 22.6; 8.4; 66.2; 88.6; 610 ❙ 442; 304; 8 ❙ −27.6; 442; 165.0

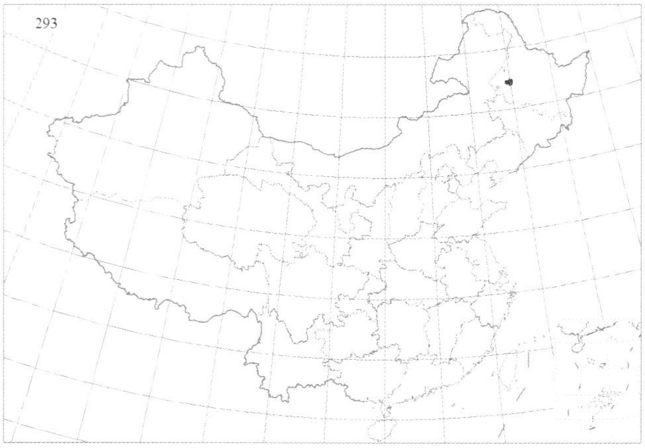

Salicaceae 杨柳科 *Populus* 杨属
Populus ciliata Wall. ex Royle 缘毛杨
Trees, deciduous

3.9 (−1.1~9.6); −4.5 (−9.1~2.4); 11.3 (6.2~15.7); 5.2 (2.0~9.6); 27.2 (2.5~60.8); 40.3 (5.4~75.6); 450 (334~600) ❙ 583 (252~1007); 357 (193~578); 20 (0~40) ❙ 23.1 (−35.9~100.7); 412 (252~600); 199.9 (79.3~351.7)

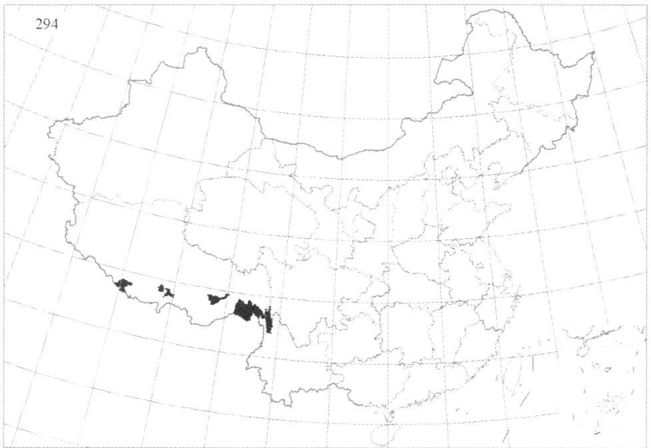

Salicaceae 杨柳科 *Populus* 杨属

Populus ciliata var. aurea C. Marquand et Airy Shaw 金色缘毛杨

Trees, deciduous

3.1；-19.6；22.6；8.4；66.2；88.6；610 | 442；304；8 | -27.6；442；166.0

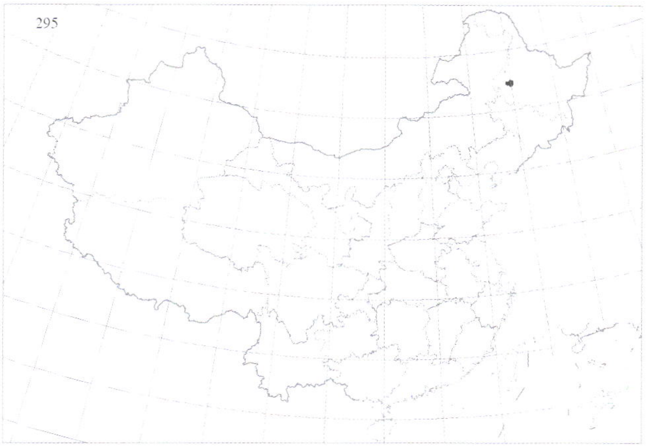

Salicaceae 杨柳科 *Populus* 杨属

Populus ciliata var. gyirongensis C. Wang et S. L. Tung 吉隆缘毛杨

Trees, deciduous

-1.1；-9.1；6.2；2.0；2.5；75.6；334 | 321；193；33 | -4.0；321；79.3

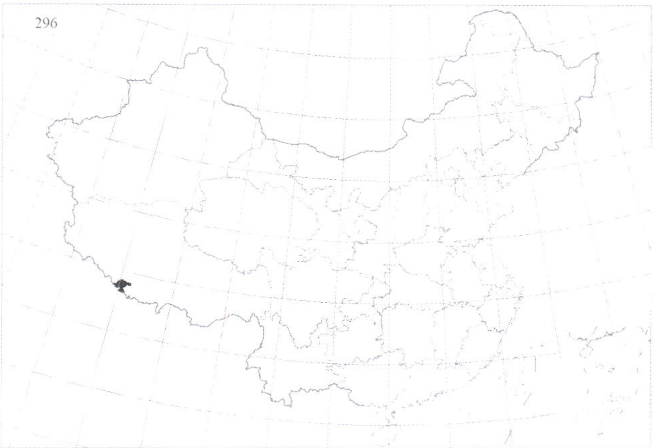

Salicaceae 杨柳科 *Populus* 杨属

Populus ciliata var. weixi C. Wang et S. L. Tung 维西缘毛杨

Trees, deciduous

9.6；2.4；15.7；9.6；60.8；5.4；600 | 1007；578；40 | 67.9；600；351.7

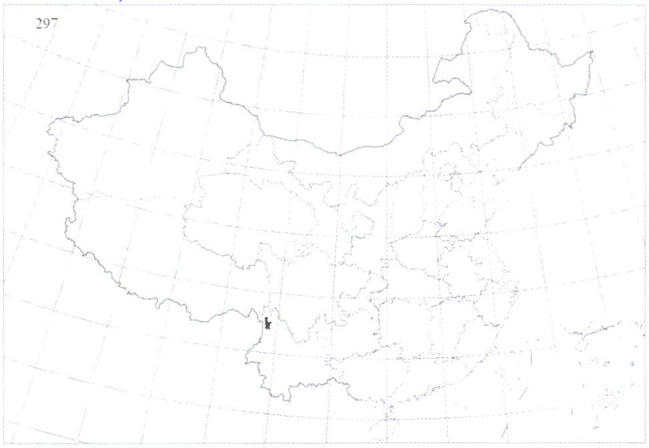

Salicaceae 杨柳科 *Populus* 杨属

Populus davidiana Dode 山杨

Trees, deciduous

8.4 (-5.8~22.2)；-6.5 (-29.8~15.1)；21.6 (7.2~29.5)；10.2 (2.0~22.3)；81.4 (3.7~207.5)；39.8 (0.0~157.4)；654 (335~1190) | 631 (58~2208)；355 (36~750)；23 (1~341) | -2.9 (-90.8~128.6)；544 (58~1190)；194.7 (7.5~659.0)

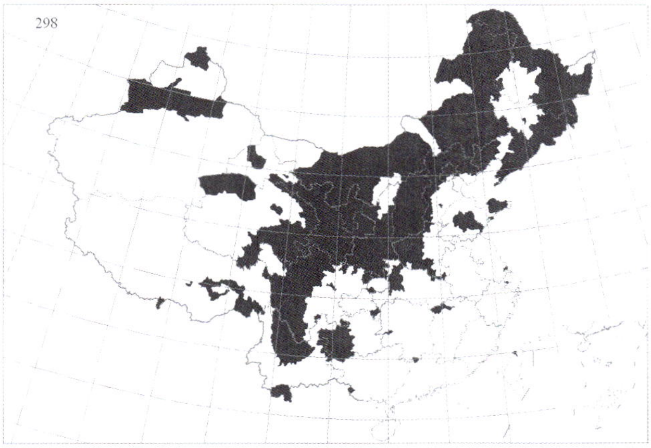

Salicaceae 杨柳科 *Populus* 杨属

Populus davidiana var. tomentella (C. K. Schneid.) Nakai 茸毛山杨

Trees, deciduous

4.8 (-0.3~13.4)；-4.9 (-10.9~2.5)；13.3 (9.0~24.0)；6.0 (3.0~13.4)；35 6 (10.4~105.4)；37.8 (4.2~73.4)；486 (378~745) | 832 (375~1208)；453 (203~606)；19 (8~33) | 73.9 (-32.2~128.6)；480 (375~745)；270.5 (95.7~357.7)

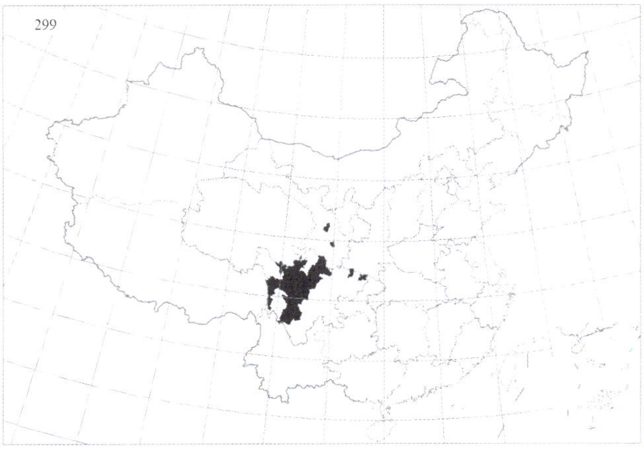

Salicaceae 杨柳科 *Populus* 杨属

Populus euphratica Olivier 胡杨

Trees, deciduous

6.2 (-4.3~12.5)；-11.5 (-20.9~-5.1)；21.0 (9.6~27.5)；9.0 (2.5~13.2)；72.6 (9.7~114.1)；57.2 (23.1~121.1)；625 (329~822) | 145 (21~495)；80 (15~319)；8 (0~51) | -74.0 (-97.3~-2.4)；145 (21~495)；54.3 (1.6~297.3)

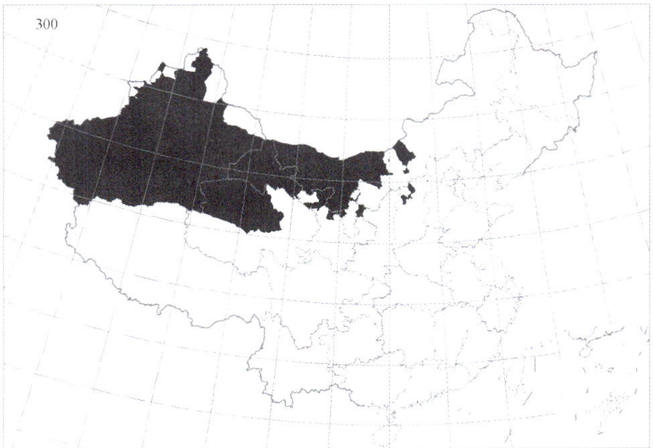

Salicaceae 杨柳科 *Populus* 杨属

Populus girinensis Skvortsov 东北杨

Trees, deciduous

-[illegible].3; -25.1; 19.4; 6.4; 48.5; 124.1; 518 ❙ 512; 334; 13 ❙ -0.9; 493; 276.8

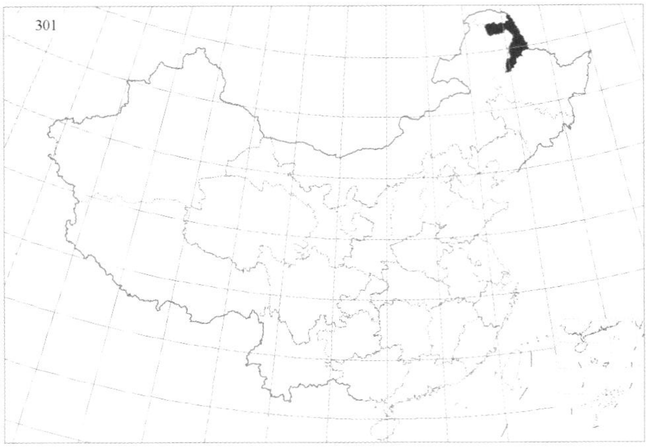

Salicaceae 杨柳科 *Populus* 杨属

Populus glauca H. H. Haines 灰背杨

Trees, deciduous

6.1; -1.5; 12.7; 6.3; 35.2; 21.8; 496 ❙ 801; 496; 21 ❙ 61.4; 496; 282.9

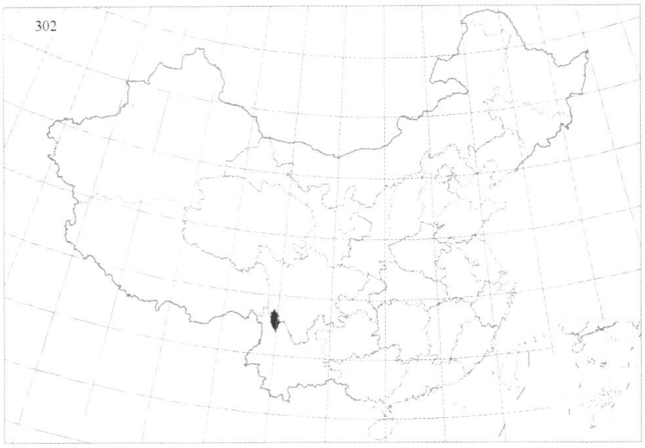

Salicaceae 杨柳科 *Populus* 杨属

Populus haoana Cheng et C. Wang 德钦杨

Trees, deciduous

5.7 (1.8~11.0); -2.3 (-7.1~4.2); 12.7 (9.9~16.6); 6.5 (3.8~11.0); 36.7 (15.5~73.1); 27.2 (0.8~53.3); 498 (404~636) ❙ 852 (583~1134); 502 (348~651); 25 (11~47) ❙ 71.1 (44.2~125.9); 498 (404~636); 292.3 (169.5~357.7)

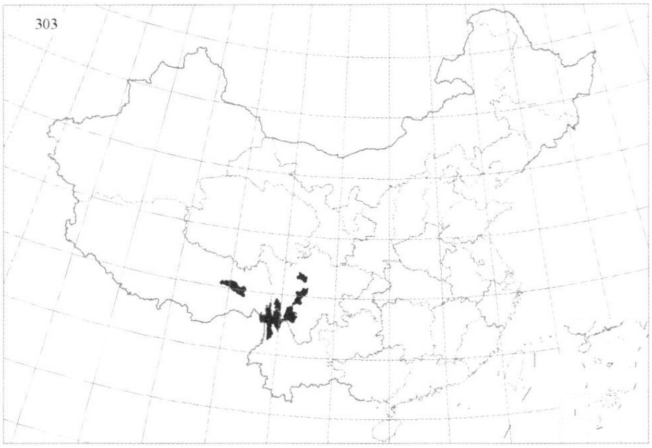

Salicaceae 杨柳科 *Populus* 杨属

Populus haoana var. macrocarpa C. Wang et S. L. Tung 大果德钦杨

Trees, deciduous

4.4; -3.9; 11.7; 5.3; 27.4; 33.8; 459 ❙ 740; 441; 19 ❙ 61.4; 459; 287.6

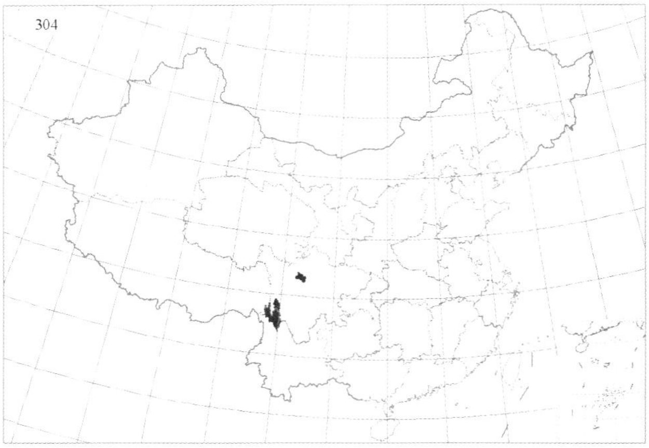

Salicaceae 杨柳科 *Populus* 杨属

Populus haoana var. megaphylla C. Wang et S. L. Tung 大叶德钦杨

Trees, deciduous

7.9; 0.5; 14.3; 8.0; 48.3; 13.3; 552 ❙ 930; 530; 38 ❙ 67.8; 552; 325.8

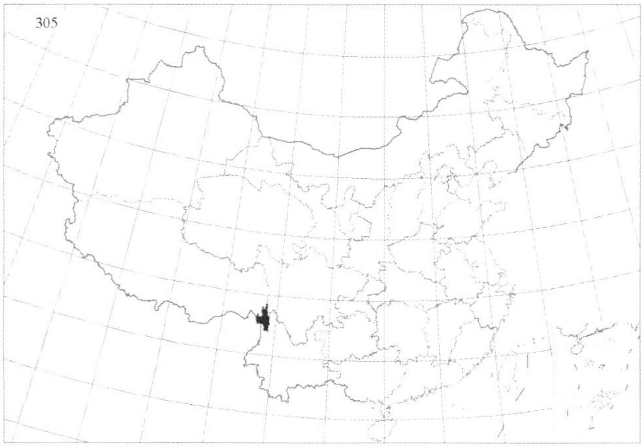

Salicaceae 杨柳科 *Populus* 杨属

Populus haoana var. microcarpa C. Wang et S. L. Tung 小果德钦杨

Trees, deciduous

6.1; -1.5; 12.7; 6.3; 35.2; 21.8; 496 ❙ 801; 496; 21 ❙ 61.4; 496; 282.9

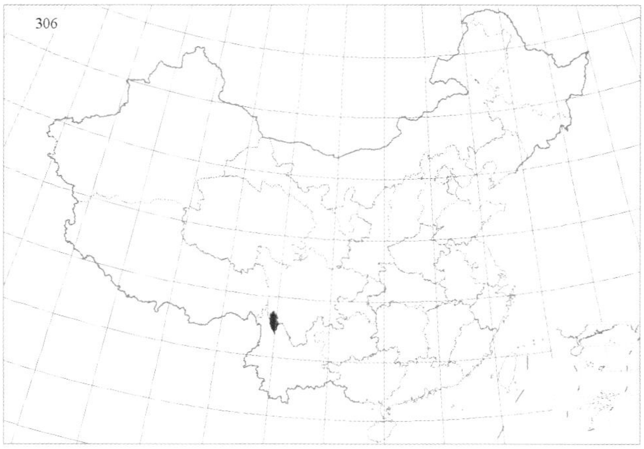

Salicaceae 杨柳科 *Populus* 杨属

Populus hsinganica C. Wang et Skvortsov 兴安杨

Trees, deciduous

2.6 (-5.8~10.2); -17.2 (-29.8~-6.2); 20.3 (15.3~24.7); 7.5 (4.4~11.4); 57.7 (28.0~95.3); 85.9 (32.8~157.4); 556 (426~722) ▎361 (111~590); 247 (74~453); 9 (2~16) ▎-33.0 (-82.3~15.9); 357 (111~590); 164.6 (13.9~309.5)

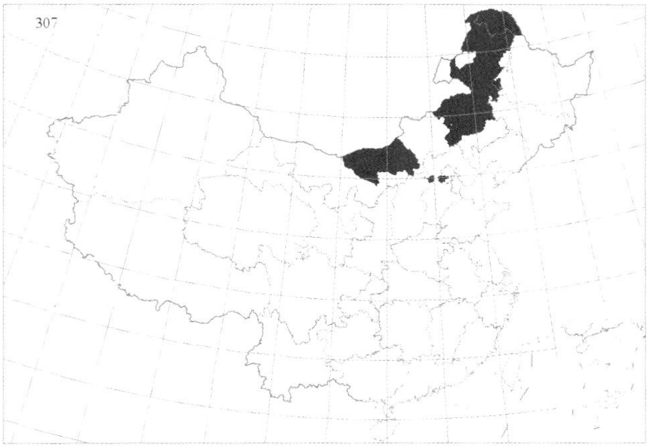

Salicaceae 杨柳科 *Populus* 杨属

Populus hsinganica var. trichorachis Z. F. Chen 毛轴兴安杨

Trees, deciduous

5.7 (2.2~9.4); -11.9 (-16.1~-7.5); 21.3 (18.6~24.3); 8.7 (6.8~10.9); 69.1 (48.4~91.7); 59.8 (38.2~81.0); 603 (516~704) ▎486 (332~569); 341 (215~406); 10 (8~11) ▎-19.7 (-35.7~-10.6); 486 (332~569); 237.0 (105.8~287.6)

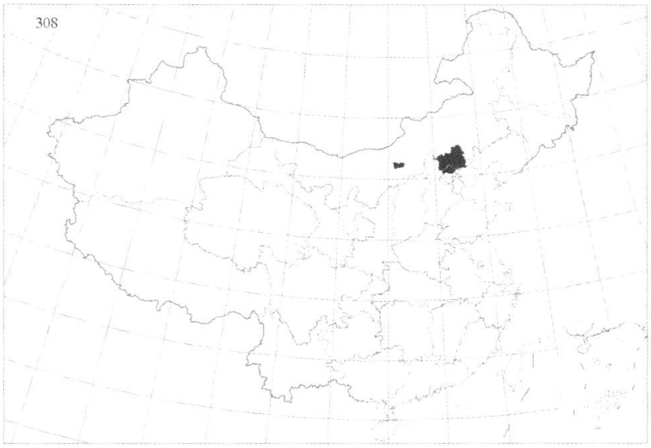

Salicaceae 杨柳科 *Populus* 杨属

Populus iliensis Drobow 伊犁杨

Trees, deciduous

2.9 (-1.3~9.0); -15.7 (-20.9~-9.3); 17.3 (13.2~23.2); 7.0 (4.6~10.7); 50.4 (26.7~88.9); 74.7 (40.7~103.9); 537 (440~703) ▎310 (225~406); 143 (65~211); 27 (12~48) ▎-40.1 (-65.1~-7.6); 310 (225~406); 216.4 (165.7~297.3)

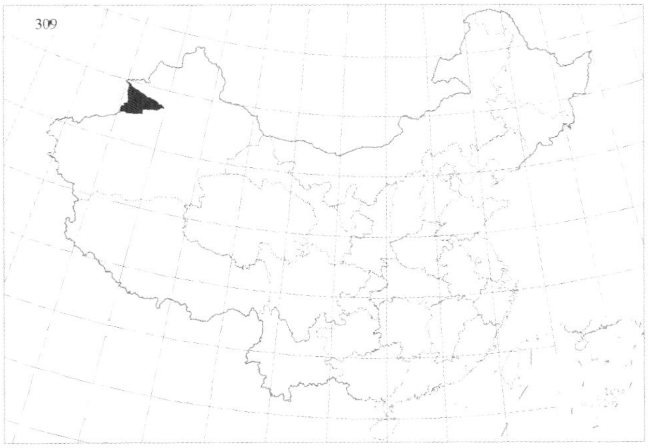

Salicaceae 杨柳科 *Populus* 杨属

Populus intramongolica T. Y. Sun et E. W. Ma 内蒙杨

Trees, deciduous

5.1 (2.2~8.0); -13.1 (-16.1~-9.0); 21.3 (18.5~24.2); 8.4 (6.7~10.1); 65.6 (48.1~84.9); 64.0 (46.5~81.1); 589 (515~671) ▎280 (111~442); 185 (74~312); 7 (2~15) ▎-51.3 (-82.3~-25.7); 280 (111~442); 91.8 (13.9~186.8)

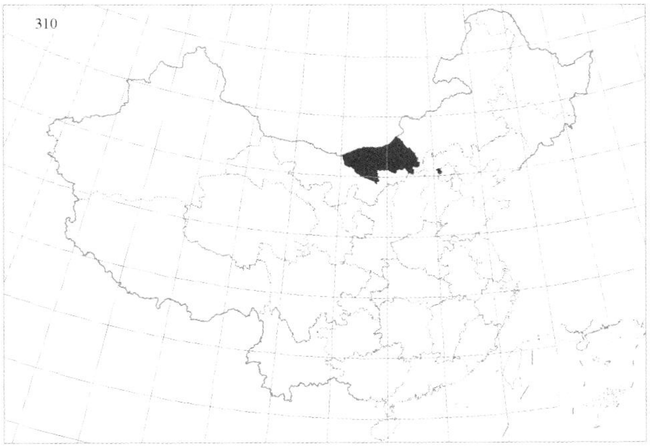

Salicaceae 杨柳科 *Populus* 杨属

Populus kangdingensis C. Wang et S. L. Tung 康定杨

Trees, deciduous

3.6 (-1.7~14.1); -5.3 (-12.3~6.5); 11.5 (8.0~20.2); 5.2 (2.4~14.2); 27.5 (6.4~109.8); 43.3 (0.0~86.1); 450 (344~729) ▎669 (256~1243); 400 (200~696); 16 (0~58) ▎46.6 (-35.1~125.9); 431 (256~707); 248.3 (93.6~384.7)

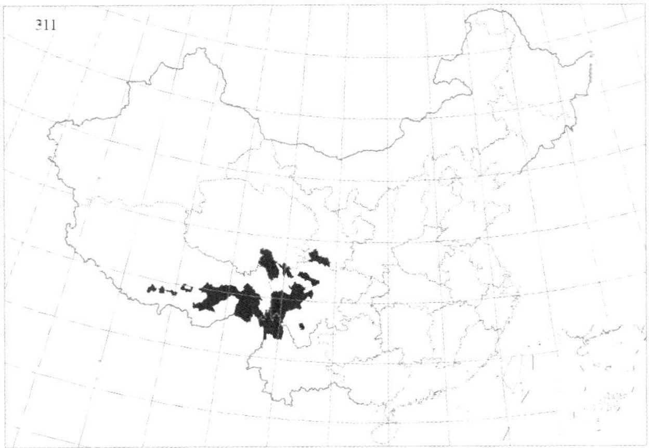

Salicaceae 杨柳科 *Populus* 杨属

Populus keerqinensis T. Y. Sun 科尔沁杨

Trees, deciduous

6.0 (0.6~7.7); -14.0 (-19.8~-11.8); 23.5 (19.0~24.8); 9.4 (6.5~10.3); 78.3 (48.2~88.1); 65.4 (54.9~101.0); 647 (514~688) ▎515 (392~715); 345 (282~451); 12 (6~22) ▎-20.6 (-40.5~8.8); 511 (392~675); 171.9 (97.2~267.9)

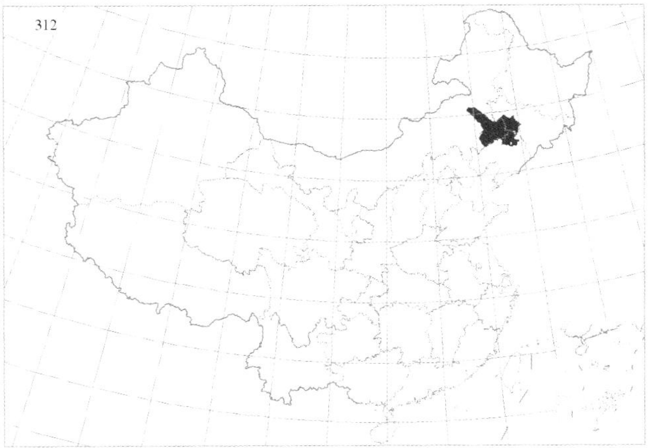

Salicaceae 杨柳科 *Populus* 杨属
Populus koreana Rehder 香杨
Trees, deciduous

4.7 (-5.8~11.6); -15.0 (-29.8~-4.8); 21.7 (15.3~26.1); 8.6 (4.4~12.5); 68.4 (28.0~105.6); 71.9 (25.4~157.4); 604 (426~782) ❙ 603 (349~1073); 392 (227~691); 18 (7~40) ❙ 0.1 (-41.6~74.1); 551 (349~690); 251.3 (105.6~346.5)

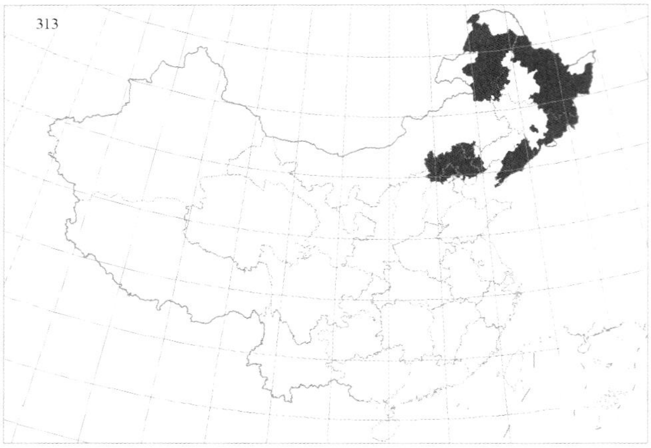

Salicaceae 杨柳科 *Populus* 杨属
Populus lancifolia N. Chao 瘦叶杨
Trees, deciduous

2.2; -6.6; 9.9; 3.8; 15.5; 49.1; 413 ❙ 932; 546; 19 ❙ 125.9; 413; 248.3

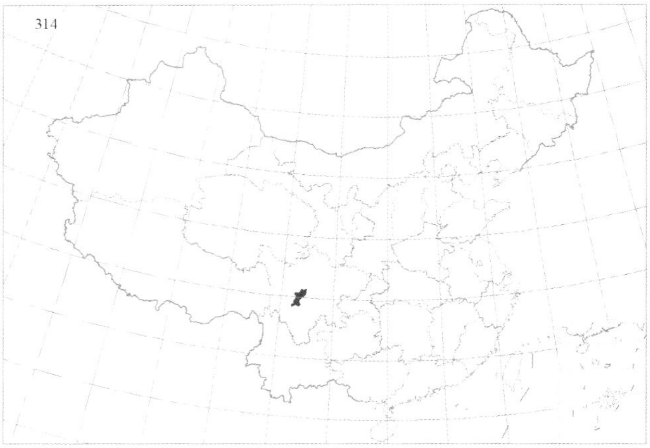

Salicaceae 杨柳科 *Populus* 杨属
Populus lasiocarpa Oliv. 大叶杨
Trees, deciduous

12.4 (2.7~17.6); 2.5 (-7.1~9.0); 21.6 (11.1~28.1); 12.6 (4.4~17.6); 96.4 (20.5~151.6); 6.8 (0.0~48.0); 708 (429~924) ❙ 1118 (731~1496); 524 (332~812); 59 (12~142) ❙ 59.0 (5.8~107.2); 708 (429~924); 244.3 (152.4~384.7)

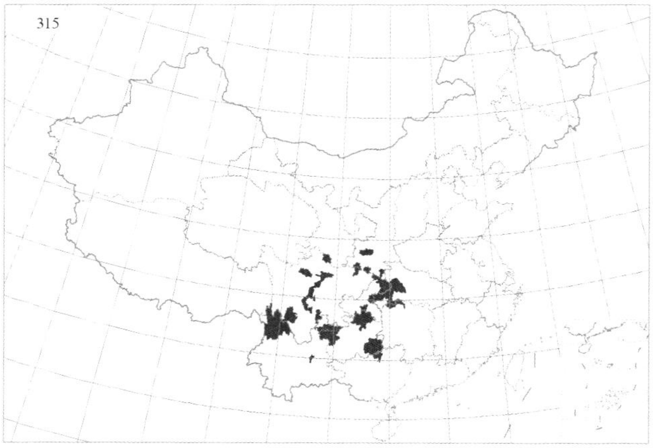

Salicaceae 杨柳科 *Populus* 杨属
Populus lasiocarpa var. longiamenta Mao et P. X. He 长序大叶杨
Trees, deciduous

12.8; 3.4; 21.3; 12.9; 96.5; 1.9; 701 ❙ 911; 480; 41 ❙ 30.0; 701; 224.5

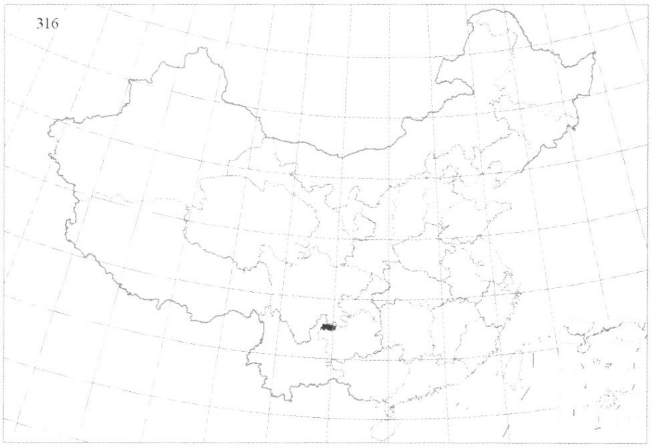

Salicaceae 杨柳科 *Populus* 杨属
Populus laurifolia Ledeb. 苦杨
Trees, deciduous

4.8 (-2.8~11.9); -13.7 (-20.9~-1.9); 20.3 (10.9~27.1); 8.4 (3.5~12.1); 65.7 (16.3~107.6); 67.0 (16.6~109.2); 600 (390~810) ❙ 251 (45~785); 123 (27~453); 18 (3~51) ❙ -56.6 (-94.0~19.6); 247 (45~711); 103.7 (7.6~297.3)

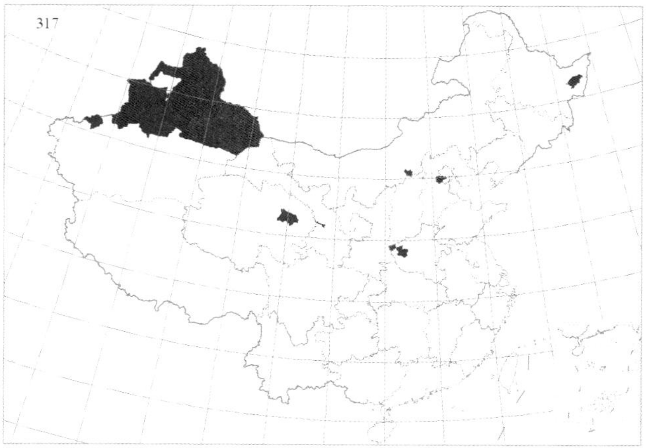

Salicaceae 杨柳科 *Populus* 杨属
Populus mainlingensis C. Wang et S. L. Tung 米林杨
Trees, deciduous

3.5; -5.2; 11.2; 4.7; 23.6; 40.5; 434 ❙ 542; 338; 9 ❙ 23.6; 434; 228.8

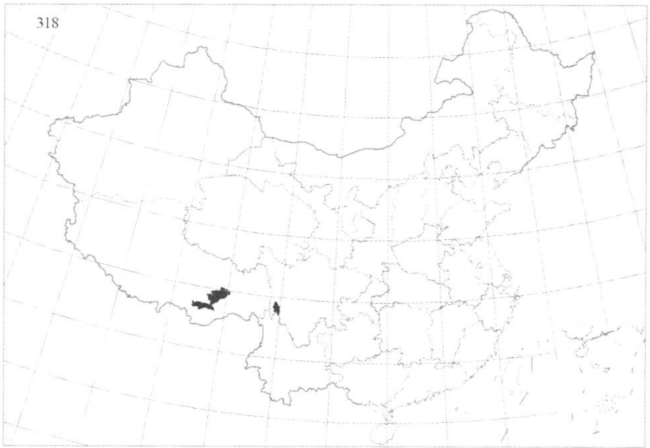

Salicaceae 杨柳科 *Populus* 杨属

Populus manshurica Nakai 热河杨

Trees, deciduous

7.5 (4.9~9.1); -10.4 (-12.8~-7.7); 23.3 (20.8~24.9); 9.8 (8.2~10.8); 81.7 (63.3~91.7); 51.0 (40.0~63.6); 656 (580~703) ❙ 539 (408~643); 372 (281~448); 9 (7~13) ❙ -17.8 (-33.0~-0.6); 539 (408~643); 173.3 (106.8~225.0)

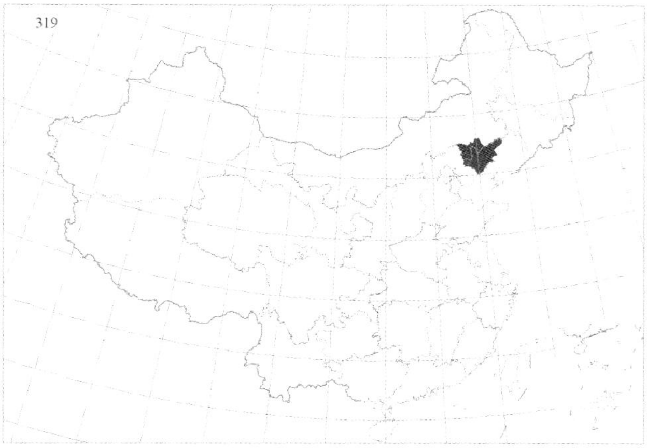

Salicaceae 杨柳科 *Populus* 杨属

Populus maximowiczii A. Henry 辽杨

Trees, deciduous

6.5 (-5.8~16.1); -11.1 (-29.8~5.7); 21.9 (15.3~25.4); 9.2 (4.4~16.1); 73.8 (28.0~133.5); 55.0 (0.0~157.4); 622 (426~847) ❙ 680 (384~1073); 424 (256~691); 22 (5~40) ❙ 9.9 (-37.6~74.1); 585 (384~845); 242.1 (60.2~346.5)

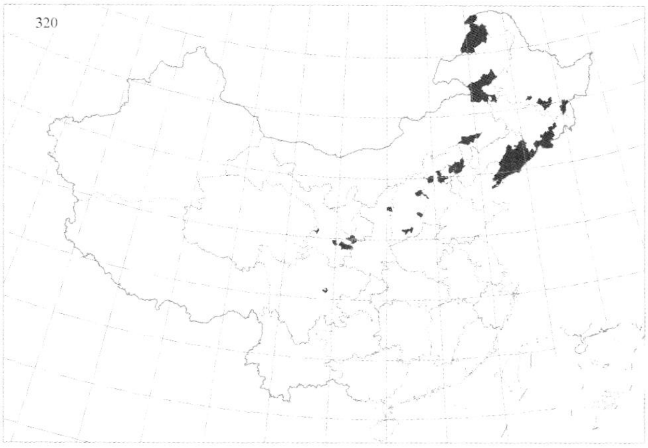

Salicaceae 杨柳科 *Populus* 杨属

Populus minhoensis S. F. Yang et H. F. Wu 民和杨

Trees, deciduous

5.2; -3.0; 16.5; 7.0; 47.4; 44.1; 518 ❙ 443; 239; 8 ❙ -14.6; 443; 141.6

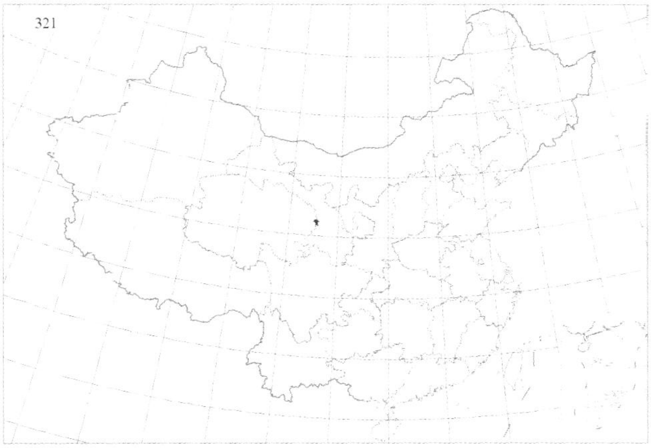

Salicaceae 杨柳科 *Populus* 杨属

Populus nakaii Skvortsov 玉泉杨

Trees, deciduous

7.6; -9.3; 22.4; 9.7; 78.4; 47.1; 645 ❙ 528; 375; 13 ❙ -18.0; 528; 187.0

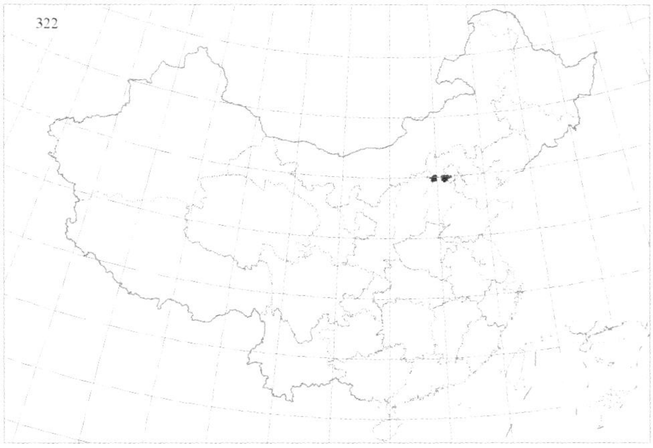

Salicaceae 杨柳科 *Populus* 杨属

Populus nigra L. 黑杨

Trees, deciduous

2.4 (-0.6~6.1); -16.6 (-19.5~-14.2); 19.1 (15.9~23.3); 7.3 (5.4~9.7); 55.2 (37.5~80.9); 85.4 (65.1~104.2); 557 (475~668) ❙ 184 (156~251); 78 (56~106); 21 (13~33) ❙ -66.5 (-76.7~-52.0); 184 (156~251); 80.8 (16.6~139.1)

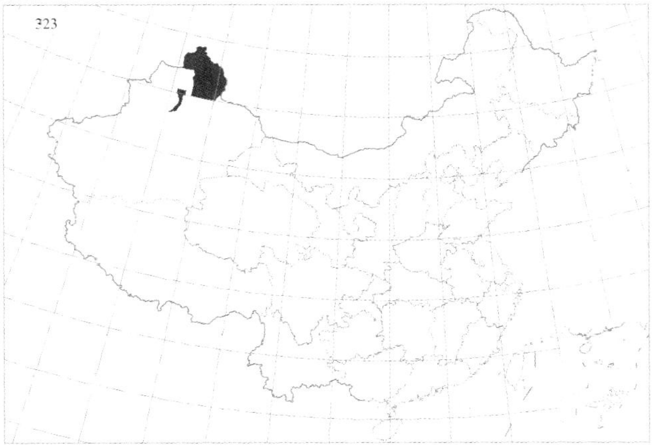

Salicaceae 杨柳科 *Populus* 杨属

Populus ningshanica C. Wang et S. L. Tung 汉白杨

Trees, deciduous

10.2; -1.2; 21.4; 10.4; 79.4; 16.1; 640 ❙ 1075; 462; 58 ❙ 68.6; 640; 279.3

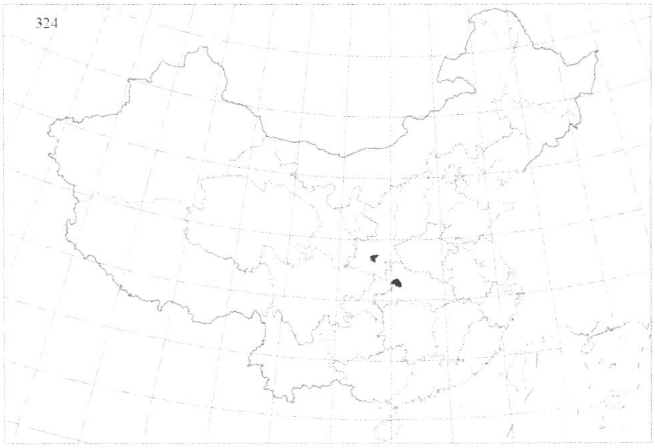

Salicaceae 杨柳科 *Populus* 杨属
Populus pamirica Kom. 帕米杨
Trees, deciduous

-0.1 (-4.3~5.9); -17.1 (-20.1~-13.8); 14.2 (9.6~23.0); 4.9 (2.5~9.4); 31.5 (9.7~77.9); 92.1 (66.6~121.1); 448 (329~655) ❙ 197 (71~406); 90 (24~211); 17 (9~26) ❙ -55.5 (-78.5~-7.6); 197 (71~406); 65.1 (15.7~261.4)

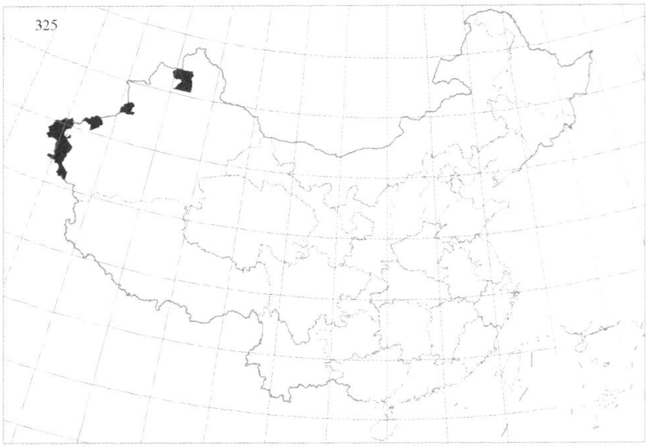

Salicaceae 杨柳科 *Populus* 杨属
Populus pilosa Rehder 柔毛杨
Trees, deciduous

3.8 (-1.3~11.4); -14.8 (-19.5~-7.2); 19.8 (12.5~25.7); 7.9 (3.9~12.7); 61.1 (21.0~109.3); 74.7 (31.6~104.2); 581 (390~793) ❙ 168 (32~315); 73 (21~124); 18 (0~51) ❙ -69.6 (-95.9~-46.3); 168 (32~312); 67.6 (1.6~189.0)

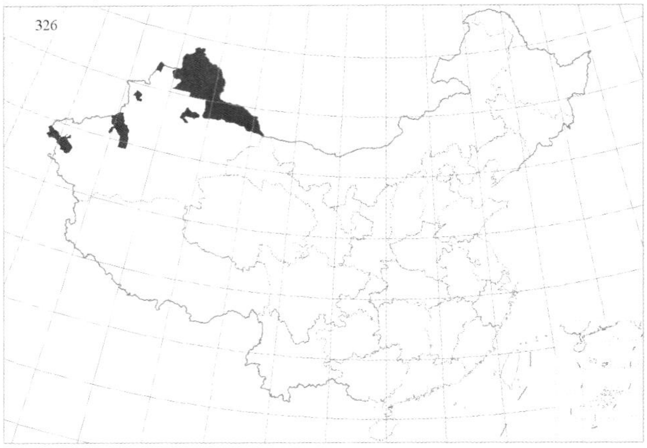

Salicaceae 杨柳科 *Populus* 杨属
Populus platyphylla T. Y. Sun 阔叶杨
Trees, deciduous

4.8 (2.2~7.9); -13.4 (-16.1~-10.4); 20.9 (18.5~24.2); 8.2 (6.7~10.1); 63.2 (48.1~84.9); 65.6 (49.6~81.1); 579 (515~671) ❙ 284 (111~407); 187 (74~270); 8 (2~15) ❙ -49.8 (-82.3~-25.7); 284 (111~407); 89.7 (13.9~145.6)

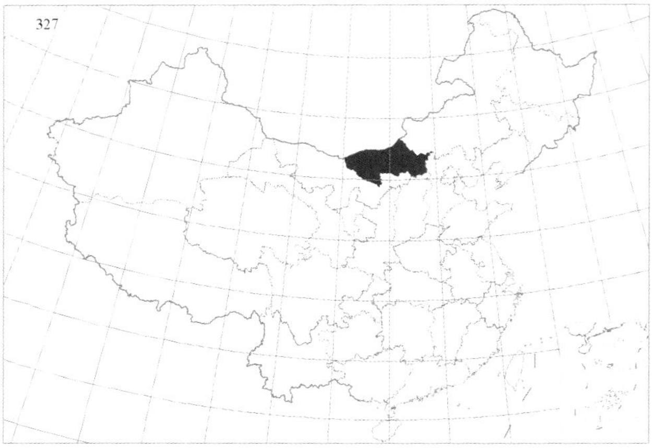

Salicaceae 杨柳科 *Populus* 杨属
Populus pruinosa Schrenk 灰胡杨
Trees, deciduous

9.4 (-4.1~12.4); -8.9 (-19.0~-5.3); 23.8 (9.8~26.5); 11.3 (2.5~13.2); 95.1 (9.9~114.1); 41.7 (24.5~117.9); 729 (330~822) ❙ 90 (21~357); 43 (15~136); 8 (0~48) ❙ -85.5 (-97.3~-34.9); 90 (21~357); 50.7 (1.6~297.3)

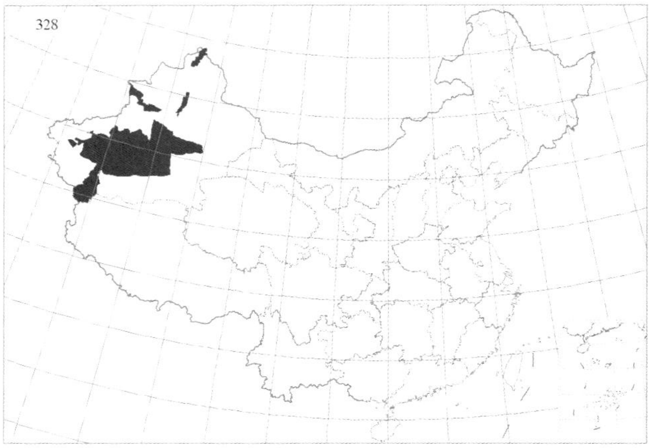

Salicaceae 杨柳科 *Populus* 杨属
Populus przewalskii Maxim. 青甘杨
Trees, deciduous

4.7 (-4.1~13.4); -9.2 (-17.3~0.9); 16.9 (7.8~25.2); 7.0 (2.1~13.4); 49.4 (5.2~108.9); 52.8 (8.5~114.2); 526 (340~764) ❙ 402 (49~1214); 220 (34~516); 10 (1~71) ❙ -20.0 (-92.8~101.4); 360 (49~764); 156.3 (5.1~341.6)

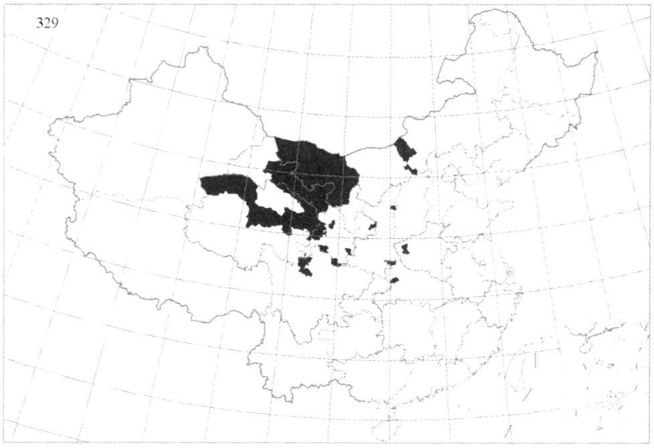

Salicaceae 杨柳科 *Populus* 杨属
Populus pseudoglauca C. Wang et P. Y. Fu 长序杨
Trees, deciduous

2.7 (0.0~5.3); -6.0 (-8.7~-2.4); 10.5 (8.1~12.2); 4.3 (2.7~5.7); 19.4 (8.0~30.5); 45.9 (26.2~67.9); 421 (363~479) ❙ 687 (341~984); 415 (226~601); 13 (4~31) ❙ 61.5 (-15.3~125.9); 416 (341~479); 238.1 (111.0~357.7)

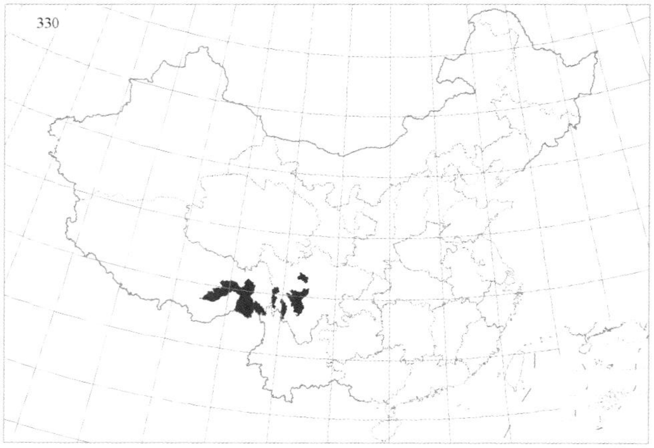

Salicaceae 杨柳科 *Populus* 杨属

Populus pseudomaximowiczii C. Wang et S. L. Tung 梧桐杨

Trees, deciduous

8.0 (2.5~12.8); -8.6 (-15.6~-3.0); 22.7 (18.6~26.5); 9.9 (6.9~13.2); 80.9 (48.9~112.9); 43.9 (18.7~78.3); 654 (523~810) ❙ 526 (397~636); 353 (254~453); 12 (9~23) ❙ -19.1 (-37.2~-0.3); 526 (397~636); 206.6 (111.2~273.5)

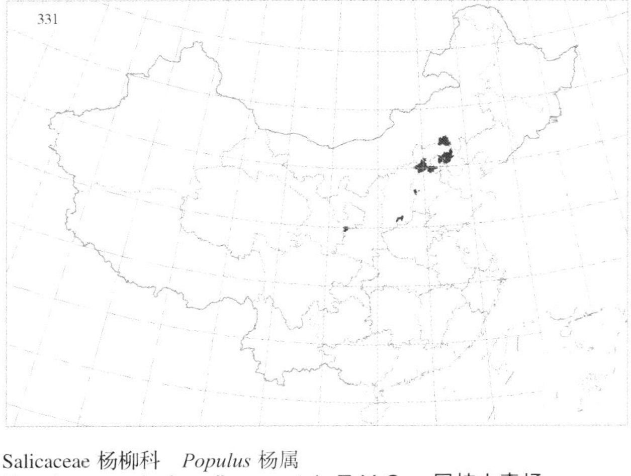

Salicaceae 杨柳科 *Populus* 杨属

Populus pseudosimonii Kitag. 小青杨

Trees, deciduous

7.7 (-1.4~15.1); -8.5 (-22.4~1.8); 22.0 (9.9~28.2); 9.7 (3.5~15.2); 77.3 (13.3~128.0); 44.4 (5.6~106.1); 641 (375~880) ❙ 547 (225~892); 320 (130~499); 16 (5~53) ❙ -13.8 (-64.1~63.7); 531 (225~817); 178.1 (44.5~349.1)

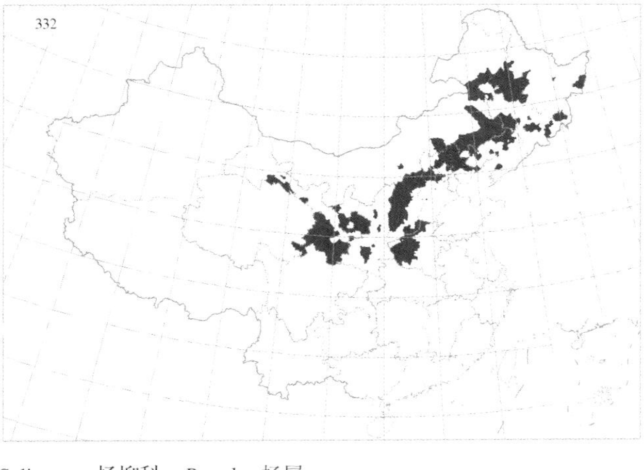

Salicaceae 杨柳科 *Populus* 杨属

Populus pseudosimonii var. patula T. Y. Sun 展枝小青杨

Trees, deciduous

2.2; -16.[illegible]; 18.6; 6.8; 48.4; 81.0; 516 ❙ 332; 215; 10 ❙ -35.7; 332; 105.8

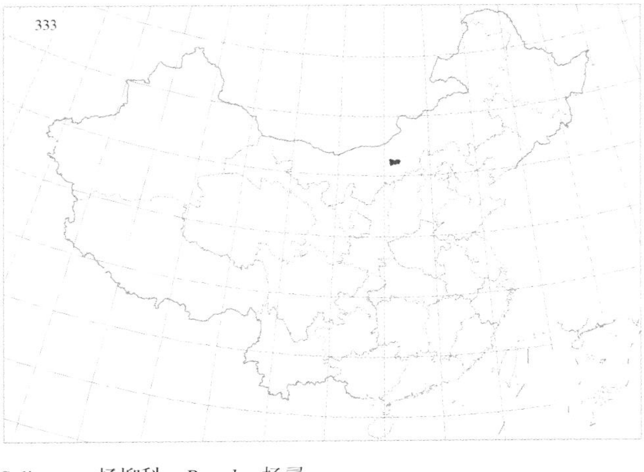

Salicaceae 杨柳科 *Populus* 杨属

Populus purdomii Rehder 冬瓜杨

Trees, deciduous

8.4 (-4.1~16.9); -3.9 (-17.3~7.0); 19.6 (7.8~27.8); 9.4 (2.1~17.0); 70.6 (5.2~143.7); 29.4 (0.0~114.2); 613 (340~881) ❙ 725 (282~1577); 369 (182~932); 23 (5~105) ❙ 20.2 (-43.4~128.6); 582 (282~881); 237.3 (81.2~357.7)

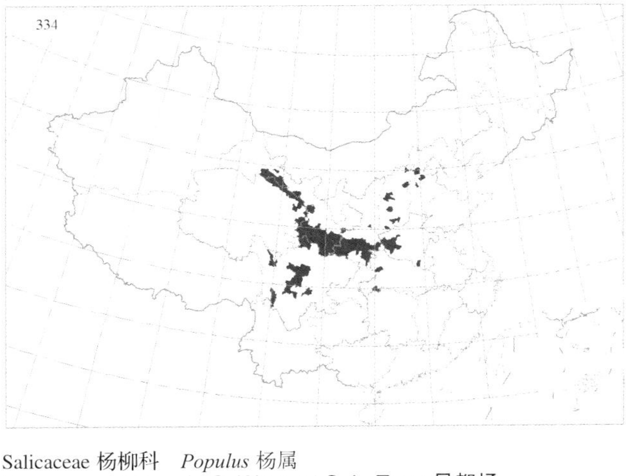

Salicaceae 杨柳科 *Populus* 杨属

Populus purdomii var. rockii (Rehder) C. F. Fang et H. L. Yang 光皮冬瓜杨

Trees, deciduous

2.9 (-4.1~9.4); -9.6 (-17.3~-2.8); 13.9 (7.8~21.1); 5.5 (2.1~9.9); 33.9 (5.2~75.0); 58.5 (21.0~114.2); 466 (340~628) ❙ 485 (361~656); 267 (219~323); 9 (5~15) ❙ 5.2 (-29.0~48.8); 435 (340~628); 214.2 (141.6~282.7)

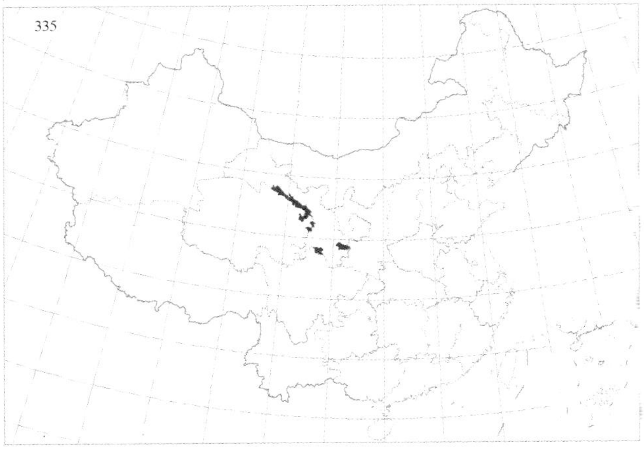

Salicaceae 杨柳科 *Populus* 杨属

Populus qamdoensis C. Wang et S. L. Tung 昌都杨

Trees, deciduous

2.8 (-1.8~5.4); -6.2 (-12.0~-2.1); 11.1 (8.6~12.3); 4.5 (2.4~5.8); 22.5 (7.9~31.5); 48.2 (25.6~88.8); 427 (333~481) ❙ 607 (337~961); 356 (233~558); 30 (1~98) ❙ 41.5 (-23.5~100.7); 409 (333~481); 194.0 (32.2~273.5)

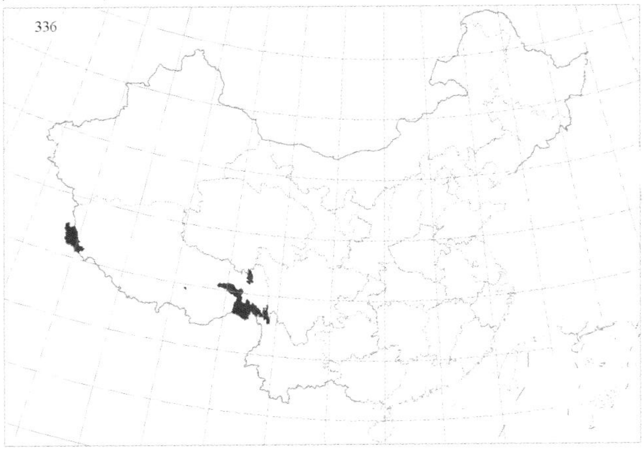

Salicaceae 杨柳科 *Populus* 杨属
Populus qiongdaoensis T. Hong et P. Luo 琼岛杨
Trees, deciduous

23.7; 18.0; 27.8; 23.7; 224.5; 0.0; 1280 ❙ 1303; 460; 46 ❙ 2.2; 1097; 477.7

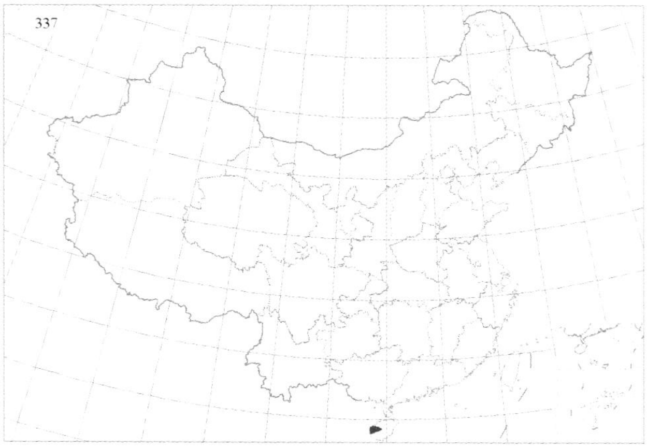

Salicaceae 杨柳科 *Populus* 杨属
Populus rotundifolia Griff. 圆叶杨
Trees, deciduous

13.7 (−4.1~21.0); 5.7 (−17.3~15.1); 20.1 (7.8~26.8); 13.9 (2.1~21.0); 112.2 (5.2~191.8); 7.3 (0.0~114.2); 735 (340~1053) ❙ 1052 (320~1751); 579 (203~996); 41 (1~212) ❙ 43.4 (−32.2~125.9); 720 (320~1053); 310.3 (95.7~659.0)

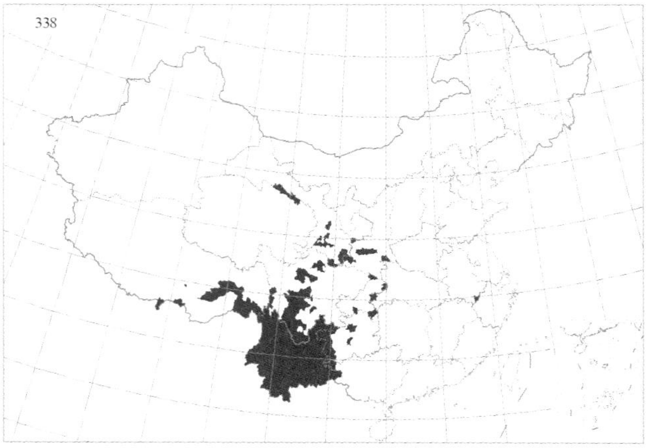

Salicaceae 杨柳科 *Populus* 杨属
Populus rotundifolia var. bonatii (H. Lév.) C. Wang et S. L. Tung 滇南山杨
Trees, deciduous

14.5; 7.7; 19.8; 14.6; 114.8; 0.0; 735 ❙ 1001; 577; 41 ❙ 36.3; 724; 297.8

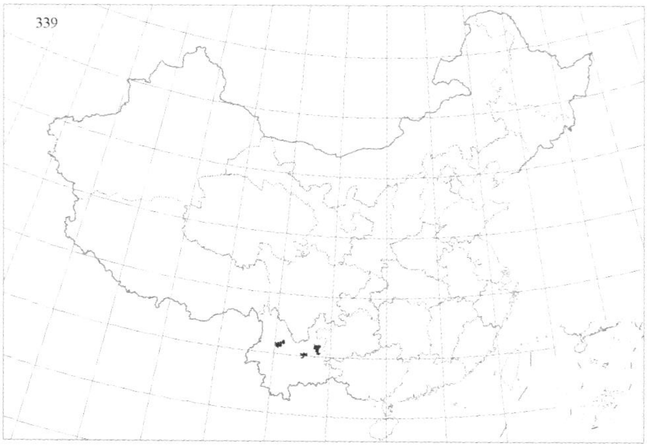

Salicaceae 杨柳科 *Populus* 杨属
Populus rotundifolia var. duclouxiana (Dode) Gombócz 清溪杨
Trees, deciduous

8.7 (−1.4~16.7); 0.1 (−15.4~9.7); 16.2 (8.1~23.2); 9.5 (2.7~16.7); 66.9 (8.0~140.9); 21.6 (0.0~92.1); 589 (363~819) ❙ 838 (282~1337); 481 (182~727); 25 (5~69) ❙ 43.3 (−32.2~125.9); 572 (282~730); 285.4 (95.7~407.1)

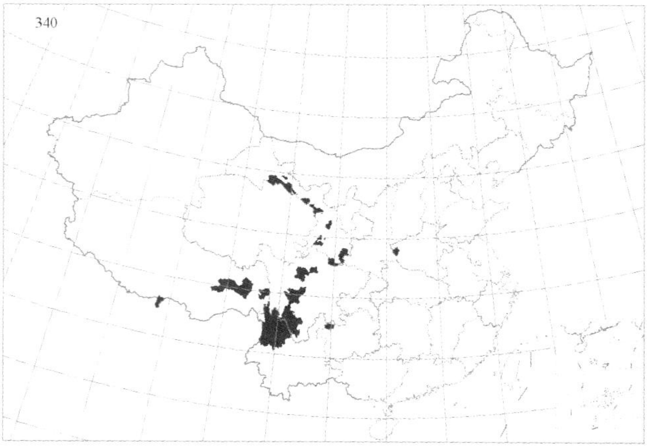

Salicaceae 杨柳科 *Populus* 杨属
Populus schneideri (Rehder) N. Chao 西南杨
Trees, deciduous

4.4 (−0.3~9.6); −4.5 (−10.9~2.4); 12.3 (9.0~18.1); 5.5 (3.0~9.6); 30.0 (10.4~64.1); 36.3 (5.4~73.4); 468 (378~600) ❙ 839 (510~1041); 474 (311~601); 21 (11~45) ❙ 80.3 (20.7~128.6); 468 (378~600); 282.9 (200.8~357.7)

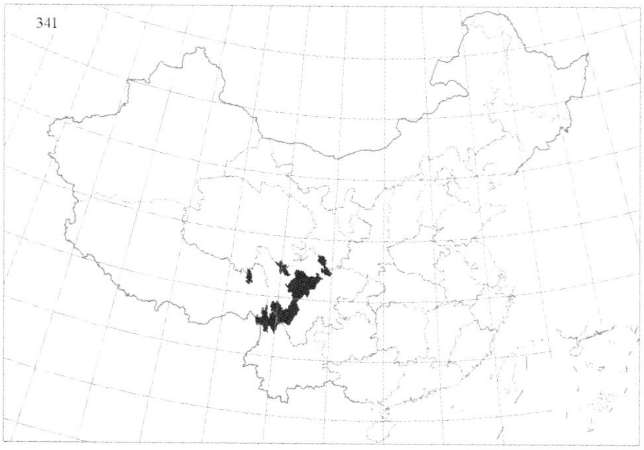

Salicaceae 杨柳科 *Populus* 杨属
Populus shanxiensis C. Wang et S. L. Tung 青毛杨
Trees, deciduous

7.6 (3.3~13.0); −8.5 (−12.9~−2.4); 21.7 (17.5~26.6); 9.4 (6.6~13.3); 74.8 (45.4~113.3); 42.9 (17.1~65.0); 625 (507~800) ❙ 479 (367~565); 287 (235~341); 13 (10~17) ❙ −22.9 (−38.1~−2.4); 479 (367~565); 140.0 (70.3~225.2)

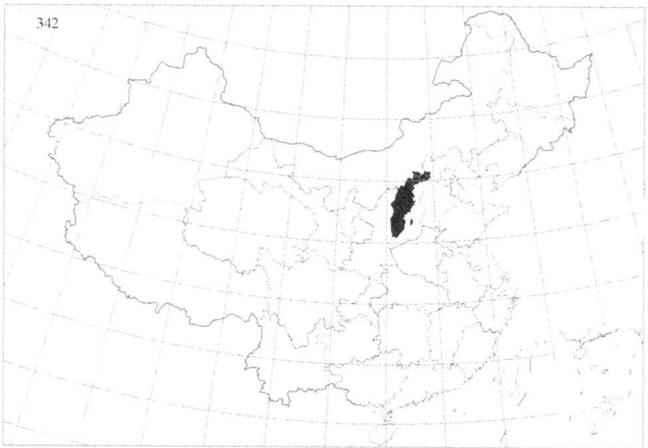

Salicaceae 杨柳科 *Populus* 杨属

Populus simonii Carrière 小叶杨

Trees, deciduous

11.2 (-4.8~17.1); -3.5 (-23.8~7.1); 24.3 (7.4~29.5); 12.1 (1.9~17.2); 99.3 (4.3~147.0); 24.8 (0.0~121.6); 740 (332~954) ❙ 686 (35~1770); 370 (23~691); 37 (1~223) ❙ -7.1 (-94.9~128.6); 607 (35~942); 172.5 (5.1~364.1)

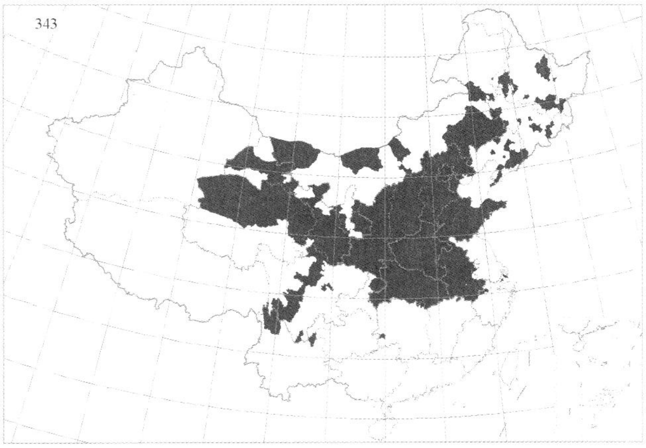

Salicaceae 杨柳科 *Populus* 杨属

Populus simonii var. liaotungensis (C. Wang et Skvortsov) C. Wang et S. L. Tung 辽东小叶杨

Trees, deciduous

5.2 (2.2~10.5); -12.9 (-16.1~-4.9); 21.3 (18.5~24.2); 8.4 (6.7~11.3); 66.0 (48.1~93.2); 63.2 (26.8~81.1); 591 (515~705) ❙ 288 (111~632); 139 (74~398); 8 (2~27) ❙ -50.3 (-82.3~-10.4); 288 (111~632); 91.5 (13.9~180.8)

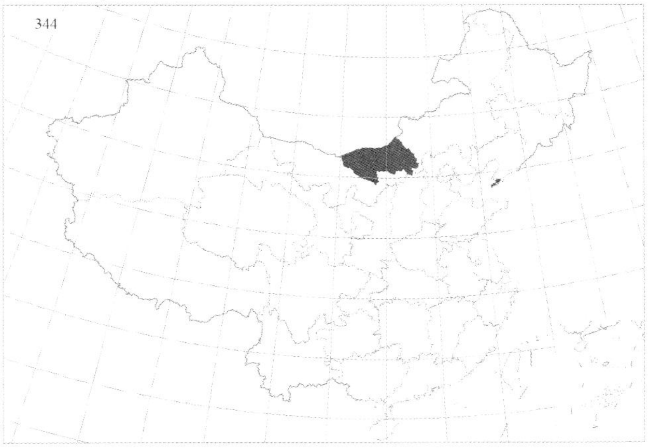

Salicaceae 杨柳科 *Populus* 杨属

Populus simonii var. rotundifolia X. C. Lu ex C. Wang et S. L. Tung 圆叶小叶杨

Trees, deciduous

6.4 (5.2~7.7); -13.7 (-16.0~-12.1); 23.9 (21.6~24.8); 9.6 (8.5~10.3); 80.6 (67.6~88.1); 62.9 (54.9~71.8); 656 (598~688) ❙ 561 (406~715); 357 (281~451); 14 (7~22) ❙ -15.2 (-38.1~8.8); 554 (406~686); 184.7 (117.7~267.9)

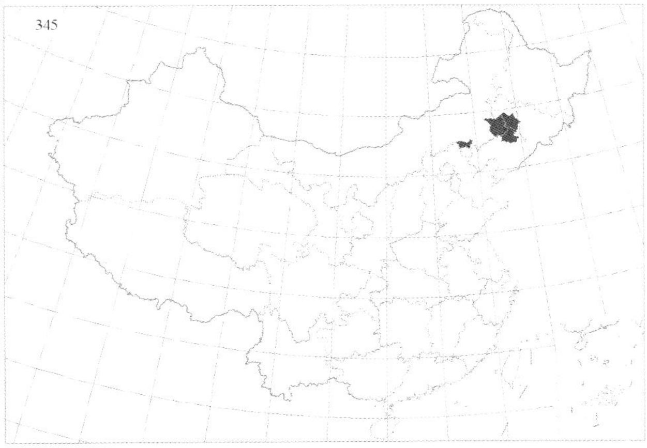

Salicaceae 杨柳科 *Populus* 杨属

Populus simonii var. tsinlingensis C. Wang et C. Y. Yu 秦岭小叶杨

Trees, deciduous

11.3 (9.2~13.5); -1.2 (-2.9~1.4); 23.3 (20.7~25.1); 11.6 (9.6~13.5); 91.8 (72.1~109.4); 15.0 (7.1~20.9); 693 (615~767) ❙ 799 (738~951); 357 (332~436); 29 (20~39) ❙ 15.4 (2.3~35.6); 693 (615~767); 262.1 (228.0~323.3)

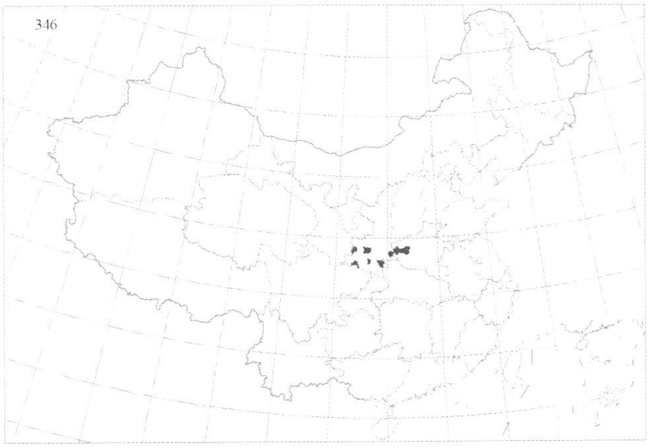

Salicaceae 杨柳科 *Populus* 杨属

Populus suaveolens Fisch. 甜杨

Trees, deciduous

3.1 (-5.8~13.9); -16.7 (-29.8~1.9); 20.5 (11.7~25.4); 8.0 (4.3~13.9); 61.2 (21.1~112.7); 83.7 (5.6~157.4); 572 (426~781) ❙ 569 (311~942); 346 (214~601); 16 (5~39) ❙ -1.1 (-40.0~56.9); 522 (311~781); 249.1 (50.7~333.3)

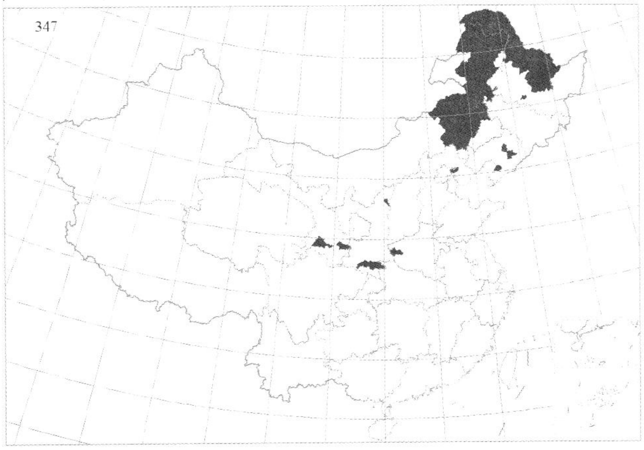

Salicaceae 杨柳科 *Populus* 杨属

Populus szechuanica C. K. Schneid. 川杨

Trees, deciduous

11.4 (-1.8~18.7); 1.1 (-12.3~9.0); 20.9 (8.6~28.9); 11.9 (2.4~18.8); 92.3 (7.9~166.0); 14.5 (0.0~88.8); 698 (333~1002) ❙ 980 (325~1736); 475 (205~1007); 42 (1~98) ❙ 43.7 (-29.3~145.8); 688 (325~1002); 220.4 (32.2~447.4)

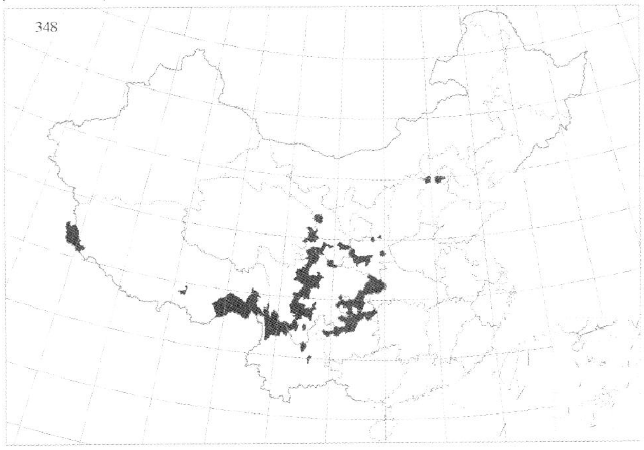

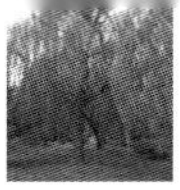

Salicaceae 杨柳科 *Populus* 杨属

Populus szechuanica var. tibetica C. K. Schneid. 藏川杨

Trees, deciduous

2.3 (−1.7~10.8); −6.8 (−11.8~3.9); 10.5 (7.9~16.6); 4.3 (2.4~10.9); 20.5 (7.0~71.8); 52.0 (1.3~87.2); 417 (335~633) ❙ 543 (227~968); 343 (180~594); 11 (0~31) ❙ 26.5 (−35.9~125.9); 366 (227~633); 198.2 (83.7~341.6)

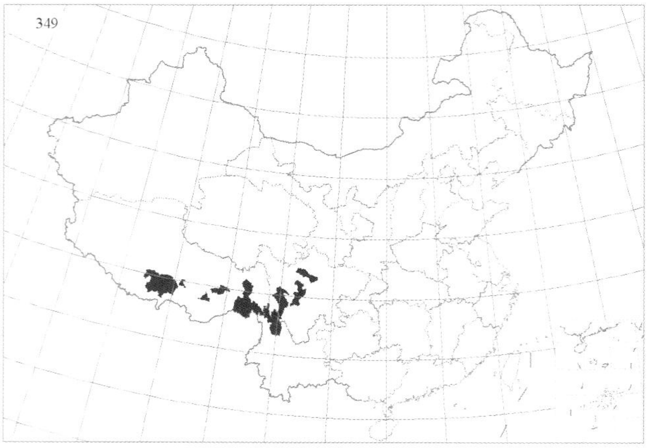

Salicaceae 杨柳科 *Populus* 杨属

Populus talassica Kom. 密叶杨

Trees, deciduous

4.8 (−2.8~11.4); −14.3 (−20.9~−7.2); 20.4 (10.9~27.1); 8.4 (3.5~12.7); 66.6 (16.3~109.3); 68.7 (31.6~109.2); 605 (390~810) ❙ 192 (32~450); 89 (21~211); 17 (0~51) ❙ −65.4 (−95.9~3.9); 191 (32~406); 86.3 (1.6~297.3)

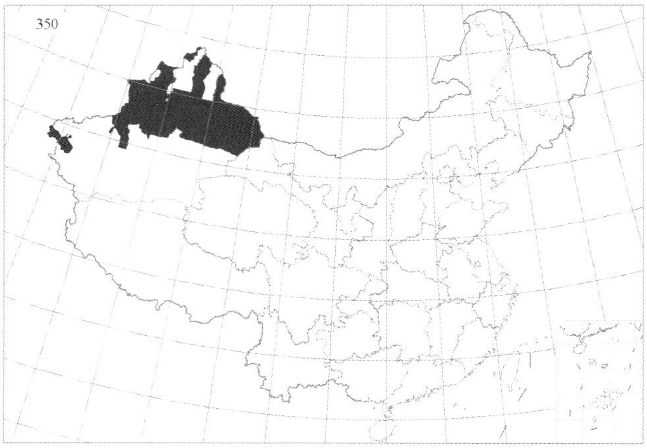

Salicaceae 杨柳科 *Populus* 杨属

Populus tomentosa Carrière 毛白杨

Trees, deciduous

12.7 (0.0~18.9); −1.5 (−11.8~12.0); 25.4 (9.7~29.6); 13.1 (3.3~19.0); 110.1 (12.2~167.8); 16.9 (0.0~72.2); 783 (395~959) ❙ 672 (225~1663); 390 (130~816); 32 (5~213) ❙ −13.5 (−64.1~101.3); 620 (225~959); 176.2 (44.9~382.6)

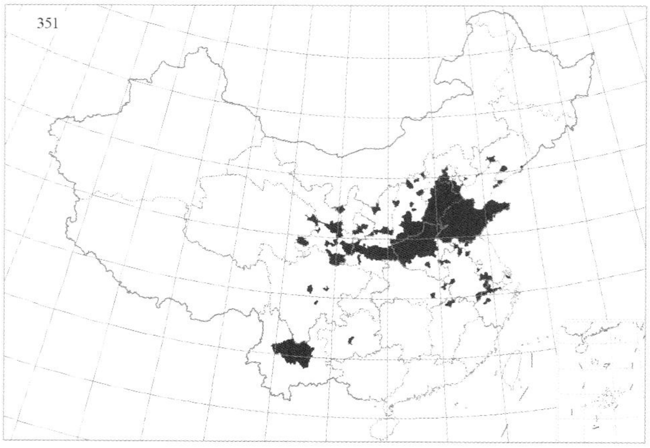

Salicaceae 杨柳科 *Populus* 杨属

Populus tomentosa var. truncata Y. C. Fu et C. H. Wang 截叶毛白杨

Trees, deciduous

13.9; −0.6; 27.0; 14.0; 119.6; 12.6; 828 ❙ 616; 338; 25 ❙ −25.6; 616; 152.6

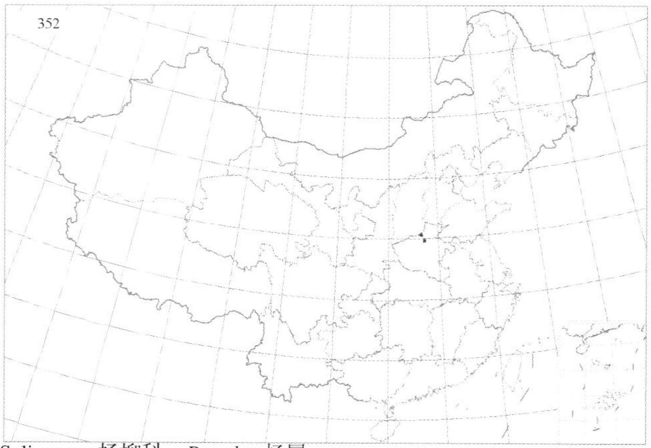

Salicaceae 杨柳科 *Populus* 杨属

Populus tremula L. 欧洲山杨

Trees, deciduous

4.8 (−2.8~16.4); −14.2 (−20.9~5.9); 20.5 (10.9~27.5); 8.5 (3.5~16.4); 67.5 (16.3~136.9); 68.9 (0.0~109.2); 609 (390~862) ❙ 215 (32~1033); 101 (21~614); 18 (0~51) ❙ −62.7 (−95.9~19.8); 210 (32~862); 91.4 (1.6~297.3)

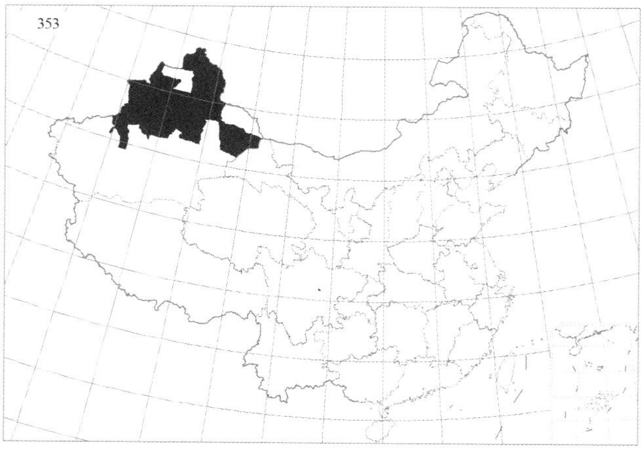

Salicaceae 杨柳科 *Populus* 杨属

Populus trinervis C. Wang et S. L. Tung 三脉青杨

Trees, deciduous

5.6 (2.0~14.8); −3.4 (−7.2~3.8); 13.5 (9.5~25.3); 6.6 (3.8~14.8); 39.5 (15.4~119.2); 32.0 (1.4~51.4); 503 (411~795) ❙ 896 (628~1358); 499 (317~606); 23 (11~88) ❙ 80.7 (28.2~128.6); 503 (411~795); 276.4 (187.9~357.7)

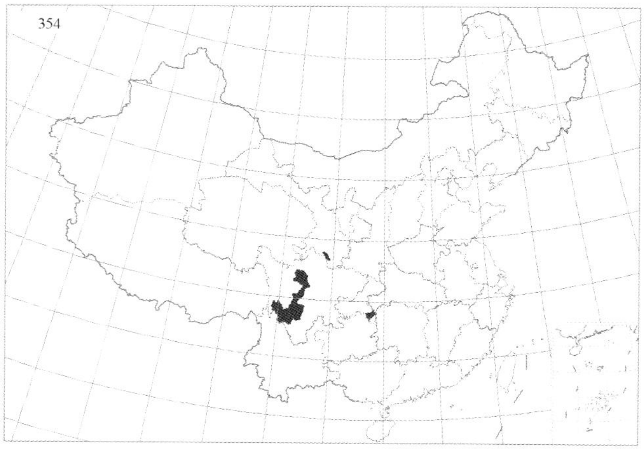

Salicaceae 杨柳科 *Populus* 杨属

Populus trinervis var. shimianica C. Wang et N. Chao 石棉杨

Trees, deciduous

10.1；1.8；17.7；10.3；70.1；7.0；621 | 1023；568；22 | 64.7；621；318.7

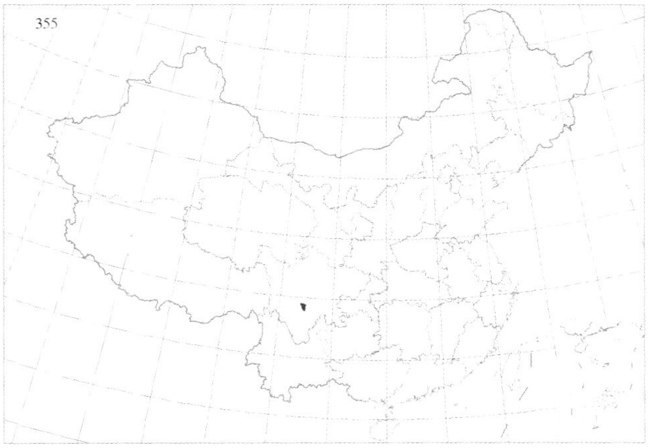

Salicaceae 杨柳科 *Populus* 杨属

Populus ussuriensis Kom. 大青杨

Trees, deciduous

3.5 (-4.0~9.5)；-17.5 (-28.2~-7.2)；21.5 (17.6~24.8)；8.2 (5.3~10.8)；64.0 (38.6~91.5)；81.5 (35.9~146.2)；587 (463~701) | 665 (451~1073)；413 (309~691)；23 (9~40) | 13.0 (-25.6~74.1)；577 (451~690)；277.1 (157.0~346.5)

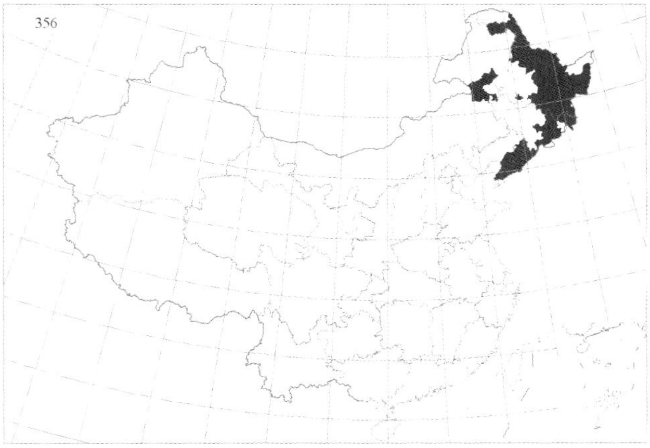

Salicaceae 杨柳科 *Populus* 杨属

Populus violascens Dode 堇柄杨

Trees, deciduous

12.6；1.6；23.2；12.7；98.8；6.9；711 | 1466；602；105 | 106.2；711；285.1

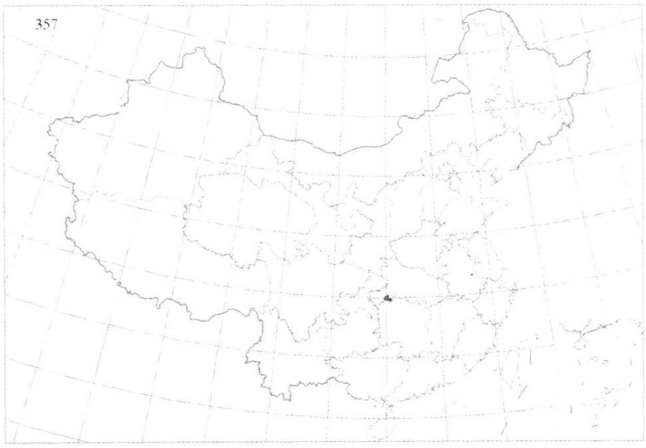

Salicaceae 杨柳科 *Populus* 杨属

Populus wenxianica Z. C. Feng et J. L. Guo ex G. H. Zhu 文县杨

Trees, deciduous

9.8 (6.0~12.3)；-1.3 (-4.0~1.1)；20.1 (14.4~22.9)；10.0 (6.6~12.3)；74.6 (39.7~96.2)；16.6 (8.1~29.1)；629 (508~709) | 706 (620~1023)；351 (301~535)；16 (13~28) | 13.9 (-8.1~101.2)；621 (508~707)；253.8 (196.0~320.6)

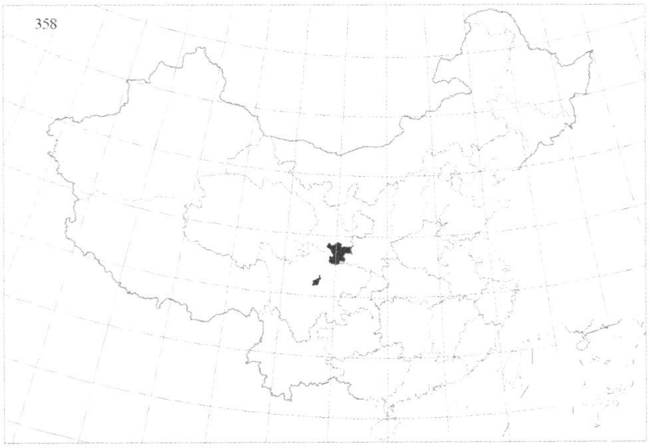

Salicaceae 杨柳科 *Populus* 杨属

Populus wilsonii C. K. Schneid. 椅杨

Trees, deciduous

10.5 (0.1~16.3)；-0.3 (-9.7~6.4)；20.5 (8.1~26.5)；10.8 (2.8~16.4)；81.9 (8.9~136.6) 15.6 (0.0~67.7)；657 (358~863) | 925 (273~1466)；443 (168~727)；38 (1~113) | 40.9 (-51.0~121.7)；646 (273~863)；258.5 (42.8~407.1)

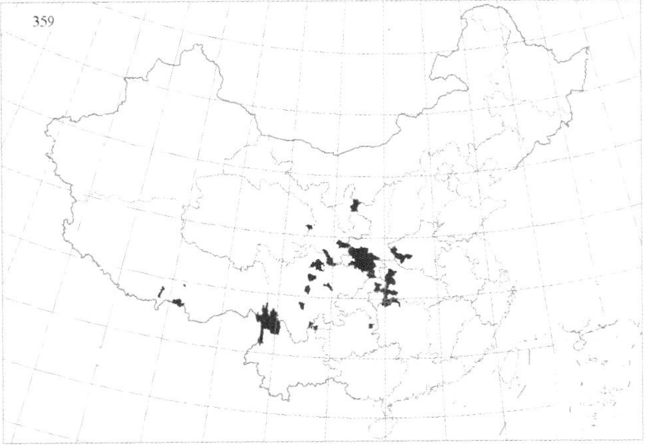

Salicaceae 杨柳科 *Populus* 杨属

Populus wuana C. Wang et S. L. Tung 长叶杨

Trees, deciduous

3.6；-4.7；11.0；4.7；23.1；39.8；441 | 772；453；22 | 72.5；441；201.8

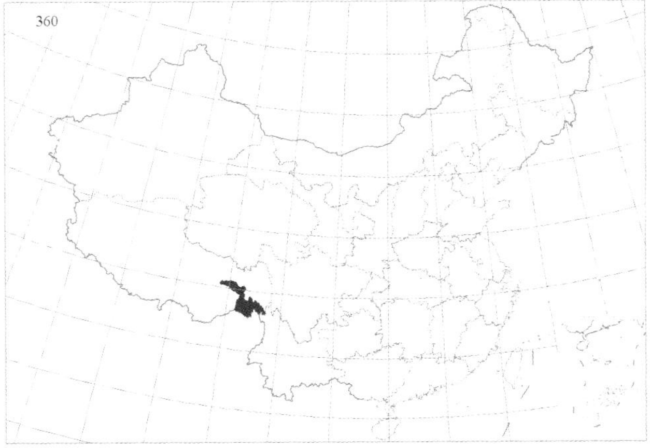

Salicaceae 杨柳科 *Populus* 杨属

Populus wulianensis S. B. Liang et X. W. Li 五莲杨

Trees, deciduous

11.7 (11.1~12.1); -2.0 (-3.3~-1.2); 23.9 (23.0~24.6); 12.0 (11.6~12.4); 98.6 (95.6~103.4); 17.1 (14.2~22.0); 722 (709~746) | 713 (670~834); 427 (392~523); 41 (31~65) | -1.3 (-9.5~14.2); 696 (670~730); 166.3 (155.2~189.3)

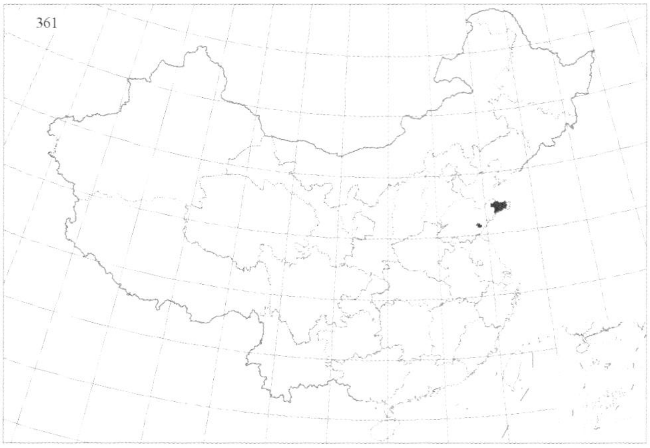

Salicaceae 杨柳科 *Populus* 杨属

Populus xiangchengensis C. Wang et S. L. Tung 乡城杨

Trees, deciduous

2.4 (0.1~3.8); -6.4 (-10.1~-4.3); 10.2 (9.5~10.8); 4.1 (3.2~4.7); 18.0 (12.3~21.6); 48.2 (35.4~71.1); 417 (377~444) | 738 (533~984); 460 (328~601); 13 (10~19) | 76.4 (38.4~125.9); 417 (377~444); 252.1 (208.8~299.8)

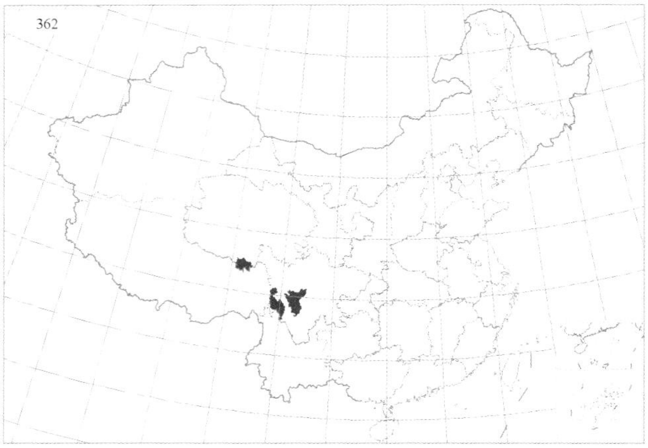

Salicaceae 杨柳科 *Populus* 杨属

Populus yatungensis (C. Wang et P. Y. Fu) C. Wang et S. L. Tung 亚东杨

Trees, deciduous

4.4 (-1.1~10.8); -3.8 (-9.1~3.9); 11.7 (6.2~16.6); 5.6 (2.0~10.9); 30.1 (2.5~71.8); 36.4 (1.3~75.6); 467 (334~633) | 716 (252~1041); 436 (193~603); 19 (0~45) | 50.6 (-35.9~125.9); 458 (252~633); 241.4 (79.3~352.3)

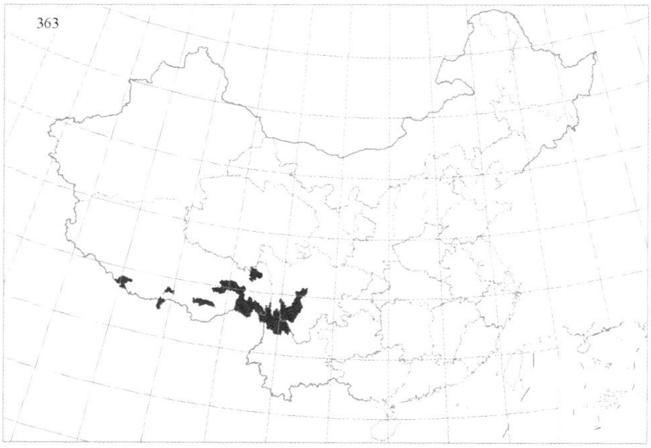

Salicaceae 杨柳科 *Populus* 杨属

Populus yuana C. Wang et S. L. Tung 五瓣杨

Trees, deciduous

7.8 (5.3~13.3); 0.1 (-2.4~5.4); 14.3 (12.2~19.6); 8.0 (5.7~13.4); 50.0 (30.5~100.4); 16.5 (0.0~26.2); 547 (479~706) | 1056 (741~1736); 614 (426~1007); 35 (21~51) | 88.7 (54.0~145.8); 547 (479~706); 318.1 (234.2~447.4)

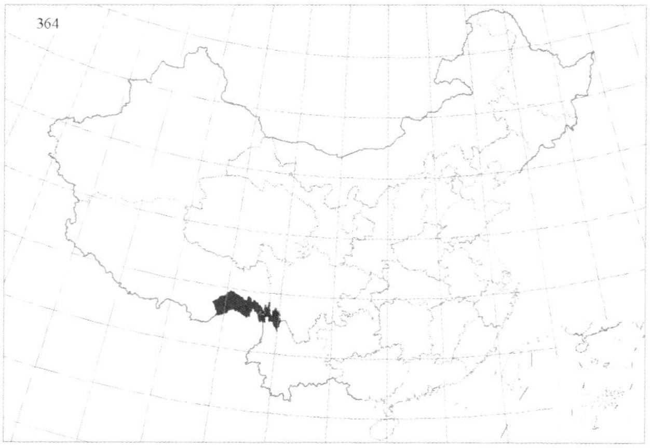

Salicaceae 杨柳科 *Populus* 杨属

Populus yunnanensis Dode 滇杨

Trees, deciduous

15.0 (5.4~19.3); 7.4 (-2.1~13.1); 21.1 (12.3~26.7); 15.1 (5.8~19.4); 121.9 (31.5~173.2); 0.8 (0.0~25.6); 769 (481~966) | 1076 (741~1456); 598 (411~836); 44 (15~69) | 40.7 (-14.2~89.0); 753 (481~966); 281.4 (153.9~516.7)

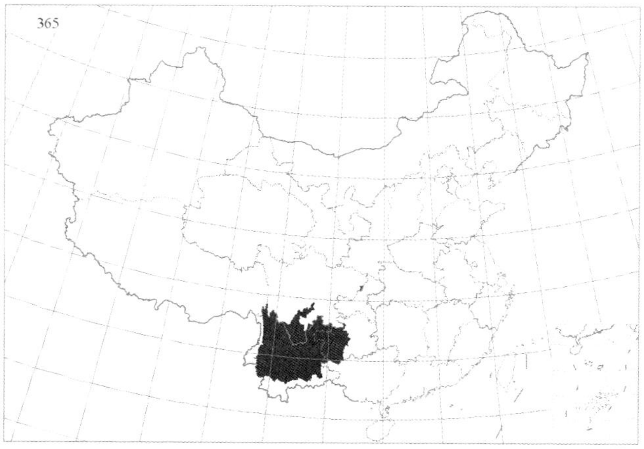

Salicaceae 杨柳科 *Populus* 杨属

Populus × gansuensis C. Wang et H. L. Yang 二白杨

Trees, deciduous

6.4 (-1.4~10.3); -8.5 (-15.4~-0.1); 19.6 (11.5~23.1); 8.4 (3.5~10.3); 63.6 (17.2~82.9); 45.6 (11.7~92.1); 579 (375~662) | 308 (58~711); 173 (36~357); 7 (2~18) | -44.8 (-90.8~48.8); 296 (58~634); 94.3 (11.3~282.7)

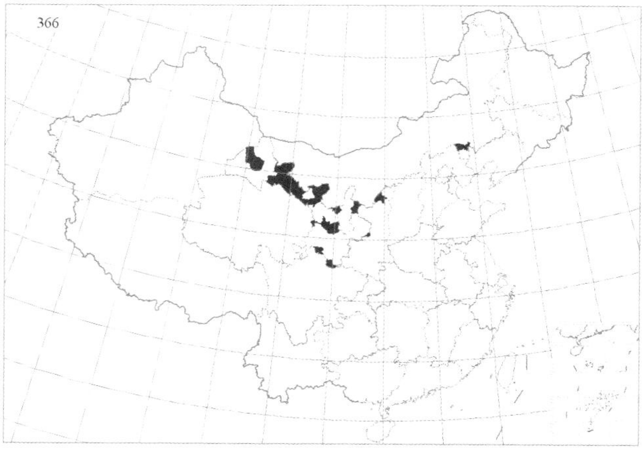

Salicaceae 杨柳科 *Populus* 杨属

Populus × hopeiensis Hu et Chow 河北杨

Trees, deciduous

8.7 (3.3~14.5); -6.9 (-12.7~0.3); 22.4 (12.2~28.2); 10.1 (5.7~14.6); 80.9 (30.5~125.3); 36.2 (10.4~65.0); 654 (479~864) ❙ 497 (58~961); 295 (36~558); 14 (4~44) ❙ -23.3 (-90.8~100.7); 493 (58~773); 139.9 (11.3~268.0)

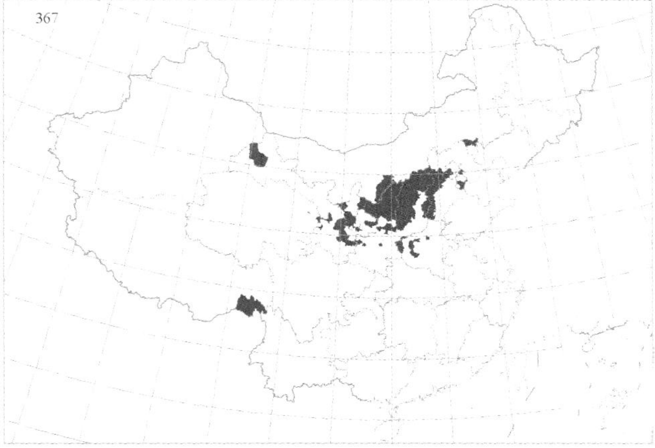

Salicaceae 杨柳科 *Populus* 杨属

Populus × jrtyschensis C. Y. Yang 额河杨

Trees, deciduous

2.1 (-0.6~5.8); -16.9 (-19.5~-14.8); 18.8 (15.9~23.3); 7.1 (5.4~9.7); 52.8 (37.5~80.9); 87.5 (71.2~104.2); 548 (475~668) ❙ 184 (156~251); 78 (56~106); 22 (15~33) ❙ -65.9 (-76.7~-52.0); 184 (156~251); 83.0 (16.6~139.1)

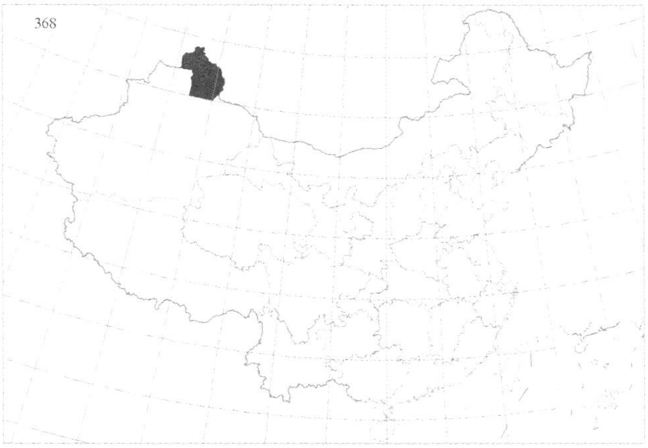

Salicaceae 杨柳科 *Populus* 杨属

Populus × pseudotomentosa C. Wang et S. L. Tung 响毛杨

Trees, deciduous

13.3; -1.7; 26.8; 13.5; 116.5; 15.9; 817 ❙ 710; 466; 27 ❙ -13.1; 710; 180.5

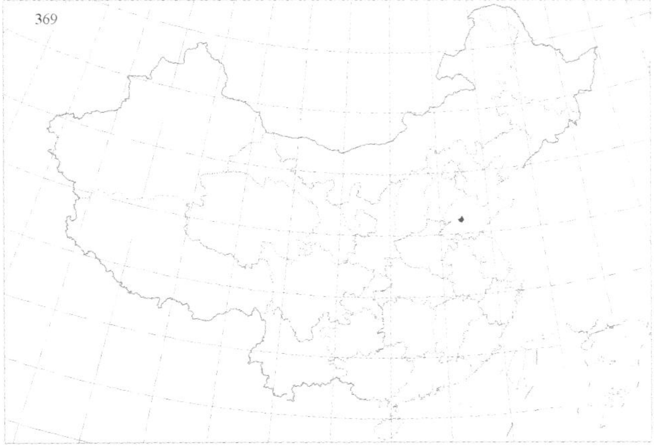

Salicaceae 杨柳科 *Salix* 柳属

Salix alatavica Kar. et Kir. ex Stschegl. 阿拉套柳

Shrubs, deciduous

5.2 (-2.8~10.2); -14.1 (-20.9~-8.5); 20.8 (10.9~27.5); 8.7 (3.5~12.1); 69.0 (16.3~107.6); 66.3 (36.6~109.2); 616 (390~813) ❙ 195 (45~450); 94 (27~211); 15 (3~48) ❙ -65.1 (-94.0~3.9); 194 (45~406); 87.0 (7.6~297.3)

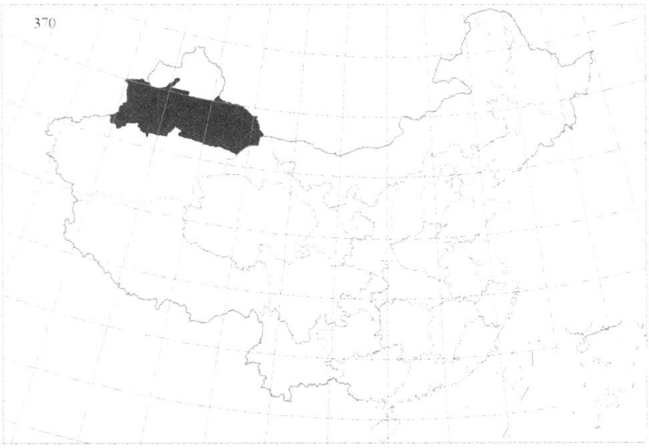

Salicaceae 杨柳科 *Salix* 柳属

Salix alba L. 白柳

Trees, deciduous

5.8 (-4.3~12.5); -11.9 (-20.9~-5.1); 20.4 (9.6~27.5); 8.8 (2.5~13.2); 69.6 (9.7~114.1); 59.6 (23.1~121.1); 616 (329~822) ❙ 153 (21~658); 79 (15~453); 11 (0~51) ❙ -71.2 (-97.3~45.4); 149 (21~658); 62.8 (1.6~297.3)

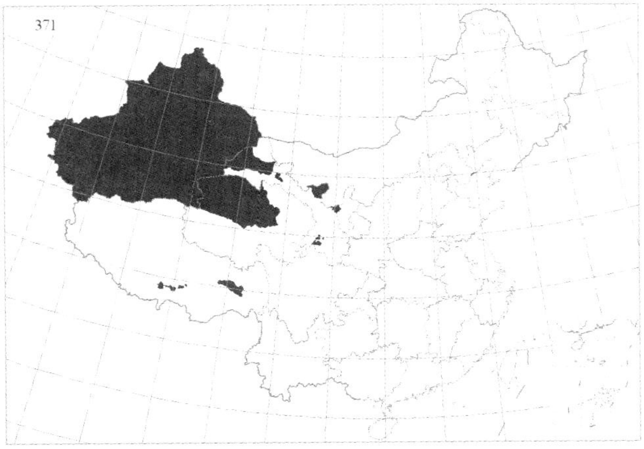

Salicaceae 杨柳科 *Salix* 柳属

Salix alfredii Goerz ex Rehder et Kobuski 秦岭柳

Shrubs or trees, deciduous

6.9 (-1.3~12.8); -6.0 (-13.5~0.7); 18.6 (9.2~27.2); 8.3 (2.9~13.3); 60.9 (10.1~114.7); 37.4 (9.4~86.3); 576 (359~827) ❙ 604 (373~1246); 3[illegible]3 (201~525); 16 (7~46) ❙ 7.5 (-47.7~93.8); 514 (359~702); 224.9 (81.2~323.3)

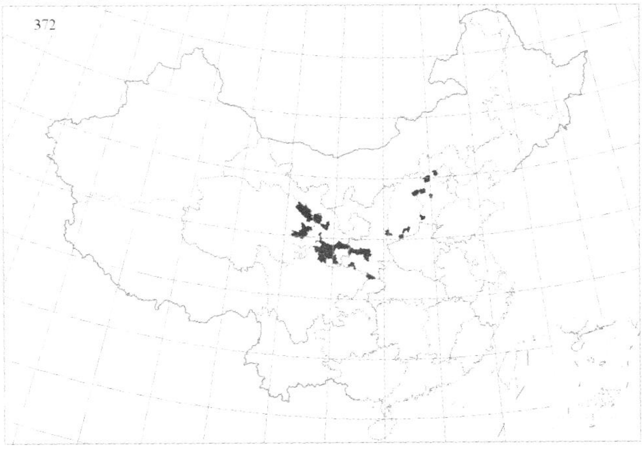

Salicaceae 杨柳科 *Salix* 柳属
Salix alfredii var. fengxianica (N. Chao) G. H. Zhu 凤县柳
Shrubs or trees, deciduous

9.2; −2.6; 20.7; 9.6; 72.1; 20.9; 615 ❙ 773; 363; 20 ❙ 25.6; 615; 323.3

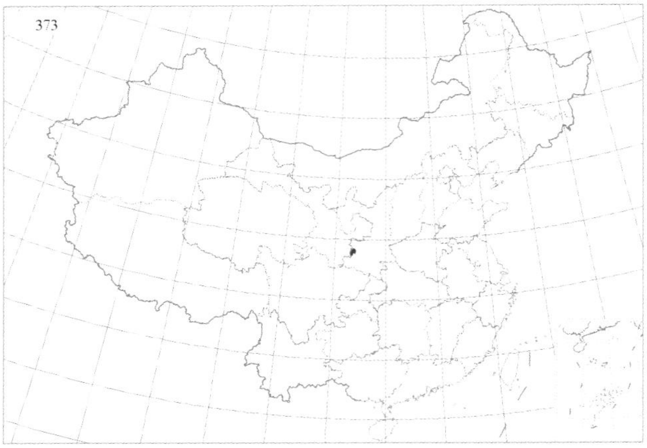

Salicaceae 杨柳科 *Salix* 柳属
Salix amphibola C. K. Schneid. 九鼎柳
Shrubs, deciduous

8.6 (2.9~15.1); −0.4 (−6.3~5.3); 16.6 (11.2~24.0); 9.0 (4.4~15.1); 60.8 (20.5~121.7); 17.4 (0.0~44.6); 583 (428~789) ❙ 969 (822~1076); 522 (420~594); 28 (19~48) ❙ 71.2 (36.4~121.7); 583 (428~789); 265.3 (200.8~319.8)

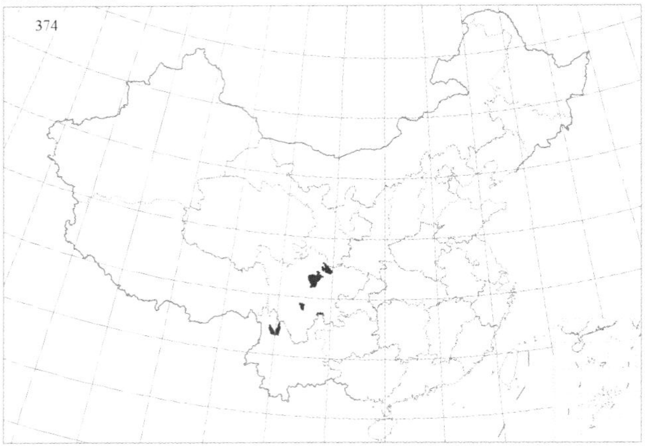

Salicaceae 杨柳科 *Salix* 柳属
Salix annulifera C. Marquand et Airy Shaw 环纹矮柳
Shrubs, deciduous

6.6 (1.2~13.3); −1.2 (−7.1~5.4); 13.5 (9.0~19.6); 7.3 (3.3~13.4); 44.2 (11.9~100.4); 24.3 (0.0~56.5); 522 (390~706) ❙ 899 (473~1736); 525 (287~1007); 28 (7~51) ❙ 66.7 (5.0~145.8); 522 (390~706); 278.8 (130.1~447.4)

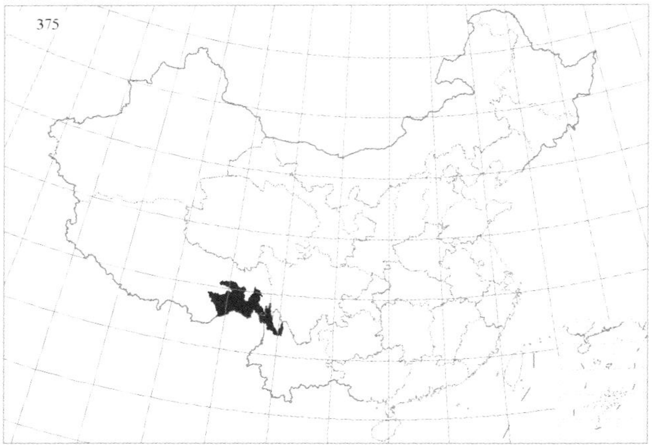

Salicaceae 杨柳科 *Salix* 柳属
Salix annulifera var. dentata S. D. Zhao 齿苞矮柳
Shrubs, deciduous

11.0 (1.2~16.7); 3.8 (−7.1~10.0); 16.8 (9.0~21.3); 11.2 (3.3~16.7); 81.6 (11.9~140.8); 9.4 (0.0~56.5); 639 (390~809) ❙ 1030 (473~1736); 587 (287~1007); 41 (7~69) ❙ 60.7 (5.0~145.8); 638 (390~809); 330.5 (130.1~447.4)

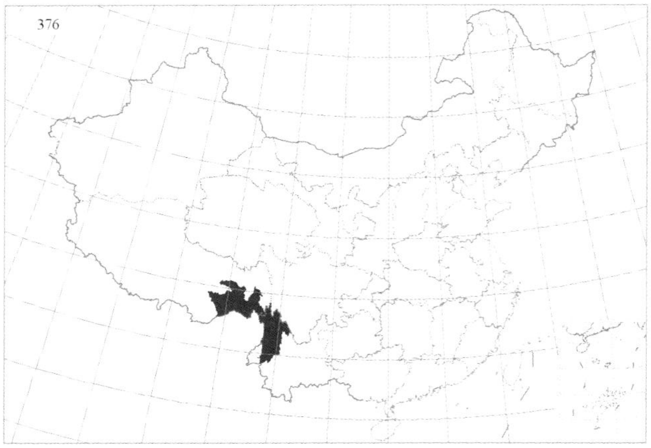

Salicaceae 杨柳科 *Salix* 柳属
Salix annulifera var. glabra P. Y. Mao et W. Z. Li 五毛矮柳
Shrubs, deciduous

5.4; −2.1; 12.3; 5.8; 31.5; 25.6; 481 ❙ 741; 426; 28 ❙ 54.0; 481; 273.5

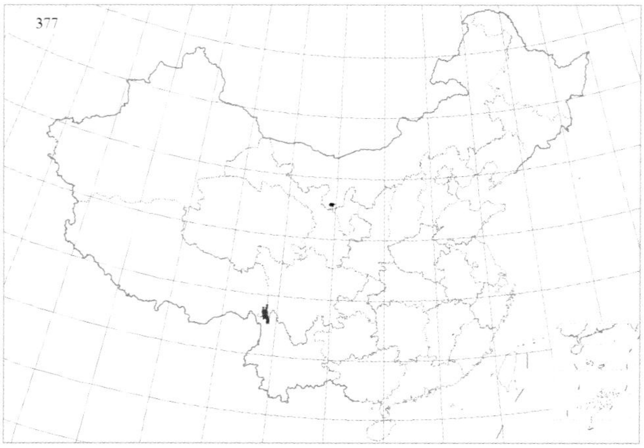

Salicaceae 杨柳科 *Salix* 柳属
Salix annulifera var. macroula C. Marquand et Airy Shaw 匙叶矮柳
Shrubs, deciduous

6.9 (1.8~13.3); −1.0 (−7.1~5.4); 13.8 (9.9~19.6); 7.5 (3.8~13.4); 46.1 (15.7~100.4); 22.8 (0.0~53.3); 529 (404~706) ❙ 1012 (583~1736); 585 (348~1007); 33 (13~51) ❙ 85.3 (44.2~145.8); 529 (404~706); 295.4 (169.5~447.4)

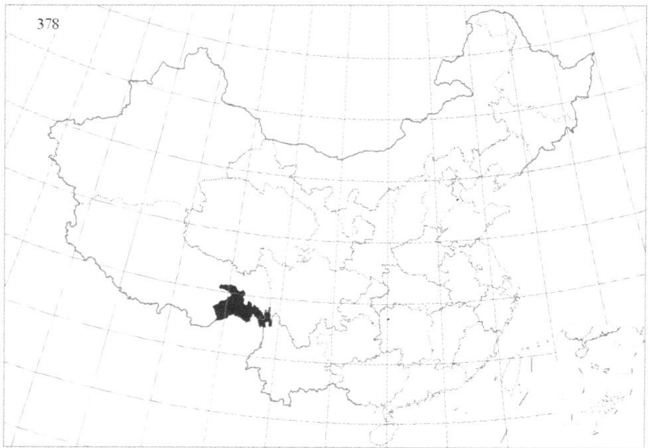

Salicaceae 杨柳科 *Salix* 柳属
Salix anticecrenata Kimura 圆齿垫柳
Shrubs, deciduous

7.4 (1.2~20.5); -0.4 (-7.1~14.9); 14.0 (9.0~23.7); 8.1 (3.3~20.5); 54.3 (11.9~185.9); 24.8 (0.0~56.5); 551 (390~992) ❙ 914 (578~1736); 535 (345~1007); 26 (11~66) ❙ 65.2 (34.2~145.8); 549 (390~959); 284.7 (130.1~570.1)

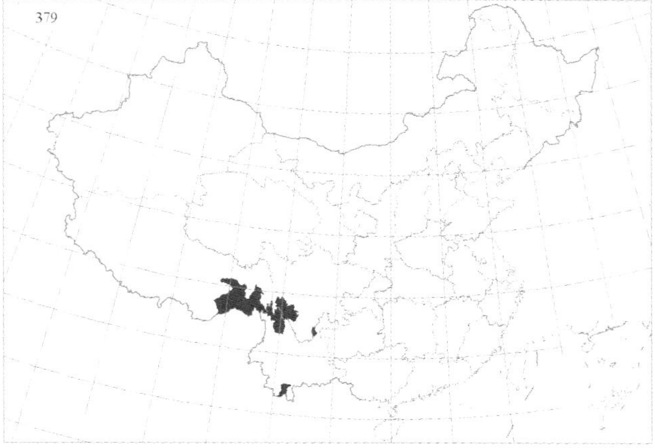

Salicaceae 杨柳科 *Salix* 柳属
Salix araeostachya C. K. Schneid. 纤序柳
Shrubs or small trees, deciduous

16.3 (6.6~21.0); 9.5 (-1.3~15.1); 21.1 (13.3~26.0); 16.3 (6.8~21.0); 136.4 (39.6~191.8); 0.3 (0.0~19.7); 807 (514~1053) ❙ 1165 (797~1751); 655 (411~996); 46 (12~77) ❙ 44.6 (-14.2~112.2); 789 (514~1053); 357.5 (215.5~659.0)

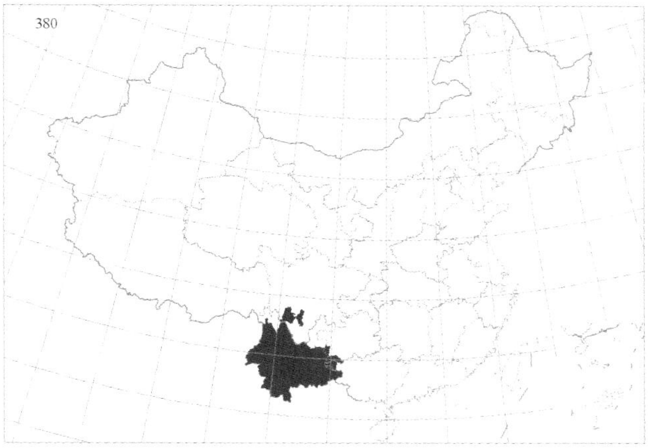

Salicaceae 杨柳科 *Salix* 柳属
Salix arctica Pall. 北极柳
Shrubs, deciduous

2.6 (-0.6~5.9); -16.4 (-19.5~-13.8); 19.3 (15.9~23.3); 7.4 (5.4~9.7); 55.7 (37.5~80.9); 84.2 (66.6~104.2); 560 (475~668) ❙ 199 (152~315); 82 (56~115); 24 (15~51) ❙ -64.1 (-76.8~-46.4); 198 (152~312); 91.5 (16.6~173.1)

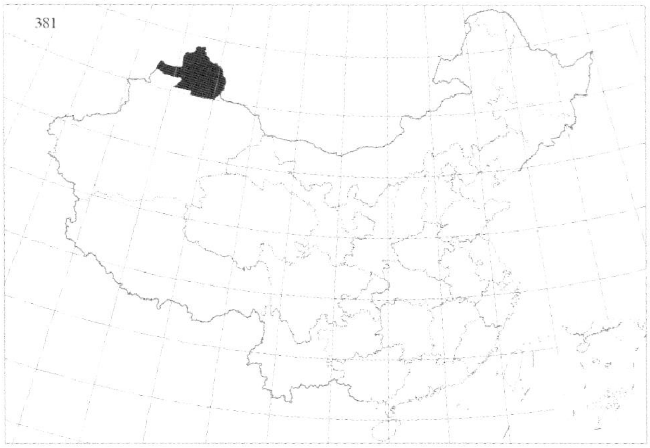

Salicaceae 杨柳科 *Salix* 柳属
Salix argyracea E. L. Wolf 银柳
Shrubs, deciduous

4.5 (-4.3~11.4); -14.4 (-20.9~-6.7); 20.1 (9.6~27.5); 8.3 (2.5~12.7); 64.8 (9.7~109.3); 69.8 (29.5~121.1); 597 (329~813) ❙ 186 (32~450); 86 (21~211); 17 (0~51) ❙ -66.2 (-95.9~3.9); 185 (32~406); 79.7 (1.6~297.3)

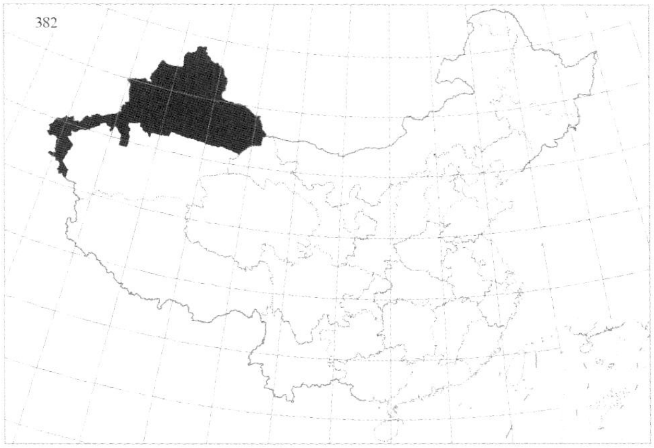

Salicaceae 杨柳科 *Salix* 柳属
Salix argyrophegga C. K. Schneid. 银光柳
Trees or shrubs, deciduous

9.1 (-1.1~15.9); -0.2 (-9.1~6.1); 17.6 (6.2~24.9); 9.7 (2.0~16.0); 69.9 (2.5~131.7); 20.1 (0.0~75.6); 609 (334~824) ❙ 1035 (321~1410); 555 (193~816); 35 (22~65) ❙ 72.5 (-4.0~128.6); 608 (321~824); 223.2 (79.3~318.7)

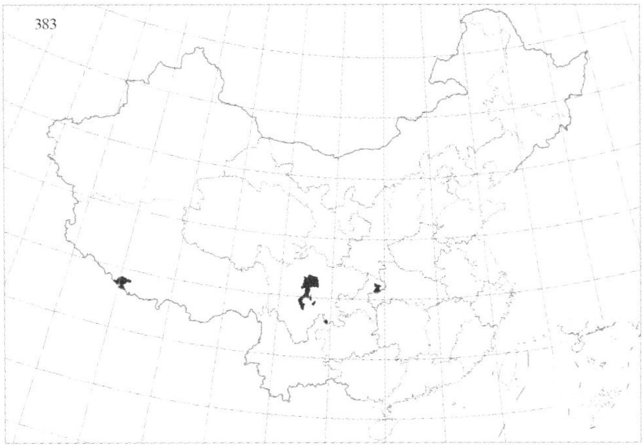

Salicaceae 杨柳科 *Salix* 柳属
Salix argyrotrichocarpa C. F. Fang 银毛果柳
Shrubs, deciduous

13.3; 5.4; 19.6; 13.4; 100.4; 0.0; 706 ❙ 1736; 1007; 51 ❙ 145.8; 706; 447.4

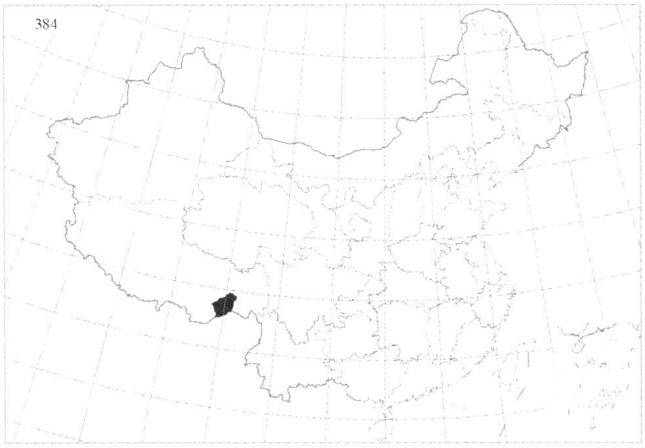

Salicaceae 杨柳科 *Salix* 柳属
Salix atopantha C. K. Schneid. 奇花柳
Shrubs, deciduous

2.4 (-4.8~13.3); -8.7 (-18.0~5.4); 12.1 (7.0~20.0); 4.8 (1.9~13.4); 26.1 (3.7~100.4); 57.0 (0.0~121.6); 441 (332~706) ❙ 618 (287~1736); 346 (188~1007); 13 (5~51) ❙ 39.4 (-37.6~145.8); 425 (287~706); 247.3 (81.2~447.4)

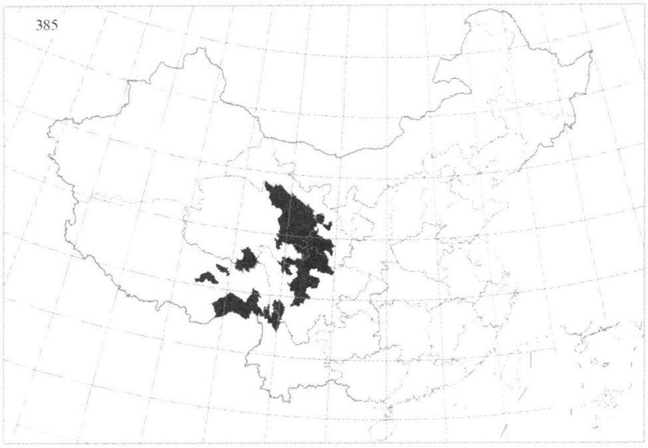

Salicaceae 杨柳科 *Salix* 柳属
Salix austrotibetica N. Chao 藏南柳
Shrubs, deciduous

13.3; 5.4; 19.6; 13.4; 100.4; 0.0; 706 ❙ 1736; 1007; 51 ❙ 145.8; 706; 447.4

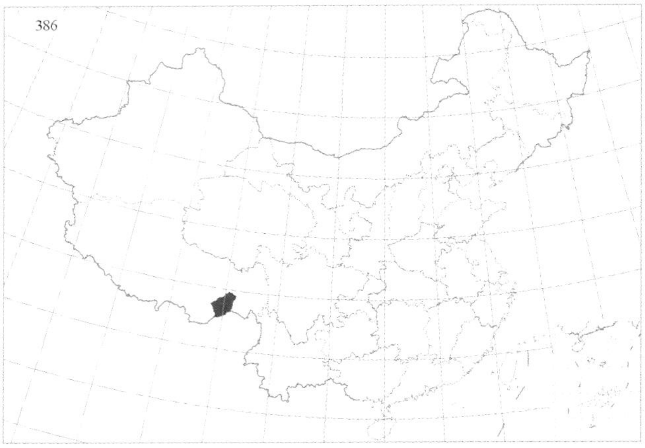

Salicaceae 杨柳科 *Salix* 柳属
Salix babylonica L. 垂柳
Trees, deciduous

14.2 (-3.6~24.8); 2.4 (-23.0~20.1); 24.7 (7.2~30.1); 14.7 (2.0~24.8); 123.1 (3.7~237.6); 12.5 (0.0~107.2); 817 (335~1385) ❙ 1077 (35~2037); 471 (23~1033); 90 (0~259) ❙ 29.0 (-94.9~147.0); 754 (35~1360); 214.2 (7.4~659.0)

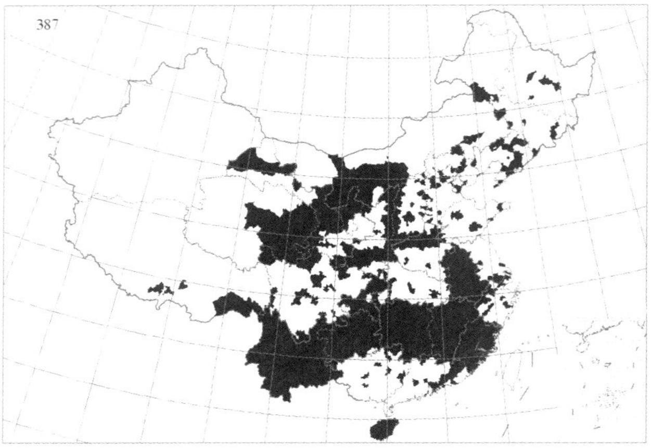

Salicaceae 杨柳科 *Salix* 柳属
Salix babylonica var. glandulipilosa P. Y. Mao et W. Z. Li 腺毛垂柳
Trees, deciduous

10.8; 3.9; 16.6; 10.9; 71.8; 1.3; 633 ❙ 968; 594; 24 ❙ 52.8; 633; 319.8

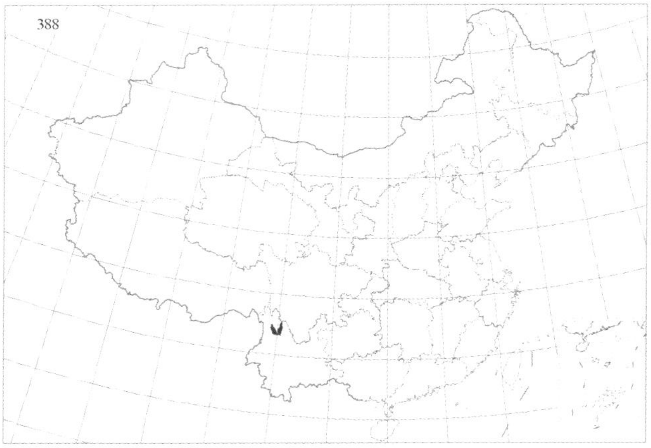

Salicaceae 杨柳科 *Salix* 柳属
Salix baileyi C. K. Schneid. 井冈柳
Trees, deciduous

14.7 (4.7~16.9); 2.4 (-6.1~4.4); 26.6 (14.7~29.2); 14.8 (6.0~16.9); 122.4 (36.3~143.8); 5.1 (0.6~39.6); 832 (482~936) ❙ 1345 (629~1621); 558 (309~666); 131 (14~195) ❙ 61.6 (22.3~108.0); 832 (482~930); 208.6 (161.2~299.5)

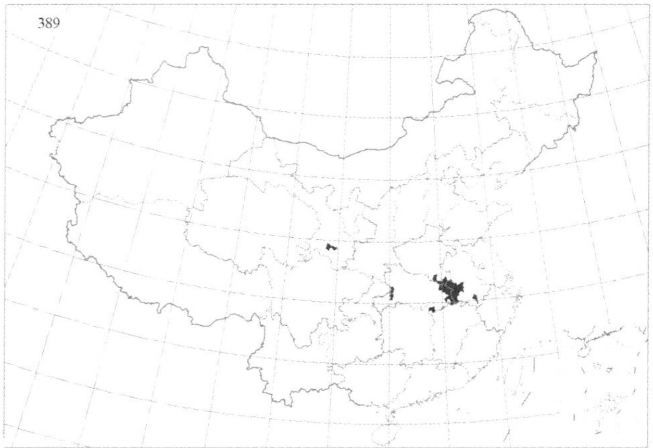

Salicaceae 杨柳科 *Salix* 柳属
Salix balansaei Seemen 中越柳
Trees

19.4; 8.9; 28.7; 19.5; 173.7; 0.0; 1038 ❙ 1776; 718; 188 ❙ 71.1; 1038; 203.9

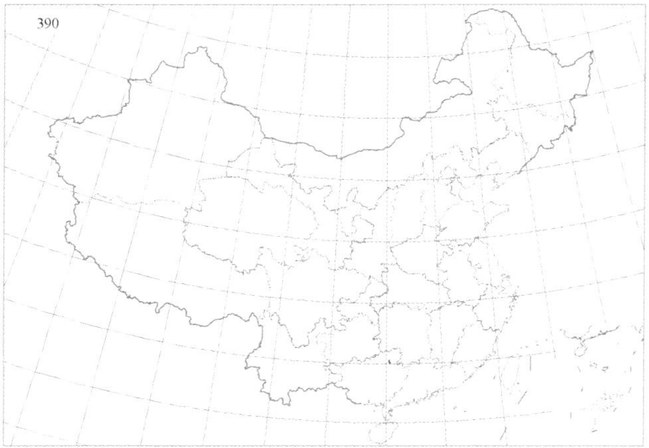

Salicaceae 杨柳科 *Salix* 柳属
Salix balansaei var. hunanensis N. Chao 湘柳
Trees, deciduous

16.2; 5.4; 26.0; 16.3; 135.4; 0.4; 853 ▎1564; 558; 184 ▎84.4; 853; 236.1

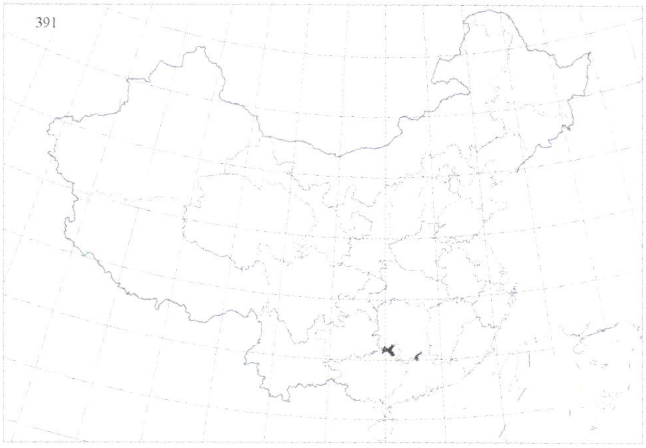

Salicaceae 杨柳科 *Salix* 柳属
Salix balfouriana C. K. Schneid. 白背柳
Shrubs or trees, deciduous

9.3 (2.9~14.5); 1.7 (-7.3~7.8); 15.8 (10.5~19.7); 9.7 (4.3~14.5); 65.3 (19.9~113.5); 12.8 (0.0~45.2); 594 (421~730) ▎953 (375~1337); 555 (203~727); 33 (8~69) ▎60.8 (-32.2~121.7); 584 (375~730); 307.8 (95.7~407.1)

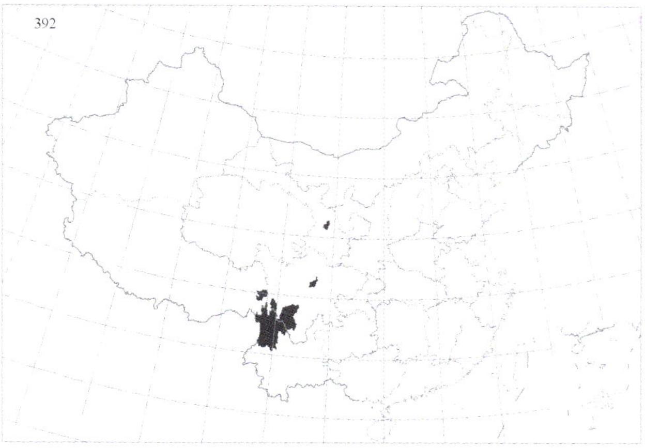

Salicaceae 杨柳科 *Salix* 柳属
Salix bangongensis C. Wang et C. F. Fang 班公柳
Trees, deciduous

-3.4 (-5.7~-1.8); -13.8 (-18.7~-10.9); 7.2 (6.0~8.6); 1.9 (1.6~2.4); 4.2 (1.5~7.9); 104.3 (88.8~133.0); 309 (294~333) ▎295 (74~620); 151 (46~306); 47 (7~103) ▎-6.3 (-75.0~91.4); 235 (74~333); 38.6 (16.7~61.9)

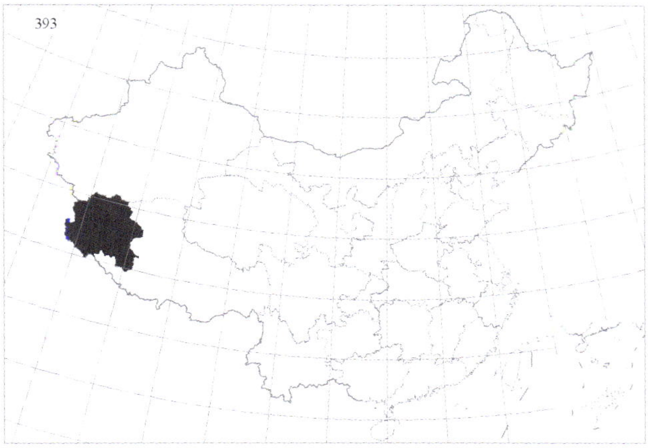

Salicaceae 杨柳科 *Salix* 柳属
Salix berberifolia Pall. 刺叶柳
Shrubs, deciduous

1.9 (-0.6~5.8); -17.1 (-19.5~-14.9); 18.7 (15.9~23.3); 7.0 (5.4~9.7); 52.4 (37.5~80.9); 88.6 (71.2~104.2); 545 (475~668) ▎184 (156~251); 77 (56~106); 22 (15~33) ▎-65.8 (-76.7~-52.0); 184 (156~251); 86.2 (16.6~139.1)

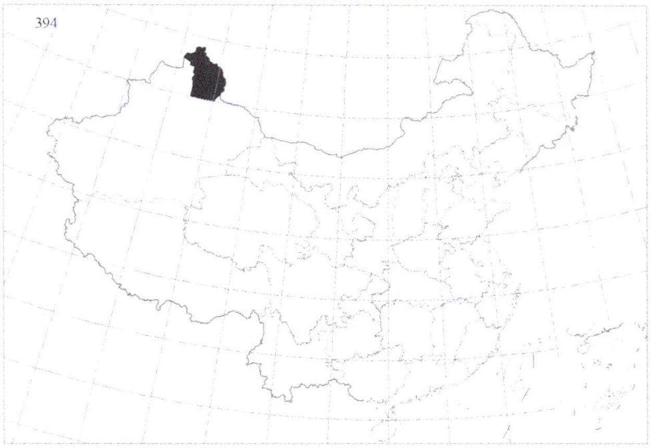

Salicaceae 杨柳科 *Salix* 柳属
Salix bhutanensis Flod. 不丹柳
Shrubs or small trees, deciduous

-1.8; -10.2; 6.1; 1.8; 1.9; 82.8; 329 ▎320; 217; 20 ▎-3.0; 320; 72.2

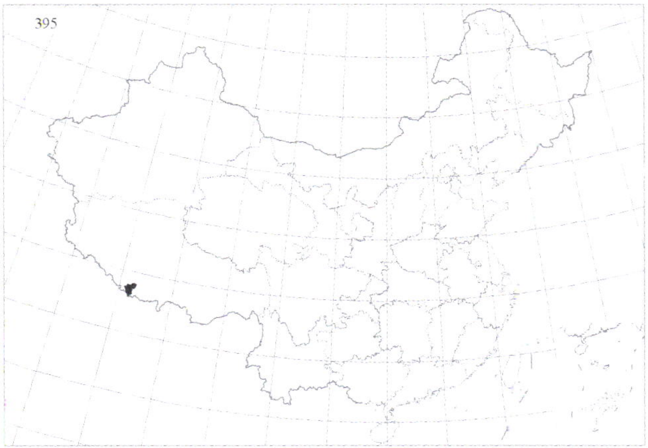

Salicaceae 杨柳科 *Salix* 柳属
Salix bikouensis Y. L. Chou 碧口柳
Shrubs or trees, deciduous

11.4 (9.3~13.5); -0.2 (-2.8~1.7); 22.5 (19.8~25.1); 11.5 (9.5~13.5); 89.6 (69.8~109.4); 12.6 (6.5~21.0); 684 (603~767) ▎899 (632~1214); 398 (306~516); 37 (15~71) ▎32.4 (0.0~101.4); 684 (603~767); 271.4 (241.9~311.0)

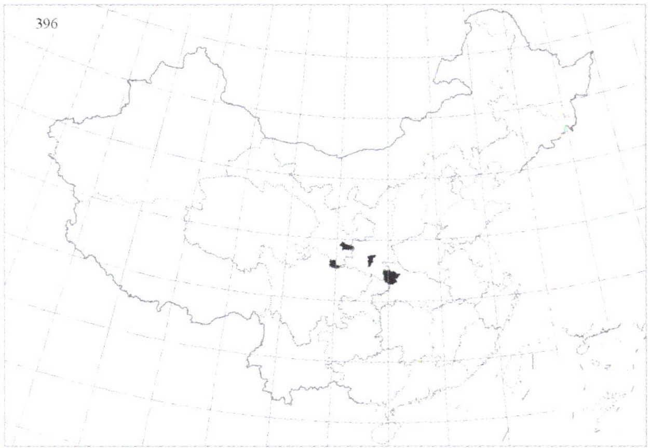

Salicaceae 杨柳科 *Salix* 柳属

Salix biondiana Seemen ex Diels 庙王柳

Shrubs or small trees, deciduous

8.2 (-1.2~14.4); -3.9 (-13.2~2.6); 19.5 (8.5~25.8); 9.3 (2.6~14.4); 70.2 (8.1~116.7); 30.8 (3.6~82.3); 609 (347~793) ❙ 751 (101~1478); 363 (61~602); 31 (5~105) ❙ 22.9 (-83.8~107.2); 581 (101~793); 237.5 (58.4~318.2)

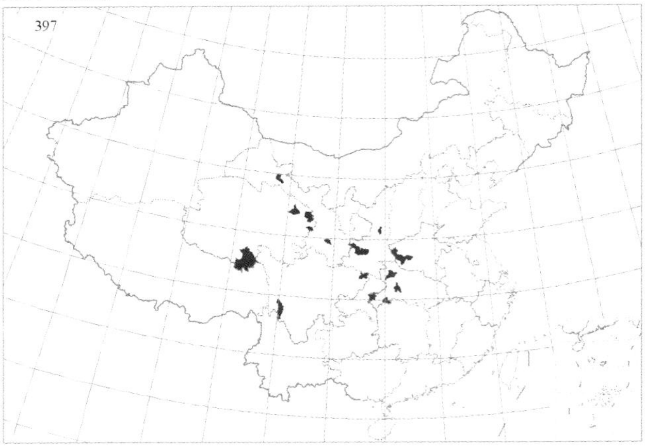

Salicaceae 杨柳科 *Salix* 柳属

Salix bistyla Hand.-Mazz. 双柱柳

Shrubs, deciduous

11.1 (-1.8~16.9); 4.1 (-10.2~10.3); 16.8 (6.1~21.4); 11.4 (1.8~16.9); 83.5 (1.9~142.7); 9.4 (0.0~82.8); 643 (329~813) ❙ 1034 (320~1454); 585 (217~792); 44 (7~70) ❙ 59.2 (-3.0~100.7); 642 (320~813); 334.8 (72.2~487.2)

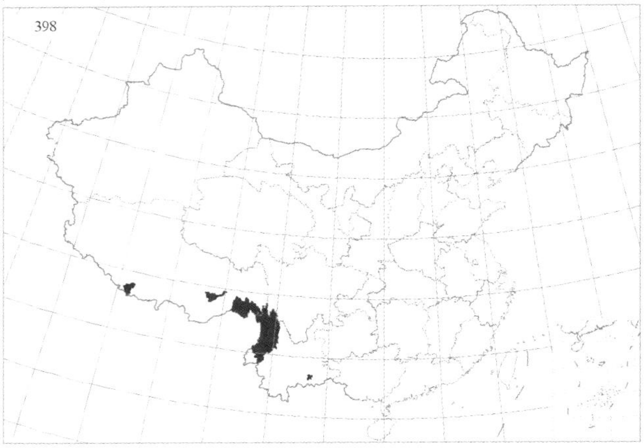

Salicaceae 杨柳科 *Salix* 柳属

Salix blakii Goerz 黄线柳

Shrubs, deciduous

4.5 (-2.8~9.7); -15.0 (-20.9~-8.1); 20.2 (10.9~27.5); 8.3 (3.5~12.1); 65.5 (16.3~107.6); 71.1 (37.2~109.2); 601 (390~813) ❙ 227 (81~450); 107 (49~211); 18 (3~46) ❙ -58.4 (-88.8~3.9); 226 (81~406); 107.4 (10.9~297.3)

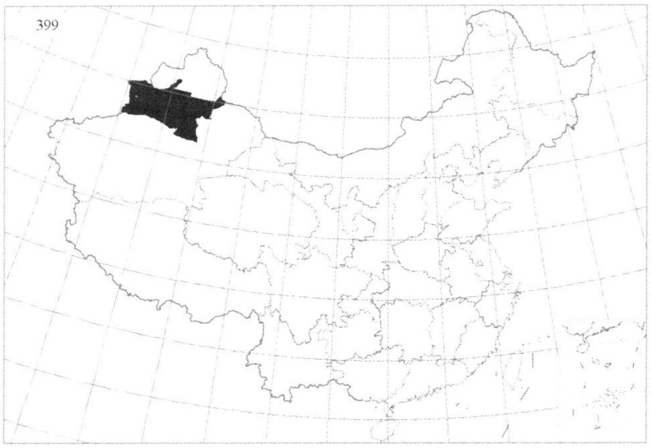

Salicaceae 杨柳科 *Salix* 柳属

Salix boseensis N. Chao 桂柳

Shrubs, deciduous

20.2; 11.8; 26.5; 20.3; 183.2; 0.0; 1038 ❙ 1336; 741; 64 ❙ 30.8; 1037; 314.7

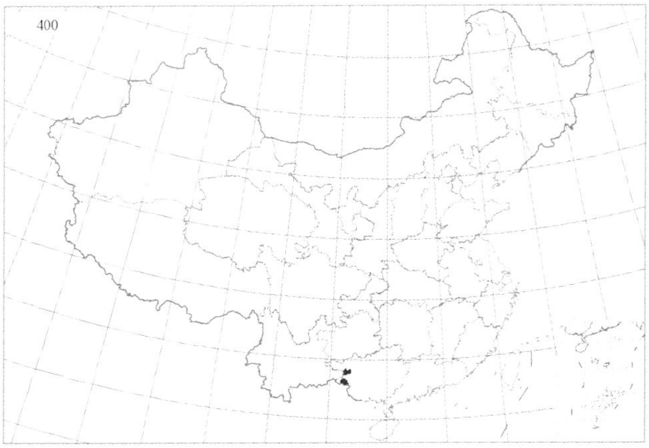

Salicaceae 杨柳科 *Salix* 柳属

Salix brachista C. K. Schneid. 小垫柳

Shrubs, deciduous

10.9 (1.8~17.2); 3.1 (-7.7~10.0); 17.5 (9.9~25.4); 11.2 (3.8~17.3); 82.4 (15.5~147.5); 11.2 (0.0~53.3); 643 (404~847) ❙ 997 (583~1410); 555 (323~816); 37 (12~69) ❙ 58.8 (26.7~125.9); 639 (404~847); 297.5 (148.9~406.0)

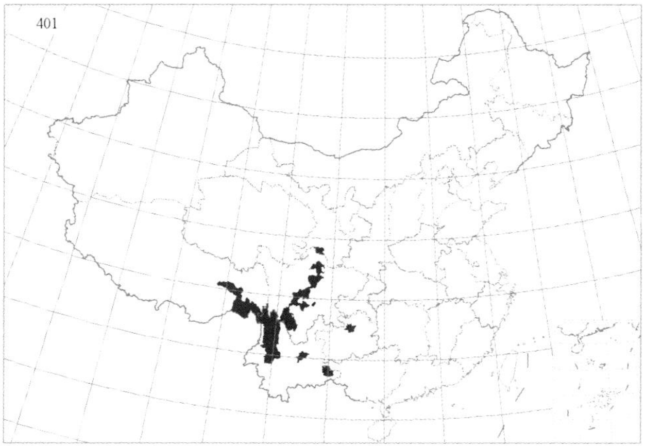

Salicaceae 杨柳科 *Salix* 柳属

Salix brachista var. integra C. Wang et C. F. Fang 全缘小垫柳

Shrubs, deciduous

10.8; 3.9; 16.6; 10.9; 71.8; 1.3; 633 ❙ 968; 594; 24 ❙ 52.8; 633; 319.8

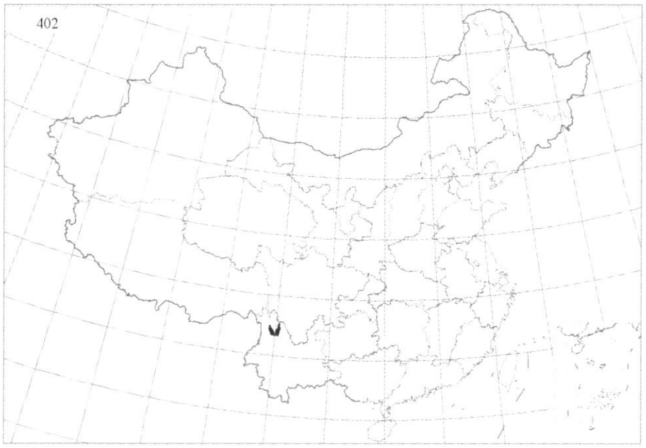

Salicaceae 杨柳科 *Salix* 柳属
Salix brachista var. pilifera N. Chao 毛果小垫柳
Shrubs, deciduous

5.3; -2.4; 12.2; 5.7; 30.5; 26.2; 479 ▎961; 558; 31 ▎100.7; 479; 234.2

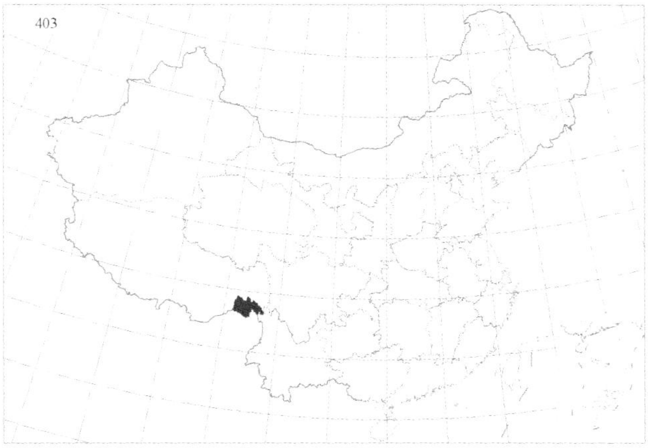

Salicaceae 杨柳科 *Salix* 柳属
Salix burqinensis C. Y. Yang 布尔津柳
Trees, deciduous

1.3; -16.5; 17.6; 6.4; 45.7; 90.0; 519 ▎244; 102; 30 ▎-53.0; 244; 137.8

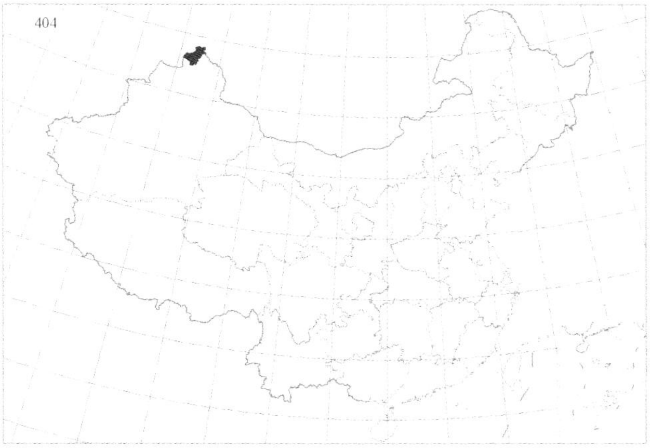

Salicaceae 杨柳科 *Salix* 柳属
Salix caesia Vill. 欧杞柳
Shrubs, deciduous

4.5 (-2.8~10.2); -14.7 (-20.9~-8.5); 20.4 (10.9~27.5); 8.3 (3.5~12.1); 65.5 (16.3~107.6); 70.9 (36.6~109.2); 601 (390~813) ▎198 (45~450); 91 (27~211); 18 (3~51) ▎-64.5 (-94.0~3.9); 197 (45~406); 88.5 (7.6~297.3)

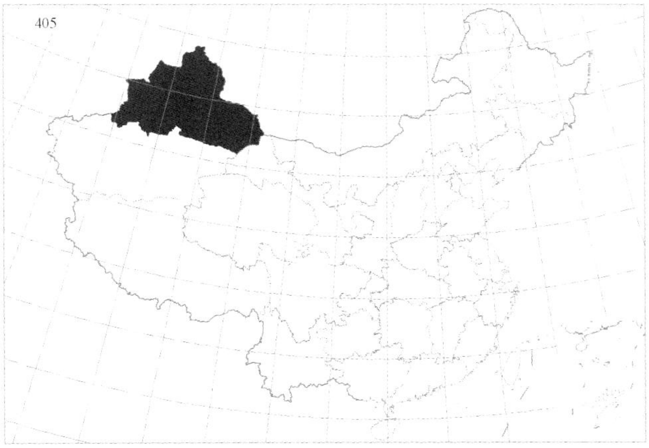

Salicaceae 杨柳科 *Salix* 柳属
Salix calyculata Hook. f. ex Andersson 长柄垫柳
Shrubs, deciduous

7.5 (1.8~13.3); -0.2 (-7.1~6.4); 14.1 (9.5~19.6); 8.0 (3.8~13.4); 50.4 (15.7~100.4); 19.9 (0.0~53.3); 543 (404~706) ▎956 (473~1736); 560 (287~1007); 33 (7~69) ▎71.7 (5.0~145.8); 543 (404~706); 302.7 (169.5~447.4)

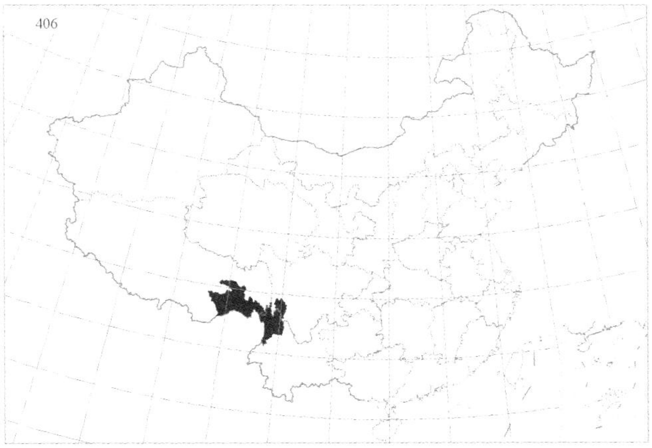

Salicaceae 杨柳科 *Salix* 柳属
Salix calyculata var. gongshanica C. Wang et C. F. Fang 贡山长柄垫柳
Shrubs, deciduous

7.7 (2.2~13.2); 0.3 (-6.3~6.4); 14.1 (9.5~18.4); 8.1 (3.9~13.3); 50.3 (15.8~99.4); 17.1 (0.0~48.6); 547 (411~695) ▎983 (729~1337); 569 (426~727); 38 (11~69) ▎78.5 (54.0~100.7); 547 (411~695); 311.9 (208.8~407.1)

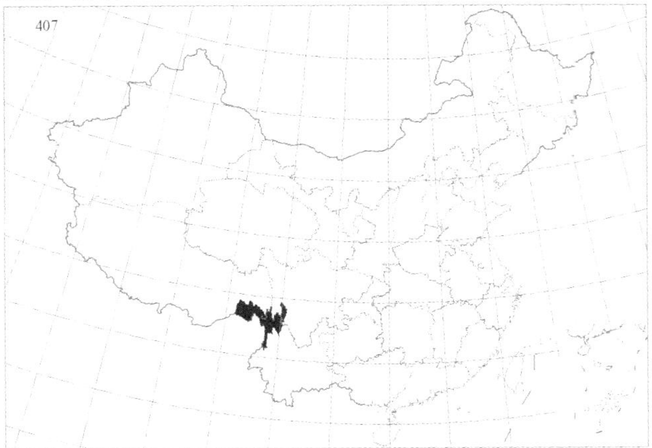

Salicaceae 杨柳科 *Salix* 柳属
Salix capitata Y. L. Chou et Skvortsov 圆头柳
Trees, deciduous

7.8 (-0.2~12.9); -9.3 (-23.8~1.1); 22.9 (18.7~25.5); 10.0 (6.3~12.9); 81.0 (50.3~106.3); 46.4 (8.0~112.6); 655 (525~773) ▎574 (408~966); 407 (247~616); 20 (6~35) ▎4.2 (-37.3~54.7); 602 (408~737); 227.9 (46.6~323.4)

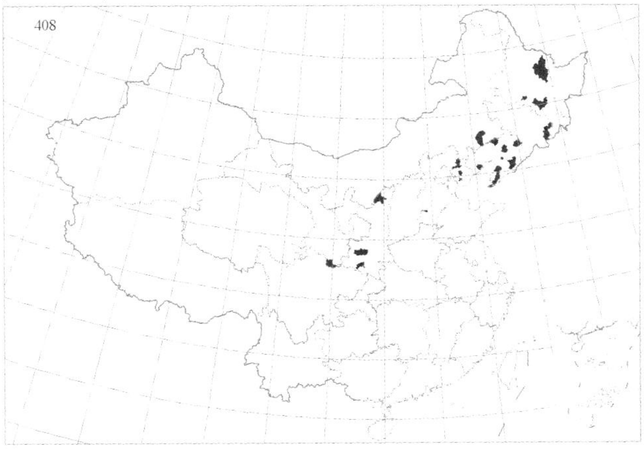

Salicaceae 杨柳科 *Salix* 柳属
Salix caprea L. 黄花柳
Shrubs or small trees, deciduous

3.6 (-2.8~9.3); -15.5 (-20.9~-8.8); 19.4 (10.9~27.1); 7.7 (3.5~12.1); 59.2 (16.3~107.6); 76.0 (40.4~109.2); 574 (390~810) ❙ 216 (75~406); 98 (39~211); 20 (3~48) ❙ -60.2 (-89.7~-7.6); 216 (75~406); 110.8 (14.6~297.3)

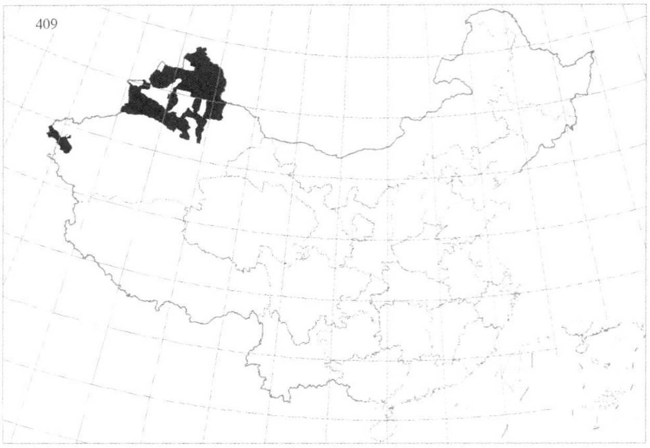

Salicaceae 杨柳科 *Salix* 柳属
Salix capusii Franch. 蓝叶柳
Shrubs, deciduous

4.8 (-4.3~12.3); -13.8 (-20.9~-5.2); 20.1 (9.6~27.5); 8.3 (2.5~13.0); 65.4 (9.7~112.3); 67.2 (24.3~121.1); 599 (329~813) ❙ 181 (42~450); 85 (17~211); 16 (3~51) ❙ -66.7 (-94.7~3.9); 181 (42~406); 78.8 (7.6~297.3)

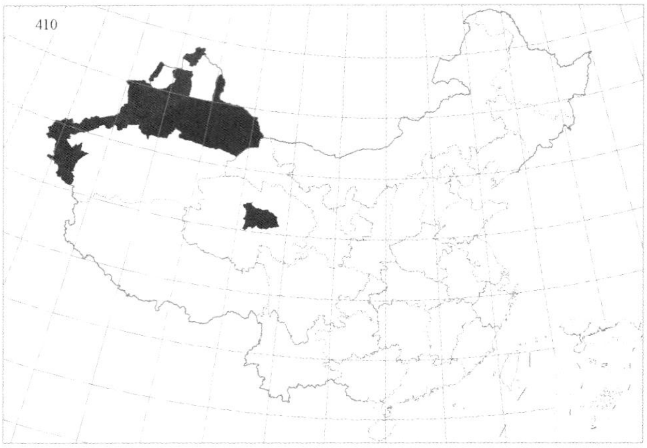

Salicaceae 杨柳科 *Salix* 柳属
Salix carmanica Bornm. 黄皮柳
Shrubs, deciduous

4.7 (-4.3~12.5); -12.9 (-20.3~-5.1); 18.7 (9.6~26.5); 8.0 (2.5~13.2); 61.3 (9.7~114.1); 64.7 (23.1~121.1); 577 (329~817) ❙ 148 (38~406); 74 (17~211); 11 (3~37) ❙ -70.5 (-95.3~-7.6); 148 (38~406); 80.7 (2.8~297.3)

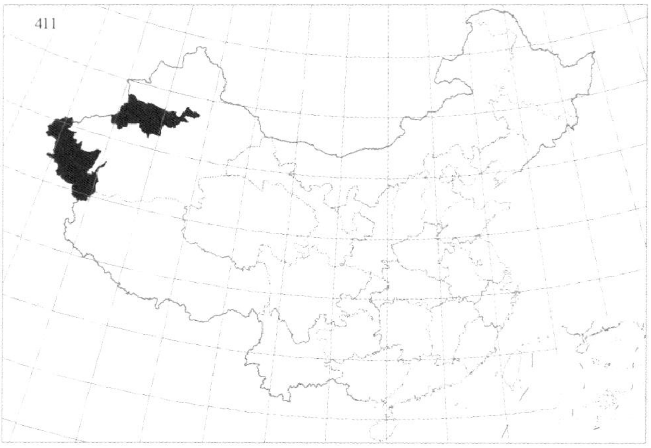

Salicaceae 杨柳科 *Salix* 柳属
Salix caspica Pall. 油柴柳
Shrubs, deciduous

4.5 (-0.6~10.2); -14.7 (-19.5~-8.5); 20.8 (15.9~27.1); 8.4 (5.4~12.1); 66.8 (37.5~107.6); 72.3 (36.6~104.2); 608 (475~810) ❙ 178 (53~325); 78 (33~135); 19 (3~51) ❙ -69.3 (-92.4~-46.4); 178 (53~325); 81.5 (10.3~233.8)

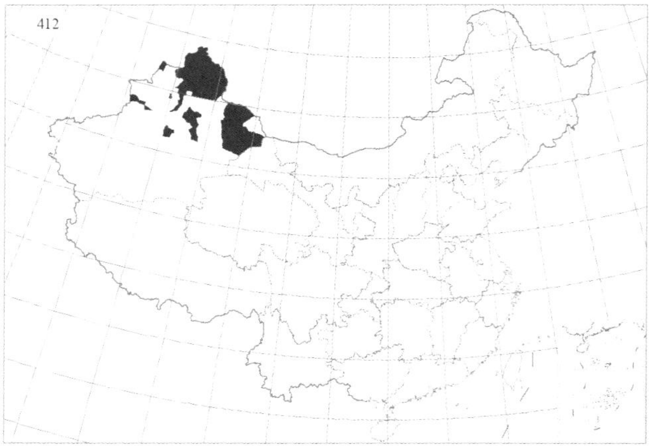

Salicaceae 杨柳科 *Salix* 柳属
Salix cathayana Diels 中华柳
Shrubs, deciduous

12.2 (-3.6~19.1); 1.4 (-15.6~12.5); 21.9 (7.2~28.9); 12.8 (2.0~19.1); 101.6 (3.7~169.2); 14.1 (0.0~107.2); 721 (335~1002) ❙ 938 (252~1526); 473 (153~816); 44 (4~181) ❙ 30.6 (-47.7~128.6); 687 (252~1002); 224.1 (46.7~487.2)

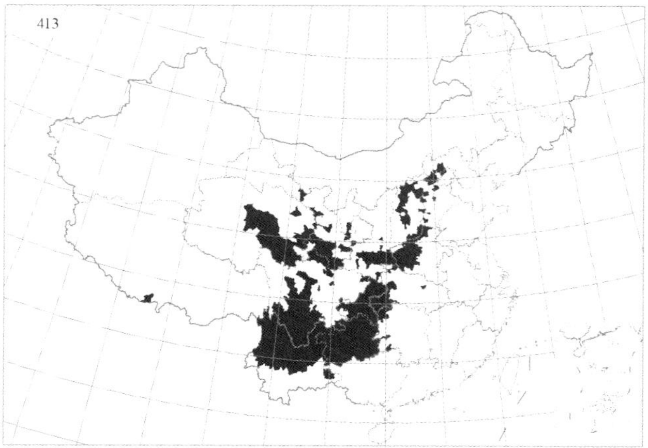

Salicaceae 杨柳科 *Salix* 柳属
Salix cavaleriei H. Lév. 滇大叶柳
Trees, deciduous

15.1 (0.5~22.3); 7.5 (-9.4~15.1); 20.9 (9.0~28.9); 15.2 (3.3~22.3); 124.0 (11.9~207.8); 2.5 (0.0~67.5); 777 (376~1193) ❙ 1119 (282~1751); 617 (216~996); 47 (1~173) ❙ 44.0 (-26.4~112.2); 761 (282~1193); 314.7 (130.1~659.0)

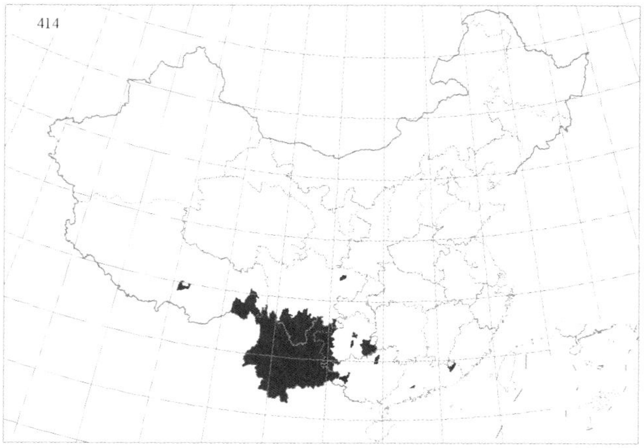

Salicaceae 杨柳科 *Salix* 柳属

Salix chaenomeloides Kimura 腺柳

Trees, deciduous

14.8 (3.3~21.0); 2.1 (-12.7~11.9); 26.6 (15.6~29.9); 14.9 (6.6~21.1); 125.1 (45.4~193.0); 7.2 (0.0~65.0); 839 (507~1114) ❙ 1053 (397~1997); 462 (245~821); 93 (10~259) ❙ 24.6 (-36.6~130.7); 803 (397~1114); 183.2 (93.5~334.4)

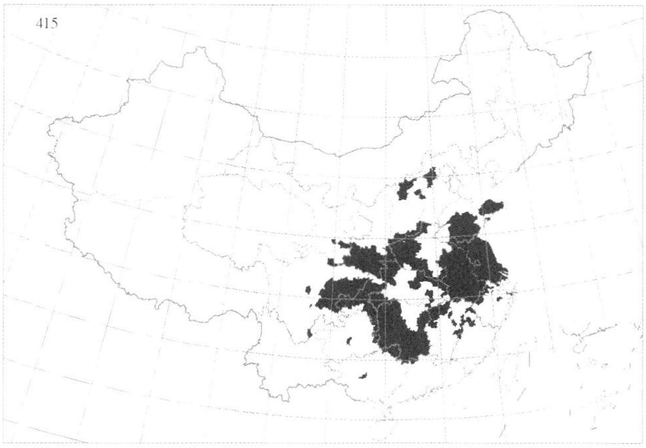

Salicaceae 杨柳科 *Salix* 柳属

Salix chaenomeloides var. glandulifolia (C. Wang et C. Y. Yu) C. F. Fang 腺叶腺柳

Trees, deciduous

12.4 (10.2~15.4); -0.5 (-2.9~3.4); 24.6 (22.4~27.7); 12.6 (10.6~15.5); 102.2 (83.1~128.9); 12.7 (1.7~20.5); 739 (656~861) ❙ 880 (622~1436); 402 (298~582); 57 (19~171) ❙ 17.1 (-13.4~68.8); 720 (622~861); 239.1 (183.4~281.2)

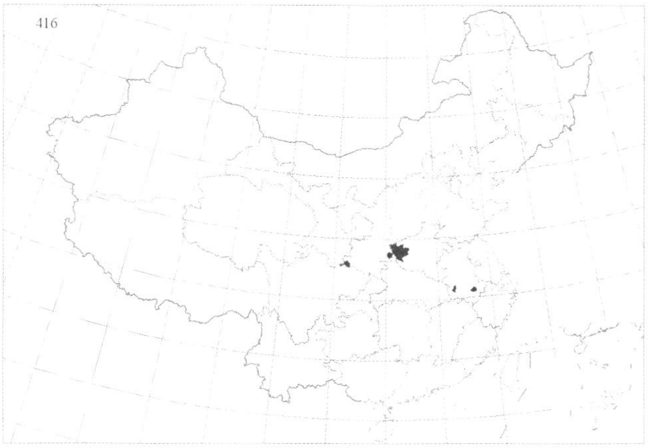

Salicaceae 杨柳科 *Salix* 柳属

Salix characta C. K. Schneid. 密齿柳

Shrubs, deciduous

7.0 (-5.8~15.1); -9.0 (-29.8~1.8); 21.2 (8.5~28.0); 9.2 (2.6~15.2); 72.7 (8.1~128.0); 48.4 (5.6~157.4); 619 (347~880) ❙ 468 (63~856); 273 (45~453); 15 (2~53) ❙ -24.3 (-84.9~52.6); 457 (63~817); 154.6 (8.4~318.2)

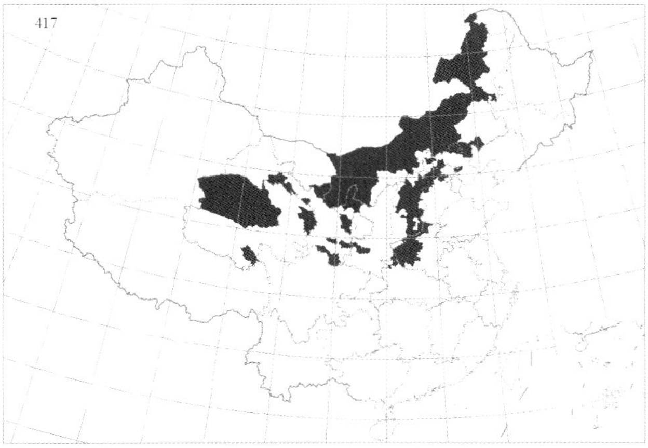

Salicaceae 杨柳科 *Salix* 柳属

Salix cheilophila C. K. Schneid. 乌柳

Shrubs or trees, deciduous

8.1 (-4.4~17.9); -6.5 (-21.5~7.6); 21.1 (6.9~29.3); 9.7 (1.7~17.9); 75.7 (3.2~154.8); 38.0 (0.0~115.9); 635 (323~961) ❙ 559 (35~1679); 314 (23~816); 20 (0~195) ❙ -10.0 (-94.9~128.6); 496 (35~961); 168.9 (5.1~364.1)

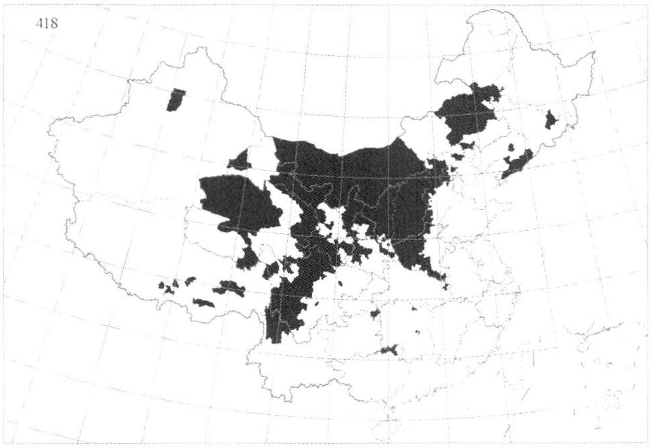

Salicaceae 杨柳科 *Salix* 柳属

Salix cheilophila var. microstachyoides (C. Wang et P. Y. Fu) C. Wang et C. F. Fang 大红柳

Shrubs or trees, deciduous

2.3; -7.7; 11.0; 4.3; 21.6; 54.4; 414 ❙ 324; 244; 2 ❙ -21.8; 324; 211.3

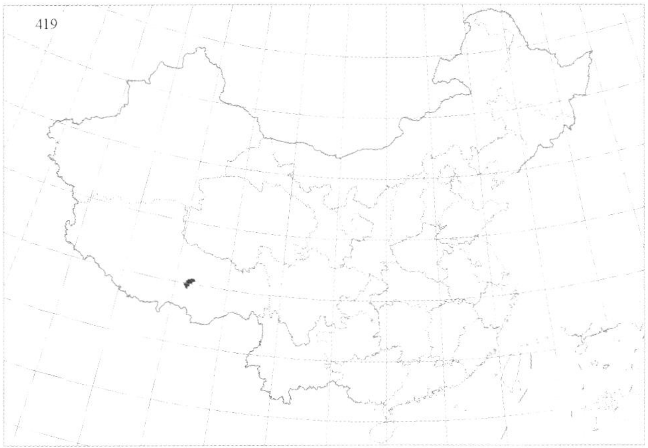

Salicaceae 杨柳科 *Salix* 柳属

Salix chekiangensis W. C. Cheng 浙江柳

Trees, deciduous

16.5; 5.1; 27.7; 16.5; 138.4; 0.0; 884 ❙ 1613; 556; 203 ❙ 82.9; 880; 233.9

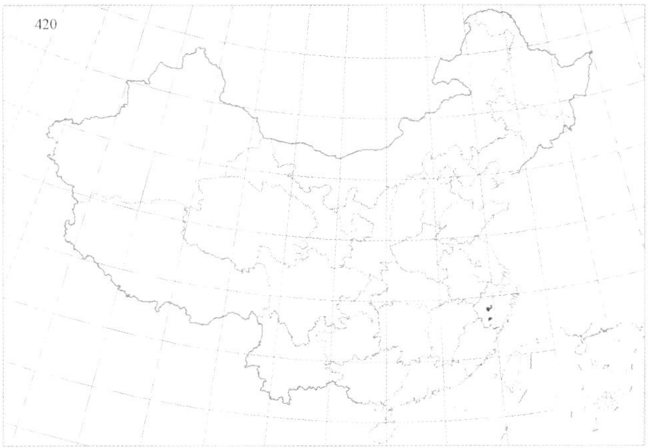

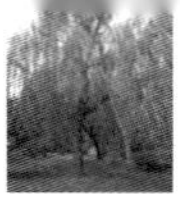

Salicaceae 杨柳科 *Salix* 柳属
Salix chienii W. C. Cheng 银叶柳
Shrubs or trees, deciduous

16.4 (12.7~19.7); 4.5 (1.5~11.5); 27.9 (23.0~29.9); 16.5 (12.8~19.7); 139.0 (99.0~176.5); 1.3 (0.0~7.3); 896 (713~1004) ❙ 1469 (879~1997); 530 (363~713); 176 (76~259) ❙ 64.6 (-1.4~147.0); 884 (713~1001); 206.4 (110.1~319.9)

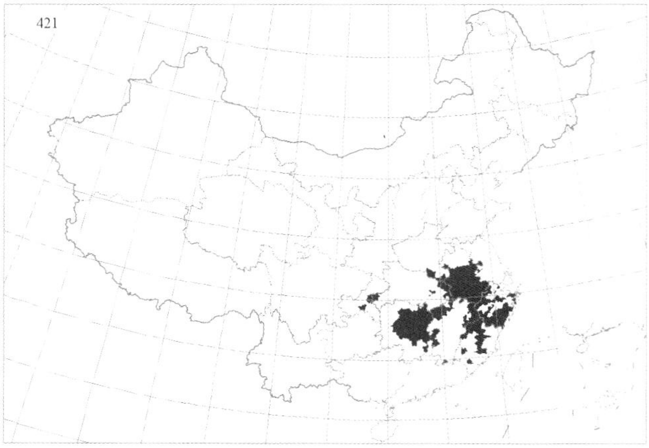

Salicaceae 杨柳科 *Salix* 柳属
Salix chienii var. pubigera N. Chao 常宁柳
Shrubs or trees, deciduous

16.8; 5.1; 27.9; 16.8; 142.3; 0.4; 903 ❙ 1444; 433; 204 ❙ 61.2; 854; 181.8

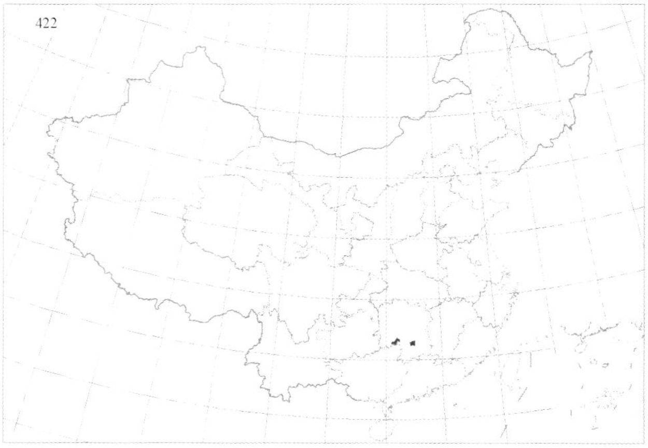

Salicaceae 杨柳科 *Salix* 柳属
Salix chikungensis C. K. Schneid. 鸡公柳
Shrubs, deciduous

15.8 (14.7~17.2); 3.0 (1.6~5.0); 27.9 (26.7~29.4); 15.8 (14.7~17.2); 132.7 (121.2~146.8); 2.8 (0.0~6.0); 881 (818~952) ❙ 1249 (845~1590); 515 (411~661); 128 (54~193) ❙ 42.0 (-3.5~88.3); 877 (818~941); 194.4 (111.9~318.0)

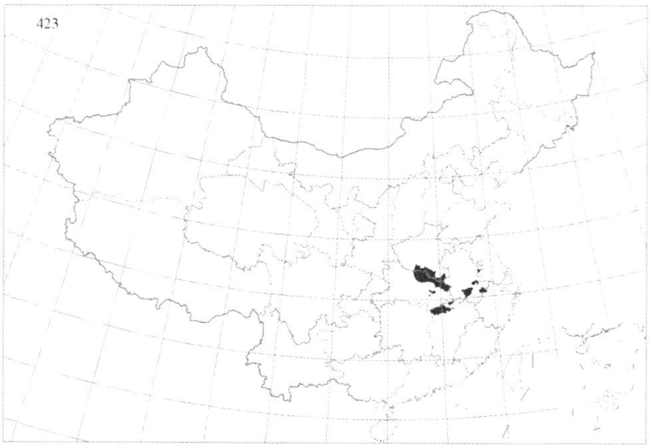

Salicaceae 杨柳科 *Salix* 柳属
Salix chingiana K. S. Hao ex C. F. Fang et A. K. Skvortsov 秦柳
Trees, deciduous

10.6 (2.5~15.1); -0.9 (-9.7~4.2); 21.4 (13.1~25.6); 11.0 (5.0~15.1); 83.7 (27.9~122.0); 16.5 (0.8~56.4); 662 (449~813) ❙ 852 (516~1246); 394 (283~532); 24 (9~46) ❙ 28.0 (-0.2~93.8); 662 (449~813); 262.6 (211.4~311.0)

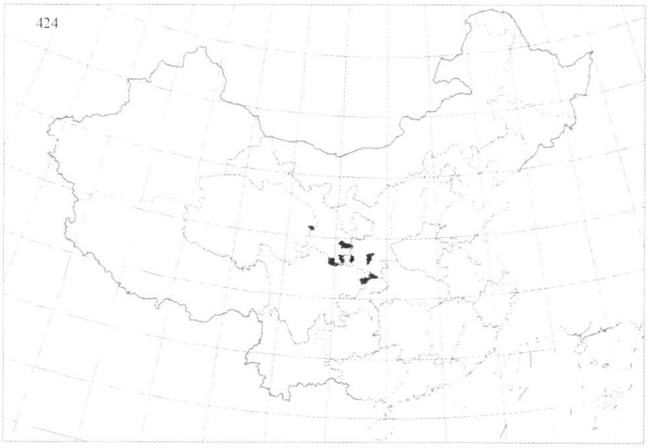

Salicaceae 杨柳科 *Salix* 柳属
Salix cinerea L. 灰柳
Shrubs, deciduous

2.6 (-1.3~6.5); -16.2 (-20.9~-11.6); 18.5 (13.2~23.3); 7.1 (4.5~9.7); 53.0 (26.7~81.0); 80.9 (53.6~104.2); 548 (433~668) ❙ 251 (152~450); 111 (56~211); 25 (12~51) ❙ -52.5 (-76.8~3.9); 249 (152~406); 130.3 (16.6~297.3)

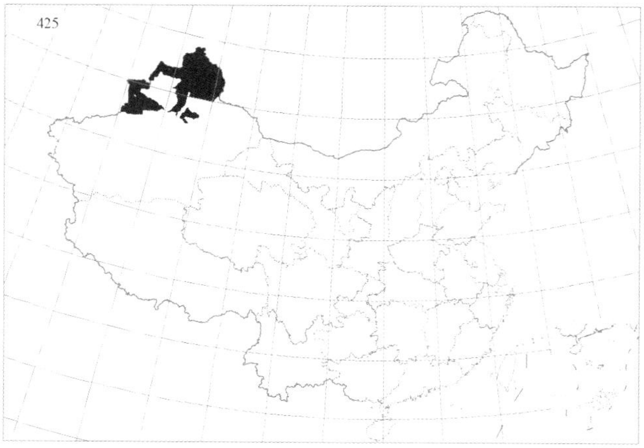

Salicaceae 杨柳科 *Salix* 柳属
Salix clathrata Hand.-Mazz. 栅枝垫柳
Shrubs, deciduous

5.8 (-2.2~12.2); -2.4 (-12.8~4.3); 12.9 (7.6~18.5); 6.8 (2.2~12.2); 39.5 (5.5~87.4); 29.6 (0.7~91.3); 504 (336~672) ❙ 859 (462~1431); 515 (299~817); 21 (9~45) ❙ 69.8 (37.7~125.9); 504 (336~672); 286.4 (130.1~363.7)

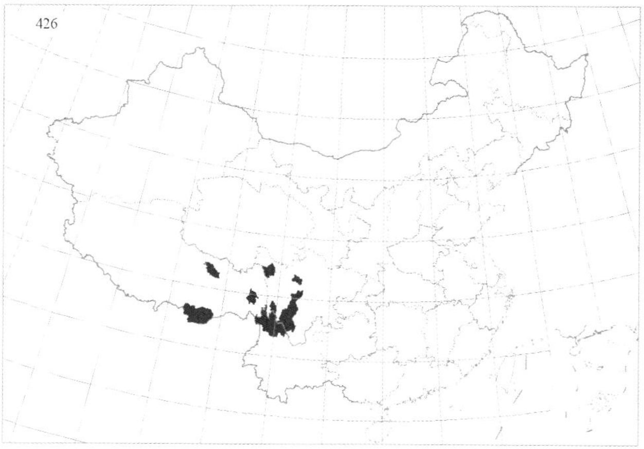

Salicaceae 杨柳科 *Salix* 柳属
Salix coggygria Hand.-Mazz. 怒江矮柳
Shrubs, deciduous

7.6 (1.8~13.3); -0.1 (-7.1~5.4); 14.2 (9.9~19.6); 8.1 (3.8~13.4); 50.2 (15.5~100.4); 19.0 (0.0~53.3); 547 (404~706) ❙ 1011 (583~1736); 588 (348~1007); 33 (13~51) ❙ 83.5 (44.2~145.8); 547 (404~706); 304.0 (169.5~447.4)

427

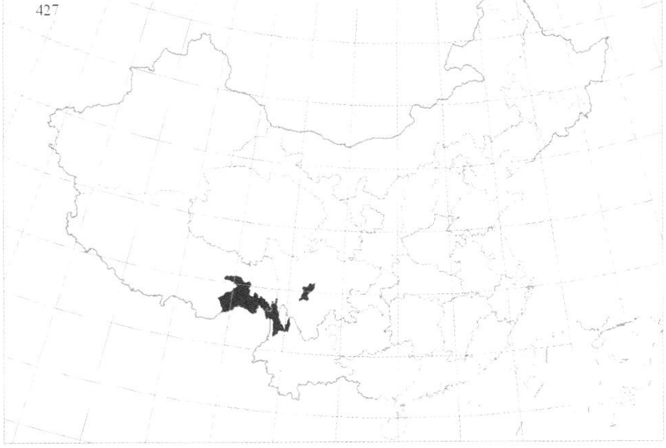

Salicaceae 杨柳科 *Salix* 柳属
Salix contortiapiculata P. Y. Mao et W. Z. Li 扭尖柳
Shrubs, deciduous

8.6; 1.3; 15.0; 8.7; 52.6; 8.8; 574 ❙ 1041; 585; 45 ❙ 81.5; 574; 352.3

428

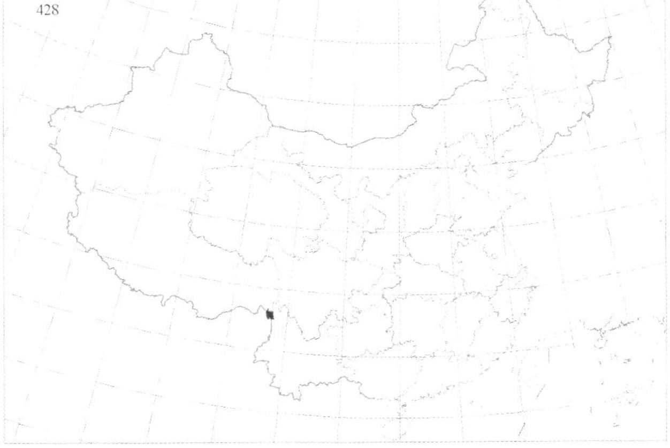

Salicaceae 杨柳科 *Salix* 柳属
Salix crenata K. S. Hao ex C. F. Fang et A. K. Skvortsov 锯齿叶垫柳
Shrubs, deciduous

12.9 (1.8~17.9); 6.0 (-7.1~11.6); 18.2 (9.9~22.1); 13.0 (3.8~17.9); 100.2 (15.7~155.1); 5.3 (0.0~53.3); 693 (404~858) ❙ 1051 (473~1454); 593 (287~792); 45 (7~70) ❙ 52.2 (5.0~99.9); 686 (404~836); 346.6 (169.5~487.2)

429

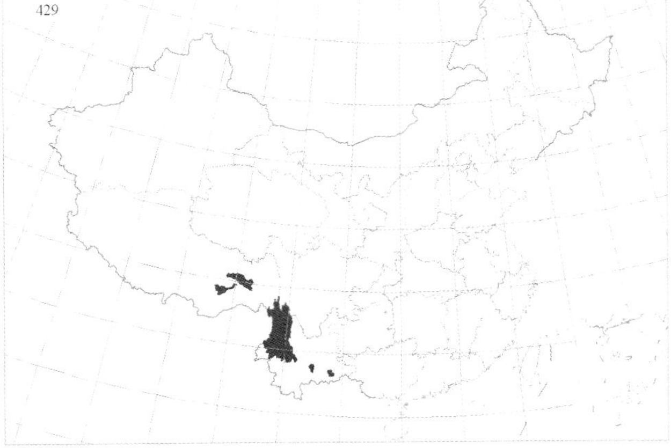

Salicaceae 杨柳科 *Salix* 柳属
Salix cupularis Rehder 杯腺柳
Shrubs, deciduous

8.4 (-0.5~14.9); -4.6 (-13.2~5.3); 20.4 (9.5~25.1); 9.5 (3.2~15.0); 72.3 (12.3~119.6); 31.3 (0.0~80.0); 616 (377~782) ❙ 638 (84~1410); 309 (53~816); 20 (1~83) ❙ 4.3 (-87.1~101.4); 509 (84~782); 218.3 (7.5~334.4)

430

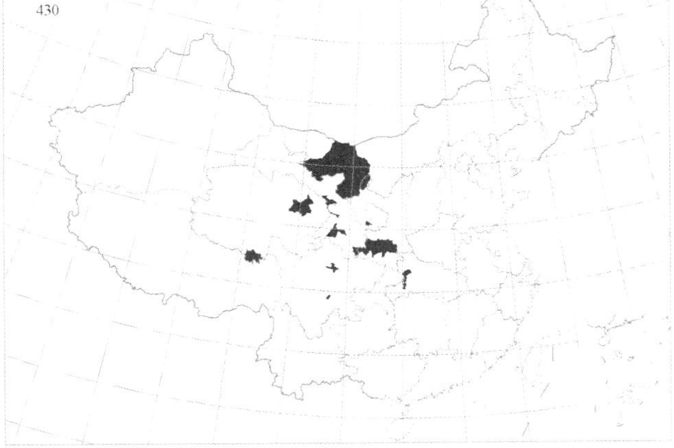

Salicaceae 杨柳科 *Salix* 柳属
Salix cupularis var. acutifolia S. Q. Zhou 尖叶杯腺柳
Shrubs, deciduous

7.7 (3.2~8.8); -9.1 (-11.4~-7.1); 22.6 (16.0~24.0); 9.6 (6.2~10.3); 77.9 (40.3~84.8); 45.0 (35.3~61.3); 637 (489~667) ❙ 208 (84~300) 124 (53~168); 4 (1~7) ❙ -66.9 (-87.1~-44.6); 208 (84~300); 59.4 (7.5~116.9)

431

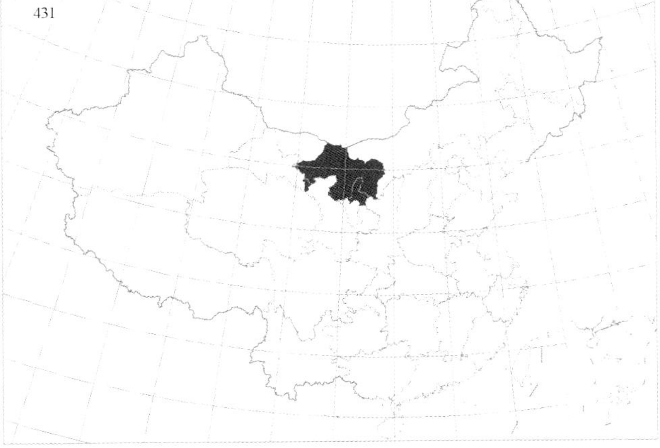

Salicaceae 杨柳科 *Salix* 柳属
Salix cyanolimnea Hance 光果乌柳
Shrubs or trees, deciduous

2.3 (-4.1~10.8); -8.4 (-17.3~3.9); 11.6 (7.8~17.3); 4.9 (2.1~10.9); 26.2 (5.2~71.8); 58.1 (1.3~114.2); 442 (340~633) ❙ 643 (361~968); 379 (219~594); 13 (5~24) ❙ 46.5 (-29.0~125.9); 422 (340~633); 259.3 (155.3~357.7)

432

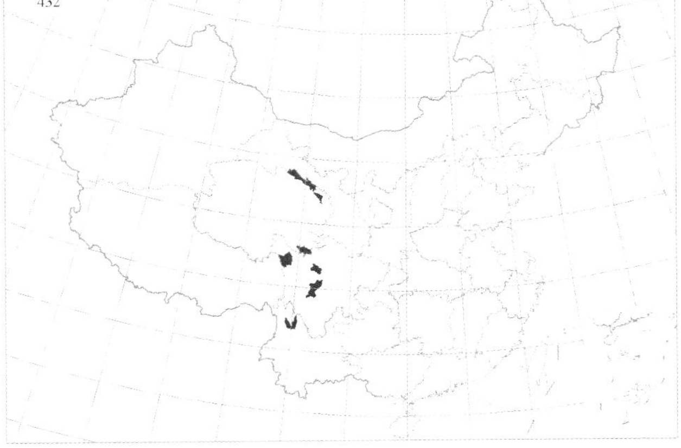

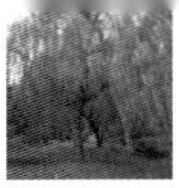

Salicaceae 杨柳科 *Salix* 柳属
Salix daguanensis P. Y. Mao et P. X. He 大关柳
Shrubs, deciduous

12.8；3.5；21.1；12.9；95.9；1.5；699 ❙ 882；476；35 ❙ 26.1；699；253.1

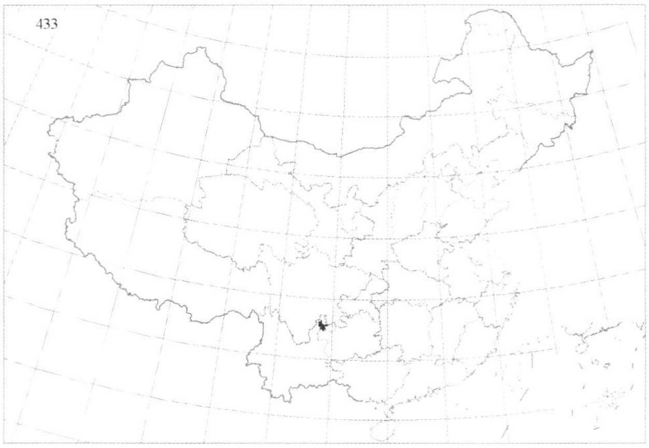

Salicaceae 杨柳科 *Salix* 柳属
Salix daliensis C. F. Fang et S. D. Zhao 大理柳
Shrubs, deciduous

14.9 (−0.7~21.0)；7.4 (−9.6~15.1)；20.8 (7.7~27.8)；15.0 (2.5~21.0)；121.6 (6.8~191.8)；2.0 (0.0~74.2)；767 (338~1053) ❙ 1095 (370~1751)；608 (262~996)；45 (10~91) ❙ 43.6 (−14.2~113.0)；753 (338~1053)；303.9 (61.6~659.0)

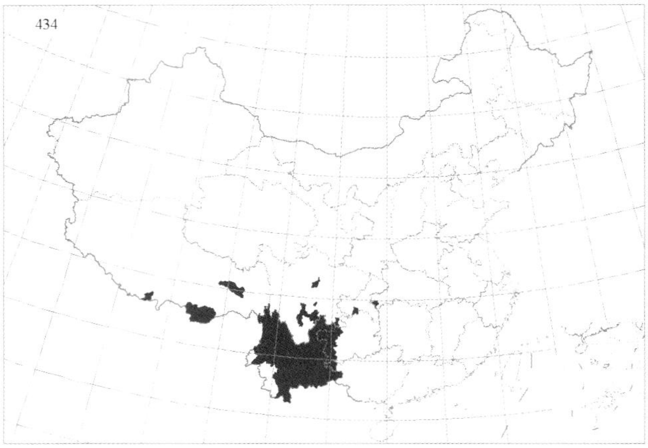

Salicaceae 杨柳科 *Salix* 柳属
Salix daltoniana Andersson 褐背柳
Shrubs or trees, deciduous

8.2 (−1.8~16.7)；0.3 (−11.7~9.7)；15.0 (6.1~24.0)；9.1 (1.8~16.7)；63.5 (1.9~140.9)；24.1 (0.0~82.8)；571 (329~833) ❙ 861 (258~1454)；495 (191~817)；32 (2~70) ❙ 44.6 (−24.7~113.0)；563 (258~833)；274.9 (60.4~487.2)

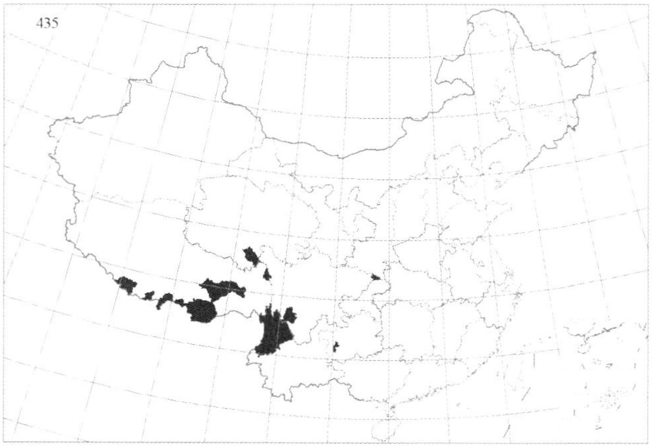

Salicaceae 杨柳科 *Salix* 柳属
Salix dalungensis C. Wang et P. Y. Fu 节枝柳
Trees, deciduous

−0.3；−9.9；8.5；2.9；10.0；72.4；358 ❙ 276；203；2 ❙ −22.9；276；161.2

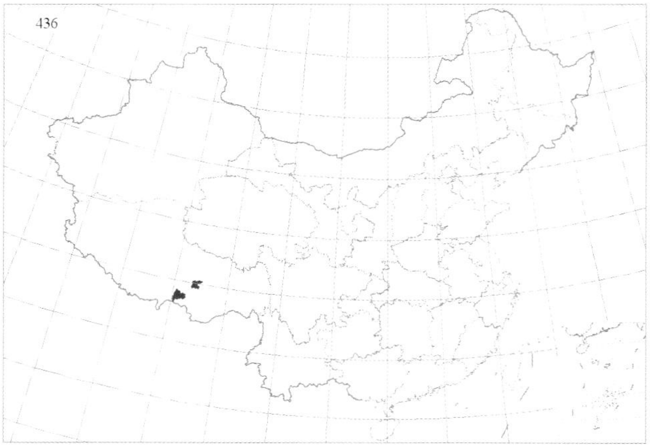

Salicaceae 杨柳科 *Salix* 柳属
Salix dasyclados Wimm. 毛枝柳
Shrubs or trees, deciduous

5.5 (−5.8~14.5)；−13.4 (−29.8~0.1)；21.4 (9.6~28.0)；9.0 (2.5~14.6)；72.0 (9.7~127.2)；66.1 (10.7~157.4)；624 (329~880) ❙ 420 (45~1018)；238 (24~648)；20 (3~51) ❙ −31.7 (−94.0~68.8)；390 (45~679)；162.0 (7.6~346.5)

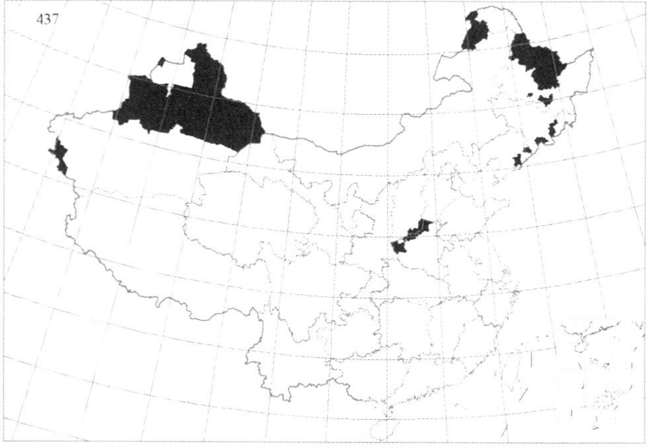

Salicaceae 杨柳科 *Salix* 柳属
Salix delavayana Hand.-Mazz. 腹毛柳
Shrubs or trees, deciduous

11.7 (−2.5~18.9)；3.4 (−12.7~12.0)；18.9 (7.0~27.7)；12.2 (2.1~19.0)；93.1 (4.0~167.8)；11.8 (0.0~93.0)；678 (332~969) ❙ 965 (382~1492)；525 (254~816)；38 (6~107) ❙ 44.7 (−14.2~128.6)；664 (332~969)；263.3 (130.1~491.5)

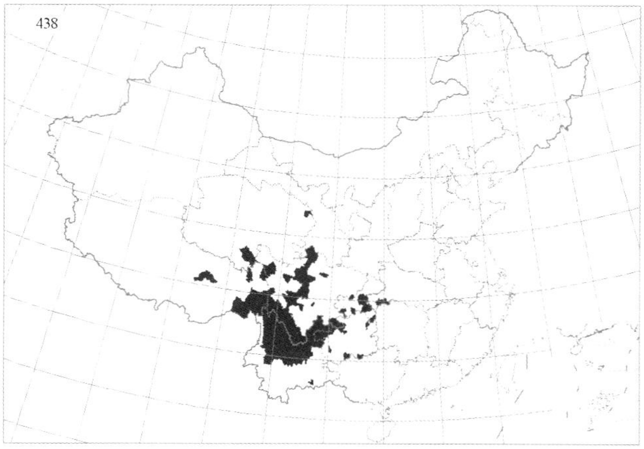

Salicaceae 杨柳科 *Salix* 柳属

Salix delavayana var. pilososuturalis Y. L. Chou et C. F. Fang 毛缝腹毛柳

Shrubs or trees, deciduous

2.2; -7.1; 10.7; 4.2; 20.0; 53.4; 416 ❙ 654; 390; 18 ❙ 54.4; 4[illegible]6; 254.0

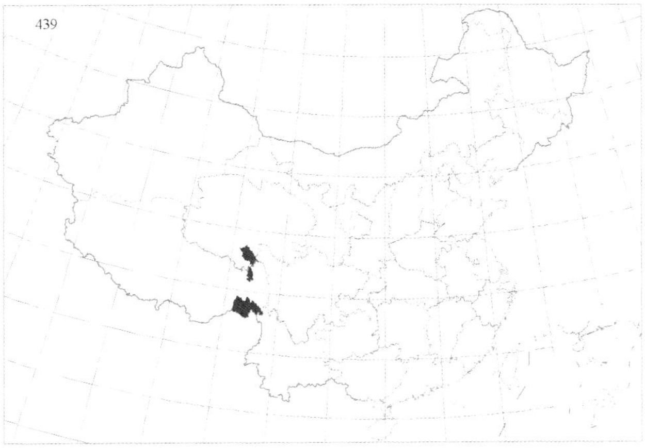

Salicaceae 杨柳科 *Salix* 柳属

Salix denticulata Andersson 齿叶柳

Shrubs, deciduous

5.0 (-4.1~14.1); -4.7 (-17.3~6.5); 13.5 (7.8~20.2); 6.9 (2.1~14.2); 43.6 (5.2~109.8); 42.8 (0.0~114.2); 506 (340~729) ❙ 771 (361~1431); 444 (235~817); 17 (5~40) ❙ 46.6 (0.0~113.0); 503 (340~707); 300.6 (208.7~363.7)

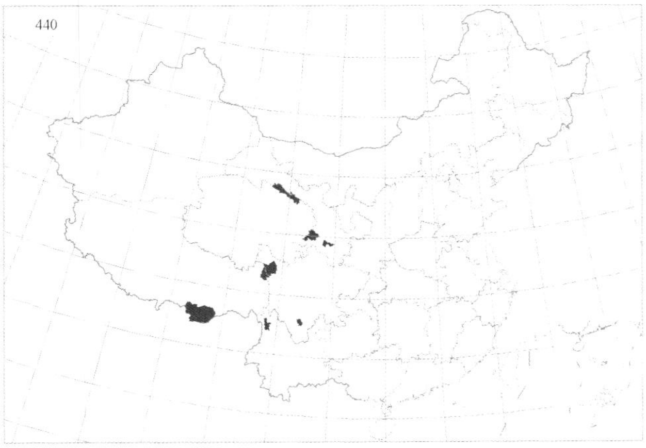

Salicaceae 杨柳科 *Salix* 柳属

Salix dibapha C. K. Schneid. 异色柳

Shrubs, deciduous

6.6 (-4.1~17.5); -2.4 (-17.3~10.5); 14.3 (7.8~22.3); 7.9 (2.1~17.5); 51.9 (5.2~149.9); 31.8 (0.0~114.2); 538 (340~846) ❙ 809 (361~1454); 481 (203~792); 21 (5~70) ❙ 50.9 (-32.2~121.7); 525 (340~823); 268.5 (95.7~487.2)

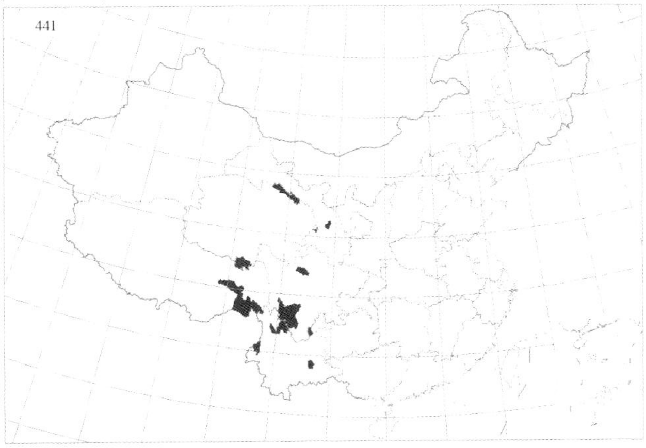

Salicaceae 杨柳科 *Salix* 柳属

Salix dissa C. K. Schneid. 异型柳

Shrubs, deciduous

10.2 (0.3~18.5); 1.4 (-10.6~11.9); 17.8 (9.3~26.3); 10.7 (3.3~18.3); 79.5 (11.9~160.0); 16.4 (0.0~67.5); 634 (391~906) ❙ 918 (471~1410); 510 (247~816); 30 (9~71) ❙ 48.3 (-16.6~128.6); 622 (391~854); 273.5 (101.3~516.7)

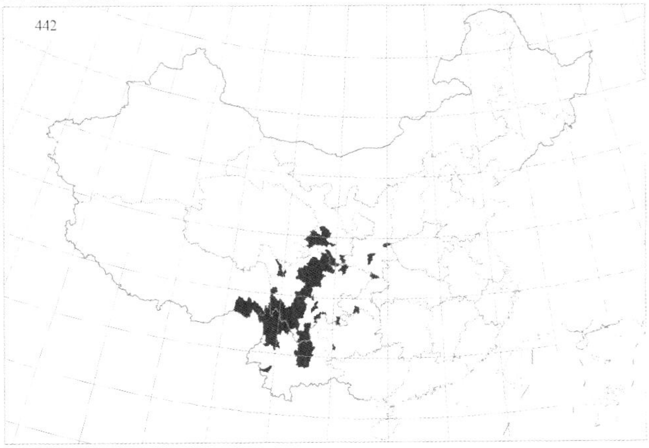

Salicaceae 杨柳科 *Salix* 柳属

Salix dissa var. cereifolia (Goerz ex Rehder et Kobuski) C F. Fang 单腺异型柳

Shrubs, deciduous

4.8 (1.9~6.4); -7.2 (-9.2~-6.1); 15.4 (11.7~17.7); 6.4 (4.3~7.7); 41.2 (21.1~55.1); 42.9 (38.1~57.8); 498 (427~547) ❙ 551 (476~629); 281 (249~318); 11 (9~14) ❙ 11.6 (-11.1~45.4); 481 (427~517); 225.6 (137.5~289.4)

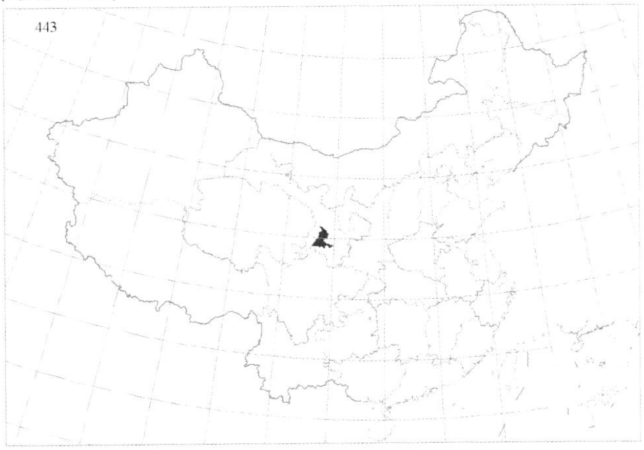

Salicaceae 杨柳科 *Salix* 柳属

Salix divaricata var. metaformosa (Nakai) Kitag. 长圆叶柳

Shrubs, deciduous

1.2 (-5.0~4.5); -19.0 (-29.4~-14.1); 19.1 (17.0~21.1); 6.8 (5.0~8.2); 50.0 (35.2~63.5); 95.4 (68.3~154.8); 521 (450~581) ❙ 535 (457~761); 358 (300~471); 19 (9~30) ❙ 11.9 (-12.6~45.2); 513 (450~566); 300.4 (247.3~322.3)

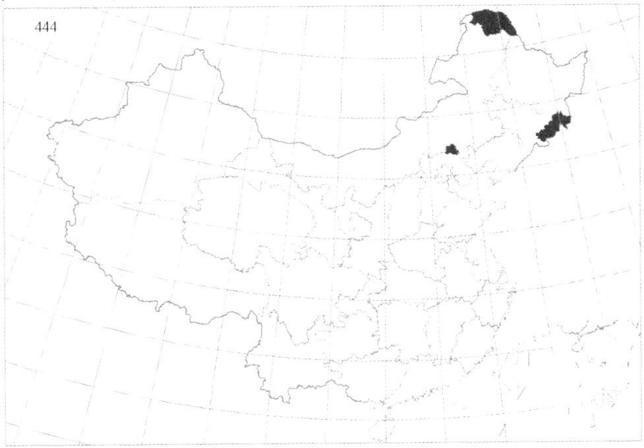

Salicaceae 杨柳科 *Salix* 柳属
Salix divergentistyla C. F. Fang 叉柱柳
Shrubs or trees, deciduous

1.6；-7.5；9.7；3.8；16.8；57.5；400 ❙ 561；356；12 ❙ 33.5；387；213.8

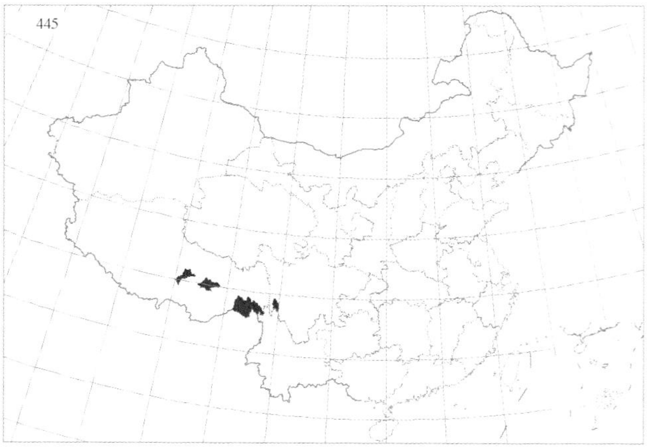

Salicaceae 杨柳科 *Salix* 柳属
Salix doii Hayata 台湾柳
Shrubs, deciduous

20.8 (16.2~23.3)；15.3 (11.0~17.7)；25.2 (20.1~28.0)；20.8 (16.2~23.3)；190.0 (134.2~219.7)；0.0 (0.0~0.0)；1060 (773~1248) ❙ 2187 (1498~2759)；1227 (859~1594)；125 (73~245) ❙ 119.7 (25.2~246.5)；1012 (773~1144)；223.5 (179.2~289.1)

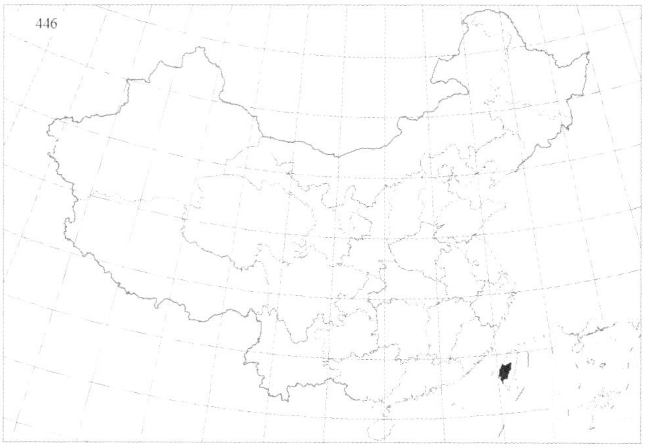

Salicaceae 杨柳科 *Salix* 柳属
Salix donggouxianica C. F. Fang 东沟柳
Shrubs, deciduous

8.7；-8.3；23.3；10.3；84.9；40.2；666 ❙ 877；537；35 ❙ 31.7；666；246.2

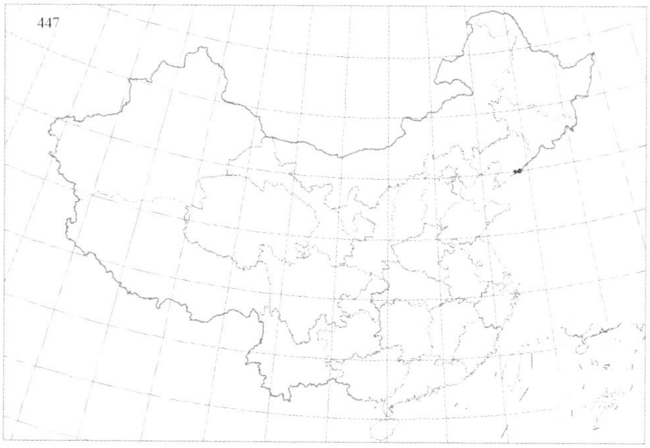

Salicaceae 杨柳科 *Salix* 柳属
Salix driophila C. K. Schneid. 林柳
Shrubs, deciduous

9.2 (1.8~16.2)；0.3 (-7.1~7.2)；17.2 (9.5~26.3)；9.7 (3.8~16.2)；67.8 (15.5~134.6)；16.6 (0.0~53.3)；605 (404~854) ❙ 969 (535~1398)；530 (326~812)；30 (10~71) ❙ 62.1 (27.1~125.9)；605 (404~854)；254.5 (169.5~372.3)

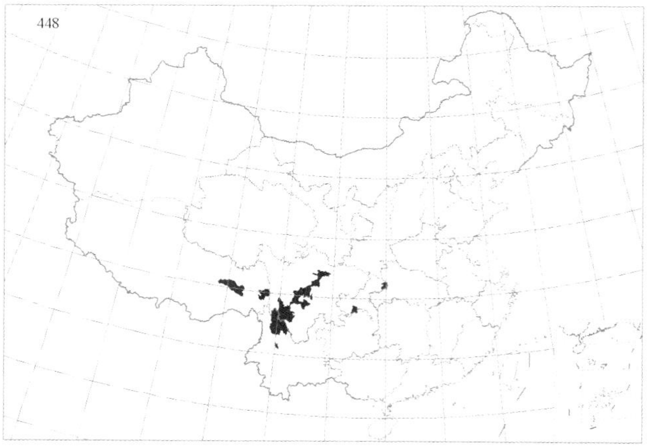

Salicaceae 杨柳科 *Salix* 柳属
Salix dunnii C. K. Schneid. 长梗柳
Shrubs or trees, deciduous

17.4 (2.2~22.3)；7.2 (-6.3~14.5)；26.8 (9.5~29.9)；17.5 (3.9~22.4)；149.8 (15.8~208.6)；0.4 (0.0~48.6)；917 (411~1193) ❙ 1495 (729~1997)；549 (299~809)；169 (11~259) ❙ 63.9 (-6.7~147.0)；906 (411~1193)；231.9 (123.9~332.5)

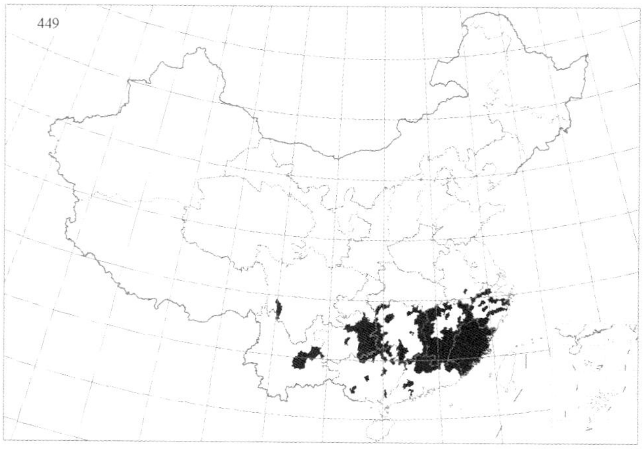

Salicaceae 杨柳科 *Salix* 柳属
Salix eriocarpa Franch. et Sav. 长柱柳
Shrubs or trees, deciduous

5.3 (0.1~10.5)；-14.4 (-20.9~-4.9)；22.2 (18.9~25.0)；8.9 (6.4~11.3)；71.1 (47.4~93.2)；66.5 (26.8~106.1)；613 (513~705) ❙ 759 (470~1073)；479 (328~691)；27 (11~40) ❙ 24.1 (-11.7~74.1)；604 (470~698)；271.9 (177.1~346.5)

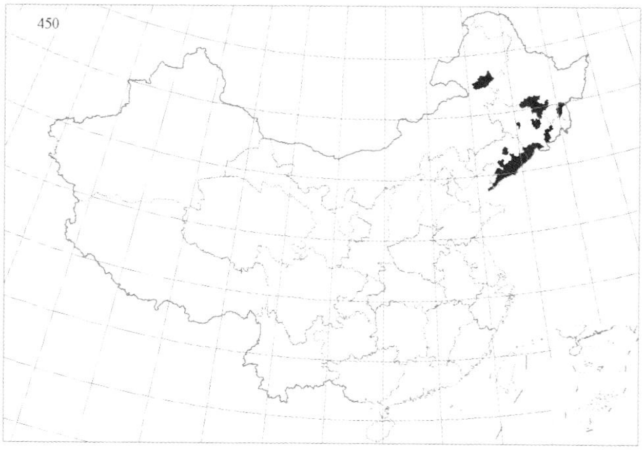

Salicaceae 杨柳科 *Salix* 柳属
Salix ericclada H. Lév. et Vaniot 绵毛柳
Shrubs or trees, deciduous

12.8 (2.2~17.2); 3.9 (-6.6~9.8); 20.7 (9.9~26.9); 13.0 (3.8~17.3); 100.3 (15.5~147.5); 6.2 (0.0~49.1); 712 (413~866) ❙ 1065 (636~1478); 552 (308~816); 46 (12~153) ❙ 52.2 (-0.3~125.9); 701 (413~866); 272.6 (148.9~382.6)

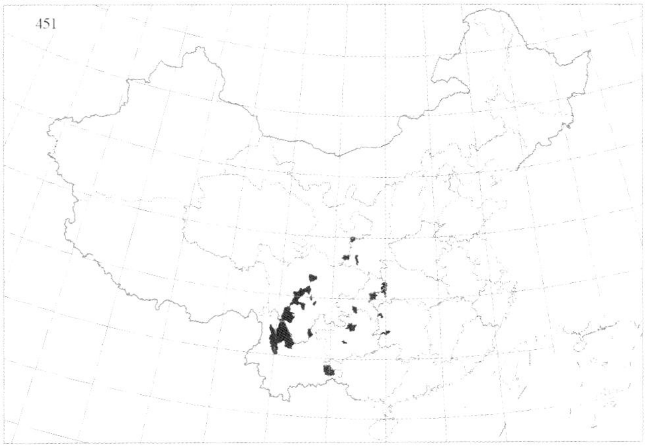

Salicaceae 杨柳科 *Salix* 柳属
Salix eriostachya Wall. ex Andersson 绵穗柳
Shrubs, deciduous

4.5 (2.0~14.8); -4.6 (-7.4~3.8); 12.6 (9.5~25.3); 5.7 (3.8~14.8); 32.1 (15.4~119.2); 37.5 (1.4~51.9); 475 (411~795) ❙ 917 (729~1358); 498 (400~601); 25 (11~88) ❙ 95.0 (67.7~128.6); 475 (411~795); 256.9 (187.9~311.5)

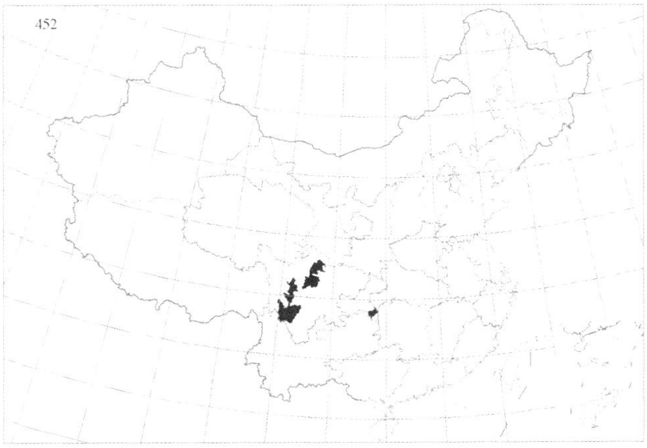

Salicaceae 杨柳科 *Salix* 柳属
Salix ernestii C. K. Schneid. 银背柳
Shrubs, deciduous

7.0 (-1.3~18.5); -2.1 (-13.5~10.5); 15.1 (8.1~25.6); 8.0 (2.7~18.3); 54.2 (8.0~160.0); 29.3 (0.0~86.3); 547 (358~906) ❙ 830 (320~1478); 462 (203~651); 24 (3~85) ❙ 54.3 (-32.2~128.6); 535 (320~845); 272.6 (95.7~384.1)

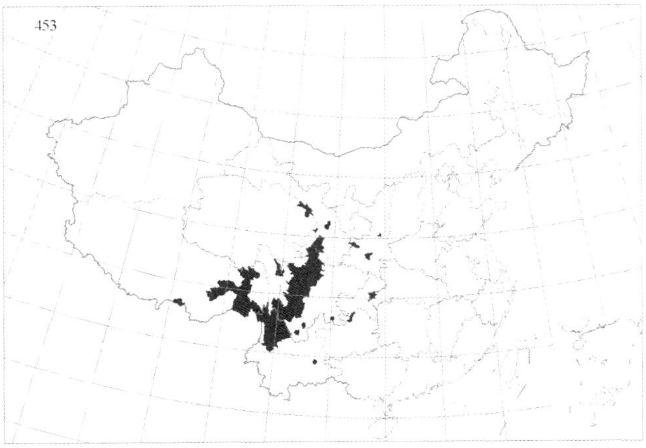

Salicaceae 杨柳科 *Salix* 柳属
Salix etosia C. K. Schneid. 巴柳
Shrubs or trees, deciduous

14.0 (2.0~18.9); 4.3 (-7.4~12.0); 22.9 (10.1~28.9); 14.2 (3.8~19.0); 113.5 (15.4~167.8); 4.6 (0.0~51.9); 773 (413~1002) ❙ 1130 (525~1526); 531 (270~816); 58 (10~181) ❙ 49.7 (-14.2~128.6); 765 (413~1002); 225.7 (108.2~366.3)

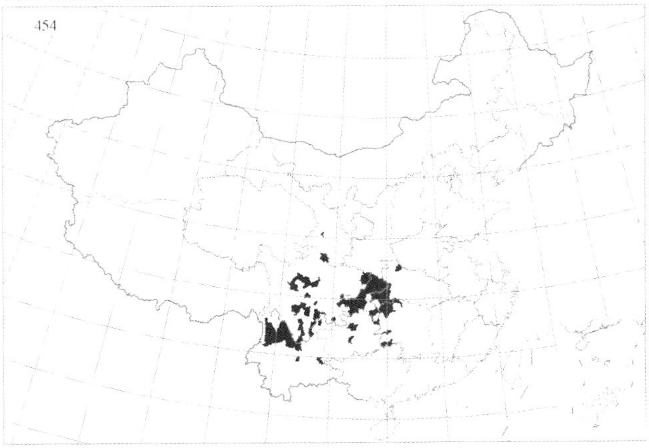

Salicaceae 杨柳科 *Salix* 柳属
Salix fargesii Burkill 川鄂柳
Shrubs or trees, deciduous

12.0 (2.9~16.9); 0.7 (-7.7~6.3); 22.8 (11.2~27.4); 12.2 (4.4~17.0); 95.8 (20.5~143.4); 10.8 (0.0~49.3); 709 (428~907) ❙ 1010 (563~1492); 452 (281~812); 45 (13~123) ❙ 42.8 (-25.4~121.7); 705 (428~907); 248.3 (124.7~334.4)

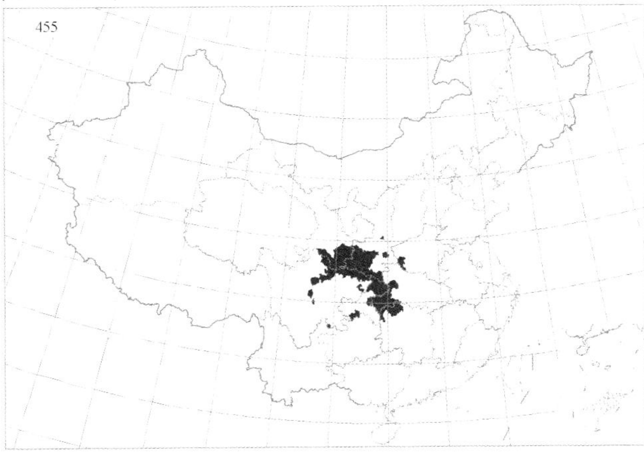

Salicaceae 杨柳科 *Salix* 柳属
Salix fargesii var. kansuensis (K. S. Hao ex C. F. Fang et A. K. Skvortsov) G. H. Zhu 甘肃柳
Shrubs or trees, deciduous

11.7 (2.9~15.8); 0.8 (-7.7~4.4); 22.0 (11.2~27.1); 12.0 (4.4~15.9); 92.9 (20.5~131.5); 11.6 (0.6~49.3); 700 (428~859) ❙ 1049 (620~1451); 466 (301~600); 55 (13~123) ❙ 49.3 (-0.2~121.7); 700 (428~859); 234.3 (182.7~284.3)

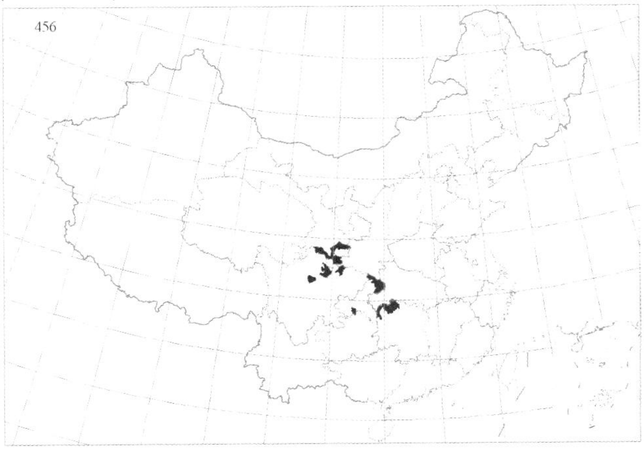

Salicaceae 杨柳科 *Salix* 柳属
Salix faxonianoides C. Wang et P. Y. Fu 藏匐柳
Shrubs, deciduous

6.1 (−1.8~13.3); −1.7 (−10.2~5.4); 12.9 (6.1~19.6); 7.0 (1.8~13.4); 42.1 (1.9~100.4); 28.8 (0.0~82.8); 510 (329~706) ❙ 868 (320~1736); 510 (193~1007); 30 (13~51) ❙ 62.8 (−4.0~145.8); 508 (320~706); 253.7 (72.2~447.4)

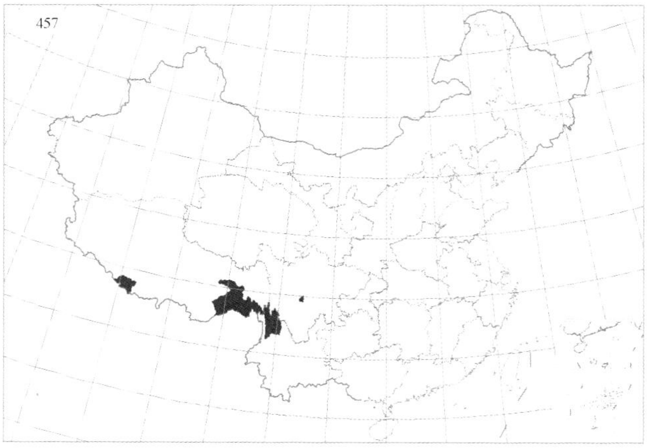

Salicaceae 杨柳科 *Salix* 柳属
Salix fedtschenkoi Goerz 山羊柳
Shrubs, deciduous

1.2 (−4.3~5.9); −16.3 (−20.1~−13.8); 16.6 (9.6~23.0); 6.0 (2.5~9.4); 42.7 (9.7~77.9); 88.2 (66.6~121.1); 495 (329~655) ❙ 190 (71~315); 77 (24~116); 23 (10~51) ❙ −62.3 (−78.5~−46.4); 190 (71~312); 75.0 (15.7~173.1)

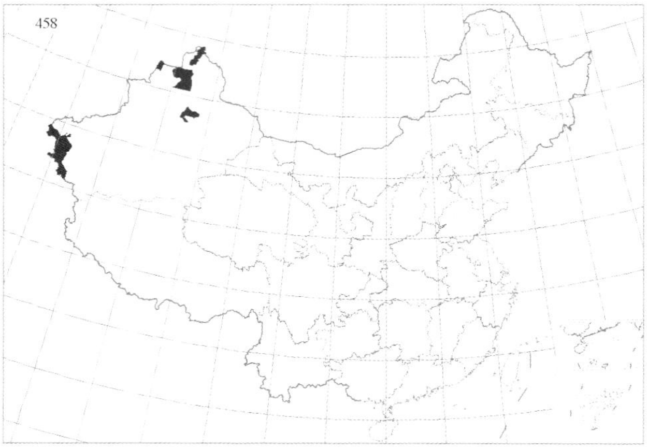

Salicaceae 杨柳科 *Salix* 柳属
Salix fengiana C. F. Fang et C. Y. Yang 贡山柳
Shrubs, deciduous

5.2 (0.0~11.0); −2.9 (−10.1~4.2); 12.4 (8.1~16.6); 6.4 (2.7~11.0); 36.3 (8.0~73.1); 33.3 (0.8~71.1); 491 (363~636) ❙ 800 (533~1134); 465 (328~651); 28 (12~47) ❙ 60.2 (41.1~81.5); 491 (363~636); 267.0 (111.0~352.3)

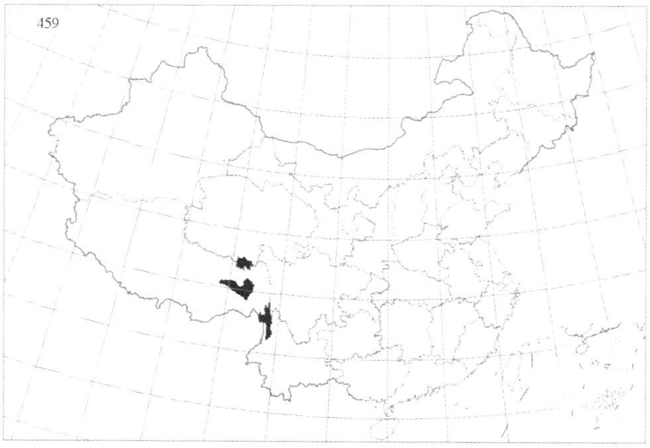

Salicaceae 杨柳科 *Salix* 柳属
Salix fengiana var. gymnocarpa Mao et W. Z. Li 裸果贡山柳
Shrubs, deciduous

11.1; 4.0; 16.8; 11.1; 76.2; 3.2; 638 ❙ 1207; 669; 57 ❙ 89.0; 638; 381.4

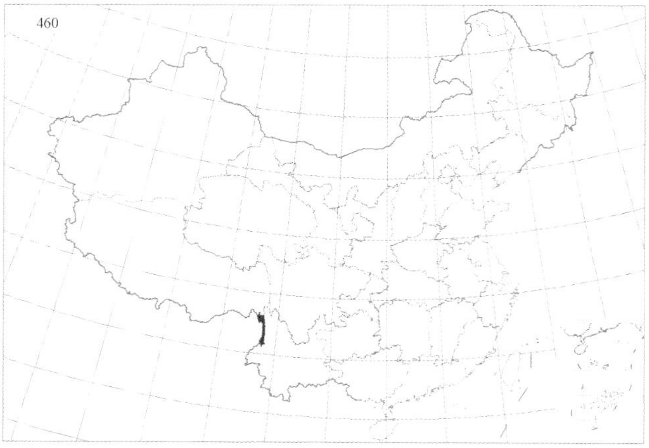

Salicaceae 杨柳科 *Salix* 柳属
Salix flabellaris Andersson 扇叶垫柳
Shrubs, deciduous

7.9 (−2.2~16.2); 0.3 (−10.2~9.4); 14.4 (5.4~20.8); 8.8 (1.6~16.3); 59.8 (0.4~135.1); 24.1 (0.0~86.4); 561 (327~790) ❙ 893 (320~1348); 507 (193~762); 37 (7~84) ❙ 56.0 (−4.0~121.7); 559 (320~790); 280.9 (61.6~407.1)

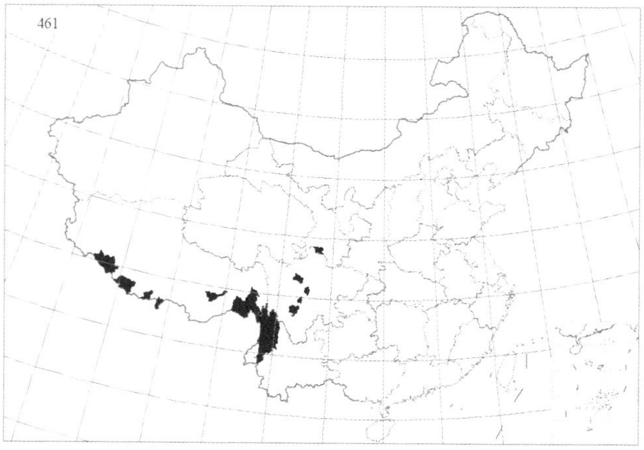

Salicaceae 杨柳科 *Salix* 柳属
Salix floccosa Burkill 丛毛矮柳
Shrubs, deciduous

8.9 (1.2~14.5); 1.6 (−7.1~7.8); 15.2 (9.0~19.7); 9.3 (3.3~14.5); 62.7 (11.9~113.5); 15.1 (0.0~56.5); 582 (390~730) ❙ 951 (473~1736); 553 (287~1007); 33 (7~54) ❙ 61.7 (5.0~145.8); 581 (390~730); 305.3 (130.1~447.4)

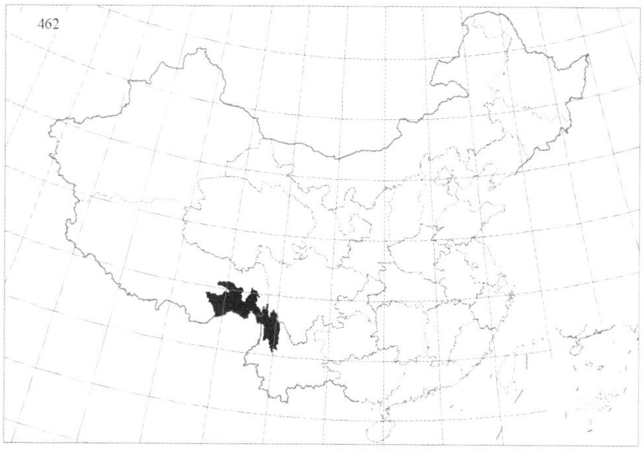

Salicaceae 杨柳科 *Salix* 柳属
Salix floderusii Nakai 崖柳
Shrubs, deciduous

5.8 (-5.8~13.5); -12.2 (-29.8~-1.6); 21.6 (15.3~27.1); 8.9 (4.4~13.7); 70.7 (28.0~117.5); 60.2 (14.6~157.4); 610 (426~820) ❙ 562 (311~1018); 346 (214~648); 18 (5~40) ❙ -7.0 (-41.6~68.8); 520 (311~698); 207.6 (70.3~346.5)

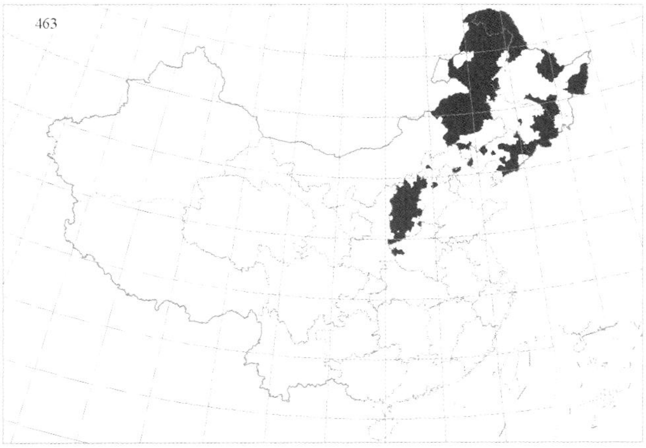

Salicaceae 杨柳科 *Salix* 柳属
Salix fulvopubescens Hayata 褐毛柳
Shrubs or small trees, deciduous

19.9 (16.2~23.3); 14.4 (11.0~17.7); 24.3 (20.1~28.0); 19.9 (16.2~23.3); 179.1 (134.2~219.7); 0.0 (0.0~0.0); 999 (773~1248) ❙ 2310 (1498~2859); 1158 (859~1594); 182 (73~402) ❙ 146.6 (25.2~246.5); 963 (773~1144); 238.8 (179.2~303.6)

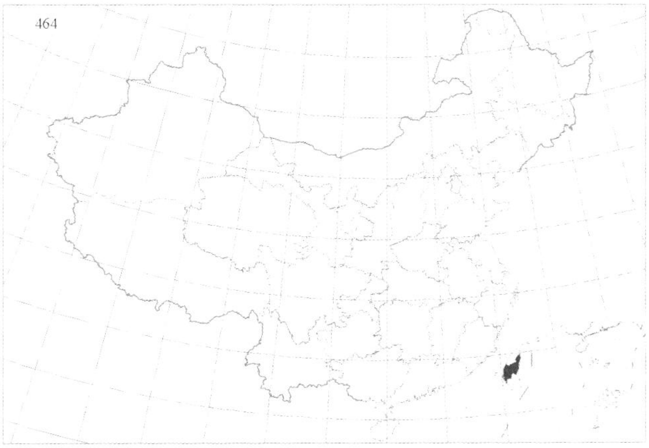

Salicaceae 杨柳科 *Salix* 柳属
Salix gilashanica C. Wang et P. Y. Fu 吉拉柳
Shrubs, deciduous

3.1 (-2.6~13.3); -6.0 (-15.3~5.4); 11.3 (8.1~19.6); 5.0 (2.5~13.4); 26.0 (7.7~100.4); 47.7 (0.0~98.0); 442 (347~706) ❙ 687 (298~1736); 409 (208~1007); 15 (1~51) ❙ 50.5 (-27.3~145.8); 433 (298~706); 254.2 (111.0~447.4)

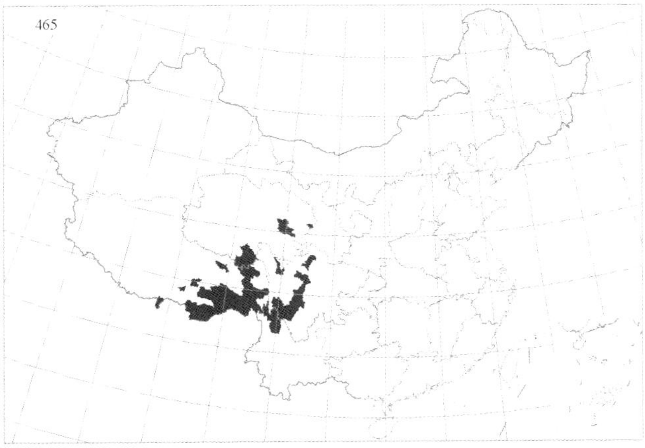

Salicaceae 杨柳科 *Salix* 柳属
Salix glareorum Mao et W. Z. Li 石流垫柳
Shrubs, deciduous

10.8; 3.9; 16.6; 10.9; 71.8; 1.3; 633 ❙ 968; 594; 24 ❙ 52.8; 633; 319.8

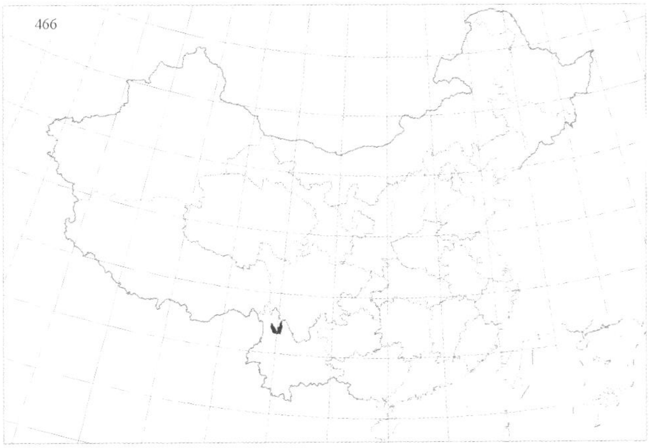

Salicaceae 杨柳科 *Salix* 柳属
Salix glauca L. 灰蓝柳
Shrubs, deciduous

1.9 (-0.6~5.3); -17.1 (-19.5~-14.9); 18.7 (15.9~23.3); 7.0 (5.4~9.7); 52.4 (37.5~80.9); 88.6 (71.2~104.2); 545 (475~668) ❙ 184 (156~251); 77 (56~106); 22 (15~33) ❙ -65.8 (-76.7~-52.0); 184 (156~251); 86.2 (16.6~139.1)

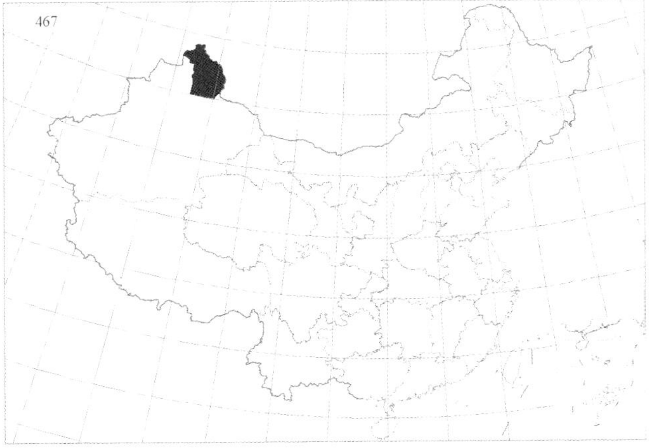

Salicaceae 杨柳科 *Salix* 柳属
Salix gordejevii Y. L. Chang et Skvortsov 黄柳
Shrubs, deciduous

3.8 (-5.8~8.9); -15.7 (-29.8~-7.5); 21.2 (12.3~25.1); 3.1 (4.1~10.8); 63.9 (22.0~92.3); 77.5 (39.0~157.4); 582 (410~706) ❙ 395 (115~718); 266 (67~446); 9 (2~29) ❙ -31.7 (-82.2~21.0); 391 (115~698); 145.2 (21.5~321.0)

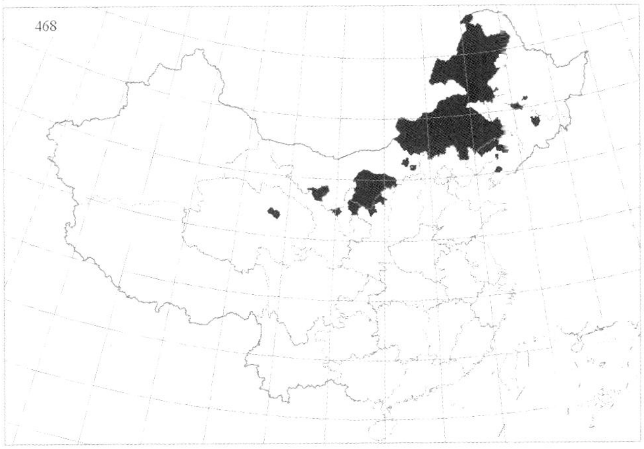

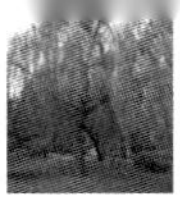

Salicaceae 杨柳科 *Salix* 柳属
Salix gracilior (Siuzew) Nakai 细枝柳
Shrubs, deciduous

5.2 (−5.8~12.1); −14.3 (−29.8~−1.6); 22.1 (13.1~25.1); 8.9 (4.4~12.4); 71.3 (27.4~103.4); 68.5 (16.7~157.4); 615 (426~746) ❙ 574 (244~1018); 371 (139~648); 17 (5~40) ❙ −6.5 (−62.1~68.8); 519 (244~698); 204.6 (46.6~346.5)

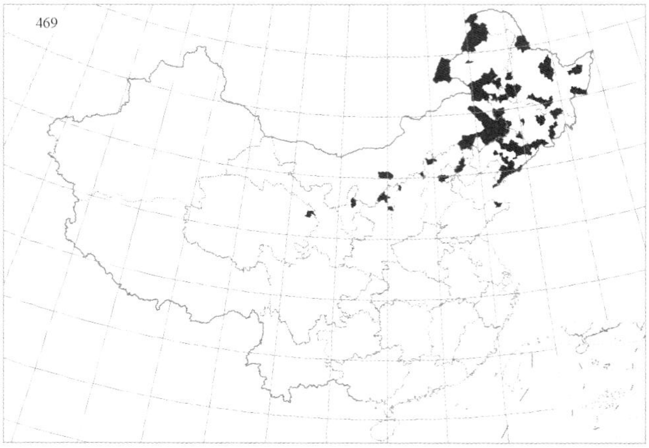

Salicaceae 杨柳科 *Salix* 柳属
Salix gracilistyla Miq. 细柱柳
Shrubs, deciduous

3.3 (−5.8~10.5); −17.8 (−29.8~−4.9); 21.3 (15.3~25.0); 8.1 (4.4~11.3); 63.0 (28.0~93.2); 83.5 (26.8~157.4); 583 (426~705) ❙ 661 (372~1073); 412 (258~691); 22 (10~40) ❙ 13.0 (−28.6~74.1); 571 (372~698); 278.3 (174.9~346.5)

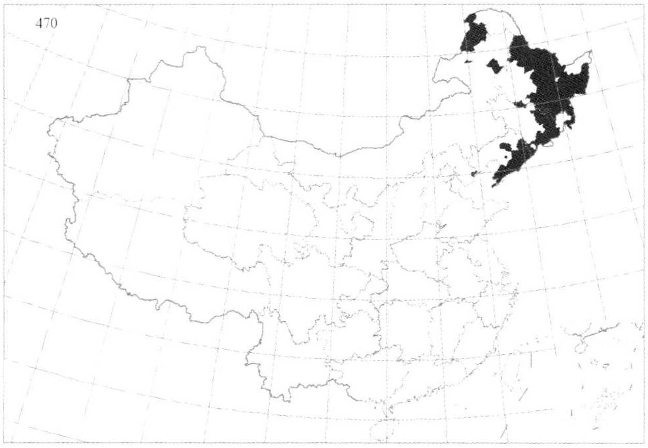

Salicaceae 杨柳科 *Salix* 柳属
Salix guebriantiana C. K. Schneid. 细序柳
Shrubs, deciduous

14.0 (1.2~19.3); 6.6 (−7.1~13.1); 19.7 (9.0~26.6); 14.1 (3.3~19.4); 112.3 (11.9~173.2); 4.1 (0.0~56.5); 734 (390~969) ❙ 998 (578~1451); 563 (345~651); 40 (11~112) ❙ 38.7 (−14.2~125.9); 713 (390~969); 283.8 (130.1~382.6)

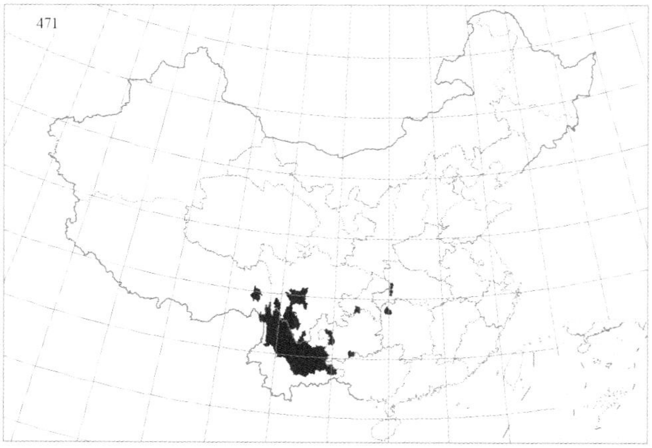

Salicaceae 杨柳科 *Salix* 柳属
Salix gyamdaensis C. F. Fang 江达柳
Shrubs, deciduous

5.1 (−0.4~16.5); −3.6 (−10.0~7.3); 12.7 (8.4~24.0); 6.3 (2.8~16.6); 37.4 (9.4~139.0); 35.6 (0.0~73.3); 489 (360~833) ❙ 772 (369~1274); 461 (243~671); 19 (5~59) ❙ 56.9 (0.0~100.7); 489 (360~833); 270.9 (205.9~357.7)

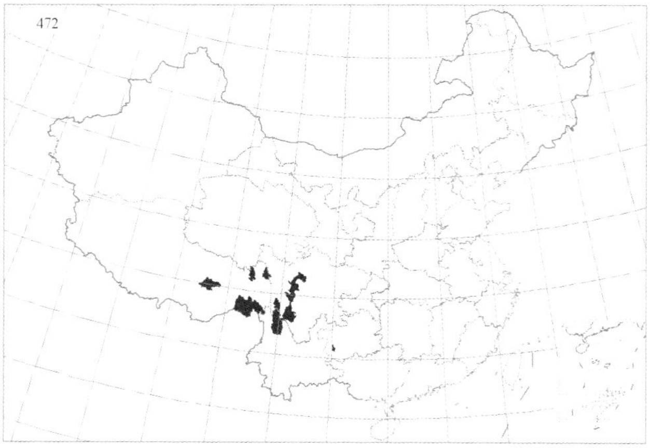

Salicaceae 杨柳科 *Salix* 柳属
Salix gyirongensis S. D. Zhao et C. F. Fang 吉隆垫柳
Shrubs, deciduous

4.2 (−1.1~10.8); −3.7 (−9.1~3.9); 11.3 (6.2~16.6); 5.5 (2.0~10.9); 29.5 (2.5~71.8); 37.9 (1.3~75.6); 461 (334~633) ❙ 699 (321~968); 423 (193~594); 21 (7~33) ❙ 48.2 (−4.0~125.9); 458 (321~633); 232.2 (79.3~319.8)

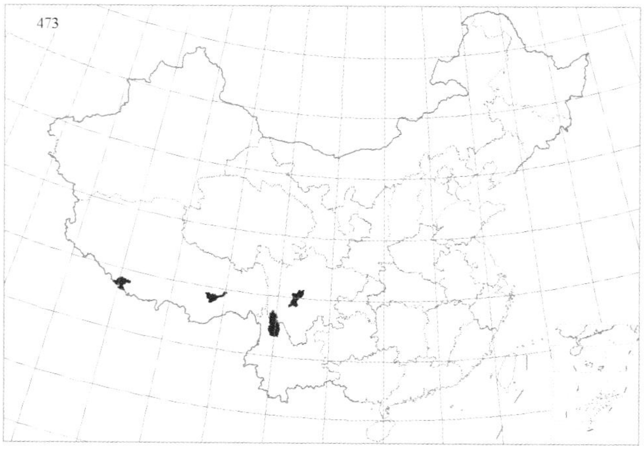

Salicaceae 杨柳科 *Salix* 柳属
Salix haoana Fang 川红柳
Shrubs, deciduous

12.7 (2.9~16.5); 1.9 (−6.3~6.0); 22.7 (11.2~28.1); 12.9 (4.4~16.5); 101.4 (20.5~138.1); 9.0 (0.0~44.6); 731 (428~891) ❙ 1162 (892~1621); 537 (389~645); 70 (25~195) ❙ 62.6 (12.1~121.7); 731 (428~891); 225.7 (133.8~311.0)

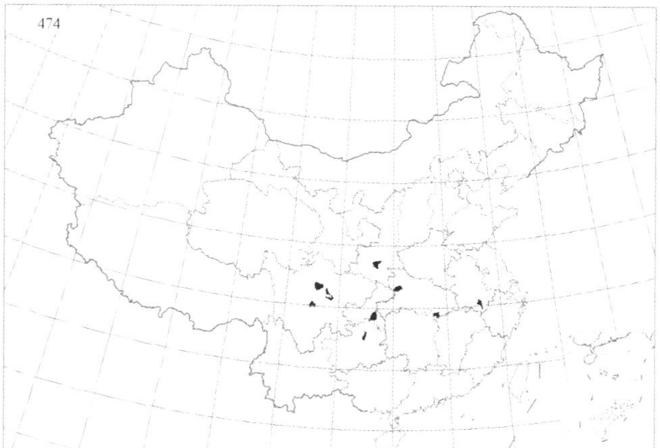

Salicaceae 杨柳科 *Salix* 柳属
Salix hastata L. 戟柳
Shrubs, deciduous

1.9 (-0.6~5.8); -17.1 (-19.5~-14.9); 18.7 (15.9~23.3); 7.0 (5.4~9.7); 52.4 (37.5~80.9); 88.6 (71.2~104.2); 545 (475~668) ❙ 184 (156~251); 77 (56~106); 22 (15~33) ❙ -65.8 (-76.7~-52.0); 184 (156~251); 36.2 (16.6~139.1)

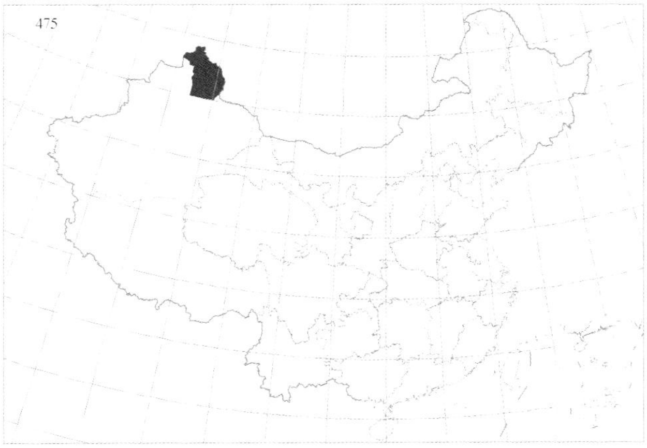

Salicaceae 杨柳科 *Salix* 柳属
Salix heishuiensis N. Chao 黑水柳
Shrubs, deciduous

8.1 (1.9~16.2); -1.6 (-8.2~6.1); 17.0 (9.9~26.3); 8.9 (3.8~16.2); 62.3 (15.4~134.6); 24.1 (0.0~53.8); 584 (413~854) ❙ 987 (590~1410); 508 (299~816); 33 (12~88) ❙ 73.9 (19.4~128.6); 584 (413~854); 247.8 (171.0~341.6)

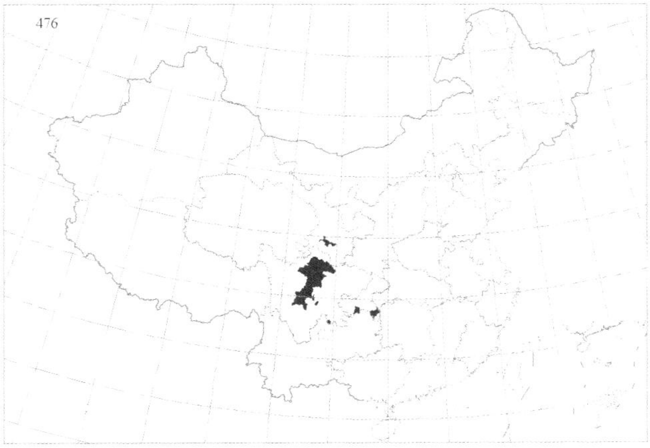

Salicaceae 杨柳科 *Salix* 柳属
Salix heterochroma Seemen 紫枝柳
Shrubs or trees, deciduous

10.4 (0.8~18.0); -1.4 (-12.7~10.4); 21.3 (9.5~28.1); 11.1 (3.8~18.0); 85.5 (15.5~156.3); 19.8 (0.0~66.0); 666 (411~958) ❙ 837 (189~1478); 420 (114~817); 32 (4~113) ❙ 25.1 (-70.2~125.9); 631 (189~958); 233.6 (33.5~344.4)

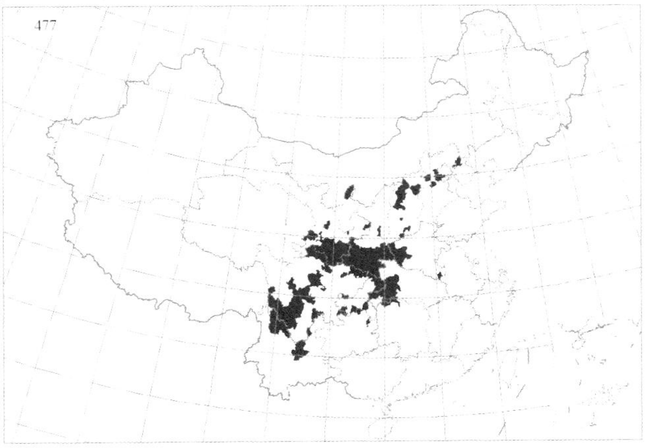

Salicaceae 杨柳科 *Salix* 柳属
Salix heteromera Hand.-Mazz. 异蕊柳
Trees, deciduous

14.8 (6.6~18.3); 8.0 (-3.2~11.8); 19.9 (13.3~22.7); 14.8 (6.6~18.3); 119.0 (39.5~159.8); 1.0 (0.0~26.7); 748 (508~882) ❙ 1037 (862~1454); 594 (521~816); 41 (12~70) ❙ 39.8 (6.5~101.2); 727 (508~834); 315.5 (200.4~487.2)

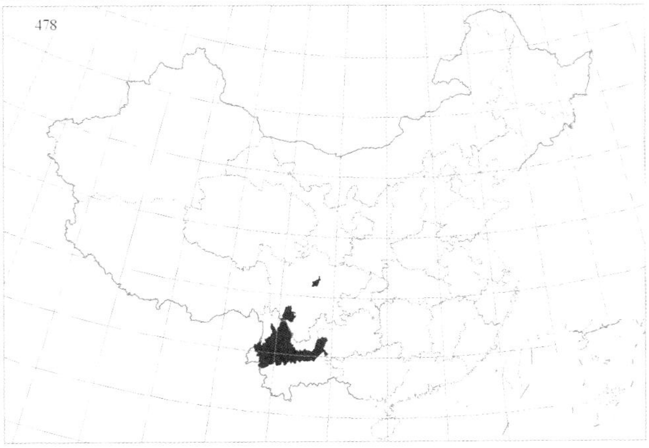

Salicaceae 杨柳科 *Salix* 柳属
Salix hirticaulis Hand.-Mazz. 毛枝垫柳
Shrubs, deciduous

12.7 (5.4~19.3); 5.7 (-4.1~13.1); 18.3 (12.3~23.6); 12.8 (5.8~19.4); 96.8 (31.5~173.2); 3.8 (0.0~29.9); 692 (481~939) ❙ 1088 (669~1428); 511 (317~835); 46 (15~69) ❙ 58.2 (21.7~93.0); 689 (481~917); 354.9 (271.9~506.2)

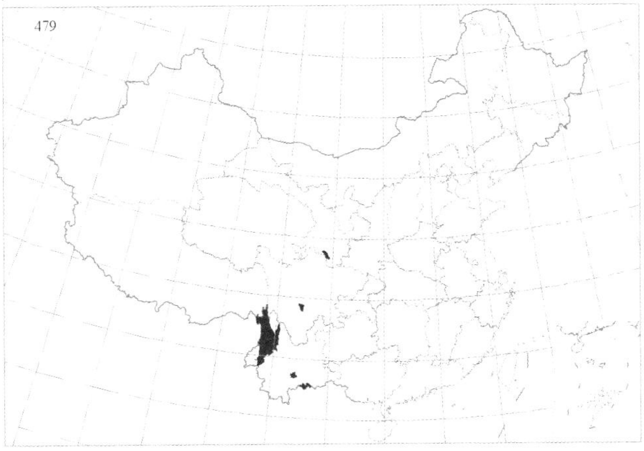

Salicaceae 杨柳科 *Salix* 柳属
Salix hsinganica Y. L. Chang et Skvortsov 兴安柳
Shrubs, deciduous

2.1 (-5.8~5.7); -19.1 (-29.8~-11.7); 20.6 (15.3~23.8); 7.5 (4.4~9.6); 57.3 (28.0~80.5); 92.1 (58.2~157.4); 559 (426~659) ❙ 464 (137~680); 301 (92~439); 13 (3~32) ❙ -16.8 (-77.5~23.8); 448 (137~611); 211.2 (25.4~333.3)

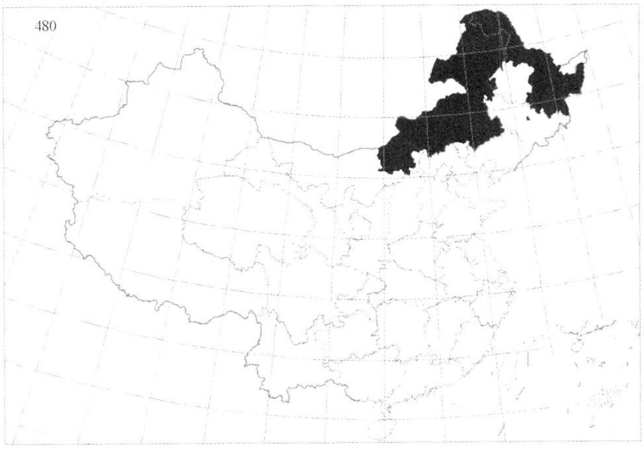

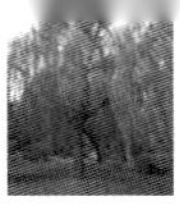

Salicaceae 杨柳科 *Salix* 柳属

Salix humaensis Y. L. Chou et R. C. Zhou 呼玛柳

Shrubs, deciduous

-4.8; -28.9; 16.5; 4.9; 33.7; 150.9; 446 ❙ 484; 317; 15 ❙ 8.6; 445; 284.3

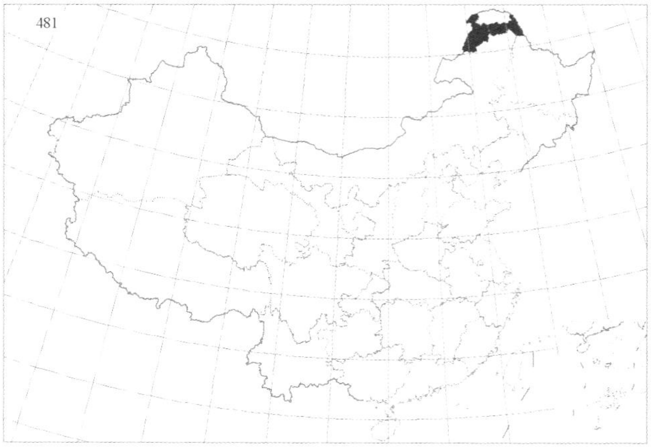

Salicaceae 杨柳科 *Salix* 柳属

Salix hupehensis K. S. Hao ex C. F. Fang et A. K. Skvortsov 湖北柳

Shrubs, deciduous

12.3 (2.2~16.2); 1.3 (-6.6~3.6); 22.7 (9.9~28.9); 12.6 (3.8~16.3); 99.1 (15.5~137.9); 10.3 (1.5~49.1); 730 (413~914) ❙ 1192 (932~1418); 539 (486~571); 82 (19~145) ❙ 68.6 (30.1~125.9); 729 (413~910); 227.8 (146.0~278.8)

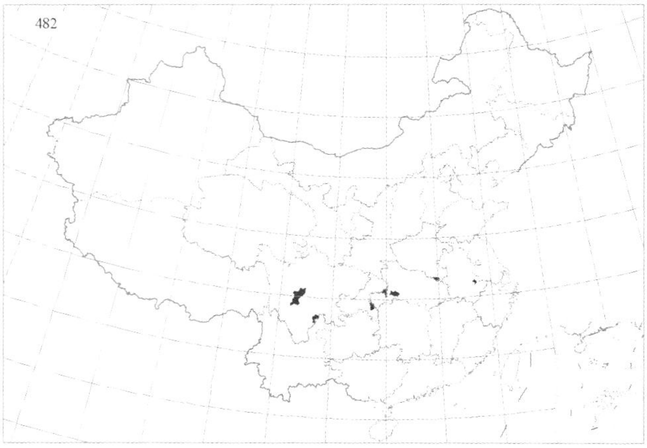

Salicaceae 杨柳科 *Salix* 柳属

Salix hylonoma C. K. Schneid. 川柳

Trees, deciduous

10.0 (-1.3~17.8); -1.8 (-13.5~9.1); 20.6 (9.4~28.5); 10.8 (2.9~17.8); 82.7 (11.3~153.7); 22.7 (0.0~86.3); 659 (359~939) ❙ 842 (373~1577); 424 (201~932); 35 (7~94) ❙ 28.4 (-37.6~101.4); 621 (359~939); 214.1 (88.1~372.3)

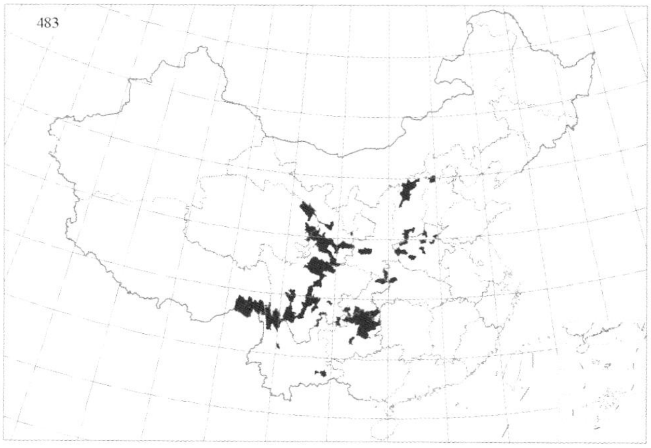

Salicaceae 杨柳科 *Salix* 柳属

Salix hypoleuca Seemen ex Diels 小叶柳

Shrubs, deciduous

11.7 (0.8~19.1); -0.2 (-12.7~10.1); 22.5 (9.5~28.5); 12.2 (3.7~19.2); 97.2 (15.1~170.1); 16.4 (0.0~66.0); 713 (409~981) ❙ 888 (341~1502); 438 (190~817); 43 (8~142) ❙ 25.1 (-40.9~128.6); 669 (341~981); 213.3 (61.2~372.3)

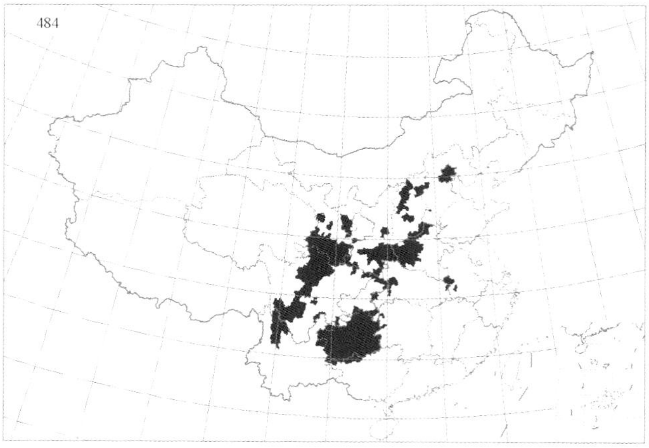

Salicaceae 杨柳科 *Salix* 柳属

Salix hypoleuca var. platyphylla C. K. Schneid. 宽叶翻白柳

Shrubs, deciduous

11.6 (9.4~13.5); -1.5 (-2.5~1.4); 23.9 (21.0~25.9); 11.8 (9.8~13.5); 95.4 (73.9~109.4); 15.7 (7.1~20.6); 711 (619~771) ❙ 734 (548~892); 337 (264~389); 27 (22~33) ❙ 4.8 (-28.9~44.0); 647 (548~767); 233.9 (153.3~311.0)

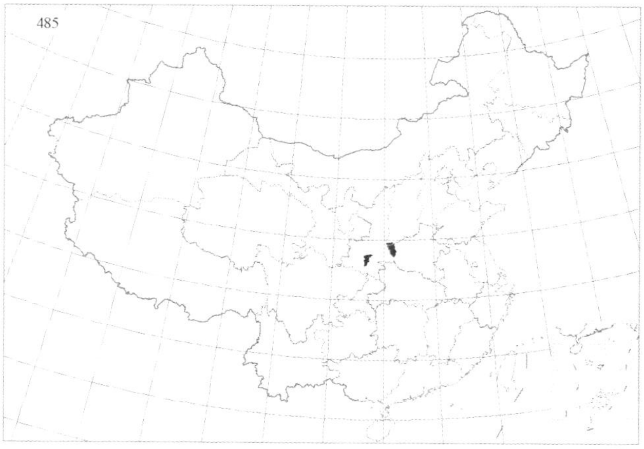

Salicaceae 杨柳科 *Salix* 柳属

Salix iliensis Regel 伊犁柳

Shrubs, deciduous

4.7 (-2.8~11.4); -14.2 (-20.9~-7.2); 20.3 (10.9~27.5); 8.4 (3.5~12.7); 66.0 (16.3~109.3); 68.7 (31.6~109.2); 603 (390~813) ❙ 183 (32~450); 86 (21~211); 16 (0~48) ❙ -66.7 (-95.9~3.9); 183 (32~406); 77.5 (1.6~297.3)

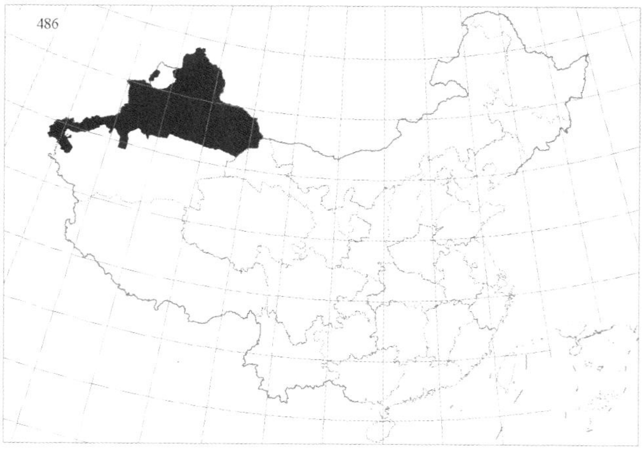

Salicaceae 杨柳科 *Salix* 柳属

Salix inamoena Hand.-Mazz. 丑柳

Shrubs, deciduous

15.3 (10.8~18.9); 8.0 (1.6~12.5); 20.8 (16.6~26.3); 15.3 (10.9~19.0); 124.3 (71.8~167.8); 0.3 (0.0~6.9); 770 (633~929) ❙ 1028 (797~1496); 576 (411~816); 43 (23~142) ❙ 34.1 (-14.2~106.2); 749 (633~909); 284.1 (170.4~487.2)

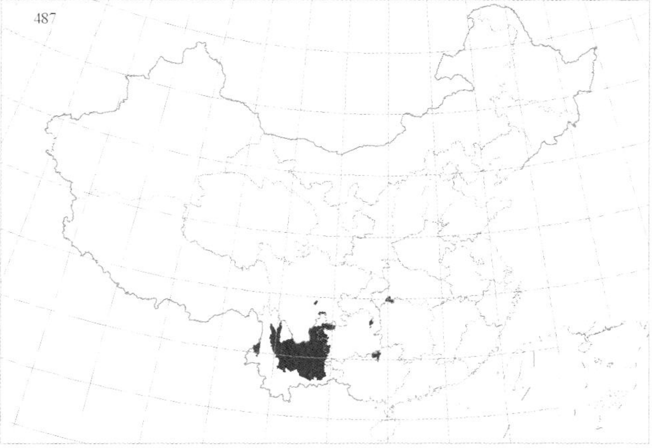

Salicaceae 杨柳科 *Salix* 柳属

Salix insignis Andersson 藏西柳

Shrubs or trees, deciduous

-0.9; -11.4; 9.1; 3.0; 12.3; 83.3; 356 ❙ 310; 190; 27 ❙ -14.0; 266; 96.3

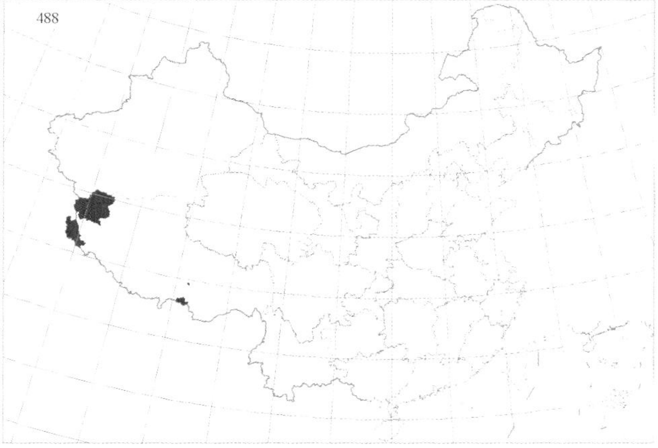

Salicaceae 杨柳科 *Salix* 柳属

Salix integra Thunb. 杞柳

Shrubs, deciduous

6.7 (-5.0~15.7); -11.9 (-29.4~2.9); 23.0 (17.0~28.6); 9.8 (5.0~15.8); 79.6 (35.2~131.4); 58.1 (2.3~154.8); 652 (450~889) ❙ 643 (349~1420); 402 (227~691); 24 (6~162) ❙ -0.6 (-41.6~74.1); 591 (349~875); 225.1 (97.2~346.5)

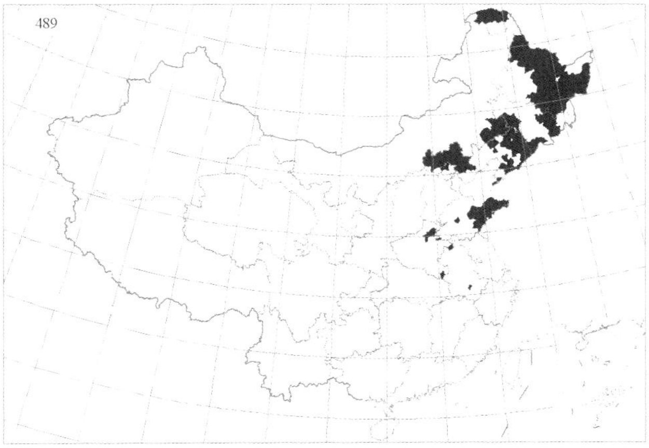

Salicaceae 杨柳科 *Salix* 柳属

Salix jinchuanica N. Chao 金川柳

Shrubs, deciduous

4.1 (0.6~15.3); -5.6 (-9.4~5.2); 12.7 (9.5~24.7); 5.8 (3.4~15.3); 34.3 (12.7~124.2); 44.6 (0.0~65.4); 478 (394~801) ❙ 846 (581~1048); 455 (347~546); 26 (12~64) ❙ 83.9 (30.8~128.6); 478 (394~801); 279.0 (184.8~341.6)

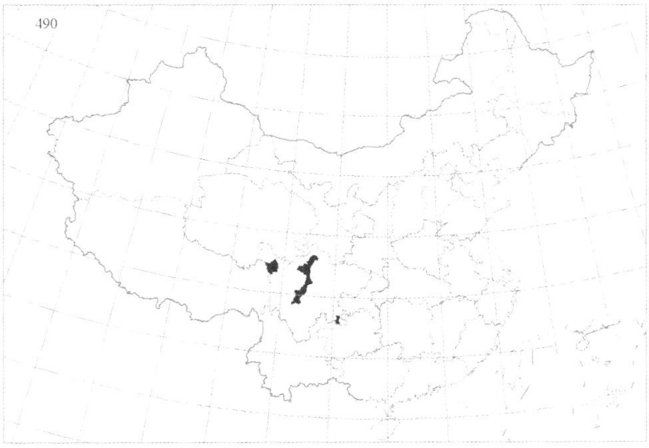

Salicaceae 杨柳科 *Salix* 柳属

Salix jingdongensis C. F. Fang 景东矮柳

Shrubs, deciduous

17.3; 11.0; 21.6; 17.3; 148.0; 0.0; 832 ❙ 1049; 598; 44 ❙ 26.1; 783; 365.3

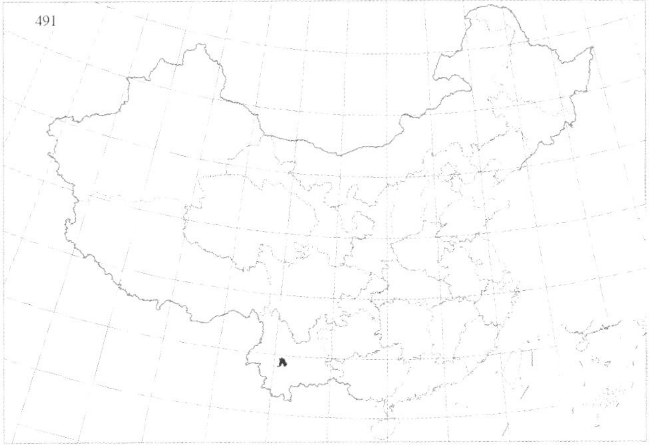

Salicaceae 杨柳科 *Salix* 柳属

Salix jishiensis C. F. Fang et J. Q. Wang 积石柳

Trees, deciduous

5.1 (0.8~9.4); -7.4 (-10.6~-2.8); 16.1 (10.7~21.1); 6.8 (3.8~9.9); 45.7 (16.2~75.0); 43.5 (21.0~66.0); 515 (411~628) ❙ 543 (476~632); 282 (249~321); 11 (9~15) ❙ 6.3 (-9.7~44.8); 501 (411~628); 224.2 (137.5~313.2)

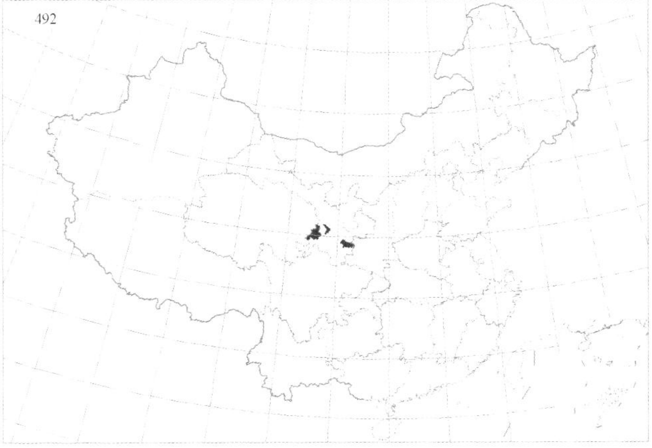

Salicaceae 杨柳科 *Salix* 柳属
Salix juparica Goerz ex Rehder et Kobuski 贵南柳
Shrubs, deciduous

0.1 (−3.6~6.4); −11.7 (−15.6~−7.1); 10.6 (7.0~18.6); 3.7 (1.9~7.9); 17.2 (3.6~58.0); 74.9 (40.5~107.2); 400 (328~554) ❙ 486 (226~800); 279 (138~427); 10 (7~18) ❙ 21.5 (−44.9~86.5); 379 (226~449); 234.4 (77.1~364.1)

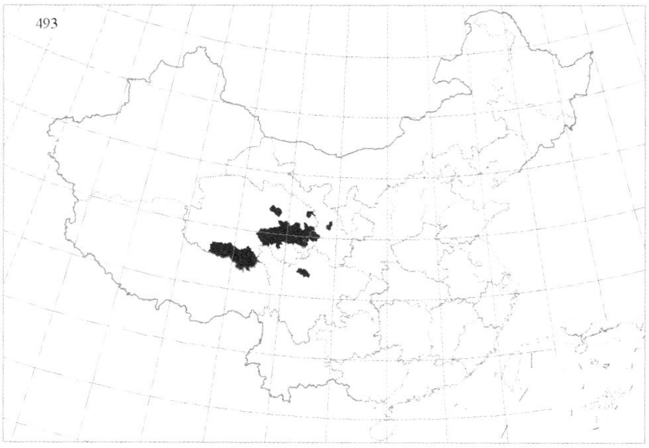

Salicaceae 杨柳科 *Salix* 柳属
Salix juparica var. tibetica (Goerz ex Rehder et Kobuski) C. F. Fang 光果贵南柳
Shrubs, deciduous

−0.3 (−3.6~2.7); −12.1 (−15.6~−7.1); 10.1 (7.0~13.1); 3.4 (1.9~5.0); 14.7 (3.6~27.9); 77.7 (48.0~107.2); 390 (328~449) ❙ 486 (226~800); 281 (138~427); 10 (7~18) ❙ 23.4 (−44.9~86.5); 378 (226~449); 238.2 (77.1~364.1)

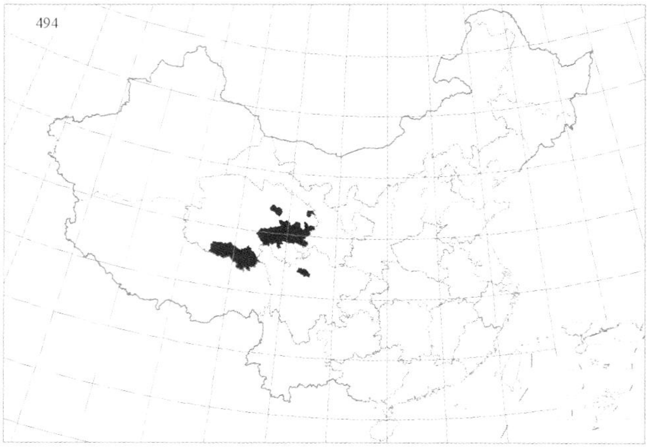

Salicaceae 杨柳科 *Salix* 柳属
Salix kamanica C. Wang et P. Y. Fu 卡马垫柳
Shrubs, deciduous

3.7 (−0.7~9.6); −4.5 (−9.9~2.4); 11.1 (7.7~15.7); 5.2 (2.5~9.6); 26.9 (6.8~60.8); 41.8 (5.4~74.2); 447 (338~600) ❙ 660 (258~1041); 397 (191~585); 21 (2~45) ❙ 39.7 (−24.7~100.7); 432 (258~600); 202.7 (60.4~352.3)

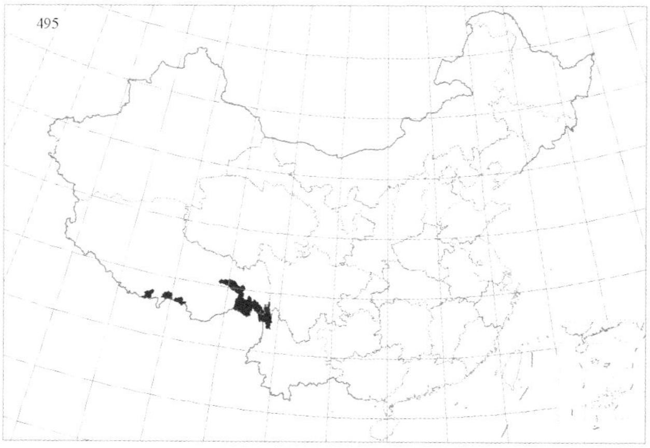

Salicaceae 杨柳科 *Salix* 柳属
Salix kangdingensis S. D. Zhao et C. F. Fang 康定垫柳
Shrubs, deciduous

2.7; −6.5; 10.7; 4.3; 19.1; 46.5; 424 ❙ 861; 489; 18 ❙ 103.7; 424; 303.0

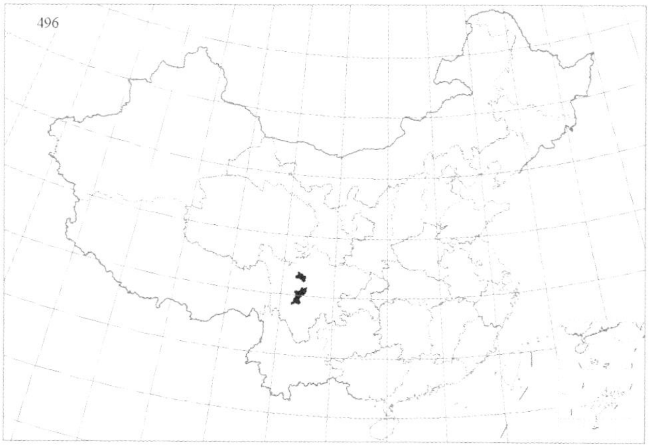

Salicaceae 杨柳科 *Salix* 柳属
Salix kangensis Nakai 江界柳
Trees, deciduous

5.1 (1.4~8.7); −13.6 (−18.0~−8.3); 21.4 (18.4~25.0); 8.6 (6.4~10.6); 67.5 (45.8~90.9); 65.3 (40.2~88.8); 596 (501~700) ❙ 776 (598~1018); 480 (350~648); 28 (18~36) ❙ 30.1 (−0.3~55.1); 596 (501~698); 281.5 (191.9~329.9)

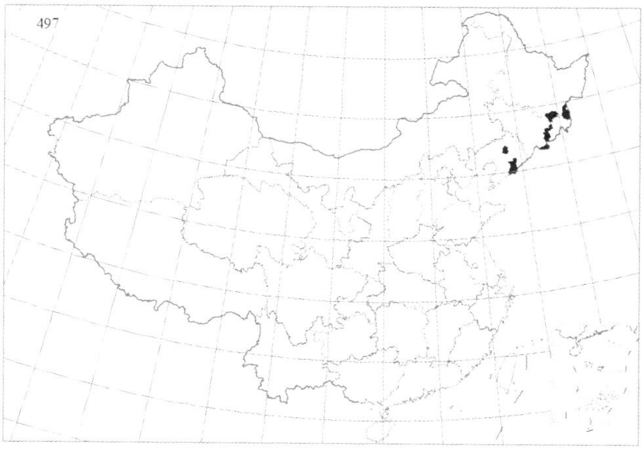

Salicaceae 杨柳科 *Salix* 柳属
Salix kangensis var. leiocarpa Kitag. 光果江界柳
Trees, deciduous

7.0; −11.0; 22.4; 9.3; 76.0; 51.4; 624 ❙ 966; 616; 35 ❙ 54.7; 624; 311.2

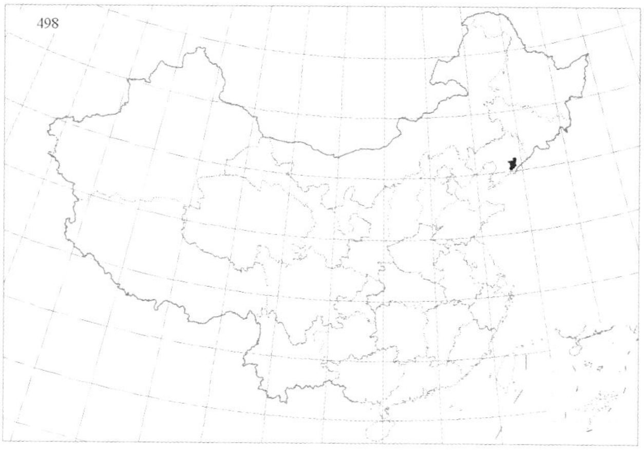

Salicaceae 杨柳科　*Salix* 柳属
Salix karelinii Turcz. 枸子叶柳
Shrubs or trees, deciduous

3.8 (-1.8~9.3)；-13.9 (-20.9~-5.5)；18.3 (6.1~27.5)；7.4 (1.8~12.1)；56.2 (1.9~107.6)；70.3 (40.7~103.9)；564 (329~813) ❙ 264 (120~527)；133 (51~317)；19 (9~48) ❙ -46.8 (-85.3~32.3)；254 (120~431)；126.9 (16.6~297.3)

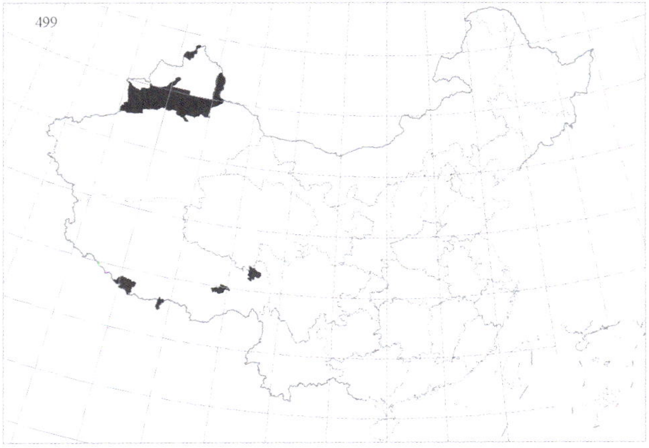

Salicaceae 杨柳科　*Salix* 柳属
Salix kirilowiana Stschegl. 天山筐柳
Shrubs or small trees, deciduous

-2.8；-20.3；10.9；3.5；16.3；109.2；390 ❙ 261；170；6 ❙ -33.2；261；111.1

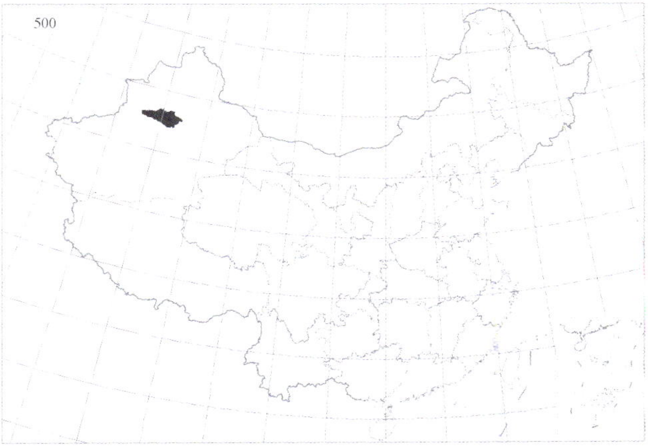

Salicaceae 杨柳科　*Salix* 柳属
Salix kochiana Trautv. 沙杞柳
Shrubs, deciduous

0.5 (-5.8~5.4)；-20.9 (-29.8~-12.5)；19.5 (15.3~23.0)；6.8 (4.4~8.7)；50.6 (28.0~69.0)；104.3 (62.5~157.4)；526 (426~611) ❙ 360 (137~507)；248 (92~339)；9 (3~16) ❙ -30.2 (-77.5~15.9)；355 (137~487)；180.1 (25.4~309.5)

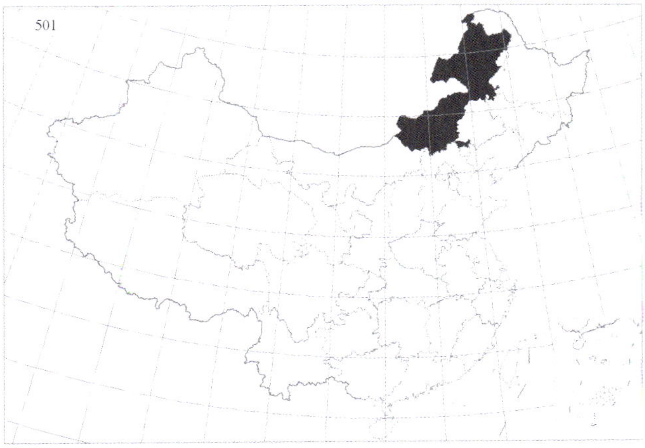

Salicaceae 杨柳科　*Salix* 柳属
Salix kongbanica C. Wang et P. Y. Fu 康巴柳
Trees, deciduous

1.2；-8.2；9.7；3.6；15.5；60.4；389 ❙ 414；269；6 ❙ 4.6；360；213.3

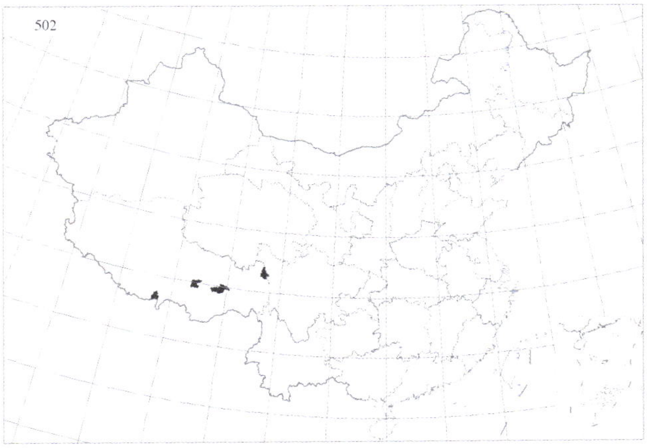

Salicaceae 杨柳科　*Salix* 柳属
Salix koreensis Andersson 朝鲜柳
Trees, deciduous

4.1 (-5.8~12.1)；-16.4 (-29.8~-1.2)；21.8 (15.3~25.0)；8.5 (4.4~12.4)；67.7 (28.0~103.4)；77.9 (14.2~157.4)；601 (426~746) ❙ 605 (311~1073)；390 (214~691)；19 (5~65) ❙ 0.7 (-40.5~74.1)；548 (311~729)；239.3 (97.2~346.5)

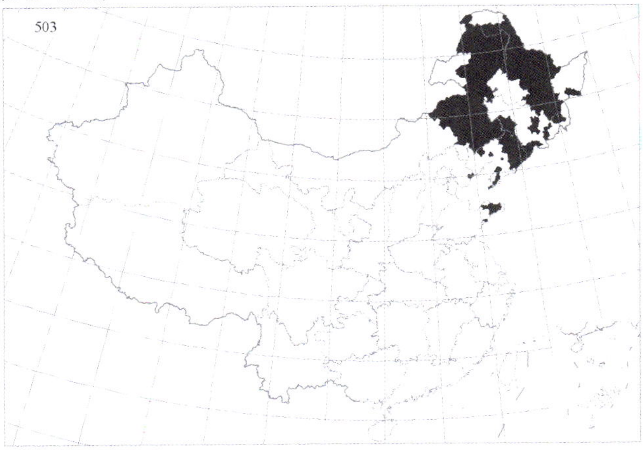

Salicaceae 杨柳科　*Salix* 柳属
Salix koreensis var. brevistyla Y. L. Chou et Skvortsov 短柱朝鲜柳
Trees, deciduous

3.4；-19.6；23.0；8.7；68.9；87.4；616 ❙ 556；367；15 ❙ -9.7；556；177.1

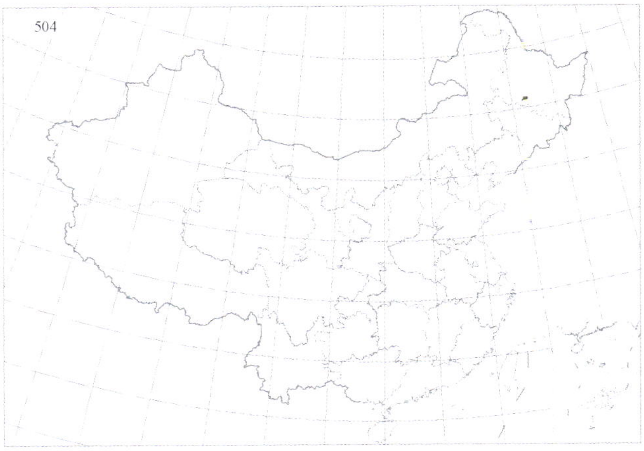

Salicaceae 杨柳科 *Salix* 柳属

Salix koreensis var. shandongensis C. F. Fang 山东柳

Trees, deciduous

11.7 (11.1~12.1); -2.1 (-3.3~-1.2); 24.0 (23.0~24.8); 12.0 (11.6~12.4); 98.5 (95.6~103.4); 17.4 (14.2~22.0); 723 (709~746) ❙ 669 (562~759); 388 (288~466); 37 (21~65) ❙ -7.5 (-23.0~3.8); 665 (562~724); 175.3 (155.2~205.7)

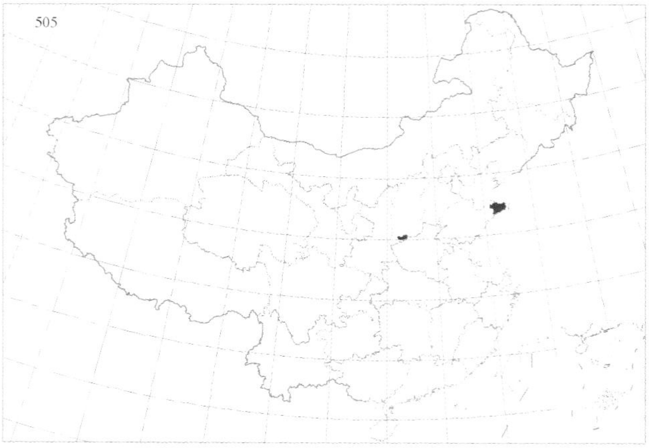

Salicaceae 杨柳科 *Salix* 柳属

Salix koriyanagi Kimura ex Goerz 尖叶紫柳

Shrubs, deciduous

6.0 (-1.5~11.7); -13.2 (-25.5~-1.6); 22.6 (18.9~25.4); 9.2 (6.4~11.9); 74.9 (47.4~99.6); 62.5 (16.7~128.7); 629 (513~745) ❙ 681 (314~1073); 438 (230~691); 23 (6~40) ❙ 9.4 (-40.5~74.1); 561 (314~698); 231.4 (97.2~346.5)

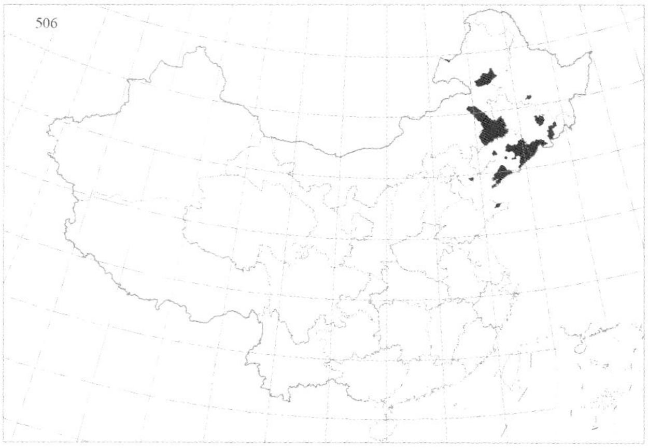

Salicaceae 杨柳科 *Salix* 柳属

Salix kouytchensis (H. Lév.) C. K. Schneid. 贵州柳

Shrubs, deciduous

11.6 (1.8~16.8); 3.0 (-9.2~7.8); 19.3 (9.9~27.5); 12.0 (3.8~16.8); 89.6 (15.7~141.5); 9.5 (0.0~57.8); 677 (404~854) ❙ 1003 (583~1334); 526 (318~720); 50 (13~145) ❙ 49.9 (13.4~94.4); 675 (404~843); 254.7 (153.9~382.6)

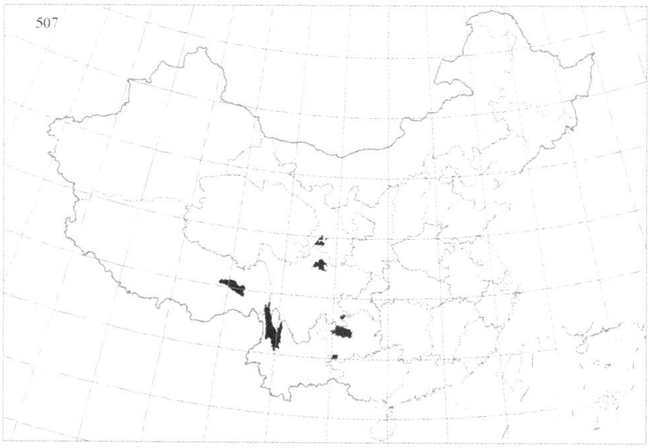

Salicaceae 杨柳科 *Salix* 柳属

Salix kungmuensis Mao et W. Z. Li 孔目矮柳

Shrubs, deciduous

8.6; 1.3; 15.0; 8.7; 52.6; 8.8; 574 ❙ 1041; 585; 45 ❙ 81.5; 574; 352.3

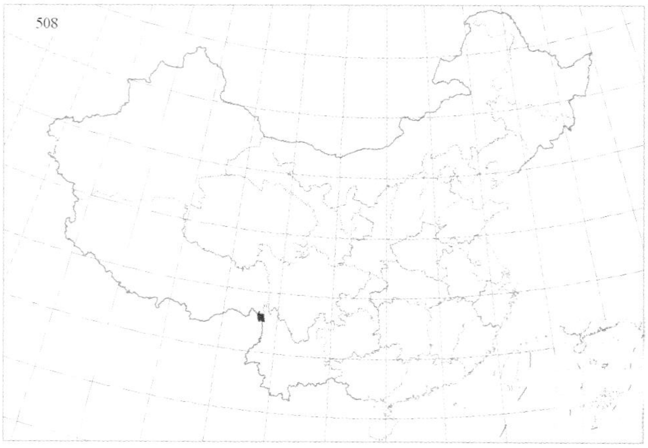

Salicaceae 杨柳科 *Salix* 柳属

Salix kusanoi (Hayata) C. K. Schneid. 水社柳

Trees, deciduous

19.8 (16.4~24.1); 14.3 (11.2~18.9); 24.3 (20.3~28.0); 19.8 (16.5~24.1); 178.1 (137.8~229.7); 0.0 (0.0~0.0); 989 (783~1322) ❙ 2509 (1751~2993); 1190 (924~1847); 237 (47~547) ❙ 164.5 (32.5~235.6); 955 (783~1123); 252.3 (154.2~303.6)

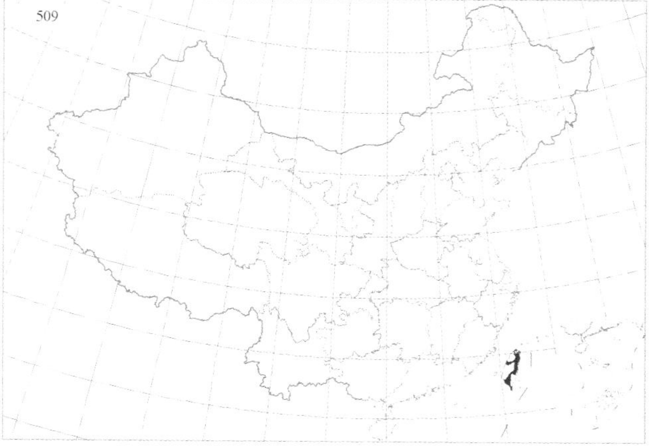

Salicaceae 杨柳科 *Salix* 柳属

Salix lamashanensis K. S. Hao ex C. F. Fang et A. K. Skvortsov 拉马山柳

Shrubs, deciduous

3.9 (-5.7~11.3); -8.5 (-18.7~-0.3); 15.1 (7.6~22.6); 6.3 (1.7~11.4); 41.4 (4.8~89.6); 53.5 (12.1~133.0); 493 (294~683) ❙ 480 (74~678); 254 (46~343); 10 (5~16) ❙ -3.7 (-75.0~61.8); 431 (74~675); 196.7 (16.7~344.4)

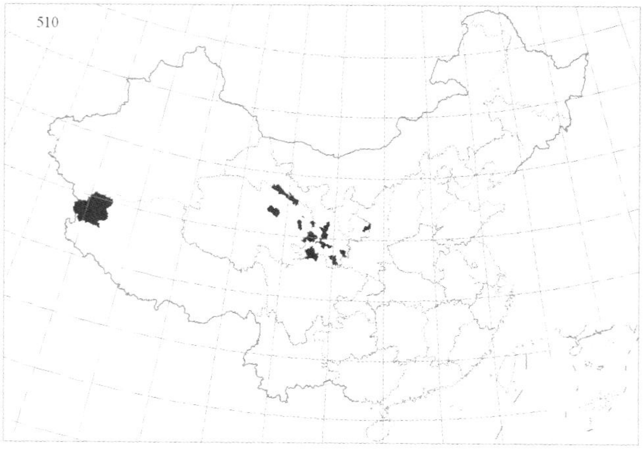

Salicaceae 杨柳科 *Salix* 柳属
Salix lanifera C. F. Fang et S. D. Zhao 白毛柳
Shrubs, deciduous

0.6; −8.5; 8.8; 3.1; 11.0; 63.5; 378 ❙ 620; 375; 13 ❙ 60.0; 378; 188.4

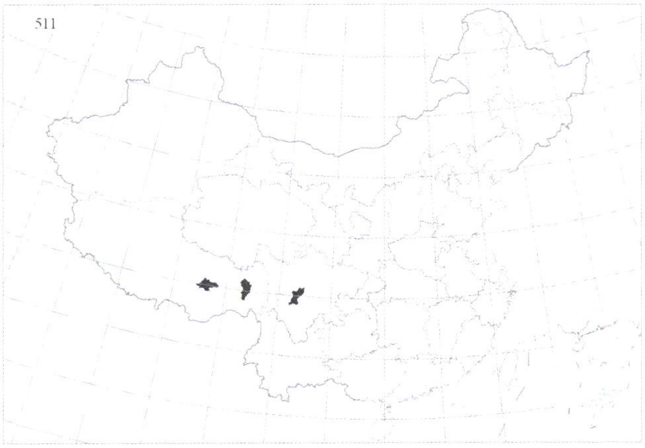

Salicaceae 杨柳科 *Salix* 柳属
Salix lasiopes C. Wang et P. Y. Fu 毛柄柳
Shrubs, deciduous

0.3; −8.6; 8.3; 2.9; 9.6; 65.4; 368 ❙ 389; 246; 11 ❙ 0.0; 368; 142.5

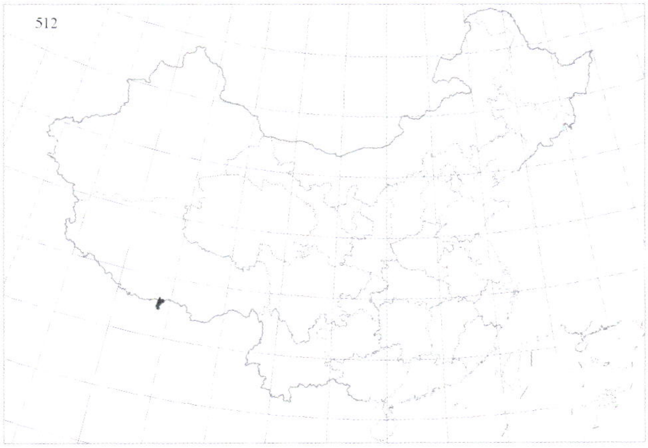

Salicaceae 杨柳科 *Salix* 柳属
Salix leveilleana C. K. Schneid. 荞麦地柳
Shrubs or trees, deciduous

16.0 (15.2~17.0); 4.3 (3.6~5.7); 27.3 (25.8~29.1); 16.0 (15.3~17.0); 133.3 (124.1~144.3); 0.9 (0.0~1.4); 865 (810~935) ❙ 1539 (1355~1671); 527 (481~559); 192 (161~221) ❙ 78.6 (61.0~106.4); 865 (810~935); 241.3 (212.3~263.0)

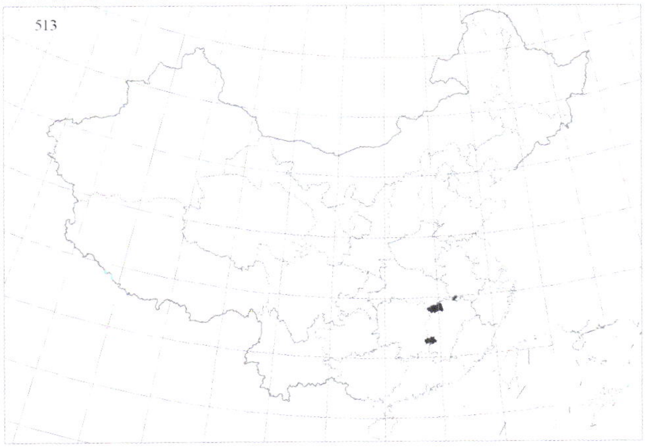

Salicaceae 杨柳科 *Salix* 柳属
Salix lindleyana Wall. ex Andersson 青藏垫柳
Shrubs, deciduous

3.4 (−4.4~12.2); −5.4 (−16.2~4.3); 11.2 (6.1~18.5); 5.0 (1.7~12.2); 26.1 (1.9~87.4); 45.0 (0.7~115.9); 445 (323~672) ❙ 728 (300~1431); 430 (198~817); 18 (3~40) ❙ 59.6 (−19.2~125.9); 442 (300~672); 246.4 (72.2~363.7)

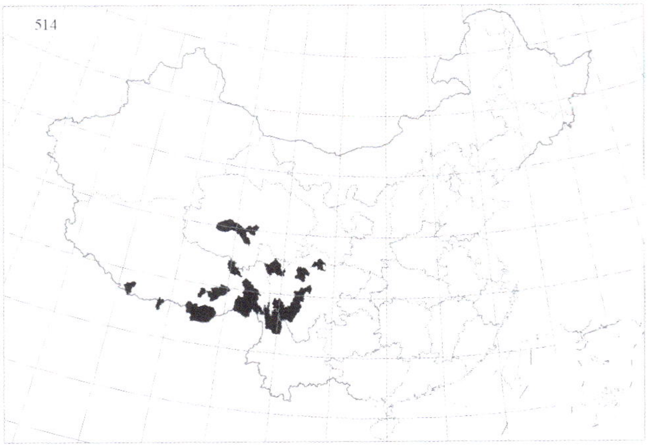

Salicaceae 杨柳科 *Salix* 柳属
Salix linearistipularis K. S. Hao 筐柳
Shrubs or trees, deciduous

9.1 (−5.8~15.8); −6.9 (−29.8~2.9); 23.4 (15.3~28.6); 10.8 (4.4~15.8); 87.8 (28.0~133.2); 38.3 (2.8~157.4); 685 (426~893) ❙ 568 (72~1246); 326 (46~648); 22 (2~109) ❙ −17.3 (−89.1~93.8); 540 (72~893); 166.8 (12.0~334.4)

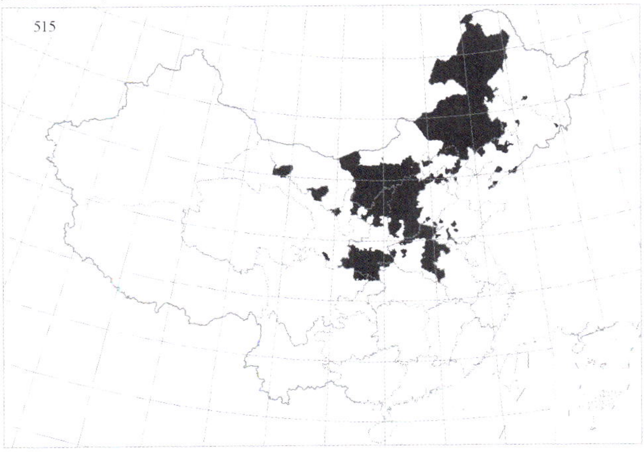

Salicaceae 杨柳科 *Salix* 柳属
Salix liouana C. Wang et C. Y. Yang 黄龙柳
Shrubs, deciduous

10.0 (1.9~13.5); −4.0 (−11.6~−0.1); 22.7 (11.7~26.8); 10.7 (4.3~13.6); 85.6 (21.1~116.5); 25.5 (11.5~57.8); 673 (427~817) ❙ 675 (336~1246); 358 (203~525); 25 (8~71) ❙ 1.5 (−45.2~101.4); 596 (336~767); 189.2 (56.4~316.6)

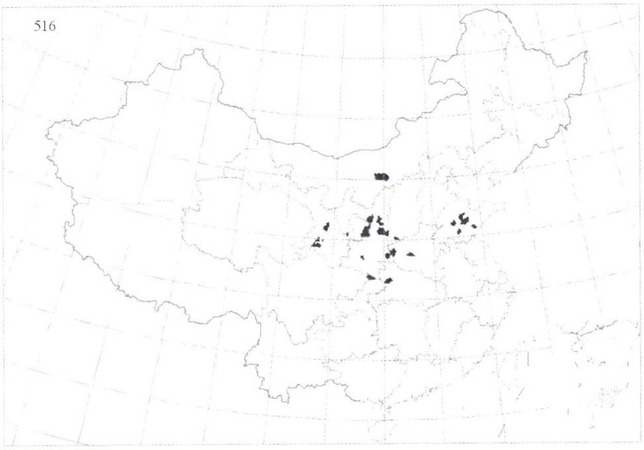

Salicaceae 杨柳科 *Salix* 柳属
Salix longiflora Wall. ex Andersson 长花柳
Shrubs, deciduous

10.3 (−5.5~19.3); 1.9 (−17.1~13.1); 17.4 (5.8~25.3); 11.2 (1.3~19.4); 84.9 (1.2~173.2); 20.5 (0.0~127.1); 642 (299~939) ❙ 917 (252~1736); 511 (153~1007); 34 (5~100) ❙ 44.9 (−37.4~145.8); 629 (252~903); 275.6 (46.7~491.5)

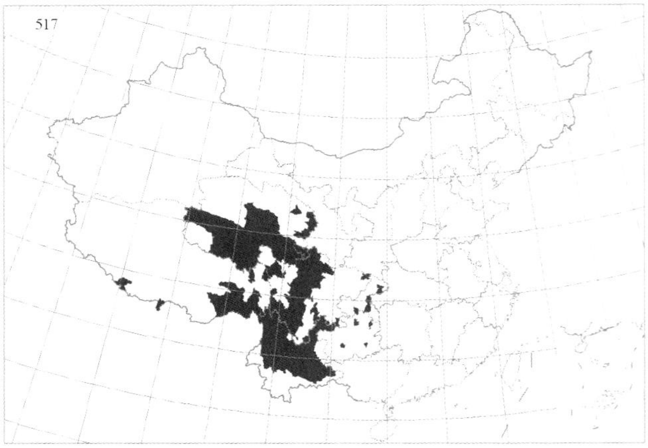

Salicaceae 杨柳科 *Salix* 柳属
Salix longissimipedicellaris N. Chao ex P. I Mao 苍山长梗柳
Shrubs, deciduous

13.6; 7.2; 18.8; 13.7; 104.1; 0.0; 706 ❙ 1042; 592; 46 ❙ 47.6; 706; 372.3

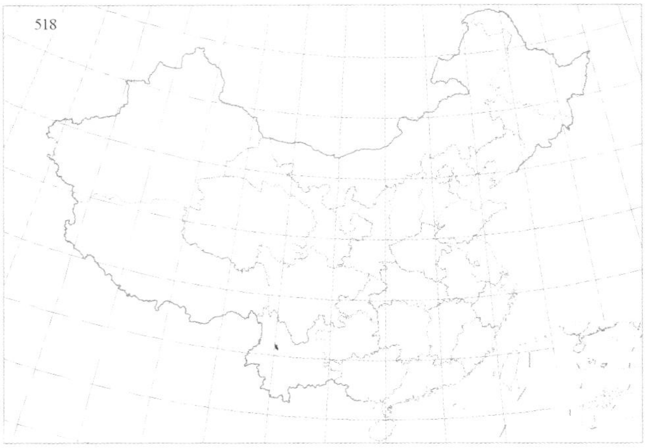

Salicaceae 杨柳科 *Salix* 柳属
Salix longistamina C. Wang et P. Y. Fu 长蕊柳
Trees, deciduous

2.4; −7.3; 10.9; 4.3; 21.7; 52.3; 417 ❙ 295; 227; 1 ❙ −29.7; 295; 162.0

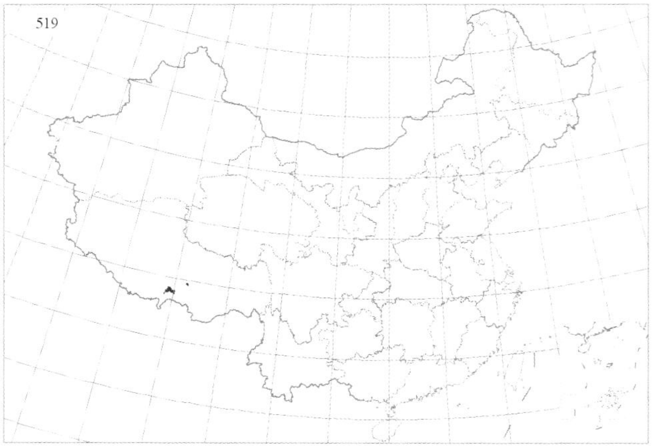

Salicaceae 杨柳科 *Salix* 柳属
Salix luctuosa H. Lév. 丝毛柳
Shrubs, deciduous

10.2 (−0.4~18.9); 0.3 (−10.0~12.5); 19.1 (8.4~26.5); 10.7 (2.8~19.0); 79.2 (9.4~167.5); 16.8 (0.0~73.3); 642 (360~914) ❙ 954 (341~1414); 492 (190~826); 35 (5~113) ❙ 51.2 (−40.9~128.6); 633 (341~888); 246.6 (61.2~372.3)

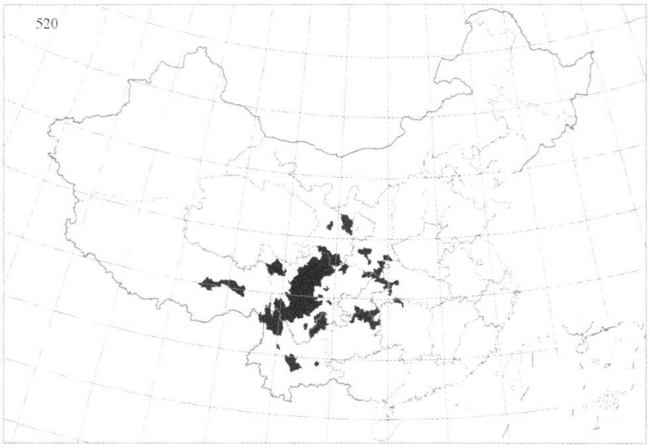

Salicaceae 杨柳科 *Salix* 柳属
Salix ludingensis T. Y. Ding et C. F. Fang 泸定垫柳
Shrubs, deciduous

6.2; −2.4; 14.1; 6.7; 39.4; 24.1; 509 ❙ 981; 539; 23 ❙ 92.9; 509; 221.3

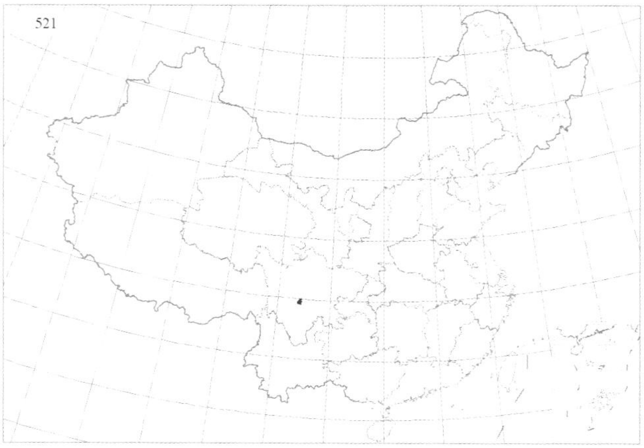

Salicaceae 杨柳科 *Salix* 柳属
Salix macroblasta C. K. Schneid. 灌西柳
Shrubs, deciduous

8.1 (2.9~14.2); −2.2 (−6.3~3.3); 17.7 (11.2~24.7); 8.8 (4.4~14.3); 62.0 (20.5~113.3); 23.8 (2.1~44.6); 581 (428~776) ❙ 865 (592~1265); 422 (295~584); 27 (13~65) ❙ 50.2 (2.4~121.7); 581 (428~776); 246.5 (214.7~267.7)

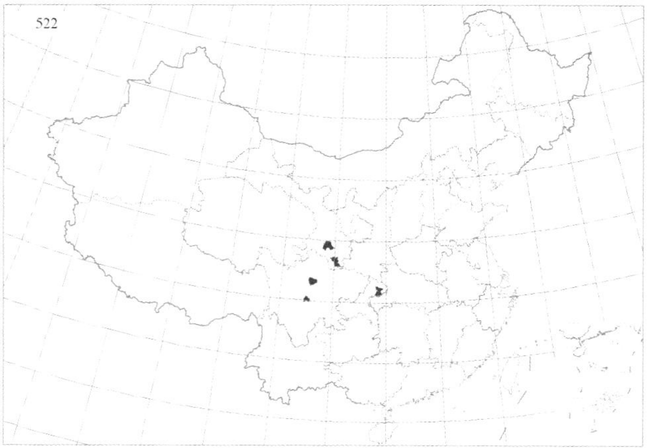

Salicaceae 杨柳科 *Salix* 柳属
Salix maerkangensis N. Chao 簇毛柳
Shrubs, deciduous

6.4 (-0.3~14.9); -3.3 (-10.9~5.3); 15.0 (9.0~23.6); 7.6 (3.0~15.0); 50.2 (10.4~119.6); 32.5 (0.0~73.4); 535 (378~782) | 938 (656~1410); 517 (382~816) 26 (15~49) | 77.6 (49.4~101.2); 535 (378~782); 279.3 (171.0~357.7)

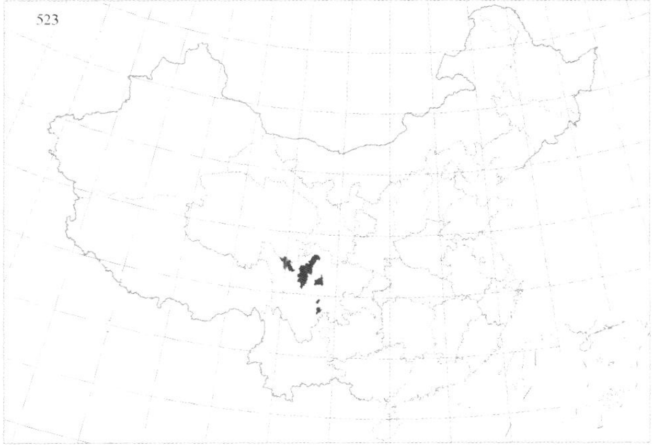

Salicaceae 杨柳科 *Salix* 柳属
Salix magnifica Hemsl. 大叶柳
Shrubs or trees, deciduous

8.9 (2.0~16.2); -0.3 (-7.2~5.8); 17.3 (10.1~26.3); 9.4 (3.8~16.2); 65.8 (15.4~134.6); 17.9 (0.0~51.4); 600 (413~854) | 1037 (791~1410); 557 (389~816); 32 (12~71) | 76.8 (32.5~128.6); 600 (413~854); 263.6 (171.0~357.7)

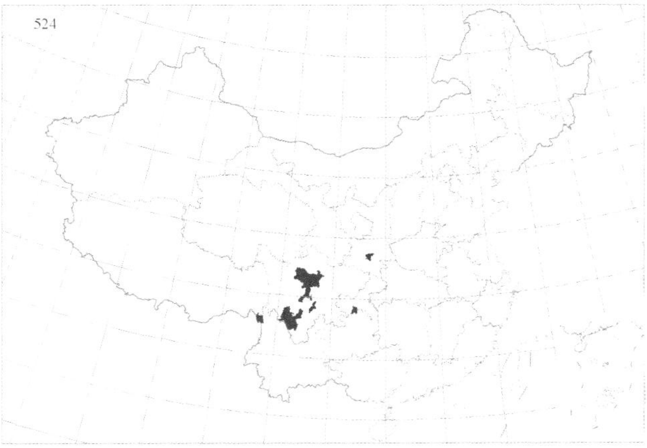

Salicaceae 杨柳科 *Salix* 柳属
Salix magnifica var. ulotricha (C. K. Schneid.) N. Chao 卷毛大叶柳
Shrubs or trees, deciduous

7.5 (2.0~14.9); -1.5 (-7.2~5.3); 15.4 (10.1~23.6); 8.1 (3.8~15.0); 53.0 (15.4~119.6); 22.7 (0.0~51.4); 553 (413~782) | 996 (791~1410); 557 (427~816); 25 (12~49) | 83.5 (49.4~128.6); 553 (413~782); 266.9 (171.0~357.7)

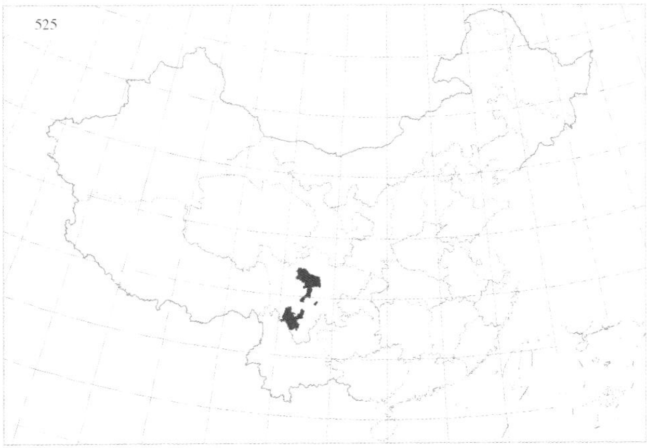

Salicaceae 杨柳科 *Salix* 柳属
Salix maizhokunggarensis N. Chao 墨竹柳
Shrubs, deciduous

3.6 (-2.5~13.3); -5.0 (-12.7~5.4); 11.3 (6.1~19.6); 5.6 (1.8~13.4); 31.9 (1.9~100.4); 47.9 (0.0~93.0); 457 (329~706) | 693 (287~1736); 433 (213~1007); 17 (2~51) | 38.7 (-18.0~145.8); 443 (287~706); 241.0 (72.2~447.4)

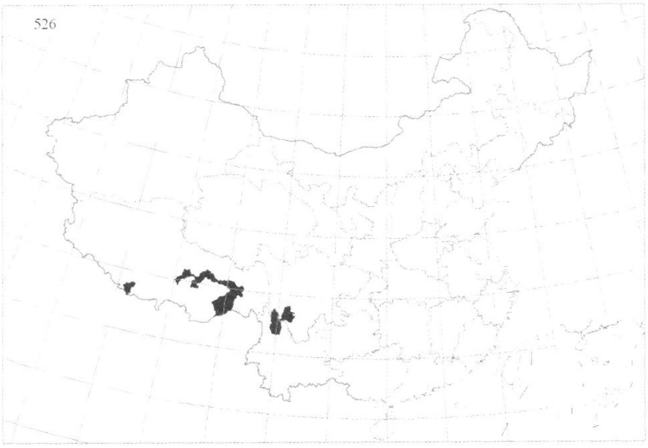

Salicaceae 杨柳科 *Salix* 柳属
Salix matsudana Koidz. 旱柳
Trees, deciduous

12.4 (-5.9~22.1); -1.7 (-25.0~13.7); 25.1 (5.4~29.9); 13.2 (1.3~22.1); 110.2 (0.4~205.6); 20.9 (0.0~130.5); 780 (300~1182) | 839 (23~1922); 412 (16~826); 62 (1~259) | 5.4 (-97.2~128.9); 664 (23~1182); 185.4 (3.0~434.4)

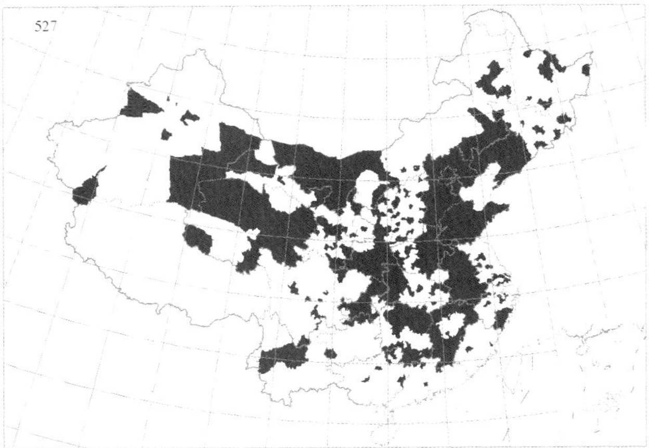

Salicaceae 杨柳科 *Salix* 柳属
Salix matsudana var. pseudomatsudana (Y. L. Chou et Skvortsov) Y. L. Chou 旱垂柳
Trees, deciduous

6.7 (3.4~8.8); -12.8 (-19.6~-8.3); 23.4 (21.8~24.9); 9.6 (8.6~10.7); 79.5 (68.0~91.3); 59.1 (40.2~87.4); 650 (595~702) | 706 (513~1005); 450 (335~635); 22 (9~39) | 9.9 (-23.4~68.8); 582 (513~666); 219.4 (141.3~322.3)

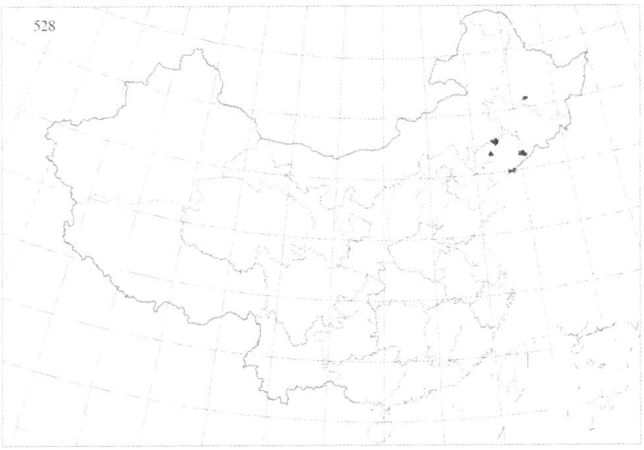

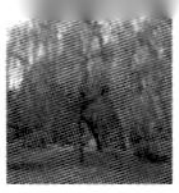

Salicaceae 杨柳科 *Salix* 柳属
Salix maximowiczii Kom. 大白柳
Trees, deciduous

2.1 (-0.9~6.5); -19.8 (-24.7~-12.0); 20.7 (19.0~22.3); 7.6 (6.5~9.1); 57.4 (48.8~73.9); 92.3 (55.1~120.7); 562 (522~616) | 644 (513~1073); 403 (330~691); 21 (11~40) | 13.9 (-6.0~74.1); 559 (513~616); 298.0 (224.5~346.5)

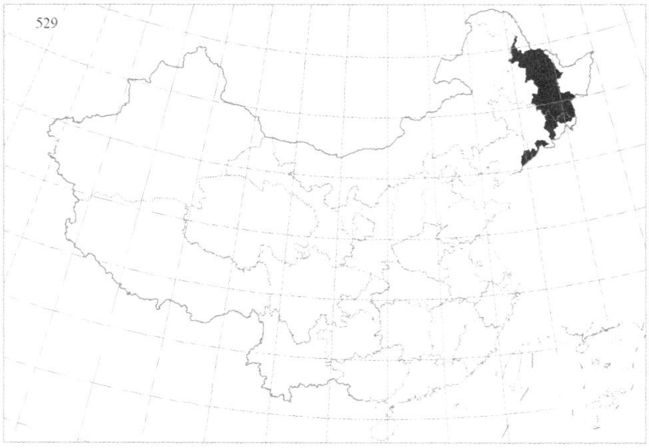

Salicaceae 杨柳科 *Salix* 柳属
Salix medogensis Y. L. Chou 墨脱柳
Trees, deciduous

5.3; -3.3; 12.9; 7.0; 43.2; 38.8; 504 | 935; 548; 29 | 70.9; 504; 283.7

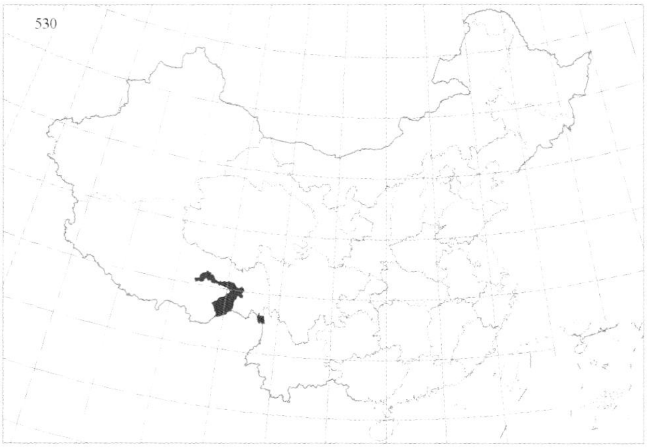

Salicaceae 杨柳科 *Salix* 柳属
Salix mesnyi Hance 粤柳
Trees, deciduous

19.1 (14.7~24.8); 9.4 (2.3~20.1); 27.8 (23.6~29.3); 19.2 (14.8~24.8); 170.3 (121.1~237.6); 0.3 (0.0~3.9); 1024 (815~1385) | 1582 (868~2037); 600 (312~1033); 161 (41~259) | 57.9 (-34.3~130.7); 1003 (815~1360); 252.2 (95.3~620.0)

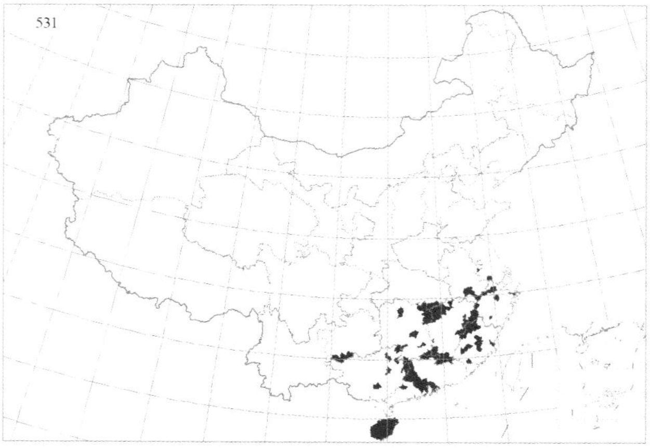

Salicaceae 杨柳科 *Salix* 柳属
Salix metaglauca C. Y. Yang 绿叶柳
Shrubs, deciduous

2.7 (-0.6~12.3); -16.0 (-19.5~-5.3); 18.9 (13.8~26.0); 7.3 (4.4~13.0); 55.4 (26.2~112.3); 82.8 (24.7~104.2); 557 (413~801) | 159 (44~251); 67 (19~106); 19 (4~33) | -70.4 (-94.5~-52.0); 159 (44~251); 73.2 (9.3~139.1)

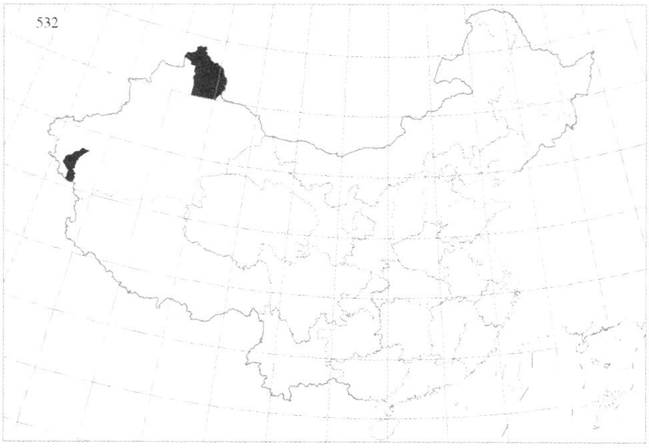

Salicaceae 杨柳科 *Salix* 柳属
Salix michelsonii Goerz ex Nasarow 米黄柳
Shrubs, deciduous

3.3 (-4.3~11.2); -15.3 (-20.9~-7.8); 18.3 (9.6~27.1); 7.4 (2.5~12.6); 55.6 (9.7~109.2); 75.4 (34.4~121.1); 558 (329~810) | 234 (23~450); 110 (16~211); 19 (1~48) | -54.3 (-97.2~3.9); 232 (23~406); 122.2 (3.0~297.3)

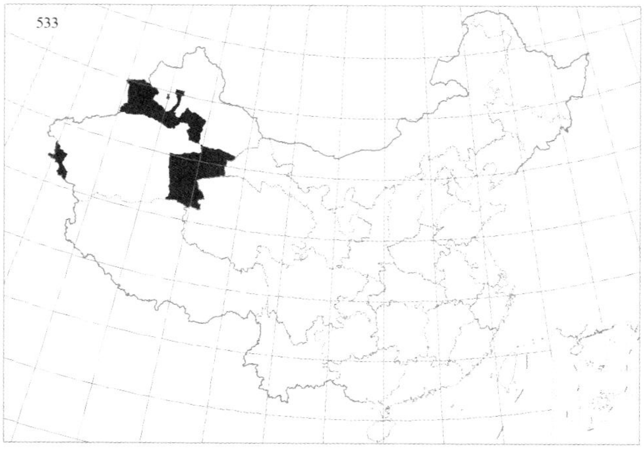

Salicaceae 杨柳科 *Salix* 柳属
Salix microphyta Franch. 宝兴矮柳
Shrubs, deciduous

8.2 (2.7~13.2); 0.0 (-7.1~6.4); 15.4 (11.1~19.4); 8.6 (4.4~13.3); 55.8 (20.5~99.4); 16.6 (0.0~48.0); 566 (429~695) | 1014 (741~1337); 558 (420~727); 36 (17~69) | 79.1 (38.2~101.2); 566 (429~695); 320.8 (200.8~407.1)

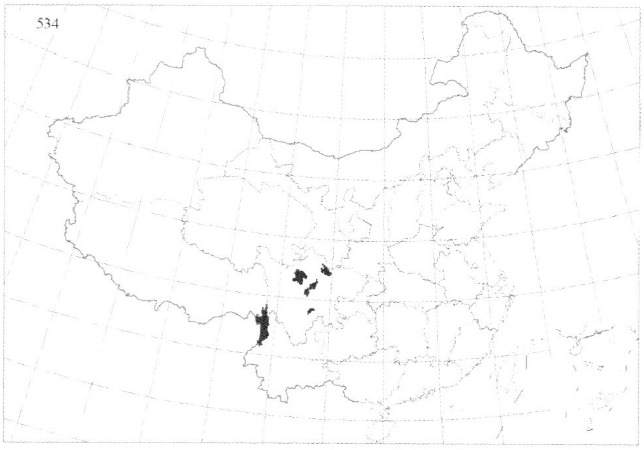

Salicaceae 杨柳科　*Salix* 柳属
Salix microstachya Turcz. ex Trautv. 小穗柳
Shrubs, deciduous

4.5 (−5.8~12.1)；−13.7 (−29.8~3.9)；20.7 (11.5~26.2)；8.2 (3.5~12.7)；63.8 (17.2~107.9)；68.9 (1.3~157.4)；582 (375~793) ❙ 349 (35~968)；232 (23~594)；8 (1~24) ❙ −38.8 (−94.9~61.4)；340 (35~698)；128.6 (5.1~319.8)

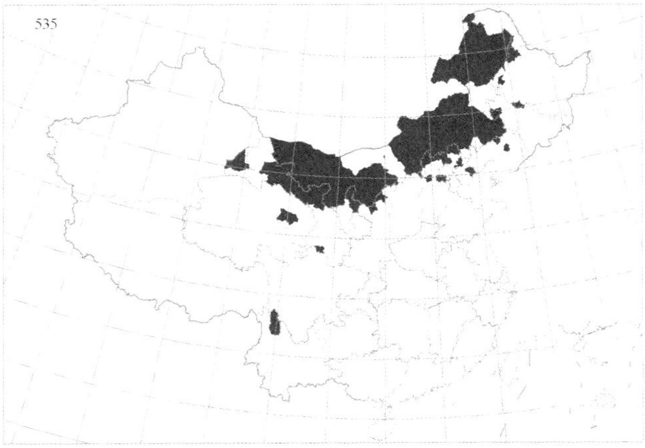

Salicaceae 杨柳科　*Salix* 柳属
Salix microstachya var. bordensis (Nakai) C. F. Fang 小红柳
Shrubs, deciduous

4.7 (−5.8~8.8)；−14.1 (−29.8~−7.1)；21.4 (11.7~24.1)；8.4 (4.0~10.3)；67.2 (20.4~84.8)；70.2 (35.3~157.4)；594 (414~670) ❙ 361 (128~694)；244 (81~499)；8 (2~16) ❙ −38.0 (−80.8~15.9)；359 (128~659)；144.4 (13.6~308.8)

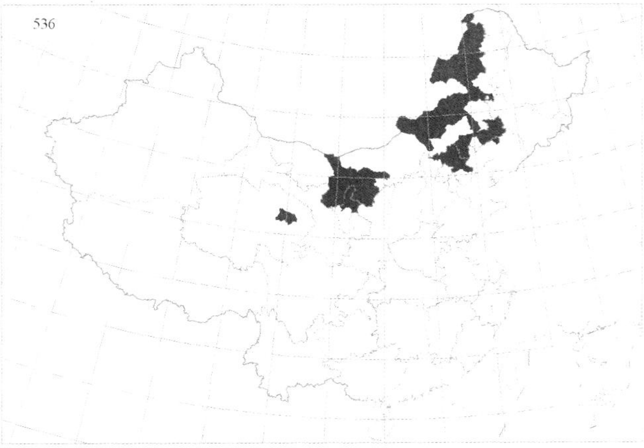

Salicaceae 杨柳科　*Salix* 柳属
Salix mictotricha C. K. Schneid. 兴山柳
Shrubs, deciduous

9.8 (0.8~17.8)；−1.0 (−10.6~7.1)；19.8 (9.9~28.0)；10.3 (3.8~17.8)；76.6 (15.5~154.1)；18.6 (0.0~66.0)；637 (411~944) ❙ 937 (519~1492)；455 (273~816)；37 (10~105) ❙ 49.3 (−18.4~125.9)；634 (411~944)；253.2 (137.9~320.9)

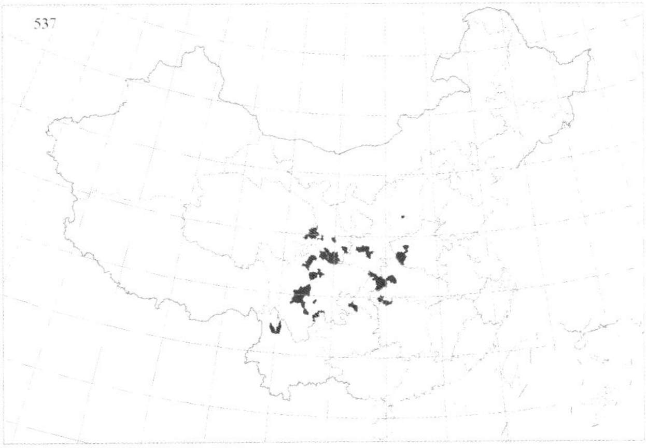

Salicaceae 杨柳科　*Salix* 柳属
Salix minjiangensis N. Chao 岷江柳
Shrubs, deciduous

2.7；−8.0；12.1；4.7；24.2；50.5；439 ❙ 733；387；17 ❙ 68.2；439；243.0

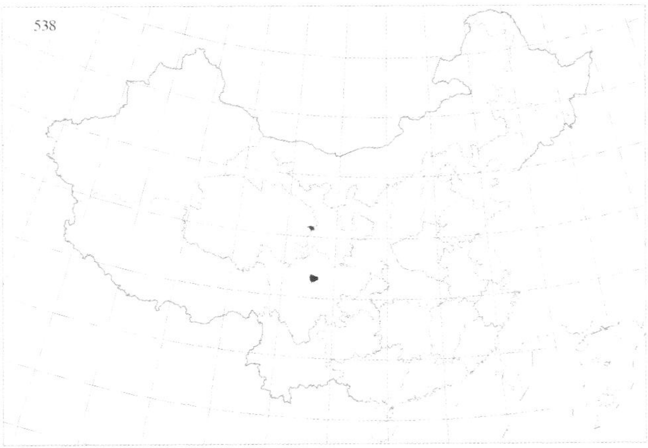

Salicaceae 杨柳科　*Salix* 柳属
Salix morrisonicola Kimura 玉山柳
Shrubs, deciduous

20.8；15.3；25.0；20.8；189.2；0.0；1019 ❙ 2445；1412；110 ❙ 140.0；1019；207.7

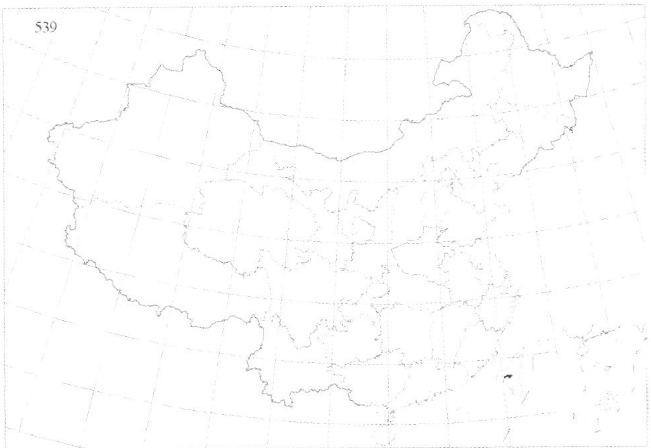

Salicaceae 杨柳科　*Salix* 柳属
Salix moupinensis Franch. 宝兴柳
Trees, deciduous

12.0 (2.9~17.3)；2.3 (−6.3~7.4)；21.0 (11.2~26.5)；12.2 (4.4~17.4)；93.3 (20.5~148.6)；8.2 (0.0~44.6)；695 (428~900) ❙ 1102 (669~1590)；571 (317~946)；42 (15~85) ❙ 60.6 (12.1~121.7)；695 (428~900)；236.0 (153.3~318.7)

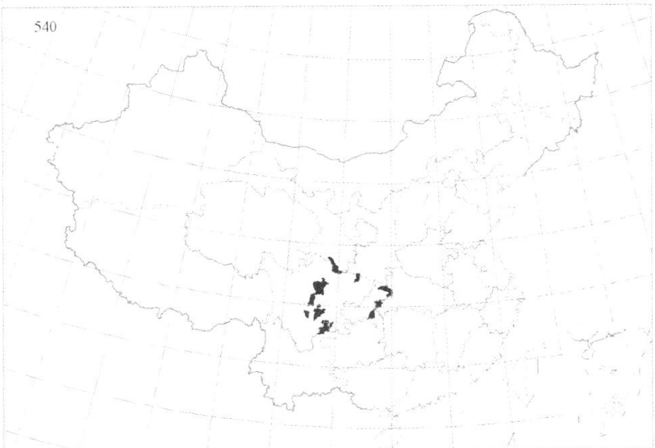

Salicaceae 杨柳科 *Salix* 柳属

Salix muliensis Goerz ex Rehder et Kobuski 木里柳

Shrubs, deciduous

5.9 (0.6~15.3); -2.9 (-9.4~5.2); 13.7 (9.0~25.3); 6.8 (3.3~15.3); 41.9 (11.9~124.2); 30.5 (0.0~65.4); 512 (390~801) ❙ 855 (578~1240); 483 (345~619); 25 (11~84) ❙ 70.0 (25.6~128.6); 511 (390~801); 264.6 (130.1~357.7)

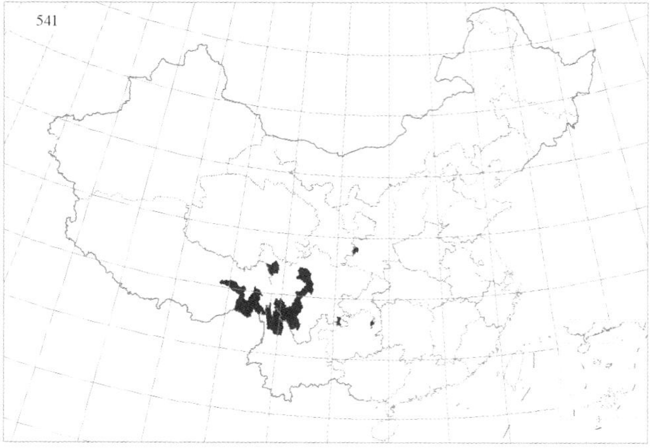

Salicaceae 杨柳科 *Salix* 柳属

Salix myrtillacea Andersson 坡柳

Shrubs, deciduous

3.6 (-4.8~14.9); -6.9 (-16.1~5.4); 12.9 (6.2~25.3); 5.6 (1.6~15.0); 32.9 (2.0~119.6); 49.0 (0.0~118.7); 468 (300~795) ❙ 671 (220~1736); 372 (138~1007); 17 (0~103) ❙ 43.0 (-44.9~145.8); 455 (220~795); 246.9 (46.9~447.4)

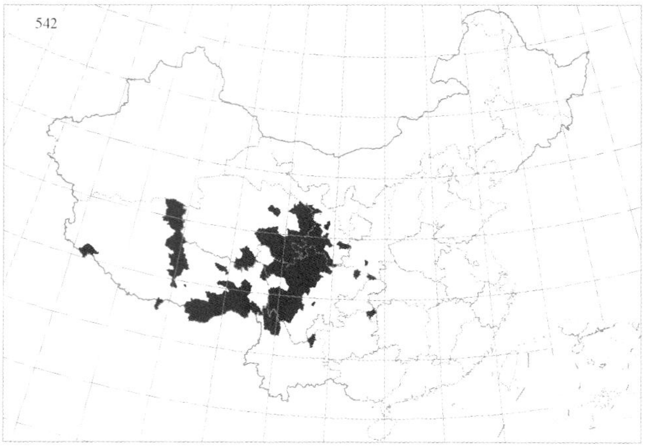

Salicaceae 杨柳科 *Salix* 柳属

Salix myrtilloides L. 越橘柳

Shrubs, deciduous

1.5 (-5.8~8.4); -20.2 (-29.8~-8.9); 20.4 (15.3~23.2); 7.3 (4.4~10.2); 55.6 (28.0~84.0); 96.8 (42.8~157.4); 552 (426~661) ❙ 553 (297~1073); 351 (203~691); 17 (5~40) ❙ -0.4 (-44.3~74.1); 508 (297~661); 267.0 (127.8~346.5)

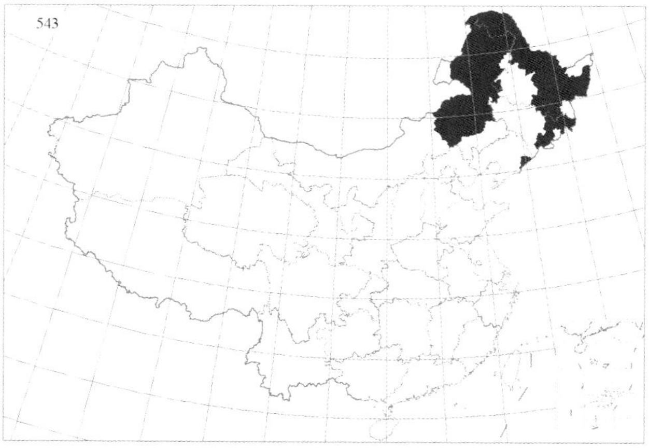

Salicaceae 杨柳科 *Salix* 柳属

Salix myrtilloides var. mandshurica Nakai 东北越橘柳

Shrubs, deciduous

1.5 (-5.0~3.9); -21.1 (-29.4~-16.9); 20.8 (17.0~22.1); 7.5 (5.0~8.4); 57.1 (35.2~65.3); 98.8 (77.6~154.8); 562 (450~603) ❙ 586 (468~669); 361 (300~423); 19 (12~32) ❙ 3.9 (-11.0~25.8); 553 (450~591); 281.7 (203.8~333.3)

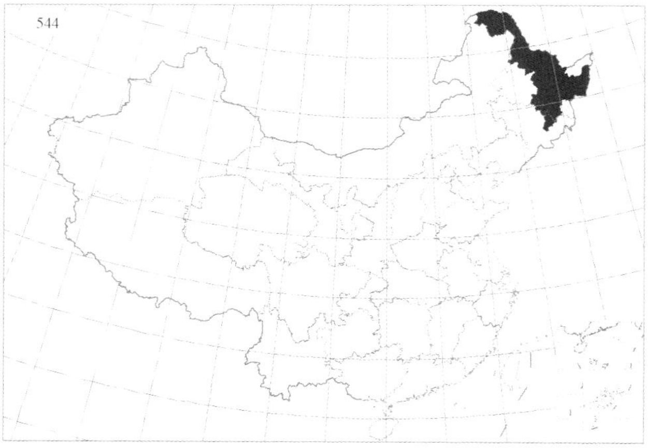

Salicaceae 杨柳科 *Salix* 柳属

Salix nankingensis C. Wang et S. L. Tung 南京柳

Shrubs or trees, deciduous

16.1; 3.6; 28.1; 16.1; 135.0; 1.7; 888 ❙ 1268; 482; 149 ❙ 42.9; 872; 173.9

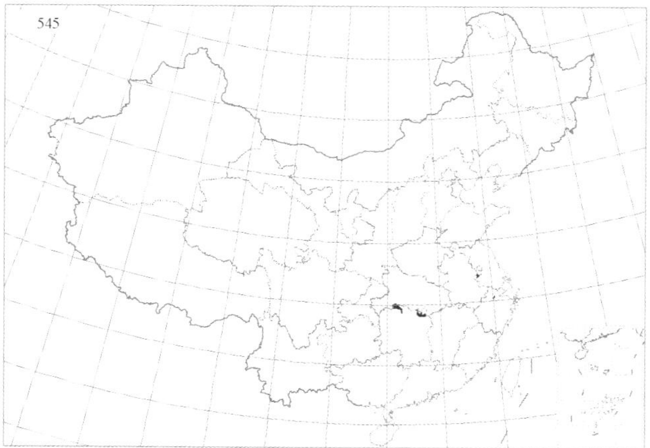

Salicaceae 杨柳科 *Salix* 柳属

Salix neoamnematchinensis T. Y. Ding et C. F. Fang 新山生柳

Shrubs, deciduous

1.0 (-4.1~6.1); -11.5 (-17.3~-7.2); 11.8 (7.2~17.3); 4.3 (2.0~7.6); 22.1 (3.7~53.6); 70.0 (39.3~114.2); 422 (339~544) ❙ 461 (317~611); 266 (196~340); 8 (5~13) ❙ 10.3 (-29.0~64.6); 403 (317~463); 226.0 (138.5~364.1)

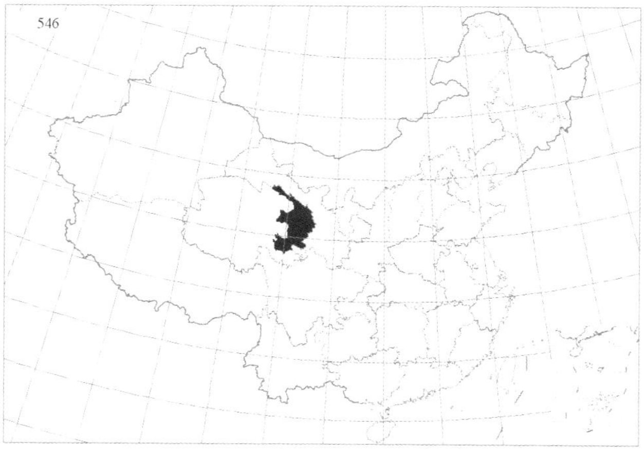

Salicaceae 杨柳科 *Salix* 柳属
Salix neclapponum C. Y. Yang 绢柳
Shrubs, deciduous

2.1 (-0.6~5.8); -16.8 (-19.5~-14.9); 18.5 (15.9~23.3); 7.0 (5.4~9.7); 51.7 (37.5~80.9); 86.3 (71.2~104.2); 542 (475~668) ❙ 196 (156~267); 86 (56~135); 21 (12~33) ❙ -63.5 (-76.7~-50.9); 196 (156~267); 98.2 (16.6~233.8)

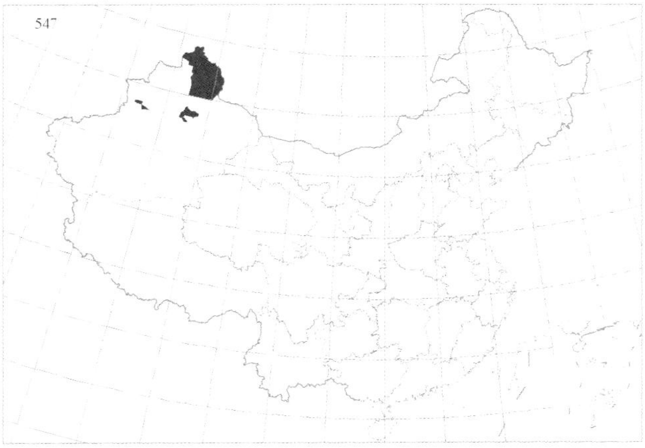

Salicaceae 杨柳科 *Salix* 柳属
Salix neowilsonii W. P. Fang 新紫柳
Trees, deciduous

14.7; 4.4; 23.8; 14.7; 117.2; 1.2; 785 ❙ 1015; 590; 30 ❙ 30.8; 784; 152.4

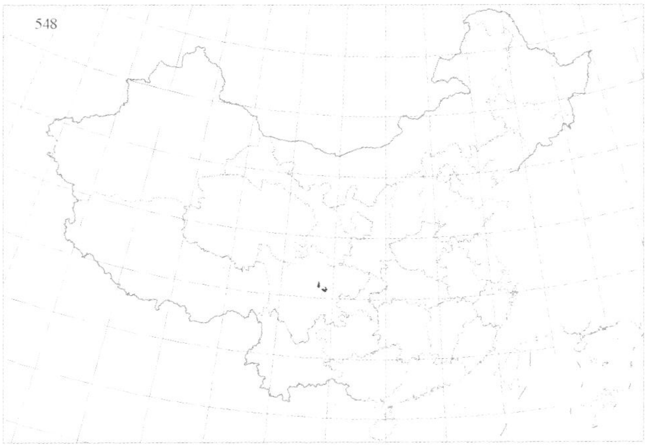

Salicaceae 杨柳科 *Salix* 柳属
Salix nippcnica Franch. et Sav. 三蕊柳
Shrubs or trees, deciduous

10.0 (-4.0~17.8); -5.9 (-28.2~5.1); 24.4 (15.9~29.9); 11.9 (5.0~17.8); 97.6 (34.9~153.8); 36.7 (0.0~146.2); 728 (448~979) ❙ 804 (152~1592); 395 (56~648); 61 (5~214) ❙ 7.7 (-76.8~101.4); 656 (152~977); 201.5 (16.6~346.5)

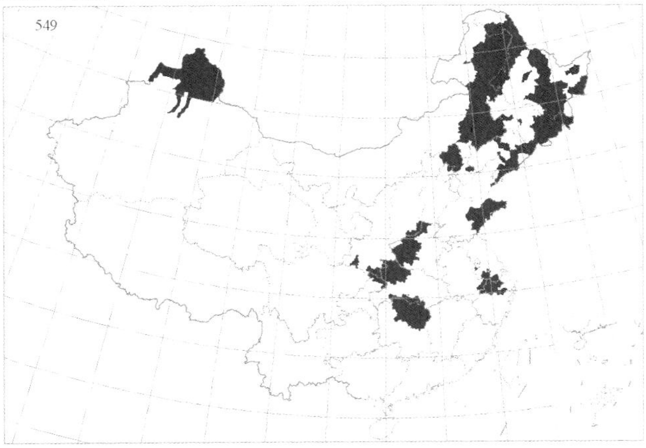

Salicaceae 杨柳科 *Salix* 柳属
Salix nipponica var. mengshanensis (S. B. Liang) G. H. Zhu 蒙山柳
Shrubs or trees, deciduous

12.6; -2.3; 25.8; 12.8; 108.7; 17.4; 775 ❙ 774; 498; 32 ❙ 0.0; 767; 136.6

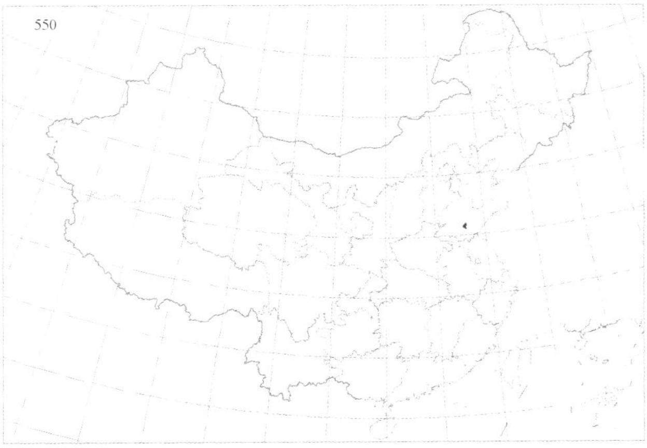

Salicaceae 杨柳科 *Salix* 柳属
Salix nujiangensis N. Chao 怒江柳
Shrubs, deciduous

8.6; 1.3; 15.0; 8.7; 52.6; 8.8; 574 ❙ 1041; 585; 45 ❙ 81.5; 574; 352.3

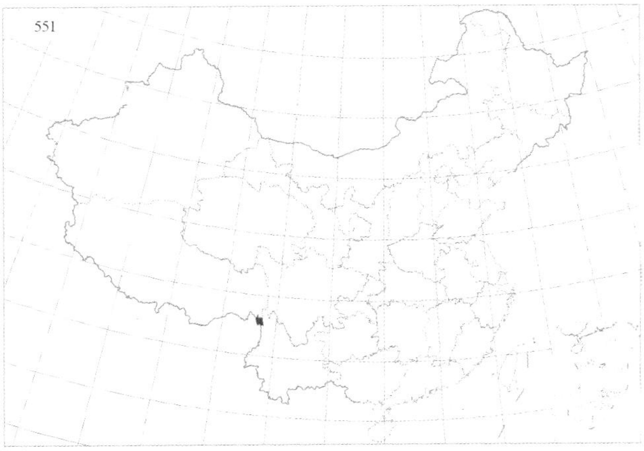

Salicaceae 杨柳科 *Salix* 柳属
Salix nummularia Andersson 多腺柳
Shrubs, deciduous

3.4 (1.4~5.2); -16.1 (-18.7~-14.1); 20.2 (18.4~22.1); 7.6 (6.4~8.7); 57.5 (45.8~69.2); 76.2 (65.0~88.8); 554 (501~601) ❙ 736 (561~1005); 449 (333~635); 27 (16~39) ❙ 32.6 (-2.6~68.8); 552 (501~601); 316.1 (289.3~330.3)

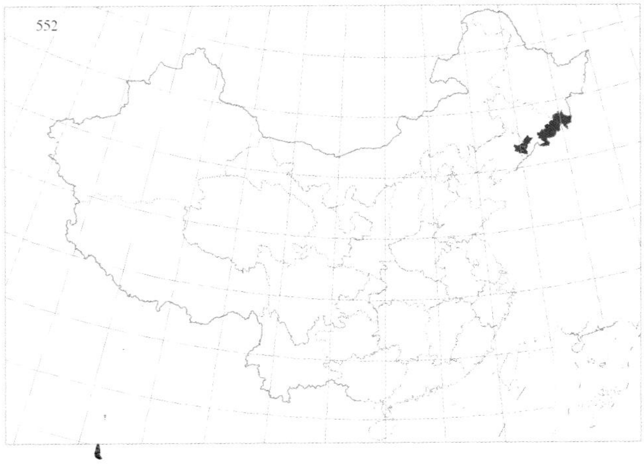

Salicaceae 杨柳科 *Salix* 柳属

Salix obscura Andersson 毛坡柳

Shrubs, deciduous

4.1 (−1.0~13.3); −5.4 (−12.0~5.4); 12.6 (8.3~19.6); 5.6 (2.7~13.4); 31.6 (8.3~100.4); 41.6 (0.0~79.5); 467 (363~706) ❙ 779 (375~1736); 432 (203~1007); 19 (7~51) ❙ 65.2 (−32.2~145.8); 458 (363~706); 254.1 (95.7~447.4)

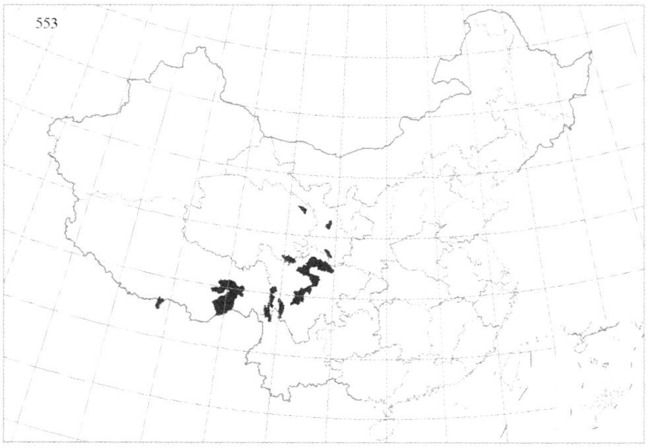

Salicaceae 杨柳科 *Salix* 柳属

Salix occidentalisinensis N. Chao 华西柳

Shrubs, deciduous

4.5 (−0.3~15.1); −5.0 (−9.9~4.4); 13.1 (8.1~25.4); 6.0 (2.7~15.2); 36.3 (9.0~122.8); 41.4 (0.6~71.8); 487 (354~817) ❙ 759 (255~1208); 429 (191~548); 17 (1~33) ❙ 59.5 (−27.8~128.6); 480 (255~817); 251.7 (101.7~341.6)

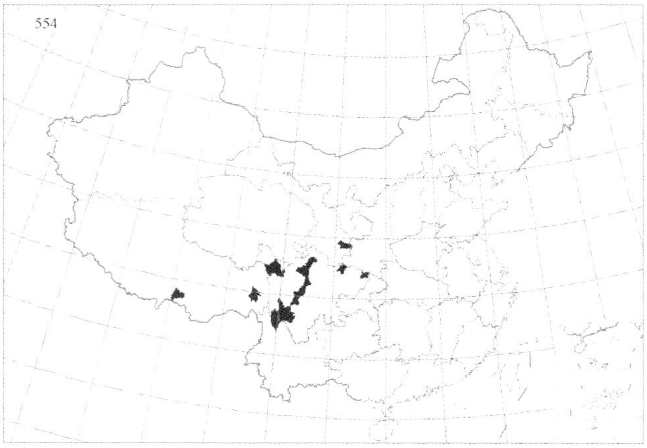

Salicaceae 杨柳科 *Salix* 柳属

Salix ochetophylla Goerz 汶川柳

Shrubs, deciduous

8.7 (2.0~15.5); −0.8 (−7.3~5.8); 17.2 (9.5~24.7); 9.3 (3.8~15.6); 64.6 (15.4~127.2); 19.8 (0.0~51.4); 593 (411~812) ❙ 1009 (375~1414); 545 (203~817); 30 (8~68) ❙ 73.1 (−32.2~128.6); 586 (375~812); 254.0 (95.7~357.7)

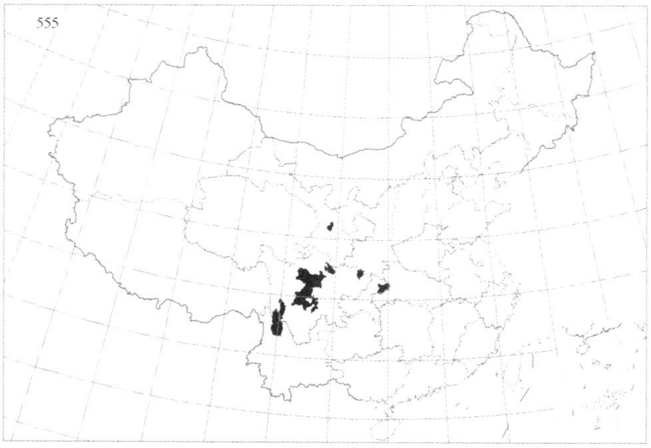

Salicaceae 杨柳科 *Salix* 柳属

Salix okamotoana Koidz. 关岭柳

Shrubs, deciduous

20.2 (16.2~24.1); 14.5 (11.0~18.9); 25.0 (20.1~28.2); 20.2 (16.2~24.1); 182.9 (134.2~229.7); 0.0 (0.0~0.0); 1019 (773~1322) ❙ 2402 (1498~3495); 1106 (750~1847); 255 (47~897) ❙ 146.2 (25.2~246.5); 983 (773~1144); 245.0 (119.4~363.7)

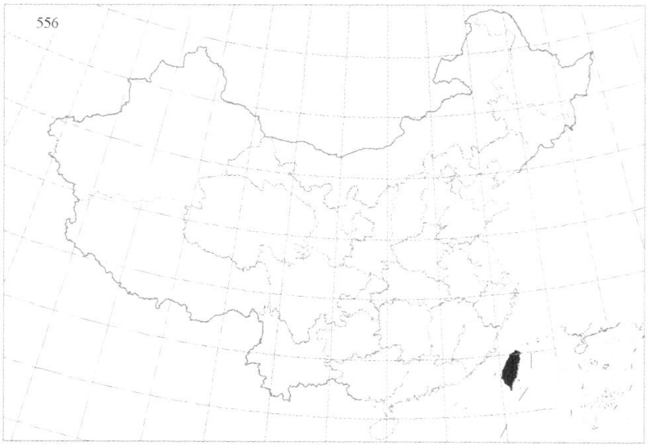

Salicaceae 杨柳科 *Salix* 柳属

Salix omeiensis C. K. Schneid. 峨眉柳

Shrubs or trees, deciduous

11.6 (5.7~15.5); 1.9 (−3.5~7.2); 20.7 (14.2~25.1); 11.8 (6.4~15.6); 89.0 (38.1~127.2); 8.8 (0.0~28.1); 679 (504~812) ❙ 1094 (711~1414); 562 (357~817); 37 (15~65) ❙ 62.4 (12.1~101.2); 679 (504~812); 256.1 (200.8~372.3)

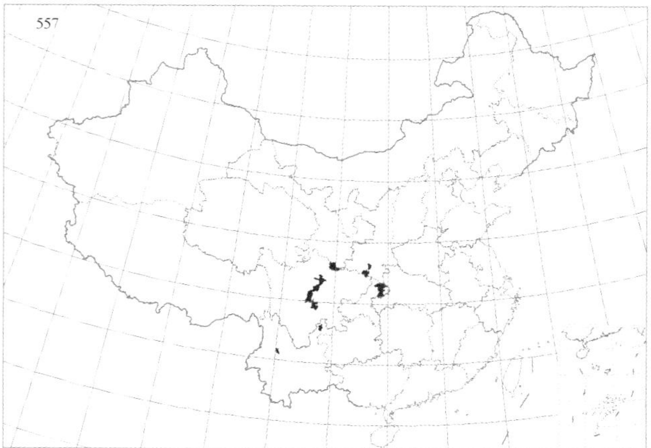

Salicaceae 杨柳科 *Salix* 柳属

Salix opsimantha C. K. Schneid. 迟花柳

Shrubs, deciduous

5.9 (0.3~17.3); −2.6 (−9.7~10.2); 13.3 (8.3~26.5); 6.9 (2.9~17.4); 42.6 (9.6~148.6); 31.4 (0.0~65.4); 510 (368~900) ❙ 821 (311~1431); 474 (232~817); 23 (2~60) ❙ 60.2 (−20.0~125.9); 506 (311~900); 260.7 (130.1~372.4)

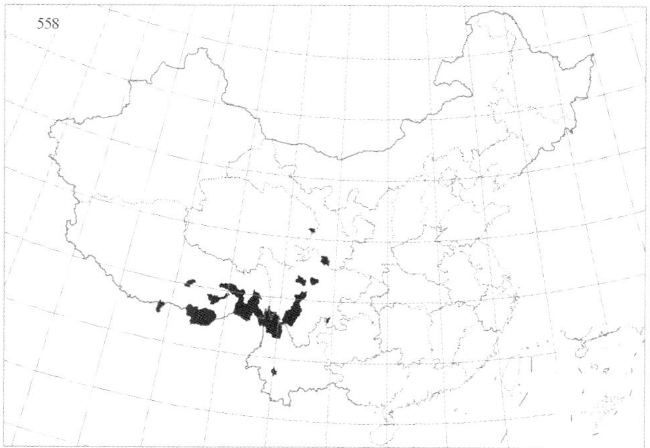

Salicaceae 杨柳科 *Salix* 柳属

Salix opsimantha var. wawashanica (P. Y. Mao et P. X. He) G. H. Zhu 娃娃山柳 Small shrubs, deciduous

15.9; 9.4; 20.6; 15.9; 130.7; 0.0; 781 ❙ 1046; 584; 49 ❙ 34.6; 756; 373.9

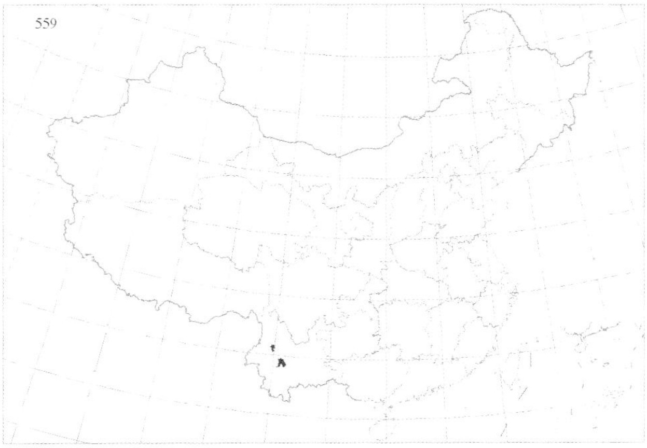

Salicaceae 杨柳科 *Salix* 柳属

Salix oreinoma C. K. Schneid. 迟花矮柳

Shrubs, deciduous

6.6 (−1.5~14.5); −1.2 (−10.5~7.8); 13.5 (7.1~19.7); 7.5 (2.2~14.5); 48.0 (4.7~113.5); 27.9 (0.0~82.2); 525 (332~730) ❙ 841 (311~1736); 499 (227~1007); 26 (2~54) ❙ 56.7 (−20.0~145.8); 519 (311~730); 271.5 (62.8~447.4)

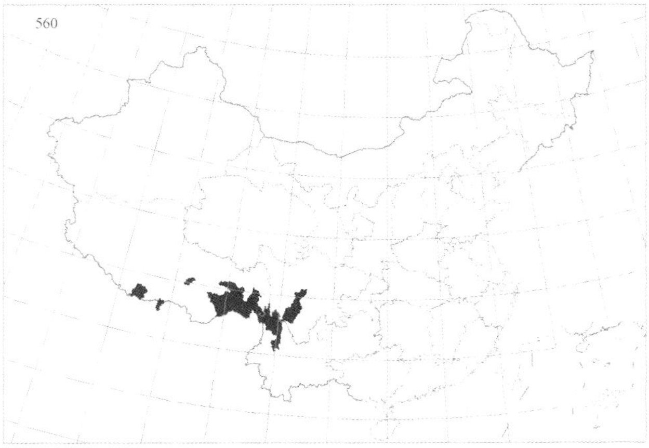

Salicaceae 杨柳科 *Salix* 柳属

Salix oreophila Hook. f. ex Andersson 尖齿叶垫柳

Shrubs, deciduous

6.0 (0.3~12.2); −2.0 (−8.6~4.3); 13.0 (8.3~18.5); 6.9 (2.9~12.2); 40.9 (9.6~87.4); 28.3 (0.7~65.4); 508 (368~672) ❙ 911 (389~1431); 527 (246~817); 29 (11~58) ❙ 74.8 (0.0~125.9); 508 (368~672); 271.0 (142.5~384.7)

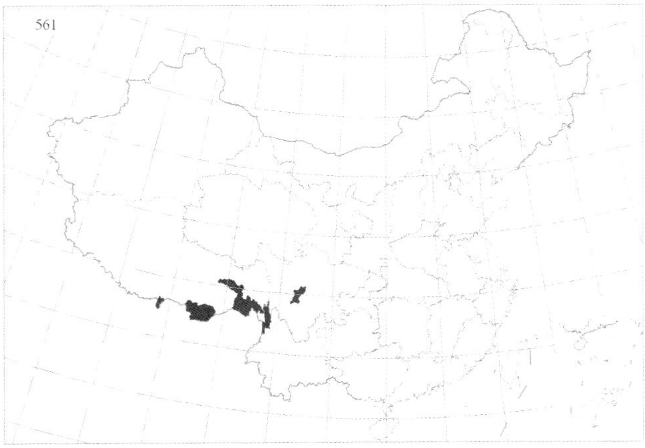

Salicaceae 杨柳科 *Salix* 柳属

Salix oreophila var. secta (Hook. f. ex Andersson) Andersson 五齿叶垫柳 Shrubs, deciduous

12.2; 4.3; 18.5; 12.2; 87.4; 0.7; 672 ❙ 1431; 817; 31 ❙ 113.0; 672; 363.7

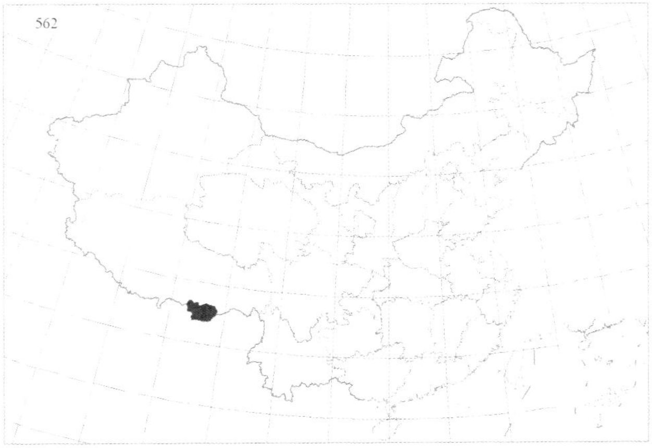

Salicaceae 杨柳科 *Salix* 柳属

Salix oritrepha C. K. Schneid. 山生柳

Shrubs, deciduous

2.7 (−5.9~14.9); −9.1 (−18.0~5.3); 13.2 (5.4~25.5); 5.3 (1.3~15.0); 32.2 (0.4~119.6); 59.2 (0.0~130.5); 460 (299~782) ❙ 511 (35~1410); 290 (23~816); 11 (2~49) ❙ 15.1 (−94.9~128.6); 385 (35~782); 205.7 (7.4~364.1)

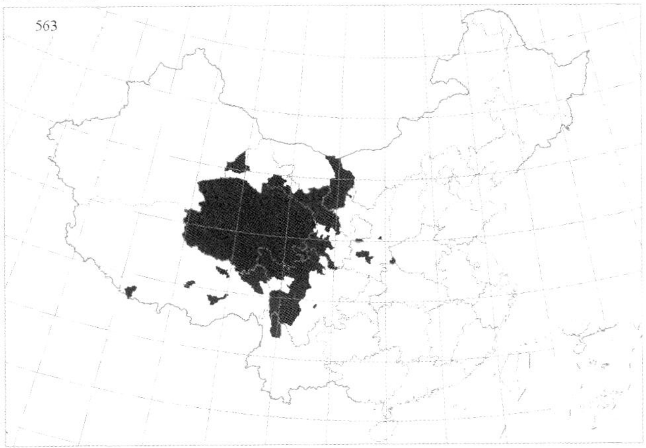

Salicaceae 杨柳科 *Salix* 柳属

Salix ovatomicrophylla K. S. Hao ex C. F. Fang et A. K. Skvortsov 卵小叶垫柳 Shrubs, deciduous

6.9 (0.0~14.5); −0.9 (−8.7~7.8); 13.7 (8.1~19.7); 7.7 (2.7~14.5); 50.1 (8.0~113.5); 26.7 (0.0~67.9); 531 (363~730) ❙ 862 (559~1057); 496 (335~610); 28 (13~54) ❙ 65.1 (36.1~125.9); 530 (363~730); 260.3 (111.0~382.6)

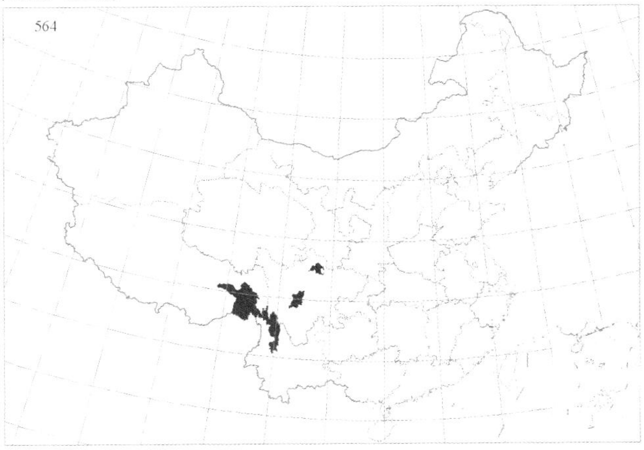

Salicaceae 杨柳科 *Salix* 柳属
Salix paraflabellaris S. D. Zhao 类扇叶垫柳
Shrubs, deciduous

10.2 (−0.7~16.2); 2.8 (−9.6~9.4); 16.4 (7.7~20.8); 10.5 (2.5~16.3); 74.7 (6.8~135.1); 11.5 (0.0~74.2); 619 (338~790) ❙ 1042 (370~1348); 588 (262~762); 42 (10~69) ❙ 66.6 (4.7~101.2); 619 (338~790); 310.9 (61.6~407.1)

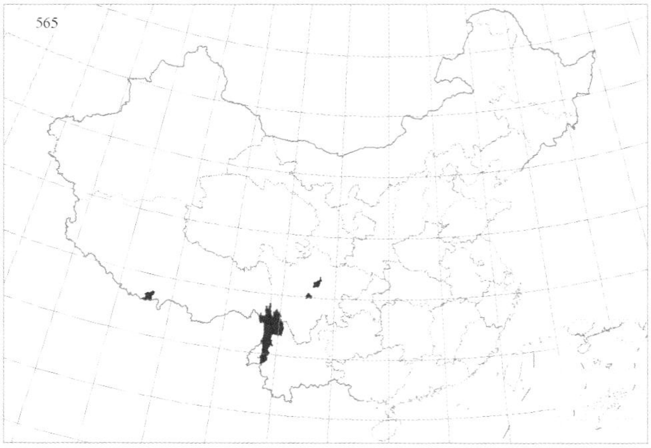

Salicaceae 杨柳科 *Salix* 柳属
Salix paraheterochroma C. Wang et P. Y. Fu 藏紫枝柳
Shrubs, deciduous

1.5; −7.1; 9.4; 3.7; 16.4; 58.3; 400 ❙ 620; 364; 19 ❙ 48.0; 397; 134.8

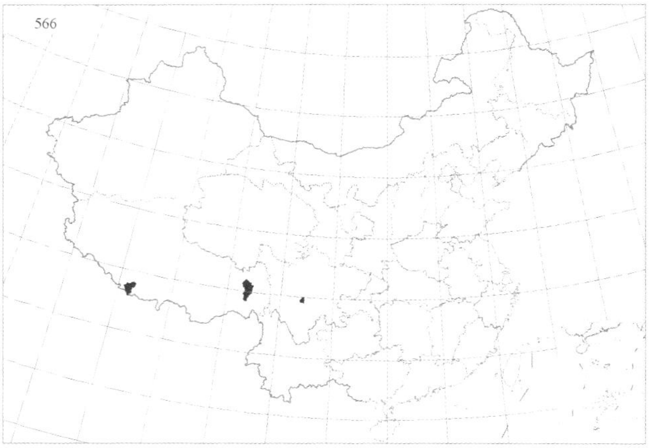

Salicaceae 杨柳科 *Salix* 柳属
Salix paraphylicifolia C. Y. Yang 光叶柳
Shrubs, deciduous

3.5 (−0.6~10.2); −15.4 (−19.5~−8.5); 19.9 (15.9~25.6); 7.9 (5.4~11.7); 60.9 (37.5~98.7); 78.3 (36.6~104.2); 581 (475~747) ❙ 160 (70~251); 71 (39~106); 18 (3~33) ❙ −71.0 (−90.6~−52.0); 160 (70~251); 70.9 (14.6~139.1)

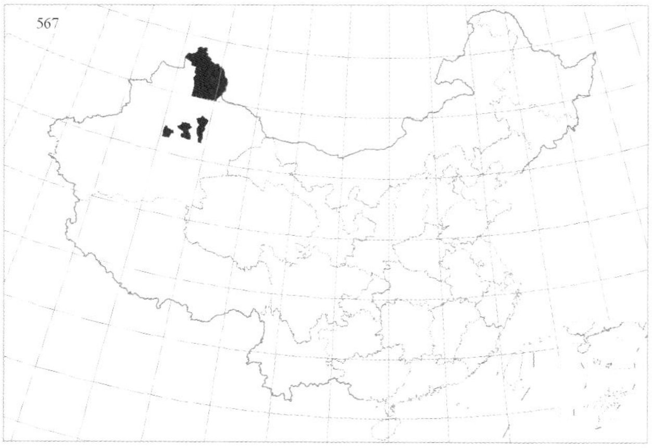

Salicaceae 杨柳科 *Salix* 柳属
Salix paraplesia C. K. Schneid. 康定柳
Trees, deciduous

6.3 (−4.1~18.5); −5.9 (−17.3~10.5); 17.3 (7.2~28.5); 7.9 (2.0~18.3); 56.1 (3.7~160.0); 39.6 (0.0~114.2); 556 (339~906) ❙ 641 (63~1441); 342 (45~688); 18 (0~94) ❙ 17.6 (−84.9~128.6); 510 (63~900); 228.7 (8.4~364.1)

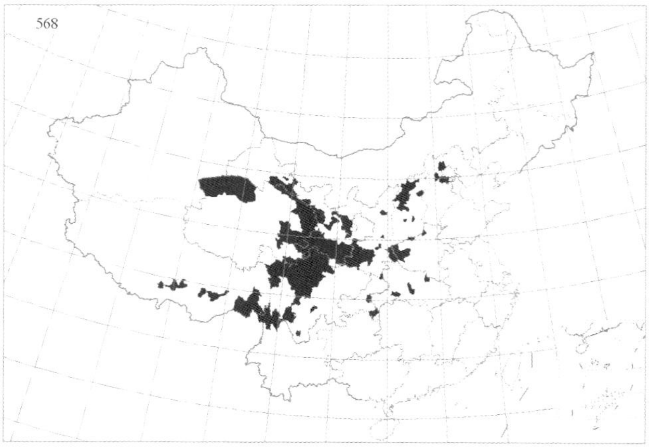

Salicaceae 杨柳科 *Salix* 柳属
Salix paraplesia var. pubescens C. Wang et C. F. Fang 毛枝康定柳
Trees, deciduous

2.5; −9.7; 13.1; 5.0; 27.9; 56.4; 449 ❙ 516; 283; 9 ❙ 14.7; 449; 218.4

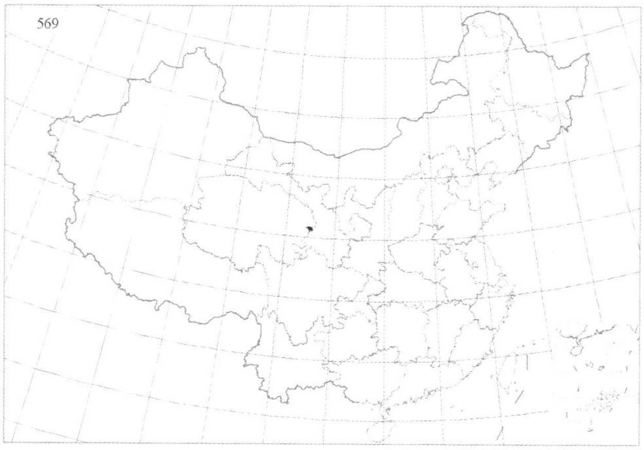

Salicaceae 杨柳科 *Salix* 柳属
Salix paraplesia var. subintegra C. Wang et P. Y. Fu 左旋康定柳
Trees, deciduous

2.6; −7.2; 11.2; 4.5; 23.0; 51.1; 422 ❙ 306; 237; 1 ❙ −27.8; 306; 163.4

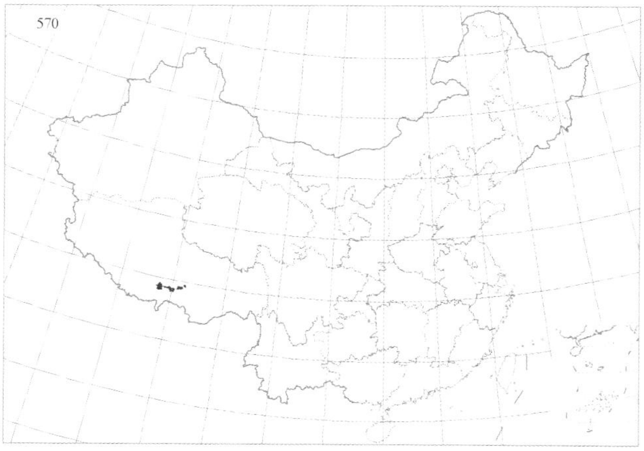

Salicaceae 杨柳科 *Salix* 柳属
Salix paratetradenia C. Wang et P. Y. Fu 类四腺柳
Shrubs, deciduous

4.9 (−2.5~12.8); −3.4 (−12.7~3.9); 12.3 (7.0~21.0); 6.1 (2.1~12.9); 35.1 (4.0~95.9); 35.6 (1.3~93.0); 481 (332~699) ❙ 683 (382~1007); 410 (246~594); 18 (6~40) ❙ 38.5 (0.0~67.9); 481 (332~699); 244.6 (130.1~351.7)

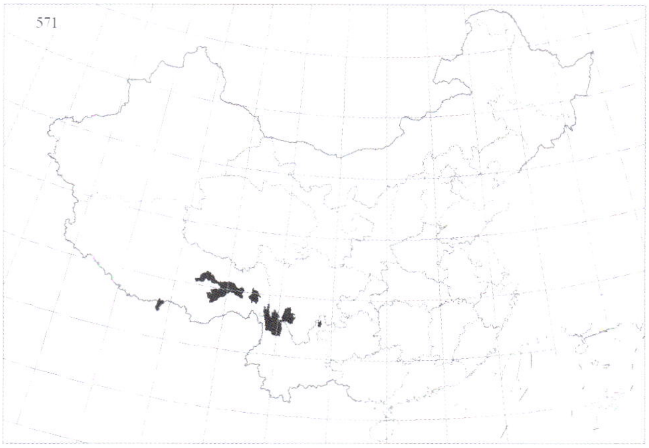

Salicaceae 杨柳科 *Salix* 柳属
Salix paratetradenia var. yatungensis C. Wang et P. Y. Fu 亚东柳
Shrubs, deciduous

0.3; −8.6; 8.3; 2.9; 9.6; 65.4; 368 ❙ 389; 246; 11 ❙ 0.0; 368; 142.5

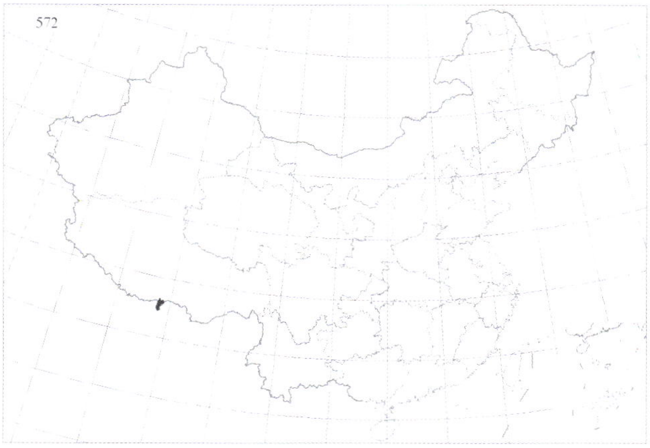

Salicaceae 杨柳科 *Salix* 柳属
Salix parvidenticulata C. F. Fang 小齿叶柳
Shrubs, deciduous

7.3 (−1.1~17.4); −2.0 (−9.1~7.4); 15.6 (6.2~25.5); 8.5 (2.0~17.4); 61.9 (2.5~148.9); 34.3 (0.0~75.6); 569 (334~886) ❙ 802 (321~1381); 397 (193~637); 47 (9~107) ❙ 33.9 (−4.0~55.9); 567 (321~886); 191.4 (79.3~280.0)

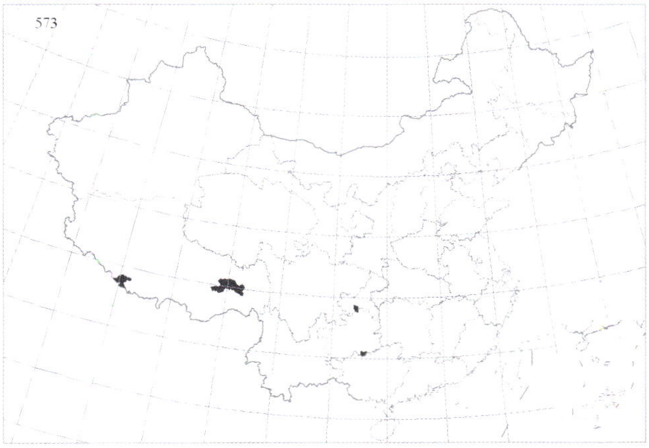

Salicaceae 杨柳科 *Salix* 柳属
Salix pella C. K. Schneid. 黑枝柳
Shrubs, deciduous

9.8 (3.0~15.5); 0.9 (−6.4~7.2); 17.9 (10.5~25.3); 10.2 (4.3~15.6); 72.9 (20.1~127.2); 14.2 (0.0~43.9); 622 (423~812) ❙ 1060 (628~1410); 575 (410~816); 36 (11~88) ❙ 71.2 (47.6~101.2); 622 (423~812); 269.3 (187.9~372.3)

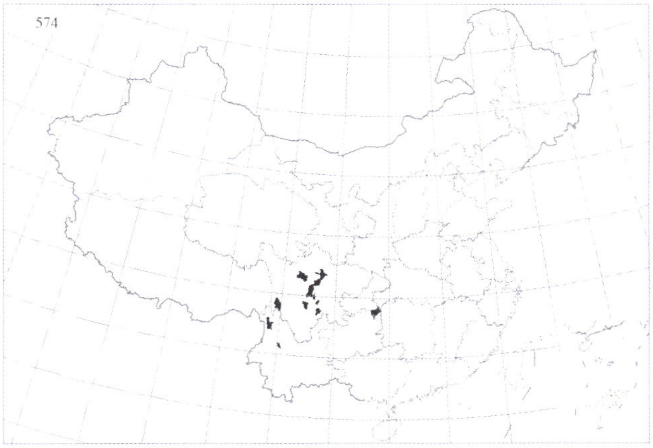

Salicaceae 杨柳科 *Salix* 柳属
Salix pentandra L. 五蕊柳
Shrubs or trees, deciduous

7.1 (−5.8~18.9); −9.3 (−29.8~12.0); 21.4 (12.7~28.0); 9.8 (4.4~19.0); 78.0 (28.0~167.8); 52.8 (0.0~157.4); 634 (426~929) ❙ 555 (111~1087); 334 (56~627); 19 (2~54) ❙ −13.4 (−82.3~67.7); 507 (111~817); 138.4 (13.9~382.6)

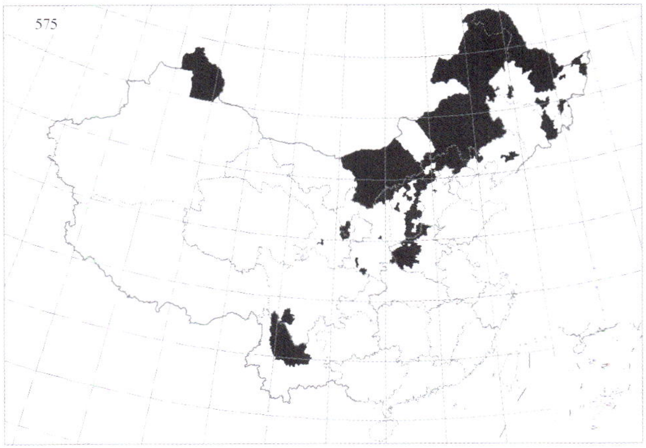

Salicaceae 杨柳科 *Salix* 柳属
Salix pentandra var. intermedia Nakai 白背五蕊柳
Shrubs or trees, deciduous

1.9 (−0.6~5.8); −17.1 (−19.5~−14.9); 18.7 (15.9~23.3); 7.0 (5.4~9.7); 52.4 (37.5~80.9); 88.6 (71.2~104.2); 545 (475~668) ❙ 184 (156~251); 77 (56~106); 22 (15~33) ❙ −65.8 (−76.7~−52.0); 184 (156~251); 86.2 (16.6~139.1)

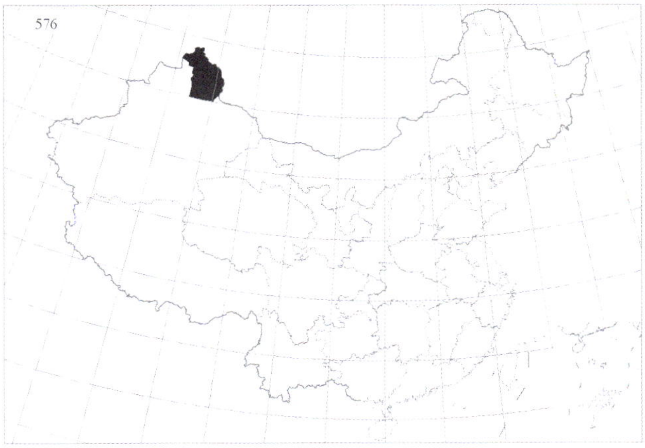

Salicaceae 杨柳科 *Salix* 柳属
Salix permollis C. Wang et C. Y. Yu 山毛柳
Trees, deciduous

10.2 (4.7~13.5); −2.1 (−6.1~2.3); 21.9 (14.7~27.1); 10.6 (6.0~13.7); 82.0 (36.3~117.5); 19.3 (4.9~39.6); 656 (482~820) ❙ 793 (533~1214); 367 (258~516); 27 (14~71) ❙ 23.3 (−35.0~101.4); 634 (482~767); 267.8 (159.3~323.3)

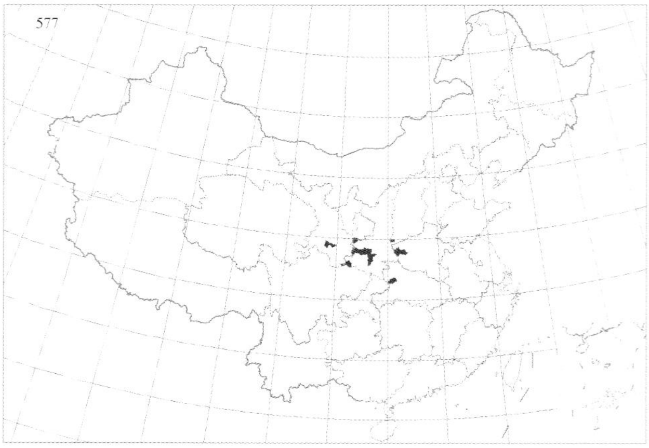

Salicaceae 杨柳科 *Salix* 柳属
Salix phaidima C. K. Schneid. 纤柳
Shrubs or trees, deciduous

9.8 (−1.8~15.5); 0.2 (−10.2~5.8); 18.6 (6.1~26.5); 10.3 (1.8~15.6); 75.2 (1.9~127.2); 17.4 (0.0~82.8); 631 (329~832) ❙ 1077 (320~1466); 544 (217~812); 50 (13~113) ❙ 70.7 (−3.0~125.9); 630 (320~832); 230.8 (72.2~318.7)

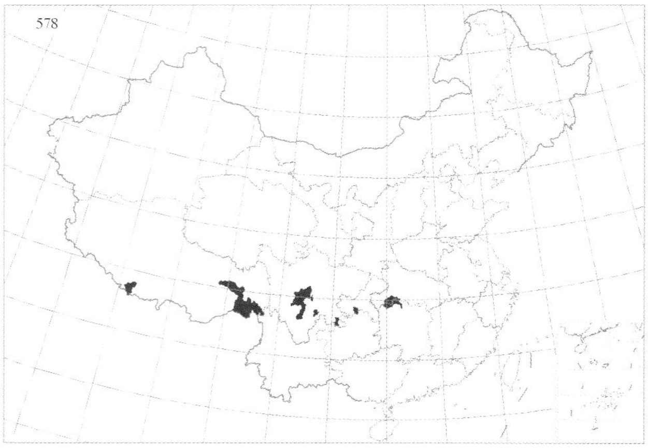

Salicaceae 杨柳科 *Salix* 柳属
Salix phanera C. K. Schneid. 长叶柳
Shrubs or trees, deciduous

8.1 (−1.5~18.5); −0.4 (−10.5~10.5); 15.6 (7.1~25.3); 8.8 (2.2~18.3); 60.4 (4.7~160.0); 22.5 (0.0~82.2); 572 (332~906) ❙ 920 (311~1358); 506 (227~651); 29 (10~88) ❙ 64.5 (−6.3~128.6); 563 (311~795); 290.7 (62.8~382.6)

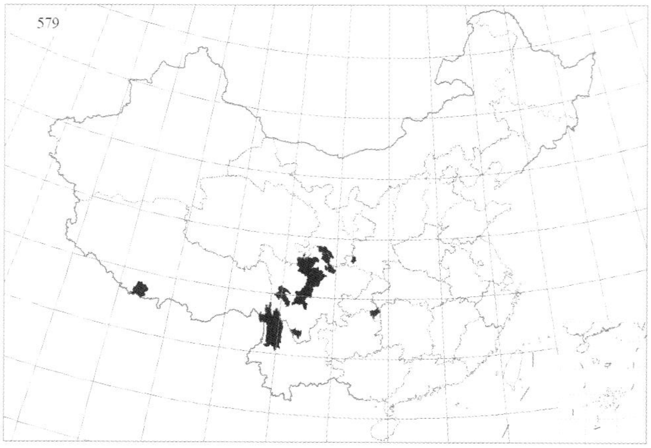

Salicaceae 杨柳科 *Salix* 柳属
Salix phanera var. weixiensis C. F. Fang 维西长叶柳
Shrubs or trees, deciduous

8.6; 1.5; 14.9; 8.8; 54.5; 10.2; 573 ❙ 981; 560; 40 ❙ 70.4; 573; 329.2

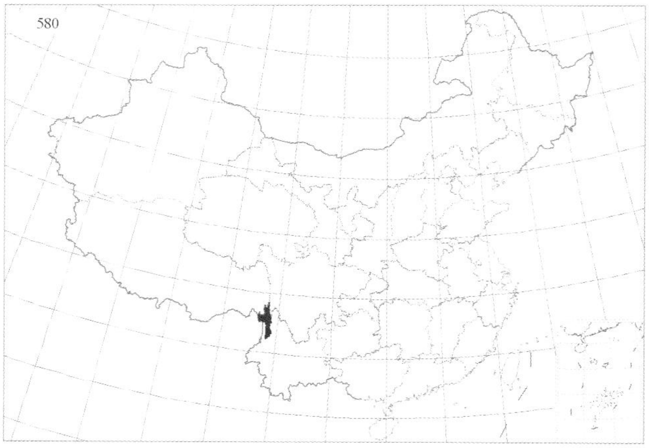

Salicaceae 杨柳科 *Salix* 柳属
Salix pierotii Miq. 白皮柳
Shrubs or trees, deciduous

5.8; −13.9; 22.7; 9.1; 73.8; 63.2; 624 ❙ 799; 505; 27 ❙ 29.1; 623; 280.8

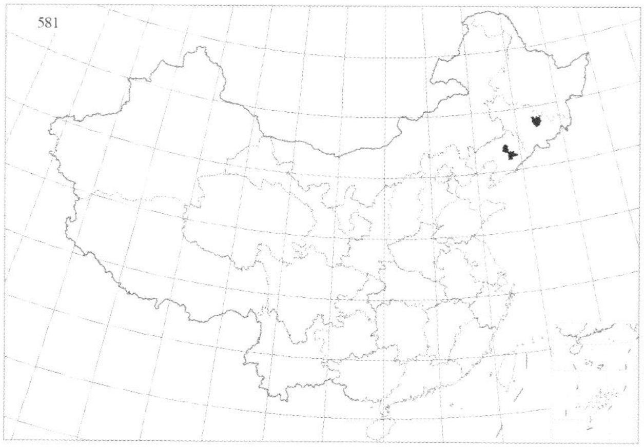

Salicaceae 杨柳科 *Salix* 柳属
Salix pilosomicrophylla C. Wang et P. Y. Fu 毛小叶垫柳
Shrubs, deciduous

9.2 (0.1~16.2); 1.7 (−8.9~9.4); 15.4 (8.1~20.8); 9.7 (2.8~16.3); 68.0 (8.9~135.1); 17.0 (0.0~67.7); 592 (358~790) ❙ 999 (320~1736); 568 (225~1007); 40 (3~69) ❙ 63.8 (−10.6~145.8); 589 (320~790); 314.5 (118.8~447.4)

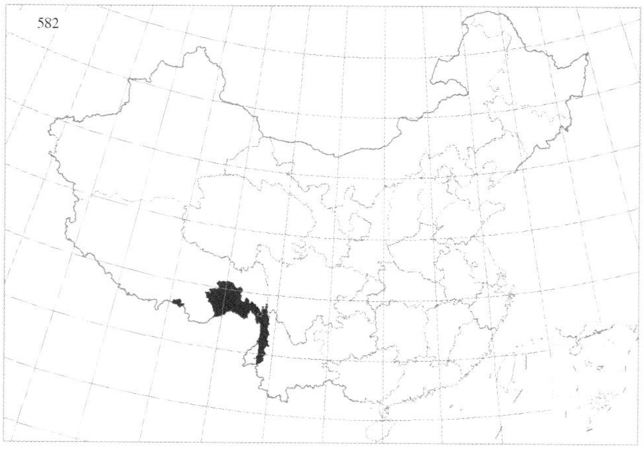

Salicaceae 杨柳科 *Salix* 柳属
Salix pingliensis Y. L. Chou 平利柳
Trees, deciduous

10.6; -2.2; 22.6; 11.1; 87.1; 19.7; 673 ▎744; 345; 22 ▎8.6; 621; 215.5

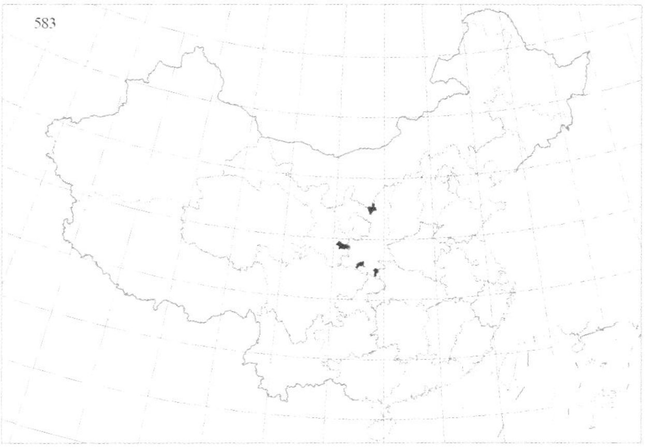

Salicaceae 杨柳科 *Salix* 柳属
Salix piptotricha Hand.-Mazz. 毛果垫柳
Shrubs, deciduous

8.1 (5.4~10.8); 0.8 (-2.1~3.9); 14.5 (12.3~16.6); 8.3 (5.8~10.9); 50.4 (31.5~71.8); 12.6 (1.3~25.6); 557 (481~633) ▎912 (741~1041); 536 (426~594); 32 (21~45) ▎63.5 (52.8~81.5); 557 (481~633); 316.0 (273.5~352.3)

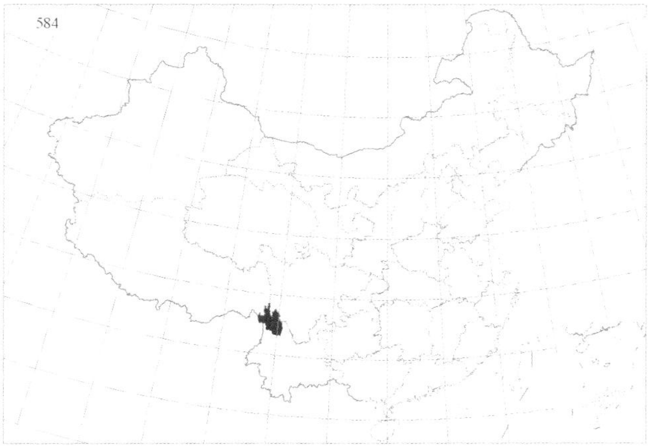

Salicaceae 杨柳科 *Salix* 柳属
Salix plocotricha C. K. Schneid. 曲毛柳
Shrubs, deciduous

6.5 (0.8~15.9); -3.2 (-10.6~6.1); 15.1 (9.5~24.9); 7.4 (3.8~16.0); 48.0 (15.5~131.7); 29.7 (0.0~66.0); 531 (411~824) ▎819 (375~1410); 457 (203~816); 21 (8~52) ▎55.2 (-32.2~125.9); 522 (375~824); 249.9 (95.7~318.7)

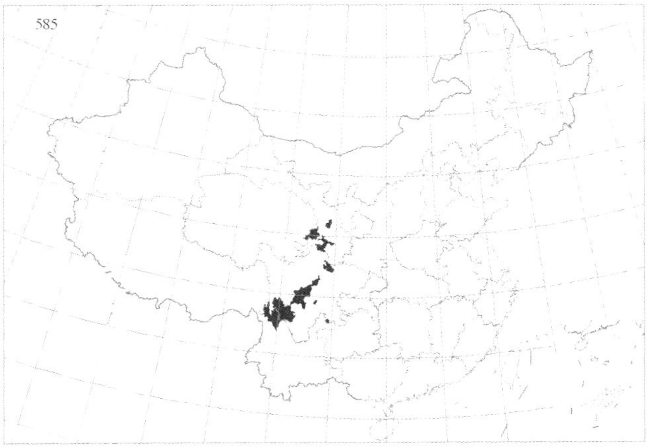

Salicaceae 杨柳科 *Salix* 柳属
Salix polyclona C. K. Schneid. 多枝柳
Shrubs, deciduous

11.2 (0.9~15.8); 0.0 (-9.5~5.8); 21.6 (9.8~26.7); 11.4 (3.6~15.8); 88.4 (14.0~130.6); 14.0 (0.0~63.1); 681 (403~851) ▎1004 (622~1466); 479 (290~816); 43 (11~123) ▎50.7 (-16.6~125.9); 673 (403~851); 245.7 (171.0~318.7)

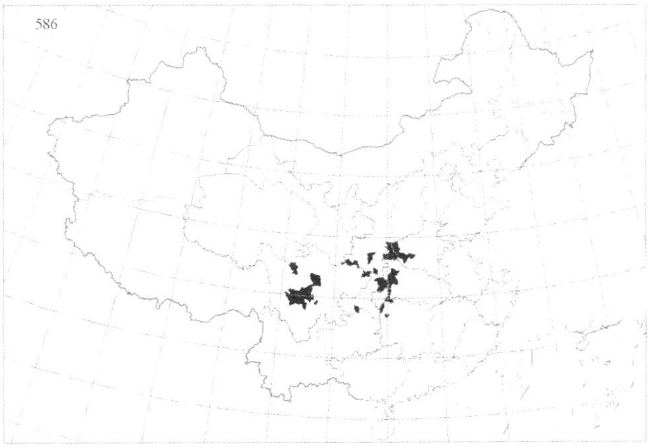

Salicaceae 杨柳科 *Salix* 柳属
Salix praticola Hand.-Mazz. ex Enander 草地柳
Shrubs, deciduous

15.3 (2.2~21.3); 6.2 (-6.6~14.2); 23.1 (9.5~28.9); 15.4 (3.8~21.3); 125.9 (15.5~196.0); 1.9 (0.0~49.1); 803 (411~1123) ▎1164 (628~1751); 573 (406~996); 63 (11~191) ▎46.5 (-14.2~125.9); 796 (411~1121); 249.0 (108.2~551.6)

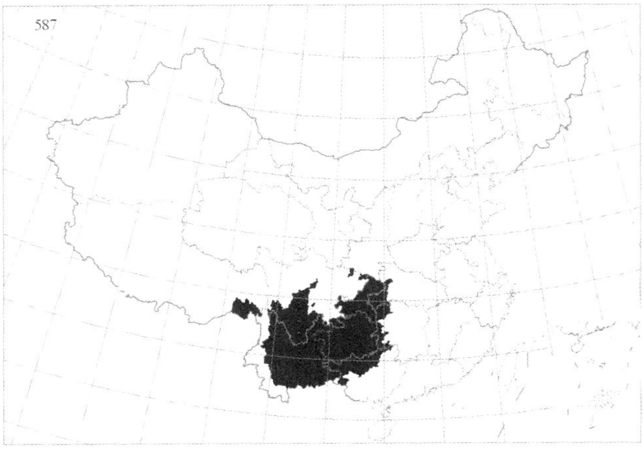

Salicaceae 杨柳科 *Salix* 柳属
Salix psammophila C. Wang et C. Y. Yang 北沙柳
Shrubs, deciduous

7.1 (1.2~9.0); -10.2 (-19.9~-7.3); 22.3 (18.6~24.0); 9.4 (6.9~10.5); 75.7 (48.9~84.9); 49.7 (36.0~97.3); 627 (523~668) ▎283 (58~491); 171 (36~309); 7 (2~19) ▎-54.3 (-90.8~-12.6); 283 (58~491); 54.9 (11.3~247.3)

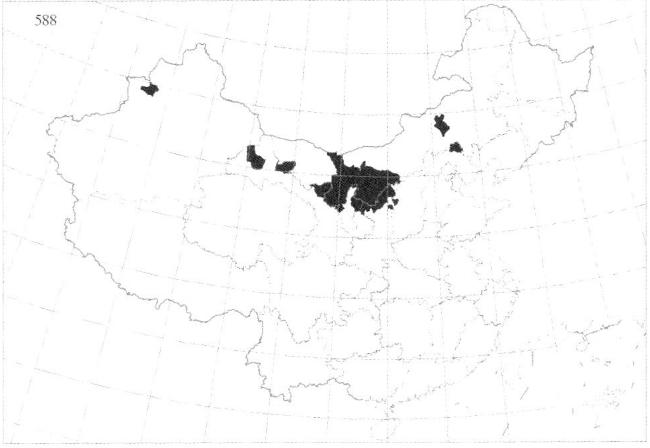

Salicaceae 杨柳科 *Salix* 柳属
Salix pseudolasiogyne H. Lév. 朝鲜垂柳
Trees, deciduous

5.7 (0.2~10.5); -13.7 (-20.8~-4.9); 22.5 (18.6~25.0); 9.1 (6.4~11.3); 73.8 (47.8~93.2); 65.1 (26.8~104.9); 626 (513~705) ❙ 569 (409~718); 374 (298~446); 17 (6~29) ❙ -9.4 (-36.8~9.1); 562 (409~698); 204.6 (113.4~279.5)

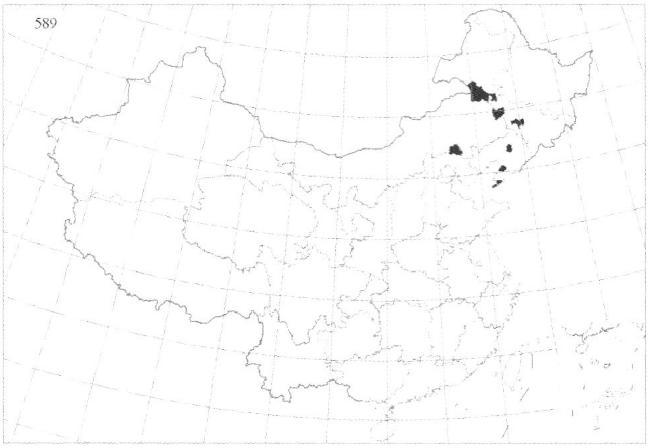

Salicaceae 杨柳科 *Salix* 柳属
Salix pseudopermollis C. Y. Yu et C. Y. Yang 小叶山毛柳
Shrubs or trees, deciduous

9.6 (-1.4~13.8); -4.4 (-15.4~1.5); 22.3 (11.5~27.3); 10.5 (3.5~14.0); 84.3 (17.2~121.1); 28.6 (7.1~92.1); 669 (375~842) ❙ 664 (282~882); 371 (182~485); 21 (5~31) ❙ 0.2 (-24.8~52.6); 611 (282~760); 209.1 (101.3~318.2)

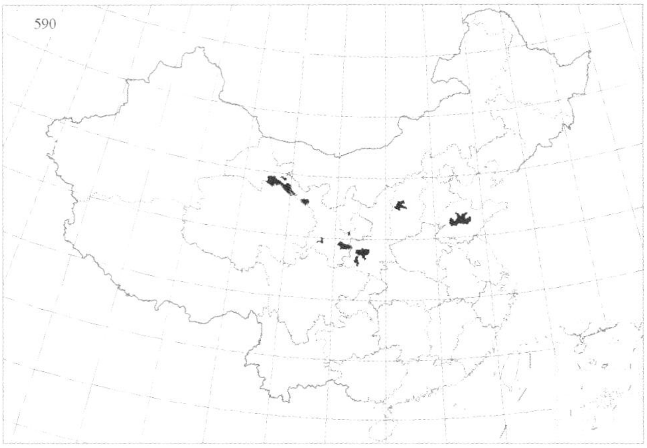

Salicaceae 杨柳科 *Salix* 柳属
Salix pseudospissa Goerz ex Rehder et Kobuski 大苞柳
Shrubs, deciduous

3.3 (-1.2~6.6); -8.2 (-11.7~-5.9); 13.5 (8.5~17.8); 5.3 (2.6~8.0); 31.3 (8.1~57.8); 51.6 (37.9~82.3); 459 (347~557) ❙ 580 (457~829); 304 (249~400); 13 (8~23) ❙ 28.3 (-11.1~94.4); 451 (347~515); 249.6 (137.5~320.9)

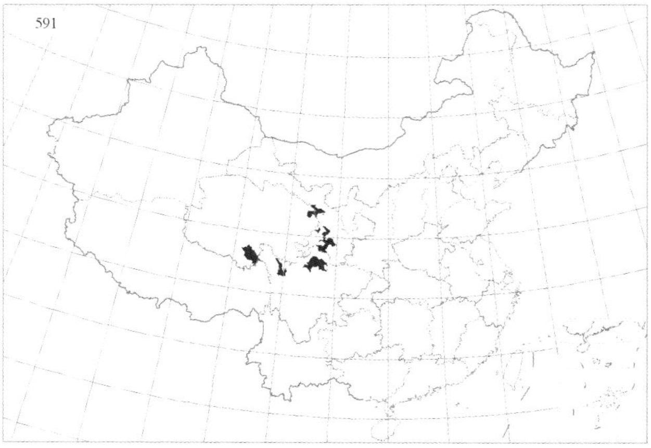

Salicaceae 杨柳科 *Salix* 柳属
Salix pseudotangii C. Wang et C. Y. Yu 山柳
Shrubs, deciduous

8.8 (3.5~11.3); -3.8 (-7.3~-0.7); 20.6 (12.6~24.3); 9.6 (5.0~11.5); 73.0 (26.3~92.4); 26.5 (13.5~43.4); 622 (446~701) ❙ 723 (543~856); 360 (331~406); 20 (12~25) ❙ 20.4 (-22.6~63.8); 594 (446~697); 275.9 (214.0~318.2)

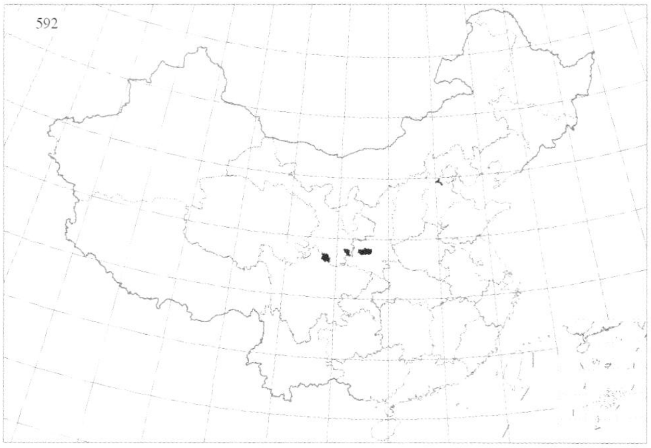

Salicaceae 杨柳科 *Salix* 柳属
Salix pseudowallichiana Goerz ex Rehder et Kobuski 青皂柳
Shrubs or small trees, deciduous

5.0 (-1.4~16.0); -6.6 (-15.4~5.3); 15.3 (9.7~26.1); 6.7 (2.9~16.1); 44.3 (11.3~132.7); 43.9 (0.0~92.1); 512 (359~856) ❙ 643 (282~1081); 355 (182~606); 14 (5~32) ❙ 28.5 (-32.2~121.7); 484 (282~856); 238.3 (95.7~357.7)

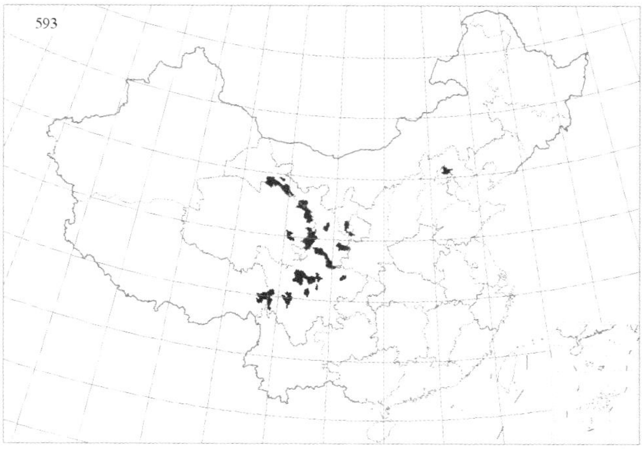

Salicaceae 杨柳科 *Salix* 柳属
Salix pseudowolohoensis K. S. Hao ex C. F. Fang et A. K. Skvortsov 西柳
Shrubs, deciduous

10.0 (2.2~16.7); 1.9 (-6.6~9.8); 17.0 (9.9~24.7); 10.2 (3.8~16.8); 70.5 (15.5~141.1); 10.4 (0.0~49.1); 619 (413~818) ❙ 1015 (741~1398); 577 (426~812); 32 (12~65) ❙ 66.4 (11.7~125.9); 612 (413~812); 290.5 (214.7~352.3)

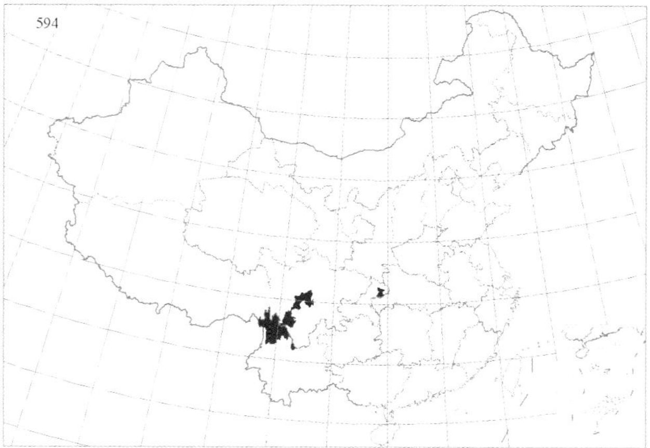

Salicaceae 杨柳科 *Salix* 柳属

Salix psilostigma Andersson 裸柱头柳

Shrubs, deciduous

13.8 (-0.7~18.9); 6.5 (-9.6~12.5); 19.6 (7.7~26.3); 14.0 (2.5~19.0); 110.5 (6.8~167.8); 4.3 (0.0~74.2); 727 (338~929) ❙ 1020 (341~1736); 574 (226~1007); 41 (4~106) ❙ 41.8 (-15.3~145.8); 705 (338~854); 291.6 (61.6~447.4)

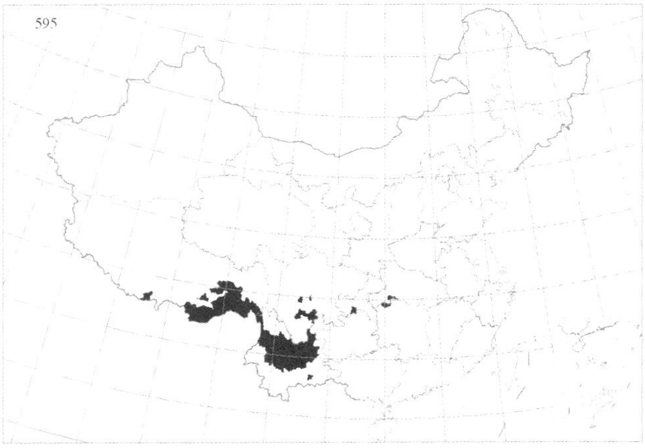

Salicaceae 杨柳科 *Salix* 柳属

Salix pycnostachya Andersson 密穗柳

Tall shrubs, deciduous

2.5 (-5.7~12.3); -13.2 (-20.1~-5.3); 15.9 (6.2~26.0); 6.4 (1.7~13.0); 46.2 (2.1~112.3); 75.9 (24.7~133.0); 507 (294~801) ❙ 186 (21~620); 93 (17~306); 22 (0~103) ❙ -53.2 (-97.3~91.4); 159 (21~337); 58.2 (1.6~297.3)

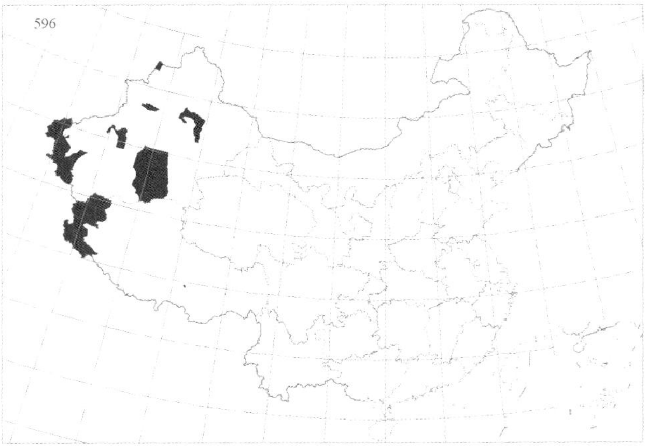

Salicaceae 杨柳科 *Salix* 柳属

Salix pyrolifolia Ledeb. 鹿蹄柳

Shrubs or trees, deciduous

3.1 (-5.8~10.2); -16.7 (-29.8~-8.5); 20.1 (10.9~27.5); 7.8 (3.5~12.1); 60.6 (16.3~107.6); 82.4 (36.6~157.4); 577 (390~813) ❙ 269 (45~510); 155 (27~339); 16 (3~51) ❙ -49.8 (-94.0~15.9); 266 (45~487); 133.0 (7.6~309.5)

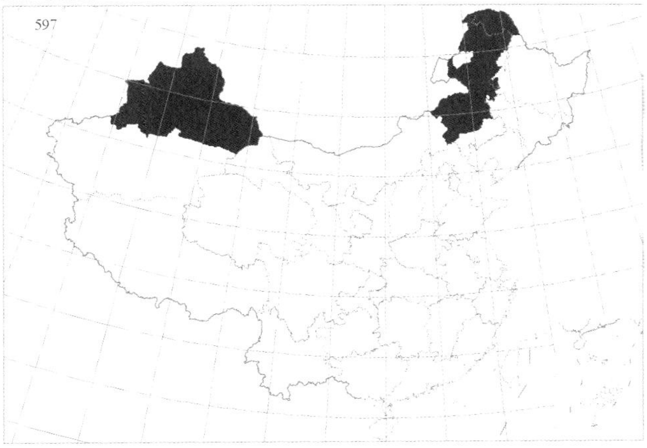

Salicaceae 杨柳科 *Salix* 柳属

Salix qinghaiensis Y. L. Chou 青海柳

Shrubs or trees, deciduous

3.7 (-4.1~10.5); -8.2 (-17.3~-0.1); 14.2 (7.8~20.4); 5.8 (2.1~10.6); 36.0 (5.2~78.8); 51.2 (11.7~114.2); 477 (340~644) ❙ 548 (361~773); 291 (219~394); 11 (5~19) ❙ 15.4 (-29.0~84.1); 460 (340~644); 235.8 (137.5~320.9)

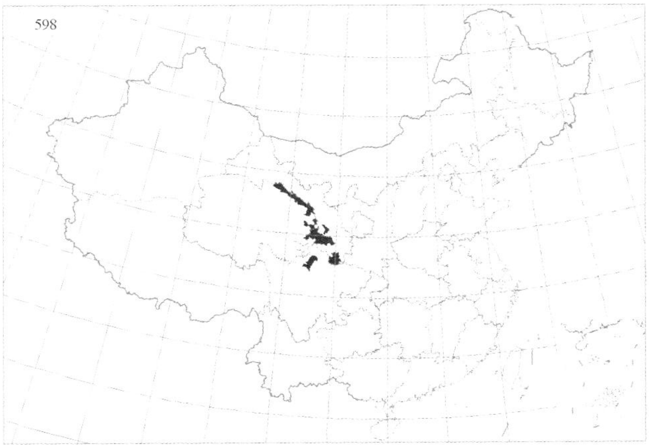

Salicaceae 杨柳科 *Salix* 柳属

Salix qinghaiensis var. microphylla Y. L. Chou 小叶青海柳

Shrubs or trees, deciduous

4.0; -7.1; 14.1; 5.6; 33.2; 45.0; 470 ❙ 608; 305; 13 ❙ 29.9; 470; 276.1

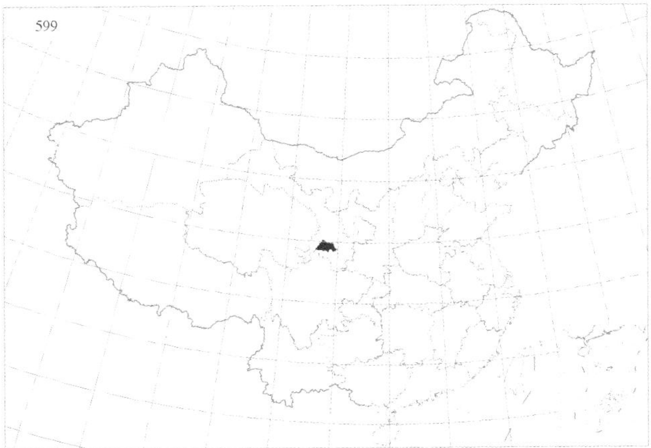

Salicaceae 杨柳科 *Salix* 柳属

Salix qinlingica C. Wang et N. Chao 陕西柳

Shrubs or small trees, deciduous

9.5; -2.6; 21.3; 9.9; 76.1; 21.3; 630 ❙ 820; 367; 25 ❙ 31.6; 630; 290.5

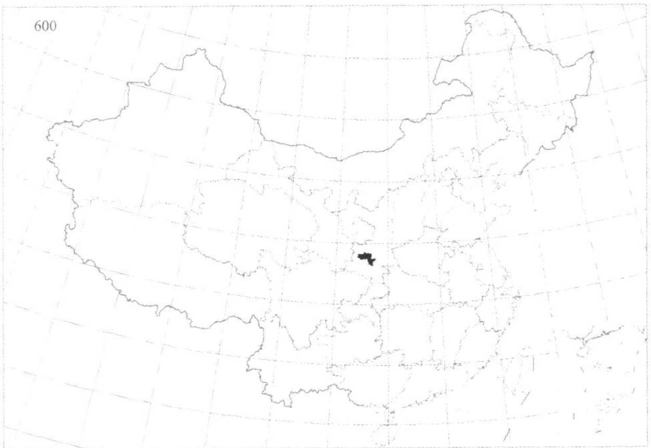

Salicaceae 杨柳科 *Salix* 柳属
Salix raddeana Lacksch. ex Nasarow 大黄柳
Shrubs or trees, deciduous

2.7 (−5.8~9.4); −18.3 (−29.8~−7.5); 20.9 (15.3~25.0); 7.8 (4.4~10.9); 60.3 (28.0~91.7); 87.8 (38.2~157.4); 571 (426~704) ❙ 567 (311~966); 363 (214~616); 17 (5~35) ❙ −0.8 (−40.0~57.4); 522 (311~698); 254.9 (106.8~346.5)

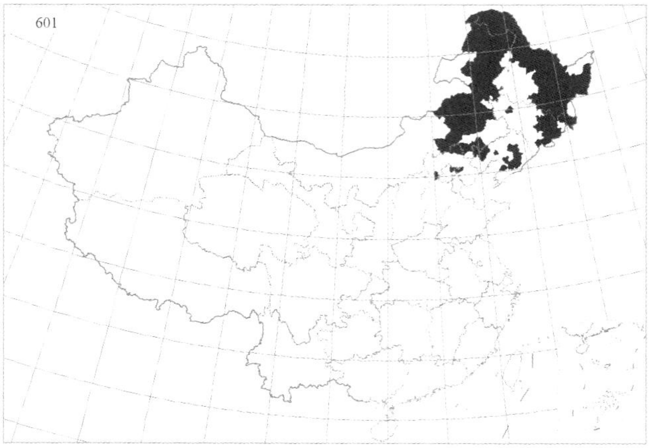

Salicaceae 杨柳科 *Salix* 柳属
Salix raddeana var. subglabra Y. L. Chang et Skvortsov 稀毛大黄柳
Shrubs or trees, deciduous

6.4; −13.8; 23.6; 9.6; 79.3; 62.2; 649 ❙ 728; 459; 24 ❙ 11.3; 649; 242.9

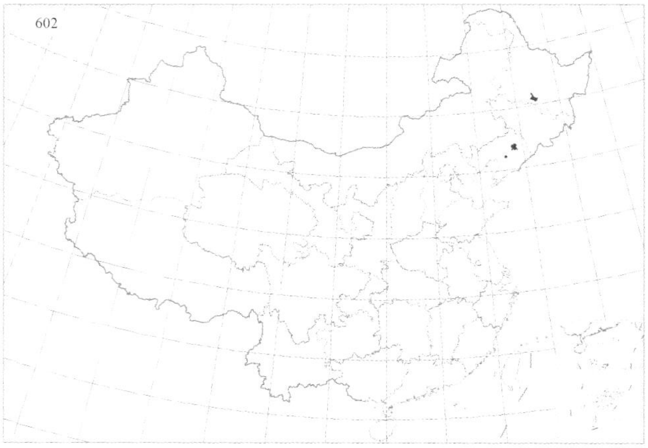

Salicaceae 杨柳科 *Salix* 柳属
Salix radinostachya C. K. Schneid. 长穗柳
Shrubs, deciduous

11.8 (0.3~18.9); 4.5 (−8.6~12.0); 17.9 (8.3~24.7); 12.1 (2.9~19.0); 90.3 (9.6~167.8); 7.7 (0.0~65.4); 668 (368~929) ❙ 974 (389~1454); 561 (246~792); 34 (11~70) ❙ 48.3 (−14.2~100.7); 644 (368~801); 301.3 (130.1~487.2)

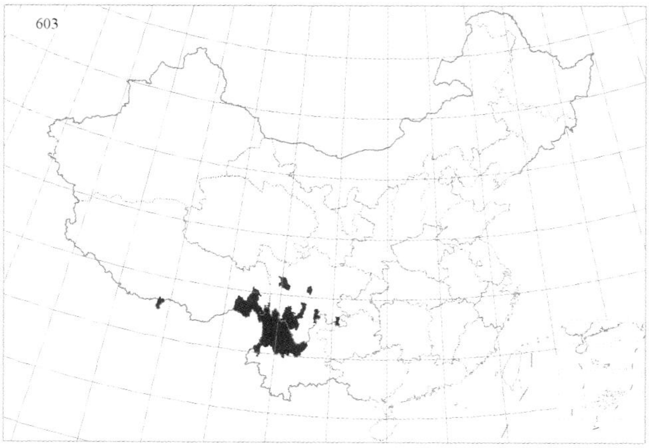

Salicaceae 杨柳科 *Salix* 柳属
Salix radinostachya var. pseudophanera C. F. Fang 绒毛长穗柳
Shrubs, deciduous

7.9; 0.5; 14.3; 8.0; 48.3; 13.3; 552 ❙ 930; 530; 38 ❙ 67.8; 552; 325.8

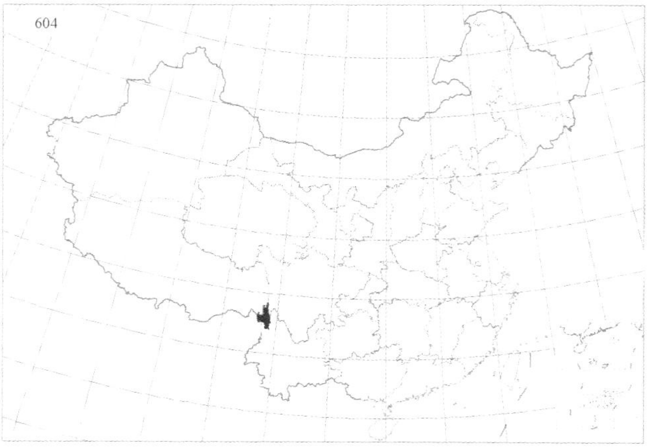

Salicaceae 杨柳科 *Salix* 柳属
Salix rectijulis Ledeb. ex Trautv. 欧越橘柳
Shrubs, deciduous

1.2 (−2.8~5.8); −17.5 (−20.3~−14.9); 17.5 (10.9~23.3); 6.5 (3.5~9.7); 46.8 (16.3~80.9); 91.5 (71.2~109.2); 521 (390~668) ❙ 215 (156~450); 95 (56~185); 23 (6~46) ❙ −56.5 (−76.7~3.9); 211 (156~404); 91.2 (16.6~139.1)

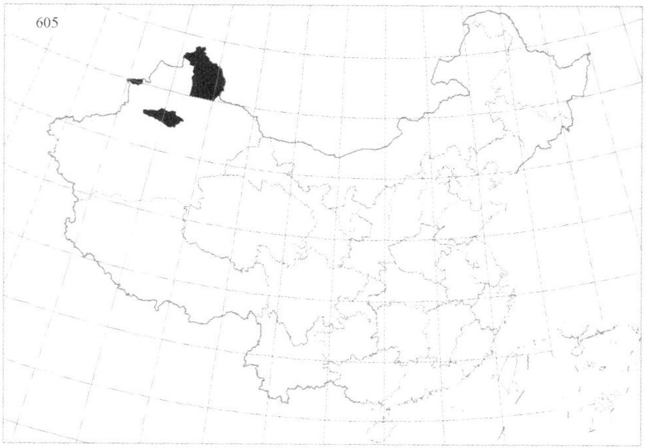

Salicaceae 杨柳科 *Salix* 柳属
Salix rehderiana C. K. Schneid. 川滇柳
Shrubs or small trees, deciduous

7.9 (−4.1~17.0); −4.9 (−17.3~10.0); 19.2 (7.2~28.8); 9.3 (2.0~17.0); 70.0 (3.7~144.0); 34.8 (0.0~114.2); 607 (339~932) ❙ 643 (63~1736); 363 (45~1007); 19 (1~115) ❙ 7.6 (−87.1~145.8); 519 (63~917); 210.6 (7.5~447.4)

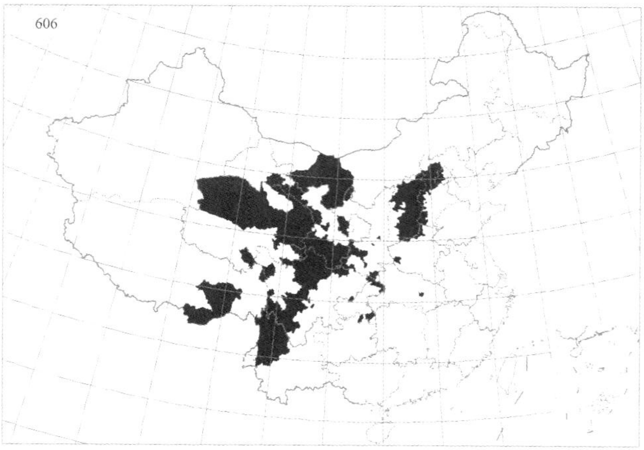

Salicaceae 杨柳科 *Salix* 柳属
Salix rehderiana var. dolia (C. K. Schneid.) N. Chao 灌柳
Shrubs or small trees, deciduous

5.7 (-1.3~15.1); -4.4 (-13.5~4.4); 14.7 (8.5~25.4); 7.1 (2.6~15.2); 45.1 (8.1~122.8); 36.2 (0.6~86.3); 517 (347~817) | 802 (386~1081); 444 (219~619); 18 (5~32) | 57.0 (-29.0~125.9); 510 (347~817); 260.3 (155.3~357.7)

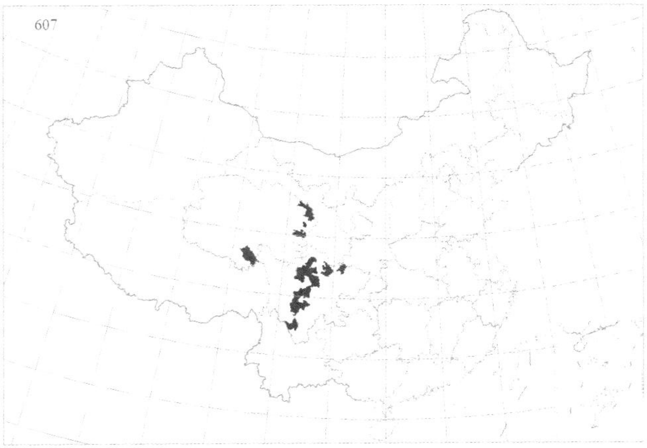

Salicaceae 杨柳科 *Salix* 柳属
Salix resecta Diels 截苞柳
Shrubs, deciduous

10.4 (2.2~15.1); 2.0 (-6.4~7.8); 17.8 (9.5~25.6); 10.7 (3.9~15.1); 76.2 (15.8~122.0); 11.4 (0.0~48.6); 634 (411~813) | 1020 (729~1441); 538 (426~627); 39 (11~84) | 62.5 (36.1~95.9); 633 (411~813); 292.6 (170.4~382.6)

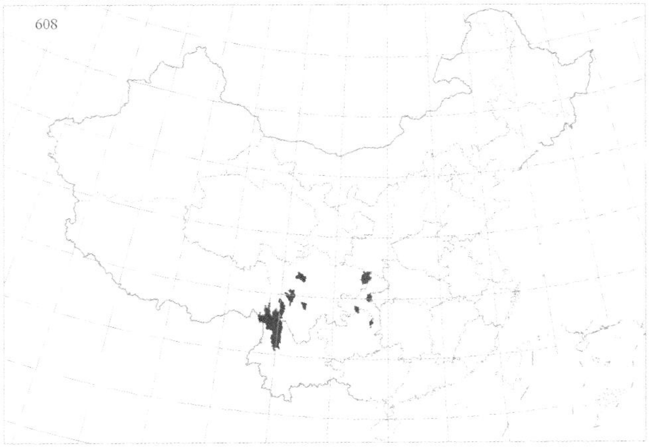

Salicaceae 杨柳科 *Salix* 柳属
Salix resectoides Hand.-Mazz. 藏截苞矮柳
Small shrubs, deciduous

5.4 (-4.8~16.2); -3.4 (-17.3~9.4); 13.0 (5.4~20.8); 7.4 (1.6~16.3); 48.2 (0.4~135.1); 42.8 (0.0~118.7); 511 (300~790) | 767 (220~1736); 446 (160~1007); 31 (1~103) | 43.1 (-27.5~145.8); 501 (220~790); 260.7 (46.9~447.4)

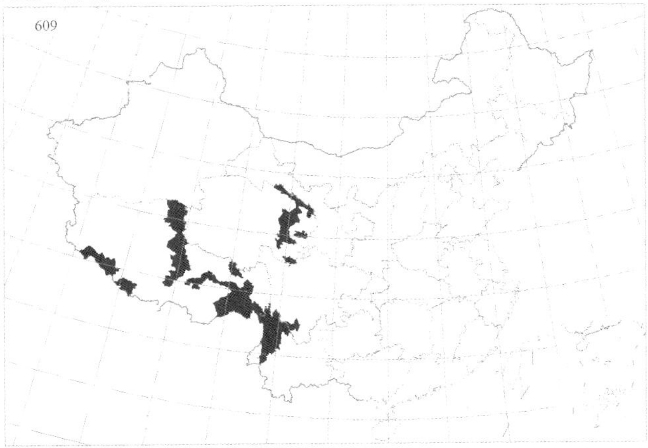

Salicaceae 杨柳科 *Salix* 柳属
Salix rhododendrifolia C. Wang et P. Y. Fu 杜鹃叶柳
Shrubs, deciduous

5.8 (-3.2~13.3); -4.8 (-14.4~5.4); 15.5 (6.1~24.7); 7.2 (1.8~13.4); 48.7 (1.9~101.9); 38.4 (0.0~101.5); 529 (329~733) | 721 (283~1736); 389 (204~1007); 23 (2~51) | 39.7 (-27.2~145.8); 520 (283~733); 254.6 (72.2~447.4)

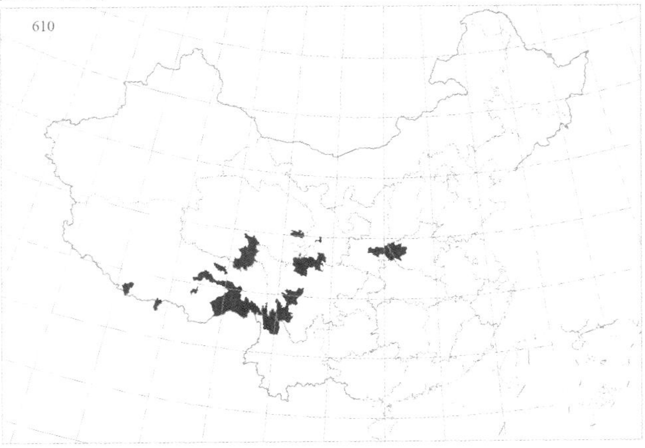

Salicaceae 杨柳科 *Salix* 柳属
Salix rhoophila C. K. Schneid. 房县柳
Shrubs, deciduous

12.3 (1.6~15.4); -0.2 (-9.3~4.3); 23.9 (10.8~27.7); 12.5 (4.0~15.4); 101.5 (17.9~129.9); 13.0 (0.7~58.4); 744 (419~875) | 865 (626~1492); 417 (316~613); 46 (15~96) | 21.1 (-22.2~121.7); 707 (419~864); 203.4 (139.1~352.3)

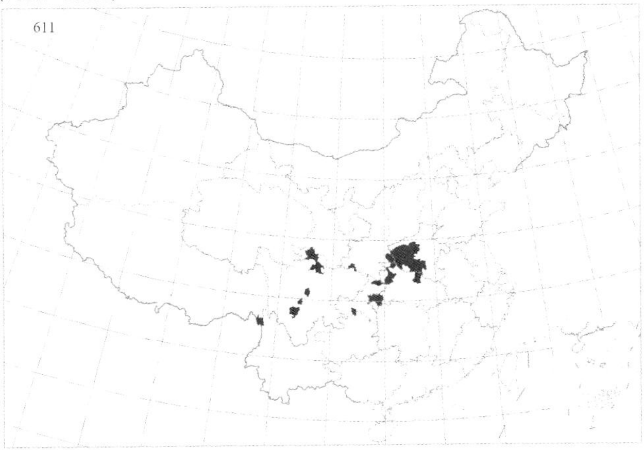

Salicaceae 杨柳科 *Salix* 柳属
Salix rockii Goerz ex Rehder et Kobuski 拉加柳
Shrubs, deciduous

0.5 (-3.3~4.7); -11.4 (-15.2~-6.1); 11.0 (7.2~14.7); 3.8 (2.0~6.0); 18.0 (3.7~36.3); 72.0 (39.6~103.2); 406 (339~482) | 500 (226~773); 274 (138~394); 12 (8~19) | 23.1 (-44.9~84.1); 369 (226~482); 235.5 (46.7~364.1)

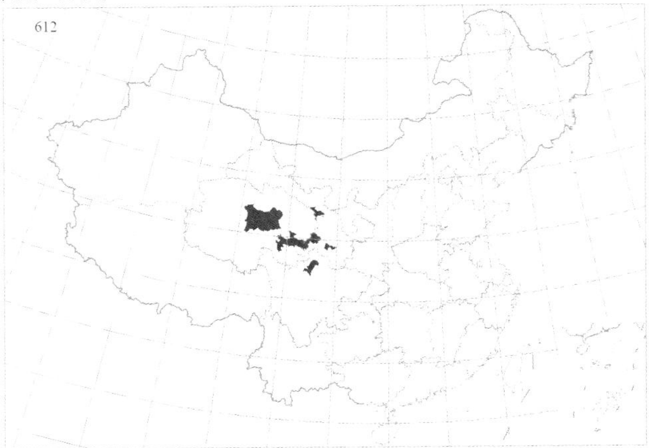

Salicaceae 杨柳科 *Salix* 柳属
Salix rorida Lacksch. 粉枝柳
Trees, deciduous

3.1 (-5.8~10.5); -17.9 (-29.8~-4.9); 21.3 (15.3~25.0); 8.0 (4.4~11.4); 62.7 (28.0~95.3); 85.5 (26.8~157.4); 581 (426~722) ❙ 560 (297~1018); 362 (203~648); 16 (4~40) ❙ -3.3 (-44.3~68.8); 510 (297~698); 240.4 (97.2~346.5)

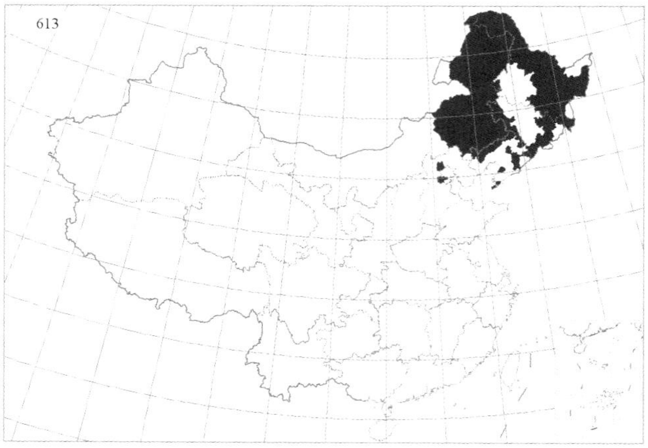

Salicaceae 杨柳科 *Salix* 柳属
Salix rorida var. roridiformis (Nakai) Ohwi 伪粉枝柳
Trees, deciduous

2.1 (-1.5~8.4); -20.1 (-25.0~-1.0); 21.1 (19.2~25.0); 7.7 (6.2~10.6); 59.5 (47.5~90.9); 93.3 (49.4~124.5); 573 (513~700) ❙ 605 (513~966); 373 (309~616); 20 (12~35) ❙ 5.2 (-11.0~54.7); 564 (513~698); 281.3 (191.9~333.3)

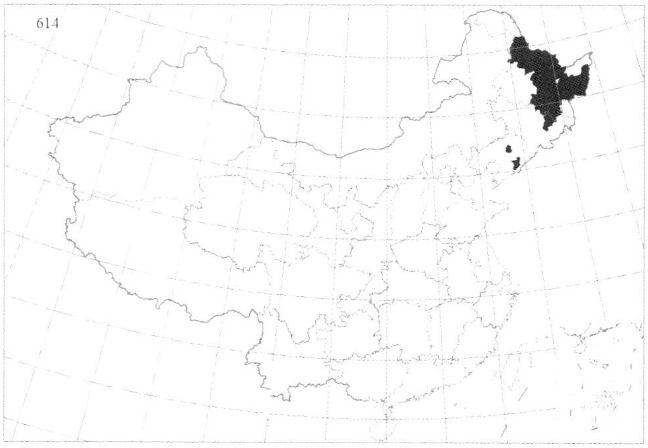

Salicaceae 杨柳科 *Salix* 柳属
Salix rosmarinifolia L. 细叶沼柳
Shrubs, deciduous

2.8 (-5.8~12.4); -17.3 (-29.8~-5.2); 20.2 (9.6~27.5); 7.7 (2.5~13.2); 58.8 (9.7~114.1); 84.3 (24.3~157.4); 567 (329~813) ❙ 410 (38~942); 249 (17~601); 16 (3~51) ❙ -26.2 (-95.3~57.4); 390 (38~658); 188.4 (2.8~346.5)

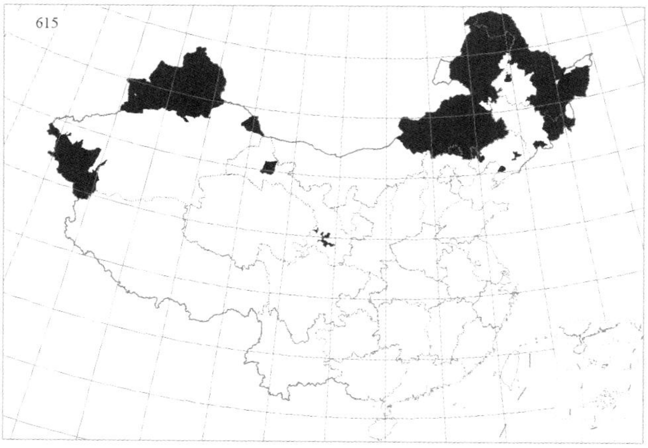

Salicaceae 杨柳科 *Salix* 柳属
Salix rosmarinifolia var. brachypoda (Trautv. et C. A. Mey.) Y. L. Chou 沼柳
Shrubs, deciduous

2.9 (-5.8~9.0); -17.4 (-29.8~-6.1); 20.6 (13.2~24.8); 7.7 (4.4~10.8); 59.3 (26.7~91.5); 84.4 (38.3~157.4); 567 (426~701) ❙ 447 (77~942); 285 (48~601); 14 (2~51) ❙ -20.6 (-87.2~56.9); 428 (77~674); 197.9 (13.9~333.3)

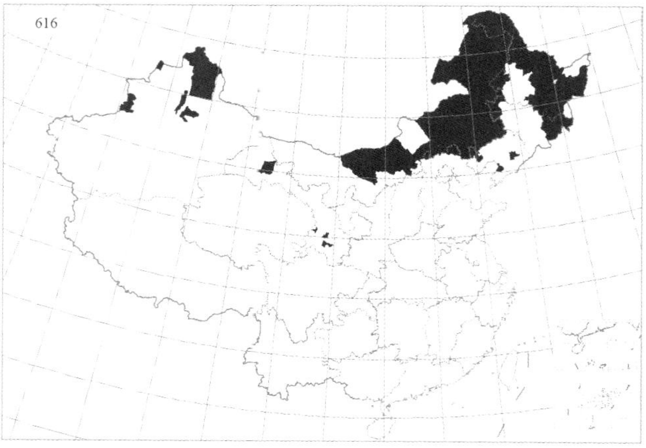

Salicaceae 杨柳科 *Salix* 柳属
Salix rosmarinifolia var. tungbeiana Y. L. Chou et Skvortsov 东北细叶沼柳
Shrubs, deciduous

0.6 (-1.5~2.7); -22.8 (-25.0~-19.9); 20.5 (19.2~21.9); 7.2 (6.2~8.1); 54.9 (47.5~62.5); 107.4 (89.5~124.5); 552 (513~590) ❙ 596 (513~657); 376 (337~416); 17 (12~21) ❙ 7.4 (-2.7~23.8); 550 (513~583); 292.9 (263.0~333.3)

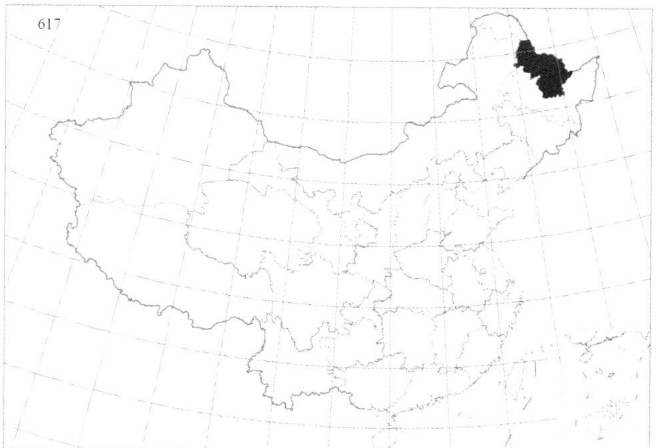

Salicaceae 杨柳科 *Salix* 柳属
Salix rosthornii Seemen 南川柳
Shrubs or trees, deciduous

16.0 (0.5~21.8); 4.3 (-11.5~13.5); 27.0 (11.8~30.1); 16.1 (3.9~21.8); 135.2 (19.7~201.6); 2.8 (0.0~73.3); 875 (406~1162) ❙ 1322 (374~1997); 504 (205~776); 137 (8~259) ❙ 50.8 (-29.3~130.7); 858 (374~1162); 204.7 (81.2~323.3)

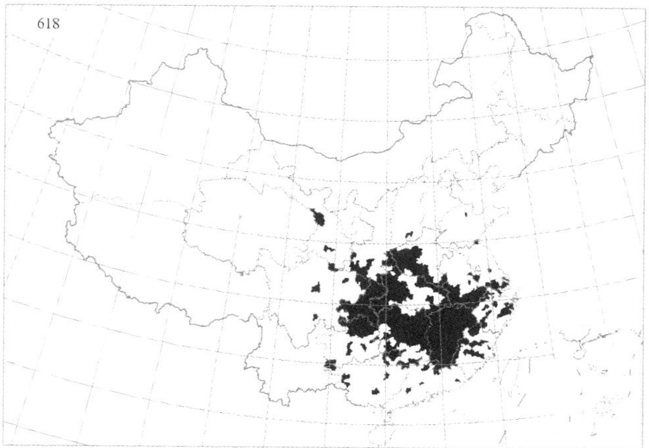

Salicaceae 杨柳科 *Salix* 柳属

Salix sachalinensis F. Schmidt 龙江柳

Shrubs or trees, deciduous

2.3 (-5.0~8.4); -18.5 (-29.4~-9.8); 20.3 (17.0~25.0); 7.5 (5.0~10.6); 57.2 (35.2~90.9); 88.7 (46.4~154.8); 553 (450~700) | 681 (468~1005); 423 (300~635); 23 (15~39) | 22.7 (-2.6~68.8); 552 (450~698); 303.0 (191.9~346.5)

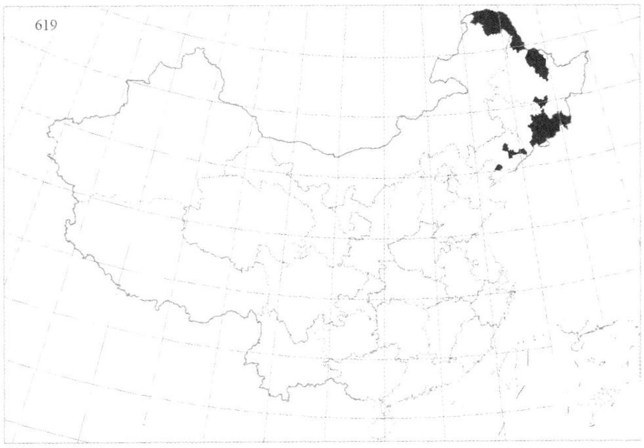

Salicaceae 杨柳科 *Salix* 柳属

Salix sajanensis Nasarow 萨彦柳

Shrubs or trees, deciduous

3.3 (-4.3~12.3); -15.2 (-20.1~-5.3); 19.1 (9.6~26.1); 7.5 (2.5~13.0); 57.9 (9.7~112.3); 78.1 (24.7~121.1); 564 (329~801) | 161 (44~315); 67 (19~127); 19 (3~51) | -70.3 (-94.5~-44.6); 161 (44~312); 71.1 (7.6~173.1)

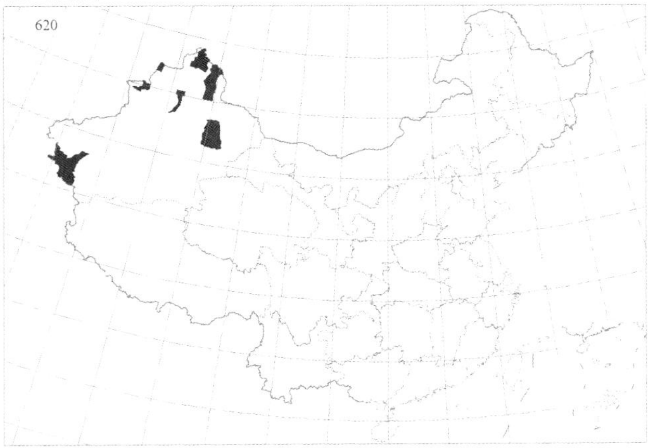

Salicaceae 杨柳科 *Salix* 柳属

Salix salwinensis Hand.-Mazz. ex Enander 对叶柳

Shrubs, deciduous

9.6 (-1.8~14.5); 2.4 (-10.2~7.8); 15.6 (6.1~19.7); 10.1 (1.8~14.5); 69.3 (1.9~113.5); 13.8 (0.0~82.8); 600 (329~730) | 985 (320~1454); 567 (217~792); 41 (10~70) | 61.0 (-3.0~100.7); 599 (320~730); 315.9 (61.6~487.2)

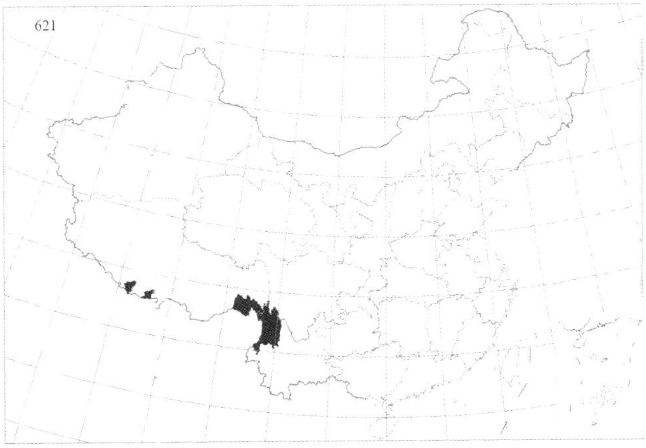

Salicaceae 杨柳科 *Salix* 柳属

Salix salwinensis var. longiamentifera C. F. Fang 长穗对叶柳

Shrubs, deciduous

9.6; 2.4; 15.7; 9.6; 60.8; 5.4; 600 | 1007; 578; 40 | 67.9; 600; 351.7

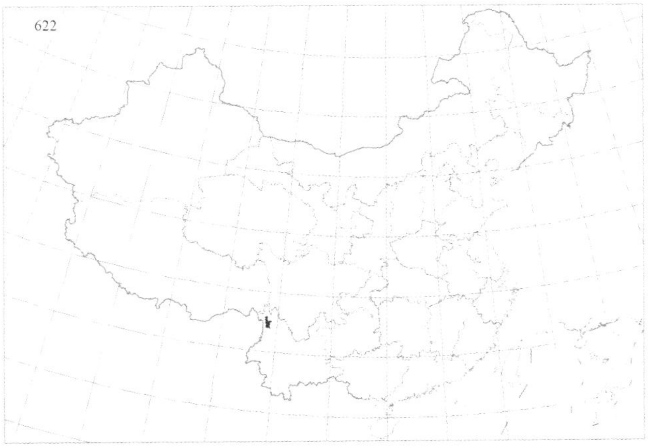

Salicaceae 杨柳科 *Salix* 柳属

Salix saposhnikovii A. K. Skvortsov 灌木柳

Shrubs, deciduous

2.5 (-1.3~6.2); -16.4 (-19.5~-13.7); 18.8 (12.5~23.6); 7.2 (3.9~9.8); 54.2 (21.0~82.7); 83.8 (65.1~104.2); 551 (390~670) | 199 (107~406); 84 (46~211); 22 (8~51) | -62.6 (-82.7~-7.6); 199 (107~406); 91.2 (16.6~261.4)

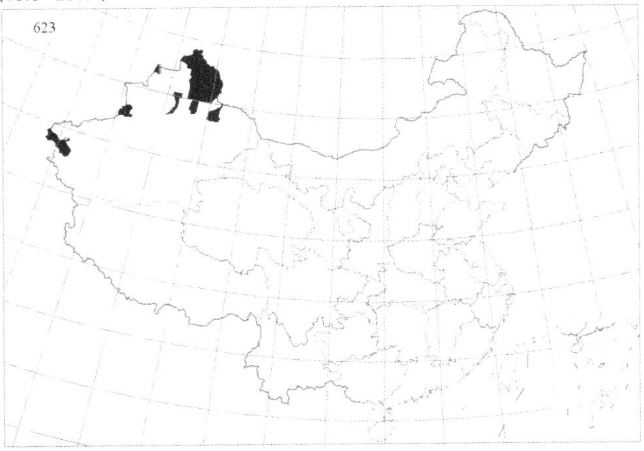

Salicaceae 杨柳科 *Salix* 柳属

Salix schugnanica Goerz 阿克苏柳

Shrubs, deciduous

11.2; -7.3; 25.3; 12.4; 106.7; 32.1; 780 | 42; 26; 1 | -94.6; 42; 17.0

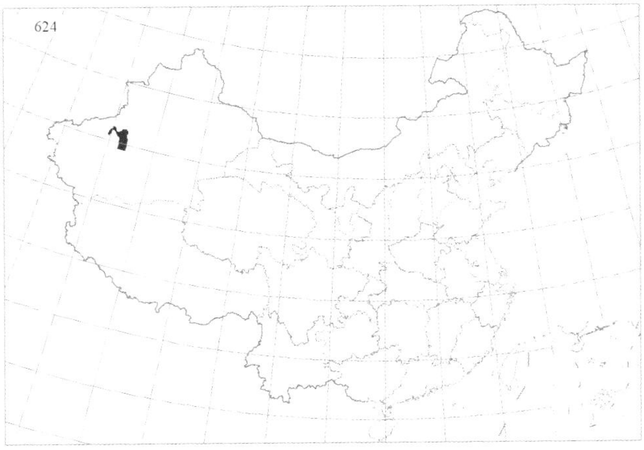

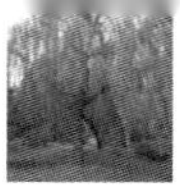

Salicaceae 杨柳科 *Salix* 柳属

Salix schwerinii E. L. Wolf 蒿柳

Shrubs or small trees, deciduous

5.3 (-5.8~15.5); -14.1 (-29.8~2.1); 22.2 (13.7~28.0); 9.1 (4.4~15.6); 73.0 (27.2~131.3); 68.8 (4.3~157.4); 627 (426~881) ❙ 544 (69~1077); 329 (44~691); 20 (4~97) ❙ -12.5 (-89.2~74.1); 505 (69~881); 202.1 (14.2~346.5)

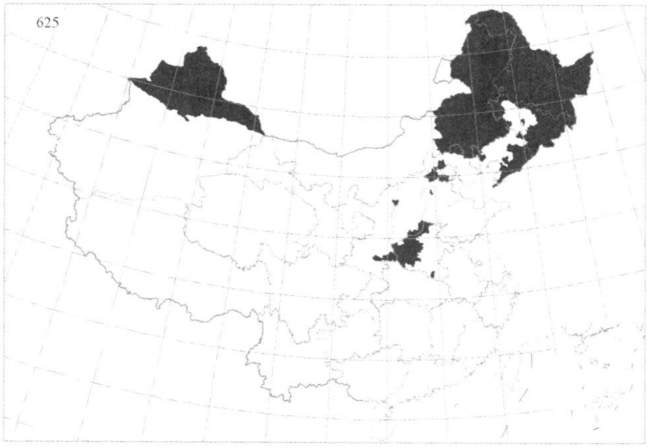

Salicaceae 杨柳科 *Salix* 柳属

Salix sclerophylla Andersson 硬叶柳

Shrubs, deciduous

1.0 (-5.5~10.8); -9.2 (-17.3~3.9); 10.2 (5.4~22.7); 3.9 (1.3~10.9); 18.4 (0.4~84.8); 65.4 (1.3~127.1); 405 (299~667) ❙ 568 (220~1007); 335 (160~601); 15 (0~103) ❙ 38.6 (-32.2~128.6); 392 (220~633); 222.8 (46.9~364.1)

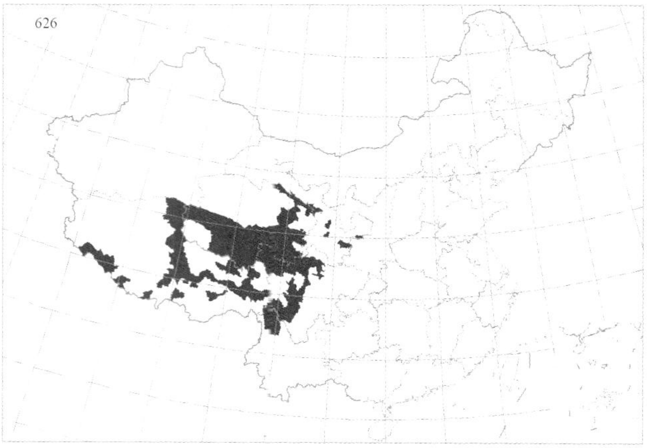

Salicaceae 杨柳科 *Salix* 柳属

Salix sclerophylla var. obtusa (C. Wang et P. Y. Fu) C. F. Fang 宽苞金背柳

Shrubs, deciduous

-1.3; -11.4; 7.9; 2.5; 7.6; 83.0; 347 ❙ 340; 235; 4 ❙ -2.9; 315; 193.3

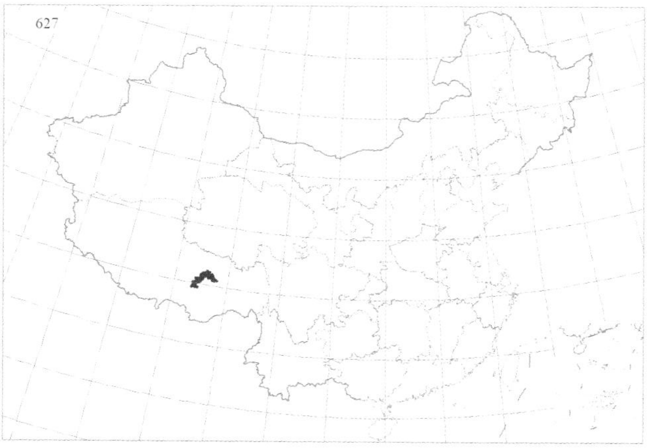

Salicaceae 杨柳科 *Salix* 柳属

Salix sclerophylla var. tibetica (Goerz ex Rehder et Kobuski) C. F. Fang 小叶硬叶柳

Shrubs, deciduous

5.3 (-1.3~11.0); -2.9 (-13.5~4.2); 12.5 (7.7~16.6); 6.5 (2.5~11.0); 37.8 (6.8~73.1); 34.0 (0.8~86.3); 492 (338~636) ❙ 761 (370~1134); 455 (262~651); 24 (8~47) ❙ 50.3 (4.7~78.3); 492 (338~636); 249.0 (61.6~351.7)

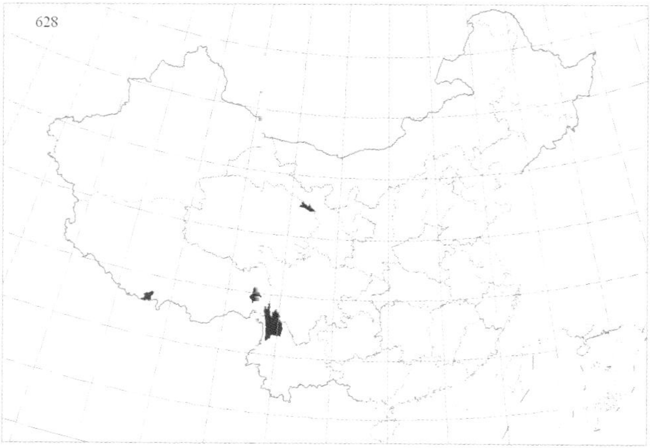

Salicaceae 杨柳科 *Salix* 柳属

Salix sclerophylloides Y. L. Chou 近硬叶柳

Shrubs, deciduous

0.9 (-4.4~11.0); -9.1 (-17.3~4.2); 10.0 (6.1~16.6); 3.9 (1.7~11.0); 18.2 (1.9~73.1); 66.2 (0.8~115.9); 405 (323~636) ❙ 599 (282~1134); 362 (182~651); 15 (2~47) ❙ 44.7 (-24.8~125.9); 400 (282~636); 238.0 (72.2~351.7)

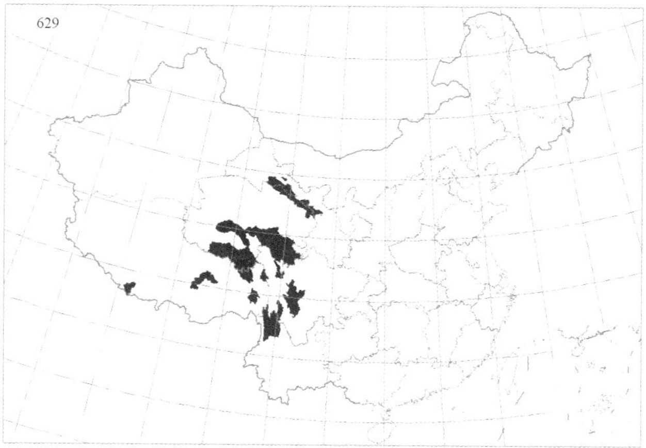

Salicaceae 杨柳科 *Salix* 柳属

Salix scopulicola Mao et W. Z. Li 岩壁垫柳

Shrubs, deciduous

8.6; 1.3; 15.0; 8.7; 52.6; 8.8; 574 ❙ 1041; 585; 45 ❙ 81.5; 574; 352.3

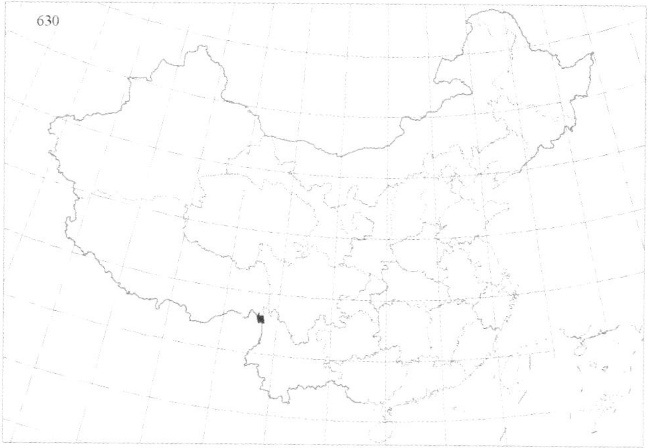

Salicaceae 杨柳科 *Salix* 柳属
Salix sericocarpa Andersson 绢果柳
Small trees, deciduous

3.8 (−5.7~18.3); −5.2 (−18.7~11.8); 12.0 (6.1~22.7); 6.3 (1.7~18.3); 41.6 (1.9~159.8); 55.2 (0.0~133.0); 473 (294~882) ❙ 532 (74~1243); 318 (46~696); 31 (0~103) ❙ 7.4 (−75.0~93.0); 403 (74~799); 161.8 (16.7~384.7)

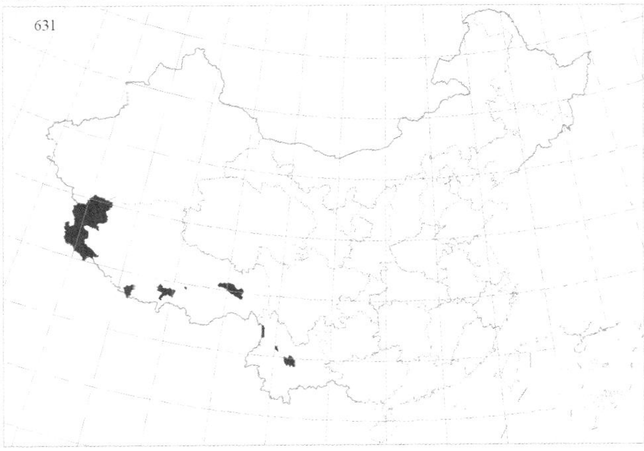

Salicaceae 杨柳科 *Salix* 柳属
Salix serpyllum Andersson 多花小垫柳
Shrubs, deciduous

3.5 (−1.8~12.2); −4.8 (−10.2~4.3); 10.8 (6.1~18.5); 5.2 (1.8~12.2); 28.3 (1.9~87.4); 46.0 (0.7~82.8); 446 (329~672) ❙ 653 (320~1431); 397 (193~817); 20 (9~33) ❙ 36.3 (−4.0~113.0); 443 (320~672); 179.0 (61.6~363.7)

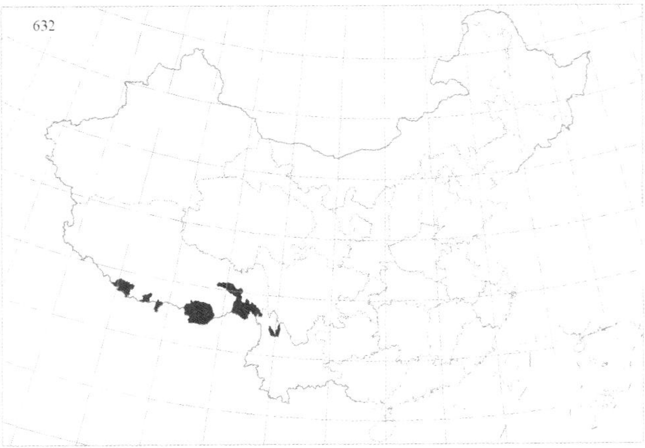

Salicaceae 杨柳科 *Salix* 柳属
Salix shandanensis C. F. Fang 山丹柳
Shrubs, deciduous

5.9 (−1.4~13.3); −8.1 (−15.6~1.1); 18.5 (9.5~26.2); 7.9 (2.9~13.4); 58.2 (11.3~111.7); 47.3 (8.1~92.1); 559 (359~787) ❙ 445 (189~773); 241 (114~363); 10 (4~20) ❙ −16.0 (−70.2~61.8); 406 (189~707); 177.3 (33.5~344.4)

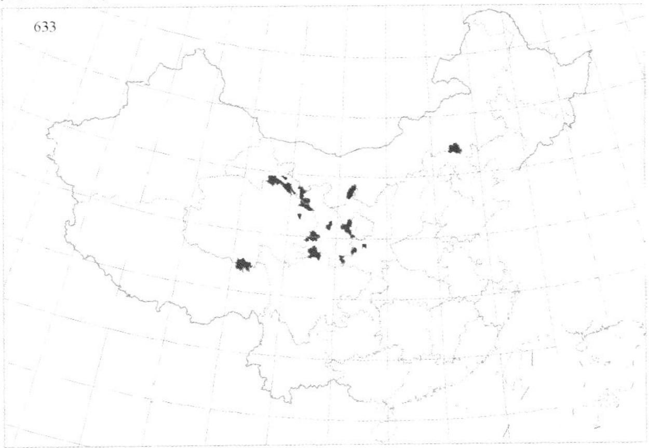

Salicaceae 杨柳科 *Salix* 柳属
Salix shihtsuanensis C. Wang et C. Y. Yu 石泉柳
Shrubs or small trees, deciduous

9.1 (−0.8~13.6); −3.2 (−13.2~1.4); 20.5 (9.2~25.8); 9.8 (2.9~13.6); 75.0 (10.1~113.4); 25.6 (7.1~80.3); 629 (374~786) ❙ 724 (467~892); 348 (268~398); 24 (8~49) ❙ 16.3 (−1.2~52.6); 629 (374~786); 266.9 (189.3~323.3)

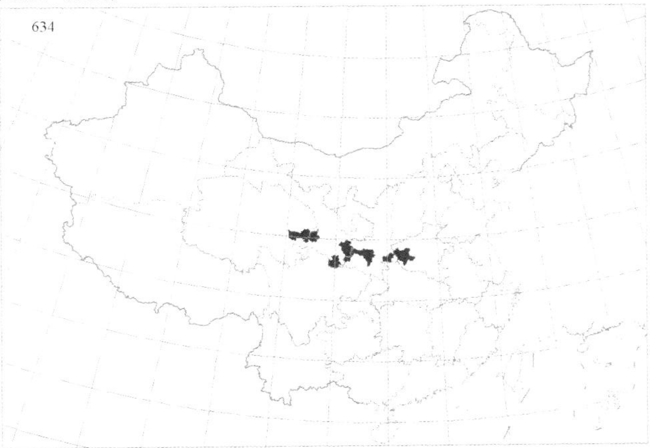

Salicaceae 杨柳科 *Salix* 柳属
Salix shihtsuanensis var. globosa C. Y. Yu 球果石泉柳
Shrubs or small trees, deciduous

9.4; −2.8; 21.1; 9.9; 75.0; 21.0; 628 ❙ 632; 306; 15 ❙ 0.0; 628; 241.9

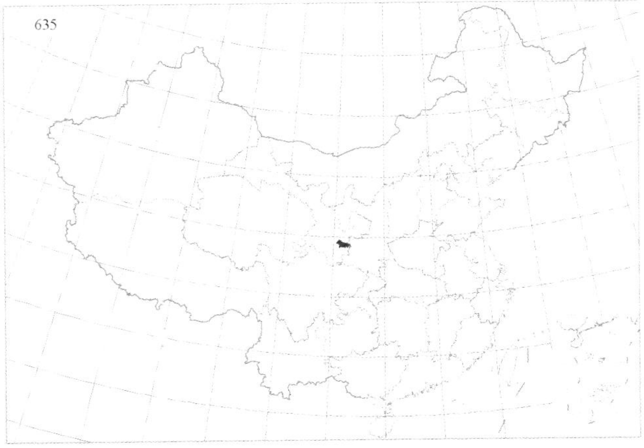

Salicaceae 杨柳科 *Salix* 柳属
Salix shihtsuanensis var. sessilis C. Y. Yu 无柄石泉柳
Shrubs or small trees, deciduous

11.5; −0.5; 23.1; 11.6; 91.6; 13.8; 693 ❙ 882; 385; 26 ❙ 28.9; 693; 286.4

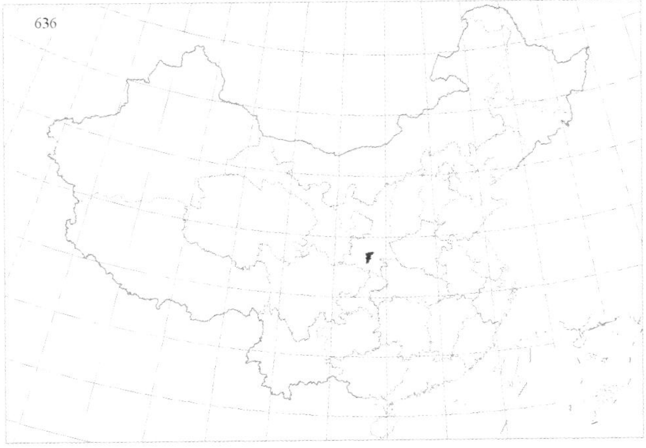

Salicaceae 杨柳科 *Salix* 柳属

Salix sikkimensis Andersson 锡金柳

Shrubs, deciduous

4.1 (−0.7~10.8); −4.2 (−9.6~3.9); 11.6 (7.7~16.6); 5.4 (2.5~10.9); 29.2 (6.8~71.8); 39.0 (1.3~74.2); 455 (338~633) ❙ 620 (370~968); 384 (262~594); 14 (7~24) ❙ 31.8 (4.7~61.4); 455 (338~633); 224.1 (61.6~319.8)

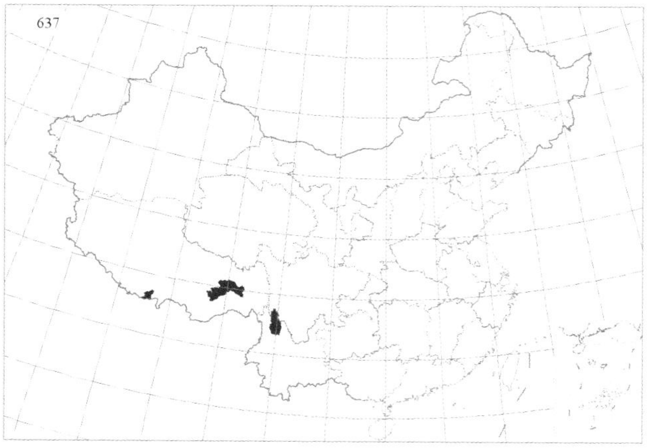

Salicaceae 杨柳科 *Salix* 柳属

Salix sinica (K. S. Hao ex C. F. Fang et A. K. Skvortsov) G. H. Zhu 中国黄花柳

Shrubs or small trees, deciduous

9.5 (−1.4~16.6); −5.1 (−16.1~5.8); 22.6 (8.5~28.8); 10.6 (2.7~16.7); 85.2 (8.3~140.6); 31.0 (0.0~92.1); 673 (359~920) ❙ 597 (84~1571); 323 (53~816); 22 (1~153) ❙ −10.9 (−87.1~125.9); 544 (84~899); 172.5 (7.5~364.1)

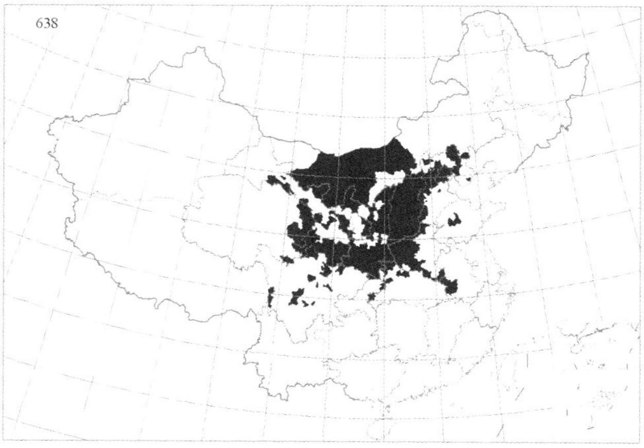

Salicaceae 杨柳科 *Salix* 柳属

Salix sinica var. dentata (K. S. Hao ex C. F. Fang et A. K. Skvortsov) G. H. Zhu 齿叶黄花柳

Shrubs or small trees, deciduous

5.8 (−1.3~10.2); −9.6 (−15.6~−6.2); 19.5 (9.7~24.7); 8.1 (2.9~11.4); 61.6 (11.3~95.3); 51.6 (32.8~86.3); 571 (359~722) ❙ 498 (341~609); 313 (190~453); 11 (7~14) ❙ −10.7 (−40.9~44.8); 479 (341~609); 199.9 (61.2~313.2)

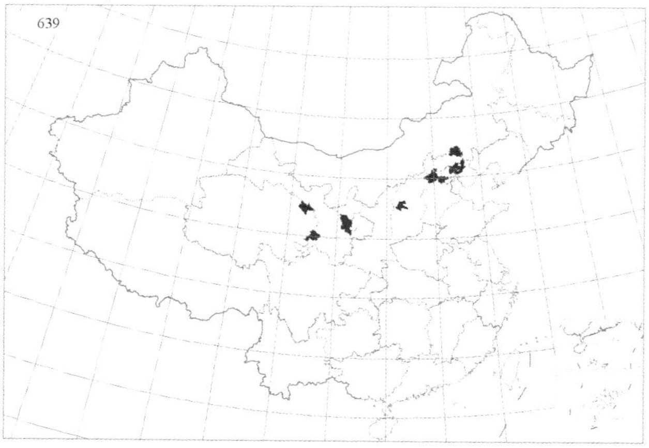

Salicaceae 杨柳科 *Salix* 柳属

Salix sinica var. subsessilis (K. S. Hao ex C. F. Fang et A. K. Skvortsov) G. Zhu 无柄黄花柳

Shrubs or small trees, deciduous

5.6; −11.5; 20.8; 8.4; 66.3; 58.2; 586 ❙ 469; 305; 13 ❙ −20.0; 469; 194.0

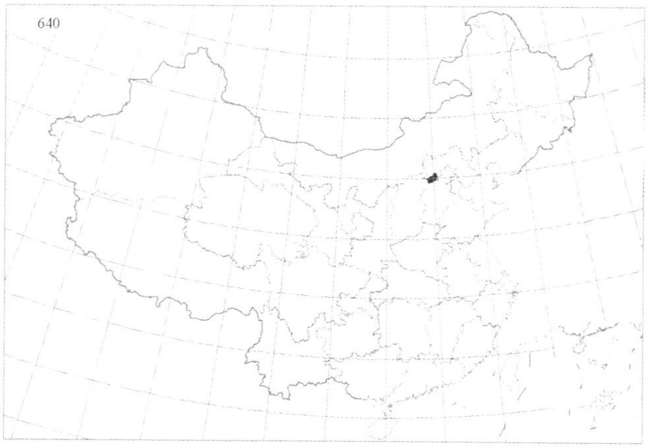

Salicaceae 杨柳科 *Salix* 柳属

Salix sinopurpurea C. Wang et C. Y. Yang 红皮柳

Shrubs, deciduous

9.9 (0.5~17.1); −4.7 (−13.9~4.2); 23.0 (10.4~29.4); 11.0 (3.7~17.2); 88.0 (15.1~147.0); 28.7 (0.8~73.3); 686 (406~954) ❙ 632 (58~1663); 337 (36~628); 31 (4~212) ❙ −8.3 (−90.8~101.4); 583 (58~921); 176.1 (11.3~346.5)

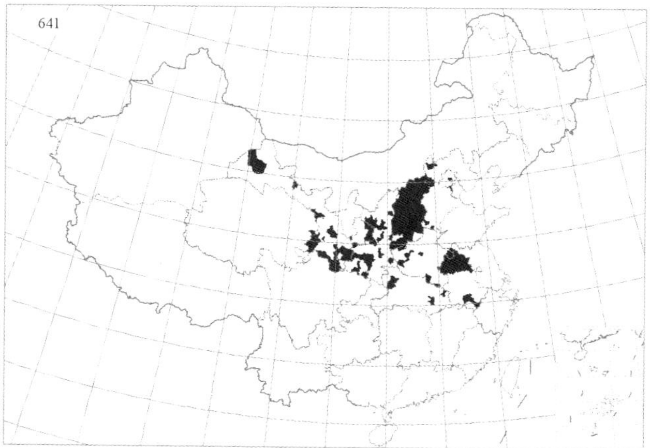

Salicaceae 杨柳科 *Salix* 柳属

Salix siuzevii Seemen 卷边柳

Shrubs or trees, deciduous

2.7 (−5.8~8.7); −18.3 (−29.8~−8.3); 20.9 (15.3~25.0); 7.8 (4.4~10.6); 60.4 (28.0~90.9); 87.5 (40.2~157.4); 572 (426~700) ❙ 558 (297~1018); 361 (203~648); 16 (5~40) ❙ −2.3 (−44.3~66.0); 510 (297~698); 251.3 (97.2~346.5)

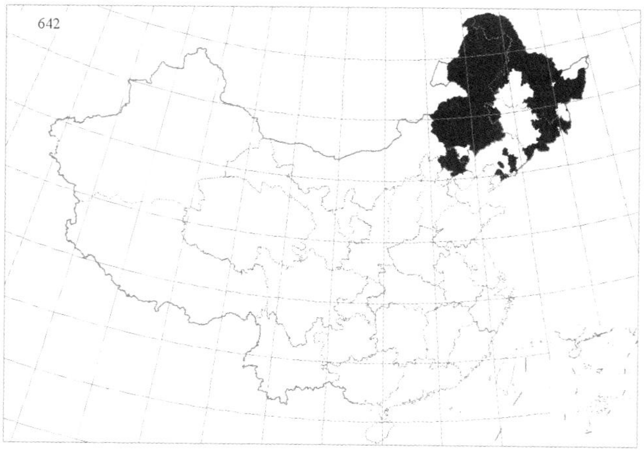

Salicaceae 杨柳科 *Salix* 柳属
Salix skvortzovii Y. L. Chang et Y. L. Chou 司氏柳
Trees, deciduous

2.4 (−1.5~8.4); −19.2 (−25.0~−9.8); 20.9 (18.5~25.0); 7.7 (6.2~10.6); 59.5 (47.5~90.9); 90.3 (46.4~124.5); 569 (513~700) ❙ 638 (513~976); 396 (315~608); 20 (12~40) ❙ 12.3 (−14.6~66.0); 562 (513~698); 288.4 (177.1~346.5)

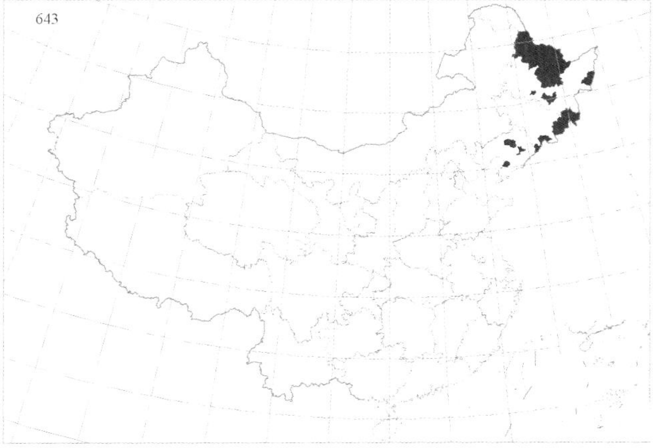

Salicaceae 杨柳科 *Salix* 柳属
Salix songarica Andersson 准噶尔柳
Small trees, deciduous

4.6 (−1.3~9.3); −15.0 (−20.9~−8.8); 20.4 (13.2~27.1); 8.4 (4.6~12.1); 66.6 (26.7~107.6); 70.3 (40.4~103.9); 605 (440~810) ❙ 229 (101~406); 107 (53~211); 18 (3~45) ❙ −59.4 (−85.2~−7.6); 229 (101~406); 120.8 (41.1~297.3)

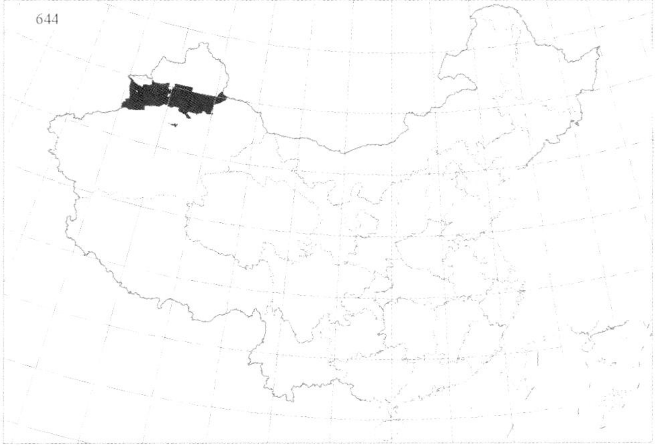

Salicaceae 杨柳科 *Salix* 柳属
Salix souliei Seemen 黄花垫柳
Shrubs, deciduous

2.9 (−4.4~18.5); −6.5 (−16.2~10.5); 11.3 (6.1~24.3); 5.0 (1.7~18.3); 27.5 (1.9~160.0); 52.2 (0.0~115.9); 445 (323~906) ❙ 690 (252~1736); 402 (153~1007); 17 (6~51) ❙ 53.9 (−37.4~145.8); 438 (252~786); 253.2 (46.7~447.4)

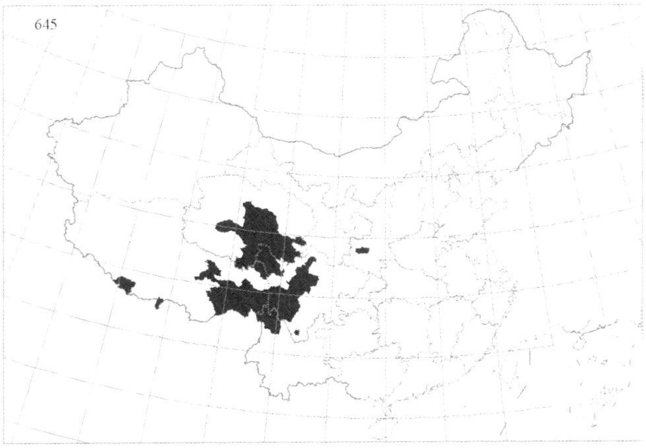

Salicaceae 杨柳科 *Salix* 柳属
Salix spathulifolia Seemen ex Diels 匙叶柳
Shrubs, deciduous

8.6 (−4.1~15.5); −3.9 (−17.3~5.3); 20.1 (7.8~29.5); 9.6 (2.1~15.5); 73.4 (5.2~136.0); 29.6 (0.0~114.2); 627 (340~932) ❙ 670 (189~1410); 329 (114~816); 19 (4~51) ❙ 10.7 (−70.2~128.6); 561 (189~795); 230.2 (33.5~349.1)

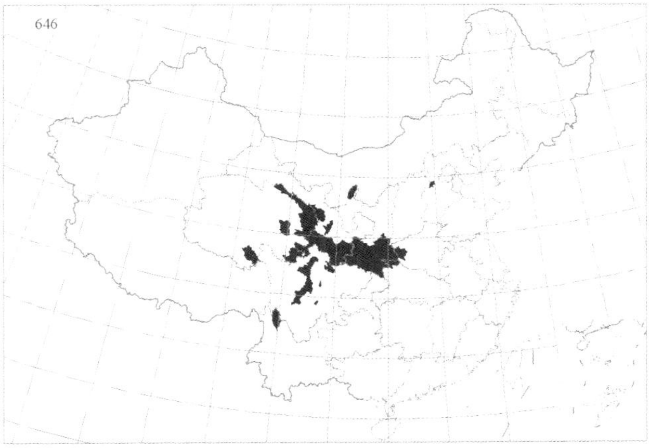

Salicaceae 杨柳科 *Salix* 柳属
Salix sphaeronymphe Goerz 巴郎柳
Shrubs or small trees, deciduous

2.9 (−5.5~14.9); −7.0 (−17.1~5.3); 11.7 (5.8~24.6); 5.0 (1.3~15.0); 27.8 (1.2~119.6); 53.1 (0.0~127.1); 447 (299~782) ❙ 712 (270~1410); 411 (183~816); 17 (5~49) ❙ 58.9 (−9.9~128.6); 447 (270~782); 254.2 (88.8~357.7)

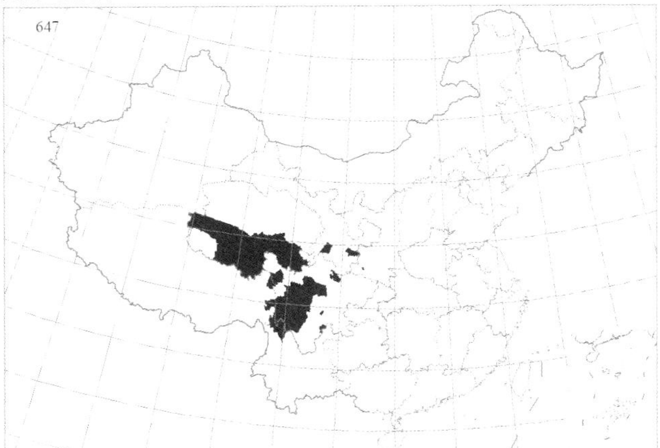

Salicaceae 杨柳科 *Salix* 柳属
Salix sphaeronymphoides Y. L. Chou 光果巴郎柳
Trees, deciduous

2.0 (−0.6~3.7); −8.2 (−10.6~−6.1); 11.1 (8.3~12.6); 4.3 (2.7~5.1); 20.9 (9.3~27.8); 56.0 (43.5~76.9); 416 (344~442) ❙ 474 (246~800); 291 (192~427); 8 (0~18) ❙ 11.3 (−28.5~86.5); 375 (246~442); 231.4 (151.1~341.6)

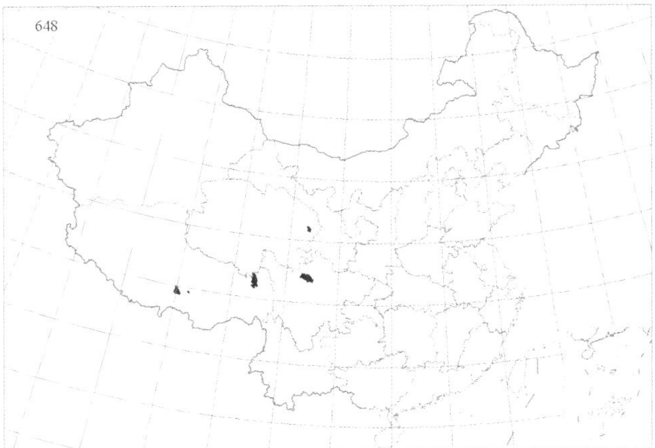

Salicaceae 杨柳科 *Salix* 柳属
Salix spodiophylla Hand.-Mazz. 灰叶柳
Shrubs, deciduous

7.7 (1.2~16.2); -0.8 (-8.2~7.8); 15.1 (9.0~26.3); 8.4 (3.3~16.3); 55.5 (11.9~135.7); 22.8 (0.0~56.5); 559 (390~856) ❙ 940 (527~1736); 523 (317~1007); 30 (9~119) ❙ 71.5 (19.6~145.8); 558 (390~856); 287.7 (130.1~447.4)

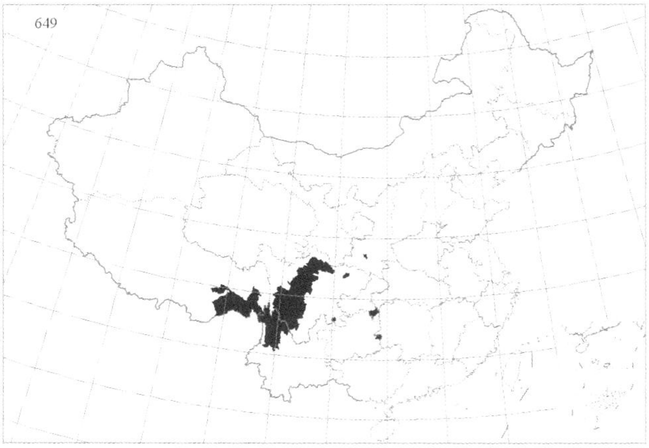

Salicaceae 杨柳科 *Salix* 柳属
Salix spodiophylla var. liocarpa (K. S. Hao) G. H. Zhu 光果灰叶柳
Shrubs, deciduous

9.5 (2.7~14.5); 2.1 (-7.1~7.8); 15.7 (11.1~19.7); 9.7 (4.4~14.5); 66.3 (20.5~113.5); 12.2 (0.0~48.0); 595 (429~730) ❙ 946 (741~1057); 549 (426~627); 32 (12~54) ❙ 60.8 (36.1~92.9); 594 (429~730); 321.3 (221.3~382.6)

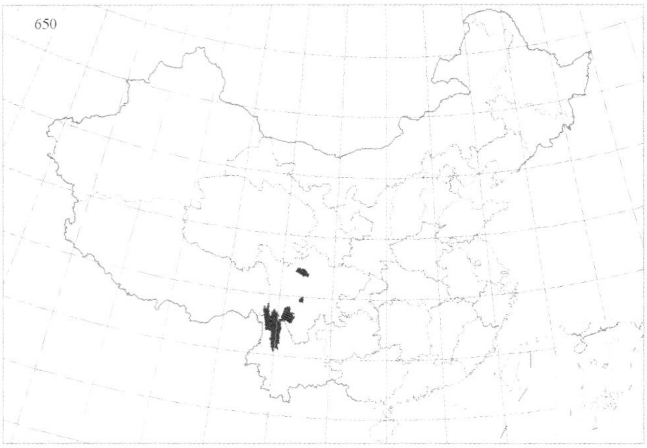

Salicaceae 杨柳科 *Salix* 柳属
Salix suchowensis W. C. Cheng ex G. Zhu 簸箕柳
Shrubs, deciduous

15.0 (10.5~17.0); 1.3 (-4.0~4.9); 27.8 (23.6~29.2); 15.1 (11.1~17.0); 127.8 (89.9~144.9); 7.2 (0.1~23.0); 867 (689~941) ❙ 912 (558~1821); 448 (256~671); 82 (17~213) ❙ 5.1 (-36.5~124.8); 779 (558~932); 168.7 (110.1~284.0)

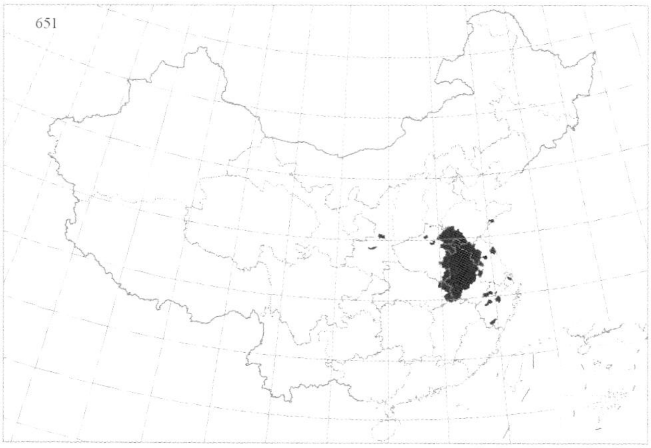

Salicaceae 杨柳科 *Salix* 柳属
Salix sungkianica Y. L. Chou et Skvortsov 松江柳
Tall shrubs, deciduous

2.5 (-5.8~8.0); -19.3 (-29.8~-9.8); 21.2 (15.3~24.1); 7.9 (4.4~10.0); 61.3 (28.0~84.2); 91.3 (46.4~157.4); 578 (426~670) ❙ 515 (366~718); 340 (246~446); 14 (6~29) ❙ -10.3 (-38.1~23.8); 494 (366~658); 223.6 (109.7~323.4)

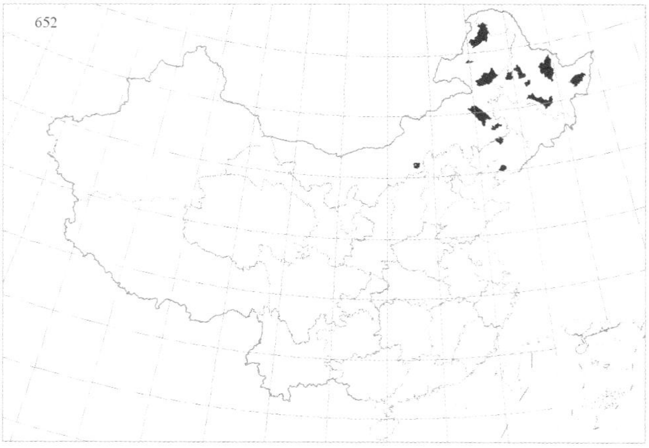

Salicaceae 杨柳科 *Salix* 柳属
Salix tagawana Koidz. 花莲柳
Shrubs, deciduous

16.4; 11.5; 20.3; 16.5; 137.8; 0.0; 783 ❙ 2499; 947; 303 ❙ 219.3; 783; 303.6

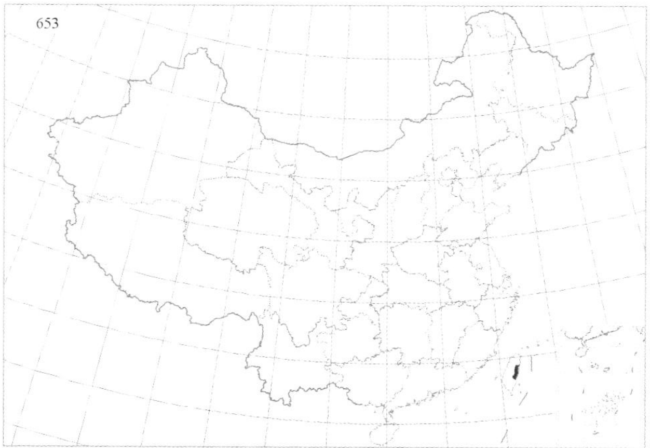

Salicaceae 杨柳科 *Salix* 柳属
Salix taipaiensis C. Y. Yu 太白柳
Shrubs, deciduous

9.5; -2.7; 21.4; 10.0; 76.8; 21.6; 633 ❙ 796; 359; 24 ❙ 27.5; 633; 283.7

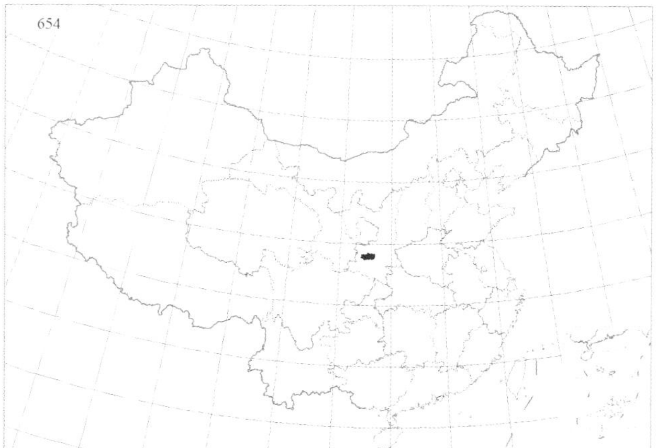

Salicaceae 杨柳科 *Salix* 柳属
Salix taishanensis C. Wang et C. F. Fang 泰山柳
Shrubs, deciduous

10.3 (4.9~14.7); -5.5 (-12.3~-0.6); 24.2 (19.5~28.5); 11.4 (7.7~14.9); 95.0 (57.9~128.7); 31.1 (11.0~61.5); 721 (553~886) ❙ 572 (457~720); 364 (254~471); 19 (12~28) ❙ -20.0 (-38.3~-8.5); 572 (457~720); 164.0 (115.9~225.2)

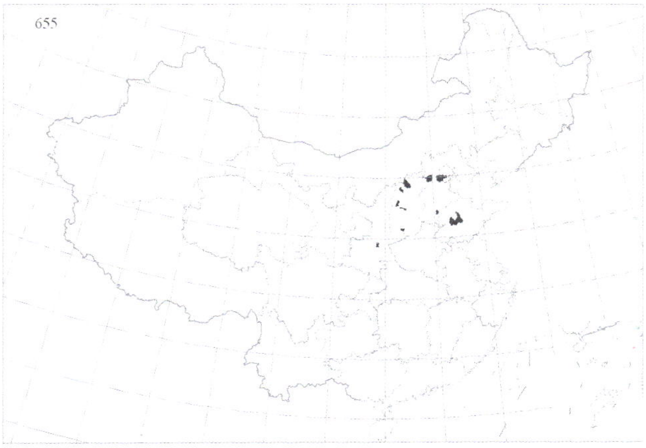

Salicaceae 杨柳科 *Salix* 柳属
Salix taishanensis var. glabra C. F. Fang et W. D. Liu 光子房泰山柳
Shrubs, deciduous

6.3; -9.3; 20.2; 8.3; 63.7; 47.4; 573 ❙ 498; 301; 14 ❙ -13.1; 498; 210.1

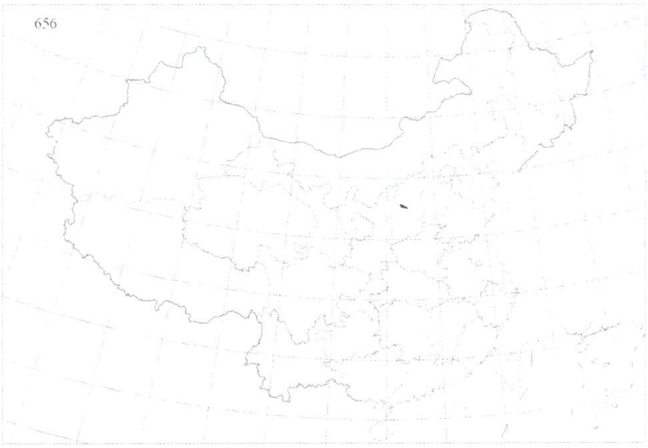

Salicaceae 杨柳科 *Salix* 柳属
Salix taishanensis var. hebeinica C. F. Fang 河北柳
Shrubs, deciduous

8.2 (2.5~13.5); -7.8 (-15.6~-1.6); 22.3 (17.5~27.1); 9.9 (6.6~13.7); 79.1 (45.4~117.5); 40.8 (14.6~78.3); 645 (507~820) ❙ 522 (397~640); 317 (254~465); 16 (9~26) ❙ -18.2 (-36.6~-2.4); 522 (397~640); 184.9 (111.2~273.5)

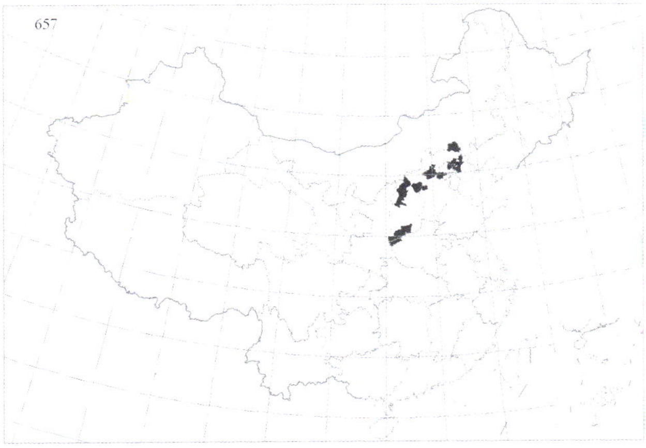

Salicaceae 杨柳科 *Salix* 柳属
Salix taiwanalpina Kimura 台湾山柳
Shrubs, deciduous

20.2 (16.2~24.1); 14.5 (11.0~18.9); 25.0 (20.1~28.2); 20.2 (16.2~24.1); 182.9 (134.2~229.7); 0.0 (0.0~0.0); 1019 (773~1322) ❙ 2402 (1498~3495); 1106 (750~1847); 255 (47~897) ❙ 146.2 (25.2~246.5); 983 (773~1144); 245.0 (119.4~363.7)

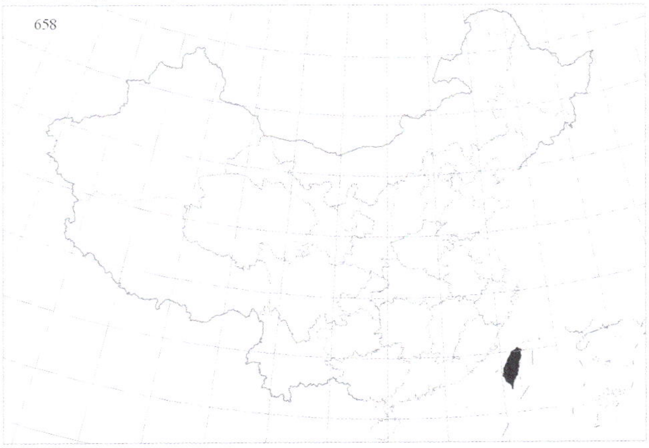

Salicaceae 杨柳科 *Salix* 柳属
Salix takasagoalpina Koidz. 高山柳
Shrubs, deciduous

17.8; 12.3; 22.3; 17.9; 154.3; 0.0; 873 ❙ 2349; 1053; 228 ❙ 180.4; 862; 269.2

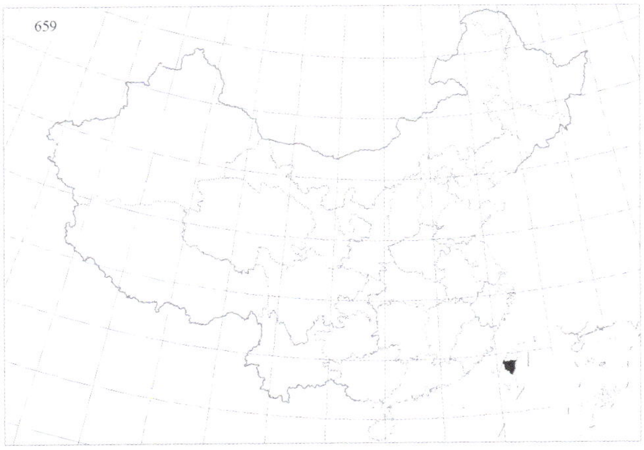

Salicaceae 杨柳科 *Salix* 柳属
Salix tangii K. S. Hao ex C. F. Fang et A. K. Skvortsov 周至柳
Shrubs, deciduous

7.7 (0.8~13.5); -5.7 (-13.9~1.4); 19.8 (10.1~26.3); 9.0 (3.8~13.5); 67.7 (15.4~114.9); 34.2 (7.1~71.1); 601 (411~800) ❙ 627 (77~1246); 333 (48~535); 18 (2~75) ❙ 7.1 (-87.2~128.6); 532 (77~767); 210.1 (20.1~357.7)

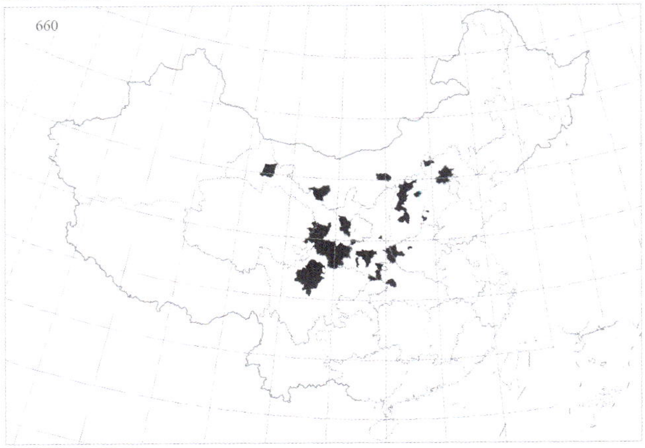

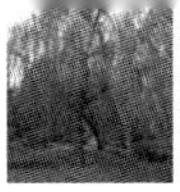

Salicaceae 杨柳科 *Salix* 柳属
Salix tangii var. angustifolia C. Y. Yu 细叶周至柳
Shrubs, deciduous

8.3; -2.2; 18.1; 8.8; 62.0; 21.0; 583 | 673; 326; 15 | 15.3; 583; 246.4

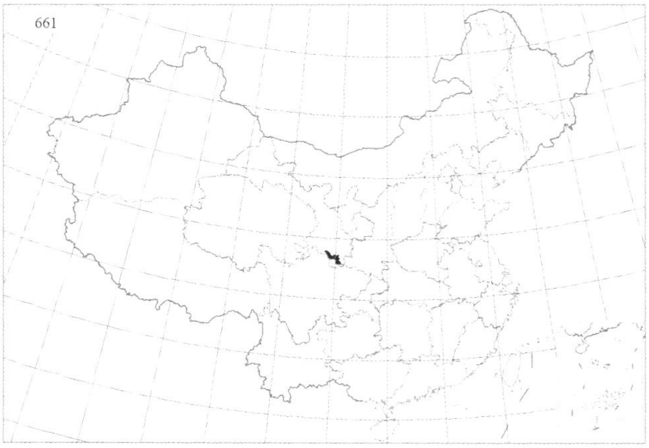

Salicaceae 杨柳科 *Salix* 柳属
Salix taoensis Goerz ex Rehder et Kobuski 洮河柳
Shrubs, deciduous

1.8 (-4.1~9.4); -10.3 (-17.3~-2.8); 12.4 (7.2~21.1); 4.7 (2.0~9.9); 25.8 (3.7~75.0); 63.7 (21.0~114.2); 436 (339~628) | 497 (63~773); 276 (45~394); 10 (2~19) | 15.8 (-84.9~87.3); 400 (63~628); 231.9 (8.4~364.1)

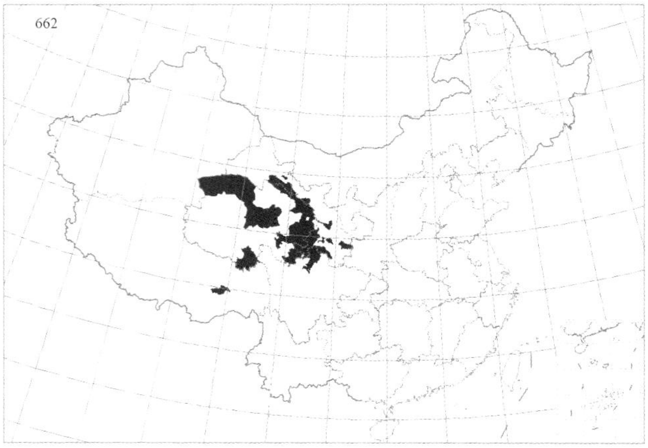

Salicaceae 杨柳科 *Salix* 柳属
lix taoensis var. leiocarpa T. Y. Ding et C. F. Fang 光果洮河柳
Shrubs, deciduous

-1.3; -13.5; 9.7; 2.9; 11.3; 86.3; 359 | 486; 289; 8 | 35.3; 359; 233.1

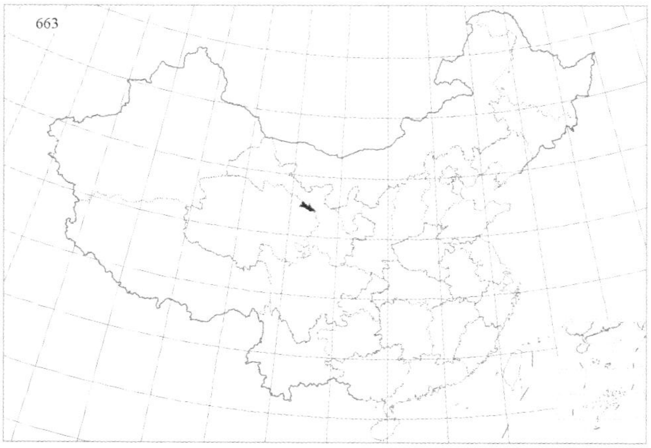

Salicaceae 杨柳科 *Salix* 柳属
Salix taoensis var. pedicellata C. F. Fang et J. Q. Wang 柄果洮河柳
Shrubs, deciduous

3.0; -7.7; 12.4; 4.8; 25.1; 49.3; 441 | 656; 323; 15 | 48.8; 441; 282.7

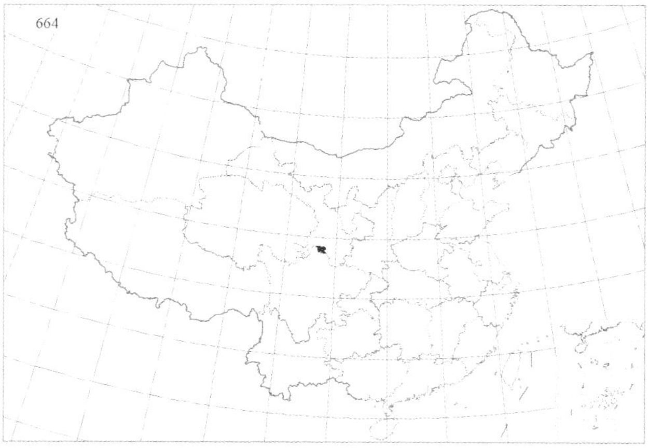

Salicaceae 杨柳科 *Salix* 柳属
Salix taraikensis Kimura 谷柳
Shrubs or small trees, deciduous

3.9 (-5.8~12.5); -15.6 (-29.8~-3.1); 20.9 (15.3~26.0); 8.1 (4.4~13.0); 63.3 (28.0~110.1); 76.2 (18.7~157.4); 582 (426~791) | 515 (44~1073); 327 (19~691); 16 (3~40) | -11.4 (-94.0~74.1); 470 (44~698); 210.1 (10.3~346.5)

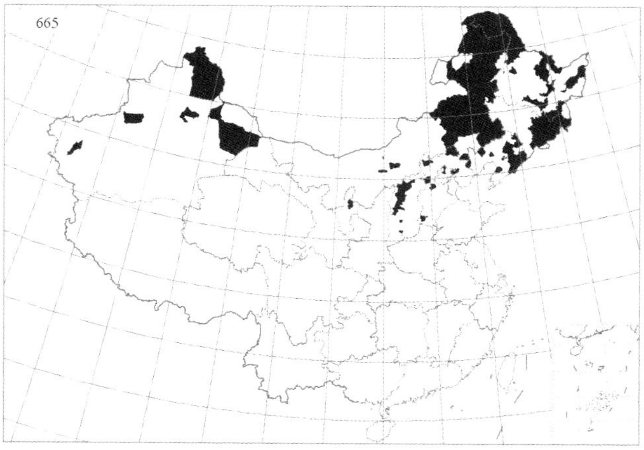

Salicaceae 杨柳科 *Salix* 柳属
Salix taraikensis var. oblanceolata C. Wang et C. F. Fang 倒披针谷柳
Shrubs or small trees, deciduous

8.4; -11.0; 25.0; 10.6; 90.9; 49.4; 700 | 698; 434; 22 | -0.3; 698; 191.9

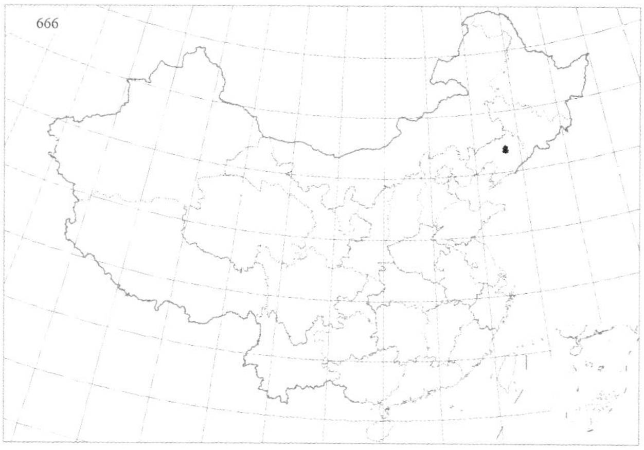

Salicaceae 杨柳科 *Salix* 柳属
Salix tarbagataica C. Y. Yang 塔城柳
Shrubs, deciduous

5.3; -13.7; 21.8; 8.9; 70.6; 66.3; 626 ❙ 231; 84; 32 ❙ -60.9; 229; 109.7

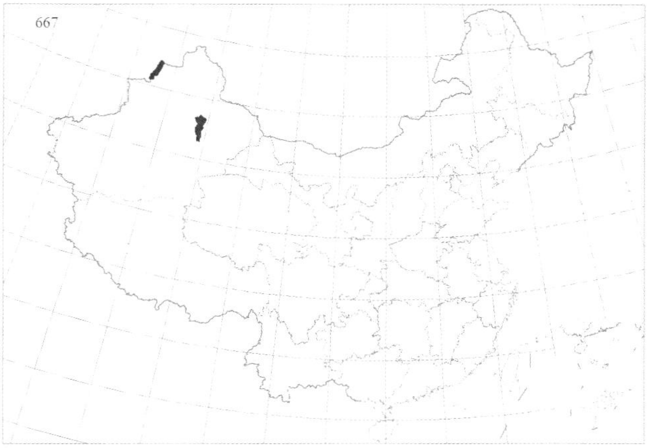

Salicaceae 杨柳科 *Salix* 柳属
Salix tenella C. K. Schneid. 光苞柳
Shrubs, deciduous

12.1 (2.2~18.7); 3.3 (-7.7~10.5); 19.9 (9.5~28.9); 12.4 (3.8~18.8); 95.7 (15.5~166.0); 9.8 (0.0~49.3); 703 (411~1002) ❙ 963 (627~1410); 516 (290~816); 36 (11~71) ❙ 41.8 (-16.6~125.9); 686 (411~1002); 261.7 (108.2~382.6)

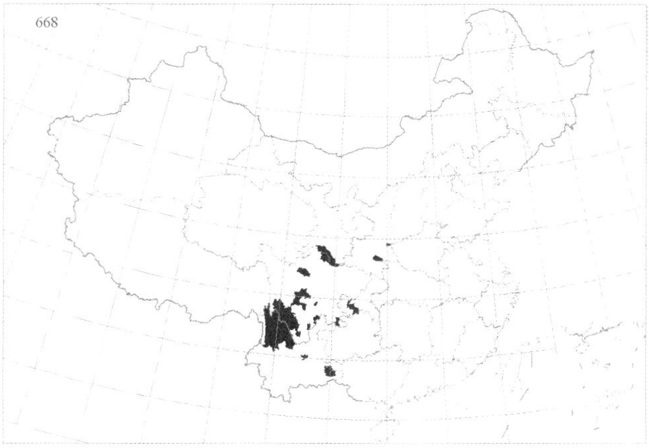

Salicaceae 杨柳科 *Salix* 柳属
Salix tenella var. trichadenia Hand.-Mazz. 基毛光苞柳
Shrubs, deciduous

10.8; 3.9; 16.6; 10.9; 71.8; 1.3; 633 ❙ 968; 594; 24 ❙ 52.8; 633; 319.8

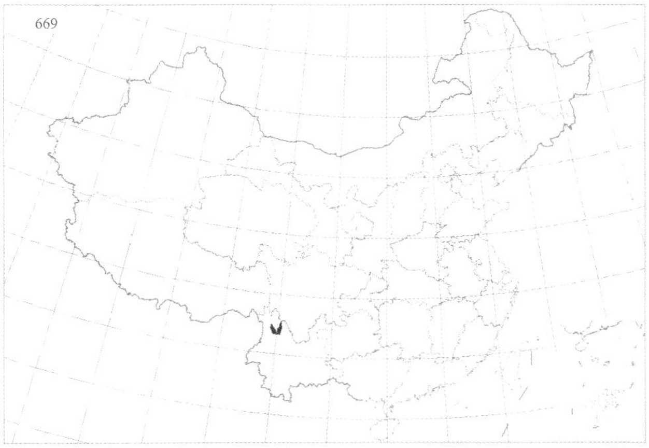

Salicaceae 杨柳科 *Salix* 柳属
Salix tengchongensis C. F. Fang 腾冲柳
Shrubs, deciduous

14.4; 7.5; 19.1; 14.4; 113.1; 0.0; 727 ❙ 1454; 792; 70 ❙ 99.9; 727; 487.2

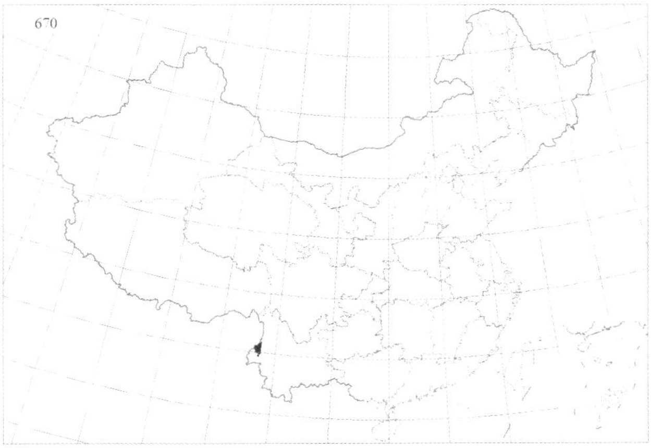

Salicaceae 杨柳科 *Salix* 柳属
Salix tenuijulis Ledeb. 细穗柳
Shrubs, deciduous

5.1 (-4.3~12.5); -13.8 (-20.9~-5.1); 20.6 (9.6~27.1); 8.6 (2.5~13.2); 68.5 (9.7~114.0); 67.3 (23.1~121.1); 613 (329~817) ❙ 183 (22~450); 83 (15~211); 17 (0~51) ❙ -66.9 (-97.2~3.9); 182 (22~406); 83.7 (1.6~297.3)

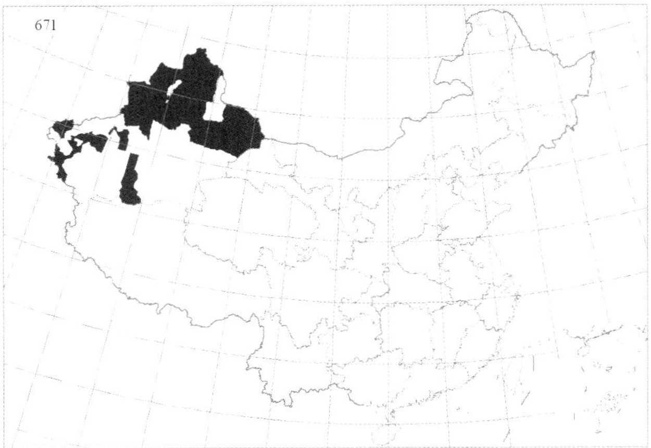

Salicaceae 杨柳科 *Salix* 柳属
Salix tetrasperma Roxb. 四子柳
Trees, deciduous

15.9 (-1.8~24.3); 8.5 (-10.5~18.7); 21.6 (6.1~29.4); 16.1 (1.8~24.3); 134.9 (1.9~231.9); 3.4 (0.0~82.8); 818 (329~1338) ❙ 1181 (311~1751); 615 (193~1007); 65 (3~217) ❙ 44.5 (-34.3~145.8); 793 (311~1153); 325.9 (61.6~659.0)

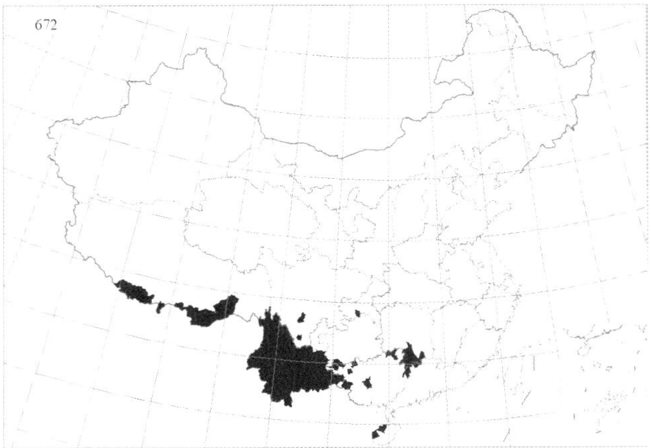

Salicaceae 杨柳科 *Salix* 柳属

Salix tianschanica Regel 天山柳

Shrubs, deciduous

4.8 (-2.8~10.2); -14.1 (-20.9~-8.5); 20.3 (10.9~27.1); 8.4 (3.5~12.1); 66.1 (16.3~107.6); 67.4 (36.6~109.2); 602 (390~810) ❙ 199 (45~450); 96 (27~211); 16 (3~48) ❙ -63.9 (-94.0~3.9); 197 (45~406); 88.2 (7.6~297.3)

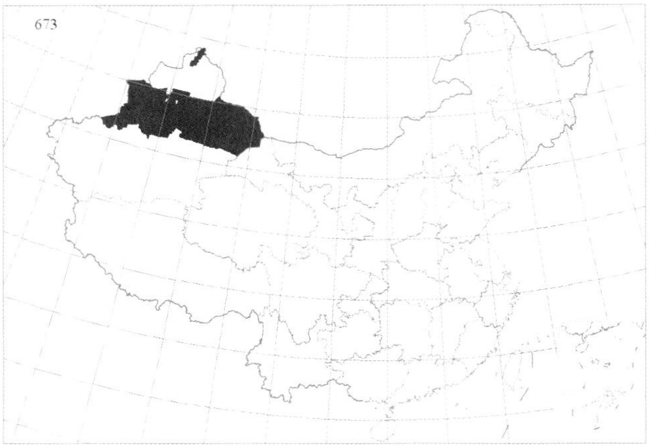

Salicaceae 杨柳科 *Salix* 柳属

Salix triandroides W. P. Fang 川三蕊柳

Shrubs or small trees, deciduous

16.3 (9.4~17.4); 4.0 (-2.8~5.7); 27.9 (21.1~29.7); 16.3 (9.9~17.5); 137.3 (75.0~150.4); 1.6 (0.0~21.0); 897 (628~964) ❙ 1250 (632~1483); 468 (306~606); 130 (15~185) ❙ 33.9 (0.0~67.0); 870 (628~927); 173.2 (133.8~260.4)

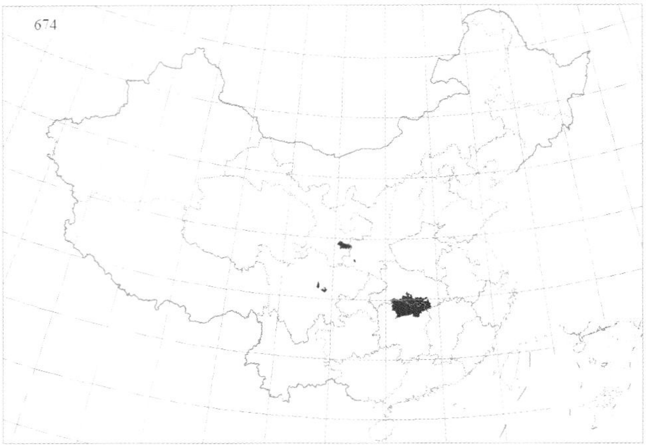

Salicaceae 杨柳科 *Salix* 柳属

Salix trichocarpa C. F. Fang 毛果柳

Small trees, deciduous

7.1 (-1.1~13.3); -1.1 (-9.1~5.4); 14.3 (6.2~22.1); 8.1 (2.0~13.4); 53.4 (2.5~100.9); 27.8 (0.0~75.6); 544 (334~723) ❙ 883 (321~1736); 524 (193~1007); 29 (11~51) ❙ 50.9 (-4.0~145.8); 541 (321~723); 240.5 (79.3~447.4)

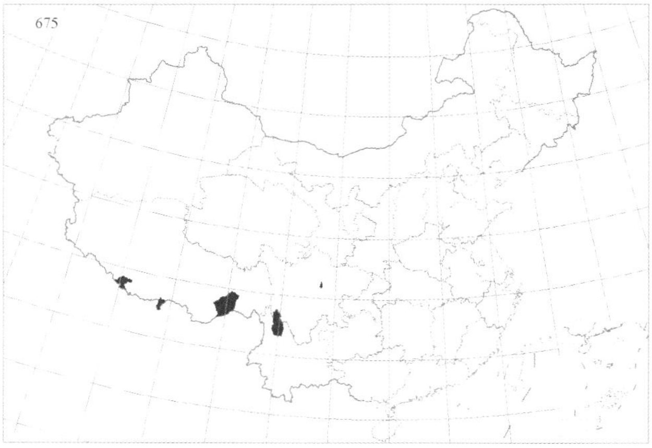

Salicaceae 杨柳科 *Salix* 柳属

Salix turanica Nasarow 吐兰柳

Shrubs, deciduous

3.2 (-4.3~12.3); -15.7 (-20.9~-5.3); 19.0 (9.6~27.5); 7.5 (2.5~13.0); 56.9 (9.7~112.3); 77.9 (24.7~121.1); 564 (329~813) ❙ 217 (44~450); 95 (19~211); 22 (4~51) ❙ -59.6 (-94.5~3.9); 216 (44~406); 104.6 (9.3~297.3)

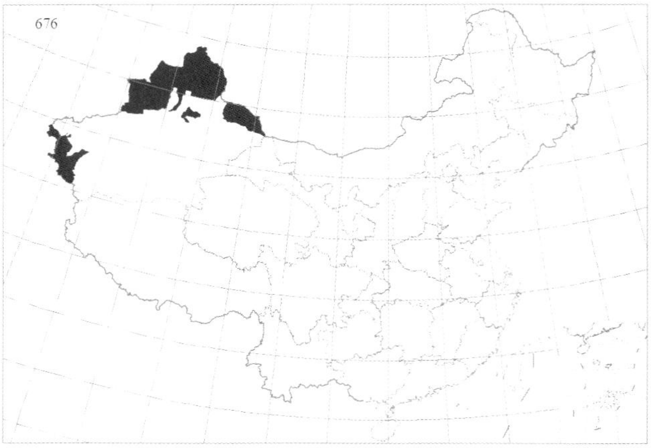

Salicaceae 杨柳科 *Salix* 柳属

Salix turczaninowii Lacksch. 蔓柳

Shrubs, deciduous

1.7 (-0.6~5.8); -17.1 (-19.5~-14.9); 18.1 (13.2~23.3); 6.8 (4.6~9.7); 49.8 (26.7~80.9); 89.0 (71.2~104.2); 535 (440~668) ❙ 206 (156~406); 91 (56~211); 22 (15~33) ❙ -60.0 (-76.7~-7.6); 206 (156~406); 103.7 (16.6~261.4)

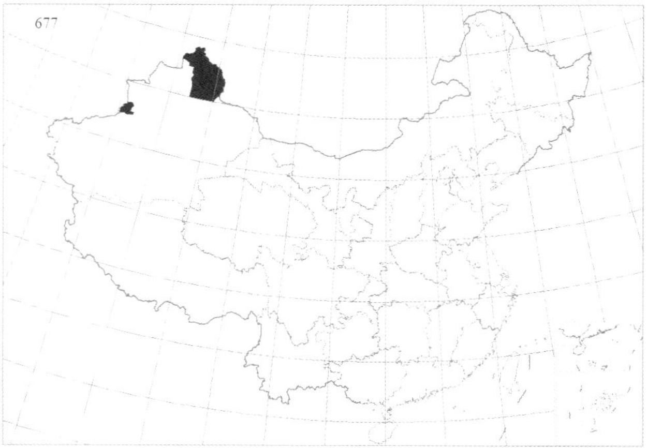

Salicaceae 杨柳科 *Salix* 柳属

Salix vaccinioides Hand.-Mazz. 乌饭叶矮柳

Shrubs, deciduous

7.2 (2.6~11.0); -0.5 (-7.0~4.2); 13.9 (11.2~16.6); 7.6 (4.3~11.0); 45.8 (20.0~73.1); 18.6 (0.8~48.2); 536 (427~636) ❙ 931 (741~1134); 539 (400~651); 28 (12~47) ❙ 75.7 (51.8~121.7); 535 (427~636); 298.7 (234.2~352.3)

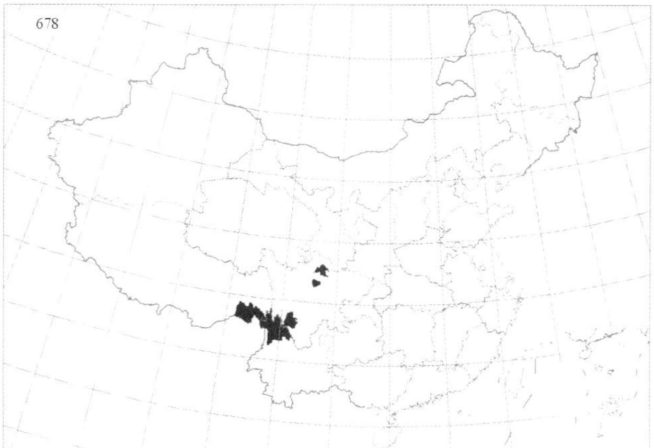

Salicaceae 杨柳科 *Salix* 柳属
Salix variegata Franch. 秋华柳
Shrubs, deciduous

14.7 (0.3~22.4); 5.4 (−11.9~14.5); 23.0 (9.3~29.2); 14.8 (3.3~22.4); 120.7 (11.9~209.2); 3.5 (0.0~69.8); 790 (391~1198) ❙ 1073 (373~1716); 555 (201~946); 48 (7~195) ❙ 37.6 (−37.6~125.9); 781 (373~1198); 226.8 (88.1~551.3)

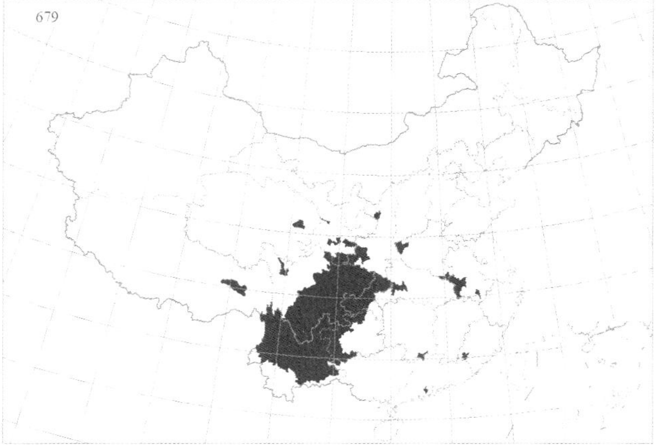

Salicaceae 杨柳科 *Salix* 柳属
Salix vestita Pursh 皱纹柳
Shrubs, deciduous

2.1 (0.4~5.8); −16.7 (−18.4~−14.9); 18.7 (16.6~23.3); 7.0 (5.8~9.7); 52.8 (40.7~80.9); 87.5 (71.2~97.3); 549 (494~668) ❙ 199 (156~251); 83 (56~106); 25 (19~33) ❙ −62.9 (−76.7~−52.0); 199 (156~251); 109.0 (27.7~139.1)

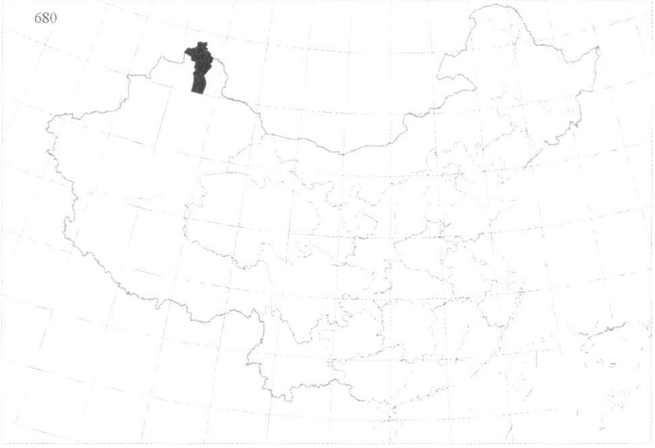

Salicaceae 杨柳科 *Salix* 柳属
Salix wallichiana Andersson 皂柳
Shrubs or small trees, deciduous

12.2 (−1.3~19.3); 0.5 (−16.1~13.1); 22.7 (8.1~28.9); 12.8 (2.6~19.4); 103.0 (8.0~173.2); 16.1 (0.0~86.3); 732 (347~1002) ❙ 891 (111~1828); 449 (74~946); 44 (0~253) ❙ 21.2 (−82.3~128.6); 678 (111~1002); 200.5 (13.6~487.2)

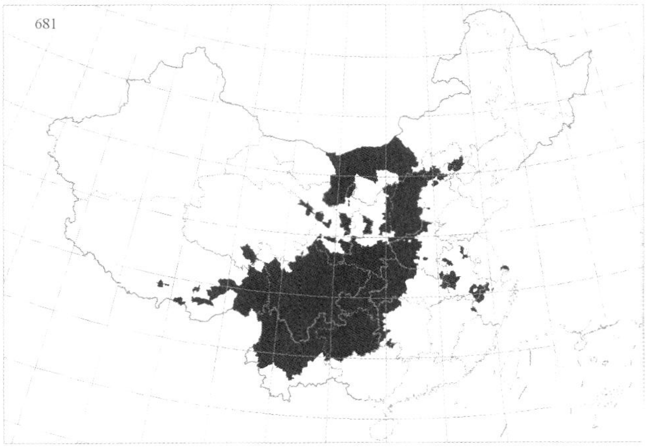

Salicaceae 杨柳科 *Salix* 柳属
Salix wallichiana var. pachyclada (H. Lév. et Vaniot) C. Wang et C. F. Fang 绒毛皂柳
Shrubs or small trees, deciduous

14.4 (9.8~15.5); 1.8 (−4.9~3.6); 26.4 (22.8~27.7); 14.5 (10.6~15.5); 118.9 (84.0~128.9); 6.0 (1.7~26.3); 809 (665~861) ❙ 1433 (624~1689); 566 (350~617); 162 (23~212) ❙ 75.9 (−6.3~105.7); 805 (624~861); 257.2 (162.7~319.9)

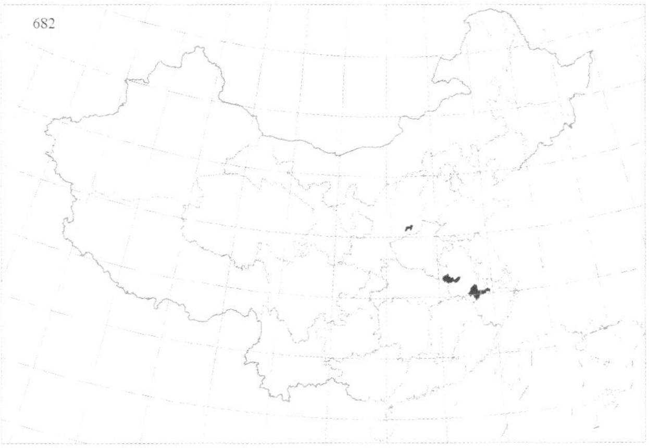

Salicaceae 杨柳科 *Salix* 柳属
Salix wangiana K. S. Hao ex C. F. Fang et A. K. Skvortsov 眉柳
Shrubs, deciduous

6.1 (−0.9~15.9); −4.6 (−12.5~6.2); 15.9 (8.1~25.5); 7.5 (2.7~15.9); 51.8 (8.0~130.8); 37.8 (0.0~79.5); 539 (354~822) ❙ 703 (300~1431); 374 (208~817); 20 (2~71) ❙ 30.4 (−25.4~113.0); 525 (300~822); 245.5 (111.0~364.1)

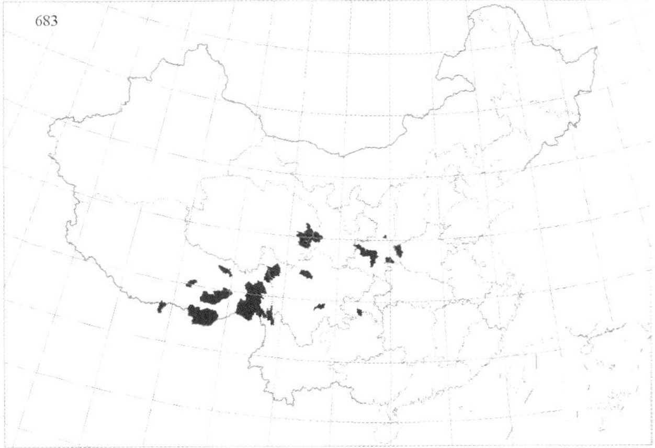

Salicaceae 杨柳科 *Salix* 柳属
Salix warburgii Seemen 水柳
Trees, deciduous

20.2 (16.2~24.1); 14.5 (11.0~18.9); 25.0 (20.1~28.2); 20.2 (16.2~24.1); 182.9 (134.2~229.7); 0.0 (0.0~0.0); 1019 (773~1322) ❙ 2402 (1498~3495); 1106 (750~1847); 255 (47~897) ❙ 146.2 (25.2~246.5); 983 (773~1144); 245.0 (119.4~363.7)

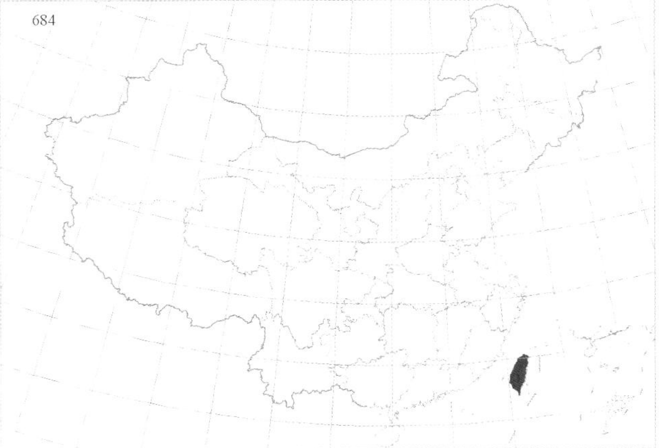

Salicaceae 杨柳科 *Salix* 柳属
Salix weixiensis Y. L. Chou 维西柳
Shrubs or small trees, deciduous

9.6; 2.4; 15.7; 9.6; 60.8; 5.4; 600 ❙ 1007; 578; 40 ❙ 67.9; 600; 351.7

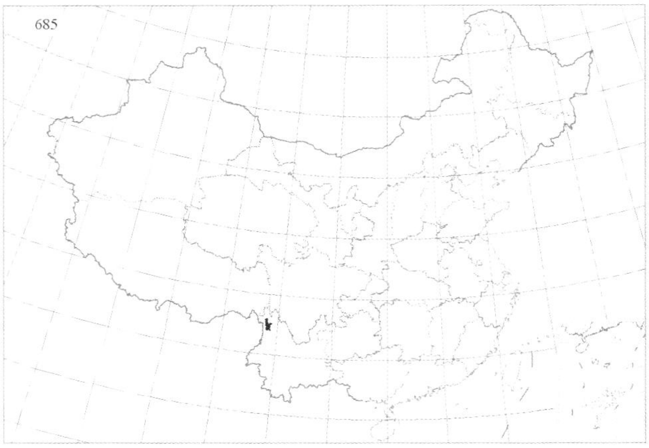

Salicaceae 杨柳科 *Salix* 柳属
Salix wilhelmsiana Bieb. 线叶柳
Shrubs or small trees, deciduous

6.3 (-4.3~12.5); -11.3 (-22.1~0.2); 21.0 (7.8~27.5); 9.0 (2.1~13.2); 72.1 (5.2~114.1); 56.1 (10.8~121.1); 623 (329~822) ❙ 185 (21~845); 103 (15~378); 8 (0~48) ❙ -67.0 (-97.3~48.8); 181 (21~717); 67.0 (1.6~297.3)

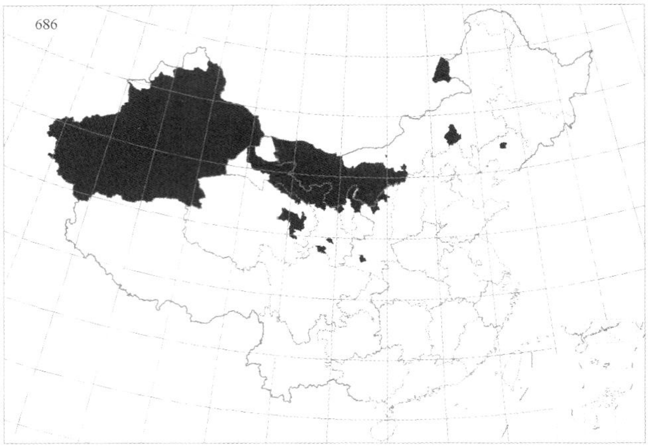

Salicaceae 杨柳科 *Salix* 柳属
Salix wilhelmsiana var. latifolia C. Y. Yang 宽线叶柳
Shrubs or small trees, deciduous

6.8 (-2.3~11.7); -10.4 (-17.2~-7.6); 20.9 (10.7~26.8); 9.3 (3.2~13.0); 75.4 (14.3~113.4); 53.2 (32.4~102.0); 642 (364~820) ❙ 88 (21~317); 57 (16~196); 2 (0~5) ❙ -82.6 (-97.3~-23.3); 88 (21~317); 32.1 (2.1~138.5)

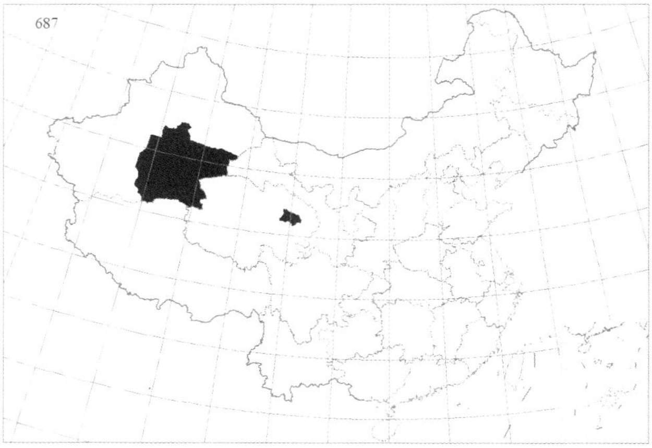

Salicaceae 杨柳科 *Salix* 柳属
Salix wilsonii Seemen ex Diels 紫柳
Trees, deciduous

15.8 (3.5~21.0); 3.9 (-6.5~1[illegible].9); 27.0 (12.6~30.1); 15.9 (5.0~21.1); 133.2 (26.3~193.0); 3.0 (0.0~43.4); 867 (446~1114) ❙ 1258 (590~1997); 501 (299~821); 130 (19~259) ❙ 44.8 (-27.6~130.7); 853 (446~1114); 201.1 (108.2~407.1)

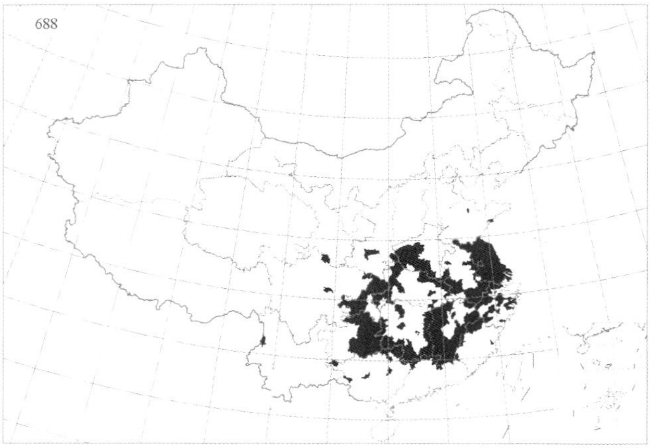

Salicaceae 杨柳科 *Salix* 柳属
Salix wolohoensis C. K. Schneid. 川南柳
Shrubs, deciduous

9.4 (2.0~15.5); 0.2 (-7.4~7.2); 17.6 (9.5~26.3); 9.8 (3.8~15.6); 70.1 (15.5~128.1); 17.1 (0.0~51.9); 612 (411~830) ❙ 978 (729~1469); 525 (332~812); 35 (11~192) ❙ 63.3 (5.8~125.9); 611 (411~830); 270.7 (182.9~372.3)

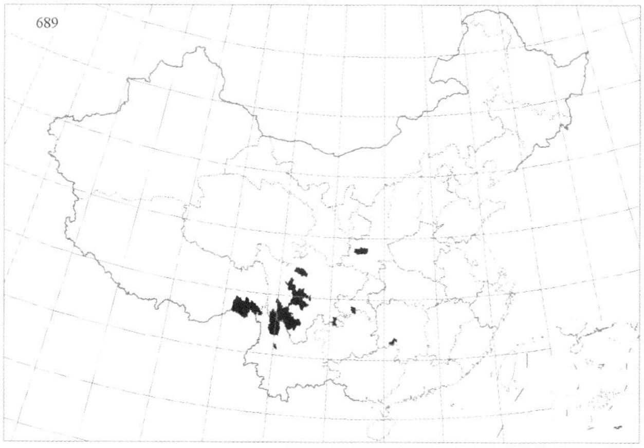

Salicaceae 杨柳科 *Salix* 柳属
Salix wuxuhaiensis N. Chao 伍须柳
Shrubs, deciduous

3.0; -5.5; 10.4; 4.3; 18.6; 42.3; 428 ❙ 958; 573; 17 ❙ 123.8; 428; 256.8

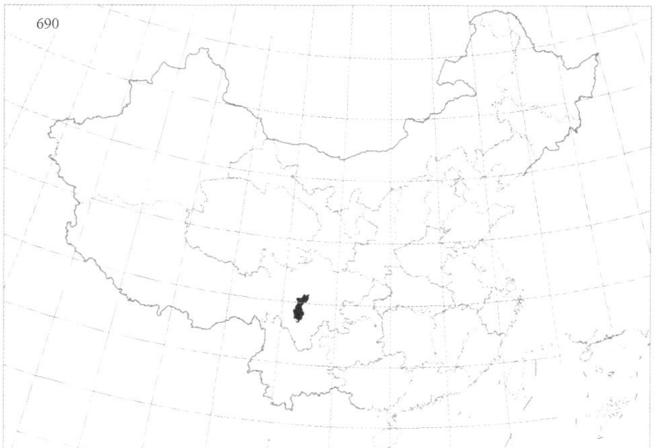

Salicaceae 杨柳科 *Salix* 柳属

Salix xiaoguangshanica Y. L. Chou et N. Chao 小光山柳

Shrubs, deciduous

14.1 (11.3~17.3); 7.2 (4.3~11.0); 19.3 (17.0~21.6); 14.2 (11.3~17.3); 110.0 (76.6~148.0); 0.1 (0.0~0.7); 726 (644~832) ❙ 1306 (1019~1736); 730 (562~1007); 57 (44~70) ❙ 82.1 (26.1~145.8); 717 (644~783); 412.2 (365.3~487.2)

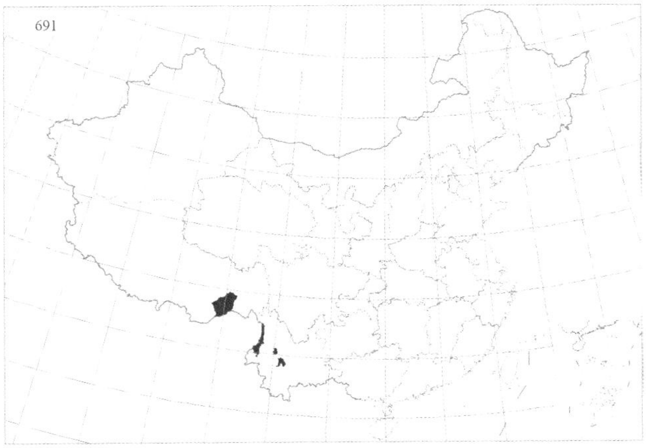

Salicaceae 杨柳科 *Salix* 柳属

Salix xizangensis Y. L. Chou 西藏柳

Shrubs, deciduous

13.3; 5.4; 19.6; 13.4; 100.4; 0.0; 706 ❙ 1736; 1007; 51 ❙ 145.8; 706; 447.4

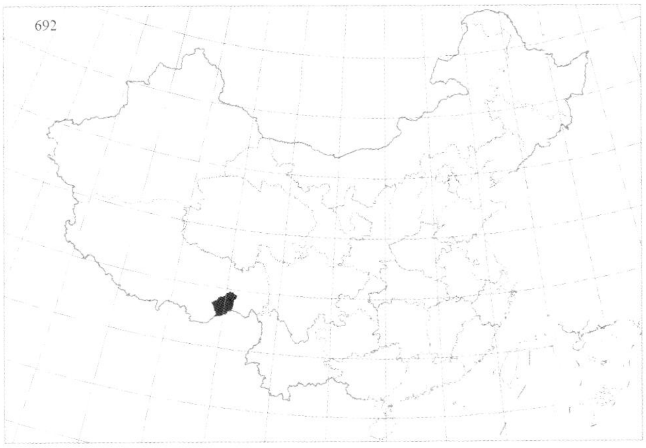

Salicaceae 杨柳科 *Salix* 柳属

Salix yadongensis N. Chao 亚东毛柳

Shrubs or small trees, deciduous

0.3; -8.6; 8.3; 2.9; 9.6; 65.4; 368 ❙ 389; 246; 11 ❙ 0.0; 368; 142.5

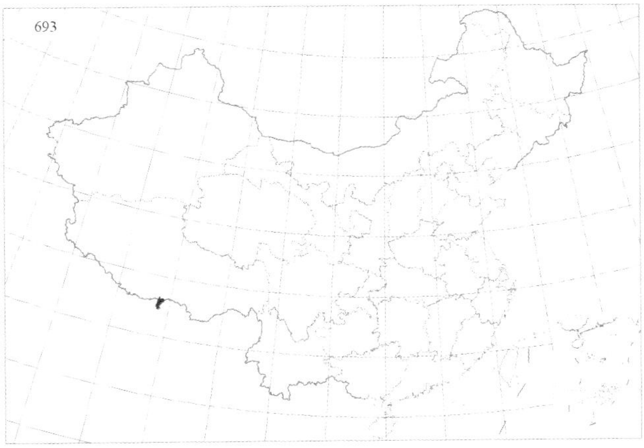

Salicaceae 杨柳科 *Salix* 柳属

Salix yanbianica C. F. Fang et C. Y. Yang 白河柳

Shrubs, deciduous

4.7 (2.2~8.4); -15.0 (-19.6~-8.9); 21.5 (19.2~25.0); 8.4 (6.9~ 0.6); 66.2 (50.3~90.9); 69.6 (42.8~87.4); 593 (525~700) ❙ 744 (556~ 018); 460 (333~648); 26 (15~40) ❙ 26.1 (-14.6~68.8); 582 (525~698); 289.9 (177.1~346.5)

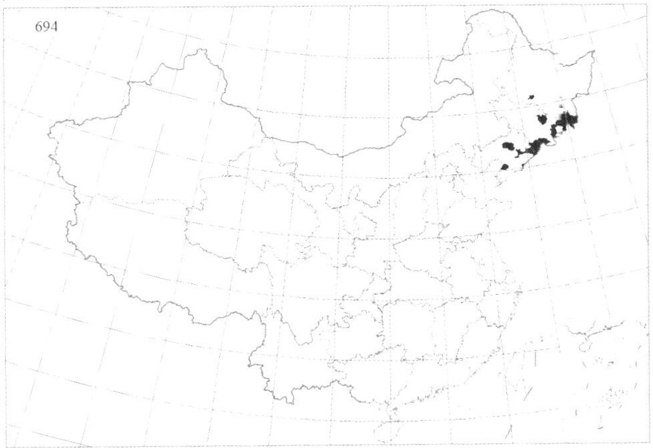

Salicaceae 杨柳科 *Salix* 柳属

Salix yuhuangshanensis C. Wang et C. Y. Yu 玉皇柳

Shrubs, deciduous

9.7 (6.2~11.5); -2.8 (-6.5~-1.5); 2 .8 (18.3~24.2); 10.2 (7.6~11.7); 79.1 (53.8~94.5); 21.4 (15.5~38.9); 642 (539~706) ❙ 736 (580~892); 336 (298~389); 22 (13~29) ❙ 14.9 (0.0~44.0); 641 (539~702); 264.7 (229.4~311.0)

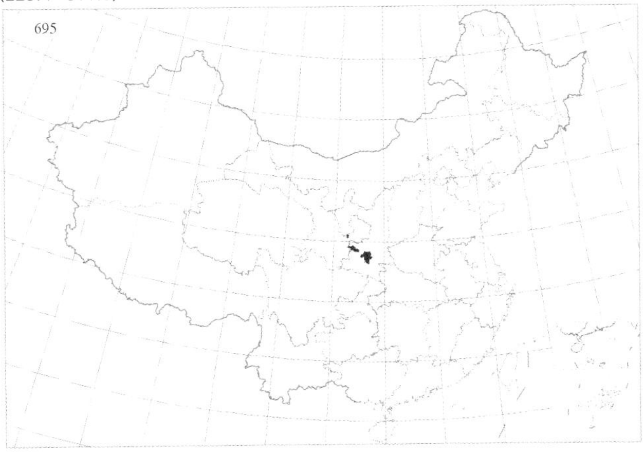

Salicaceae 杨柳科 *Salix* 柳属

Salix yumenensis H. L. Yang 玉门柳

Shrubs, deciduous

5.5; -10.1; 19.5; 8.0; 61.7; 55.1; 566 ❙ 131; 79; 3 ❙ -76.5; 131; 26.7

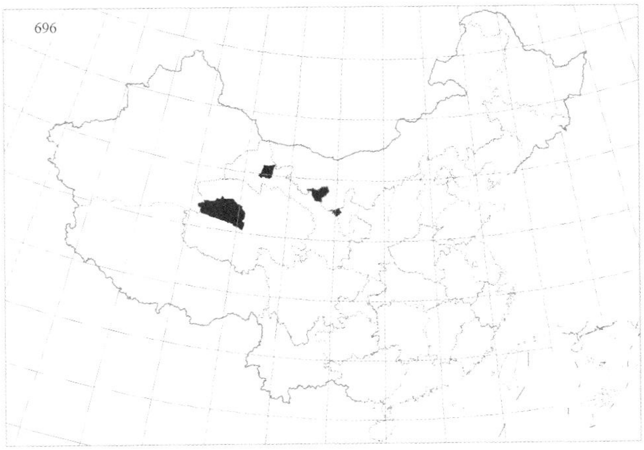

Salicaceae 杨柳科 *Salix* 柳属

Salix zangica N. Chao 藏柳

Shrubs, deciduous

2.5 (−0.9~8.6); −6.5 (−11.3~1.3); 10.6 (8.7~15.0); 4.5 (2.7~8.7); 21.7 (9.4~52.6); 51.3 (8.8~79.1); 425 (354~574) ❙ 608 (311~1041); 372 (232~585); 17 (2~45) ❙ 39.4 (−20.0~81.5); 410 (311~574); 233.8 (130.1~352.3)

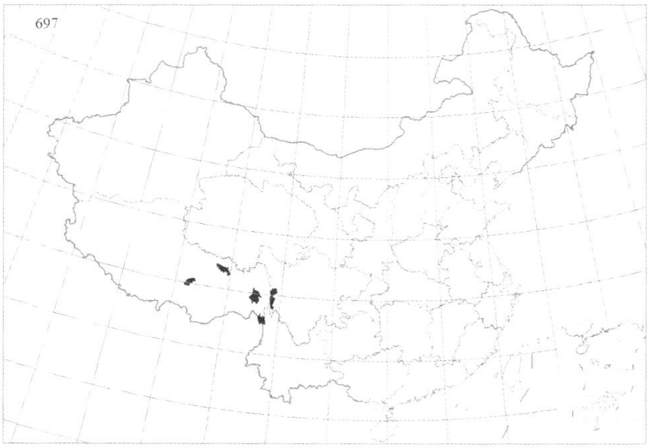

Salicaceae 杨柳科 *Salix* 柳属

Salix zayulica C. Wang et C. F. Fang 察隅矮柳

Shrubs, deciduous

4.6; −3.1; 11.6; 6.2; 37.7; 41.5; 479 ❙ 761; 427; 43 ❙ 58.1; 479; 210.5

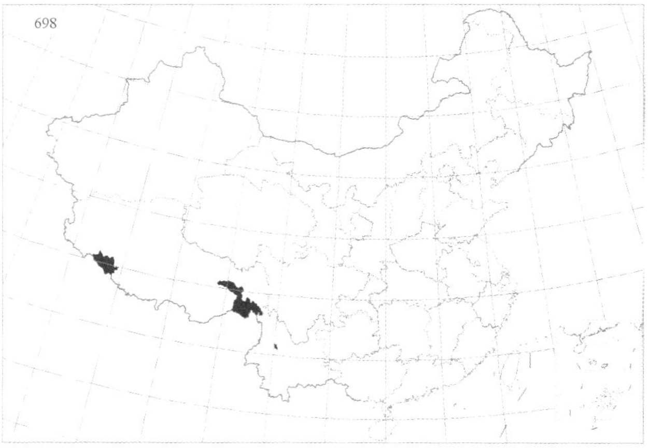

Salicaceae 杨柳科 *Salix* 柳属

Salix zhegushanica N. Chao 鹧鸪柳

Shrubs, deciduous

5.7 (0.6~15.4); −4.7 (−9.4~3.7); 15.1 (9.5~28.0); 6.9 (3.4~15.5); 44.9 (12.7~128.4); 35.6 (1.8~65.4); 520 (394~864) ❙ 776 (341~1076); 419 (190~619); 23 (8~135) ❙ 55.0 (−40.9~128.6); 498 (341~864); 244.1 (61.2~357.7)

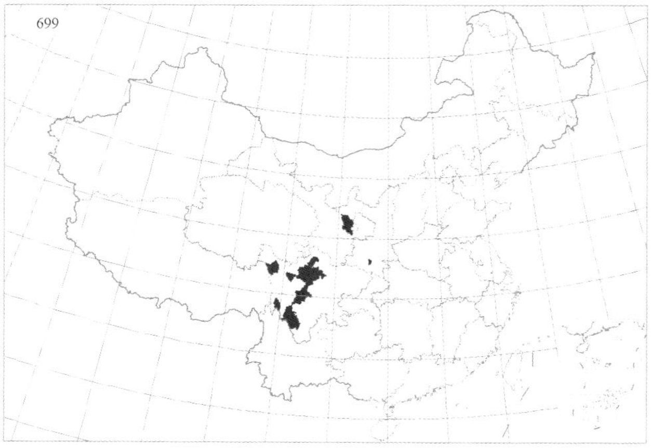

Salicaceae 杨柳科 *Salix* 柳属

Salix zhouquensis X. G. Sun 舟曲柳

Shrubs, deciduous

6.2; −4.1; 15.8; 7.0; 45.2; 29.9; 522 ❙ 669; 317; 15 ❙ 28.2; 522; 271.9

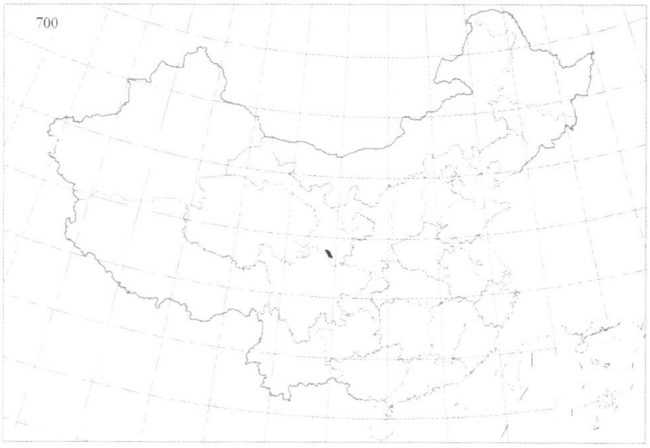

Myricaceae 杨梅科 *Myrica* 杨梅属

Myrica adenophora Hance 青杨梅

Shrubs, evergreen

21.3 (17.6~24.7); 13.3 (7.1~19.8); 27.8 (22.8~29.0); 21.4 (17.6~24.7); 196.5 (151.6~236.5); 0.0 (0.0~0.0); 1133 (909~1376) ❙ 1661 (868~2992); 737 (312~1847); 134 (42~189) ❙ 49.6 (−34.3~202.5); 1098 (868~1360); 276.2 (154.2~620.0)

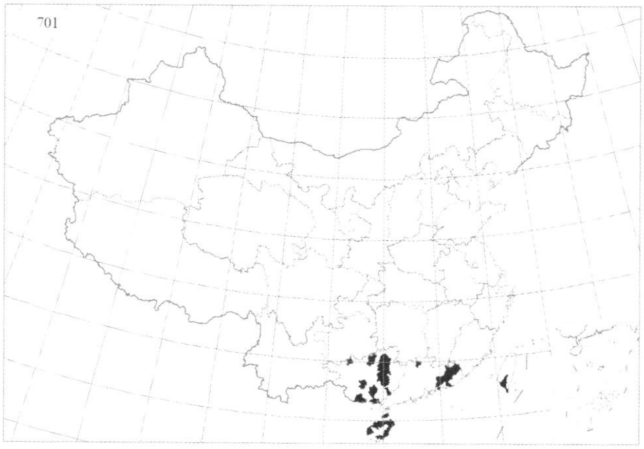

Myricaceae 杨梅科 *Myrica* 杨梅属

Myrica esculenta Buch.-Ham. ex D. Don 毛杨梅

Trees, evergreen

16.8 (5.3~23.2); 8.9 (−2.4~15.8); 22.9 (12.2~29.8); 16.8 (5.7~23.2); 142.6 (30.5~218.9); 0.6 (0.0~26.2); 857 (479~1257) ❙ 1243 (741~2035); 638 (411~996); 70 (15~196) ❙ 45.4 (−14.2~112.2); 845 (479~1257); 303.4 (108.2~659.0)

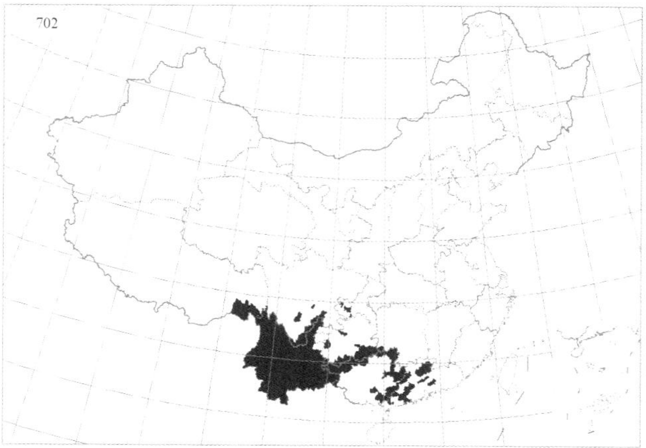

M

Myricaceae 杨梅科　*Myrica* 杨梅属
Myrica nana A. Chev. 云南杨梅
Shrubs, evergreen

15.1 (5.3~21.0); 7.8 (-2.4~14.2); 20.8 (12.2~26.6); 15.2 (5.7~21.0); 123.0 (30.5~191.8); 1.0 (0.0~26.2); 769 (479~1053) ❙ 1058 (741~1751); 599 (411~996); 42 (15~77) ❙ 38.4 (-14.2~100.7); 751 (479~1053); 298.5 (142.9~551.6)

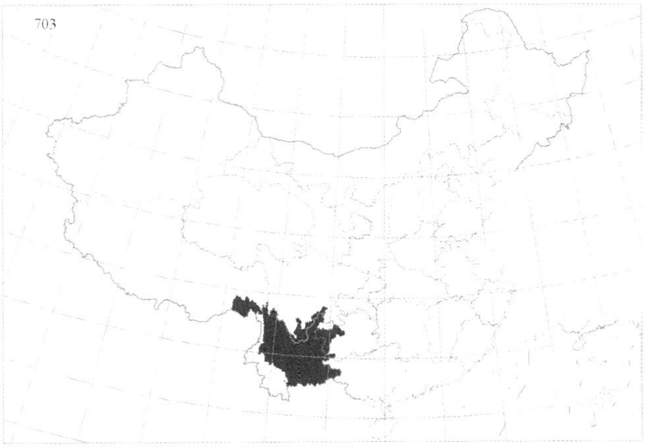

Myricaceae 杨梅科　*Myrica* 杨梅属
Myrica rubra (Lour.) Siebold et Zucc. 杨梅
Trees, evergreen

17.6 (5.3~23.0); 8.0 (-2.4~15.5); 26.2 (12.2~30.1); 17.7 (5.7~23.0); 152.3 (30.5~216.2); 0.3 (0.0~26.2); 926 (479~1247) ❙ 1446 (741~3495); 597 (299~1428); 140 (15~897) ❙ 56.9 (-14.2~246.5); 914 (479~1247); 242.2 (95.3~659.0)

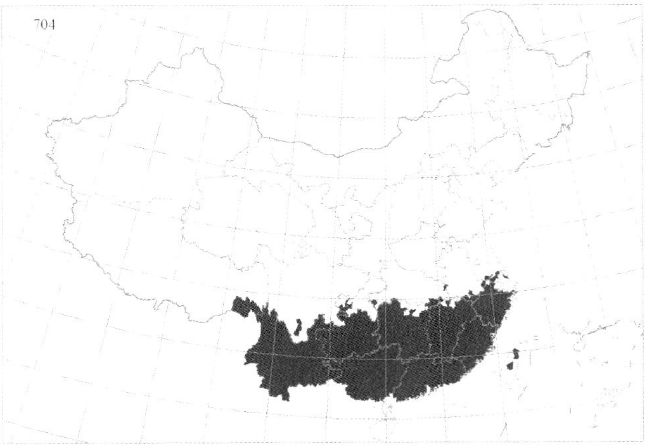

Juglandaceae 胡桃科　*Annamocarya* 喙核桃属
Annamocarya sinensis (Dode) J. -F. Leroy 喙核桃
Trees, deciduous

18.0 (15.5~22.2); 9.3 (4.8~14.2); 24.7 (20.3~28.3); 18.0 (15.6~22.3); 156.2 (127.2~207.5); 0.0 (0.0~0.2); 906 (789~1190) ❙ 1365 (1107~1758); 690 (518~860); 84 (42~189) ❙ 51.7 (15.4~93.2); 904 (780~1190); 317.3 (214.2~551.6)

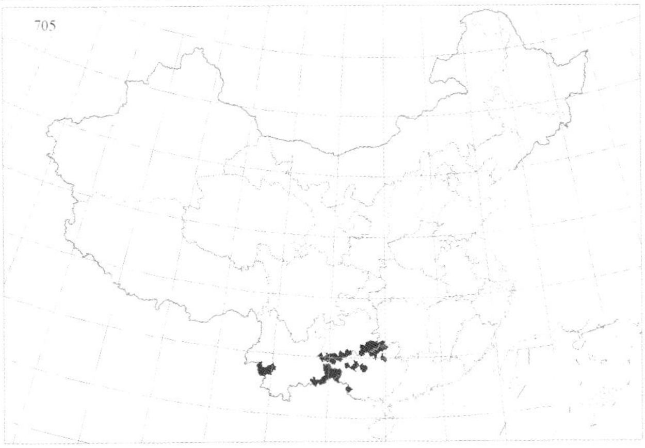

Juglandaceae 胡桃科　*Carya* 山核桃属
Carya cathayensis Sarg. 山核桃
Trees, deciduous

16.2 (14.1~19.2); 4.7 (-0.4~11.8); 26.8 (22.5~29.1); 16.2 (14.1~19.2); 136.8 (116.7~170.8); 2.2 (0.0~12.5); 872 (796~992) ❙ 1334 (610~1689); 563 (312~723); 135 (28~220) ❙ 53.5 (-27.3~105.7); 861 (610~992); 230.8 (139.3~411.1)

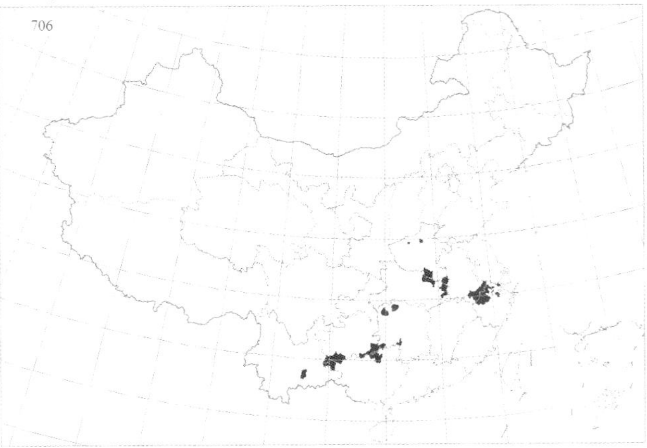

Juglandaceae 胡桃科　*Carya* 山核桃属
Carya hunanensis W. C. Cheng et R. H. Chang ex Chang et Lu 湖南山核桃
Trees, deciduous

16.0 (14.3~17.9); 4.8 (3.3~7.3); 26.4 (24.3~28.1); 16.1 (14.3~17.9); 133.3 (114.1~154.8); 0.6 (0.0~2.3); 852 (765~937) ❙ 1394 (1206~1526); 516 (419~639); 143 (75~198) ❙ 63.9 (45.6~99.4); 850 (765~937); 209.7 (152.4~257.[illegible])

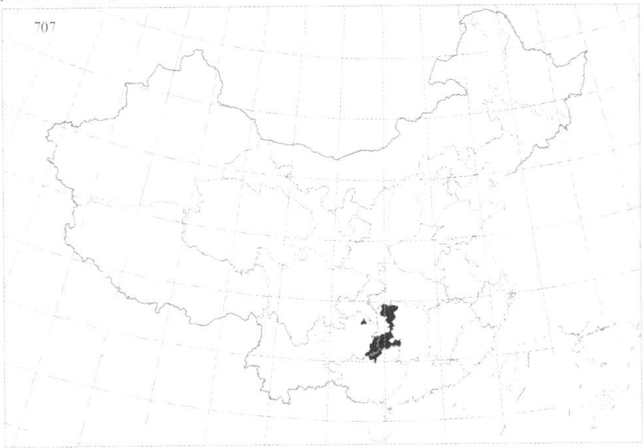

Juglandaceae 胡桃科　*Carya* 山核桃属
Carya kweichowensis Kuang et A. M. Lu 贵州山核桃
Trees, deciduous

16.5 (14.3~19.1); 7.1 (4.6~10.1); 24.3 (22.8~26.0); 16.5 (14.3~19.1); 138.5 (112.3~169.2); 0.0 (0.0~0.4); 842 (751~966) ❙ 1265 (1194~1381); 643 (561~720); 65 (51~107) ❙ 50.9 (30.0~62.6); 842 (751~966); 217.8 (160.8~283.9)

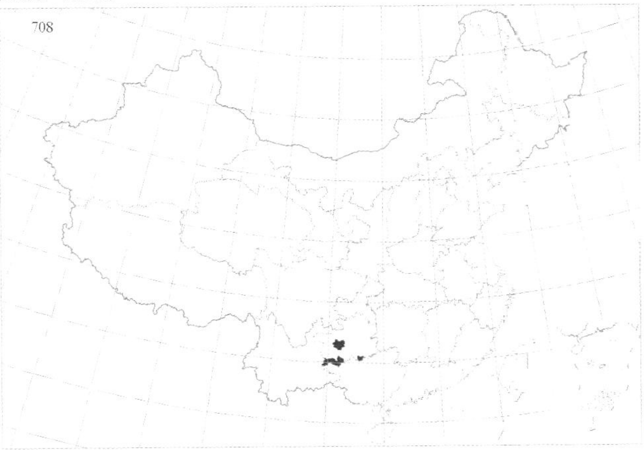

Juglandaceae 胡桃科 *Carya* 山核桃属
Carya tonkinensis Lecomte 越南山核桃
Trees, deciduous

16.6 (8.6~21.0)；9.6 (1.3~14.2)；21.6 (15.0~27.0)；16.6 (8.7~21.0)；139.9 (52.6~191.8)；0.4 (0.0~8.8)；827 (574~1053) ❙ 1257 (1041~1522)；696 (553~835)；58 (44~96) ❙ 53.9 (21.3~99.9)；821 (574~1053)；391.0 (288.8~551.6)

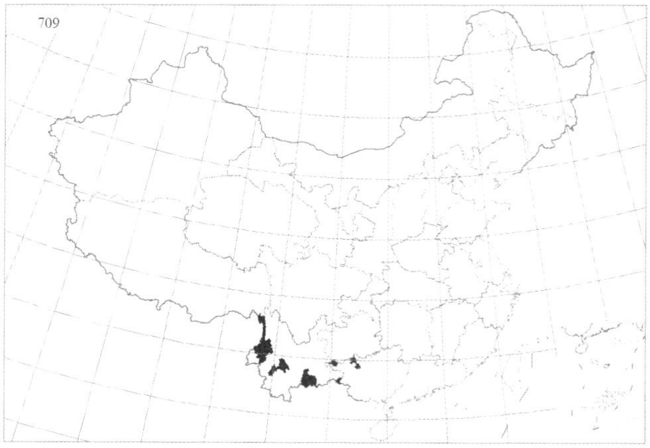

Juglandaceae 胡桃科 *Cyclocarya* 青钱柳属
Cyclocarya paliurus (Batalin) Iljinsk. 青钱柳
Trees, deciduous

15.7 (7.4~21.0)；4.2 (-6.5~12.2)；26.5 (18.7~29.9)；15.8 (8.2~21.1)；132.3 (58.8~193.0)；3.3 (0.0~32.5)；857 (561~1114) ❙ 1332 (517~1997)；506 (290~821)；135 (12~259) ❙ 54.4 (-29.6~147.0)；845 (517~1114)；220.8 (126.1~354.5)

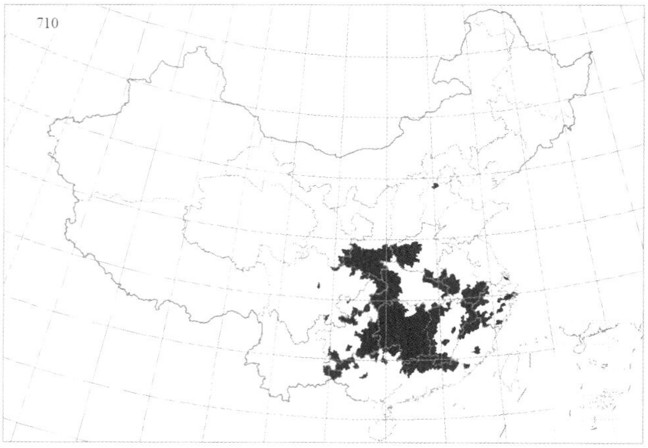

Juglandaceae 胡桃科 *Engelhardia* 黄杞属
Engelhardia hainanensis P. Y. Chen 海南黄杞
Trees

23.9；18.4；27.9；23.9；227.3；0.0；1301 ❙ 1267；430；45 ❙ -2.5；1086；450.1

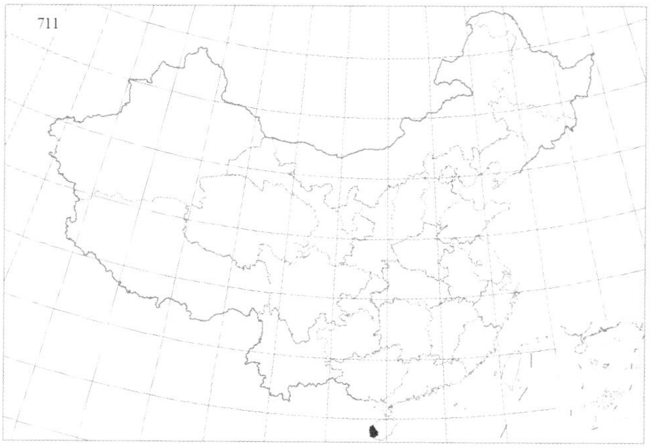

Juglandaceae 胡桃科 *Engelhardia* 黄杞属
Engelhardia roxburghiana Wall. 黄杞
Trees, semievergreen

18.2 (9.3~24.8)；8.6 (-1.7~20.1)；26.7 (16.7~30.1)；18.2 (9.5~24.8)；158.9 (69.8~237.6)；0.2 (0.0~18.0)；958 (603~1385) ❙ 1504 (868~3495)；604 (312~1847)；143 (20~897) ❙ 58.9 (-34.3~246.5)；946 (603~1360)；235.1 (95.3~659.0)

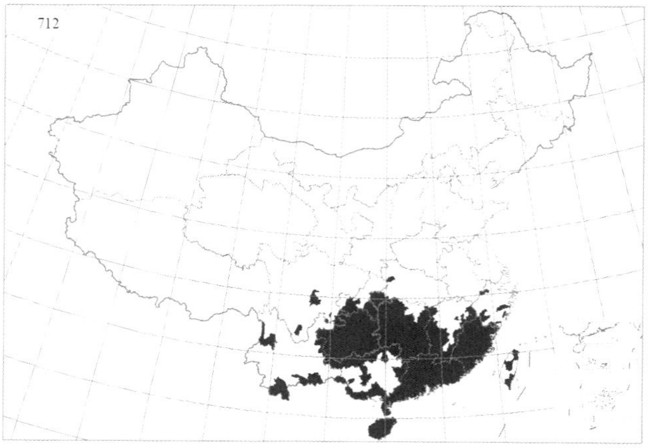

Juglandaceae 胡桃科 *Engelhardia* 黄杞属
Engelhardia serrata var. cambodica W. E. Manning 齿叶黄杞
Trees

18.9 (16.3~20.5)；12.6 (9.5~15.1)；23.0 (20.3~26.0)；18.9 (16.4~20.5)；167.0 (136.6~185.9)；0.0 (0.0~0.0)；915 (789~992) ❙ 1403 (1232~1607)；699 (655~783)；58 (45~77) ❙ 53.5 (32.2~64.9)；900 (780~986)；482.4 (262.1~659.0)

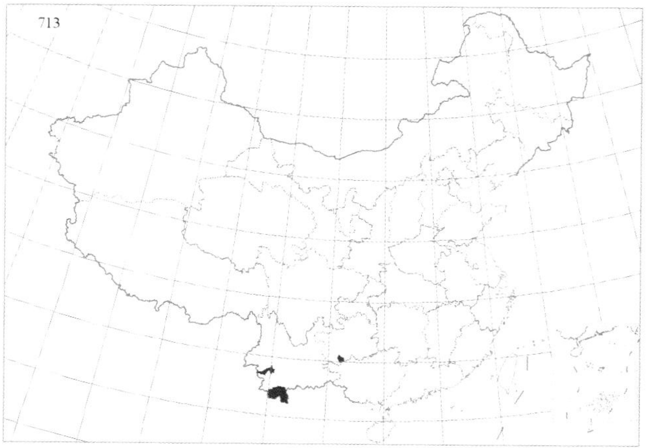

Juglandaceae 胡桃科 *Engelhardia* 黄杞属
Engelhardia spicata Lesch. ex Blume 云南黄杞
Trees

15.7 (5.4~24.3)；8.4 (-2.1~18.7)；21.3 (12.3~28.7)；15.7 (5.8~24.3)；129.5 (31.5~231.9)；0.9 (0.0~25.6)；803 (481~1338) ❙ 1138 (741~1751)；630 (312~1007)；45 (15~89) ❙ 43.8 (-34.3~145.8)；777 (481~1139)；332.7 (142.9~659.0)

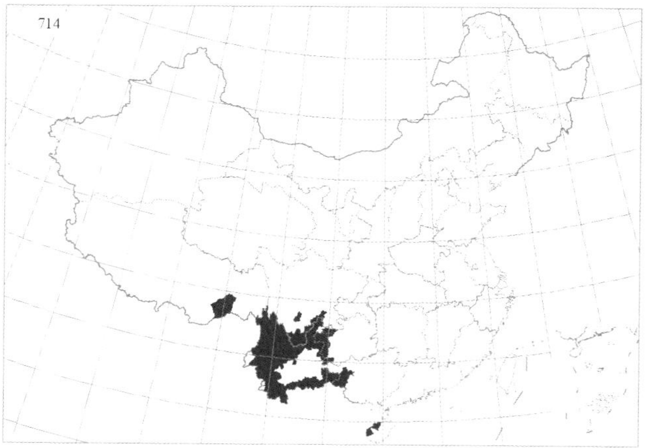

Juglandaceae 胡桃科 *Engelhardia* 黄杞属

Engelhardia spicata var. aceriflora (Reinw.) Koord. et Valeton 爪哇黄杞

Trees

18.5 (16.9~21.0); 12.0 (9.9~14.2); 23.0 (21.4~25.6); 18.6 (16.9~21.0) 163.0 (142.7~191.8); 0.0 (0.0~0.0); 899 (813~1053) | 1403 (1178~1547); 795 (646~911); 58 (51~68) | 56.7 (36.8~80.6); 897 (813~1053); 478.5 (383.4~551.6)

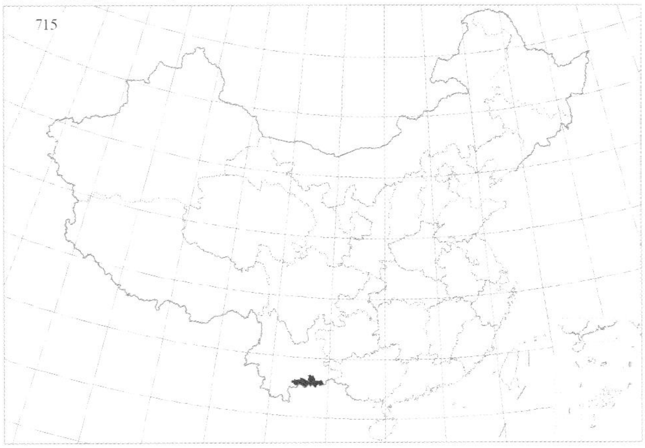

Juglandaceae 胡桃科 *Engelhardia* 黄杞属

Engelhardia spicata var. colebrookeana (Lindl. ex Wall.) Koord. et Valeton 毛叶黄杞

Small trees

17.7 (10.2~24.8); 10.3 (2.4~20.1); 23.4 (16.7~28.7); 17.8 (10.3~24.8); 153.3 (68.2~237.6); 0.1 (0.0~5.1); 899 (620~1385) | 1232 (338~1751); 644 (312~996); 54 (15~130) | 40.0 (-34.3~99.9); 867 (620~1254); 341.8 (184.6~659.0)

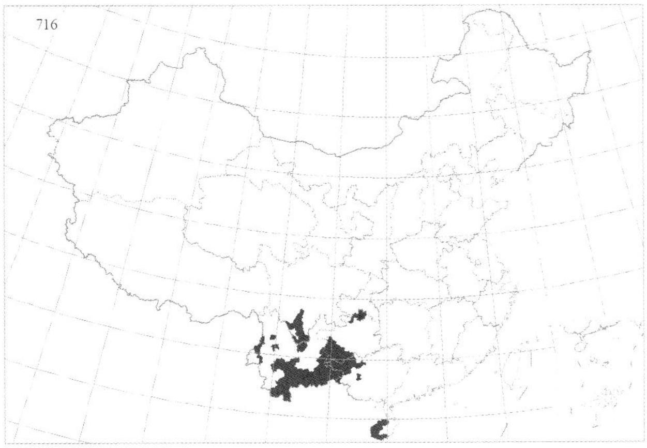

Juglandaceae 胡桃科 *Juglans* 胡桃属

Juglans mandshurica Maxim. 胡桃楸

Shrubs or trees, deciduous

12.0 (-1.5~21.0); -1.8 (-25.0~12.0); 24.3 (10.8~29.9); 12.9 (4.3~21.1); 106.2 (20.0~193.0); 22.1 (0.0~124.5); 753 (427~1114) | 954 (341~2499); 448 (190~947); 69 (7~303) | 24.0 (-48.6~219.3); 700 (341~1114); 211.2 (58.7~491.5)

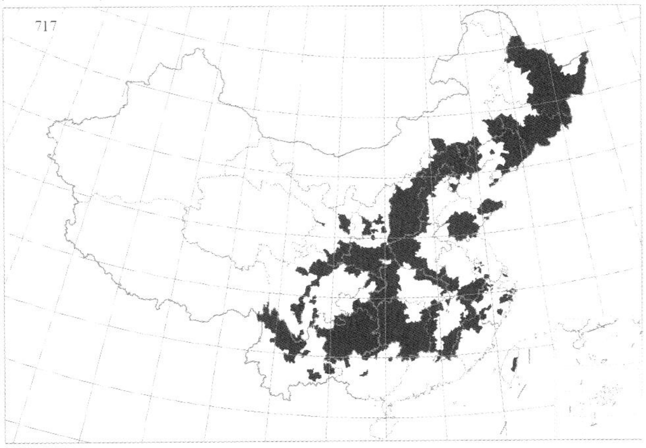

Juglandaceae 胡桃科 *Juglans* 胡桃属

Juglans regia L. 胡桃

Trees, deciduous

12.8 (-1.8~21.0); 1.6 (-19.0~15.1); 22.6 (6.1~29.8); 13.4 (1.8~21.0); 109.4 (1.9~192.6); 14.9 (0.0~90.3); 749 (329~1113) | 929 (225~1852); 472 (65~1007); 49 (0~203) | 22.1 (-65.1~145.8); 693 (225~1107); 233.7 (32.2~659.0)

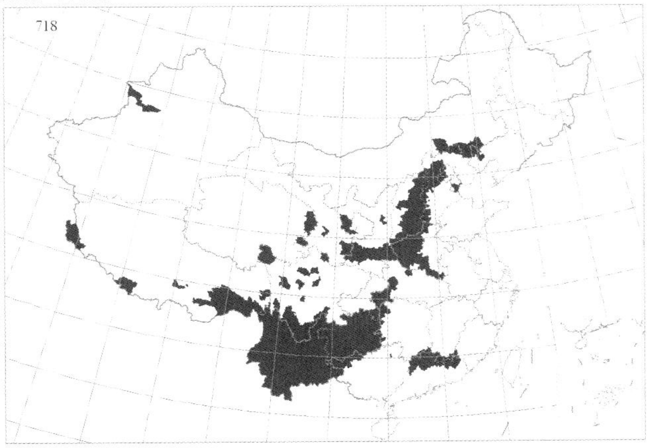

Juglandaceae 胡桃科 *Juglans* 胡桃属

Juglans sigillata Dode 泡核桃

Trees, deciduous

13.6 (-1.8~20.5); 6.3 (-10.2~15.1); 19.5 (6.1~25.5); 13.8 (1.8~20.5); 108.8 (1.9~185.9); 5.0 (0.0~82.8); 726 (329~992) | 1068 (320~1736); 592 (193~1007); 41 (12~88) | 48.6 (-14.2~145.8); 703 (320~986); 315.5 (72.2~659.0)

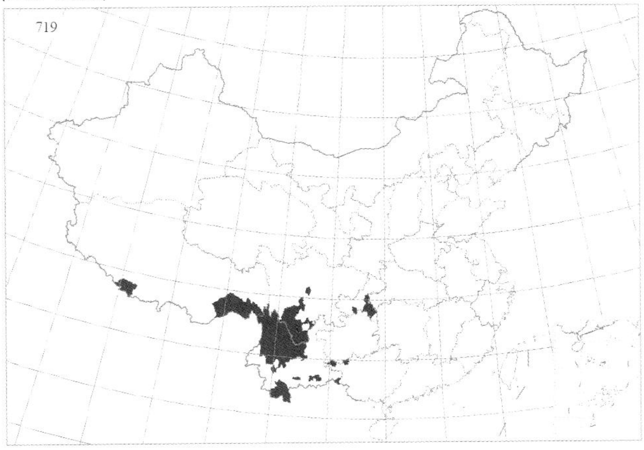

Juglandaceae 胡桃科 *Platycarya* 化香树属

Platycarya strobilacea Siebold et Zucc. 化香树

Small trees, deciduous

15.3 (3.3~23.4); 3.5 (-13.7~16.1); 26.0 (10.8~29.9); 15.5 (4.6~23.5); 130.1 (21.6~221.7); 6.4 (0.0~71.7); 850 (434~1273) | 1208 (367~2680); 510 (235~1186); 110 (8~308) | 40.3 (-40.9~246.5); 817 (367~1273); 207.9 (95.3~551.6)

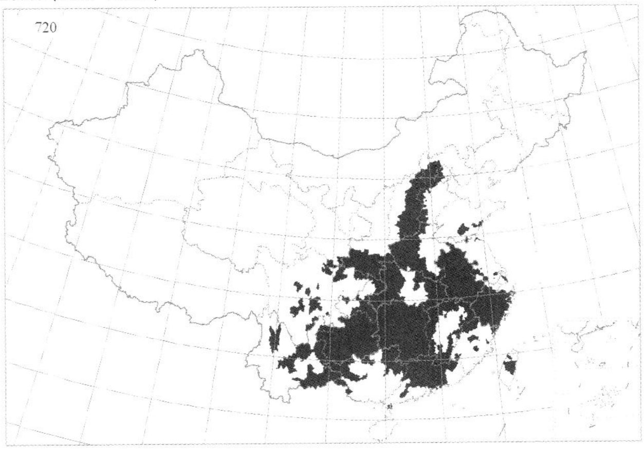

Juglandaceae 胡桃科 *Pterocarya* 枫杨属

Pterocarya hupehensis Skan 湖北枫杨

Trees, deciduous

11.8 (3.9~17.4)；-1.7 (-13.7~7.4)；24.2 (15.6~29.4)；12.4 (7.3~17.4)；100.9 (48.2~148.9)；18.4 (0.0~71.7)；734 (535~952) ❙ 826 (367~1571)；404 (235~816)；45 (8~161) ❙ 11.2 (-40.9~108.0)；664 (367~921)；195.3 (101.9~334.4)

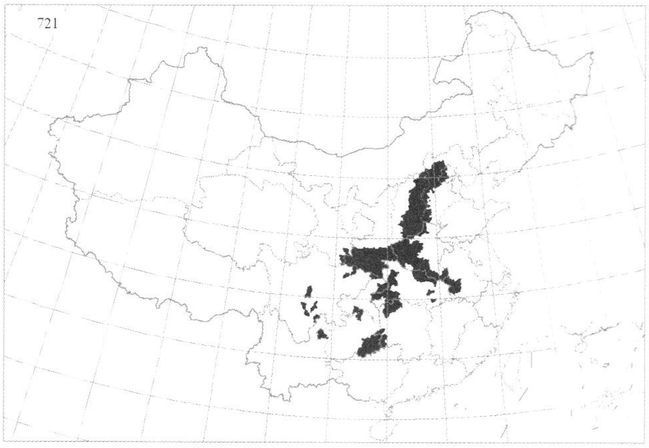

Juglandaceae 胡桃科 *Pterocarya* 枫杨属

Pterocarya macroptera Batalin 甘肃枫杨

Trees, deciduous

9.7 (5.7~13.5)；-1.7 (-6.7~2.9)；20.3 (14.2~23.7)；10.0 (6.4~13.5)；75.0 (38.1~105.3)；18.1 (3.0~35.7)；630 (504~744) ❙ 707 (373~922)；346 (201~460)；18 (7~35) ❙ 13.2 (-37.6~83.0)；613 (373~744)；254.1 (88.1~320.6)

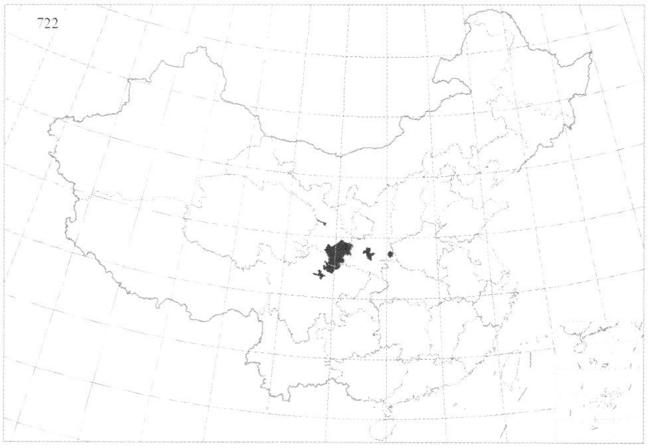

Juglandaceae 胡桃科 *Pterocarya* 枫杨属

Pterocarya macroptera var. delavayi (Franch.) W. E. Manning 云南枫杨

Trees, deciduous

12.6 (2.2~18.3)；4.8 (-6.6~11.8)；19.2 (9.9~23.9)；12.7 (3.8~18.3)；96.5 (15.5~159.8)；4.2 (0.0~49.1)；692 (413~882) ❙ 1047 (741~1410)；581 (420~816)；42 (15~83) ❙ 53.7 (6.5~125.9)；679 (413~845)；291.5 (185.6~407.1)

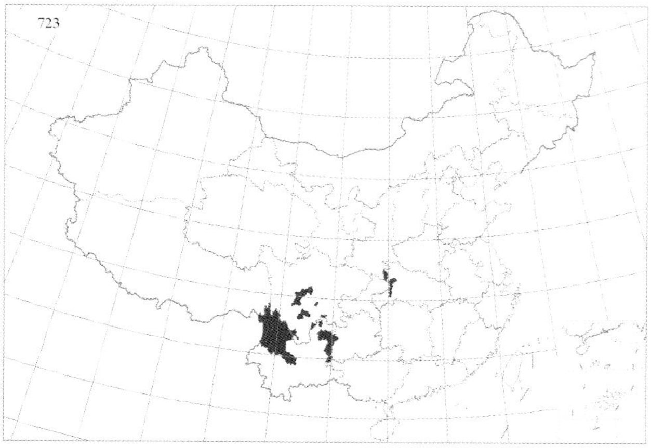

Juglandaceae 胡桃科 *Pterocarya* 枫杨属

Pterocarya macroptera var. insignis (Rehder et E. H. Wilson) W. E. Manning 华西枫杨

Trees, deciduous

11.6 (2.2~16.8)；1.4 (-6.6~7.3)；21.1 (9.9~26.8)；11.8 (3.8~16.9)；89.5 (15.5~142.7)；9.2 (0.0~49.1) 681 (413~869) ❙ 1058 (632~1893)；514 (306~816)；53 (15~227) ❙ 56.0 (-1.2~129.9)；681 (413~869)；257.7 (182.9~382.6)

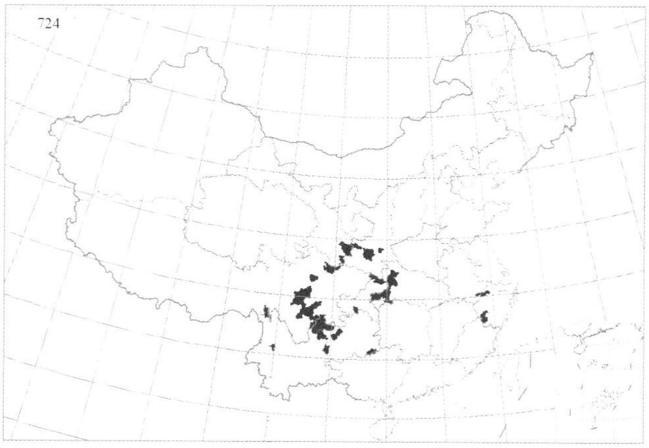

Juglandaceae 胡桃科 *Pterocarya* 枫杨属

Pterocarya rhoifolia Siebold et Zucc. 水胡桃

Trees, deciduous

11.9；-2.0；24.1；12.2；100.7；17.1；729 ❙ 747；444；34 ❙ 0.0；729；136.4

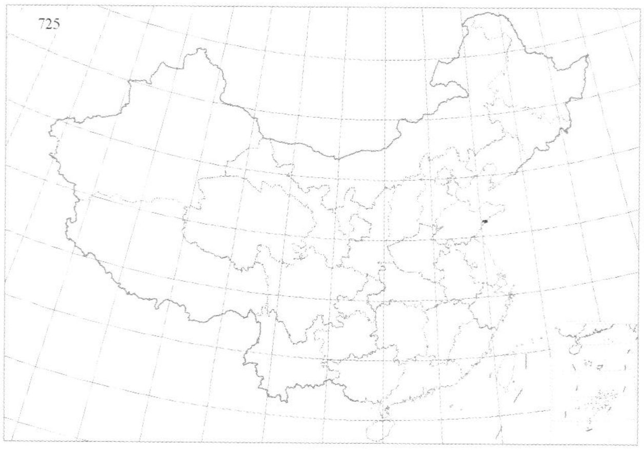

Juglandaceae 胡桃科 *Pterocarya* 枫杨属

Pterocarya stenoptera C. DC. 枫杨

Trees, deciduous

15.1 (2.6~22.4)；3.3 (-14.9~15.3)；26.1 (11.2~30.1)；15.4 (4.3~22.7)；129.6 (20.0~212.7)；7.2 (0.0~71.7)；848 (427~1216) ❙ 1176 (367~2680)；498 (235~1186)；107 (8~308) ❙ 36.9 (-40.9~246.5)；813 (367~1216)；205.1 (95.3~659.0)

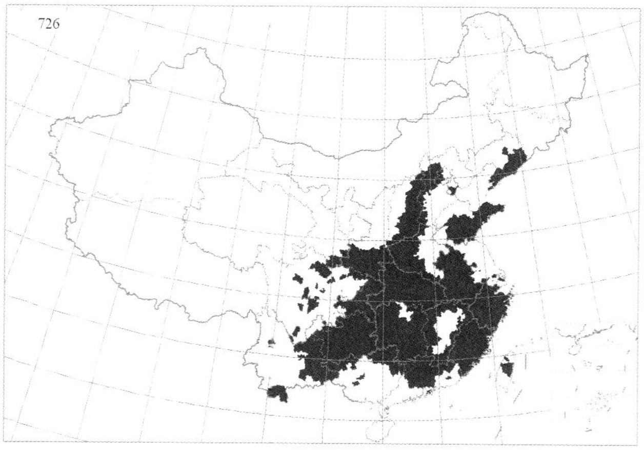

Juglandaceae 胡桃科 *Pterocarya* 枫杨属
Pterocarya tonkinensis (Franch.) Dode 越南枫杨
Trees, deciduous

17.3 (11.1~21.0); 9.9 (1.8~15.1); 22.9 (19.4~25.6); 17.4 (11.1~21.0); 148.8 (79.8~191.8); 0.5 (0.0~6.9); 860 (648~1053) ‖ 1315 (1014~1607); 704 (519~860); 51 (20~77) ‖ 53.3 (21.7~80.6); 849 (648~1053); 393.9 (214.5~659.0)

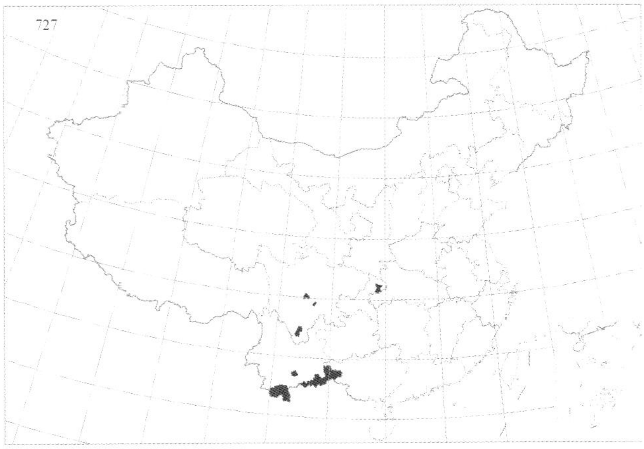

Betulaceae 桦木科 *Alnus* 桤木属
Alnus cremastogyne Burkill 桤木
Trees, deciduous

13.1 (2.2~18.7); 3.0 (-7.0~8.1); 22.4 (9.9~28.9); 13.3 (3.8~18.8); 104.6 (15.5~166.0); 6.7 (0.0~49.1); 742 (413~1002) ‖ 1120 (677~1478); 537 (336~817); 54 (14~122) ‖ 54.8 (2.4~125.9); 742 (413~1002); 222.8 (108.2~351.6)

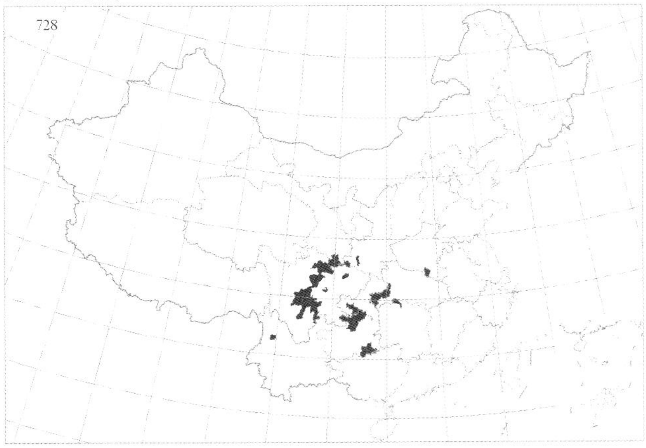

Betulaceae 桦木科 *Alnus* 桤木属
Alnus ferdinandi-coburgii C. K. Schneid. 川滇桤木
Trees, deciduous

13.8 (5.4~18.9); 6.1 (-2.1~12.0); 20.2 (12.3~27.1); 13.9 (5.8~19.0); 107.8 (31.5~167.8); 1.6 (0.0~25.6); 728 (481~969) ‖ 982 (741~1454); 561 (411~792); 35 (15~70) ‖ 36.3 (-14.2~99.9); 706 (481~969); 266.0 (127.1~487.2)

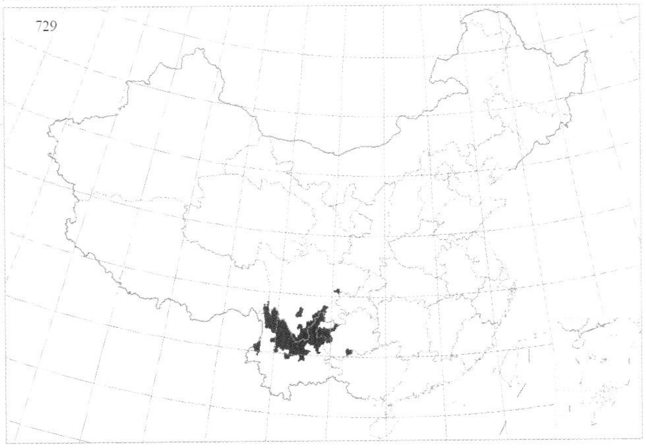

Betulaceae 桦木科 *Alnus* 桤木属
Alnus formosana (Burkill) Makino 台湾桤木
Trees, deciduous

20.5 (17.1~24.1); 14.8 (11.2~18.9); 25.1 (21.8~28.0); 20.5 (17.1~24.1); 185.9 (145.0~229.7); 0.0 (0.0~0.0); 1029 (816~1322) ‖ 2361 (1751~2993); 1196 (924~1594); 199 (47~547) ‖ 139.0 (32.5~203.4); 984 (816~1095); 226.6 (154.2~281.4)

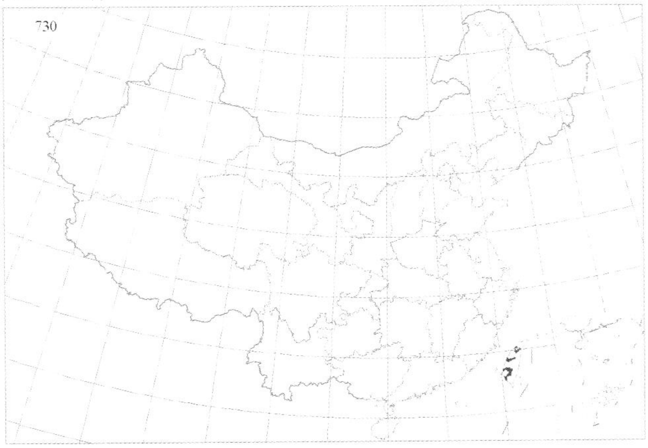

Betulaceae 桦木科 *Alnus* 桤木属
Alnus henryi C. K. Schneid. 台北桤木
Trees, deciduous

20.0; 14.3; 24.9; 20.0; 180.5; 0.0; 991 ‖ 2780; 1224; 321 ‖ 180.8; 991; 280.9

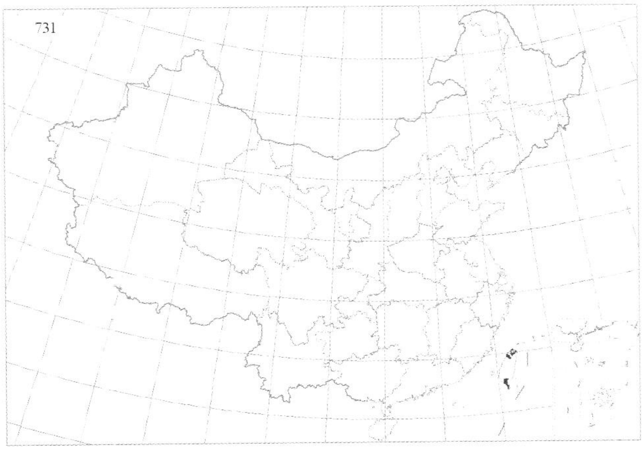

Betulaceae 桦木科 *Alnus* 桤木属
Alnus hirsuta Turcz. ex Rupr. 辽东桤木
Trees, deciduous

3.7 (-5.8~13.8); -17.0 (-29.8~-1.2); 21.4 (15.3~27.3); 8.3 (4.4~14.0); 65.2 (28.0~121.1); 80.8 (14.4~157.4); 592 (426~842) ‖ 661 (404~1073); 414 (276~691); 23 (6~40) ‖ 11.9 (-28.7~74.1); 574 (404~800); 272.2 (136.4~346.5)

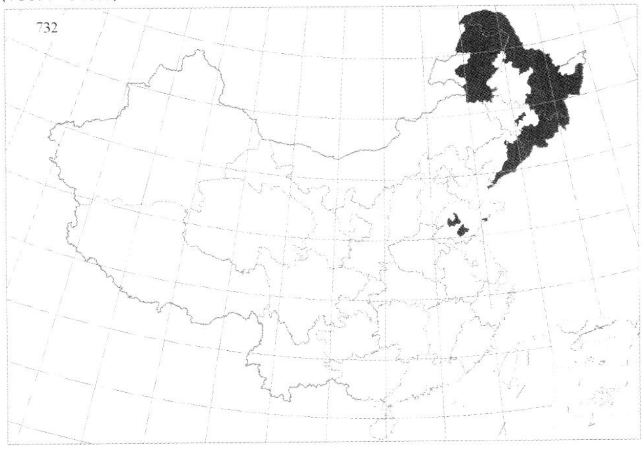

Betulaceae 桦木科 *Alnus* 桤木属
Alnus japonica (Thunb.) Steud. 日本桤木
Trees, deciduous

13.2 (2.2~19.1); -0.8 (-17.4~10.0); 25.9 (18.5~28.8); 13.9 (6.7~19.1); 116.2 (47.6~169.1); 17.5 (0.0~83.8); 802 (513~973) ❙ 1110 (561~1997); 515 (340~669); 96 (16~254) ❙ 37.3 (-23.9~130.7); 784 (513~973); 210.9 (134.1~337.1)

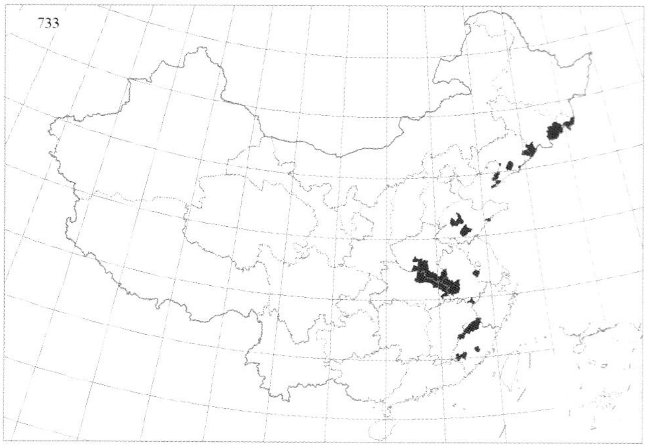

Betulaceae 桦木科 *Alnus* 桤木属
Alnus lanata Duthie ex Bean 毛桤木
Trees, deciduous

8.4; 0.6; 14.8; 9.2; 65.7; 24.6; 574 ❙ 970; 571; 29 ❙ 81.5; 572; 263.3

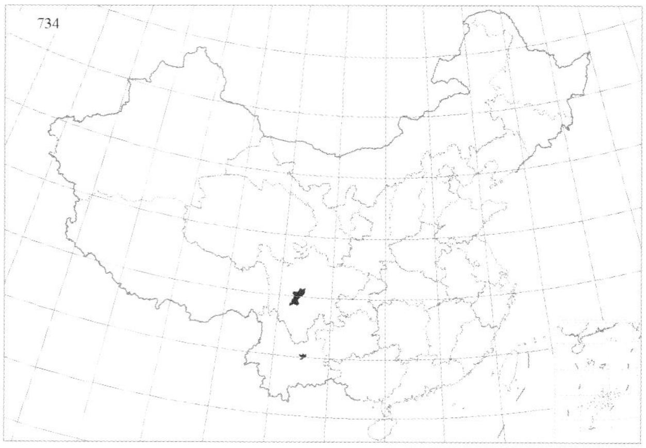

Betulaceae 桦木科 *Alnus* 桤木属
Alnus mandshurica (Callier ex C. K. Schneid.) Hand.-Mazz. 东北桤木
Shrubs or small trees, deciduous

1.1 (-5.8~7.9); -19.7 (-29.8~-10.4); 19.4 (15.3~24.3); 6.9 (4.4~10.2); 52.0 (28.0~85.9); 98.5 (50.4~157.4); 530 (426~679) ❙ 571 (49~1073); 365 (34~691); 18 (1~40) ❙ 8.1 (-92.8~74.1); 485 (49~624); 286.2 (5.1~335.3)

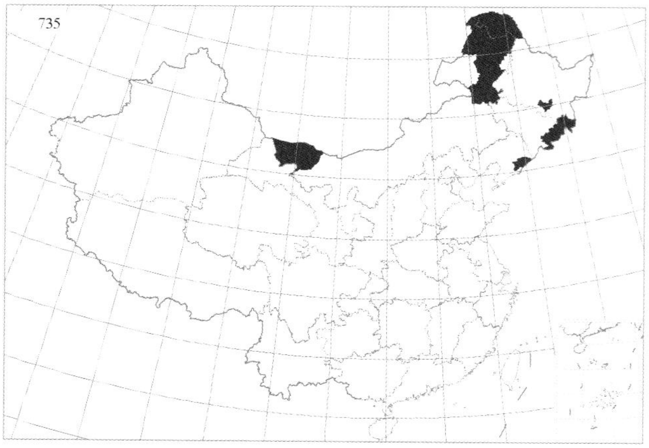

Betulaceae 桦木科 *Alnus* 桤木属
Alnus nepalensis D. Don 尼泊尔桤木
Trees, deciduous

15.0 (-1.8~21.7); 6.7 (-10.2~15.1); 21.7 (6.1~28.5); 15.1 (1.8~21.8); 122.7 (1.9~201.3); 2.4 (0.0~82.8); 784 (329~1162) ❙ 1132 (320~1736); 604 (217~1007); 52 (9~173) ❙ 45.4 (-14.2~145.8); 773 (320~1162); 278.4 (72.2~659.0)

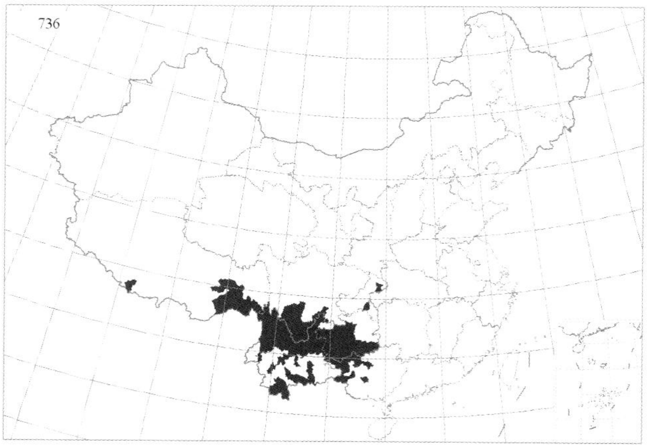

Betulaceae 桦木科 *Alnus* 桤木属
Alnus trabeculosa Hand.-Mazz. 江南桤木
Trees, deciduous

16.5 (12.6~21.8); 4.9 (1.1~13.1); 27.4 (23.2~29.8); 16.6 (12.7~21.8); 139.7 (98.8~201.6); 1.1 (0.0~7.9); 892 (711~1160) ❙ 1459 (753~1997); 531 (384~821); 168 (43~259) ❙ 64.0 (-12.2~147.0); 883 (711~1160); 211.7 (131.2~354.5)

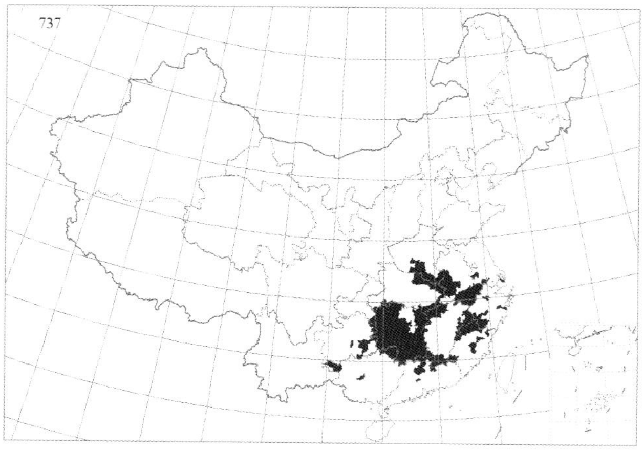

Betulaceae 桦木科 *Betula* 桦木属
Betula albosinensis Burkill 红桦
Trees, deciduous

9.2 (-3.3~16.8); -4.0 (-15.3~7.8); 21.0 (7.2~28.0); 10.2 (2.0~16.8); 79.1 (3.7~141.5); 28.6 (0.0~103.2); 648 (339~880) ❙ 694 (341~1583); 362 (190~938); 24 (5~113) ❙ 10.3 (-40.9~128.6); 579 (339~857); 207.9 (61.2~364.1)

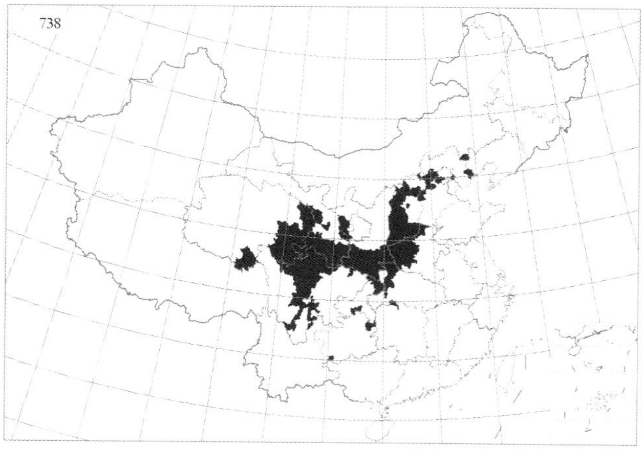

Betulaceae 桦木科 *Betula* 桦木属
Betula alnoides Buch.-Ham. ex D. Don 西桦
Trees, deciduous

18.3 (13.2~23.9); 10.6 (3.8~18.7); 24.0 (18.4~28.5); 18.3 (13.3~23.9); 159.7 (99.4~226.8); 0.0 (0.0~1.5); 915 (695~1294) ❙ 1390 (1014~1740); 710 (444~1007); 79 (42~191) ❙ 54.3 (-2.2~145.8); 904 (695~1153); 373.8 (208.7~559.0)

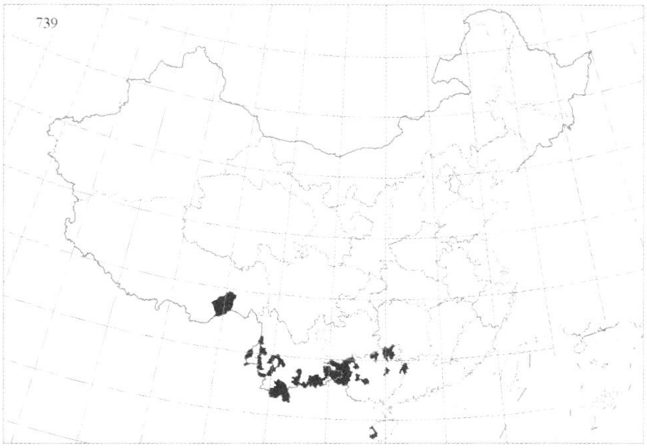

Betulaceae 桦木科 *Betula* 桦木属
Betula austrosinensis Chun ex P. C. Li 华南桦
Trees, deciduous

17.0 (11.1~21.6); 6.6 (1.8~13.6); 26.2 (19.4~29.8); 17.0 (11.1~21.7); 144.5 (79.8~199.8); 0.4 (0.0~6.9); 892 (648~1153) ❙ 1422 (942~1829); 561 (403~850); 141 (30~230) ❙ 59.7 (21.3~128.9); 888 (648~1153); 227.9 (148.9~506.2)

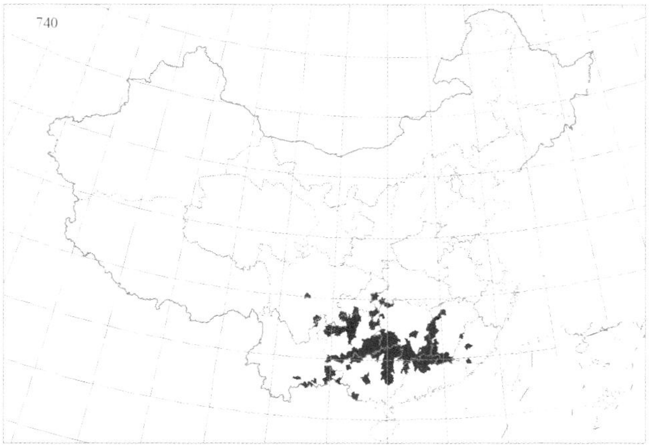

Betulaceae 桦木科 *Betula* 桦木属
Betula calcicola (W. W. Sm.) P. C. Li 岩桦
Shrubs, deciduous

7.3 (2.2~13.7); -0.5 (-6.3~6.2); 13.9 (9.5~19.7); 7.8 (3.9~13.7); 48.5 (15.8~104.7); 20.5 (0.0~48.6); 537 (411~714) ❙ 823 (628~968); 517 (406~603); 18 (11~26) ❙ 54.6 (34.8~77.4); 534 (411~696); 266.6 (201.9~319.8)

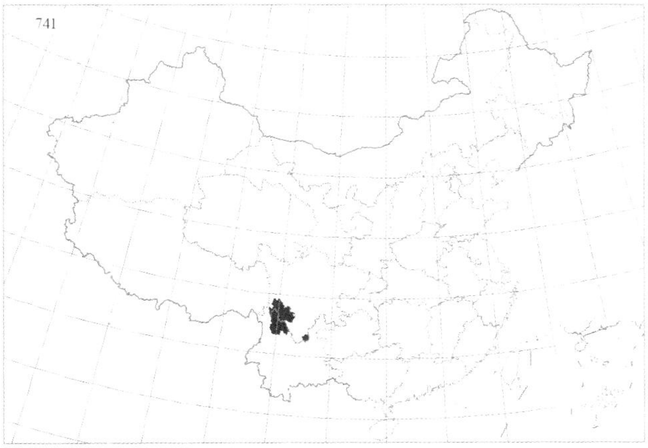

Betulaceae 桦木科 *Betula* 桦木属
Betula chinensis Maxim. 坚桦
Shrubs, deciduous

10.3 (0.1~15.5); -4.8 (-15.9~4.0); 23.7 (9.5~28.0); 11.3 (3.2~15.6); 92.1 (12.3~131.3); 28.5 (1.0~80.4); 702 (377~881) ❙ 648 (367~1451); 374 (235~691); 24 (7~120) ❙ -7.0 (-47.7~121.7); 600 (367~881); 186.4 (70.3~335.3)

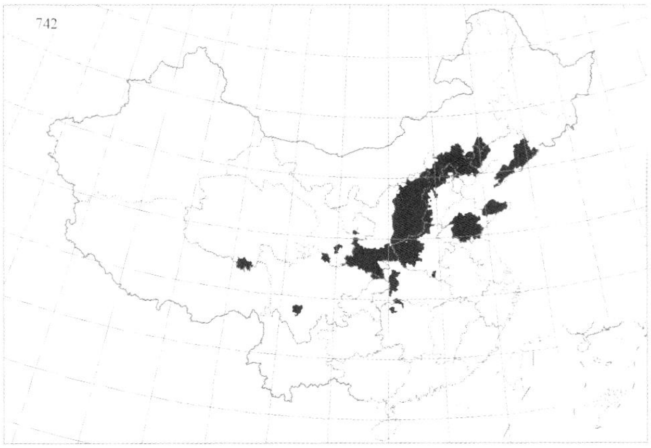

Betulaceae 桦木科 *Betula* 桦木属
Betula costata Trautv. 硕桦
Trees, deciduous

3.3 (-5.8~10.5); -17.1 (-29.8~-4.9); 21.1 (15.3~25.2); 8.0 (4.4~11.7); 62.4 (28.0~99.1); 81.8 (26.8~157.4); 579 (426~743) ❙ 589 (249~1073); 375 (179~691); 18 (6~40) ❙ 1.9 (-54.1~74.1); 532 (249~681); 263.3 (111.2~346.5)

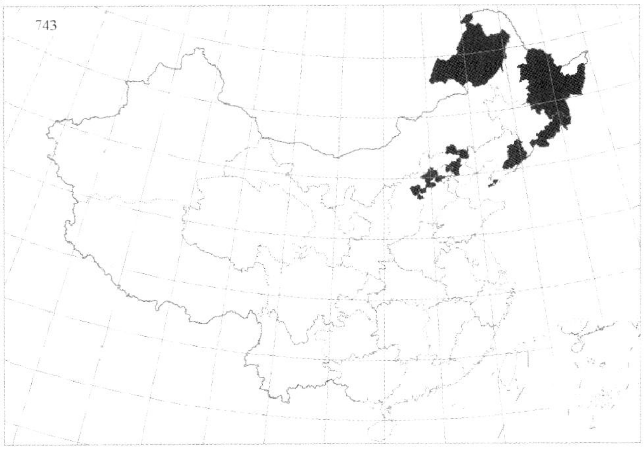

Betulaceae 桦木科 *Betula* 桦木属
Betula cylindrostachya Lindl. 长穗桦
Trees, deciduous

10.9 (1.8~17.9); 2.4 (-7.1~7.4); 18.4 (9.9~27.1); 11.2 (3.8~18.0); 82.0 (15.7~155.6); 10.2 (0.0~53.3); 661 (404~938) ❙ 1199 (583~1740); 579 (348~727); 88 (13~191) ❙ 79.1 (44.2~101.8); 661 (404~938); 288.1 (169.5~407.1)

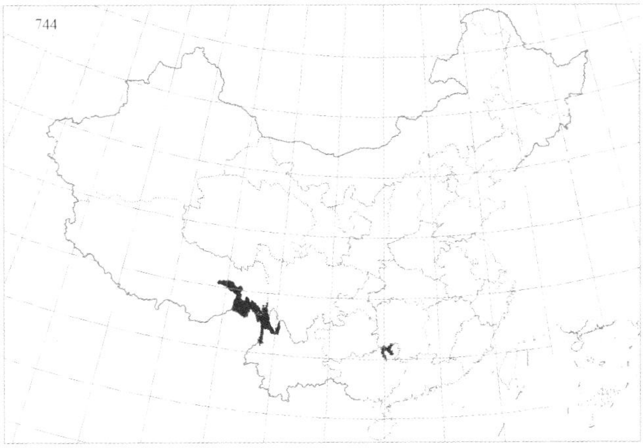

Betulaceae 桦木科 *Betula* 桦木属

Betula dahurica Pall. 黑桦

Trees, deciduous

4.0 (-5.8~12.0); -15.7 (-29.8~-3.6); 21.1 (15.3~25.4); 8.2 (4.4~12.5); 64.0 (28.0~104.9); 75.5 (20.3~157.4); 585 (426~768) ❙ 579 (297~1073); 370 (203~691); 18 (5~40) ❙ -0.7 (-44.3~74.1); 524 (297~681); 245.3 (104.4~346.5)

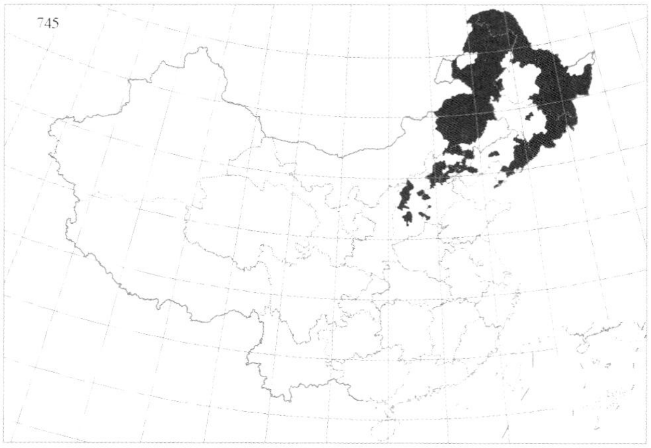

Betulaceae 桦木科 *Betula* 桦木属

Betula delavayi Franch. 高山桦

Trees or shrubs, deciduous

6.8 (0.3~14.9); -1.5 (-9.9~7.6); 14.0 (9.3~23.6); 7.6 (3.3~15.0); 47.6 (11.9~119.6); 25.7 (0.0~67.5); 531 (391~782) ❙ 879 (527~1410); 513 (324~816); 24 (8~49) ❙ 66.8 (26.6~128.6); 530 (391~782); 286.8 (201.9~372.3)

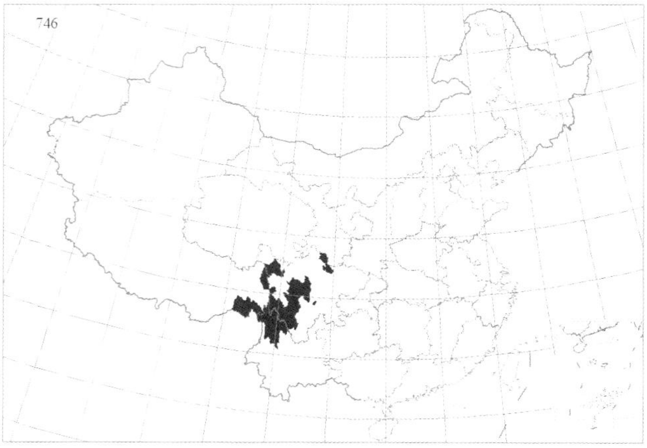

Betulaceae 桦木科 *Betula* 桦木属

Betula delavayi var. microstachya P. C. Li 细穗高山桦

Shrubs, deciduous

2.4 (0.3~5.3); -6.8 (-9.9~-2.4); 10.6 (9.3~12.6); 4.2 (3.3~5.7); 19.1 (11.9~30.5); 49.9 (26.2~67.5); 421 (391~479) ❙ 696 (527~961); 404 (324~558); 15 (8~31) ❙ 64.6 (26.6~125.9); 421 (391~479); 266.5 (169.5~337.3)

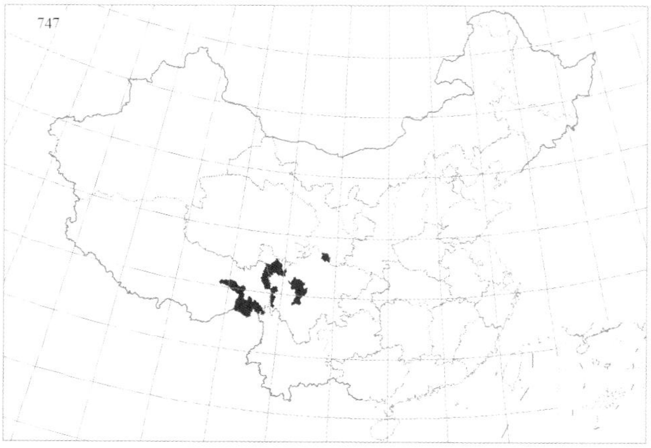

Betulaceae 桦木科 *Betula* 桦木属

Betula delavayi var. polyneura Hu ex P. C. Li 多脉高山桦

Trees or shrubs, deciduous

10.8; 3.9; 16.6; 10.9; 71.8; 1.3; 633 ❙ 968; 594; 24 ❙ 52.8; 633; 319.8

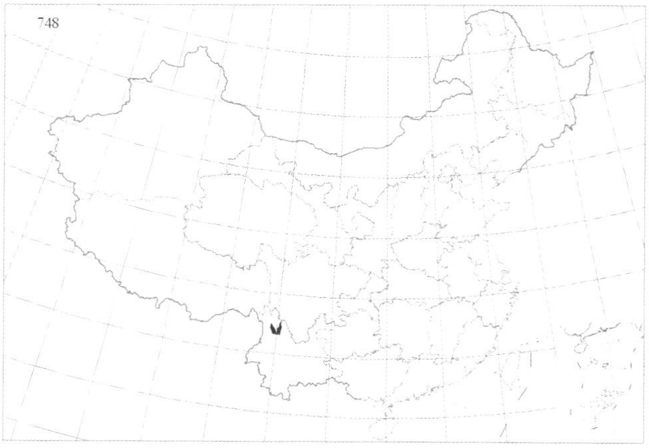

Betulaceae 桦木科 *Betula* 桦木属

Betula ermanii Cham. 岳桦

Trees, deciduous

2.0 (-5.8~8.0); -19.2 (-29.8~-9.8); 20.5 (15.3~23.5); 7.5 (4.4~10.0); 56.9 (28.0~83.4); 91.9 (46.4~157.4); 557 (426~658) ❙ 590 (311~1073); 373 (214~691); 19 (5~40) ❙ 5.6 (-40.0~74.1); 520 (311~658); 274.3 (148.8~346.5)

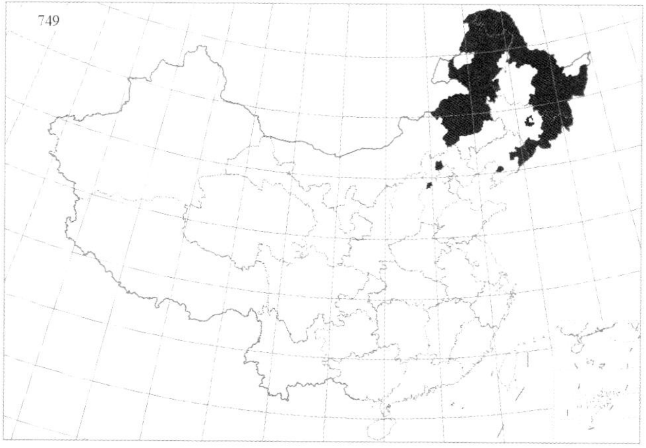

Betulaceae 桦木科 *Betula* 桦木属

Betula ermanii var. macrostrobila Liou 帽儿山岳桦

Trees, deciduous

10.8; 3.9; 16.6; 10.9; 71.8; 1.3; 633 ❙ 968; 594; 24 ❙ 52.8; 633; 319.8

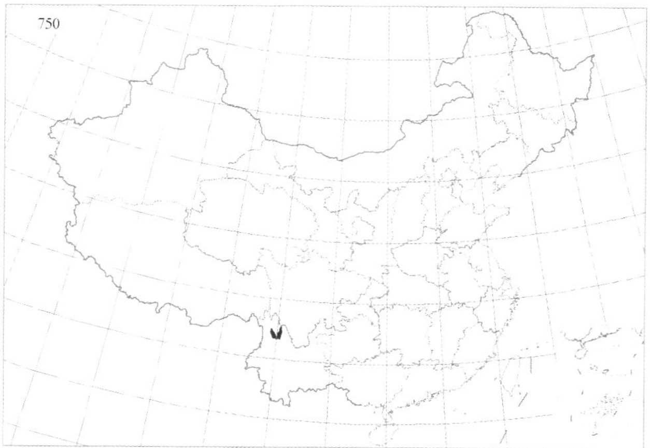

Betulaceae 桦木科 *Betula* 桦木属
Betula ermanii var. yingkiliensis Liou et Z. Wang 英吉里岳桦
Trees, deciduous

-4.4 (-5.8~-2.9); -28.4 (-29.8~-26.4); 17.1 (15.3~18.0); 5.1 (4.4~5.4); 36.3 (28.0~40.2); 148.0 (134.6~157.4); 455 (426~471) | 494 (468~510); 319 (300~338); 15 (13~16) | 6.8 (0.9~15.9); 455 (426~471); 287.7 (275.4~306.9)

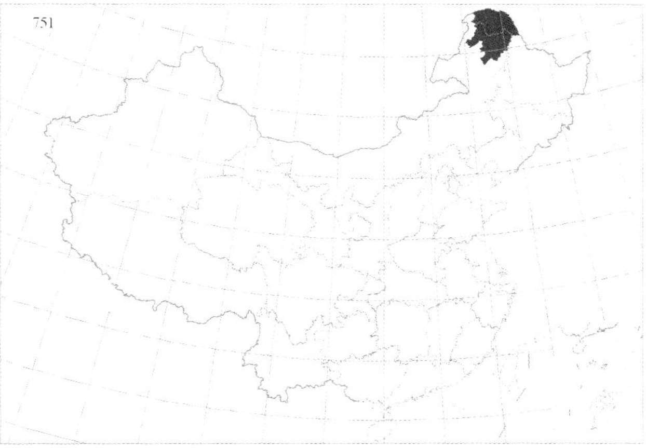

Betulaceae 桦木科 *Betula* 桦木属
Betula fargesii Franch. 狭翅桦
Shrubs or small trees, deciduous

13.0 (3.5~17.9); 2.0 (-6.5~6.0); 23.5 (10.8~29.6); 13.2 (4.7~18.0); 105.4 (21.6~155.8); 8.5 (0.0~43.4); 753 (444~972) | 1255 (731~1602); 507 (346~816); 100 (16~204) | 69.0 (38.2~121.7); 741 (444~972); 230.6 (142.8~284.3)

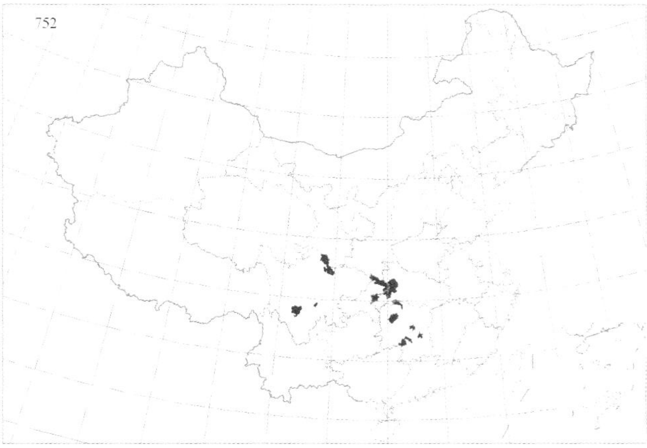

Betulaceae 桦木科 *Betula* 桦木属
Betula fruticosa Pall. 柴桦
Shrubs, deciduous

0.5 (-5.8~4.8); -21.3 (-29.8~-14.0); 19.7 (15.3~23.0); 6.9 (4.4~8.7); 51.7 (28.0~68.9); 105.1 (69.7~157.4); 534 (426~616) | 505 (311~761); 332 (214~471); 14 (5~32) | -5.1 (-40.0~45.2); 472 (311~591); 253.9 (148.8~322.3)

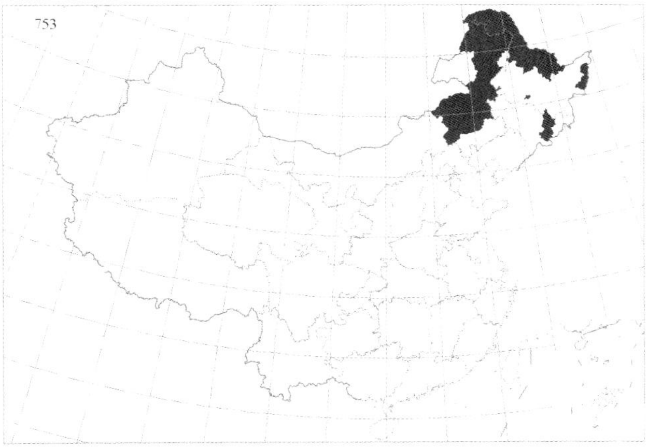

Betulaceae 桦木科 *Betula* 桦木属
Betula gmelinii Bunge 砂生桦
Shrubs, deciduous

1.4 (-5.8~10.5); -19.7 (-29.8~-4.9); 20.1 (15.3~23.8); 7.3 (4.4~11.3); 56.2 (28.0~93.2); 99.1 (26.8~157.4); 550 (426~705) | 443 (297~632); 303 (203~398); 11 (5~27) | -17.9 (-44.3~15.9); 434 (297~632); 208.3 (97.2~309.5)

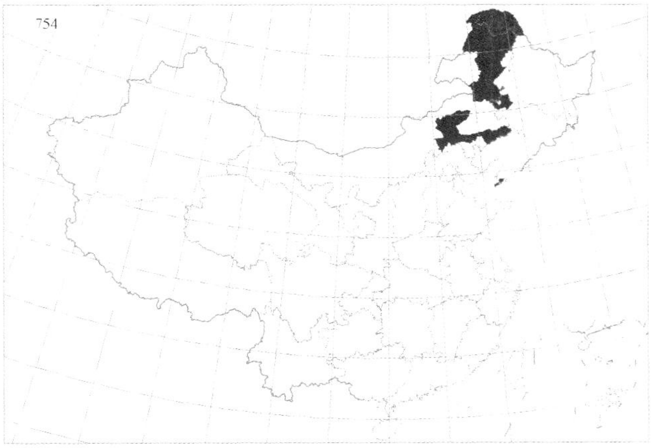

Betulaceae 桦木科 *Betula* 桦木属
Betula gynoterminalis Y. C. Hsu et C. J. Wang 贡山桦
Trees, deciduous

8.6; 1.3; 15.0; 8.7; 52.6; 8.8; 574 | 1041; 585; 45 | 81.5; 574; 352.3

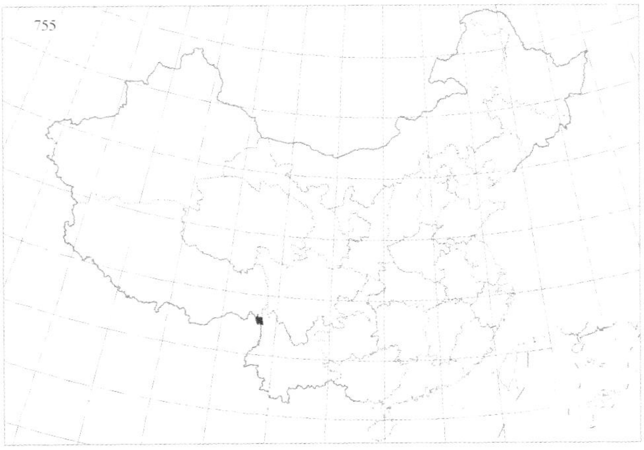

Betulaceae 桦木科 *Betula* 桦木属
Betula halophila Ching 盐桦
Shrubs, deciduous

1.2; -17.1; 17.6; 6.4; 45.8; 91.5; 519 | 210; 90; 26 | -59.5; 210; 129.3

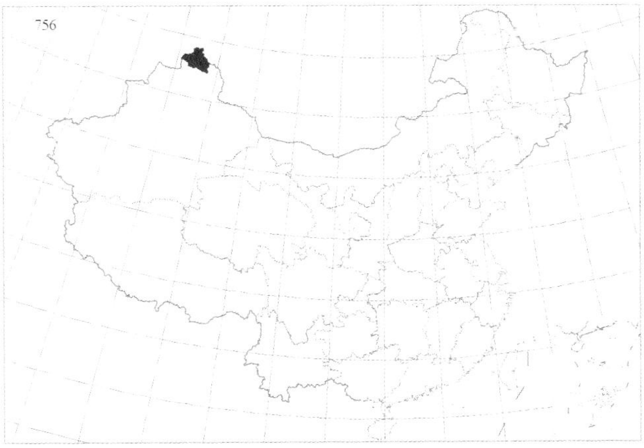

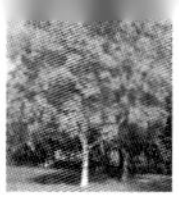

Betulaceae 桦木科 *Betula* 桦木属

Betula humilis Schrank 甸生桦

Shrubs, deciduous

0.8 (-0.7~2.8); -20.5 (-25.0~-16.0); 19.2 (15.9~22.0); 6.8 (5.4~8.2); 50.8 (37.5~63.3); 100.5 (85.3~118.4); 537 (475~595) ❙ 435 (161~657); 253 (74~416); 21 (15~33) ❙ -20.4 (-68.1~23.8); 399 (161~583); 216.1 (74.6~333.3)

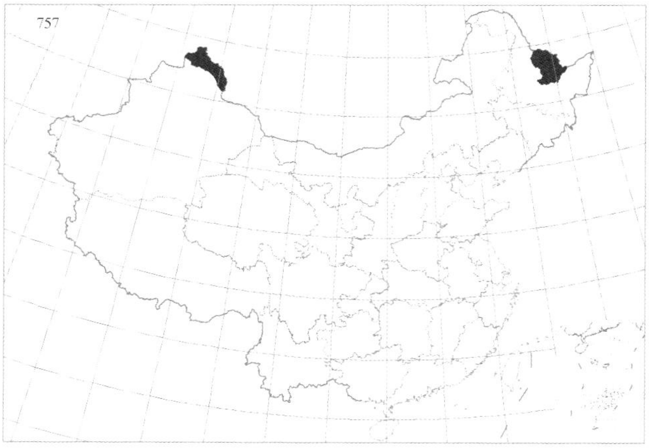

Betulaceae 桦木科 *Betula* 桦木属

Betula insignis Franch. 香桦

Trees, deciduous

13.9 (2.2~19.4); 3.4 (-6.6~9.4); 23.7 (9.9~29.0); 14.1 (3.8~19.4); 112.9 (15.5~173.4); 5.1 (0.0~49.1); 775 (413~1029) ❙ 1279 (707~1745); 536 (362~817); 100 (15~222) ❙ 65.9 (-0.2~125.9); 773 (413~1029); 236.9 (170.4~318.7)

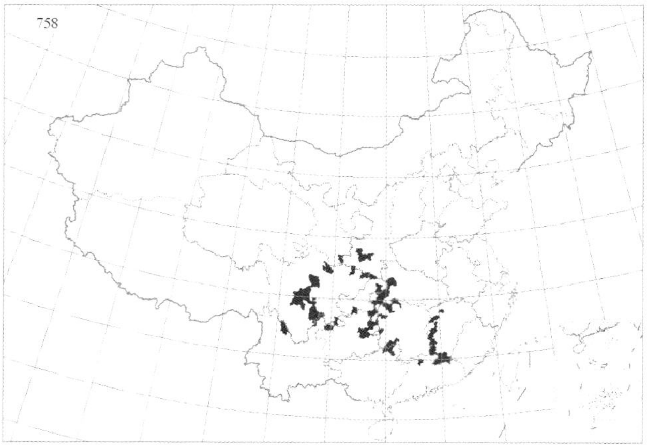

Betulaceae 桦木科 *Betula* 桦木属

Betula jinpingensis P. C. Li 金平桦

Trees, deciduous

19.1; 13.0; 23.3; 19.1; 169.7; 0.0; 917 ❙ 1428; 835; 56 ❙ 55.7; 917; 506.2

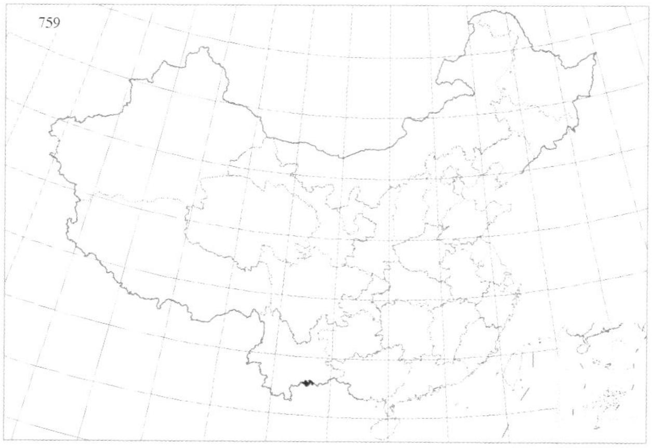

Betulaceae 桦木科 *Betula* 桦木属

Betula jiulungensis Hu ex P. C. Li 九龙桦

Trees, deciduous

3.8; -4.3; 10.8; 4.7; 21.6; 35.4; 444 ❙ 984; 601; 16 ❙ 121.7; 444; 265.2

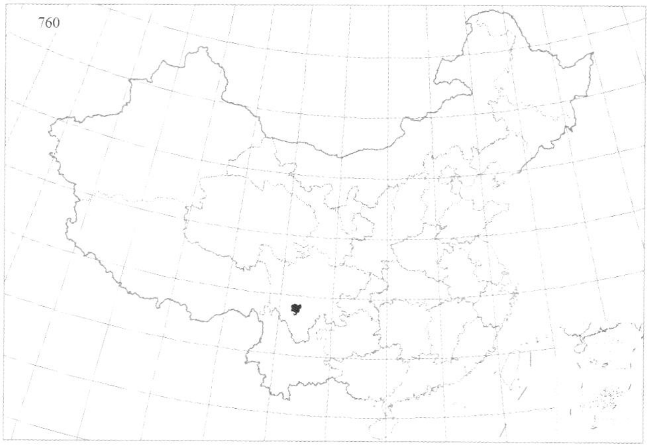

Betulaceae 桦木科 *Betula* 桦木属

Betula luminifera H. J. P. Winkl. 亮叶桦

Trees, deciduous

15.8 (2.2~23.9); 5.3 (-6.6~18.7); 25.4 (9.9~29.9); 15.9 (3.8~23.9); 132.9 (15.5~226.8); 2.3 (0.0~49.1); 848 (413~1294) ❙ 1312 (626~2435); 551 (316~1428); 114 (15~259) ❙ 54.6 (-22.2~147.0); 840 (413~1221); 236.2 (108.2~659.0)

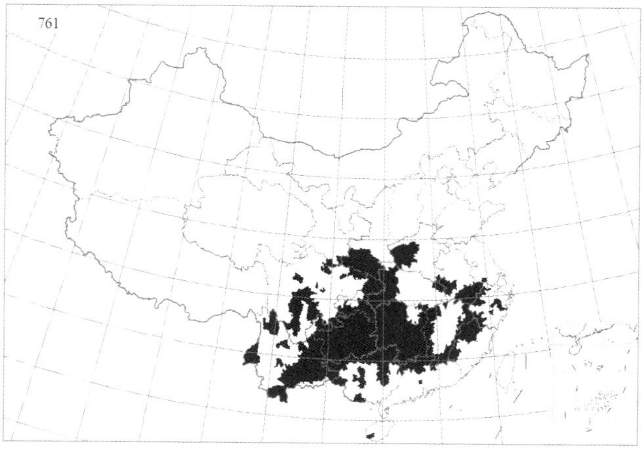

Betulaceae 桦木科 *Betula* 桦木属

Betula microphylla Bunge 小叶桦

Small trees, deciduous

2.3 (-4.3~10.7); -15.8 (-20.9~-6.7); 17.4 (9.6~24.4); 6.7 (2.5~11.8); 48.2 (9.7~98.3); 80.3 (29.5~121.1); 524 (329~736) ❙ 217 (45~450); 95 (22~211); 22 (4~51) ❙ -58.0 (-90.8~3.9); 215 (45~406); 104.1 (12.6~297.3)

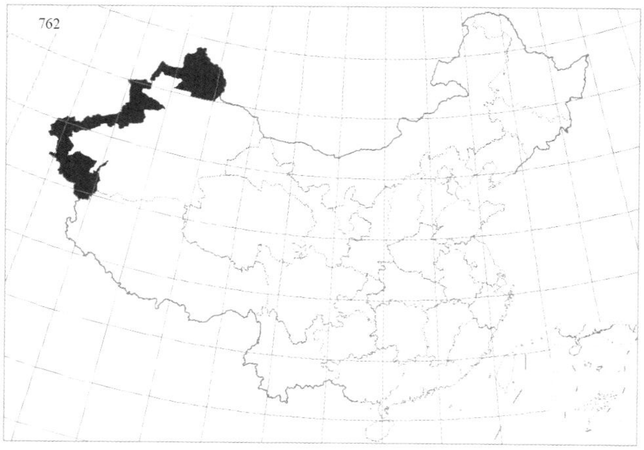

Betulaceae 桦木科 *Betula* 桦木属

Betula middendorfii Trautv. et C. A. Mey. 扇叶桦

Shrubs, deciduous

-1.0 (-5.8~9.3); -23.2 (-29.8~-7.4); 18.7 (15.3~23.7); 6.3 (4.4~10.7); 47.1 (28.0~88.9); 118.6 (36.2~157.4); 508 (426~686) ❙ 489 (429~684); 328 (300~426); 14 (6~30) ❙ -3.6 (-28.7~15.9); 471 (426~684); 273.2 (185.5~309.5)

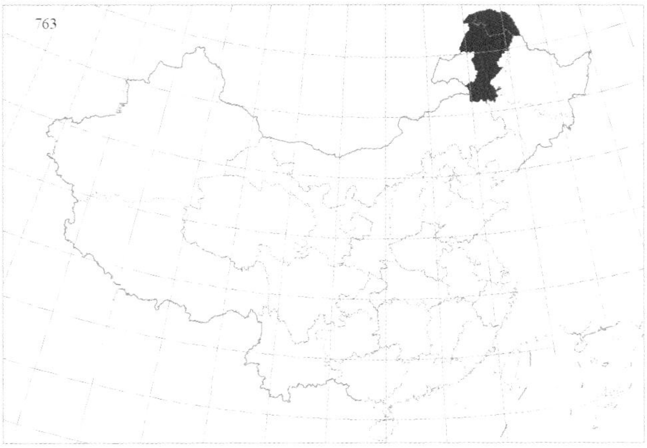

Betulaceae 桦木科 *Betula* 桦木属

Betula ovalifolia Rupr. 油桦

Shrubs, deciduous

1.9 (-3.1~4.5); -18.4 (-26.4~-14.1); 19.8 (16.5~23.0); 7.2 (5.0~8.7); 53.5 (34.9~69.0); 89.8 (68.3~134.6); 539 (448~611) ❙ 472 (137~808); 297 (92~500); 15 (3~32) ❙ -12.3 (-77.5~45.2); 426 (137~591); 221.0 (25.4~338.4)

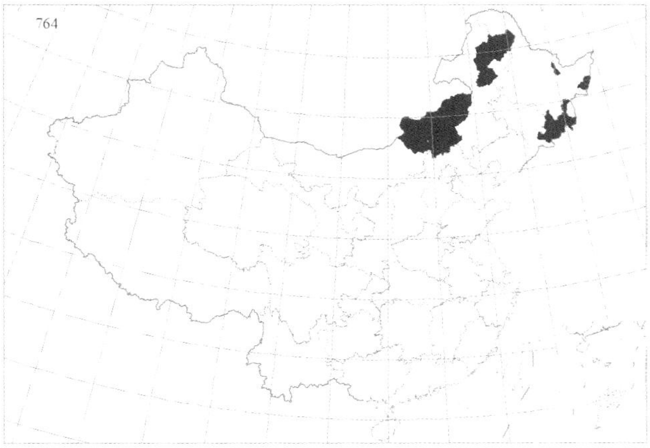

Betulaceae 桦木科 *Betula* 桦木属

Betula pendula Roth 垂枝桦

Trees, deciduous

5.4 (-4.6~14.8); -8.9 (-28.8~4.3); 18.1 (7.0~27.2); 7.8 (2.0~14.9); 56.7 (3.7~119.2); 51.0 (0.7~149.1); 560 (332~827) ❙ 563 (45~1197); 321 (27~691); 18 (2~71) ❙ 5.1 (-94.0~128.6); 459 (45~791); 214.4 (7.6~357.7)

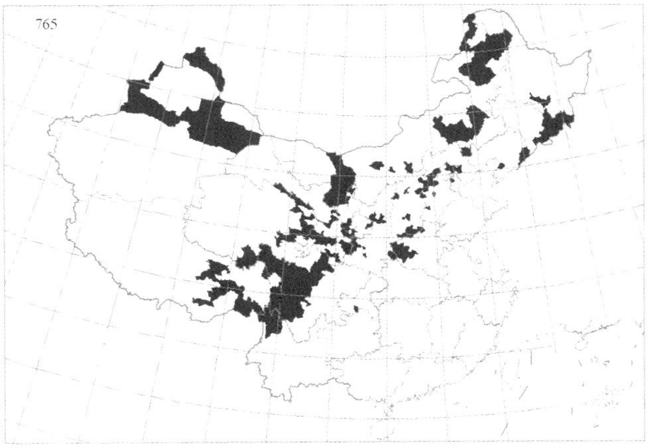

Betulaceae 桦木科 *Betula* 桦木属

Betula platyphylla Sukaczev 白桦

Trees, deciduous

6.9 (-5.8~15.5); -8.7 (-29.8~4.2); 20.7 (7.2~28.0); 9.2 (2.0~15.6); 71.1 (3.7~131.3); 47.3 (0.8~157.4); 617 (339~881) ❙ 594 (132~1358); 345 (81~691); 18 (1~97) ❙ -0.8 (-80.1~128.6); 525 (132~881); 207.4 (13.6~357.7)

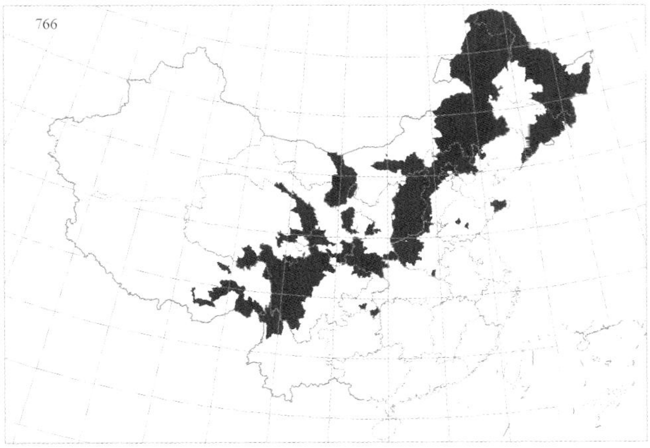

Betulaceae 桦木科 *Betula* 桦木属

Betula potaninii Batalin 矮桦

Shrubs or small trees, deciduous

6.9 (-1.2~14.9); -2.6 (-11.7~5.3); 15.5 (8.5~24.1); 7.8 (2.6~15.0); 51.0 (8.1~119.6); 27.6 (0.0~82.3); 543 (347~782) ❙ 914 (491~1410); 492 (301~816); 25 (11~49) ❙ 72.0 (11.5~125.9); 543 (347~782); 268.0 (171.0~352.3)

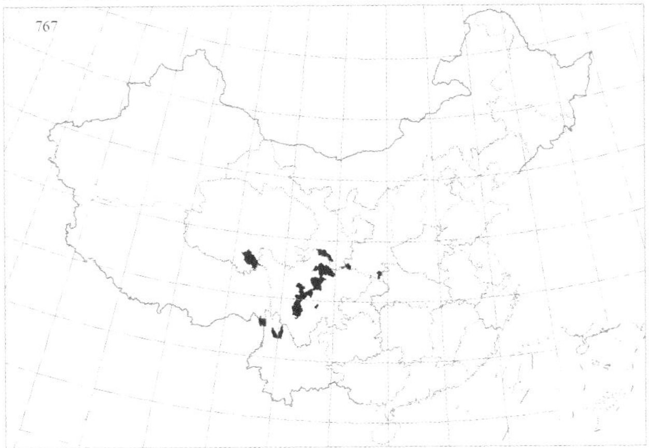

Betulaceae 桦木科 *Betula* 桦木属

Betula rhombibracteata P. C. Li 菱苞桦

Trees, deciduous

5.4; -2.1; 12.3; 5.8; 31.5; 25.6; 481 ❙ 741; 426; 28 ❙ 54.0; 48[illegible]; 273.5

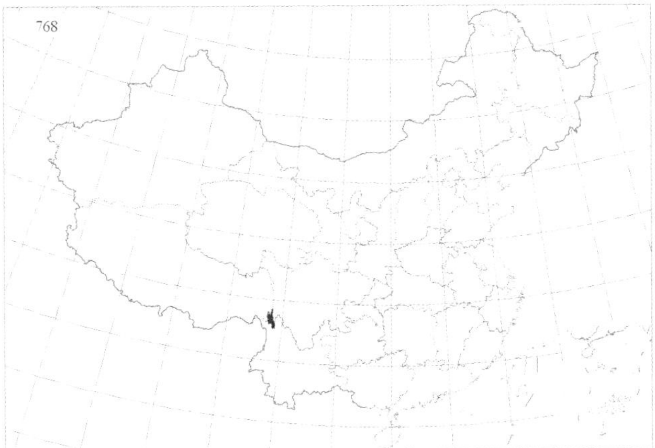

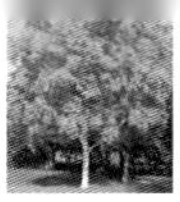

Betulaceae 桦木科 *Betula* 桦木属
Betula rotundifolia Spach 圆叶桦
Shrubs, deciduous

0.8 (-0.6~2.1); -17.8 (-19.5~-16.0); 17.2 (15.9~18.7); 6.2 (5.4~6.9); 44.1 (37.5~50.8); 94.7 (85.3~104.2); 509 (475~544) ❙ 194 (161~251); 85 (74~106); 23 (15~33) ❙ -61.9 (-68.1~-52.0); 194 (161~251); 118.5 (74.6~139.1)

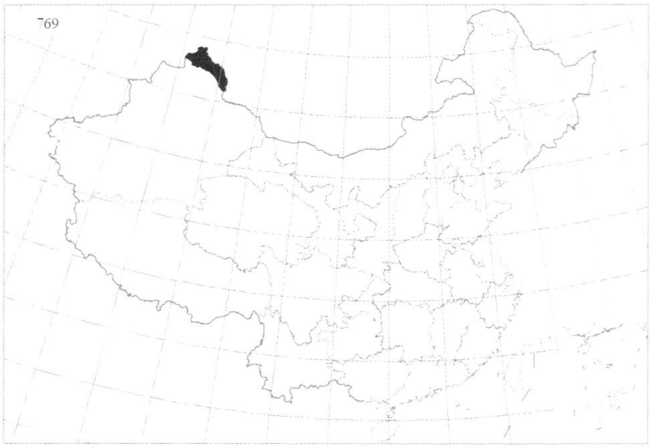

Betulaceae 桦木科 *Betula* 桦木属
Betula schmidtii Regel 赛黑桦
Trees, deciduous

4.5 (2.2~7.3); -14.4 (-18.8~-10.4); 20.9 (18.5~22.8); 8.1 (6.7~9.4); 63.2 (47.6~77.5); 68.2 (49.7~84.3); 576 (513~631) ❙ 790 (561~1073); 488 (333~691); 28 (16~40) ❙ 36.4 (-2.6~74.1); 574 (513~631); 315.2 (287.1~346.5)

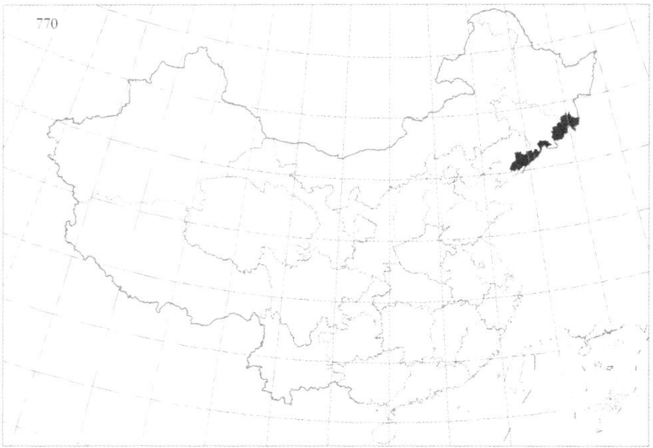

Betulaceae 桦木科 *Betula* 桦木属
Betula tianschanica Rupr. 天山桦
Small trees, deciduous

4.4 (-4.3~10.2); -14.3 (-20.9~-8.5); 19.6 (9.6~26.5); 8.0 (2.5~11.8); 62.5 (9.7~104.4); 69.2 (36.6~121.1); 584 (329~786) ❙ 192 (45~450); 92 (24~211); 15 (3~48) ❙ -64.3 (-94.0~3.9); 191 (45~406); 81.3 (7.6~297.3)

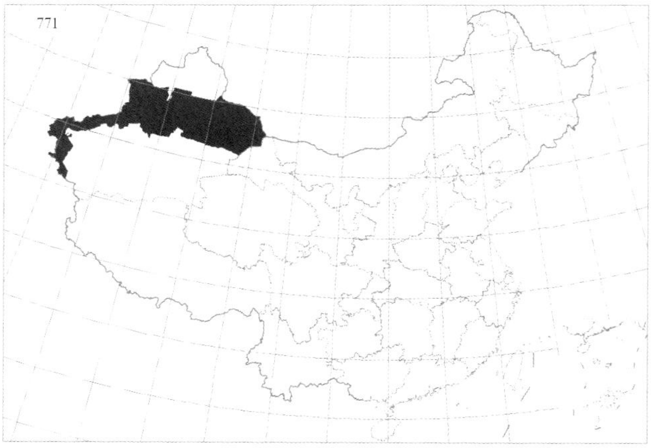

Betulaceae 桦木科 *Betula* 桦木属
Betula trichogemma (Hu ex P. C. Li) T. Hong 峨眉矮桦
Shrubs or small trees, deciduous

13.0; 3.8; 21.5; 13.1; 100.2; 2.6; 717 ❙ 1315; 752; 43 ❙ 83.8; 717; 253.6

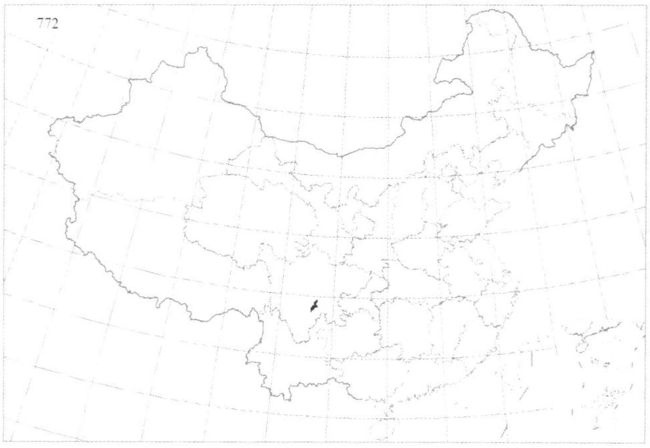

Betulaceae 桦木科 *Betula* 桦木属
Betula utilis D. Don 糙皮桦
Trees, deciduous

8.5 (-1.8~21.6); -3.6 (-15.9~12.6); 19.5 (6.1~28.5); 9.6 (1.8~21.7); 72.8 (1.9~199.8); 30.0 (0.0~86.3); 622 (329~1153) ❙ 765 (320~1740); 404 (190~1007); 28 (5~194) ❙ 24.9 (-41.6~145.8); 589 (320~1153); 235.7 (61.2~447.4)

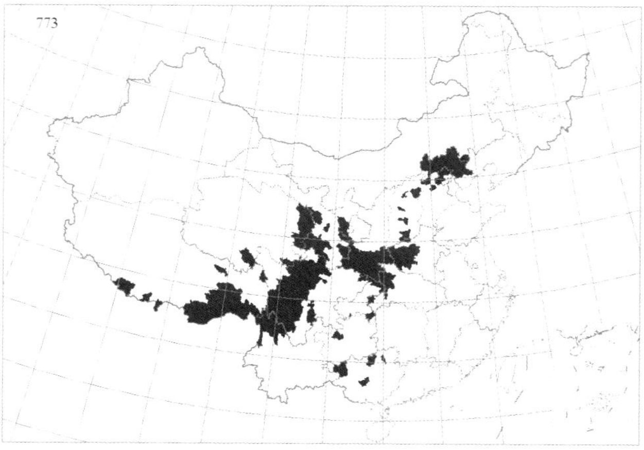

Betulaceae 桦木科 *Carpinus* 鹅耳枥属
Carpinus chuniana Hu 粤北鹅耳枥
Trees, deciduous

16.9 (13.5~21.2); 6.4 (2.6~12.5); 26.5 (22.4~29.0); 16.9 (13.6~21.2); 143.7 (106.7~194.6); 0.4 (0.0~4.1); 889 (744~1114) ❙ 1433 (1089~1829); 545 (403~860); 145 (55~230) ❙ 61.7 (14.8~104.6); 882 (744~1114); 225.5 (151.0~491.5)

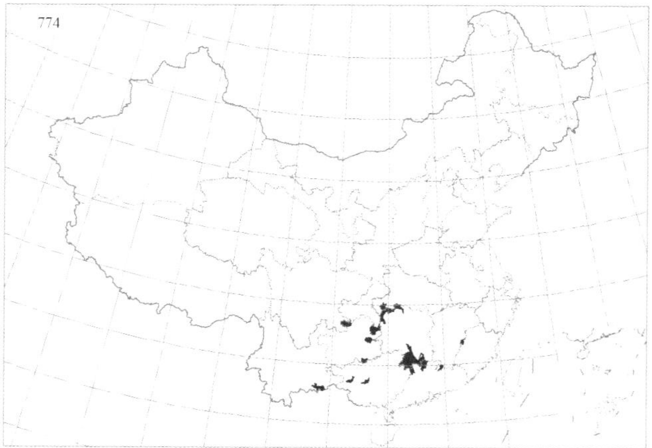

Betulaceae 桦木科 *Carpinus* 鹅耳枥属

Carpinus cordata Blume 千金榆

Trees, deciduous

11.0 (1.9~17.[illegible]); -3.3 (-20.1~5.0); 24.0 (14.2~29.3); 11.9 (6.4~17.1); 96.5 (38.1~145.7); 23.6 (0.0~91.9); 720 (504~941) ❙ 778 (367~1828); 396 (201~691); 43 (7~253) ❙ 7.2 (-40.9~112.2); 645 (367~941); 201.9 (88.1~346.5)

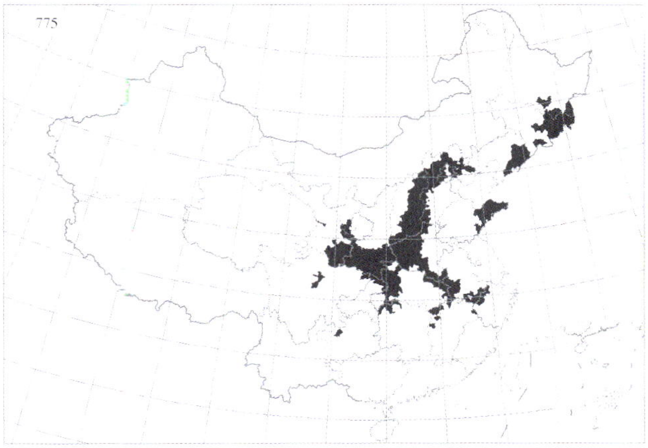

Betulaceae 桦木科 *Carpinus* 鹅耳枥属

Carpinus cordata var. chinensis Franch. 华千金榆

Trees, deciduous

14.0 (9.8~16.2); 2.1 (-4.9~4.4); 25.5 (22.6~27.6); 14.1 (10.6~16.2); 114.5 (84.0~135.6); 5.4 (0.6~26.3); 786 (665~877) ❙ 1343 (624~1689); 544 (350~666); 125 (23~212) ❙ 70.1 (-6.3~108.0); 785 (624~877); 251.4 (162.7~319.9)

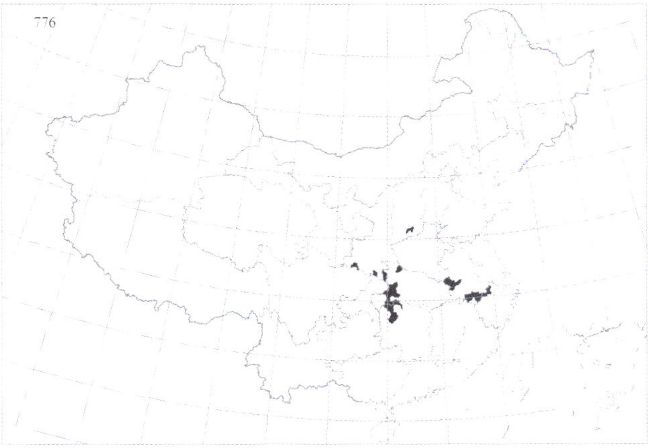

Betulaceae 桦木科 *Carpinus* 鹅耳枥属

Carpinus cordata var. mollis (Rehder) W. C. Cheng ex Chun 毛叶千金榆

Trees, deciduous

12.2 (5.7~16.6); -0.5 (-4.9~4.4); 24.2 (14.2~28.8); 12.4 (6.4~16.7); 100.4 (38.1~140.6); 13.2 (0.6~29.9); 736 (504~920) ❙ 840 (620~1571); 393 (290~666); 46 (13~153) ❙ 14.6 (-22.2~108.0); 708 (504~899); 217.6 (139.1~334.4)

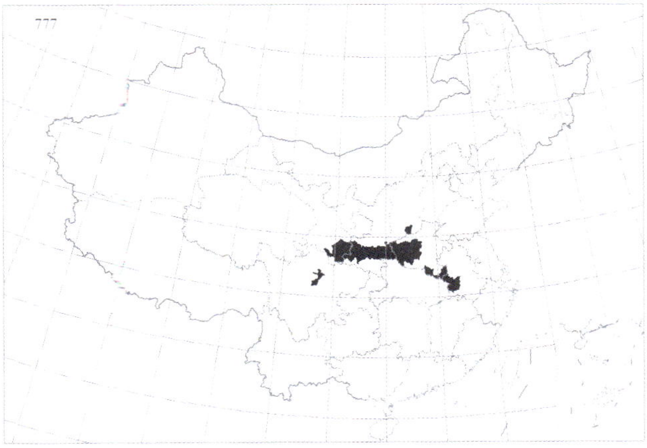

Betulaceae 桦木科 *Carpinus* 鹅耳枥属

Carpinus dayongina K. W. Liu et Q. Z. Lin 大庸鹅耳枥

Trees, deciduous

15.7; 4.0; 26.7; 15.7; 129.4; 1.0; 847 ❙ 1421; 566; 121 ❙ 67.8; 847; 208.0

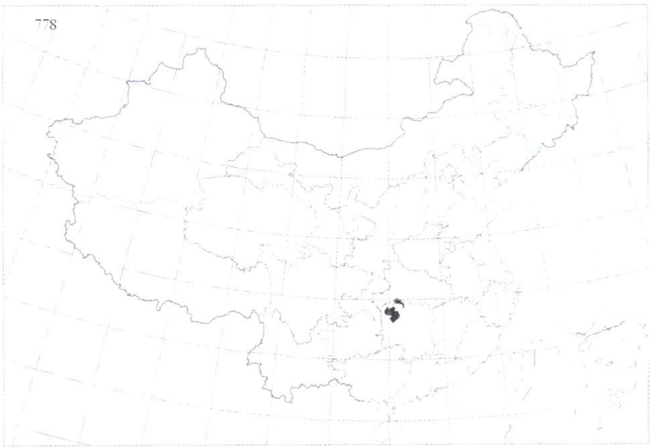

Betulaceae 桦木科 *Carpinus* 鹅耳枥属

Carpinus fangiana Hu 川黔千金榆

Trees, deciduous

14.7 (2.6~17.9); 4.3 (-7.0~10.9); 24.2 (11.2~28.7); 14.8 (4.3~18.0); 119.3 (20.0~155.6); 2.6 (0.0~48.2); 792 (427~948) ❙ 1264 (822~1740); 540 (400~946); 97 (19~221) ❙ 60.4 (19.4~106.4); 790 (427~938); 225.8 (166.3~491.5)

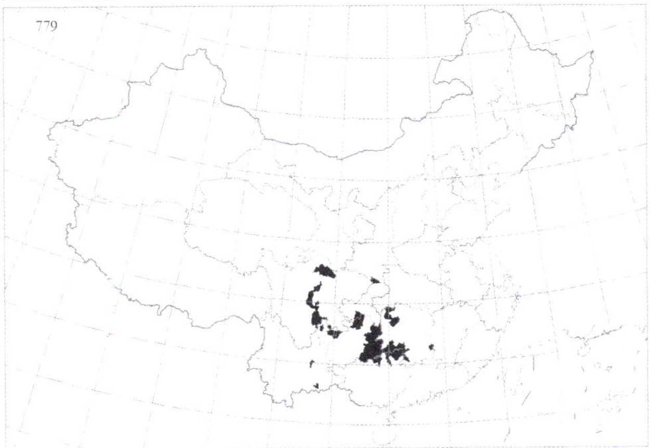

Betulaceae 桦木科 *Carpinus* 鹅耳枥属

Carpinus fargesiana H. J. P. Winkl. 川陕鹅耳枥

Trees, deciduous

12.3 (2.0~16.6); 0.4 (-7.7~8.2); 23.4 (9.9~28.8); 12.5 (3.8~16.7); 100.2 (15.4~140.6); 12.3 (0.0~51.4); 735 (413~920) ❙ 997 (520~1663); 460 (258~688); 61 (10~212) ❙ 38.5 (-35.9~128.6); 706 (413~910); 221.4 (128.6~357.7)

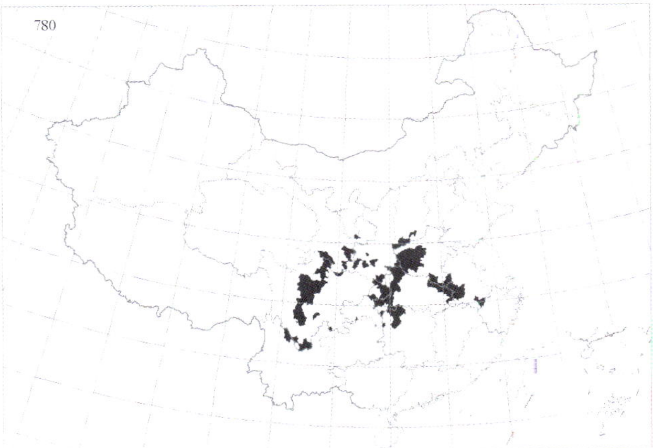

Betulaceae 桦木科 *Carpinus* 鹅耳枥属

Carpinus fargesiana var. hwai (Hu et W. C. Cheng) P. C. Li 狭叶鹅耳枥

Trees, deciduous

14.9 (13.4~16.1); 3.9 (2.7~5.1); 25.5 (23.5~27.1); 15.0 (13.4~16.2); 121.3 (104.5~134.2); 1.4 (0.0~3.7); 808 (735~864) ❙ 1341 (1197~1441); 544 (489~592); 83 (64~123) ❙ 66.6 (51.2~95.9); 808 (735~864); 196.5 (173.7~220.9)

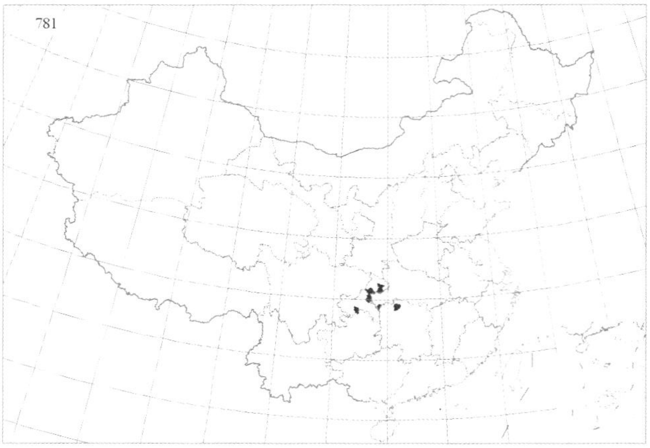

Betulaceae 桦木科 *Carpinus* 鹅耳枥属

Carpinus firmifolia (H. J. P. Winkl.) Hu 厚叶鹅耳枥

Trees, deciduous

15.6; 5.5; 24.6; 15.6; 127.5; 0.0; 808 ❙ 1206; 546; 69 ❙ 49.2; 808; 189.4

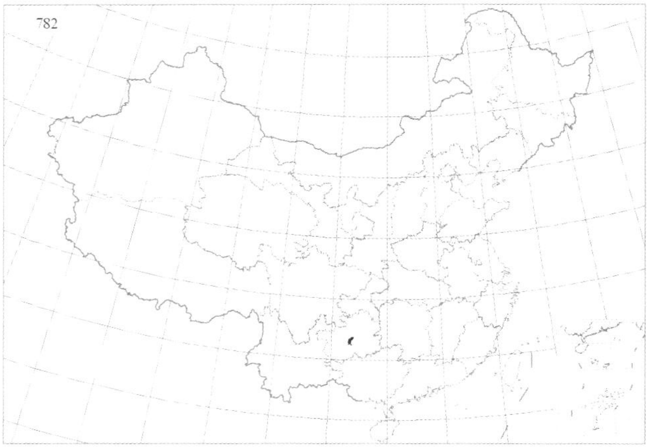

Betulaceae 桦木科 *Carpinus* 鹅耳枥属

Carpinus hebestroma Yamam. 太鲁阁鹅耳枥

Trees, deciduous

16.4; 11.5; 20.3; 16.5; 137.8; 0.0; 783 ❙ 2499; 947; 303 ❙ 219.3; 783; 303.6

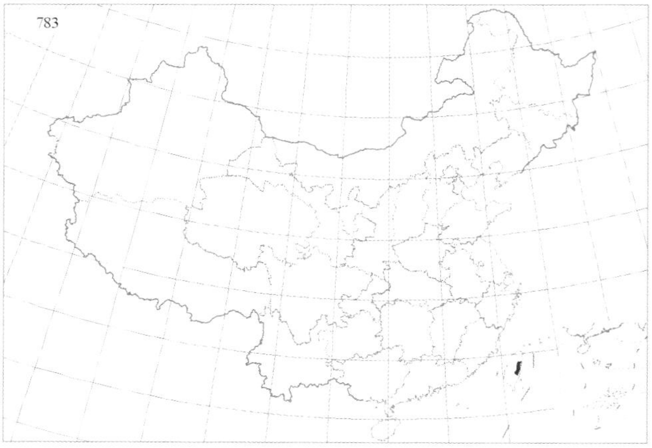

Betulaceae 桦木科 *Carpinus* 鹅耳枥属

Carpinus henryana (H. J. P. Winkl.) H. J. P. Winkl. 川鄂鹅耳枥

Trees, deciduous

11.7 (2.2~18.5); 0.2 (-7.0~8.6); 22.4 (9.9~27.5); 11.9 (3.8~18.5); 93.5 (15.5~162.1); 13.1 (0.0~49.1); 701 (413~953) ❙ 1015 (632~1590); 453 (306~600); 58 (15~182) ❙ 45.9 (-13.5~125.9); 696 (413~953); 243.6 (141.2~334.4)

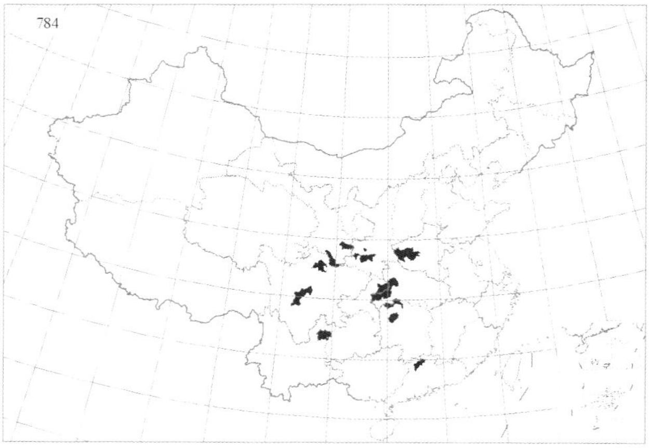

Betulaceae 桦木科 *Carpinus* 鹅耳枥属

Carpinus hupeana Hu 湖北鹅耳枥

Trees, deciduous

13.9 (9.3~16.7); 1.7 (-2.9~5.0); 25.6 (20.1~28.3); 14.0 (9.5~16.8); 114.5 (69.8~141.2); 6.6 (0.0~20.5); 788 (603~903) ❙ 1239 (622~1828); 507 (298~666); 108 (26~253) ❙ 56.2 (-13.5~108.0); 781 (603~903); 231.8 (141.2~319.9)

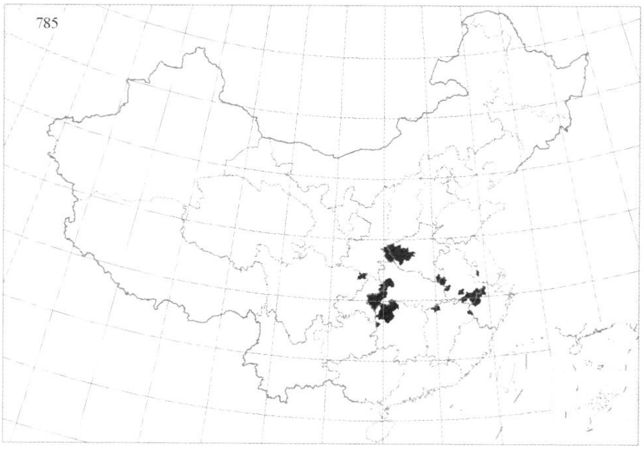

Betulaceae 桦木科 *Carpinus* 鹅耳枥属

Carpinus kawakamii Hayata 阿里山鹅耳枥

Trees, deciduous

19.0 (16.2~22.2); 12.4 (8.5~17.7); 24.4 (20.1~26.8); 19.0 (16.2~22.2); 168.4 (134.2~206.7); 0.0 (0.0~0.0); 943 (773~1123) ❙ 2316 (1716~2992); 1107 (617~1847); 184 (107~246) ❙ 148.3 (76.3~246.5); 943 (773~1123); 272.3 (207.7~298.9)

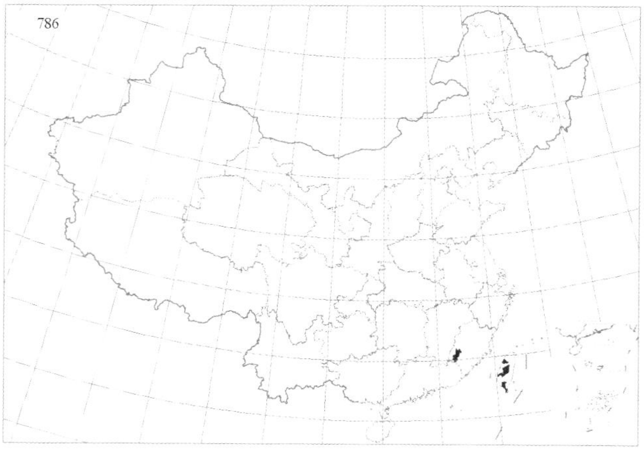

Betulaceae 桦木科 *Carpinus* 鹅耳枥属

Carpinus kweichowensis Hu 贵州鹅耳枥

Trees, deciduous

16.7 (13.4~20.5); 7.9 (3.6~15.1); 24.0 (21.4~25.9); 16.7 (13.4~20.5); 140.9 (102.7~185.9); 0.3 (0.0~1.8); 844 (721~992) ▎ 1273 (1049~1607); 600 (473~720); 70 (44~107) ▎ 50.8 (26.1~64.9); 837 (721~986); 294.0 (170.4~659.0)

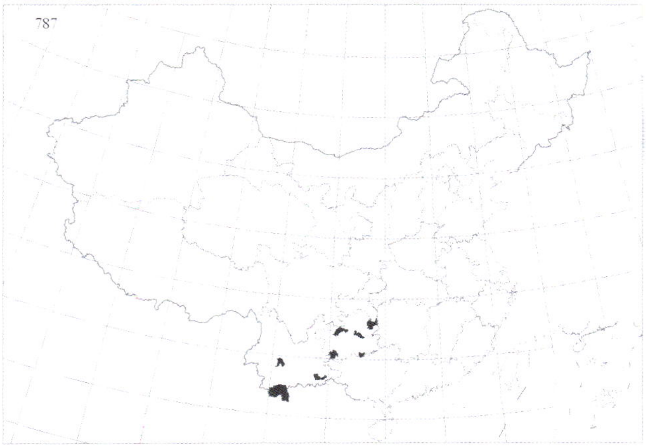

Betulaceae 桦木科 *Carpinus* 鹅耳枥属

Carpinus londoniana H. J. P. Winkl. 短尾鹅耳枥

Trees, deciduous

17.4 (11.1~24.1); 7.4 (1.8~18.7); 26.4 (19.1~29.3); 17.5 (11.1~24.2); 150.3 (79.8~230.1); 0.4 (0.0~6.9); 916 (648~1326) ▎ 1529 (976~1997); 592 (403~959); 158 (30~259) ▎ 68.2 (6.7~130.7); 908 (648~1326); 259.4 (151.0~627.0)

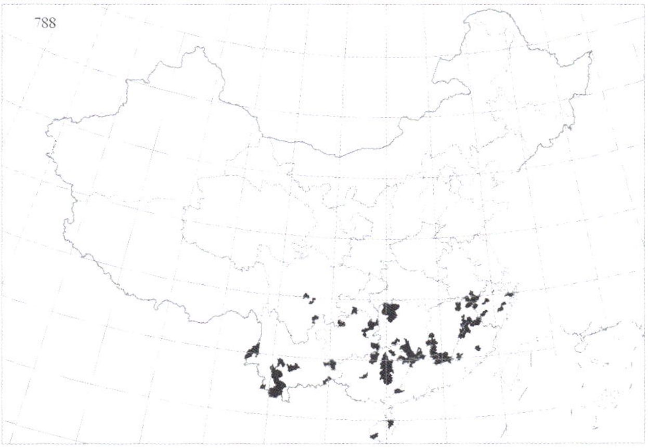

Betulaceae 桦木科 *Carpinus* 鹅耳枥属

Carpinus londoniana var. lanceolata (Hand.-Mazz.) P. C. Li 海南鹅耳枥

Trees, deciduous

11.9; 2.9; 20.0; 12.0; 87.2; 3.4; 671 ▎ 1001; 550; 35 ▎ 49.1; 671; 278.8

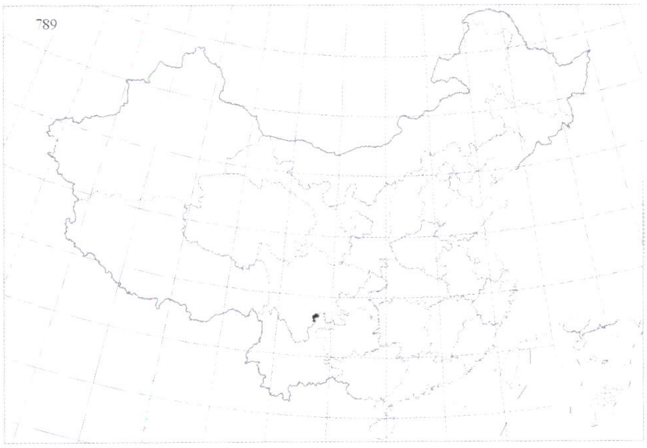

Betulaceae 桦木科 *Carpinus* 鹅耳枥属

Carpinus londoniana var. latifolia P. C. Li 宽叶鹅耳枥

Trees, deciduous

16.5; 5.3; 27.7; 16.5; 138.0; 0.0; 888 ▎ 1363; 481; 183 ▎ 53.5; 888; 222.6

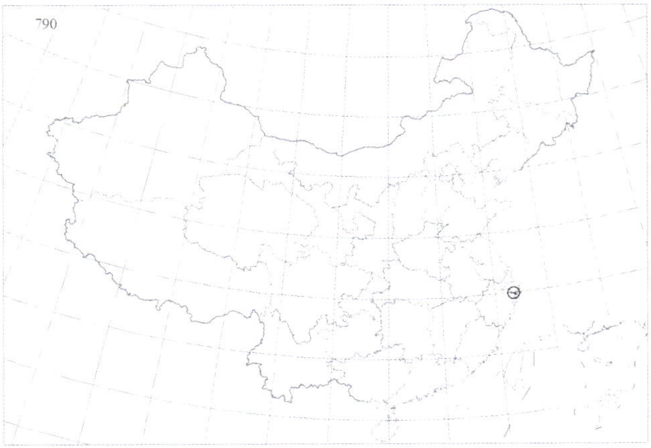

Betulaceae 桦木科 *Carpinus* 鹅耳枥属

Carpinus londoniana var. xiphobracteata P. C. Li 剑苞鹅耳枥

Trees, deciduous

16.2; 4.9; 27.6; 16.2; 134.7; 0.1; 874 ▎ 1413; 505; 185 ▎ 61.6; 874; 260.0

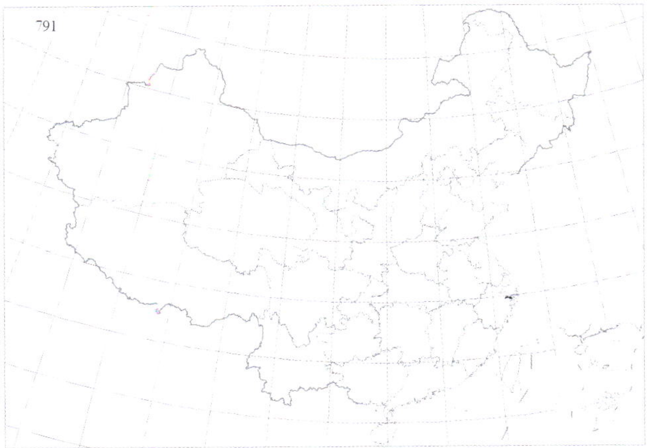

Betulaceae 桦木科 *Carpinus* 鹅耳枥属

Carpinus microphylla Z. C. Chen ex Y. S. Wang et J. P. Huang 田阳鹅耳枥

Shrubs, deciduous

11.9 (7.4~21.2); 0.0 (−4.2~12.5); 23.3 (18.7~27.7); 12.1 (8.2~2[illegible].2); 97.6 (58.8~194.6); 13.8 (0.0~29.4); 716 (561~1114) ▎ 907 (634~1335); 408 (290~676); 41 (16~86) ▎ 28.6 (−16.6~101.4); 707 (561~1114); 268.2 (212.0~318.2)

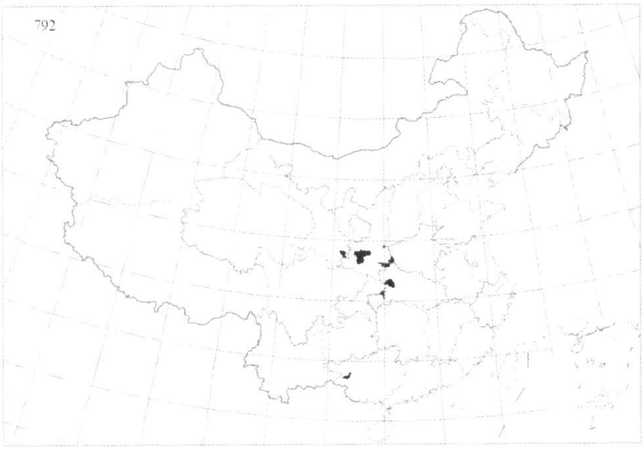

Betulaceae 桦木科 *Carpinus* 鹅耳枥属
Carpinus minutiserrata Hayata 细齿鹅耳枥
Trees, deciduous

17.0；11.3；21.7；17.1；144.9；0.0；819 ❙ 2517；1043；277 ❙ 209.5；819；286.8

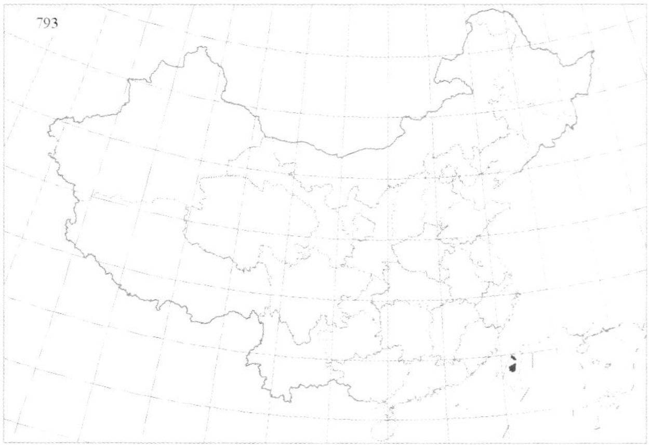

Betulaceae 桦木科 *Carpinus* 鹅耳枥属
Carpinus mollicoma Hu 软毛鹅耳枥
Trees, deciduous

14.1 (1.8~17.6)；5.2 (-7.1~9.9)；21.7 (9.9~25.9)；14.4 (3.8~17.7)；118.2 (15.7~152.4)；8.9 (0.0~53.3)；764 (404~881) ❙ 1261 (583~1577)；727 (348~932)；45 (13~60) ❙ 63.2 (44.2~80.4)；764 (404~881)；239.4 (167.5~359.7)

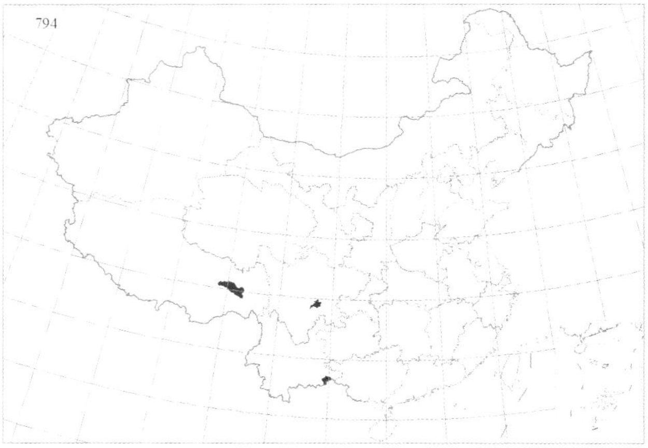

Betulaceae 桦木科 *Carpinus* 鹅耳枥属
Carpinus monbeigiana Hand.-Mazz. 云南鹅耳枥
Trees, deciduous

13.0 (1.8~18.9)；5.4 (-7.1~12.0)；19.3 (9.9~26.6)；13.1 (3.8~19.0)；99.6 (15.7~167.8)；3.4 (0.0~53.3)；703 (404~929) ❙ 985 (583~1736)；572 (348~1007)；32 (13~69) ❙ 42.5 (-14.2~145.8)；677 (404~912)；279.4 (142.9~447.4)

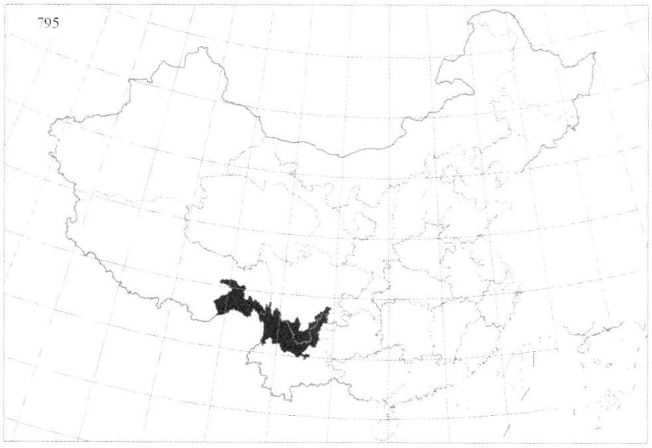

Betulaceae 桦木科 *Carpinus* 鹅耳枥属
Carpinus oblongifolia (Hu) Hu et W. C. Cheng 宝华鹅耳枥
Trees, deciduous

15.5；2.3；28.1；15.6；130.7；3.9；879 ❙ 1002；448；114 ❙ 14.0；879；134.1

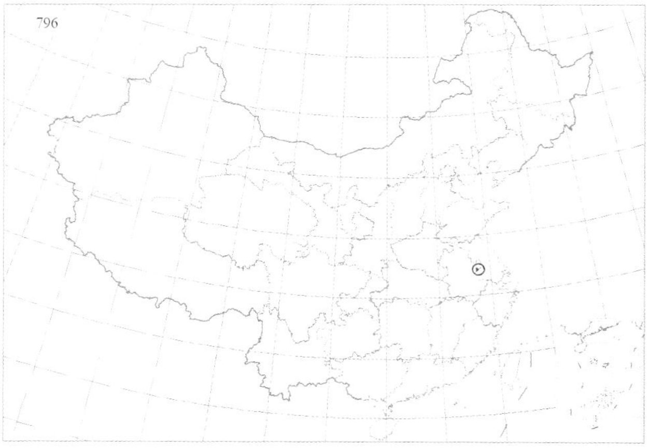

Betulaceae 桦木科 *Carpinus* 鹅耳枥属
Carpinus omeiensis Hu et D. Fang 峨眉鹅耳枥
Small trees, deciduous

15.4；5.3；24.8；15.4；125.1；0.1；806 ❙ 1162；578；60 ❙ 44.5；806；188.0

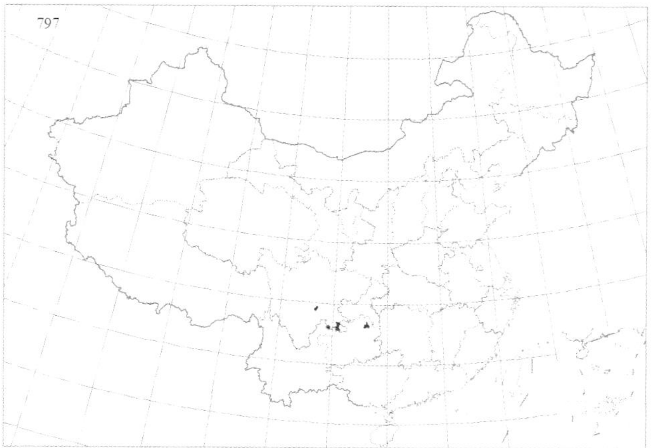

Betulaceae 桦木科 *Carpinus* 鹅耳枥属
Carpinus polyneura Franch. 多脉鹅耳枥
Trees, deciduous

15.0 (2.6~23.6)；4.0 (-7.0~16.7)；25.2 (11.2~29.3)；15.0 (4.3~23.6)；123.5 (20.0~223.5)；3.5 (0.0~48.2)；818 (427~1281) ❙ 1314 (624~1997)；526 (317~818)；114 (14~259) ❙ 60.7 (-8.1~130.7)；814 (427~1273)；225.1 (134.6~551.6)

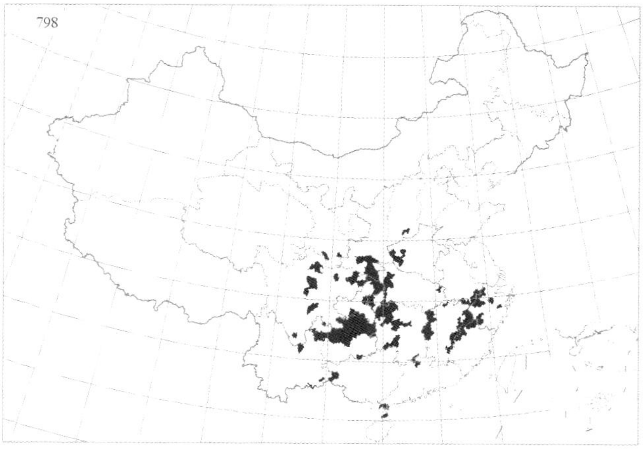

Betulaceae 桦木科 *Carpinus* 鹅耳枥属
Carpinus pubescens Burkill 云贵鹅耳枥
Trees, deciduous

14.9 (3.5~19.1); 4.7 (−6.8~11.0); 24.1 (10.8~27.7); 15.0 (4.7~19.2); 122.7 (21.6~170.1); 3.1 (0.0~43.4); 797 (444~981) ❙ 1184 (557~1590); 538 (286~860); 74 (10~182) ❙ 48.7 (−10.0~121.7); 796 (444~981); 223.8 (132.4~491.5)

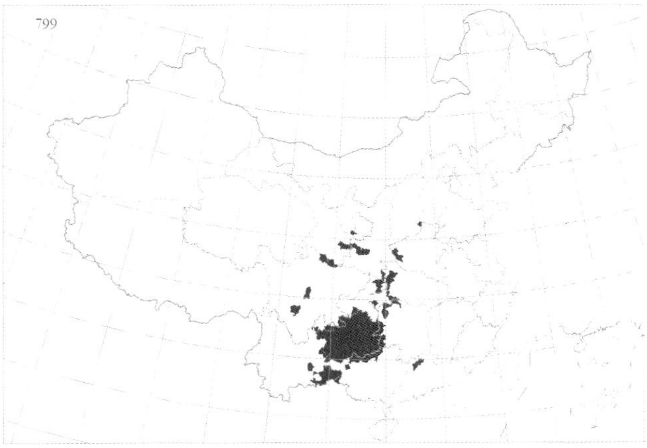

Betulaceae 桦木科 *Carpinus* 鹅耳枥属
Carpinus purpurinervis Hu 紫脉鹅耳枥
Small trees, deciduous

18.7 (16.2~20.5); 9.2 (6.3~10.9); 26.4 (23.4~28.3); 18.8 (16.2~20.5); 165.2 (134.5~186.1); 0.0 (0.0~0.0); 967 (826~1083) ❙ 1411 (1232~1567); 682 (611~775); 108 (59~141) ❙ 46.3 (30.9~62.5); 967 (826~1083); 253.2 (215.4~322.1)

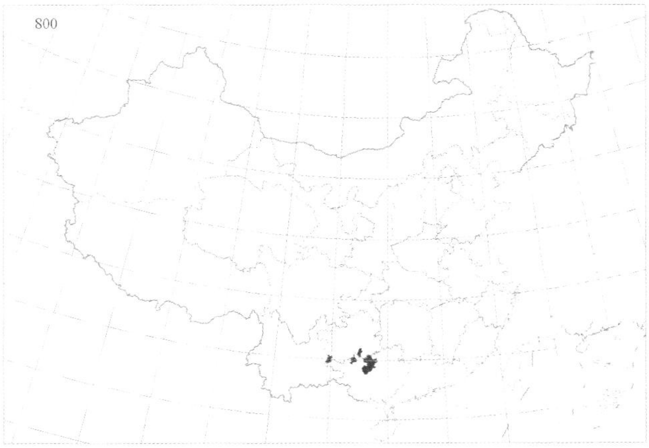

Betulaceae 桦木科 *Carpinus* 鹅耳枥属
Carpinus putoensis W. C. Cheng 普陀鹅耳枥
Trees, deciduous

16.3 (15.9~16.5); 4.9 (4.2~5.3); 27.8 (27.6~28.2); 16.3 (15.9~16.5); 136.2 (131.9~138.3); 0.2 (0.0~0.8); 883 (866~897) ❙ 1354 (1274~1413); 484 (454~505); 184 (181~186) ❙ 53.5 (41.9~61.6); 883 (866~897); 229.7 (196.9~260.0)

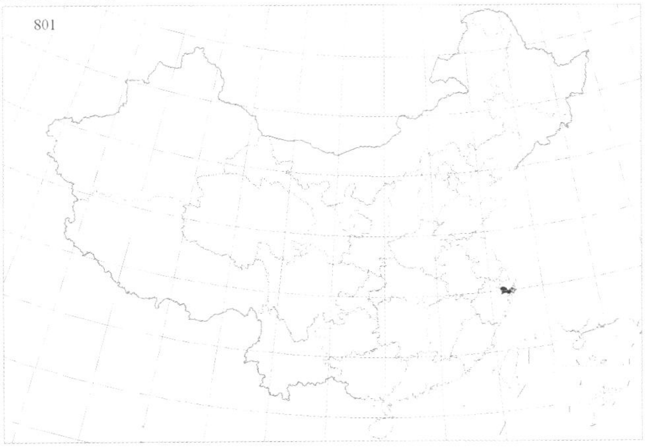

Betulaceae 桦木科 *Carpinus* 鹅耳枥属
Carpinus rankanensis Hayata 兰邯千金榆
Trees, deciduous

18.3; 13.4; 22.0; 18.3; 159.6; 0.0; 893 ❙ 2724; 1327; 218 ❙ 210.8; 893; 296.5

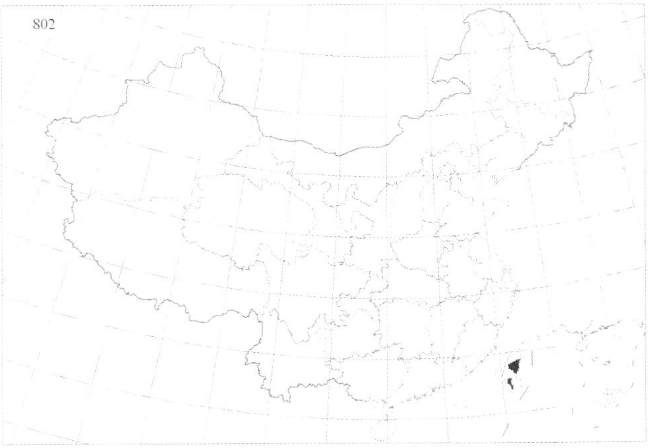

Betulaceae 桦木科 *Carpinus* 鹅耳枥属
Carpinus rankanensis var. matsudae Yamamoto 细叶兰邯千金榆
Trees, deciduous

21.7; 16.7; 25.4; 21.7; 200.8; 0.0; 1117 ❙ 2255; 1381; 87 ❙ 117.5; 1003; 203.3

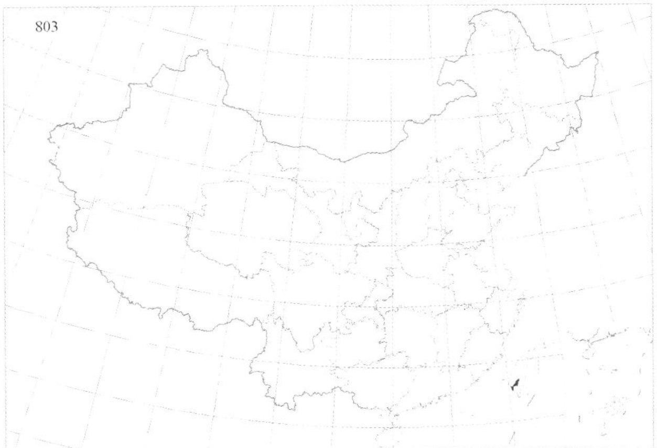

Betulaceae 桦木科 *Carpinus* 鹅耳枥属
Carpinus rupestris A. Camus 岩生鹅耳枥
Small trees, deciduous

16.5 (12.5~22.2); 7.1 (3.6~13.9); 24.6 (19.6~28.6); 16.6 (12.5~22.3); 139.3 (92.0~207.5); 0.4 (0.0~1.7); 864 (685~1190) ❙ 1288 (1071~1736); 625 (471~1007); 77 (45~140) ❙ 51.3 (12.9~145.8); 864 (685~1190); 246.8 (153.9~447.4)

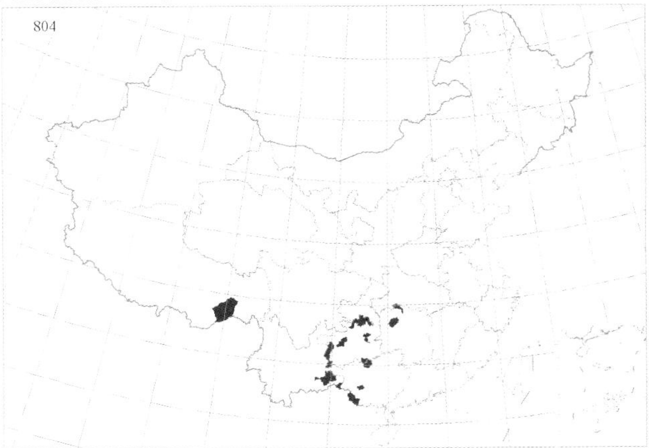

Betulaceae 桦木科 *Carpinus* 鹅耳枥属
Carpinus shensiensis Hu 陕西鹅耳枥
Trees, deciduous

11.8 (7.4~16.2); -0.2 (-4.4~4.4); 23.2 (18.7~27.5); 12.0 (8.2~16.2); 94.8 (58.8~135.6); 13.0 (0.6~29.4); 706 (561~877) ❙ 976 (588~1448); 423 (294~600); 51 (15~140) ❙ 37.1 (-12.5~83.5); 693 (561~877); 249.1 (162.9~318.2)

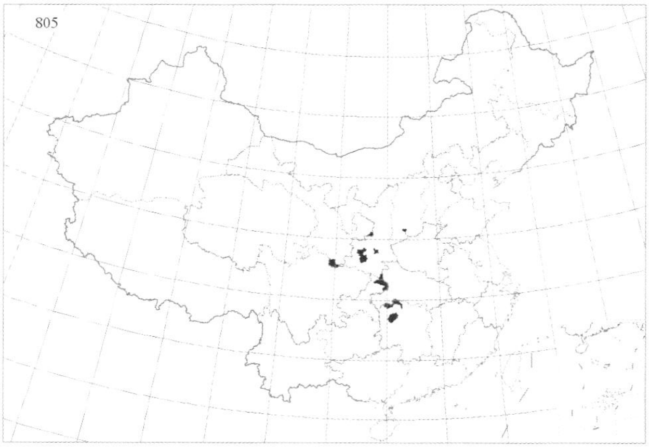

Betulaceae 桦木科 *Carpinus* 鹅耳枥属
Carpinus stipulata H. Winkler 小叶鹅耳枥
Trees, deciduous

12.4 (8.4~16.2); -1.3 (-6.4~4.4); 24.9 (21.6~27.5); 12.7 (9.7~16.2); 104.1 (74.2~135.6); 15.2 (0.6~32.8); 749 (624~877) ❙ 825 (583~1448); 413 (290~600); 45 (23~140) ❙ 9.0 (-23.0~79.3); 711 (583~877); 199.1 (136.6~260.3)

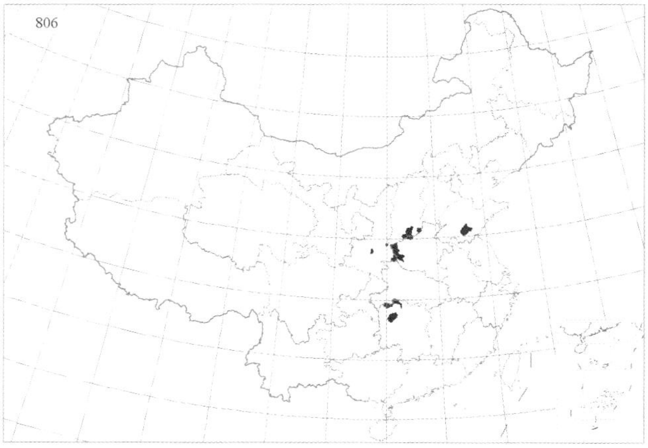

Betulaceae 桦木科 *Carpinus* 鹅耳枥属
Carpinus sungpanensis W. Y. Hsia 松潘鹅耳枥
Trees, deciduous

2.6; -7.0; 11.2; 4.3; 20.0; 48.2; 427 ❙ 829; 400; 23 ❙ 94.4; 427; 252.2

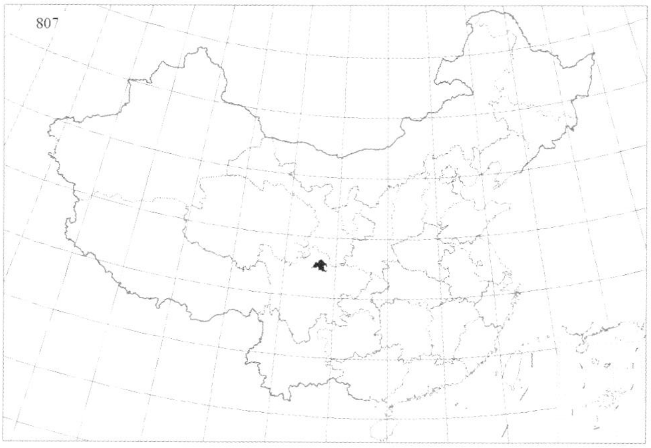

Betulaceae 桦木科 *Carpinus* 鹅耳枥属
Carpinus tientaiensis W. C. Cheng 天台鹅耳枥
Trees, deciduous

15.6; 4.4; 27.1; 15.7; 128.8; 0.6; 843 ❙ 1581; 587; 185 ❙ 87.6; 843; 236.5

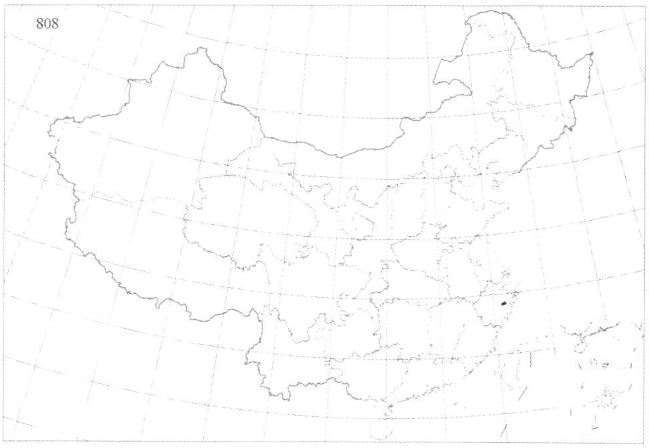

Betulaceae 桦木科 *Carpinus* 鹅耳枥属
Carpinus tsaiana Hu 宽苞鹅耳枥
Trees, deciduous

17.2 (15.3~19.1); 8.7 (4.2~11.0); 23.9 (22.5~25.9); 17.2 (15.4~19.1); 147.0 (125.2~169.6); 0.1 (0.0~0.8); 857 (822~945) ❙ 1344 (1232~1439); 709 (592~820); 64 (48~92) ❙ 57.1 (45.8~75.0); 857 (822~945); 315.9 (185.4~491.5)

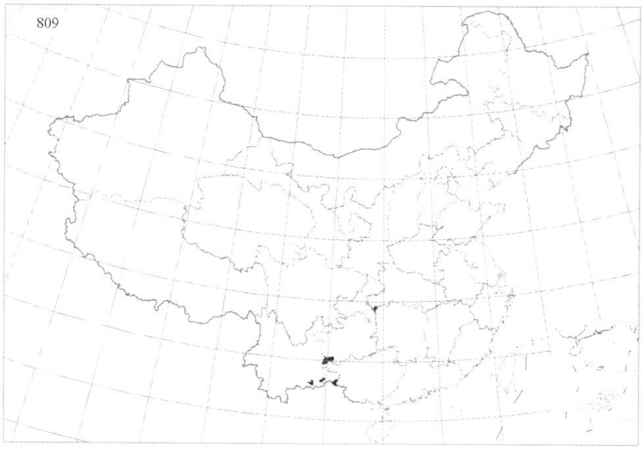

Betulaceae 桦木科 *Carpinus* 鹅耳枥属
Carpinus tschonoskii Maxim. 昌化鹅耳枥
Trees, deciduous

14.3 (7.3~20.5); 2.4 (-2.9~14.9); 25.4 (15.6~29.2); 14.4 (7.5~20.5); 117.7 (48.2~185.9); 5.8 (0.0~20.6); 803 (537~992) ❙ 1151 (512~1828); 494 (258~727); 96 (15~253) ❙ 43.3 (-35.3~112.2); 780 (512~959); 217.2 (121.6~570.1)

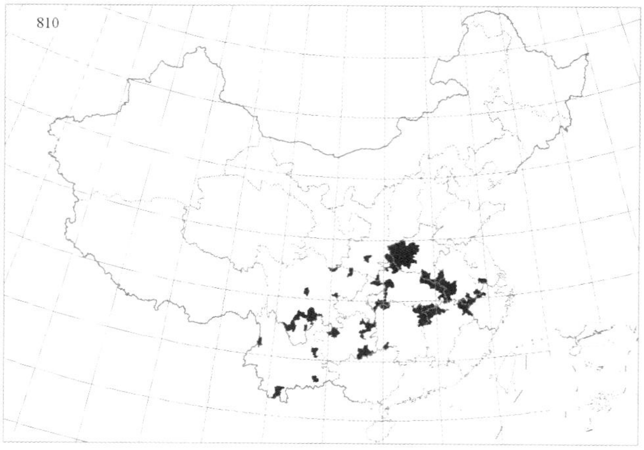

Betulaceae 桦木科 *Carpinus* 鹅耳枥属
Carpinus tsunyihensis Hu 遵义鹅耳枥
Trees, deciduous

15.4; 4.7; 25.3; 15.4; 125.2; 0.3; 812 ❙ 1106; 481; 69 ❙ 36.3; 812; 164.2

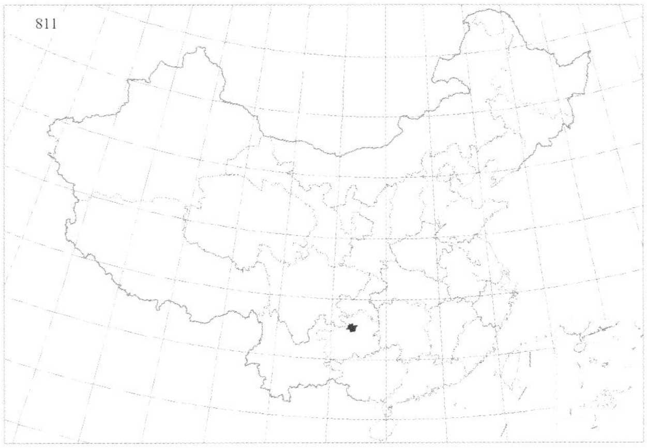

Betulaceae 桦木科 *Carpinus* 鹅耳枥属
Carpinus turczaninowii Hance 鹅耳枥
Trees, deciduous

11.0 (2.2~17.1); -3.6 (-17.9~8.0); 24.1 (12.4~29.4); 11.7 (4.8~17.1); 96.2 (25.1~146.1); 24.3 (0.0~83.9); 721 (441~952) ❙ 668 (341~1663); 374 (190~661); 27 (7~212) ❙ -7.1 (-40.9~101.3); 619 (341~921); 187.4 (61.2~334.4)

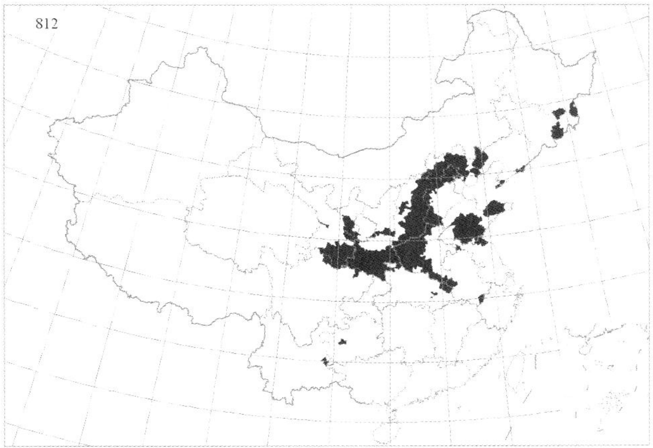

Betulaceae 桦木科 *Carpinus* 鹅耳枥属
Carpinus viminea Lindl. 雷公鹅耳枥
Trees, deciduous

15.5 (-1.8~21.0); 4.7 (-10.2~13.0); 25.7 (6.1~29.8); 15.6 (1.8~21.1); 129.3 (1.9~193.0); 2.3 (0.0~82.8); 839 (329~1114) ❙ 1356 (320~1997); 532 (193~835); 135 (13~259) ❙ 60.9 (-22.2~147.0); 828 (320~1114); 223.4 (72.2~506.2)

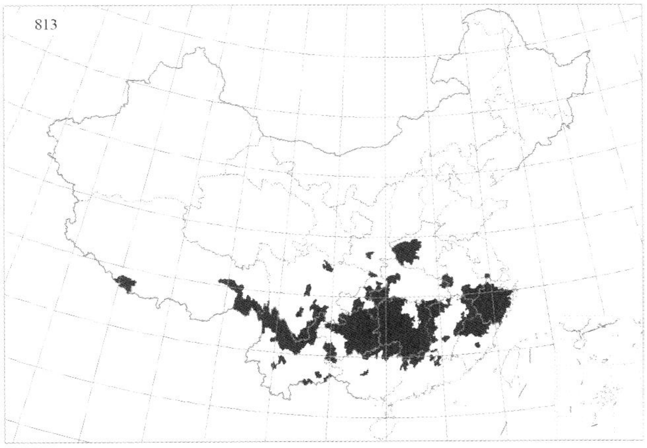

Betulaceae 桦木科 *Carpinus* 鹅耳枥属
Carpinus viminea var. chiukiangensis Hu 贡山鹅耳枥
Trees, deciduous

17.1 (8.6~20.5); 10.9 (1.3~15.1); 21.3 (15.0~23.7); 17.1 (8.7~20.5); 146.6 (52.6~185.9); 1.5 (0.0~8.8); 849 (574~992) ❙ 1308 (1041~1607); 647 (585~702); 57 (44~77) ❙ 55.4 (26.1~81.5); 830 (574~986); 468.0 (347.3~659.0)

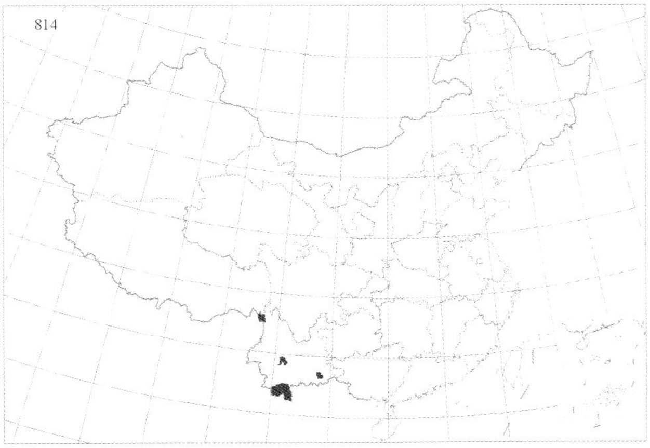

Betulaceae 桦木科 *Corylus* 榛属
Corylus chinensis Franch. 华榛
Trees, deciduous

13.1 (5.4~18.9); 2.5 (-4.2~12.0); 22.8 (12.3~28.1); 13.2 (5.8~19.0); 104.6 (31.5~167.8); 6.9 (0.0~29.9); 737 (481~929) ❙ 985 (512~1736); 479 (258~1007); 48 (12~221) ❙ 34.6 (-35.3~145.8); 713 (481~924); 242.2 (121.6~447.4)

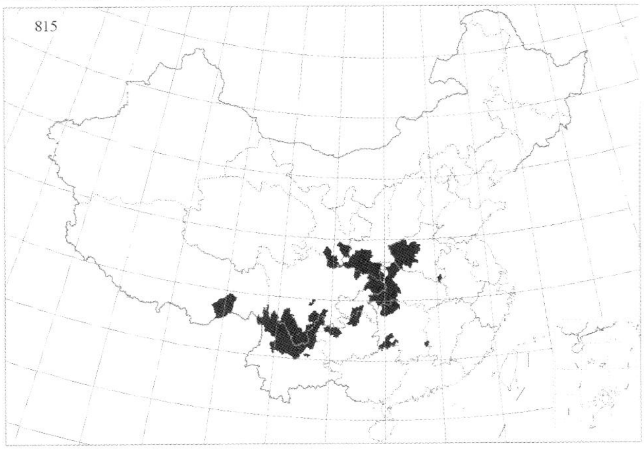

Betulaceae 桦木科 *Corylus* 榛属
Corylus fargesii C. K. Schneid. 披针叶榛
Small trees, deciduous

11.4 (2.6~16.2); 0.2 (-7.0~5.5); 22.0 (11.2~27.1); 11.7 (4.3~16.2); 90.0 (20.0~134.6); 12.4 (0.0~48.2); 687 (427~859) ❙ 963 (373~1478); 445 (201~817); 42 (7~192) ❙ 40.7 (-37.6~107.2); 679 (373~859); 244.0 (88.1~319.8)

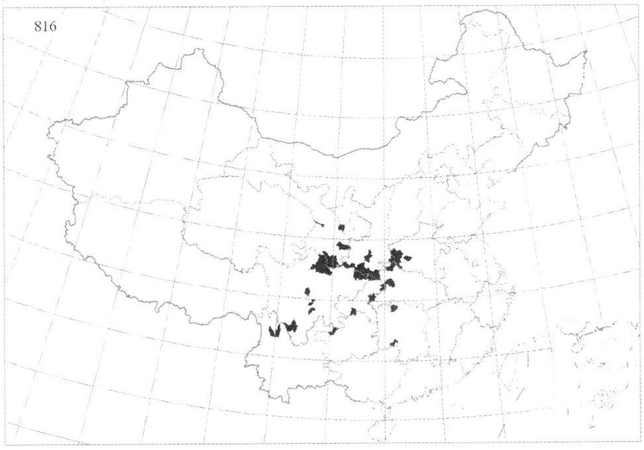

Betulaceae 桦木科 *Corylus* 榛属
Corylus ferox Wall. 刺榛
Trees, deciduous

11.5 (−1.8~17.1); 1.9 (−10.2~9.0); 20.2 (6.1~26.8); 11.8 (1.8~17.1); 89.0 (1.9~145.2); 10.6 (0.0~82.8); 680 (329~889) ❙ 1095 (320~1736); 561 (193~1007); 47 (10~113) ❙ 62.0 (−4.0~145.8); 677 (320~889); 255.2 (61.6~487.2)

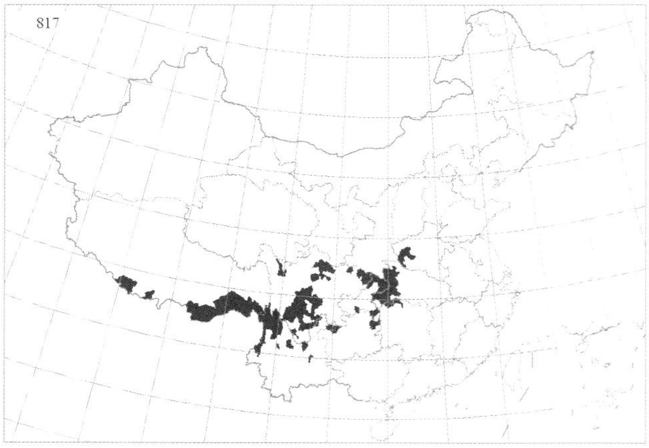

Betulaceae 桦木科 *Corylus* 榛属
Corylus ferox var. thibetica (Batalin) Franch. 藏刺榛
Trees, deciduous

11.2 (2.2~15.3); 0.3 (−7.7~4.7); 21.3 (9.9~26.8); 11.5 (3.8~15.4); 87.7 (15.5~125.7); 13.0 (0.3~49.3); 677 (413~834) ❙ 992 (620~1492); 456 (301~613); 49 (13~113) ❙ 48.1 (−8.1~125.9); 675 (413~832); 233.4 (151.3~311.0)

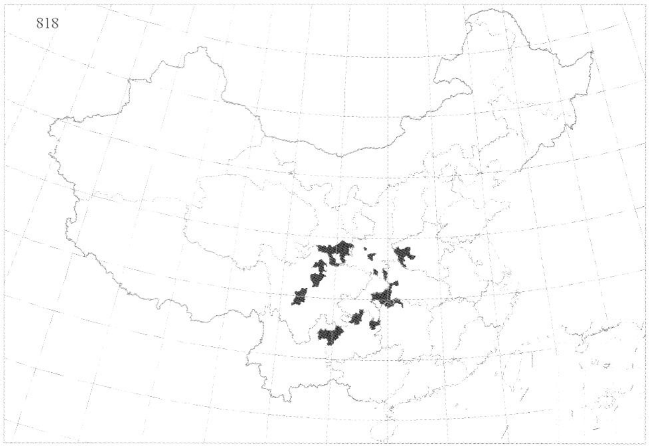

Betulaceae 桦木科 *Corylus* 榛属
Corylus heterophylla Fisch. ex Trautv. 榛
Shrubs or small trees, deciduous

9.4 (−5.8~17.7); −6.4 (−29.8~7.0); 23.5 (12.4~29.3); 11.1 (4.4~17.7); 90.3 (25.1~152.4); 36.8 (0.0~157.4); 695 (426~964) ❙ 743 (309~1679); 400 (190~691); 38 (5~220) ❙ 6.2 (−47.7~108.0); 630 (309~961); 210.1 (61.2~346.5)

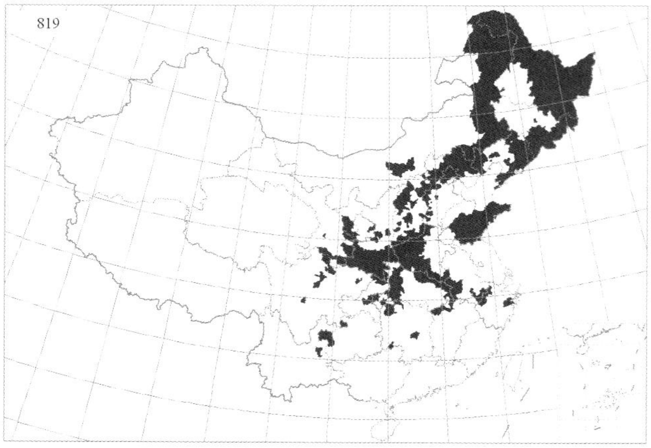

Betulaceae 桦木科 *Corylus* 榛属
Corylus heterophylla var. sutchuanensis Franch. 川榛
Shrubs or small trees, deciduous

14.0 (0.0~19.1); 2.3 (−11.4~10.1); 24.9 (9.7~29.3); 14.1 (3.3~19.1); 115.1 (12.2~169.2); 7.0 (0.0~72.2); 784 (395~969) ❙ 1086 (341~1770); 478 (190~720); 78 (8~223) ❙ 37.4 (−40.9~112.2); 771 (341~969); 212.4 (61.2~364.1)

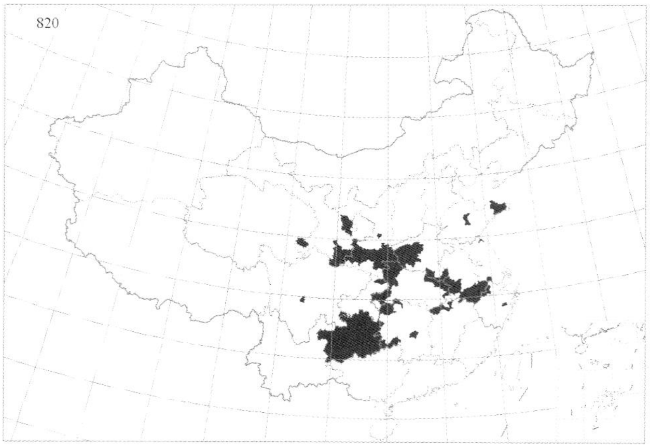

Betulaceae 桦木科 *Corylus* 榛属
Corylus mandshurica Maxim. 毛榛
Shrubs, deciduous

7.5 (−1.9~15.5); −8.9 (−25.0~2.1); 22.0 (10.7~28.0); 9.7 (3.8~15.6); 76.6 (16.2~131.3); 45.8 (4.3~125.0); 638 (411~881) ❙ 589 (307~1246); 345 (173~691); 19 (7~97) ❙ −6.0 (−49.4~121.7); 548 (307~881); 201.1 (42.6~346.5)

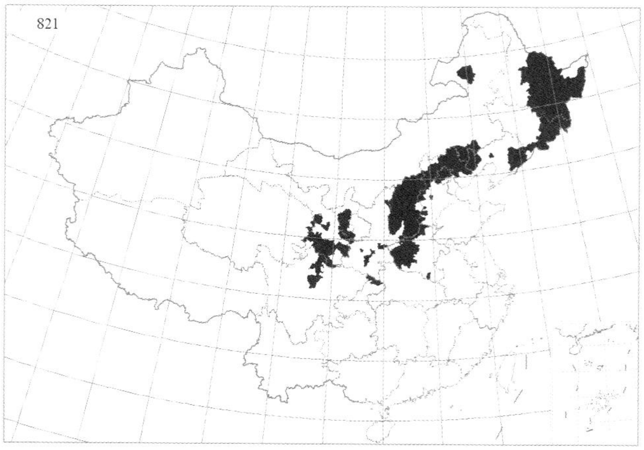

Betulaceae 桦木科 *Corylus* 榛属
Corylus wangii Hu 维西榛
Small trees, deciduous

9.6; 2.4; 15.7; 9.6; 60.8; 5.4; 600 ❙ 1007; 578; 40 ❙ 67.9; 600; 351.7

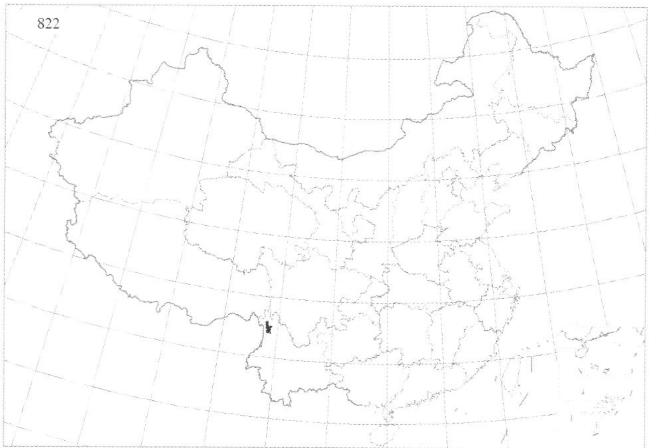

Betulaceae 桦木科 *Corylus* 榛属

Corylus yunnanensis (Franch.) A. Camus 滇榛

Shrubs or small trees, deciduous

13.0 (2.2~18.9); 5.1 (-6.6~12.4); 19.5 (9.9~27.1); 13.1 (3.8~19.0); 100.3 (15.5~167.8); 4.6 (0.0~49.1); 704 (413~929) ❙ 1009 (741~1500); 573 (411~792); 35 (12~106) ❙ 46.7 (-14.2~125.9); 687 (413~915); 280.3 (127.1~487.2)

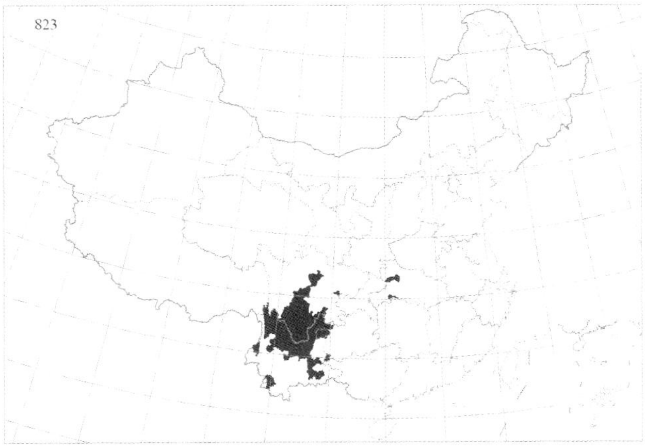

Betulaceae 桦木科 *Ostrya* 铁木属

Ostrya japonica Sarg. 铁木

Trees, deciduous

10.9 (2.2~15.5); -2.9 (-12.5~2.1); 23.5 (9.9~27.8); 11.5 (3.8~15.6); 93.3 (15.5~131.3); 21.3 (4.3~62.3); 709 (413~881) ❙ 695 (373~1478); 364 (201~616); 27 (7~140) ❙ 1.1 (-43.4~125.9); 608 (373~881); 201.0 (88.1~357.7)

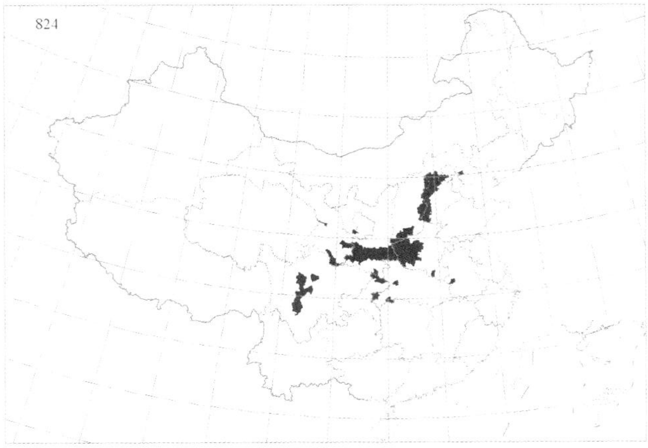

Betulaceae 桦木科 *Ostrya* 铁木属

Ostrya multinervis Rehder 多脉铁木

Trees, deciduous

16.6 (10.8~18.7); 5.9 (1.6~8.1); 26.7 (16.6~28.9); 16.7 (10.9~18.8); 140.4 (71.8~166.0); 0.5 (0.0~6.9); 886 (633~1002) ❙ 1201 (968~1893); 509 (442~677); 76 (24~227) ❙ 37.5 (8.0~129.9); 886 (633~1002); 165.9 (108.2~319.8)

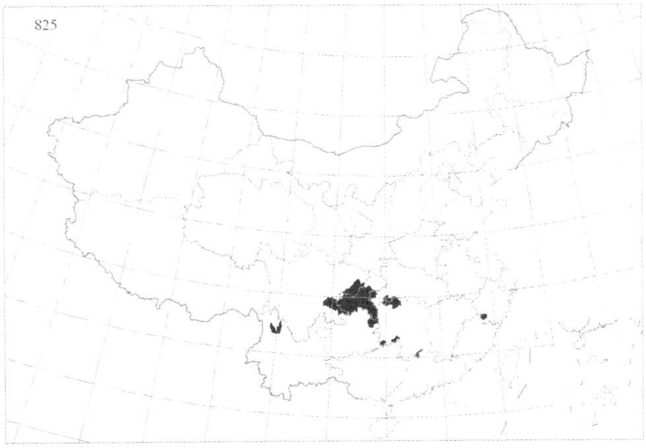

Betulaceae 桦木科 *Ostrya* 铁木属

Ostrya rehderiana Chun 天目铁木

Trees, deciduous

15.2 (14.4~16.1); 2.9 (2.1~4.1); 27.2 (26.3~28.3); 15.2 (14.5~16.1); 125.7 (118.7~134.7); 3.0 (0.9~4.9); 839 (801~887) ❙ 1490 (1296~1663); 562 (517~599); 195 (177~220) ❙ 78.1 (46.2~101.3); 839 (801~887); 257.8 (198.0~284.0)

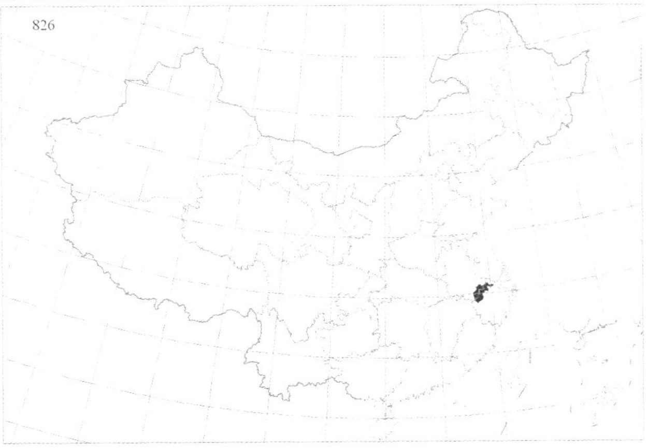

Betulaceae 桦木科 *Ostrya* 铁木属

Ostrya trichocarpa D. Fang et Y. S. Wang 毛果铁木

Trees, deciduous

18.5; 9.8; 25.2; 18.6; 163.0; 0.0; 928 ❙ 1434; 768; 79 ❙ 54.7; 928; 310.2

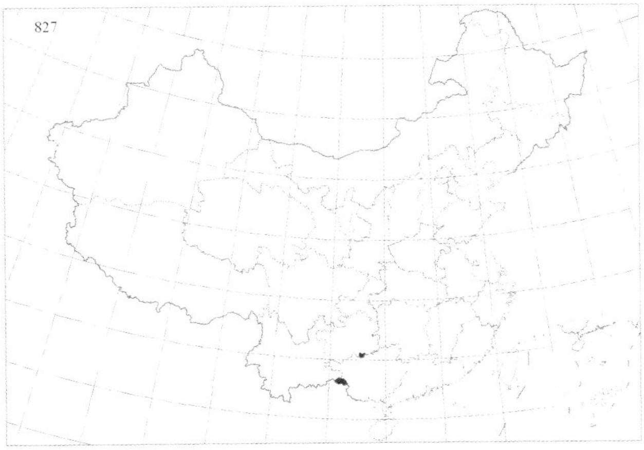

Betulaceae 桦木科 *Ostrya* 铁木属

Ostrya yunnanensis Hu ex P. C. Li 云南铁木

Trees, deciduous

13.8; 6.7; 19.3; 13.8; 105.9; 0.0; 714 ❙ 963; 592; 29 ❙ 35.0; 690; 266.2

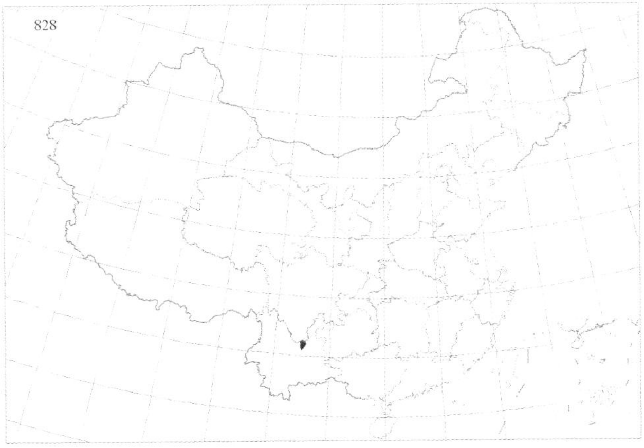

Betulaceae 桦木科 *Ostryopsis* 虎榛子属
Ostryopsis davidiana Decne. 虎榛子
Shrubs, deciduous

8.1 (−1.3~15.5); −6.7 (−20.8~6.0); 21.3 (8.3~28.0); 9.7 (2.5~15.6); 75.8 (7.0~131.3); 38.1 (0.0~104.9); 634 (354~881) ❙ 554 (128~1358); 308 (81~610); 16 (2~97) ❙ −11.1 (−80.8~121.7); 508 (128~881); 173.8 (13.6~357.7)

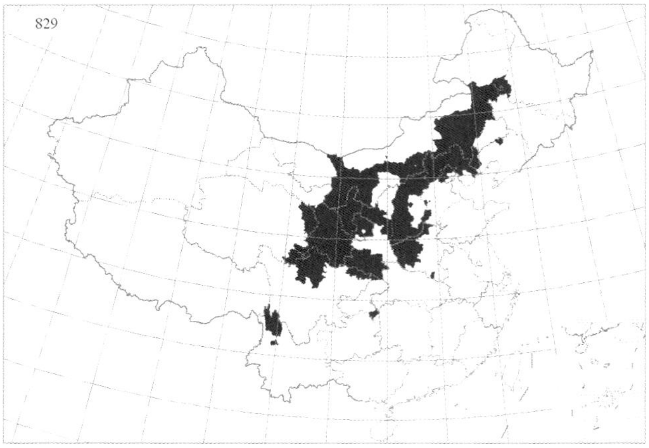

Betulaceae 桦木科 *Ostryopsis* 虎榛子属
Ostryopsis nobilis I. B. Balfour et W. W. Sm. 滇虎榛
Shrubs, deciduous

13.0 (3.0~18.9); 5.2 (−5.6~12.0); 19.5 (10.5~26.6); 13.1 (4.3~19.0); 100.2 (20.1~167.8); 3.9 (0.0~43.9); 705 (423~929) ❙ 939 (628~1141); 546 (406~627); 30 (11~64) ❙ 35.3 (−14.2~70.8); 679 (423~912); 262.1 (142.9~351.7)

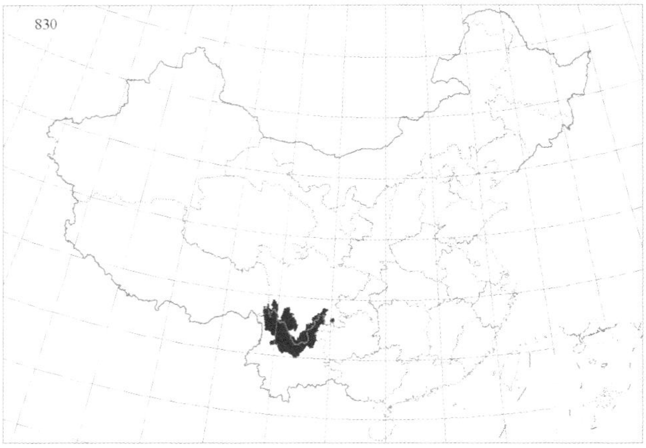

Fagaceae 壳斗科 *Castanea* 栗属
Castanea henryi (Skan) Rehder et E. H. Wilson 锥栗
Trees, deciduous

16.4 (9.3~22.1); 5.3 (−1.7~13.7); 26.8 (19.4~29.9); 16.4 (9.5~22.1); 138.5 (69.8~205.6); 1.1 (0.0~18.0); 880 (603~1182) ❙ 1431 (842~1997); 547 (370~848); 153 (19~259) ❙ 63.2 (5.8~147.0); 875 (603~1182); 224.2 (131.2~411.1)

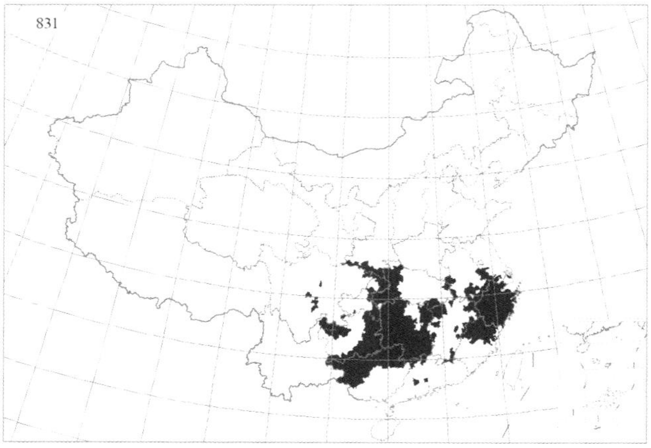

Fagaceae 壳斗科 *Castanea* 栗属
Castanea mollissima Blume 栗
Trees, deciduous

15.3 (2.6~24.8); 3.9 (−15.2~20.1); 25.8 (11.2~30.1); 15.6 (4.3~24.8); 131.8 (20.0~237.6); 7.2 (0.0~71.7); 856 (427~1385) ❙ 1217 (203~2993); 515 (118~1428); 109 (4~547) ❙ 40.7 (−69.4~246.5); 821 (203~1360); 222.0 (101.9~627.0)

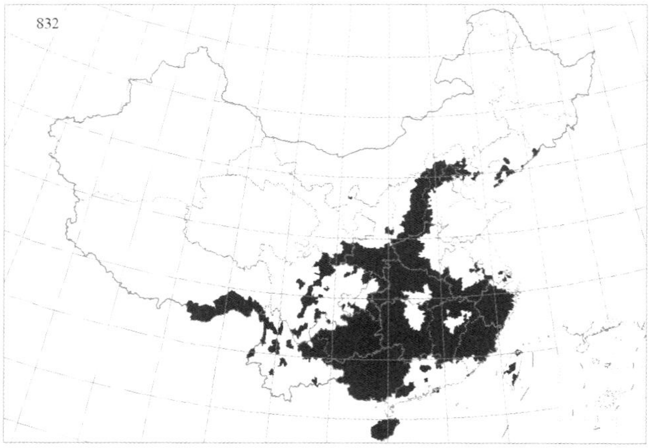

Fagaceae 壳斗科 *Castanea* 栗属
Castanea seguinii Dode 茅栗
Shrubs or small trees, deciduous

14.7 (3.9~22.3); 2.5 (−13.7~14.0); 26.0 (18.7~29.9); 14.9 (7.3~22.4); 124.0 (52.8~208.3); 7.6 (0.0~71.7); 826 (535~1200) ❙ 1178 (367~1997); 482 (235~821); 113 (8~259) ❙ 40.0 (−40.9~147.0); 794 (367~1200); 203.9 (101.9~432.9)

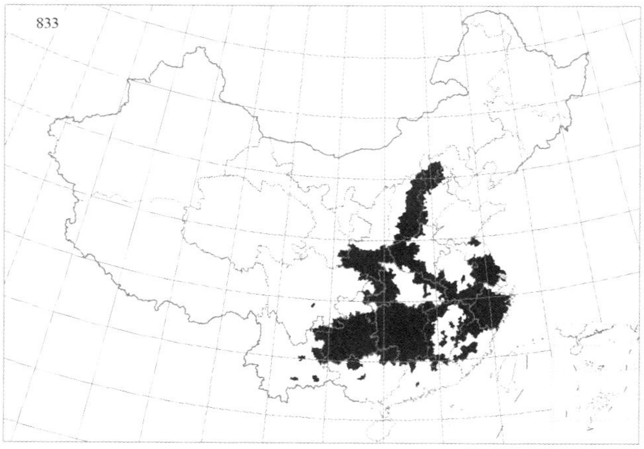

Fagaceae 壳斗科 *Castanopsis* 锥属
Castanopsis amabilis W. C. Cheng et C. S. Chao 南宁锥
Trees, evergreen

21.3 (17.4~22.7); 13.1 (7.4~15.0); 27.5 (22.5~28.9); 21.3 (17.4~22.8); 196.0 (148.9~213.1); 0.0 (0.0~0.0); 1123 (854~1221) ❙ 1663 (1302~2435); 895 (637~1428); 111 (59~136) ❙ 48.5 (12.9~99.5); 1123 (854~1221); 294.4 (201.4~491.5)

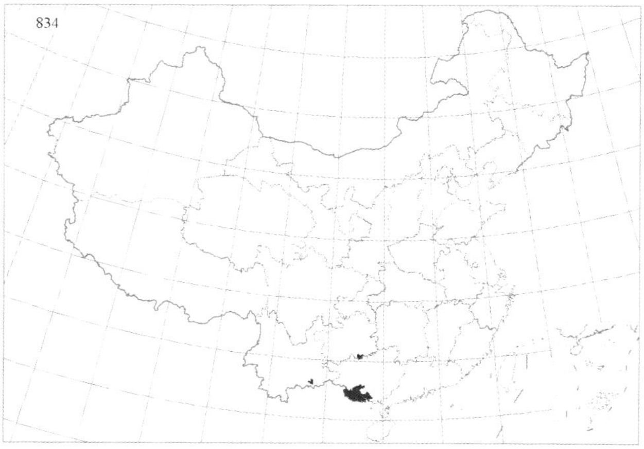

Fagaceae 壳斗科 *Castanopsis* 锥属
Castanopsis argyrophylla King ex Hook. f. 银叶锥
Trees, evergreen

18.2 (14.4~20.5); 12.0 (7.5~15.1); 22.2 (19.1~23.7); 18.2 (14.4~20.5); 158.9 (113.1~185.9); 0.0 (0.0~0.0); 876 (727~992) ❙ 1395 (1049~1607); 743 (598~835); 57 (42~77) ❙ 59.6 (26.1~99.9); 860 (727~986); 460.4 (354.7~659.0)

835
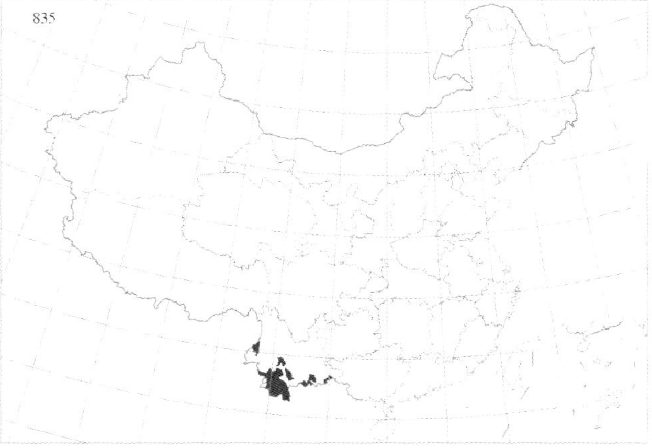

Fagaceae 壳斗科 *Castanopsis* 锥属
Castanopsis boisii Hickel et A. Camus 榄壳锥
Trees, evergreen

21.8 (18.9~24.3); 15.0 (10.6~19.1); 26.9 (23.6~28.6); 21.9 (19.0~24.3); 202.7 (167.5~231.9); 0.0 (0.0~0.0); 1148 (939~1338) ❙ 1595 (1179~2435); 758 (392~1428); 77 (42~136) ❙ 40.8 (-11.9~39.5); 1077 (939~1221); 421.2 (223.8~659.0)

836
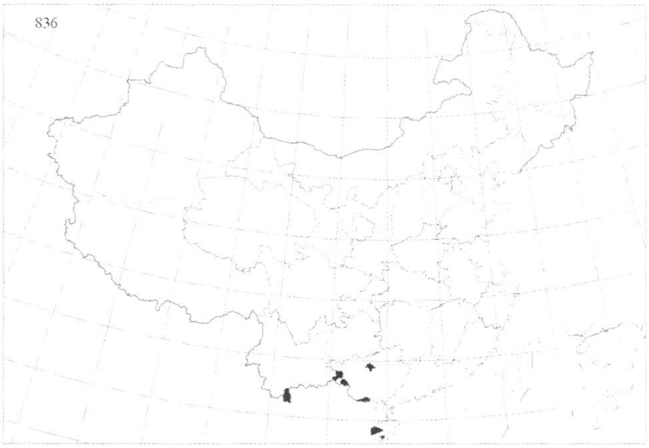

Fagaceae 壳斗科 *Castanopsis* 锥属
Castanopsis calathiformis (Skan) Rehder et E. H. Wilson 枹丝锥
Trees, evergreen

19.0 (13.3~22.7); 12.0 (5.4~15.1); 23.9 (19.6~28.6); 19.1 (13.4~22.8); 168.7 (100.4~213.1); 0.0 (0.0~0.0); 948 (706~1221) ❙ 1467 (1028~2435); 795 (566~1428); 74 (42~191) ❙ 55.8 (12.5~145.8); 940 (706~1221); 393.1 (235.1~659.0)

837
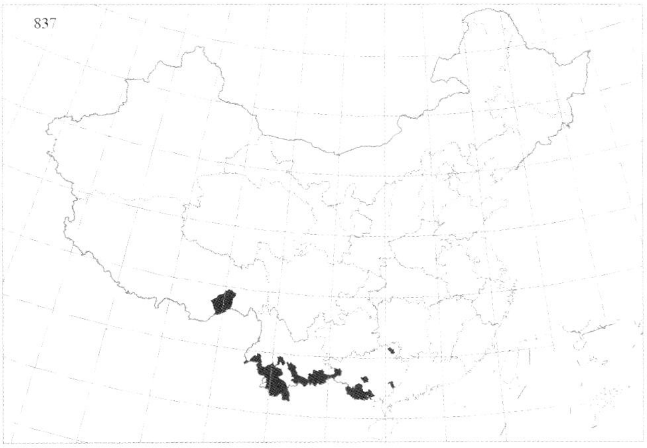

Fagaceae 壳斗科 *Castanopsis* 锥属
Castanopsis carlesii (Hemsl.) Hayata 米槠
Trees, evergreen

18.0 (10.6~24.7); 8.0 (-0.2~19.8); 27.0 (19.4~29.8); 18.0 (10.6~24.7); 156.8 (79.4~236.5); 0.3 (0.0~12.1); 950 (643~1376) ❙ 1505 (868~2993); 598 (312~1847); 150 (30~551) ❙ 60.1 (-34.3~246.5); 940 (543~1239); 228.0 (108.2~659.0)

838
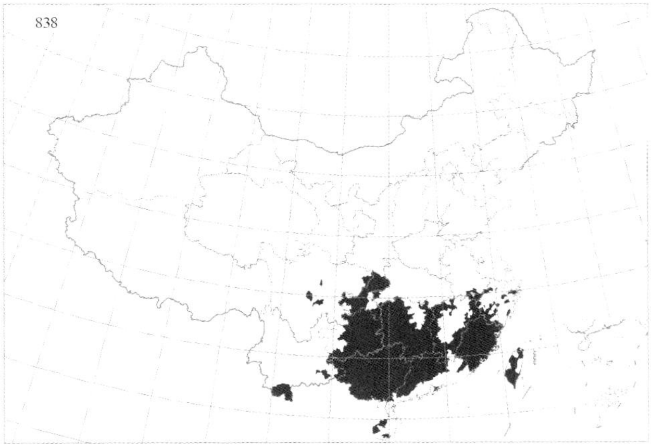

Fagaceae 壳斗科 *Castanopsis* 锥属
Castanopsis carlesii var. spinulosa W. C. Cheng et C. S. Chao 短刺米槠
Trees, evergreen

18.8 (11.1~22.7); 10.6 (1.8~15.1); 25.3 (19.4~28.6); 18.8 (11.1~22.8); 166.3 (79.8~213.1); 0.7 (0.0~6.9); 976 (648~1221) ❙ 1487 (1076~2435); 771 (458~1428); 83 (30~182) ❙ 53.1 (12.9~99.5); 974 (648~1221); 320.0 (126.1~659.0)

839
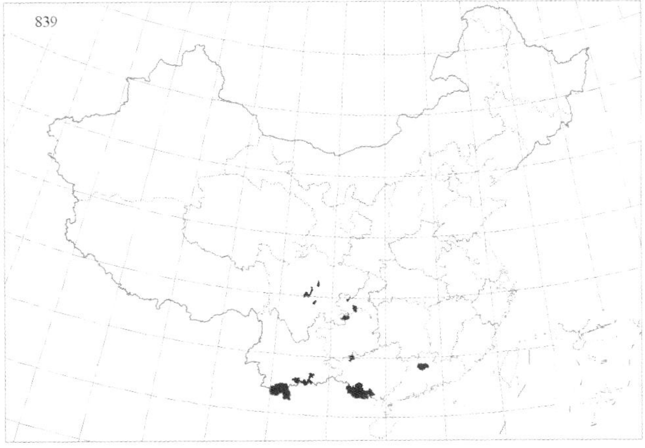

Fagaceae 壳斗科 *Castanopsis* 锥属
Castanopsis ceratacantha Rehder et E. H. Wilson 瓦山锥
Trees, evergreen

16.9 (5.3~22.7); 8.0 (-2.4~15.1); 24.2 (12.2~29.0); 16.9 (5.7~22.8); 144.3 (30.5~213.1); 1.4 (0.0~26.2); 875 (479~1221) ❙ 1353 (961~2435); 676 (448~1428); 76 (23~177) ❙ 56.8 (12.9~112.2); 872 (479~1221); 296.5 (149.6~659.0)

840
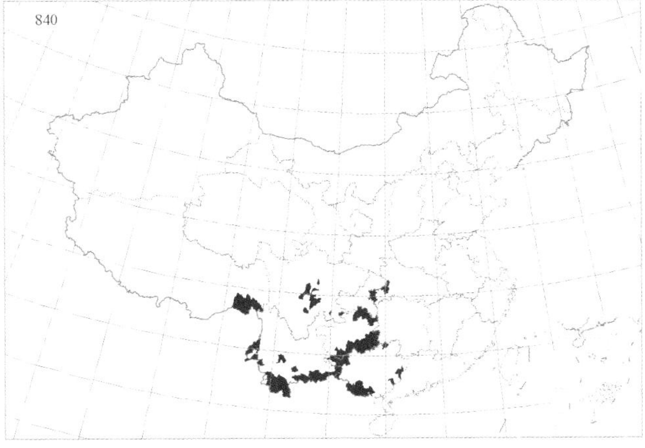

Fagaceae 壳斗科　*Castanopsis* 锥属

Castanopsis cerebrina (Hickel et A. Camus) Barnett 毛叶杯锥

Trees, evergreen

19.4；12.6；24.1；19.4；172.6；0.0；954 ❙ 1445；787；64 ❙ 52.3；954；521.5

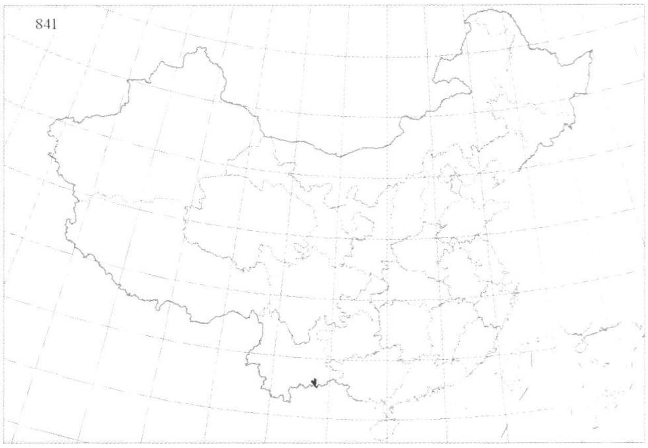

Fagaceae 壳斗科　*Castanopsis* 锥属

Castanopsis chinensis (Spreng.) Hance 锥

Trees, evergreen

19.1 (14.6~23.5)；9.5 (3.8~16.4)；27.3 (21.9~29.8)；19.2 (14.7~23.5)；170.2 (117.7~222.3)；0.0 (0.0~1.5)；1005 (779~1276) ❙ 1538 (1074~2005)；641 (403~891)；147 (43~211) ❙ 54.3 (-6.9~101.8)；999 (779~1264)；230.0 (151.0~354.5)

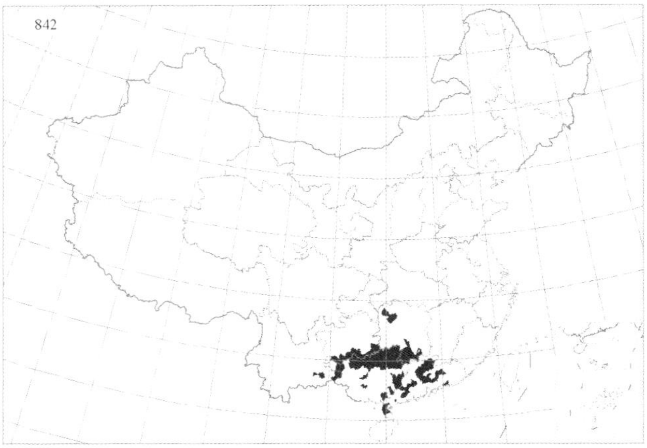

Fagaceae 壳斗科　*Castanopsis* 锥属

Castanopsis choboensis Hickel et A. Camus 窄叶锥

Trees, evergreen

19.9 (18.3~21.6)；10.9 (9.4~13.2)；27.0 (25.5~27.9)；19.9 (18.4~21.6)；179.3 (160.9~199.6)；0.0 (0.0~0.0)；1030 (924~1143) ❙ 1418 (1236~1900)；737 (617~1056)；93 (54~141) ❙ 37.4 (24.6~71.0)；1030 (924~1143)；279.7 (231.5~335.1)

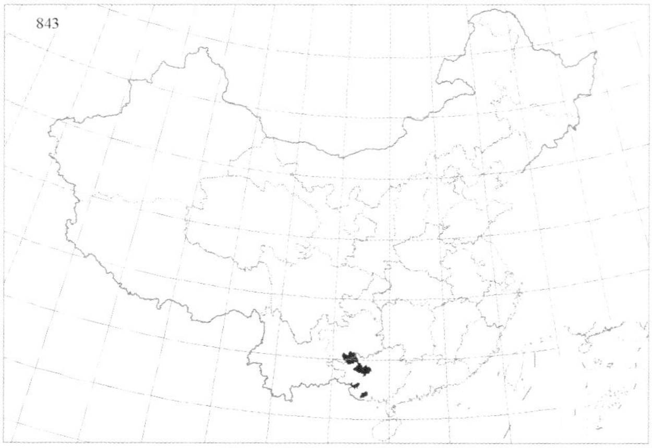

Fagaceae 壳斗科　*Castanopsis* 锥属

Castanopsis chunii W. C. Cheng 厚皮锥

Trees, evergreen

16.9 (11.1~21.6)；6.7 (2.2~14.1)；26.2 (19.4~29.4)；17.0 (11.3~21.6)；143.8 (80.8~199.4)；0.4 (0.0~5.3)；890 (652~1128) ❙ 1418 (999~1991)；557 (435~902)；135 (30~230) ❙ 59.4 (11.7~104.6)；889 (652~1128)；214.2 (128.7~316.1)

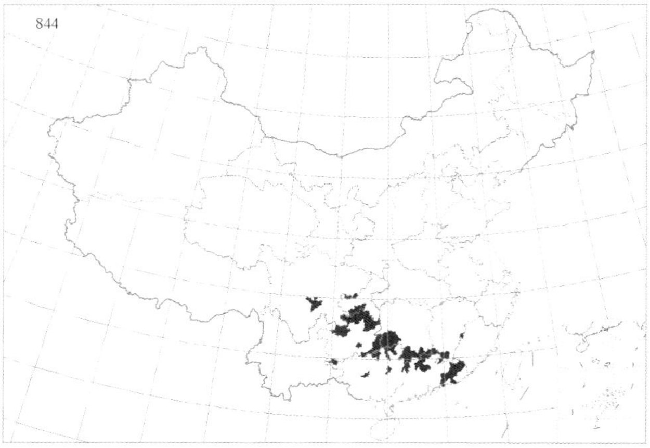

Fagaceae 壳斗科　*Castanopsis* 锥属

Castanopsis clarkei King ex Hook. f. 棱刺锥

Trees, evergreen

18.4；12.3；22.3；18.4；160.7；0.0；899 ❙ 1576；762；63 ❙ 80.2；883；547.6

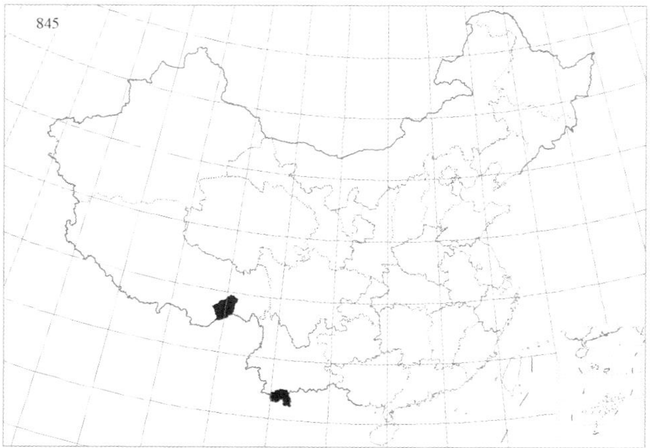

Fagaceae 壳斗科　*Castanopsis* 锥属

Castanopsis concinna (Champ. ex Benth.) A. DC. 华南锥

Trees, evergreen

21.2 (17.9~23.4)；12.8 (7.7~16.1)；27.8 (25.5~29.2)；21.3 (17.9~23.5)；195.3 (155.4~221.7)；0.0 (0.0~0.0)；1120 (918~1273) ❙ 1668 (1250~2435)；819 (507~1428)；120 (58~188) ❙ 49.1 (11.9~99.5)；1120 (918~1273)；255.4 (157.7~327.7)

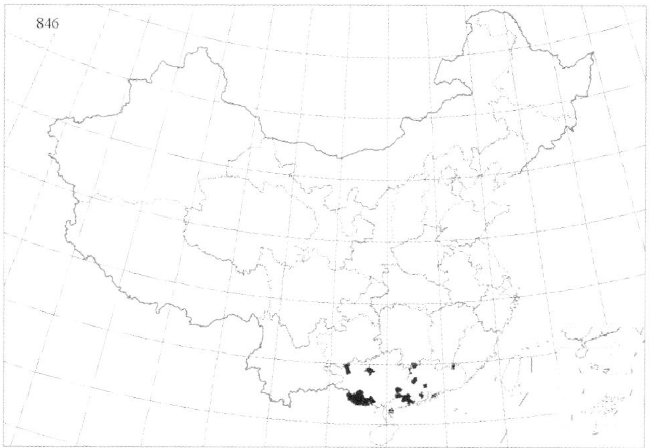

Fagaceae 壳斗科 *Castanopsis* 锥属

Castanopsis crassifolia Hickel et A. Camus 厚叶锥

Trees, evergreen

21.9 (21.3~22.7); 13.8 (13.1~15.0); 28.0 (27.3~28.6); 21.9 (21.3~22.8); 203.3 (195.9~213.1); 0.0 (0.0~0.0); 1161 (1111~1221) ❙ 1806 (1451~2435); 996 (766~1428); 119 (102~130) ❙ 55.4 (27.0~99.5); 1161 (1111~1221); 294.7 (256.6~335.1)

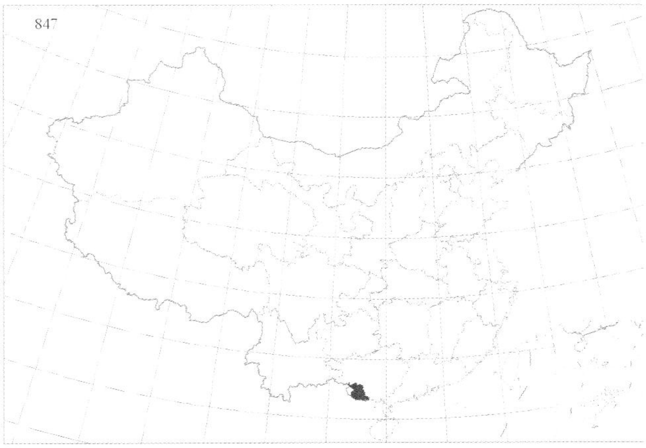

Fagaceae 壳斗科 *Castanopsis* 锥属

Castanopsis damingshanensis S. L. Mo ex C. C. Huang et Y. T. Chang 大明山锥

Trees, evergreen

21.3; 12.2; 28.3; 21.3; 196.0; 0.0; 1131 ❙ 1636; 788; 173 ❙ 44.7; 1131; 219.4

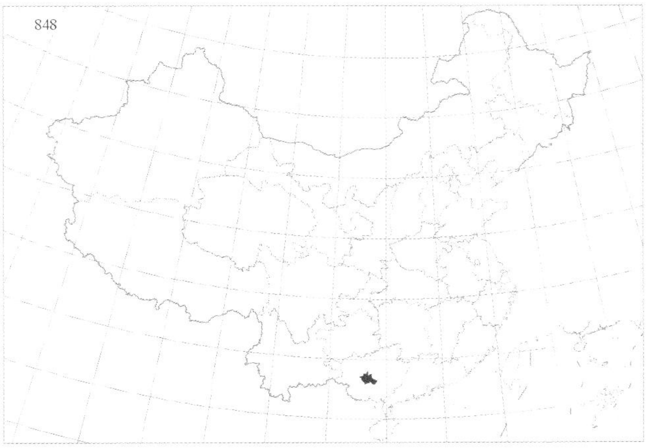

Fagaceae 壳斗科 *Castanopsis* 锥属

Castanopsis delavayi Franch. 高山锥

Trees, evergreen

15.7 (6.1~21.6); 8.3 (-1.5~13.3); 21.2 (12.7~28.5); 15.7 (6.3~21.7); 129.5 (35.2~199.8); 0.9 (0.0~21.8); 796 (496~1153) ❙ 1118 (797~1625); 633 (411~911); 44 (12~173) ❙ 41.8 (-14.2~93.0); 777 (496~1153); 320.8 (189.5~551.3)

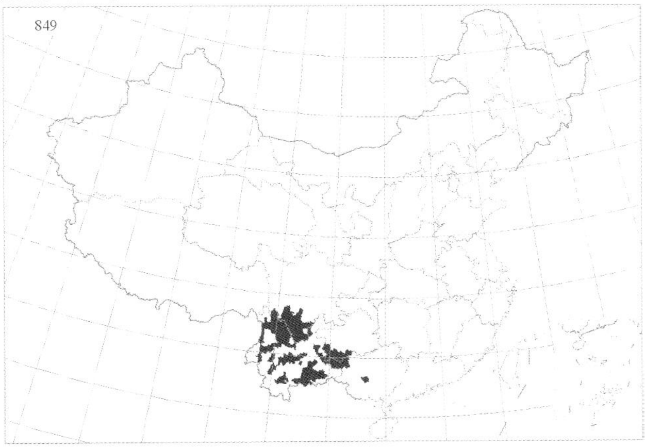

Fagaceae 壳斗科 *Castanopsis* 锥属

Castanopsis densispinosa Y. C. Hsu et H. W. Jen 密刺锥

Trees, evergreen

19.1; 13.0; 23.3; 19.1; 169.7; 0.0; 917 ❙ 1428; 835; 56 ❙ 55.7; 917; 506.2

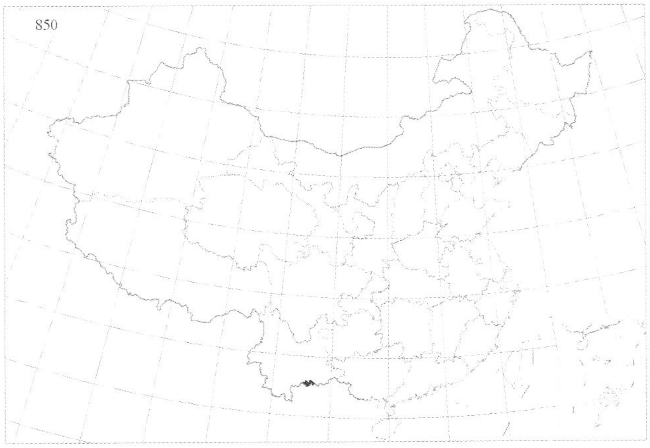

Fagaceae 壳斗科 *Castanopsis* 锥属

Castanopsis echinocarpa Hook. f. et Thomson ex Miq. 短刺锥

Trees, evergreen

17.4 (11.3~21.0); 10.8 (4.3~15.1); 22.0 (17.0~25.6); 17.4 (11.3~21.0); 149.4 (76.6~191.8); 0.0 (0.0~0.7); 855 (644~1053) ❙ 1349 (956~1736); 738 (531~1007); 54 (20~77) ❙ 59.5 (5.5~145.8); 838 (644~1053); 444.7 (311.9~659.0)

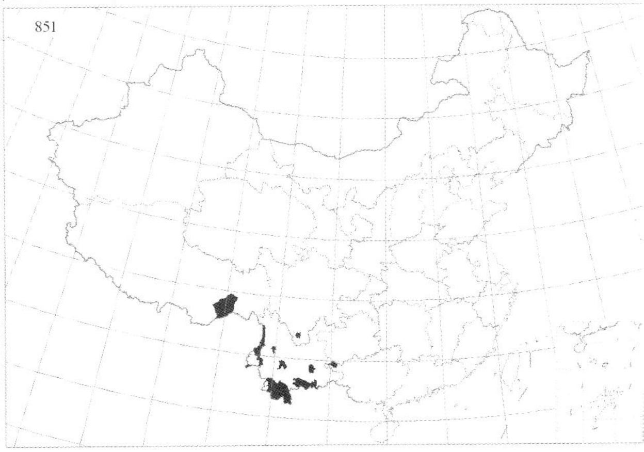

Fagaceae 壳斗科 *Castanopsis* 锥属

Castanopsis eyrei (Champ. ex Benth.) Tutcher 甜槠

Trees, evergreen

17.4 (9.3~23.0); 6.8 (-1.7~15.2); 27.0 (18.8~29.8); 17.4 (9.5~23.0); 149.8 (69.8~216.2); 0.8 (0.0~18.0); 926 (603~1247) ❙ 1444 (607~2680); 570 (349~1428); 147 (31~259) ❙ 57.0 (-7.5~246.5); 919 (603~1247); 220.4 (111.9~411.1)

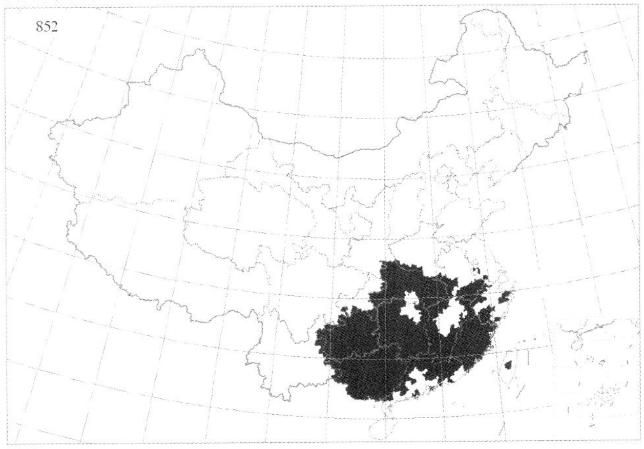

Fagaceae 壳斗科 *Castanopsis* 锥属
Castanopsis fabri Hance 罗浮锥
Trees, evergreen

18.9 (12.5~24.0); 9.6 (3.1~19.1); 27.1 (20.1~29.8); 19.0 (12.5~24.0); 167.9 (92.0~228.1); 0.1 (0.0~2.3); 988 (685~1302) ❙ 1583 (1070~2992); 642 (356~1847); 162 (43~303) ❙ 62.3 (−5.0~219.3); 984 (685~1247); 252.0 (123.9~620.0)

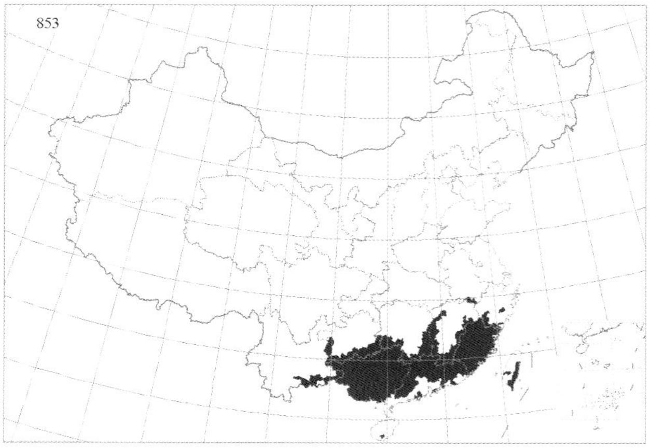

Fagaceae 壳斗科 *Castanopsis* 锥属
Castanopsis fargesii Franch. 栲
Trees, evergreen

17.5 (7.3~24.3); 7.4 (−2.0~18.7); 26.8 (15.6~29.9); 17.6 (7.5~24.3); 151.6 (48.2~231.9); 0.5 (0.0~20.4); 928 (537~1338) ❙ 1453 (862~2680); 581 (299~1428); 142 (21~259) ❙ 58.1 (−34.3~246.5); 919 (537~1247); 226.0 (108.2~551.6)

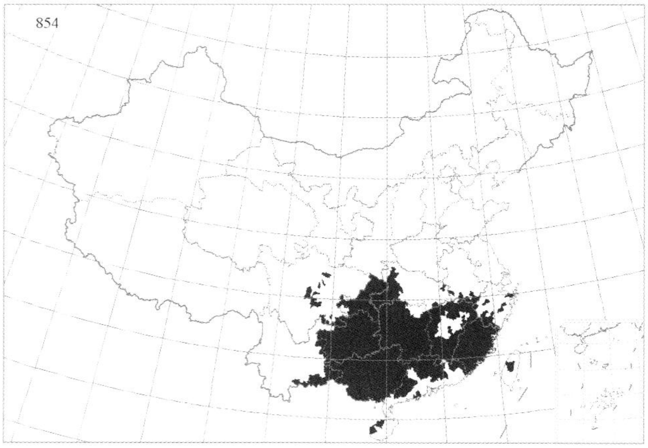

Fagaceae 壳斗科 *Castanopsis* 锥属
Castanopsis ferox (Roxb.) Spach 思茅锥
Trees, evergreen

18.0 (13.3~20.5); 11.6 (5.4~15.1); 22.0 (19.1~23.7); 18.0 (13.4~20.5); 156.0 (100.4~185.9); 0.0 (0.0~0.0); 870 (706~992) ❙ 1472 (1049~1736); 806 (598~1007); 56 (42~77) ❙ 71.0 (26.1~145.8); 858 (706~986); 490.7 (365.3~659.0)

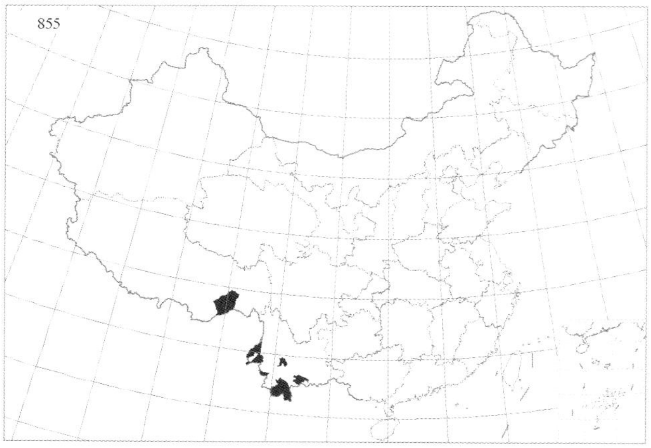

Fagaceae 壳斗科 *Castanopsis* 锥属
Castanopsis fissa (Champ. ex Benth.) Rehder et E. H. Wilson 黧蒴锥
Trees, evergreen

19.6 (14.3~24.8); 10.4 (3.3~20.1); 27.3 (21.6~29.8); 19.7 (14.3~24.8); 176.0 (113.2~237.6); 0.0 (0.0~2.3); 1030 (761~1385) ❙ 1570 (868~2435); 650 (312~1428); 150 (41~225) ❙ 54.6 (−34.3~128.9); 1018 (761~1360); 261.6 (160.7~620.0)

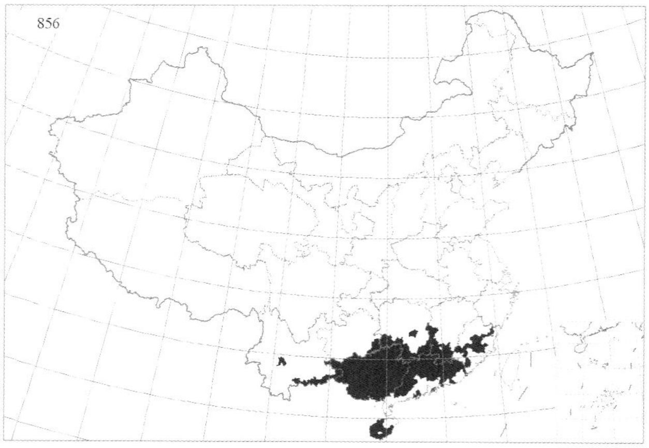

Fagaceae 壳斗科 *Castanopsis* 锥属
Castanopsis fleuryi Hickel et A. Camus 小果锥
Trees, evergreen

18.3 (16.1~20.5); 12.3 (9.6~15.1); 22.2 (20.3~23.7); 18.4 (16.2~20.5); 160.7 (133.9~185.9); 0.0 (0.0~0.0); 880 (785~992) ❙ 1372 (1049~1607); 749 (598~911); 54 (42~77) ❙ 56.0 (21.7~78.9); 865 (779~986); 464.4 (346.9~659.0)

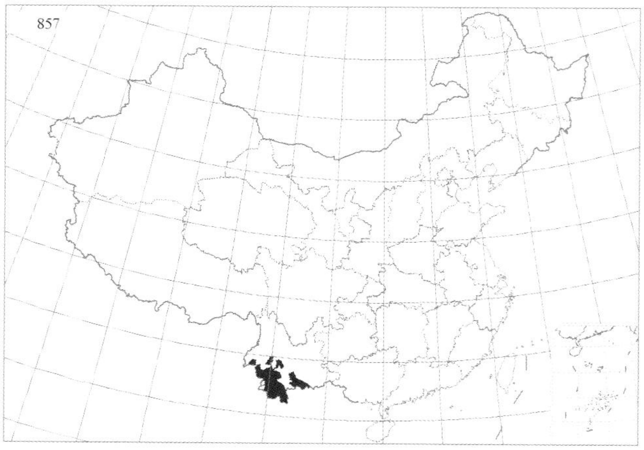

Fagaceae 壳斗科 *Castanopsis* 锥属
Castanopsis fordii Hance 毛锥
Trees, evergreen

18.7 (14.3~23.2); 9.0 (3.1~15.8); 27.3 (24.2~30.1); 18.7 (14.3~23.2); 164.5 (113.2~218.9); 0.1 (0.0~2.2); 975 (761~1257) ❙ 1627 (1070~2024); 601 (356~902); 184 (60~259) ❙ 68.9 (−5.0~147.0); 968 (761~1257); 243.0 (123.9~332.5)

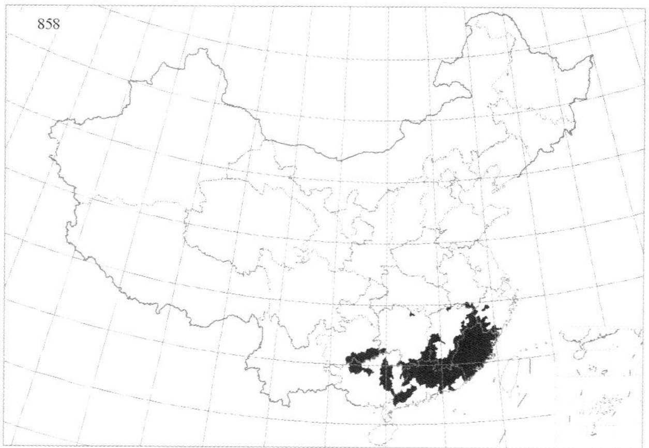

Fagaceae 壳斗科 *Castanopsis* 锥属

Castanopsis globigemmata Chun et C. C. Huang 圆芽锥

Trees, evergreen

19.3; 12.9; 23.7; 19.3; 171.9; 0.0; 939 ❙ 1398; 788; 59 ❙ 49.3; 937; 500.3

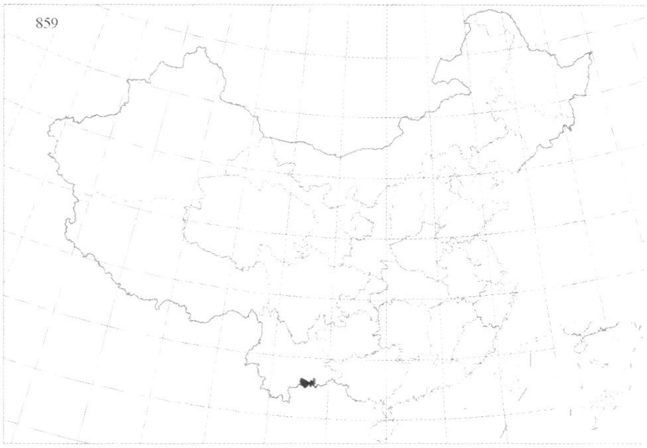

Fagaceae 壳斗科 *Castanopsis* 锥属

Castanopsis hainanensis Merr. 海南锥

Trees, evergreen

23.7 (22.2~24.8); 18.0 (13.9~20.1); 27.7 (26.7~28.4); 23.7 (22.3~24.8); 224.8 (207.5~237.6); 0.0 (0.0~0.0); 1288 (1190~1385) ❙ 1436 (1179~1725); 530 (392~713); 56 (41~87) ❙ 12.1 (-11.9~41.8); 1134 (1056~1190); 493.1 (327.7~620.0)

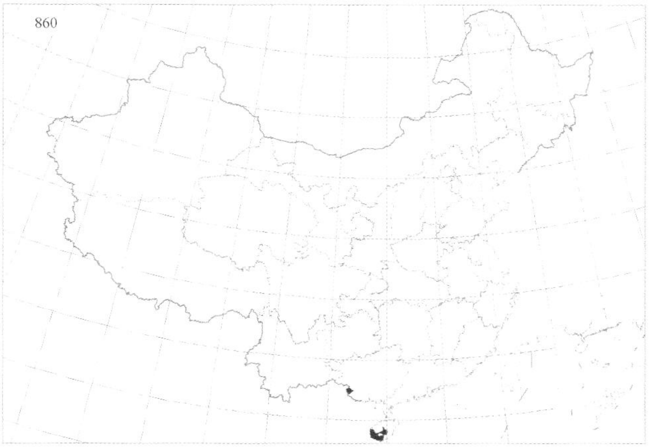

Fagaceae 壳斗科 *Castanopsis* 锥属

Castanopsis hupehensis C. S. Chao 湖北锥

Trees, evergreen

15.6 (12.7~19.3); 5.3 (2.1~13.1); 25.2 (22.5~26.7); 15.7 (12.8~19.4); 129.0 (99.0~173.2); 0.9 (0.0~5.6); 822 (713~939) ❙ 1276 (1132~1547); 539 (448~911); 89 (48~145) ❙ 55.6 (21.7~107.2); 821 (713~903); 229.2 (152.4~551.3)

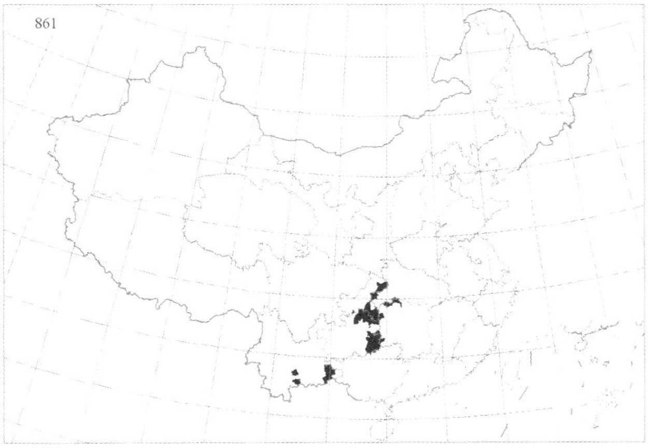

Fagaceae 壳斗科 *Castanopsis* 锥属

Castanopsis hystrix Hook. f. et Thomson ex A. DC. 红锥

Trees, evergreen

19.4 (13.3~24.3); 10.7 (3.3~18.7); 26.6 (19.1~29.8); 19.5 (13.4~24.3); 173.6 (100.4~231.9); 0.0 (0.0~2.3); 1008 (706~1338) ❙ 1548 (868~2992); 695 (312~1847); 132 (42~341) ❙ 55.7 (-34.3~245.5); 1000 (706~1257); 271.0 (130.1~659.0)

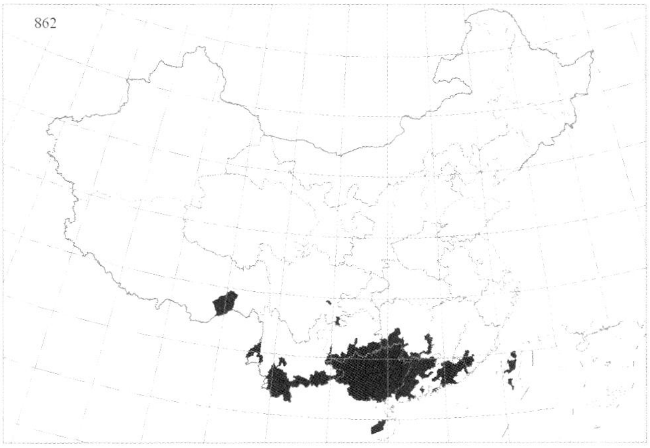

Fagaceae 壳斗科 *Castanopsis* 锥属

Castanopsis indica (Roxb. ex Lindl.) A. DC. 印度锥

Trees, evergreen

20.5 (13.3~24.8); 13.8 (5.4~20.1); 25.4 (19.6~28.7); 20.6 (13.4~24.8); 187.1 (100.4~237.6); 0.0 (0.0~0.0); 1056 (706~1385) ❙ 1551 (868~2992); 782 (312~1847); 77 (41~188) ❙ 50.9 (-34.3~214.4); 1013 (706~1299); 411.2 (226.9~659.0)

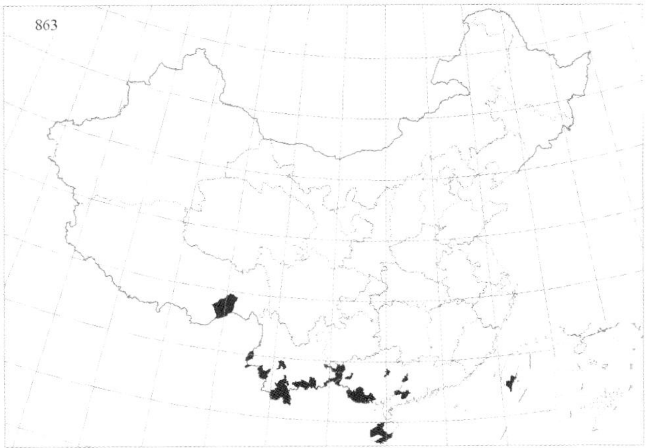

Fagaceae 壳斗科 *Castanopsis* 锥属

Castanopsis jianfenglingensis Duanmu 尖峰岭锥

Trees, evergreen

24.3; 18.7; 28.4; 24.3; 231.9; 0.0; 1338 ❙ 1179; 392; 42 ❙ -11.9; 1056; 455.2

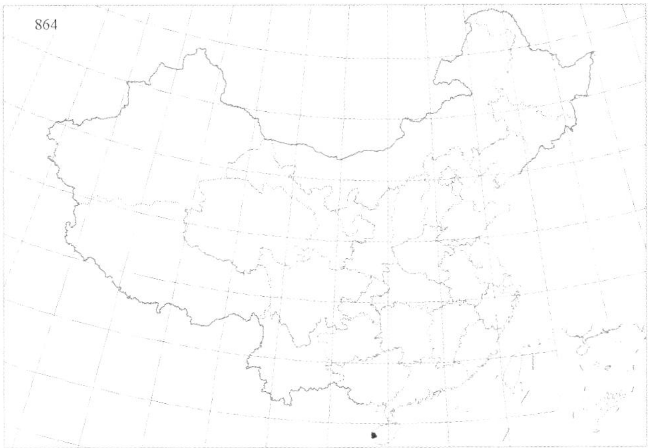

Fagaceae 壳斗科 *Castanopsis* 锥属
Castanopsis jucunda Hance 秀丽锥
Trees, evergreen

18.3 (14.3~24.8); 8.2 (2.6~20.1); 27.5 (21.7~30.1); 18.4 (14.3~24.8); 160.7 (113.2~237.6); 0.2 (0.0~3.5); 975 (761~1385) ❙ 1601 (868~2992); 593 (299~1847); 173 (41~402) ❙ 67.1 (-34.3~235.6); 952 (755~1360); 252.0 (142.8~620.0)

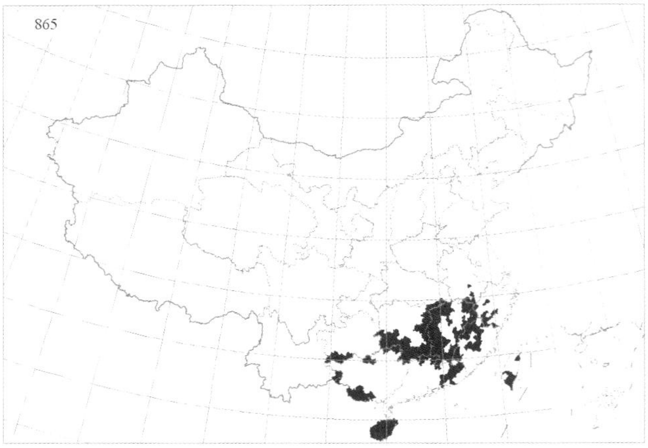

Fagaceae 壳斗科 *Castanopsis* 锥属
Castanopsis kawakamii Hayata 吊皮锥
Trees, evergreen

19.1 (14.3~22.8); 9.9 (3.7~15.3); 27.2 (20.1~29.8); 19.2 (14.3~22.8); 170.7 (113.2~213.6); 0.0 (0.0~1.4); 1000 (761~1230) ❙ 1683 (1284~2859); 661 (488~1412); 174 (73~402) ❙ 71.1 (12.9~246.5); 998 (761~1230); 248.2 (174.9~327.7)

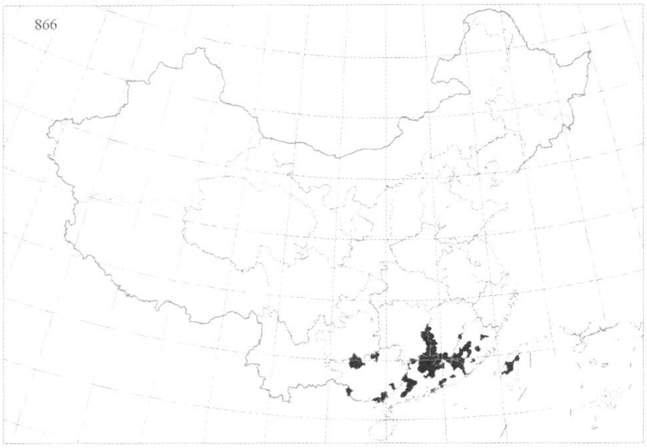

Fagaceae 壳斗科 *Castanopsis* 锥属
Castanopsis kweichowensis Hu 贵州锥
Trees, evergreen

16.5 (14.3~19.2); 6.3 (4.0~9.9); 25.5 (23.5~27.5); 16.6 (14.3~19.2); 139.0 (112.6~170.8); 0.1 (0.0~1.1); 859 (756~1002) ❙ 1271 (1072~1496); 561 (476~683); 96 (64~142) ❙ 48.2 (30.1~71.9); 859 (756~1002); 227.6 (162.8~322.1)

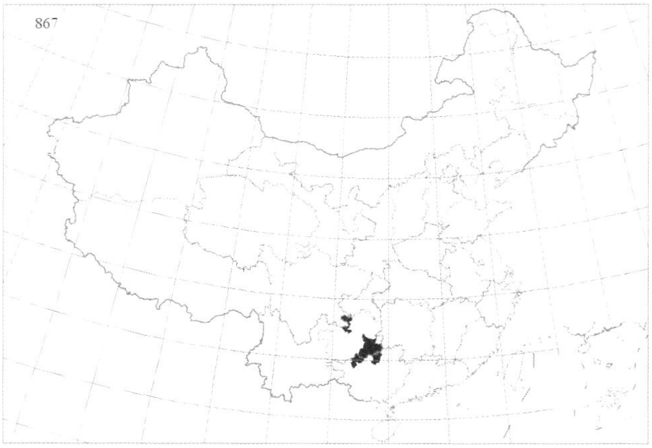

Fagaceae 壳斗科 *Castanopsis* 锥属
Castanopsis lamontii Hance 鹿角锥
Trees, evergreen

19.3 (14.3~23.0); 9.9 (3.3~15.2); 27.1 (21.4~29.8); 19.3 (14.3~23.0); 171.8 (112.6~216.2); 0.0 (0.0~2.3); 1006 (756~1247) ❙ 1574 (1172~2148); 662 (411~1234); 152 (43~254) ❙ 58.0 (5.8~130.7); 1004 (756~1247); 250.1 (160.7~551.6)

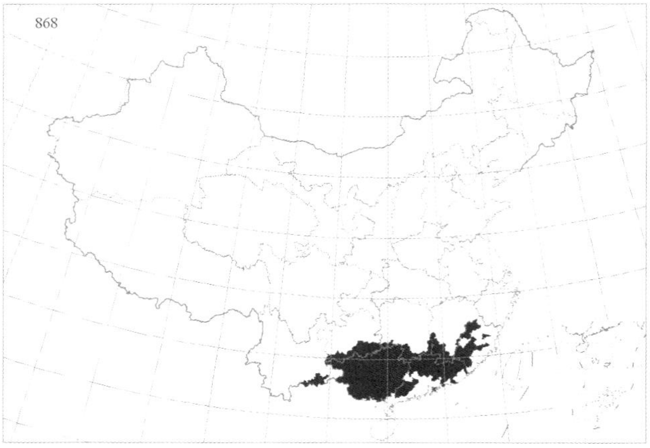

Fagaceae 壳斗科 *Castanopsis* 锥属
Castanopsis ledongensis C. C. Huang et Y. T. Chang 乐东锥
Trees, evergreen

23.9; 18.9; 27.4; 24.0; 227.5; 0.0; 1298 ❙ 1466; 497; 48 ❙ 12.9; 1130; 537.6

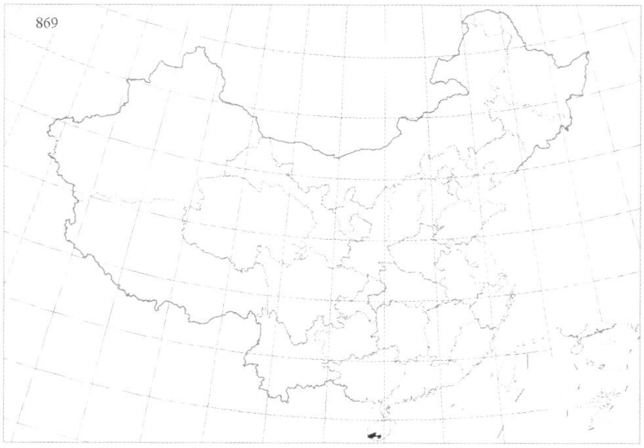

Fagaceae 壳斗科 *Castanopsis* 锥属
Castanopsis longispina (King ex Hook. f.) C. C. Huang et Y. T. Chang 长刺锥
Trees, evergreen

13.3; 5.4; 19.6; 13.4; 100.4; 0.0; 706 ❙ 1736; 1007; 51 ❙ 145.8; 706; 447.4

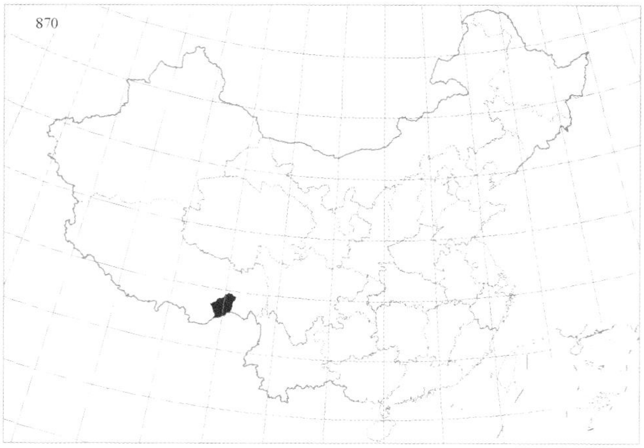

Fagaceae 壳斗科 *Castanopsis* 锥属
Castanopsis longzhouica C. C. Huang et Y. T. Chang 龙州锥
Trees, evergreen

22.1; 13.6; 28.4; 22.1; 205.6; 0.0; 1183 | 1446; 743; 112 | 22.2; 1183; 300.0

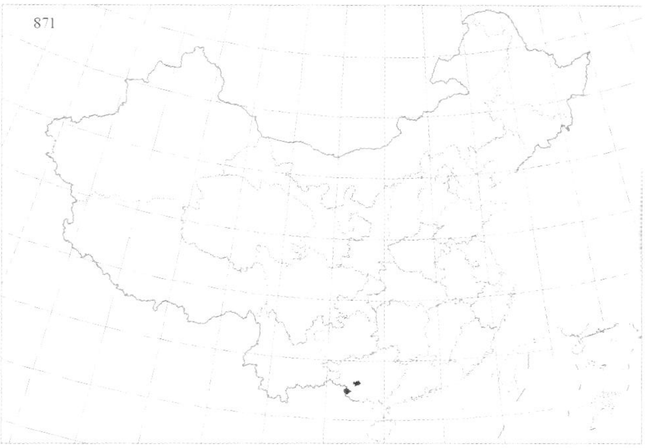

Fagaceae 壳斗科 *Castanopsis* 锥属
Castanopsis megaphylla Hu 大叶锥
Trees, evergreen

19.6; 13.3; 23.8; 19.6; 175.2; 0.0; 953 | 1481; 778; 65 | 55.7; 953; 552.1

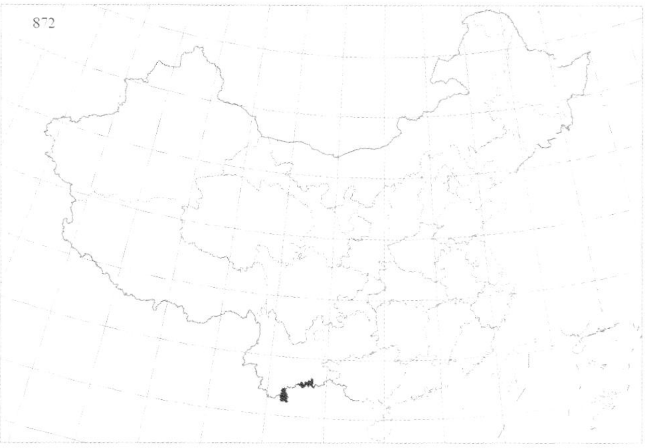

Fagaceae 壳斗科 *Castanopsis* 锥属
Castanopsis mekongensis A. Camus 湄公锥
Trees, evergreen

19.5 (17.7~21.0); 13.6 (10.9~15.1); 23.2 (21.6~25.6); 19.5 (17.8~21.0); 174.3 (153.4~191.8); 0.0 (0.0~0.0); 943 (854~1053) | 1493 (1368~1607); 734 (655~818); 64 (52~77) | 59.0 (44.5~74.2); 923 (839~1053); 537.6 (439.5~659.0)

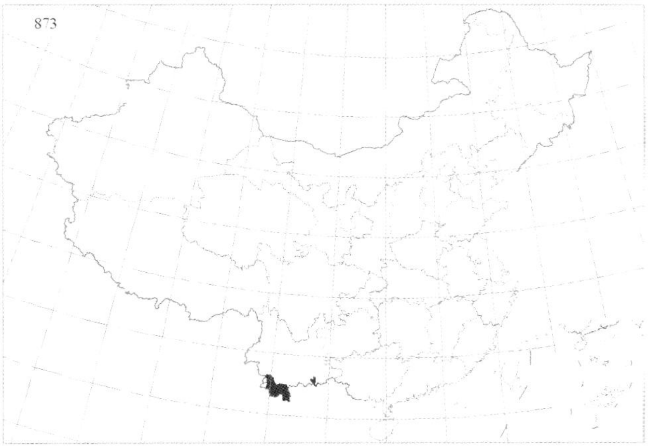

Fagaceae 壳斗科 *Castanopsis* 锥属
Castanopsis nigrescens Chun et C. C. Huang 黑叶锥
Trees, evergreen

18.5 (14.3~22.0); 8.5 (3.3~13.5); 27.4 (24.2~29.7); 18.5 (14.3~22.0); 162.2 (113.2~203.8); 0.1 (0.0~2.3); 969 (761~1170) | 1728 (1371~2005); 611 (513~827); 202 (132~258) | 80.5 (17.4~130.7); 963 (761~1170); 247.2 (147.5~305.7)

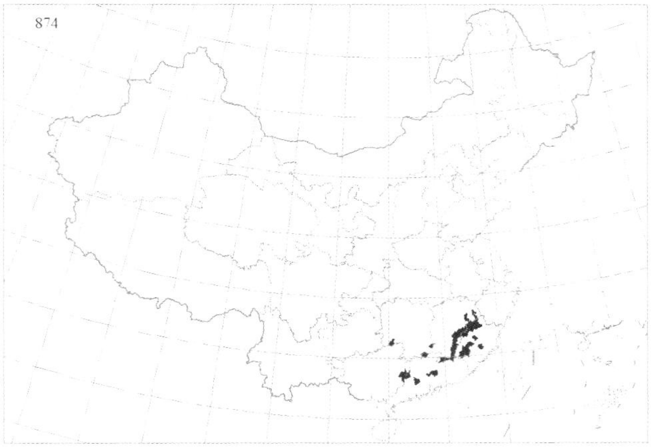

Fagaceae 壳斗科 *Castanopsis* 锥属
Castanopsis oblonga Y. C. Hsu et H. W. Jen 矩叶锥
Trees, evergreen

21.1 (18.6~22.2); 13.5 (12.7~13.9); 26.7 (22.5~28.5); 21.2 (18.7~22.3); 194.3 (163.9~207.5); 0.0 (0.0~0.0); 1098 (888~1190) | 1497 (1143~1900); 809 (637~1056); 95 (50~128) | 37.3 (12.9~74.2); 1094 (888~1190); 329.8 (256.6~551.3)

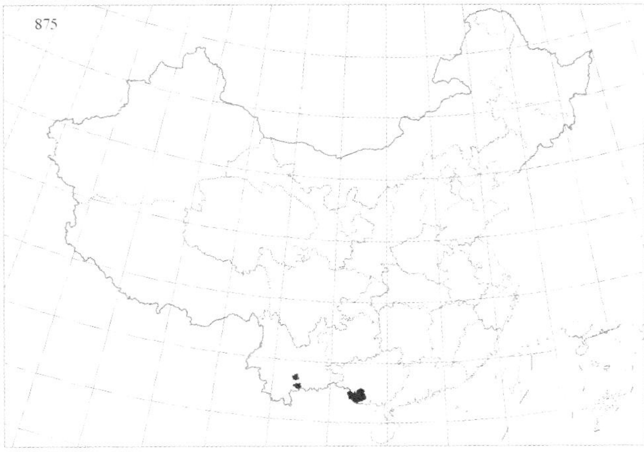

Fagaceae 壳斗科 *Castanopsis* 锥属
Castanopsis orthacantha Franch. 元江锥
Trees, evergreen

15.6 (5.4~21.0); 8.4 (-2.1~14.2); 21.1 (12.3~26.9); 15.7 (5.8~21.0); 128.9 (31.5~191.8); 0.7 (0.0~25.6); 787 (481~1053) | 1098 (741~1751); 619 (411~996); 43 (15~77) | 39.9 (-14.2~99.9); 769 (481~1053); 317.4 (142.9~551.6)

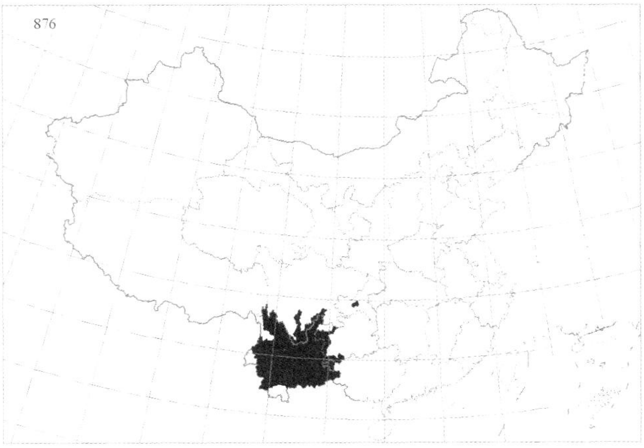

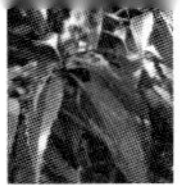

Fagaceae 壳斗科 *Castanopsis* 锥属
Castanopsis ouonbiensis Hickel et A. Camus 屏边锥
Trees, evergreen

17.7；10.4；22.9；17.7；152.9；0.0；857 ❙ 1395；785；56 ❙ 62.8；857；425.6

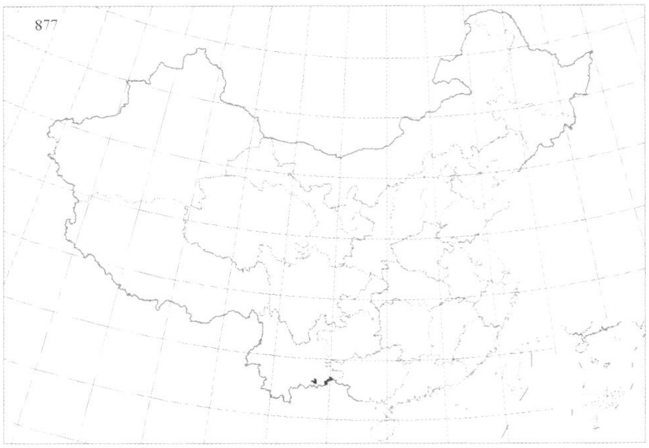

Fagaceae 壳斗科 *Castanopsis* 锥属
Castanopsis platyacantha Rehder et E. H. Wilson 扁刺锥
Trees, evergreen

16.6 (11.1~21.3)；7.3 (2.2~14.2)；24.4 (19.4~28.7)；16.7 (11.3~21.3)；140.6 (80.8~196.0)；0.4 (0.0~5.3)；863 (652~1123) ❙ 1236 (879~1620)；622 (453~911)；72 (33~166) ❙ 43.9 (5.8~87.2)；863 (652~1121)；256.2 (128.7~551.6)

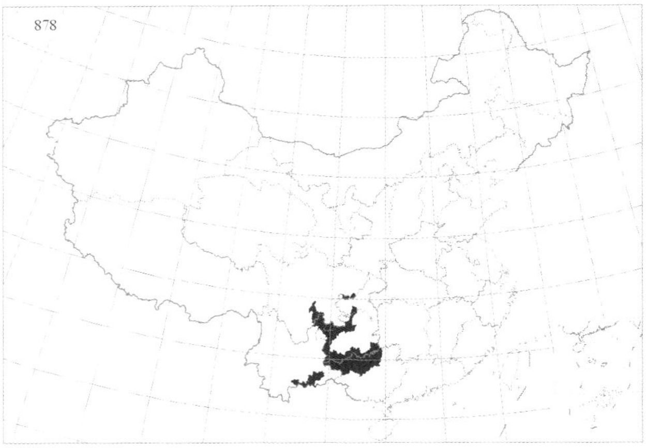

Fagaceae 壳斗科 *Castanopsis* 锥属
Castanopsis remotidenticulata Hu 疏齿锥
Trees, evergreen

18.2 (16.3~19.3)；11.7 (9.1~13.1)；22.6 (21.4~23.6)；18.3 (16.4~19.4)；159.3 (136.4~173.2)；0.0 (0.0~0.0)；879 (797~939) ❙ 1311 (1028~1547)；748 (566~911)；54 (45~63) ❙ 49.9 (12.5~80.6)；865 (797~917)；419.7 (339.5~551.3)

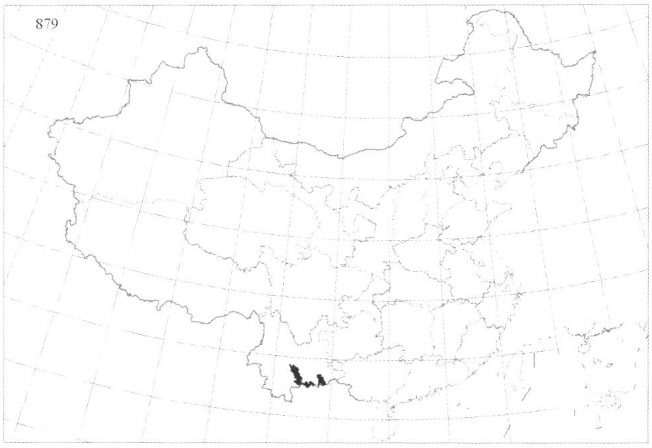

Fagaceae 壳斗科 *Castanopsis* 锥属
Castanopsis rockii A. Camus 龙陵锥
Trees, evergreen

16.4 (13.3~19.1)；9.8 (5.4~13.0)；21.0 (19.1~23.3)；16.5 (13.4~19.1)；137.4 (100.4~169.7)；0.0 (0.0~0.0)；806 (706~917) ❙ 1477 (1348~1736)；846 (762~1007)；57 (47~70) ❙ 85.7 (55.7~145.8)；803 (706~917)；471.1 (395.4~551.3)

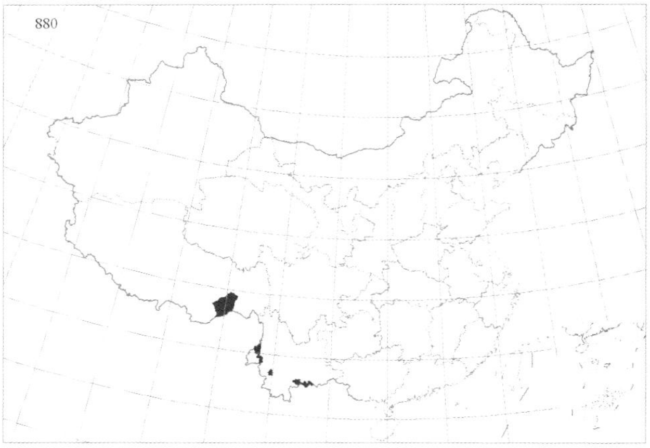

Fagaceae 壳斗科 *Castanopsis* 锥属
Castanopsis rufotomentosa Hu 红壳锥
Trees, evergreen

17.2；9.5；22.7；17.3；147.0；0.0；838 ❙ 1254；699；48 ❙ 49.7；838；336.3

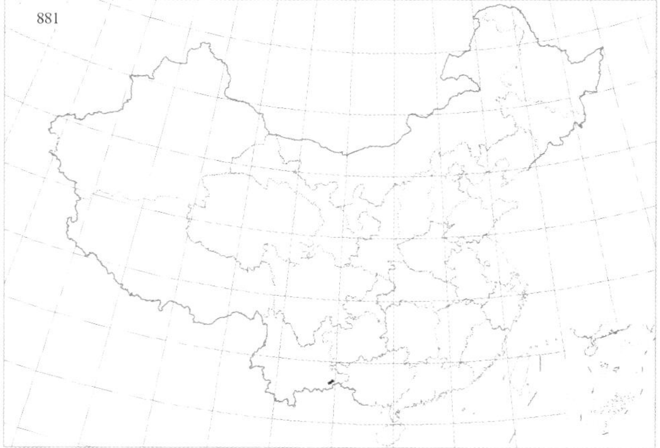

Fagaceae 壳斗科 *Castanopsis* 锥属
Castanopsis sclerophylla (Lindl.) Schottky 苦槠
Trees, evergreen

16.7 (9.3~21.8)；5.3 (−1.7~15.2)；27.6 (20.1~29.9)；16.7 (9.5~22.7)；141.7 (69.8~212.7)；0.8 (0.0~18.0)；900 (603~1216) ❙ 1452 (871~1997)；528 (299~994)；166 (49~259) ❙ 62.0 (−6.7~147.0)；891 (603~1216)；206.6 (108.2~371.7)

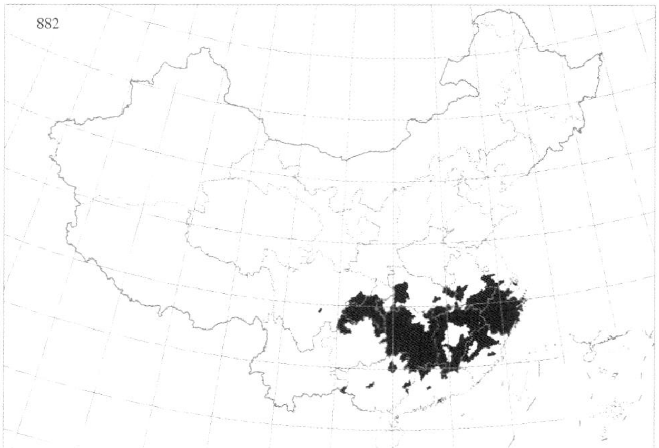

Fagaceae 壳斗科 *Castanopsis* 锥属
Castanopsis subuliformis Chun et C. C. Huang 钻刺锥
Trees, evergreen

21.8; 12.8; 28.7; 21.8; 201.6; 0.0; 1164 ❙ 1454; 654; 139 ❙ 25.1; 1164; 214.2

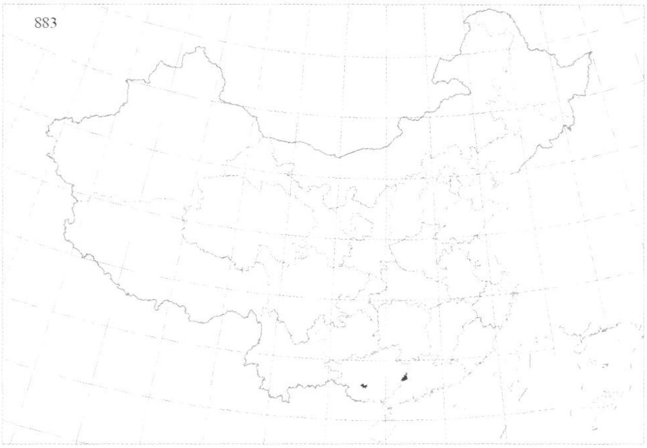

Fagaceae 壳斗科 *Castanopsis* 锥属
Castanopsis tcheponensis Hickel et A. Camus 薄叶锥
Trees, evergreen

19.1; 12.9; 23.1; 19.1; 169.7; 0.0; 923 ❙ 1416; 706; 63 ❙ 53.9; 907; 530.8

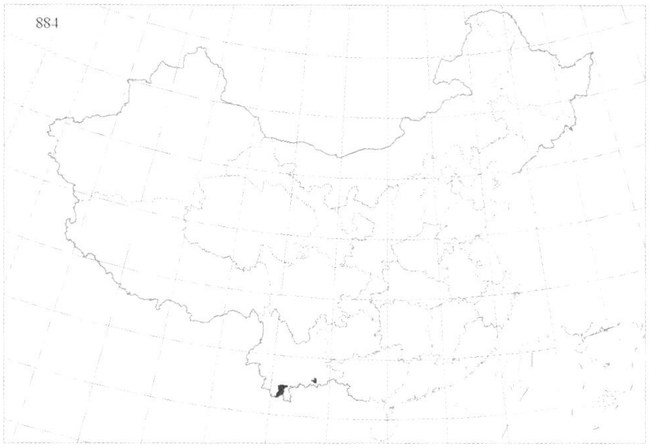

Fagaceae 壳斗科 *Castanopsis* 锥属
Castanopsis tessellata Hickel et A. Camus 棕毛锥
Trees, evergreen

19.4; 12.6; 24.1; 19.4; 172.6; 0.0; 954 ❙ 1445; 787; 64 ❙ 52.3; 954; 521.5

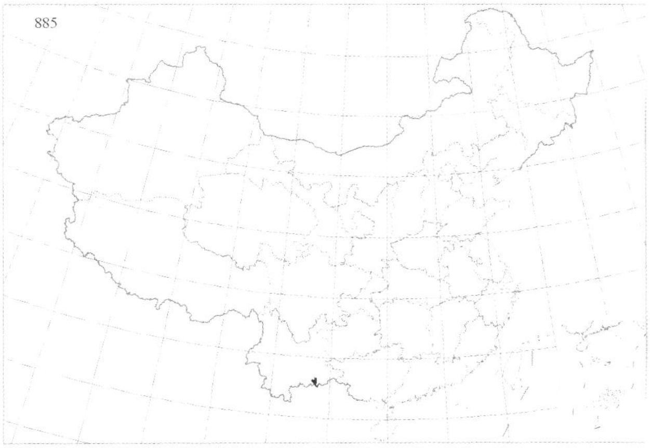

Fagaceae 壳斗科 *Castanopsis* 锥属
Castanopsis tibetana Hance 钩锥
Trees, evergreen

17.4 (12.1~21.8); 6.9 (1.1~14.1); 27.1 (22.9~29.9); 17.4 (12.2~21.8); 149.6 (95.0~202.0); 0.4 (0.0~8.5); 919 (696~1168) ❙ 1497 (931~1997); 555 (323~902); 163 (43~259) ❙ 63.8 (-5.0~147.0); 911 (696~1139); 225.8 (123.9~411.1)

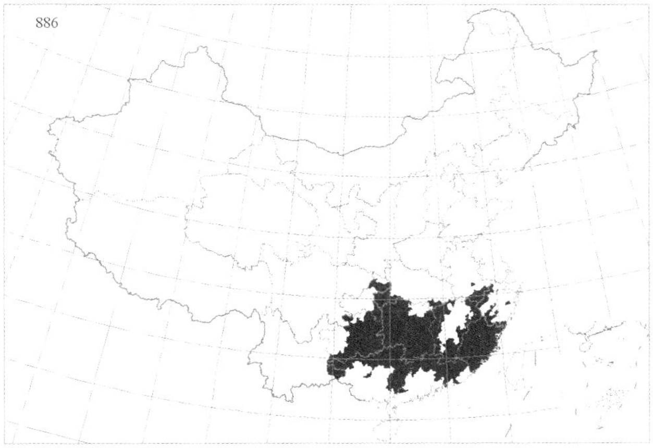

Fagaceae 壳斗科 *Castanopsis* 锥属
Castanopsis tonkinensis Seemen 公孙锥
Trees, evergreen

22.5 (16.3~24.8); 16.0 (8.5~20.1); 27.3 (21.4~28.9); 22.6 (16.4~24.8); 210.7 (136.4~237.6); 0.0 (0.0~0.0); 1204 (797~1385) ❙ 1534 (868~1900); 645 (312~1056); 86 (41~169) ❙ 29.9 (-34.3~74.2); 1136 (797~1360); 391.2 (194.5~620.0)

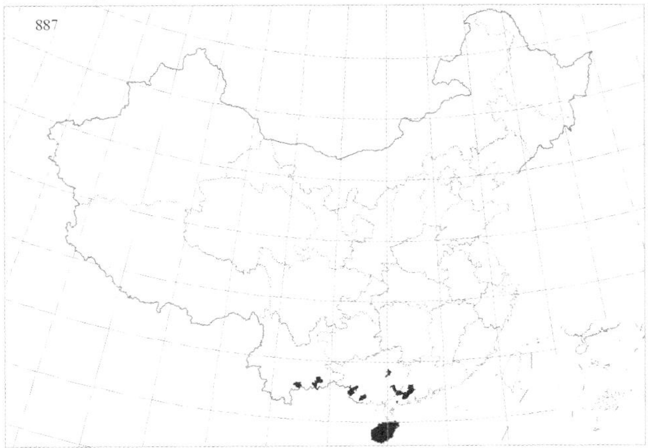

Fagaceae 壳斗科 *Castanopsis* 锥属
Castanopsis tribuloides (Sm.) A. DC. 蒺藜锥
Trees, evergreen

19.2 (13.3~22.2); 11.8 (5.4~15.2); 24.9 (19.6~29.8); 19.3 (13.4~22.7); 172.0 (100.4~212.7); 0.0 (0.0~0.0); 978 (706~1216) ❙ 1581 (1044~1965); 775 (508~1007); 97 (42~194) ❙ 63.5 (26.1~145.8); 969 (706~1216); 381.0 (174.9~659.0)

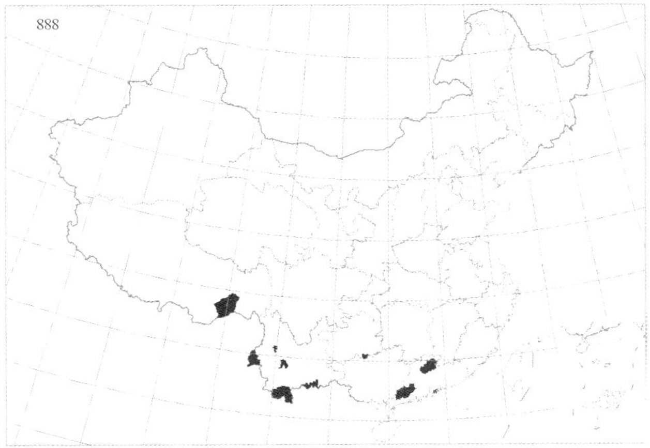

Fagaceae 壳斗科 *Castanopsis* 锥属
Castanopsis undulatifolia G. A. Fu 波叶锥
Trees, evergreen

24.1；18.0；28.5；24.2；230.2；0.0；1329 ❙ 1813；712；113 ❙ 36.4；1329；248.2

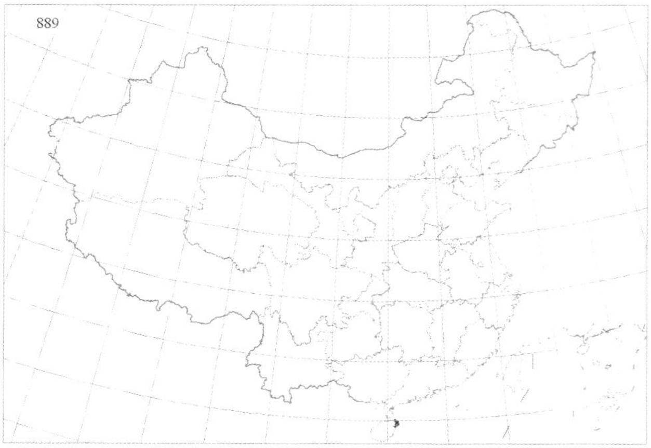

Fagaceae 壳斗科 *Castanopsis* 锥属
Castanopsis uraiana (Hayata) Kanehira et Hatusima 淋漓锥
Trees, evergreen

18.1 (16.2~22.2)；9.4 (5.0~17.7)；26.0 (20.1~29.0)；18.2 (16.2~22.2)；158.3 (134.2~206.7)；0.0 (0.0~0.0)；926 (773~1123) ❙ 1895 (1381~2993)；743 (469~1847)；207 (107~547) ❙ 105.3 (55.9~246.5)；922 (773~1123)；260.5 (188.8~363.7)

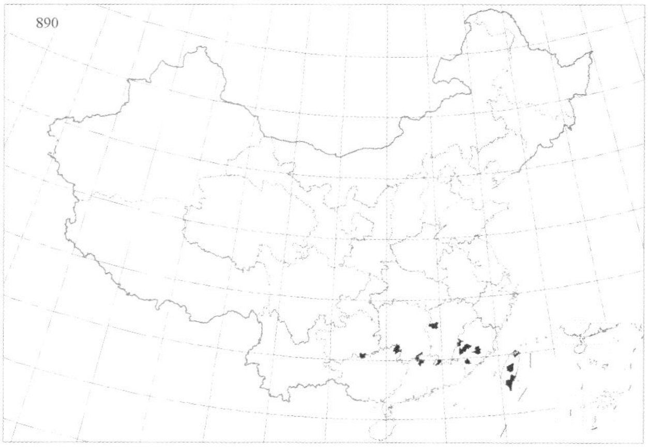

Fagaceae 壳斗科 *Castanopsis* 锥属
Castanopsis wattii (King ex Hook. f.) A. Camus 变色锥
Trees, evergreen

17.0 (13.3~19.3)；10.5 (5.4~13.8)；21.4 (19.1~23.6)；17.1 (13.4~19.4)；144.8 (100.4~173.2)；0.0 (0.0~0.0)；832 (706~939) ❙ 1370 (1049~1736)；746 (598~1007)；57 (44~70) ❙ 67.2 (21.7~145.8)；817 (706~917)；425.0 (335.1~513.9)

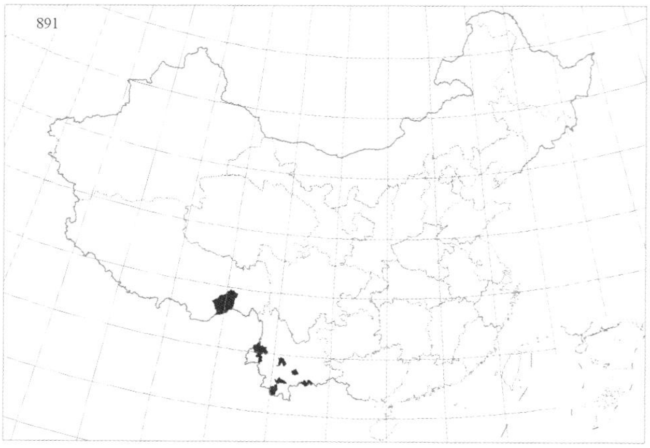

Fagaceae 壳斗科 *Castanopsis* 锥属
Castanopsis wenchangensis G. A. Fu et C. C. Huang 文昌锥
Trees, evergreen

24.1；18.0；28.5；24.2；230.2；0.0；1329 ❙ 1813；712；113 ❙ 36.4；1329；248.2

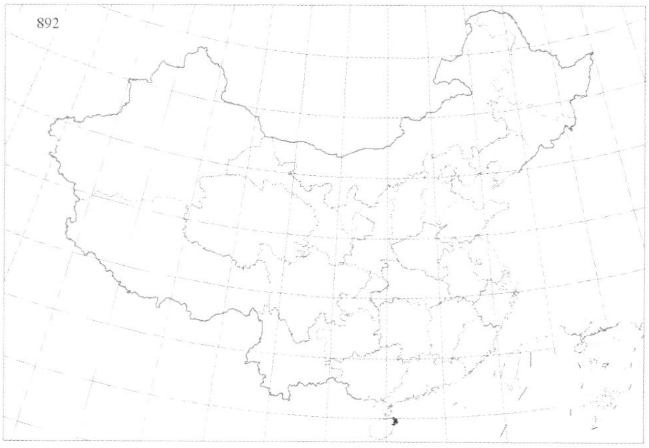

Fagaceae 壳斗科 *Castanopsis* 锥属
Castanopsis xichouensis C. C. Huang et Y. T. Chang 西畴锥
Trees, evergreen

17.2；9.5；22.7；17.3；147.0；0.0；838 ❙ 1254；699；48 ❙ 49.7；838；336.3

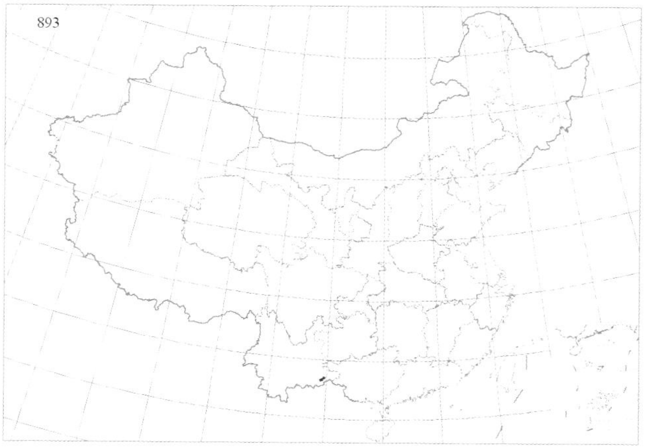

Fagaceae 壳斗科 *Cyclobalanopsis* 青冈属
Cyclobalanopsis albicaulis (Chun et W. C. Ko) Y. C. Hsu et H. W. Jen 白枝青冈
Trees, evergreen

24.8；20.1；28.0；24.8；237.6；0.0；1385 ❙ 1405；473；41 ❙ 1.5；1137；527.5

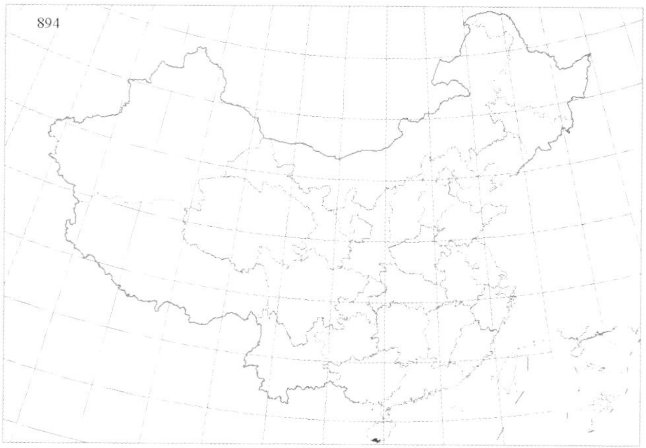

Fagaceae 壳斗科 *Cyclobalanopsis* 青冈属

Cyclobalanopsis annulata (Sm.) Oerst. 环青冈

Trees, evergreen

18.8 (-1.8~22.2); 10.5 (-10.2~13.9); 25.2 (6.1~28.8); 19.1 (1.8~22.3); 173.1 (1.9~207.5); 7.5 (0.0~82.8); 1020 (329~1130) | 1366 (320~1647); 690 (217~860); 95 (20~142) | 33.2 (-3.0~80.6); 1019 (320~1190); 268.7 (72.2~413.5)

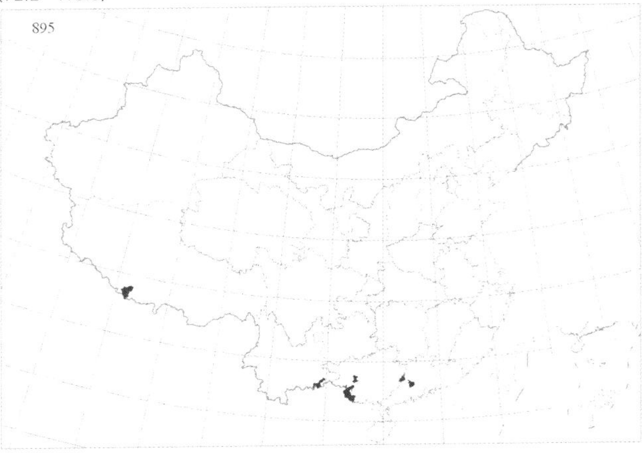

Fagaceae 壳斗科 *Cyclobalanopsis* 青冈属

Cyclobalanopsis argyrotricha (A. Camus) Chun et Y. T. Chang ex Y. C. Hsu et H. W. Jen 贵州青冈

Trees, evergreen

16.6 (13.3~21.0); 7.2 (3.6~14.2); 24.7 (21.6~27.6); 16.6 (13.3~21.0); 139.9 (101.4~191.8); 0.5 (0.0~1.7); 864 (715~1053) | 1351 (999~1697); 595 (501~818); 106 (53~194) | 55.8 (39.0~71.6); 864 (715~1053); 267.4 (183.0~551.6)

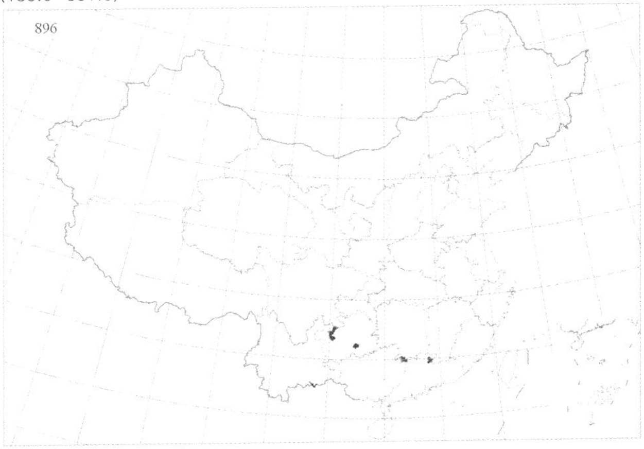

Fagaceae 壳斗科 *Cyclobalanopsis* 青冈属

Cyclobalanopsis augustinii (Skan) Schottky 窄叶青冈

Trees, evergreen

16.8 (13.3~22.7); 8.4 (3.6~15.0); 23.7 (18.8~28.6); 16.9 (13.3~22.8); 142.5 (101.9~213.1); 0.2 (0.0~1.8); 859 (706~1221) | 1288 (951~2435); 672 (481~1428); 67 (26~130) | 49.3 (12.9~112.2); 854 (689~1221); 288.5 (152.4~627.0)

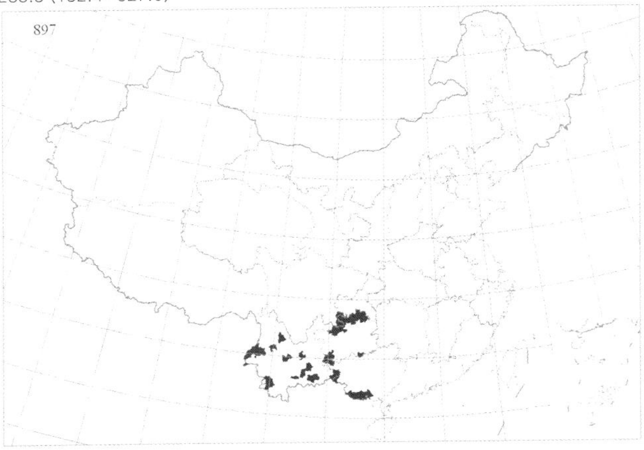

Fagaceae 壳斗科 *Cyclobalanopsis* 青冈属

Cyclobalanopsis austrocochinchinensis (Hickel et A. Camus) Hjelmq. 越南青冈

Trees, evergreen

19.9; 14.2; 23.3; 19.9; 178.5; 0.0; 956 | 1539 756; 69 | 61.5; 944; 593.5

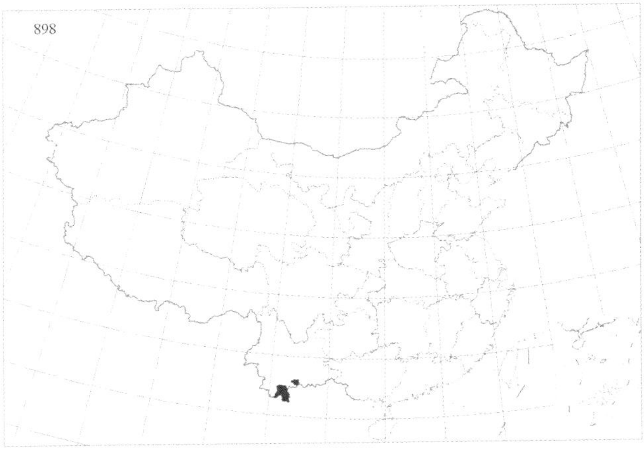

Fagaceae 壳斗科 *Cyclobalanopsis* 青冈属

Cyclobalanopsis austroglauca Y. T. Chang ex Y. C. Hsu et H. W. Jen 滇南青冈

Trees, evergreen

18.7 (17.2~21.0); 10.7 (8.8~14.2); 24.7 (22.5~26.9); 18.8 (17.3~21.0); 165.3 (147.0~191.8); 0.0 (0.0~0.0); 931 (838~1053) | 1349 (1250~1522); 734 (670~848); 66 (48~117) | 45.1 (30.5~60.1); 930 (838~1053); 367.7 (238.3~551.6)

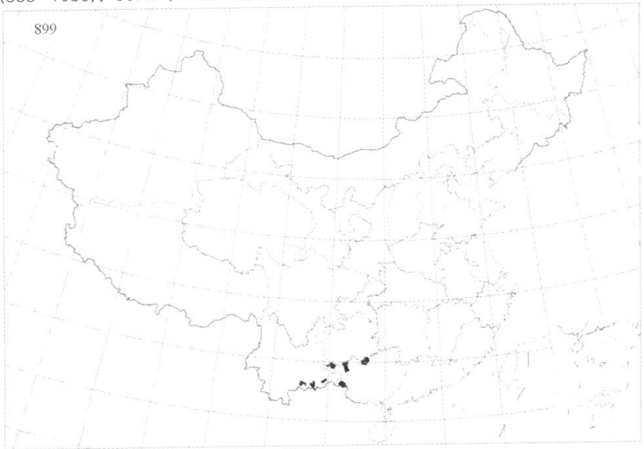

Fagaceae 壳斗科 *Cyclobalanopsis* 青冈属

Cyclobalanopsis bella (Chun et Tsiang) Chun ex Y. C. Hsu et H. W. Jen 槟榔青冈

Trees, evergreen

21.4 (17.7~24.3); 13.3 (7.7~19.1); 27.9 (26.3~28.8); 21.4 (17.8~24.3); 197.1 (153.6~231.9); 0.0 (0.0~0.0); 1137 (916~1338) | 1540 (868~2005); 607 (312~827); 130 (42~191) | 39.2 (-34.3~95.5); 1063 (868~1170); 295.4 (203.4~620.0)

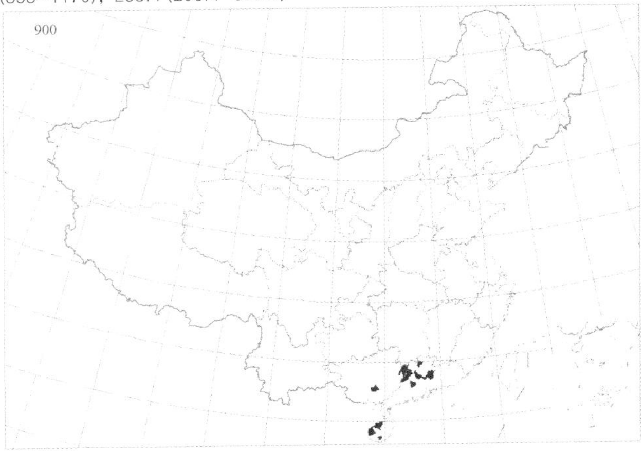

Fagaceae 壳斗科 *Cyclobalanopsis* 青冈属
Cyclobalanopsis blakei (Skan) Schottky 栎子青冈
Trees, evergreen

20.9 (15.9~24.7); 13.1 (6.5~19.8); 26.9 (22.3~28.9); 20.9 (16.0~24.7); 191.1 (131.6~236.5); 0.0 (0.0~0.0); 1094 (802~1376) ❙ 1578 (868~2435); 724 (312~1428); 103 (42~181) ❙ 46.1 (-34.3~99.5); 1057 (802~1257); 312.8 (163.7~620.0)

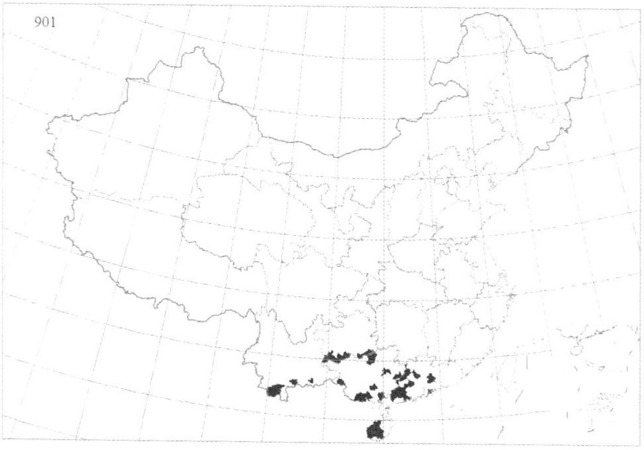

Fagaceae 壳斗科 *Cyclobalanopsis* 青冈属
Cyclobalanopsis camusiae (Trel. ex Hickel et A. Camus) Y. C. Hsu et H. W. Jen 法斗青冈
Trees, evergreen

19.7; 11.8; 25.7; 19.7; 176.9; 0.0; 1007 ❙ 1476; 726; 93 ❙ 47.7; 1007; 309.5

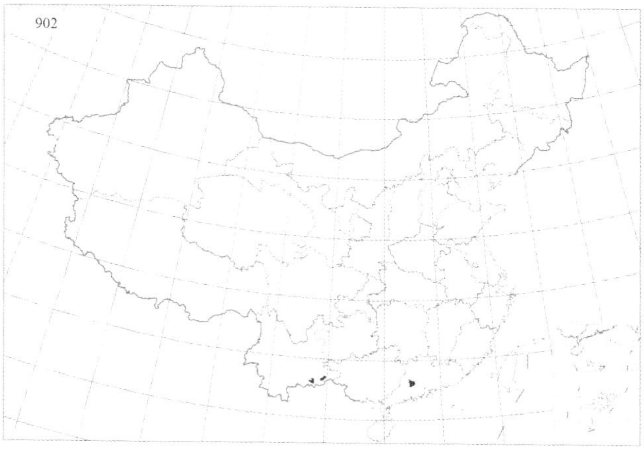

Fagaceae 壳斗科 *Cyclobalanopsis* 青冈属
Cyclobalanopsis championii (Benth.) Oerst. 岭南青冈
Trees, evergreen

20.5 (17.3~24.7); 12.3 (6.5~19.8); 27.3 (21.7~28.8); 20.6 (17.3~24.7); 187.5 (147.9~236.5); 0.0 (0.0~0.0); 1076 (863~1376) ❙ 1687 (1284~2992); 727 (448~1847); 137 (48~235) ❙ 58.4 (12.9~214.4); 1064 (863~1299); 292.1 (163.7~620.0)

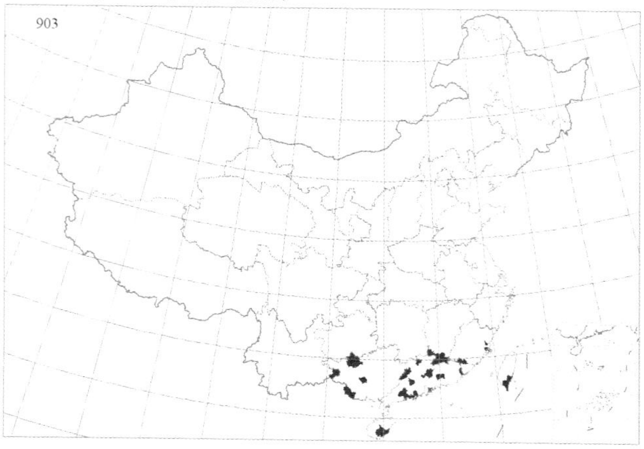

Fagaceae 壳斗科 *Cyclobalanopsis* 青冈属
Cyclobalanopsis chapensis (Hickel et A. Camus) Y. C. Hsu et H. W. Jen 扁果青冈
Trees, evergreen

18.8 (17.2~20.5); 12.5 (9.5~15.1); 23.0 (22.3~23.7); 18.8 (17.3~20.5); 166.1 (147.0~185.9); 0.0 (0.0~0.0); 906 (838~992) ❙ 1449 (1254~1607); 757 (655~911); 60 (48~77) ❙ 60.1 (47.6~74.2); 898 (838~986); 498.5 (336.3~659.0)

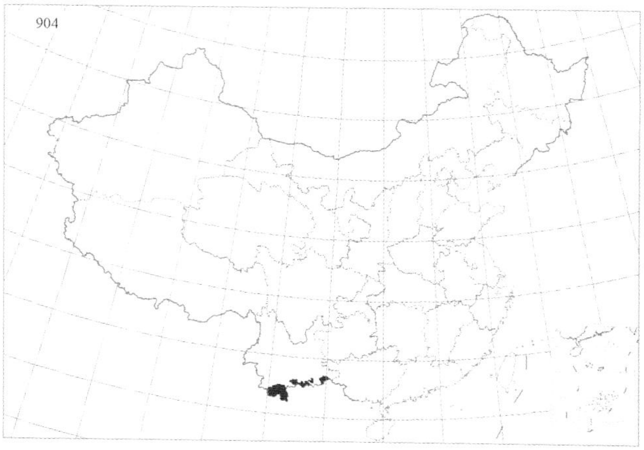

Fagaceae 壳斗科 *Cyclobalanopsis* 青冈属
Cyclobalanopsis chevalieri (Hickel et A. Camus) Y. C. Hsu et H. W. Jen 黑果青冈
Trees, evergreen

18.4; 10.5; 24.0; 18.4; 160.7; 0.0; 905 ❙ 1274; 714; 59 ❙ 40.2; 900; 314.1

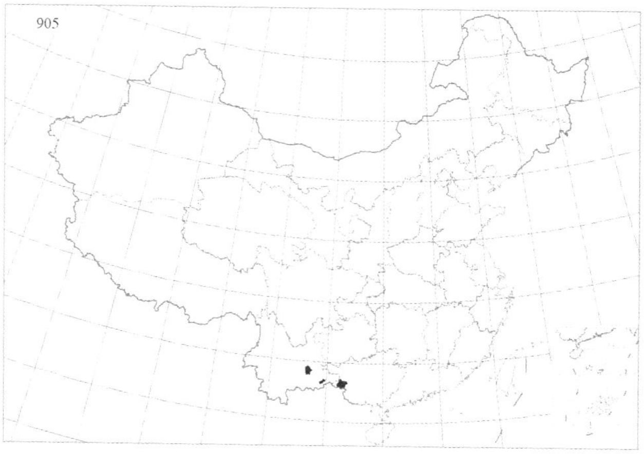

Fagaceae 壳斗科 *Cyclobalanopsis* 青冈属
Cyclobalanopsis chingsiensis (Y. T. Chang) Y. T. Chang 靖西青冈
Trees, evergreen

19.2; 11.0; 25.2; 19.2; 170.5; 0.0; 952 ❙ 1485; 848; 72 ❙ 55.9; 952; 313.7

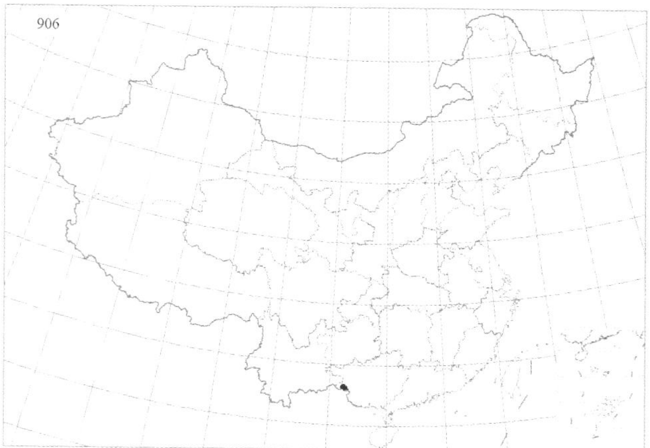

Fagaceae 壳斗科 *Cyclobalanopsis* 青冈属

Cyclobalanopsis chungii (F. P. Metcalf) Y. C. Hsu et H. W. Jen 福建青冈

Trees, evergreen

17.7 (14.3~21.7); 7.2 (3.3~13.2); 27.3 (24.2~29.4); 17.7 (14.3~21.7); 153.1 (113.2~200.8); 0.2 (0.0~2.3); 933 (761~1151) ❙ 1559 (1168~1953); 547 (411~858); 183 (58~259) ❙ 68.0 (27.0~128.9); 929 (761~1151); 228.5 (159.0~316.1)

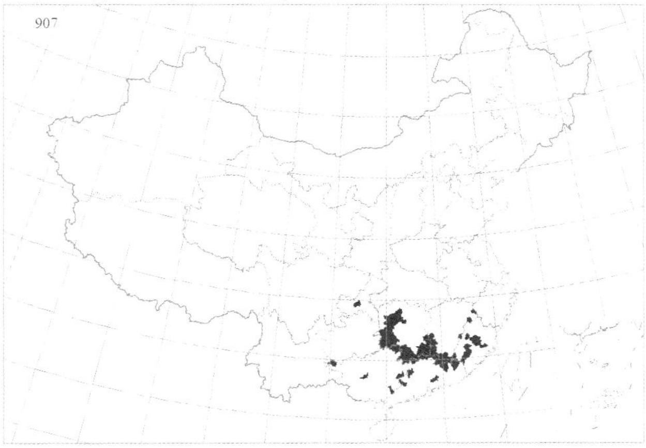

Fagaceae 壳斗科 *Cyclobalanopsis* 青冈属

Cyclobalanopsis daimingshanensis S. Lee 大明山青冈

Trees, evergreen

21.2; 12.1; 28.3; 21.3; 195.2; 0.0; 1126 ❙ 1663; 808; 180 ❙ 47.7; 1126; 223.8

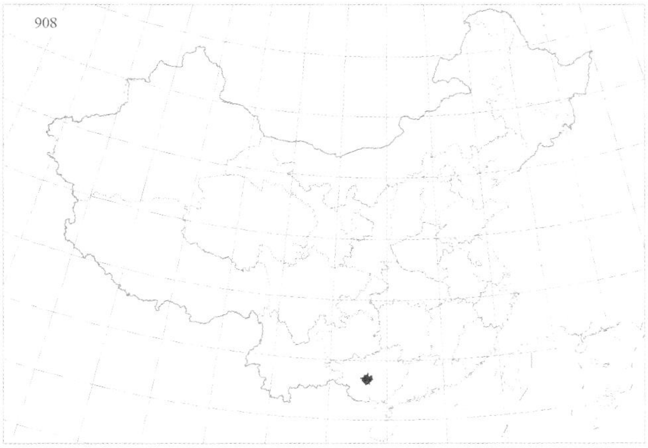

Fagaceae 壳斗科 *Cyclobalanopsis* 青冈属

Cyclobalanopsis delavayi (Franch.) Schottky 黄毛青冈

Trees, evergreen

14.9 (5.4~19.8); 6.8 (-2.1~13.1); 21.6 (12.3~28.4); 15.0 (5.8~19.8); 120.7 (31.5~177.9); 1.3 (0.0~25.6); 774 (481~1010) ❙ 1073 (741~1492); 584 (411~774); 45 (15~146) ❙ 39.7 (-14.2~107.2); 753 (481~1010); 273.5 (142.9~432.9)

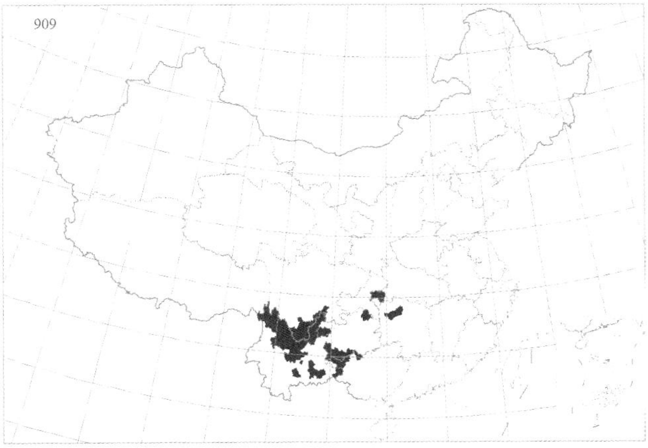

Fagaceae 壳斗科 *Cyclobalanopsis* 青冈属

Cyclobalanopsis delicatula (Chun et Tsiang) Y. C. Hsu et H. W. Jen 上思青冈

Trees, evergreen

20.0 (16.4~22.7); 10.8 (5.4~15.0); 27.4 (26.3~28.6); 20.0 (16.5~22.8); 180.3 (137.7~213.1); 0.0 (0.0~0.0); 1053 (864~1221) ❙ 1818 (1425~2435); 916 (518~1428); 141 (124~189) ❙ 71.9 (49.1~99.5); 1053 (864~1221); 262.3 (224.9~318.8)

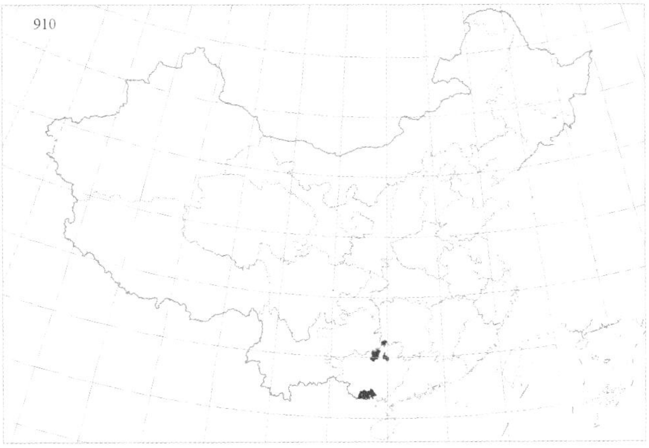

Fagaceae 壳斗科 *Cyclobalanopsis* 青冈属

Cyclobalanopsis dinghuensis (C. C. Huang) Y. C. Hsu et H. W. Jen 鼎湖青冈

Trees, evergreen

21.7; 13.1; 28.7; 21.7; 200.8; 0.0; 1154 ❙ 1582; 682; 135 ❙ 37.0; 1154; 212.3

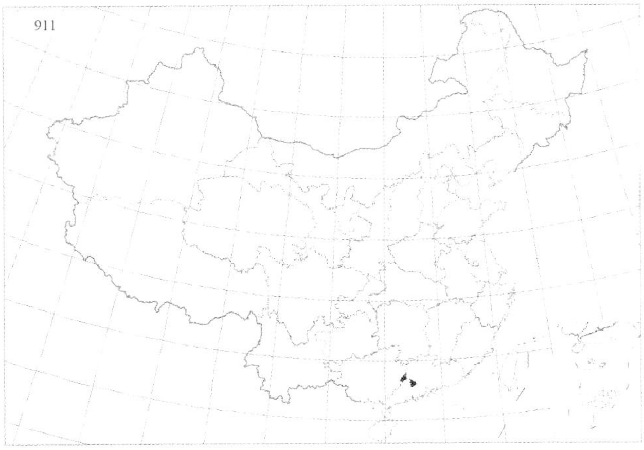

Fagaceae 壳斗科 *Cyclobalanopsis* 青冈属

Cyclobalanopsis disciformis (Chun et Tsiang) Y. C. Hsu et H. W. Jen 碟斗青冈

Trees, evergreen

18.4 (15.0~23.1); 8.9 (4.5~17.5); 26.5 (23.0~28.3); 18.4 (15.0~23.2); 161.4 (120.8~218.4); 0.0 (0.0~0.5); 958 (794~1232) ❙ 1478 (1089~1776); 629 (499~828); 129 (53~187) ❙ 54.9 (20.7~85.5); 952 (794~1147); 246.0 (181.2~538.2)

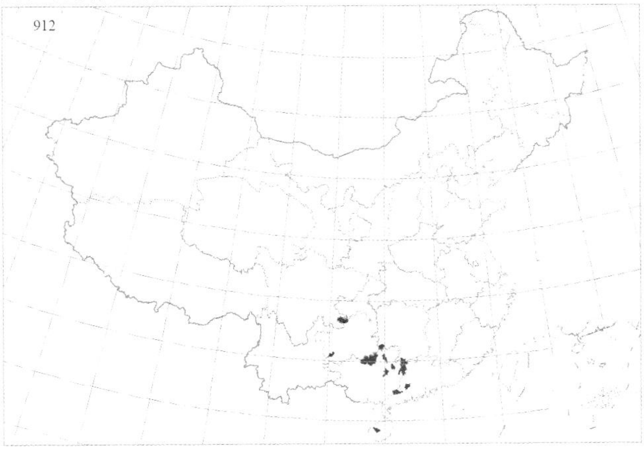

Fagaceae 壳斗科 *Cyclobalanopsis* 青冈属

Cyclobalanopsis edithiae (Skan) Schottky 华南青冈

Trees, evergreen

22.8 (21.3~24.6); 15.9 (13.2~19.5); 28.1 (27.2~28.7); 22.8 (21.3~24.7); 214.1 (195.9~235.9); 0.0 (0.0~0.0); 1225 (1111~1372) ❙ 1932 (1551~2435); 954 (539~1428); 102 (48~130) ❙ 58.9 (19.1~99.5); 1193 (1111~1299); 346.9 (139.6~620.0)

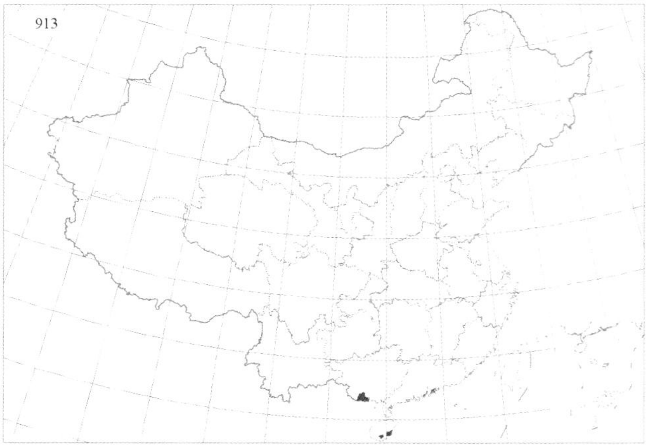

Fagaceae 壳斗科 *Cyclobalanopsis* 青冈属

Cyclobalanopsis elevaticostata Q. F. Zheng 突脉青冈

Trees, evergreen

17.8; 8.3; 26.5; 17.9; 154.4; 0.0; 916 ❙ 1732; 600; 202 ❙ 89.6; 916; 275.0

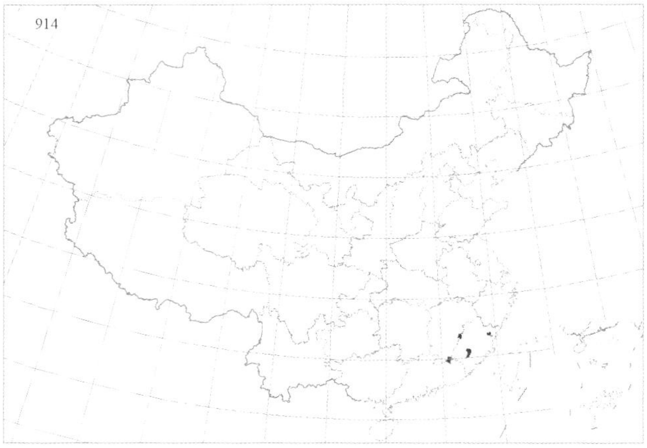

Fagaceae 壳斗科 *Cyclobalanopsis* 青冈属

Cyclobalanopsis fleuryi (Hickel et A. Camus) Chun ex Q. F. Zheng 饭甑青冈

Trees, evergreen

19.1 (8.6~24.7); 10.3 (1.3~19.8); 26.4 (15.0~29.8); 19.1 (8.7~24.7); 169.6 (52.6~236.5); 0.1 (0.0~8.8); 995 (574~1376) ❙ 1531 (868~2035); 642 (312~1044); 131 (42~221) ❙ 56.6 (-34.3~145.8); 977 (574~1326); 284.5 (160.7~620.0)

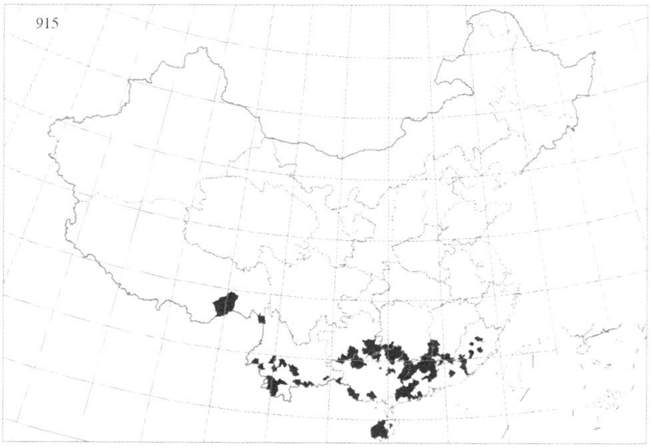

Fagaceae 壳斗科 *Cyclobalanopsis* 青冈属

Cyclobalanopsis gambleana (A. Camus) Y. C. Hsu et H. W. Jen 毛曼青冈

Trees, evergreen

12.9 (-1.8~19.1); 5.0 (-10.2~13.0); 19.7 (6.1~26.6); 13.2 (1.8~19.1); 101.7 (1.9~169.7); 5.8 (0.0~82.8); 710 (329~929) ❙ 1020 (320~1736); 565 (217~1007); 40 (10~122) ❙ 44.9 (-14.2~145.8); 690 (320~917); 273.7 (61.6~506.2)

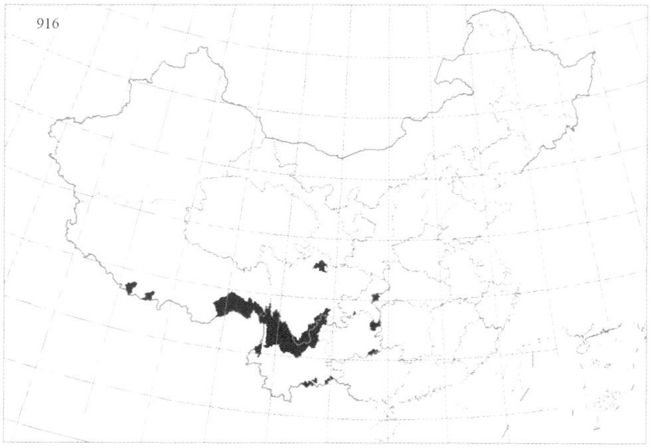

Fagaceae 壳斗科 *Cyclobalanopsis* 青冈属

Cyclobalanopsis gilva (Blume) Oerst. 赤皮青冈

Trees, evergreen

16.7 (14.3~21.1); 5.9 (3.3~14.6); 26.6 (20.1~29.4); 16.7 (14.3~21.1); 140.7 (112.6~193.1); 0.4 (0.0~2.3); 880 (756~1077) ❙ 1548 (1101~3495); 543 (392~1186); 187 (51~897) ❙ 75.4 (29.7~246.5); 872 (756~1077); 215.5 (111.9~383.4)

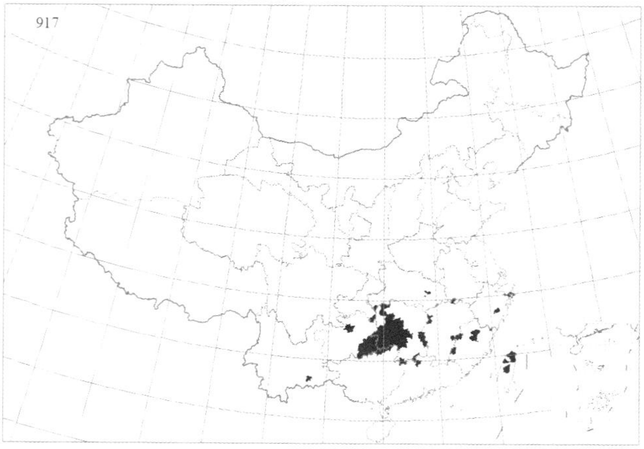

Fagaceae 壳斗科 *Cyclobalanopsis* 青冈属

Cyclobalanopsis glauca (Thunb.) Oerst. 青冈

Trees, evergreen

15.9 (5.3~24.1); 5.3 (-4.2~18.9); 25.8 (12.2~29.8); 16.0 (5.7~24.1); 134.4 (30.5~229.7); 2.4 (0.0~29.4); 860 (479~1322) ❙ 1344 (620~3495); 541 (290~1847); 131 (12~897) ❙ 55.9 (-22.2~246.5); 848 (479~1216); 229.1 (111.9~551.6)

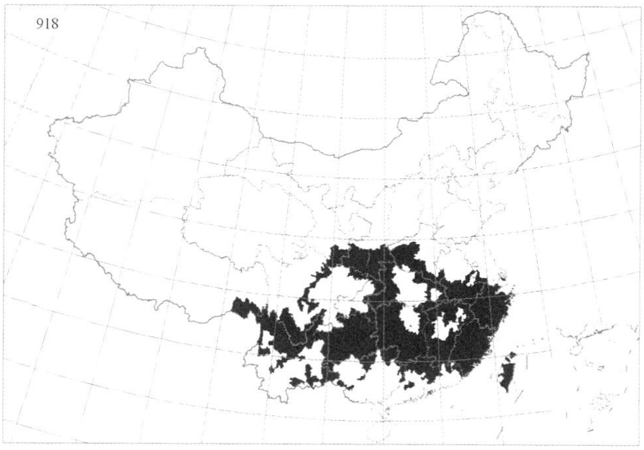

Fagaceae 壳斗科 *Cyclobalanopsis* 青冈属
Cyclobalanopsis glaucoides Schottky 滇青冈
Trees, evergreen

13.5 (−1.2~19.2); 5.9 (−11.7~13.1); 19.7 (8.5~26.0); 13.7 (2.6~19.2); 108.2 (8.1~170.8); 5.3 (0.0~82.3); 726 (347~966) ❙ 1036 (491~1495); 589 (301~835); 36 (11~97) ❙ 45.8 (−14.2~125.9); 704 (347~966); 287.9 (185.6~506.2)

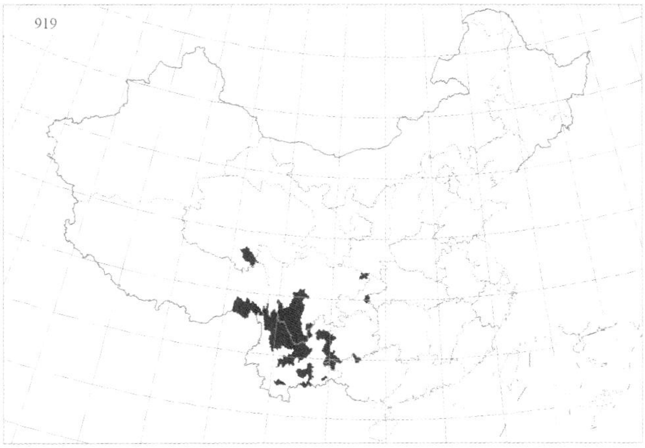

Fagaceae 壳斗科 *Cyclobalanopsis* 青冈属
Cyclobalanopsis gracilis (Rehder et E. H. Wilson) W. C. Cheng et T. Hong 细叶青冈
Trees, evergreen

15.8 (2.6~22.0); 4.5 (−7.0~13.5); 26.6 (11.2~29.8); 15.9 (4.3~22.0); 132.6 (20.0~203.8); 2.2 (0.0~48.2); 860 (427~1170) ❙ 1429 (677~1997); 528 (336~821); 155 (12~259) ❙ 66.1 (−0.2~147.0); 854 (427~1170); 224.0 (134.6~319.9)

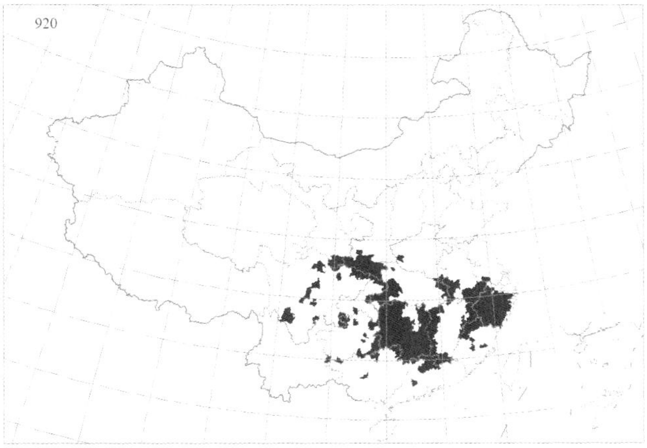

Fagaceae 壳斗科 *Cyclobalanopsis* 青冈属
Cyclobalanopsis helferiana (A. DC.) Oerst. 毛枝青冈
Trees, evergreen

21.1 (17.7~24.8); 13.9 (9.4~20.1); 26.2 (21.6~28.9); 21.1 (17.8~24.8); 193.0 (153.4~237.6); 0.0 (0.0~0.0); 1094 (854~1385) ❙ 1314 (868~1776); 621 (312~893); 61 (20~125) ❙ 23.3 (−34.3~74.2); 1011 (786~1230); 385.3 (229.5~570.1)

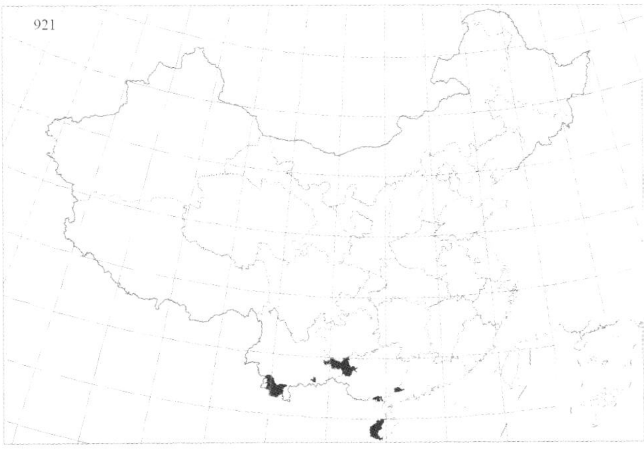

Fagaceae 壳斗科 *Cyclobalanopsis* 青冈属
Cyclobalanopsis hui (Chun) Chun ex Y. C. Hsu et H. W. Jen 雷公青冈
Trees, evergreen

20.5 (16.5~24.7); 11.9 (5.3~19.8); 27.8 (26.2~29.8); 20.6 (16.5~24.7); 187.0 (138.2~236.5); 0.0 (0.0~0.0); 1084 (861~1376) ❙ 1621 (868~2035); 637 (312~1033); 149 (42~215) ❙ 51.8 (−34.3~95.5); 1054 (861~1212); 264.0 (157.7~620.0)

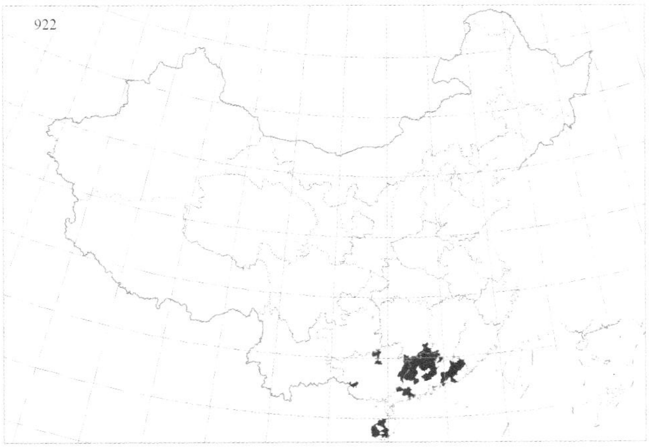

Fagaceae 壳斗科 *Cyclobalanopsis* 青冈属
Cyclobalanopsis hypophaea (Hayata) Kudo 绒毛青冈
Trees, evergreen

18.5; 14.1; 21.7; 18.5; 162.1; 0.0; 863 ❙ 2713; 1388; 188 ❙ 214.4; 863; 363.7

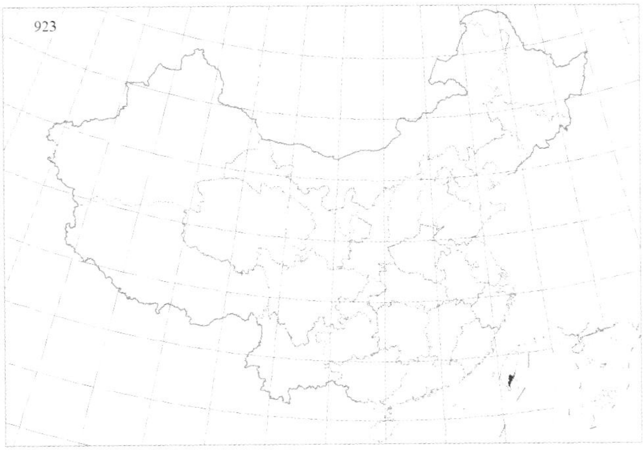

Fagaceae 壳斗科 *Cyclobalanopsis* 青冈属
Cyclobalanopsis jenseniana (Hand.-Mazz.) W. C. Cheng et T. Hong ex Q. F. Zheng 大叶青冈
Trees, evergreen

16.9 (12.7~21.3); 6.2 (2.1~14.2); 26.8 (22.5~29.8); 17.0 (12.8~21.3); 143.9 (99.0~196.0); 0.3 (0.0~5.6); 894 (713~1123) ❙ 1468 (1089~1997); 549 (392~835); 154 (51~254) ❙ 64.9 (5.8~147.0); 888 (713~1121); 227.8 (134.6~551.6)

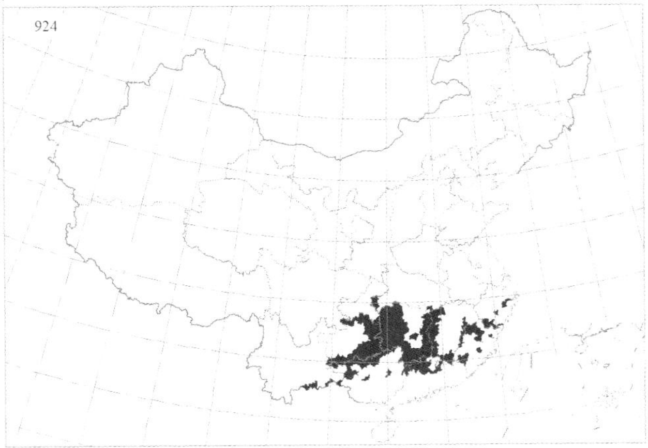

Fagaceae 壳斗科 *Cyclobalanopsis* 青冈属
Cyclobalanopsis jinpinensis Y. C. Hsu et H. W. Jen 金平青冈
Trees, evergreen

19.1；13.0；23.3；19.1；169.7；0.0；917 | 1428；835；56 | 55.7；917；506.2

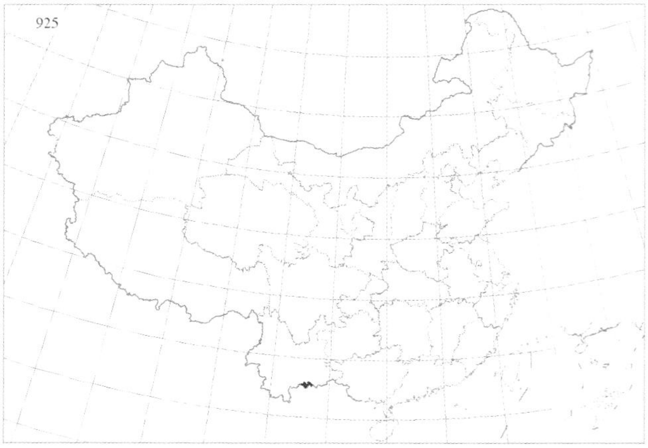

Fagaceae 壳斗科 *Cyclobalanopsis* 青冈属
Cyclobalanopsis kerrii (Craib) Hu 毛叶青冈
Trees, evergreen

19.0 (12.6~24.8)；12.1 (6.0~20.1)；23.9 (17.9~28.7)；19.0 (12.6~24.8)；167.9 (91.6~237.6)；0.0 (0.0~0.0)；960 (678~1385) | 1246 (868~1607)；628 (312~835)；51 (18~77) | 33.8 (-34.3~74.2)；900 (678~1142)；375.8 (189.5~659.0)

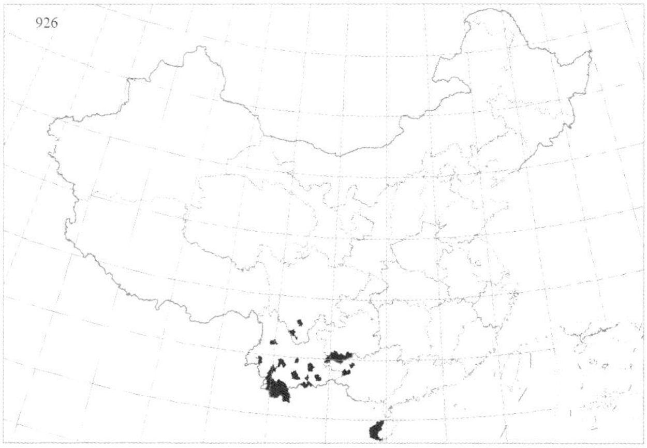

Fagaceae 壳斗科 *Cyclobalanopsis* 青冈属
Cyclobalanopsis kiukiangensis Y. T. Chang ex Y. C. Hsu et H. W. Jen 俅江青冈
Trees, evergreen

13.5；6.5；19.0；13.6；105.6；2.9；723 | 1441；834；53 | 100.5；723；450.3

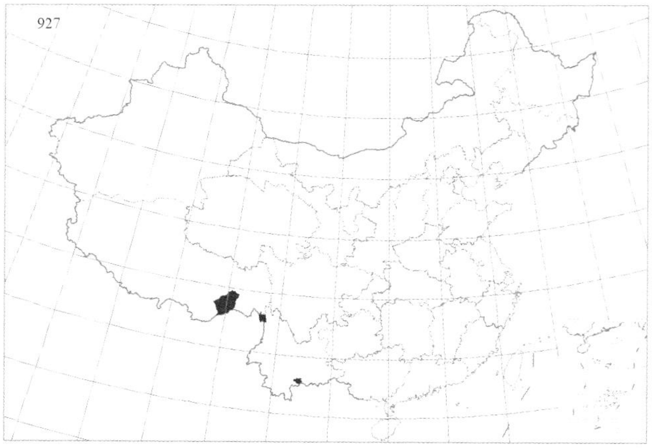

Fagaceae 壳斗科 *Cyclobalanopsis* 青冈属
Cyclobalanopsis kouangsiensis (A. Camus) Y. C. Hsu et H. W. Jen 广西青冈
Trees, evergreen

21.4 (17.7~24.3)；14.5 (8.5~19.1)；26.7 (22.5~28.6)；21.5 (17.8~24.3)；197.6 (153.4~231.9)；0.0 (0.0~0.0)；1134 (854~1338) | 1413 (1179~1551)；571 (392~756)；88 (42~169) | 29.2 (-11.9~69.8)；1030 (854~1147)；408.6 (185.5~620.0)

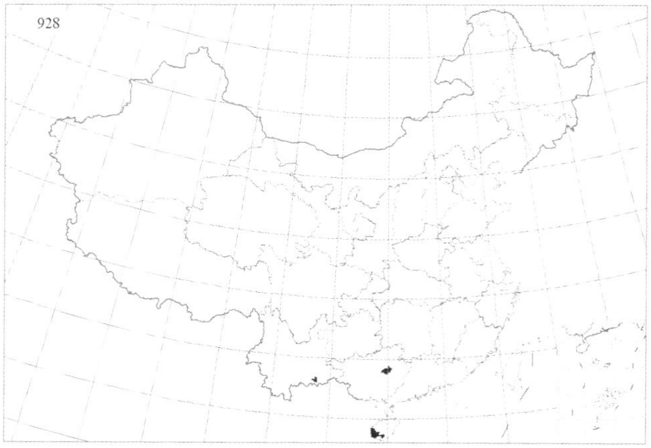

Fagaceae 壳斗科 *Cyclobalanopsis* 青冈属
Cyclobalanopsis lamellosa (Sm.) Oerst. 薄片青冈
Trees, evergreen

14.4 (-1.8~19.8)；7.1 (-10.2~13.0)；19.9 (6.1~26.4)；14.7 (1.8~19.8)；119.3 (1.9~177.9)；6.2 (0.0~82.8)；762 (329~1010) | 1234 (320~1736)；699 (217~1007)；51 (20~70) | 63.2 (-3.0~145.8)；751 (320~1010)；387.2 (72.2~627.0)

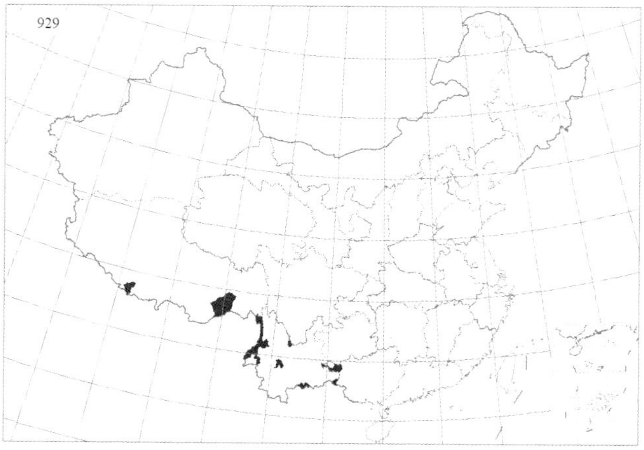

Fagaceae 壳斗科 *Cyclobalanopsis* 青冈属
Cyclobalanopsis litoralis Chun et P. C. Tam ex Y. C. Hsu et H. W. Jen 尖峰青冈
Trees, evergreen

24.0 (23.6~24.3)；18.2 (17.8~18.7)；28.2 (27.5~28.7)；24.0 (23.6~24.3)；228.5 (223.3~231.9)；0.0 (0.0~0.0)；1312 (1271~1338) | 1169 (868~1380)；410 (312~455)；53 (42~71) | -10.7 (-34.3~6.7)；1053 (868~1139)；410.0 (288.3~455.2)

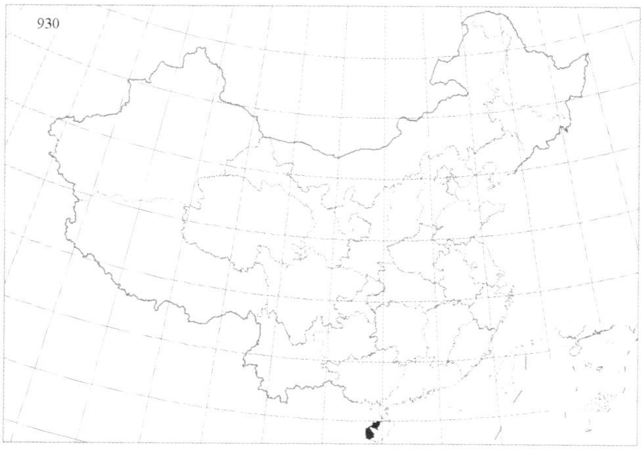

Fagaceae 壳斗科 *Cyclobalanopsis* 青冈属

Cyclobalanopsis litseoides (Dunn) Schottky 木姜叶青冈

Trees, evergreen

22.2 (21.8~22.4)；14.7 (14.1~15.2)；28.3 (27.8~28.7)；22.3 (21.9~22.7)；207.5 (202.6~212.7)；0.0 (0.0~0.0)；1183 (1150~1216) | 1993 (1876~2037)；978 (891~1033)；117 (104~142) | 68.5 (61.6~72.1)；1183 (1150~1216)；214.2 (163.7~273.3)

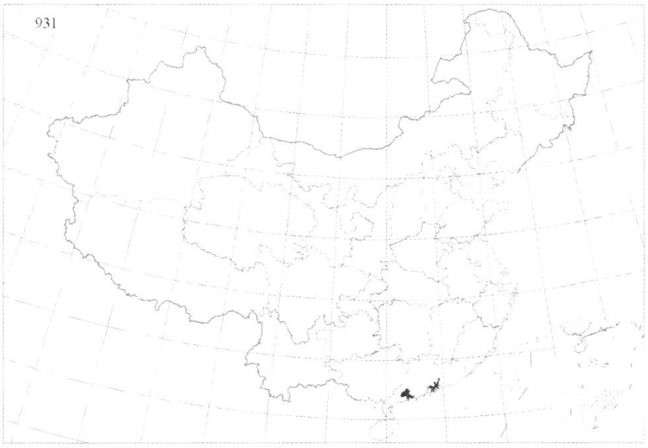

Fagaceae 壳斗科 *Cyclobalanopsis* 青冈属

Cyclobalanopsis lobbii (Hook. f. et Thomson ex Wenz.) Y. C. Hsu et H. W. Jen 滇西青冈

Trees, evergreen

8.9 (5.4~14.4)；1.6 (-2.1~7.5)；15.1 (12.3~19.1)；9.1 (5.8~14.4)；59.5 (31.5~113.1)；11.9 (0.0~25.6)；579 (481~727) | 1021 (741~1454)；586 (426~792)；39 (12~70) | 74.2 (54.0~99.9)；579 (481~727)；355.2 (273.5~487.2)

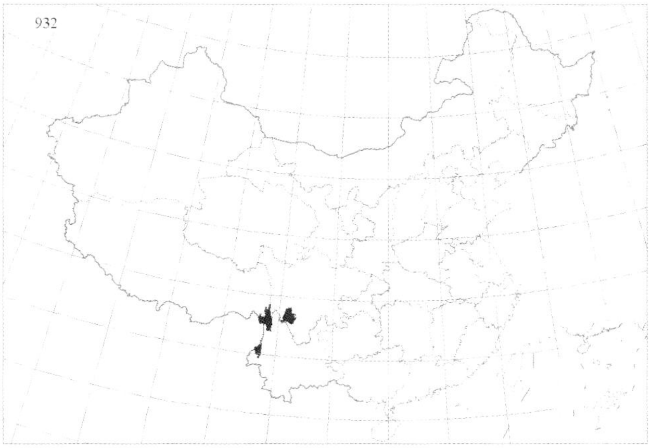

Fagaceae 壳斗科 *Cyclobalanopsis* 青冈属

Cyclobalanopsis longinux Schottky 长果青冈

Trees, evergreen

19.5 (16.2~24.1)；13.9 (11.0~18.9)；24.1 (20.1~28.0)；19.5 (16.2~24.1)；174.2 (134.2~229.7)；0.0 (0.0~0.0)；964 (773~1322) | 2503 (1751~2993)；1163 (750~1847)；243 (47~547) | 168.7 (32.5~246.5)；940 (773~1123)；264.8 (154.2~363.7)

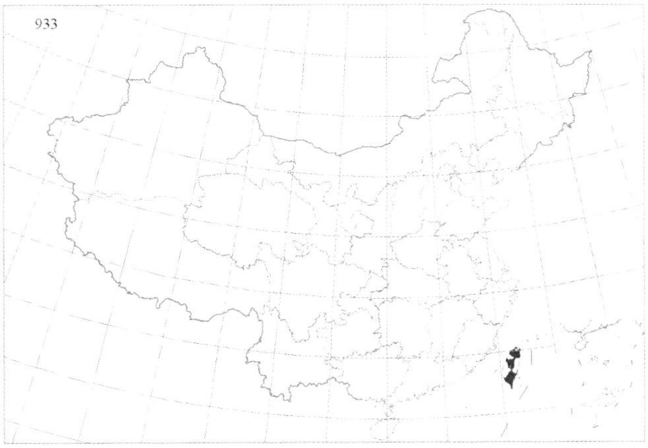

Fagaceae 壳斗科 *Cyclobalanopsis* 青冈属

Cyclobalanopsis lungmaiensis Hu 龙迈青冈

Trees, evergreen

15.7 (8.6~18.9)；7.7 (1.3~10.6)；21.8 (15.0~25.5)；15.7 (8.7~19.0)；130.7 (52.6~167.5)；1.8 (0.0~8.8)；806 (574~939) | 1331 (1041~1688)；722 (585~959)；62 (45~107) | 67.4 (37.6~112.2)；806 (574~939)；380.6 (245.3~627.0)

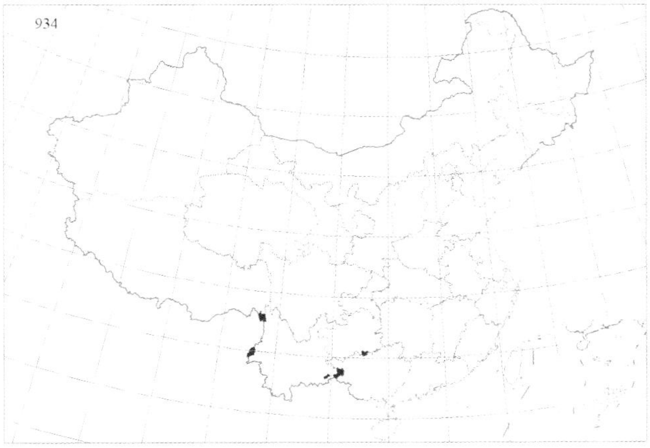

Fagaceae 壳斗科 *Cyclobalanopsis* 青冈属

Cyclobalanopsis morii (Hayata) Schottky 台湾青冈

Trees, evergreen

18.4 (16.2~21.7)；12.8 (11.0~15.3)；23.1 (20.1~27.0)；18.5 (16.2~21.7)；161.8 (134.2~200.2)；0.0 (0.0~0.0)；896 (773~1118) | 2442 (1836~2859)；1069 (750~1412)；251 (110~402) | 178.9 (64.2~246.5)；892 (773~1075)；273.2 (202.8~363.7)

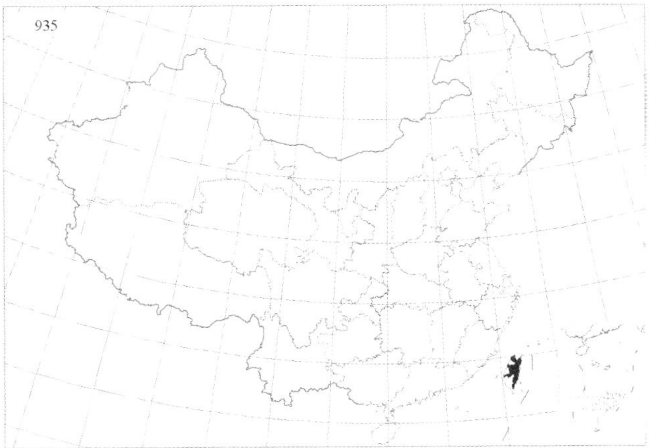

Fagaceae 壳斗科 *Cyclobalanopsis* 青冈属

Cyclobalanopsis motuoensis (C. C. Huang) Y. C. Hsu et H. W. Jen 墨脱青冈

Trees, evergreen

13.3；5.4；19.6；13.4；100.4；0.0；706 | 1736；1007；51 | 145.8；706；447.4

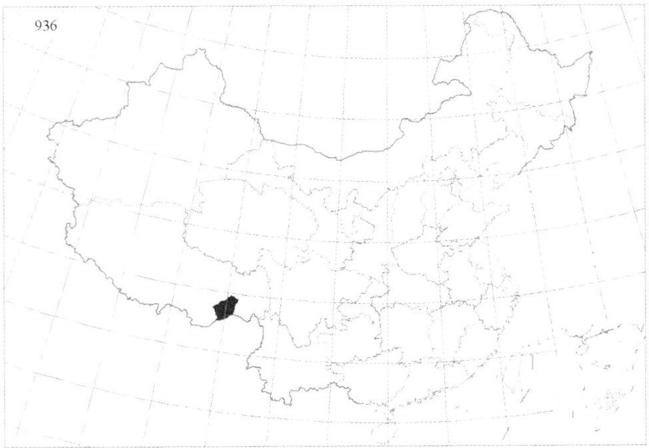

Fagaceae 壳斗科 *Cyclobalanopsis* 青冈属
Cyclobalanopsis multinervis W. C. Cheng et T. Hong 多脉青冈
Trees, evergreen

15.7 (9.1~21.0)；4.6 (-2.5~11.9)；26.2 (17.7~29.8)；15.8 (9.2~21.1)；131.0 (64.1~193.0)；1.8 (0.0~20.6)；847 (595~1114) | 1393 (806~1997)；515 (352~821)；141 (16~259) | 64.0 (7.2~147.0)；842 (595~1114)；223.3 (140.5~318.7)

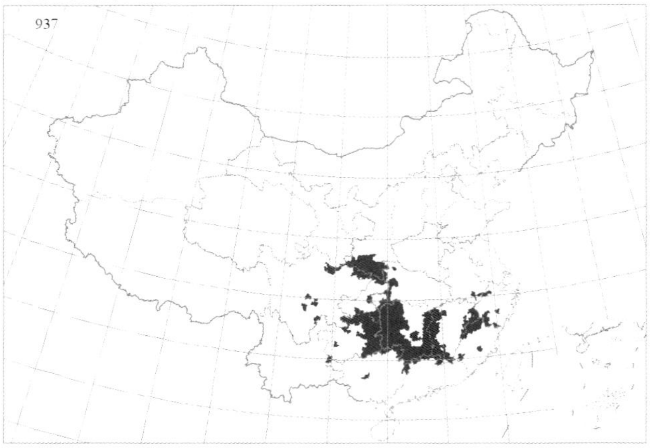

Fagaceae 壳斗科 *Cyclobalanopsis* 青冈属
Cyclobalanopsis myrsinifolia (Blume) Oerst. 小叶青冈
Trees, evergreen

16.3 (5.4~22.5)；5.6 (-2.9~17.7)；26.2 (12.3~29.9)；16.4 (5.8~22.7)；138.1 (31.5~212.7)；1.9 (0.0~25.6)；877 (481~1216) | 1376 (626~2993)；549 (316~1847)；134 (15~547) | 56.4 (-22.2~235.6)；865 (481~1216)；224.1 (108.2~659.0)

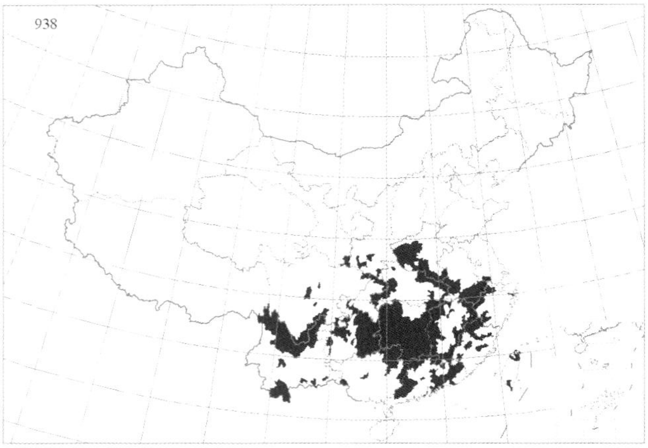

Fagaceae 壳斗科 *Cyclobalanopsis* 青冈属
Cyclobalanopsis neglecta Schottky 竹叶青冈
Trees, evergreen

21.7 (17.6~24.7)；14.1 (7.3~19.8)；27.7 (25.7~28.8)；21.7 (17.6~24.7)；200.8 (151.6~236.5)；0.0 (0.0~0.0)；1146 (909~1376) | 1718 (1179~2435)；775 (392~1428)；118 (42~182) | 51.0 (-11.9~99.5)；1116 (909~1221)；286.2 (139.6~590.7)

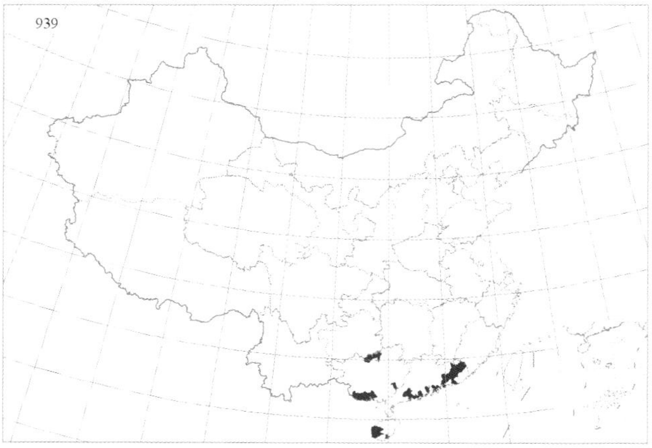

Fagaceae 壳斗科 *Cyclobalanopsis* 青冈属
Cyclobalanopsis ningangensis W. C. Cheng et Y. C. Hsu 宁冈青冈
Trees, evergreen

16.4 (14.3~17.8)；4.5 (3.3~7.0)；27.7 (24.3~29.3)；16.4 (14.3~17.8)；138.1 (114.1~154.1)；0.7 (0.0~2.3)；888 (765~964) | 1498 (1257~1679)；501 (427~591)；188 (112~228) | 69.2 (40.5~106.4)；877 (765~961)；218.4 (158.9~268.8)

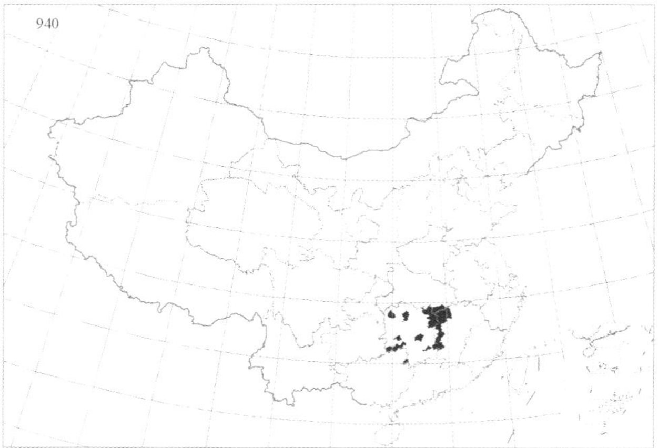

Fagaceae 壳斗科 *Cyclobalanopsis* 青冈属
Cyclobalanopsis obovatifolia (C. C. Huang) Q. F. Zheng 倒卵叶青冈
Trees, evergreen

18.9 (17.7~20.7)；9.6 (6.9~12.0)；27.1 (26.5~27.7)；19.0 (17.7~20.7)；167.5 (152.4~189.0)；0.0 (0.0~0.0)；976 (916~1083) | 1672 (1486~2005)；613 (505~824)；176 (151~193) | 70.9 (60.2~85.1)；976 (916~1083)；242.1 (188.8~273.2)

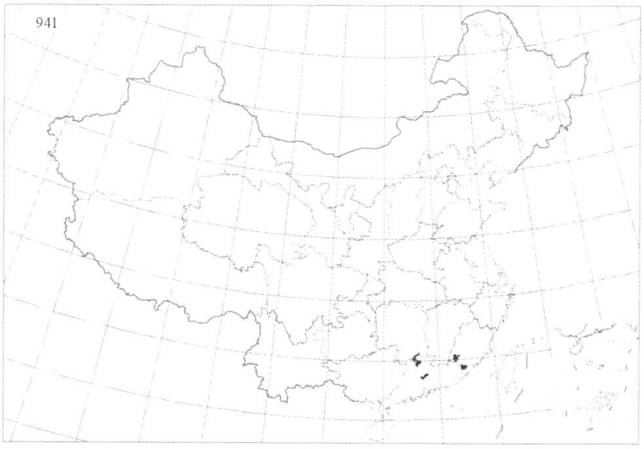

Fagaceae 壳斗科 *Cyclobalanopsis* 青冈属
Cyclobalanopsis oxyodon (Miq.) Oerst. 曼青冈
Trees, evergreen

14.2 (-1.8~19.4)；3.5 (-10.2~10.2)；24.1 (6.1~29.2)；14.3 (1.8~19.4)；115.2 (1.9~173.4)；4.5 (0.0~82.8)；786 (329~1029) | 1277 (320~1922)；528 (217~1007)；103 (15~259) | 63.0 (-3.0~145.8)；785 (320~1029)；229.4 (72.2~447.4)

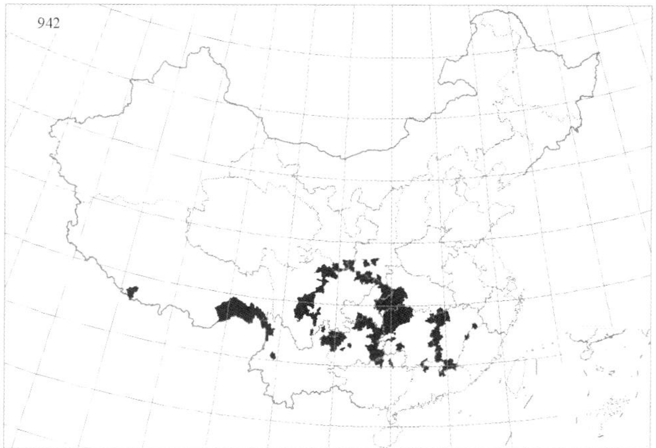

Fagaceae 壳斗科 *Cyclobalanopsis* 青冈属

Cyclobalanopsis pachyloma (Seemen) Schottky 毛果青冈

Trees, evergreen

18.2 (12.4~22.5); 8.9 (4.0~17.7); 26.4 (18.6~29.8); 18.3 (12.4~22.6); 159.3 (89.1~210.9); 0.0 (0.0~1.1); 946 (675~1206) | 1607 (896~2992); 652 (473~1847); 154 (31~259) | 70.6 (29.9~246.5); 944 (675~1206); 251.5 (139.6~551.6)

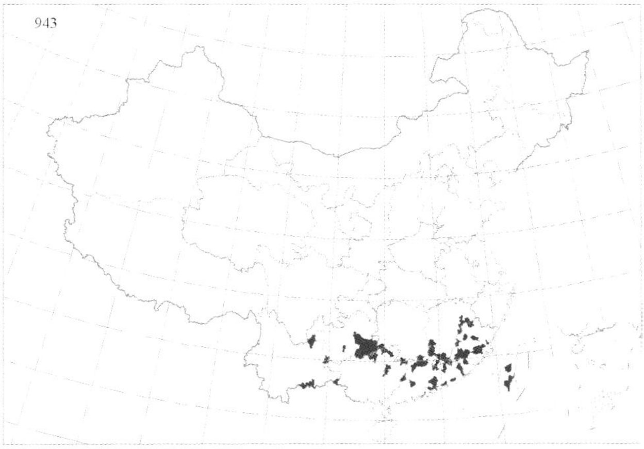

Fagaceae 壳斗科 *Cyclobalanopsis* 青冈属

Cyclobalanopsis patelliformis (Chun) Y. C. Hsu et H. W. Jen 托盘青冈

Trees, evergreen

21.3 (17.3~24.8); 13.5 (6.4~20.1); 27.7 (26.6~28.7); 21.4 (17.3~24.8); 196.6 (147.9~237.6); 0.0 (0.0~0.0); 1141 (907~1385 | 1548 (868~1876); 579 (312~891); 125 (41~197) | 40.0 (-34.3~78.8); 1062 (868~1299); 337.8 (208.1~620.0)

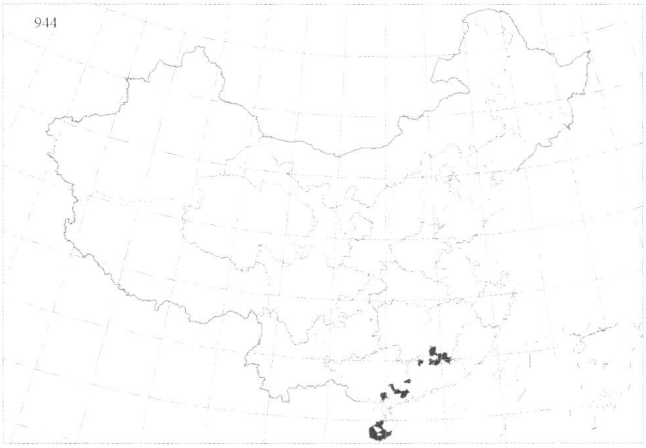

Fagaceae 壳斗科 *Cyclobalanopsis* 青冈属

Cyclobalanopsis pentacycla (Y. T. Chang) Y. T. Chang ex Y. C. Hsu et H. W. Jen 五环青冈

Trees, evergreen

16.7 (15.7~17.6); 6.8 (5.2~9.9); 25.2 (23.2~25.7); 16.8 (15.7~17.7); 141.0 (128.4~152.4); 0.0 (0.0~0.0); 857 (824~886) | 1327 (1187~1422); 595 (486~813); 100 (52~129) | 54.8 (44.1~65.6); 857 (824~886); 250.2 (203.1~359.7)

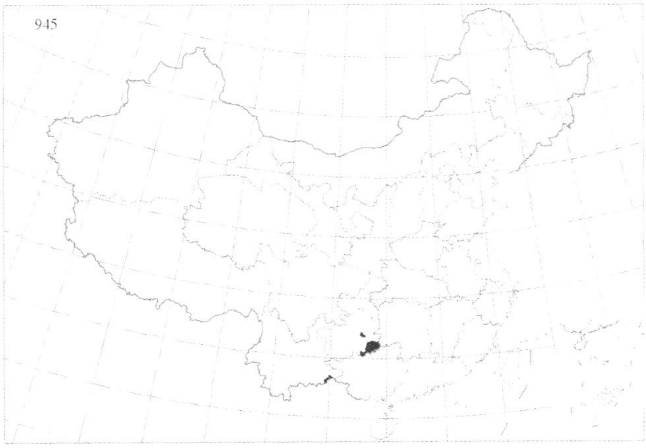

Fagaceae 壳斗科 *Cyclobalanopsis* 青冈属

Cyclobalanopsis phanera (Chun) Y. C. Hsu et H. W. Jen 亮叶青冈

Trees, evergreen

24.5 (24.0~24.8); 19.4 (18.7~20.1); 28.0 (27.2~28.4); 24.5 (24.0~24.8); 234.0 (228.1~237.6); 0.0 (0.0~0.0); 1355 (1302~1385) | 1498 (1179~1767); 501 (392~567); 55 (41~88) | 10.6 (-11.9~23.8); 1170 (1056~1299); 529.1 (455.2~620.0)

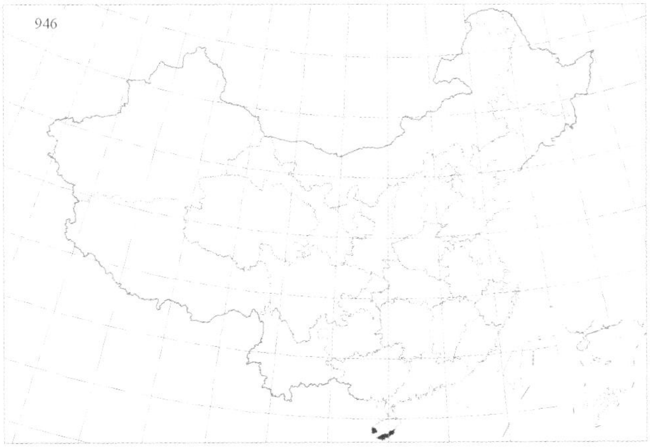

Fagaceae 壳斗科 *Cyclobalanopsis* 青冈属

Cyclobalanopsis poilanei (Hickel et A. Camus) Hjelmq. 黄背青冈

Trees, evergreen

21.7 (21.0~22.7); 13.1 (11.8~15.0); 28.2 (27.3~28.6); 21.7 (21.1~22.8); 200.5 (192.6~213.1); 0.0 (0.0~0.0); 1151 (1111~1221) | 1864 (1554~2435); 989 (727~1428); 152 (124~194) | 61.6 (35.6~99.5); 1151 (1111~1221); 253.6 (206.0~318.8)

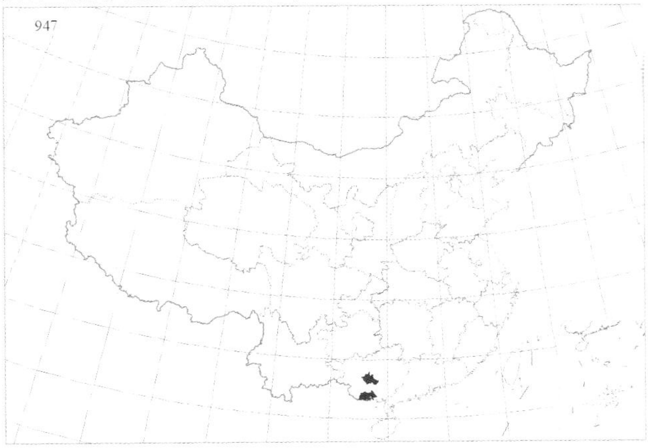

Fagaceae 壳斗科 *Cyclobalanopsis* 青冈属

Cyclobalanopsis rex (Hemsl.) Schottky 大果青冈

Trees, evergreen

18.5 (14.4~20.5); 12.5 (7.5~15.1); 22.3 (19.1~23.7); 18.5 (14.4~20.5); 162.5 (113.1~185.9); 0.0 (0.0~0.0); 895 (727~992) | 1453 (1323~1607); 724 (655~792); 64 (51~77) | 63.7 (47.1~99.9); 879 (727~986); 520.8 (403.0~659.0)

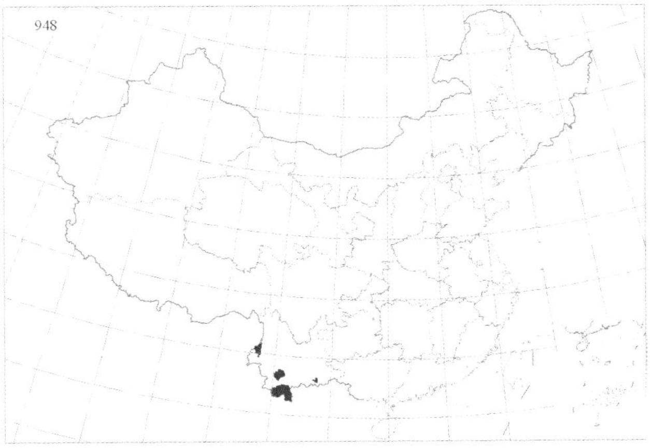

Fagaceae 壳斗科 *Cyclobalanopsis* 青冈属
Cyclobalanopsis saravanensis (A. Camus) Hjelmq. 薄叶青冈
Trees, evergreen

17.2; 9.1; 23.1; 17.3; 147.5; 0.0; 847 ❙ 1161; 633; 50 ❙ 37.1; 847; 314.4

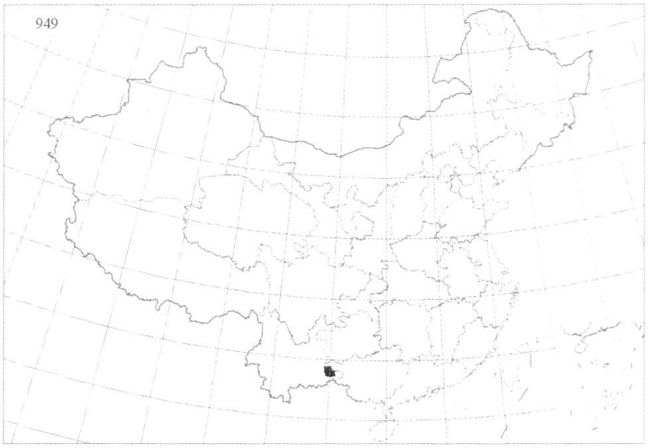

Fagaceae 壳斗科 *Cyclobalanopsis* 青冈属
Cyclobalanopsis semiserrata (Roxb.) Oerst. 无齿青冈
Trees, evergreen

17.8; 10.9; 22.8; 17.8; 153.8; 0.0; 883 ❙ 1513; 854; 59 ❙ 76.5; 883; 499.2

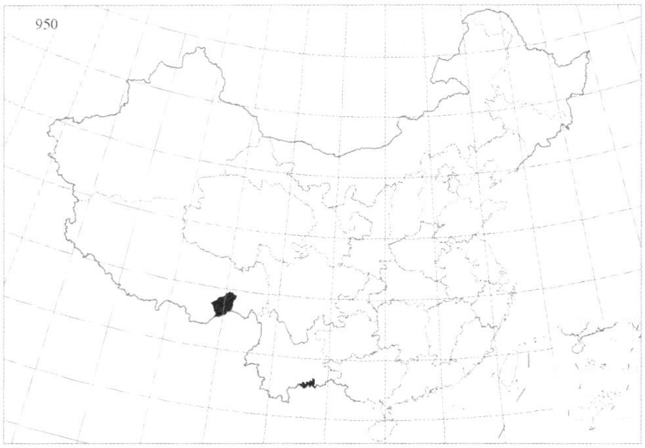

Fagaceae 壳斗科 *Cyclobalanopsis* 青冈属
Cyclobalanopsis sessilifolia (Blume) Schottky 云山青冈
Trees, evergreen

16.9 (11.9~22.2); 6.0 (2.1~17.7); 27.0 (20.0~29.9); 16.9 (12.0~22.2); 143.6 (87.2~206.7); 0.4 (0.0~4.9); 897 (671~1170) ❙ 1535 (1001~2993); 557 (392~1847); 174 (35~551) ❙ 71.2 (16.9~203.4); 890 (671~1170); 221.1 (140.5~354.5)

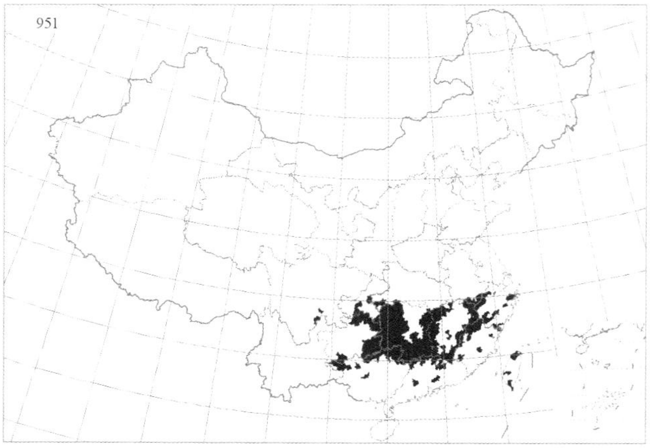

Fagaceae 壳斗科 *Cyclobalanopsis* 青冈属
Cyclobalanopsis sichourensis Hu 西畴青冈
Trees, evergreen

18.1; 10.1; 23.9; 18.1; 157.3; 0.0; 888 ❙ 1273; 714; 49 ❙ 43.7; 888; 339.4

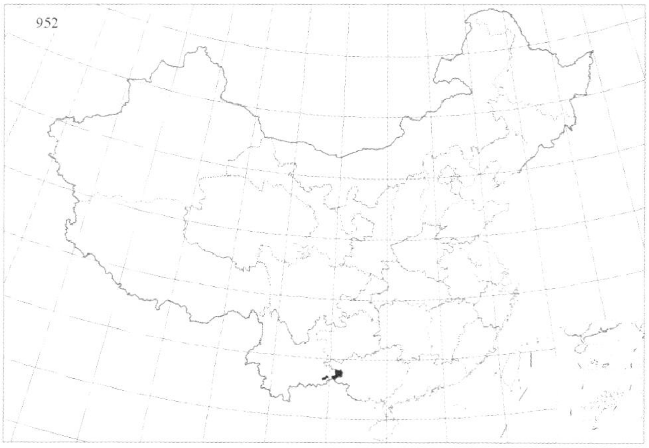

Fagaceae 壳斗科 *Cyclobalanopsis* 青冈属
Cyclobalanopsis stenophylloides (Hayata) Kudo et Masam. 台湾窄叶青冈
Trees, evergreen

19.7 (16.2~24.1); 14.3 (11.0~18.9); 23.8 (20.1~28.0); 19.7 (16.2~24.1); 176.0 (134.2~229.7); 0.0 (0.0~0.0); 975 (773~1322) ❙ 2367 (1751~2759); 1261 (1033~1594); 154 (47~245) ❙ 156.0 (32.5~246.5); 936 (773~1095); 250.2 (154.2~363.7)

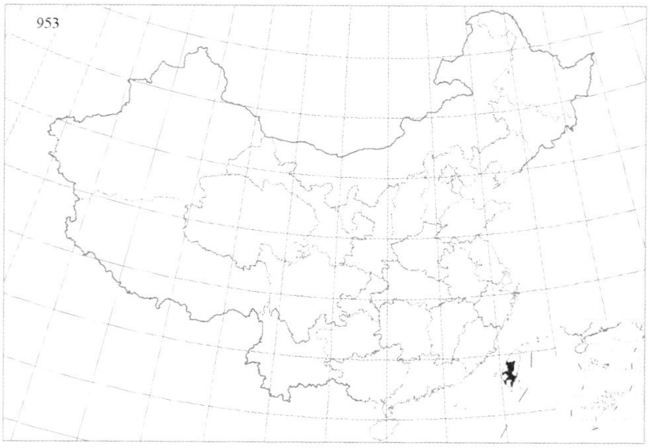

Fagaceae 壳斗科 *Cyclobalanopsis* 青冈属
Cyclobalanopsis stewardiana (A. Camus) Y. C. Hsu et H. W. Jen 褐叶青冈
Trees, evergreen

16.2 (8.6~21.0); 5.1 (1.1~11.9); 26.5 (15.0~29.4); 16.2 (8.7~21.1); 135.3 (52.6~193.0); 0.8 (0.0~8.8); 863 (574~1114) ❙ 1453 (976~1997); 532 (392~821); 156 (44~259) ❙ 68.6 (20.8~130.7); 857 (574~1114); 217.8 (134.6~438.9)

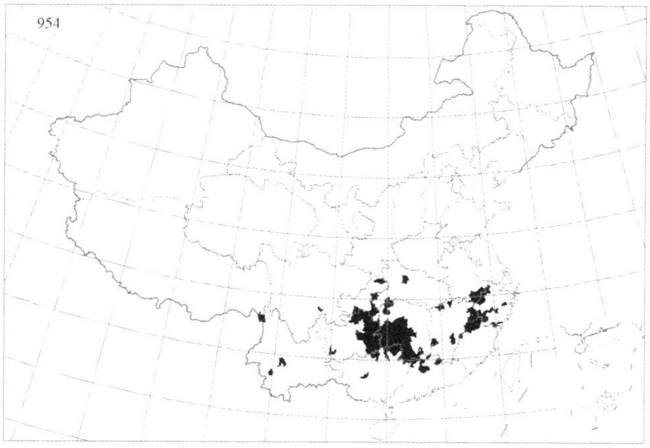

Fagaceae 壳斗科 *Cyclobalanopsis* 青冈属
Cyclobalanopsis subhinoidea (Chun et W. C. Ko) Y. C. Hsu et H. W. Jen 鹿茸青冈 Trees, evergreen

23.5; 18.1; 27.3; 23.5; 222.4; 0.0; 1260 ❙ 1449; 503; 53 ❙ 15.4; 1122; 489.3

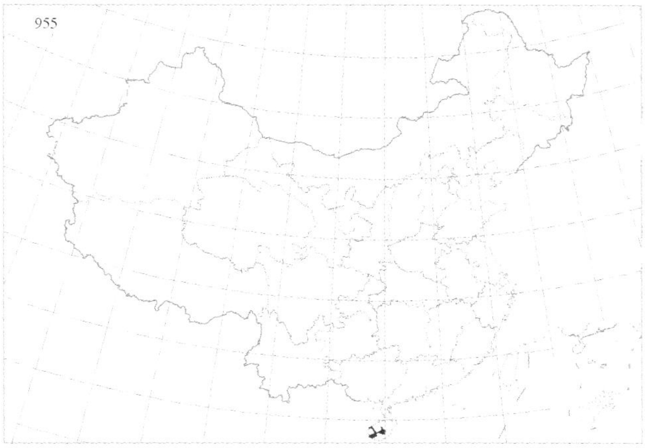

Fagaceae 壳斗科 *Cyclobalanopsis* 青冈属
Cyclobalanopsis tenuicupula Y. C. Hsu et H. W. Jen 薄斗青冈
Trees, evergreen

19.1; 13.0; 23.3; 19.1; 169.7; 0.0; 917 ❙ 1428; 835; 56 ❙ 55.7; 917; 506.2

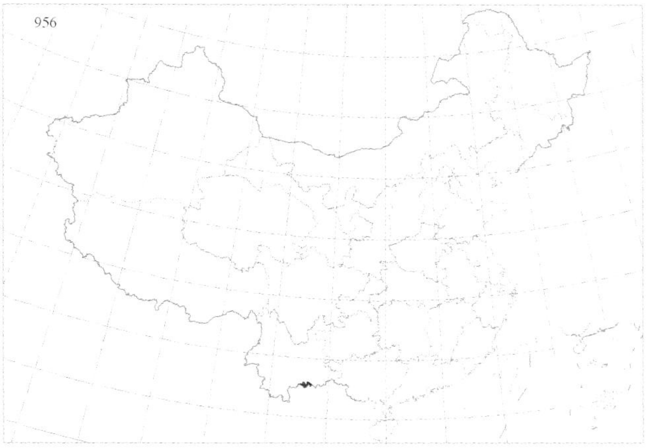

Fagaceae 壳斗科 *Cyclobalanopsis* 青冈属
Cyclobalanopsis thorelii (Hickel et A. Camus) Hu 厚缘青冈
Trees, evergreen

19.2 (17.4~20.5); 12.0 (7.4~15.1); 24.5 (22.3~26.6); 19.3 (17.4~20.5); 171.5 (148.9~185.9); 0.0 (0.0~0.0); 949 (886~992) ❙ 1445 (1236~1607); 691 (617~848); 73 (58~107) ❙ 52.4 (27.5~64.9); 939 (880~986); 426.1 (245.3~659.0)

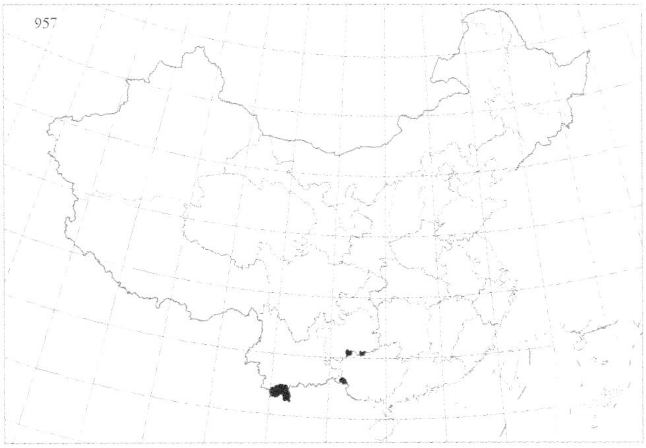

Fagaceae 壳斗科 *Cyclobalanopsis* 青冈属
Cyclobalanopsis tiaoloshanica (Chun et W. C. Ko) Y. C. Hsu et H. W. Jen 吊罗山青冈 Trees, evergreen

24.0; 19.1; 27.2; 24.0; 228.1; 0.0; 1302 ❙ 1551; 539; 48 ❙ 19.1; 1147; 620.0

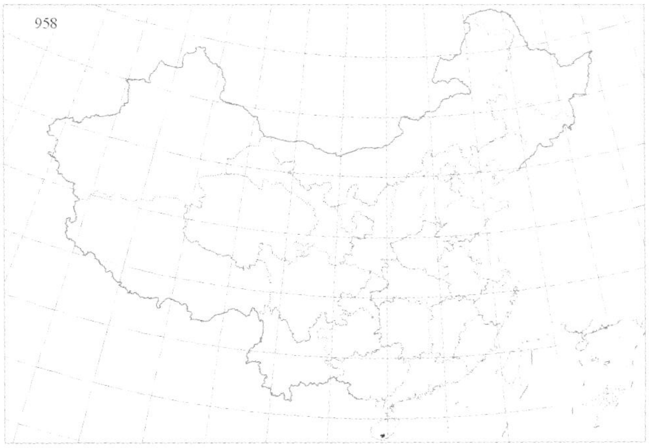

Fagaceae 壳斗科 *Cyclobalanopsis* 青冈属
Cyclobalanopsis tomentosinervis Y. C. Hsu et H. W. Jen 毛脉青冈
Trees, evergreen

19.1; 13.0; 23.3; 19.1; 169.7; 0.0; 917 ❙ 1428; 835; 56 ❙ 55.7; 917; 506.2

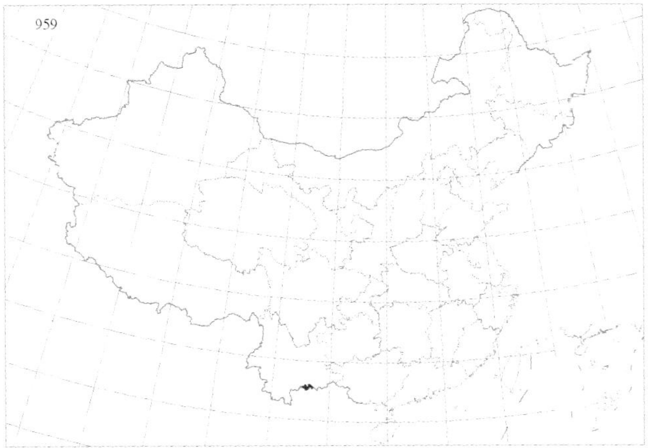

Fagaceae 壳斗科 *Cyclobalanopsis* 青冈属
Cyclobalanopsis xanthotricha (A. Camus) Y. C. Hsu et H. W. Jen 思茅青冈 Trees, evergreen

18.6; 12.8; 22.1; 18.6; 163.7; 0.0; 885 ❙ 1497; 780; 57 ❙ 69.3; 865; 433.7

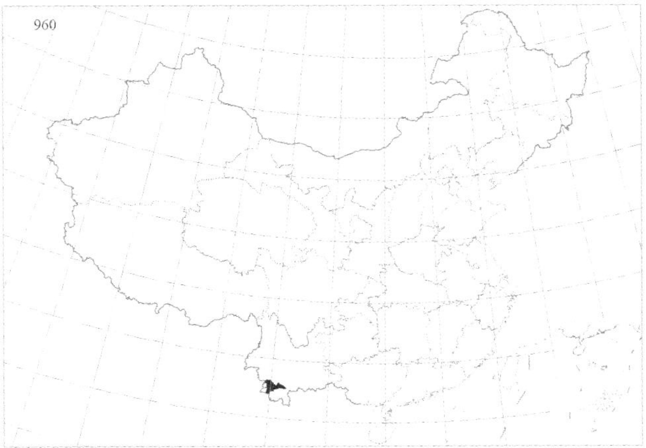

Fagaceae 壳斗科 *Cyclobalanopsis* 青冈属
Cyclobalanopsis yingjiangensis Y. C. Hsu et Q. Z. Dong 盈江青冈
Trees, evergreen

16.4; 9.6; 20.6; 16.4; 137.3; 0.0; 795 ❙ 1688; 959; 62 ❙ 112.2; 795; 627.0

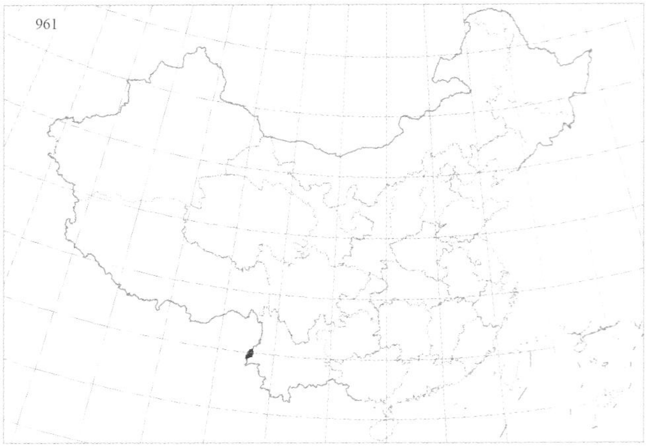

Fagaceae 壳斗科 *Cyclobalanopsis* 青冈属
Cyclobalanopsis yonganensis (L. G. Lin et C. C. Huang) Y. C. Hsu et H. W. Jen 永安青冈
Trees, evergreen

18.0; 8.8; 26.3; 18.0; 156.0; 0.0; 916 ❙ 1711; 583; 202 ❙ 86.9; 916; 285.2

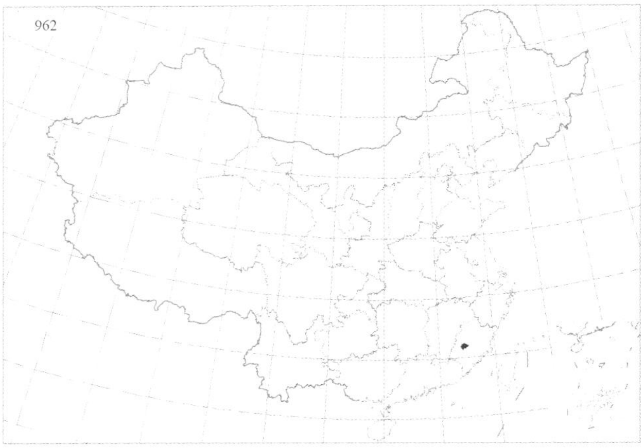

Fagaceae 壳斗科 *Fagus* 水青冈属
Fagus engleriana Seemen 米心水青冈
Trees, deciduous

13.3 (5.7~17.9); 2.0 (-3.5~7.1); 24.0 (14.2~28.3); 13.3 (6.4~17.9); 106.1 (38.1~154.8); 6.3 (0.0~28.1); 749 (504~937) ❙ 1146 (622~1950); 502 (298~817); 76 (15~246) ❙ 52.6 (-13.4~147.0); 748 (504~937); 240.3 (139.5~334.4)

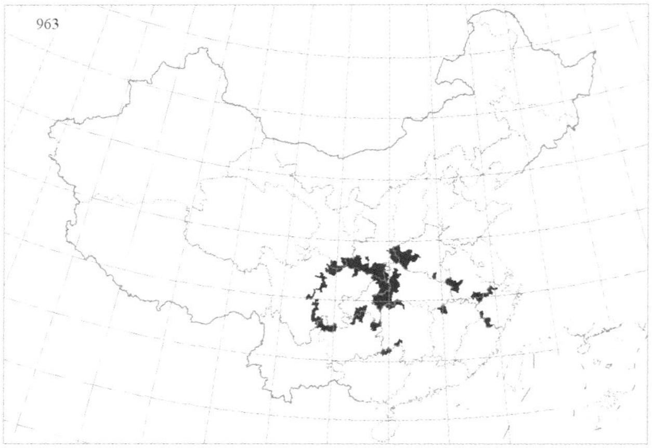

Fagaceae 壳斗科 *Fagus* 水青冈属
Fagus hayatae Palib. 台湾水青冈
Trees, deciduous

15.0 (9.3~21.1); 5.1 (-1.7~14.6); 24.6 (20.1~27.4); 15.1 (9.5~21.1); 125.0 (69.8~193.1); 4.1 (0.0~18.0); 811 (603~1077) ❙ 1673 (806~3495); 630 (436~959); 201 (16~897) ❙ 100.4 (8.2~235.6); 811 (603~1077); 254.4 (171.5~285.1)

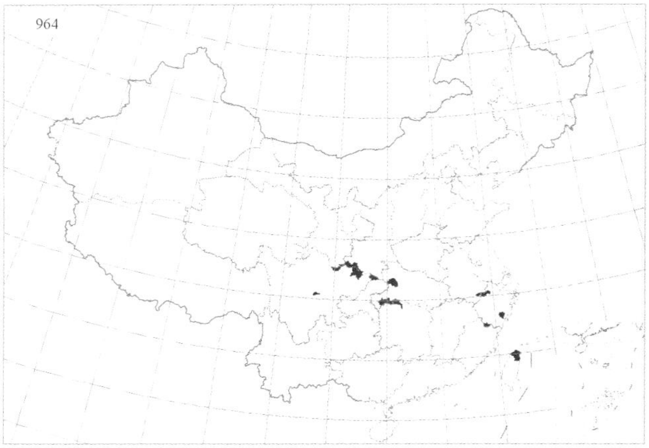

Fagaceae 壳斗科 *Fagus* 水青冈属
Fagus longipetiolata Seemen 水青冈
Trees, deciduous

16.4 (2.6~21.9); 5.9 (-7.0~13.2); 26.0 (11.2~29.8); 16.5 (4.3~22.0); 138.5 (20.0~203.8); 1.1 (0.0~48.2); 871 (427~1176) ❙ 1420 (711~1997); 566 (352~911); 136 (15~259) ❙ 63.4 (5.8~147.0); 868 (427~1176); 232.0 (134.6~551.3)

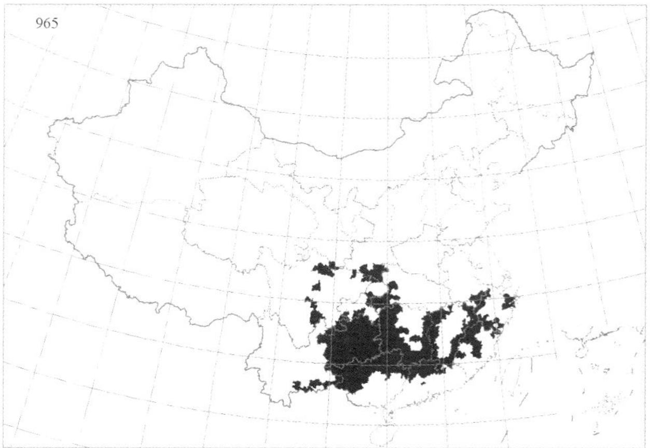

Fagaceae 壳斗科 *Fagus* 水青冈属
Fagus lucida Rehder et E. H. Wilson 光叶水青冈
Trees, deciduous

16.1 (9.1~21.0); 5.1 (-1.7~11.9); 26.4 (17.7~29.8); 16.1 (9.2~21.1); 134.6 (64.1~193.0); 1.0 (0.0~18.0); 861 (595~1114) ❙ 1420 (822~1997); 528 (376~821); 146 (19~259) ❙ 65.2 (7.8~147.0); 855 (595~1114); 214.7 (134.6~354.5)

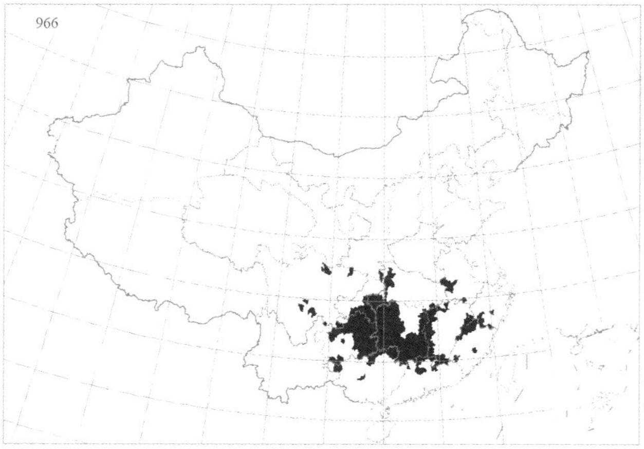

Fagaceae 壳斗科 *Formanodendron* 三棱栎属

Formanodendron doichangensis (A. Camus) Nixon et Crepet 三棱栎

Trees, evergreen

18.7 (14.4~20.5); 12.9 (7.5~15.1); 22.2 (19.1~23.7); 18.7 (14.4~20.5); 164.6 (113.1~185.9); 0.0 (0.0~0.0); 897 (727~992) ❙ 1483 (1434~1607); 733 (655~803); 57 (41~77) ❙ 66.4 (47.6~99.9); 873 (727~986); 511.9 (427.9~659.0)

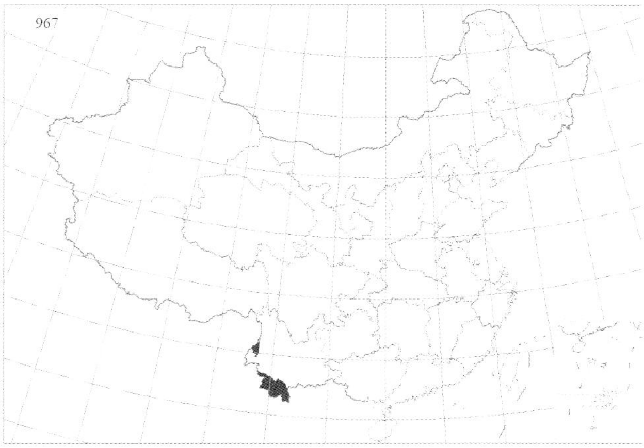

Fagaceae 壳斗科 *Lithocarpus* 石栎属

Lithocarpus amoenus Chun et C. C. Huang 愉柯

Trees, evergreen

20.5 (16.2~22.7); 11.9 (5.4~15.0); 27.5 (24.4~28.8); 20.6 (16.2~22.8); 186.8 (134.5~213.1); 0.0 (0.0~0.0); 1079 (826~1221) ❙ 1715 (1302~2435); 828 (518~1428); 134 (86~178) ❙ 59.3 (12.9~99.5); 1079 (826~1221); 257.0 (203.4~318.8)

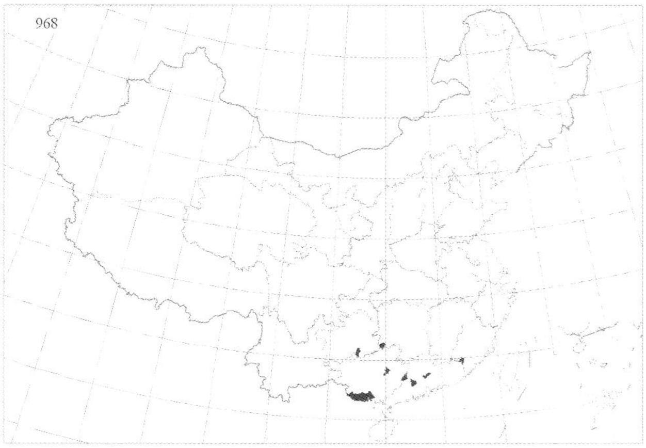

Fagaceae 壳斗科 *Lithocarpus* 石栎属

Lithocarpus amygdalifolius (Skan) Hayata 杏叶柯

Trees, evergreen

21.5 (16.2~24.7); 14.1 (6.5~19.8); 27.2 (20.1~28.9); 21.5 (16.2~24.7); 198.1 (134.2~236.5); 0.0 (0.0~0.0); 1130 (773~1376) ❙ 1700 (868~2992); 797 (312~1847); 114 (42~245) ❙ 55.4 (-34.3~246.5); 1087 (773~1299); 316.6 (145.3~620.0)

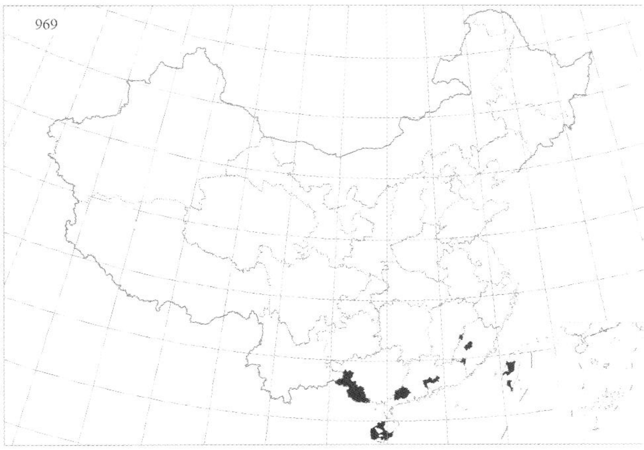

Fagaceae 壳斗科 *Lithocarpus* 石栎属

Lithocarpus apricus C. C. Huang et Y. T. Chang 向阳柯

Trees, evergreen

18.1; 11.7; 22.4; 18.1; 157.8; 0.0; 873 ❙ 1039; 582; 45 ❙ 13.3; 817; 352.4

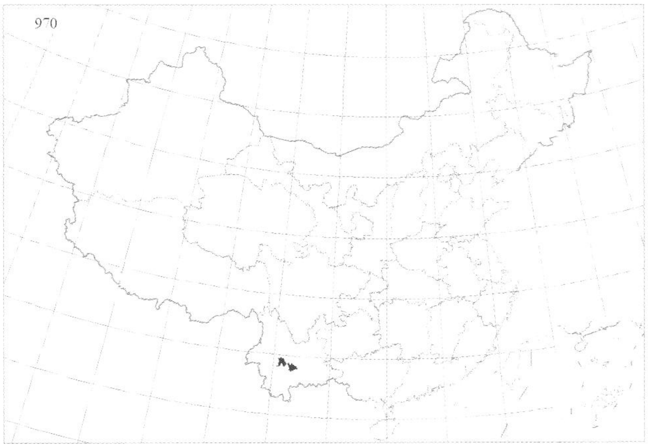

Fagaceae 壳斗科 *Lithocarpus* 石栎属

Lithocarpus arcaulus (Buch.-Ham. ex Spreng.) C. C. Huang et Y. T. Chang 小箱柯

Trees, evergreen

-1.8; -10.2; 6.1; 1.8; 1.9; 82.8; 329 ❙ 320; 217; 20 ❙ -3.0; 320; 72.2

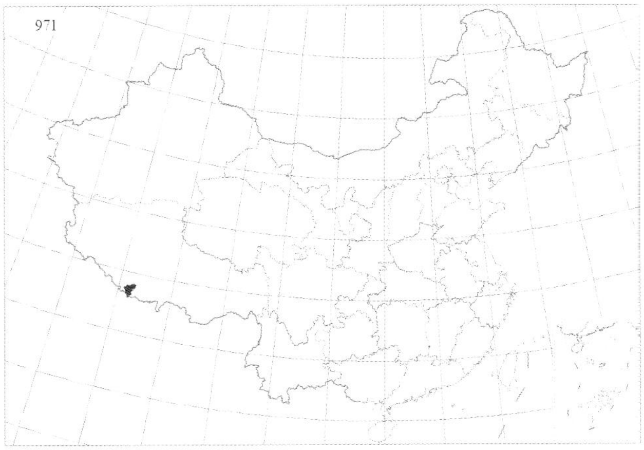

Fagaceae 壳斗科 *Lithocarpus* 石栎属

Lithocarpus areca (Hickel et A. Camus) A. Camus 槟榔柯

Trees, evergreen

20.2 (17.6~22.2); 11.7 (9.9~13.9); 26.5 (23.2~28.5); 20.2 (17.7~22.3); 182.7 (152.4~207.5); 0.0 (0.0~0.0); 1043 (859~1190) ❙ 1469 (1374~1600); 760 (614~839); 98 (52~165) ❙ 42.9 (15.4~65.6); 1043 (859~1190); 313.7 (253.0~371.7)

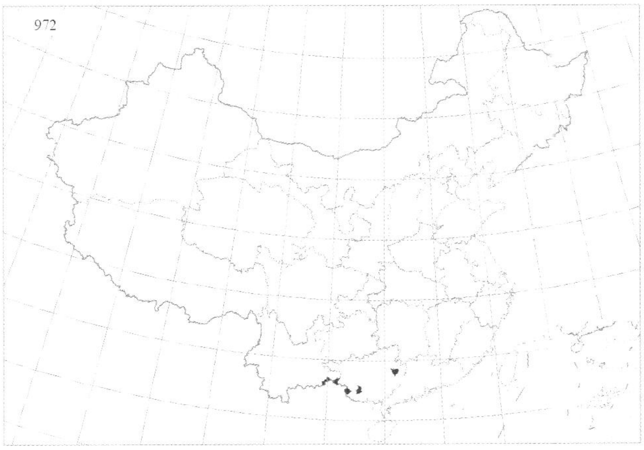

Fagaceae 壳斗科 *Lithocarpus* 石栎属

Lithocarpus attenuatus (Skan) Rehder 尖叶柯

Trees, evergreen

22.2 (21.3~23.2); 14.4 (13.2~15.8); 28.2 (27.3~28.9); 22.3 (21.3~23.2); 207.4 (195.9~218.9); 0.0 (0.0~0.0); 1185 (1111~1257) ❙ 1797 (1302~2435); 938 (670~1428); 110 (75~130) ❙ 51.7 (12.9~99.5); 1185 (1111~1257); 263.4 (194.5~327.7)

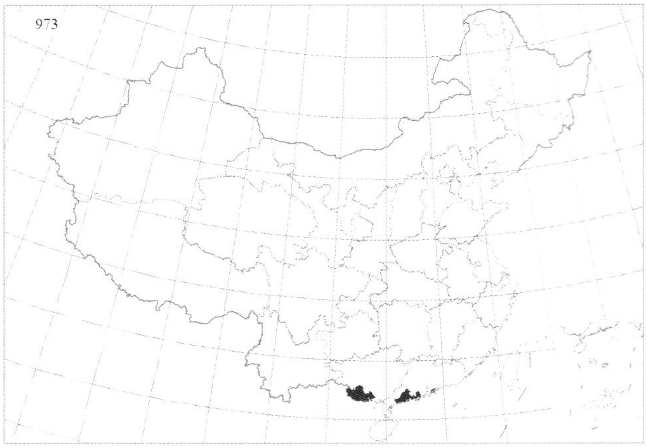

Fagaceae 壳斗科 *Lithocarpus* 石栎属

Lithocarpus bacgiangensis (Hickel et A. Camus) A. Camus 茸果柯

Trees, evergreen

22.1 (16.3~24.8); 15.8 (7.3~20.1); 26.6 (21.4~28.8); 22.2 (16.4~24.8); 206.2 (136.4~237.6); 0.0 (0.0~0.0); 1173 (797~1385) ❙ 1475 (868~1876); 614 (312~891); 73 (41~142) ❙ 28.7 (-34.3~64.9); 1099 (797~1360); 426.6 (225.9~620.0)

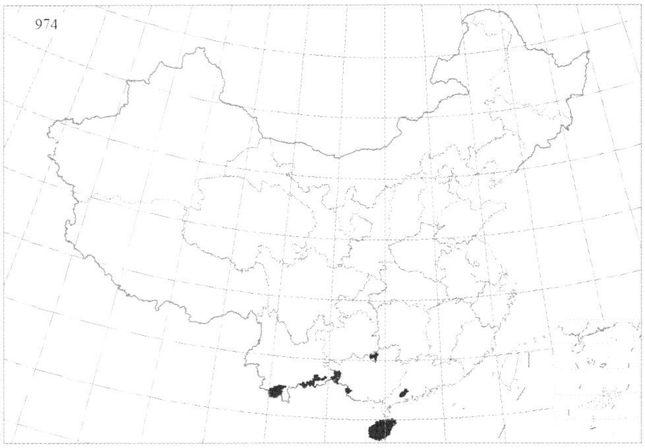

Fagaceae 壳斗科 *Lithocarpus* 石栎属

Lithocarpus balansae (Drake) A. Camus 猴面柯

Trees, evergreen

18.2 (17.2~21.0); 11.5 (9.5~14.2); 22.9 (21.6~25.6); 18.2 (17.3~21.0); 159.0 (147.0~191.8); 0.0 (0.0~0.0); 882 (832~1053) ❙ 1381 (1049~1547); 785 (598~911); 56 (44~68) ❙ 56.8 (26.1~80.6); 876 (783~1053); 445.4 (336.3~551.6)

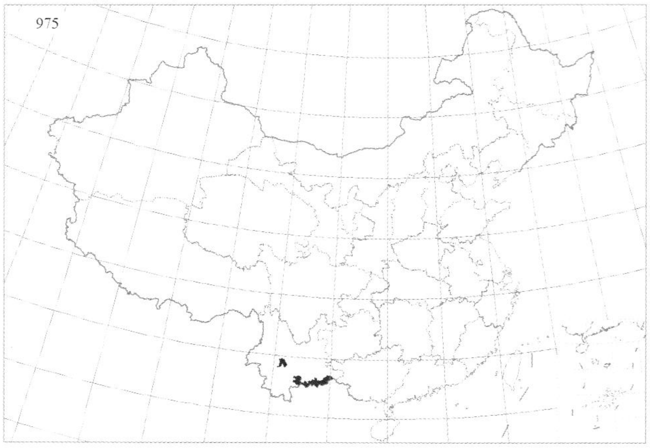

Fagaceae 壳斗科 *Lithocarpus* 石栎属

Lithocarpus bonnetii (Hickel et A. Camus) A. Camus 帽柯

Trees, evergreen

21.6; 15.5; 25.9; 21.6; 199.6; 0.0; 1131 ❙ 1487; 632; 51 ❙ 35.1; 1050; 506.0

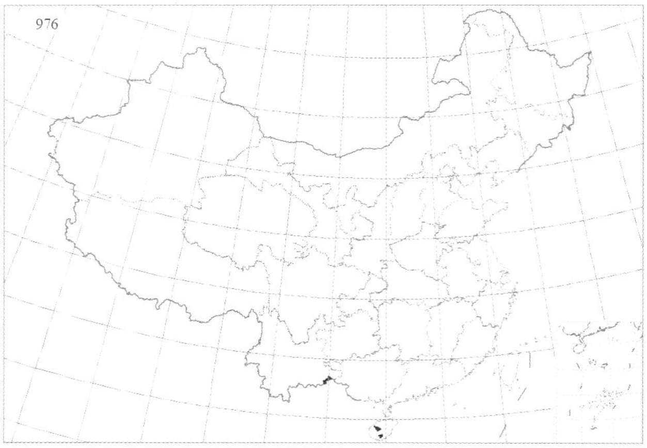

Fagaceae 壳斗科 *Lithocarpus* 石栎属

Lithocarpus brachystachyus Chun 短穗柯

Trees, evergreen

23.7 (21.8~24.6); 18.0 (14.2~19.5); 27.8 (27.2~28.4); 23.8 (21.9~24.7); 225.1 (202.6~235.9); 0.0 (0.0~0.0); 1291 (1150~1372) ❙ 1550 (1179~1876); 569 (392~891); 69 (42~116) ❙ 21.2 (-11.9~63.2); 1153 (1056~1299); 470.3 (273.3~620.0)

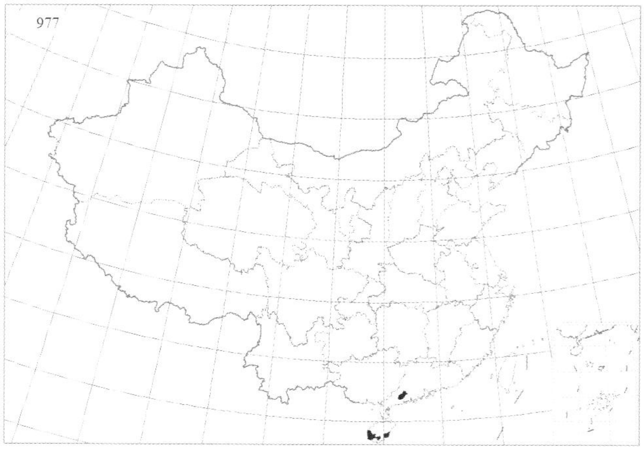

Fagaceae 壳斗科 *Lithocarpus* 石栎属

Lithocarpus brevicaudatus (Skan) Hayata 岭南柯

Trees, evergreen

17.1 (13.5~24.3); 6.9 (2.6~18.7); 26.3 (20.1~29.7); 17.1 (13.6~24.3); 145.6 (106.7~231.9); 0.3 (0.0~4.1); 897 (734~1338) ❙ 1477 (868~2993); 568 (312~1847); 143 (42~547) ❙ 65.6 (-34.3~246.5); 887 (734~1163); 228.1 (140.5~572.9)

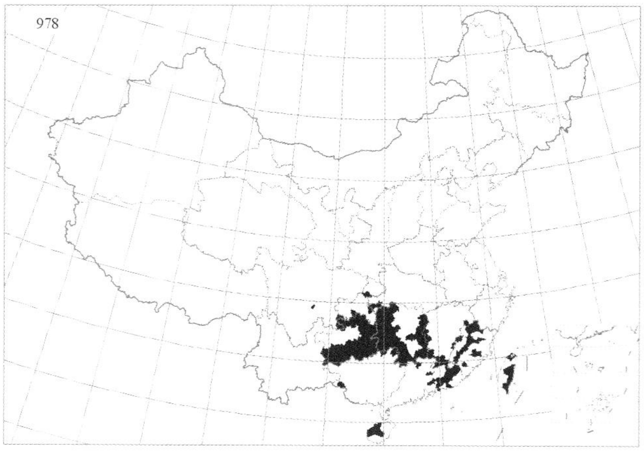

Fagaceae 壳斗科 *Lithocarpus* 石栎属
Lithocarpus calolepis Y. C. Hsu et H. W. Jen 美苞柯
Trees, evergreen

17.1; 9.8; 22.2; 17.1; 145.6; 0.0; 830 ❙ 1271; 704; 53 ❙ 53.1; 830; 391.7

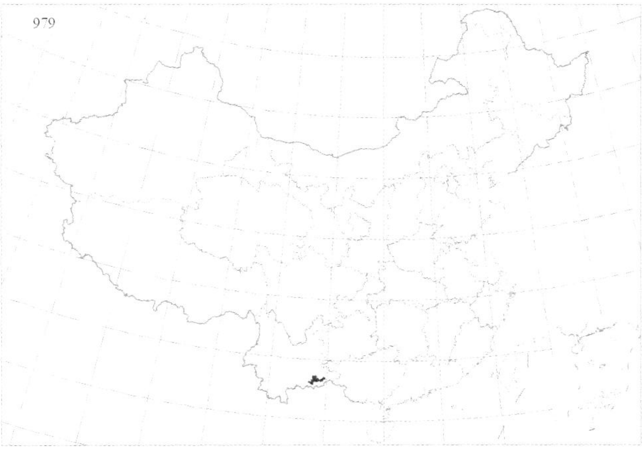

Fagaceae 壳斗科 *Lithocarpus* 石栎属
Lithocarpus calophyllus Chun ex C. C. Huang et Y. T. Chang 美叶柯
Trees, evergreen

17.6 (14.3~22.2); 7.0 (3.3~14.7); 27.3 (24.2~29.8); 17.7 (14.3~22.3); 152.5 (113.2~207.2); 0.3 (0.0~2.3); 934 (761~1186) ❙ 1529 (1171~2014); 561 (392~954); 168 (73~230) ❙ 64.2 (30.9~128.9); 926 (761~1186); 219.5 (134.6~322.1)

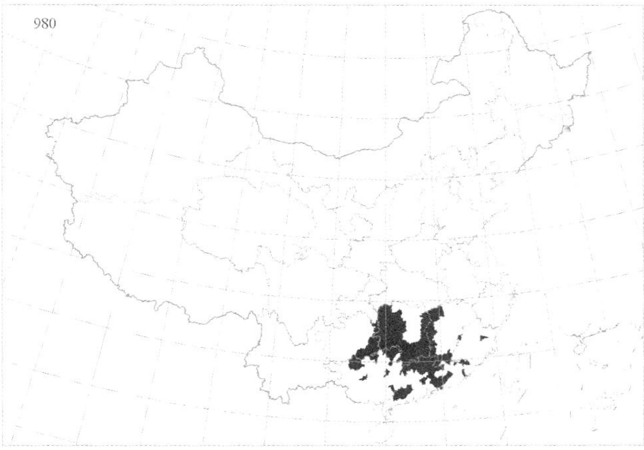

Fagaceae 壳斗科 *Lithocarpus* 石栎属
Lithocarpus carolineae (Skan) Rehder 红心柯
Trees, evergreen

19.3; 12.7; 23.8; 19.3; 171.6; 0.0; 942 ❙ 1439; 803; 61 ❙ 53.4; 942; 516.4

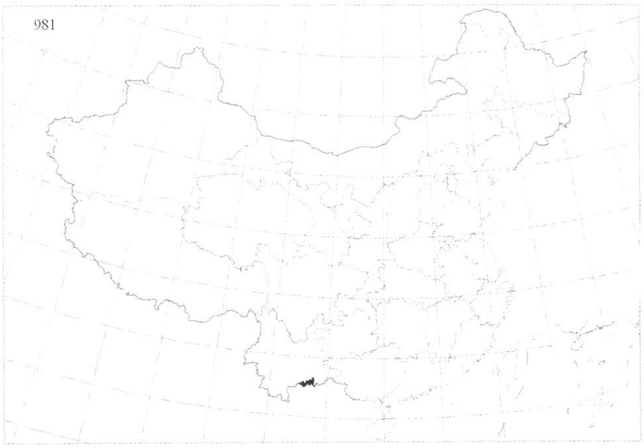

Fagaceae 壳斗科 *Lithocarpus* 石栎属
Lithocarpus caudatilimbus (Merr.) A. Camus 尾叶柯
Trees, evergreen

24.7; 19.8; 28.1; 24.7; 236.8; 0.0; 1378 ❙ 1586; 520; 65 ❙ 15.1; 1218; 537.7

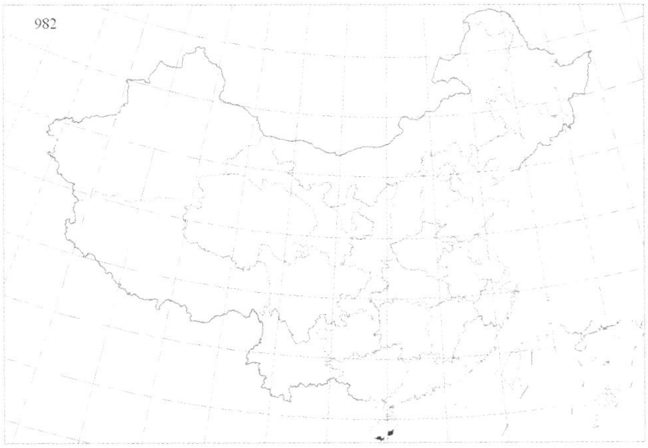

Fagaceae 壳斗科 *Lithocarpus* 石栎属
Lithocarpus chifui Chun et Tsiang 粤北柯
Trees, evergreen

17.7; 7.7; 26.6; 17.8; 153.6; 0.0; 916 ❙ 1564; 556; 182 ❙ 70.7; 916; 232.5

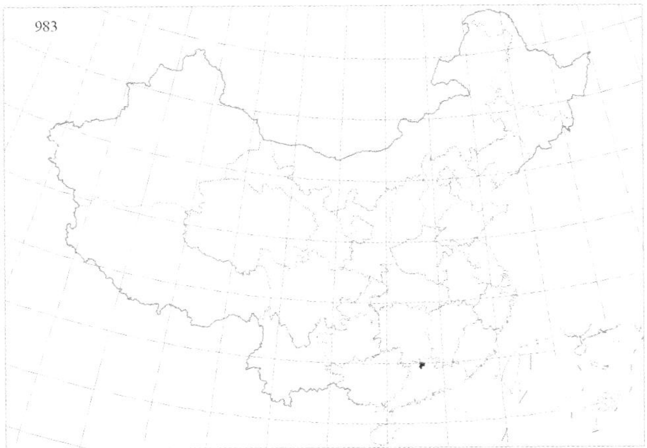

Fagaceae 壳斗科 *Lithocarpus* 石栎属
Lithocarpus chiungchungensis Chun et P. C. Tam 琼中柯
Trees, evergreen

23.0; 17.7; 26.7; 23.1; 217.0; 0.0; 1217 ❙ 1725; 610; 68 ❙ 41.8; 1163; 572.9

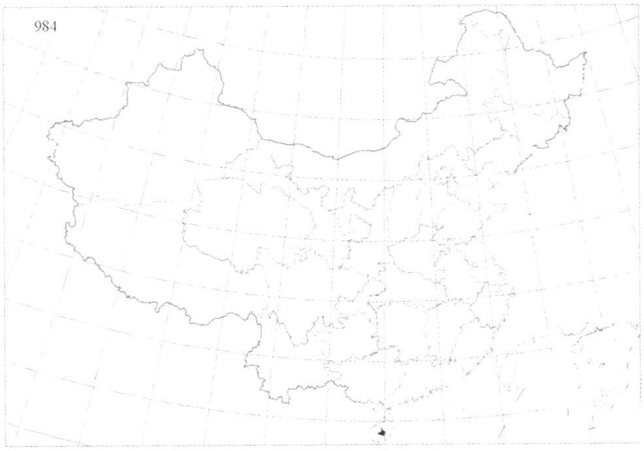

Fagaceae 壳斗科 *Lithocarpus* 石栎属
Lithocarpus chrysocomus Chun et Tsiang 金毛柯
Trees, evergreen

17.8 (14.3~21.6); 7.5 (3.7~13.2); 26.8 (24.2~28.5); 17.8 (14.3~21.6); 153.7 (113.2~199.2); 0.1 (0.0~1.4); 926 (761~1139) ❙ 1576 (1187~1953); 574 (465~858); 178 (84~218) ❙ 70.7 (44.1~128.9); 926 (761~1139); 234.0 (188.8~300.5)

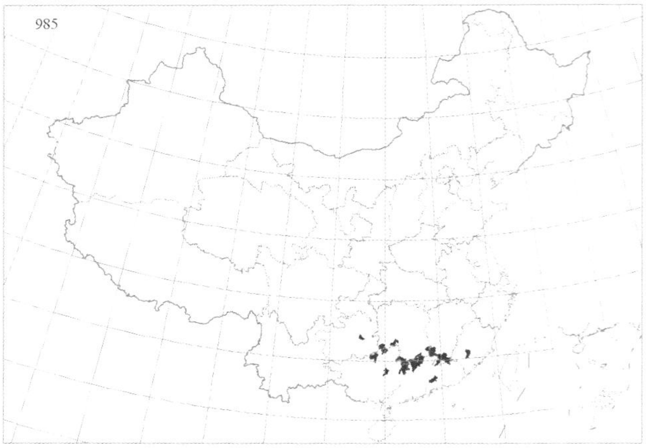

Fagaceae 壳斗科 *Lithocarpus* 石栎属
Lithocarpus cinereus Chun et C. C. Huang 炉灰柯
Trees, evergreen

21.2 (17.7~22.7); 13.4 (10.9~15.0); 26.9 (22.5~28.6); 21.2 (17.8~22.8); 194.7 (153.4~213.1); 0.0 (0.0~0.0); 1108 (854~1221) ❙ 1948 (1368~2435); 1104 (756~1428); 112 (59~130) ❙ 74.8 (56.8~99.5); 1108 (854~1221); 328.6 (235.1~491.5)

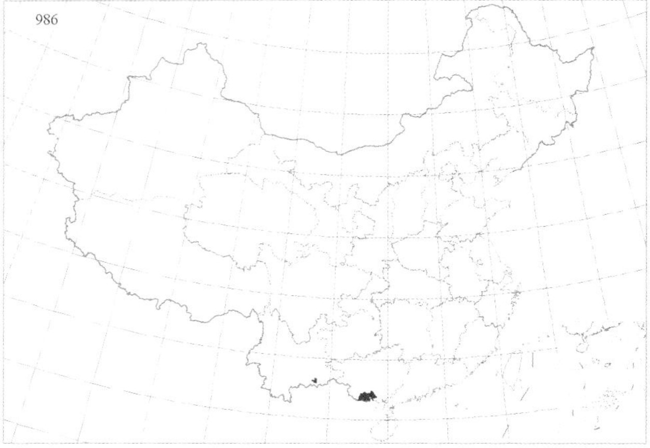

Fagaceae 壳斗科 *Lithocarpus* 石栎属
Lithocarpus cleistocarpus (Seemen) Rehder et E. H. Wilson 包果柯
Trees, evergreen

15.8 (6.2~22.2); 5.1 (-2.4~14.7); 25.7 (14.1~29.8); 15.8 (6.7~22.2); 131.2 (39.4~206.9); 1.4 (0.0~24.1); 848 (509~1177) ❙ 1380 (711~1997); 541 (357~891); 132 (15~259) ❙ 63.0 (4.4~147.0); 841 (509~1177); 222.5 (130.1~506.2)

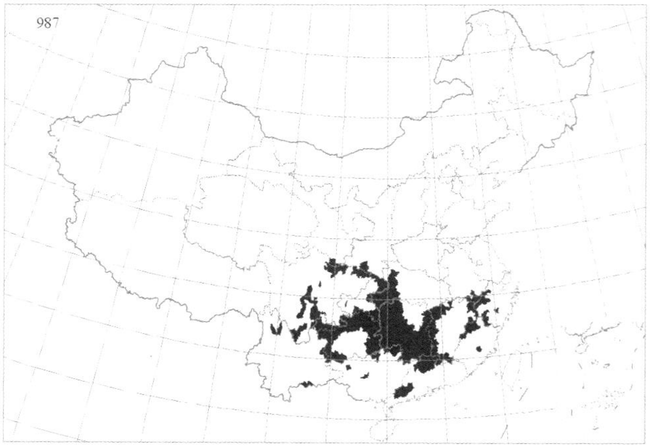

Fagaceae 壳斗科 *Lithocarpus* 石栎属
Lithocarpus cleistocarpus var. omeiensis W. P. Fang 峨眉包果柯
Trees, evergreen

14.1 (10.2~18.5); 6.6 (2.4~10.5); 20.1 (16.6~24.3); 14.1 (10.3~18.3); 110.4 (68.2~160.0); 1.1 (0.0~5.1); 740 (620~906) ❙ 1000 (956~1042); 611 (584~641); 20 (18~24) ❙ 37.7 (5.5~67.1); 691 (620~786); 318.7 (298.0~331.6)

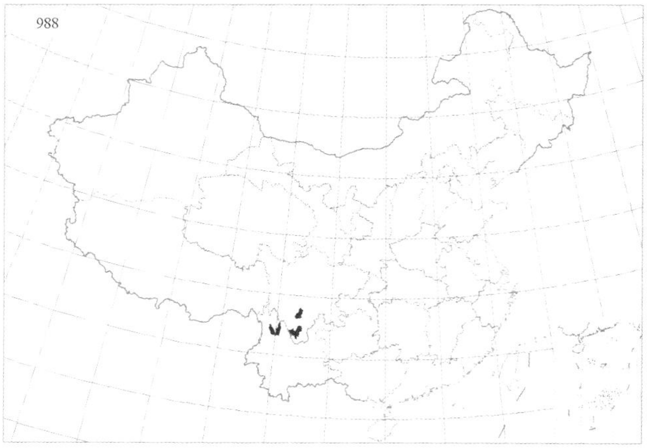

Fagaceae 壳斗科 *Lithocarpus* 石栎属
Lithocarpus collettii (King ex Hook. f.) A. Camus 格林柯
Trees, evergreen

13.3; 5.4; 19.6; 13.4; 100.4; 0.0; 706 ❙ 1736; 1007; 51 ❙ 145.8; 706; 447.4

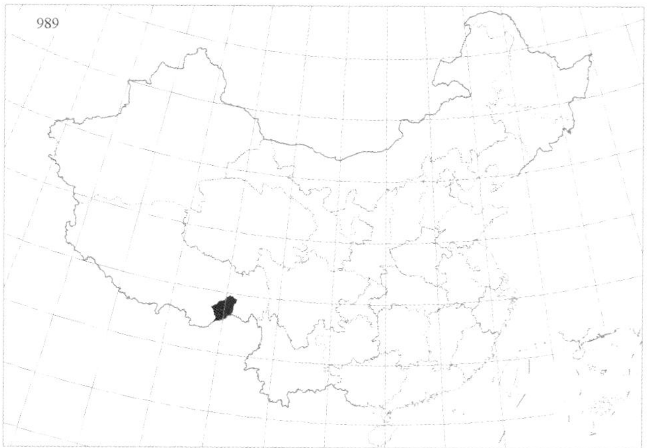

Fagaceae 壳斗科 *Lithocarpus* 石栎属
Lithocarpus confinis C. C. Huang ex Y. C. Hsu et H. W. Jen 窄叶柯
Trees, evergreen

15.8 (11.7~19.3); 7.7 (2.7~13.1); 22.2 (18.4~26.0); 15.8 (11.7~19.4); 130.3 (82.6~173.2); 0.2 (0.0~3.8); 800 (658~966) ❙ 1145 (916~1422); 625 (486~813); 49 (19~70) ❙ 43.6 (13.3~92.4); 790 (658~966); 283.4 (181.2~491.5)

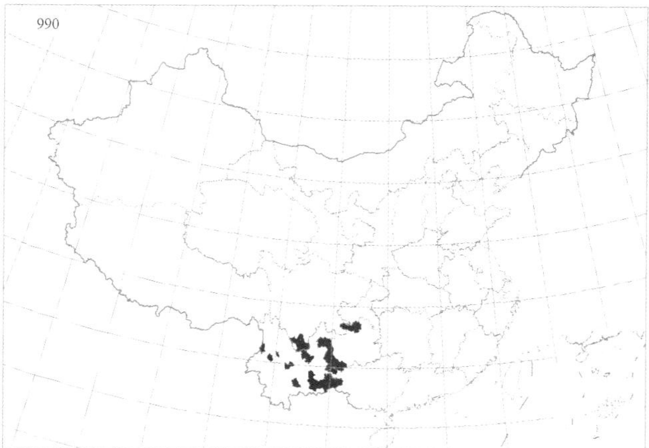

Fagaceae 壳斗科 *Lithocarpus* 石栎属
Lithocarpus corneus (Lour.) Rehder 烟斗柯
Trees, evergreen

20.1 (14.6~24.8); 11.5 (3.8~20.1); 27.2 (20.1~29.8); 20.2 (14.7~24.8); 182.3 (117.7~237.6); 0.0 (0.0~1.5); 1057 (773~1385) ❙ 1586 (868~2992); 696 (312~1847); 134 (41~303) ❙ 52.3 (-34.3~246.5); 1045 (773~1360); 262.7 (139.6~620.0)

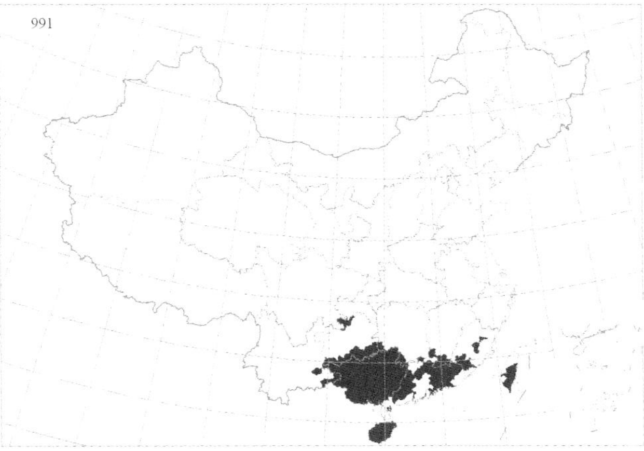

Fagaceae 壳斗科 *Lithocarpus* 石栎属
Lithocarpus corneus var. angustifolius C. C. Huang et Y. T. Chang 窄叶烟斗柯
Trees, evergreen

19.2 (17.7~21.3); 10.5 (9.1~12.6); 25.9 (22.5~27.9); 19.2 (17.8~21.3); 170.7 (153.4~196.0); 0.0 (0.0~0.0); 973 (854~1123) ❙ 1331 (1172~1609); 695 (634~820); 77 (50~177) ❙ 37.5 (5.8~60.1); 973 (854~1121); 335.2 (238.7~491.5)

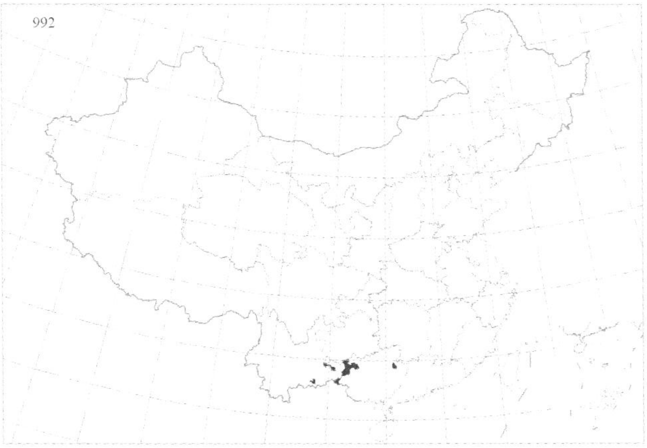

Fagaceae 壳斗科 *Lithocarpus* 石栎属
Lithocarpus corneus var. zonatus C. C. Huang et Y. T. Chang 环鳞烟斗柯
Trees, evergreen

21.4; 12.7; 28.4; 21.4; 197.2; 0.0; 1132 ❙ 1707; 729; 154 ❙ 50.7; 1132; 233.5

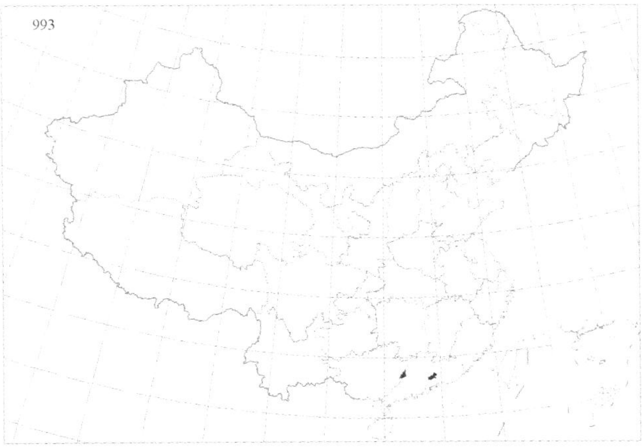

Fagaceae 壳斗科 *Lithocarpus* 石栎属
Lithocarpus craibianus Barnett 白穗柯
Trees, evergreen

13.9 (6.1~19.2); 7.0 (-1.5~13.8); 19.2 (12.7~23.5); 14.0 (6.3~19.4); 110.1 (35.2~172.6); 2.6 (0.0~21.8); 732 (496~932) ❙ 1077 (801~1499); 617 (496~774); 38 (18~58) ❙ 49.0 (13.3~81.5); 714 (496~923); 353.3 (265.0~513.9)

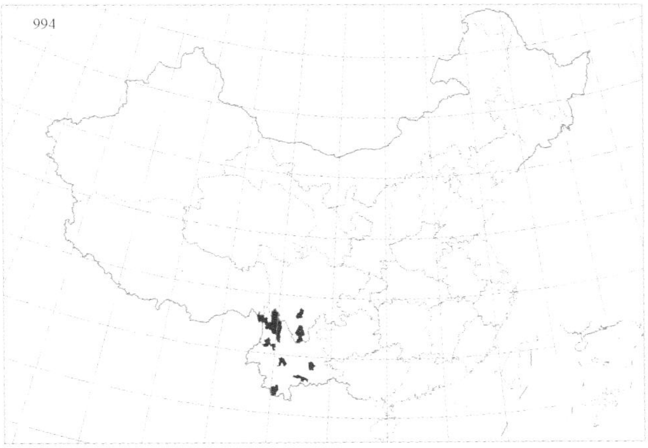

Fagaceae 壳斗科 *Lithocarpus* 石栎属
Lithocarpus crassifolius A. Camus 硬叶柯
Trees, evergreen

18.1; 11.8; 22.3; 18.1; 157.6; 0.0; 869 ❙ 1138; 646; 48 ❙ 31.2; 832; 379.2

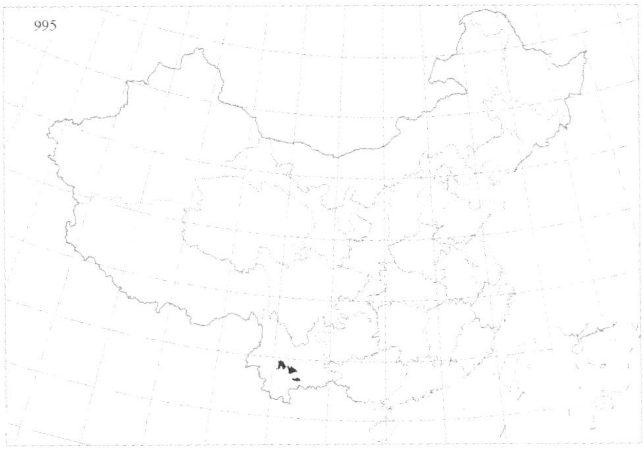

Fagaceae 壳斗科 *Lithocarpus* 石栎属
Lithocarpus cryptocarpus A. Camus 闭壳柯
Trees, evergreen

21.0; 14.2; 25.6; 21.0; 191.8; 0.0; 1053 ❙ 1522; 818; 68 ❙ 44.5; 1053; 551.6

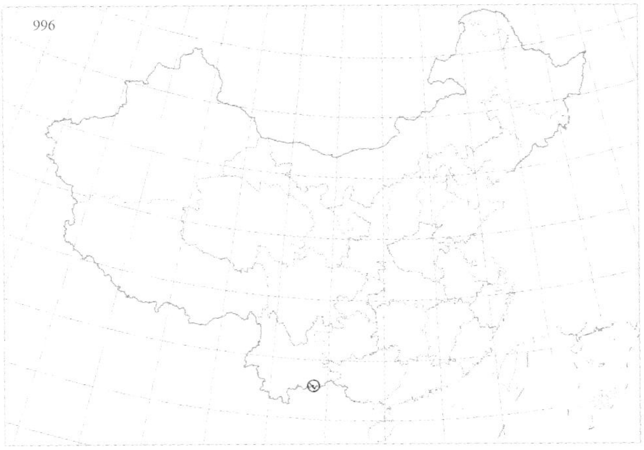

Fagaceae 壳斗科 *Lithocarpus* 石栎属
Lithocarpus cucullatus C. C. Huang et Y. T. Chang 风兜柯
Trees, evergreen

17.9；7.6；27.1；17.9；155.4；0.3；942 ❙ 1597；609；159 ❙ 69.8；942；218.1

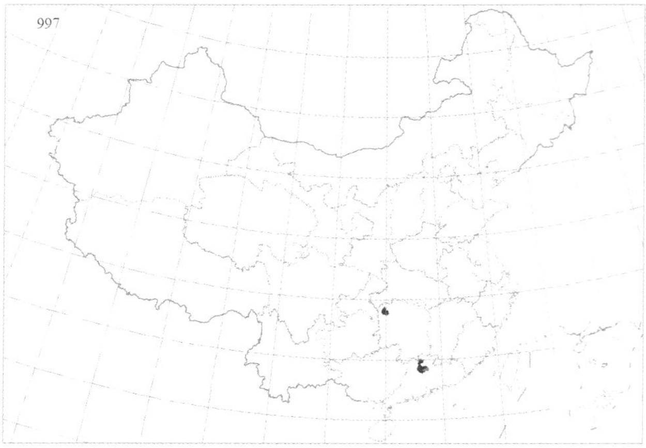

Fagaceae 壳斗科 *Lithocarpus* 石栎属
Lithocarpus cyrtocarpus (Drake) A. Camus 鱼篮柯
Trees, evergreen

20.9 (16.9~23.2)；13.4 (10.3~15.8)；26.5 (21.4~28.9)；21.0 (16.9~23.2)；191.8 (142.7~218.9)；0.0 (0.0~0.0)；1087 (813~1257) ❙ 1606 (1049~2435)；867 (598~1428)；94 (44~130) ❙ 48.1 (12.9~99.5)；1084 (783~1257)；335.4 (194.5~551.6)

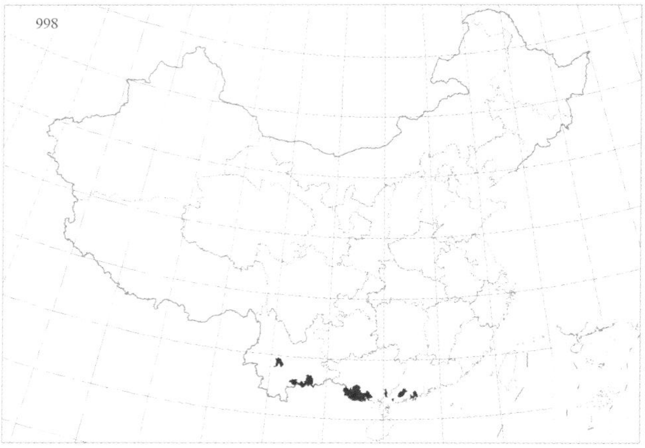

Fagaceae 壳斗科 *Lithocarpus* 石栎属
Lithocarpus damiaoshanicus C. C. Huang et Y. T. Chang 大苗山柯
Trees, evergreen

18.5；8.4；26.9；18.5；162.5；0.0；959 ❙ 1476；662；131 ❙ 54.1；959；240.2

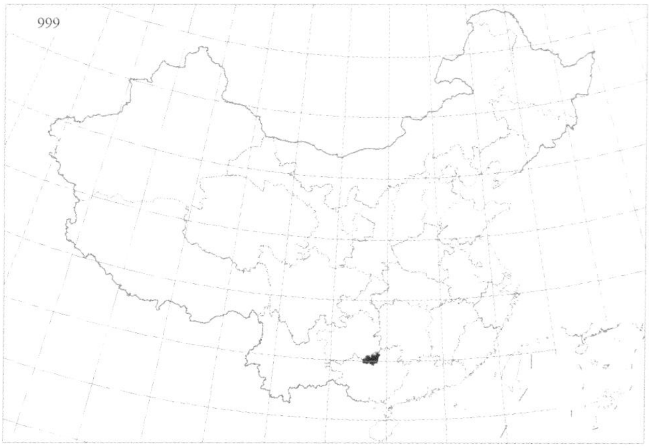

Fagaceae 壳斗科 *Lithocarpus* 石栎属
Lithocarpus dealbatus (Hook. f. et Thomson ex Miq.) Rehder 白柯
Trees, evergreen

14.3 (5.4~19.2)；6.7 (−2.1~13.3)；20.4 (12.3~26.6)；14.3 (5.8~19.4)；113.1 (31.5~172.6)；1.3 (0.0~25.6)；743 (481~932) ❙ 1045 (741~1688)；595 (411~959)；38 (15~69) ❙ 41.9 (−14.2~112.2)；721 (481~923)；294.7 (142.9~627.0)

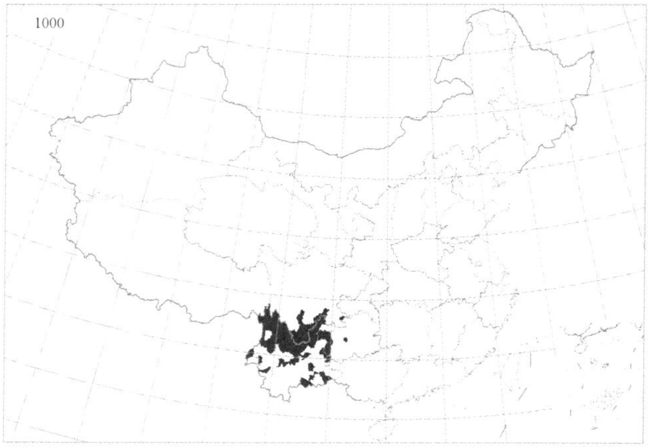

Fagaceae 壳斗科 *Lithocarpus* 石栎属
Lithocarpus dodonaeifolius (Hayata) Hayata 柳叶柯
Trees, evergreen

20.3；15.9；23.6；20.4；184.4；0.0；993 ❙ 2853；1618；147 ❙ 190.4；993；330.3

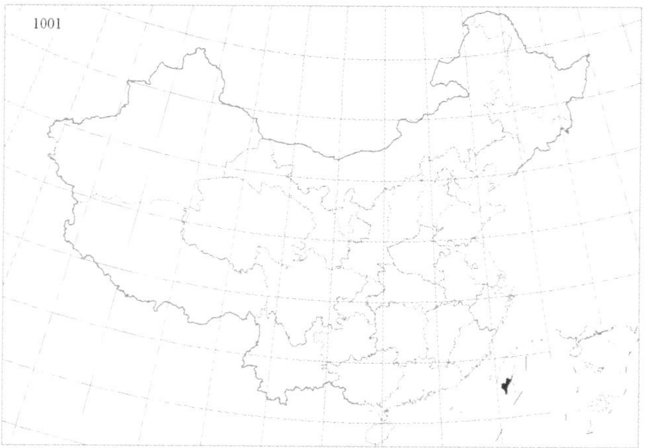

Fagaceae 壳斗科 *Lithocarpus* 石栎属
Lithocarpus echinophorus Hu ex A. Camus 壶壳柯
Trees, evergreen

18.2 (17.3~19.3)；11.9 (10.5~13.1)；22.6 (21.6~23.6)；18.3 (17.3~19.4)；159.3 (148.0~173.2)；0.0 (0.0~0.0)；877 (832~939) ❙ 1261 (958~1547)；720 (531~911)；53 (44~63) ❙ 43.7 (13.3~74.2)；862 (783~917)；429.4 (311.9~551.3)

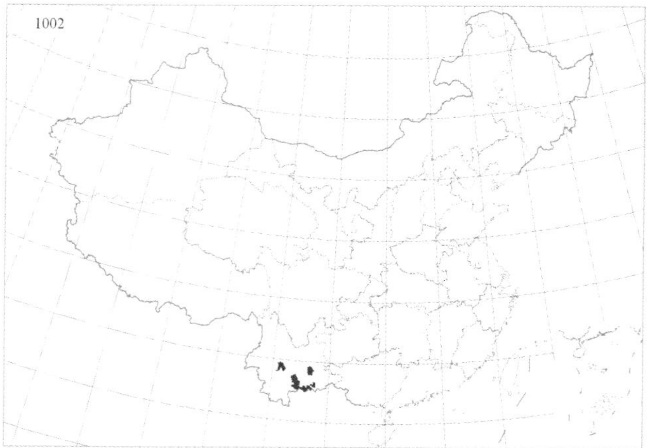

Fagaceae 壳斗科 *Lithocarpus* 石栎属
Lithocarpus echinophorus var. bidoupensis A. Camus 金平柯
Trees, evergreen

18.5; 12.2; 22.8; 18.5; 162.3; 0.0; 886 | 1448; 834; 59 | 63.4; 886; 516.3

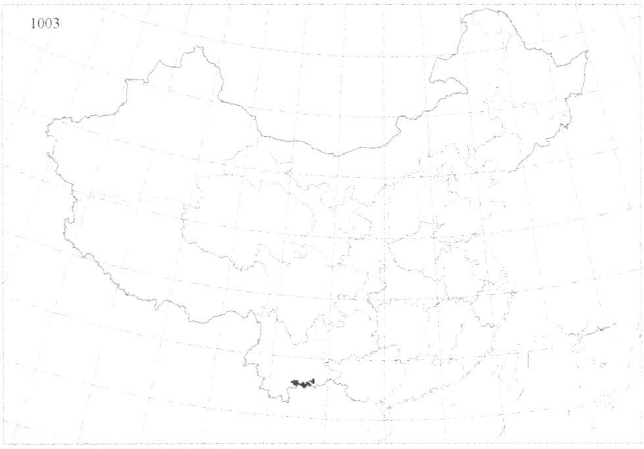

Fagaceae 壳斗科 *Lithocarpus* 石栎属
Lithocarpus echinophorus var. chapensis A. Camus 沙坝柯
Trees, evergreen

17.7; 10.9; 22.5; 17.8; 153.4; 0.0; 854 | 1368; 756; 59 | 60.1; 854; 491.5

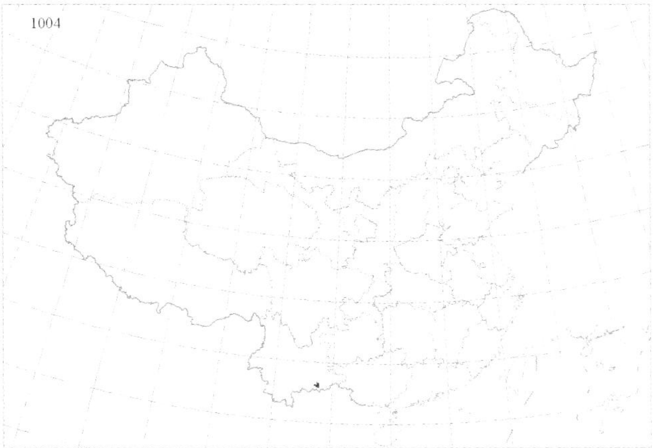

Fagaceae 壳斗科 *Lithocarpus* 石栎属
Lithocarpus echinotholus Hu ex A. Camus 刺壳柯
Trees, evergreen

18.4 (16.9~21.0); 11.7 (9.9~14.2); 23.0 (21.4~25.6); 18.4 (16.9~21.0); 160.9 (142.7~191.8); 0.0 (0.0~0.0); 894 (813~1053) | 1400 (1178~1522); 783 (646~860); 58 (51~68) | 57.2 (44.5~80.6); 894 (813~1053); 469.2 (383.4~551.6)

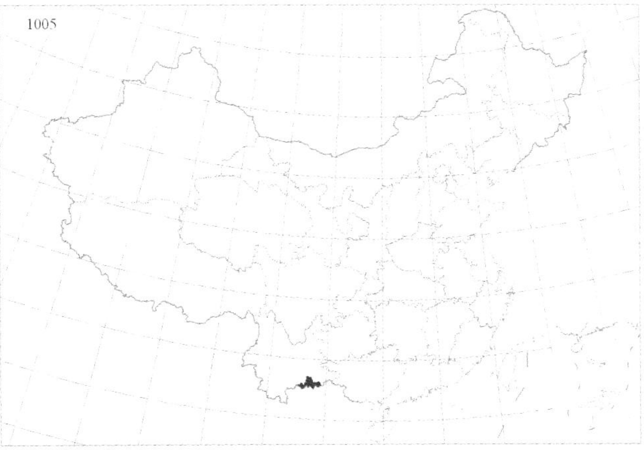

Fagaceae 壳斗科 *Lithocarpus* 石栎属
Lithocarpus elaeagnifolius (Seemen) Chun 胡颓子叶柯
Trees, evergreen

24.2; 18.8; 27.9; 24.2; 230.3; 0.0; 1323 | 1491; 502; 63 | 12.6; 1169; 489.8

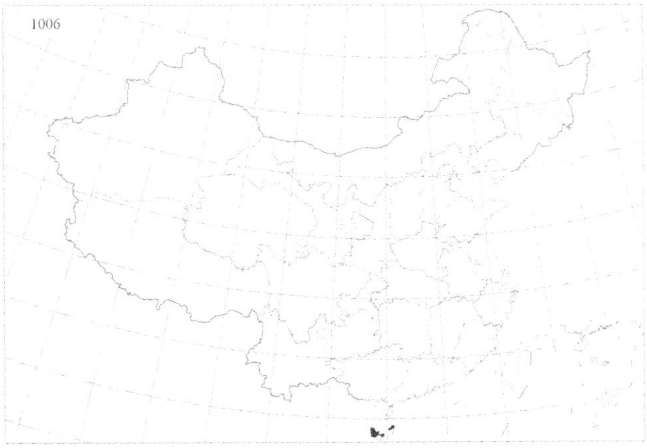

Fagaceae 壳斗科 *Lithocarpus* 石栎属
Lithocarpus elizabethiae (Tutcher) Rehder 厚斗柯
Trees, evergreen

18.3 (13.3~22.7); 8.8 (3.6~15.2); 26.5 (22.2~28.7); 18.4 (13.3~22.8); 160.9 (101.9~213.1); 0.1 (0.0~1.7); 957 (718~1221) | 1506 (999~2435); 670 (473~1428); 129 (50~216) | 57.5 (12.9~103.1); 957 (718~1221); 245.0 (162.8~411.1)

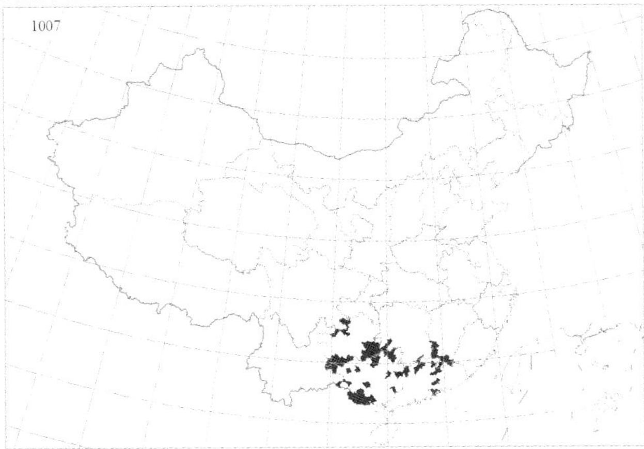

Fagaceae 壳斗科 *Lithocarpus* 石栎属
Lithocarpus elmerrillii Chun 万宁柯
Trees, evergreen

24.0; 18.4; 27.9; 24.0; 228.6; 0.0; 1311 | 1603; 588; 78 | 22.2; 1234; 451.9

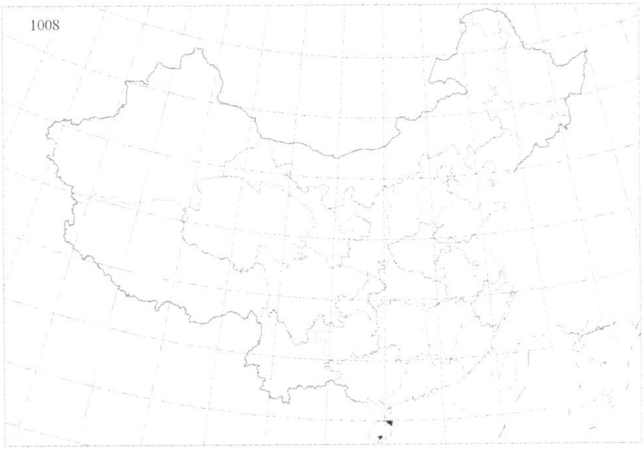

Fagaceae 壳斗科 *Lithocarpus* 石栎属
Lithocarpus eriobotryoides C. C. Huang et Y. T. Chang 枇杷叶柯
Trees, evergreen

15.2 (7.3~17.4); 4.0 (-2.0~6.6); 25.8 (15.6~29.2); 15.2 (7.5~17.5); 124.3 (48.2~149.6); 1.6 (0.0~20.4); 825 (537~937) ❙ 1328 (1005~1526); 544 (457~817); 101 (27~181) ❙ 62.3 (19.5~107.2); 824 (537~930); 193.2 (141.4~272.3)

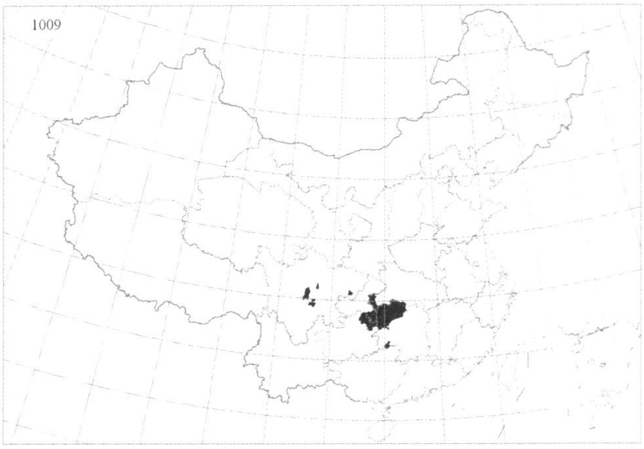

Fagaceae 壳斗科 *Lithocarpus* 石栎属
Lithocarpus fangii (Hu et W. C. Cheng) C. C. Huang et Y. T. Chang 川柯
Trees, evergreen

16.8; 6.5; 26.4; 16.8; 141.8; 0.0; 886 ❙ 1202; 584; 60 ❙ 38.1; 886; 161.0

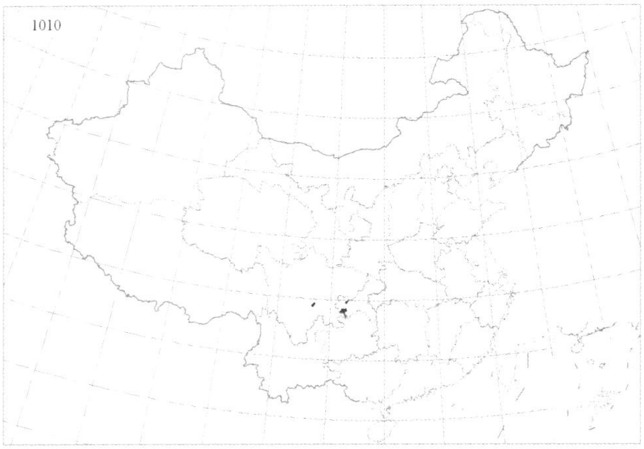

Fagaceae 壳斗科 *Lithocarpus* 石栎属
Lithocarpus farinulentus (Hance) A. Camus 易武柯
Trees, evergreen

20.0; 14.6; 23.2; 20.1; 180.8; 0.0; 963 ❙ 1523; 681; 67 ❙ 58.4; 942; 581.0

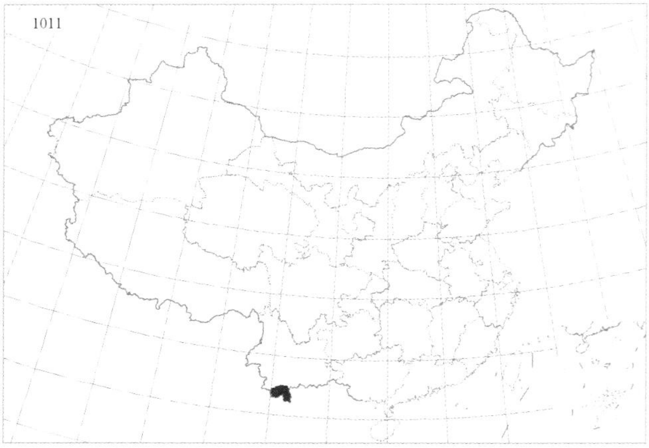

Fagaceae 壳斗科 *Lithocarpus* 石栎属
Lithocarpus fenestratus (Roxb.) Rehder 泥柯
Trees, evergreen

18.3 (6.1~24.8); 10.2 (-1.5~20.1); 24.8 (12.7~29.8); 18.4 (6.3~24.8); 161.4 (35.2~237.6); 0.4 (0.0~21.8); 945 (496~1385) ❙ 1494 (801~2435); 708 (473~1428); 105 (21~194) ❙ 59.8 (1.5~145.8); 934 (496~1221); 333.2 (174.9~659.0)

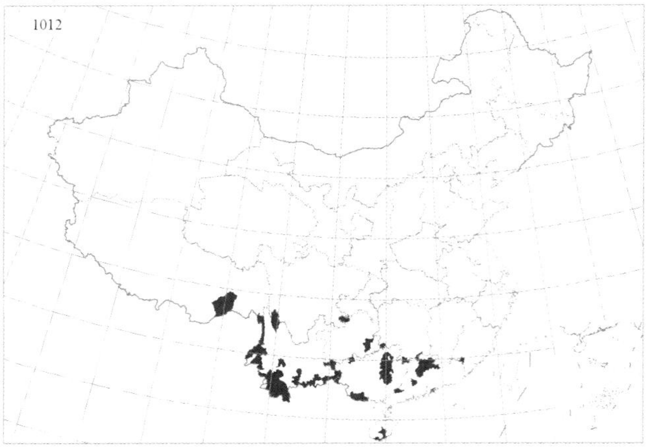

Fagaceae 壳斗科 *Lithocarpus* 石栎属
Lithocarpus fenzelianus A. Camus 红柯
Trees, evergreen

23.8; 18.5; 27.6; 23.9; 226.3; 0.0; 1292 ❙ 1399; 482; 48 ❙ 8.7; 1114; 517.2

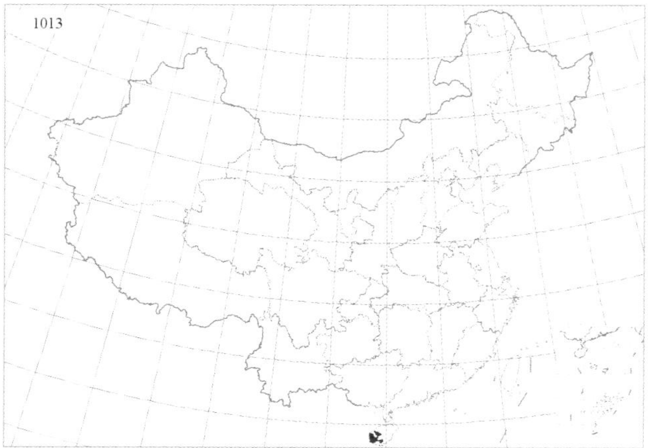

Fagaceae 壳斗科 *Lithocarpus* 石栎属
Lithocarpus floccosus C. C. Huang et Y. T. Chang 卷毛柯
Trees, evergreen

19.7 (16.9~22.2); 10.5 (6.4~14.1); 27.7 (25.8~28.8); 19.8 (17.0~22.3); 177.5 (143.6~207.2); 0.0 (0.0~0.0); 1034 (862~1186) ❙ 1676 (1462~2014); 644 (539~954); 173 (132~200) ❙ 63.2 (29.9~100.8); 1034 (862~1186); 236.2 (187.2~302.2)

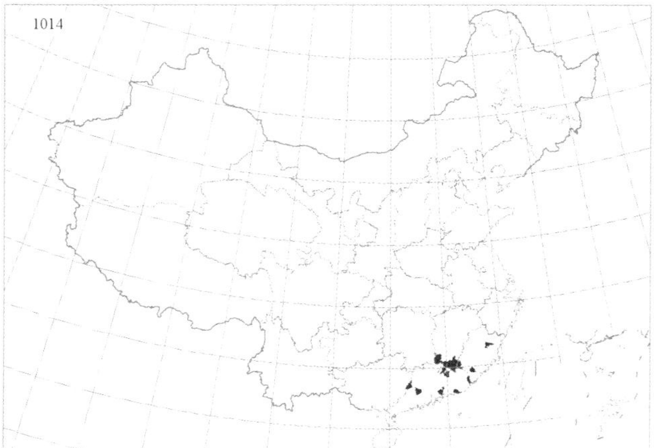

Fagaceae 壳斗科 *Lithocarpus* 石栎属
Lithocarpus fohaiensis (Hu) A. Camus 勐海柯
Trees, evergreen

18.5; 12.7; 22.1; 18.6; 162.9; 0.0; 889 ❙ 1390; 707; 54 ❙ 56.5; 867; 493.3

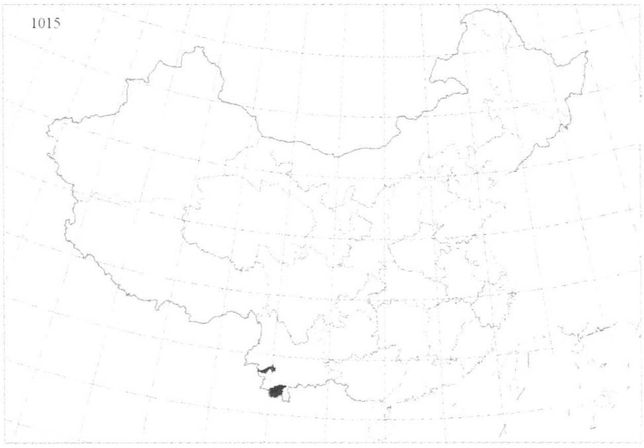

Fagaceae 壳斗科 *Lithocarpus* 石栎属
Lithocarpus fordianus (Hemsl.) Chun 密脉柯
Trees, evergreen

17.9 (14.3~21.9); 9.2 (4.0~15.1); 24.8 (20.8~28.6); 17.9 (14.3~22.0); 154.8 (112.6~203.8); 0.1 (0.0~1.1); 900 (756~1176) ❙ 1369 (1172~1758); 654 (473~835); 90 (42~189) ❙ 52.6 (21.9~82.1); 896 (756~1176); 325.4 (215.4~659.0)

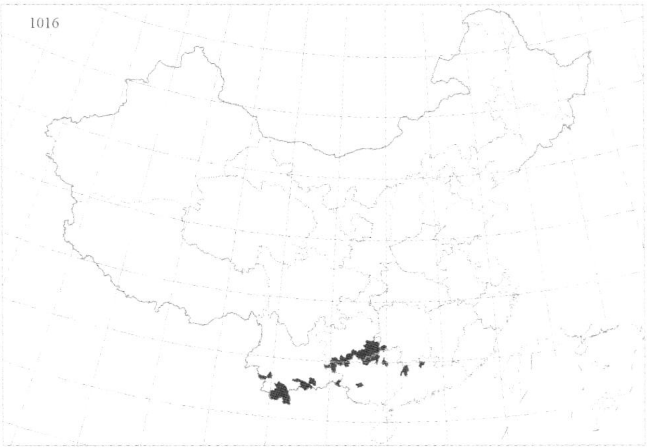

Fagaceae 壳斗科 *Lithocarpus* 石栎属
Lithocarpus formosanus (Skan) Hayata 台湾柯
Trees, evergreen

20.3; 15.9; 23.6; 20.4; 184.4; 0.0; 993 ❙ 2853; 1618; 147 ❙ 190.4; 993; 330.3

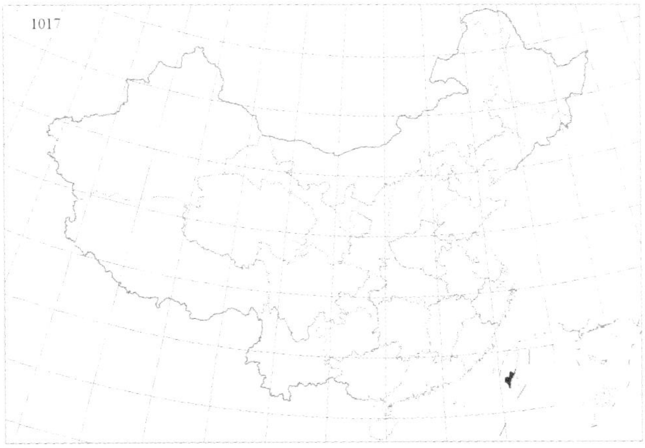

Fagaceae 壳斗科 *Lithocarpus* 石栎属
Lithocarpus gaoligongensis C. C. Huang et Y. T. Chang 高黎贡柯
Trees, evergreen

13.8; 7.0; 18.8; 13.9; 106.3; 0.0; 711 ❙ 1396; 760; 70 ❙ 96.1; 711; 447.2

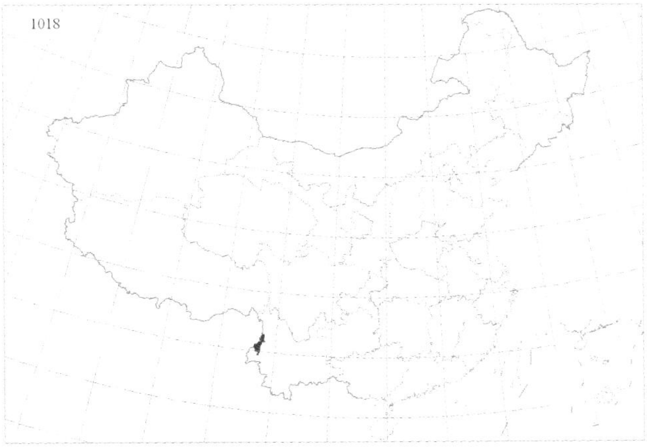

Fagaceae 壳斗科 *Lithocarpus* 石栎属
Lithocarpus garrettianus (Craib) A. Camus 望楼柯
Trees, evergreen

17.3; 11.0; 21.6; 17.3; 148.0; 0.0; 832 ❙ 1049; 598; 44 ❙ 26.1; 783; 365.3

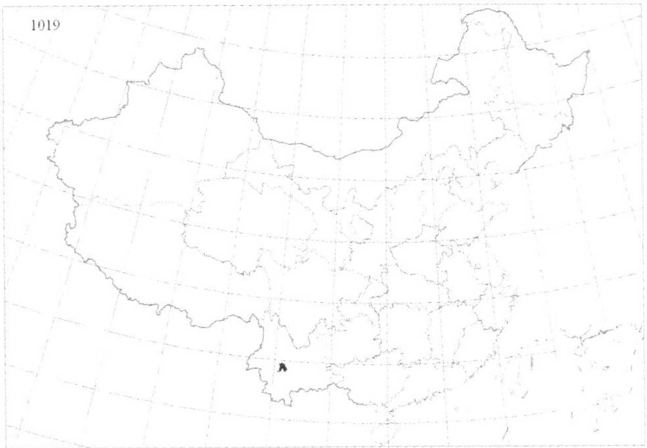

Fagaceae 壳斗科 *Lithocarpus* 石栎属
Lithocarpus glaber (Thunb.) Nakai 柯
Trees, evergreen

17.5 (10.6~24.1); 6.8 (-0.2~18.9); 27.5 (20.1~29.9); 17.6 (10.6~24.1); 151.5 (79.4~229.7); 0.5 (0.0~12.1); 934 (643~1322) ❙ 1523 (871~2759); 564 (299~1594); 173 (46~308) ❙ 64.4 (-6.7~246.5); 924 (643~1273); 223.2 (111.9~411.1)

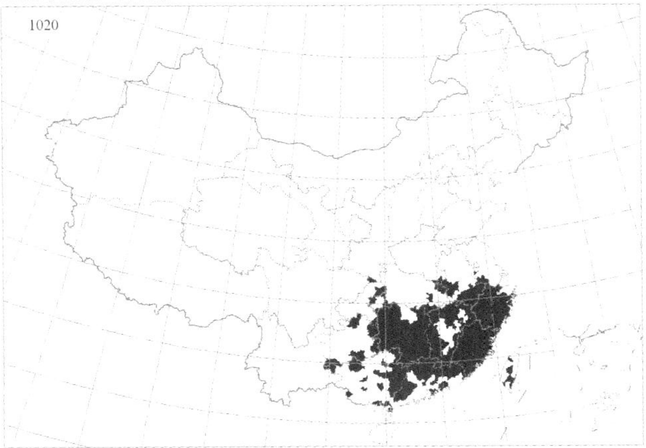

Fagaceae 壳斗科 *Lithocarpus* 石栎属
Lithocarpus grandifolius (D. Don) S. N. Biswas 耳叶柯
Trees, evergreen

16.6 (12.8~21.0); 8.6 (3.3~15.1); 23.2 (18.4~29.8); 16.7 (12.9~21.0); 140.4 (96.0~191.8); 0.3 (0.0~2.2); 847 (692~1113) ❙ 1350 (879~1803); 648 (472~1007); 80 (26~194) ❙ 59.9 (21.3~145.8); 838 (692~1072); 343.5 (174.9~659.0)

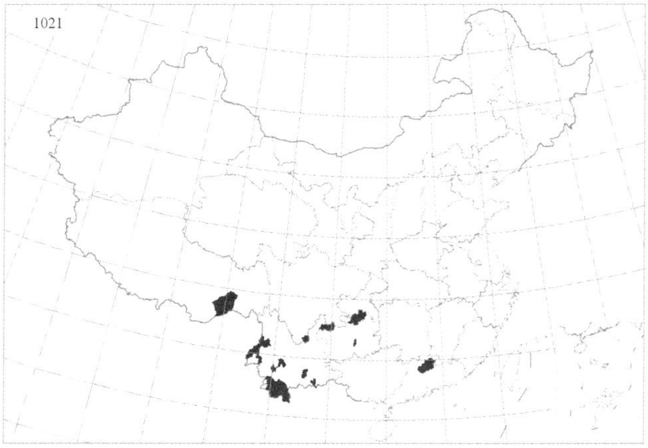

Fagaceae 壳斗科 *Lithocarpus* 石栎属
Lithocarpus gymnocarpus A. Camus 假鱼篮柯
Trees, evergreen

21.1; 13.6; 26.8; 21.1; 193.3; 0.0; 1097 ❙ 1682; 848; 93 ❙ 53.9; 1097; 291.9

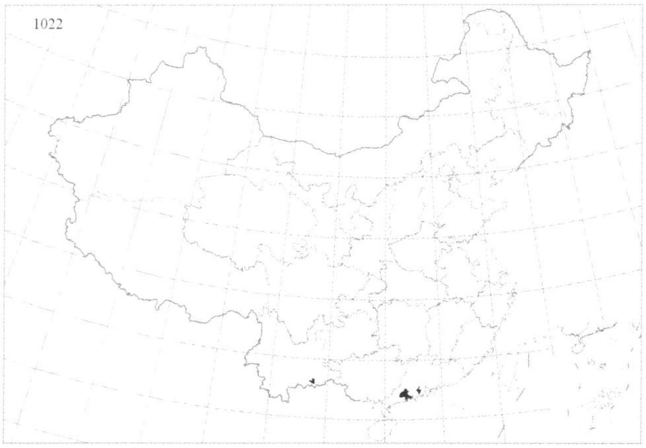

Fagaceae 壳斗科 *Lithocarpus* 石栎属
Lithocarpus haipinii Chun 瘰耳柯
Trees, evergreen

18.4 (14.3~22.0); 8.9 (4.0~15.2); 26.9 (23.5~28.8); 18.6 (14.3~22.7); 163.2 (112.6~212.7); 0.1 (0.0~1.1); 972 (756~1216) ❙ 1529 (1159~1991); 624 (476~994); 147 (77~191) ❙ 57.7 (22.7~95.5); 972 (756~1216); 225.9 (185.5~277.1)

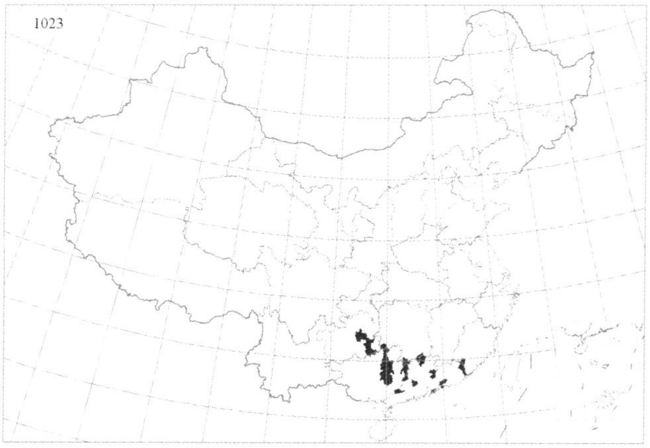

Fagaceae 壳斗科 *Lithocarpus* 石栎属
Lithocarpus hancei (Benth.) Rehder 硬壳柯
Trees, evergreen

17.4 (7.3~24.7); 7.7 (-2.0~19.8); 26.1 (15.0~29.8); 17.5 (7.5~24.7); 149.9 (48.2~236.5); 0.4 (0.0~20.4); 915 (537~1376) ❙ 1476 (868~2993); 606 (312~1847); 136 (18~547) ❙ 62.3 (-34.3~246.5); 905 (537~1221); 241.9 (126.1~620.0)

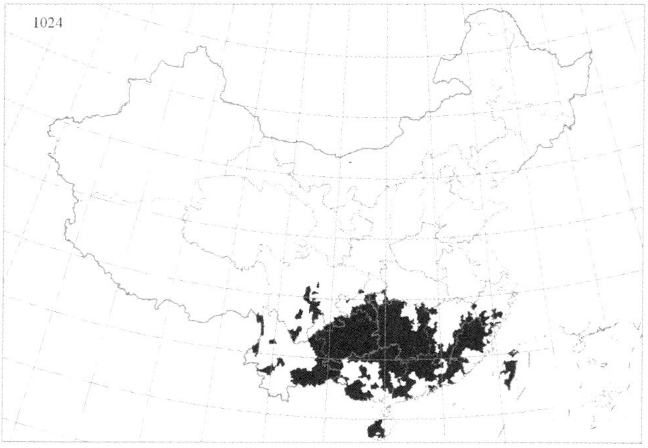

Fagaceae 壳斗科 *Lithocarpus* 石栎属
Lithocarpus handelianus A. Camus 瘤果柯
Trees, evergreen

22.0 (19.1~24.8); 14.5 (11.0~20.1); 27.5 (25.0~28.6); 22.0 (19.1~24.8); 204.0 (169.6~237.6); 0.0 (0.0~0.0); 1165 (945~1385) ❙ 1566 (1188~2435); 780 (473~1428); 87 (41~130) ❙ 35.4 (1.5~99.5); 1132 (945~1326); 353.9 (256.6~620.0)

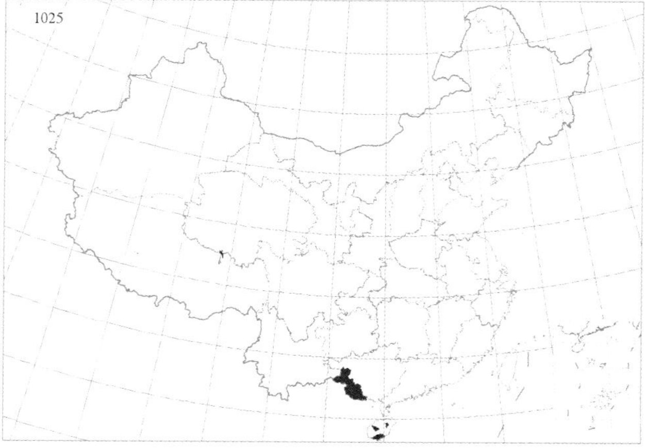

Fagaceae 壳斗科 *Lithocarpus* 石栎属
Lithocarpus harlandii (Hance ex Walp.) Rehder 港柯
Trees, evergreen

16.9 (12.6~23.9); 6.5 (1.6~18.7); 26.7 (20.1~29.8); 17.0 (12.7~23.9); 144.3 (98.8~226.8); 0.4 (0.0~6.9); 895 (711~1294) ❙ 1565 (999~2993); 588 (403~1847); 171 (43~547) ❙ 75.5 (6.7~246.5); 890 (711~1216); 242.4 (151.0~551.6)

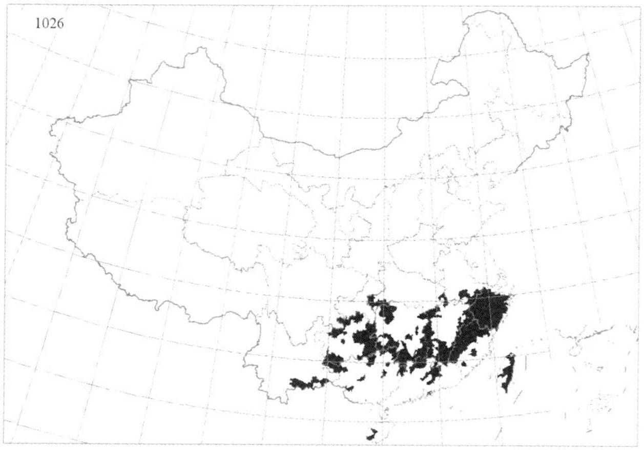

Fagaceae 壳斗科 *Lithocarpus* 石栎属
Lithocarpus henryi (Seemen) Rehder et E. H. Wilson 灰柯
Trees, evergreen

15.9 (10.2~19.7); 4.5 (-0.7~10.5); 26.5 (21.0~29.4); 15.9 (10.4~19.8); 132.2 (77.2~177.1); 1.4 (0.0~13.6); 855 (634~1029) ❙ 1391 (964~1922); 531 (415~816); 136 (29~259) ❙ 63.1 (14.0~112.2); 852 (634~1029); 219.4 (111.9~322.1)

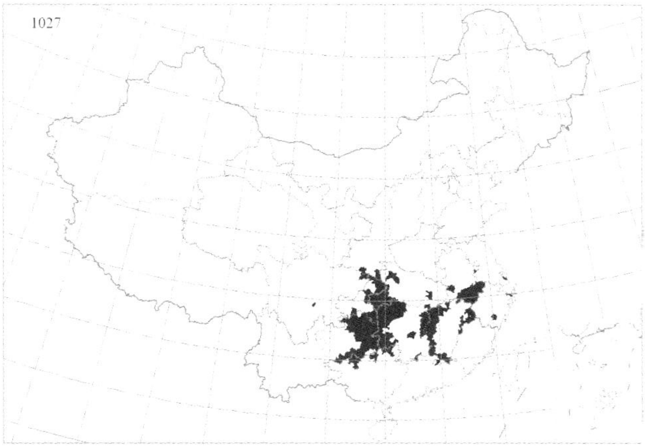

Fagaceae 壳斗科 *Lithocarpus* 石栎属
Lithocarpus howii Chun 梨果柯
Trees, evergreen

24.2 (23.9~24.6); 18.8 (18.2~19.5); 28.0 (27.2~28.4); 24.2 (23.9~24.7); 230.6 (226.8~235.9); 0.0 (0.0~0.0); 1326 (1294~1372) ❙ 1520 (1179~1767); 514 (392~619); 67 (42~110) ❙ 14.5 (-11.9~29.9); 1188 (1056~1326); 486.3 (352.9~620.0)

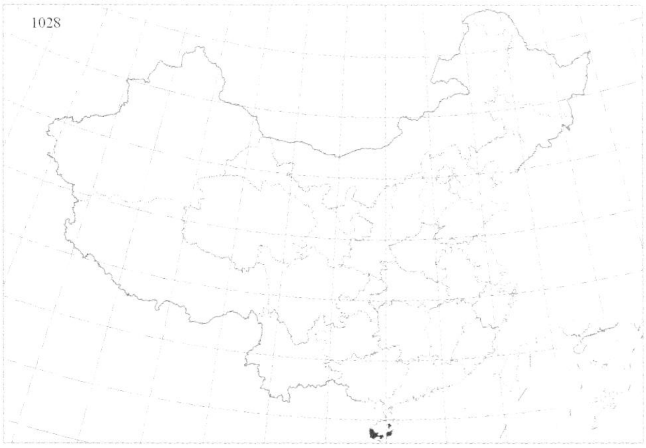

Fagaceae 壳斗科 *Lithocarpus* 石栎属
Lithocarpus hypoglaucus (Hu) C. C. Huang ex Y. C. Hsu et H. W. Jen 灰背叶柯
Trees, evergreen

13.7 (5.4~19.2); 6.2 (-2.1~13.8); 19.8 (12.3~26.6); 13.7 (5.8~19.2); 106.5 (31.5~170.5); 1.8 (0.0~25.6); 724 (481~929) ❙ 991 (741~1499); 574 (411~771); 32 (15~58) ❙ 38.5 (-14.2~93.0); 696 (481~912); 290.1 (142.9~513.9)

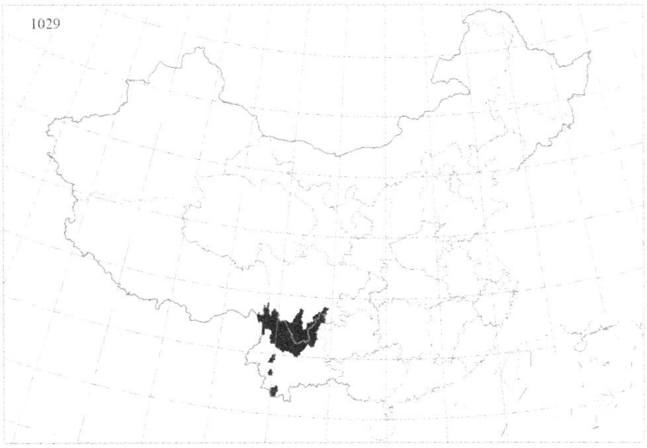

Fagaceae 壳斗科 *Lithocarpus* 石栎属
Lithocarpus irwinii (Hance) Rehder 广南柯
Trees, evergreen

21.2 (19.0~22.2); 12.8 (9.1~15.2); 28.2 (26.9~29.8); 21.3 (19.0~22.7); 196.2 (168.2~212.7); 0.0 (0.0~0.0); 1127 (989~1216) ❙ 1683 (1463~1965); 746 (508~994); 137 (94~194) ❙ 49.9 (24.9~76.3); 1121 (989~1216); 226.6 (174.9~273.3)

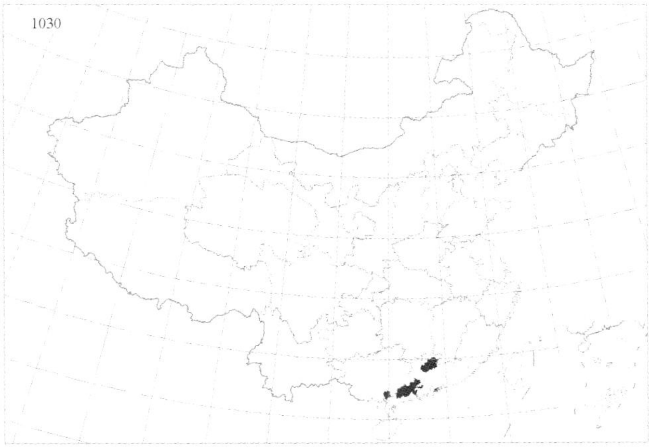

Fagaceae 壳斗科 *Lithocarpus* 石栎属
Lithocarpus iteaphyllus (Hance) Rehder 鼠刺叶柯
Trees, evergreen

18.8 (15.3~22.5); 9.1 (4.2~15.2); 27.6 (25.2~29.8); 18.9 (15.4~22.7); 167.2 (125.8~212.7); 0.0 (0.0~0.8); 992 (818~1216) ❙ 1647 (1462~1965); 612 (499~994); 181 (97~216) ❙ 67.5 (29.9~103.1); 989 (818~1216); 232.3 (139.6~305.7)

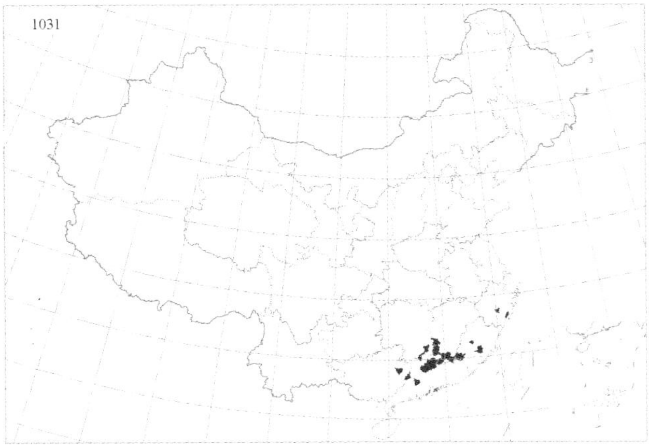

Fagaceae 壳斗科 *Lithocarpus* 石栎属
Lithocarpus ithyphyllus Chun ex H. T. Chang 挺叶柯
Trees, evergreen

20.7 (19.5~21.7); 12.3 (10.6~14.0); 27.5 (26.9~27.8); 20.7 (19.6~21.7); 188.5 (175.1~200.2); 0.0 (0.0~0.0); 1074 (1003~1134) ❙ 1887 (1711~1990); 778 (682~827); 168 (129~191) ❙ 76.1 (60.2~95.5); 1074 (1003~1134); 255.7 (233.6~275.2)

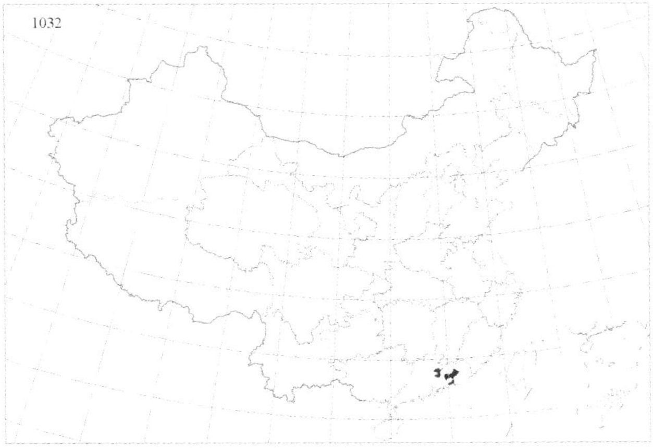

Fagaceae 壳斗科 *Lithocarpus* 石栎属
Lithocarpus jenkinsii (Benth.) C. C. Huang et Y. T. Chang 盈江柯
Trees, evergreen

16.4；9.6；20.6；16.4；137.3；0.0；795 | 1688；959；62 | 112.2；795；627.0

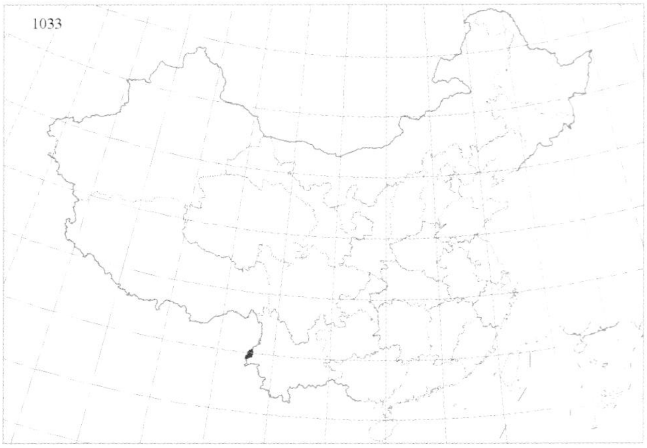

Fagaceae 壳斗科 *Lithocarpus* 石栎属
Lithocarpus kawakamii (Hayata) Hayata 齿叶柯
Trees, evergreen

18.7 (16.2~21.1)；13.0 (11.0~15.3)；23.6 (20.1~27.4)；18.7 (16.2~21.1)；164.6 (134.2~192.9)；0.0 (0.0~0.0)；915 (773~1077) | 2712 (2445~2993)；1046 (844~1412)；360 (110~551) | 200.7 (140.0~246.5)；915 (773~1077)；261.0 (207.7~303.6)

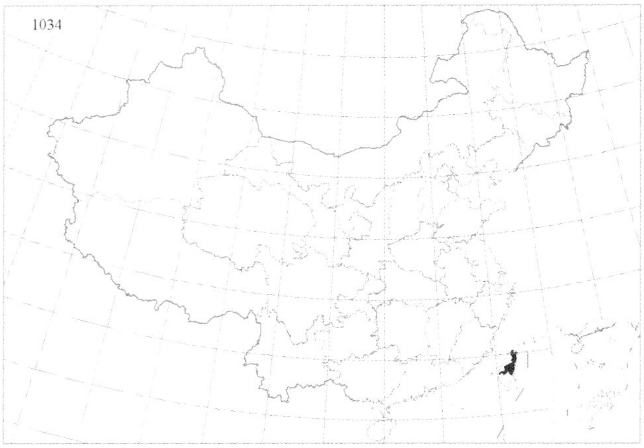

Fagaceae 壳斗科 *Lithocarpus* 石栎属
Lithocarpus konishii (Hayata) Hayata 油叶柯
Trees, evergreen

22.8 (16.2~24.8)；17.2 (11.0~20.1)；26.8 (20.1~28.8)；22.8 (16.2~24.8)；213.9 (134.2~237.6)；0.0 (0.0~0.0)；1220 (773~1385) | 1738 (868~2992)；693 (312~1847)；103 (41~303) | 53.5 (-34.3~246.5)；1130 (773~1360)；402.3 (139.6~620.0)

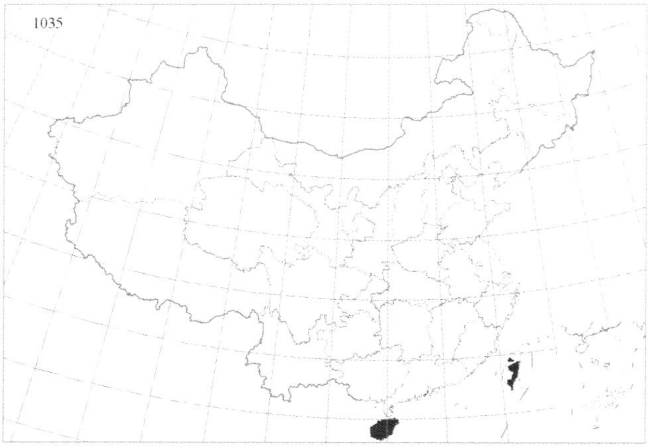

Fagaceae 壳斗科 *Lithocarpus* 石栎属
Lithocarpus laetus Chun et C. C. Huang ex Y. C. Hsu et H. W. Jen 屏边柯
Trees, evergreen

17.7；10.9；22.5；17.8；153.4；0.0；854 | 1368；756；59 | 60.1；854；491.5

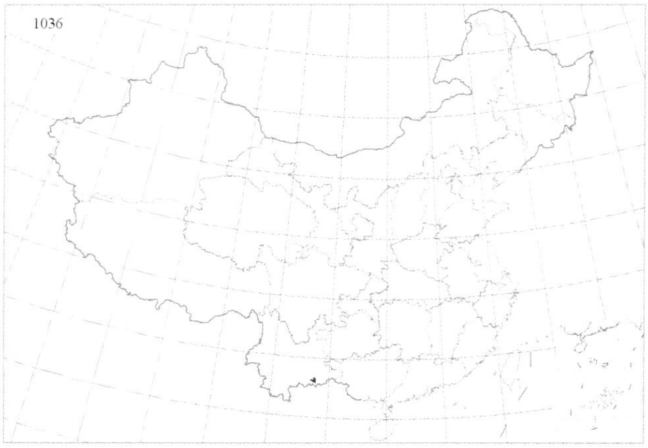

Fagaceae 壳斗科 *Lithocarpus* 石栎属
Lithocarpus laoticus (Hickel et A. Camus) A. Camus 老挝柯
Trees, evergreen

17.4 (13.3~19.1)；10.8 (5.4~13.0)；22.0 (19.6~23.3)；17.4 (13.4~19.1)；148.9 (100.4~169.7)；0.0 (0.0~0.0)；846 (706~917) | 1483 (1336~1736)；857 (756~1007)；57 (51~63) | 78.1 (54.8~145.8)；846 (706~917)；485.9 (432.9~551.3)

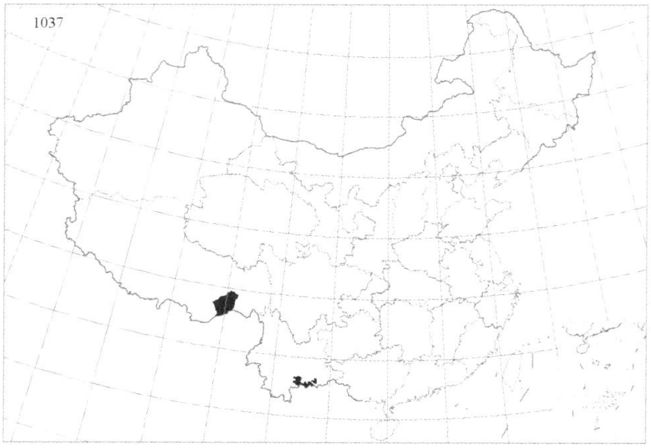

Fagaceae 壳斗科 *Lithocarpus* 石栎属
Lithocarpus lepidocarpus (Hayata) Hayata 鬼石柯
Trees, evergreen

18.8 (16.2~21.7)；13.4 (11.0~15.3)；23.1 (20.1~27.0)；18.8 (16.2~21.7)；166.1 (134.2~200.2)；0.0 (0.0~0.0)；918 (773~1118) | 2411 (1836~2713)；1213 (1033~1412)；181 (110~245) | 171.4 (64.2~246.5)；909 (773~1075)；269.0 (202.8~363.7)

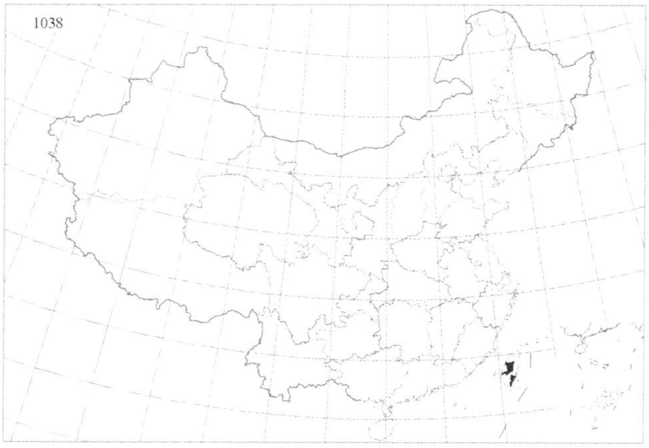

Fagaceae 壳斗科 *Lithocarpus* 石栎属
Lithocarpus leucodermis Chun et C. C. Huang 白枝柯
Trees, evergreen

17.7；10.9；22.5；17.8；153.4；0.0；854 ❙ 1368；756；59 ❙ 60.1；854；491.5

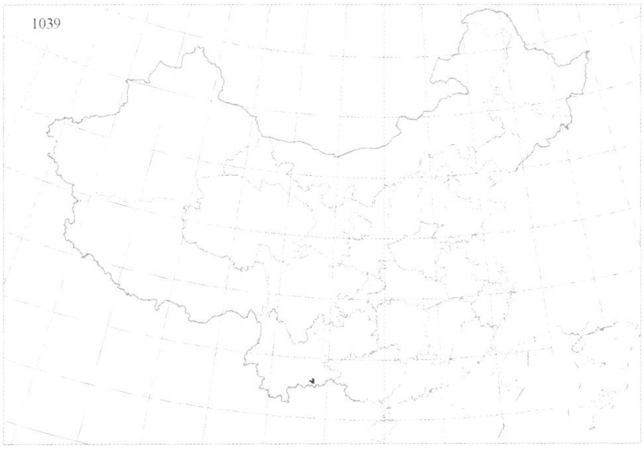

Fagaceae 壳斗科 *Lithocarpus* 石栎属
Lithocarpus levis Chun et C. C. Huang 滑壳柯
Trees, evergreen

15.5 (14.3~16.7)；5.3 (4.0~6.9)；24.7 (23.0~25.7)；15.6 (14.3~16.7)；127.4 (112.6~140.5)；0.3 (0.0~1.1)；811 (756~863) ❙ 1256 (1211~1321)；538 (476~696)；94 (64~106) ❙ 55.3 (45.6~71.9)；811 (756~863)；219.6 (188.8~246.1)

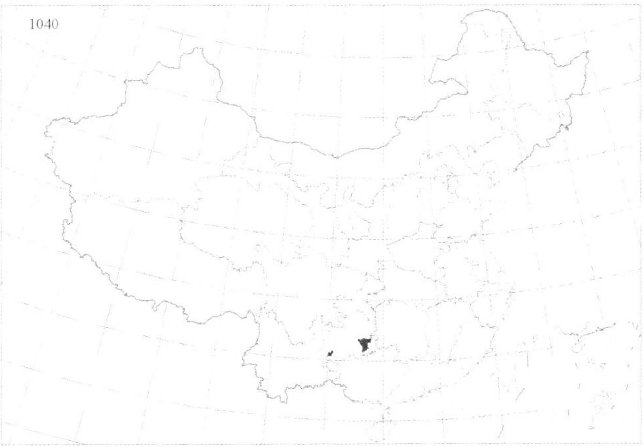

Fagaceae 壳斗科 *Lithocarpus* 石栎属
Lithocarpus listeri (King) Grierson et D. G. Long 谊柯
Trees, evergreen

13.3；5.4；19.6；13.4；100.4；0.0；706 ❙ 1736；1007；51 ❙ 145.8；706；447.4

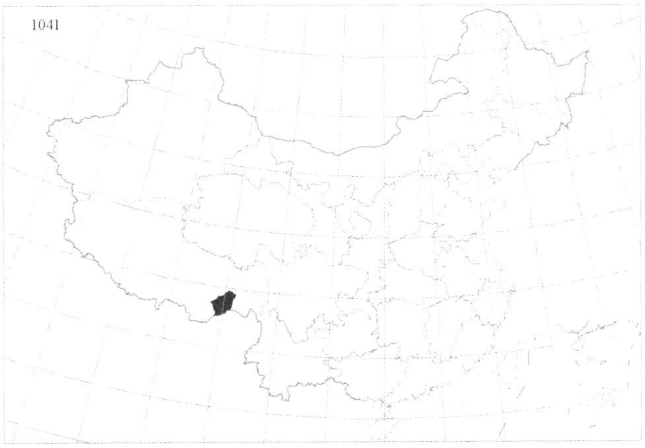

Fagaceae 壳斗科 *Lithocarpus* 石栎属
Lithocarpus litseifolius (Hance) Chun 木姜叶柯
Trees, evergreen

17.6 (10.2~24.8)；7.5 (-0.7~20.1)；26.7 (16.7~29.8)；17.6 (10.3~24.8)；152.0 (68.2~237.6)；0.4 (0.0~13.6)；929 (620~1385) ❙ 1483 (899~2435)；584 (383~1428)；144 (18~259) ❙ 60.6 (1.5~132.1)；920 (620~1326)；237.2 (108.2~620.0)

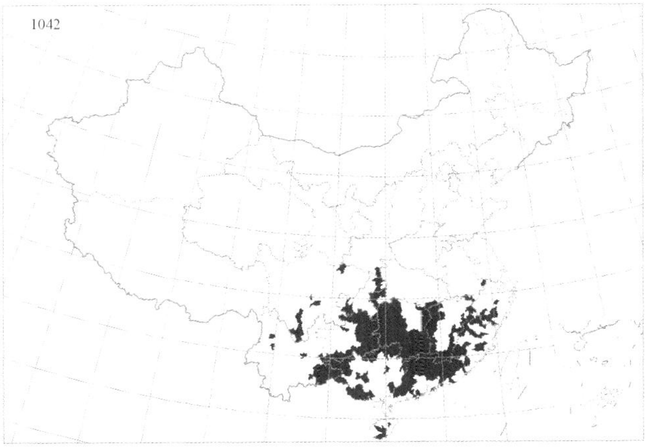

Fagaceae 壳斗科 *Lithocarpus* 石栎属
Lithocarpus litseifolius var. pubescens C. C. Huang et Y. T. Chang 毛枝木姜叶柯
Trees, evergreen

19.1；9.8；26.6；19.2；170.1；0.0；981 ❙ 1284；651；73 ❙ 30.9；981；322.1

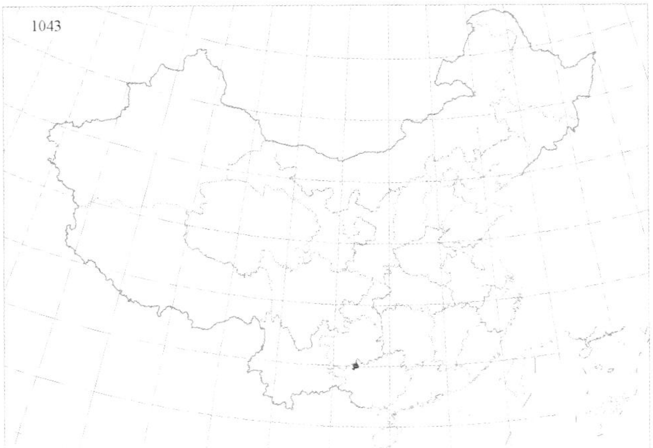

Fagaceae 壳斗科 *Lithocarpus* 石栎属
Lithocarpus longanoides C. C. Huang et Y. T. Chang 龙眼柯
Trees, evergreen

20.4 (18.0~22.0)；11.7 (8.5~14.2)；27.3 (25.0~28.8)；20.4 (18.0~22.0)；185.0 (156.4~203.8)；0.0 (0.0~0.0)；1059 (913~1170) ❙ 1598 (1250~1990)；723 (599~891)；130 (49~181) ❙ 50.9 (29.9~83.4)；1059 (913~1170)；257.2 (185.5~371.7)

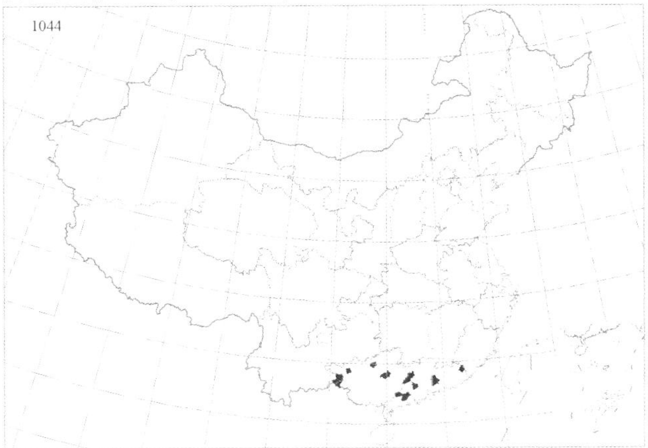

Fagaceae 壳斗科 *Lithocarpus* 石栎属
Lithocarpus longipedicellatus (Hickel et A. Camus) A. Camus 柄果柯
Trees, evergreen

21.2 (16.3~24.8); 13.5 (8.2~20.1); 27.1 (21.4~28.7); 21.3 (16.4~24.8); 195.0 (136.4~237.6); 0.0 (0.0~0.0); 1113 (797~1385) ❙ 1531 (868~2435); 717 (312~1428); 101 (41~189) ❙ 39.5 (-34.3~99.5); 1072 (797~1326); 325.5 (201.9~620.0)

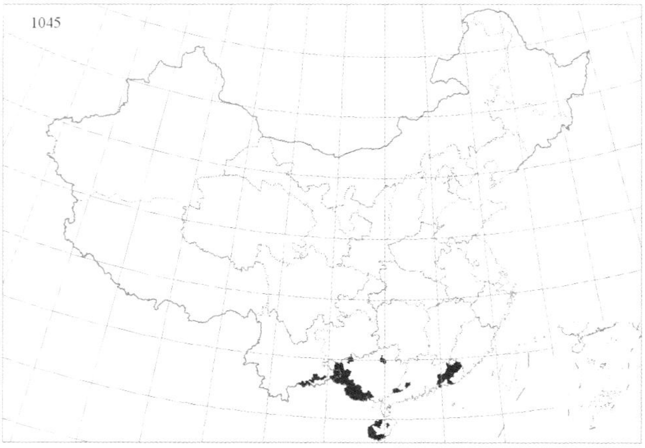

Fagaceae 壳斗科 *Lithocarpus* 石栎属
Lithocarpus lycoperdon (Skan) A. Camus 香菌柯
Trees, evergreen

17.9; 11.4; 22.4; 17.9; 155.3; 0.0; 862 ❙ 1325; 746; 55 ❙ 53.6; 862; 460.4

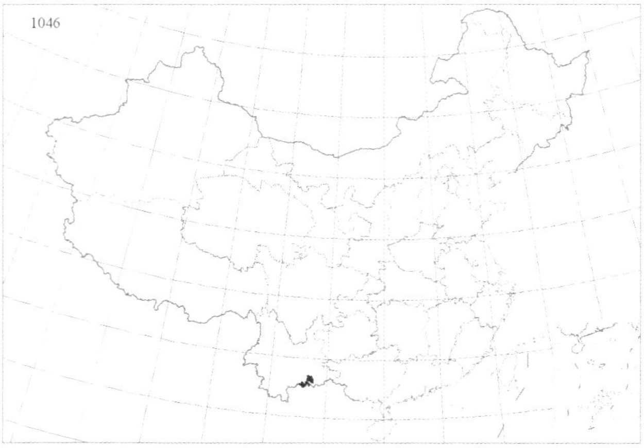

Fagaceae 壳斗科 *Lithocarpus* 石栎属
Lithocarpus macilentus Chun et C. C. Huang 粉叶柯
Trees, evergreen

21.6 (21.1~22.0); 13.1 (11.9~15.2); 28.7 (28.4~29.0); 21.8 (21.1~22.7); 201.2 (193.2~212.7); 0.0 (0.0~0.0); 1157 (1115~1216) ❙ 1579 (1371~1965); 693 (565~994); 132 (104~144) ❙ 36.2 (17.4~61.6); 1151 (1107~1216); 215.5 (203.4~232.5)

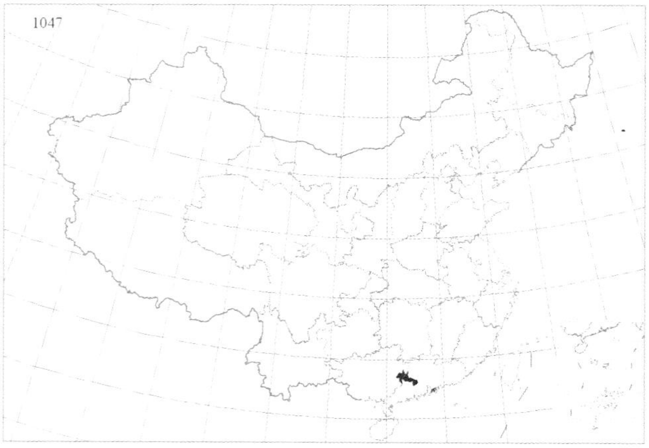

Fagaceae 壳斗科 *Lithocarpus* 石栎属
Lithocarpus magneinii (Hickel et A. Camus) A. Camus 黑家柯
Trees, evergreen

19.4 (17.7~21.0); 13.2 (10.9~14.9); 23.5 (22.5~25.6); 19.4 (17.8~21.0); 172.9 (153.4~191.8); 0.0 (0.0~0.0); 941 (854~1053) ❙ 1466 (1368~1547); 795 (655~911); 62 (56~68) ❙ 56.4 (44.5~74.2); 934 (854~1053); 534.1 (491.5~570.1)

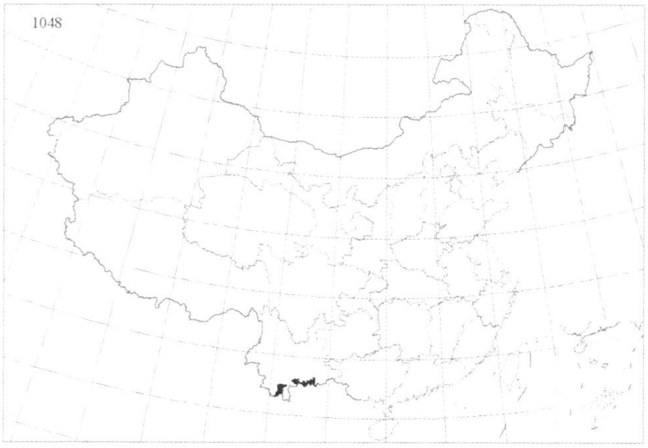

Fagaceae 壳斗科 *Lithocarpus* 石栎属
Lithocarpus mairei (Schottky) Rehder 光叶柯
Trees, evergreen

16.0 (10.8~19.3); 9.4 (3.9~13.1); 20.7 (16.6~23.6); 16.0 (10.9~19.4); 132.6 (71.8~173.2); 0.1 (0.0~1.3); 795 (633~939) ❙ 1081 (968~1454); 612 (539~792); 45 (24~70) ❙ 37.7 (12.5~99.9); 776 (633~903); 342.9 (278.4~487.2)

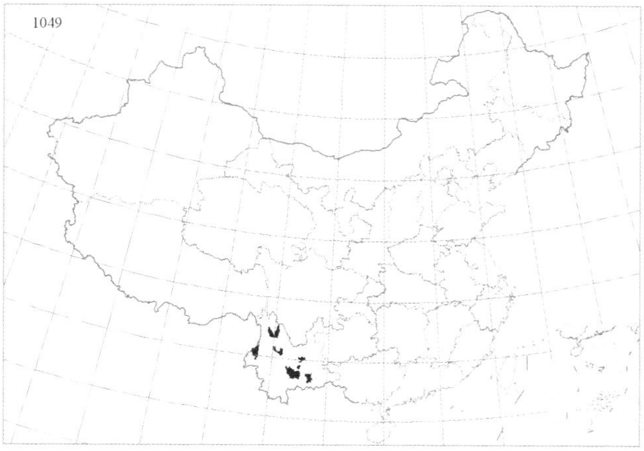

Fagaceae 壳斗科 *Lithocarpus* 石栎属
Lithocarpus megalophyllus Rehder et E. H. Wilson 大叶柯
Trees, evergreen

16.7 (11.1~21.3); 7.7 (1.8~14.2); 24.4 (19.4~27.3); 16.8 (11.1~21.3); 141.9 (79.8~195.9); 0.6 (0.0~6.9); 862 (648~1111) ❙ 1303 (985~1900); 682 (471~1056); 66 (30~177) ❙ 52.3 (16.9~94.5); 862 (648~1111); 284.2 (142.9~551.6)

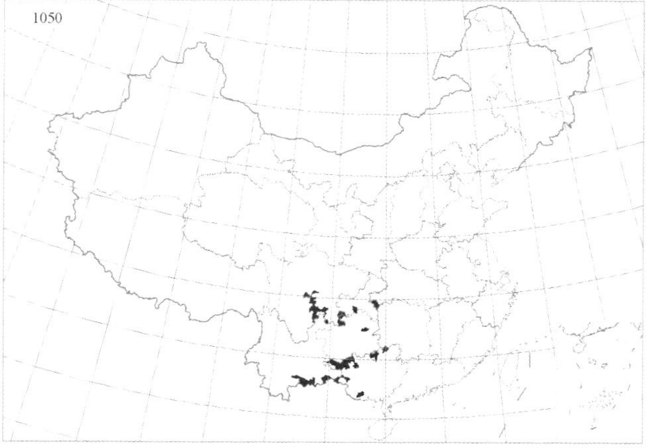

Fagaceae 壳斗科 *Lithocarpus* 石栎属

Lithocarpus mekongensis (A. Camus) C. C. Huang et Y. T. Chang 澜沧柯 Trees, evergreen

20.0; 14.5; 23.2; 20.0; 180.5; 0.0; 961 ❙ 1520; 682; 59 ❙ 58.1; 934; 594.5

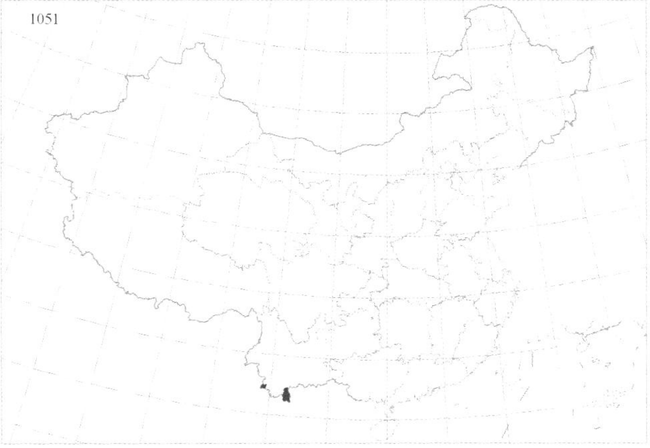

Fagaceae 壳斗科 *Lithocarpus* 石栎属

Lithocarpus melanochromus Chun et Tsiang ex C. C. Huang et Y. T. Chang 黑柯 Trees, evergreen

21.7 (21.0~22.7); 13.3 (11.8~15.0); 28.0 (27.3~28.6); 21.7 (21.1~22.8); 200.7 (192.6~213.1); 0.0 (0.0~0.0); 1149 (1111~1221) ❙ 1788 (1302~2435); 939 (670~1428); 138 (86~194) ❙ 55.3 (12.9~99.5); 1[illegible]49 (1111~1221); 262.6 (206.0~318.8)

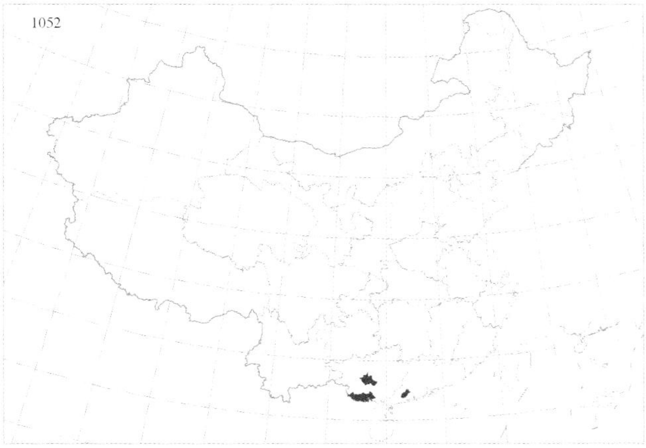

Fagaceae 壳斗科 *Lithocarpus* 石栎属

Lithocarpus mianningensis Hu 缅宁柯
Trees, evergreen

15.2 (10.2~17.3); 8.5 (2.4~11.0); 20.0 (16.9~21.6); 15.2 (10.3~17.3); 123.8 (68.2~148.0); 1.0 (0.0~5.1); 761 (620~832) ❙ 1150 (1036~1348); 654 (598~762); 43 (18~56) ❙ 52.0 (26.1~72.8); 748 (620~783); 367.2 (298.0~405.0)

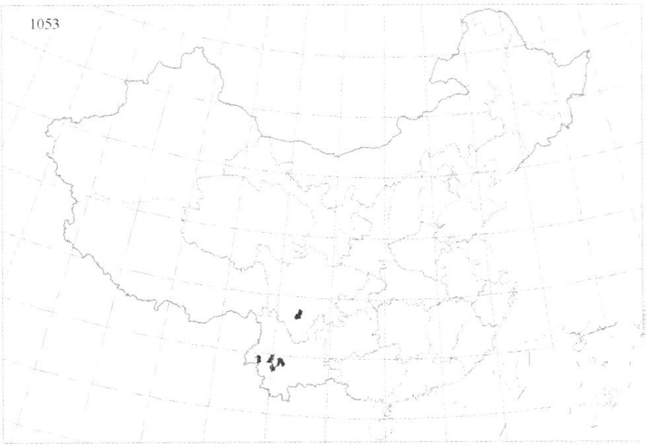

Fagaceae 壳斗科 *Lithocarpus* 石栎属

Lithocarpus microspermus A. Camus 小果柯
Trees, evergreen

19.5 (17.7~21.0); 13.4 (10.9~14.9); 23.5 (22.3~25.6); 19.5 (17.8~21.0); 174.3 (153.4~191.8); 0.0 (0.0~0.0); 945 (854~1053) ❙ 1456 (1368~1522); 750 (655~835); 61 (56~68) ❙ 54.6 (44.5~64.9); 933 (854~1053); 526.6 (491.5~570.1)

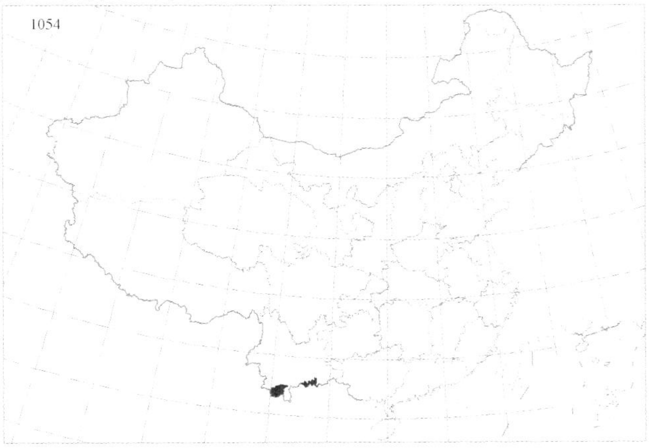

Fagaceae 壳斗科 *Lithocarpus* 石栎属

Lithocarpus naiadarum (Hance) Chun 水仙柯
Trees, evergreen

21.6 (13.8~24.8); 15.0 (6.7~20.1); 26.5 (19.3~28.8); 21.7 (13.8~24.8); 200.2 (105.9~237.6); 0.0 (0.0~0.0); 1156 (714~1385) ❙ 1475 (868~2435); 646 (312~1428); 78 (27~144) ❙ 29.0 (-34.3~99.5); 1089 (683~1360); 368.6 (157.7~620.0)

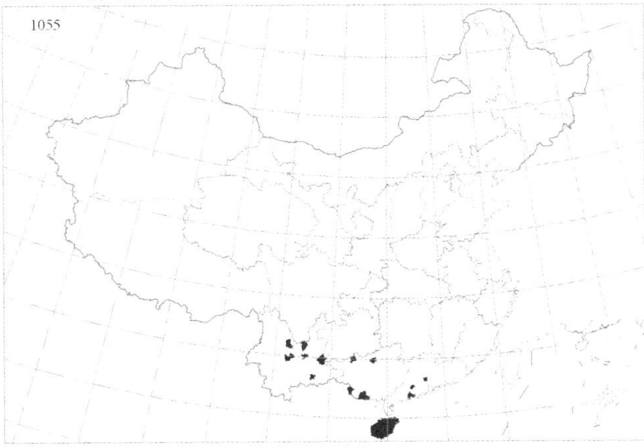

Fagaceae 壳斗科 *Lithocarpus* 石栎属

Lithocarpus nantoensis (Hayata) Hayata 南投柯
Trees, evergreen

17.3; 12.6; 20.9; 17.3; 148.1; 0.0; 818 ❙ 2697; 1287; 216 ❙ 230.4; 818; 326.4

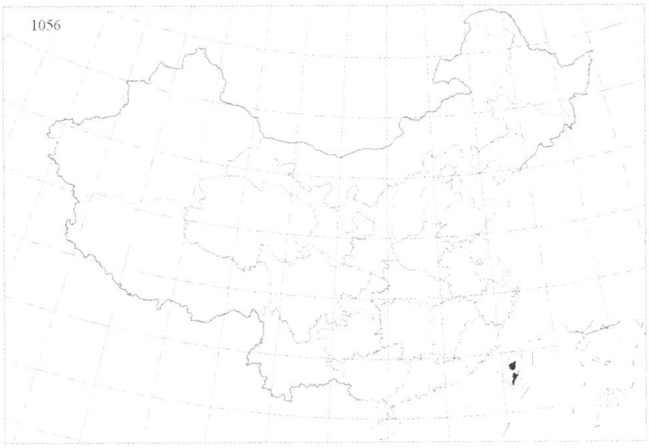

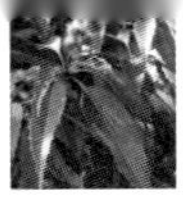

Fagaceae 壳斗科 *Lithocarpus* 石栎属
Lithocarpus nitidinux (Hu) Chun ex C. C. Huang et Y. T. Chang 光果柯
Trees, evergreen

18.1; 9.5; 24.6; 18.1; 157.2; 0.0; 904 ❙ 1245; 658; 54 ❙ 38.6; 904; 295.3

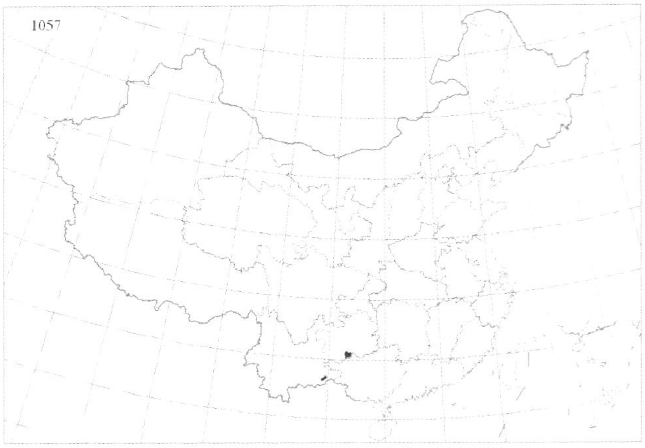

Fagaceae 壳斗科 *Lithocarpus* 石栎属
Lithocarpus oblanceolatus C. C. Huang et Y. T. Chang 峨眉柯
Trees, evergreen

13.0; 3.8; 21.5; 13.1; 100.2; 2.6; 717 ❙ 1315; 752; 43 ❙ 83.8; 717; 253.6

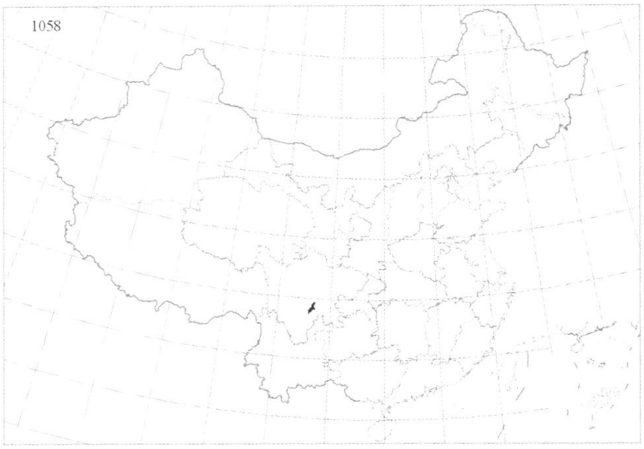

Fagaceae 壳斗科 *Lithocarpus* 石栎属
Lithocarpus obovatilimbus Chun 卵叶柯
Trees, evergreen

24.7; 19.8; 27.9; 24.7; 236.5; 0.0; 1376 ❙ 1588; 534; 58 ❙ 15.4; 1212; 494.7

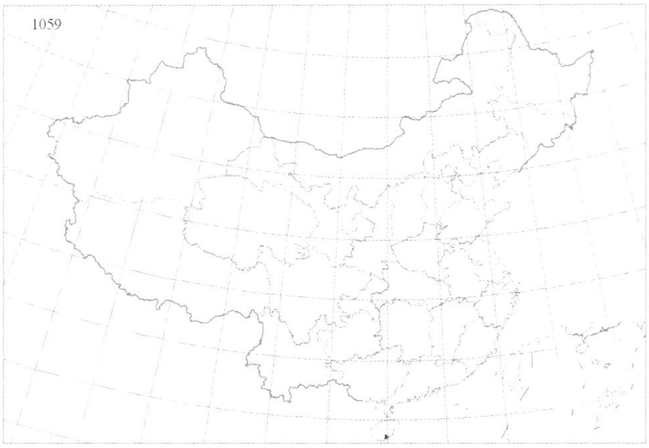

Fagaceae 壳斗科 *Lithocarpus* 石栎属
Lithocarpus obscurus C. C. Huang et Y. T. Chang 墨脱柯
Trees, evergreen

13.3; 5.4; 19.6; 13.4; 100.4; 0.0; 706 ❙ 1736; 1007; 51 ❙ 145.8; 706; 447.4

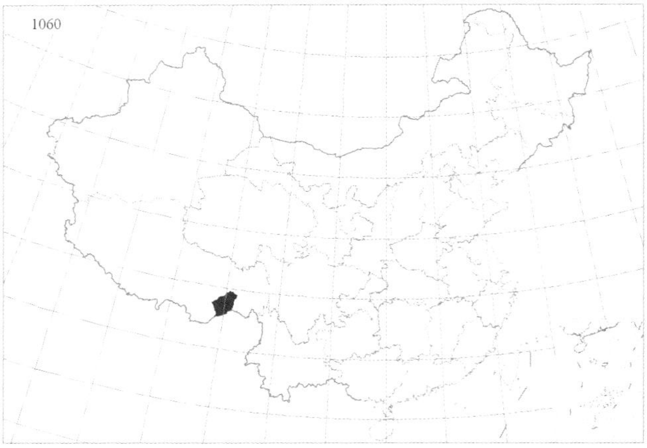

Fagaceae 壳斗科 *Lithocarpus* 石栎属
Lithocarpus oleifolius A. Camus 榄叶柯
Trees, evergreen

17.8 (14.3~21.6); 7.5 (3.3~13.2); 27.1 (24.2~29.8); 17.9 (14.3~21.6); 154.7 (113.2~199.2); 0.2 (0.0~2.3); 934 (761~1139) ❙ 1600 (1251~1961); 565 (411~858); 188 (54~230) ❙ 72.0 (32.2~128.9); 930 (761~1139); 231.9 (169.7~305.7)

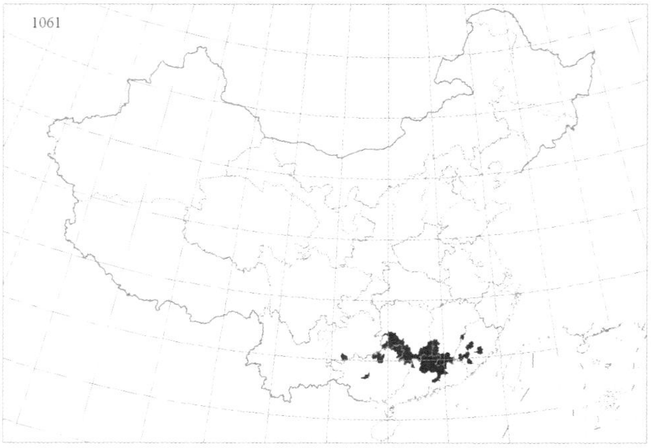

Fagaceae 壳斗科 *Lithocarpus* 石栎属
Lithocarpus pachylepis A. Camus 厚鳞柯
Trees, evergreen

18.7 (17.2~21.0); 10.9 (8.7~14.2); 24.4 (22.1~27.7); 18.7 (17.3~21.0); 164.8 (147.0~191.8); 0.0 (0.0~0.0); 926 (833~1053) ❙ 1380 (1161~1547); 759 (633~911); 66 (48~111) ❙ 49.7 (30.9~80.6); 925 (833~1053); 381.5 (248.4~551.6)

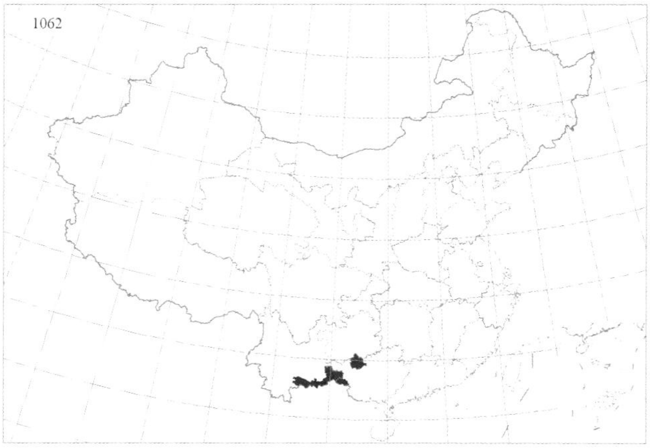

Fagaceae 壳斗科 *Lithocarpus* 石栎属

Lithocarpus pachyphyllus (Kurz) Rehder 厚叶柯

Trees, evergreen

15.6 (8.6~19.1); 8.9 (1.3~13.0); 20.3 (15.0~23.3); 15.6 (8.7~19.1); 127.9 (52.6~169.7); 0.6 (0.0~8.8); 778 (574~917) | 1169 (939~1454); 648 (524~835); 54 (39~70) | 52.5 (6.5~99.9); 763 (574~917); 385.6 (316.6~506.2)

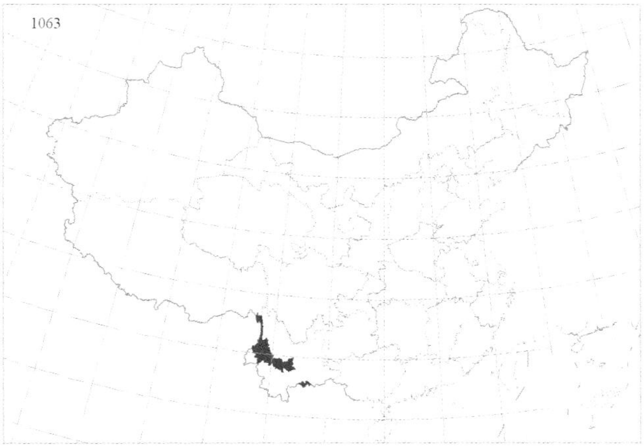

Fagaceae 壳斗科 *Lithocarpus* 石栎属

Lithocarpus pachyphyllus var. fruticosus (Wall. ex King) A. Camus 顺宁厚叶柯

Trees, evergreen

13.0; 6.1; 18.3; 13.0; 100.3; 4.4; 703 | 1045; 591; 44 | 53.8; 678; 358.8

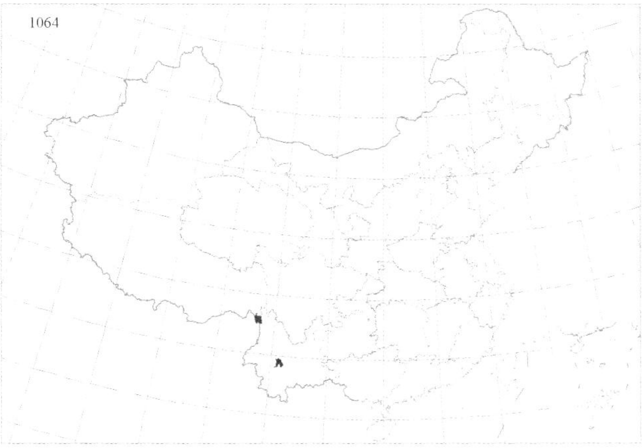

Fagaceae 壳斗科 *Lithocarpus* 石栎属

Lithocarpus paihengii Chun et Tsiang 大叶苦柯

Trees, evergreen

19.3 (16.7~22.3); 9.7 (5.6~14.0); 27.5 (26.2~29.0); 19.3 (16.8~22.4); 171.6 (141.2~208.2); 0.0 (0.0~0.0); 1005 (885~1198) | 1662 (1486~2005); 638 (497~827); 180 (132~197) | 65.9 (38.8~85.1); 1005 (885~1198); 234.9 (111.7~300.5)

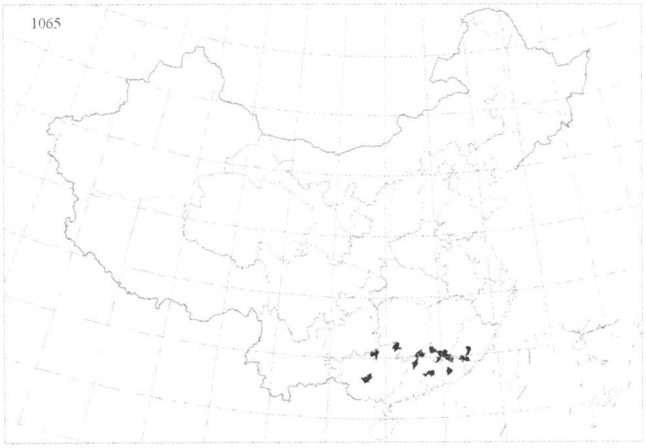

Fagaceae 壳斗科 *Lithocarpus* 石栎属

Lithocarpus pakhaensis A. Camus 滇南柯

Trees, evergreen

19.4; 12.6; 24.1; 19.4; 172.6; 0.0; 954 | 1445; 787; 64 | 52.3; 954; 521.5

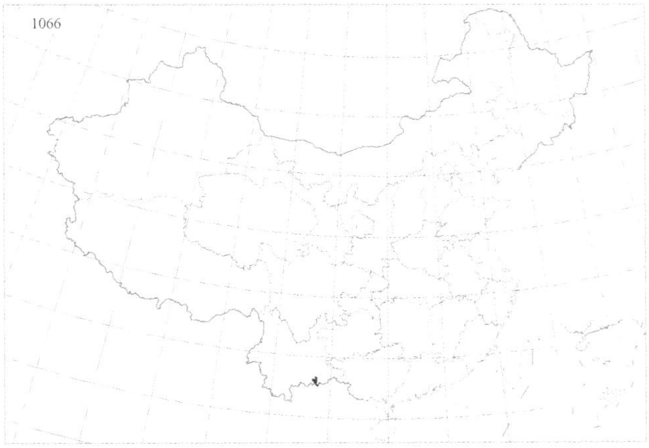

Fagaceae 壳斗科 *Lithocarpus* 石栎属

Lithocarpus paniculatus Hand.-Mazz. 圆锥柯

Trees, evergreen

16.8 (9.3~22.4); 5.7 (-1.7~14.8); 27.1 (20.1~29.8); 16.8 (9.5~22.4); 142.4 (69.8~208.9); 0.6 (0.0~18.0); 896 (603~1191) | 1459 (999~2037); 515 (392~1033); 167 (53~230) | 63.6 (30.9~128.9); 884 (603~1191); 202.6 (134.6~285.1)

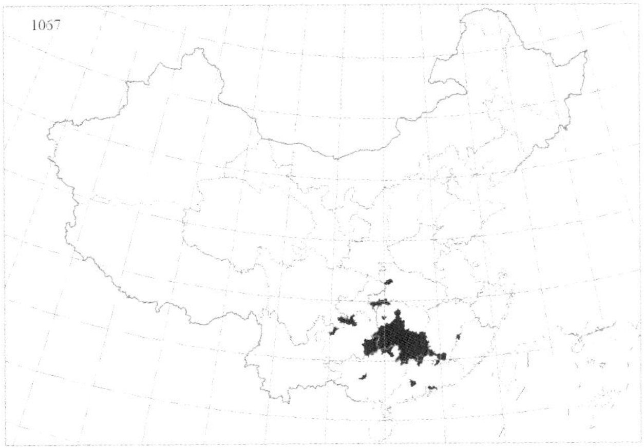

Fagaceae 壳斗科 *Lithocarpus* 石栎属

Lithocarpus pasania C. C. Huang et Y. T. Chang 石柯

Trees, evergreen

13.3; 5.4; 19.6; 13.4; 100.4; 0.0; 706 | 1736; 1007; 51 | 145.8; 706; 447.4

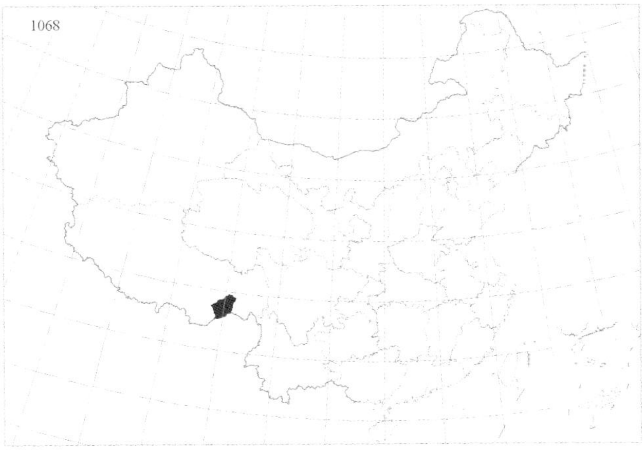

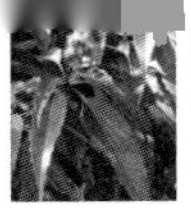

Fagaceae 壳斗科 *Lithocarpus* 石栎属
Lithocarpus petelotii A. Camus 星毛柯
Trees, evergreen

16.7 (14.9~21.0); 7.3 (3.6~14.2); 24.8 (21.4~27.5); 16.8 (15.0~21.0); 141.4 (121.2~191.8); 0.3 (0.0~1.6); 856 (791~1053) ❙ 1344 (1159~1522); 633 (454~860); 85 (48~174) ❙ 57.3 (37.6~80.6); 856 (791~1053); 293.9 (189.0~551.6)

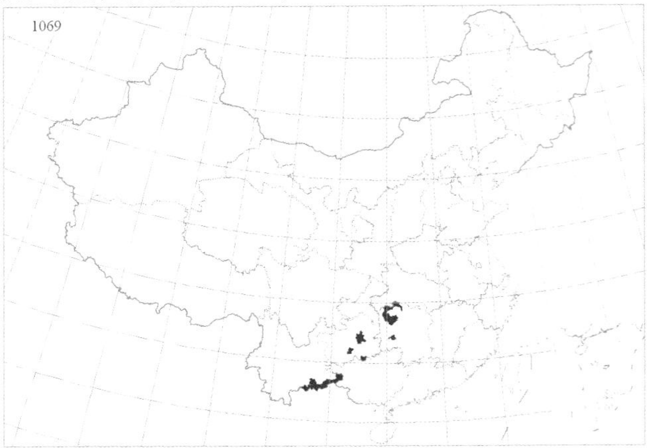

Fagaceae 壳斗科 *Lithocarpus* 石栎属
Lithocarpus phansipanensis A. Camus 桂南柯
Trees, evergreen

21.3; 13.2; 27.3; 21.3; 195.9; 0.0; 1111 ❙ 1900; 1056; 125 ❙ 71.0; 1111; 297.2

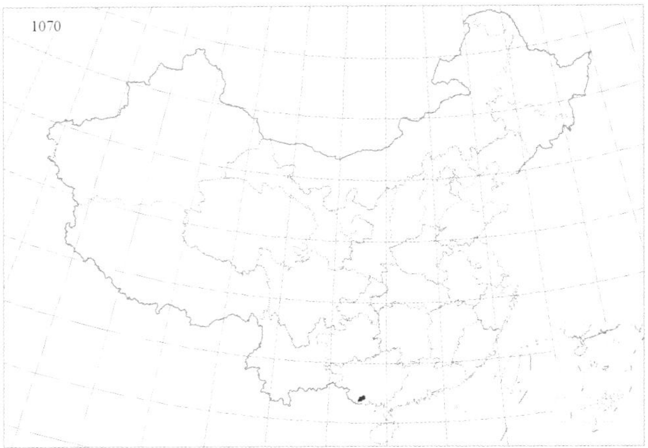

Fagaceae 壳斗科 *Lithocarpus* 石栎属
Lithocarpus propinquus C. C. Huang et Y. T. Chang 三柄果柯
Trees, evergreen

17.9 (16.3~21.0); 10.4 (9.1~14.2); 23.3 (21.4~25.6); 18.0 (16.4~21.0); 155.4 (136.4~191.8); 0.0 (0.0~0.0); 877 (797~1053) ❙ 1339 (1161~1522); 746 (633~860); 54 (48~68) ❙ 53.1 (37.1~80.6); 877 (797~1053); 394.6 (314.4~551.6)

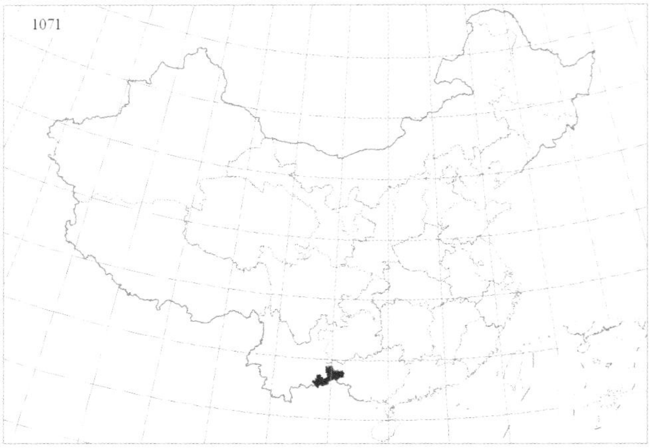

Fagaceae 壳斗科 *Lithocarpus* 石栎属
Lithocarpus pseudoreinwardtii A. Camus 单果柯
Trees, evergreen

18.7 (16.3~21.0); 12.3 (9.1~15.1); 23.0 (21.4~25.6); 18.8 (16.4~21.0); 165.1 (136.4~191.8); 0.0 (0.0~0.0); 905 (797~1053) ❙ 1465 (1190~1751); 782 (655~996); 61 (48~77) ❙ 62.2 (44.5~88.8); 899 (797~1053); 480.2 (336.3~659.0)

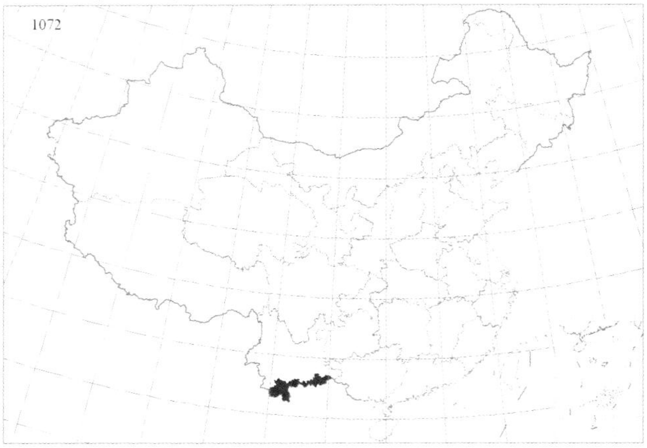

Fagaceae 壳斗科 *Lithocarpus* 石栎属
Lithocarpus pseudovestitus A. Camus 毛果柯
Trees, evergreen

20.9 (17.2~24.7); 13.5 (9.1~19.8); 26.3 (22.4~28.6); 20.9 (17.3~24.7); 191.2 (147.0~236.5); 0.0 (0.0~0.0); 1085 (833~1376) ❙ 1508 (1161~2435); 782 (392~1428); 78 (42~130) ❙ 41.1 (-11.9~99.5); 1054 (833~1221); 367.3 (235.1~620.0)

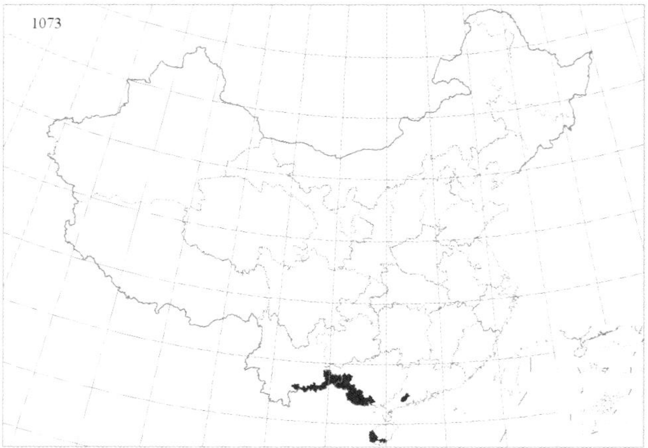

Fagaceae 壳斗科 *Lithocarpus* 石栎属
Lithocarpus pseudoxizangensis Z. K. Zhou et H. Sun 假西藏柯
Trees, evergreen

13.3; 5.4; 19.6; 13.4; 100.4; 0.0; 706 ❙ 1736; 1007; 51 ❙ 145.8; 706; 447.4

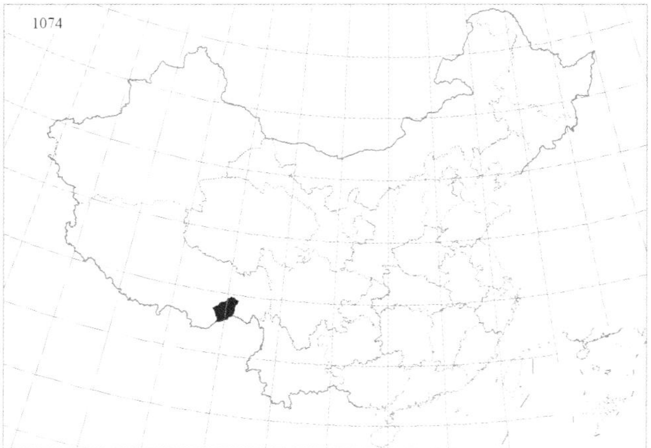

Fagaceae 壳斗科 *Lithocarpus* 石栎属

Lithocarpus qinzhouicus C. C. Huang et Y. T. Chang 钦州柯

Trees, evergreen

19.9; 10.7; 27.1; 19.9; 179.0; 0.0; 1045 ❙ 1634; 840; 116 ❙ 56.4; 1045; 240.2

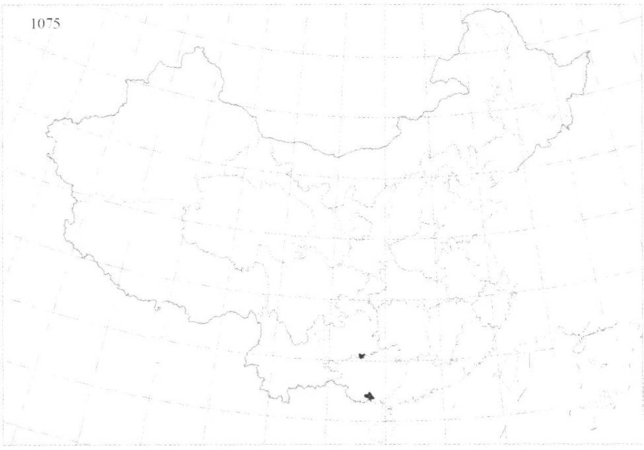

Fagaceae 壳斗科 *Lithocarpus* 石栎属

Lithocarpus quercifolius C. C. Huang et Y. T. Chang 栎叶柯

Trees, evergreen

19.8 (15.3~22.4); 10.7 (4.2~14.8); 27.7 (26.0~28.4); 19.9 (15.4~22.4); 178.4 (125.8~208.9); 0.2 (0.0~0.8); 1053 (818~1191) ❙ 1860 (1561~2037); 812 (512~1033); 157 (110~216) ❙ 78.1 (70.4~103.1); 1053 (818~1191); 200.9 (163.7~222.5)

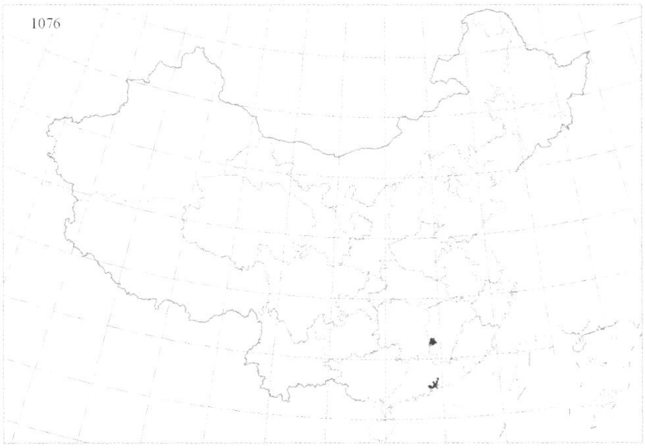

Fagaceae 壳斗科 *Lithocarpus* 石栎属

Lithocarpus rhabdostachyus subsp. dakhaensis A. Camus 毛枝柯

Trees, evergreen

18.5 (17.2~21.0); 11.1 (9.5~14.2); 23.7 (22.4~25.6); 18.5 (17.3~21.0); 162.3 (147.0~191.8); 0.0 (0.0~0.0); 905 (833~1053) ❙ 1403 (1254~1522); 791 (699~860); 56 (48~68) ❙ 55.7 (37.6~80.6); 905 (833~1053); 421.5 (336.3~551.6)

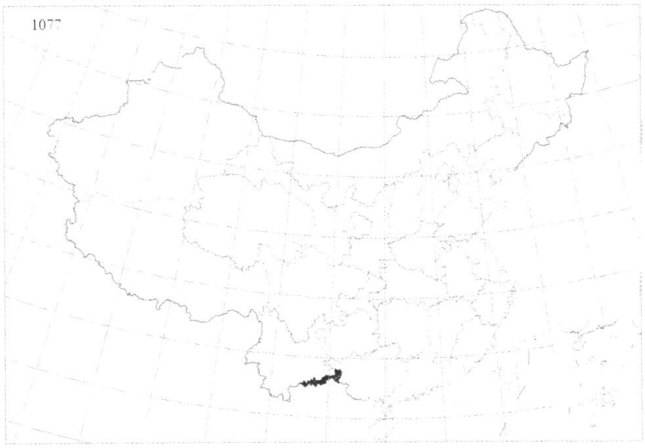

Fagaceae 壳斗科 *Lithocarpus* 石栎属

Lithocarpus rosthornii (Schottky) Barnett 南川柯

Trees, evergreen

18.3 (11.9~22.0); 8.6 (2.9~13.5); 26.7 (20.0~28.8); 18.3 (12.0~22.0); 160.4 (87.2~203.8); 0.3 (0.0~3.4); 963 (671~1170) ❙ 1519 (999~2005); 654 (489~902); 137 (35~218) ❙ 58.4 (21.3~128.9); 963 (671~1170); 221.1 (158.7~294.6)

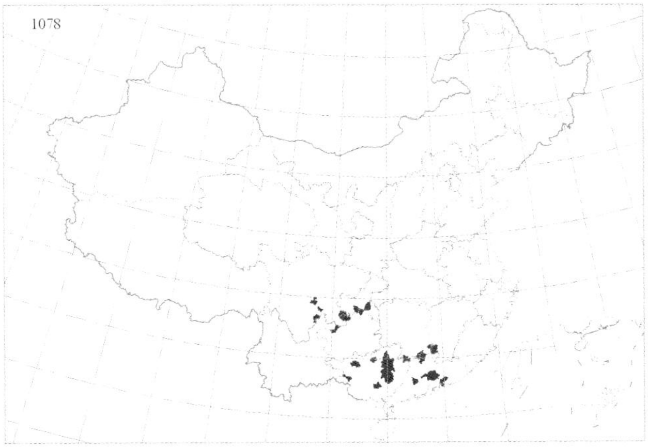

Fagaceae 壳斗科 *Lithocarpus* 石栎属

Lithocarpus shinsuiensis Hayata et Kaneh. 浸水营柯

Trees, evergreen

20.3; 15.9; 23.6; 20.4; 184.4; 0.0; 993 ❙ 2853; 1618; 147 ❙ 190.4; 993; 330.3

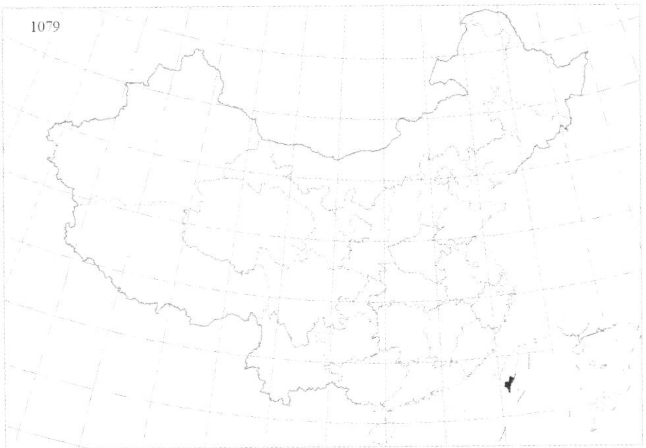

Fagaceae 壳斗科 *Lithocarpus* 石栎属

Lithocarpus silvicolarum (Hance) Chun 犁耙柯

Trees, evergreen

21.3 (16.3~24.8); 14.2 (7.3~20.1); 26.4 (21.4~28.8); 21.3 (16.4~24.8); 196.0 (136.4~237.6); 0.0 (0.0~0.0); 1118 (797~1385) ❙ 1481 (868~2435); 695 (312~1428); 79 (41~142) ❙ 35.1 (-34.3~99.5); 1074 (797~1360); 376.6 (225.9~620.0)

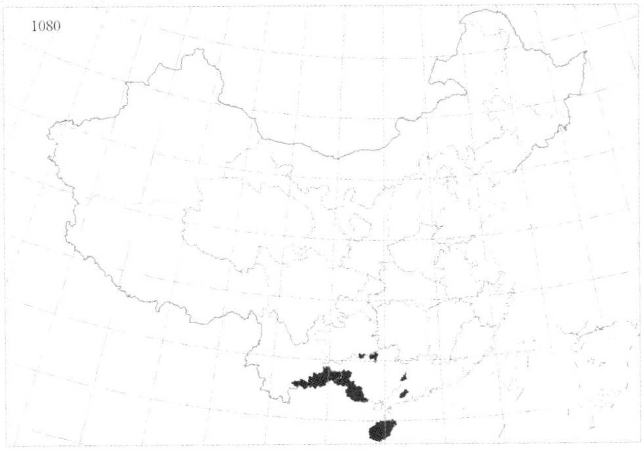

Fagaceae 壳斗科 *Lithocarpus* 石栎属
Lithocarpus skanianus (Dunn) Rehder 滑皮柯
Trees, evergreen

17.7 (14.3~24.8); 7.5 (3.3~20.1); 27.0 (21.4~29.4); 17.8 (14.3~24.8); 153.3 (113.2~237.6); 0.1 (0.0~2.3); 930 (761~1385) ❙ 1549 (1161~1900); 568 (392~1056); 173 (41~230) ❙ 67.8 (1.5~128.9); 917 (761~1147); 258.8 (145.3~620.0)

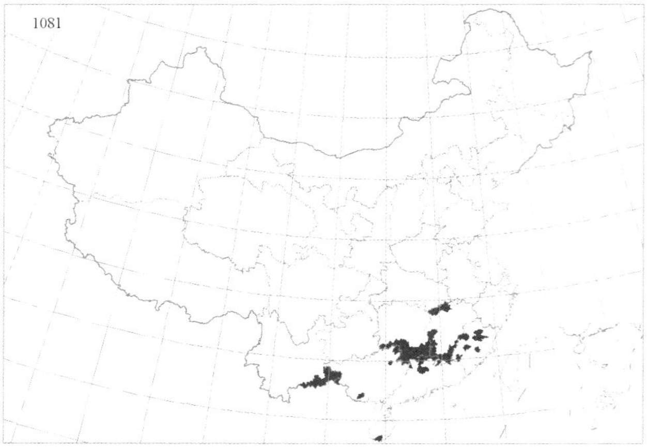

Fagaceae 壳斗科 *Lithocarpus* 石栎属
Lithocarpus sphaerocarpus (Hickel et A. Camus) A. Camus 球壳柯
Trees, evergreen

20.0 (17.2~22.7); 12.5 (9.1~15.1); 25.4 (22.3~28.6); 20.1 (17.3~22.8); 180.8 (147.0~213.1); 0.0 (0.0~0.0); 1014 (833~1221) ❙ 1499 (1161~2435); 814 (633~1428); 78 (48~130) ❙ 48.7 (5.8~99.5); 1011 (833~1221); 380.3 (256.6~659.0)

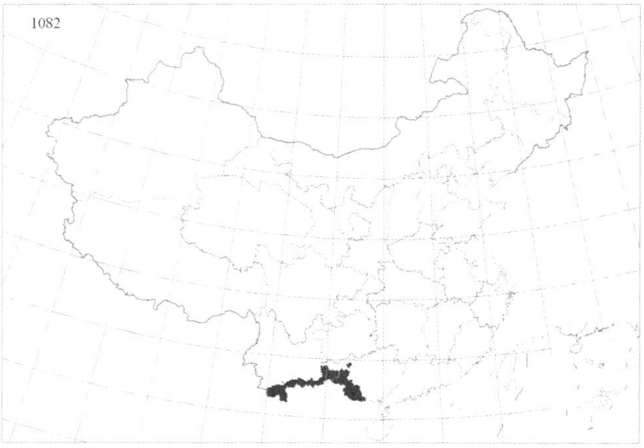

Fagaceae 壳斗科 *Lithocarpus* 石栎属
Lithocarpus tabularis Y. C. Hsu et H. Wei Jen 平头柯
Trees, evergreen

18.9 (17.2~21.0); 12.5 (9.5~15.1); 23.1 (22.1~25.6); 18.9 (17.3~21.0); 166.9 (147.0~191.8); 0.0 (0.0~0.0); 912 (833~1053) ❙ 1460 (1254~1751); 788 (655~996); 60 (48~77) ❙ 60.5 (36.8~88.8); 906 (833~1053); 488.6 (336.3~659.0)

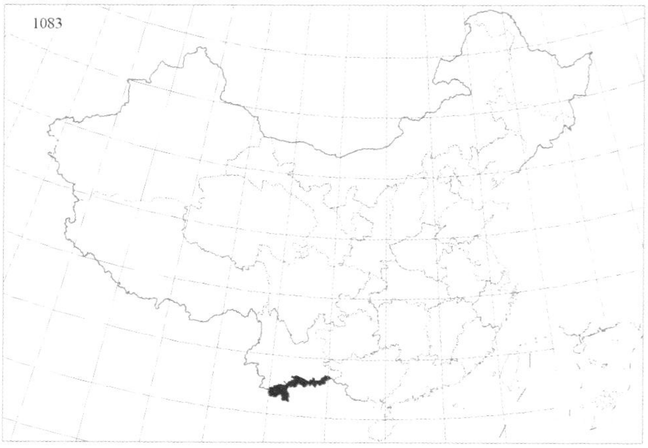

Fagaceae 壳斗科 *Lithocarpus* 石栎属
Lithocarpus taitoensis (Hayata) Hayata 菱果柯
Trees, evergreen

19.1 (16.2~22.2); 13.8 (11.0~17.7); 24.4 (20.1~28.7); 19.7 (16.2~22.7); 176.2 (134.2~212.7); 0.0 (0.0~0.0); 987 (773~1216) ❙ 2357 (1836~2992); 1178 (960~1847); 178 (104~246) ❙ 148.7 (61.6~246.5); 980 (773~1216); 262.3 (202.8~298.9)

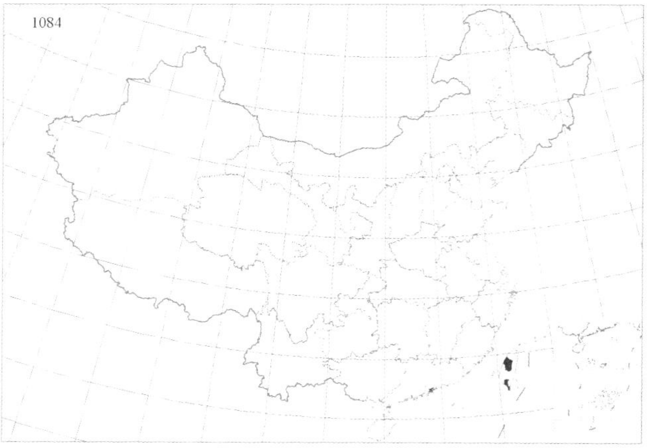

Fagaceae 壳斗科 *Lithocarpus* 石栎属
Lithocarpus talangensis C. C. Huang et Y. T. Chang 石屏柯
Trees, evergreen

18.7; 12.5; 22.8; 18.8; 165.0; 0.0; 899 ❙ 1283; 738; 54 ❙ 42.9; 880; 387.1

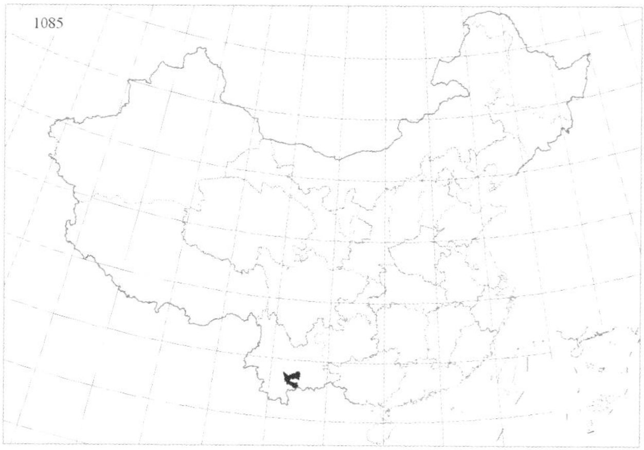

Fagaceae 壳斗科 *Lithocarpus* 石栎属
Lithocarpus tenuilimbus H. T. Chang 薄叶柯
Trees, evergreen

18.7 (16.3~21.8); 9.6 (6.5~14.2); 26.3 (21.4~29.8); 18.8 (16.4~21.9); 165.5 (136.4~202.6); 0.0 (0.0~0.0); 965 (797~1150) ❙ 1582 (1190~1961); 657 (508~891); 147 (48~197) ❙ 64.4 (35.6~95.5); 959 (797~1150); 278.5 (174.9~491.5)

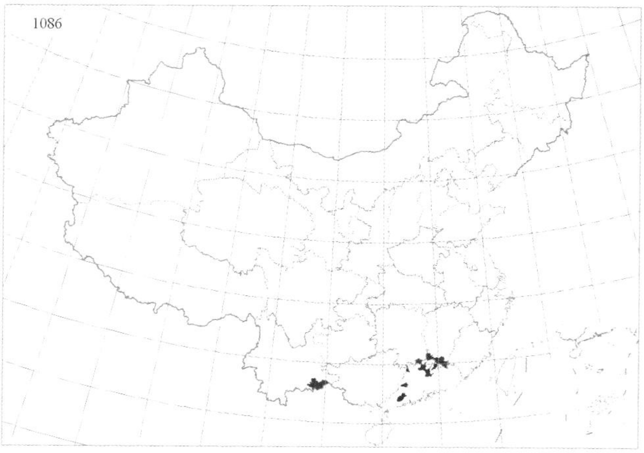

Fagaceae 壳斗科 *Lithocarpus* 石栎属
Lithocarpus tephrocarpus (Drake) A. Camus 灰壳柯
Trees, evergreen

17.6; 9.9; 23.2; 17.7; 152.4; 0.0; 859 ❙ 1422; 813; 52 ❙ 65.6; 859; 359.7

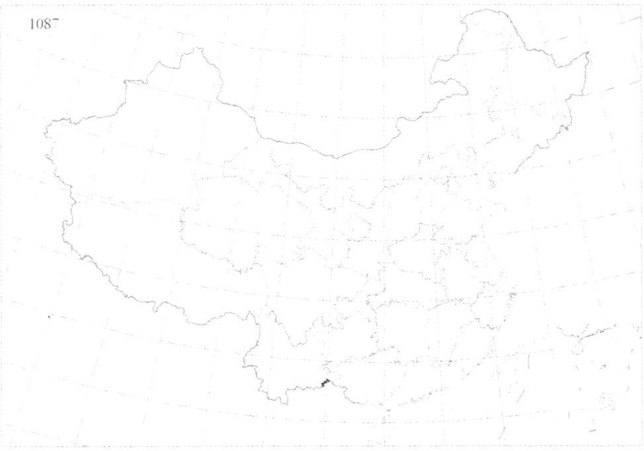

Fagaceae 壳斗科 *Lithocarpus* 石栎属
Lithocarpus thomsonii (Miq.) Rehder 潞西柯
Trees, evergreen

16.2 (8.6~19.6); 9.5 (1.3~12.8); 20.7 (15.0~23.5); 16.2 (8.7~19.6); 134.9 (52.6~175.7); 0.4 (0.0~8.8); 798 (574~953) ❙ 1331 (1041~1736); 756 (585~1007); 52 (42~70) ❙ 68.0 (26.1~145.8); 790 (574~932); 428.5 (335.1~589.2)

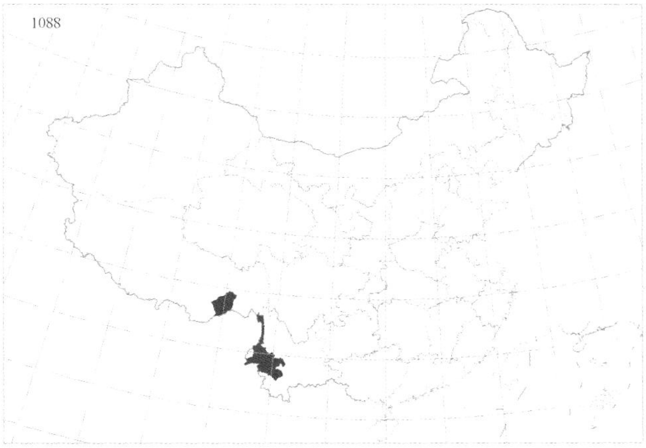

Fagaceae 壳斗科 *Lithocarpus* 石栎属
Lithocarpus trachycarpus (Hickel et A. Camus) A. Camus 糙果柯
Trees, evergreen

19.4; 13.7; 22.8; 19.4; 172.9; 0.0; 931 ❙ 1494; 715; 62 ❙ 60.9; 915; 544.4

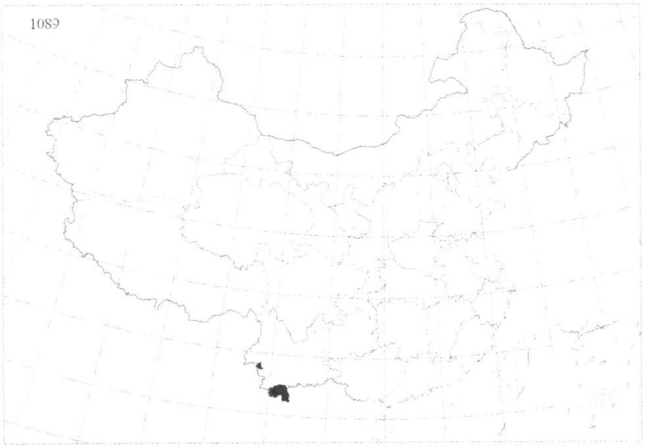

Fagaceae 壳斗科 *Lithocarpus* 石栎属
Lithocarpus triqueter (Hickel et A. Camus) A. Camus 棱果柯
Trees, evergreen

19.4; 12.6; 24.1; 19.4; 172.6; 0.0; 954 ❙ 1445; 787; 64 ❙ 52.3; 954; 521.5

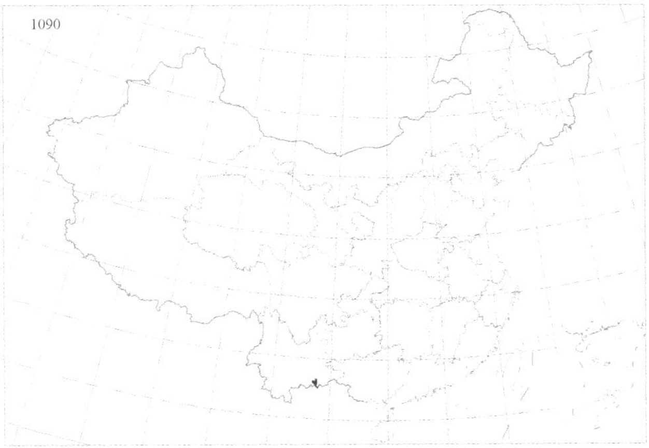

Fagaceae 壳斗科 *Lithocarpus* 石栎属
Lithocarpus truncatus (King ex Hook. f.) Rehder et E. H. Wilson 截果柯
Trees, evergreen

17.8 (13.3~21.0); 11.4 (5.4~15.1); 22.1 (19.1~25.6); 17.9 (13.4~21.0); 154.6 (100.4~191.8); 0.0 (0.0~0.0); 862 (706~1053) ❙ 1356 (1007~1751); 751 (539~1007); 53 (41~77) ❙ 57.9 (12.5~145.8); 849 (706~1053); 438.7 (294.5~659.0)

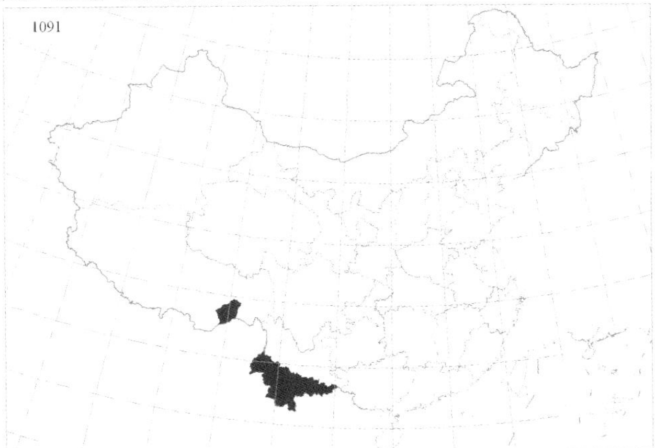

Fagaceae 壳斗科 *Lithocarpus* 石栎属
Lithocarpus tubulosus (Hickel et A. Camus) A. Camus 壶嘴柯
Trees, evergreen

21.0; 14.2; 25.6; 21.0; 191.8; 0.0; 1053 ❙ 1522; 818; 68 ❙ 44.5; 1053; 551.6

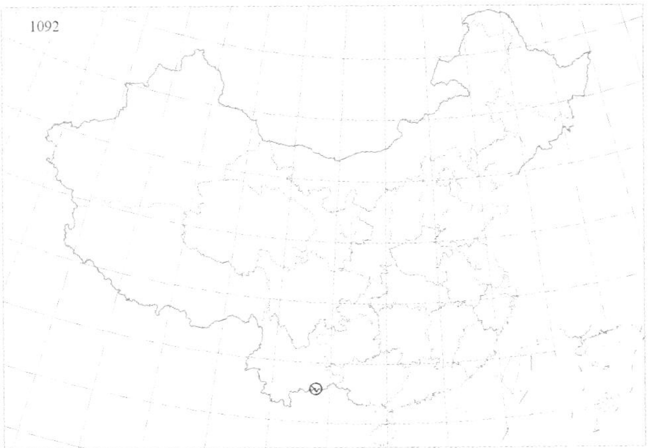

Fagaceae 壳斗科 *Lithocarpus* 石栎属

Lithocarpus uvariifolius (Hance) Rehder 紫玉盘柯

Trees, evergreen

19.9 (16.9~22.4); 11.1 (7.1~15.2); 27.4 (25.8~28.7); 20.0 (17.0~22.7); 180.0 (143.6~212.7); 0.0 (0.0~0.0); 1039 (862~1216) ❙ 1682 (1284~2037); 696 (507~1033); 155 (73~221) ❙ 62.5 (29.9~100.8); 1039 (862~1216); 252.9 (163.7~332.5)

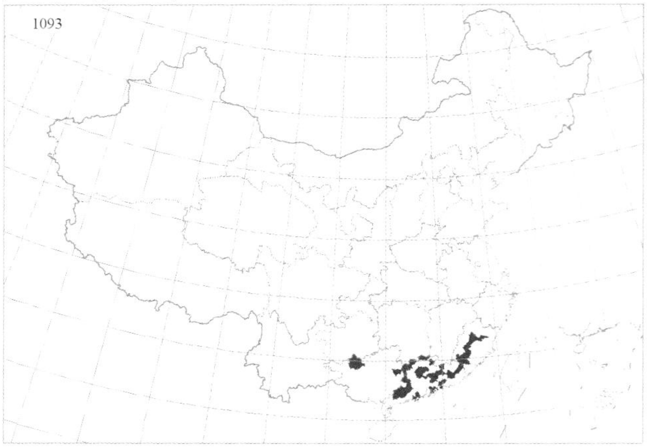

Fagaceae 壳斗科 *Lithocarpus* 石栎属

Lithocarpus uvariifolius var. ellipticus (F. P. Metcalf) C. C. Huang et Y. T. Chang 卵叶玉盘柯

Shrubs or small trees, evergreen

20.6 (17.7~22.4); 11.8 (7.1~14.8); 27.9 (26.3~28.8); 20.6 (17.8~22.4); 187.2 (153.6~208.9); 0.0 (0.0~0.0); 1079 (916~1191) ❙ 1709 (1462~2037); 699 (507~1033); 159 (75~194) ❙ 58.7 (29.9~85.1); 1079 (916~1191); 232.1 (157.7~270.0)

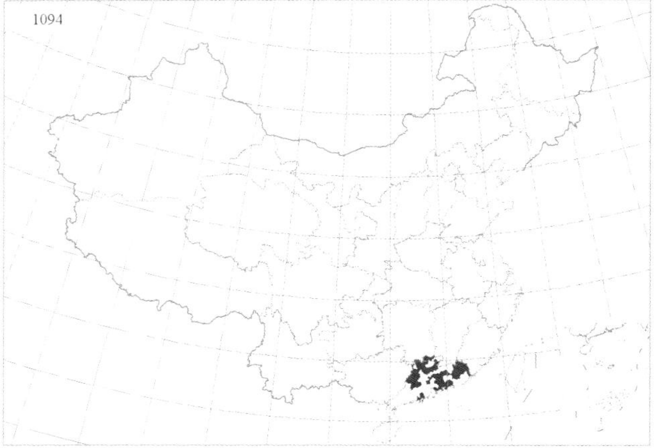

Fagaceae 壳斗科 *Lithocarpus* 石栎属

Lithocarpus variolosus (Franch.) Chun 麻子壳柯

Trees, evergreen

15.0 (5.4~20.4); 7.8 (-2.4~15.1); 20.5 (12.3~26.6); 15.1 (5.8~20.5); 121.9 (31.5~185.8); 1.1 (0.0~25.6); 765 (481~989) ❙ 1089 (741~1688); 616 (411~959); 41 (12~77) ❙ 43.2 (-14.2~112.2); 747 (481~986); 315.6 (142.9~659.0)

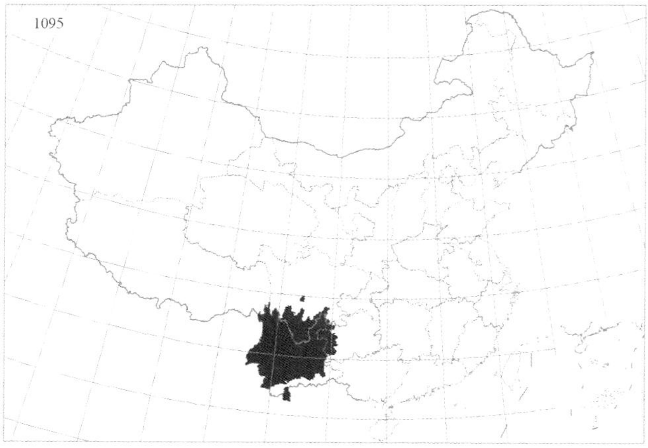

Fagaceae 壳斗科 *Lithocarpus* 石栎属

Lithocarpus xizangensis C. C. Huang et Y. T. Chang 西藏柯

Trees, evergreen

13.3; 5.4; 19.6; 13.4; 100.4; 0.0; 706 ❙ 1736; 1007; 51 ❙ 145.8; 706; 447.4

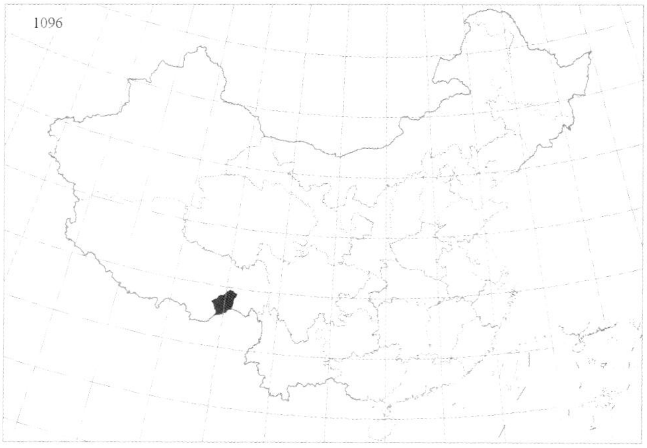

Fagaceae 壳斗科 *Lithocarpus* 石栎属

Lithocarpus xylocarpus (Kurz) Markgr. 木果柯

Trees, evergreen

17.7 (15.2~21.0); 11.4 (8.7~14.2); 21.9 (20.1~25.6); 17.8 (15.2~21.0); 153.2 (122.5~191.8); 0.0 (0.0~0.0); 855 (753~1053) ❙ 1239 (939~1547); 698 (524~911); 50 (37~68) ❙ 44.8 (6.5~74.2); 833 (717~1053); 407.4 (289.9~551.6)

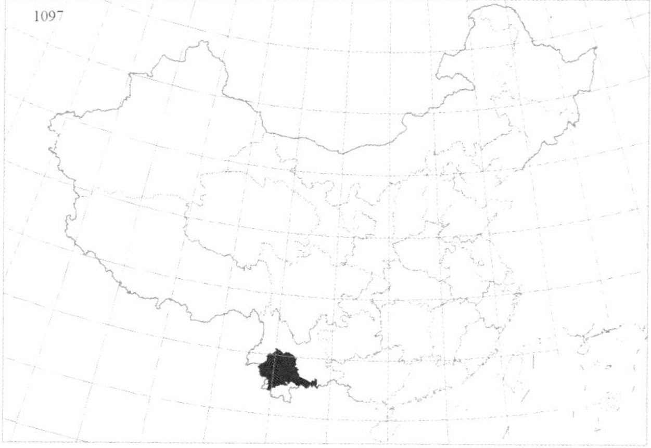

Fagaceae 壳斗科 *Lithocarpus* 石栎属

Lithocarpus yongfuensis Q. F. Zheng 永福柯

Trees, evergreen

18.2; 9.3; 26.2; 18.2; 158.7; 0.0; 922 ❙ 1693; 598; 193 ❙ 83.7; 922; 292.8

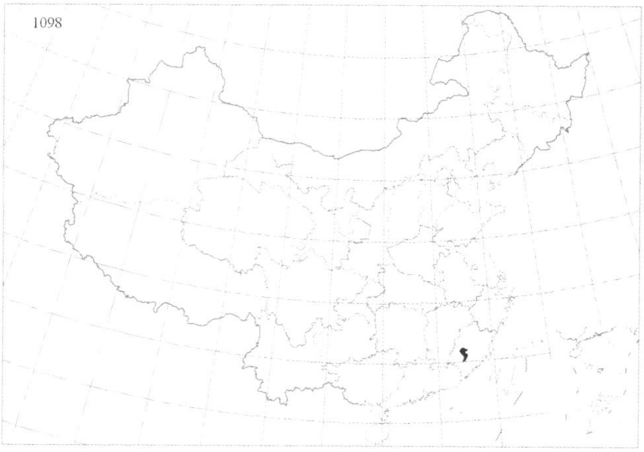

Fagaceae 壳斗科 *Quercus* 栎属
Quercus acrodonta Seemen 岩栎
Trees, evergreen

13.8 (2.6~19.2); 2.4 (-7.0~10.9); 24.4 (11.2~29.1); 13.9 (4.3~19.2); 112.4 (20.0~170.8); 6.7 (0.0~48.2); 775 (427~961) ❙ 1075 (626~1574); 465 (290~727); 76 (14~209) ❙ 38.8 (-22.2~108.0); 763 (427~961); 233.1 (139.1~411.1)

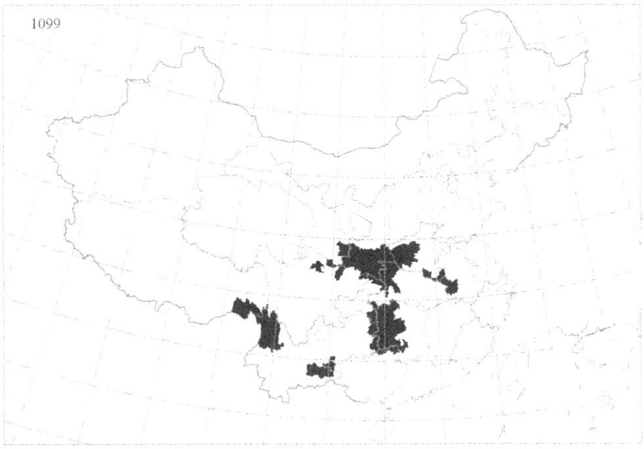

Fagaceae 壳斗科 *Quercus* 栎属
Quercus acutissima Carruth. 麻栎
Trees, deciduous

15.4 (2.2~24.8); 4.2 (-15.9~20.1); 25.7 (9.5~29.9); 15.6 (3.8~24.8); 131.4 (15.5~237.6); 5.5 (0.0~80.4); 851 (411~1385) ❙ 1154 (409~3495); 528 (258~1428); 93 (8~897) ❙ 35.0 (-37.2~224.4); 823 (409~1247); 217.5 (108.2~659.0)

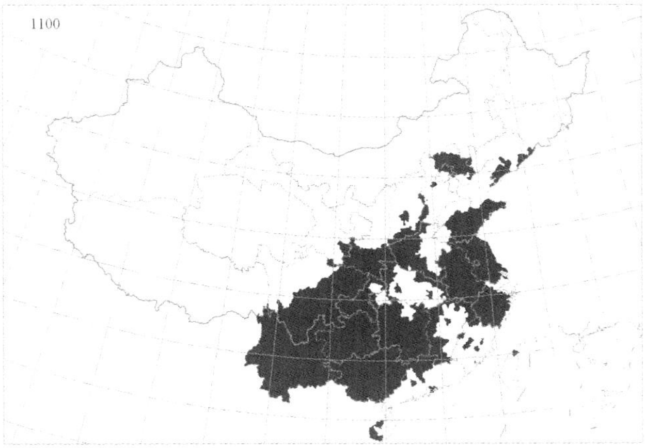

Fagaceae 壳斗科 *Quercus* 栎属
Quercus aliena Blume 槲栎
Trees, deciduous

13.9 (2.2~21.0); 1.8 (-15.9~15.1); 24.8 (9.5~29.9); 14.2 (3.9~21.0); 117.8 (15.8~191.8); 10.7 (0.0~80.4); 795 (411~1053) ❙ 10[illegible]9 (377~2355); 480 (241~996); 75 (8~308) ❙ 26.6 (-43.4~172.5); 746 (377~1053); 218.3 (70.3~659.0)

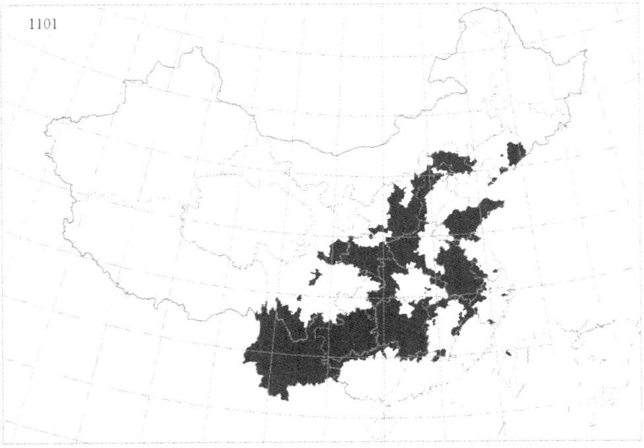

Fagaceae 壳斗科 *Quercus* 栎属
Quercus aliena var. acutiserrata Maxim. ex Wenz. 锐齿槲栎
Trees, deciduous

14.2 (5.4~21.0); 3.7 (-13.5~15.1); 23.5 (12.3~29.6); 14.4 (5.8~21.0); 118.4 (31.5~191.8); 6.9 (0.0~61.8); 781 (481~1053) ❙ 1029 (517~1751); 531 (288~996); 55 (11~209) ❙ 31.5 (-29.6~112.2); 756 (481~1053); 258.7 (134.1~659.0)

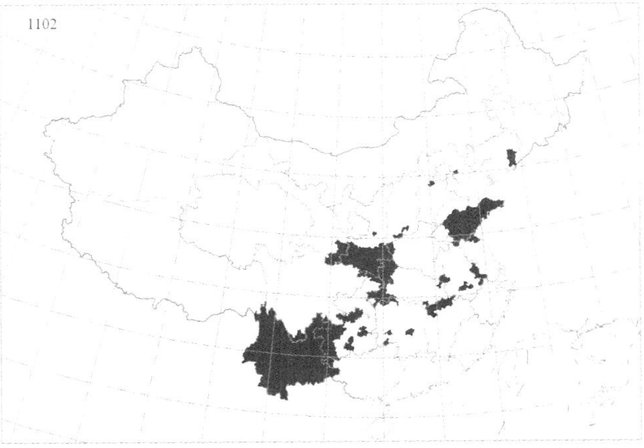

Fagaceae 壳斗科 *Quercus* 栎属
Quercus aliena var. pekingensis Schottky 北京槲栎
Trees, deciduous

12.0 (6.6~14.4); -3.1 (-10.5~0.0); 25.6 (21.4~28.0); 12.6 (8.9~14.5); 106.0 (71.3~124.7); 21.2 (11.2~52.1); 766 (604~858) ❙ 683 (455~917); 438 (285~560); 27 (9~44) ❙ -10.5 (-35.6~17.2); 671 (455~812); 168.5 (131.6~273.5)

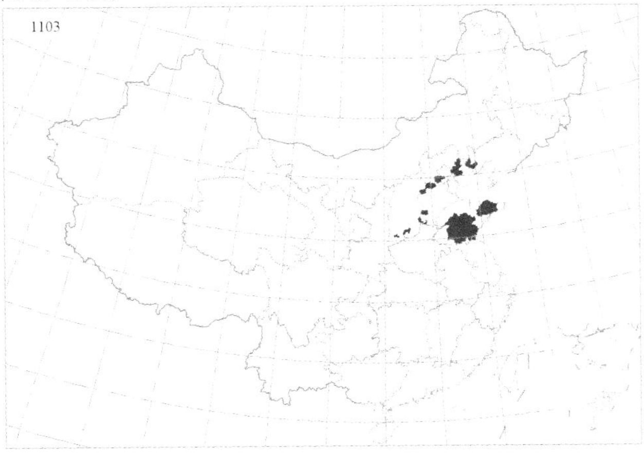

Fagaceae 壳斗科 *Quercus* 栎属
Quercus aquifolioides Rehder et E. H. Wilson 川滇高山栎
Trees, evergreen

12.2 (-0.4~19.6); 4.5 (-10.9~13.9); 18.5 (8.4~25.9); 12.6 (2.8~19.6); 98.4 (9.4~175.7); 10.9 (0.0~73.4); 688 (360~953) ❙ 1006 (369~1736); 567 (243~1007); 36 (5~88) ❙ 48.9 (-14.2~145.8); 672 (360~932); 302.6 (161.9~627.0)

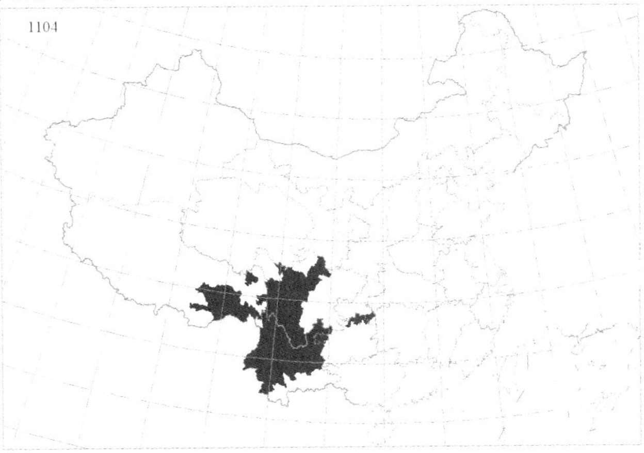

Fagaceae 壳斗科 *Quercus* 栎属
Quercus baronii Skan 橿子栎
Shrubs or trees, semievergreen

12.2 (2.6~16.1); -0.5 (-7.7~5.3); 24.1 (11.1~28.1); 12.5 (4.3~16.1); 100.7 (20.0~135.3); 13.7 (0.0~49.3); 737 (427~902) | 831 (503~1492); 395 (258~816); 39 (14~120) | 15.1 (-36.5~121.7); 681 (427~884); 210.8 (120.1~341.6)

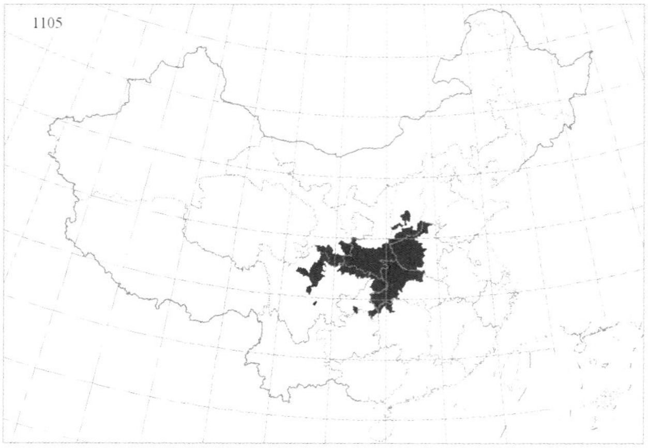

Fagaceae 壳斗科 *Quercus* 栎属
Quercus bawanglingensis C. C. Huang 坝王栎
Trees, evergreen

23.6; 17.8; 27.8; 23.6; 223.3; 0.0; 1271 | 1242; 444; 44 | -2.2; 1091; 439.7

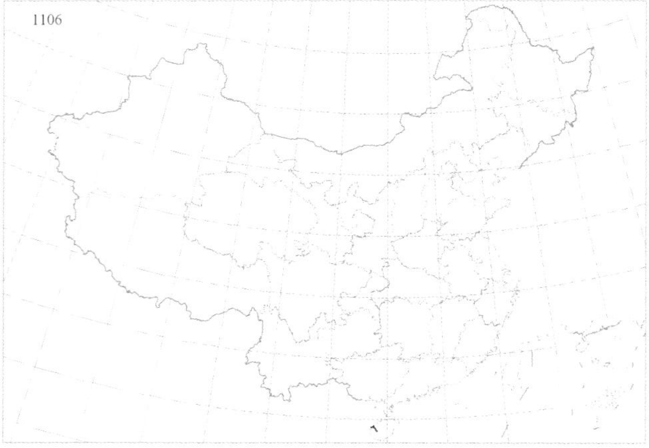

Fagaceae 壳斗科 *Quercus* 栎属
Quercus chenii Nakai 小叶栎
Trees, deciduous

16.1 (10.2~18.8); 3.7 (-2.9~9.2); 27.9 (22.4~30.1); 16.1 (10.6~18.9); 136.0 (83.1~166.5); 2.7 (0.0~20.5); 888 (656~1024) | 1341 (626~1950); 497 (299~685); 155 (29~259) | 50.3 (-22.2~121.9); 866 (626~977); 200.2 (111.9~319.9)

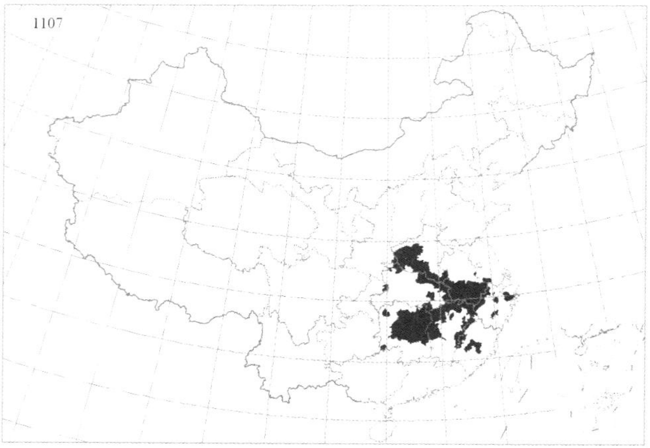

Fagaceae 壳斗科 *Quercus* 栎属
Quercus cocciferoides Hand.-Mazz. 铁橡栎
Trees, evergreen or semievergreen

15.0 (2.2~21.0); 7.4 (-7.0~15.1); 20.9 (9.5~26.6); 15.1 (3.9~21.0); 122.8 (15.8~191.8); 2.3 (0.0~48.6); 770 (411~1053) | 1111 (729~1751); 613 (390~996); 43 (11~77) | 45.3 (-14.2~121.7); 755 (411~1053); 324.0 (142.9~659.0)

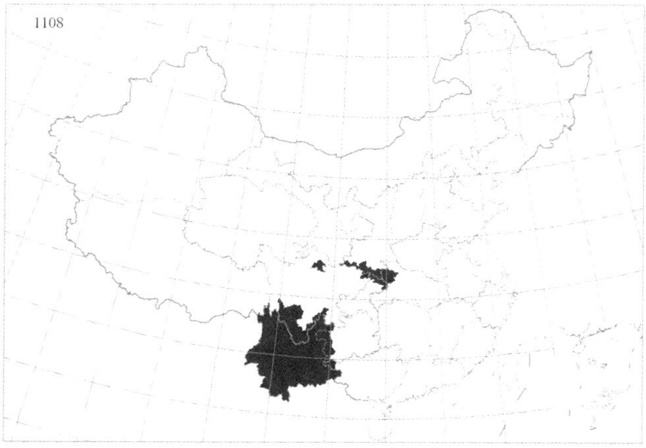

Fagaceae 壳斗科 *Quercus* 栎属
Quercus dentata Thunb. 槲树
Trees, deciduous

12.2 (1.4~21.7); -1.1 (-20.6~15.3); 24.0 (9.9~29.9); 13.1 (3.8~21.7); 107.5 (15.5~200.2); 20.2 (0.0~93.8); 755 (413~1123) | 872 (397~2382); 465 (254~1045); 48 (7~244) | 15.1 (-43.4~191.8); 700 (397~1121); 225.9 (70.3~627.0)

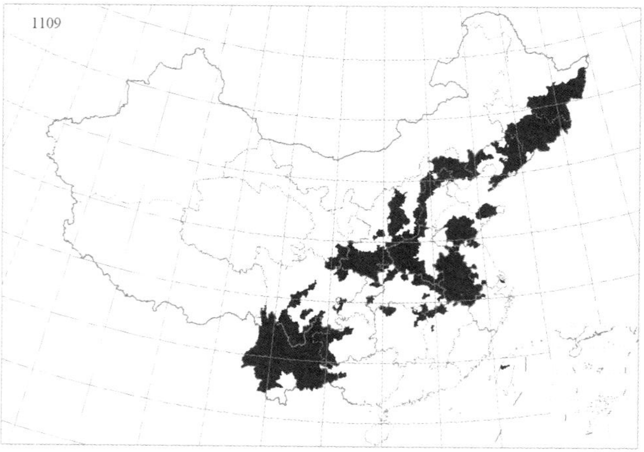

Fagaceae 壳斗科 *Quercus* 栎属
Quercus dolicholepis A. Camus 匙叶栎
Trees, evergreen

12.7 (2.2~19.2); 1.6 (-7.0~10.7); 22.8 (9.9~28.1); 12.9 (3.8~19.2); 103.0 (15.5~170.8); 10.0 (0.0~49.1); 734 (413~961) | 919 (512~1494); 455 (258~723); 44 (12~197) | 27.5 (-35.9~125.9); 694 (413~961); 238.5 (121.6~411.1)

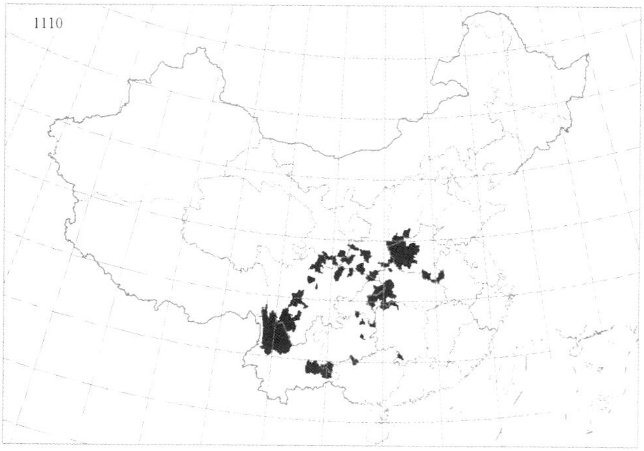

Fagaceae 壳斗科 *Quercus* 栎属

Quercus engleriana Seemen 巴东栎

Trees, evergreen or semievergreen

15.0 (5.3~19.4); 4.5 (-2.5~10.9); 24.6 (12.2~29.5); 15.0 (5.7~19.4); 122.5 (30.5~172.7); 2.3 (0.0~26.2); 807 (479~1002) ❙ 1275 (689~1997); 529 (316~1007); 106 (15~258) ❙ 57.9 (-2.0~147.0); 803 (479~1002); 227.8 (140.5~447.4)

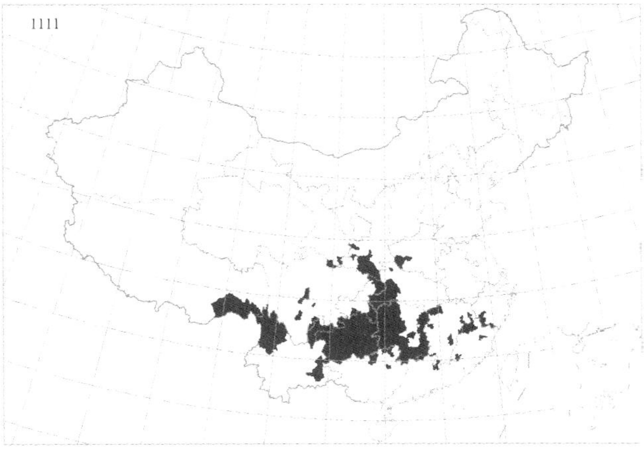

Fagaceae 壳斗科 *Quercus* 栎属

Quercus fabri Hance 白栎

Trees, deciduous

16.0 (2.6~21.3); 4.7 (-7.0~15.2); 26.5 (11.2~30.1); 16.0 (4.3~22.7); 134.3 (20.0~212.7); 1.9 (0.0~48.2); 868 (427~1216) ❙ 1316 (685~1997); 525 (299~994); 134 (16~259) ❙ 52.0 (-23.1~147.0); 853 (427~1216); 210.7 (108.2~411.1)

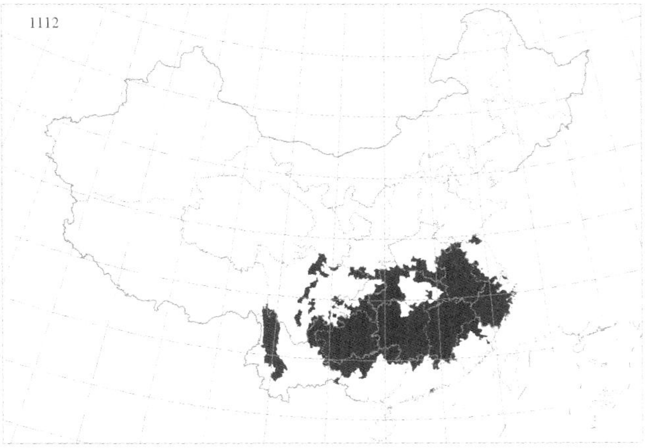

Fagaceae 壳斗科 *Quercus* 栎属

Quercus franchetii Skan 锥连栎

Trees, evergreen

14.5 (3.8~19.3); 7.0 (-4.3~13.1); 20.4 (10.8~26.6); 14.6 (4.7~19.4); 115.8 (21.6~173.2); 1.4 (0.0~35.4); 748 (444~939) ❙ 1003 (741~1307); 573 (411~700); 38 (15~60) ❙ 35.8 (-14.2~121.7); 727 (444~912); 277.2 (142.9~383.4)

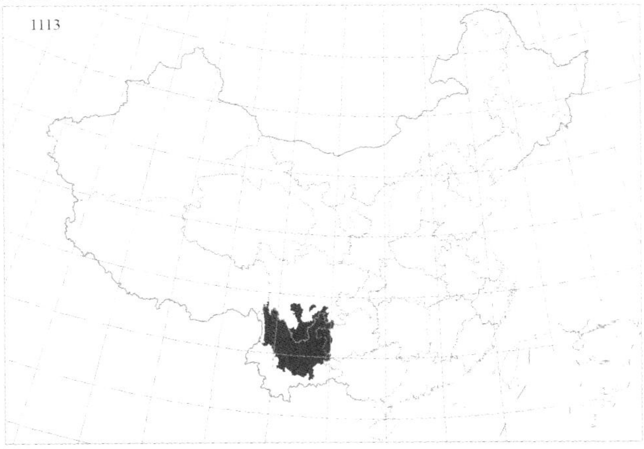

Fagaceae 壳斗科 *Quercus* 栎属

Quercus griffithii Hook. f. et Thomson ex Miq. 大叶栎

Trees, deciduous

15.2 (5.4~20.5); 7.5 (-2.1~15.1); 21.3 (12.3~27.1); 15.2 (5.8~20.5); 123.9 (31.5~185.9); 1.0 (0.0~25.6); 779 (481~992) ❙ 1104 (741~1736); 597 (411~1007); 50 (12~196) ❙ 42.8 (-14.2~145.8); 762 (481~986); 295.7 (142.9~659.0)

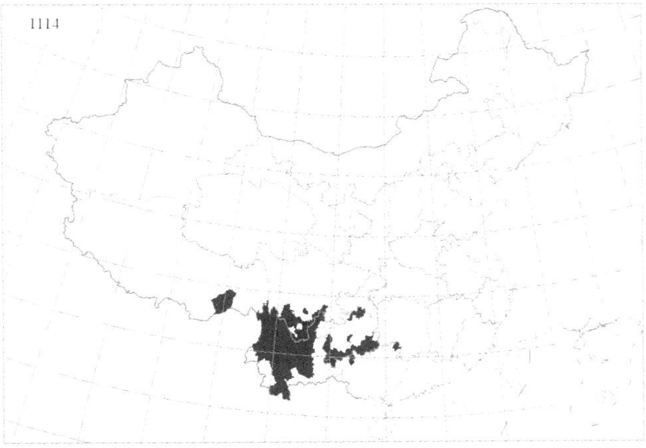

Fagaceae 壳斗科 *Quercus* 栎属

Quercus guajavifolia H. Lév. 帽斗栎

Shrubs or small trees, evergreen

12.7 (2.2~18.9); 5.0 (-7.1~12.0); 19.0 (9.5~26.6); 12.9 (3.8~19.0); 98.4 (15.5~167.8); 5.6 (0.0~49.1); 695 (411~929) ❙ 977 (579~1356); 560 (365~727); 34 (10~88) ❙ 43.9 (-14.2~125.9); 674 (411~912); 276.7 (142.9~407.1)

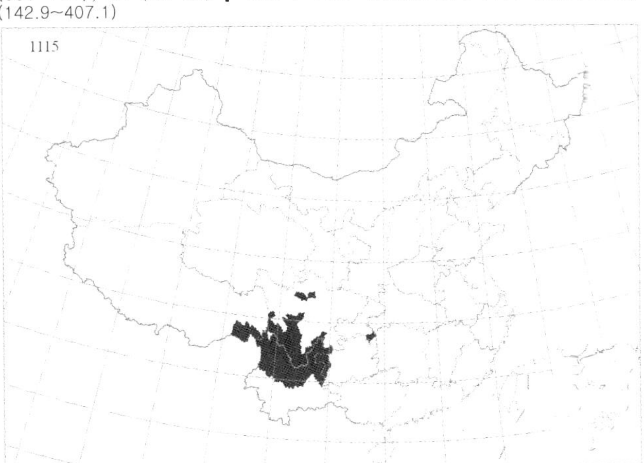

Fagaceae 壳斗科 *Quercus* 栎属

Quercus kingiana Craib 澜沧栎

Trees, evergreen

17.6; 11.6; 21.3; 17.6; 151.6; 0.0; 841 ❙ 1352; 753; 49 ❙ 60.9; 822; 413.4

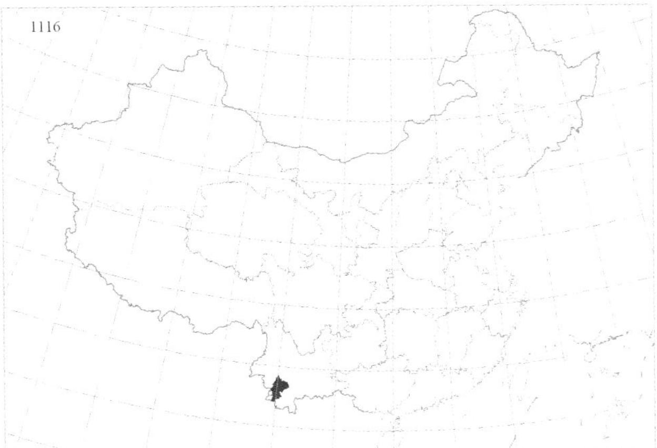

Fagaceae 壳斗科 *Quercus* 栎属
Quercus lanata Sm. 通麦栎
Trees, evergreen

8.4 (1.8~17.6); 0.3 (−7.1~9.9); 15.2 (9.9~23.2); 9.0 (3.8~17.7); 62.5 (15.7~152.4); 21.7 (0.0~53.3); 575 (404~859) ❙ 1045 (527~1736); 605 (317~1007); 33 (9~52) ❙ 76.7 (22.4~145.8); 575 (404~859); 307.2 (169.5~447.4)

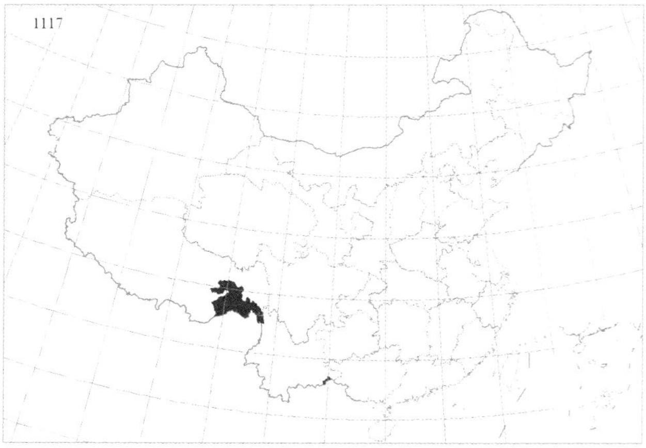

Fagaceae 壳斗科 *Quercus* 栎属
Quercus lodicosa O. E. Warb. et E. F. Warb. 西藏栎
Trees, evergreen

8.4; 0.0; 15.5; 9.0; 61.9; 20.9; 568 ❙ 1132; 662; 30 ❙ 84.1; 568; 363.7

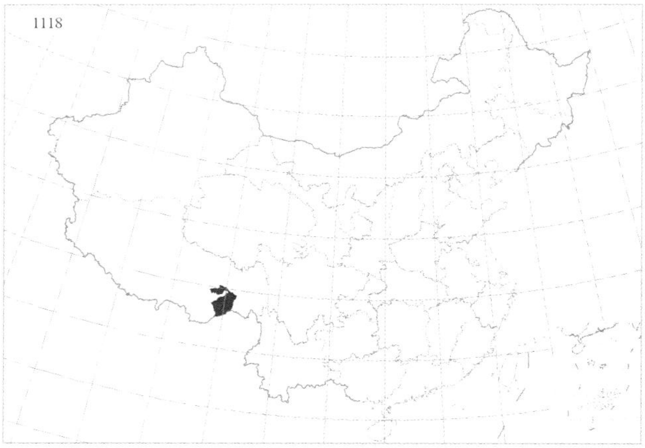

Fagaceae 壳斗科 *Quercus* 栎属
Quercus marlipoensis Hu et W. C. Cheng 麻栗坡栎
Trees, evergreen

17.6; 9.9; 23.2; 17.7; 152.4; 0.0; 859 ❙ 1422; 813; 52 ❙ 65.6; 859; 359.7

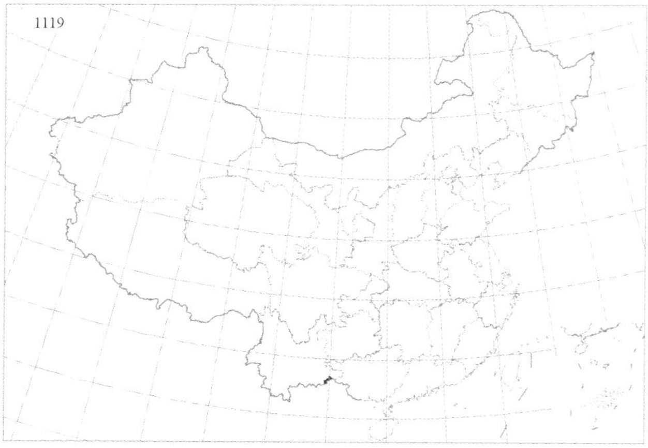

Fagaceae 壳斗科 *Quercus* 栎属
Quercus mongolica Fisch. ex Ledeb. 蒙古栎
Trees, deciduous

7.7 (−5.8~15.5); −9.1 (−29.8~2.1); 22.5 (10.7~28.0); 9.9 (3.8~15.6); 80.0 (16.2~131.3); 46.9 (4.3~157.4); 653 (411~881) ❙ 576 (132~1077); 347 (81~691); 19 (2~97) ❙ −10.2 (−80.1~121.7); 545 (132~881); 195.1 (13.6~337.1)

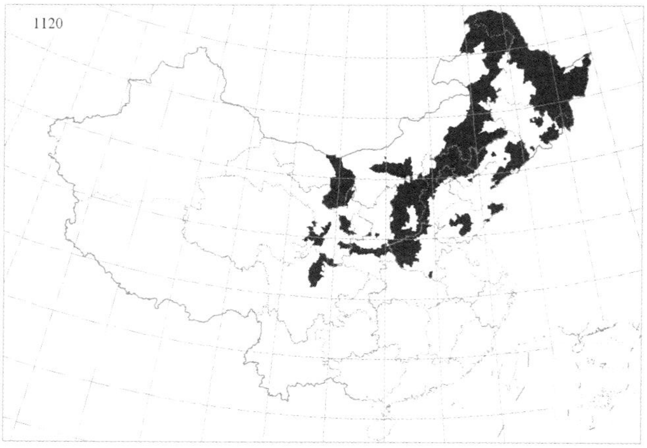

Fagaceae 壳斗科 *Quercus* 栎属
Quercus monimotricha Hand.-Mazz. 矮高山栎
Shrubs, evergreen

12.2 (2.0~18.9); 4.8 (−7.4~12.0); 18.1 (9.5~24.3); 12.4 (3.8~19.0); 94.7 (15.4~167.8); 8.3 (0.0~51.9); 678 (411~929) ❙ 976 (729~1454); 565 (411~792); 32 (11~70) ❙ 49.3 (−14.2~128.6); 652 (411~799); 299.8 (205.2~487.2)

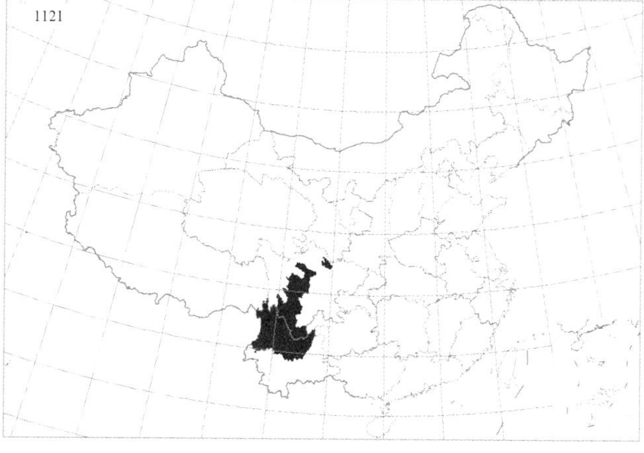

Fagaceae 壳斗科 *Quercus* 栎属
Quercus oxyphylla (E. H. Wilson) Hand.-Mazz. 尖叶栎
Trees, evergreen

14.1 (2.2~19.2); 2.7 (−6.3~9.1); 24.8 (9.5~28.5); 14.1 (3.9~19.2); 115.4 (15.8~170.8); 6.2 (0.0~48.6); 785 (411~1004) ❙ 1241 (677~1997); 497 (336~709); 103 (11~259) ❙ 56.6 (−0.2~130.7); 784 (411~1004); 251.7 (149.5~334.4)

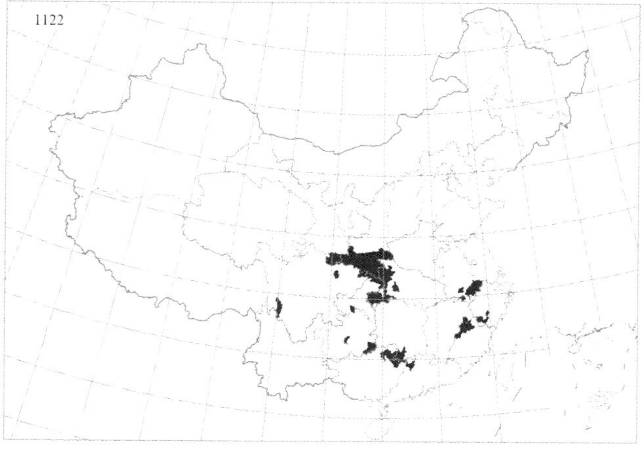

Fagaceae 壳斗科 *Quercus* 栎属
Quercus phillyreoides A. Gray 乌冈栎
Trees, evergreen

15.7 (7.4~21.8); 4.8 (-4.2~13.5); 25.8 (18.7~29.7); 15.8 (8.2~21.8); 131.5 (58.8~201.6); 2.6 (0.0~29.4); 846 (561~1162) ❙ 1350 (634~2993); 528 (290~924); 129 (15~547) ❙ 58.8 (-16.6~203.4); 841 (561~1162); 227.6 (126.1~411.1)

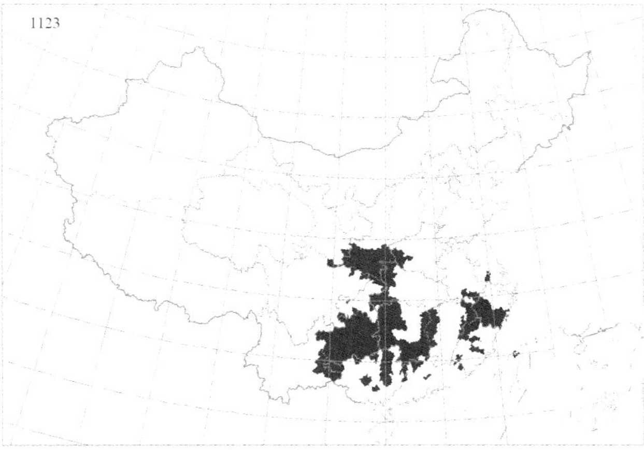

Fagaceae 壳斗科 *Quercus* 栎属
Quercus rehderiana Hand.-Mazz. 毛脉高山栎
Trees, evergreen

12.6 (-1.8~18.9); 4.8 (-10.2~12.5); 19.0 (6.1~26.6); 12.8 (1.8~19.0); 98.7 (1.9~167.8); 7.3 (0.0~82.8); 694 (329~929) ❙ 987 (320~1736); 559 (193~1007); 37 (10~70) ❙ 44.9 (-14.2~145.8); 676 (320~912); 274.5 (72.2~487.2)

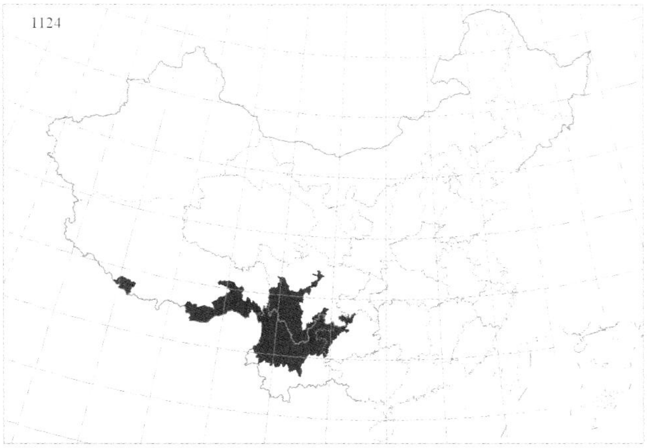

Fagaceae 壳斗科 *Quercus* 栎属
Quercus semecarpifolia Sm. 高山栎
Trees, evergreen

3.3 (-1.8~12.2); -4.9 (-10.2~4.3); 10.6 (6.1~18.5); 5.1 (1.8~12.2); 27.6 (1.9~87.4); 47.7 (0.7~82.8); 444 (329~672) ❙ 723 (320~1431); 427 (193~817); 26 (13~33) ❙ 50.2 (-4.0~113.0); 439 (320~672); 183.8 (72.2~363.7)

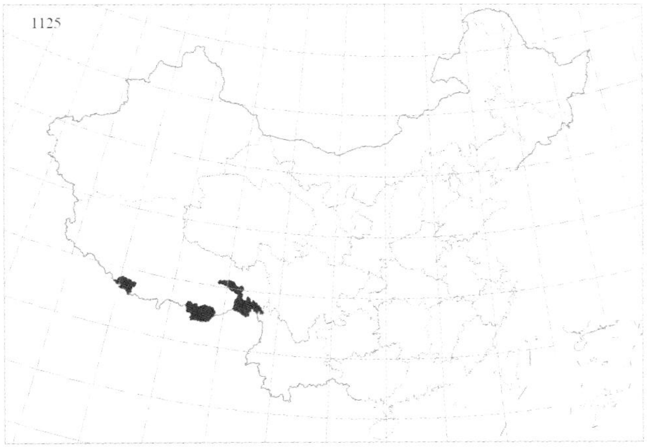

Fagaceae 壳斗科 *Quercus* 栎属
Quercus senescens Hand.-Mazz. 灰背栎
Shrubs or trees, evergreen

13.0 (1.8~18.9); 5.2 (-7.1~12.0); 19.4 (9.5~26.6); 13.1 (3.8~19.0); 101.0 (15.5~167.8); 4.4 (0.0~53.3); 705 (404~929) ❙ 1002 (583~1431); 564 (348~817); 38 (11~76) ❙ 45.2 (-14.2~125.9); 688 (404~912); 270.8 (142.9~407.1)

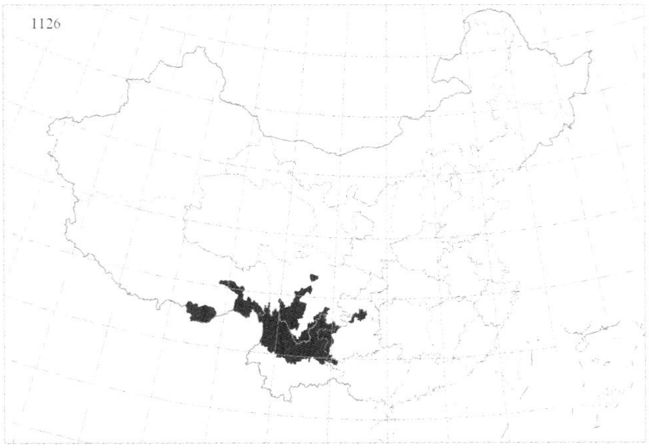

Fagaceae 壳斗科 *Quercus* 栎属
Quercus serrata Thunb. 枹栎
Trees, deciduous

14.6 (2.6~21.7); 2.5 (-15.5~15.3); 25.7 (10.8~30.1); 14.8 (4.3~21.7); 122.4 (20.0~200.2); 7.1 (0.0~68.8); 819 (427~1123) ❙ 1174 (473~2332); 496 (258~1045); 103 (7~259) ❙ 42.5 (-35.9~191.8); 798 (427~1121); 208.2 (108.2~411.1)

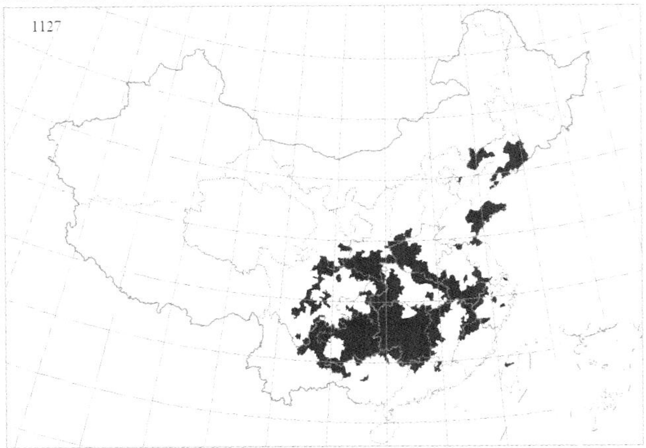

Fagaceae 壳斗科 *Quercus* 栎属
Quercus setulosa Hickel et A. Camus 富宁栎
Trees, evergreen

19.3 (16.2~22.2); 10.3 (6.3~13.9); 26.4 (24.4~28.3); 19.4 (16.2~22.3); 172.2 (134.5~207.5); 0.0 (0.0~0.0); 993 (826~1190) ❙ 1367 (1236~1567); 705 (611~848); 84 (54~141) ❙ 38.9 (14.8~62.5); 993 (826~1190); 284.0 (215.4~371.7)

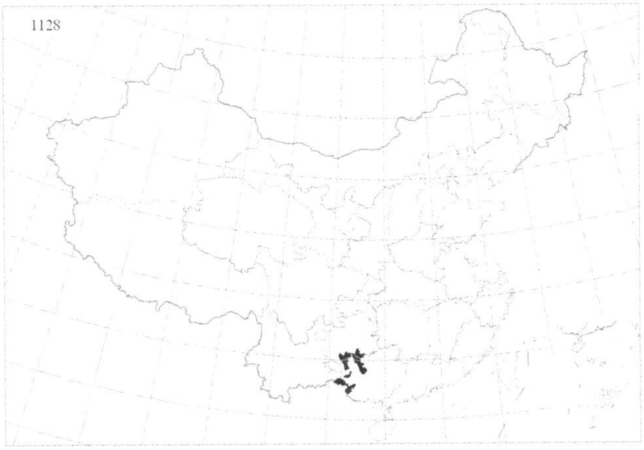

Fagaceae 壳斗科 *Quercus* 栎属
Quercus spinosa David ex Franch. 刺叶高山栎
Trees, evergreen

13.2 (-0.7~20.8); 3.6 (-9.6~15.3); 21.9 (7.7~28.6); 13.4 (2.5~20.8); 106.3 (6.8~189.2); 7.0 (0.0~74.2); 735 (338~1019) ❙ 1092 (370~2713); 517 (262~1412); 65 (10~341) ❙ 48.8 (-16.6~246.5); 726 (338~1019); 258.7 (61.6~487.2)

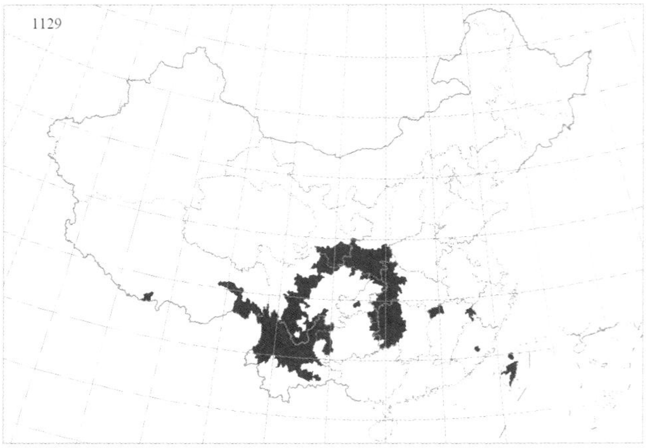

Fagaceae 壳斗科 *Quercus* 栎属
Quercus tarokoensis Hayata 太鲁阁栎
Trees, evergreen

17.4; 12.8; 21.0; 17.5; 150.0; 0.0; 823 ❙ 2606; 1168; 246 ❙ 216.8; 823; 333.6

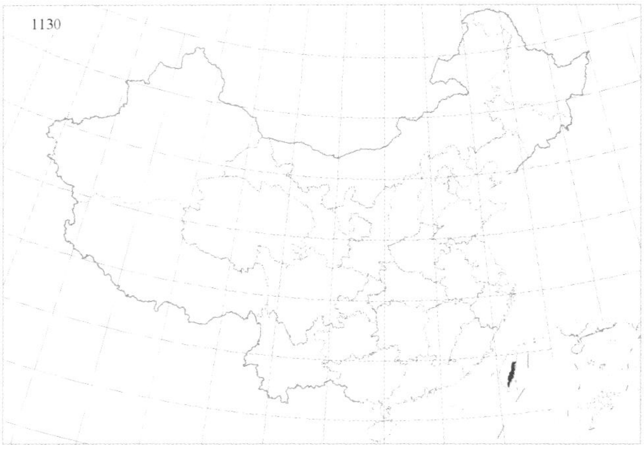

Fagaceae 壳斗科 *Quercus* 栎属
Quercus utilis Hu et W. C. Cheng 炭栎
Trees, evergreen

17.8 (14.3~21.0); 10.2 (4.6~14.2); 23.4 (22.4~25.6); 17.8 (14.3~21.0); 154.1 (112.3~191.8); 0.0 (0.0~0.4); 875 (751~1053) ❙ 1402 (1196~1547); 770 (597~911); 63 (48~107) ❙ 60.6 (44.5~80.6); 875 (751~1053); 403.9 (179.5~551.6)

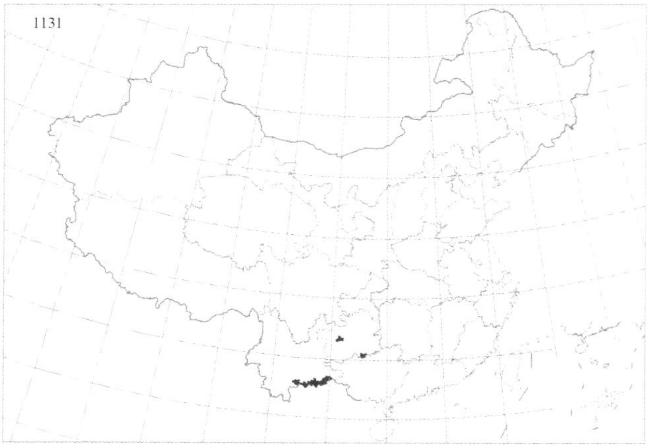

Fagaceae 壳斗科 *Quercus* 栎属
Quercus variabilis Blume 栓皮栎
Trees, deciduous

14.7 (2.6~22.2); 2.9 (-15.9~17.7); 25.3 (11.2~29.8); 14.9 (4.3~22.2); 124.0 (20.0~206.7); 7.6 (0.0~80.4); 820 (427~1123) ❙ 1088 (397~3495); 502 (254~1847); 87 (8~897) ❙ 31.5 (-43.4~246.5); 783 (397~1123); 219.6 (108.2~659.0)

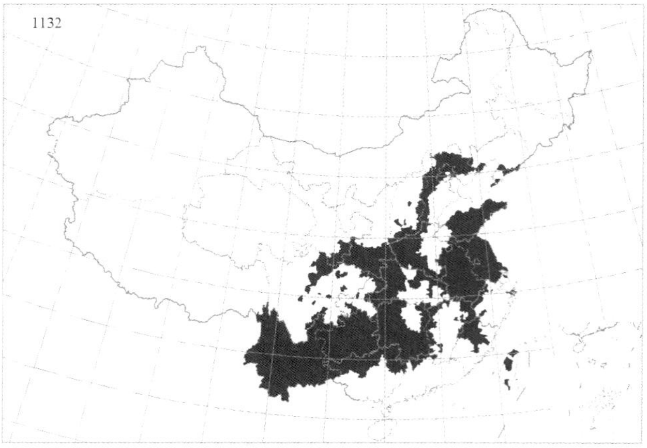

Fagaceae 壳斗科 *Quercus* 栎属
Quercus yunnanensis Franch. 云南波罗栎
Trees, deciduous

14.9 (5.4~19.6); 7.4 (-2.1~13.0); 20.8 (12.3~27.4); 14.9 (5.8~19.6); 120.4 (31.5~175.7); 1.1 (0.0~25.6); 764 (481~961) ❙ 1086 (741~1688); 608 (411~959); 44 (12~180) ❙ 42.9 (-14.2~112.2); 747 (481~961); 304.9 (142.9~627.0)

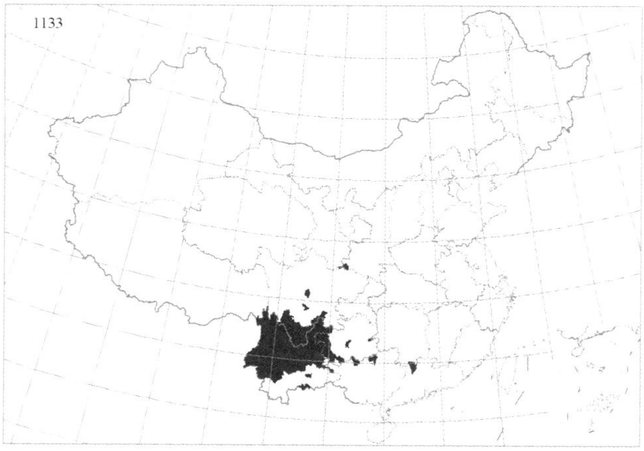

Ulmaceae 榆科 *Aphananthe* 糙叶树属
Aphananthe aspera (Thunb.) Planch. 糙叶树
Trees, deciduous

16.9 (6.0~24.1); 6.9 (-4.9~18.9); 25.9 (14.4~30.1); 17.0 (6.6~24.1); 144.8 (39.7~229.7); 1.3 (0.0~26.7); 889 (508~1322) ❙ 1391 (520~3495); 570 (258~1594); 137 (18~897) ❙ 56.6 (-35.9~246.5); 873 (508~1257); 243.9 (123.9~659.0)

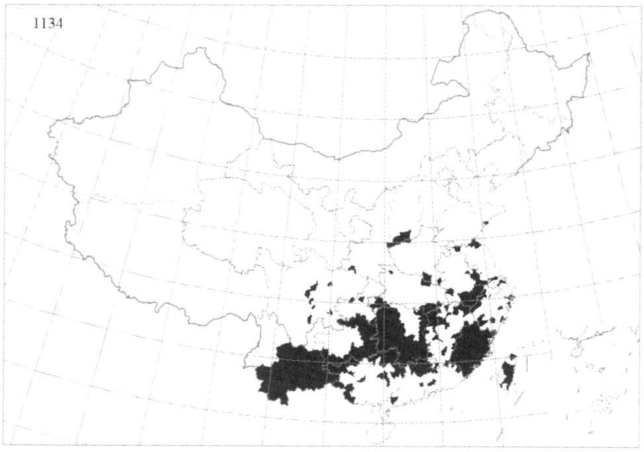

U

Ulmaceae 榆科 *Aphananthe* 糙叶树属
Aphananthe aspera var. pubescens C. J. Chen 柔毛糙叶树
Trees, deciduous

20.0 (16.3~22.2); 12.0 (5.5~14.2); 26.1 (22.3~28.3); 20.0 (16.4~22.3); 180.5 (136.4~207.5); 0.0 (0.0~0.0); 1024 (854~1190) | 1397 (1163~1693); 742 (633~850); 81 (43~188) | 38.4 (5.8~95.9); 1023 (854~1190); 355.0 (234.1~551.6)

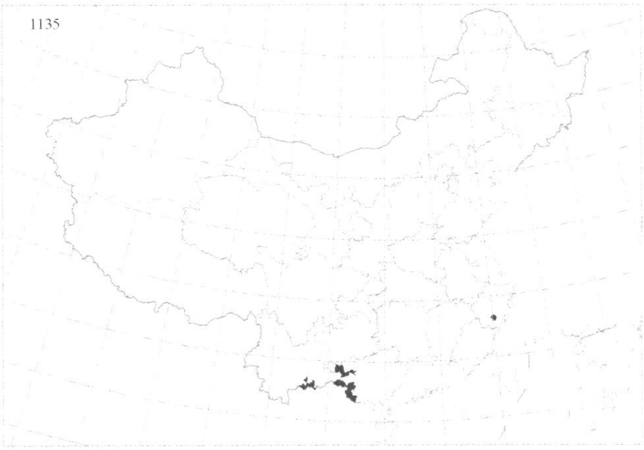

Ulmaceae 榆科 *Aphananthe* 糙叶树属
Aphananthe cuspidata (Blume) Planch. 滇糙叶树
Trees

22.8 (17.4~24.8); 16.5 (11.0~20.1); 27.4 (21.4~29.0); 22.8 (17.4~24.8); 213.9 (149.3~237.6); 0.0 (0.0~0.0); 1218 (834~1385) | 1546 (868~2037); 647 (312~1033); 80 (41~118) | 28.9 (-34.3~71.8); 1150 (834~1360); 388.0 (163.7~659.0)

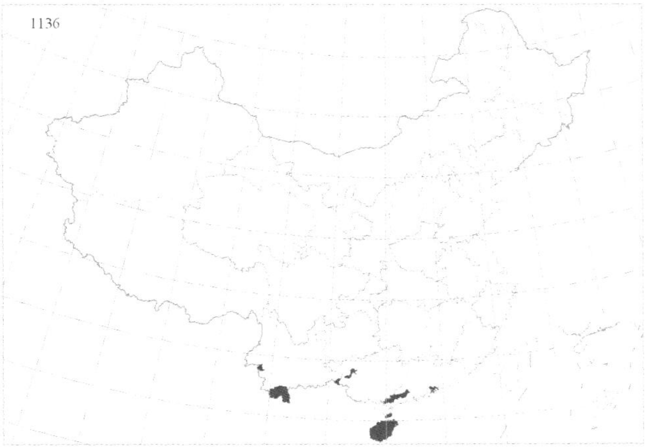

Ulmaceae 榆科 *Celtis* 朴属
Celtis biondii Pamp. 紫弹树
Trees, deciduous

16.2 (2.2~24.1); 5.8 (-6.6~18.9); 25.7 (9.9~30.1); 16.2 (3.8~24.1); 136.8 (15.5~229.7); 2.0 (0.0~49.1); 864 (413~1322) | 1320 (512~2992); 549 (258~1847); 123 (14~259) | 52.5 (-35.3~246.5); 851 (413~1206); 235.9 (110.1~627.0)

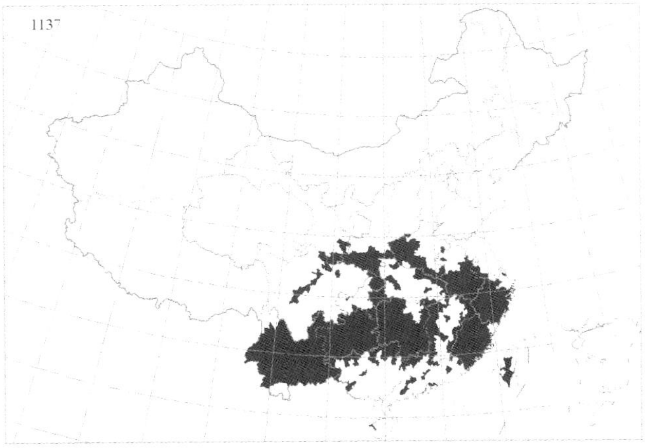

Ulmaceae 榆科 *Celtis* 朴属
Celtis bungeana Blume 黑弹树
Trees, deciduous

13.1 (0.6~21.3); 0.0 (-20.7~15.1); 24.8 (9.0~29.8); 13.6 (3.3~21.3); 112.6 (11.9~196.0); 15.4 (0.0~101.4); 781 (389~1123) | 869 (132~1776); 439 (81~996); 58 (2~228) | 10.1 (-80.1~128.9); 689 (132~1121); 200.1 (13.6~659.0)

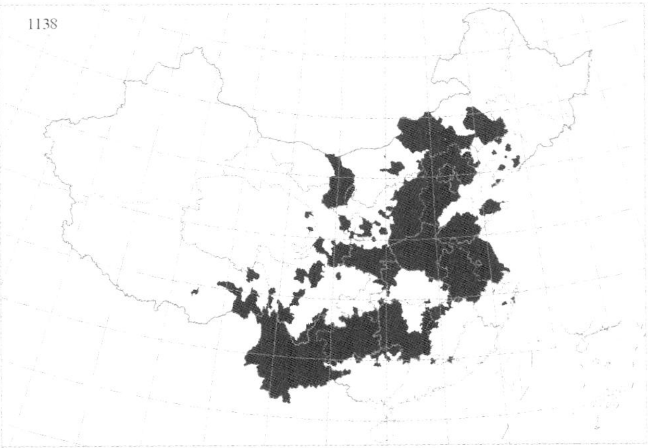

Ulmaceae 榆科 *Celtis* 朴属
Celtis cerasifera C. K. Schneid. 小果朴
Trees, deciduous

14.6 (1.8~19.1); 5.9 (-7.6~12.0); 21.7 (9.9~28.1); 14.7 (3.8~19.1); 119.1 (15.5~169.2); 3.9 (0.0~53.3); 766 (404~988) | 1082 (428~1719); 560 (266~860); 54 (9~191) | 41.9 (-35.6~125.9); 752 (404~988); 278.8 (125.5~491.5)

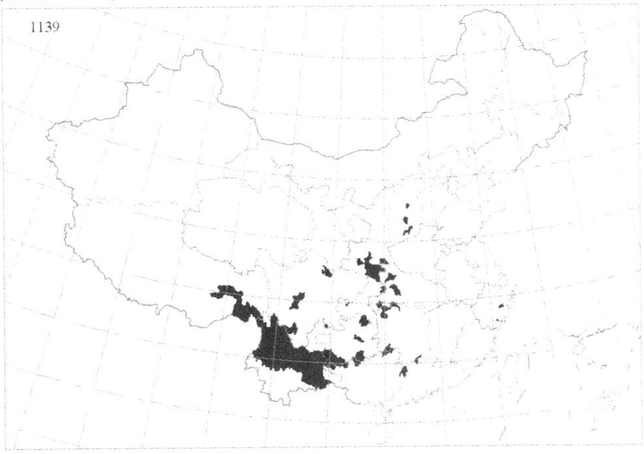

Ulmaceae 榆科 *Celtis* 朴属
Celtis chekiangensis W. C. Cheng 天目朴树
Trees, deciduous

15.7 (14.4~19.2); 3.7 (2.1~9.2); 27.1 (25.1~28.3); 15.7 (14.5~19.2); 131.2 (118.7~170.8); 2.3 (0.0~4.9); 857 (801~1002) | 1530 (1331~1663); 575 (494~678); 189 (135~222) | 79.3 (49.1~101.3); 856 (801~1002); 264.8 (226.0~299.5)

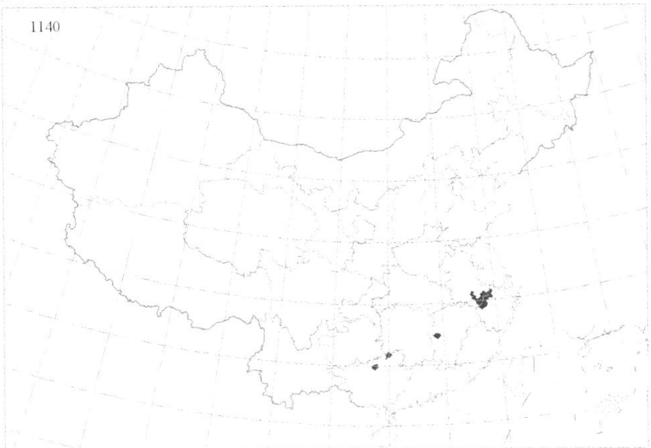

Ulmaceae 榆科 *Celtis* 朴属

Celtis julianae C. K. Schneid. 珊瑚朴

Trees, deciduous

15.2 (5.7~21.1); 3.6 (-4.2~13.2); 26.0 (14.2~29.9); 15.2 (6.4~21.1); 126.4 (38.1~193.7); 4.0 (0.0~29.4); 832 (504~1104) ❙ 1237 (512~1829); 489 (258~720); 114 (13~253) ❙ 47.8 (-35.3~128.9); 819 (504~1057); 214.9 (111.9~334.4)

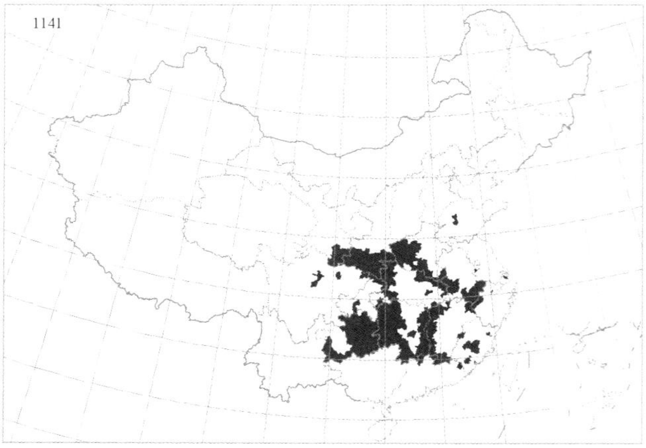

Ulmaceae 榆科 *Celtis* 朴属

Celtis koraiensis Nakai 大叶朴

Trees, deciduous

11.3 (3.9~16.6); -3.4 (-13.7~4.4); 24.6 (18.7~28.8); 12.0 (7.3~16.7); 99.4 (52.8~140.6); 23.1 (0.6~71.7); 734 (535~920) ❙ 695 (367~1571); 388 (235~666); 31 (8~157) ❙ -5.7 (-47.7~108.0); 641 (367~910); 185.0 (101.9~334.4)

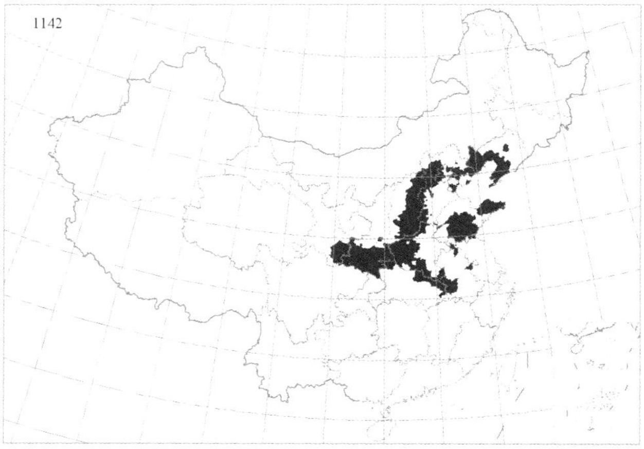

Ulmaceae 榆科 *Celtis* 朴属

Celtis philippensis Blanco 大果油朴

Trees, evergreen

21.6 (18.1~24.3); 14.5 (9.5~18.7); 26.9 (21.7~28.8); 21.7 (18.2~24.3); 200.5 (158.1~231.9); 0.0 (0.0~0.0); 1139 (863~1338) ❙ 1462 (868~2713); 678 (312~1388); 86 (42~188) ❙ 32.0 (-34.3~214.4); 1092 (863~1331); 350.2 (204.4~659.0)

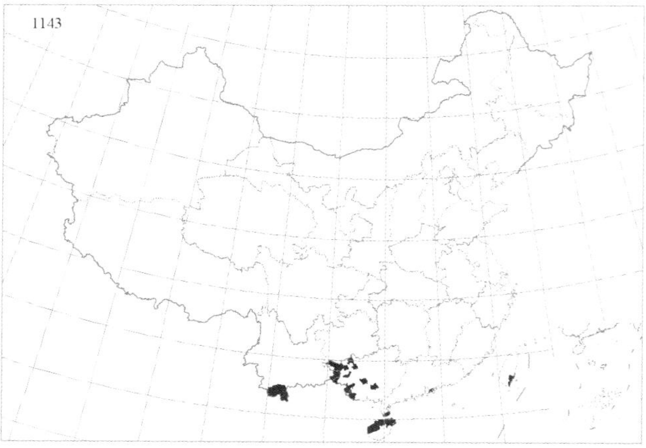

Ulmaceae 榆科 *Celtis* 朴属

Celtis philippensis var. wightii (Planch.) Soepadmo 铁灵花

Trees

24.3; 18.9; 28.1; 24.3; 231.9; 0.0; 1338 ❙ 1285; 442; 48 ❙ -3.9; 1111; 462.4

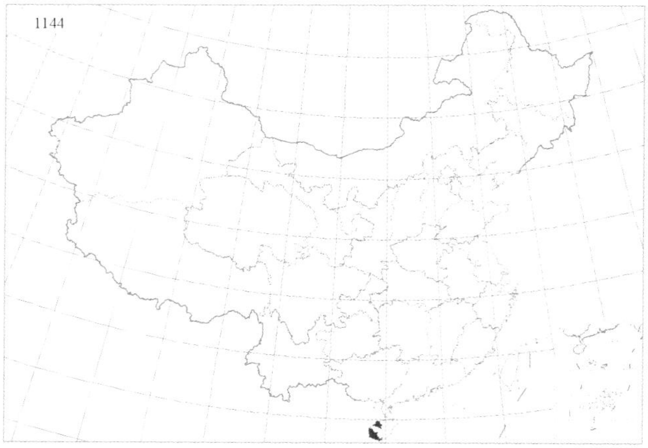

Ulmaceae 榆科 *Celtis* 朴属

Celtis sinensis Pers. 朴树

Trees

15.7 (2.2~24.7); 3.9 (-13.7~19.8); 26.5 (9.5~29.9); 15.9 (3.9~24.7); 136.1 (15.8~236.5); 7.1 (0.0~71.7); 878 (411~1376) ❙ 1199 (367~3495); 509 (235~1594); 112 (8~897) ❙ 34.4 (-43.4~224.4); 834 (367~1326); 199.8 (95.3~538.2)

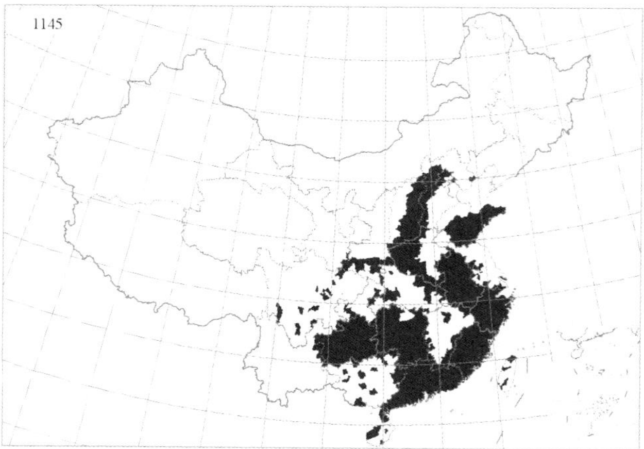

Ulmaceae 榆科 *Celtis* 朴属

Celtis tetrandra Roxb. 四蕊朴

Trees

16.0 (-1.1~24.7); 5.9 (-9.1~19.8); 24.9 (6.2~29.8); 16.1 (2.0~24.7); 136.3 (2.5~236.5); 3.4 (0.0~75.6); 858 (334~1376) ❙ 1213 (321~2993); 566 (193~1847); 90 (15~547) ❙ 40.7 (-36.5~246.5); 832 (321~1269); 246.0 (79.3~659.0)

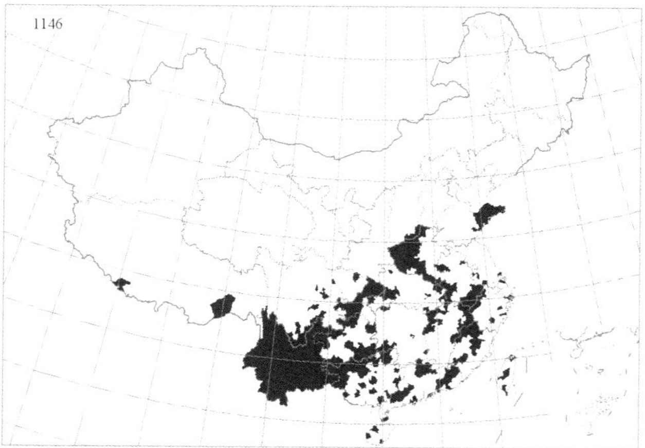

Ulmaceae 榆科 *Celtis* 朴属

Celtis timorensis Span. 假玉桂

Trees, evergreen

20.0 (13.3~24.8); 11.9 (3.6~20.1); 26.5 (19.6~29.2); 20.1 (13.3~24.8); 181.1 (100.4~237.6); 0.0 (0.0~1.7); 1045 (706~1385) ❙ 1477 (939~2037); 696 (392~1044); 99 (37~235) ❙ 43.0 (−11.9~145.8); 1034 (706~1331); 283.6 (95.3~659.0)

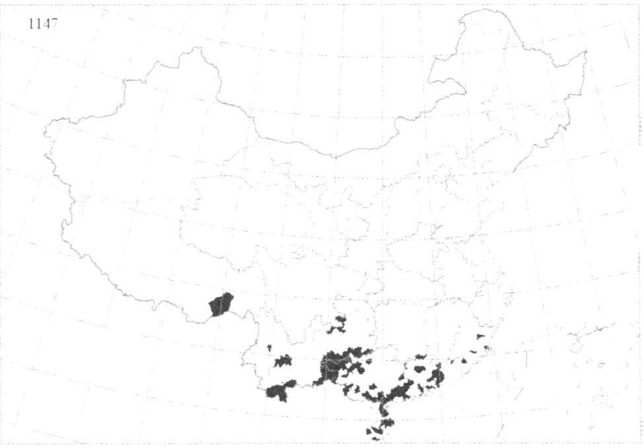

Ulmaceae 榆科 *Celtis* 朴属

Celtis vandervoetiana C. K. Schneid. 西川朴

Trees, deciduous

17.3 (9.3~23.4); 7.8 (−1.7~16.1); 25.6 (16.6~29.8); 17.3 (9.5~23.5); 148.2 (69.8~221.7); 0.3 (0.0~18.0); 900 (603~1273) ❙ 1426 (797~1997); 583 (392~996); 136 (23~259) ❙ 58.5 (−14.2~130.7); 890 (603~1273); 254.4 (134.6~551.6)

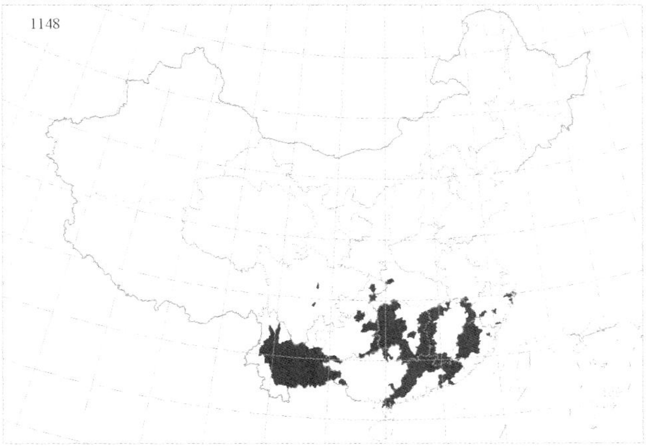

Ulmaceae 榆科 *Gironniera* 白颜树属

Gironniera subaequalis Planch. 白颜树

Trees, evergreen

21.2 (16.7~24.5); 13.7 (9.5~19.1); 26.9 (21.3~29.1); 21.2 (16.8~24.5); 194.5 (141.5~234.4); 0.0 (0.0~0.0); 1102 (811~1360) ❙ 1670 (868~2435); 770 (312~1428); 107 (37~194) ❙ 47.0 (−34.3~99.5); 1085 (744~1360); 285.9 (95.3~659.0)

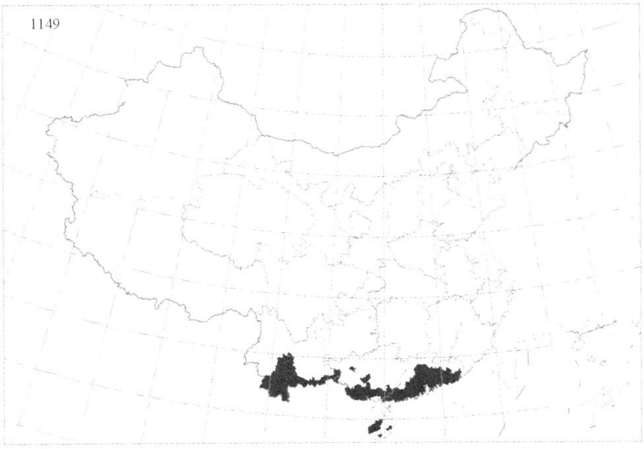

Ulmaceae 榆科 *Hemiptelea* 刺榆属

Hemiptelea davidii (Hance) Planch. 刺榆

Small trees, deciduous

11.9 (−0.3~17.6); −2.6 (−25.0~5.6); 25.1 (12.4~29.3); 12.7 (4.8~17.6); 104.7 (25.1~151.4); 21.5 (0.0~117.1); 757 (441~953) ❙ 817 (341~1393); 409 (190~691); 61 (7~250) ❙ 5.8 (−47.7~129.9); 687 (341~941); 186.0 (61.2~337.1)

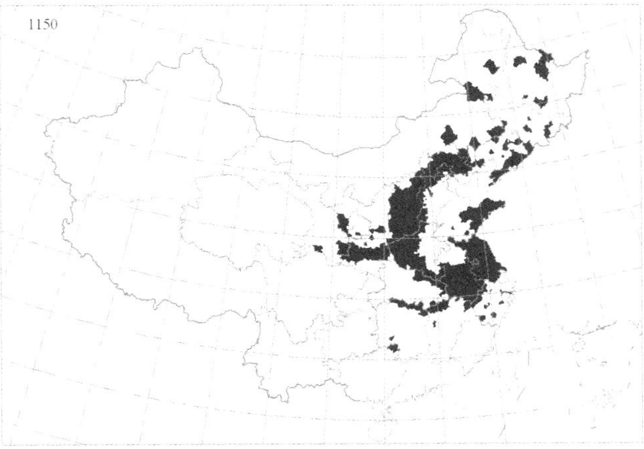

Ulmaceae 榆科 *Pteroceltis* 青檀属

Pteroceltis tatarinowii Maxim. 青檀

Trees, deciduous

14.4 (2.2~22.3); 2.1 (−12.5~13.9); 25.7 (9.9~29.5); 14.7 (3.8~22.3); 122.5 (15.5~208.1); 8.7 (0.0~62.6); 819 (413~1202) ❙ 1098 (377~1997); 476 (241~1056); 90 (8~259) ❙ 32.3 (−43.4~130.7); 777 (377~1202); 204.9 (108.2~371.7)

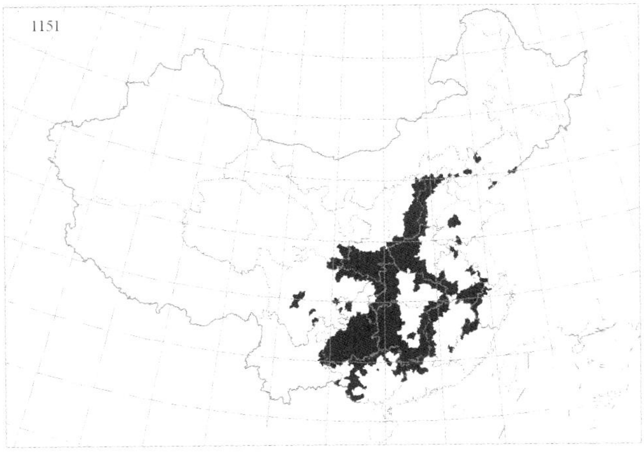

Ulmaceae 榆科 *Trema* 山黄麻属

Trema angustifolia (Planch.) Blume 狭叶山黄麻

Shrubs or small trees, evergreen

17.3 (10.8~23.9); 10.0 (2.7~18.7); 22.8 (16.6~28.9); 17.4 (10.9~23.9); 148.7 (71.8~226.8); 0.2 (0.0~3.8); 877 (633~1294) ❙ 1256 (797~2435); 672 (411~1428); 62 (22~192) ❙ 42.9 (−14.2~112.2); 862 (633~1257); 321.4 (166.0~659.0)

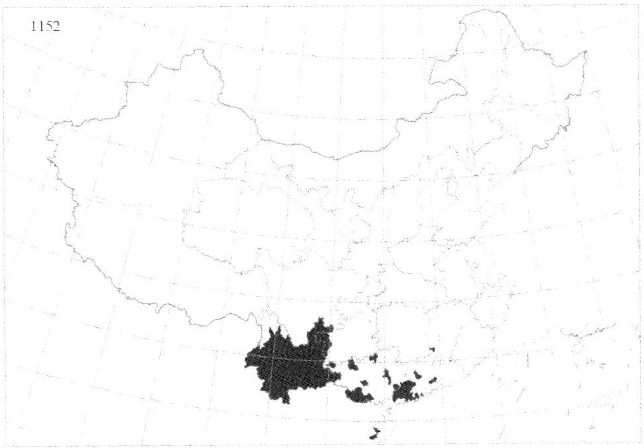

Ulmaceae 榆科 *Trema* 山黄麻属
Trema cannabina Lour. 光叶山黄麻
Shrubs or small trees, evergreen

19.2 (12.6~24.8); 9.9 (1.6~20.1); 27.4 (20.1~29.8); 19.2 (12.7~24.8); 170.7 (98.8~237.6); 0.2 (0.0~6.9); 1011 (711~1385) ❙ 1600 (868~3495); 622 (312~1428); 163 (41~897) ❙ 61.5 (-34.3~246.5); 998 (711~1360); 244.3 (95.3~620.0)

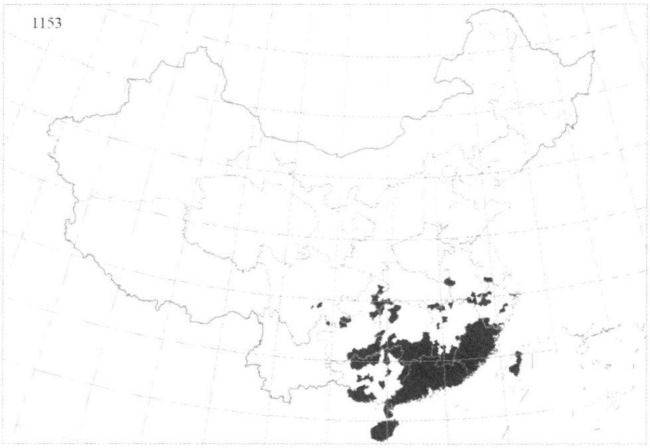

Ulmaceae 榆科 *Trema* 山黄麻属
Trema cannabina var. dielsiana (Hand.-Mazz.) C. J. Chen 山油麻
Shrubs or small trees, evergreen

16.9 (12.6~22.4); 5.6 (1.6~14.8); 27.4 (23.0~30.1); 16.9 (12.7~22.4); 143.4 (98.8~208.9); 0.6 (0.0~6.9); 904 (711~1191) ❙ 1493 (1089~2037); 532 (392~1033); 170 (65~259) ❙ 65.8 (28.8~130.7); 892 (711~1191); 213.7 (134.6~319.9)

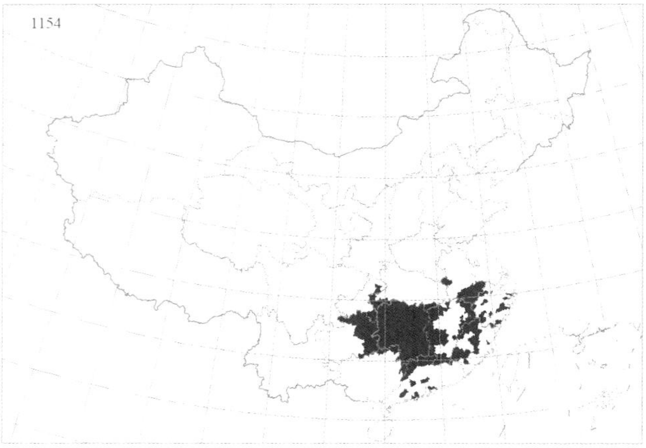

Ulmaceae 榆科 *Trema* 山黄麻属
Trema levigata Hand.-Mazz. 羽脉山黄麻
Small trees, evergreen

15.4 (5.4~21.0); 7.5 (-2.1~14.2); 21.8 (12.3~27.8); 15.5 (5.8~21.0); 126.5 (31.5~191.8); 0.9 (0.0~25.6); 789 (481~1053) ❙ 1087 (741~1547); 598 (411~911); 46 (15~105) ❙ 38.8 (-14.2~107.2); 774 (481~1053); 283.1 (118.7~551.6)

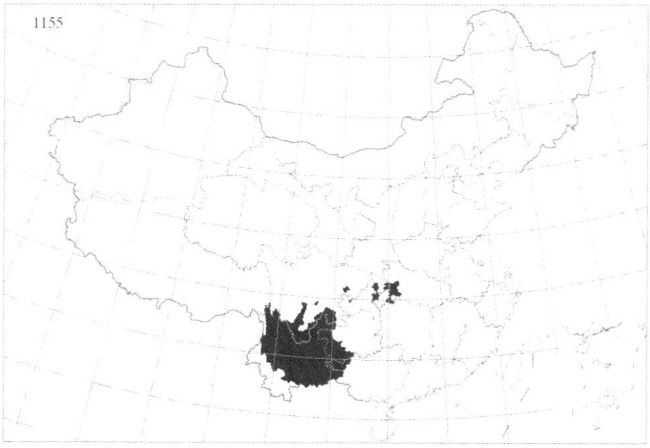

Ulmaceae 榆科 *Trema* 山黄麻属
Trema nitida C. J. Chen 银毛叶山黄麻
Small trees, evergreen

17.8 (10.6~22.2); 8.8 (-0.2~15.1); 25.5 (20.0~28.9); 17.9 (10.6~22.2); 154.6 (79.4~206.9); 0.3 (0.0~12.1); 917 (643~1177) ❙ 1340 (999~1991); 629 (442~996); 84 (35~191) ❙ 46.7 (8.0~93.8); 914 (643~1177); 266.4 (108.2~659.0)

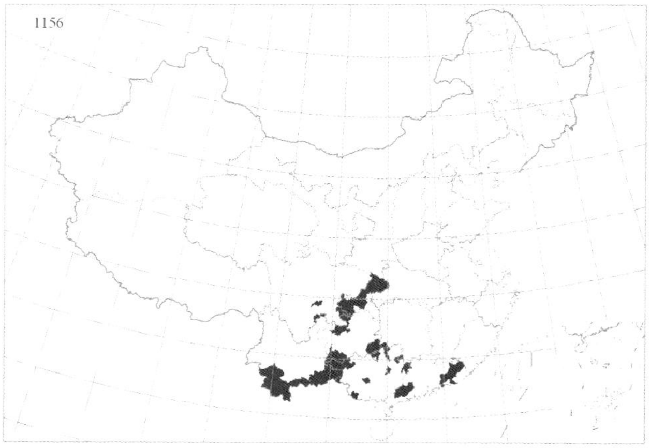

Ulmaceae 榆科 *Trema* 山黄麻属
Trema orientalis (L.) Blume 异色山黄麻
Trees, evergreen

18.5 (3.8~24.8); 10.3 (-4.3~20.1); 25.4 (10.8~29.8); 18.6 (4.7~24.8); 163.1 (21.6~237.6); 0.2 (0.0~35.4); 957 (444~1385) ❙ 1460 (741~2993); 652 (356~1594); 121 (16~547) ❙ 53.9 (-14.2~203.4); 945 (444~1299); 276.6 (95.3~659.0)

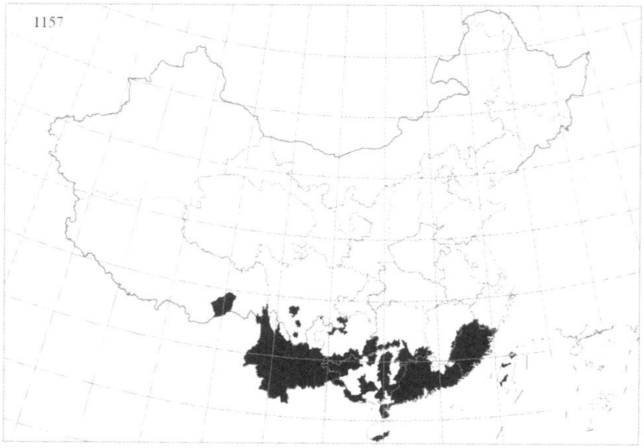

Ulmaceae 榆科 *Trema* 山黄麻属
Trema tomentosa (Roxb.) H. Hara 山黄麻
Small trees, evergreen

18.1 (3.8~24.8); 10.1 (-4.3~20.1); 24.5 (10.8~29.8); 18.1 (4.7~24.8); 157.8 (21.6~237.6); 0.4 (0.0~35.4); 934 (444~1385) ❙ 1436 (741~3495); 667 (411~1847); 108 (15~897) ❙ 54.4 (-14.2~246.5); 921 (444~1299); 280.2 (95.3~659.0)

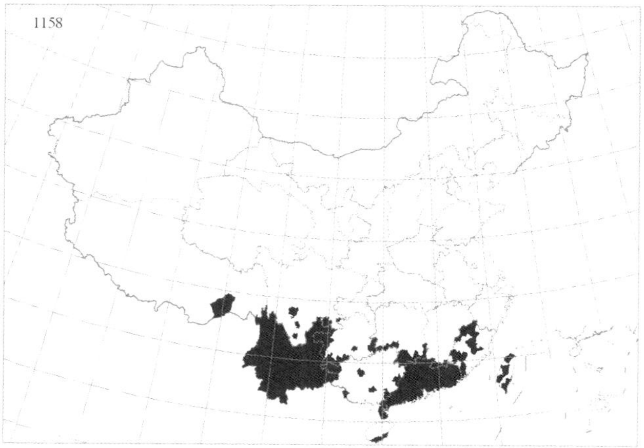

Ulmaceae 榆科 *Ulmus* 榆属

Ulmus androssowii var. subhirsuta (C. K. Schneid.) P. H. Huang 毛枝榆
Trees, deciduous or semievergreen

14.1 (−1.1~18.9); 5.1 (−9.1~12.5); 20.7 (6.2~29.1); 14.2 (2.0~19.0); 112.4 (2.5~167.8); 2.6 (0.0~75.6); 743 (334~930) ❙ 1043 (321~1502); 568 (193~792); 47 (12~172) ❙ 41.6 (−14.2~107.2); 723 (321~930); 281.3 (79.3~487.2)

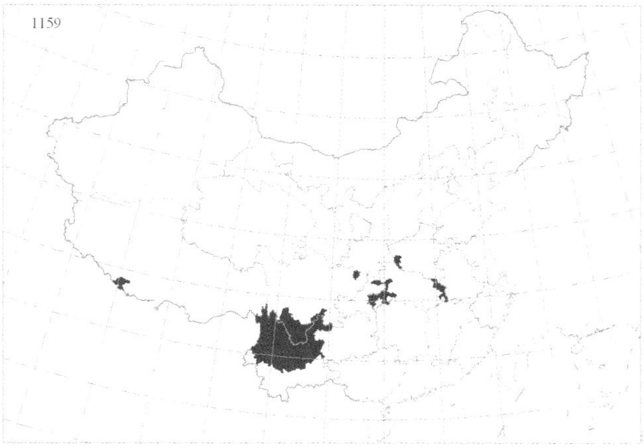

Ulmaceae 榆科 *Ulmus* 榆属

Ulmus bergmanniana C. K. Schneid. 兴山榆
Trees, deciduous

13.9 (2.2~19.4); 2.1 (−7.7~10.2); 25.0 (9.9~29.3); 14.0 (3.8~19.4); 114.6 (15.5~172.7); 7.2 (0.0~49.3); 789 (413~994) ❙ 1152 (512~1922); 474 (258~817); 97 (13~259) ❙ 44.9 (−35.3~128.9); 772 (413~994); 226.8 (118.7~382.6)

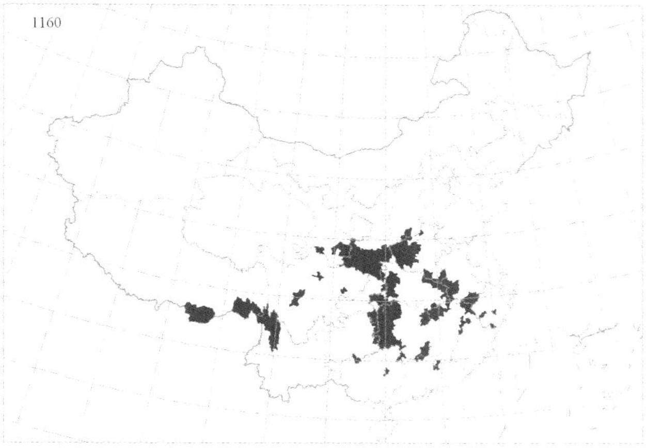

Ulmaceae 榆科 *Ulmus* 榆属

Ulmus bergmanniana var. lasiophylla C. K. Schneid. 蜀榆
Trees, deciduous

13.0 (2.2~21.3); 3.7 (−6.6~12.3); 21.1 (9.9~28.6); 13.2 (3.8~21.3); 105.5 (15.5~196.1); 8.9 (0.0~49.1); 752 (413~1130) ❙ 1179 (701~1758); 569 (303~817); 80 (15~189) ❙ 59.4 (−2.2~125.9); 751 (413~1130); 272.5 (185.5~382.6)

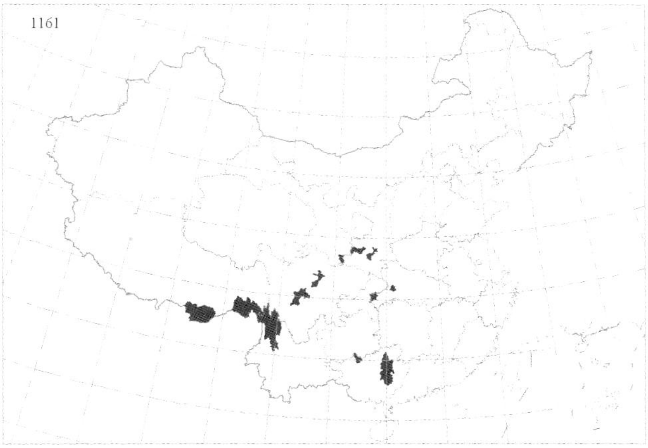

Ulmaceae 榆科 *Ulmus* 榆属

Ulmus castaneifolia Hemsl. 多脉榆
Trees, deciduous

16.4 (12.7~21.0); 5.4 (1.5~11.9); 26.6 (23.0~29.5); 16.5 (12.8~21.1); 138.1 (99.0~193.0); 0.7 (0.0~7.3); 872 (713~1114) ❙ 1423 (1082~1997); 537 (392~821); 146 (50~259) ❙ 63.7 (8.0~137.5); 867 (713~1114); 216.5 (108.2~322.1)

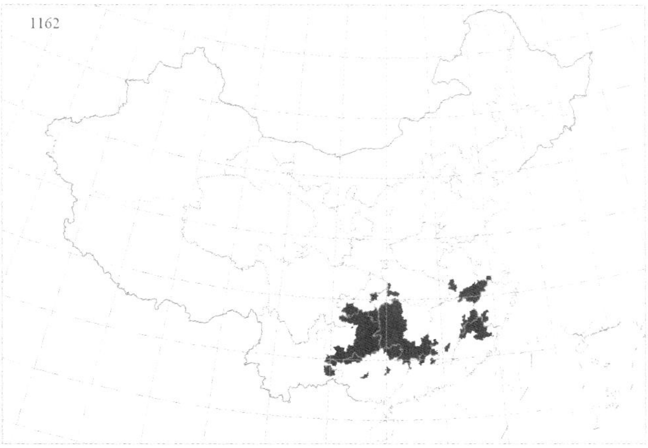

Ulmaceae 榆科 *Ulmus* 榆属

Ulmus changii W. C. Cheng 杭州榆
Trees, deciduous

16.3 (11.0~23.6); 5.3 (1.5~17.8); 26.6 (16.6~29.9); 16.4 (11.0~23.3); 137.5 (73.1~223.3); 0.8 (0.0~7.3); 874 (636~1271) ❙ 1476 (916~1997); 548 (411~82); 162 (18~259) ❙ 69.4 (−2.2~132.1); 867 (636~1114); 231.4 (141.7~439.7)

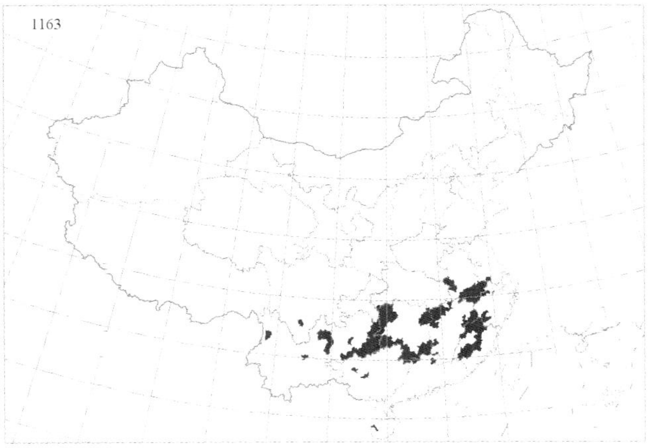

Ulmaceae 榆科 *Ulmus* 榆属

Ulmus changii var. kunmingensis (W. C. Cheng) W. C. Cheng et L. K. Fu 昆明榆
Trees, deciduous

14.6 (11.0~18.9); 5.7 (3.2~9.5); 22.3 (16.6~26.6); 14.7 (11.0~18.9); 116.6 (73.1~167.3); 0.7 (0.0~2.5); 779 (636~969) ❙ 1146 (1022~1259); 592 (473~651); 56 (18~123) ❙ 49.1 (27.5~78.3); 775 (636~969); 257.5 (171.0~339.5)

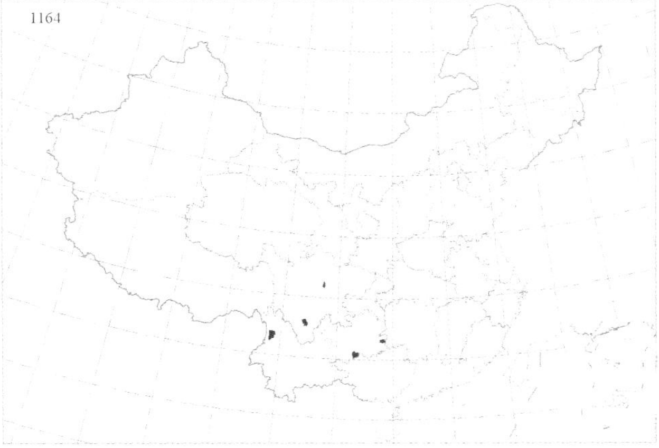

Ulmaceae 榆科 *Ulmus* 榆属
Ulmus chenmoui W. C. Cheng 琅琊榆
Trees, deciduous

15.4; 2.2; 28.1; 15.5; 130.1; 4.3; 877 ❙ 1018; 460; 113 ❙ 16.2; 877; 146.5

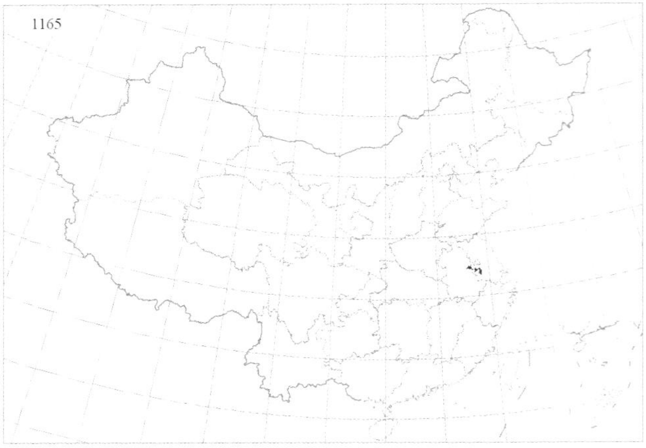

Ulmaceae 榆科 *Ulmus* 榆属
Ulmus davidiana Planch. 黑榆
Shrubs or trees, deciduous

9.2 (1.2~15.5); -6.8 (-20.1~2.1); 23.4 (17.5~28.0); 10.7 (6.4~15.6); 87.6 (45.4~131.3); 36.4 (4.3~93.6); 684 (506~881) ❙ 570 (309~1077); 341 (199~616); 19 (8~97) ❙ -15.9 (-43.4~54.7); 556 (309~881); 178.9 (70.3~333.1)

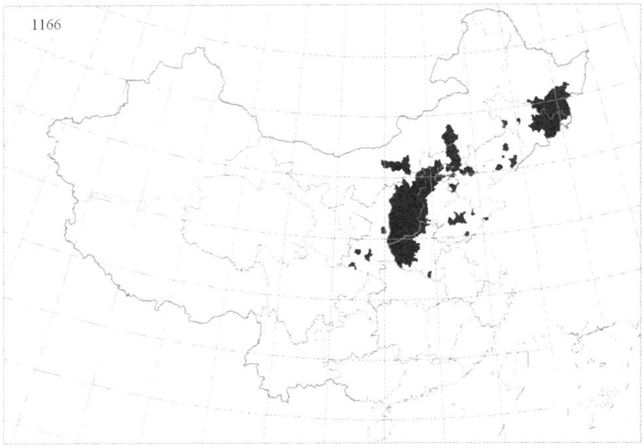

Ulmaceae 榆科 *Ulmus* 榆属
Ulmus davidiana var. japonica (Rehder) Nakai 春榆
Shrubs or trees, deciduous

2.5 (-0.7~8.4); -19.5 (-25.0~-8.9); 21.2 (19.0~23.2); 7.9 (6.6~10.2); 60.3 (50.1~84.0); 89.8 (42.8~118.4); 577 (527~661) ❙ 618 (537~1018); 379 (309~648); 20 (11~36) ❙ 6.5 (-11.0~56.9); 567 (527~661); 279.9 (203.8~333.3)

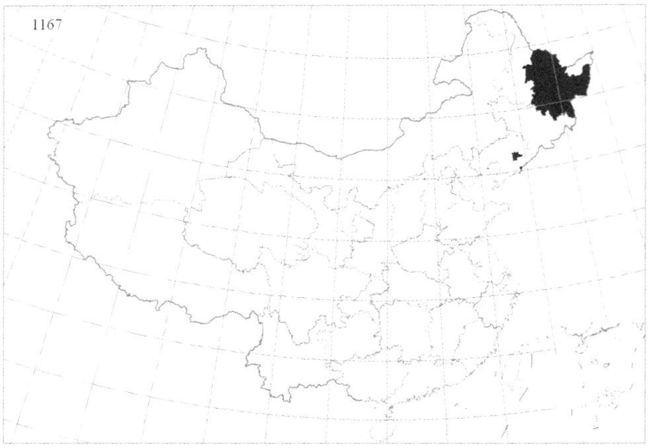

Ulmaceae 榆科 *Ulmus* 榆属
Ulmus elongata L. K. Fu et C. S. Ding 长序榆
Trees, deciduous

16.4 (14.4~18.6); 4.9 (2.1~9.3); 27.4 (24.7~29.1); 16.5 (14.5~18.7); 138.5 (118.7~164.0); 1.0 (0.0~4.9); 885 (789~984) ❙ 1715 (1436~1950); 584 (467~713); 220 (169~259) ❙ 94.4 (56.7~147.0); 879 (789~965); 250.6 (183.9~300.0)

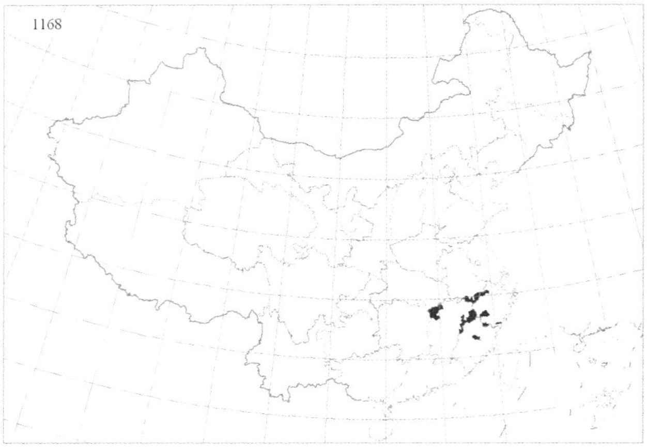

Ulmaceae 榆科 *Ulmus* 榆属
Ulmus gaussenii W. C. Cheng 醉翁榆
Trees, deciduous

15.4; 1.8; 28.0; 15.4; 130.0; 5.0; 875 ❙ 994; 464; 102 ❙ 13.5; 875; 151.6

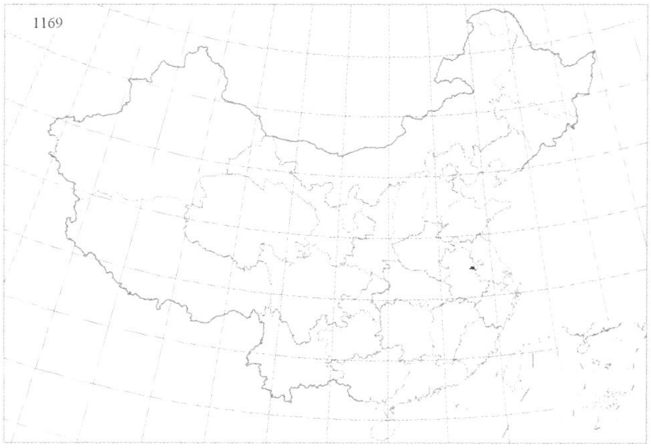

Ulmaceae 榆科 *Ulmus* 榆属
Ulmus glaucescens Franch. 旱榆
Shrubs or trees, deciduous

8.6 (-4.1~15.5); -7.1 (-17.3~2.1); 22.6 (7.8~28.5); 10.2 (2.1~15.6); 81.9 (5.2~131.3); 37.5 (4.3~114.2); 658 (340~886) ❙ 481 (63~1077); 283 (45~635); 14 (1~97) ❙ -26.7 (-87.1~68.8); 472 (63~881); 140.6 (7.5~323.3)

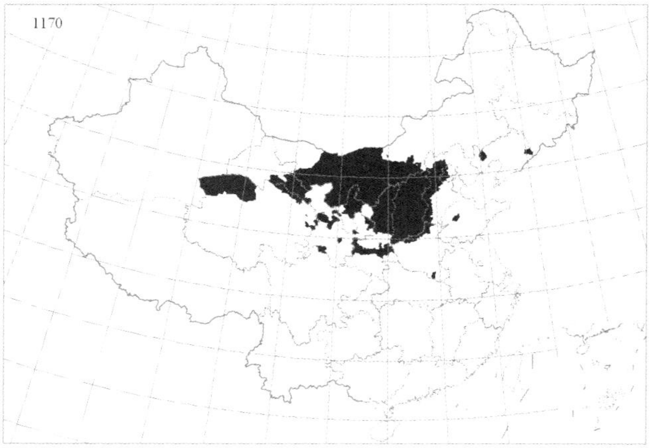

Ulmaceae 榆科 *Ulmus* 榆属

Ulmus glaucescens var. lasiocarpa Rehder 毛果旱榆

Shrubs or trees, deciduous

7.7 (2.0~9.8); -9.0 (-12.9~-4.9); 22.4 (12.6~24.0); 9.6 (4.8~10.6); 77.2 (25.6~85.4); 44.7 (26.3~63.1); 635 (442~671) ❙ 255 (128~624); 152 (81~350); 6 (2~23) ❙ -59.1 (-80.8~1.2); 254 (128~624); 75.9 (13.6~209.6)

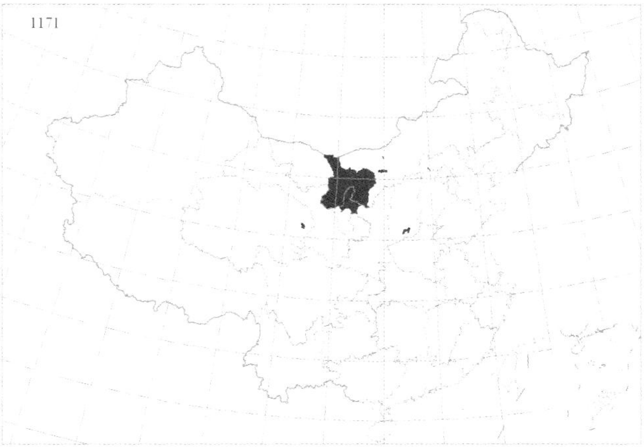

Ulmaceae 榆科 *Ulmus* 榆属

Ulmus harbinensis S. Q. Nie et K. Q. Huang 哈尔滨榆

Trees, deciduous

3.4; -19.6; 23.0; 8.7; 68.9; 87.4; 616 ❙ 556; 367; 15 ❙ -9.7; 556; 177.1

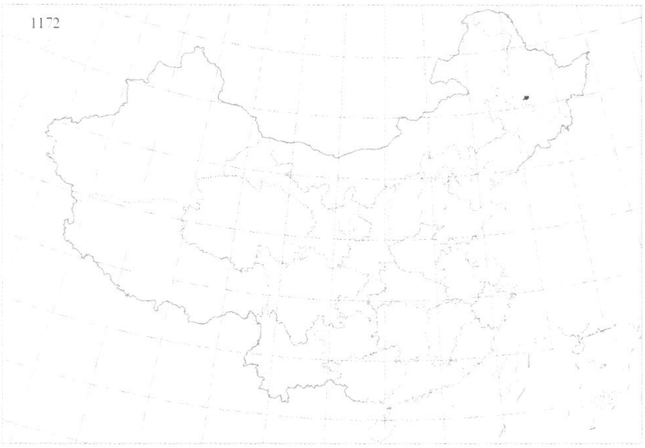

Ulmaceae 榆科 *Ulmus* 榆属

Ulmus laciniata (Trautv.) Mayr 裂叶榆

Trees, deciduous

6.6 (-0.2~13.8); -11.1 (-23.8~-1.2); 22.1 (17.5~27.3); 9.3 (6.4~14.0); 74.0 (45.4~121.1); 54.1 (14.4~112.6); 626 (501~842) ❙ 546 (202~1073); 340 (134~691); 17 (6~40) ❙ -11.9 (-65.5~74.1); 515 (202~720); 195.1 (52.7~338.4)

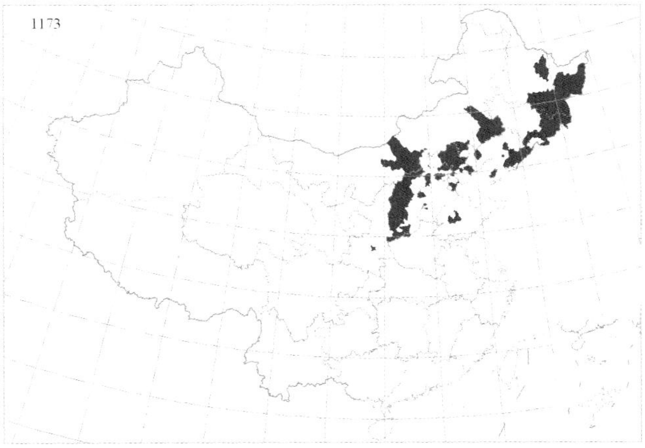

Ulmaceae 榆科 *Ulmus* 榆属

Ulmus lamellosa C. Wang et S. L. Chang 脱皮榆

Small trees, deciduous

11.1 (3.7~13.5); -4.0 (-14.4~-1.3); 24.7 (19.7~27.1); 11.9 (7.6~13.7); 98.3 (56.3~117.5); 24.4 (14.6~71.0); 729 (553~820) ❙ 556 (304~640); 309 (195~465); 20 (8~26) ❙ -23.5 (-48.6~-6.9); 556 (304~640); 172.2 (87.0~256.0)

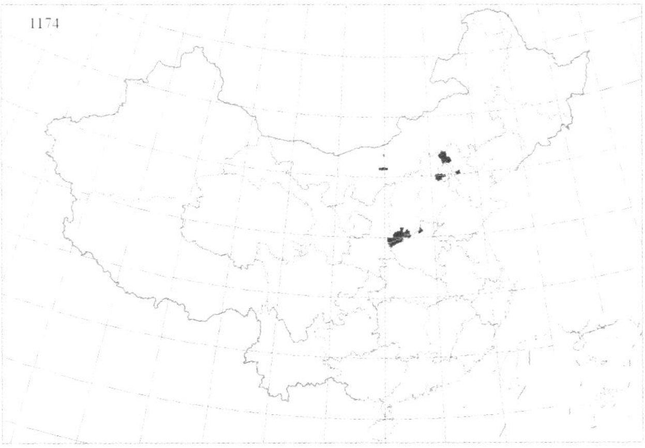

Ulmaceae 榆科 *Ulmus* 榆属

Ulmus lanceifolia Roxb. 常绿榆

Small trees, evergreen

18.2 (8.6~24.8); 11.2 (1.3~20.1); 23.1 (15.0~29.0); 18.2 (8.7~24.8); 158.5 (52.6~237.6); 0.2 (0.0~8.8); 906 (574~1385) ❙ 1340 (1028~1776); 714 (473~996); 66 (41~188) ❙ 49.7 (1.5~93.0); 892 (574~1190); 388.1 (160.7~659.0)

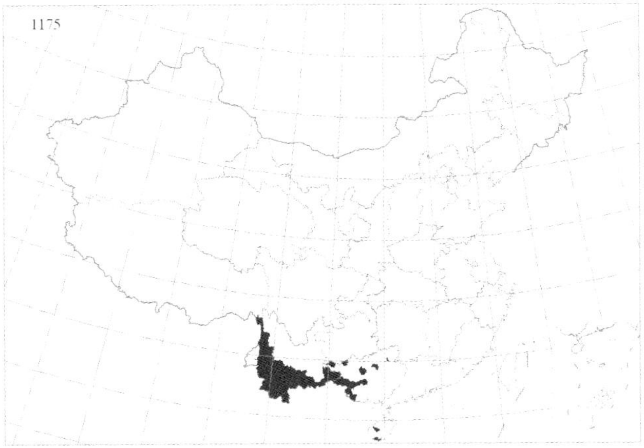

Ulmaceae 榆科 *Ulmus* 榆属

Ulmus macrocarpa Hance 大果榆

Shrubs or trees, deciduous

8.8 (-5.8~17.6); -7.4 (-29.8~5.6); 23.2 (11.0~29.0); 10.6 (3.8~17.6); 86.0 (17.4~151.4); 40.1 (0.0~157.4); 677 (407~953) ❙ 624 (195~1621); 359 (133~691); 26 (4~228) ❙ -8.3 (-67.0~106.2); 574 (195~915); 185.7 (43.7~346.5)

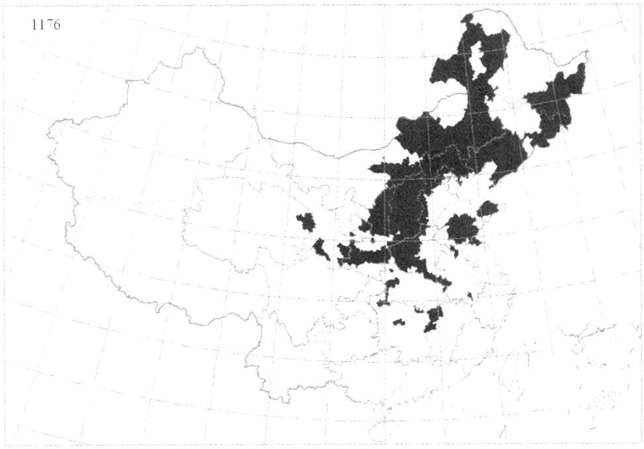

Ulmaceae 榆科 *Ulmus* 榆属
Ulmus macrocarpa var. glabra S. Q. Nie et K. Q. Huang 光秃大果榆
Trees, deciduous

3.4; -19.6; 23.0; 8.7; 68.9; 87.4; 616 ❙ 556; 367; 15 ❙ -9.7; 556; 177.1

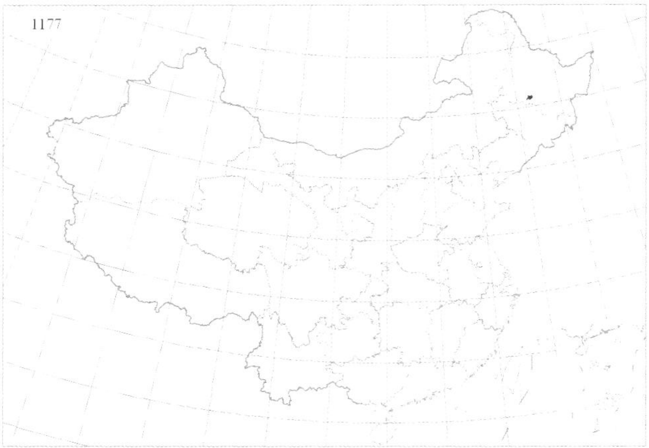

Ulmaceae 榆科 *Ulmus* 榆属
Ulmus microcarpa L. K. Fu 小果榆
Trees, deciduous

5.3; -2.4; 12.2; 5.7; 30.5; 26.2; 479 ❙ 961; 558; 31 ❙ 100.7; 479; 234.2

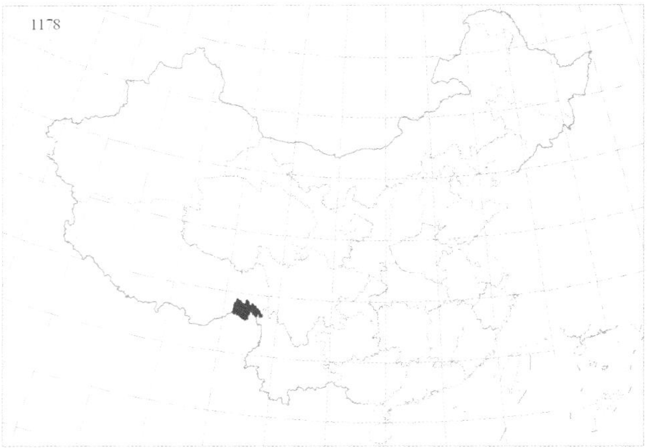

Ulmaceae 榆科 *Ulmus* 榆属
Ulmus parvifolia Jacquem. 榔榆
Trees, deciduous

15.0 (3.9~24.1); 2.7 (-13.7~18.9); 26.6 (18.7~29.9); 15.3 (7.3~24.1); 128.8 (52.8~229.7); 7.9 (0.0~71.7); 849 (535~1322) ❙ 1135 (367~3495); 478 (235~1847); 110 (8~897) ❙ 31.4 (-51.1~224.4); 804 (367~1205); 197.3 (101.9~538.2)

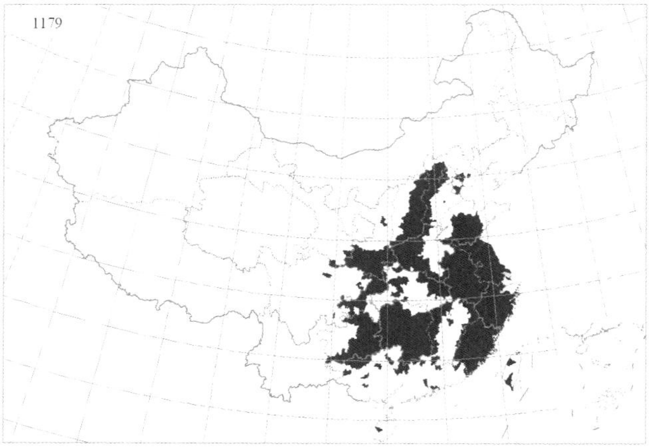

Ulmaceae 榆科 *Ulmus* 榆属
Ulmus prunifolia W. C. Cheng et L. K. Fu 李叶榆
Trees, deciduous

14.2; 3.3; 24.7; 14.3; 113.9; 2.9; 778 ❙ 1303; 533; 70 ❙ 68.6; 778; 220.3

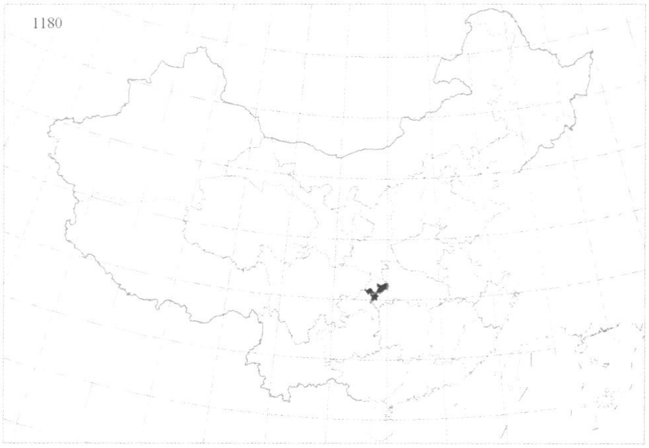

Ulmaceae 榆科 *Ulmus* 榆属
Ulmus pseudopropinqua Wang et Li 假春榆
Trees, deciduous

7.0; -12.2; 23.3; 10.0; 81.1; 57.1; 661 ❙ 594; 383; 21 ❙ -10.1; 594; 179.0

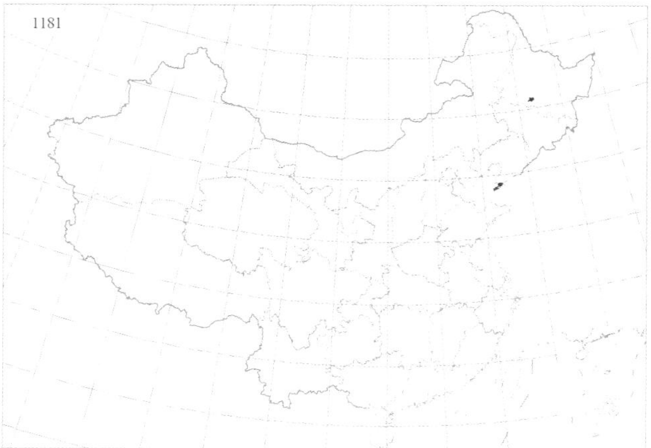

Ulmaceae 榆科 *Ulmus* 榆属
Ulmus pumila L. 榆树
Trees, deciduous

10.4 (-5.8~21.6); -4.8 (-29.8~13.2); 23.8 (8.3~29.4); 11.9 (2.5~21.7); 97.9 (9.3~199.8); 32.3 (0.0~157.4); 726 (329~1153) ❙ 672 (21~1770); 372 (15~959); 37 (0~228) ❙ -8.9 (-97.3~112.2); 581 (21~1153); 179.8 (1.6~627.0)

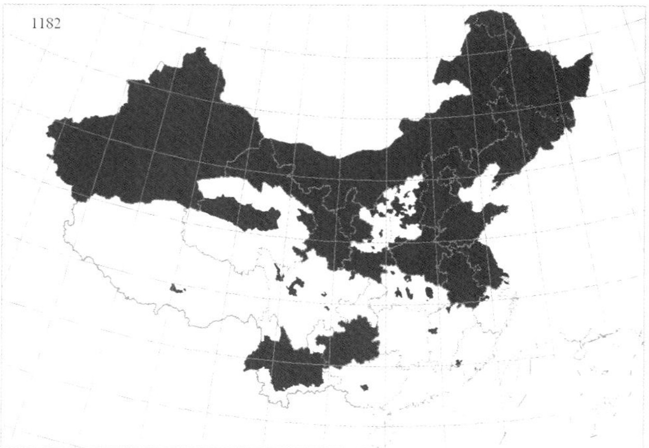

Ulmaceae 榆科 *Ulmus* 榆属

Ulmus szechuanica W. P. Fang 红果榆

Trees, deciduous

16.1 (14.1~18.2); 4.1 (1.6~6.9); 27.6 (25.4~29.1); 16.1 (14.2~18.2); 135.4 (116.7~158.8); 1.7 (0.0~6.5); 880 (796~984) ❙ 1553 (950~1922); 562 (448~632); 194 (27~259) ❙ 76.5 (12.1~123.6); 877 (796~961); 235.0 (133.8~299.5)

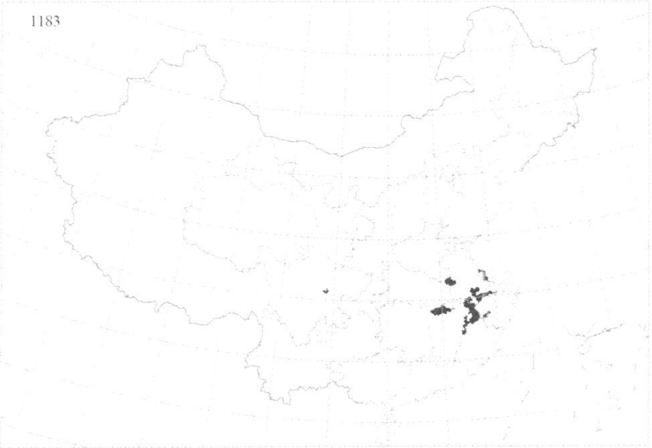

Ulmaceae 榆科 *Ulmus* 榆属

Ulmus uyematsui Hayata 阿里山榆

Trees, deciduous

18.5; 13.2; 22.5; 18.5; 161.7; 0.0; 896 ❙ 2563; 1299; 178 ❙ 193.3; 896; 248.4

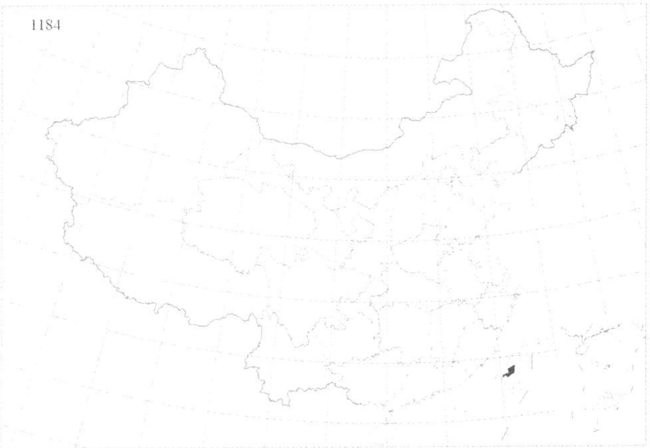

Ulmaceae 榆科 *Zelkova* 榉属

Zelkova schneideriana Hand.-Mazz. 大叶榉树

Trees, deciduous

15.5 (5.3~21.3); 4.4 (-2.9~14.2); 25.7 (12.2~29.9); 15.5 (5.7~21.3); 12[illegible].4 (30.5~196.0); 3.0 (0.0~26.2); 841 (479~1123) ❙ 1146 (512~1893); 51[illegible] (258~860); 100 (15~231) ❙ 36.2 (-35.3~129.9); 820 (479~1121); 21[illegible].7 (110.1~551.6)

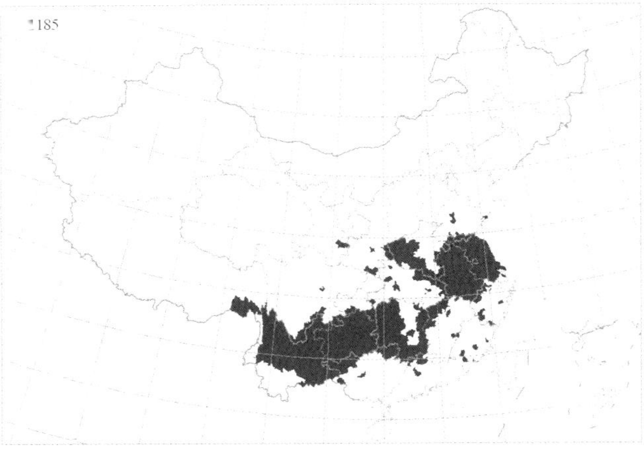

Ulmaceae 榆科 *Zelkova* 榉属

Zelkova serrata (Thunb.) Makino 榉树

Trees, deciduous

14.6 (7.4~21.7); 2.8 (-6.9~15.3); 25.6 (18.7~29.9); 14.7 (8.2~21.7); 122.0 (58.8~200.2); 5.8 (0.0~34.4); 813 (561~1118) ❙ 1145 (512~2713); 482 (258~1412); 95 (14~308) ❙ 39.7 (-36.5~246.5); 783 (512~1075); 209.9 (120.1~363.7)

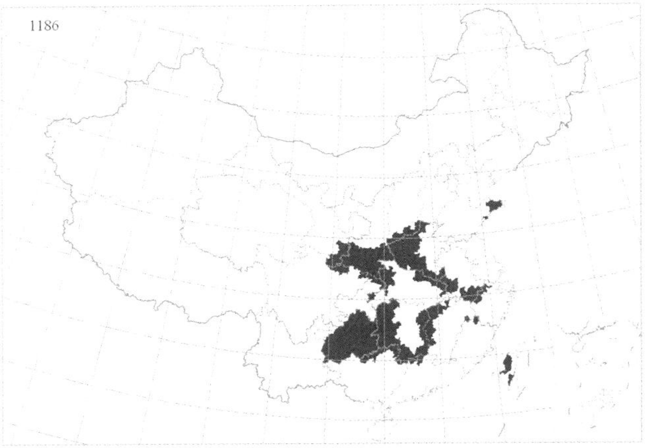

Ulmaceae 榆科 *Zelkova* 榉属

Zelkova sinica C. K. Schneid. 大果榉

Trees, deciduous

12.9 (2.6~19.7); -0.4 (-10.1~10.5); 25.2 (11.2~28.8); 13.2 (4.3~19.8); 109.2 (20.[illegible]~177.1); 13.6 (0.0~50.4); 776 (427~1018) ❙ 833 (373~1694); 402 (201~[illegible]23); 51 (7~209) ❙ 8.1 (-37.6~108.0); 694 (373~1018); 187.9 (88.1~322.1)

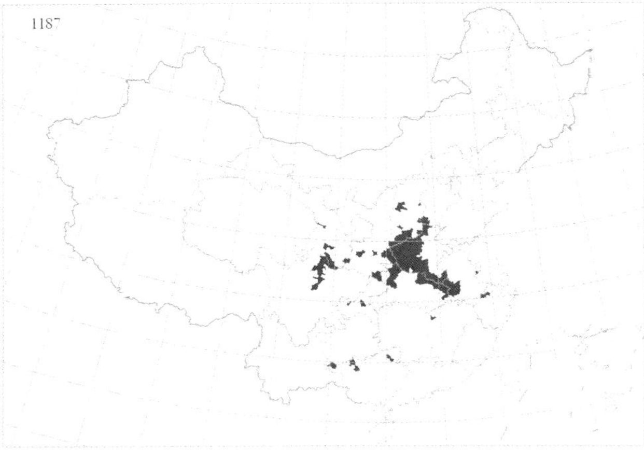

Rhoipteleaceae 马尾树科 *Rhoiptelea* 马尾树属

Rhoiptelea chiliantha Diels et Hand.-Mazz. 马尾树

Trees, deciduous

17.5 (14.3~21.3); 8.5 (3.8~14.2); 24.9 (21.4~28.7); 17.6 (14.3~21.3); 150.9 (112.6~196.0); 0.1 (0.0~1.5); 890 (756~1123) ❙ 1348 (1007~1776); 647 (476~860); 93 (43~191) ❙ 52.1 (5.8~101.8); 890 (756~1121); 294.8 (188.8~551.6)

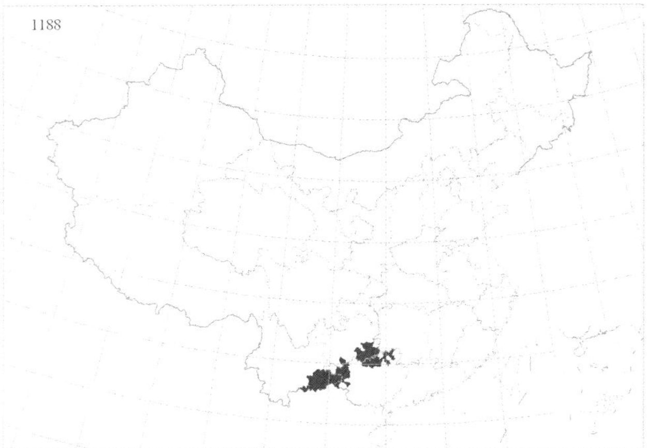

R

Moraceae 桑科 *Antiaris* 见血封喉属
Antiaris toxicaria (Pers.) Lesch. 见血封喉
Trees, evergreen

21.6 (17.2~24.7); 14.6 (9.5~19.8); 26.8 (21.6~29.2); 21.6 (17.3~24.7); 199.5 (147.0~236.5); 0.0 (0.0~0.0); 1130 (833~1376) ❙ 1617 (1254~2435); 791 (480~1428); 89 (48~137) ❙ 45.2 (-1.4~99.5); 1117 (833~1331); 338.6 (178.7~659.0)

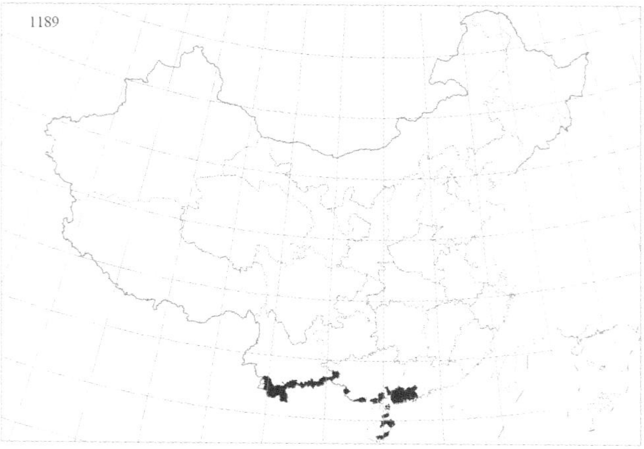

Moraceae 桑科 *Artocarpus* 波罗密属
Artocarpus chama Buch.-Ham. 野树波罗
Trees, deciduous

19.5 (18.2~21.0); 13.7 (12.4~15.1); 23.2 (21.6~25.6); 19.6 (18.2~21.0); 174.6 (158.6~191.8); 0.0 (0.0~0.0); 941 (861~1053) ❙ 1510 (1276~1751); 792 (655~996); 63 (51~77) ❙ 61.0 (36.8~88.8); 930 (839~1053); 528.6 (439.5~659.0)

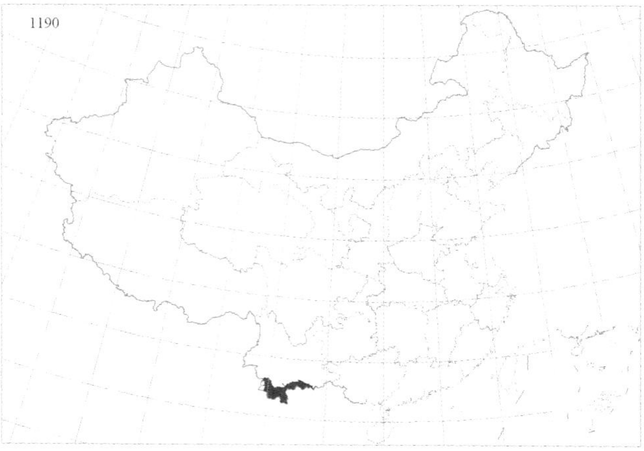

Moraceae 桑科 *Artocarpus* 波罗密属
Artocarpus gongshanensis S. K. Wu ex C. Y. Wu et S. S. Chang 贡山波罗蜜
Trees, evergreen

8.6; 1.3; 15.0; 8.7; 52.6; 8.8; 574 ❙ 1041; 585; 45 ❙ 81.5; 574; 352.3

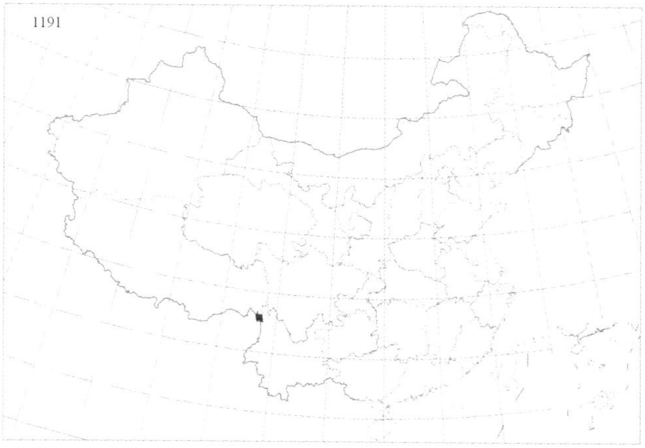

Moraceae 桑科 *Artocarpus* 波罗密属
Artocarpus heterophyllus Lam. 波罗蜜
Trees, evergreen

21.4 (13.3~24.8); 14.1 (5.4~20.1); 26.9 (19.6~29.2); 21.4 (13.4~24.8); 197.1 (100.4~237.6); 0.0 (0.0~0.0); 1120 (706~1385) ❙ 1599 (868~2037); 739 (312~1033); 101 (41~188) ❙ 45.1 (-34.3~145.8); 1098 (706~1360); 311.7 (95.3~659.0)

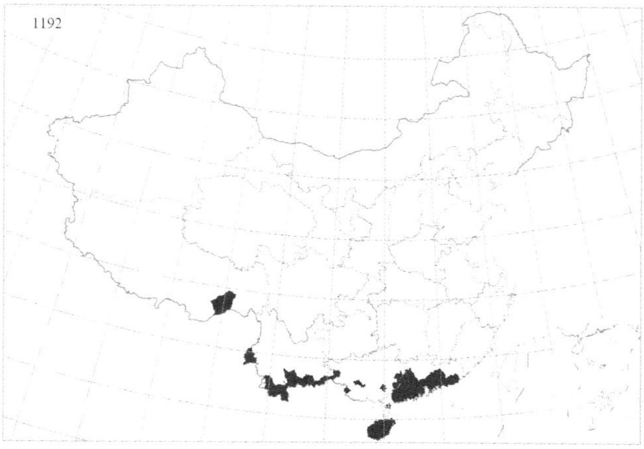

Moraceae 桑科 *Artocarpus* 波罗密属
Artocarpus hypargyreus Hance 白桂木
Trees, evergreen

19.2 (10.6~24.5); 10.0 (-0.2~18.9); 27.3 (21.1~29.8); 19.3 (10.6~24.5); 171.5 (79.4~234.4); 0.2 (0.0~12.1); 1013 (643~1360) ❙ 1483 (868~2037); 619 (312~1033); 123 (19~235) ❙ 47.6 (-34.3~128.9); 1005 (643~1360); 228.4 (95.3~551.6)

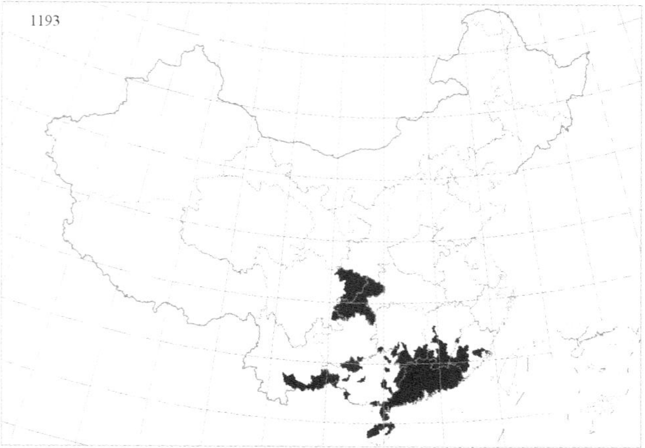

Moraceae 桑科 *Artocarpus* 波罗密属
Artocarpus lakoocha Wall. ex Roxb. 野波罗蜜
Trees, evergreen

17.9 (15.2~21.0); 11.6 (8.7~15.1); 22.0 (20.1~25.6); 18.0 (15.2~21.0); 155.7 (122.5~191.8); 0.0 (0.0~0.0); 865 (753~1053) ❙ 1332 (939~1751); 738 (524~996); 52 (37~77) ❙ 53.9 (6.5~112.2); 846 (717~1053); 440.3 (289.9~659.0)

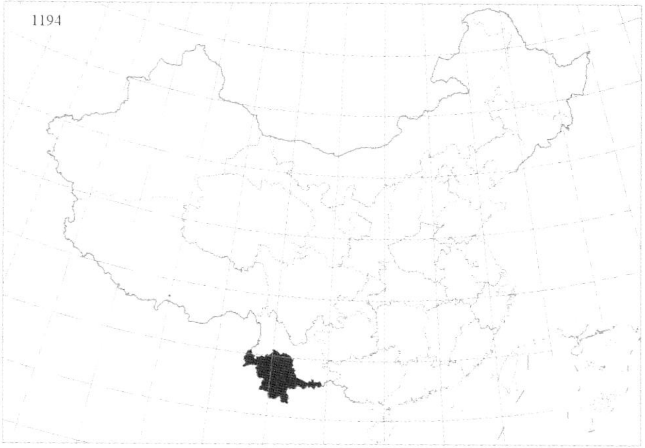

Moraceae 桑科 *Artocarpus* 波罗密属

Artocarpus nanchuanensis S. S. Chang, S. H. Tan et Z. Y. Liu 南川木波罗

Trees, evergreen

14.8; 4.3; 24.9; 14.9; 119.2; 0.7; 791 ❙ 1197; 489; 71 ❙ 51.2; 791; 182.9

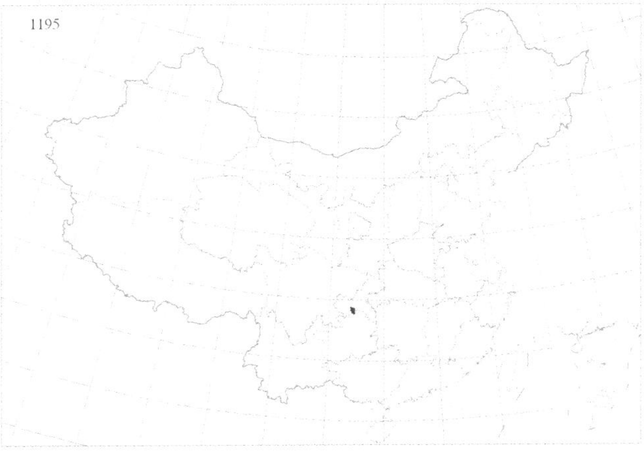

Moraceae 桑科 *Artocarpus* 波罗密属

Artocarpus nigrifolius C. Y. Wu 牛李

Trees, evergreen

20.0; 13.6; 24.4; 20.1; 180.7; 0.0; 985 ❙ 1475; 827; 62 ❙ 50.1; 985; 528.9

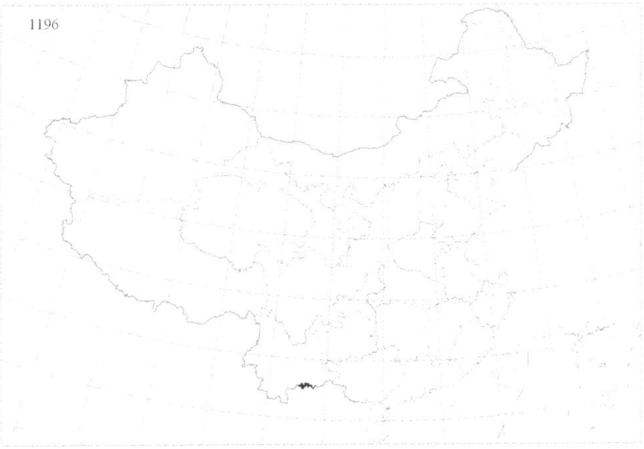

Moraceae 桑科 *Artocarpus* 波罗密属

Artocarpus nitidus subsp. griffithii (King ex Hook. f.) F. M. Jarrett 披针叶桂木

Small trees, evergreen

19.2; 13.8; 22.3; 19.2; 170.5; 0.0; 909 ❙ 1499; 686; 58 ❙ 64.9; 880; 513.9

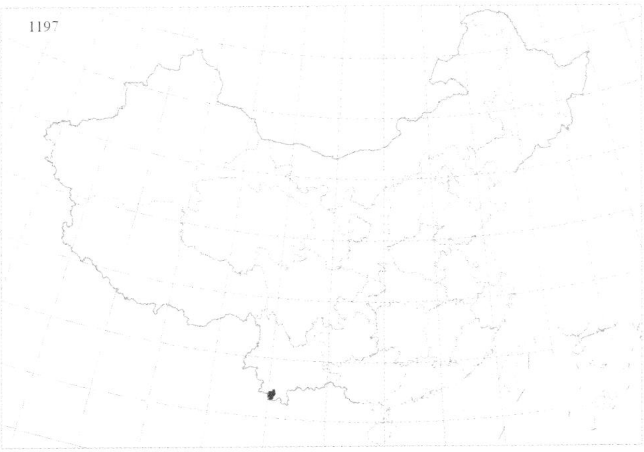

Moraceae 桑科 *Artocarpus* 波罗密属

Artocarpus nitidus subsp. lingnanensis (Merr.) F. M. Jarrett 桂木

Trees, evergreen

22.0 (16.4~24.8); 14.3 (5.4~20.1); 28.1 (22.3~29.2); 22.1 (16.5~24.8); 205.0 (137.7~237.6); 0.0 (0.0~0.0); 1175 (854~1385) ❙ 1629 (868~2035); 709 (312~994); 118 (41~191) ❙ 40.4 (-34.3~95.5); 1147 (854~1360); 262.7 (95.3~659.0)

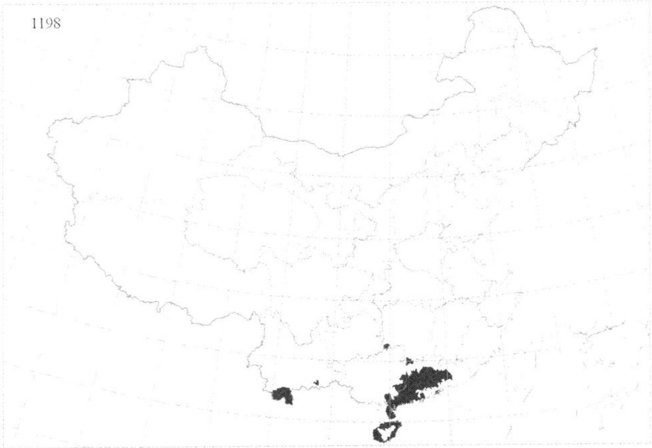

Moraceae 桑科 *Artocarpus* 波罗密属

Artocarpus petelotii Gagnep. 短绢毛波罗蜜

Trees, evergreen

17.7 (16.3~21.0); 10.5 (9.0~14.2); 22.7 (21.4~25.6); 17.7 (16.4~21.0); 152.8 (136.4~191.8); 0.0 (0.0~0.0); 862 (797~1053) ❙ 1236 (958~1522); 686 (531~860); 51 (44~68) ❙ 43.4 (13.3~80.6); 860 (797~1053); 379.6 (277.0~551.6)

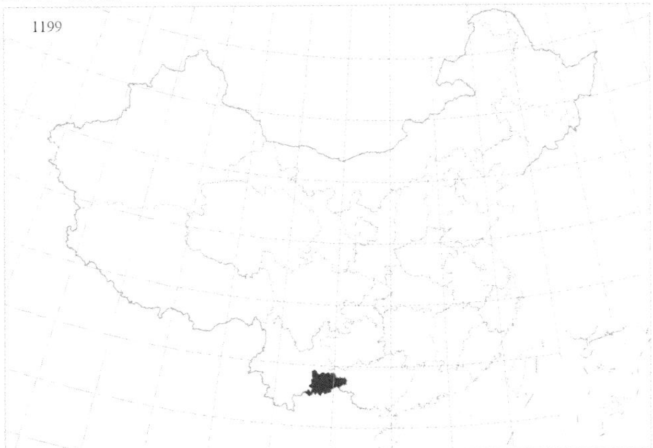

Moraceae 桑科 *Artocarpus* 波罗密属

Artocarpus pithecogallus C. Y. Wu 猴子瘿袋

Trees, evergreen

20.0; 14.6; 23.2; 20.1; 180.8; 0.0; 963 ❙ 1523; 681; 67 ❙ 58.4; 942; 581.0

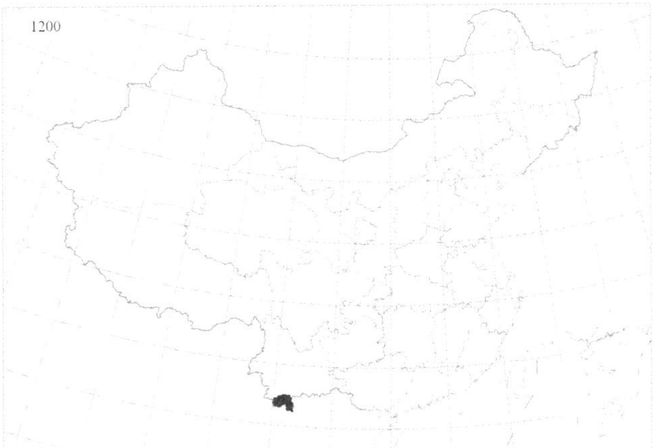

Moraceae 桑科 *Artocarpus* 波罗密属
Artocarpus styracifolius Pierre 二色波罗蜜
Trees, evergreen

20.7 (15.7~24.8); 12.1 (4.1~20.1); 27.8 (22.4~29.8); 20.7 (15.7~24.8); 188.7 (129.7~237.6); 0.0 (0.0~0.9); 1091 (833~1385) ❙ 1612 (868~2435); 685 (312~1428); 140 (41~230) ❙ 49.4 (-34.3~104.6); 1076 (833~1273); 241.9 (95.3~620.0)

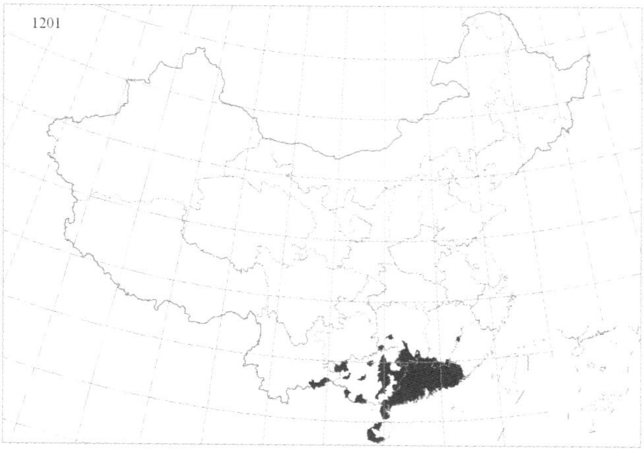

Moraceae 桑科 *Artocarpus* 波罗密属
Artocarpus tonkinensis A. Chev. ex Gagnep. 胭脂
Trees, evergreen

21.1 (16.3~24.8); 13.3 (7.3~20.1); 27.2 (21.4~29.2); 21.2 (16.4~24.8); 194.0 (136.4~237.6); 0.0 (0.0~0.0); 1108 (797~1385) ❙ 1602 (868~2435); 744 (312~1428); 111 (41~205) ❙ 46.2 (-34.3~99.5); 1091 (797~1331); 282.6 (95.3~659.0)

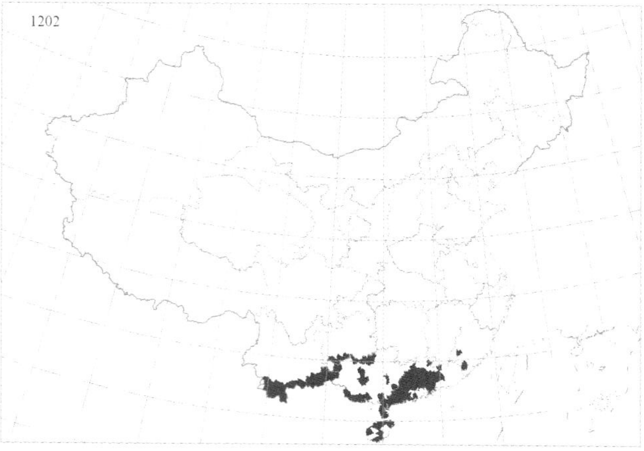

Moraceae 桑科 *Artocarpus* 波罗密属
Artocarpus xanthocarpus Merr. 黄果波罗蜜
Trees, evergreen

18.5; 14.1; 21.7; 18.5; 162.1; 0.0; 863 ❙ 2713; 1388; 188 ❙ 214.4; 863; 363.7

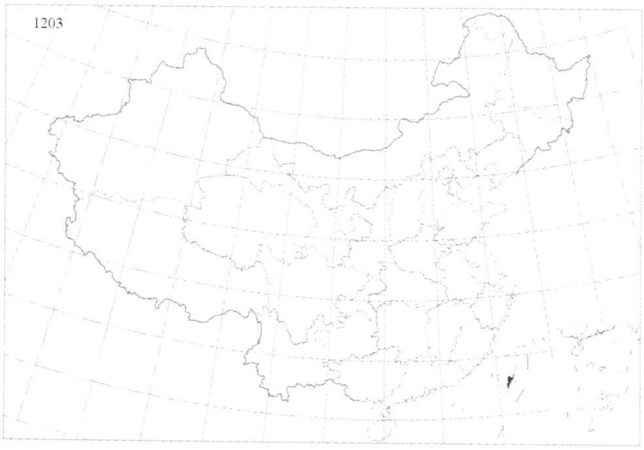

Moraceae 桑科 *Broussonetia* 构属
Broussonetia kaempferi var. australis T. Suzuki 藤构
Scandent shrubs, deciduous

17.3 (5.4~23.6); 7.0 (-3.2~16.7); 26.7 (12.3~30.1); 17.3 (5.8~23.6); 148.8 (31.5~223.5); 0.7 (0.0~26.7); 920 (481~1281) ❙ 1412 (741~3495); 584 (349~1428); 132 (20~897) ❙ 54.1 (-6.9~224.4); 914 (481~1273); 220.7 (95.3~627.0)

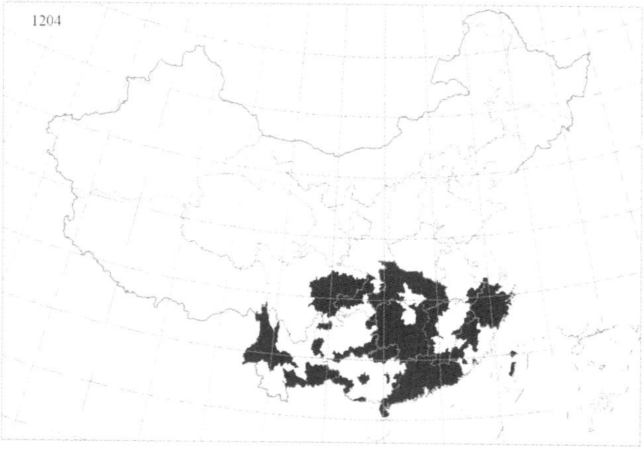

Moraceae 桑科 *Broussonetia* 构属
Broussonetia kazinoki Siebold et Zucc. 楮
Shrubs, deciduous

17.0 (5.3~24.7); 6.7 (-2.9~19.8); 26.4 (12.2~30.1); 17.1 (5.7~24.7); 146.6 (30.5~236.5); 1.7 (0.0~26.2); 912 (479~1376) ❙ 1401 (512~3495); 569 (258~1428); 138 (15~897) ❙ 54.0 (-35.3~246.5); 901 (479~1360); 231.0 (95.3~551.6)

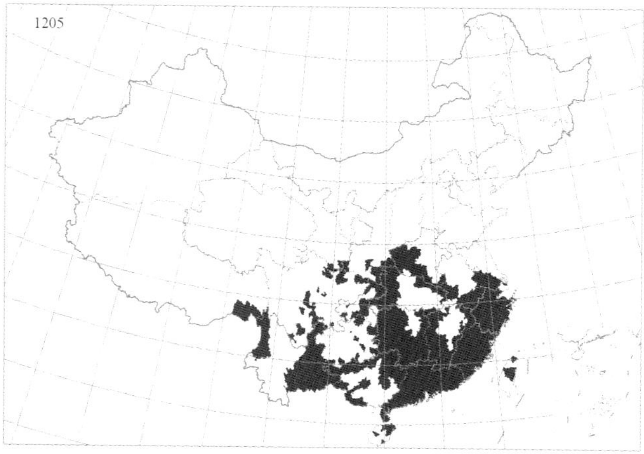

Moraceae 桑科 *Broussonetia* 构属
Broussonetia kurzii (Hook. f.) Corner 落叶花桑
Scandent shrubs, deciduous

18.9 (16.9~21.0); 12.8 (10.3~15.1); 22.8 (21.4~25.6); 18.9 (16.9~21.0); 167.0 (142.7~191.8); 0.0 (0.0~0.0); 907 (813~1053) ❙ 1384 (1059~1751); 744 (604~996); 56 (41~77) ❙ 52.6 (21.3~88.8); 892 (813~1053); 454.3 (294.5~659.0)

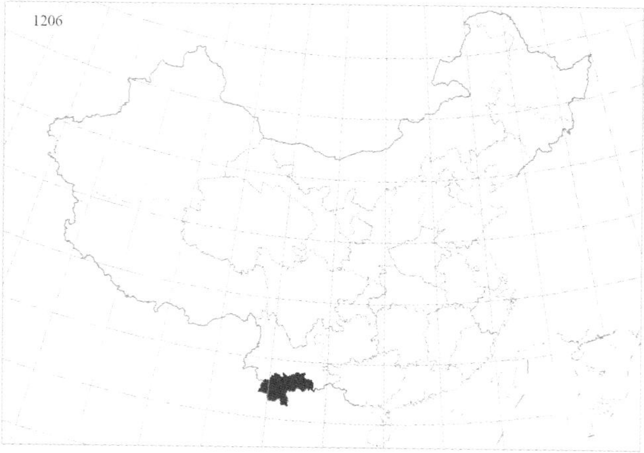

Moraceae 桑科 *Broussonetia* 构属

Broussonetia papyrifera (L.) L'Hér. ex Vent. 构树

Trees, deciduous

16.5 (2.6~24.8); 6.2 (-11.4~20.1); 25.7 (10.8~29.9); 16.6 (4.3~24.8); 141.4 (20.0~237.6); 2.9 (0.0~57.0); 888 (427~1385) ❙ 1296 (373~3495); 566 (201~1594); 109 (7~897) ❙ 46.0 (-48.6~246.5); 869 (373~1331); 229.9 (88.1~659.0)

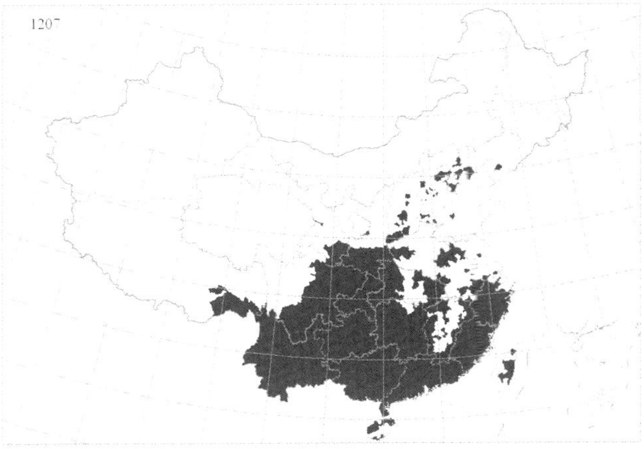

Moraceae 桑科 *Ficus* 榕属

Ficus abelii Miq. 石榕树

Shrubs

18.3 (11.2~24.8); 9.5 (2.7~20.1); 25.6 (16.7~29.9); 18.3 (11.3~24.8); 159.8 (75.8~237.6); 0.1 (0.0~3.7); 948 (642~1385) ❙ 1432 (797~2037); 632 (403~1033); 119 (18~259) ❙ 51.6 (-14.2~128.9); 938 (642~1299); 265.0 (95.3~659.0)

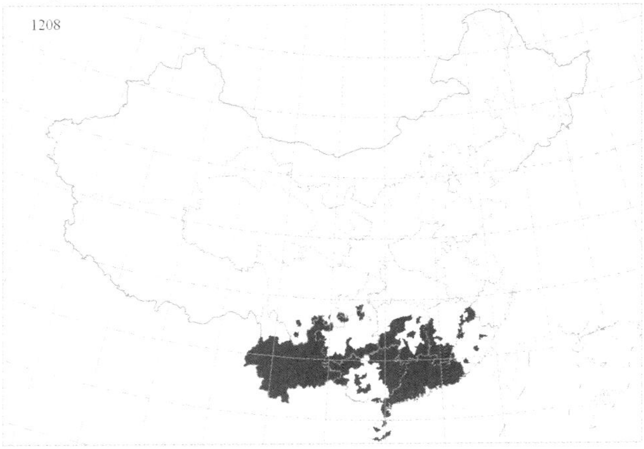

Moraceae 桑科 *Ficus* 榕属

Ficus altissima Blume 高山榕

Trees, evergreen

20.1 (12.6~24.8); 12.5 (5.1~20.1); 25.9 (17.9~29.2); 20.1 (12.6~24.8); 181.5 (91.6~237.6); 0.0 (0.0~0.0); 1038 (678~1385) ❙ 1475 (868~2435); 712 (312~1428); 91 (39~197) ❙ 43.2 (-34.3~112.2); 1022 (678~1360); 300.9 (95.3~659.0)

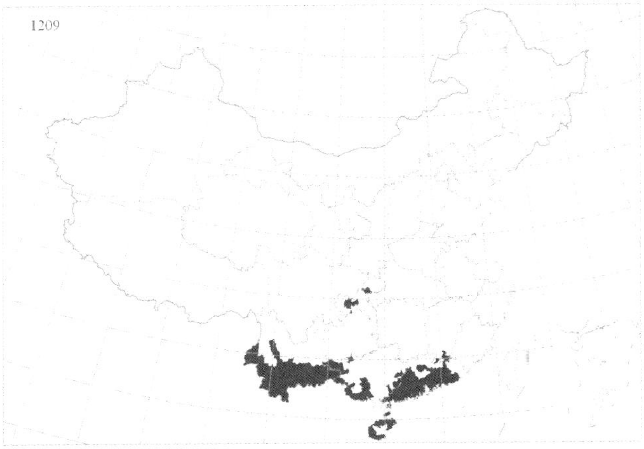

Moraceae 桑科 *Ficus* 榕属

Ficus ampelos Burm. f. 菲律宾榕

Small trees, evergreen

18.6 (16.2~22.2); 13.1 (11.0~17.7); 23.2 (20.1~25.9); 18.6 (16.2~22.2); 163.4 (134.2~206.7); 0.0 (0.0~0.0); 903 (773~1123) ❙ 2662 (2208~2993); 1113 (750~1847); 305 (107~547) ❙ 198.0 (125.7~246.5); 903 (773~1123); 290.7 (260.5~363.7)

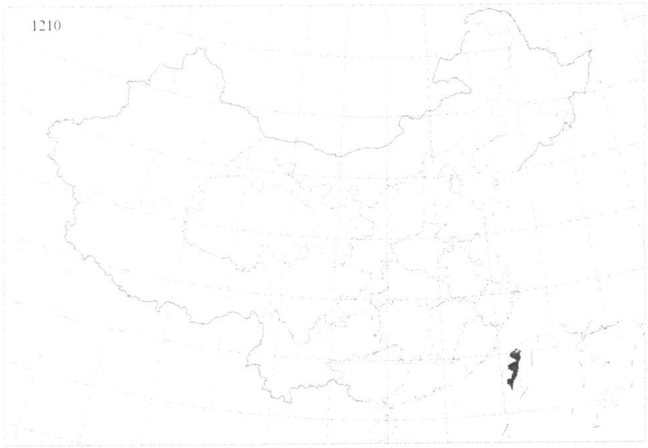

Moraceae 桑科 *Ficus* 榕属

Ficus annulata Blume 环纹榕

Trees, evergreen

18.5 (16.8~21.0); 12.4 (10.3~15.1); 22.5 (20.8~25.6); 18.6 (16.9~21.0); 162.8 (142.6~191.8); 0.0 (0.0~0.0); 890 (811~1053) ❙ 1362 (1007~1751); 746 (539~996); 55 (42~77) ❙ 53.0 (19.0~88.8); 878 (802~1053); 447.9 (294.5~659.0)

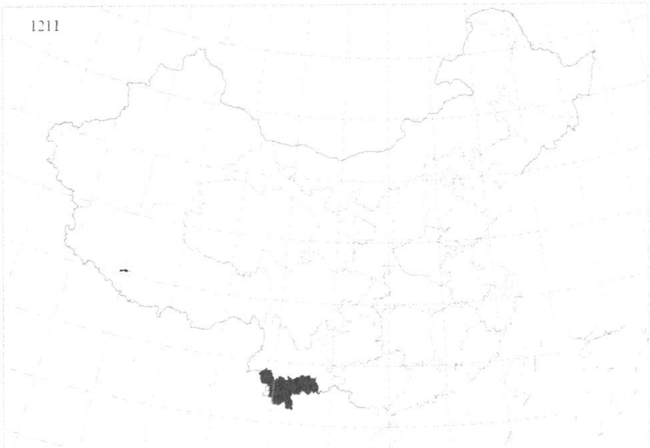

Moraceae 桑科 *Ficus* 榕属

Ficus aurantiaca Griff. 橙黄榕

Scandent shrubs, evergreen

20.3; 15.9; 23.6; 20.4; 184.4; 0.0; 993 ❙ 2853; 1618; 147 ❙ 190.4; 993; 330.3

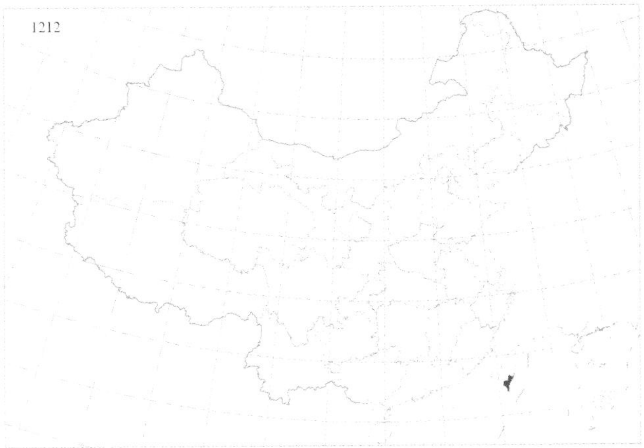

Moraceae 桑科 *Ficus* 榕属
Ficus auriculata Lour. 大果榕
Trees, evergreen

18.3 (8.6~24.7)；11.0 (1.3~19.8)；23.7 (15.0~29.2)；18.4 (8.7~24.7)；160.5 (52.6~236.5)；0.1 (0.0~8.8)；934 (574~1376) ❙ 1300 (797~2435)；681 (312~1428)；68 (19~177) ❙ 40.6 (-34.3~112.2)；912 (574~1273)；320.8 (139.6~659.0)

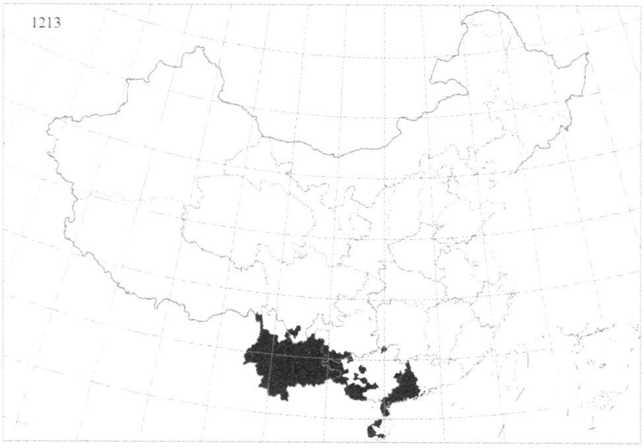

Moraceae 桑科 *Ficus* 榕属
Ficus beipeiensis S. S. Chang 北碚榕
Trees, evergreen

17.7；6.9；27.8；17.7；152.2；0.0；937 ❙ 1105；458；63 ❙ 18.0；937；126.1

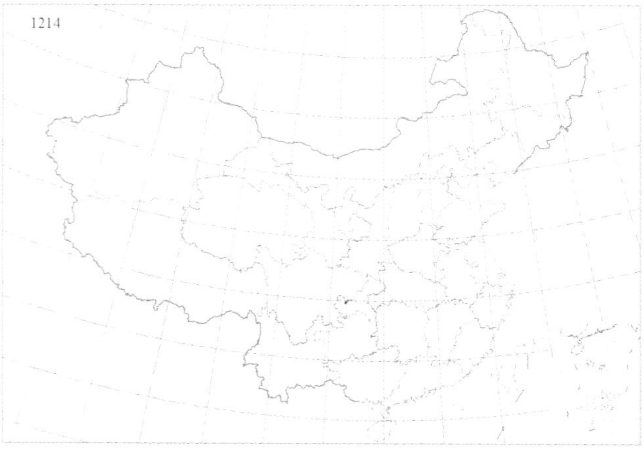

Moraceae 桑科 *Ficus* 榕属
Ficus benguetensis Merr. 黄果榕
Trees, evergreen

20.2 (16.2~24.1)；14.5 (11.0~18.9)；25.0 (20.1~28.2)；20.2 (16.2~24.1)；182.9 (134.2~229.7)；0.0 (0.0~0.0)；1019 (773~1322) ❙ 2402 (1498~3495)；1106 (750~1847)；255 (47~897) ❙ 146.2 (25.2~246.5)；983 (773~1144)；245.0 (119.4~363.7)

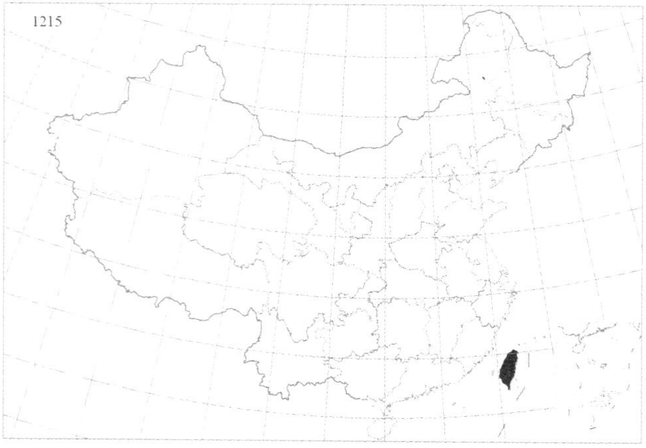

Moraceae 桑科 *Ficus* 榕属
Ficus benjamina L. 垂叶榕
Trees, evergreen

19.8 (15.1~24.8)；12.8 (7.8~20.1)；24.8 (20.1~29.2)；19.8 (15.2~24.8)；177.9 (121.9~237.6)；0.0 (0.0~0.0)；1010 (753~1385) ❙ 1405 (868~2992)；702 (312~1847)；76 (37~188) ❙ 41.1 (-34.3~214.4)；984 (717~1331)；335.7 (152.9~659.0)

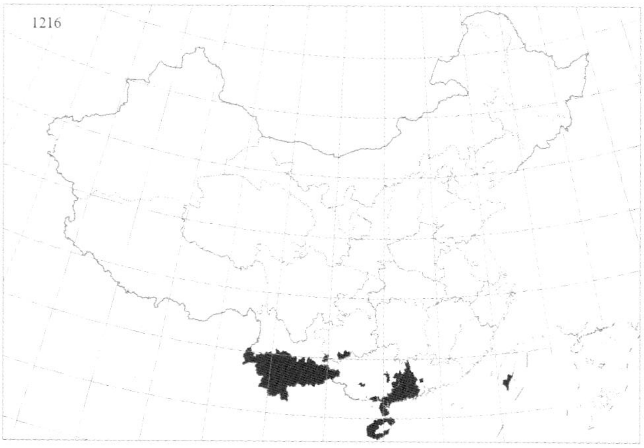

Moraceae 桑科 *Ficus* 榕属
Ficus benjamina var. nuda (Miq.) Barrett 丛毛垂叶榕
Trees, evergreen

19.0 (18.1~19.3)；12.9 (12.0~13.5)；23.0 (22.1~23.6)；19.0 (18.1~19.4)；168.2 (157.3~173.2)；0.0 (0.0~0.0)；911 (863~939) ❙ 1404 (1143~1751)；809 (637~996)；59 (50~77) ❙ 54.2 (21.7~88.8)；902 (863~927)；451.2 (346.9~551.3)

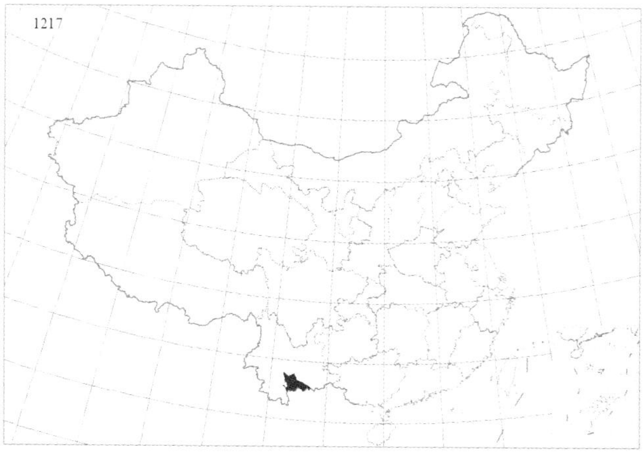

Moraceae 桑科 *Ficus* 榕属
Ficus callosa Willd. 硬皮榕
Trees, evergreen

19.1 (18.1~20.5)；13.2 (12.0~15.1)；22.8 (21.6~23.7)；19.1 (18.1~20.5)；169.7 (157.3~185.9)；0.0 (0.0~0.0)；915 (861~992) ❙ 1442 (1143~1751)；766 (637~996)；58 (41~77) ❙ 57.7 (21.7~88.8)；897 (839~986)；473.2 (346.9~659.0)

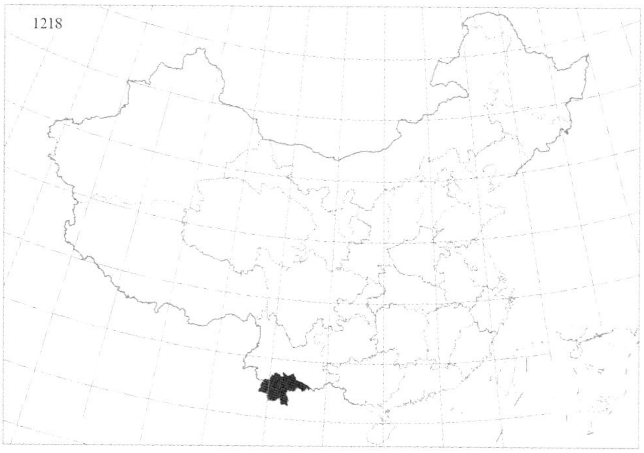

Moraceae 桑科 *Ficus* 榕属

Ficus cardiophylla Merr. 龙州榕

Trees, evergreen

21.9; 13.5; 28.1; 22.0; 203.5; 0.0; 1167 ❙ 1413; 739; 94 ❙ 21.2; 1167; 331.4

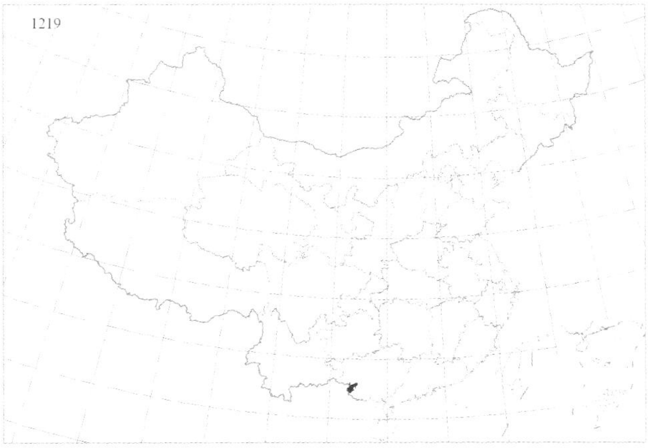

Moraceae 桑科 *Ficus* 榕属

Ficus caulocarpa (Miq.) Miq. 大叶赤榕

Trees, deciduous

19.9; 14.6; 24.1; 20.0; 179.4; 0.0; 972 ❙ 2532; 1249; 224 ❙ 160.0; 972; 357.0

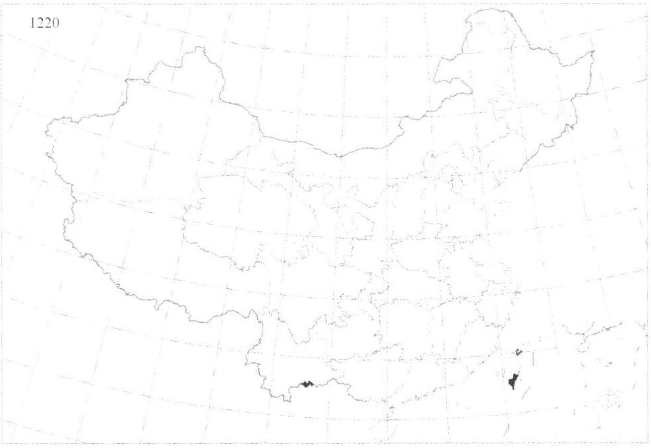

Moraceae 桑科 *Ficus* 榕属

Ficus chapaensis Gagnep. 沙坝榕

Shrubs or small trees, evergreen

16.5 (8.6~21.0); 9.8 (1.3~15.1); 21.2 (15.0~25.6); 16.6 (8.7~21.0); 138.9 (52.6~191.8); 0.2 (0.0~8.8); 814 (574~1053) ❙ 1183 (907~1751); 662 (489~996); 49 (20~77) ❙ 45.7 (5.5~112.2); 797 (574~1053); 370.6 (219.7~659.0)

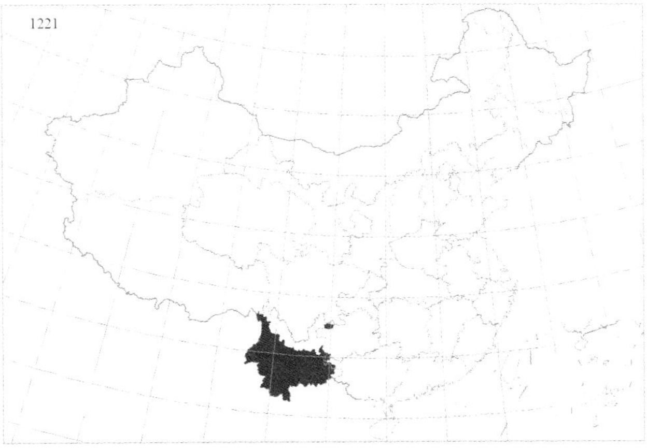

Moraceae 桑科 *Ficus* 榕属

Ficus chartacea Wall. ex King 纸叶榕

Shrubs, evergreen

16.3 (5.4~22.2); 8.8 (-2.1~14.2); 22.1 (12.3~29.0); 16.3 (5.8~22.3); 136.7 (31.5~207.5); 0.7 (0.0~25.6); 827 (481~1190) ❙ 1148 (741~1700); 600 (411~911); 68 (15~211) ❙ 39.3 (-14.2~81.5); 805 (481~1190); 303.9 (160.7~551.6)

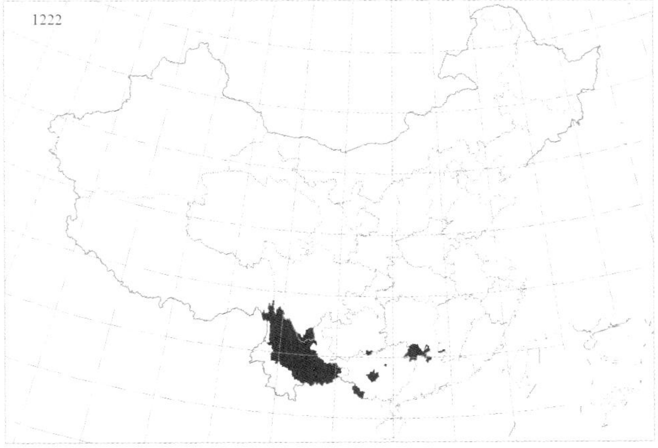

Moraceae 桑科 *Ficus* 榕属

Ficus chartacea var. torulosa King 无柄纸叶榕

Shrubs, evergreen

13.9 (2.2~21.0); 6.2 (-6.6~14.2); 20.0 (9.5~27.5); 14.0 (3.8~21.0); 113.3 (15.5~191.8); 6.4 (0.0~49.1); 748 (411~1053) ❙ 1099 (685~1564); 621 (406~911); 40 (11~182) ❙ 51.3 (4.4~125.9); 732 (411~1053); 323.3 (188.8~551.6)

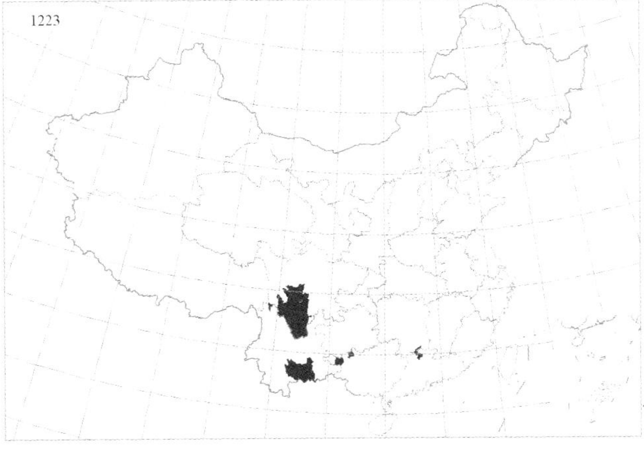

Moraceae 桑科 *Ficus* 榕属

Ficus concinna (Miq.) Miq. 雅榕

Trees, evergreen

17.4 (12.4~22.4); 9.8 (4.7~15.1); 23.2 (18.6~29.8); 17.5 (12.4~22.4); 149.4 (89.1~209.1); 0.0 (0.0~0.3); 870 (675~1204) ❙ 1294 (797~1893); 649 (411~1044); 81 (23~231) ❙ 48.4 (-14.2~129.9); 855 (675~1204); 324.5 (174.9~659.0)

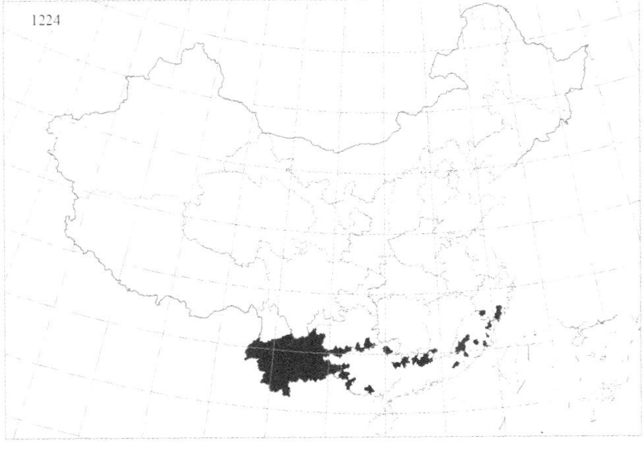

Moraceae 桑科 *Ficus* 榕属
Ficus cumingii Miq. 糙毛榕
Shrubs or small trees, evergreen

18.5; 14.1; 21.7; 18.5; 162.1; 0.0; 863 ❙ 2713; 1388; 188 ❙ 214.4; 863; 363.7

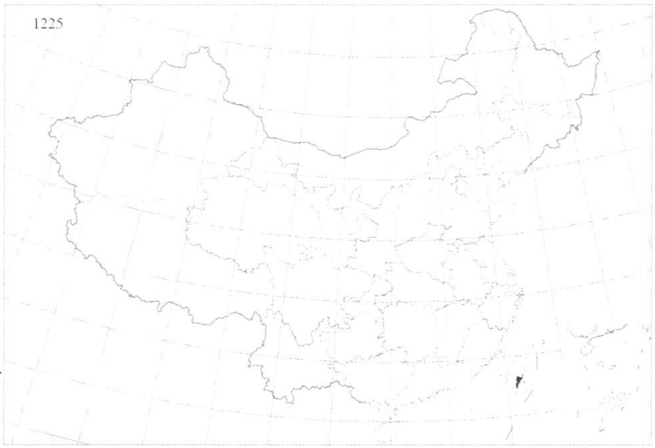

Moraceae 桑科 *Ficus* 榕属
Ficus curtipes Corner 钝叶榕
Trees, evergreen

18.6 (16.3~20.5); 12.5 (9.6~15.1); 22.4 (20.3~23.7); 18.6 (16.4~20.5); 163.7 (136.6~185.9); 0.0 (0.0~0.0); 892 (789~992) ❙ 1418 (1028~1751); 772 (566~996); 55 (41~77) ❙ 59.5 (12.5~112.2); 875 (780~986); 465.9 (295.5~659.0)

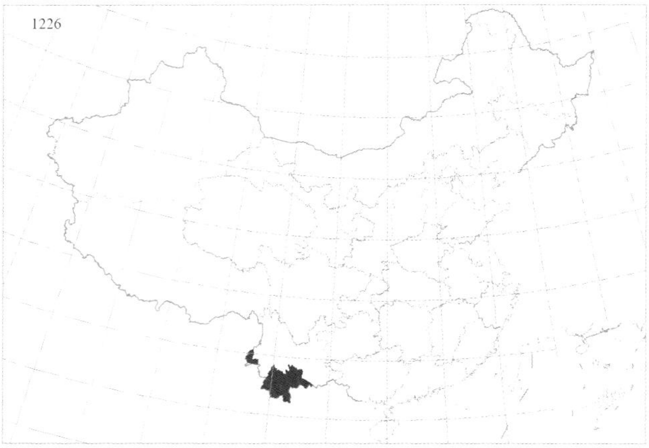

Moraceae 桑科 *Ficus* 榕属
Ficus cyrtophylla (Wall. ex Miq.) Miq. 歪叶榕
Shrubs or small trees, evergreen

17.1 (8.6~22.2); 9.8 (1.3~15.1); 22.5 (15.0~28.5); 17.2 (8.7~22.3); 146.3 (52.6~207.5); 0.1 (0.0~8.8); 853 (574~1190) ❙ 1242 (907~1758); 672 (521~1007); 60 (27~189) ❙ 46.1 (5.8~145.8); 841 (574~1190); 347.1 (189.5~659.0)

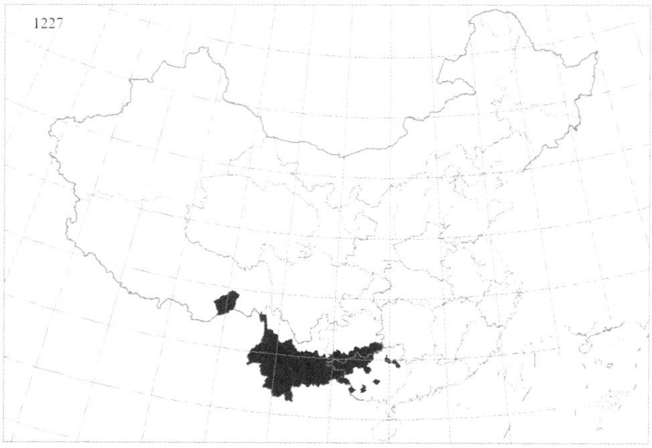

Moraceae 桑科 *Ficus* 榕属
Ficus daimingshanensis S. S. Chang 大明山榕
Shrubs, evergreen

21.3 (17.7~22.7); 12.7 (6.9~15.0); 28.0 (27.3~28.6); 21.3 (17.7~22.8); 196.2 (152.4~213.1); 0.0 (0.0~0.0); 1129 (927~1221) ❙ 1753 (1302~2435); 903 (505~1428); 144 (86~194) ❙ 55.1 (12.9~99.5); 1129 (927~1221); 254.9 (188.8~318.8)

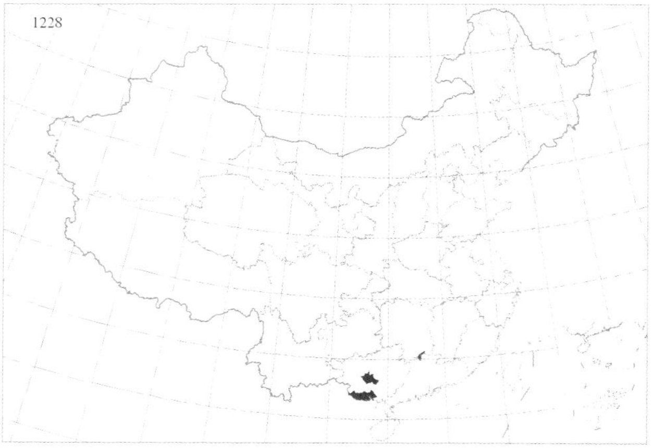

Moraceae 桑科 *Ficus* 榕属
Ficus dinganensis S. S. Chang 定安榕
Scandent shrubs, evergreen

24.1; 18.2; 28.4; 24.2; 230.1; 0.0; 1326 ❙ 1723; 619; 110 ❙ 29.9; 1326; 352.9

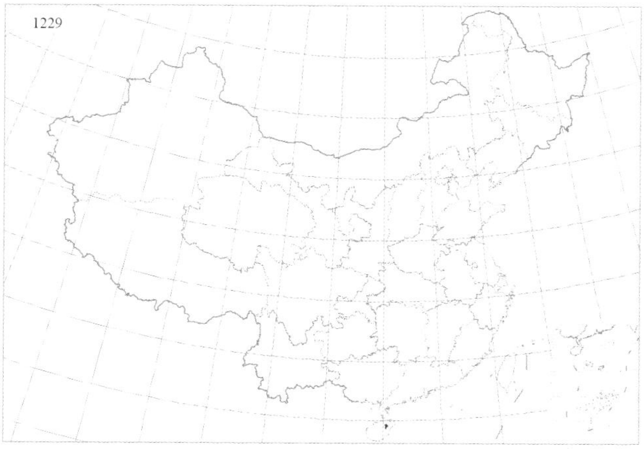

Moraceae 桑科 *Ficus* 榕属
Ficus drupacea Thunb. 枕果榕
Trees, evergreen

22.2 (18.2~24.8); 16.5 (12.4~20.1); 26.1 (21.6~28.8); 22.3 (18.2~24.8); 207.1 (158.6~237.6); 0.0 (0.0~0.0); 1164 (861~1385) ❙ 1500 (868~1854); 627 (312~996); 68 (41~115) ❙ 32.7 (-34.3~88.8); 1085 (839~1360); 469.5 (248.2~659.0)

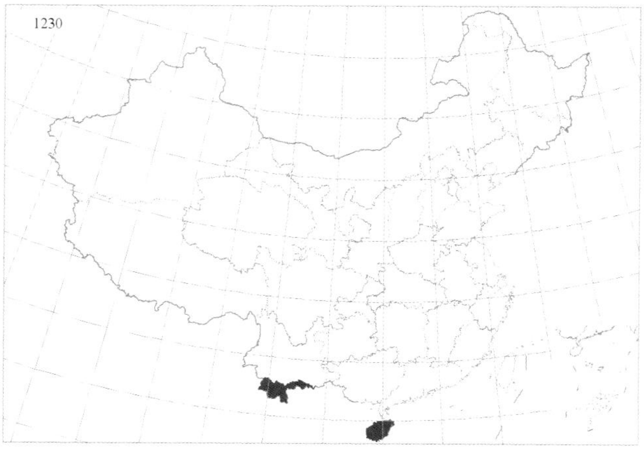

Moraceae 桑科 *Ficus* 榕属

Ficus drupacea var. pubescens (Roem. et Schult.) Corner 毛枕果榕

Trees, evergreen

21.0; 14.2; 25.6; 21.0; 191.8; 0.0; 1053 ❙ 1522; 818; 68 ❙ 44.5; 1053; 551.6

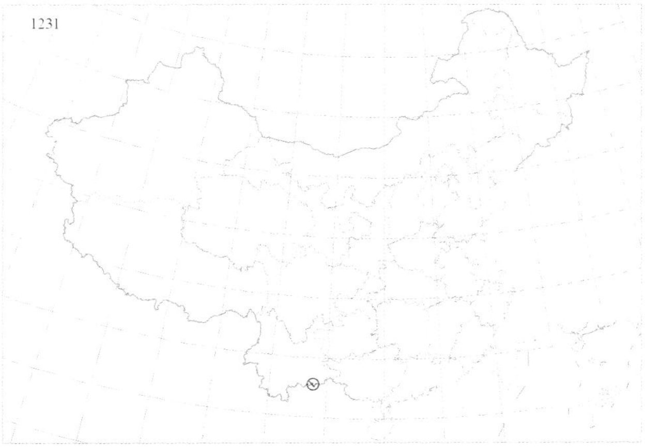

Moraceae 桑科 *Ficus* 榕属

Ficus elastica Roxb. 印度榕

Trees, evergreen

19.3 (11.2~24.8); 12.3 (4.6~20.1); 24.5 (16.7~29.1); 19.4 (11.3~24.8); 172.5 (75.8~237.6); 0.0 (0.0~0.4); 990 (642~1385) ❙ 1400 (868~2037); 693 (312~1033); 77 (29~188) ❙ 42.4 (-34.3~112.2); 967 (642~1360); 327.9 (95.3~659.0)

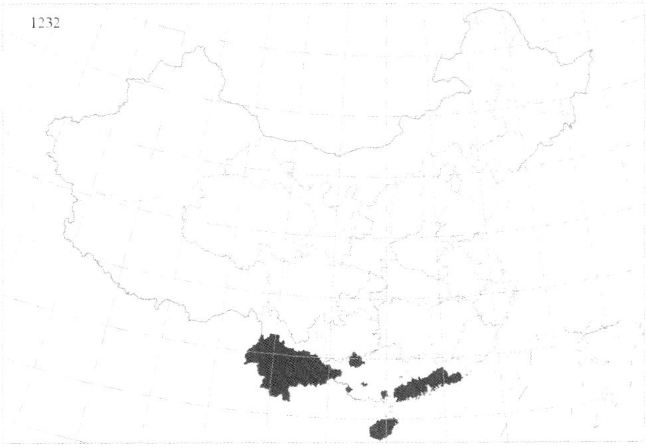

Moraceae 桑科 *Ficus* 榕属

Ficus erecta Thunb. 矮小天仙果

Shrubs or small trees, deciduous

18.3 (13.4~24.1); 8.1 (1.5~18.9); 27.5 (20.1~29.9); 18.3 (13.4~24.1); 160.1 (108.3~229.7); 0.3 (0.0~7.3); 966 (755~1322) ❙ 1591 (862~3495); 612 (356~1847); 176 (47~897) ❙ 66.5 (-5.0~246.5); 956 (755~1216); 226.2 (111.9~506.2)

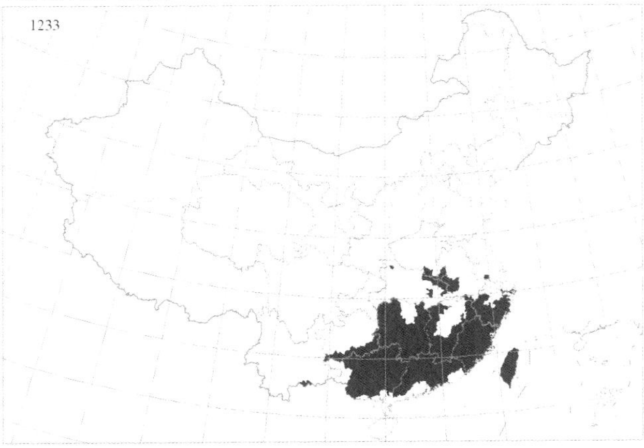

Moraceae 桑科 *Ficus* 榕属

Ficus esquiroliana H. Lév. 黄毛榕

Shrubs or small trees, evergreen

19.1 (8.6~24.1); 11.1 (1.3~19.1); 25.5 (15.0~29.8); 19.2 (8.7~24.1); 170.1 (52.6~229.7); 0.1 (0.0~8.8); 985 (574~1322) ❙ 1451 (911~2992); 694 (474~1847); 102 (31~203) ❙ 47.7 (-6.9~166.5); 978 (574~1273); 279.9 (95.3~659.0)

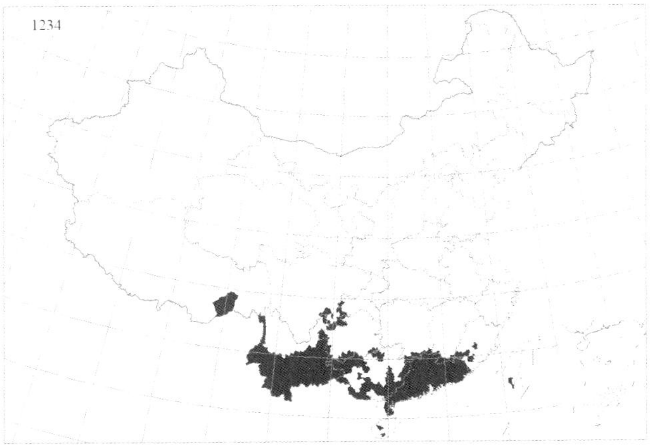

Moraceae 桑科 *Ficus* 榕属

Ficus filicauda Hand.-Mazz. 线尾榕

Trees, evergreen

16.5 (1.8~21.0); 9.9 (-7.1~14.9); 21.0 (9.9~25.6); 16.6 (3.8~21.0); 140.9 (15.7~191.8); 2.7 (0.0~53.3); 824 (404~1053) ❙ 1375 (583~1751); 767 (348~1007); 55 (13~77) ❙ 68.6 (21.7~145.8); 816 (404~1053); 438.6 (169.5~627.0)

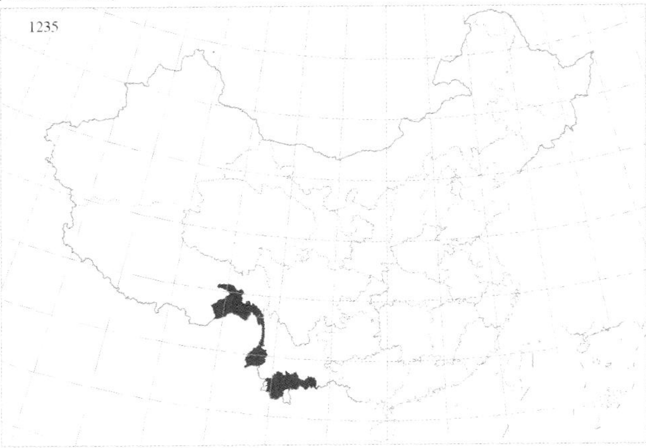

Moraceae 桑科 *Ficus* 榕属

Ficus filicauda var. longipes S. S. Chang 长柄线尾榕

Trees, evergreen

13.3; 5.4; 19.6; 13.4; 100.4; 0.0; 706 ❙ 1736; 1007; 51 ❙ 145.8; 706; 447.4

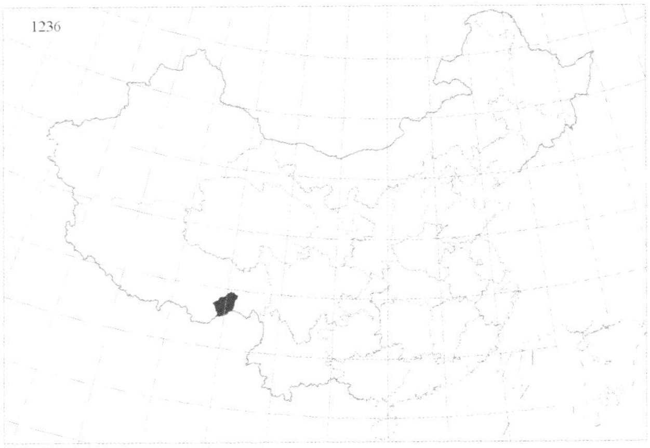

Moraceae 桑科 *Ficus* 榕属
Ficus fistulosa Reinw. ex Blume 水同木
Small trees, evergreen

18.9 (11.2~24.8); 11.7 (4.6~20.1); 24.5 (16.7~29.0); 19.0 (11.3~24.8); 168.1 (75.8~237.6); 0.0 (0.0~0.4); 966 (642~1385) ❙ 1443 (797~3495); 697 (312~1847); 98 (23~897) ❙ 49.4 (-34.3~224.4); 947 (642~1326); 314.2 (139.6~659.0)

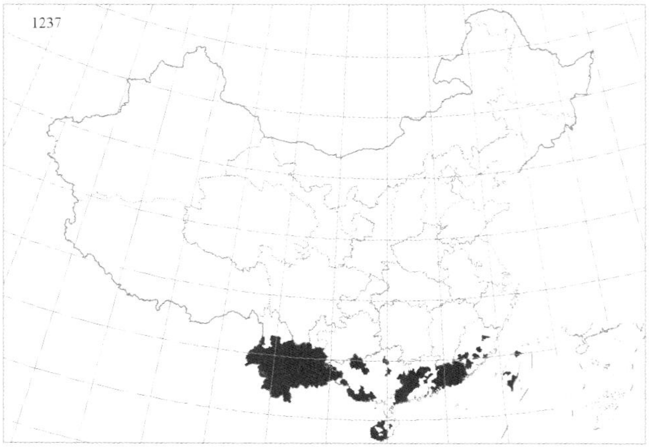

Moraceae 桑科 *Ficus* 榕属
Ficus formosana Maxim. 台湾榕
Shrubs, evergreen

19.0 (12.7~24.8); 9.3 (2.1~20.1); 27.7 (20.1~30.1); 19.0 (12.8~24.8); 168.3 (99.0~237.6); 0.1 (0.0~5.6); 1005 (713~1385) ❙ 1606 (868~3495); 621 (299~1847); 171 (41~897) ❙ 62.6 (-34.3~246.5); 987 (713~1360); 232.3 (95.3~620.0)

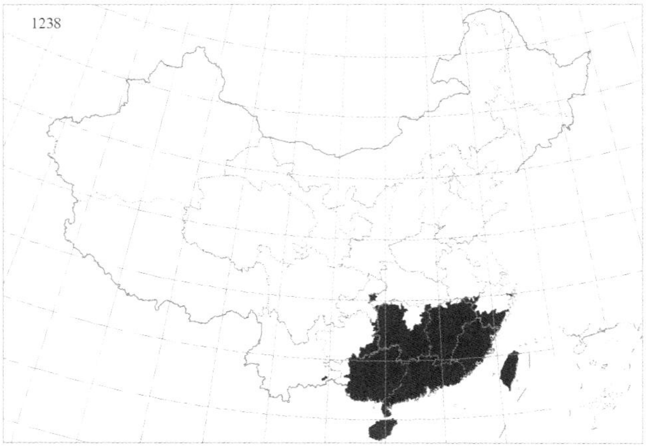

Moraceae 桑科 *Ficus* 榕属
Ficus fulva Reinwardt ex Blume 金毛榕
Small trees, evergreen

20.6 (13.3~24.8); 12.5 (4.5~20.1); 27.0 (19.6~29.8); 20.6 (13.4~24.8); 187.4 (100.4~237.6); 0.0 (0.0~0.5); 1074 (706~1385) ❙ 1541 (868~2435); 699 (312~1428); 114 (41~197) ❙ 44.6 (-34.3~145.8); 1062 (706~1360); 277.1 (95.3~659.0)

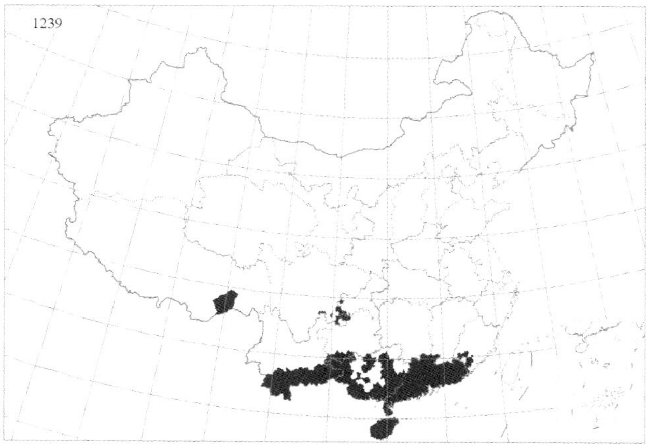

Moraceae 桑科 *Ficus* 榕属
Ficus fusuiensis S. S. Chang 扶绥榕
Shrubs, evergreen

20.6; 12.4; 26.7; 20.7; 188.1; 0.0; 1066 ❙ 1519; 829; 93 ❙ 43.5; 1066; 314.1

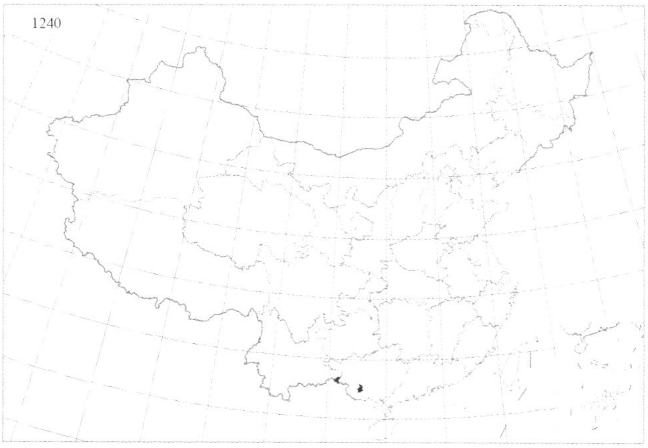

Moraceae 桑科 *Ficus* 榕属
Ficus gasparriniana Miq. 冠毛榕
Shrubs, evergreen

16.7 (8.6~22.7); 8.2 (-0.2~15.1); 23.8 (15.0~28.9); 16.8 (8.7~22.8); 141.5 (52.6~213.1); 0.4 (0.0~12.1); 856 (574~1221) ❙ 1227 (797~2435); 613 (411~1428); 71 (22~255) ❙ 43.8 (-14.2~112.2); 848 (574~1221); 278.6 (108.2~659.0)

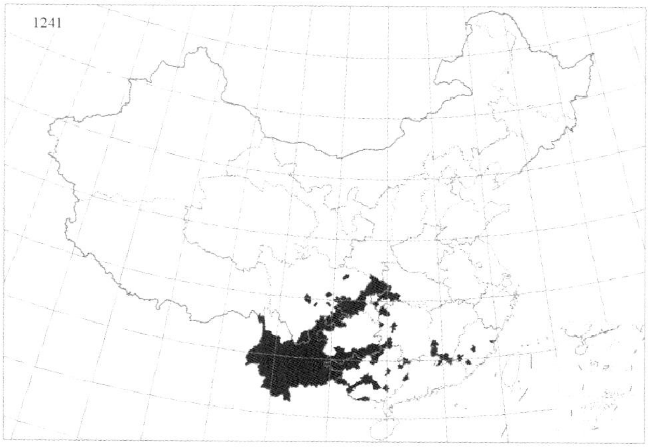

Moraceae 桑科 *Ficus* 榕属
Ficus gasparriniana var. esquirolii (Lév. et Vaniot) Corner 长叶冠毛榕
Shrubs, evergreen

18.3 (15.0~21.6); 8.7 (3.8~14.2); 26.8 (23.4~28.6); 18.3 (15.1~21.7); 159.9 (122.0~199.8); 0.0 (0.0~1.2); 948 (810~1153) ❙ 1582 (1097~1997); 619 (445~821); 167 (43~255) ❙ 68.0 (16.9~130.7); 948 (810~1153); 259.0 (158.7~551.6)

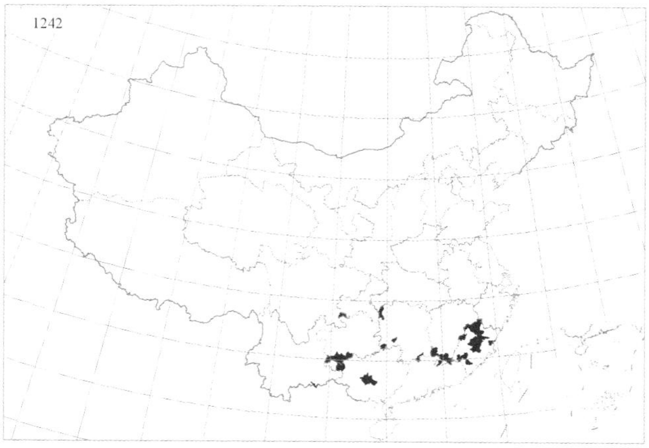

Moraceae 桑科 *Ficus* 榕属
Ficus gasparriniana var. laceratifolia (Lév. et Vaniot) Corner 菱叶冠毛榕
Shrubs, evergreen

17.1 (11.3~21.9); 7.9 (3.2~15.1); 25.1 (17.0~28.9); 17.2 (11.3~22.0); 146.1 (76.6~203.8); 0.1 (0.0~2.5); 882 (644~1176) ❙ 1167 (899~1751); 592 (442~996); 56 (19~173) ❙ 32.9 (5.6~93.0); 878 (644~1176); 228.1 (108.2~659.0)

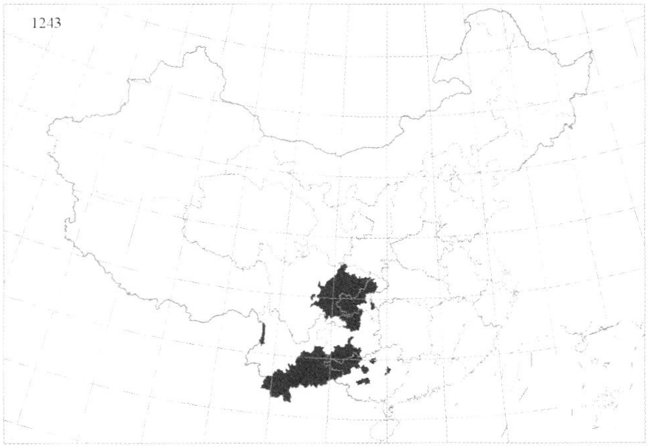

Moraceae 桑科 *Ficus* 榕属
Ficus geniculata Kurz 曲枝榕
Trees, evergreen

15.0; 8.4; 20.2; 15.1; 121.2; 0.0; 753 ❙ 950; 570; 27 ❙ 26.3; 683; 307.7

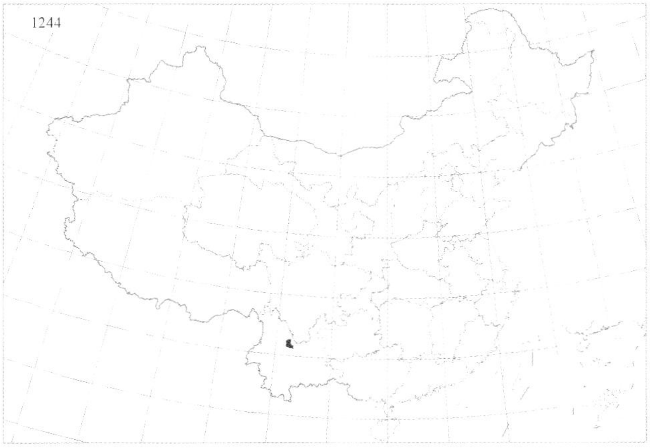

Moraceae 桑科 *Ficus* 榕属
Ficus glaberrima Blume 大叶水榕
Trees, evergreen

19.5 (13.3~24.8); 12.6 (5.4~20.1); 24.4 (19.1~28.8); 19.5 (13.4~24.8); 174.5 (100.4~237.6); 0.0 (0.0~0.0); 989 (706~1385) ❙ 1386 (868~2435); 706 (312~1428); 66 (37~194) ❙ 42.9 (-34.3~145.8); 959 (706~1360); 386.0 (189.5~659.0)

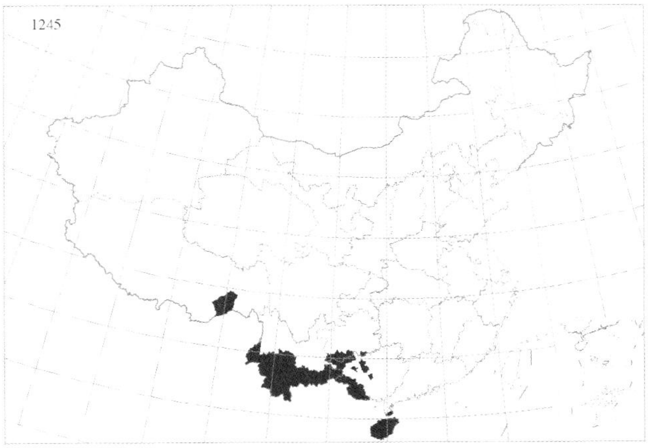

Moraceae 桑科 *Ficus* 榕属
Ficus guangxiensis S. S. Chang 广西榕
Vines, evergreen

19.1; 11.0; 25.1; 19.2; 170.0; 0.0; 949 ❙ 1461; 834; 65 ❙ 54.0; 949; 342.7

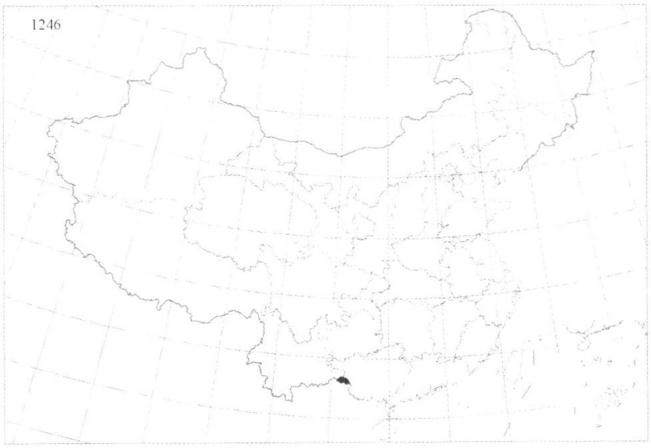

Moraceae 桑科 *Ficus* 榕属
Ficus guizhouensis S. S. Chang 贵州榕
Scandent shrubs, evergreen

17.1 (14.3~19.1); 7.7 (4.0~11.0); 24.9 (22.7~26.9); 17.2 (14.3~19.1); 146.3 (112.6~169.6); 0.2 (0.0~1.1); 873 (756~968) ❙ 1344 (1254~1438); 640 (531~820); 92 (48~122) ❙ 54.4 (47.0~71.9); 873 (756~968); 273.9 (225.1~371.7)

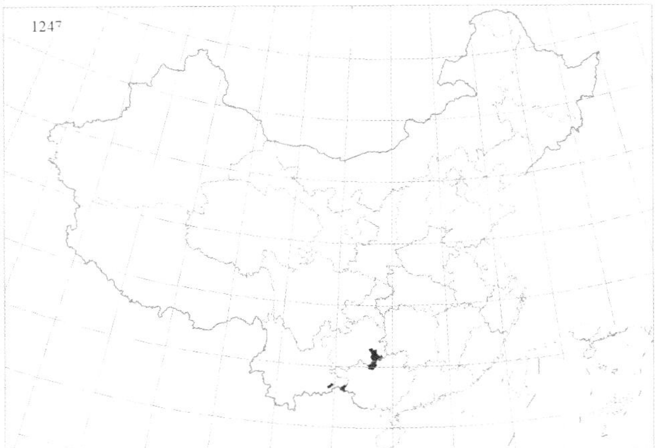

Moraceae 桑科 *Ficus* 榕属
Ficus hederacea Roxb. 藤榕
Scandent shrubs, evergreen

19.3 (8.6~24.8); 12.0 (1.3~20.1); 24.6 (15.0~29.0); 19.3 (8.7~24.8); 172.0 (52.6~237.6); 0.1 (0.0~8.8); 983 (574~1385) ❙ 1428 (958~2035); 715 (447~996); 81 (41~194) ❙ 47.3 (-12.0~112.2); 970 (574~1331); 344.9 (157.7~659.0)

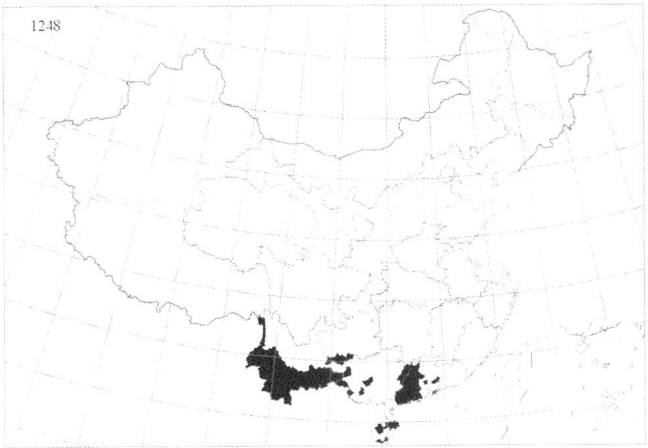

Moraceae 桑科 *Ficus* 榕属
Ficus henryi Warb. 尖叶榕
Small trees, evergreen

16.1 (5.3~21.6); 6.0 (−3.2~14.2); 25.3 (12.2~28.9); 16.2 (5.7~21.7); 135.3 (30.5~199.8); 1.2 (0.0~26.7); 852 (479~1153) ❙ 1199 (711~1751); 577 (357~1007); 64 (15~174) ❙ 42.6 (5.6~145.8); 851 (479~1153); 206.0 (108.2~551.6)

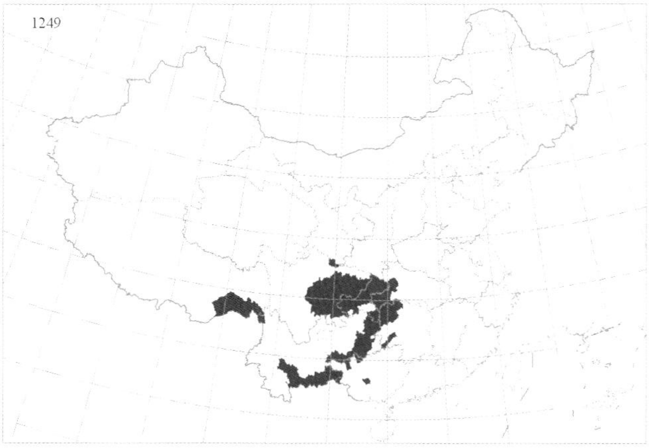

Moraceae 桑科 *Ficus* 榕属
Ficus heteromorpha Hemsl. 异叶榕
Trees, deciduous

15.8 (5.7~23.6); 5.2 (−4.9~16.7); 25.5 (14.2~29.8); 15.9 (6.4~23.6); 133.3 (38.1~223.5); 2.8 (0.0~29.4); 851 (504~1281) ❙ 1288 (512~1997); 538 (258~1056); 112 (14~259) ❙ 50.6 (−35.3~147.0); 839 (504~1269); 231.0 (111.9~551.6)

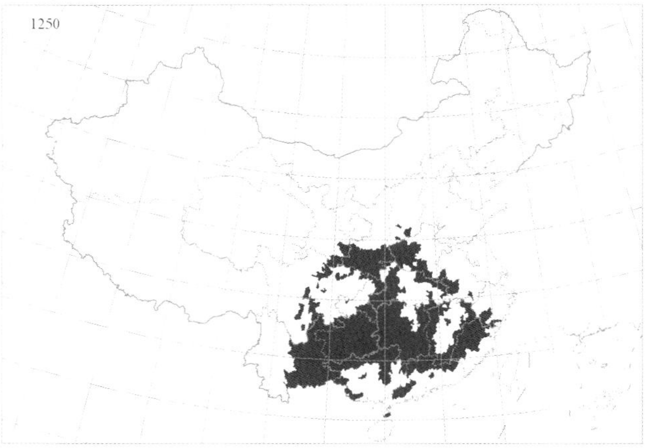

Moraceae 桑科 *Ficus* 榕属
Ficus heterophylla L. f. 山榕
Shrubs, evergreen

23.0 (19.2~24.8); 17.1 (13.5~20.1); 27.2 (22.3~28.8); 23.1 (19.2~24.8); 216.8 (170.5~237.6); 0.0 (0.0~0.0); 1232 (909~1385) ❙ 1522 (1176~1813); 590 (444~726); 79 (41~132) ❙ 25.7 (−12.0~64.9); 1155 (880~1331); 431.0 (203.4~659.0)

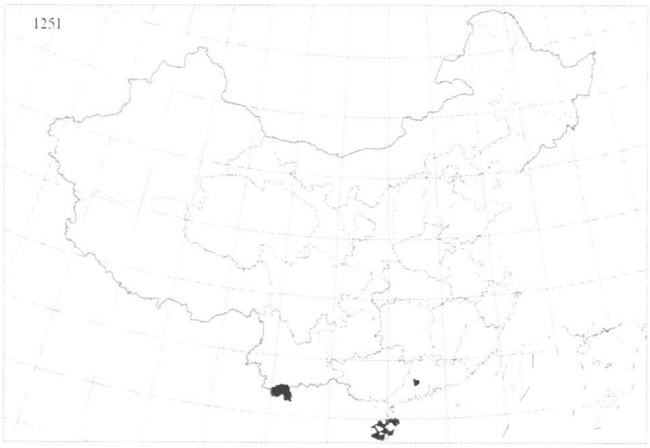

Moraceae 桑科 *Ficus* 榕属
Ficus heteropleura Blume 尾叶榕
Scandent shrubs, evergreen

22.6 (18.5~24.8); 15.9 (11.0~20.1); 27.5 (21.7~28.8); 22.7 (18.5~24.8); 212.2 (162.1~237.6); 0.0 (0.0~0.0); 1212 (863~1385) ❙ 1566 (868~2713); 697 (312~1428); 86 (41~188) ❙ 32.0 (−34.3~214.4); 1148 (863~1360); 379.3 (248.2~620.0)

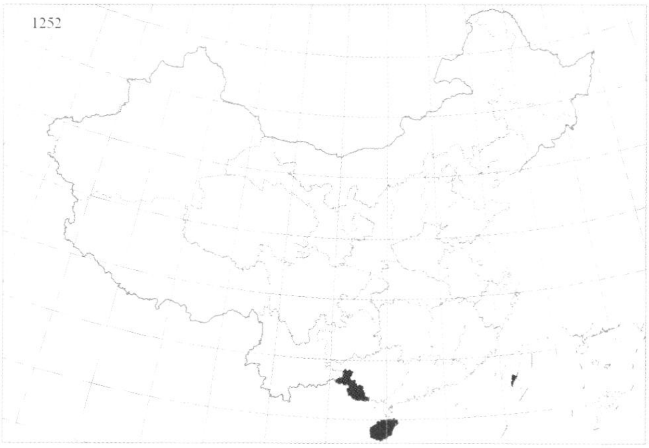

Moraceae 桑科 *Ficus* 榕属
Ficus hirta Vahl 粗叶榕
Shrubs or small trees, evergreen

19.5 (13.2~24.8); 10.8 (3.3~20.1); 26.7 (18.4~29.8); 19.5 (13.3~24.8); 174.1 (99.4~237.6); 0.0 (0.0~2.3); 1012 (695~1385) ❙ 1534 (868~2435); 651 (312~1428); 138 (41~257) ❙ 53.7 (−34.3~137.5); 1000 (695~1360); 272.0 (95.3~659.0)

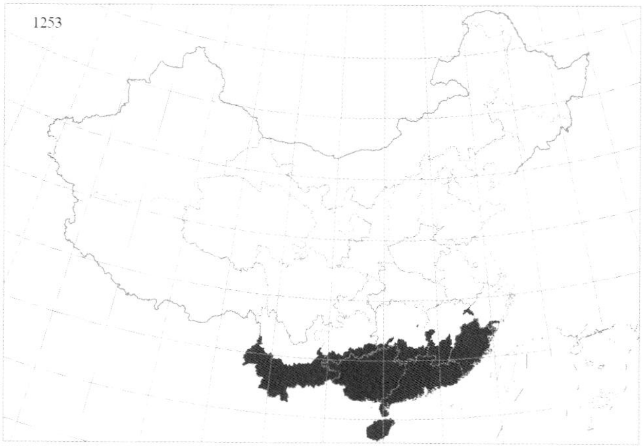

Moraceae 桑科 *Ficus* 榕属
Ficus hispida L. f. 对叶榕
Shrubs or small trees, evergreen

20.2 (14.4~24.8); 12.1 (3.8~20.1); 26.6 (19.1~29.8); 20.3 (14.4~24.8); 183.3 (113.1~237.6); 0.0 (0.0~1.5); 1052 (727~1385) ❙ 1533 (868~2435); 704 (312~1428); 116 (39~197) ❙ 47.3 (−34.3~112.2); 1041 (727~1360); 285.3 (95.3~659.0)

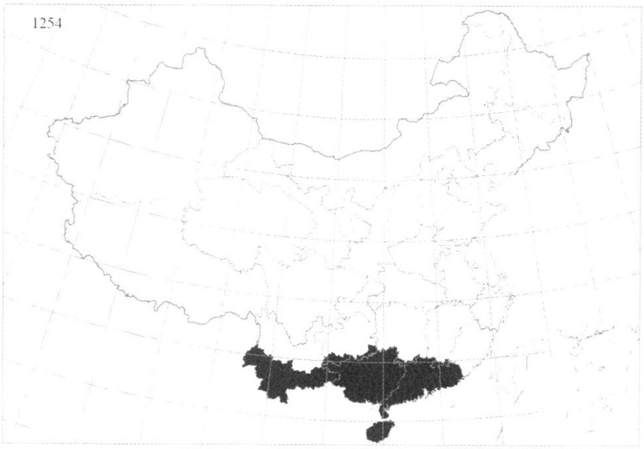

Moraceae 桑科 *Ficus* 榕属
Ficus hookeriana Corner 大青树
Trees, evergreen

17.1 (13.6~21.0); 10.4 (7.2~15.1); 21.8 (18.8~25.6); 17.2 (13.7~21.0); 146.2 (104.1~191.8); 0.0 (0.0~0.0); 836 (706~1053) ❙ 1199 (797~1751) 669 (411~996); 49 (23~77) ❙ 43.2 (-14.2~112.2); 818 (690~1053); 369.5 (189.5~659.0)

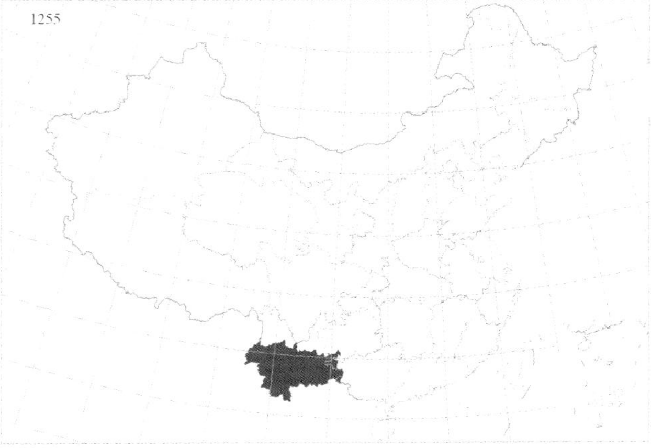

Moraceae 桑科 *Ficus* 榕属
Ficus irisana Elmer 糙叶榕
Trees, evergreen

19.0 (16.2~22.2); 13.5 (11.0~17.7); 23.6 (20.1~27.0); 19.1 (16.2~22.2); 168.6 (134.2~206.7); 0.0 (0.0~0.0); 931 (773~1123) ❙ 256 (1836~2993); 1149 (750~1847); 261 (107~547) ❙ 180.7 (64.2~246.5); 927 (773~1123); 273.3 (202.8~363.7)

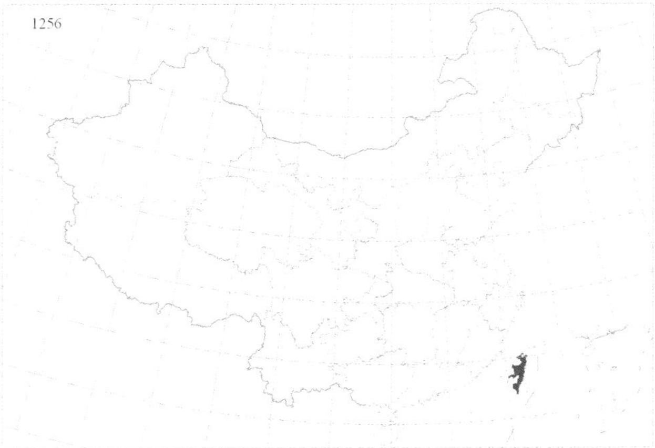

Moraceae 桑科 *Ficus* 榕属
Ficus ischnopoda Miq. 壶托榕
Small trees, evergreen

17.0 (8.6~21.3); 10.0 (1.3~15.1); 21.9 (15.0~28.4); 17.0 (8.7~21.3); 144.4 (52.6~195.9); 0.1 (0.0~8.8); 837 (574~1115) ❙ 1228 (896~1900); 679 (524~1056); 53 (31~144) ❙ 46.8 (6.5~112.2); 827 (574~1111); 363.0 (193.3~659.0)

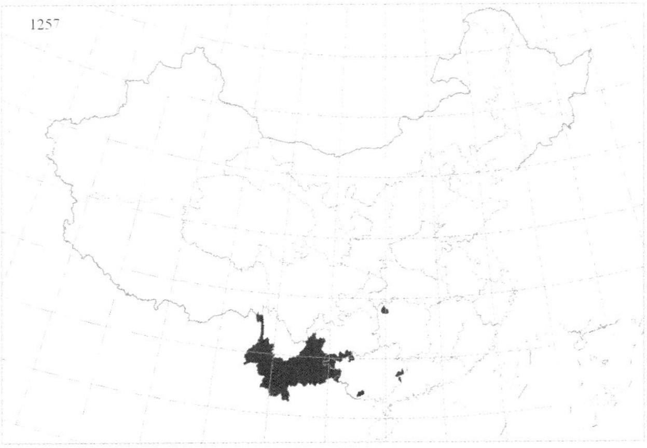

Moraceae 桑科 *Ficus* 榕属
Ficus kurzii King 滇缅榕
Trees

20.0; 14.6; 23.2; 20.1; 180.8; 0.0; 963 ❙ 1523; 681; 67 ❙ 58.4; 942; 581.0

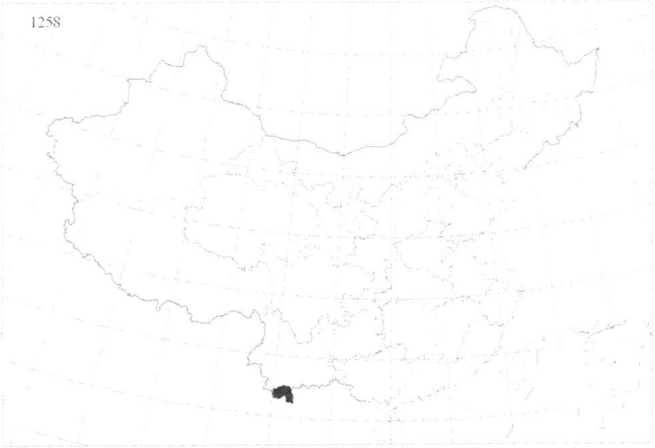

Moraceae 桑科 *Ficus* 榕属
Ficus laevis Blume 光叶榕
Scandent shrubs, evergreen

18.8 (16.3~22.7); 11.6 (7.3~15.1); 23.8 (20.3~28.6); 18.8 (16.4~22.8); 165.6 (136.4~213.1); 0.0 (0.0~0.0); 926 (789~1221) ❙ 1382 (1007~2435); 736 (539~1428); 70 (41~169) ❙ 49.2 (12.5~99.5); 916 (780~1221); 375.7 (218.0~659.0)

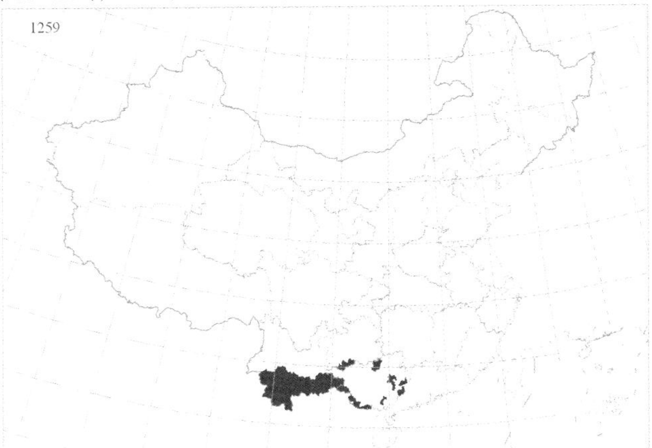

Moraceae 桑科 *Ficus* 榕属
Ficus langkokensis Drake 青藤公
Trees, evergreen

20.4 (14.3~24.7); 12.3 (3.3~19.8); 26.8 (20.7~29.8); 20.4 (14.3~24.7); 184.9 (114.1~236.5); 0.0 (0.0~2.3); 1060 (765~1376) ❙ 1562 (858~2435); 695 (312~1428); 120 (41~199) ❙ 48.9 (-34.3~99.5); 1047 (735~1331); 282.0 (95.3~659.0)

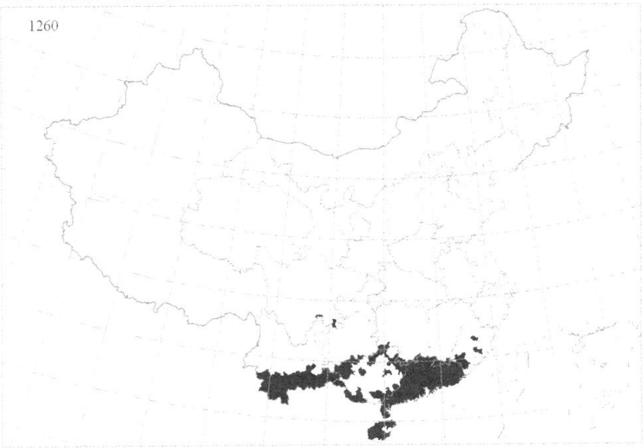

Moraceae 桑科 *Ficus* 榕属

Ficus maclellandii King 瘤枝榕

Trees, evergreen

17.9 (13.2~21.0); 11.6 (6.4~15.1); 22.1 (18.4~25.6); 18.0 (13.3~21.0); 155.8 (99.4~191.8); 0.0 (0.0~0.0); 867 (695~1053) ❙ 1378 (1028~1751); 762 (566~996); 54 (41~77) ❙ 59.5 (12.5~112.2); 853 (695~1053); 449.7 (295.5~659.0)

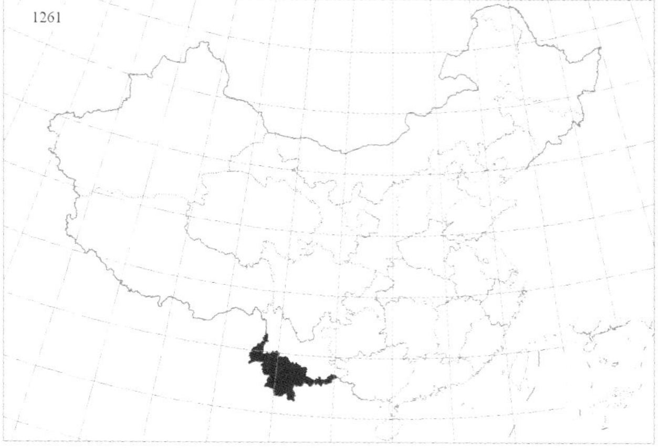

Moraceae 桑科 *Ficus* 榕属

Ficus microcarpa L. f. 榕树

Trees, evergreen

19.3 (13.4~24.8); 11.3 (3.3~20.1); 25.7 (18.8~29.8); 19.3 (13.5~24.8); 172.2 (102.1~237.6); 0.0 (0.0~2.1); 995 (706~1385) ❙ 1475 (797~3495); 686 (386~1847); 114 (22~897) ❙ 48.8 (-14.2~224.4); 983 (683~1326); 278.5 (95.3~659.0)

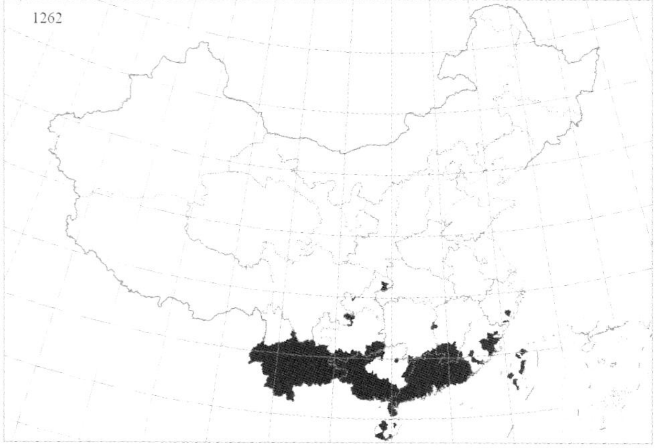

Moraceae 桑科 *Ficus* 榕属

Ficus napoensis S. S. Chang 那坡榕

Scandent shrubs, evergreen

20.7; 12.5; 26.7; 20.7; 188.5; 0.0; 1068 ❙ 1406; 766; 72 ❙ 33.8; 1068; 349.7

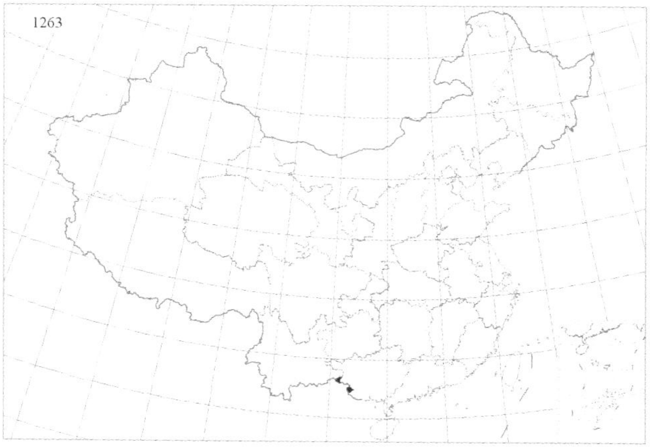

Moraceae 桑科 *Ficus* 榕属

Ficus neriifolia Sm. 森林榕

Trees, evergreen

16.4 (-1.1~21.0); 9.9 (-9.1~15.1); 20.8 (6.2~25.6); 16.5 (2.0~21.0); 139.7 (2.5~191.8); 2.9 (0.0~75.6); 815 (334~1053) ❙ 1271 (321~1751); 704 (193~1007); 51 (13~77) ❙ 56.2 (-4.0~145.8); 799 (321~1053); 417.1 (79.3~659.0)

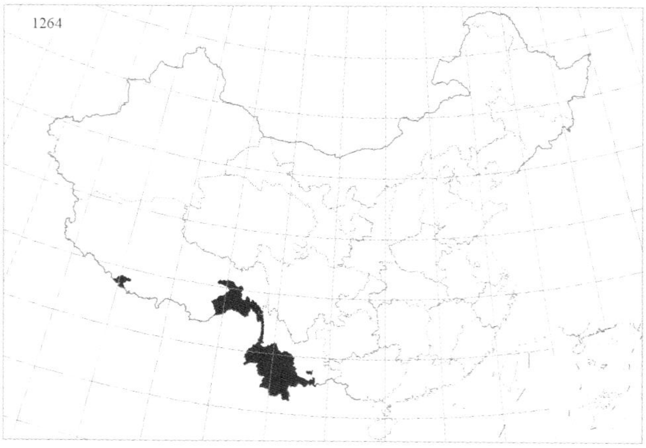

Moraceae 桑科 *Ficus* 榕属

Ficus nervosa B. Heyne ex Roth 九丁榕

Trees, evergreen

20.5 (11.6~24.8); 12.7 (3.1~20.1); 26.8 (18.6~29.0); 20.6 (11.5~24.8); 187.0 (80.9~237.6); 0.0 (0.0~2.8); 1068 (654~1385) ❙ 1690 (988~3495); 749 (455~1847); 140 (28~897) ❙ 60.8 (-1.4~246.5); 1060 (654~1331); 283.2 (126.1~659.0)

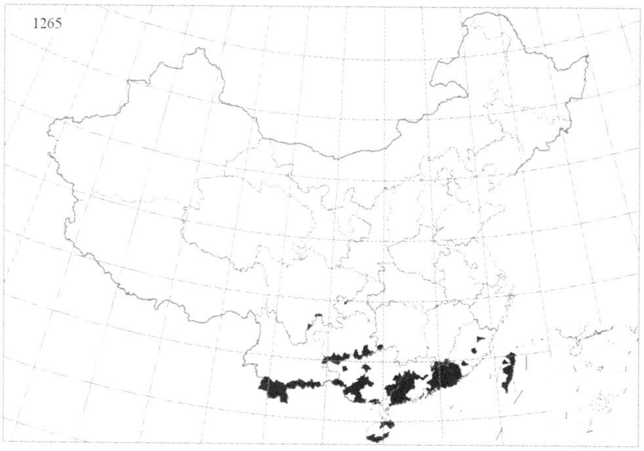

Moraceae 桑科 *Ficus* 榕属

Ficus oligodon Miq. 苹果榕

Small trees, evergreen

19.0 (8.6~24.8); 12.1 (1.3~20.1); 23.8 (15.0~28.8); 19.0 (8.7~24.8); 168.0 (52.6~237.6); 0.1 (0.0~8.8); 964 (574~1385) ❙ 1352 (868~1888); 684 (312~1044); 62 (37~154) ❙ 44.0 (-34.3~145.8); 932 (574~1360); 393.3 (149.6~659.0)

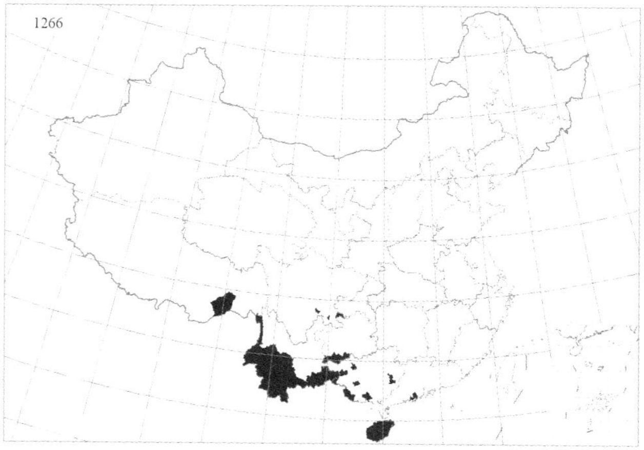

Moraceae 桑科 *Ficus* 榕属

Ficus orthoneura H. Lév. et Vaniot 直脉榕

Trees, evergreen

17.5 (14.2~22.2); 10.5 (6.2~15.1); 22.4 (19.1~28.3); 17.6 (14.2~22.3); 150.7 (110.5~207.5); 0.0 (0.0~0.0); 858 (724~1190) ❙ 1250 (946~1751); 694 (531~996); 52 (34~102) ❙ 45.9 (12.2~112.2); 848 (716~1190); 366.8 (189.5~659.0)

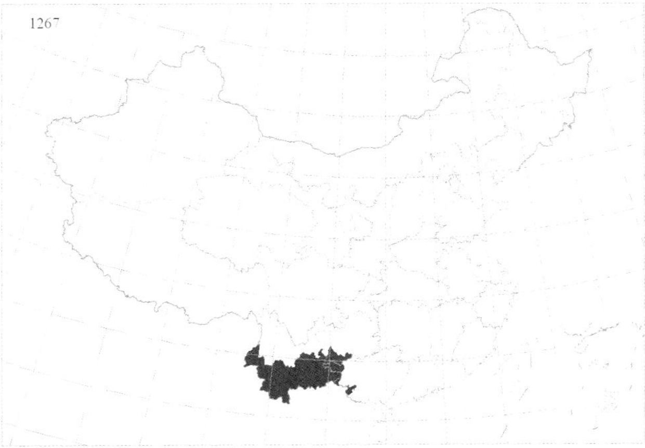

Moraceae 桑科 *Ficus* 榕属

Ficus ovatifolia S. S. Chang 卵叶榕

Shrubs or trees, evergreen

18.5 (17.3~21.0); 11.6 (9.9~14.2); 23.4 (22.4~25.6); 18.6 (17.3~21.0); 162.9 (147.1~191.8); 0.0 (0.0~0.0); 903 (833~1053) ❙ 1449 (1368~1522); 817 (756~860); 58 (52~68) ❙ 61.3 (44.5~80.6); 903 (833~1053); 464.5 (359.7~551.6)

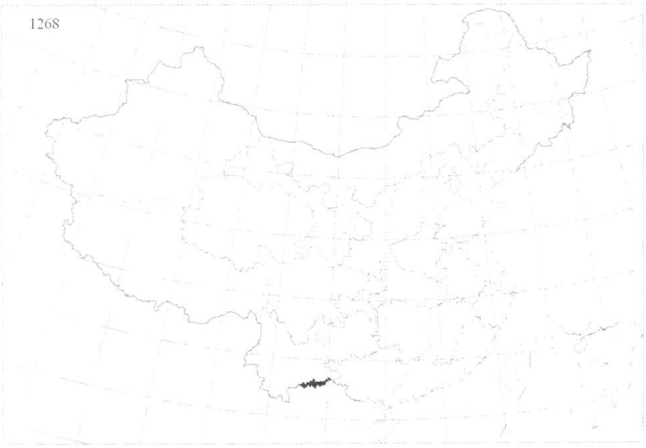

Moraceae 桑科 *Ficus* 榕属

Ficus pandurata Hance 琴叶榕

Small shrubs, evergreen

18.6 (14.2~24.8); 9.2 (2.1~20.1); 26.8 (19.1~30.1); 18.7 (14.2~24.8); 164.1 (110.5~237.6); 0.2 (0.0~4.9); 977 (724~1385) ❙ 1519 (868~2037); 610 (299~1033); 149 (37~259) ❙ 57.3 (-34.3~147.0); 960 (717~1360); 259.3 (95.3~659.0)

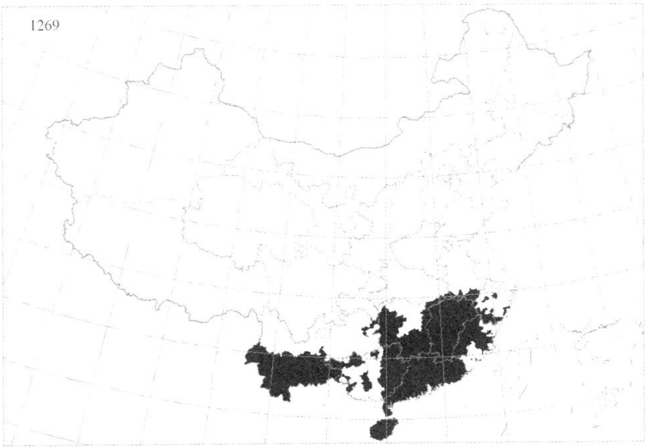

Moraceae 桑科 *Ficus* 榕属

Ficus pedunculosa Miq. 蔓榕

Shrubs, evergreen

20.3; 15.9; 23.6; 20.4; 184.4; 0.0; 993 ❙ 2853; 1618; 147 ❙ 190.4; 993; 330.3

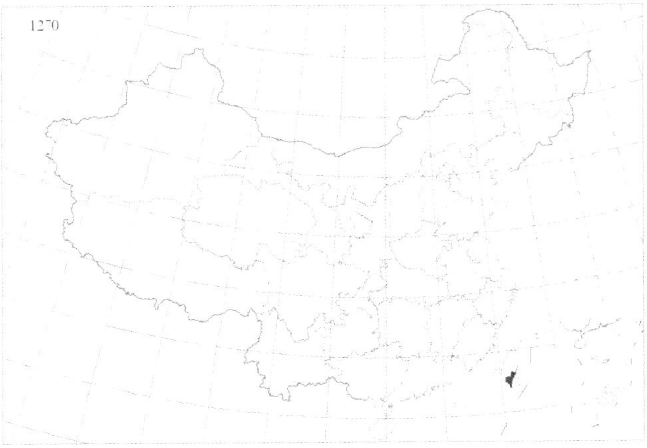

Moraceae 桑科 *Ficus* 榕属

Ficus pisocarpa Blume 豆果榕

Trees, evergreen

18.7 (16.3~21.0); 12.5 (9.1~15.1); 22.8 (21.4~25.6); 18.8 (16.4~21.0); 165.1 (136.4~191.8); 0.0 (0.0~0.0); 901 (797~1053) ❙ 1405 (1143~1751); 762 (637~996); 59 (48~77) ❙ 56.1 (21.7~88.8); 891 (797~1053); 458.0 (346.9~659.0)

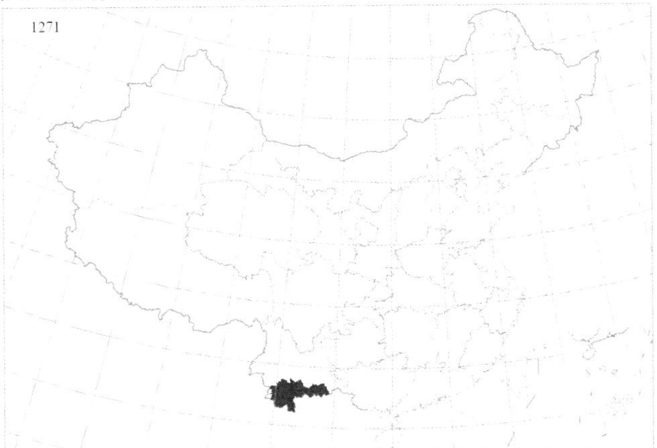

Moraceae 桑科 *Ficus* 榕属

Ficus polynervis S. S. Chang 多脉榕

Scandent shrubs, evergreen

18.2 (17.2~21.0); 10.9 (9.5~14.2); 23.3 (22.4~25.6); 18.2 (17.3~21.0); 158.4 (147.0~191.8); 0.0 (0.0~0.0); 887 (833~1053) ❙ 1414 (1254~1522); 789 (699~860); 57 (48~68) ❙ 60.1 (44.5~80.6); 887 (833~1053); 430.5 (336.3~551.6)

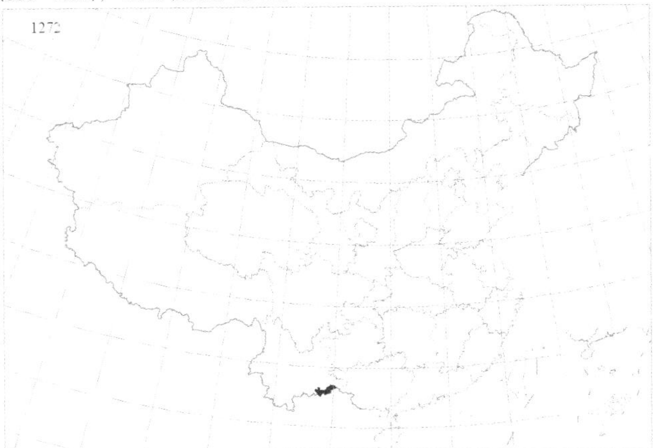

Moraceae 桑科 *Ficus* 榕属
Ficus praetermissa Corner 钩毛榕
Shrubs, evergreen

18.9 (16.9~21.0); 12.8 (9.9~15.1); 22.8 (21.4~25.6); 18.9 (16.9~21.0); 167.3 (142.7~191.8); 0.0 (0.0~0.0); 908 (813~1053) ❙ 1447 (1178~1751); 774 (646~996); 58 (41~77) ❙ 59.6 (36.8~88.8); 895 (813~1053); 477.0 (354.7~659.0)

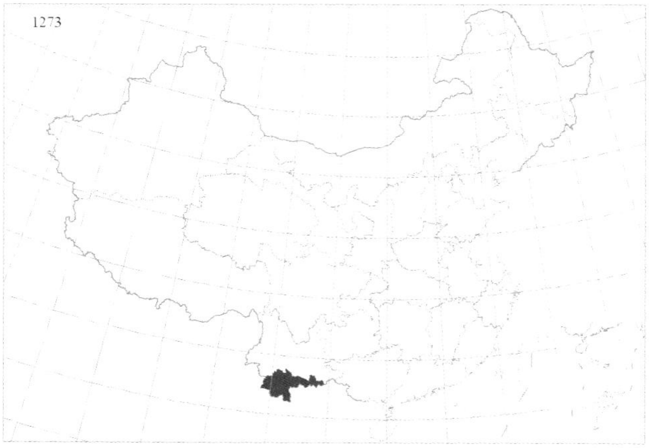

Moraceae 桑科 *Ficus* 榕属
Ficus prostrata (Wall. ex Miq.) Miq. 平枝榕
Small trees, evergreen

8.6; 1.3; 15.0; 8.7; 52.6; 8.8; 574 ❙ 1041; 585; 45 ❙ 81.5; 574; 352.3

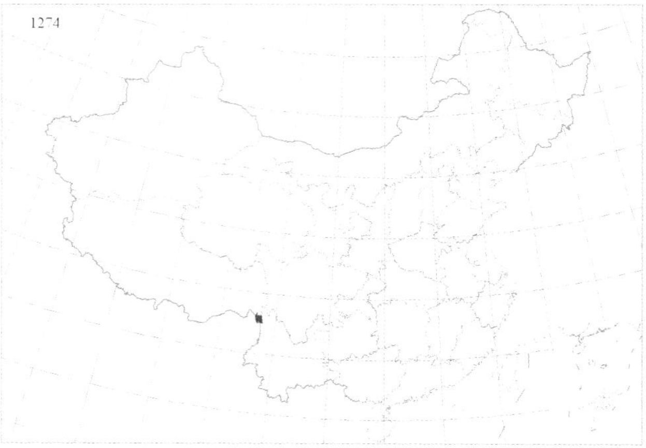

Moraceae 桑科 *Ficus* 榕属
Ficus pubigera (Wall. ex Miq.) Kurz 褐叶榕
Scandent shrubs, evergreen

19.9 (13.2~24.8); 12.9 (5.4~20.1); 25.0 (18.4~28.9); 20.0 (13.3~24.8); 179.7 (99.4~237.6); 0.0 (0.0~0.0); 1020 (695~1385) ❙ 1429 (868~2435); 704 (312~1428); 71 (41~145) ❙ 43.2 (-34.3~145.8); 986 (695~1360); 382.6 (189.5~659.0)

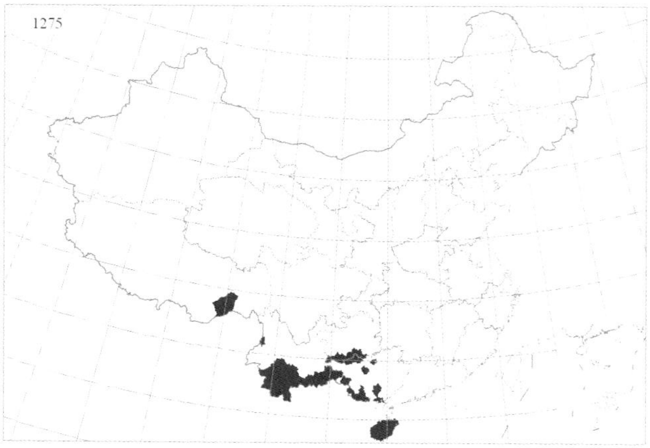

Moraceae 桑科 *Ficus* 榕属
Ficus pubigera var. anserina Corner 鳞果褐叶榕
Tall shrubs, evergreen

19.1; 13.0; 23.3; 19.1; 169.7; 0.0; 917 ❙ 1428; 835; 56 ❙ 55.7; 917; 506.2

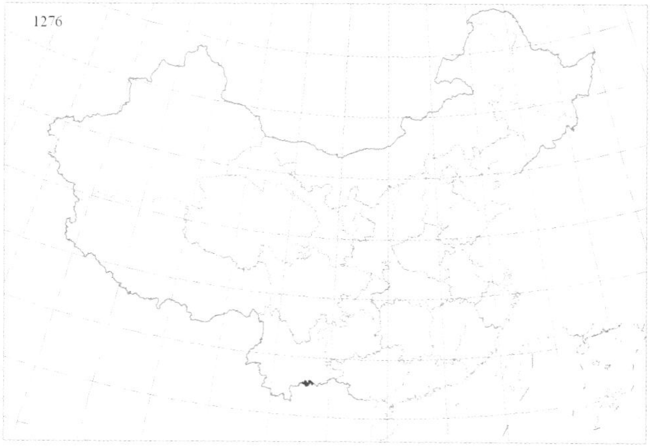

Moraceae 桑科 *Ficus* 榕属
Ficus pubigera var. maliformis (King) Corner 大果褐叶榕
Scandent shrubs, evergreen

17.0; 10.4; 21.8; 17.1; 144.7; 0.0; 837 ❙ 1570; 918; 57 ❙ 91.9; 837; 501.6

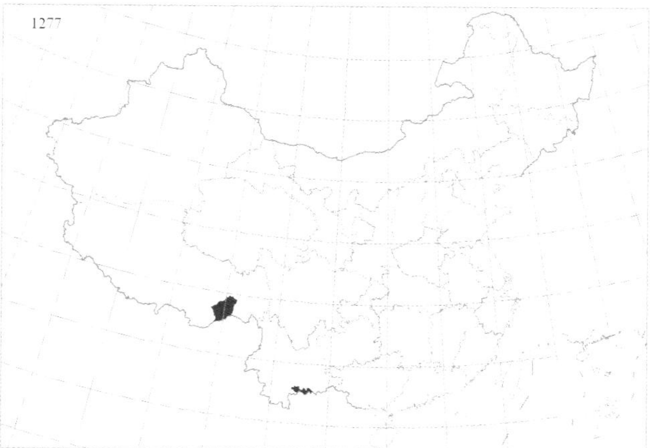

Moraceae 桑科 *Ficus* 榕属
Ficus pubigera var. reticulata S. S. Chang 网果褐叶榕
Scandent shrubs, evergreen

20.0; 14.6; 23.2; 20.1; 180.8; 0.0; 963 ❙ 1523; 681; 67 ❙ 58.4; 942; 581.0

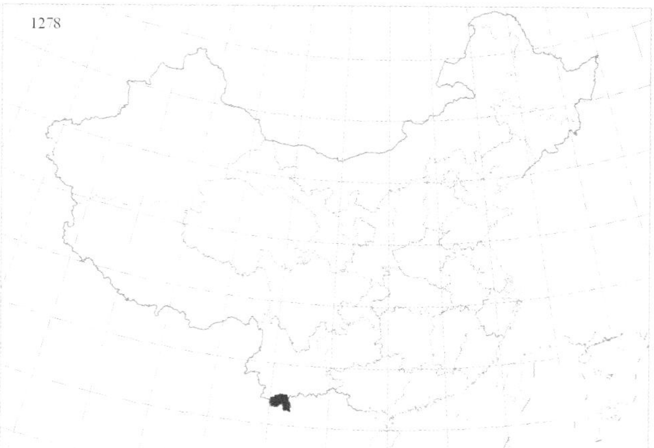

Moraceae 桑科 *Ficus* 榕属
Ficus pubilimba Merr. 球果山榕
Trees, evergreen

23.7 (19.1~24.8); 18.0 (11.0~20.1); 27.8 (25.0~28.8); 23.8 (19.1~24.8); 225.3 (169.6~237.6); 0.0 (0.0~0.0); 1291 (945~1385) ❙ 1498 (868~1854); 548 (312~820); 73 (41~115) ❙ 16.8 (-34.3~52.2); 1179 (868~1360); 436.9 (248.2~620.0)

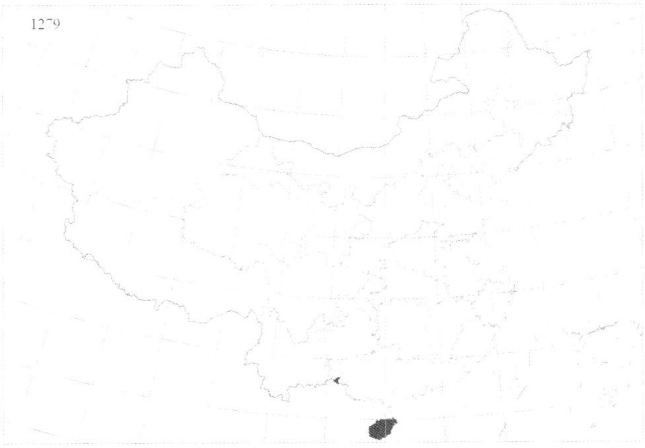

Moraceae 桑科 *Ficus* 榕属
Ficus pubinervis Blume 绿岛榕
Shrubs, evergreen

20.2 (16.2~24.1); 14.5 (11.0~18.9); 25.0 (20.1~28.2); 20.2 (16.2~24.1); 182.9 (134.2~229.7); 0.0 (0.0~0.0); 1019 (773~1322) ❙ 2402 (1498~3495); 1106 (750~1847); 255 (47~897) ❙ 146.2 (25.2~246.5); 983 (773~1144); 245.0 (119.4~363.7)

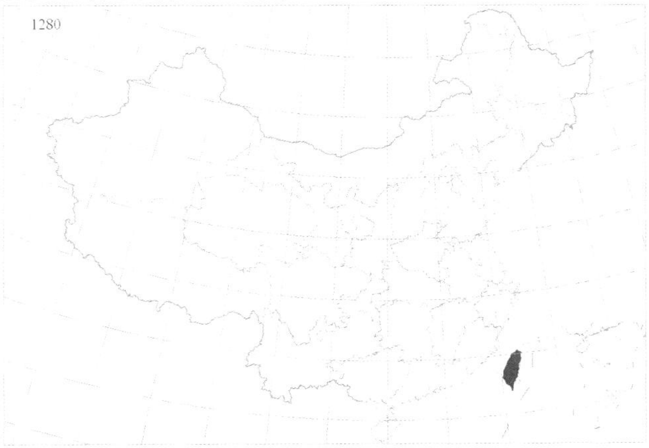

Moraceae 桑科 *Ficus* 榕属
Ficus pumila L. 薜荔
Shrubs, evergreen

17.3 (2.2~24.6); 6.7 (-6.6~19.5); 27.0 (9.9~30.1); 17.3 (3.8~24.7); 149.3 (15.5~235.9); 1.5 (0.0~49.1); 926 (413~1372) ❙ 1375 (626~3495); 569 (299~1847); 134 (16~897) ❙ 49.0 (-23.1~246.5); 914 (413~1331); 214.2 (95.3~620.0)

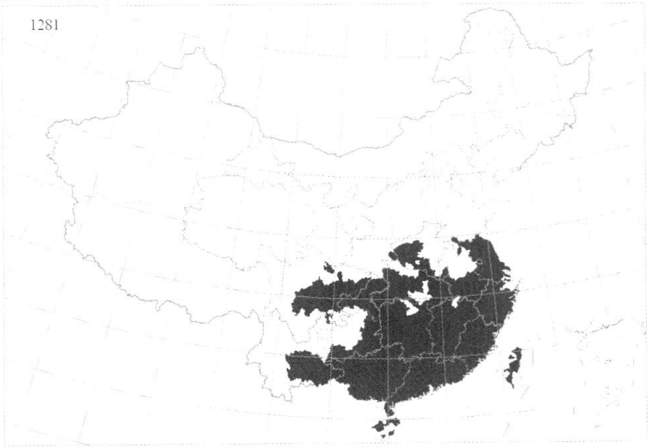

Moraceae 桑科 *Ficus* 榕属
Ficus pumila var. awkeotsang (Makino) Corner 爱玉子
Shrubs, evergreen

18.7; 9.8; 27.5; 18.7; 164.5; 0.0; 959 ❙ 1485; 522; 231 ❙ 54.8; 959; 219.3

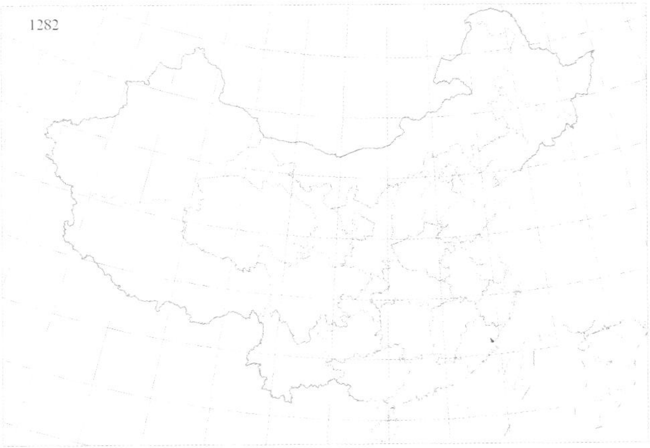

Moraceae 桑科 *Ficus* 榕属
Ficus pyriformis Hook. et Arn. 舶梨榕
Shrubs, evergreen

19.6 (13.6~24.8); 11.6 (3.8~20.1); 25.8 (18.8~29.8); 19.6 (13.6~24.8); 175.5 (103.6~237.6); 0.0 (0.0~1.5); 1010 (706~1385) ❙ 1489 (868~2435); 690 (312~1428); 110 (29~221) ❙ 48.3 (-34.3~112.2); 997 (690~1326); 288.0 (95.3~659.0)

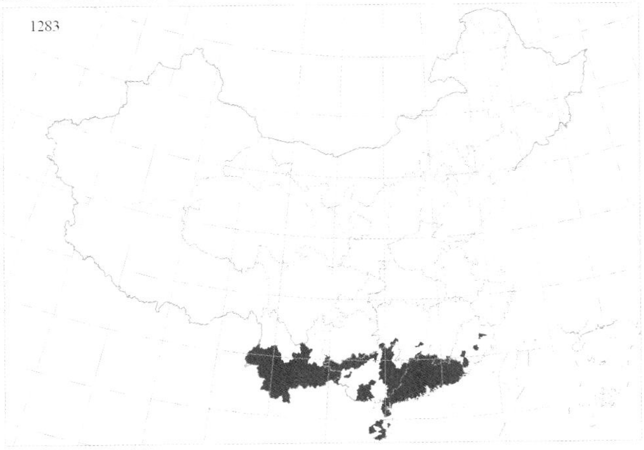

Moraceae 桑科 *Ficus* 榕属
Ficus racemosa L. 聚果榕
Trees, evergreen

17.2 (8.6~21.6); 10.3 (1.3~15.1); 22.1 (15.0~28.5); 17.3 (8.7~21.7); 147.5 (52.6~199.8); 0.1 (0.0~8.8); 848 (574~1153) ❙ 1223 (797~1751); 679 (411~996); 52 (23~173) ❙ 44.6 (-14.2~112.2); 833 (574~1153); 36[illegible].3 (189.5~659.0)

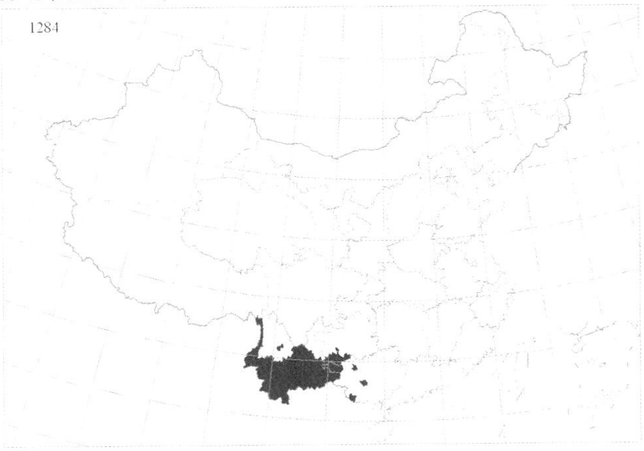

Moraceae 桑科 *Ficus* 榕属
Ficus racemosa var. miquelli (King) Corner 柔毛聚果榕
Trees, evergreen

18.3 (16.9~21.0); 11.7 (9.9~14.2); 22.8 (21.4~25.6); 18.3 (16.9~21.0); 159.9 (142.7~191.8); 0.0 (0.0~0.0); 885 (813~1053) ❙ 1295 (958~1547); 728 (531~911); 54 (44~68) ❙ 46.4 (13.3~80.6); 880 (813~1053); 440.9 (311.9~551.6)

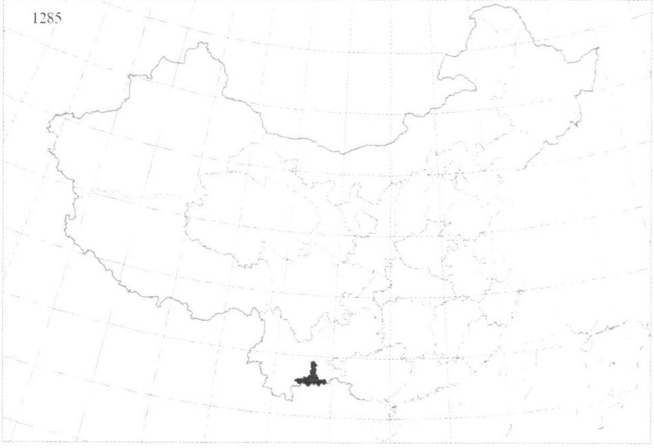

Moraceae 桑科 *Ficus* 榕属
Ficus religiosa L. 菩提树
Trees, evergreen

20.3 (16.3~22.7); 12.8 (7.1~15.2); 26.0 (20.3~28.8); 20.3 (16.4~22.7); 183.9 (136.6~212.3); 0.0 (0.0~0.0); 1040 (789~1214) ❙ 1592 (1232~2014); 741 (425~959); 105 (42~185) ❙ 54.7 (14.0~112.2); 1030 (780~1214); 330.1 (116.6~659.0)

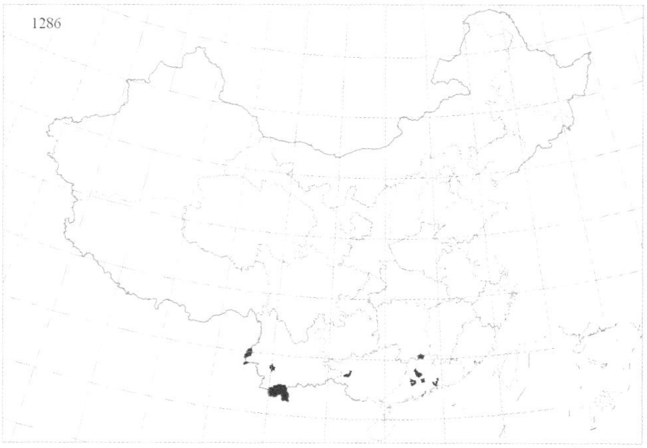

Moraceae 桑科 *Ficus* 榕属
Ficus ruficaulis Merr. 红茎榕
Small trees, deciduous

20.3; 15.9; 23.6; 20.4; 184.4; 0.0; 993 ❙ 2853; 1618; 147 ❙ 190.4; 993; 330.3

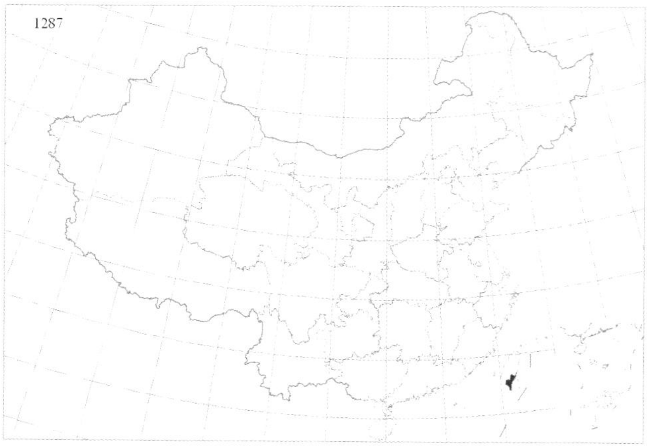

Moraceae 桑科 *Ficus* 榕属
Ficus rumphii Blume 心叶榕
Trees, evergreen

15.3; 8.6; 19.9; 15.3; 123.5; 0.0; 756 ❙ 1268; 696; 60 ❙ 68.9; 753; 446.1

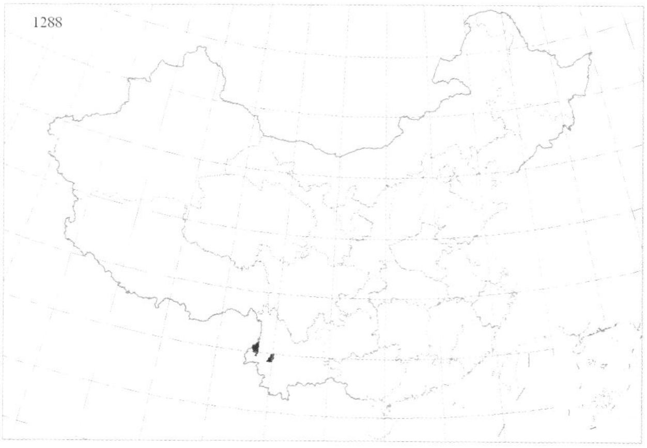

Moraceae 桑科 *Ficus* 榕属
Ficus ruyuanensis S. S. Chang 乳源榕
Shrubs or trees, evergreen

17.7; 7.5; 26.5; 17.7; 152.6; 0.0; 912 ❙ 1530; 597; 162 ❙ 67.7; 912; 229.2

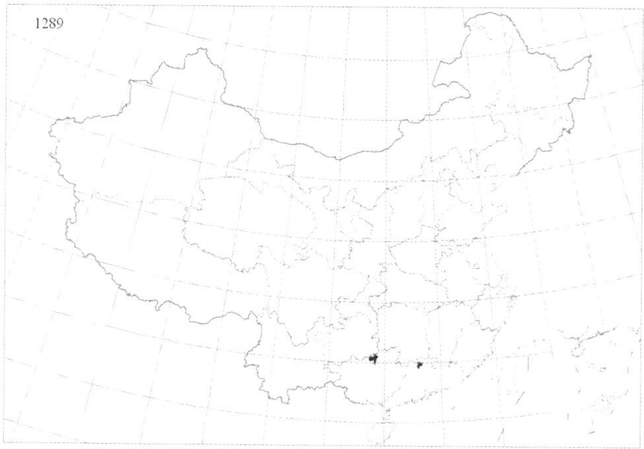

Moraceae 桑科 *Ficus* 榕属
Ficus sagittata Vahl 羊乳榕
Trees, evergreen

22.3 (19.5~24.8); 14.7 (10.6~20.1); 28.3 (23.6~29.2); 22.4 (19.6~24.8); 208.6 (175.1~237.6); 0.0 (0.0~0.0); 1195 (989~1385) ❙ 1649 (868~2435); 721 (312~1428); 117 (41~197) ❙ 39.3 (-34.3~99.5); 1167 (868~1360); 271.0 (95.3~659.0)

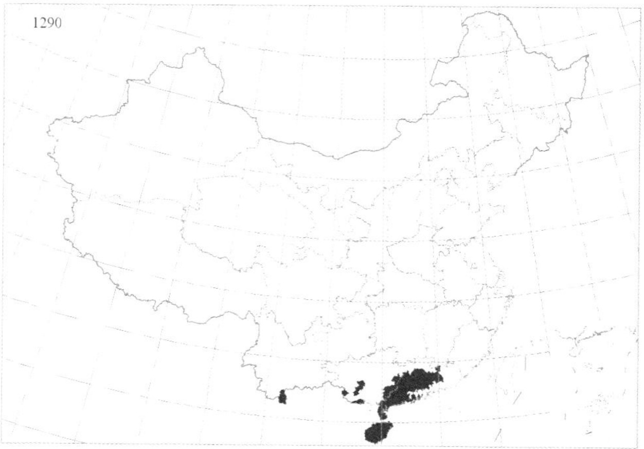

Moraceae 桑科 *Ficus* 榕属
Ficus sarmentosa Buch.-Ham. ex Sm. 匍茎榕
Scandent shrubss or vines, evergreen

16.0 (−1.8~24.3); 6.0 (−10.2~19.1); 24.9 (6.1~29.4); 16.0 (1.8~24.3); 134.7 (1.9~231.9); 2.4 (0.0~82.8); 851 (329~1338) ❙ 1301 (320~2993); 569 (217~1847); 108 (10~547) ❙ 52.4 (−22.2~246.5); 841 (320~1326); 246.0 (61.6~659.0)

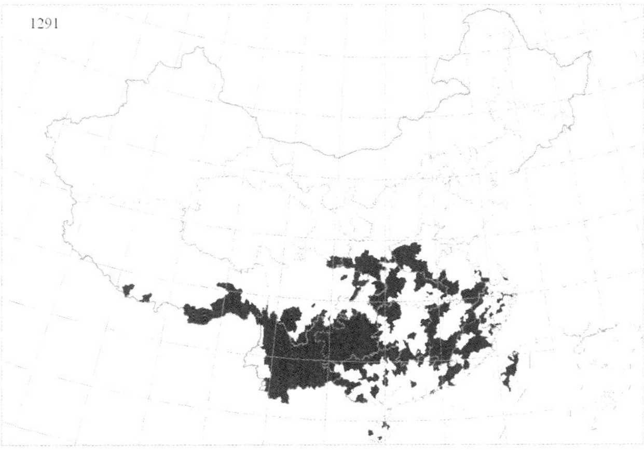

Moraceae 桑科 *Ficus* 榕属
Ficus sarmentosa var. duclouxii (Lév. et Vaniot) Corner 大果藤爬榕
Scandent shrubs, evergreen

15.3 (6.1~19.8); 6.4 (−1.5~13.3); 22.7 (12.7~26.6); 15.4 (6.3~19.8); 126.8 (35.2~177.9); 2.3 (0.0~21.8); 804 (496~1010) ❙ 1168 (801~1334); 612 (452~741); 55 (12~106) ❙ 47.0 (16.9~70.6); 804 (496~1010); 253.7 (181.2~452.0)

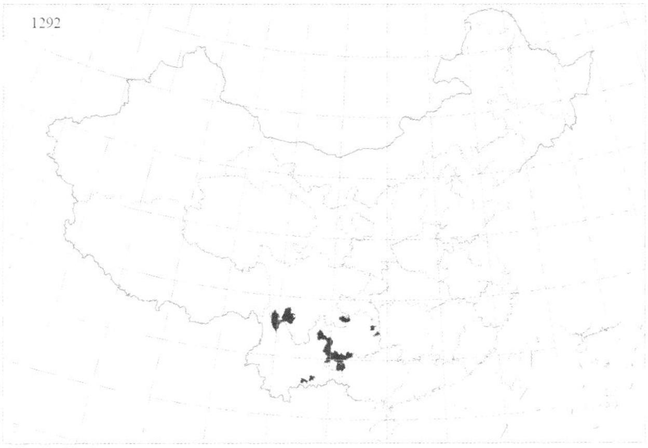

Moraceae 桑科 *Ficus* 榕属
Ficus sarmentosa var. henryi (King ex Oliv.) Corner 珍珠莲
Scandent shrubs, evergreen

17.0 (2.2~23.6); 6.5 (−6.6~16.7); 26.5 (9.9~29.9); 17.0 (3.8~23.6); 145.5 (15.5~223.5); 1.5 (0.0~49.1); 907 (413~1281) ❙ 1367 (626~2037); 570 (316~1033); 123 (14~259) ❙ 51.4 (−22.2~147.0); 900 (413~1273); 212.9 (95.3~551.6)

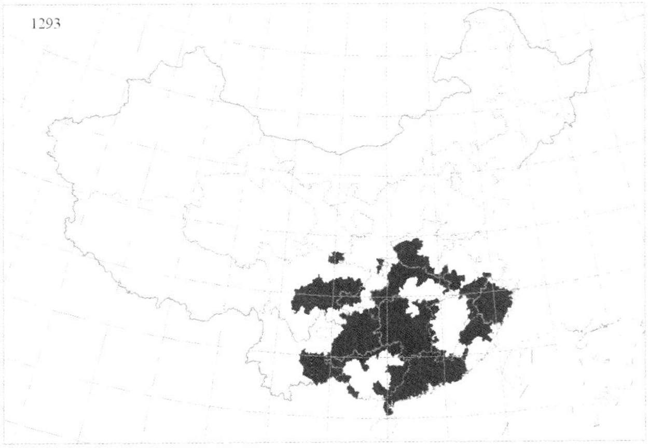

Moraceae 桑科 *Ficus* 榕属
Ficus sarmentosa var. impressa (Champ. ex Benth.) Corner 爬藤榕
Scandent shrubs, evergreen

17.1 (2.2~23.3); 6.5 (−6.6~16.0); 26.7 (9.9~29.4); 17.2 (3.8~23.3); 148.0 (15.5~220.1); 2.1 (0.0~49.1); 922 (413~1264) ❙ 1374 (626~2435); 589 (316~1428); 122 (14~259) ❙ 49.8 (−22.2~147.0); 917 (413~1264); 216.7 (108.2~491.5)

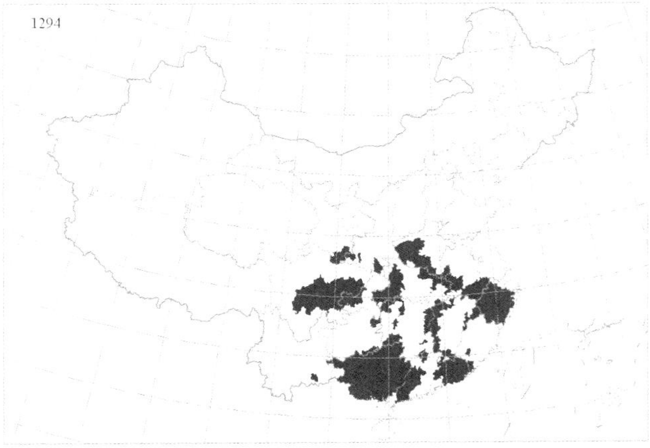

Moraceae 桑科 *Ficus* 榕属
Ficus sarmentosa var. lacrymans (Lév.) Corner 尾尖爬藤榕
Scandent shrubs, evergreen

17.0 (2.2~23.2); 6.9 (−6.6~15.8); 26.1 (9.9~29.8); 17.1 (3.8~23.2); 145.8 (15.5~218.9); 1.2 (0.0~49.1); 904 (413~1257) ❙ 1335 (922~2035); 592 (442~969); 99 (16~259) ❙ 49.2 (6.6~125.9); 903 (413~1257); 206.2 (108.2~371.7)

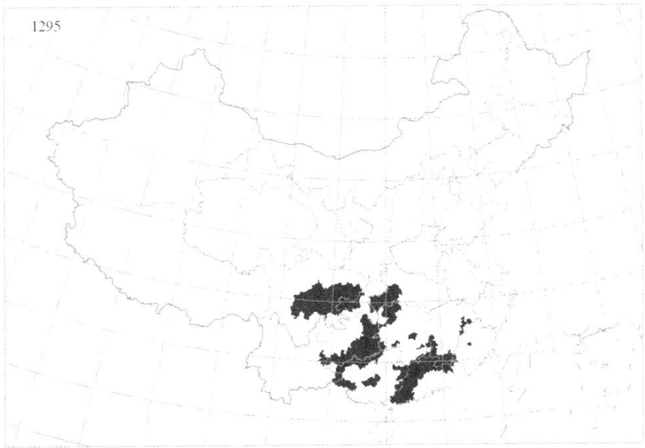

Moraceae 桑科 *Ficus* 榕属
Ficus sarmentosa var. luducca Corner 长柄爬藤榕
Vines, evergreen

17.4 (12.0~22.2); 7.7 (0.6~14.1); 26.0 (21.4~29.8); 17.5 (12.1~22.3); 150.8 (94.7~207.2); 1.4 (0.0~9.7); 917 (697~1186) ❙ 1520 (707~2014); 631 (362~954); 129 (15~216) ❙ 66.4 (−0.2~107.2); 915 (697~1186); 251.4 (166.9~551.3)

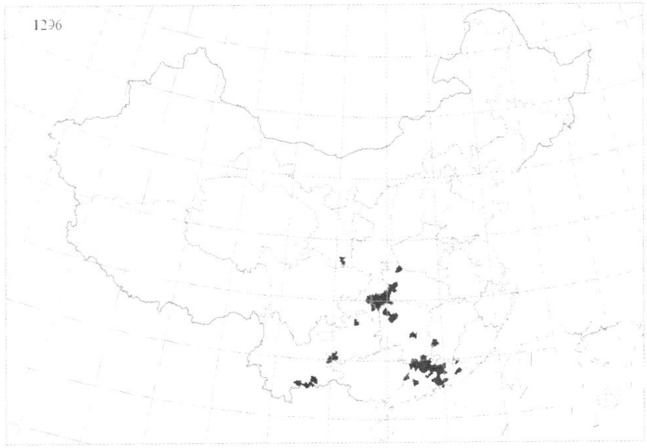

Moraceae 桑科 *Ficus* 榕属

Ficus sarmentosa var. nipponica (Franch. et Sav.) Corner 白背爬藤榕

Scandent shrubs, evergreen

16.2 (1.8~22.3); 5.4 (-7.1~14.9); 26.3 (9.9~29.3); 16.3 (3.8~22.3); 136.7 (15.7~208.2); 2.2 (0.0~53.3); 870 (404~1185) ❙ 1495 (583~2035); 591 (348~969); 158 (13~242) ❙ 72.3 (19.5~147.0); 867 (404~1185); 253.1 (149.5~551.3)

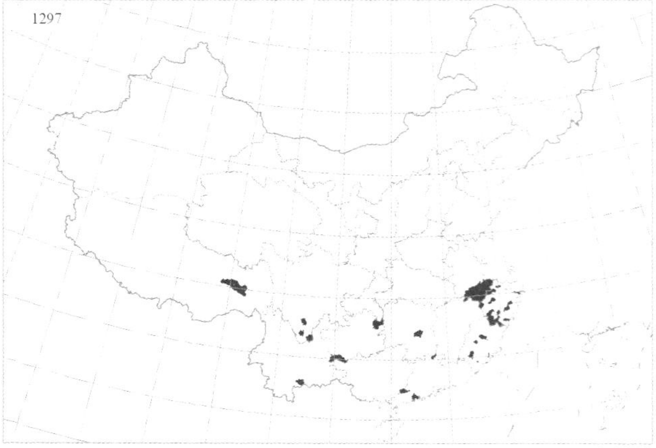

Moraceae 桑科 *Ficus* 榕属

Ficus semicordata Buch.-Ham. ex Sm. 鸡嗉子榕

Small trees, evergreen

17.6 (8.6~22.3); 10.5 (1.3~15.1); 22.6 (15.0~28.9); 17.6 (8.7~22.3); 151.2 (52.6~208.1); 0.1 (0.0~8.8); 869 (574~1202) ❙ 1251 (797~1900); 690 (411~1056); 57 (23~177) ❙ 44.7 (-14.2~145.8); 855 (574~1202); 357.2 (189.5~659.0)

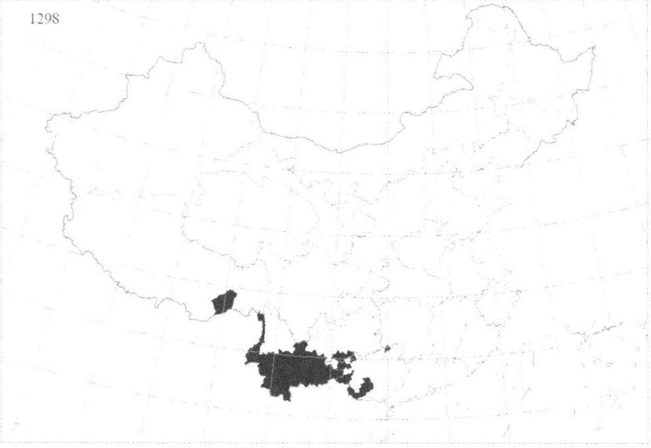

Moraceae 桑科 *Ficus* 榕属

Ficus septica Burm. f. 棱果榕

Trees, evergreen

19.6; 14.3; 24.0; 19.6; 175.6; 0.0; 956 ❙ 2889; 1280; 311 ❙ 205.0; 956; 297.0

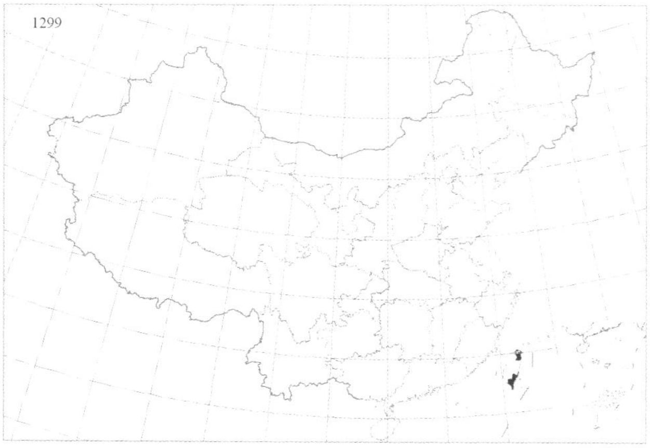

Moraceae 桑科 *Ficus* 榕属

Ficus simplicissima Lour. 极简榕

Shrubs, evergreen

20.6 (15.5~24.8); 13.7 (4.8~20.1); 25.9 (21.6~28.7); 20.8 (15.6~24.8); 189.3 (127.2~237.6); 0.0 (0.0~0.2); 1073 (815~1385) ❙ 1552 (1049~2037); 714 (455~1033); 86 (41~180) ❙ 47.1 (1.5~93.2); 1037 (783~1216); 392.1 (139.6~659.0)

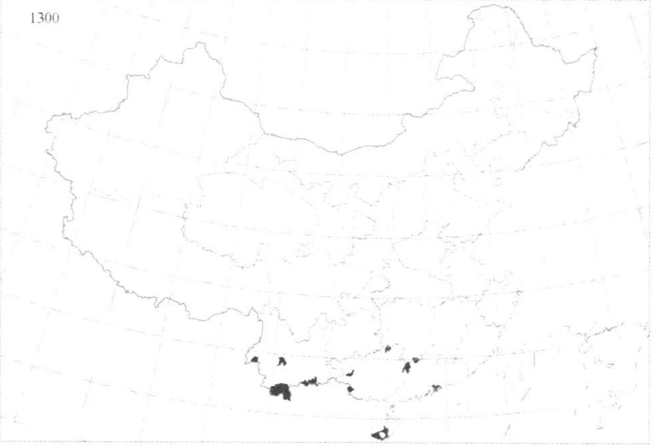

Moraceae 桑科 *Ficus* 榕属

Ficus sinociliata Z. K. Zhou et M. G. Gilbert 缘毛榕

Shrubs, evergreen

20.8; 13.1; 26.6; 20.9; 190.5; 0.0; 1070 ❙ 1577; 691; 130 ❙ 48.2; 1069; 381.9

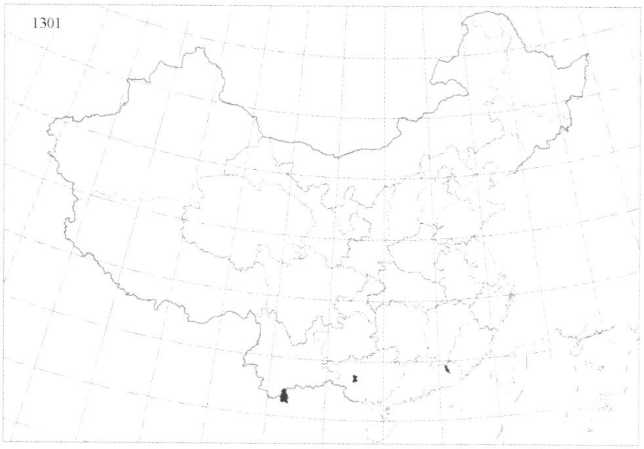

Moraceae 桑科 *Ficus* 榕属

Ficus squamosa Roxb. 肉托榕

Shrubs, evergreen

18.5 (16.3~20.5); 12.5 (10.2~15.1); 22.3 (20.3~23.7); 18.5 (16.4~20.5); 162.1 (136.6~185.9); 0.0 (0.0~0.0); 885 (789~992) ❙ 1358 (1028~1751); 739 (566~996); 55 (42~77) ❙ 53.4 (12.5~88.8); 863 (780~986); 445.2 (339.5~659.0)

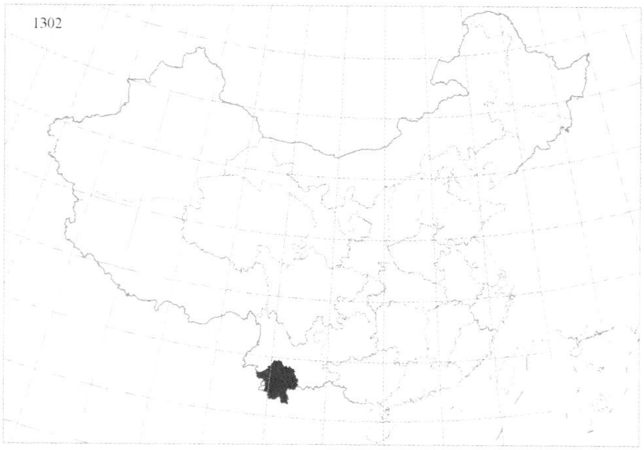

Moraceae 桑科 *Ficus* 榕属

Ficus stenophylla Hemsl. 竹叶榕

Small shrubs, evergreen

18.4 (12.0~24.8); 8.9 (1.1~20.1); 26.6 (19.4~30.1); 18.4 (12.0~24.8); 161.4 (92.3~237.6); 0.2 (0.0~8.5); 962 (690~1385) ❙ 1502 (868~2435); 622 (312~1428); 137 (19~259) ❙ 57.2 (-34.3~147.0); 953 (690~1326); 250.6 (111.7~659.0)

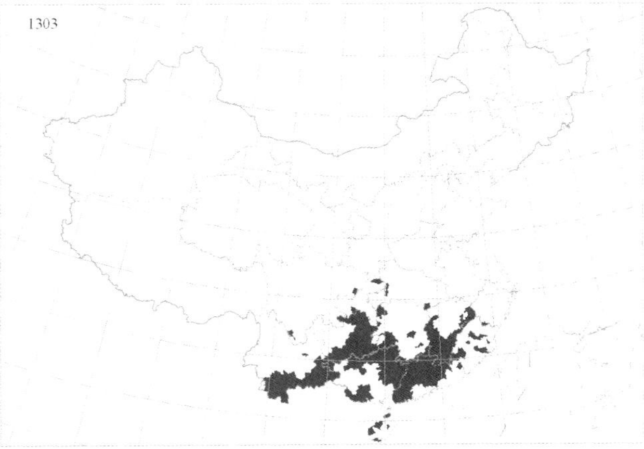

Moraceae 桑科 *Ficus* 榕属

Ficus stricta (Miq.) Miq. 劲直榕

Trees, evergreen

13.0 (14.4~20.5); 11.8 (7.5~15.1); 22.0 (19.1~23.7); 18.1 (14.4~20.5); 156.8 (113.1~185.9); 0.0 (0.0~0.0); 867 (727~992) ❙ 1391 (1049~1751); 769 (594~996); 54 (41~77) ❙ 60.7 (21.7~112.2); 853 (727~986); 460.9 (346.6~659.0)

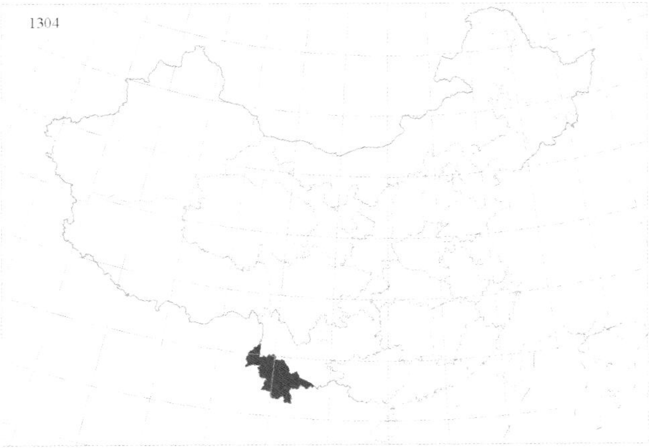

Moraceae 桑科 *Ficus* 榕属

Ficus subincisa Buch.-Ham. ex Sm. 棒果榕

Shrubs, evergreen

17.1 (5.3~22.2); 10.3 (-2.4~15.1); 21.8 (12.2~28.5); 17.1 (5.7~22.3); 146.0 (30.5~207.5); 0.4 (0.0~26.2); 840 (479~1190) ❙ 1223 (911~1751); 683 (524~996); 52 (31~173) ❙ 46.2 (6.5~112.2); 827 (479~1190); 373.0 (234.2~659.0)

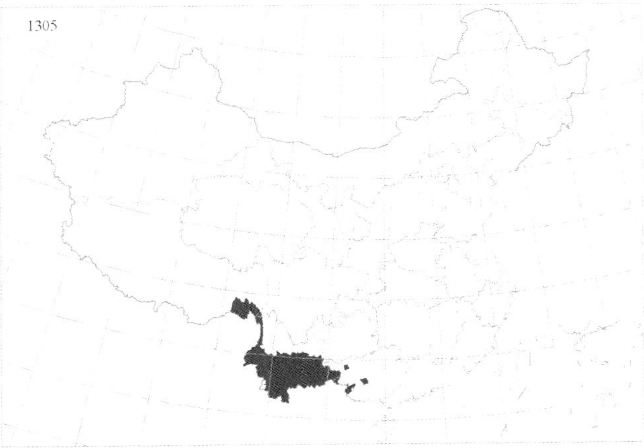

Moraceae 桑科 *Ficus* 榕属

Ficus subpisocarpa Gagnep. 笔管榕

Trees, deciduous

20.6 (12.4~24.8); 12.6 (4.7~20.1); 27.2 (18.6~29.8); 20.7 (12.4~24.8); 188.1 (89.1~237.6); 0.0 (0.0~0.3); 1086 (675~1385) ❙ 1622 (797~3495); 698 (312~1428); 139 (23~897) ❙ 50.8 (-34.3~246.5); 1067 (675~1360); 249.0 (95.3~572.9)

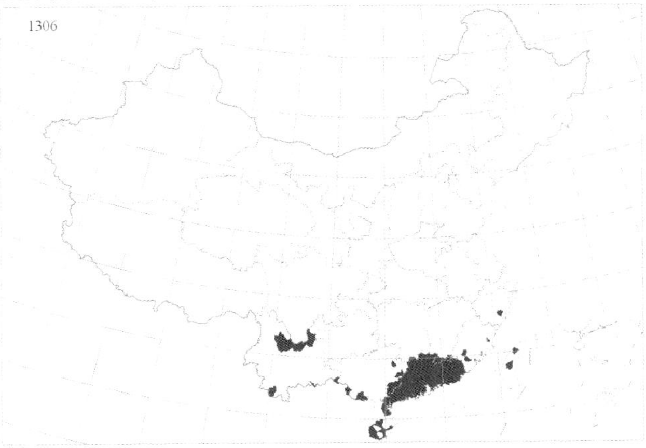

Moraceae 桑科 *Ficus* 榕属

Ficus subulata Blume 假斜叶榕

Scandent shrubs, evergreen

20.5 (13.3~24.8); 13.0 (5.4~20.1); 26.1 (19.1~29.2); 20.5 (13.4~24.8); 186.3 (100.4~237.6); 0.0 (0.0~0.0); 1060 (706~1385) ❙ 1523 (868~2037); 735 (312~1033); 93 (41~197) ❙ 45.9 (-34.3~145.8); 1042 (706~1331); 317.2 (95.3~659.0)

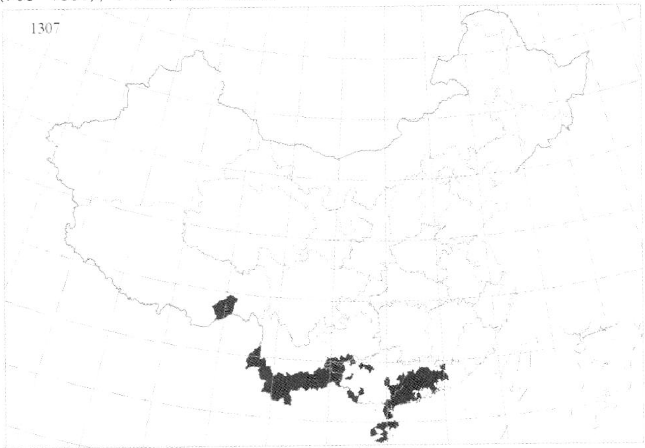

Moraceae 桑科 *Ficus* 榕属

Ficus tannoensis Hayata 滨榕

Scandent shrubs, evergreen

18.9 (16.4~22.2); 13.7 (11.5~17.7); 23.2 (20.3~25.7); 19.0 (16.5~22.2); 167.7 (137.8~206.7); 0.0 (0.0~0.0); 920 (783~1123) ❙ 2654 (2203~2992); 1178 (750~1847); 268 (107~402) ❙ 192.3 (125.7~235.6); 920 (783~1123); 298.1 (260.5~363.7)

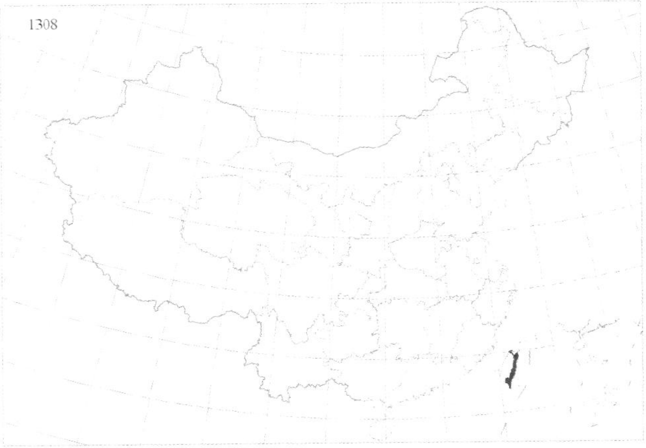

Moraceae 桑科 *Ficus* 榕属
Ficus tikoua Bureau 地果
Vines, evergreen

15.6 (2.2~22.7); 6.1 (-6.6~15.1); 23.9 (9.5~29.4); 15.6 (3.8~22.8); 129.4 (15.5~213.1); 1.8 (0.0~49.1); 823 (411~1221) ❙ 1175 (677~2435); 574 (336~1428); 66 (11~209) ❙ 44.0 (-14.2~125.9); 815 (411~1221); 239.9 (108.2~659.0)

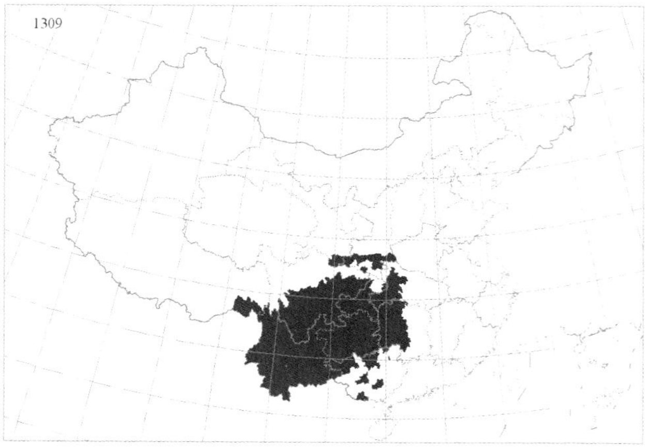

Moraceae 桑科 *Ficus* 榕属
Ficus tinctoria G. Forst. 梁料榕
Small trees, evergreen

18.2 (11.2~24.8); 10.9 (2.7~20.1); 23.7 (16.7~28.9); 18.3 (11.3~24.8); 159.5 (75.8~237.6); 0.1 (0.0~3.7); 930 (642~1385) ❙ 1322 (797~2992); 678 (312~1847); 70 (22~194) ❙ 43.6 (-34.3~214.4); 906 (642~1360); 331.0 (154.2~659.0)

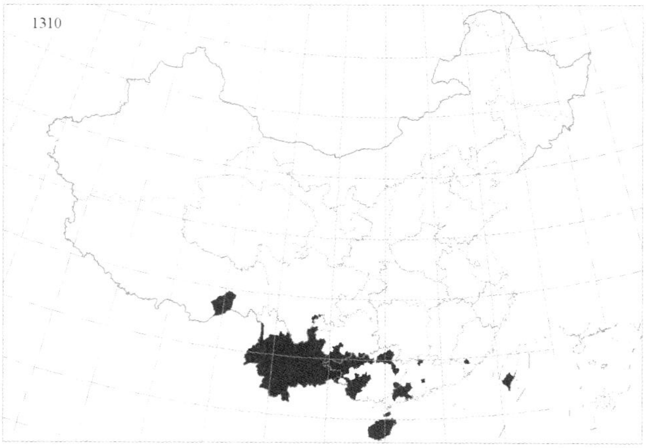

Moraceae 桑科 *Ficus* 榕属
Ficus tinctoria subsp. gibbosa (Blume) Corner 斜叶榕
Trees or shrubs, evergreen

20.3 (11.6~23.6); 11.6 (2.9~16.7); 27.3 (18.6~29.2); 20.4 (11.5~23.6); 184.7 (80.9~223.5); 0.1 (0.0~3.4); 1069 (654~1281) ❙ 1571 (988~2435); 733 (497~1428); 127 (28~196) ❙ 48.3 (-6.9~145.8); 1068 (654~1273); 251.1 (95.3~659.0)

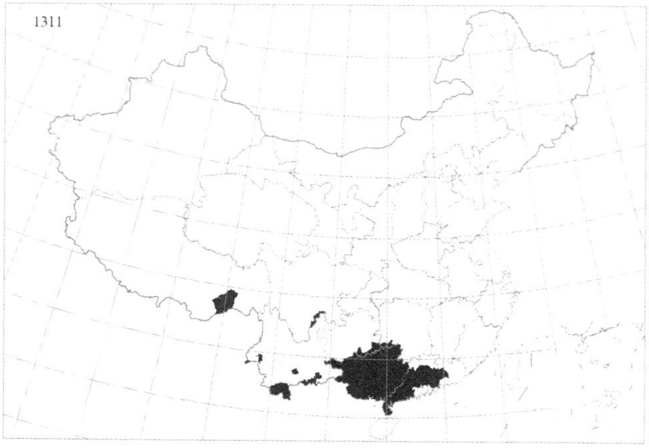

Moraceae 桑科 *Ficus* 榕属
Ficus tinctoria subsp. swinhoei (King) Corner 匍匐斜叶榕
Shrubs or small trees

20.3; 15.9; 23.6; 20.4; 184.4; 0.0; 993 ❙ 2853; 1618; 147 ❙ 190.4; 993; 330.3

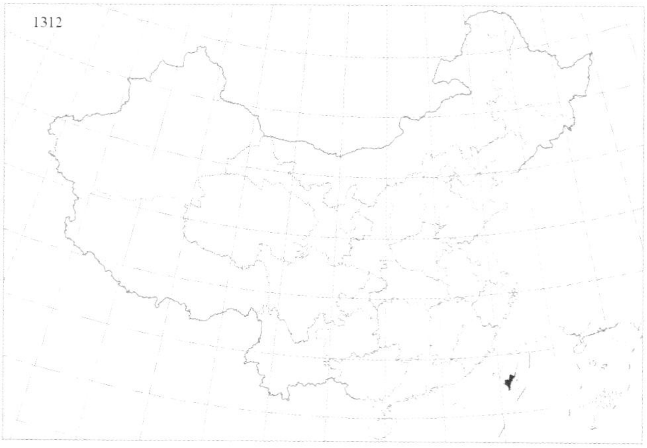

Moraceae 桑科 *Ficus* 榕属
Ficus trichocarpa var. obtusa (Hassk.) Corner 钝叶毛果榕
Scandent shrubs, evergreen

18.5; 14.1; 21.7; 18.5; 162.1; 0.0; 863 ❙ 2713; 1388; 188 ❙ 214.4; 863; 363.7

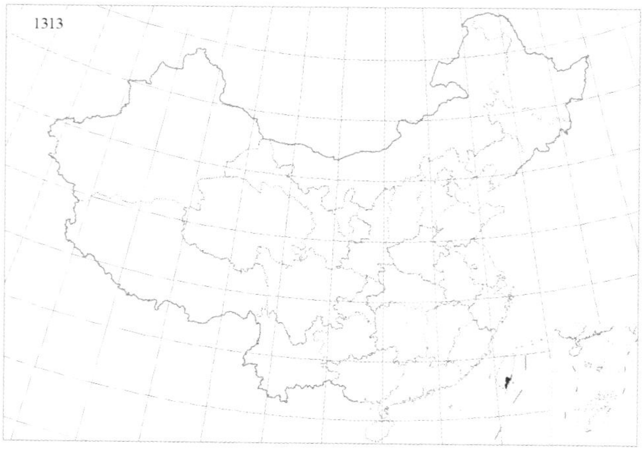

Moraceae 桑科 *Ficus* 榕属
Ficus trivia Corner 楔叶榕
Shrubs or small trees, evergreen

18.4 (15.0~22.2); 9.3 (4.6~14.2); 26.0 (21.4~29.8); 18.5 (15.1~22.3); 161.7 (121.2~207.5); 0.0 (0.0~0.4); 945 (795~1190) ❙ 1462 (1007~1852); 626 (494~860); 128 (44~197) ❙ 54.8 (15.4~93.2); 943 (795~1190); 272.4 (174.9~551.6)

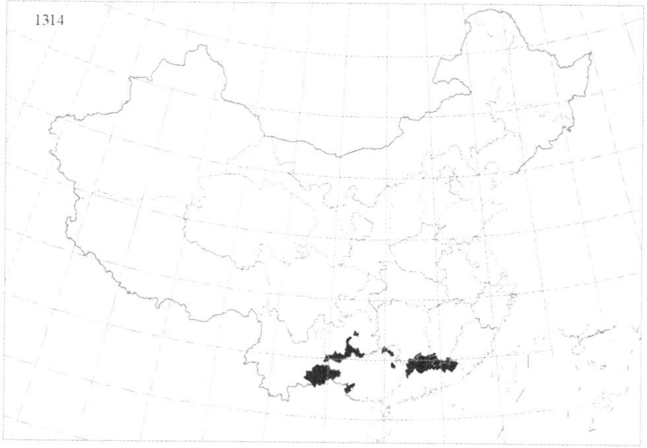

Moraceae 桑科 *Ficus* 榕属
Ficus trivia var. laevigata S. S. Chang 光叶楔叶榕
Shrubs or small trees, evergreen

19.1; 9.7; 26.6; 19.1; 169.3; 0.0; 995 ❙ 1390; 680; 104 ❙ 41.1; 995; 251.5

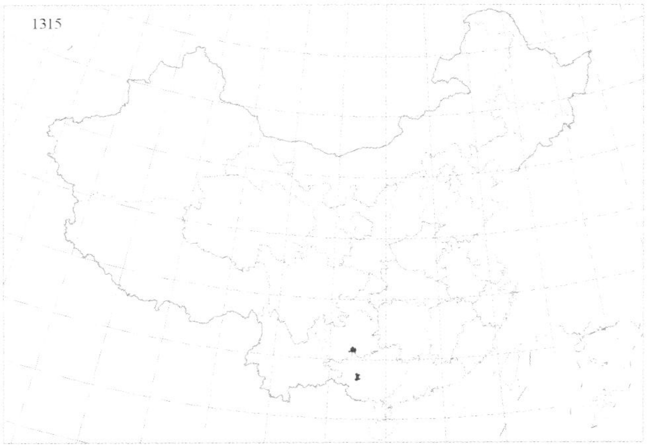

Moraceae 桑科 *Ficus* 榕属
Ficus tsiangii Merr. ex Corner 岩木瓜
Shrubs or trees, evergreen

17.6 (13.0~22.2); 8.6 (1.9~14.2); 25.0 (21.4~29.0); 17.6 (13.1~22.3); 151.5 (102.9~207.5); 0.2 (0.0~6.1); 893 (730~1190) ❙ 1324 (939~1758); 629 (403~996); 92 (39~211) ❙ 48.8 (6.5~100.5); 885 (730~1190); 280.6 (130.1~551.6)

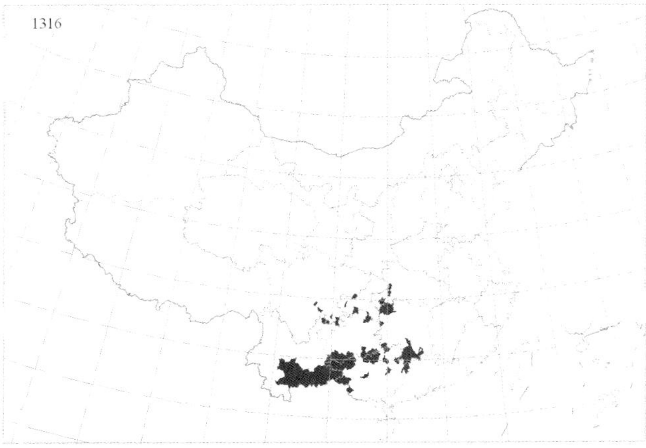

Moraceae 桑科 *Ficus* 榕属
Ficus tuphapensis Drake 平塘榕
Shrubs, evergreen

20.9 (16.2~24.8); 14.0 (6.3~20.1); 26.0 (21.4~28.8); 21.0 (16.2~24.8); 191.5 (134.5~237.6); 0.0 (0.0~0.0); 1099 (797~1385) ❙ 1410 (868~1854); 617 (312~860); 70 (41~177) ❙ 31.9 (-34.3~80.6); 1040 (797~1360); 387.9 (215.4~620.0)

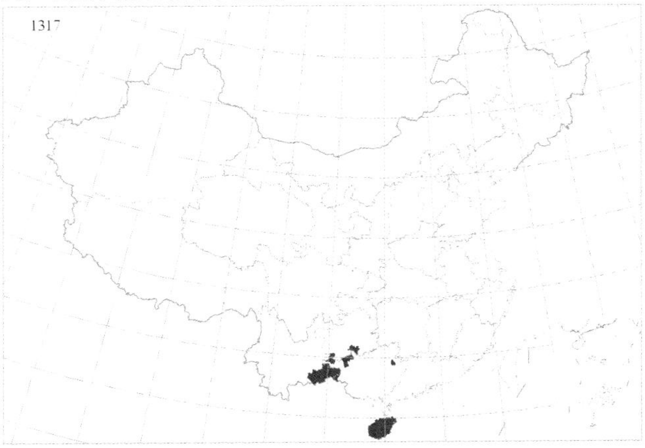

Moraceae 桑科 *Ficus* 榕属
Ficus undulata S. S. Chang 波缘榕
Shrubs, evergreen

21.2; 12.2; 28.5; 21.3; 195.2; 0.0; 1126 ❙ 1462; 599; 142 ❙ 29.9; 1126; 226.9

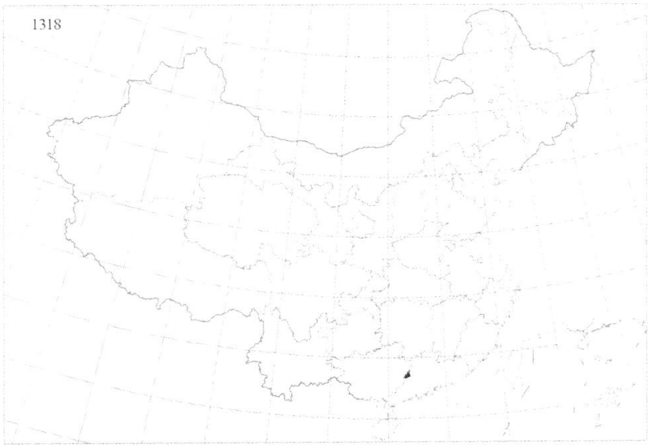

Moraceae 桑科 *Ficus* 榕属
Ficus vaccinioides Hemsl. ex King 越橘榕
Small shrubs, evergreen

18.9 (16.4~22.2); 13.8 (11.5~17.7); 23.2 (20.3~25.9); 19.0 (16.5~22.2); 168.0 (137.8~206.7); 0.0 (0.0~0.0); 921 (783~1123) ❙ 2811 (2499~2993); 1213 (924~1847); 309 (107~547) ❙ 207.8 (166.5~235.6); 921 (783~1123); 298.3 (261.4~363.7)

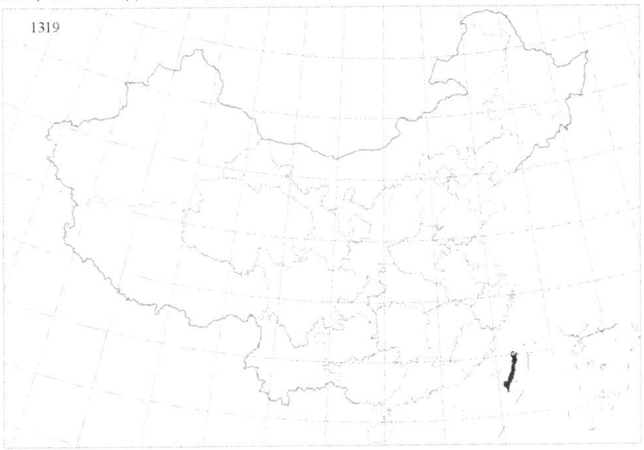

Moraceae 桑科 *Ficus* 榕属
Ficus variegata Blume 杂色榕
Trees, evergreen

20.8 (16.1~24.8); 13.3 (5.3~20.1); 26.8 (20.3~29.2); 20.9 (16.2~24.8); 190.5 (133.9~237.6); 0.0 (0.0~0.0); 1086 (785~1385) ❙ 1602 (868~2993); 714 (312~1847); 120 (41~547) ❙ 49.7 (-34.3~214.4); 1066 (779~1360); 296.5 (95.3~659.0)

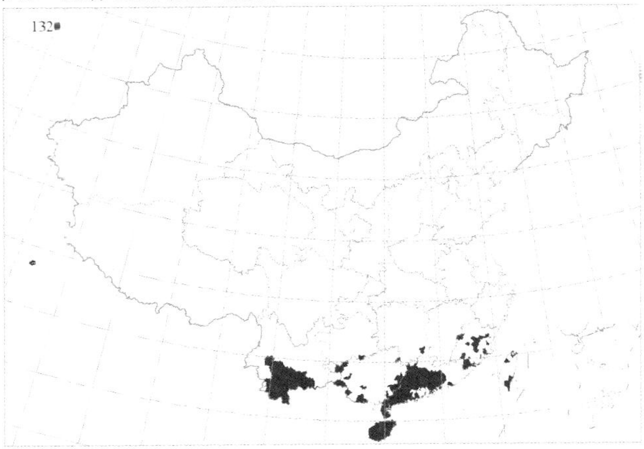

Moraceae 桑科 *Ficus* 榕属

Ficus variolosa Lindl. ex Benth. 变叶榕

Shrubs or small trees, evergreen

19.5 (14.3~24.7); 10.5 (3.3~19.8); 27.2 (21.4~29.8); 19.6 (14.3~24.7); 174.9 (112.6~236.5); 0.0 (0.0~2.3); 1023 (756~1376) ❙ 1569 (868~2435); 646 (312~1428); 150 (42~259) ❙ 55.8 (-34.3~147.0); 1013 (756~1331); 252.2 (95.3~659.0)

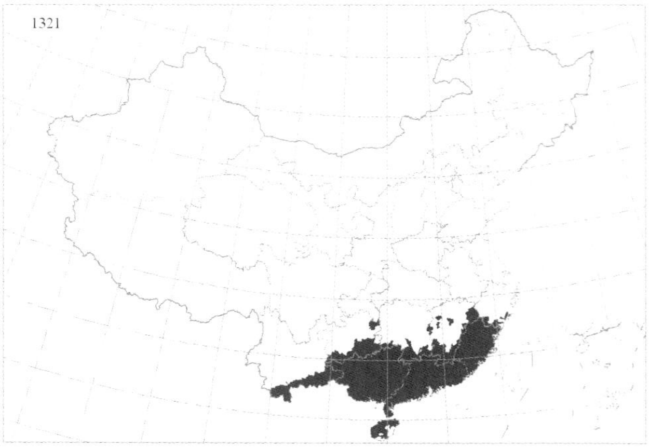

Moraceae 桑科 *Ficus* 榕属

Ficus vasculosa Wall. ex Miq. 白肉榕

Trees, evergreen

19.6 (14.7~24.8); 11.1 (3.8~20.1); 26.6 (22.2~29.0); 19.7 (14.8~24.8); 176.6 (118.7~237.6); 0.1 (0.0~1.5); 1031 (772~1385) ❙ 1498 (868~2435); 678 (312~1428); 103 (41~194) ❙ 46.8 (-34.3~99.5); 1014 (772~1360); 270.7 (139.6~659.0)

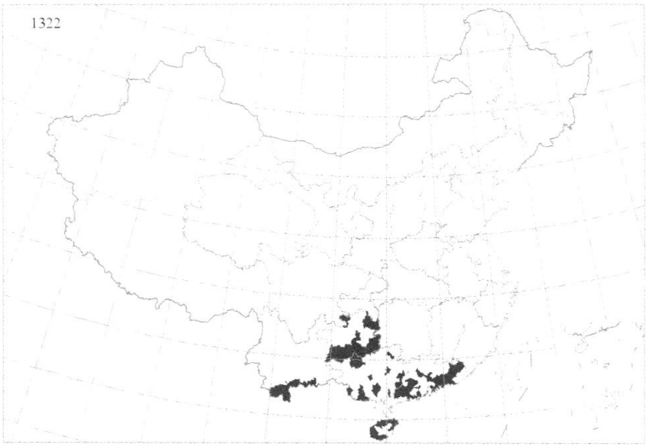

Moraceae 桑科 *Ficus* 榕属

Ficus virens Aiton 黄葛树

Trees, deciduous or semideciduous

18.5 (3.8~24.8); 10.2 (-4.3~20.1); 25.4 (10.8~29.8); 18.6 (4.7~24.8); 163.1 (21.6~237.6); 0.2 (0.0~35.4); 961 (444~1385) ❙ 1373 (797~2993); 649 (312~1847); 93 (16~547) ❙ 43.4 (-34.3~214.4); 949 (444~1360); 260.2 (95.3~659.0)

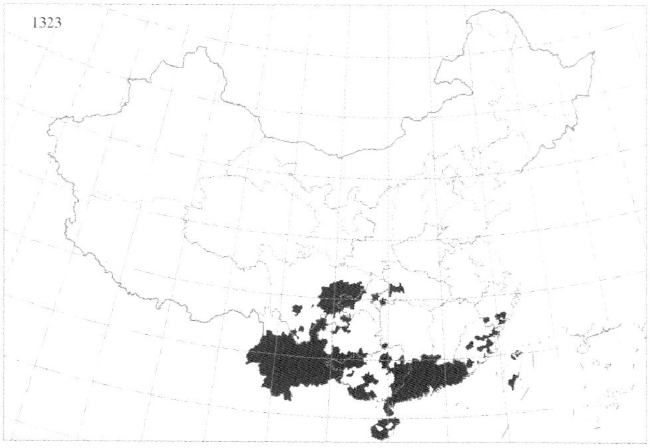

Moraceae 桑科 *Ficus* 榕属

Ficus virgata Reinw. ex Blume 岛榕

Trees, evergreen

19.4; 14.3; 25.0; 20.3; 183.2; 0.0; 1025 ❙ 2657; 1238; 251 ❙ 169.5; 1025; 263.0

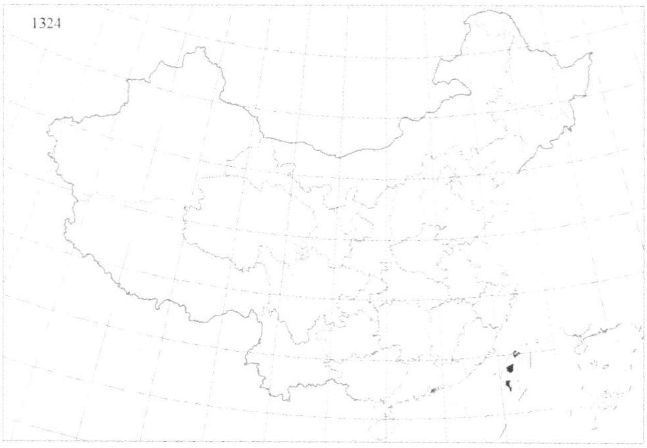

Moraceae 桑科 *Ficus* 榕属

Ficus yunnanensis S. S. Chang 云南榕

Trees, evergreen

16.6 (8.6~19.3); 10.1 (1.3~13.1); 21.2 (15.0~23.6); 16.6 (8.7~19.4); 140.6 (52.6~173.2); 0.9 (0.0~8.8); 821 (574~939) ❙ 1106 (939~1454); 624 (524~792); 47 (37~70) ❙ 37.3 (6.5~99.9); 781 (574~903); 351.7 (289.9~487.2)

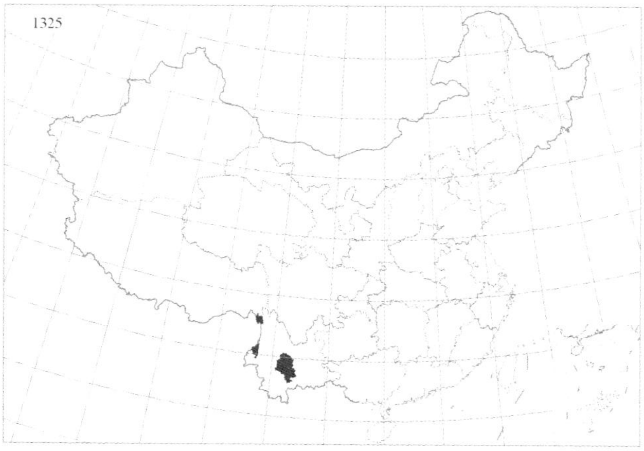

Moraceae 桑科 *Maclura* 柘属

Maclura amboinensis Blume 景东柘

Scandent shrubs

17.5 (13.3~19.3); 10.8 (5.4~13.1); 22.1 (19.6~23.6); 17.5 (13.4~19.4); 150.3 (100.4~173.2); 0.0 (0.0~0.0); 851 (706~939) ❙ 1196 (939~1736); 674 (524~1007); 49 (39~59) ❙ 42.7 (6.5~145.8); 819 (706~903); 371.8 (298.7~491.5)

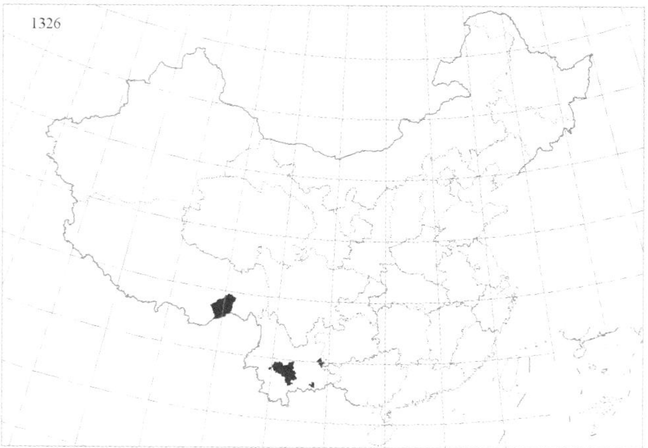

Moraceae 桑科 *Maclura* 柘属
Maclura cochinchinensis (Lour.) Corner 构棘
Shrubs, evergreen

18.4 (5.3~24.8); 8.7 (−2.4~20.1); 26.9 (12.2~30.1); 18.4 (5.7~24.8); 161.2 (30.5~237.6); 0.4 (0.0~26.2); 967 (479~1385) ❙ 1525 (711~3495); 620 (312~1847); 150 (15~897) ❙ 59.8 (−34.3~246.5); 956 (479~1360); 242.9 (95.3~659.0)

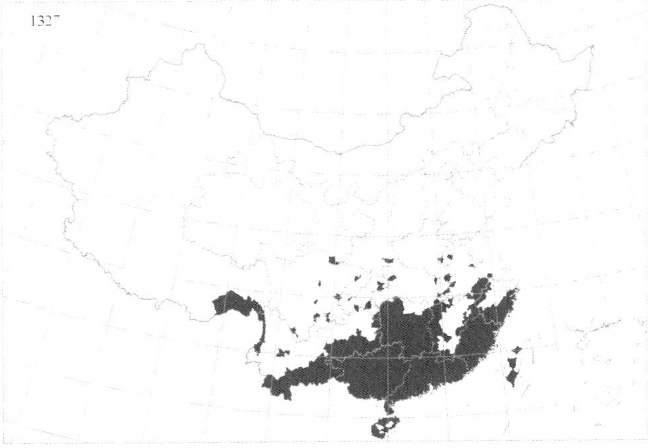

Moraceae 桑科 *Maclura* 柘属
Maclura fruticosa (Roxb.) Corner 柘藤
Vines

17.5 (10.2~22.0); 10.7 (2.4~15.1); 22.2 (16.6~29.0); 17.5 (10.3~22.0); 150.1 (68.2~203.8); 0.1 (0.0~5.1); 860 (620~1170) ❙ 1311 (939~1776); 711 (524~996); 60 (18~188) ❙ 53.1 (6.5~112.2); 848 (620~1170); 396.5 (160.7~659.0)

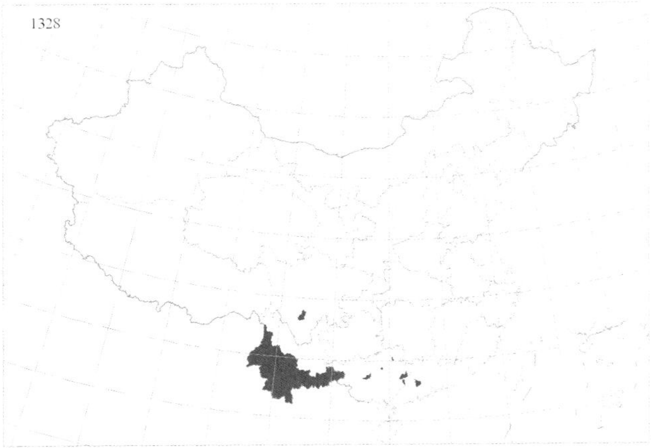

Moraceae 桑科 *Maclura* 柘属
Maclura pubescens (Trécul) Z. K. Zhou et M. G. Gilbert 毛柘藤
Vines

19.9 (13.2~23.6); 11.5 (5.1~16.7); 26.8 (18.4~29.8); 19.9 (13.3~23.6); 179.3 (99.4~223.5); 0.0 (0.0~0.0); 1033 (695~1281) ❙ 1572 (1007~2037); 679 (403~1033); 134 (41~254) ❙ 53.5 (−6.9~130.7); 1027 (695~1273); 263.1 (95.3~659.0)

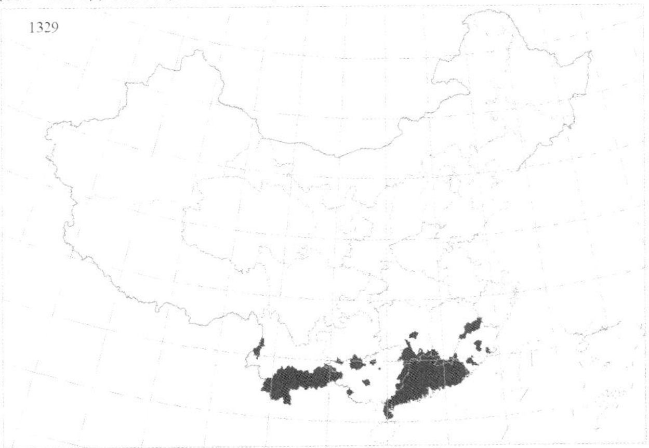

Moraceae 桑科 *Maclura* 柘属
Maclura tricuspidata Carrière 柘
Trees or shrubs, deciduous

15.0 (3.8~23.0); 2.8 (−12.7~15.2); 26.0 (10.8~30.1); 15.2 (4.7~23.0); 128.5 (21.6~216.2); 8.0 (0.0~65.5); 846 (444~1247) ❙ 1063 (367~2435); 493 (235~1428); 88 (8~259) ❙ 24.4 (−51.1~147.0); 776 (367~1247); 210.4 (76.1~659.0)

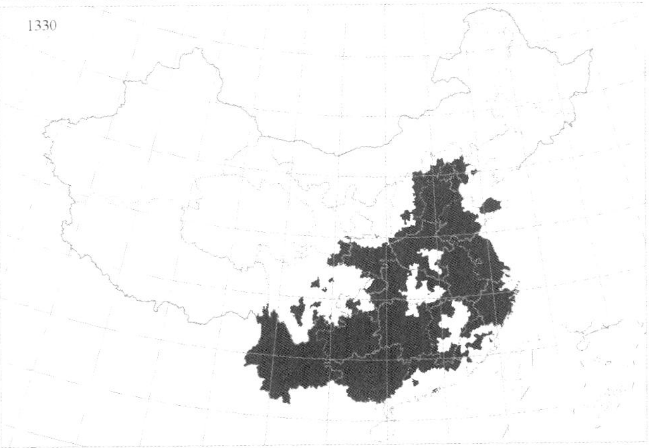

Moraceae 桑科 *Malaisia* 牛筋藤属
Malaisia scandens (Lour.) Planch. 牛筋藤
Scandent shrubs

22.1 (16.2~24.8); 14.7 (9.2~20.1); 27.7 (20.1~29.0); 22.1 (16.2~24.8); 205.4 (134.2~237.6); 0.0 (0.0~0.0); 1175 (773~1385) ❙ 1650 (868~2993); 767 (312~1847); 121 (41~547) ❙ 44.2 (−34.3~246.5); 1140 (773~1360); 295.2 (139.6~548.0)

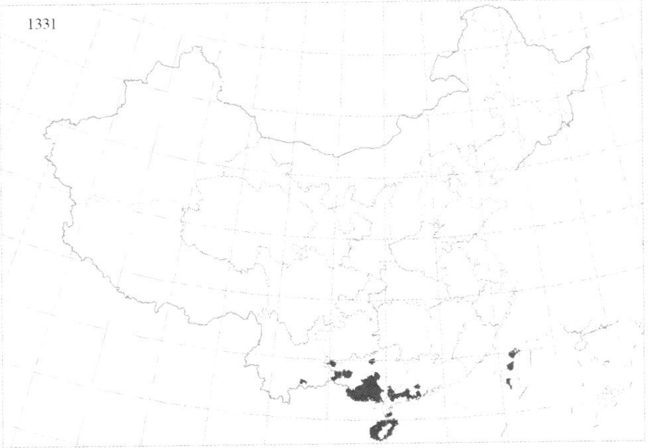

Moraceae 桑科 *Morus* 桑属
Morus alba L. 桑
Shrubs or trees, deciduous

14.9 (1.0~24.8); 2.8 (−22.4~20.1); 25.8 (9.5~29.9); 15.3 (3.8~24.8); 130.0 (15.5~237.6); 10.2 (0.0~104.9); 853 (411~1385) ❙ 1069 (379~2435); 508 (219~1428); 81 (5~259) ❙ 23.7 (−51.1~145.8); 788 (379~1273); 208.6 (76.1~659.0)

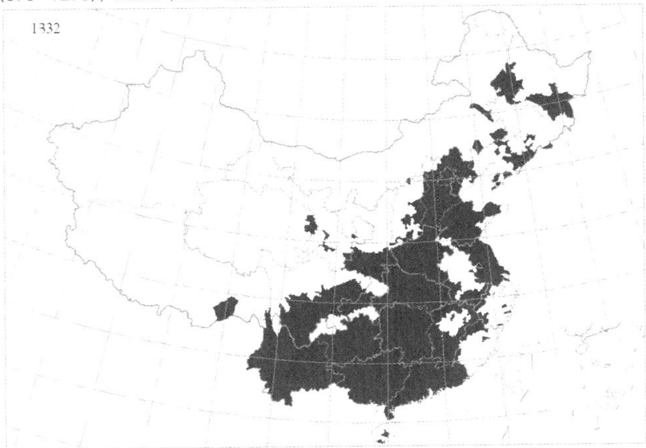

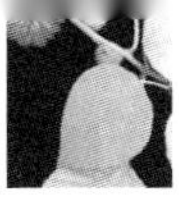

Moraceae 桑科 *Morus* 桑属
Morus australis Poir. 鸡桑
Shrubs or small trees, deciduous

14.8 (−1.8~23.0); 3.5 (−13.7~17.7); 25.0 (6.1~29.8); 15.1 (1.8~23.0); 127.3 (1.9~216.2); 8.7 (0.0~82.8); 830 (329~1247) ▌ 1127 (255~3495); 518 (191~1847); 87 (1~897) ▌ 33.4 (−43.4~246.5); 792 (255~1247); 221.0 (60.4~659.0)

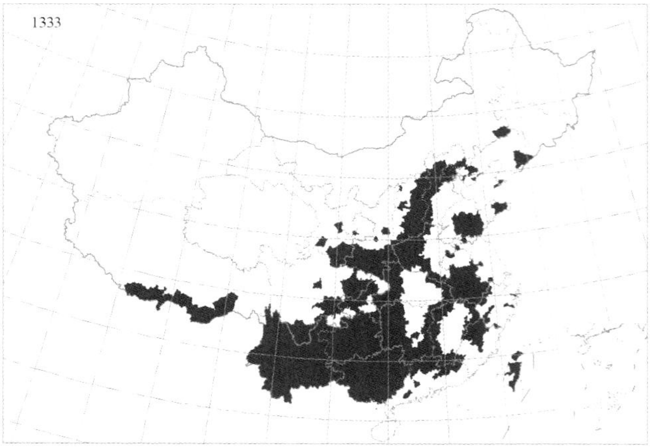

Moraceae 桑科 *Morus* 桑属
Morus cathayana Hemsl. 华桑
Shrubs or small trees, deciduous

15.0 (5.7~23.6); 3.4 (−5.8~17.8); 25.7 (14.2~29.4); 15.1 (6.4~23.6); 125.9 (38.1~223.3); 5.4 (0.0~30.6); 828 (504~1271) ▌ 1209 (512~1997); 510 (258~996); 104 (11~259) ▌ 45.1 (−35.9~130.7); 805 (504~1091); 240.2 (117.3~659.0)

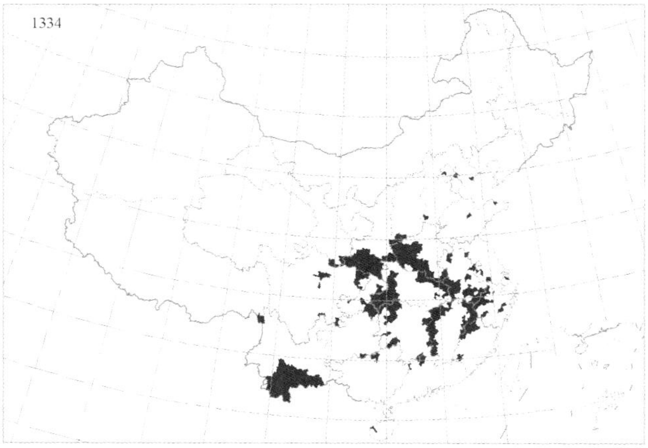

Moraceae 桑科 *Morus* 桑属
Morus cathayana var. gongshanensis (Z. Y. Cao) Z. Y. Cao 贡山桑
Shrubs or small trees

8.6; 1.3; 15.0; 8.7; 52.6; 8.8; 574 ▌ 1041; 585; 45 ▌ 81.5; 574; 352.3

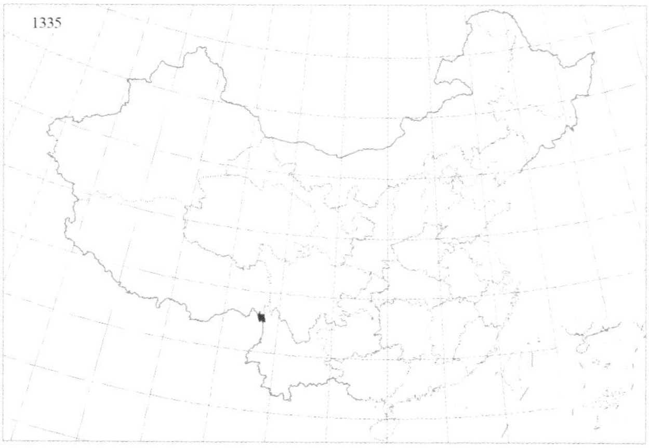

Moraceae 桑科 *Morus* 桑属
Morus liboensis S. S. Chang 荔波桑
Trees, deciduous

18.1; 8.1; 26.2; 18.1; 157.0; 0.0; 927 ▌ 1409; 653; 112 ▌ 52.2; 927; 241.8

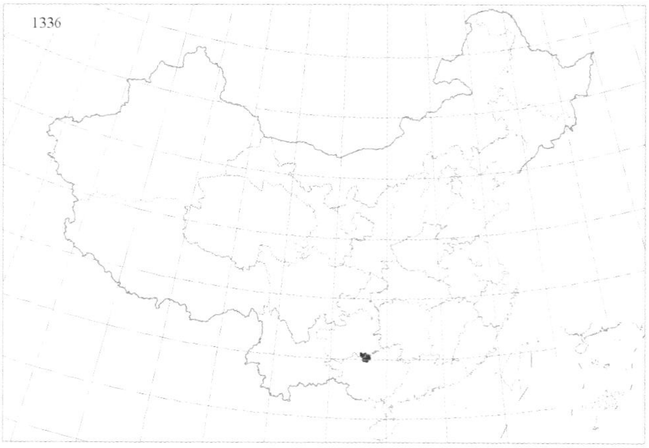

Moraceae 桑科 *Morus* 桑属
Morus macroura Miq. 奶桑
Small trees, deciduous

16.4 (1.8~22.2); 9.7 (−7.1~15.1); 21.1 (9.5~28.3); 16.5 (3.8~22.3); 139.0 (15.7~207.5); 1.8 (0.0~53.3); 815 (404~1190) ▌ 1193 (583~1751); 667 (348~996); 49 (11~89) ▌ 46.9 (6.5~112.2); 801 (404~1190); 374.2 (169.5~659.0)

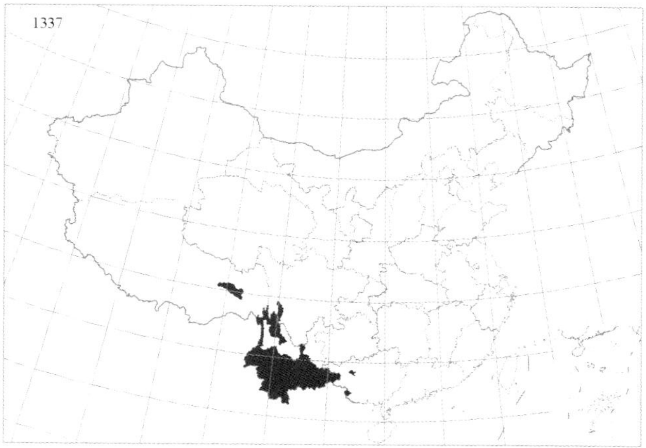

Moraceae 桑科 *Morus* 桑属
Morus mongolica (Bureau) C. K. Schneid. 蒙桑
Shrubs or small trees, deciduous

12.1 (0.2~21.6); −1.7 (−20.8~13.1); 24.3 (9.5~28.7); 12.8 (3.8~21.6); 105.9 (15.5~199.3); 20.3 (0.0~104.9); 756 (411~1150) ▌ 718 (187~1776); 416 (127~816); 31 (5~197) ▌ −4.4 (−67.8~125.9); 626 (187~1148); 197.5 (35.7~516.7)

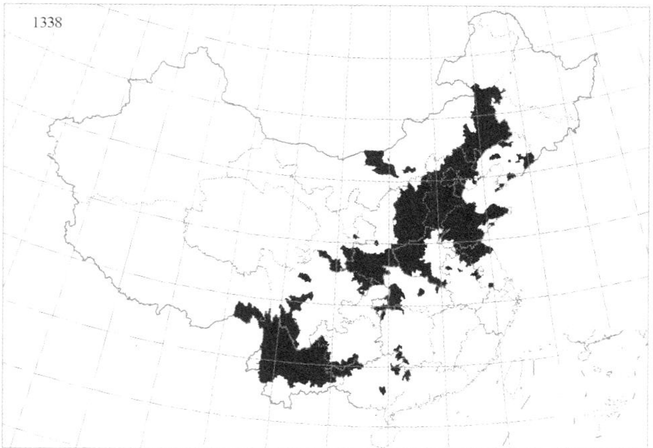

Moraceae 桑科 *Morus* 桑属
Morus notabilis C. K. Schneid. 川桑
Trees, deciduous

14.4 (5.4~21.0); 5.7 (-2.1~14.2); 21.9 (12.3~27.5); 14.5 (5.8~21.0); 115.5 (31.5~191.8); 2.2 (0.0~25.6); 764 (481~1053) ❙ 1215 (741~1522); 645 (426~860); 57 (28~140) ❙ 59.9 (19.4~93.0); 764 (481~1053); 303.3 (166.0~551.6)

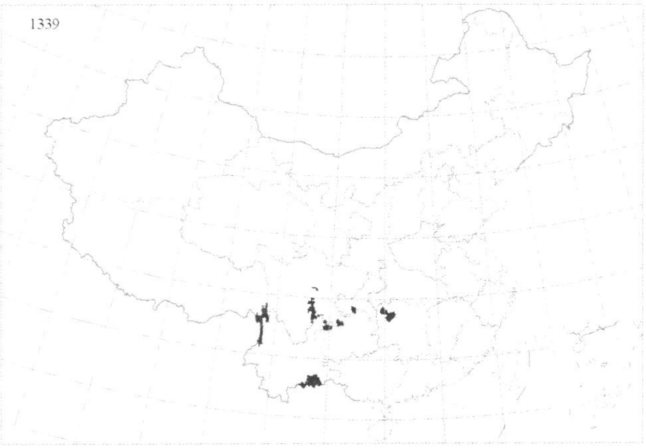

Moraceae 桑科 *Morus* 桑属
Morus serrata Roxb. 吉隆桑
Small trees, deciduous

-1.1; -9.1; 6.2; 2.0; 2.5; 75.6; 334 ❙ 321; 193; 33 ❙ -4.0; 321; 79.3

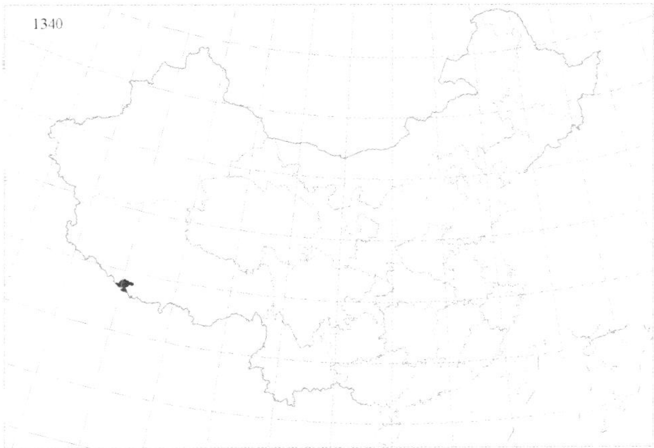

Moraceae 桑科 *Morus* 桑属
Morus trilobata (S. S. Chang) Z. Y. Cao 裂叶桑
Trees, deciduous

15.8 (14.3~17.7); 6.0 (4.0~10.9); 24.6 (22.5~25.7); 15.9 (14.3~17.8); 131.0 (112.6~153.4); 0.3 (0.0~1.1); 820 (756~863) ❙ 1264 (1211~1368); 548 (476~756); 93 (59~106) ❙ 54.5 (45.6~71.9); 820 (756~863); 267.6 (188.8~491.5)

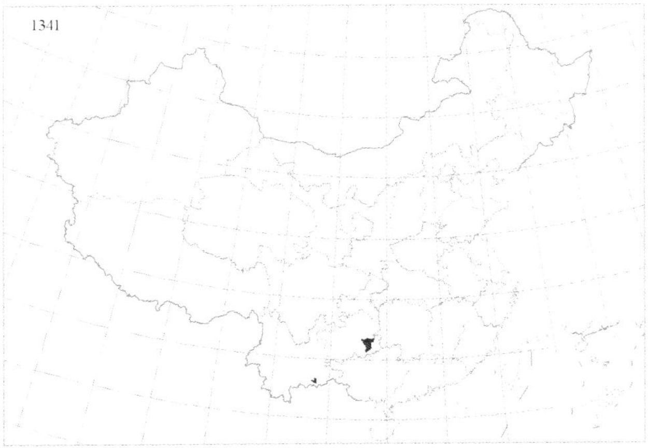

Moraceae 桑科 *Morus* 桑属
Morus wittiorum Hand.-Mazz. 长穗桑
Trees, deciduous

17.5 (14.3~22.2); 7.4 (3.3~14.9); 26.5 (20.4~29.4); 17.6 (14.3~22.3); 151.2 (112.6~207.5); 0.2 (0.0~2.3); 918 (756~1190) ❙ 1462 (1159~1776); 586 (392~911); 146 (51~218) ❙ 60.2 (15.4~128.9); 912 (756~1190); 241.7 (145.3~570.1)

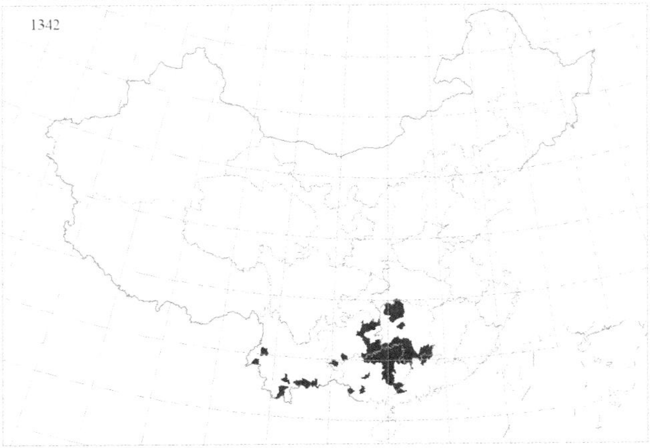

Moraceae 桑科 *Streblus* 鹊肾树属
Streblus asper Lour. 鹊肾树
Shrubs or trees, evergreen

21.3 (17.3~24.8); 14.0 (7.8~20.1); 26.7 (21.6~29.2); 21.3 (17.3~24.8); 195.9 (147.1~237.6); 0.0 (0.0~0.0); 1111 (833~1385) ❙ 1490 (868~2035); 708 (312~997); 92 (41~177) ❙ 36.3 (-34.3~88.8); 1082 (816~1331); 318.5 (152.9~659.0)

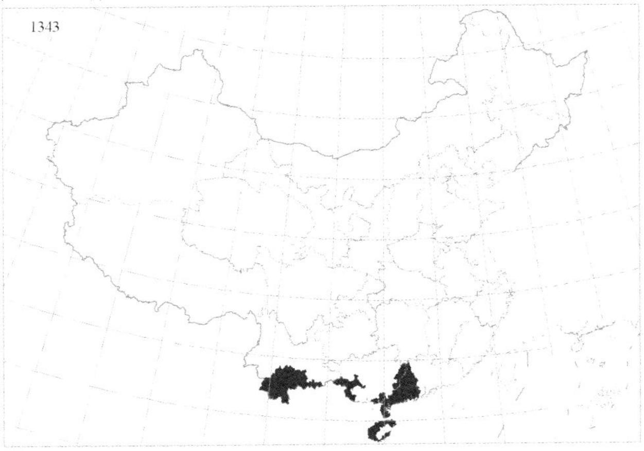

Moraceae 桑科 *Streblus* 鹊肾树属
Streblus ilicifolius (Vidal) Corner 刺桑
Shrubs or trees

22.5 (16.9~24.8); 16.3 (10.3~20.1); 26.8 (21.4~28.8); 22.5 (16.9~24.8); 210.0 (142.7~237.6); 0.0 (0.0~0.0); 1195 (813~1385) ❙ 1506 (868~2435); 656 (312~1428); 74 (41~130) ❙ 28.3 (-34.3~99.5); 1121 (813~1360); 428.9 (248.2~620.0)

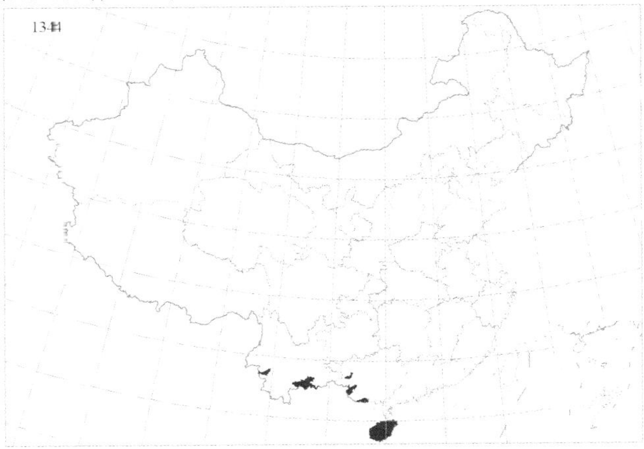

Moraceae 桑科 *Streblus* 鹊肾树属
Streblus indicus (Bureau) Corner 假鹊肾树
Trees

20.0 (14.4~24.8); 13.8 (7.5~20.1); 24.2 (19.1~28.9); 20.0 (14.4~24.8); 180.4 (113.1~237.6); 0.0 (0.0~0.0); 1018 (727~1385) ❙ 1422 (868~2435); 700 (312~1428); 64 (37~136) ❙ 43.4 (-34.3~112.2); 972 (717~1360); 426.0 (201.4~659.0)

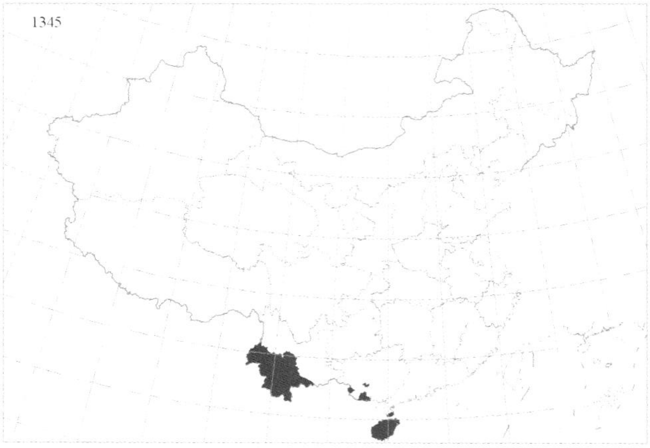

Moraceae 桑科 *Streblus* 鹊肾树属
Streblus macrophyllus Blume 双果桑
Scandent shrubs

18.9 (16.3~23.6); 11.9 (9.1~16.7); 23.8 (21.4~28.6); 18.9 (16.4~23.6); 167.2 (136.4~223.5); 0.0 (0.0~0.0); 933 (797~1281) ❙ 1373 (1163~1751); 762 (551~996); 60 (48~87) ❙ 49.3 (-1.0~88.8); 932 (797~1269); 411.9 (235.0~551.6)

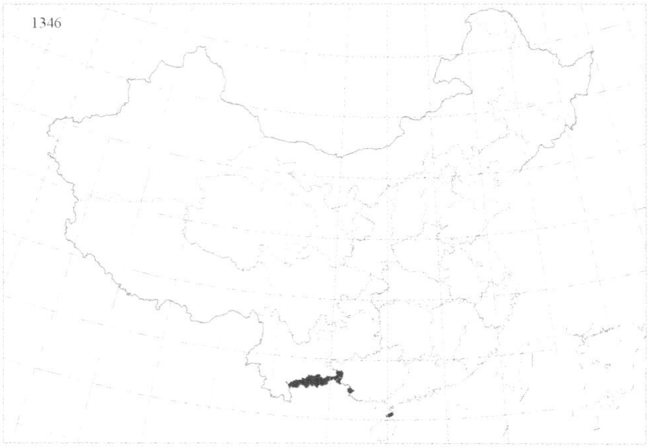

Moraceae 桑科 *Streblus* 鹊肾树属
Streblus taxoides (Roth) Kurz 叶被木
Shrubs, evergreen

24.4; 19.2; 28.0; 24.4; 233.1; 0.0; 1347 ❙ 1433; 472; 55 ❙ 6.3; 1151; 496.5

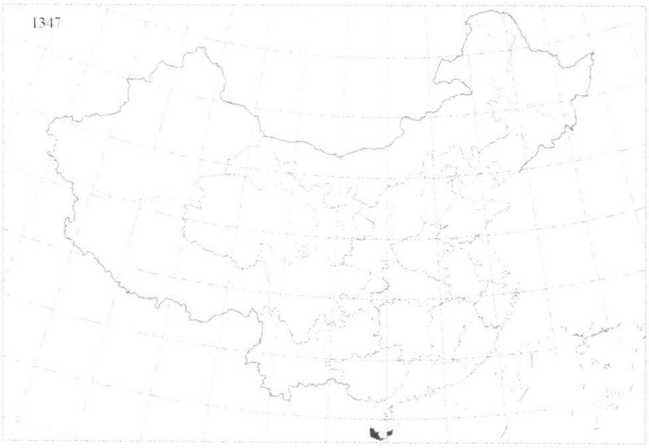

Moraceae 桑科 *Streblus* 鹊肾树属
Streblus tonkinensis (Dubard et Eberh.) Corner 米扬
Trees, evergreen

19.6 (15.4~24.3); 12.2 (8.1~18.7); 24.9 (20.7~28.8); 19.6 (15.5~24.3); 175.5 (125.7~231.9); 0.0 (0.0~0.0); 997 (767~1338) ❙ 1267 (868~1600); 662 (312~860); 67 (42~171) ❙ 30.0 (-34.3~80.6); 959 (767~1190); 351.0 (188.6~551.6)

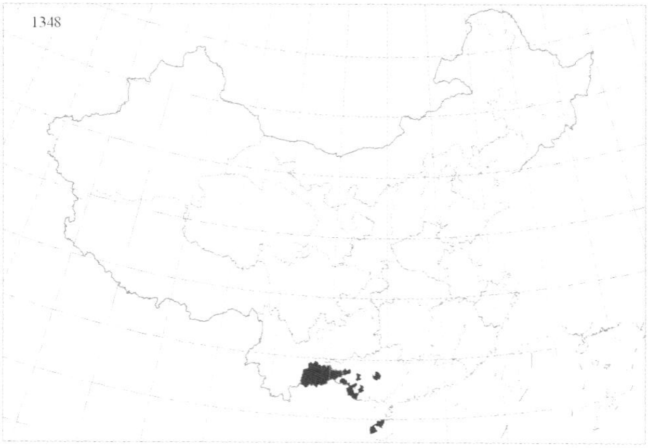

U

Moraceae 桑科 *Streblus* 鹊肾树属
Streblus zeylanicus (Thwaites) Kurz 尾叶刺桑
Shrubs

20.2 (18.6~22.2); 13.5 (12.7~14.2); 25.0 (22.5~28.5); 20.3 (18.7~22.3); 183.5 (163.9~207.5); 0.0 (0.0~0.0); 1014 (888~1190) ❙ 1500 (1276~1751); 836 (713~996); 76 (51~128) ❙ 50.0 (15.4~88.8); 1012 (888~1190); 451.3 (256.6~551.6)

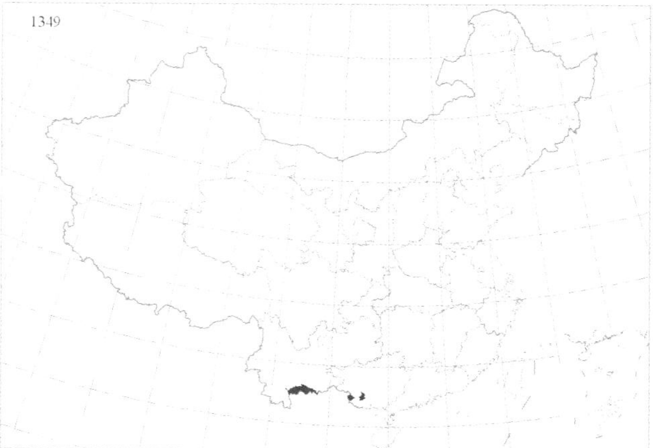

Urticaceae 荨麻科 *Archiboehmeria* 舌柱麻属
Archiboehmeria atrata (Gagnep.) C. J. Chen 舌柱麻
Shrubs or subshrubs

19.8 (14.6~24.7); 10.8 (3.8~19.8); 27.3 (24.6~28.8); 19.9 (14.7~24.7); 178.5 (117.7~236.5); 0.0 (0.0~1.5); 1041 (779~1376) ❙ 1641 (1182~2435); 721 (445~1428); 139 (43~202) ❙ 58.6 (15.4~101.8); 1035 (779~1299); 258.6 (163.7~548.0)

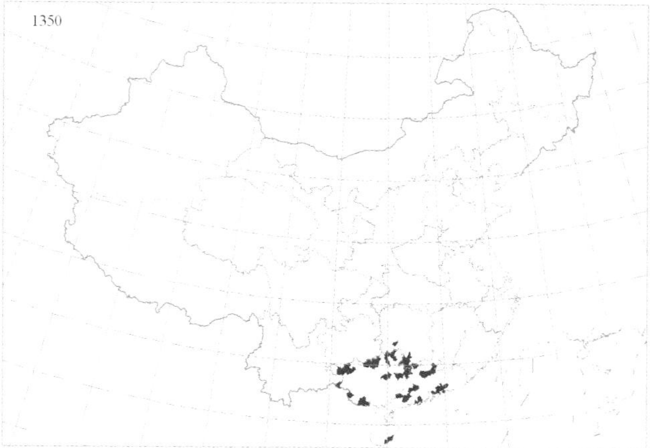

Urticaceae 荨麻科 *Boehmeria* 苎麻属
Boehmeria clidemioides Miq. 白面苎麻
Herbs or subshrubs

15.8 (1.8~23.0); 6.8 (-7.1~17.7); 23.5 (9.9~29.3); 15.9 (3.8~23.1); 132.2 (15.7~217.0); 1.6 (0.0~53.3); 827 (404~1217) ❙ 1300 (583~1997); 595 (348~996); 99 (12~259) ❙ 56.9 (-14.2~130.7); 817 (404~1163); 285.4 (130.1~659.0)

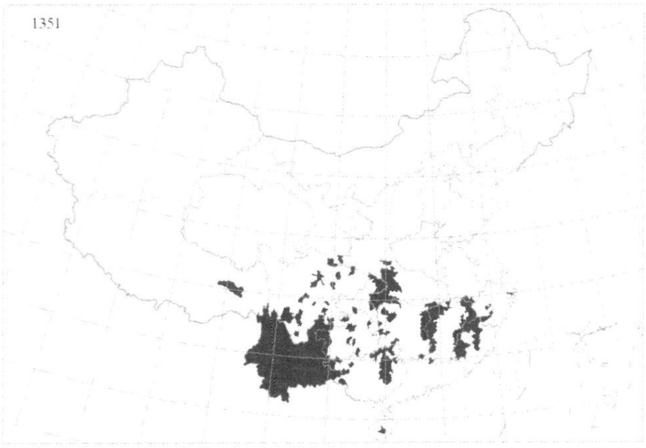

Urticaceae 荨麻科 *Boehmeria* 苎麻属
Boehmeria clidemioides var. diffusa (Wedd.) Hand.-Mazz. 序叶苎麻
Herbs or subshrubs

15.6 (5.7~21.8); 6.4 (-3.5~14.2); 23.9 (12.7~28.3); 15.7 (6.3~21.9); 131.0 (35.2~202.6); 2.7 (0.0~28.1); 828 (496~1150) ❙ 1393 (801~1997); 599 (445~911); 113 (12~254) ❙ 68.7 (19.0~130.7); 825 (496~1150); 281.6 (152.4~551.6)

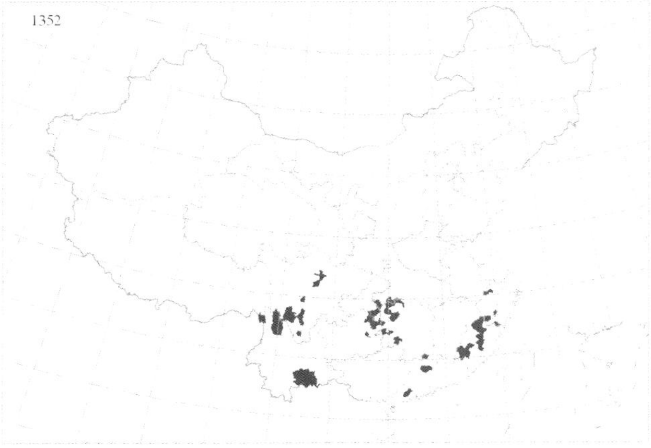

Urticaceae 荨麻科 *Boehmeria* 苎麻属
Boehmeria conica C. J. Chen 锥序苎麻
Shrubs

19.0; 13.1; 22.6; 19.1; 168.8; 0.0; 909 ❙ 1495; 775; 62 ❙ 64.5; 891; 427.9

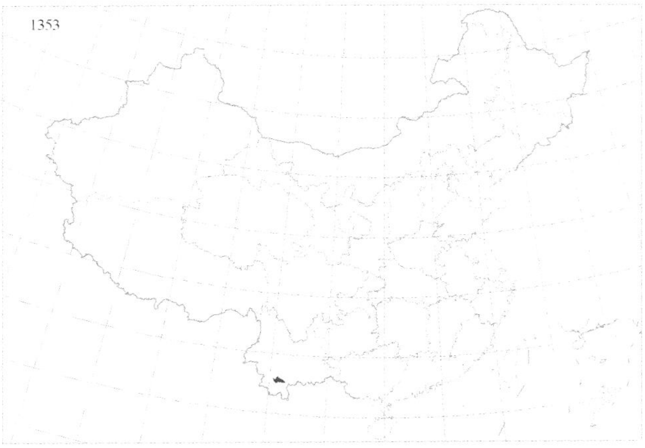

Urticaceae 荨麻科 *Boehmeria* 苎麻属
Boehmeria densiflora Hook. et Arn. 密花苎麻
Shrubs

20.2 (16.2~23.8); 14.4 (11.0~18.1); 25.2 (20.1~28.7); 20.4 (16.2~23.8); 185.1 (134.2~225.8); 0.0 (0.0~0.0); 1030 (773~1297) ❙ 2331 (1485~2993); 1167 (750~1847); 190 (58~547) ❙ 134.6 (35.5~246.5); 1004 (773~1216); 256.7 (119.4~363.7)

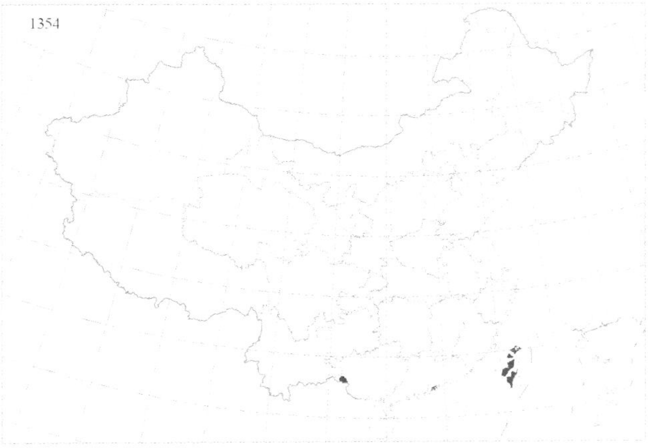

Urticaceae 荨麻科 *Boehmeria* 苎麻属
Boehmeria dolichostachya W. T. Wang 长序苎麻
Shrubs or subshrubs

18.9 (15.9~22.3); 9.5 (6.3~13.9); 26.4 (23.0~29.0); 18.9 (16.0~22.3); 166.9 (131.6~208.1); 0.0 (0.0~0.0); 974 (802~1202) ❙ 1416 (1232~1776); 676 (556~820); 106 (51~189) ❙ 46.6 (15.4~82.1); 974 (802~1202); 247.9 (160.7~371.7)

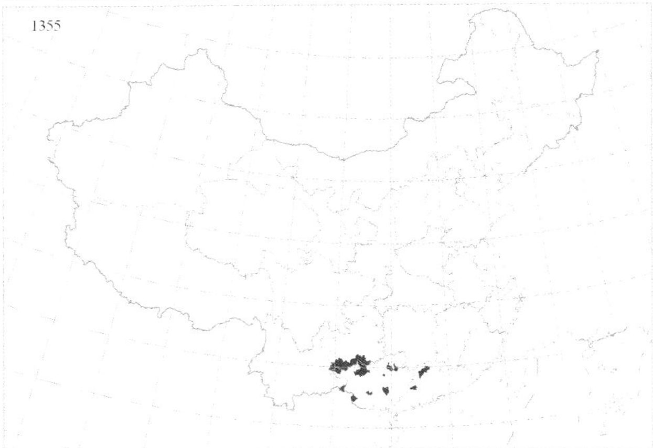

Urticaceae 荨麻科 *Boehmeria* 苎麻属
Boehmeria dolichostachya var. mollis (W. T. Wang) W. T. Wang et C. J. Chen 柔毛苎麻
Shrubs or subshrubs

21.9; 13.5; 28.6; 21.9; 202.6; 0.0; 1161 ❙ 1634; 730; 129 ❙ 40.7; 1161; 206.4

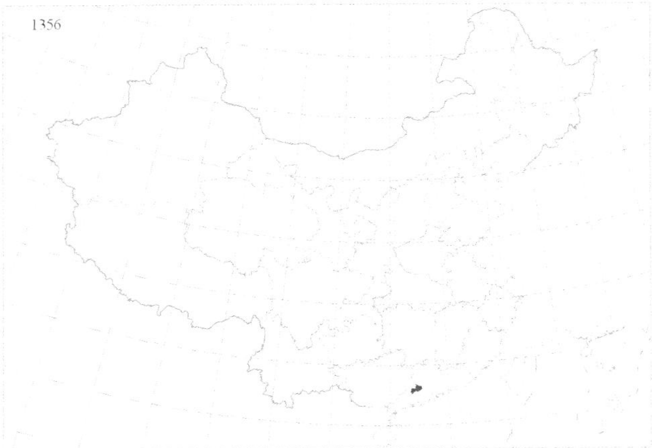

Urticaceae 荨麻科 *Boehmeria* 苎麻属
Boehmeria glomerulifera Miq. 腋球苎麻
Shrubs or small trees

17.3 (5.3~22.2); 10.7 (-2.4~15.1); 21.9 (12.2~28.3); 17.3 (5.7~22.3); 150.1 (30.5~207.5); 2.4 (0.0~26.2); 877 (479~1190) ❙ 1459 (961~1736); 765 (558~1007); 63 (31~87) ❙ 72.1 (15.4~145.8); 869 (479~1190); 464.0 (234.2~659 0)

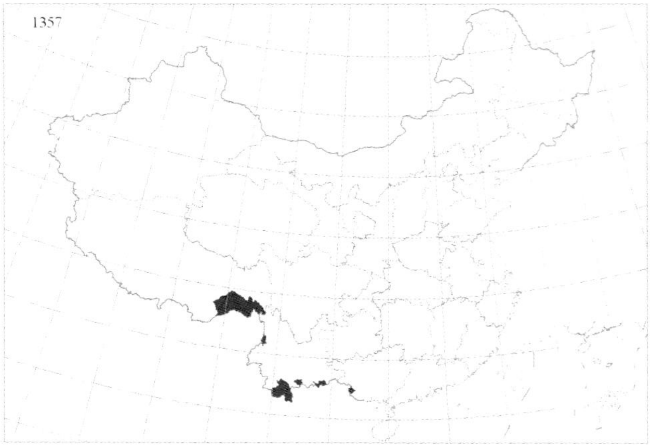

Urticaceae 荨麻科 *Boehmeria* 苎麻属
Boehmeria hamiltoniana Wedd. 细序苎麻
Shrubs

18.8 (16.7~20.5); 12.8 (10.3~15.1); 22.6 (21.3~23.7); 18.8 (16.8~20.5); 165.5 (141.5~185.9); 0.0 (0.0~0.0); 899 (811~992) ❙ 1340 (939~1751); 714 (524~996); 54 (37~77) ❙ 48.8 (6.5~88.8); 869 (744~986); 436.7 (301.1~659.0)

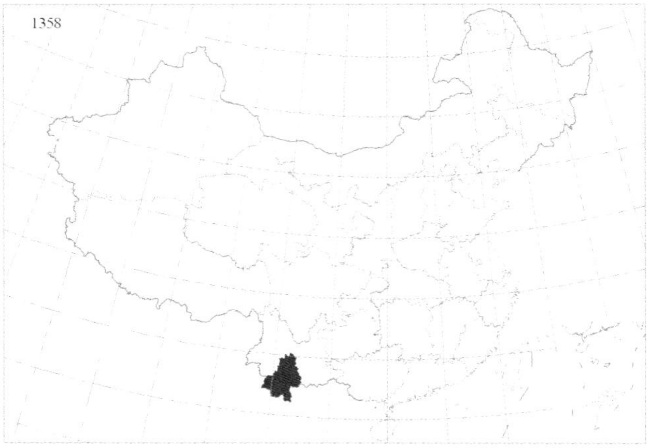

Urticaceae 荨麻科 *Boehmeria* 苎麻属
Boehmeria ingjiangensis W. T. Wang 盈江苎麻
Shrubs

16.4; 9.6; 20.6; 16.4; 137.3; 0.0; 795 ❙ 1688; 959; 62 ❙ 112.2; 795; 627.0

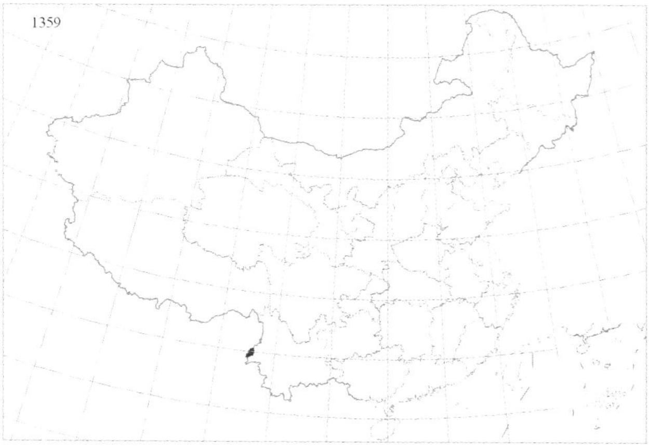

Urticaceae 荨麻科 *Boehmeria* 苎麻属
Boehmeria japonica (L. f.) Miq. 野线麻
Herbs or subshrubs

16.6 (9.4~22.2); 5.7 (-3.3~15.3); 26.7 (20.1~29.4); 16.7 (9.8~22.2); 141.6 (73.9~206.9); 1.8 (0.0~20.6); 890 (619~1177) ❙ 1456 (585~3495); 564 (316~1412); 153 (20~897) ❙ 62.9 (-25.6~246.5); 882 (585~1177); 222.7 (95.9~659.0)

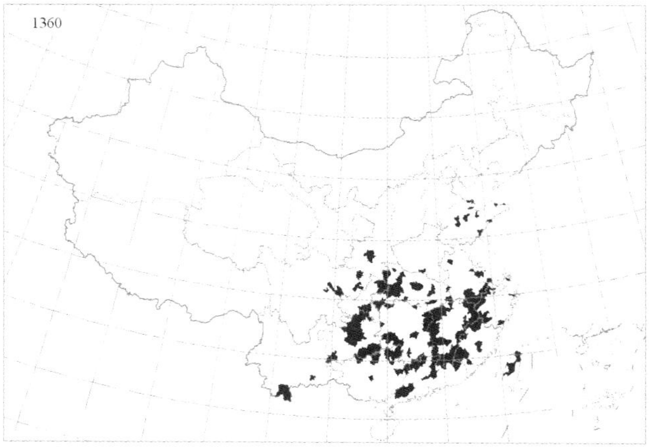

Urticaceae 荨麻科 *Boehmeria* 苎麻属
Boehmeria lanceolata Ridl. 北越苎麻
Small shrubs

21.9 (18.6~24.2); 15.1 (12.6~18.1); 26.7 (22.5~28.7); 21.9 (18.7~24.3); 202.4 (163.9~231.2); 0.0 (0.0~0.0); 1144 (888~1335) ❙ 1260 (868~1547); 624 (312~911); 63 (56~71) ❙ 15.7 (-34.3~74.2); 1014 (868~1139); 423.7 (288.3~551.6)

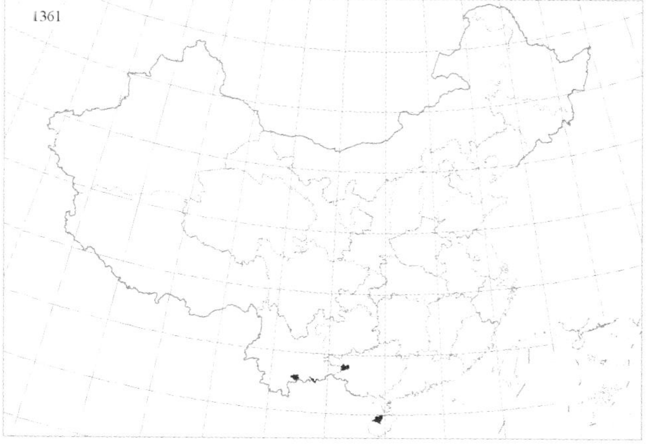

Urticaceae 荨麻科 *Boehmeria* 苎麻属
Boehmeria lohuiensis S. S. Chien 琼海苎麻
Small shrubs

24.2; 18.8; 28.0; 24.2; 230.6; 0.0; 1327 ❙ 1617; 538; 82 ❙ 21.5; 1236; 441.9

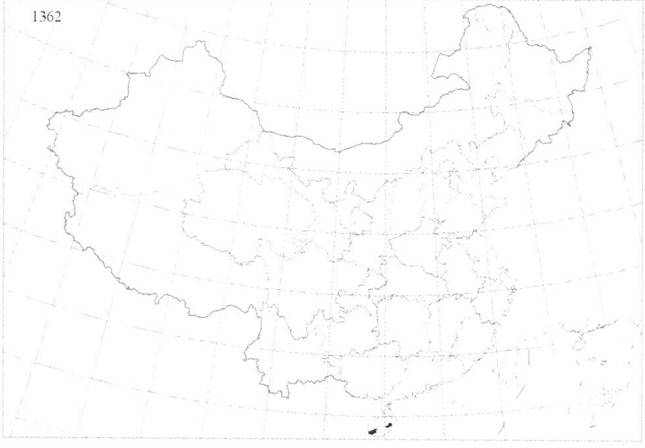

Urticaceae 荨麻科 *Boehmeria* 苎麻属

Boehmeria macrophylla Hornem. 水苎麻

Herbs or subshrubs

16.7 (5.4~22.7); 9.2 (-2.1~15.1); 22.4 (12.3~29.0); 16.8 (5.8~22.8); 141.7 (31.5~213.1); 0.6 (0.0~25.6); 851 (481~1221) ❙ 1221 (741~2435); 660 (411~1428); 61 (12~191) ❙ 43.9 (-14.2~145.8); 837 (481~1221); 319.8 (185.6~659.0)

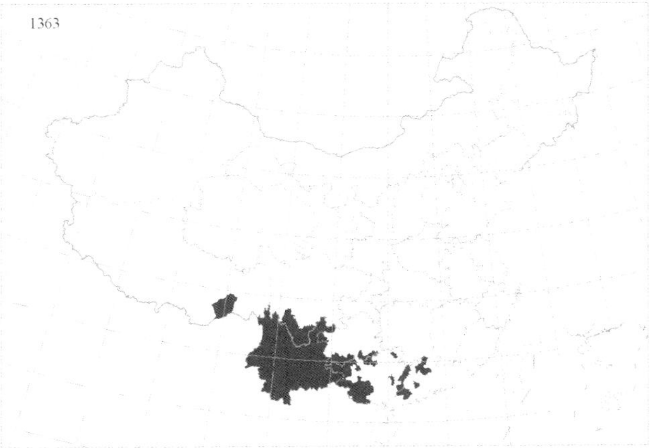

Urticaceae 荨麻科 *Boehmeria* 苎麻属

Boehmeria macrophylla var. canescens (Wedd.) D. G. Long 灰绿水苎麻

Herbs or subshrubs

16.5; 9.8; 21.4; 16.5; 140.4; 2.2; 842 ❙ 1370; 767; 59 ❙ 65.1; 842; 486.7

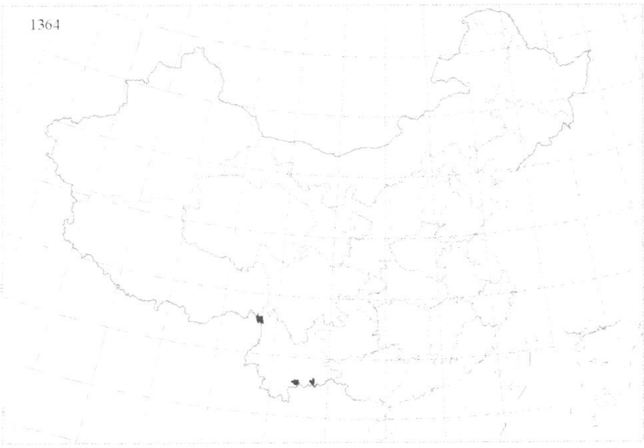

Urticaceae 荨麻科 *Boehmeria* 苎麻属

Boehmeria macrophylla var. scabrella (Roxb.) D. G. Long 糙叶苎麻

Herbs or subshrubs

20.4 (16.9~23.2); 12.9 (10.3~15.8); 26.2 (21.4~28.9); 20.5 (16.9~23.2); 185.8 (142.7~218.9); 0.0 (0.0~0.0); 1050 (813~1257) ❙ 1596 (1007~2005); 765 (539~911); 109 (44~192) ❙ 51.9 (19.0~85.1); 1049 (813~1257); 311.4 (157.7~551.6)

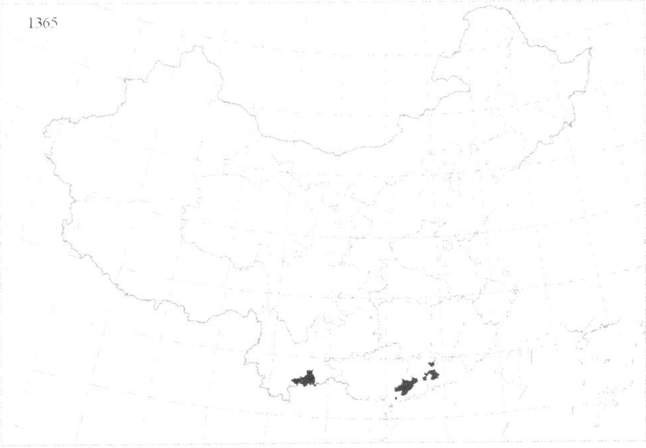

Urticaceae 荨麻科 *Boehmeria* 苎麻属

Boehmeria nivea (L.) Gaudich. 苎麻

Shrubs or subshrubs

16.5 (5.4~24.5); 6.3 (-4.2~19.1); 25.7 (12.3~30.1); 16.6 (5.8~24.5); 141.6 (31.5~234.4); 2.7 (0.0~29.4); 884 (481~1360) ❙ 1323 (604~2993); 5[illegible]2 (316~1847); 116 (14~547) ❙ 49.5 (-23.0~246.5); 869 (481~1360); 242.9 (120.3~659.0)

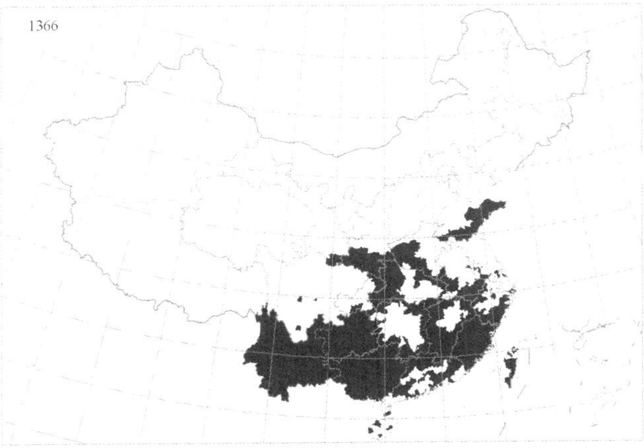

Urticaceae 荨麻科 *Boehmeria* 苎麻属

Boehmeria nivea var. tenacissima (Gaudich.) Miq. 青叶苎麻

Shrubs or subshrubs

20.6 (17.6~24.0); 12.2 (7.7~19.1); 27.4 (22.5~28.8); 20.6 (17.7~24.0); 187.1 (152.5~228.1); 0.0 (0.0~0.0); 1073 (888~1302) ❙ 1696 (1428~2037); 739 (507~1033); 138 (48~192) ❙ 58.7 (19.1~85.1); 1068 (888~1191); 267.1 (157.7~620.0)

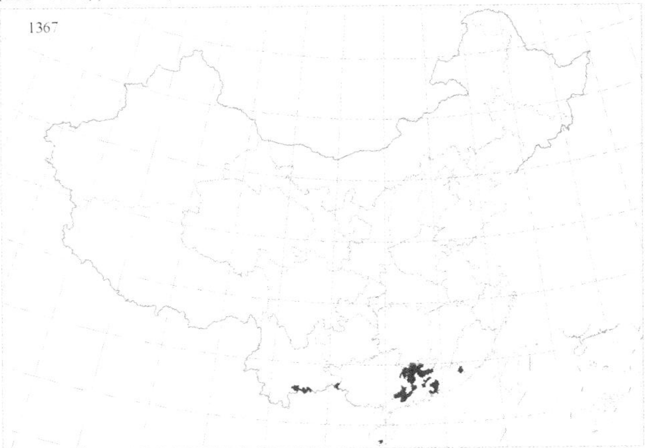

Urticaceae 荨麻科 *Boehmeria* 苎麻属

Boehmeria penduliflora Weddell ex D. G. Long 长叶苎麻

Shrubs

16.2 (5.4~22.7); 9.1 (-2.1~15.1); 21.5 (12.3~28.6); 16.3 (5.8~22.8); 136.0 (31.5~213.1); 0.6 (0.0~25.6); 817 (481~1221) ❙ 1168 (741~2435); 656 (411~1428); 48 (15~173) ❙ 43.1 (-14.2~145.8); 800 (481~1221); 338.2 (185.6~659.0)

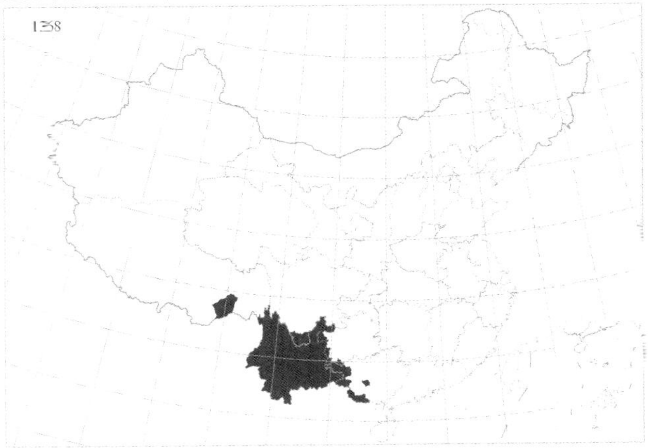

Urticaceae 荨麻科 *Boehmeria* 苎麻属
Boehmeria pilosiuscula (Blume) Hassk. 疏毛苎麻
Herbs or subshrubs

18.0 (-1.8~24.8); 12.1 (-10.2~20.1); 22.2 (6.1~28.7); 18.2 (1.8~24.8); 161.1 (1.9~237.6); 4.4 (0.0~82.8); 932 (329~1385) ❙ 1597 (320~2992); 821 (193~1847); 81 (20~308) ❙ 74.2 (-12.0~246.5); 890 (320~1326); 384.2 (72.2~659.0)

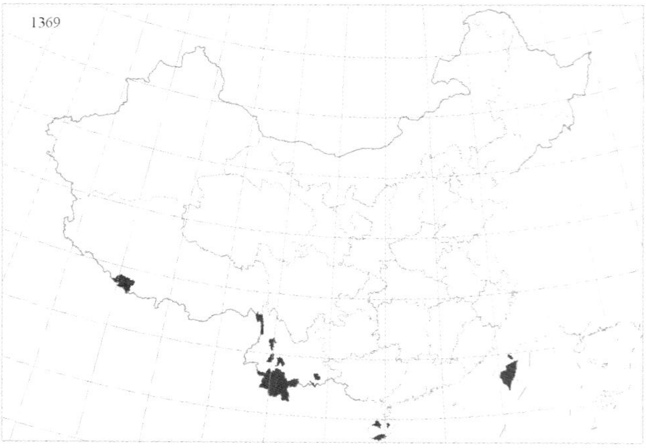

Urticaceae 荨麻科 *Boehmeria* 苎麻属
Boehmeria polystachya Wedd. 歧序苎麻
Shrubs

17.6 (5.3~21.0); 11.1 (-2.4~15.1); 21.9 (12.2~25.6); 17.6 (5.7~21.0); 151.9 (30.5~191.8); 0.5 (0.0~26.2); 853 (479~1053) ❙ 1328 (939~1751); 736 (524~1007); 52 (31~77) ❙ 56.5 (6.5~145.8); 839 (479~1053); 427.1 (234.2~659.0)

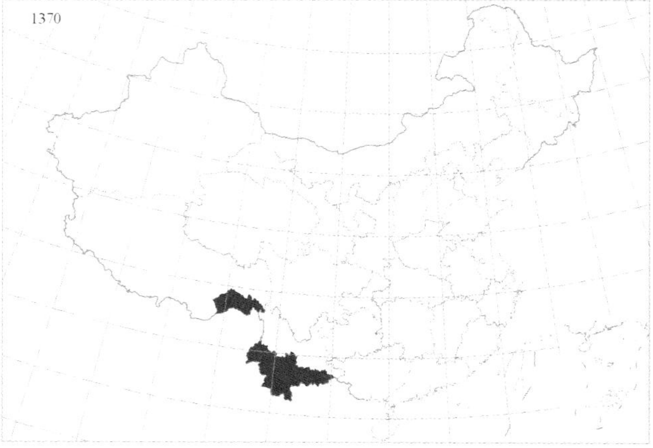

Urticaceae 荨麻科 *Boehmeria* 苎麻属
Boehmeria siamensis Craib 八棱麻
Shrubs

17.0 (11.0~22.2); 10.1 (4.2~15.1); 21.9 (16.6~28.5); 17.0 (11.0~22.3); 144.4 (73.1~207.5); 0.0 (0.0~0.8); 840 (636~1190) ❙ 1192 (797~1751); 665 (411~996); 51 (23~130) ❙ 42.4 (-14.2~112.2); 822 (636~1190); 357.0 (207.5~659.0)

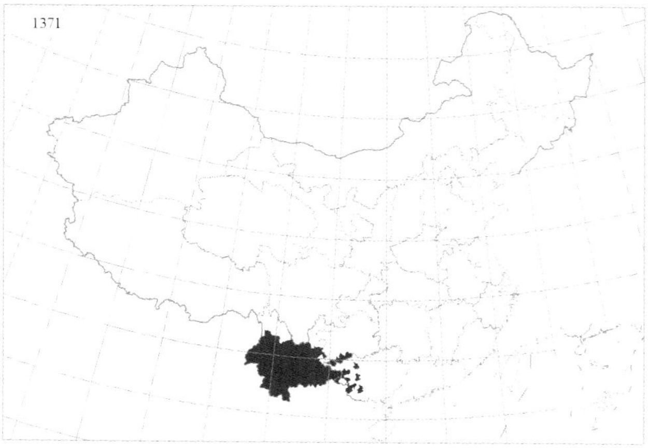

Urticaceae 荨麻科 *Boehmeria* 苎麻属
Boehmeria tomentosa Wedd. 密毛苎麻
Shrubs

16.4 (12.6~19.6); 9.8 (6.0~13.9); 21.1 (17.9~23.9); 16.5 (12.6~19.6); 137.5 (91.6~175.1); 0.0 (0.0~0.0); 805 (678~939) ❙ 1113 (797~1751); 630 (411~996); 45 (18~77) ❙ 38.4 (-14.2~99.9); 780 (678~927); 343.3 (218.0~551.3)

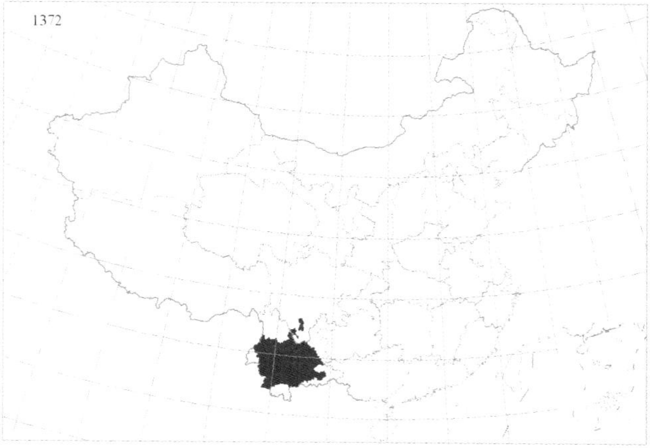

Urticaceae 荨麻科 *Boehmeria* 苎麻属
Boehmeria tricuspis (Hance) Makino 八角麻
Shrubs

14.1 (2.1~21.8); 1.5 (-17.9~13.1); 25.9 (14.2~30.1); 14.5 (6.4~21.8); 120.8 (38.1~201.6); 10.5 (0.0~84.9); 815 (504~1160) ❙ 1122 (367~1997); 469 (235~848); 109 (8~259) ❙ 34.4 (-40.9~147.0); 777 (367~1160); 207.9 (101.9~337.1)

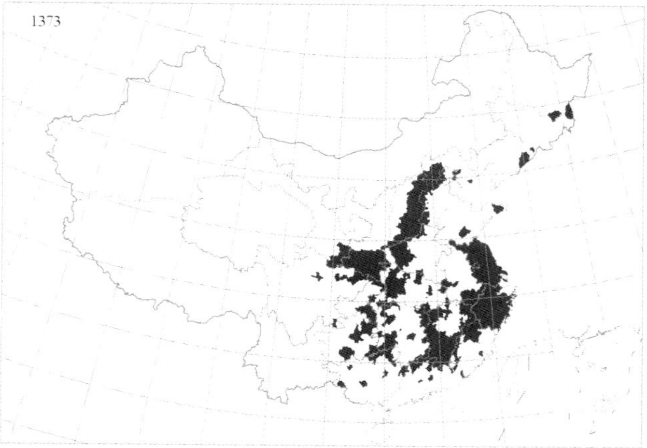

Urticaceae 荨麻科 *Boehmeria* 苎麻属
Boehmeria zollingeriana Wedd. 帚序苎麻
Shrubs

19.1 (16.2~23.8); 13.3 (10.3~18.1); 23.1 (20.1~28.2); 19.2 (16.2~23.8); 170.1 (134.2~225.8); 0.0 (0.0~0.0); 931 (773~1297) ❙ 1696 (939~2993); 883 (524~1847); 102 (37~547) ❙ 83.2 (6.5~246.5); 902 (744~1144); 383.8 (119.4~659.0)

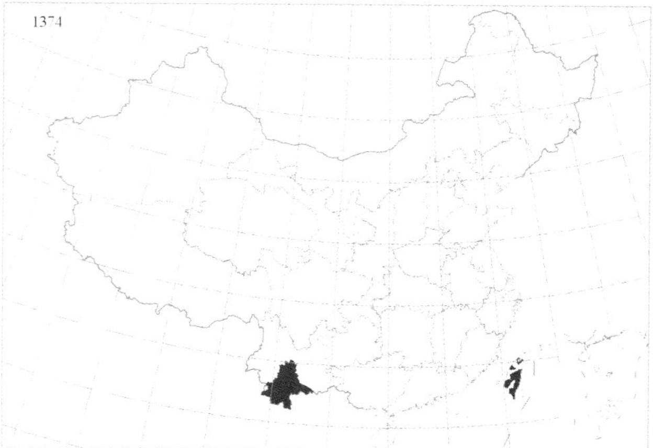

Urticaceae 荨麻科 *Boehmeria* 苎麻属
Boehmeria zollingeriana var. blinii (H. Lév.) C. J. Chen 黔桂苎麻
Shrubs

19.1 (17.0~21.3); 9.7 (7.1~12.6); 26.6 (24.6~28.7); 19.1 (17.0~21.3); 169.5 (143.7~196.0); 0.0 (0.0~0.0); 983 (854~1123) ❙ 1359 (1172~1620); 679 (588~775); 92 (43~166) ❙ 38.7 (5.8~64.7); 983 (854~1121); 264.9 (189.5~411.1)

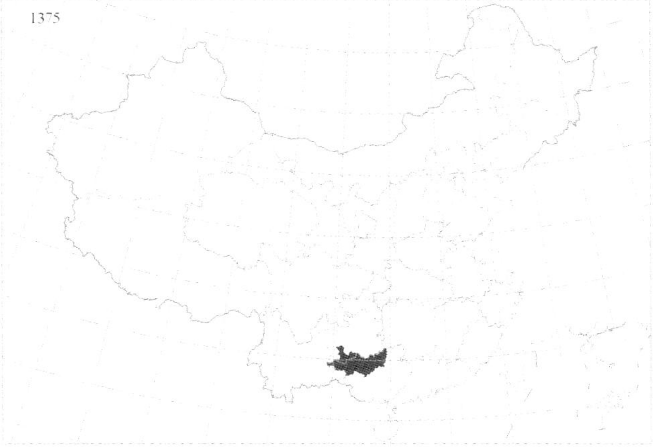

Urticaceae 荨麻科 *Cypholophus* 瘤冠麻属
Cypholophus moluccanus (Blume) Miq. 瘤冠麻
Tall shrubs

18.5; 14.1; 21.7; 18.5; 162.1; 0.0; 863 ❙ 2713; 1388; 188 ❙ 214.4; 863; 363.7

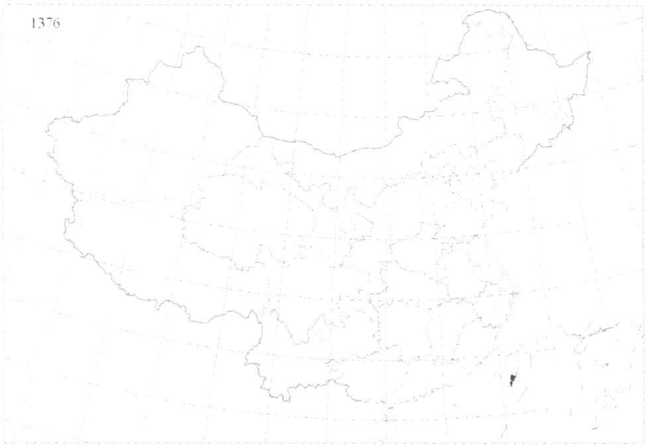

Urticaceae 荨麻科 *Debregeasia* 水麻属
Debregeasia elliptica C. J. Chen 椭圆叶水麻
Shrubs or small trees, deciduous

18.7 (17.2~22.2); 11.0 (9.1~14.2); 24.2 (22.4~28.3); 18.7 (17.3~22.3); 164.9 (147.0~207.5); 0.0 (0.0~0.0); 929 (833~1190) ❙ 1370 (1161~1522); 760 (633~860); 59 (48~87) ❙ 49.2 (15.4~80.6); 929 (833~1190); 389.9 (314.4~551.6)

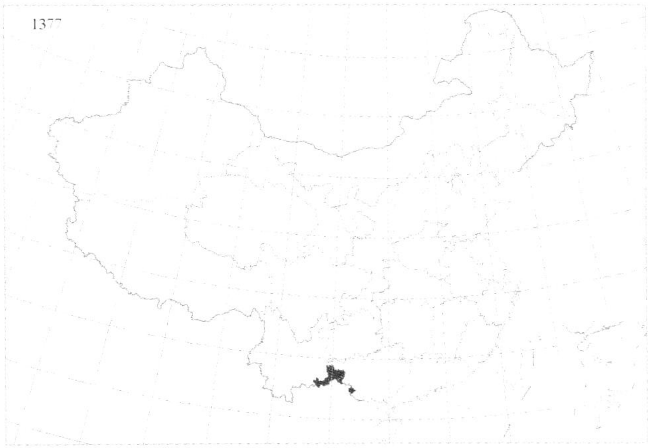

Urticaceae 荨麻科 *Debregeasia* 水麻属
Debregeasia longifolia (Burm. f.) Wedd. 长叶水麻
Shrubs or small trees

15.7 (−1.8~22.3); 7.0 (−10.2~15.1); 23.1 (6.1~28.9); 15.8 (1.8~22.3); 130.7 (1.9~208.1); 1.5 (0.0~82.8); 815 (329~1202) ❙ 1169 (320~1900); 587 (217~1056); 60 (10~161) ❙ 44.2 (−14.2~145.8); 807 (320~1202); 268.1 (72.2~659.0)

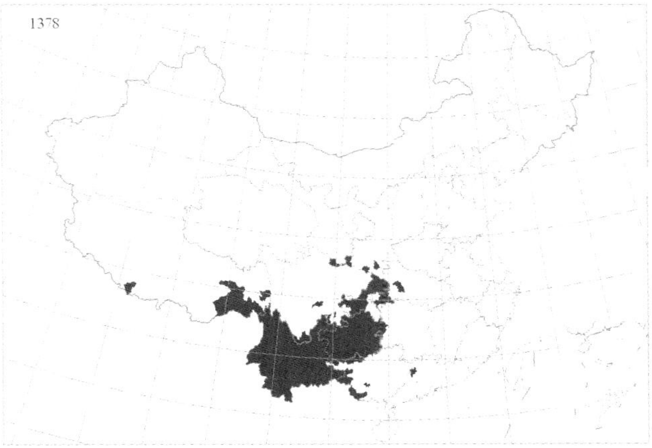

Urticaceae 荨麻科 *Debregeasia* 水麻属
Debregeasia orientalis C. J. Chen 水麻
Shrubs

14.5 (0.3~24.1); 5.3 (−9.9~18.9); 22.5 (9.0~29.9); 14.7 (3.3~24.1); 119.8 (11.9~229.7); 5.2 (0.0~67.5); 783 (389~1322) ❙ 1166 (514~3495); 572 (314~1594); 67 (9~897) ❙ 49.7 (−22.2~246.5); 772 (389~1170); 260.7 (108.2~659.0)

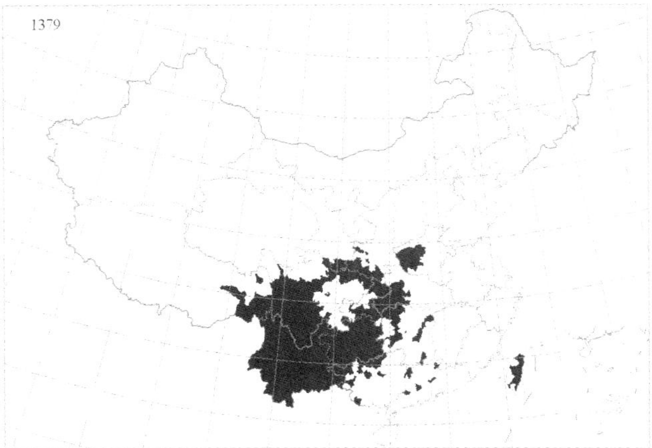

Urticaceae 荨麻科 *Debregeasia* 水麻属
Debregeasia saeneb (Forssk.) Hepper et Wood 柳叶水麻
Shrubs or small trees

14.4 (−1.8~18.9); 7.6 (−10.2~12.5); 19.3 (6.1~23.9); 14.5 (1.8~19.0); 116.5 (1.9~167.8); 3.5 (0.0~82.8); 741 (329~929) ❙ 965 (320~1164); 559 (193~651); 38 (15~59) ❙ 31.0 (−14.2~78.3); 711 (320~851); 293.0 (72.2~406.0)

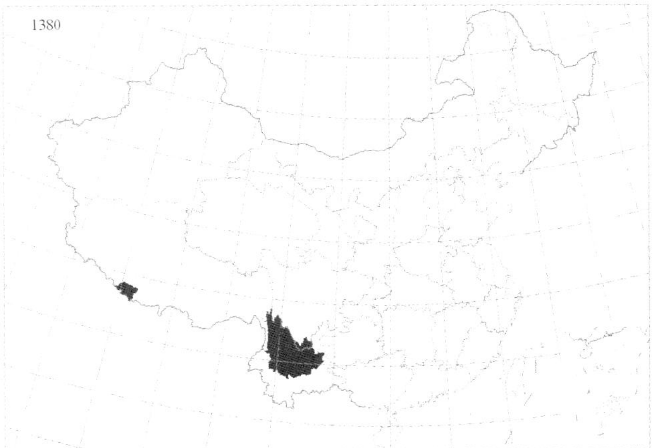

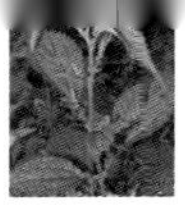

Urticaceae 荨麻科 *Debregeasia* 水麻属
Debregeasia squamata King ex Hook. f. 鳞片水麻
Dwarf shrubs, deciduous

20.5 (17.2~24.8); 12.4 (7.3~20.1); 27.0 (22.4~28.8); 20.6 (17.3~24.8); 186.8 (147.0~237.6); 0.0 (0.0~0.0); 1073 (833~1385) ❙ 1574 (1179~2148); 699 (392~1234); 119 (41~197) ❙ 48.8 (-11.9~86.7); 1047 (833~1326); 318.2 (185.5~659.0)

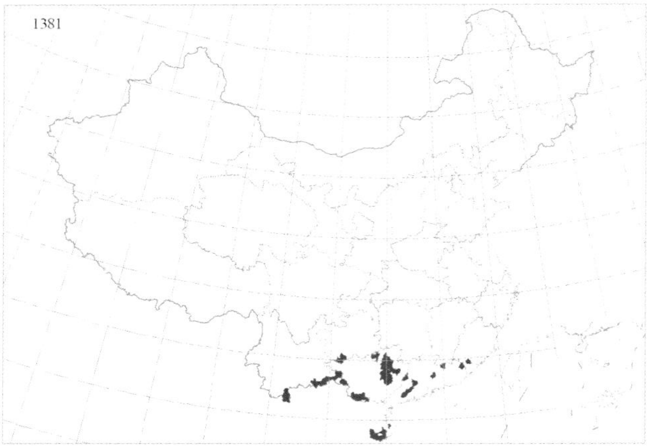

Urticaceae 荨麻科 *Dendrocnide* 火麻树属
Dendrocnide basirotunda (C. Y. Wu) Chew 圆基火麻树
Small trees

20.5; 15.0; 23.7; 20.5; 185.9; 0.0; 990 ❙ 1536; 678; 72 ❙ 55.1; 973; 614.6

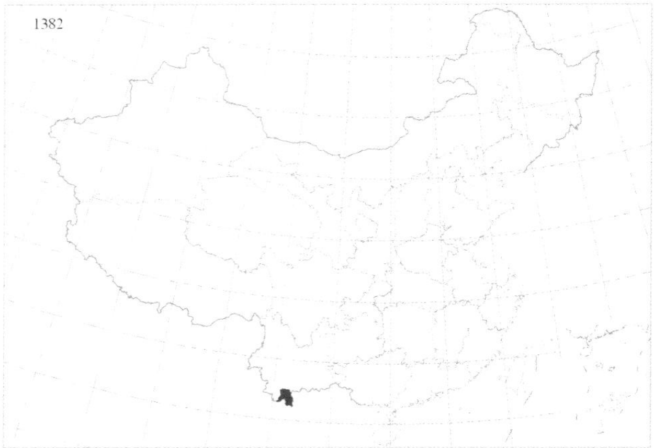

Urticaceae 荨麻科 *Dendrocnide* 火麻树属
Dendrocnide meyeniana (Walp.) Chew 咬人狗
Trees, evergreen

20.6 (16.2~23.8); 15.3 (11.0~18.1); 24.8 (20.1~28.2); 20.7 (16.2~23.8); 188.0 (134.2~225.8); 0.0 (0.0~0.0); 1048 (773~1297) ❙ 2542 (1757~2993); 1312 (924~1847); 203 (58~547) ❙ 155.9 (35.5~246.5); 1002 (773~1144); 256.5 (119.4~363.7)

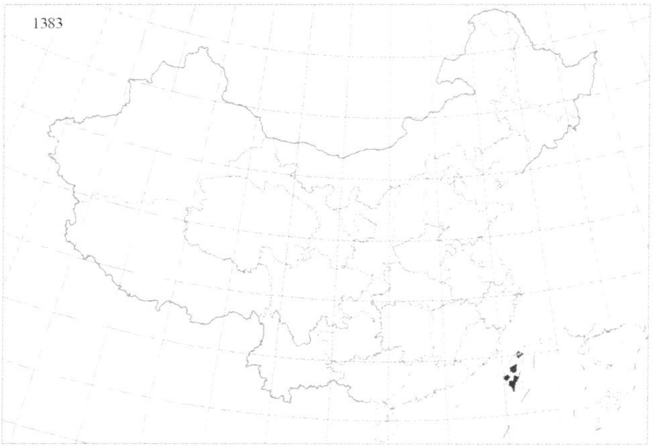

Urticaceae 荨麻科 *Dendrocnide* 火麻树属
Dendrocnide sinuata (Blume) Chew 全缘火麻树
Shrubs or small trees, evergreen

20.0 (13.3~24.5); 12.9 (5.4~18.9); 25.0 (19.6~28.8); 20.0 (13.4~24.5); 179.9 (100.4~234.4); 0.0 (0.0~0.0); 1012 (706~1360) ❙ 1528 (1176~1876); 779 (447~1007); 76 (48~132) ❙ 54.7 (-12.0~145.8); 1000 (706~1360); 408.8 (203.4~659.0)

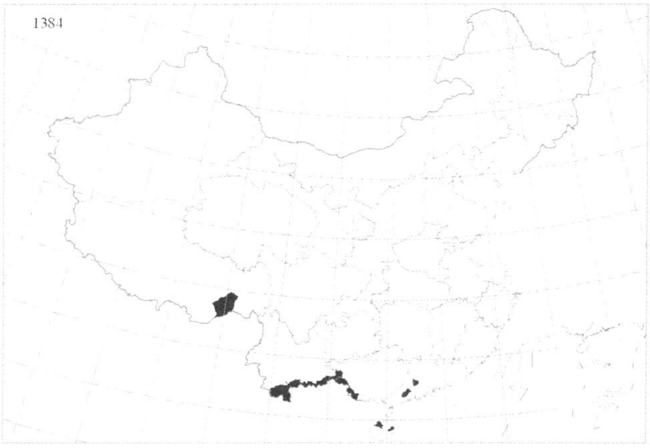

Urticaceae 荨麻科 *Dendrocnide* 火麻树属
Dendrocnide stimulans (L. f.) Chew 海南火树麻
Shrubs or small trees

23.5 (21.3~24.2); 17.0 (13.5~18.0); 28.4 (27.5~28.8); 23.6 (21.3~24.2); 223.1 (195.6~230.4); 0.0 (0.0~0.0); 1283 (1107~1331) ❙ 1649 (1309~1991); 662 (480~902); 110 (88~142) ❙ 30.2 (-1.4~79.9); 1268 (1107~1331); 285.4 (248.2~351.4)

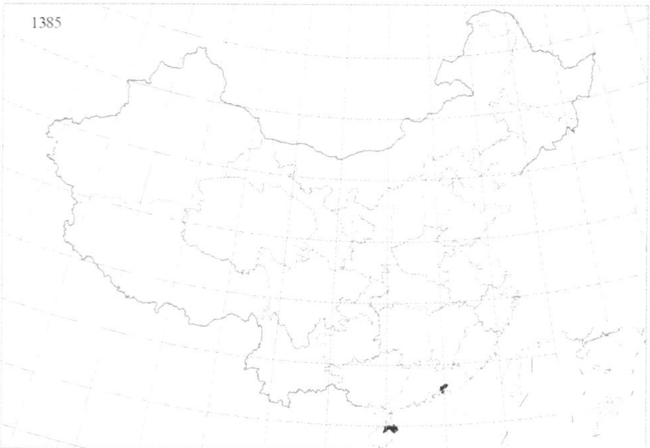

Urticaceae 荨麻科 *Dendrocnide* 火麻树属
Dendrocnide urentissima (Gagnep.) Chew 火麻树
Trees

19.1 (17.2~22.2); 11.8 (9.5~15.1); 24.2 (22.1~28.3); 19.1 (17.3~22.3); 169.4 (147.0~207.5); 0.0 (0.0~0.0); 940 (833~1190) ❙ 1395 (1143~1751); 765 (637~996); 63 (48~87) ❙ 49.3 (15.4~88.8); 935 (833~1190); 409.1 (289.7~659.0)

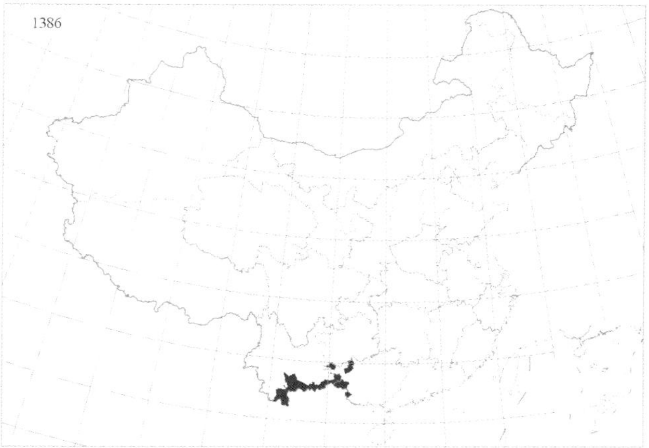

Urticaceae 荨麻科 *Laportea* 艾麻属
Laportea violacea Gagnep. 葡萄叶艾麻
Shrubs or subshrubs

19.1 (17.8~22.2); 9.7 (7.1~13.9); 26.6 (24.6~28.3); 19.1 (17.8~22.3); 169.3 (153.8~207.5); 0.0 (0.0~0.0); 983 (901~1190) ❙ 1492 (1250~1740); 711 (651~848); 119 (58~189) ❙ 53.1 (15.4~85.5); 983 (901~1190); 277.3 (208.7~327.7)

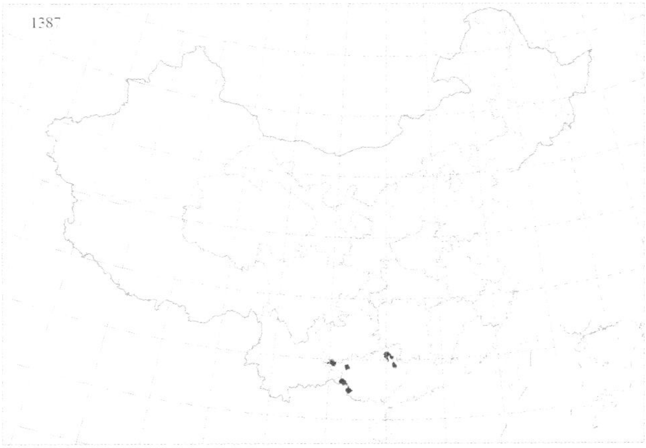

Urticaceae 荨麻科 *Leucosyke* 四脉麻属
Leucosyke quadrinervia C. B. Rob. 四脉麻
Shrubs or small trees, evergreen

18.2; 12.9; 22.5; 18.2; 158.9; 0.0; 864 ❙ 2534; 1144; 248 ❙ 93.4; 864; 324.1

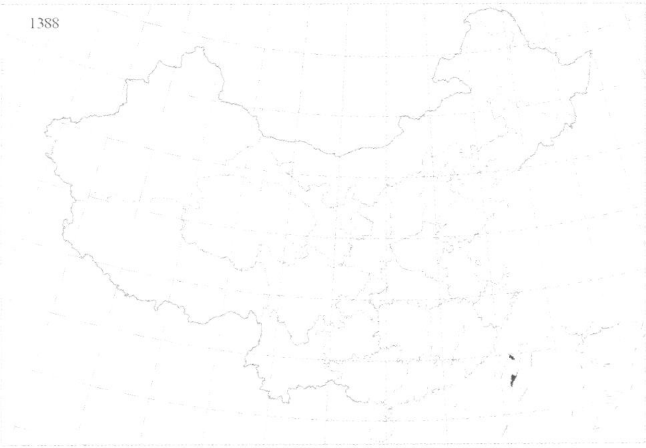

Urticaceae 荨麻科 *Maoutia* 水丝麻属
Maoutia puya (Hook.) Wedd. 水丝麻
Shrubs

16.9 (5.3~22.2); 9.9 (-2.4~15.1); 21.9 (12.2~28.3); 17.0 (5.7~22.3); 144.0 (30.5~207.5); 0.5 (0.0~26.2); 840 (479~1190) ❙ 1200 (797~1751); 671 (411~1007); 49 (12~141) ❙ 43.9 (-14.2~145.8); 821 (479~1190); 357.7 (189.5~659.0)

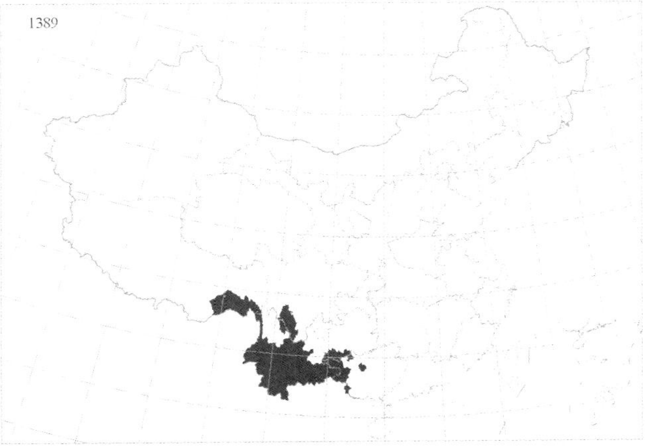

Urticaceae 荨麻科 *Oreocnide* 紫麻属
Oreocnide boniana (Gagnep.) Hand.-Mazz. 膜叶紫麻
Shrubs

21.0; 14.2; 25.6; 21.0; 191.8; 0.0; 1053 ❙ 1522; 818; 68 ❙ 44.5; 1053; 551.6

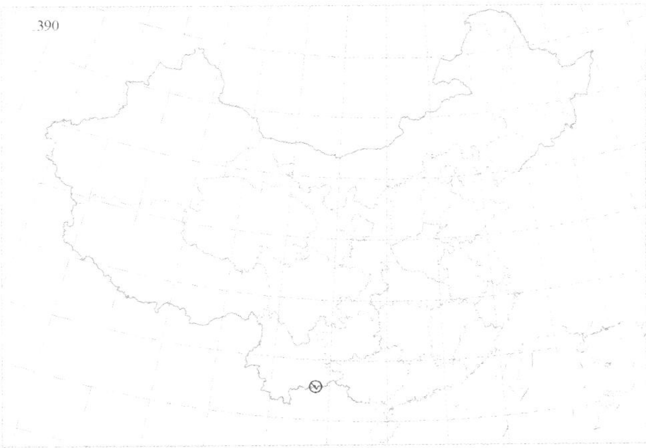

Urticaceae 荨麻科 *Oreocnide* 紫麻属
Oreocnide frutescens (Thunb.) Miq. 紫麻
Shrubs or small trees

16.5 (-1.8~24.7); 7.1 (-10.2~19.8); 24.8 (6.1~29.9); 16.6 (1.8~24.7); 140.3 (1.9~236.5); 1.3 (0.0~82.8); 869 (329~1376) ❙ 1359 (320~2037); 591 (217~1056); 111 (11~259) ❙ 56.6 (-14.2~145.8); 860 (320~1216); 261.5 (72.2~659.0)

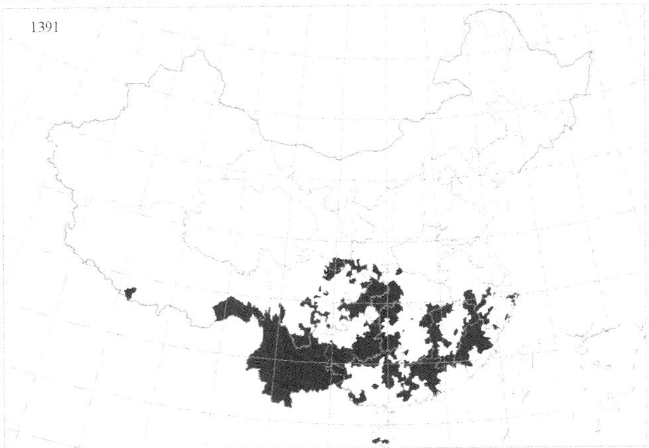

Urticaceae 荨麻科 *Oreocnide* 紫麻属
Oreocnide frutescens subsp. insignis C. J. Chen 细梗紫麻
Shrubs or small trees

21.0; 13.2; 26.9; 21.0; 192.3; 0.0; 1081 ❙ 1776; 828; 125 ❙ 64.3; 1081; 254.1

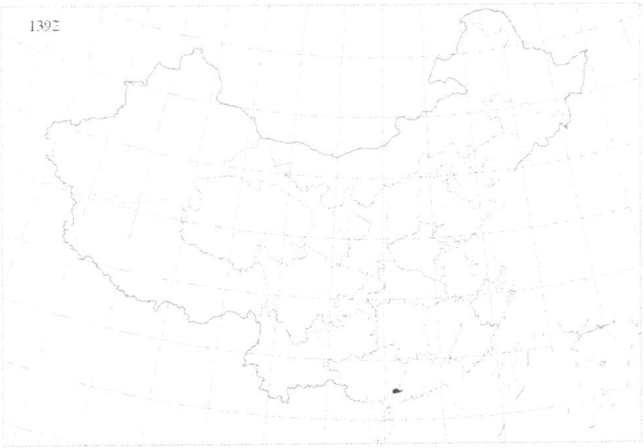

Urticaceae 荨麻科 *Oreocnide* 紫麻属
Oreocnide frutescens subsp. occidentalis C. J. Chen 滇藏紫麻
Small trees

18.2；11.8；22.5；18.2；158.6；0.0；871 ❙ 1457；834；61 ❙ 67.2；871；521.4

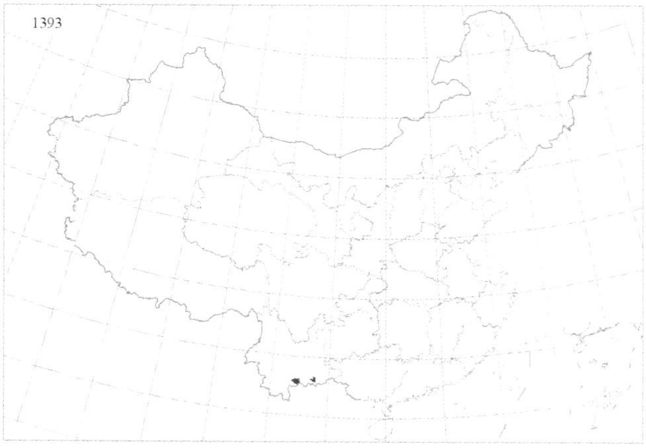

Urticaceae 荨麻科 *Oreocnide* 紫麻属
Oreocnide integrifolia (Gaudich.) Miq. 全缘叶紫麻
Shrubs or small trees, evergreen

20.1 (8.6~24.3)；13.9 (1.3~18.7)；24.7 (15.0~28.7)；20.2 (8.7~24.3)；182.7 (52.6~231.9)；0.9 (0.0~8.8)；1077 (574~1338) ❙ 1438 (1007~1736)；645 (392~1007)；63 (40~110) ❙ 42.1 (-12.0~145.8)；1014 (574~1326)；447.4 (283.8~572.9)

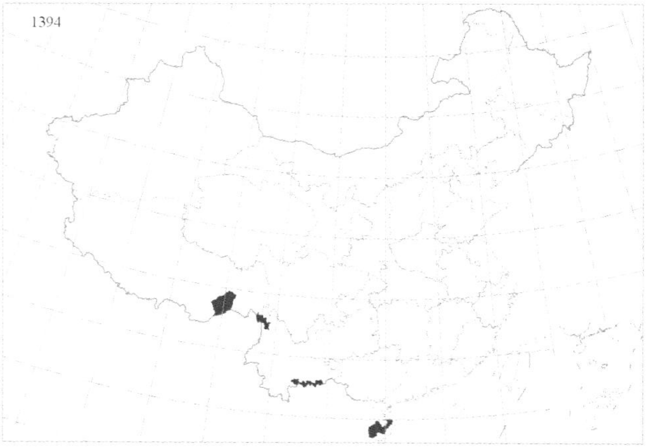

Urticaceae 荨麻科 *Oreocnide* 紫麻属
Oreocnide kwangsiensis Hand.-Mazz. 广西紫麻
Shrubs

20.0 (18.7~22.2)；10.9 (8.8~13.9)；27.0 (25.0~28.3)；20.0 (18.8~22.3)；180.4 (165.2~207.5)；0.0 (0.0~0.0)；1035 (945~1190) ❙ 1439 (1250~1662)；736 (651~848)；104 (57~173) ❙ 39.4 (15.4~55.9)；1035 (945~1190)；279.7 (204.5~371.7)

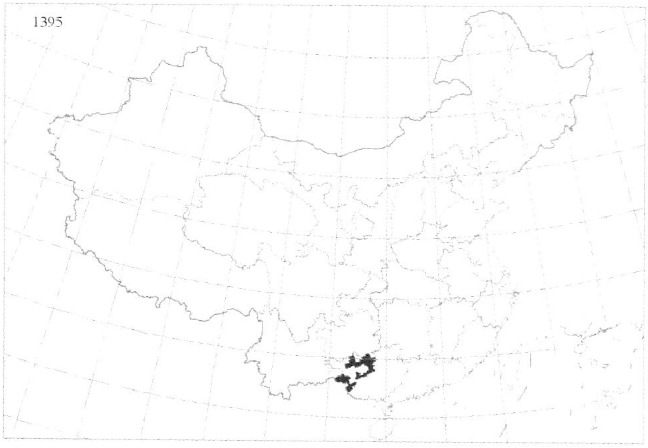

Urticaceae 荨麻科 *Oreocnide* 紫麻属
Oreocnide obovata (C. H. Wright) Merr. 倒卵叶紫麻
Shrubs

18.9 (14.3~22.7)；10.9 (3.3~15.2)；25.0 (20.8~29.0)；19.0 (14.3~22.8)；167.5 (114.1~213.1)；0.1 (0.0~2.3)；956 (765~1221) ❙ 1485 (1059~2435)；724 (445~1428)；101 (41~191) ❙ 56.1 (15.4~101.8)；950 (765~1221)；335.6 (169.7~659.0)

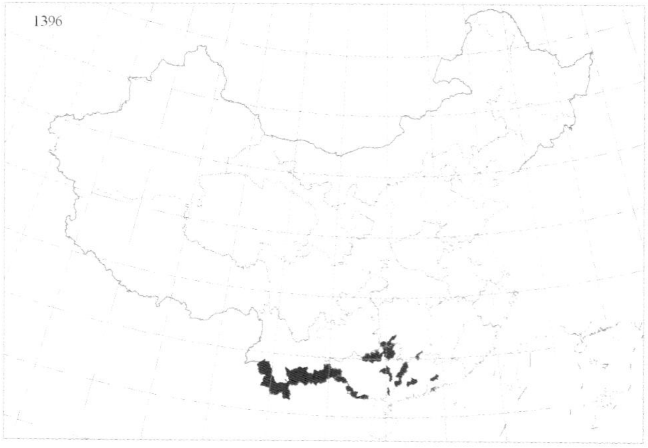

Urticaceae 荨麻科 *Oreocnide* 紫麻属
Oreocnide obovata var. paradoxa (Gagnep.) C. J. Chen 凹尖紫麻
Shrubs

16.4；5.4；26.3；16.5；137.7；0.0；864 ❙ 1425；518；161 ❙ 64.9；864；239.6

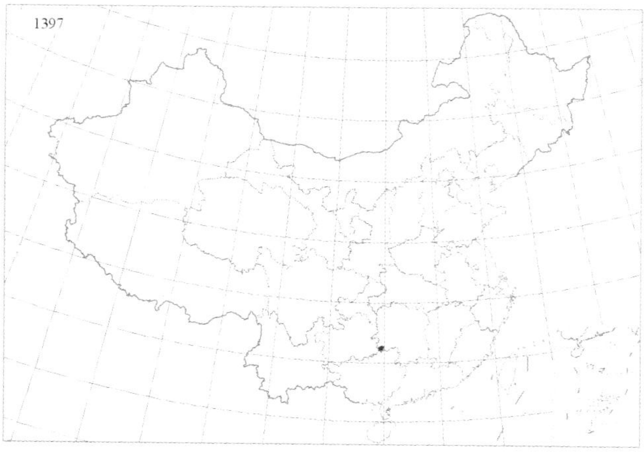

Urticaceae 荨麻科 *Oreocnide* 紫麻属
Oreocnide pedunculata (Shirai) Masam. 长梗紫麻
Shrubs or small trees

19.7 (16.2~24.1)；14.1 (11.0~18.9)；24.5 (20.1~28.0)；19.8 (16.2~24.1)；177.2 (134.2~229.7)；0.0 (0.0~0.0)；983 (773~1322) ❙ 2641 (1751~3495)；1146 (821~1847)；317 (47~897) ❙ 175.4 (32.5~246.5)；962 (773~1123)；255.8 (154.2~303.6)

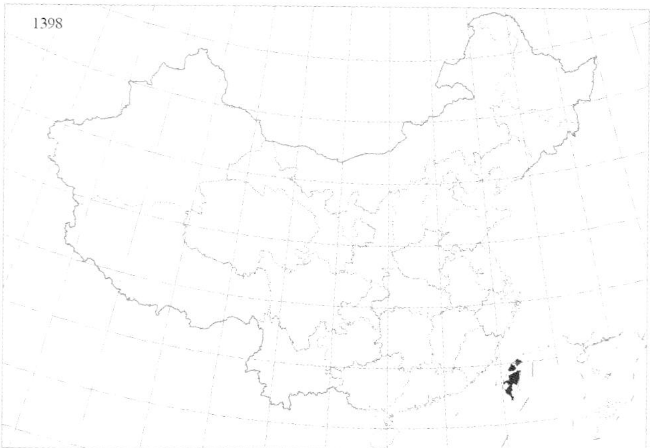

Urticaceae 荨麻科 *Oreocnide* 紫麻属
Oreocnide rubescens (Blume) Miq. 红紫麻
Shrubs or small trees, evergreen

19.6 (16.3~24.3); 12.8 (9.1~18.7); 24.3 (21.4~28.7); 19.6 (16.4~24.3); 175.7 (136.4~231.9); 0.0 (0.0~0.0); 981 (797~1338) ❙ 1383 (1007~2435); 717 (392~1428); 63 (41~171) ❙ 42.4 (-12.0~99.5); 955 (783~1326); 399.0 (206.0~659.0)

1399
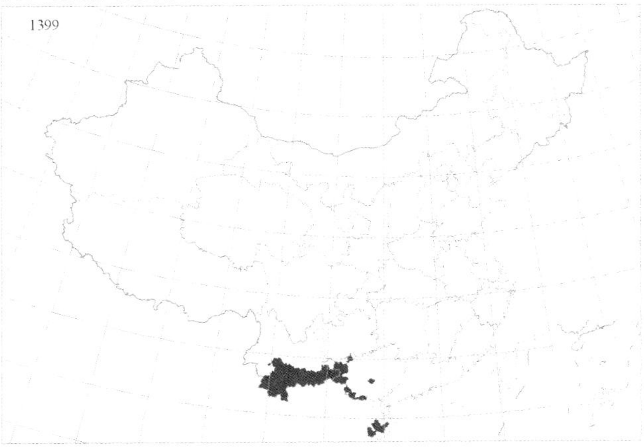

Urticaceae 荨麻科 *Oreocnide* 紫麻属
Oreocnide serrulata C. J. Chen 细齿紫麻
Shrubs

19.6 (17.2~22.2); 11.6 (9.5~14.2); 25.3 (22.4~28.3); 19.6 (17.3~22.3); 175.2 (147.0~207.5); 0.0 (0.0~0.0); 988 (833~1190) ❙ 1380 (1188~1522); 762 (634~860); 64 (48~87) ❙ 42.0 (5.8~80.6); 988 (833~1190); 357.4 (294.6~551.6)

1400
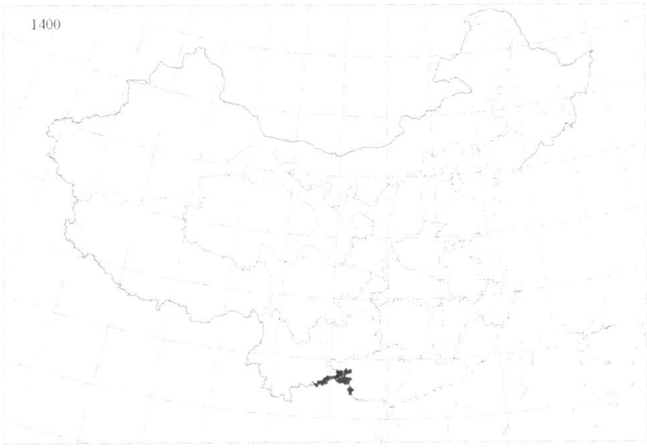

Urticaceae 荨麻科 *Oreocnide* 紫麻属
Oreocnide tonkinensis (Gagnep.) Merr. et Chun 宽叶紫麻
Shrubs

18.8 (16.3~21.8); 11.5 (9.1~14.2); 23.9 (21.4~28.5); 18.8 (16.4~21.8); 165.6 (136.4~201.6); 0.0 (0.0~0.0); 926 (797~1162) ❙ 1409 (1074~1751); 774 (582~996); 68 (47~145) ❙ 53.0 (30.7~88.8); 926 (797~1162); 396.0 (208.5~551.6)

1401
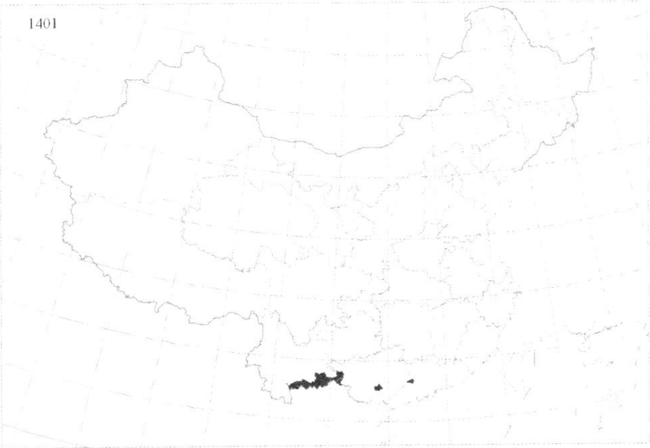

Urticaceae 荨麻科 *Pipturus* 落尾麻属
Pipturus arborescens (Link) C. B. Rob. 落尾木
Shrubs or small trees, evergreen

17.4; 12.8; 21.0; 17.5; 150.0; 0.0; 823 ❙ 2606; 1168; 246 ❙ 216.8; 823; 333.6

1402
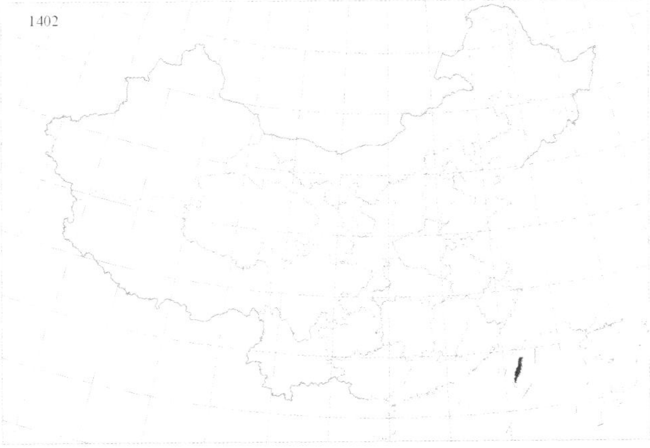

Urticaceae 荨麻科 *Poikilospermum* 锥头麻属
Poikilospermum lanceolatum (Trécul) Merr. 毛叶锥头麻
Scandent shrubs

17.7 (5.3~21.0); 11.4 (-2.4~15.1); 22.0 (12.2~25.6); 17.7 (5.7~21.0); 154.2 (30.5~191.8); 1.7 (0.0~26.2); 871 (479~1053) ❙ 1403 (961~1751); 755 (558~1007); 57 (31~77) ❙ 63.5 (26.1~145.8); 862 (479~1053); 471.6 (234.2~659.0)

1403
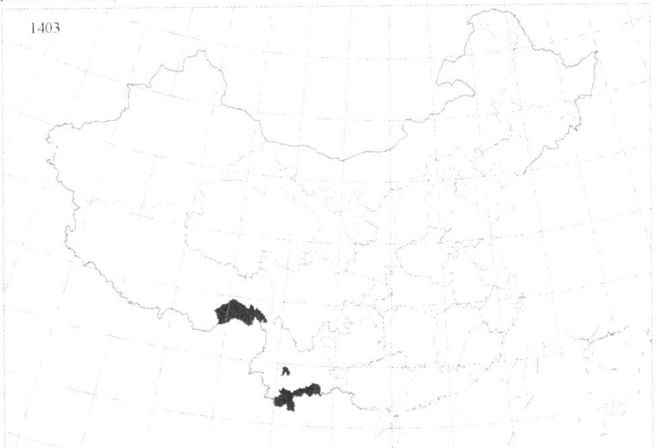

Urticaceae 荨麻科 *Poikilospermum* 锥头麻属
Poikilospermum naucleiflorum (Roxb. ex Lindl.) Chew 大序锥头麻
Shrubs or vines

18.9 (16.9~21.0); 12.8 (10.3~15.1); 23.1 (21.4~25.6); 19.0 (16.9~21.0); 167.7 (142.7~191.8); 0.0 (0.0~0.0); 915 (813~1053) ❙ 1427 (1163~1751); 783 (641~996); 61 (48~77) ❙ 55.3 (34.8~88.8); 914 (813~1053); 501.8 (383.4~659.0)

1404
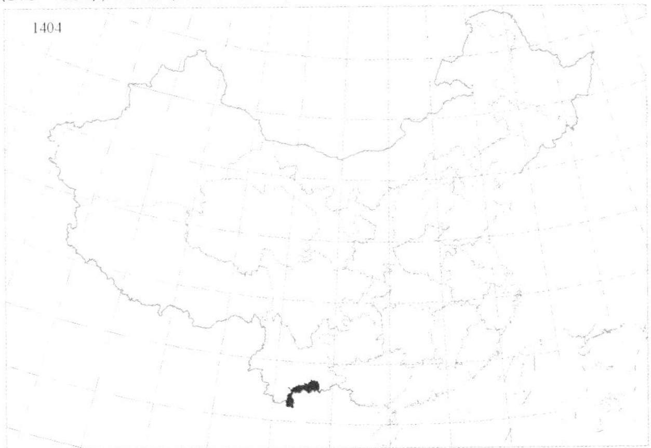

Urticaceae 荨麻科 *Pouzolzia* 雾水葛属
Pouzolzia calophylla W. T. Wang et C. J. Chen 美叶雾水葛
Shrubs

11.9 (5.3~19.3); 4.7 (-2.4~13.1); 17.6 (12.2~23.6); 12.0 (5.7~19.4); 88.8 (30.5~173.2); 5.9 (0.0~26.2); 673 (479~939) ❙ 1244 (961~1736); 702 (558~1007); 51 (31~69) ❙ 89.2 (21.7~145.8); 667 (479~903); 362.1 (234.2~447.4)

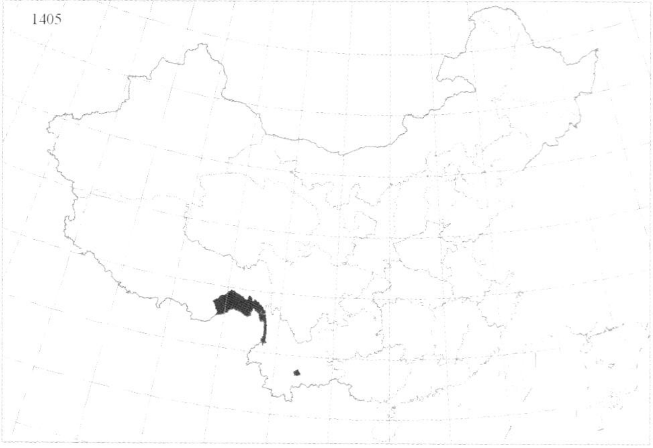

Urticaceae 荨麻科 *Pouzolzia* 雾水葛属
Pouzolzia niveotomentosa W. T. Wang 雪毡雾水葛
Shrubs

13.6 (10.1~19.2); 5.5 (2.3~13.3); 20.4 (16.9~24.0); 13.7 (10.2~19.4); 105.8 (68.1~172.6); 1.9 (0.0~5.2); 730 (618~932) ❙ 1060 (963~1276); 608 (550~741); 35 (26~51) ❙ 46.5 (34.8~70.8); 725 (618~923); 291.4 (212.0~452.0)

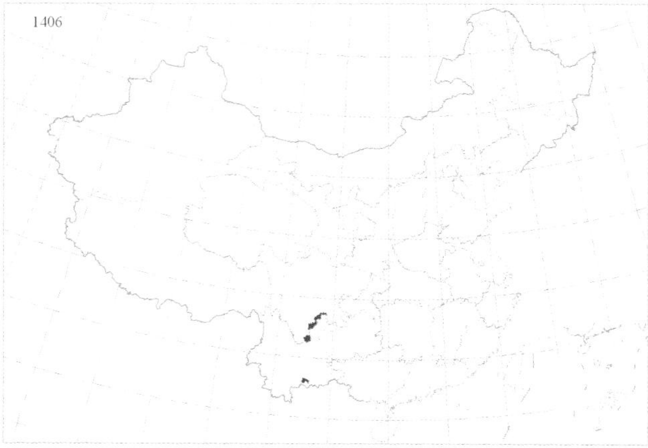

Urticaceae 荨麻科 *Pouzolzia* 雾水葛属
Pouzolzia sanguinea (Blume) Merr. 红雾水葛
Shrubs

16.5 (-1.8~24.8); 8.3 (-10.2~20.1); 23.2 (6.1~28.9); 16.6 (1.8~24.8); 139.9 (1.9~237.6); 1.3 (0.0~82.8); 854 (329~1385) ❙ 1201 (320~2435); 612 (193~1428); 61 (13~182) ❙ 41.6 (-34.3~145.8); 839 (320~1221); 285.1 (72.2~659.0)

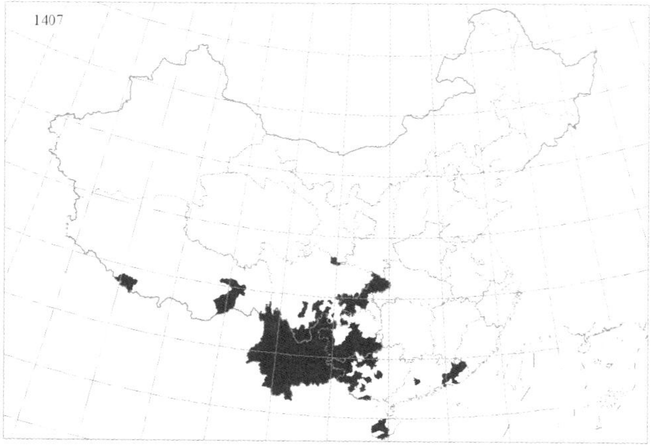

Urticaceae 荨麻科 *Pouzolzia* 雾水葛属
Pouzolzia sanguinea var. elegans (Wedd.) Friis, Wilmot-Dear et C. J. Chen 雅致雾水葛
Shrubs

15.7 (1.8~24.1); 8.5 (-7.1~18.9); 21.3 (9.9~28.5); 15.8 (3.8~24.1); 130.4 (15.5~229.7); 1.7 (0.0~53.3); 799 (404~1322) ❙ 1243 (583~3495); 669 (348~1847); 66 (12~897) ❙ 55.0 (-14.2~246.5); 781 (404~1153); 313.2 (119.4~659.0)

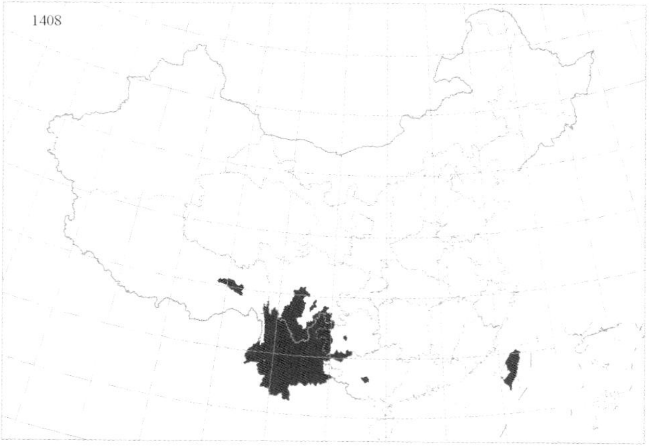

P

Urticaceae 荨麻科 *Sarcochlamys* 肉被麻属
Sarcochlamys pulcherrima Gaudich. 肉被麻
Shrubs or small trees, evergreen

11.0; 3.4; 17.3; 11.0; 76.5; 4.4; 640 ❙ 1389; 796; 48 ❙ 113.7; 640; 399.9

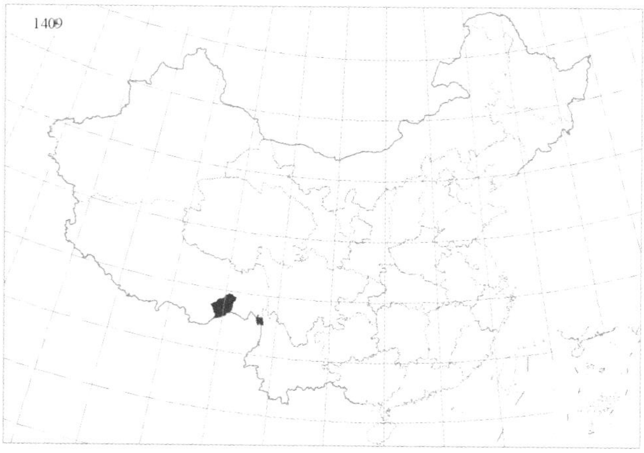

Proteaceae 山龙眼科 *Helicia* 山龙眼属
Helicia clivicola W. W. Sm. 山地山龙眼
Shrubs or small trees, evergreen

17.5 (14.4~19.6); 11.1 (7.5~13.1); 21.8 (19.1~23.6); 17.6 (14.4~19.6); 150.9 (113.1~175.7); 0.0 (0.0~0.0); 849 (727~953) ❙ 1275 (939~1688); 731 (524~959); 50 (37~70) ❙ 50.6 (6.5~112.2); 822 (717~932); 416.9 (289.9~627.0)

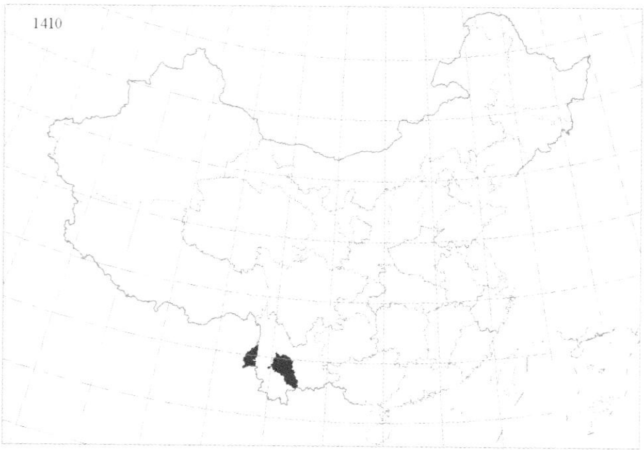

Proteaceae 山龙眼科 *Helicia* 山龙眼属
Helicia cochinchinensis Lour. 小果山龙眼
Shrubs or trees, evergreen

18.6 (9.3~24.8); 8.8 (-1.7~20.1); 27.3 (19.1~30.1); 18.6 (9.5~24.8); 164.0 (69.8~237.6); 0.2 (0.0~18.0); 983 (603~1385) ❙ 1527 (868~3495); 607 (299~1594); 151 (32~897) ❙ 57.5 (-34.3~246.5); 968 (603~1360); 238.3 (95.3~659.0)

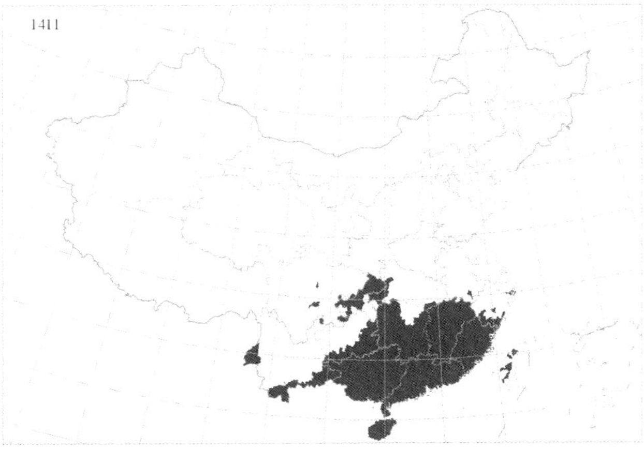

Proteaceae 山龙眼科 *Helicia* 山龙眼属
Helicia dongxingensis H. S. Kiu 东兴山龙眼
Trees, evergreen

21.9; 14.1; 27.9; 22.0; 203.7; 0.0; 1161 ❙ 2161; 1240; 126 ❙ 85.7; 1161; 305.5

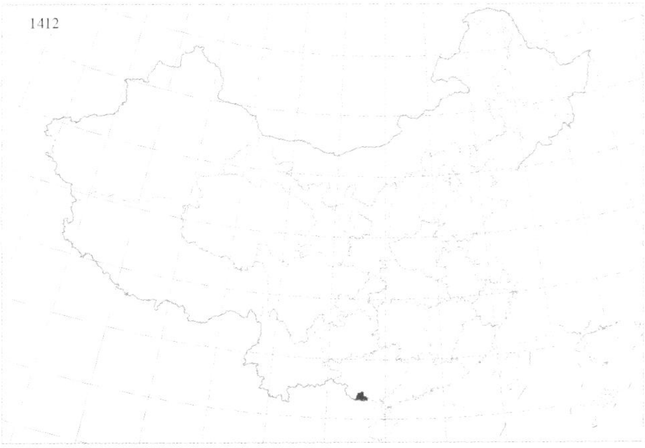

Proteaceae 山龙眼科 *Helicia* 山龙眼属
Helicia falcata C. Y. Wu 镰叶山龙眼
Trees, evergreen

19.2 (17.3~22.8); 12.1 (9.9~14.7); 24.3 (22.4~28.9); 19.3 (17.3~22.8); 171.3 (147.1~213.6); 0.0 (0.0~0.0); 958 (833~1230) ❙ 1482 (1368~1647); 829 (756~893); 66 (52~106) ❙ 56.7 (33.9~80.6); 958 (833~1230); 425.3 (229.5~551.6)

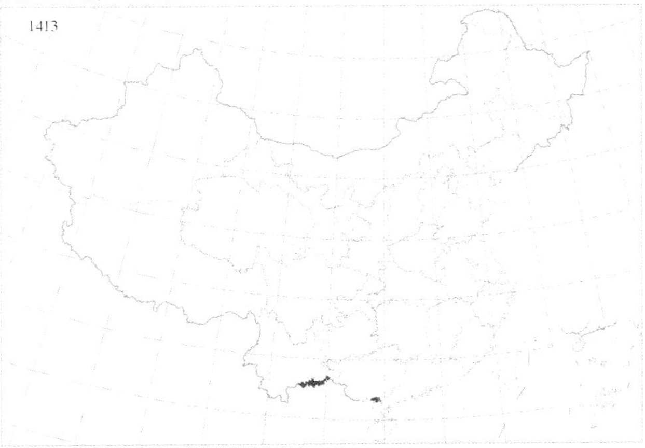

Proteaceae 山龙眼科 *Helicia* 山龙眼属
Helicia formosana Hemsl. 山龙眼
Trees, evergreen

21.3 (14.6~24.7); 13.9 (3.8~19.8); 27.1 (20.1~28.9); 21.3 (14.7~24.7); 196.2 (117.7~236.5); 0.0 (0.0~1.5); 1125 (773~1376) ❙ 1749 (868~2993); 816 (312~1847); 124 (42~547) ❙ 61.2 (-34.3~246.5); 1079 (773~1360); 301.7 (145.3~620.0)

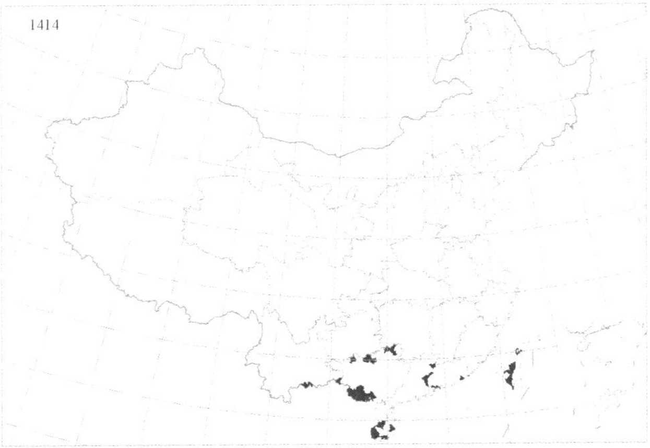

Proteaceae 山龙眼科 *Helicia* 山龙眼属
Helicia grandis Hemsl. 大山龙眼
Trees, evergreen

18.0 (16.3~21.0); 11.0 (9.0~14.2); 22.8 (21.4~25.6); 18.0 (16.4~21.0); 156.3 (136.4~191.8); 0.0 (0.0~0.0); 873 (797~1053) ❙ 1323 (1018~1751); 744 (554~996); 55 (47~77) ❙ 51.3 (25.7~88.8); 873 (797~1053); 411.3 (277.0~551.6)

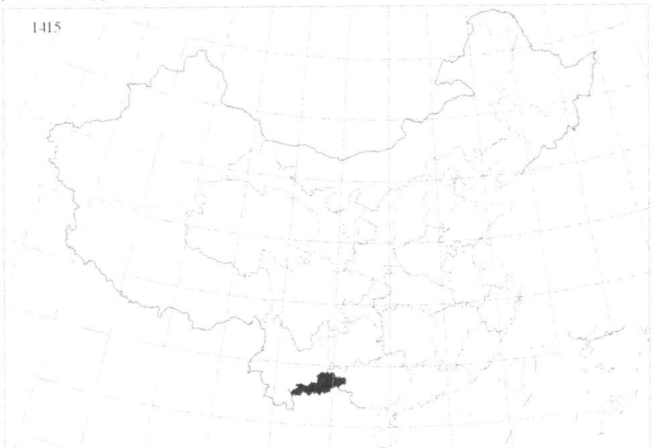

Proteaceae 山龙眼科 *Helicia* 山龙眼属
Helicia hainanensis Hayata 海南山龙眼
Shrubs or trees, evergreen

21.6 (16.3~24.8); 14.5 (9.1~20.1); 26.7 (21.4~28.8); 21.6 (16.4~24.8); 199.3 (136.4~237.6); 0.0 (0.0~0.0); 1136 (797~1385) ❙ 1508 (868~2435); 708 (312~1428); 81 (41~131) ❙ 34.7 (-34.3~99.5); 1093 (797~1360); 363.0 (209.2~620.0)

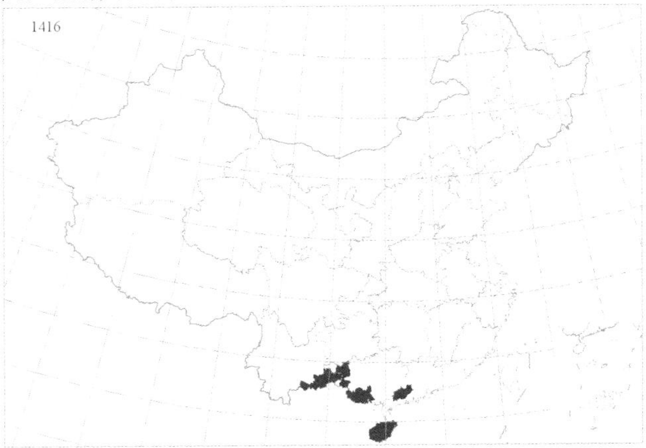

Proteaceae 山龙眼科 *Helicia* 山龙眼属
Helicia kwangtungensis W. T. Wang 广东山龙眼
Trees, evergreen

20.2 (16.5~22.4); 11.1 (5.8~14.7); 27.8 (25.7~29.8); 20.2 (16.5~22.5); 182.4 (138.2~209.6); 0.0 (0.0~0.0); 1056 (861~1205) ❙ 1659 (1403~2024); 664 (505~891); 163 (95~205) ❙ 58.1 (24.9~98.6); 1055 (861~1205); 231.0 (95.3~284.3)

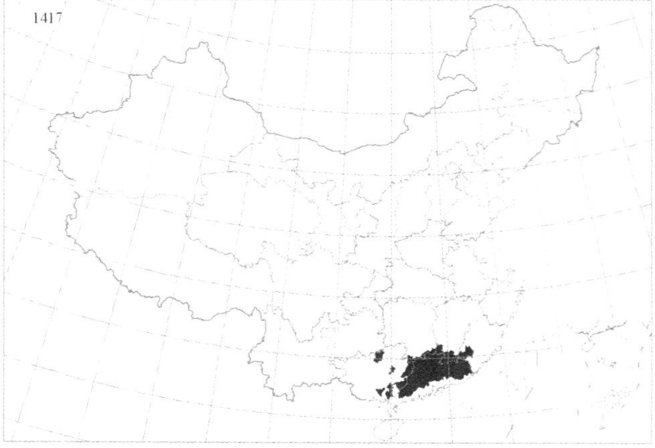

Proteaceae 山龙眼科 *Helicia* 山龙眼属
Helicia longipetiolata Merr. et Chun 长柄山龙眼
Trees, evergreen

21.5 (18.1~24.8); 13.8 (9.4~20.1); 27.3 (24.6~28.9); 21.6 (18.2~24.8); 198.8 (158.1~237.6); 0.0 (0.0~0.0); 1135 (901~1385) ❙ 1539 (1172~2435); 755 (473~1428); 89 (41~130) ❙ 36.2 (-1.4~99.5); 1116 (901~1326); 322.9 (194.5~620.0)

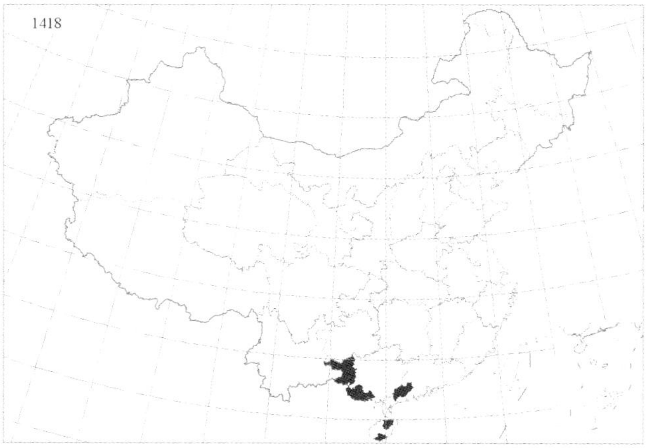

Proteaceae 山龙眼科 *Helicia* 山龙眼属
Helicia nilagirica Bedd. 深绿山龙眼
Trees, evergreen

17.8 (14.4~21.0); 11.1 (7.5~15.1); 22.2 (19.1~25.6); 17.8 (14.4~21.0); 153.6 (113.1~191.8); 0.0 (0.0~0.0); 860 (727~1053) ❙ 1304 (958~1751); 720 (531~996); 53 (41~77) ❙ 51.7 (12.5~112.2); 849 (727~1053); 417.5 (274.9~659.0)

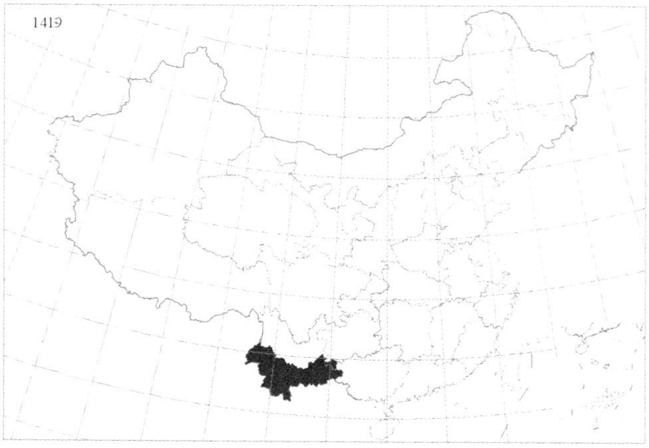

Proteaceae 山龙眼科 *Helicia* 山龙眼属
Helicia obovatifolia Merr. et Chun 倒卵叶山龙眼
Trees, evergreen

22.5 (18.0~24.8); 15.5 (8.5~20.1); 27.7 (25.0~28.8); 22.6 (18.0~24.8); 210.8 (156.4~237.6); 0.0 (0.0~0.0); 1206 (913~1385) ❙ 1556 (868~2435); 698 (312~1428); 90 (41~169) ❙ 30.6 (-34.3~99.5); 1149 (868~1360); 359.8 (209.2~620.0)

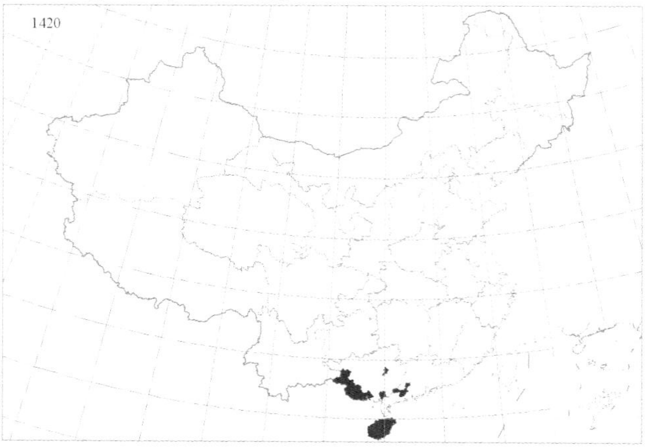

Proteaceae 山龙眼科 *Helicia* 山龙眼属
Helicia obovatifolia var. mixta (H. L. Li) Sleumer 枇杷叶山龙眼
Trees, evergreen

22.2 (17.9~24.8); 14.7 (7.4~20.1); 27.9 (25.7~28.8); 22.2 (18.0~24.8); 206.5 (155.6~237.6); 0.0 (0.0~0.0); 1189 (913~1385) ❙ 1588 (868~2435); 681 (312~1428); 111 (41~189) ❙ 35.7 (-34.3~99.5); 1137 (868~1360); 334.2 (185.5~620.0)

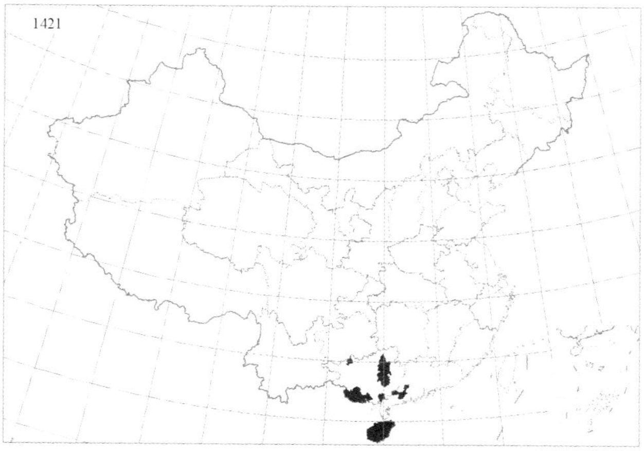

Proteaceae 山龙眼科 *Helicia* 山龙眼属
Helicia pyrrhobotrya Kurz 焰序山龙眼
Trees, evergreen

21.0 (16.3~24.8); 14.7 (9.1~20.1); 25.3 (21.4~28.8); 21.0 (16.4~24.8); 192.1 (136.4~237.6); 0.0 (0.0~0.0); 1083 (797~1385) ❙ 1450 (868~1854); 661 (312~996); 68 (41~194) ❙ 37.9 (-34.3~88.8); 1027 (797~1360); 432.2 (213.5~659.0)

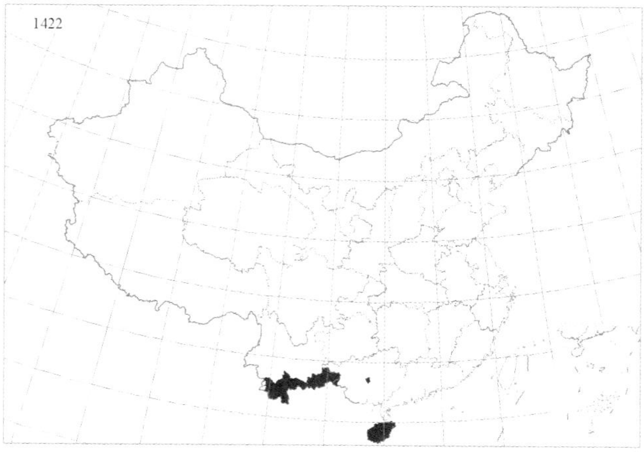

Proteaceae 山龙眼科 *Helicia* 山龙眼属
Helicia reticulata W. T. Wang 网脉山龙眼
Shrubs or trees, evergreen

19.7 (14.3~23.6)；10.8 (3.3~16.7)；27.1 (21.4~29.8)；19.8 (14.3~23.6)；177.2 (112.6~223.5)；0.0 (0.0~2.3)；1031 (756~128) | 1544 (1007~2435)；684 (444~1428)；132 (43~207) | 50.7 (-6.9~101.8)；1030 (756~1273)；243.2 (95.3~551.6)

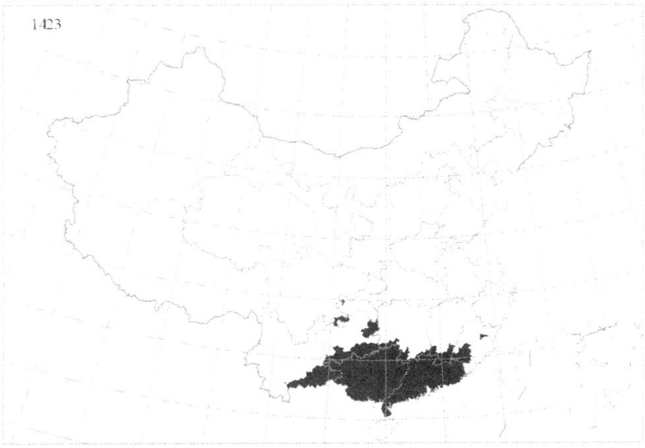

Proteaceae 山龙眼科 *Helicia* 山龙眼属
Helicia shweliensis W. W. Sm. 瑞丽山龙眼
Shrubs or trees, evergreen

17.3 (14.4~19.6)；10.5 (7.5~12.8)；21.5 (19.1~23.5)；17.3 (14.4~19.6)；147.7 (113.1~175.7)；0.0 (0.0~0.0)；839 (727~953) | 1482 (1049~1688)；854 (598~959)；54 (42~70) | 77.6 (26.1~112.2)；829 (727~932)；491.2 (365.3~627.0)

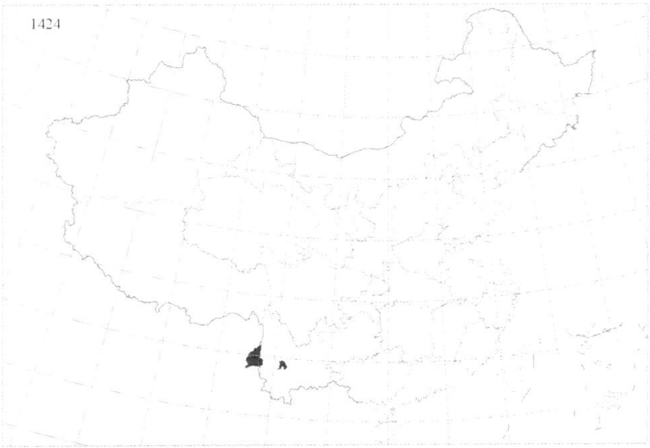

Proteaceae 山龙眼科 *Helicia* 山龙眼属
Helicia silvicola W. W. Sm. 林地山龙眼
Trees, evergreen

18.5 (16.3~21.0)；11.8 (9.1~15.1)；23.0 (21.4~25.6)；18.5 (16.4~21.0)；162.1 (136.4~191.8)；0.0 (0.0~0.0)；894 (797~1053) | 1401 (1074~1751)；759 (582~996)；58 (47~77) | 56.7 (32.4~88.8)；889 (797~1053)；438.9 (277.0~659.0)

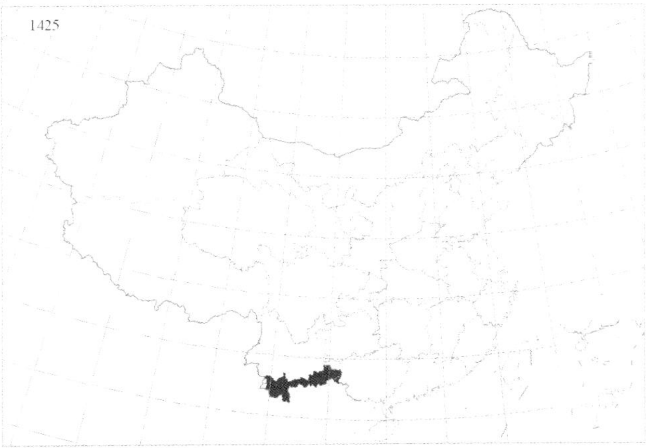

Proteaceae 山龙眼科 *Helicia* 山龙眼属
Helicia tibetensis H. X. Kiu 西藏山龙眼
Trees, evergreen

11.0；3.4；17.3；11.0；76.5；4.4；640 | 1389；796；48 | 113.7；640；399.9

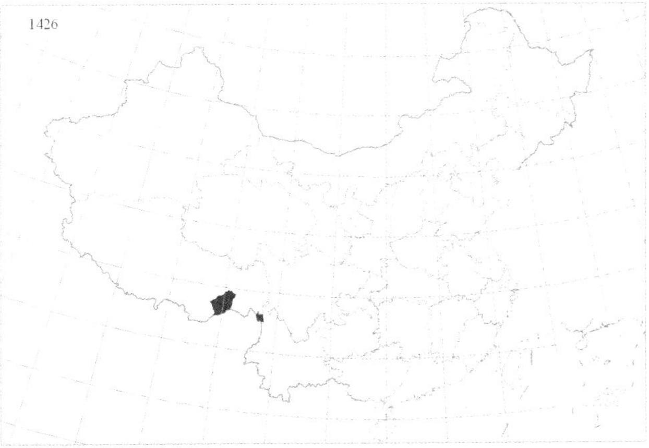

Proteaceae 山龙眼科 *Helicia* 山龙眼属
Helicia tsaii W. T. Wang 潞西山龙眼
Trees, evergreen

17.6 (16.0~19.6)；11.2 (9.3~13.9)；21.6 (20.3~23.5)；17.6 (16.0~19.6)；151.6 (132.4~175.7)；0.0 (0.0~0.0)；846 (780~953) | 1394 (1083~1683)；773 (594~959)；50 (41~62) | 64.7 (33.9~112.2)；833 (779~932)；464.9 (346.6~627.0)

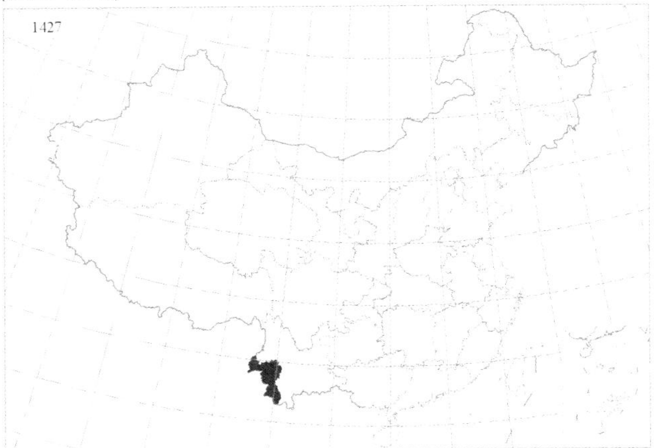

Proteaceae 山龙眼科 *Helicia* 山龙眼属
Helicia vestita W. W. Sm. 浓毛山龙眼
Trees, evergreen

18.5 (13.3~20.5)；12.5 (5.4~15.1)；22.4 (19.6~23.7)；18.5 (13.4~20.5)；162.3 (100.4~185.9)；0.0 (0.0~0.0)；901 (706~992) | 1560 (1464~1736)；765 (655~1007)；63 (51~77) | 77.1 (47.6~145.8)；885 (706~986)；523.7 (427.9~659.0)

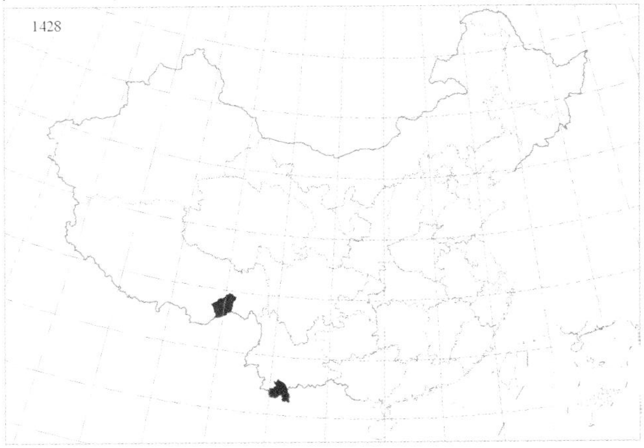

Proteaceae 山龙眼科 *Helicia* 山龙眼属
Helicia vestita var. longipes W. T. Wang 锈毛山龙眼
Trees, evergreen

13.3; 5.4; 19.6; 13.4; 100.4; 0.0; 706 ❙ 1736; 1007; 51 ❙ 145.8; 706; 447.4

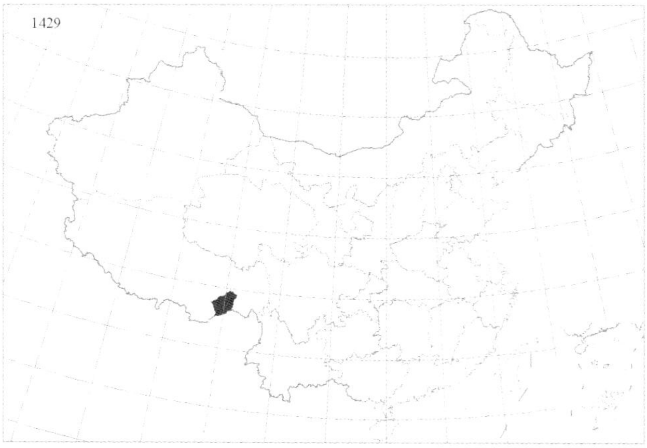

Proteaceae 山龙眼科 *Helicia* 山龙眼属
Helicia yangchunensis H. S. Kiu 阳春山龙眼
Trees, evergreen

21.8; 14.2; 27.8; 21.9; 202.6; 0.0; 1150 ❙ 1876; 891; 116 ❙ 63.2; 1150; 273.3

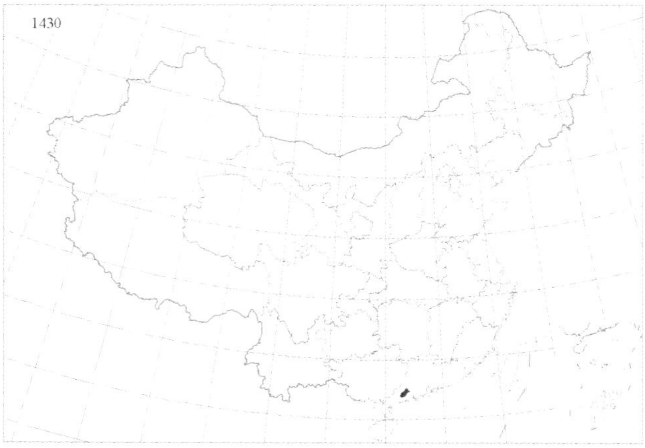

Proteaceae 山龙眼科 *Heliciopsis* 假山龙眼属
Heliciopsis henryi (Diels) W. T. Wang 假山龙眼
Trees

21.5 (19.1~22.7); 13.3 (11.0~15.0); 27.6 (25.0~28.6); 21.5 (19.1~22.8); 198.0 (169.6~213.1); 0.0 (0.0~0.0); 1128 (945~1221) ❙ 1648 (1188~2435); 884 (634~1428); 106 (56~137) ❙ 46.1 (5.8~99.5); 1128 (945~1221); 290.4 (223.4~371.7)

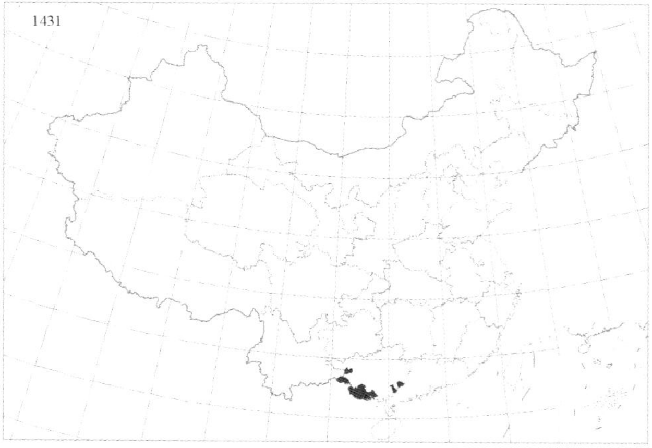

Proteaceae 山龙眼科 *Heliciopsis* 假山龙眼属
Heliciopsis lobata (Merr.) Sleumer 调羹树
Trees, evergreen

20.7 (17.2~24.7); 13.7 (7.1~19.8); 25.6 (21.6~28.7); 20.7 (17.3~24.7); 188.4 (147.0~236.5); 0.0 (0.0~0.0); 1058 (833~1376) ❙ 1573 (1176~2435); 789 (447~1428); 80 (41~182) ❙ 50.9 (-12.0~99.5); 1034 (833~1360); 411.8 (197.5~659.0)

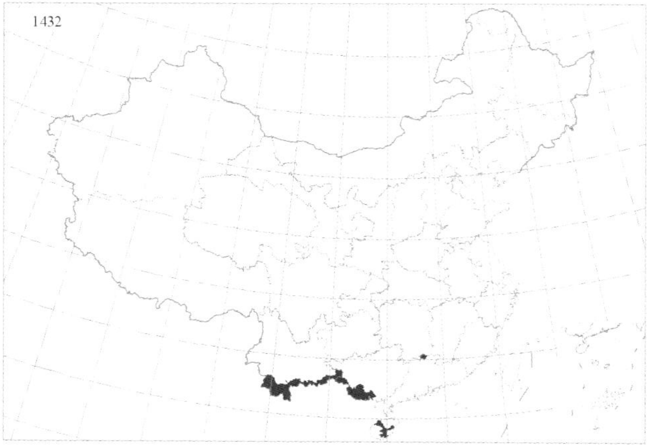

Proteaceae 山龙眼科 *Heliciopsis* 假山龙眼属
Heliciopsis terminalis (Kurz) Sleumer 痄腮树
Trees

19.9 (16.0~24.8); 13.0 (9.3~20.1); 24.7 (20.3~28.9); 20.0 (16.0~24.8); 179.7 (132.4~237.6); 0.0 (0.0~0.0); 1009 (780~1385) ❙ 1508 (1178~2435); 788 (455~1428); 74 (41~146) ❙ 52.1 (1.5~112.2); 990 (780~1299); 407.2 (194.5~659.0)

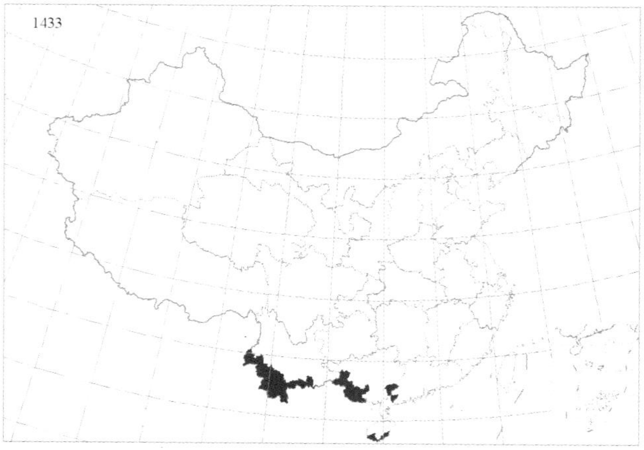

Olacaceae 铁青树科 *Erythropalum* 赤苍藤属
Erythropalum scandens Blume 赤苍藤
Vines, evergreen

20.0 (14.9~24.8); 12.3 (4.6~20.1); 25.7 (20.3~29.0); 20.0 (15.0~24.8); 179.9 (119.9~237.6); 0.0 (0.0~0.4); 1028 (781~1385) ❙ 1430 (868~2435); 704 (312~1428); 81 (41~173) ❙ 41.4 (-34.3~99.5); 1005 (780~1331); 333.1 (160.7~659.0)

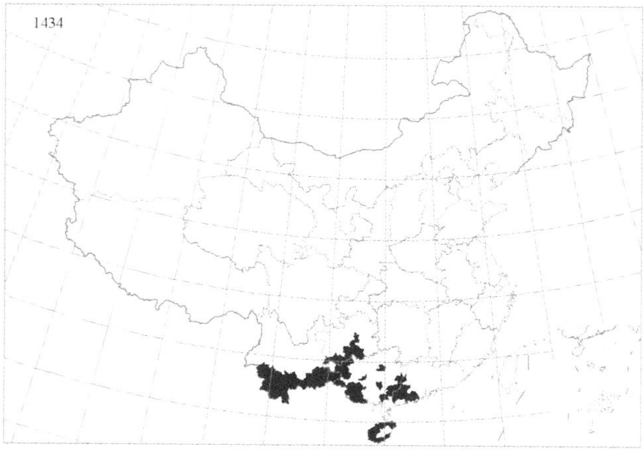

Olacaceae 铁青树科 *Malania* 蒜头果属
Malania oleifera Chun et S. K. Lee 蒜头果
Trees

20.0 (17.2~22.2); 11.3 (8.8~13.9); 26.5 (23.1~28.6); 20.0 (17.3~22.3); 180.1 (147.5~207.5); 0.0 (0.0~0.0); 1028 (847~1190) ❙ 1335 (1161~1518); 713 (633~848); 78 (43~138) ❙ 30.6 (5.8~55.9); 1028 (847~1190); 305.9 (238.3~411.1)

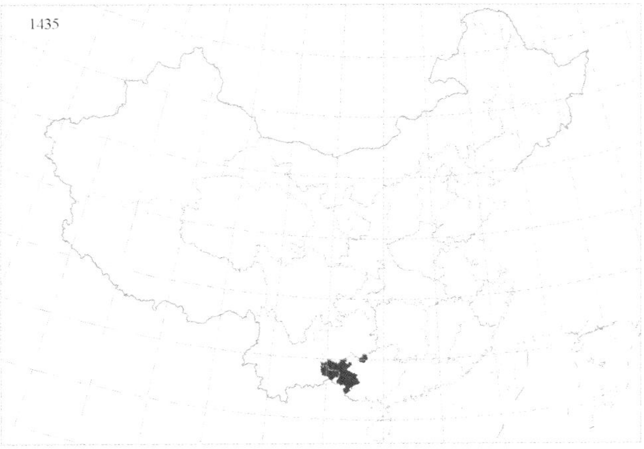

Olacaceae 铁青树科 *Olax* 铁青树属
Olax austrosinensis Y. R. Ling 疏花铁青树
Shrubs

22.9 (21.3~24.8); 16.0 (12.9~20.1); 28.0 (27.2~28.6); 22.9 (21.3~24.8); 215.2 (195.9~237.6); 0.0 (0.0~0.0); 1234 (1111~1385) ❙ 1748 (1380~2435); 821 (455~1428); 101 (41~145) ❙ 43.1 (1.5~99.5); 1176 (1111~1360); 370.1 (208.5~538.2)

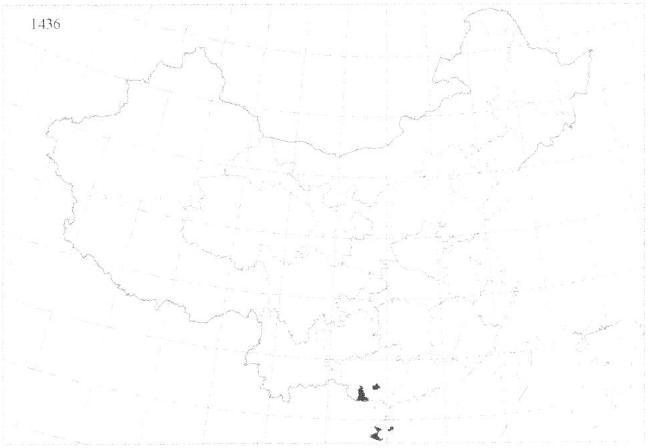

Olacaceae 铁青树科 *Olax* 铁青树属
Olax imbricata Roxb. 铁青树
Shrubs

23.2 (21.3~24.8); 16.6 (12.9~20.1); 28.0 (27.2~28.7); 23.2 (21.3~24.8); 218.6 (195.9~237.6); 0.0 (0.0~0.0); 1254 (1111~1385) ❙ 1610 (1176~2435); 722 (392~1428); 88 (41~145) ❙ 30.1 (-12.0~99.5); 1156 (1056~1360); 386.5 (208.5~538.2)

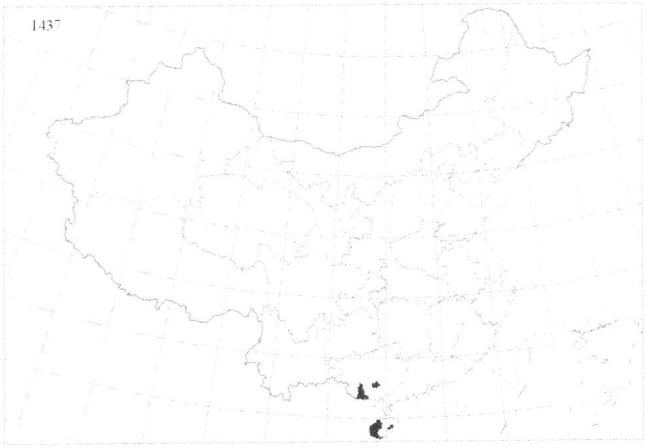

Olacaceae 铁青树科 *Schoepfia* 青皮木属
Schoepfia chinensis Gardner et Champ. 华南青皮木
Small trees, deciduous

18.0 (6.6~24.3); 8.8 (-1.3~18.7); 26.0 (13.3~29.8); 18.1 (6.8~24.3); 157.5 (39.6~231.9); 0.5 (0.0~19.7); 954 (514~1338) ❙ 1488 (862~2435); 621 (312~1428); 137 (12~254) ❙ 57.1 (-34.3~130.7); 940 (514~1232); 243.0 (148.9~538.2)

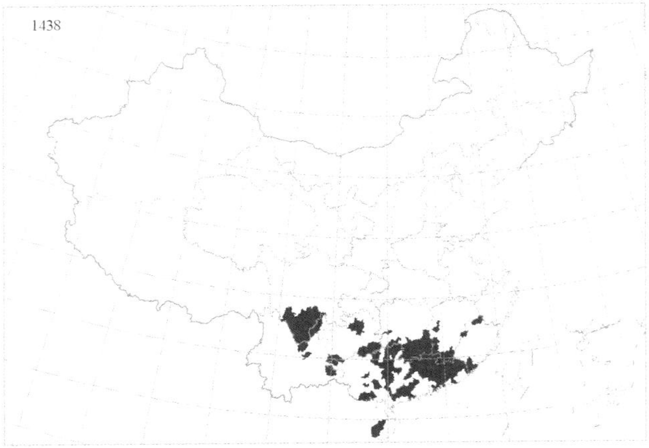

Olacaceae 铁青树科 *Schoepfia* 青皮木属
Schoepfia fragrans Wall. 香芙木
Small trees, evergreen

16.8 (5.3~20.5); 10.3 (-2.4~15.1); 21.2 (12.2~23.9); 16.8 (5.7~20.5); 141.9 (30.5~185.9); 0.4 (0.0~26.2); 819 (479~992) ❙ 1180 (797~1751); 662 (411~996); 46 (22~77) ❙ 44.1 (-14.2~112.2); 792 (479~986); 374.3 (218.0~659.0)

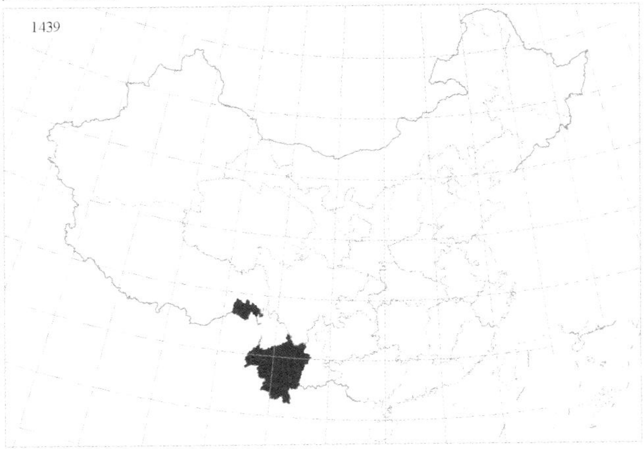

Olacaceae 铁青树科 *Schoepfia* 青皮木属
Schoepfia jasminodora Siebold et Zucc. 青皮木
Shrubs or small trees, deciduous

16.0 (2.6~22.7); 5.7 (-7.0~17.7); 25.5 (11.2~30.1); 16.1 (4.3~22.8); 134.9 (20.0~213.1); 1.8 (0.0~48.2); 859 (427~1221) ❙ 1385 (707~2992); 570 (350~1847); 132 (15~303) ❙ 61.1 (-14.2~246.5); 851 (427~1221); 239.3 (133.8~551.6)

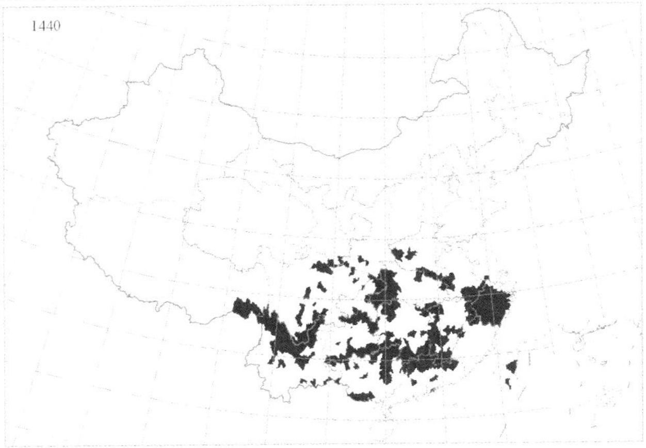

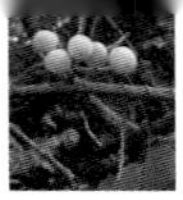

Olacaceae 铁青树科 *Schoepfia* 青皮木属

Schoepfia jasminodora var. malipoensis Y. R. Ling 大果青皮木

Trees, deciduous

19.0; 10.2; 25.8; 19.0; 168.4; 0.0; 958 ❙ 1250; 675; 60 ❙ 30.5; 958; 291.2

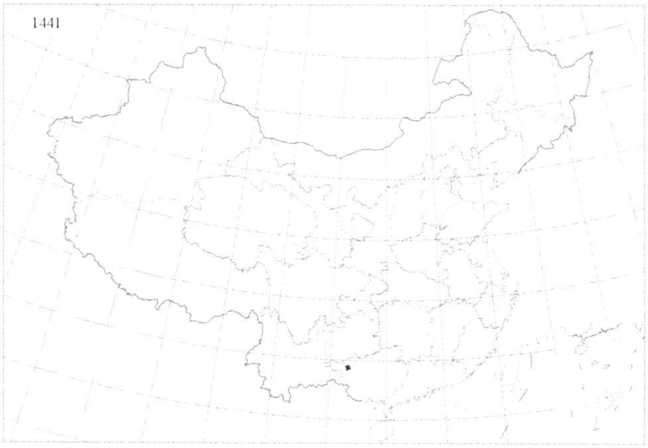

Olacaceae 铁青树科 *Ximenia* 海檀木属

Ximenia americana L. 海檀木

Shrubs or small trees

24.0 (22.2~24.8); 18.9 (17.2~20.1); 27.5 (25.7~28.5); 24.0 (22.3~24.8); 228.4 (207.2~237.6); 0.0 (0.0~0.0); 1307 (1127~1385) ❙ 1604 (1380~1854); 542 (455~622); 66 (41~115) ❙ 23.4 (1.5~47.8); 1185 (1055~1360); 528.3 (428.6~620.0)

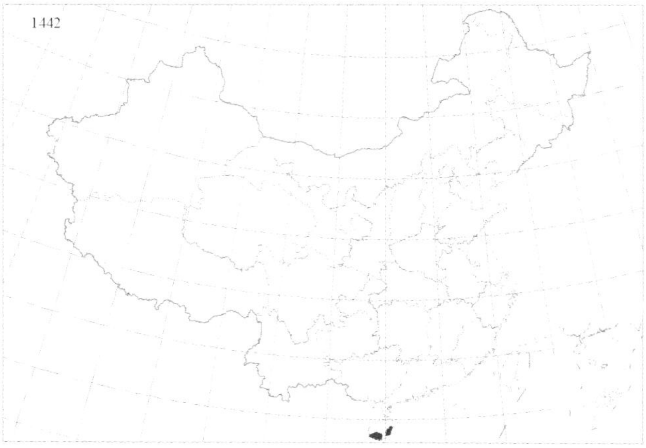

Opiliaceae 山柚子科 *Cansjera* 山柑藤属

Cansjera rheedei J. F. Gmel. 山柑藤

Vines or shrubs, evergreen

20.7 (16.3~24.8); 13.5 (9.0~20.1); 26.2 (21.4~29.0); 20.8 (16.4~24.8); 189.8 (136.4~237.6); 0.0 (0.0~0.0); 1080 (797~1385) ❙ 1468 (958~2037); 721 (447~1056); 84 (41~165) ❙ 37.1 (-12.0~80.6); 1056 (797~1326); 320.2 (122.7~620.0)

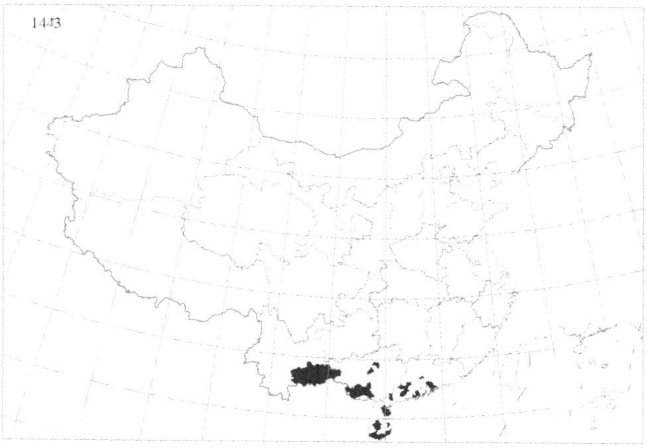

Opiliaceae 山柚子科 *Champereia* 台湾山柚属

Champereia manillana (Blume) Merr. 台湾山柚

Shrubs or trees, evergreen

22.2 (19.3~24.1); 16.8 (14.4~18.9); 26.4 (22.8~28.0); 22.2 (19.3~24.1); 206.7 (171.9~229.7); 0.0 (0.0~0.0); 1152 (912~1322) ❙ 2125 (1498~2992); 1284 (859~1847); 95 (47~127) ❙ 92.2 (25.2~202.5); 1070 (912~1131); 217.5 (154.2~296.9)

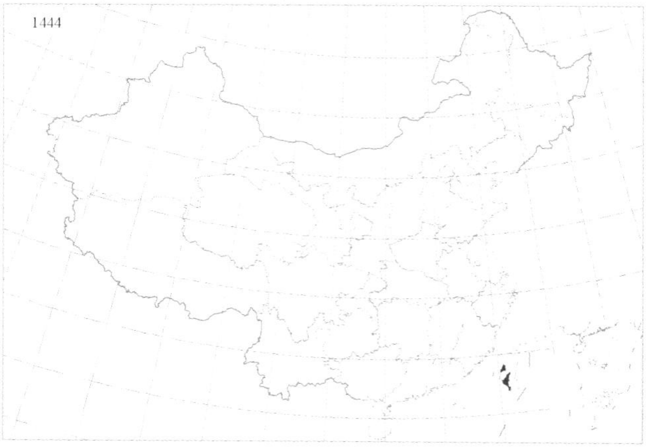

Opiliaceae 山柚子科 *Lepionurus* 鳞尾木属

Lepionurus sylvestris Blume 鳞尾木

Shrubs, evergreen

18.6 (16.9~21.0); 12.3 (9.9~14.2); 22.9 (21.4~25.6); 18.7 (16.9~21.0); 163.9 (142.7~191.8); 0.0 (0.0~0.0); 899 (813~1053) ❙ 1423 (1163~1751); 798 (641~996); 59 (48~77) ❙ 58.5 (34.8~88.8); 896 (813~1053); 469.8 (383.4~551.6)

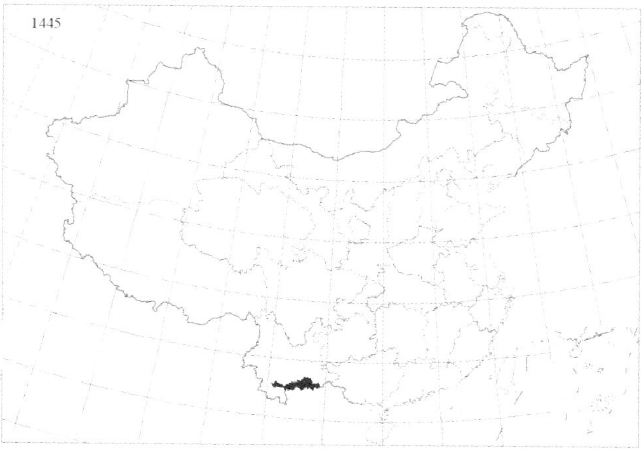

Opiliaceae 山柚子科 *Opilia* 山柚子属

Opilia amentacea Roxb. 山柚子

Vines, evergreen

20.4; 15.1; 23.6; 20.5; 185.8; 0.0; 989 ❙ 1607; 702; 77 ❙ 62.5; 986; 659.0

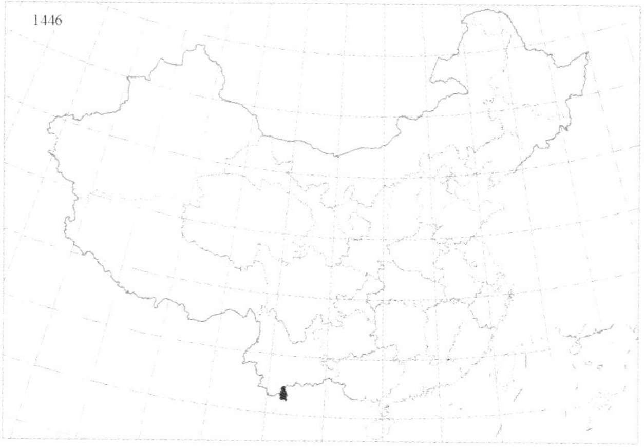

Opiliaceae 山柚子科 *Urobotrya* 尾球木属
Urobotrya latisquama (Gagnep.) Hiepko 尾球木
Shrubs, evergreen

19.3 (17.2~22.2); 11.7 (8.8~15.1); 24.7 (22.4~28.5); 19.3 (17.3~22.3); 171.8 (147.0~207.5); 0.0 (0.0~0.0); 963 (833~1190) ❙ 1431 (1161~1751); 774 (633~996); 71 (48~128) ❙ 50.2 (14.8~88.8); 963 (833~1190); 397.7 (238.3~659.0)

1447

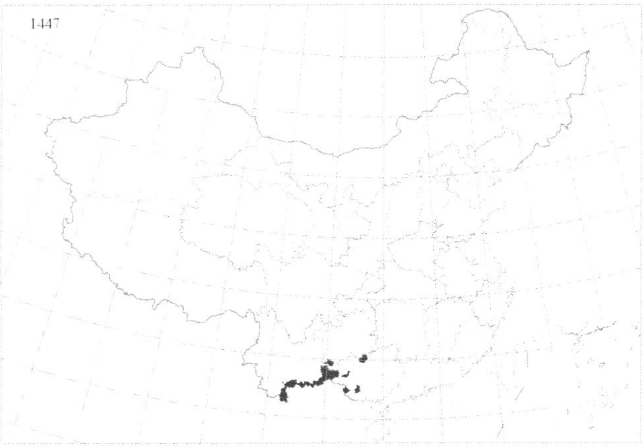

Santalaceae 檀香科 *Buckleya* 米面蓊属
Buckleya graebneriana Diels 秦岭米面蓊
Shrubs, deciduous

11.8 (7.4~15.2); -1.1 (-4.2~2.3); 24.0 (18.7~27.4); 12.0 (8.2~15.3); 96.8 (58.8~127.7); 14.7 (4.3~29.4); 717 (561~858) ❙ 760 (512~1214); 354 (258~536); 32 (14~109) ❙ 7.0 (-35.3~101.4); 678 (512~858); 232.6 (121.6~334.4)

1448

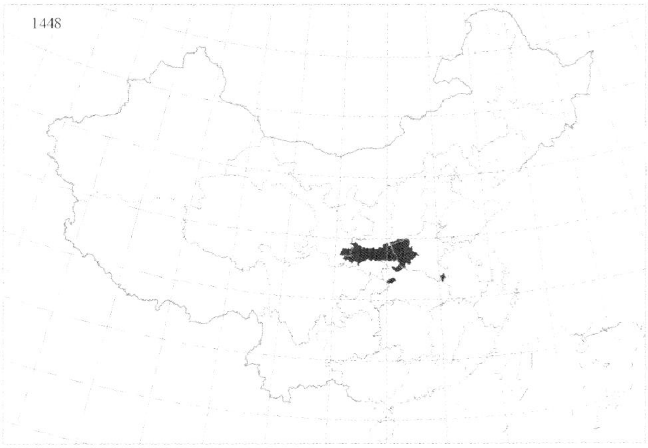

Santalaceae 檀香科 *Buckleya* 米面蓊属
Buckleya henryi Diels 米面蓊
Shrubs, deciduous

12.5 (4.7~17.0); -0.5 (-8.6~5.9); 24.7 (14.7~29.2); 12.8 (6.0~17.0); 104.3 (36.3~144.0); 13.3 (0.0~41.2); 751 (482~936) ❙ 877 (201~1743); 396 (119~666); 56 (4~220) ❙ 15.8 (-69.9~108.0); 693 (201~930); 208.8 (47.6~334.4)

1449

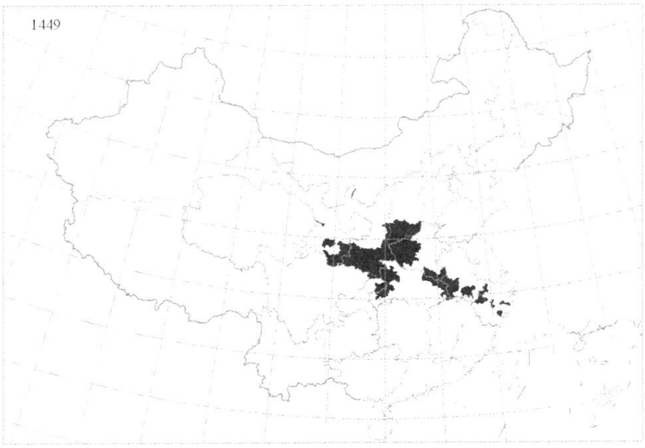

Santalaceae 檀香科 *Dendrotrophe* 寄生藤属
Dendrotrophe buxifolia (Blume) Miq. 黄杨叶寄生藤
Vines or shrubs, evergreen

19.3; 12.1; 24.4; 19.3; 171.9; 0.0; 971 ❙ 1474; 827; 84 ❙ 48.6; 947; 331.2

1450

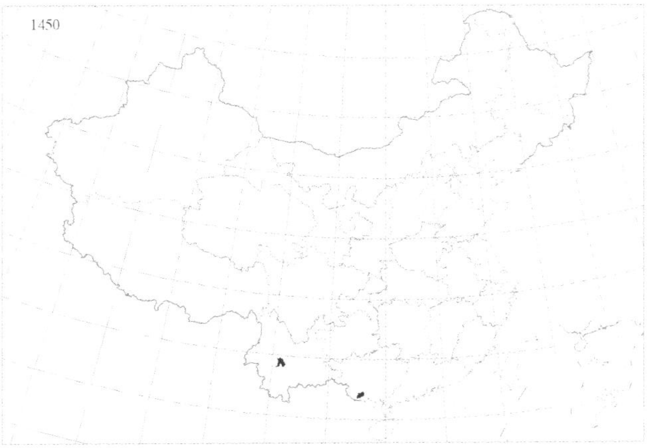

Santalaceae 檀香科 *Dendrotrophe* 寄生藤属
Dendrotrophe granulata (Hook. f. et Thomson ex A. DC.) A. N. Henry et B. Roy 疣枝寄生藤 Vines, evergreen

15.3; 7.5; 21.2; 15.3; 123.7; 0.0; 772 ❙ 1495; 853; 50 ❙ 97.8; 772; 391.8

1451

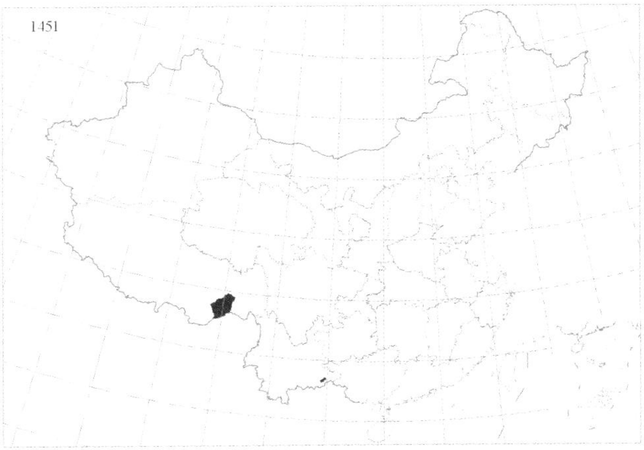

Santalaceae 檀香科 *Dendrotrophe* 寄生藤属
Dendrotrophe platyphylla (Sprengel) N. H. Xia et M. G. Gilbert 异花寄生藤 Vines, evergreen

8.6; 1.3; 15.0; 8.7; 52.6; 8.8; 574 ❙ 1041; 585; 45 ❙ 31.5; 574; 352.3

1452

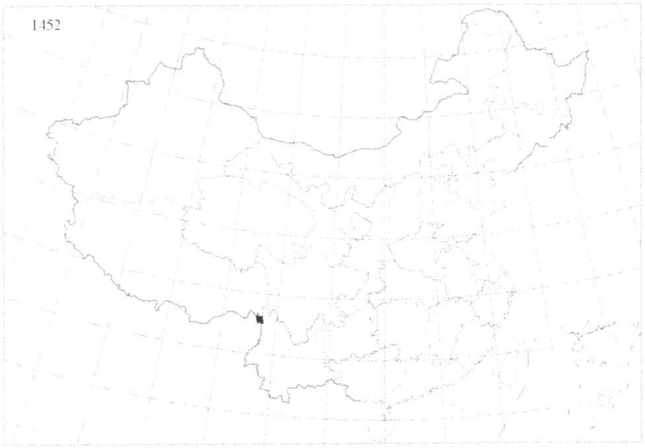

Santalaceae 檀香科 *Dendrotrophe* 寄生藤属
Dendrotrophe polyneura (Hu) D. D. Tao ex P. C. Tam 多脉寄生藤
Vines, evergreen

17.7 (8.6~21.0); 11.3 (1.3~15.1); 22.0 (15.0~25.6); 17.8 (8.7~21.0); 153.2 (52.6~191.8); 0.2 (0.0~8.8); 858 (574~1053) ❙ 1314 (939~1751); 728 (524~996); 52 (37~77) ❙ 53.6 (6.5~112.2); 841 (574~1053); 425.4 (277.0~659.0)

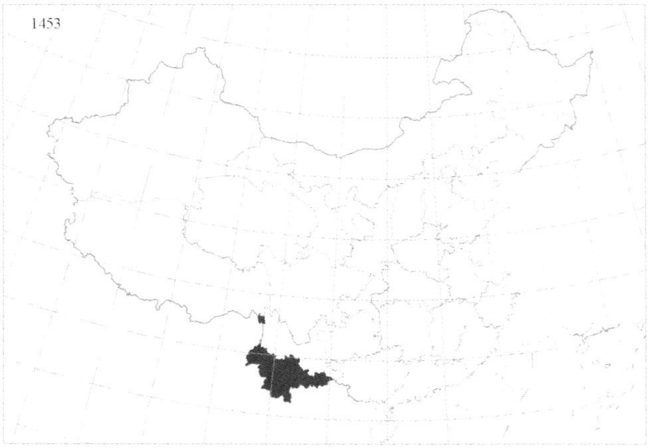

Santalaceae 檀香科 *Dendrotrophe* 寄生藤属
Dendrotrophe umbellata (Blume) Miq. 伞花寄生藤
Vines, evergreen

24.0 (22.2~24.8); 18.9 (17.2~20.1); 27.5 (25.7~28.7); 24.0 (22.3~24.8); 227.9 (207.2~237.6); 0.0 (0.0~0.0); 1303 (1127~1385) ❙ 1461 (1176~1666); 508 (447~597); 51 (41~59) ❙ 13.1 (-12.0~47.8); 1134 (1055~1212); 516.7 (411.8~620.0)

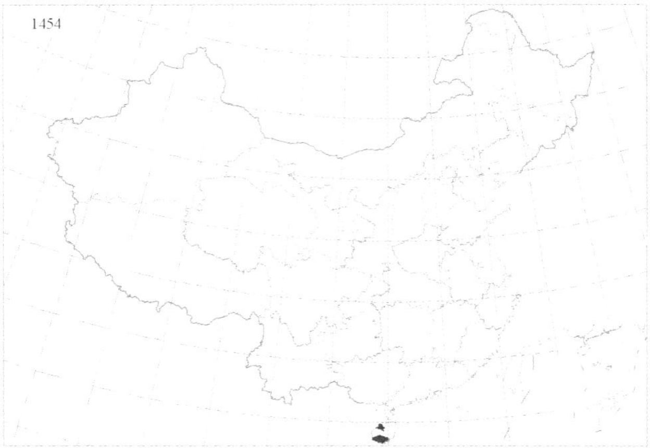

Santalaceae 檀香科 *Dendrotrophe* 寄生藤属
Dendrotrophe varians (Blume) Miq. 寄生藤
Vines, evergreen

20.0 (14.4~24.8); 11.9 (5.6~20.1); 26.7 (19.1~29.8); 20.1 (14.4~24.8); 180.9 (113.1~237.6); 0.0 (0.0~0.0); 1039 (727~1385) ❙ 1575 (868~2435); 671 (312~1428); 135 (41~257) ❙ 54.1 (-34.3~137.5); 1025 (727~1360); 289.0 (95.3~659.0)

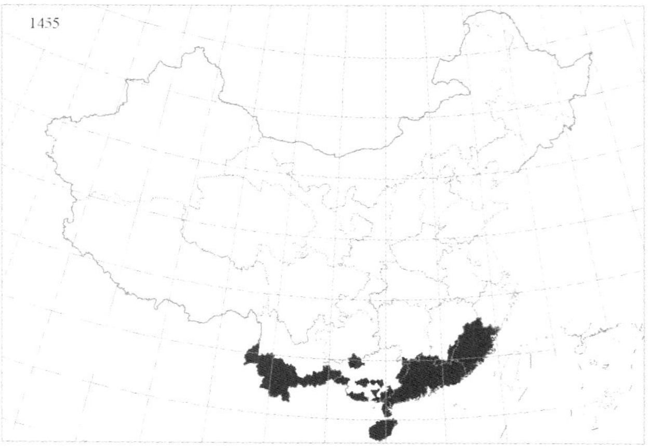

Santalaceae 檀香科 *Osyris* 沙针属
Osyris quadripartita Salzmann ex Decne. 沙针
Shrubs or small trees, evergreen

14.9 (-1.1~22.3); 7.6 (-9.1~15.1); 20.6 (6.2~28.9); 15.0 (2.0~22.3); 123.1 (2.5~208.1); 3.5 (0.0~75.6); 776 (334~1202) ❙ 1120 (321~1751); 628 (193~996); 46 (11~186) ❙ 46.2 (-14.2~128.6); 760 (321~1202); 323.3 (79.3~659.0)

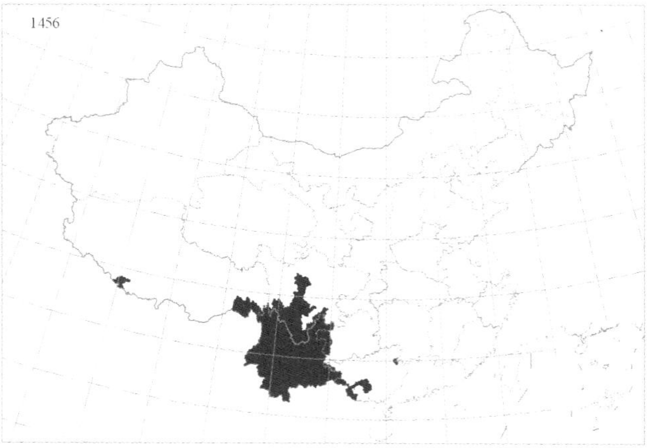

Santalaceae 檀香科 *Pyrularia* 檀梨属
Pyrularia edulis (Wall.) A. DC. 檀梨
Shrubs or small trees, deciduous

17.1 (10.6~21.8); 7.9 (-0.2~15.1); 25.1 (16.6~29.5); 17.2 (10.6~21.8); 146.1 (73.1~201.6); 0.3 (0.0~12.1); 882 (636~1160) ❙ 1363 (797~1997); 578 (392~1007); 116 (23~259) ❙ 54.8 (-14.2~145.8); 873 (636~1160); 264.8 (108.2~659.0)

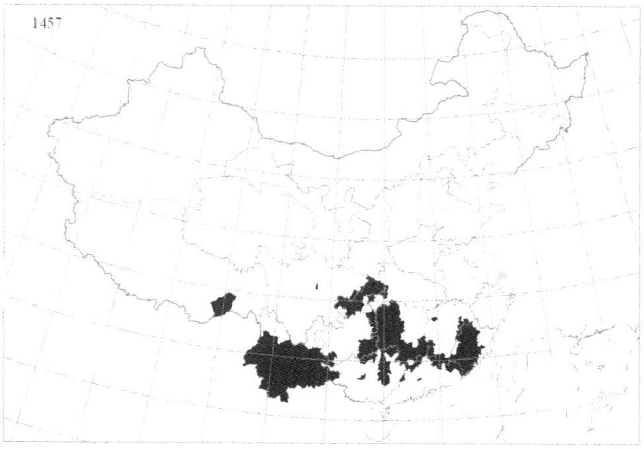

Santalaceae 檀香科 *Scleropyrum* 硬核属
Scleropyrum wallichianum (Wight et Arn.) Arn. 硬核
Trees, evergreen

20.0 (14.4~24.8); 13.7 (7.5~20.1); 24.1 (19.1~28.6); 20.0 (14.4~24.8); 180.1 (113.1~237.6); 0.0 (0.0~0.0); 1005 (727~1385) ❙ 1536 (1028~2435); 766 (444~1428); 66 (41~182) ❙ 56.1 (-2.2~112.2); 961 (727~1221); 466.2 (232.5~659.0)

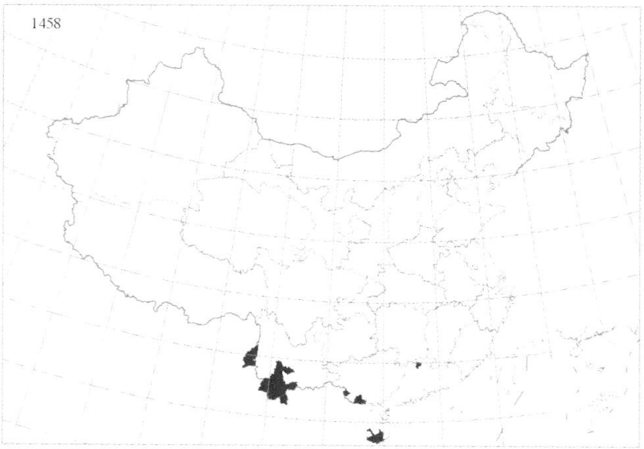

Santalaceae 檀香科 *Scleropyrum* 硬核属

Scleropyrum wallichianum var. mekongense (Gagnep.) Lecomte 无刺硬核

Trees, evergreen

19.1; 13.0; 23.3; 19.1; 169.7; 0.0; 917 ❙ 1428; 835; 56 ❙ 55.7; 917; 506.2

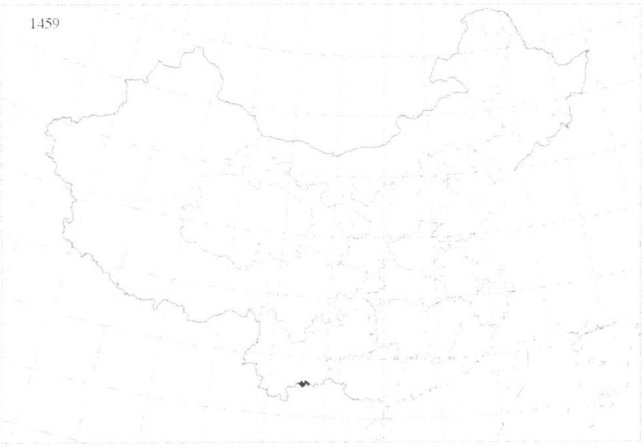

Loranthaceae 桑寄生科 *Dendrophthoe* 五蕊寄生属

Dendrophthoe pentandra (L.) Miq. 五蕊寄生

Shrubs, evergreen

19.4 (8.6~24.8); 12.2 (1.3~20.1); 24.7 (15.0~29.2); 19.5 (8.7~24.8); 173.5 (52.6~237.6); 0.1 (0.0~8.8); 995 (574~1385) ❙ 1421 (907~2037); 717 (447~1056); 80 (29~175) ❙ 43.9 (-12.0~112.2); 979 (574~1299); 317.6 (95.3~659.0)

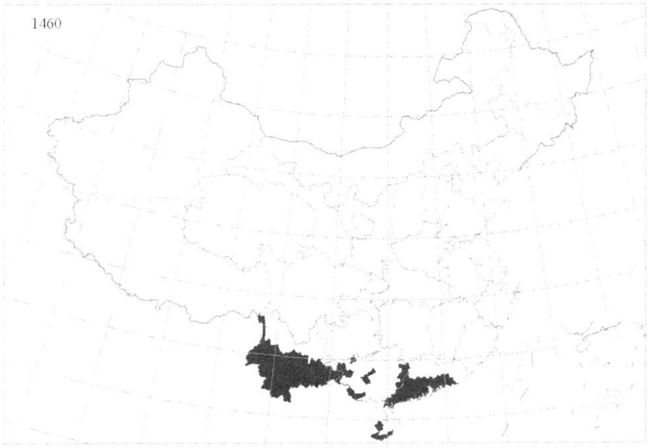

Loranthaceae 桑寄生科 *Elytranthe* 大苞鞘花属

Elytranthe albida (Blume) Blume 大苞鞘花

Shrubs

17.7 (13.1~21.0); 11.3 (6.4~15.1); 21.9 (18.4~25.6); 17.7 (13.2~21.0); 152.6 (98.1~191.8); 0.0 (0.0~0.0); 856 (692~1053) ❙ 1346 (939~1751); 742 (524~996); 54 (37~77) ❙ 57.8 (6.5~112.2); 840 (692~1053); 441.4 (301.1~659.0)

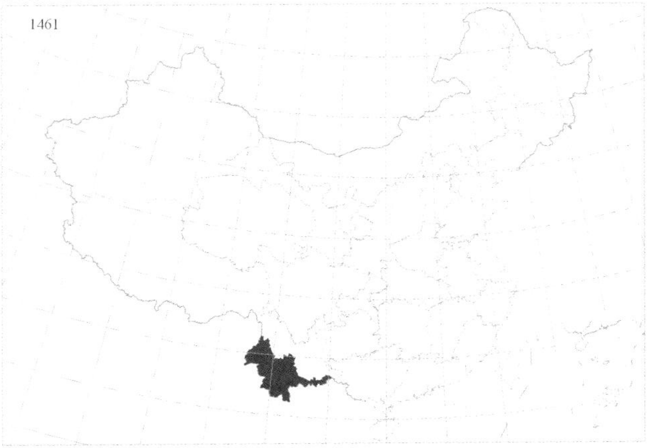

Loranthaceae 桑寄生科 *Elytranthe* 大苞鞘花属

Elytranthe parasitica (L.) Danser 墨脱大苞鞘花

Shrubs

13.3; 5.4; 19.6; 13.4; 100.4; 0.0; 706 ❙ 1736; 1007; 51 ❙ 145.8; 705; 447.4

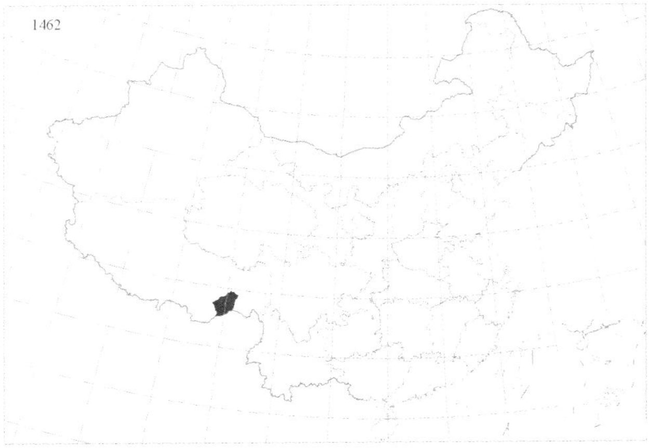

Loranthaceae 桑寄生科 *Helixanthera* 离瓣寄生属

Helixanthera coccinea (Jack) Danser 景洪离瓣寄生

Shrubs

20.5; 14.9; 23.7; 20.5; 185.9; 0.0; 992 ❙ 1464; 655; 66 ❙ 47.6; 959; 570.1

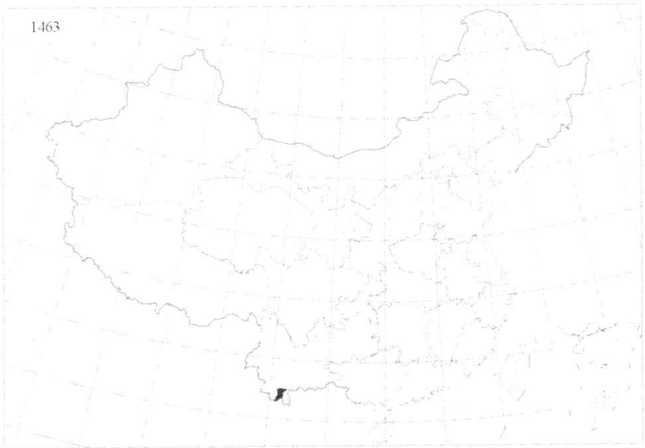

Loranthaceae 桑寄生科 *Helixanthera* 离瓣寄生属

Helixanthera guangxiensis H. S. Kiu 广西离瓣寄生

Shrubs

22.2 (21.0~24.3); 14.8 (13.1~18.7); 27.9 (26.9~28.4); 22.3 (21.0~24.3); 207.2 (192.3~231.9); 0.0 (0.0~0.0); 1181 (1081~1338) ❙ 1608 (1179~2148); 734 (392~1234); 103 (42~137) ❙ 37.5 (-11.9~86.7); 1132 (1056~1177); 296.2 (209.2~455.2)

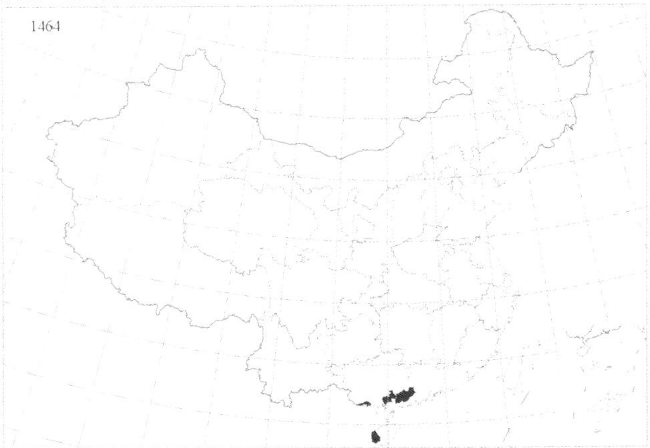

Loranthaceae 桑寄生科 *Helixanthera* 离瓣寄生属
Helixanthera parasitica Lour. 离瓣寄生
Shrubs

19.7 (5.4~24.8); 11.7 (-2.1~20.1); 26.0 (12.3~29.8); 19.7 (5.8~24.8); 177.1 (31.5~237.6); 0.2 (0.0~25.6); 1024 (481~1385) ❙ 1509 (741~2435); 709 (426~1428); 111 (21~220) ❙ 49.0 (-12.0~145.8); 1016 (481~1289); 289.0 (95.3~659.0)

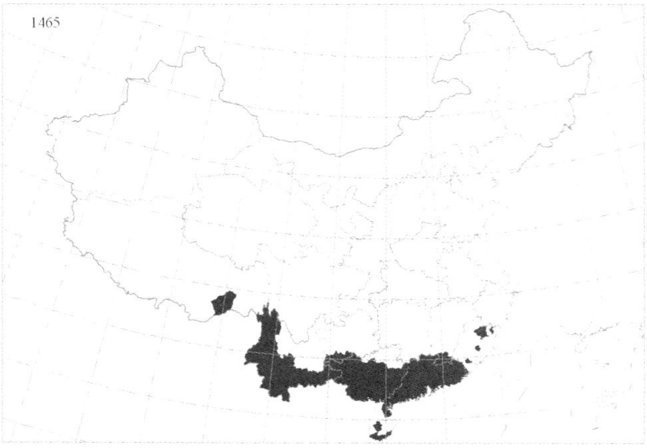

Loranthaceae 桑寄生科 *Helixanthera* 离瓣寄生属
Helixanthera pulchra (DC.) Danser 密花离瓣寄生
Shrubs

20.0; 14.6; 23.2; 20.1; 180.8; 0.0; 963 ❙ 1523; 681; 67 ❙ 58.4; 942; 581.0

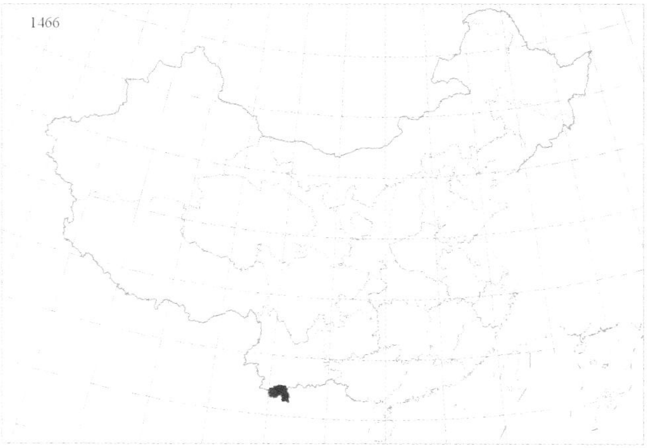

Loranthaceae 桑寄生科 *Helixanthera* 离瓣寄生属
Helixanthera sampsonii (Hance) Danser 油茶离瓣寄生
Shrubs

20.5 (8.6~23.6); 12.1 (1.3~17.5); 27.4 (15.0~29.8); 20.6 (8.7~23.6); 187.0 (52.6~223.5); 0.1 (0.0~8.8); 1077 (574~1281) ❙ 1631 (1041~2148); 715 (503~1234); 134 (45~208) ❙ 52.9 (-6.9~145.8); 1075 (574~1273); 246.7 (95.3~659.0)

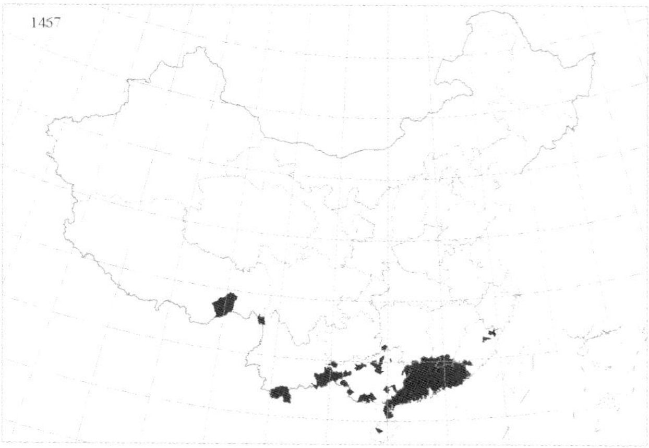

Loranthaceae 桑寄生科 *Helixanthera* 离瓣寄生属
Helixanthera scoriarum (W. W. Sm.) Danser 滇西离瓣寄生
Shrubs

16.1; 9.6; 20.4; 16.1; 133.8; 0.0; 786 ❙ 1356; 763; 56 ❙ 73.6; 772; 463.0

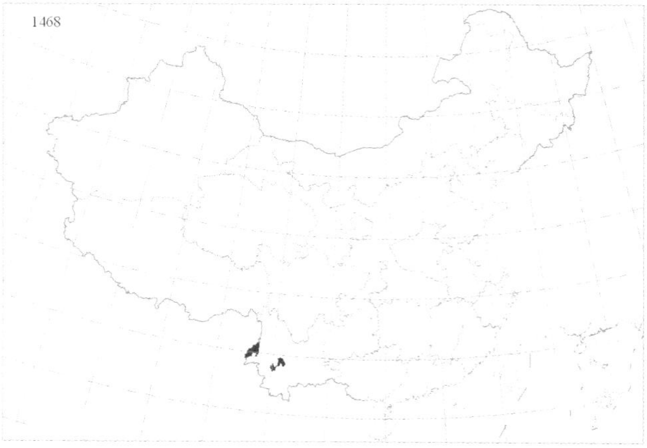

Loranthaceae 桑寄生科 *Helixanthera* 离瓣寄生属
Helixanthera terrestris (Hook. f.) Danser 林地离瓣寄生
Shrubs

13.3; 5.4; 19.6; 13.4; 100.4; 0.0; 706 ❙ 1736; 1007; 51 ❙ 145.8; 705; 447.4

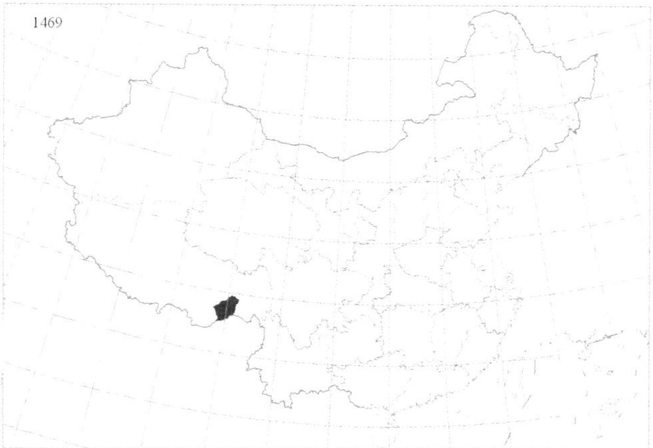

Loranthaceae 桑寄生科 *Loranthus* 桑寄生属
Loranthus delavayi Tiegh. 周树桑寄生
Shrubs

16.7 (5.3~24.1); 7.6 (-2.4~18.9); 24.5 (12.2~29.8); 16.7 (5.7~24.1); 141.6 (30.5~229.7); 1.1 (0.0~26.2); 867 (479~1322) ❙ 1384 (685~3495); 607 (357~1847); 114 (15~897) ❙ 59.3 (-14.2~246.5); 858 (479~1190); 261.1 (146.0~551.6)

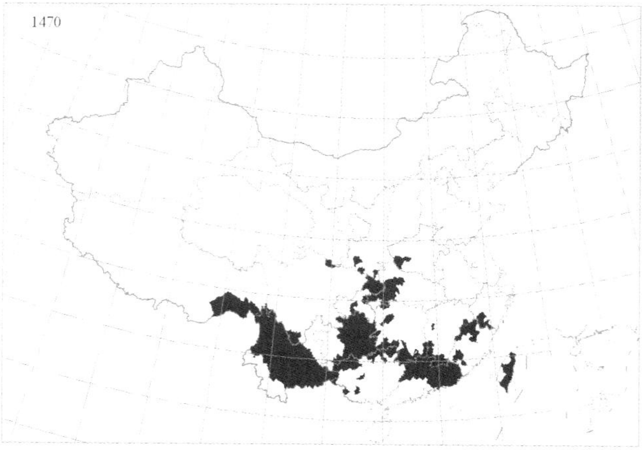

Loranthaceae 桑寄生科 *Loranthus* 桑寄生属
Loranthus guizhouensis H. S. Kiu 南桑寄生
Shrubs, deciduous

17.3 (14.9~21.2); 7.8 (4.2~12.2); 25.3 (23.0~28.5); 17.4 (15.0~21.3); 148.6 (119.9~195.2); 0.0 (0.0~0.8); 887 (781~1126) ❙ 1294 (1136~1660); 617 (448~820); 83 (43~216) ❙ 46.7 (16.9~103.1); 887 (781~1126); 249.4 (160.1~411.1)

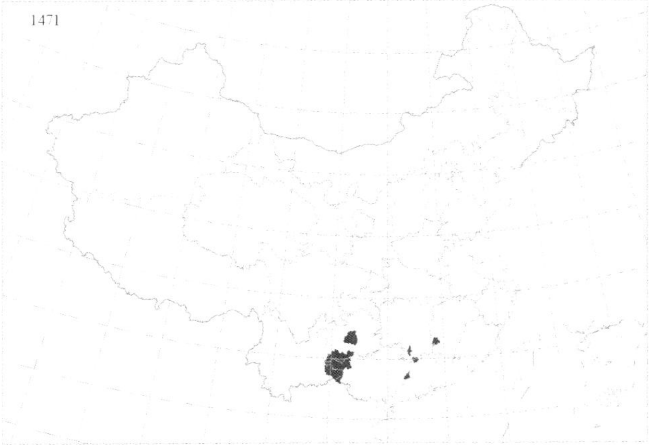

Loranthaceae 桑寄生科 *Loranthus* 桑寄生属
Loranthus kaoi (J. M. Chao) H. S. Kiu 台中桑寄生
Shrubs

9.6 (16.2~24.1); 14.0 (11.0~18.9); 23.9 (20.1~28.0); 19.6 (16.2~24.1); 74.9 (134.2~229.7); 0.0 (0.0~0.0); 973 (773~1322) ❙ 2387 (1751~2859); 1200 (959~1594); 185 (47~402) ❙ 159.0 (32.5~246.5); 935 (773~1095); 236.2 (154.2~289.1)

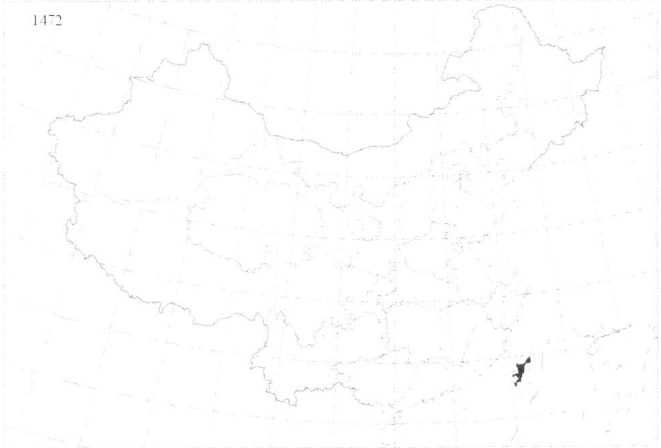

Loranthaceae 桑寄生科 *Loranthus* 桑寄生属
Loranthus lambertianus Schult. f. 吉隆桑寄生
Shrubs, deciduous

6.1; -1.8; 12.9; 7.7; 51.5; 37.8; 520 ❙ 1028; 600; 42 ❙ 70.9; 514; 263.3

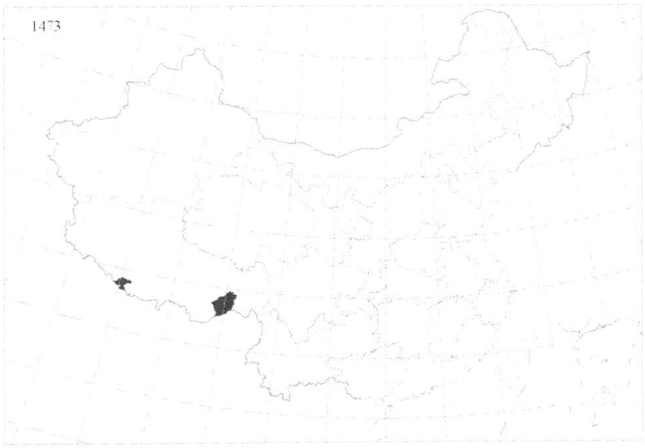

Loranthaceae 桑寄生科 *Loranthus* 桑寄生属
Loranthus pseudo-odoratus Lingelsh. 华中桑寄生
Shrubs

14.9; 4.4; 25.1; 15.0; 120.5; 0.7; 796 ❙ 1703; 636; 172 ❙ 113.3; 796; 256.0

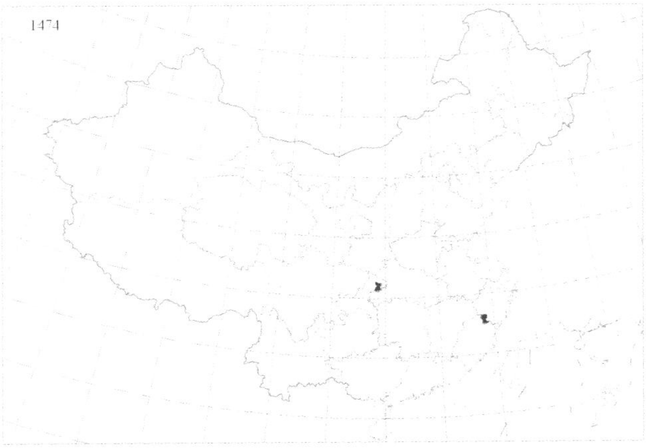

Loranthaceae 桑寄生科 *Loranthus* 桑寄生属
Loranthus tanakae Franch. et Sav. 北桑寄生
Shrubs, deciduous

9.6 (1.9~15.5); -5.5 (-14.4~2.1); 23.1 (11.7~28.0); 10.7 (4.3~15.6); 87.0 (21.1~131.3); 31.0 (4.3~71.0); 680 (427~881) ❙ 571 (229~1246); 325 (148~535); 19 (5~97) ❙ -15.2 (-63.6~101.2); 552 (229~881); 164.0 (33.5~323.3)

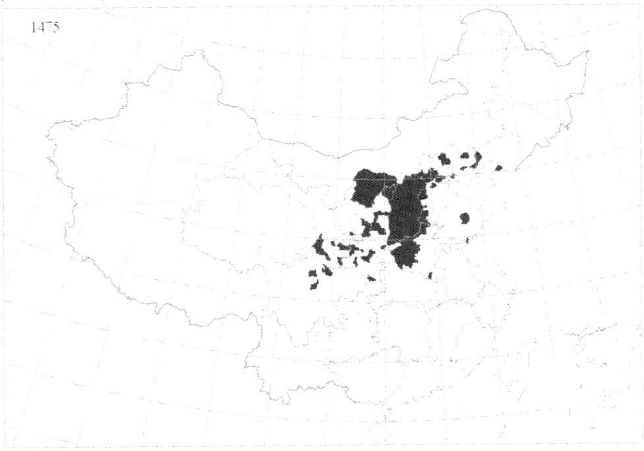

Loranthaceae 桑寄生科 *Macrosolen* 鞘花属
Macrosolen bibracteolatus (Hance) Danser 双花鞘花
Shrubs

20.1 (15.2~24.7); 11.8 (6.3~19.8); 26.7 (21.4~29.8); 20.1 (16.2~24.7); 181.6 (134.5~236.5); 0.0 (0.0~0.0); 1042 (797~1376) ❙ 1556 (868~2 48); 677 (312~1234); 127 (42~197) ❙ 51.1 (-34.3~95.5); 1022 (797~1326); 303.8 (174.9~659.0)

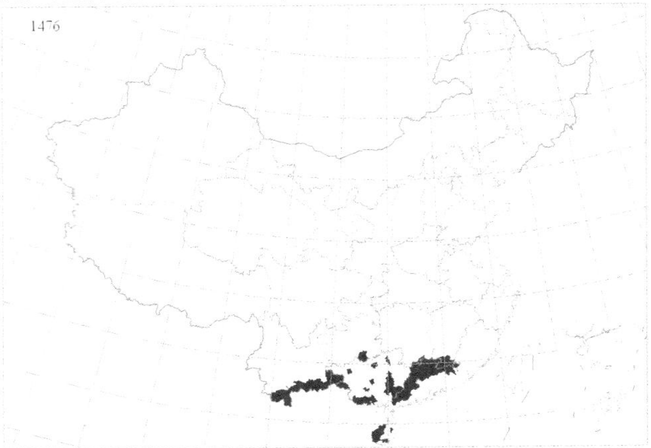

Loranthaceae 桑寄生科 *Macrosolen* 鞘花属
Macrosolen cochinchinensis (Lour.) Tiegh. 鞘花
Shrubs

19.3 (3.2~24.8); 11.0 (-6.4~20.1); 26.0 (11.4~29.8); 19.4 (4.7~24.8); 172.8 (22.7~237.6); 0.2 (0.0~43.8); 1001 (436~1385) ❙ 1482 (791~2148); 685 (403~1234); 111 (17~230) ❙ 49.2 (-6.9~145.8); 994 (436~1273); 277.2 (95.3~659.0)

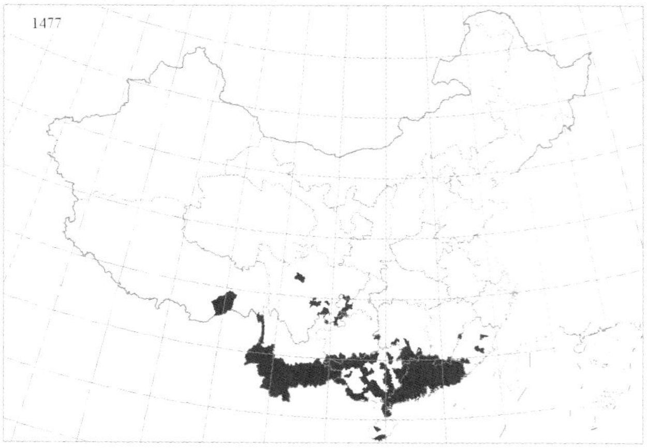

Loranthaceae 桑寄生科 *Macrosolen* 鞘花属
Macrosolen geminatus (Merr.) Danser 勐腊鞘花
Shrubs

19.7; 14.1; 23.5; 19.8; 177.8; 0.0; 953 ❙ 1517; 768; 66 ❙ 59.1; 952; 582.6

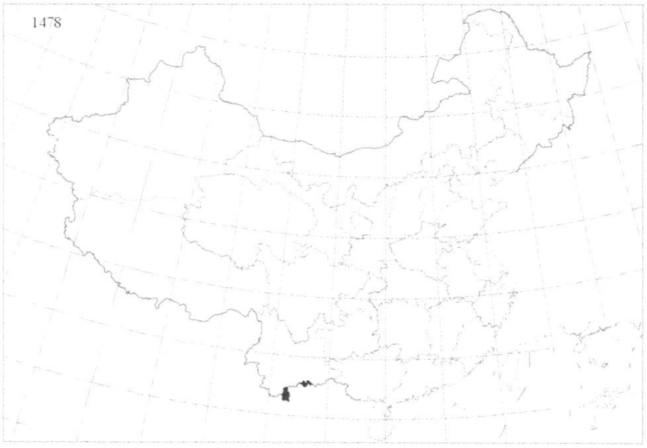

Loranthaceae 桑寄生科 *Macrosolen* 鞘花属
Macrosolen robinsonii (Gamble) Danser 短序鞘花
Shrubs

20.5; 14.9; 23.7; 20.5; 185.9; 0.0; 992 ❙ 1464; 655; 66 ❙ 47.6; 959; 570.1

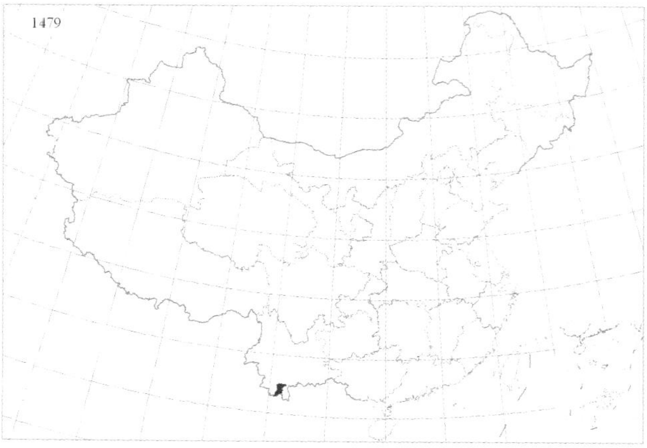

Loranthaceae 桑寄生科 *Macrosolen* 鞘花属
Macrosolen tricolor (Lecomte) Danser 三色鞘花
Shrubs

23.3 (21.3~24.8); 16.9 (13.2~20.1); 28.1 (25.7~29.2); 23.4 (21.3~24.8); 220.5 (195.9~237.6); 0.0 (0.0~0.0); 1262 (1111~1385) ❙ 1624 (1179~2435); 710 (392~1428); 84 (41~130) ❙ 29.8 (-11.9~99.5); 1193 (1055~1360); 378.6 (178.7~620.0)

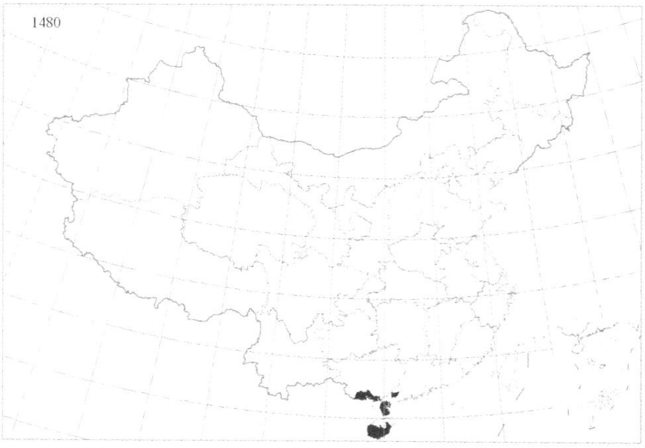

Loranthaceae 桑寄生科 *Scurrula* 梨果寄生属
Scurrula atropurpurea (Blume) Danser 梨果寄生
Shrubs

16.0 (-1.8~21.0); 9.0 (-10.2~15.1); 21.0 (6.1~26.0); 16.1 (1.8~21.0); 133.6 (1.9~191.8); 1.4 (0.0~82.8); 800 (329~1053) ❙ 1152 (320~1751); 646 (217~996); 46 (15~77) ❙ 44.4 (-14.2~112.2); 781 (320~1053); 344.3 (72.2~659.0)

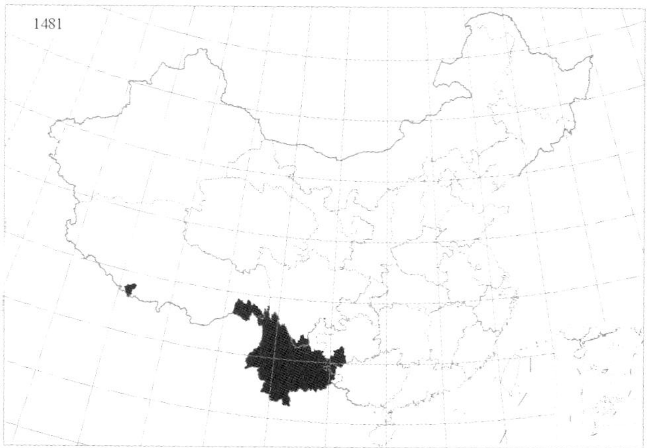

Loranthaceae 桑寄生科 *Scurrula* 梨果寄生属
Scurrula buddleioides (Desr.) G. Don 滇藏梨果寄生
Shrubs

14.4 (1.8~20.5); 7.4 (-7.1~15.1); 19.6 (9.9~24.3); 14.5 (3.8~20.5); 118.1 (15.7~185.9); 5.2 (0.0~53.3); 754 (404~992) ❙ 1171 (583~1736); 640 (348~1007); 45 (13~77) ❙ 58.4 (5.5~145.8); 731 (404~986); 369.3 (169.5~659.0)

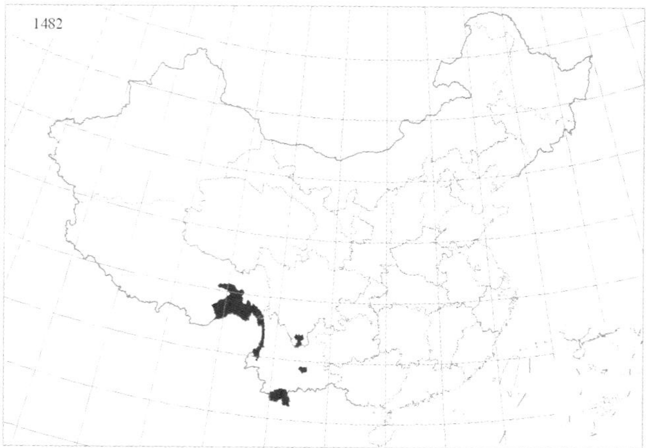

Loranthaceae 桑寄生科 *Scurrula* 梨果寄生属

Scurrula chingii (W. C. Cheng) H. S. Kiu 卵叶梨果寄生

Shrubs

18.4 (8.6~23.6); 11.3 (1.3~17.8); 23.4 (15.0~28.6); 18.5 (8.7~23.6); 161.9 (52.6~223.3); 0.2 (0.0~8.8); 920 (574~1271) | 1422 (1011~2148); 765 (444~1234); 70 (41~194) | 56.7 (-2.2~112.2); 911 (574~1190); 403.1 (211.1~659.0)

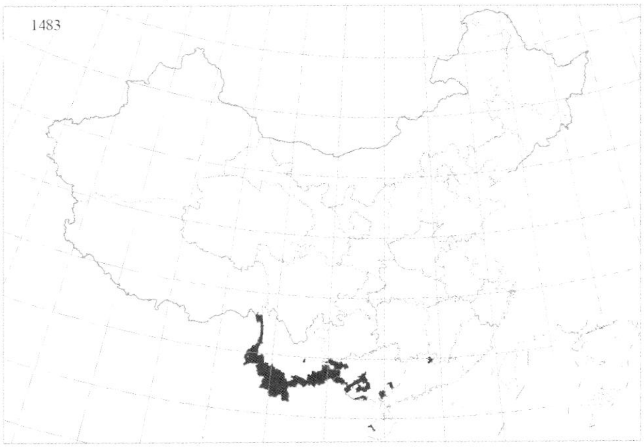

Loranthaceae 桑寄生科 *Scurrula* 梨果寄生属

Scurrula chingii var. yunnanensis H. S. Kiu 短柄梨果寄生

Shrubs

19.5 (18.2~20.5); 13.9 (12.4~15.1); 22.7 (21.6~23.7); 19.5 (18.2~20.5); 173.9 (158.6~185.9); 0.0 (0.0~0.0); 931 (861~992) | 1494 (1434~1607); 705 (655~785); 56 (41~77) | 60.9 (47.6~74.2); 901 (839~986); 531.1 (439.5~659.0)

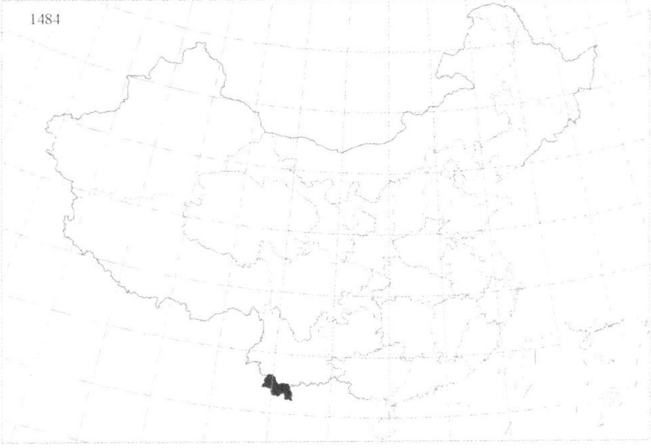

Loranthaceae 桑寄生科 *Scurrula* 梨果寄生属

Scurrula elata (Edgew.) Danser 高山寄生

Shrubs

6.3 (-1.8~16.0); -1.7 (-10.2~9.3); 13.1 (6.1~20.4); 8.0 (1.8~16.0); 55.3 (1.9~132.4); 38.9 (0.0~82.8); 527 (329~780) | 921 (320~1736); 543 (193~1007); 33 (10~56) | 54.9 (-4.0~145.8); 523 (320~780); 236.6 (61.6~447.4)

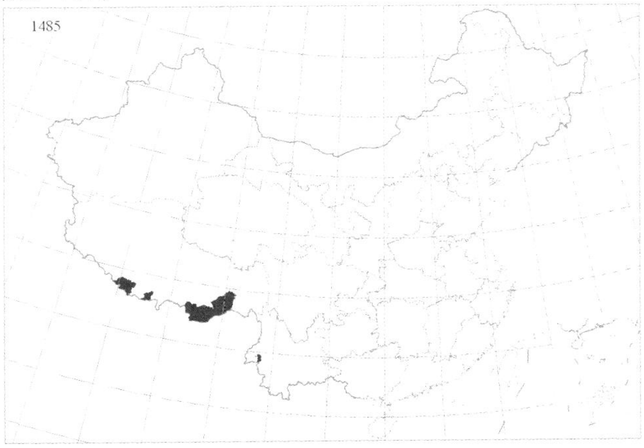

Loranthaceae 桑寄生科 *Scurrula* 梨果寄生属

Scurrula ferruginea (Jack) Danser 锈毛梨果寄生

Shrubs

18.1 (14.4~21.0); 11.5 (7.5~15.1); 22.4 (19.1~25.6); 18.1 (14.4~21.0); 157.6 (113.1~191.8); 0.0 (0.0~0.0); 876 (727~1053) | 1391 (958~1688); 771 (531~959); 55 (42~77) | 59.5 (13.3~112.2); 867 (727~1053); 459.6 (311.9~659.0)

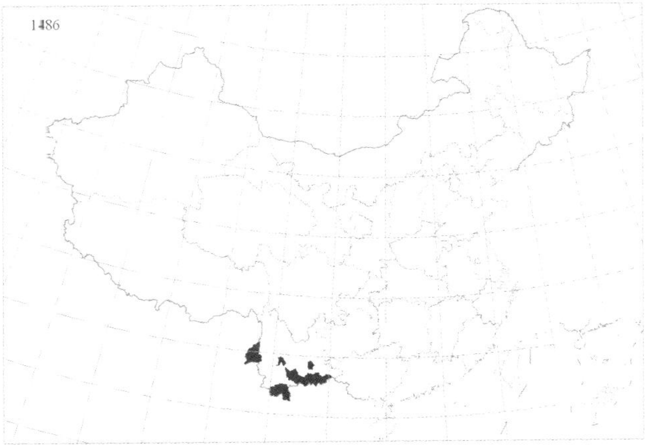

Loranthaceae 桑寄生科 *Scurrula* 梨果寄生属

Scurrula gongshanensis H. S. Kiu 贡山梨果寄生

Shrubs

11.1; 4.0; 16.8; 11.1; 76.2; 3.2; 638 | 1207; 669; 57 | 89.0; 638; 381.4

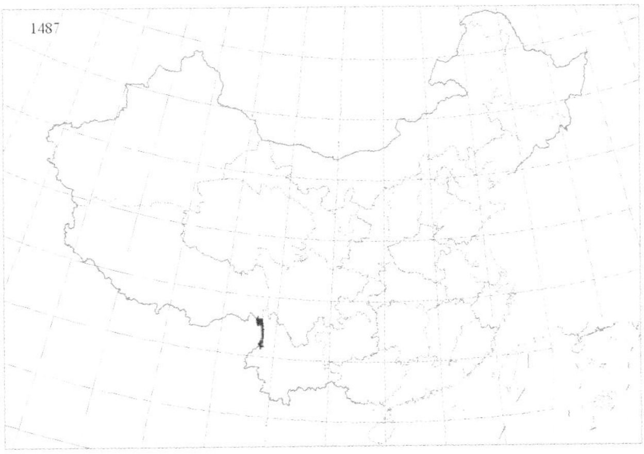

Loranthaceae 桑寄生科 *Scurrula* 梨果寄生属

Scurrula notothixoides (Hance) Danser 小叶梨果寄生

Shrubs

23.9 (22.2~24.8); 18.0 (15.9~20.1); 28.1 (25.7~29.2); 23.9 (22.3~24.8); 227.2 (207.2~237.6); 0.0 (0.0~0.0); 1303 (1127~1385) | 1469 (868~1354); 546 (312~712); 75 (41~115) | 12.9 (-34.3~47.8); 1201 (868~1360); 401.1 (178.7~620.0)

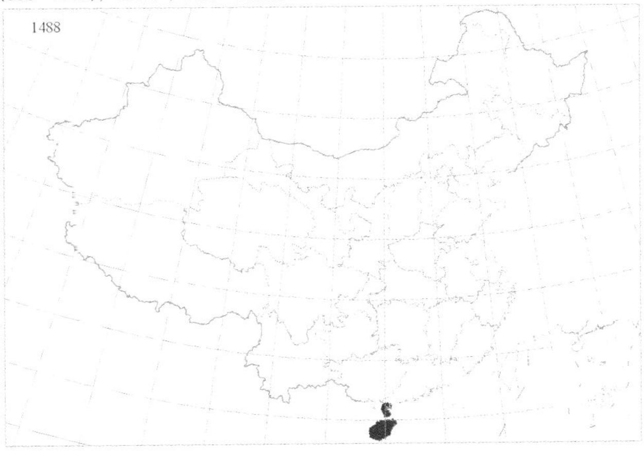

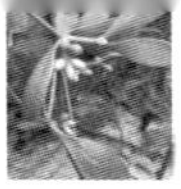

Loranthaceae 桑寄生科 *Scurrula* 梨果寄生属

Scurrula parasitica L. 红花寄生

Shrubs, evergreen

18.5 (6.6~24.8); 10.2 (-1.3~20.1); 25.3 (13.3~29.8); 18.5 (6.8~24.8); 162.5 (39.6~237.6); 0.2 (0.0~19.7); 957 (514~1385) ❙ 1420 (797~3495); 655 (312~1847); 106 (12~897) ❙ 49.1 (-34.3~224.4); 943 (514~1326); 273.0 (95.3~659.0)

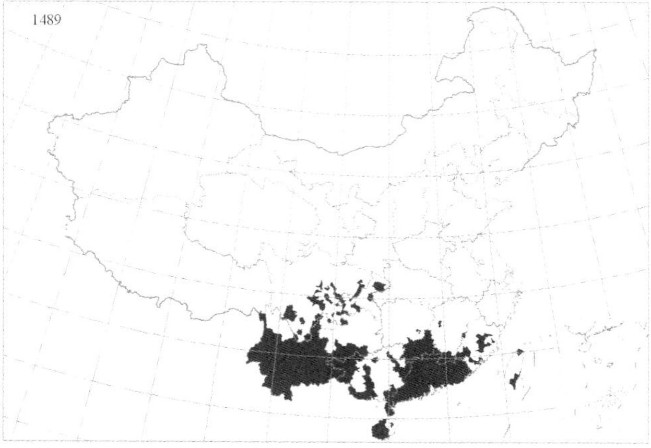

Loranthaceae 桑寄生科 *Scurrula* 梨果寄生属

Scurrula parasitica var. graciliflora (Roxb. ex Schult.) H. S. Kiu 小红花寄生

Shrubs

16.7 (6.6~21.0); 9.1 (-1.3~14.9); 22.3 (13.3~26.0); 16.7 (6.8~21.0); 141.7 (39.6~191.8); 0.8 (0.0~19.7); 839 (514~1053) ❙ 1239 (862~1530); 690 (539~909); 50 (12~68) ❙ 48.9 (19.0~93.0); 833 (514~1053); 348.1 (189.5~570.1)

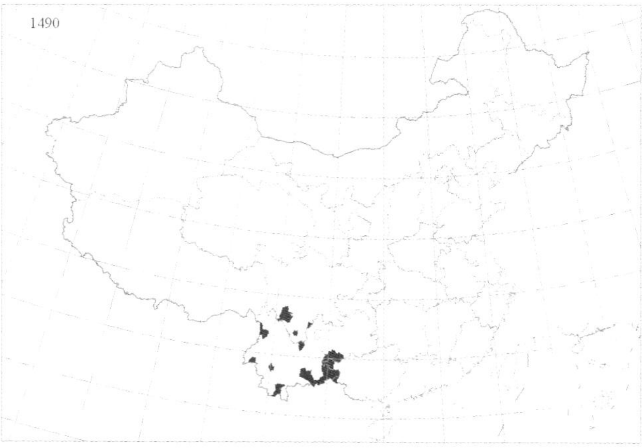

Loranthaceae 桑寄生科 *Scurrula* 梨果寄生属

Scurrula pulverulenta (Wall.) G. Don 白花梨果寄生

Shrubs

17.5 (16.3~19.2); 11.3 (9.6~13.8); 21.3 (20.3~22.3); 17.5 (16.4~19.2); 150.5 (136.6~170.5); 0.0 (0.0~0.0); 839 (789~909) ❙ 1438 (1232~1688); 789 (586~959); 52 (45~62) ❙ 71.9 (56.2~112.2); 827 (780~880); 492.9 (372.4~627.0)

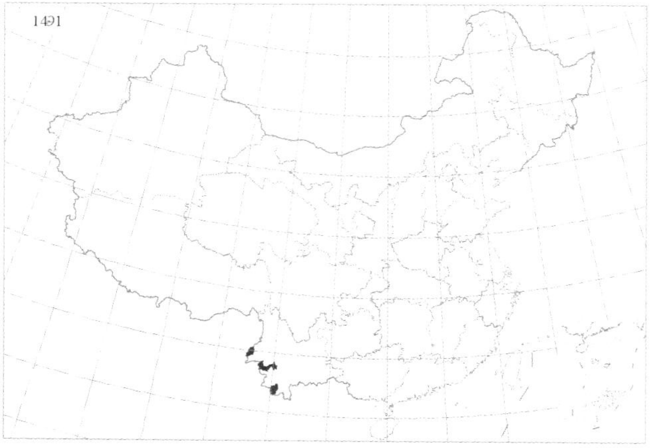

Loranthaceae 桑寄生科 *Taxillus* 钝果寄生属

Taxillus balansae (Lecomte) Danser 栗毛钝果寄生

Shrubs

19.7 (16.3~22.7); 11.8 (9.1~15.0); 25.5 (21.4~28.6); 19.8 (16.4~22.8); 177.2 (136.4~213.1); 0.0 (0.0~0.0); 1005 (797~1221) ❙ 1541 (1161~2435); 851 (633~1428); 84 (48~194) ❙ 53.1 (12.9~99.5); 1005 (797~1221); 335.1 (213.5~551.6)

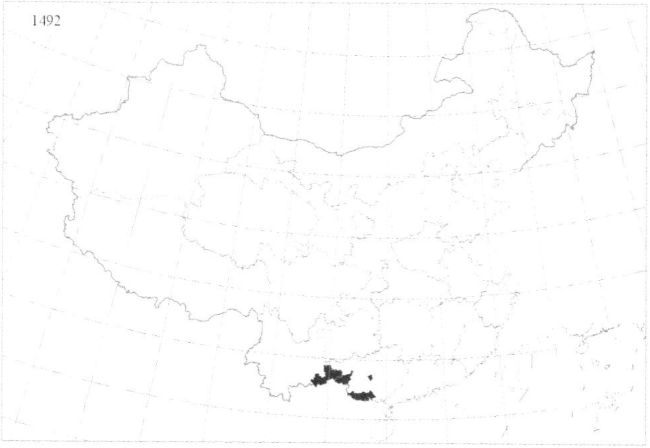

Loranthaceae 桑寄生科 *Taxillus* 钝果寄生属

Taxillus caloreas (Diels) Danser 松柏钝果寄生

Shrubs

14.5 (0.3~24.1); 6.4 (-8.6~18.3); 21.3 (8.3~28.6); 14.6 (2.9~24.1); 118.4 (9.3~229.4); 4.1 (0.0~65.4); 773 (368~1321) ❙ 1215 (389~2859); 594 (246~1388); 81 (10~402) ❙ 57.0 (-14.2~246.5); 759 (368~1294); 270.1 (142.5~572.9)

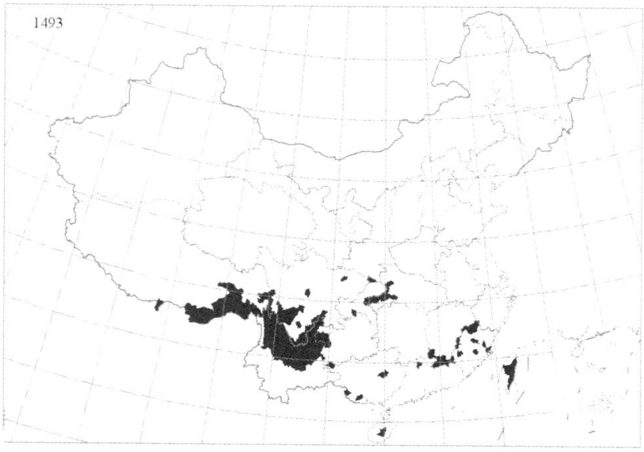

Loranthaceae 桑寄生科 *Taxillus* 钝果寄生属

Taxillus caloreas var. fargesii (Lecomte) H. S. Kiu 显脉钝果寄生

Shrubs

13.6 (2.2~18.9); 6.6 (-6.3~12.0); 19.2 (9.5~23.9); 13.8 (3.9~19.0); 108.6 (15.8~167.8); 4.5 (0.0~48.6); 719 (411~929) ❙ 963 (535~1441); 553 (326~627); 35 (10~84) ❙ 36.2 (-14.2~95.9); 691 (411~815); 288.6 (201.9~382.6)

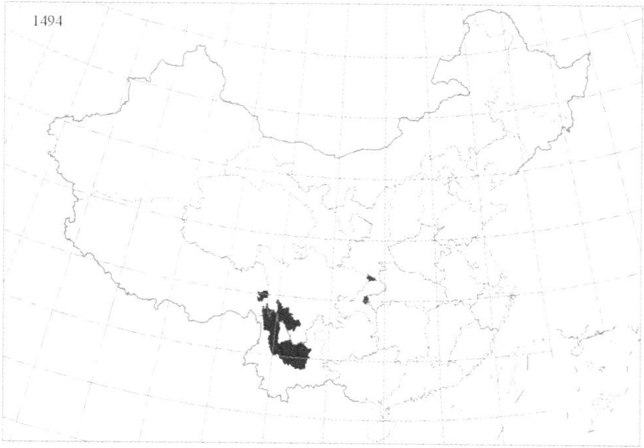

Loranthaceae 桑寄生科 *Taxillus* 钝果寄生属
Taxillus chinensis (DC.) Danser 广寄生
Shrubs

21.1 (15.5~24.8); 12.7 (4.2~20.1); 27.9 (24.6~29.8); 21.1 (15.6~24.8); 193.5 (128.1~237.6); 0.0 (0.0~0.8); 1114 (830~1385) ❙ 1646 (1174~2435); 712 (386~1428); 138 (41~230) ❙ 49.1 (-6.9~104.6); 1105 (830~1326); 242.0 (95.3~620.0)

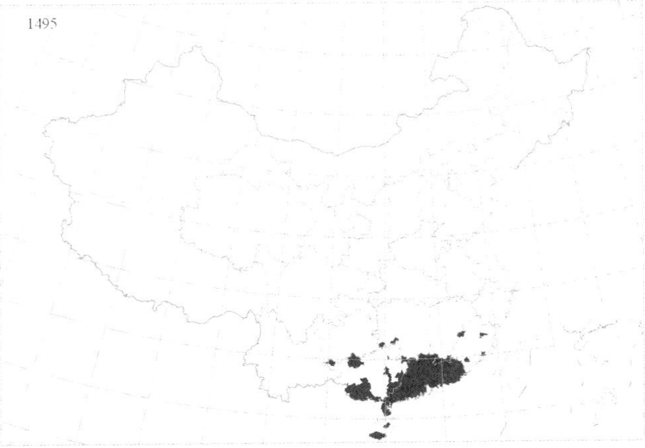

Loranthaceae 桑寄生科 *Taxillus* 钝果寄生属
Taxillus delavayi (Tiegh.) Danser 柳树寄生
Shrubs

12.6 (-1.1~21.0); 4.9 (-9.1~14.2); 19.0 (6.2~26.6); 12.9 (2.0~21.0); 100.2 (2.5~191.8); 8.0 (0.0~75.6); 698 (334~1053) ❙ 1010 (321~1522); 572 (193~818); 37 (9~106) ❙ 47.8 (-14.2~128.6); 683 (321~1053); 286.6 (79.3~551.6)

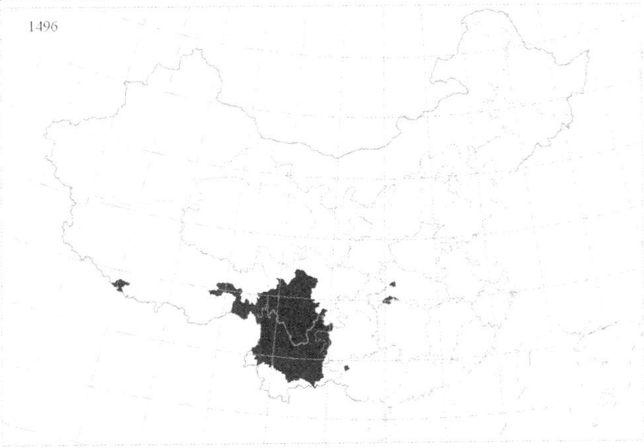

Loranthaceae 桑寄生科 *Taxillus* 钝果寄生属
Taxillus kaempferi (DC.) Danser 小叶钝果寄生
Shrubs

16.0 (3.3~18.7); 5.0 (-6.1~9.8); 26.4 (11.4~29.1); 16.1 (4.6~18.7); 134.8 (22.5~164.9); 2.2 (0.0~42.7); 866 (434~984) ❙ 1691 (812~1997); 589 (452~713); 209 (17~259) ❙ 95.7 (51.2~147.0); 861 (434~981); 255.4 (182.9~337.3)

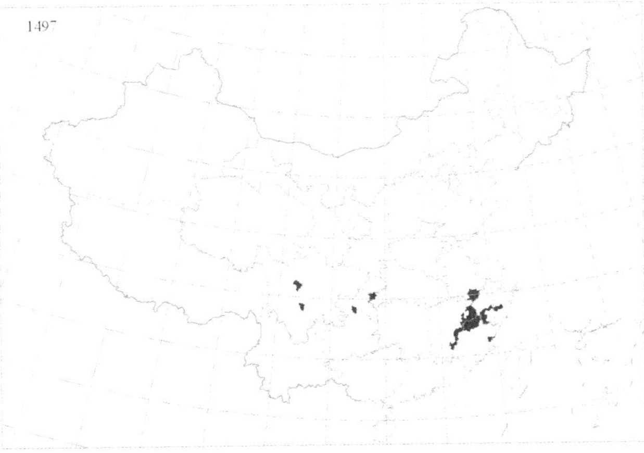

Loranthaceae 桑寄生科 *Taxillus* 钝果寄生属
Taxillus levinei (Merr.) H. S. Kiu 锈毛钝果寄生
Shrubs

18.0 (9.3~22.2); 7.7 (-1.7~14.3); 27.3 (20.1~29.9); 18.1 (9.5~22.3); 157.4 (69.8~207.2); 0.6 (0.0~18.0); 953 (603~1186) ❙ 1598 (964~2024); 602 (421~954); 174 (43~259) ❙ 68.9 (16.9~130.7); 950 (603~1186); 240.1 (140.2~371.7)

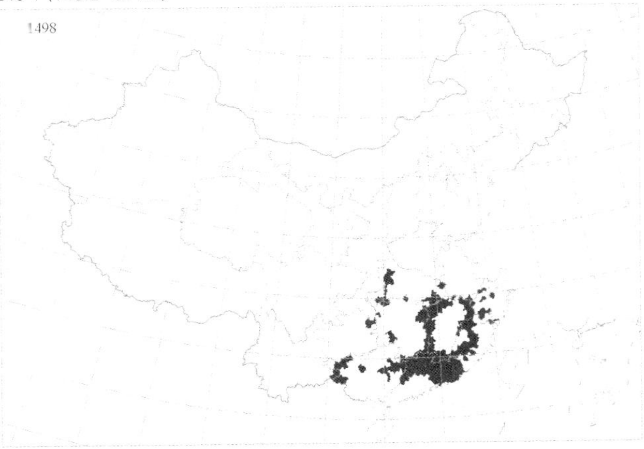

Loranthaceae 桑寄生科 *Taxillus* 钝果寄生属
Taxillus limprichtii (Grüning) H. S. Kiu 木兰寄生
Shrubs

18.8 (14.2~24.8); 9.9 (3.3~20.1); 26.4 (20.1~29.8); 18.8 (14.2~24.8); 166.0 (112.5~237.6); 0.1 (0.0~2.3); 976 (756~1385) ❙ 1562 (868~2993); 639 (312~1594); 144 (41~547) ❙ 62.9 (-34.3~246.5); 957 (756~1360); 279.7 (119.4~659.0)

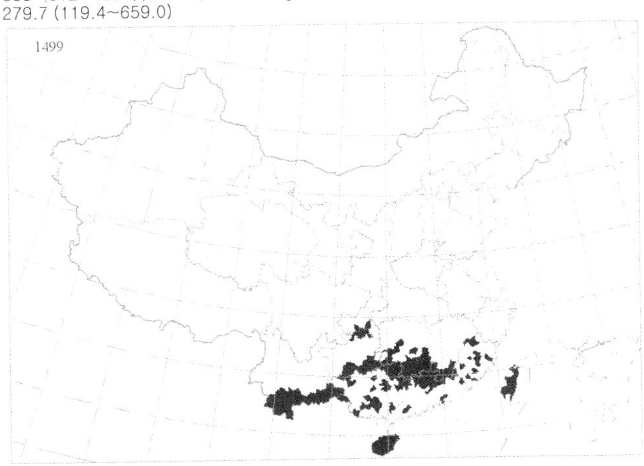

Loranthaceae 桑寄生科 *Taxillus* 钝果寄生属
Taxillus limprichtii var. longiflorus (Lecomte) H. S. Kiu 亮叶木兰寄生
Shrubs

19.0 (16.9~21.0); 13.0 (10.3~15.1); 22.9 (21.4~25.6); 19.1 (16.9~21.0); 168.7 (142.7~191.8); 0.0 (0.0~0.0); 916 (813~1053) ❙ 1446 (1163~1751); 765 (641~996); 61 (48~77) ❙ 58.0 (34.8~88.8); 907 (813~1053); 497.5 (383.4~659.0)

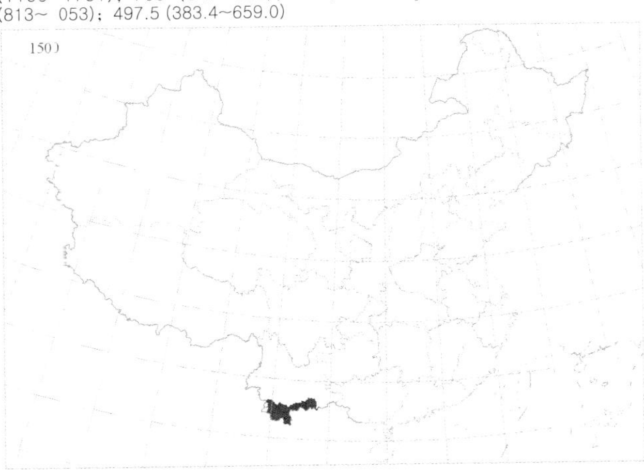

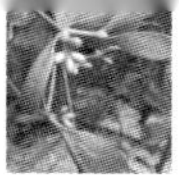

Loranthaceae 桑寄生科 *Taxillus* 钝果寄生属

Taxillus liquidambaricola (Hayata) Hosok. 阆阚果寄生

Shrubs

23.4 (16.2~24.8); 17.7 (11.0~20.1); 27.6 (20.1~28.8); 23.5 (16.2~24.8); 221.8 (-34.2~237.6); 0.0 (0.0~0.0); 1272 (773~1385) ❙ 1562 (868~2680); 568 (312~1186); 86 (41~245) ❙ 27.7 (-34.3~246.5); 1165 (773~1360); 424.9 (248.2~620.0)

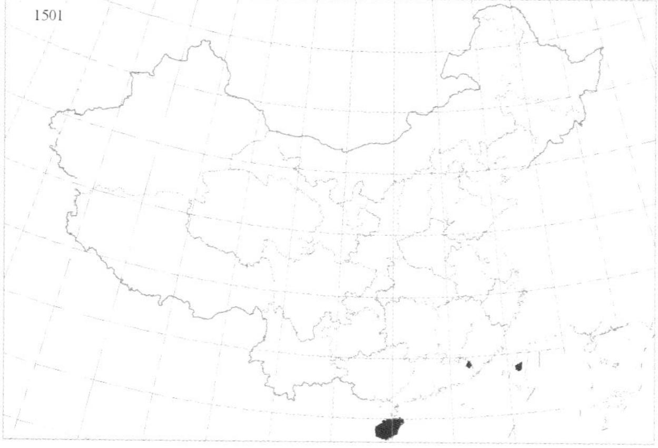

Loranthaceae 桑寄生科 *Taxillus* 钝果寄生属

Taxillus nigrans (Hance) Danser 毛叶钝果寄生

Shrubs, evergreen

15.7 (0.3~21.7); 5.2 (-9.9~15.3); 25.1 (9.3~29.3); 15.7 (3.3~21.7); 131.1 (11.9~200.2); 2.7 (0.0~67.5); 837 (391~1118) ❙ 1278 (626~3495); 550 (316~1412); 107 (14~897) ❙ 52.3 (-22.2~246.5); 824 (391~1114); 225.6 (123.1~551.6)

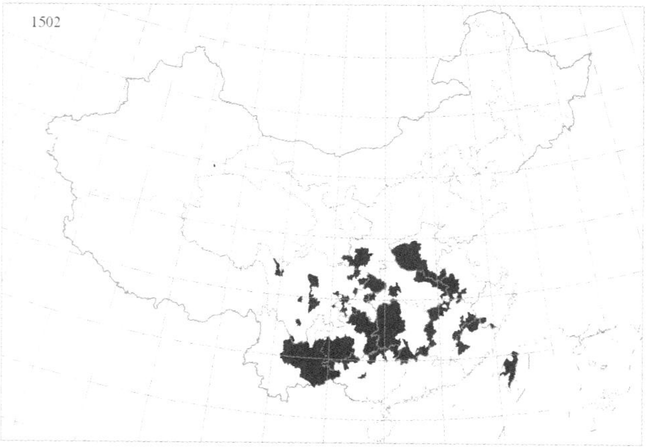

Loranthaceae 桑寄生科 *Taxillus* 钝果寄生属

Taxillus pseudochinensis (Yamam.) Danser 高雄钝果寄生

Shrubs

21.7; 16.7; 25.4; 21.7; 200.8; 0.0; 1117 ❙ 2255; 1381; 87 ❙ 117.5; 1003; 203.3

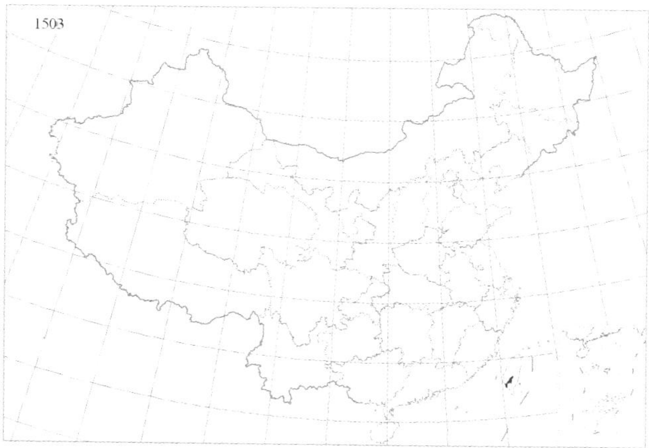

Loranthaceae 桑寄生科 *Taxillus* 钝果寄生属

Taxillus sericus Danser 龙陵钝果寄生

Shrubs

14.4 (8.6~17.8); 7.6 (1.3~11.1); 19.3 (15.0~22.0); 14.5 (8.7~17.8); 114.8 (52.6~154.1); 1.2 (0.0~8.8); 740 (574~855) ❙ 1320 (1041~1561); 746 (585~909); 57 (44~70) ❙ 79.5 (26.1~99.9); 734 (574~855); 426.8 (352.3~539.7)

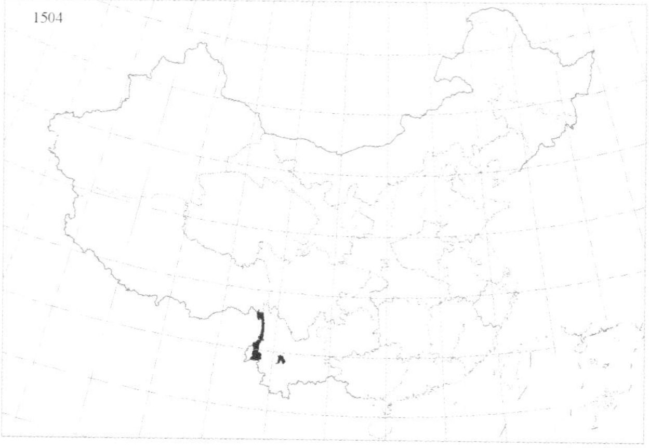

Loranthaceae 桑寄生科 *Taxillus* 钝果寄生属

Taxillus sutchuenensis (Lecomte) Danser 桑寄生

Shrubs

16.6 (2.2~22.8); 6.4 (-6.6~15.0); 25.8 (9.9~29.9); 16.7 (3.8~22.8); 141.3 (15.5~213.6); 1.5 (0.0~49.1); 884 (413~1230) ❙ 1342 (548~2680); 562 (264~1428); 117 (12~259) ❙ 52.1 (-28.9~246.5); 877 (413~1230); 227.1 (118.7~551.6)

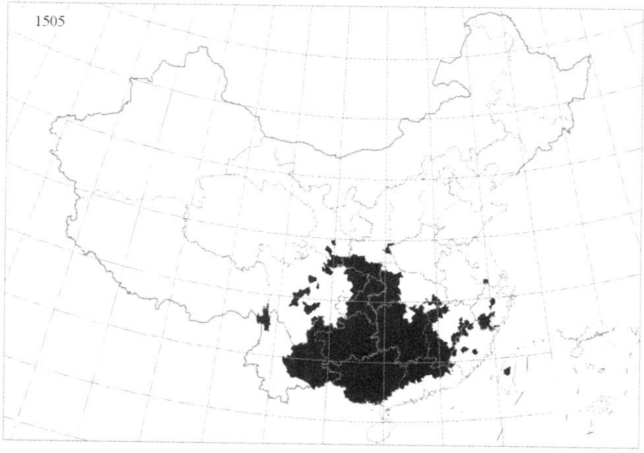

Loranthaceae 桑寄生科 *Taxillus* 钝果寄生属

Taxillus sutchuenensis var. duclouxii (Lecomte) H. S. Kiu 灰毛桑寄生

Shrubs

15.0 (5.4~18.2); 4.6 (-2.1~9.0); 24.4 (12.3~29.9); 15.0 (5.8~18.2); 121.4 (31.5~158.8); 1.2 (0.0~25.6); 802 (481~976) ❙ 1223 (741~1526); 530 (420~816); 89 (19~221) ❙ 52.9 (9.2~99.4); 799 (481~951); 207.3 (133.8~382.6)

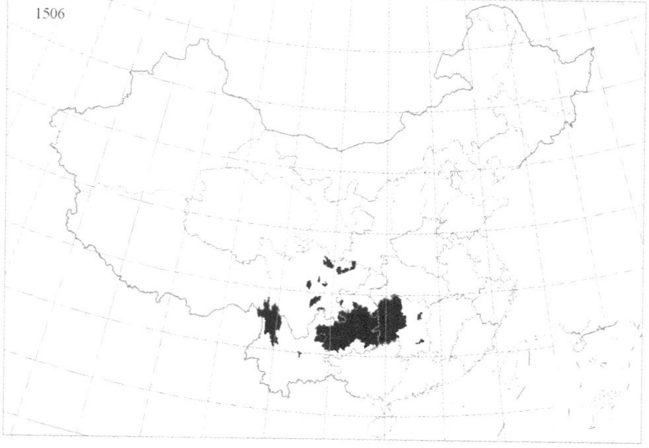

Loranthaceae 桑寄生科 *Taxillus* 钝果寄生属
Taxillus theifer (Hayata) H. S. Kiu 台湾钝果寄生
Shrubs

20.9; 15.2; 25.6; 20.9; 191.3; 0.0; 1061 ❙ 2282; 1202; 172 ❙ 121.4; 1026; 262.0

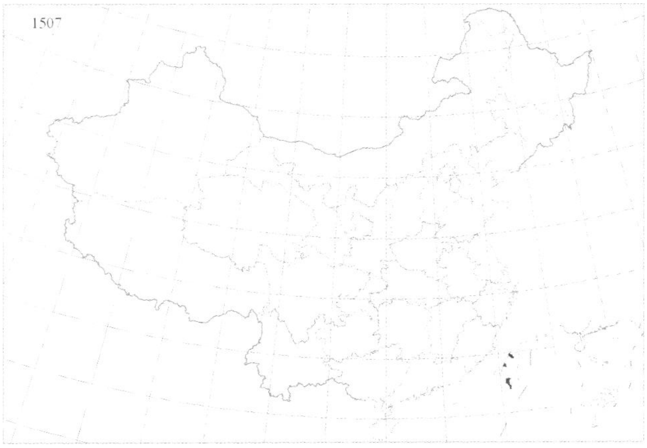

Loranthaceae 桑寄生科 *Taxillus* 钝果寄生属
Taxillus thibetensis (Lecomte) Danser 滇藏钝果寄生
Shrubs

13.6 (1.2~18.9); 5.9 (-7.1~12.5); 19.9 (9.0~26.6); 13.7 (3.3~19.0); 107.8 (11.9~167.8); 3.7 (0.0~56.5); 725 (390~929) ❙ 1003 (578~1583); 572 (345~938); 36 (12~79) ❙ 41.0 (-14.2~121.7); 704 (390~912); 264.0 (130.1~352.3)

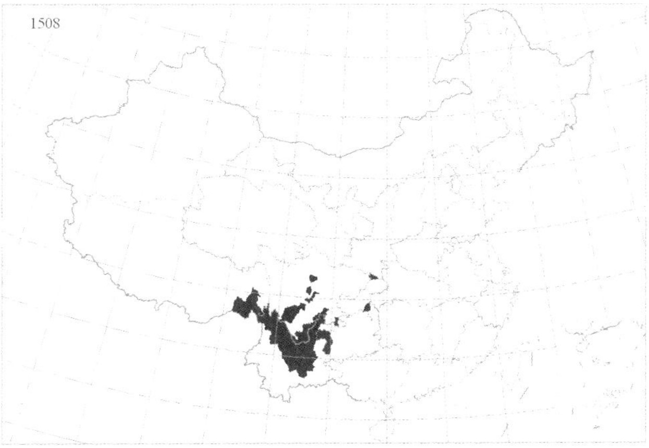

Loranthaceae 桑寄生科 *Taxillus* 钝果寄生属
Taxillus umbellifer (Schult. f.) Danser 伞花钝果寄生
Shrubs

13.3; 5.4; 19.6; 13.4; 100.4; 0.0; 706 ❙ 1736; 1007; 51 ❙ 145.8; 706; 447.4

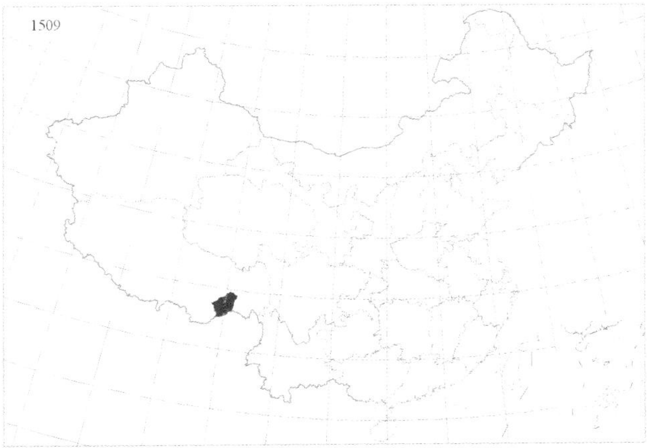

Viscaceae 槲寄生科 *Arceuthobium* 油杉寄生属
Arceuthobium chinense Lecomte 油杉寄生
Subshrubs

12.7 (-4.1~19.3); 5.3 (-17.3~13.1); 18.5 (7.6~24.3); 13.2 (2.1~19.4); 103.8 (5.2~173.2); 11.2 (0.0~114.2); 701 (336~939) ❙ 946 (361~1736); 550 (235~1007); 34 (5~69) ❙ 37.7 (-14.2~145.8); 675 (336~903); 297.2 (181.4~447.4)

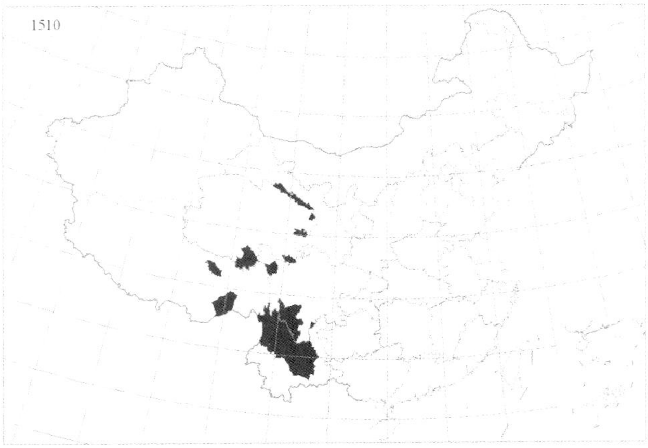

Viscaceae 槲寄生科 *Arceuthobium* 油杉寄生属
Arceuthobium oxycedri (DC.) Bieb. 圆柏寄生
Subshrubs

0.9; -8.3; 9.2; 3.3; 12.7; 61.9; 383 ❙ 480; 300; 9 ❙ 24.0; 371; 174.8

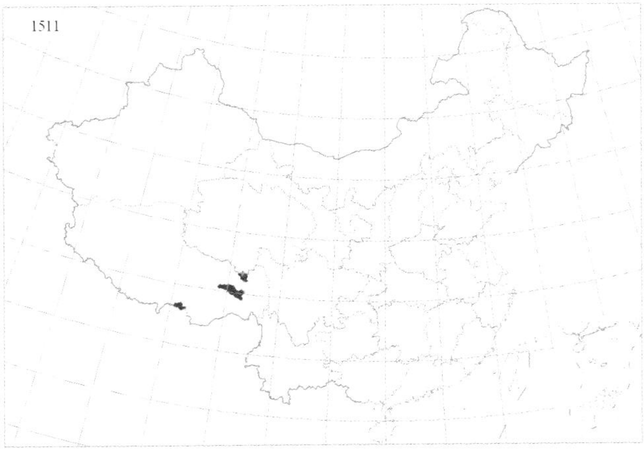

Viscaceae 槲寄生科 *Arceuthobium* 油杉寄生属
Arceuthobium pini Hawksw. et Wiens 高山松寄生
Subshrubs

5.4 (-2.2~14.8); -3.1 (-12.8~6.5); 12.9 (7.6~25.3); 6.6 (2.2~14.8); 39.9 (5.5~119.2); 34.4 (0.0~91.3); 498 (336~795) ❙ 803 (389~1736); 471 (246~1007); 22 (7~88) ❙ 57.8 (0.0~145.8); 497 (336~795); 261.2 (142.5~447.4)

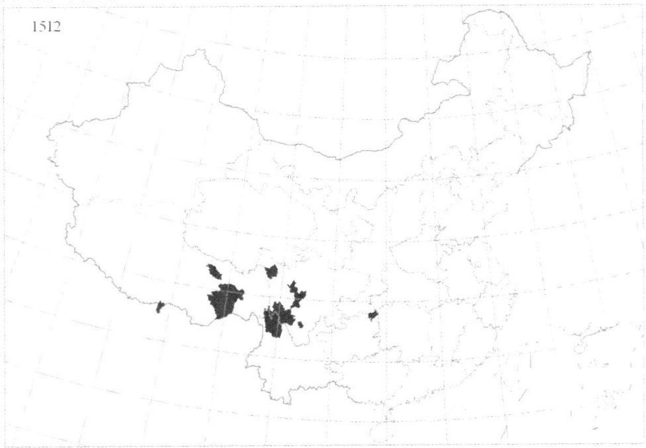

Viscaceae 槲寄生科 *Arceuthobium* 油杉寄生属

Arceuthobium sichuanense (H. S. Kiu) Hawksw. et Wiens 云杉寄生

Subshrubs

0.4; -9.1; 8.9; 3.2; 12.0; 66.5; 380 ❙ 515; 326; 11 ❙ 33.5; 380; 218.0

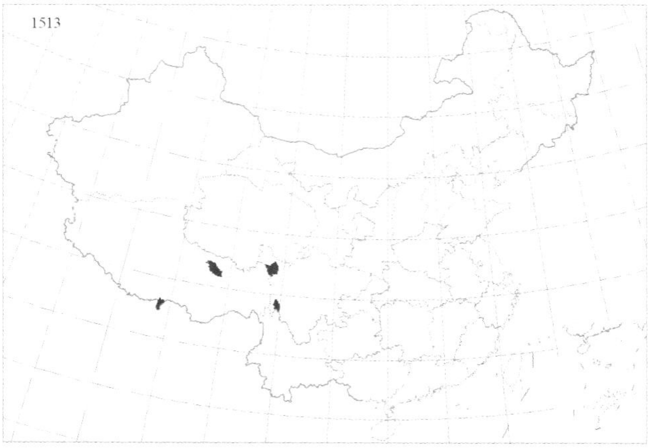

Viscaceae 槲寄生科 *Arceuthobium* 油杉寄生属

Arceuthobium tibetense H. S. Kiu et W. Ren 冷杉寄生

Subshrubs

3.4; -5.4; 11.2; 4.6; 22.8; 41.8; 429 ❙ 500; 302; 8 ❙ 13.7; 429; 255.3

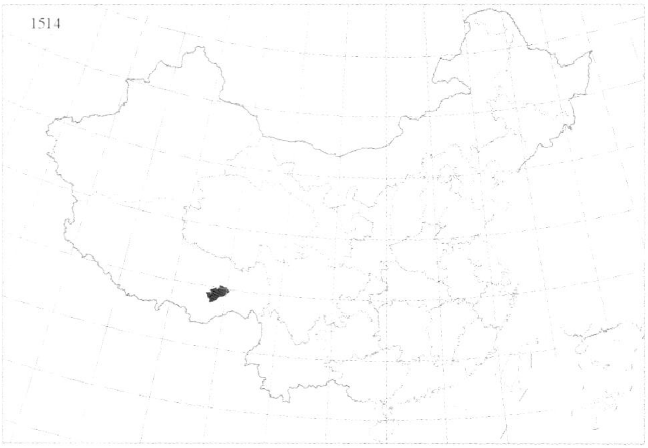

Viscaceae 槲寄生科 *Korthalsella* 栗寄生属

Korthalsella japonica (Thunb.) Engl. 栗寄生

Subshrubs, evergreen

16.7 (-1.3~24.8); 6.6 (-13.5~20.1); 25.8 (9.7~29.9); 16.8 (2.9~24.8); 144.2 (11.3~237.6); 2.8 (0.0~86.3); 902 (359~1385) ❙ 1353 (486~3495); 566 (289~1412); 116 (8~897) ❙ 50.3 (-34.3~246.5); 887 (359~1326); 222.7 (95.3~659.0)

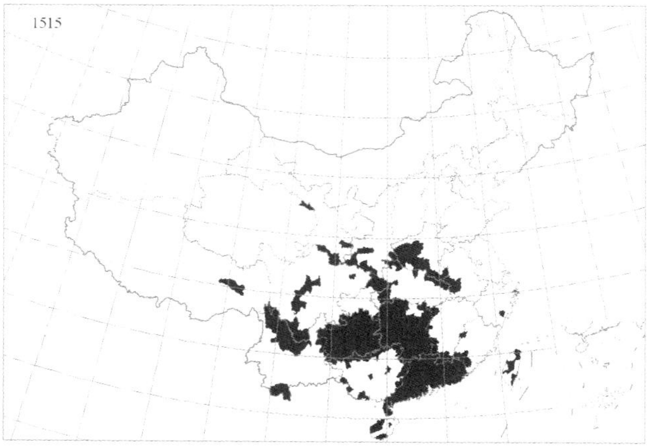

Viscaceae 槲寄生科 *Viscum* 槲寄生属

Viscum album subsp. meridianum (Danser) D. G. Long 卵叶槲寄生

Shrubs, evergreen

15.4 (-1.1~21.0); 7.4 (-9.1~14.2); 21.9 (6.2~26.0); 15.6 (2.0~21.0); 129.3 (2.5~191.8); 3.5 (0.0~75.6); 800 (334~1053) ❙ 1201 (321~1547); 638 (193~911); 58 (31~77) ❙ 50.7 (-4.0~100.7); 798 (321~1053); 309.0 (79.3~551.6)

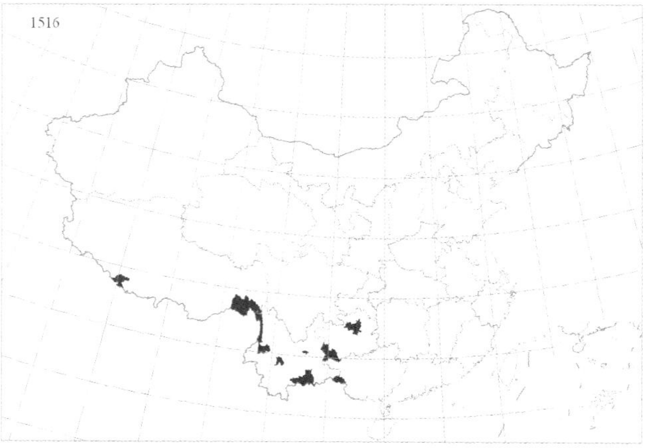

Viscaceae 槲寄生科 *Viscum* 槲寄生属

Viscum articulatum Burm. f. 扁枝槲寄生

Subshrubs, evergreen

18.6 (1.8~24.8); 10.1 (-7.1~20.1); 25.6 (9.9~29.9); 18.7 (3.8~24.8); 164.7 (15.7~237.6); 0.9 (0.0~53.3); 972 (404~1385) ❙ 1509 (527~3495); 692 (312~1847); 121 (9~897) ❙ 57.2 (-34.3~246.5); 960 (404~1257); 267.7 (95.3~659.0)

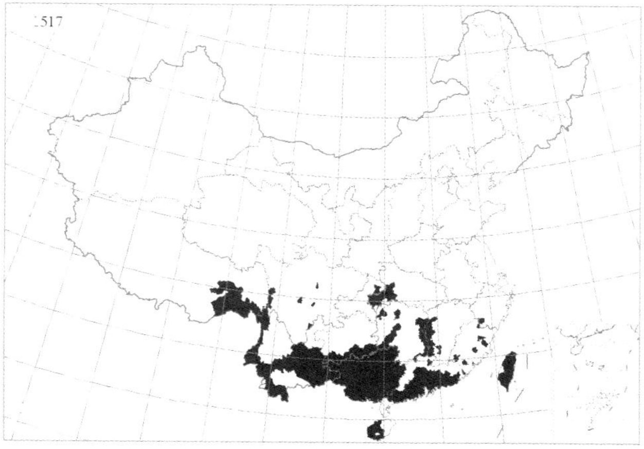

Viscaceae 槲寄生科 *Viscum* 槲寄生属

Viscum coloratum (Kom.) Nakai 槲寄生

Shrubs, evergreen

12.3 (-0.8~22.3); -2.2 (-25.0~17.7); 25.3 (11.7~29.9); 13.2 (4.3~22.3); 110.1 (21.1~208.1); 22.5 (0.0~118.4); 779 (427~1202) ❙ 909 (111~2992); 424 (74~1847); 76 (2~308) ❙ 13.6 (-82.3~246.5); 678 (111~1202); 194.5 (13.9~346.5)

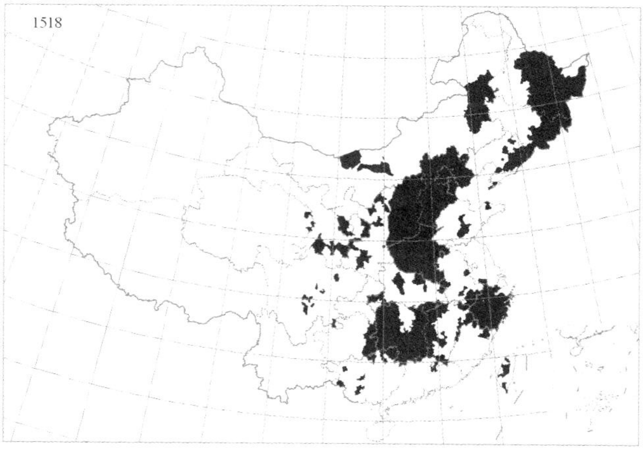

Viscaceae 槲寄生科 *Viscum* 槲寄生属

Viscum diospyrosicola Hayata 柿寄生

Subshrubs, evergreen

17.9 (3.8~24.5); 8.0 (-4.3~18.9); 26.8 (10.8~29.9); 18.0 (4.7~24.5); 156.3 (21.6~234.4); 0.8 (0.0~35.4); 952 (444~1360) ❙ 1475 (711~2680); 589 (312~1186); 140 (15~245) ❙ 56.7 (-34.3~246.5); 942 (444~1360); 221.4 (95.3~455.2)

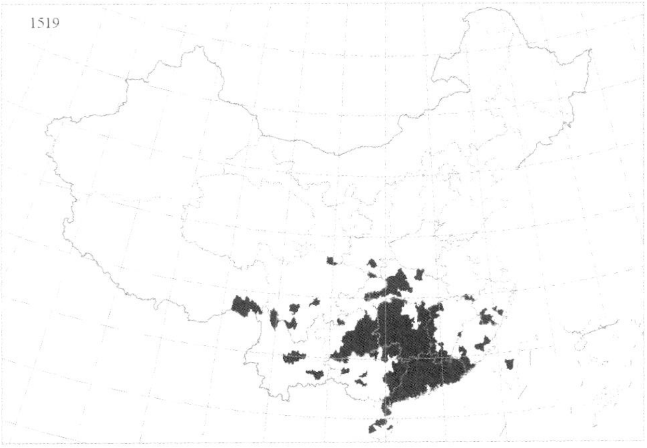

Viscaceae 槲寄生科 *Viscum* 槲寄生属

Viscum fargesii Lecomte 线叶槲寄生

Shrubs, evergreen

9.4 (1.6~15.1); -3.3 (-10.2~4.4); 21.2 (10.9~28.2); 10.2 (4.0~15.2); 78.8 (17.9~126.9); 25.0 (0.6~60.7); 647 (419~874) ❙ 677 (373~1246); 328 (201~525); 20 (7~46) ❙ 6.6 (-38.3~93.8); 591 (373~817); 227.5 (88.1~344.4)

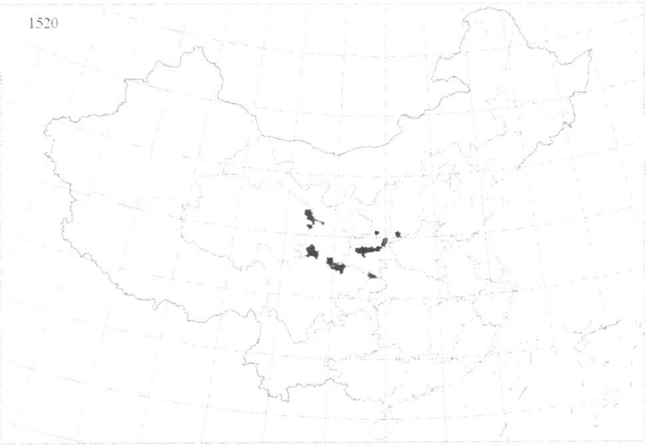

Viscaceae 槲寄生科 *Viscum* 槲寄生属

Viscum liquidambaricola Hayata 枫寄生

Shrubs, evergreen

17.2 (-1.1~24.8); 8.2 (-9.6~20.1); 24.9 (6.2~29.8); 17.3 (2.0~24.8); 148.5 (2.5~237.6); 1.5 (0.0~75.6); 910 (334~1385) ❙ 1359 (321~2759); 613 (193~1594); 102 (10~230) ❙ 50.8 (-34.3~202.5); 899 (321~1360); 256.1 (61.6~659.0)

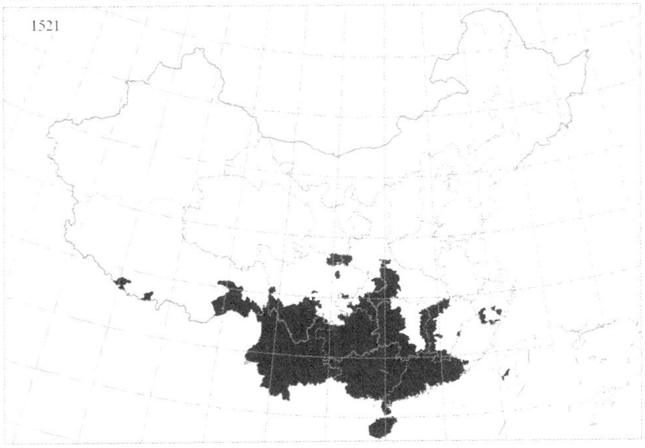

Viscaceae 槲寄生科 *Viscum* 槲寄生属

Viscum loranthi Elmer 聚花槲寄生

Subshrubs, evergreen

12.6 (6.1~19.2); 6.0 (-1.5~13.8); 17.8 (12.7~22.3); 12.6 (6.3~19.2); 94.6 (35.2~170.5); 3.3 (0.0~21.8); 685 (496~909) ❙ 1066 (801~1499); 597 (496~686); 40 (21~58) ❙ 55.7 (43.0~64.9); 681 (496~880); 367.5 (282.9~513.9)

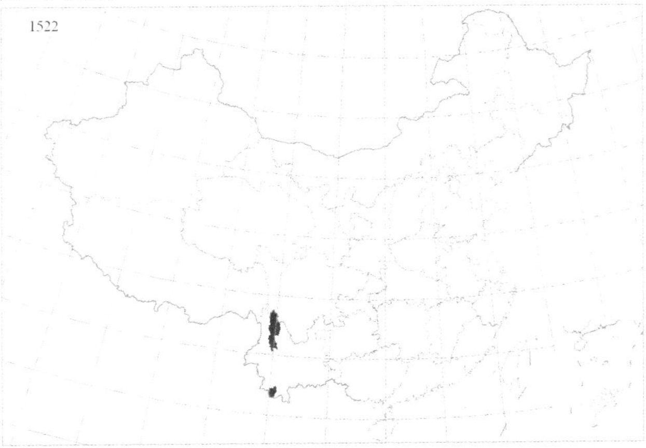

Viscaceae 槲寄生科 *Viscum* 槲寄生属

Viscum monoicum Roxb. ex DC. 五脉槲寄生

Shrubs, evergreen

18.9 (17.2~21.3); 11.7 (9.0~15.1); 24.1 (21.6~27.9); 19.0 (17.3~21.3); 167.8 (147.1~196.0); 0.0 (0.0~0.0); 931 (833~1123) ❙ 1400 (1161~1751); 757 (617~996); 59 (49~77) ❙ 51.3 (5.8~88.8); 927 (833~1121); 413.7 (189.5~659.0)

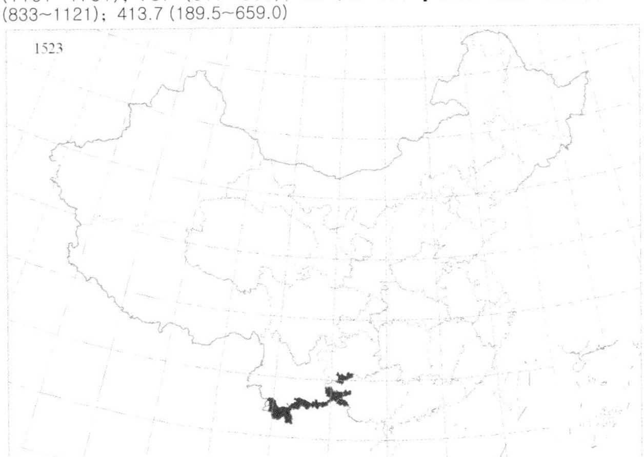

Viscaceae 槲寄生科 *Viscum* 槲寄生属

Viscum multinerve (Hayata) Hayata 柄果槲寄生

Shrubs, evergreen

20.6 (15.1~24.8); 12.5 (6.2~20.1); 27.1 (20.1~29.2); 20.6 (15.2~24.8); 187.5 (122.6~237.6); 0.0 (0.0~0.0); 1076 (772~1385) ❙ 1588 (868~2759); 711 (312~1594); 121 (41~257) ❙ 49.7 (-34.3~246.5); 1061 (772~1273); 253.8 (95.3~659.0)

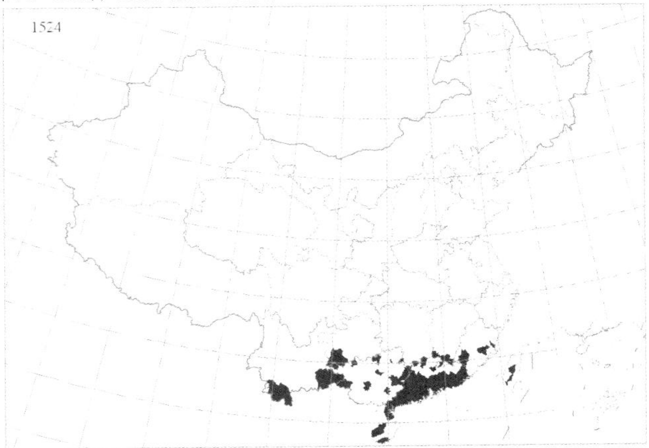

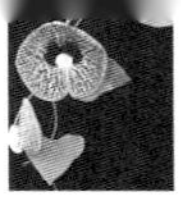

Viscaceae 槲寄生科 *Viscum* 槲寄生属
Viscum nudum Danser 绿茎槲寄生
Shrubs, evergreen

7.8 (-0.3~17.5); -1.0 (-10.9~10.9); 15.5 (9.0~24.6); 8.7 (3.0~17.6); 59.2 (10.4~151.2); 24.6 (0.0~73.4); 566 (378~846) ❙ 951 (579~1414); 542 (347~817); 27 (9~51) ❙ 70.5 (19.0~128.6); 566 (378~836); 270.6 (185.6~383.4)

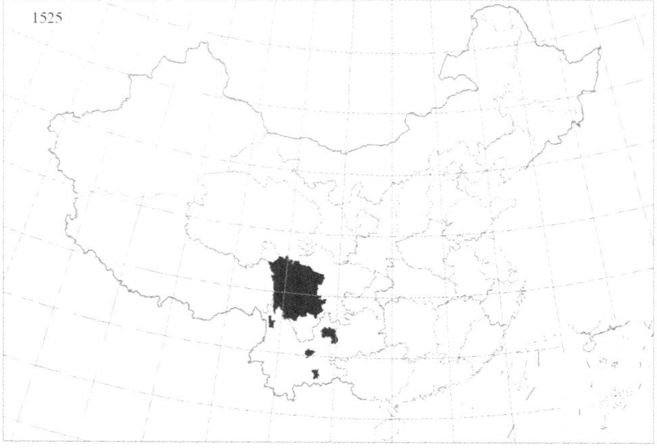

Viscaceae 槲寄生科 *Viscum* 槲寄生属
Viscum ovalifolium DC. 瘤果槲寄生
Shrubs, evergreen

21.8 (17.9~24.8); 14.2 (7.4~20.1); 27.7 (21.6~29.2); 21.9 (18.0~24.8); 202.6 (155.6~237.6); 0.0 (0.0~0.0); 1155 (861~1385) ❙ 1641 (868~2148); 753 (312~1234); 112 (41~188) ❙ 43.4 (-34.3~88.8); 1133 (839~1360); 277.1 (95.3~659.0)

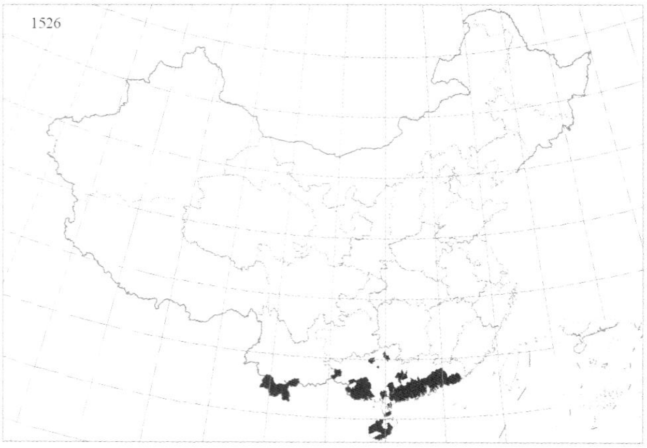

Viscaceae 槲寄生科 *Viscum* 槲寄生属
Viscum yunnanense H. S. Kiu 云南槲寄生
Subshrubs, evergreen

20.4; 15.1; 23.6; 20.5; 185.8; 0.0; 989 ❙ 1607; 702; 77 ❙ 62.5; 986; 659.0

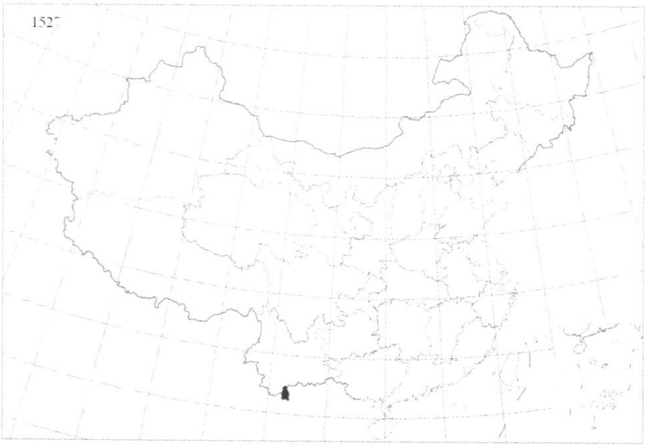

Aristolochiaceae 马兜铃科 *Aristolochia* 马兜铃属
Aristolochia caulialata C. Y. Wu ex J. S. Ma et C. Y. Cheng 翅茎马兜铃
Vines, deciduous

19.0 (16.3~21.3); 11.2 (8.7~14.2); 24.7 (20.3~27.9); 19.0 (16.4~21.3); 168.5 (136.6~196.0); 0.0 (0.0~0.0); 951 (789~1123) ❙ 1387 (1182~1688); 754 (634~959); 67 (43~111) ❙ 47.8 (5.8~112.2); 949 (780~1121); 374.8 (248.4~627.0)

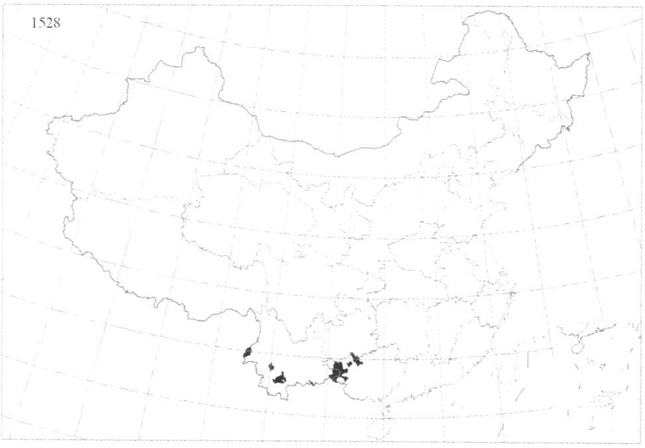

Aristolochiaceae 马兜铃科 *Aristolochia* 马兜铃属
Aristolochia championii Merr. et W. Y. Chun 长叶马兜铃
Vines, deciduous

17.5 (10.6~22.3); 7.7 (-0.2~15.1); 26.2 (20.6~28.9); 17.6 (10.6~22.3); 151.4 (79.4~208.1); 0.4 (0.0~12.1); 918 (643~1202) ❙ 1309 (1028~1776); 589 (442~959); 90 (46~194) ❙ 43.7 (8.0~112.2); 917 (643~1202); 223.2 (108.2~659.0)

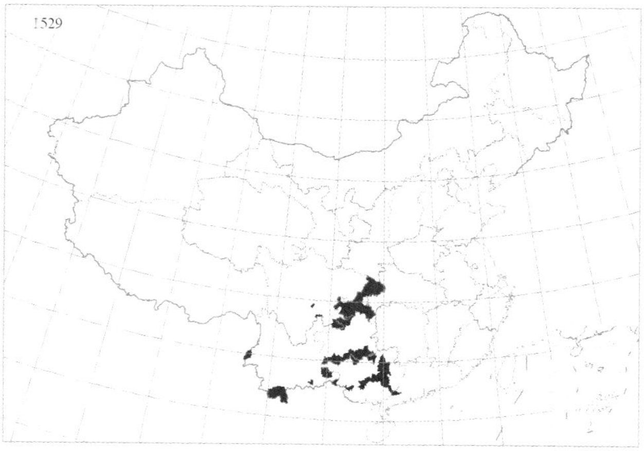

Aristolochiaceae 马兜铃科 *Aristolochia* 马兜铃属
Aristolochia fangchi Y. C. Wu ex L. D. Chow et S. M. Hwang 广防己
Vines, deciduous

21.0 (14.6~23.4); 12.7 (3.8~16.1); 27.7 (22.3~29.2); 21.0 (14.7~23.5); 192.3 (117.7~221.7); 0.0 (0.0~1.5); 1104 (779~1273) ❙ 1650 (1188~2037); 746 (544~1033); 128 (48~191) ❙ 50.3 (5.8~101.8); 1103 (779~1273); 230.2 (95.3~659.0)

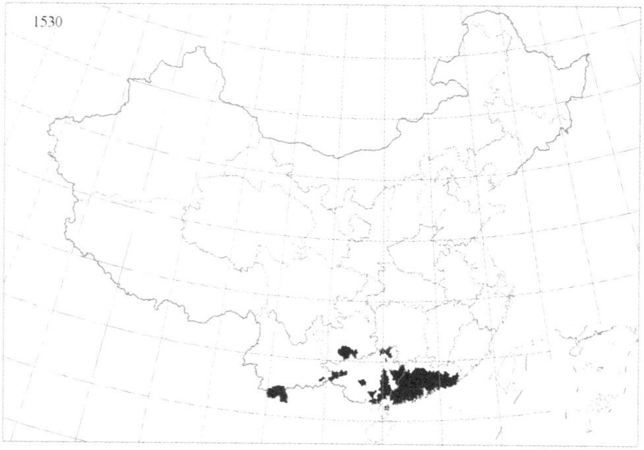

Aristolochiaceae 马兜铃科 *Aristolochia* 马兜铃属
Aristolochia fulvicoma Merr. et Chun 黄毛马兜铃
Climbing shrubs, deciduous

24.2 (23.0~24.8); 19.1 (17.7~20.1); 27.7 (26.7~28.4); 24.2 (23.1~24.8); 230.9 (217.0~237.6); 0.0 (0.0~0.0); 1330 (1217~1385) ❙ 1626 (1405~1767); 557 (473~619); 69 (41~110) ❙ 22.7 (1.5~41.8); 1214 (1137~1326); 519.3 (352.9~620.0)

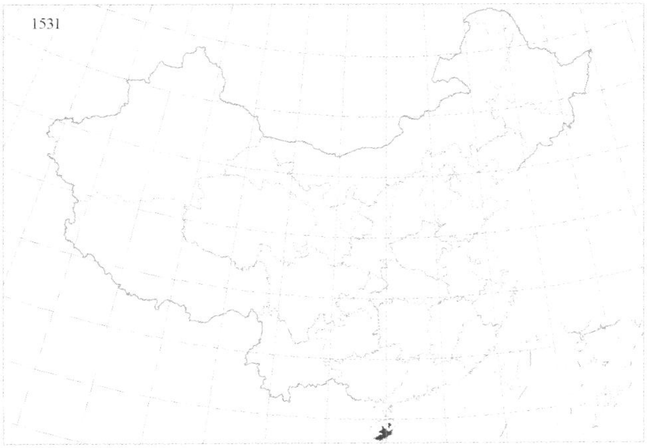

Aristolochiaceae 马兜铃科 *Aristolochia* 马兜铃属
Aristolochia griffithii Hook. f. et Thomson ex Duch. 西藏马兜铃
Climbing shrubs, deciduous

8.6 (-1.8~17.5); 1.0 (-10.2~10.5); 15.1 (6.1~22.3); 9.4 (1.8~17.5); 65.9 (1.9~149.9); 21.9 (0.0~82.8); 582 (329~846) ❙ 909 (320~1736); 533 (217~1007); 31 (9~51) ❙ 51.4 (-3.0~145.8); 575 (320~823); 278.1 (61.6~447.4)

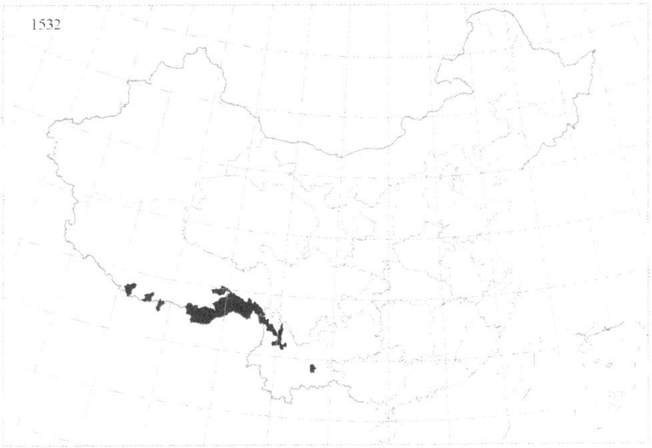

Aristolochiaceae 马兜铃科 *Aristolochia* 马兜铃属
Aristolochia hainanensis Merr. 海南马兜铃
Climbing shrubs, deciduous

22.8 (19.1~24.5); 16.1 (11.0~19.1); 27.8 (25.0~28.7); 22.9 (19.1~24.5); 214.3 (169.6~234.4); 0.0 (0.0~0.0); 1224 (945~1360) ❙ 1586 (868~2435); 721 (312~1428); 85 (42~130) ❙ 31.1 (-34.3~99.5); 1140 (868~1360); 382.8 (235.1~620.0)

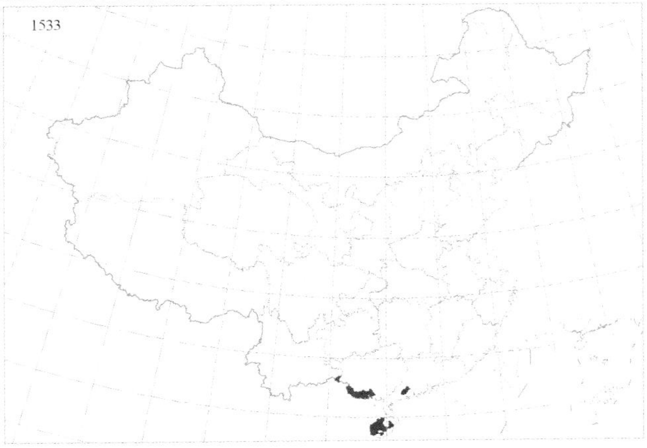

Aristolochiaceae 马兜铃科 *Aristolochia* 马兜铃属
Aristolochia howii Merr. et Chun 南粤马兜铃
Climbing shrubs, deciduous

24.1; 19.0; 27.6; 24.1; 229.3; 0.0; 1316 ❙ 1634; 561; 68 ❙ 24.7; 1202; 537.3

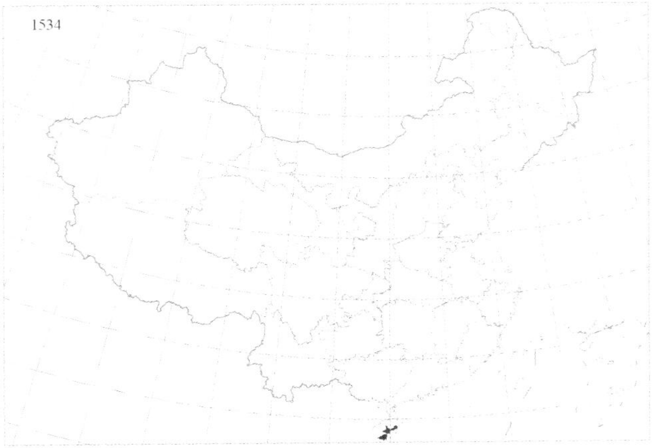

Aristolochiaceae 马兜铃科 *Aristolochia* 马兜铃属
Aristolochia kunmingensis C. Y. Cheng et J. S. Ma 昆明马兜铃
Climbing shrubs

15.8 (11.3~22.3); 8.8 (4.3~14.1); 20.9 (17.0~28.9); 15.8 (11.3~22.3); 129.9 (76.6~207.8); 0.1 (0.0~0.7); 804 (644~1193) ❙ 1253 (952~1688); 694 (576~959); 58 (28~11[illegible]) ❙ 58.6 (26.1~112.2); 791 (644~1193); 360.4 (197.7~627.0)

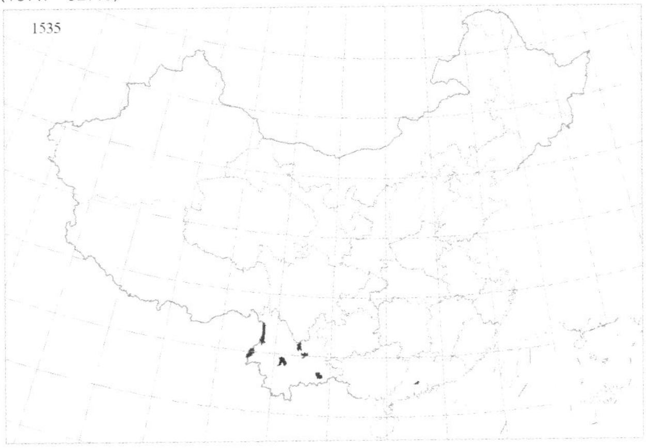

Aristolochiaceae 马兜铃科 *Aristolochia* 马兜铃属
Aristolochia kwangsiensis Chun et F. C. How 广西马兜铃
Climbing shrubs, deciduous

18.3 (11.1~22.7); 9.1 (2.1~15.0); 26.0 (19.4~28.7); 18.3 (11.3~22.8); 160.4 (80.8~213.1); 0.4 (0.0~5.6); 954 (652~1221) ❙ 1403 (985~2435); 683 (475~1428); 94 (37~190) ❙ 48.1 (5.8~107.2); 954 (652~1221); 258.8 (154.4~506.2)

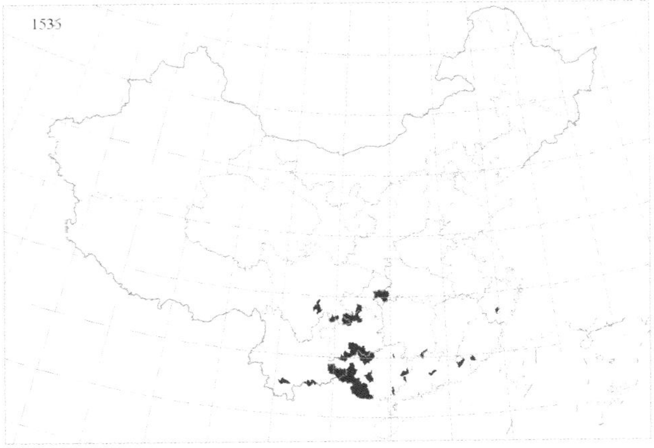

Aristolochiaceae 马兜铃科 *Aristolochia* 马兜铃属
Aristolochia manshuriensis Kom. 关木通
Vines, deciduous

7.9 (1.4~15.1); -9.1 (-20.6~2.9); 22.8 (18.4~27.5); 10.2 (6.4~15.2); 81.3 (45.8~128.0); 46.4 (3.0~93.8); 657 (501~865) ❙ 714 (503~1246); 389 (258~691); 29 (15~83) ❙ 10.3 (-35.9~101.4); 616 (501~844); 244.0 (121.6~346.5)

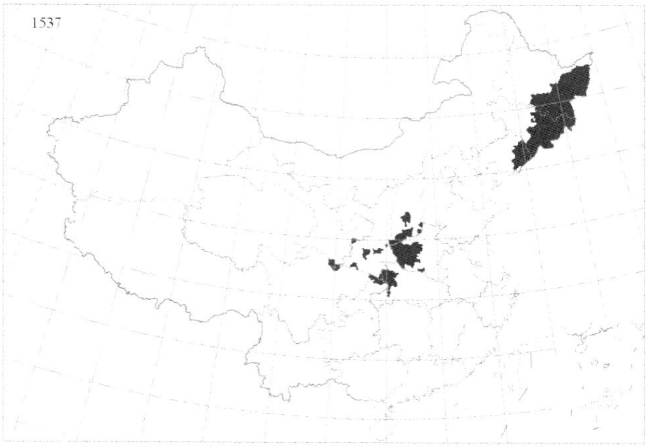

Aristolochiaceae 马兜铃科 *Aristolochia* 马兜铃属
Aristolochia mollissima Hance 寻骨风
Climbing shrubs, deciduous

14.8 (8.2~18.1); 1.5 (-6.9~6.9); 27.3 (21.6~29.9); 14.9 (9.6~18.1); 125.7 (73.3~157.7); 7.4 (0.0~34.4); 850 (622~979) ❙ 1031 (558~1922); 462 (302~685); 97 (19~259) ❙ 20.3 (-36.5~112.2); 800 (558~977); 178.0 (110.1~319.9)

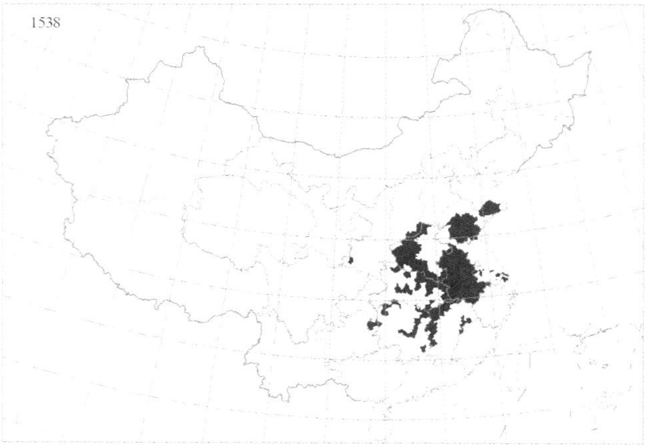

Aristolochiaceae 马兜铃科 *Aristolochia* 马兜铃属
Aristolochia moupinensis Franch. 淮通
Vines, deciduous

14.9 (2.2~18.9); 4.6 (-6.6~12.0); 24.7 (9.9~29.3); 15.0 (3.8~19.0); 121.8 (15.5~167.8); 2.0 (0.0~49.1); 811 (413~984) ❙ 1325 (741~1959); 549 (356~817); 127 (15~259) ❙ 63.2 (-14.2~147.0); 803 (413~965); 239.1 (142.9~384.7)

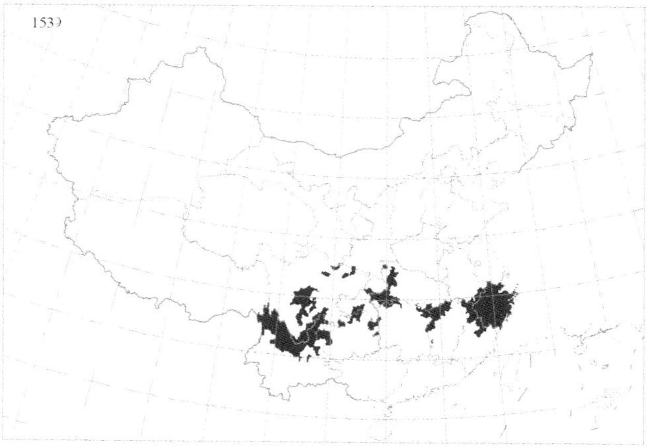

Aristolochiaceae 马兜铃科 *Aristolochia* 马兜铃属
Aristolochia ovatifolia S. M. Hwang 卵叶马兜铃
Climbing shrubs, deciduous

13.5 (6.6~21.0); 4.8 (-1.3~11.9); 20.9 (13.3~28.2); 13.6 (6.8~21.1); 105.0 (39.6~193.0); 2.1 (0.0~19.7); 732 (514~1114) ❙ 1153 (862~1662); 604 (479~821); 61 (12~191) ❙ 57.6 (26.5~101.8); 728 (514~1114); 252.0 (185.6~352.3)

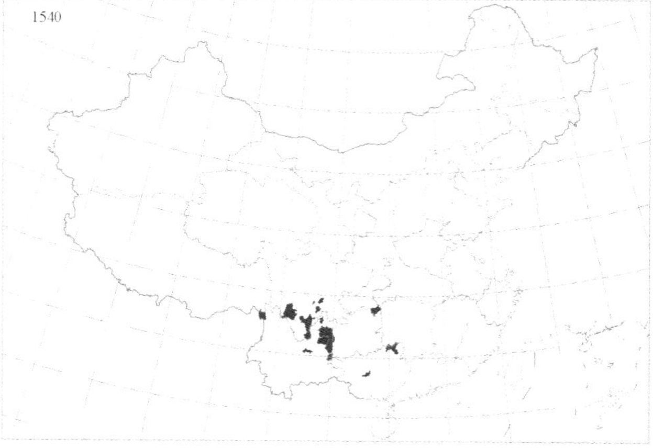

Aristolochiaceae 马兜铃科 *Aristolochia* 马兜铃属
Aristolochia petelotii O. C. Schmidt 滇南马兜铃
Climbing shrubs, deciduous

18.7 (17.3~21.0); 11.9 (9.9~14.2); 23.5 (22.4~25.6); 18.8 (17.3~21.0); 165.0 (147.1~191.8); 0.0 (0.0~0.0); 910 (833~1053) ❙ 1438 (1276~1547); 819 (741~911); 58 (51~68) ❙ 58.7 (36.8~80.6); 909 (833~1053); 462.2 (359.7~551.6)

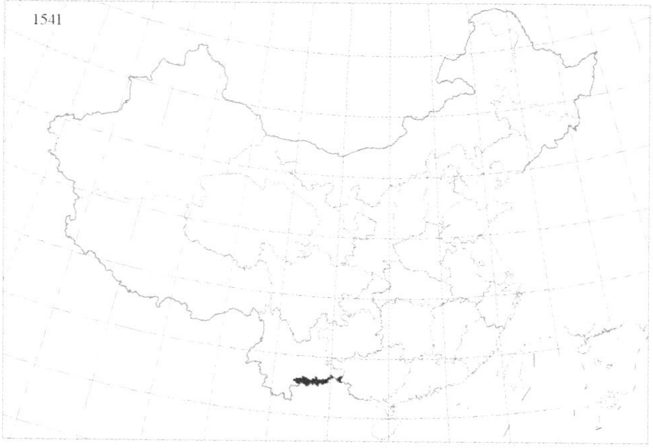

Aristolochiaceae 马兜铃科 *Aristolochia* 马兜铃属
Aristolochia saccata Wall. 管兰香
Climbing shrubs

16.3 (8.6~19.1); 9.9 (1.3~13.1); 20.7 (15.0~23.3); 16.3 (8.7~19.1); 137.5 (52.6~169.7); 1.5 (0.0~8.8); 809 (574~917) ❙ 1390 (1041~1688); 782 (585~959); 56 (45~62) ❙ 72.8 (55.7~112.2); 804 (574~917); 446.8 (352.3~627.0)

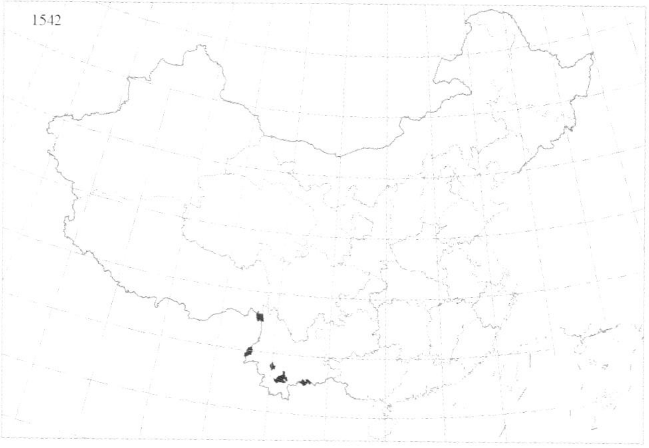

Aristolochiaceae 马兜铃科 *Aristolochia* 马兜铃属
Aristolochia scytophylla S. M. Hwang et D. Y. Chen 革叶马兜铃
Climbing shrubs, deciduous

16.6; 7.2; 24.4; 16.6; 139.7; 0.3; 844 ❙ 1263; 619; 63 ❙ 50.4; 844; 269.2

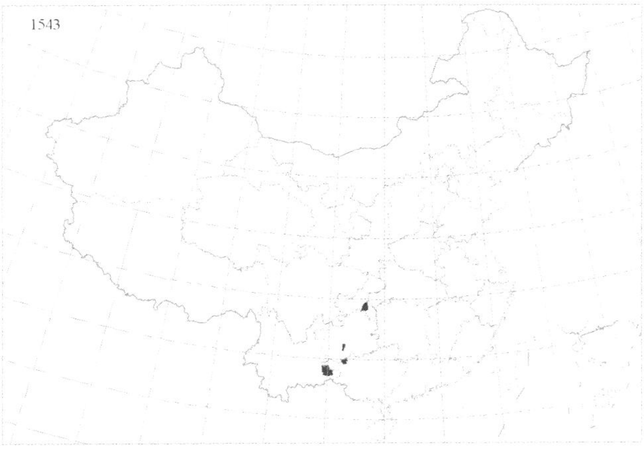

Aristolochiaceae 马兜铃科 *Aristolochia* 马兜铃属
Aristolochia thwaitesii Hook. 海边马兜铃
Subshrubs, deciduous

22.2 (21.3~22.8); 14.6 (13.5~15.5); 28.3 (27.4~28.9); 22.2 (21.3~22.8); 206.9 (195.6~214.1); 0.0 (0.0~0.0); 1180 (1107~1225) ❙ 1775 (1447~2037); 849 (620~1033); 114 (75~152) ❙ 50.5 (20.8~79.9); 1180 (1107~1225); 191.3 (116.6~259.2)

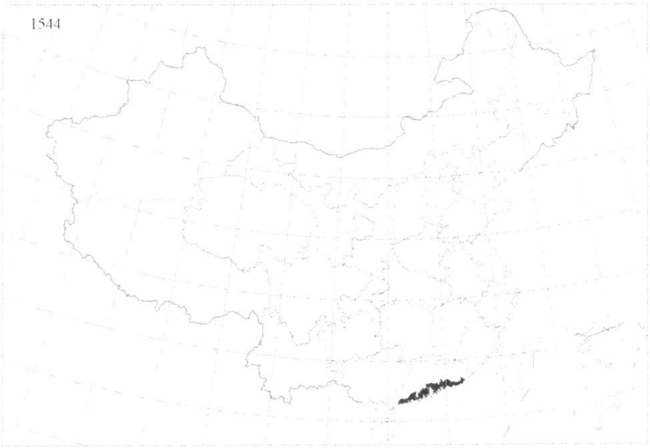

Aristolochiaceae 马兜铃科 *Aristolochia* 马兜铃属
Aristolochia transsecta (Chatterjee) C. Y. Wu ex S. M. Hwang 粉花马兜铃
Climbing shrubs, deciduous

13.4 (8.6~16.2); 6.5 (1.3~9.4); 18.6 (15.0~20.8); 13.5 (8.7~16.3); 103.1 (52.6~135.1); 1.4 (0.0~8.8); 706 (574~790) ❙ 1215 (1041~1348); 669 (585~762); 58 (45~69) ❙ 73.2 (50.8~93.0); 706 (574~790); 369.6 (335.1~407.1)

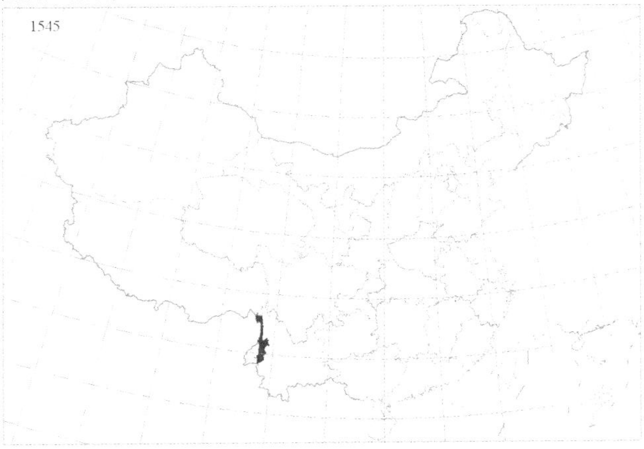

Aristolochiaceae 马兜铃科 *Aristolochia* 马兜铃属
Aristolochia versicolor S. M. Hwang 过石珠
Climbing shrubs, deciduous

19.4 (11.4~22.3); 10.6 (2.7~15.1); 26.5 (18.8~29.8); 19.5 (11.4~22.3); 173.5 (80.7~208.1); 0.1 (0.0~3.7); 1003 (652~1202) ❙ 1572 (874~1991); 673 (494~959); 143 (25~197) ❙ 57.6 (15.4~112.2); 999 (652~1202); 289.3 (157.7~659.0)

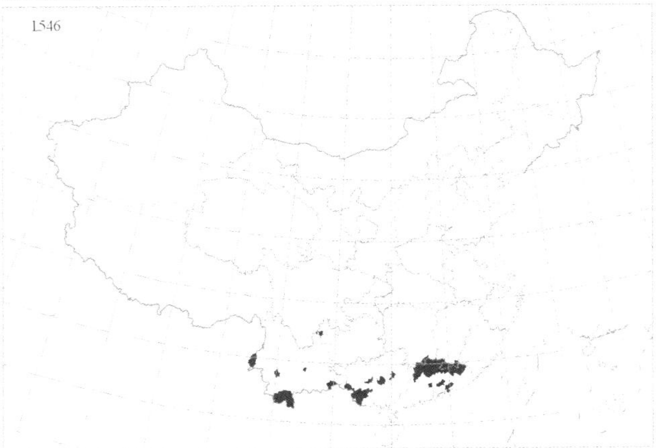

Aristolochiaceae 马兜铃科 *Thottea* 阿柏麻属
Thottea hainanensis (Merr. et Chun) D. Hou 海南线果兜铃
Subshrubs

24.3; 19.3; 27.7; 24.3; 231.8; 0.0; 1337 ❙ 1621; 546; 64 ❙ 21.9; 1203; 535.8

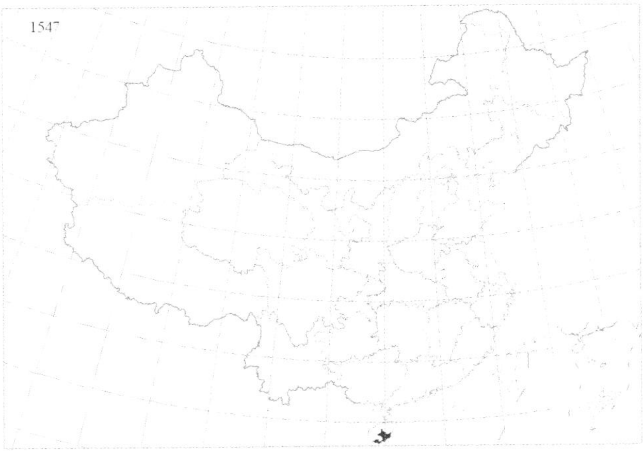

Polygonaceae 蓼科 *Atraphaxis* 针枝蓼属
Atraphaxis bracteata Losinsk. 沙木蓼
Shrubs deciduous

5.1 (-1.4~8.3); -12.5 (-20.7~-7.5); 20.8 (11.5~24.3); 8.3 (3.5~10.2); 64.3 (17.2~85.9); 62.5 (39.0~101.4); 582 (375~679) ❙ 261 (49~437); 168 (34~270); 6 (1~12) ❙ -53.9 (-92.8~-24.8); 261 (49~437); 62.8 (5.1~145.6)

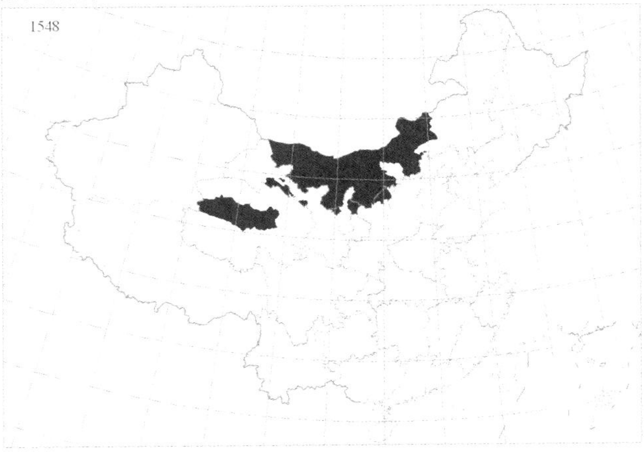

P

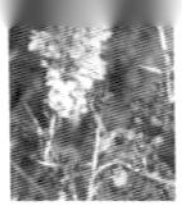

Polygonaceae 蓼科 *Atraphaxis* 针枝蓼属
Atraphaxis canescens Bunge 糙叶木蓼
Shrubs, deciduous

1.9；-16.2；18.3；6.8；49.3；86.5；536 ❙ 216；91；27 ❙ -59.4；216；112.8

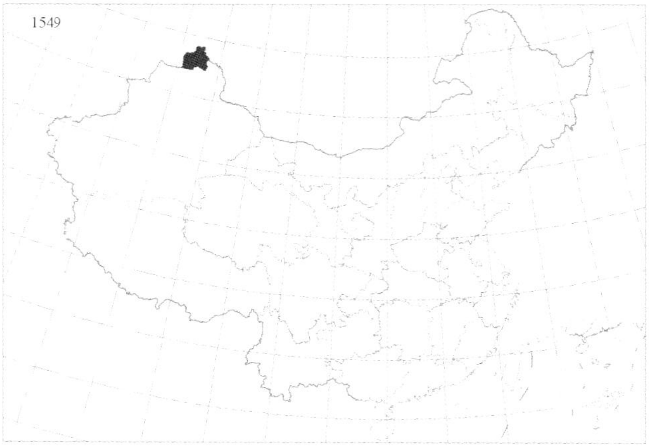

Polygonaceae 蓼科 *Atraphaxis* 针枝蓼属
Atraphaxis compacta Ledeb. 拳木蓼
Shrubs, deciduous

5.1 (-2.1~11.2)；-14.4 (-19.5~-7.8)；21.6 (10.7~27.5)；8.9 (3.2~12.6)；72.1 (14.3~109.2)；69.8 (34.4~104.2)；630 (364~813) ❙ 167 (23~308)；75 (16~127)；16 (1~33) ❙ -72.2 (-97.2~-44.6)；167 (23~308)；61.1 (3.0~139.1)

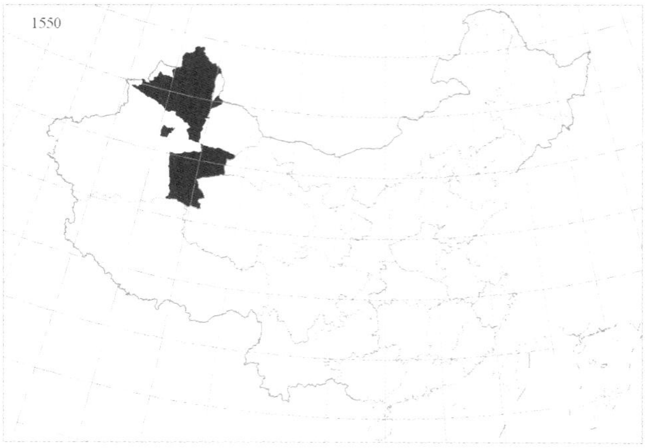

Polygonaceae 蓼科 *Atraphaxis* 针枝蓼属
Atraphaxis decipiens Jaub. et Spach 细枝木蓼
Shrubs, deciduous

3.9 (-2.8~9.3)；-15.7 (-20.9~-11.1)；20.3 (10.9~27.5)；8.1 (3.5~12.1)；63.9 (16.3~107.6)；76.3 (51.6~109.2)；595 (390~813) ❙ 203 (120~315)；91 (51~171)；19 (3~51) ❙ -64.2 (-85.3~-29.7)；203 (120~312)；93.9 (16.6~297.3)

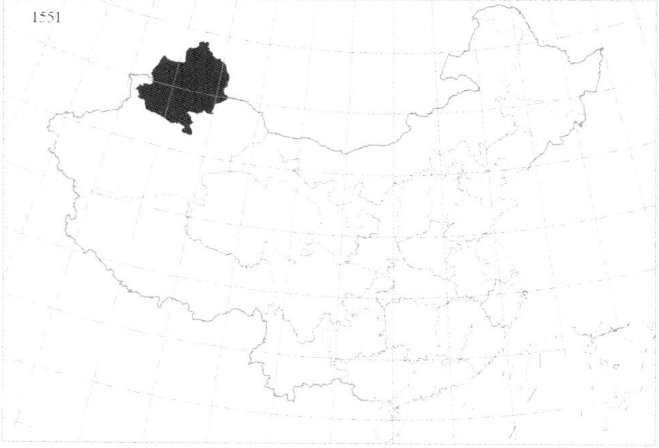

Polygonaceae 蓼科 *Atraphaxis* 针枝蓼属
Atraphaxis frutescens (L.) Eversm. 木蓼
Shrubs, deciduous

5.4 (-2.8~11.8)；-13.1 (-20.9~-6.7)；20.9 (10.9~27.5)；8.7 (3.5~13.0)；69.6 (16.3~113.8)；63.7 (30.1~109.2)；616 (390~822) ❙ 174 (28~450)；88 (20~211)；13 (0~51) ❙ -68.9 (-96.6~3.9)；174 (28~406)；71.5 (1.6~297.3)

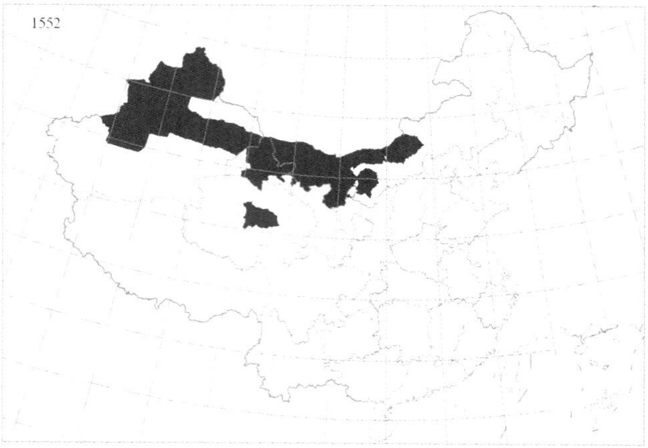

Polygonaceae 蓼科 *Atraphaxis* 针枝蓼属
Atraphaxis irtyschensis C. Y. Yang et Y. L. Han 额河木蓼
Shrubs, deciduous

1.3；-15.9；18.3；6.8；49.3；85.7；536 ❙ 226；95；27 ❙ -57.5；226；110.1

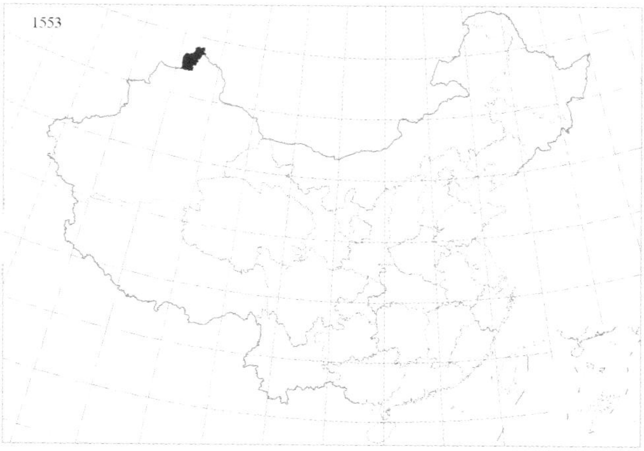

Polygonaceae 蓼科 *Atraphaxis* 针枝蓼属
Atraphaxis laetevirens (Ledeb.) Jaub. et Spach 绿叶木蓼
Shrubs, deciduous

2.7 (-1.3~9.0)；-16.1 (-20.9~-9.3)；18.5 (13.2~23.3)；7.2 (4.6~10.7)；53.5 (26.7~88.9)；80.3 (40.7~104.2)；551 (440~703) ❙ 244 (152~406)；107 (56~211)；25 (12~51) ❙ -54.3 (-76.8~-7.6)；244 (152~406)；142.6 (16.6~297.3)

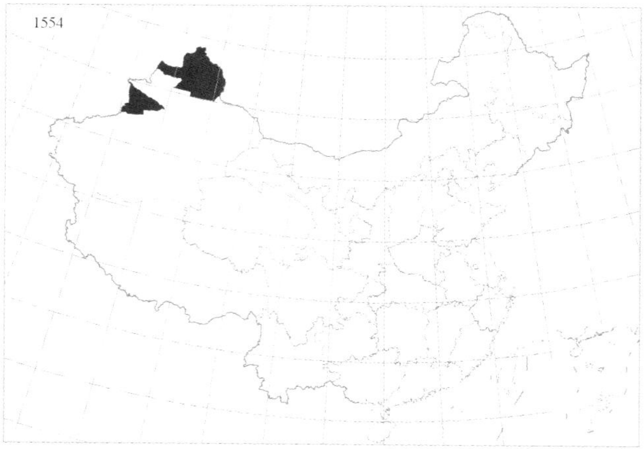

Polygonaceae 蓼科 *Atraphaxis* 针枝蓼属
Atraphaxis manshurica Kitag. 东北木蓼
Shrubs, deciduous

4.9 (-4.6~11.4); -14.0 (-28.8~-6.5); 21.2 (10.9~26.5); 8.5 (3.5~12.7); 67.7 (16.3~109.3); 68.4 (31.6~149.1); 601 (390~793) ❙ 322 (32~694); 207 (21~499); 11 (0~51) ❙ -44.8 (-95.9~5.4); 321 (32~659); 130.1 (1.6~309.5)

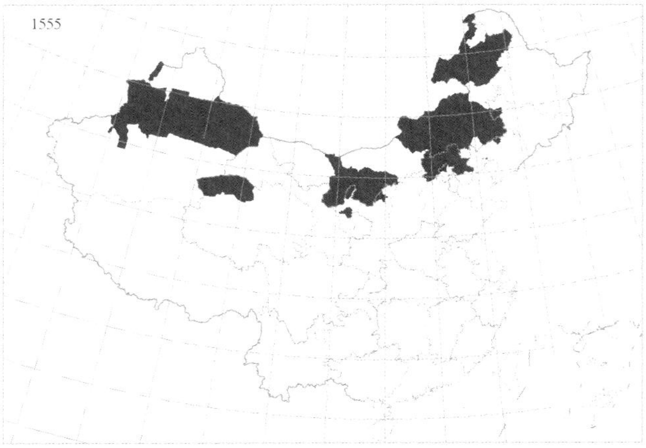

Polygonaceae 蓼科 *Atraphaxis* 针枝蓼属
Atraphaxis pungens (Bieb.) Jaub. et Spach 锐枝木蓼
Shrubs, deciduous

5.2 (-2.5~8.7); -12.1 (-19.5~-8.1); 20.5 (10.1~24.3); 8.3 (2.9~10.3); 64.6 (12.0~85.9); 61.7 (39.2~104.2); 584 (358~679) ❙ 182 (39~416); 109 (29~265); 7 (1~25) ❙ -67.8 (-92.8~-24.8); 182 (39~416); 52.6 (5.1~147.7)

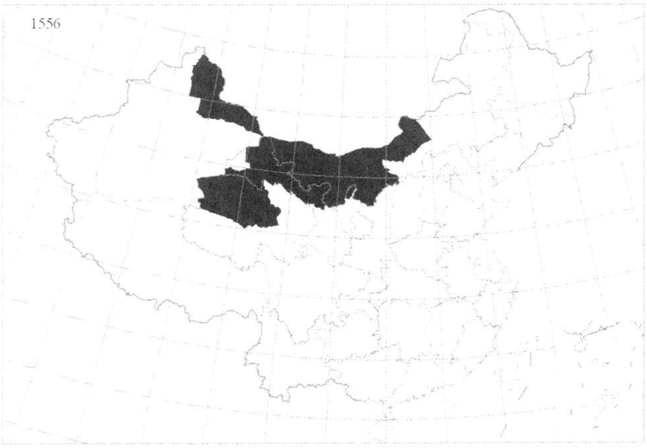

Polygonaceae 蓼科 *Atraphaxis* 针枝蓼属
Atraphaxis pyrifolia Bunge 梨叶木蓼
Shrubs, deciduous

2.2 (-1.3~9.0); -16.5 (-20.9~-9.3); 17.4 (13.2~23.2); 6.8 (4.6~10.7); 48.8 (26.7~88.9); 81.6 (40.7~104.2); 529 (440~703) ❙ 251 (158~406); 113 (62~211); 24 (12~48) ❙ -51.4 (-72.3~-7.6); 251 (158~406); 153.1 (16.6~297.3)

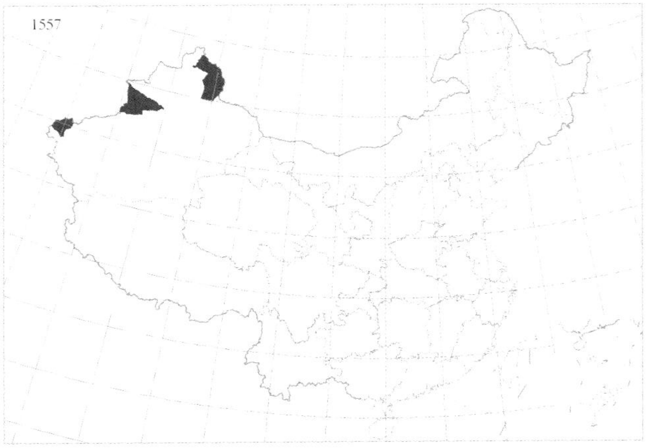

Polygonaceae 蓼科 *Atraphaxis* 针枝蓼属
Atraphaxis spinosa L. 刺木蓼
Shrubs, deciduous

4.9 (-2.8~10.2); -14.3 (-20.9~-8.1); 20.6 (10.9~27.5); 8.5 (3.5~12.1); 67.3 (16.3~107.6); 68.2 (36.6~109.2); 609 (390~813) ❙ 200 (45~450); 93 (27~211); 18 (3~51) ❙ -64.3 (-94.0~3.9); 199 (45~406); 90.6 (7.6~297.3)

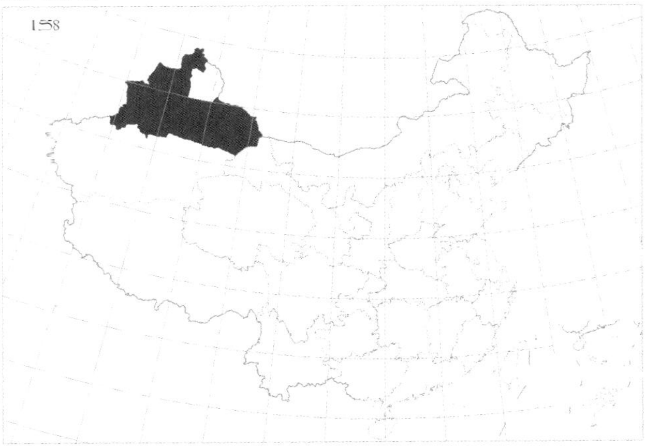

Polygonaceae 蓼科 *Atraphaxis* 针枝蓼属
Atraphaxis virgata (Regel) Krassn. 帚枝木蓼
Shrubs, deciduous

3.6 (-1.1~9.0); -15.7 (-19.5~-10.7); 20.3 (13.7~27.5); 8.0 (4.5~12.0); 62.4 (27.2~107.3); 78.2 (50.7~104.2); 589 (433~813) ❙ 209 (95~450); 86 (49~185); 23 (5~51) ❙ -62.9 (-86.2~3.9); 207 (95~404); 78.2 (10.9~173.1)

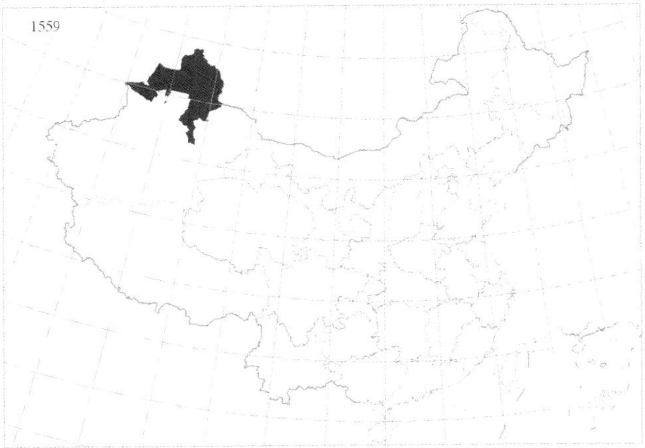

Polygonaceae 蓼科 *Calligonum* 沙拐枣属
Calligonum alashanicum Losinsk. 阿拉善沙拐枣
Shrubs, deciduous

6.8 (2.5~8.0); -10.1 (-11.9~-7.5); 21.8 (15.1~24.0); 9.0 (5.7~10.1); 72.4 (36.0~83.7); 50.3 (39.0~64.8); 613 (472~663) ❙ 260 (115~416); 163 (67~265); 6 (2~12) ❙ -56.8 (-82.2~-31.0); 260 (115~416); 52.1 (13.6~147.7)

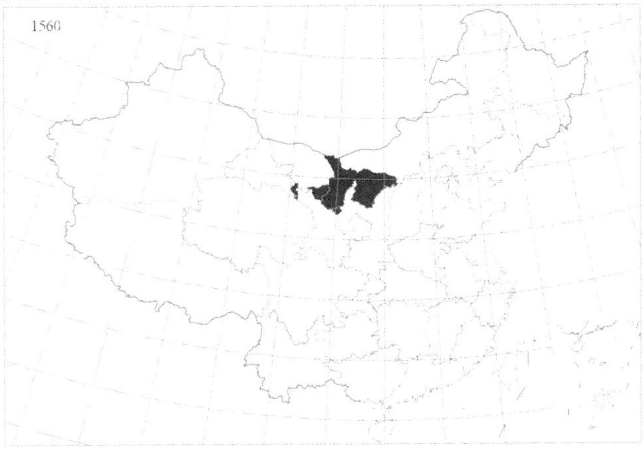

Polygonaceae 蓼科 *Calligonum* 沙拐枣属

Calligonum aphyllum (Pall.) Guerke 无叶沙拐枣

Shrubs, deciduous

6.4; -11.5; 20.6; 8.9; 70.1; 53.2; 620 | 309; 110; 44 | -48.9; 309; 174.4

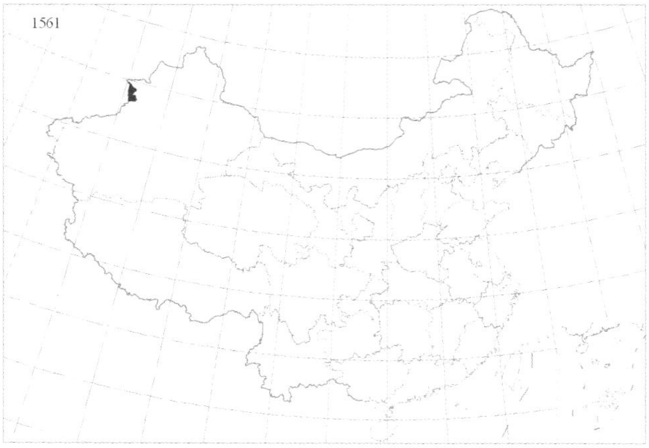

Polygonaceae 蓼科 *Calligonum* 沙拐枣属

Calligonum arborescens Litv. 乔木沙拐枣

Shrubs, deciduous

7.3 (4.7~8.6); -10.6 (-16.5~-7.5); 22.7 (21.2~25.6); 9.7 (8.7~11.0); 79.0 (69.8~93.7); 50.4 (39.0~72.0); 647 (614~725) | 148 (75~225); 80 (39~130); 7 (2~19) | -76.7 (-89.7~-64.1); 148 (75~225); 39.0 (14.6~57.8)

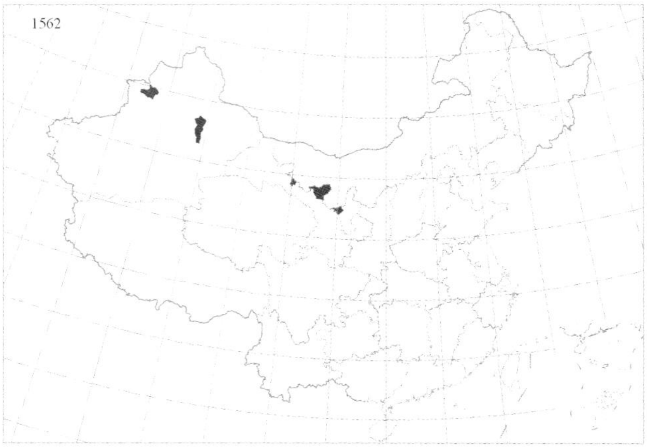

Polygonaceae 蓼科 *Calligonum* 沙拐枣属

Calligonum calliphysa Bunge 泡果沙拐枣

Shrubs, deciduous

4.7 (-0.6~9.3); -14.8 (-19.5~-8.8); 21.3 (15.9~27.1); 8.6 (5.4~12.1); 69.4 (37.5~107.6); 72.3 (40.4~104.2); 618 (475~810) | 162 (45~308); 71 (27~127); 16 (3~30) | -72.4 (-94.0~-44.6); 162 (45~308); 59.4 (7.6~139.1)

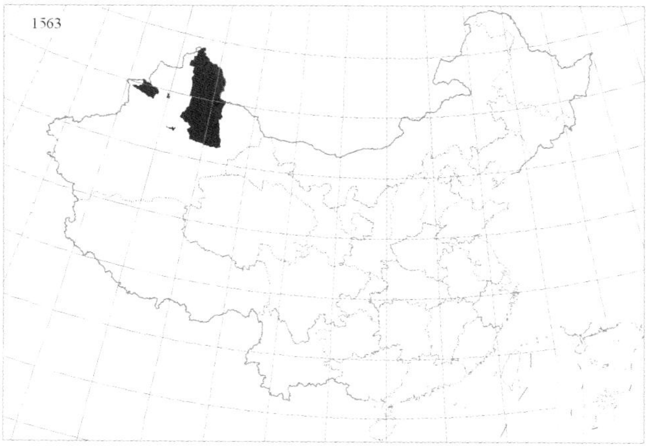

Polygonaceae 蓼科 *Calligonum* 沙拐枣属

Calligonum caput-medusae Schrenk 头状沙拐枣

Shrubs, deciduous

7.3 (4.7~8.6); -11.2 (-16.5~-9.1); 23.1 (21.2~25.6); 9.8 (8.7~11.0); 80.4 (69.8~93.7); 52.6 (44.9~72.0); 654 (614~725) | 128 (75~209); 71 (39~94); 6 (2~19) | -80.0 (-89.7~-65.9); 128 (75~209); 38.3 (14.6~57.8)

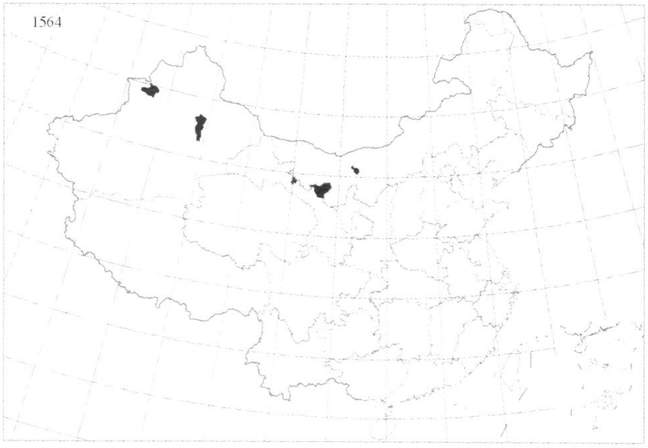

Polygonaceae 蓼科 *Calligonum* 沙拐枣属

Calligonum chinense Losinsk. 甘肃沙拐枣

Shrubs, deciduous

6.2 (-2.1~11.2); -10.9 (-16.2~-7.8); 21.2 (10.7~26.8); 8.8 (3.2~12.6); 70.2 (14.3~109.2); 55.5 (34.4~99.3); 608 (364~800) | 143 (23~416); 93 (16~265); 4 (1~12) | -74.4 (-97.2~-24.8); 143 (23~416); 38.0 (3.0~101.3)

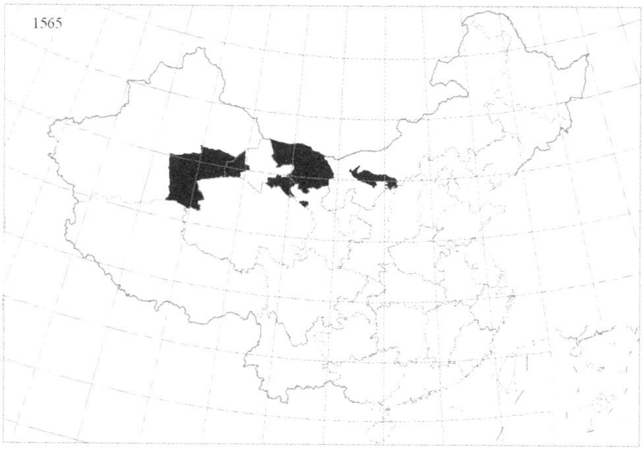

Polygonaceae 蓼科 *Calligonum* 沙拐枣属

Calligonum colubrinum E. Borszcow 褐色沙拐枣

Shrubs, deciduous

5.1; -14.9; 22.3; 9.1; 73.9; 71.5; 631 | 159; 66; 15 | -74.8; 159; 45.4

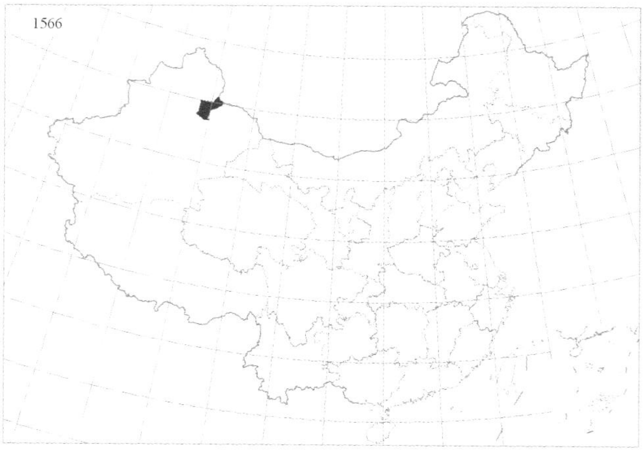

Polygonaceae 蓼科 *Calligonum* 沙拐枣属
Calligonum cordatum Korovin ex N. Pavlov 心形沙拐枣
Shrubs, deciduous

8.6; -10.8; 25.6; 11.0; 93.7; 50.3; 725 | 75; 39; 5 | -89.7; 75; 14.6

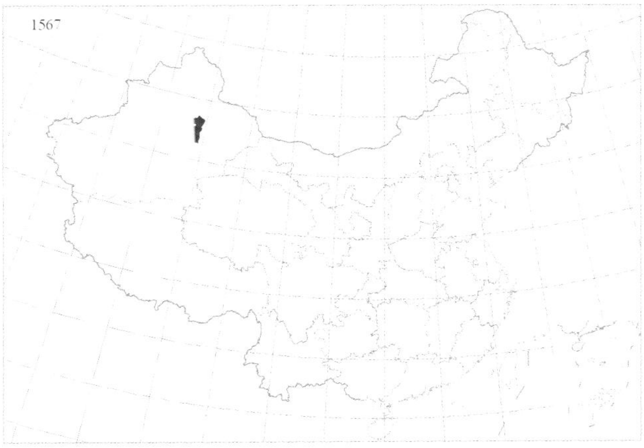

Polygonaceae 蓼科 *Calligonum* 沙拐枣属
Calligonum densum E. Borszczow 密刺沙拐枣
Shrubs, deciduous

6.4; -11.5; 20.6; 8.9; 70.1; 53.2; 620 | 309; 110; 44 | -48.9; 309; 174.4

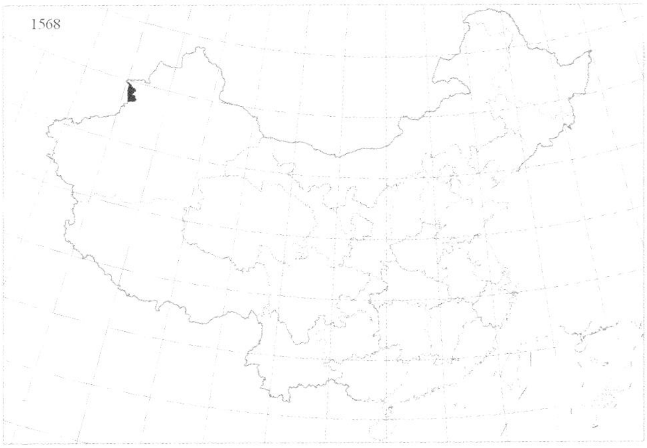

Polygonaceae 蓼科 *Calligonum* 沙拐枣属
Calligonum ebinuricum N. A. Ivanova ex Soskov 艾比湖沙拐枣
Shrubs, deciduous

6.6 (2.7~9.3); -14.5 (-18.4~-10.8); 23.5 (19.2~27.1); 10.0 (7.6~12.1); 83.8 (57.1~107.6); 63.9 (50.3~83.0); 685 (562~810) | 164 (75~224); 77 (39~117); 13 (5~19) | -74.8 (-89.7~-60.2); 164 (75~224); 60.6 (14.6~108.3)

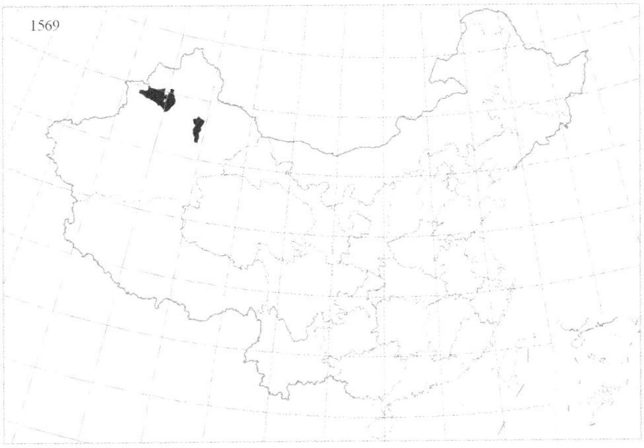

Polygonaceae 蓼科 *Calligonum* 沙拐枣属
Calligonum gobicum (Bunge ex Meisn.) Losinsk. 戈壁沙拐枣
Shrubs, deciduous

6.7 (-2.1~11.2); -11.7 (-16.2~-7.8); 22.6 (10.7~27.5); 9.5 (3.2~12.6); 78.6 (14.3~109.2); 57.1 (34.4~99.3); 653 (364~813) | 95 (23~175); 51 (16~71); 7 (1~17) | -84.8 (-97.2~-72.8); 95 (23~175); 24.2 (3.0~55.9)

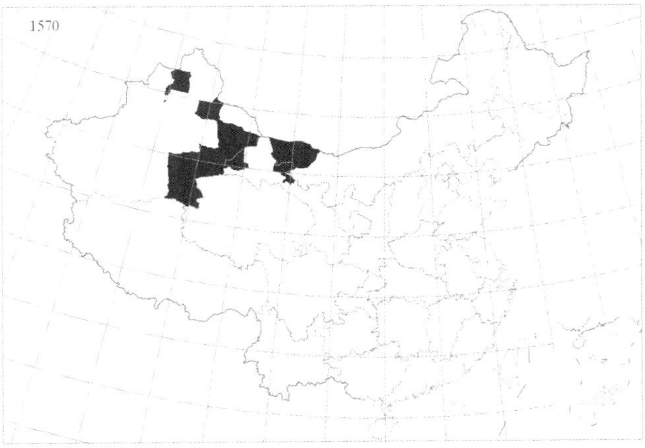

Polygonaceae 蓼科 *Calligonum* 沙拐枣属
Calligonum jeminaicum Z. M. Mao 吉木乃沙拐枣
Shrubs, deciduous

3.2; -14.8; 19.7; 7.6; 56.5; 77.1; 570 | 190; 80; 21 | -66.6; 190; 54.8

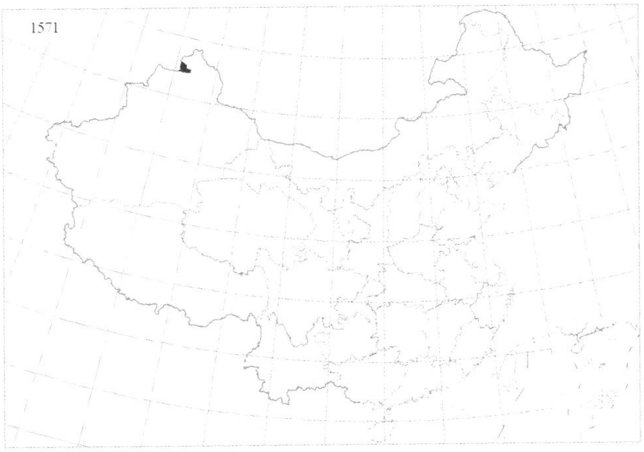

Polygonaceae 蓼科 *Calligonum* 沙拐枣属
Calligonum klementzii Losinsk. 奇台沙拐枣
Shrubs, deciduous

6.1 (4.9~8.9); -13.3 (-15.0~-8.8); 22.8 (21.9~24.0); 9.5 (8.8~10.6); 78.0 (71.0~88.2); 63.6 (41.2~71.6); 649 (619~692) | 129 (35~186); 56 (23~69); 12 (3~20) | -79.9 (-94.9~-72.2); 129 (35~186); 35.0 (7.4~49.6)

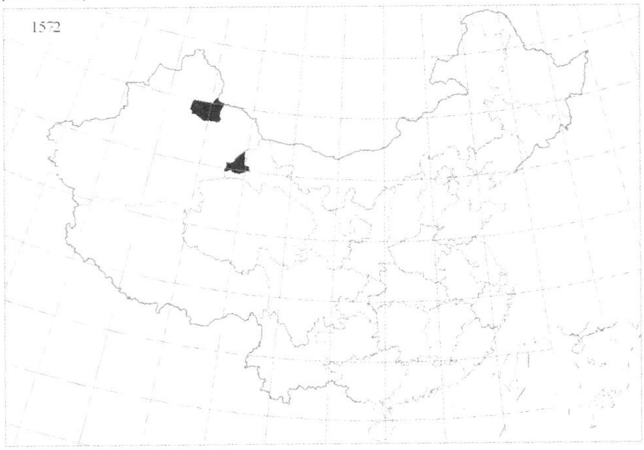

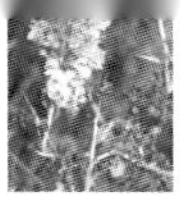

Polygonaceae 蓼科 *Calligonum* 沙拐枣属
Calligonum korlaense Z. M. Mao 库尔勒沙拐枣
Shrubs, deciduous

8.9；-10.2；23.3；10.8；90.5；43.8；705 ❙ 83；53；4 ❙ -88.0；83；28.9

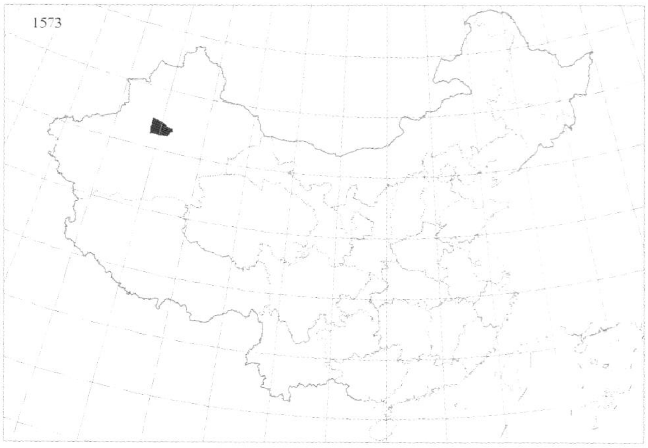

Polygonaceae 蓼科 *Calligonum* 沙拐枣属
Calligonum leucocladum (Schrenk) Bunge 淡枝沙拐枣
Shrubs, deciduous

5.3 (-0.6~9.3)；-14.7 (-19.5~-9.3)；21.9 (15.9~27.5)；9.0 (5.4~12.1)；73.3 (37.5~107.6)；69.6 (40.7~104.2)；635 (475~813) ❙ 187 (75~357)；79 (39~136)；19 (5~48) ❙ -69.6 (-89.7~-34.9)；187 (75~357)；69.8 (14.6~190.1)

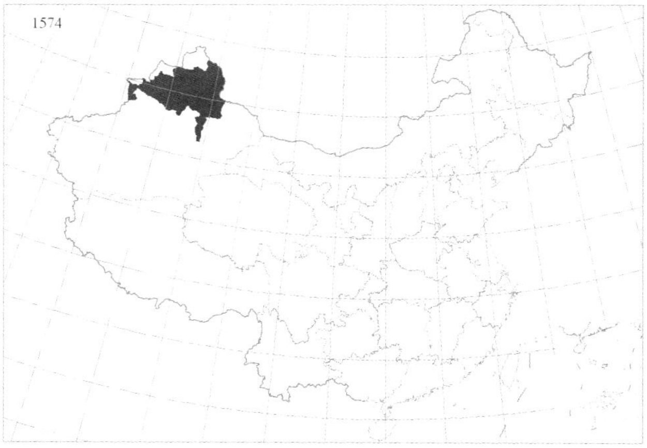

Polygonaceae 蓼科 *Calligonum* 沙拐枣属
Calligonum mongolicum Turcz. 沙拐枣
Shrubs, deciduous

5.7 (-4.1~12.4)；-11.9 (-18.9~-5.4)；21.0 (9.8~26.8)；8.6 (2.5~13.2)；68.1 (9.9~114.1)；59.5 (25.2~117.9)；602 (330~820) ❙ 173 (23~416)；111 (16~270)；5 (1~14) ❙ -69.5 (-97.2~-25.7)；173 (23~416)；44.4 (2.1~145.6)

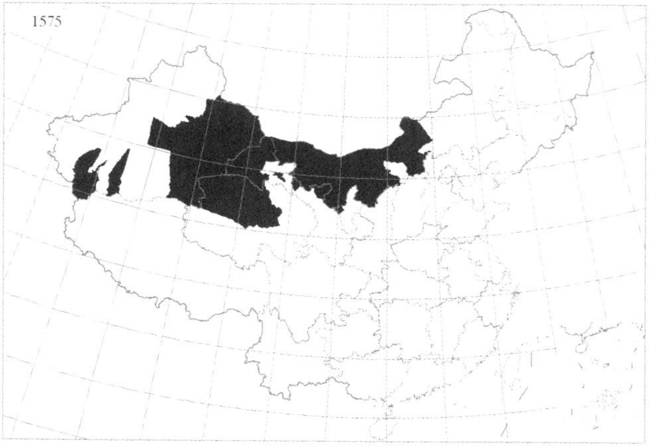

Polygonaceae 蓼科 *Calligonum* 沙拐枣属
Calligonum pumilum Losinsk. 小沙拐枣
Shrubs, deciduous

9.1 (-2.3~12.5)；-8.8 (-17.2~-5.1)；23.4 (10.7~26.8)；10.9 (3.2~13.2)；91.7 (14.3~114.0)；42.2 (23.1~102.0)；712 (364~822) ❙ 55 (21~131)；32 (15~71)；3 (0~11) ❙ -91.4 (-97.3~-76.9)；55 (21~131)；19.8 (1.6~67.2)

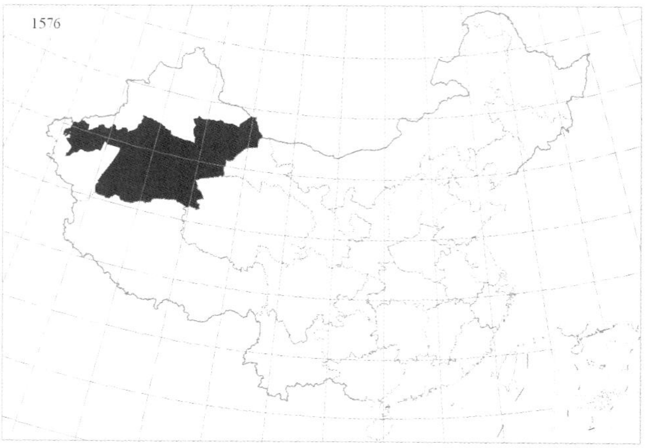

Polygonaceae 蓼科 *Calligonum* 沙拐枣属
Calligonum roborowskii Losinsk. 塔里木沙拐枣
Shrubs, deciduous

8.2 (-4.1~12.4)；-9.5 (-20.3~-5.2)；22.5 (9.8~26.8)；10.3 (2.5~13.2)；85.4 (9.9~114.1)；46.5 (24.3~117.9)；683 (330~822) ❙ 62 (21~261)；37 (15~170)；3 (0~11) ❙ -89.5 (-97.3~-33.2)；62 (21~261)；22.0 (1.6~120.2)

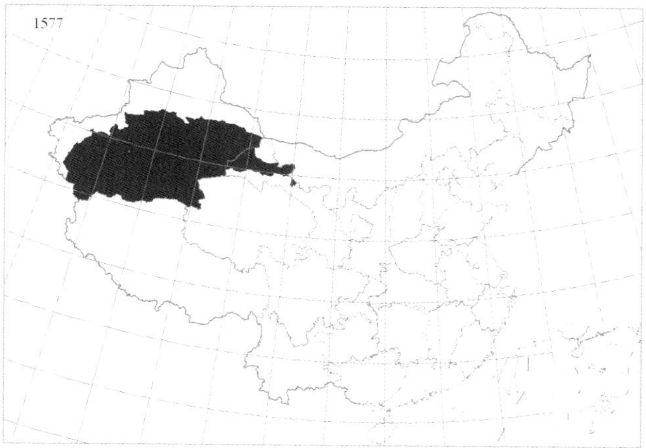

Polygonaceae 蓼科 *Calligonum* 沙拐枣属
Calligonum rubicundum Bunge 红果沙拐枣
Shrubs, deciduous

2.4 (-0.6~5.9)；-16.6 (-19.5~-13.8)；19.2 (15.9~23.3)；7.3 (5.4~9.7)；55.1 (37.5~80.9)；85.6 (66.6~104.2)；557 (475~668) ❙ 181 (152~251)；77 (56~106)；21 (15~33) ❙ -66.9 (-76.8~-52.0)；181 (152~251)；78.0 (16.6~139.1)

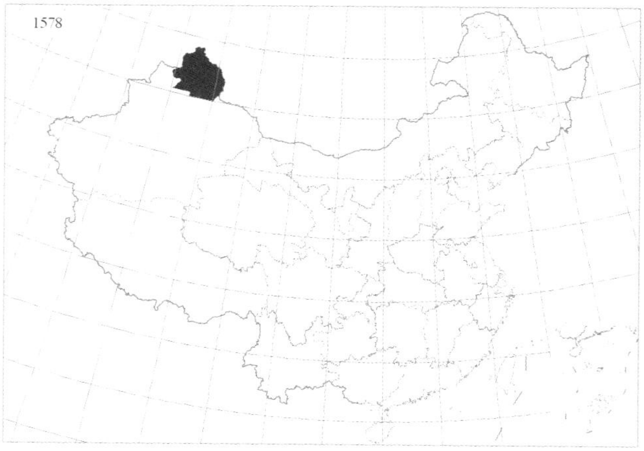

Polygonaceae 蓼科 *Calligonum* 沙拐枣属

Calligonum squarrosum N. Pavlov 粗糙沙拐枣

Shrubs, deciduous

4.9; -14.8; 21.9; 8.3; 71.1; 71.6; 620 ❙ 144; 63; 14 ❙ -76.8; 144; 41.1

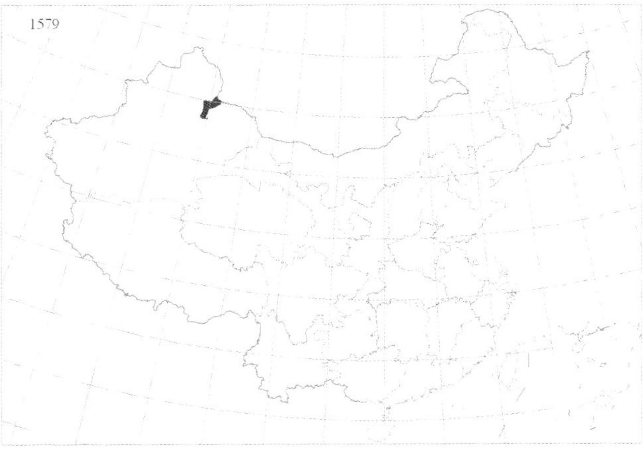

Polygonaceae 蓼科 *Calligonum* 沙拐枣属

Calligonum trifarium Z. M. Mao 三裂沙拐枣

Shrubs, deciduous

4.7; -16.5; 21.2; 8.7; 69.8; 72.0; 614 ❙ 209; 94; 19 ❙ -65.9; 209; 57.8

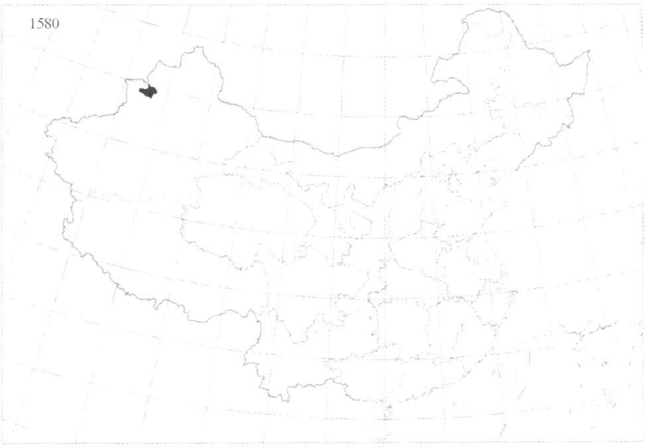

Polygonaceae 蓼科 *Calligonum* 沙拐枣属

Calligonum yengisaricum Z. M. Mao 英吉沙沙拐枣

Shrubs, deciduous

9.2 (-0.2~12.4); -7.9 (-16.4~-5.2); 23.1 (13.8~26.1); 10.9 (4.4~13.2); 90.4 (26.2~114.1); 38.8 (24.3~88.3); 703 (413~811) ❙ 45 (38~53); 20 (17~23); 5 (3~8) ❙ -93.2 (-95.3~-88.0); 45 (38~53); 25.3 (2.8~65.1)

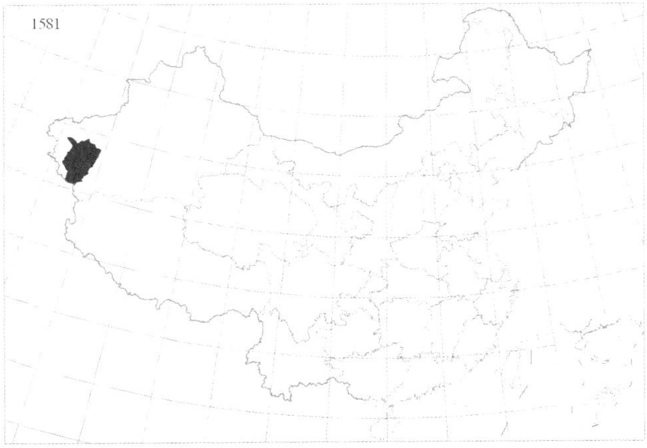

Polygonaceae 蓼科 *Calligonum* 沙拐枣属

Calligonum zaidamense Losinsk. 柴达木沙拐枣

Shrubs, deciduous

1.7 (-2.5~7.4); -12.7 (-16.1~-9.1); 14.8 (10.1~22.2); 5.4 (2.9~9.3); 35.8 (12.0~74.9); 74.7 (44.9~101.3); 461 (358~628) ❙ 147 (58~252); 93 (36~153); 6 (2~12) ❙ -64.7 (-90.8~-37.4); 147 (58~252); 36.0 (8.4~77.1)

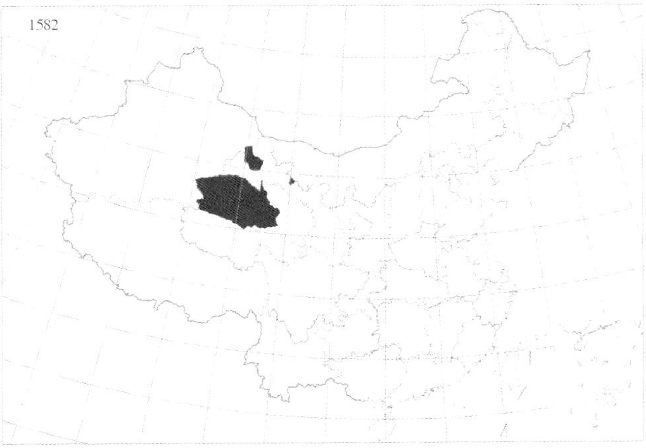

Polygonaceae 蓼科 *Fagopyrum* 荞麦属

Fagopyrum urophyllum (Bureau et Franch.) H. Gross 万年乔

Subshrubs, deciduous

13.3 (2.0~18.5); 5.6 (-7.4~11.6); 19.3 (10.1~24.3); 13.4 (4.0~18.3); 105.3 (16.9~160.0); 5.9 (0.0~51.9); 711 (416~906) ❙ 957 (677~1076); 553 (336~630); 33 (12~48) ❙ 38.1 (2.4~121.7); 697 (416~815); 268.0 (207.5~330.0)

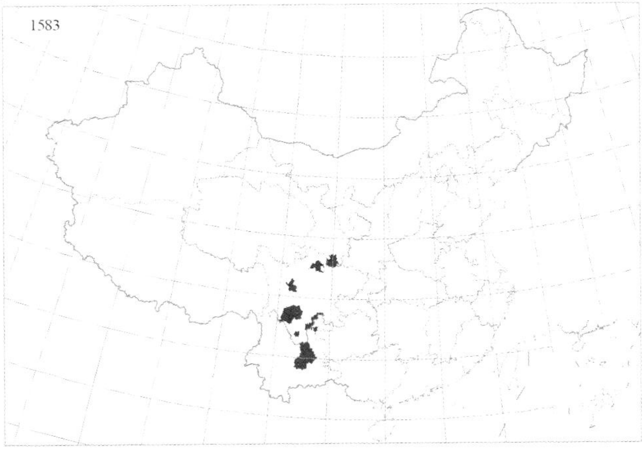

Polygonaceae 蓼科 *Fallopia* 何首乌属

Fallopia aubertii (L. Henry) Holub 木藤首乌

Subshrubs, deciduous

10.8 (0.4~18.9); 0.4 (-11.3~12.0); 20.0 (9.9~27.1); 11.3 (3.6~19.0); 86.9 (15.4~167.8); 17.1 (0.0~69.4); 668 (406~929) ❙ 811 (132~1358); 440 (81~627); 29 (2~88) ❙ 25.9 (-80.1~128.6); 617 (132~912); 234.0 (13.6~351.7)

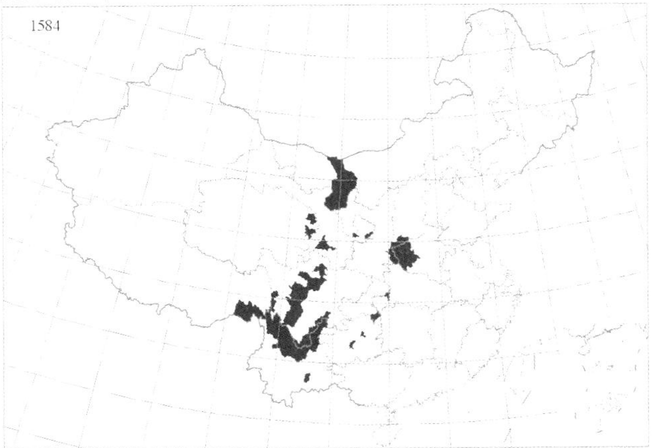

Polygonaceae 蓼科 *Parapteropyrum* 翅果蓼属
Parapteropyrum tibeticum A. J. Li 翅果蓼
Shrubs, deciduous

5.3; -3.9; 13.6; 6.8; 44.5; 40.1; 500 ❙ 590; 330; 19 ❙ 6.6; 467; 207.4

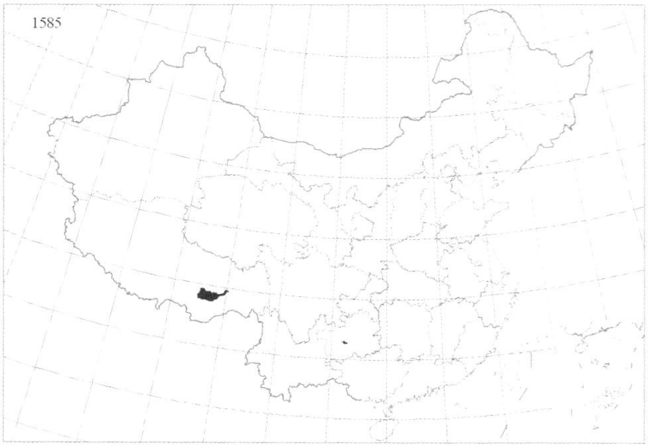

Polygonaceae 蓼科 *Polygonum* 蓼属
Polygonum emodi Meisn. 竹叶舒筋
Shrubs, deciduous

13.9 (-1.8~18.9); 7.0 (-10.2~12.0); 19.2 (6.1~23.9); 14.1 (1.8~19.0); 111.4 (1.9~167.8); 3.8 (0.0~82.8); 726 (329~929) ❙ 1001 (320~1736); 581 (217~1007); 38 (1~69) ❙ 38.9 (-27.3~145.8); 697 (320~834); 308.1 (72.2~447.4)

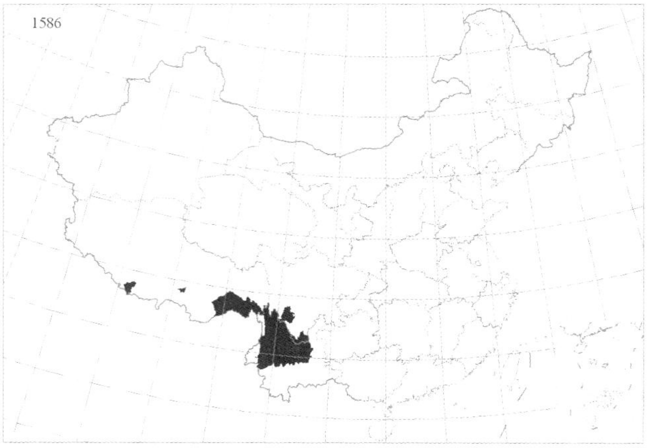

Polygonaceae 蓼科 *Polygonum* 蓼属
Polygonum emodi var. dependens Diels 宽竹叶舒筋
Shrubs, deciduous

14.0 (5.3~18.9); 7.1 (-2.4~12.0); 19.3 (12.2~23.9); 14.0 (5.7~19.0); 110.0 (30.5~167.8); 1.8 (0.0~26.2); 726 (479~929) ❙ 998 (741~1337); 578 (411~727); 38 (15~69) ❙ 40.0 (-14.2~100.7); 702 (479~815); 296.8 (218.0~407.1)

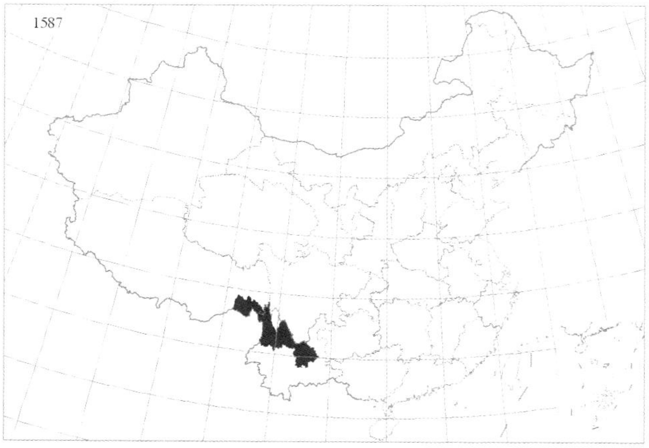

Polygonaceae 蓼科 *Polygonum* 蓼属
Polygonum intramongolicum Borodina 圆叶萹蓄
Shrubs, deciduous

4.6 (1.2~7.9); -13.6 (-17.8~-10.4); 20.8 (18.2~24.2); 8.1 (6.4~10.1); 62.3 (45.9~84.9); 66.8 (49.6~90.7); 576 (506~671) ❙ 282 (111~407); 187 (74~275); 7 (2~15) ❙ -49.6 (-82.3~-20.8); 282 (111~407); 91.6 (13.9~200.5)

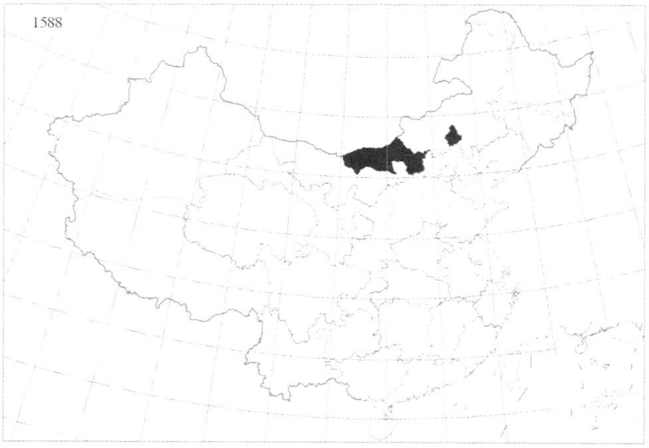

Polygonaceae 蓼科 *Polygonum* 蓼属
Polygonum molle D. Don 绢毛神血宁
Subshrubs, deciduous

15.5 (-1.8~21.0); 8.6 (-10.2~15.1); 20.6 (6.1~26.4); 15.7 (1.8~21.0); 129.9 (1.9~191.8); 3.2 (0.0~82.8); 787 (329~1053) ❙ 1157 (320~1751); 648 (193~1007); 47 (11~77) ❙ 47.1 (-4.0~145.8); 774 (320~1053); 353.6 (72.2~659.0)

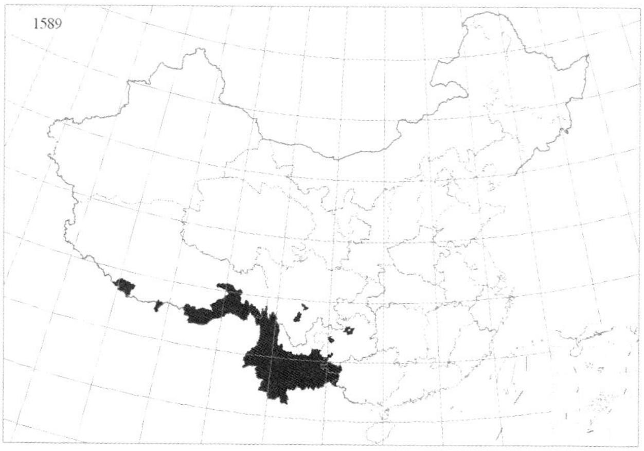

Polygonaceae 蓼科 *Polygonum* 蓼属
Polygonum molle var. frondosum (Meisn.) A. J. Li 光叶神血宁
Subshrubs, deciduous

12.5; 5.5; 18.0; 12.6; 96.8; 6.6; 684 ❙ 1177; 658; 51 ❙ 75.2; 684; 376.3

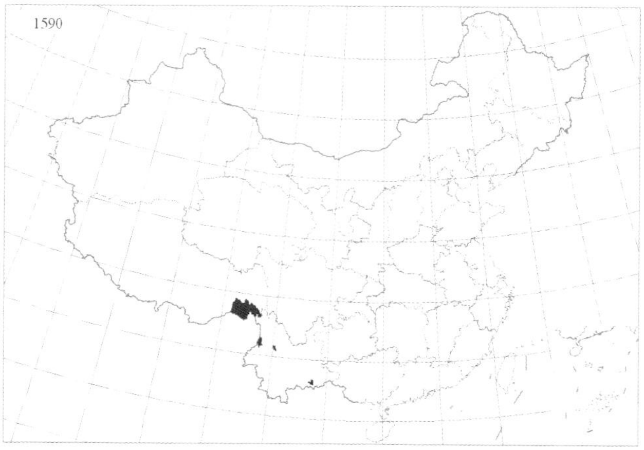

Polygonaceae 蓼科 *Polygonum* 蓼属
Polygonum molle var. rude (Meisn.) A. J. Li 倒毛神血宁
Subshrubs, deciduous

13.7 (−1.8~21.0); 6.8 (−10.2~14.2); 19.0 (6.1~25.6); 14.2 (1.8~21.0); 116.7 (1.9~191.8); 11.7 (0.0~82.8); 747 (329~1053) | 1146 (320~1547); 653 (217~911); 47 (11~68) | 50.6 (−3.0~113.0); 746 (320~1053); 388.6 (72.2~551.6)

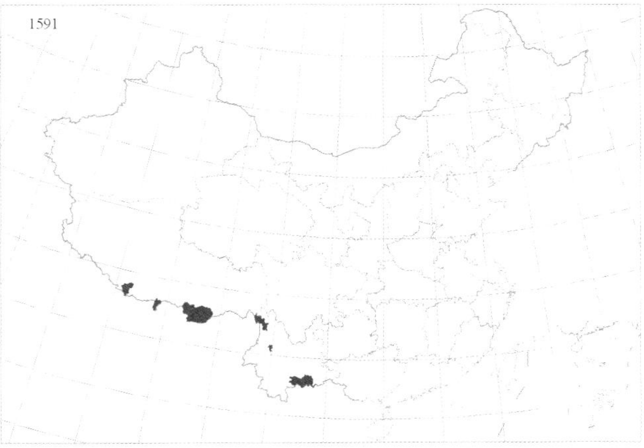

Polygonaceae 蓼科 *Polygonum* 蓼属
Polygonum paronychioides C. A. Mey. 线叶萹蓄
Subshrubs, deciduous

−2.1; −11.5; 7.4; 2.1; 5.0; 89.2; 328 | 564; 269; 101 | 72.2; 328; 39.6

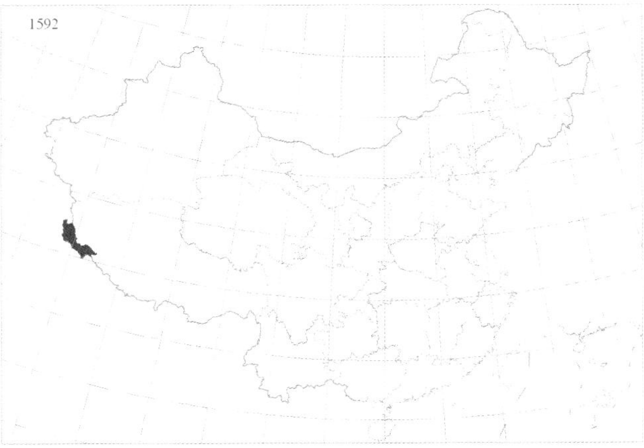

Polygonaceae 蓼科 *Polygonum* 蓼属
Polygonum polystachyum Wall. ex Meisn. 多穗神血宁
Subshrubs, deciduous

9.7 (0.3~20.5); 2.1 (−9.9~14.9); 16.0 (8.3~23.7); 10.4 (2.9~20.5); 75.4 (9.6~185.9); 18.8 (0.0~67.5); 613 (368~992) | 962 (389~1736); 541 (246~1007); 36 (7~59) | 54.4 (0.0~145.8); 604 (368~959); 331.7 (142.5~570.1)

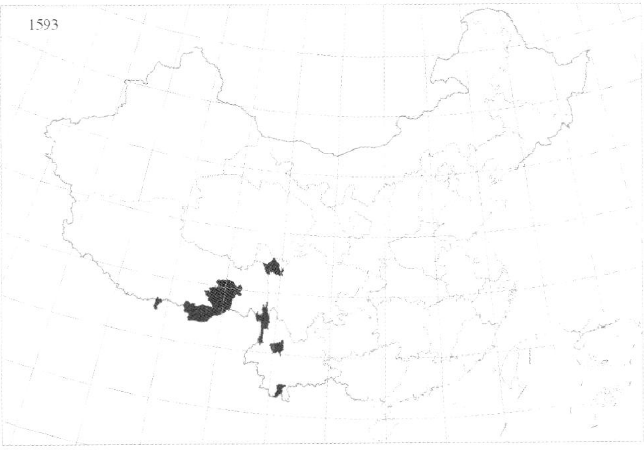

Polygonaceae 蓼科 *Polygonum* 蓼属
Polygonum popovii A. J. Li ex Borodina 库车萹蓄
Shrubs, deciduous

3.6; −10.5; 22.9; 10.6; 88.2; 45.2; 694 | 91; 57; 4 | −86.6; 91; 33.8

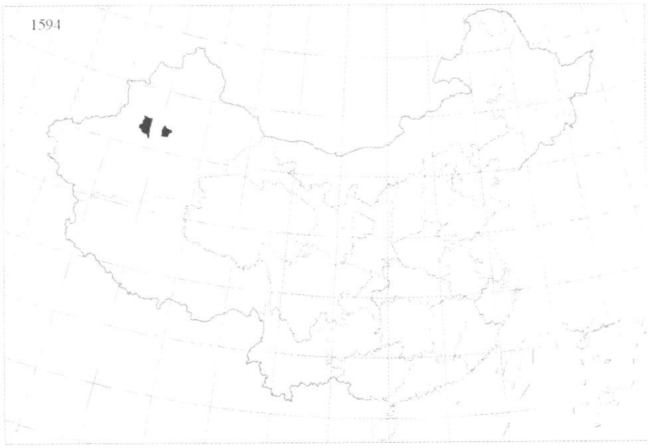

Polygonaceae 蓼科 *Polygonum* 蓼属
Polygonum tibeticum Hemsl. 西藏神血宁
Subshrubs, deciduous

0.7 (−2.2~3.2); −8.5 (−12.8~−5.8); 9.0 (6.1~11.3); 3.3 (1.8~4.5); 12.7 (1.9~21.4); 63.4 (42.6~91.3); 383 (329~430) | 541 (250~780); 349 (198~487); 11 (1~20) | 38.6 (−27.5~81.6); 370 (250~430); 188.0 (72.2~299.8)

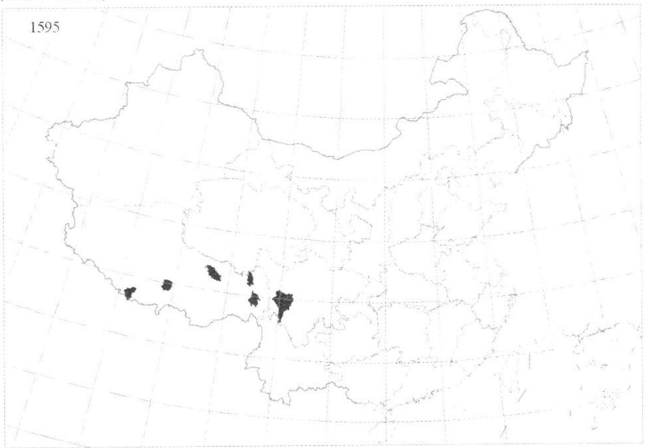

Polygonaceae 蓼科 *Polygonum* 蓼属
Polygonum tortuosum D. Don 叉枝神血宁
Subshrubs, deciduous

−0.6 (−5.7~3.7); −10.0 (−18.7~−5.5); 8.4 (5.4~12.1); 2.8 (1.6~5.1); 10.0 (0.4~27.8); 76.4 (41.8~133.0); 359 (294~441) | 425 (74~943); 253 (46~507); 30 (0~103) | 16.8 (−75.0~128.6); 318 (74~431); 128.0 (16.7~280.0)

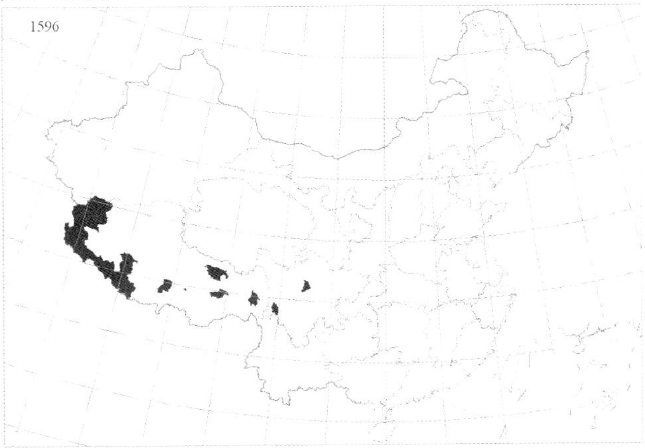

Polygonaceae 蓼科 *Polygonum* 蓼属

Polygonum vacciniifolium Wall. ex Meisn. 乌饭树叶蓼

Shrubs, deciduous

3.7; -4.9; 11.3; 5.8; 33.9; 49.4; 457 ❙ 710; 430; 18 ❙ 35.6; 450; 189.7

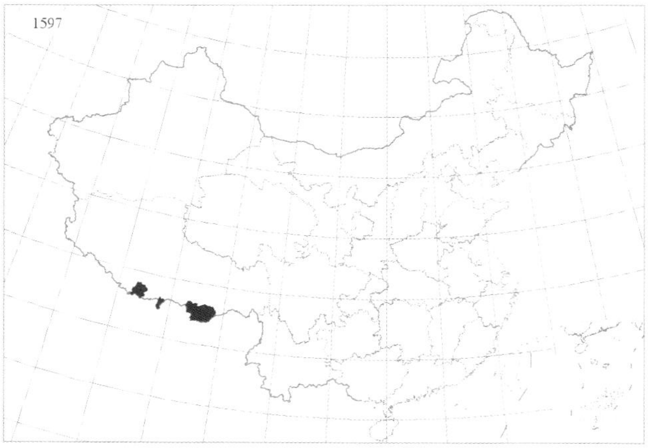

Polygonaceae 蓼科 *Rumex* 酸模属

Rumex hastatus D. Don 戟叶酸模

Shrubs, deciduous

13.2 (2.2~18.9); 5.7 (-7.1~12.4); 19.1 (9.5~24.0); 13.3 (3.8~19.0); 103.4 (15.5~167.8); 4.7 (0.0~51.0); 706 (411~929) ❙ 993 (535~1736); 571 (326~1007); 36 (9~69) ❙ 43.0 (-14.2~145.8); 688 (411~839); 287.4 (185.6~447.4)

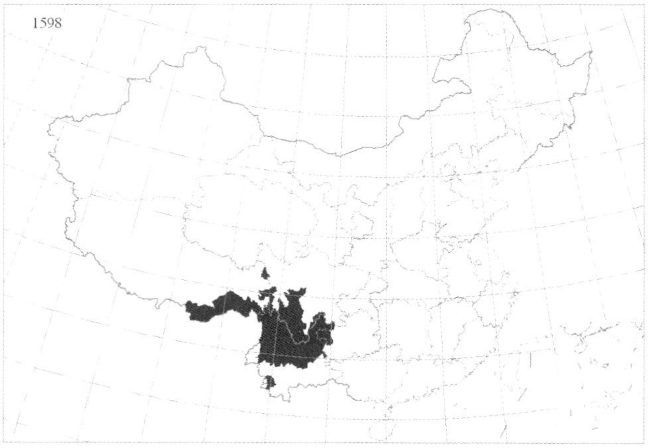

Chenopodiaceae 藜科 *Atriplex* 滨藜属

Atriplex maximowicziana Makino 海滨藜

Herbs or subshrubs

21.4 (19.5~23.8); 14.2 (10.9~18.1); 28.0 (27.7~28.6); 21.5 (19.6~23.8); 197.5 (174.7~225.8); 0.0 (0.0~0.0); 1128 (1007~1297) ❙ 1498 (1258~2114); 725 (425~1357); 148 (58~235) ❙ 32.2 (14.0~69.4); 1065 (1007~1144); 198.2 (119.4~231.6)

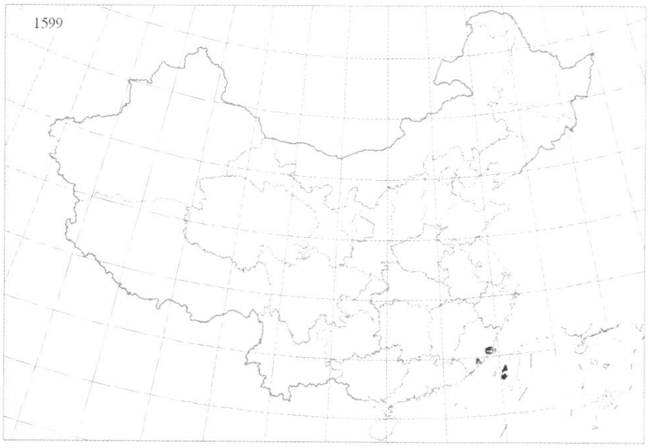

Chenopodiaceae 藜科 *Halostachys* 盐穗木属

Halostachys caspica C. A. Mey. ex Schrenk 盐穗木

Shrubs

6.9 (-4.1~12.5); -11.5 (-20.9~-5.1); 21.9 (9.8~27.5); 9.7 (2.5~13.2); 79.0 (9.9~114.1); 55.5 (23.1~117.9); 657 (330~822) ❙ 113 (21~344); 57 (15~188); 8 (0~34) ❙ -80.6 (-97.3~-23.3); 113 (21~344); 46.6 (1.6~297.3)

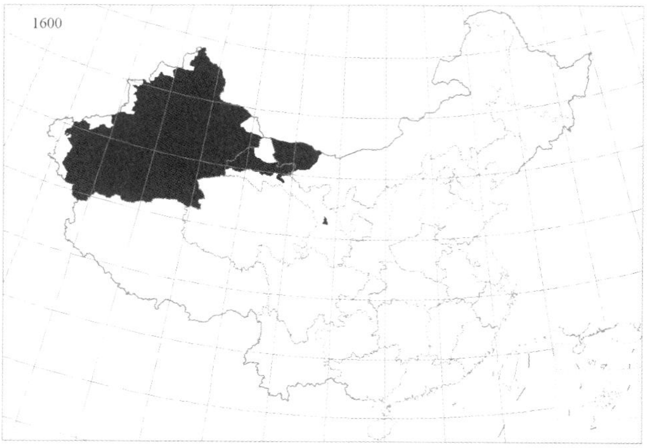

Chenopodiaceae 藜科 *Haloxylon* 琐琐属

Haloxylon ammodendron (C. A. Mey.) Bunge 琐琐

Small trees, deciduous

6.3 (-2.8~12.5); -11.6 (-20.3~-5.1); 21.3 (10.1~27.5); 9.1 (2.9~13.2); 73.2 (12.0~114.0); 57.6 (23.1~109.2); 629 (358~822) ❙ 137 (21~450); 71 (15~185); 9 (0~51) ❙ -75.6 (-97.3~3.9); 137 (21~404); 48.5 (1.6~190.1)

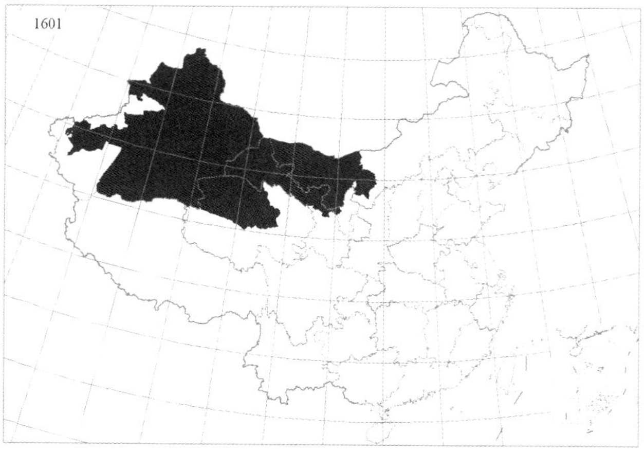

Chenopodiaceae 藜科 *Haloxylon* 琐琐属

Haloxylon persicum Bunge ex Boiss. et Buhse 白琐琐

Small trees, deciduous

4.5 (-0.6~9.3); -15.3 (-19.5~-10.8); 21.3 (15.9~27.5); 8.6 (5.4~12.1); 69.0 (37.5~107.6); 74.0 (50.3~104.2); 618 (475~813) ❙ 190 (75~315); 81 (39~127); 20 (5~51) ❙ -68.2 (-89.7~-44.6); 190 (75~312); 73.7 (14.6~173.1)

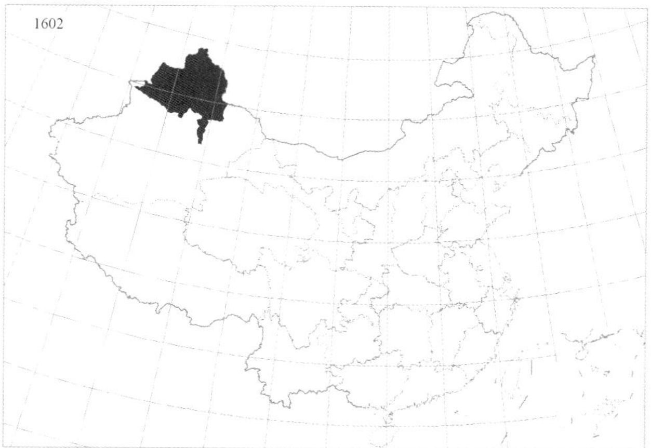

Chenopodiaceae 藜科 *Kalidium* 盐爪爪属
Kalidium caspicum (L.) Ung.-Sternb. 里海盐爪爪
Small shrubs

3.7 (-4.3~9.3); -15.1 (-20.9~-11.6); 19.3 (9.6~27.5); 7.7 (2.5~12.1); 60.1 (9.7~107.6); 74.4 (51.6~121.1); 576 (329~813) ❙ 206 (71~357); 96 (24~171); 18 (8~45) ❙ -61.8 (-85.3~-29.7); 206 (71~357); 87.6 (15.7~297.3)

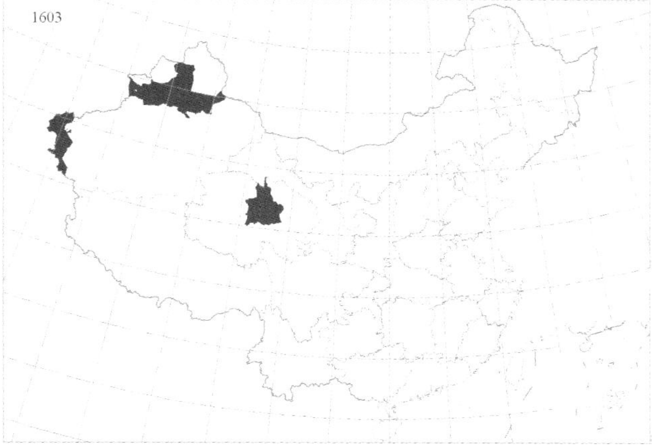

Chenopodiaceae 藜科 *Kalidium* 盐爪爪属
Kalidium cuspidatum (Ung.-Sternb.) Grubov 尖叶盐爪爪
Small shrubs

5.2 (-4.3~11.9); -12.6 (-24.4~-5.6); 20.8 (9.6~27.5); 8.4 (2.5~12.7); 66.2 (9.7~107.9); 62.7 (25.2~122.1); 592 (329~813) ❙ 258 (49~516); 156 (23~313); 9 (1~30) ❙ -55.1 (-92.8~14.7); 258 (49~471); 80.2 (5.1~235.8)

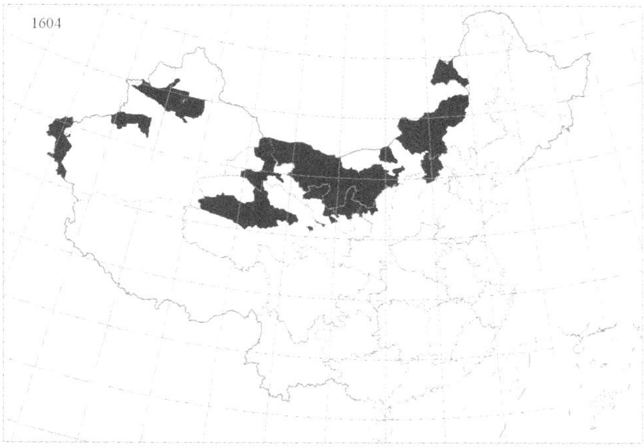

Chenopodiaceae 藜科 *Kalidium* 盐爪爪属
Kalidium foliatum (Pall.) Moq. 盐爪爪
Small shrubs

6.5 (-2.8~12.9); -11.6 (-24.4~-4.0); 21.9 (10.1~27.6); 9.3 (2.9~13.4); 75.1 (12.0~115.6); 56.6 (20.3~122.1); 634 (358~836) ❙ 232 (21~694); 142 (16~499); 9 (0~48) ❙ -61.5 (-97.3~5.4); 231 (21~659); 85.4 (1.6~297.3)

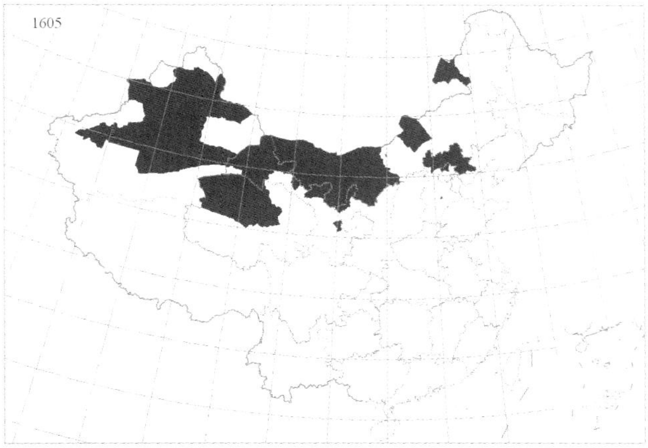

Chenopodiaceae 藜科 *Kalidium* 盐爪爪属
Kalidium gracile Fenzl 细枝盐爪爪
Small shrubs

6.9 (-4.1~12.5); -10.7 (-22.1~-5.1); 21.7 (9.8~27.5); 9.4 (2.5~13.2); 76.4 (9.9~114.1); 52.9 (23.1~117.9); 640 (330~822) ❙ 139 (21~443); 82 (15~287); 6 (0~21) ❙ -75.8 (-97.3~-22.1); 139 (21~443); 40.9 (1.6~143.7)

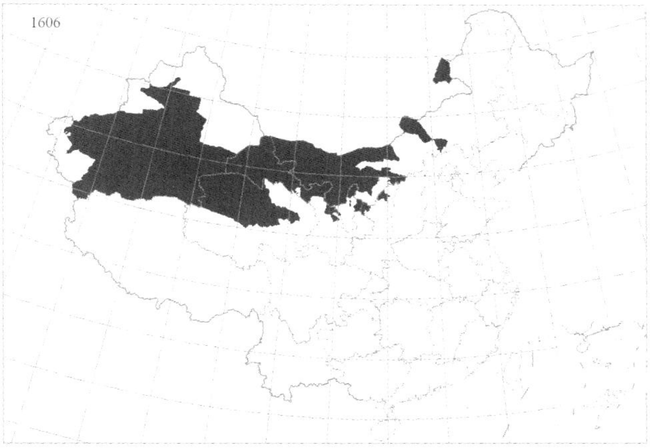

Chenopodiaceae 藜科 *Krascheninnikovia* 驼绒藜属
Krascheninnikovia arborescens (Losinsk.) Mosyakin 华北驼绒藜
Shrubs

4.9 (-4.8~12.1); -11.5 (-20.7~-2.5); 19.4 (7.4~25.2); 7.8 (1.9~12.4); 59.1 (4.3~103.1); 59.8 (17.8~121.6); 562 (332~747) ❙ 395 (101~829); 245 (61~453); 9 (4~23) ❙ -27.6 (-83.8~94.4); 380 (101~611); 130.3 (32.7~344.4)

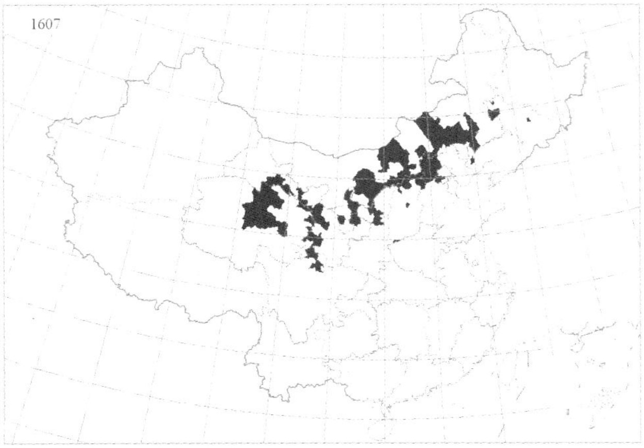

Chenopodiaceae 藜科 *Krascheninnikovia* 驼绒藜属
Krascheninnikovia ceratoides (L.) Gueldenst. 驼绒藜
Shrubs, evergreen

4.8 (-5.7~12.5); -12.0 (-22.1~-2.0); 19.2 (6.2~27.5); 8.0 (1.7~13.2); 61.4 (2.1~114.1); 62.8 (20.4~133.0); 578 (294~822) ❙ 227 (21~1005); 129 (15~531); 11 (0~103) ❙ -54.0 (-97.3~91.4); 211 (21~632); 87.4 (1.6~364.1)

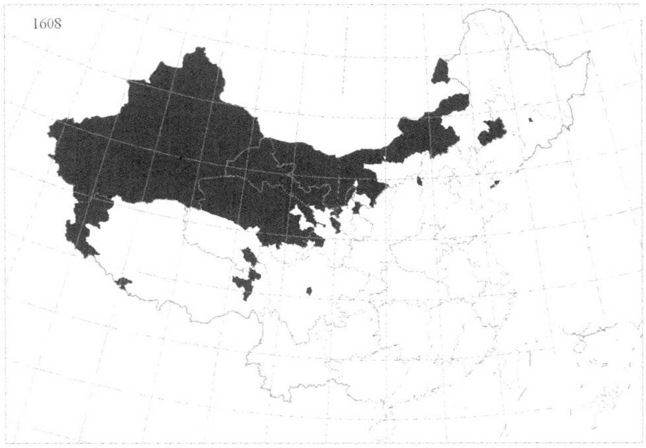

Chenopodiaceae 藜科 *Krascheninnikovia* 驼绒藜属
Krascheninnikovia compacta (Losina-Losinskaja) Grubov 垫状驼绒藜
Small shrubs

-1.3 (-5.9~11.6); -15.1 (-20.1~-5.7); 11.3 (5.4~25.1); 3.8 (1.3~12.5); 20.9 (0.4~106.3); 95.8 (26.7~133.0); 390 (294~766) ❙ 155 (34~372); 96 (16~235); 8 (1~38) ❙ -55.8 (-95.6~0.0); 153 (34~340); 44.2 (7.7~208.7)

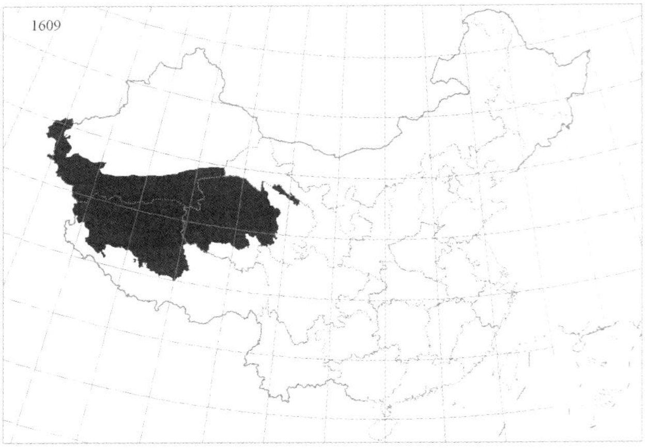

Chenopodiaceae 藜科 *Krascheninnikovia* 驼绒藜属
Krascheninnikovia compacta var. longipilosa (C. P. Tsien et C. G. Ma) Mosyakin 长毛驼绒藜
Small shrubs

-2.5 (-5.9~3.2); -15.0 (-18.7~-11.5); 9.4 (5.4~17.4); 2.8 (1.3~6.5); 12.0 (0.4~45.1); 101.0 (65.3~133.0); 354 (294~502) ❙ 242 (39~457); 153 (29~259); 9 (1~38) ❙ -30.9 (-92.3~12.5); 235 (39~406); 79.6 (10.8~208.7)

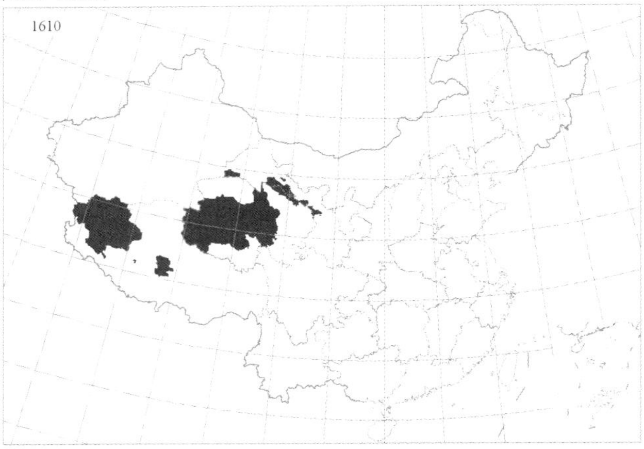

Chenopodiaceae 藜科 *Krascheninnikovia* 驼绒藜属
Krascheninnikovia ewersmannia (Stschegl. ex Losinsk.) Grubov 心叶驼绒藜
Shrubs

5.0 (-2.8~11.6); -14.1 (-20.9~-6.7); 20.8 (10.9~27.5); 8.6 (3.5~12.7); 68.8 (16.3~109.3); 67.7 (30.1~109.2); 615 (390~813) ❙ 186 (32~450); 86 (21~211); 17 (0~51) ❙ -66.9 (-95.9~3.9); 185 (32~406); 82.9 (1.6~297.3)

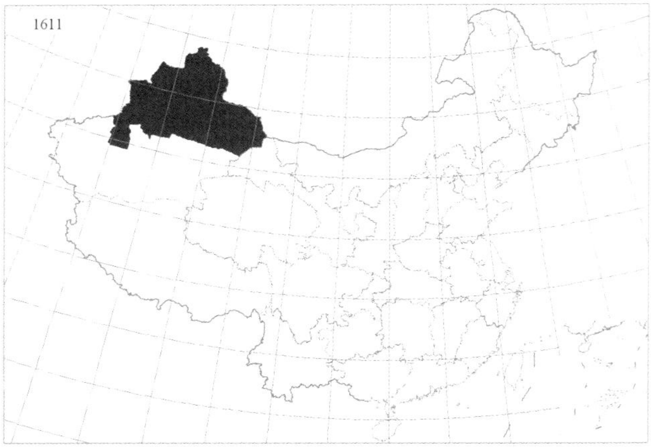

Chenopodiaceae 藜科 *Salsola* 猪毛菜属
Salsola arbuscula Pall. 木本猪毛菜
Small shrubs, deciduous

4.5 (-4.3~11.2); -13.7 (-20.9~-6.7); 20.0 (9.6~27.5); 8.1 (2.5~12.6); 63.1 (9.7~109.2); 68.9 (34.4~121.1); 585 (329~813) ❙ 188 (23~450); 90 (16~201); 15 (1~51) ❙ -66.1 (-97.2~3.9); 188 (23~404); 74.1 (3.0~297.3)

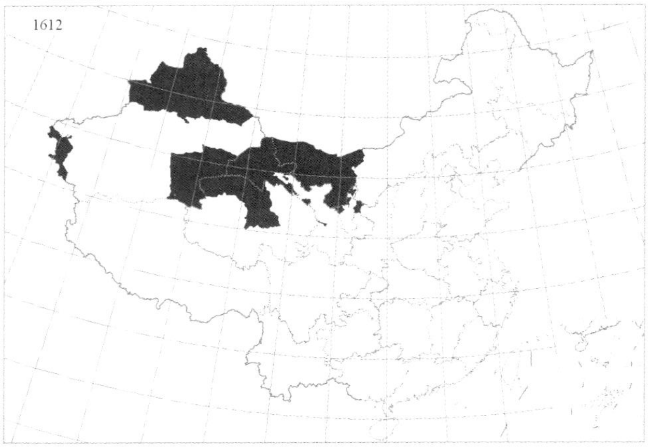

Chenopodiaceae 藜科 *Salsola* 猪毛菜属
Salsola laricifolia Turcz. ex Litv. 松叶猪毛菜
Small shrubs, deciduous

6.0 (-1.3~9.3); -11.9 (-20.9~-7.0); 21.5 (13.4~27.5); 8.9 (4.5~12.1); 70.4 (27.2~107.6); 57.9 (35.3~103.9); 611 (433~813) ❙ 228 (49~450); 126 (34~265); 11 (1~51) ❙ -61.5 (-92.8~3.9); 228 (49~416); 69.2 (5.1~213.5)

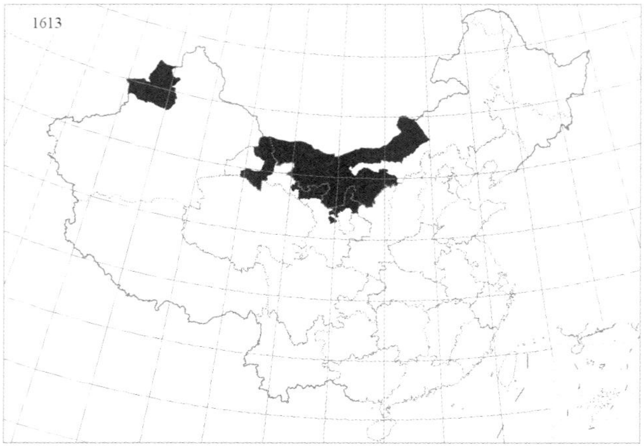

Chenopodiaceae 藜科 *Suaeda* 碱蓬属
Suaeda australis (R. Br.) Moq. 南方碱蓬
Small shrubs

20.8 (14.7~24.8); 12.7 (2.2~20.1); 28.0 (26.1~29.0); 20.8 (14.8~24.8); 189.9 (122.0~237.6); 0.1 (0.0~4.5); 1101 (824~1385) ❙ 1507 (1024~2435); 600 (356~1428); 151 (41~257) ❙ 38.4 (-5.0~99.5); 1068 (824~1329); 259.9 (123.9~572.9)

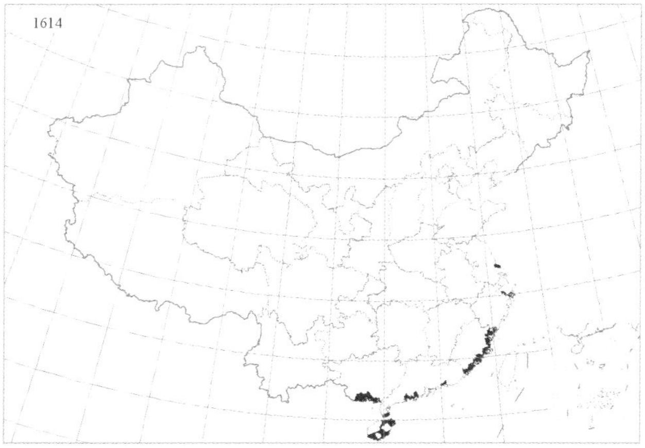

Chenopodiaceae 藜科 *Suaeda* 碱蓬属
Suaeda microphylla Pall. 小叶碱蓬
Subshrubs

5.2 (-1.3~9.3); -14.6 (-20.9~-9.3); 21.4 (13.4~27.5); 8.9 (4.5~12.1); 71.5 (27.2~107.6); 68.3 (40.7~103.9); 627 (433~813) ▌ 224 (107~450); 96 (51~185); 22 (8~48) ▌ -61.6 (-85.3~3.9); 222 (107~404); 90.7 (21.1~213.5)

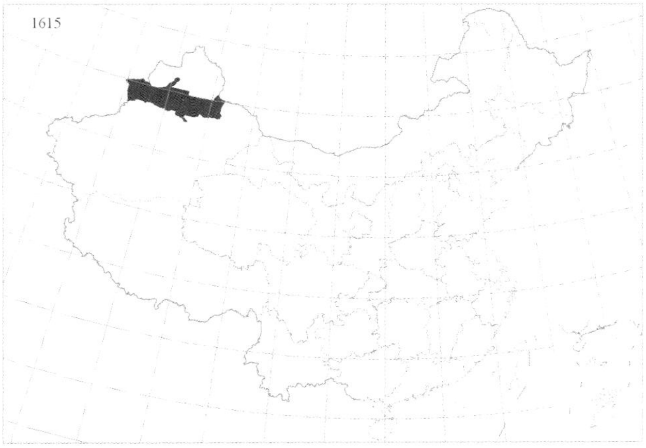

Chenopodiaceae 藜科 *Suaeda* 碱蓬属
Suaeda physophora Pall. 囊果碱蓬
Subshrubs

4.7 (-2.8~11.7); -14.5 (-20.9~-7.6); 20.6 (10.9~27.5); 8.5 (3.5~13.0); 66.9 (16.3~113.4); 69.9 (32.4~109.2); 607 (390~820) ▌ 196 (30~450); 90 (21~211); 18 (1~51) ▌ -64.9 (-96.3~3.9); 195 (30~406); 89.6 (2.1~297.3)

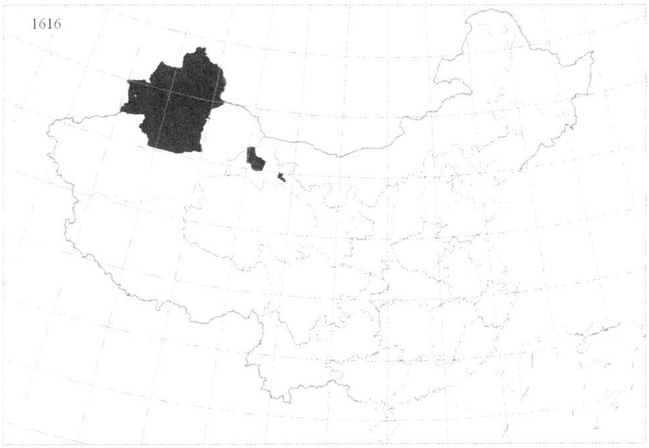

Amaranthaceae 苋科 *Deeringia* 浆果苋属
Deeringia amaranthoides (Lam.) Merr. 浆果苋
Shrubs

16.9 (5.4~24.8); 9.7 (-2.1~20.1); 22.3 (12.3~29.2); 16.9 (5.8~24.8); 143.8 (31.5~237.6); 0.5 (0.0~25.6); 862 (481~1385) ▌ 1193 (741~2992); 641 (312~1847); 51 (15~194) ▌ 40.4 (-34.3~166.5); 835 (481~1360); 338.1 (142.9~659.0)

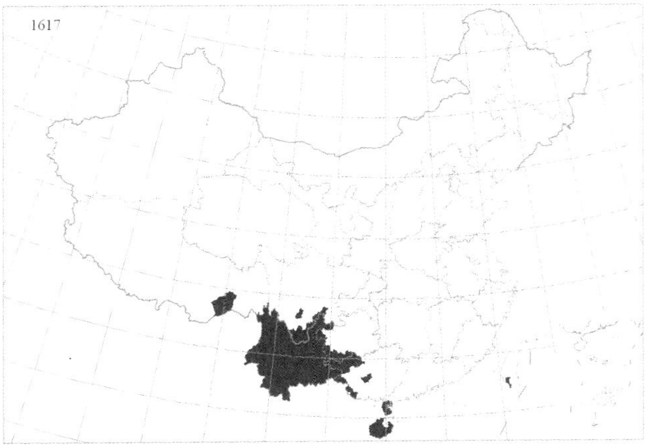

Amaranthaceae 苋科 *Stilbanthus* 巨苋藤属
Stilbanthus scandens Hook. f. 巨苋藤
Large climbers

19. ; 11.0; 25.0; 19.1; 169.6; 0.0; 945 ▌ 1438; 820; 57 ▌ 52.2; 945; 371.7

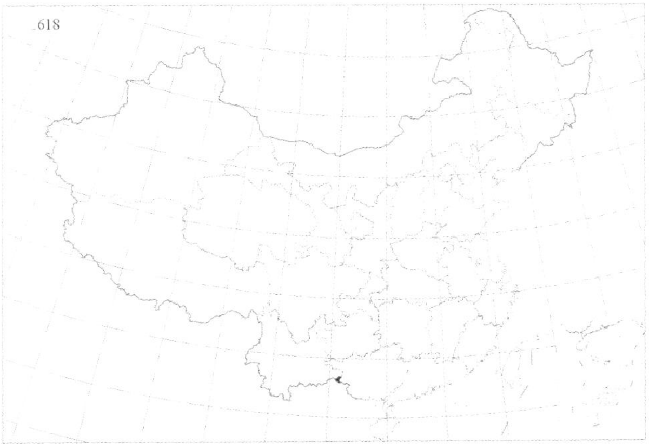

Nyctaginaceae 紫茉莉科 *Bougainvillea* 叶子花属
Bougainvillea glabra Choisy 光叶子花
Scandent shrubs

21.2 (13.6~24.8); 13.4 (6.1~20.1); 27.5 (19.6~29.8); 21.2 (13.7~24.8); 194.4 (103.9~237.6); 0.0 (0.0~0.0); 1115 (713~1385) ▌ 1717 (868~3495); 736 (312~1847); 144 (18~897) ▌ 57.7 (-34.3~246.5); 1092 (691~1360); 251.1 (95.3~620.0)

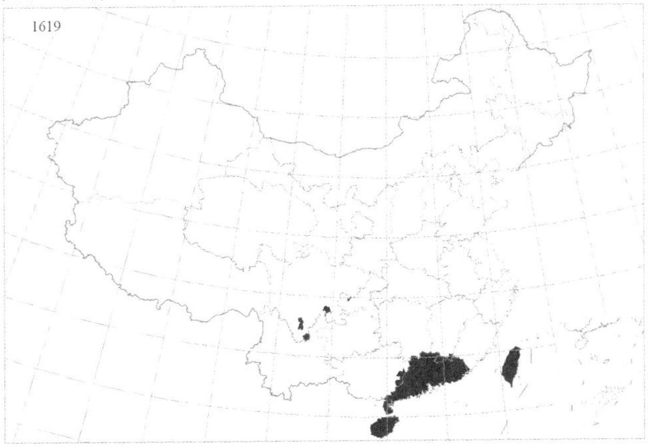

Nyctaginaceae 紫茉莉科 *Commicarpus* 黏腺果属
Commicarpus lantsangensis D. Q. Lu 澜沧黏腺果
Subshrubs

4.0; -4.2; 11.3; 5.0; 24.9; 36.5; 450 ▌ 723; 426; 20 ▌ 62.2; 450; 239.6

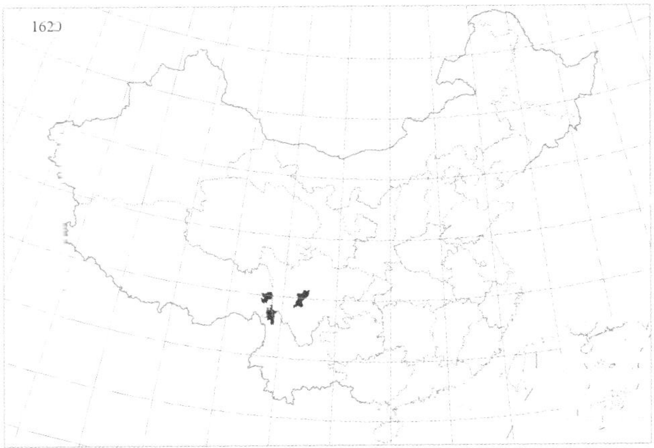

Nyctaginaceae 紫茉莉科 *Pisonia* 腺果藤属
Pisonia aculeata L. 腺果藤
Scandent shrubs

23.1 (16.4~24.8); 17.6 (11.5~20.1); 27.0 (20.3~28.8); 23.1 (16.5~24.8); 217.5 (137.8~237.6); 0.0 (0.0~0.0); 1237 (783~1385) ❙ 1697 (868~2992); 705 (312~1847); 90 (41~303) ❙ 45.4 (-34.3~219.3); 1141 (783~1360); 408.9 (154.2~659.0)

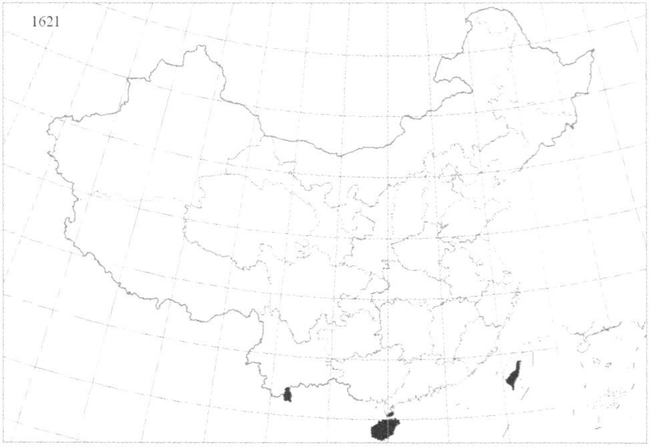

Nyctaginaceae 紫茉莉科 *Pisonia* 腺果藤属
Pisonia grandis R. Br. 抗风桐
Trees, evergreen

23.3 (14.6~24.8); 17.5 (3.8~20.1); 27.5 (21.7~28.8); 23.3 (14.7~24.8); 219.9 (117.7~237.6); 0.1 (0.0~1.5); 1263 (779~1385) ❙ 1562 (868~2713); 575 (312~1388); 85 (41~191) ❙ 28.6 (-34.3~214.4); 1156 (779~1360); 428.0 (248.2~620.0)

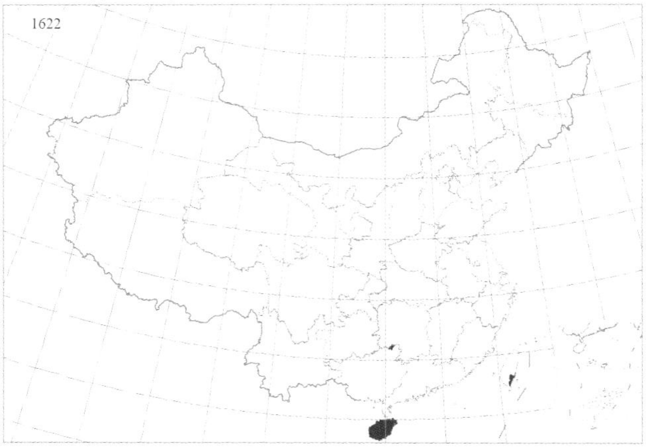

Nyctaginaceae 紫茉莉科 *Pisonia* 腺果藤属
Pisonia umbellifera (J. R. Forst. et G. Forst.) Seem. 胶果木
Trees

23.5; 18.6; 26.8; 23.5; 222.1; 0.0; 1254 ❙ 1964; 883; 70 ❙ 60.7; 1161; 496.1

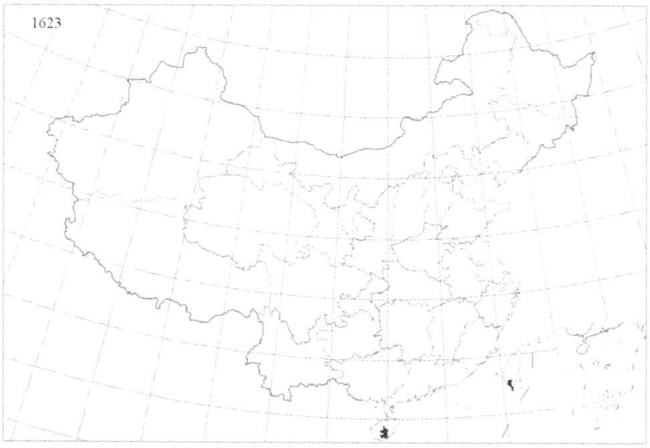

Caryophyllaceae 石竹科 *Gymnocarpos* 裸果木属
Gymnocarpos przewalskii Bunge ex Maxim. 裸果木
Subshrubs

5.9 (-4.3~12.5); -11.4 (-20.1~-5.1); 20.7 (9.6~26.8); 8.6 (2.5~13.2); 68.3 (9.7~114.0); 57.3 (23.1~121.1); 602 (329~817) ❙ 138 (23~317); 81 (16~196); 6 (1~26) ❙ -75.2 (-97.2~-23.3); 138 (23~317); 43.2 (3.0~138.5)

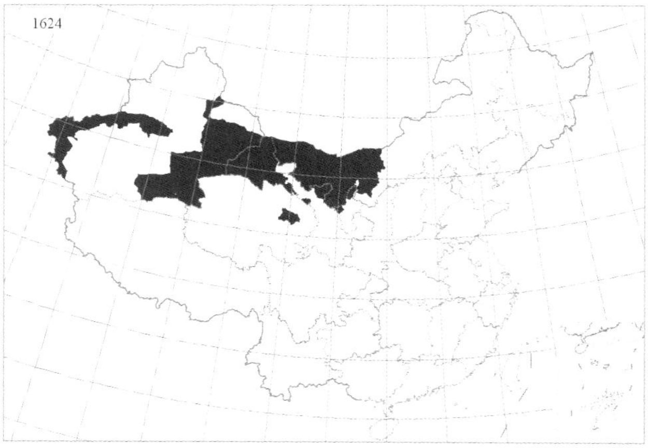

Eupteleaceae 领春木科 *Euptelea* 领春木属
Euptelea pleiosperma Hook. f. et Thomson 领春木
Shrubs or small trees, deciduous

11.8 (1.8~21.0); -0.5 (-13.7~14.2); 22.9 (9.9~29.1); 12.3 (3.8~21.0); 98.3 (15.5~191.8); 16.2 (0.0~71.7); 718 (404~1053) ❙ 860 (367~1922); 435 (235~1007); 43 (8~259) ❙ 19.9 (-40.9~145.8); 653 (367~1053); 216.6 (101.9~551.6)

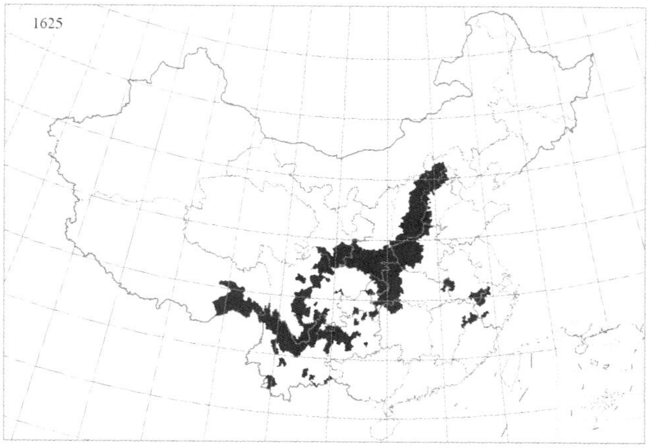

Trochodendraceae 昆栏树科 *Trochodendron* 昆栏树属
Trochodendron aralioides Siebold et Zucc. 昆栏树
Shrubs or small trees, evergreen

19.0 (16.2~21.7); 13.3 (11.0~15.3); 23.9 (20.1~27.4); 19.1 (16.2~21.7); 168.8 (134.2~200.2); 0.0 (0.0~0.0); 937 (773~1118) ❙ 2670 (1836~3495); 1056 (821~1412); 361 (110~897) ❙ 189.9 (64.2~246.5); 932 (773~1077); 264.1 (202.8~363.7)

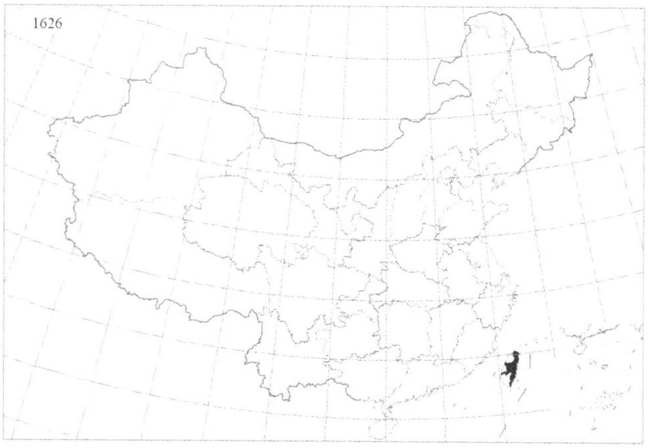

Tetracentraceae 水青树科 *Tetracentron* 水青树属
Tetracentron sinense Oliv. 水青树
Trees, evergreen

13.1 (-0.7~21.0); 3.3 (-9.6~14.9); 22.0 (7.7~28.1); 13.3 (2.5~21.0); 105.6 (6.8~191.8); 7.3 (0.0~74.2); 733 (338~1053) ❙ 1093 (370~1751); 527 (262~1007); 56 (9~198) ❙ 49.6 (-8.1~145.8); 730 (338~1053); 272.9 (61.6~570.1)

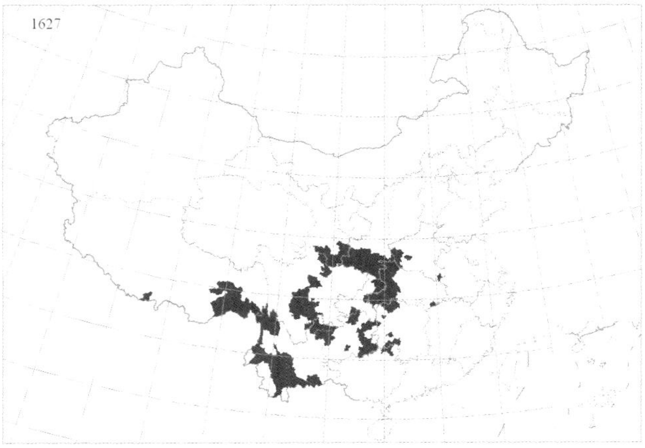

Cercidiphyllaceae 连香树科 *Cercidiphyllum* 连香树属
Cercidiphyllum japonicum Siebold et Zucc. 连香树
Trees, deciduous

11.5 (2.0~18.2); -1.9 (-13.7~6.6); 23.7 (9.9~29.3); 12.1 (3.8~18.2); 97.5 (15.4~158.8); 19.2 (0.0~71.7); 723 (413~984) ❙ 860 (367~1922); 412 (235~817); 53 (8~259) ❙ 18.2 (-40.9~128.6); 646 (367~961); 205.9 (101.9~337.3)

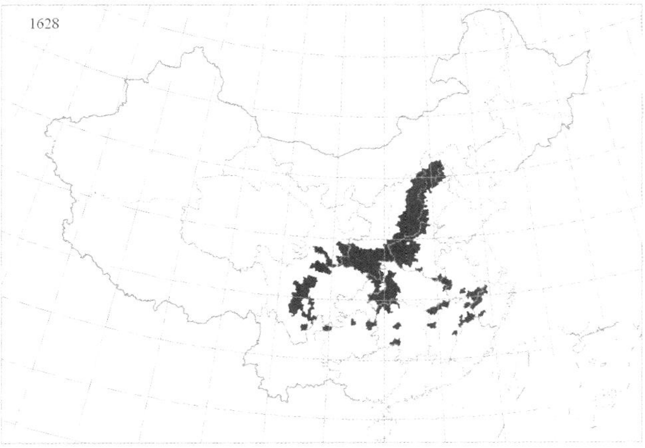

Paeoniaceae 芍药科 *Paeonia* 芍药属
Paeonia decomposita Hand.-Mazz. 四川牡丹
Shrubs, deciduous

2.8; -6.5; 10.9; 4.4; 20.3; 45.9; 428 ❙ 834; 464; 18 ❙ 95.2; 428; 321.2

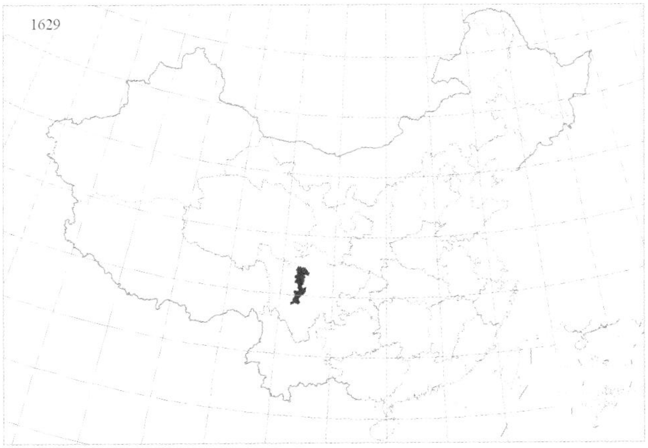

Paeoniaceae 芍药科 *Paeonia* 芍药属
Paeonia delavayi Franch. 滇牡丹
Shrubs, deciduous

12.2 (1.8~18.9); 4.5 (-7.4~12.5); 18.4 (9.4~26.6); 12.4 (3.7~19.0); 95.4 (14.2~167.8); 9.0 (0.0~53.3); 681 (402~929) ❙ 921 (307~1197); 534 (231~651); 31 (1~71) ❙ 37.9 (-28.5~125.9); 658 (307~912); 270.0 (142.9~382.6)

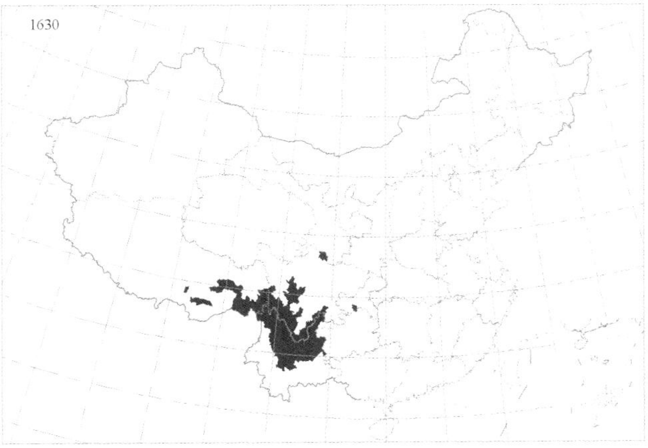

Paeoniaceae 芍药科 *Paeonia* 芍药属
Paeonia jishanensis T. Hong et W. Z. Zhao 矮牡丹
Shrubs, deciduous

10.9 (2.6~17.0); -2.2 (-10.6~6.3); 22.9 (11.2~28.2); 11.5 (4.3~17.1); 91.4 (20.0~145.1); 20.0 (0.0~50.1); 695 (427~908) ❙ 713 (409~1214); 357 (235~577); 22 (8~71) ❙ 4.0 (-38.3~121.7); 621 (409~908); 191.5 (44.7~334.4)

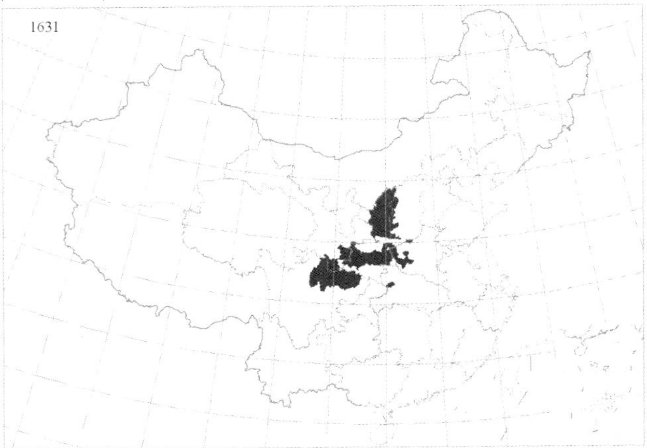

Paeoniaceae 芍药科 *Paeonia* 芍药属
Paeonia ludlowii (Stern et Taylor) D. Y. Hong 大花黄牡丹
Shrubs, deciduous

3.7; -5.1; 11.5; 4.8; 24.7; 39.4; 437 ❙ 513; 314; 8 ❙ 15.2; 437; 226.3

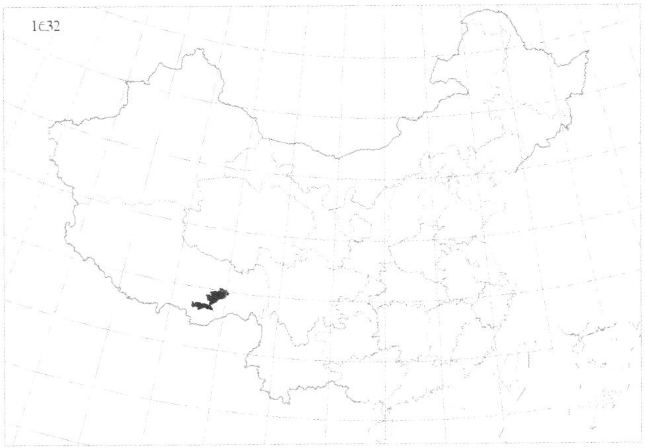

Paeoniaceae 芍药科 *Paeonia* 芍药属
Paeonia ostii T. Hong et J. X. Zhang 凤丹
Shrubs, deciduous

11.8; -1.8; 24.3; 12.0; 98.4; 16.4; 723 ❙ 693; 344; 32 ❙ -4.0; 664; 219.4

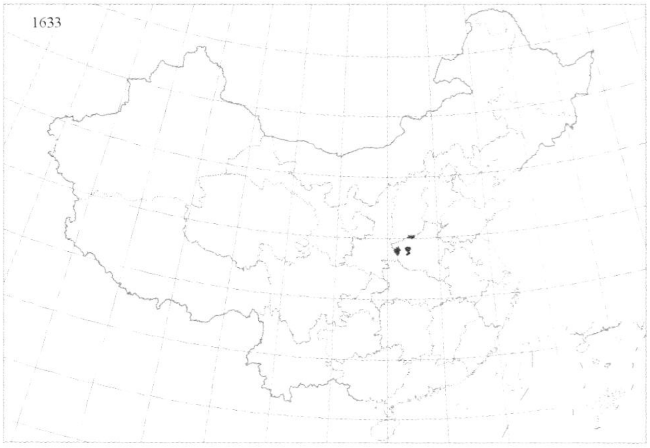

Paeoniaceae 芍药科 *Paeonia* 芍药属
Paeonia qiui Y. L. Pei et D. Y. Hong 卵叶牡丹
Shrubs, deciduous

11.4; -0.6; 22.8; 11.5; 90.4; 13.6; 684 ❙ 1016; 452; 60 ❙ 51.5; 684; 261.9

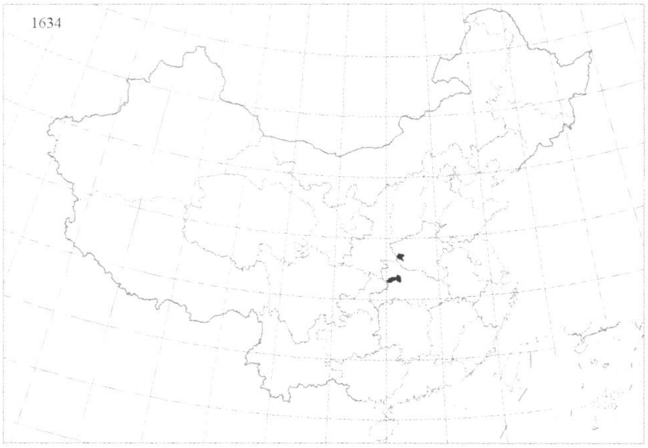

Paeoniaceae 芍药科 *Paeonia* 芍药属
Paeonia rockii (S. G. Haw et Lauener) T. Hong et J. J. Li 紫斑牡丹
Shrubs, deciduous

9.6 (1.9~13.6); -2.3 (-9.2~1.0); 20.7 (11.7~25.8); 10.2 (4.3~13.6); 77.4 (21.1~113.4); 21.2 (8.6~57.8); 638 (427~786) ❙ 740 (375~1214); 356 (203~516); 28 (8~71) ❙ 16.4 (-32.2~101.4); 615 (375~786); 242.4 (95.7~294.3)

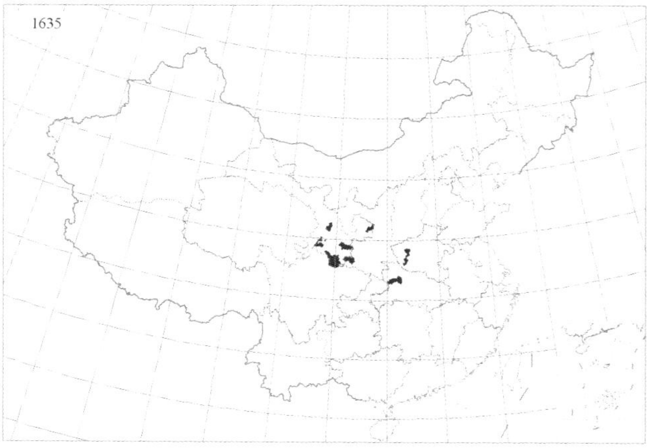

Paeoniaceae 芍药科 *Paeonia* 芍药属
Paeonia suffruticosa Andr. 牡丹
Shrubs, deciduous

10.8 (5.9~17.1); -3.1 (-10.8~4.0); 23.5 (16.6~29.4); 11.5 (7.3~17.1); 93.0 (51.7~146.1); 22.8 (1.0~53.6); 707 (529~952) ❙ 747 (341~1478); 372 (190~601); 45 (8~171) ❙ 3.8 (-40.9~107.2); 616 (341~921); 182.7 (61.2~319.8)

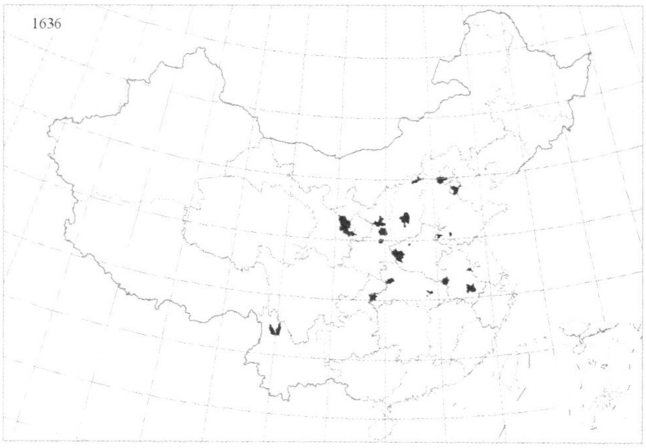

R

Ranunculaceae 毛茛科 *Clematis* 铁线莲属
Clematis acerifolia Maxim. 槭叶铁线莲
Shrubs, deciduous

7.8 (2.6~11.0); -9.4 (-15.9~-5.7); 23.0 (19.1~25.5); 9.9 (7.2~12.0); 81.6 (52.0~101.7); 47.2 (29.5~80.4); 657 (536~757) ❙ 554 (409~694); 401 (258~499); 11 (8~12) ❙ -15.7 (-32.6~5.4); 552 (409~659); 232.9 (174.5~287.6)

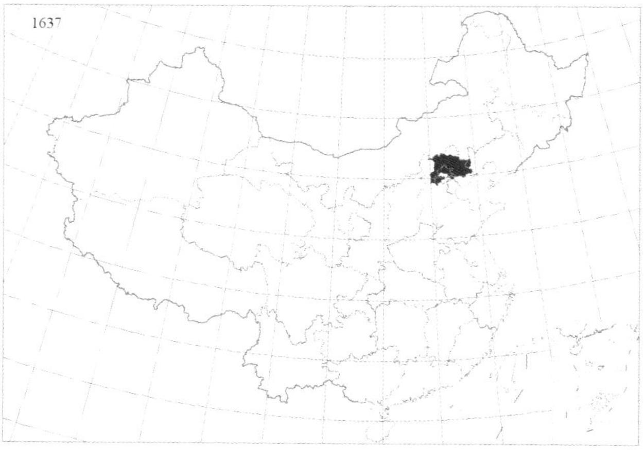

Ranunculaceae 毛茛科 *Clematis* 铁线莲属
Clematis acuminata var. longicaudata W. T. Wang 长尾尖铁线莲
Vines, deciduous

16.4 (5.3~19.3); 9.9 (-2.4~13.3); 20.9 (12.2~23.6); 16.5 (5.7~19.4); 138.7 (30.5~173.2); 1.0 (0.0~26.2); 810 (479~939) ❙ 1236 (961~1736); 695 (539~1007); 51 (31~70) ❙ 54.8 (12.5~145.8); 796 (479~923); 391.0 (234.2~516.7)

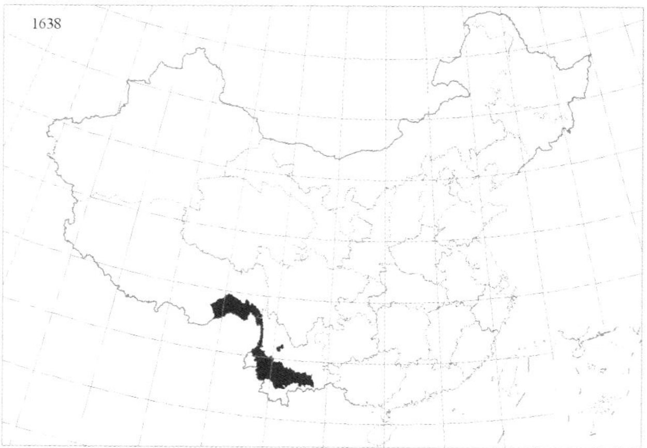

Ranunculaceae 毛茛科 *Clematis* 铁线莲属
Clematis akebioides (Maxim.) H. J. Veitch 甘川铁线莲
Vines, deciduous

3.9 (-4.1~11.0); -6.5 (-17.3~3.9); 13.2 (7.8~24.3); 5.7 (2.1~11.1); 33.3 (5.2~85.9); 45.4 (1.3~114.2); 474 (340~679) ❙ 663 (49~1023); 376 (34~619); 15 (1~31) ❙ 45.1 (-92.8~128.6); 411 (49~638); 233.7 (5.1~357.7)

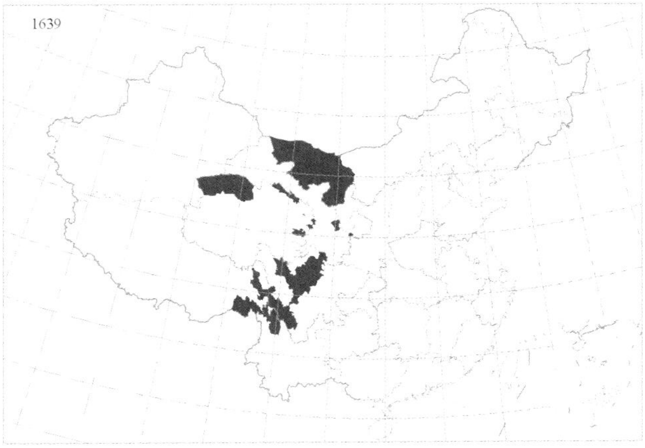

Ranunculaceae 毛茛科 *Clematis* 铁线莲属
Clematis akoensis Hayata 屏东铁线莲
Vines, deciduous

19.5; 14.5; 23.5; 19.6; 174.8; 0.0; 950 ❙ 2687; 1378; 201 ❙ 184.5; 950; 315.0

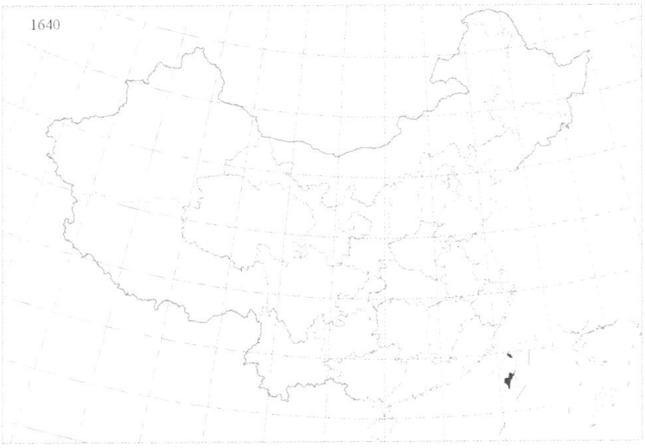

Ranunculaceae 毛茛科 *Clematis* 铁线莲属
Clematis alternata Kitam. et Tamura 互叶铁线莲
Vines, deciduous

-1.1; -9.1; 6.2; 2.0; 2.5; 75.6; 334 ❙ 321; 193; 33 ❙ -4.0; 321; 79.3

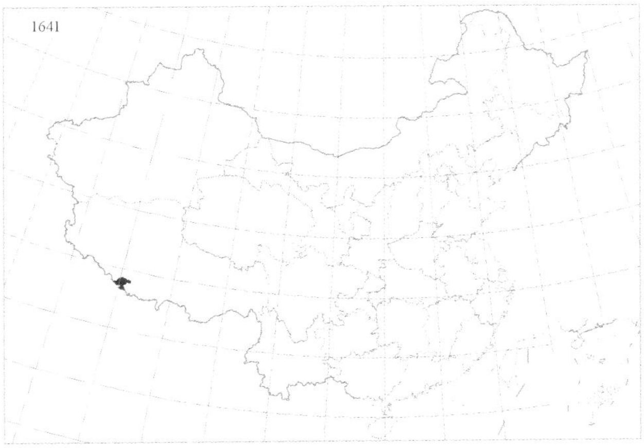

Ranunculaceae 毛茛科 *Clematis* 铁线莲属
Clematis apiifolia DC. 女萎
Vines, deciduous

15.9 (3.2~21.1); 4.5 (-6.4~12.0); 26.8 (11.4~30.1); 16.0 (4.7~21.2); 133.3 (22.7~194.2); 1.7 (0.0~43.8); 864 (436~1121) ❙ 1373 (711~1997); 538 (357~812); 146 (15~259) ❙ 59.2 (-1.4~147.0); 860 (436~1121); 211.6 (110.1~357.7)

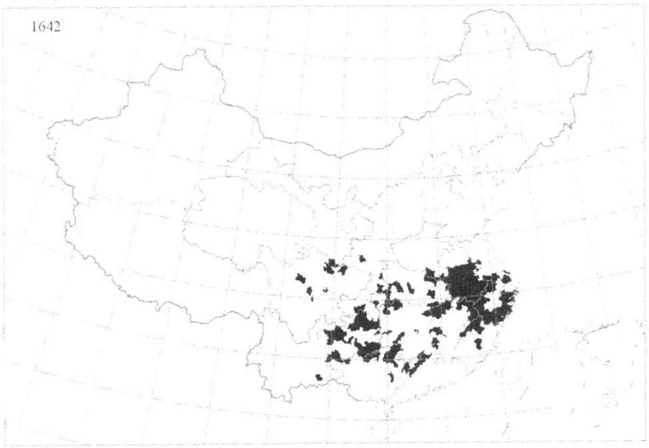

Ranunculaceae 毛茛科 *Clematis* 铁线莲属
Clematis armandii Franch. 小木通
Vines, deciduous

16.1 (2.6~22.5); 6.7 (-7.0~15.1); 24.3 (11.2~29.8); 16.2 (4.3~22.6); 135.7 (20.0~210.9); 1.8 (0.0~48.2); 849 (427~1206) ❙ 1283 (527~2035); 588 (317~1033); 91 (9~259) ❙ 51.3 (-14.2~130.7); 843 (427~1206); 256.5 (118.7~659.0)

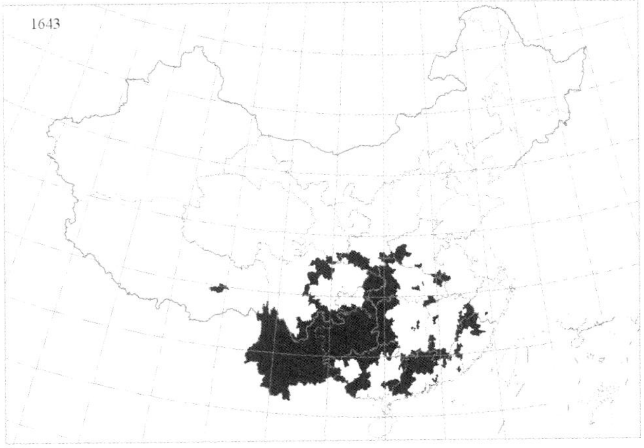

Ranunculaceae 毛茛科 *Clematis* 铁线莲属
Clematis armandii var. hefengensis (G. F. Tao) W. T. Wang 鹤峰铁线莲
Vines, deciduous

14.0; 3.1; 24.5; 14.0; 111.9; 3.7; 765 ❙ 1336; 541; 90 ❙ 76.8; 765; 218.8

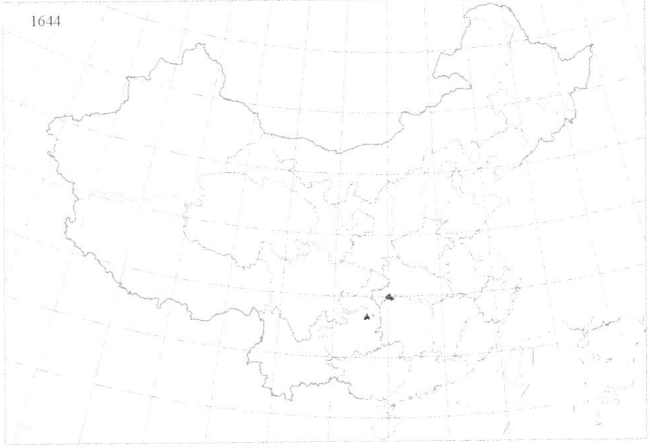

Ranunculaceae 毛茛科 *Clematis* 铁线莲属
Clematis baominiana W. T. Wang 多毛铁线莲
Vines, deciduous

16.9; 4.3; 29.0; 17.0; 144.4; 0.7; 930 ❙ 1404; 467; 176 ❙ 51.0; 900; 178.7

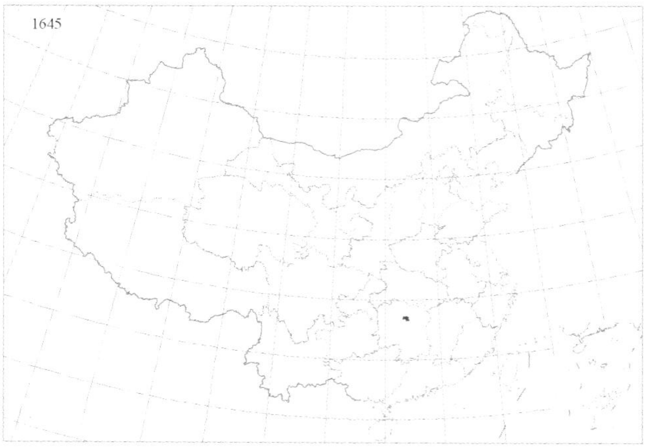

Ranunculaceae 毛茛科 *Clematis* 铁线莲属
Clematis barbellata Kitam. et Tamura 吉隆铁线莲
Vines, deciduous

-1.1; -9.1; 6.2; 2.0; 2.5; 75.6; 334 ❙ 321; 193; 33 ❙ -4.0; 321; 79.3

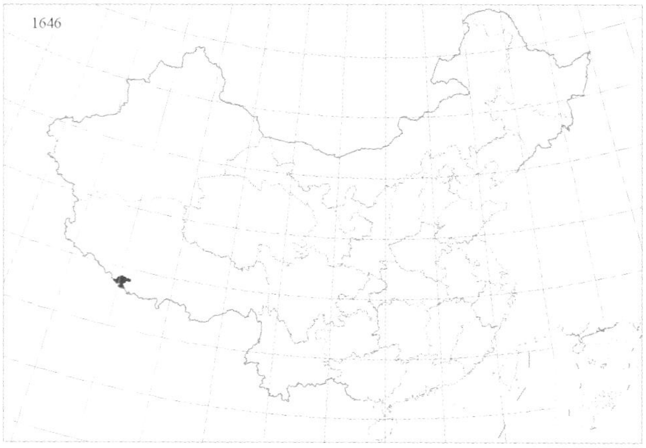

Ranunculaceae 毛茛科 *Clematis* 铁线莲属
Clematis brevicaudata DC. 短尾铁线莲
Vines, deciduous

9.8 (-3.1~17.7); -5.7 (-26.4~7.0); 23.5 (9.2~29.4); 11.2 (2.9~17.7); 91.5 (10.1~152.4); 33.3 (0.0~134.6); 708 (359~961) ❙ 616 (189~1742); 359 (114~666); 27 (4~221) ❙ -11.5 (-70.2~128.9); 553 (189~961); 173.7 (33.5~357.7)

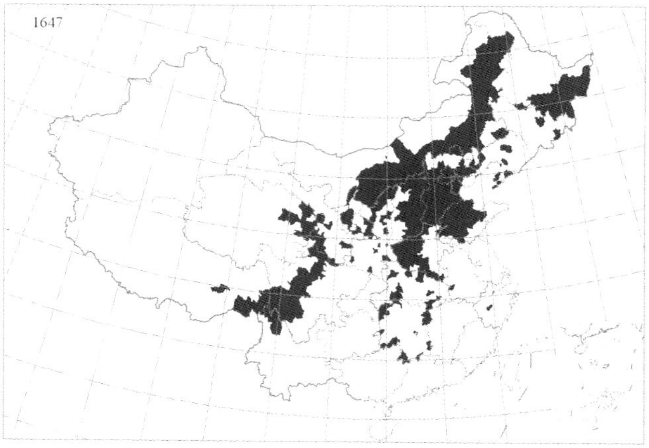

Ranunculaceae 毛茛科 *Clematis* 铁线莲属
Clematis brevipes Rehder 短梗铁线莲
Vines, deciduous

4.7; -6.1; 14.7; 6.0; 36.3; 39.6; 482 ❙ 629; 309; 14 ❙ 30.6; 482; 267.7

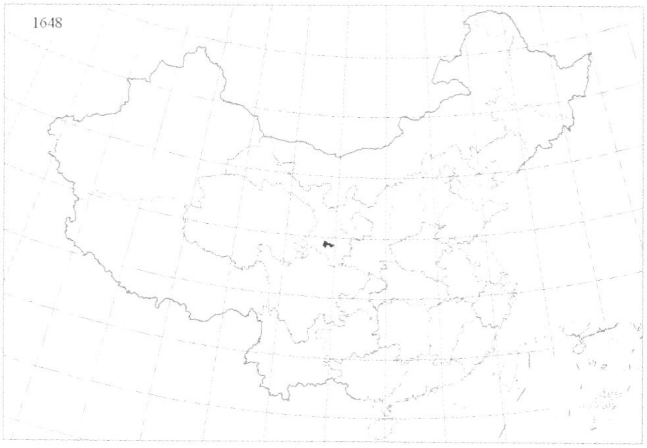

Ranunculaceae 毛茛科 *Clematis* 铁线莲属
Clematis buchananiana DC. 毛木通
Vines, deciduous

15.4 (-1.8~21.0); 8.1 (-10.2~14.2); 21.0 (6.1~26.4); 15.5 (1.8~21.0); 127.7 (1.9~191.8); 2.1 (0.0~82.8); 785 (329~1053) ❙ 1112 (320~1751); 625 (193~1007); 45 (10~77) ❙ 42.2 (-14.2~145.8); 771 (320~1053); 314.3 (72.2~551.6)

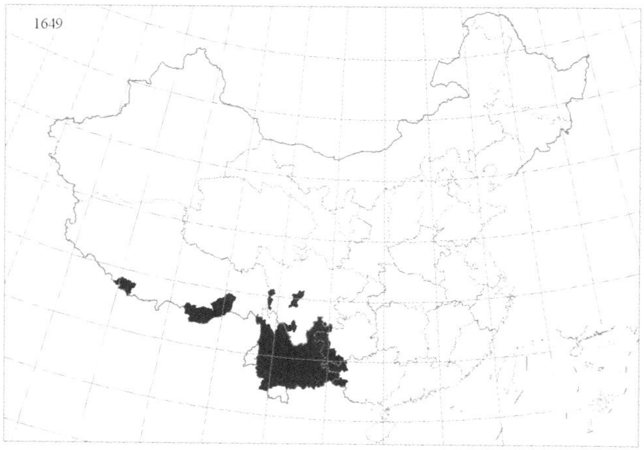

Ranunculaceae 毛茛科 *Clematis* 铁线莲属
Clematis caudigera W. T. Wang 尾尖铁线莲
Vines

2.3; -13.3; 15.9; 5.9; 38.5; 70.2; 471 ❙ 46; 29; 2 ❙ -90.3; 46; 13.0

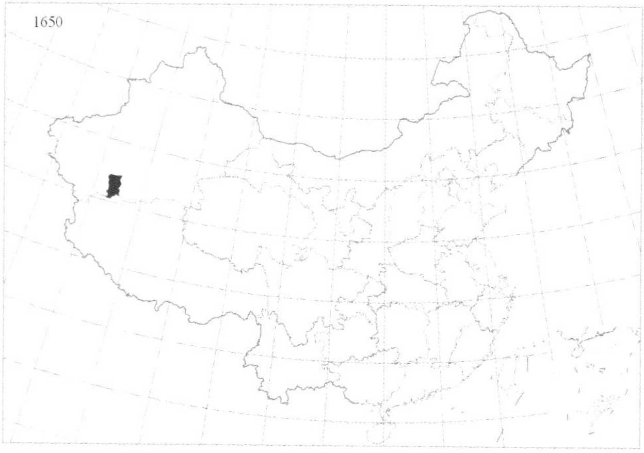

Ranunculaceae 毛茛科 *Clematis* 铁线莲属
Clematis chekiangensis C. P'ei 浙江山木通
Vines, deciduous

15.0；5.0；24.7；15.1；120.8；0.0；789 ❙ 1950；713；223 ❙ 147.0；789；283.9

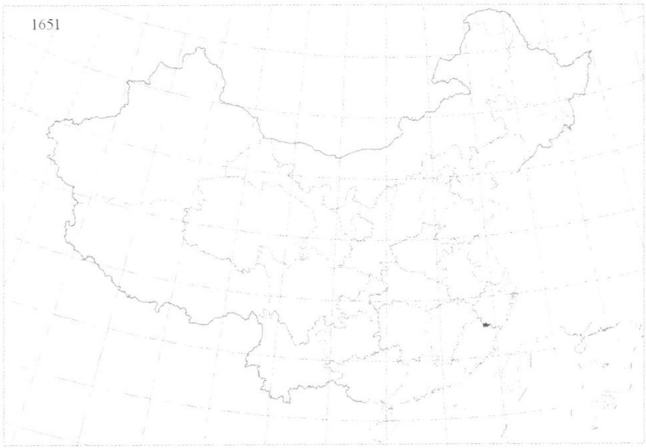

Ranunculaceae 毛茛科 *Clematis* 铁线莲属
Clematis chinensis (Rehder et E. H. Wilson) W. T. Wang 威灵仙
Vines, deciduous

16.8 (3.8~23.6)；5.9 (-4.3~17.8)；26.9 (10.8~30.1)；16.9 (4.7~23.6)；144.8 (21.6~223.3)；2.1 (0.0~35.4)；907 (444~1271) ❙ 1351 (481~2992)；548 (258~1847)；135 (13~259) ❙ 49.0 (-42.2~214.4)；885 (444~1247)；220.1 (110.1~659.0)

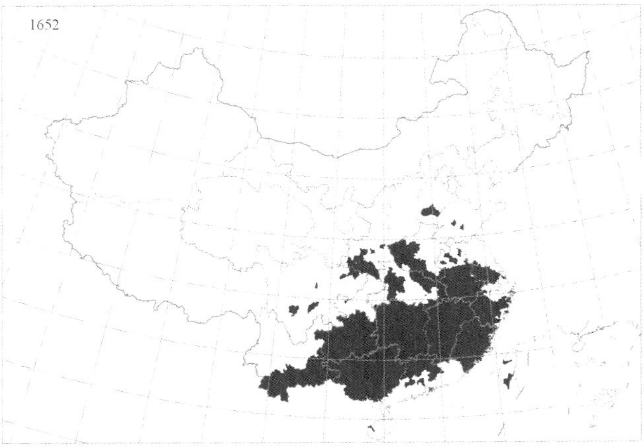

Ranunculaceae 毛茛科 *Clematis* 铁线莲属
Clematis chinensis var. anhweiensis (M. C. Chang) W. T. Wang 安徽铁线莲
Vines, deciduous

15.8；3.3；27.8；15.8；131.7；2.0；871 ❙ 1504；545；185 ❙ 73.2；871；245.5

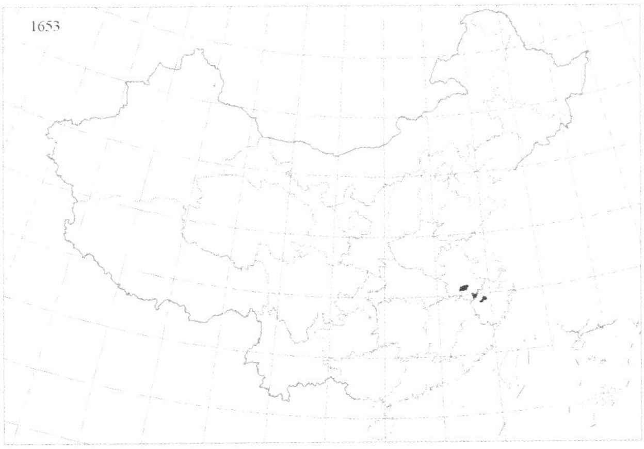

Ranunculaceae 毛茛科 *Clematis* 铁线莲属
Clematis chinensis var. vestita (Rehder et E. H. Wilson) W. T. Wang 毛叶威灵仙
Vines, deciduous

15.7 (13.4~17.0)；2.9 (0.7~4.5)；28.0 (25.2~29.2)；15.7 (13.4~17.0)；131.7 (108.3~144.9)；2.9 (0.5~8.5)；877 (755~941) ❙ 1277 (722~1801)；514 (363~666)；146 (60~242) ❙ 46.3 (-16.6~112.2)；869 (722~935)；191.3 (110.1~319.9)

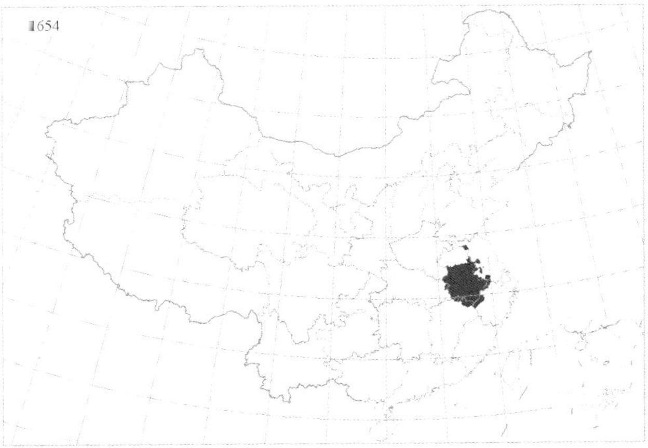

Ranunculaceae 毛茛科 *Clematis* 铁线莲属
Clematis chingii W. T. Wang 两广铁线莲
Vines, deciduous

18.2 (15.5~21.3)；9.1 (4.2~12.6)；25.4 (22.7~28.7)；18.2 (15.6~21.3)；158.5 (128.1~196.0)；0.1 (0.0~0.8)；921 (802~1123) ❙ 1341 (1161~1776)；678 (465~848)；80 (48~192) ❙ 46.3 (5.8~76.9)；921 (802~1121)；276.9 (189.5~371.7)

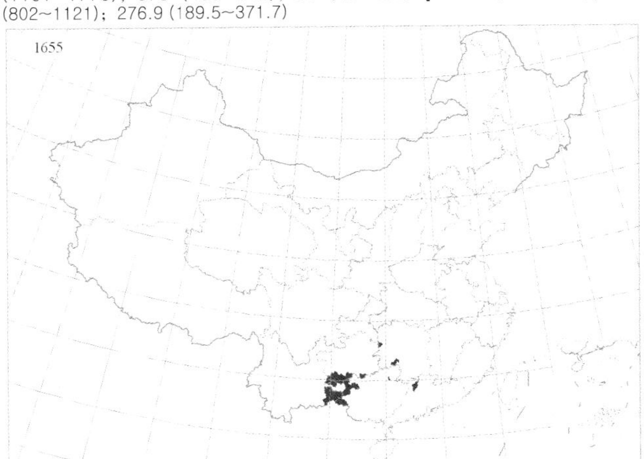

Ranunculaceae 毛茛科 *Clematis* 铁线莲属
Clematis chiupehensis M. Y. Fang 邱北铁线莲
Vines, deciduous

17.5；10.6；22.2；17.5；150.1；0.0；846 ❙ 988；539；44 ❙ 17.1；815；322.9

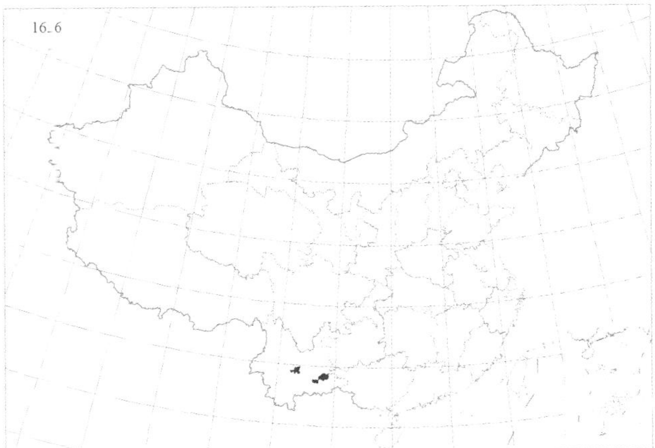

Ranunculaceae 毛茛科　*Clematis* 铁线莲属

Clematis chrysocoma Franch. 金毛铁线莲

Vines, deciduous

13.8 (0.3~19.1); 6.1 (-9.9~12.0); 20.0 (9.3~26.6); 13.9 (3.3~19.1); 109.1 (11.9~169.2); 3.1 (0.0~67.5); 728 (391~966) ❙ 1015 (632~1478); 573 (372~727); 39 (12~85) ❙ 41.8 (-14.2~125.9); 712 (391~966); 274.5 (142.9~407.1)

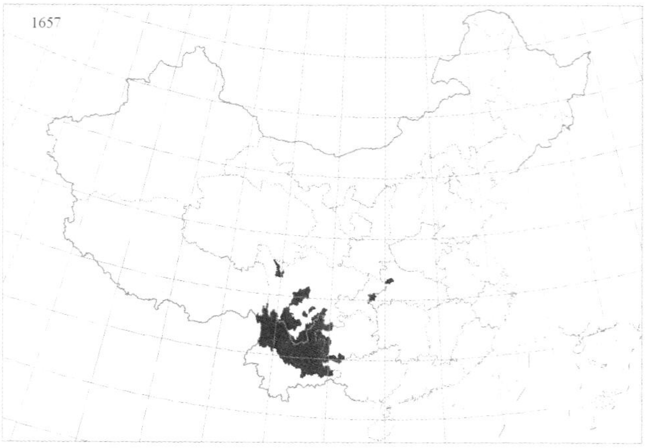

Ranunculaceae 毛茛科　*Clematis* 铁线莲属

Clematis clarkeana H. Lév. et Vaniot 平坝铁线莲

Vines, deciduous

16.4 (14.3~19.1); 7.1 (4.6~10.1); 24.2 (22.8~26.0); 16.5 (14.3~19.1); 137.9 (112.3~169.2); 0.0 (0.0~0.4); 837 (751~966) ❙ 1269 (1194~1381); 649 (561~723); 65 (51~107) ❙ 52.1 (30.0~62.6); 837 (751~966); 223.2 (160.8~316.1)

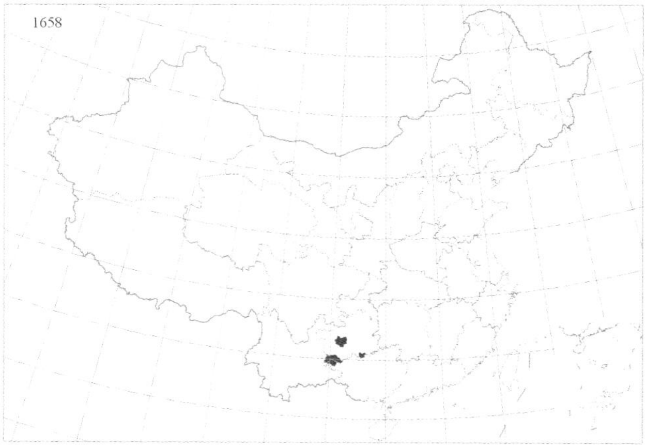

Ranunculaceae 毛茛科　*Clematis* 铁线莲属

Clematis connata DC. 合柄铁线莲

Vines, deciduous

12.2 (-1.8~18.9); 4.7 (-10.2~12.0); 18.4 (6.1~26.6); 12.5 (1.8~19.0); 95.1 (1.9~167.8); 7.7 (0.0~82.8); 682 (329~929) ❙ 944 (320~1736); 549 (193~1007); 32 (7~58) ❙ 40.2 (-14.2~145.8); 660 (320~912); 276.1 (72.2~447.4)

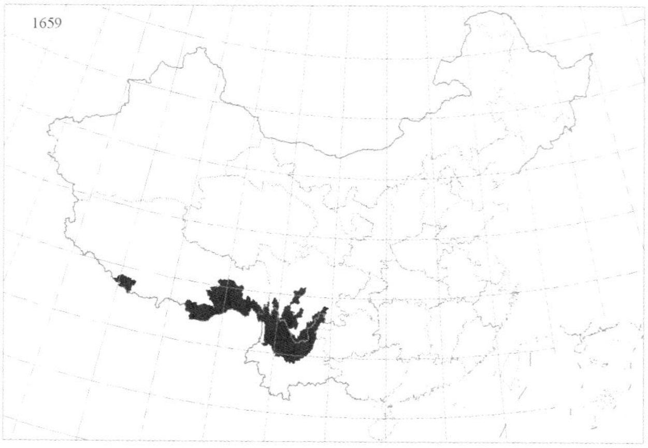

Ranunculaceae 毛茛科　*Clematis* 铁线莲属

Clematis connata var. pseudoconnata (Kuntze) W. T. Wang 川藏铁线莲

Vines, deciduous

5.0; -3.0; 12.1; 6.3; 36.6; 35.9; 488 ❙ 739; 457; 17 ❙ 44.0; 485; 227.2

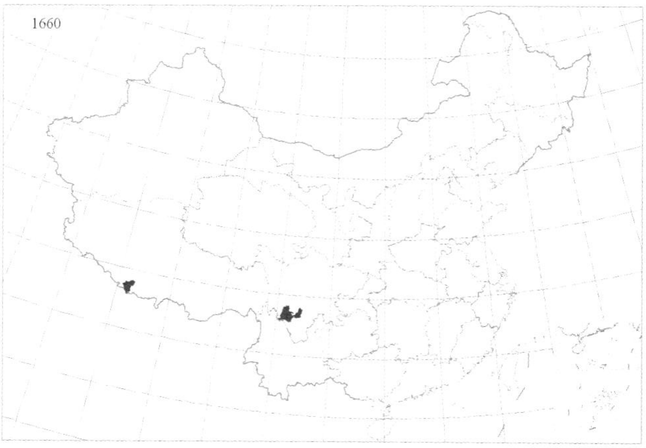

Ranunculaceae 毛茛科　*Clematis* 铁线莲属

Clematis connata var. trullifera (Franch.) W. T. Wang 杯柄铁线莲

Vines, deciduous

9.7 (2.2~15.3); 1.5 (-6.6~7.1); 16.8 (9.9~26.5); 10.0 (3.8~15.4); 68.2 (15.5~125.7); 11.6 (0.0~49.1); 609 (413~832) ❙ 954 (579~1337); 544 (365~651); 33 (10~113) ❙ 58.2 (37.0~125.9); 606 (413~832); 262.3 (183.0~351.7)

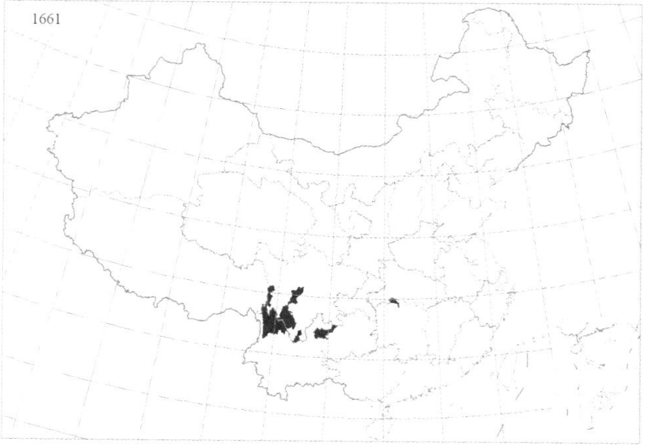

Ranunculaceae 毛茛科　*Clematis* 铁线莲属

Clematis corniculata W. T. Wang 角萼铁线莲

Vines, deciduous

6.2; -10.6; 20.1; 8.9; 70.6; 55.0; 614 ❙ 41; 21; 3 ❙ -92.3; 41; 11.8

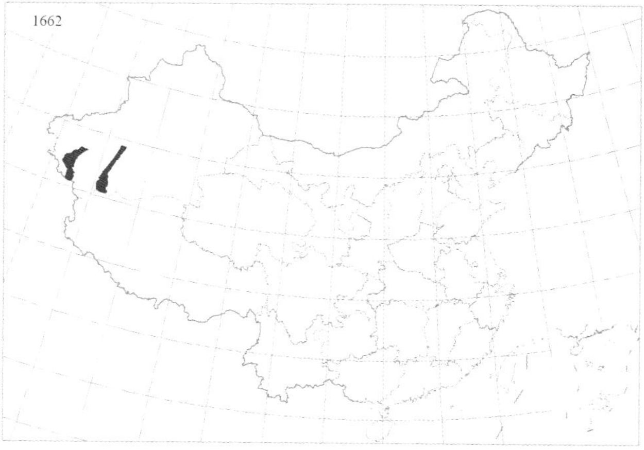

Ranunculaceae 毛茛科 *Clematis* 铁线莲属
Clematis courtoisii Hand.-Mazz. 大花威灵仙
Vines, deciduous

15.5 (14.1~17.6); 3.0 (1.6~5.1); 27.6 (26.2~29.3); 15.6 (14.2~17.6); 129.5 (116.7~151.3); 2.8 (0.0~6.5); 861 (796~961) ❙ 1442 (1117~1770); 558 (478~661); 170 (107~223) ❙ 68.1 (28.5~112.2); 860 (796~961); 236.3 (149.5~319.9)

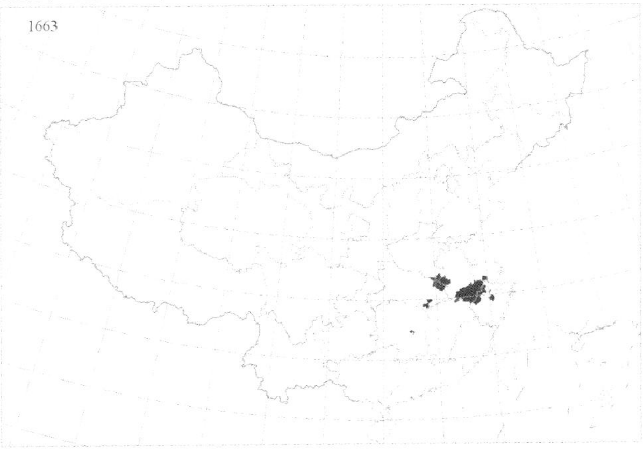

Ranunculaceae 毛茛科 *Clematis* 铁线莲属
Clematis crassifolia Benth. 厚叶铁线莲
Vines, deciduous

19.5 (14.6~24.3); 10.3 (3.4~18.7); 27.4 (20.1~29.0); 19.5 (14.7~24.3); 174.5 (117.7~231.9); 0.0 (0.0~2.1); 1023 (773~1338) ❙ 1703 (868~3495); 668 (312~1428); 184 (42~897) ❙ 69.6 (-34.3~246.5); 1006 (773~1254); 254.6 (151.0~538.2)

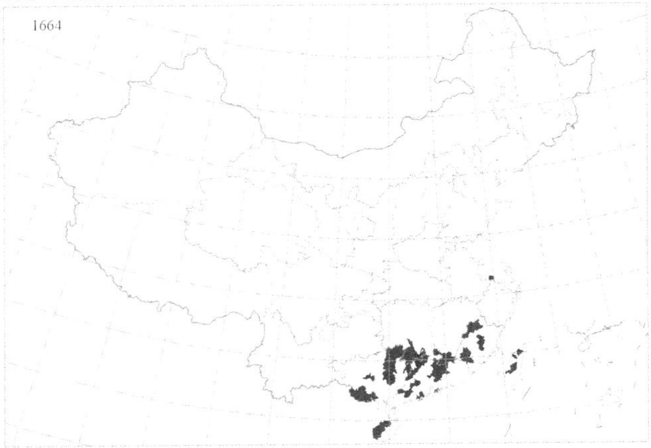

Ranunculaceae 毛茛科 *Clematis* 铁线莲属
Clematis crassipes Chun et F. C. How 粗柄铁线莲
Vines, deciduous

22.9 (21.0~24.8); 16.3 (13.2~20.1); 27.7 (25.7~28.7); 23.0 (21.0~24.8); 215.6 (192.3~237.6); 0.0 (0.0~0.0); 1231 (1081~1385) ❙ 1592 (868~2435); 722 (312~1428); 85 (41~130) ❙ 31.2 (-34.3~99.5); 1137 (868~1254); 390.1 (235.1~620.0)

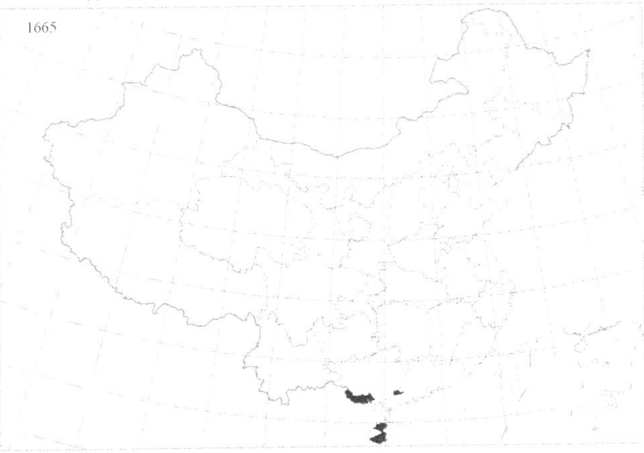

Ranunculaceae 毛茛科 *Clematis* 铁线莲属
Clematis delavayi Franch. 银叶铁线莲
Shrubs, deciduous

9.5 (1.6~18.9); 1.1 (-8.2~12.0); 16.6 (9.5~25.9); 10.1 (3.7~19.0); 74.0 (15.2~167.8); 19.7 (0.0~56.2); 614 (408~929) ❙ 844 (535~1352); 497 (264~771); 24 (10~47) ❙ 44.5 (-28.9~125.9); 576 (408~802); 272.0 (153.3~438.9)

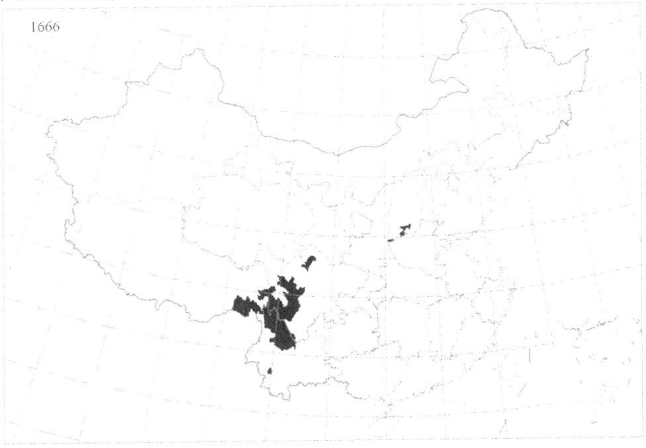

Ranunculaceae 毛茛科 *Clematis* 铁线莲属
Clematis delavayi var. calvescens C. K. Schneid. 疏毛银叶铁线莲
Shrubs, deciduous

6.3; -1.4; 13.1; 6.8; 39.9; 23.6; 508 ❙ 747; 456; 19 ❙ 45.8; 508; 259.9

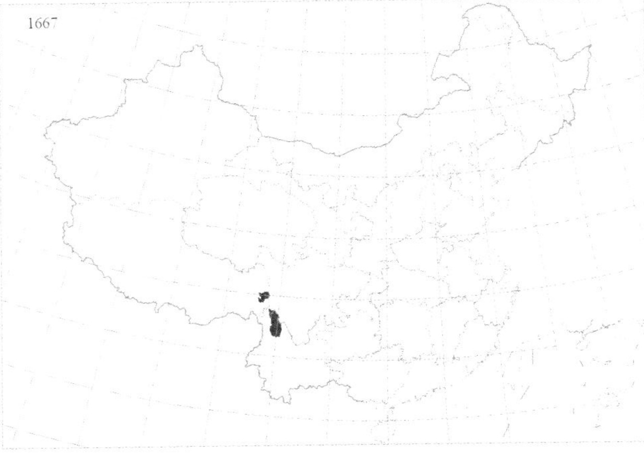

Ranunculaceae 毛茛科 *Clematis* 铁线莲属
Clematis delavayi var. limprichtii (Ulbr.) M. C. Chang 裂银叶铁线莲
Shrubs, deciduous

2.4 (1.6~3.2); -6.6 (-8.0~-5.6); 10.1 (9.5~10.8); 4.1 (3.7~4.5); 17.9 (15.2~21.4); 48.7 (42.6~56.2); 417 (408~430) ❙ 720 (628~780); 451 (410~487); 12 (11~15) ❙ 72.6 (48.6~86.8); 417 (408~430); 253.6 (208.8~299.8)

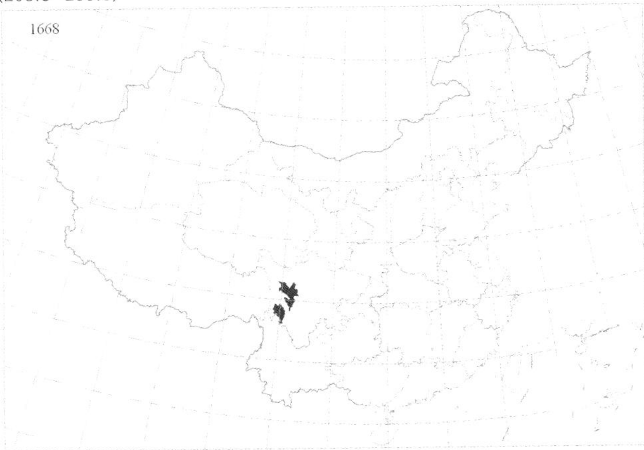

Ranunculaceae 毛茛科 *Clematis* 铁线莲属

Clematis delavayi var. spinescens Balf. f. ex Diels 刺铁线莲

Shrubs, deciduous

5.2 (2.7~10.8); -2.7 (-6.0~3.9); 12.2 (10.4~16.6); 5.9 (4.2~10.9); 32.6 (18.9~71.8); 29.6 (1.3~46.9); 479 (419~633) ❙ 737 (535~968); 448 (326~594); 20 (10~31) ❙ 53.1 (27.1~100.7); 479 (419~633); 252.2 (201.9~319.8)

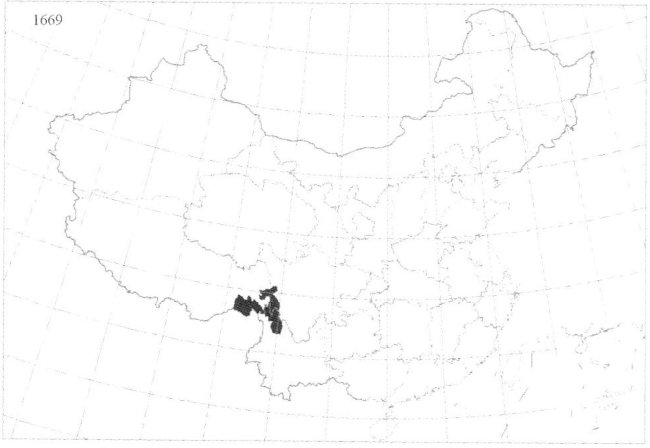

Ranunculaceae 毛茛科 *Clematis* 铁线莲属

Clematis dilatata C. P'ei 舟柄铁线莲

Vines, deciduous

16.0 (15.1~17.0); 5.0 (4.0~6.7); 27.0 (25.6~27.9); 16.1 (15.2~17.0); 133.4 (122.9~144.4); 0.3 (0.0~1.0); 856 (810~896) ❙ 1679 (1544~1821); 602 (525~671); 198 (188~226) ❙ 96.8 (74.1~124.8); 856 (810~896); 239.2 (213.9~273.1)

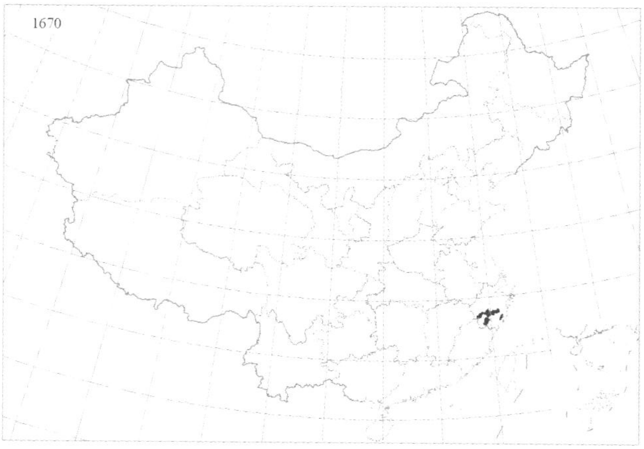

Ranunculaceae 毛茛科 *Clematis* 铁线莲属

Clematis fasciculiflora Franch. 滑叶藤

Vines, deciduous

15.5 (5.4~21.0); 8.4 (-2.4~15.1); 20.9 (12.3~26.6); 15.6 (5.8~21.0); 127.8 (31.5~191.8); 1.0 (0.0~25.6); 783 (481~1053) ❙ 1122 (741~1751); 631 (411~996); 44 (12~77) ❙ 43.5 (-14.2~112.2); 766 (481~1053); 331.0 (142.9~659.0)

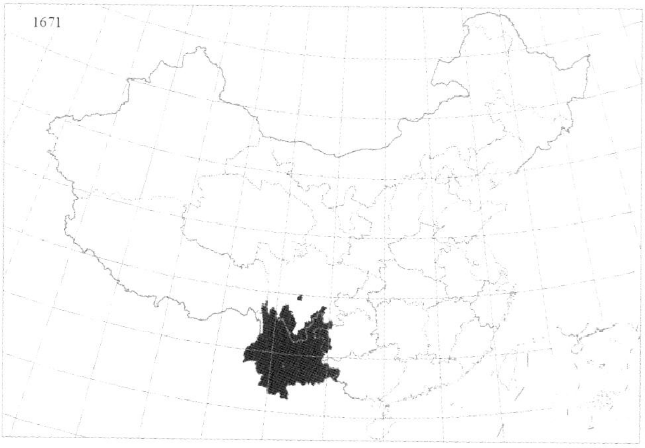

Ranunculaceae 毛茛科 *Clematis* 铁线莲属

Clematis fasciculiflora var. angustifolia H. F. Comber 狭叶滑叶藤

Vines, deciduous

6.4; -1.4; 13.0; 6.5; 37.4; 20.7; 505 ❙ 832; 522; 16 ❙ 64.6; 505; 297.2

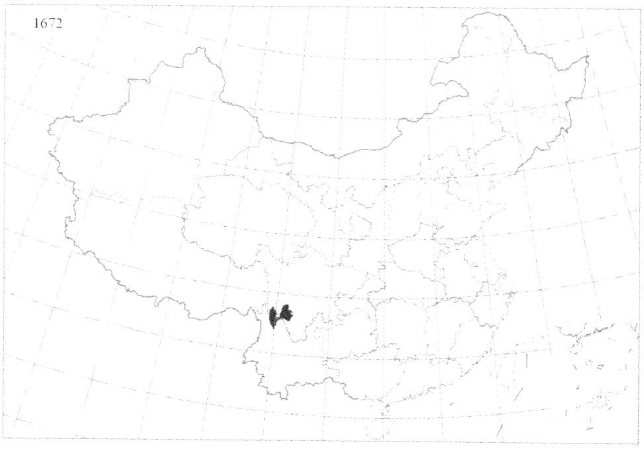

Ranunculaceae 毛茛科 *Clematis* 铁线莲属

Clematis finetiana H. Lév. et Vaniot 山木通

Vines, deciduous

16.9 (9.3~22.4); 6.0 (-2.9~14.7); 27.1 (20.1~30.1); 17.0 (9.5~22.5); 145.1 (69.8~209.7); 1.3 (0.0~20.5); 908 (603~1205) ❙ 1421 (512~2024); 542 (258~954); 151 (16~259) ❙ 56.6 (-35.3~147.0); 894 (512~1205); 213.5 (95.3~335.1)

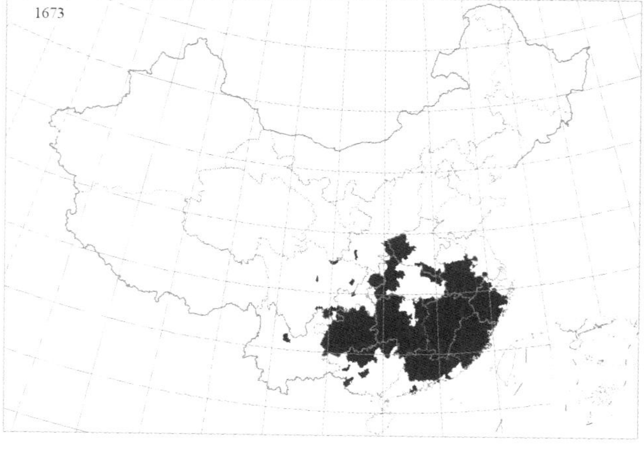

Ranunculaceae 毛茛科 *Clematis* 铁线莲属

Clematis finetiana var. pedata W. T. Wang 鸟足叶铁线莲

Vines, deciduous

15.4; 3.9; 26.5; 15.5; 126.9; 1.2; 835 ❙ 1428; 579; 116 ❙ 71.2; 835; 210.9

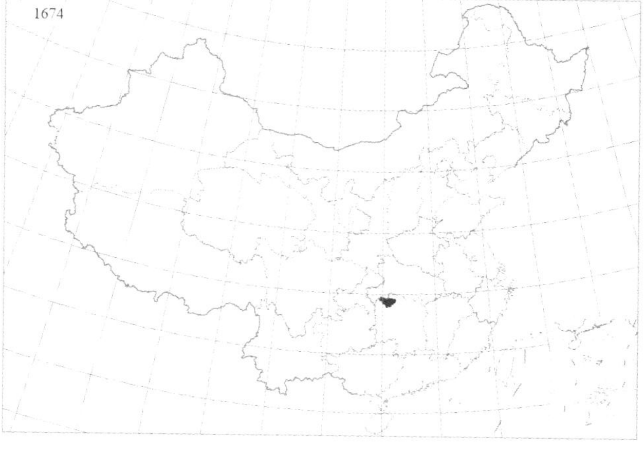

Ranunculaceae 毛茛科 *Clematis* 铁线莲属
Clematis formosana Kuntze 宝岛铁线莲
Vines, deciduous

20.0; 15.4; 23.3; 20.0; 180.2; 0.0; 966 | 2821; 1610; 140 | 194.5; 966; 304.3

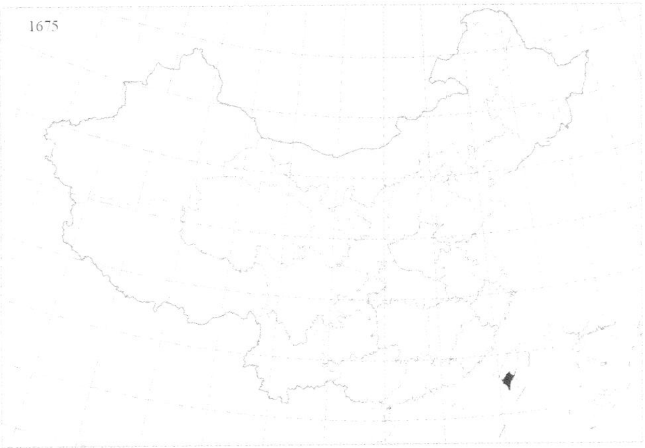

Ranunculaceae 毛茛科 *Clematis* 铁线莲属
Clematis fruticosa Turcz. 灌木铁线莲
Shrubs, deciduous

7.0 (0.1~14.3); -9.4 (-21.5~0.1); 21.6 (12.4~27.2); 9.2 (4.8~14.3); 72.8 (25.1~122.8); 47.7 (10.7~108.6); 618 (441~845) | 419 (77~677); 249 (48~453); 11 (1~26) | -31.9 (-87.2~48.8); 415 (77~644); 123.1 (7.5~282.7)

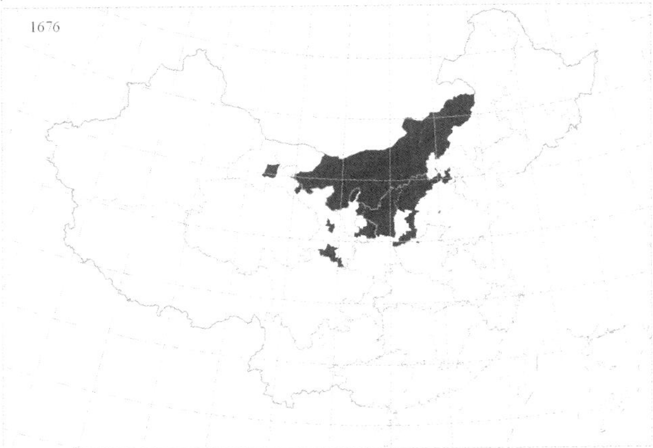

Ranunculaceae 毛茛科 *Clematis* 铁线莲属
Clematis fruticosa var. canescens Turcz. 毛灌木铁线莲
Shrubs, deciduous

5.8 (-0.3~8.7); -9.9 (-16.0~-2.1); 19.7 (9.0~24.3); 8.1 (3.0~10.3); 62.2 (10.4~85.9); 52.5 (25.6~77.9); 578 (378~679) | 282 (49~943); 167 (34~507); 7 (1~28) | -43.5 (-92.8~128.6); 227 (49~481); 94.6 (5.1~341.6)

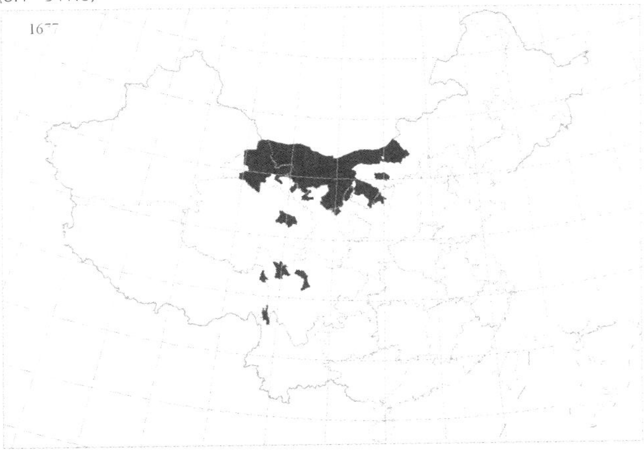

Ranunculaceae 毛茛科 *Clematis* 铁线莲属
Clematis fulvicoma Rehder et E. H. Wilson 滇南铁线莲
Vines, deciduous

18.2 (16.3~21.0); 11.7 (9.1~15.2); 22.9 (20.6~28.7); 18.5 (16.4~22.7); 161.7 (136.4~212.7); 0.0 (0.0~0.0); 900 (795~1216) | 1473 (1049~1965); 801 (598~994); 58 (42~104) | 64.3 (26.1~112.2); 891 (783~1216); 464.9 (204.4~659.0)

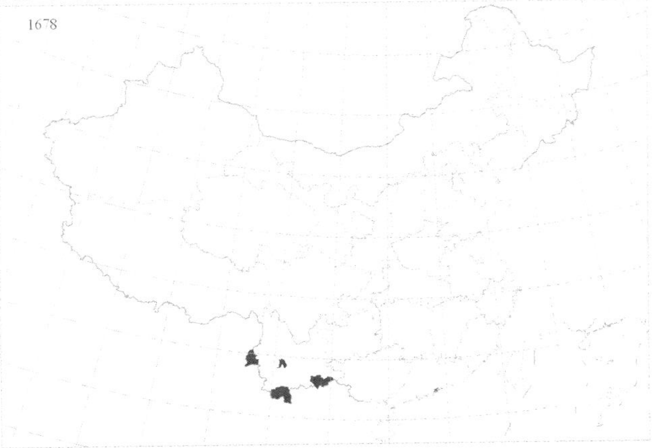

Ranunculaceae 毛茛科 *Clematis* 铁线莲属
Clematis gouriana Roxb. ex DC. 小蓑衣藤
Vines, deciduous

15.2 (2.9~21.3); 5.2 (-6.5~14.9); 24.1 (11.2~28.7); 15.3 (4.4~21.3); 125.9 (20.5~196.0); 2.8 (0.0~44.6); 812 (428~1123) | 1266 (731~1777); 571 (346~959); 84 (12~223) | 57.4 (5.8~121.7); 811 (428~1121); 237.8 (126.1~627.0)

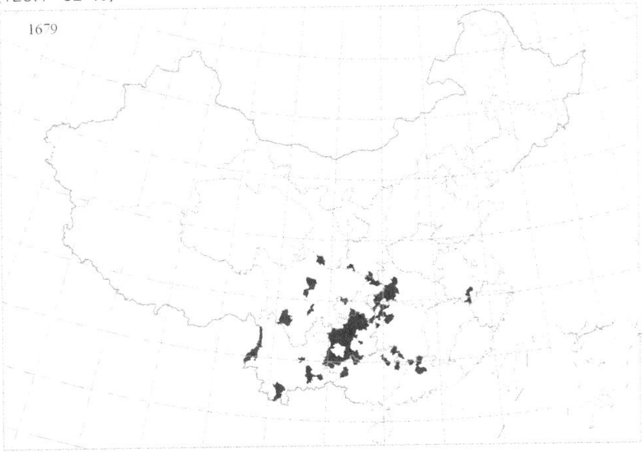

Ranunculaceae 毛茛科 *Clematis* 铁线莲属
Clematis gracilifolia Roxb. ex DC. 薄叶铁线莲
Vines, deciduous

7.4 (-0.9~18.9); -1.2 (-11.3~12.0); 14.9 (8.7~25.3); 8.4 (2.7~19.0); 57.7 (9.4~167.8); 28.1 (0.0~79.1); 555 (354~929) | 840 (473~1736); 477 (270~1007); 22 (7~88) | 55.8 (-14.2~145.8); 538 (354~795); 279.3 (169.5~447.4)

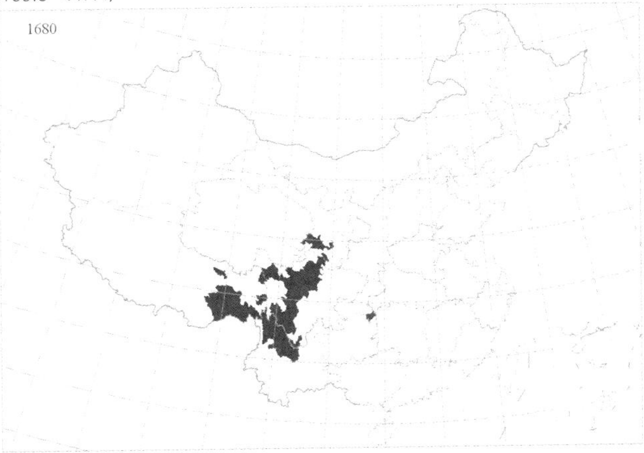

Ranunculaceae 毛茛科 *Clematis* 铁线莲属

Clematis gracilifolia var. dissectifolia W. T. Wang et M. C. Chang 狭裂薄叶铁线莲 Vines, deciduous

5.4 (−1.8~11.0); −2.6 (−10.2~3.9); 12.4 (6.1~17.0); 6.3 (1.8~11.1); 35.9 (1.9~74.6); 30.1 (1.3~82.8); 492 (329~640) ❙ 824 (320~1041); 498 (217~619); 20 (11~45) ❙ 67.6 (−3.0~125.9); 490 (320~638); 275.4 (72.2~357.7)

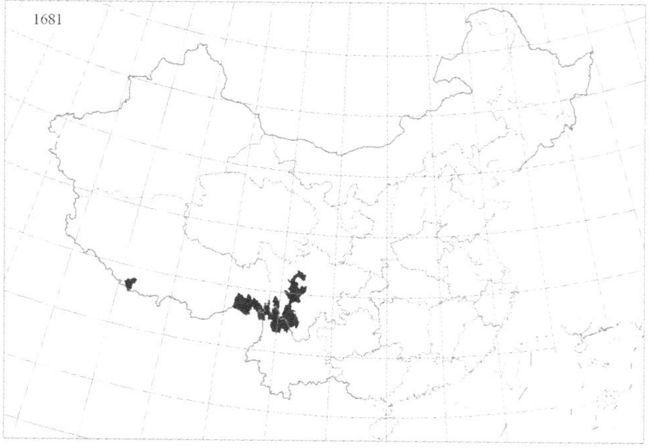

Ranunculaceae 毛茛科 *Clematis* 铁线莲属

Clematis gracilifolia var. macrantha W. T. Wang et M. C. Chang 大花薄叶铁线莲 Vines, deciduous

7.3; −2.0; 15.6; 7.5; 48.2; 20.4; 537 ❙ 1005; 531; 27 ❙ 87.3; 537; 272.3

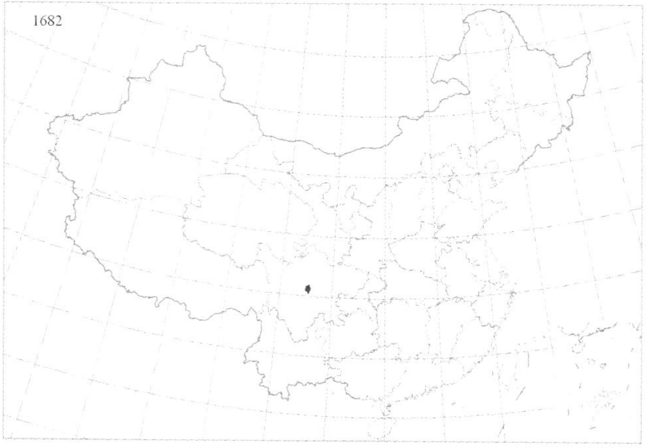

Ranunculaceae 毛茛科 *Clematis* 铁线莲属

Clematis grandidentata (Rehder et E. H. Wilson) W. T. Wang 粗齿铁线莲 Vines, deciduous

13.5 (−2.6~21.6); 1.8 (−15.3~14.2); 24.0 (8.5~29.4); 13.9 (2.5~21.7); 113.8 (7.7~199.8); 11.5 (0.0~98.0); 775 (349~1153) ❙ 1026 (341~1997); 476 (190~860); 73 (8~254) ❙ 30.5 (−40.9~130.7); 729 (341~1153); 214.7 (61.2~551.6)

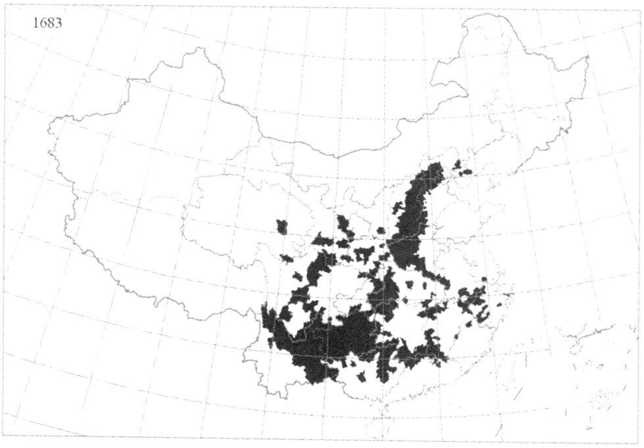

Ranunculaceae 毛茛科 *Clematis* 铁线莲属

Clematis grandidentata var. likiangensis (Rehder) W. T. Wang 丽江铁线莲 Vines, deciduous

11.1 (2.6~15.5); 1.5 (−7.0~7.9); 19.5 (11.2~26.8); 11.3 (4.3~15.6); 84.0 (20.0~127.1); 10.8 (0.0~48.2); 662 (427~826) ❙ 1014 (596~1663); 537 (350~816); 46 (12~212) ❙ 54.9 (−20.9~101.4); 651 (427~826); 265.8 (117.3~382.6)

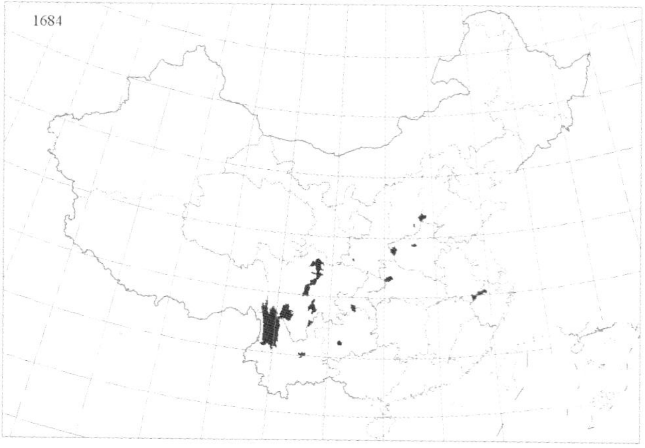

Ranunculaceae 毛茛科 *Clematis* 铁线莲属

Clematis gratopsis W. T. Wang 金佛铁线莲

Vines, deciduous

14.3 (9.1~21.3); 3.4 (−2.5~12.3); 24.5 (18.1~28.6); 14.4 (9.2~21.3); 117.9 (64.1~196.1); 5.3 (0.0~20.6); 795 (595~1130) ❙ 1210 (675~1758); 520 (331~728); 80 (15~189) ❙ 52.4 (−1.2~107.2); 795 (595~1130); 243.0 (164.4~334.4)

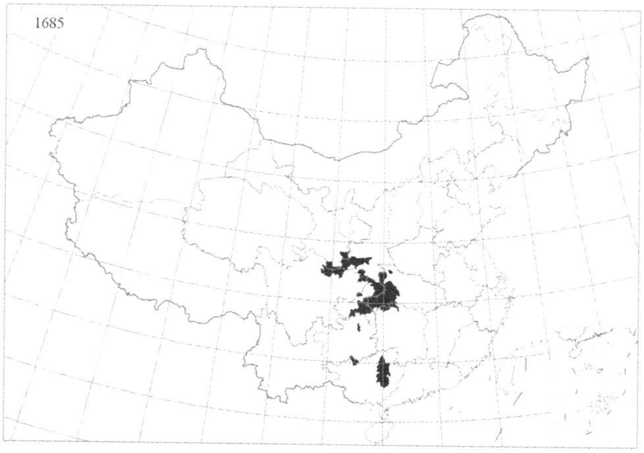

Ranunculaceae 毛茛科 *Clematis* 铁线莲属

Clematis grewiiflora DC. 黄毛铁线莲

Vines, deciduous

−1.1; −9.1; 6.2; 2.0; 2.5; 75.6; 334 ❙ 321; 193; 33 ❙ −4.0; 321; 79.3

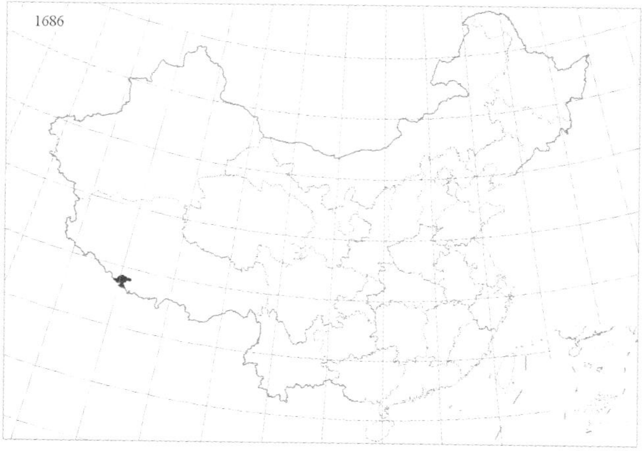

Ranunculaceae 毛茛科 *Clematis* 铁线莲属

Clematis hainanensis W. T. Wang 海南铁线莲

Vines, deciduous

24.0 (22.2~24.8); 18.4 (17.2~20.1); 28.0 (25.7~28.8); 24.0 (22.3~24.8); 228.3 (207.2~237.6); 0.0 (0.0~0.0); 1310 (1127~1385) | 1501 (868~1854); 534 (312~712); 74 (41~115) | 15.0 (-34.3~47.8); 119 (868~1360); 440.3 (248.2~620.0)

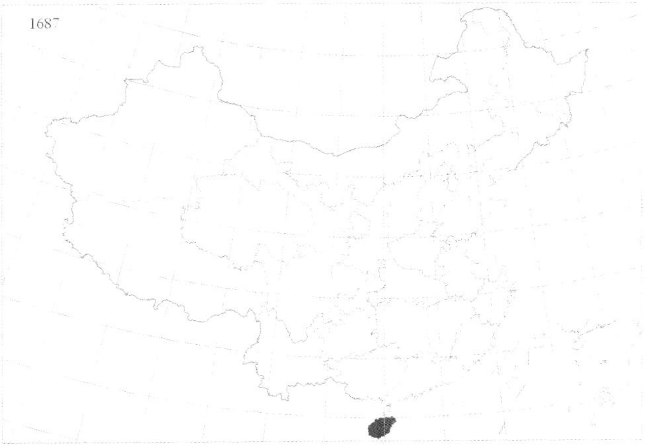

Ranunculaceae 毛茛科 *Clematis* 铁线莲属

Clematis hancockiana Maxim. 毛萼铁线莲

Vines, deciduous

15.6 (14.4~16.7); 3.2 (2.1~5.3); 27.7 (26.3~29.0); 15.7 (14.5~16.8); 130.8 (118.7~142.4); 2.4 (0.0~4.9); 867 (801~936) | 1410 (1077~1770); 544 (476~628); 169 (97~223) | 63.3 (22.3~112.2); 867 (801~931); 230.8 (140.7~319.9)

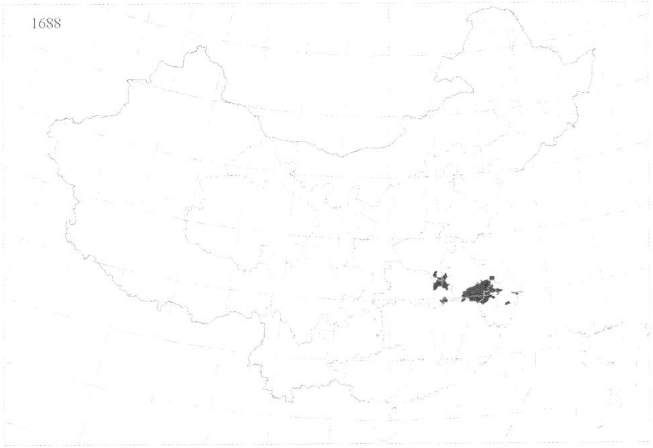

Ranunculaceae 毛茛科 *Clematis* 铁线莲属

Clematis henryi Oliv. 单叶铁线莲

Vines, deciduous

16.6 (5.4~21.7); 6.0 (-2.1~15.3); 26.5 (12.3~30.1); 16.6 (5.8~21.7); 140.8 (31.5~200.2); 1.2 (0.0~25.6); 885 (481~1153) | 1452 (741~3495); 546 (323~1412); 160 (17~897) | 64.2 (-3.3~246.5); 874 (481~1153); 233.2 (131.2~551.6)

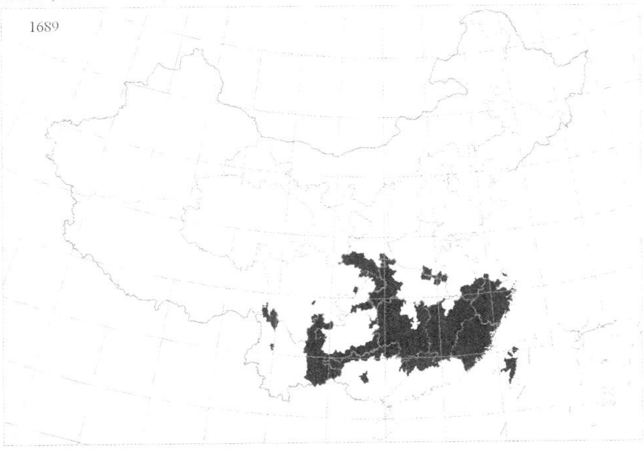

Ranunculaceae 毛茛科 *Clematis* 铁线莲属

Clematis henryi var. mollis W. T. Wang 毛单叶铁线莲

Vines, deciduous

14.1; 2.9; 24.8; 14.1; 113.5; 3.9; 775 | 1459; 596; 108 | 89.6; 775; 239.2

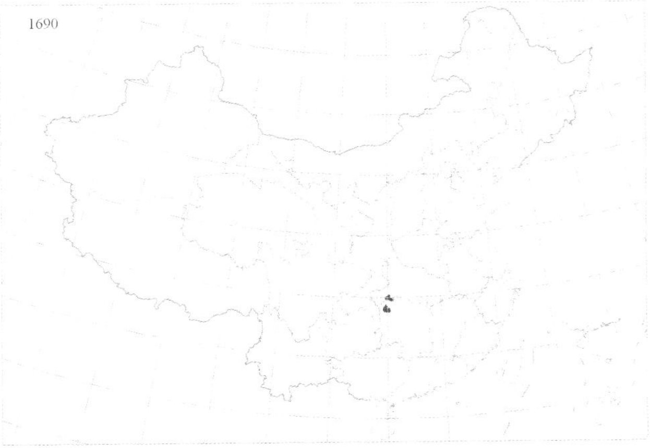

Ranunculaceae 毛茛科 *Clematis* 铁线莲属

Clematis iliensis Y. S. Hou et W. H. Hou 伊犁铁线莲

Vines, deciduous

3.9 (-1.3~9.0); -14.9 (-20.9~-9.3); 18.5 (13.4~23.2); 7.6 (4.7~10.7); 56.9 (28.1~88.9); 69.4 (40.7~103.9); 563 (443~703) | 292 (225~357); 127 (65~171); 30 (12~48) | -47.2 (-65.1~-29.7); 292 (225~357); 208.1 (165.7~297.3)

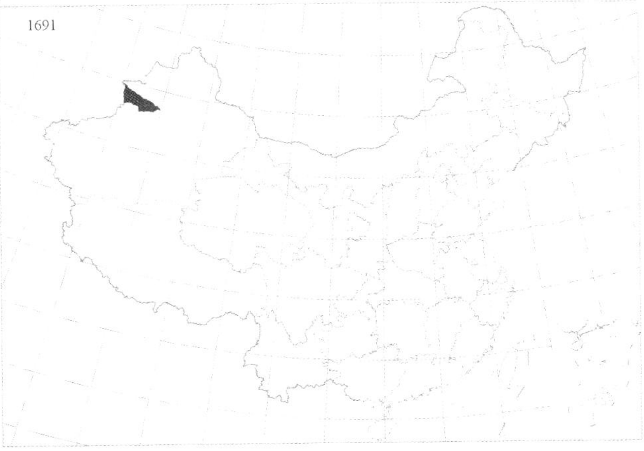

Ranunculaceae 毛茛科 *Clematis* 铁线莲属

Clematis intricata Bunge 黄花铁线莲

Vines, deciduous

6.7 (-4.1~13.6); -9.3 (-19.6~-0.3); 20.7 (7.3~27.5); 8.8 (2.0~13.8); 69.1 (4.1~119.7); 48.7 (12.1~114.2); 607 (335~841) | 464 (84~1073); 290 (53~691); 11 (1~40) | -21.7 (-87.1~74.1); 441 (84~644); 154.4 (7.5~335.3)

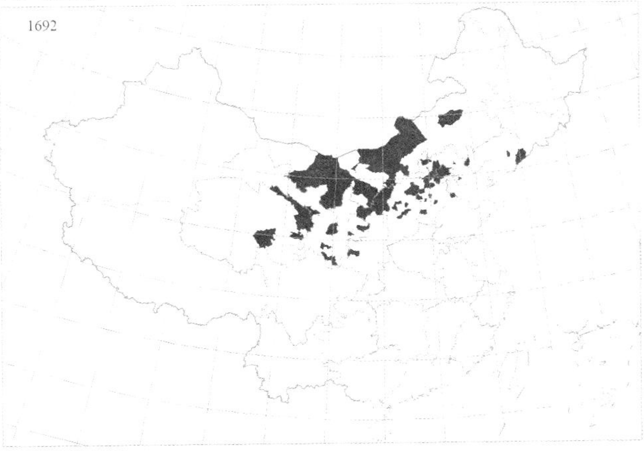

Ranunculaceae 毛茛科 *Clematis* 铁线莲属
Clematis javana DC. 台湾铁线莲
Vines, deciduous

20.5 (16.2~24.1); 14.7 (11.0~18.9); 25.4 (20.1~28.2); 20.5 (16.2~24.1); 186.3 (134.2~229.7); 0.0 (0.0~0.0); 1042 (773~1322) ❙ 2467 (1751~3495); 1110 (750~1847); 287 (47~897) ❙ 146.6 (32.5~246.5); 999 (773~1144); 245.2 (119.4~363.7)

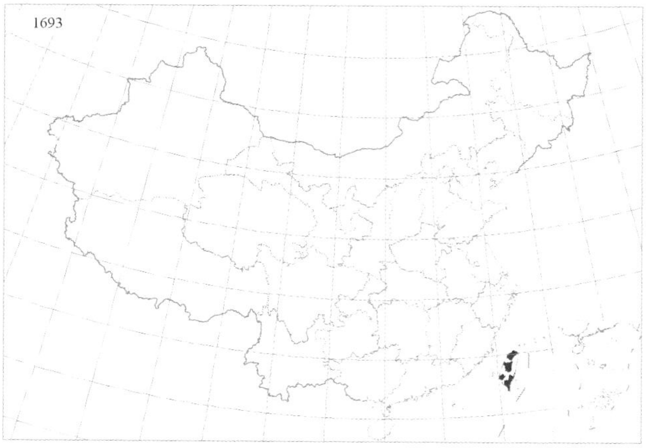

Ranunculaceae 毛茛科 *Clematis* 铁线莲属
Clematis jialasaensis W. T. Wang 加拉萨铁线莲
Vines, deciduous

13.3; 5.4; 19.6; 13.4; 100.4; 0.0; 706 ❙ 1736; 1007; 51 ❙ 145.8; 706; 447.4

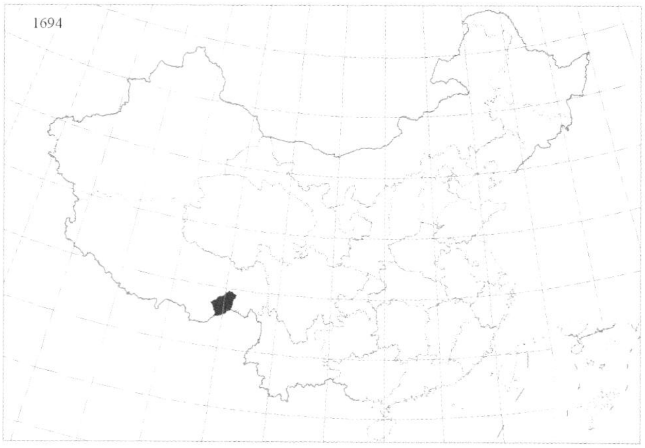

Ranunculaceae 毛茛科 *Clematis* 铁线莲属
Clematis jingdungensis W. T. Wang 多花铁线莲
Vines, deciduous

17.3; 11.0; 21.6; 17.3; 148.0; 0.0; 832 ❙ 1049; 598; 44 ❙ 26.1; 783; 365.3

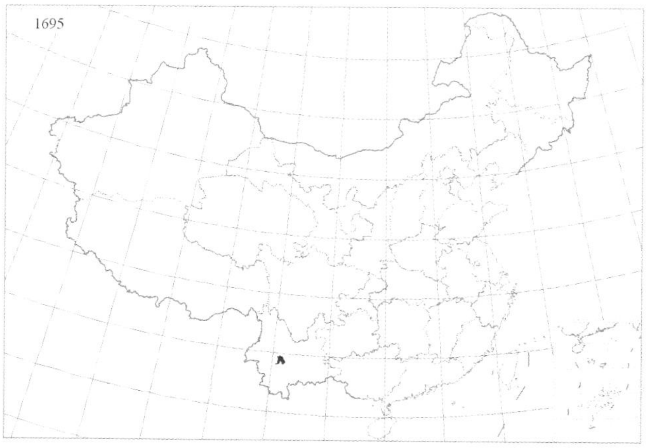

Ranunculaceae 毛茛科 *Clematis* 铁线莲属
Clematis jinzhaiensis Z. W. Xue et Z. W. Wang 金寨铁线莲
Vines, deciduous

14.1; 1.6; 26.2; 14.2; 116.7; 6.5; 796 ❙ 1364; 601; 129 ❙ 71.2; 796; 243.1

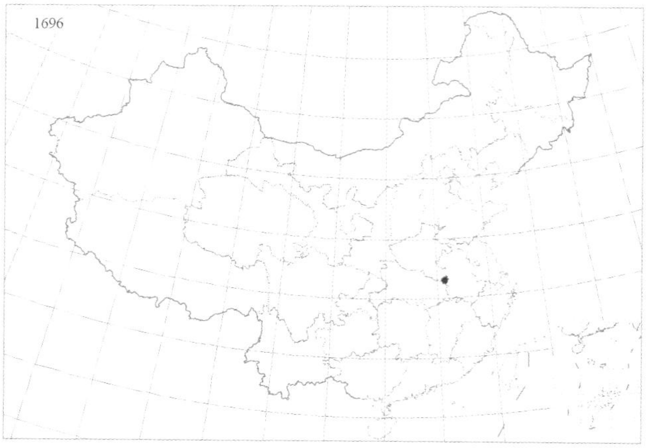

Ranunculaceae 毛茛科 *Clematis* 铁线莲属
Clematis kirilowii Maxim. 太行铁线莲
Vines, deciduous

10.4 (2.6~16.5); -5.1 (-15.9~5.3); 24.1 (18.7~28.7); 11.5 (7.2~16.5); 94.4 (52.0~139.4); 29.3 (0.0~80.4); 714 (535~921) ❙ 610 (367~1581); 370 (235~587); 26 (8~213) ❙ -15.1 (-47.7~87.6); 587 (367~921); 172.6 (101.9~287.6)

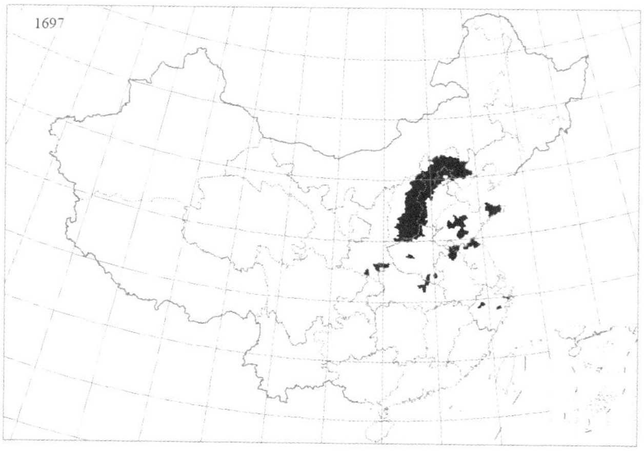

Ranunculaceae 毛茛科 *Clematis* 铁线莲属
Clematis kirilowii var. chanetii (H. Lév.) Hand.-Mazz. 狭裂太行铁线莲
Vines, deciduous

10.0 (4.2~13.8); -5.0 (-12.5~0.1); 23.4 (18.7~27.3); 11.0 (7.3~14.0); 89.6 (52.8~121.1); 29.3 (10.8~62.6); 690 (535~842) ❙ 622 (377~917); 385 (241~554); 26 (8~65) ❙ -10.5 (-37.2~16.3); 615 (377~812); 165.3 (109.4~228.3)

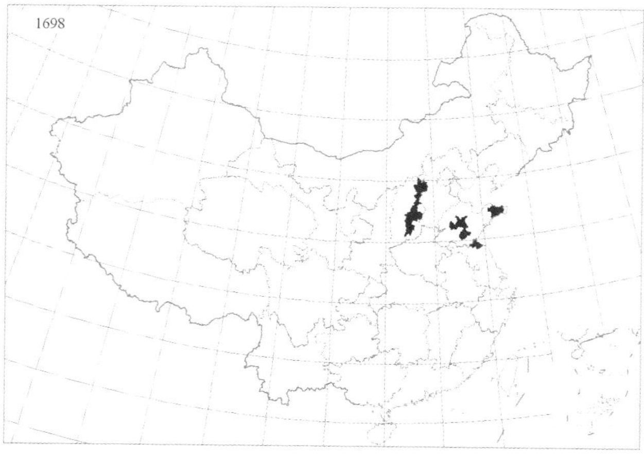

Ranunculaceae 毛茛科 *Clematis* 铁线莲属
Clematis koreana Kom. 朝鲜铁线莲
Subshrubs or vines, deciduous

4.4 (1.8~8.7); -14.4 (-18.7~-8.3); 20.8 (18.5~23.4); 8.1 (6.7~10.4); 62.3 (47.6~85.5); 69.0 (40.2~87.9); 573 (513~668) ❙ 749 (561~1073); 460 (333~691); 26 (16~40) ❙ 30.2 (-2.6~74.1); 571 (513~668); 307.8 (232.8~335.3)

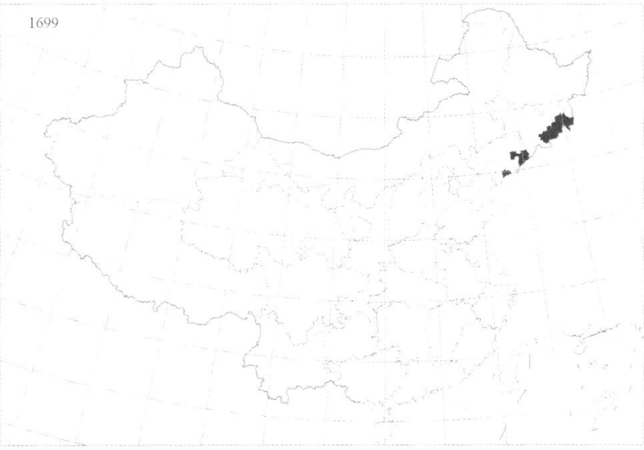

Ranunculaceae 毛茛科 *Clematis* 铁线莲属
Clematis kweichowensis C. P'ei 贵州铁线莲
Vines, deciduous

15.4 (12.8~18.1); 5.0 (0.9~9.5); 25.2 (24.0~27.0); 15.5 (12.8~18.2); 127.2 (102.5~158.1); 1.3 (0.0~8.7); 818 (732~901) ❙ 1111 (862~1451); 515 (376~723); 64 (47~112) ❙ 36.0 (7.8~72.9); 818 (732~901); 203.9 (154.4~316.1)

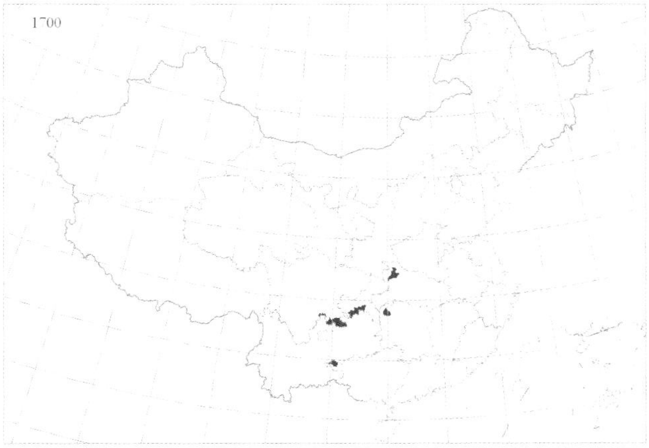

Ranunculaceae 毛茛科 *Clematis* 铁线莲属
Clematis lancifolia Bureau et Franch. 披针叶铁线莲
Shrubs, deciduous

12.8 (2.2~18.9); 5.0 (-6.6~12.0); 19.0 (9.9~24.3); 13.1 (3.8~19.0); 103.3 (15.5~167.8); 10.1 (0.0~49.1); 707 (413~929) ❙ 904 (473~1042); 536 (287~641); 21 (7~38) ❙ 33.1 (-14.2~125.9); 656 (413~786); 281.1 (218.0~331.6)

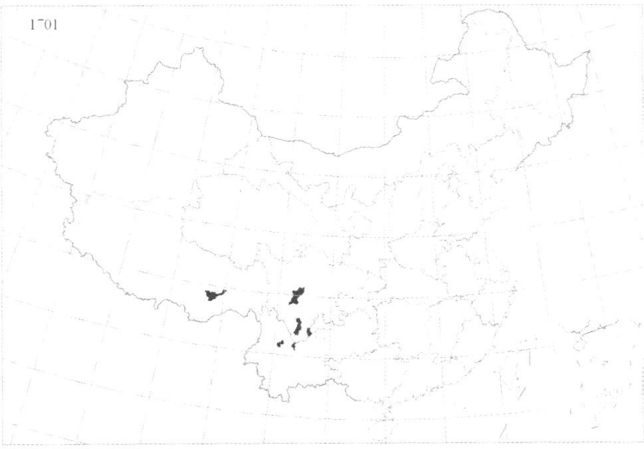

Ranunculaceae 毛茛科 *Clematis* 铁线莲属
Clematis lancifolia var. ternata W. T. Wang et M. C. Chang 竹叶铁线莲
Shrubs, deciduous

16.1; 8.6; 21.9; 16.1; 133.0; 0.0; 797 ❙ 1014; 630; 20 ❙ 27.3; 719; 330.0

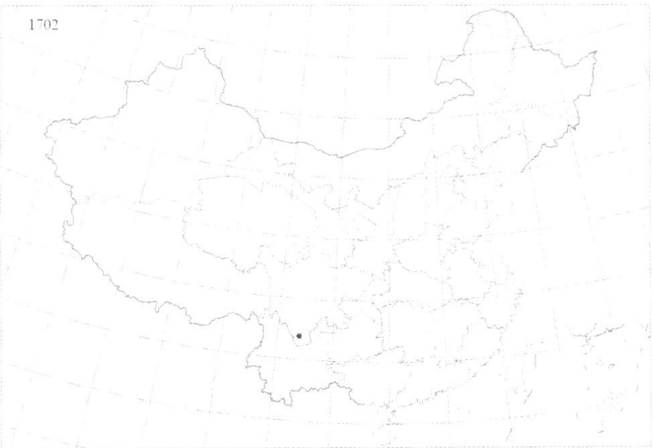

Ranunculaceae 毛茛科 *Clematis* 铁线莲属
Clematis laxistrigosa (W. T. Wang et M. C. Chang) W. T. Wang 糙毛铁线莲
Vines, deciduous

11.5; 2.8; 19.3; 11.5; 82.3; 3.8; 659 ❙ 1086; 602; 27 ❙ 65.1; 659; 316.6

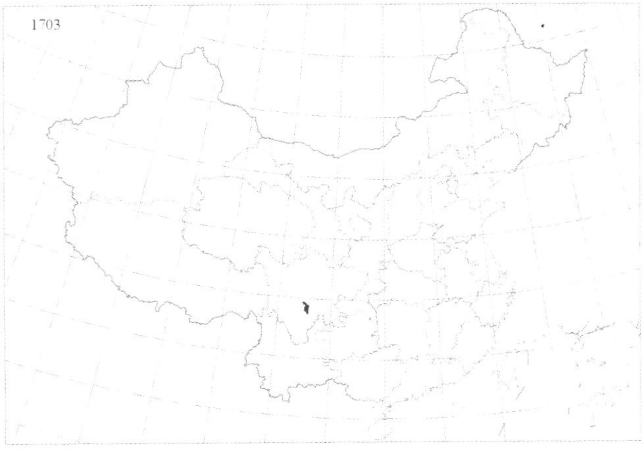

Ranunculaceae 毛茛科 *Clematis* 铁线莲属
Clematis leschenaultiana DC. 锈毛铁线莲
Vines, deciduous

18.8 (12.1~24.8); 9.8 (1.1~20.1); 26.5 (19.1~29.8); 18.9 (12.2~24.8); 166.8 (95.0~237.6); 0.2 (0.0~8.5); 983 (696~1385) ❙ 1571 (868~3495); 641 (312~1847); 150 (41~897) ❙ 62.6 (-34.3~246.5); 969 (696~1360); 274.6 (133.6~620.0)

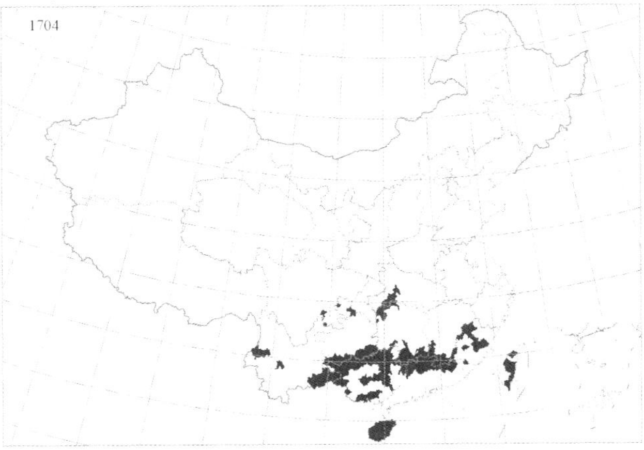

Ranunculaceae 毛茛科 *Clematis* 铁线莲属
Clematis lingyunensis W. T. Wang 凌云铁线莲
Vines, deciduous

19.0；10.2；25.8；19.0；168.4；0.0；958 ❙ 1250；675；60 ❙ 30.5；958；291.2

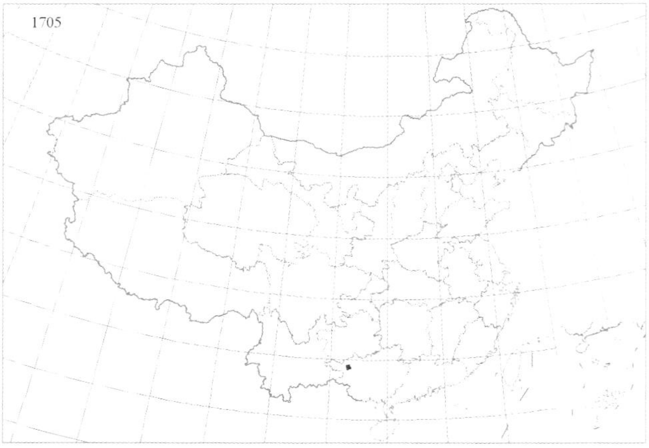

Ranunculaceae 毛茛科 *Clematis* 铁线莲属
Clematis longistyla Hand.-Mazz. 光柱铁线莲
Vines, deciduous

15.9 (15.1~17.1)；2.9 (1.7~4.0)；28.2 (27.3~29.4)；16.0 (15.2~17.1)；134.8 (126.5~146.1)；3.0 (1.0~5.4)；896 (848~952) ❙ 1195 (1042~1429)；517 (480~628)；111 (89~134) ❙ 33.8 (21.1~68.5)；888 (848~931)；156.6 (111.9~202.8)

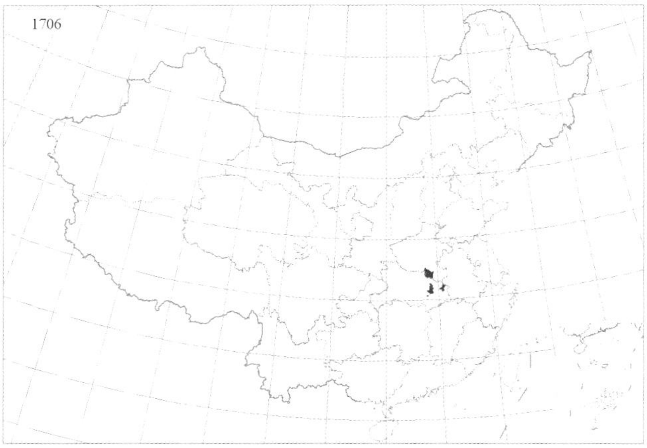

Ranunculaceae 毛茛科 *Clematis* 铁线莲属
Clematis loureiroana DC. 丝铁线莲
Vines, deciduous

21.4 (14.6~24.8)；13.3 (3.8~20.1)；27.9 (23.4~29.8)；21.4 (14.7~24.8)；197.2 (117.7~237.6)；0.0 (0.0~1.5)；1134 (779~1385) ❙ 1616 (868~2037)；698 (312~1044)；129 (41~197) ❙ 44.1 (-34.3~101.8)；1117 (779~1360)；256.0 (95.3~620.0)

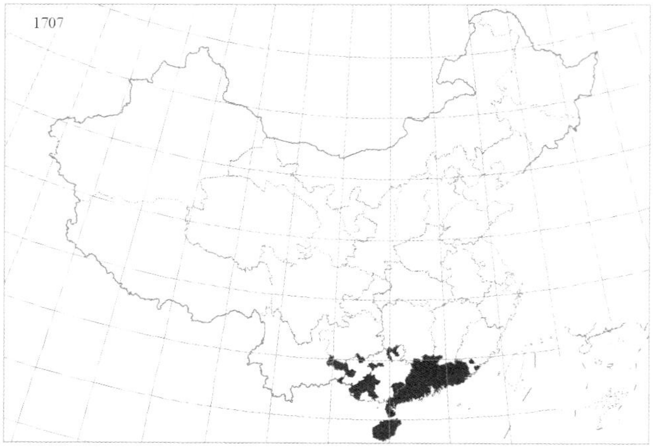

Ranunculaceae 毛茛科 *Clematis* 铁线莲属
Clematis macropetala Ledeb. 长瓣铁线莲
Vines, deciduous

8.0 (-4.0~15.1)；-7.6 (-28.2~1.8)；22.1 (9.7~28.0)；9.9 (2.9~15.2)；78.7 (11.3~128.0)；41.7 (5.6~146.2)；646 (359~880) ❙ 509 (84~882)；296 (53~499)；16 (1~53) ❙ -20.3 (-87.1~52.6)；496 (84~817)；167.8 (7.5~323.3)

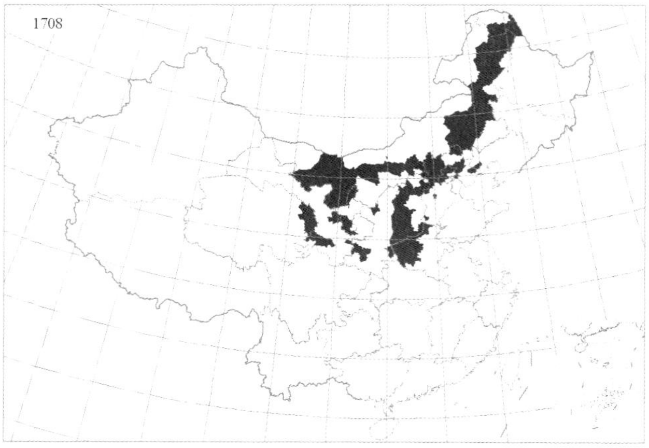

Ranunculaceae 毛茛科 *Clematis* 铁线莲属
Clematis macropetala var. albiflora (Maxim. ex Kuntze) Hand.-Mazz. 白花长瓣铁线莲
Vines, deciduous

6.3；-9.6；20.3；8.4；64.7；48.2；577 ❙ 502；301；13 ❙ -13.1；502；196.5

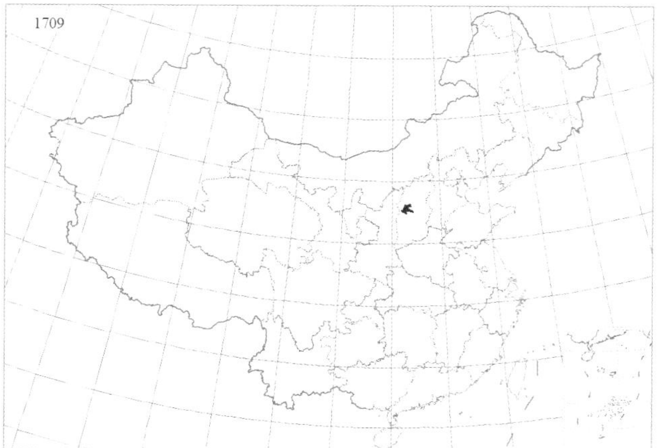

Ranunculaceae 毛茛科 *Clematis* 铁线莲属
Clematis mashanensis W. T. Wang 马山铁线莲
Vines, deciduous

21.0；11.9；28.2；21.1；193.0；0.0；1114 ❙ 1662；821；173 ❙ 49.2；1114；220.1

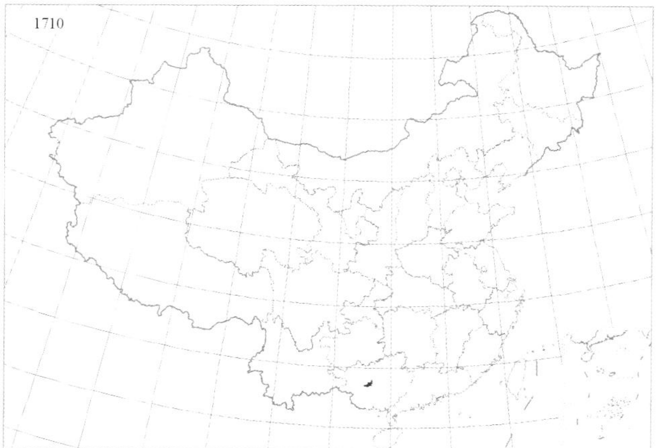

Ranunculaceae 毛茛科 *Clematis* 铁线莲属

Clematis menglaensis M. C. Chang 勐腊铁线莲

Vines, deciduous

19.1；13.0；23.1；19.1；169.6；0.0；922 ❙ 1487；729；68 ❙ 61.3；920；575.3

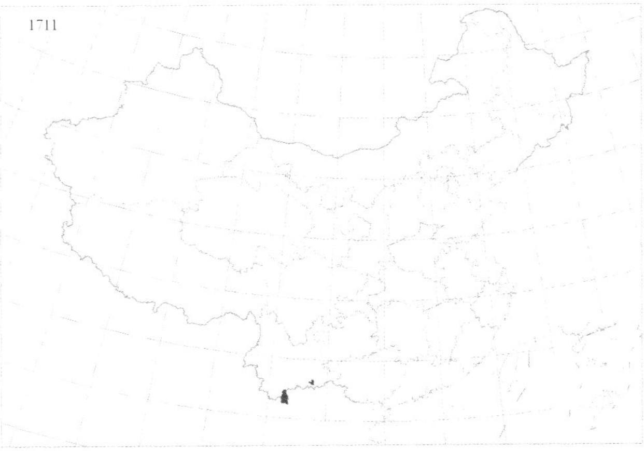

Ranunculaceae 毛茛科 *Clematis* 铁线莲属

Clematis metuoensis M. Y. Fang 墨脱铁线莲

Vines, deciduous

13.3；5.4；19.6；13.4；100.4；0.0；706 ❙ 1736；1007；51 ❙ 145.8；706；447.4

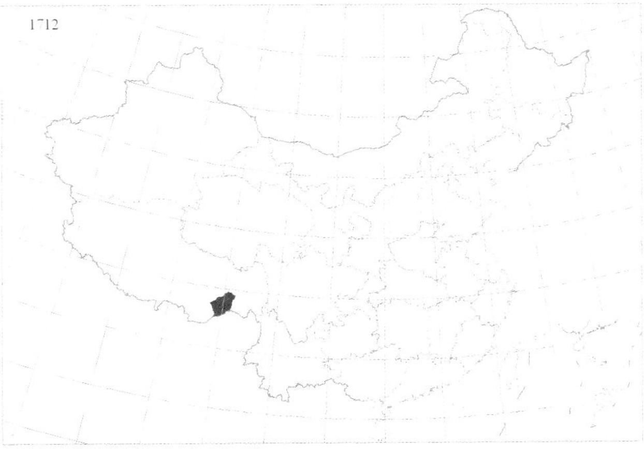

Ranunculaceae 毛茛科 *Clematis* 铁线莲属

Clematis meyeniana Walp. 毛柱铁线莲

Vines, deciduous

18.5 (11.1~24.8)；9.0 (1.1~20.1)；26.8 (19.4~29.8)；18.6 (11.1~24.8)；163.1 (79.8~237.6)；0.2 (0.0~8.5)；970 (648~1385) ❙ 1565 (955~3495)；637 (392~1847)；154 (30~897) ❙ 62.8 (-1.0~224.4)；964 (648~1273)；252.6 (126.1~659.0)

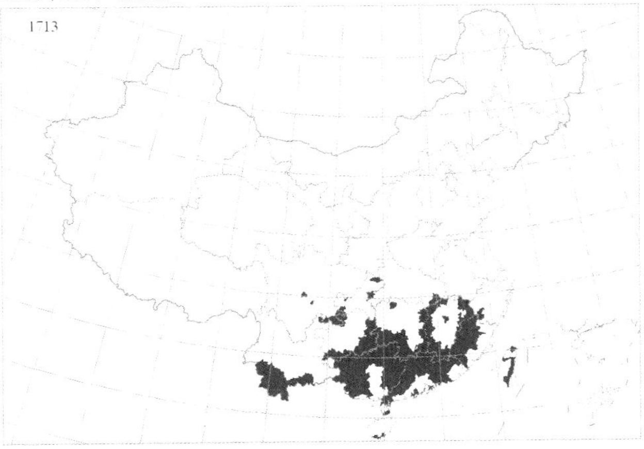

Ranunculaceae 毛茛科 *Clematis* 铁线莲属

Clematis meyeniana var. granulata Finet et Gagnep. 沙叶铁线莲

Vines, deciduous

2 .5 (17.9~24 8)；13.3 (7.4~20.1)；28.0 (25.0~29.1)；21.6 (18.0~24.8)；198.6 (155.6~237.6)；0.0 (0.0~0.0)；1138 (913~1385) ❙ 1676 (1188~2037)；771 (473~1056)；129 (41~194) ❙ 47.9 (1.5~85.5)；1133 (913~1257)；229.5 (95.3~551.6)

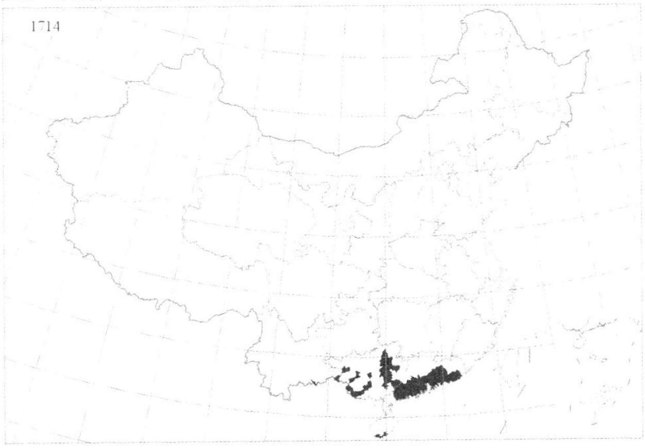

Ranunculaceae 毛茛科 *Clematis* 铁线莲属

Clematis montana Buch.-Ham. ex DC. 绣球藤

Vines, deciduous

14.0 (-1.8~21.7)；3.7 (-12.4~15.3)；23.2 (6.1~29.3)；14.2 (1.8~21.7)；115.0 (1.9~200.2)；6.8 (0.0~82.8)；774 (329~1118) ❙ 1145 (320~2859)；519 (190~1412)；85 (6~402) ❙ 47.1 (-40.9~235.6)；759 (320~1114)；248.0 (61.2~491.5)

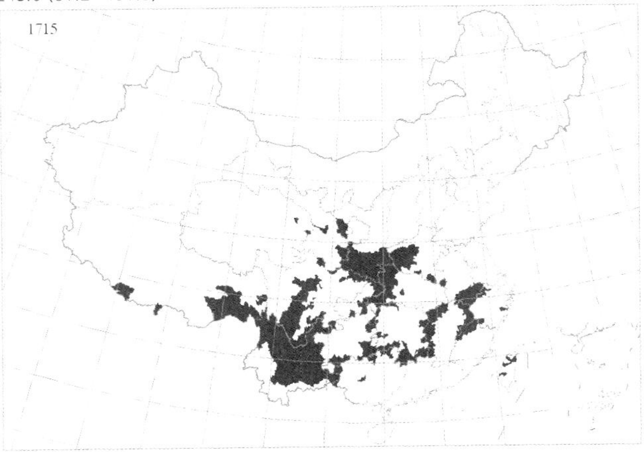

Ranunculaceae 毛茛科 *Clematis* 铁线莲属

Clematis montana var. glabrescens (H. F. Comber) W. T. Wang et M. C. Chang 毛果绣球藤

Vines, deciduous

7.1 (1.4~13.2)；-1.1 (-7.6~6.4)；14.3 (9.2~21.1)；7.8 (3.5~13.3)；48.4 (13.4~99.4)；22.1 (0.0~55.6)；538 (402~702) ❙ 927 (579~1337)；540 (348~727)；25 (10~69) ❙ 72.8 (38.4~125.9)；537 (402~702)；286.4 (169.5~407.1)

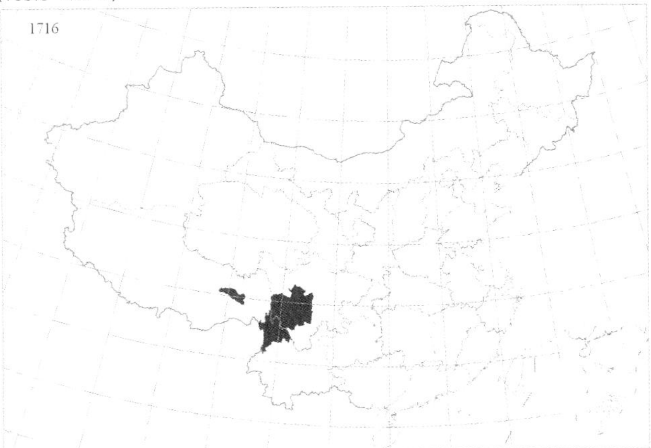

Ranunculaceae 毛茛科 *Clematis* 铁线莲属
Clematis montana var. longipes W. T. Wang 大花绣球藤
Vines, deciduous

11.8 (2.2~19.1); 2.6 (-6.6~10.1); 20.0 (9.9~26.5); 12.0 (3.8~19.1); 91.3 (15.5~169.2); 9.2 (0.0~49.1); 692 (413~966) ❙ 1156 (741~1492); 584 (426~816); 55 (15~113) ❙ 70.6 (30.0~125.9); 691 (413~966); 251.3 (170.4~352.3)

1717

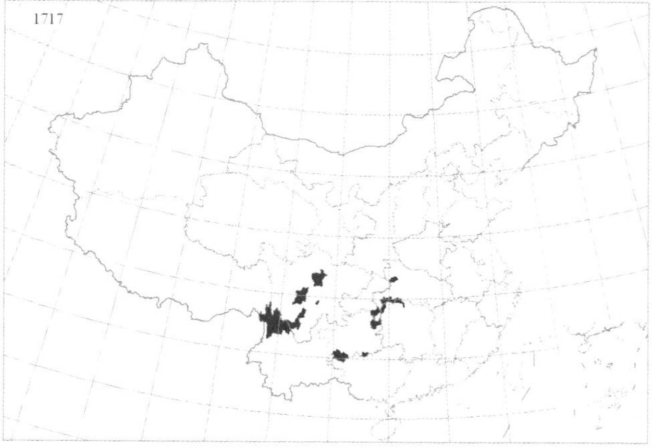

Ranunculaceae 毛茛科 *Clematis* 铁线莲属
Clematis montana var. sterilis Hand.-Mazz. 小叶绣球藤
Vines, deciduous

8.4 (2.2~15.4); 0.6 (-9.7~8.2); 15.0 (9.9~21.1); 8.8 (3.8~15.4); 57.1 (15.5~125.4); 16.0 (0.0~56.4); 570 (413~770) ❙ 969 (516~1337); 569 (283~727); 29 (9~69) ❙ 71.7 (14.7~125.9); 565 (413~696); 303.9 (200.8~407.1)

1718

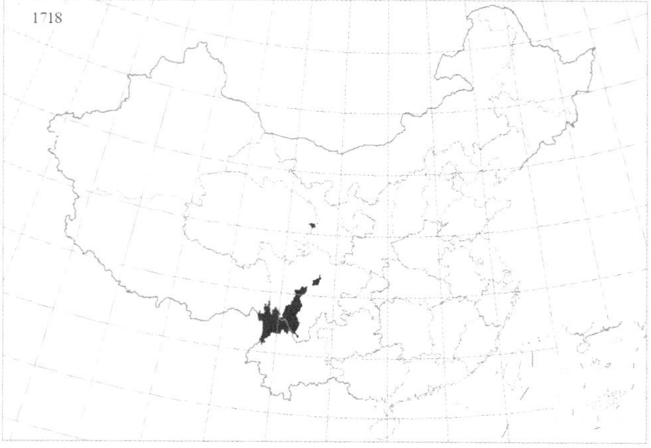

Ranunculaceae 毛茛科 *Clematis* 铁线莲属
Clematis morii Hayata 森氏铁线莲
Vines, deciduous

20.2 (16.2~24.1); 14.5 (11.0~18.9); 25.0 (20.1~28.2); 20.2 (16.2~24.1); 182.9 (134.2~229.7); 0.0 (0.0~0.0); 1019 (773~1322) ❙ 2402 (1498~3495); 1106 (750~1847); 255 (47~897) ❙ 146.2 (25.2~246.5); 983 (773~1144); 245.0 (119.4~363.7)

1719

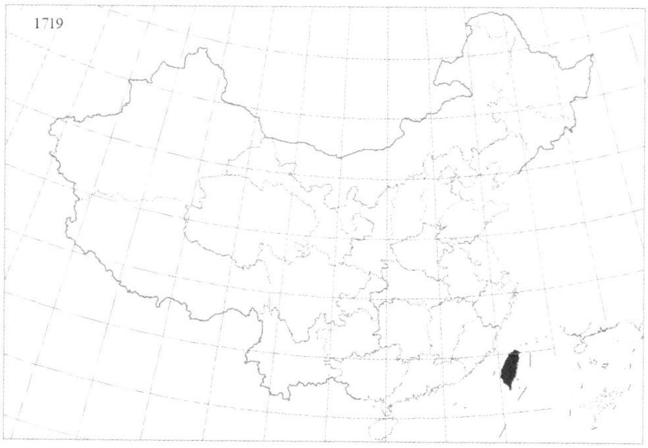

Ranunculaceae 毛茛科 *Clematis* 铁线莲属
Clematis nannophylla Maxim. 小叶铁线莲
Shrubs, deciduous

3.7 (-2.5~10.5); -9.3 (-16.1~-0.3); 15.2 (9.7~23.8); 6.1 (2.9~10.6); 39.9 (11.3~83.7); 54.6 (12.1~101.3); 488 (358~663) ❙ 409 (63~677); 225 (45~336); 8 (2~15) ❙ -14.5 (-84.9~45.4); 374 (63~644); 150.4 (8.4~313.2)

1720

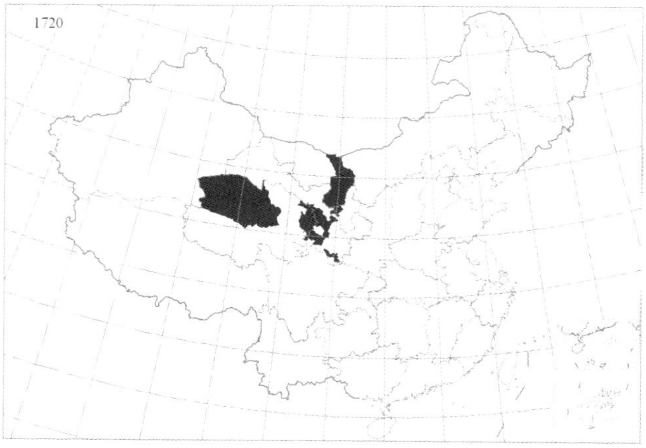

Ranunculaceae 毛茛科 *Clematis* 铁线莲属
Clematis nannophylla var. foliosa Maxim. 多叶铁线莲
Shrubs, deciduous

6.7; -4.7; 17.1; 7.7; 53.0; 32.3; 545 ❙ 612; 305; 13 ❙ 14.3; 514; 219.5

1721

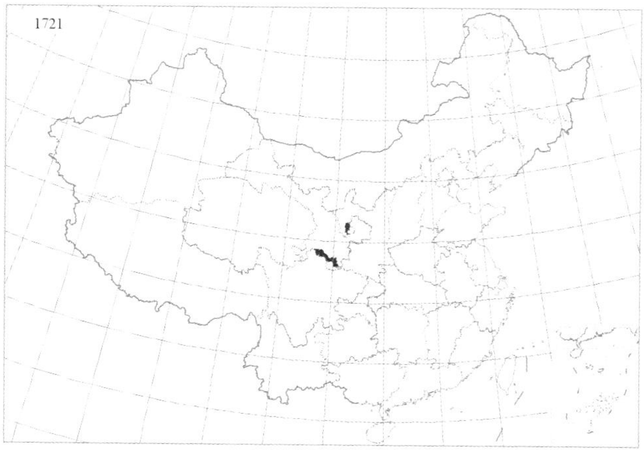

Ranunculaceae 毛茛科 *Clematis* 铁线莲属
Clematis napaulensis DC. 合苞铁线莲
Vines, deciduous

14.7 (-1.8~18.9); 7.7 (-10.2~12.5); 19.9 (6.1~25.0); 15.0 (1.8~19.0); 123.6 (1.9~167.5); 6.4 (0.0~82.8); 758 (329~939) ❙ 1145 (320~1736); 644 (217~1007); 48 (13~70) ❙ 51.7 (-3.0~145.8); 749 (320~939); 357.3 (72.2~539.7)

1722

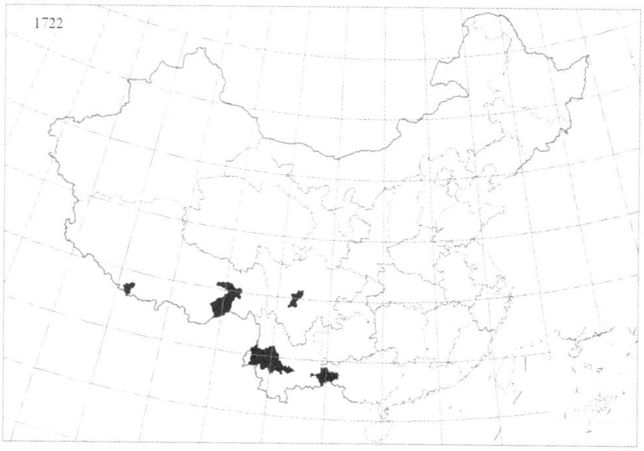

Ranunculaceae 毛茛科 *Clematis* 铁线莲属
Clematis nukiangensis M. Y. Fang 怒江铁线莲
Vines, deciduous

14.0 (8.6~17.9); 7.2 (1.3~11.6); 19.1 (15.0~22.1); 14.0 (8.7~17.9); 109.6 (52.6~155.1); 1.2 (0.0~8.8); 725 (574~858) ❙ 1201 (1041~1348); 665 (585~762); 57 (45~69) ❙ 67.5 (27.9~93.0); 723 (574~836); 360.4 (295.5~407.1)

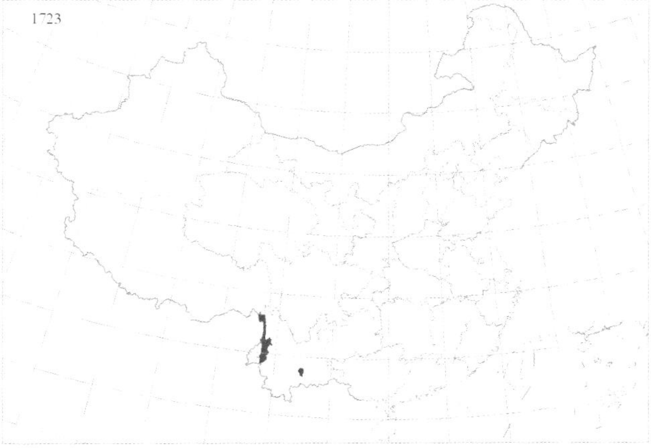

Ranunculaceae 毛茛科 *Clematis* 铁线莲属
Clematis obscura Maxim. 秦岭铁线莲
Vines, deciduous

10.6 (2.6~18.1); -2.2 (-7.6~7.5); 22.6 (11.2~28.6); 11.1 (4.3~18.2); 87.8 (20.0~158.0); 19.7 (0.0~48.2); 681 (427~962) ❙ 744 (483~1214); 351 (245~516); 27 (10~104) ❙ 11.8 (-36.2~121.7); 628 (427~962); 227.3 (126.1~334.4)

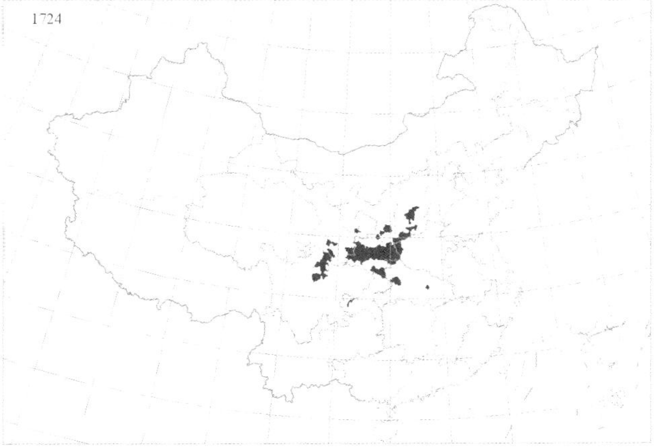

Ranunculaceae 毛茛科 *Clematis* 铁线莲属
Clematis parviloba Gardner et Champ. 裂叶铁线莲
Vines, deciduous

16.4 (5.4~21.9); 7.8 (-2.4~15.2); 23.6 (12.3~28.7); 16.5 (5.8~22.7); 138.5 (31.5~212.7); 1.0 (0.0~25.6); 850 (481~1216) ❙ 1348 (741~1965); 628 (426~994); 101 (12~235) ❙ 59.1 (12.5~147.0); 845 (481~1216); 287.1 (160.1~530.1)

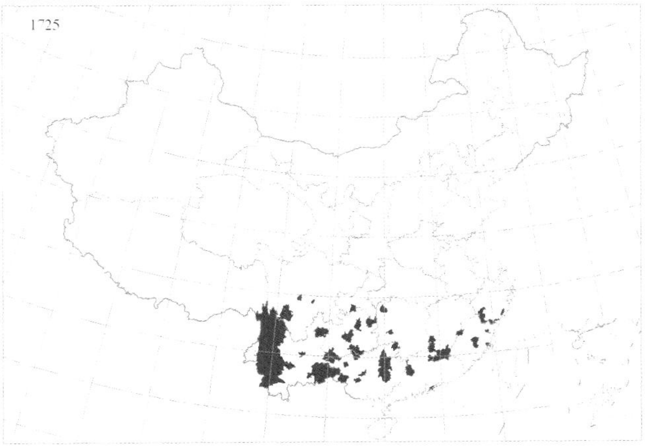

Ranunculaceae 毛茛科 *Clematis* 铁线莲属
Clematis parviloba var. bartletti (Yamam.) W. T. Wang 巴氏铁线莲
Vines, deciduous

17.2; 12.2; 21.2; 17.3; 147.1; 0.0; 818 ❙ 2688; 1120; 284 ❙ 228.9; 818; 305.6

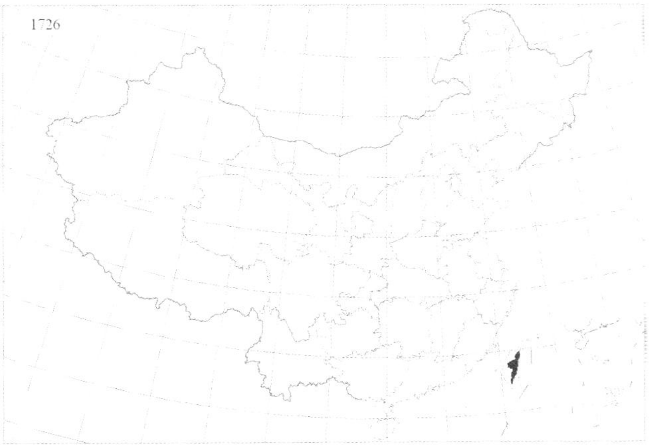

Ranunculaceae 毛茛科 *Clematis* 铁线莲属
Clematis pashanensis (M. C. Chang) W. T. Wang 巴山铁线莲
Vines

17.1 (15.1~18.7); 6.3 (4.1~7.9); 27.4 (25.6~28.8); 17.2 (15.1~18.7); 146.3 (122.0~164.3); 0.1 (0.0~0.9); 915 (813~994) ❙ 1132 (1020~1320); 478 (442~546); 55 (36~74) ❙ 24.3 (9.3~58.1); 915 (813~992); 146.3 (112.1~211.4)

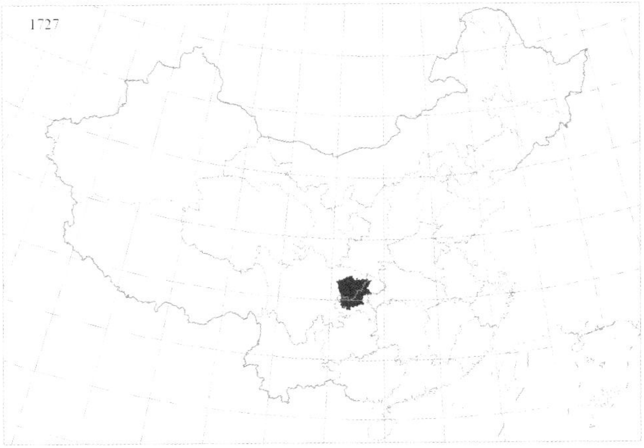

Ranunculaceae 毛茛科 *Clematis* 铁线莲属
Clematis peterae Hand.-Mazz. 钝萼铁线莲
Vines, deciduous

13.0 (2.2~20.5); 2.6 (-7.7~14.9); 22.4 (9.9~28.9); 13.2 (3.8~20.5); 105.5 (15.5~185.9); 8.5 (0.0~49.3); 736 (413~1038) ❙ 987 (491~1776); 488 (245~817); 46 (12~188) ❙ 35.4 (-36.2~125.9); 716 (413~1038); 239.6 (108.2~570.1)

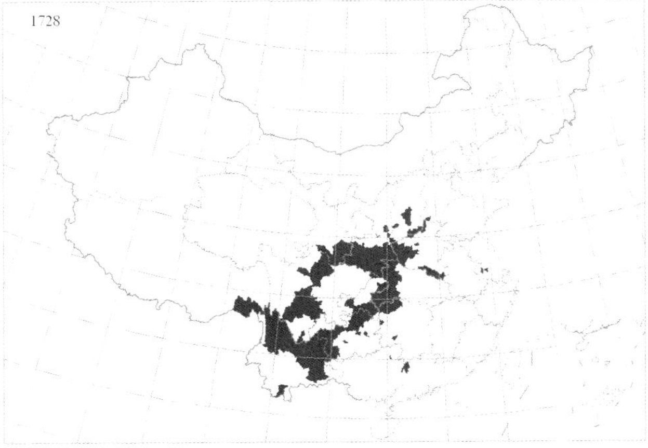

Ranunculaceae 毛茛科 *Clematis* 铁线莲属
Clematis pianmaensis W. T. Wang 片马铁线莲
Vines, deciduous

13.2; 6.4; 18.4; 13.3; 99.4; 0.0; 695 ❙ 1337; 727; 69 ❙ 92.4; 695; 407.1

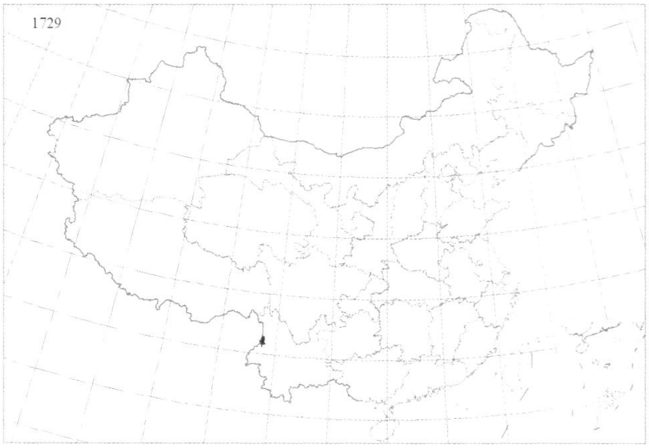

Ranunculaceae 毛茛科 *Clematis* 铁线莲属
Clematis pinnata Maxim. 羽叶铁线莲
Vines, deciduous

8.2 (5.0~11.0); -8.8 (-12.3~-5.7); 23.2 (20.2~25.5); 10.1 (8.0~12.0); 83.7 (61.6~101.7); 44.6 (29.5~61.5); 665 (568~757) ❙ 530 (397~694); 382 (254~499); 11 (10~13) ❙ -20.3 (-36.6~5.4); 527 (397~659); 218.5 (111.2~273.5)

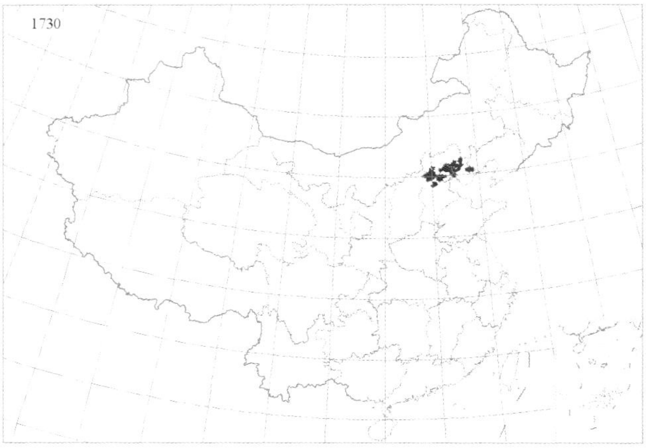

Ranunculaceae 毛茛科 *Clematis* 铁线莲属
Clematis potaninii Maxim. 美花铁线莲
Vines, deciduous

9.3 (-1.0~14.5); -2.4 (-11.8~2.3); 20.2 (8.5~28.2); 9.9 (2.7~14.6); 74.7 (8.3~125.3); 22.7 (4.9~79.5); 631 (363~864) ❙ 790 (516~1164); 379 (245~601); 25 (9~51) ❙ 31.1 (-36.2~128.6); 621 (363~786); 260.5 (131.4~357.7)

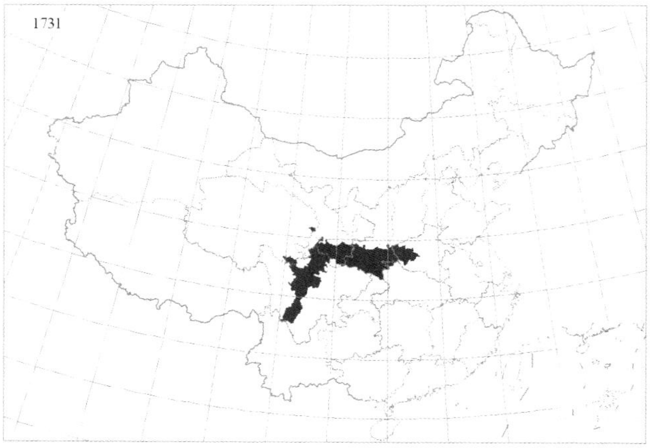

Ranunculaceae 毛茛科 *Clematis* 铁线莲属
Clematis pseudootophora M. Y. Fang 华中铁线莲
Vines, deciduous

10.2; -2.9; 22.4; 10.6; 83.1; 20.5; 656 ❙ 785; 381; 39 ❙ 19.6; 656; 274.1

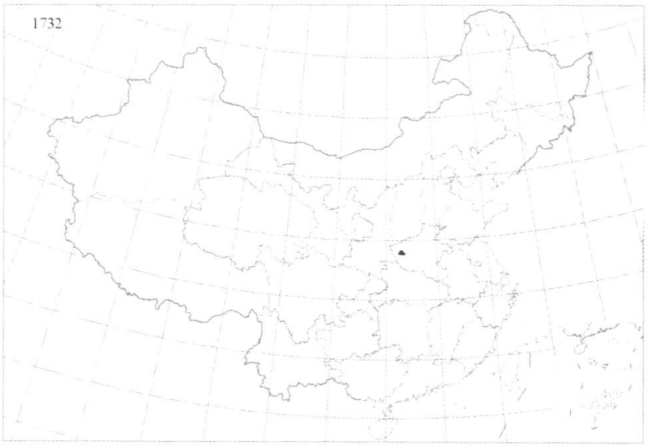

Ranunculaceae 毛茛科 *Clematis* 铁线莲属
Clematis pseudopogonandra Finet et Gagnep. 西南铁线莲
Vines, deciduous

6.4 (-0.3~14.6); -1.7 (-10.9~7.8); 13.4 (9.0~20.0); 7.3 (3.0~14.6); 44.6 (10.4~115.2); 27.1 (0.0~73.4); 519 (378~738) ❙ 878 (473~1243); 509 (287~696); 25 (7~58) ❙ 70.7 (5.0~125.9); 516 (378~707); 297.5 (169.5~384.7)

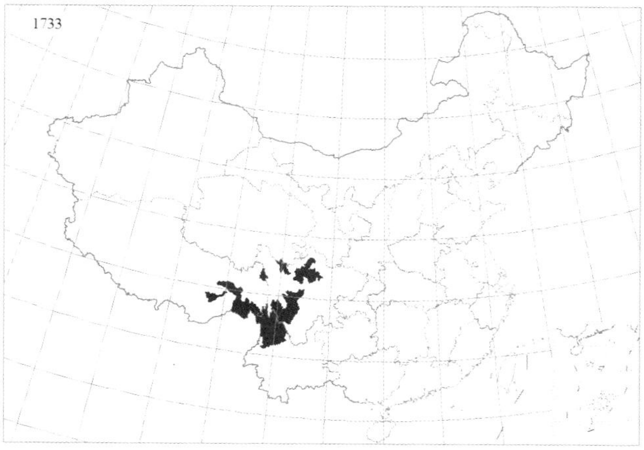

Ranunculaceae 毛茛科 *Clematis* 铁线莲属
Clematis psilandra Kitag. 光蕊铁线莲
Vines, deciduous

22.2; 17.7; 25.5; 22.2; 206.7; 0.0; 1123 ❙ 2992; 1847; 107 ❙ 166.5; 1123; 296.9

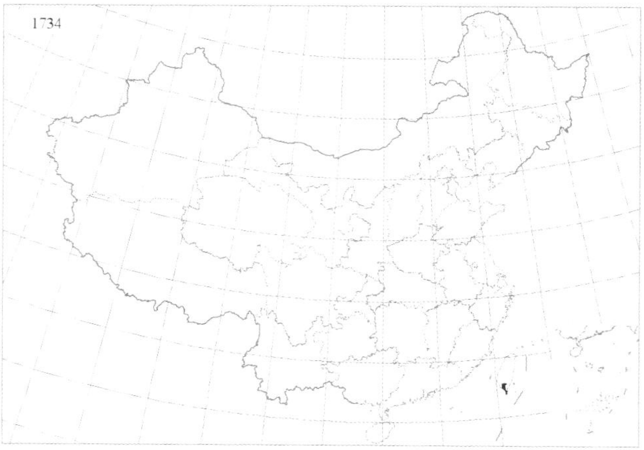

Ranunculaceae 毛茛科 *Clematis* 铁线莲属

Clematis puberula var. ganpiniana (H. Lév. et Vaniot) W. T. Wang 扬子铁线莲

Vines, deciduous

15.3 (1.8~21.0); 4.3 (-7.6~13.2); 25.4 (9.9~30.1); 15.4 (3.8~21.0); 128.3 (15.4~192.3); 3.9 (0.0~53.3); 838 (404~1081) ❙ 1319 (503~1997); 535 (299~828); 124 (9~259) ❙ 57.5 (-23.9~130.7); 822 (404~1081); 222.2 (127.2~491.5)

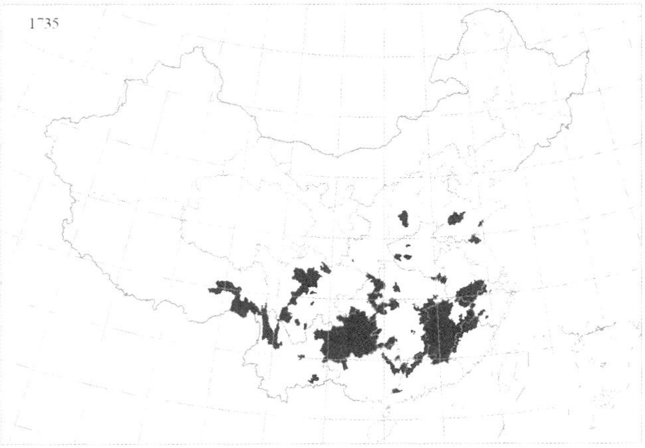

Ranunculaceae 毛茛科 *Clematis* 铁线莲属

Clematis qingchengshanica W. T. Wang 青城山铁线莲

Vines, deciduous

13.3; 3.2; 22.1; 13.2; 100.9; 2.5; 723 ❙ 1081; 606; 32 ❙ 49.4; 723; 171.0

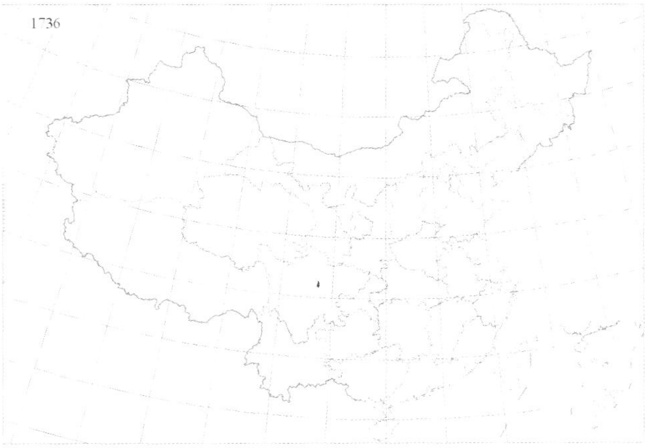

Ranunculaceae 毛茛科 *Clematis* 铁线莲属

Clematis quinquefoliolata Hutch. 五叶铁线莲

Vines, deciduous

14.3 (9.3~16.7); 3.3 (-1.7~7.3); 24.6 (19.4~27.1); 14.3 (9.5~16.7); 115.1 (69.8~140.6); 3.6 (0.0~18.0); 779 (603~872) ❙ 1194 (862~1466); 526 (376~816); 75 (33~123) ❙ 54.2 (7.8~106.2); 779 (603~872); 217.7 (143.4~285.1)

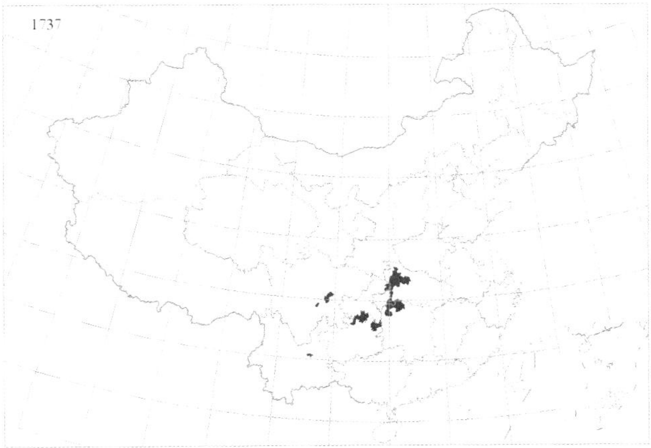

Ranunculaceae 毛茛科 *Clematis* 铁线莲属

Clematis ranunculoides Franch. 毛茛铁线莲

Vines, deciduous

7.2 (1.8~14.3); -0.7 (-7.6~7.6); 13.9 (9.5~20.2); 7.8 (3.8~14.3); 50.0 (15.5~111.9); 23.4 (0.0~54.8); 537 (411~729) ❙ 888 (591~1057); 533 (362~615); 22 (9~42) ❙ 69.2 (36.1~125.9); 534 (411~707); 281.6 (208.8~349.8)

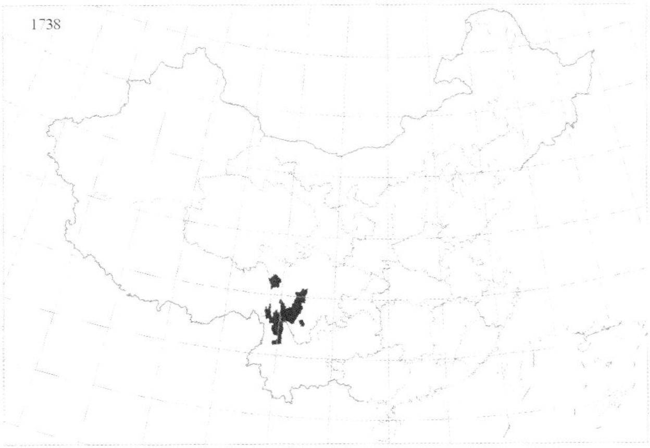

Ranunculaceae 毛茛科 *Clematis* 铁线莲属

Clematis rehderiana Craib 长花铁线莲

Vines, deciduous

3.4 (-3.3~14.8); -6.2 (-15.2~7.8); 12.0 (6.2~25.3); 5.3 (2.0~14.8); 29.3 (2.5~119.2); 48.0 (0.0~103.2); 455 (334~795) ❙ 735 (321~1478); 407 (193~594); 22 (4~88) ❙ 59.1 (-15.3~128.6); 452 (321~795); 251.5 (79.3~357.7)

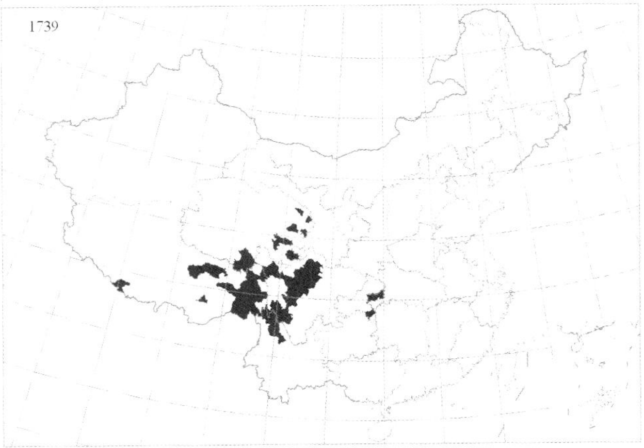

Ranunculaceae 毛茛科 *Clematis* 铁线莲属

Clematis rubifolia C. H. Wright 莓叶铁线莲

Vines, deciduous

18.0 (15.0~21.0); 11.3 (8.2~15.1); 22.5 (20.0~25.6); 18.0 (15.0~21.0); 155.8 (119.9~191.8); 0.0 (0.0~0.0); 872 (748~1053) ❙ 1307 (1007~1607); 686 (539~860); 56 (44~77) ❙ 49.7 (19.0~80.6); 866 (748~1053); 431.1 (277.0~659.0)

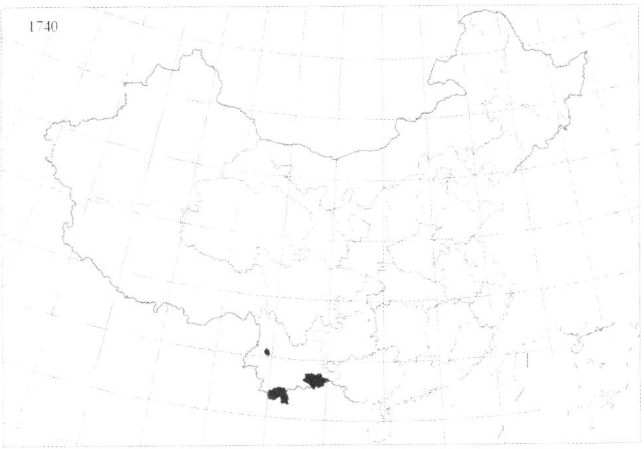

Ranunculaceae 毛茛科 *Clematis* 铁线莲属

Clematis serratifolia Rehder 齿叶铁线莲

Vines, deciduous

5.5 (1.9~8.7); -13.4 (-18.6~-8.3); 21.8 (18.5~24.6); 8.7 (6.7~10.4); 69.6 (47.6~87.1); 63.2 (40.2~88.8); 603 (513~681) ❙ 798 (566~1073); 500 (333~691); 27 (9~40) ❙ 32.4 (-4.6~74.1); 601 (513~681); 293.2 (201.5~335.3)

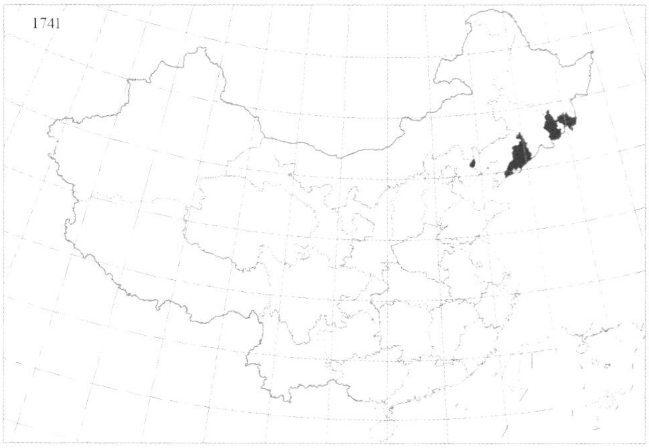

Ranunculaceae 毛茛科 *Clematis* 铁线莲属

Clematis shenlungchiaensis M. Y. Fang 神农架铁线莲

Vines, deciduous

10.7; -0.6; 21.6; 10.8; 82.2; 13.9; 650 ❙ 1166; 499; 73 ❙ 80.9; 650; 263.4

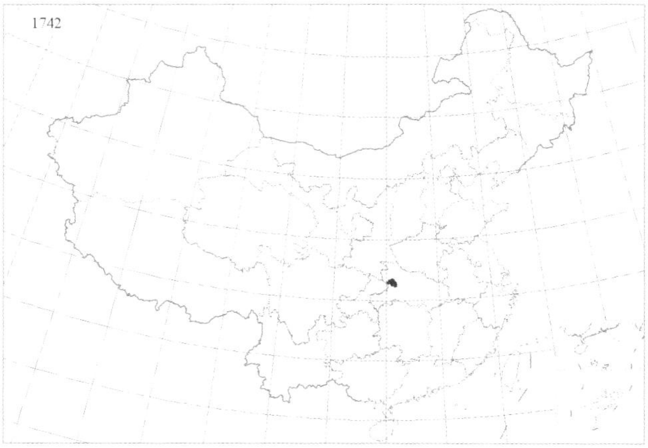

Ranunculaceae 毛茛科 *Clematis* 铁线莲属

Clematis shensiensis W. T. Wang 陕西铁线莲

Vines, deciduous

11.6 (7.2~14.8); -2.5 (-7.6~1.8); 24.4 (18.7~28.2); 12.0 (8.2~14.9); 98.3 (58.8~125.3); 18.9 (5.7~39.8); 727 (561~864) ❙ 651 (503~927); 325 (245~447); 26 (12~68) ❙ -9.7 (-36.7~52.6); 625 (503~844); 187.9 (111.8~318.2)

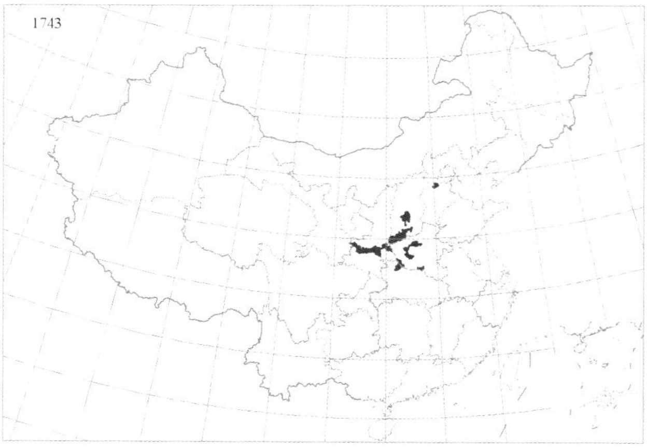

Ranunculaceae 毛茛科 *Clematis* 铁线莲属

Clematis siamensis Drumm. et Craib 锡金铁线莲

Vines, deciduous

17.2 (14.4~19.6); 10.5 (7.5~13.1); 21.5 (19.1~23.5); 17.2 (14.4~19.6); 146.5 (113.1~175.7); 0.0 (0.0~0.0); 832 (727~953) ❙ 1384 (1049~1688); 782 (594~959); 54 (42~70) ❙ 66.5 (26.1~112.2); 824 (727~932); 456.4 (346.6~627.0)

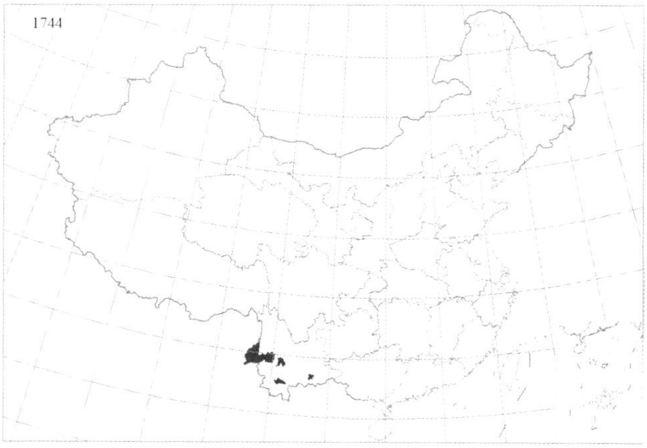

Ranunculaceae 毛茛科 *Clematis* 铁线莲属

Clematis sibirica (L.) Mill. 西伯利亚铁线莲

Subshrubs, deciduous

4.6 (-5.8~12.3); -14.1 (-29.8~-4.9); 20.4 (9.6~27.5); 8.3 (2.5~13.0); 65.4 (9.7~113.4); 69.3 (24.7~157.4); 595 (329~820) ❙ 237 (30~657); 128 (19~416); 14 (1~51) ❙ -56.7 (-96.3~35.3); 229 (30~632); 101.6 (2.1~323.4)

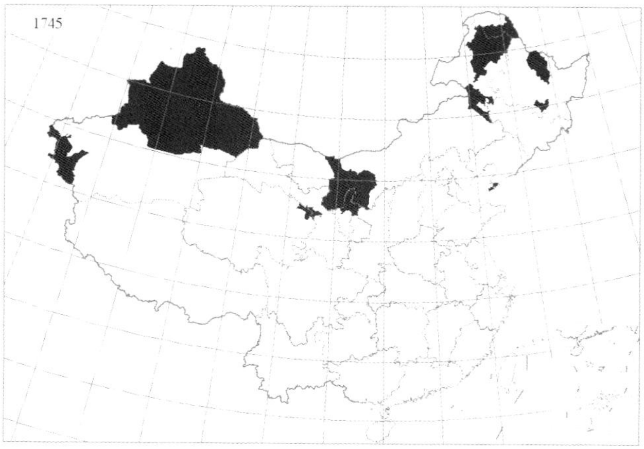

Ranunculaceae 毛茛科 *Clematis* 铁线莲属

Clematis sibirica var. ochotensis (Pall.) S. H. Li et Y. H. Huang 半钟铁线莲

Vines, deciduous

7.1 (-4.0~12.0); -9.9 (-28.2~-3.6); 22.2 (17.6~26.4); 9.4 (5.3~12.6); 75.9 (38.6~107.0); 49.6 (20.3~146.2); 631 (463~786) ❙ 431 (128~694); 282 (81~499); 11 (2~24) ❙ -30.9 (-80.8~27.4); 425 (128~659); 162.5 (13.6~322.3)

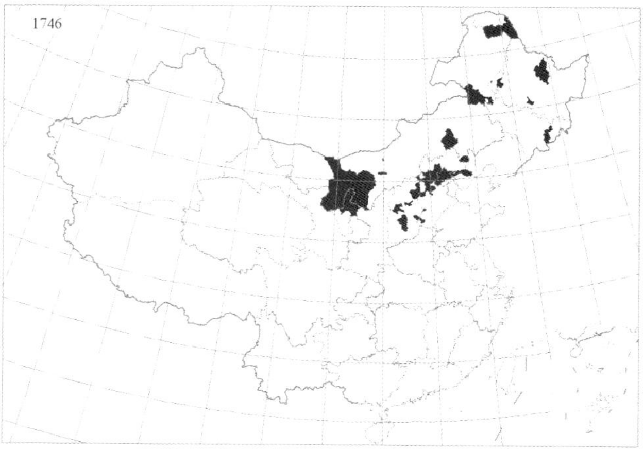

Ranunculaceae 毛茛科 *Clematis* 铁线莲属
Clematis smilacifolia Wall. 菝葜叶铁线莲
Vines, deciduous

18.8 (8.6~24.8); 11.7 (1.3~20.1); 23.9 (15.0~28.9); 18.8 (8.7~24.8); 165.9 (52.6~237.6); 0.1 (0.0~8.8); 939 (574~1385) ❙ 1356 (958~2435); 721 (473~1428); 66 (41~173) ❙ 44.7 (1.5~112.2); 923 (574~1257); 358.9 (186.7~659.0)

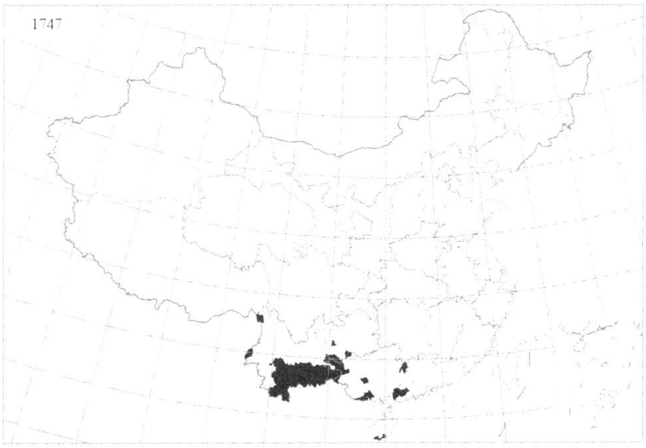

Ranunculaceae 毛茛科 *Clematis* 铁线莲属
Clematis smilacifolia var. peltata (W. T. Wang) W. T. Wang 盾叶铁线莲
Vines, deciduous

20.5; 15.0; 23.7; 20.5; 185.9; 0.0; 990 ❙ 1536; 678; 72 ❙ 55.1; 973; 614.6

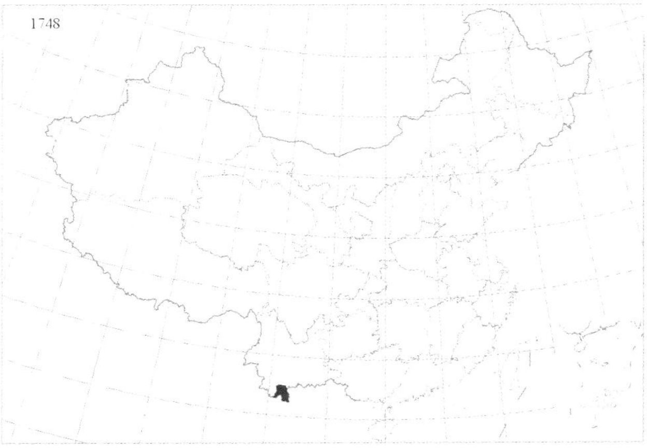

Ranunculaceae 毛茛科 *Clematis* 铁线莲属
Clematis songorica Bunge 准噶尔铁线莲
Shrubs, deciduous

4.6 (-4.3~12.3); -14.1 (-20.9~-5.3); 20.1 (9.6~27.5); 8.3 (2.5~13.0); 65.0 (9.7~112.3); 68.9 (24.7~121.1); 597 (329~813) ❙ 178 (44~450); 82 (19~211); 16 (1~51) ❙ -67.7 (-94.5~3.9); 177 (44~406); 76.3 (5.1~297.3)

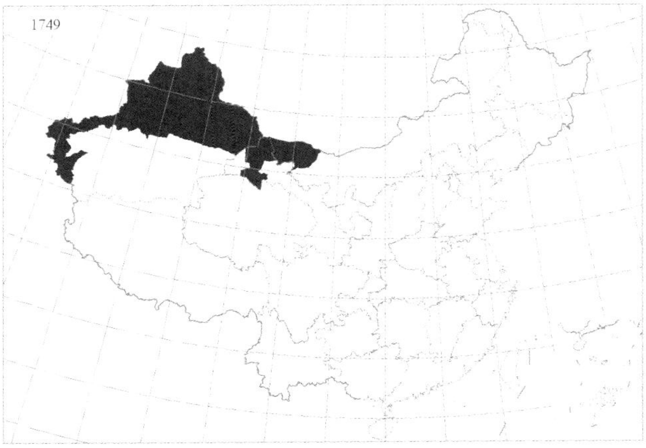

Ranunculaceae 毛茛科 *Clematis* 铁线莲属
Clematis songorica var. aspleniifolia (Schrenk) Trautv. 蕨叶铁线莲
Shrubs, deciduous

7.1; -11.5; 21.2; 9.6; 77.3; 51.5; 647 ❙ 126; 71; 7 ❙ -78.6; 126; 56.0

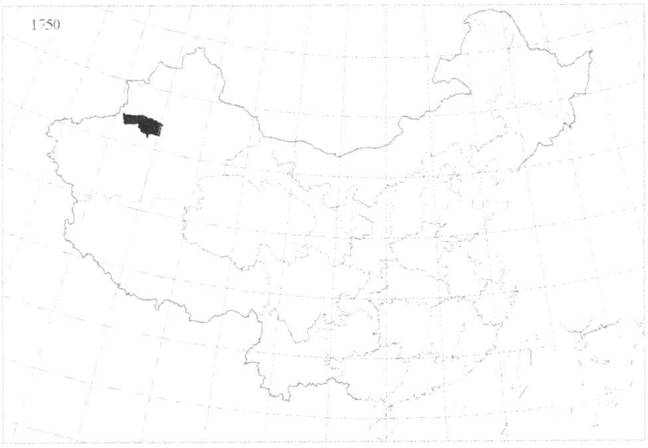

Ranunculaceae 毛茛科 *Clematis* 铁线莲属
Clematis tangutica (Maxim.) Korsh. 甘青铁线莲
Vines, deciduous

3.0 (-5.7~13.3); -10.3 (-20.1~3.2); 14.8 (5.8~26.8); 5.9 (1.3~13.2); 39.9 (1.2~114.1); 62.8 (2.5~133.0); 489 (294~811) ❙ 380 (21~1081); 216 (15~606); 9 (0~41) ❙ -12.5 (-97.3~128.6); 294 (21~723); 152.3 (2.8~364.1)

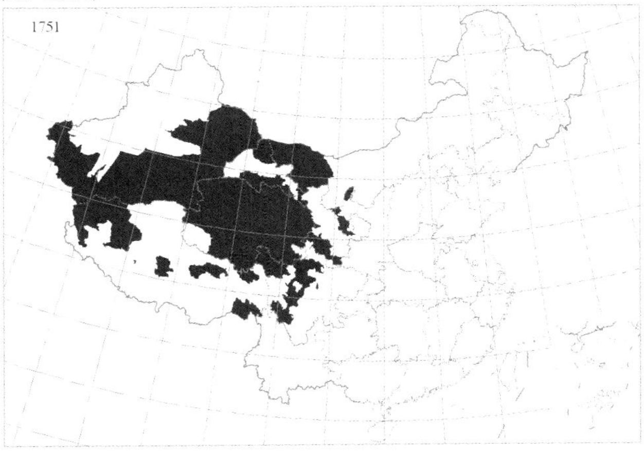

Ranunculaceae 毛茛科 *Clematis* 铁线莲属
Clematis tashiroi Maxim. 长萼铁线莲
Vines, deciduous

19.9 (16.4~24.1); 14.2 (11.2~18.9); 24.7 (20.3~28.0); 19.9 (16.5~24.1); 178.8 (137.8~229.7); 0.0 (0.0~0.0); 992 (783~1322) ❙ 2593 (1751~3495); 1107 (821~1847); 319 (47~897) ❙ 168.7 (32.5~235.6); 969 (783~1123); 257.3 (154.2~363.7)

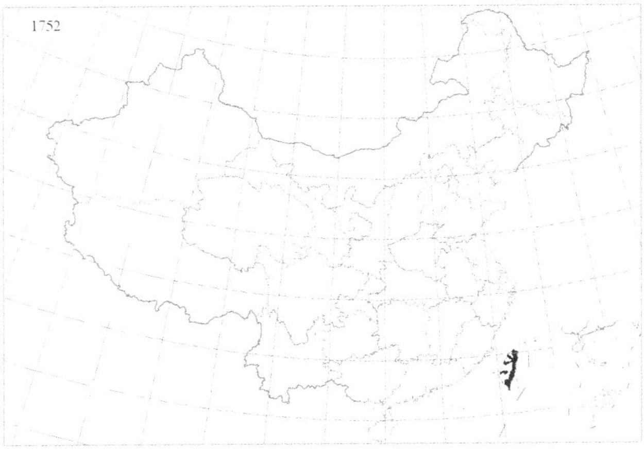

Ranunculaceae 毛茛科 *Clematis* 铁线莲属

Clematis tatarinowii Maxim. 细花铁线莲

Subshrubs or vines, deciduous

10.3 (7.7~12.1); -6.3 (-9.2~-4.1); 24.9 (22.8~26.2); 11.5 (9.7~12.7); 96.6 (80.8~107.9); 33.0 (22.7~47.4); 731 (648~793) ❙ 557 (492~606); 425 (359~479); 11 (10~12) ❙ -23.7 (-29.5~-18.2); 557 (492~606); 218.8 (167.2~263.7)

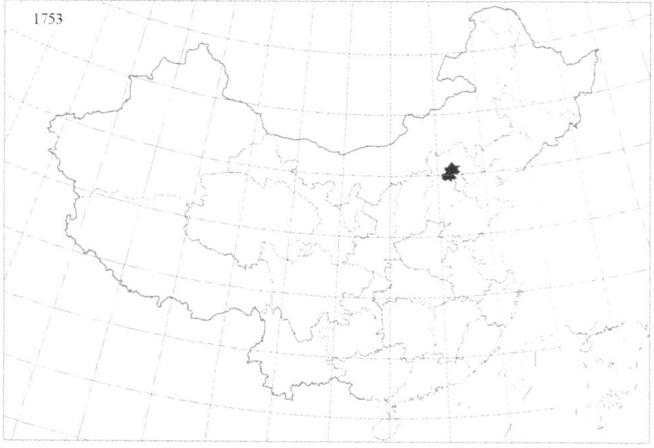

Ranunculaceae 毛茛科 *Clematis* 铁线莲属

Clematis terniflora DC. 圆锥铁线莲

Vines, deciduous

14.4 (-0.2~24.1); 1.4 (-23.8~18.9); 26.6 (18.4~29.7); 14.7 (6.4~24.1); 123.4 (45.8~229.7); 10.6 (0.0~112.6); 833 (501~1322) ❙ 1132 (397~3495); 496 (254~1847); 111 (9~897) ❙ 34.2 (-43.4~246.5); 802 (397~1144); 200.1 (110.1~363.7)

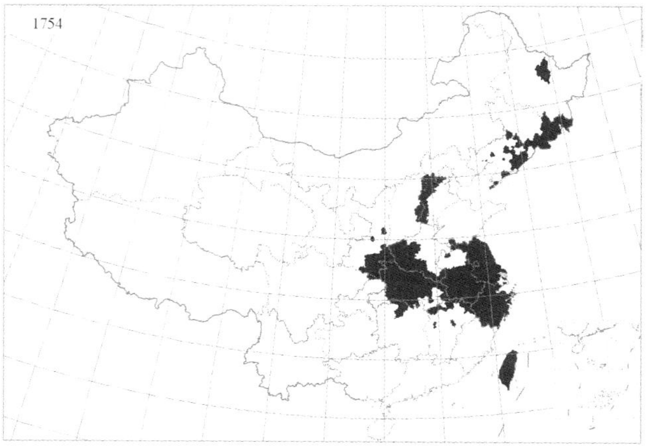

Ranunculaceae 毛茛科 *Clematis* 铁线莲属

Clematis tibetana var. vernayi (C. E. C. Fisch.) W. T. Wang 厚萼中印铁线莲

Vines, deciduous

0.5 (-2.3~3.7); -8.8 (-11.8~-5.5); 9.0 (6.1~12.1); 3.2 (1.7~5.1); 12.8 (1.9~27.8); 66.0 (41.8~91.0); 376 (324~441) ❙ 355 (212~620); 239 (156~348); 11 (0~103) ❙ -5.1 (-35.9~91.4); 313 (212~431); 146.3 (46.9~280.0)

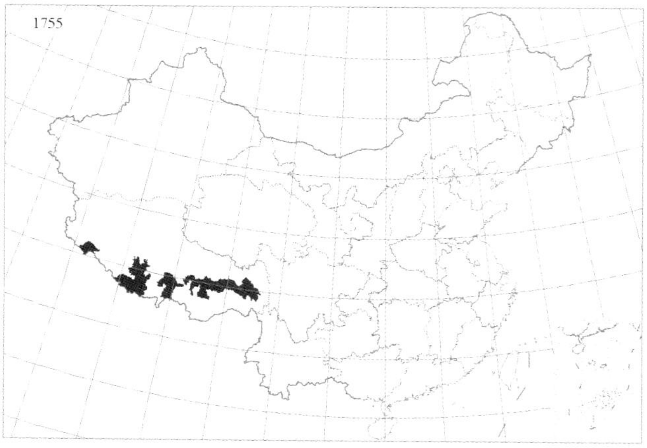

Ranunculaceae 毛茛科 *Clematis* 铁线莲属

Clematis tsugetorum Ohwi 高山铁线莲

Shrubs, deciduous

17.3; 11.6; 22.1; 17.3; 148.1; 0.0; 830 ❙ 2631; 1015; 318 ❙ 218.2; 830; 279.8

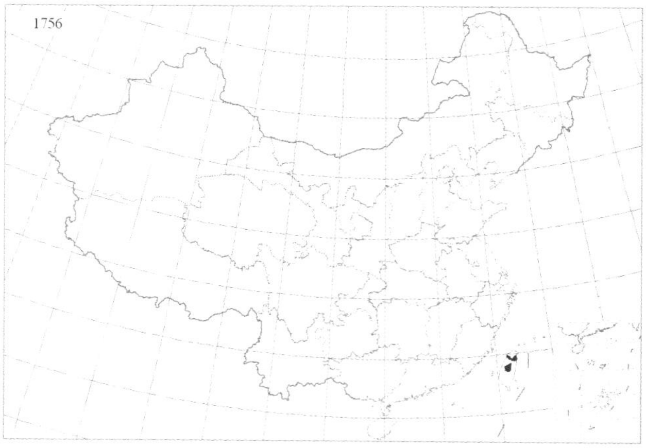

Ranunculaceae 毛茛科 *Clematis* 铁线莲属

Clematis uncinata Turcz. 柱果铁线莲

Vines, deciduous

17.2 (9.3~24.1); 6.7 (-2.9~18.9); 26.8 (19.8~30.1); 17.3 (9.5~24.1); 148.3 (69.8~229.7); 1.4 (0.0~20.6); 918 (603~1322) ❙ 1463 (644~3495); 589 (299~1847); 147 (14~897) ❙ 60.0 (-12.2~246.5); 909 (603~1247); 233.4 (108.2~551.6)

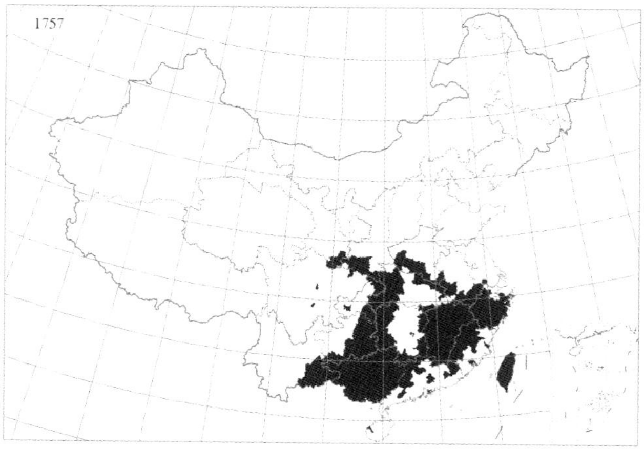

Ranunculaceae 毛茛科 *Clematis* 铁线莲属

Clematis uncinata var. coriacea Pamp. 皱叶铁线莲

Vines, deciduous

12.2 (5.7~16.3); 0.2 (-3.5~4.8); 23.5 (14.2~27.5); 12.3 (6.4~16.3); 98.0 (38.1~136.2); 11.5 (0.2~28.1); 719 (504~877) ❙ 911 (738~1444); 404 (350~551); 43 (17~140) ❙ 26.9 (5.0~83.0); 719 (504~877); 255.1 (186.2~334.4)

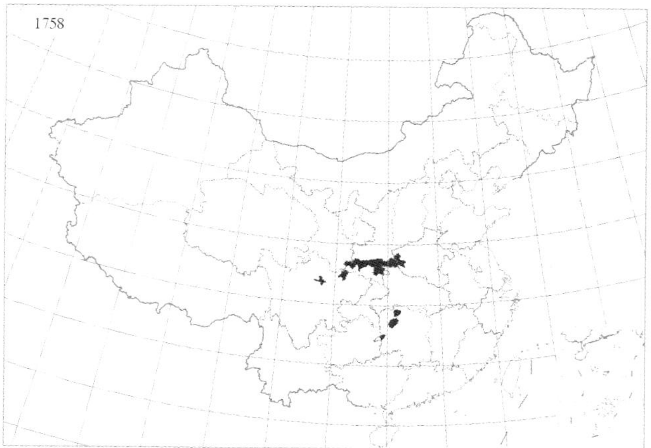

Ranunculaceae 毛茛科 *Clematis* 铁线莲属
Clematis uncinata var. okinawensis (Ohwi) Ohwi 毛柱果铁线莲
Vines, deciduous

14.8 (2.7~24.1); 4.8 (-6.3~18.9); 24.1 (10.4~29.6); 14.9 (4.2~24.1); 125.3 (18.9~229.7); 7.2 (0.0~46.9); 821 (419~1322) ❙ 1421 (579~3495); 634 (301~1847); 124 (10~897) ❙ 69.4 (-8.1~246.5); 809 (419~1144); 245.3 (119.4~363.7)

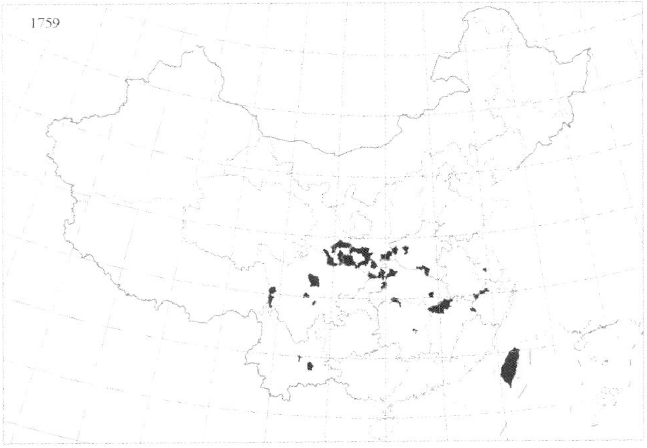

Ranunculaceae 毛茛科 *Clematis* 铁线莲属
Clematis urophylla Franch. 尾叶铁线莲
Vines, deciduous

15.9 (2.2~21.3); 6.3 (-6.6~14.2); 24.1 (9.9~28.6); 15.9 (3.8~21.3); 133.5 (15.5~196.1); 2.7 (0.0~49.1); 834 (413~1130) ❙ 1355 (932~1758); 606 (450~860); 102 (19~221) ❙ 64.7 (13.3~125.9); 834 (413~1130); 263.2 (70.4~551.6)

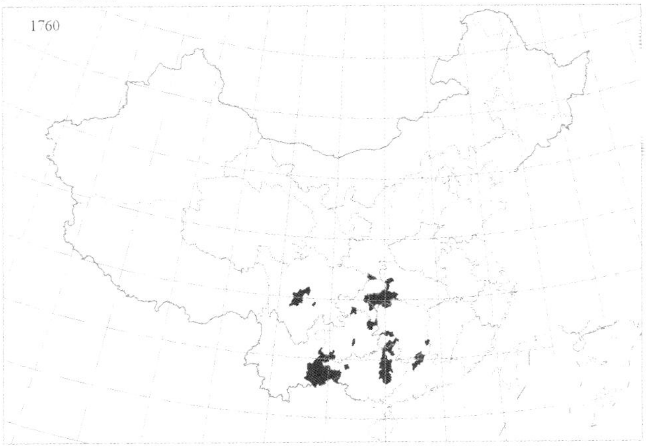

Ranunculaceae 毛茛科 *Clematis* 铁线莲属
Clematis vaniotii H. Lév. 云贵铁线莲
Vines, deciduous

17.2 (15.1~18.9); 8.5 (5.0~10.6); 24.0 (21.4~26.6); 17.2 (15.2~19.0); 146.6 (121.9~167.5); 0.0 (0.0~0.0); 859 (788~969) ❙ 1246 (1018~1504); 652 (507~860); 60 (47~107) ❙ 45.3 (25.7~80.6); 859 (788~969); 284.6 (184.8~413.5)

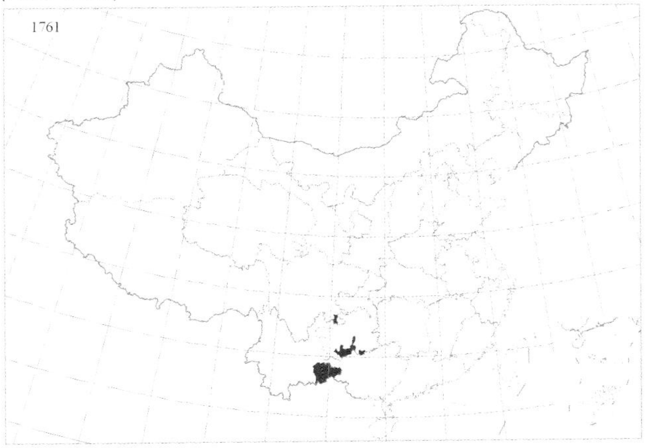

Ranunculaceae 毛茛科 *Clematis* 铁线莲属
Clematis venusta M. C. Chang 丽叶铁线莲
Vines, deciduous

8.5 (5.4~10.8); 1.2 (-2.1~3.9); 14.7 (12.3~16.6); 8.6 (5.8~10.9); 53.5 (31.5~71.8); 11.3 (1.3~25.6); 567 (481~633) ❙ 893 (741~1007); 539 (426~603); 26 (15~40) ❙ 57.6 (51.8~67.9); 565 (481~633); 305.6 (273.5~351.7)

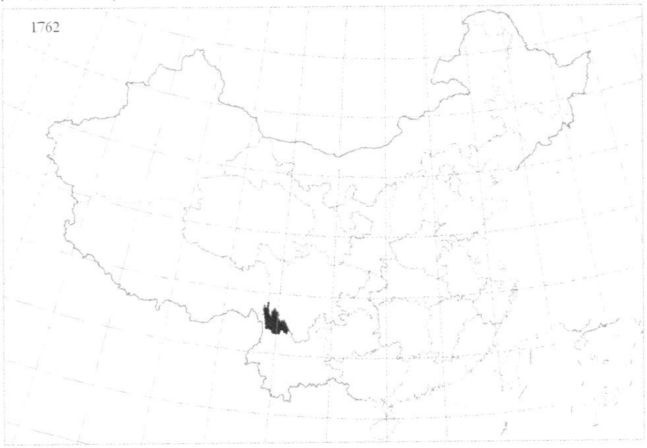

Ranunculaceae 毛茛科 *Clematis* 铁线莲属
Clematis viridis (W. T. Wang et M. C. Chang) W. T. Wang 绿叶铁线莲
Shrubs, deciduous

1.4 (-0.3~2.7); -8.4 (-10.9~-7.1); 10.1 (9.0~11.1); 3.8 (3.0~4.4); 15.7 (10.4~20.5); 58.3 (48.0~73.4); 407 (378~429) ❙ 715 (543~943); 404 (330~507); 16 (9~23) ❙ 75.8 (28.5~128.6); 407 (378~429); 303.4 (270.7~341.6)

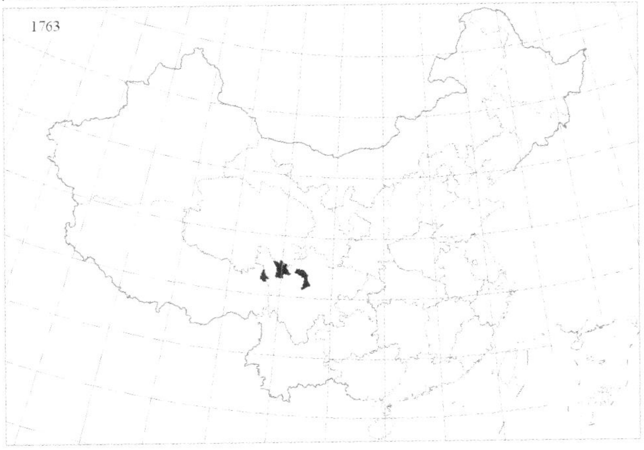

Ranunculaceae 毛茛科 *Clematis* 铁线莲属
Clematis yuanjiangensis W. T. Wang 元江铁线莲
Vines, deciduous

19.3; 13.1; 23.6; 19.4; 173.2; 0.0; 939 ❙ 1143; 637; 50 ❙ 21.7; 903; 346.9

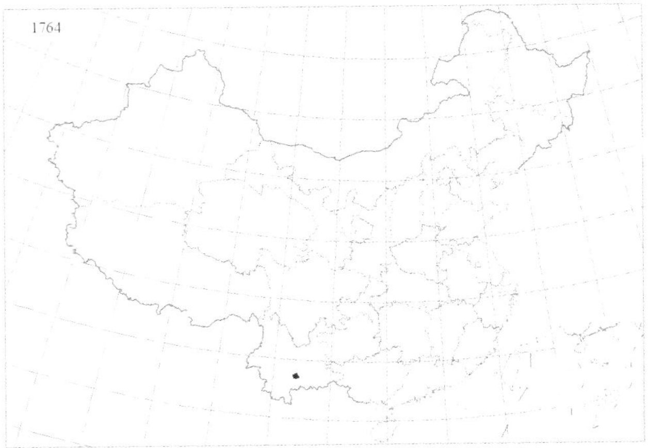

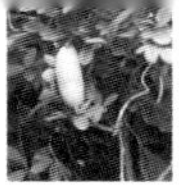

Ranunculaceae 毛茛科 *Clematis* 铁线莲属

Clematis yui W. T. Wang 俞氏铁线莲

Vines, deciduous

8.6；1.3；15.0；8.7；52.6；8.8；574 ❙ 1041；585；45 ❙ 81.5；574；352.3

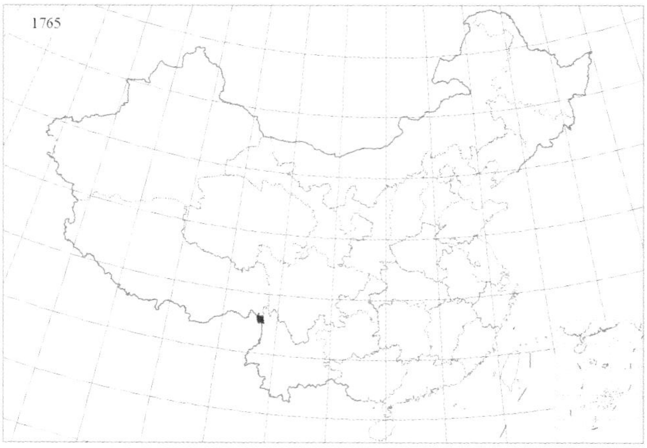

Ranunculaceae 毛茛科 *Clematis* 铁线莲属

Clematis yunnanensis Franch. 云南铁线莲

Vines, deciduous

14.0 (1.4~21.0)；6.4 (-7.6~14.2)；20.0 (9.2~26.6)；14.1 (3.5~21.0)；112.1 (13.4~191.8)；3.9 (0.0~55.6)；737 (402~1053) ❙ 1056 (579~1736)；602 (348~1007)；39 (10~70) ❙ 45.5 (-14.2~145.8)；722 (402~1053)；300.1 (142.9~551.6)

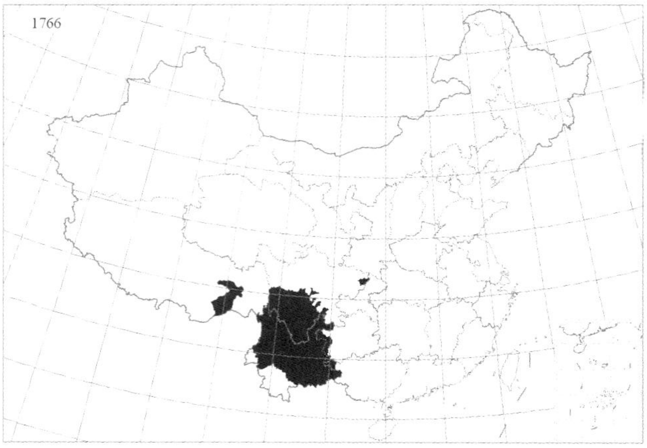

Ranunculaceae 毛茛科 *Naravelia* 锡兰莲属

Naravelia pilulifera Hance 两广锡兰莲

Vines

22.2 (20.1~24.8)；14.5 (10.7~20.1)；28.1 (27.1~28.8)；22.3 (20.1~24.8)；207.2 (181.1~237.6)；0.0 (0.0~0.0)；1188 (1048~1385) ❙ 1447 (868~1876)；656 (312~891)；97 (41~168) ❙ 23.4 (-34.3~63.2)；1117 (868~1190)；304.9 (203.4~527.5)

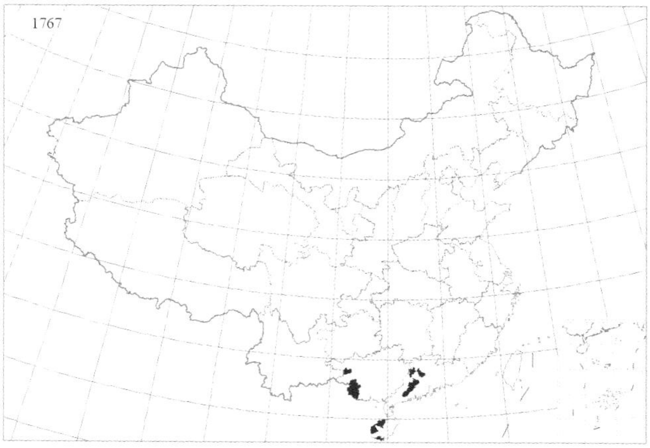

Ranunculaceae 毛茛科 *Naravelia* 锡兰莲属

Naravelia zeylanica (L.) DC. 锡兰莲

Vines

20.5；15.0；23.7；20.5；185.9；0.0；990 ❙ 1536；678；72 ❙ 55.1；973；614.6

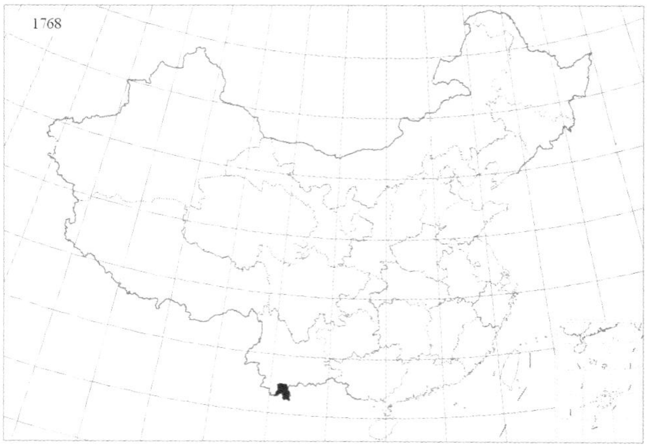

Lardizabalaceae 木通科 *Akebia* 木通属

Akebia chingshuiensis T. Shimizu 清水山木通

Vines, deciduous or subevergreen

16.4；11.5；20.3；16.5；137.8；0.0；783 ❙ 2499；947；303 ❙ 219.3；783；303.6

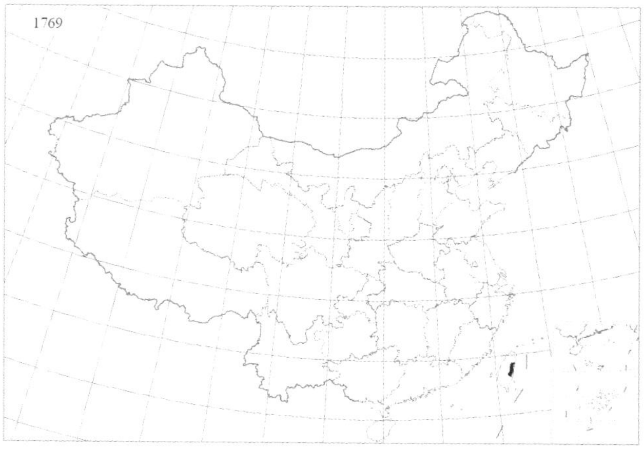

Lardizabalaceae 木通科 *Akebia* 木通属

Akebia longeracemosa Matsum. 长序木通

Vines, subvergreen

18.4 (17.1~21.7)；9.7 (6.9~15.3)；26.1 (21.8~27.5)；18.4 (17.1~21.7)；161.0 (145.0~200.2)；0.0 (0.0~0.0)；942 (816~1118) ❙ 1754 (1486~2382)；730 (505~1045)；181 (120~244) ❙ 89.6 (60.2~191.8)；934 (816~1075)；220.6 (188.8~281.4)

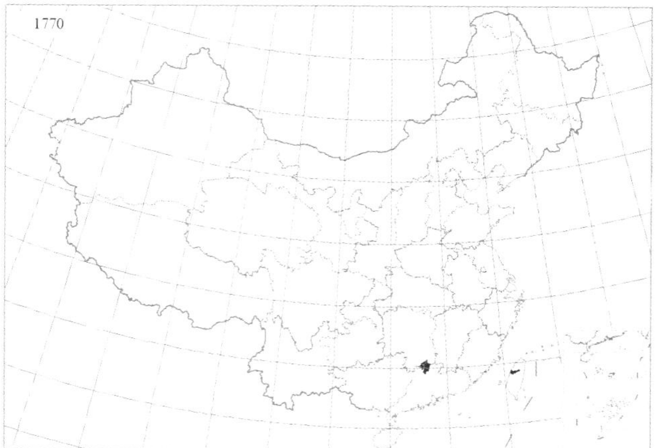

Lardizabalaceae 木通科 *Akebia* 木通属
Akebia quinata (Houtt.) Decne. 木通
Vines, deciduous

15.4 (6.2~21.8); 3.0 (-3.7~14.2); 27.0 (14.1~29.9); 15.5 (6.7~21.9); 130.1 (39.4~202.6); 5.1 (0.0~24.1); 857 (509~1150) ❙ 1207 (512~1997); 489 (258~891); 127 (17~259) ❙ 39.5 (-35.3~147.0); 831 (509~1150); 198.0 (108.2~334.4)

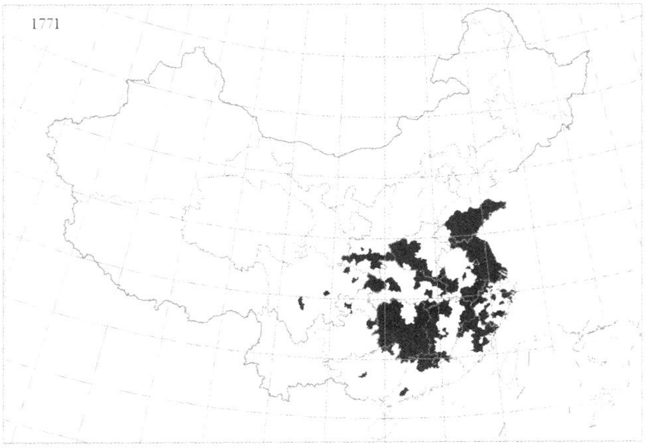

Lardizabalaceae 木通科 *Akebia* 木通属
Akebia trifoliata (Thunb.) Koidz. 三叶木通
Vines, deciduous

15.5 (2.6~22.7); 3.8 (-10.1~15.0); 26.3 (11.2~30.1); 15.6 (4.3~22.8); 130.8 (20.0~213.1); 4.3 (0.0~50.4); 853 (427~1221) ❙ 1228 (509~2435); 499 (245~1423); 112 (13~259) ❙ 42.8 (-36.5~130.7); 826 (427~1221); 207.7 (108.2~411.1)

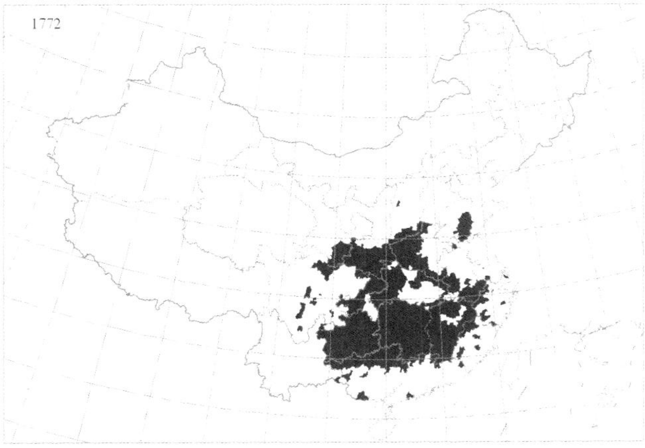

Lardizabalaceae 木通科 *Akebia* 木通属
Akebia trifoliata subsp. australis (Diels) T. Shimizu 白木通
Vines, deciduous

16.0 (5.4~21.3); 4.9 (-4.9~14.2); 26.2 (12.3~30.1); 16.1 (5.8~21.3); 135.2 (31.5~196.1); 2.6 (0.0~26.7); 865 (481~1130) ❙ 1311 (512~1997); 522 (258~848); 125 (18~259) ❙ 51.1 (-35.9~130.7); 845 (481~1130); 219.3 (121.6~551.6)

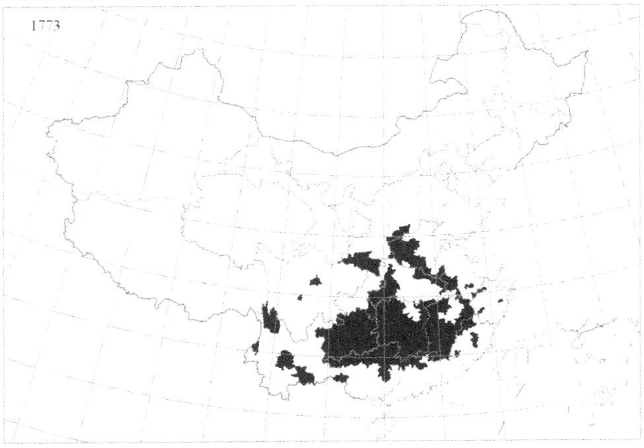

Lardizabalaceae 木通科 *Archakebia* 长萼木通属
Archakebia apetala (Q. Xia, J. Z. Suen et Z. X. Peng) Q. Xia et al. 长萼木通
Vines, deciduous

14.8 (10.3~16.6); 4.0 (-0.1~6.1); 24.8 (19.8~26.6); 14.8 (10.3~16.7); 120.8 (75.8~140.1); 3.0 (0.0~11.7); 807 (634~885) ❙ 907 (711~1135); 484 (357~562); 25 (15~39) ❙ 12.1 (5.0~31.0); 802 (634~866); 198.1 (130.0~294.3)

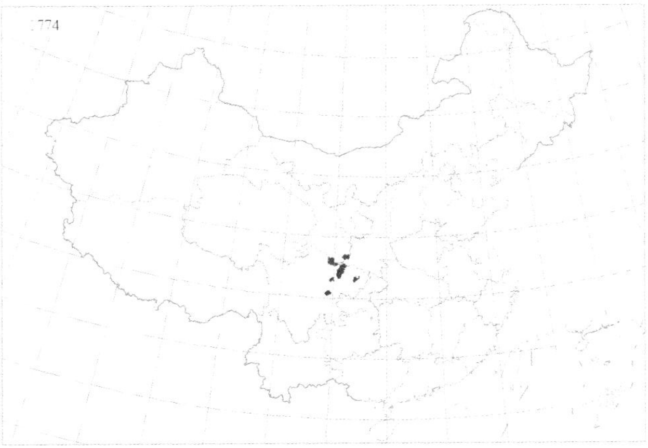

Lardizabalaceae 木通科 *Decaisnea* 猫儿子属
Decaisnea insignis (Griff.) Hook. f. et Thomson 猫儿屎
Shrubs, deciduous

14.6 (2.2~21.0); 4.8 (-7.0~14.2); 23.3 (9.9~29.3); 14.7 (3.8~21.0); 119.6 (15.5~191.8); 3.8 (0.0~49.1); 783 (413~1053) ❙ 1136 (551~1922); 537 (245~1007); 70 (12~259) ❙ 45.2 (-36.2~145.8); 772 (413~1053); 258.5 (131.4~551.6)

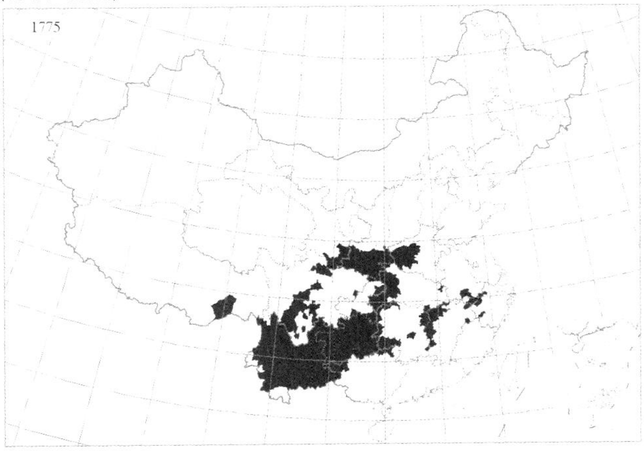

Lardizabalaceae 木通科 *Holboellia* 八月瓜属
Holboellia angustifolia (Gagnep.) H. N. Qin 五月瓜藤
Vines, evergreen

15.1 (-1.8~21.0); 5.3 (-10.2~14.2); 23.7 (6.1~29.8); 15.1 (1.8~21.1); 124.1 (1.9~193.0); 2.9 (0.0~82.8); 803 (329~1114) ❙ 1233 (320~1997); 549 (93~996); 91 (10~259) ❙ 53.3 (-14.2~147.0); 796 (320~1114); 249.2 (61.6~551.6)

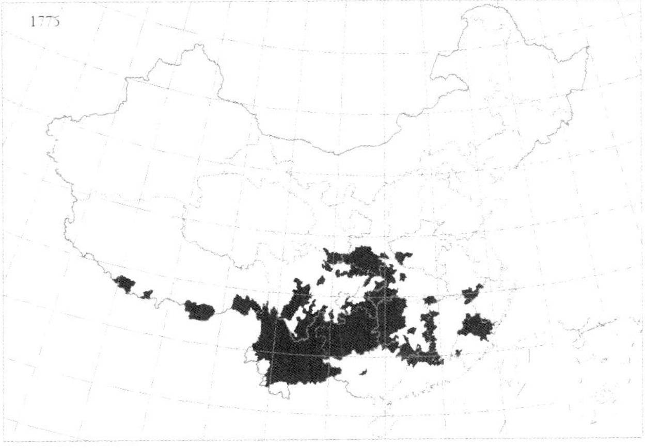

Lardizabalaceae 木通科 *Holboellia* 八月瓜属

Holboellia chapaensis Gagnep. 沙坝八月瓜

Vines, evergreen

17.9 (14.4~21.0); 10.9 (7.5~14.2); 22.7 (19.1~25.6); 17.9 (14.4~21.0); 154.8 (113.1~191.8); 0.0 (0.0~0.0); 869 (727~1053) ❙ 1297 (939~1688); 732 (524~959); 53 (37~70) ❙ 50.1 (6.5~112.2); 853 (727~1053); 412.5 (301.1~627.0)

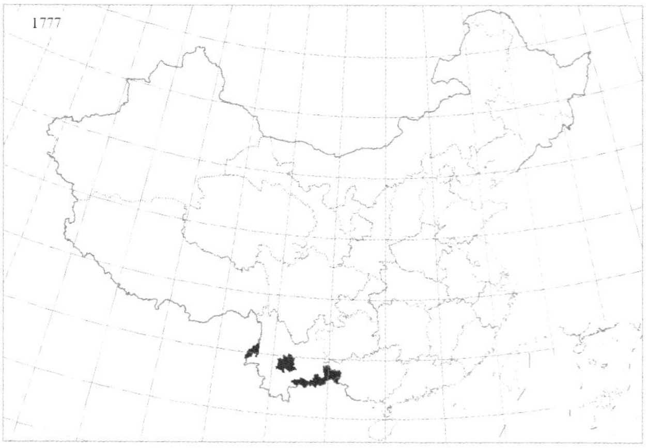

Lardizabalaceae 木通科 *Holboellia* 八月瓜属

Holboellia coriacea Diels 鹰爪枫

Vines, evergreen

15.5 (9.2~19.4); 4.0 (-2.6~10.2); 26.2 (16.7~29.8); 15.5 (9.5~19.4); 128.4 (62.7~172.7); 2.3 (0.0~20.9); 840 (600~994) ❙ 1304 (677~1997); 512 (316~816); 122 (14~254) ❙ 55.0 (-2.0~130.7); 835 (600~994); 213.5 (134.6~334.4)

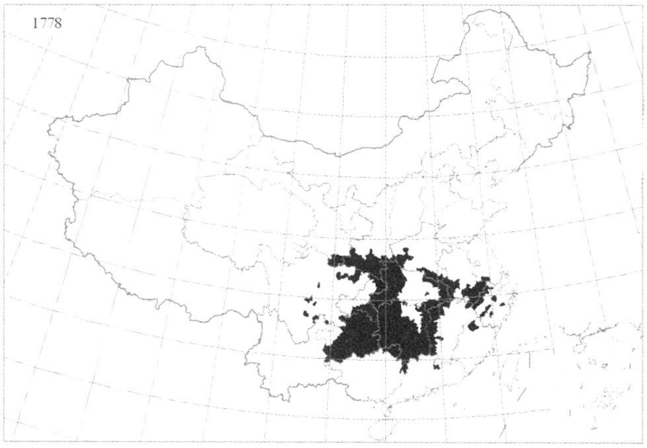

Lardizabalaceae 木通科 *Holboellia* 八月瓜属

Holboellia grandiflora Réaub. 牛姆瓜

Vines, evergreen

13.7 (2.2~18.5); 2.2 (-6.6~11.6); 24.6 (9.9~28.9); 13.8 (3.8~18.3); 111.9 (15.5~160.0); 6.4 (0.0~49.1); 775 (413~933) ❙ 1101 (626~1770); 493 (316~817); 74 (14~223) ❙ 42.7 (-22.2~125.9); 762 (413~927); 225.2 (139.1~352.3)

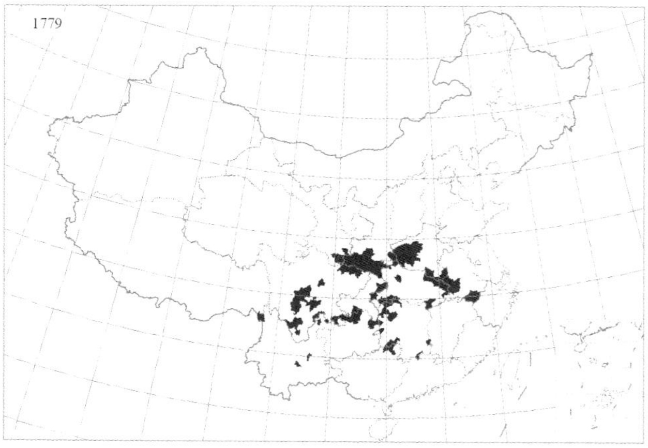

Lardizabalaceae 木通科 *Holboellia* 八月瓜属

Holboellia latifolia C. Y. Wu et S. H. Huang ex H. N. Qin 八月瓜

Vines, evergreen

14.6 (-1.8~21.0); 7.2 (-10.2~14.2); 20.2 (6.1~25.6); 14.7 (1.8~21.0); 118.7 (1.9~191.8); 3.5 (0.0~82.8); 754 (329~1053) ❙ 1076 (320~1751); 604 (193~1007); 43 (9~107) ❙ 43.1 (-14.2~145.8); 739 (320~1053); 309.8 (72.2~551.6)

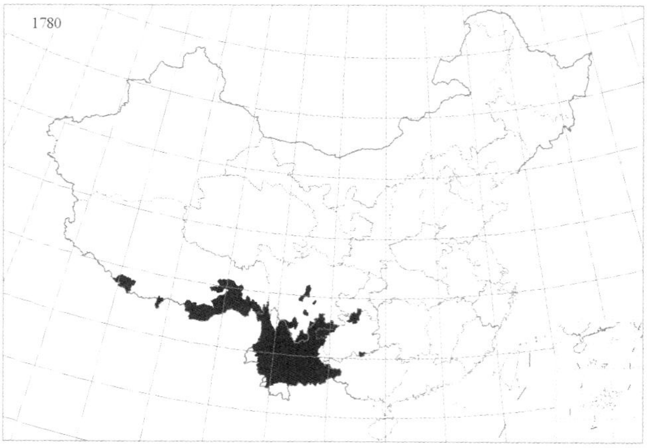

Lardizabalaceae 木通科 *Holboellia* 八月瓜属

Holboellia parviflora (Hemsl.) Gagnep. 小花鹰爪枫

Vines, evergreen

17.0 (14.0~21.0); 8.7 (4.3~14.2); 23.5 (21.4~26.3); 17.0 (14.1~21.0); 144.3 (109.4~191.8); 0.1 (0.0~0.7); 848 (748~1053) ❙ 1251 (958~1522); 674 (475~860); 56 (44~107) ❙ 47.8 (13.3~89.0); 846 (748~1053); 310.5 (176.4~551.6)

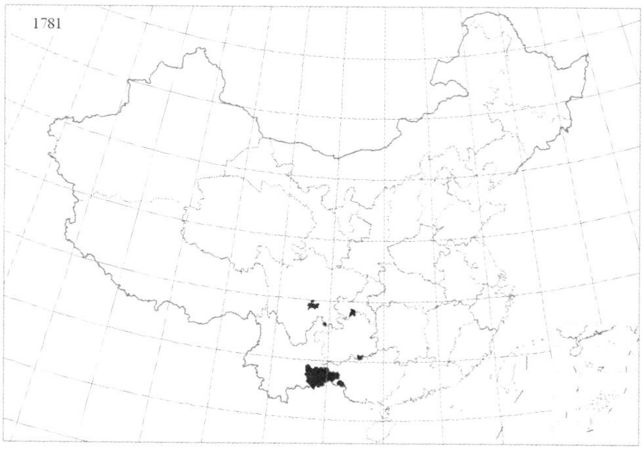

Lardizabalaceae 木通科 *Holboellia* 八月瓜属

Holboellia pterocaulis T. Chen et Q. H. Chen 棱茎八月瓜

Vines, evergreen

15.5 (13.4~17.4); 5.0 (2.7~7.4); 25.0 (23.5~25.7); 15.5 (13.4~17 4); 126.9 (104.5~148.9); 0.6 (0.0~3.7); 814 (735~886) ❙ 1247 (1064~1441); 526 (476~637); 89 (64~122) ❙ 53.6 (30.1~95.9); 814 (735~886); 209.9 (148.9~246.1)

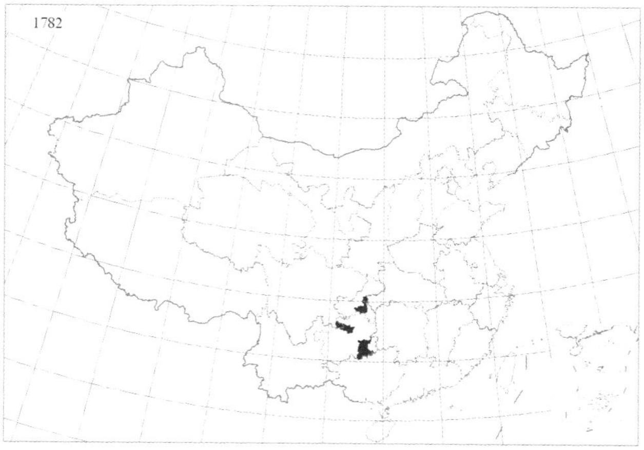

Lardizabalaceae 木通科 *Sargentodoxa* 大血藤属
Sargentodoxa cuneata (Oliv.) Rehder et E. H. Wilson 大血藤
Vines, deciduous

16.7 (2.2~23.0); 5.9 (-6.6~17.7); 26.6 (9.9~30.1); 16.7 (3.8~23.1); 142.5 (15.5~217.0); 1.7 (0.0~49.1); 893 (413~1221) ❙ 1369 (626~2435); 546 (299~1428); 133 (15~259) ❙ 53.6 (-22.2~147.0); 881 (413~1221); 222.5 (108.2~590.7)

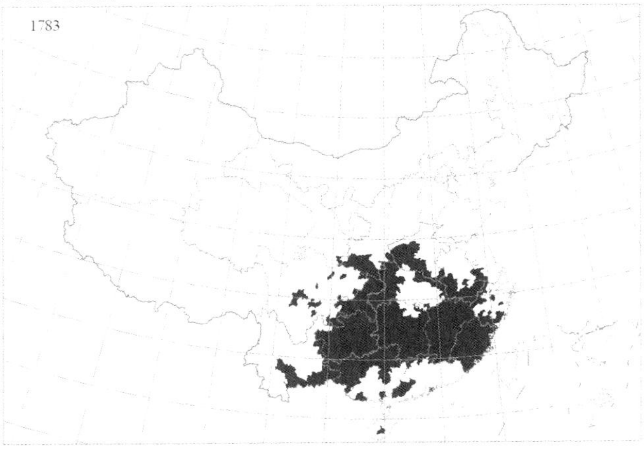

Lardizabalaceae 木通科 *Stauntonia* 野木瓜属
Stauntonia brachyanthera Hand.-Mazz. 黄蜡果
Large climbers, evergreen

17.2 (14.3~21.3); 6.1 (2.1~12.3); 27.3 (24.2~29.3); 17.2 (14.3~21.3); 147.0 (113.2~196.1); 0.6 (0.0~4.9); 914 (761~1132) ❙ 1578 (1117~1893); 573 (445~728); 188 (84~242) ❙ 73.7 (22.7~129.9); 912 (761~1132); 231.2 (149.5~319.9)

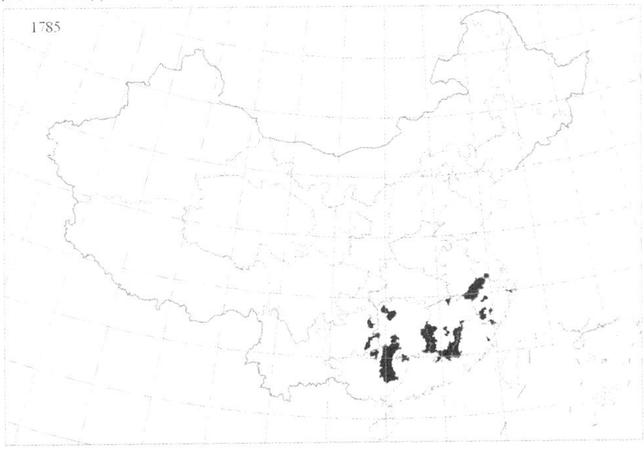

Lardizabalaceae 木通科 *Stauntonia* 野木瓜属
Stauntonia cavalerieana Gagnep. 西南野木瓜
Vines, evergreen

16.9 (12.7~22.7); 6.7 (2.[illegible]~15.0); 26.1 (23.0~28.7); 17.0 (12.8~22.8); 144.2 (99.0~213.1); 0.3 (0.0~5.6); 889 (713~1221) ❙ 1406 (985~2435); 593 (445~1428); 122 (43~197) ❙ 58.1 (16.9~107.2); 889 (713~1221); 222.6 (146.0~354.5)

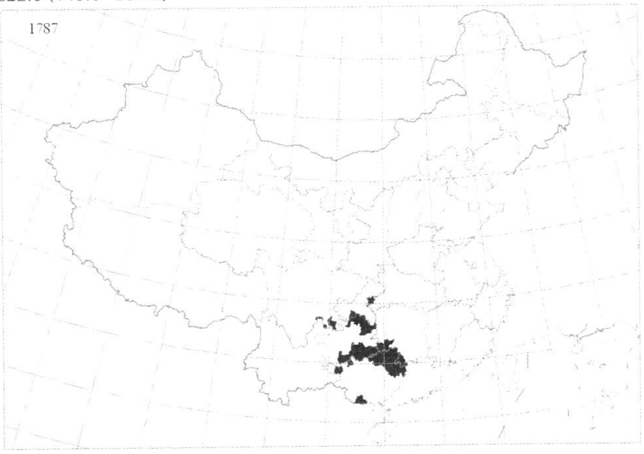

Lardizabalaceae 木通科 *Sinofranchetia* 串果藤属
Sinofranchetia chinensis (Franch.) Hemsl. 串果藤
Vines, deciduous

12.5 (2.2~19.0); 1.3 (-6.6~9.3); 23.1 (9.9~29.7); 12.7 (3.8~19.1); 99.9 (15.5~168.8); 9.0 (0.0~49.1); 727 (413~986) ❙ 1010 (534~1745); 467 (283~817); 53 (14~194) ❙ 39.6 (-18.9~125.9); 720 (413~986); 243.2 (111.9~334.4)

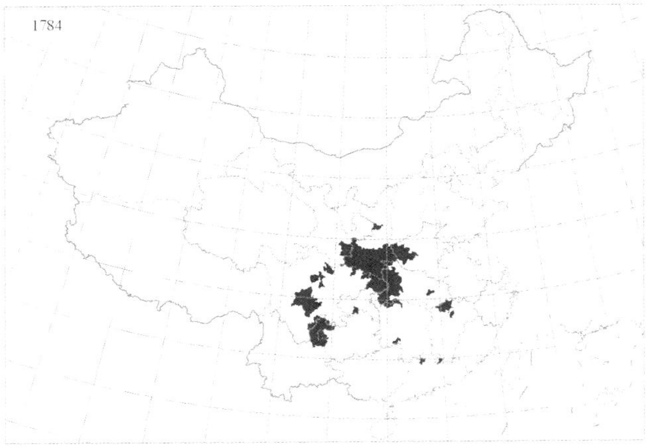

Lardizabalaceae 木通科 *Stauntonia* 野木瓜属
Stauntonia brunoniana Wall. ex Hemsl. 三叶野木瓜
Large climbers, evergreen

18.3 (14.9~21.0); 11.5 (5.3~15.1); 23.2 (20.6~25.9); 18.4 (15.0~21.0); 160.4 (119.6~191.8); 0.0 (0.0~0.0); 895 (782~1053) ❙ 1518 (1318~1751); 832 (655~996); 60 (42~77) ❙ 70.5 (44.5~112.2); 889 (782~1053); 462.6 (167.5~659.0)

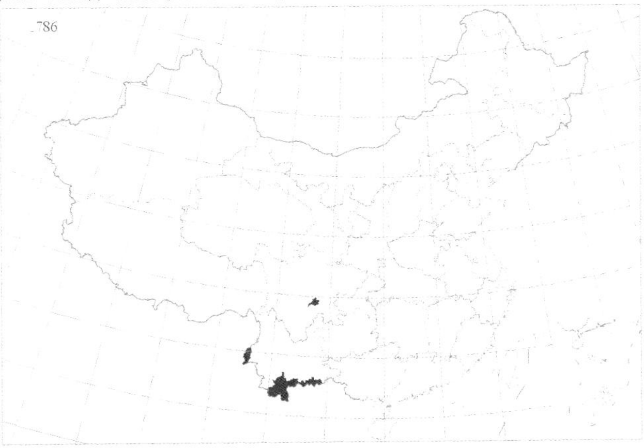

Lardizabalaceae 木通科 *Stauntonia* 野木瓜属
Stauntonia chinensis DC. 野木瓜
Vines, evergreen

17.9 (3.6~24.8); 7.8 (1.1~20.1); 27.2 (15.0~30.1); 18.0 (8.7~24.8); 156.1 (52.6~237.6); 0.3 (0.0~8.8); 946 (574~1385) ❙ 1537 (871~2030); 567 (299~1033); 169 (41~259) ❙ 64.0 (-6.7~147.0); 931 (574~1360); 243.7 (123.9~659.0)

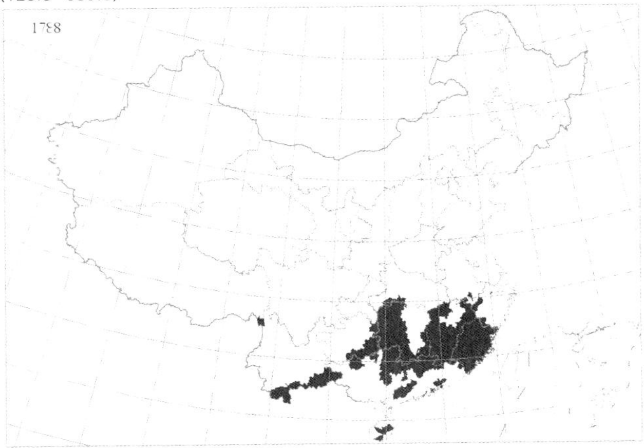

Lardizabalaceae 木通科 *Stauntonia* 野木瓜属
Stauntonia conspicua R. H. Chang 腺脉野木瓜
Vines, evergreen

17.1 (15.0~20.7); 6.2 (4.0~10.6); 27.2 (24.3~29.8); 17.1 (15.1~20.8); 145.5 (120.8~189.3); 0.1 (0.0~1.0); 904 (789~1113) ❙ 1625 (1373~1997); 541 (392~713); 206 (176~259) ❙ 82.1 (35.6~147.0); 884 (789~1036); 220.7 (145.3~303.4)

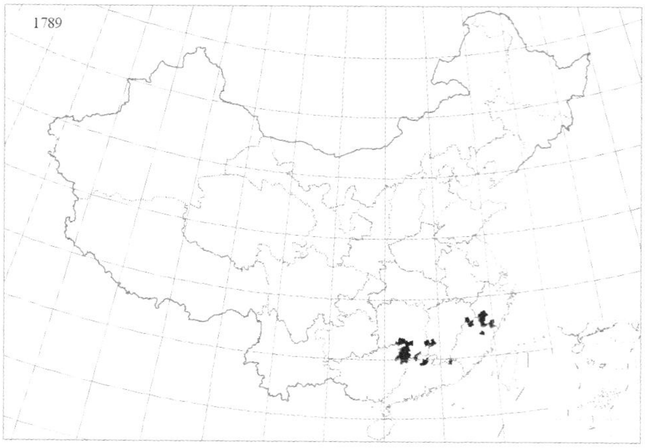

Lardizabalaceae 木通科 *Stauntonia* 野木瓜属
Stauntonia decora (Dunn) C. Y. Wu ex S. H. Huang 翅野木瓜
Vines, evergreen

19.5 (16.9~21.8); 11.1 (7.3~14.2); 26.2 (21.4~28.6); 19.5 (16.9~21.8); 174.5 (142.7~201.9); 0.0 (0.0~0.0); 999 (813~1158) ❙ 1535 (1007~1990); 713 (539~911); 127 (44~191) ❙ 54.0 (5.8~95.5); 998 (813~1158); 301.4 (185.5~551.6)

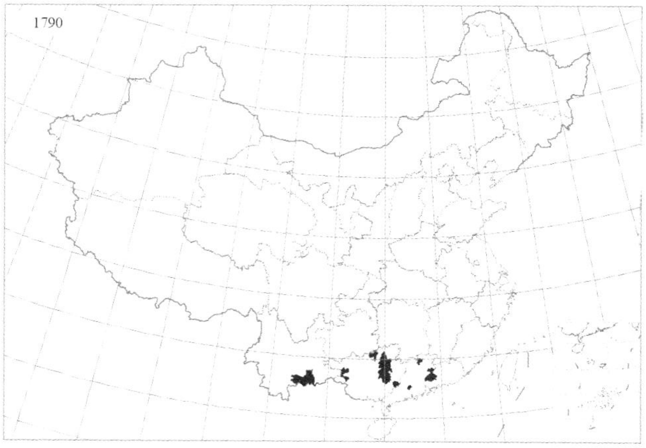

Lardizabalaceae 木通科 *Stauntonia* 野木瓜属
Stauntonia duclouxii Gagnep. 羊瓜藤
Large climbers, evergreen

15.1 (6.0~21.0); 4.6 (-3.2~14.2); 25.0 (14.4~28.9); 15.2 (6.6~21.0); 124.3 (39.7~191.8); 2.0 (0.0~26.7); 817 (508~1053) ❙ 1185 (711~1526); 526 (357~818); 70 (15~181) ❙ 46.9 (8.0~107.2); 815 (508~1053); 205.6 (108.2~551.6)

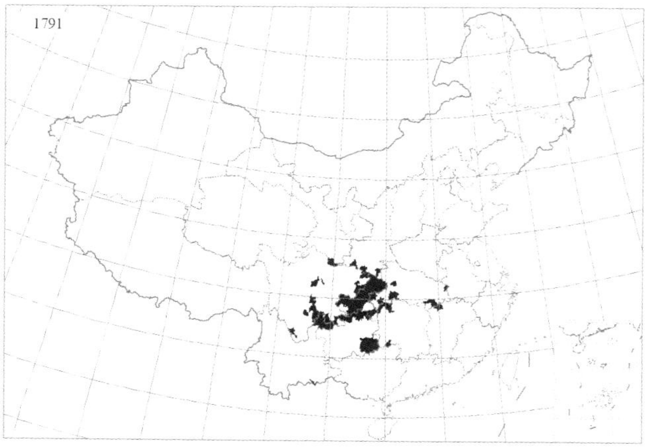

Lardizabalaceae 木通科 *Stauntonia* 野木瓜属
Stauntonia elliptica Hemsl. 牛藤果
Vines, evergreen

16.8 (3.5~22.4); 5.8 (-6.5~14.8); 26.9 (12.6~29.7); 16.8 (5.0~22.4); 142.4 (26.3~208.9); 0.7 (0.0~43.4); 894 (446~1191) ❙ 1449 (731~2030); 523 (346~1033); 158 (19~221) ❙ 62.2 (16.1~128.9); 885 (446~1191); 204.9 (134.6~284.1)

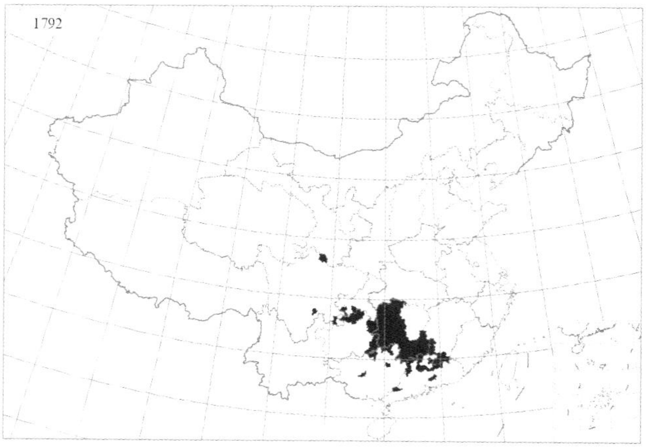

Lardizabalaceae 木通科 *Stauntonia* 野木瓜属
Stauntonia glauca Merr. et F. P. Metcalf 粉叶野木瓜
Vines, evergreen

20.6 (18.2~22.2); 12.1 (8.7~14.7); 27.3 (25.9~28.3); 20.7 (18.3~22.2); 188.2 (159.3~206.9); 0.0 (0.0~0.0); 1075 (928~1177) ❙ 1588 (1284~1900); 780 (651~1056); 107 (73~125) ❙ 47.3 (30.9~71.0); 1075 (928~1177); 262.7 (209.2~322.1)

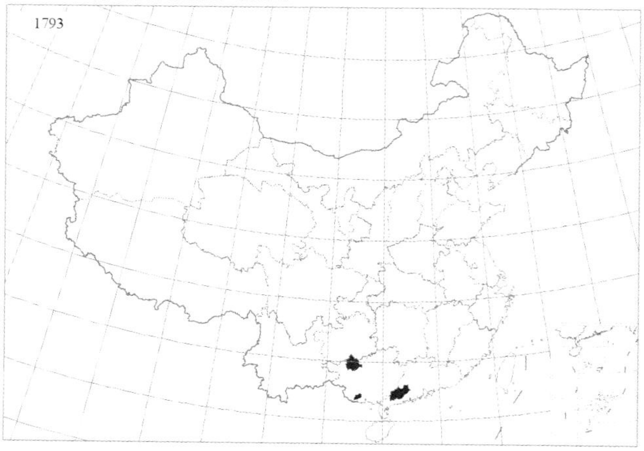

Lardizabalaceae 木通科 *Stauntonia* 野木瓜属
Stauntonia leucantha Y. C. Wu 钝药野木瓜
Vines, evergreen

17.6 (12.0~22.3); 6.8 (1.1~13.8); 27.6 (22.5~30.1); 17.7 (12.0~22.3); 152.2 (92.3~208.1); 0.3 (0.0~8.5); 937 (690~1202) ❙ 1537 (871~1997); 560 (299~1056); 179 (48~254) ❙ 65.4 (-6.7~130.7); 922 (690~1202); 224.1 (126.1~371.7)

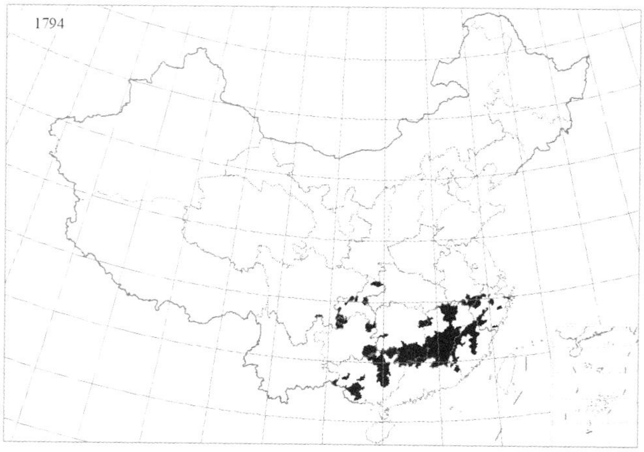

Lardizabalaceae 木通科 *Stauntonia* 野木瓜属

Stauntonia maculata Merr. 斑叶野木瓜

Vines, evergreen

19.5 (15.2~21.8); 9.8 (4.1~13.4); 27.9 (25.8~29.8); 19.5 (15.3~21.8); 174.2 (124.1~201.9); 0.1 (0.0~0.9); 1028 (810~1158) ❙ 1712 (1509~1990); 640 (508~827); 187 (163~221) ❙ 68.3 (35.6~106.4); 1014 (810~1158); 226.7 (174.9~260.7)

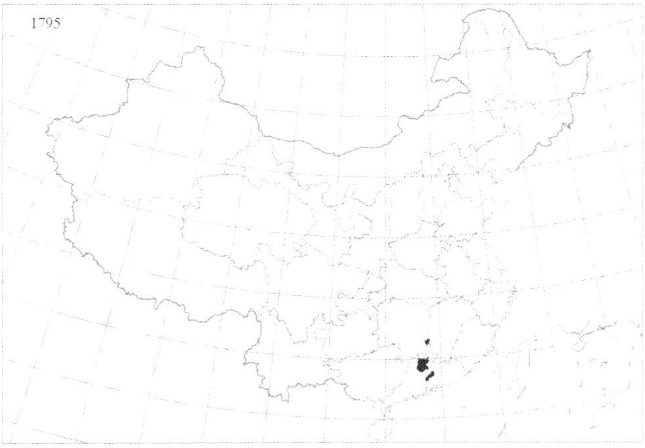

Lardizabalaceae 木通科 *Stauntonia* 野木瓜属

Stauntonia obcordatilimba C. Y. Wu et S. H. Huang 倒心叶野木瓜

Vines, evergreen

18.9; 10.6; 25.0; 19.0; 167.5; 0.0; 939 ❙ 1292; 729; 49 ❙ 37.6; 939; 342.5

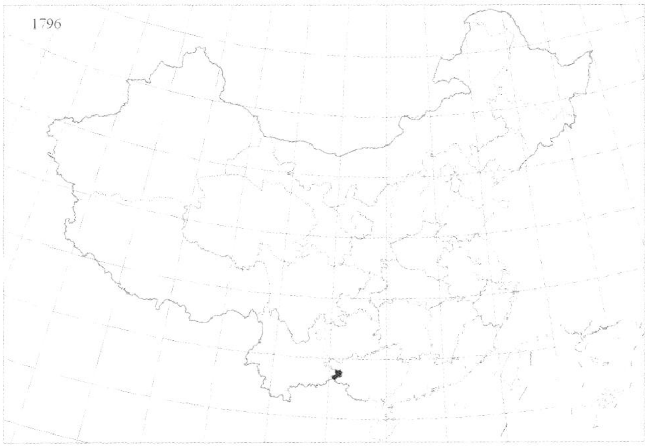

Lardizabalaceae 木通科 *Stauntonia* 野木瓜属

Stauntonia obovata Hemsl. 倒卵叶野木瓜

Vines, evergreen

17.4 (13.4~24.0); 7.0 (2.5~19.1); 27.0 (20.1~29.4); 17.4 (13.4~24.0); 149.6 (104.5~228.1); 0.4 (0.0~3.9); 920 (735~1302) ❙ 1552 (1064~3495); 593 (403~1412); 169 (43~897) ❙ 69.5 (8.0~246.5); 916 (735~1216); 228.1 (108.2~620.0)

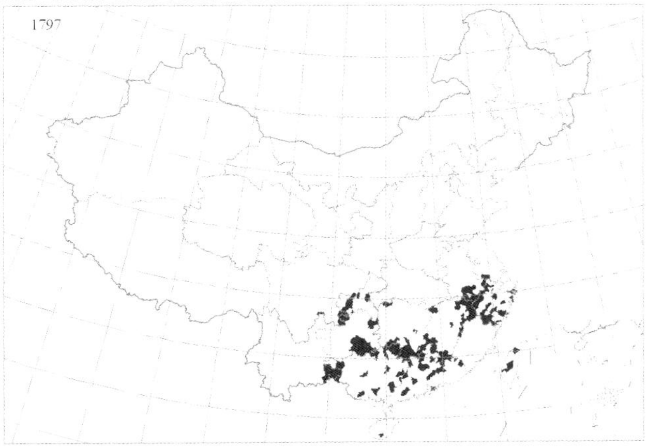

Lardizabalaceae 木通科 *Stauntonia* 野木瓜属

Stauntonia obovatifoliola Hemsl. 石月

Vines, evergreen

17.8 (12.9~24.1); 7.6 (1.5~18.9); 27.1 (20.1~29.8); 17.9 (12.9~24.1); 155.0 (101.2~229.7); 0.4 (0.0~7.3); 942 (723~1322) ❙ 1653 (1117~3495); 624 (392~1847); 188 (47~897) ❙ 76.9 (17.4~246.5); 932 (723~1216); 231.9 (119.4~363.7)

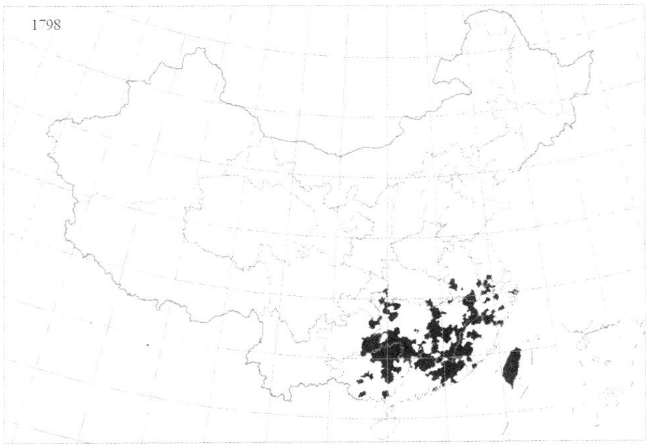

Lardizabalaceae 木通科 *Stauntonia* 野木瓜属

Stauntonia obovatifoliola subsp. urophylla (Hand.-Mazz.) H. N. Qin 尾叶那藤

Vines, evergreen

17.0 (14.3~21.6); 6.3 (2.5~14.1); 26.8 (23.5~29.7); 17.0 (14.3~21.6); 144.8 (112.6~199.4); 0.4 (0.0~3.9); 899 (756~1128) ❙ 1502 (1175~1991); 543 (403~902); 167 (84~242) ❙ 67.6 (41.1~147.0); 893 (756~1128); 221.0 (151.0~302.2)

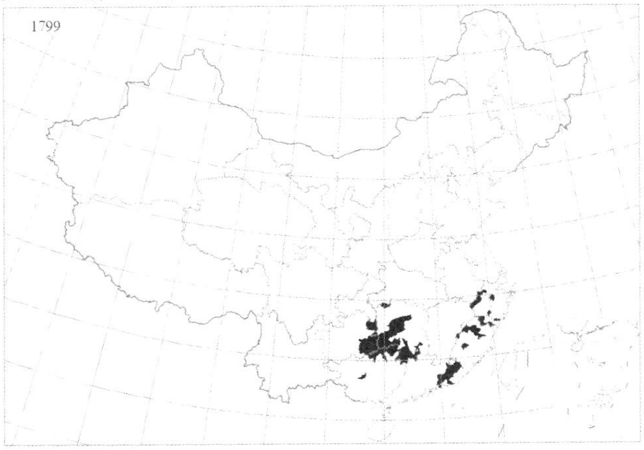

Lardizabalaceae 木通科 *Stauntonia* 野木瓜属

Stauntonia oligophylla Merr. et Chun 少叶野木瓜

Vines, evergreen

24.0; 19.1; 27.2; 24.0; 228.1; 0.0; 1302 ❙ 1551; 539; 48 ❙ 19.1; 1147; 620.0

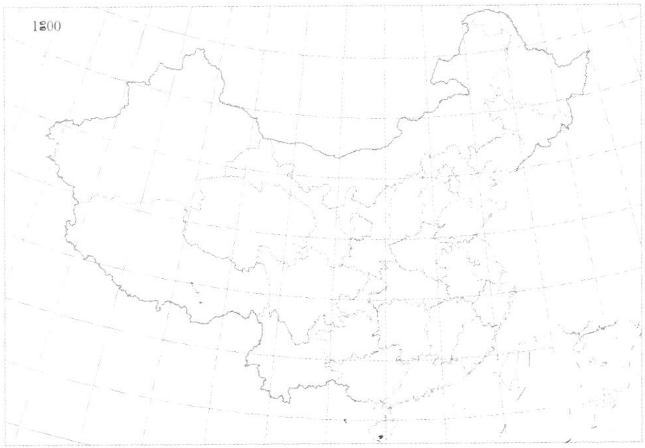

Lardizabalaceae 木通科 *Stauntonia* 野木瓜属

Stauntonia pseudomaculata C. Y. Wu et S. H. Huang 假斑叶野木瓜

Vines, evergreen

18.0 (16.9~21.0); 10.5 (9.1~14.2); 23.2 (21.4~25.6); 18.0 (16.9~21.0); 156.2 (142.7~191.8); 0.0 (0.0~0.0); 879 (813~1053) ❙ 1338 (1161~1522); 744 (633~860); 54 (48~68) ❙ 52.5 (37.1~80.6); 879 (813~1053); 399.1 (314.4~551.6)

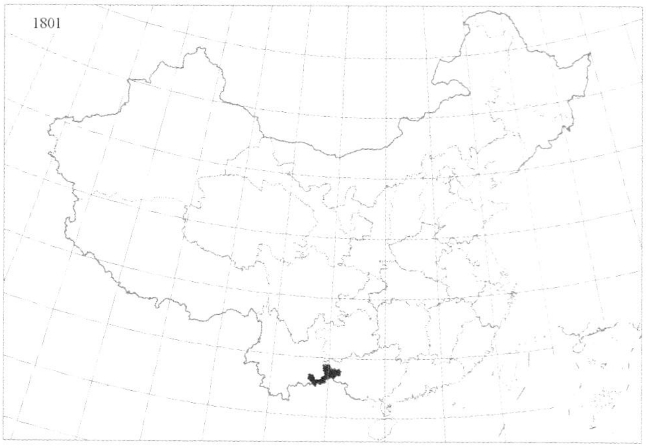

Lardizabalaceae 木通科 *Stauntonia* 野木瓜属

Stauntonia trinervia Merr. 三脉野木瓜

Vines, evergreen

19.8 (15.5~21.8); 11.0 (4.8~13.4); 27.3 (25.1~28.5); 19.8 (15.6~21.8); 178.0 (127.2~201.9); 0.0 (0.0~0.2); 1029 (815~1158) ❙ 1707 (1494~2005); 682 (539~1056); 167 (125~192) ❙ 66.6 (39.6~93.2); 1029 (815~1158); 255.4 (213.4~297.2)

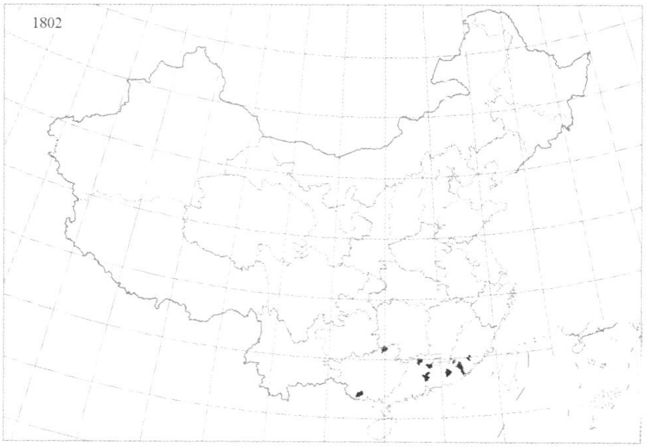

Lardizabalaceae 木通科 *Stauntonia* 野木瓜属

Stauntonia yaoshanensis F. N. Wei et S. L. Mo 瑶山野木瓜

Large climbers, evergreen

19.6 (17.9~21.3); 9.8 (7.4~12.3); 27.8 (25.8~28.6); 19.7 (18.0~21.3); 176.1 (155.6~196.1); 0.0 (0.0~0.0); 1032 (913~1130) ❙ 1596 (1371~1758); 669 (562~728); 175 (161~189) ❙ 56.0 (22.7~85.5); 1031 (913~1130); 220.0 (185.5~243.9)

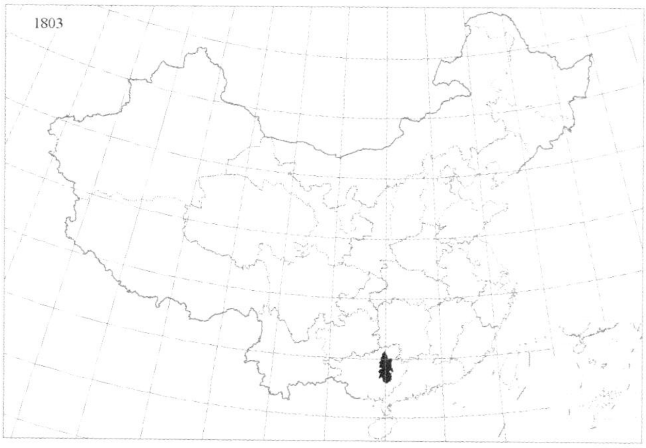

Menispermaceae 防已科 *Albertisia* 崖藤属

Albertisia laurifolia Yamam. 崖藤

Vines

22.9 (17.6~24.8); 16.4 (9.9~20.1); 27.8 (23.2~29.0); 23.0 (17.7~24.8); 215.9 (152.4~237.6); 0.0 (0.0~0.0); 1235 (859~1385) ❙ 1588 (868~2435); 711 (312~1428); 83 (41~130) ❙ 30.3 (-34.3~99.5); 1145 (859~1360); 387.9 (199.6~620.0)

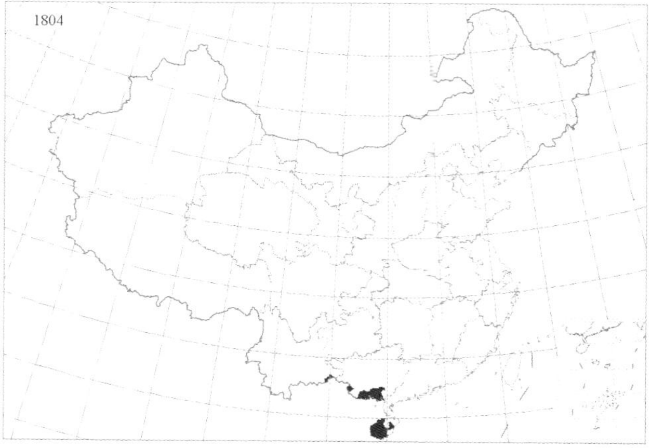

Menispermaceae 防已科 *Arcangelisia* 古山龙属

Arcangelisia gusanlung H. S. Lo 古山龙

Large climbers

24.0 (22.2~24.8); 18.4 (17.2~20.1); 28.0 (25.7~28.8); 24.0 (22.3~24.8); 228.3 (207.2~237.6); 0.0 (0.0~0.0); 1310 (1127~1385) ❙ 1501 (868~1854); 534 (312~712); 74 (41~115) ❙ 15.0 (-34.3~47.8); 1191 (868~1360); 440.3 (248.2~620.0)

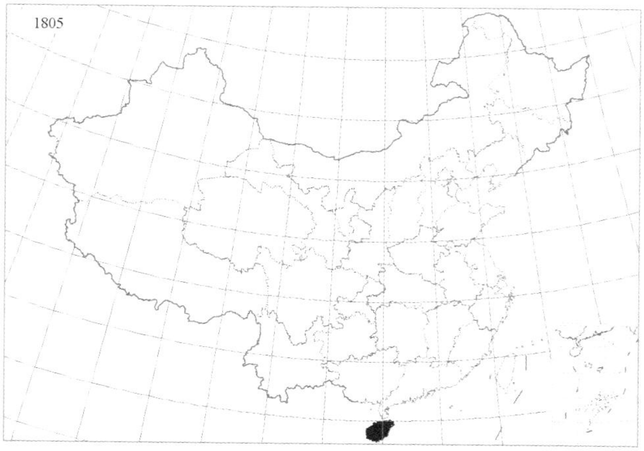

Menispermaceae 防已科 *Aspidocarya* 球果藤属

Aspidocarya uvifera Hook. f. et Thomson 球果藤

Vines

17.3 (13.3~21.0); 10.0 (3.6~15.1); 22.8 (20.3~25.9); 17.4 (13.3~21.0); 148.5 (101.9~191.8); 0.1 (0.0~1.8); 851 (718~1053) ❙ 1278 (958~1751); 685 (475~996); 56 (42~77) ❙ 50.2 (12.5~112.2); 842 (718~1053); 373.7 (148.9~659.0)

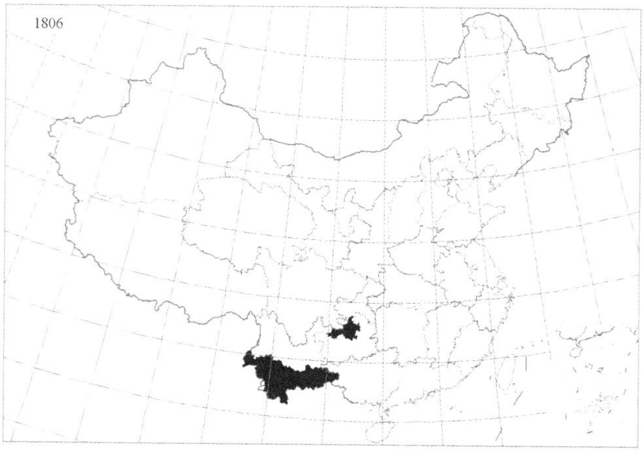

Menispermaceae 防己科 *Cissampelos* 锡生藤属

Cissampelos pareira var. hirsuta (Buch. ex DC.) Forman 锡生藤

Vines

17.9 (−1.8~21.0); 11.0 (−10.2~15.1); 22.7 (6.1~26.6); 18.1 (1.8~21.0); 158.2 (1.9~191.8); 3.2 (0.0~82.8); 887 (329~1053) ▎1346 (320~1751); 728 (217~996); 58 (20~77) ▎51.0 (−3.0~112.2); 879 (320~1053); 412.7 (72.2~659.0)

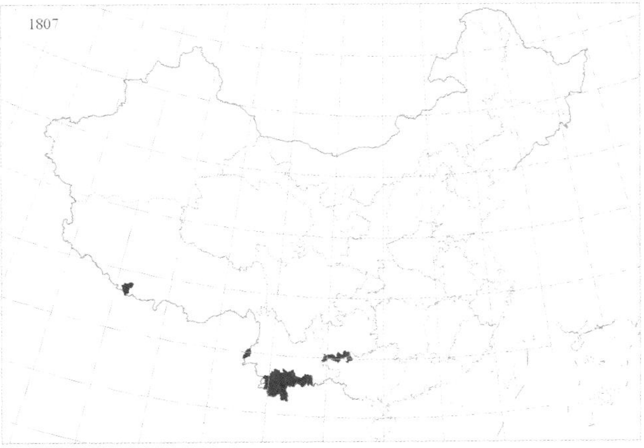

Menispermaceae 防己科 *Cocculus* 木防己属

Cocculus laurifolius DC. 樟叶木防己

Shrubs or small trees, evergreen

19.0 (−1.1~24.3); 10.1 (−9.1~19.1); 26.3 (6.2~29.8); 19.1 (2.0~24.3); 169.2 (2.5~231.9); 0.5 (0.0~75.6); 987 (334~1338) ▎1464 (321~2992); 672 (193~1847); 120 (33~230) ▎49.4 (−11.9~166.5); 979 (321~1247); 277.9 (79.3~659.0)

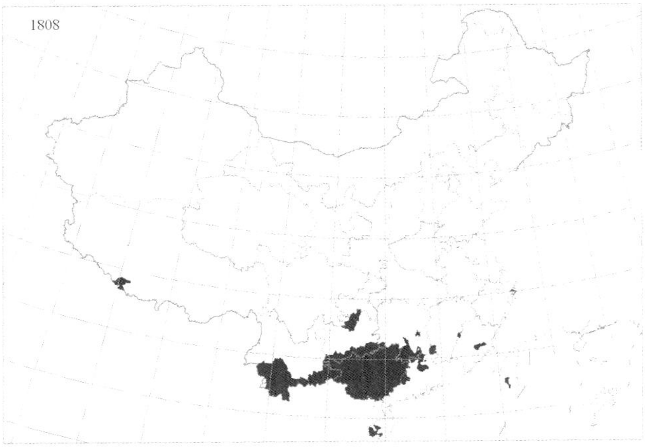

Menispermaceae 防己科 *Cocculus* 木防己属

Cocculus orbiculatus (L.) DC. 木防己

Vines

16.8 (5.4~24.7); 6.4 (−7.4~19.8); 26.1 (12.3~30.1); 16.9 (5.8~24.7); 144.4 (31.5~236.5); 2.3 (0.0~36.2); 901 (481~1376) ▎1320 (512~2993); 567 (258~1847); 120 (12~547) ▎46.4 (−36.5~246.5); 881 (481~1299); 232.5 (95.3~659.0)

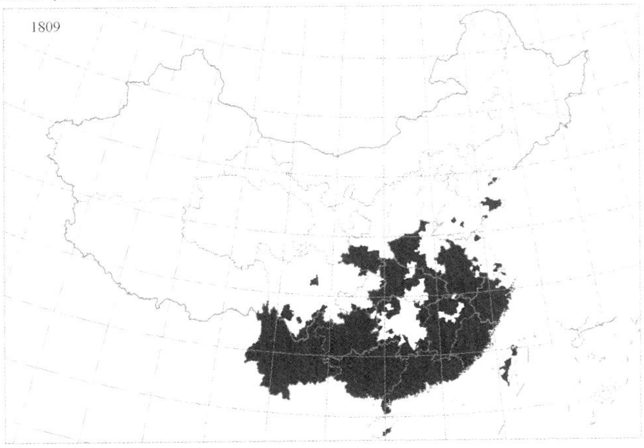

Menispermaceae 防己科 *Cocculus* 木防己属

Cocculus orbiculatus var. mollis (Wall. ex Hook. f. et Thomson) Hara 毛木防己

Vines

16.4 (5.4~21.3); 8.5 (−2.5~15.1); 22.5 (12.3~29.9); 16.5 (5.8~21.3); 138.2 (31.5~196.0); 0.9 (0.0~25.6); 832 (481~1123) ▎1200 (711~1751); 632 (357~996); 62 (15~215) ▎44.5 (−14.2~112.2); 818 (481~1121); 321.2 (141.7~659.0)

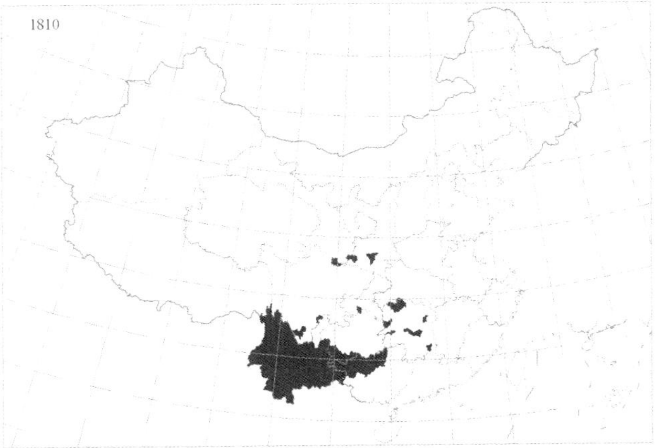

Menispermaceae 防己科 *Cyclea* 轮环藤属

Cyclea debiliflora Miers 纤花轮环藤

Vines

16.4; 9.6; 20.6; 16.4; 137.3; 0.0; 795 ▎1688; 959; 62 ▎112.2; 795; 627.0

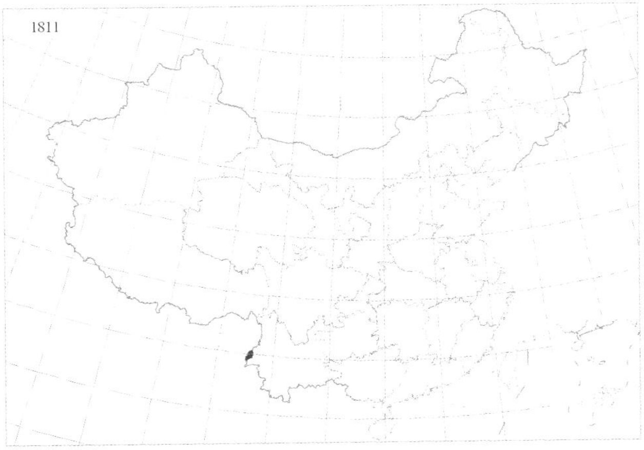

Menispermaceae 防己科 *Cyclea* 轮环藤属

Cyclea hypoglauca (Schauer) Diels 粉叶轮环藤

Vines

20.1 (14.3~24.1); 11.0 (3.7~19.1); 27.6 (21.4~29.8); 20.1 (14.3~24.2); 181.4 (113.2~230.2); 0.0 (0.0~1.5); 1055 (761~1327) ▎1579 (868~2435); 674 (312~1428); 146 (43~230) ▎51.1 (−34.3~128.9); 1049 (761~1273); 236.2 (95.3~620.0)

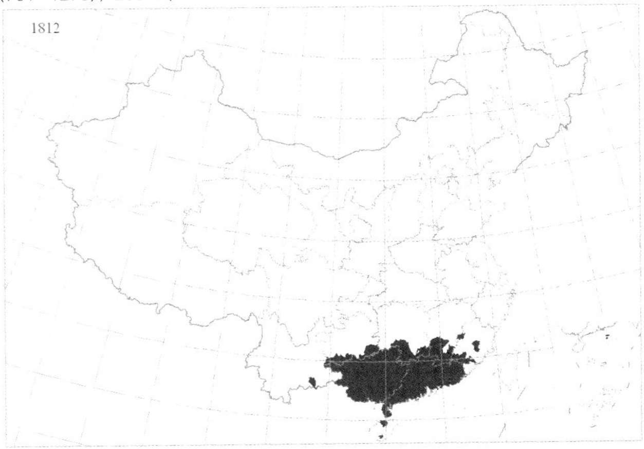

Menispermaceae 防己科 *Cyclea* 轮环藤属
Cyclea insularis (Makino) Hatsus. 海岛轮环藤
Vines

19.5 (16.2~24.1); 11.9 (7.8~18.9); 25.4 (20.1~28.0); 19.5 (16.2~24.1); 174.1 (134.2~229.7); 0.0 (0.0~0.0); 989 (773~1322) ❙ 1673 (1172~2759); 913 (617~1594); 92 (47~245) ❙ 73.3 (21.9~246.5); 962 (773~1144); 260.2 (154.2~411.1)

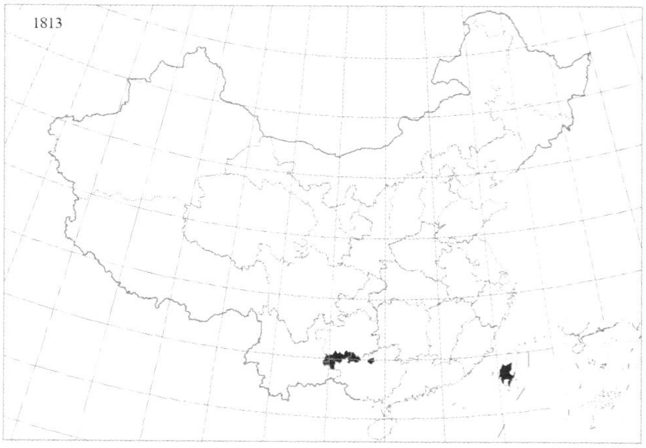

Menispermaceae 防己科 *Cyclea* 轮环藤属
Cyclea longgangensis J. Y. Luo 弄岗轮环藤
Vines

22.2; 13.9; 28.3; 22.3; 207.5; 0.0; 1190 ❙ 1374; 713; 87 ❙ 15.4; 1190; 327.7

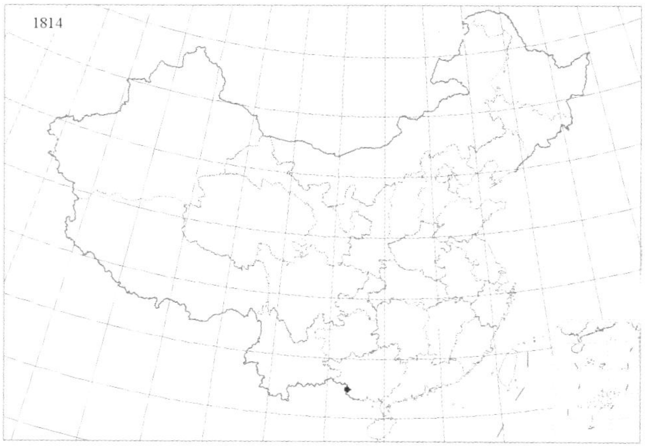

Menispermaceae 防己科 *Cyclea* 轮环藤属
Cyclea meeboldii Diels 云南轮环藤
Scandent shrubs

20.3; 14.5; 23.8; 20.3; 183.5; 0.0; 986 ❙ 1523; 715; 67 ❙ 54.9; 970; 573.6

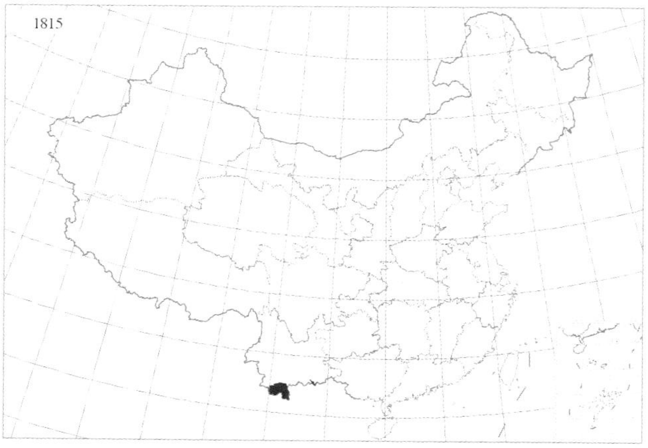

Menispermaceae 防己科 *Cyclea* 轮环藤属
Cyclea ochiaiana (Yamamoto) S. F. Huang et T. C. Huang 台湾轮环藤
Vines

21.4; 16.2; 25.3; 21.4; 196.9; 0.0; 1084 ❙ 2318; 1392; 95 ❙ 125.0; 1008; 204.7

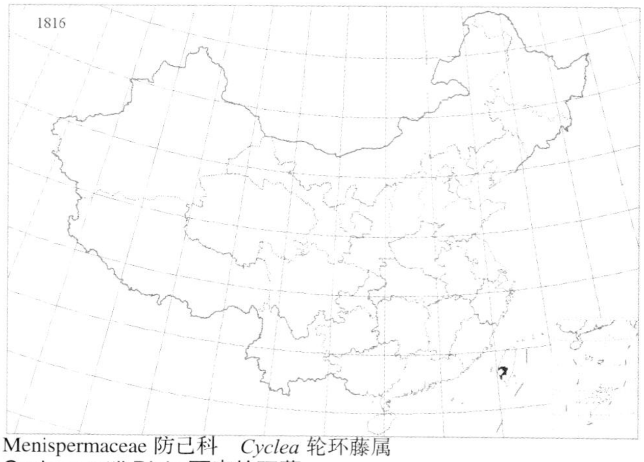

Menispermaceae 防己科 *Cyclea* 轮环藤属
Cyclea polypetala Dunn 铁藤
Large climbers

19.4 (8.6~24.8); 13.2 (1.3~20.1); 23.6 (15.0~28.7); 19.5 (8.7~24.8); 173.7 (52.6~237.6); 0.1 (0.0~8.8); 977 (574~1385) ❙ 1408 (868~2148); 710 (312~1234); 58 (41~125) ❙ 47.9 (-34.3~112.2); 933 (574~1360); 449.6 (288.3~659.0)

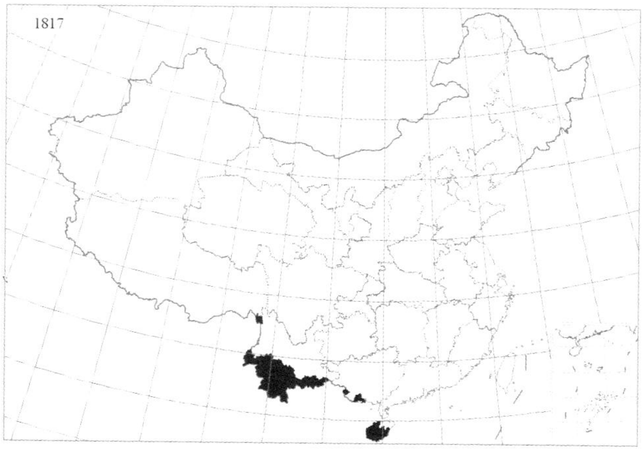

Menispermaceae 防己科 *Cyclea* 轮环藤属
Cyclea wattii Diels 西南轮环藤
Vines

17.3 (12.8~21.0); 10.3 (3.5~15.1); 22.3 (18.4~26.6); 17.4 (12.9~21.0); 148.6 (95.9~191.8); 0.0 (0.0~1.5); 850 (692~1053) ❙ 1277 (885~1751); 696 (475~996); 55 (34~107) ❙ 50.5 (6.5~112.2); 838 (692~1053); 385.8 (176.4~659.0)

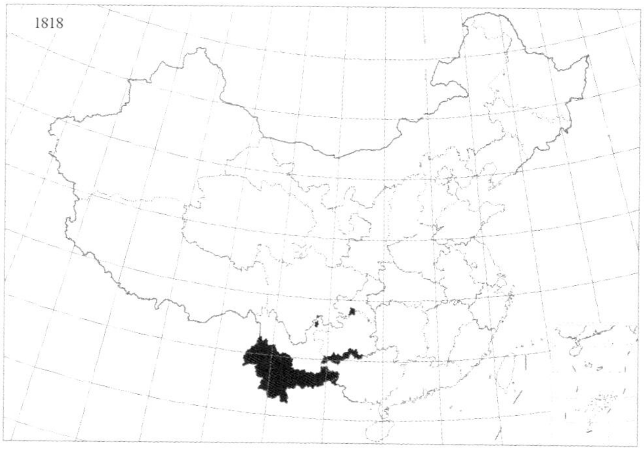

Menispermaceae 防已科 *Diploclisia* 秤钩风属
Diploclisia affinis (Oliv.) Diels 秤钩风
Vines

17.5 (10.6~22.3); 6.9 (-0.2~14.7); 27.1 (21.1~30.1); 17.5 (10.6~22.4); 150.9 (79.4~208.7); 0.7 (0.0~12.1); 929 (643~1200) ❙ 1506 (1035~2037); 586 (403~1026); 153 (31~259) ❙ 63.2 (5.8~130.7); 925 (643~1200); 224.9 (108.2~354.5)

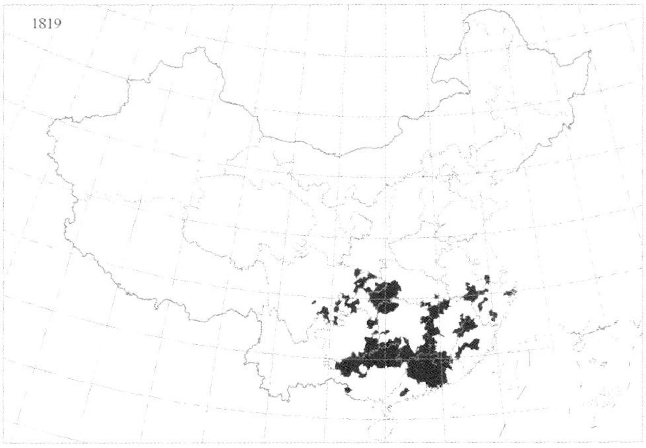

Menispermaceae 防已科 *Diploclisia* 秤钩风属
Diploclisia glaucescens (Blume) Diels 苍白秤钩风
Large climbers, evergreen

20.3 (14.3~24.8); 11.9 (3.6~20.1); 27.0 (20.6~29.1); 20.3 (14.3~24.8); 184.0 (112.6~237.6); 0.0 (0.0~1.6); 1063 (756~1385) ❙ 1548 (368~2435); 696 (312~1428); 118 (41~230) ❙ 47.3 (-34.3~112.2); 1050 (756~1360); 277.4 (95.3~659.0)

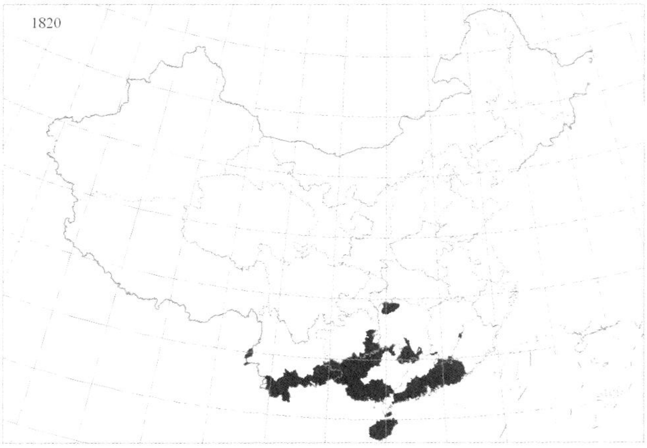

Menispermaceae 防已科 *Eleutharrhena* 藤枣属
Eleutharrhena macrocarpa (Diels) Forman 藤枣
Vines

19.1 (17.7~21.0); 13.0 (10.9~15.1); 23.1 (21.6~25.6); 19.2 (17.8~21.0); 169.9 (153.4~191.8); 0.0 (0.0~0.0); 921 (854~1053) ❙ 1441 (1143~1751); 780 (637~996); 60 (50~77) ❙ 56.7 (21.7~88.8); 909 (839~1053); 469.5 (346.9~659.0)

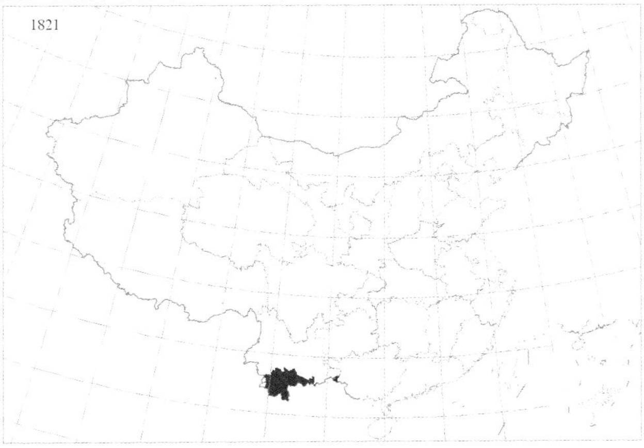

Menispermaceae 防已科 *Fibraurea* 天仙藤属
Fibraurea recisa Pierre 天仙藤
Large climbers

20.3 (16.3~23.6); 13.1 (9.1~16.7); 26.4 (20.6~29.2); 20.8 (16.4~23.6); 189.7 (136.4~223.5); 0.0 (0.0~0.0); 1076 (795~1281) ❙ 1546 (1074~2435); 795 (545~1428); 97 (43~173) ❙ 45.1 (-6.9~112.2); 1072 (795~1273); 314.3 (139.6~659.0)

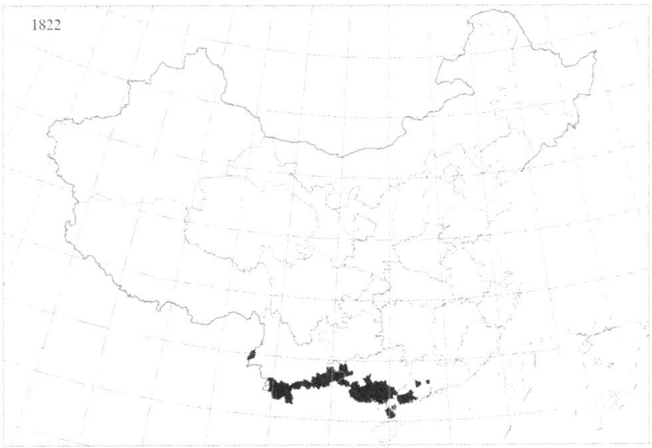

Menispermaceae 防已科 *Hypserpa* 夜花藤属
Hypserpa nitida Miers 夜花藤
Vines

21.0 (14.6~24.8); 12.9 (3.8~20.1); 27.4 (21.6~29.2); 21.0 (14.7~24.8); 192.4 (117.7~237.6); 0.0 (0.0~1.5); 1102 (779~1385) ❙ 1580 (868~2435); 716 (312~1428); 117 (41~202) ❙ 45.0 (-34.3~101.8); 1089 (779~1360); 275.1 (95.3~659.0)

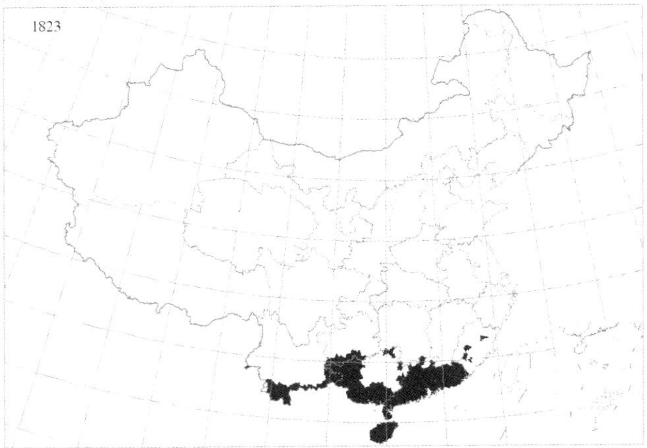

Menispermaceae 防已科 *Pachygone* 粉绿藤属
Pachygone sinica Diels 粉绿藤
Vines

19.6 (14.6~23.0); 10.0 (3.8~15.1); 27.6 (24.6~29.8); 19.6 (14.7~23.0); 175.[illegible] (117.7~216.2); 0.0 (0.0~1.5); 1027 (779~1247) ❙ 1595 (1250~2014); 667 (476~1044); 158 (58~197) ❙ 56.7 (17.4~101.8); 1025 (779~1247); 231.9 (174.9~318.5)

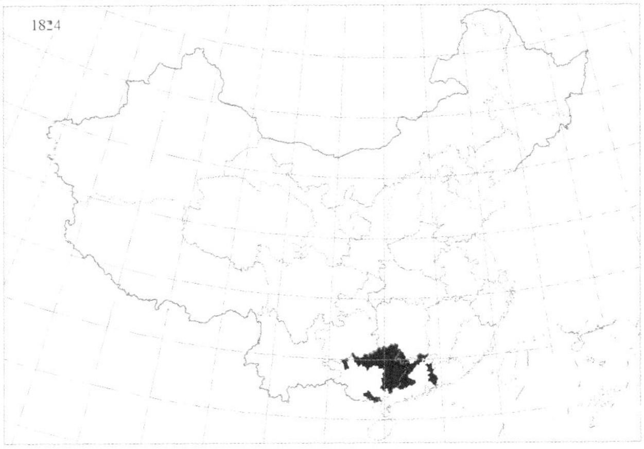

Menispermaceae 防己科 *Pachygone* 粉绿藤属
Pachygone valida Diels 肾子藤
Vines

19.8 (16.9~22.8); 12.9 (10.2~15.1); 24.7 (21.4~28.9); 19.9 (16.9~22.8); 178.6 (142.7~213.6); 0.0 (0.0~0.0); 990 (813~1230) ❙ 1531 (1178~2148); 828 (634~1234); 74 (51~125) ❙ 55.1 (5.8~88.8); 985 (813~1230); 421.3 (229.5~659.0)

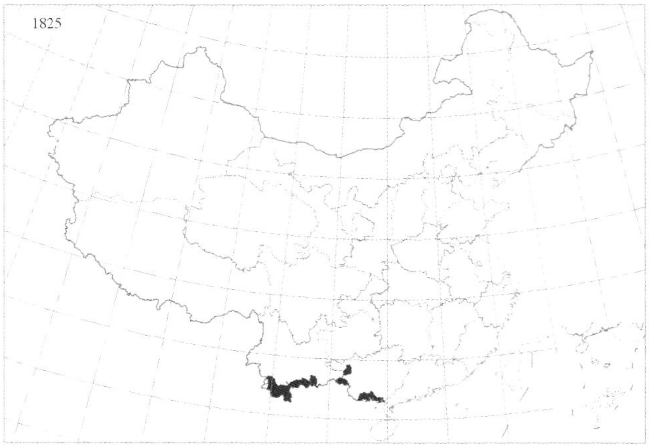

Menispermaceae 防己科 *Pericampylus* 细圆藤属
Pericampylus glaucus (Lam.) Merr. 细圆藤
Vines, evergreen

19.2 (10.6~24.8); 10.3 (-0.2~20.1); 26.7 (19.1~29.8); 19.2 (10.6~24.8); 171.0 (79.4~237.6); 0.2 (0.0~12.1); 1005 (643~1385) ❙ 1513 (868~2993); 659 (312~1428); 123 (29~547) ❙ 52.7 (-34.3~246.5); 996 (643~1360); 259.4 (95.3~659.0)

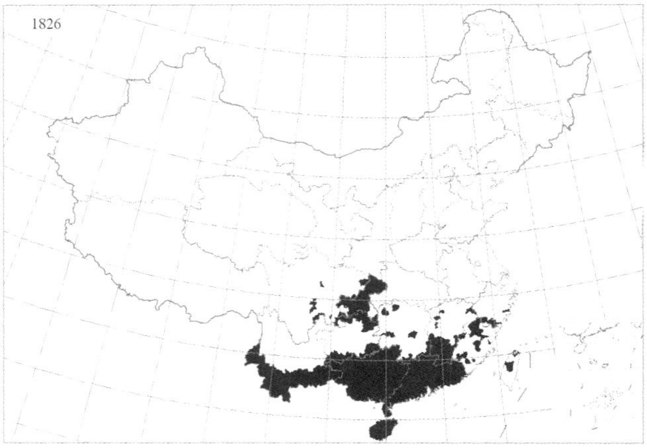

Menispermaceae 防己科 *Pycnarrhena* 密花藤属
Pycnarrhena lucida (Teijsm. et Binn.) Miq. 密花藤
Vines

23.9 (22.2~24.8); 18.6 (17.2~20.1); 27.6 (25.7~28.5); 23.9 (22.3~24.8); 227.0 (207.2~237.6); 0.0 (0.0~0.0); 1298 (1127~1385) ❙ 1531 (1179~1854); 525 (392~622); 60 (41~115) ❙ 18.5 (-11.9~47.8); 1161 (1055~1360); 515.5 (428.6~620.0)

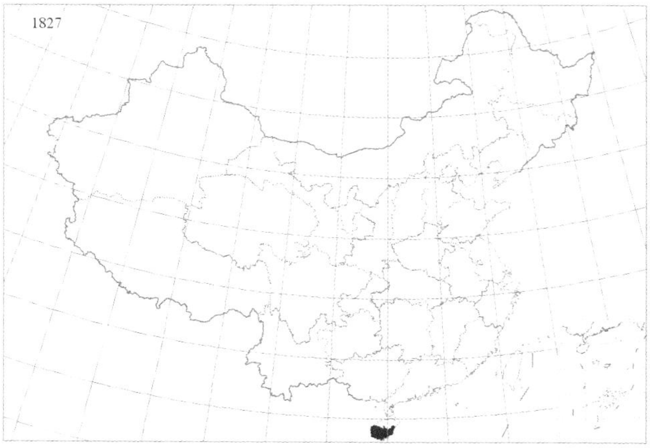

Menispermaceae 防己科 *Pycnarrhena* 密花藤属
Pycnarrhena poilanei (Gagnep.) Forman 硬骨藤
Vines or shrubs

21.6 (17.2~24.8); 15.6 (9.1~20.1); 25.8 (21.6~28.8); 21.6 (17.3~24.8); 199.8 (147.0~237.6); 0.0 (0.0~0.0); 1127 (833~1385) ❙ 1475 (868~1854); 644 (312~996); 67 (41~115) ❙ 35.1 (-34.3~88.8); 1059 (833~1360); 451.7 (248.2~659.0)

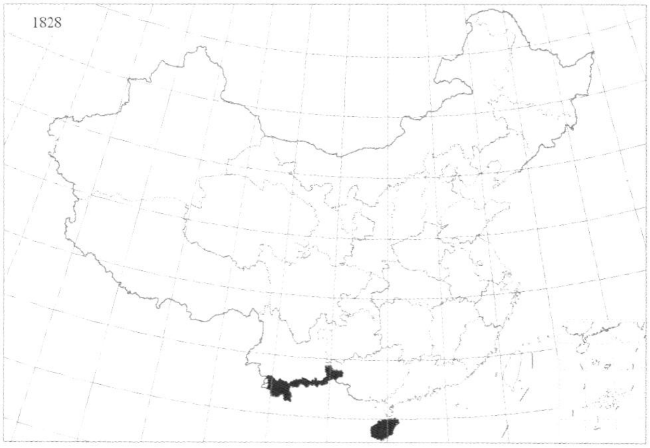

Menispermaceae 防己科 *Sinomenium* 风龙属
Sinomenium acutum (Thunb.) Rehder et E. H. Wilson 风龙
Large climbers

15.6 (2.6~22.2); 4.5 (-7.0~14.7); 25.9 (11.2~29.8); 15.7 (4.3~22.2); 130.9 (20.0~206.9); 3.0 (0.0~48.2); 847 (427~1177) ❙ 1322 (512~1997); 540 (258~959); 122 (15~259) ❙ 55.7 (-35.3~147.0); 836 (427~1177); 233.4 (121.6~627.0)

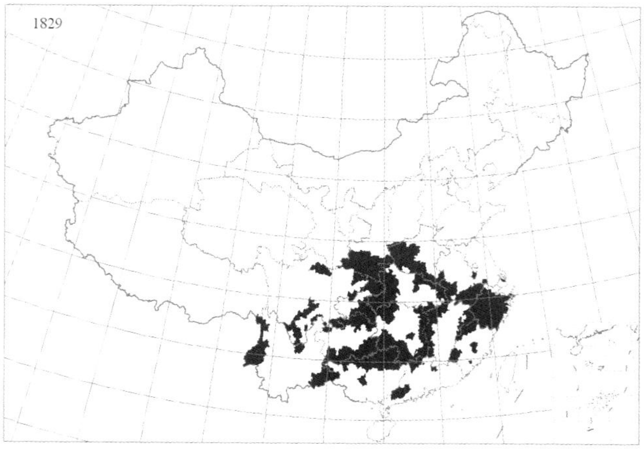

Menispermaceae 防己科 *Stephania* 千金藤属
Stephania merrillii Diels 台湾千近藤
Vines

18.5; 14.1; 21.7; 18.5; 162.1; 0.0; 863 ❙ 2713; 1388; 188 ❙ 214.4; 863; 363.7

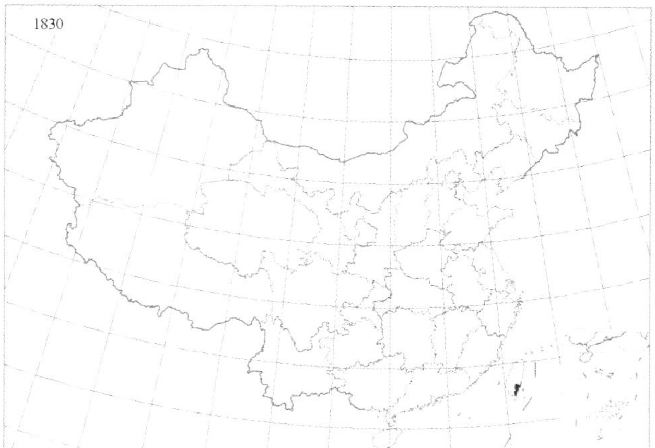

Menispermaceae 防已科 *Tinomiscium* 大叶藤属

Tinomiscium petiolare Hook. f. et Thomson 大叶藤

Vines

19.0 (14.6~22.2); 12.0 (7.9~15.1); 23.9 (19.6~28.6); 19.0 (14.6~22.3); 168.5 (115.8~207.5); 0.0 (0.0~0.0); 940 (736~1190) ❙ 1390 (1007~2148); 752 (539~1234); 69 (39~173) ❙ 48.0 (15.4~88.8); 932 (732~1190); 383.8 (220.1~659.0)

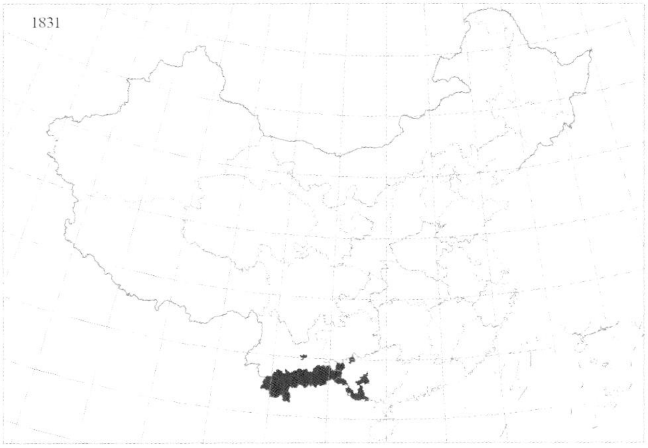

Menispermaceae 防已科 *Tinospora* 青牛胆属

Tinospora crispa (L.) Hook. f. et Thomson 波叶青牛胆

Vines, deciduous

19.9 (16.3~23.0); 12.5 (9.0~15.1); 25.3 (21.4~29.0); 20.0 (16.4~23.0); 179.7 (136.4~216.2); 0.0 (0.0~0.0); 1012 (797~1247) ❙ 1455 (1007~2435); 772 (539~1428); 86 (44~171) ❙ 44.5 (12.9~99.5); 1007 (797~1247); 340.9 (160.7~659.0)

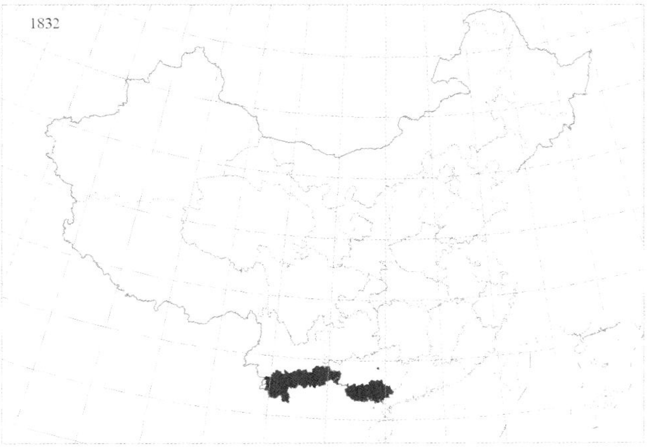

Menispermaceae 防已科 *Tinospora* 青牛胆属

Tinospora hainanensis H. S. Lo et Z. X. Li 海南青牛胆

Large climbers, deciduous

24.4 (24.1~24.8); 18.7 (17.8~20.1); 28.4 (28.0~28.7); 24.4 (24.1~24.8); 232.9 (229.4~237.6); 0.0 (0.0~0.0); 1348 (1320~1385) ❙ 1406 (868~1813); 502 (312~712); 75 (41~113) ❙ 4.1 (-34.3~36.4); 1154 (868~1329); 404.7 (248.2~548.0)

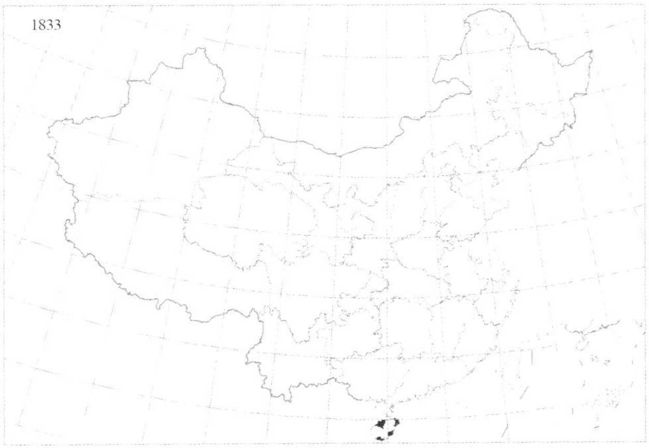

Menispermaceae 防已科 *Tinospora* 青牛胆属

Tinospora sinensis (Lour.) Merr. 中华青牛胆

Vines, deciduous

20.6 (13.3~24.8); 12.5 (5.4~20.1); 27.0 (19.6~29.8); 20.6 (13.4~24.8); 187.4 (100.4~237.6); 0.0 (0.0~0.0); 1070 (706~1385) ❙ 1606 (1007~2435); 740 (473~1428); 123 (41~197) ❙ 51.0 (-1.4~145.8); 1065 (706~1254); 271.4 (95.3~659.0)

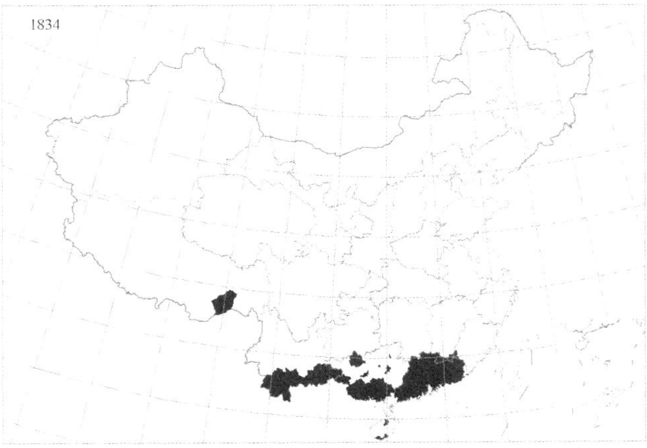

Illiciaceae 八角科 *Illicium* 八角属

Illicium angustisepalum A. C. Sm. 大屿八角

Trees, evergreen

17.6 (14.3~22.4); 7.1 (2.8~15.2); 27.2 (24.3~29.3); 17.7 (14.3~22.7); 152.6 (114.1~212.7); 0.3 (0.0~2.7); 934 (765~1216) ❙ 1683 (1284~2037); 621 (465~1033); 195 (73~259) ❙ 81.9 (27.0~147.0); 932 (765~1216); 242.1 (163.7~335.1)

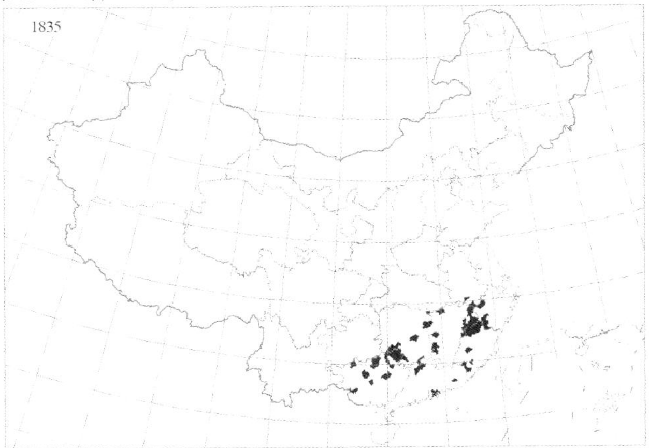

Illiciaceae 八角科 *Illicium* 八角属

Illicium arborescens Hayata 台湾八角

Trees, evergreen

20.4 (16.2~23.3); 14.9 (11.0~17.7); 24.9 (20.1~28.0); 20.4 (16.2~23.3); 185.2 (134.2~219.7); 0.0 (0.0~0.0); 1026 (773~1248) ❙ 2353 (1493~2993); 1209 (859~1594); 186 (73~547) ❙ 142.0 (25.2~246.5); 990 (773~1144); 245.8 (179.2~363.7)

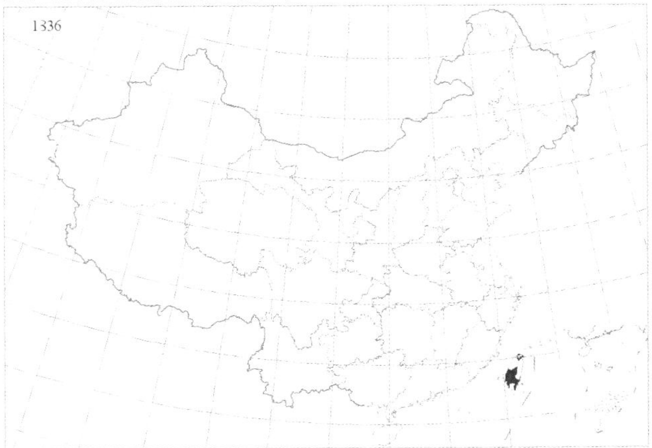

Illiciaceae 八角科 *Illicium* 八角属

Illicium brevistylum A. C. Sm. 短柱八角

Shrubs or trees, evergreen

18.5 (12.7~21.3); 8.4 (2.1~12.3); 27.2 (22.5~29.8); 18.5 (12.8~21.3); 162.5 (99.0~196.1); 0.2 (0.0~5.6); 967 (713~1130) ❙ 1602 (1292~2024); 610 (500~806); 177 (49~216) ❙ 66.9 (22.7~107.2); 965 (713~1130); 237.7 (174.9~491.5)

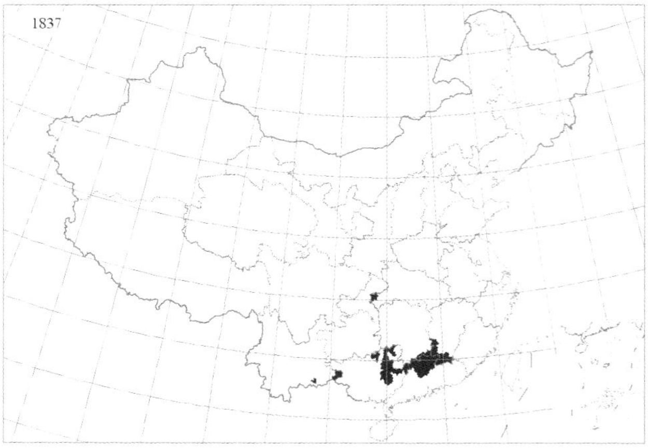

Illiciaceae 八角科 *Illicium* 八角属

Illicium burmanicum Wils. 中缅八角

Shrubs or trees, evergreen

16.8 (8.6~21.0); 10.2 (1.3~14.2); 21.3 (15.0~25.6); 16.8 (8.7~21.0); 141.8 (52.6~191.8); 0.3 (0.0~8.8); 823 (574~1053) ❙ 1292 (1028~1736); 726 (553~1007); 54 (44~70) ❙ 58.5 (12.5~145.8); 313 (574~1053); 412.3 (335.1~551.6)

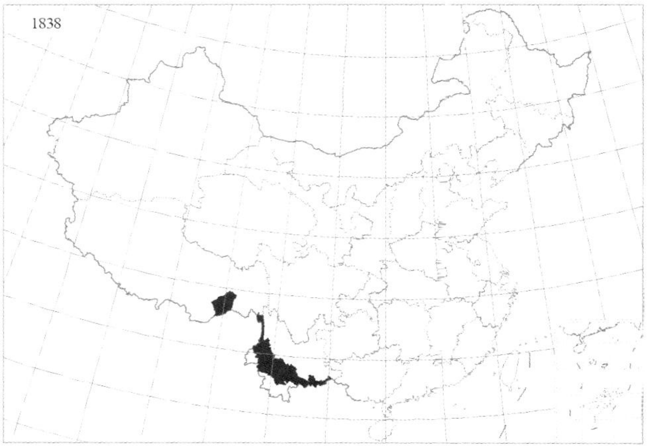

Illiciaceae 八角科 *Illicium* 八角属

Illicium difengpi K. I. B. et K. I. M. ex B. N. Chang 地枫皮

Shrubs, evergreen

20.3 (17.2~22.7); 12.0 (8.8~15.0); 26.6 (22.4~28.9); 20.4 (17.3~22.8); 184.5 (147.0~213.1); 0.0 (0.0~0.0); 1051 (833~1221) ❙ 1502 (1161~2435); 796 (633~1428); 100 (48~173) ❙ 43.7 (5.8~99.5); 1051 (833~1221); 309.3 (201.4~551.6)

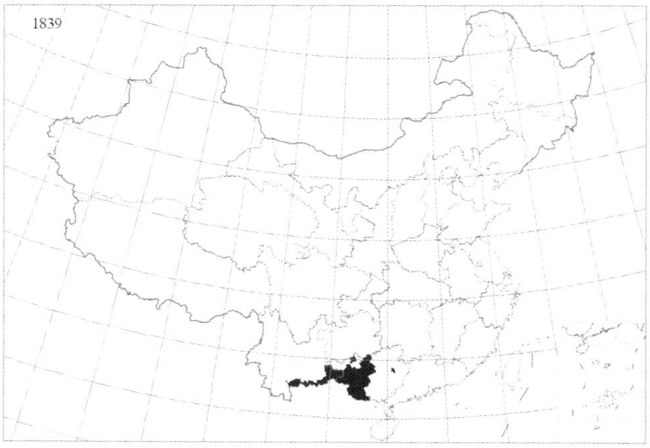

Illiciaceae 八角科 *Illicium* 八角属

Illicium dunnianum Tutcher 红花八角

Shrubs, evergreen

18.0 (11.7~22.7); 8.7 (1.1~15.2); 26.0 (18.8~29.0); 18.1 (11.7~22.8); 157.8 (82.6~213.1); 0.3 (0.0~8.5); 947 (658~1221) ❙ 1463 (916~2435); 659 (454~1428); 117 (31~254) ❙ 55.1 (12.9~130.7); 946 (658~1221); 229.3 (139.6~354.5)

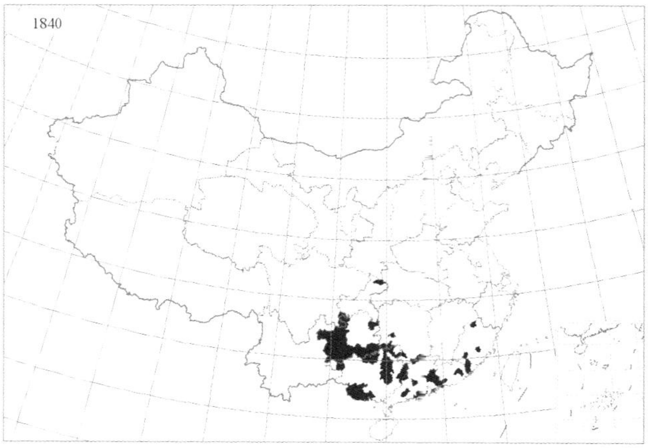

Illiciaceae 八角科 *Illicium* 八角属

Illicium griffithii Hook. f. et Thomson 西藏八角

Trees, evergreen

1.7 (-1.8~13.3); -7.3 (-10.5~5.4); 9.7 (6.1~19.6); 4.3 (1.8~13.4); 21.8 (1.9~100.4); 60.8 (0.0~82.8); 405 (329~706) ❙ 500 (255~1736); 324 (191~1007); 11 (1~51) ❙ 7.3 (-31.1~145.8); 358 (255~706); 137.2 (60.4~447.4)

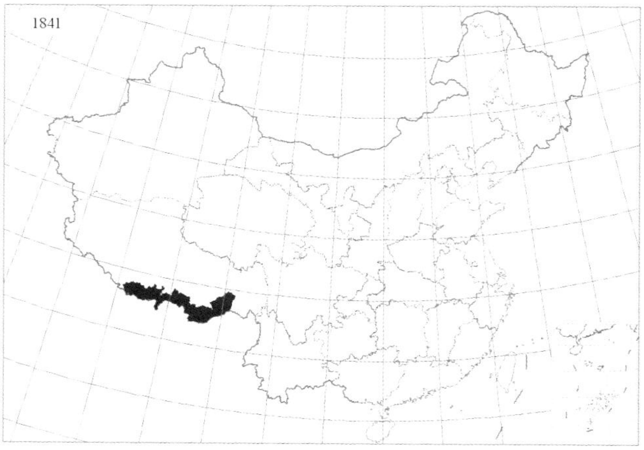

Illiciaceae 八角科 *Illicium* 八角属

Illicium henryi Diels 红茴香

Shrubs or trees, evergreen

15.6 (7.3~19.4); 4.2 (-4.2~9.4); 26.4 (15.6~30.1); 15.7 (7.5~19.4); 130.8 (48.2~173.4); 2.6 (0.0~29.4); 853 (537~1037) ❙ 1321 (707~1997); 513 (299~816); 126 (15~259) ❙ 54.6 (-6.7~130.7); 842 (537~1029); 218.2 (108.2~334.4)

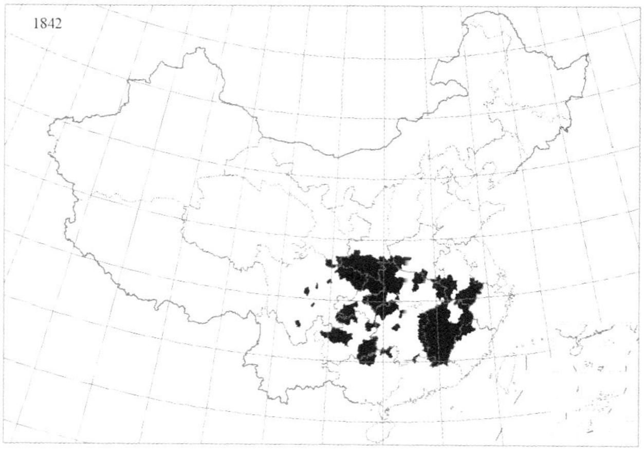

Illiciaceae 八角科 *Illicium* 八角属
Illicium jiadifengpi B. N. Chang 假地枫皮
Trees, evergreen

16.0 (11.1~19.5); 4.8 (1.1~11.0); 26.5 (19.4~29.3); 16.0 (11.3~19.5); 133.6 (80.8~174.1); 1 1 (0.0~8.5); 859 (652~991) ❙ 1523 (1197~1997); 541 (419~713); 172 (37~254) ❙ 78.4 (28.8~147.0); 855 (652~991); 232.7 (177.4~303.4)

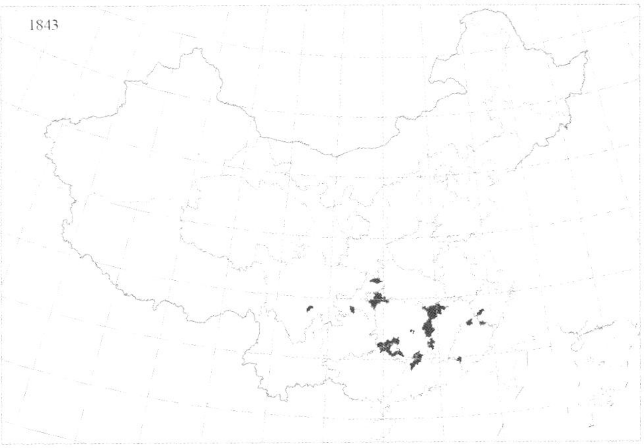

Illiciaceae 八角科 *Illicium* 八角属
Illicium lanceolatum A. C. Sm. 红毒茴
Shrubs or trees, evergreen

15.6 (12.3~21.0); 5.4 (-0.8~13.0); 27.2 (21.6~30.1); 16.7 (12.3~21.1); 140.8 (98.8~193.0); 0.8 (0.0~13.4); 891 (711~1114) ❙ 1456 (793~1997); 532 (299~821); 162 (42~259) ❙ 63.7 (-6.7~147.0); 880 (711~1114); 2 4.8 (131.2~332.5)

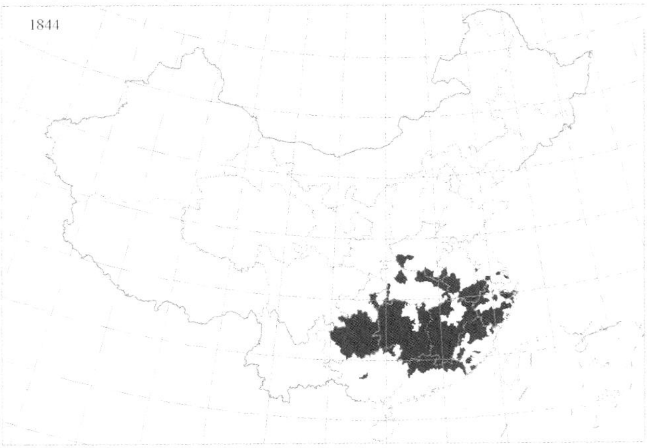

Illiciaceae 八角科 *Illicium* 八角属
Illicium leiophyllum A. C. Sm. 平滑八角
Shrubs, evergreen

22.3; 14.9; 28.5; 22.5; 210.5; 0.0; 1202 ❙ 1770; 919; 89 ❙ 47.0; 1202; 209.9

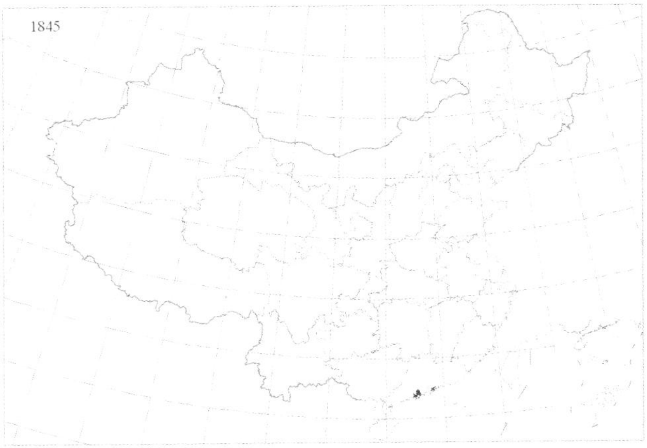

Illiciaceae 八角科 *Illicium* 八角属
Illicium macranthum A. C. Sm. 大花八角
Shrubs or trees, evergreen

14.6 (5.4~19.1); 7.8 (-2.1~13.0); 19.7 (12.3~23.9); 14.6 (5.8~19.1); 117.1 (31.5~169.7); 1.6 (0.0~25.6); 748 (481~929) ❙ 1065 (741~1504); 611 (411~860); 41 (15~70) ❙ 44.2 (-14.2~99.9); 716 (481~917); 339.0 (218.0~506.2)

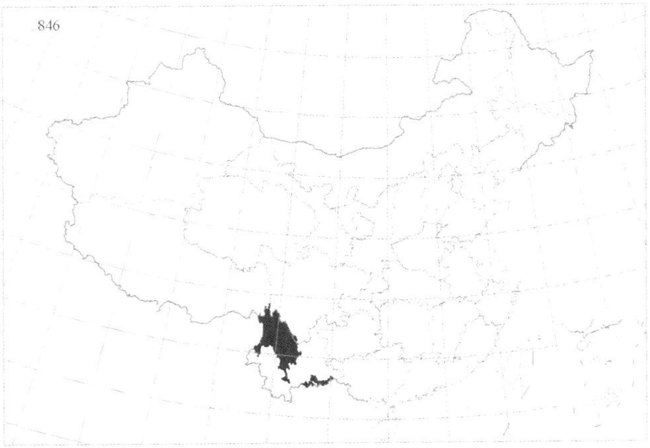

Illiciaceae 八角科 *Illicium* 八角属
Illicium majus Hook. f. et Thomson 大八角
Trees, evergreen

17.7 (9.5~22.5); 7.8 (1.1~15.2); 26.3 (16.7~29.8); 17.7 (9.6~22.7); 153.1 (62.7~212.7); 0.3 (0.0~8.5); 928 (600~1216) ❙ 1456 (985~2037); 610 (392~1026); 128 (27~221) ❙ 57.6 (8.0~128.9); 924 (600~1216); 238.6 (108.2~627.0)

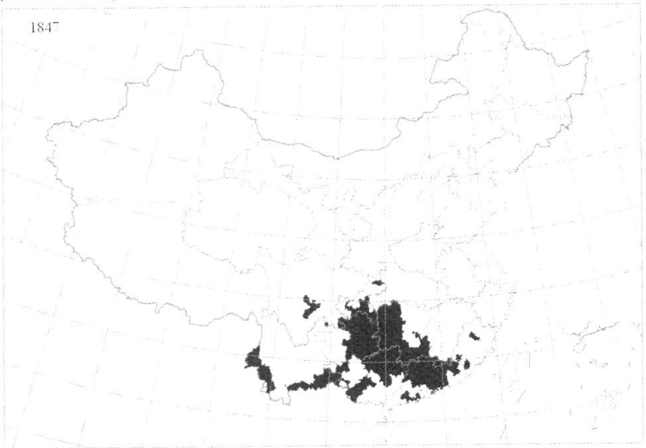

Illiciaceae 八角科 *Illicium* 八角属
Illicium merrillianum A. C. Sm. 滇西八角
Small trees, evergreen

16.3 (11.0~19.5); 9.8 (4.2~13.8); 20.6 (16.6~23.3); 16.3 (11.0~19.5); 135.9 (73.1~173.8); 0.0 (0.0~0.8); 799 (636~942) ❙ 1352 (1090~1688); 746 (594~959); 54 (42~70) ❙ 69.7 (34.7~112.2); 792 (536~921); 438.8 (335.1~627.0)

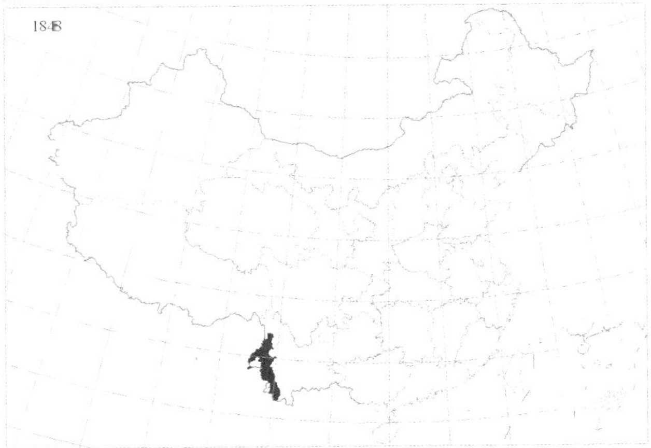

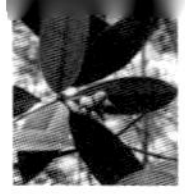

Illiciaceae 八角科 *Illicium* 八角属
Illicium micranthum Dunn 小花八角
Shrubs or small trees, evergreen

17.2 (8.6~23.4); 8.2 (1.3~16.1); 24.8 (15.0~29.3); 17.3 (8.7~23.5); 147.5 (52.6~221.7); 0.4 (0.0~8.8); 889 (574~1273) ❙ 1312 (896~2035); 615 (419~996); 89 (18~204) ❙ 48.3 (6.5~107.2); 884 (574~1273); 257.6 (108.2~659.0)

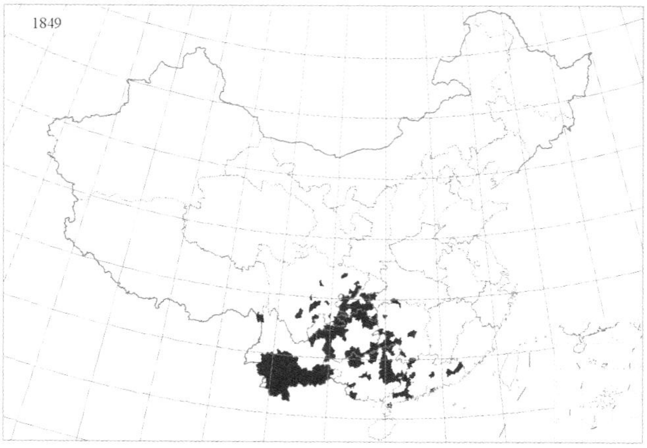

Illiciaceae 八角科 *Illicium* 八角属
Illicium modestum A. C. Sm. 滇南八角
Shrubs or small trees, evergreen

18.3; 10.6; 23.9; 18.3; 159.7; 0.0; 897 ❙ 1440; 662; 83 ❙ 60.4; 883; 379.6

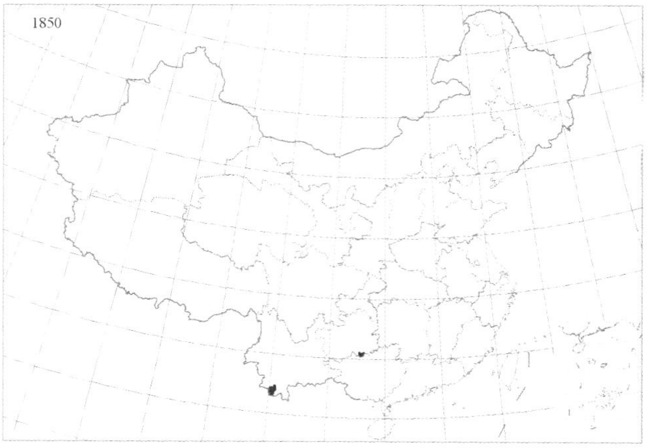

Illiciaceae 八角科 *Illicium* 八角属
Illicium oligandrum Merr. et Chun 少药八角
Small trees, evergreen

21.3 (17.9~24.1); 12.8 (7.4~19.1); 27.9 (25.8~28.9); 21.3 (18.0~24.2); 195.7 (155.6~230.1); 0.0 (0.0~0.0); 1128 (913~1326) ❙ 1647 (1302~2435); 759 (455~1428); 141 (48~194) ❙ 47.0 (6.7~99.5); 1116 (913~1326); 274.7 (185.5~620.0)

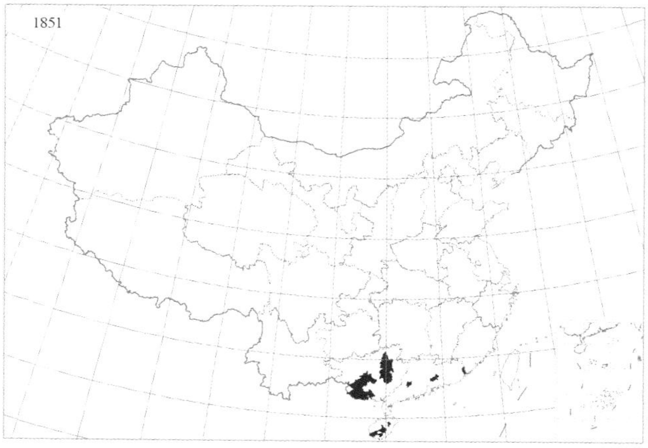

Illiciaceae 八角科 *Illicium* 八角属
Illicium pachyphyllum A. C. Sm. 短梗八角
Shrubs, evergreen

19.1 (14.2~22.7); 9.5 (3.3~15.0); 27.0 (20.8~28.6); 19.1 (14.2~22.8); 169.6 (110.9~213.1); 0.2 (0.0~2.3); 1003 (734~1221) ❙ 1548 (1168~2435); 699 (445~1428); 141 (53~194) ❙ 55.7 (12.9~101.8); 1003 (734~1221); 244.4 (169.7~371.7)

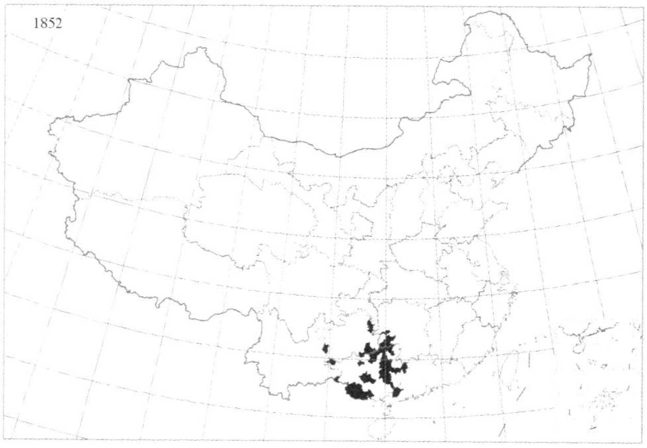

Illiciaceae 八角科 *Illicium* 八角属
Illicium petelotii A. C. Sm. 少果八角
Shrubs or small trees, evergreen

18.7; 11.5; 23.6; 18.7; 164.1; 0.0; 915 ❙ 1381; 758; 59 ❙ 51.4; 915; 459.8

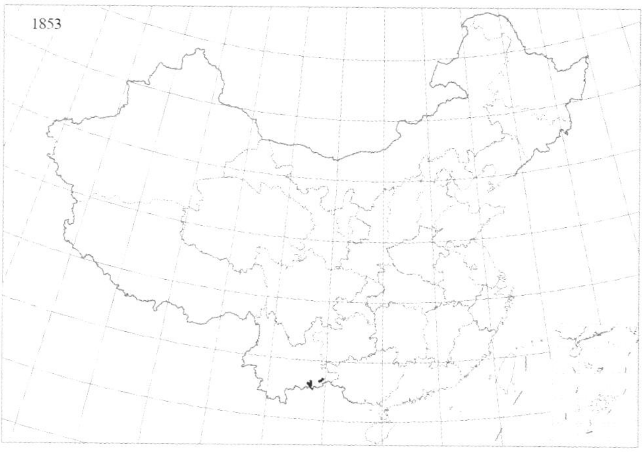

Illiciaceae 八角科 *Illicium* 八角属
Illicium philippinense Merr. 白花八角
Shrubs or trees, evergreen

20.3; 15.0; 24.4; 20.3; 183.4; 0.0; 998 ❙ 2765; 1406; 206 ❙ 180.7; 998; 256.8

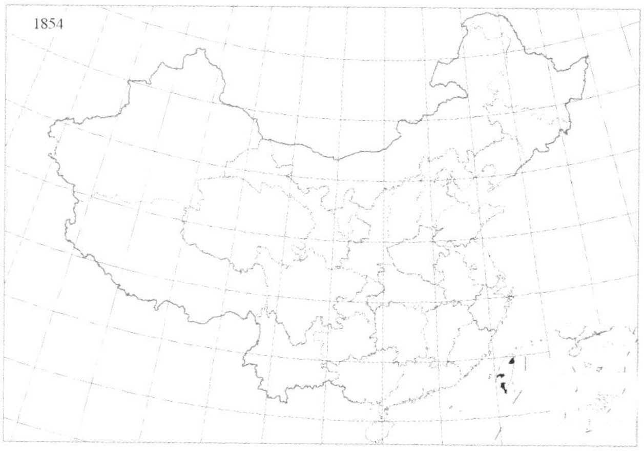

Illiciaceae 八角科 *Illicium* 八角属
Illicium simonsii Maxim. 野八角
Trees, evergreen

15.0 (5.4~21.0)；6.6 (−2.1~14.2)；22.0 (12.3~28.9)；15.0 (5.8~21.0)；121.4 (31.5~191.8)；1.1 (0.0~25.6)；777 (481~1053) | 1121 (741~1624)；585 (411~911)；55 (15~196) | 45.3 (−14.2~106.2)；766 (481~1053)；269.3 (108.2~551.6)

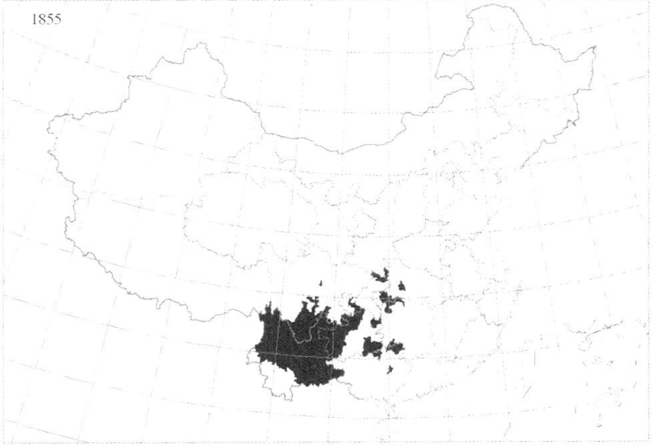

Illiciaceae 八角科 *Illicium* 八角属
Illicium tashiroi Maxim. 峦大八角
Shrubs or small trees, evergreen

19.0 (16.2~22.2)；13.5 (11.0~17.7)；23.5 (20.1~27.0)；19.0 (16.2~22.2)；158.1 (134.2~206.7)；0.0 (0.0~0.0)；937 (773~1123) | 2550 (1836~2992)；1214 (959~1847)；223 (107~402) | 180.9 (64.2~246.5)；928 (773~1123)；267.2 (202.8~296.9)

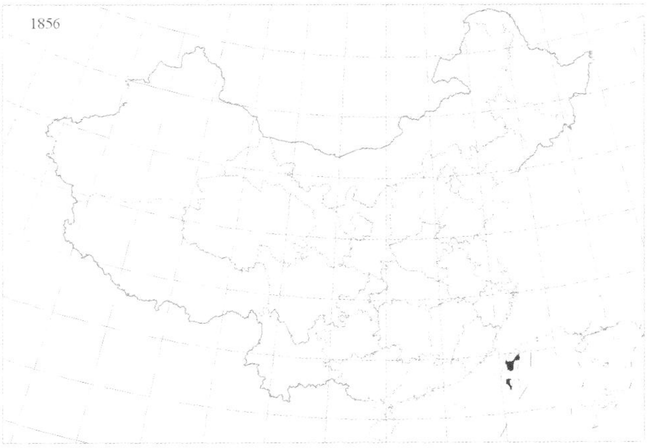

Illiciaceae 八角科 *Illicium* 八角属
Illicium ternstroemioides A. C. Sm. 厚皮香八角
Trees, evergreen

20.1 (14.6~24.8)；11.4 (3.8~20.1)；27.5 (24.6~29.8)；20.1 (14.7~24.8)；181.7 (117.7~237.6)；0.0 (0.0~1.5)；1063 (779~1385) | 1622 (1176~1997)；579 (392~818)；157 (41~258) | 56.5 (−12.0~130.7)；1028 (779~1360)；315.8 (174.9~620.0)

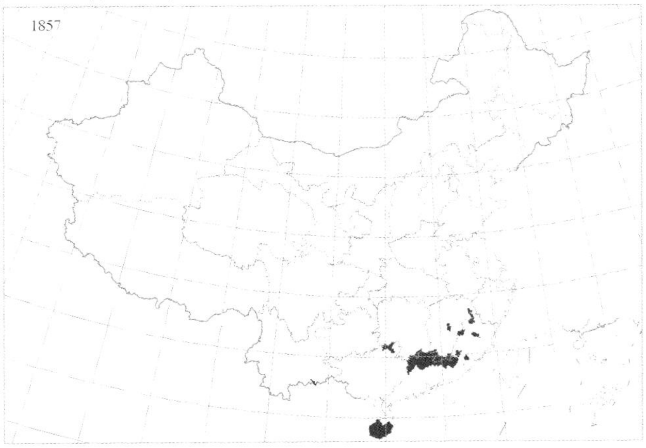

Illiciaceae 八角科 *Illicium* 八角属
Illicium tsaii A. C. Sm. 文山八角
Shrubs or trees, evergreen

18.1 (16.3~21.0)；11.2 (9.1~14.2)；22.9 (21.4~25.6)；18.2 (16.4~21.0)；157.9 (136.4~191.8)；0.0 (0.0~0.0)；880 (797~1053) | 1375 (1074~1751)；775 (582~996)；56 (47~77) | 56.3 (32.4~88.8)；879 (797~1053)；426.2 (277.0~551.6)

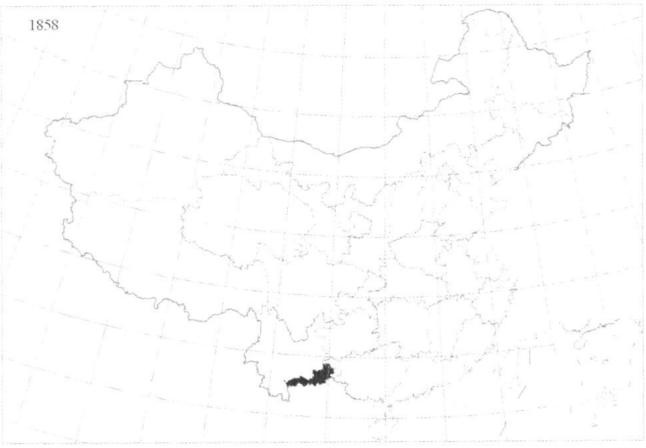

Illiciaceae 八角科 *Illicium* 八角属
Illicium tsangii A. C. Sm. 粤中八角
Shrubs, evergreen

19.9 (15.4~21.8)；10.9 (4.1~14.0)；27.5 (25.8~28.5)；19.9 (15.4~21.8)；179.3 (126.2~201.9)；0.1 (0.0~0.9)；1041 (820~1158) | 1733 (1423~2005)；687 (479~858)；163 (129~181) | 66.5 (39.6~85.1)；1041 (820~1158)；233.8 (188.8~263.8)

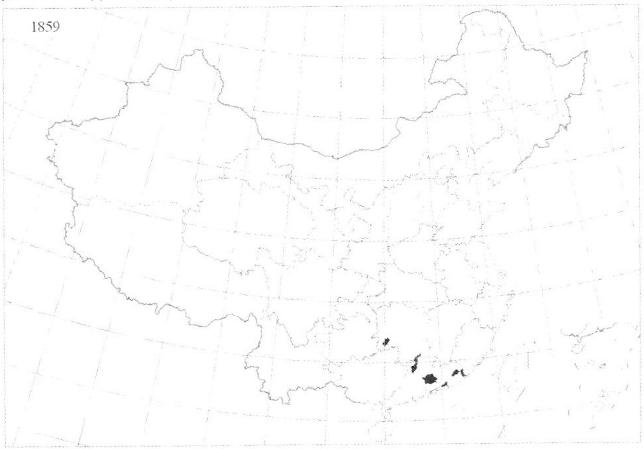

Illiciaceae 八角科 *Illicium* 八角属
Illicium verum Hook. f. 八角
Trees, evergreen

18.1 (12.6~24.6)；10.0 (3.8~19.5)；24.4 (17.9~29.3)；18.1 (12.6~24.7)；157.6 (91.6~235.9)；0.0 (0.0~1.5)；924 (678~1372) | 1308 (797~2435)；658 (411~1428)；84 (23~204) | 41.9 (−14.2~99.9)；915 (678~1299)；288.4 (157.7~659.0)

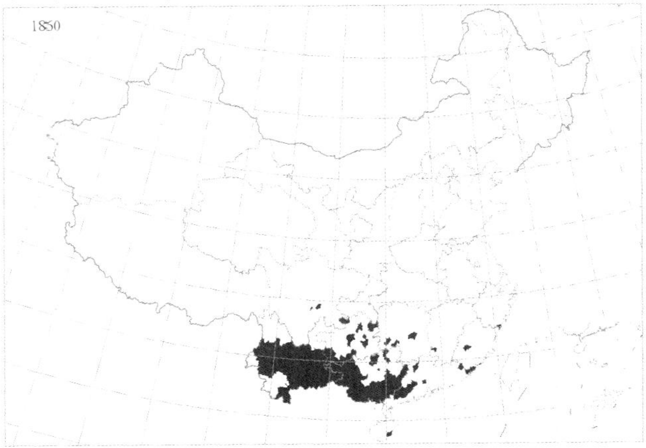

Illiciaceae 八角科 *Illicium* 八角属

Illicium wardii A. C. Sm. 贡山八角

Small trees, evergreen

12.7 (8.6~16.0); 5.8 (1.3~9.3); 18.0 (15.0~20.4); 12.7 (8.7~16.0); 94.8 (52.6~132.4); 1.9 (0.0~8.8); 684 (574~780) ❙ 1285 (1041~1454); 712 (585~792); 59 (45~70) ❙ 87.9 (72.8~99.9); 684 (574~780); 405.3 (352.3~487.2)

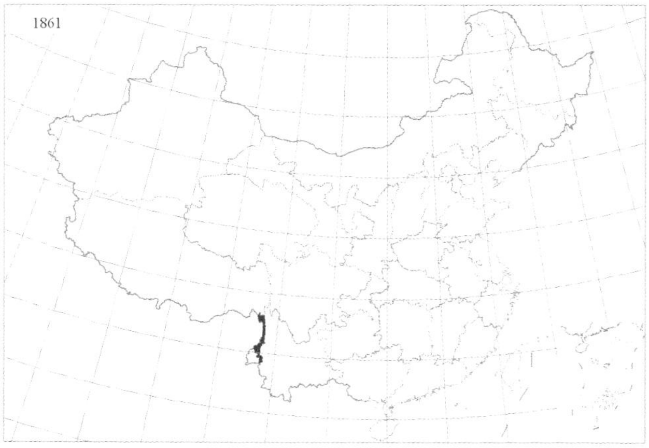

Schisandraceae 五味子科 *Kadsura* 南五味子属

Kadsura angustifolia A. C. Sm. 狭叶南五味子

Vines, evergreen

17.3 (14.6~21.0); 8.6 (3.8~14.2); 24.4 (21.4~29.3); 17.4 (14.7~21.0); 148.8 (117.7~191.8); 0.2 (0.0~1.5); 873 (779~1053) ❙ 1428 (1161~1679); 684 (454~860); 101 (48~195) ❙ 64.3 (37.1~101.8); 873 (779~1053); 343.9 (193.4~551.6)

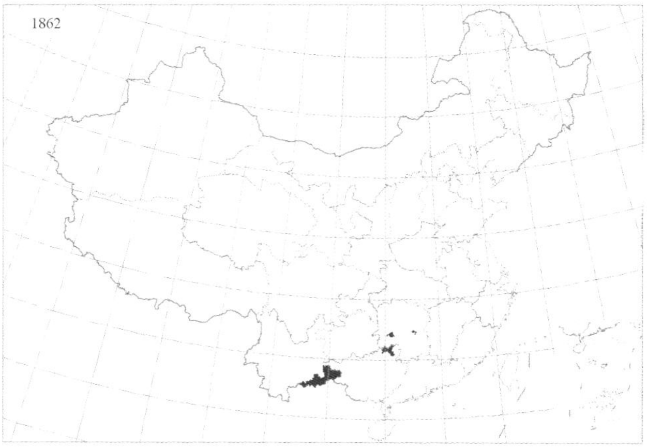

Schisandraceae 五味子科 *Kadsura* 南五味子属

Kadsura coccinea (Lem.) A. C. Sm. 黑老虎

Vines, evergreen

18.0 (10.3~24.8); 8.4 (-0.1~20.1); 26.3 (18.6~30.1); 18.0 (10.3~24.8); 156.7 (75.8~237.6); 0.2 (0.0~11.7); 944 (634~1385) ❙ 1431 (707~2435); 605 (299~1428); 128 (15~259) ❙ 52.2 (-34.3~128.9); 932 (634~1326); 241.5 (135.1~659.0)

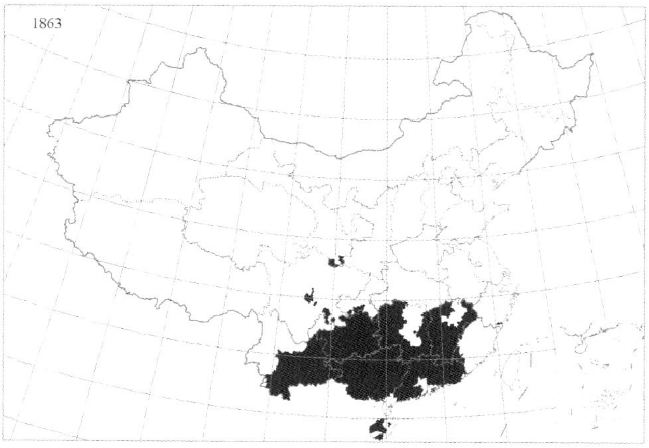

Schisandraceae 五味子科 *Kadsura* 南五味子属

Kadsura heteroclita (Roxb.) Craib 异形南五味子

Vines, evergreen

18.8 (8.6~24.8); 9.7 (-1.7~20.1); 26.5 (15.0~29.8); 18.9 (8.7~24.8); 166.7 (52.6~237.6); 0.4 (0.0~18.0); 989 (574~1385) ❙ 1488 (868~2435); 650 (312~1428); 123 (39~258) ❙ 52.6 (-34.3~145.8); 979 (574~1360); 260.6 (95.3~659.0)

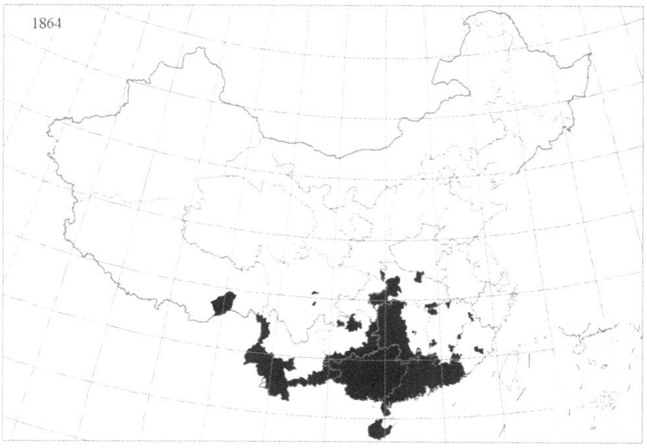

Schisandraceae 五味子科 *Kadsura* 南五味子属

Kadsura induta A. C. Sm. 毛南五味子

Vines, evergreen

20.2 (17.2~22.7); 12.2 (9.1~15.0); 26.1 (22.4~28.6); 20.3 (17.3~22.8); 183.2 (147.0~213.1); 0.0 (0.0~0.0); 1039 (833~1221) ❙ 1490 (1161~2435); 820 (633~1428); 81 (43~130) ❙ 44.0 (5.8~99.5); 1039 (833~1221); 335.2 (256.6~551.6)

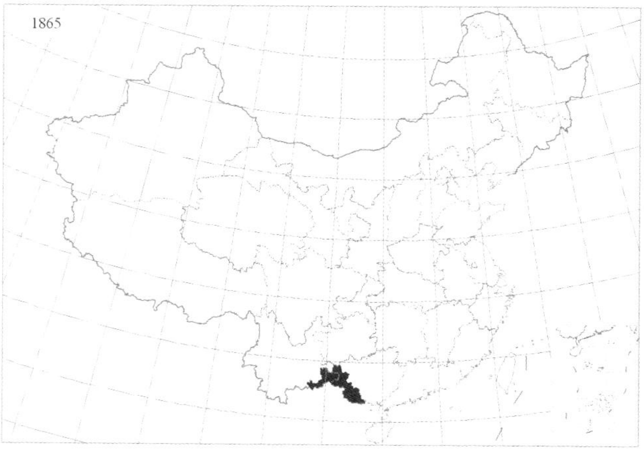

Schisandraceae 五味子科 *Kadsura* 南五味子属

Kadsura japonica (L.) Dunal 日本南五味子

Vines, evergreen

18.9 (16.2~22.2); 11.4 (7.2~17.7); 25.5 (20.1~27.4); 18.9 (16.2~22.2); 166.8 (134.2~206.7); 0.0 (0.0~0.0); 949 (773~1123) ❙ 2241 (1427~3495); 857 (495~1847); 284 (107~897) ❙ 138.2 (50.3~246.5); 946 (773~1123); 278.0 (202.8~363.7)

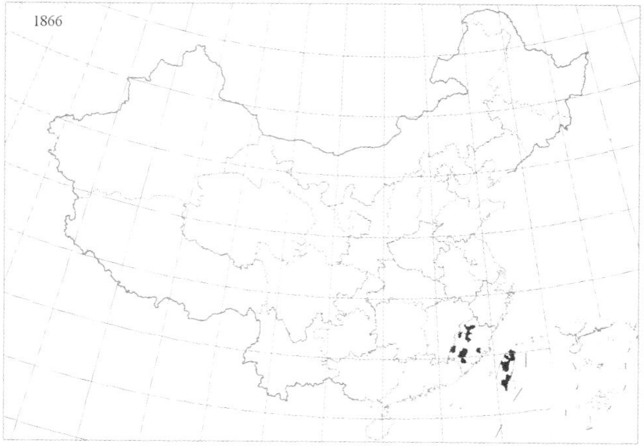

Schisandraceae 五味子科 *Kadsura* 南五味子属
Kadsura longipedunculata Finet et Gagnep. 南五味子
Vines, evergreen

17.5 (7.3~23.6); 7.0 (-2.9~16.7); 27.2 (15.6~30.1); 17.6 (7.5~23.6); 152.2 (48.2~223.5); 1.2 (0.0~20.6); 938 (537~1281) ❙ 1443 (512~2435); 575 (258~1428); 145 (16~259) ❙ 54.6 (-35.3~147.0); 926 (512~1273); 217.3 (95.3~447.4)

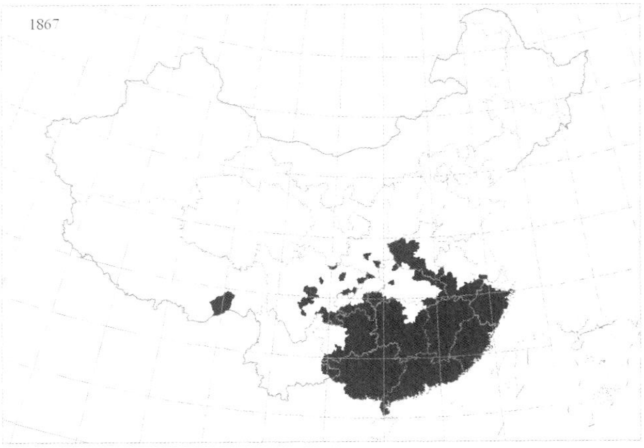

Schisandraceae 五味子科 *Kadsura* 南五味子属
Kadsura oblongifolia Merr. 冷饭藤
Vines, evergreen

21.0 (14.3~24.5); 12.4 (3.3~19.1); 27.9 (24.3~29.1); 21.0 (14.3~24.5); 192.5 (114.1~234.4); 0.0 (0.0~2.3); 1114 (765~1360) ❙ 1639 (868~2435); 737 (312~1428); 138 (48~196) ❙ 48.8 (-34.3~101.8); 1105 (765~1360); 233.1 (95.3~620.0)

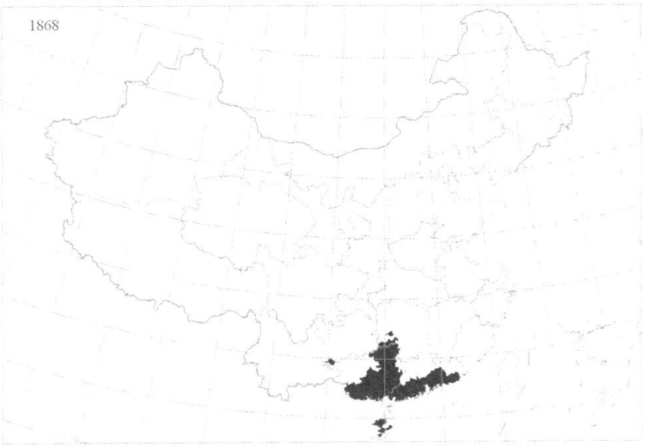

Schisandraceae 五味子科 *Kadsura* 南五味子属
Kadsura renchangiana S. F. Lan 仁昌南五味子
Vines, evergreen

17.2 (14.6~19.2); 7.1 (3.8~10.1); 25.7 (23.4~27.7); 17.2 (14.7~19.2); 146.7 (117.7~170.8); 0.3 (0.0~1.5); 887 (779~1002) ❙ 1407 (1221~1624); 621 (473~723); 117 (51~191) ❙ 59.2 (30.0~101.8); 887 (779~1002); 243.9 (170.4~318.5)

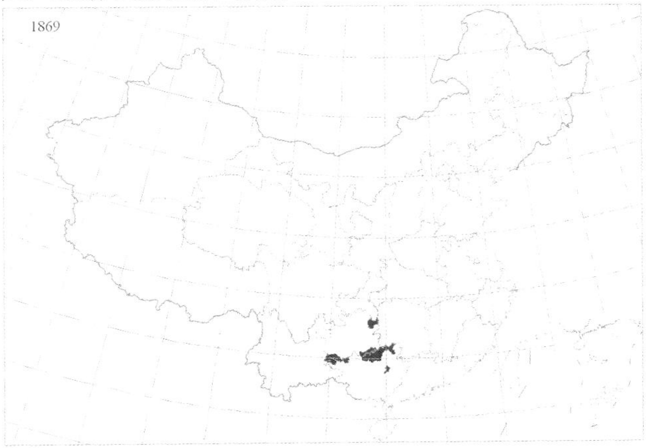

Schisandraceae 五味子科 *Schisandra* 五味子属
Schisandra arisanensis Hayata 阿里山五味子
Vines, evergreen

20.0 (16.2~23.3); 14.3 (11.0~17.7); 24.6 (20.1~28.0); 20.0 (16.2~23.3); 180.2 (134.2~219.7); 0.0 (0.0~0.0); 1006 (773~1248) ❙ 2283 (1498~2993); 1131 (859~1594); 196 (73~547) ❙ 139.7 (25.2~246.5); 973 (773~1144); 239.0 (179.2~303.6)

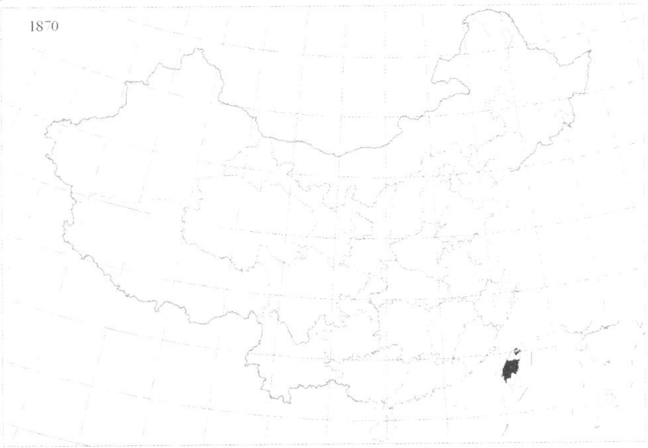

Schisandraceae 五味子科 *Schisandra* 五味子属
Schisandra arisanensis subsp. viridis (A. C. Smith) R. M. K. Saunders 绿叶五味子
Vines, deciduous

17.9 (12.5~22.3); 7.7 (1.6~14.2); 26.9 (20.1~29.8); 18.0 (12.5~22.3); 155.9 (92.0~208.1); 0.3 (0.0~6.5); 942 (685~1202) ❙ 1502 (1071~1997); 599 (392~835); 155 (43~254) ❙ 60.7 (5.8~130.7); 938 (685~1202); 240.1 (145.3~551.6)

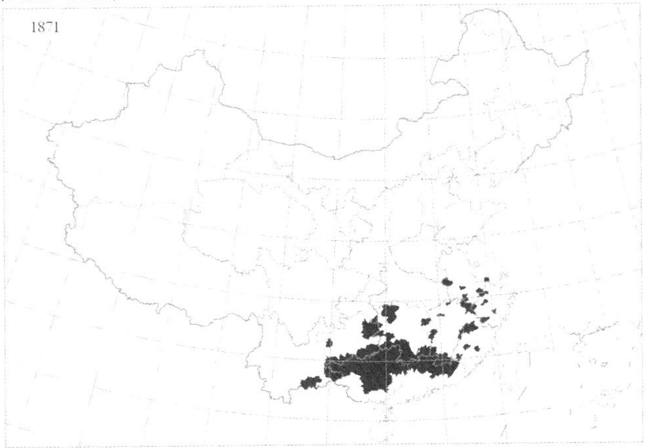

Schisandraceae 五味子科 *Schisandra* 五味子属
Schisandra bicolor Cheng 二色五味子
Vines, deciduous

15.7 (6.1~18.7); 4.2 (-1.5~8.6); 26.6 (12.7~29.2); 15.8 (6.3~18.8); 130.8 (35.2~165.2); 1.8 (0.0~21.8); 855 (496~976) ❙ 1523 (801~1793); 544 (445~666); 185 (21~238) ❙ 78.6 (36.1~128.9); 852 (496~976); 240.7 (169.7~319.9)

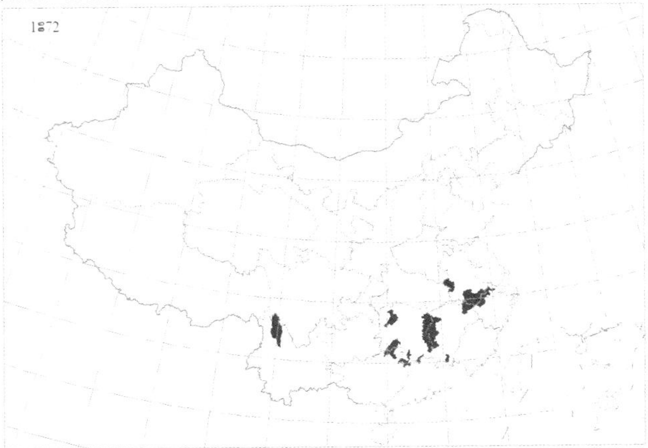

Schisandraceae 五味子科 *Schisandra* 五味子属

Schisandra chinensis (Turcz.) Baill. 五味子

Vines, deciduous

9.1 (−4.0~16.9); −7.5 (−28.2~4.4); 23.8 (16.5~29.2); 11.0 (5.0~16.9); 90.4 (34.9~143.8); 41.2 (0.6~146.2); 698 (448~936) ❙ 684 (196~1571); 390 (117~691); 33 (4~175) ❙ −2.1 (−69.4~108.0); 612 (196~930); 203.9 (60.5~346.5)

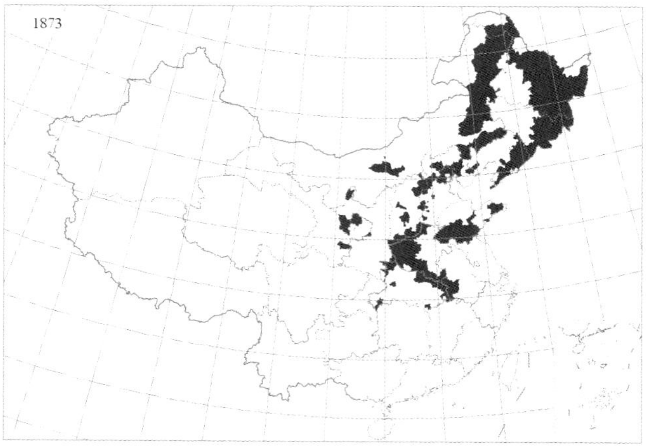

Schisandraceae 五味子科 *Schisandra* 五味子属

Schisandra glaucescens Diels 金山五味子

Vines, deciduous

13.1 (9.3~16.2); 2.0 (−1.7~5.5); 23.9 (20.1~27.9); 13.2 (9.5~16.2); 104.5 (69.8~134.8); 6.2 (0.0~18.0); 740 (603~883) ❙ 1194 (862~1495); 502 (376~602); 64 (26~173) ❙ 62.3 (7.8~107.2); 740 (603~883); 247.3 (172.9~305.6)

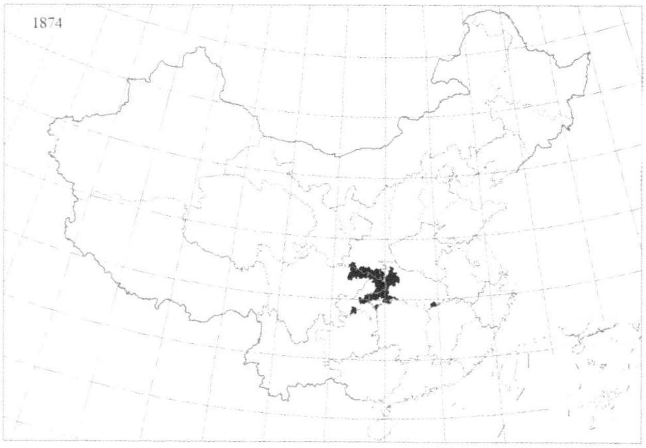

Schisandraceae 五味子科 *Schisandra* 五味子属

Schisandra grandiflora (Wall.) Hook. f. et Thoms. 大花五味子

Vines, deciduous

15.8 (−1.8~20.5); 9.1 (−10.2~15.1); 20.7 (6.1~26.5); 16.1 (1.8~20.5); 135.1 (1.9~185.9); 4.6 (0.0~82.8); 799 (329~992) ❙ 1228 (320~1751); 675 (193~1007); 51 (10~113) ❙ 51.9 (−4.0~145.8); 783 (320~986); 394.8 (61.6~659.0)

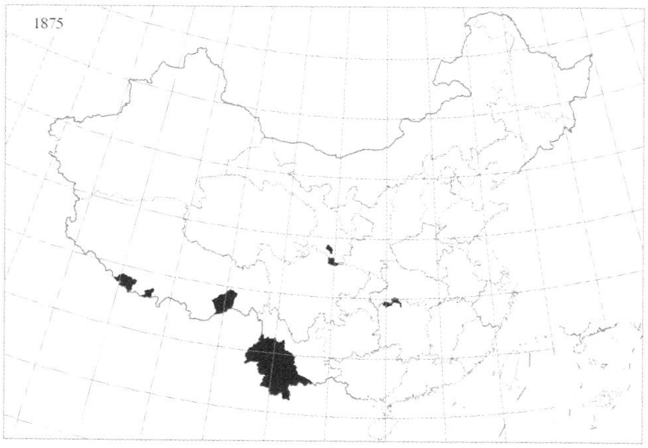

Schisandraceae 五味子科 *Schisandra* 五味子属

Schisandra henryi C. B. Clarke 翼梗五味子

Vines, deciduous

16.8 (8.6~22.3); 6.7 (−0.1~15.1); 25.7 (15.0~30.1); 16.8 (8.7~22.3); 142.0 (52.6~208.1); 0.4 (0.0~11.7); 879 (574~1202) ❙ 1370 (711~1997); 570 (299~1007); 124 (15~259) ❙ 56.2 (−14.2~147.0); 867 (574~1202); 241.5 (130.1~659.0)

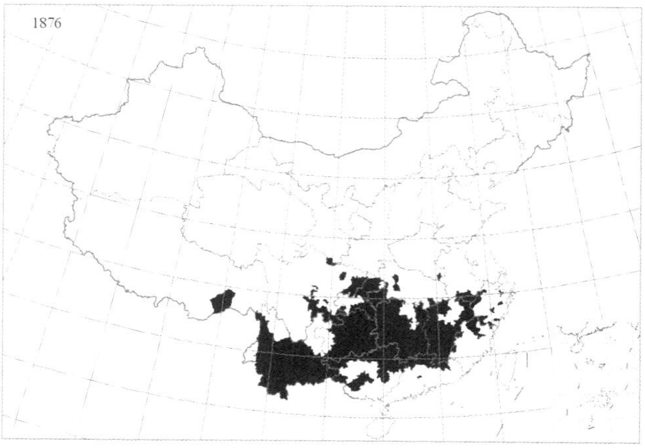

Schisandraceae 五味子科 *Schisandra* 五味子属

Schisandra incarnata Stapf 兴山五味子

Vines, deciduous

13.7 (9.3~16.8); 1.9 (−1.7~4.3); 25.1 (20.1~29.1); 13.8 (9.5~16.9); 111.7 (69.8~143.0); 6.3 (0.7~18.0); 780 (603~930) ❙ 1162 (854~1439); 493 (389~579); 88 (46~172) ❙ 51.7 (−5.3~101.4); 776 (603~930); 227.6 (133.5~279.5)

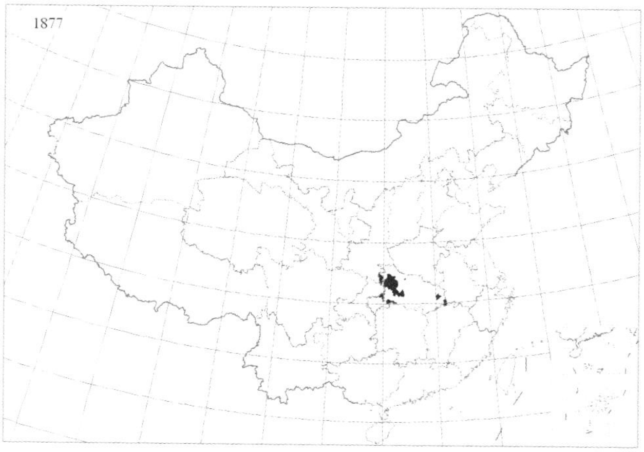

Schisandraceae 五味子科 *Schisandra* 五味子属

Schisandra lancifolia (Rehder et E. H. Wilson) A. C. Sm. 狭叶五味子

Vines, deciduous

14.0 (2.0~21.0); 6.5 (−7.2~14.2); 19.9 (9.9~28.4); 14.1 (3.8~21.0); 112.3 (15.4~191.8); 4.2 (0.0~51.4); 740 (413~1053) ❙ 1101 (741~1751); 519 (389~996); 39 (12~225) ❙ 51.1 (−14.2~128.6); 722 (413~1053); 323.6 (142.9~551.6)

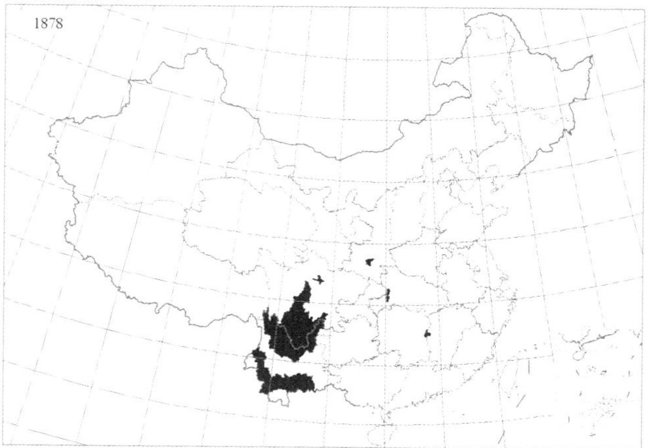

Schisandraceae 五味子科 *Schisandra* 五味子属

Schisandra longipes (Merr. et Chun) R. M. K. Saunders 长柄五味子

Vines, deciduous

17.5 (15.5~18.5); 7.2 (4.8~8.6); 26.7 (25.1~27.5); 17.5 (15.6~18.5); 150.5 (127.2~162.1); 0.0 (0.0~0.2); 911 (815~953) ❙ 1594 (1503~1740); 589 (513~709); 181 (176~187) ❙ 75.5 (61.1~93.2); 911 (815~953); 224.8 (197.5~256.8)

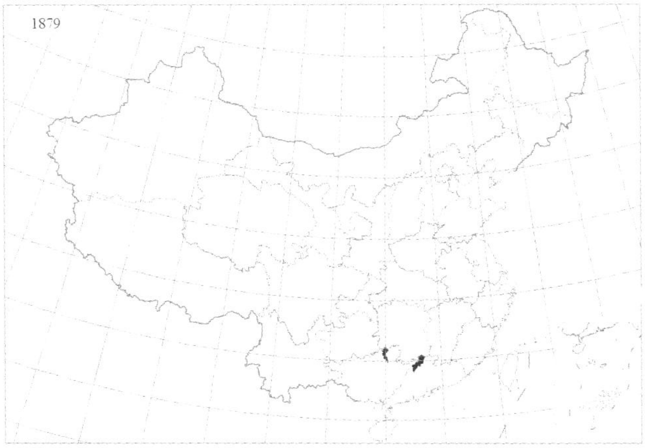

Schisandraceae 五味子科 *Schisandra* 五味子属

Schisandra micrantha A. C. Sm. 小花五味子

Vines, deciduous

16.5 (11.7~21.6); 8.7 (2.7~14.2); 22.5 (18.6~27.9); 16.6 (11.7~21.6); 138.8 (84.6~199.6); 0.2 (0.0~3.8); 827 (661~1143) ❙ 1167 (896~1551); 632 (471~911); 56 (31~169) ❙ 41.3 (12.2~80.6); 821 (661~1143); 287.4 (170.4~551.6)

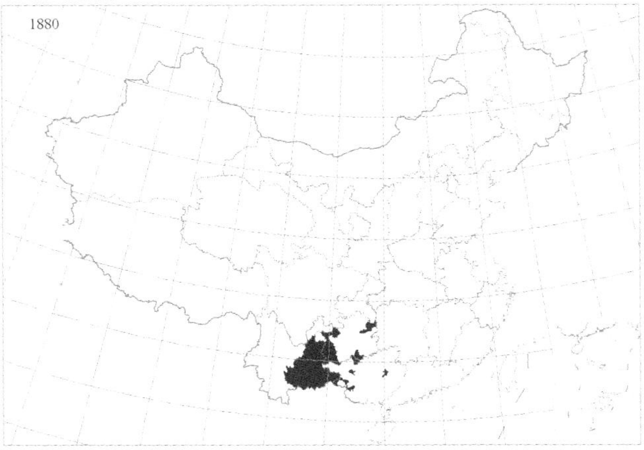

Schisandraceae 五味子科 *Schisandra* 五味子属

Schisandra neglecta A. C. Sm. 滇藏五味子

Vines, deciduous

14.1 (-0.7~19.3); 6.8 (-9.6~13.8); 19.7 (7.7~26.6); 14.2 (2.5~19.4); 112.4 (6.8~173.2); 3.1 (0.0~74.2); 735 (338~939) ❙ 1049 (370~1736); 601 (262~1007); 37 (10~70) ❙ 44.2 (-14.2~145.8); 712 (338~912); 312.1 (61.6~516.7)

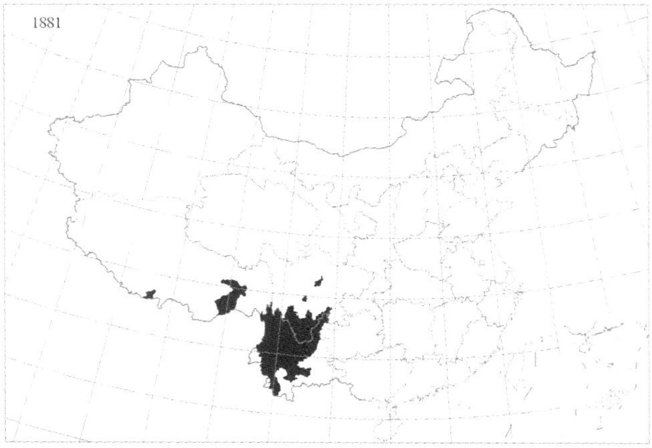

Schisandraceae 五味子科 *Schisandra* 五味子属

Schisandra plena A. C. Sm. 重瓣五味子

Vines, evergreen

19.2; 13.3; 22.8; 19.3; 171.1; 0.0; 923 ❙ 1442; 738; 58 ❙ 56.5; 899; 504.9

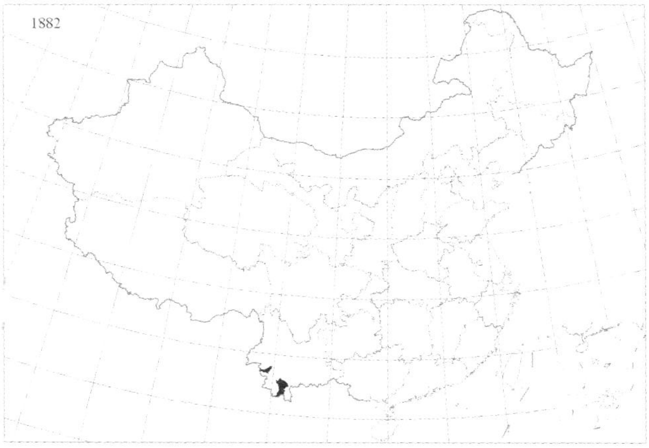

Schisandraceae 五味子科 *Schisandra* 五味子属

Schisandra propinqua (Wall.) Baill. 合蕊五味子

Vines, deciduous

14.7 (-1.1~21.0); 5.0 (-9.1~14.2); 23.4 (6.2~30.1); 14.8 (2.0~21.0); 120.4 (2.5~191.8); 3.2 (0.0~75.6); 794 (334~1053) ❙ 1166 (321~1820); 543 (193~911); 80 (12~250) ❙ 47.2 (-14.2~125.9); 779 (321~1053); 254.9 (79.3~551.6)

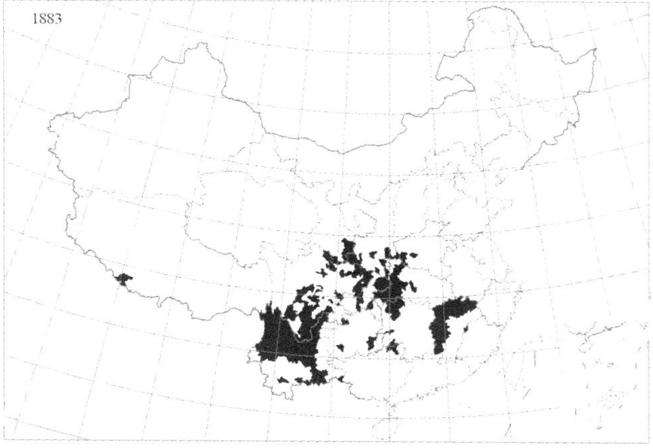

Schisandraceae 五味子科 *Schisandra* 五味子属

Schisandra pubescens Hemsl. et E. H. Wilson 毛叶五味子

Vines, deciduous

14.8 (2.0~19.1); 4.6 (-7.2~13.0); 24.1 (10.1~29.3); 14.9 (3.8~19.1); 120.5 (15.4~169.7); 2.6 (0.0~51.4); 795 (413~1002) ❙ 1226 (943~1679); 566 (445~911); 74 (22~221) ❙ 56.3 (8.0~128.6); 794 (413~1002); 216.0 (108.2~551.3)

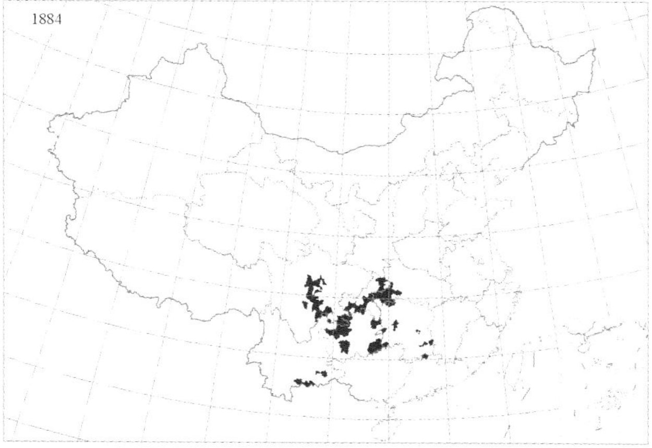

Schisandraceae 五味子科 *Schisandra* 五味子属
Schisandra pubinervis (Rehder et E. H. Wilson) R. M. K. Saunders 毛脉五味子
Vines

10.7 (2.0~14.9); 1.3 (−7.2~5.3); 19.4 (10.1~24.9); 11.1 (3.8~15.0); 81.8 (15.4~119.6); 12.2 (0.0~51.4); 653 (413~791) ❙ 1181 (943~1414); 633 (489~817); 41 (23~71) ❙ 84.2 (51.2~128.6); 653 (413~791); 242.4 (182.9~292.6)

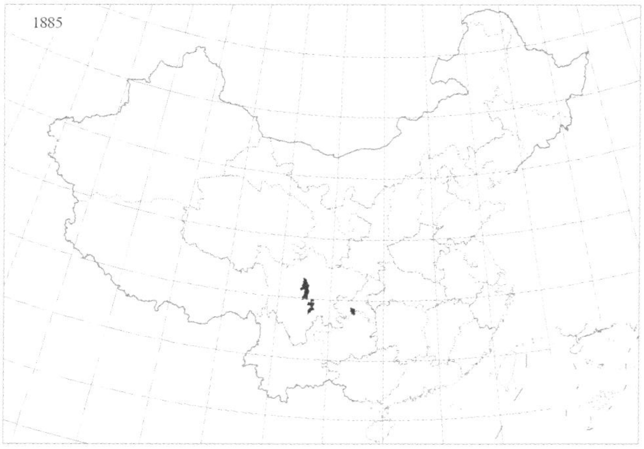

Schisandraceae 五味子科 *Schisandra* 五味子属
Schisandra rubriflora (Franch.) Rehd. et Wils. 红花五味子
Vines, deciduous

11.2 (0.3~18.9); 2.1 (−10.7~12.0); 19.2 (8.3~26.6); 11.5 (2.9~19.0); 86.4 (9.6~167.8); 11.3 (0.0~68.1); 664 (368~929) ❙ 971 (389~1736); 530 (246~1007); 31 (11~106) ❙ 48.3 (−14.2~145.8); 651 (368~912); 265.3 (113.6~487.2)

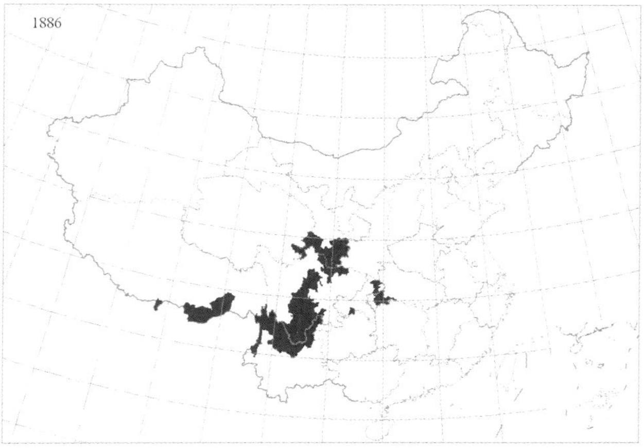

Schisandraceae 五味子科 *Schisandra* 五味子属
Schisandra sphaerandra Stapf 球蕊五味子
Vines, deciduous

12.0 (−1.8~18.9); 4.6 (−10.2~12.0); 18.0 (6.1~24.3); 12.3 (1.8~19.0); 92.7 (1.9~167.8); 8.6 (0.0~82.8); 673 (329~929) ❙ 963 (320~1454); 560 (193~817); 31 (9~70) ❙ 45.4 (−14.2~121.7); 649 (320~809); 296.9 (72.2~487.2)

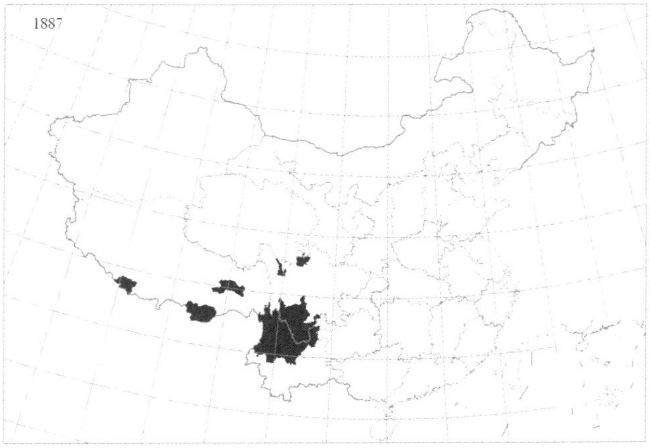

Schisandraceae 五味子科 *Schisandra* 五味子属
Schisandra sphenanthera Rehd. et Wils. 华中五味子
Vines, deciduous

14.2 (2.2~21.0); 2.3 (−13.7~14.2); 25.1 (9.9~30.1); 14.5 (3.8~21.1); 119.5 (15.5~193.0); 8.7 (0.0~71.7); 803 (413~1114) ❙ 1095 (367~1997); 473 (235~860); 90 (8~259) ❙ 34.4 (−40.9~147.0); 761 (367~1114); 212.2 (101.9~551.6)

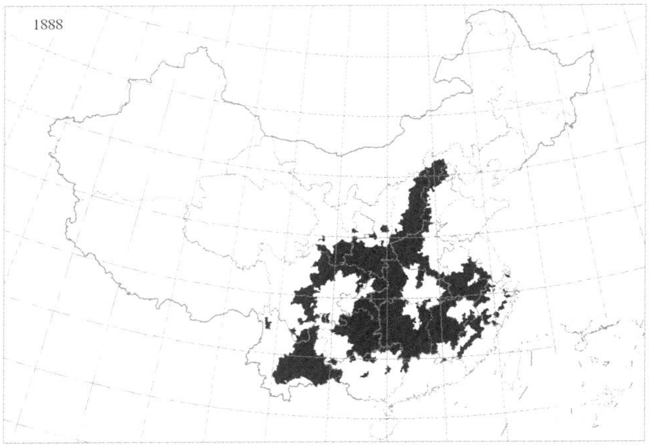

Schisandraceae 五味子科 *Schisandra* 五味子属
Schisandra tomentella A. C. Sm. 柔毛五味子
Vines, deciduous

12.0; 3.0; 20.3; 12.1; 88.7; 3.3; 677 ❙ 1106; 616; 37 ❙ 63.8; 677; 275.5

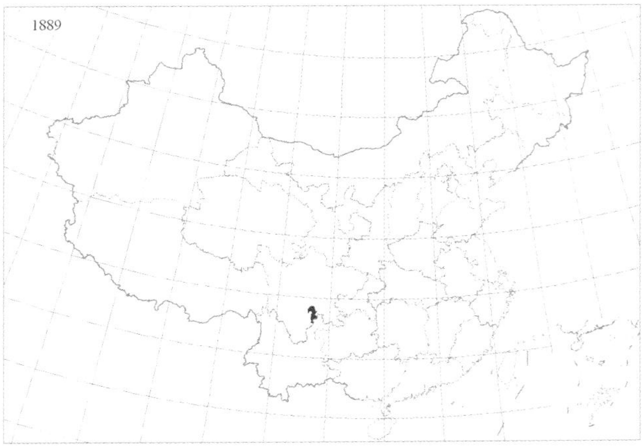

Magnoliaceae 木兰科 *Alcimandra* 长蕊木兰属
Alcimandra cathcartii (Hook. f. et Thomson) Dandy 长蕊木兰
Trees, deciduous

13.8 (−1.8~21.0); 6.7 (−10.5~15.1); 19.0 (6.1~25.6); 14.3 (1.8~21.0); 117.9 (1.9~191.8); 12.2 (0.0~82.8); 742 (329~1053) ❙ 1153 (255~1751); 648 (191~1007); 45 (1~77) ❙ 50.4 (−31.1~145.8); 725 (255~1053); 368.0 (60.4~659.0)

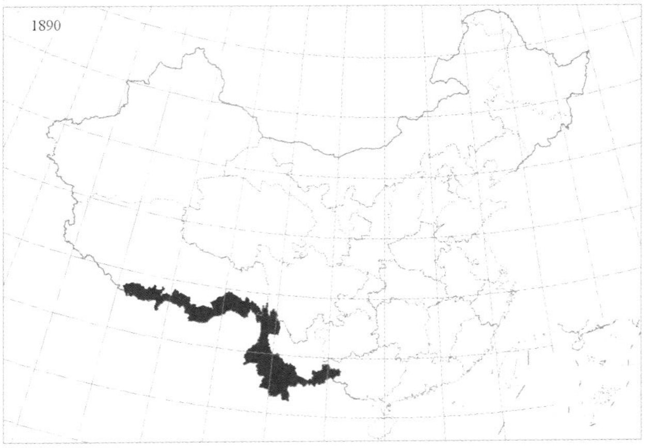

Magnoliaceae 木兰科 *Houpoea* 厚朴属

Houpoea officinalis (Rehder et E. H. Wilson) N. H. Xia et C. Y. Wu 厚朴

Trees, deciduous

15.6 (2.2~21.3); 4.9 (-6.6~12.3); 25.6 (9.9~30.1); 15.7 (3.8~21.3); 130.3 (15.5~196.1); 2.2 (0.0~49.1); 844 (413~1130) ❙ 1351 (619~1997); 544 (299~946); 124 (14~259) ❙ 60.1 (-8.1~147.0); 841 (413~1130); 222.6 (108.2~332.5)

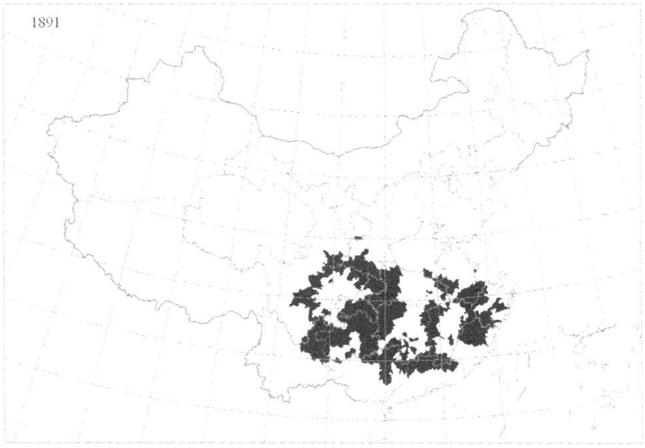

Magnoliaceae 木兰科 *Houpoea* 厚朴属

Houpoea rostrata (W. W. Smith) N. H. Xia et C. Y. Wu 长喙厚朴

Trees, deciduous

11.9 (6.2~14.9); 4.3 (-2.4~7.5); 18.1 (14.1~23.6); 12.0 (6.7~15.0); 87.4 (39.4~119.6); 4.2 (0.0~24.1); 666 (509~782) ❙ 1296 (981~1736); 725 (539~1007); 53 (23~70) ❙ 94.2 (68.1~145.8); 666 (509~782); 360.1 (214.5~487.2)

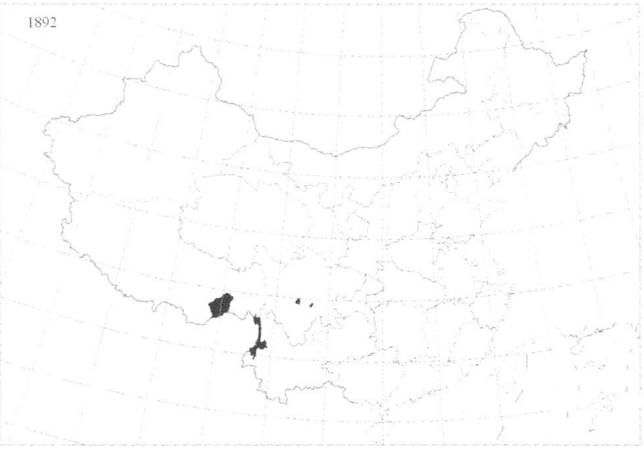

Magnoliaceae 木兰科 *Lirianthe* 长喙木兰属

Lirianthe albosericea (Chun et C. Tsoong) N. H. Xia et C. Y. Wu 绢毛木兰

Small trees, evergreen

22.9 (21.3~24.7); 16.2 (13.2~19.8); 27.7 (27.2~28.6); 22.9 (21.3~24.7); 215.2 (195.9~236.5); 0.0 (0.0~0.0); 1232 (1111~1376) ❙ 1924 (1551~2435); 958 (534~1428); 97 (48~130) ❙ 58.3 (15.4~99.5); 1168 (1111~1221); 406.2 (297.2~620.0)

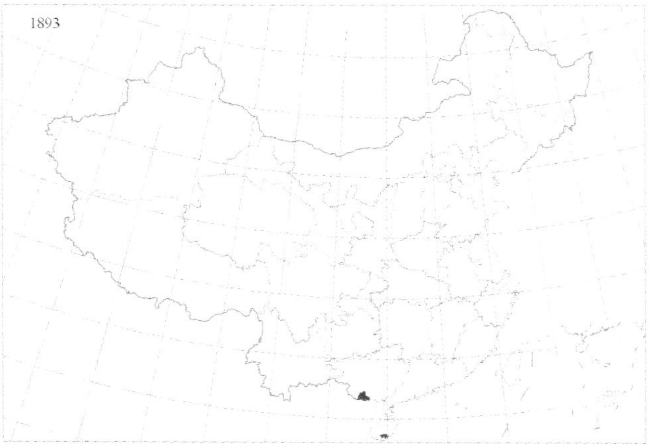

Magnoliaceae 木兰科 *Lirianthe* 长喙木兰属

Lirianthe championii (Bentham) N. H. Xia et C. Y. Wu 香港木兰

Shrubs or small trees, evergreen

20.9 (14.3~24.8); 12.7 (4.0~20.1); 27.6 (23.4~29.8); 20.9 (14.3~24.8); 191.2 (112.6~237.6); 0.0 (0.0~1.1); 1107 (756~1385) ❙ 1579 (868~2435); 700 (312~1428); 117 (41~194) ❙ 44.3 (-34.3~99.5); 1090 (756~1360); 253.9 (95.3~620.0)

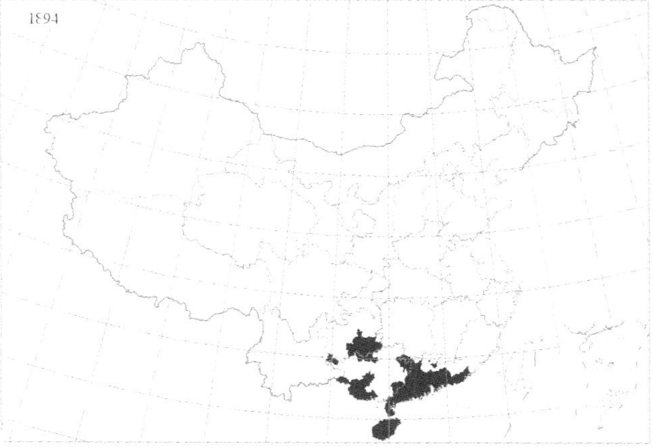

Magnoliaceae 木兰科 *Lirianthe* 长喙木兰属

Lirianthe coco (Lour.) N. H. Xia et C. Y. Wu 夜香木兰

Shrubs or small trees, evergreen

19.4 (13.3~22.8); 10.6 (3.2~15.2); 27.1 (22.1~28.9); 19.6 (13.2~22.8); 175.3 (100.9~213.6); 0.1 (0.0~2.5); 1033 (723~1230) ❙ 1561 (950~1965); 733 (458~994); 114 (27~197) ❙ 51.7 (12.1~80.4); 1033 (723~1230); 214.5 (126.1~290.3)

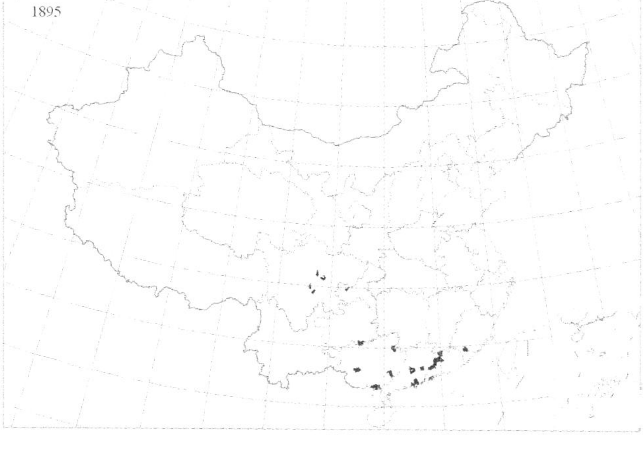

Magnoliaceae 木兰科 *Lirianthe* 长喙木兰属

Lirianthe delavayi (Franch.) N. H. Xia et C. Y. Wu 山玉兰

Trees, evergreen

16.0 (5.4~22.0); 6.4 (-2.1~15.1); 24.4 (12.3~29.7); 16.0 (5.6~22.0); 133.1 (31.5~203.8); 1.0 (0.0~25.6); 839 (481~1170) ❙ 1254 (741~1893); 567 (389~996); 101 (15~253) ❙ 51.1 (-14.2~129.9); 827 (481~1170); 255.3 (108.2~659.0)

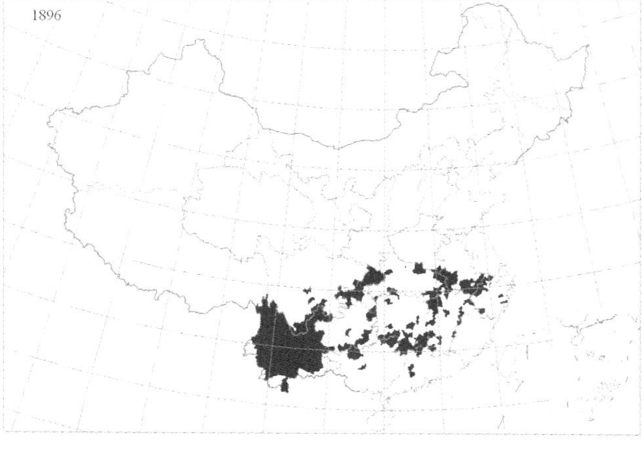

Magnoliaceae 木兰科 *Lirianthe* 长喙木兰属
Lirianthe fistulosa (Finet et Gagnepain) N. H. Xia et C. Y. Wu 显脉木兰
Shrubs or small trees, evergreen

21.4 (14.3~24.8); 13.8 (4.0~20.1); 27.4 (21.4~29.8); 21.5 (14.3~24.8); 197.9 (112.6~237.6); 0.0 (0.0~1.1); 1135 (756~1385) ❙ 1614 (868~2435); 718 (312~1428); 113 (41~194) ❙ 44.0 (-34.3~99.5); 1108 (756~1360); 298.1 (157.7~620.0)

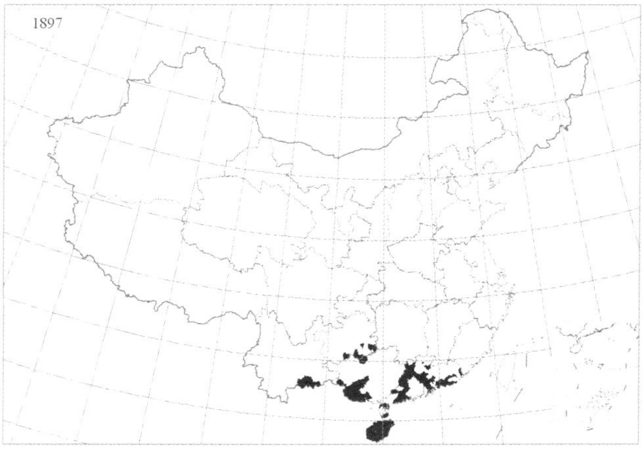

Magnoliaceae 木兰科 *Lirianthe* 长喙木兰属
Lirianthe fujianensis N. H. Xia et C. Y. Wu 福建木兰
Trees, evergreen

18.6; 9.3; 27.6; 18.7; 164.0; 0.0; 965 ❙ 1702; 591; 199 ❙ 76.4; 965; 300.0

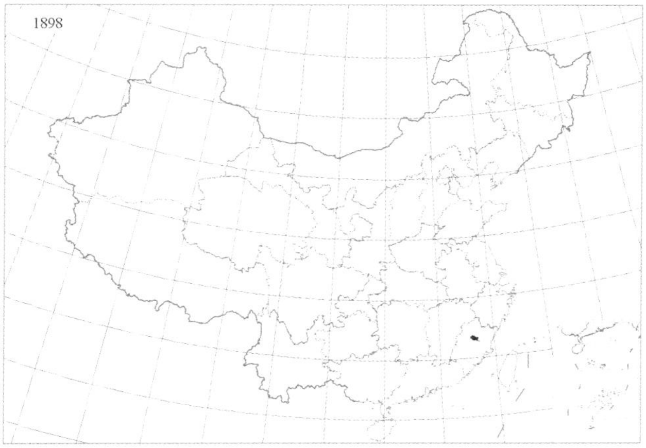

Magnoliaceae 木兰科 *Lirianthe* 长喙木兰属
Lirianthe henryi Dunn 大叶木兰
Trees, evergreen

18.1 (14.4~21.0); 11.8 (7.5~15.1); 22.2 (19.1~25.6); 18.1 (14.4~21.0); 157.5 (113.1~191.8); 0.0 (0.0~0.0); 873 (727~1053) ❙ 1364 (1049~1751); 742 (594~996); 55 (41~77) ❙ 56.4 (26.1~99.9); 859 (727~1053); 450.4 (313.7~659.0)

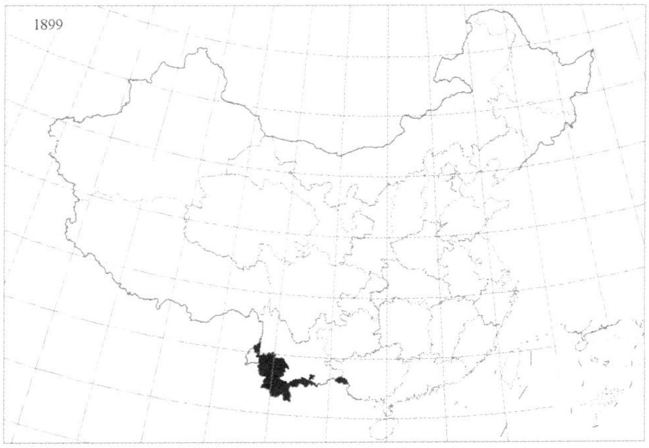

Magnoliaceae 木兰科 *Lirianthe* 长喙木兰属
Lirianthe odoratissima (Y. W. Law et R. Z. Zhou) N. H. Xia et C. Y. Wu 馨香木兰
Trees, evergreen

18.5 (16.5~21.8); 10.7 (9.0~14.2); 24.2 (21.9~27.8); 18.6 (16.6~21.9); 163.1 (139.2~202.6); 0.0 (0.0~0.0); 928 (810~1150) ❙ 1409 (1018~2148); 773 (554~1234); 66 (47~124) ❙ 50.0 (25.7~86.7); 928 (810~1150); 322.7 (273.3~371.7)

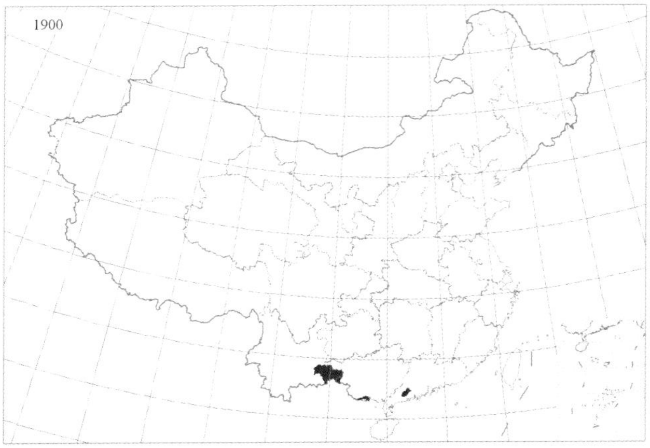

Magnoliaceae 木兰科 *Liriodendron* 鹅掌楸属
Liriodendron chinense (Hemsl.) Sarg. 鹅掌楸
Trees, deciduous

15.7 (9.3~22.1); 4.9 (-3.6~14.9); 25.9 (19.4~29.4); 15.8 (9.5~22.1); 131.2 (69.8~205.6); 1.9 (0.0~21.0); 844 (603~1182) ❙ 1325 (739~1997); 550 (376~860); 120 (26~259) ❙ 57.6 (-2.1~147.0); 842 (603~1182); 229.7 (108.2~570.1)

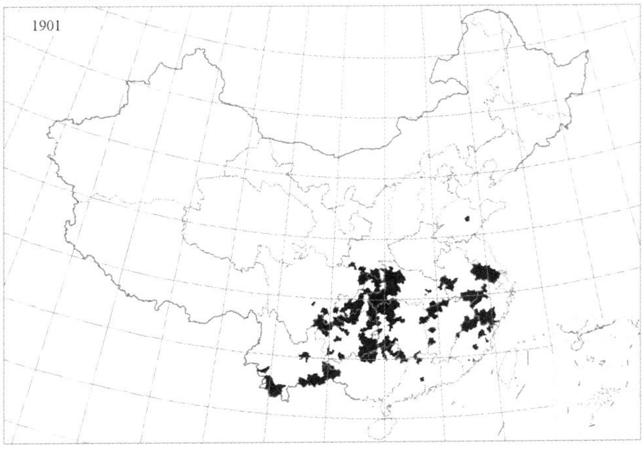

Magnoliaceae 木兰科 *Manglietia* 木莲属
Manglietia aromatica Dandy 香木莲
Trees, evergreen

18.5 (15.0~22.2); 9.7 (4.5~14.2); 25.4 (21.4~28.3); 18.5 (15.0~22.3); 162.3 (120.8~207.5); 0.0 (0.0~0.5); 940 (794~1190) ❙ 1314 (1072~1581); 677 (488~860); 76 (43~165) ❙ 40.7 (5.8~80.6); 940 (794~1190); 296.7 (162.8~551.6)

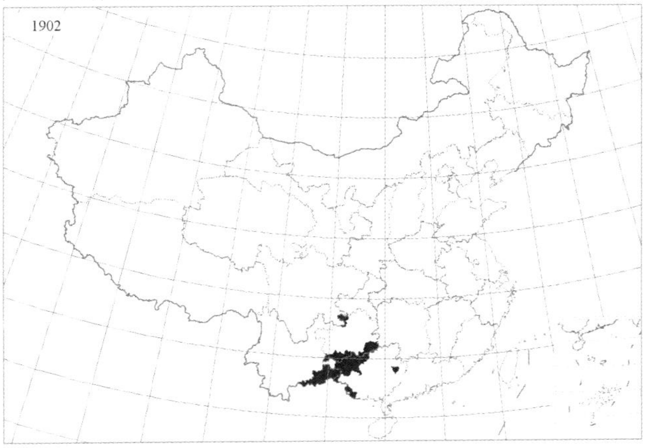

Magnoliaceae 木兰科 *Manglietia* 木莲属
Manglietia calcarea X. H. Song 石山木莲
Trees, evergreen

17.4；7.4；25.5；17.4；148.9；0.0；886 ❙ 1381；637；107 ❙ 55.9；886；245.3

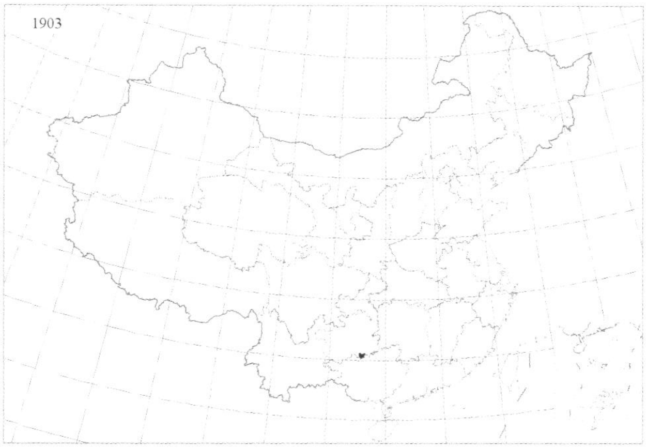

Magnoliaceae 木兰科 *Manglietia* 木莲属
Manglietia caveana Hook. f. et Thoms. 西藏木莲
Trees, evergreen

13.3；5.4；19.6；13.4；100.4；0.0；706 ❙ 1736；1007；51 ❙ 145.8；706；447.4

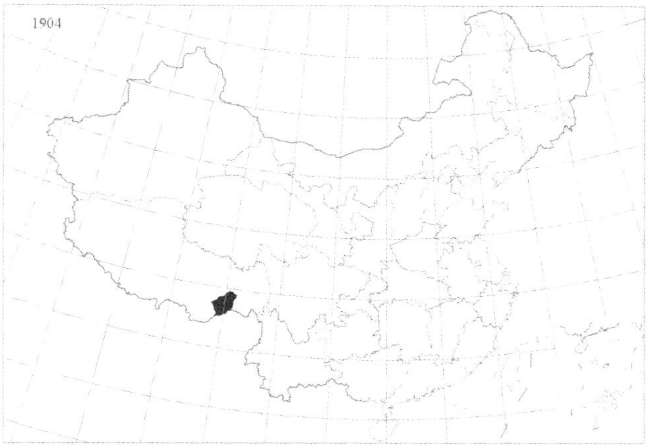

Magnoliaceae 木兰科 *Manglietia* 木莲属
Manglietia chevalieri Dandy 睦南木莲
Trees, evergreen

20.4；15.1；23.6；20.5；185.8；0.0；989 ❙ 1607；702；77 ❙ 62.5；986；659.0

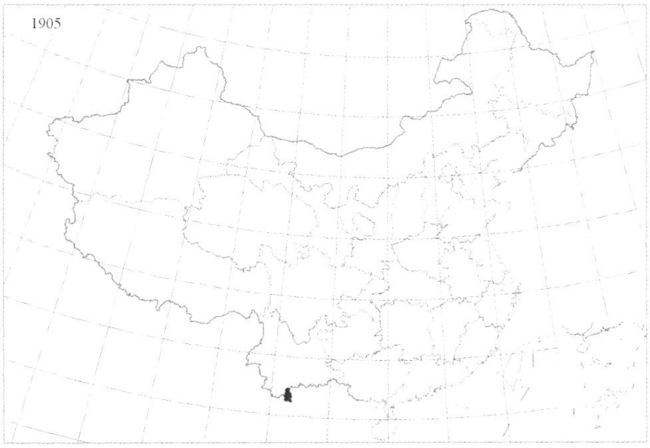

Magnoliaceae 木兰科 *Manglietia* 木莲属
Manglietia conifera Dandy 桂南木莲
Trees, evergreen

18.3 (14.2~22.7)；8.6 (3.3~15.0)；26.6 (21.4~29.0)；18.4 (14.2~22.8)；160.5 (112.5~213.1)；0.1 (0.0~2.3)；956 (756~1221) ❙ 1494 (1072~2435)；639 (403~1428)；140 (43~230) ❙ 57.4 (5.8~128.9)；954 (756~1221)；246.7 (151.0~491.5)

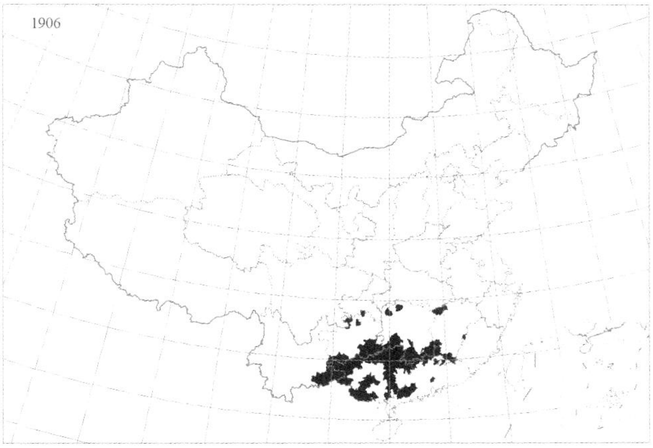

Magnoliaceae 木兰科 *Manglietia* 木莲属
Manglietia crassipes Y. W. Law 粗梗木莲
Shrubs or trees, evergreen

18.0；8.5；25.8；18.0；156.4；0.0；913 ❙ 1551；672；169 ❙ 69.8；913；243.9

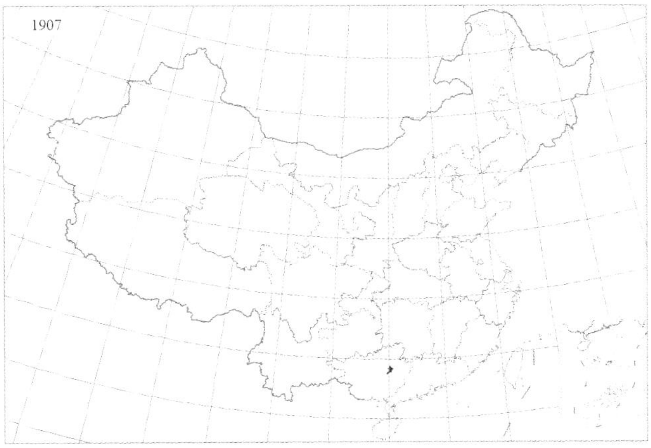

Magnoliaceae 木兰科 *Manglietia* 木莲属
Manglietia dandyi (Gagnep.) Dandy 大叶木莲
Trees, evergreen

18.4 (16.3~21.3)；10.6 (9.1~13.8)；24.1 (21.4~27.9)；18.5 (16.4~21.3)；151.9 (136.4~196.0)；0.0 (0.0~0.0)；914 (797~1123) ❙ 1342 (1161~1504)；734 (633~860)；58 (48~96) ❙ 48.1 (5.8~80.6)；911 (797~1121)；360.8 (282.2~513.9)

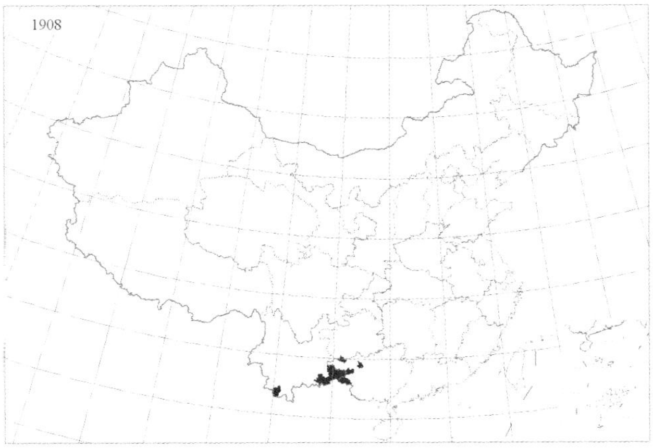

Magnoliaceae 木兰科 *Manglietia* 木莲属

Manglietia decidua Q. Y. Zheng 落叶木莲

Trees, deciduous

17.0; 5.1; 28.4; 17.1; 145.0; 0.0; 921 ▎1595; 493; 225 ▎73.2; 911; 220.5

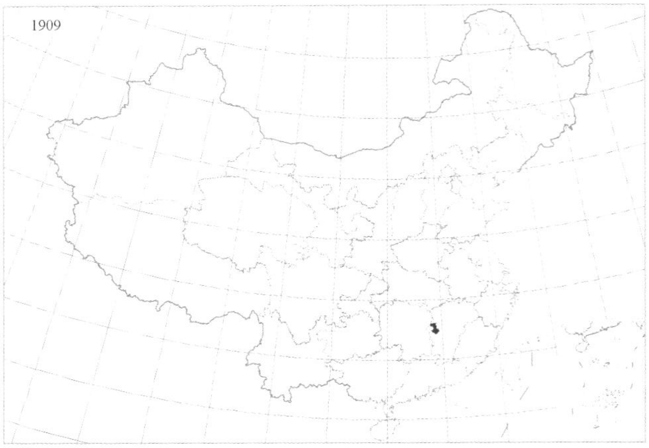

Magnoliaceae 木兰科 *Manglietia* 木莲属

Manglietia duclouxii Finet et Gagnepain 川滇木莲

Trees, evergreen

17.1 (12.8~21.0); 9.3 (3.5~14.2); 23.3 (21.0~27.0); 17.2 (12.9~21.0); 146.2 (95.9~191.8); 0.1 (0.0~1.5); 852 (699~1053) ▎1197 (885~1547); 650 (479~911); 55 (34~76) ▎40.2 (19.0~80.6); 850 (699~1053); 345.4 (158.7~551.6)

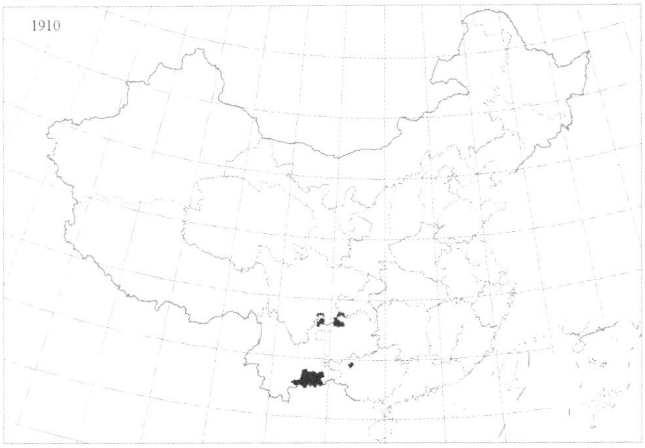

Magnoliaceae 木兰科 *Manglietia* 木莲属

Manglietia fordiana Oliver 木莲

Trees, evergreen

18.0 (11.7~23.6); 7.9 (2.1~17.8); 26.9 (18.8~29.9); 18.0 (11.7~23.6); 156.7 (82.6~223.5); 0.3 (0.0~4.9); 950 (658~1281) ▎1487 (916~2435); 603 (392~1428); 142 (31~259) ▎57.5 (-6.9~147.0); 942 (658~1273); 227.1 (95.3~659.0)

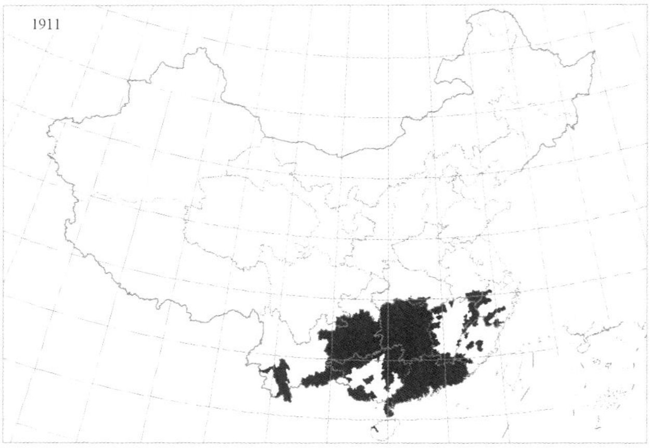

Magnoliaceae 木兰科 *Manglietia* 木莲属

Manglietia fordiana var. hainarensis (Dandy) N. H. Xia 海南木莲

Trees, evergreen

23.5 (19.1~24.8); 17.4 (11.0~20.1); 27.9 (25.0~29.0); 23.6 (19.1~24.8); 222.8 (169.6~237.6); 0.0 (0.0~0.0); 1277 (945~1385) ▎1553 (868~2035); 613 (312~997); 79 (41~116) ▎22.7 (-34.3~71.8); 1183 (868~1360); 403.3 (199.6~620.0)

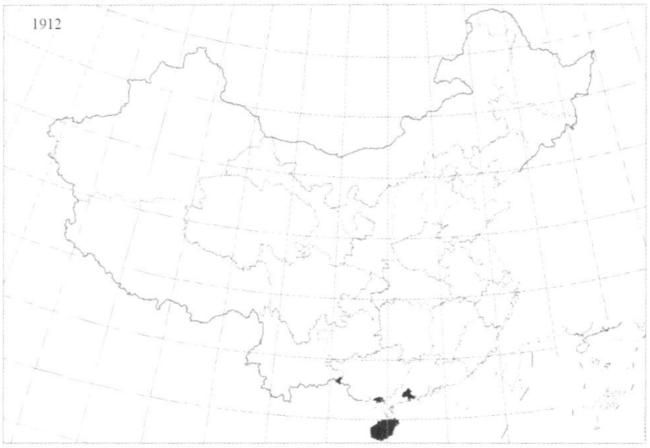

Magnoliaceae 木兰科 *Manglietia* 木莲属

Manglietia forrestii W. W. Sm. ex Dandy 滇桂木莲

Trees, evergreen

18.5 (8.6~22.7); 11.7 (1.3~15.1); 23.3 (15.0~28.6); 18.5 (8.7~22.8); 162.8 (52.6~213.1); 0.2 (0.0~8.8); 925 (574~1221) ▎1448 (963~2435); 793 (566~1428); 69 (26~142) ▎58.0 (12.5~99.9); 914 (574~1221); 399.8 (225.9~659.0)

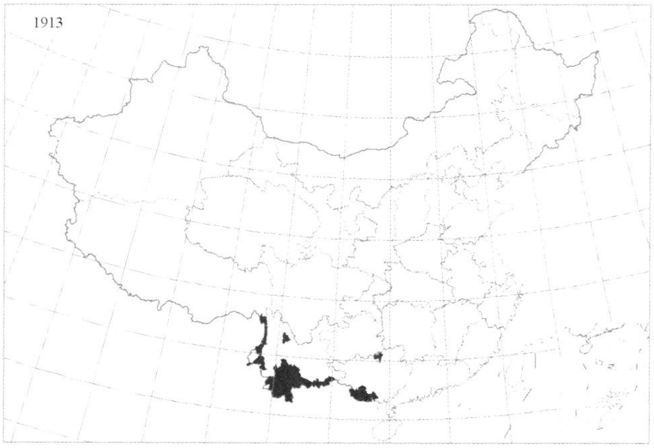

Magnoliaceae 木兰科 *Manglietia* 木莲属

Manglietia garrettii Craib 泰国木莲

Trees, evergreen

19.2; 13.8; 22.3; 19.2; 170.5; 0.0; 909 ▎1499; 686; 58 ▎64.9; 880; 513.9

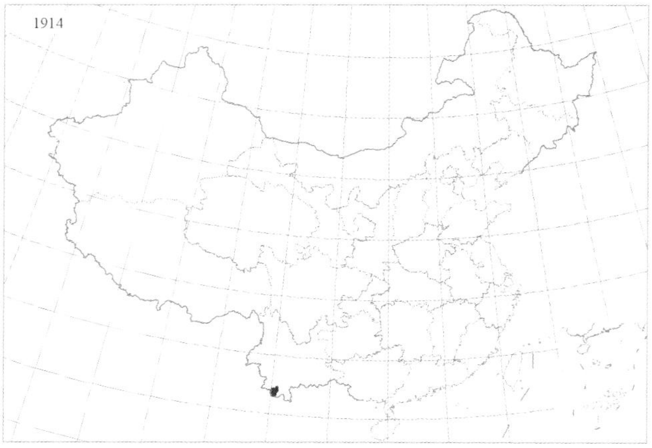

Magnoliaceae 木兰科 *Manglietia* 木莲属

Manglietia glaucifolia Y. W. Law et Y. F. Wu 苍背木莲

Trees, evergreen

16.7; 6.3; 25.7; 16.7; 140.5; 0.0; 863 ▎1268; 531; 105 ▎47.0; 863; 246.1

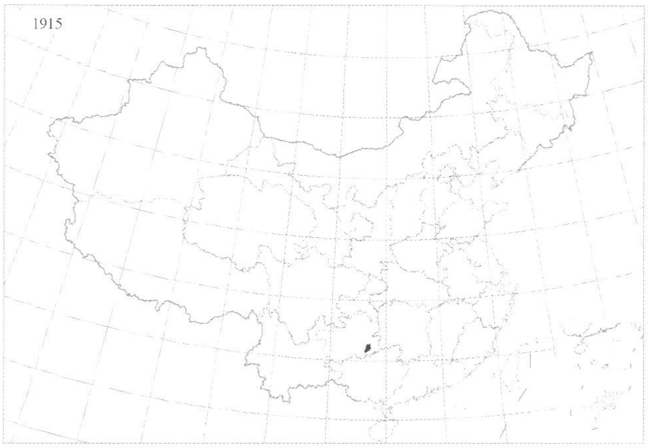

Magnoliaceae 木兰科 *Manglietia* 木莲属

Manglietia grandis Hu et W. C. Cheng 大果木莲

Trees, evergreen

18.4 (17.2~21.0); 10.9 (9.1~14.2); 23.8 (22.4~25.6); 18.5 (17.3~21.0); 161.7 (147.0~191.8); 0.0 (0.0~0.0); 904 (833~1053) ▎1387 (116[illegible]~1522); 781 (633~860); 57 (48~72) ▎53.9 (37.1~80.6); 904 (833~1053); 400.1 (313.7~551.6)

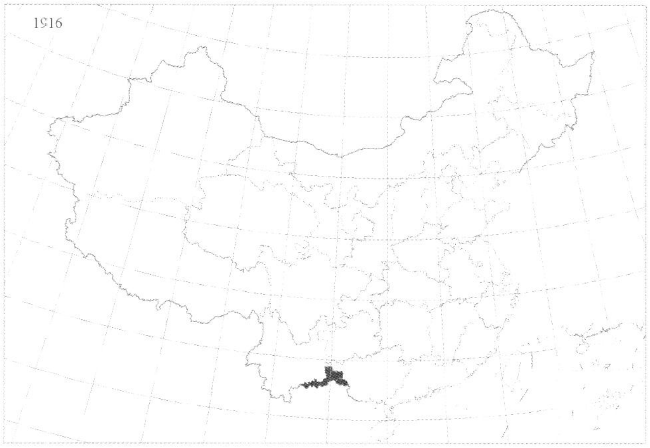

Magnoliaceae 木兰科 *Manglietia* 木莲属

Manglietia hongheensis Y. M. Shui et W. H. Chen 红河木莲

Trees, evergreen

18.6; 12.6; 22.7; 18.7; 164.6; 0.0; 894 ▎1386; 808; 56 ▎55.3; 891; 478.7

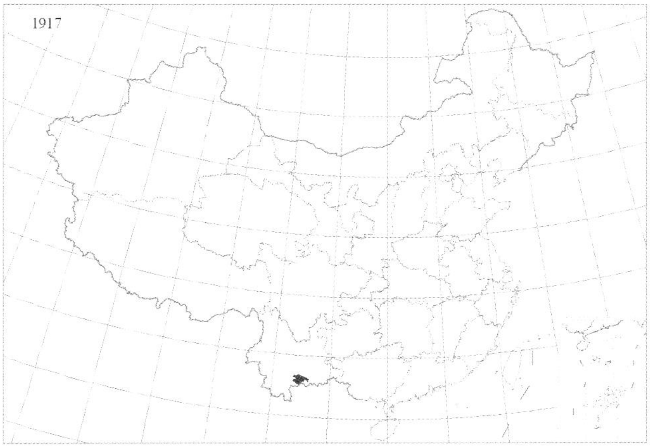

Magnoliaceae 木兰科 *Manglietia* 木莲属

Manglietia hookeri Cubitt et W. W. Smith 中缅木莲

Trees, evergreen

19.2 (17.3~20.5); 12.9 (9.5~15.1); 23.4 (21.6~26.0); 19.2 (17.3~20.5); 170.9 (148.0~185.9); 0.0 (0.0~0.0); 933 (832~992) ▎1374 (1049~1607); 662 (598~702); 60 (44~77) ▎46.7 (26.1~64.9); 911 (783~986); 474.1 (262.1~659.0)

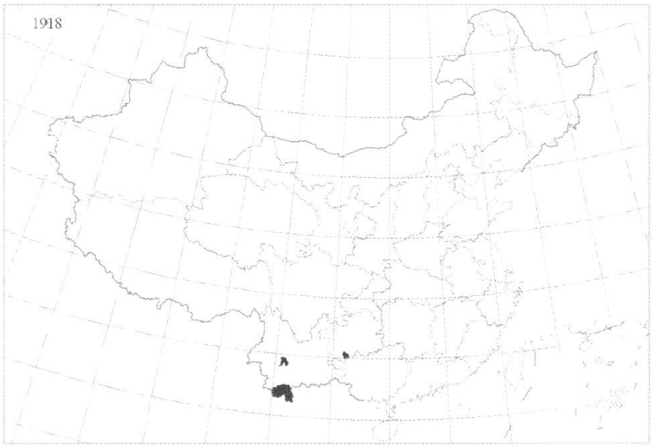

Magnoliaceae 木兰科 *Manglietia* 木莲属

Manglietia insignis Wall. Tent. Fl. Nepal. 红花木莲

Trees, deciduous

16.8 (5.3~22.7); 8.0 (-2.4~15.1); 24.1 (12.2~28.6); 16.9 (5.7~22.8); 143.0 (30.5~213.1); 0.7 (0.0~26.2); 863 (479~1221) ▎1334 (939~2435); 641 (419~1428); 87 (20~218) ▎55.4 (5.5~128.9); 857 (479~1221); 296.4 (127.1~659.0)

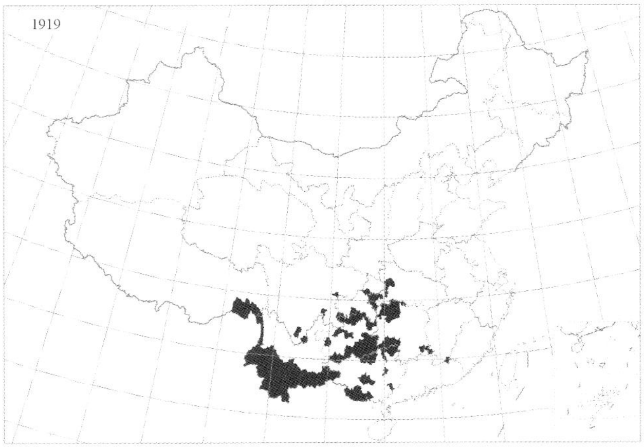

Magnoliaceae 木兰科 *Manglietia* 木莲属

Manglietia kwangtungensis (Merr.) Dandy 毛桃木莲

Trees, evergreen

19.9 (14.6~22.9); 10.6 (3.8~15.2); 27.7 (24.6~29.8); 20.0 (14.7~22.9); 179.6 (117.7~214.8); 0.0 (0.0~1.5); 1048 (779~1232) ▎1660 (1250~2435); 688 (465~1428); 162 (60~215) ▎59.5 (12.9~101.8); 1046 (779~1232); 233.4 (111.7~327.7)

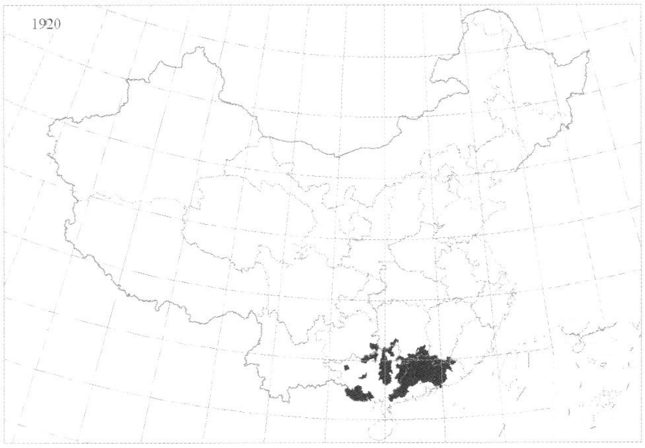

Magnoliaceae 木兰科 *Manglietia* 木莲属
Manglietia lucida B. L. Chen et S. C. Yang 亮叶木莲
Trees, evergreen

21.0; 14.2; 25.6; 21.0; 191.8; 0.0; 1053 ❙ 1522; 818; 68 ❙ 44.5; 1053; 551.6

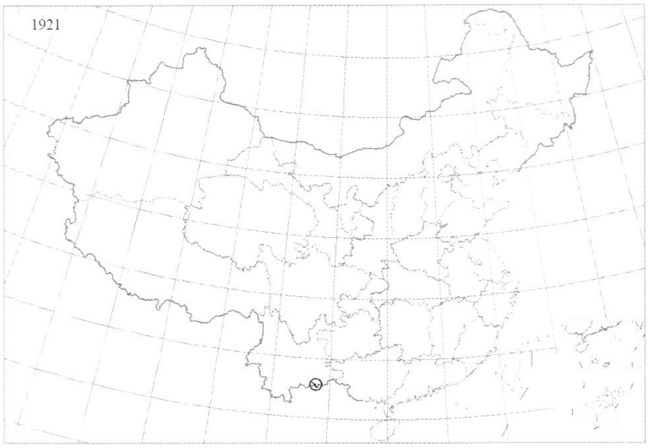

Magnoliaceae 木兰科 *Manglietia* 木莲属
Manglietia oblonga Y. W. Law et al. 椭圆叶木莲
Trees, evergreen

17.5; 6.8; 26.9; 17.5; 150.0; 0.0; 914 ❙ 1682; 652; 189 ❙ 83.9; 914; 224.7

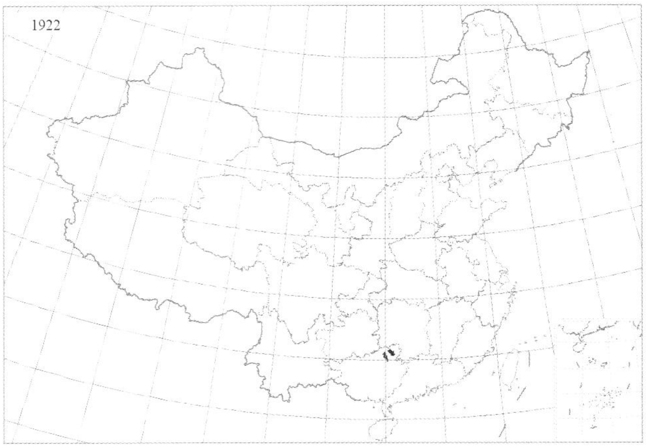

Magnoliaceae 木兰科 *Manglietia* 木莲属
Manglietia obovalifolia C. Y. Wu et Y. W. Law 倒卵叶木莲
Trees, evergreen

17.5; 8.5; 25.0; 17.5; 150.2; 0.3; 895 ❙ 1406; 645; 105 ❙ 58.3; 885; 372.5

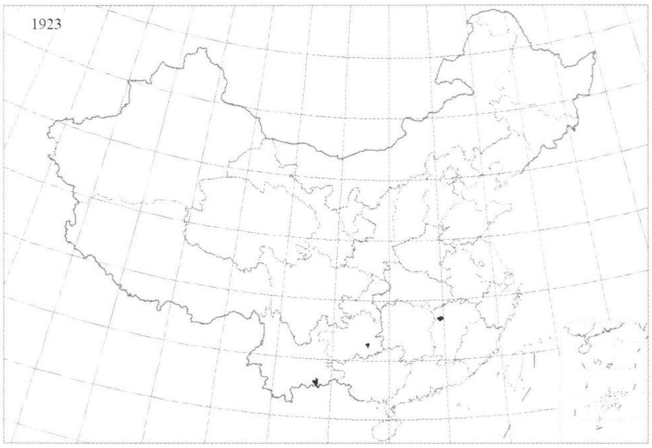

Magnoliaceae 木兰科 *Manglietia* 木莲属
Manglietia ovoidea Hung T. Chang et B. L. Chen 卵果木莲
Trees, evergreen

17.3; 9.9; 22.4; 17.3; 147.1; 0.0; 833 ❙ 1504; 860; 55 ❙ 80.6; 833; 413.5

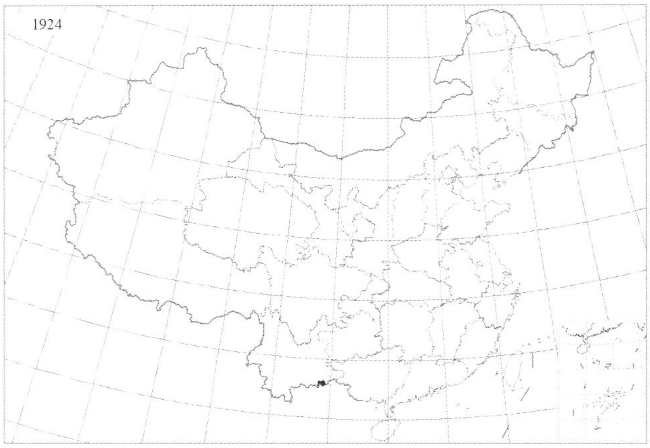

Magnoliaceae 木兰科 *Manglietia* 木莲属
Manglietia pachyphylla Chang 厚叶木莲
Trees, evergreen

21.6 (15.8~22.8); 13.6 (4.4~15.5); 28.2 (26.7~29.1); 21.7 (15.9~22.8); 200.1 (131.1~214.1); 0.0 (0.0~0.6); 1145 (852~1225) ❙ 1749 (1384~2037); 790 (553~1033); 131 (75~191) ❙ 53.3 (20.8~95.5); 1145 (852~1225); 205.1 (95.3~275.2)

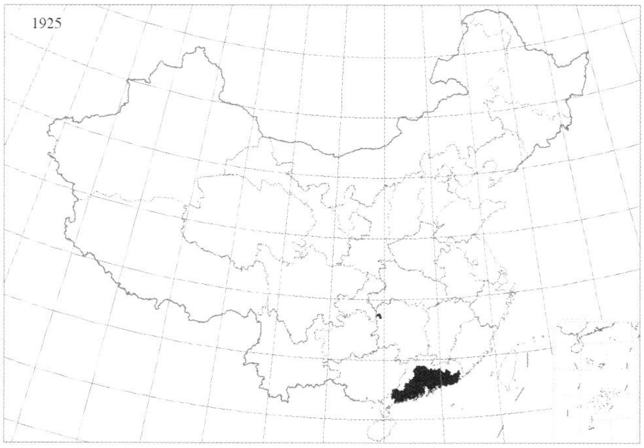

Magnoliaceae 木兰科 *Manglietia* 木莲属
Manglietia patungensis Hu 巴东木莲
Trees, evergreen

14.5; 3.8; 24.7; 14.5; 117.1; 3.1; 785 ❙ 1244; 520; 78 ❙ 61.1; 785; 209.6

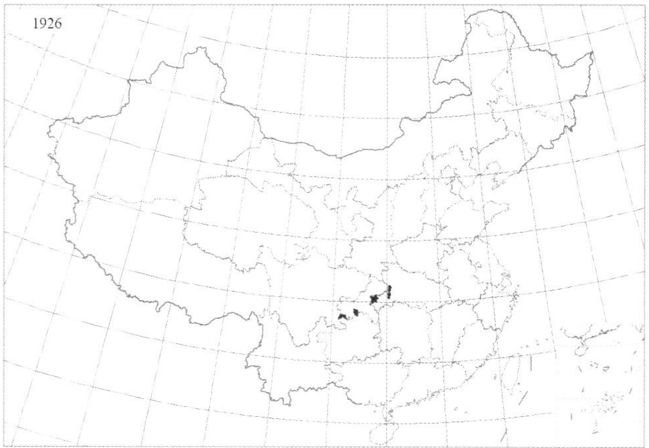

Magnoliaceae 木兰科 *Manglietia* 木莲属
Manglietia rufibarbata Dandy 毛瓣木莲
Trees, evergreen

18.2; 10.0; 24.3; 18.2; 158.9; 0.0; 905 ❙ 1354; 739; 66 ❙ 50.4; 905; 328.2

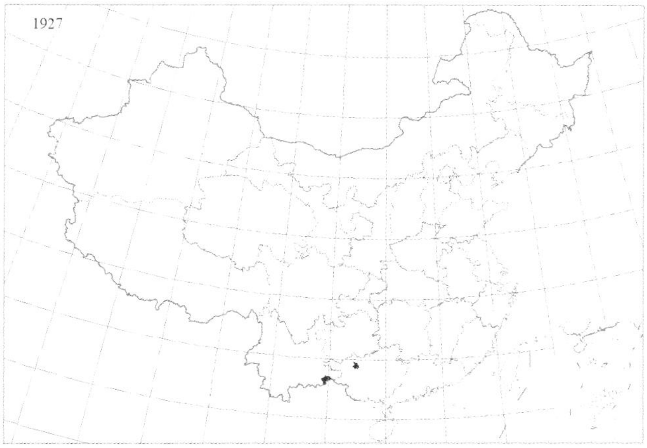

Magnoliaceae 木兰科 *Manglietia* 木莲属
Manglietia sinoconifera F. N. Wei 那坡木莲
Trees, evergreen

19.1; 11.0; 25.0; 19.1; 169.6; 0.0; 945 ❙ 1438; 820; 57 ❙ 52.2; 945; 371.7

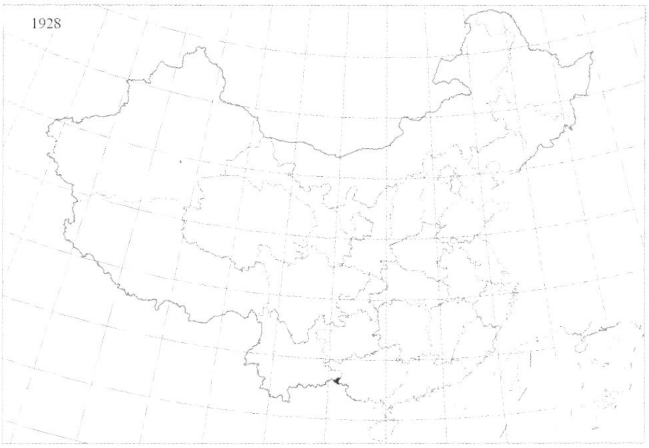

Magnoliaceae 木兰科 *Manglietia* 木莲属
Manglietia szechuanica Hu 四川木莲
Trees, evergreen

14.4 (11.1~17.8); 5.1 (1.8~7.9); 22.8 (19.3~27.7); 14.4 (11.1~17.8); 114.4 (79.8~153.7); 1.3 (0.0~6.9); 766 (648~939) ❙ 1098 (952~1410); 588 (478~816); 47 (28~72) ❙ 44.9 (16.1~87.2); 760 (648~939); 224.9 (142.5~292.6)

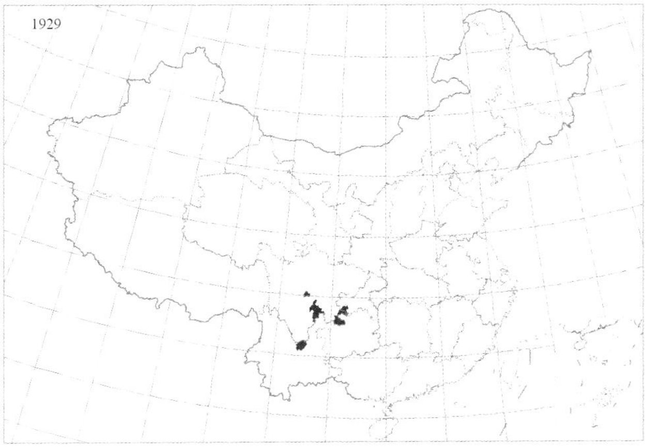

Magnoliaceae 木兰科 *Manglietia* 木莲属
Manglietia ventii Tiep 毛果木莲
Trees, evergreen

19.3; 12.7; 23.8; 19.3; 171.6; 0.0; 942 ❙ 1439; 803; 61 ❙ 53.4; 942; 516.4

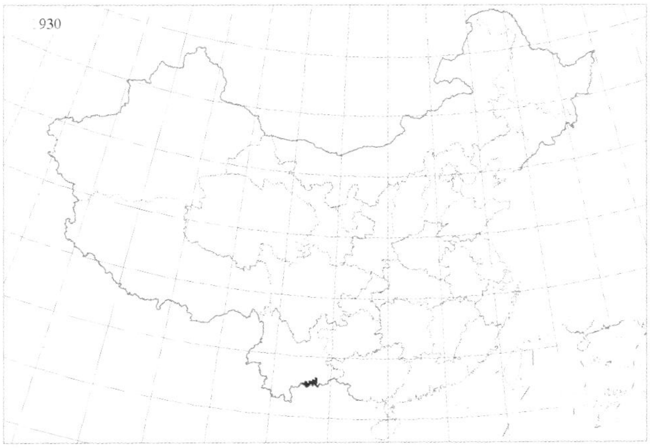

Magnoliaceae 木兰科 *Manglietia* 木莲属
Manglietia zhengyiana N. H. Xia 锈毛木莲
Trees, evergreen

17.2; 9.7; 22.5; 17.3; 147.1; 0.0; 835 ❙ 1379; 780; 52 ❙ 65.1; 835; 374.9

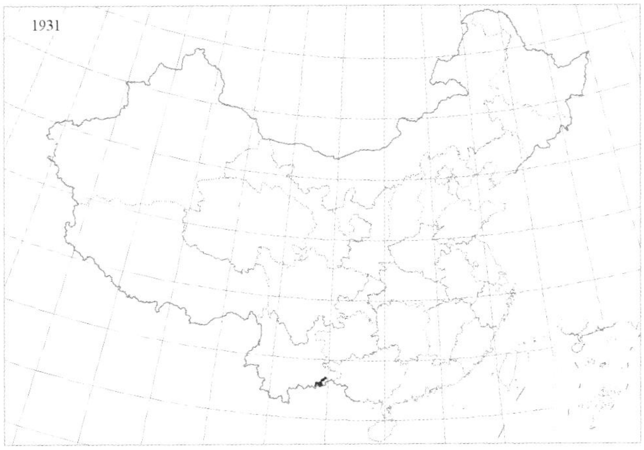

Magnoliaceae 木兰科 *Michelia* 含笑属
Michelia angustioblonga Y. W. Law 狭叶含笑
Small trees, evergreen

17.4; 7.4; 25.5; 17.4; 148.9; 0.0; 886 ❙ 1381; 637; 107 ❙ 55.9; 886; 245.3

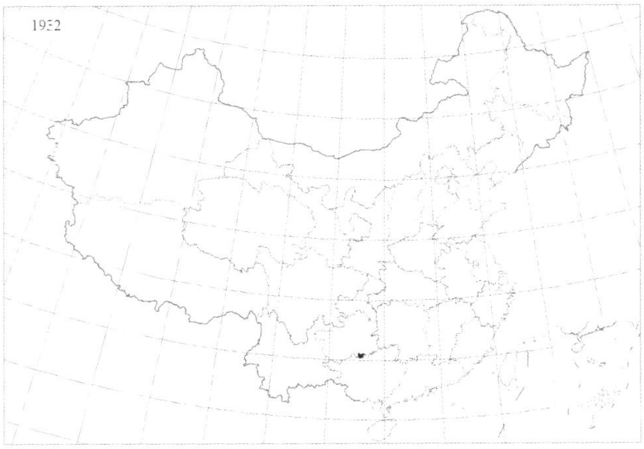

Magnoliaceae 木兰科 *Michelia* 含笑属
Michelia baillonii (Pierre) Finet et Gagnep. 合果木
Trees, evergreen

18.0 (11.3~21.0); 11.7 (4.3~15.1); 22.0 (17.0~25.6); 18.0 (11.3~21.0); 156.1 (76.6~191.8); 0.0 (0.0~0.7); 868 (644~1053) ❙ 1399 (1049~1751); 772 (598~996); 55 (41~77) ❙ 61.8 (21.7~112.2); 856 (644~1053); 461.2 (346.6~659.0)

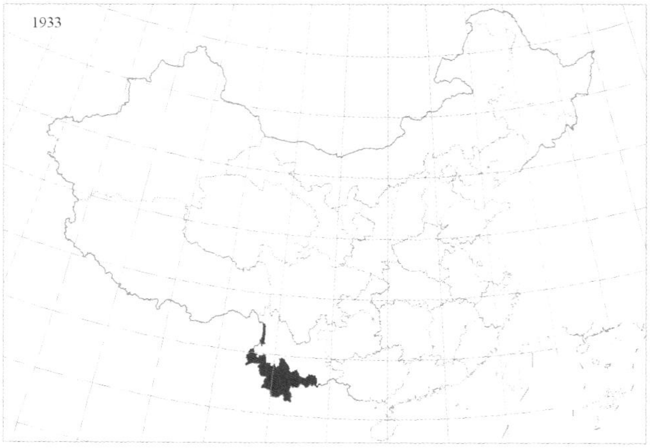

Magnoliaceae 木兰科 *Michelia* 含笑属
Michelia balansae (Aug. DC.) Dandy 苦梓含笑
Trees, evergreen

20.7 (14.6~24.8); 12.3 (3.8~20.1); 27.4 (21.4~29.2); 20.7 (14.7~24.8); 188.7 (117.7~237.6); 0.0 (0.0~1.5); 1087 (779~1385) ❙ 1539 (868~2435); 693 (312~1428); 122 (41~221) ❙ 43.3 (-34.3~101.8); 1074 (779~1360); 273.0 (152.9~620.0)

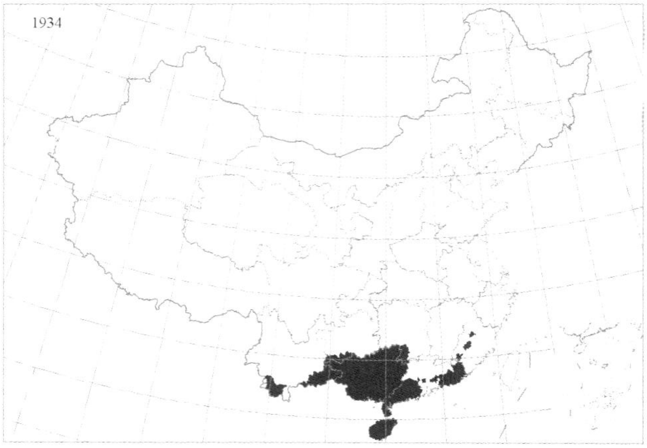

Magnoliaceae 木兰科 *Michelia* 含笑属
Michelia cavaleriei Finet et Gagnep. 平伐含笑
Trees, evergreen

18.8 (13.4~23.0); 9.1 (2.7~15.1); 27.1 (20.6~29.0); 18.8 (13.4~23.0); 165.9 (104.5~216.2); 0.1 (0.0~3.7); 988 (735~1247) ❙ 1378 (1017~2148); 626 (442~1234); 108 (43~207) ❙ 40.5 (5.8~101.8); 987 (735~1247); 221.1 (108.2~551.6)

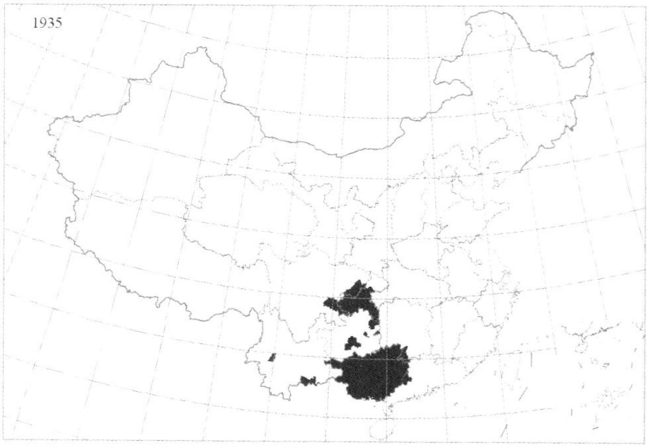

Magnoliaceae 木兰科 *Michelia* 含笑属
Michelia cavaleriei var. platypetala (Handel-Mazzetti) N. H. Xia 阔瓣含笑
Trees, evergreen

17.8 (12.7~22.4); 7.6 (2.1~14.3); 26.9 (23.0~29.7); 17.8 (12.8~22.4); 154.1 (99.0~209.1); 0.3 (0.0~5.6); 935 (713~1204) ❙ 1494 (1064~1907); 573 (392~1044); 154 (43~248) ❙ 60.5 (16.9~107.2); 930 (713~1204); 221.9 (134.6~354.5)

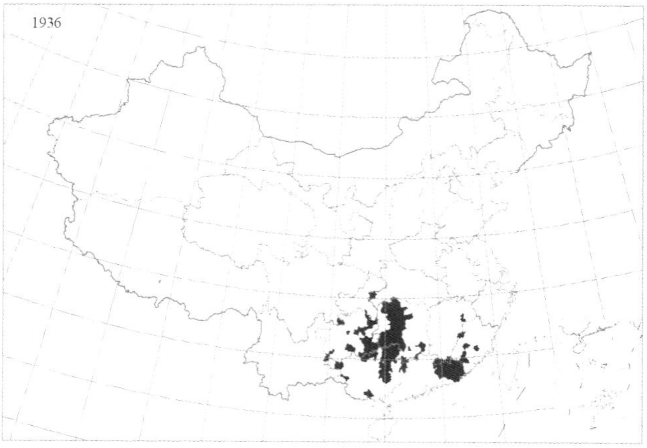

Magnoliaceae 木兰科 *Michelia* 含笑属
Michelia champaca L. 黄兰
Trees, evergreen

20.3 (0.3~23.6); 12.0 (-8.6~16.7); 27.1 (8.3~29.8); 20.4 (2.9~23.6); 184.7 (9.6~223.5); 0.7 (0.0~65.4); 1065 (368~1281) ❙ 1618 (389~2037); 704 (246~1033); 132 (11~197) ❙ 53.2 (-6.9~112.2); 1062 (368~1273); 247.5 (95.3~659.0)

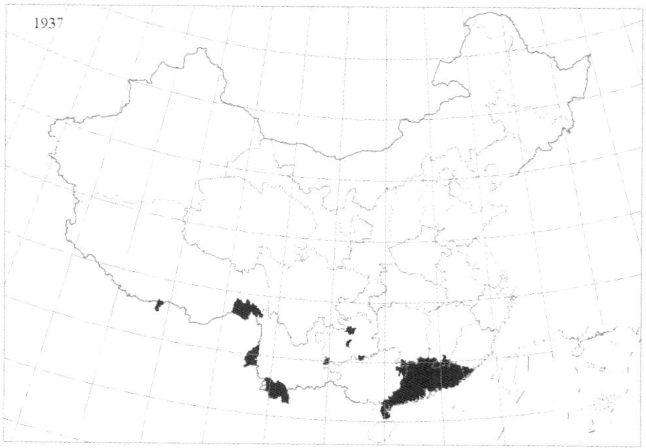

Magnoliaceae 木兰科 *Michelia* 含笑属
Michelia champaca var. pubinervia (Blume) Miquel 毛叶脉黄兰
Trees, evergreen

19.2; 13.8; 22.3; 19.2; 170.5; 0.0; 909 ❙ 1499; 686; 58 ❙ 64.9; 830; 513.9

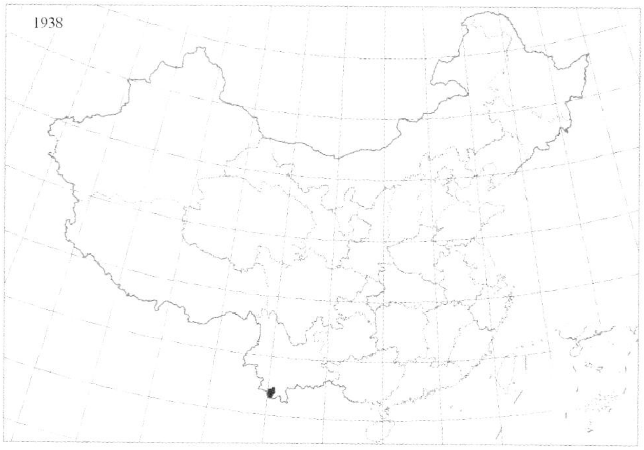

Magnoliaceae 木兰科 *Michelia* 含笑属
Michelia shiluensis Chun et Y. F. Wu 石碌含笑
Trees, evergreen

23.7; 18.3; 27.5; 23.8; 225.1; 0 0; 1282 ▌ 1424; 496; 50 ▌ 11.7; 1115; 522.0

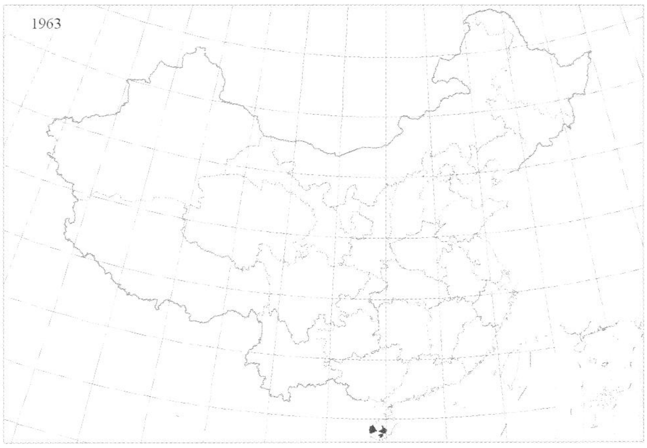

Magnoliaceae 木兰科 *Michelia* 含笑属
Michelia skinneriana Dunn 野含笑
Trees, evergreen

18 5 (14.3~24.8); 8.6 (2.3~20.1); 27.5 (23.4~29.9); 18.6 (14.3~24.8); 162.9 (112.6~237.6); 0.2 (0.0~3.9); 979 (756~1385) ▌ 1578 (868~2024); 567 (312~862); 178 (41~259) ▌ 63.9 (-34.3~147.0); 961 (756~1360); 245.2 (123.9~620.0)

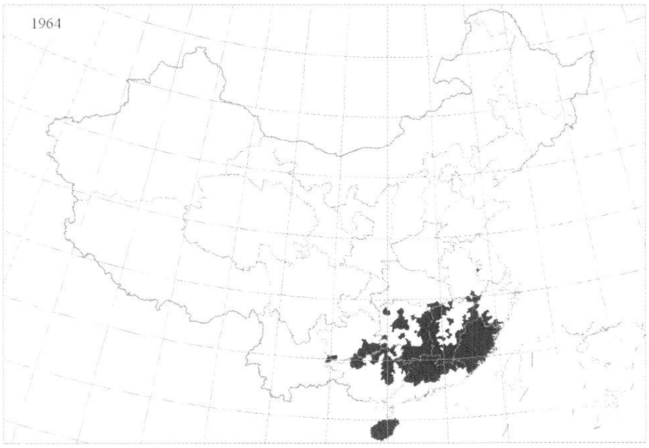

Magnoliaceae 木兰科 *Michelia* 含笑属
Michelia sphaerantha C. Y. Wu ex Y. W. Law et Y. F. Wu 球花含笑
Trees, evergreen

17.5 (15.9~18.9); 11.1 (9.5~12.5); 21.9 (20.6~23.2); 17.5 (16.0~19.0); 150.5 (131.6~167.5); 0.0 (0.0~0.0); 845 (779~914) ▌ 1020 (939~1139); 576 (524~651); 43 (37~47) ▌ 21.0 (6.5~33.4); 791 (744~851); 334.1 (301.1~365.3)

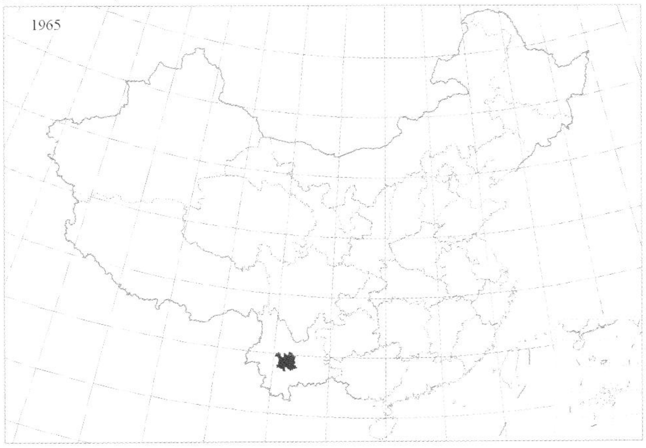

Magnoliaceae 木兰科 *Michelia* 含笑属
Michelia velutina DC. 绒毛含笑
Trees, evergreen

13 8 (1.8~19.2); 6.7 (-7.1~13.3); 19.2 (9.9~23.5); 14.1 (3.8~19.4); 113.8 (15.7~172.6); 7.2 (0.0~53.3); 735 (404~932) ▌ 1134 (527~1736); 647 (3 7~1007); 43 (9~69) ▌ 57.0 (6.5~145.8); 717 (404~923); 343.7 (169.5~452.0)

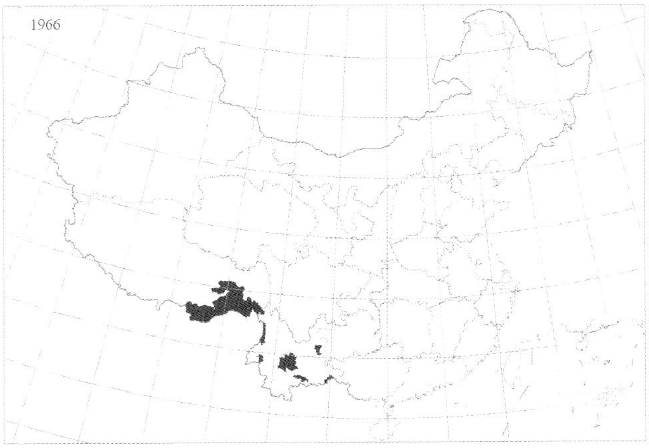

Magnoliaceae 木兰科 *Michelia* 含笑属
Michelia wilsonii Finet et Gagnep. 峨眉含笑
Trees, evergreen

14.6 (11.1~17.6); 4.6 (0.7~9.9); 23.9 (19.4~28.4); 14.7 (11.3~17.7); 117.8 (80.8~152.4); 1.4 (0.0~9.4); 780 (652~917) ▌ 1187 (968~1478); 560 (462~817); 64 (20~190) ▌ 52.6 (27.3~107.2); 776 (652~875); 219.2 (139.5~359.7)

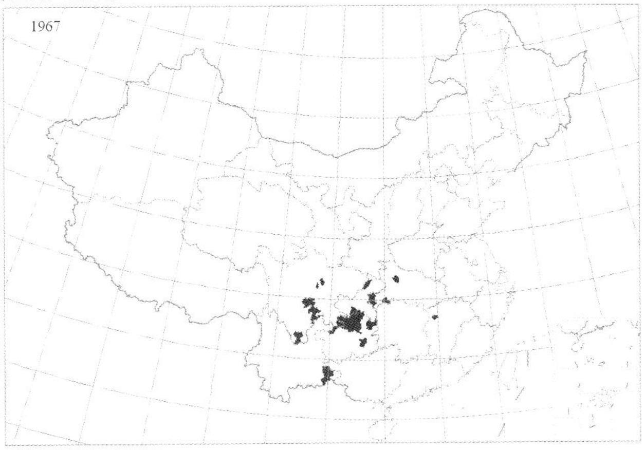

Magnoliaceae 木兰科 *Michelia* 含笑属
M chelia wilsonii subsp. szechuanica (Dandy) J. Li 川含笑
Trees, evergreen

14.4 (6.6~18.7); 4.8 (-1.3~8.4); 22.9 (13.3~28.9); 14.4 (6.8~18.8); 114.4 (39.6~166.0); 1.7 (0.0~19.7); 776 (514~1002) ▌ 1091 (862~1478); 532 (442~663); 51 (12~112) ▌ 42.8 (8.0~107.2); 774 (514~1002); 211.2 (108.2~331.6)

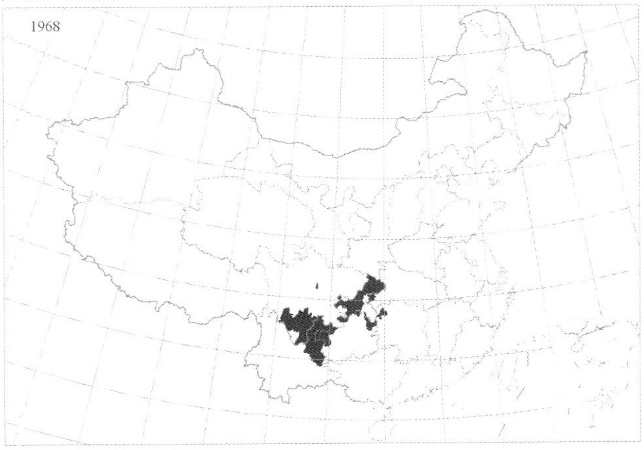

Magnoliaceae 木兰科 *Michelia* 含笑属

Michelia xanthantha C. Y. Wu ex Y. W. Law et Y. F. Wu 黄花含笑

Trees, evergreen

20.0; 14.6; 23.2; 20.1; 180.8; 0.0; 963 ❙ 1523; 681; 67 ❙ 58.4; 942; 581.0

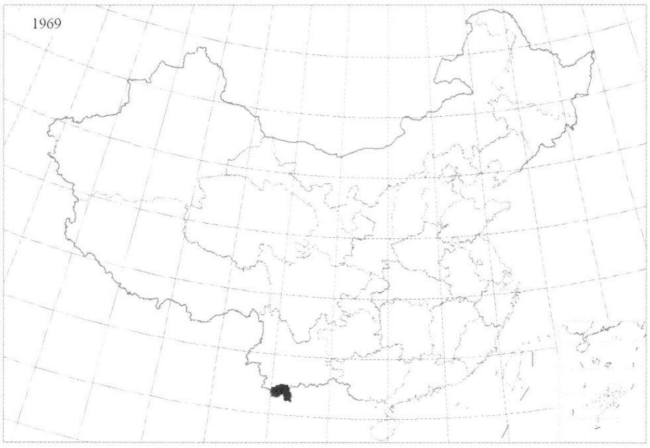

Magnoliaceae 木兰科 *Michelia* 含笑属

Michelia yunnanensis Franch. ex Finet et Gagnep. 云南含笑

Shrubs, evergreen

15.6 (5.4~21.0); 8.5 (-2.1~15.1); 20.8 (12.3~26.0); 15.6 (5.8~21.0); 128.0 (31.5~191.8); 0.7 (0.0~25.6); 783 (481~1053) ❙ 1078 (741~1751); 614 (411~996); 43 (15~77) ❙ 38.2 (-14.2~99.9); 763 (481~1053); 311.9 (185.6~659.0)

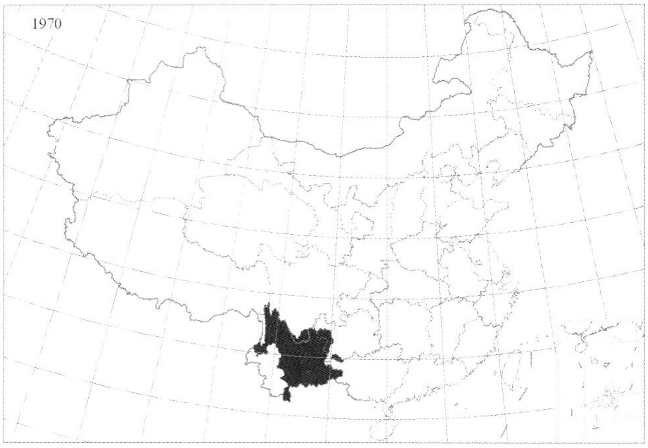

Magnoliaceae 木兰科 *Oyama* 天女花属

Oyama globosa (Hook. f. et Thoms.) N. H. Xia et C. Y. Wu 毛叶天女花

Small trees, deciduous

12.2 (-0.7~18.9); 4.4 (-9.9~12.5); 18.7 (7.7~26.6); 12.4 (2.5~19.0); 94.4 (6.8~167.8); 7.5 (0.0~74.2); 683 (338~929) ❙ 1003 (370~1736); 580 (262~1007); 32 (10~70) ❙ 49.8 (-14.2~145.8); 663 (338~912); 282.7 (61.6~487.2)

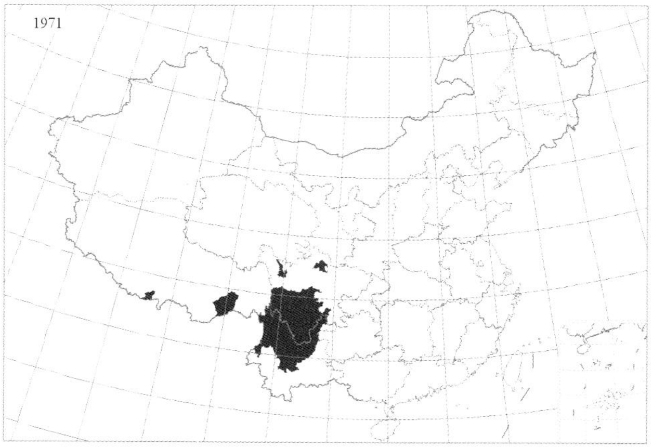

Magnoliaceae 木兰科 *Oyama* 天女花属

Oyama sieboldii (K. Koch) N. H. Xia et C. Y. Wu 天女花

Small trees, deciduous

12.8 (1.4~17.6); 0.4 (-18.8~11.0); 24.0 (11.2~29.3); 13.5 (4.3~17.6); 109.4 (20.0~151.7); 15.7 (0.0~88.8); 759 (427~941) ❙ 1273 (632~1997); 562 (398~816); 113 (11~259) ❙ 65.8 (-10.4~147.0); 755 (427~941); 254.8 (136.4~372.3)

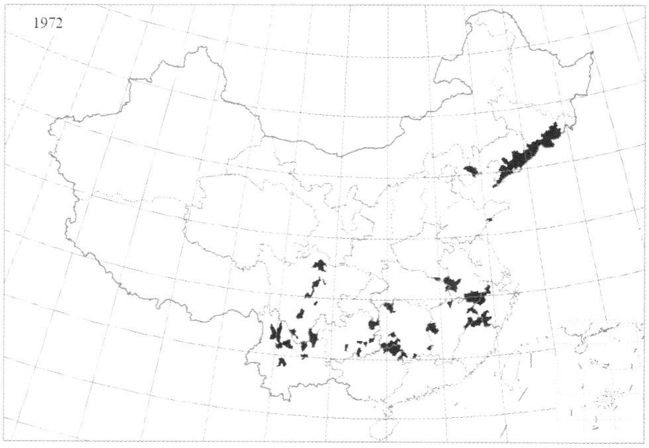

Magnoliaceae 木兰科 *Oyama* 天女花属

Oyama sinensis (Rehder et E. H. Wilson) N. H. Xia et C. Y. Wu 圆叶天女花

Shrubs, deciduous

11.9 (2.6~16.2); 2.7 (-7.0~8.2); 20.0 (11.2~25.3); 12.1 (4.3~16.2); 91.5 (20.0~134.5); 8.7 (0.0~48.2); 682 (427~826) ❙ 1136 (829~1410); 590 (400~816); 47 (18~100) ❙ 68.7 (27.6~101.2); 674 (427~826); 239.4 (187.9~310.2)

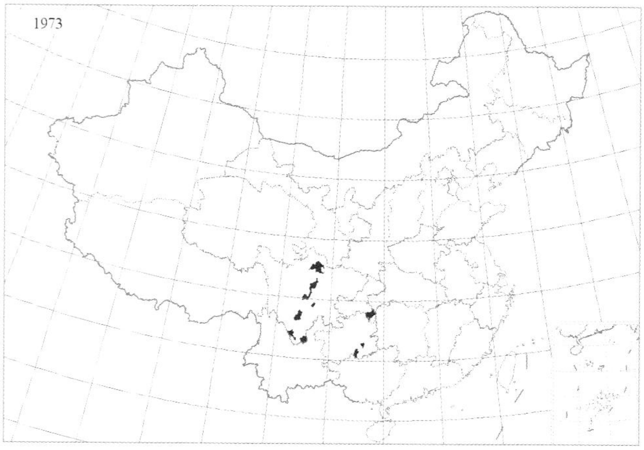

Magnoliaceae 木兰科 *Oyama* 天女花属

Oyama wilsonii (Finet et Gagnep.) N. H. Xia et C. Y. Wu 西康天女花

Shrubs or small trees, deciduous

13.3 (2.2~18.9); 5.5 (-7.0~12.5); 19.7 (9.9~24.9); 13.4 (3.8~19.0); 104.4 (15.5~167.8); 4.0 (0.0~49.1); 716 (413~929) ❙ 1028 (711~1414); 582 (357~817); 35 (15~71) ❙ 46.2 (-14.2~125.9); 691 (413~851); 272.3 (182.9~372.3)

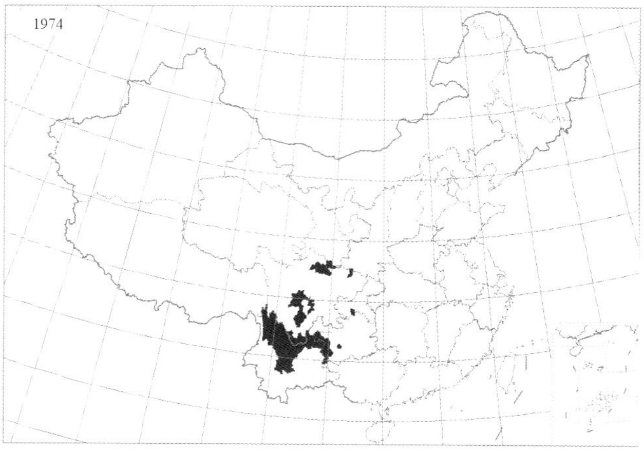

Magnoliaceae 木兰科 *Pachylarnax* 厚壁木属
Pachylarnax sinica (Law) N. H Xia et C. Y. Wu 华盖木
Trees, evergreen

18.5 (17.2~21.0); 11.2 (9.5~14.2); 23.7 (22.4~25.6); 18.6 (17.3~21.0); 162.6 (147.0~191.8); 0.0 (0.0~0.0); 906 (833~1053) ❙ 1404 (1254~1522); 793 (699~860); 55 (48~68) ❙ 55.6 (37.6~80.6); 906 (833~1053); 418.3 (336.3~551.6)

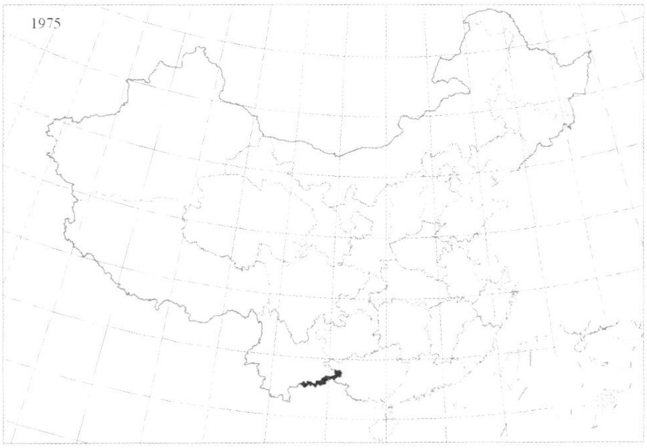

Magnoliaceae 木兰科 *Parakmeria* 拟单性木兰属
Parakmeria kachirachirai (Kaneh. et Yamam.) Y. W. Law 恒春拟单性木兰
Trees, evergreen

20.3; 15.9; 23.6; 20.4; 184.4; 0.0; 993 ❙ 2853; 1618; 147 ❙ 190.4; 993; 330.3

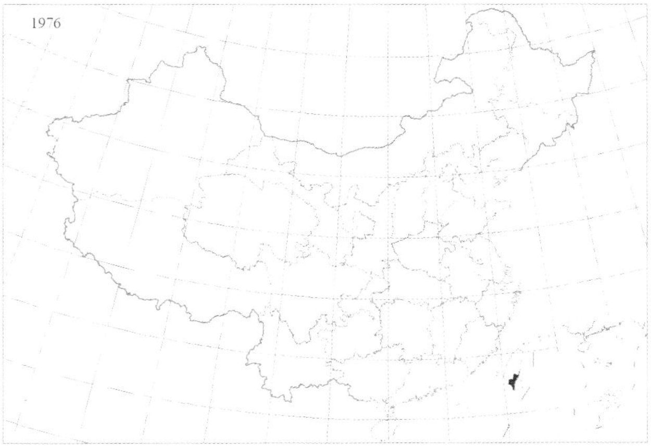

Magnoliaceae 木兰科 *Parakmeria* 拟单性木兰属
Parakmeria lotungensis (Chun et C. H. Tsoong) Y. W. Law 乐东拟单性木兰
Trees, evergreen

18.3 (14.3~24.8); 8.5 (3.3~20.1); 27.0 (23.5~29.4); 18.3 (14.3~24.8); 159.7 (112.6~237.6); 0.2 (0.0~2.3); 964 (756~1385) ❙ 1560 (868~1997); 589 (312~891); 159 (41~254) ❙ 65.4 (-34.3~147.0); 945 (756~1326); 258.5 (151.0~620.0)

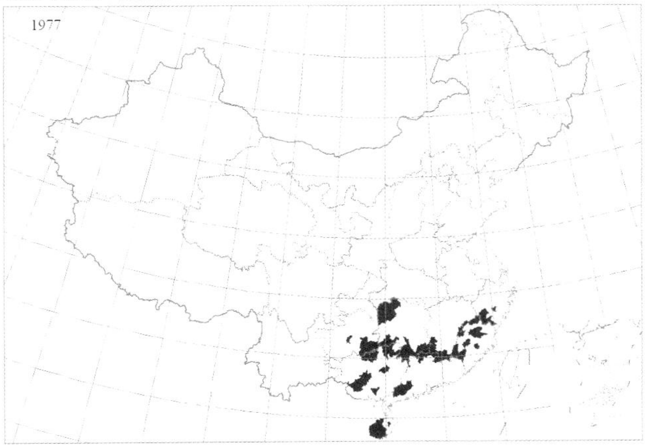

Magnoliaceae 木兰科 *Parakmeria* 拟单性木兰属
Parakmeria nitida (W. W. Sm.) Y. W. Law 光叶拟单性木兰
Trees, evergreen

17.3 (5.3~23.0); 9.0 (-2.4~17.7); 25.1 (12.2~29.8); 17.9 (5.7~23.1); 156.0 (30.5~217.0); 1.5 (0.0~26.2); 946 (479~1221) ❙ 1508 (741~2435); 708 (426~1428); 125 (21~196) ❙ 62.1 (12.9~145.8); 944 (479~1221); 285.4 (174.9~572.9)

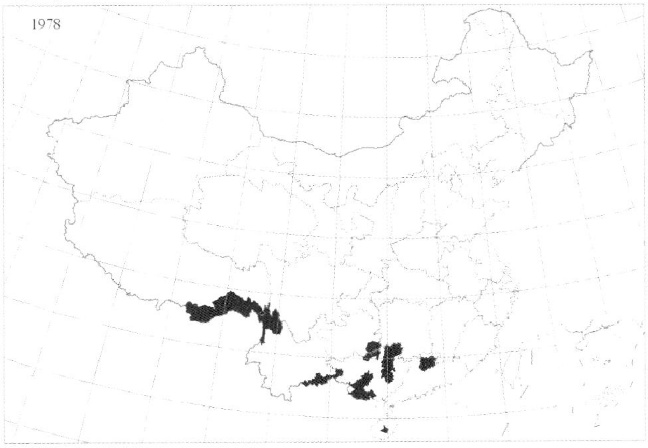

Magnoliaceae 木兰科 *Parakmeria* 拟单性木兰属
Parakmeria omeiensis W. C. Cheng 峨眉拟单性木兰
Trees, evergreen

14.9; 5.3; 23.6; 15.0; 119.6; 0.0; 782 ❙ 1410; 816; 49 ❙ 80.4; 782; 214.5

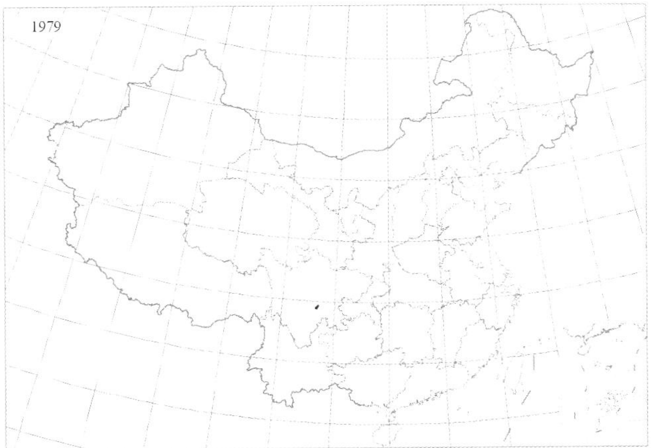

Magnoliaceae 木兰科 *Parakmeria* 拟单性木兰属
Parakmeria yunnanensis Hu 云南拟单性木兰
Trees, evergreen

17.4 (12.5~21.6); 9.3 (3.6~14.2); 23.6 (20.1~28.6); 17.5 (12.5~21.7); 149.8 (92.0~199.8); 0.1 (0.0~1.4); 877 (685~1153) ❙ 1316 (1014~1703); 683 (531~911); 80 (44~194) ❙ 50.4 (26.1~80.6); 873 (685~1153); 326.9 (185.5~551.6)

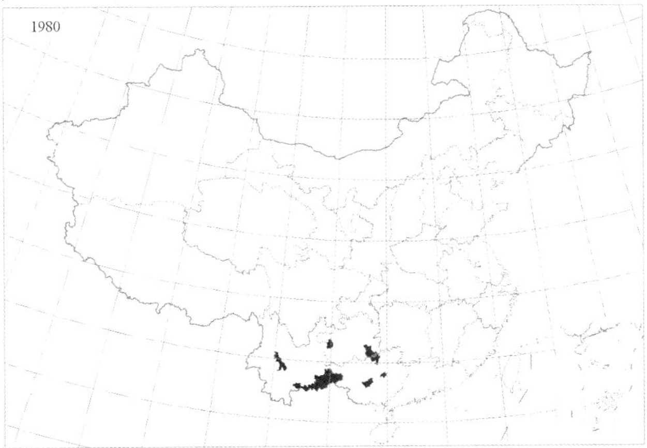

Magnoliaceae 木兰科 *Talauma* 盖裂木属
Talauma hodgsonii Hook. f. et Thomson 盖裂木
Trees, evergreen

16.0；9.1；21.1；16.0；132.1；0.0；797 ❙ 1641；959；57 ❙ 110.0；797；499.4

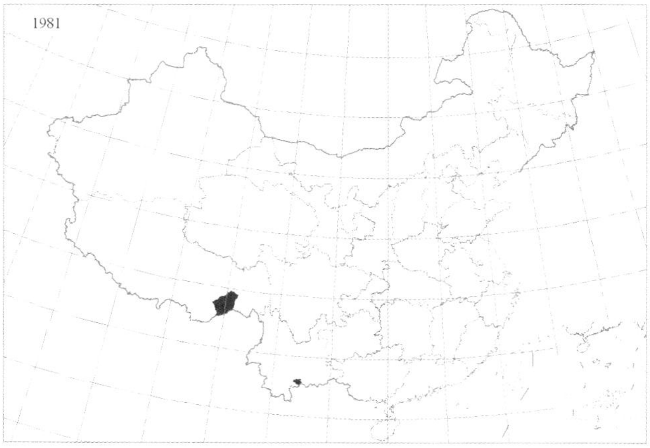

Magnoliaceae 木兰科 *Woonyoungia* 焕镛木属
Woonyoungia septentrionalis (Dandy) Y. W. Law 焕镛木
Trees, deciduous

18.4 (17.3~21.0)；9.5 (7.3~14.2)；25.3 (22.4~27.5)；18.4 (17.3~21.0)；161.1 (147.1~191.8)；0.0 (0.0~0.0)；930 (833~1053) ❙ 1465 (1381~1522)；731 (637~860)；97 (52~142) ❙ 58.4 (44.5~80.6)；930 (833~1053)；327.2 (225.9~551.6)

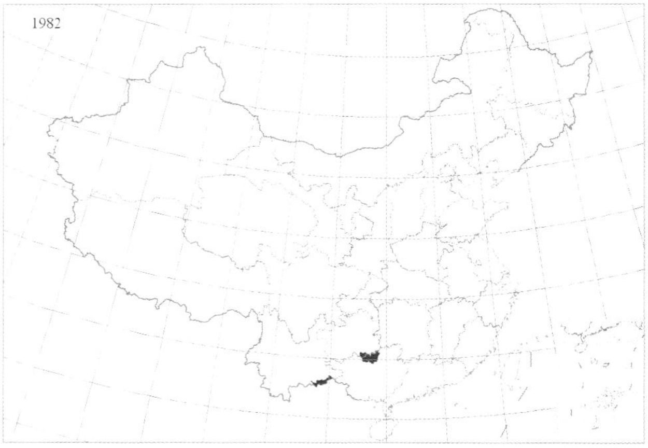

Magnoliaceae 木兰科 *Yulania* 玉兰属
Yulania amoena (W. C. Cheng) D. L. Fu 天目玉兰
Trees, deciduous

15.7 (13.4~18.2)；3.8 (1.5~6.5)；27.2 (24.6~29.5)；15.8 (13.4~18.2)；131.3 (108.3~158.8)；2.0 (0.0~7.3)；859 (755~984) ❙ 1535 (838~1922)；560 (366~677)；189 (45~259) ❙ 79.1 (5.4~129.9)；854 (755~927)；239.4 (149.5~319.9)

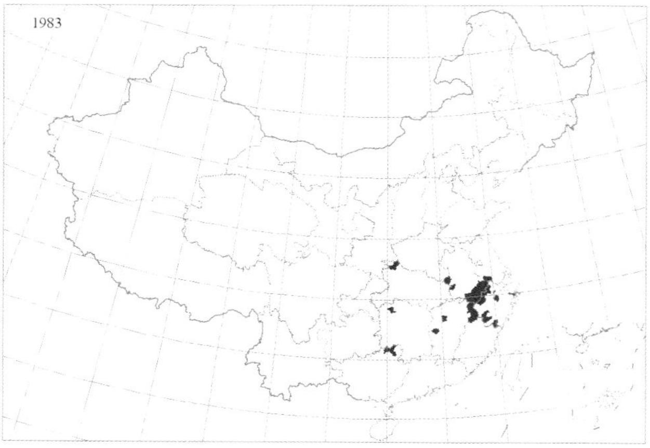

Magnoliaceae 木兰科 *Yulania* 玉兰属
Yulania biondii (Pamp.) D. L. Fu 望春玉兰
Trees, deciduous

13.8 (6.2~17.5)；1.5 (-4.2~5.4)；25.4 (15.8~29.7)；13.9 (7.0~17.5)；113.3 (45.2~150.6)；7.3 (0.0~29.9)；787 (522~964) ❙ 1085 (626~1770)；456 (303~666)；85 (14~223) ❙ 37.3 (-22.2~112.2)；774 (522~924)；221.8 (139.1~334.4)

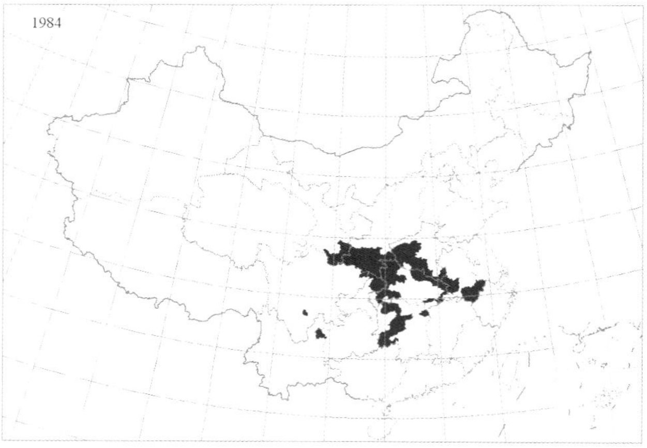

Magnoliaceae 木兰科 *Yulania* 玉兰属
Yulania campbellii (Hook. f. et Thoms.) D. L. Fu 滇藏玉兰
Trees, deciduous

9.0 (-1.8~19.6)；1.0 (-10.5~12.8)；15.6 (6.1~23.5)；9.6 (1.8~19.6)；66.7 (1.9~175.7)；18.7 (0.0~82.8)；593 (329~953) ❙ 959 (311~1736)；570 (217~1007)；25 (10~69) ❙ 61.4 (-6.3~145.8)；582 (311~932)；282.0 (61.6~482.8)

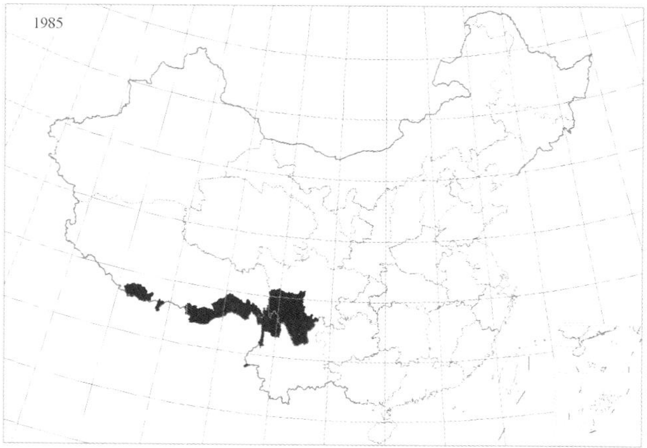

Magnoliaceae 木兰科 *Yulania* 玉兰属
Yulania cylindrica (E. H. Wilson) D. L. Fu 黄山玉兰
Trees, deciduous

16.2 (12.7~19.1)；4.6 (1.5~10.0)；27.3 (23.0~29.7)；16.3 (12.8~19.1)；136.6 (99.0~169.1)；1.2 (0.0~7.3)；879 (713~1006) ❙ 1578 (862~1997)；566 (376~713)；187 (47~259) ❙ 80.5 (7.8~147.0)；871 (713~980)；236.4 (161.2~319.9)

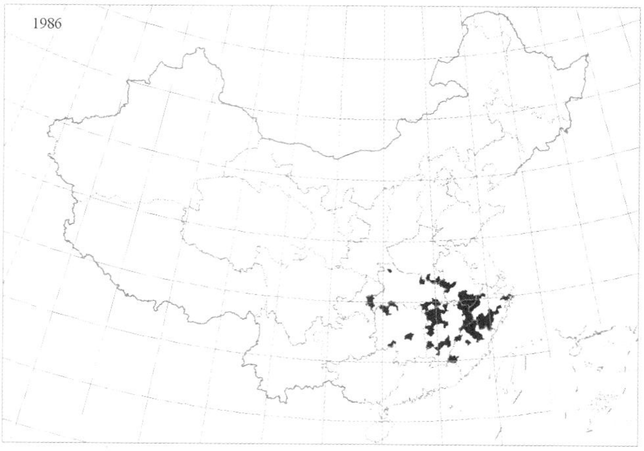

Magnoliaceae 木兰科 *Yulania* 玉兰属

Yulania dawsoniana (Rehder et E. H. Wilson) D. L. Fu 光叶玉兰

Trees, deciduous

11.5 (2.2~16.1); 1.8 (-6.6~6.4); 20.2 (9.9~25.7); 11.7 (3.8~16.1); 87.3 (15.5~133.6); 9.0 (0.0~49.1); 676 (413~822) ❙ 1076 (806~1448); 562 (448~688); 39 (16~105) ❙ 62.9 (8.2~125.9); 676 (413~822); 244.3 (181.8~318.7)

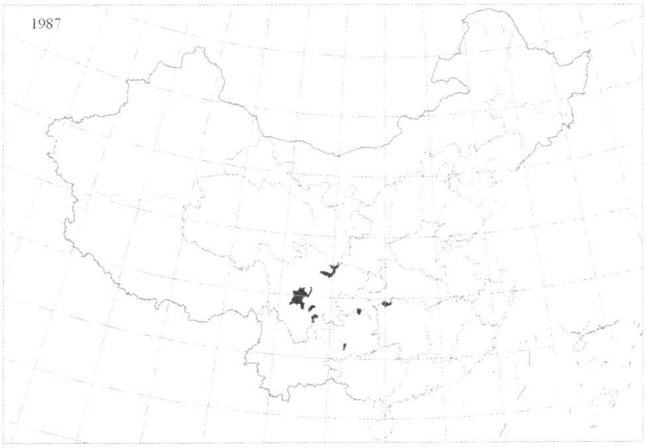

Magnoliaceae 木兰科 *Yulania* 玉兰属

Yulania denudata (Desr.) D. L. Fu 玉兰

Trees, deciduous

15.7 (5.4~21.8); 3.8 (-4.2~13.1); 26.9 (12.3~30.1); 15.8 (5.8~21.8); 132.2 (31.5~201.6); 3.4 (0.0~29.4); 865 (481~1160) ❙ 1337 (626~1997); 492 (299~770); 146 (14~259) ❙ 53.6 (-25.5~147.0); 845 (481~1160); 214.2 (133.5~351.7)

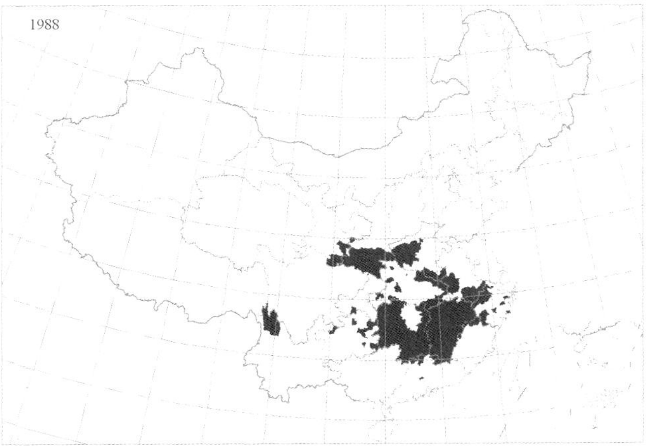

Magnoliaceae 木兰科 *Yulania* 玉兰属

Yulania elliptigemmata (C. L. Guo et L. L. Huang) N. H. Xia 椭蕾玉兰

Trees, deciduous

14.9; 2.8; 26.3; 14.9; 122.2; 3.0; 822 ❙ 1054; 482; 78 ❙ 28.2; 822; 219.8

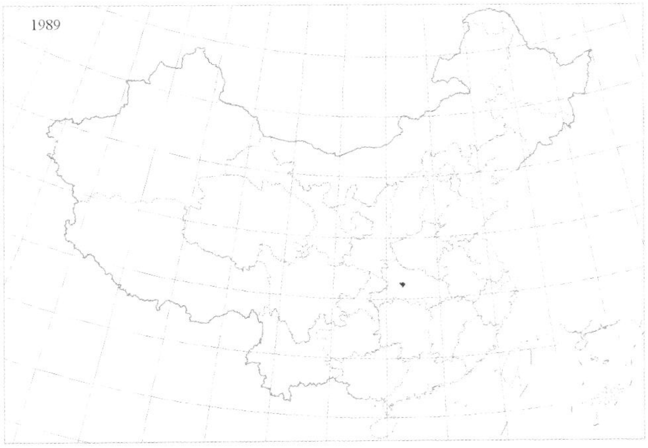

Magnoliaceae 木兰科 *Yulania* 玉兰属

Yulania jigongshanensis (T. B. Chao et al.) D. L. Fu 鸡公山玉兰

Trees, deciduous

15.[illegible]; 1.7; 27.4; 15.2; 127.7; 5.4; 860 ❙ 1042; 480; 89 ❙ 21.1; 860; 166.6

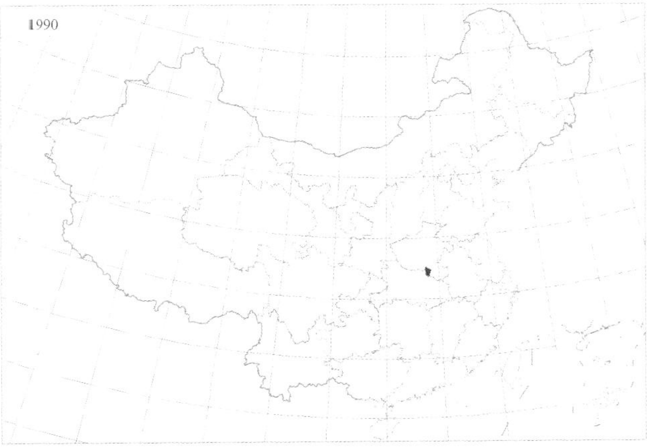

Magnoliaceae 木兰科 *Yulania* 玉兰属

Yulania liliiflora (Desr.) D. C. Fu 紫玉兰

Shrubs, deciduous

14.7 (5.4~22.0); 2.9 (-6.2~13.5); 25.9 (12.3~29.9); 14.8 (5.8~22.0); 121.9 (31.5~203.8); 4.8 (0.0~32.8); 823 (481~1170) ❙ 1156 (590~1997); 485 (303~938); 94 (12~254) ❙ 40.3 (-31.4~130.7); 810 (481~1170); 212.4 (111.9~352.3)

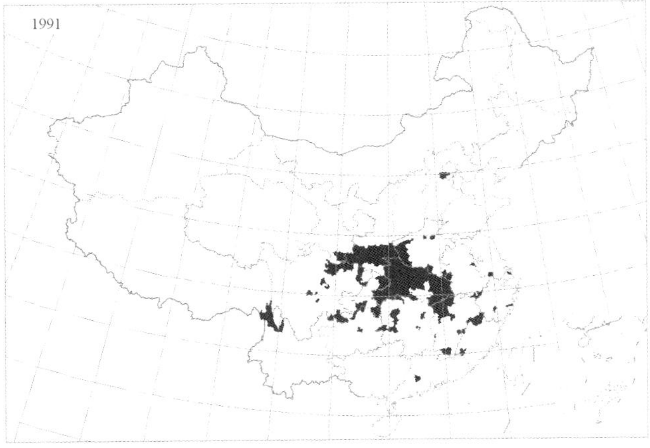

Magnoliaceae 木兰科 *Yulania* 玉兰属

Yulania mirifolia D. L. Fu et al. 奇叶玉兰

Trees, deciduous

15.1 1.7; 27.4; 15.2; 127.7; 5.4; 860 ❙ 1042; 480; 89 ❙ 21.1; 860; 166.6

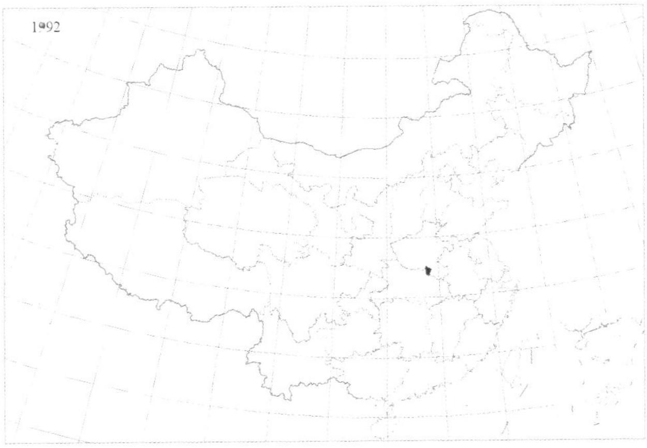

Magnoliaceae 木兰科 *Yulania* 玉兰属

Yulania multiflora (M. C. Wang et C. L. Min) D. L. Fu 多花玉兰 Trees, deciduous

9.4; -2.5; 21.0; 9.8; 73.9; 20.6; 619 ❙ 892; 389; 29 ❙ 44.0; 619; 311.0

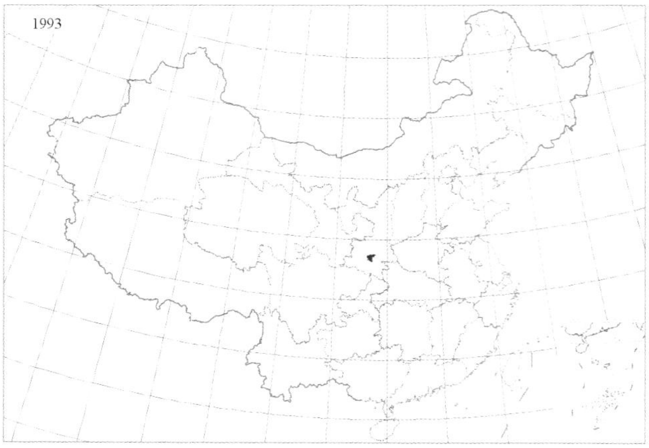

Magnoliaceae 木兰科 *Yulania* 玉兰属

Yulania pilocarpa (Z. Z. Zhao et Z. W. Xie) D. L. Fu 罗田玉兰

Trees, deciduous

15.3; 2.8; 27.3; 15.3; 126.5; 3.1; 848 ❙ 1429; 628; 134 ❙ 68.5; 848; 202.8

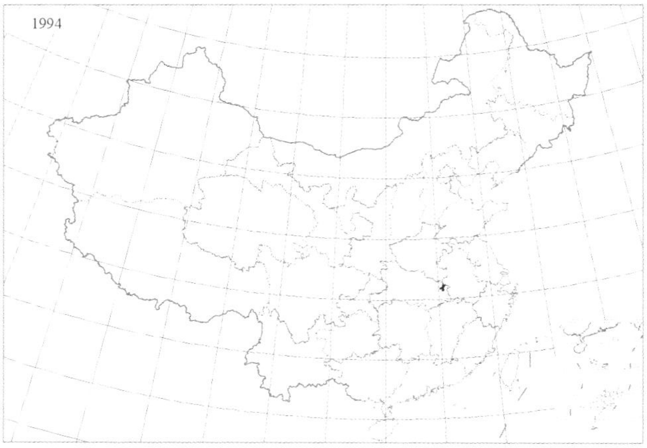

Magnoliaceae 木兰科 *Yulania* 玉兰属

Yulania sargentiana (Rehder et E. H. Wilson) D. L. Fu 凹叶玉兰 Trees, deciduous

12.0 (7.3~15.5); 2.8 (-2.0~5.8); 20.4 (15.6~24.4); 12.1 (7.5~15.6); 90.1 (48.2~127.2); 5.1 (0.0~20.4); 684 (537~812) ❙ 1174 (1001~1414); 656 (531~817); 37 (22~49) ❙ 72.5 (36.4~101.6); 684 (537~812); 256.7 (211.6~318.7)

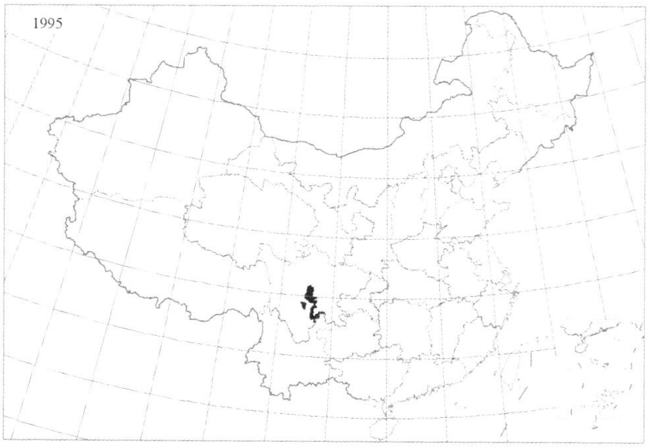

Magnoliaceae 木兰科 *Yulania* 玉兰属

Yulania sprengeri (Pamp.) D. L. Fu 武当玉兰

Trees, deciduous

13.4 (3.0~18.3); 1.7 (-7.7~8.1); 24.5 (12.4~28.9); 13.5 (4.8~18.4); 109.0 (25.1~160.7); 7.9 (0.0~49.3); 765 (441~965) ❙ 1016 (555~1562); 450 (277~816); 61 (12~202) ❙ 33.1 (-17.7~107.2); 758 (441~965); 219.5 (113.6~334.4)

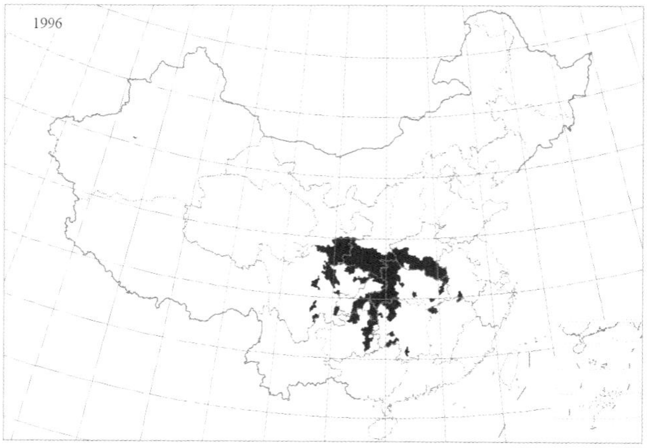

Magnoliaceae 木兰科 *Yulania* 玉兰属

Yulania stellata (Maxim.) N. H. Xia 星花玉兰

Shrubs, deciduous

15.4; 4.9; 25.6; 15.4; 124.9; 0.1; 810 ❙ 1821; 671; 204 ❙ 124.8; 810; 273.1

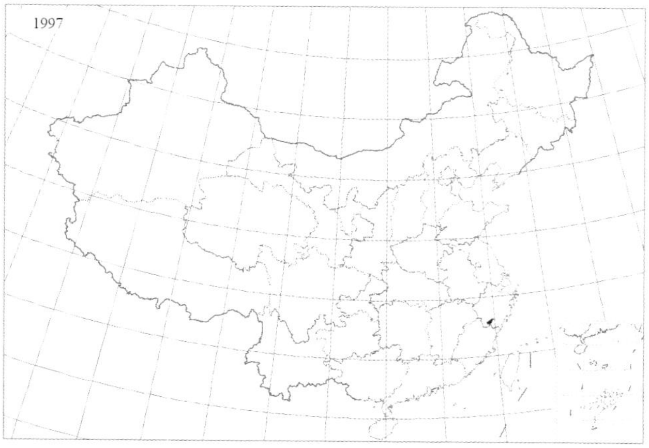

Magnoliaceae 木兰科 *Yulania* 玉兰属

Yulania viridula D. L. Fu et al. 青皮玉兰

Trees, deciduous

13.4; 1.5; 24.6; 13.4; 108.0; 7.1; 760 ❙ 864; 393; 23 ❙ 13.7; 760; 260.4

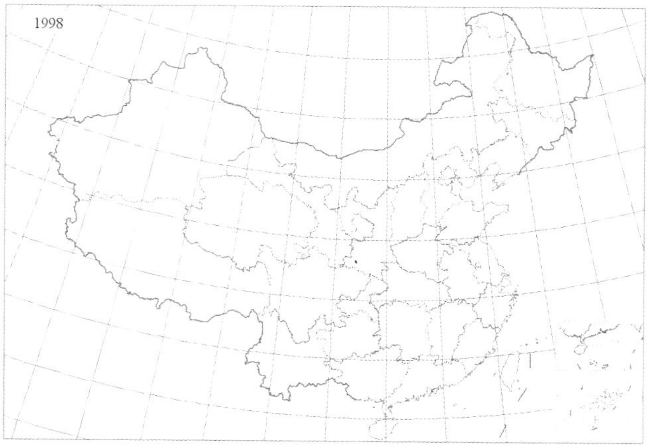

Magnoliaceae 木兰科 *Yulania* 玉兰属

Yulania zenii (W. C. Cheng) D. L. Fu 宝华玉兰

Trees, deciduous

15.3 (13.2~16.4); 2.3 (-0.2~4.0); 28.0 (25.6~28.5); 15.4 (13.2~16.4); 128.9 (110.4~138.2); 4.3 (1.0~11.7); 869 (773~901) ❙ 1066 (793~1463); 469 (398~563); 117 (44~165) ❙ 21.9 (0.0~62.3); 869 (773~901); 164.5 (134.1~223.4)

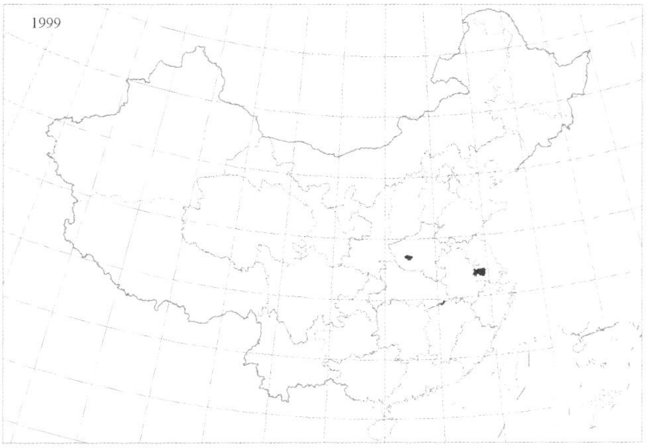

Magnoliaceae 木兰科 *Yulania* 玉兰属

Yulania × soulangeana (Soul.-Bod.) D. L. Fu 二乔玉兰

Trees, deciduous

15.2; 28.7; 22.7; 212.7; 0.0; 1216 ❙ 1965; 994; 104 ❙ 61.6; 1216; 204.4

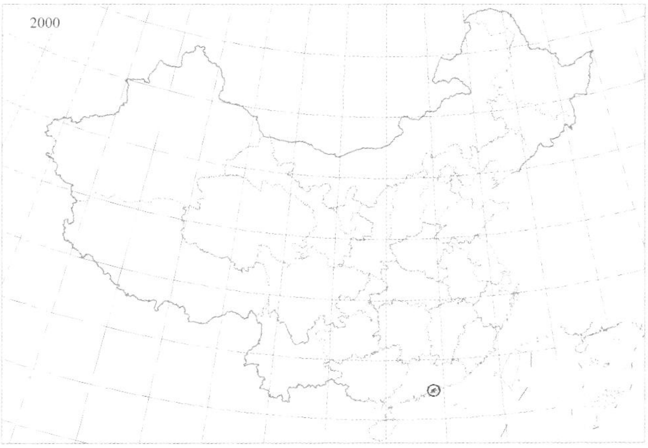

Calycanthaceae 蜡梅科 *Calycanthus* 夏蜡梅属

Calycanthus chinensis (Cheng et S. Y. Chang) Cheng et S. Y. Chang 夏蜡梅

Shrubs, deciduous

15.2; 3.5; 27.0; 15.2; 125.0; 2.2; 829 ❙ 1535; 576; 191 ❙ 85.1; 829; 260.2

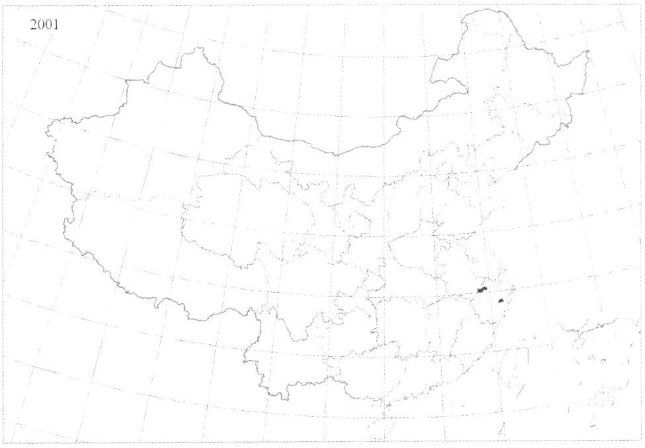

Calycanthaceae 蜡梅科 *Chimonanthus* 蜡梅属

Chimonanthus campanulatus R. H. Chang et C. S. Ding 西南蜡梅

Shrubs, evergreen

16.6; 8.7; 22.6; 16.7; 140.2; 0.0; 834 ❙ 1211; 692; 49 ❙ 44.3; 827; 307.3

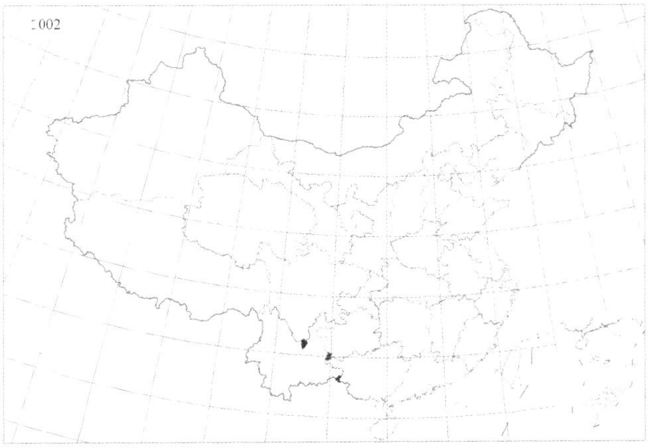

Calycanthaceae 蜡梅科 *Chimonanthus* 蜡梅属

Chimonanthus grammatus M. C. Liu 突托蜡梅

Shrubs or trees, evergreen

18.7; 8.6; 27.4; 18.8; 165.2; 0.0; 976 ❙ 1672; 587; 196 ❙ 71.3; 976; 254.7

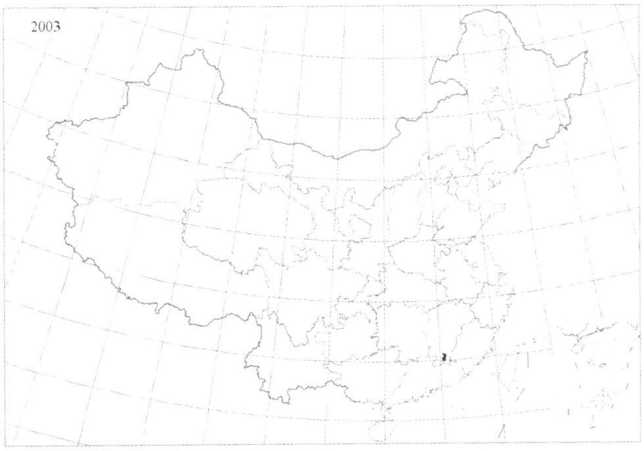

Calycanthaceae 蜡梅科 *Chimonanthus* 蜡梅属

Chimonanthus nitens Oliv. 山蜡梅

Shrubs, evergreen

15.0 (5.4~19.6); 4.5 (-4.2~12.0); 24.6 (12.3~29.8); 15.0 (5.8~19.6); 123.0 (31.5~175.5); 2.7 (0.0~29.4); 807 (481~1002) ❙ 1232 (627~1997); 524 (290~813); 99 (15~259) ❙ 51.7 (-16.6~130.7); 796 (481~1002); 235.0 (142.9~372.3)

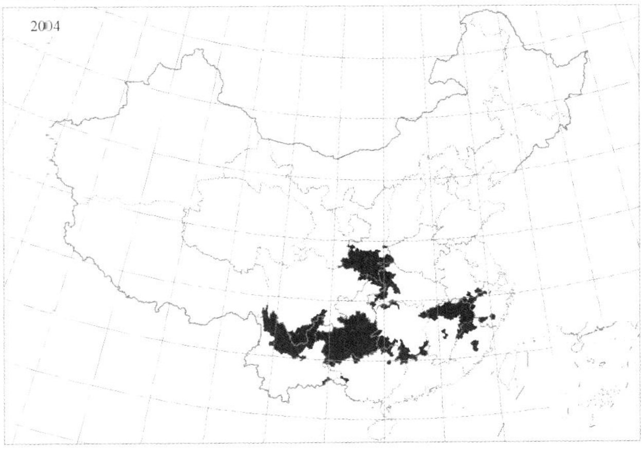

C

Calycanthaceae 蜡梅科 *Chimonanthus* 蜡梅属
Chimonanthus praecox (L.) Link 蜡梅
Shrubs, deciduous

14.7 (6.6~21.3); 2.2 (-7.3~12.6); 26.0 (13.3~29.9); 14.8 (6.8~21.3); 123.8 (39.6~196.0); 7.3 (0.0~37.2); 829 (514~1123) ❙ 1016 (411~1997); 474 (258~1007); 71 (10~254) ❙ 22.7 (-51.1~145.8); 739 (411~1121); 193.0 (108.2~491.5)

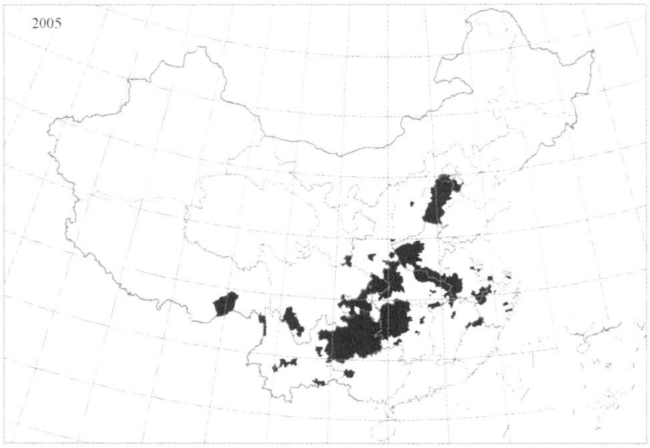

Calycanthaceae 蜡梅科 *Chimonanthus* 蜡梅属
Chimonanthus salicifolius Hu 柳叶蜡梅
Shrubs, semievergreen

16.4 (15.0~18.1); 4.5 (1.5~6.9); 27.7 (25.9~29.1); 16.4 (15.0~18.1); 138.3 (123.7~157.7); 1.0 (0.0~6.1); 889 (824~965) ❙ 1633 (828~1922); 560 (427~677); 211 (81~259) ❙ 83.9 (-6.0~129.9); 884 (824~965); 240.7 (172.8~299.5)

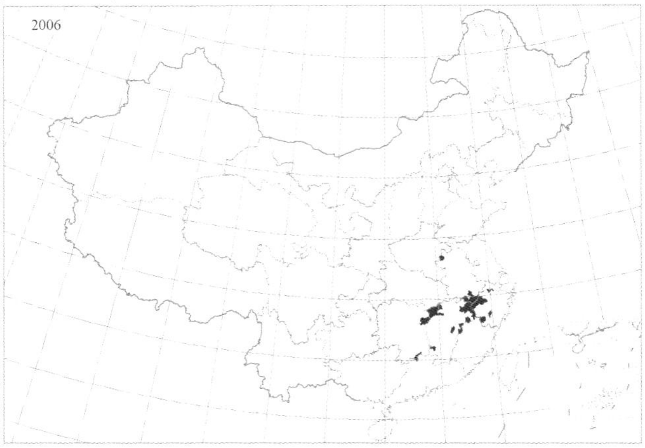

Calycanthaceae 蜡梅科 *Chimonanthus* 蜡梅属
Chimonanthus zhejiangensis M. C. Liu 浙江蜡梅
Shrubs, evergreen

16.2 (15.1~17.6); 5.2 (3.9~7.4); 26.9 (25.9~27.9); 16.2 (15.2~17.6); 134.9 (122.9~151.2); 0.4 (0.0~1.1); 861 (811~919) ❙ 1745 (1573~1893); 615 (555~677); 211 (179~242) ❙ 103.4 (75.7~129.9); 861 (811~919); 243.7 (216.1~269.3)

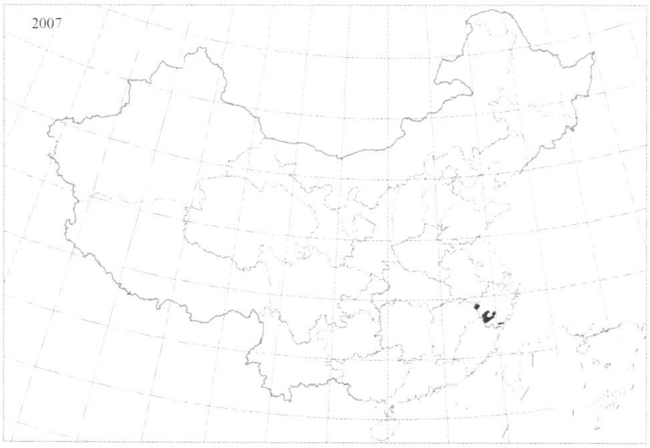

Myristicaceae 肉豆蔻科 *Horsfieldia* 风吹楠属
Horsfieldia amygdalina (Wall. ex Hook. f. et Thomson) Warb. 风吹楠
Trees, evergreen

21.4 (16.0~24.8); 15.0 (9.3~20.1); 25.7 (20.4~29.0); 21.4 (16.0~24.8); 197.0 (132.4~237.6); 0.0 (0.0~0.0); 1112 (780~1385) ❙ 1546 (868~2435); 734 (312~1428); 75 (41~130) ❙ 43.4 (-34.3~112.2); 1062 (780~1360); 435.0 (203.1~659.0)

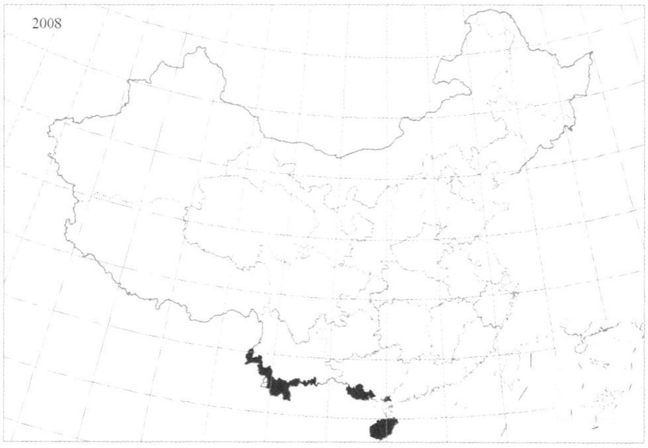

Myristicaceae 肉豆蔻科 *Horsfieldia* 风吹楠属
Horsfieldia kingii (Hook. f.) Warb. 大叶风吹楠
Trees, evergreen

19.8 (16.0~24.8); 12.8 (8.0~20.1); 24.6 (20.3~28.6); 19.8 (16.0~24.8); 177.8 (132.4~237.6); 0.0 (0.0~0.0); 996 (780~1385) ❙ 1511 (1182~2435); 791 (444~1428); 71 (41~184) ❙ 54.4 (-2.2~112.2); 973 (780~1221); 432.7 (215.8~659.0)

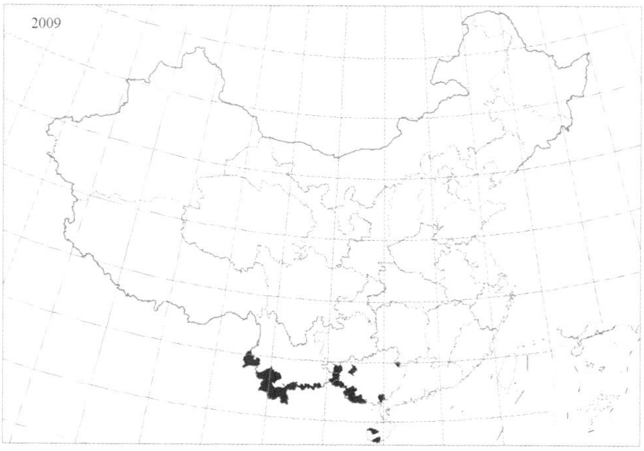

Myristicaceae 肉豆蔻科 *Horsfieldia* 风吹楠属
Horsfieldia prainii (King) Warburg 云南风吹楠
Trees, evergreen

18.6; 12.7; 22.2; 18.6; 163.5; 0.0; 898 ❙ 1528; 772; 63 ❙ 72.1; 885; 573.7

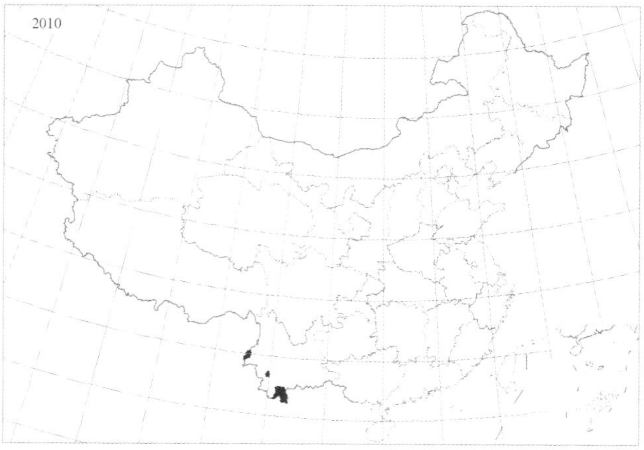

M

Myristicaceae 肉豆蔻科 *Knema* 红光树属
Knema elegans Warburg 假广子
Small trees, evergreen

18.2 (16.4~20.5); 12.0 (9.6~15.1); 22.0 (20.6~23.7); 18.3 (16.4~20.5); 159.2 (137.3~185.9); 0.0 (0.0~0.0); 875 (795~992) ❙ 1506 (1352~1688); 817 (655~959); 54 (42~77) ❙ 72.9 (47.6~112.2); 863 (795~986); 523.0 (434.4~659.0)

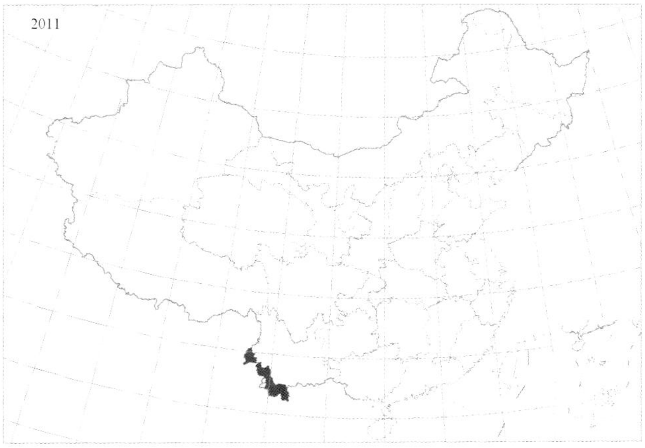

Myristicaceae 肉豆蔻科 *Knema* 红光树属
Knema globularia (Lam.) Warb. 小叶红光树
Small trees, evergreen

19.6 (16.4~22.2); 12.9 (9.6~15.1); 24.0 (20.6~28.3); 19.6 (16.4~22.3); 175.4 (137.3~207.5); 0.0 (0.0~0.0); 969 (795~1190) ❙ 1568 (1302~2148); 855 (655~1234); 70 (42~125) ❙ 63.7 (12.9~112.2); 961 (795~1190); 473.2 (281.7~659.0)

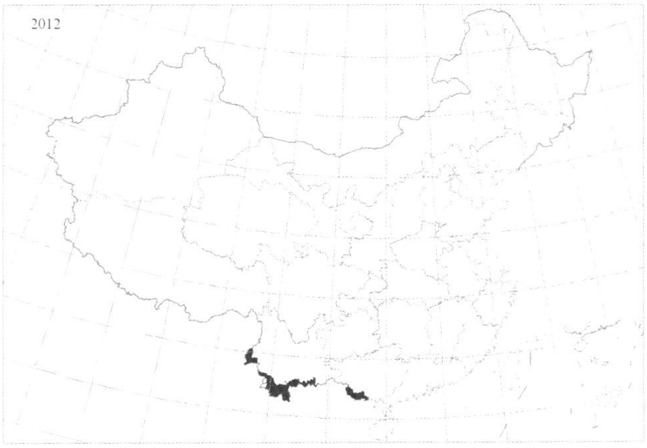

Myristicaceae 肉豆蔻科 *Knema* 红光树属
Knema lenta Pierre ex Warb. 狭叶红光树
Small trees, evergreen

17.4 (13.3~20.4); 10.7 (5.4~15.1); 21.7 (19.6~23.6); 17.5 (13.4~20.5); 149.5 (100.4~185.8); 0.0 (0.0~0.0); 850 (706~989) ❙ 1614 (1530~1736); 903 (702~1007); 57 (42~77) ❙ 92.4 (62.5~145.8); 847 (706~986); 546.8 (447.4~659.0)

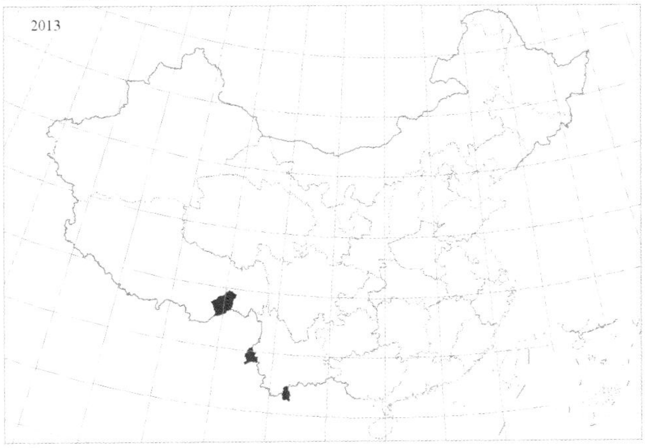

Myristicaceae 肉豆蔻科 *Knema* 红光树属
Knema linifolia (Roxb.) Warb. 大叶红光树
Trees, evergreen

17.9; 11.9; 21.6; 18.0; 155.8; 0.0; 854 ❙ 1435; 803; 42 ❙ 68.0; 833; 505.1

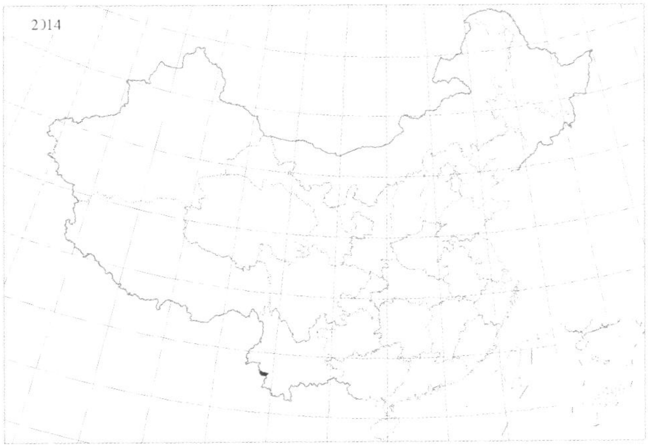

Myristicaceae 肉豆蔻科 *Knema* 红光树属
Knema tenuinervia W. J. de Wilde 红光树
Trees, evergreen

18.8 (16.4~21.0); 12.5 (9.6~15.1); 22.8 (20.6~25.6); 18.8 (16.4~21.0); 166.2 (137.3~191.8); 0.0 (0.0~0.0); 910 (795~1053) ❙ 1510 (1049~1751); 842 (598~996); 58 (42~77) ❙ 66.9 (26.1~112.2); 900 (783~1053); 524.4 (365.3~659.0)

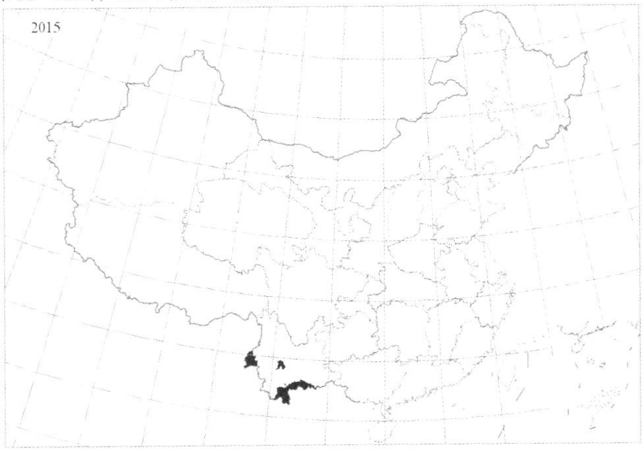

Myristicaceae 肉豆蔻科 *Knema* 红光树属
Knema tonkinensis (Warburg) W. J. de Wilde 密花红光树
Small trees, evergreen

17.9; 11.9; 21.6; 18.0; 155.8; 0.0; 854 ❙ 1435; 803; 42 ❙ 68.0; 833; 505.1

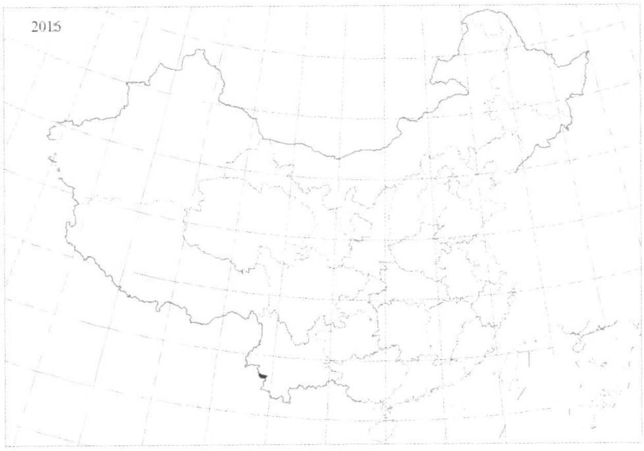

Myristicaceae 肉豆蔻科 *Myristica* 肉豆蔻属
Myristica yunnanensis Y. H. Li 云南肉豆蔻
Trees, evergreen

18.9 (17.1~20.5); 13.0 (11.0~15.1); 22.6 (20.8~23.7); 18.9 (17.1~20.5); 167.3 (145.0~185.9); 0.0 (0.0~0.0); 904 (815~992) ❙ 1453 (1276~1751); 776 (655~996); 57 (41~77) ❙ 60.9 (36.8~88.8); 887 (802~986); 480.4 (354.7~659.0)

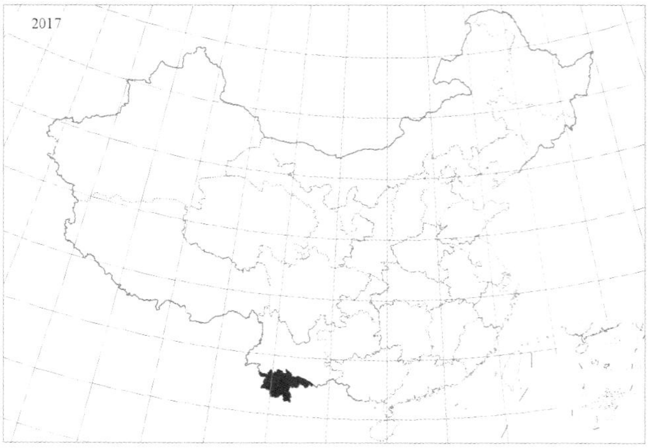

Lauraceae 樟科 *Actinodaphne* 黄肉楠属
Actinodaphne acuminata (Blume) Meisner 南投黄肉楠
Trees, evergreen

20.3 (15.5~22.7); 12.0 (4.8~15.3); 26.8 (20.1~28.8); 20.3 (15.6~22.8); 183.6 (127.2~213.1); 0.0 (0.0~0.2); 1058 (773~1221) ❙ 1778 (1250~2680); 860 (497~1428); 143 (60~303) ❙ 73.9 (12.9~246.5); 1058 (773~1221); 264.0 (203.4~327.7)

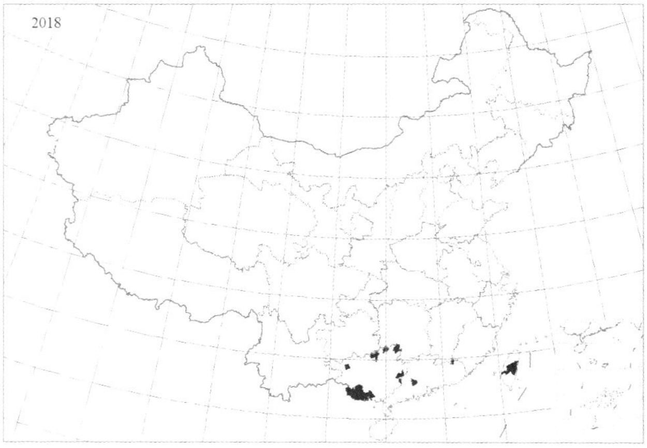

Lauraceae 樟科 *Actinodaphne* 黄肉楠属
Actinodaphne cupularis (Hemsl.) Gamble 红果黄肉楠
Shrubs or small trees, evergreen

15.5 (10.2~21.0); 4.7 (-2.9~14.2); 25.3 (21.0~28.9); 15.5 (10.4~21.0); 129.2 (77.2~191.8); 2.8 (0.0~20.5); 829 (634~1053) ❙ 1190 (626~1852); 539 (316~996); 79 (25~216) ❙ 43.9 (-22.2~108.0); 820 (626~1053); 221.7 (108.2~551.6)

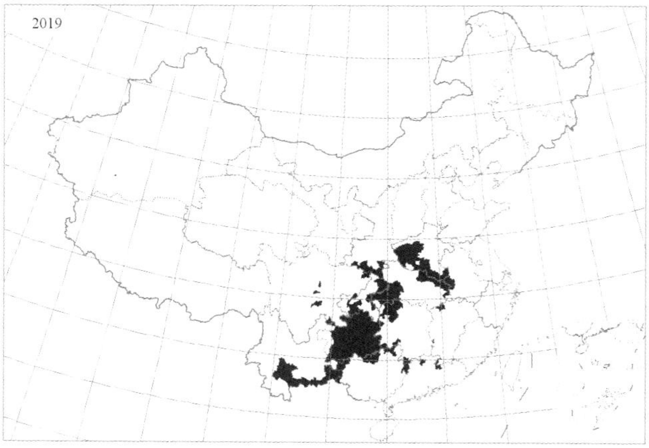

Lauraceae 樟科 *Actinodaphne* 黄肉楠属
Actinodaphne forrestii (C. K. Allen) Kosterm. 毛尖树
Trees, evergreen

17.7 (5.4~22.7); 10.5 (-2.1~15.0); 22.8 (12.3~28.6); 17.7 (5.8~22.8); 153.1 (31.5~213.1); 0.6 (0.0~25.6); 886 (481~1221) ❙ 1343 (741~2435); 747 (426~1428); 63 (28~130) ❙ 52.8 (5.8~112.2); 877 (481~1221); 373.9 (189.5~627.0)

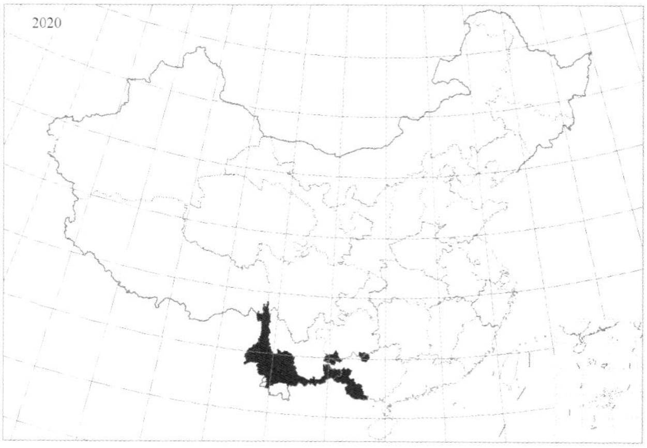

Lauraceae 樟科 *Actinodaphne* 黄肉楠属
Actinodaphne glaucina C. K. Allen 白背黄肉楠
Trees, evergreen

24.0 (22.2~24.8); 18.4 (17.2~20.1); 28.0 (25.7~28.8); 24.0 (22.3~24.8); 228.3 (207.2~237.6); 0.0 (0.0~0.0); 1310 (1127~1385) ❙ 1501 (868~1854); 534 (312~712); 74 (41~115) ❙ 15.0 (-34.3~47.8); 1191 (868~1360); 440.3 (248.2~620.0)

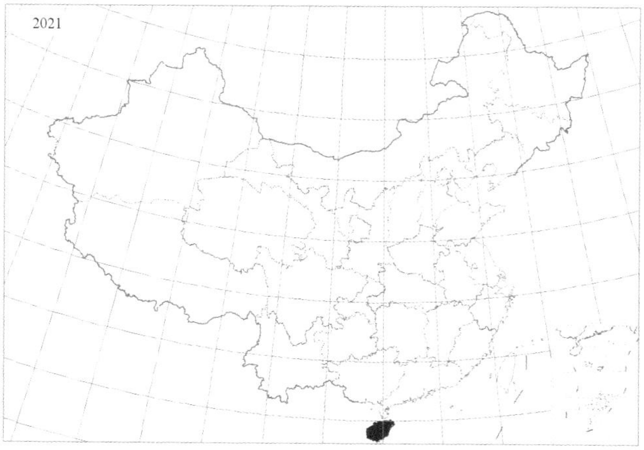

Lauraceae 樟科 *Actinodaphne* 黄肉楠属
Actinodaphne henryi Gamble 思茅黄肉楠
Trees, evergreen

19.0 (16.4~20.5); 13.1 (9.6~15.1); 22.5 (20.6~23.7); 19.1 (16.4~20.5); 168.7 (137.3~185.9); 0.0 (0.0~0.0); 914 (795~992) ❙ 1544 (1464~1688); 781 (655~959); 60 (42~77) ❙ 70.2 (47.6~112.2); 896 (795~986); 531.5 (427.9~659.0)

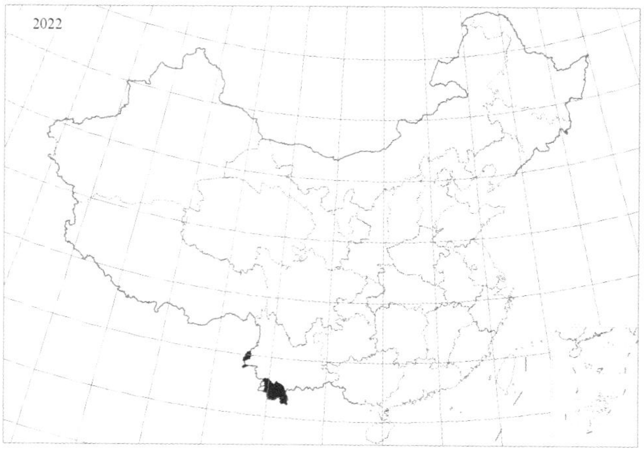

Lauraceae 樟科　*Actinodaphne* 黄肉楠属

Actinodaphne koshepangii Chun ex H. T. Chang 广东黄肉楠

Trees, evergreen

17.1 (14.6~21.0); 6.2 (3.6~11.9); 27.1 (24.6~29.0); 17.2 (14.7~21.1); 146.2 (117.7~193.0); 0.2 (0.0~1.6); 905 (779~1114) ▌1506 (1384~1662); 525 (403~821); 178 (105~211) ▌67.1 (47.4~101.8); 896 (779~1114); 210.7 (151.0~260.5)

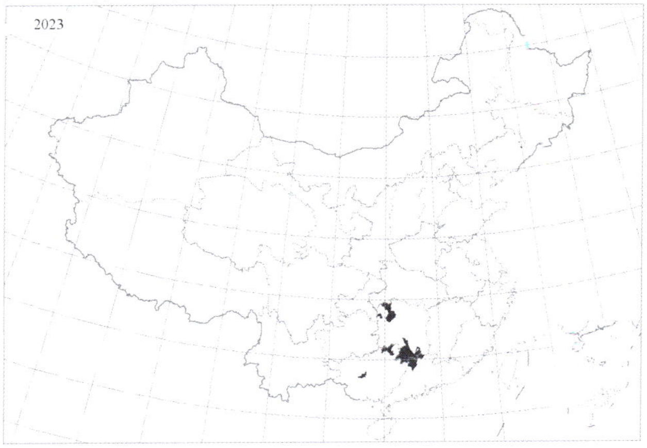

Lauraceae 樟科　*Actinodaphne* 黄肉楠属

Actinodaphne kweichowensis Yang et P. H. Huang 黔桂黄肉楠

Trees, evergreen

18.3 (15.9~22.2); 9.3 (5.3~13.9); 25.4 (23.4~28.3); 18.3 (15.9~22.3); 160.2 (131.1~207.5); 0.0 (0.0~0.0); 931 (833~1190) ▌1322 (1089~1485); 695 (503~848); 72 (51~129) ▌43.0 (15.4~59.6); 931 (833~1190); 269.3 (181.2~371.7)

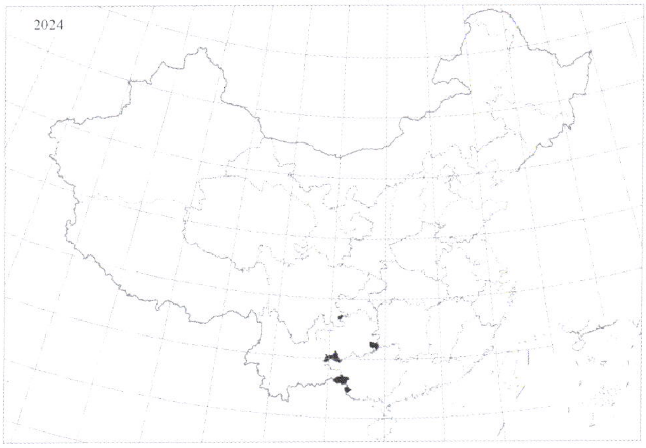

Lauraceae 樟科　*Actinodaphne* 黄肉楠属

Actinodaphne lecomtei C. K. Allen 柳叶黄肉楠

Trees, evergreen

18.7 (11.9~23.4); 9.3 (2.5~16.1); 26.6 (20.0~29.8); 18.7 (12.0~23.5); 165.0 (87.2~221.7); 0.2 (0.0~4.2); 981 (671~1273) ▌1483 (999~1875); 632 (475~891); 127 (33~197) ▌52.5 (9.0~101.8); 979 (671~1273); 228.0 (153.9~322.1)

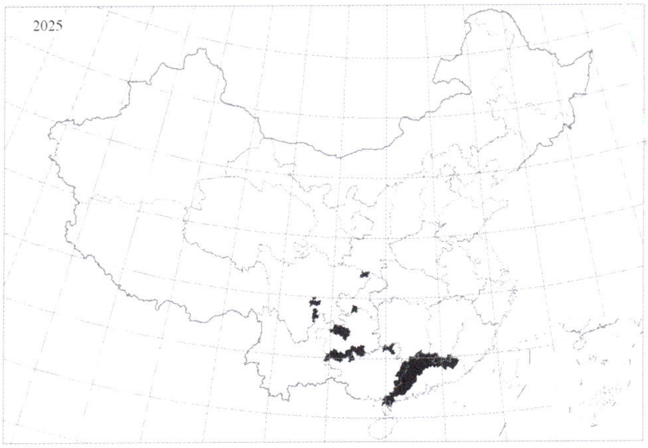

Lauraceae 樟科　*Actinodaphne* 黄肉楠属

Actinodaphne menghaiensis J. Li 勐海黄肉楠

Trees, evergreen

19.2; 13.8; 22.3; 19.2; 170.5; 0.0; 909 ▌1499; 686; 58 ▌64.9; 880; 513.9

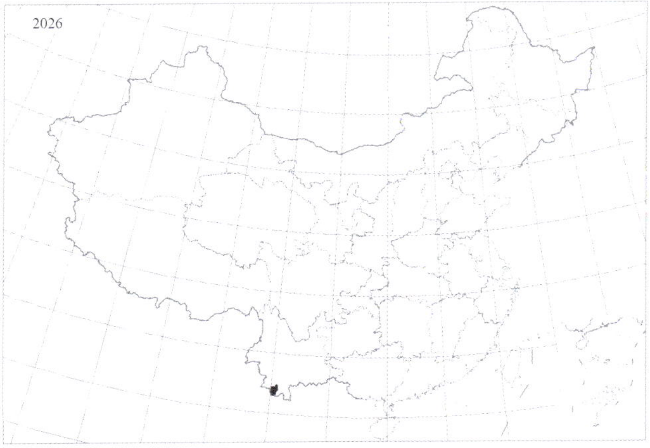

Lauraceae 樟科　*Actinodaphne* 黄肉楠属

Actinodaphne mushaensis (Hayata) Hayata 雾社黄肉楠

Trees, evergreen

19.1 (16.2~21.7); 13.3 (11.0~15.3); 24.0 (20.1~27.0); 19.1 (16.2~21.7); 169.6 (134.2~200.2); 0.0 (0.0~0.0); 943 (773~1118) ▌2467 (1836~2993); 1120 (924~1412); 253 (110~547) ▌169.2 (64.2~246.5); 934 (773~1075); 248.5 (202.8~289.1)

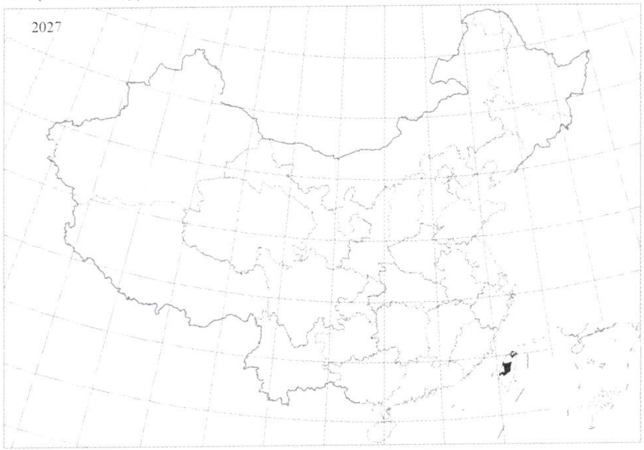

Lauraceae 樟科　*Actinodaphne* 黄肉楠属

Actinodaphne obovata (Nees) Blume 倒卵叶黄肉楠

Trees, evergreen

17.5 (1.8~21.0); 11.1 (-7.1~15.1); 21.9 (9.9~25.6); 17.7 (3.8~21.0); 154.2 (15.7~191.8); 3.3 (0.0~53.3); 870 (404~1053) ▌1460 (583~1751); 796 (348~1007); 57 (13~77) ▌68.9 (36.8~145.8); 863 (404~1053); 488.2 (169.5~659.0)

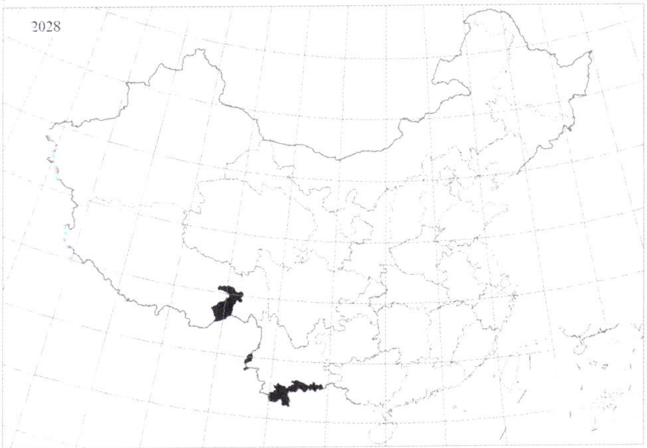

Lauraceae 樟科 *Actinodaphne* 黄肉楠属
Actinodaphne obscurinervia Yang et P. H. Huang 隐脉黄肉楠
Trees, evergreen

14.3; 3.0; 25.0; 14.3; 114.9; 3.5; 785 ❙ 1263; 514; 78 ❙ 61.2; 785; 228.0

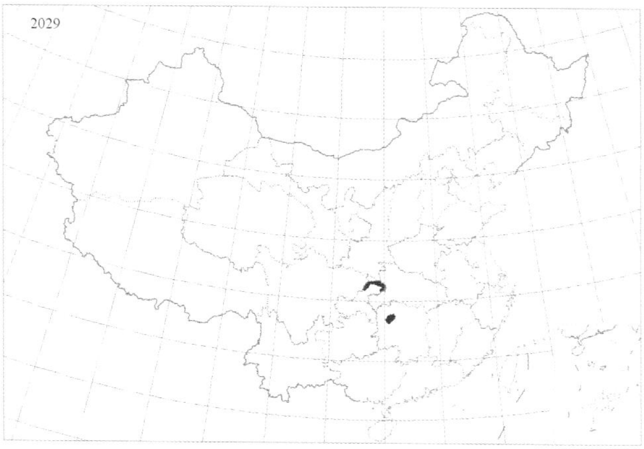

Lauraceae 樟科 *Actinodaphne* 黄肉楠属
Actinodaphne omeiensis (H. Liu) C. K. Allen 峨眉黄肉楠
Shrubs or trees, evergreen

15.5 (13.0~17.8); 5.4 (3.7~7.8); 24.9 (21.4~27.0); 15.6 (13.1~17.9); 127.2 (98.2~154.2); 0.4 (0.0~1.5); 818 (708~929) ❙ 1169 (985~1414); 578 (473~817); 64 (38~94) ❙ 44.1 (18.3~89.0); 818 (708~929); 189.0 (142.9~255.2)

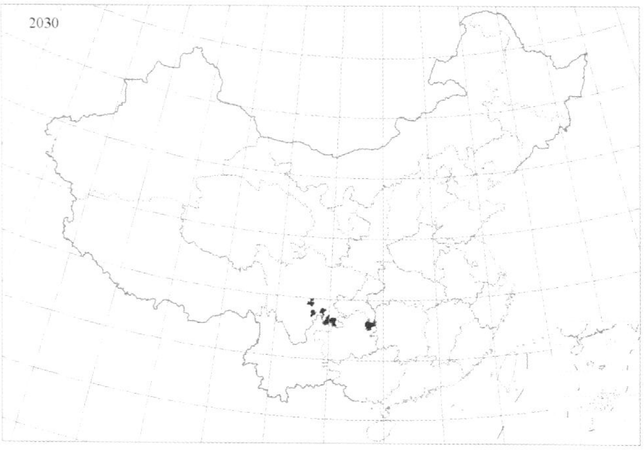

Lauraceae 樟科 *Actinodaphne* 黄肉楠属
Actinodaphne paotingensis Yen C. Yang et P. H. Huang 保亭黄肉楠
Shrubs or trees, evergreen

23.5; 18.4; 27.0; 23.5; 222.6; 0.0; 1259 ❙ 1638; 575; 58 ❙ 30.4; 1155; 596.5

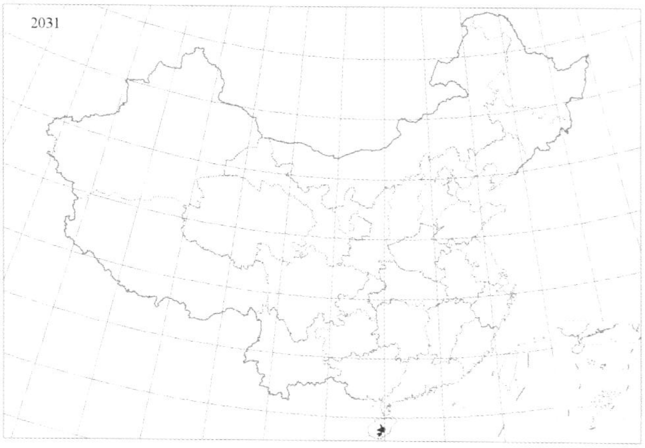

Lauraceae 樟科 *Actinodaphne* 黄肉楠属
Actinodaphne pilosa (Lour.) Merr. 毛黄肉楠
Trees, evergreen

21.7 (18.2~24.8); 14.3 (8.6~20.1); 27.3 (21.6~29.2); 21.8 (18.2~24.8); 201.2 (158.6~237.6); 0.0 (0.0~0.0); 1146 (861~1385) ❙ 1606 (868~2435); 730 (312~1428); 112 (41~197) ❙ 42.5 (-34.3~99.5); 1113 (839~1360); 333.8 (157.7~659.0)

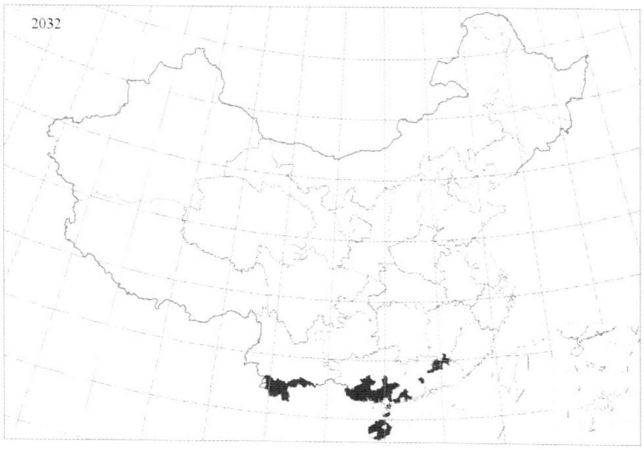

Lauraceae 樟科 *Actinodaphne* 黄肉楠属
Actinodaphne trichocarpa C. K. Allen 毛果黄肉楠
Shrubs or trees, evergreen

15.0 (5.4~20.5); 7.4 (-2.1~15.1); 20.8 (12.3~25.8); 15.0 (5.8~20.5); 121.0 (31.5~185.9); 1.1 (0.0~25.6); 766 (481~992) ❙ 1153 (741~1688); 633 (426~959); 47 (21~86) ❙ 50.6 (6.5~112.2); 756 (481~986); 329.1 (150.8~659.0)

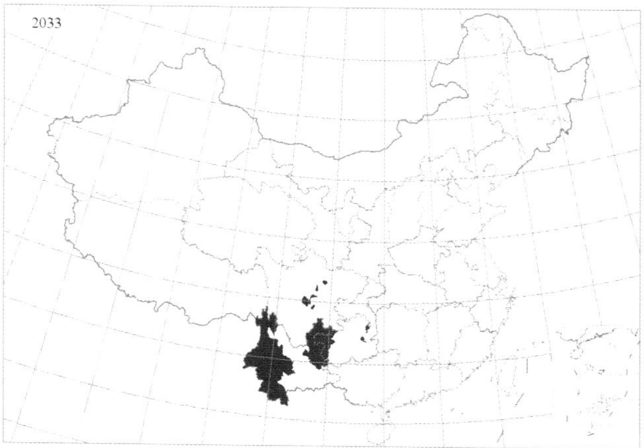

Lauraceae 樟科 *Actinodaphne* 黄肉楠属
Actinodaphne tsaii Hu 马关黄肉楠
Trees, evergreen

19.9 (17.2~21.6); 12.2 (9.5~14.2); 25.9 (22.1~28.4); 19.9 (17.3~21.6); 178.8 (147.0~199.4); 0.0 (0.0~0.0); 1007 (833~1128) ❙ 1580 (1254~1991); 726 (584~902); 119 (48~182) ❙ 57.1 (36.8~80.6); 1007 (833~1128); 322.9 (201.9~551.6)

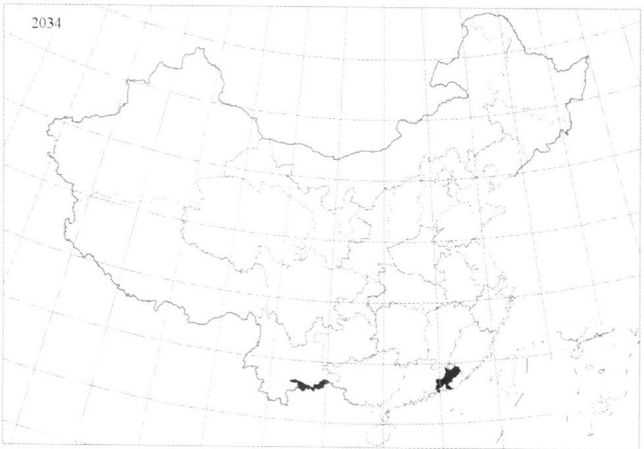

Lauraceae 樟科 *Alseodaphne* 油丹属
Alseodaphne andersonii (King ex Hook. f.) Kosterm. 毛叶油丹
Trees, evergreen

18.2 (13.3~21.0); 11.5 (5.4~15.1); 22.8 (19.6~25.6); 18.3 (13.4~21.0); 159.2 (100.4~191.8); 0.0 (0.0~0.0); 886 (706~1053) ❙ 1428 (1074~1751); 780 (582~1007); 57 (42~77) ❙ 62.3 (32.4~145.8); 879 (706~1053); 449.3 (277.0~659.0)

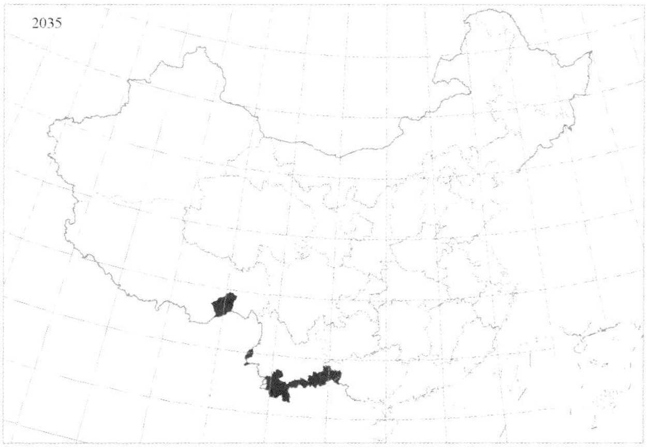

Lauraceae 樟科 *Alseodaphne* 油丹属
Alseodaphne gracilis Kosterm. 细梗油丹
Trees, evergreen

21.0; 14.2; 25.6; 21.0; 191.8; 0.0; 1053 ❙ 1522; 818; 68 ❙ 44.5; 1053; 551.6

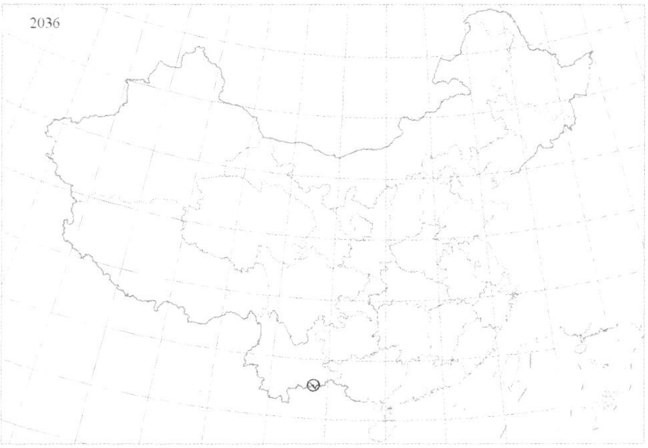

Lauraceae 樟科 *Alseodaphne* 油丹属
Alseodaphne hainanensis Merr. 油丹
Trees, evergreen

23.1 (17.4~24.8); 17.3 (7.4~20.1); 27.4 (23.3~28.7); 23.2 (17.4~24.8); 218.1 (148.9~237.6); 0.0 (0.0~0.0); 1252 (886~1385) ❙ 1371 (868~1767); 515 (312~835); 60 (41~107) ❙ 12.0 (-34.3~55.9); 1084 (868~1299); 460.8 (245.3~620.0)

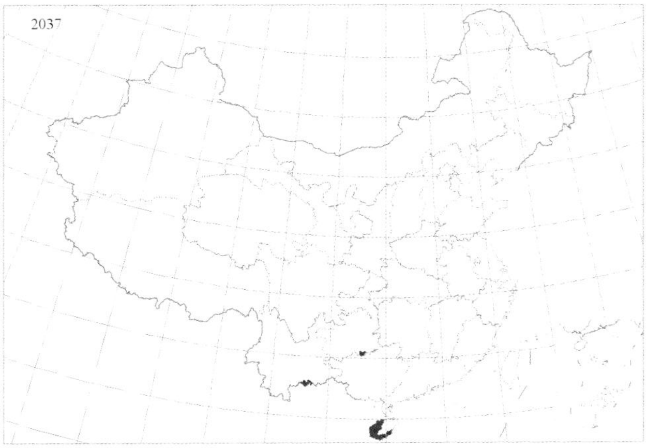

Lauraceae 樟科 *Alseodaphne* 油丹属
Alseodaphne hokouensis H. W. Li 河口油丹
Trees, evergreen

21.0; 14.2; 25.6; 21.0; 191.8; 0.0; 1053 ❙ 1522; 818; 68 ❙ 44.5; 1053; 551.6

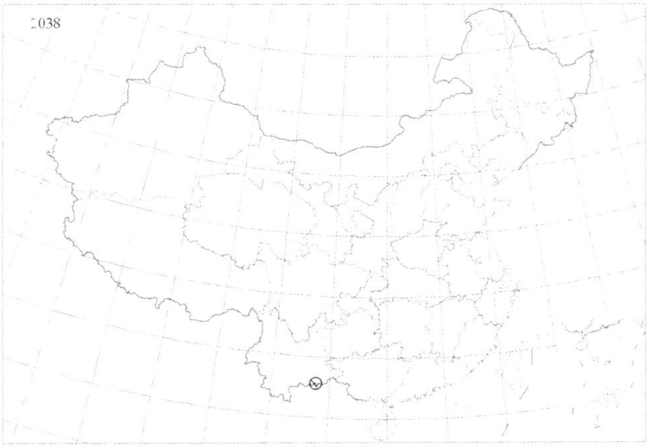

Lauraceae 樟科 *Alseodaphne* 油丹属
Alseodaphne huanglianshanensis H. W. Li et Y. M. Shui 黄莲山油丹
Trees, evergreen

18.6; 12.7; 22.5; 18.7; 163.9; 0.0; 888 ❙ 1547; 911; 63 ❙ 74.2; 888; 551.3

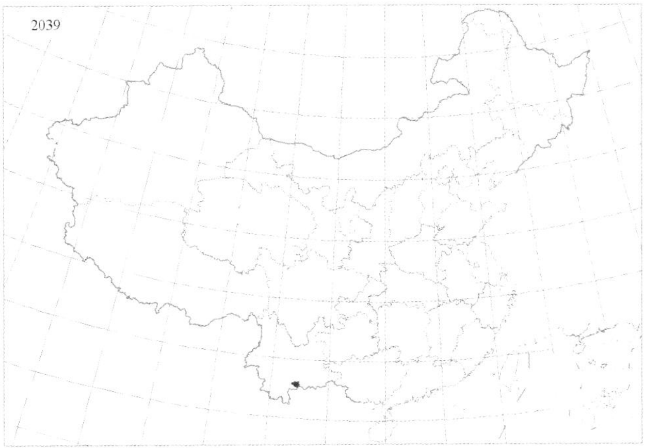

Lauraceae 樟科 *Alseodaphne* 油丹属
Alseodaphne marlipoensis (H. W. Li) H. W. Li 麻栗坡油丹
Trees evergreen

18.3; 10.9; 23.5; 18.3; 159.6; 0.0; 896 ❙ 1425; 798; 56 ❙ 60.1; 896; 415.3

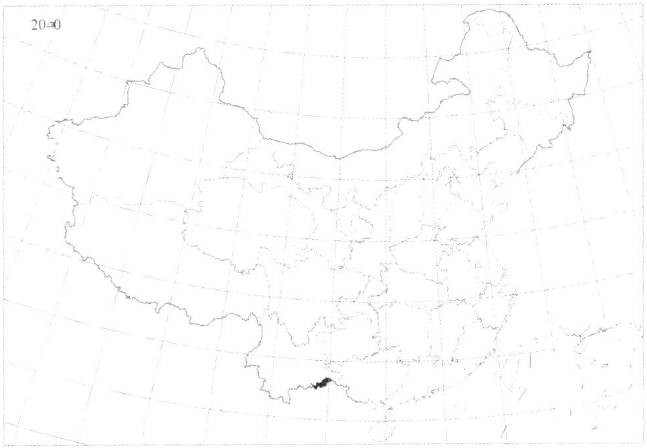

Lauraceae 樟科 *Alseodaphne* 油丹属
Alseodaphne petiolaris (Meisn.) Hook. f. 长柄油丹
Trees, evergreen

18.1 (13.3~21.0); 11.5 (5.4~15.1); 22.3 (19.6~25.8); 18.1 (13.4~21.0); 157.4 (100.4~191.8); 0.0 (0.0~0.0); 875 (706~1053) ❙ 1446 (1049~1751); 800 (598~1007); 55 (41~77) ❙ 66.2 (26.1~145.8); 865 (706~1053); 462.2 (294.6~659.0)

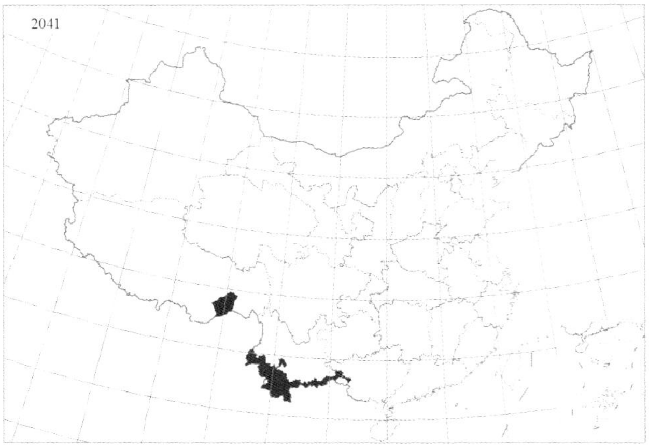

Lauraceae 樟科 *Alseodaphne* 油丹属
Alseodaphne rugosa Merr. et Chun 皱皮油丹
Trees, evergreen

23.9; 18.9; 27.3; 23.9; 227.2; 0.0; 1298 ❙ 1621; 561; 58 ❙ 25.4; 1174; 562.6

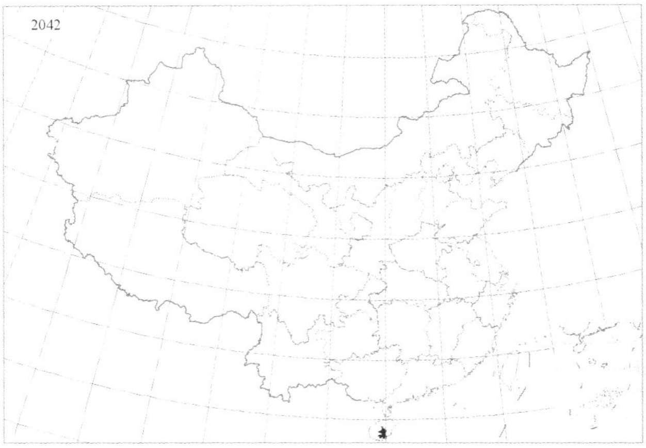

Lauraceae 樟科 *Alseodaphne* 油丹属
Alseodaphne sichourensis H. W. Li 西畴油丹
Trees, evergreen

19.4; 11.2; 25.4; 19.5; 173.5; 0.0; 986 ❙ 1403; 764; 81 ❙ 43.7; 986; 326.7

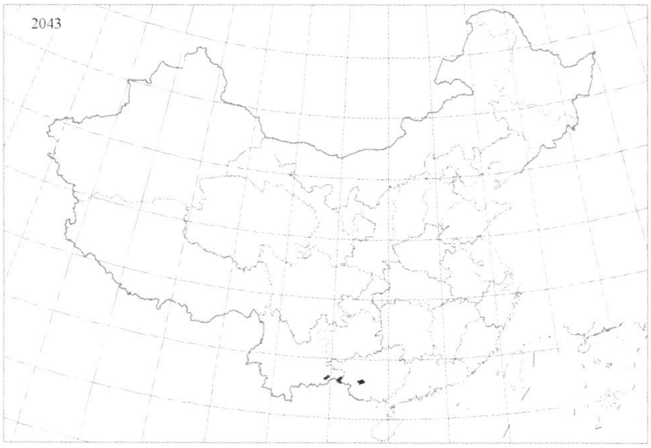

Lauraceae 樟科 *Alseodaphne* 油丹属
Alseodaphne yunnanensis Kosterm. 云南油丹
Trees, evergreen

18.9; 12.9; 22.9; 18.9; 166.8; 0.0; 902 ❙ 1487; 873; 59 ❙ 65.0; 902; 528.7

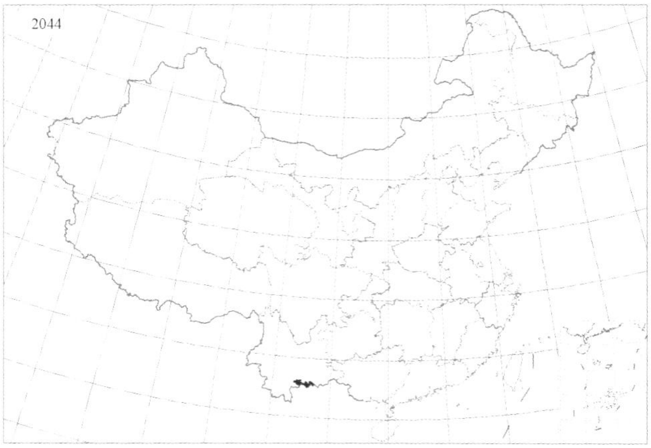

Lauraceae 樟科 *Beilschmiedia* 琼楠属
Beilschmiedia appendiculata (C. K. Allen) S. K. Lee et Y. T. Wei 山潺
Trees, evergreen

23.7 (22.3~24.8); 18.1 (14.7~20.1); 27.8 (26.7~28.4); 23.8 (22.4~24.8); 225.5 (208.3~237.6); 0.0 (0.0~0.0); 1291 (1188~1385) ❙ 1471 (1179~1725); 541 (392~845); 58 (41~92) ❙ 14.5 (-11.9~41.8); 1155 (1056~1294); 463.7 (215.3~572.9)

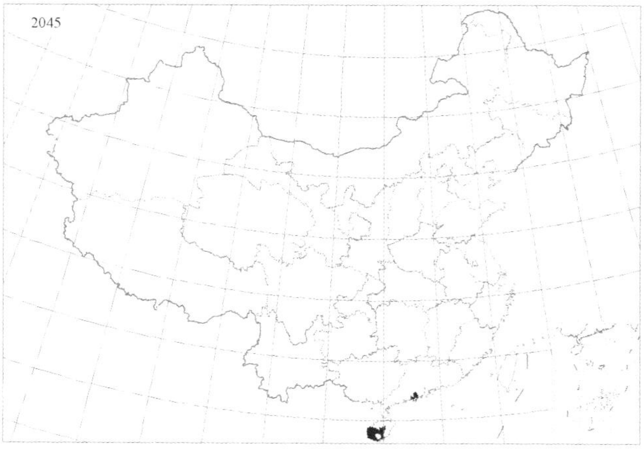

Lauraceae 樟科 *Beilschmiedia* 琼楠属
Beilschmiedia baotingensis S. K. Lee et Y. T. Wei 保亭琼楠
Trees, evergreen

24.0; 19.1; 27.2; 24.0; 228.1; 0.0; 1302 ❙ 1551; 539; 48 ❙ 19.1; 1147; 620.0

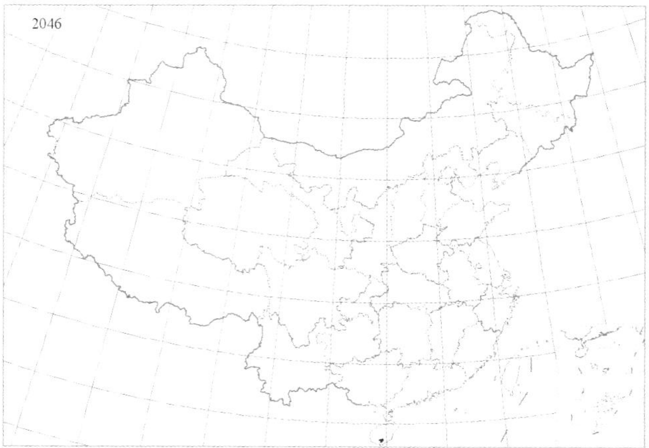

Lauraceae 樟科 *Beilschmiedia* 琼楠属
Beilschmiedia brachythyrsa H. W. Li 勐仑琼楠
Trees, evergreen

20.0；14.6；23.2；20.1；180.8；0.0；963 ❙ 1523；681；67 ❙ 58.4；942；581.0

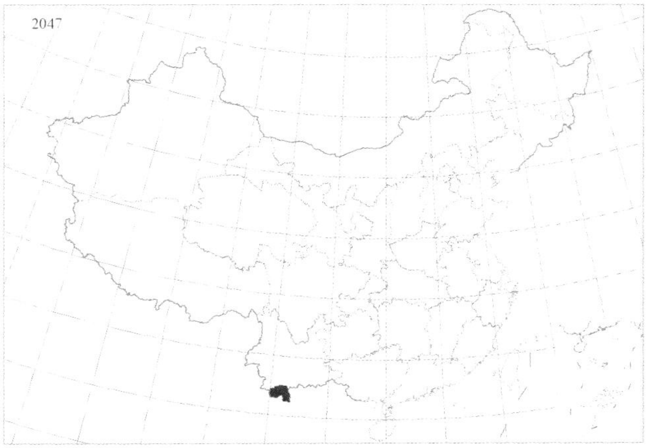

Lauraceae 樟科 *Beilschmiedia* 琼楠属
Beilschmiedia brevifolia Y. T. Wei 短叶琼楠
Trees, evergreen

24.0；19.1；27.2；24.0；228.1；0.0；1302 ❙ 1551；539；48 ❙ 19.1；1147；620.0

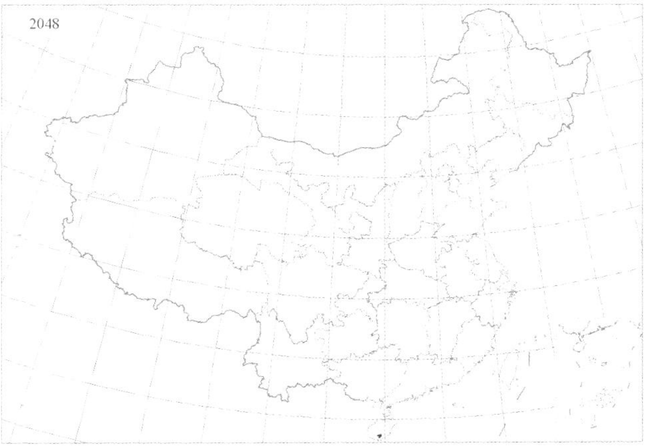

Lauraceae 樟科 *Beilschmiedia* 琼楠属
Beilschmiedia brevipaniculata C. K. Allen 短序琼楠
Trees, evergreen

22.1 (19.1~24.0)；14.3 (9.3~19.1)；28.0 (27.2~28.7)；22.1 (19.1~24.0)；205.3 (169.2~228.1)；0.0 (0.0~0.0)；1175 (988~1302) ❙ 1696 (1302~2435)；849 (455~1428)；112 (48~173) ❙ 44.9 (6.7~99.5)；1153 (988~1221)；304.0 (139.6~620.0)

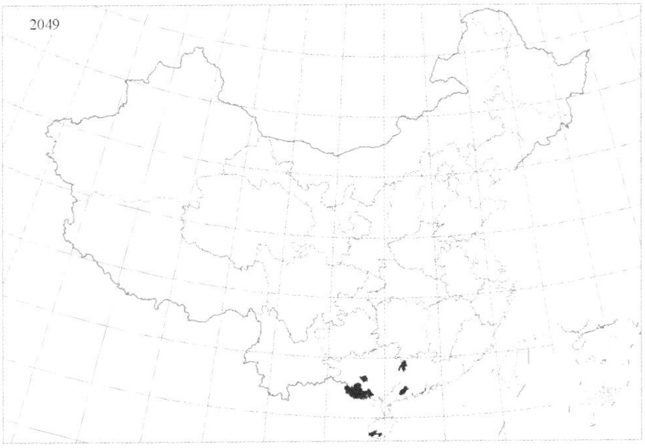

Lauraceae 樟科 *Beilschmiedia* 琼楠属
Beilschmiedia cylindrica S. K. Lee et Y. T. Wei 柱果琼楠
Shrubs or trees, evergreen

18.3 (17.2~21.0)；11.2 (9.5~14.2)；23.3 (22.4~25.6)；18.4 (17.3~21.0)；160.2 (147.0~191.8)；0.0 (0.0~0.0)；892 (833~1053) ❙ 1416 (1254~1522)；797 (699~860)；57 (48~68) ❙ 59.4 (44.5~80.6)；892 (833~1053)；443.1 (336.3~551.6)

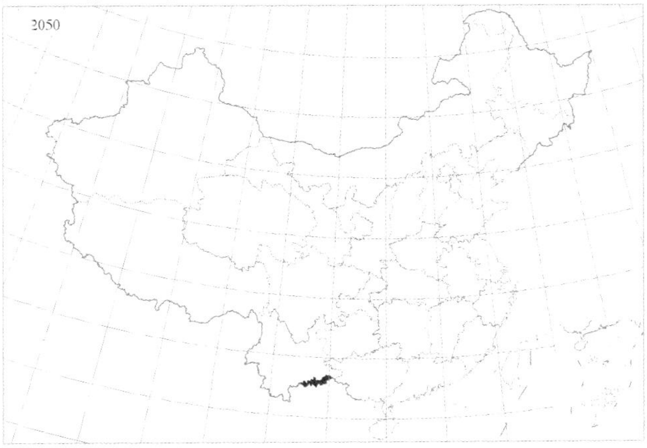

Lauraceae 樟科 *Beilschmiedia* 琼楠属
Beilschmiedia delicata S. K. Lee et Y. T. Wei 美脉琼楠
Shrubs or trees, evergreen

19.0 (14.0~22.4)；9.9 (4.1~14.8)；26.4 (21.4~28.9)；19.1 (14.1~22.4)；168.7 (109.8~208.9)；0.0 (0.0~0.9)；985 (747~1202) ❙ 1473 (988~2030)；684 (499~1033)；124 (47~194) ❙ 50.3 (14.8~93.2)；985 (747~1202)；264.4 (163.7~551.6)

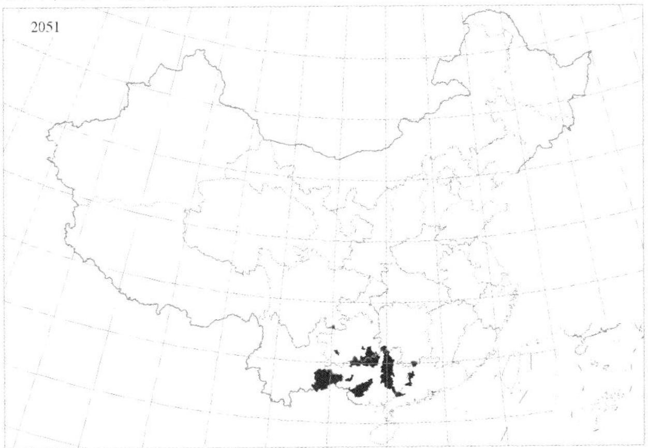

Lauraceae 樟科 *Beilschmiedia* 琼楠属
Beilschmiedia erythrophloia Hayata 台琼楠
Trees, evergreen

19.2 (16.2~22.2)；13.9 (11.0~17.7)；23.5 (20.1~25.9)；19.2 (16.2~22.2)；171.0 (134.2~206.7)；0.0 (0.0~0.0)；936 (773~1123) ❙ 2780 (2445~2993)；1286 (924~1847)；266 (107~547) ❙ 201.1 (140.0~246.5)；936 (773~1123)；280.8 (207.7~363.7)

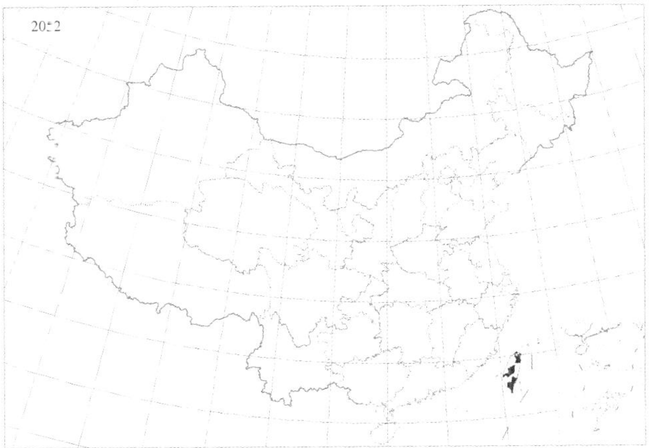

Lauraceae 樟科 *Beilschmiedia* 琼楠属

Beilschmiedia fasciata H. W. Li 白柴果

Trees, evergreen

18.5 (17.2~21.0)；11.5 (9.5~14.2)；23.3 (22.4~25.6)；18.5 (17.3~21.0)；162.0 (147.0~191.8)；0.0 (0.0~0.0)；898 (833~1053) | 1396 (1254~1522)；789 (699~860)；56 (48~68) | 56.1 (36.8~80.6)；897 (833~1053)；444.4 (336.3~551.6)

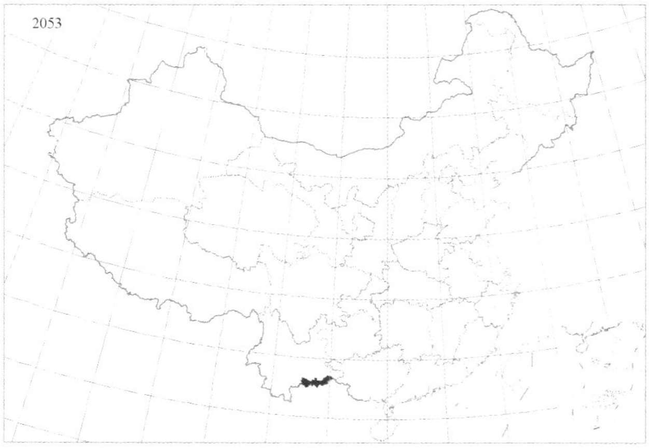

Lauraceae 樟科 *Beilschmiedia* 琼楠属

Beilschmiedia fordii Dunn 广东琼楠

Trees, evergreen

18.5 (14.3~21.8)；8.7 (3.3~15.2)；27.0 (23.6~29.8)；18.6 (14.3~22.7)；162.8 (114.1~212.7)；0.1 (0.0~2.3)；966 (765~1216) | 1557 (1089~2148)；625 (403~1234)；156 (51~230) | 61.5 (27.5~104.6)；962 (765~1216)；237.6 (151.0~322.1)

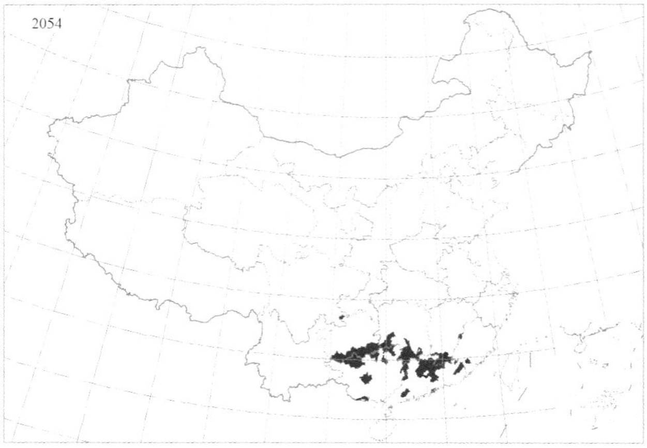

Lauraceae 樟科 *Beilschmiedia* 琼楠属

Beilschmiedia furfuracea Chun ex H. T. Chang 糠秕琼楠

Trees, evergreen

22.0 (21.0~23.2)；13.8 (12.6~15.8)；28.2 (26.9~28.9)；22.0 (21.0~23.2)；204.3 (192.3~218.9)；0.0 (0.0~0.0)；1169 (1081~1257) | 1723 (1302~2435)；895 (670~1428)；121 (86~173) | 47.6 (12.9~99.5)；1169 (1081~1257)；263.6 (194.5~327.7)

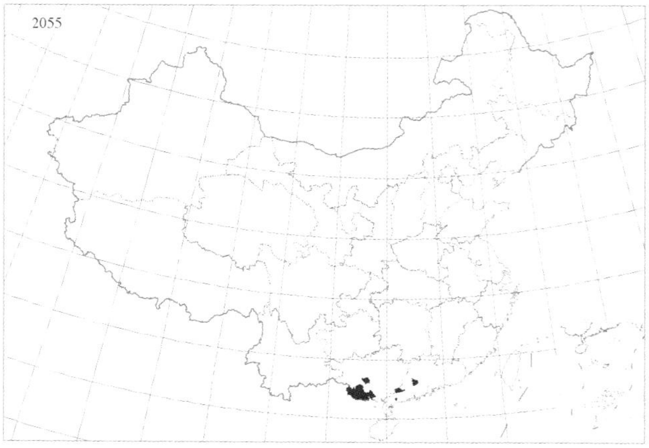

Lauraceae 樟科 *Beilschmiedia* 琼楠属

Beilschmiedia glauca S. K. Lee et L. F. Lau 粉背琼楠

Trees, evergreen

21.6 (17.2~24.8)；14.8 (7.8~20.1)；26.9 (22.7~28.4)；21.7 (17.3~24.8)；199.8 (147.0~237.6)；0.0 (0.0~0.0)；1154 (838~1385) | 1353 (1104~1725)；539 (392~699)；57 (41~80) | 19.1 (-11.9~49.7)；1050 (838~1163)；419.3 (151.7~620.0)

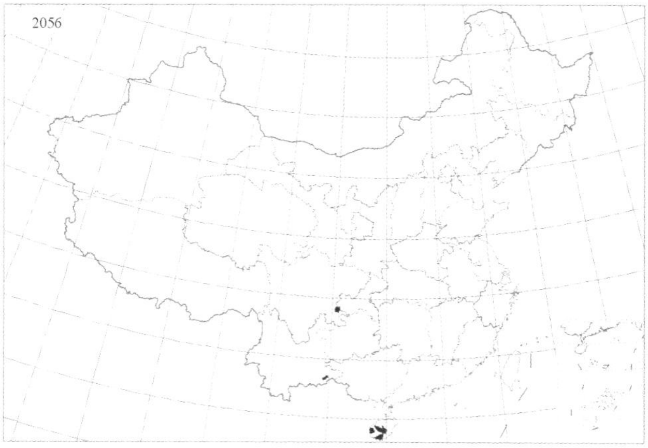

Lauraceae 樟科 *Beilschmiedia* 琼楠属

Beilschmiedia glauca var. glaucoides H. W. Li 顶序琼楠

Trees, evergreen

18.3；10.9；23.5；18.3；159.6；0.0；896 | 1425；798；56 | 60.1；896；415.3

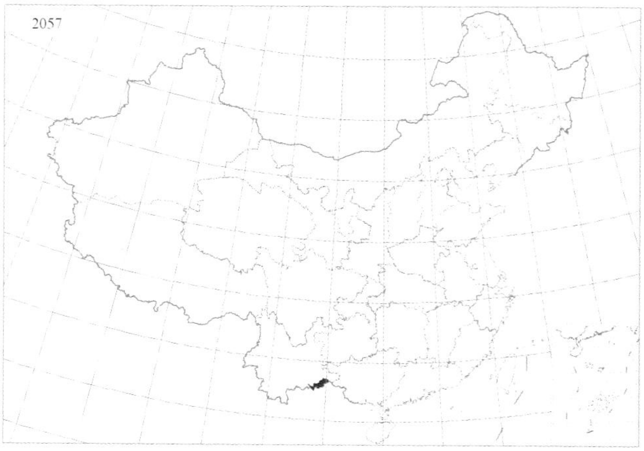

Lauraceae 樟科 *Beilschmiedia* 琼楠属

Beilschmiedia henghsienensis S. K. Lee et Y. T. Wei 横县琼楠

Trees, evergreen

21.8；13.4；28.1；21.8；201.9；0.0；1156 | 1833；984；134 | 58.7；1156；263.6

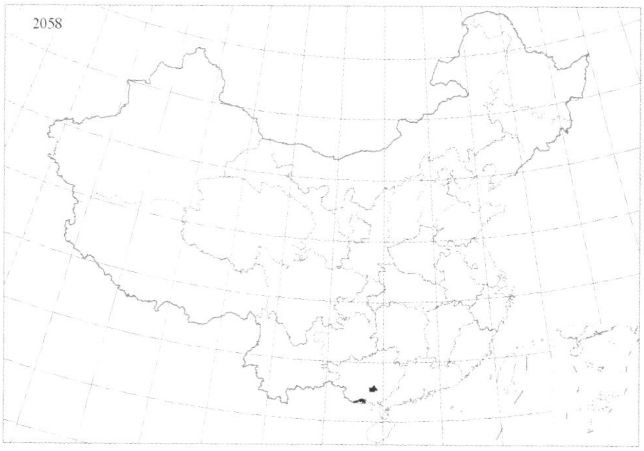

Lauraceae 樟科 *Beilschmiedia* 琼楠属

Beilschmiedia intermedia C. K. Allen 琼楠

Trees, evergreen

22.8 (21.0~24.8); 15.8 (13.2~20.1); 28.1 (25.7~29.2); 22.9 (21.0~24.8) 214.2 (192.3~237.6); 0.0 (0.0~0.0); 1226 (1081~1385) ❙ 1624 (868~2435); 724 (312~1428); 99 (41~167) ❙ 33.8 (-34.3~99.5); 1168 (868~1331); 327.5 (157.7~620.0)

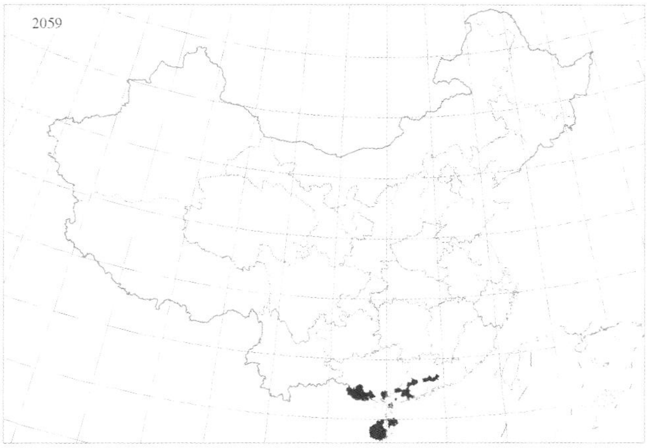

Lauraceae 樟科 *Beilschmiedia* 琼楠属

Beilschmiedia kweichowensis Cheng 贵州琼楠

Trees, evergreen

16.7 (10.6~19.2); 6.2 (-0.2~10.2); 26.3 (20.0~28.9); 16.7 (10.6~19.2); 141.1 (79.4~170.8); 0.8 (0.0~12.1); 884 (643~1002) ❙ 1192 (985~1496); 533 (442~816); 67 (35~142) ❙ 36.7 (8.0~93.8); 884 (643~1002); 186.0 (108.2~322.1)

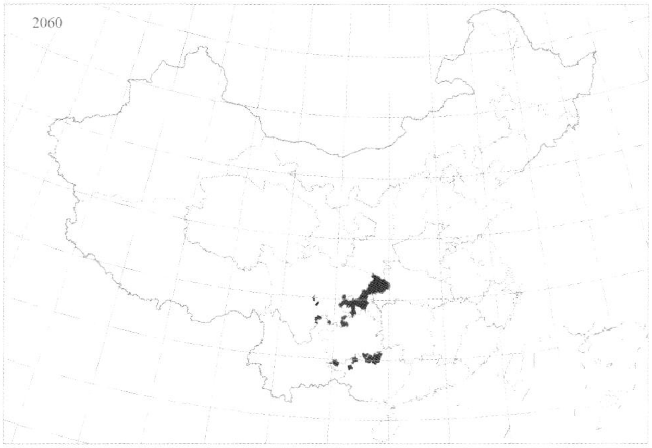

Lauraceae 樟科 *Beilschmiedia* 琼楠属

Beilschmiedia laevis C. K. Allen 红枝琼楠

Trees, evergreen

23.8 (21.8~24.8); 17.9 (13.7~20.1); 28.1 (27.2~28.7); 23.9 (21.9~24.8); 226.3 (202.7~237.6); 0.0 (0.0~0.0); 1301 (1159~1385) ❙ 1278 (868~1551); 474 (312~708); 61 (41~124) ❙ -1.1 (-34.3~31.9); 1085 (868~1159); 426.8 (245.3~620.0)

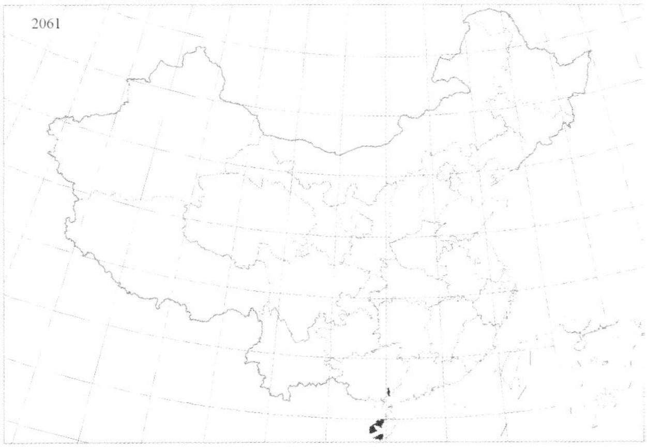

Lauraceae 樟科 *Beilschmiedia* 琼楠属

Beilschmiedia linocieroides H. W. Li 李榄琼楠

Trees, evergreen

18.6 (16.3~21.0); 12.3 (9.5~15.1); 22.5 (20.3~25.6); 18.6 (16.4~21.0); 163.4 (136.6~191.8); 0.0 (0.0~0.0); 896 (789~1053) ❙ 1441 (1232~1607); 759 (655~907); 52 (41~77) ❙ 61.2 (44.5~80.6); 878 (780~1053); 477.6 (336.3~659.0)

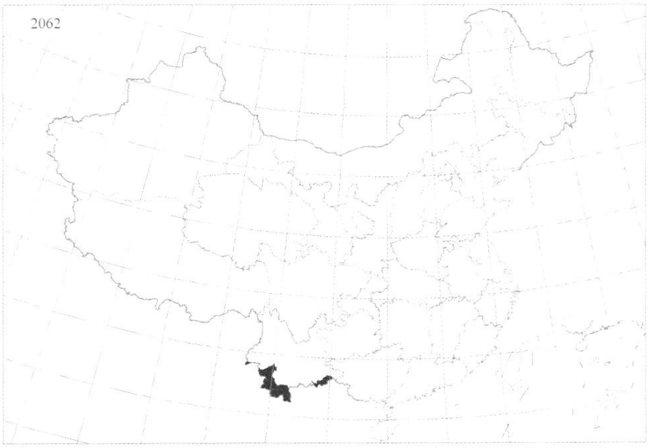

Lauraceae 樟科 *Beilschmiedia* 琼楠属

Beilschmiedia longepetiolata C. K. Allen 长柄琼楠

Trees, evergreen

21.8; 14.8; 27.1; 21.9; 202.3; 0.0; 1172 ❙ 1404; 581; 67 ❙ 22.8; 1071; 417.8

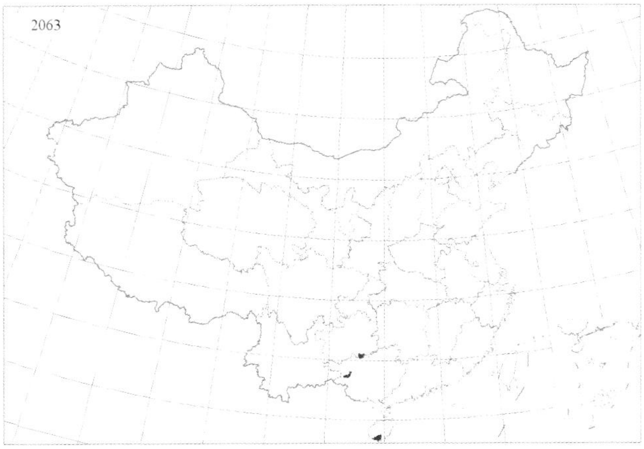

Lauraceae 樟科 *Beilschmiedia* 琼楠属

Beilschmiedia macropoda C. K. Allen 肉柄琼楠

Trees, evergreen

24.5; 19.4; 28.1; 24.6; 234.7; 0.0; 1362 ❙ 1572; 542; 70 ❙ 15.6; 1225; 458.4

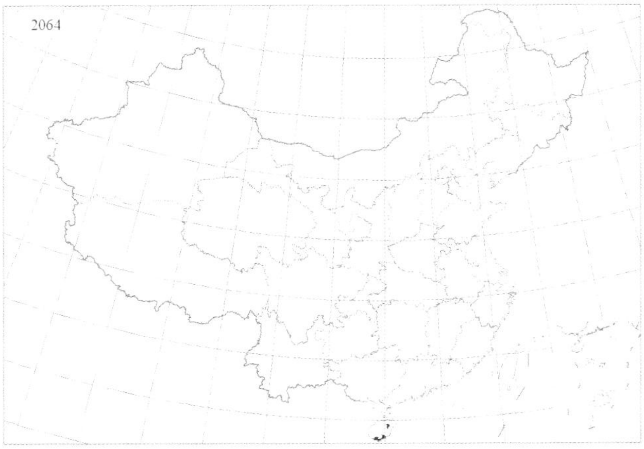

Lauraceae 樟科 *Beilschmiedia* 琼楠属
Beilschmiedia muricata H. T. Chang 瘤果琼楠
Shrubs or trees, evergreen

22.0 (21.3~22.7); 13.9 (13.2~15.0); 28.1 (27.3~28.6); 22.1 (21.3~22.8); 204.7 (195.9~213.1); 0.0 (0.0~0.0); 1171 (1111~1221) ❙ 1751 (1302~2435); 959 (670~1428); 114 (86~130) ❙ 49.5 (12.9~99.5); 1171 (1111~1221); 288.3 (235.1~327.7)

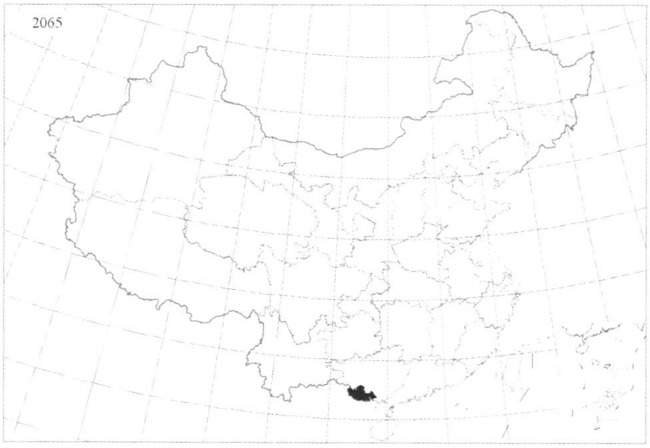

Lauraceae 樟科 *Beilschmiedia* 琼楠属
Beilschmiedia ningmingensis S. K. Lee et Y. T. Wei 宁明琼楠
Trees, evergreen

21.5; 13.6; 27.4; 21.6; 199.1; 0.0; 1131 ❙ 1581; 850; 106 ❙ 39.9; 1131; 281.7

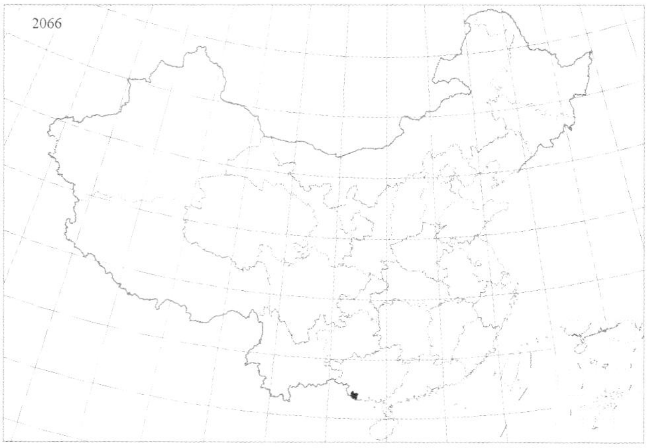

Lauraceae 樟科 *Beilschmiedia* 琼楠属
Beilschmiedia obconica C. K. Allen 锈叶琼楠
Trees, evergreen

24.3; 18.7; 28.4; 24.3; 231.9; 0.0; 1338 ❙ 1179; 392; 42 ❙ -11.9; 1056; 455.2

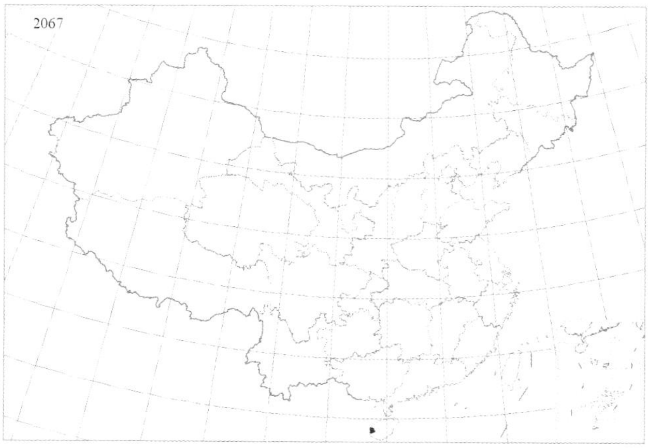

Lauraceae 樟科 *Beilschmiedia* 琼楠属
Beilschmiedia obscurinervia H. T. Chang 隐脉琼楠
Trees, evergreen

21.9 (21.2~22.7); 13.8 (12.2~15.0); 28.1 (27.3~28.6); 22.0 (21.3~22.8); 203.6 (195.2~213.1); 0.0 (0.0~0.0); 1165 (1111~1221) ❙ 1736 (1302~2435); 920 (599~1428); 117 (86~142) ❙ 49.0 (12.9~99.5); 1165 (1111~1221); 281.3 (226.9~327.7)

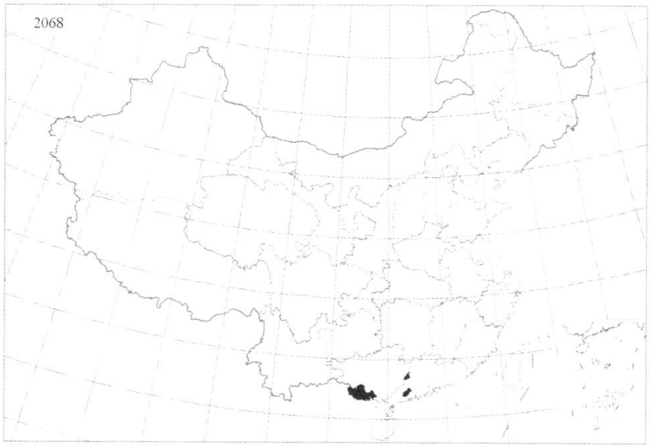

Lauraceae 樟科 *Beilschmiedia* 琼楠属
Beilschmiedia ovoidea F. N. Wei 卵果琼楠
Trees, evergreen

18.7; 8.8; 26.9; 18.8; 165.2; 0.0; 968 ❙ 1437; 670; 117 ❙ 48.5; 968; 238.3

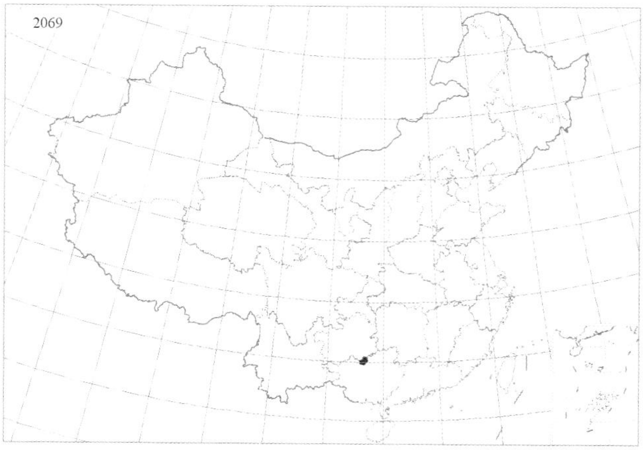

Lauraceae 樟科 *Beilschmiedia* 琼楠属
Beilschmiedia pauciflora H. W. Li 少花琼楠
Trees, evergreen

20.5; 14.9; 23.7; 20.5; 185.9; 0.0; 992 ❙ 1464; 655; 66 ❙ 47.6; 959; 570.1

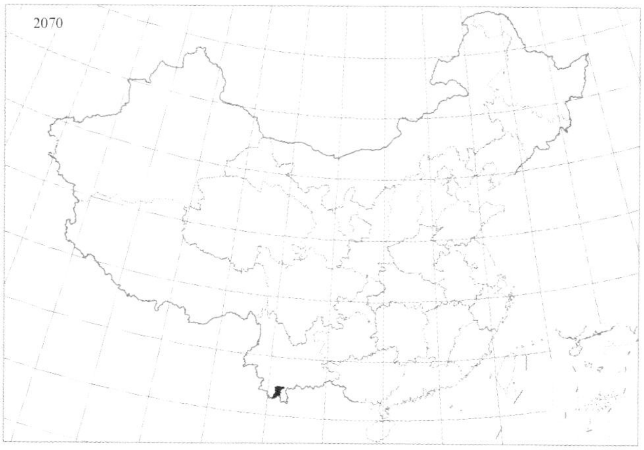

Lauraceae 樟科 *Beilschmiedia* 琼楠属
Beilschmiedia percoriacea C. K. Allen 厚叶琼楠
Trees, evergreen

21.5 (17.7~24.8)；14.0 (7.4~20.1)；27.3 (22.3~28.9)；21.6 (17.8~24.8)；198.7 (153.4~237.6)；0.0 (0.0~0.0)；1136 (854~1385) ▌1590 (868~2148)；706 (3[illegible]2~1234)；115 (41~194) ▌42.7 (-34.3~88.8)；1095 (854~1331)；332.9 (185.5~659.0)

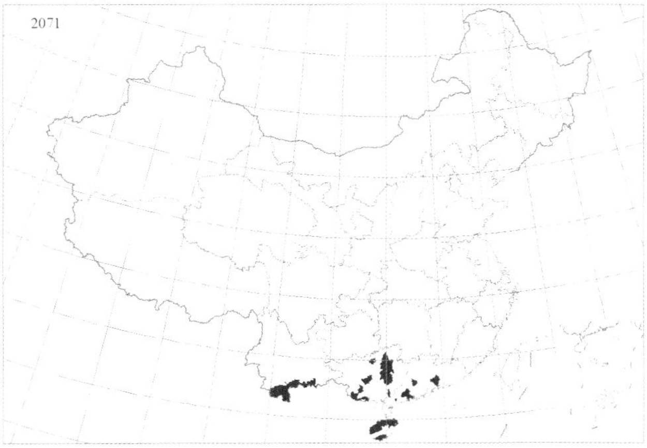

Lauraceae 樟科 *Beilschmiedia* 琼楠属
Beilschmiedia percoriacea var. ciliata H. W. Li 缘毛琼楠
Trees, evergreen

18.9 (17.1~21.0)；13.0 (10.9~15.1)；22.7 (20.8~25.6)；19.0 (17.1~21.0)；167.8 (145.0~191.8)；0.0 (0.0~0.0)；909 (815~1053) ▌1453 (1276~1751)；777 (655~996)；57 (41~77) ▌60.0 (36.8~88.8)；894 (802~1053)；484.8 (354.7~659.0)

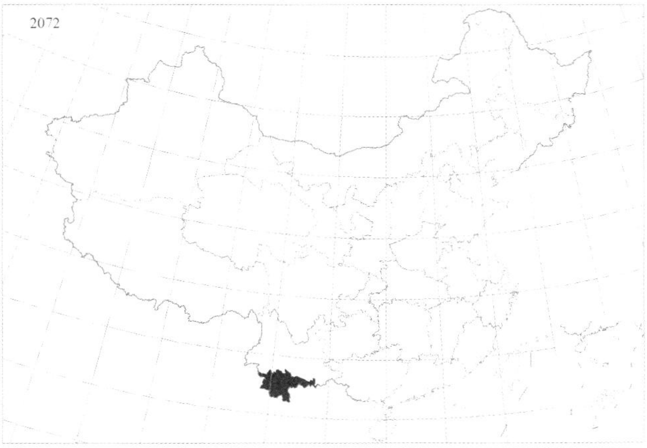

Lauraceae 樟科 *Beilschmiedia* 琼楠属
Beilschmiedia pergamentacea C. K. Allen 纸叶琼楠
Trees, evergreen

21.8 (17.4~24.1)；14.4 (7.4~19.1)；27.2 (22.5~28.6)；21.8 (17.4~24.2)；201.9 (148.9~230.1)；0.0 (0.0~0.0)；1150 (854~1326) ▌1657 (1302~2435)；822 (539~1428)；96 (48~130) ▌44.9 (12.9~99.5)；1132 (854~1326)；372.9 (235.1~620.0)

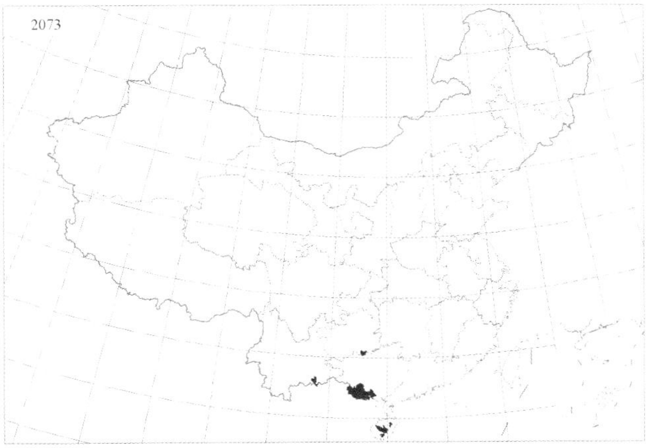

Lauraceae 樟科 *Beilschmiedia* 琼楠属
Beilschmiedia punctilimba H. W. Li 点叶琼楠
Trees, evergreen

17.7；10.9；22.5；17.8；153.4；0.0；854 ▌1368；756；59 ▌60.1；854；491.5

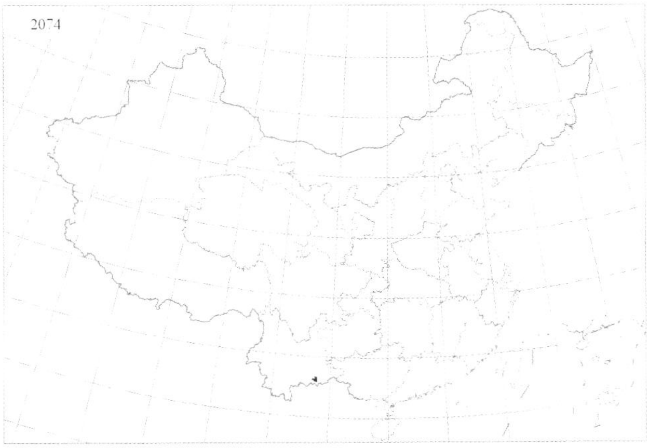

Lauraceae 樟科 *Beilschmiedia* 琼楠属
Beilschmiedia purpurascens H. W. Li 紫叶琼楠
Trees, evergreen

19.0 (17.7~20.5)；13.1 (10.9~15.1)；22.5 (21.6~23.7)；19.0 (17.7~20.5)；168.4 (152.7~185.9)；0.0 (0.0~0.0)；910 (849~992) ▌1494 (1366~1616)；767 (655~946)；52 (41~77) ▌64.7 (47.6~90.4)；886 (833~986)；528.0 (439.5~659.0)

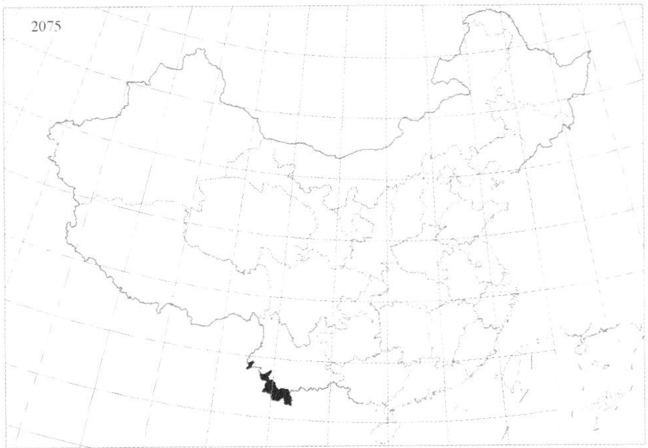

Lauraceae 樟科 *Beilschmiedia* 琼楠属
Beilschmiedia robusta C. K. Allen 粗壮琼楠
Trees, evergreen

18.4 (13.3~21.0)；11.3 (5.4~15.1)；23.5 (19.6~26.9)；18.5 (13.4~21.0)；161.6 (100.4~191.8)；0.0 (0.0~0.0)；903 (706~1053) ▌1397 (1028~1751)；754 (566~1007)；61 (42~117) ▌55.7 (12.5~145.8)；892 (706~1053)；406.8 (189.5~659.0)

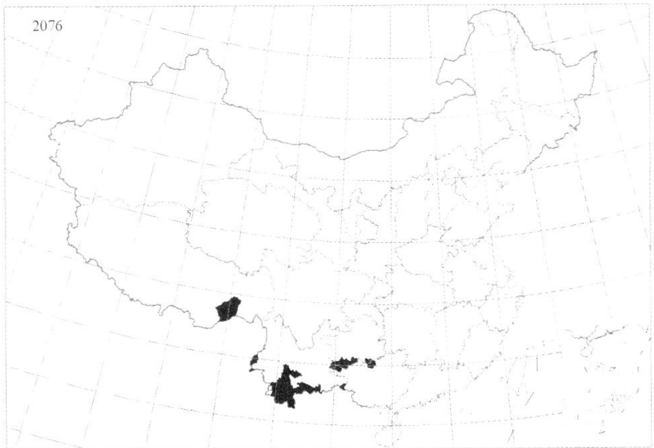

Lauraceae 樟科 *Beilschmiedia* 琼楠属
Beilschmiedia roxburghiana Nees 桐琼楠
Trees, evergreen

19.1 (13.3~22.2); 11.1 (5.4~15.2); 25.7 (19.6~28.7); 19.3 (13.4~22.7); 172.2 (100.4~212.7); 0.0 (0.0~0.0); 988 (706~1216) | 1541 (1049~1965); 750 (562~1007); 120 (44~189) | 58.2 (15.4~145.8); 984 (706~1216); 330.2 (185.5~551.6)

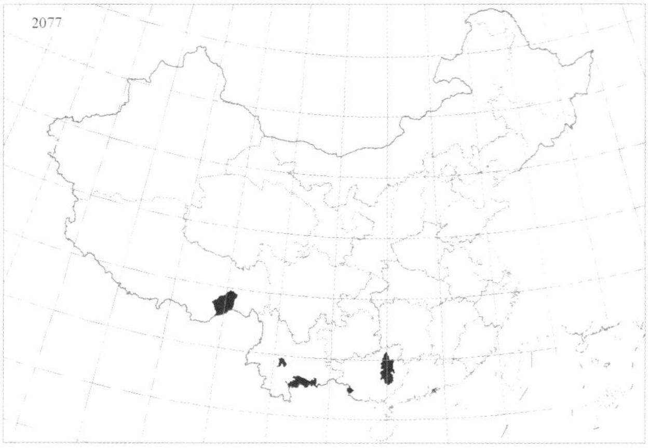

Lauraceae 樟科 *Beilschmiedia* 琼楠属
Beilschmiedia rufohirtella H. W. Li 红毛琼楠
Trees, evergreen

18.2 (17.2~21.0); 10.9 (9.5~14.2); 23.3 (22.4~25.6); 18.2 (17.3~21.0); 158.4 (147.0~191.8); 0.0 (0.0~0.0); 887 (833~1053) | 1414 (1254~1522); 789 (699~860); 57 (48~68) | 60.1 (44.5~80.6); 887 (833~1053); 430.5 (336.3~551.6)

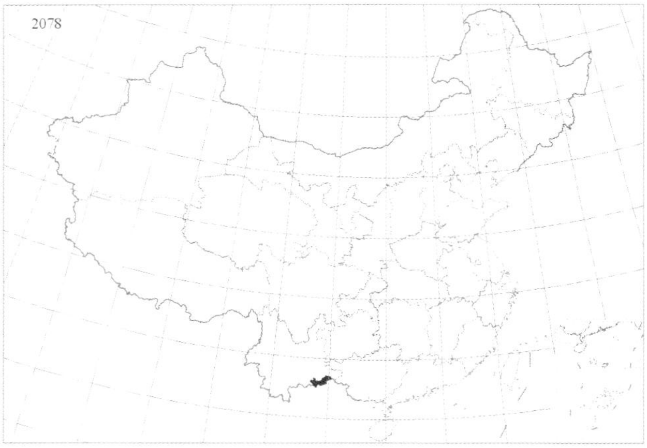

Lauraceae 樟科 *Beilschmiedia* 琼楠属
Beilschmiedia shangsiensis Y. T. Wei 上思琼楠
Trees, evergreen

21.9 (21.3~22.7); 13.9 (13.2~15.0); 27.9 (27.3~28.6); 22.0 (21.3~22.8); 203.6 (195.9~213.1); 0.0 (0.0~0.0); 1162 (1111~1221) | 1876 (1302~2435); 1047 (670~1428); 116 (86~130) | 61.1 (12.9~99.5); 1162 (1111~1221); 289.6 (235.1~318.8)

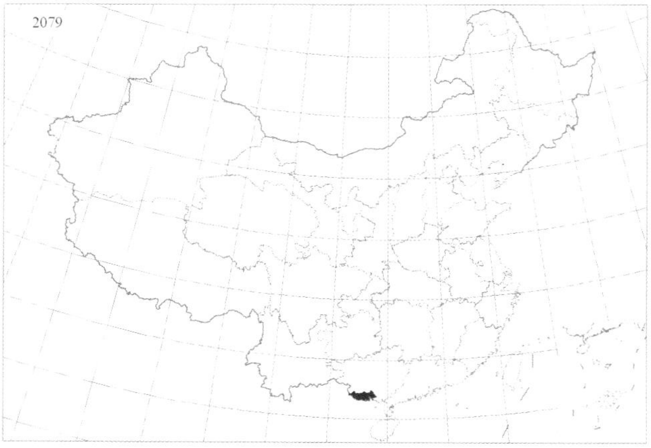

Lauraceae 樟科 *Beilschmiedia* 琼楠属
Beilschmiedia sichourensis H. W. Li 西畴琼楠
Trees, evergreen

17.2; 9.5; 22.7; 17.3; 147.0; 0.0; 838 | 1254; 699; 48 | 49.7; 838; 336.3

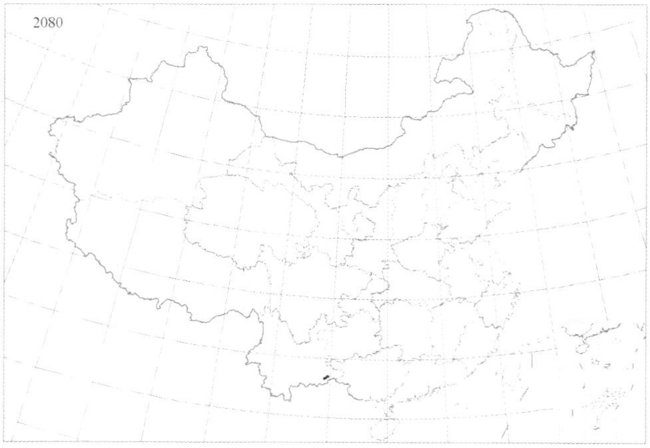

Lauraceae 樟科 *Beilschmiedia* 琼楠属
Beilschmiedia tsangii Merr. 网脉琼楠
Trees, evergreen

21.0 (14.0~24.8); 13.1 (3.8~20.1); 27.3 (21.4~29.8); 21.1 (14.1~24.8); 193.1 (109.8~237.6); 0.0 (0.0~1.5); 1114 (747~1385) | 1605 (868~2992); 711 (312~1847); 119 (41~194) | 46.8 (-34.3~166.5); 1077 (747~1331); 281.6 (154.2~620.0)

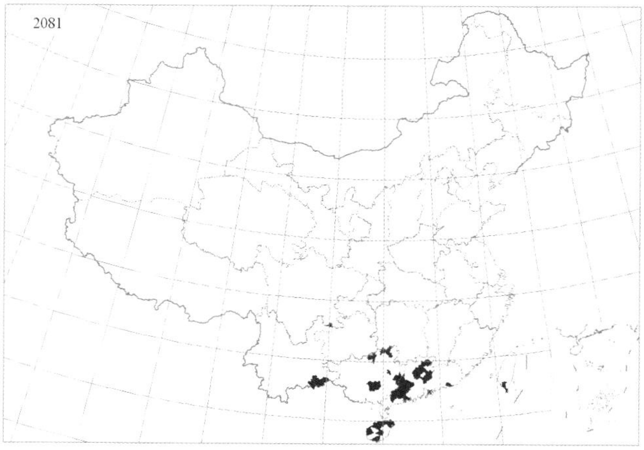

Lauraceae 樟科 *Beilschmiedia* 琼楠属
Beilschmiedia tungfangensis S. K. Lee et L. F. Lau 东方琼楠
Trees, evergreen

24.3; 18.7; 28.4; 24.3; 231.9; 0.0; 1338 | 1179; 392; 42 | -11.9; 1056; 455.2

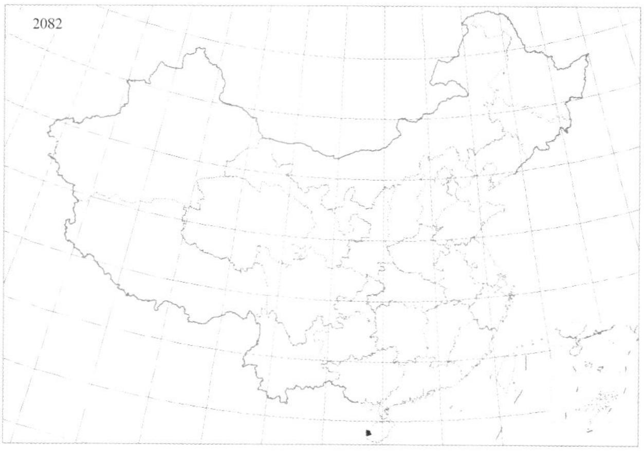

Lauraceae 樟科 *Beilschmiedia* 琼楠属
Beilschmiedia wangii C. K. Allen 海南琼楠
Trees, evergreen

21.5 (16.3~24.3); 13.9 (9.1~19.1); 27.4 (21.4~29.8); 21.6 (16.4~24.3); 198.7 (136.4~231.9); 0.0 (0.0~0.0); 1137 (797~1338) ❙ 1603 (868~2435); 732 (312~1428); 107 (42~197) ❙ 43.1 (-34.3~99.5); 1097 (797~1326); 317.0 (122.7~620.0)

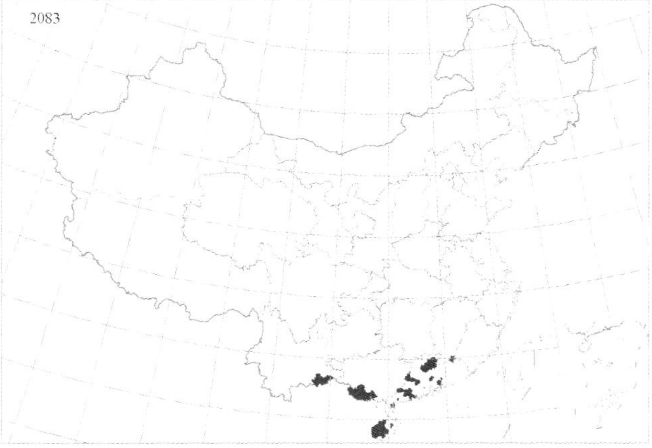

Lauraceae 樟科 *Beilschmiedia* 琼楠属
Beilschmiedia yunnanensis Hu 滇琼楠
Trees, evergreen

20.4 (16.3~24.3); 12.3 (7.3~18.7); 26.6 (21.4~29.2); 20.4 (16.4~24.3); 184.8 (136.4~231.9); 0.0 (0.0~0.0); 1058 (797~1338) ❙ 1533 (868~2435); 739 (312~1428); 110 (42~189) ❙ 47.1 (-34.3~99.5); 1032 (797~1273); 301.1 (122.7~551.6)

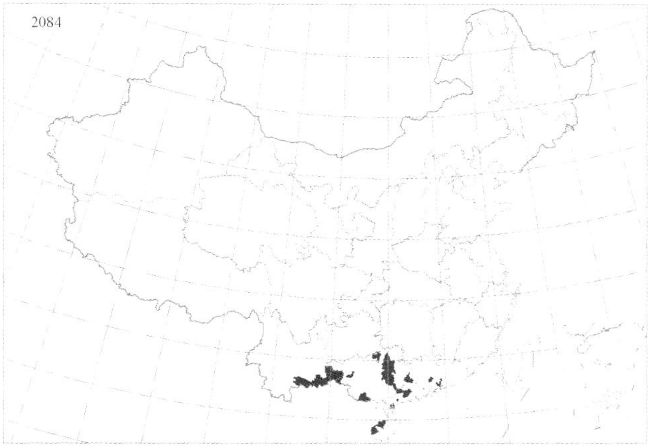

Lauraceae 樟科 *Caryodaphnopsis* 檬果樟属
Caryodaphnopsis henryi Airy Shaw 小花檬果樟
Trees

18.3 (16.9~21.0); 11.6 (9.5~14.2); 22.9 (21.4~25.6); 18.3 (16.9~21.0); 159.8 (142.7~191.8); 0.0 (0.0~0.0); 885 (813~1053) ❙ 1383 (1178~1547); 785 (646~911); 56 (48~68) ❙ 56.7 (36.8~80.6); 884 (813~1053); 447.8 (336.3~551.6)

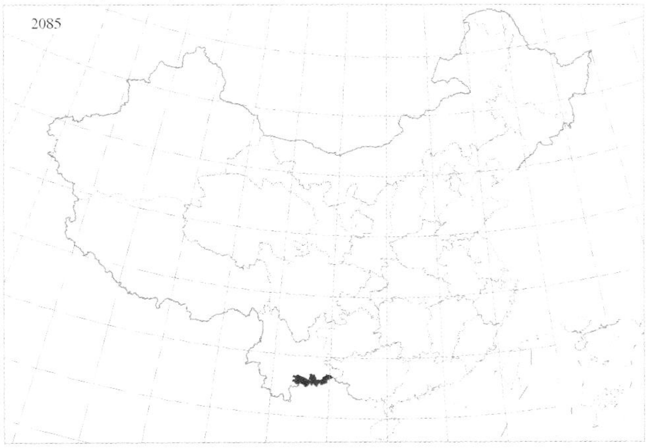

Lauraceae 樟科 *Caryodaphnopsis* 檬果樟属
Caryodaphnopsis laotica Airy Shaw 老挝檬果樟
Trees

19.3; 12.7; 23.8; 19.3; 171.6; 0.0; 942 ❙ 1439; 803; 61 ❙ 53.4; 942; 516.4

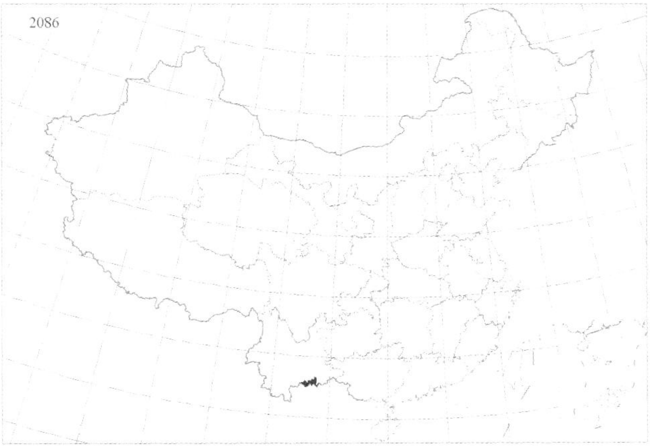

Lauraceae 樟科 *Caryodaphnopsis* 檬果樟属
Caryodaphnopsis tonkinensis (Lecomte) Airy Shaw 檬果樟
Trees

19.4 (17.7~22.2); 12.6 (10.9~14.2); 23.9 (21.6~28.3); 19.4 (17.8~22.3); 172.8 (153.4~207.5); 0.0 (0.0~0.0); 956 (854~1190) ❙ 1416 (1366~1522); 785 (713~835); 60 (42~87) ❙ 50.2 (15.4~68.0); 949 (833~1190); 483.1 (327.7~551.6)

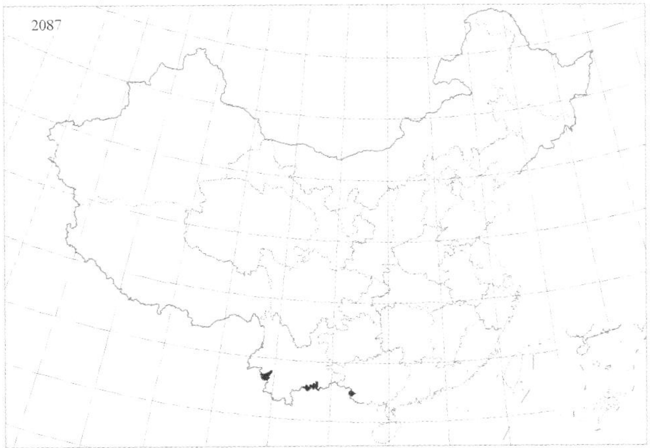

Lauraceae 樟科 *Cinnamomum* 樟属
Cinnamomum appelianum Schewe 毛桂
Trees, evergreen

18.0 (10.6~22.8); 7.9 (-0.2~15.5); 27.0 (21.1~29.8); 18.0 (10.6~22.8); 156.8 (79.4~214.1); 0.3 (0.0~12.1); 950 (643~1225) ❙ 1483 (1028~2035); 594 (411~1033); 138 (43~259) ❙ 56.9 (5.8~128.9); 948 (643~1225); 220.2 (108.2~551.6)

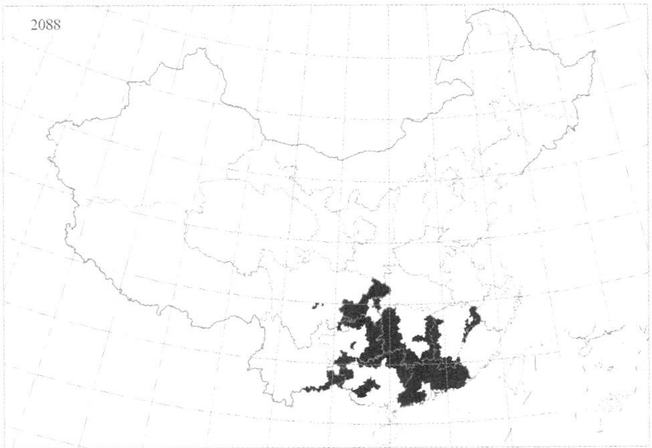

Lauraceae 樟科 *Cinnamomum* 樟属
Cinnamomum austrosinense H. T. Chang 华南桂
Trees, evergreen

17.9 (14.3~22.7); 7.7 (3.1~15.0); 27.0 (22.8~29.3); 17.9 (14.3~22.8); 155.0 (112.6~213.1); 0.2 (0.0~2.3); 940 (756~1221) ❙ 1623 (1165~3495); 611 (403~1428); 182 (75~897) ❙ 73.5 (12.9~235.6); 938 (756~1221); 234.6 (146.0~327.7)

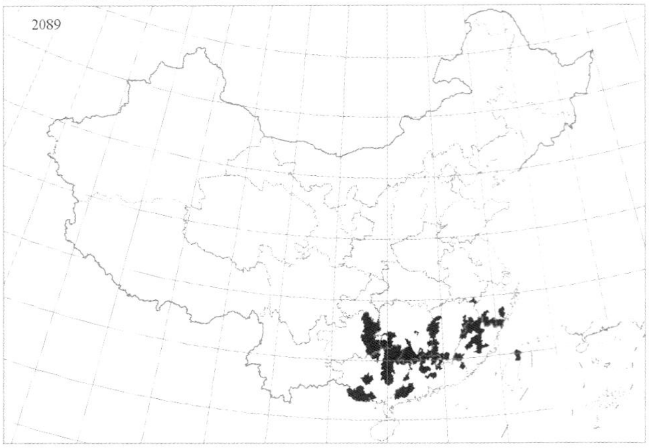

Lauraceae 樟科 *Cinnamomum* 樟属
Cinnamomum austroyunnanense H. W. Li 滇南桂
Trees, evergreen

18.5 (15.2~21.0); 11.4 (4.1~15.1); 23.6 (22.3~25.8); 18.5 (15.2~21.0); 161.9 (123.4~191.8); 0.1 (0.0~0.9); 904 (812~1053) ❙ 1456 (1221~1751); 748 (473~996); 67 (48~94) ❙ 61.3 (44.5~88.8); 898 (812~1053); 451.2 (211.8~659.0)

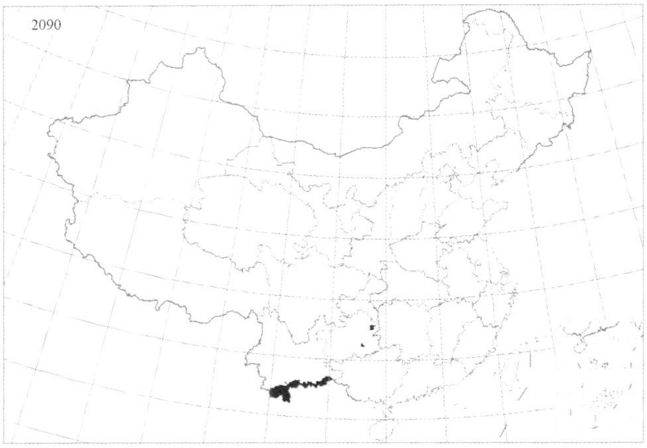

Lauraceae 樟科 *Cinnamomum* 樟属
Cinnamomum bejolghota (Buch.-Ham.) Sweet 钝叶桂
Trees, evergreen

19.8 (8.6~24.7); 12.8 (1.3~19.8); 24.8 (15.0~29.2); 19.8 (8.7~24.7); 178.0 (52.6~236.5); 0.1 (0.0~8.8); 1011 (574~1376) ❙ 1468 (868~1876); 712 (312~996); 83 (42~189) ❙ 49.0 (-34.3~112.2); 979 (574~1360); 397.0 (178.7~659.0)

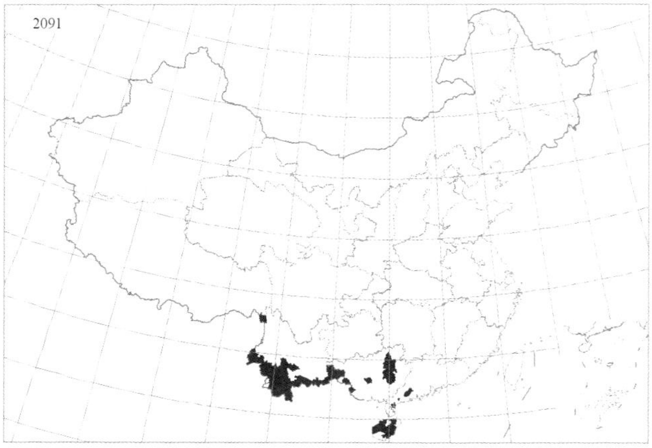

Lauraceae 樟科 *Cinnamomum* 樟属
Cinnamomum bodinieri H. Lév. 猴樟
Trees, evergreen

16.1 (9.3~21.6); 5.5 (-1.7~14.1); 25.7 (20.1~29.2); 16.1 (9.5~21.6); 134.4 (69.8~199.4); 1.0 (0.0~18.0); 852 (603~1128) ❙ 1332 (862~1991); 546 (376~902); 108 (33~221) ❙ 56.7 (7.8~107.2); 851 (603~1128); 213.4 (128.7~342.5)

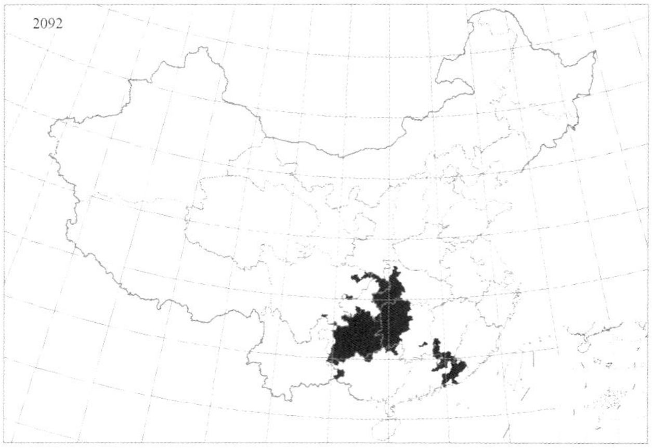

Lauraceae 樟科 *Cinnamomum* 樟属
Cinnamomum brachythyrsum J. Li 短序樟
Trees, evergreen

20.0; 14.6; 23.2; 20.1; 180.8; 0.0; 963 ❙ 1523; 681; 67 ❙ 58.4; 942; 581.0

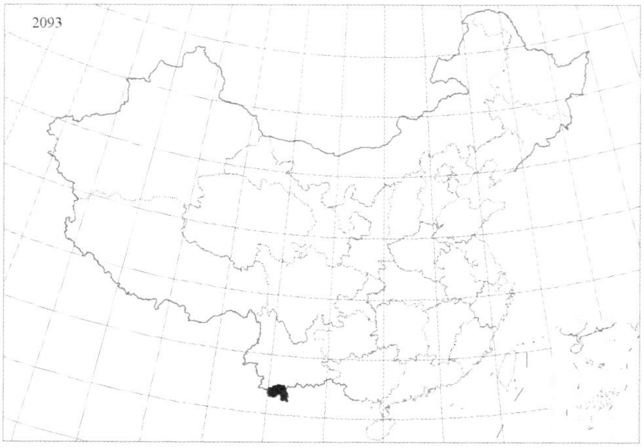

Lauraceae 樟科 *Cinnamomum* 樟属
Cinnamomum burmannii (Nees et T. Nees) Blume 阴香
Trees, evergreen

18.5 (11.7~24.8); 9.4 (2.6~20.1); 26.3 (17.9~29.8); 18.6 (11.7~24.8); 162.9 (82.6~237.6); 0.2 (0.0~4.0); 972 (658~1385) ❙ 1470 (868~2435); 629 (312~1428); 120 (31~259) ❙ 52.8 (-34.3~128.9); 963 (658~1360); 249.5 (95.3~627.0)

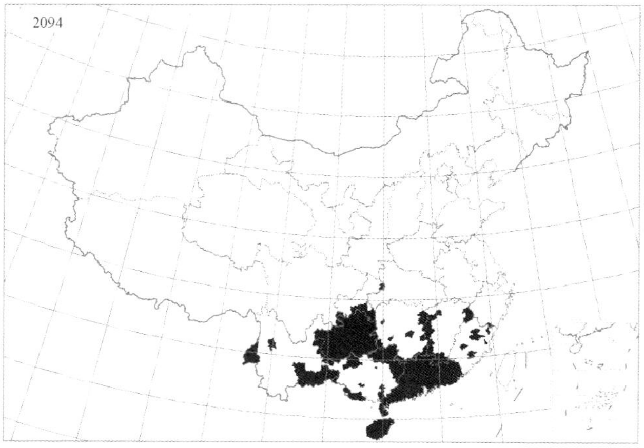

Lauraceae 樟科 *Cinnamomum* 樟属
Cinnamomum camphora (L.) Presl 樟
Trees, evergreen

17.6 (5.4~24.8); 7.7 (-2.1~20.1); 26.4 (12.3~30.1); 17.6 (5.8~24.8); 151.9 (31.5~237.6); 0.5 (0.0~25.6); 929 (481~1385) ❙ 1403 (741~3495); 584 (312~1847); 130 (15~897) ❙ 51.8 (-34.3~224.4); 918 (481~1360); 234.1 (95.3~659.0)

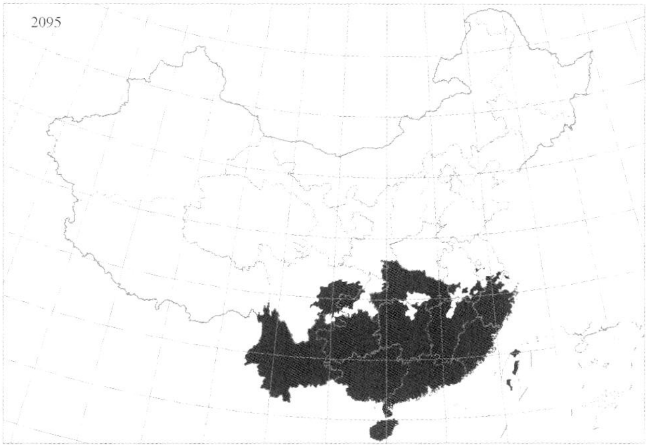

Lauraceae 樟科 *Cinnamomum* 樟属
Cinnamomum cassia (L.) D. Don 肉桂
Trees, evergreen

20.5 (15.5~24.6); 12.2 (4.8~19.5); 27.3 (22.3~29.2); 20.6 (15.6~24.7); 186.8 (127.2~235.9); 0.0 (0.0~0.2); 1075 (815~1372) ❙ 1638 (1176~3495); 703 (425~1428); 151 (58~897) ❙ 54.0 (-12.0~224.4); 1069 (815~1360); 275.6 (157.7~659.0)

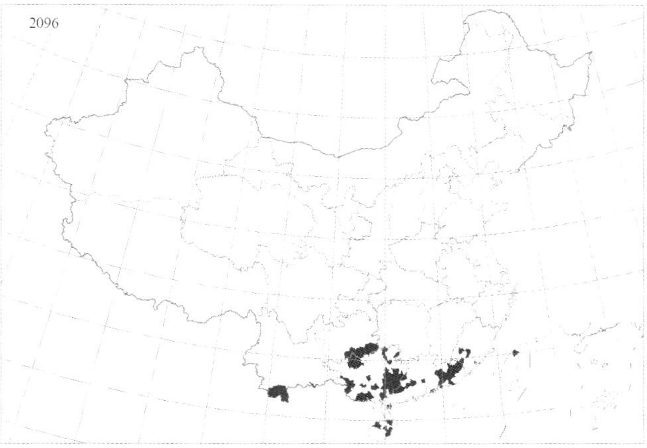

Lauraceae 樟科 *Cinnamomum* 樟属
Cinnamomum chartophyllum H. W. Li 坚叶樟
Trees, evergreen

18.8 (15.4~21.0); 12.5 (8.1~15.1); 23.0 (20.7~25.6); 18.8 (15.5~21.0); 166.2 (125.7~191.8); 0.0 (0.0~0.0); 914 (767~1053) ❙ 1461 (1007~1751); 797 (561~996); 64 (49~77) ❙ 59.6 (31.4~88.8); 913 (767~1053); 506.9 (274.9~659.0)

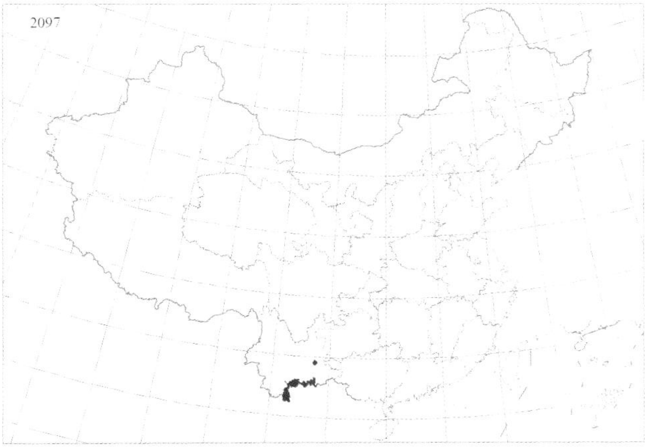

Lauraceae 樟科 *Cinnamomum* 樟属
Cinnamomum contractum H. W. Li 聚花桂
Trees, evergreen

13.0 (3.4~18.9); 6.0 (-5.5~12.0); 18.6 (11.4~23.9); 13.2 (4.7~19.0); 102.0 (23.3~167.8); 5.1 (0.0~41.8); 702 (431~929) ❙ 955 (527~1736); 561 (317~1007); 30 (9~51) ❙ 39.4 (-14.2~145.8); 658 (431~760); 296.6 (218.0~447.4)

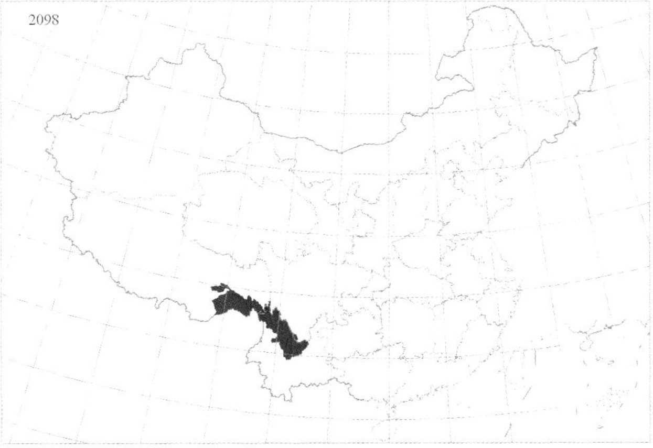

Lauraceae 樟科 *Cinnamomum* 樟属
Cinnamomum foveolatum (Merr.) H. W. Li et J. Li 尾叶樟
Trees, evergreen

17.1 (14.3~21.0); 7.9 (4.0~14.2); 24.8 (22.4~26.9); 17.2 (14.3~21.0); 146.1 (112.6~191.8); 0.1 (0.0~1.1); 869 (756~1053) ❙ 1314 (1161~1547); 640 (473~911); 79 (48~129) ❙ 51.6 (27.5~80.6); 869 (756~1053); 280.5 (188.8~551.6)

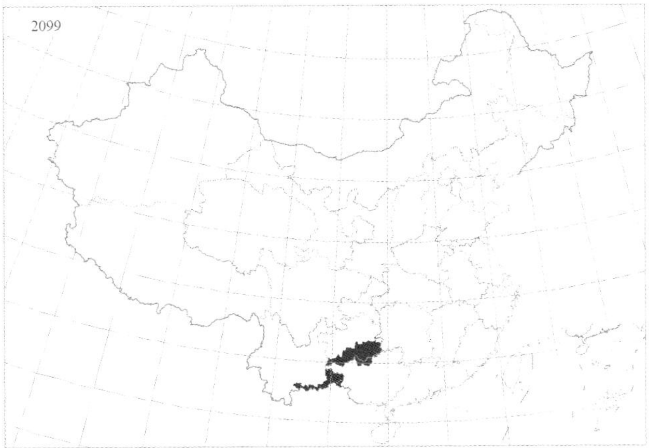

Lauraceae 樟科 *Cinnamomum* 樟属
Cinnamomum glanduliferum (Wall.) Nees 云南樟
Trees, evergreen

15.8 (-0.7~22.4); 6.7 (-9.6~14.0); 23.7 (7.7~30.1); 15.9 (2.5~22.4); 131.7 (6.8~209.1); 1.4 (0.0~74.2); 829 (338~1204) ❙ 1226 (370~1888); 569 (262~1044); 89 (10~228) ❙ 48.3 (-14.2~112.2); 815 (338~1204); 248.8 (61.6~627.0)

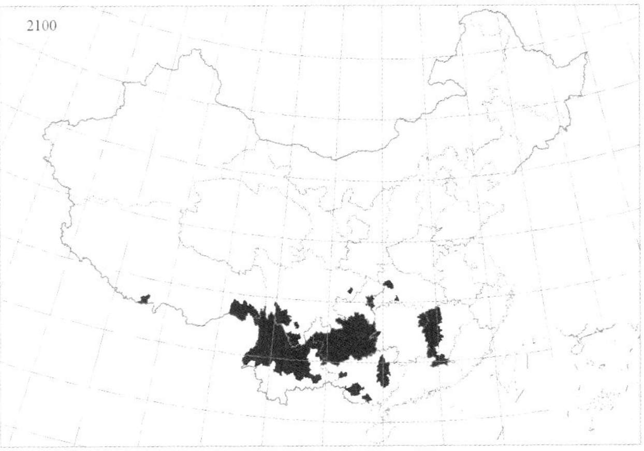

Lauraceae 樟科 *Cinnamomum* 樟属
Cinnamomum heyneanum Nees 狭叶桂
Trees, evergreen

17.3 (13.0~22.0); 7.9 (1.9~13.5); 25.2 (23.0~28.8); 17.3 (13.1~22.0); 149.0 (102.9~203.8); 1.1 (0.0~6.1); 894 (730~1170) ❙ 1286 (1064~1647); 639 (475~835); 72 (50~132) ❙ 45.1 (21.9~92.3); 894 (730~1170); 269.2 (187.6~506.2)

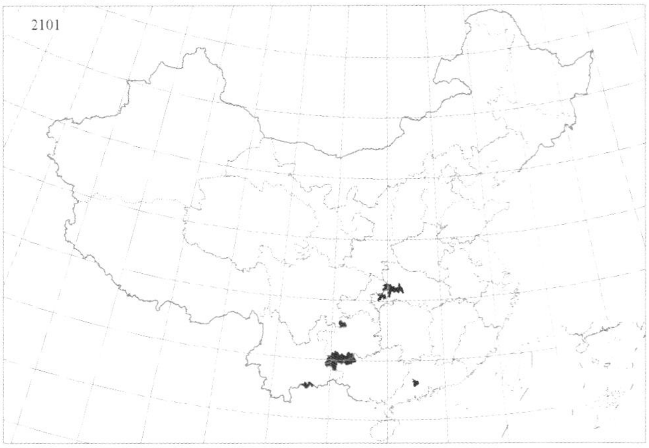

Lauraceae 樟科 *Cinnamomum* 樟属
Cinnamomum illicioides A. Chev. 八角樟
Trees, evergreen

20.5 (16.3~24.2); 12.1 (7.4~19.1); 27.1 (21.4~28.9); 20.6 (16.4~24.3); 186.7 (136.4~231.2); 0.0 (0.0~0.0); 1075 (797~1335) ❙ 1598 (1176~2435); 777 (447~1428); 119 (48~189) ❙ 50.0 (-12.0~99.5); 1064 (797~1257); 285.2 (185.5~620.0)

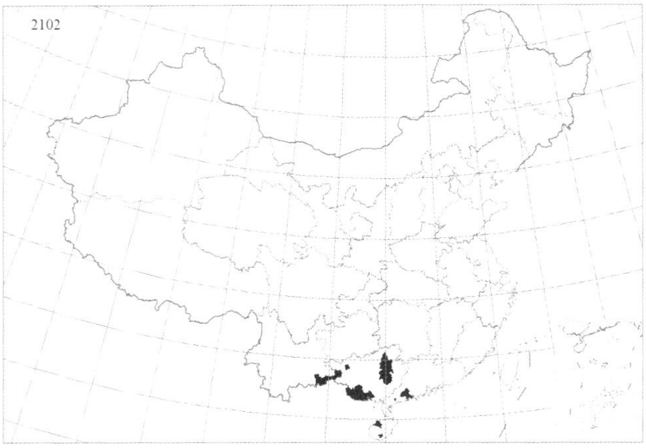

Lauraceae 樟科 *Cinnamomum* 樟属
Cinnamomum iners Reinwardt ex Blume 大叶桂
Trees, evergreen

19.1 (8.6~22.7); 11.8 (1.3~15.1); 24.3 (15.0~28.6); 19.2 (8.7~22.8); 170.2 (52.6~213.1); 0.2 (0.0~8.8); 965 (574~1221) ❙ 1477 (1041~2435); 806 (582~1428); 72 (42~130) ❙ 55.2 (5.8~145.8); 961 (574~1221); 392.2 (256.6~659.0)

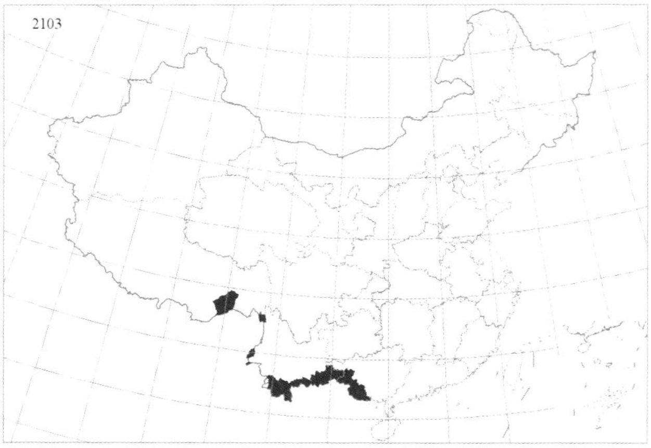

Lauraceae 樟科 *Cinnamomum* 樟属
Cinnamomum japonicum Siebold 天竺桂
Trees, evergreen

16.1 (10.2~22.2); 4.5 (-2.9~17.7); 27.1 (20.1~29.7); 16.2 (10.6~22.7); 136.8 (83.1~212.7); 2.7 (0.0~20.5); 878 (656~1216) ❙ 1472 (626~2992); 559 (316~1847); 166 (29~402) ❙ 66.9 (-22.2~246.5); 865 (626~1216); 230.2 (139.1~551.6)

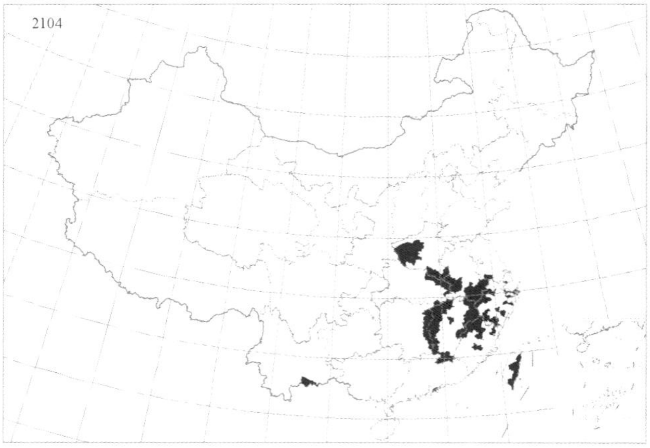

Lauraceae 樟科 *Cinnamomum* 樟属
Cinnamomum javanicum Blume 爪哇肉桂
Trees, evergreen

18.7; 11.7; 23.5; 18.7; 164.1; 0.0; 914 ❙ 1465; 812; 61 ❙ 61.7; 914; 485.5

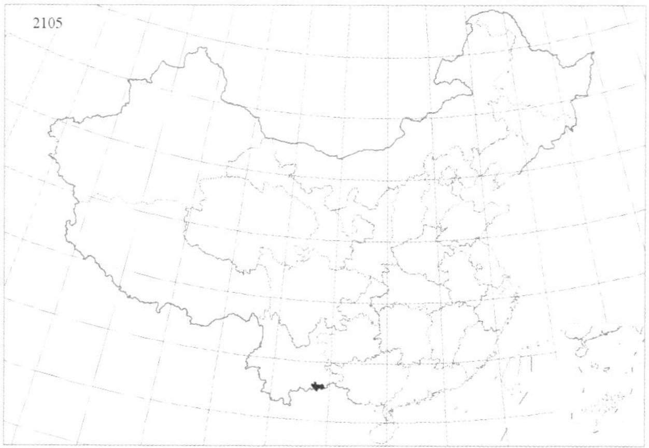

Lauraceae 樟科 *Cinnamomum* 樟属
Cinnamomum jensenianum Hand.-Mazz. 野黄桂
Trees, evergreen

17.1 (9.3~21.3); 6.0 (-1.7~12.3); 27.5 (20.1~30.1); 17.2 (9.5~21.3); 147.1 (69.8~196.1); 0.7 (0.0~18.0); 919 (603~1130) ❙ 1529 (871~1997); 552 (299~827); 175 (49~259) ❙ 67.3 (-6.7~130.7); 903 (603~1130); 221.1 (111.9~305.7)

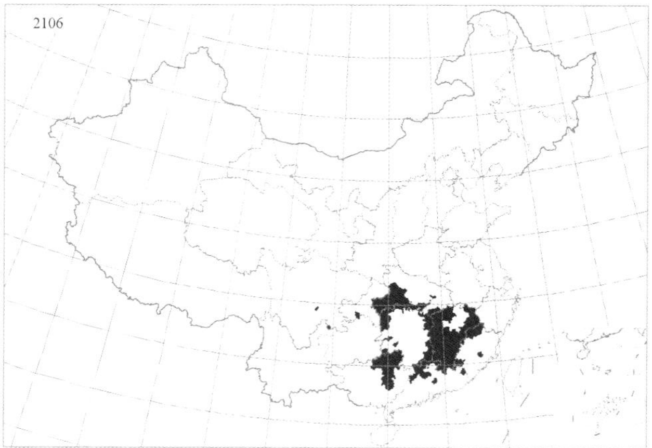

Lauraceae 樟科 *Cinnamomum* 樟属
Cinnamomum kotoense Kaneh. et Sasaki 兰屿肉桂
Trees, evergreen

18.5; 14.1; 21.7; 18.5; 162.1; 0.0; 863 ❙ 2713; 1388; 188 ❙ 214.4; 863; 363.7

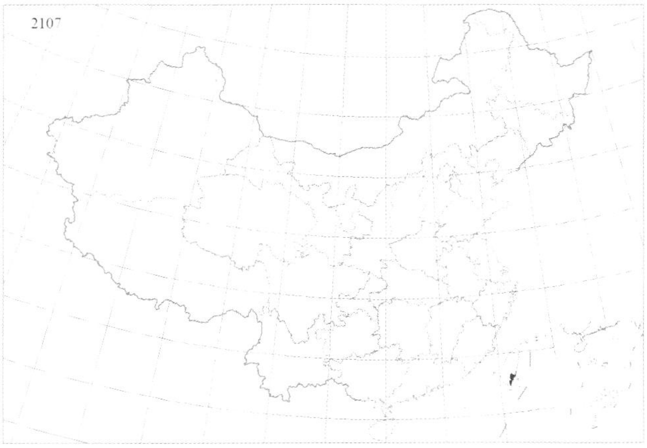

Lauraceae 樟科 *Cinnamomum* 樟属
Cinnamomum kwangtungense Merr. 红辣槁树
Trees, evergreen

21.7; 13.3; 28.5; 21.8; 201.3; 0.0; 1155 ❙ 1880; 823; 156 ❙ 63.2; 1155; 195.2

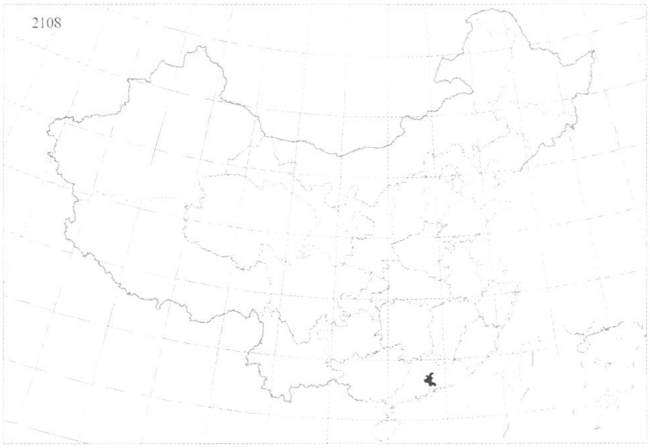

Lauraceae 樟科 *Cinnamomum* 樟属
Cinnamomum liangii C. K. Allen 软皮桂
Trees, evergreen

21.7 (17.6~24.7); 13.7 (7.3~19.8); 28.1 (25.8~29.1); 21.8 (17.6~24.7); 201.0 (151.6~236.5); 0.0 (0.0~0.0); 1150 (909~1376) ❙ 1714 (1374~2037); 780 (534~1033); 125 (53~188) ❙ 49.6 (15.4~83.4); 1145 (909~1225); 220.5 (95.3~538.2)

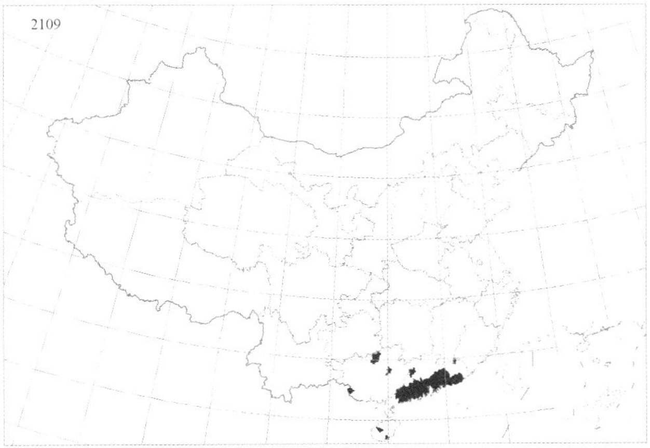

Lauraceae 樟科 *Cinnamomum* 樟属
Cinnamomum longepaniculatum (Gamble) N. Chao ex H. W. Li 油樟
Trees, evergreen

15.4 (2.2~20.1); 5.1 (-6.3~10.5); 24.7 (9.5~29.3); 15.4 (3.9~20.1); 128.8 (15.8~181.6); 3.8 (0.0~48.6); 837 (411~1052) ❙ 1295 (707~1745); 561 (357~817); 102 (11~224) ❙ 54.7 (-0.2~103.1); 830 (411~1052); 240.4 (135.1~322.5)

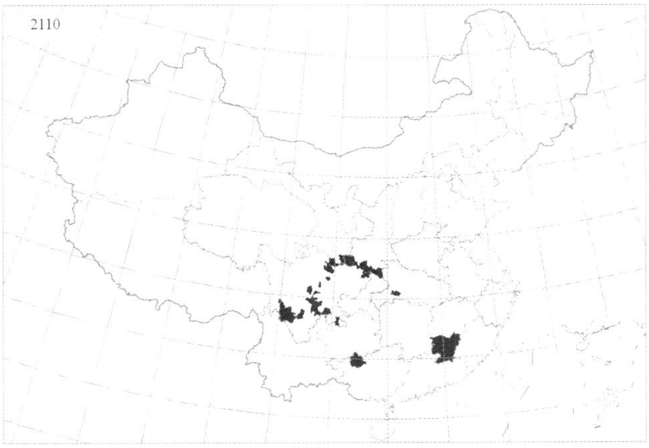

Lauraceae 樟科 *Cinnamomum* 樟属
Cinnamomum longipetiolatum H. W. Li 长柄樟
Trees, evergreen

18.2; 12.0; 22.4; 18.2; 158.9; 0.0; 874 ❙ 1238; 717; 50 ❙ 40.9; 850; 435.7

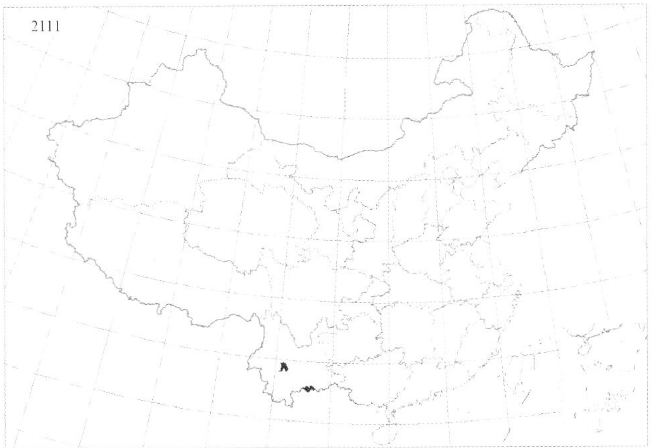

Lauraceae 樟科 *Cinnamomum* 樟属
Cinnamomum mairei H. Lév. 银叶桂
Trees, evergreen

14.9 (2.2~21.0); 6.3 (-6.3~15.1); 22.2 (9.5~27.3); 15.0 (3.9~21.0); 121.0 (15.8~191.8); 2.0 (0.0~48.6); 781 (411~1053) ❙ 1128 (729~1751); 605 (420~996); 46 (11~92) ❙ 44.6 (5.5~94.5); 776 (411~1053); 273.8 (135.1~659.0)

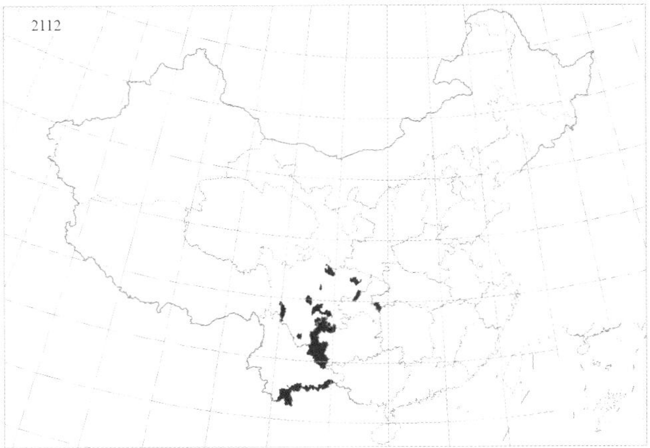

Lauraceae 樟科 *Cinnamomum* 樟属
Cinnamomum micranthum (Hayata) Hayata 沉水樟
Trees, evergreen

17.7 (14.3~22.0); 7.5 (3.3~15.3); 26.9 (20.1~30.1); 17.7 (14.3~22.0); 152.5 (114.1~203.8); 0.1 (0.0~2.3); 925 (765~1170) ❙ 1655 (1168~3495); 589 (403~1412); 199 (84~897) ❙ 79.8 (34.4~246.5); 917 (765~1170); 233.0 (148.9~305.7)

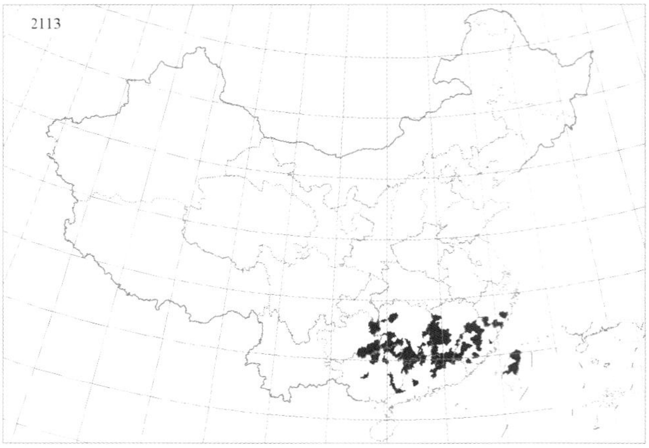

Lauraceae 樟科 *Cinnamomum* 樟属
Cinnamomum migao H. W. Li 米槁
Trees, evergreen

18.2 (16.1~20.5); 9.1 (5.3~11.5); 25.4 (22.7~27.7); 18.2 (16.1~20.6); 159.0 (133.1~187.0); 0.0 (0.0~0.0); 923 (836~1072) ❙ 1341 (1161~1758); 682 (523~848); 82 (43~189) ❙ 45.7 (16.9~82.1); 923 (836~1072); 279.3 (189.5~371.7)

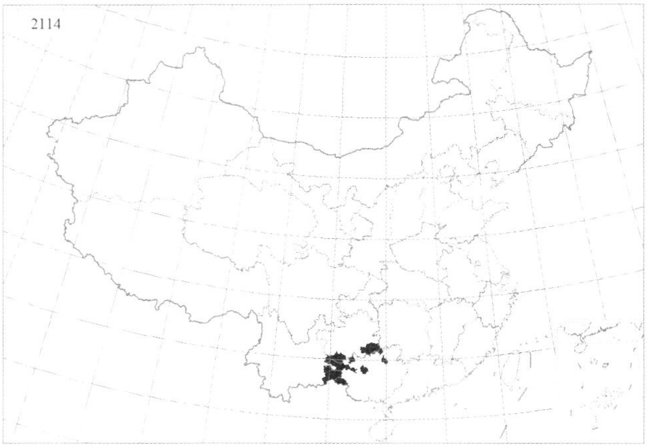

Lauraceae 樟科 *Cinnamomum* 樟属
Cinnamomum mollifolium H. W. Li 毛叶樟
Trees, evergreen

19.9; 14.1; 23.2; 19.9; 179.0; 0.0; 958 ❙ 1532; 737; 61 ❙ 60.1; 937; 556.5

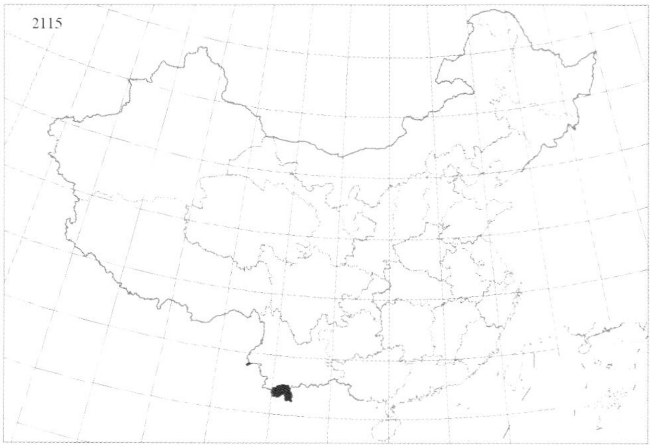

Lauraceae 樟科 *Cinnamomum* 樟属
Cinnamomum osmophloeum Kaneh. 土肉桂
Trees, evergreen

19.9; 14.5; 24.2; 19.9; 178.7; 0.0; 998 ❙ 2546; 1218; 241 ❙ 171.2; 942; 239.3

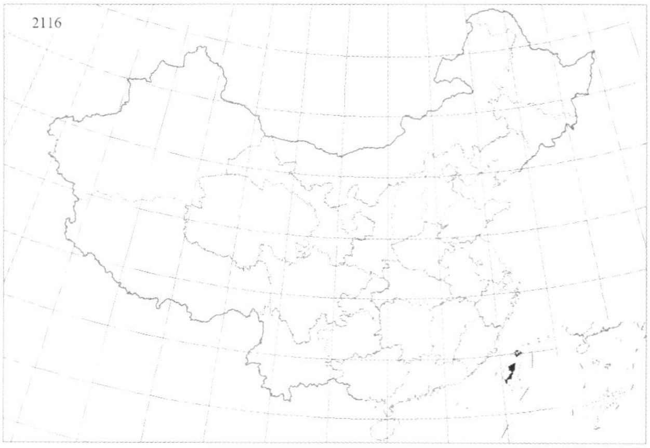

Lauraceae 樟科 *Cinnamomum* 樟属
Cinnamomum parthenoxylon (Jack) Meisner 黄樟
Trees, evergreen

18.3 (11.7~24.8); 8.8 (1.6~20.1); 26.6 (18.4~29.8); 18.4 (11.7~24.8); 161.0 (82.6~237.6); 0.2 (0.0~6.9); 966 (658~1385) ❙ 1480 (868~2435); 614 (312~1428); 130 (31~259) ❙ 54.8 (−34.3~128.9); 954 (658~1360); 247.0 (95.3~659.0)

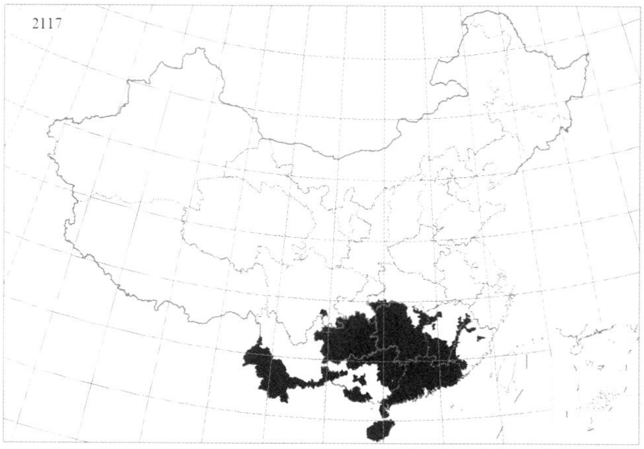

Lauraceae 樟科 *Cinnamomum* 樟属
Cinnamomum pauciflorum Nees 少花桂
Trees, evergreen

16.8 (9.3~22.2); 6.5 (−1.7~14.7); 26.1 (18.6~29.8); 16.9 (9.5~22.3); 142.9 (69.8~207.5); 0.6 (0.0~18.0); 889 (603~1190) ❙ 1376 (797~2014); 550 (388~954); 132 (23~254) ❙ 54.7 (−14.2~130.7); 879 (603~1190); 224.7 (133.5~551.6)

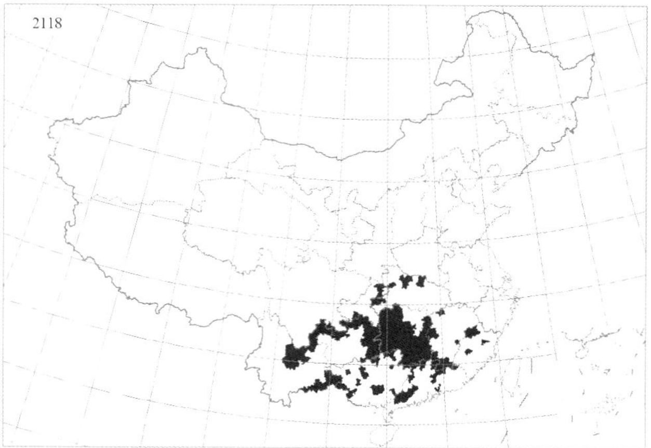

Lauraceae 樟科 *Cinnamomum* 樟属
Cinnamomum philippinense (Merr.) C. E. Chang 菲律宾樟树
Trees, evergreen

20.7; 15.7; 24.4; 20.7; 188.2; 0.0; 1029 ❙ 2417; 1391; 118 ❙ 147.3; 972; 244.5

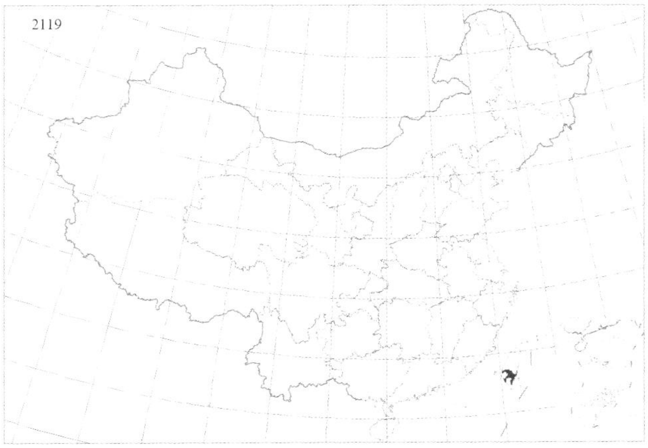

Lauraceae 樟科 *Cinnamomum* 樟属
Cinnamomum pingbienense H. W. Li 屏边桂
Trees, evergreen

18.7 (14.3~22.7); 10.3 (4.0~15.0); 25.2 (21.4~28.6); 18.8 (14.3~22.8); 165.1 (112.6~213.1); 0.0 (0.0~1.1); 953 (756~1221) ❙ 1407 (1161~2435); 741 (490~1428); 81 (48~130) ❙ 48.5 (5.8~99.5); 953 (756~1221); 315.6 (188.8~551.6)

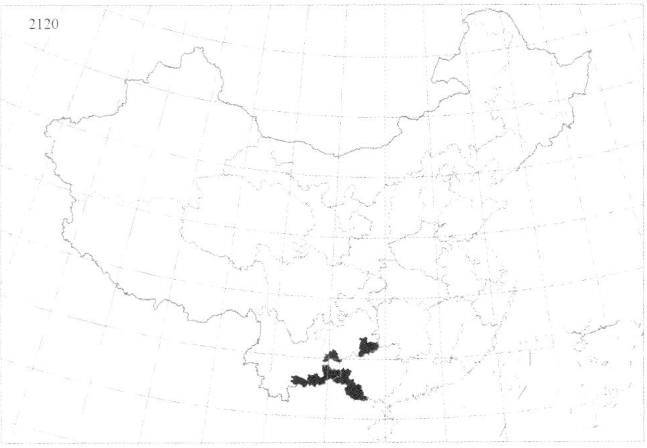

Lauraceae 樟科 *Cinnamomum* 樟属
Cinnamomum pittosporoides Hand.-Mazz. 刀把木
Trees, evergreen

16.7 (14.1~21.0); 9.9 (5.3~14.2); 21.8 (19.1~25.6); 16.7 (14.2~21.0); 141.0 (109.8~191.8); 0.0 (0.0~0.0); 824 (727~1053) ❙ 1066 (797~1522); 611 (411~835); 38 (18~70) ❙ 29.7 (-14.2~99.9); 769 (683~1053); 338.0 (212.0~551.6)

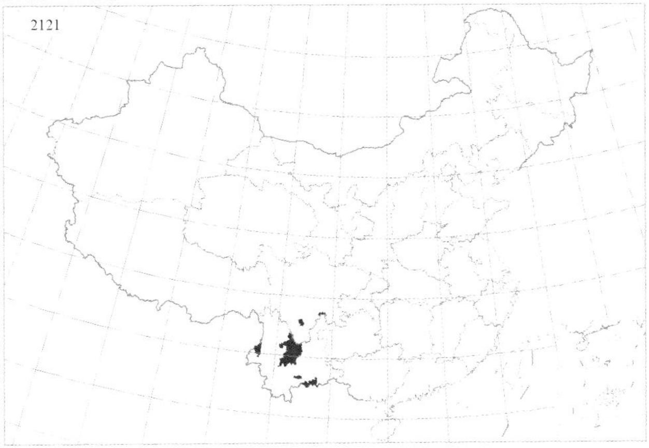

Lauraceae 樟科 *Cinnamomum* 樟属
Cinnamomum platyphyllum (Diels) C. K. Allen 阔叶樟
Trees, evergreen

13.8 (10.6~16.2); 2.8 (-0.7~5.5); 24.3 (21.1~26.4); 13.8 (10.6~16.2); 110.6 (79.4~134.6); 4.7 (0.0~13.5); 765 (643~866) ❙ 1094 (675~1478); 484 (331~592); 44 (16~85) ❙ 43.7 (-1.2~107.2); 764 (643~866); 229.6 (164.0~306.1)

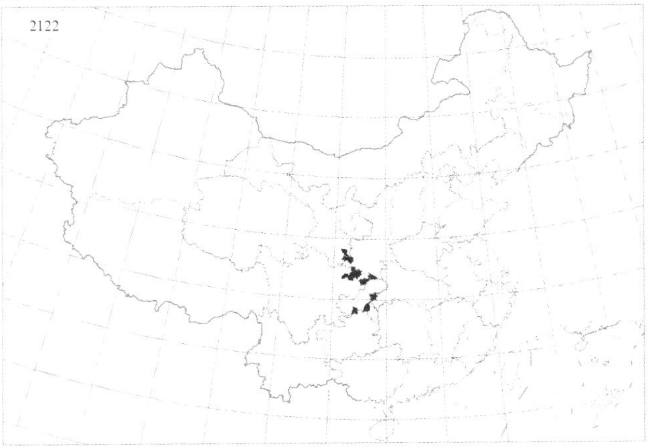

Lauraceae 樟科 *Cinnamomum* 樟属
Cinnamomum reticulatum Hayata 网脉桂
Trees, evergreen

22.2; 17.7; 25.5; 22.2; 206.7; 0.0; 1123 ❙ 2992; 1847; 107 ❙ 166.5; 1123; 296.9

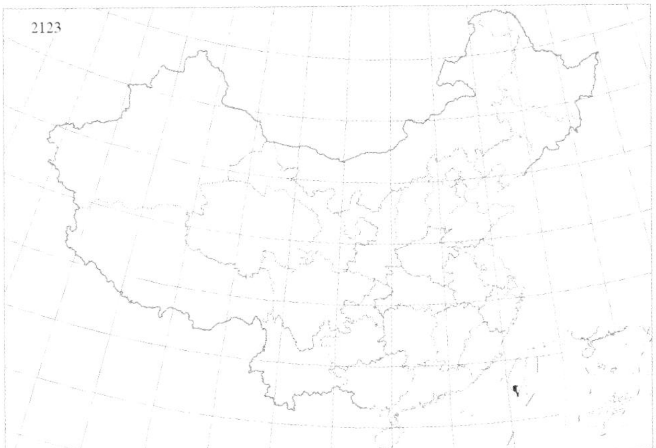

Lauraceae 樟科 *Cinnamomum* 樟属
Cinnamomum rigidissimum H. T. Chang 卵叶桂
Trees, evergreen

19.8 (14.6~24.3); 10.5 (3.8~19.1); 27.5 (21.7~29.0); 19.9 (14.7~24.3); 178.6 (117.7~231.9); 0.0 (0.0~1.5); 1042 (779~1338) ❙ 1551 (1172~2992); 592 (392~1847); 143 (42~211) ❙ 50.6 (-11.9~214.4); 1036 (779~1247); 249.5 (151.0~620.0)

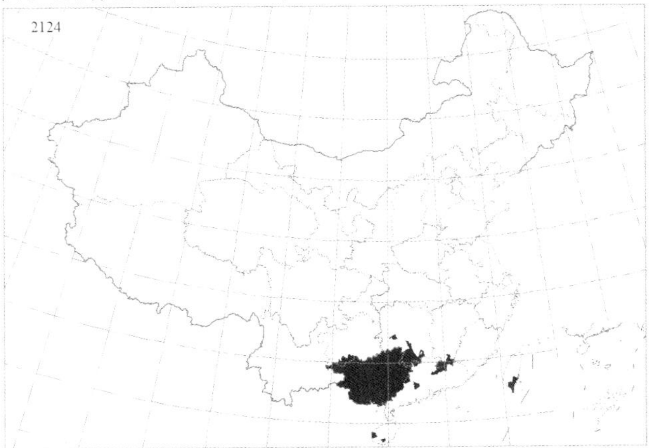

Lauraceae 樟科 *Cinnamomum* 樟属

Cinnamomum rufotomentosum K. M. Lan 绒毛樟

Trees, evergreen

17.1; 8.5; 23.4; 17.1; 145.1; 0.0; 845 ❙ 1232; 666; 59 ❙ 45.8; 845; 283.9

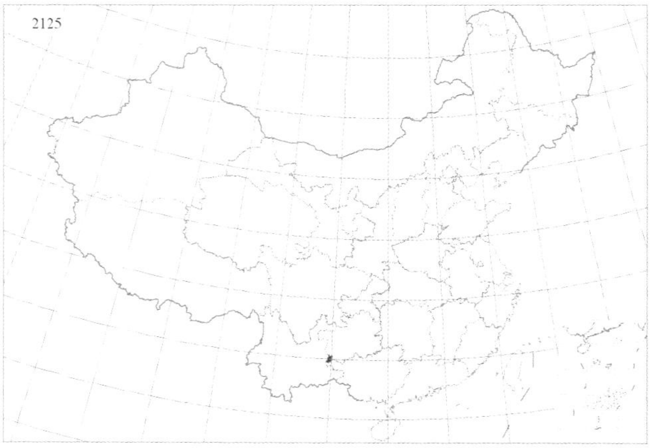

Lauraceae 樟科 *Cinnamomum* 樟属

Cinnamomum saxatile H. W. Li 岩樟

Trees, evergreen

19.5 (16.2~22.2); 11.0 (6.3~14.2); 25.8 (22.4~28.5); 19.5 (16.2~22.3); 174.2 (134.5~207.5); 0.0 (0.0~0.0); 994 (826~1190) ❙ 1448 (1161~1710); 757 (611~860); 96 (48~194) ❙ 46.8 (15.4~80.6); 994 (826~1190); 321.0 (213.5~551.6)

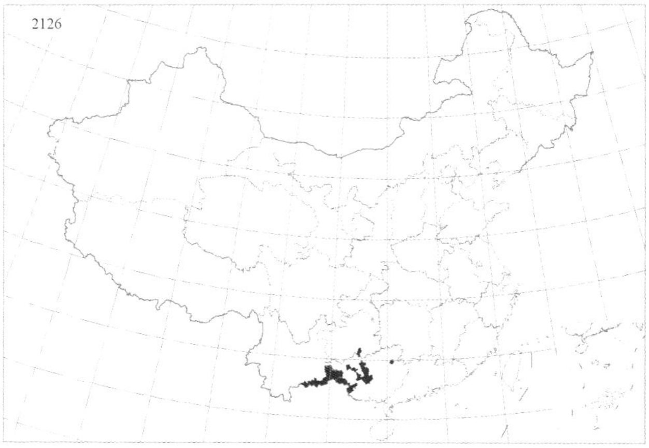

Lauraceae 樟科 *Cinnamomum* 樟属

Cinnamomum septentrionale Hand.-Mazz. 银木

Trees, evergreen

12.8 (10.2~15.9); 1.4 (-0.7~4.2); 23.7 (19.8~28.0); 12.9 (10.3~16.0); 102.0 (75.8~134.0); 7.5 (0.8~13.6); 735 (634~897) ❙ 1000 (644~1478); 445 (321~599); 44 (14~92) ❙ 36.6 (-8.1~107.2); 729 (634~849); 252.3 (141.3~306.1)

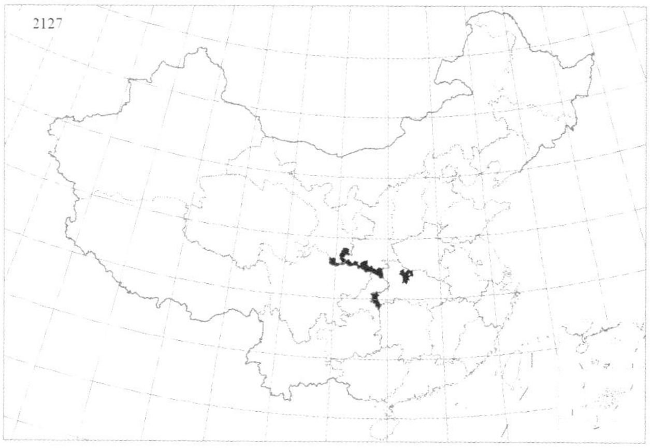

Lauraceae 樟科 *Cinnamomum* 樟属

Cinnamomum subavenium Miq. 香桂

Trees, evergreen

18.1 (8.6~24.6); 8.2 (-0.2~19.5); 26.9 (15.0~29.9); 18.1 (8.7~24.7); 158.2 (52.6~235.9); 0.4 (0.0~12.1); 956 (574~1372) ❙ 1504 (849~3495); 620 (299~1412); 145 (36~897) ❙ 59.4 (-34.3~246.5); 946 (574~1299); 249.0 (108.2~659.0)

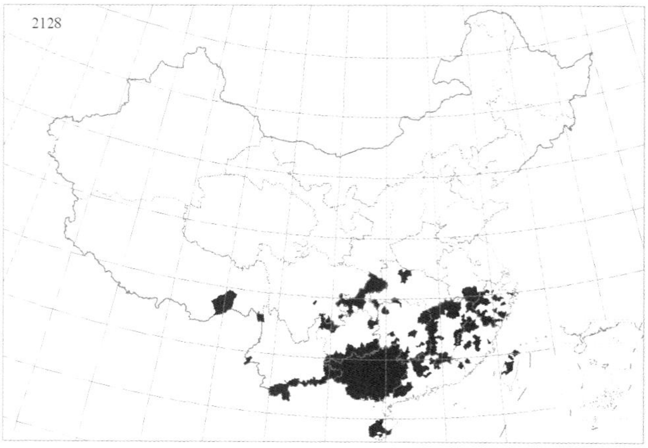

Lauraceae 樟科 *Cinnamomum* 樟属

Cinnamomum tamala (Buchanan-Hamilton) T. Nees et Nees 柴桂

Trees, evergreen

15.9 (-1.8~20.5); 9.4 (-10.2~15.1); 20.5 (6.1~23.7); 16.1 (1.8~20.5); 135.7 (1.9~185.9); 4.1 (0.0~82.8); 800 (329~992) ❙ 1261 (320~1751); 702 (193~1007); 51 (20~77) ❙ 57.3 (-4.0~145.8); 785 (320~986); 411.4 (72.2~659.0)

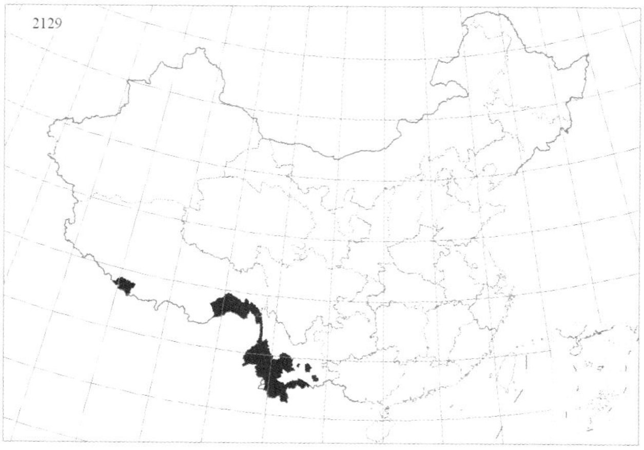

Lauraceae 樟科 *Cinnamomum* 樟属

Cinnamomum tenuipile Kosterm. 细毛樟

Trees, evergreen

17.7 (13.2~20.5); 11.3 (6.4~15.1); 21.7 (18.4~23.7); 17.7 (13.3~20.5); 152.5 (99.4~185.9); 0.0 (0.0~0.0); 856 (695~992) ❙ 1485 (1180~1751); 814 (622~996); 58 (42~77) ❙ 74.3 (47.6~112.2); 846 (695~986); 496.8 (335.1~659.0)

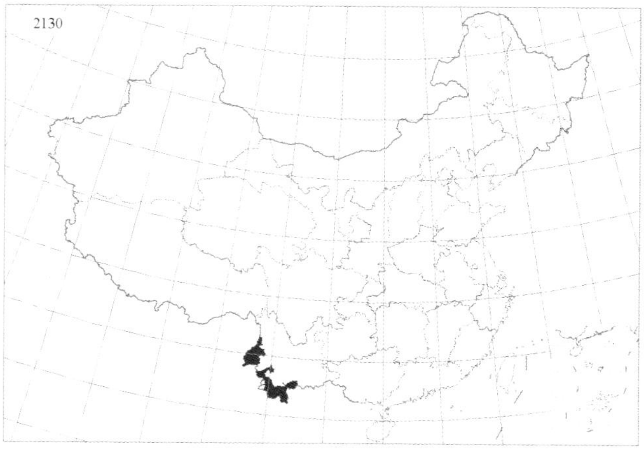

Lauraceae 樟科 *Cinnamomum* 樟属
Cinnamomum tonkinense (Lecomte) A. Chev. 假桂皮树
Trees, evergreen

18.2 (16.7~21.0); 11.5 (9.5~14.2); 22.8 (21.3~25.6); 18.3 (16.8~21.0); 159.3 (141.5~191.8); 0.0 (0.0~0.0); 882 (811~1053) ❙ 1275 (939~1559); 712 (524~907); 50 (37~68) ❙ 44.5 (6.5~80.6); 859 (744~1053); 407.8 (301.1~551.6)

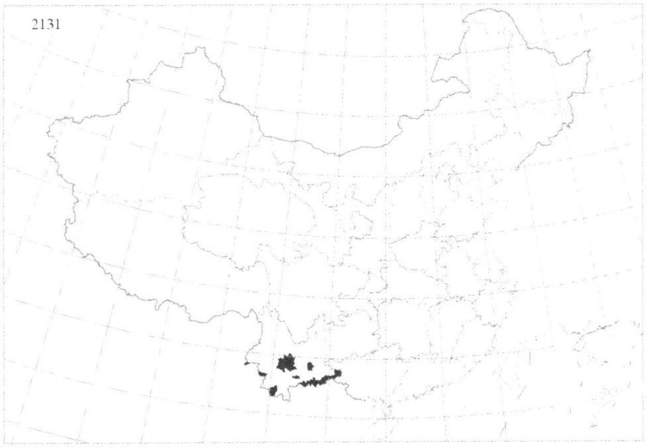

Lauraceae 樟科 *Cinnamomum* 樟属
Cinnamomum tsangii Merr. 辣汁树
Trees, evergreen

18.1 (14.3~24.3); 7.6 (3.3~18.7); 27.7 (24.2~30.1); 18.2 (14.3~24.3); 158.3 (113.2~231.9); 0.1 (0.0~2.3); 963 (761~1338) ❙ 1573 (868~2014); 538 (312~954); 187 (42~230) ❙ 65.6 (-34.3~128.9); 936 (761~1170); 231.2 (148.9~538.2)

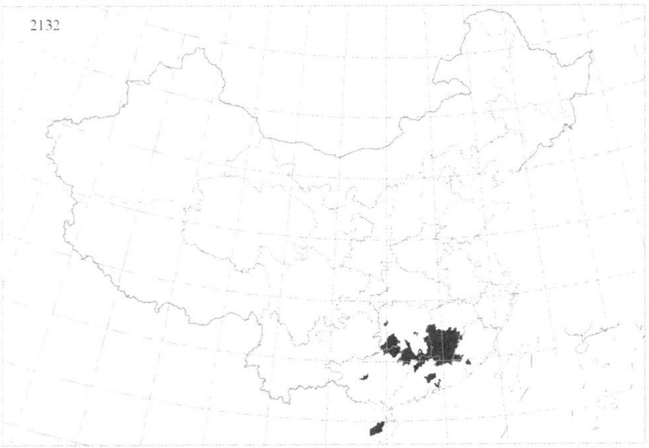

Lauraceae 樟科 *Cinnamomum* 樟属
Cinnamomum tsoi C. K. Allen 平托桂
Trees, evergreen

21.9; 15.0; 27.1; 21.9; 203.0; 0.0; 1153 ❙ 1599; 607; 96 ❙ 39.8; 1105; 449.1

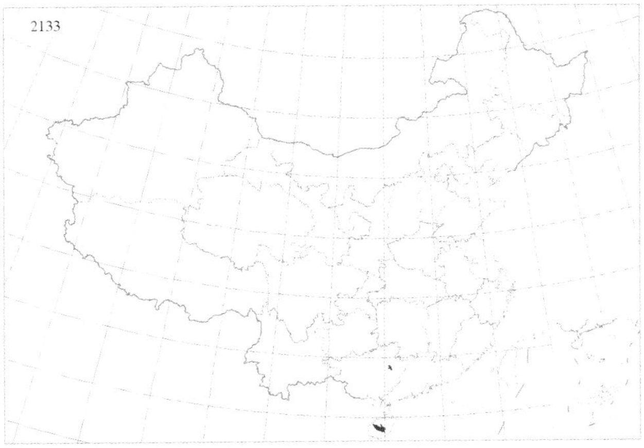

Lauraceae 樟科 *Cinnamomum* 樟属
Cinnamomum validinerve Hance 粗脉桂
Trees, evergreen

20.2 (14.6~22.7); 11.2 (3.8~15.2); 27.6 (24.6~28.8); 20.3 (14.7~22.8); 183.6 (117.7~213.1); 0.0 (0.0~1.5); 1067 (779~1221) ❙ 1649 (1293~2435); 759 (505~1428); 143 (57~191) ❙ 55.9 (12.9~101.8); 1067 (779~1221); 243.6 (163.7~371.7)

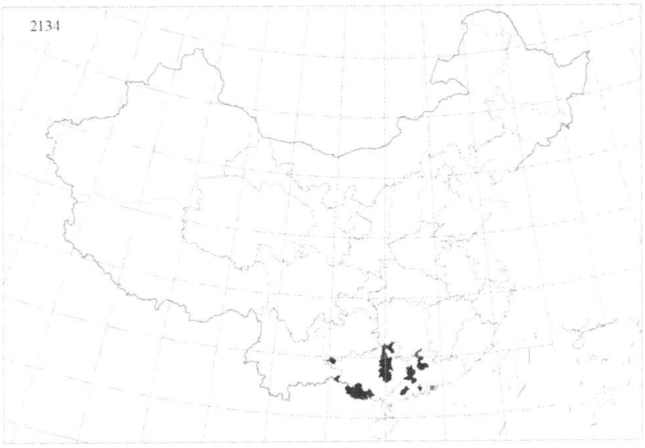

Lauraceae 樟科 *Cinnamomum* 樟属
Cinnamomum wilsonii Gamble 川桂
Trees, evergreen

16.3 (6.1~22.4); 5.4 (-2.9~14.3); 26.2 (12.7~29.8); 16.3 (6.3~22.5); 138.0 (35.2~209.7); 2.3 (0.0~21.8); 877 (496~1205) ❙ 1345 (626~2024); 545 (316~1007); 122 (15~230) ❙ 53.2 (-22.2~145.8); 867 (496~1205); 217.2 (95.3~447.4)

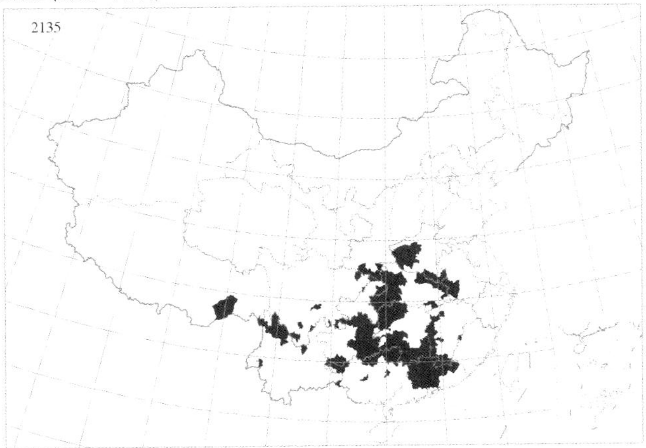

Lauraceae 樟科 *Cryptocarya* 厚壳桂属
Cryptocarya acutifolia H. W. Li 尖叶厚壳桂
Trees, evergreen

19.8 (18.6~21.8); 13.9 (12.7~15.1); 23.7 (22.3~27.7); 19.9 (18.7~21.9); 178.7 (163.9~202.2); 0.0 (0.0~0.0); 967 (888~1150) ❙ 1635 (1428~2148); 860 (655~1234); 74 (56~124) ❙ 68.7 (47.6~88.8); 958 (880~1150); 519.0 (318.8~659.0)

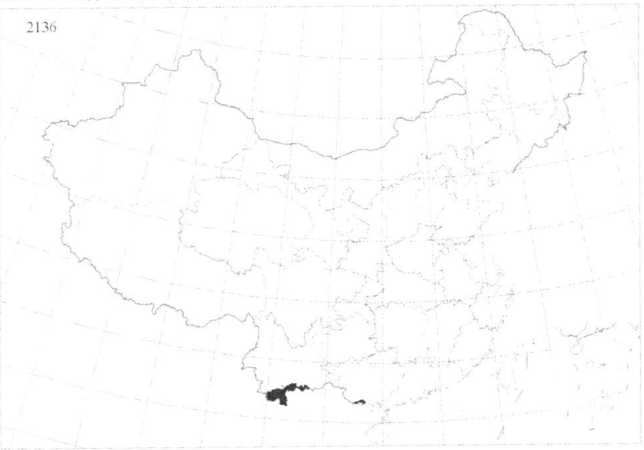

Lauraceae 樟科 *Cryptocarya* 厚壳桂属
Cryptocarya amygdalina Nees 杏仁厚壳桂
Trees, evergreen

13.3; 5.4; 19.6; 13.4; 100.4; 0.0; 706 ❙ 1736; 1007; 51 ❙ 145.8; 706; 447.4

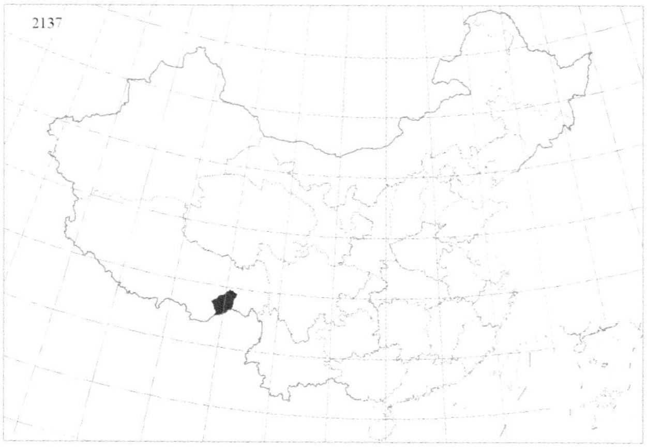

Lauraceae 樟科 *Cryptocarya* 厚壳桂属
Cryptocarya brachythyrsa H. W. Li 短序厚壳桂
Trees, evergreen

19.3 (18.6~20.5); 13.5 (12.7~14.9); 23.0 (22.3~23.7); 19.4 (18.7~20.5); 172.6 (163.9~185.9); 0.0 (0.0~0.0); 928 (888~992) ❙ 1494 (1276~1751); 804 (655~996); 62 (51~77) ❙ 61.4 (36.8~88.8); 916 (880~959); 517.9 (452.0~570.1)

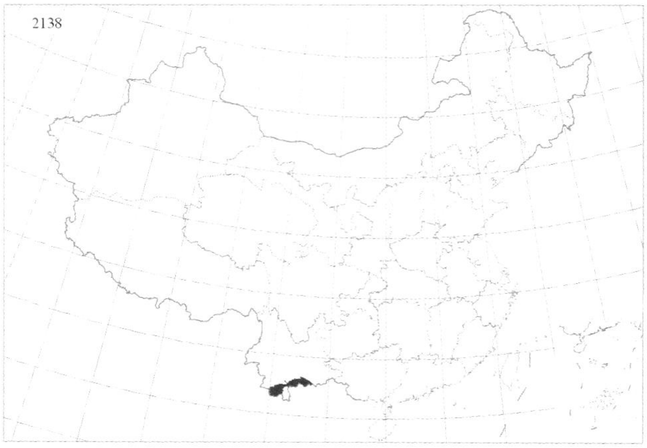

Lauraceae 樟科 *Cryptocarya* 厚壳桂属
Cryptocarya calcicola H. W. Li 岩生厚壳桂
Trees, evergreen

20.2 (15.0~24.8); 12.2 (4.5~20.1); 26.2 (22.1~28.7); 20.2 (15.0~24.8); 182.6 (120.8~237.6); 0.0 (0.0~0.5); 1042 (794~1385) ❙ 1405 (868~2435); 721 (312~1428); 77 (41~169) ❙ 36.9 (-34.3~99.5); 1012 (794~1221); 342.2 (181.2~659.0)

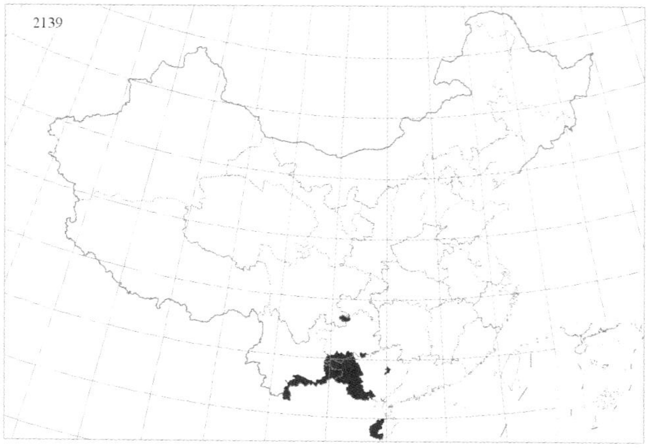

Lauraceae 樟科 *Cryptocarya* 厚壳桂属
Cryptocarya chinensis (Hance) Hemsl. 厚壳桂
Trees, evergreen

20.5 (14.6~24.8); 12.0 (3.8~20.1); 27.6 (20.1~29.8); 20.6 (14.7~24.8); 186.6 (117.7~237.6); 0.0 (0.0~1.5); 1079 (773~1385) ❙ 1685 (868~3495); 703 (312~1847); 155 (41~897) ❙ 59.1 (-34.3~246.5); 1066 (773~1326); 242.8 (95.3~620.0)

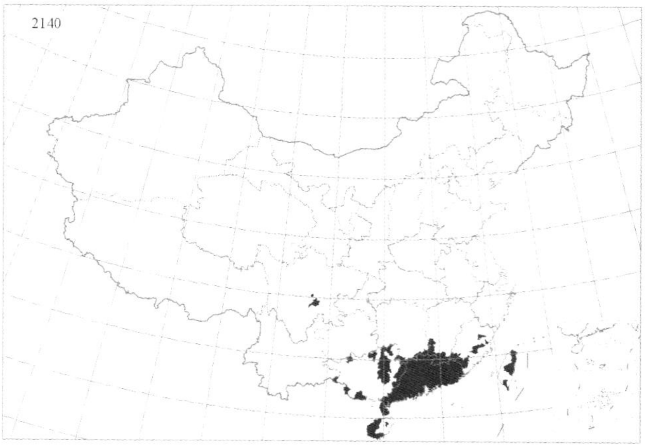

Lauraceae 樟科 *Cryptocarya* 厚壳桂属
Cryptocarya chingii W. C. Cheng 硬壳桂
Trees, evergreen

19.8 (14.6~24.8); 10.8 (3.8~20.1); 27.6 (24.3~29.8); 19.8 (14.7~24.8); 177.9 (117.7~237.6); 0.0 (0.0~1.5); 1038 (779~1385) ❙ 1631 (868~2435); 639 (312~1428); 169 (41~259) ❙ 59.9 (-34.3~137.5); 1025 (779~1257); 247.8 (95.3~620.0)

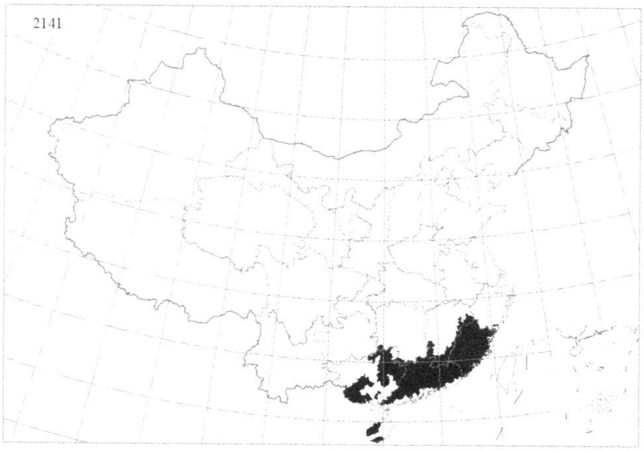

Lauraceae 樟科 *Cryptocarya* 厚壳桂属
Cryptocarya concinna Hance 黄果厚壳桂
Trees, evergreen

20.0 (14.3~24.1); 11.1 (3.3~19.1); 27.4 (20.1~29.8); 20.0 (14.3~24.2); 180.6 (114.1~230.2); 0.0 (0.0~2.3); 1051 (765~1329) ❙ 1709 (1268~2993); 734 (445~1847); 161 (48~547) ❙ 65.3 (-1.0~246.5); 1047 (765~1329); 243.3 (139.6~620.0)

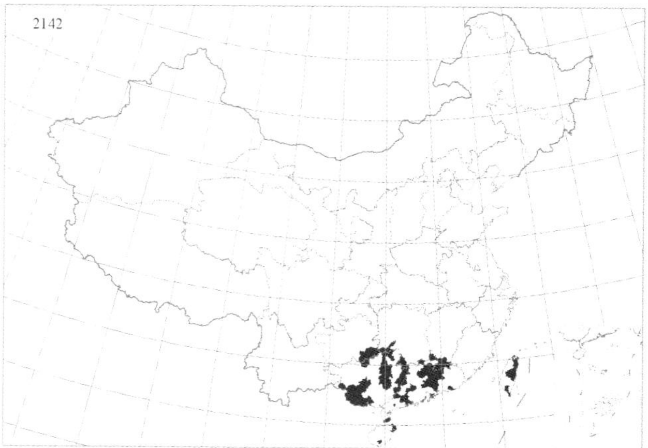

Lauraceae 樟科 *Cryptocarya* 厚壳桂属
Cryptocarya densiflora Blume 丛花厚壳桂
Trees, evergreen

19.6 (14.6~24.0); 11.0 (3.8~19.1); 26.6 (21.4~29.0); 19.6 (14.7~24.0); 175.7 (117.7~228.1); 0.0 (0.0~1.5); 1012 (779~1302) ❙ 1600 (1074~2435); 699 (425~1428); 138 (47~230) ❙ 59.1 (12.9~104.6); 1008 (779~1257); 289.8 (163.7~659.0)

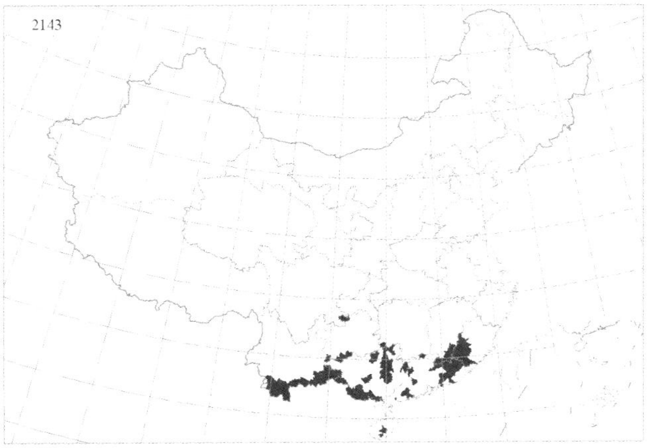

Lauraceae 樟科 *Cryptocarya* 厚壳桂属
Cryptocarya depauperata H. W. Li 贫花厚壳桂
Trees, evergreen

18.5 (17.2~21.0); 11.7 (9.5~14.2); 23.2 (22.4~25.6); 18.5 (17.3~21.0); 162.3 (147.0~191.8); 0.0 (0.0~0.0); 896 (833~1053) ❙ 1474 (1254~1751); 836 (699~996); 60 (48~77) ❙ 64.9 (44.5~88.8); 896 (833~1053); 465.5 (336.3~551.6)

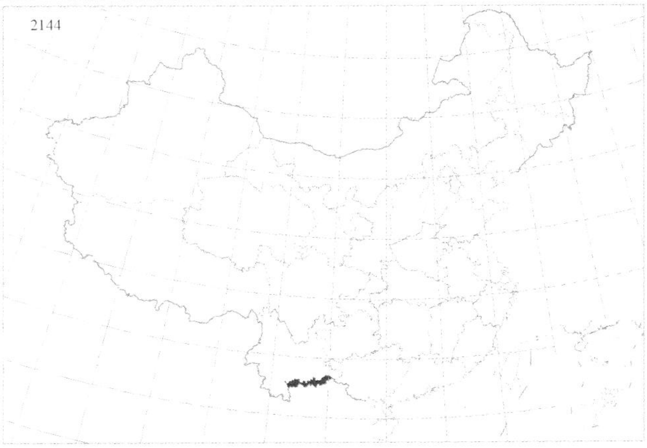

Lauraceae 樟科 *Cryptocarya* 厚壳桂属
Cryptocarya elliptifolia Merr. 菲岛厚壳桂
Trees

18.5; 14.1; 21.7; 18.5; 162.1; 0.0; 863 ❙ 2713; 1388; 188 ❙ 214.4; 863; 363.7

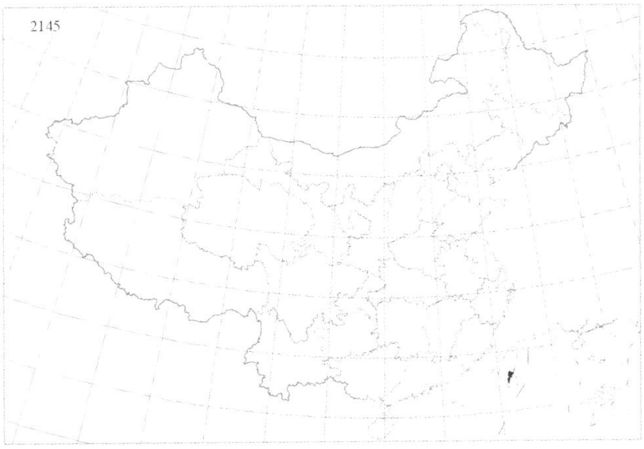

Lauraceae 樟科 *Cryptocarya* 厚壳桂属
Cryptocarya hainanensis Merr. 海南厚壳桂
Trees, evergreen

20.1 (17.2~24.3); 12.9 (9.1~20.1); 25.2 (22.3~28.3); 20.1 (17.3~24.8); 181.5 (147.0~237.6); 0.0 (0.0~0.0); 1020 (833~1385) ❙ 1441 (1161~1900); 738 (473~1056); 66 (41~125) ❙ 43.8 (1.5~88.8); 995 (833~1212); 418.7 (281.7~659.0)

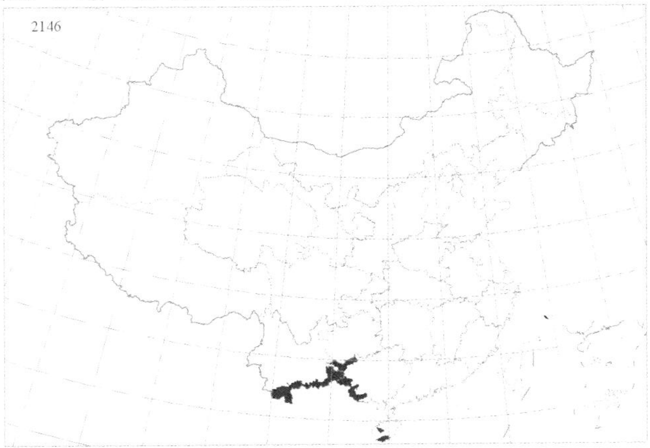

Lauraceae 樟科 *Cryptocarya* 厚壳桂属
Cryptocarya impressinervia H. W. Li 钝叶厚壳桂
Trees, evergreen

23.3 (21.8~24.7); 17.2 (14.0~19.8); 27.7 (26.7~28.4); 23.3 (21.9~24.7); 219.7 (202.2~236.5); 0.0 (0.0~0.0); 1253 (1150~1376) ❙ 1563 (1179~2148); 658 (392~1234); 66 (42~124) ❙ 26.1 (-11.9~86.7); 1143 (1056~1212); 433.6 (215.3~572.9)

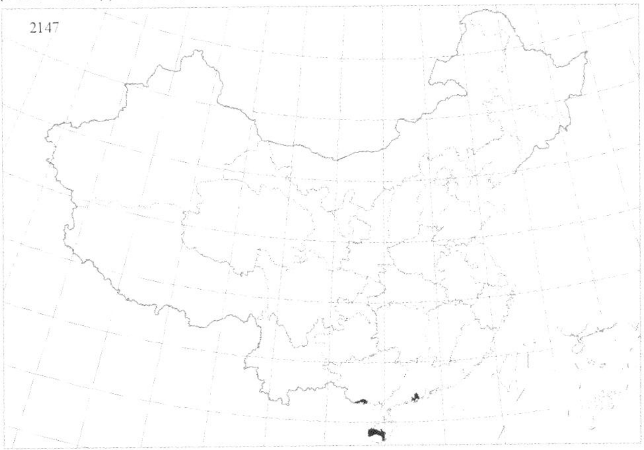

Lauraceae 樟科 *Cryptocarya* 厚壳桂属
Cryptocarya kwangtungensis H. T. Chang 广东厚壳桂
Trees, evergreen

19.6 (17.8~22.2); 10.0 (7.1~13.9); 27.7 (26.9~29.8); 19.7 (17.8~22.3); 176.0 (154.0~207.5); 0.0 (0.0~0.0); 1027 (933~1190) ❙ 1692 (1374~1961); 544 (508~809); 185 (87~197) ❙ 65.7 (15.4~95.5); 1021 (933~1190); 245.0 (174.9~327.7)

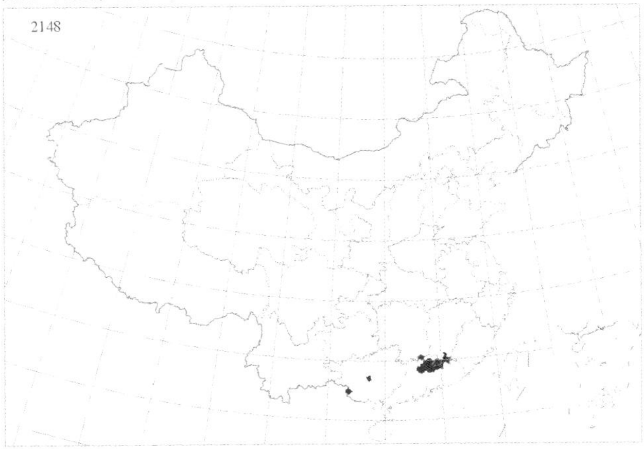

Lauraceae 樟科 *Cryptocarya* 厚壳桂属
Cryptocarya leiana C. K. Allen 鸡卵槁
Trees, evergreen

24.1; 18.0; 28.6; 24.2; 230.2; 0.0; 1327 ❙ 1309; 480; 88 ❙ -1.4; 1254; 351.4

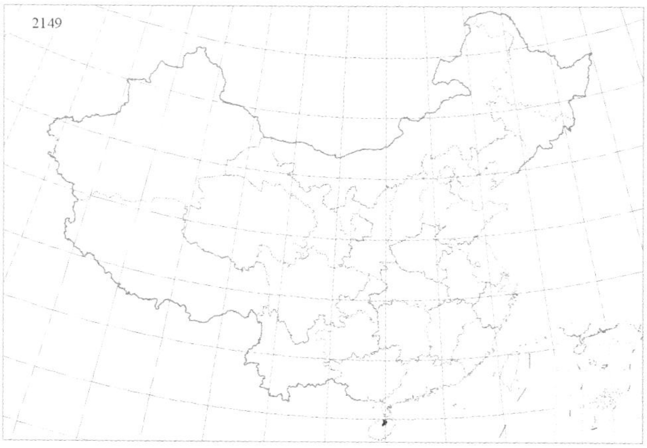

Lauraceae 樟科 *Cryptocarya* 厚壳桂属
Cryptocarya lyoniifolia S. K. Lee et F. N. Wei 南烛厚壳桂
Trees, evergreen

20.7; 12.4; 26.9; 20.8; 189.3; 0.0; 1074 ❙ 1437; 779; 86 ❙ 35.2; 1074; 305.7

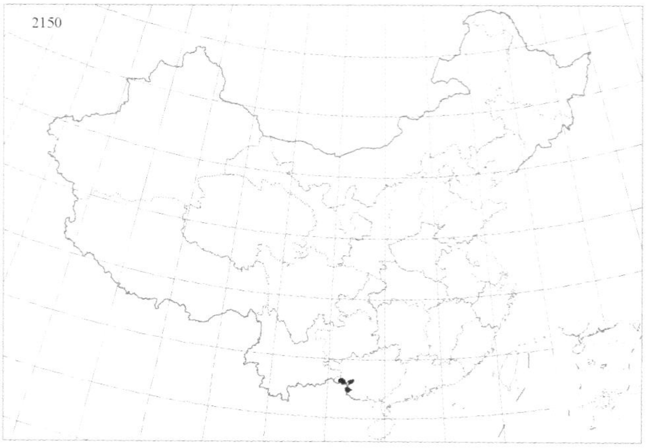

Lauraceae 樟科 *Cryptocarya* 厚壳桂属
Cryptocarya maclurei Merr. 白背厚壳桂
Trees, evergreen

21.6 (17.6~24.8); 13.4 (7.4~20.1); 28.1 (25.8~29.2); 21.6 (17.7~24.8); 199.7 (152.5~237.6); 0.0 (0.0~0.0); 1146 (906~1385) ❙ 1674 (1302~2435); 760 (455~1428); 130 (41~194) ❙ 46.9 (1.5~99.5); 1138 (906~1273); 235.6 (95.3~620.0)

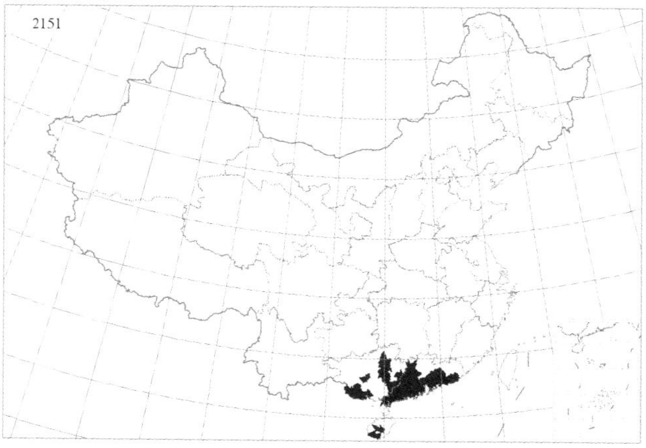

Lauraceae 樟科 *Cryptocarya* 厚壳桂属
Cryptocarya maculata H. W. Li 斑果厚壳桂
Trees, evergreen

19.4; 12.6; 24.1; 19.4; 172.6; 0.0; 954 ❙ 1445; 787; 64 ❙ 52.3; 954; 521.5

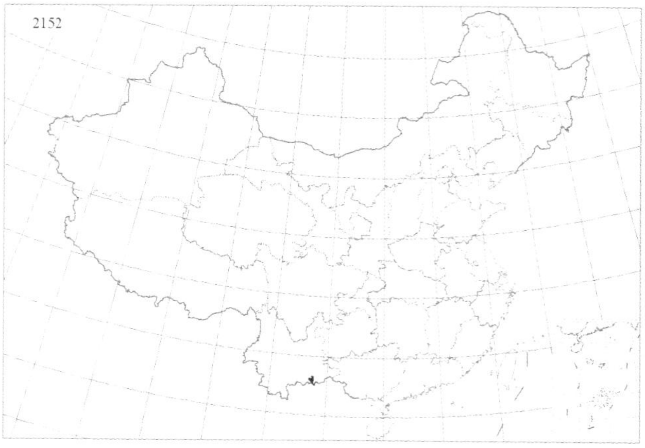

Lauraceae 樟科 *Cryptocarya* 厚壳桂属
Cryptocarya metcalfiana C. K. Allen 长序厚壳桂
Trees, evergreen

23.8 (21.2~24.8); 18.1 (12.2~20.1); 27.9 (27.2~28.5); 23.8 (21.3~24.8); 225.6 (195.2~237.6); 0.0 (0.0~0.0); 1299 (1126~1385) ❙ 1401 (1179~1588); 491 (392~599); 60 (41~142) ❙ 8.3 (-11.9~29.9); 1126 (1056~1212); 459.9 (226.9~620.0)

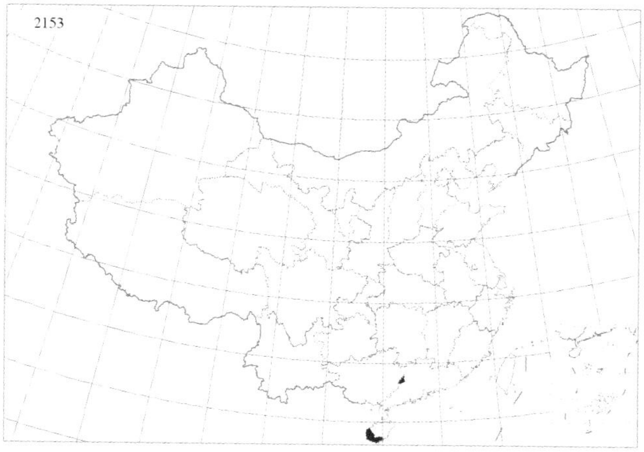

Lauraceae 樟科 *Cryptocarya* 厚壳桂属
Cryptocarya tsangii Nakai 红柄厚壳桂
Trees, evergreen

23.0 (17.6~24.8); 16.6 (7.1~20.1); 27.8 (25.7~28.8); 23.0 (17.7~24.8); 216.5 (152.5~237.6); 0.0 (0.0~0.0); 1247 (906~1385) ❙ 1508 (868~1854); 537 (312~712); 91 (41~182) ❙ 23.2 (-34.3~71.7); 1150 (868~1360); 405.4 (197.5~620.0)

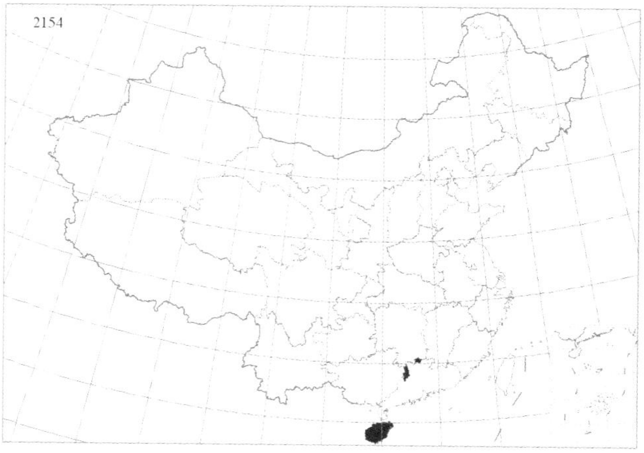

Lauraceae 樟科 *Cryptocarya* 厚壳桂属

Cryptocarya yaanica N. Chao ex H. W. Li et al. 雅安厚壳桂

Trees, evergreen

16.6 (10.6~18.7); 5.8 (-0.2~8.1); 26.8 (21.1~28.9); 16.6 (10.6~18.8); 140.3 (79.4~166.0); 0.9 (0.0~12.1); 887 (643~1002) ❙ 1167 (1028~1398); 496 (442~812); 62 (46~76) ❙ 33.6 (8.0~93.8); 887 (643~1002); 161.6 (108.2~284.3)

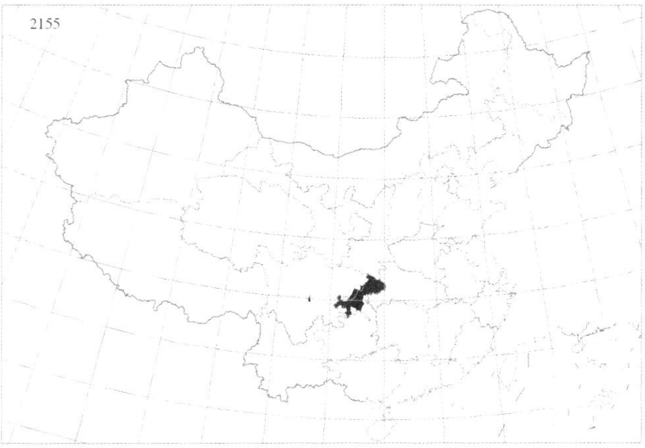

Lauraceae 樟科 *Cryptocarya* 厚壳桂属

Cryptocarya yunnanensis H. W. Li 云南厚壳桂

Trees, evergreen

20.1; 14.3; 23.5; 20.2; 181.9; 0.0; 974 ❙ 1543; 755; 62 ❙ 58.5; 956; 570.7

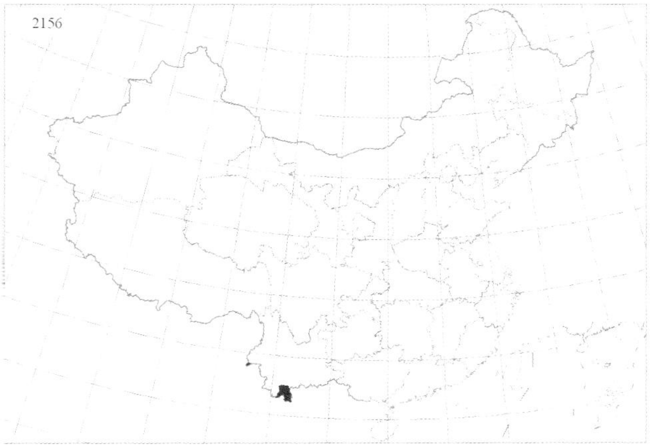

Lauraceae 樟科 *Dehaasia* 莲桂属

Dehaasia hainanensis Kosterm. 莲桂

Shrubs or trees, evergreen

24.1 (23.6~24.8); 18.5 (17.8~20.1); 28.2 (27.5~28.7); 24.2 (23.6~24.8); 230.1 (223.3~237.6); 0.0 (0.0~0.0); 1324 (1271~1385) ❙ 1208 (868~1405); 420 (312~473); 51 (41~71) ❙ -8.7 (-34.3~6.7); 1067 (868~1139); 429.6 (288.3~527.5)

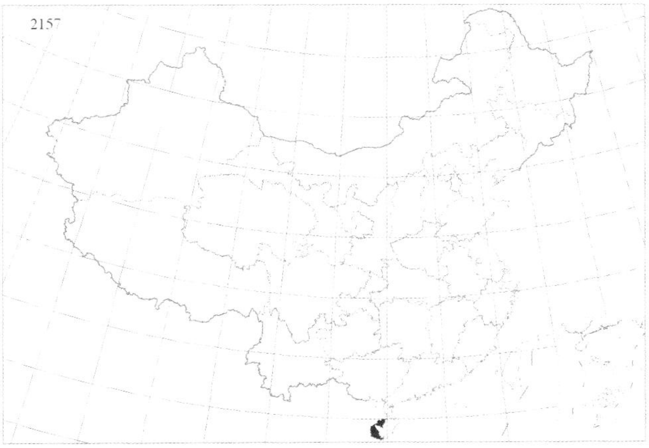

Lauraceae 樟科 *Dehaasia* 莲桂属

Dehaasia incrassata (Jack) Kosterm. 腰果楠

Trees, evergreen

18.5; 14.1; 21.7; 18.5; 162.1; 0.0; 863 ❙ 2713; 1388; 188 ❙ 214.4; 863; 363.7

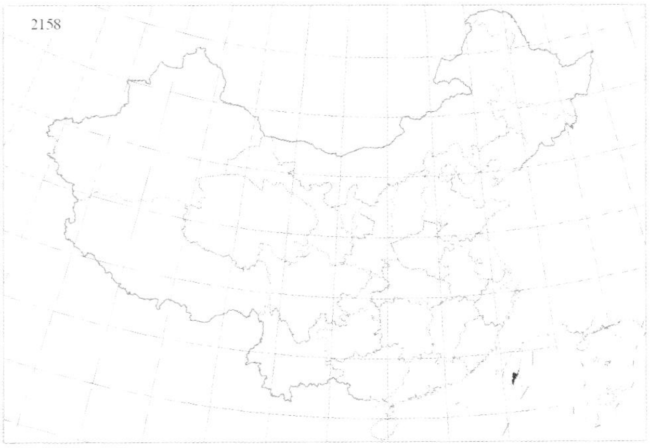

Lauraceae 樟科 *Dehaasia* 莲桂属

Dehaasia kwangtungensis Kosterm. 广东莲桂

Trees, evergreen

23.2; 15.8; 28.9; 23.2; 218.9; 0.0; 1257 ❙ 1686; 824; 102 ❙ 34.1; 1257; 194.5

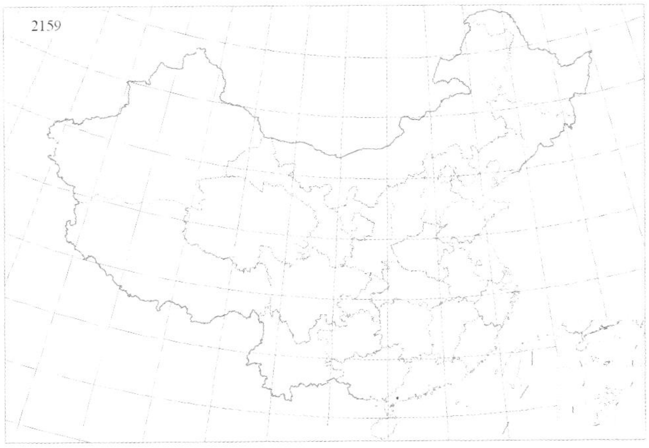

Lauraceae 樟科 *Dodecadenia* 单花木姜子属

Dodecadenia grandiflora (Hook. f.) D. G. Long 单花木姜子

Trees, evergreen

13.0 (-1.8~18.9); 6.0 (-10.2~12.5); 18.3 (6.1~23.2); 13.2 (1.8~19.0); 105.5 (1.9~167.5); 9.5 (0.0~82.8); 704 (329~914) ❙ 1038 (320~1736); 604 (217~1007); 35 (11~51) ❙ 49.1 (-3.0~145.8); 679 (320~851); 332.5 (72.2~516.7)

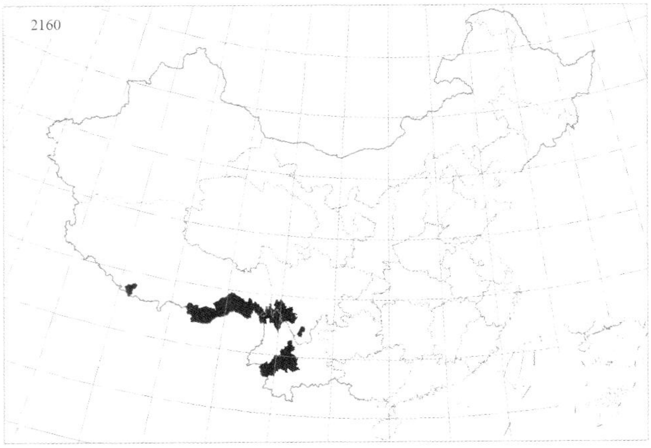

Lauraceae 樟科 *Dodecadenia* 单花木姜子属
Dodecadenia grandiflora var. griffithii (Hook. f.) D. G. Long 无毛单花木姜子
Trees, evergreen

8.6；1.3；15.0；8.7；52.6；8.8；574 ❙ 1041；585；45 ❙ 81.5；574；352.3

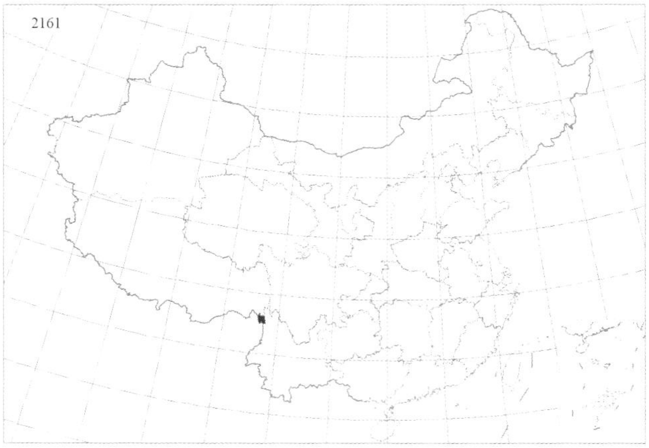

Lauraceae 樟科 *Endiandra* 土楠属
Endiandra coriacea Merr. 革叶土楠
Trees

18.5；14.1；21.7；18.5；162.1；0.0；863 ❙ 2713；1388；188 ❙ 214.4；863；363.7

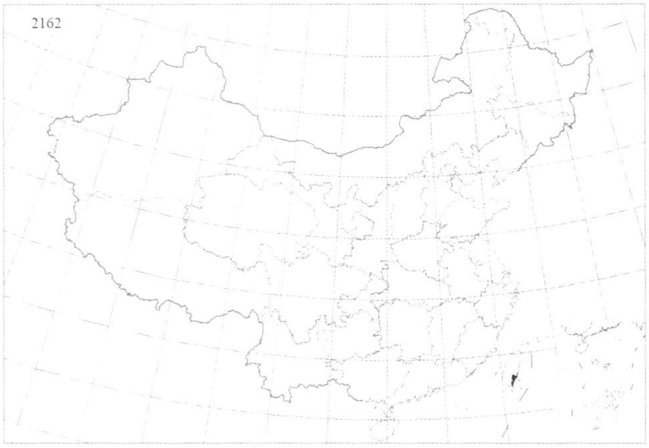

Lauraceae 樟科 *Endiandra* 土楠属
Endiandra dolichocarpa S. K. Lee et Y. T. Wei 长果土楠
Trees

21.7；13.2；28.0；21.8；201.0；0.0；1152 ❙ 1326；694；80 ❙ 15.1；1152；303.1

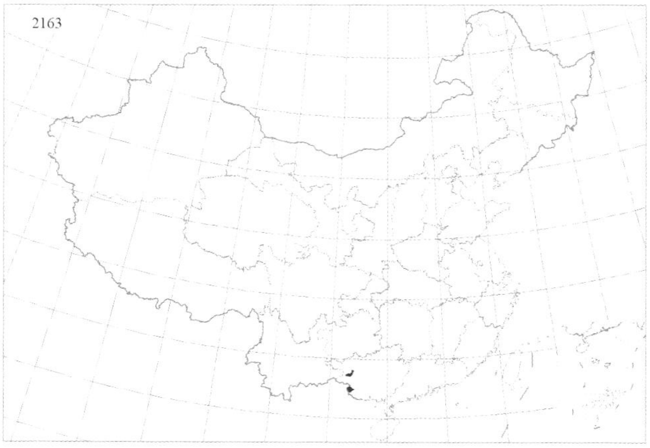

Lauraceae 樟科 *Endiandra* 土楠属
Endiandra hainanensis Merr. et Metcalf ex Allen 土楠
Trees

23.6 (21.0~24.2)；17.5 (14.2~19.1)；27.9 (25.6~28.8)；23.6 (21.0~24.3)；223.5 (191.8~231.2)；0.0 (0.0~0.0)；1278 (1053~1335) ❙ 1448 (1176~1655)；584 (447~818)；79 (48~109) ❙ 14.4 (-12.0~44.5)；1208 (1053~1331)；417.2 (283.8~620.0)

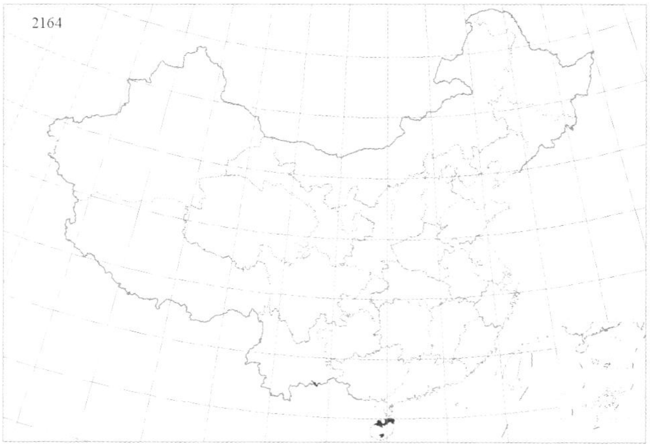

Lauraceae 樟科 *Iteadaphne* 香面叶属
Iteadaphne caudata (Nees) H. W. Li 香面叶
Shrubs or small trees, evergreen

18.2 (14.4~22.7)；11.4 (7.5~15.1)；22.8 (19.1~28.6)；18.2 (14.4~22.8)；158.8 (113.1~213.1)；0.0 (0.0~0.0)；891 (727~1221) ❙ 1320 (939~2435)；735 (524~1428)；57 (37~130) ❙ 47.8 (6.5~112.2)；877 (717~1221)；385.5 (235.1~659.0)

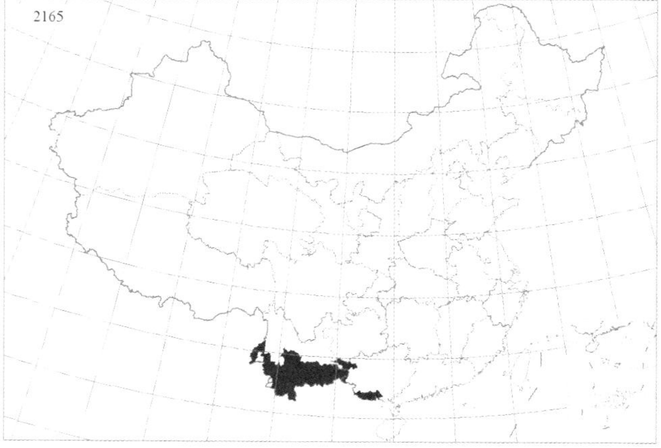

Lauraceae 樟科 *Lindera* 山胡椒属
Lindera aggregata (Sims) Kosterm. 乌药
Shrubs or trees, evergreen

17.5 (6.2~24.6)；6.8 (-4.1~19.5)；27.4 (15.8~30.1)；17.6 (7.0~24.7)；152.5 (45.2~235.9)；1.8 (0.0~29.9)；943 (522~1372) ❙ 1465 (626~2680)；560 (299~1186)；155 (14~259) ❙ 55.8 (-22.2~246.5)；929 (522~1299)；220.0 (95.3~548.0)

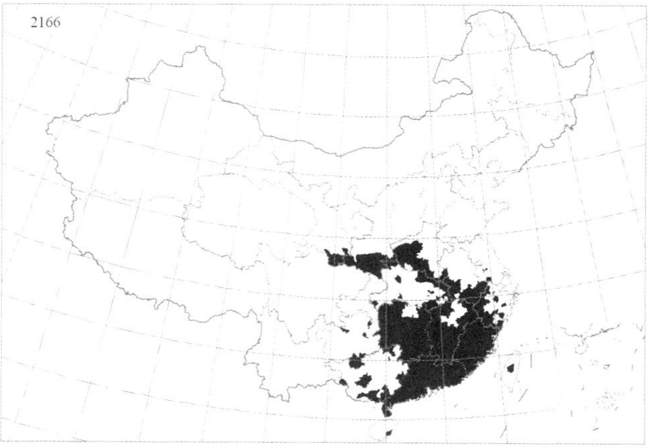

Lauraceae 樟科 *Lindera* 山胡椒属
Lindera aggregata var. playfairii (Hemsl.) H. B. Cui 小叶乌药
Shrubs or trees, evergreen

21.2 (17.6~24.6); 12.8 (6.0~19.5); 28.1 (26.2~29.8); 21.2 (17.7~24.7); 194.7 (152.5~235.9); 0.0 (0.0~0.0); 1119 (906~1372) ▎1664 (1187~2037); 712 (445~1044); 142 (75~221) ▎49.6 (-6.9~95.5); 1116 (906~1299); 222.6 (95.3~548.0)

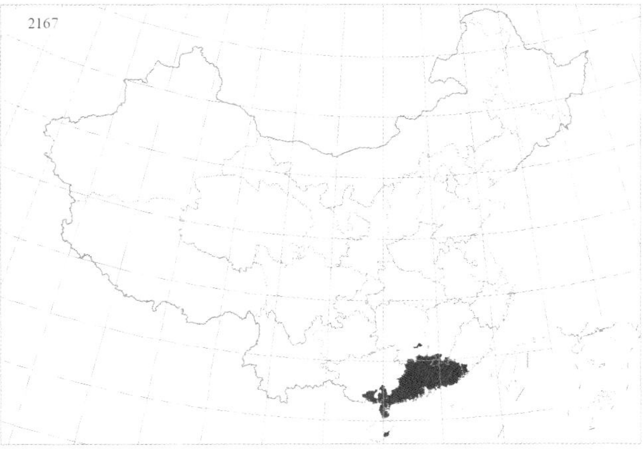

Lauraceae 樟科 *Lindera* 山胡椒属
Lindera akoensis Hayata 台湾香叶树
Shrubs or trees, evergreen

20.1 (17.8~22.2); 14.3 (12.1~17.7); 25.0 (22.8~25.9); 20.1 (17.9~22.2); 181.4 (154.4~206.7); 0.0 (0.0~0.0); 992 (852~1123) ▎2699 (2208~2993); 1178 (750~1847); 301 (107~547) ▎174.2 (125.7~235.6); 992 (852~1123); 258.5 (207.7~296.9)

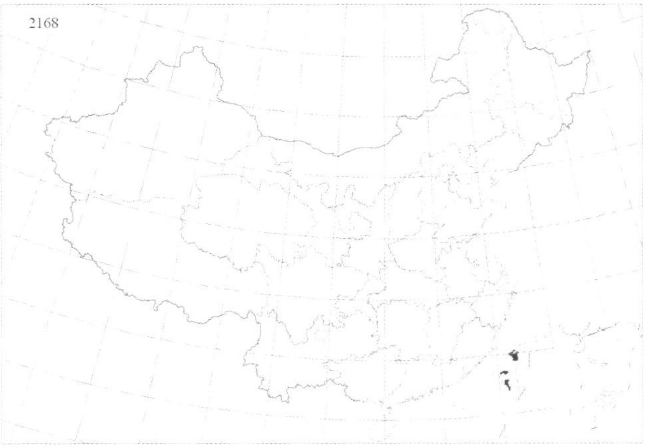

Lauraceae 樟科 *Lindera* 山胡椒属
Lindera angustifolia W. C. Cheng 狭叶山胡椒
Shrubs or trees, deciduous

15.5 (3.9~22.8); 3.0 (-13.7~15.5); 26.9 (18.7~30.1); 15.7 (7.3~22.8); 134.4 (52.8~214.1); 8.3 (0.0~71.7); 876 (535~1225) ▎1192 (367~2035); 483 (235~969); 122 (8~259) ▎32.6 (-43.4~130.7); 821 (367~1225); 192.6 (101.9~319.9)

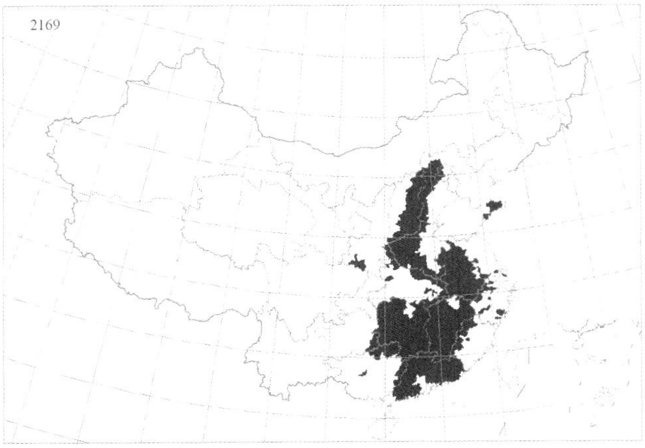

Lauraceae 樟科 *Lindera* 山胡椒属
Lindera chienii W. C. Cheng 江浙山胡椒
Shrubs or trees, deciduous

15.6 (13.4~17.6); 3.0 (1.1~5.6); 27.8 (25.2~29.3); 15.7 (13.4~17.6); 131.0 (108.3~151.4); 3.0 (0.0~7.9); 871 (755~953) ▎1290 (753~1828); 521 (376~685); 141 (46~253) ▎48.2 (-12.2~112.2); 866 (753~941); 199.9 (123.9~319.9)

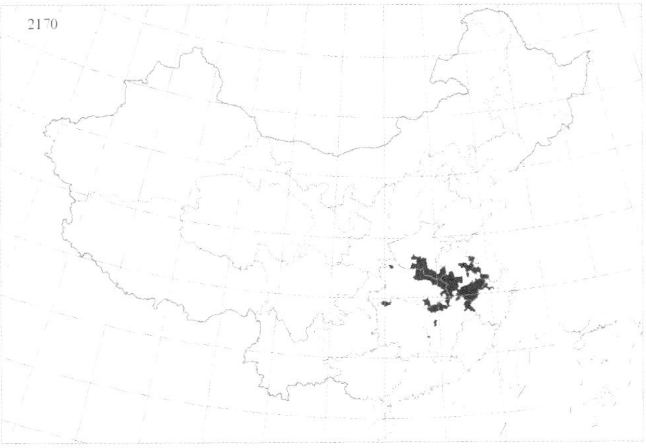

Lauraceae 樟科 *Lindera* 山胡椒属
Lindera chunii Merr. 鼎湖钓樟
Shrubs or trees

20.8 (14.9~23.1); 12.1 (3.6~17.5); 27.8 (25.7~29.0); 20.8 (15.0~23.2); 190.1 (121.2~218.4); 0.0 (0.0~1.6); 1096 (807~1232) ▎1664 (1284~2435); 749 (544~1428); 139 (53~197) ▎52.6 (12.9~99.5); 1094 (807~1221); 250.2 (139.6~538.2)

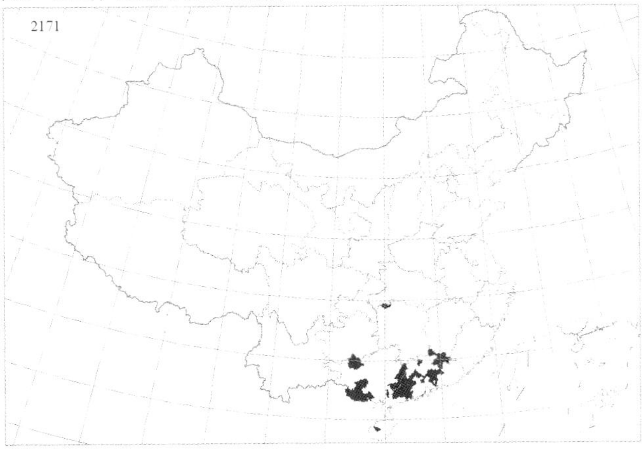

Lauraceae 樟科 *Lindera* 山胡椒属
Lindera communis Hemsl. 香叶树
Shrubs or trees, evergreen

16.6 (2.0~23.6); 6.6 (-7.2~17.7); 25.4 (9.5~30.1); 16.6 (3.8~23.6); 141.3 (15.4~223.5); 2.1 (0.0~51.4); 884 (411~1281) ▎1317 (512~2993); 580 (258~1847); 105 (11~547) ▎49.5 (-35.3~246.5); 872 (411~1273); 230.2 (95.3~659.0)

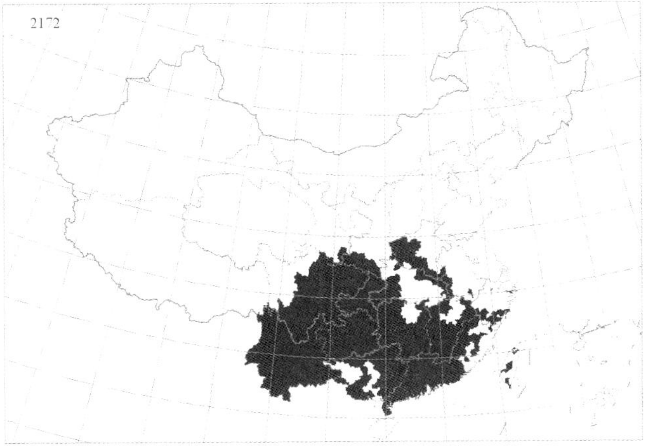

Lauraceae 樟科 *Lindera* 山胡椒属
Lindera doniana C. K. Allen 贡山山胡椒
Trees, evergreen

12.6; 5.7; 18.0; 12.6; 94.1; 2.4; 686 ❙ 1168; 651; 54 ❙ 73.3; 674; 377.3

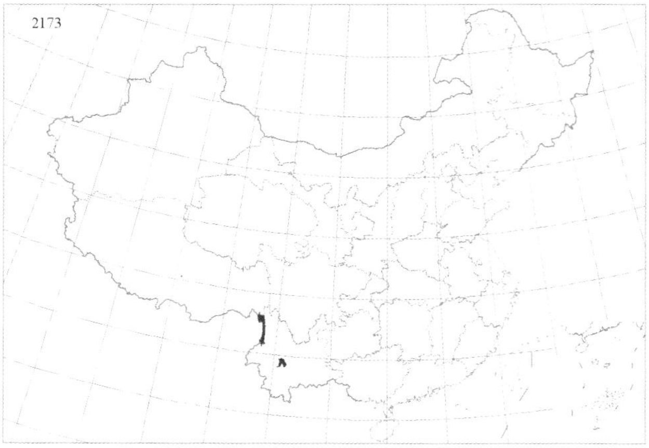

Lauraceae 樟科 *Lindera* 山胡椒属
Lindera erythrocarpa Makino 红果山胡椒
Shrubs or trees, deciduous

16.8 (2.5~22.8); 5.4 (-9.7~15.5); 27.3 (12.4~30.1); 16.8 (4.8~22.8); 143.9 (25.1~214.1); 2.3 (0.0~56.4); 909 (441~1225) ❙ 1362 (512~2859); 532 (258~969); 138 (9~402) ❙ 49.2 (-35.9~235.6); 891 (441~1225); 199.0 (108.2~319.9)

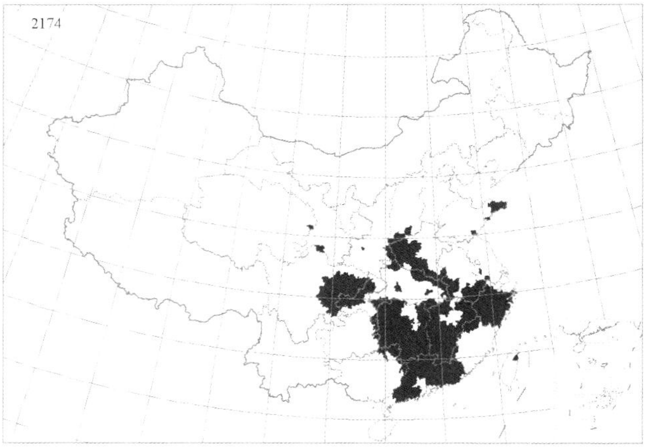

Lauraceae 樟科 *Lindera* 山胡椒属
Lindera floribunda (C. K. Allen) H. B. Cui 绒毛钓樟
Trees, evergreen

15.6 (5.7~19.1); 5.4 (-3.5~12.7); 24.7 (14.2~29.4); 15.6 (6.4~19.1); 128.9 (38.1~169.2); 1.6 (0.0~28.1); 824 (504~972) ❙ 1285 (711~1624); 549 (357~816); 101 (15~209) ❙ 56.5 (12.1~106.2); 817 (504~966); 241.3 (133.8~516.7)

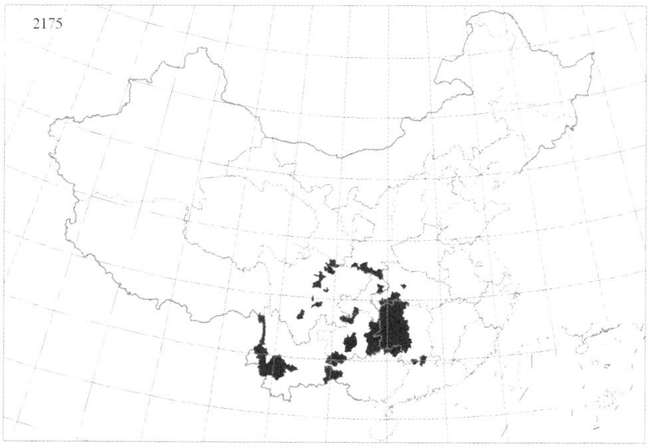

Lauraceae 樟科 *Lindera* 山胡椒属
Lindera foveolata H. W. Li 蜂房叶山胡椒
Trees, evergreen

18.6 (17.2~21.0); 11.6 (9.5~14.2); 23.4 (22.4~25.6); 18.6 (17.3~21.0); 163.4 (147.0~191.8); 0.0 (0.0~0.0); 905 (833~1053) ❙ 1401 (1254~1522); 794 (699~860); 55 (48~68) ❙ 55.5 (36.8~80.6); 904 (833~1053); 436.5 (336.3~551.6)

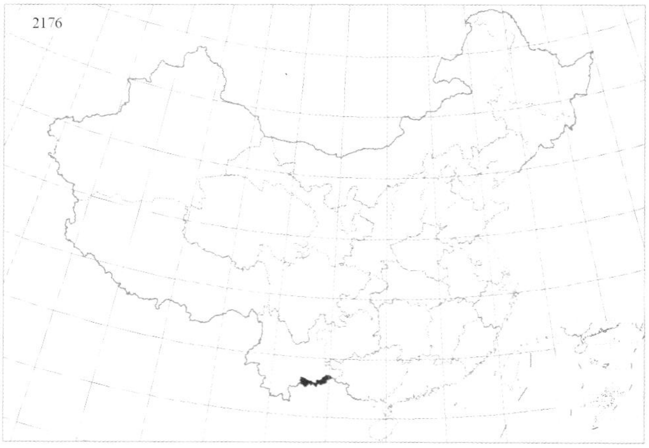

Lauraceae 樟科 *Lindera* 山胡椒属
Lindera fragrans Oliv. 香叶子
Trees, evergreen

16.2 (2.6~23.0); 5.4 (-7.0~15.1); 25.8 (11.2~29.0); 16.2 (4.3~23.0); 137.9 (20.0~216.2); 3.3 (0.0~48.2); 876 (427~1247) ❙ 1224 (626~2148); 556 (316~1234); 89 (15~225) ❙ 40.0 (-22.2~108.0); 869 (427~1247); 220.2 (134.1~411.1)

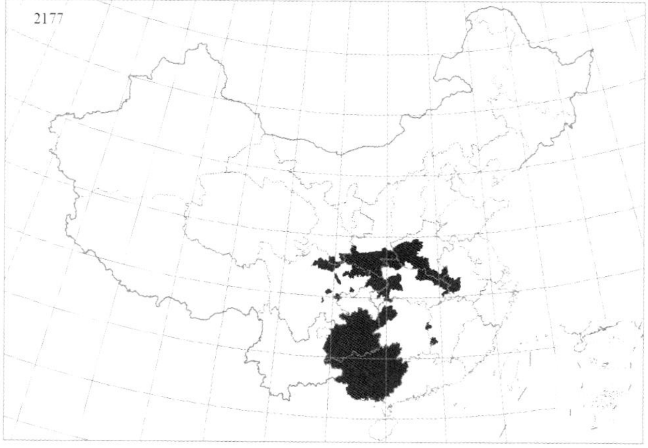

Lauraceae 樟科 *Lindera* 山胡椒属
Lindera glauca (Sieb. et Zucc.) Bl. 山胡椒
Shrubs or trees, deciduous

16.6 (7.4~23.6); 5.4 (-4.9~16.7); 27.0 (18.1~30.1); 16.7 (8.2~23.6); 142.2 (58.8~223.5); 2.4 (0.0~29.4); 900 (561~1281) ❙ 1307 (512~2680); 531 (258~1186); 124 (14~308) ❙ 44.8 (-35.9~246.5); 884 (512~1273); 200.2 (95.3~334.4)

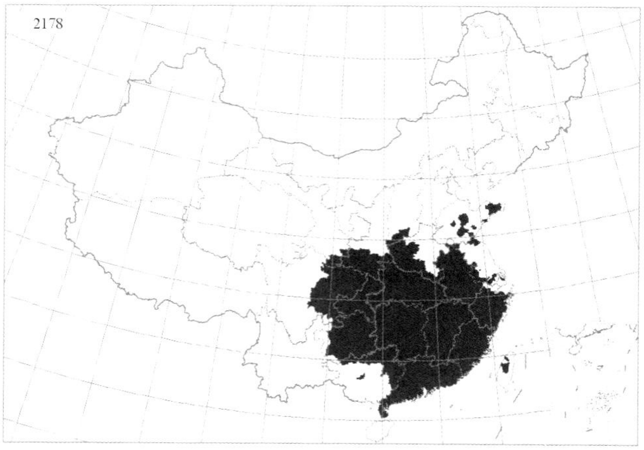

Lauraceae 樟科 *Lindera* 山胡椒属

Lindera gracilipes H. W. Li 纤梗山胡椒

Shrubs or trees

13.2 (5.3~21.0); 5.9 (−2.4~14.2); 19.0 (12.2~25.6); 13.3 (5.7~21.0); 105.7 (30.5~191.8); 7.0 (0.0~26.2); 733 (479~1053) | 1326 (961~1736); 745 (558~1007); 51 (31~68) | 86.5 (44.5~145.8); 733 (479~1053); 415.4 (234.2~551.6)

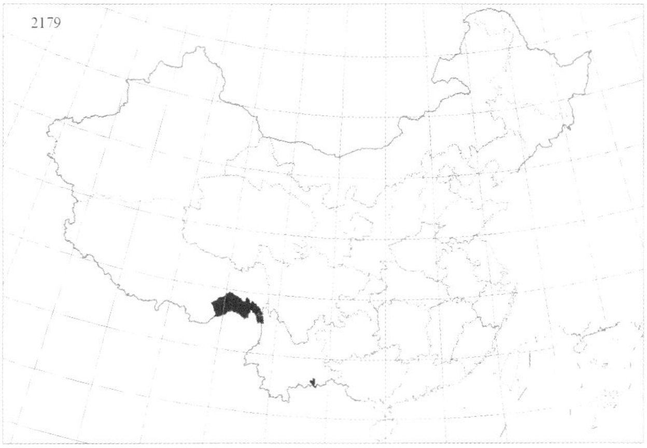

Lauraceae 樟科 *Lindera* 山胡椒属

Lindera guangxiensis H. B. Cui 广西钓樟

Trees, evergreen

17.4 (14.3~22.2); 7.4 (3.3~13.9); 26.1 (24.3~28.3); 17.5 (14.3~22.3); 149.8 (114.1~207.5); 0.4 (0.0~2.3); 907 (765~1190) | 1522 (1250~1740); 637 (465~848); 148 (57~196) | 70.6 (15.4~101.8); 907 (765~1190); 265.5 (205.4~371.7)

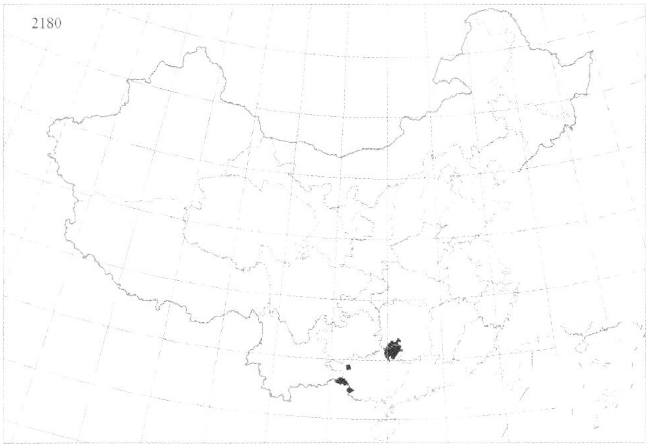

Lauraceae 樟科 *Lindera* 山胡椒属

Lindera kariensis W. W. Sm. 更里山胡椒

Shrubs or trees, deciduous

12.7 (5.3~16.7); 5.9 (−2.4~10.0); 18.2 (12.2~22.1); 12.8 (5.7~16.7); 96.8 (30.5~140.9); 3.3 (0.0~26.2); 687 (479~819) | 1050 (741~1454); 599 (426~792); 43 (15~70) | 54.7 (14.4~100.7); 675 (479~809); 343.4 (234.2~487.2)

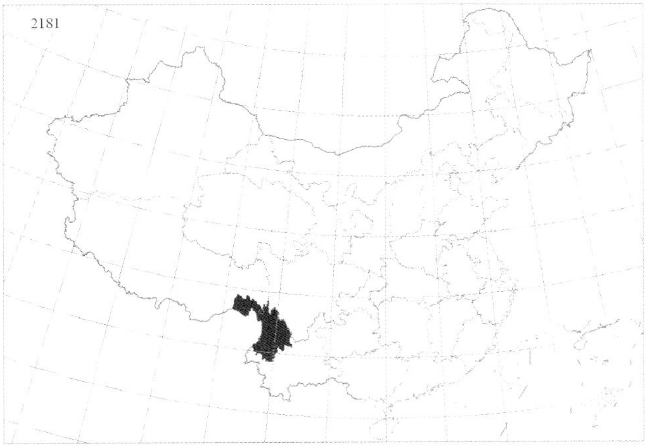

Lauraceae 樟科 *Lindera* 山胡椒属

Lindera kwangtungensis (H. Liu) C. K. Allen 广东山胡椒

Trees, evergreen

19.0 (9.0~24.7); 9.7 (−0.2~19.8); 27.0 (16.5~29.8); 19.0 (9.0~24.7); 168.4 (58.7~236.5); 0.3 (0.0~12.1); 1002 (584~1376) | 1475 (868~2435); 638 (312~1428); 119 (22~216) | 48.4 (−34.3~103.1); 994 (584~1331); 229.5 (108.2~620.0)

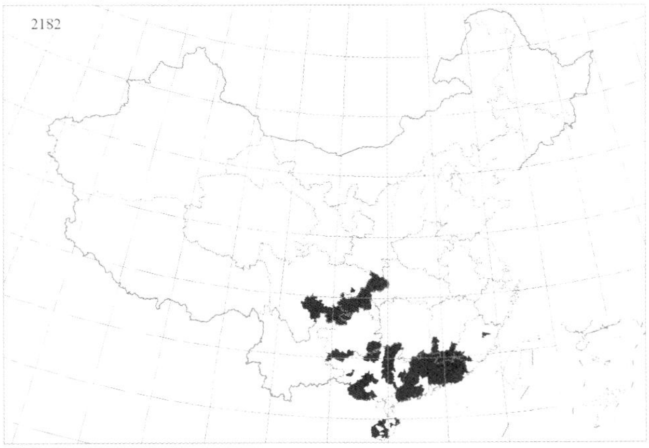

Lauraceae 樟科 *Lindera* 山胡椒属

Lindera latifolia Hook. f. 团香果

Trees, evergreen

16.2 (5.4~21.0); 9.6 (−2.1~15.1); 21.0 (12.3~25.6); 16.3 (5.8~21.0); 136.0 (31.5~191.8); 0.6 (0.0~25.6); 804 (481~1053) | 1170 (741~1751); 658 (411~1007); 47 (21~77) | 45.9 (−14.2~145.8); 785 (481~1053); 366.6 (218.0~659.0)

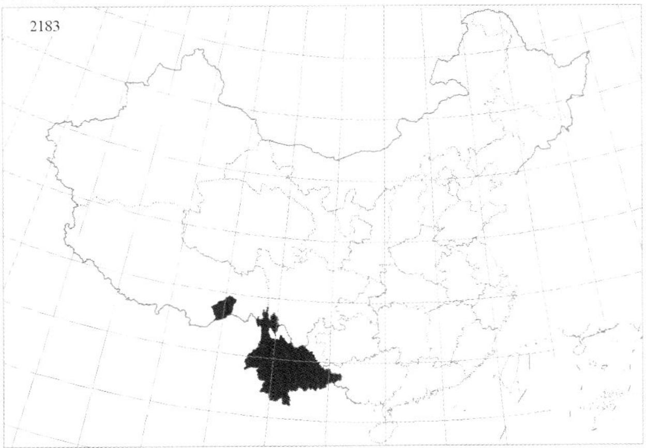

Lauraceae 樟科 *Lindera* 山胡椒属

Lindera limprichtii H. Winkl. 卵叶钓樟

Trees, evergreen

9.5 (2.0~16.2); −0.9 (−7.2~5.2); 19.2 (10.1~26.4); 10.0 (3.8~16.2); 73.1 (15.4~134.4); 18.0 (0.0~51.4); 626 (413~866) | 903 (644~1208); 445 (317~584); 23 (14~36) | 50.4 (−8.1~128.6); 623 (413~866); 263.1 (172.9~294.3)

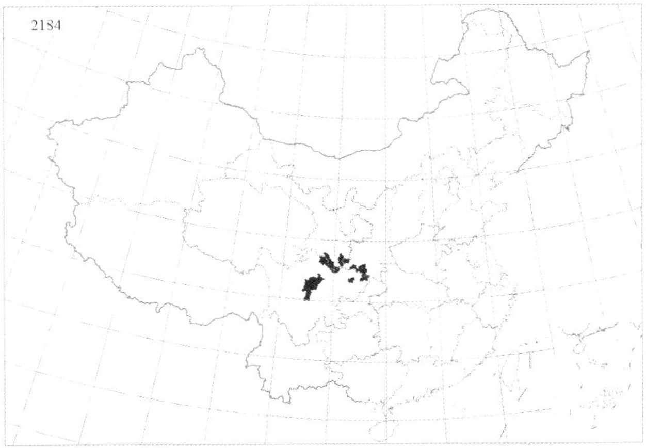

Lauraceae 樟科 *Lindera* 山胡椒属

Lindera longipedunculata C. K. Allen 山柿子果

Trees, evergreen

13.5 (8.6~16.9); 6.4 (1.3~10.0); 18.7 (15.0~21.2); 13.5 (8.7~16.9); 103.1 (52.6~142.6); 1.4 (0.0~8.8); 708 (574~814) ❙ 1437 (1041~1736); 809 (585~1007); 59 (45~70) ❙ 102.4 (81.5~145.8); 708 (574~814); 463.6 (352.3~627.0)

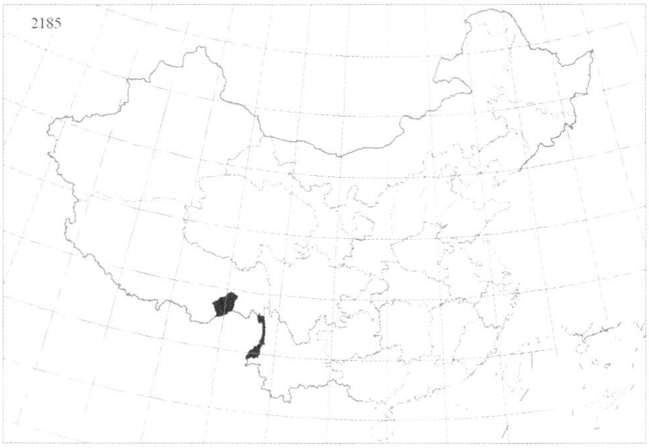

Lauraceae 樟科 *Lindera* 山胡椒属

Lindera lungshengensis S. K. Lee 龙胜钓樟

Shrubs or trees, evergreen

15.9 (14.3~17.9); 5.1 (3.3~7.4); 25.6 (24.3~27.1); 15.9 (14.3~18.0); 131.8 (114.1~155.6); 0.8 (0.0~2.3); 838 (765~938) ❙ 1607 (1526~1740); 594 (530~709); 185 (177~191) ❙ 92.5 (82.3~101.8); 838 (765~938); 244.3 (208.7~258.0)

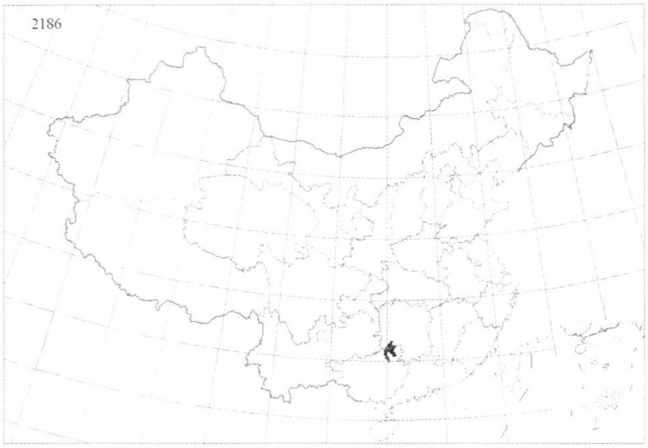

Lauraceae 樟科 *Lindera* 山胡椒属

Lindera megaphylla Hemsl. 黑壳楠

Trees, evergreen

16.6 (5.4~23.4); 6.1 (−2.9~16.1); 26.1 (12.3~30.1); 16.6 (5.8~23.5); 141.2 (31.5~221.7); 1.5 (0.0~25.6); 887 (481~1273) ❙ 1322 (626~2993); 551 (299~1412); 115 (14~547) ❙ 49.0 (−22.2~235.6); 875 (481~1273); 216.6 (95.3~551.6)

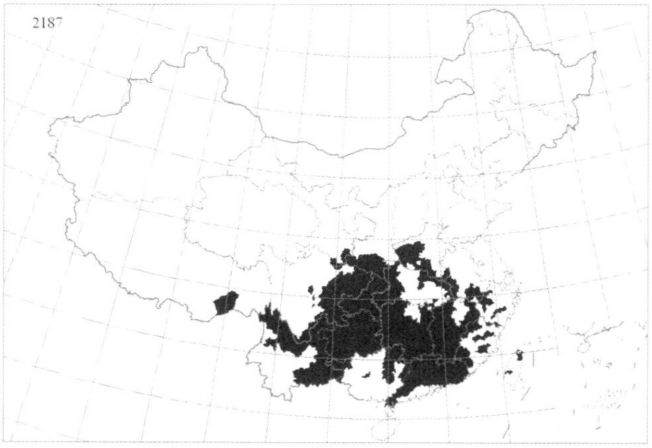

Lauraceae 樟科 *Lindera* 山胡椒属

Lindera menghaiensis H. W. Li 勐海山胡椒

Trees

19.2; 13.8; 22.3; 19.2; 170.5; 0.0; 909 ❙ 1499; 686; 58 ❙ 64.9; 880; 513.9

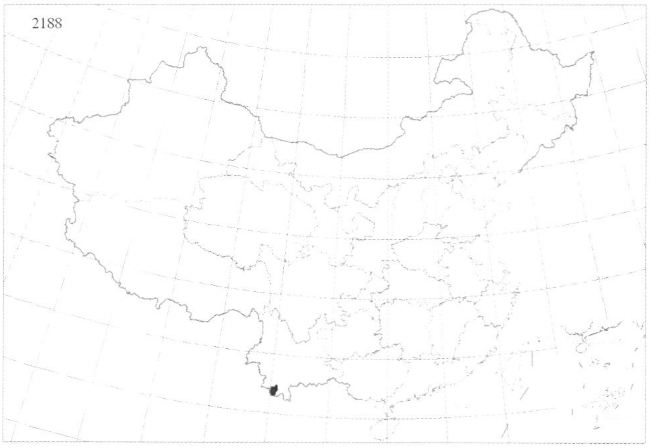

Lauraceae 樟科 *Lindera* 山胡椒属

Lindera metcalfiana C. K. Allen 滇粤山胡椒

Shrubs or trees

19.8 (8.6~24.7); 11.6 (1.3~19.8); 26.3 (15.0~29.8); 19.9 (8.7~24.7); 178.5 (52.6~236.5); 0.1 (0.0~8.8); 1030 (574~1376) ❙ 1575 (868~2435); 708 (312~1428); 125 (42~200) ❙ 54.8 (−34.3~112.2); 1019 (574~1326); 297.6 (161.4~659.0)

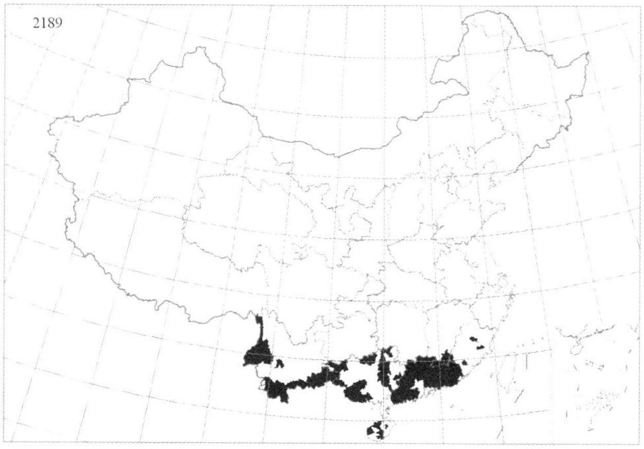

Lauraceae 樟科 *Lindera* 山胡椒属

Lindera metcalfiana var. dictyophylla (C. K. Allen) H. B. Cui 网叶山胡椒

Shrubs or trees

17.8 (14.2~21.0); 11.0 (6.2~15.1); 22.3 (19.1~27.1); 17.8 (14.2~21.0); 153.7 (110.5~191.8); 0.0 (0.0~0.0); 863 (727~1053) ❙ 1332 (958~1751); 729 (531~996); 55 (41~163) ❙ 54.5 (13.3~112.2); 856 (727~1053); 417.8 (248.4~659.0)

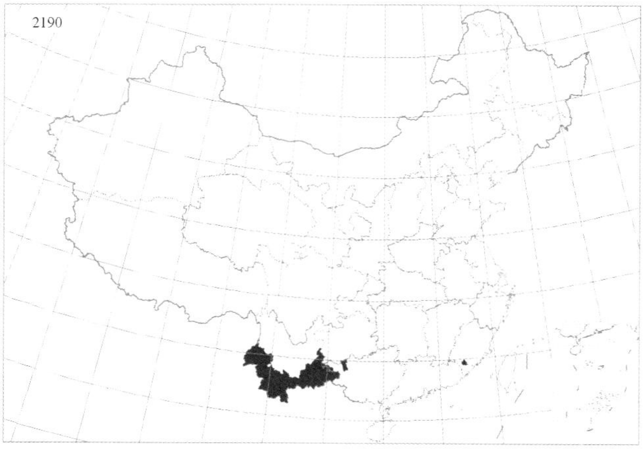

Lauraceae 樟科 *Lindera* 山胡椒属
Lindera motuoensis H. B. Cui 西藏山胡椒
Trees, evergreen

13.3；5.4；19.6；13.4；100.4；0.0；706 ❙ 1736；1007；51 ❙ 145.8；706；447.4

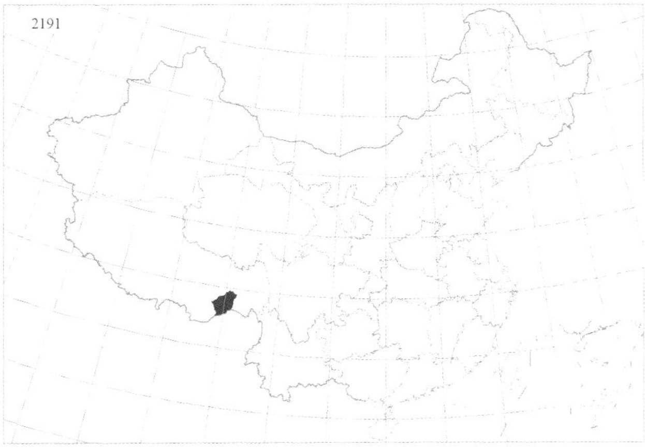

Lauraceae 樟科 *Lindera* 山胡椒属
Lindera nacusua (D. Don) Merr. 绒毛山胡椒
Shrubs or trees, evergreen

18.2 (1.8~24.3)；9.4 (-7.1~19.1)；25.6 (9.9~29.8)；18.2 (3.8~24.3)；159.3 (15.7~231.9)；0.6 (0.0~53.3)；954 (404~1338) ❙ 1458 (583~2037)；641 (312~1033)；119 (13~254) ❙ 54.1 (-34.3~145.8)；943 (404~1273)；258.0 (95.3~659.0)

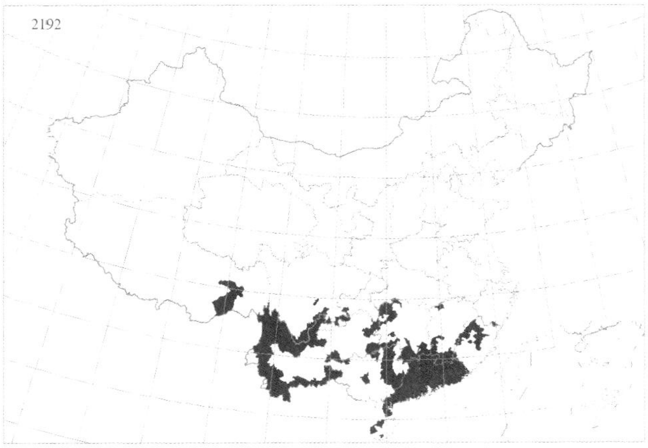

Lauraceae 樟科 *Lindera* 山胡椒属
Lindera nacusua var. menglungensis H. B. Cui 勐仑山胡椒
Shrubs or trees, evergreen

18.5；12.1；22.3；18.5；162.4；0.0；894 ❙ 1617；878；58 ❙ 82.7；888；589.5

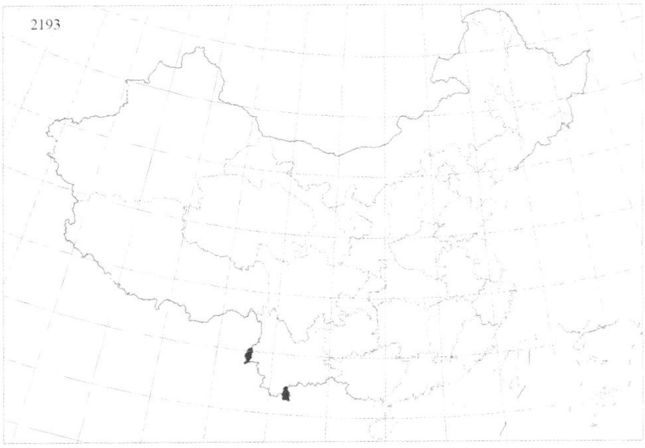

Lauraceae 樟科 *Lindera* 山胡椒属
Lindera neesiana (Wall. ex Nees) Kurz 绿叶甘橿
Shrubs or trees, deciduous

15.3 (1.8~20.5)；4.2 (-7.1~15.1)；25.6 (9.5~29.9)；15.4 (3.8~20.5)；127.8 (15.7~185.9)；3.6 (0.0~53.3)；834 (404~992) ❙ 1268 (512~1997)；513 (258~1007)；116 (10~254) ❙ 51.8 (-35.3~145.8)；815 (404~986)；235.1 (121.6~659.0)

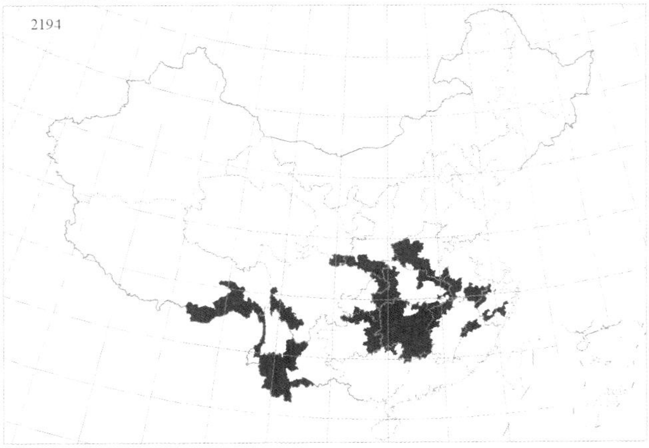

Lauraceae 樟科 *Lindera* 山胡椒属
Lindera obtusiloba Blume 三桠乌药
Shrubs or trees

12.8 (-1.4~19.4)；0.5 (-13.7~12.0)；24.1 (7.7~30.1)；13.2 (2.4~19.4)；108.1 (6.6~173.4)；13.9 (0.0~82.3)；763 (338~1037) ❙ 998 (367~1997)；448 (235~1007)；77 (7~259) ❙ 28.6 (-40.9~145.8)；711 (338~1029)；211.4 (61.6~447.4)

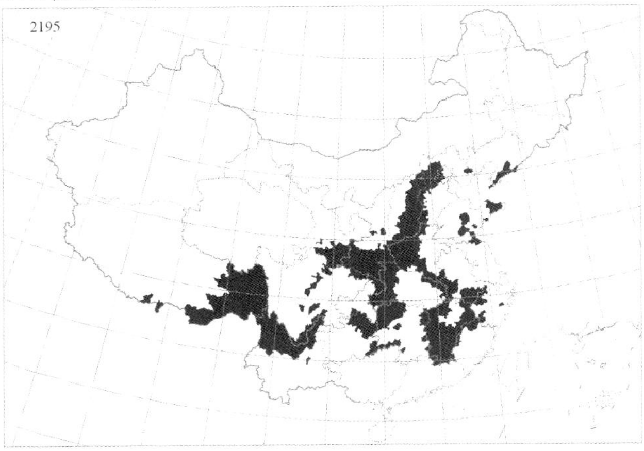

Lauraceae 樟科 *Lindera* 山胡椒属
Lindera obtusiloba var. heterophylla (Meisn.) H. P. Tsui 滇藏钓樟
Shrubs or trees, deciduous

9.0 (-0.7~13.9)；0.8 (-9.6~6.4)；16.2 (7.7~24.3)；9.5 (2.5~14.0)；65.0 (6.8~110.5)；16.1 (0.0~74.2)；593 (338~760) ❙ 1068 (370~1736)；579 (246~1007)；44 (10~105) ❙ 73.1 (0.0~145.8)；593 (338~760)；290.6 (61.6~447.4)

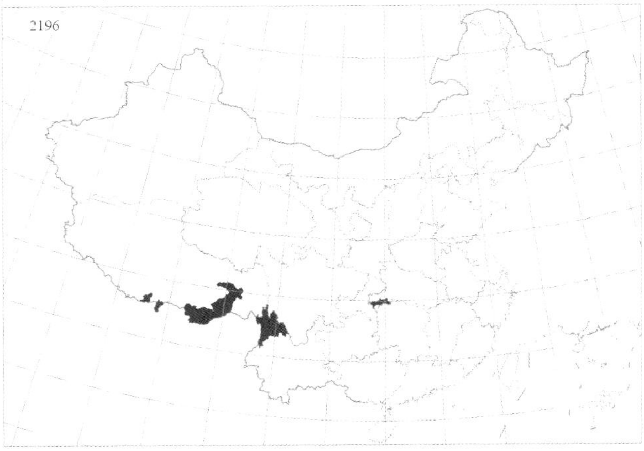

Lauraceae 樟科 *Lindera* 山胡椒属
Lindera praecox (Siebold et Zucc.) Blume 大果山胡椒
Shrubs, deciduous

15.7 (13.4~18.2)；3.6 (1.5~6.5)；27.5 (25.2~29.7)；15.8 (13.4~18.2)；131.4 (108.3~158.8)；2.2 (0.0~7.3)；864 (755~985) | 1556 (1157~1922)；574 (425~666)；191 (122~259) | 80.4 (29.5~112.2)；858 (755~927)；240.8 (158.1~319.9)

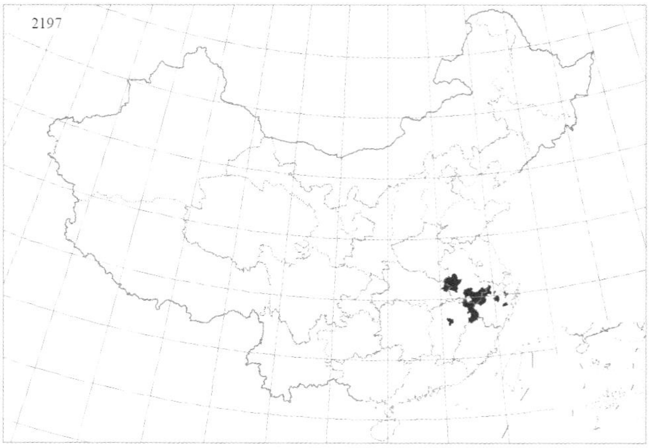

Lauraceae 樟科 *Lindera* 山胡椒属
Lindera prattii Gamble 峨眉钓樟
Trees, evergreen

14.7 (7.3~16.7)；4.2 (-2.0~6.2)；24.6 (15.0~27.7)；14.8 (7.5~16.7)；118.4 (48.2~140.6)；1.4 (0.0~20.4)；791 (537~900) | 1257 (885~1574)；545 (465~817)；83 (27~181) | 59.3 (26.5~99.4)；791 (537~900)；204.2 (146.0~352.3)

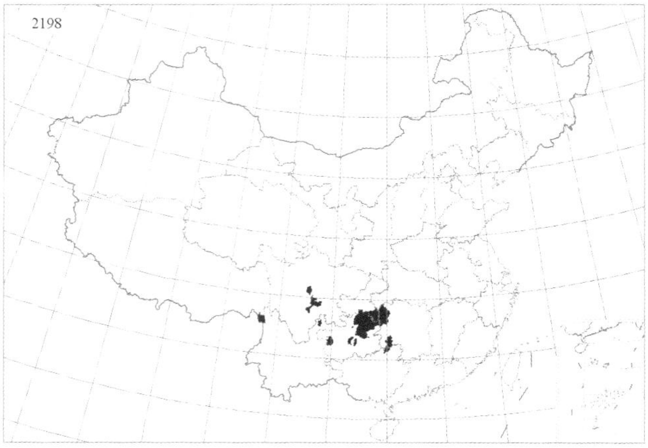

Lauraceae 樟科 *Lindera* 山胡椒属
Lindera pulcherrima C. K. Allen 西藏钓樟
Trees, evergreen

15.6 (-1.8~22.5)；4.9 (-10.2~15.0)；25.4 (6.1~29.4)；15.7 (1.8~22.6)；131.2 (1.9~210.9)；3.3 (0.0~82.8)；846 (329~1206) | 1388 (320~1829)；549 (217~1007)；129 (20~230) | 65.4 (-3.0~145.8)；841 (320~1206)；226.3 (72.2~447.4)

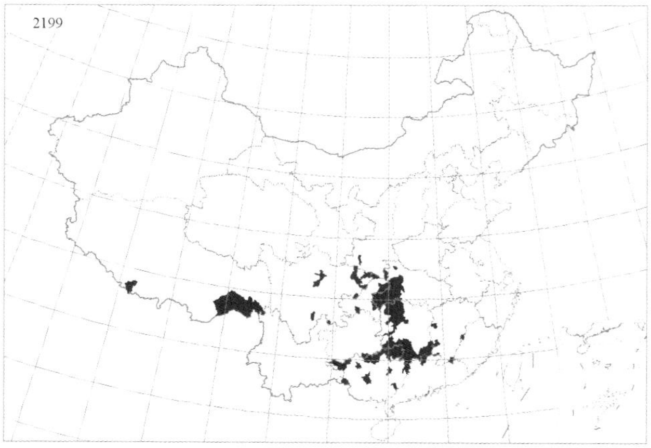

Lauraceae 樟科 *Lindera* 山胡椒属
Lindera pulcherrima var. attenuata C. K. Allen 香粉叶
Trees, evergreen

16.5 (10.6~21.6)；5.3 (-0.2~13.2)；27.0 (18.8~29.9)；16.5 (10.6~21.6)；139.2 (79.4~199.2)；0.7 (0.0~12.1)；884 (643~1139) | 1423 (916~2005)；512 (392~858)；154 (31~225) | 61.5 (30.9~128.9)；872 (643~1139)；203.1 (134.6~318.5)

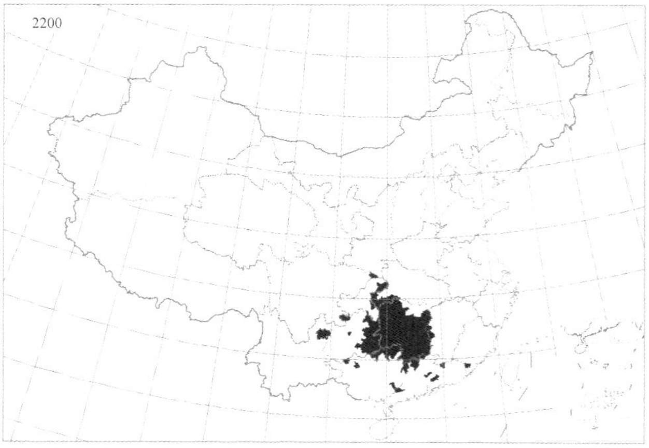

Lauraceae 樟科 *Lindera* 山胡椒属
Lindera pulcherrima var. hemsleyana (Diels) H. B. Cui 川钓樟
Trees, evergreen

14.7 (5.3~19.2)；3.9 (-4.1~12.0)；24.6 (12.2~29.2)；14.7 (5.7~19.2)；120.2 (30.5~170.8)；3.8 (0.0~29.9)；799 (479~1002) | 1215 (644~1526)；527 (317~848)；84 (14~181) | 52.8 (-8.1~106.2)；798 (479~1002)；225.6 (135.3~432.9)

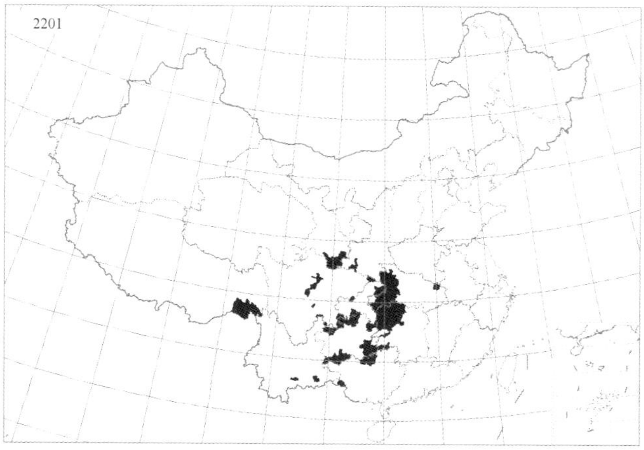

Lauraceae 樟科 *Lindera* 山胡椒属
Lindera reflexa Hemsl. 山橿
Shrubs or trees, deciduous

16.8 (9.3~22.8)；5.5 (-3.8~15.5)；27.3 (15.7~30.1)；16.9 (9.5~22.8)；144.3 (60.8~214.1)；2.1 (0.0~22.5)；910 (600~1225) | 1423 (512~2035)；538 (258~969)；152 (18~259) | 56.1 (-35.9~130.7)；890 (512~1225)；214.4 (121.6~516.7)

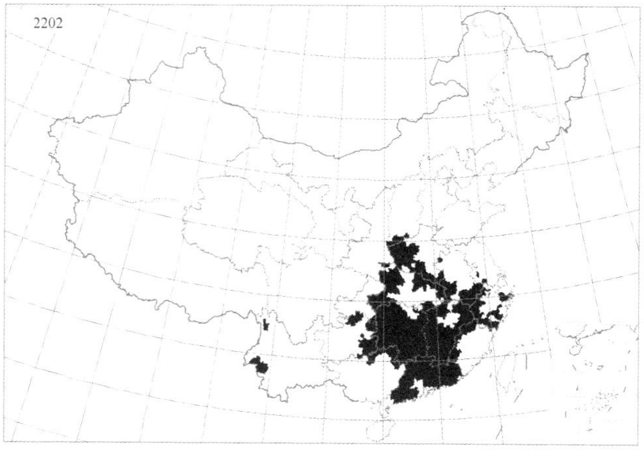

Lauraceae 樟科 *Lindera* 山胡椒属
Lindera robusta (C. K. Allen) H. B. Cui 海南山胡椒
Trees, evergreen

24.0 (23.0~24.7); 18.7 (17.7~19.8); 27.7 (26.7~28.4); 24.1 (23.1~24.7); 228.7 (217.0~236.5); 0.0 (0.0~0.0); 1312 (1217~1376) ▎1553 (1179~1725); 539 (392~619); 65 (42~110) ▎18.9 (-11.9~41.8); 1181 (1056~1326); 499.2 (352.9~620.0)

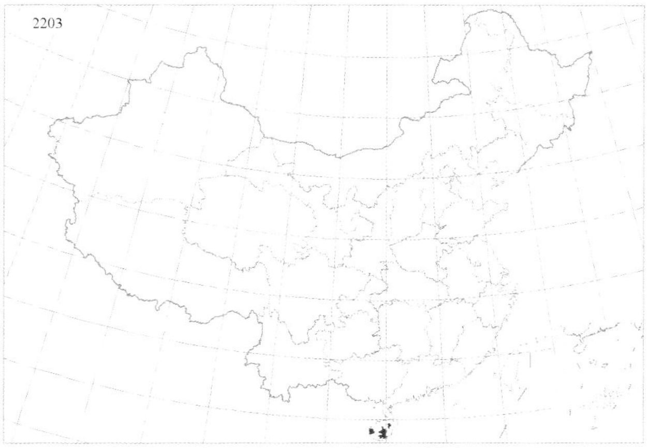

Lauraceae 樟科 *Lindera* 山胡椒属
Lindera rubronervia Gamble 红脉钓樟
Shrubs or trees, deciduous

15.4 (10.2~18.9); 2.9 (-2.9~9.2); 27.3 (22.4~29.6); 15.5 (10.6~18.9); 129.5 (83.1~166.7); 4.3 (0.0~20.5); 859 (656~1006) ▎1263 (512~1922); 490 (258~685); 139 (21~259) ▎45.7 (-35.3~123.6); 837 (512~980); 205.0 (121.6~319.9)

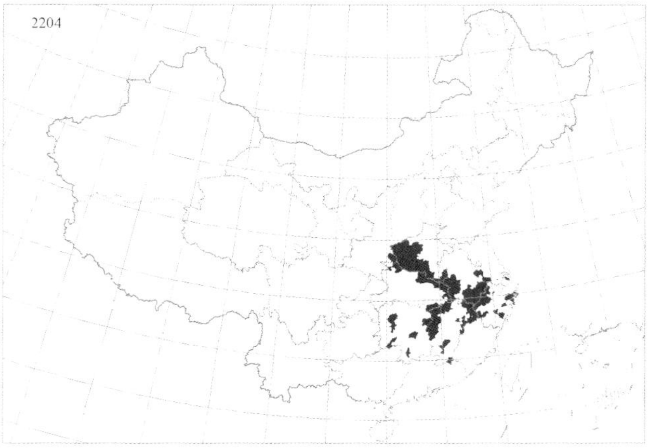

Lauraceae 樟科 *Lindera* 山胡椒属
Lindera setchuenensis Gamble 四川山胡椒
Shrubs, evergreen

14.6 (12.6~17.5); 4.4 (1.6~7.6); 24.1 (21.6~26.6); 14.7 (12.7~17.6); 117.9 (98.8~150.7); 1.7 (0.0~6.9); 783 (711~912) ▎1264 (999~1577); 614 (486~932); 66 (32~105) ▎62.5 (22.7~107.2); 783 (711~912); 197.8 (130.1~285.1)

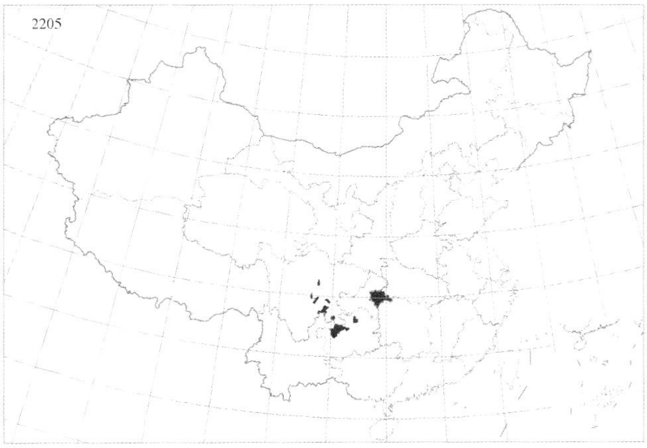

Lauraceae 樟科 *Lindera* 山胡椒属
Lindera supracostata Lecomte 菱叶钓樟
Shrubs or trees, evergreen

14.3 (2.0~19.3); 5.3 (-7.2~13.1); 22.1 (9.9~29.2); 14.3 (3.8~19.4); 114.5 (15.4~173.2); 2.8 (0.0~51.4); 765 (413~939) ▎1110 (741~1494); 549 (411~727); 61 (15~158) ▎46.9 (-14.2~128.6); 750 (413~937); 252.1 (142.9~407.1)

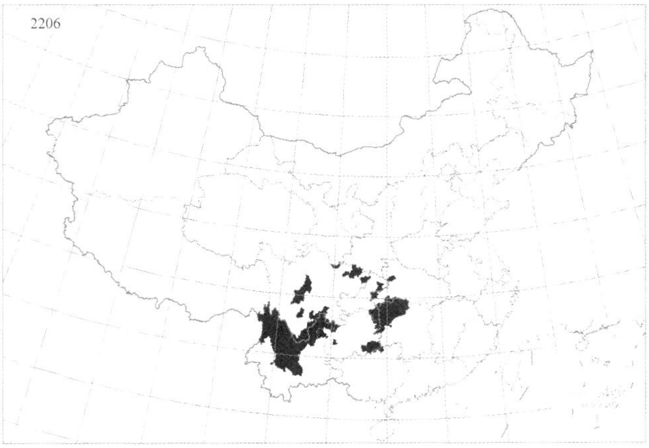

Lauraceae 樟科 *Lindera* 山胡椒属
Lindera thomsonii C. K. Allen 三股筋香
Trees, evergreen

16.5 (5.3~21.0); 9.4 (-2.4~14.2); 21.7 (12.2~26.6); 16.6 (5.7~21.0); 139.6 (30.5~191.8); 1.0 (0.0~26.2); 827 (479~1053) ▎1236 (939~1561); 696 (524~911); 51 (24~70) ▎51.6 (6.5~100.7); 817 (479~1053); 377.1 (189.5~551.6)

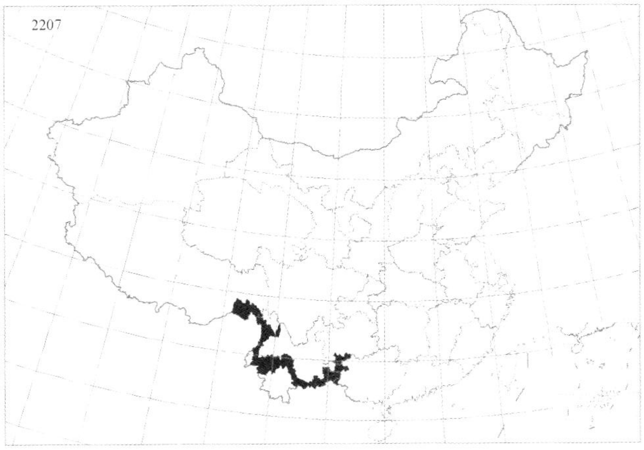

Lauraceae 樟科 *Lindera* 山胡椒属
Lindera thomsonii var. velutina (Forrest) L. C. Wang 长尾钓樟
Trees, evergreen

16.6 (13.1~19.6); 10.1 (6.4~13.1); 21.2 (18.4~23.9); 16.7 (13.2~19.6); 140.1 (98.1~175.7); 0.0 (0.0~0.0); 813 (692~953) ▎1170 (797~1688); 666 (411~959); 46 (23~70) ▎44.3 (-14.2~112.2); 783 (683~932); 375.0 (218.0~627.0)

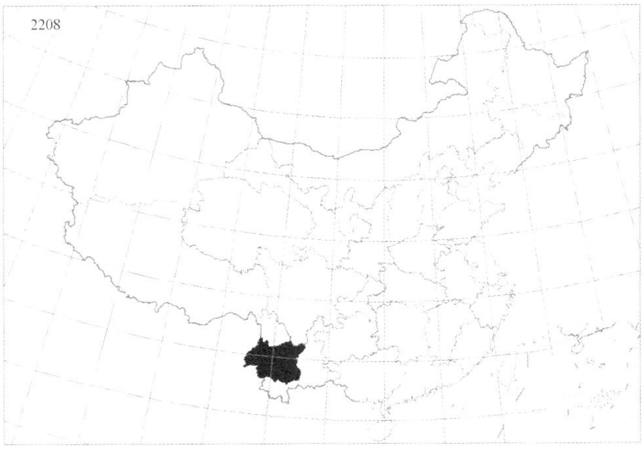

Lauraceae 樟科 *Lindera* 山胡椒属

Lindera tienchuanensis Fang et Kung 天全钓樟

Shrubs or trees, evergreen

9.0; 0.1; 16.9; 9.4; 64.5; 15.4; 592 ❙ 1189; 647; 34 ❙ 99.0; 592; 289.3

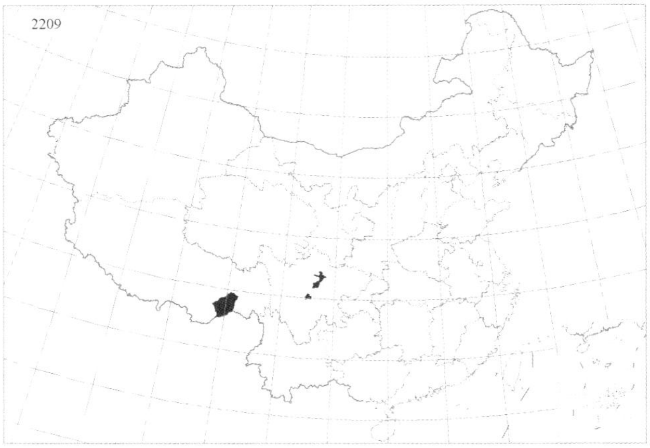

Lauraceae 樟科 *Lindera* 山胡椒属

Lindera tonkinensis Lecomte 假桂钓樟

Trees, evergreen

19.2 (14.4~24.7); 12.6 (7.5~19.8); 23.7 (19.1~28.6); 19.2 (14.4~24.7); 170.7 (113.1~236.5); 0.0 (0.0~0.0); 954 (727~1376) ❙ 1406 (939~2435); 749 (524~1428); 63 (37~130) ❙ 48.1 (6.5~99.9); 932 (717~1299); 405.1 (235.1~659.0)

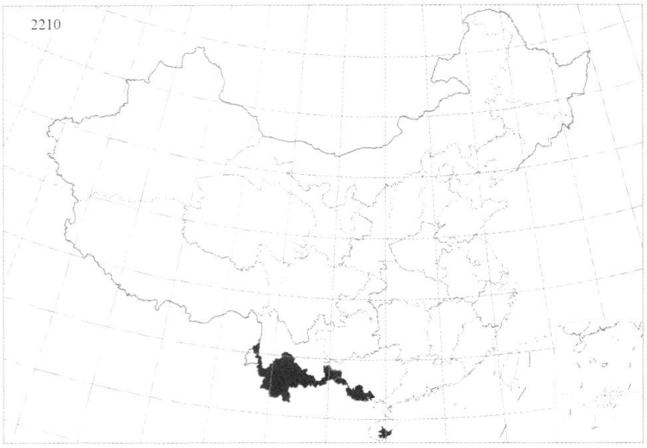

Lauraceae 樟科 *Lindera* 山胡椒属

Lindera tonkinensis var. subsessilis H. W. Li 无梗钓樟

Trees, evergreen

16.3 (5.4~21.0); 9.8 (-2.1~15.1); 20.8 (12.3~25.6); 16.3 (5.8~21.0); 136.1 (31.5~191.8); 0.6 (0.0~25.6); 806 (481~1053) ❙ 1166 (741~1751); 651 (411~996); 47 (23~77) ❙ 45.3 (-14.2~99.9); 780 (481~1053); 377.2 (218.0~659.0)

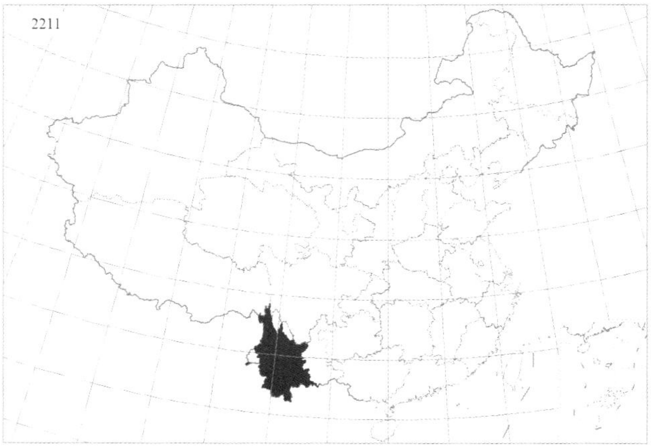

Lauraceae 樟科 *Lindera* 山胡椒属

Lindera villipes H. B. Cui 毛柄钓樟

Trees, evergreen

14.7 (8.6~17.8); 7.7 (1.3~11.1); 19.7 (15.0~22.0); 14.7 (8.7~17.8); 117.4 (52.6~154.1); 0.8 (0.0~8.8); 746 (574~855) ❙ 1316 (1041~1736); 737 (585~1007); 57 (45~70) ❙ 77.8 (34.7~145.8); 745 (574~855); 415.8 (335.1~539.7)

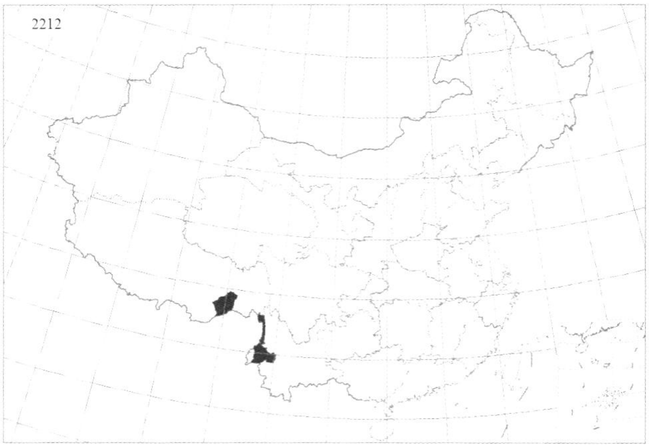

Lauraceae 樟科 *Litsea* 木姜子属

Litsea acutivena Hayata 尖叶木姜子

Trees, evergreen

19.9 (14.3~24.6); 11.3 (4.0~19.5); 27.2 (20.1~29.8); 20.0 (14.3~24.7); 179.8 (112.6~235.9); 0.0 (0.0~1.1); 1043 (756~1372) ❙ 1703 (1211~2992); 719 (425~1847); 154 (48~254) ❙ 65.7 (6.7~246.5); 1035 (756~1299); 258.4 (163.7~620.0)

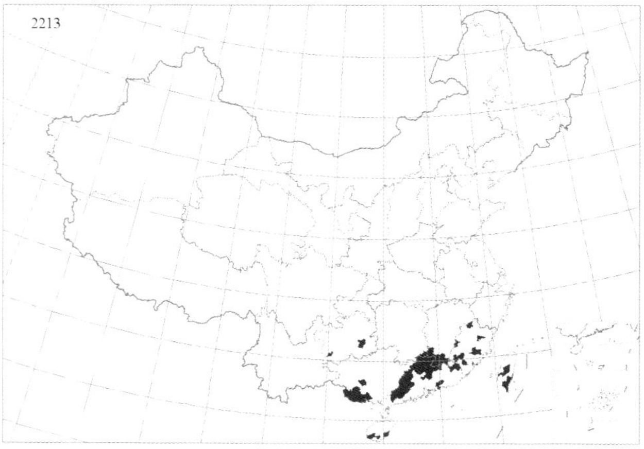

Lauraceae 樟科 *Litsea* 木姜子属

Litsea akoensis Hayata 屏东木姜子

Trees, evergreen

19.4 (16.2~22.2); 13.7 (11.0~17.7); 24.0 (20.1~27.0); 19.4 (16.2~22.2); 172.7 (134.2~206.7); 0.0 (0.0~0.0); 955 (773~1123) ❙ 2598 (1836~2993); 1201 (924~1847); 254 (107~547) ❙ 178.3 (64.2~246.5); 949 (773~1123); 257.9 (202.8~296.9)

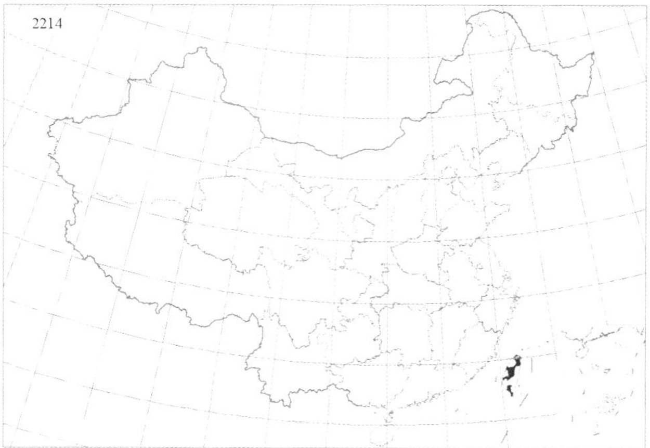

Lauraceae 樟科 *Litsea* 木姜子属
Litsea akoensis var. sasakii (Kamikoti) J. C. Liao 浸水营木姜子
Trees, evergreen

18.5; 14.1; 21.7; 18.5; 162.1; 0.0; 863 ❙ 2713; 1388; 188 ❙ 214.4; 863; 363.7

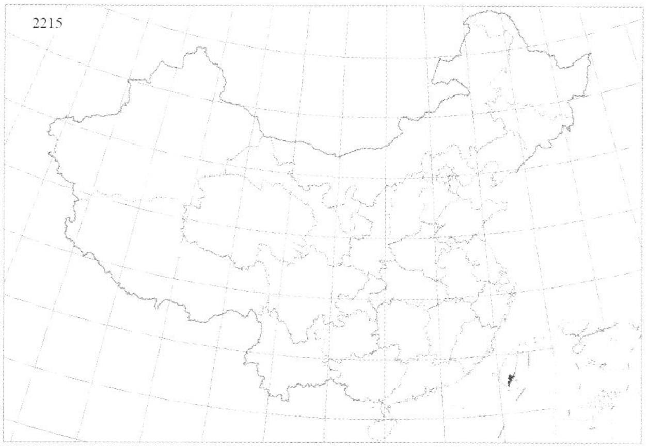

Lauraceae 樟科 *Litsea* 木姜子属
Litsea albescens (Hook. f.) D. G. Long 白叶木姜子
Shrubs or small trees, evergreen

13.3; 5.4; 19.6; 13.4; 100.4; 0.0; 706 ❙ 1736; 1007; 51 ❙ 145.8; 706; 447.4

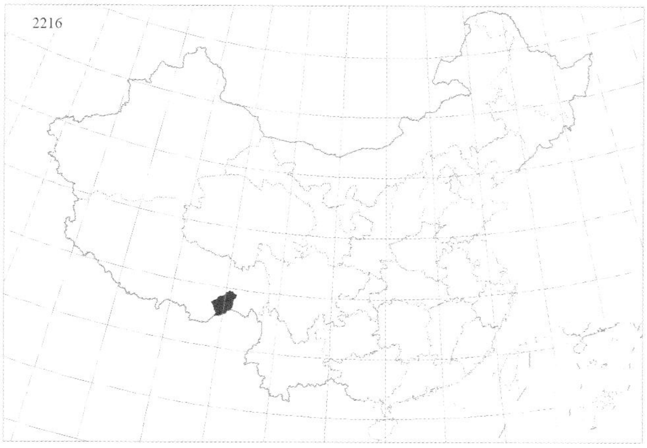

Lauraceae 樟科 *Litsea* 木姜子属
Litsea auriculata S. S. Chien et W. C. Cheng 天目木姜子
Trees, deciduous

15.6 (13.2~18.5); 3.4 (-0.2~7.8); 27.2 (24.7~29.1); 15.6 (13.2~18.5); 130.5 (108.3~162.3); 2.9 (0.0~11.7); 860 (755~984) ❙ 1404 (734~1950); 538 (376~713); 159 (41~259) ❙ 63.1 (-13.8~147.0); 852 (734~975); 222.0 (140.4~284.0)

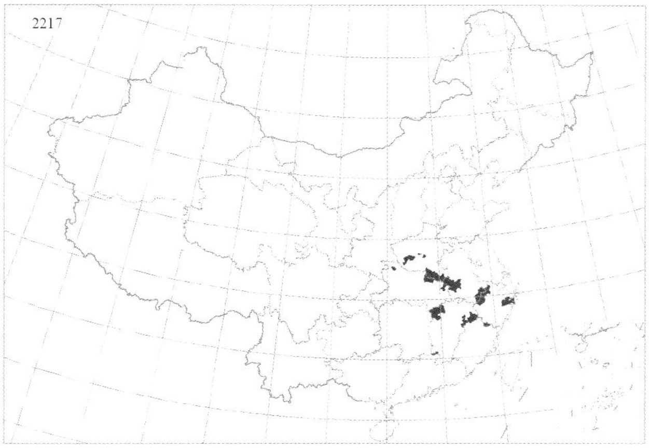

Lauraceae 樟科 *Litsea* 木姜子属
Litsea balansae Lecomte 假辣子
Shrubs or trees, evergreen

17.4 (13.3~21.0); 10.2 (5.4~14.2); 22.5 (19.6~25.6); 17.4 (13.4~21.0); 149.0 (100.4~191.8); 0.0 (0.0~0.0); 852 (706~1053) ❙ 1400 (1178~1736); 788 (646~1007); 55 (48~68) ❙ 66.3 (44.5~145.8); 852 (706~1053); 426.3 (336.3~551.6)

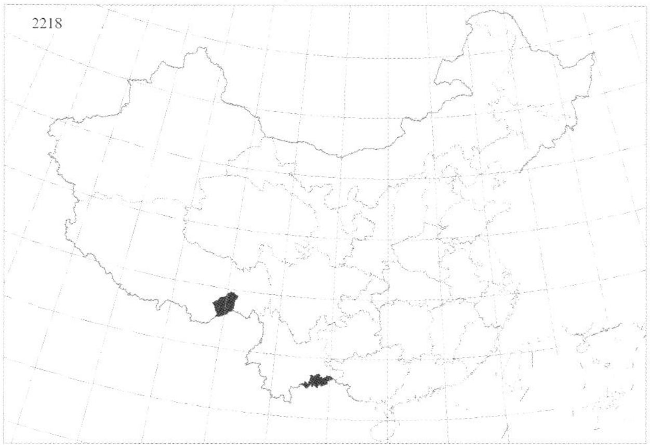

Lauraceae 樟科 *Litsea* 木姜子属
Litsea baviensis Lecomte 大萼木姜子
Trees, evergreen

20.4 (17.2~24.8); 12.7 (7.4~20.1); 26.3 (22.4~28.7); 20.5 (17.3~24.8); 185.7 (147.0~237.6); 0.0 (0.0~0.0); 1060 (833~1385) ❙ 1495 (1161~1777); 678 (444~996); 104 (41~194) ❙ 43.9 (-12.0~88.8); 1025 (833~1326); 365.0 (185.5~659.0)

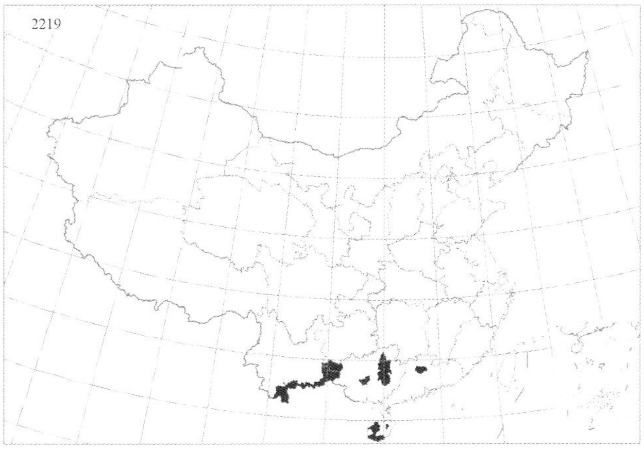

Lauraceae 樟科 *Litsea* 木姜子属
Litsea beilschmiediifolia H. W. Li 琼楠叶木姜子
Trees, evergreen

18.8; 12.0; 23.5; 18.8; 165.5; 0.0; 914 ❙ 1455; 818; 60 ❙ 60.2; 914; 490.7

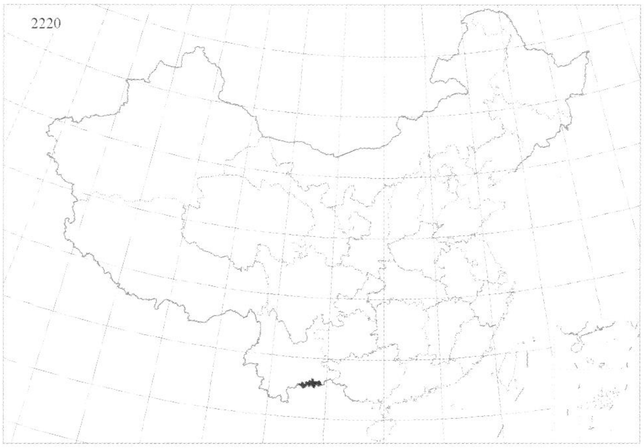

Lauraceae 樟科 *Litsea* 木姜子属
Litsea biflora H. B. Cui 少花木姜子
Trees, evergreen

13.3; 5.4; 19.6; 13.4; 100.4; 0.0; 706 ❙ 1736; 1007; 51 ❙ 145.8; 706; 447.4

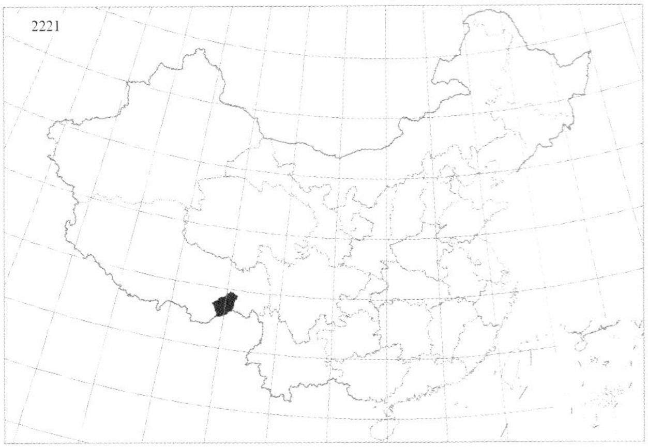

Lauraceae 樟科 *Litsea* 木姜子属
Litsea cangyuanensis J. Li et H. W. Li 沧源木姜子
Trees, evergreen

17.9; 11.9; 21.6; 18.0; 155.8; 0.0; 854 ❙ 1435; 803; 42 ❙ 68.0; 833; 505.1

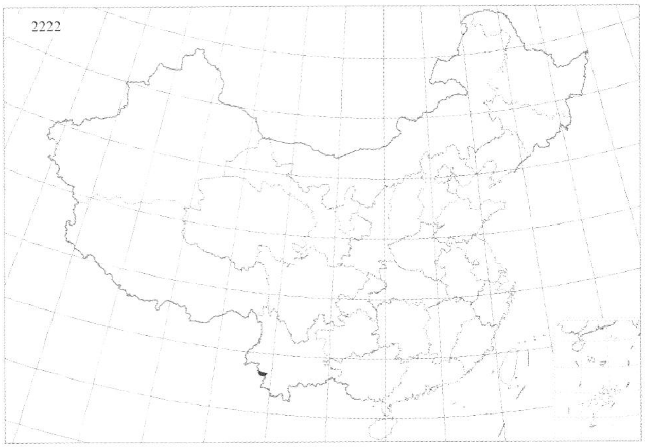

Lauraceae 樟科 *Litsea* 木姜子属
Litsea chengshuzhii H. B. Cui 树志木姜子
Trees, evergreen

13.3; 5.4; 19.6; 13.4; 100.4; 0.0; 706 ❙ 1736; 1007; 51 ❙ 145.8; 706; 447.4

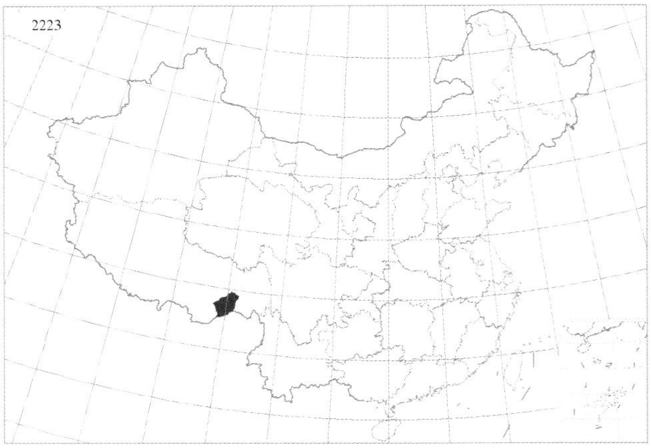

Lauraceae 樟科 *Litsea* 木姜子属
Litsea chinpingensis Yang et P. H. Huang 金平木姜子
Trees, evergreen

17.4 (-1.8~21.0); 11.1 (-10.2~15.1); 21.6 (6.1~25.6); 17.5 (1.8~21.0); 152.2 (1.9~191.8); 3.3 (0.0~82.8); 857 (329~1053) ❙ 1319 (320~1751); 727 (217~996); 54 (20~77) ❙ 53.3 (-3.0~112.2); 838 (320~1053); 434.3 (72.2~659.0)

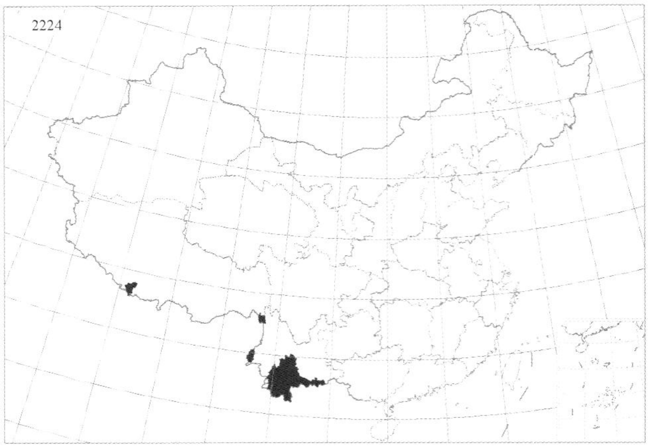

Lauraceae 樟科 *Litsea* 木姜子属
Litsea chunii W. C. Cheng 高山木姜子
Shrubs, deciduous

7.6 (0.9~16.4); -1.4 (-9.5~8.6); 15.5 (9.2~26.7); 8.3 (3.5~16.4); 55.7 (13.4~137.2); 24.3 (0.0~63.1); 562 (402~869) ❙ 957 (579~1583); 530 (346~938); 27 (9~88) ❙ 72.8 (8.2~128.6); 560 (402~869); 272.8 (165.6~407.1)

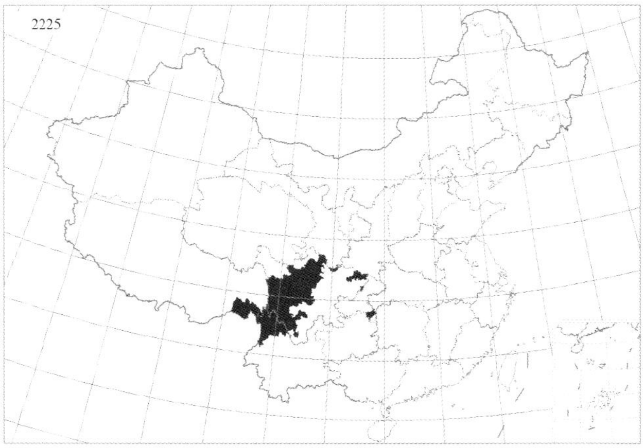

Lauraceae 樟科 *Litsea* 木姜子属
Litsea chunii var. likiangensis Yang et P. H. Huang 丽江木姜子
Shrubs, deciduous

7.7 (5.4~10.8); 0.3 (-2.1~3.9); 14.1 (12.3~16.6); 7.9 (5.8~10.9); 47.8 (31.5~71.8); 14.8 (1.3~25.6); 545 (481~633) ❙ 876 (741~1007); 529 (426~594); 25 (12~40) ❙ 60.8 (52.8~67.9); 545 (481~633); 307.9 (273.5~351.7)

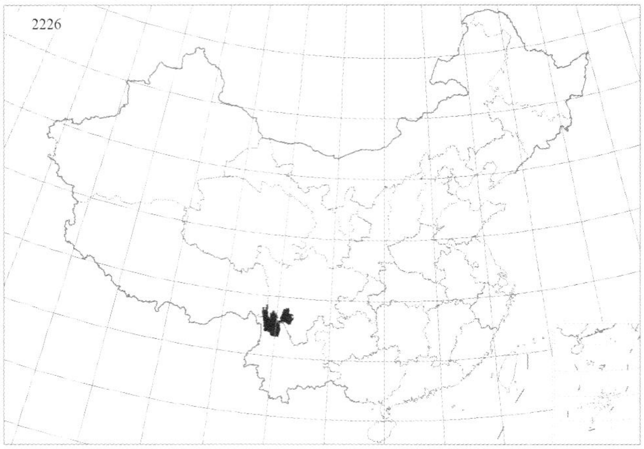

Lauraceae 樟科 *Litsea* 木姜子属

Litsea coelestis H. B. Cui 蓝叶木姜子

Trees, evergreen

13.3; 5.4; 19.6; 13.4; 100.4; 0.0; 706 ❙ 1736; 1007; 51 ❙ 145.8; 706; 447.4

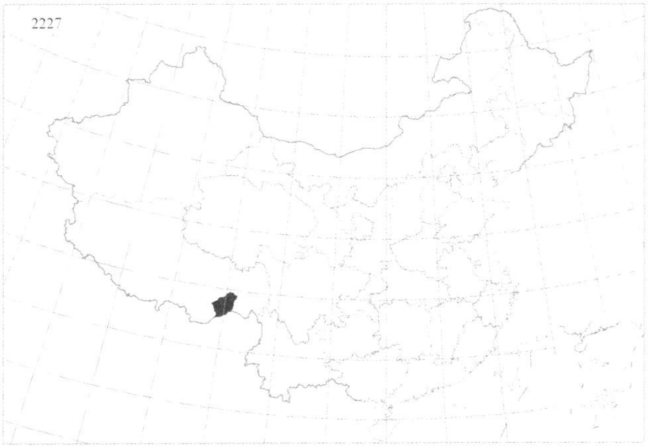

Lauraceae 樟科 *Litsea* 木姜子属

Litsea coreana H. Lév. 朝鲜木姜子

Trees, evergreen

16.4 (10.6~23.0); 5.4 (-0.2~15.3); 26.8 (19.4~29.9); 16.5 (10.6~23.0); 138.7 (79.4~216.2); 1.1 (0.0~12.1); 881 (643~1247) ❙ 1435 (828~2859); 555 (403~1045); 149 (32~402) ❙ 63.7 (-6.0~235.6); 873 (643~1247); 217.2 (108.2~365.3)

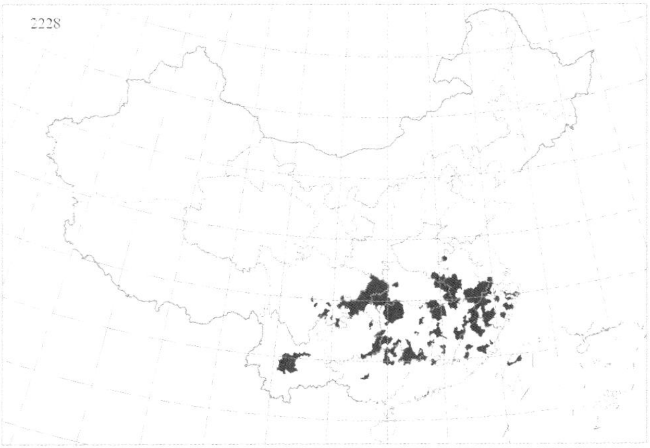

Lauraceae 樟科 *Litsea* 木姜子属

Litsea coreana var. lanuginosa (Migo) Yang et P. H. Huang 毛豹皮樟

Trees, evergreen

16.7 (12.3~19.4); 5.1 (1.1~9.4); 27.7 (20.8~30.1); 16.8 (12.3~19.4); 142.2 (96.2~173.4); 0.7 (0.0~8.1); 903 (709~1037) ❙ 1461 (707~1997); 519 (299~816); 170 (15~259) ❙ 62.0 (-6.7~130.7); 884 (707~1029); 204.6 (126.1~319.9)

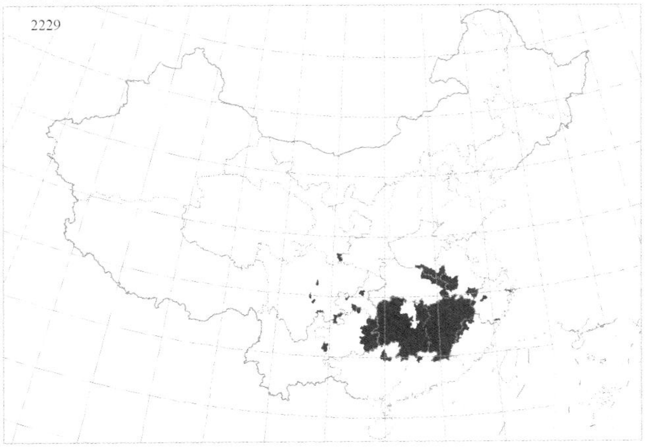

Lauraceae 樟科 *Litsea* 木姜子属

Litsea coreana var. sinensis (C. K. Allen) Yang et P. H. Huang 豹皮樟

Trees, evergreen

16.0 (10.2~20.0); 4.1 (-2.9~10.6); 27.3 (22.3~29.7); 16.1 (10.6~20.0); 135.8 (83.1~180.0); 2.8 (0.0~20.5); 878 (656~1040) ❙ 1410 (626~1997); 520 (316~778); 161 (29~259) ❙ 59.6 (-22.2~147.0); 862 (626~1040); 217.3 (139.1~319.9)

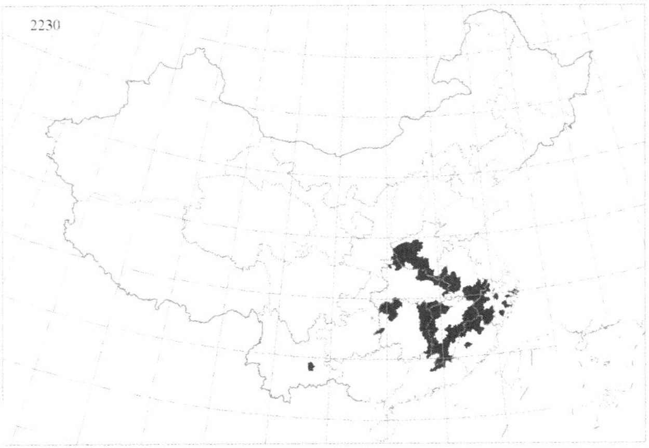

Lauraceae 樟科 *Litsea* 木姜子属

Litsea cubeba Yang et P. C. Huang 山鸡椒

Shrubs or trees, deciduous

17.4 (-1.5~24.8); 7.9 (-10.5~20.1); 25.9 (7.1~30.1); 17.5 (2.2~24.8); 151.2 (4.7~237.6); 1.2 (0.0~82.2); 924 (332~1385) ❙ 1421 (311~3495); 594 (227~1428); 129 (11~897) ❙ 55.1 (-34.3~246.5); 911 (311~1360); 247.2 (62.8~659.0)

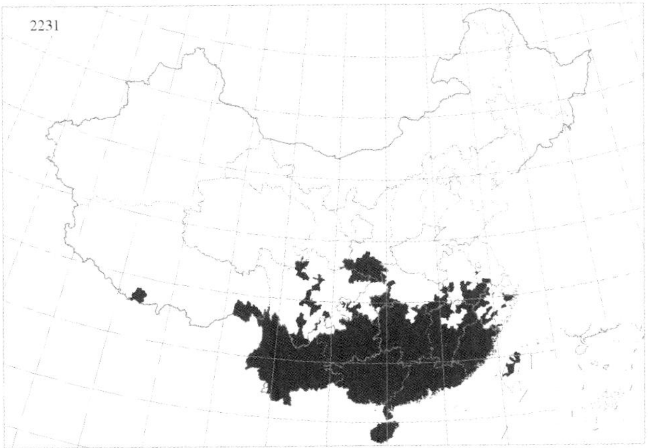

Lauraceae 樟科 *Litsea* 木姜子属

Litsea cubeba var. formosana (Nakai) Yang et P. H. Huang 毛山鸡椒

Shrubs or trees, deciduous

18.0 (14.4~21.4); 8.4 (2.1~13.6); 27.1 (24.3~29.0); 18.1 (14.5~21.5); 157.0 (118.7~197.7); 0.2 (0.0~4.9); 938 (789~1128) ❙ 1622 (1070~1997); 570 (356~713); 196 (130~257) ❙ 75.5 (-5.0~147.0); 931 (789~1094); 252.6 (123.9~332.5)

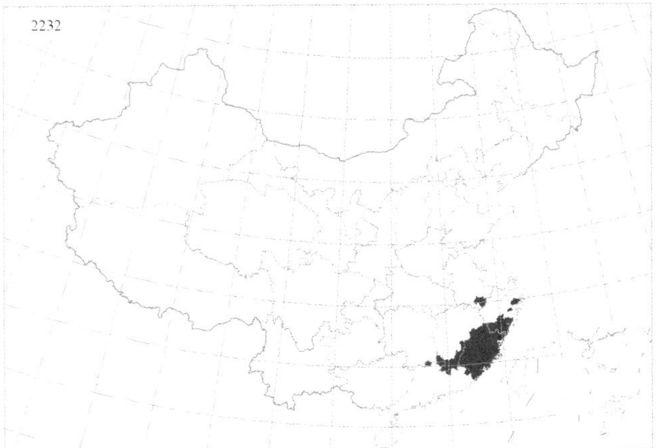

Lauraceae 樟科 *Litsea* 木姜子属
Litsea depressa H. P. Tsui 扁果木姜子
Trees, evergreen

13.3；5.4；19.6；13.4；100.4；0.0；706 ❙ 1736；1007；51 ❙ 145.8；706；447.4

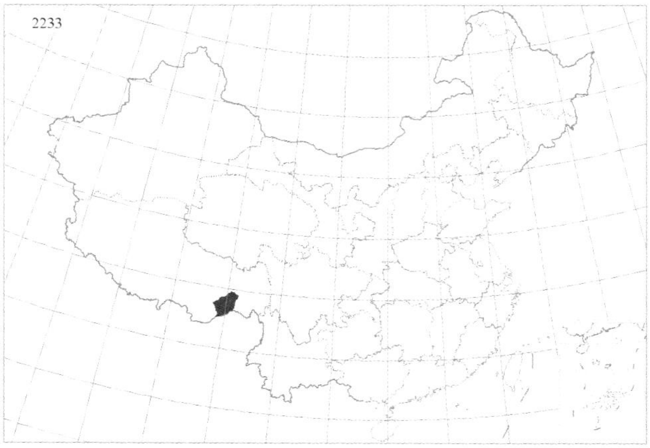

Lauraceae 樟科 *Litsea* 木姜子属
Litsea dilleniifolia P. Y. Pai et P. H. Huang 五桠果叶木姜子
Trees, evergreen

19.0 (16.8~22.2)；12.3 (9.9~15.1)；23.5 (20.8~28.4)；19.0 (16.9~22.3)；167.9 (142.6~207.5)；0.0 (0.0~0.0)；932 (811~1190) ❙ 1426 (1163~1751)；789 (641~996)；63 (42~116) ❙ 54.6 (15.4~88.8)；928 (802~1190)；447.2 (273.1~659.0)

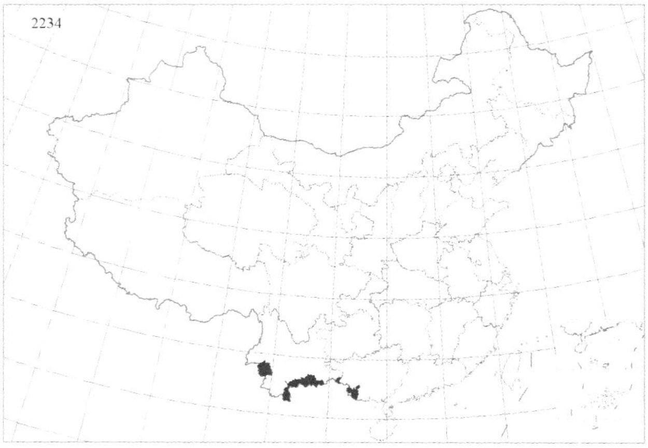

Lauraceae 樟科 *Litsea* 木姜子属
Litsea elongata (Nees) Benth. et Hook. f. 黄丹木姜子
Trees, evergreen

16.8 (5.3~24.3)；6.8 (-2.9~19.1)；25.8 (12.2~30.1)；16.9 (5.7~24.3)；144.0 (30.5~231.9)；1.3 (0.0~26.2)；892 (479~1338) ❙ 1354 (512~2435)；574 (258~1428)；119 (19~259) ❙ 52.0 (-35.3~147.0)；877 (479~1247)；241.9 (108.2~659.0)

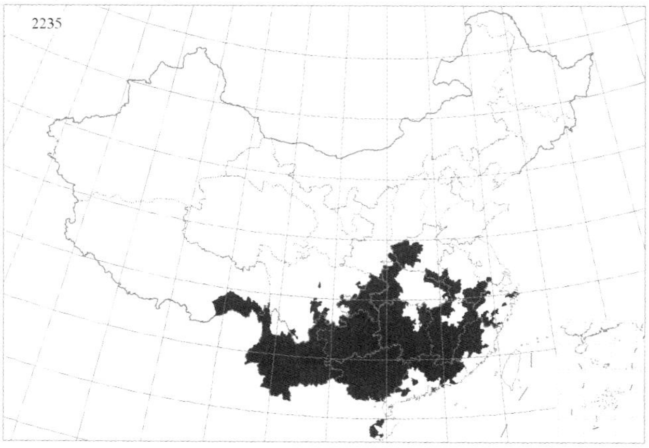

Lauraceae 樟科 *Litsea* 木姜子属
Litsea elongata var. faberi (Hemsl.) Yang et P. H. Huang 石木姜子
Trees, evergreen

15.9 (9.5~19.1)；5.0 (1.4~10.0)；25.9 (16.7~29.8)；15.9 (9.6~19.1)；131.5 (62.7~169.1)；0.6 (0.0~8.0)；847 (600~986) ❙ 1352 (874~1997)；527 (392~817)；133 (25~254) ❙ 59.6 (18.3~147.0)；838 (600~977)；202.9 (134.6~303.4)

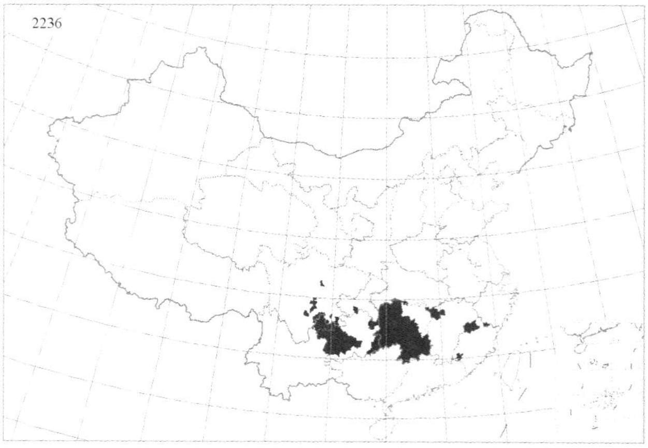

Lauraceae 樟科 *Litsea* 木姜子属
Litsea elongata var. subverticillata (Y. C. Yang) Yang et P. H. Huang 近轮叶木姜子
Trees, evergreen

16.6 (14.2~21.0)；6.2 (3.3~14.2)；26.1 (21.4~29.4)；16.7 (14.2~21.0)；140.5 (112.5~191.8)；0.4 (0.0~2.3)；873 (765~1108) ❙ 1307 (1074~1547)；540 (392~911)；108 (43~209) ❙ 50.3 (9.3~99.4)；867 (765~1108)；226.4 (112.1~551.6)

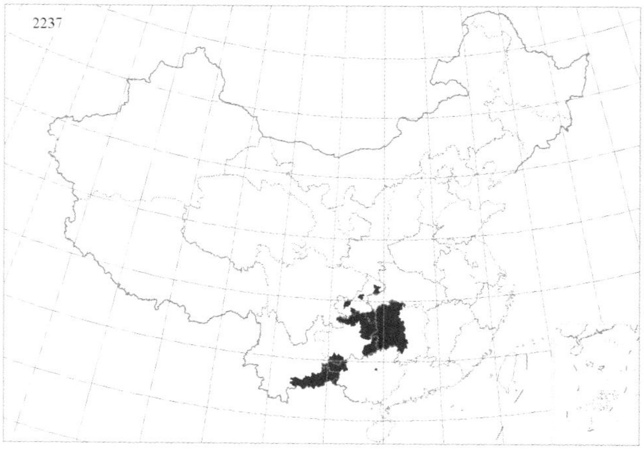

Lauraceae 樟科 *Litsea* 木姜子属
Litsea foveola Kosterm. 蜂窝木姜子
Shrubs or trees, evergreen

21.8；13.2；28.2；21.9；202.3；0.0；1162 ❙ 1483；758；121 ❙ 27.8；1162；300.2

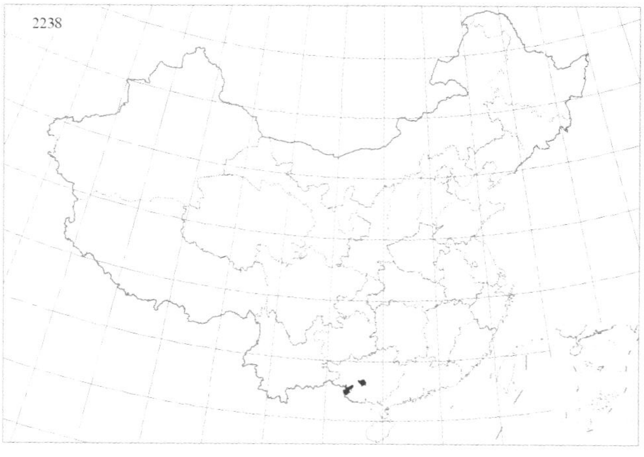

Lauraceae 樟科 *Litsea* 木姜子属

Litsea garciae Vidal 兰屿木姜子

Trees, evergreen

18.5; 14.1; 21.7; 18.5; 162.1; 0.0; 863 ❙ 2713; 1388; 188 ❙ 214.4; 863; 363.7

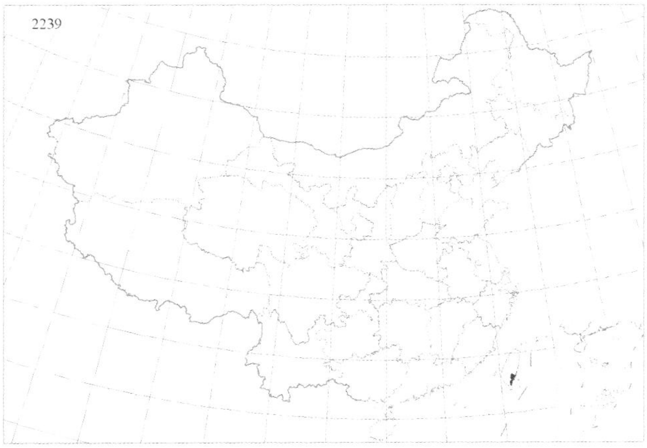

Lauraceae 樟科 *Litsea* 木姜子属

Litsea glutinosa (Lour.) C. B. Rob. 潺槁木姜子

Trees, evergreen or deciduous

20.1 (8.6~24.8); 12.0 (1.3~20.1); 26.6 (15.0~29.8); 20.2 (8.7~24.8); 182.3 (52.6~237.6); 0.0 (0.0~8.8); 1050 (574~1385) ❙ 1564 (868~2435); 693 (312~1428); 125 (37~259) ❙ 51.1 (-34.3~130.7); 1036 (574~1360); 289.1 (95.3~659.0)

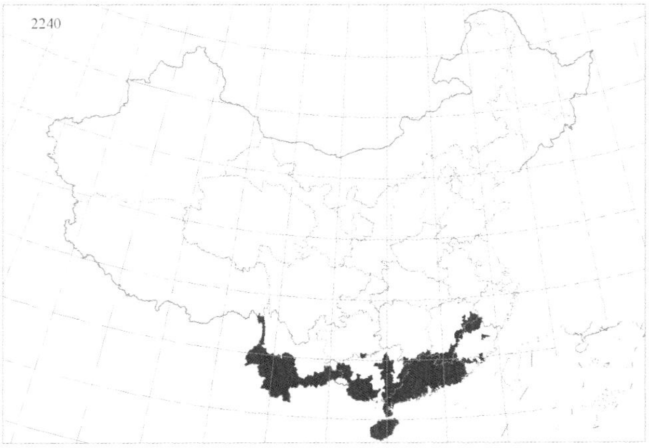

Lauraceae 樟科 *Litsea* 木姜子属

Litsea glutinosa var. brideliifolia (Hayata) Merr. 白野槁树

Trees, evergreen or deciduous

22.4 (21.0~23.5); 14.5 (13.0~16.4); 28.6 (26.9~29.0); 22.4 (21.0~23.5); 208.9 (192.3~222.3); 0.0 (0.0~0.0); 1197 (1081~1276) ❙ 1613 (1187~1776); 746 (545~828); 115 (79~151) ❙ 35.8 (-6.9~64.3); 1182 (1081~1257); 191.4 (120.3~254.1)

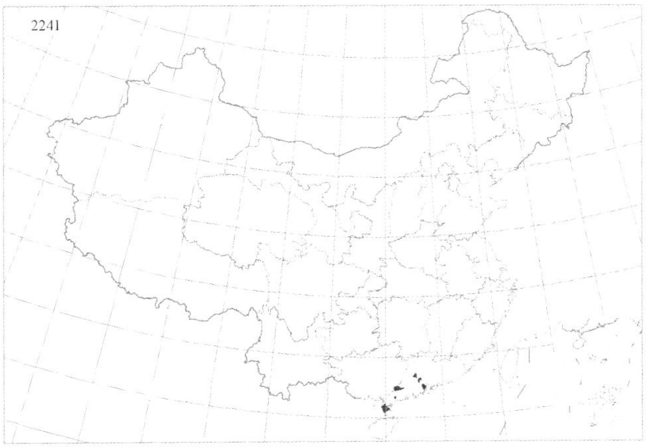

Lauraceae 樟科 *Litsea* 木姜子属

Litsea gongshanensis H. W. Li 贡山木姜子

Shrubs or trees, evergreen

11.0; 3.4; 17.3; 11.0; 76.5; 4.4; 640 ❙ 1389; 796; 48 ❙ 113.7; 640; 399.9

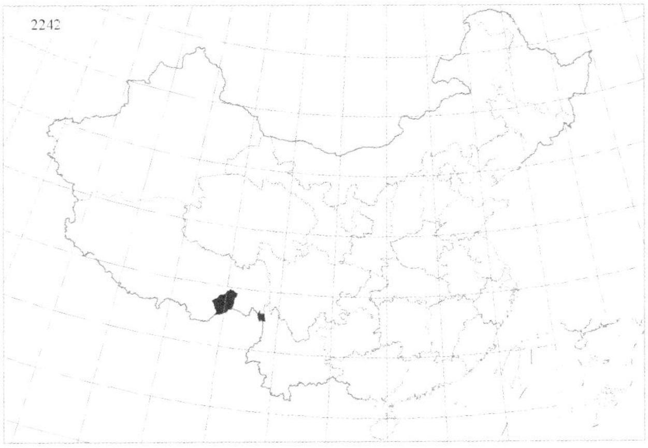

Lauraceae 樟科 *Litsea* 木姜子属

Litsea greenmaniana C. K. Allen 华南木姜子

Trees, evergreen

19.0 (8.6~22.2); 9.8 (1.3~15.1); 26.7 (15.0~29.8); 19.0 (8.7~22.2); 168.2 (52.6~206.9); 0.1 (0.0~8.8); 982 (574~1177) ❙ 1587 (1041~1997); 648 (421~996); 154 (45~254) ❙ 62.6 (22.7~130.7); 979 (574~1177); 275.5 (163.7~659.0)

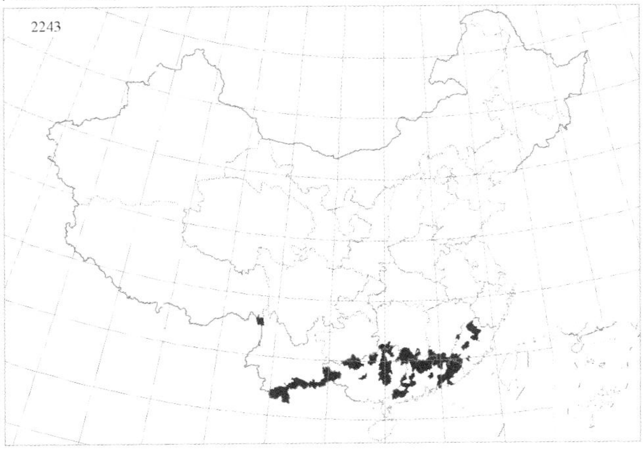

Lauraceae 樟科 *Litsea* 木姜子属

Litsea honghoensis H. Liu 红河木姜子

Trees, evergreen

17.9 (16.3~21.0); 11.0 (9.0~14.2); 22.7 (21.4~25.6); 17.9 (16.4~21.0); 155.0 (136.4~191.8); 0.0 (0.0~0.0); 867 (797~1053) ❙ 1243 (958~1547); 698 (531~911); 51 (44~68) ❙ 43.3 (13.3~80.6); 862 (797~1053); 386.0 (277.0~551.6)

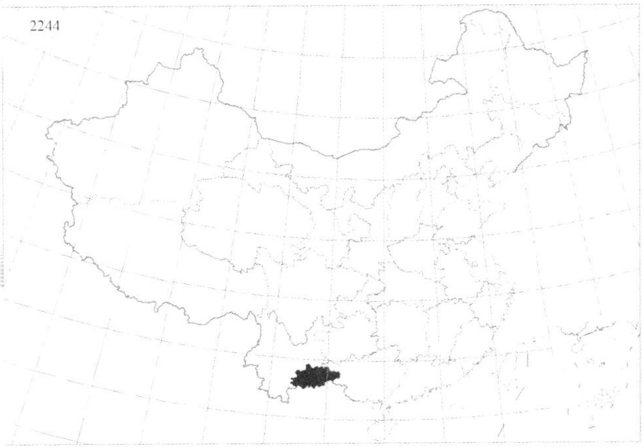

Lauraceae 樟科 *Litsea* 木姜子属

Litsea hunanensis Yang et P. H. Huang 湖南木姜子

Trees, evergreen

15.9 (14.3~17.3); 5.1 (3.3~6.9); 25.5 (24.3~26.7); 15.9 (14.3~17.3); 131.4 (114.1~148.1); 0.6 (0.0~2.3); 833 (765~896) ❙ 1495 (1227~1624); 544 (517~594); 167 (85~191) ❙ 80.0 (48.8~101.8); 833 (765~896); 238.5 (203.1~258.0)

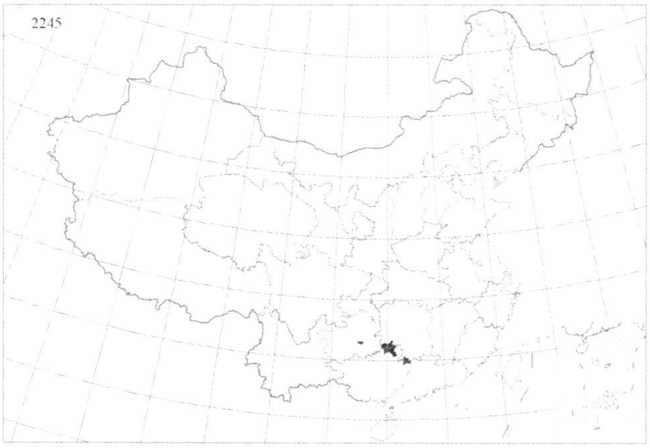

Lauraceae 樟科 *Litsea* 木姜子属

Litsea hupehana Hemsl. 湖北木姜子

Trees, evergreen

16.8 (9.3~18.7); 6.2 (-1.7~8.1); 26.7 (20.1~28.8); 16.9 (9.5~18.7); 142.9 (69.8~164.3); 0.4 (0.0~18.0); 894 (603~994) ❙ 1113 (922~1492); 533 (421~826); 54 (24~105) ❙ 25.5 (5.6~107.2); 893 (603~992); 152.7 (112.1~279.5)

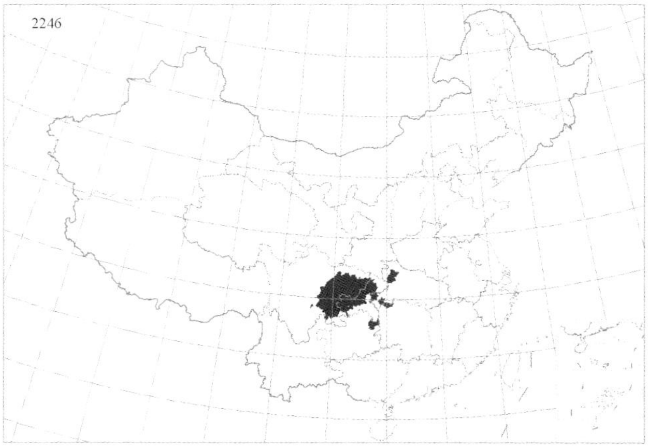

Lauraceae 樟科 *Litsea* 木姜子属

Litsea hypophaea Hayata 黄肉树

Trees

19.2; 14.1; 23.3; 19.2; 170.5; 0.0; 936 ❙ 2845; 1336; 272 ❙ 207.7; 936; 302.8

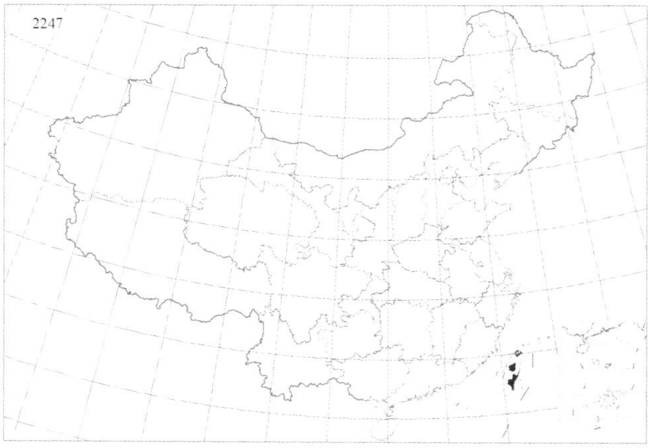

Lauraceae 樟科 *Litsea* 木姜子属

Litsea ichangensis Gamble 宜昌木姜子

Shrubs or trees, deciduous

14.1 (9.3~16.9); 3.2 (-1.7~6.3); 24.6 (19.8~28.4); 14.2 (9.5~16.9); 113.4 (69.8~143.4); 3.4 (0.0~18.0); 773 (603~924) ❙ 1233 (711~1526); 514 (357~613); 82 (15~181) ❙ 60.4 (12.1~107.2); 773 (603~924); 220.8 (161.9~285.1)

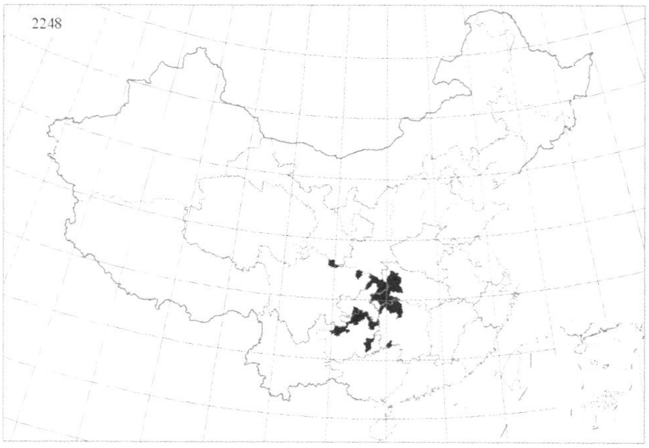

Lauraceae 樟科 *Litsea* 木姜子属

Litsea kingii Hook. f. 秃净木姜子

Shrubs or small trees, deciduous

9.8; 2.4; 16.1; 10.0; 67.0; 8.4; 605 ❙ 1151; 637; 49 ❙ 90.0; 605; 341.4

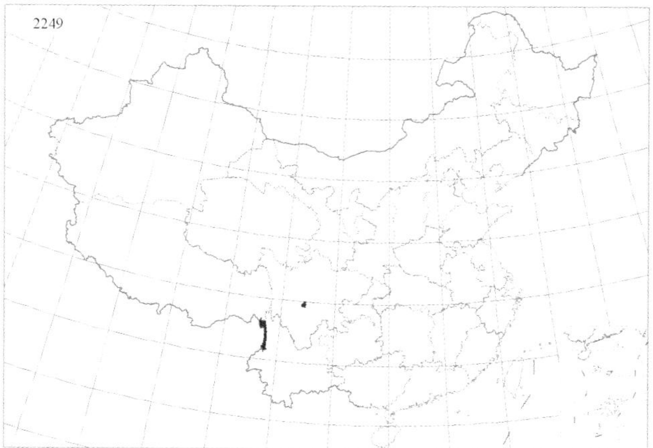

Lauraceae 樟科 *Litsea* 木姜子属

Litsea kobuskiana C. K. Allen 安顺木姜子

Small trees, evergreen

18.3 (16.8~19.8); 8.4 (6.5~11.1); 26.6 (25.7~27.5); 18.4 (16.9~19.8); 160.6 (142.7~177.9); 0.0 (0.0~0.0); 949 (869~1010) ❙ 1453 (1182~1740); 658 (579~709); 124 (43~187) ❙ 53.7 (16.9~85.5); 949 (869~1010); 251.5 (208.7~354.5)

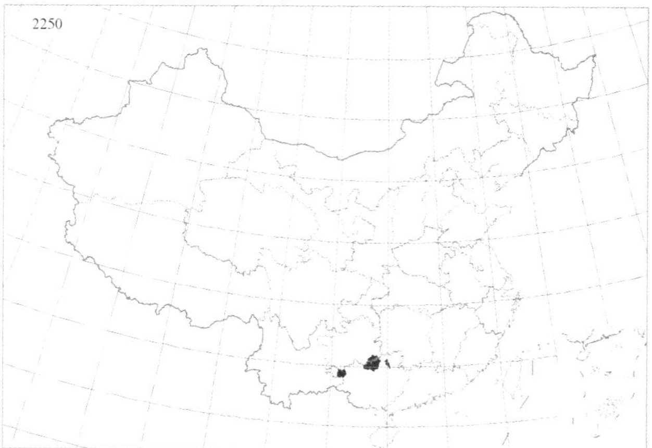

Lauraceae 樟科 *Litsea* 木姜子属
Litsea kwangsiensis Yang et P. H. Huang 红楠刨
Trees, evergreen

21.2 (19.1~22.7); 13.0 (11.0~15.0); 27.4 (25.0~28.6); 21.3 (19.1~22.8); 195.4 (169.6~213.1); 0.0 (0.0~0.0); 1113 (945~1221) ❙ 1739 (1302~2435); 934 (670~1428); 122 (57~194) ❙ 55.8 (12.9~99.5); 1113 (945~1221); 283.4 (213.5~371.7)

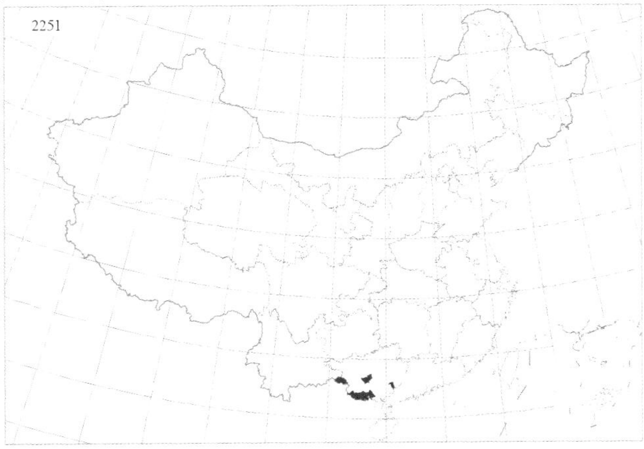

Lauraceae 樟科 *Litsea* 木姜子属
Litsea kwangtungensis H. T. Chang 广东木姜子
Shrubs, evergreen

21.3 (20.7~21.8); 13.0 (12.0~14.2); 27.8 (27.3~28.5); 21.3 (20.7~21.9); 195.7 (189.0~202.6); 0.0 (0.0~0.0); 1117 (1083~1158) ❙ 1879 (1621~2005); 855 (720~1056); 152 (116~181) ❙ 68.4 (45.4~85.1); 1117 (1083~1158); 256.0 (213.4~297.2)

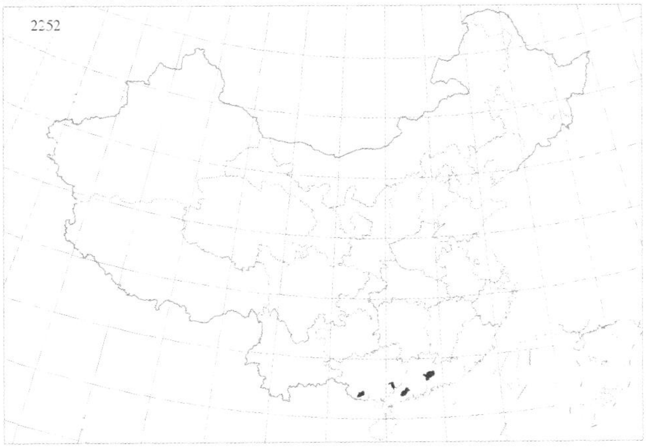

Lauraceae 樟科 *Litsea* 木姜子属
Litsea lancifolia Roxb. ex Nees 剑叶木姜子
Shrubs, evergreen

20.7 (14.4~24.8); 14.0 (7.5~20.1); 25.4 (19.1~28.8); 20.7 (14.4~24.8); 188.5 (113.1~237.6); 0.0 (0.0~0.0); 1067 (727~1385) ❙ 1451 (868~2435); 705 (312~1428); 72 (37~130) ❙ 38.9 (-34.3~99.9); 1024 (727~1360); 396.0 (163.7~659.0)

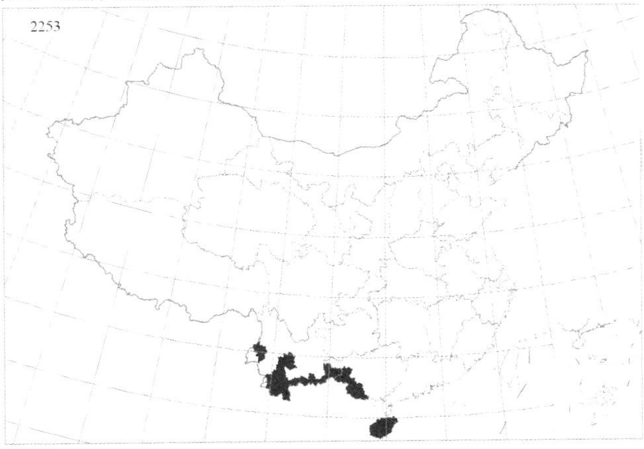

Lauraceae 樟科 *Litsea* 木姜子属
Litsea lancifolia var. ellipsoidea Yang et P. H. Huang 椭圆果木姜子
Shrubs, evergreen

18.9 (17.7~21.0); 12.5 (10.9~14.2); 23.2 (21.8~25.6); 18.9 (17.7~21.0); 167.3 (152.7~191.8); 0.0 (0.0~0.0); 916 (849~1053) ❙ 1459 (1276~1616); 834 (741~946); 58 (51~68) ❙ 60.3 (36.8~90.4); 914 (849~1053); 523.6 (452.0~589.2)

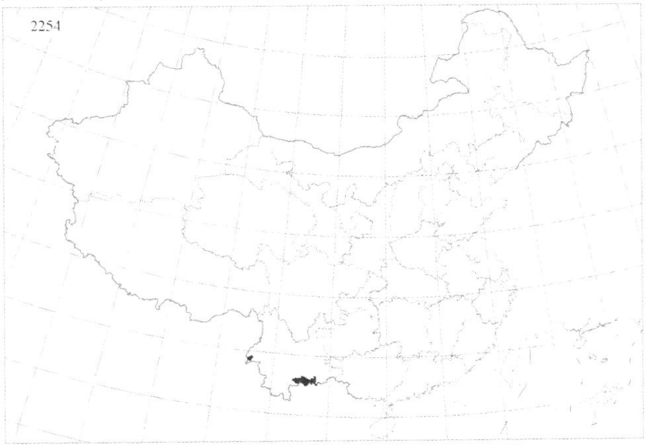

Lauraceae 樟科 *Litsea* 木姜子属
Litsea lancifolia var. pedicellata Hook. f. 有梗木姜子
Shrubs, evergreen

19.3 (17.3~21.0); 13.3 (10.9~15.1); 23.2 (21.6~25.6); 19.3 (17.3~21.0); 172.2 (148.0~191.8); 0.0 (0.0~0.0); 935 (832~1053) ❙ 1420 (1049~1607); 722 (598~835); 61 (44~77) ❙ 51.7 (26.1~64.9); 919 (783~1053); 522.5 (365.3~659.0)

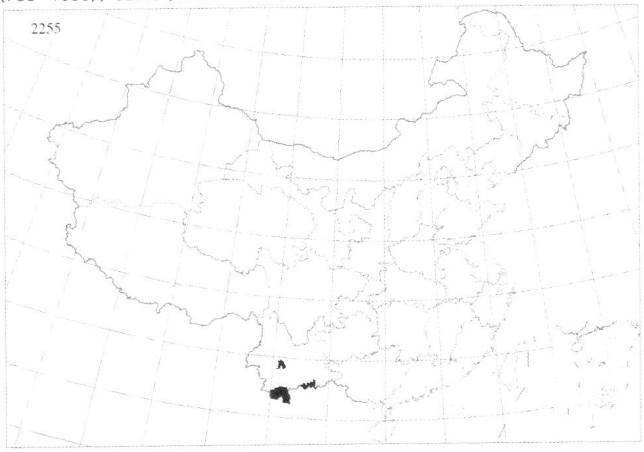

Lauraceae 樟科 *Litsea* 木姜子属
Litsea lancilimba Merr. 大果木姜子
Trees, evergreen

19.3 (14.3~24.7); 9.7 (3.4~19.8); 27.6 (22.5~29.8); 19.3 (14.3~24.7); 171.8 (113.2~236.5); 0.1 (0.0~2.1); 1021 (761~1376) ❙ 1586 (1117~2435); 629 (392~1428); 158 (42~230) ❙ 57.6 (-11.9~128.9); 1004 (761~1326); 260.1 (151.0~620.0)

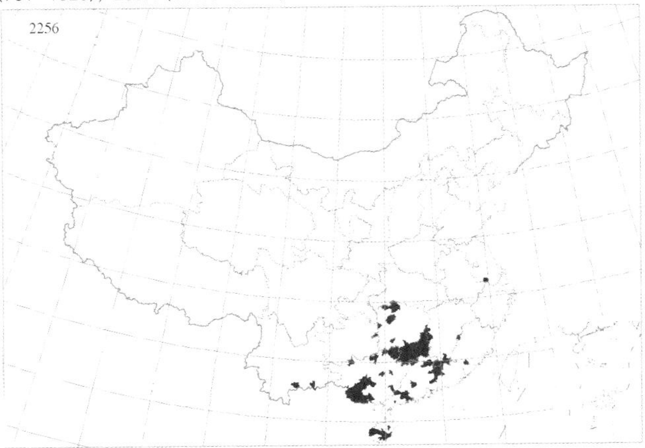

Lauraceae 樟科 *Litsea* 木姜子属
Litsea liboshengii H. B. Cui 勃生木姜子
Trees, evergreen

13.3; 5.4; 19.6; 13.4; 100.4; 0.0; 706 ❙ 1736; 1007; 51 ❙ 145.8; 706; 447.4

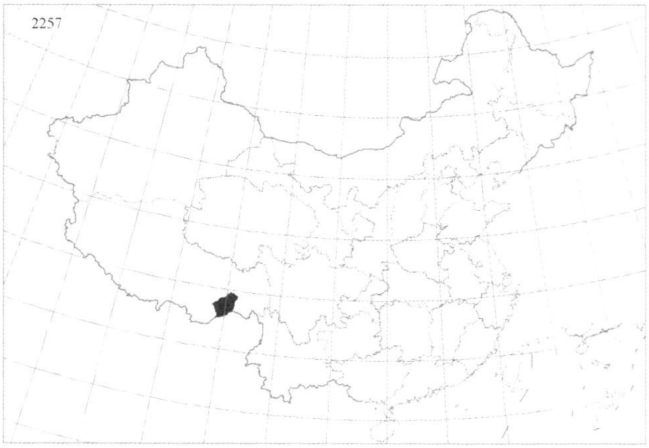

Lauraceae 樟科 *Litsea* 木姜子属
Litsea litseifolia (C. K. Allen) Yang et P. H. Huang 海南木姜子
Shrubs or small trees, evergreen

24.3; 19.3; 27.7; 24.3; 232.0; 0.0; 1337 ❙ 1659; 553; 68 ❙ 23.9; 1223; 584.0

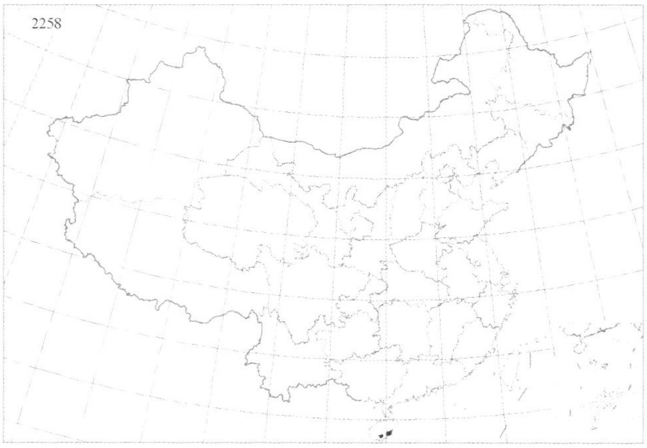

Lauraceae 樟科 *Litsea* 木姜子属
Litsea liyuyingi H. Liu 圆锥木姜子
Trees, evergreen

19.2 (16.4~20.5); 13.2 (9.6~15.1); 22.7 (20.6~23.7); 19.2 (16.4~20.5); 170.7 (137.3~185.9); 0.0 (0.0~0.0); 925 (795~992) ❙ 1563 (1464~1688); 782 (655~959); 61 (42~77) ❙ 70.6 (47.6~112.2); 908 (795~986); 570.6 (482.8~659.0)

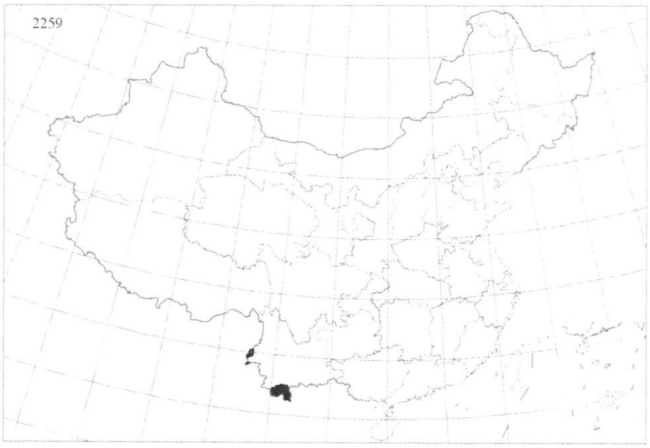

Lauraceae 樟科 *Litsea* 木姜子属
Litsea longistaminata (H. Liu) Kosterm. 长蕊木姜子
Trees, evergreen

17.9 (13.3~21.0); 11.5 (5.4~15.1); 21.9 (19.1~25.6); 17.9 (13.4~21.0); 154.8 (100.4~191.8); 0.0 (0.0~0.0); 863 (706~1053) ❙ 1439 (1049~1751); 799 (598~1007); 55 (42~77) ❙ 67.7 (26.1~145.8); 852 (706~1053); 473.2 (346.6~659.0)

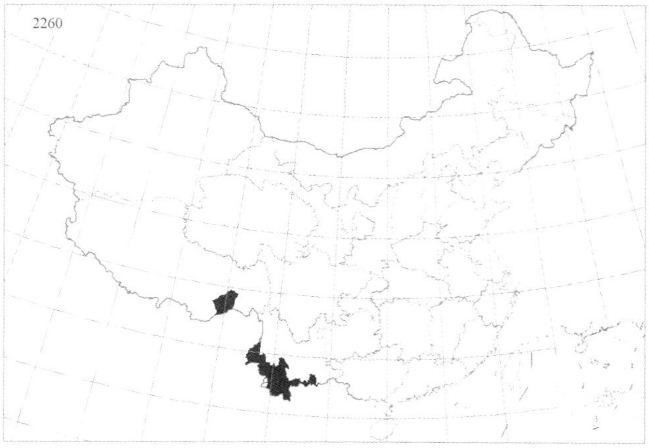

Lauraceae 樟科 *Litsea* 木姜子属
Litsea machiloides Yang et P. H. Huang 润楠叶木姜子
Trees, evergreen

21.1; 11.8; 28.9; 21.1; 193.2; 0.0; 1122 ❙ 1597; 630; 160 ❙ 42.5; 1098; 202.9

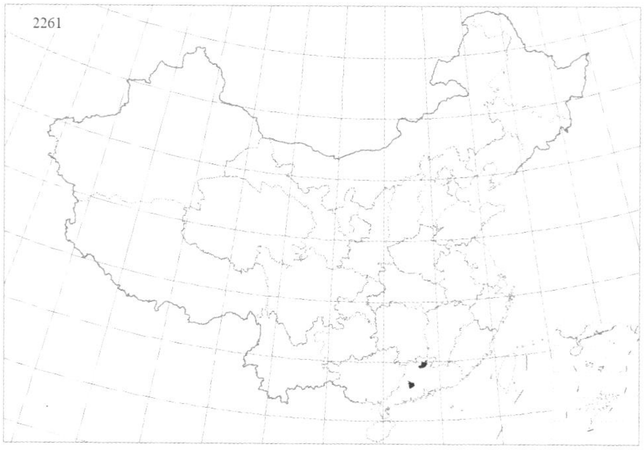

Lauraceae 樟科 *Litsea* 木姜子属
Litsea martabanica (Kurz) Hook. f. 滇南木姜子
Trees, evergreen

18.3 (16.3~20.5); 12.2 (9.6~15.1); 22.1 (20.3~23.7); 18.3 (16.4~20.5); 159.8 (136.6~185.9); 0.0 (0.0~0.0); 876 (789~992) ❙ 1434 (1049~1751); 787 (598~996); 55 (42~77) ❙ 63.8 (26.1~112.2); 860 (780~986); 473.4 (354.7~659.0)

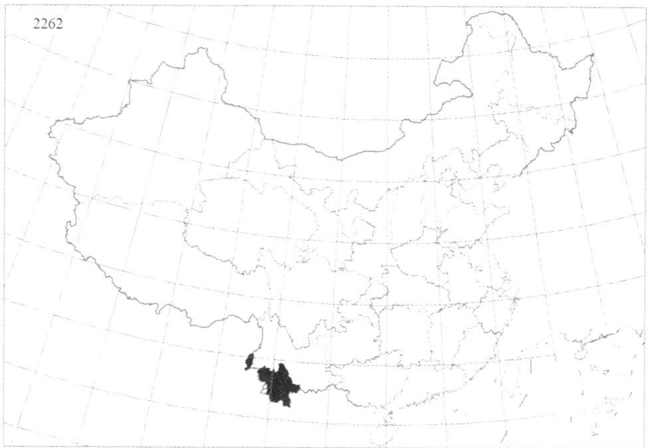

Lauraceae 樟科 *Litsea* 木姜子属

Litsea mishmiensis Hook. f. 察南木姜子

Trees

5.3; -2.4; 12.2; 5.7; 30.5; 26.2; 479 ❙ 961; 558; 31 ❙ 100.7; 479; 234.2

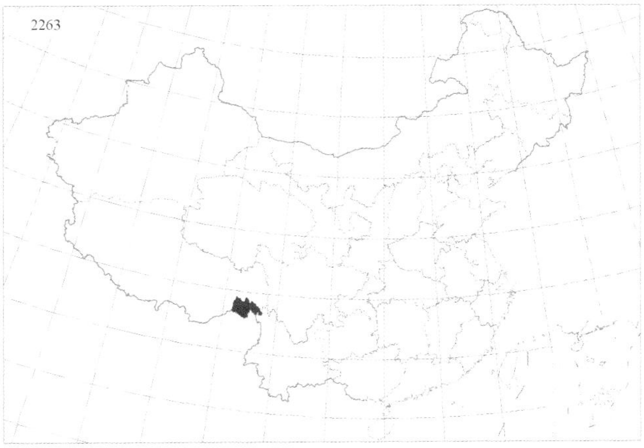

Lauraceae 樟科 *Litsea* 木姜子属

Litsea mollis Hemsl. 毛叶木姜子

Shrubs or small trees, deciduous

16.7 (6.6~22.7); 7.0 (-1.7~15.1); 25.2 (13.3~29.9); 16.8 (6.8~22.8); 141.6 (39.6~213.1); 0.6 (0.0~19.7); 872 (514~1221) ❙ 1338 (797~2435); 578 (376~1428); 112 (12~233) ❙ 53.5 (-14.2~145.8); 861 (514~1221); 252.7 (108.2~659.0)

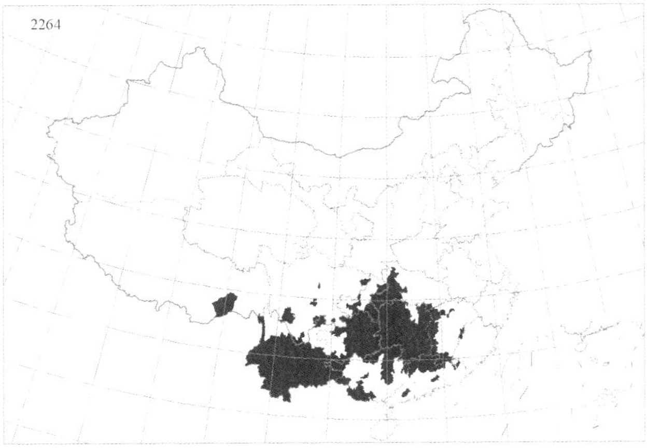

Lauraceae 樟科 *Litsea* 木姜子属

Litsea monopetala (Roxb.) Pers. 假柿木姜子

Trees, evergreen

20.4 (8.6~24.8); 12.5 (1.3~20.1); 26.6 (15.0~29.8); 20.5 (8.7~24.8); 185.6 (52.6~237.6); 0.0 (0.0~8.8); 1066 (574~1385) ❙ 1546 (868~2148); 709 (312~1234); 113 (41~197) ❙ 47.0 (-34.3~145.8); 1053 (574~1360); 288.8 (95.3~659.0)

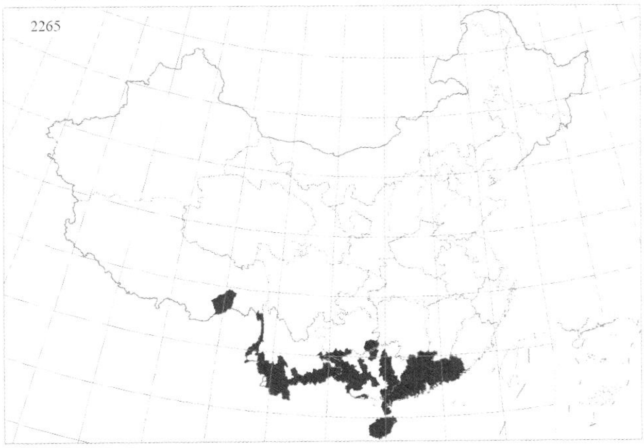

Lauraceae 樟科 *Litsea* 木姜子属

Litsea morrisonensis Hayata 玉山木姜子

Small trees, evergreen

17.8; 12.5; 22.1; 17.8; 153.9; 0.0; 857 ❙ 2621; 1126; 265 ❙ 210.3; 857; 266.6

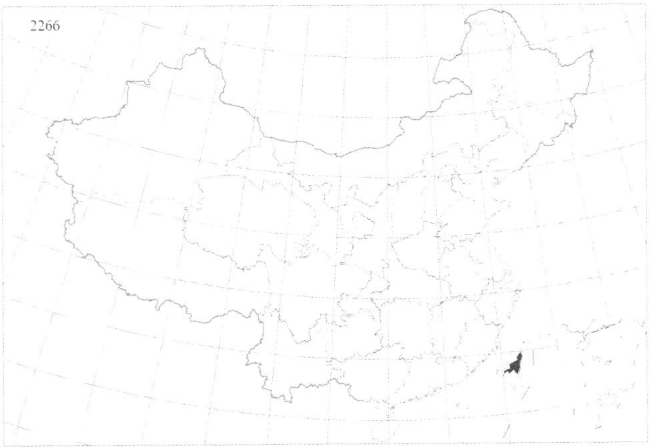

Lauraceae 樟科 *Litsea* 木姜子属

Litsea moupinensis Lecomte 宝兴木姜子

Trees, deciduous

14.5 (2.6~21.0); 6.6 (-7.0~15.1); 21.0 (11.2~27.0); 14.5 (4.3~21.0); 116.5 (20.0~191.8); 2.2 (0.0~48.2); 758 (427~1053) ❙ 1084 (741~1736); 600 (400~1007); 41 (15~77) ❙ 44.5 (-14.2~145.8); 738 (427~1053); 297.5 (142.9~659.0)

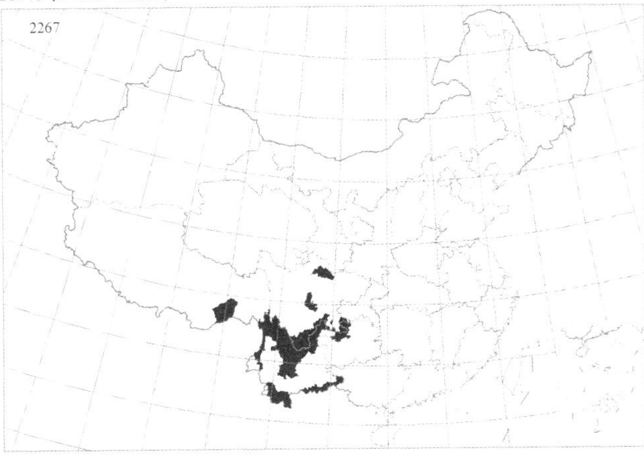

Lauraceae 樟科 *Litsea* 木姜子属

Litsea moupinensis var. szechuanica (C. K. Allen) Yang et P. H. Huang 四川木姜子

Trees, deciduous

13.8; 4.0; 22.7; 13.8; 107.4; 1.4; 741 ❙ 1163; 641; 45 ❙ 56.3; 741; 189.5

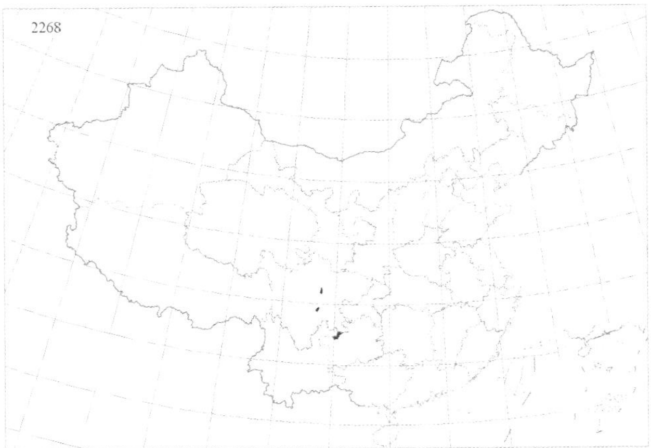

Lauraceae 樟科 *Litsea* 木姜子属
Litsea oligophlebia H. T. Chang 少脉木姜子
Trees, evergreen

22.0 (21.3~22.7); 14.1 (13.2~15.0); 27.9 (27.3~28.6); 22.0 (21.3~22.8); 204.0 (195.9~213.1); 0.0 (0.0~0.0); 1163 (1111~1221) ❙ 1896 (1302~2435); 1018 (670~1428); 116 (86~130) ❙ 62.7 (12.9~99.5); 1163 (1111~1221); 281.1 (235.1~318.8)

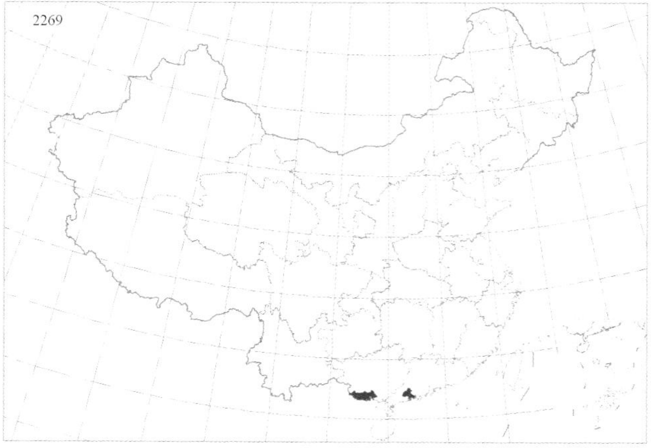

Lauraceae 樟科 *Litsea* 木姜子属
Litsea panamanja (Nees) Hook. f. 香花木姜子
Trees, evergreen

19.7 (6.6~22.7); 12.4 (-1.3~15.1); 24.9 (13.3~28.6); 19.7 (6.8~22.8); 177.5 (39.6~213.1); 0.8 (0.0~19.7); 1005 (514~1221) ❙ 1527 (862~2435); 825 (548~1428); 79 (12~130) ❙ 53.9 (5.8~112.2); 1000 (514~1221); 389.0 (256.6~659.0)

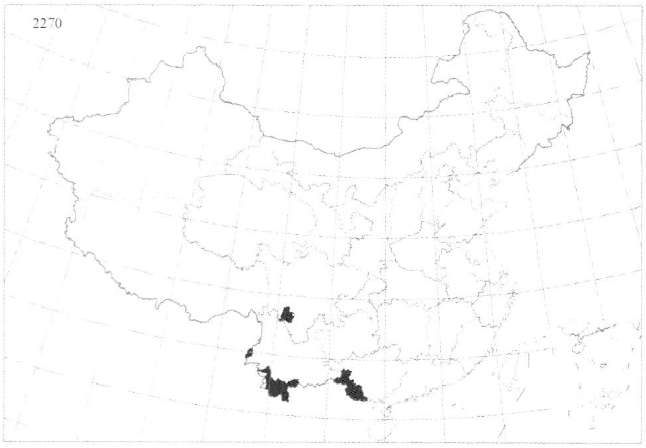

Lauraceae 樟科 *Litsea* 木姜子属
Litsea pedunculata (Diels) Yang et P. H. Huang 红皮木姜子
Shrubs or small trees, evergreen

16.3 (12.6~21.0); 6.2 (1.6~14.2); 25.3 (21.4~27.6); 16.3 (12.7~21.0); 136.6 (98.8~191.8); 0.6 (0.0~6.9); 848 (711~1053) ❙ 1397 (1064~1751); 588 (445~996); 119 (50~216) ❙ 65.2 (32.5~106.2); 848 (711~1053); 255.8 (169.7~551.6)

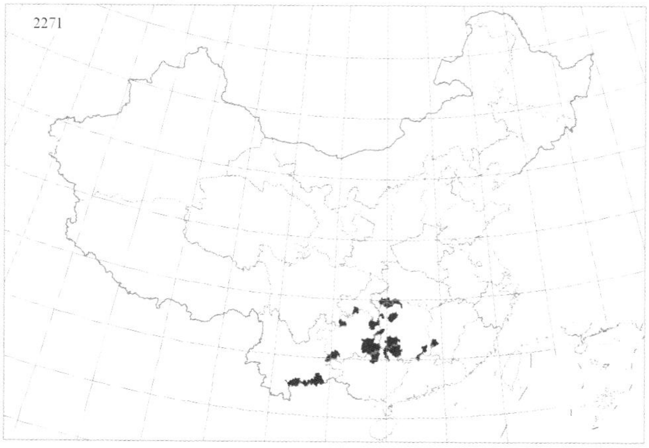

Lauraceae 樟科 *Litsea* 木姜子属
Litsea pedunculata var. pubescens Yang et P. H. Huang 毛红皮木姜子
Shrubs or small trees, evergreen

16.5; 7.7; 23.8; 16.5; 138.7; 0.2; 830 ❙ 1281; 654; 64 ❙ 54.2; 830; 302.6

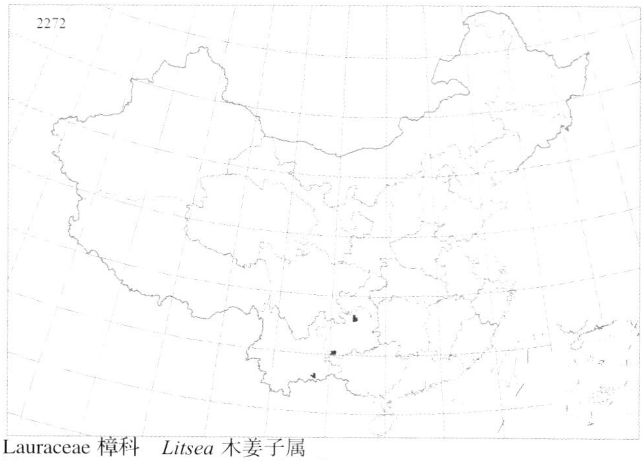

Lauraceae 樟科 *Litsea* 木姜子属
Litsea pittosporifolia Yang et P. H. Huang 海桐叶木姜子
Shrubs, evergreen

22.0; 13.5; 28.8; 22.0; 203.6; 0.0; 1169 ❙ 1641; 724; 132 ❙ 40.5; 1169; 205.0

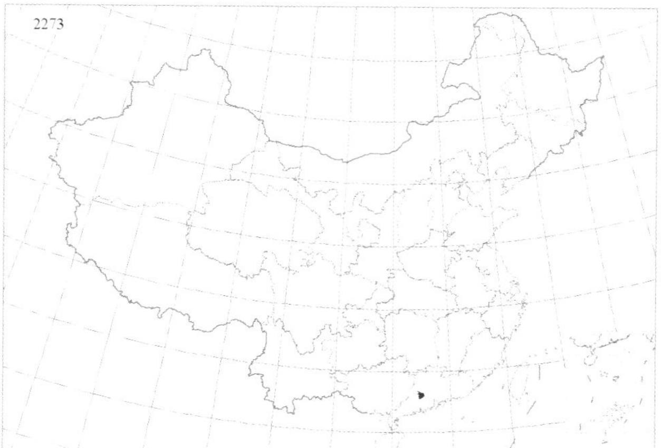

Lauraceae 樟科 *Litsea* 木姜子属
Litsea populifolia Hemsl 杨叶木姜子
Small trees, deciduous

13.5 (1.8~18.1); 3.3 (-7.1~8.1); 23.1 (9.9~27.3); 13.6 (3.8~18.2); 107.7 (15.7~158.1); 4.5 (0.0~53.3); 748 (404~949) ❙ 1129 (583~1577); 559 (348~932); 56 (13~106) ❙ 52.3 (5.4~113.0); 748 (404~949); 228.1 (135.1~363.7)

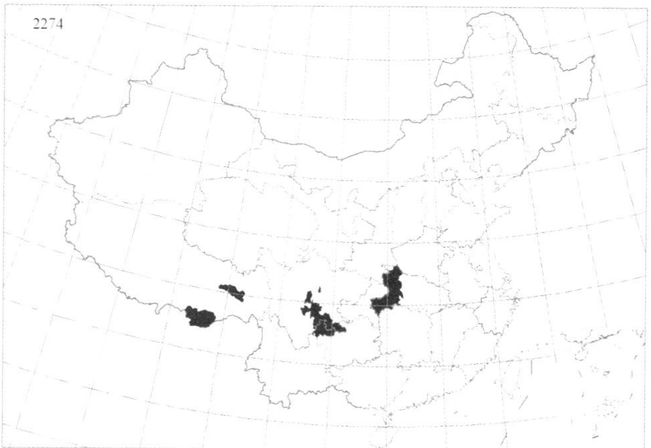

Lauraceae 樟科 *Litsea* 木姜子属
Litsea pseudoelongata H. Liou 竹叶木姜子
Small trees, evergreen

22.2 (19.1~24.6); 14.7 (9.3~19.5); 28.0 (25.5~29.2); 22.2 (19.1~24.7); 206.4 (169.2~235.9); 0.0 (0.0~0.0); 1181 (988~1372) ❙ 1860 (1187~2992); 914 (545~1847); 122 (79~182) ❙ 59.2 (-6.9~166.5); 1169 (988~1329); 280.1 (178.7~548.0)

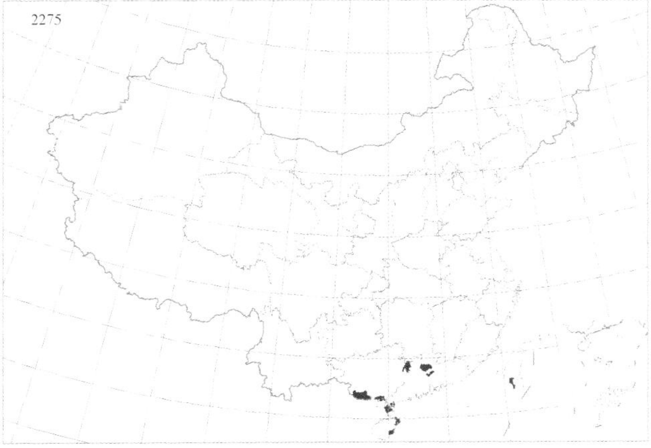

Lauraceae 樟科 *Litsea* 木姜子属
Litsea pungens Hemsl. 木姜子
Small trees, deciduous

14.1 (-0.7~22.7); 3.9 (-9.6~15.0); 23.3 (7.7~29.8); 14.3 (2.5~22.8); 117.5 (6.8~213.1); 7.2 (0.0~74.2); 783 (338~1221) ❙ 1097 (370~2435); 516 (245~1428); 70 (7~254) ❙ 39.7 (-36.2~145.8); 764 (338~1221); 249.6 (61.6~487.2)

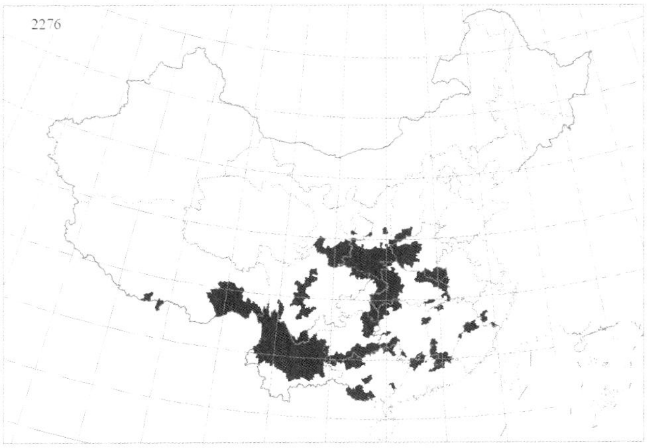

Lauraceae 樟科 *Litsea* 木姜子属
Litsea rotundifolia (Nees) Hemsl. 圆叶豺皮樟
Shrubs or small trees, evergreen

19.8 (14.3~24.1); 10.6 (3.5~19.1); 27.7 (20.1~29.8); 19.8 (14.3~24.2); 178.1 (113.2~230.2); 0.0 (0.0~1.5); 1045 (761~1329) ❙ 1654 (1182~2680); 667 (421~1428); 162 (43~259) ❙ 60.9 (-6.9~246.5); 1040 (761~1329); 233.6 (95.3~620.0)

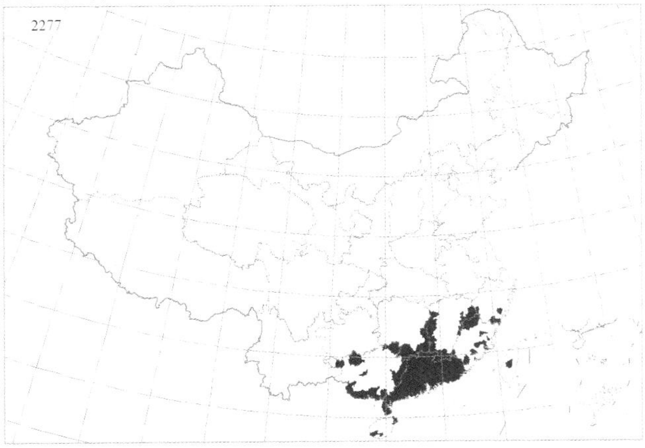

Lauraceae 樟科 *Litsea* 木姜子属
Litsea rotundifolia var. oblongifolia (Nees) C. K. Allen 豺皮樟
Shrubs or small trees, evergreen

20.4 (15.3~24.2); 11.6 (4.1~19.1); 27.8 (20.1~29.8); 20.4 (15.4~24.2); 184.8 (125.8~230.4); 0.0 (0.0~0.9); 1074 (773~1331) ❙ 1649 (1187~2680); 678 (411~1428); 153 (48~259) ❙ 55.8 (-6.9~246.5); 1069 (773~1331); 233.4 (95.3~620.0)

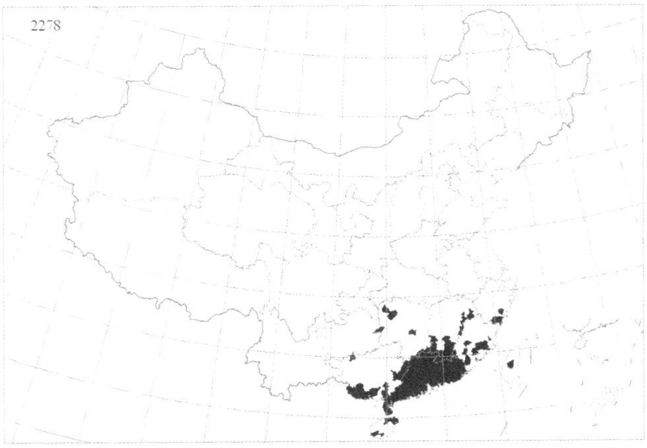

Lauraceae 樟科 *Litsea* 木姜子属
Litsea rotundifolia var. ovatifolia Yang et P. H. Huang 卵叶豺皮樟
Shrubs or small trees, evergreen

20.9 (17.3~23.6); 12.4 (6.5~16.7); 28.0 (25.9~29.8); 21.0 (17.3~23.6); 191.5 (147.9~223.5); 0.0 (0.0~0.0); 1101 (906~1281) ❙ 1671 (1187~2037); 701 (476~1033); 147 (75~197) ❙ 52.9 (-6.9~95.5); 1099 (906~1273); 223.0 (95.3~300.5)

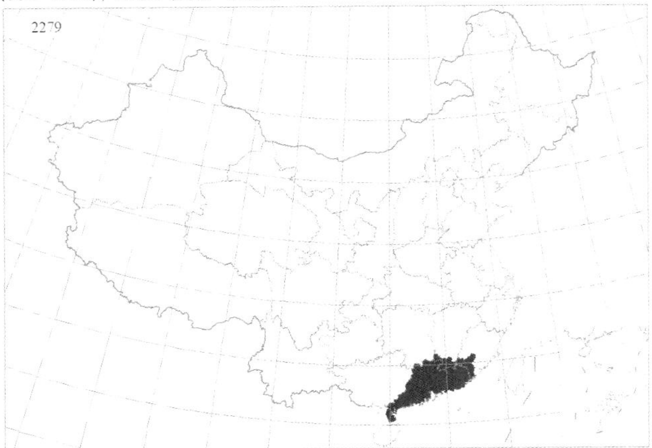

Lauraceae 樟科 *Litsea* 木姜子属
Litsea rubescens Lecomte 红叶木姜子
Shrubs or small trees, deciduous

15.4 (0.3~21.0); 6.4 (-8.6~15.1); 23.2 (8.3~28.9); 15.5 (2.9~21.0); 127.3 (9.6~191.8); 1.9 (0.0~65.4); 808 (368~1053) ❙ 1155 (389~1751); 587 (246~1007); 57 (11~196) ❙ 44.5 (-14.2~145.8); 800 (368~1053); 253.9 (108.2~659.0)

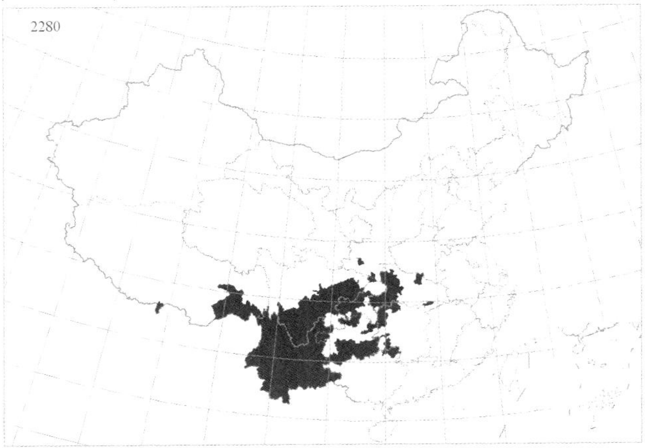

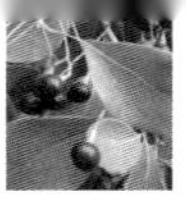

Lauraceae 樟科 *Litsea* 木姜子属

Litsea rubescens var. yunnanensis Lecomte 滇木姜子

Shrubs or small trees, deciduous

13.3 (5.3~19.2); 5.3 (-2.4~13.3); 20.0 (12.2~26.6); 13.4 (5.7~19.4); 102.9 (30.5~172.6); 2.4 (0.0~26.2); 716 (479~969) ❙ 1033 (741~1337); 569 (411~741); 43 (15~94) ❙ 45.9 (-14.2~100.7); 704 (479~969); 266.2 (170.4~452.0)

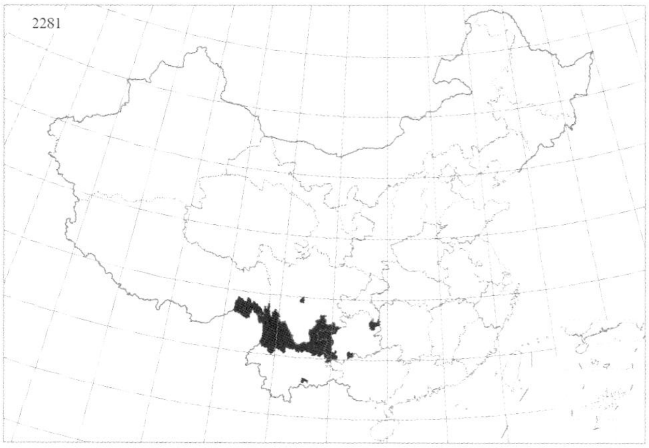

Lauraceae 樟科 *Litsea* 木姜子属

Litsea salicifolia (Roxburgh ex Nees) Hook. f. 黑木姜子

Trees, evergreen

19.6 (16.8~24.0); 11.3 (7.4~19.1); 25.9 (22.4~28.6); 19.6 (16.8~24.0); 175.2 (141.5~228.1); 0.0 (0.0~0.0); 999 (833~1302) ❙ 1464 (1161~1876); 721 (539~996); 98 (48~189) ❙ 47.7 (15.4~88.8); 992 (833~1190); 330.4 (185.5~659.0)

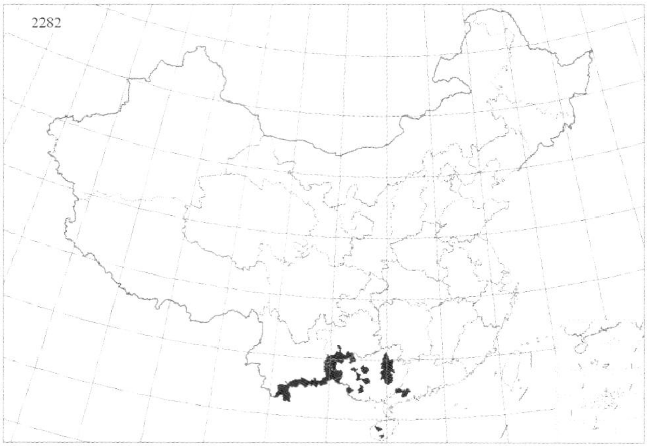

Lauraceae 樟科 *Litsea* 木姜子属

Litsea semecarpifolia (Wallich ex Nees) Hook. f. 玉兰叶木姜子

Trees, evergreen

19.3; 13.4; 22.8; 19.4; 172.3; 0.0; 930 ❙ 1415; 719; 56 ❙ 52.5; 903; 543.4

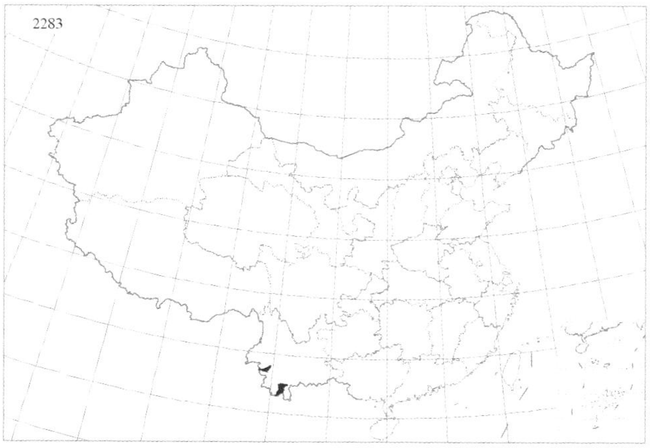

Lauraceae 樟科 *Litsea* 木姜子属

Litsea sericea (Wall. ex Nees) Hook. f. 绢毛木姜子

Shrubs or small trees, deciduous

11.4 (-1.8~19.0); 3.6 (-10.2~13.1); 18.0 (6.1~25.6); 11.8 (1.8~19.1); 88.8 (1.9~168.8); 11.2 (0.0~82.8); 663 (329~929) ❙ 993 (320~1736); 572 (193~1007); 32 (7~69) ❙ 52.3 (-14.2~145.8); 645 (320~891); 280.5 (72.2~447.4)

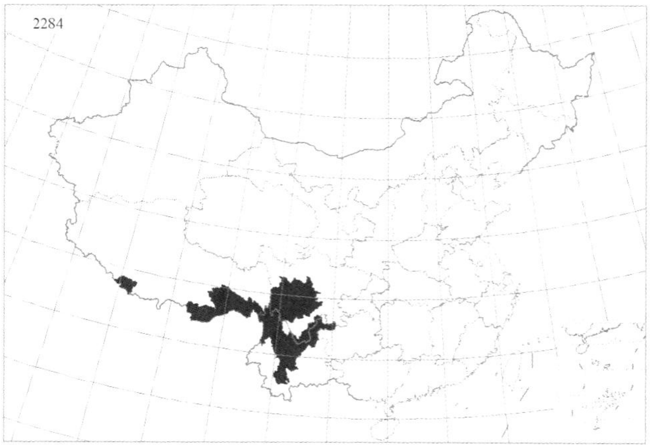

Lauraceae 樟科 *Litsea* 木姜子属

Litsea sinoglobosa Yang et P. H. Huang 圆果木姜子

Shrubs or trees, evergreen

17.8 (14.6~21.0); 7.3 (3.8~11.9); 27.2 (24.6~29.0); 17.9 (14.7~21.1); 154.3 (117.7~193.0); 0.1 (0.0~1.5); 933 (779~1114) ❙ 1558 (1284~1829); 548 (403~821); 186 (73~230) ❙ 67.5 (30.9~104.6); 926 (779~1114); 226.5 (151.0~322.1)

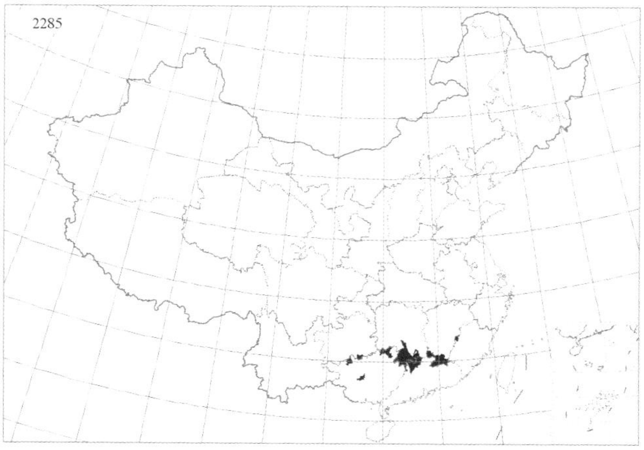

Lauraceae 樟科 *Litsea* 木姜子属

Litsea subcoriacea Yang et P. H. Huang 桂北木姜子

Trees, evergreen

17.2 (9.3~22.7); 6.8 (-1.7~15.0); 26.6 (20.1~29.8); 17.2 (9.5~22.8); 147.2 (69.8~213.1); 0.8 (0.0~18.0); 907 (603~1221) ❙ 1458 (1049~2435); 566 (392~1428); 145 (44~254) ❙ 61.5 (21.9~130.7); 897 (603~1221); 233.6 (134.6~551.6)

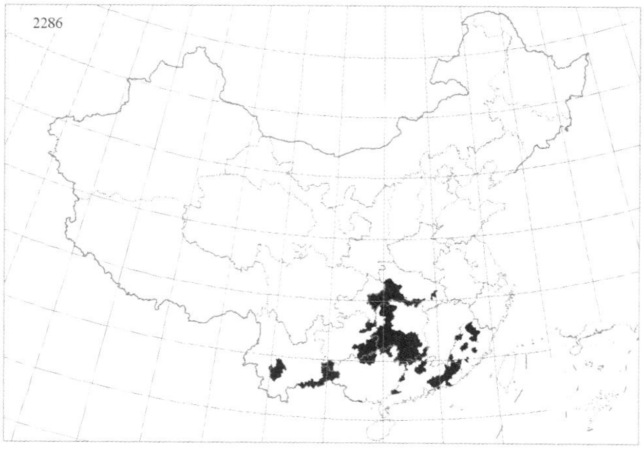

Lauraceae 樟科 *Litsea* 木姜子属
Litsea suberosa Yang et P. H. Huang 栓皮木姜子
Small trees, evergreen

15.8 (13.5~18.8); 4.9 (2.6~8.8); 26.0 (22.7~29.3); 15.8 (13.6~18.8); 130.7 (106.7~165.8); 0.8 (0.0~4.1); 842 (744~991) ❙ 1373 (1089~1679); 527 (419~817); 126 (47~204) ❙ 63.5 (30.9~101.8); 840 (744~991); 204.8 (146.0~258.0)

2287

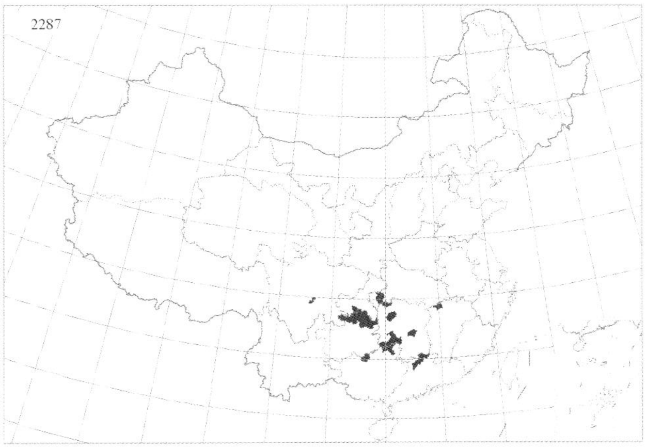

Lauraceae 樟科 *Litsea* 木姜子属
Litsea szemaois (H. Liu) J. Li et H. W. Li 思茅木姜子
Trees, evergreen

19.4 (18.2~20.5); 13.9 (12.4~15.1); 22.8 (21.6~23.7); 19.5 (18.2~20.5); 173.9 (158.6~185.9); 0.0 (0.0~0.0); 932 (861~992) ❙ 1513 (1464~1607); 721 (655~785); 63 (52~77) ❙ 62.8 (47.6~74.2); 911 (839~986); 522.1 (427.9~659.0)

2288

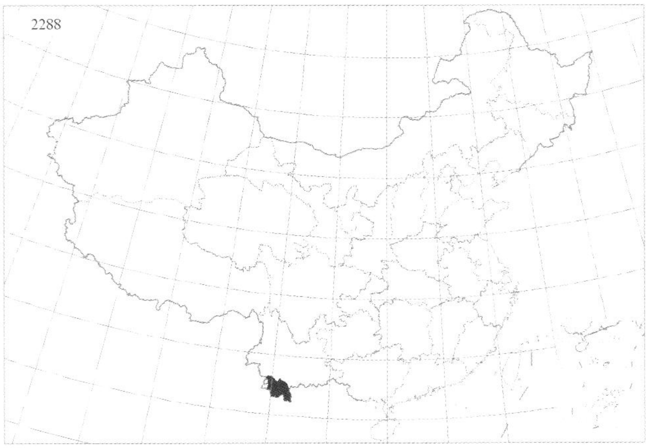

Lauraceae 樟科 *Litsea* 木姜子属
Litsea taronensis H. W. Li 独龙木姜子
Trees, deciduous

14.5; 6.4; 21.0; 14.6; 117.6; 2.9; 811 ❙ 1467; 795; 89 ❙ 89.4; 811; 345.8

2289

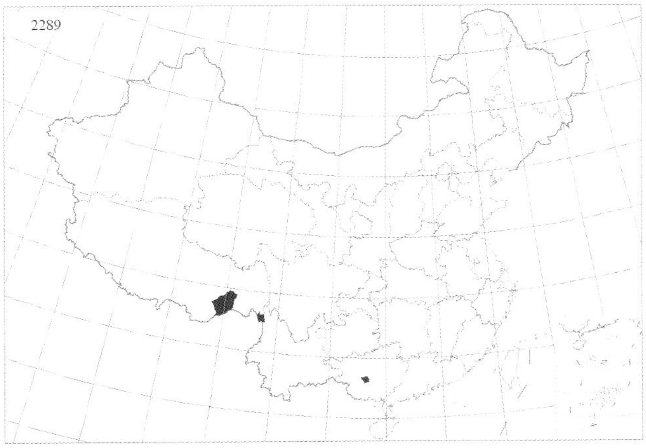

Lauraceae 樟科 *Litsea* 木姜子属
Litsea tibetana Yang et P. H. Huang 西藏木姜子
Shrubs, evergreen

13.3; 5.4; 19.6; 13.4; 100.4; 0.0; 706 ❙ 1736; 1007; 51 ❙ 145.8; 706; 447.4

2290

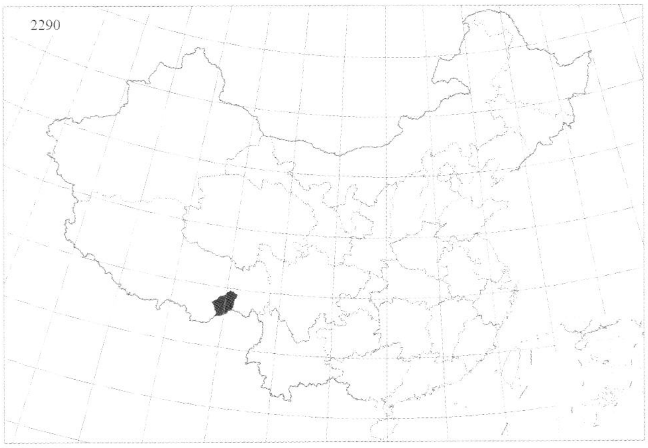

Lauraceae 樟科 *Litsea* 木姜子属
Litsea tsinlingensis Yang et P. H. Huang 秦岭木姜子
Shrubs or small trees, deciduous

11.9 (7.4~15.5); -0.3 (-4.2~2.7); 23.5 (18.7~29.5); 12.0 (8.2~15.5); 95.4 (58.8~136.0); 12.5 (3.8~29.4); 709 (561~932) ❙ 838 (511~1164); 380 (258~504); 28 (15~51) ❙ 19.0 (-45.2~83.5); 696 (511~786); 258.9 (172.7~334.4)

2291

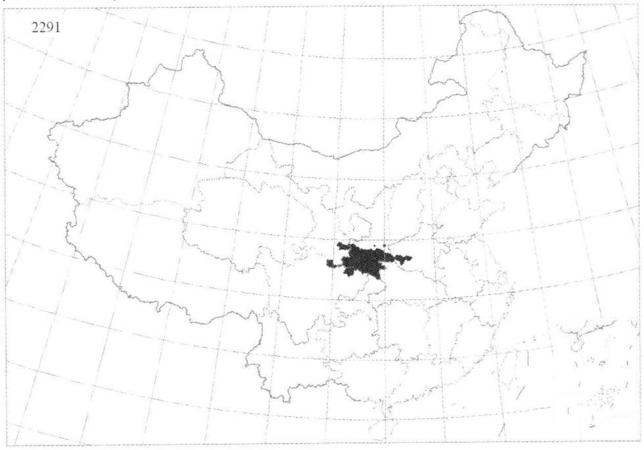

Lauraceae 樟科 *Litsea* 木姜子属
Litsea umbellata (Lour.) Merr. 伞花木姜子
Shrubs or small trees, evergreen

19.6 (17.2~22.2); 12.2 (9.1~15.1); 24.9 (22.3~28.6); 19.6 (17.3~22.3); 175.8 (147.0~207.5); 0.0 (0.0~0.0); 982 (833~1190) ❙ 1453 (1161~1751); 778 (633~996); 75 (42~174) ❙ 49.6 (12.9~88.8); 978 (833~1190); 397.8 (200.0~659.0)

2292

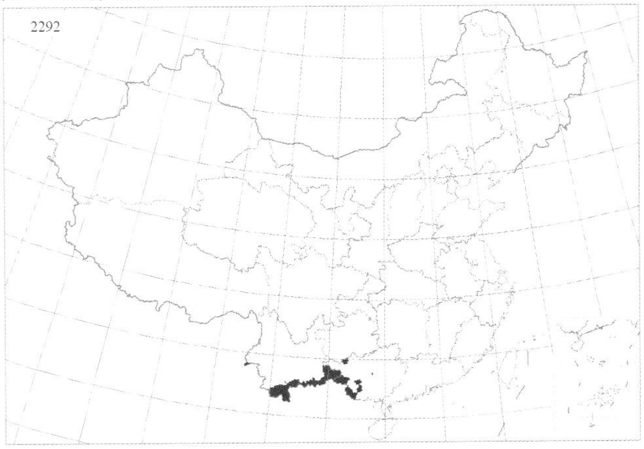

Lauraceae 樟科 *Litsea* 木姜子属

Litsea vang Lecomte var. lobata Lecomte 沧源薄托木姜子

Trees, evergreen

15.6; 8.6; 20.6; 15.7; 128.1; 0.0; 780 ❙ 1585; 905; 46 ❙ 106.9; 770; 476.3

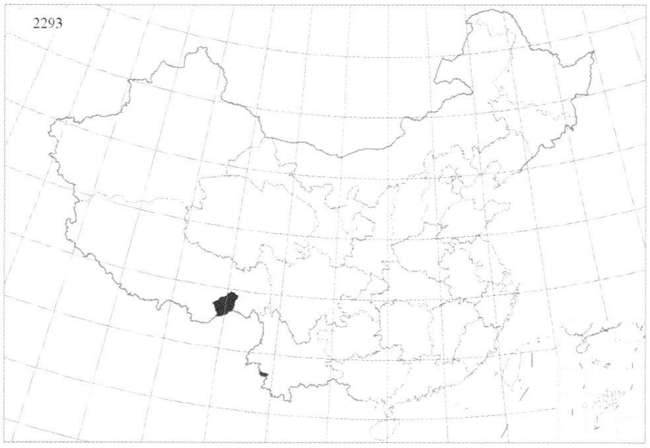

Lauraceae 樟科 *Litsea* 木姜子属

Litsea variabilis Hemsl. 黄椿木姜子

Shrubs or trees, evergreen

21.2 (16.5~24.8); 13.0 (5.8~20.1); 27.7 (22.1~29.8); 21.2 (16.5~24.8); 194.5 (138.2~237.6); 0.0 (0.0~0.0); 1117 (833~1385) ❙ 1621 (868~2435); 712 (312~1428); 131 (41~203) ❙ 46.8 (-34.3~99.5); 1104 (833~1360); 263.2 (95.3~620.0)

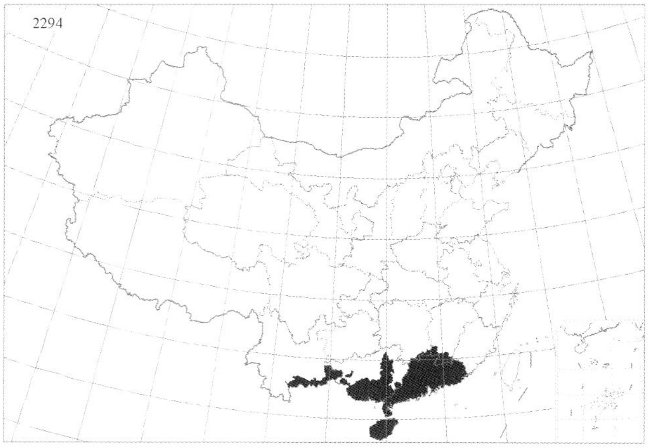

Lauraceae 樟科 *Litsea* 木姜子属

Litsea variabilis var. oblonga Lecomte 毛黄椿木姜子

Shrubs or trees, evergreen

20.6 (18.9~22.2); 12.1 (9.8~13.9); 27.0 (25.0~28.3); 20.6 (19.0~22.3); 187.2 (167.5~207.5); 0.0 (0.0~0.0); 1066 (939~1190) ❙ 1399 (1279~1630); 740 (651~848); 81 (49~107) ❙ 32.0 (14.8~55.9); 1066 (939~1190); 306.1 (223.2~342.5)

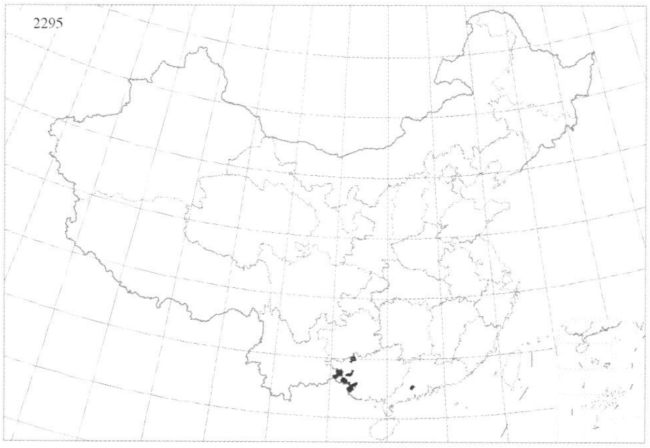

Lauraceae 樟科 *Litsea* 木姜子属

Litsea veitchiana Gamble 钝叶木姜子

Shrubs or small trees, deciduous

14.6 (3.5~21.0); 4.2 (-6.5~14.2); 24.2 (12.6~29.3); 14.7 (5.0~21.0); 118.8 (26.3~191.8); 3.2 (0.0~43.4); 791 (446~1053) ❙ 1263 (731~1745); 547 (346~1007); 92 (15~221) ❙ 59.9 (5.6~145.8); 789 (446~1053); 245.7 (147.3~551.6)

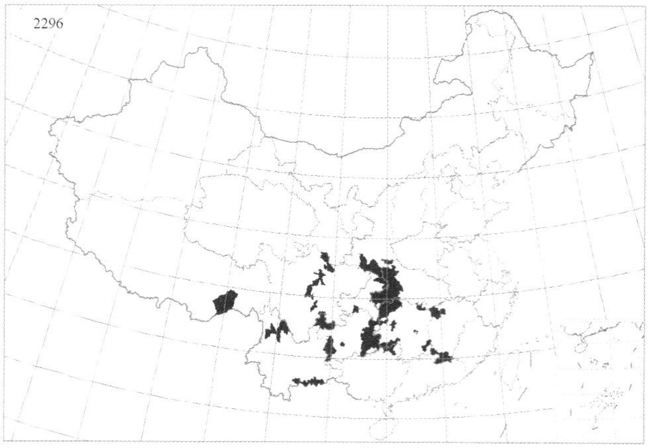

Lauraceae 樟科 *Litsea* 木姜子属

Litsea veitchiana var. trichocarpa (Yang) H. S. Kung 毛果木姜子

Shrubs or small trees, deciduous

10.2; 0.5; 18.8; 10.3; 74.1; 11.7; 628 ❙ 1013; 554; 28 ❙ 64.6; 628; 209.8

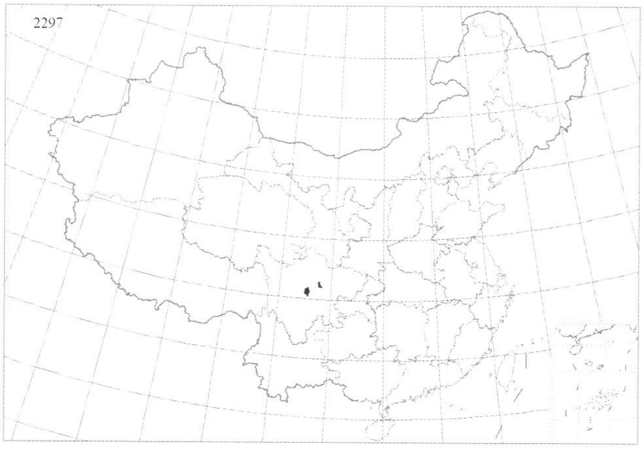

Lauraceae 樟科 *Litsea* 木姜子属

Litsea verticillata Hance 轮叶木姜子

Shrubs or small trees, evergreen

20.6 (14.6~24.8); 12.0 (3.5~20.1); 27.7 (22.4~29.8); 20.7 (14.7~24.8); 188.3 (117.7~237.6); 0.0 (0.0~1.5); 1091 (779~1385) ❙ 1634 (1161~2435); 714 (427~1428); 138 (41~218) ❙ 51.4 (-2.2~101.8); 1083 (779~1360); 250.3 (95.3~620.0)

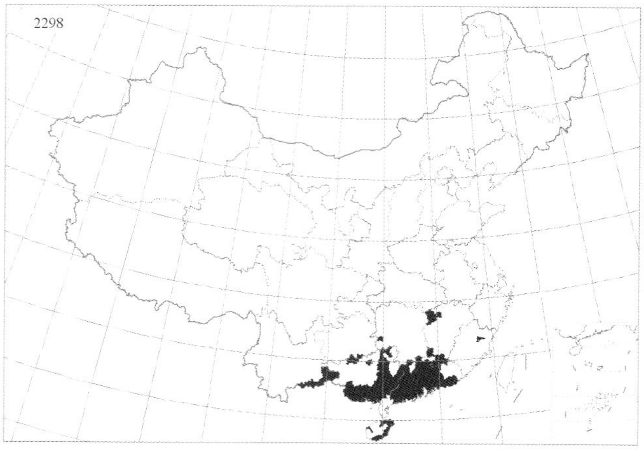

Lauraceae 樟科 *Litsea* 木姜子属
Litsea verticillifolia Yang et P. H. Huang 琼南木姜子
Shrubs, evergreen

23.9 (22.2~24.8); 18.3 (17.2~20.1); 27.6 (25.7~28.5); 23.9 (22.3~24.8); 227.0 (207.2~237.5); 0.0 (0.0~0.0); 1298 (1127~1385) ❙ 1531 (1179~1854); 525 (392~622); 60 (41~115) ❙ 18.5 (-11.9~47.8); 1161 (1055~1360); 515.5 (428.6~620.0)

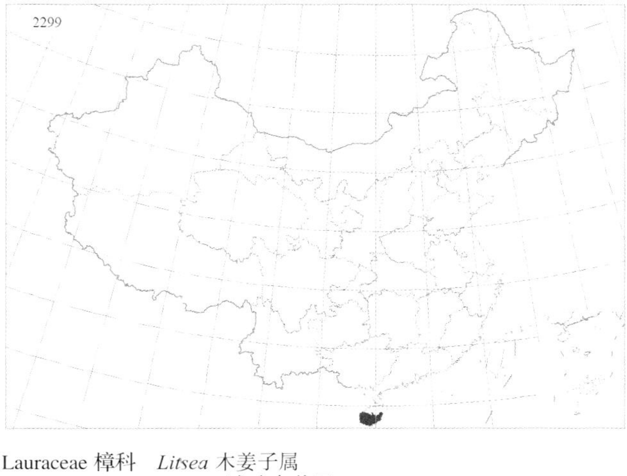

Lauraceae 樟科 *Litsea* 木姜子属
Litsea viridis H. Liu 干香柴
Small trees, evergreen

17.7; 10.9; 22.5; 17.8; 153.4; 0.0; 854 ❙ 1368; 756; 59 ❙ 60.1; 854; 491.5

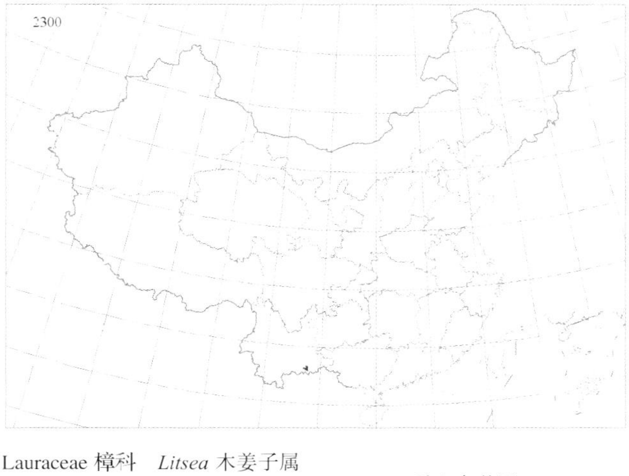

Lauraceae 樟科 *Litsea* 木姜子属
Litsea wilsonii Gamble 绒叶木姜子
Trees, evergreen

16.0 (10.2~21.8); 5.5 (-0.7~14.0); 25.9 (19.4~28.9); 16.1 (10.3~21.9); 134.4 (75.8~202.2); 1.5 (0.0~13.6); 861 (634~1150) ❙ 1179 (711~2148); 540 (357~1234); 59 (15~124) ❙ 38.9 (8.0~93.8); 861 (634~1150); 177.6 (108.2~318.8)

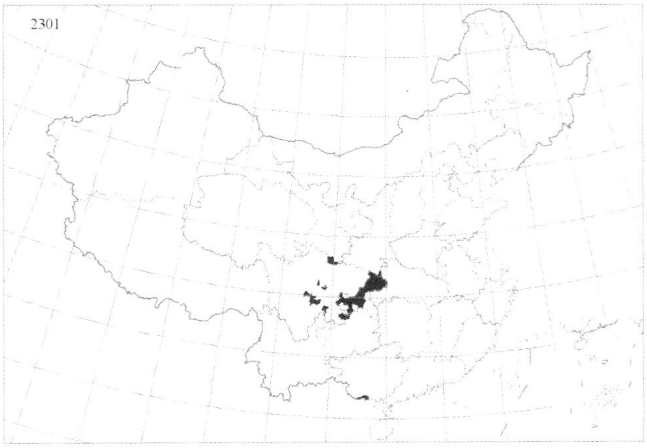

Lauraceae 樟科 *Litsea* 木姜子属
Litsea yaoshanensis Yang et P. H. Huang 瑶山木姜子
Trees, evergreen

19.6 (17.9~21.3); 9.8 (7.4~12.3); 27.8 (25.8~28.6); 19.7 (18.0~21.3); 176.1 (155.6~196.1); 0.0 (0.0~0.0); 1032 (913~1130) ❙ 1596 (1371~1758); 669 (562~728); 175 (161~189) ❙ 56.0 (22.7~85.5); 1031 (913~1130); 220.0 (185.5~243.9)

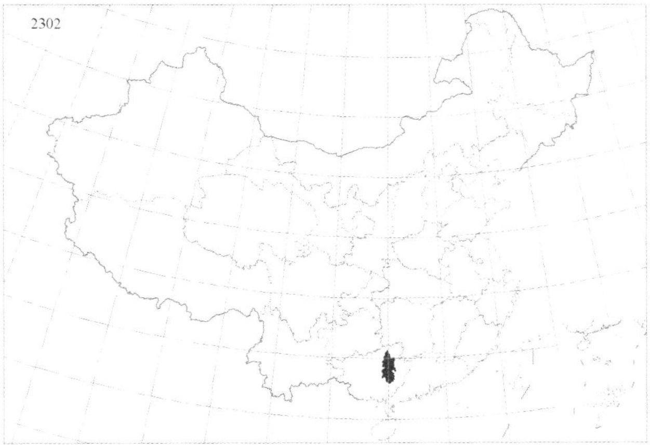

Lauraceae 樟科 *Litsea* 木姜子属
Litsea yunnanensis Yang et P. H. Huang 云南木姜子
Trees, evergreen

18.4 (15.6~21.0); 11.8 (9.1~15.1); 22.9 (20.3~26.4); 18.4 (15.6~21.0); 161.1 (127.6~191.8); 0.0 (0.0~0.0); 892 (767~1053) ❙ 1321 (939~1751); 723 (524~996); 55 (37~77) ❙ 48.0 (6.5~88.8); 872 (717~1053); 413.5 (289.9~659.0)

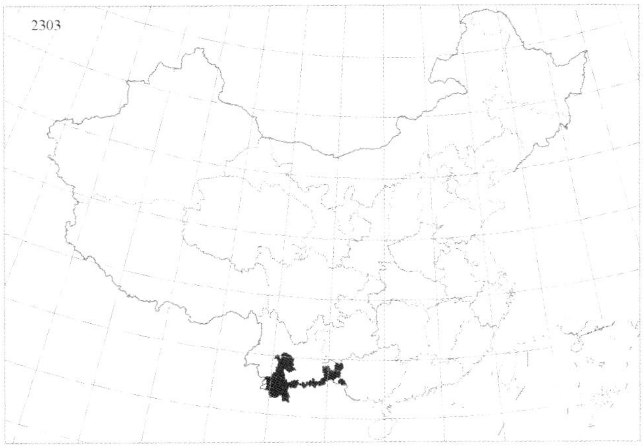

Lauraceae 樟科 *Machilus* 润楠属
Machilus austroguizhouensis (S. K. Lee et F. N. Wei) Kosterm. 黔南润楠
Trees, evergreen

17.5 (15.0~19.2); 8.0 (4.5~11.0); 25.4 (23.4~27.5); 17.6 (15.0~19.2); 151.0 (120.8~170.8); 0.0 (0.0~0.5); 895 (794~1002) ❙ 1297 (1089~1495); 642 (499~820); 77 (51~135) ❙ 45.2 (27.5~62.5); 895 (794~1002); 249.4 (181.2~371.7)

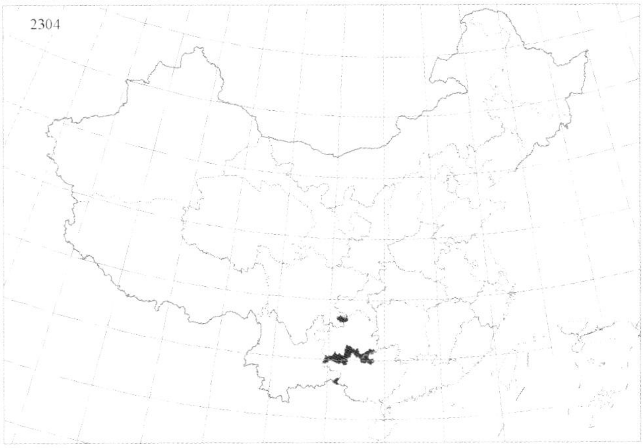

Lauraceae 樟科 *Machilus* 润楠属
Machilus bonii Lecomte 枇杷叶润楠
Trees, evergreen

18.9 (17.4~22.2); 10.2 (7.4~14.2); 25.7 (22.5~28.3); 19.0 (17.4~22.3); 167.7 (148.9~207.5); 0.0 (0.0~0.0); 962 (854~1190) ❙ 1332 (1236~1522); 696 (617~818); 71 (51~117) ❙ 39.4 (15.4~60.1); 962 (854~1190); 316.1 (189.5~551.6)

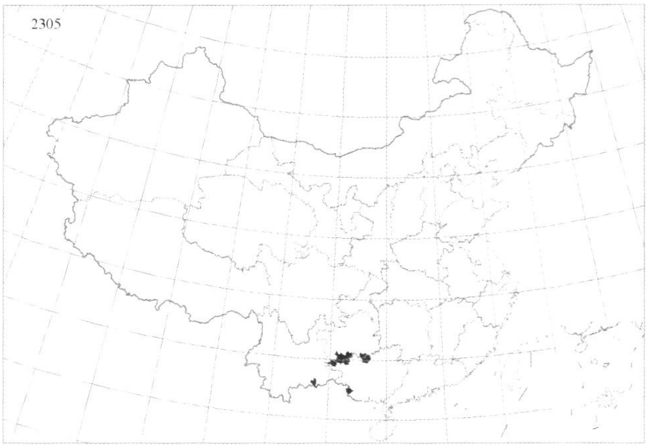

Lauraceae 樟科 *Machilus* 润楠属
Machilus breviflora (Benth.) Hemsl. 短序润楠
Trees, evergreen

21.3 (17.9~24.0); 13.1 (8.1~19.1); 27.9 (24.6~28.8); 21.4 (17.9~24.0); 196.9 (155.4~228.1); 0.0 (0.0~0.0); 1127 (901~1302) ❙ 1707 (1279~2435); 808 (539~1428); 124 (48~197) ❙ 51.7 (12.9~99.5); 1124 (901~1221); 254.4 (139.6~620.0)

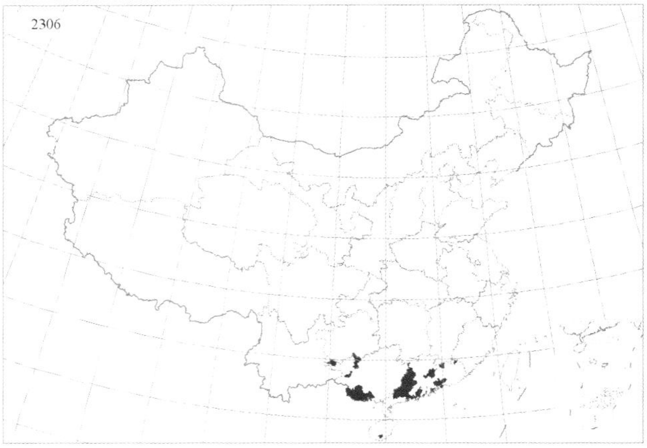

Lauraceae 樟科 *Machilus* 润楠属
Machilus calcicola C. J. Qi 灰岩润楠
Trees, evergreen

17.6 (14.6~22.2); 7.0 (3.8~13.9); 27.2 (24.6~29.0); 17.7 (14.7~22.3); 152.1 (117.7~207.5); 0.1 (0.0~1.5); 931 (779~1190) ❙ 1484 (1171~1776); 543 (403~821); 171 (58~211) ❙ 60.7 (15.4~101.8); 922 (779~1190); 216.3 (151.0~327.7)

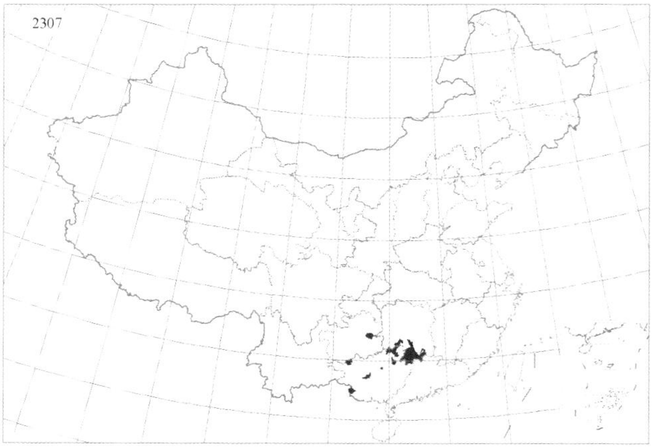

Lauraceae 樟科 *Machilus* 润楠属
Machilus cavaleriei H. Lév. 安顺润楠
Shrubs or small trees, evergreen

17.1 (15.5~19.2); 7.6 (5.5~10.7); 24.7 (23.6~26.0); 17.1 (15.6~19.2); 145.0 (127.1~170.8); 0.0 (0.0~0.0); 866 (798~961) ❙ 1293 (1172~1381); 641 (552~720); 74 (50~107) ❙ 50.3 (21.9~62.6); 866 (798~961); 254.9 (194.4~411.1)

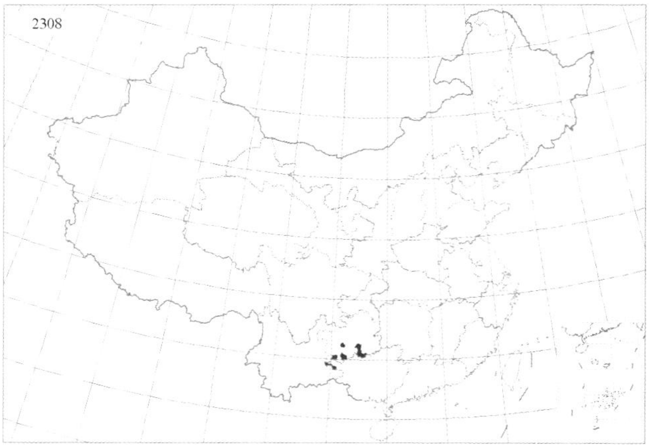

Lauraceae 樟科 *Machilus* 润楠属
Machilus chayuensis S. K. Lee 察隅润楠
Trees, evergreen

3.6; -4.7; 11.0; 4.7; 23.1; 39.8; 441 ❙ 772; 453; 22 ❙ 72.5; 441; 201.8

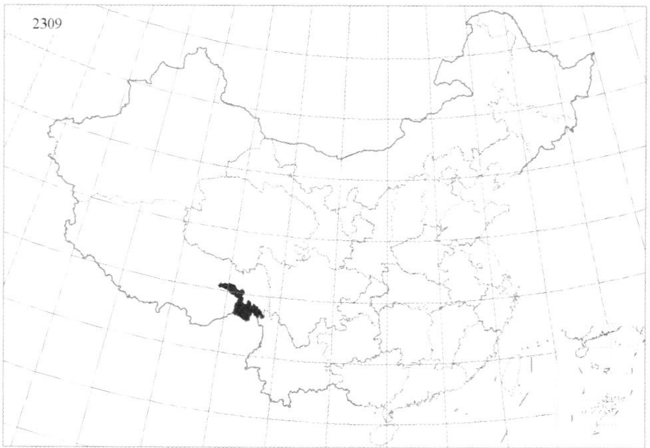

Lauraceae 樟科 *Machilus* 润楠属
Machilus chekiangensis S. K. Lee 浙江润楠
Trees, evergreen

18.5 (15.9~22.4); 8.7 (5.9~15.2); 27.4 (25.2~28.8); 18.6 (16.0~22.7); 163.5 (131.9~212.7); 0.0 (0.0~0.0); 972 (831~1216) ❙ 1733 (1485~2030); 628 (522~1033); 197 (104~259) ❙ 79.2 (41.1~121.9); 969 (831~1216); 245.1 (163.7~305.7)

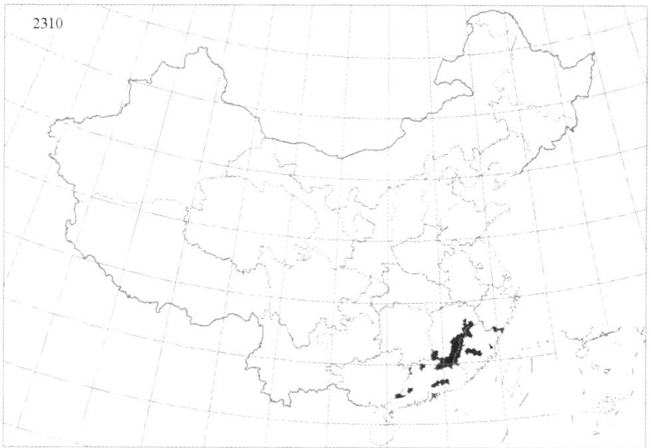

Lauraceae 樟科 *Machilus* 润楠属
Machilus chienkweiensis S. K. Lee 黔桂润楠
Trees, evergreen

16.3 (14.3~19.2); 5.5 (3.3~9.2); 26.0 (23.5~28.6); 16.3 (14.3~19.2); 136.2 (112.6~170.8); 0.4 (0.0~2.3); 856 (756~1002) ❙ 1384 (1211~1624); 534 (419~678); 137 (84~197) ❙ 62.2 (44.3~101.8); 850 (756~1002); 217.8 (134.6~258.0)

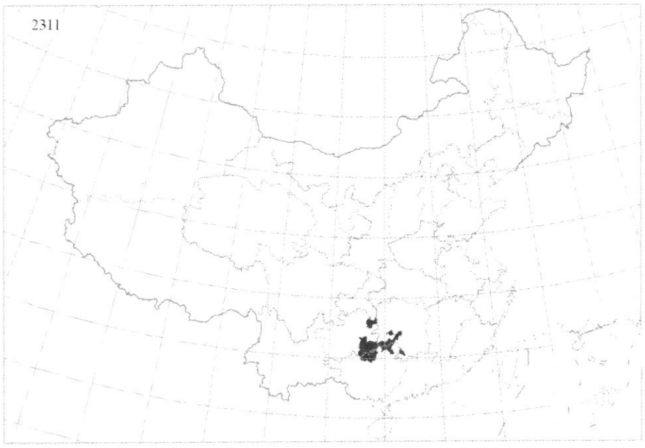

Lauraceae 樟科 *Machilus* 润楠属
Machilus chinensis (Champ. ex Benth.) Hemsl. 华润楠
Trees, evergreen

21.0 (12.7~24.8); 12.6 (2.1~20.1); 27.9 (23.0~29.8); 21.1 (12.8~24.8); 192.9 (99.0~237.6); 0.1 (0.0~5.6); 1113 (713~1385) ❙ 1637 (868~2435); 701 (312~1428); 138 (41~197) ❙ 49.0 (-34.3~107.2); 1100 (713~1331); 240.9 (95.3~620.0)

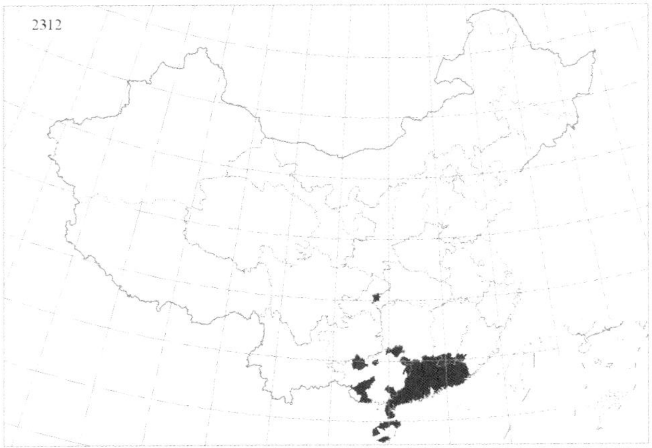

Lauraceae 樟科 *Machilus* 润楠属
Machilus chrysotricha H. W. Li 黄毛润楠
Trees, evergreen

14.4 (5.4~18.9); 7.6 (-2.1~12.0); 19.7 (12.3~23.9); 14.5 (5.8~19.0); 115.6 (31.5~167.8); 2.0 (0.0~25.6); 742 (481~929) ❙ 943 (741~1057); 555 (411~610); 31 (15~43) ❙ 29.4 (-14.2~67.9); 693 (481~799); 289.8 (218.0~351.7)

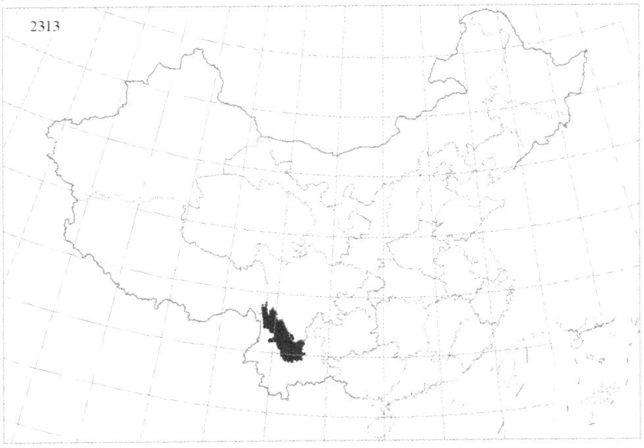

Lauraceae 樟科 *Machilus* 润楠属
Machilus chuanchienensis S. K. Lee 川黔润楠
Trees, evergreen

15.2 (14.2~17.4); 4.5 (3.6~7.4); 25.4 (24.4~26.3); 15.2 (14.2~17.4); 123.7 (112.5~148.9); 0.8 (0.0~1.7); 809 (765~886) ❙ 1220 (1132~1381); 500 (473~637); 78 (65~107) ❙ 50.9 (32.5~62.4); 809 (765~886); 187.0 (152.4~245.3)

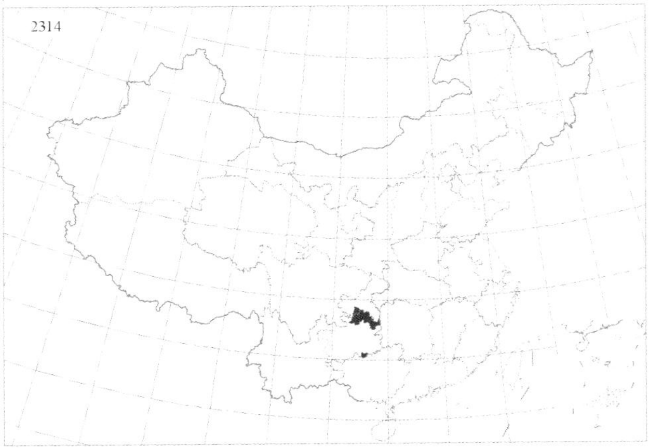

Lauraceae 樟科 *Machilus* 润楠属
Machilus cicatricosa S. K. Lee 刻节润楠
Trees, evergreen

23.5; 18.4; 27.0; 23.5; 222.6; 0.0; 1259 ❙ 1638; 575; 58 ❙ 30.4; 1155; 596.5

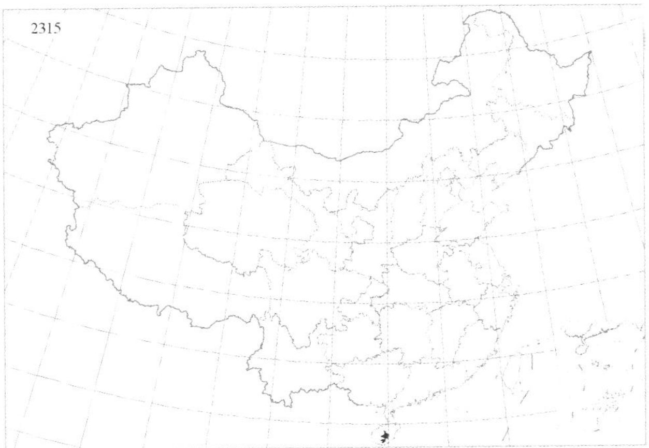

Lauraceae 樟科 *Machilus* 润楠属
Machilus daozhenensis Y. K. Li 道真润楠
Trees, evergreen

14.6; 3.6; 25.0; 14.6; 116.9; 1.6; 786 ❙ 1346; 551; 90 ❙ 70.9; 786; 206.6

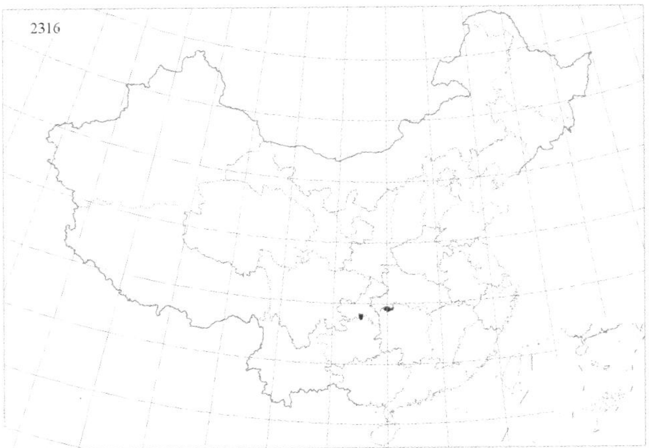

Lauraceae 樟科 *Machilus* 润楠属
Machilus decursinervis Chun 基脉润楠
Trees, evergreen

18.1 (14.3~22.7); 8.3 (3.3~15.0); 26.4 (22.4~28.7); 18.2 (14.3~22.8); 158.0 (112.6~213.1); 0.1 (0.0~2.3); 943 (756~1221) | 1502 (1182~2435); 646 (419~1428); 140 (43~221) | 60.3 (5.8~106.4); 942 (756~1221); 258.3 (177.4~551.6)

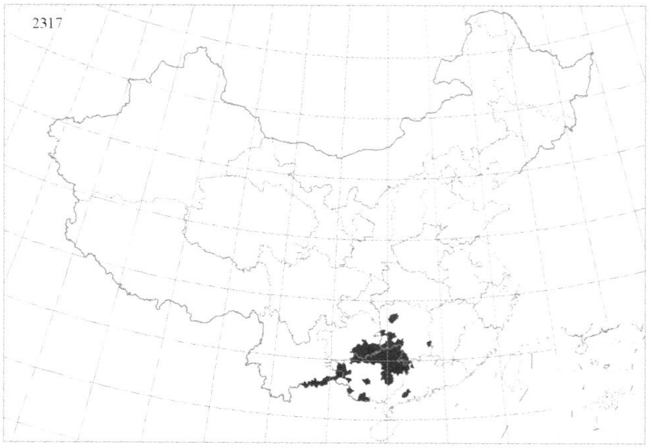

Lauraceae 樟科 *Machilus* 润楠属
Machilus dinganensis (S. K. Lee et F. N. Wei) Kosterm. 定安润楠
Trees, evergreen

24.1; 18.2; 28.4; 24.2; 230.1; 0.0; 1326 | 1723; 619; 110 | 29.9; 1326; 352.9

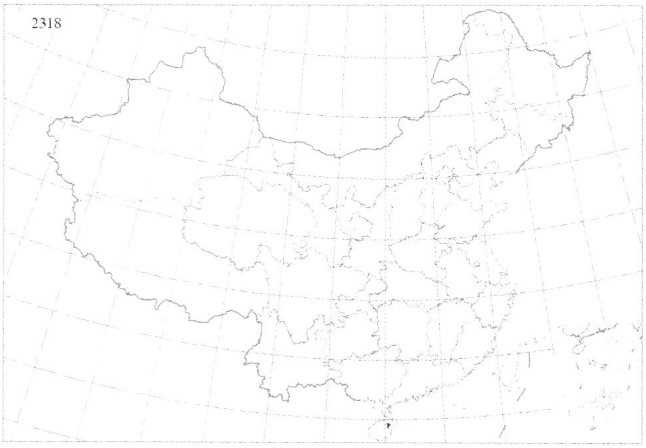

Lauraceae 樟科 *Machilus* 润楠属
Machilus dumicola (W. W. Sm.) H. W. Li 灌丛润楠
Small trees, evergreen

16.7 (13.2~19.6); 9.9 (6.4~12.8); 21.1 (18.4~23.5); 16.7 (13.3~19.6); 140.3 (99.4~175.7); 0.0 (0.0~0.0); 815 (695~953) | 1455 (1180~1688); 826 (622~959); 57 (42~70) | 79.0 (50.8~112.2); 812 (695~932); 467.6 (335.1~627.0)

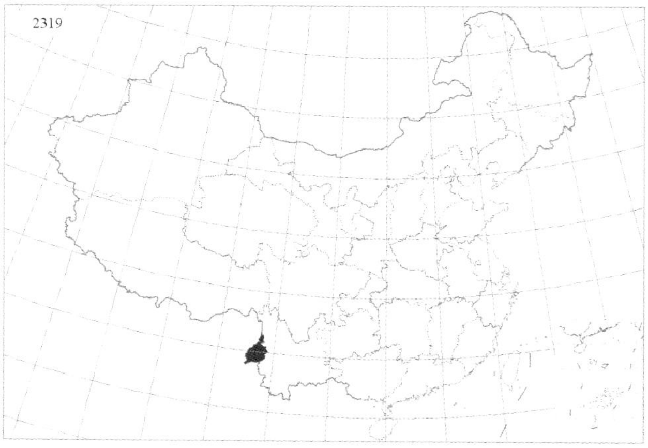

Lauraceae 樟科 *Machilus* 润楠属
Machilus duthiei King ex Hook. f. 长梗润楠
Trees, evergreen

15.7 (−1.8~22.2); 8.6 (−10.2~14.7); 21.1 (6.1~28.3); 15.8 (1.8~22.2); 130.5 (1.9~206.9); 1.6 (0.0~82.8); 803 (329~1177) | 1100 (320~1876); 607 (217~1007); 52 (19~177) | 37.6 (−14.2~145.8); 774 (320~1177); 300.6 (72.2~506.2)

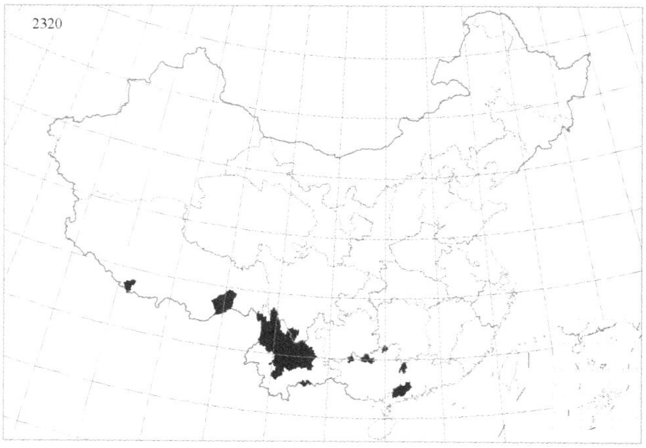

Lauraceae 樟科 *Machilus* 润楠属
Machilus fasciculata H. W. Li 簇序润楠
Shrubs or small trees, evergreen

18.1 (17.2~19.1); 9.7 (8.2~11.0); 24.4 (22.7~27.3); 18.1 (17.3~19.1); 157.7 (147.0~169.6); 0.0 (0.0~0.0); 899 (838~966) | 1388 (1161~1758); 737 (633~820); 74 (48~189) | 54.0 (37.1~82.1); 899 (838~966); 324.9 (224.9~371.7)

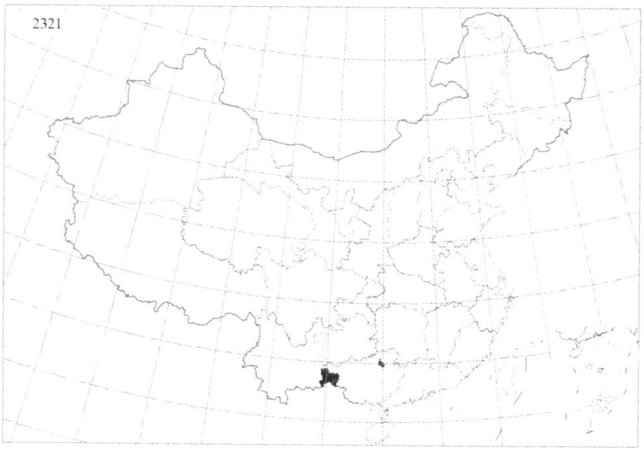

Lauraceae 樟科 *Machilus* 润楠属
Machilus foonchewii S. K. Lee 琼桂润楠
Trees, evergreen

23.2 (21.8~24.8); 16.7 (14.0~20.1); 28.0 (27.2~28.6); 23.3 (21.9~24.8); 219.3 (202.6~237.6); 0.0 (0.0~0.0); 1260 (1150~1385) | 1825 (1405~2435); 840 (473~1428); 90 (41~130) | 46.7 (1.5~99.5); 1179 (1137~1221); 384.2 (235.1~620.0)

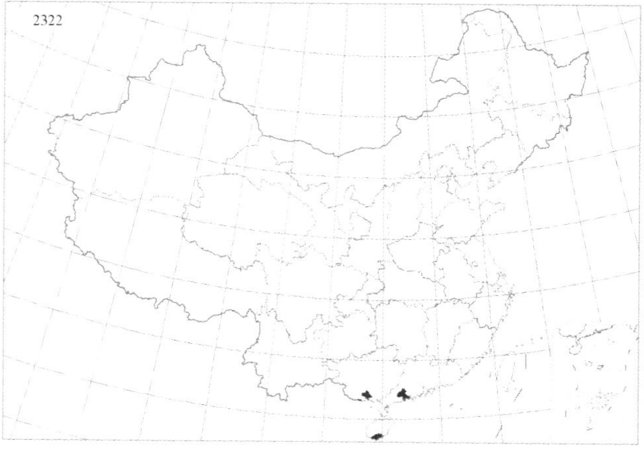

Lauraceae 樟科 *Machilus* 润楠属
Machilus fukienensis H. T. Chang 闽润楠
Small trees, evergreen

19.1 (18.1~20.5); 10.4 (8.3~12.0); 27.7 (26.5~28.9); 19.2 (18.2~20.5); 169.9 (158.2~186.5); 0.0 (0.0~0.0); 986 (920~1073) ❙ 1431 (1244~1821); 494 (411~614); 214 (172~257) ❙ 46.2 (15.9~92.7); 978 (920~1012); 238.9 (191.1~332.5)

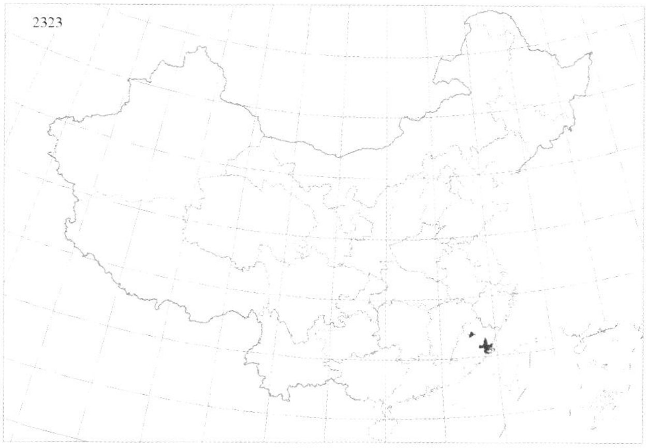

Lauraceae 樟科 *Machilus* 润楠属
Machilus gamblei King ex Hook. f. 黄心树
Trees, evergreen

19.0 (8.6~24.3); 11.2 (1.3~18.7); 25.1 (15.0~29.0); 19.1 (8.7~24.3); 169.0 (52.6~231.9); 0.1 (0.0~8.8); 973 (574~1338) ❙ 1470 (939~2435); 716 (392~1428); 96 (37~197) ❙ 52.0 (−11.9~99.5); 960 (574~1326); 315.2 (139.6~659.0)

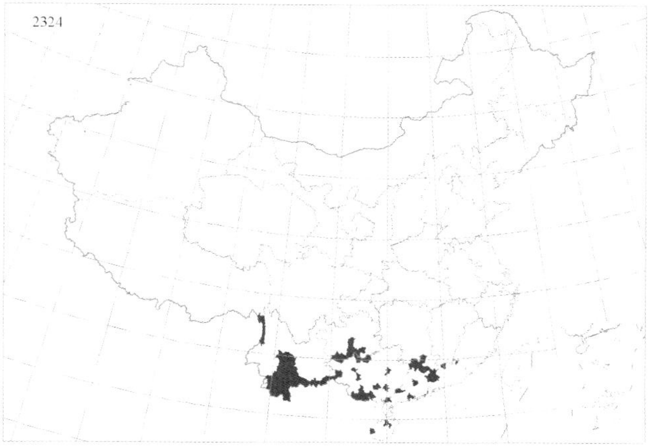

Lauraceae 樟科 *Machilus* 润楠属
Machilus glabrophylla J. F. Zuo 光叶润楠
Trees, evergreen

21.5 (19.2~22.7); 13.4 (11.0~15.0); 27.6 (25.2~28.8); 21.5 (19.2~22.8); 198.0 (170.5~213.1); 0.0 (0.0~0.0); 1128 (952~1221) ❙ 1755 (1302~2435); 935 (617~1428); 116 (72~144) ❙ 55.2 (12.9~99.5); 1128 (952~1221); 274.9 (157.7~318.8)

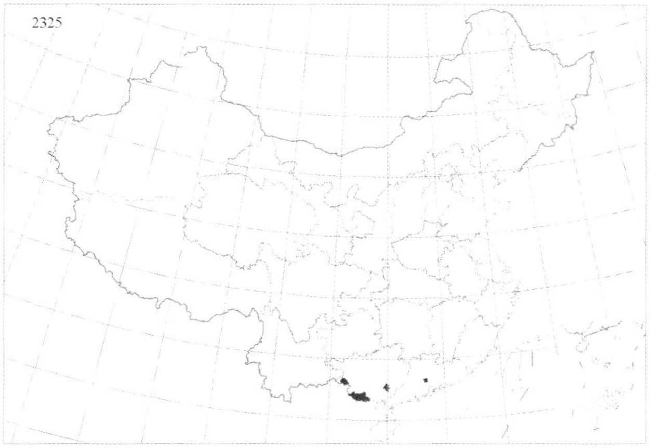

Lauraceae 樟科 *Machilus* 润楠属
Machilus glaucescens (Nees) Wight 柔毛润楠
Trees, evergreen

17.5; 11.1; 21.7; 17.6; 150.7; 0.0; 841 ❙ 1272; 715; 48 ❙ 51.1; 830; 450.1

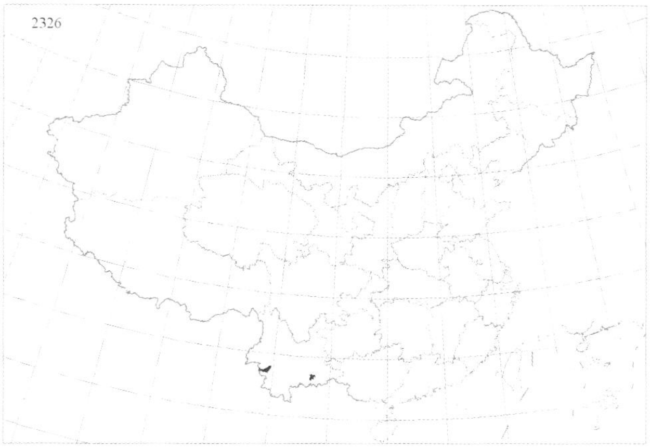

Lauraceae 樟科 *Machilus* 润楠属
Machilus glaucifolia (S. K. Lee et F. N. Wei) Kosterm. 粉叶润楠
Trees, evergreen

18.5 (17.4~19.8); 9.3 (7.4~11.1); 25.8 (24.6~26.6); 18.5 (17.4~19.8); 162.5 (148.9~177.9); 0.0 (0.0~0.0); 938 (886~1010) ❙ 1295 (1182~1381); 672 (637~723); 73 (43~107) ❙ 38.6 (16.9~55.9); 938 (886~1010); 306.2 (245.3~354.5)

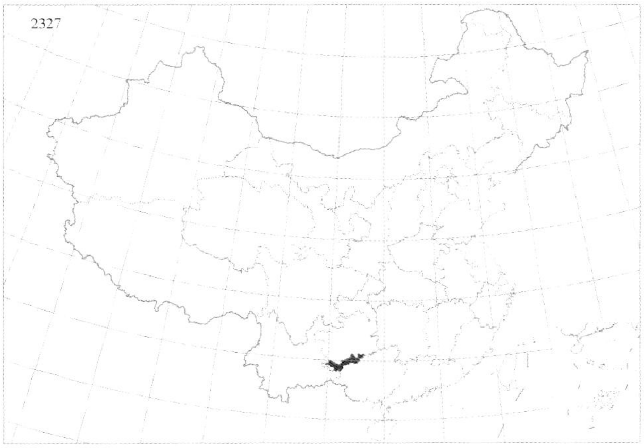

Lauraceae 樟科 *Machilus* 润楠属
Machilus gongshanensis H. W. Li 贡山润楠
Trees, evergreen

9.5 (5.4~13.2); 2.4 (−2.1~6.4); 15.5 (12.3~18.4); 9.6 (5.8~13.3); 62.6 (31.5~99.4); 8.0 (0.0~25.6); 595 (481~695) ❙ 1034 (741~1337); 594 (426~727); 42 (21~69) ❙ 72.7 (52.8~93.0); 595 (481~695); 338.9 (273.5~407.1)

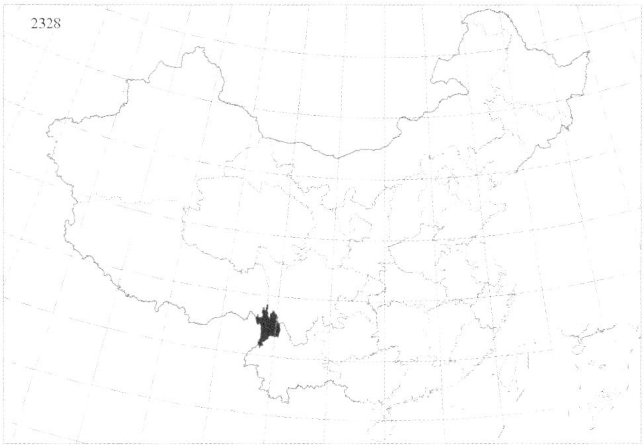

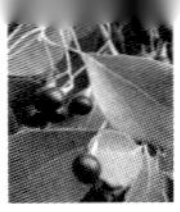

Lauraceae 樟科 *Machilus* 润楠属
Machilus gracillima Chun 柔弱润楠
Small trees, evergreen

17.9; 8.0; 26.1; 18.0; 155.5; 0.0; 918 ❙ 1434; 654; 119 ❙ 56.3; 918; 253.4

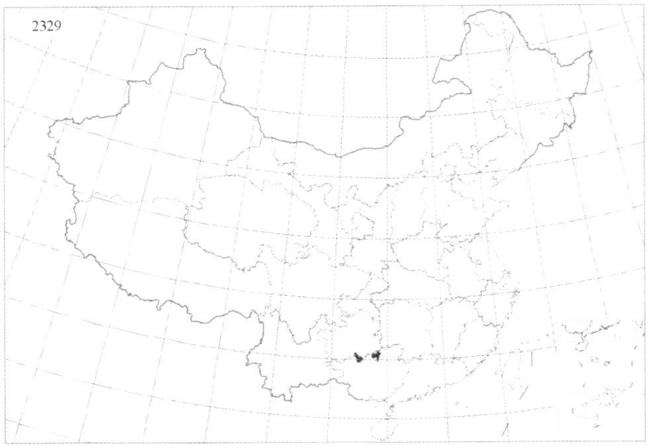

Lauraceae 樟科 *Machilus* 润楠属
Machilus grandibracteata (S. K. Lee et F. N. Wei) Kosterm. 大苞润楠
Trees, evergreen

22.2; 13.9; 28.3; 22.3; 207.5; 0.0; 1190 ❙ 1374; 713; 87 ❙ 15.4; 1190; 327.7

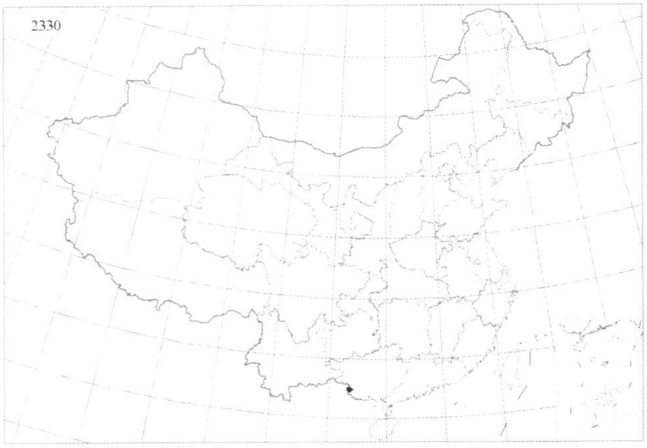

Lauraceae 樟科 *Machilus* 润楠属
Machilus grijsii Hance 黄绒润楠
Trees, evergreen

19.2 (14.3~24.8); 9.7 (3.7~20.1); 27.7 (24.2~29.8); 19.3 (14.3~24.8); 171.5 (113.2~237.6); 0.1 (0.0~1.5); 1022 (761~1385) ❙ 1658 (868~2035); 616 (299~1033); 171 (41~259) ❙ 65.9 (-34.3~130.7); 999 (755~1360); 266.8 (148.6~620.0)

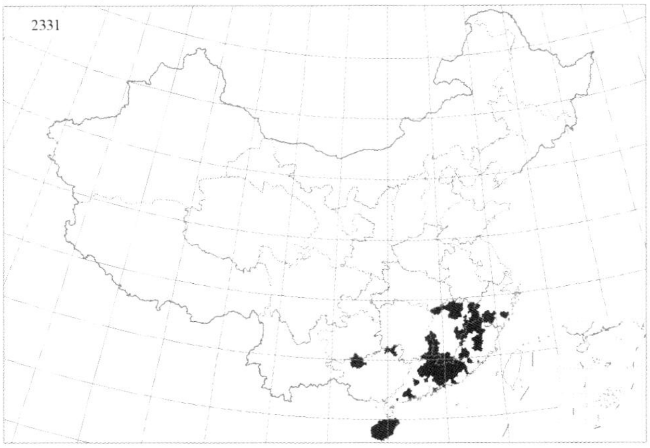

Lauraceae 樟科 *Machilus* 润楠属
Machilus ichangensis Hand.-Mazz. 宜昌润楠
Trees, evergreen

15.9 (7.3~22.4); 4.5 (-4.2~15.2); 26.4 (15.6~30.1); 15.9 (7.5~22.7); 134.2 (48.2~212.7); 3.1 (0.0~29.4); 866 (537~1216) ❙ 1325 (626~1991); 521 (299~994); 126 (14~259) ❙ 52.4 (-22.2~128.9); 851 (537~1216); 218.6 (126.1~334.4)

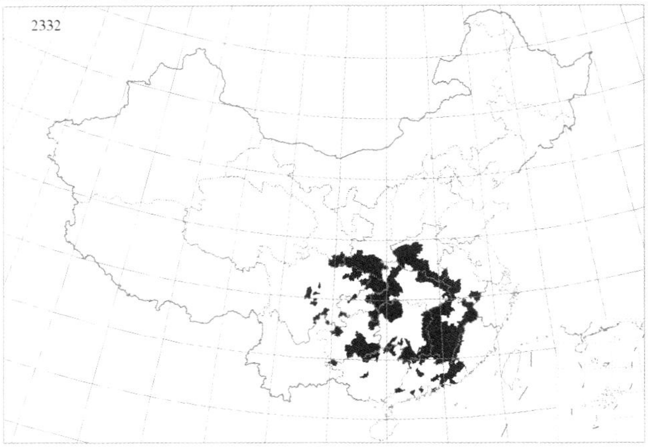

Lauraceae 樟科 *Machilus* 润楠属
Machilus ichangensis var. leiophylla Hand.-Mazz. 滑叶润楠
Trees, evergreen

17.4 (14.3~20.5); 6.6 (3.7~11.0); 27.1 (23.5~29.7); 17.4 (14.3~20.5); 149.0 (112.6~186.1); 0.2 (0.0~1.5); 916 (756~1083) ❙ 1504 (1218~1776); 552 (400~820); 173 (57~225) ❙ 64.9 (36.3~128.9); 907 (756~1083); 221.2 (151.0~371.7)

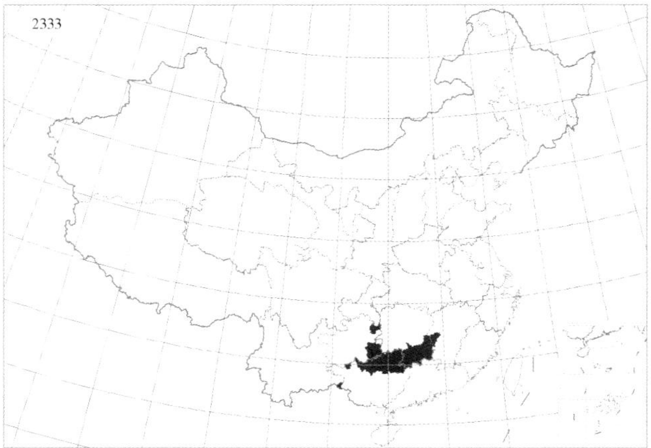

Lauraceae 樟科 *Machilus* 润楠属
Machilus kurzii King ex Hook. f. 秃枝润楠
Trees, evergreen

8.6; 1.3; 15.0; 8.7; 52.6; 8.8; 574 ❙ 1041; 585; 45 ❙ 81.5; 574; 352.3

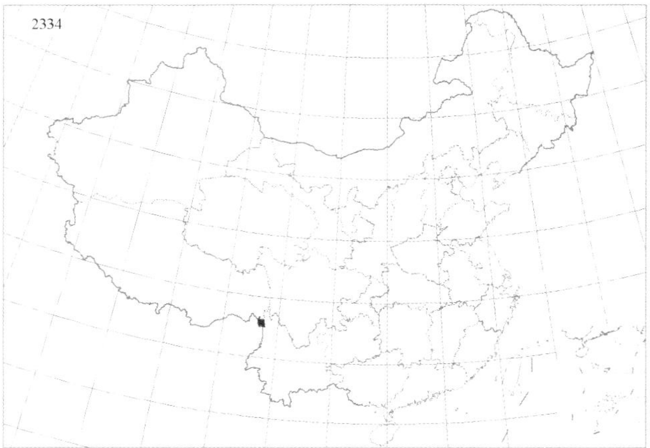

Lauraceae 樟科 *Machilus* 润楠属

Machilus kwangtungensis Y. K. Li 广东润楠

Trees, evergreen

19.1 (14.6~23.4); 9.5 (3.8~16.1); 27.3 (24.6~29.8); 19.1 (14.7~23.5); 169.2 (117.7~221.7); 0.0 (0.0~1.5); 997 (779~1273) ❙ 1644 (1268~2035); 654 (476~969); 163 (73~199) ❙ 66.0 (11.9~115.7); 996 (779~1273); 237.3 (164.9~302.2)

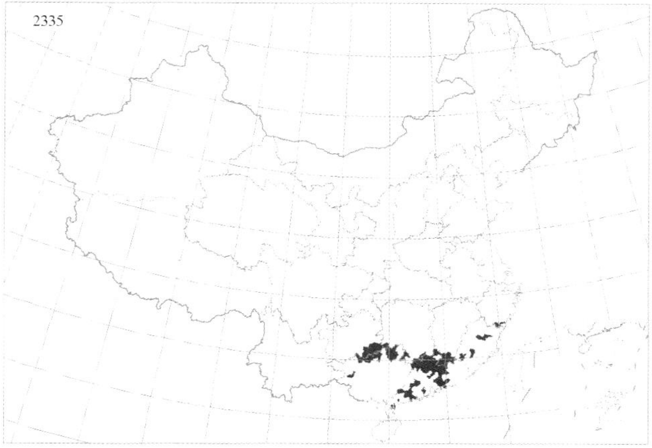

Lauraceae 樟科 *Machilus* 润楠属

Machilus lenticellata S. K. Lee et F. N. Wei 疣序润楠

Trees, evergreen

19.0; 9.5; 26.7; 19.1; 169.0; 0.0; 980 ❙ 1389; 645; 112 ❙ 41.7; 980; 285.3

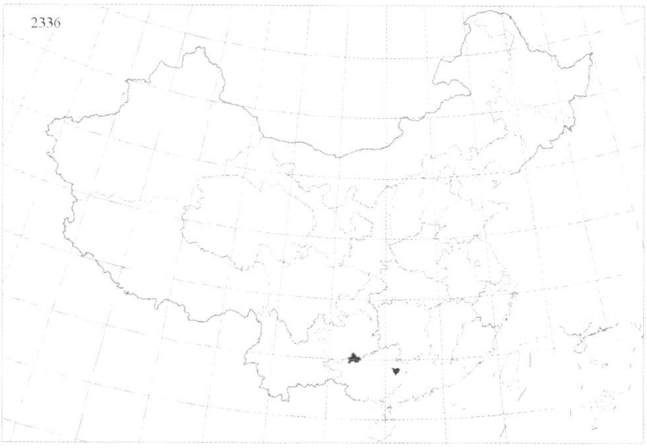

Lauraceae 樟科 *Machilus* 润楠属

Machilus leptophylla Hand.-Mazz. 薄叶润楠

Trees, evergreen

17.3 (13.3~22.7); 6.3 (1.6~15.2); 27.4 (22.2~29.9); 17.3 (13.3~22.8); 148.4 (101.9~213.1); 0.5 (0.0~6.5); 921 (718~1221) ❙ 1527 (871~2435); 566 (299~1428); 172 (43~259) ❙ 66.3 (-6.7~130.7); 913 (718~1221); 223.4 (134.6~371.7)

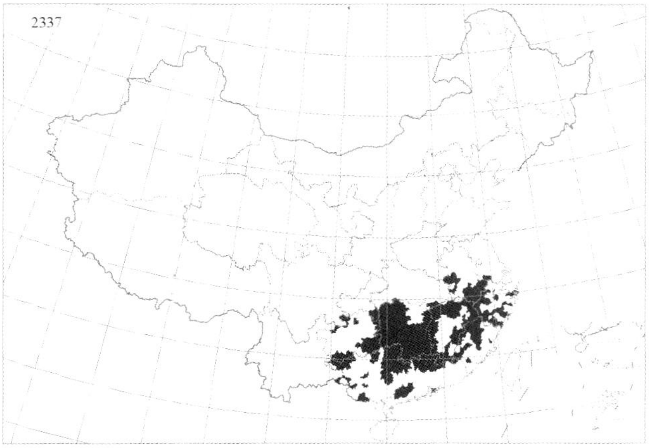

Lauraceae 樟科 *Machilus* 润楠属

Machilus lichuanensis W. C. Cheng ex S. K. Lee 利川润楠

Trees, evergreen

16.5 (12.7~18.7); 5.8 (2.1~8.1); 26.5 (23.0~29.0); 16.5 (12.8~18.8); 133.7 (99.0~156.0); 0.5 (0.0~5.6); 875 (713~1002) ❙ 1195 (985~1671); 519 (437~816); 82 (40~221) ❙ 38.0 (8.0~107.2); 873 (713~1002); 168.7 (108.2~257.1)

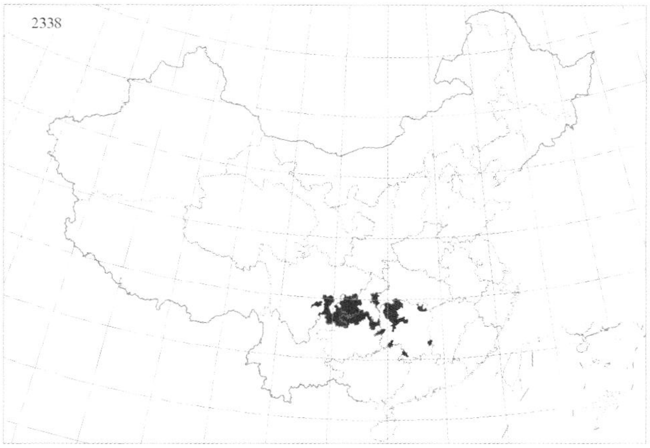

Lauraceae 樟科 *Machilus* 润楠属

Machilus litseifolia S. K. Lee 木姜润楠

Trees, evergreen

17.5 (14.3~21.0); 7.1 (3.3~12.0); 26.8 (23.5~29.0); 17.5 (14.3~21.1); 150.8 (112.6~193.0); 0.2 (0.0~2.3); 919 (756~1114) ❙ 1526 (1211~2005); 589 (403~824); 161 (57~230) ❙ 67.0 (30.9~129.9); 916 (756~1114); 230.7 (151.0~371.7)

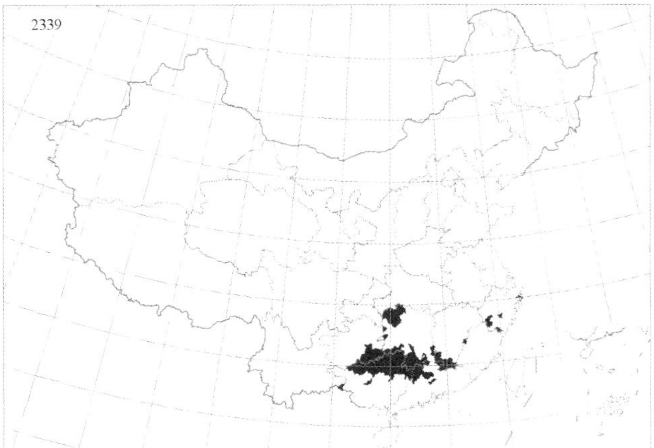

Lauraceae 樟科 *Machilus* 润楠属

Machilus lohuiensis S. K. Lee 乐会润楠

Trees, evergreen

24.2; 18.8; 28.0; 24.2; 230.6; 0.0; 1327 ❙ 1617; 538; 82 ❙ 21.5; 1236; 441.9

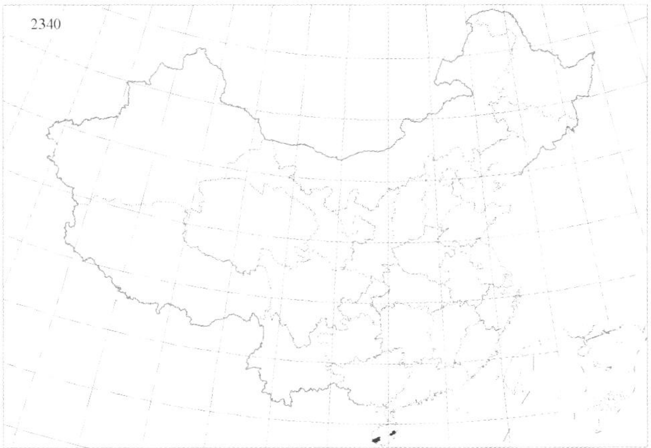

Lauraceae 樟科 *Machilus* 润楠属
Machilus longipes H. T. Chang 东莞润楠
Shrubs, evergreen

22.3；14.3；28.8；22.4；208.7；0.0；1197 ❙ 1848；878；138 ❙ 54.4；1197；161.4

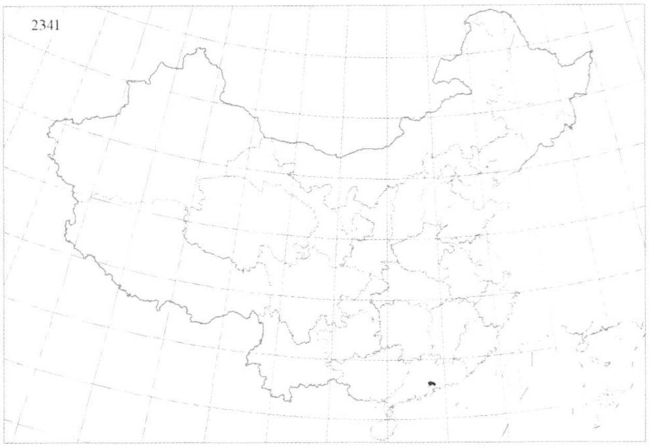

Lauraceae 樟科 *Machilus* 润楠属
Machilus mangdangshanensis Q. F. Zheng 茫荡山润楠
Shrubs or small trees, evergreen

18.6；9.3；27.6；18.7；164.0；0.0；965 ❙ 1702；591；199 ❙ 76.4；965；300.0

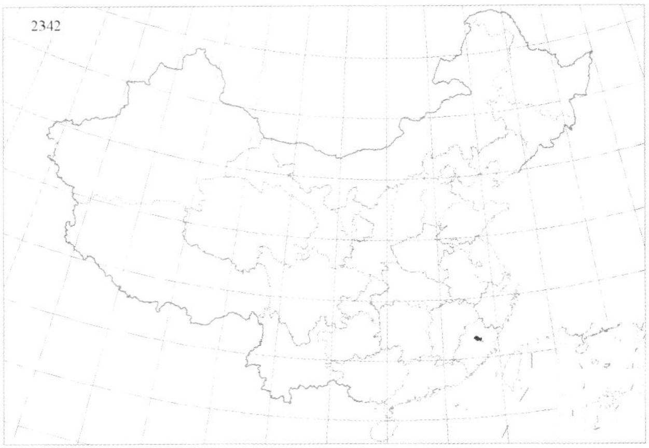

Lauraceae 樟科 *Machilus* 润楠属
Machilus melanophylla H. W. Li 暗叶润楠
Trees, evergreen

20.4；15.1；23.6；20.5；185.8；0.0；989 ❙ 1607；702；77 ❙ 62.5；986；659.0

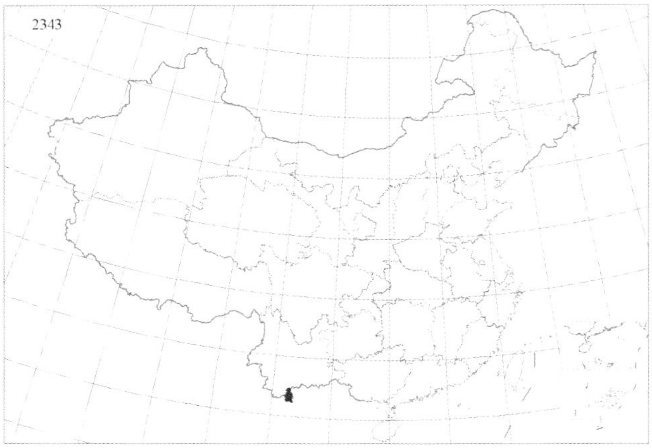

Lauraceae 樟科 *Machilus* 润楠属
Machilus miaoshanensis F. N. Wei et C. Q. Lin 苗山润楠
Trees, evergreen

17.6；7.3；26.3；17.6；151.6；0.0；909 ❙ 1496；639；142 ❙ 64.7；909；225.9

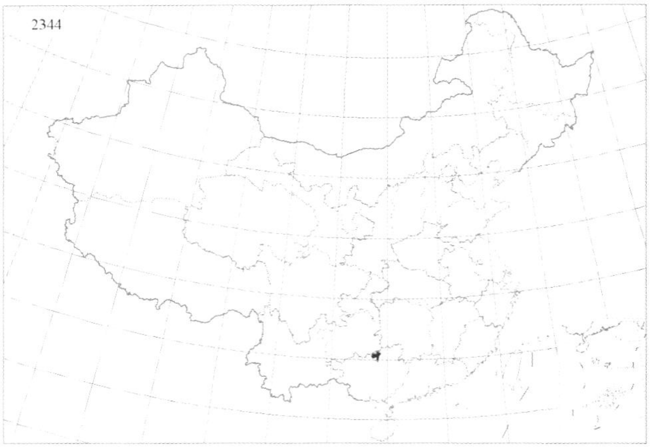

Lauraceae 樟科 *Machilus* 润楠属
Machilus microcarpa S. K. Lee 小果润楠
Trees, evergreen

14.6 (7.3~21.8)；2.6 (−2.9~14.0)；25.7 (15.6~28.8)；14.6 (7.5~21.9)；120.9 (48.2~202.2)；5.4 (0.0~20.5)；816 (537~1150) ❙ 1119 (626~2148)；504 (316~1234)；81 (26~197) ❙ 37.4 (−22.2~108.0)；797 (537~1150)；199.1 (133.8~318.8)

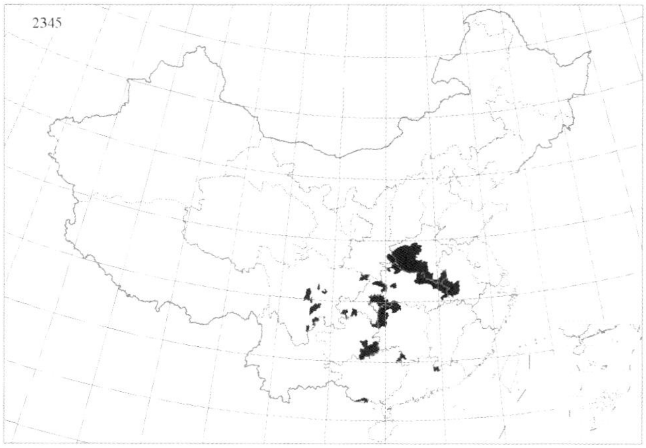

Lauraceae 樟科 *Machilus* 润楠属
Machilus microcarpa var. omeiensis S. K. Lee 峨眉润楠
Trees, evergreen

14.9；5.3；23.6；15.0；119.6；0.0；782 ❙ 1410；816；49 ❙ 80.4；782；214.5

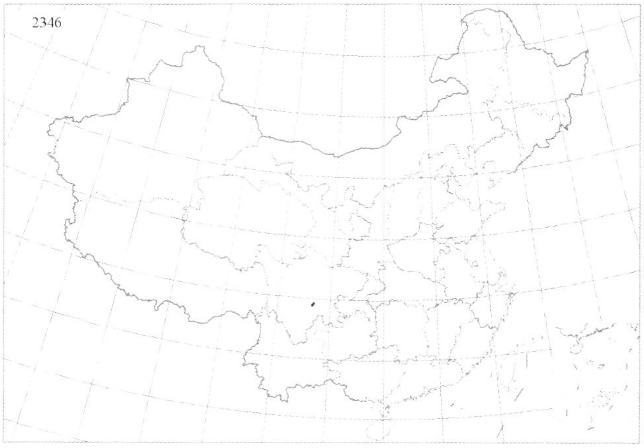

Lauraceae 樟科 *Machilus* 润楠属

Machilus minkweiensis S. K. Lee 闽桂润楠

Trees, evergreen

21.9 (19.7~24.1); 14.3 (11.5~18.2); 27.8 (27.1~28.5); 22.0 (19.7~24.2); 203.7 (176.5~23C.1); 0.0 (0.0~0.0); 1164 (1001~1326) ❙ 1485 (1188~1723); 648 (544~839); 97 (53~163) ❙ 28.1 (5.8~62.9); 1149 (1001~1326); 338.7 (256.6~538.2)

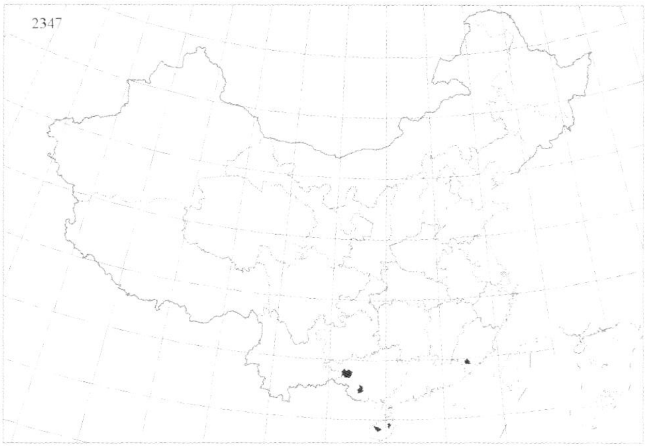

Lauraceae 樟科 *Machilus* 润楠属

Machilus minutiloba S. K. Lee 雁荡润楠

Small trees, evergreen

16.7 (15.8~18.2); 6.2 (4.5~7.8); 27.3 (25.8~28.7); 16.8 (15.8~18.3); 141.5 (130.0~159.6); 0.1 (0.0~0.5); 884 (831~965) ❙ 1687 (1585~1791); 620 (559~659); 186 (176~195) ❙ 91.3 (72.4~115.7); 884 (831~965); 225.8 (189.8~257.7)

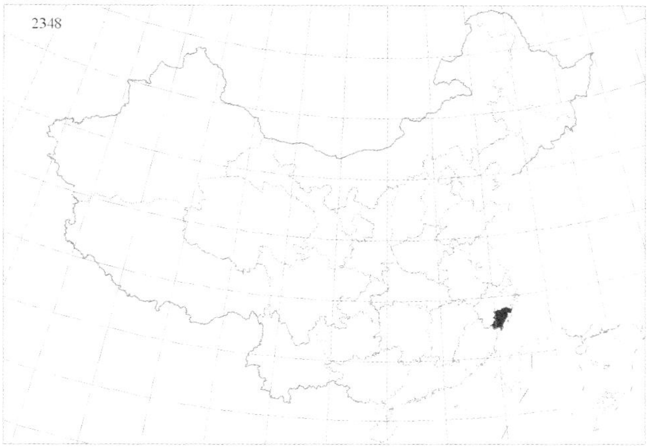

Lauraceae 樟科 *Machilus* 润楠属

Machilus monticola S. K. Lee 尖峰润楠

Trees, evergreen

23.9 (23.0~24.3); 18.5 (17.7~19.1); 27.7 (26.7~28.4); 23.9 (23.1~24.3); 226.8 (217.0~231.9); 0.0 (0.0~0.0); 1295 (1217~1338) ❙ 1512 (1179~1725); 523 (392~619); 63 (42~110) ❙ 17.1 (-11.9~41.8); 1161 (1056~1326); 491.3 (352.9~620.0)

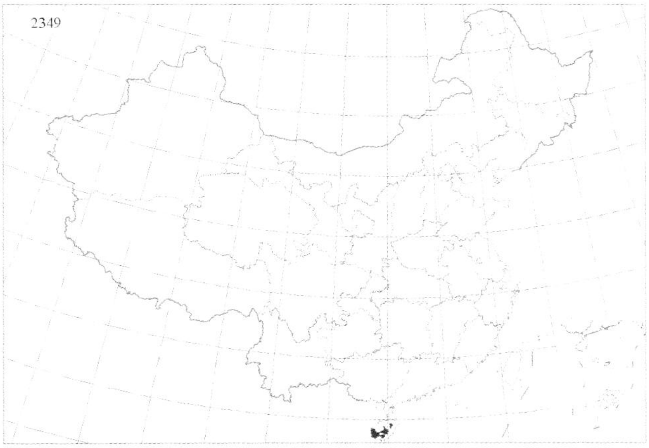

Lauraceae 樟科 *Machilus* 润楠属

Machilus multinervia H. Liu 多脉润楠

Trees, evergreen

19.2 (16.9~22.2); 10.3 (6.8~13.9); 26.0 (23.4~28.3); 19.2 (17.0~22.3); 170.4 (143.8~207.5); 0.0 (0.0~0.0); 984 (845~1190) ❙ 1301 (1172~1437); 666 (566~723); 80 (50~117) ❙ 33.8 (12.9~48.8); 984 (845~1190); 302.6 (236.8~411.1)

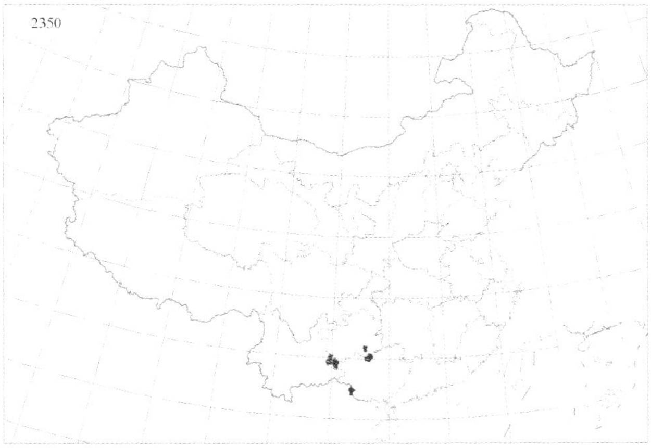

Lauraceae 樟科 *Machilus* 润楠属

Machilus nakao S. K. Lee 纳槁润楠

Trees, evergreen

23.6 (21.8~24.7); 17.3 (13.7~19.8); 28.1 (26.7~28.8); 23.6 (21.9~24.7); 223.4 (202.7~236.5); 0.0 (0.0~0.0); 1283 (1159~1376) ❙ 1583 (1309~1813); 627 (480~839); 93 (48~128) ❙ 23.8 (-1.4~41.8); 1234 (1147~1331); 373.0 (245.3~620.0)

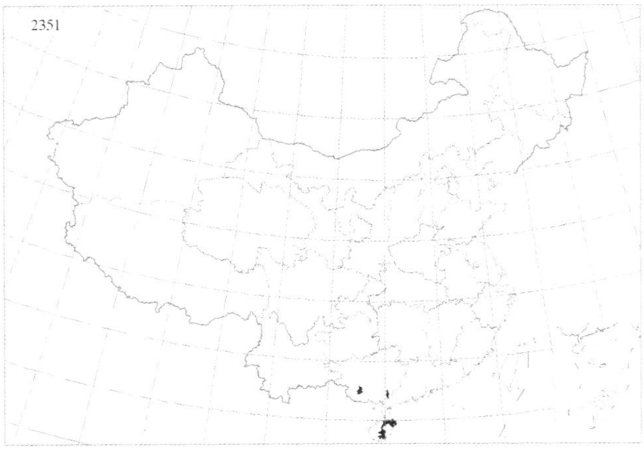

Lauraceae 樟科 *Machilus* 润楠属

Machilus nanchuanensis N. Chao ex S. K. Lee 南川润楠

Trees, evergreen

14.8; 4.3; 24.9; 14.9; 119.2; 0.7; 791 ❙ 1197; 489; 71 ❙ 51.2; 791; 182.9

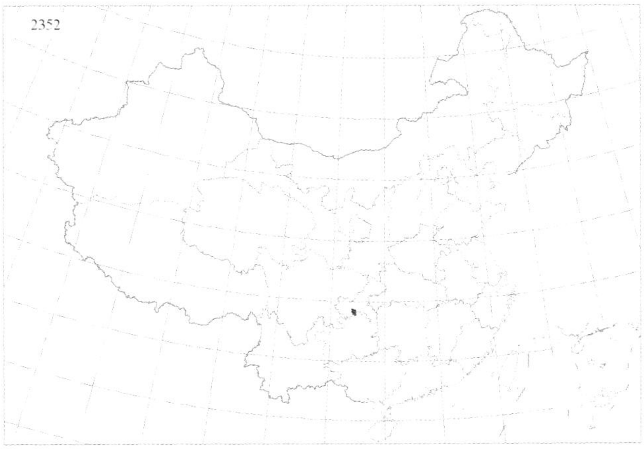

Lauraceae 樟科 *Machilus* 润楠属

Machilus nanmu (Oliv.) Hemsl. 润楠

Trees, evergreen

15.7 (10.2~19.5); 5.6 (-0.7~13.8); 25.1 (19.6~28.9); 15.8 (10.4~19.5); 130.7 (77.2~173.8); 1.6 (0.0~13.6); 836 (634~1002) ❙ 1209 (950~1736); 567 (420~1007); 60 (27~105) ❙ 47.4 (8.0~145.8); 835 (634~1002); 217.8 (108.2~627.0)

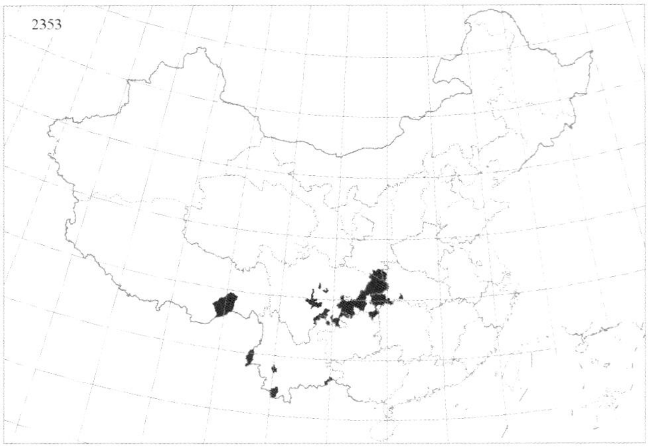

Lauraceae 樟科 *Machilus* 润楠属

Machilus obovatifolia (Hayata) Kaneh. et Sasaki 倒卵叶润楠

Small trees, evergreen

22.2; 17.7; 25.5; 22.2; 206.7; 0.0; 1123 ❙ 2992; 1847; 107 ❙ 166.5; 1123; 296.9

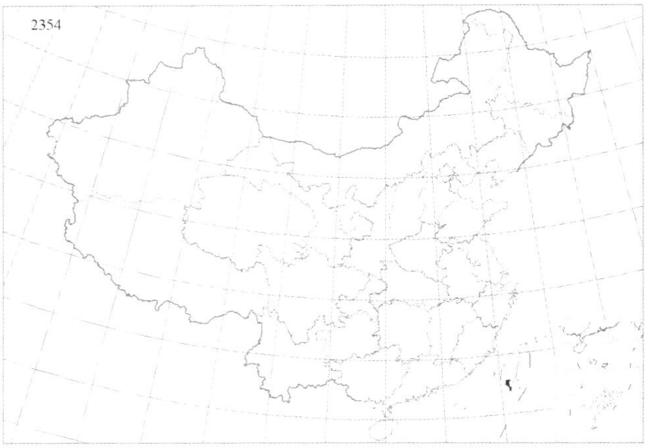

Lauraceae 樟科 *Machilus* 润楠属

Machilus obscurinervis S. K. Lee 隐脉润楠

Trees, evergreen

13.3; 5.4; 19.6; 13.4; 100.4; 0.0; 706 ❙ 1736; 1007; 51 ❙ 145.8; 706; 447.4

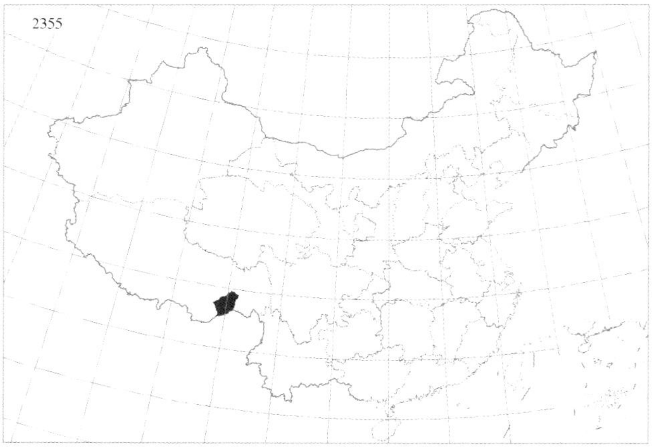

Lauraceae 樟科 *Machilus* 润楠属

Machilus oculodracontis Chun 龙眼润楠

Trees, evergreen

19.1 (17.3~22.1); 9.2 (6.4~13.7); 27.7 (26.9~28.8); 19.2 (17.3~22.1); 169.9 (147.9~205.6); 0.0 (0.0~0.0); 1001 (907~1182) ❙ 1682 (1579~1784); 629 (541~828); 183 (125~197) ❙ 68.6 (50.9~78.8); 1001 (907~1182); 232.1 (157.7~261.3)

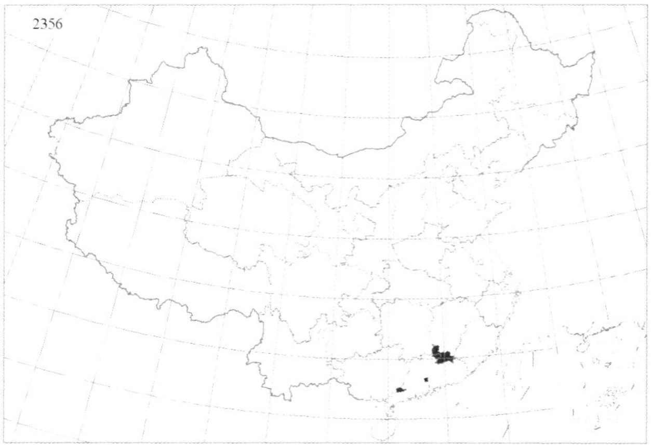

Lauraceae 樟科 *Machilus* 润楠属

Machilus oreophila Hance 建润楠

Shrubs or trees, evergreen

18.7 (14.3~22.7); 9.1 (3.3~15.2); 27.1 (23.0~29.8); 18.8 (14.3~22.8); 165.8 (114.1~213.1); 0.1 (0.0~2.3); 981 (765~1221) ❙ 1602 (1215~2435); 634 (403~1428); 165 (58~235) ❙ 64.4 (12.9~129.9); 978 (765~1221); 243.5 (151.0~570.1)

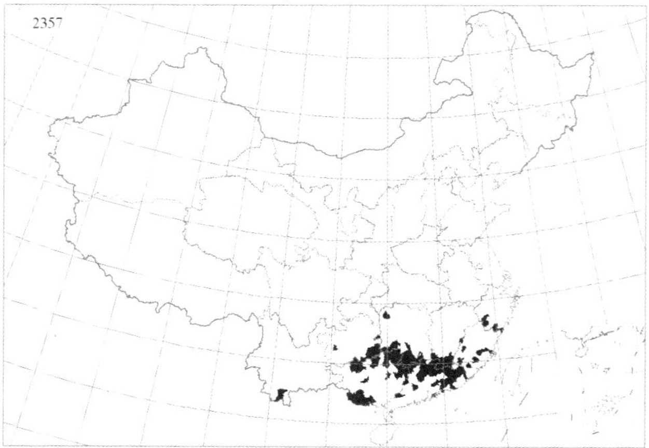

Lauraceae 樟科 *Machilus* 润楠属

Machilus ovatiloba S. K. Lee 糙枝润楠

Trees, evergreen

13.3; 5.4; 19.6; 13.4; 100.4; 0.0; 706 ❙ 1736; 1007; 51 ❙ 145.8; 706; 447.4

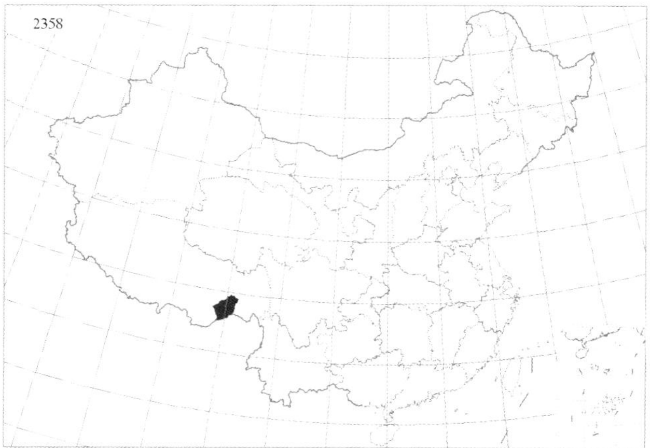

Lauraceae 樟科 *Machilus* 润楠属

Machilus parabreviflora H. T. Chang 赛短花润楠

Shrubs, evergreen

21.7 (19.1~22.7); 13.5 (9.8~15.0); 27.9 (26.6~28.6); 21.8 (19.2~22.8); 201.2 (170.1~210.1); 0.0 (0.0~0.0); 1152 (981~1221) ❙ 1704 (1284~2435); 929 (551~1428); 110 (73~130) ❙ 47.7 (12.9~99.5); 1152 (981~1221); 291.7 (235.1~327.7)

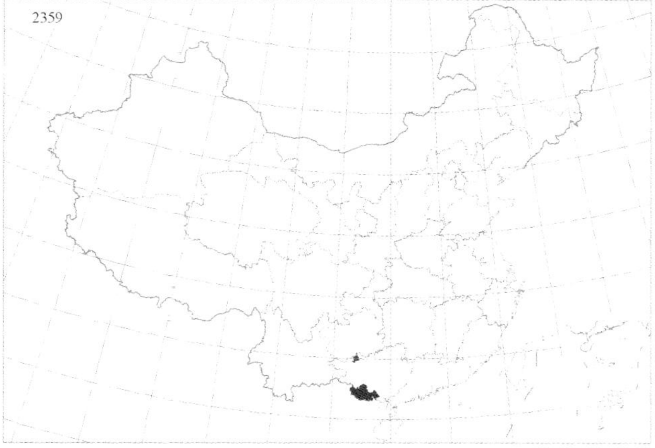

Lauraceae 樟科 *Machilus* 润楠属

Machilus pauhoi Kaneh. 刨花润楠

Trees, evergreen

18.6 (14.3~22.7); 8.8 (2.5~15.2); 27.5 (24.2~29.8); 18.7 (14.3~22.8); 164.7 (113.2~213.1); 0.1 (0.0~3.9); 983 (761~1221) ❙ 1651 (1215~2435); 633 (445~1428); 178 (57~259) ❙ 69.8 (17.4~130.7); 980 (761~1221); 238.3 (174.9~371.7)

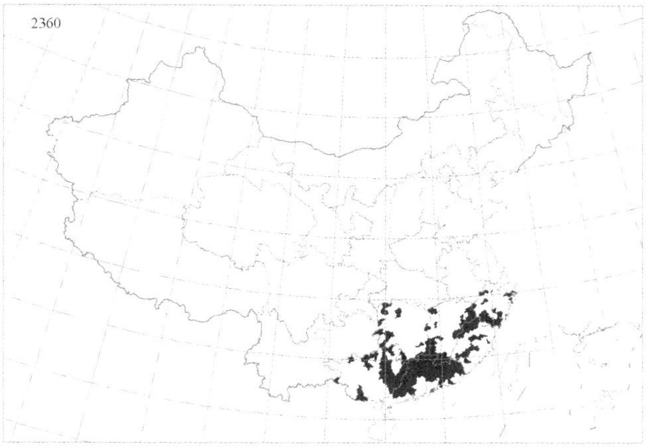

Lauraceae 樟科 *Machilus* 润楠属

Machilus phoenicis Dunn 凤凰润楠

Trees, evergreen

18.2 (14.3~22.2); 8.2 (3.3~14.7); 27.3 (24.2~29.8); 18.3 (14.3~22.2); 159.4 (113.2~206.9); 0.1 (0.0~2.3); 958 (761~1177) ❙ 1638 (1221~1997); 612 (403~902); 181 (57~258) ❙ 72.7 (22.7~147.0); 954 (761~1177); 232.8 (151.0~371.7)

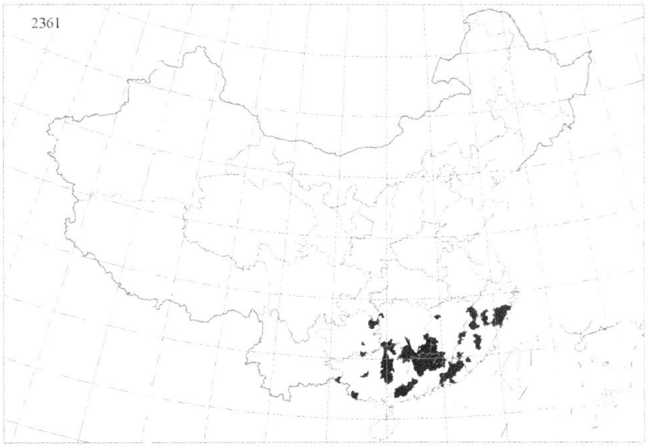

Lauraceae 樟科 *Machilus* 润楠属

Machilus platycarpa Chun 扁果润楠

Trees, evergreen

21.6 (17.0~23.2); 13.4 (6.2~15.8); 27.9 (26.7~28.9); 21.6 (17.0~23.2); 199.5 (144.4~218.9); 0.0 (0.0~0.0); 1143 (891~1257) ❙ 1721 (1302~2435); 862 (594~1428); 122 (86~191) ❙ 51.0 (12.9~99.5); 1143 (891~1257); 260.6 (194.5~327.7)

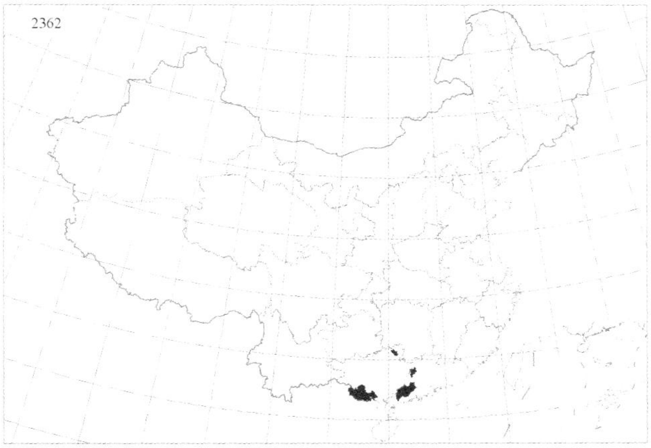

Lauraceae 樟科 *Machilus* 润楠属

Machilus pomifera (Kosterm.) S. K. Lee 梨润楠

Trees, evergreen

23.7 (21.8~24.8); 18.2 (14.2~20.1); 27.5 (26.7~28.0); 23.7 (21.9~24.8); 224.4 (202.6~237.6); 0.0 (0.0~0.0); 1286 (1150~1385) ❙ 1629 (1405~1876); 609 (473~891); 66 (41~116) ❙ 28.2 (1.5~63.2); 1162 (1137~1212); 497.7 (273.3~620.0)

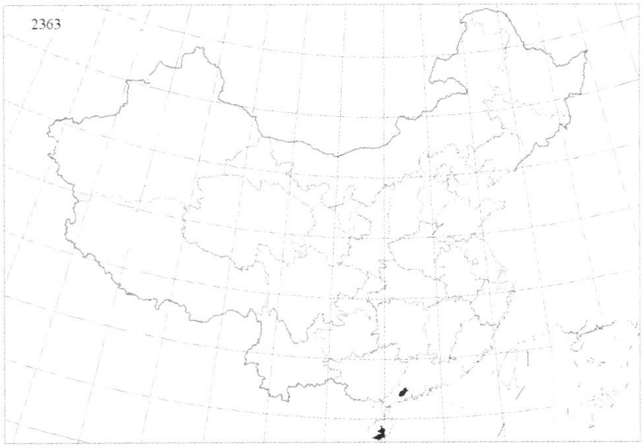

Lauraceae 樟科 *Machilus* 润楠属

Machilus pyramidalis H. W. Li 塔序润楠

Shrubs or small trees, evergreen

18.3 (17.2~21.0); 11.2 (9.5~14.2); 23.3 (22.4~25.6); 18.4 (17.3~21.0); 160.2 (147.0~191.8); 0.0 (0.0~0.0); 892 (833~1053) ❙ 1416 (1254~1522); 797 (699~860); 57 (48~68) ❙ 59.4 (44.5~80.6); 892 (833~1053); 443.1 (336.3~551.6)

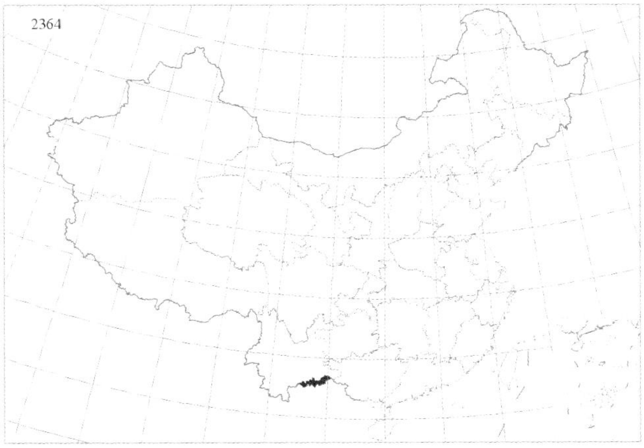

Lauraceae 樟科 *Machilus* 润楠属
Machilus rehderi C. K. Allen 狭叶润楠
Small trees, evergreen

17.0 (14.3~21.3); 6.5 (3.3~12.3); 26.3 (23.0~29.1); 17.0 (14.3~21.3); 144.7 (112.6~196.1); 0.2 (0.0~2.3); 891 (756~1130) ❙ 1393 (1165~1776); 563 (403~821); 130 (51~221) ❙ 56.7 (27.5~101.8); 888 (756~1130); 218.1 (146.0~318.5)

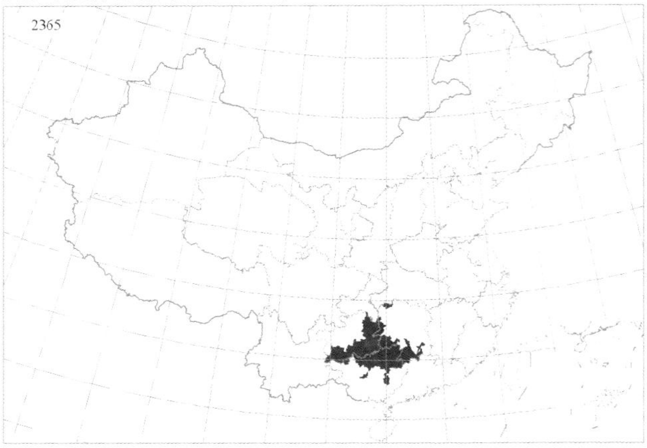

Lauraceae 樟科 *Machilus* 润楠属
Machilus reticulata K. M. Lan 网脉润楠
Trees, evergreen

17.1; 8.5; 23.4; 17.1; 145.1; 0.0; 845 ❙ 1232; 666; 59 ❙ 45.8; 845; 283.9

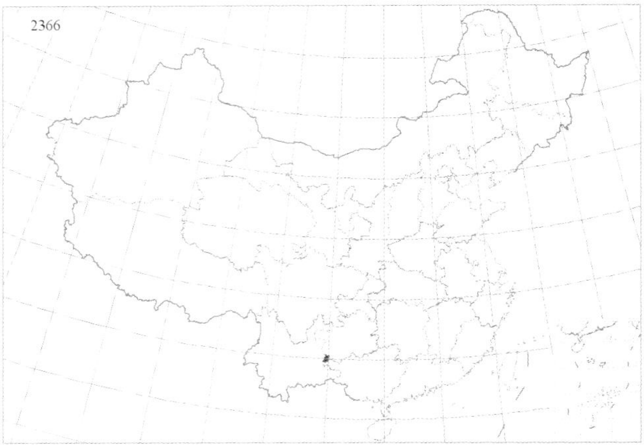

Lauraceae 樟科 *Machilus* 润楠属
Machilus robusta W. W. Sm. 粗壮润楠
Trees, evergreen

18.4 (8.6~24.7); 10.7 (1.3~19.8); 24.1 (15.0~29.0); 18.4 (8.7~24.7); 161.3 (52.6~236.5); 0.1 (0.0~8.8); 931 (574~1376) ❙ 1399 (797~2005); 707 (411~1007); 85 (23~194) ❙ 51.4 (-14.2~145.8); 919 (574~1326); 341.2 (157.7~659.0)

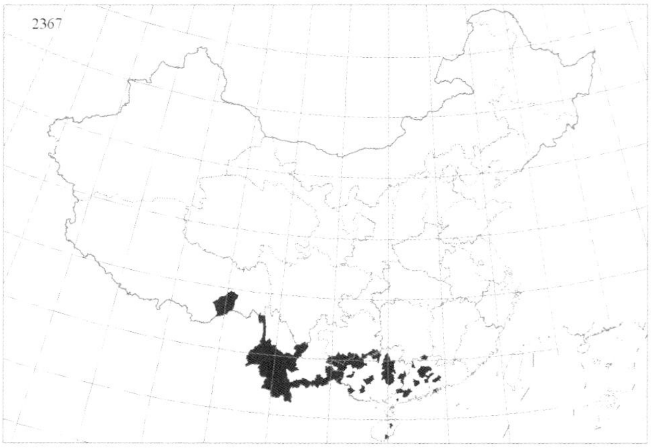

Lauraceae 樟科 *Machilus* 润楠属
Machilus rufipes H. W. Li 红梗润楠
Trees, evergreen

18.3 (13.3~21.0); 11.8 (5.4~15.1); 22.8 (19.6~25.6); 18.4 (13.4~21.0); 160.5 (100.4~191.8); 0.0 (0.0~0.0); 889 (706~1053) ❙ 1339 (939~1751); 738 (524~1007); 55 (37~77) ❙ 51.2 (6.5~145.8); 871 (706~1053); 420.0 (301.1~659.0)

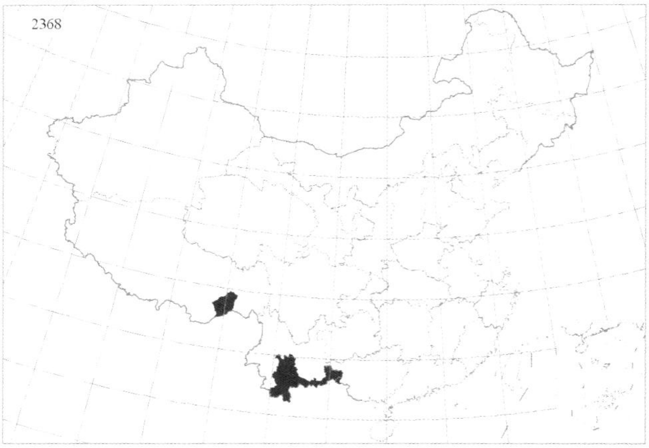

Lauraceae 樟科 *Machilus* 润楠属
Machilus salicina Hance 柳叶润楠
Shrubs, evergreen

19.0 (8.6~24.6); 10.6 (1.3~19.5); 25.6 (15.0~29.8); 19.0 (8.7~24.7); 168.4 (52.6~235.9); 0.1 (0.0~8.8); 976 (574~1372) ❙ 1505 (939~2435); 688 (411~1428); 117 (37~204) ❙ 55.6 (6.5~112.2); 967 (574~1299); 305.1 (157.7~659.0)

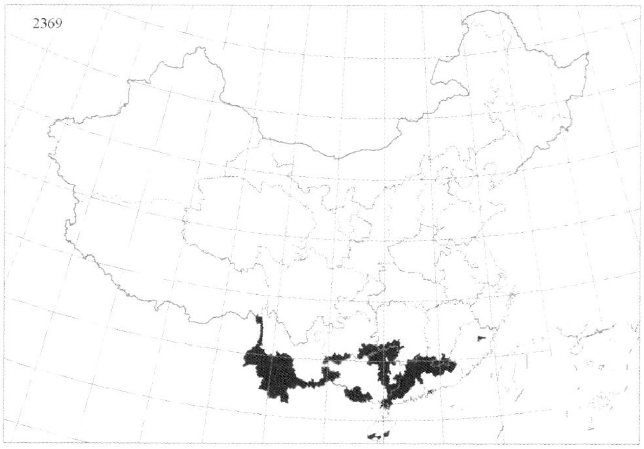

Lauraceae 樟科 *Machilus* 润楠属
Machilus salicoides S. K. Lee 华蓥润楠
Trees, evergreen

17.4; 6.6; 27.6; 17.5; 149.6; 0.0; 928 ❙ 1114; 459; 60 ❙ 20.2; 928; 132.5

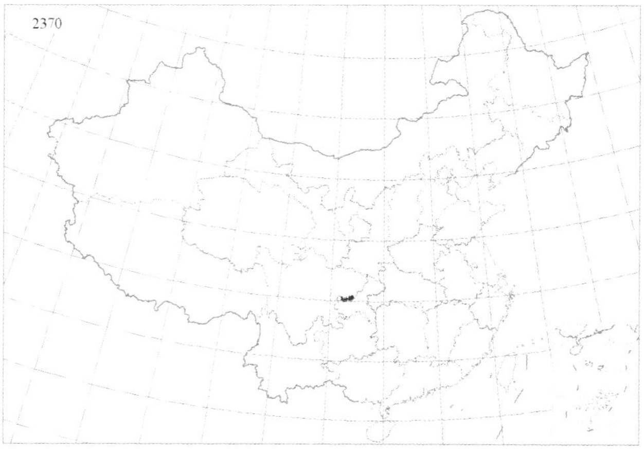

Lauraceae 樟科 *Machilus* 润楠属
Machilus shiwandashanica H. T. Chang 十万大山润楠
Shrubs, evergreen

22.0 (21.3~22.7); 13.9 (13.2~15.0); 27.9 (27.3~28.6); 22.0 (21.3~22.8); 204.1 (195.9~213.1); 0.0 (0.0~0.0); 1166 (1111~1221) ❙ 1804 (1302~2435); 999 (670~1428); 112 (86~130) ❙ 54.6 (12.9~99.5); 1166 (1111~1221); 295.0 (235.1~327.7)

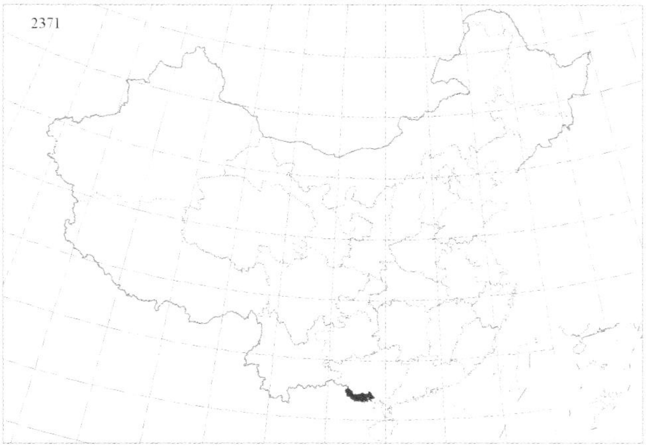

Lauraceae 樟科 *Machilus* 润楠属
Machilus shweliensis W. W. Sm. 瑞丽润楠
Shrubs or trees, evergreen

17.5; 10.8; 22.1; 17.5; 150.3; 0.0; 857 ❙ 1515; 881; 51 ❙ 80.4; 852; 457.3

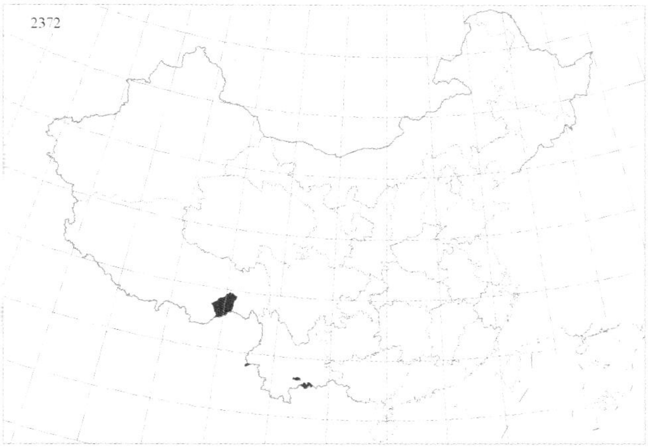

Lauraceae 樟科 *Machilus* 润楠属
Machilus sichourensis H. W. Li 西畴润楠
Trees, evergreen

18.4 (17.2~21.0); 11.7 (9.5~14.2); 23.1 (22.1~25.6); 18.5 (17.3~21.0); 161.7 (147.0~191.8); 0.0 (0.0~0.0); 893 (833~1053) ❙ 1406 (1254~1547); 801 (699~911); 56 (48~68) ❙ 58.0 (36.8~80.6); 892 (833~1053); 455.0 (336.3~551.6)

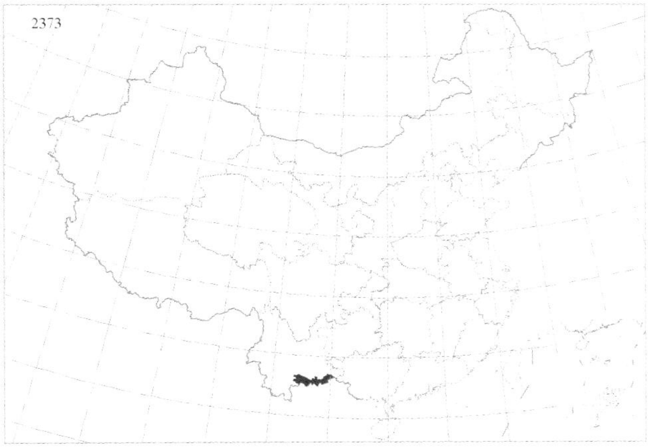

Lauraceae 樟科 *Machilus* 润楠属
Machilus sichuanensis N. Chao 四川润楠
Trees, evergreen

13.3; 3.2; 22.1; 13.2; 100.9; 2.5; 723 ❙ 1081; 606; 32 ❙ 49.4; 723; 171.0

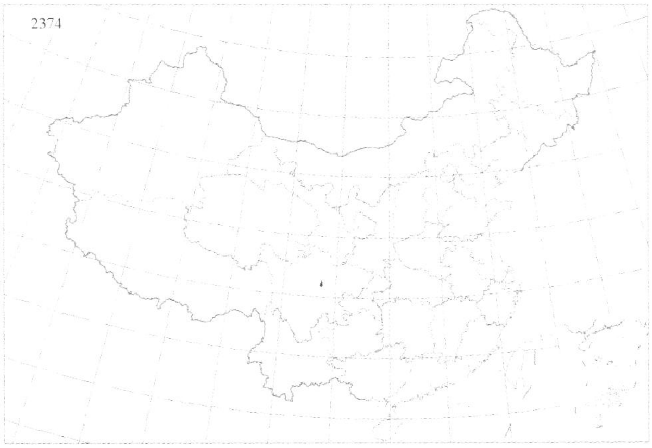

Lauraceae 樟科 *Machilus* 润楠属
Machilus submultinervia Y. K. Li 册亨润楠
Trees, evergreen

19.1; 10.6; 25.5; 19.1; 169.4; 0.0; 955 ❙ 1347; 761; 54 ❙ 41.1; 955; 327.0

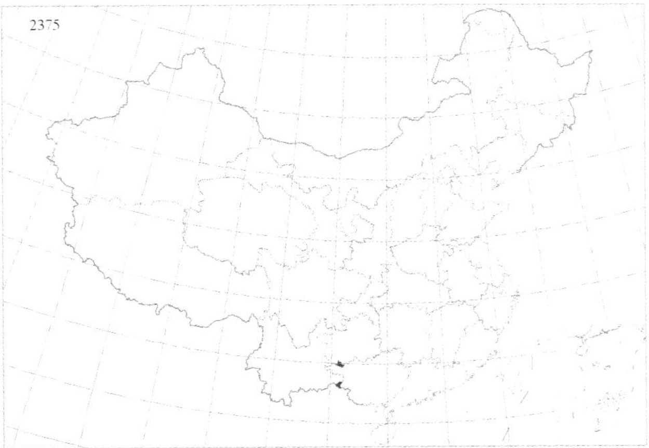

Lauraceae 樟科 *Machilus* 润楠属
Machilus tenuipilis (H. W. Li) Kosterm. 细毛润楠
Trees, evergreen

17.8 (8.6~20.5); 11.5 (1.3~15.1); 21.8 (15.0~23.7); 17.8 (8.7~20.5); 154.1 (52.6~185.9); 0.4 (0.0~8.8); 860 (574~992) ❙ 1427 (958~1751); 789 (531~996); 54 (42~77) ❙ 66.5 (13.3~112.2); 850 (574~986); 479.3 (311.9~659.0)

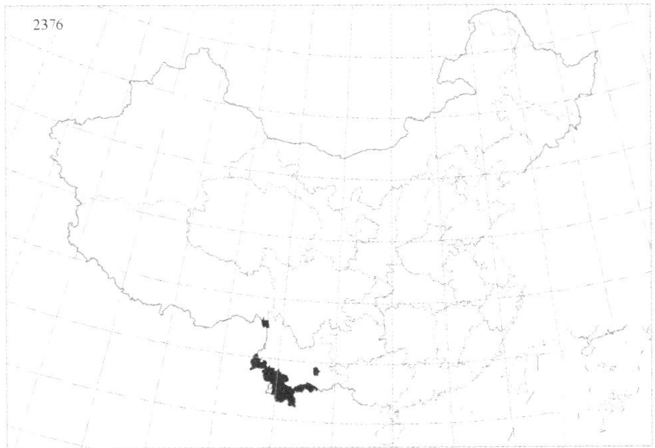

Lauraceae 樟科 *Machilus* 润楠属
Machilus thunbergii Siebold et Zucc. 红楠
Trees, evergreen

18.2 (11.9~24.1); 7.8 (-2.0~18.9); 27.8 (20.1~30.1); 18.3 (12.2~24.1); 159.5 (100.7~229.7); 0.3 (0.0~17.1); 971 (729~1322) ❙ 1614 (747~3495); 602 (299~1594); 183 (34~897) ❙ 67.9 (-6.9~246.5); 959 (729~1273); 220.6 (95.3~363.7)

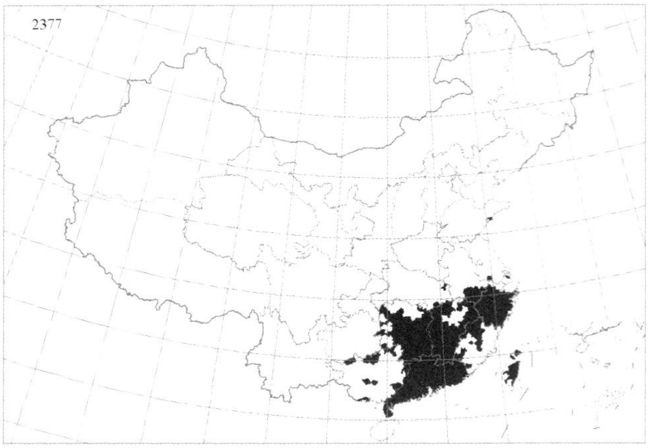

Lauraceae 樟科 *Machilus* 润楠属
Machilus velutina Champ. ex Benth. 绒毛润楠
Trees, evergreen

19.3 (14.3~24.3); 9.8 (3.3~19.1); 27.6 (23.5~29.8); 19.4 (14.3~24.3); 172.4 (112.6~231.9); 0.1 (0.0~2.3); 1021 (756~1338) ❙ 1618 (868~2435); 642 (312~1428); 163 (42~259) ❙ 61.0 (-34.3~147.0); 1011 (756~1273); 234.4 (95.3~620.0)

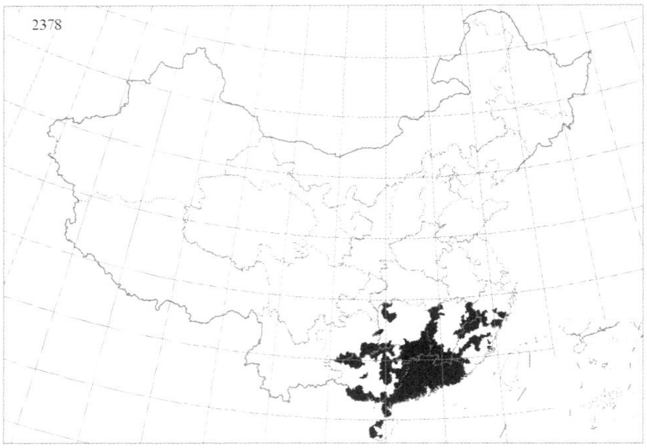

Lauraceae 樟科 *Machilus* 润楠属
Machilus velutinoides (S. K. Lee et F. N. Wei) Kosterm. 东兴润楠
Shrubs, evergreen

22.3; 14.5; 28.1; 22.3; 207.7; 0.0; 1186 ❙ 2291; 1331; 127 ❙ 93.1; 1186; 309.7

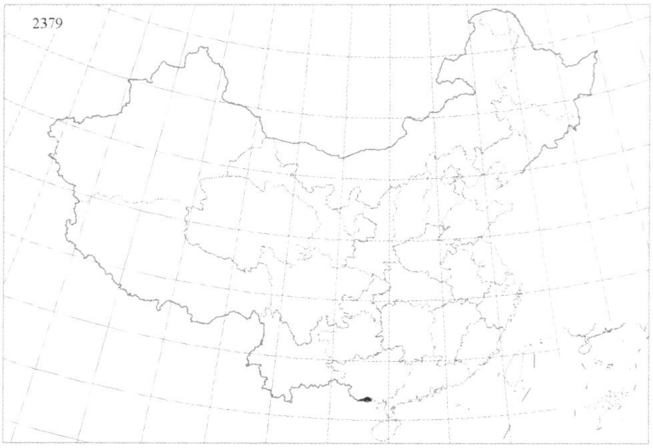

Lauraceae 樟科 *Machilus* 润楠属
Machilus verruculosa H. W. Li 疣枝润楠
Trees, evergreen

18.0 (16.3~21.0); 10.8 (9.1~14.2); 23.0 (21.4~25.6); 18.0 (16.4~21.0); 156.2 (136.4~191.8); 0.0 (0.0~0.0); 879 (797~1053) ❙ 1401 (1190~1522); 781 (656~860); 57 (50~68) ❙ 60.0 (44.5~80.6); 879 (797~1053); 432.7 (347.3~551.6)

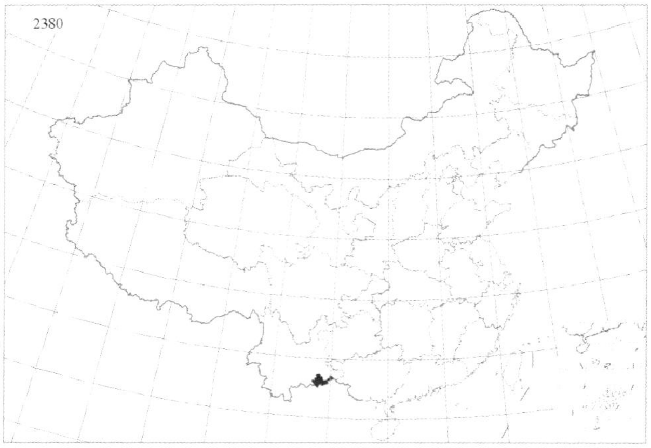

Lauraceae 樟科 *Machilus* 润楠属
Machilus viridis Hand.-Mazz. 绿叶润楠
Trees, evergreen

11.7 (5.3~18.2); 4.8 (-2.4~12.4); 17.2 (12.2~22.0); 11.9 (5.7~18.2); 88.7 (30.5~158.7); 7.3 (0.0~26.2); 663 (479~868) ❙ 1108 (741~1530); 636 (426~909); 41 (12~69) ❙ 67.8 (36.1~100.7); 658 (479~855); 360.6 (234.2~516.7)

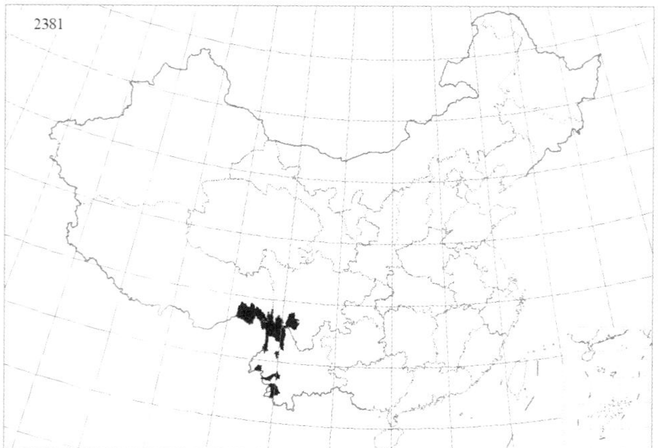

Lauraceae 樟科 *Machilus* 润楠属
Machilus wangchiana Chun 信宜润楠
Trees, evergreen

19.8 (17.6~21.3); 11.5 (7.1~15.2); 27.4 (26.3~28.7); 20.3 (17.6~22.7); 183.6 (151.6~212.7); 0.0 (0.0~0.0); 1061 (909~1216) ❙ 1710 (1496~1965); 792 (513~1056); 137 (104~182) ❙ 61.3 (45.4~71.0); 1061 (909~1216); 237.0 (197.5~297.2)

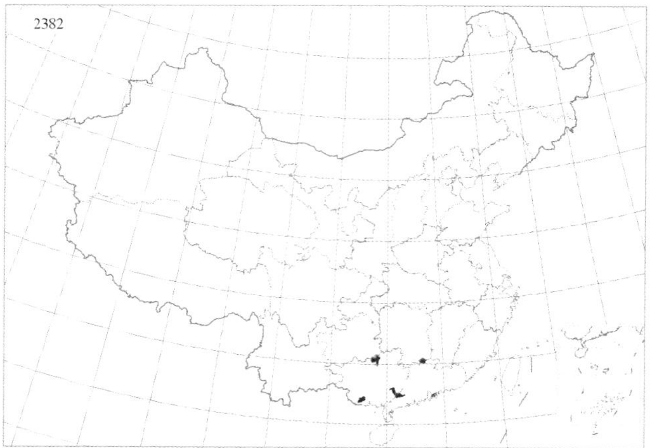

Lauraceae 樟科 *Machilus* 润楠属

Machilus wenshanensis H. W. Li 文山润楠

Trees, evergreen

18.3 (16.3~21.2); 10.0 (7.8~12.5); 24.5 (21.4~27.7); 18.3 (16.4~21.2); 160.0 (136.4~194.5); 0.0 (0.0~0.0); 919 (797~1114) ❙ 1307 (1190~1438); 719 (656~820); 61 (50~73) ❙ 43.9 (14.8~59.6); 919 (797~1114); 308.1 (225.6~371.7)

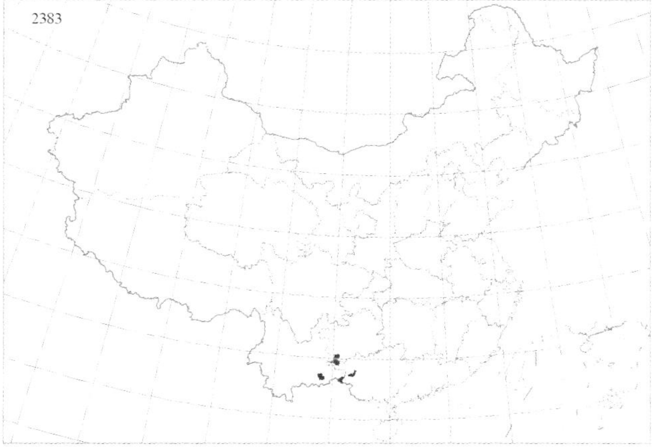

Lauraceae 樟科 *Machilus* 润楠属

Machilus yunnanensis Lecomte 滇润楠

Trees, evergreen

14.3 (−1.8~18.9); 7.9 (−10.2~12.0); 20.0 (6.1~24.6); 15.0 (1.8~19.0); 120.9 (1.9~167.8); 2.4 (0.0~82.8); 754 (329~929) ❙ 1012 (320~1454); 584 (193~816); 39 (12~70) ❙ 34.5 (−14.2~99.9); 728 (320~901); 304.1 (72.2~516.7)

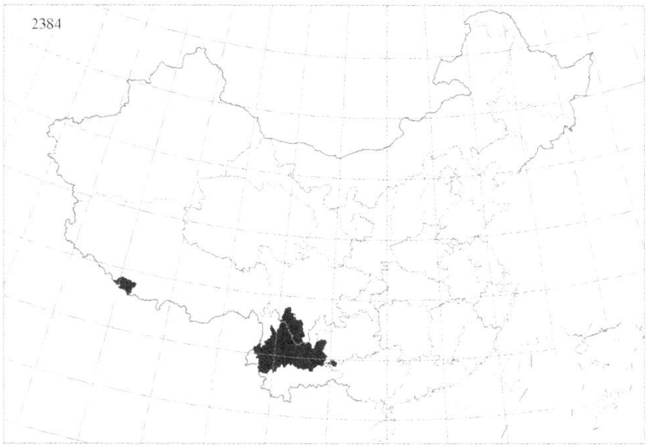

Lauraceae 樟科 *Machilus* 润楠属

Machilus zuihoensis Hayata 香润楠

Trees, evergreen

20.2 (16.2~23.8); 14.5 (11.0~18.1); 25.0 (20.1~28.2); 20.3 (16.2~23.8); 183.0 (134.2~225.8); 0.0 (0.0~0.0); 1021 (773~1297) ❙ 2602 (1757~3495); 1143 (821~1847); 312 (58~897) ❙ 164.7 (35.5~246.5); 992 (773~1144); 250.4 (119.4~363.7)

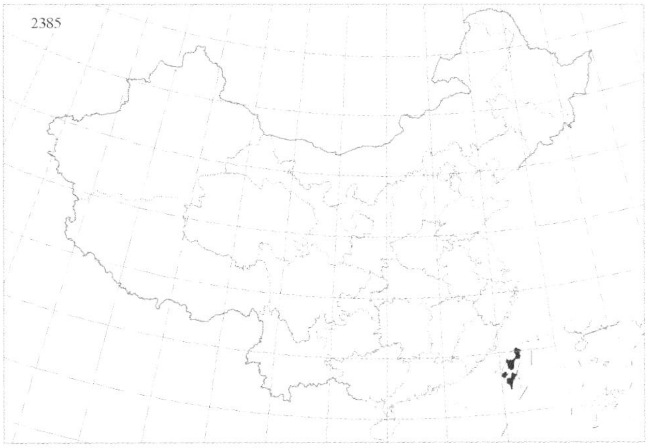

Lauraceae 樟科 *Machilus* 润楠属

Machilus zuihoensis var. mushaensis (F. Y. Lu) Y. C. Liu 青叶润楠

Trees, evergreen

17.8; 12.[illegible]; 22.8; 17.9; 154.4; 0.0; 852 ❙ 2859; 959; 402 ❙ 235.6; 852; 265.8

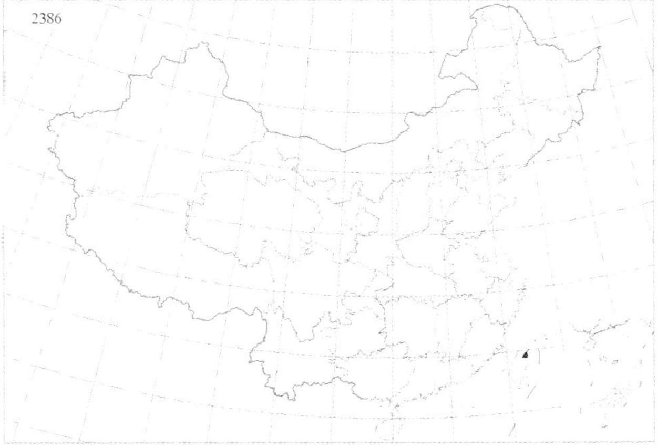

Lauraceae 樟科 *Neocinnamomum* 新樟属

Neocinnamomum caudatum (Nees) Merr. 滇新樟

Trees, evergreen

16.7 (10.4~22.2); 9.9 (3.4~15.1); 21.7 (16.3~28.3); 16.8 (10.5~22.3); 141.5 (68.0~207.5); 0.0 (0.0~2.6); 827 (623~1190) ❙ 1164 (797~1751); 651 (411~996); 48 (15~87) ❙ 41.1 (−14.2~112.2); 809 (615~1190); 350.8 (207.5~659.0)

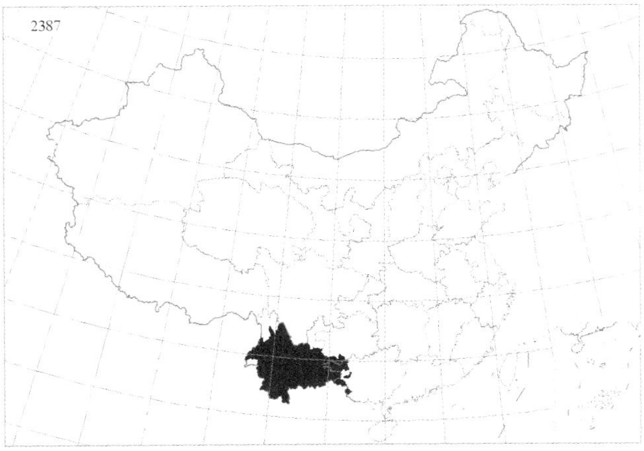

Lauraceae 樟科 *Neocinnamomum* 新樟属

Neocinnamomum delavayi (Lecomte) H. Liu 新樟

Shrubs or small trees, evergreen

15.2 (5.4~21.0); 7.9 (−2.1~14.2); 20.7 (12.3~26.6); 15.2 (5.8~21.0); 123.2 (31.5~191.8); 0.8 (0.0~25.6); 769 (481~1053) ❙ 1076 (741~1688); 613 (411~959); 42 (15~70) ❙ 40.7 (−14.2~112.2); 751 (481~1053); 314.0 (142.9~627.0)

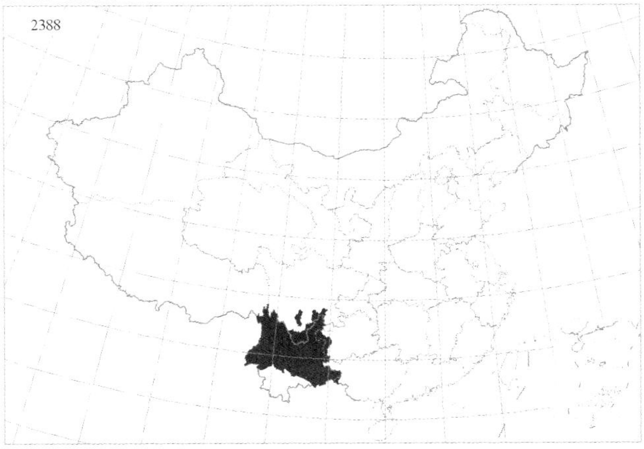

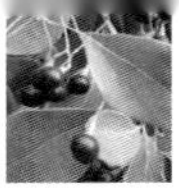

Lauraceae 樟科 *Neocinnamomum* 新樟属

Neocinnamomum fargesii (Lecomte) Kosterm. 川鄂新樟

Shrubs or small trees, evergreen

14.4 (10.6~19.1); 3.9 (-0.2~11.0); 24.1 (21.1~26.3); 14.4 (10.6~19.1); 116.8 (79.4~169.6); 3.9 (0.0~12.1); 778 (643~945) ❙ 1231 (1076~1478); 556 (475~820); 64 (42~85) ❙ 59.8 (32.5~107.2); 778 (643~945); 246.8 (176.4~371.7)

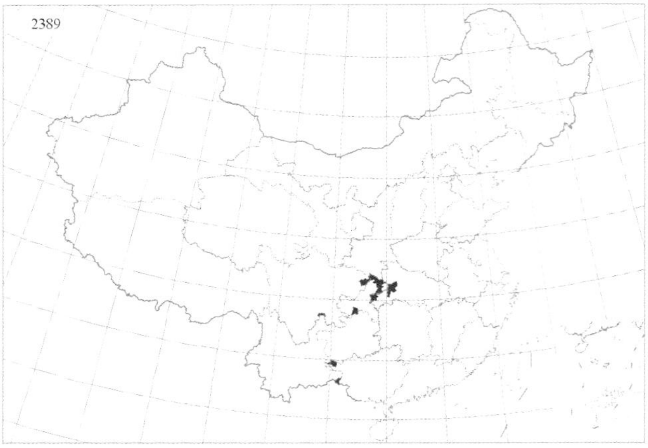

Lauraceae 樟科 *Neocinnamomum* 新樟属

Neocinnamomum lecomtei H. Liu 海南新樟

Shrubs, evergreen

20.0 (16.9~24.0); 12.0 (7.1~19.1); 26.1 (21.4~28.5); 20.0 (16.9~24.0); 180.6 (142.7~228.1); 0.0 (0.0~0.0); 1031 (813~1302) ❙ 1393 (1178~1900); 701 (455~1056); 75 (48~158) ❙ 36.6 (6.7~71.0); 1011 (813~1190); 351.9 (189.5~620.0)

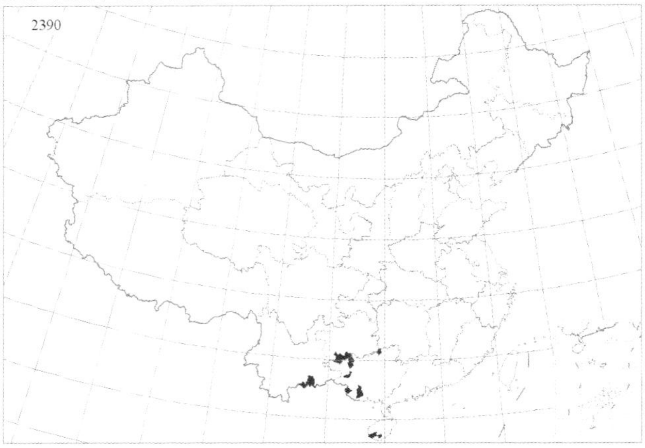

Lauraceae 樟科 *Neocinnamomum* 新樟属

Neocinnamomum mekongense (Hand.-Mazz.) Kosterm. 沧江新樟

Shrubs or small trees, evergreen

12.5 (5.3~21.6); 5.0 (-2.4~12.6); 18.5 (12.2~28.5); 12.6 (5.7~21.7); 95.3 (30.5~199.8); 5.2 (0.0~26.2); 716 (479~1153) ❙ 1223 (961~1625); 676 (558~820); 65 (31~173) ❙ 75.9 (40.9~100.7); 716 (479~1153); 334.9 (234.2~407.1)

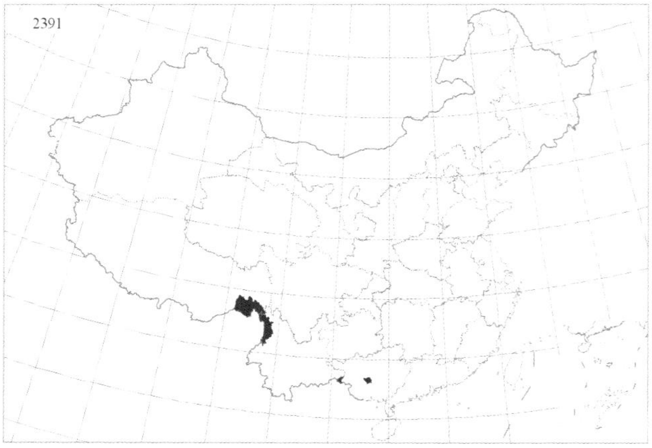

Lauraceae 樟科 *Neolitsea* 新木姜子属

Neolitsea aciculata (Blume) Koidzumi 台湾新木姜子

Trees, evergreen

18.2; 12.8; 22.6; 18.3; 159.3; 0.0; 881 ❙ 2661; 1186; 252 ❙ 207.4; 881; 254.2

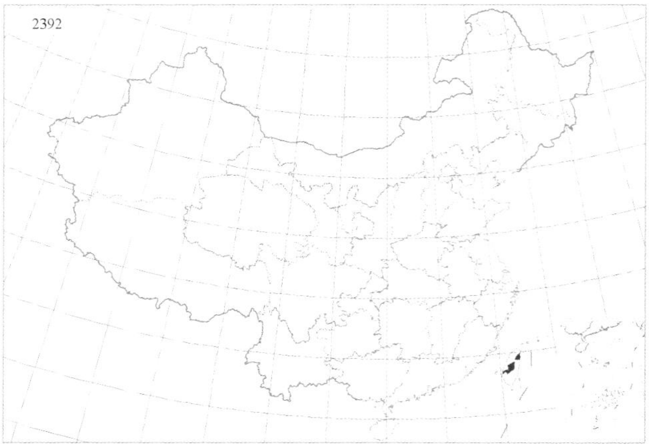

Lauraceae 樟科 *Neolitsea* 新木姜子属

Neolitsea acuminatissima (Hayata) Kaneh. et Sasaki 尖叶新木姜子

Small trees, evergreen

18.9 (16.2~21.7); 13.0 (11.0~15.3); 23.9 (20.1~27.0); 18.9 (16.2~21.7); 167.2 (134.2~200.2); 0.0 (0.0~0.0); 930 (773~1118) ❙ 2448 (1836~2993); 1083 (900~1412); 262 (110~547) ❙ 169.7 (64.2~246.5); 922 (773~1075); 254.5 (202.8~289.1)

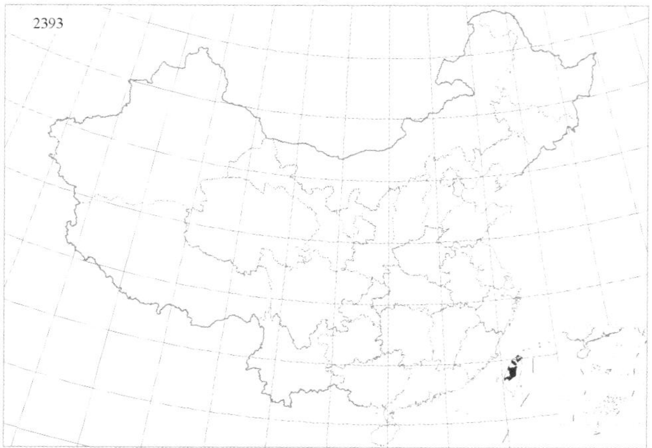

Lauraceae 樟科 *Neolitsea* 新木姜子属

Neolitsea alongensis Lecomte 下龙新木姜子

Small trees, evergreen

19.4 (17.2~21.8); 12.8 (9.5~15.1); 24.0 (22.1~27.7); 19.5 (17.3~21.9); 173.7 (147.0~202.2); 0.0 (0.0~0.0); 957 (838~1150) ❙ 1603 (1254~2148); 871 (655~1234); 76 (48~125) ❙ 66.8 (47.6~88.8); 953 (838~1150); 448.8 (297.2~659.0)

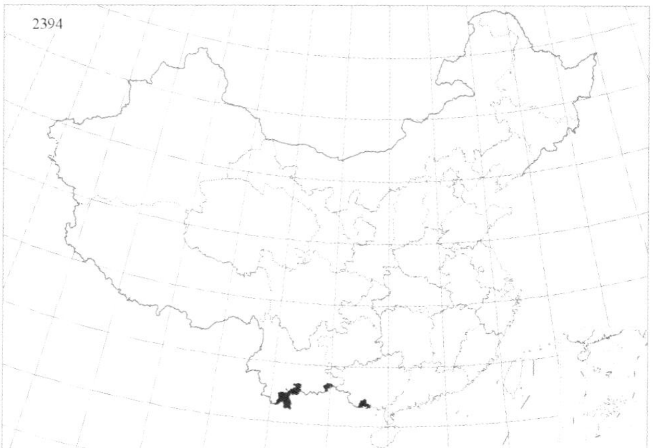

Lauraceae 樟科 *Neolitsea* 新木姜子属
Neolitsea aurata (Hayata) Koidz. 新木姜子
Trees, evergreen

18.0 (10.6~23.4); 7.8 (−0.2~16.1); 27.2 (20.0~30.1); 18.1 (10.6~23.5); 157.2 (79.4~221.7); 0.4 (0.0~12.1); 955 (643~1273) ❙ 1501 (885~2713); 598 (392~1388); 149 (33~259) ❙ 58.6 (5.8~214.4); 951 (643~1273); 224.4 (95.3~411.1)

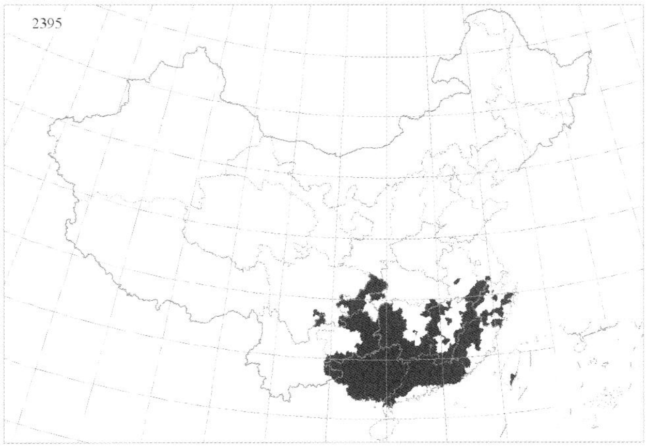

Lauraceae 樟科 *Neolitsea* 新木姜子属
Neolitsea aurata var. chekiangensis (Nakai) Yang et P. H. Huang 浙江新木姜子
Trees, evergreen

17.1 (14.3~19.4); 5.7 (2.1~9.4); 27.9 (24.2~30.1); 17.2 (14.3~19.4); 146.4 (113.2~173.4); 0.4 (0.0~4.9); 919 (761~1037) ❙ 1611 (871~1997); 549 (299~713); 204 (107~259) ❙ 76.3 (−6.7~147.0); 899 (755~1029); 230.4 (148.9~319.9)

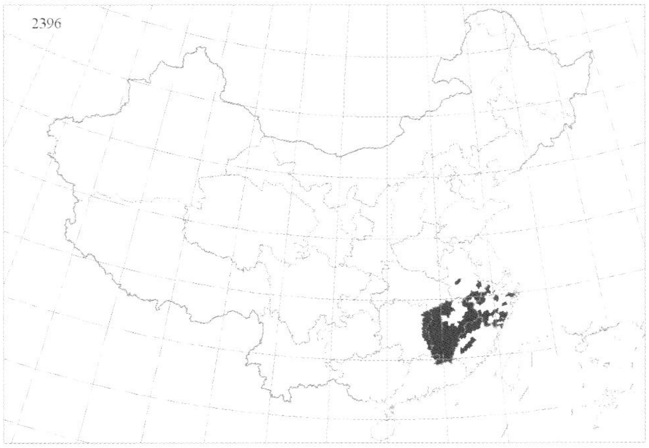

Lauraceae 樟科 *Neolitsea* 新木姜子属
Neolitsea aurata var. glauca Y. C. Yang 粉叶新木姜子
Trees, evergreen

15.6 (14.2~17.7); 4.9 (3.3~6.9); 25.7 (24.4~27.8); 15.7 (14.3~17.7); 128.5 (113.3~152.2); 0.7 (0.0~2.1); 831 (776~937) ❙ 1168 (1064~1448); 507 (458~600); 71 (63~105) ❙ 41.5 (18.0~79.3); 831 (776~937); 183.9 (126.1~223.0)

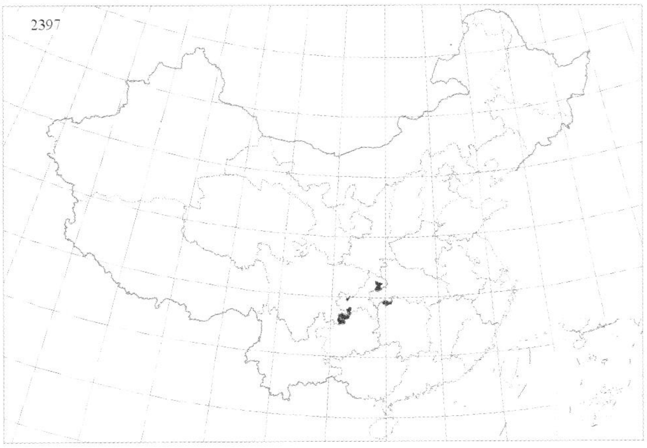

Lauraceae 樟科 *Neolitsea* 新木姜子属
Neolitsea aurata var. paraciculata (Nakai) Yang et P. H. Huang 云和新木姜子
Trees, evergreen

17.1 (14.3~20.8); 5.8 (2.5~12.2); 27.5 (24.2~29.8); 17.1 (14.3~20.8); 145.5 (113.2~189.7); 0.3 (0.0~3.9); 910 (761~1113) ❙ 1542 (1215~1997); 530 (392~827); 184 (92~259) ❙ 69.9 (35.6~130.7); 899 (761~1085); 210.9 (134.6~303.4)

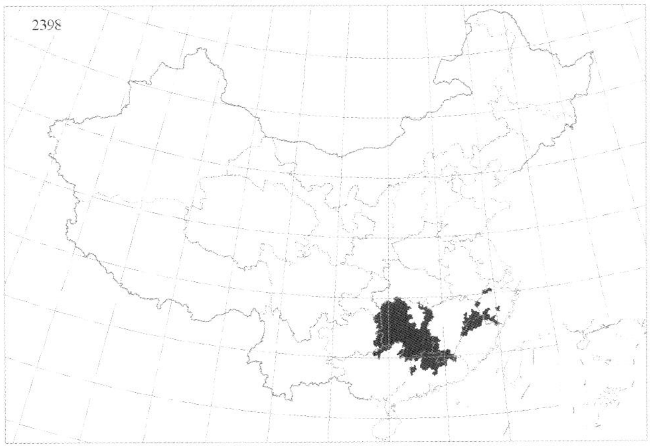

Lauraceae 樟科 *Neolitsea* 新木姜子属
Neolitsea aurata var. undulatula Yang et P. H. Huang 浙闽新木姜子
Trees, evergreen

16.8 (14.7~18.2); 5.8 (2.5~7.2); 27.1 (24.7~29.1); 16.8 (14.8~18.2); 142.2 (120.8~158.8); 0.3 (0.0~3.9); 892 (789~984) ❙ 1825 (1489~1997); 608 (531~713); 234 (191~259) ❙ 105.5 (75.9~147.0); 884 (789~944); 255.4 (183.9~303.4)

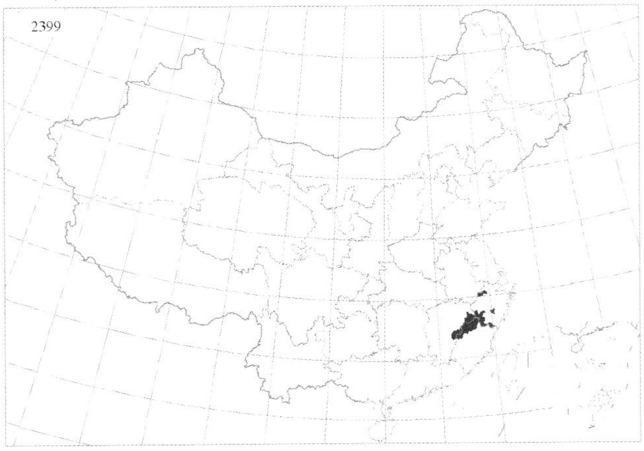

Lauraceae 樟科 *Neolitsea* 新木姜子属
Neolitsea brevipes H. W. Li 短梗新木姜子
Small trees, evergreen

18.0 (14.6~22.0); 8.4 (3.8~14.2); 26.3 (22.1~29.4); 18.0 (14.7~22.0); 156.3 (117.7~203.8); 0.0 (0.0~1.5); 926 (779~1170) ❙ 1529 (1161~1883); 612 (421~911); 155 (48~240) ❙ 65.7 (30.5~104.6); 922 (779~1170); 275.5 (163.7~551.6)

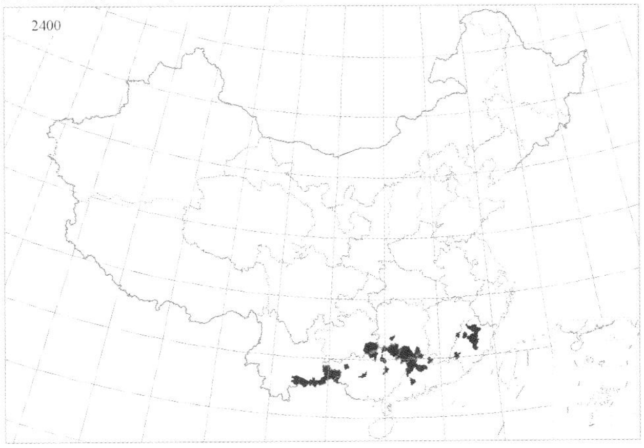

Lauraceae 樟科 *Neolitsea* 新木姜子属
Neolitsea buisanensis Yamamoto et S. Kamikoti 武威山新木姜子
Shrubs or small trees, evergreen

22.8 (21.3~24.8); 15.8 (12.4~20.1); 28.1 (25.5~29.0); 22.9 (21.3~24.8); 214.7 (195.7~237.6); 0.0 (0.0~0.0); 1230 (1111~1385) ❙ 1599 (868~2992); 726 (312~1847); 101 (41~171) ❙ 31.1 (-34.3~166.5); 1178 (868~1360); 337.4 (199.4~620.0)

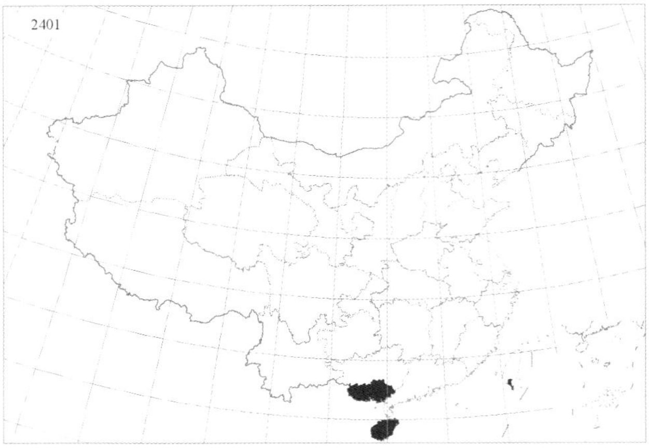

Lauraceae 樟科 *Neolitsea* 新木姜子属
Neolitsea cambodiana Lecomte 锈叶新木姜子
Trees, evergreen

18.8 (14.3~24.6); 9.0 (3.3~19.5); 27.5 (24.2~29.8); 18.9 (14.3~24.7); 166.4 (113.2~235.9); 0.1 (0.0~2.3); 991 (761~1372) ❙ 1609 (1179~2005); 604 (392~1056); 175 (42~254) ❙ 64.2 (-11.9~130.7); 980 (761~1299); 241.2 (134.6~620.0)

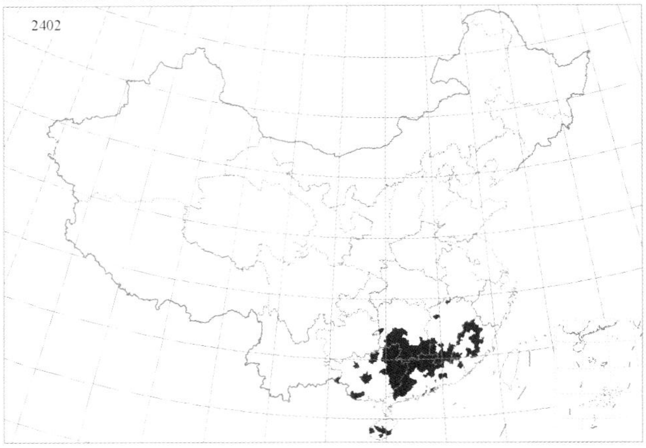

Lauraceae 樟科 *Neolitsea* 新木姜子属
Neolitsea cambodiana var. glabra C. K. Allen 香港新木姜子
Trees, evergreen

19.7 (16.6~22.5); 10.7 (6.0~15.0); 27.4 (25.0~28.7); 19.8 (16.6~22.6); 177.0 (138.9~210.9); 0.0 (0.0~0.0); 1028 (866~1206) ❙ 1663 (1188~2030); 688 (584~1033); 159 (56~254) ❙ 63.5 (5.8~130.7); 1028 (866~1206); 262.2 (139.6~371.7)

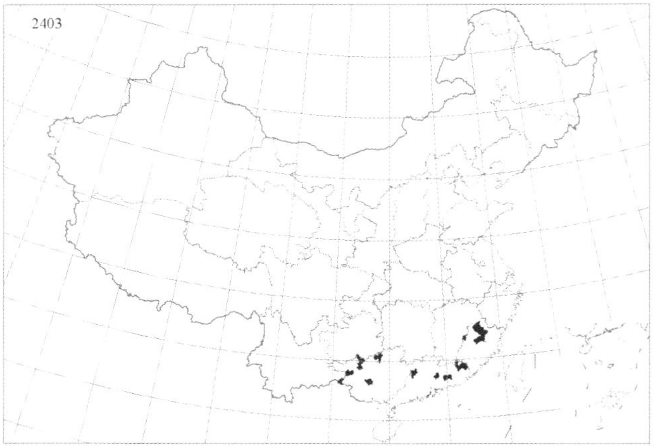

Lauraceae 樟科 *Neolitsea* 新木姜子属
Neolitsea chrysotricha H. W. Li 金毛新木姜子
Small trees, evergreen

15.1; 8.3; 20.0; 15.2; 122.0; 0.0; 759 ❙ 1386; 759; 66 ❙ 84.1; 759; 461.9

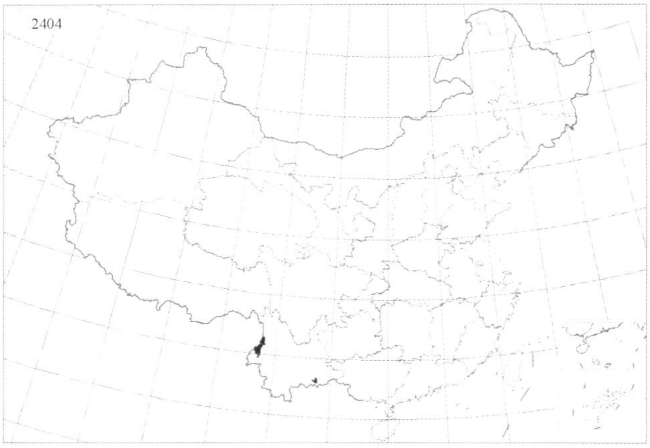

Lauraceae 樟科 *Neolitsea* 新木姜子属
Neolitsea chuii Merr. 鸭公树
Trees, evergreen

18.7 (14.3~22.7); 9.0 (3.3~15.0); 27.0 (21.4~29.8); 18.7 (14.3~22.8); 164.8 (112.6~213.1); 0.1 (0.0~2.3); 976 (756~1221) ❙ 1569 (1074~2435); 627 (392~1428); 159 (47~230) ❙ 61.6 (12.9~128.9); 971 (756~1221); 242.5 (134.6~551.6)

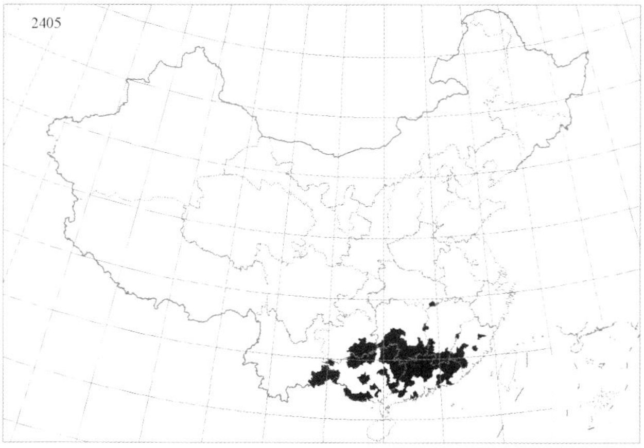

Lauraceae 樟科 *Neolitsea* 新木姜子属
Neolitsea confertifolia (Hemsl.) Merr. 簇叶新木姜子
Small trees, evergreen

15.7 (9.3~21.7); 4.2 (-2.9~13.4); 26.5 (19.4~29.8); 15.8 (9.5~21.7); 131.9 (69.8~200.7); 2.7 (0.0~20.5); 856 (603~1147) ❙ 1322 (626~1829); 498 (316~828); 132 (15~230) ❙ 54.0 (-22.2~128.9); 842 (603~1147); 208.1 (127.1~305.6)

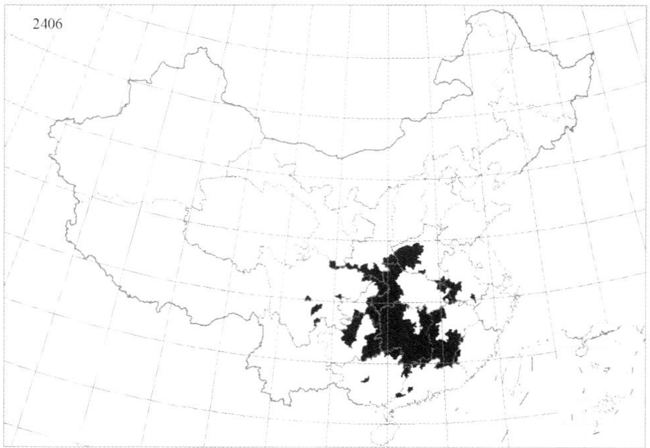

Lauraceae 樟科 *Neolitsea* 新木姜子属
Neolitsea daibuensis Kamik. 大武山新木姜子
Trees, evergreen

21.7；16.7；25.4；21.7；200.8；0.0；1117 ❙ 2255；1381；87 ❙ 117.5；1003；203.3

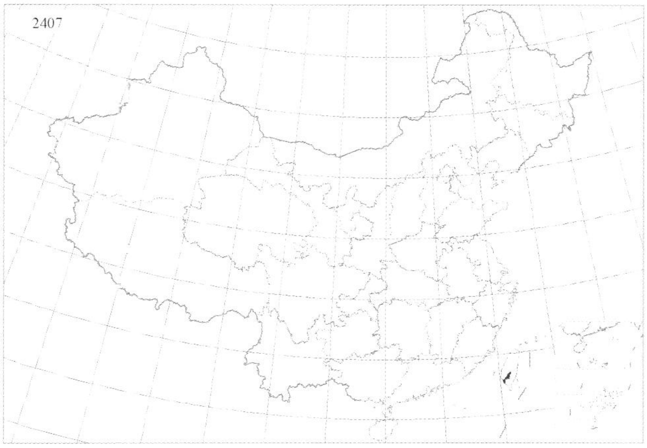

Lauraceae 樟科 *Neolitsea* 新木姜子属
Neolitsea ellipsoidea C. K. Allen 香果新木姜子
Trees, evergreen

21.7 (19.0~24.8)；14.0 (9.1~20.1)；27.9 (26.7~29.8)；21.8 (19.0~24.8)；201.1 (168.2~237.6)；0.0 (0.0~0.0)；1163 (989~1385) ❙ 1620 (1179~1852)；581 (392~711)；132 (41~194) ❙ 42.7 (-11.9~86.3)；1088 (989~1326)；357.6 (174.9~620.0)

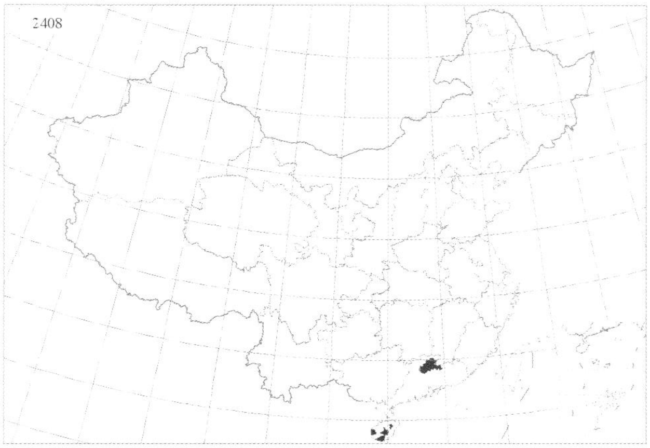

Lauraceae 樟科 *Neolitsea* 新木姜子属
Neolitsea hainanensis Yang et P. H. Huang 海南新木姜子
Trees, evergreen

22.1 (19.0~24.3)；14.8 (10.2~20.1)；27.5 (25.2~28.6)；22.1 (19.0~24.8)；205.4 (168.4~237.6)；0.0 (0.0~0.0)；1175 (952~1385) ❙ 1558 (1179~2435)；770 (392~1428)；81 (41~130) ❙ 33.9 (-11.9~99.5)；1114 (952~1221)；331.3 (235.1~620.0)

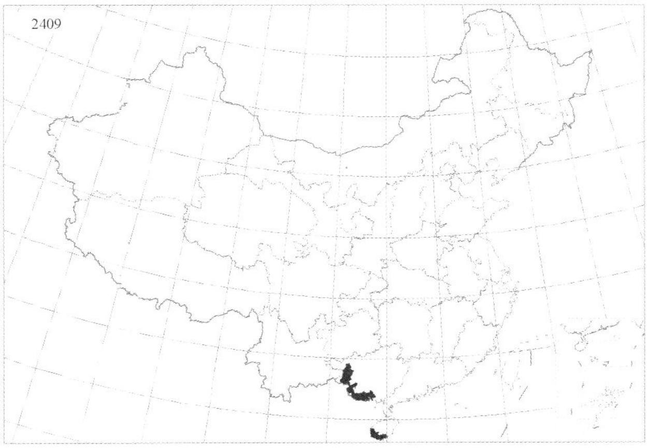

Lauraceae 樟科 *Neolitsea* 新木姜子属
Neolitsea hiiranensis Liu et J. C. Liao 南仁山新木姜子
Small trees, evergreen

22.2；17.7；25.5；22.2；206.7；0.0；1123 ❙ 2992；1847；107 ❙ 166.5；1123；296.9

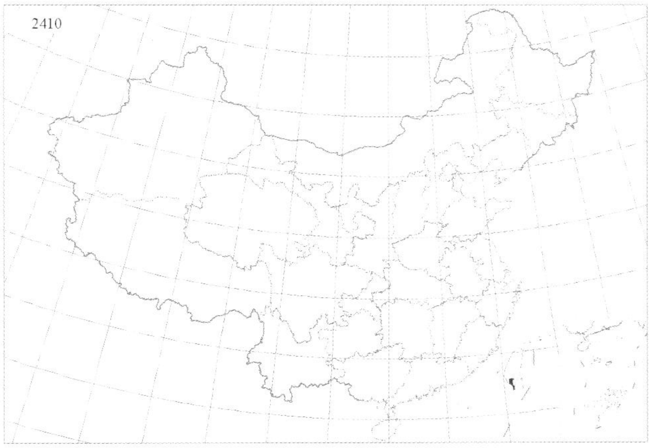

Lauraceae 樟科 *Neolitsea* 新木姜子属
Neolitsea homilantha C. K. Allen 团花新木姜子
Shrubs or small trees, evergreen

15.4 (-1.1~21.0)；8.5 (-9.1~14.2)；20.5 (6.2~25.6)；15.5 (2.0~21.0)；126.8 (2.5~19[illegible].8)；1.6 (0.0~75.6)；777 (334~1053) ❙ 1092 (321~1751)；621 (193~996)；44 (15~77) ❙ 41.1 (-14.2~100.7)；760 (321~1053)；326.8 (79.3~551.6)

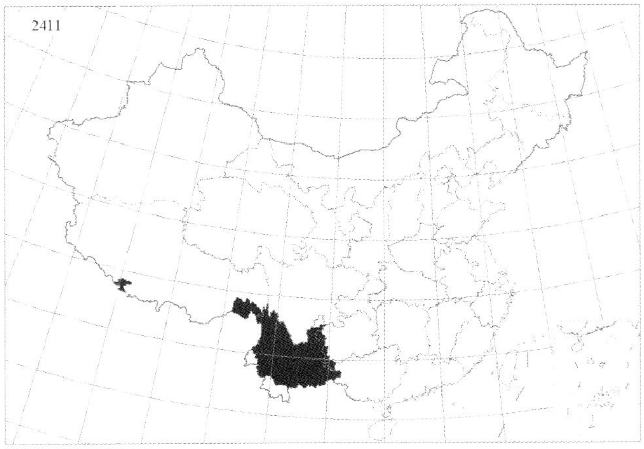

Lauraceae 樟科 *Neolitsea* 新木姜子属
Neolitsea howii C. K. Allen 保亭新木姜子
Shrubs or trees, evergreen

24.0 (22.2~24.8)；18.4 (17.2~20.1)；28.0 (25.7~28.8)；24.0 (22.3~24.8)；228.3 (207.2~237.6)；0.0 (0.0~0.0)；1310 (1127~1385) ❙ 1501 (868~1854)；534 (312~712)；74 (41~115) ❙ 15.0 (-34.3~47.8)；1191 (868~1360)；440.3 (248.2~620.0)

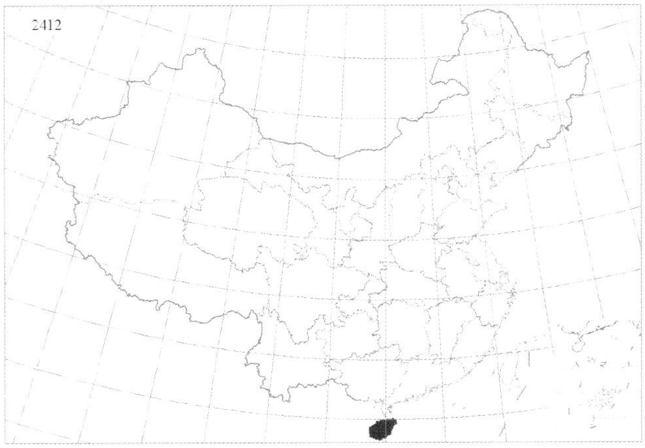

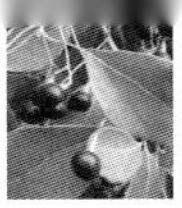

Lauraceae 樟科 *Neolitsea* 新木姜子属

Neolitsea hsiangkweiensis Yang et P. H. Huang 湘桂新木 姜子

Trees, evergreen

16.7 (14.6~19.2); 6.4 (3.6~11.0); 25.8 (24.6~26.7); 16.8 (14.7~19.2); 141.9 (117.7~170.5); 0.4 (0.0~1.5); 872 (779~958) ❙ 1439 (1250~1625); 619 (518~848); 123 (58~191) ❙ 66.0 (30.5~101.8); 872 (779~958); 253.0 (182.7~318.5)

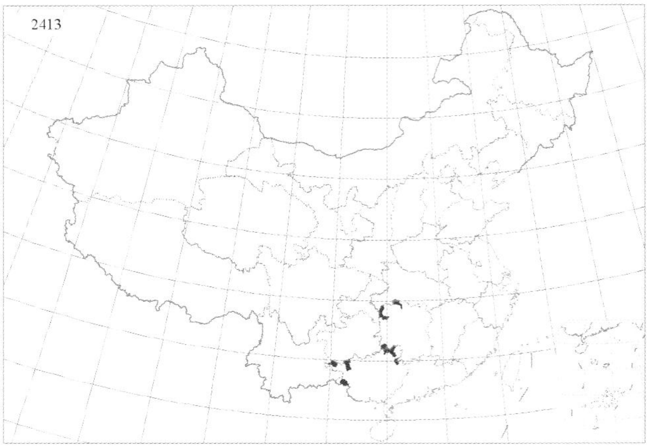

Lauraceae 樟科 *Neolitsea* 新木姜子属

Neolitsea impressa Y. C. Yang 凹脉新木姜子

Shrubs or small trees, evergreen

16.6 (10.6~18.7); 5.8 (-0.2~8.1); 26.8 (21.1~28.9); 16.6 (10.6~18.8); 140.4 (79.4~166.0); 0.9 (0.0~12.1); 889 (643~1002) ❙ 1162 (1028~1320); 488 (442~546); 61 (42~76) ❙ 32.7 (8.0~93.8); 889 (643~1002); 161.2 (108.2~284.3)

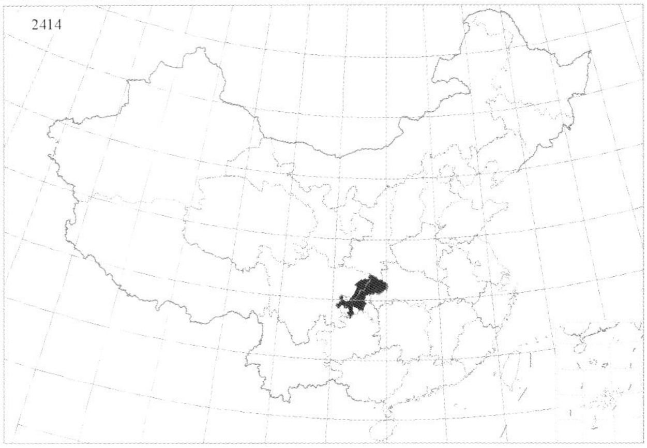

Lauraceae 樟科 *Neolitsea* 新木姜子属

Neolitsea konishii (Hayata) Kaneh. et Sasaki 五掌楠

Trees, evergreen

19.5 (16.2~24.1); 13.9 (11.0~18.9); 24.3 (20.1~28.0); 19.6 (16.2~24.1); 175.0 (134.2~229.7); 0.0 (0.0~0.0); 971 (773~1322) ❙ 2599 (1751~3495); 1098 (821~1594); 320 (47~897) ❙ 176.2 (32.5~246.5); 948 (773~1095); 254.2 (154.2~363.7)

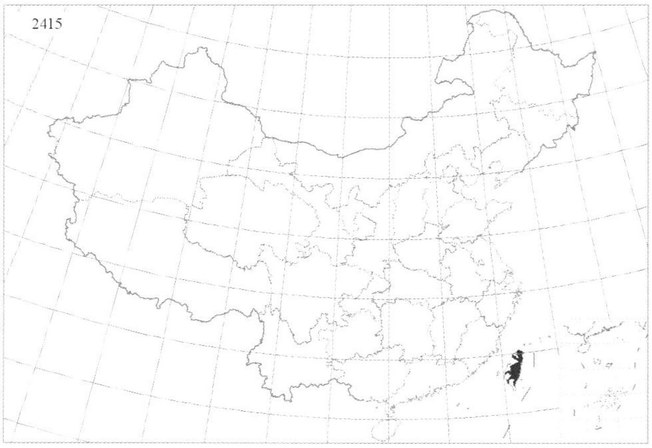

Lauraceae 樟科 *Neolitsea* 新木姜子属

Neolitsea kwangsiensis H. Liu 广西新木姜子

Shrubs or small trees, evergreen

20.7 (17.6~22.0); 12.0 (6.6~14.3); 27.7 (25.8~28.8); 20.7 (17.6~22.1); 188.5 (151.3~204.9); 0.0 (0.0~0.0); 1084 (928~1170) ❙ 1617 (1250~1876); 723 (476~891); 127 (60~197) ❙ 49.3 (30.5~64.3); 1083 (922~1170); 233.5 (180.3~291.2)

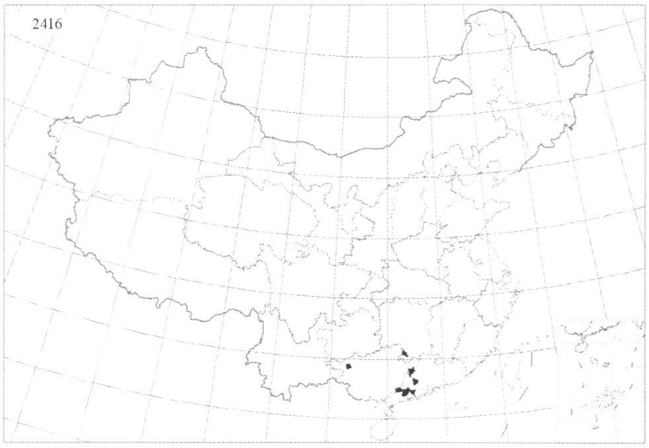

Lauraceae 樟科 *Neolitsea* 新木姜子属

Neolitsea levinei Merr. 大叶新木姜子

Trees, evergreen

17.0 (7.3~23.2); 6.7 (-2.0~15.8); 26.4 (15.6~29.8); 17.1 (7.5~23.2); 145.4 (48.2~218.9); 0.6 (0.0~20.4); 900 (537~1257) ❙ 1435 (916~2030); 570 (392~1033); 134 (27~259) ❙ 60.0 (8.0~145.8); 895 (537~1257); 218.4 (108.2~551.6)

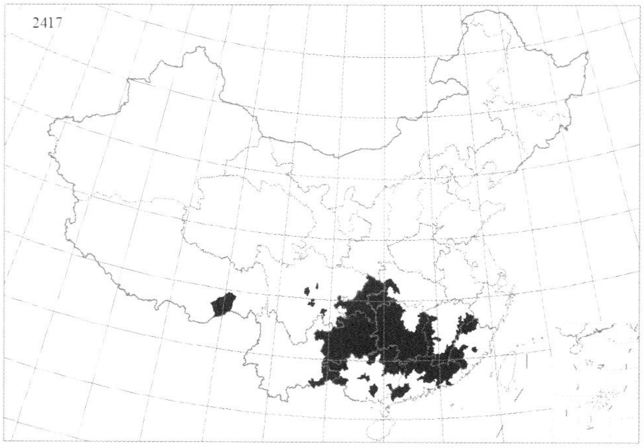

Lauraceae 樟科 *Neolitsea* 新木姜子属

Neolitsea longipedicellata Yang et P. H. Huang 长梗新木 姜子

Trees, evergreen

18.6; 8.8; 26.8; 18.7; 164.2; 0.0; 964 ❙ 1425; 656; 117 ❙ 48.2; 964; 268.1

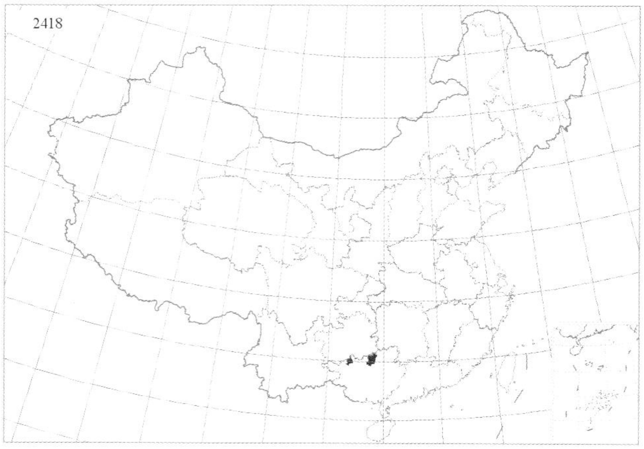

Lauraceae 樟科 *Neolitsea* 新木姜子属
Neolitsea lunglingensis H. W. Li 龙陵新木姜子
Small trees, evergreen

15.2；8.4；19.8；15.2；122.8；0.0；754 ❙ 1401；777；63 ❙ 86.3；754；441.3

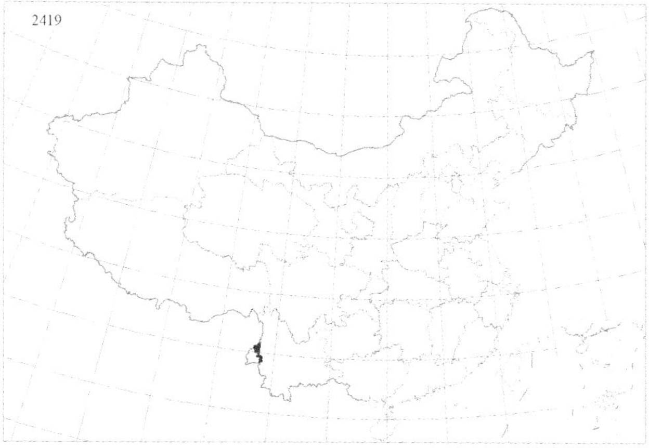

Lauraceae 樟科 *Neolitsea* 新木姜子属
Neolitsea menglaensis Yang et P. H. Huang 勐腊新木姜子
Trees, evergreen

20.4；15.1；23.6；20.5；185.8；0.0；989 ❙ 1607；702；77 ❙ 62.5；986；659.0

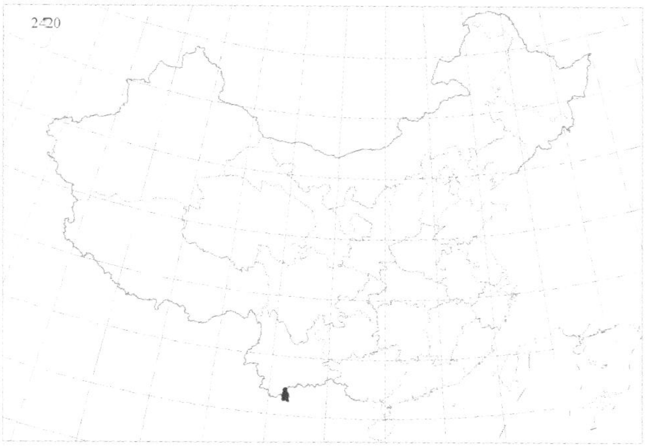

Lauraceae 樟科 *Neolitsea* 新木姜子属
Neolitsea oblongifolia Merr. et Chun 长圆叶新木姜子
Trees, evergreen

23.4 (21.5~24.7)；17.4 (13.6~19.8)；27.7 (26.7~28.4)；23.4 (21.6~24.7)；221.0 (199.1~236.5)；0.0 (0.0~0.0)；1264 (1131~1376) ❙ 1483 (1179~1725)；585 (392~850)；65 (42~106) ❙ 18.1 (-11.9~41.8)；1145 (1056~1212)；458.2 (281.7~620.0)

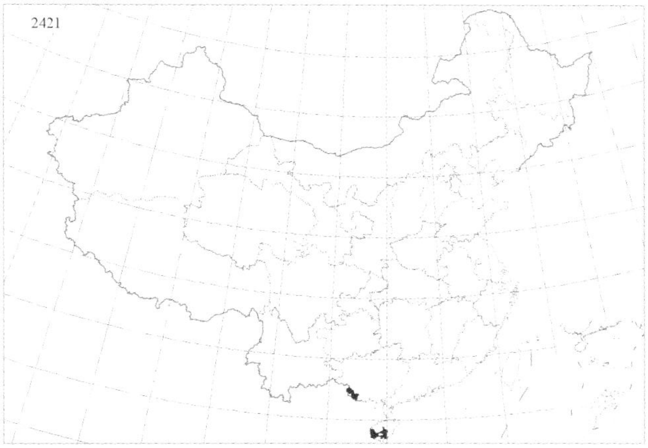

Lauraceae 樟科 *Neolitsea* 新木姜子属
Neolitsea obtusifolia Merr. 钝叶新木姜子
Trees, evergreen

24.2；19.2；27.7；24.3；231.4；0.0；1333 ❙ 1446；497；48 ❙ 8.4；1147；520.5

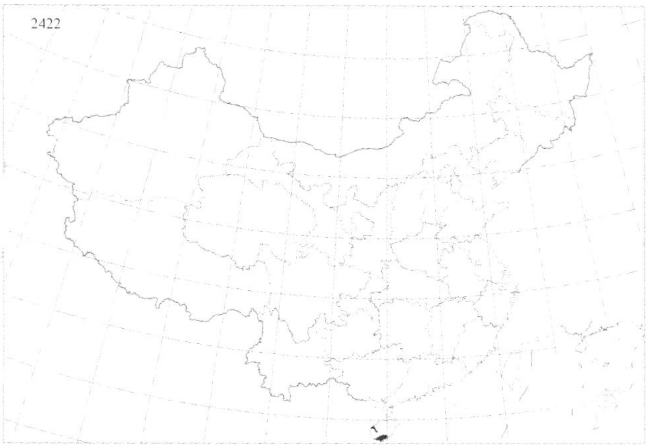

Lauraceae 樟科 *Neolitsea* 新木姜子属
Neolitsea ovatifolia Yang et P. H. Huang 卵叶新木姜子
Small shrubs, evergreen

20.4 (14.3~24.8)；12.4 (4.0~20.1)；27.0 (22.5~29.8)；20.5 (14.3~24.8)；185.9 (112.6~237.6)；0.0 (0.0~1.1)；1077 (756~1385) ❙ 1541 (1179~2035)；656 (392~969)；110 (41~194) ❙ 45.6 (-11.9~76.3)；1036 (756~1257)；314.9 (174.9~620.0)

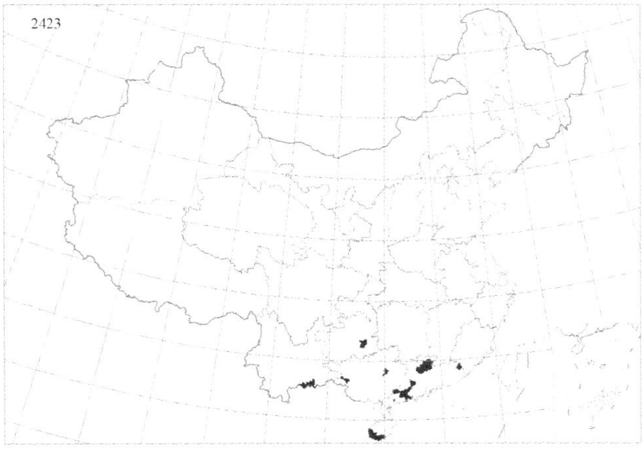

Lauraceae 樟科 *Neolitsea* 新木姜子属
Neolitsea pallens (D. Don) Momiy. et H. Hara 灰白新木姜子
Small trees, evergreen

6.3；-2.1；13.9；8.0；56.1；39.6；538 ❙ 947；558；38 ❙ 54.8；532；203.3

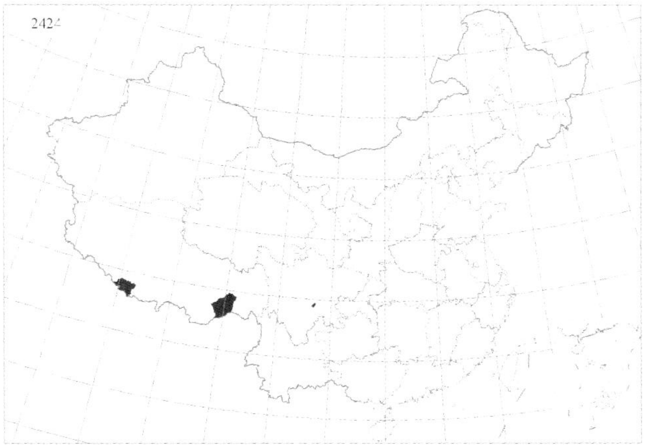

Lauraceae 樟科 *Neolitsea* 新木姜子属

Neolitsea parvigemma (Hayata) Kaneh. et Sasaki 小芽新木姜子

Small trees, evergreen

20.5; 15.5; 24.1; 20.5; 185.6; 0.0; 1032 ▎2546; 1449; 131 ▎162.0; 976; 248.1

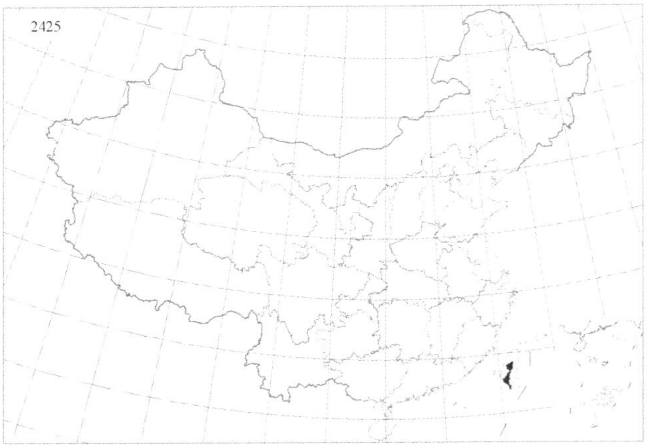

Lauraceae 樟科 *Neolitsea* 新木姜子属

Neolitsea phanerophlebia Merr. 显脉新木姜子

Small trees, evergreen

19.0 (14.6~22.7); 9.4 (3.8~15.2); 27.5 (24.6~29.8); 19.1 (14.7~22.8); 169.6 (117.7~213.1); 0.0 (0.0~1.5); 1002 (779~1221) ▎1642 (1293~2435); 649 (403~1428); 171 (57~230) ▎64.6 (25.0~104.6); 997 (779~1221); 233.1 (151.0~371.7)

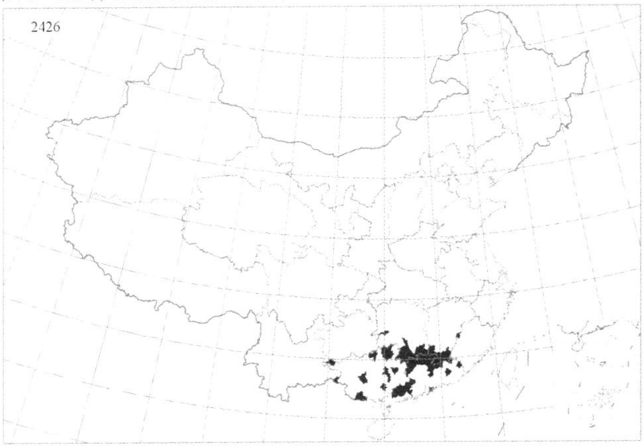

Lauraceae 樟科 *Neolitsea* 新木姜子属

Neolitsea pingbienensis Yang et P. H. Huang 屏边新木姜子

Shrubs, evergreen

18.0 (16.3~21.0); 10.9 (9.1~14.2); 23.0 (21.4~25.6); 18.1 (16.4~21.0); 156.8 (136.4~191.8); 0.0 (0.0~0.0); 879 (797~1053) ▎1384 (1190~1522); 777 (656~860); 56 (48~68) ▎57.9 (44.5~80.6); 879 (797~1053); 429.4 (336.3~551.6)

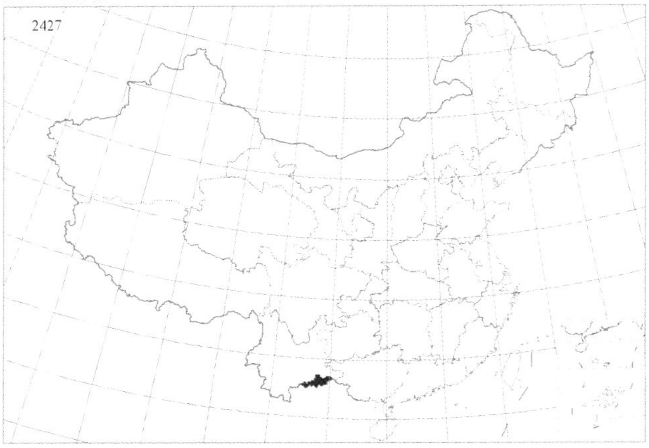

Lauraceae 樟科 *Neolitsea* 新木姜子属

Neolitsea pinninervis Yang et P. H. Huang 羽脉新木姜子

Shrubs or small trees, evergreen

17.8 (14.3~21.0); 7.5 (3.8~11.9); 26.8 (23.5~29.4); 17.9 (14.3~21.1); 154.3 (112.6~193.0); 0.1 (0.0~1.5); 930 (756~1114) ▎1505 (1211~1961); 571 (403~821); 162 (58~230) ▎62.2 (30.9~104.6); 925 (756~1114); 232.4 (151.0~322.1)

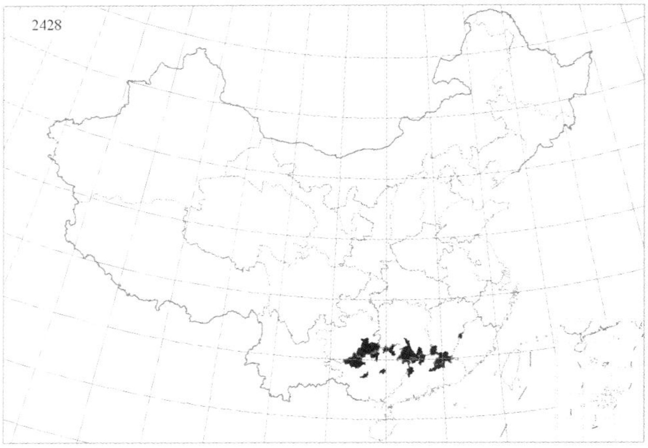

Lauraceae 樟科 *Neolitsea* 新木姜子属

Neolitsea polycarpa H. Liu 多果新木姜子

Trees, evergreen

18.0 (16.3~21.0); 11.1 (9.1~14.2); 22.8 (21.4~25.6); 18.0 (16.4~21.0); 156.2 (136.4~191.8); 0.0 (0.0~0.0); 872 (797~1053) ▎1279 (1059~1522); 718 (604~860); 52 (44~68) ▎46.8 (21.3~80.6); 868 (797~1053); 396.4 (294.5~551.6)

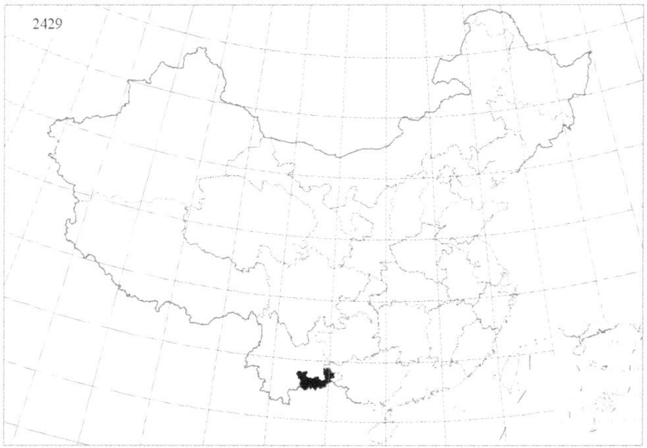

Lauraceae 樟科 *Neolitsea* 新木姜子属

Neolitsea pulchella (Meisn.) Merr. 美丽新木姜子

Small trees, evergreen

19.2 (14.6~24.0); 9.7 (3.8~19.1); 27.5 (24.6~29.8); 19.3 (14.7~24.0); 172.1 (117.7~228.1); 0.1 (0.0~1.5); 1015 (779~1302) ▎1653 (1348~2005); 645 (448~1056); 166 (48~215) ▎64.2 (15.4~101.8); 1009 (779~1216); 242.9 (139.6~620.0)

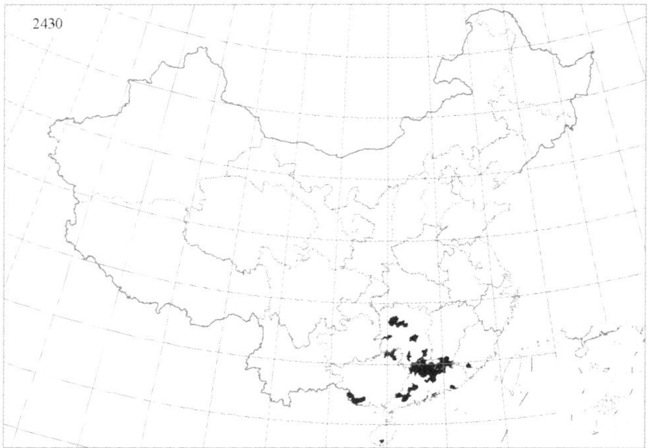

Lauraceae 樟科 *Neolitsea* 新木姜子属
Neolitsea purpurascens Yang 紫新木姜子
Small trees, evergreen

13.7；4.1；22.1；13.7；107.2；2.6；745 ❙ 1160；568；59 ❙ 56.0；745；232.4

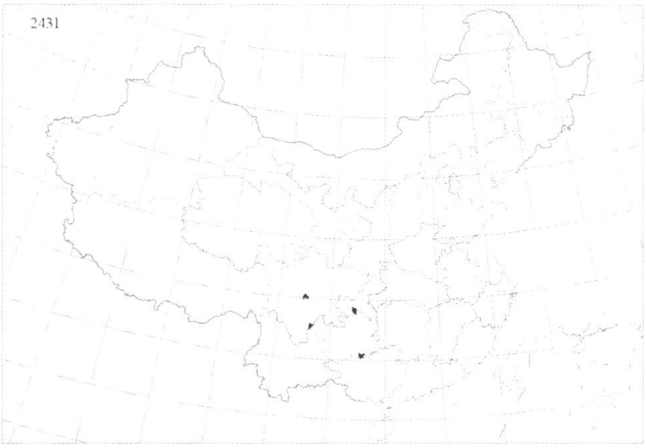

Lauraceae 樟科 *Neolitsea* 新木姜子属
Neolitsea sericea (Blume) Koidz. 舟山新木姜子
Trees, evergreen

17.0；7.9；25.8；17.0；144.6；0.3；877 ❙ 1712；761；170 ❙ 96.1；877；255.9

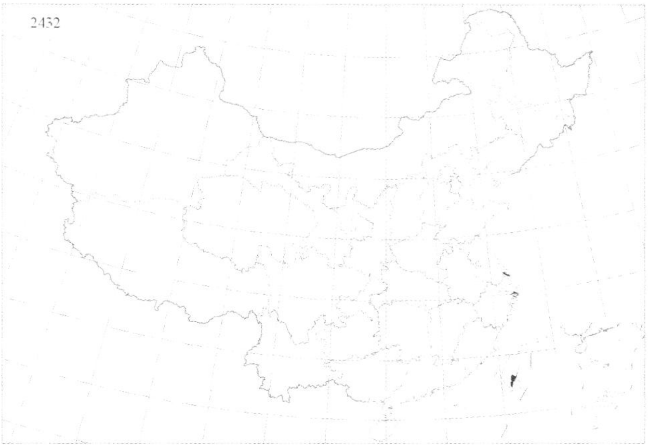

Lauraceae 樟科 *Neolitsea* 新木姜子属
Neolitsea shingningensis Yang et P. H. Huang 新宁新木 姜子
Shrubs or small trees, evergreen

16.4 (14.3~21.2)；5.5 (3.3~12.2)；26.4 (24.3~28.5)；16.4 (14.3~21.3) 137.4 (114.1~195.2)；0.6 (0.0~2.3)；867 (765~1126) ❙ 1464 (1221~1671)；524 (419~639)；163 (84~221) ❙ 70.2 (29.9~106.4)；864 (765~1126)；219.7 (170.4~258.0)

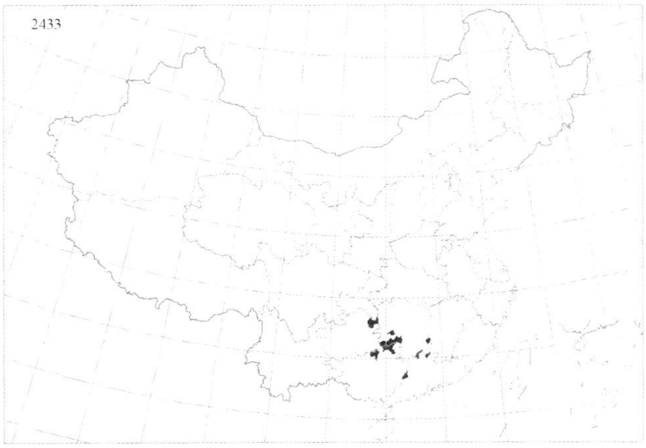

Lauraceae 樟科 *Neolitsea* 新木姜子属
Neolitsea sutchuanensis Y. C. Yang 四川新木姜子
Small trees, evergreen

14.6 (1.8~19.2)；5.1 (-7.1~11.0)；22.9 (9.9~27.5)；14.8 (3.8~19.2)；119.5 (15.7~170.8)；3.2 (0.0~53.3)；785 (404~1002) ❙ 1240 (527~1624)；598 (317~946)；82 (9~196) ❙ 58.1 (22.4~101.8)；785 (404~1002)；241.6 (153.9~407.1)

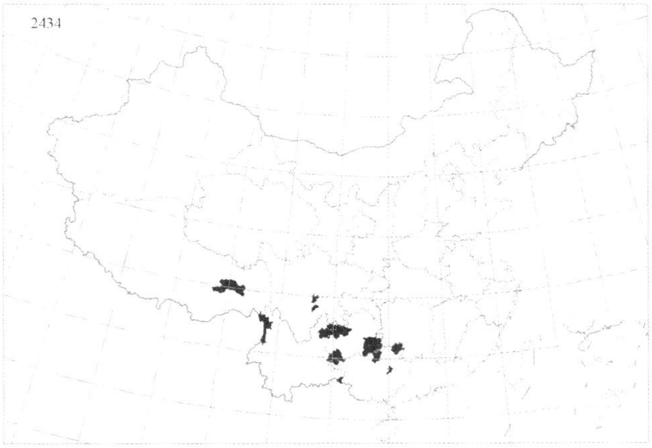

Lauraceae 樟科 *Neolitsea* 新木姜子属
Neolitsea tomentosa H. W. Li 绒毛新木姜子
Small trees, evergreen

15.8；8.8；21.0；15.8；132.6；2.9；827 ❙ 1310；720；57 ❙ 62.0；827；465.[illegible]

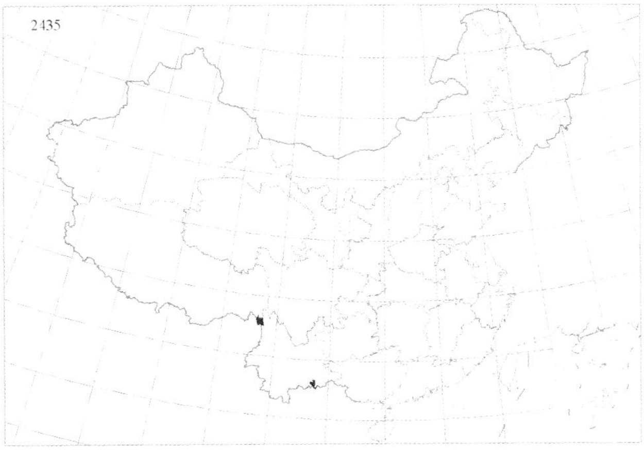

Lauraceae 樟科 *Neolitsea* 新木姜子属
Neolitsea undulatifolia (H. Lév.) C. K. Allen 波叶新木姜子
Shrubs or small trees, evergreen

18.1 (16.1~21.5)；8.8 (5.3~13.6)；25.6 (22.5~27.8)；18.1 (16.1~21.6)；157.3 (133.1~199.1)；0.0 (0.0~0.0)；924 (826~1131) ❙ 1408 (1268~1621)；667 (523~850)；104 (57~146) ❙ 52.9 (39.9~64.7)；924 (826~1131)；274.9 (215.4~491.5)

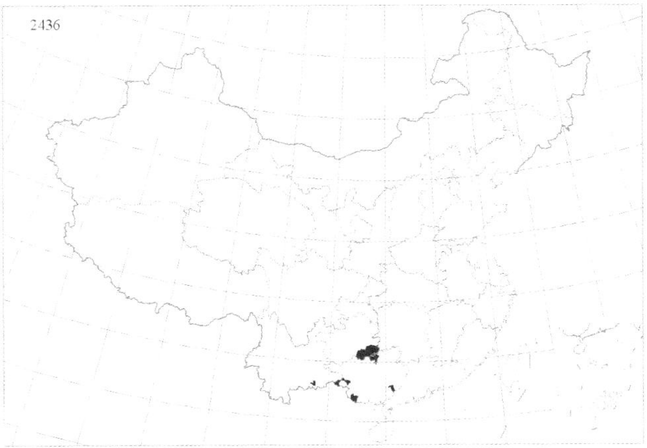

Lauraceae 樟科 *Neolitsea* 新木姜子属
Neolitsea variabillima (Hayata) Kaneh. et Sasaki 变叶新木姜子
Trees, evergreen

18.9 (16.2~21.7); 12.9 (11.0~15.3); 23.8 (20.1~27.0); 18.9 (16.2~21.7); 167.0 (134.2~200.2); 0.0 (0.0~0.0); 928 (773~1118) ▎2318 (1836~2680); 1054 (750~1412); 228 (110~341) ▎156.8 (64.2~246.5); 921 (773~1075); 254.3 (202.8~289.1)

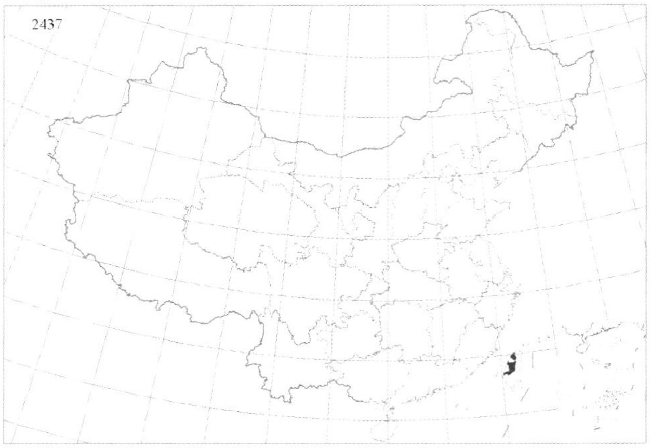

Lauraceae 樟科 *Neolitsea* 新木姜子属
Neolitsea velutina W. T. Wang 毛叶新木姜子
Small trees, evergreen

19.9 (17.2~22.0); 11.4 (7.3~13.6); 26.4 (22.7~28.8); 19.9 (17.3~22.0); 178.8 (147.0~203.8); 0.0 (0.0~0.0); 1022 (838~1170) ▎1477 (1254~1647); 733 (639~850); 98 (48~146) ▎45.3 (30.9~65.6); 1022 (838~1170); 275.0 (203.4~359.7)

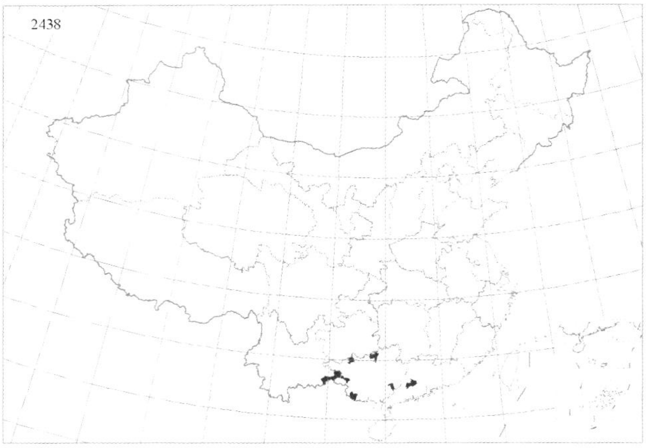

Lauraceae 樟科 *Neolitsea* 新木姜子属
Neolitsea villosa (Blume) Merr. 兰屿新木姜子
Trees, evergreen

18.5; 14.1; 21.7; 18.5; 162.1; 0.0; 863 ▎2713; 1388; 188 ▎214.4; 863; 363.7

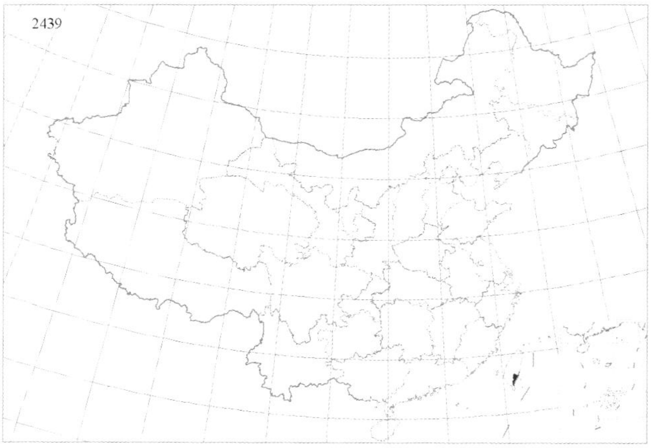

Lauraceae 樟科 *Neolitsea* 新木姜子属
Neolitsea wushanica (Chun) Merr. 巫山新木姜子
Small trees, evergreen

15.1 (9.3~19.1); 4.3 (−1.7~10.0); 25.3 (20.1~28.9); 15.1 (9.5~19.1); 125.1 (69.8~169.1); 3.4 (0.0~18.0); 819 (603~1002) ▎1307 (931~1777); 520 (411~816); 103 (25~205) ▎61.4 (8.0~107.2); 818 (603~1002); 226.5 (108.2~305.6)

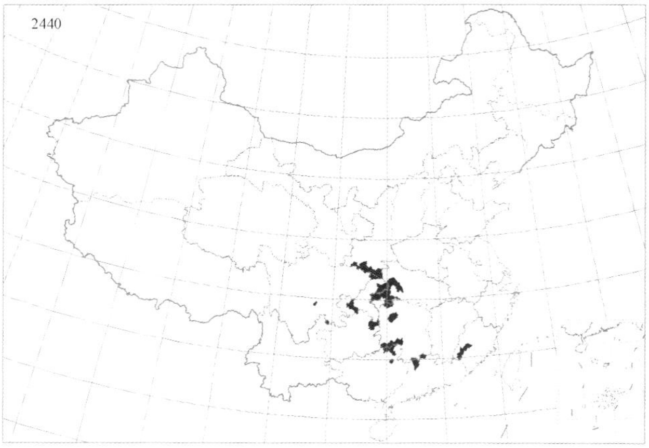

Lauraceae 樟科 *Neolitsea* 新木姜子属
Neolitsea zeylanica (Nees et T. Nees) Merr. 南亚新木姜子
Trees, evergreen

18.7 (14.9~22.2); 8.9 (3.6~13.9); 27.1 (25.7~28.3); 18.7 (15.0~22.3); 165.0 (121.2~207.5); 0.4 (0.0~1.6); 989 (807~1190) ▎1535 (1337~1900); 702 (497~1056); 124 (87~196) ▎57.0 (15.4~79.3); 989 (807~1190); 251.2 (201.5~327.7)

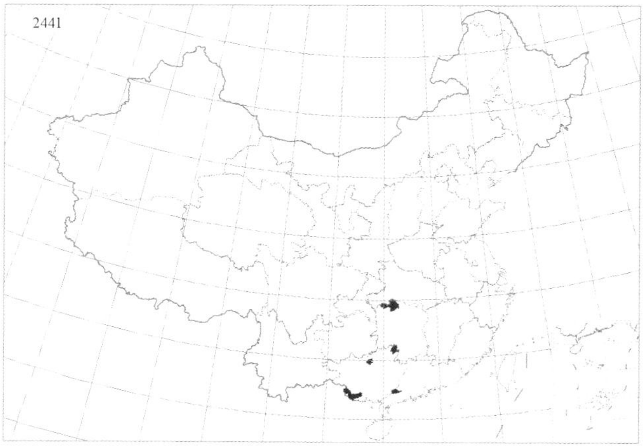

Lauraceae 樟科 *Nothaphoebe* 赛楠属
Nothaphoebe cavaleriei (Lév.) Yang 赛楠
Trees, evergreen

14.7 (11.1~18.1); 4.8 (2.2~8.1); 23.8 (19.4~27.3); 14.7 (11.3~18.2); 117.9 (80.8~158.1); 0.9 (0.0~5.3); 782 (652~949) ▎1163 (879~1448); 589 (472~817); 55 (23~105) ▎49.1 (18.6~89.0); 782 (652~949); 211.5 (135.1~298.1)

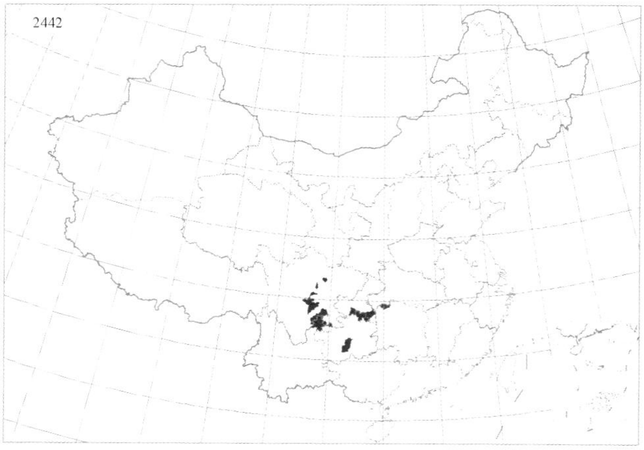

Lauraceae 樟科 *Nothaphoebe* 赛楠属
Nothaphoebe konishii (Hayata) Hayata 台湾赛楠
Trees, evergreen

19.9 (16.2~24.1)；14.7 (11.0~18.9)；23.8 (20.1~28.0)；19.9 (16.2~24.1)；178.5 (134.2~229.7)；0.0 (0.0~0.0)；990 (773~1322) ❙ 2445 (1751~2992)；1323 (1033~1847)；154 (47~245) ❙ 159.8 (32.5~246.5)；951 (773~1123)；262.9 (154.2~363.7)

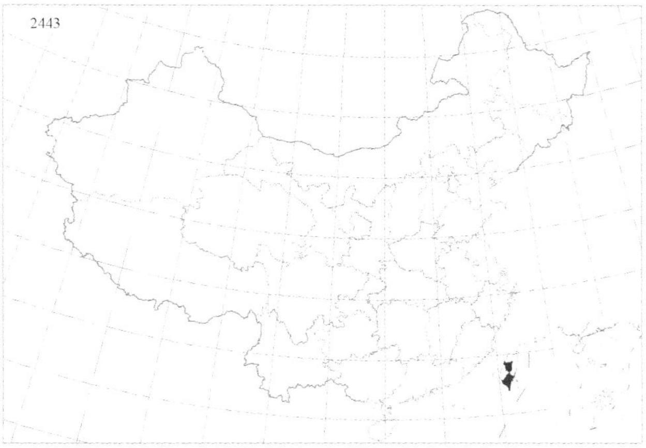

Lauraceae 樟科 *Parasassafras* 密花檫属
Parasassafras confertiflorum (Meisner) D. G. Long 拟檫木
Shrubs or small trees, evergreen

17.2 (14.4~19.6)；10.5 (7.5~12.8)；21.4 (19.1~23.5)；17.2 (14.4~19.6)；146.8 (113.1~175.7)；0.0 (0.0~0.0)；834 (727~953) ❙ 1493 (1232~1688)；861 (703~959)；53 (42~70) ❙ 79.7 (56.2~112.2)；829 (727~932)；486.2 (372.4~627.0)

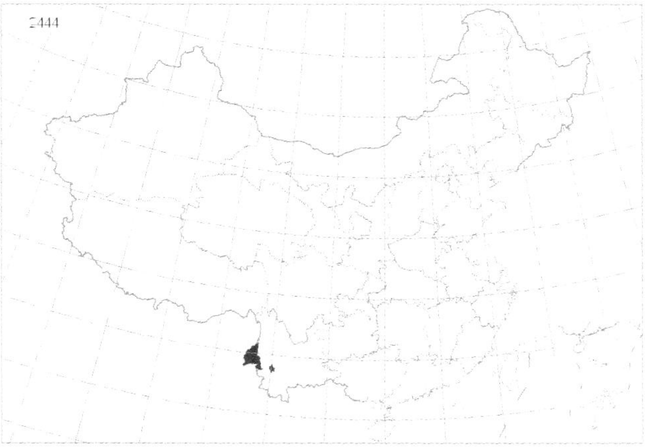

Lauraceae 樟科 *Phoebe* 楠木属
Phoebe angustifolia Meisn. 沼楠
Small shrubs, evergreen

18.7 (17.3~21.0)；12.2 (9.9~14.2)；23.2 (22.1~25.6)；18.8 (17.3~21.0)；165.3 (147.1~191.8)；0.0 (0.0~0.0)；909 (833~1053) ❙ 1406 (1276~1522)；797 (741~860)；57 (51~68) ❙ 55.4 (36.8~80.6)；907 (833~1053)；474.6 (413.5~551.6)

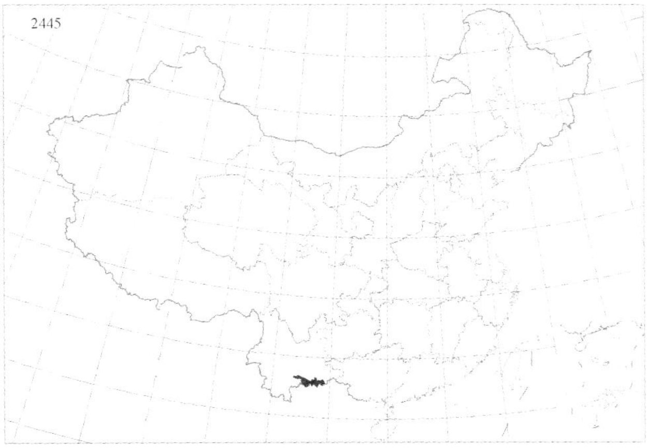

Lauraceae 樟科 *Phoebe* 楠木属
Phoebe bournei (Hemsl.) Yang 闽楠
Trees, evergreen

17.1 (12.0~21.5)；6.2 (-0.8~12.8)；27.2 (23.0~29.9)；17.2 (12.1~21.5)；146.6 (94.7~198.2)；0.5 (0.0~13.4)；910 (697~1138) ❙ 1515 (793~1997)；544 (376~821)；171 (42~259) ❙ 66.9 (-2.0~130.7)；902 (697~1138)；221.3 (134.6~354.5)

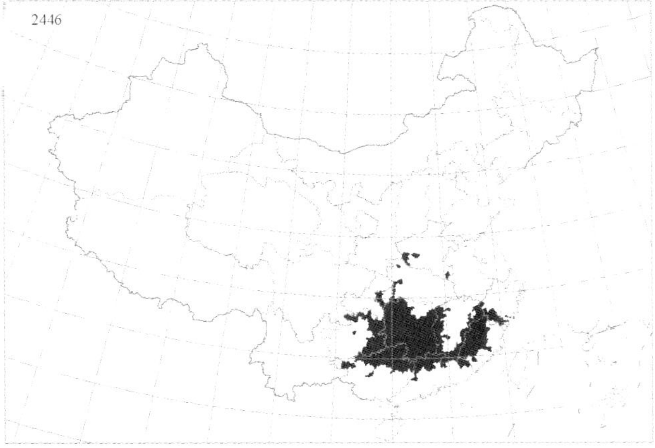

Lauraceae 樟科 *Phoebe* 楠木属
Phoebe brachythyrsa H. W. Li 短序楠
Small shrubs, evergreen

12.8；3.5；21.1；12.9；96.0；1.5；699 ❙ 879；472；36 ❙ 25.7；699；229.2

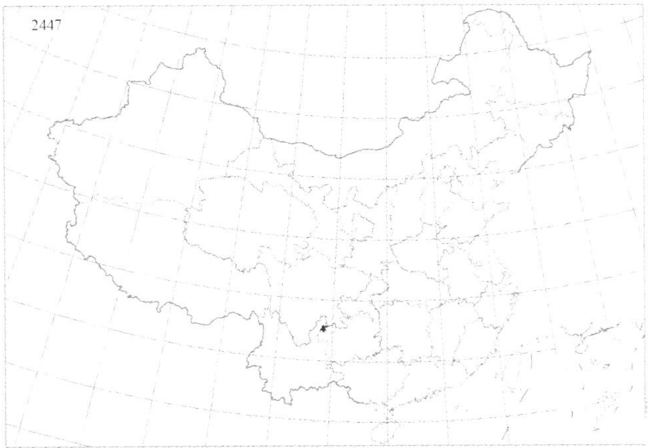

Lauraceae 樟科 *Phoebe* 楠木属
Phoebe calcarea S. K. Lee et F. N. Wei 石山楠
Trees, evergreen

19.6 (17.4~22.2)；9.9 (7.3~13.9)；27.6 (25.5~29.0)；19.6 (17.4~22.3)；175.5 (148.9~207.5)；0.0 (0.0~0.0)；1027 (886~1190) ❙ 1499 (1279~1710)；658 (561~775)；143 (73~186) ❙ 47.2 (14.8~70.6)；1027 (886~1190)；247.9 (160.7~335.1)

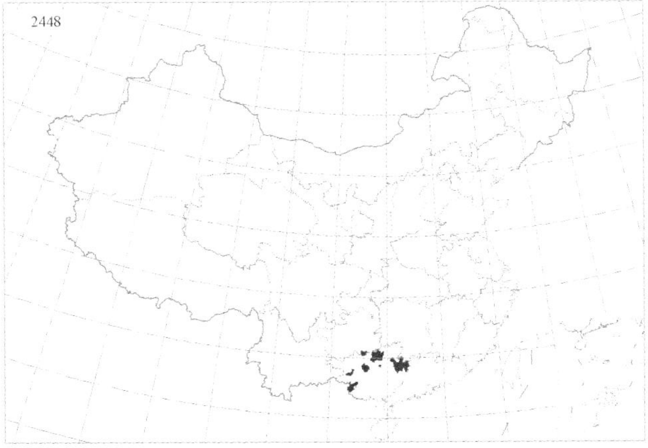

Lauraceae 樟科 *Phoebe* 楠木属

Phoebe chekiangensis C. B. Shang 浙江楠

Trees, evergreen

16.7 (14.7~19.1); 5.5 (2.3~10.0); 27.4 (24.7~28.9); 16.8 (14.8~19.1); 141.8 (120.8~169.1); 0.7 (0.0~3.9); 895 (789~984) ❙ 1653 (1002~1997); 577 (448~713); 209 (114~259) ❙ 85.2 (14.0~147.0); 892 (789~981); 251.0 (134.1~319.9)

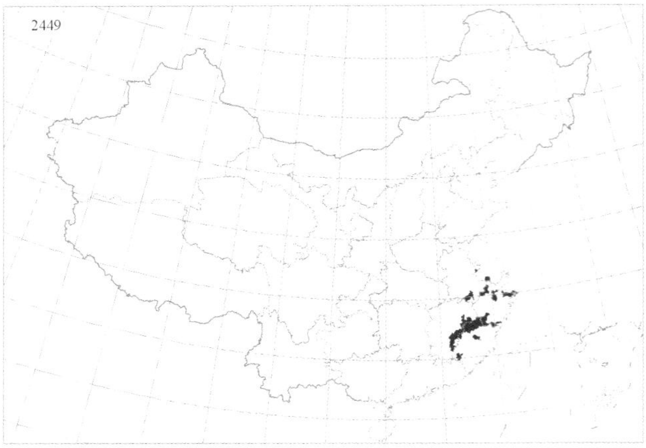

Lauraceae 樟科 *Phoebe* 楠木属

Phoebe chinensis Chun 山楠

Trees, evergreen

14.1 (5.3~19.4); 1.9 (-2.9~9.4); 25.4 (12.2~29.2); 14.1 (5.7~19.4); 116.3 (30.5~173.4); 7.1 (0.0~26.2); 799 (479~1029) ❙ 1071 (512~1745); 472 (258~817); 79 (15~216) ❙ 34.0 (-35.3~108.0); 774 (479~1029); 210.5 (121.6~316.6)

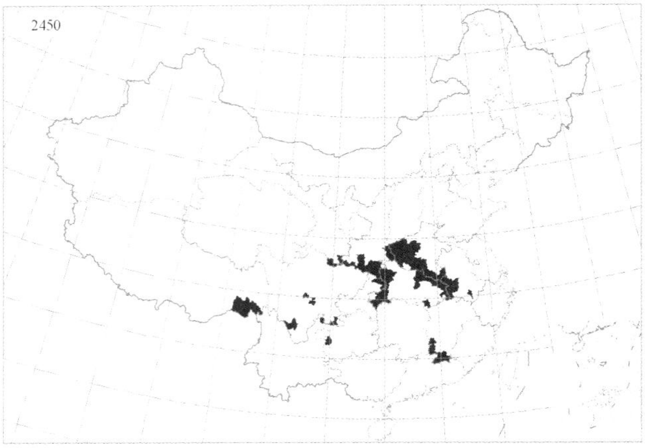

Lauraceae 樟科 *Phoebe* 楠木属

Phoebe crassipedicella S. K. Lee et F. N. Wei 粗柄楠

Trees, evergreen

18.1 (16.7~19.1); 8.4 (6.3~9.8); 26.0 (25.5~26.9); 18.1 (16.7~19.2); 157.7 (140.5~170.1); 0.0 (0.0~0.0); 926 (863~981) ❙ 1317 (1236~1437); 633 (531~678); 88 (54~122) ❙ 42.6 (27.5~57.4); 926 (863~981); 265.9 (225.1~322.1)

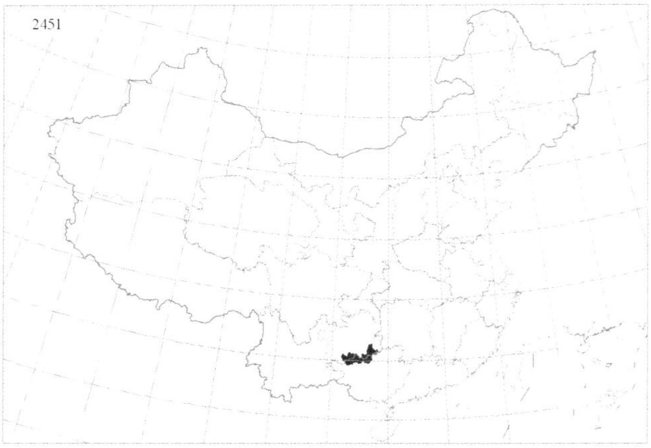

Lauraceae 樟科 *Phoebe* 楠木属

Phoebe faberi (Hemsl.) Chun 竹叶楠

Trees, evergreen

14.2 (6.0~19.0); 3.0 (-3.2~13.1); 24.7 (14.4~28.8); 14.3 (6.6~19.1); 116.6 (39.7~168.8); 5.3 (0.0~26.7); 789 (508~949) ❙ 1052 (626~1571); 497 (316~817); 62 (19~157) ❙ 34.2 (-22.2~108.0); 768 (508~949); 217.6 (135.1~427.9)

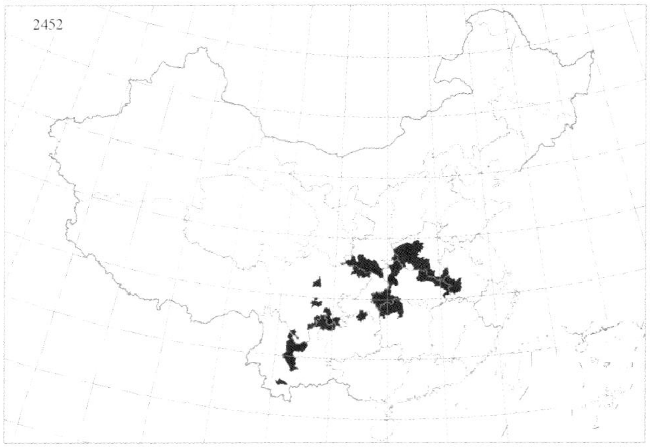

Lauraceae 樟科 *Phoebe* 楠木属

Phoebe formosana (Matsum. et Hay.) Hay. 台楠

Trees, evergreen

16.8 (14.9~21.7); 7.4 (2.6~15.3); 25.4 (20.1~28.6); 16.8 (14.9~21.7); 143.0 (122.6~200.2); 1.2 (0.0~3.5); 871 (773~1118) ❙ 2001 (1396~2993); 764 (529~1186); 243 (120~547) ❙ 132.3 (55.4~246.5); 868 (773~1075); 272.1 (202.8~319.9)

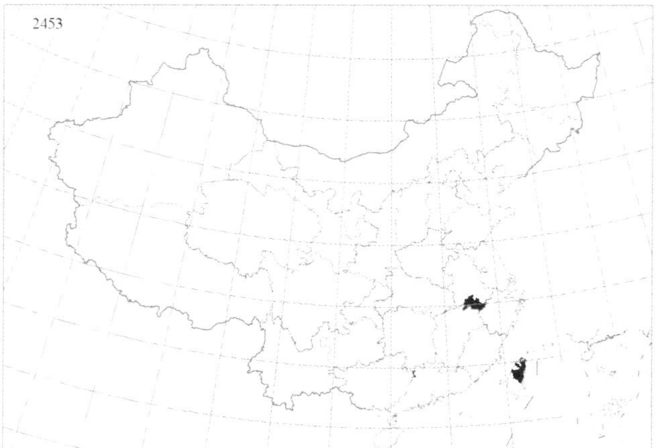

Lauraceae 樟科 *Phoebe* 楠木属

Phoebe forrestii W. W. Sm. 长毛楠

Trees, evergreen

16.0 (5.3~21.0); 9.4 (-2.4~14.2); 20.9 (12.2~25.6); 16.1 (5.7~21.0); 133.5 (30.5~191.8); 0.5 (0.0~26.2); 796 (479~1053) ❙ 1083 (797~1751); 619 (411~996); 44 (18~77) ❙ 37.0 (-14.2~100.7); 768 (479~1053); 336.4 (207.5~551.6)

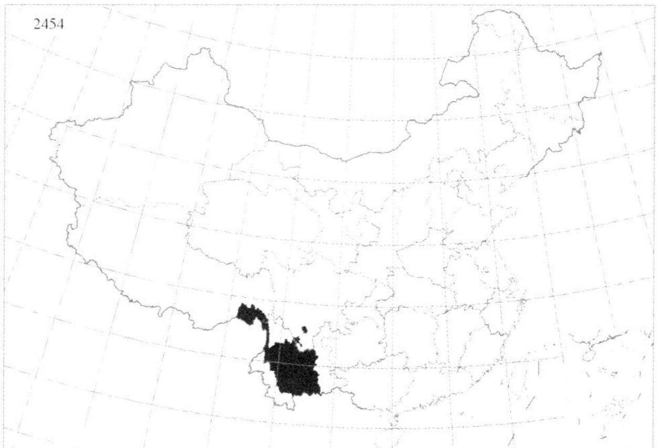

Lauraceae 樟科 *Phoebe* 楠木属
Phoebe glaucifolia S. K. Lee et F. N. Wei 白背楠
Trees, evergreen

13.3; 5.4; 19.6; 13.4; 100.4; 0.0; 706 ❙ 1736; 1007; 51 ❙ 145.8; 706; 447.4

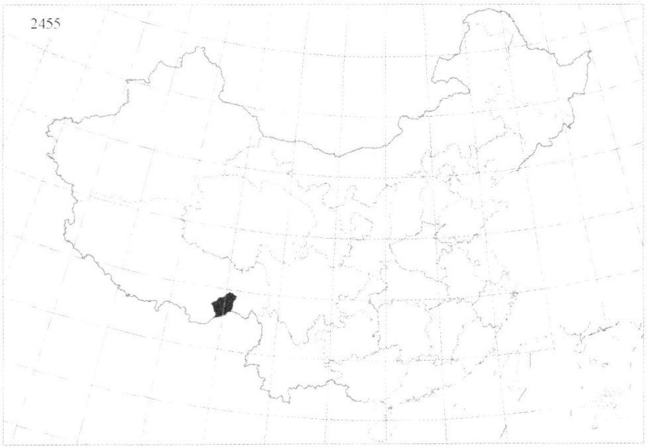

Lauraceae 樟科 *Phoebe* 楠木属
Phoebe glaucophylla H. W. Li 粉叶楠
Trees, evergreen

17.6; 9.9; 23.2; 17.7; 152.4; 0.0; 859 ❙ 1422; 813; 52 ❙ 65.6; 859; 359.7

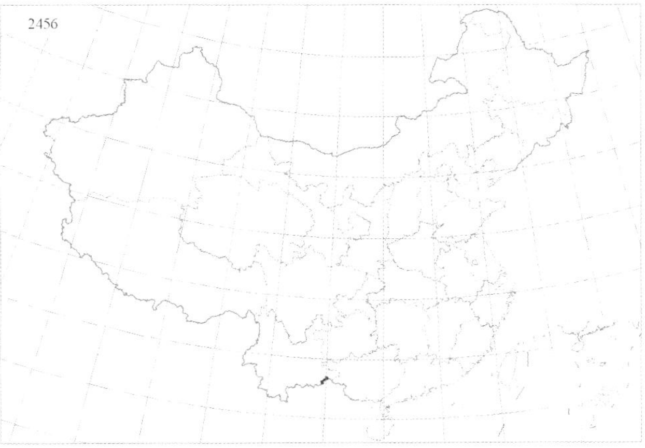

Lauraceae 樟科 *Phoebe* 楠木属
Phoebe hainanensis Merr. 茶槁楠
Trees, evergreen

23.6; 17.9; 27.7; 23.7; 224.1; 0.0; 1276 ❙ 1450; 529; 64 ❙ 14.9; 1151; 492.3

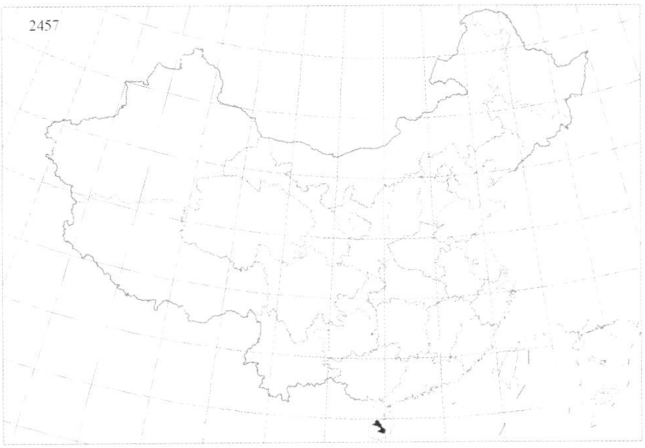

Lauraceae 樟科 *Phoebe* 楠木属
Phoebe hui Cheng 细叶楠
Trees, evergreen

14.0 (10.1~16.3); 4.0 (1.1~5.8); 23.1 (17.7~25.7); 14.0 (10.3~16.3); 110.3 (70.1~136.2); 2.2 (0.0~8.0); 758 (621~859) ❙ 1129 (931~1446); 597 (411~816); 43 (22~100) ❙ 49.8 (10.5~80.4); 757 (621~848); 222.1 (133.8~318.7)

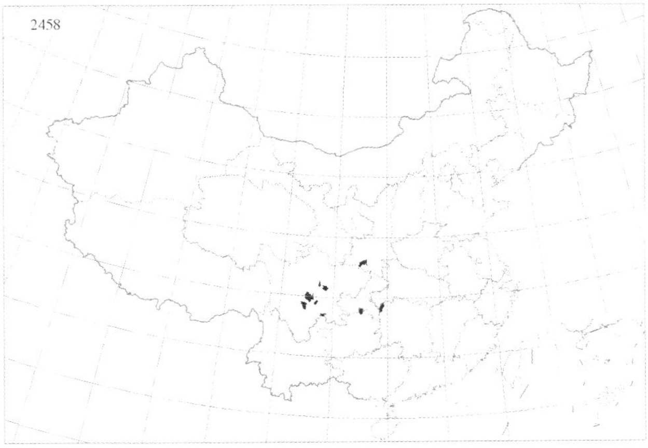

Lauraceae 樟科 *Phoebe* 楠木属
Phoebe hunanensis Hand.-Mazz. 湘楠
Shrubs or small trees, evergreen

15.4 (7.4~21.0); 3.6 (-4.2~11.9); 26.6 (18.7~29.8); 15.5 (8.2~21.1); 129.4 (58.8~193.0); 3.6 (0.0~29.4); 850 (561~1114) ❙ 1271 (626~1829); 477 (316~821); 128 (15~230) ❙ 48.7 (-22.2~128.9); 835 (561~1114); 208.6 (134.6~334.4)

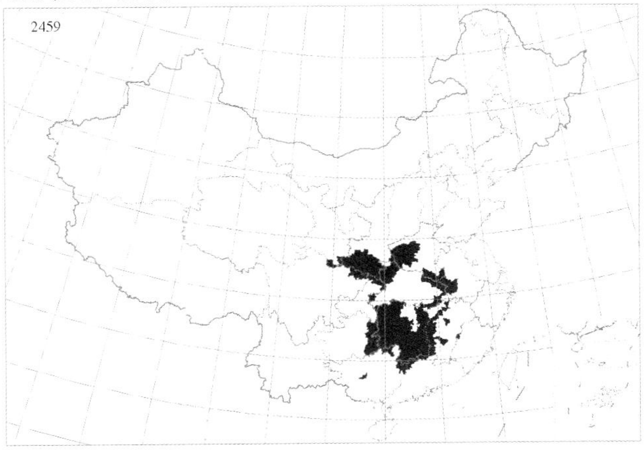

Lauraceae 樟科 *Phoebe* 楠木属
Phoebe hungmoensis S. K. Lee 红毛山楠
Trees, evergreen

23.6 (21.4~24.7); 17.5 (12.5~19.8); 28.0 (27.2~28.7); 23.6 (21.5~24.7); 223.4 (197.8~236.5); 0.0 (0.0~0.0); 1285 (1140~1376) ❙ 1411 (1176~1588); 546 (444~766); 68 (44~130) ❙ 10.3 (-12.0~31.4); 1146 (1091~1212); 442.9 (248.2~620.0)

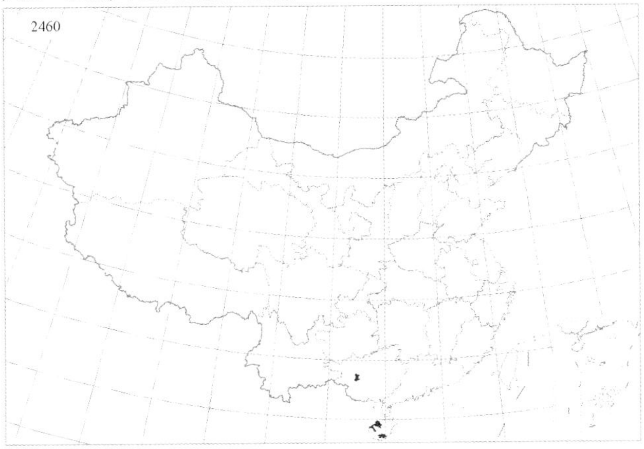

Lauraceae 樟科 *Phoebe* 楠木属
Phoebe kwangsiensis H. Liu 桂楠
Small trees, evergreen

19.4 (17.4~21.3); 10.0 (7.3~12.6); 26.7 (25.5~27.9); 19.4 (17.4~21.3); 172.8 (148.9~196.0); 0.0 (0.0~0.0); 999 (886~1123) ❙ 1351 (1182~1510); 678 (634~765); 91 (43~142) ❙ 36.1 (5.8~64.7); 999 (886~1121); 277.5 (225.9~354.5)

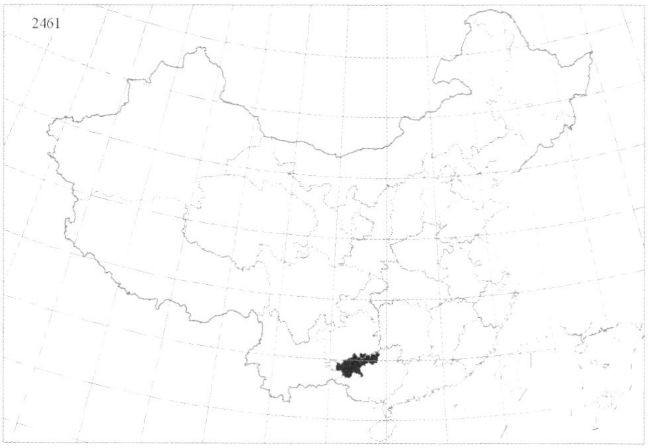

Lauraceae 樟科 *Phoebe* 楠木属
Phoebe lanceolata (Nees) Nees 披针叶楠
Trees, evergreen

18.9 (16.3~22.2); 12.5 (9.1~15.1); 23.0 (20.6~28.3); 18.9 (16.4~22.3); 166.7 (136.4~207.5); 0.0 (0.0~0.0); 916 (795~1190) ❙ 1481 (1190~1751); 796 (655~996); 61 (42~87) ❙ 63.1 (15.4~112.2); 906 (795~1190); 486.2 (327.7~659.0)

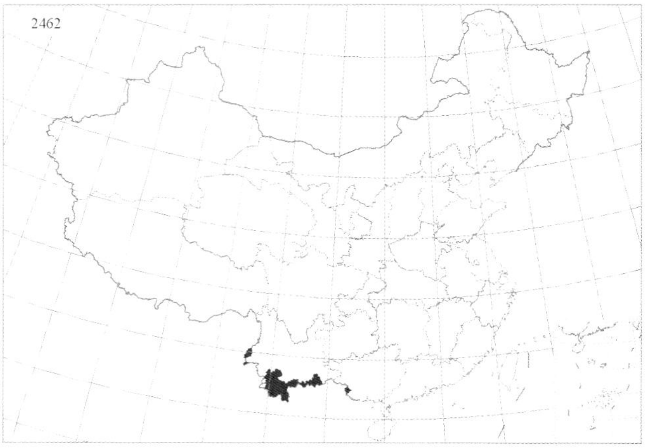

Lauraceae 樟科 *Phoebe* 楠木属
Phoebe legendrei Lecomte 雅砻江楠
Trees, evergreen

13.9 (5.4~19.0); 6.8 (-2.1~13.1); 19.5 (12.3~24.3); 14.0 (5.8~19.1); 109.9 (31.5~168.8); 2.4 (0.0~25.6); 729 (481~929) ❙ 972 (741~1495); 574 (411~792); 29 (12~70) ❙ 35.4 (-14.2~99.9); 686 (481~891); 296.1 (205.2~487.2)

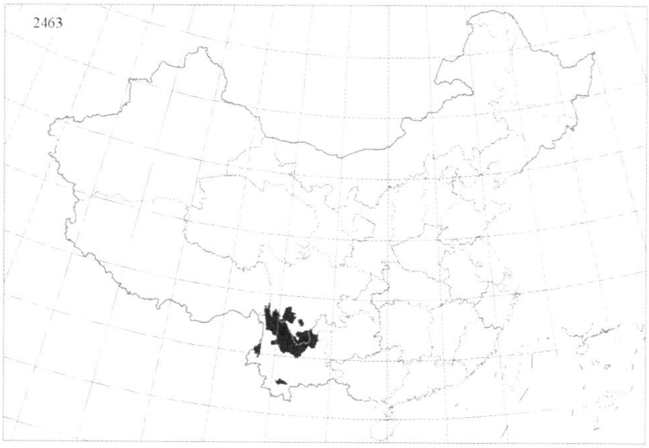

Lauraceae 樟科 *Phoebe* 楠木属
Phoebe lichuanensis S. K. Lee 利川楠
Trees, evergreen

12.7; 2.1; 23.0; 12.8; 99.0; 5.6; 713 ❙ 1478; 592; 85 ❙ 107.2; 713; 234.7

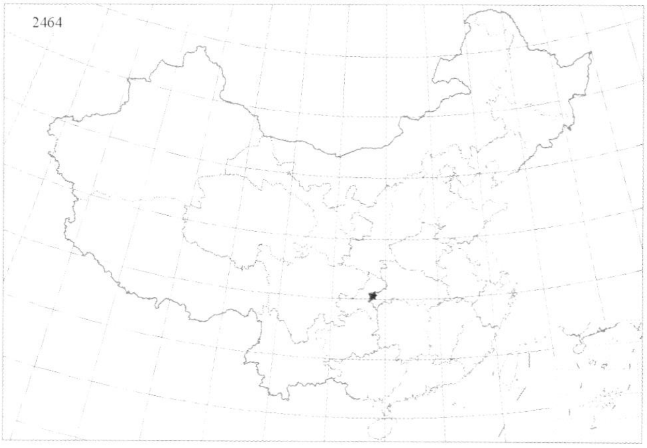

Lauraceae 樟科 *Phoebe* 楠木属
Phoebe macrocarpa C. Y. Wu 大果楠
Trees, evergreen

17.4 (8.6~21.0); 10.6 (1.3~14.2); 22.2 (15.0~25.6); 17.4 (8.7~21.0); 150.0 (52.6~191.8); 0.6 (0.0~8.8); 854 (574~1053) ❙ 1349 (1041~1751); 762 (582~996); 56 (45~77) ❙ 58.5 (32.4~88.8); 854 (574~1053); 426.4 (277.0~551.6)

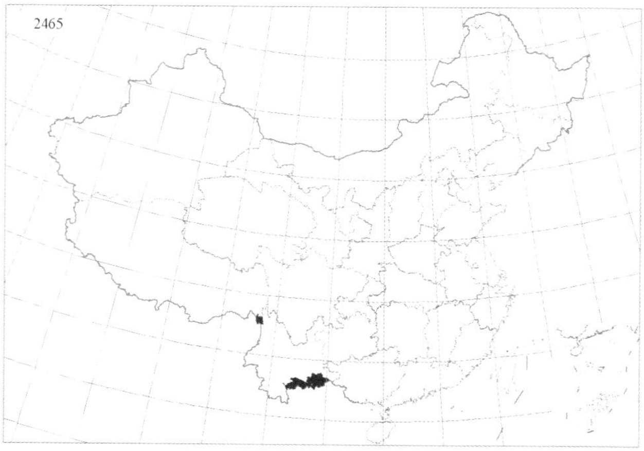

Lauraceae 樟科 *Phoebe* 楠木属
Phoebe megacalyx H. W. Li 大萼楠
Small trees, evergreen

19.2 (17.3~21.0); 13.1 (9.9~15.1); 23.4 (22.4~25.6); 19.3 (17.3~21.0); 171.4 (147.1~191.8); 0.0 (0.0~0.0); 932 (833~1053) ❙ 1496 (1276~1751); 808 (655~996); 64 (51~77) ❙ 61.2 (36.8~88.8); 927 (833~1053); 523.3 (413.5~659.0)

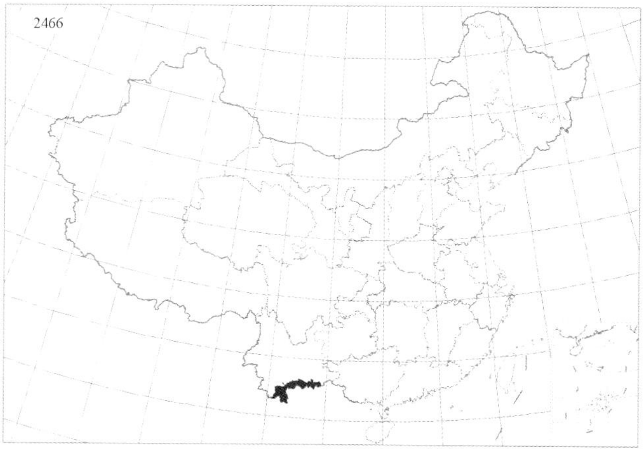

Lauraceae 樟科 *Phoebe* 楠木属
Phoebe microphylla H. W. Li 小叶楠
Trees, evergreen

19.3; 12.7; 23.8; 19.3; 171.6; 0.0; 942 | 1439; 803; 61 | 53.4; 942; 516.4

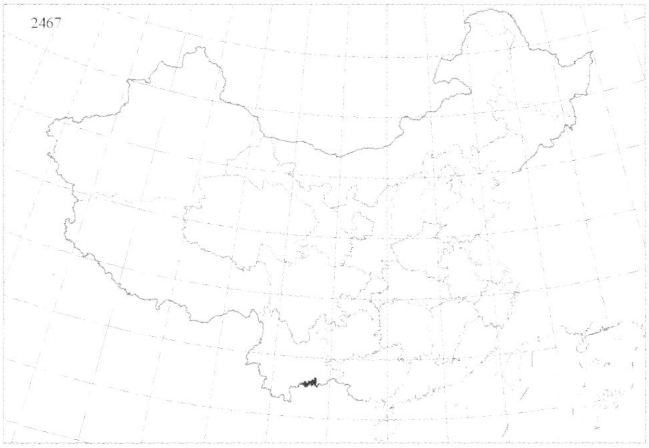

Lauraceae 樟科 *Phoebe* 楠木属
Phoebe minutiflora H. W. Li 小花楠
Trees, evergreen

18.9 (16.4~21.0); 12.6 (9.6~15.1); 22.9 (20.6~25.6); 18.9 (16.4~21.0); 167.4 (137.3~191.8); 0.0 (0.0~0.0); 915 (795~1053) | 1507 (958~1751); 833 (531~996); 58 (42~77) | 65.4 (13.3~112.2); 907 (795~1053); 520.7 (311.9~659.0)

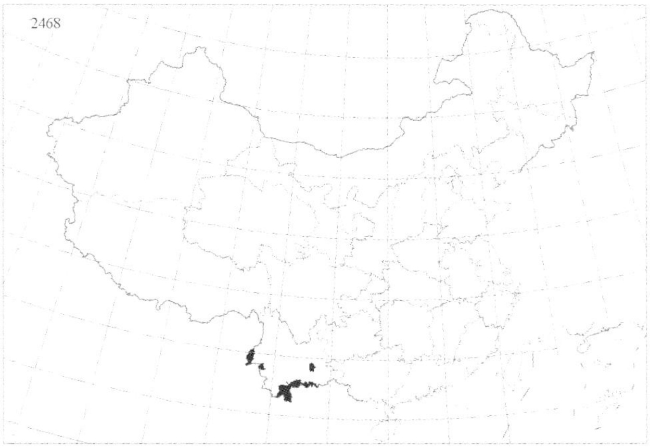

Lauraceae 樟科 *Phoebe* 楠木属
Phoebe motuonan S. K. Lee et F. N. Wei 墨脱楠
Trees, evergreen

13.3; 5.4; 19.6; 13.4; 100.4; 0.0; 706 | 1736; 1007; 51 | 145.8; 706; 447.4

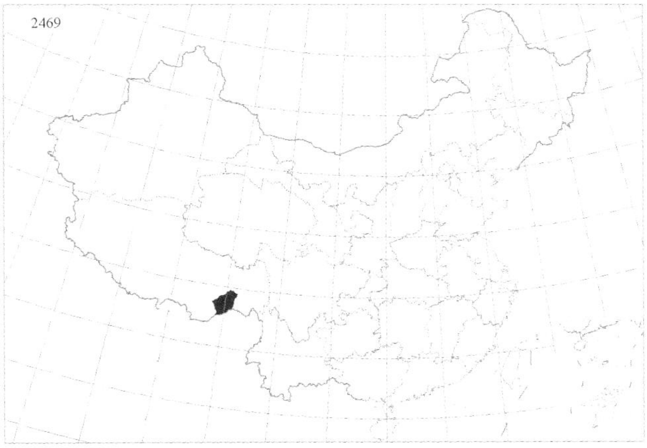

Lauraceae 樟科 *Phoebe* 楠木属
Phoebe neurantha (Hemsl.) Gamble 白楠
Shrubs or small trees, evergreen

15.3 (5.4~22.2); 5.3 (-2.9~14.2); 24.1 (12.3~29.3); 15.3 (5.8~22.3); 126.8 (31.5~207.5); 3.0 (0.0~25.6); 815 (481~1190) | 1162 (626~1922); 544 (316~996); 80 (12~259) | 42.5 (-22.2~111.3); 800 (481~1190); 255.0 (133.8~551.6)

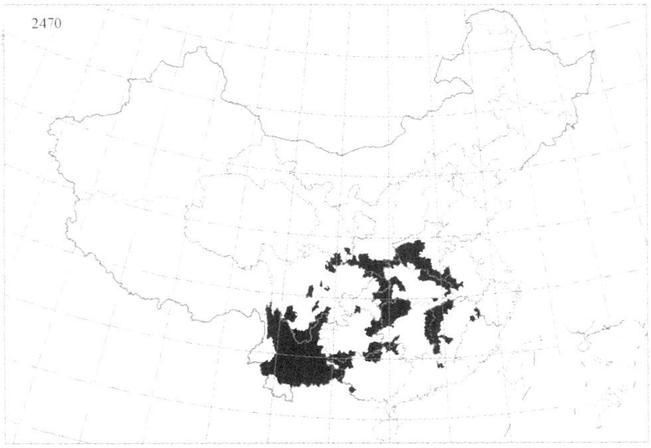

Lauraceae 樟科 *Phoebe* 楠木属
Phoebe neurantha var. brevifolia H. W. Li 短叶白楠
Shrubs, evergreen

17.7 (14.2~22.2); 10.4 (6.2~14.2); 22.9 (20.1~28.3); 17.7 (14.2~22.3); 152.7 (110.5~207.5); 0.0 (0.0~0.0); 870 (729~1190) | 1206 (958~1522); 669 (531~860); 52 (43~87) | 39.1 (13.3~80.6); 867 (729~1190); 359.4 (248.4~551.6)

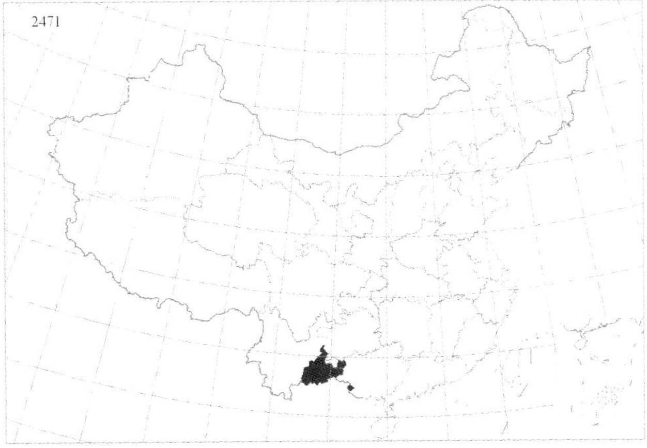

Lauraceae 樟科 *Phoebe* 楠木属
Phoebe neurantha var. cavaleriei H. Liu 兴义白楠
Shrubs or small trees, evergreen

17.1; 8.5; 23.4; 17.1; 145.1; 0.0; 845 | 1232; 666; 59 | 45.8; 845; 283.9

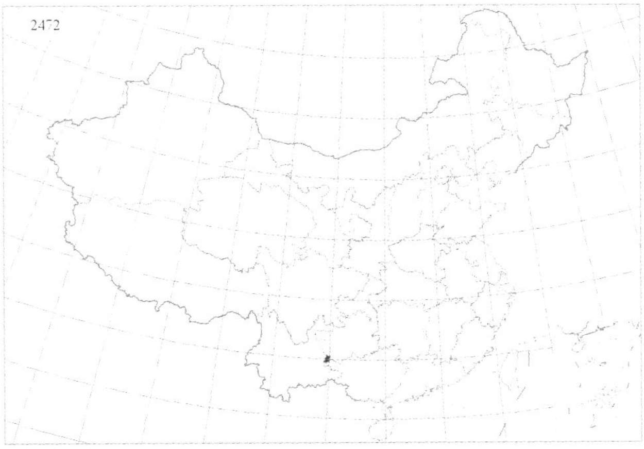

Lauraceae 樟科 *Phoebe* 楠木属
Phoebe neuranthoides S. K. Lee et F. N. Wei 光枝楠
Shrubs or small trees, evergreen

15.7 (9.1~19.2); 5.0 (-0.7~10.1); 25.6 (18.1~29.2); 15.8 (9.2~19.2); 130.1 (64.1~170.8); 1.0 (0.0~14.1); 835 (595~1002) ❙ 1274 (806~1496); 534 (375~720); 93 (16~192) ❙ 53.0 (8.2~107.2); 834 (595~1002); 210.7 (146.0~286.7)

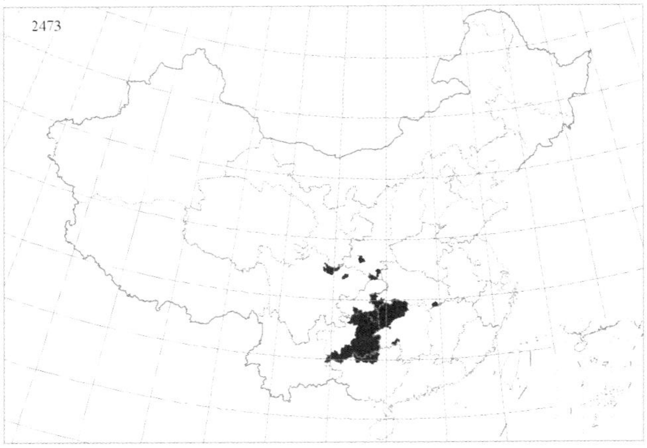

Lauraceae 樟科 *Phoebe* 楠木属
Phoebe nigrifolia S. K. Lee et F. N. Wei 黑叶楠
Shrubs or small trees, evergreen

20.2 (17.6~22.3); 11.5 (7.3~13.9); 26.9 (25.2~28.3); 20.3 (17.6~22.3); 183.2 (151.6~207.5); 0.0 (0.0~0.0); 1049 (909~1190) ❙ 1414 (1279~1496); 736 (639~848); 92 (72~142) ❙ 36.6 (14.8~64.7); 1049 (909~1190); 295.9 (225.9~335.1)

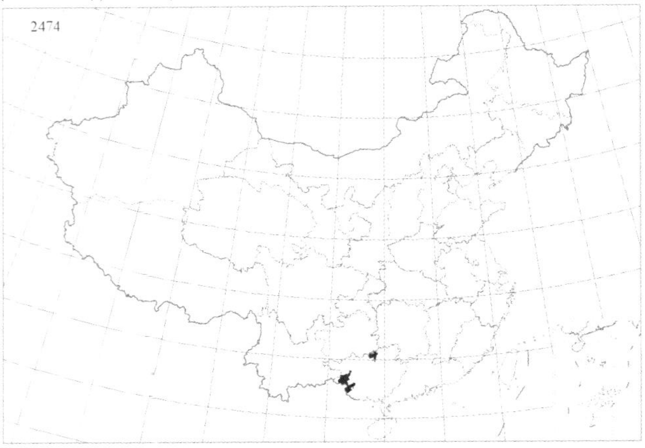

Lauraceae 樟科 *Phoebe* 楠木属
Phoebe puwenensis Cheng 普文楠
Trees, evergreen

18.5 (16.0~20.5); 12.5 (9.3~15.1); 22.4 (20.3~23.7); 18.6 (16.0~20.5); 163.0 (132.4~185.9); 0.0 (0.0~0.0); 891 (780~992) ❙ 1421 (1028~1751); 769 (566~996); 55 (41~77) ❙ 59.9 (12.5~112.2); 871 (780~986); 467.4 (339.5~659.0)

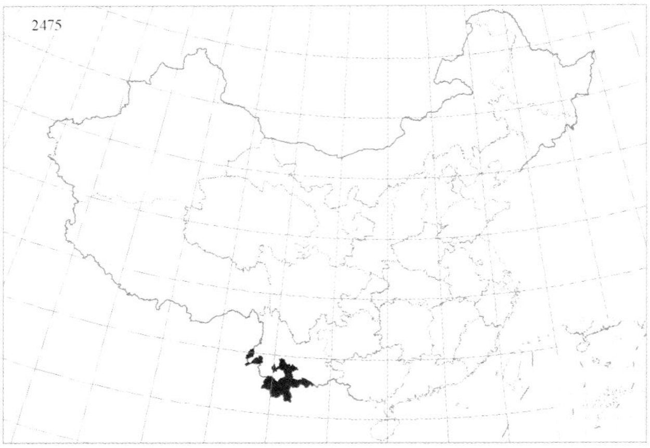

Lauraceae 樟科 *Phoebe* 楠木属
Phoebe rufescens H. W. Li 红梗楠
Trees, evergreen

17.6 (15.4~19.6); 11.3 (8.1~13.8); 21.7 (20.3~23.5); 17.7 (15.5~19.6); 152.1 (125.7~175.7); 0.0 (0.0~0.0); 849 (767~953) ❙ 1295 (939~1688); 729 (524~959); 43 (37~62) ❙ 52.2 (6.5~112.2); 823 (717~932); 414.7 (274.9~627.0)

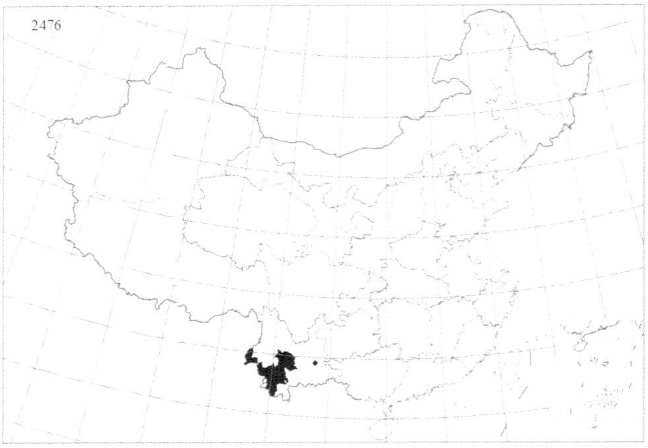

Lauraceae 樟科 *Phoebe* 楠木属
Phoebe sheareri (Hemsl.) Gamble 紫楠
Shrubs or small trees, evergreen

16.6 (10.2~21.6); 5.6 (-2.9~14.9); 26.8 (16.9~29.9); 16.6 (10.3~21.6); 141.2 (68.2~199.4); 1.5 (0.0~20.5); 890 (620~1128) ❙ 1401 (626~2005); 541 (316~902); 143 (15~259) ❙ 57.4 (-22.2~137.5); 880 (620~1128); 218.2 (108.2~570.1)

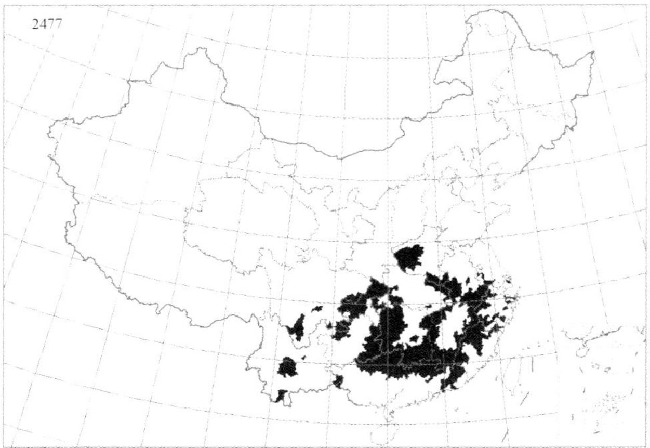

Lauraceae 樟科 *Phoebe* 楠木属
Phoebe sheareri var. omeiensis (Yang) N. Chao 峨眉楠
Shrubs or small trees evergreen

16.6 (13.3~19.1); 6.8 (3.6~10.1); 25.1 (22.2~27.8); 16.6 (13.3~19.1); 139.4 (101.9~169.2); 0.1 (0.0~1.7); 855 (718~969) ❙ 1187 (985~1410); 603 (458~816); 61 (49~107) ❙ 39.2 (18.0~80.4); 855 (718~969); 208.9 (126.1~283.9)

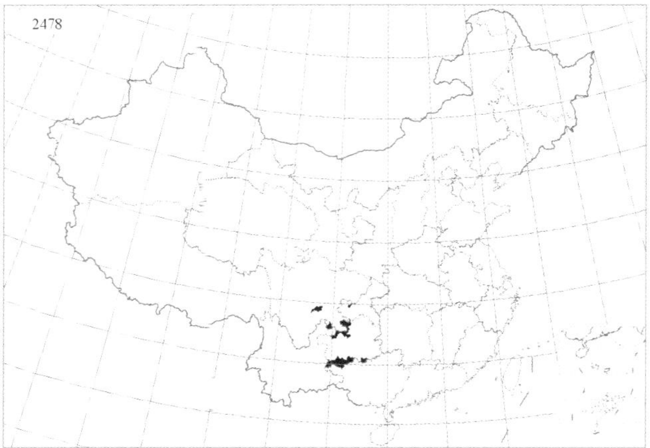

Lauraceae 樟科 *Phoebe* 楠木属
Phoebe tavoyana (Meisn.) Hook. f. 乌心楠
Trees, evergreen

23.5 (21.5~24.8)；17.3 (13.6~20.1)；28.1 (25.7~29.2)；23.5 (21.6~24.8)；222.2 (199.1~237.6)；0.0 (0.0~0.0)；1271 (1127~1385) ❙ 1443 (868~1965)；580 (312~994)；73 (41~115) ❙ 14.3 (-34.3~61.6)；1163 (868~1326)；384.3 (178.7~620.0)

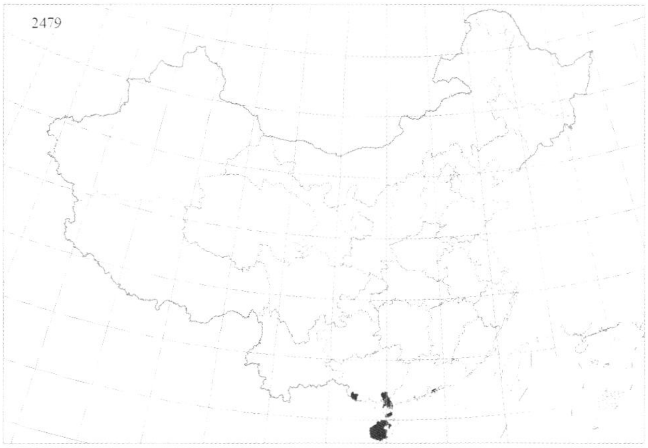

Lauraceae 樟科 *Phoebe* 楠木属
Phoebe yaiensis S. K. Lee 崖楠
Trees, evergreen

21.0；14.0；26.0；21.0；192.5；0.0；1094 ❙ 1442；714；57 ❙ 36.5；1011；404.3

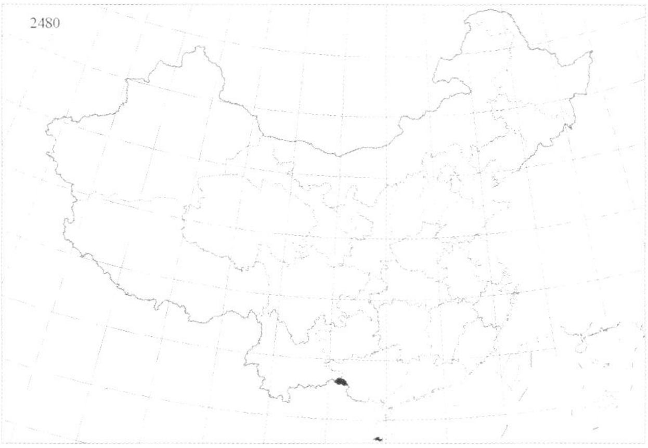

Lauraceae 樟科 *Phoebe* 楠木属
Phoebe yunnanensis H. W. Li 景东楠
Trees, evergreen

17.1 (13.6~19.6)；10.5 (7.2~12.8)；21.4 (18.8~23.5)；17.1 (13.7~19.6)；145.0 (104.1~175.7)；0.0 (0.0~0.0)；830 (706~953) ❙ 1276 (986~1688)；730 (556~959)；49 (42~62) ❙ 53.5 (12.5~112.2)；810 (706~932)；426.8 (303.3~627.0)

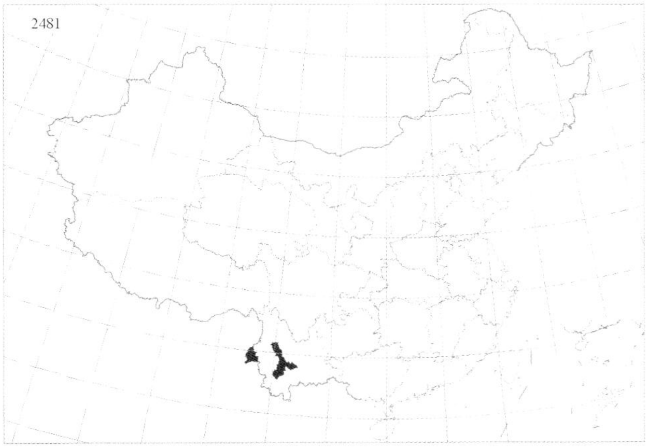

Lauraceae 樟科 *Phoebe* 楠木属
Phoebe zhennan S. K. Lee et F. N. Wei 楠木
Trees, evergreen

15.0 (7.3~18.7)；4.1 (-2.0~8.1)；25.3 (15.6~28.9)；15.0 (7.5~18.8)；122.8 (48.2~166.0)；2.3 (0.0~20.4)；813 (537~1002) ❙ 1200 (798~1495)；530 (377~817)；79 (26~175) ❙ 49.2 (1.9~107.2)；812 (537~1002)；195.9 (108.2~298.1)

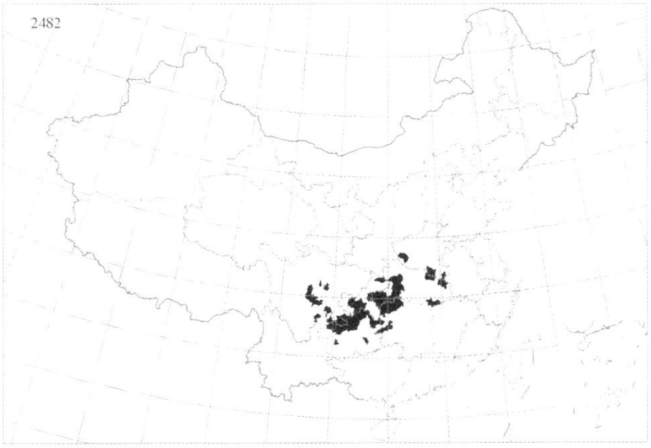

Lauraceae 樟科 *Sassafras* 檫木属
Sassafras randaiense (Hayata) Rehder 台湾檫木
Trees, deciduous

20.8；15.3；25.0；20.8；189.2；0.0；1019 ❙ 2445；1412；110 ❙ 140.0；1019；207.7

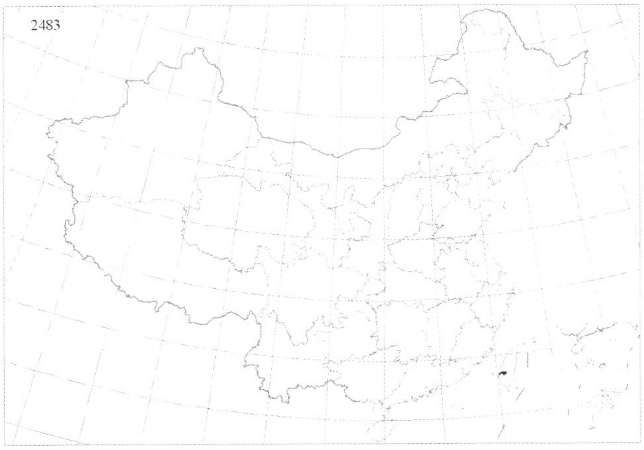

Lauraceae 樟科 *Sassafras* 檫木属
Sassafras tzumu (Hemsl.) Hemsl. 檫木
Trees, deciduous

16.7 (9.3~22.8)；5.7 (-3.3~14.7)；26.8 (18.6~30.1)；16.7 (9.5~22.8)；141.5 (69.8~213.6)；1.0 (0.0~22.0)；891 (603~1230) ❙ 1431 (670~2148)；550 (299~1234)；150 (25~259) ❙ 60.7 (-9.5~130.7)；880 (603~1230)；220.4 (131.2~506.2)

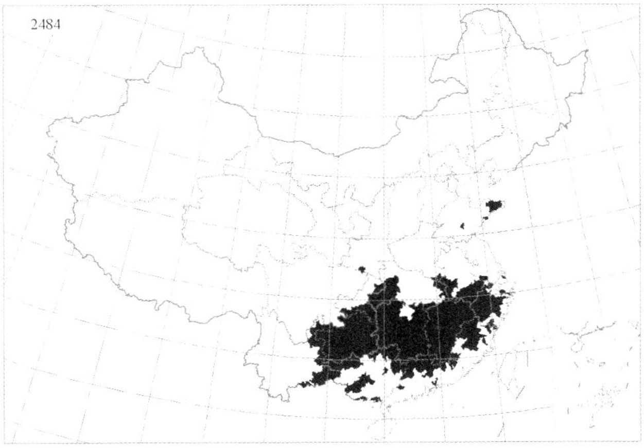

Lauraceae 樟科 *Sinosassafras* 华檫木属
Sinosassafras flavinervium (C. K. Allen) H. W. Li 华檫木
Trees, evergreen

16.1 (8.6~19.6); 9.6 (1.3~13.9); 20.5 (15.0~22.7); 16.1 (8.7~19.6); 133.9 (52.6~175.1); 0.4 (0.0~8.8); 793 (574~933) ❙ 1281 (1014~1736); 707 (568~1007); 51 (41~70) ❙ 63.0 (26.1~145.8); 778 (574~882); 414.7 (318.7~530.1)

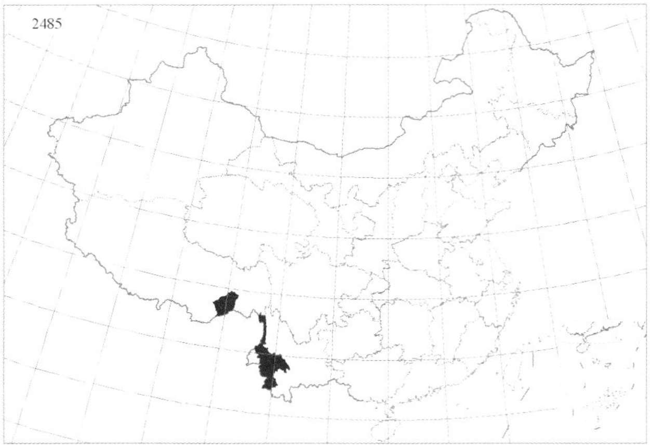

Lauraceae 樟科 *Syndiclis* 油果樟属
Syndiclis anlungensis H. W. Li 安龙油果樟
Trees, evergreen

17.4; 8.6; 24.1; 17.5; 149.8; 0.0; 869 ❙ 1314; 722; 61 ❙ 51.5; 869; 271.4

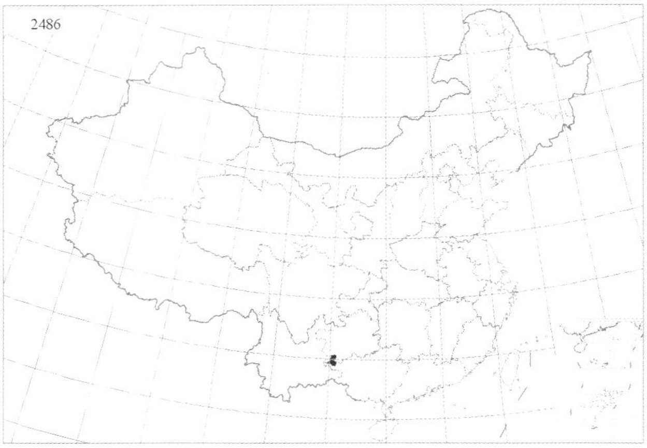

Lauraceae 樟科 *Syndiclis* 油果樟属
Syndiclis chinensis C. K. Allen 油果樟
Trees, evergreen

23.9 (22.2~24.8); 18.2 (15.0~20.1); 28.0 (25.7~28.8); 24.0 (22.3~24.8); 227.5 (207.2~237.6); 0.0 (0.0~0.0); 1305 (1127~1385) ❙ 1548 (868~2435); 578 (312~1428); 76 (41~130) ❙ 19.2 (-34.3~99.5); 1193 (868~1360); 433.3 (248.2~620.0)

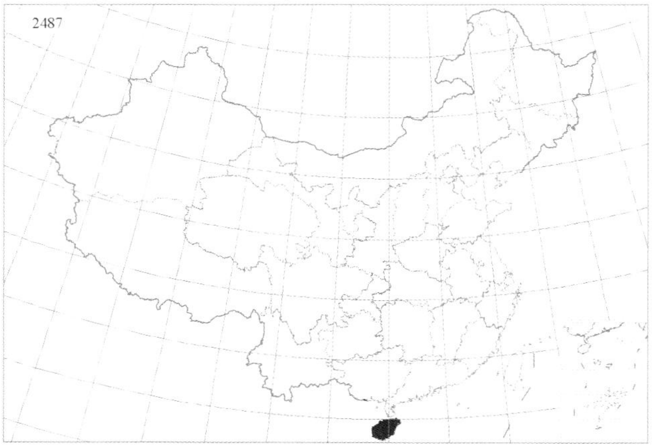

Lauraceae 樟科 *Syndiclis* 油果樟属
Syndiclis fooningensis H. W. Li 富宁油果樟
Small trees, evergreen

18.9; 10.6; 25.0; 19.0; 167.5; 0.0; 939 ❙ 1292; 729; 49 ❙ 37.6; 939; 342.5

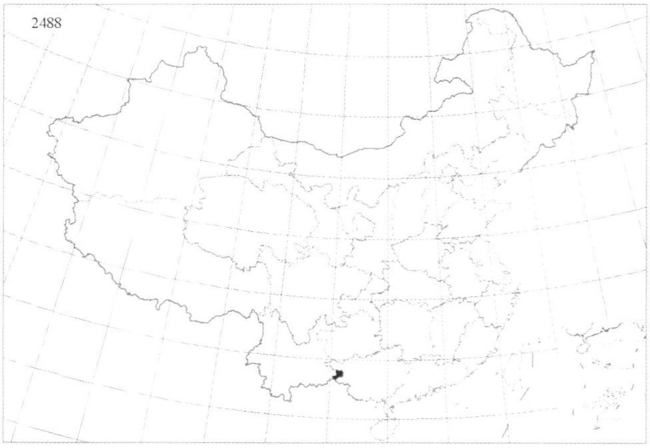

Lauraceae 樟科 *Syndiclis* 油果樟属
Syndiclis furfuracea H. W. Li 鳞秕油果樟
Trees, evergreen

19.4; 12.6; 24.1; 19.4; 172.6; 0.0; 954 ❙ 1445; 787; 64 ❙ 52.3; 954; 521.5

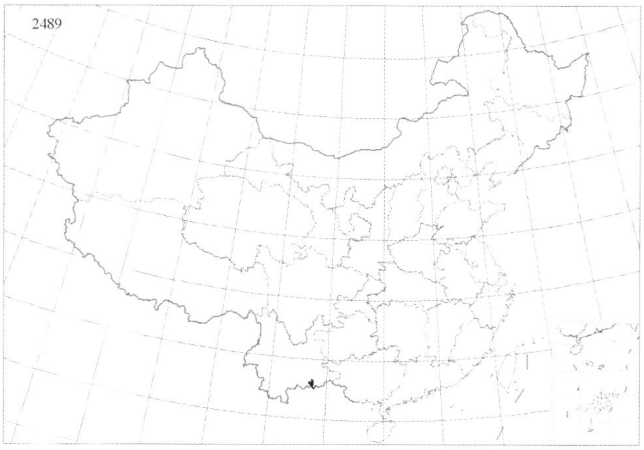

Lauraceae 樟科 *Syndiclis* 油果樟属
Syndiclis kwangsiensis (Kosterm.) H. W. Li 广西油果樟
Trees, evergreen

22.0 (21.3~22.7); 13.9 (13.2~15.0); 28.1 (27.3~28.9); 22.1 (21.3~22.8); 205.0 (195.9~213.1); 0.0 (0.0~0.0); 1174 (1111~1221) ❙ 1720 (1302~2435); 934 (670~1428); 116 (86~136) ❙ 46.6 (12.9~99.5); 1174 (1111~1221); 279.6 (201.4~327.7)

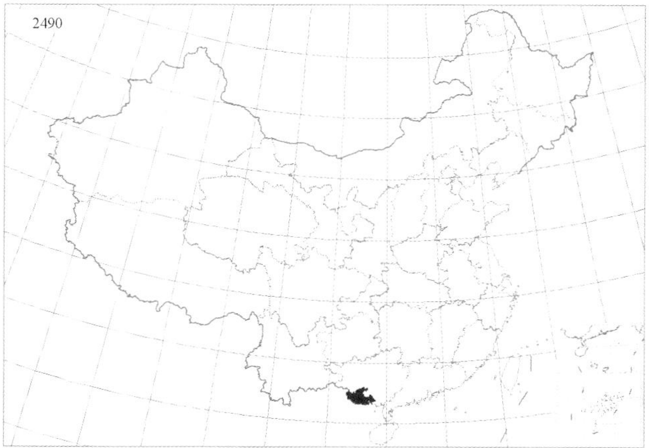

Lauraceae 樟科 *Syndiclis* 油果樟属
Syndiclis lotungensis S. K. Lee 乐东油果樟
Trees, evergreen

23.9; 18.7; 27.5; 23.9; 226.8; 0.0; 1294 ▎1380; 455; 48 ▎6.7; 1112; 455.2

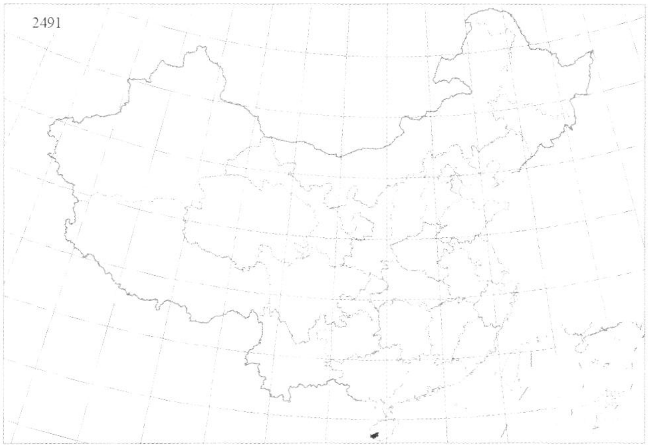

Lauraceae 樟科 *Syndiclis* 油果樟属
Syndiclis marlipoensis H. W. Li 麻栗坡油果樟
Shrubs, evergreen

17.6; 9.9; 23.2; 17.7; 152.4; 0.0; 859 ▎1422; 813; 52 ▎65.6; 859; 359.7

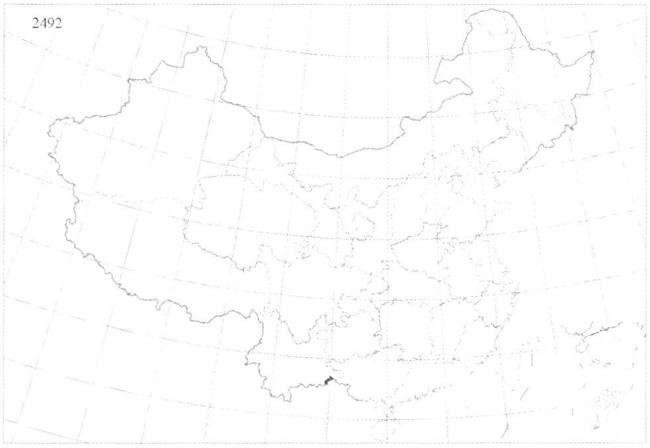

Lauraceae 樟科 *Syndiclis* 油果樟属
Syndiclis pingbienensis H. W. Li 屏边油果樟
Trees, evergreen

19.1; 12.6; 23.6; 19.1; 169.7; 0.0; 932 ▎1479; 828; 63 ▎59.6; 932; 531.5

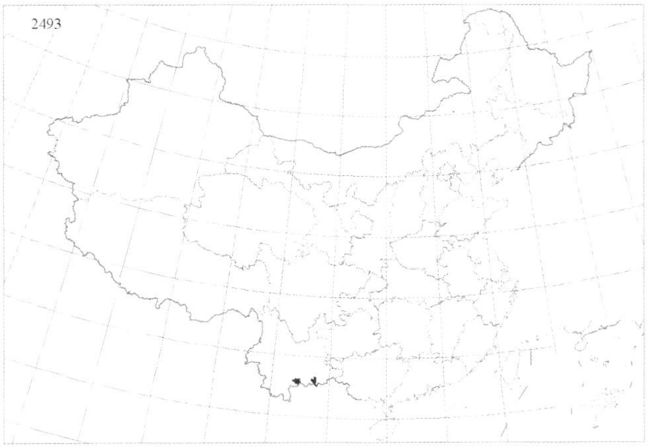

Lauraceae 樟科 *Syndiclis* 油果樟属
Syndiclis sichourensis H. W. Li 西畴油果樟
Trees, evergreen

17.2; 9.5; 22.7; 17.3; 147.0; 0.0; 838 ▎1254; 699; 48 ▎49.7; 838; 336.3

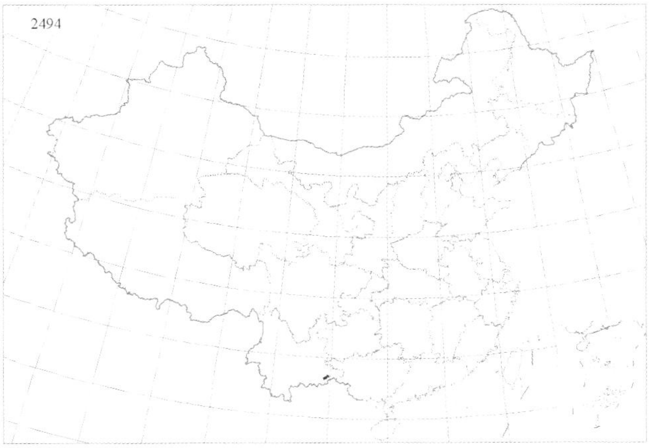

Hernandiaceae 莲叶桐科 *Hernandia* 莲叶桐属
Hernandia nymphaeifolia (Presl) Kubitzki 莲叶桐
Trees, evergreen

22.5 (16.2~24.8); 17.0 (11.0~20.1); 26.5 (20.1~28.8); 22.5 (16.2~24.8); 210.5 (134.2~237.6); 0.0 (0.0~0.0); 1201 (773~1385) ▎1988 (1309~2992); 881 (473~1847); 123 (41~245) ▎81.8 (-1.4~246.5); 1166 (773~1360); 341.5 (248.2~527.5)

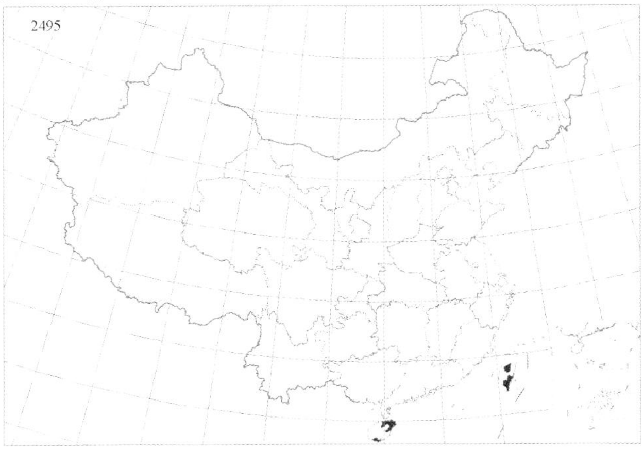

Hernandiaceae 莲叶桐科 *Illigera* 青藤属
Illigera aromatica S. Z. Huang et S. L. Mo 香青藤
Vines, evergreen

20.9 (19.1~22.2); 12.7 (11.0~13.9); 27.1 (25.0~28.6); 21.0 (19.1~22.3); 191.8 (169.6~207.5); 0.0 (0.0~0.0); 1090 (945~1190) ▎1478 (1374~1581); 795 (713~850); 93 (57~131) ▎36.8 (15.4~55.9); 1090 (945~1190); 306.8 (211.1~371.7)

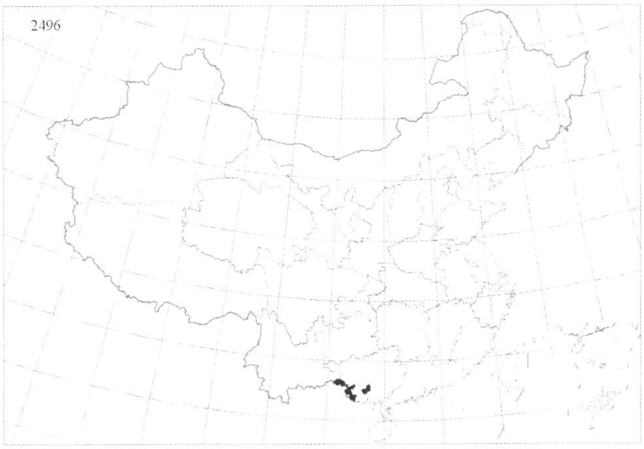

H

Hernandiaceae 莲叶桐科 *Illigera* 青藤属
Illigera brevistaminata Y. R. Li 短蕊青藤
Vines, evergreen

17.6 (14.6~20.3); 7.7 (3.8~10.9); 26.1 (23.6~27.7); 17.7 (14.7~20.4); 152.3 (117.7~184.4); 0.1 (0.0~1.5); 911 (779~1065) | 1452 (1236~1758); 626 (505~775); 130 (51~191) | 60.2 (27.5~101.8); 911 (779~1065); 236.9 (188.8~288.8)

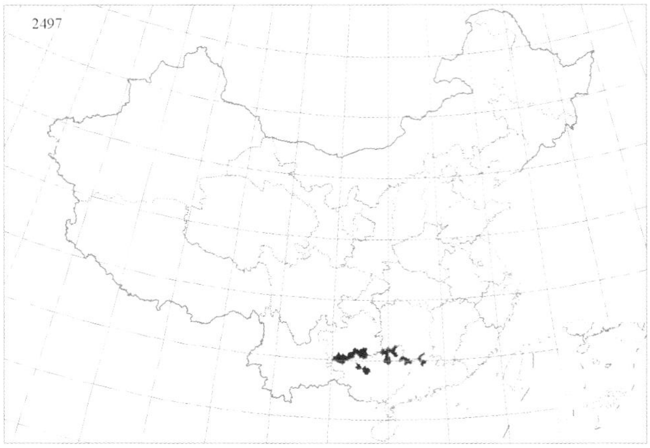

Hernandiaceae 莲叶桐科 *Illigera* 青藤属
Illigera celebica Miq. 宽药青藤
Vines, evergreen

20.7 (17.2~24.8); 13.1 (7.3~20.1); 26.7 (22.3~28.8); 20.8 (17.3~24.8); 189.9 (147.0~237.6); 0.0 (0.0~0.0); 1082 (833~1385) | 1582 (1161~2435); 744 (455~1428); 110 (41~194) | 48.0 (-1.4~99.5); 1067 (833~1331); 329.8 (163.7~659.0)

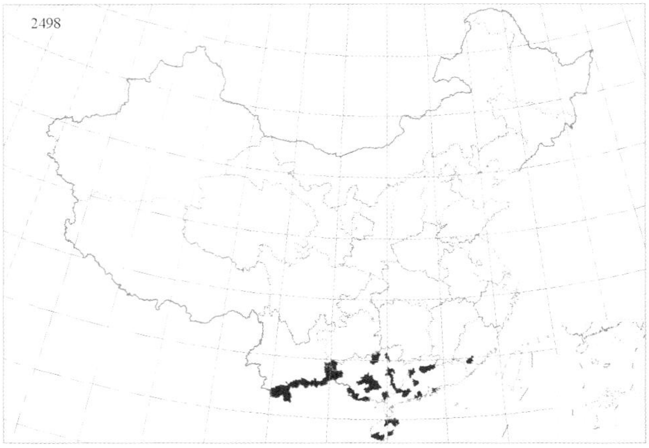

Hernandiaceae 莲叶桐科 *Illigera* 青藤属
Illigera cordata Dunn 心叶青藤
Vines, evergreen

17.0 (5.4~22.7); 9.5 (-2.1~15.1); 22.7 (12.3~28.6); 17.1 (5.8~22.8); 145.6 (31.5~213.1); 1.0 (0.0~25.6); 867 (481~1221) | 1214 (741~2435); 656 (411~1428); 62 (12~194) | 40.2 (-14.2~99.5); 847 (481~1221); 316.3 (189.5~659.0)

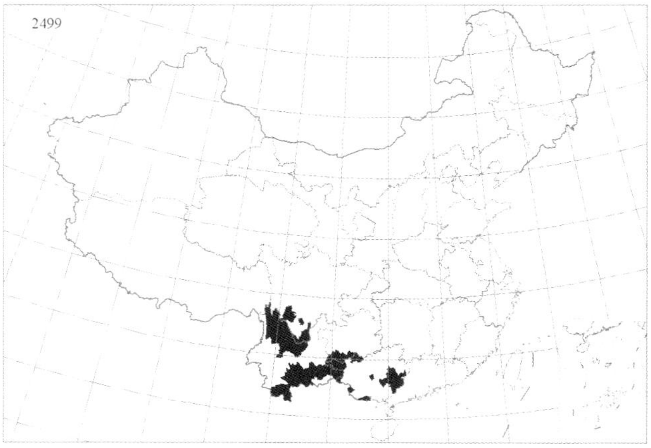

Hernandiaceae 莲叶桐科 *Illigera* 青藤属
Illigera cordata var. mollissima (W. W. Sm.) Kubitzki 多毛青藤
Vines, evergreen

16.5 (10.8~18.9); 9.4 (3.9~12.0); 21.8 (16.6~24.3); 16.5 (10.9~19.0); 138.4 (71.8~167.8); 0.3 (0.0~1.3); 826 (633~929) | 928 (797~1007); 530 (411~594); 27 (20~44) | 15.0 (-14.2~52.8); 745 (633~836); 292.2 (218.0~340.7)

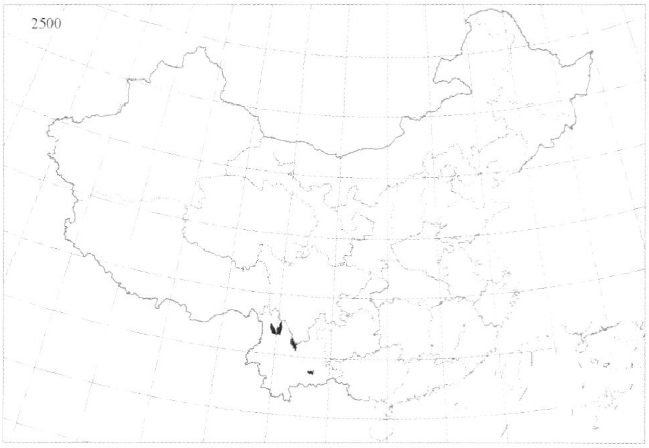

Hernandiaceae 莲叶桐科 *Illigera* 青藤属
Illigera glabra Y. R. Li 无毛青藤
Vines, evergreen

16.9 (14.4~18.2); 10.4 (7.5~11.9); 21.1 (19.1~22.0); 17.0 (14.4~18.2); 143.7 (113.1~158.7); 0.0 (0.0~0.0); 819 (727~868) | 1443 (1348~1561); 823 (762~909); 53 (42~70) | 76.8 (57.4~99.9); 813 (727~855); 480.2 (395.4~539.7)

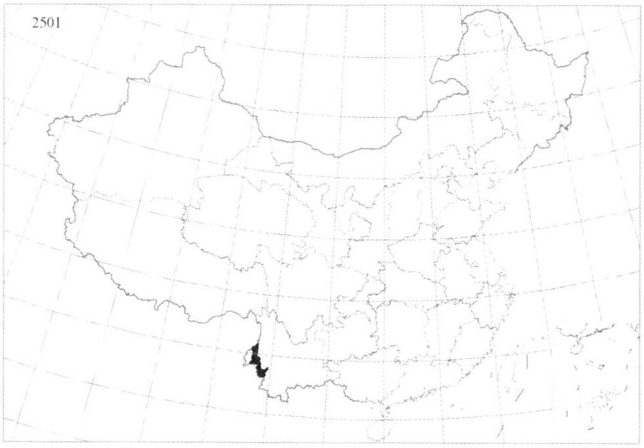

Hernandiaceae 莲叶桐科 *Illigera* 青藤属
Illigera grandiflora W. W. Sm. et Jeffrey 大花青藤
Vines, evergreen

16.2 (5.4~21.6); 9.4 (-2.1~15.1); 21.2 (12.3~28.5); 16.3 (5.8~21.7); 136.2 (31.5~199.8); 0.8 (0.0~25.6); 811 (481~1153) | 1222 (741~1751); 683 (426~996); 50 (15~173) | 51.1 (4.4~112.2); 794 (481~1153); 381.1 (189.5~659.0)

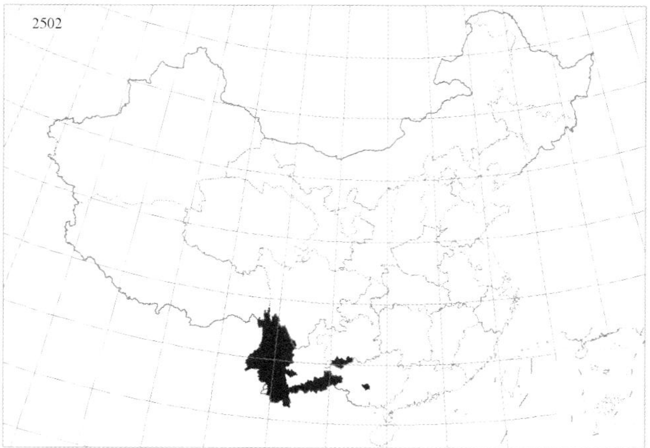

Hernandiaceae 莲叶桐科 *Illigera* 青藤属
Illigera luzonensis (C. Presl) Merr. 台湾青藤
Vines, evergreen

21.7；16.7；25.4；21.7；200.8；0.0；1117 ❙ 2255；1381；87 ❙ 117.5；1003；203.3

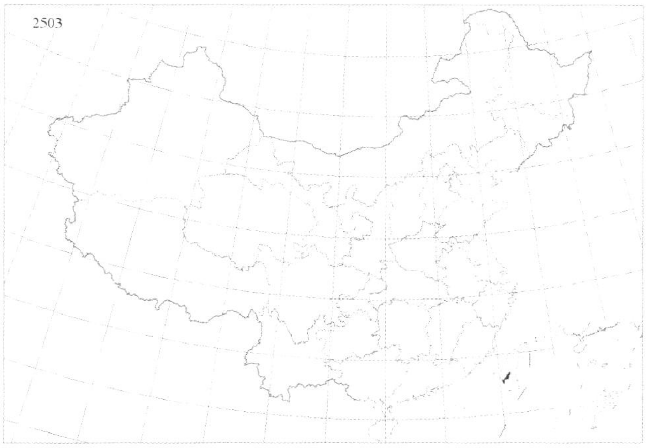

Hernandiaceae 莲叶桐科 *Illigera* 青藤属
Illigera nervosa Merr. 显脉青藤
Vines, evergreen

17.1 (14.4~18.9)；10.6 (7.5~12.5)；21.4 (19.1~23.2)；17.2 (14.4~19.0)；146.2 (113.1~167.5)；0.0 (0.0~0.0)；829 (727~914) ❙ 1225 (958~1454)；692 (531~816)；50 (44~70) ❙ 49.0 (12.5~99.9)；808 (727~851)；403.6 (311.9~516.7)

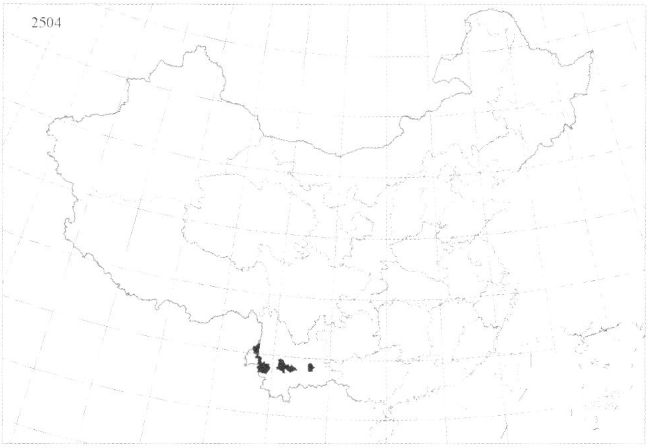

Hernandiaceae 莲叶桐科 *Illigera* 青藤属
Illigera orbiculata C. Y. Wu 圆叶青藤
Vines, evergreen

19.9；13.6；24.1；20.0；179.9；0.0；977 ❙ 1489；687；74 ❙ 52.8；961；507.9

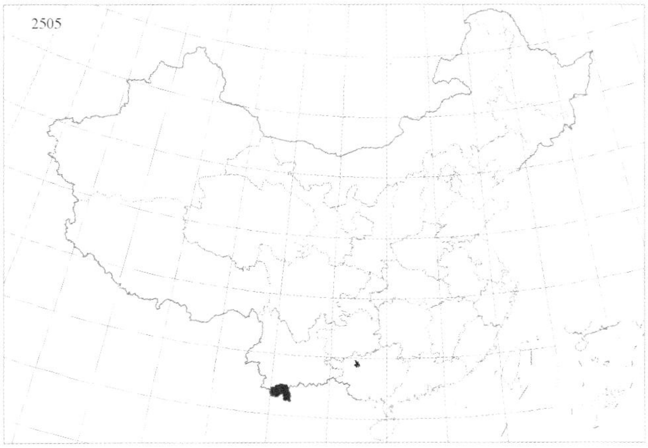

Hernandiaceae 莲叶桐科 *Illigera* 青藤属
Illigera parviflora Dunn 小花青藤
Vines, evergreen

19.6 (11.3~24.0)；10.6 (3.3~19.1)；27.2 (17.0~29.8)；19.7 (11.3~24.1)；176.2 (76.6~229.0)；0.1 (0.0~2.3)；1027 (644~1321) ❙ 1593 (1161~2435)；669 (419~1428)；148 (48~230) ❙ 56.5 (12.9~128.9)；1022 (644~1321)；261.5 (152.9~659.0)

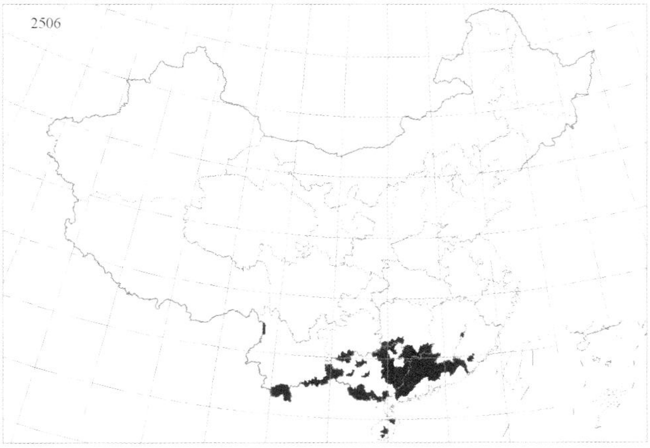

Hernandiaceae 莲叶桐科 *Illigera* 青藤属
Illigera pseudoparviflora Y. R. Li 尾叶青藤
Vines, evergreen

18.4 (17.4~19.1)；9.1 (7.4~10.1)；25.9 (25.5~26.6)；18.5 (17.4~19.1)；161.7 (148.9~169.2)；0.0 (0.0~0.0)；937 (886~969) ❙ 1280 (1236~1381)；660 (617~701)；66 (51~107) ❙ 36.9 (27.5~55.9)；937 (886~969)；246.7 (189.5~282.2)

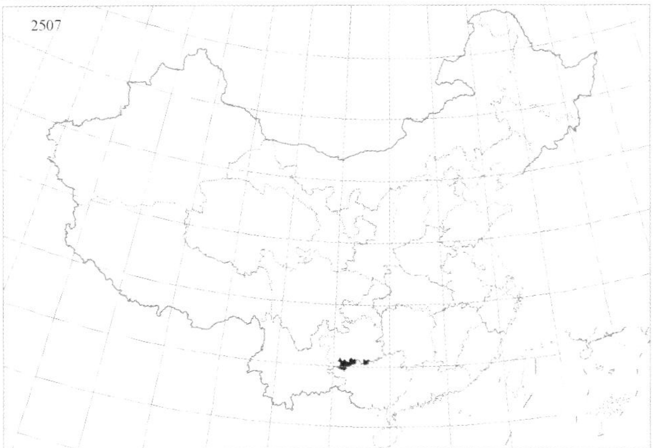

Hernandiaceae 莲叶桐科 *Illigera* 青藤属
Illigera rhodantha Hance 红花青藤
Vines, evergreen

20.6 (15.9~24.8)；12.4 (6.9~20.1)；26.9 (21.4~29.0)；20.7 (16.0~24.8)；188.1 (131.6~237.6)；0.0 (0.0~0.0)；1078 (797~1385) ❙ 1494 (868~2435)；709 (312~1428)；104 (41~194) ❙ 40.0 (-34.3~99.5)；1058 (797~1360)；295.1 (160.7~659.0)

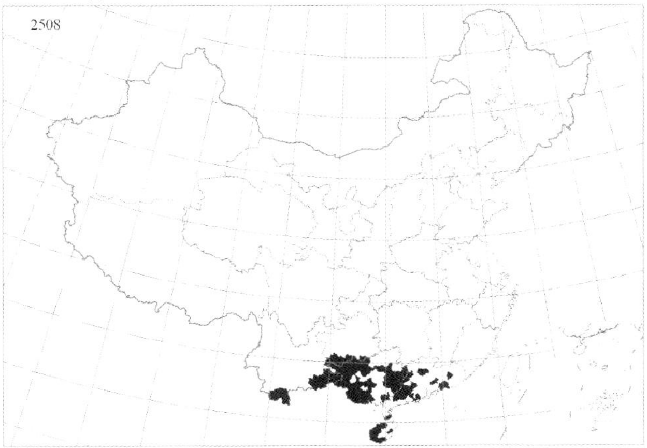

Hernandiaceae 莲叶桐科 *Illigera* 青藤属
Illigera rhodantha var. dunniana (H. Lév.) Kubitzki 锈毛青藤
Vines, evergreen

19.4 (17.7~21.3); 11.0 (9.5~14.2); 25.8 (22.5~27.9); 19.5 (17.8~21.3); 173.6 (153.4~196.0); 0.0 (0.0~0.0); 983 (854~1123) ❙ 1332 (1182~1522); 725 (617~848); 64 (43~111) ❙ 36.3 (5.8~60.1); 983 (854~1121); 340.9 (248.4~551.6)

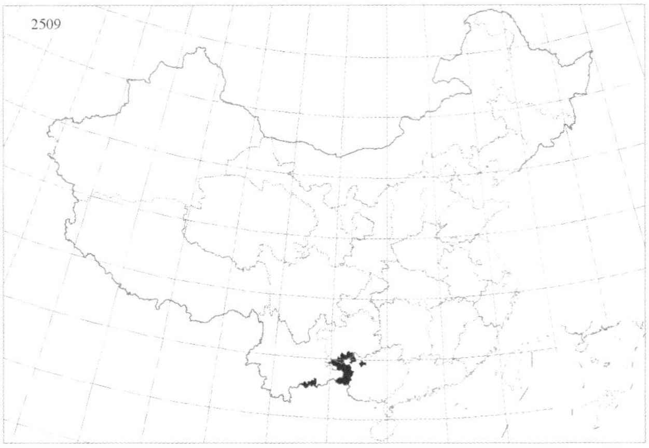

Hernandiaceae 莲叶桐科 *Illigera* 青藤属
Illigera trifoliata subsp. cucullata (Merr.) Kubitzki 兜状青藤
Vines, evergreen

17.3; 10.6; 22.0; 17.3; 148.1; 0.0; 834 ❙ 1273; 701; 55 ❙ 52.5; 834; 437.5

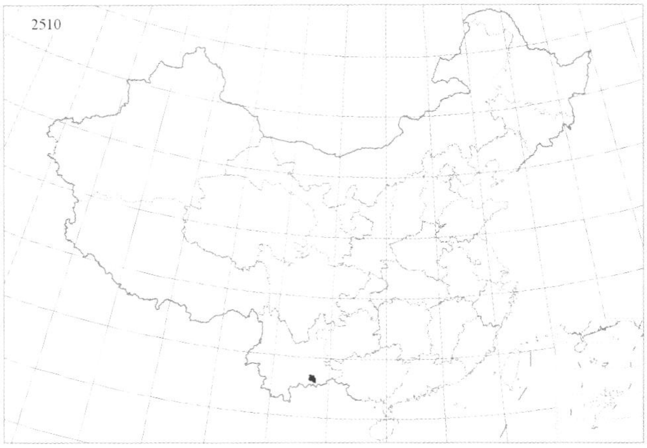

Capparaceae 山柑科 *Borthwickia* 节蒴木属
Borthwickia trifoliata W. W. Sm. 节蒴木
Shrubs or small trees, evergreen

19.6 (17.7~21.0); 13.5 (10.9~15.1); 23.5 (22.5~25.6); 19.6 (17.8~21.0); 175.1 (153.4~191.8); 0.0 (0.0~0.0); 949 (854~1053) ❙ 1489 (1368~1607); 780 (655~911); 65 (56~77) ❙ 57.5 (44.5~74.2); 943 (854~1053); 555.0 (491.5~659.0)

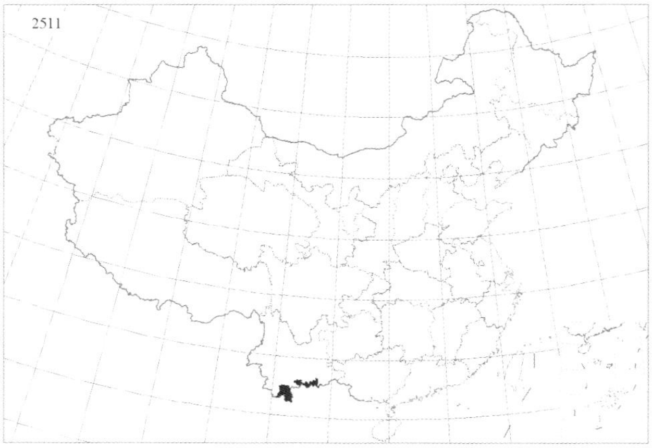

Capparaceae 山柑科 *Capparis* 山柑属
Capparis acutifolia Sweet 独行千里
Vines or shrubs, evergreen

20.8 (14.6~24.8); 12.4 (3.8~20.1); 27.7 (20.1~29.8); 20.8 (14.7~24.8); 189.7 (117.7~237.6); 0.0 (0.0~1.5); 1097 (773~1385) ❙ 1660 (868~2992); 693 (312~1847); 143 (41~259) ❙ 54.3 (-34.3~246.5); 1081 (773~1360); 253.8 (95.3~620.0)

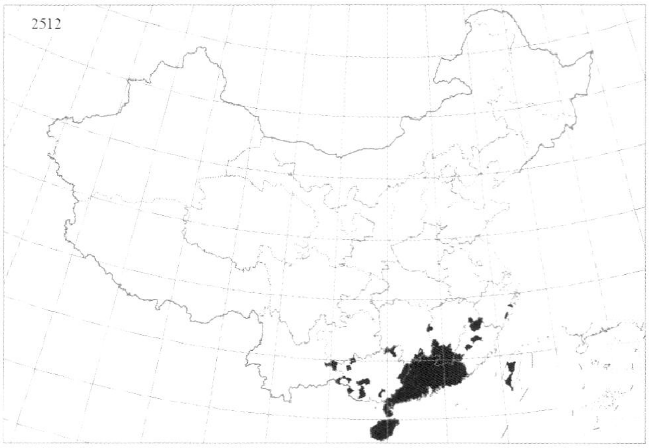

Capparaceae 山柑科 *Capparis* 山柑属
Capparis assamica Hook. f. et Thomson 总序山柑
Shrubs, evergreen

19.1 (17.2~21.0); 12.5 (9.5~14.9); 23.6 (22.5~25.6); 19.1 (17.3~21.0); 169.6 (147.0~191.8); 0.0 (0.0~0.0); 931 (838~1053) ❙ 1407 (1254~1522); 753 (655~835); 60 (48~68) ❙ 51.5 (44.5~60.1); 924 (838~1053); 491.1 (336.3~570.1)

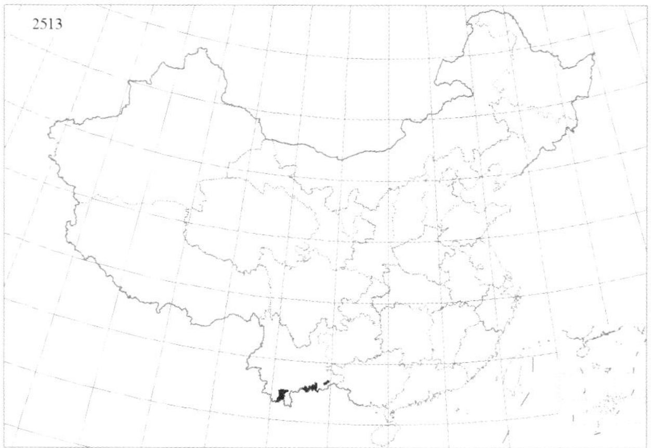

Capparaceae 山柑科 *Capparis* 山柑属
Capparis bodinieri H. Lév. 野香橼花
Shrubs or small trees, evergreen

15.7 (5.4~21.6); 8.2 (-2.4~15.1); 21.5 (12.3~28.5); 15.7 (5.8~21.7); 129.6 (31.5~199.8); 1.1 (0.0~25.6); 800 (481~1153) ❙ 1119 (741~1703); 617 (411~959); 48 (15~194) ❙ 40.8 (-14.2~112.2); 781 (481~1153); 312.8 (142.9~659.0)

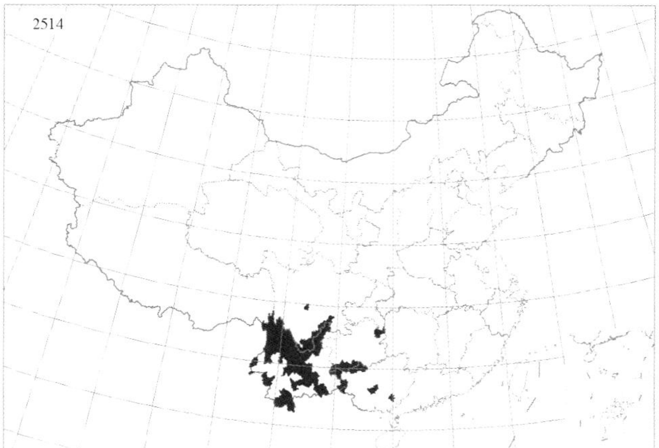

Capparaceae 山柑科 *Capparis* 山柑属

Capparis cantoniensis Lour. 广州山柑

Scandent shrubs, evergreen

20.9 (13.2~24.8); 12.7 (3.8~20.1); 27.6 (18.4~29.8); 21.0 (13.3~24.8); 191.9 (99.4~237.6); 0.0 (0.0~1.5); 1105 (695~1385) ❙ 1614 (868~2435); 692 (312~1428); 134 (41~257) ❙ 47.9 (-34.3~130.7); 1090 (695~1360); 260.8 (95.3~659.0)

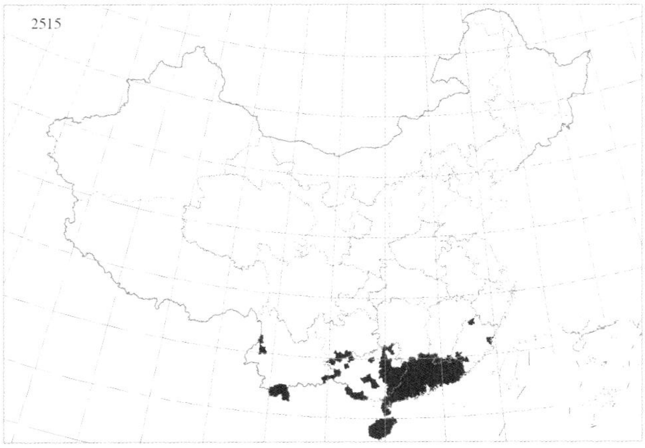

Capparaceae 山柑科 *Capparis* 山柑属

Capparis chingiana B. S. Sun 野槟榔

Shrubs, evergreen

18.8 (14.4~20.7); 9.9 (7.5~11.2); 25.9 (19.1~29.0); 18.9 (14.4~20.8); 166.7 (113.1~189.3); 0.0 (0.0~0.0); 969 (727~1108) ❙ 1426 (1083~1620); 709 (600~848); 104 (48~166) ❙ 48.5 (28.9~99.9); 968 (727~1108); 283.1 (160.7~487.2)

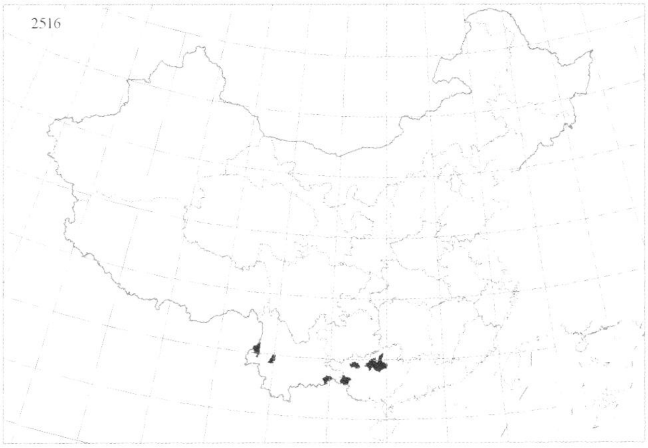

Capparaceae 山柑科 *Capparis* 山柑属

Capparis dasyphylla Merr. et F. P. Metcalf 多毛山柑

Scandent shrubs, evergreen

23.9; 18.8; 27.5; 24.0; 227.6; 0.0; 1304 ❙ 1424; 491; 47 ❙ 9.6; 1130; 507.0

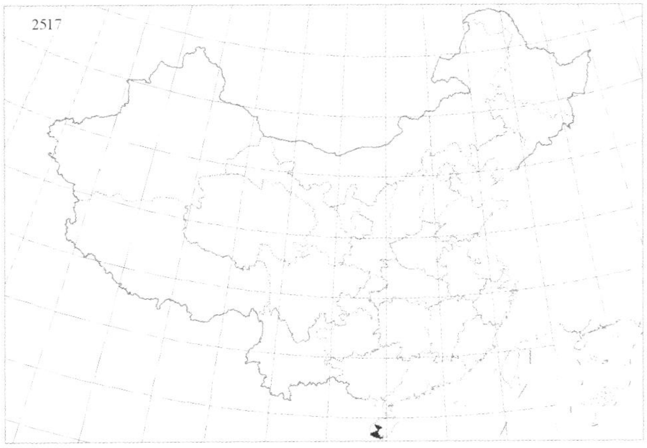

Capparaceae 山柑科 *Capparis* 山柑属

Capparis fengii B. S. Sun 文山山柑

Scandent shrubs, evergreen

17.0; 10.0; 22.0; 17.1; 144.9; 0.0; 826 ❙ 1279; 706; 55 ❙ 54.8; 826; 419.4

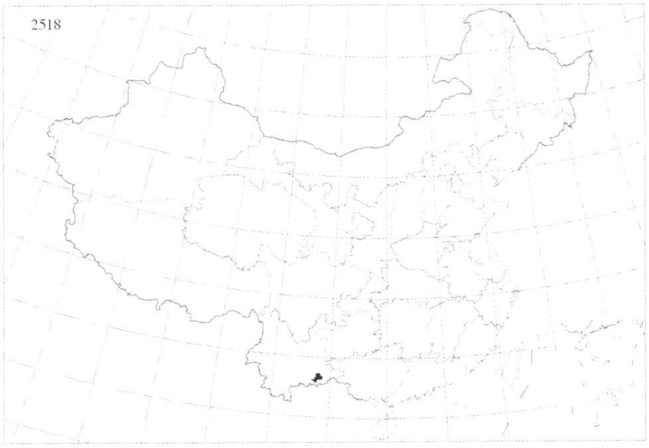

Capparaceae 山柑科 *Capparis* 山柑属

Capparis floribunda Wight 少蕊山柑

Shrubs or vines, evergreen

22.2; 17.7; 25.5; 22.2; 206.7; 0.0; 1123 ❙ 2992; 1847; 107 ❙ 166.5; 1123; 296.9

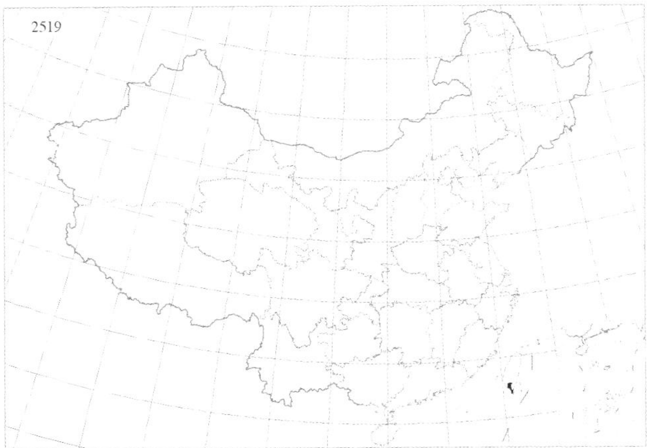

Capparaceae 山柑科 *Capparis* 山柑属

Capparis fohaiensis B. S. Sun 勐海山柑

Vines, evergreen

19.8; 14.4; 23.0; 19.8; 178.2; 0.0; 949 ❙ 1553; 694; 68 ❙ 63.7; 933; 586.4

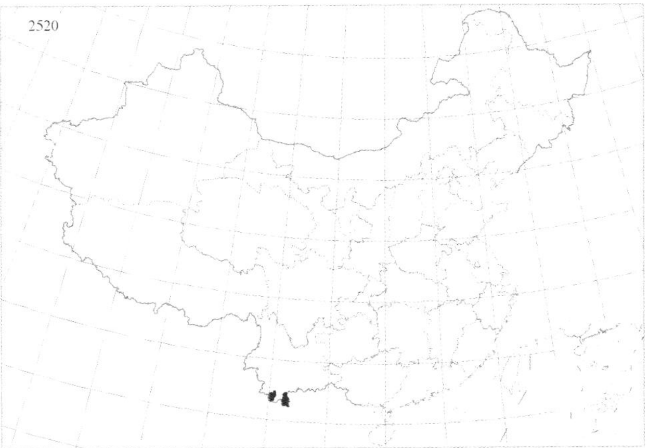

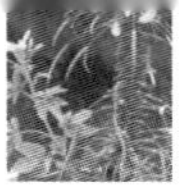

Capparaceae 山柑科 *Capparis* 山柑属
Capparis formosana Hemsl. 台湾山柑
Vines, evergreen

23.1 (18.5~24.8); 17.9 (14.1~20.1); 26.7 (21.7~28.6); 23.1 (18.5~24.8); 217.1 (162.1~237.6); 0.0 (0.0~0.0); 1231 (863~1385) ❙ 1910 (1309~2992); 882 (455~1847); 90 (41~188) ❙ 66.0 (-1.4~214.4); 1142 (863~1326); 399.7 (154.2~620.0)

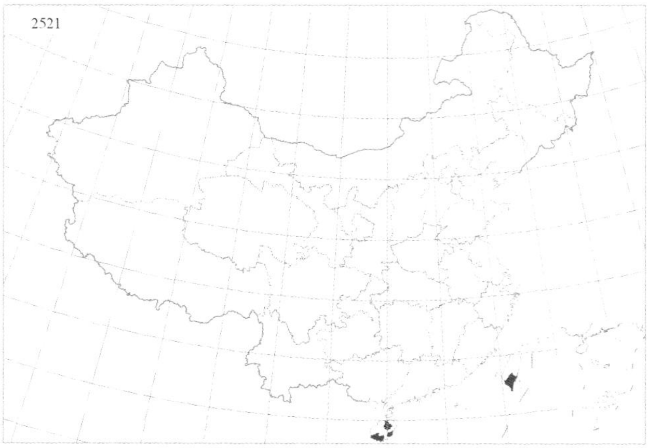

Capparaceae 山柑科 *Capparis* 山柑属
Capparis hainanensis Oliv. 海南山柑
Shrubs, evergreen

22.2; 13.9; 28.3; 22.3; 207.5; 0.0; 1190 ❙ 1374; 713; 87 ❙ 15.4; 1190; 327.7

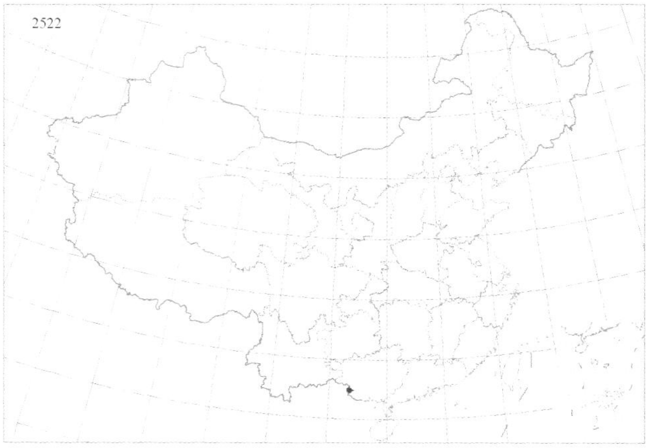

Capparaceae 山柑科 *Capparis* 山柑属
Capparis henryi Matsum. 长刺山柑
Scandent shrubs, evergreen

18.5; 14.1; 21.7; 18.5; 162.1; 0.0; 863 ❙ 2713; 1388; 188 ❙ 214.4; 863; 363.7

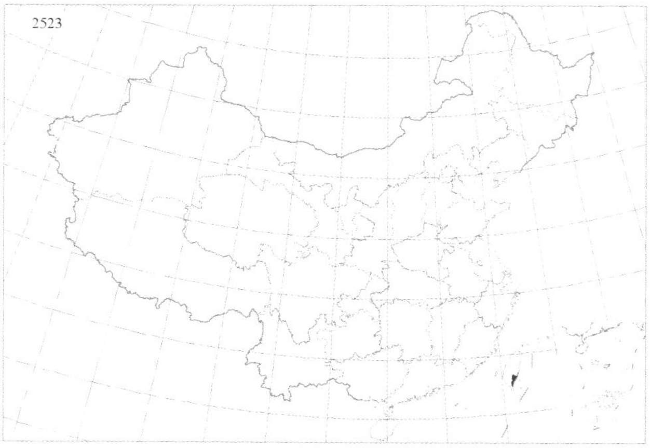

Capparaceae 山柑科 *Capparis* 山柑属
Capparis himalayensis Jafri 爪钾山柑
Shrubs, evergreen

6.7 (-3.4~12.5); -11.7 (-20.9~-5.1); 21.5 (8.4~27.5); 9.5 (2.1~13.2); 77.0 (6.6~114.0); 55.6 (23.1~109.2); 650 (300~822) ❙ 157 (28~509); 78 (20~233); 13 (0~98) ❙ -70.8 (-96.6~52.9); 154 (28~406); 68.2 (1.6~297.3)

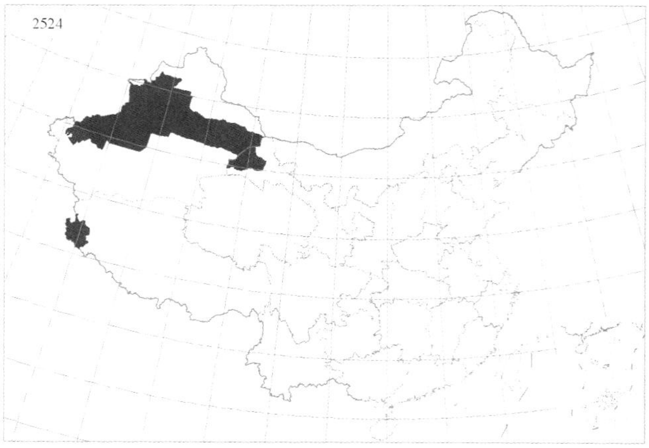

Capparaceae 山柑科 *Capparis* 山柑属
Capparis khuamak Gagnep. 屏边山柑
Scandent shrubs, evergreen

19.4; 12.6; 24.1; 19.4; 172.6; 0.0; 954 ❙ 1445; 787; 64 ❙ 52.3; 954; 521.5

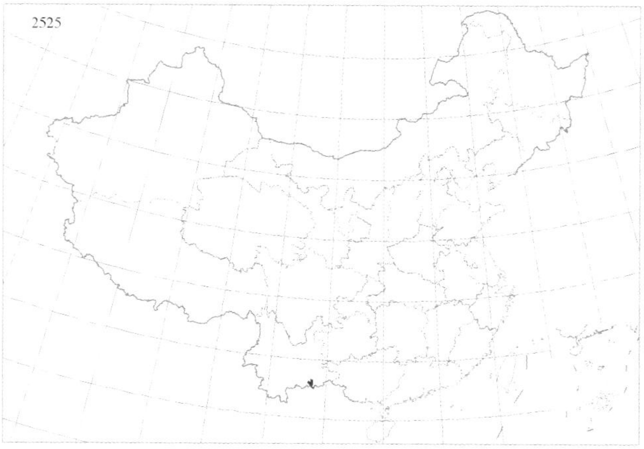

Capparaceae 山柑科 *Capparis* 山柑属
Capparis lanceolaris DC. 兰屿山柑
Scandent shrubs, evergreen

18.5; 14.1; 21.7; 18.5; 162.1; 0.0; 863 ❙ 2713; 1388; 188 ❙ 214.4; 863; 363.7

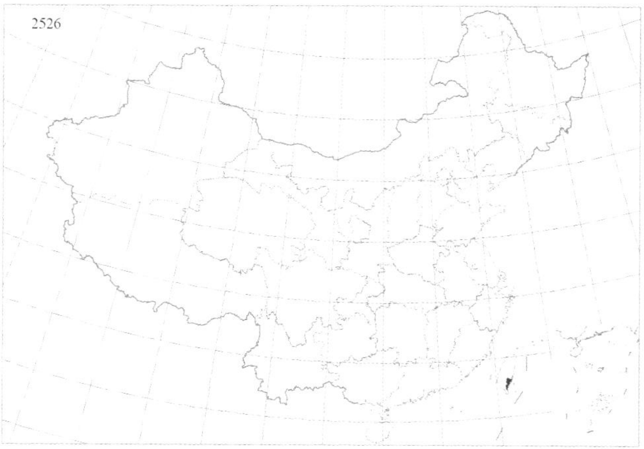

Capparaceae 山柑科 *Capparis* 山柑属
Capparis masaikai H. Lév. 马槟榔
Shrubs, evergreen

19.7 (16.3~22.7); 11.2 (8.7~15.0); 26.1 (21.4~28.9); 19.7 (16.4~22.8); 176.5 (136.4~213.1); 0.0 (0.0~0.0); 1008 (797~1221) ❙ 1443 (1182~2435); 769 (617~1428); 84 (43~177) ❙ 43.5 (5.8~99.5); 1008 (797~1221); 308.6 (189.5~551.6)

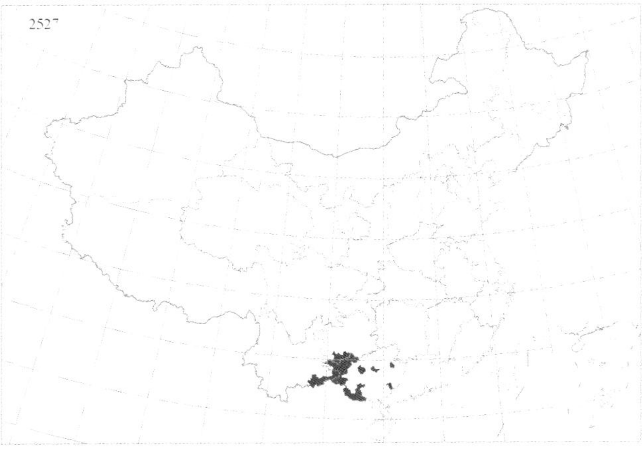

Capparaceae 山柑科 *Capparis* 山柑属
Capparis membranifolia Kurz 雷公橘
Vines or shrubs, evergreen

20.6 (13.2~24.8); 12.7 (6.4~20.1); 26.7 (18.4~28.9); 20.7 (13.3~24.8); 188.4 (99.4~237.6); 0.0 (0.0~0.0); 1082 (695~1385) ❙ 1470 (868~2435); 696 (312~1428); 96 (41~182) ❙ 37.6 (-34.3~99.5); 1057 (695~1360); 323.9 (189.5~659.0)

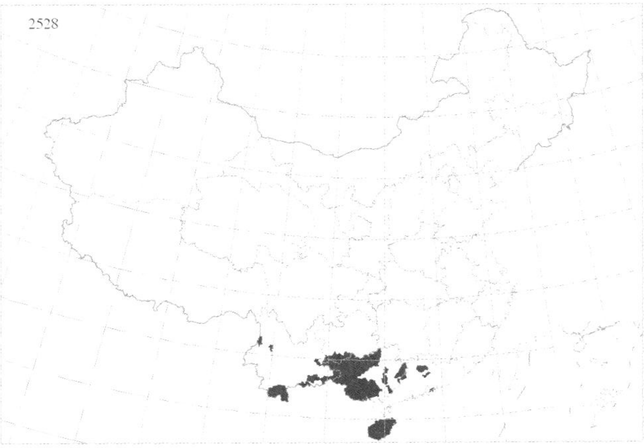

Capparaceae 山柑科 *Capparis* 山柑属
Capparis micracantha DC. 小刺山柑
Shrubs or small trees, evergreen

22.5 (15.8~24.8); 16.3 (9.0~20.1); 27.0 (20.6~29.2); 22.5 (15.8~24.8); 210.5 (129.7~237.6); 0.0 (0.0~0.0); 1201 (776~1385) ❙ 1572 (868~2992); 715 (312~1847); 78 (41~127) ❙ 33.6 (-34.3~202.5); 1136 (772~1360); 359.8 (154.2~620.0)

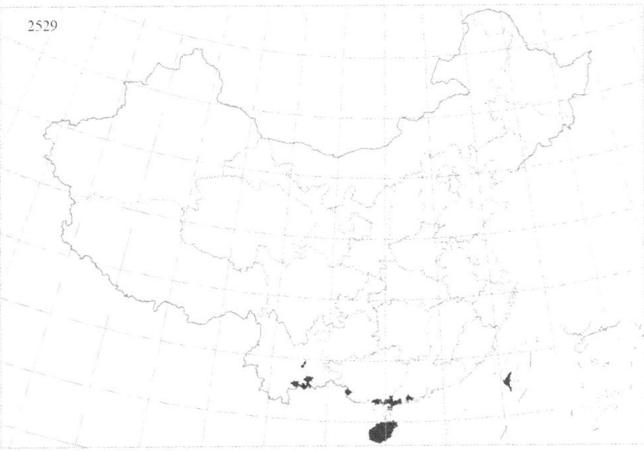

Capparaceae 山柑科 *Capparis* 山柑属
Capparis multiflora Hook. f. et Thomson 多花山柑
Shrubs or small trees, evergreen

18.4 (16.4~21.0); 11.9 (9.6~14.2); 22.6 (20.6~25.6); 18.4 (16.4~21.0); 161.0 (137.3~191.8); 0.0 (0.0~0.0); 890 (795~1053) ❙ 1421 (1178~1688); 807 (646~959); 54 (42~68) ❙ 60.6 (36.8~112.2); 884 (795~1053); 494.0 (383.4~627.0)

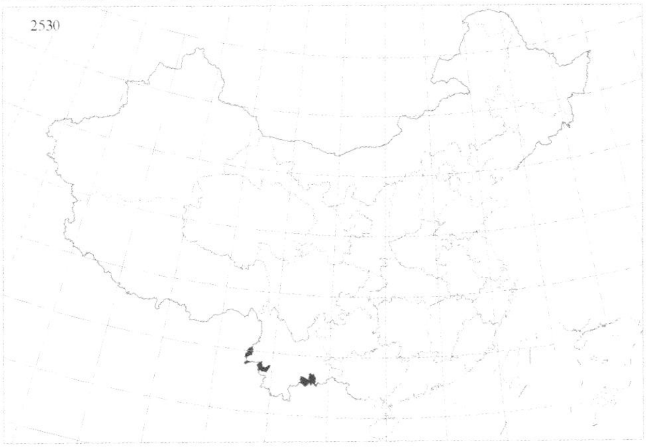

Capparaceae 山柑科 *Capparis* 山柑属
Capparis pubiflora DC. 毛蕊山柑
Shrubs or small trees, evergreen

22.0 (18.5~24.8); 14.3 (10.2~20.1); 28.1 (21.7~29.1); 22.0 (18.5~24.8); 204.3 (162.1~237.6); 0.0 (0.0~0.0); 1166 (863~1385) ❙ 1676 (868~2713); 770 (312~1388); 117 (41~188) ❙ 45.4 (-34.3~214.4); 1141 (863~1225). 236.4 (95.3~527.5)

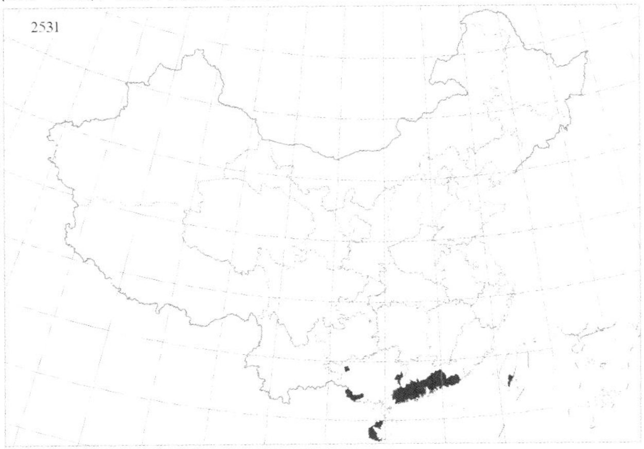

Capparaceae 山柑科 *Capparis* 山柑属
Capparis pubifolia B. S. Sun 毛叶山柑
Vines, evergreen

19.6; 11.4; 25.8; 19.7; 176.1; 0.0; 993 ❙ 1390; 769; 68 ❙ 40.0; 993; 321.8

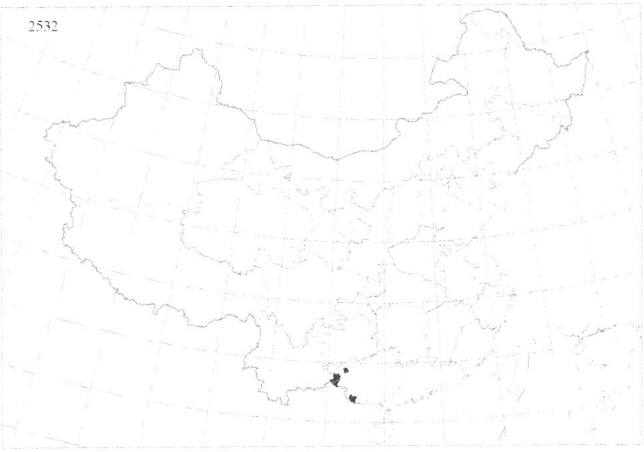

Capparaceae 山柑科 *Capparis* 山柑属
Capparis sabiifolia Hook. f. et Thomson 黑叶山柑
Shrubs or small trees, evergreen

19.3 (17.7~21.0); 12.6 (9.5~15.1); 23.9 (22.5~25.6); 19.3 (17.8~21.0); 171.8 (153.4~191.8); 0.0 (0.0~0.0); 943 (854~1053) | 1444 (1293~1607); 767 (702~835); 64 (56~77) | 53.3 (43.5~62.5); 942 (854~1053); 504.9 (316.1~659.0)

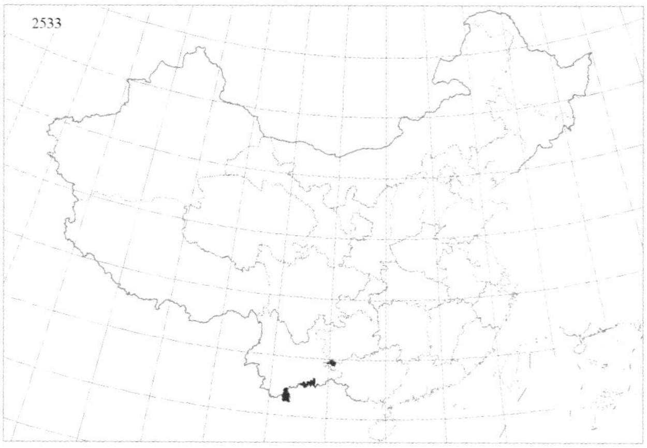

Capparaceae 山柑科 *Capparis* 山柑属
Capparis sepiaria L. 青皮刺
Shrubs, evergreen

22.3 (19.2~24.8); 14.8 (11.9~20.1); 28.1 (23.5~29.2); 22.3 (19.4~24.8); 207.8 (172.6~237.6); 0.0 (0.0~0.0); 1186 (932~1385) | 1691 (1179~2435); 771 (392~1428); 111 (41~188) | 43.4 (-11.9~99.5); 1163 (923~1360); 264.3 (95.3~620.0)

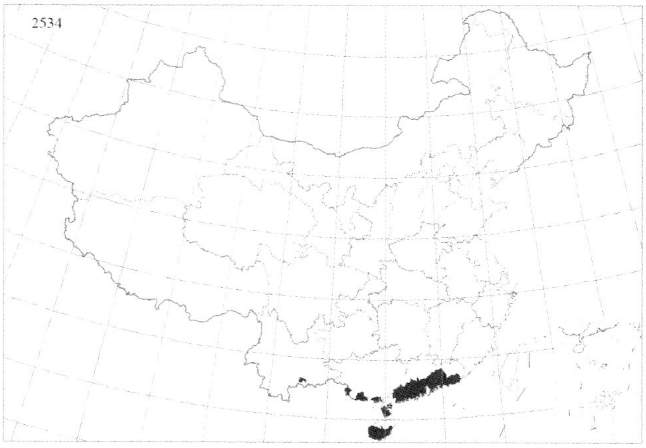

Capparaceae 山柑科 *Capparis* 山柑属
Capparis spinosa L. 山柑
Shrubs, evergreen

6.9 (-2.8~12.5); -11.8 (-20.3~-5.1); 22.1 (10.9~27.5); 9.7 (3.5~13.2); 79.2 (16.3~114.0); 55.5 (23.1~109.2); 660 (390~822) | 152 (43~450); 71 (19~185); 13 (1~48) | -74.3 (-94.7~3.9); 151 (43~404); 58.5 (7.6~190.1)

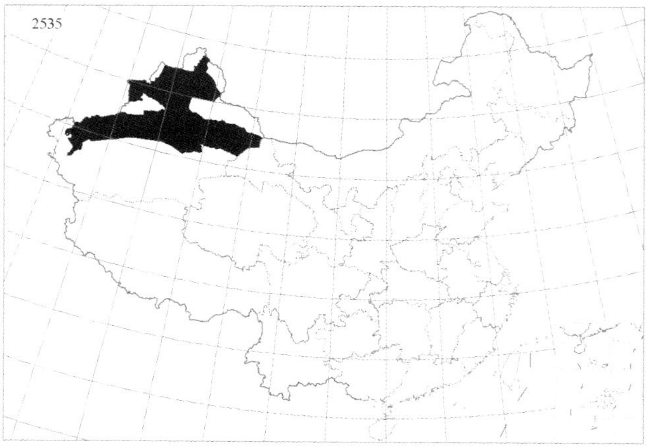

Capparaceae 山柑科 *Capparis* 山柑属
Capparis subsessilis B. S. Sun 无柄山柑
Shrubs, evergreen

20.8 (17.4~22.2); 11.6 (5.2~13.9); 28.2 (27.4~29.1); 20.9 (17.4~22.3); 190.2 (148.8~207.5); 0.0 (0.0~0.0); 1107 (947~1190) | 1368 (931~1581); 674 (323~850); 110 (86~136) | 23.0 (-1.6~40.2); 1081 (790~1190); 264.0 (198.4~327.7)

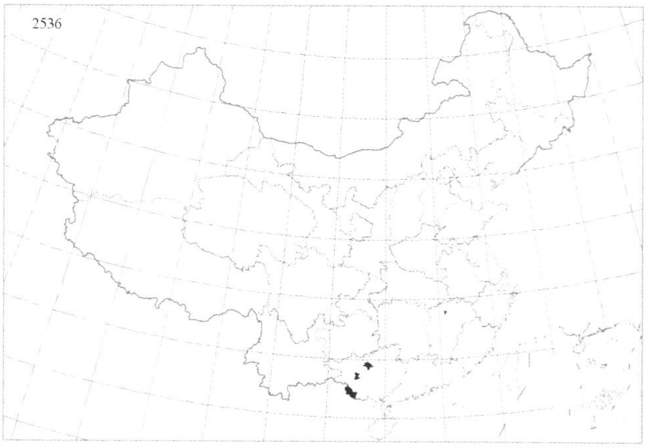

Capparaceae 山柑科 *Capparis* 山柑属
Capparis trichocarpa B. S. Sun 毛果山柑
Vines, evergreen

19.2; 13.8; 22.3; 19.2; 170.5; 0.0; 909 | 1499; 686; 58 | 64.9; 880; 513.9

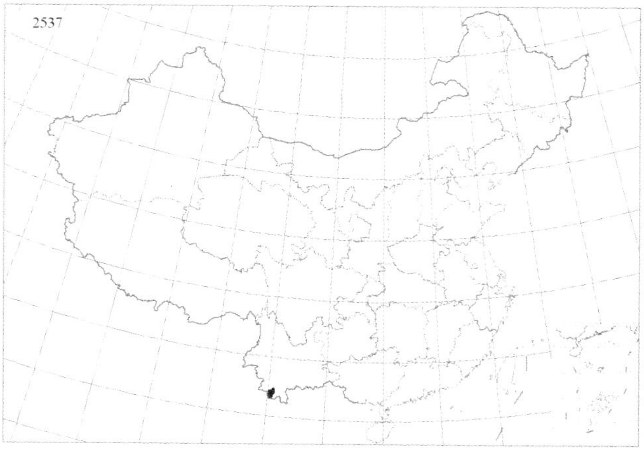

Capparaceae 山柑科 *Capparis* 山柑属
Capparis urophylla F. Chun 小绿刺
Shrubs or small trees, evergreen

19.0 (14.4~23.4); 10.4 (3.8~16.1); 25.9 (19.1~29.2); 19.0 (14.4~23.5); 168.6 (113.1~221.7); 0.0 (0.0~1.5); 976 (727~1273) | 1510 (1049~1900); 692 (411~1056); 128 (44~204) | 56.5 (11.9~101.8); 970 (727~1273); 302.3 (178.7~659.0)

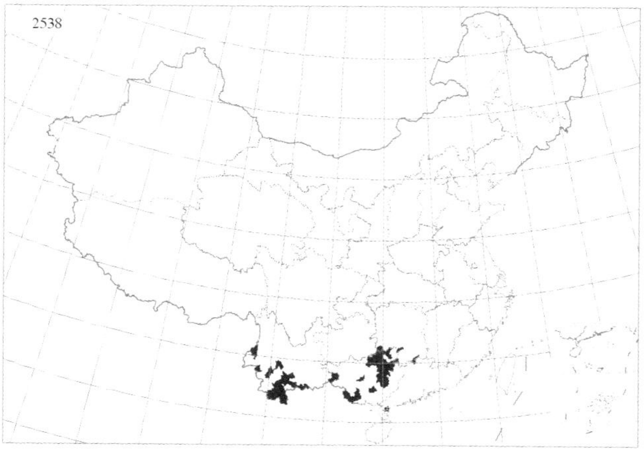

Capparaceae 山柑科 *Capparis* 山柑属
Capparis versicolor Griff. 屈头鸡
Vines or shrubs, evergreen

21.4 (14.6~24.8); 13.2 (3.8~20.1); 28.0 (24.6~29.8); 21.4 (14.7~24.8); 196.9 (117.7~237.6); 0.0 (0.0~1.5); 1133 (779~1385) ❙ 1666 (1187~2435); 714 (455~1428); 137 (41~197) ❙ 48.6 (-6.9~101.8); 1123 (779~1360); 248.3 (95.3~620.0)

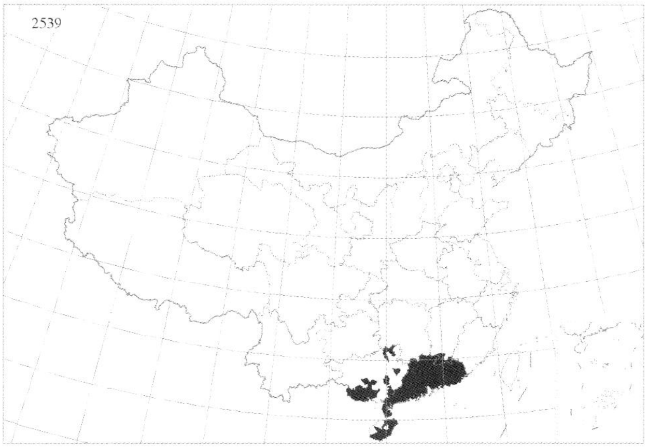

Capparaceae 山柑科 *Capparis* 山柑属
Capparis viburnifolia Gagnep. 荚蒾叶山柑
Shrubs or vines, evergreen

19.1; 13.4; 22.5; 19.1; 169.7; 0.0; 909 ❙ 1497; 730; 60 ❙ 64.7; 886; 470.9

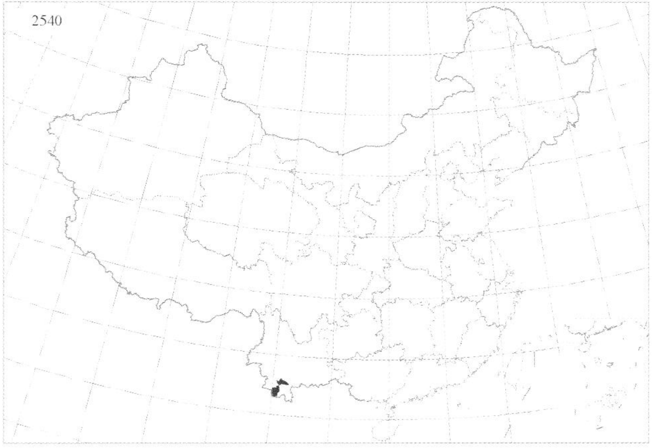

Capparaceae 山柑科 *Capparis* 山柑属
Capparis wui B. S. Sun 元江山柑
Shrubs, evergreen

19.3; 13.1; 23.6; 19.4; 173.2; 0.0; 939 ❙ 1143; 637; 50 ❙ 21.7; 903; 346.9

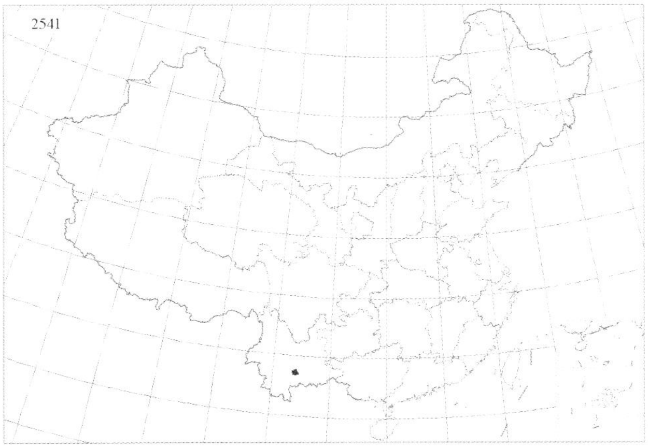

Capparaceae 山柑科 *Capparis* 山柑属
Capparis zeylanica L. 牛眼睛
Shrubs, evergreen

22.2 (19.0~24.8); 14.5 (9.1~20.1); 28.3 (26.7~29.8); 22.2 (19.0~24.8); 206.6 (168.2~237.6); 0.0 (0.0~0.0); 1184 (989~1385) ❙ 1662 (868~2037); 738 (312~1033); 118 (41~194) ❙ 41.8 (-34.3~83.4); 1153 (868~1331); 241.3 (95.3~548.0)

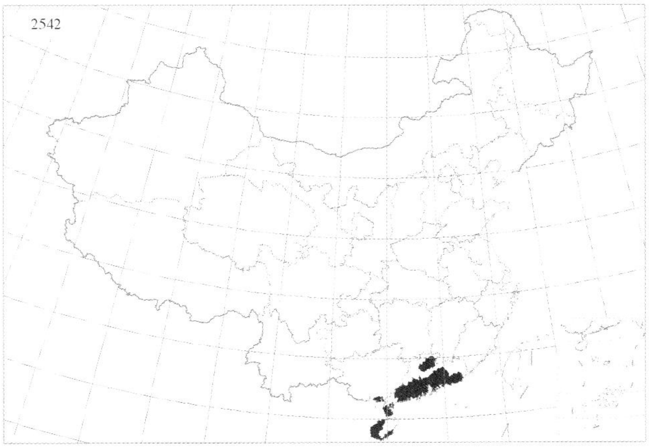

Capparaceae 山柑科 *Crateva* 鱼木属
Crateva formosensis (M. Jacobs) B. S. Sun 台湾鱼木
Shrubs or trees, deciduous

20.2 (14.1~23.6); 12.5 (3.5~17.7); 26.7 (22.8~29.0); 20.3 (14.2~23.6); 183.3 (111.5~223.5); 0.1 (0.0~1.7); 1050 (762~1281) ❙ 1866 (1097~3495); 833 (499~1847); 196 (57~897) ❙ 81.9 (-6.9~235.6); 1045 (762~1269); 251.5 (142.9~327.7)

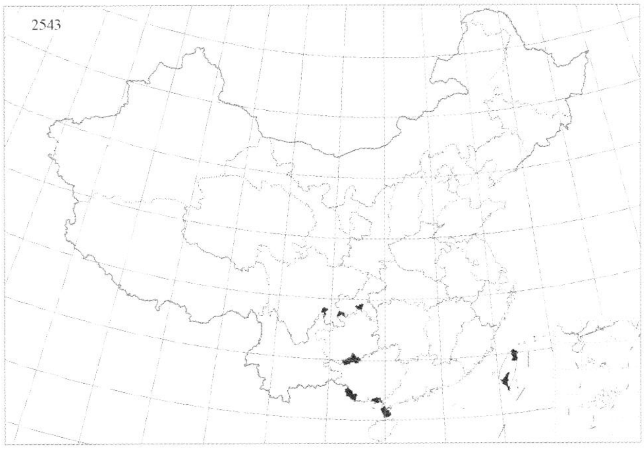

Capparaceae 山柑科 *Crateva* 鱼木属
Crateva magna (Lour.) Candolle 沙梨木
Trees

22.1 (17.3~24.8); 15.3 (9.9~20.1); 27.0 (22.4~29.0); 22.1 (17.3~24.8); 205.2 (147.1~237.6); 0.0 (0.0~0.0); 1166 (833~1385) ❙ 1550 (868~2435); 695 (312~1428); 83 (41~165) ❙ 35.6 (-34.3~99.5); 1092 (833~1299); 390.2 (95.3~659.0)

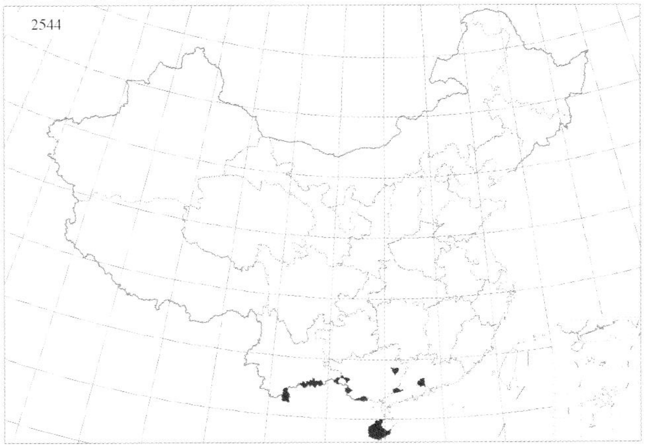

Capparaceae 山柑科 *Crateva* 鱼木属
Crateva trifoliata (Roxb.) B. S. Sun 钝叶鱼木
Shrubs or trees, deciduous

22.0 (17.9~24.8); 14.4 (7.4~20.1); 28.0 (22.6~29.1); 22.1 (18.0~24.8); 204.9 (155.6~237.6); 0.0 (0.0~0.0); 1172 (909~1385) ❙ 1666 (868~2037); 743 (312~1033); 115 (41~189) ❙ 43.7 (-34.3~85.5); 1140 (868~1299); 255.9 (95.3~620.0)

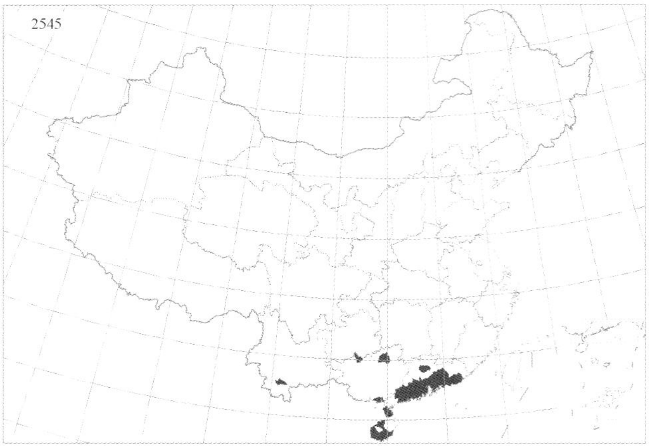

Capparaceae 山柑科 *Crateva* 鱼木属
Crateva unilocularis Buch.-Ham. 树头菜
Trees

21.0 (13.2~24.8); 13.6 (6.4~20.1); 26.7 (18.4~29.2); 21.0 (13.3~24.8); 192.3 (99.4~237.6); 0.0 (0.0~0.0); 1095 (695~1385) ❙ 1601 (976~2435); 757 (448~1428); 107 (38~257) ❙ 47.3 (-6.9~99.5); 1082 (695~1360); 290.3 (95.3~659.0)

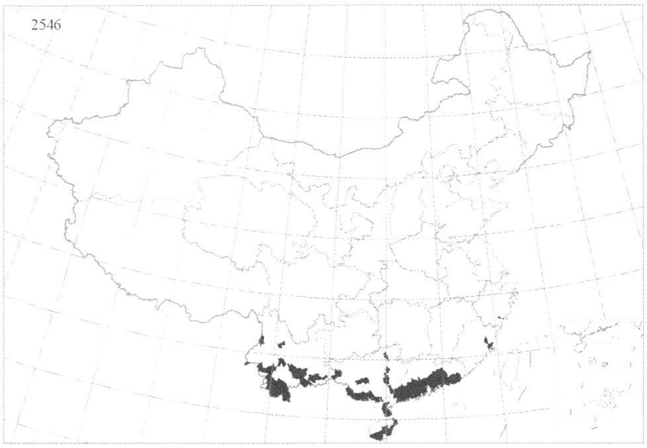

Capparaceae 山柑科 *Stixis* 斑果藤属
Stixis scandens Lour. 和闭脉斑果藤
Vines

20.4; 15.1; 23.6; 20.5; 185.8; 0.0; 989 ❙ 1607; 702; 77 ❙ 62.5; 986; 659.0

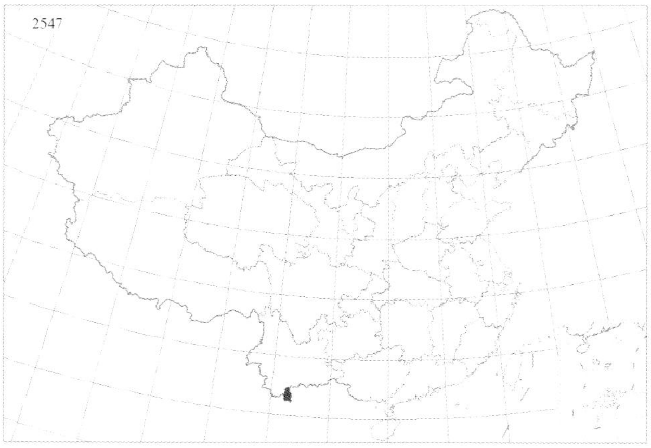

Capparaceae 山柑科 *Stixis* 斑果藤属
Stixis suaveolens (Roxb.) Pierre 斑果藤
Vines

21.2 (16.3~24.8); 15.0 (9.1~20.1); 25.5 (20.6~28.8); 21.2 (16.4~24.8); 194.8 (136.4~237.6); 0.0 (0.0~0.0); 1103 (795~1385) ❙ 1477 (1176~1813); 676 (447~959); 66 (41~113) ❙ 39.4 (-12.0~112.2); 1048 (795~1331); 460.2 (248.2~659.0)

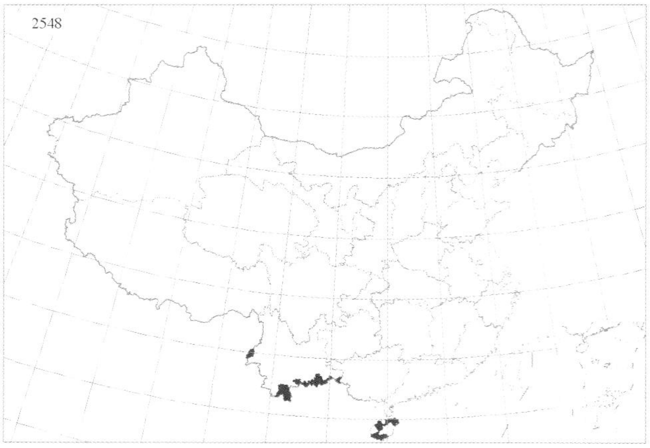

Moringaceae 辣木科 *Moringa* 辣木属
Moringa oleifera Lam. 辣木
Trees

23.7; 17.3; 28.5; 23.7; 224.8; 0.0; 1301 ❙ 1455; 566; 82 ❙ 13.5; 1153; 365.7

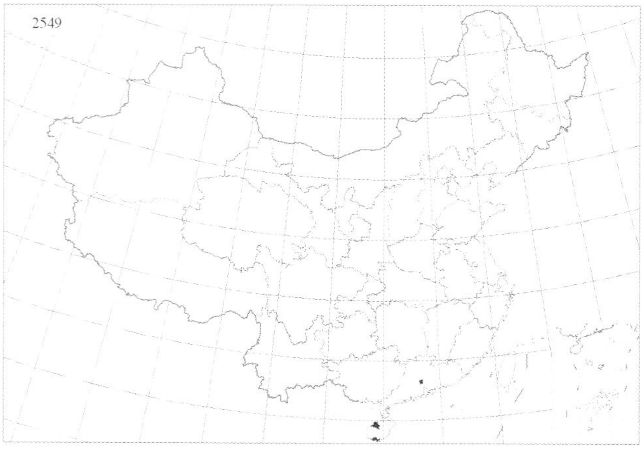

Bretschneideraceae 伯乐树科 *Bretschneidera* 伯乐树属
Bretschneidera sinensis Hemsl. 伯乐树
Trees

17.8 (9.3~23.4); 7.7 (-1.7~16.1); 26.7 (19.4~29.8); 17.8 (9.5~23.5); 154.0 (69.8~221.7); 0.4 (0.0~18.0); 935 (603~1273) ❙ 1503 (931~3495); 596 (323~996); 146 (34~897) ❙ 61.7 (-1.6~224.4); 931 (603~1273); 232.3 (95.3~659.0)

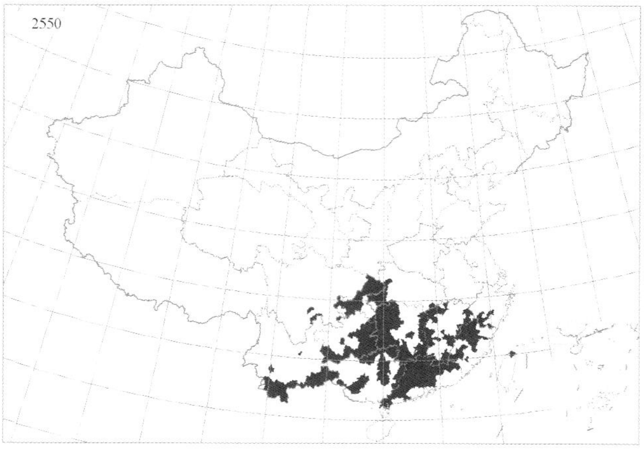

B

Saxifragaceae 虎耳草科 *Decumaria* 赤壁藤属
Decumaria sinensis Oliv. 赤壁木
Scandent shrubs, evergreen

13.7 (2.6~16.6); 1.5 (-7.0~5.5); 25.3 (11.2~28.8); 13.8 (4.3~16.7); 112.6 (20.0~140.6); 7.5 (0.0~48.2); 784 (427~920) | 1002 (626~1571); 447 (316~666); 62 (15~173) | 28.7 (-26.1~108.0); 766 (427~910); 211.0 (133.5~334.4)

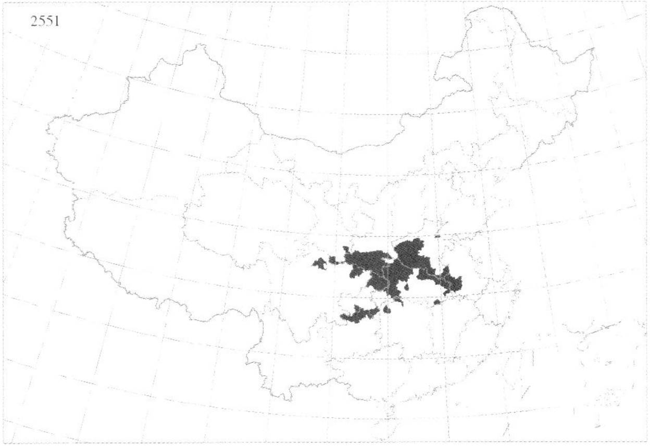

Saxifragaceae 虎耳草科 *Deutzia* 溲疏属
Deutzia albida Batalin 白溲疏
Shrubs, deciduous

9.1 (3.0~13.5); -2.0 (-7.7~2.3); 19.5 (12.4~24.1); 9.6 (4.8~13.6); 70.5 (25.1~107.5); 20.4 (4.9~49.3); 613 (441~755) | 685 (620~842); 339 (301~441); 15 (13~19) | 12.7 (-1.2~48.8); 612 (441~755); 257.7 (196.0~306.1)

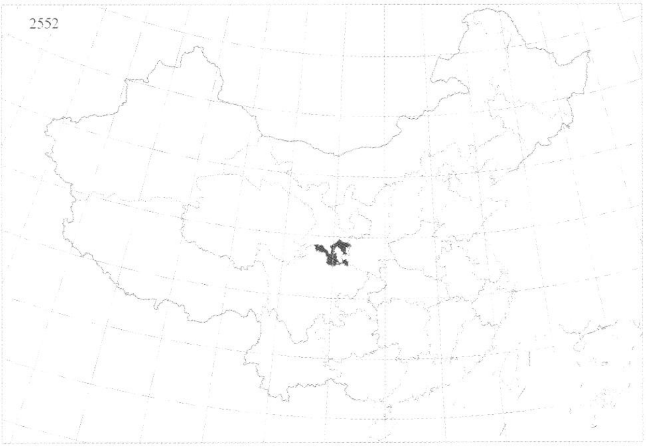

Saxifragaceae 虎耳草科 *Deutzia* 溲疏属
Deutzia aspera Rehder 马桑溲疏
Shrubs, deciduous

16.7 (11.2~21.0); 10.2 (4.6~15.1); 21.3 (16.7~25.6); 16.8 (11.3~21.0); 141.4 (75.8~191.8); 0.0 (0.0~0.4); 819 (642~1053) | 1143 (797~1751); 639 (411~996); 46 (23~77) | 39.3 (-14.2~88.8); 797 (642~1053); 357.8 (218.0~659.0)

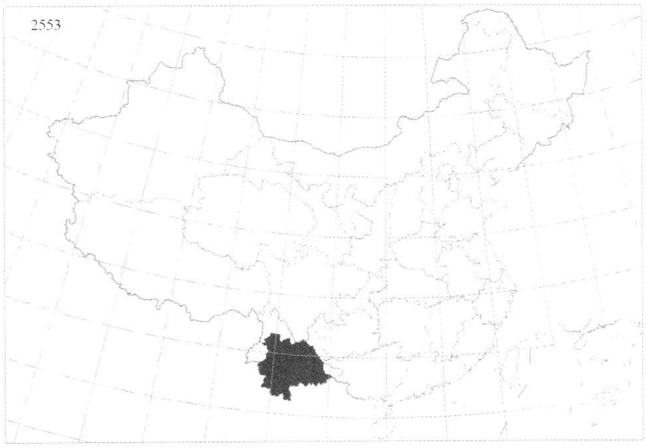

Saxifragaceae 虎耳草科 *Deutzia* 溲疏属
Deutzia baroniana Diels 钩齿溲疏
Shrubs, deciduous

9.2 (1.4~15.1); -6.9 (-20.3~1.8); 23.4 (18.4~27.8); 10.7 (6.4~15.2); 87.5 (45.8~128.0); 36.8 (5.6~93.6); 683 (501~865) | 609 (367~1073); 361 (235~691); 22 (8~53) | -9.5 (-40.9~74.1); 574 (367~817); 193.0 (101.9~346.5)

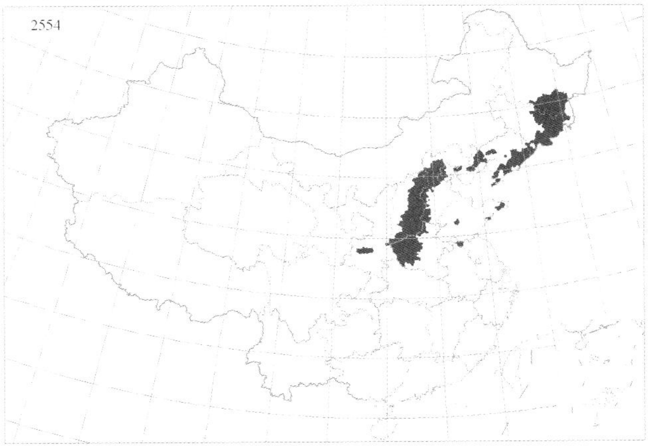

Saxifragaceae 虎耳草科 *Deutzia* 溲疏属
Deutzia bomiensis S. M. Hwang 波密溲疏
Shrubs, deciduous

5.4 (-0.7~13.3); -3.1 (-9.6~5.4); 12.8 (7.7~19.6); 6.5 (2.5~13.4); 39.4 (6.8~100.4); 34.1 (0.0~74.2); 489 (338~706) | 827 (370~1736); 492 (262~1007); 20 (7~51) | 56.8 (4.7~145.8); 489 (338~706); 244.4 (61.6~447.4)

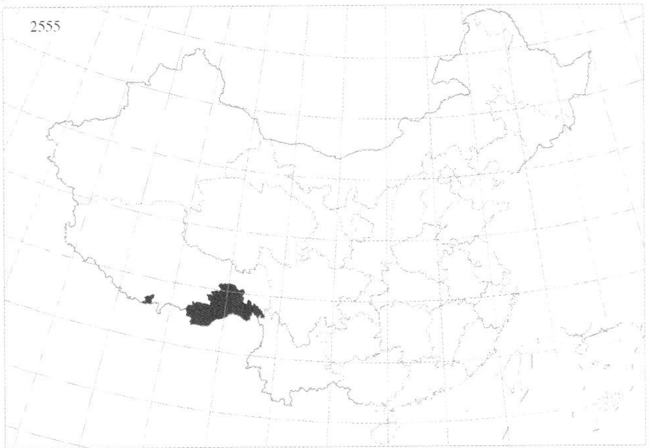

Saxifragaceae 虎耳草科 *Deutzia* 溲疏属
Deutzia breviloba S. M. Hwang 短裂溲疏
Shrubs, deciduous

6.7; -1.5; 13.8; 7.1; 42.2; 21.7; 522 | 1060; 622; 21 | 105.4; 522; 267.5

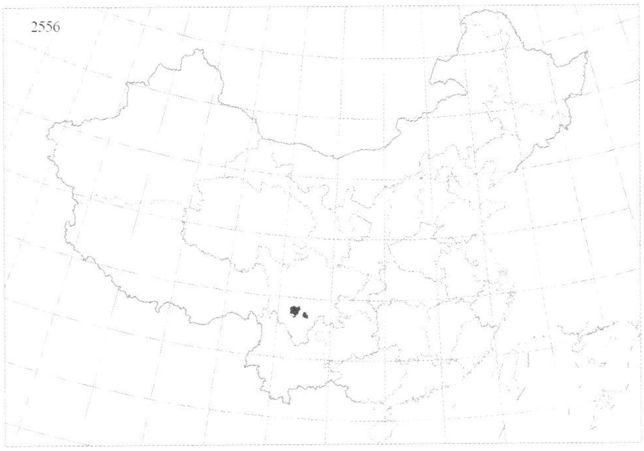

Saxifragaceae 虎耳草科 *Deutzia* 溲疏属

Deutzia calycosa Rehder 大萼溲疏

Shrubs, deciduous

11.4 (2.2~17.4); 3.7 (-6.6~8.2); 18.0 (9.9~26.4); 11.6 (3.8~17.4); 83.8 (15.5~149.1); 6.4 (0.0~49.1); 659 (413~903) ❙ 1064 (741~1410); 619 (426~816); 35 (11~69) ❙ 63.3 (30.5~125.9); 655 (413~903); 302.5 (152.1~407.1)

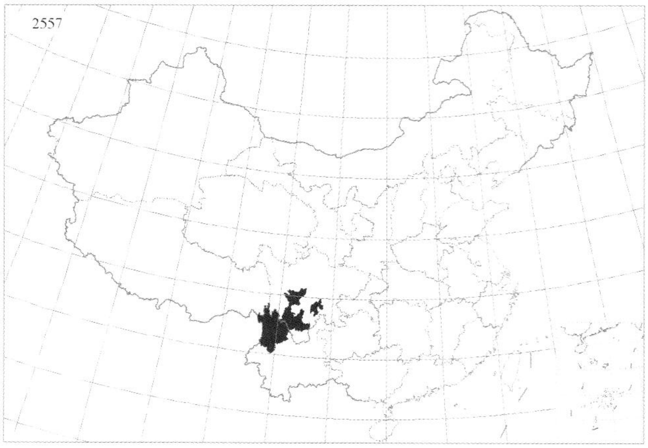

Saxifragaceae 虎耳草科 *Deutzia* 溲疏属

Deutzia cinerascens Rehder 灰叶溲疏

Shrubs, deciduous

16.2; 6.1; 25.5; 16.3; 135.6; 0.4; 846 ❙ 1175; 566; 66 ❙ 39.4; 846; 180.5

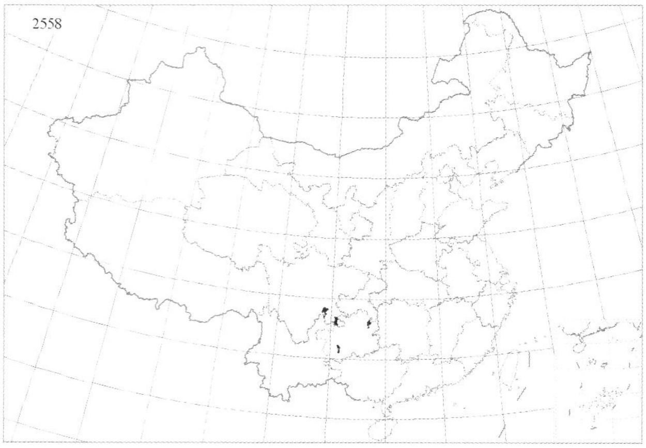

Saxifragaceae 虎耳草科 *Deutzia* 溲疏属

Deutzia compacta Craib 密序溲疏

Shrubs, deciduous

7.4 (-1.1~13.3); -0.3 (-9.1~5.4); 14.0 (6.2~19.6); 8.0 (2.0~13.4); 50.0 (2.5~100.4); 21.2 (0.0~75.6); 542 (334~706) ❙ 932 (321~1736); 547 (193~1007); 30 (9~58) ❙ 64.7 (-4.0~145.8); 540 (321~706); 293.3 (79.3~447.4)

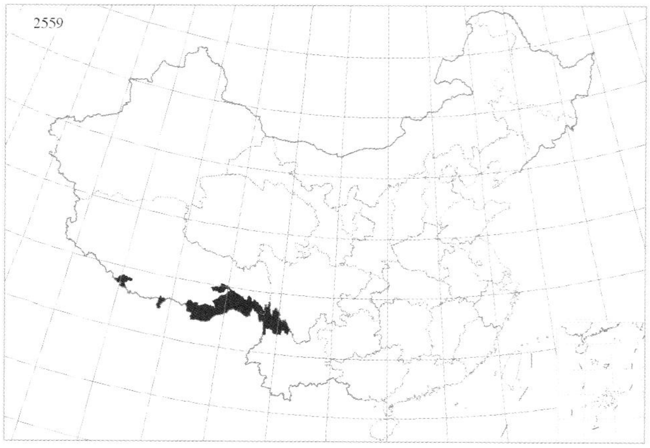

Saxifragaceae 虎耳草科 *Deutzia* 溲疏属

Deutzia crassidentata S. M. Hwang 粗齿溲疏

Shrubs, deciduous

8.8; 1.2; 15.2; 8.9; 57.1; 10.8; 577 ❙ 926; 583; 13 ❙ 61.2; 576; 303.7

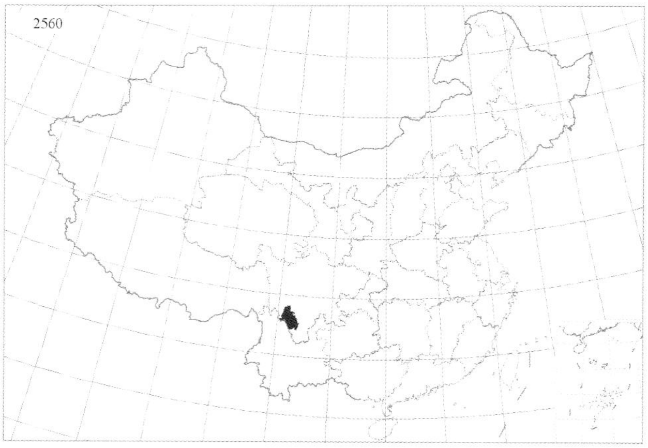

Saxifragaceae 虎耳草科 *Deutzia* 溲疏属

Deutzia crassifolia Rehder 厚叶溲疏

Shrubs, semievergreen

16.2 (10.4~19.3); 9.5 (3.4~13.3); 20.9 (16.3~23.9); 16.2 (10.5~19.4); 134.5 (68.0~173.2); 0.1 (0.0~2.6); 797 (623~939) ❙ 1044 (797~1495); 598 (411~836); 41 (15~62) ❙ 31.3 (-14.2~66.9); 768 (615~923); 319.0 (218.0~491.5)

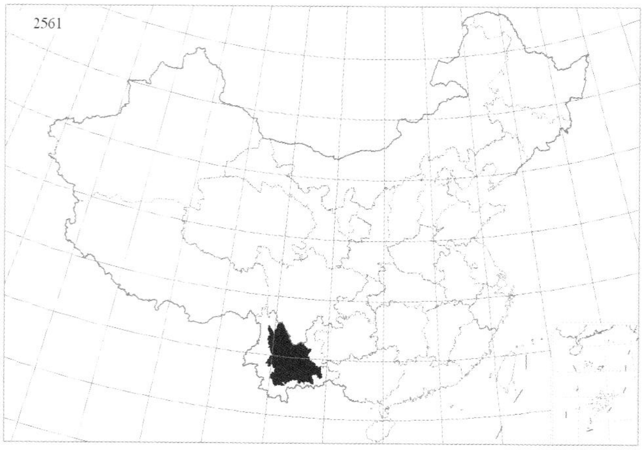

Saxifragaceae 虎耳草科 *Deutzia* 溲疏属

Deutzia crenata Siebold et Zucc. 齿叶溲疏

Shrubs, deciduous

16.0; 4.3; 27.9; 16.1; 133.8; 0.8; 879 ❙ 1059; 415; 139 ❙ 20.5; 879; 131.2

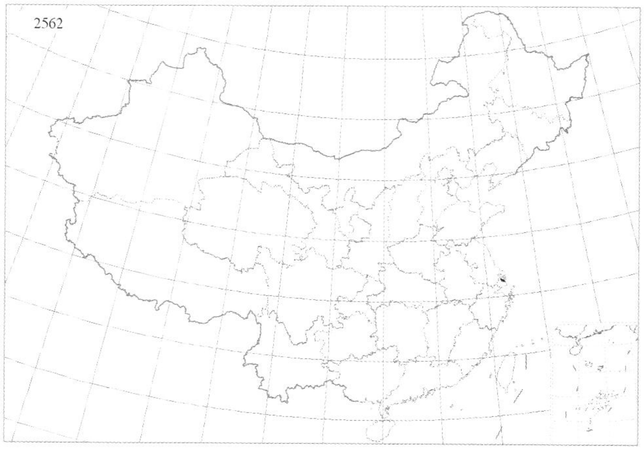

Saxifragaceae 虎耳草科 *Deutzia* 溲疏属

Deutzia cymuligera S. M. Hwang 小聚花溲疏

Shrubs, deciduous

12.0；1.1；22.5；12.0；92.3；8.5；690 ❙ 1207；501；56 ❙ 74.9；690；258.2

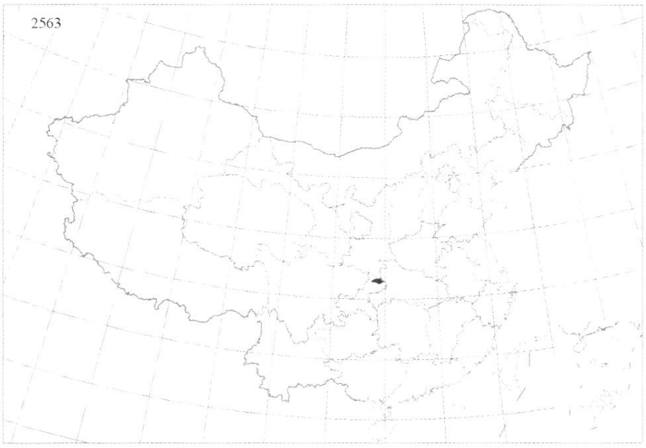

Saxifragaceae 虎耳草科 *Deutzia* 溲疏属

Deutzia discolor Hemsl. 异色溲疏

Shrubs, deciduous

12.0 (2.0~16.2)；0.1 (-7.2~6.1)；23.3 (10.1~28.1)；12.2 (3.8~16.2)；97.0 (15.4~136.2)；11.9 (0.0~51.4)；718 (413~891) ❙ 883 (620~1471)；408 (301~641)；39 (13~175) ❙ 25.0 (-22.2~128.6)；700 (413~891)；239.8 (139.1~334.4)

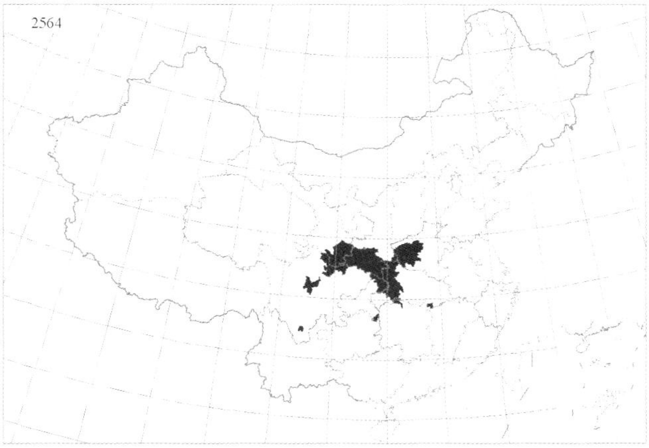

Saxifragaceae 虎耳草科 *Deutzia* 溲疏属

Deutzia esquirolii (H. Lév.) Rehder 狭叶溲疏

Shrubs, deciduous

15.5 (14.9~16.3)；5.4 (4.8~6.6)；24.4 (23.9~24.9)；15.5 (15.0~16.3)；126.1 (119.9~136.1)；0.0 (0.0~0.2)；803 (781~833) ❙ 1215 (1136~1284)；557 (501~627)；69 (64~78) ❙ 51.3 (41.2~58.8)；803 (781~833)；196.4 (160.1~243.4)

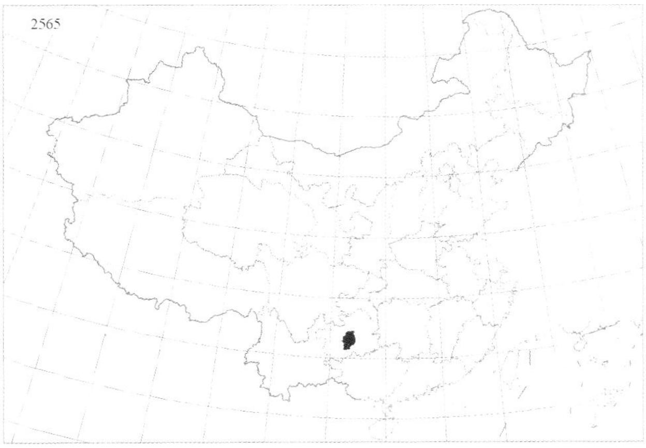

Saxifragaceae 虎耳草科 *Deutzia* 溲疏属

Deutzia faberi Rehder 浙江溲疏

Shrubs, deciduous

16.7 (14.8~21.1)；6.3 (3.4~14.4)；27.1 (20.1~29.2)；16.8 (14.8~21.1)；141.8 (120.2~192.9)；0.4 (0.0~2.1)；889 (773~1077) ❙ 1802 (1360~2993)；628 (481~1186)；244 (182~551) ❙ 102.5 (53.5~246.5)；885 (773~1077)；232.2 (176.3~289.1)

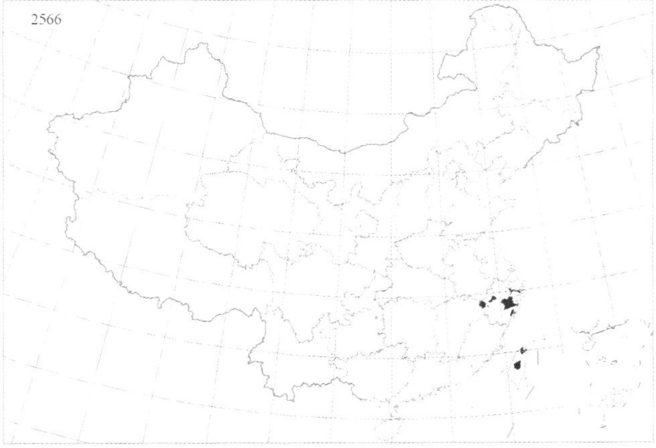

Saxifragaceae 虎耳草科 *Deutzia* 溲疏属

Deutzia glabrata Kom. 光萼溲疏

Shrubs, deciduous

6.6 (-5.0~14.2)；-11.0 (-29.4~1.8)；22.1 (17.0~26.8)；9.5 (5.0~14.2)；75.2 (35.2~116.7)；54.9 (5.8~154.8)；629 (450~811) ❙ 683 (341~1073)；400 (190~691)；27 (8~65) ❙ 9.7 (-40.9~74.1)；598 (341~800)；247.4 (61.2~346.5)

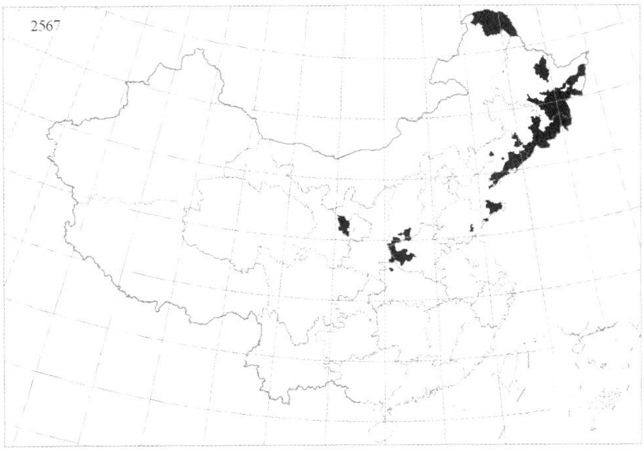

Saxifragaceae 虎耳草科 *Deutzia* 溲疏属

Deutzia glauca Kom. 黄山溲疏

Shrubs, deciduous

15.8 (13.4~18.2)；3.6 (1.5~8.4)；27.6 (25.2~29.3)；15.9 (13.4~18.3)；132.5 (108.3~159.1)；2.3 (0.0~7.3)；871 (755~941) ❙ 1447 (1042~1829)；564 (474~685)；161 (89~242) ❙ 66.6 (20.9~109.2)；871 (755~941)；223.0 (157.1~319.9)

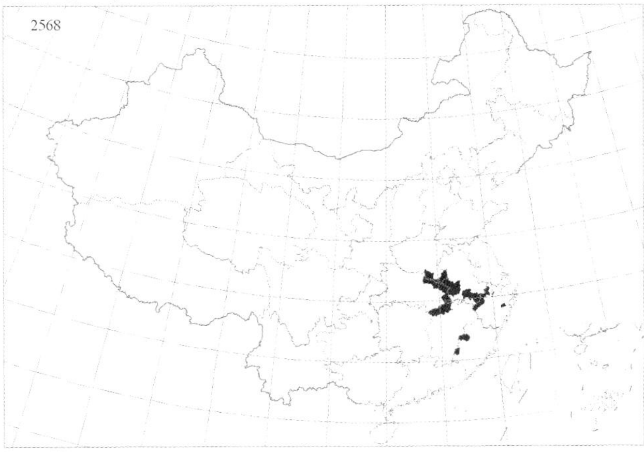

Saxifragaceae 虎耳草科 *Deutzia* 溲疏属
Deutzia glomeruliflora Franch. 球花溲疏
Shrubs, deciduous

11.4 (2.0~18.9); 3.7 (-7.2~12.0); 17.8 (9.5~25.3); 11.7 (3.8~19.0); 87.7 (15.4~167.8); 10.1 (0.0~51.4); 658 (411~929) ❙ 962 (579~1398); 553 (365~812); 30 (10~88) ❙ 51.0 (-14.2~128.6); 637 (411~812); 278.9 (187.9~366.3)

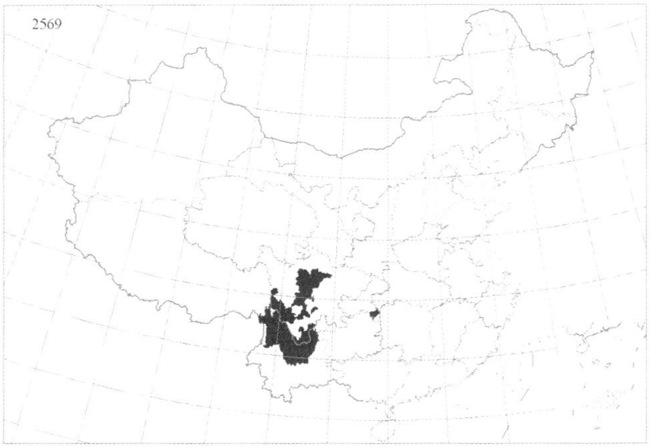

Saxifragaceae 虎耳草科 *Deutzia* 溲疏属
Deutzia gracilis Siebold et Zucc. 细梗溲疏
Shrubs, deciduous

20.8 (16.2~23.3); 15.3 (11.0~17.7); 25.2 (20.1~28.0); 20.8 (16.2~23.3); 190.0 (134.2~219.7); 0.0 (0.0~0.0); 1060 (773~1248) ❙ 2187 (1498~2759); 1227 (859~1594); 125 (73~245) ❙ 119.7 (25.2~246.5); 1012 (773~1144); 223.5 (179.2~289.1)

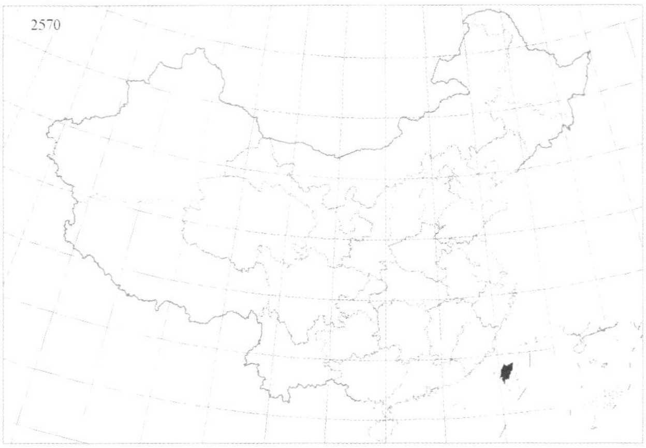

Saxifragaceae 虎耳草科 *Deutzia* 溲疏属
Deutzia grandiflora Bunge 大花溲疏
Shrubs, deciduous

11.1 (2.6~16.6); -3.6 (-15.9~4.4); 24.4 (15.8~28.8); 11.9 (6.6~16.7); 98.1 (45.2~140.6); 24.0 (0.6~80.4); 728 (507~920) ❙ 678 (367~1571); 376 (235~666); 31 (7~157) ❙ -7.2 (-43.4~108.0); 627 (367~910); 178.9 (70.3~334.4)

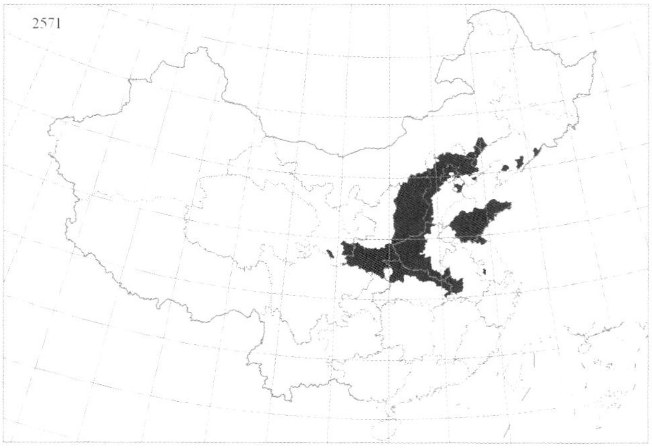

Saxifragaceae 虎耳草科 *Deutzia* 溲疏属
Deutzia heterophylla S. M. Hwang 异叶溲疏
Shrubs, deciduous

6.6; -1.3; 13.3; 6.8; 39.6; 19.7; 514 ❙ 862; 548; 12 ❙ 67.7; 514; 311.5

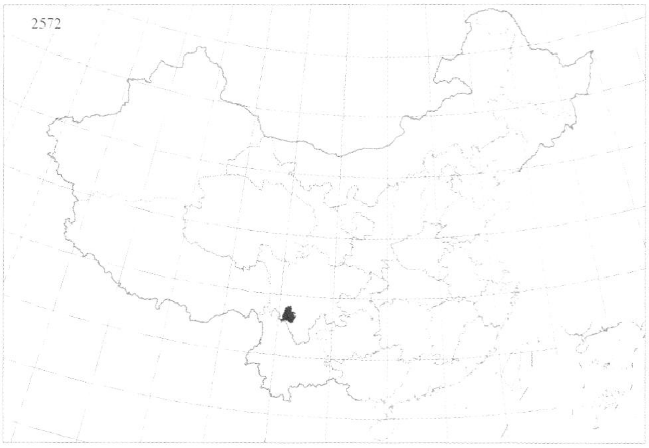

Saxifragaceae 虎耳草科 *Deutzia* 溲疏属
Deutzia hookeriana (C. K. Schneid.) Airy Shaw 西藏溲疏
Shrubs, deciduous

6.4 (1.8~13.3); -1.7 (-7.1~5.4); 13.4 (9.9~19.6); 7.0 (3.8~13.4); 42.1 (15.7~100.4); 25.4 (0.0~53.3); 513 (404~706) ❙ 884 (473~1736); 513 (287~1007); 28 (7~51) ❙ 65.2 (5.0~145.8); 513 (404~706); 292.4 (169.5~447.4)

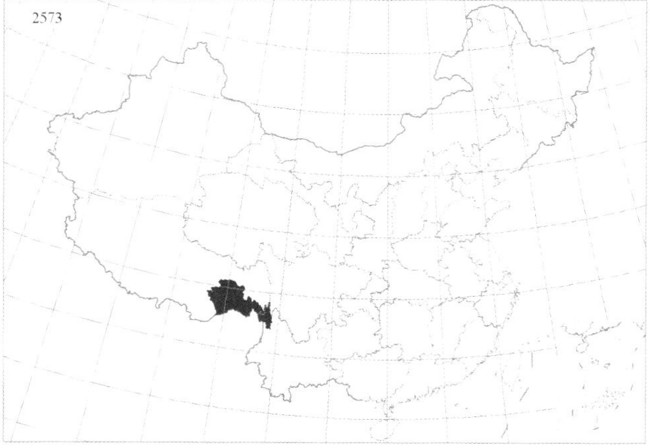

Saxifragaceae 虎耳草科 *Deutzia* 溲疏属
Deutzia hypoglauca Rehder 粉背溲疏
Shrubs, deciduous

11.8 (6.2~15.1); -0.9 (-4.2~1.8); 23.7 (15.8~27.5); 12.0 (7.0~15.2); 96.3 (45.2~128.0); 14.6 (5.6~29.9); 717 (522~865) ❙ 784 (614~1246); 373 (301~525); 34 (14~71) ❙ 11.3 (-22.2~101.4); 685 (522~817); 223.6 (139.1~334.4)

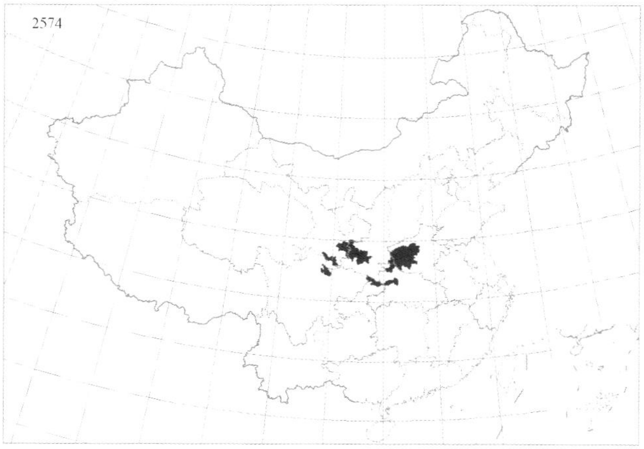

Saxifragaceae 虎耳草科 *Deutzia* 溲疏属
Deutzia hypoglauca var. shawiana (Zaik.) Zaik. 青城溲疏
Shrubs, deciduous

7.4; −2.1; 16.2; 7.8; 51.1; 21.1; 549 ❙ 872; 440; 22 ❙ 60.6; 549; 269.0

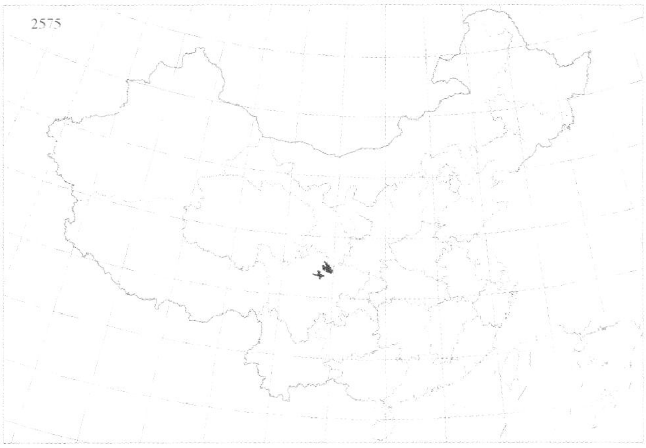

Saxifragaceae 虎耳草科 *Deutzia* 溲疏属
Deutzia longifolia Franch. 长叶溲疏
Shrubs, deciduous

12.7 (2.0~18.9); 4.7 (−7.2~12.5); 19.3 (9.9~28.1); 12.9 (3.8~19.0); 99.4 (15.4~167.8); 6.8 (0.0~51.4); 698 (413~929) ❙ 986 (628~1414); 553 (336~817); 36 (11~88) ❙ 45.2 (−14.2~128.6); 680 (413~924); 270.6 (139.5~382.6)

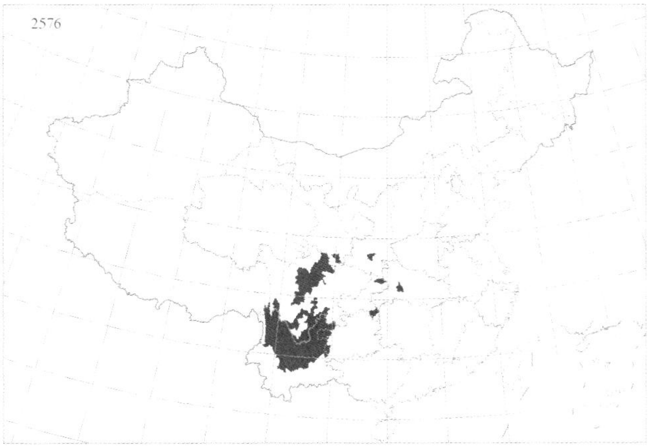

Saxifragaceae 虎耳草科 *Deutzia* 溲疏属
Deutzia longifolia var. pingwuensis S. M. Hwang 平武溲疏
Shrubs, deciduous

6.3; −3.6; 15.4; 7.1; 45.2; 28.8; 521 ❙ 777; 383; 19 ❙ 51.0; 521; 279.8

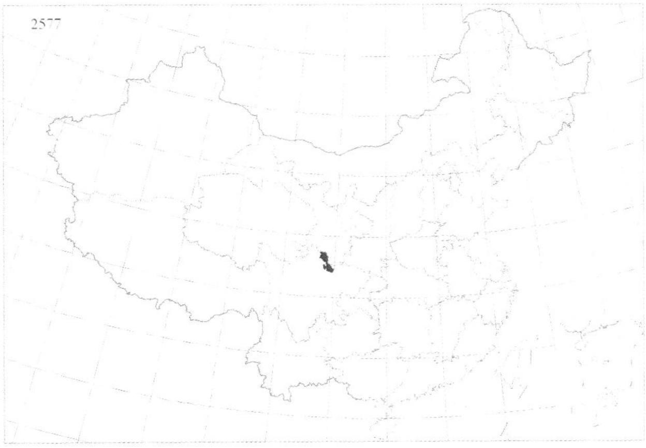

Saxifragaceae 虎耳草科 *Deutzia* 溲疏属
Deutzia mollis Duthie 钻丝溲疏
Shrubs, deciduous

12.5 (9.3~14.5); 0.6 (−1.7~1.8); 23.9 (20.1~26.4); 12.6 (9.5~14.6); 100.5 (69.8~120.7); 9.7 (5.7~18.0); 729 (603~821) ❙ 1022 (862~1214); 447 (376~516); 64 (47~75) ❙ 44.0 (7.2~101.4); 729 (603~821); 233.9 (174.3~279.5)

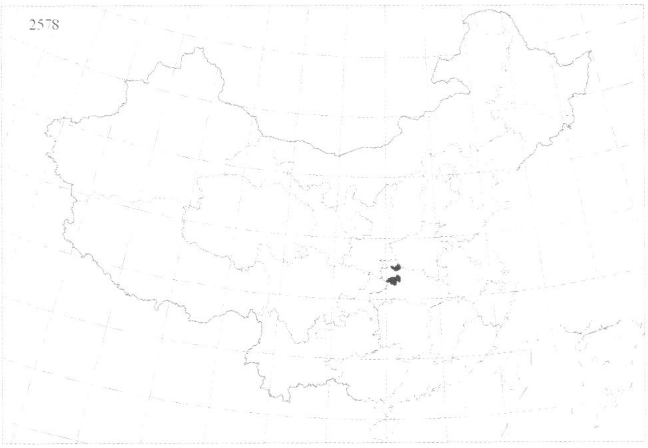

Saxifragaceae 虎耳草科 *Deutzia* 溲疏属
Deutzia monbeigii W. W. Sm. 维西溲疏
Shrubs, deciduous

6.5 (2.7~11.0); −1.2 (−6.0~4.2); 13.2 (10.4~16.6); 7.0 (4.2~11.0); 40.7 (18.9~73.1); 21.8 (0.8~46.9); 515 (419~636) ❙ 866 (579~1134); 518 (365~651); 26 (10~47) ❙ 67.9 (38.4~121.7); 515 (419~636); 283.6 (201.9~352.3)

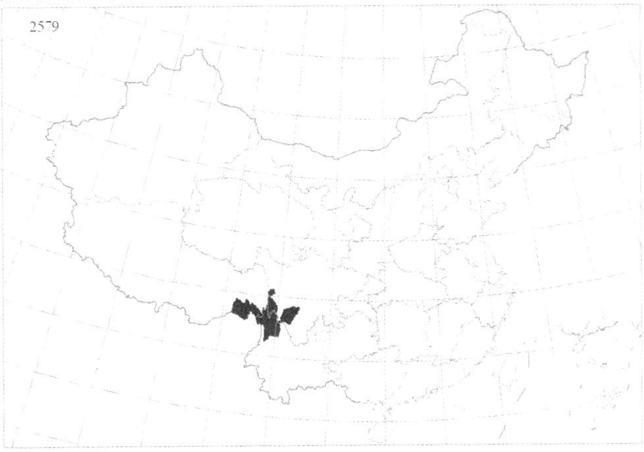

Saxifragaceae 虎耳草科 *Deutzia* 溲疏属
Deutzia muliensis S. M. Hwang 木里溲疏
Shrubs, deciduous

8.4; 0.5; 15.1; 8.5; 53.9; 12.4; 567 ❙ 949; 577; 15 ❙ 67.4; 567; 304.7

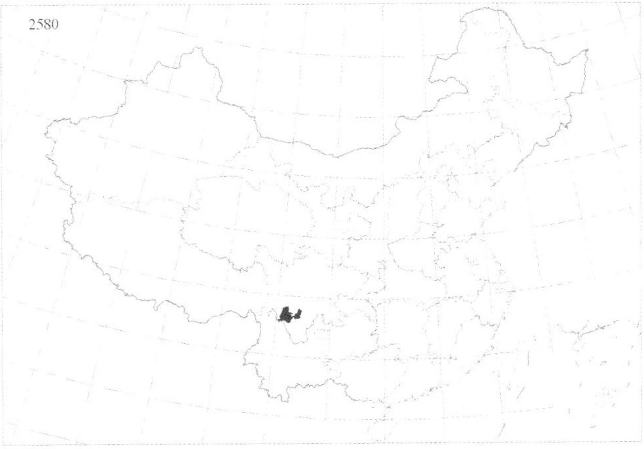

Saxifragaceae 虎耳草科 *Deutzia* 溲疏属
Deutzia multiradiata W. T. Wang 多辐线溲疏
Shrubs, deciduous

14.8; 4.3; 24.9; 14.9; 119.2; 0.7; 791 | 1197; 489; 71 | 51.2; 791; 182.9

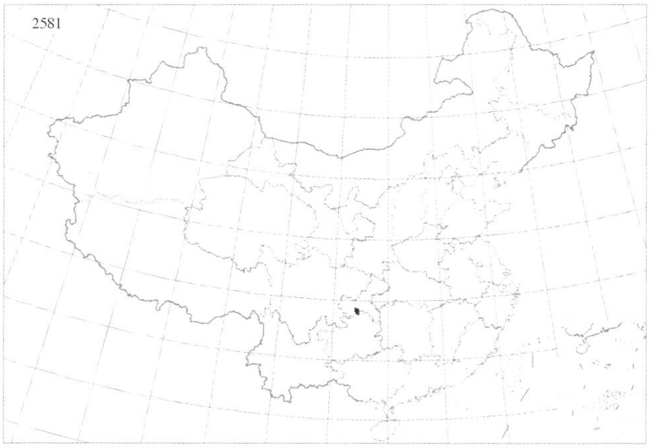

Saxifragaceae 虎耳草科 *Deutzia* 溲疏属
Deutzia nanchuanensis W. T. Wang 南川溲疏
Shrubs, deciduous

12.7; 3.3; 21.1; 12.7; 96.0; 3.9; 701 | 1098; 542; 43 | 57.7; 696; 248.5

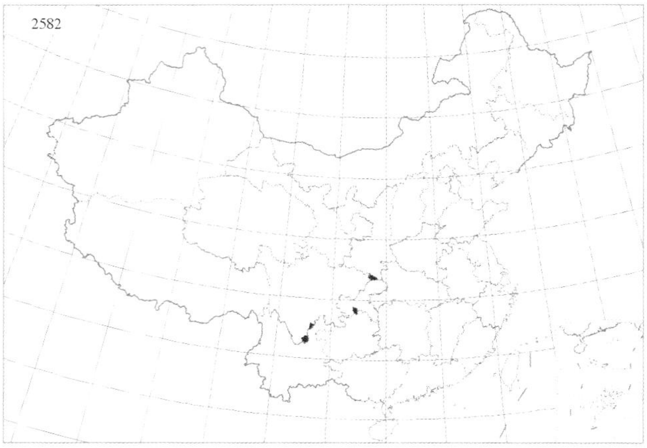

Saxifragaceae 虎耳草科 *Deutzia* 溲疏属
Deutzia ningpoensis Rehder 宁波溲疏
Shrubs, deciduous

16.0 (9.3~18.2); 4.1 (-1.7~8.4); 27.4 (20.1~29.3); 16.0 (9.5~18.3); 134.0 (69.8~159.1); 1.7 (0.0~18.0); 873 (603~984) | 1538 (849~1997); 563 (388~713); 180 (46~259) | 76.7 (-5.3~147.0); 868 (603~965); 230.4 (141.3~319.9)

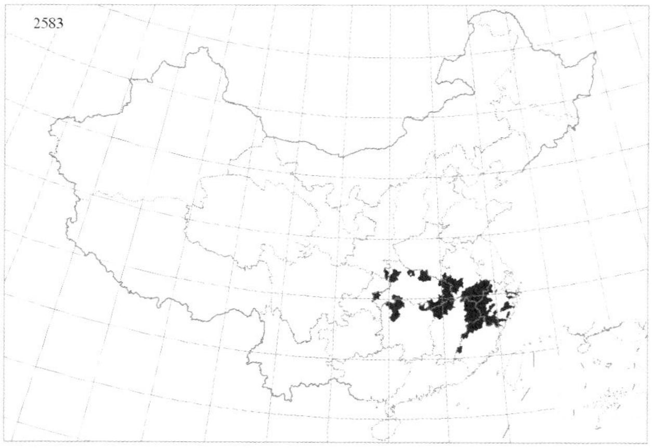

Saxifragaceae 虎耳草科 *Deutzia* 溲疏属
Deutzia obtusilobata S. M. Hwang 钝裂溲疏
Shrubs, deciduous

3.8; -4.3; 10.8; 4.7; 21.6; 35.4; 444 | 984; 601; 16 | 121.7; 444; 265.2

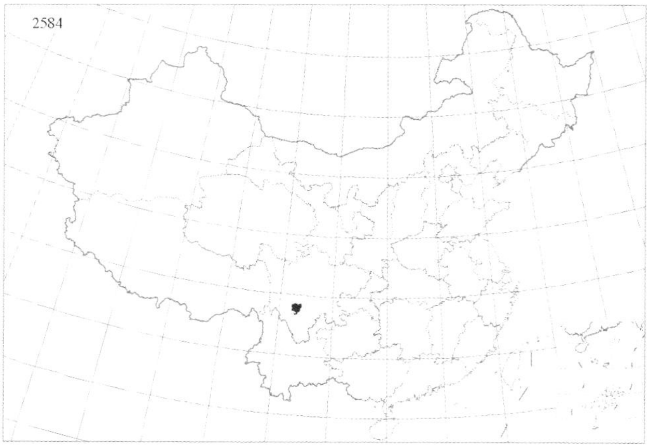

Saxifragaceae 虎耳草科 *Deutzia* 溲疏属
Deutzia parviflora Bunge 小花溲疏
Shrubs, deciduous

10.2 (-0.2~16.9); -5.0 (-23.8~5.2); 23.8 (16.6~28.3); 11.3 (6.4~16.9); 92.1 (45.4~143.4); 29.7 (0.0~112.6); 703 (501~924) | 635 (349~1466); 359 (227~691); 23 (7~109) | -9.1 (-47.7~106.2); 593 (349~924); 188.7 (70.3~346.5)

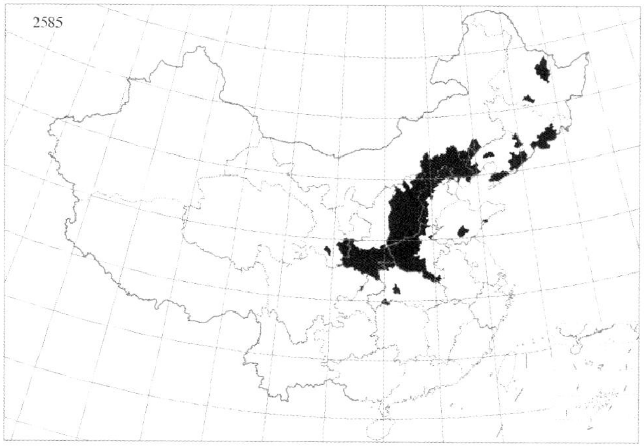

Saxifragaceae 虎耳草科 *Deutzia* 溲疏属
Deutzia parviflora var. amurensis Regel 东北溲疏
Shrubs, deciduous

4.4 (1.4~9.8); -15.4 (-20.1~-4.9); 21.3 (18.4~24.4); 8.3 (6.4~10.6); 64.7 (45.8~89.4); 71.4 (26.3~93.6); 586 (501~692) | 740 (473~1073); 464 (332~691); 25 (8~40) | 26.6 (-25.0~74.1); 576 (473~668); 291.5 (136.6~346.5)

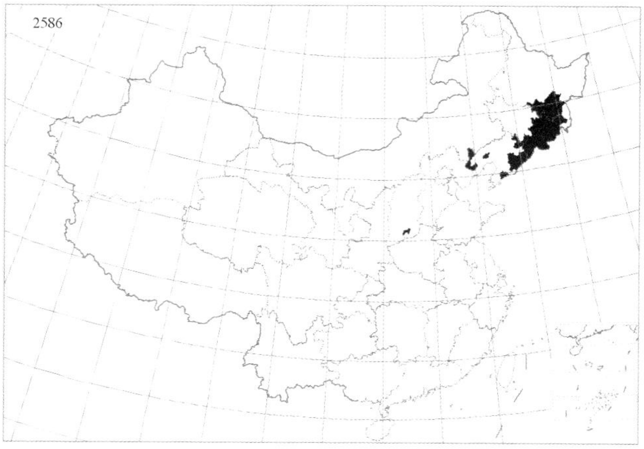

Saxifragaceae 虎耳草科 *Deutzia* 溲疏属
Deutzia parviflora var. micrantha (Engl.) Rehder 碎花溲疏
Shrubs, deciduous

10.8 (5.2~13.4); -2.7 (-10.7~1.2); 23.4 (18.7~26.6); 11.3 (7.7~13.4); 90.5 (57.5~114.9); 20.1 (7.6~54.8); 692 (550~804) ❙ 683 (478~856); 336 (290~391); 24 (14~39) ❙ 0.1 (-24.6~52.6); 632 (478~761); 230.1 (141.3~320.6)

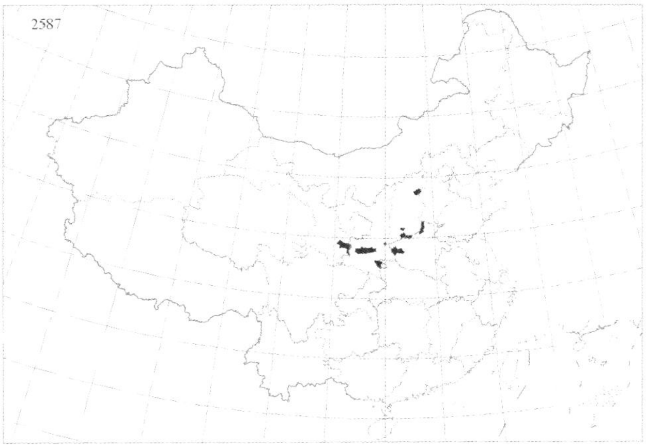

Saxifragaceae 虎耳草科 *Deutzia* 溲疏属
Deutzia pilosa Rehder 褐毛溲疏
Shrubs, deciduous

13.0 (2.2~16.3); 3.2 (-6.6~8.6); 21.9 (9.9~25.2); 13.1 (3.8~16.3); 102.5 (15.5~136.1); 5.9 (0.0~49.1); 723 (413~833) ❙ 1113 (677~1410); 557 (336~816); 48 (14~74) ❙ 56.1 (2.4~125.9); 720 (413~833); 215.9 (147.3~330.0)

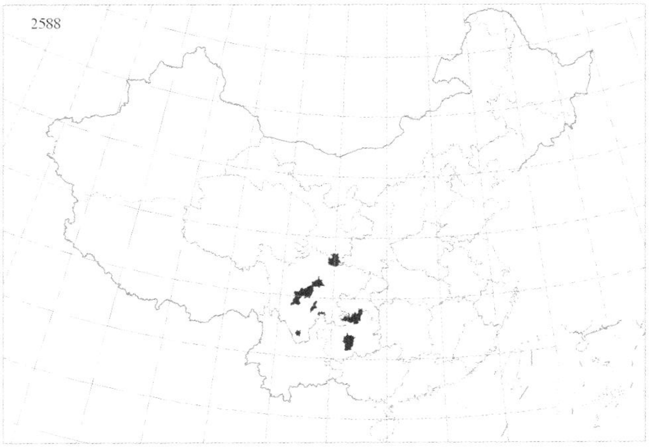

Saxifragaceae 虎耳草科 *Deutzia* 溲疏属
Deutzia pulchra S. Vidal 美丽溲疏
Shrubs or small trees, deciduous

18.9 (16.2~22.2); 13.1 (11.0~17.7); 23.8 (20.1~27.4); 18.9 (16.2~22.2); 167.0 (134.2~206.7); 0.0 (0.0~0.0); 926 (773~1123) ❙ 2550 (1836~2993); 1065 (750~1847); 300 (107~551) ❙ 180.1 (64.2~246.5); 922 (773~1123); 278.9 (202.8~363.7)

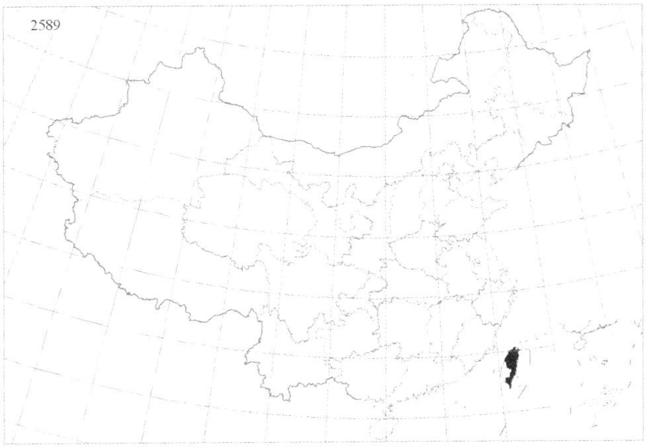

Saxifragaceae 虎耳草科 *Deutzia* 溲疏属
Deutzia purpurascens (Franch. ex L. Henry) Rehder 紫花溲疏
Shrubs, deciduous

11.2 (-0.3~17.3); 4.0 (-9.6~11.0); 17.1 (8.1~21.6); 11.5 (2.7~17.3); 84.2 (9.0~148.0); 9.0 (0.0~71.8); 646 (354~832) ❙ 999 (255~1736); 572 (191~1007); 39 (1~69) ❙ 55.2 (-27.8~145.8); 633 (255~783); 317.7 (101.7~447.4)

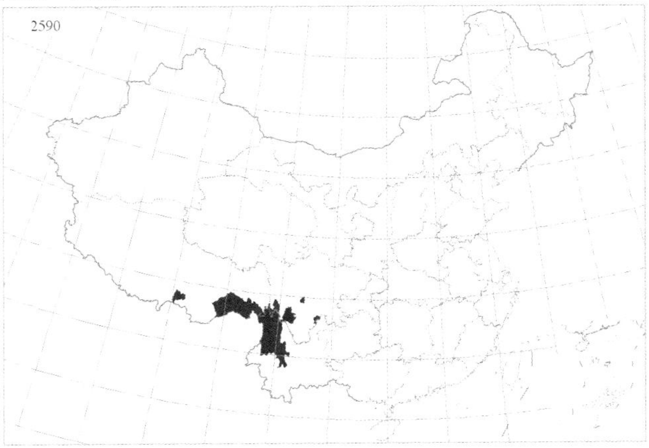

Saxifragaceae 虎耳草科 *Deutzia* 溲疏属
Deutzia rehderiana C. K. Schneid. 灌丛溲疏
Shrubs, deciduous

15.3 (3.8~18.9); 7.9 (-4.3~12.5); 21.3 (10.8~26.8); 15.4 (4.7~19.0); 125.2 (21.6~167.8); 0.8 (0.0~35.4); 777 (444~929) ❙ 1015 (797~1240); 556 (411~651); 44 (16~90) ❙ 32.1 (-14.2~121.7); 747 (444~859); 275.8 (146.0~332.6)

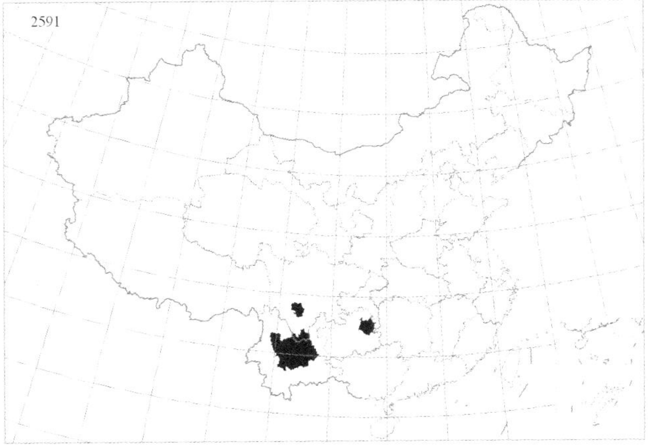

Saxifragaceae 虎耳草科 *Deutzia* 溲疏属
Deutzia rubens Rehder 粉红溲疏
Shrubs, deciduous

12.2 (2.9~17.1); 0.7 (-7.7~4.5); 23.1 (11.2~29.4); 12.5 (4.4~17.1); 98.7 (20.5~146.5); 11.5 (0.5~49.3); 730 (428~951) ❙ 1077 (632~1502); 490 (306~817); 70 (15~183) ❙ 50.0 (-2.0~121.7); 729 (428~932); 231.4 (140.2~357.7)

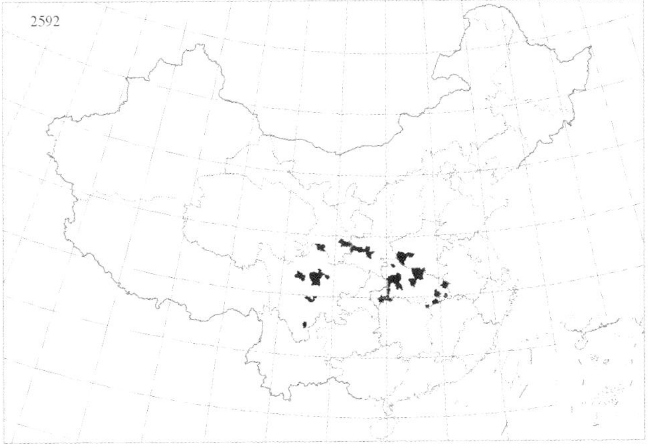

Saxifragaceae 虎耳草科 *Deutzia* 溲疏属
Deutzia schneideriana Rehder 长江溲疏
Shrubs, deciduous

15.6 (9.1~18.2); 3.6 (−0.7~6.5); 27.1 (18.1~29.7); 15.6 (9.2~18.2); 129.8 (64.1~158.8); 2.2 (0.0~14.1); 857 (595~984) ❙ 1457 (822~1922); 553 (403~685); 160 (19~259) ❙ 70.4 (13.2~123.6); 851 (595~941); 226.8 (149.5~319.9)

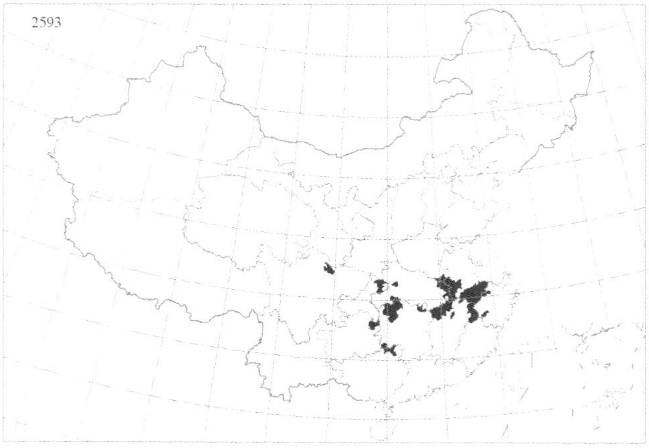

Saxifragaceae 虎耳草科 *Deutzia* 溲疏属
Deutzia setchuenensis Franch. 四川溲疏
Shrubs, deciduous

16.4 (5.7~21.6); 5.6 (−3.5~15.2); 26.5 (14.1~30.1); 16.5 (6.4~22.7); 139.0 (38.1~212.7); 1.0 (0.0~28.1); 879 (504~1216) ❙ 1393 (862~1965); 531 (376~1056); 137 (20~259) ❙ 58.9 (5.5~128.9); 871 (504~1216); 206.9 (108.2~331.6)

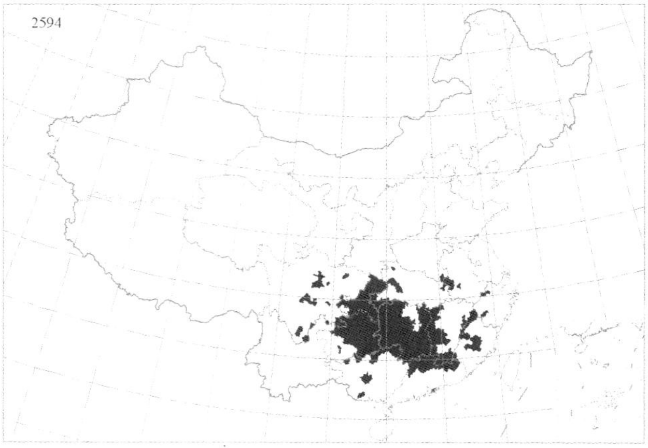

Saxifragaceae 虎耳草科 *Deutzia* 溲疏属
Deutzia setchuenensis var. corymbiflora (Lemoine ex André) Rehder 多花溲疏
Shrubs, deciduous

14.8 (10.6~17.7); 4.0 (−0.2~6.9); 25.1 (21.1~27.8); 14.9 (10.6~17.7); 120.9 (79.4~152.2); 2.6 (0.0~12.1); 808 (643~937) ❙ 1128 (899~1278); 509 (458~533); 47 (19~71) ❙ 41.8 (9.2~93.8); 808 (643~937); 196.6 (126.1~284.3)

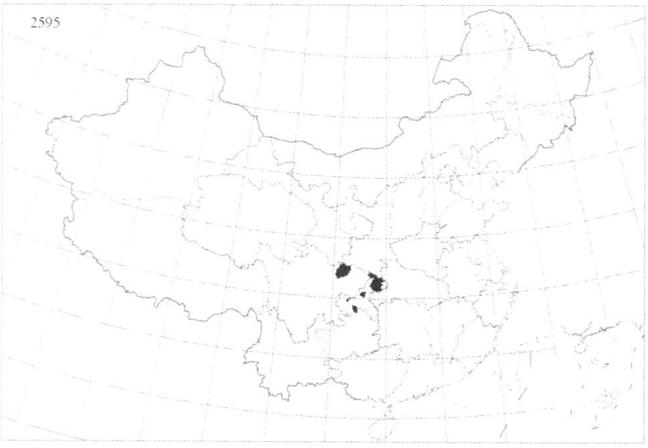

Saxifragaceae 虎耳草科 *Deutzia* 溲疏属
Deutzia setchuenensis var. longidentata Rehder 长齿溲疏
Shrubs, deciduous

14.7; 4.4; 23.8; 14.7; 117.2; 1.2; 785 ❙ 1015; 590; 30 ❙ 30.8; 784; 152.4

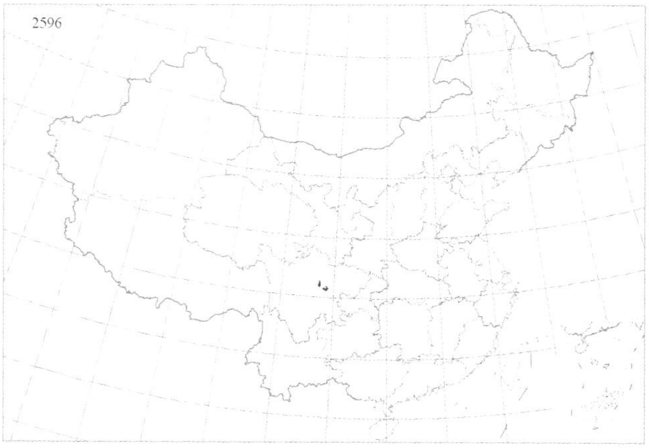

Saxifragaceae 虎耳草科 *Deutzia* 溲疏属
Deutzia silvestrii Pamp. 红花溲疏
Shrubs, deciduous

15.9; 2.9; 28.0; 16.0; 134.0; 2.4; 897 ❙ 849; 388; 65 ❙ −5.3; 849; 141.3

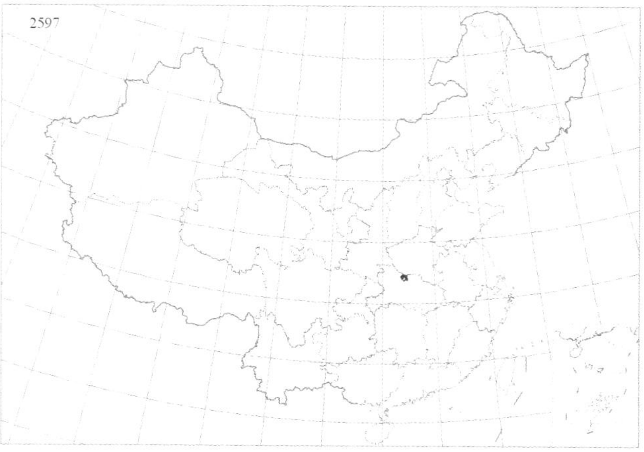

Saxifragaceae 虎耳草科 *Deutzia* 溲疏属
Deutzia squamosa S. M. Hwang 鳞毛溲疏
Shrubs, deciduous

11.9; 2.9; 20.0; 12.0; 87.2; 3.4; 671 ❙ 1001; 550; 35 ❙ 49.1; 671; 278.8

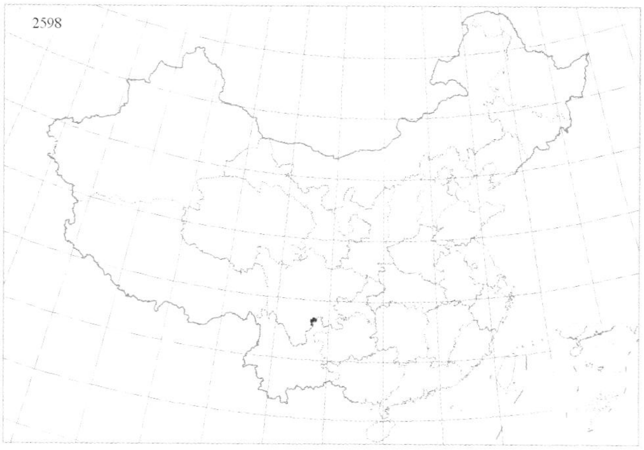

Saxifragaceae 虎耳草科 *Deutzia* 溲疏属
Deutzia staminea R. Br. ex Wall. 长柱溲疏
Shrubs, deciduous

4.4 (-1.8~13.3); -3.6 (-10.2~5.4); 11.6 (6.1~19.6); 5.6 (1.8~13.4); 31.2 (1.9~100.4); 37.5 (0.0~82.8); 465 (329~706) ❙ 767 (320~1736); 463 (193~1007); 24 (10~51) ❙ 55.9 (-4.0~145.8); 463 (320~706); 237.6 (61.6~447.4)

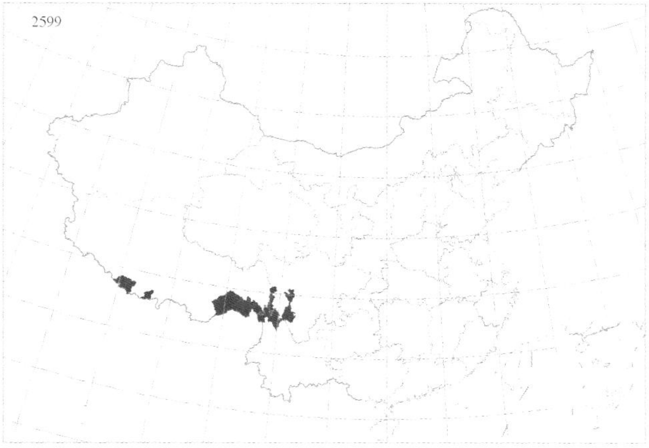

Saxifragaceae 虎耳草科 *Deutzia* 溲疏属
Deutzia subulata Hand.-Mazz. 钻齿溲疏
Shrubs, deciduous

8.2 (2.0~14.1); 0.6 (-7.2~6.5); 14.7 (10.1~20.2); 8.6 (3.8~14.2); 54.7 (15.4~109.8); 15.4 (0.0~51.4); 564 (413~729) ❙ 937 (628~1134); 563 (410~651); 23 (11~47) ❙ 68.5 (40.1~128.6); 562 (413~707); 300.2 (236.3~351.7)

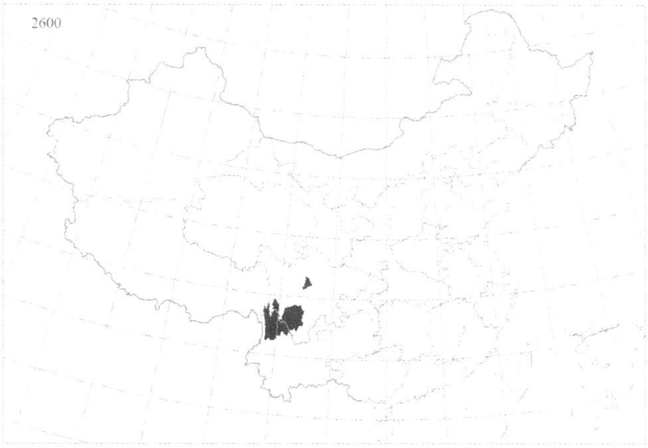

Saxifragaceae 虎耳草科 *Deutzia* 溲疏属
Deutzia taibaiensis W. T. Wang ex S. M. Hwang 太白溲疏
Shrubs, deciduous

10.0; -2.3; 21.8; 10.4; 79.9; 19.0; 647 ❙ 714; 334; 19 ❙ 10.5; 647; 271.9

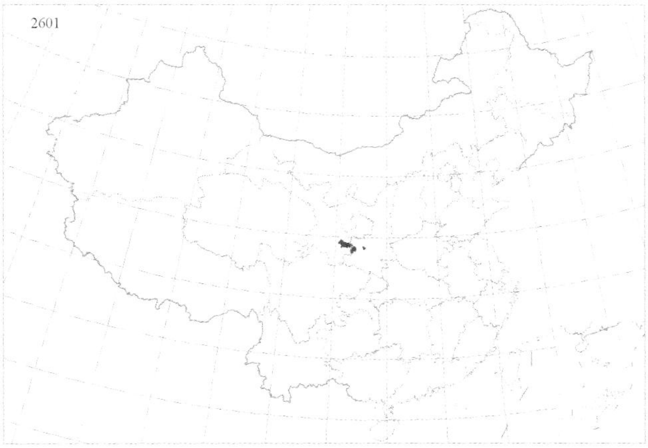

Saxifragaceae 虎耳草科 *Deutzia* 溲疏属
Deutzia taiwanensis (Maxim.) C. K. Schneid. 台湾溲疏
Shrubs, deciduous

18.9 (16.2~22.2); 13.4 (11.0~17.7); 23.4 (20.1~27.0); 18.9 (16.2~22.2); 166.7 (134.2~206.7); 0.0 (0.0~0.0); 921 (773~1123) ❙ 2569 (1836~2993); 1170 (924~1847); 251 (107~547) ❙ 184.3 (64.2~246.5); 917 (773~1123); 277.1 (202.8~363.7)

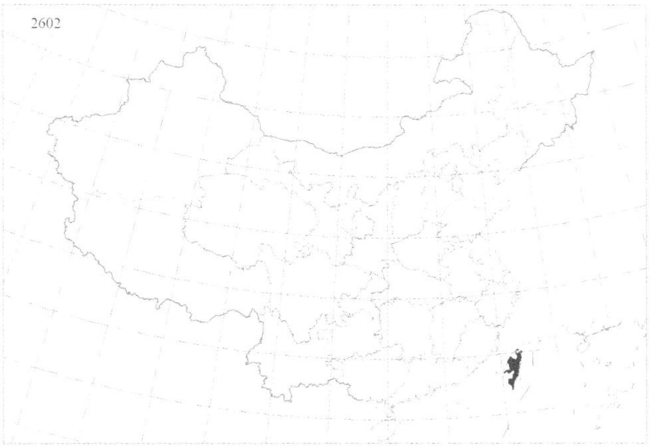

Saxifragaceae 虎耳草科 *Dichroa* 常山属
Dichroa daimingshanensis Y. C. Wu 大明常山
Subshrubs, deciduous

20.4 (16.1~22.4); 11.4 (5.3~14.3); 27.6 (25.0~28.9); 20.5 (16.1~22.5); 185.8 (133.1~209.6); 0.0 (0.0~0.0); 1075 (838~1205) ❙ 1558 (1268~1965); 726 (523~911); 140 (57~194) ❙ 45.9 (15.4~85.5); 1075 (838~1205); 239.2 (95.3~371.7)

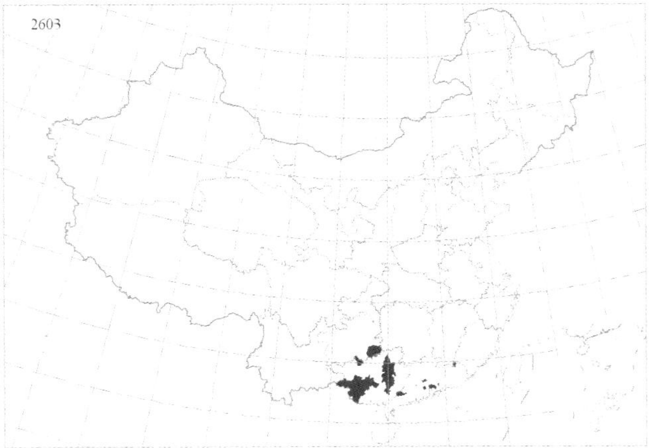

Saxifragaceae 虎耳草科 *Dichroa* 常山属
Dichroa febrifuga Lour. 常山
Shrubs, deciduous

17.4 (3.2~24.5); 7.9 (-5.8~19.1); 25.8 (10.8~30.1); 17.5 (4.5~24.5); 150.7 (21.4~234.4); 0.8 (0.0~42.6); 920 (430~1360) ❙ 1389 (622~2435); 601 (298~1428); 116 (11~259) ❙ 51.9 (-14.2~145.8); 909 (430~1360); 244.8 (95.3~659.0)

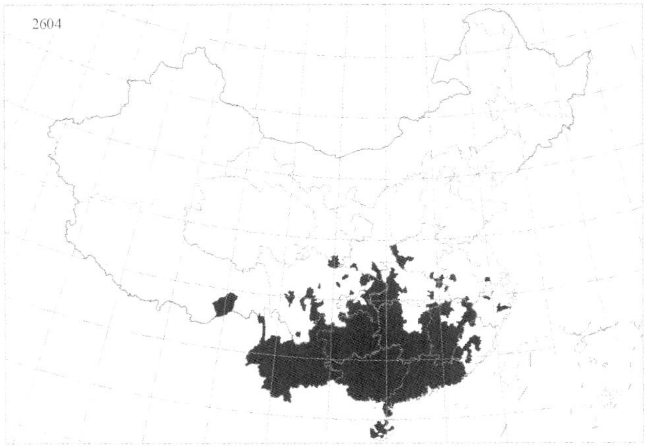

Saxifragaceae 虎耳草科 *Dichroa* 常山属

Dichroa hirsuta Gagnep. 硬毛常山

Shrubs, deciduous

19.9 (17.7~21.5); 11.5 (8.7~14.2); 26.2 (22.5~28.2); 19.9 (17.8~21.6); 178.6 (153.4~199.1); 0.0 (0.0~0.0); 1017 (854~1140) ❙ 1431 (1182~1581); 745 (670~850); 91 (43~135) ❙ 41.4 (16.9~60.1); 1017 (854~1140); 352.1 (248.2~551.6)

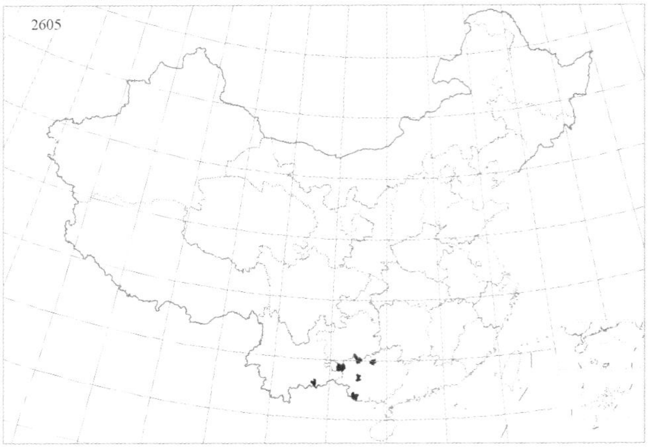

Saxifragaceae 虎耳草科 *Dichroa* 常山属

Dichroa mollissima Merr. 海南常山

Shrubs, deciduous

23.8 (22.2~24.8); 18.7 (17.2~20.1); 27.3 (25.7~28.2); 23.8 (22.3~24.8); 225.8 (207.2~237.6); 0.0 (0.0~0.0); 1287 (1127~1385) ❙ 1598 (1405~1767); 552 (473~610); 58 (41~88) ❙ 25.0 (1.5~47.8); 1165 (1055~1299); 556.0 (494.7~620.0)

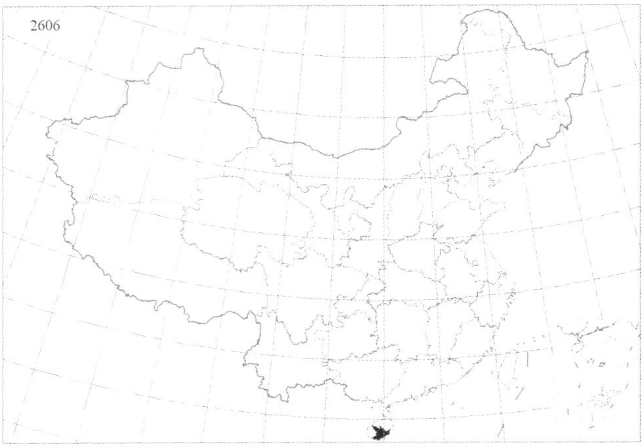

Saxifragaceae 虎耳草科 *Dichroa* 常山属

Dichroa yaoshanensis Y. C. Wu 罗蒙常山

Subshrubs, deciduous

18.7 (14.6~21.4); 9.2 (3.8~14.2); 26.4 (22.4~28.6); 18.7 (14.7~21.5); 164.5 (117.7~197.8); 0.0 (0.0~1.5); 963 (779~1140) ❙ 1498 (1161~1961); 659 (505~860); 137 (43~191) ❙ 56.8 (5.8~101.8); 963 (779~1140); 271.6 (185.5~551.6)

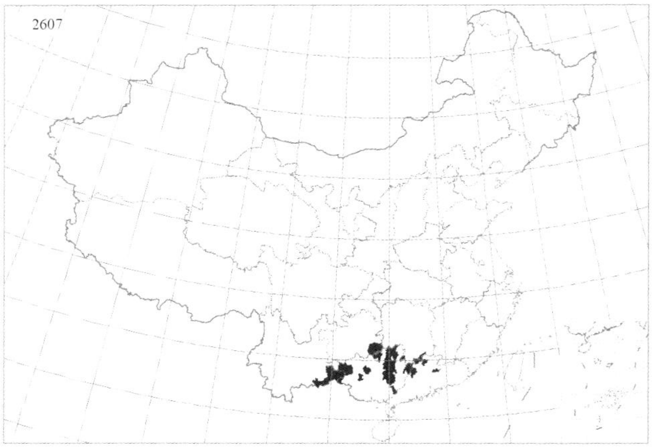

Saxifragaceae 虎耳草科 *Dichroa* 常山属

Dichroa yunnanensis S. M. Hwang 云南常山

Shrubs, deciduous

8.6; 1.3; 15.0; 8.7; 52.6; 8.8; 574 ❙ 1041; 585; 45 ❙ 81.5; 574; 352.3

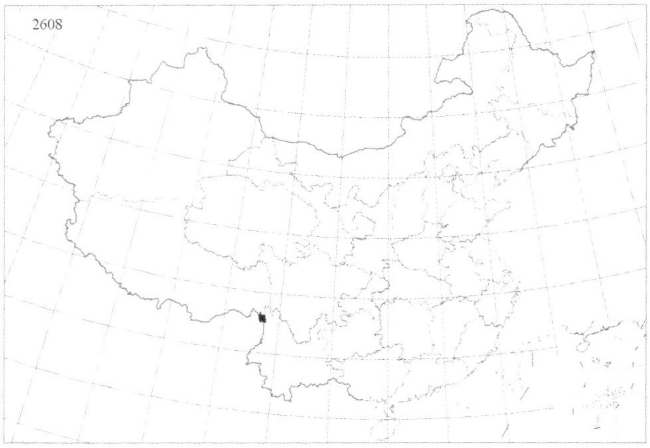

Saxifragaceae 虎耳草科 *Hydrangea* 绣球花属

Hydrangea anomala D. Don 冠盖绣球

Climbing shrubs, deciduous

15.1 (−1.8~21.7); 4.0 (−10.2~15.3); 25.3 (6.1~29.8); 15.2 (1.8~21.7); 125.5 (1.9~200.5); 4.2 (0.0~82.8); 828 (329~1140) ❙ 1306 (320~3495); 521 (217~1412); 127 (10~897) ❙ 57.1 (−22.2~246.5); 816 (320~1140); 229.1 (61.6~487.2)

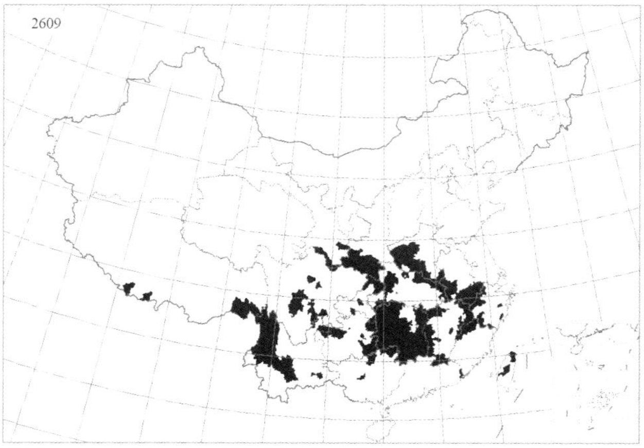

Saxifragaceae 虎耳草科 *Hydrangea* 绣球花属

Hydrangea aspera D. Don 马桑绣球

Shrubs or small trees, deciduous

15.2 (2.2~24.1); 6.5 (−6.6~18.9); 22.7 (9.9~29.3); 15.2 (3.8~24.1); 124.9 (15.5~229.7); 2.2 (0.0~49.1); 795 (413~1322) ❙ 1241 (711~3495); 608 (349~1847); 75 (15~897) ❙ 56.2 (−14.2~246.5); 786 (413~1123); 274.6 (131.4~659.0)

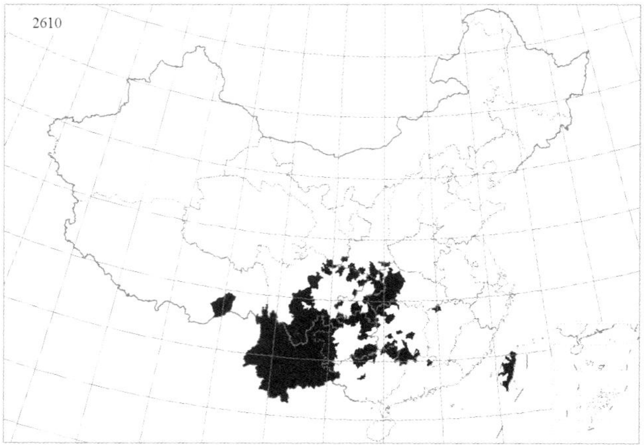

Saxifragaceae 虎耳草科 *Hydrangea* 绣球花属
Hydrangea bretschneideri Dippel 东陵绣球
Shrubs, deciduous

8.6 (-1.3~17.9); -5.5 (-18.7~11.0); 21.1 (9.7~27.8); 9.9 (2.9~17.9); 76.8 (11.3~154.8); 33.4 (0.0~90.7); 638 (359~937) ❙ 622 (256~1494); 346 (169~651); 20 (6~158) ❙ -0.6 (-57.7~128.6); 545 (256~937); 193.6 (70.3~383.4)

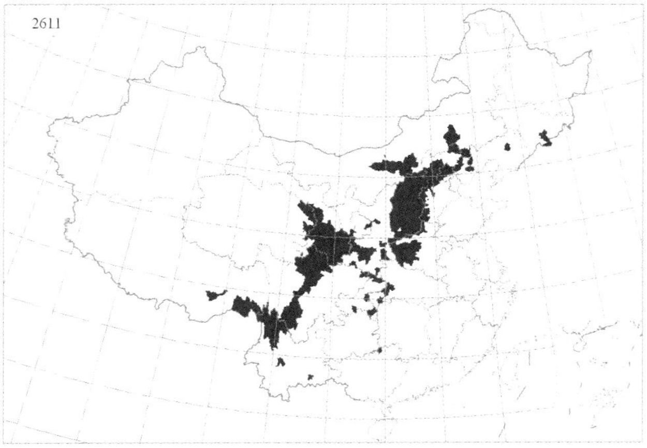

Saxifragaceae 虎耳草科 *Hydrangea* 绣球花属
Hydrangea candida Chun 珠光绣球
Shrubs, deciduous

21.9 (21.3~22.7); 13.9 (13.2~15.0); 27.9 (27.3~28.6); 22.0 (21.3~22.8); 203.6 (195.9~213.1); 0.0 (0.0~0.0); 1162 (1111~1221) ❙ 1876 (1302~2435); 1047 (670~1428); 116 (86~130) ❙ 61.1 (12.9~99.5); 1162 (1111~1221); 289.6 (235.1~318.8)

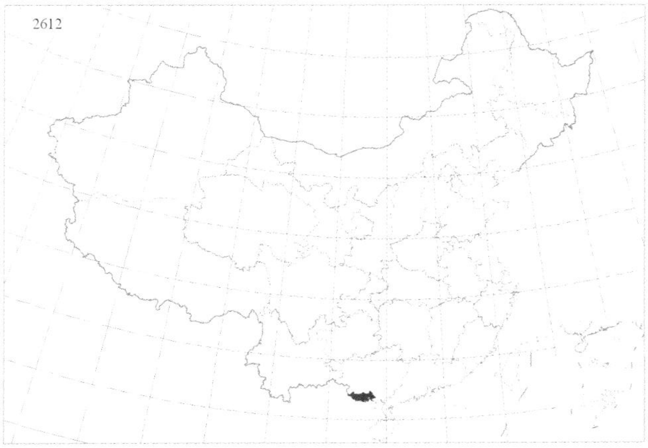

Saxifragaceae 虎耳草科 *Hydrangea* 绣球花属
Hydrangea caudatifolia W. T. Wang et M. X. Nie 尾叶绣球
Shrubs, deciduous

17.1 (16.5~18.1); 6.2 (5.0~7.2); 27.3 (26.3~28.5); 17.2 (16.6~18.1); 146.0 (138.6~157.7); 0.0 (0.0~0.0); 906 (866~965) ❙ 1890 (1773~1997); 618 (552~669); 246 (233~256) ❙ 109.1 (83.7~130.7); 903 (866~944); 266.7 (223.7~303.4)

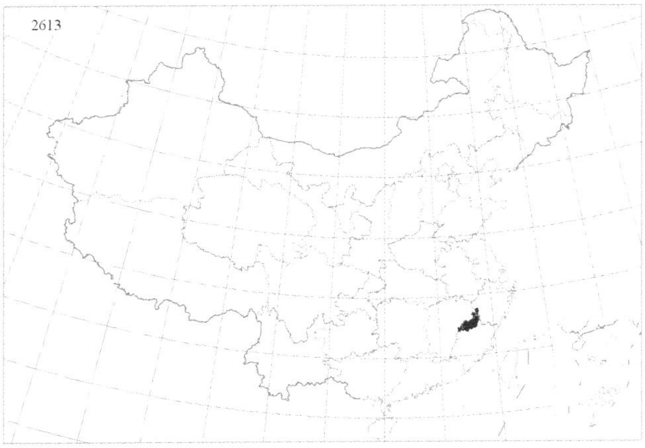

Saxifragaceae 虎耳草科 *Hydrangea* 绣球花属
Hydrangea chinensis Maxim. 中国绣球
Shrubs, deciduous

16.8 (8.6~24.2); 6.4 (0.2~18.9); 26.5 (15.0~30.1); 16.9 (8.7~24.3); 143.5 (52.6~231.2); 0.8 (0.0~9.2); 896 (574~1335) ❙ 1583 (706~3495); 611 (381~1847); 172 (20~897) ❙ 77.4 (-17.1~246.5); 888 (574~1170); 237.8 (111.9~487.2)

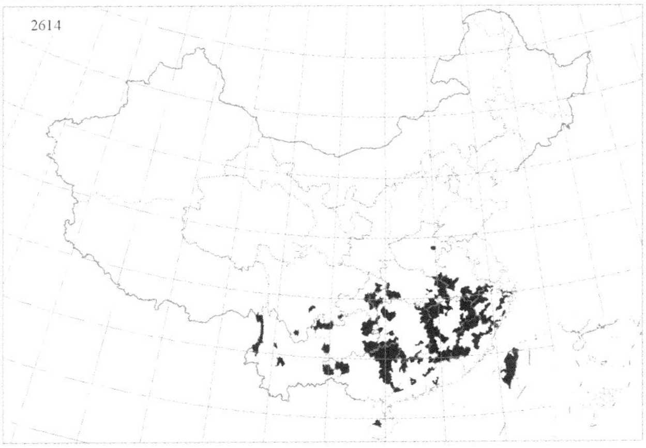

Saxifragaceae 虎耳草科 *Hydrangea* 绣球花属
Hydrangea chungii Rehder 福建绣球
Shrubs, deciduous

17.4 (16.6~18.6); 7.2 (5.9~9.3); 27.0 (26.3~27.6); 17.5 (16.6~18.7); 149.8 (138.9~164.0); 0.0 (0.0~0.0); 911 (866~965) ❙ 1861 (1702~1997); 623 (587~669) 232 (199~254) ❙ 104.8 (76.4~130.7); 911 (866~965); 288.4 (264.8~332.5)

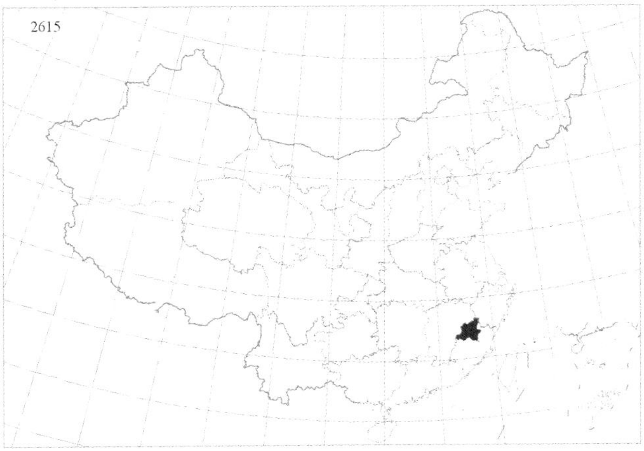

Saxifragaceae 虎耳草科 *Hydrangea* 绣球花属
Hydrangea coacta C. F. Wei 毡毛绣球
Shrubs, deciduous

12.0; 0.7; 22.9; 12.1; 94.6; 9.4; 701 ❙ 951; 436; 26 ❙ 35.6; 701; 276.4

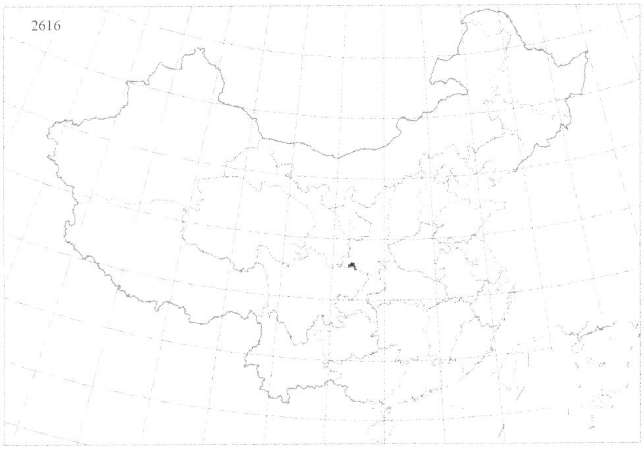

Saxifragaceae 虎耳草科 *Hydrangea* 绣球花属
Hydrangea coenobialis Chun 酥醪绣球
Shrubs, deciduous

18.3 (14.3~22.0); 8.5 (3.8~14.1); 26.7 (23.5~28.6); 18.3 (14.3~22.1); 159.6 (112.6~205.1); 0.1 (0.0~1.5); 954 (756~1170) ❙ 1593 (1182~2014); 637 (476~954); 156 (43~249) ❙ 67.5 (5.8~121.9); 954 (756~1170); 245.1 (185.5~354.5)

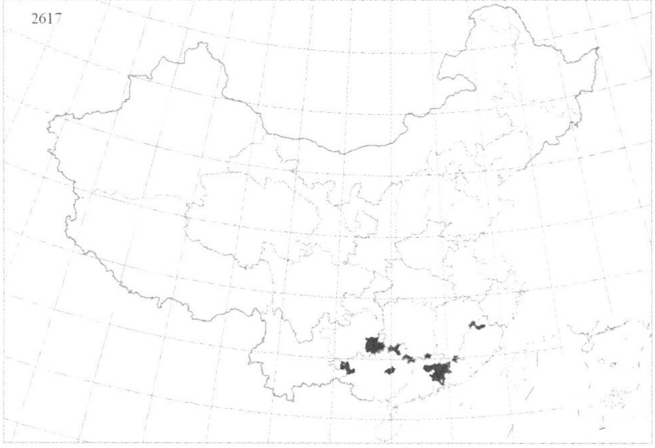

Saxifragaceae 虎耳草科 *Hydrangea* 绣球花属
Hydrangea davidii Franch. 西南绣球
Shrubs, deciduous

15.5 (5.4~22.7); 7.1 (-2.1~15.0); 22.4 (12.3~28.6); 15.5 (5.8~22.8); 127.0 (31.5~213.1); 0.8 (0.0~25.6); 797 (481~1221) ❙ 1168 (741~2435); 608 (411~1428); 58 (15~189) ❙ 46.9 (-14.2~107.2); 786 (481~1221); 275.6 (146.0~551.6)

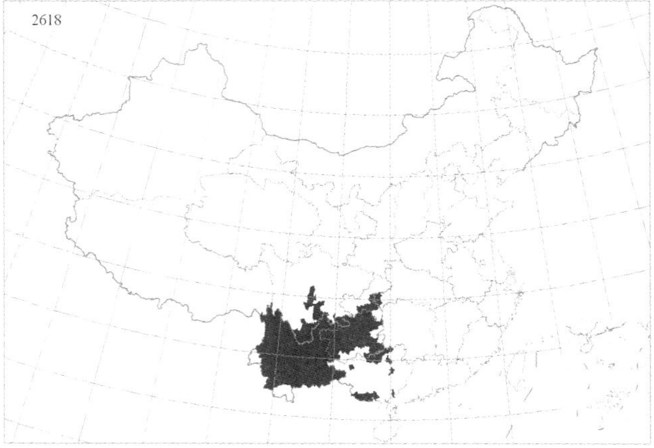

Saxifragaceae 虎耳草科 *Hydrangea* 绣球花属
Hydrangea dumicola W. W. Sm. 银针绣球
Shrubs, deciduous

10.1 (5.4~14.4); 2.9 (-2.1~7.5); 15.9 (12.3~19.1); 10.1 (5.8~14.4); 68.2 (31.5~113.1); 7.1 (0.0~25.6); 610 (481~727) ❙ 1081 (741~1454); 616 (426~792); 45 (21~70) ❙ 75.7 (52.8~99.9); 610 (481~727); 355.4 (273.5~487.2)

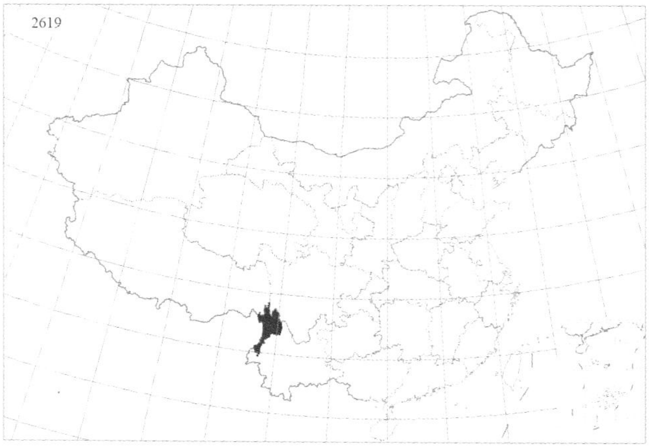

Saxifragaceae 虎耳草科 *Hydrangea* 绣球花属
Hydrangea gracilis W. T. Wang et M. X. Nie 细枝绣球
Small shrubs, deciduous

16.6 (14.3~18.5); 5.7 (3.7~7.8); 26.8 (24.2~28.1); 16.7 (14.3~18.5); 140.7 (113.2~162.3); 0.4 (0.0~1.4); 881 (761~975) ❙ 1629 (1561~1742); 551 (512~581); 202 (189~218) ❙ 86.6 (65.4~128.9); 881 (761~975); 220.6 (208.1~236.3)

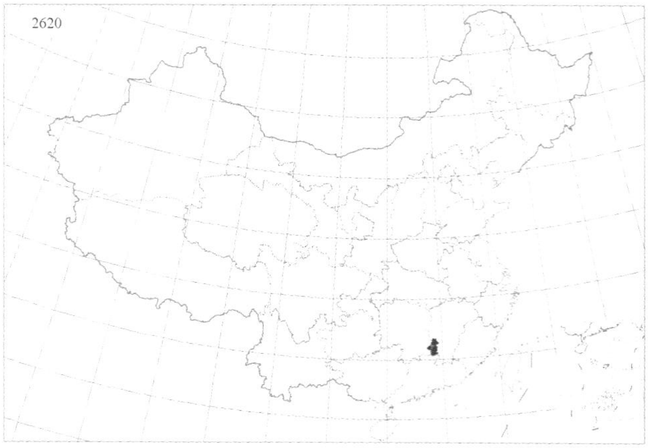

Saxifragaceae 虎耳草科 *Hydrangea* 绣球花属
Hydrangea heteromalla D. Don 微绒绣球
Shrubs or small trees, deciduous

11.9 (-1.8~18.9); 4.1 (-10.2~12.7); 18.4 (6.1~25.6); 12.2 (1.8~19.0); 92.4 (1.9~167.8); 9.1 (0.0~82.8); 673 (329~929) ❙ 958 (320~1736); 544 (193~1007); 36 (7~81) ❙ 43.3 (-14.2~145.8); 656 (320~888); 277.5 (61.6~551.3)

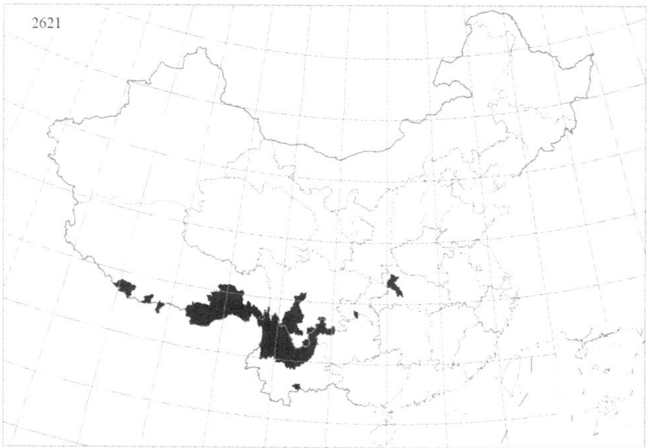

Saxifragaceae 虎耳草科 *Hydrangea* 绣球花属
Hydrangea hypoglauca Rehder 白背绣球
Shrubs, deciduous

12.3 (2.0~21.0); 2.8 (-7.2~14.2); 20.8 (9.9~28.1); 12.5 (3.8~21.0); 95.8 (15.4~191.8); 7.8 (0.0~51.4); 702 (413~1053) ❙ 1126 (731~1547); 555 (332~911); 55 (12~153) ❙ 62.9 (5.8~128.6); 700 (413~1053); 264.6 (147.3~551.6)

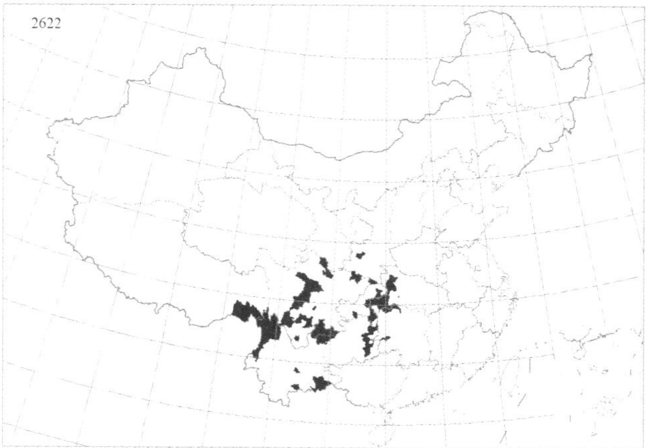

Saxifragaceae 虎耳草科 *Hydrangea* 绣球花属
Hydrangea integrifolia Hayata 全缘绣球
Climbing shrubs, deciduous

19.5 (16.2~24.1); 13.6 (11.0~18.9); 24.4 (20.1~28.0); 19.5 (16.2~24.1); 173.9 (134.2~229.7); 0.0 (0.0~0.0); 968 (773~1322) ❙ 2543 (1751~3495); 1045 (750~1594); 326 (47~897) ❙ 170.6 (32.5~246.5); 947 (773~1095); 248.3 (154.2~303.6)

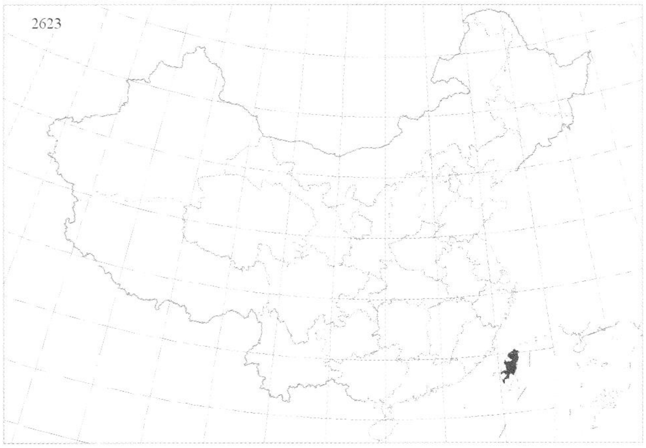

Saxifragaceae 虎耳草科 *Hydrangea* 绣球花属
Hydrangea kawakamii Hayata 蝶萼绣球
Climbing shrubs, deciduous

19.8; 14.0; 24.6; 19.8; 178.1; 0.0; 984 ❙ 2221; 1163; 158 ❙ 132.0; 970; 230.6

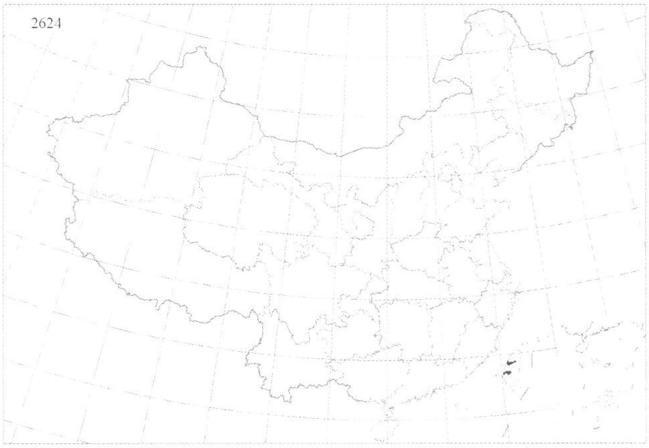

Saxifragaceae 虎耳草科 *Hydrangea* 绣球花属
Hydrangea kwangsiensis Hu 粤西绣球
Shrubs, deciduous

19.0 (14.6~22.6); 9.3 (3.8~14.9); 27.4 (24.6~29.8); 19.1 (14.7~22.6); 169.1 (117.7~211.7); 0.0 (0.0~1.5); 997 (779~1213) ❙ 1603 (1381~1876); 616 (474~891); 170 (82~221) ❙ 62.1 (20.8~106.4); 996 (779~1213); 232.6 (164.9~273.3)

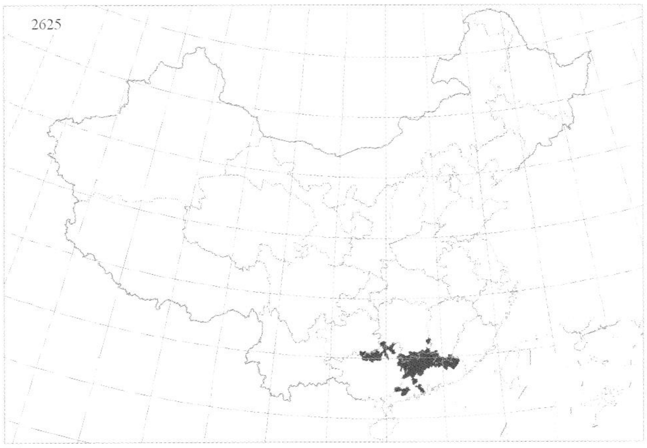

Saxifragaceae 虎耳草科 *Hydrangea* 绣球花属
Hydrangea kwangtungensis Merr. 广东绣球
Shrubs or subshrubs, deciduous

21.0 (18.2~22.7); 12.3 (8.7~15.0); 27.8 (25.9~28.6); 21.0 (18.3~22.8); 192.1 (159.3~213.1); 0.0 (0.0~0.0); 1104 (928~1221) ❙ 1718 (1302~2435); 827 (607~1428); 139 (86~194) ❙ 56.1 (12.9~99.5); 1104 (928~1221); 273.1 (228.5~327.7)

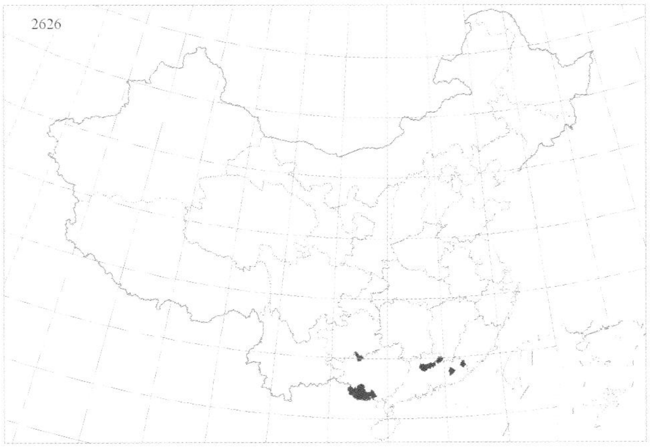

Saxifragaceae 虎耳草科 *Hydrangea* 绣球花属
Hydrangea lingii G. Hoo 狭叶绣球
Shrubs, deciduous

18.6 (16.6~22.0); 8.8 (5.9~13.5); 27.2 (25.2~28.8); 18.6 (16.6~22.0); 163.5 (138.9~203.8); 0.0 (0.0~0.0); 967 (864~1170) ❙ 1686 (1296~1997); 606 (474~778); 193 (99~254) ❙ 75.5 (40.1~130.7); 966 (864~1170); 258.7 (180.3~332.5)

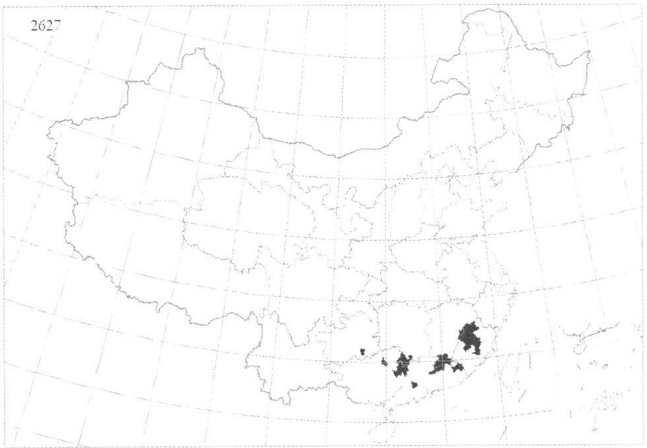

Saxifragaceae 虎耳草科 *Hydrangea* 绣球花属
Hydrangea linkweiensis Chun 临桂绣球
Shrubs, deciduous

16.1 (8.6~19.0); 4.9 (1.3~9.3); 26.5 (15.0~29.3); 16.2 (8.7~19.1); 134.5 (52.6~168.8); 0.7 (0.0~8.8); 865 (574~992) ❙ 1479 (1001~1997); 538 (444~709); 161 (35~259) ❙ 71.2 (28.8~130.7); 863 (574~986); 225.5 (170.4~384.7)

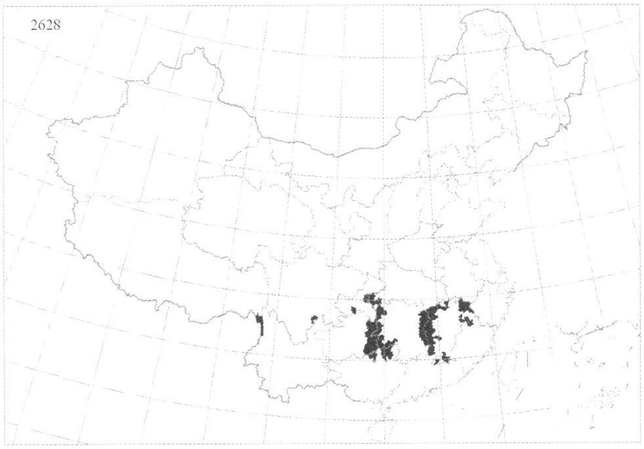

Saxifragaceae 虎耳草科 *Hydrangea* 绣球花属
Hydrangea longifolia Hayata 长叶绣球
Shrubs, deciduous

17.6; 12.6; 21.6; 17.6; 151.4; 0.0; 833 ❙ 2691; 1098; 298 ❙ 223.1; 833; 311.0

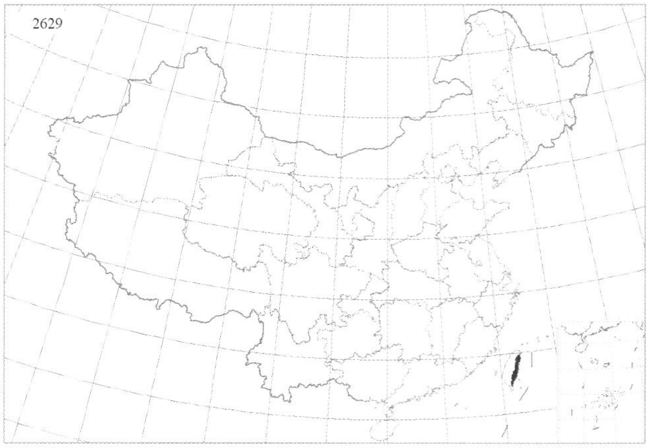

Saxifragaceae 虎耳草科 *Hydrangea* 绣球花属
Hydrangea longipes Franch. 莼兰绣球
Shrubs, deciduous

12.5 (2.2~17.7); 1.3 (-7.7~11.0); 23.0 (9.9~28.1); 12.7 (3.8~17.8); 100.1 (15.5~153.4); 9.7 (0.0~49.3); 725 (413~924) ❙ 1052 (528~1770); 467 (270~816); 58 (11~223) ❙ 45.7 (-16.6~125.9); 722 (413~924); 241.8 (141.3~491.5)

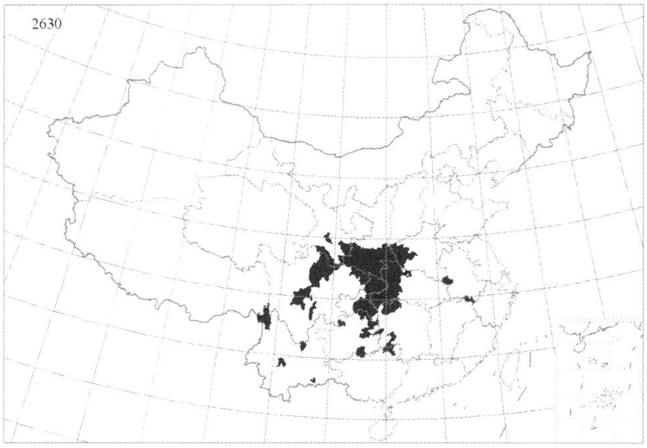

Saxifragaceae 虎耳草科 *Hydrangea* 绣球花属
Hydrangea longipes var. fulvescens (Rehder) W. T. Wang 绣毛绣球
Shrubs or small trees, deciduous

12.7; 0.9; 24.0; 12.7; 101.5; 8.7; 729 ❙ 967; 417; 53 ❙ 33.8; 729; 237.7

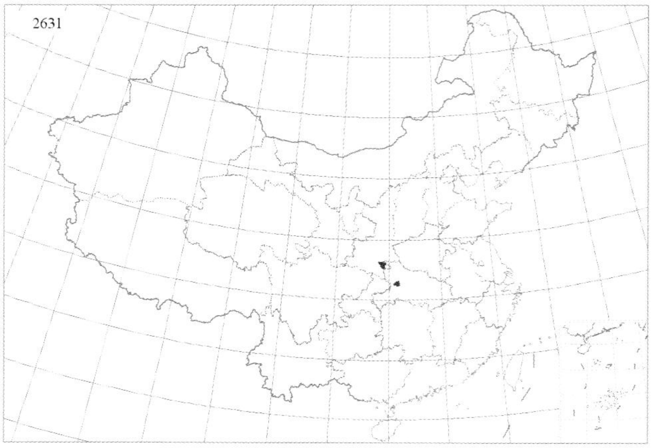

Saxifragaceae 虎耳草科 *Hydrangea* 绣球花属
Hydrangea longipes var. lanceolata Hemsley 披针绣球
Shrubs or small trees, deciduous

14.1 (12.6~16.9); 2.5 (1.0~4.8); 25.1 (23.5~28.1); 14.2 (12.6~16.9); 114.5 (99.9~143.4); 4.7 (0.2~8.6); 786 (724~924) ❙ 1102 (760~1267); 494 (382~542); 71 (17~94) ❙ 40.8 (5.0~67.3); 786 (724~924); 238.3 (166.0~294.3)

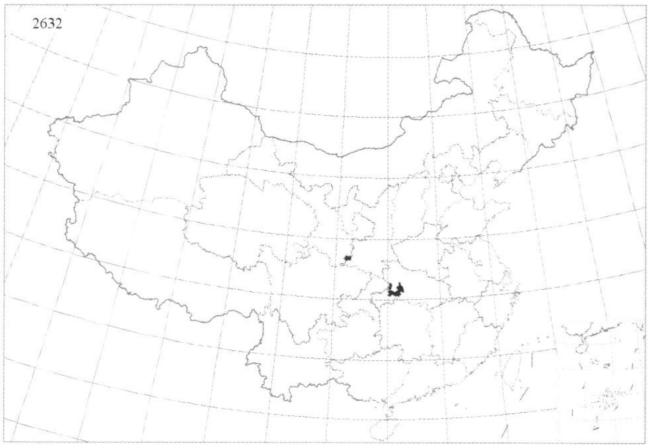

Saxifragaceae 虎耳草科 *Hydrangea* 绣球花属
Hydrangea macrocarpa Hand.-Mazz. 大果绣球
Shrubs or small trees, deciduous

12.8 (5.4~18.9); 5.4 (-2.1~12.0); 19.0 (12.3~27.3); 12.9 (5.8~19.0); 97.8 (31.5~167.8); 3.2 (0.0~25.6); 696 (481~929) ❙ 1007 (741~1466); 574 (411~651); 38 (12~134) ❙ 46.5 (-14.2~106.2); 673 (481~848); 300.5 (202.8~384.1)

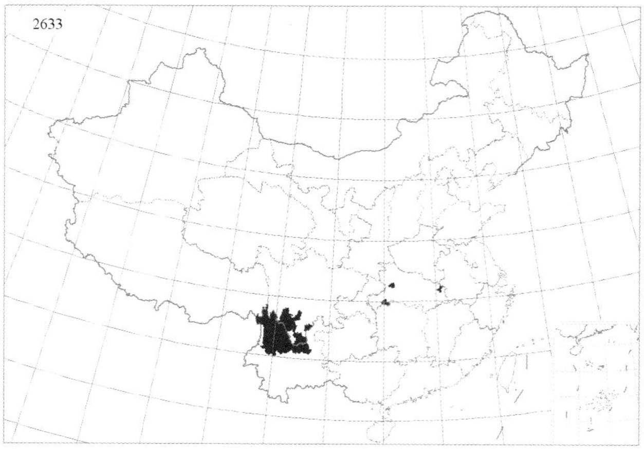

Saxifragaceae 虎耳草科 *Hydrangea* 绣球花属
Hydrangea mangshanensis C. F. Wei 莽山绣球
Shrubs, deciduous

18.3 (17.3~20.7); 8.1 (6.9~11.7); 27.3 (26.5~28.2); 18.3 (17.3~20.8); 159.7 (148.1~189.5); 0.0 (0.0~0.0); 953 (896~1094) ❙ 1565 (1486~1781); 554 (505~707); 178 (168~185) ❙ 64.4 (59.3~70.7); 953 (896~1094); 216.0 (188.8~244.8)

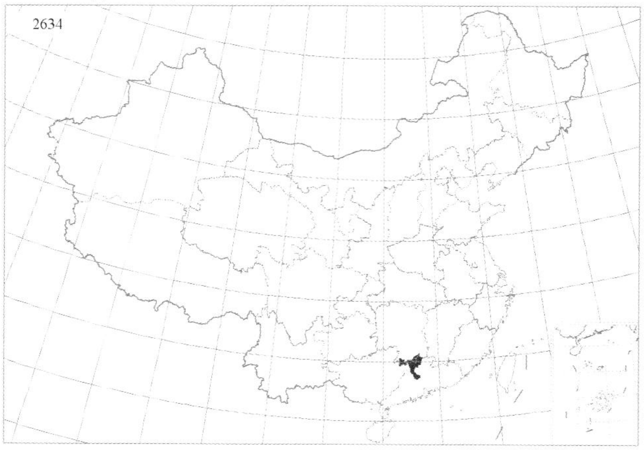

Saxifragaceae 虎耳草科 *Hydrangea* 绣球花属
Hydrangea paniculata Siebold 圆锥绣球
Shrubs or small trees, deciduous

17.0 (8.7~22.3); 6.2 (-4.6~15.3); 27.0 (19.6~30.1); 17.0 (9.6~22.3); 145.4 (73.2~208.1); 1.2 (0.0~28.1); 905 (619~1202) ❙ 1454 (557~3495); 545 (286~1045); 160 (10~897) ❙ 60.7 (-22.2~235.6); 892 (557~1202); 219.2 (123.9~335.1)

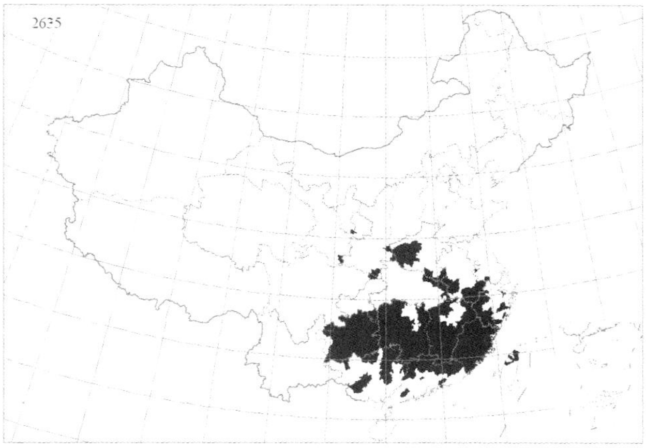

Saxifragaceae 虎耳草科 *Hydrangea* 绣球花属
Hydrangea robusta Hook. f. et Thomson 粗枝绣球
Shrubs or small trees, deciduous

14.7 (1.8~22.2); 6.1 (-7.7~15.1); 21.8 (9.9~29.3); 14.8 (3.8~22.3); 120.2 (15.5~207.5); 3.6 (0.0~53.3); 773 (404~1190) ❙ 1150 (527~1776); 590 (306~996); 59 (9~221) ❙ 49.6 (-14.2~125.9); 763 (404~1190); 294.5 (118.7~659.0)

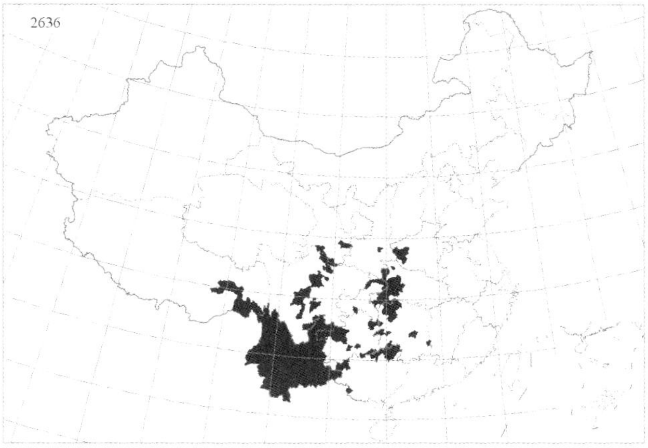

Saxifragaceae 虎耳草科 *Hydrangea* 绣球花属
Hydrangea sargentiana Rehder 紫彩绣球
Shrubs, deciduous

12.0; 0.6; 23.1; 12.1; 94.7; 9.7; 697 ❙ 1118; 482; 75 ❙ 60.4; 697; 247.3

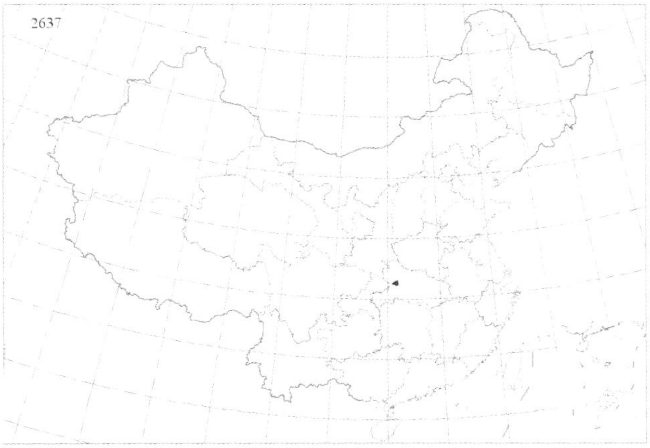

Saxifragaceae 虎耳草科 *Hydrangea* 绣球花属
Hydrangea stenophylla Merr. et Chun 柳叶绣球
Shrubs, deciduous

17.9 (14.3~22.0); 7.7 (4.0~13.5); 26.9 (23.5~28.8); 18.0 (14.3~22.0); 155.8 (112.6~203.8); 0.1 (0.0~1.1); 941 (756~1170) ❙ 1537 (1211~1745); 576 (496~726); 164 (89~221) ❙ 64.3 (29.9~106.4); 941 (756~1170); 228.2 (188.8~261.3)

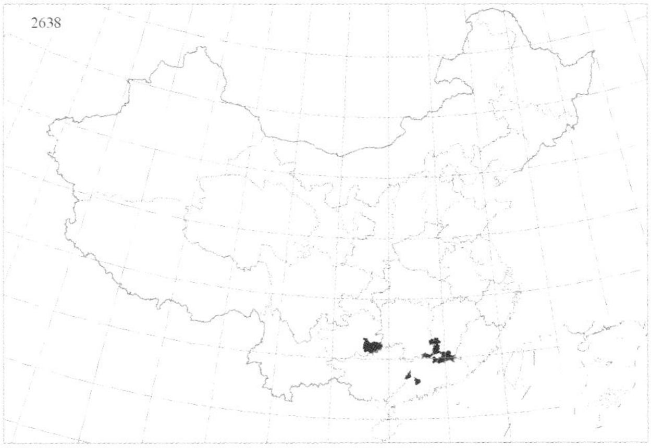

Saxifragaceae 虎耳草科 *Hydrangea* 绣球花属
Hydrangea strigosa Rehder 蜡莲绣球
Shrubs, deciduous

15.5 (2.2~21.0); 5.1 (-7.7~14.2); 25.0 (9.9~29.9); 15.6 (3.8~21.1); 129.1 (15.5~193.0); 2.6 (0.0~49.3); 832 (413~1114) ❙ 1268 (626~1997); 535 (306~996); 109 (14~259) ❙ 52.0 (-22.2~147.0); 822 (413~1114); 236.7 (118.7~551.6)

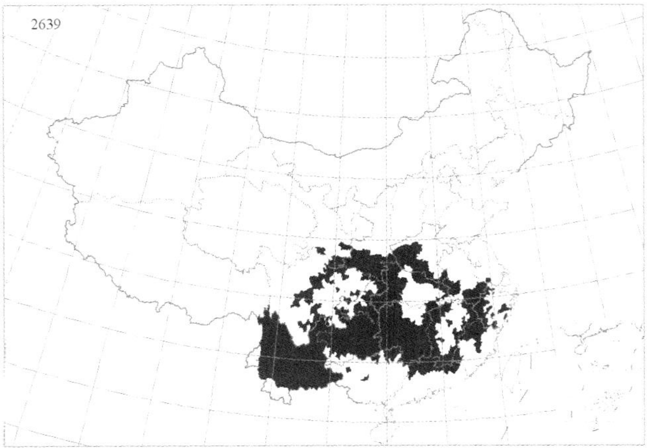

Saxifragaceae 虎耳草科 *Hydrangea* 绣球花属
Hydrangea stylosa Hook. f. et Thomson 长柱绣球
Shrubs, deciduous

14.8 (8.6~19.1); 8.0 (1.3~13.0); 20.0 (15.0~23.3); 14.9 (8.7~19.1); 119.9 (52.6~169.7); 1.4 (0.0~8.8); 756 (574~917) ❙ 1130 (944~1428); 639 (520~835); 47 (33~56) ❙ 50.1 (26.1~81.5); 750 (574~917); 383.5 (249.8~506.2)

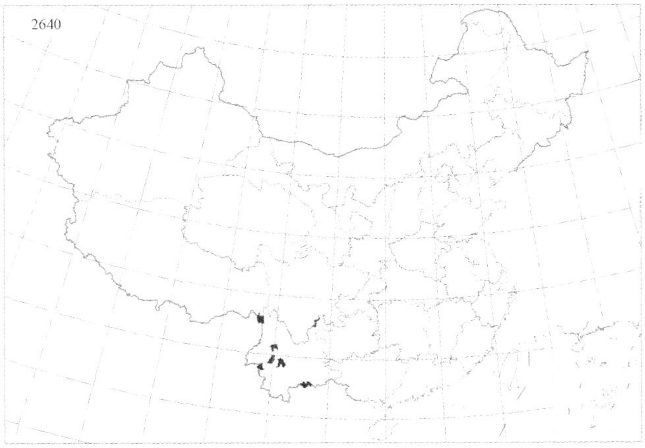

Saxifragaceae 虎耳草科 *Hydrangea* 绣球花属
Hydrangea sungpanensis Hand.-Mazz. 松潘绣球
Shrubs or small trees, deciduous

16.3 (2.6~21.0); 9.4 (-7.0~14.2); 21.3 (11.2~25.6); 16.4 (4.3~21.0); 138.6 (20.0~191.8); 2.0 (0.0~48.2); 815 (427~1053) ❙ 1251 (829~1751); 702 (400~996); 49 (15~84) ❙ 54.4 (12.5~95.9); 805 (427~1053); 385.4 (208.2~551.6)

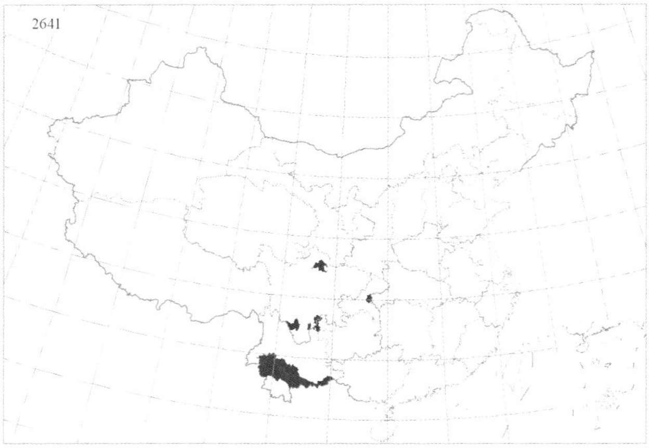

Saxifragaceae 虎耳草科 *Hydrangea* 绣球花属
Hydrangea xanthoneura Diels 挂苦绣球
Shrubs or small trees, deciduous

10.9 (1.6~19.2); 0.0 (-9.3~13.3); 20.9 (9.9~29.1); 11.3 (3.8~19.4); 86.4 (15.4~172.6); 15.2 (0.0~58.4); 674 (413~932) ❙ 938 (341~1547); 468 (190~911); 39 (8~165) ❙ 41.8 (-40.9~128.6); 657 (341~932); 249.8 (61.2~551.3)

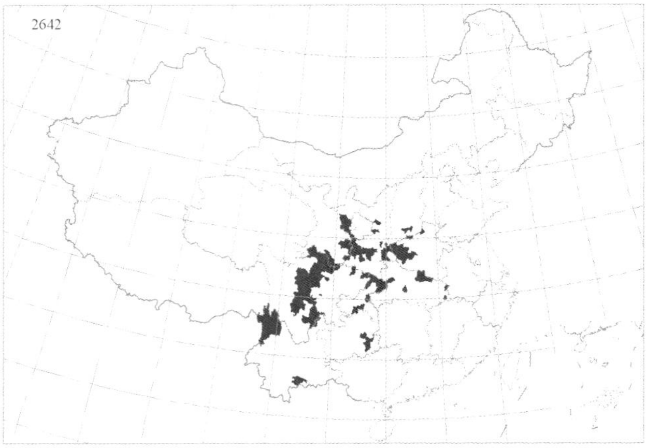

Saxifragaceae 虎耳草科 *Hydrangea* 绣球花属
Hydrangea zhewanensis P. S. Hsu et X. P. Zhang 浙皖绣球
Shrubs, deciduous

15.8 (14.7~16.8); 4.1 (2.5~5.6); 27.5 (26.8~28.8); 15.8 (14.8~16.8); 131.2 (121.1~141.7); 1.3 (0.0~3.9); 860 (815~912) ❙ 1459 (1296~1663); 536 (486~595); 189 (177~212) ❙ 70.0 (46.2~101.3); 860 (815~912); 238.6 (198.0~284.0)

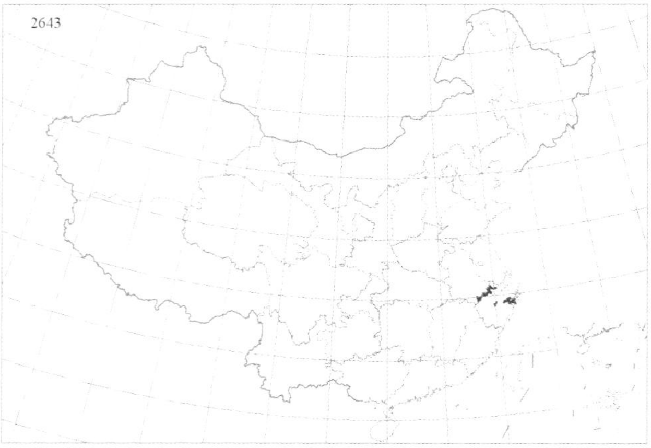

Saxifragaceae 虎耳草科 *Itea* 鼠刺属
Itea amoena Chun 秀丽鼠刺
Shrubs, evergreen

21.9 (21.3~22.7); 13.9 (13.2~15.0); 27.9 (27.3~28.6); 22.0 (21.3~22.8); 203.6 (195.9~213.1); 0.0 (0.0~0.0); 1162 (1111~1221) ❙ 1876 (1302~2435); 1047 (670~1428); 116 (86~130) ❙ 61.1 (12.9~99.5); 1162 (1111~1221); 289.6 (235.1~318.8)

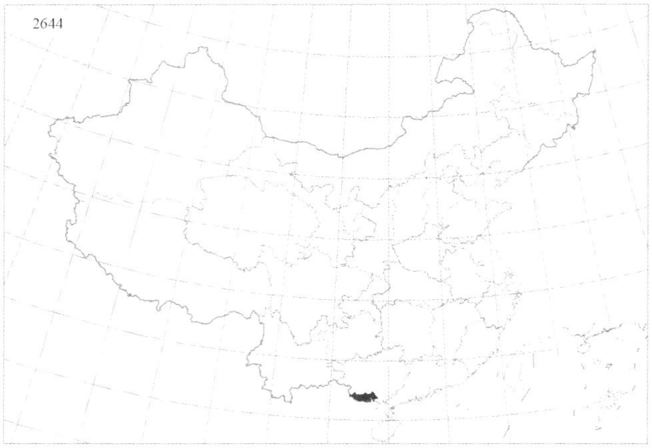

Saxifragaceae 虎耳草科 *Itea* 鼠刺属
Itea chinensis Hook. et Arn. 鼠刺
Shrubs or small trees, evergreen

17.7 (8.6~24.2); 8.3 (1.3~18.3); 25.8 (15.0~30.1); 17.7 (8.7~24.2); 153.0 (52.6~230.4); 0.2 (0.0~8.8); 922 (574~1331) ❙ 1456 (797~2435); 607 (299~1428); 137 (23~259) ❙ 58.5 (-34.3~145.8); 906 (574~1331); 267.4 (139.6~659.0)

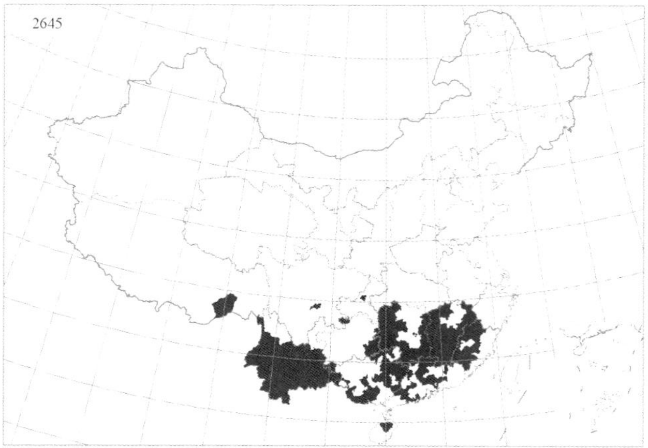

Saxifragaceae 虎耳草科 *Itea* 鼠刺属
Itea coriacea Y. C. Wu 厚叶鼠刺
Shrubs or small trees

18.9 (14.3~22.4); 9.1 (3.3~14.8); 27.2 (23.5~28.9); 19.0 (14.3~22.4); 167.6 (112.6~208.9); 0.1 (0.0~2.3); 996 (756~1202) ❙ 1580 (1211~2030); 655 (476~1033); 160 (72~254) ❙ 60.5 (20.3~130.7); 996 (756~1202); 229.8 (157.7~335.1)

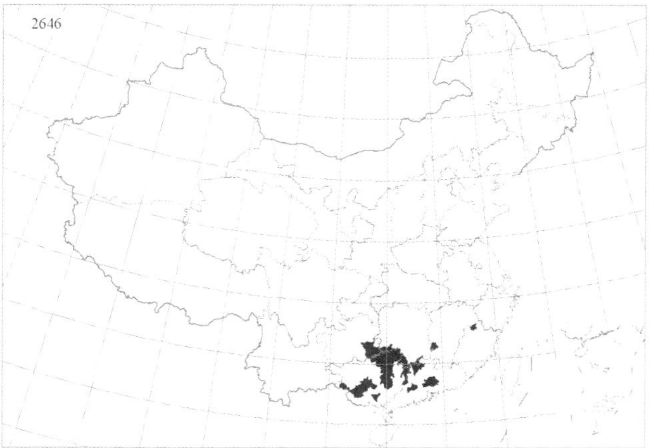

Saxifragaceae 虎耳草科 *Itea* 鼠刺属
Itea glutinosa Hand.-Mazz. 腺鼠刺
Shrubs or small trees

17.2 (14.3~21.6); 6.7 (3.3~12.6); 26.7 (23.5~28.7); 17.3 (14.3~21.7); 147.6 (112.6~199.8); 0.3 (0.0~2.3); 906 (756~1153) | 1490 (1061~1777); 564 (419~809); 156 (59~205) | 64.8 (10.5~101.8); 905 (756~1153); 224.6 (128.7~300.5)

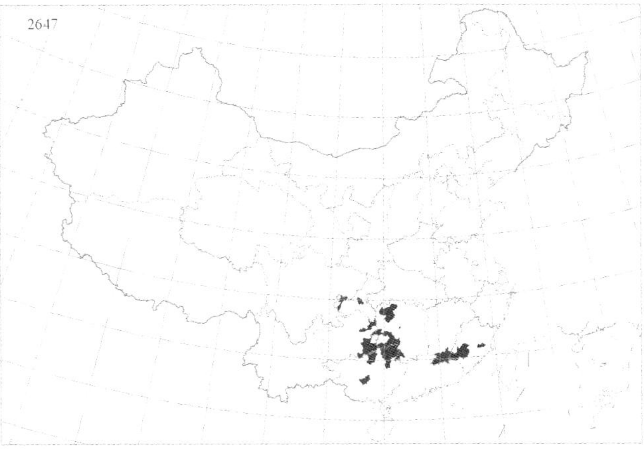

Saxifragaceae 虎耳草科 *Itea* 鼠刺属
Itea ilicifolia Oliv. 冬青叶鼠刺
Shrubs

14.4 (2.2~19.1); 4.4 (-6.3~10.7); 23.5 (9.5~28.1); 14.5 (3.9~19.2); 116.3 (15.8~170.1); 3.0 (0.0~48.6); 774 (411~981) | 1165 (729~1492); 546 (349~723); 62 (11~107) | 51.9 (4.4~121.7); 771 (411~981); 225.0 (146.0~352.3)

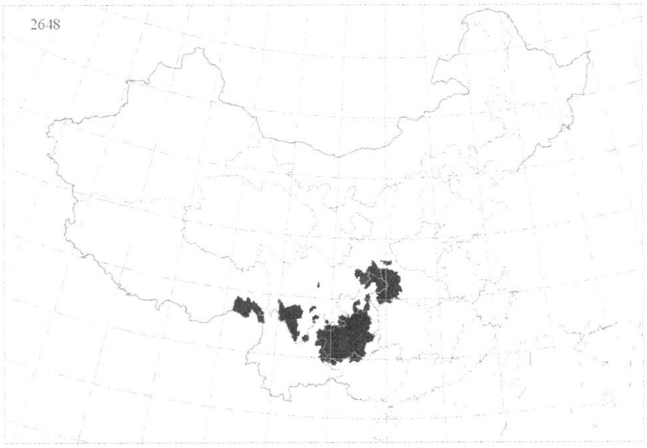

Saxifragaceae 虎耳草科 *Itea* 鼠刺属
Itea indochinensis Merr. 毛鼠刺
Shrubs or small trees

13.5 (8.6~22.2); 9.7 (1.3~14.2); 25.6 (15.0~28.7); 18.6 (8.7~22.3); 163.2 (52.6~207.5); 0.2 (0.0~8.8); 951 (574~1190) | 1435 (1041~2005); 690 (453~996); 108 (44~191) | 52.0 (15.4~101.8); 950 (574~1190); 292.0 (123.1~551.6)

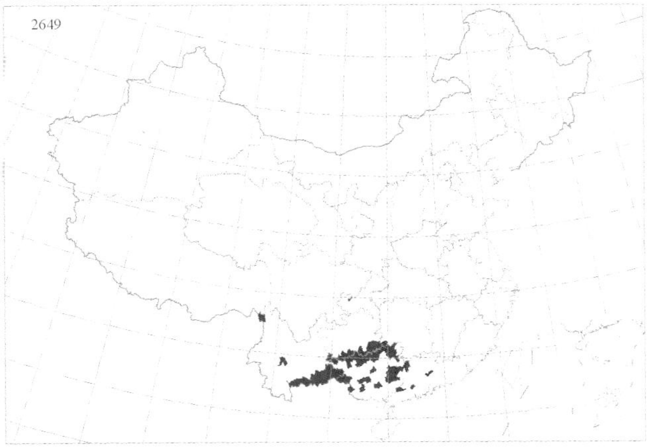

Saxifragaceae 虎耳草科 *Itea* 鼠刺属
Itea indochinensis var. pubinervia (H. T. Chang) C. Y. Wu 毛脉鼠刺
Shrubs or small trees

18.8 (8.6~23.2); 11.3 (1.3~15.8); 24.3 (15.0~28.9); 18.8 (8.7~23.2); 165.9 (52.6~218.9); 0.3 (0.0~8.8); 947 (574~1257) | 1462 (1041~2035); 757 (582~996); 79 (45~191) | 54.8 (32.2~95.5); 947 (574~1257); 346.2 (194.5~551.6)

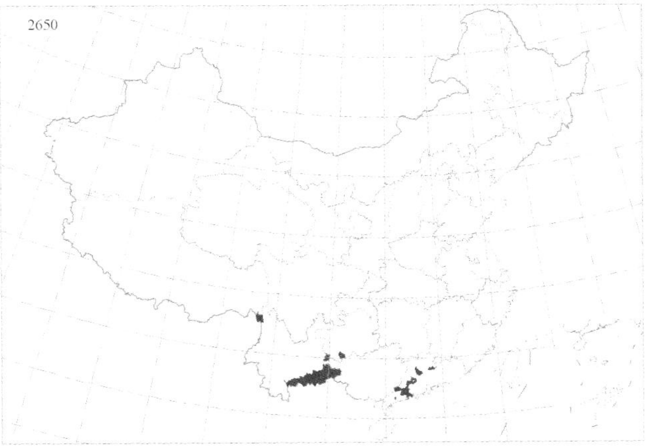

Saxifragaceae 虎耳草科 *Itea* 鼠刺属
Itea kiukiangensis C. C. Huang et S. C. Huang ex H. Chuang 俅江鼠刺
Trees

12.0 (5.3~16.7); 5.1 (-2.4~9.7); 17.7 (12.2~22.1); 12.1 (5.7~16.7); 88.6 (30.5~140.9); 4.1 (0.0~26.2); 666 (479~819) | 1049 (741~1736); 606 (426~1007); 39 (15~69) | 59.3 (14.4~145.8); 651 (479~748); 334.8 (234.2~447.4)

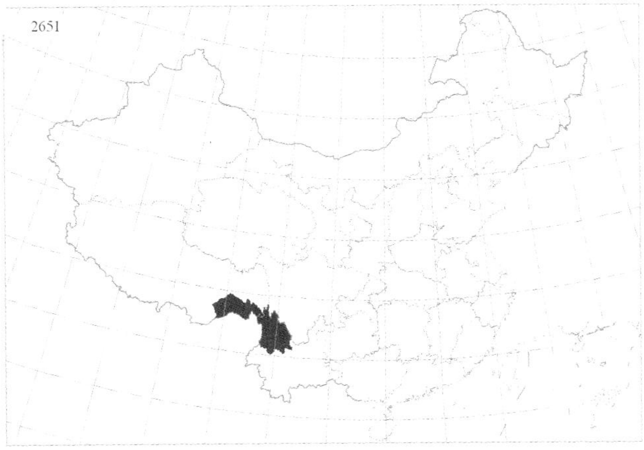

Saxifragaceae 虎耳草科 *Itea* 鼠刺属
Itea kwangsiensis H. T. Chang 子农鼠刺
Shrubs

19.2; 9.2; 27.5; 19.2; 170.8; 0.0; 1002 | 1495; 678; 135 | 49.1; 1002; 256.4

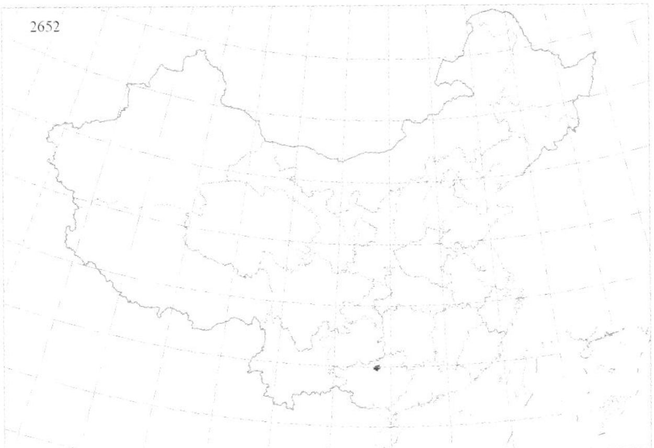

Saxifragaceae 虎耳草科 *Itea* 鼠刺属

Itea macrophylla Wall. 大叶鼠刺

Small trees

19.1 (14.3~24.8); 11.9 (4.0~20.1); 24.3 (20.3~28.9); 19.1 (14.3~24.8); 169.8 (112.6~237.6); 0.0 (0.0~1.1); 962 (756~1385) ❙ 1366 (868~1751); 697 (312~996); 70 (41~194) ❙ 45.2 (-34.3~112.2); 931 (756~1254); 385.1 (189.5~659.0)

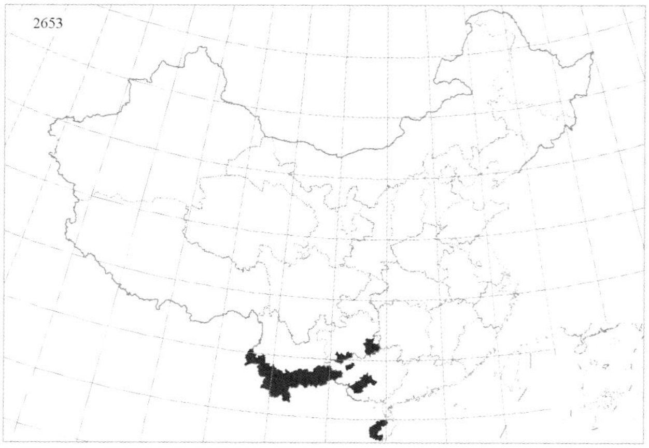

Saxifragaceae 虎耳草科 *Itea* 鼠刺属

Itea oldhamii C. K. Schneid. 台湾鼠刺

Shrubs, evergreen

19.9; 13.6; 25.9; 20.0; 179.9; 0.0; 998 ❙ 3036; 887; 599 ❙ 205.7; 998; 248.1

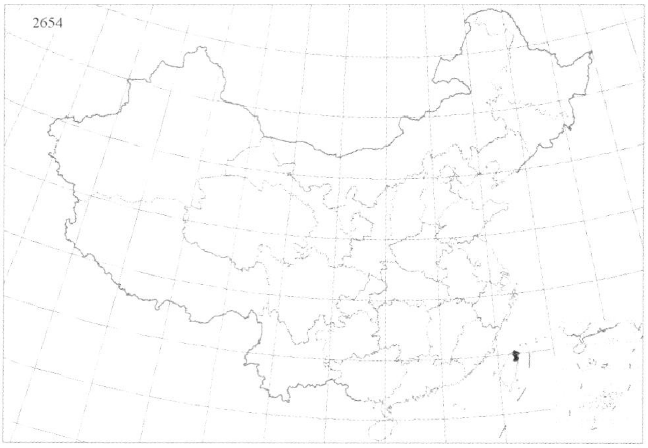

Saxifragaceae 虎耳草科 *Itea* 鼠刺属

Itea omeiensis C. K. Schneid. 峨眉鼠刺

Shrubs or small trees, deciduous

18.0 (8.6~23.6); 7.8 (1.3~16.7); 27.2 (15.0~29.8); 18.0 (8.7~23.6); 156.7 (52.6~223.5); 0.3 (0.0~8.8); 953 (574~1281) ❙ 1517 (871~2435); 595 (299~1428); 153 (20~259) ❙ 60.2 (-6.9~147.0); 945 (574~1273); 224.9 (95.3~551.6)

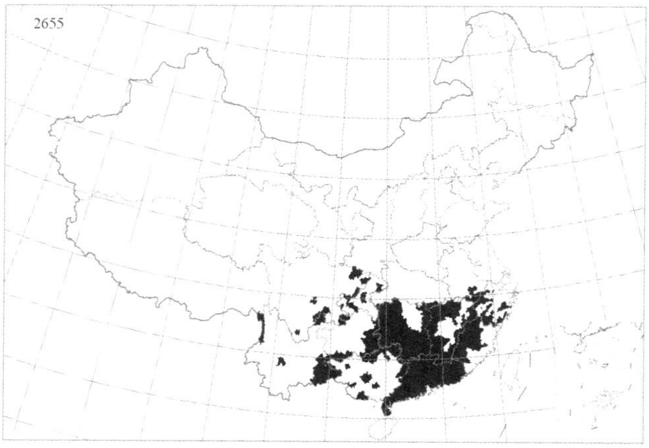

Saxifragaceae 虎耳草科 *Itea* 鼠刺属

Itea parviflora Hemsl. 小花鼠刺

Shrubs or small trees, evergreen

19.8 (17.1~24.1); 14.3 (11.2~18.9); 24.3 (21.7~28.0); 19.9 (17.1~24.1); 178.3 (145.0~229.7); 0.0 (0.0~0.0); 984 (816~1322) ❙ 2516 (1751~2993); 1207 (900~1847); 232 (47~547) ❙ 164.8 (32.5~235.6); 954 (816~1123); 262.6 (154.2~363.7)

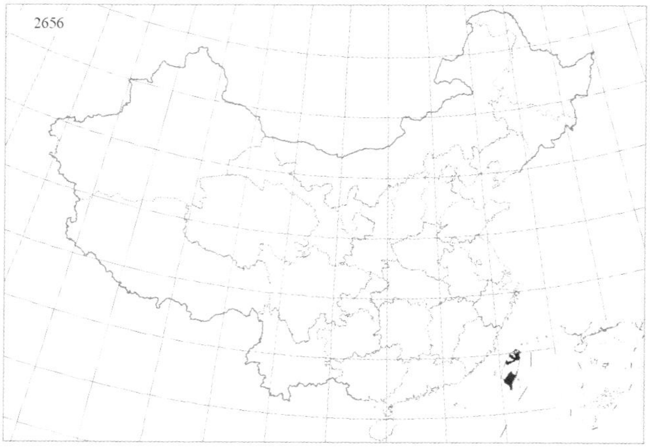

Saxifragaceae 虎耳草科 *Itea* 鼠刺属

Itea riparia Collett et Hemsl. 河岸鼠刺

Shrubs

18.3 (16.3~20.5); 11.4 (9.0~15.1); 23.0 (21.4~25.0); 18.3 (16.4~20.5); 159.7 (136.4~185.9); 0.0 (0.0~0.0); 887 (797~992) ❙ 1377 (1018~1751); 742 (554~996); 57 (47~77) ❙ 54.9 (25.7~88.8); 882 (797~986); 425.9 (277.0~659.0)

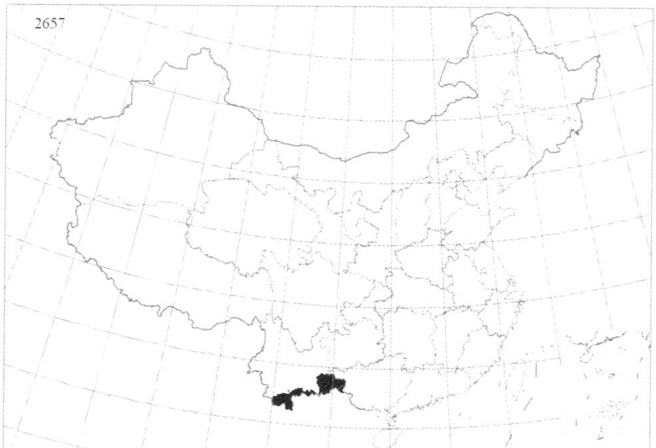

Saxifragaceae 虎耳草科 *Itea* 鼠刺属

Itea yangchunensis S. Y. Jin 阳春鼠刺

Shrubs or small trees

21.8; 14.2; 27.8; 21.9; 202.6; 0.0; 1150 ❙ 1876; 891; 116 ❙ 63.2; 1150; 273.3

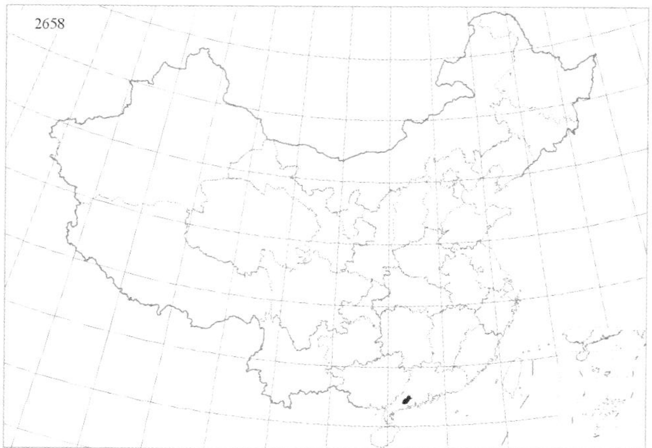

Saxifragaceae 虎耳草科　*Itea* 鼠刺属
Itea yunnanensis Franch. 滇鼠刺
Shrubs or small trees

15.2 (2.2~21.0); 7.7 (−6.3~14.2); 21.0 (9.5~27.3); 15.2 (3.9~21.0); 124.1 (15.8~19[illegible].8); 1.5 (0.0~48.6); 775 (411~1053) ❙ 1101 (729~1751); 614 (411~996); 44 (11~107) ❙ 43.0 (−14.2~121.7); 759 (411~1053); 304.5 (142.9~551.6)

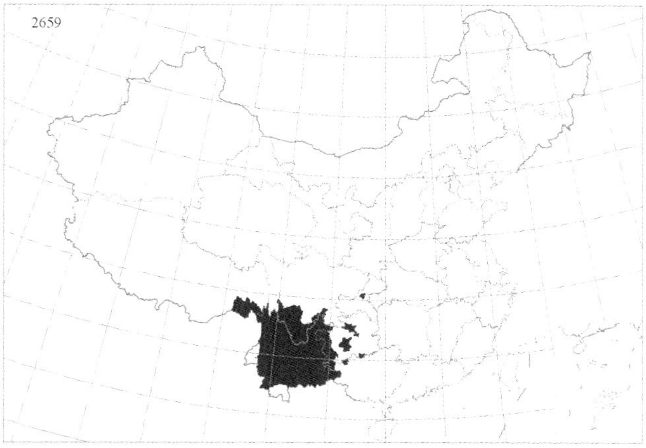

Saxifragaceae 虎耳草科　*Philadelphus* 山梅花属
Philadelphus brachybotrys (Koehne) Koehne 短序山梅花
Shrubs, deciduous

15.6 (9.4~18.2); 4.0 (−2.5~8.4); 26.9 (21.0~29.1); 15.7 (9.8~18.3); 130.1 (73.9~159.1); 2.2 (0.0~20.6); 851 (619~938) ❙ 1551 (892~1893); 561 (389~677); 188 (29~242) ❙ 82.6 (40.8~129.9); 846 (619~938); 241.8 (181.3~311.0)

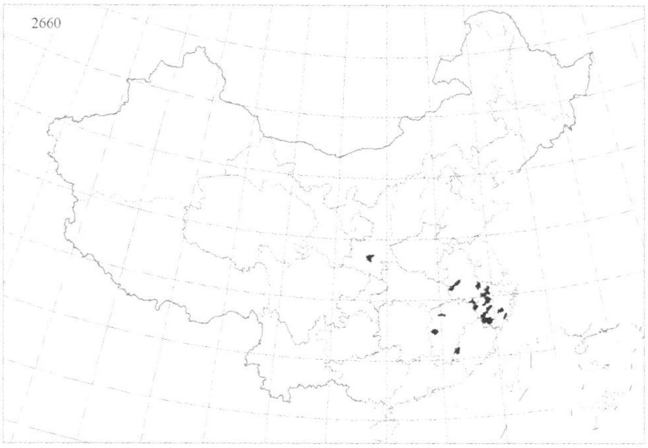

Saxifragaceae 虎耳草科　*Philadelphus* 山梅花属
Philadelphus calvescens (Rehder) S. M. Hwang 丽江山梅花
Shrubs. deciduous

10.6 (5.4~16.9); 3.5 (−2.1~10.3); 16.4 (12.3~21.4); 10.7 (5.8~16.9); 74.4 (31.5~142.7); 7.2 (0.0~25.6); 626 (481~813) ❙ 971 (741~1178); 572 (426~651); 33 (15~51) ❙ 55.9 (36.1~78.3); 623 (481~813); 318.5 (273.5~383.4)

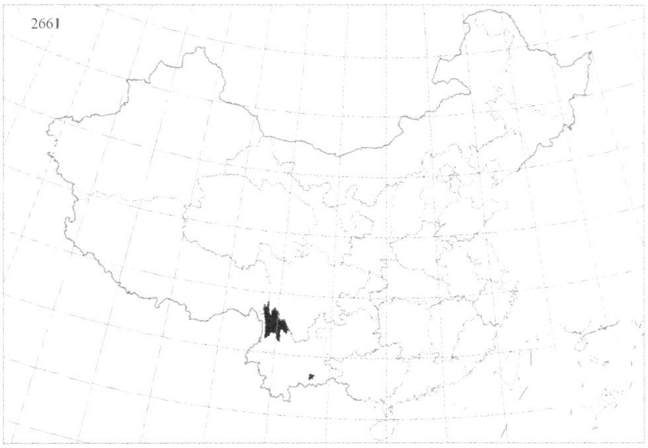

Saxifragaceae 虎耳草科　*Philadelphus* 山梅花属
Philadelphus caudatus S. M. Hwang 尾萼山梅花
Shrubs, deciduous

14.6; 7.9; 19.6; 14.6; 115.8; 0.0; 736 ❙ 1009; 596; 39 ❙ 37.1; 732; 278.4

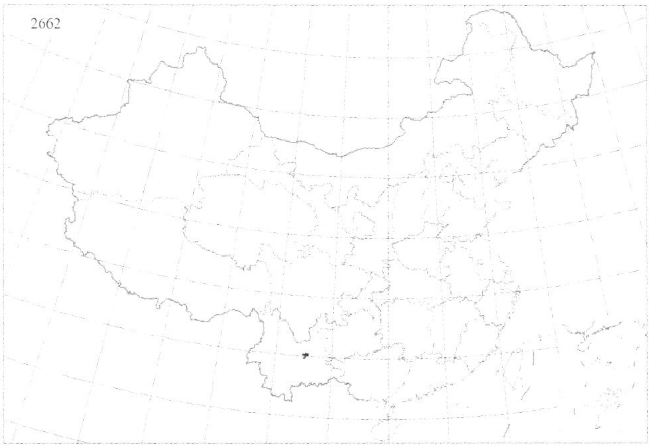

Saxifragaceae 虎耳草科　*Philadelphus* 山梅花属
Philadelphus dasycalyx (Rehder) S. Y. Hu 毛萼山梅花
Shrubs, deciduous

8.8 ([illegible].5~14.2); −6.7 (−13.7~3.3); 22.6 (11.8~27.8); 10.2 (3.9~14.4); 82.2 (19.7~124.3); 35.8 (2.1~73.3); 659 (406~859) ❙ 525 (341~1265); 313 (190~519); 16 (8~65) ❙ −19.7 (−40.9~63.0); 517 (341~776); 156.9 (61.2~318.2)

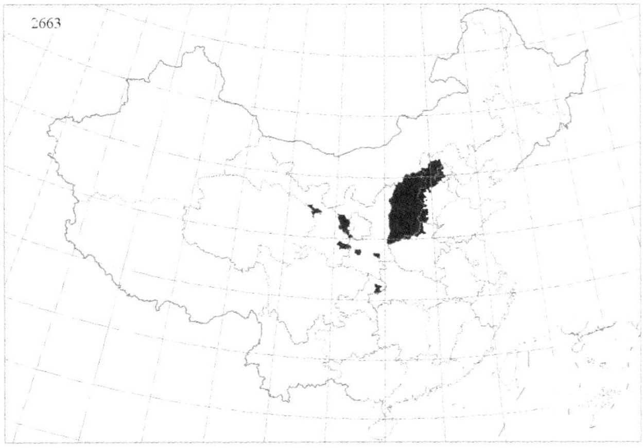

Saxifragaceae 虎耳草科　*Philadelphus* 山梅花属
Philadelphus delavayi L. Henry 云南山梅花
Shrubs, deciduous

8.2 (−1.1~15.0); 0.5 (−9.1~8.2); 14.7 (6.2~20.0); 8.8 (2.0~15.0); 58.3 (2.5~119.9); 19.7 (0.0~75.6); 563 (334~748) ❙ 932 (321~1736); 541 (193~1007); 32 (9~69) ❙ 64.1 (−4.0~145.8); 561 (321~748); 296.6 (79.3~447.4)

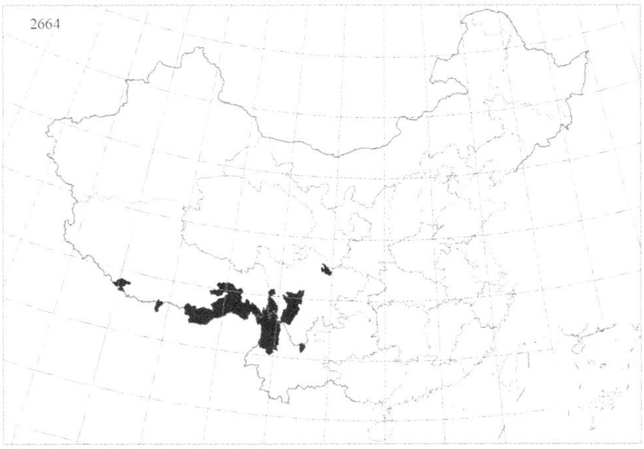

Saxifragaceae 虎耳草科 *Philadelphus* 山梅花属
Philadelphus delavayi var. melanocalyx Lemoine ex L. Henry 黑萼山梅花
Shrubs, deciduous

8.5; 1.2; 14.7; 8.6; 53.5; 11.5; 565 ❙ 884; 545; 23 ❙ 57.1; 565; 301.3

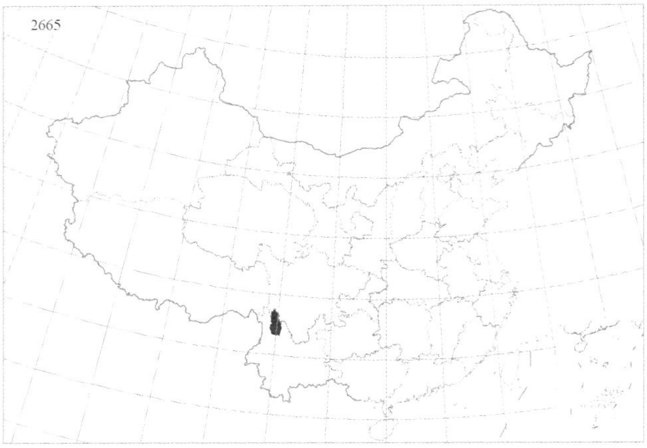

Saxifragaceae 虎耳草科 *Philadelphus* 山梅花属
Philadelphus delavayi var. trichocladus Hand.-Mazz. 毛枝山梅花
Shrubs, deciduous

6.9 (5.3~9.6); -0.6 (-2.4~2.4); 13.5 (12.2~15.7); 7.2 (5.7~9.6); 41.7 (30.5~60.8); 17.9 (5.4~26.2); 524 (479~600) ❙ 902 (741~1041); 532 (426~585); 29 (12~45) ❙ 72.2 (54.0~100.7); 524 (479~600); 301.0 (234.2~352.3)

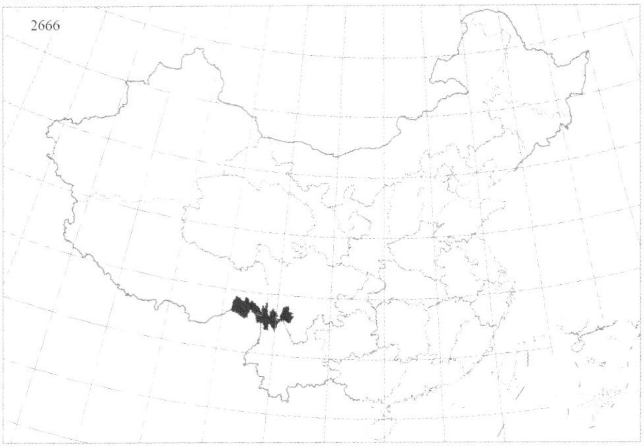

Saxifragaceae 虎耳草科 *Philadelphus* 山梅花属
Philadelphus henryi Koehne 滇南山梅花
Shrubs, deciduous

13.6 (2.0~18.9); 6.0 (-7.2~12.5); 19.7 (9.9~26.6); 13.7 (3.8~19.0); 107.9 (15.4~167.8); 4.4 (0.0~51.4); 722 (413~929) ❙ 986 (741~1410); 560 (411~816); 36 (15~71) ❙ 39.9 (-14.2~128.6); 701 (413~912); 280.9 (142.9~491.5)

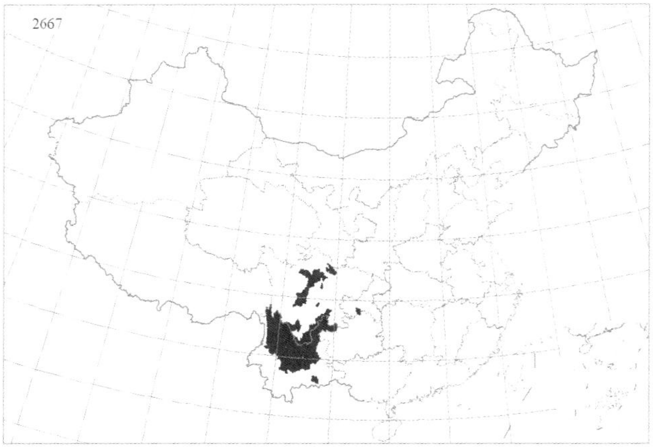

Saxifragaceae 虎耳草科 *Philadelphus* 山梅花属
Philadelphus incanus Koehne 山梅花
Shrubs, deciduous

10.8 (-1.3~17.7); -2.2 (-13.5~8.5); 22.9 (8.5~29.3); 11.5 (2.7~17.7); 91.7 (8.3~152.6); 21.0 (0.0~86.3); 699 (359~941) ❙ 797 (373~1828); 374 (201~685); 44 (5~253) ❙ 12.9 (-37.6~108.0); 646 (359~941); 202.5 (70.3~334.4)

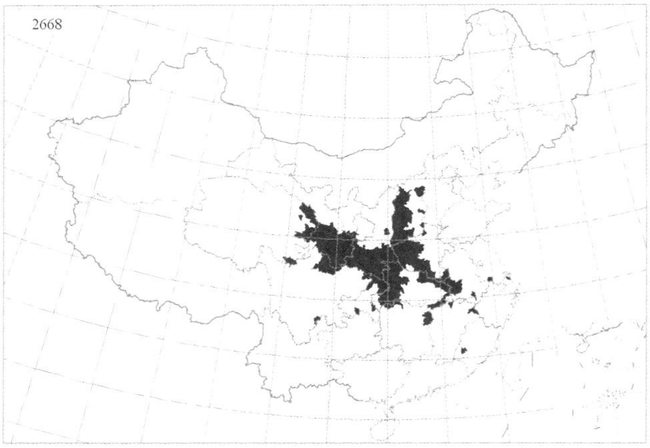

Saxifragaceae 虎耳草科 *Philadelphus* 山梅花属
Philadelphus incanus var. mitsai (S. Y. Hu) S. M. Hwang 米柴山梅花
Shrubs, deciduous

8.1; -3.2; 18.1; 9.2; 66.2; 28.5; 591 ❙ 770; 430; 20 ❙ 22.4; 552; 217.5

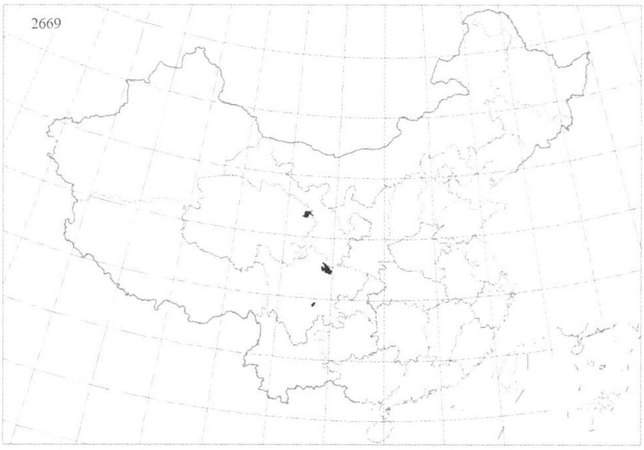

Saxifragaceae 虎耳草科 *Philadelphus* 山梅花属
Philadelphus kansuensis (Rehder) S. Y. Hu 甘肃山梅花
Shrubs, deciduous

4.8 (0.6~15.1); -6.4 (-12.0~4.4); 14.8 (9.9~25.4); 6.5 (3.8~15.2); 41.1 (15.4~122.8); 42.4 (0.6~72.5); 501 (413~817) ❙ 671 (443~950); 359 (239~546); 15 (7~25) ❙ 37.2 (-14.6~128.6); 489 (413~817); 230.5 (101.3~289.4)

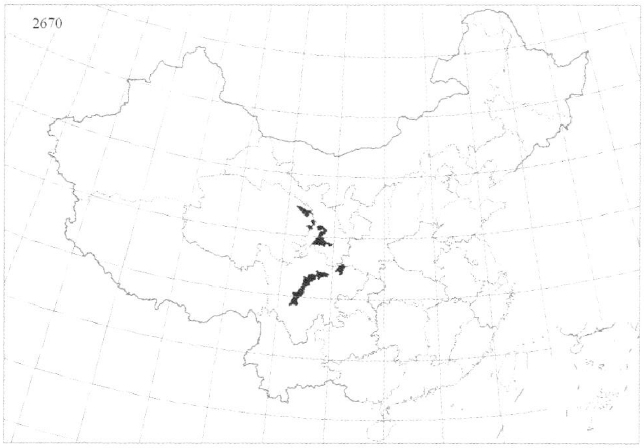

Saxifragaceae 虎耳草科 *Philadelphus* 山梅花属
Philadelphus kunmingensis S. M. Hwang 昆明山梅花
Shrubs, deciduous

14.6; 7.9; 19.6; 14.6; 115.8; 0.0; 736 ❙ 1009; 596; 39 ❙ 37.1; 732; 278.4

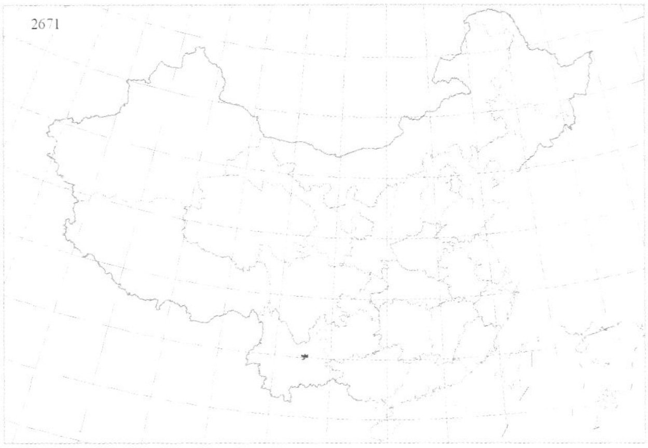

Saxifragaceae 虎耳草科 *Philadelphus* 山梅花属
Philadelphus laxiflorus Rehder 疏花山梅花
Shrubs, deciduous

11.3 (6.0~16.2); -1.9 (-10.1~3.7); 23.6 (18.3~28.1); 11.8 (7.6~16.2); 94.7 (53.8~136.2); 18.9 (1.3~50.4); 709 (539~891) ❙ 849 (447~1495); 388 (243~601); 53 (10~175) ❙ 17.2 (-22.1~71.2); 685 (447~891); 226.9 (102.7~320.6)

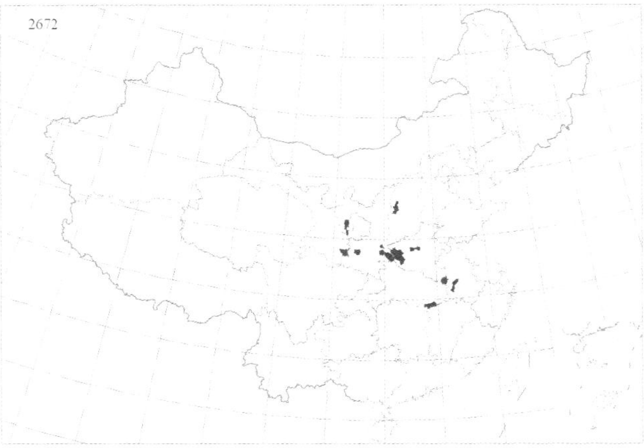

Saxifragaceae 虎耳草科 *Philadelphus* 山梅花属
Philadelphus lushuiensis T. C. Ku et S. M. Hwang 泸水山梅花
Shrubs, deciduous

13.2; 5.4; 18.4; 13.3; 99.4; 0.0; 695 ❙ 1337; 727; 69 ❙ 92.4; 695; 407.1

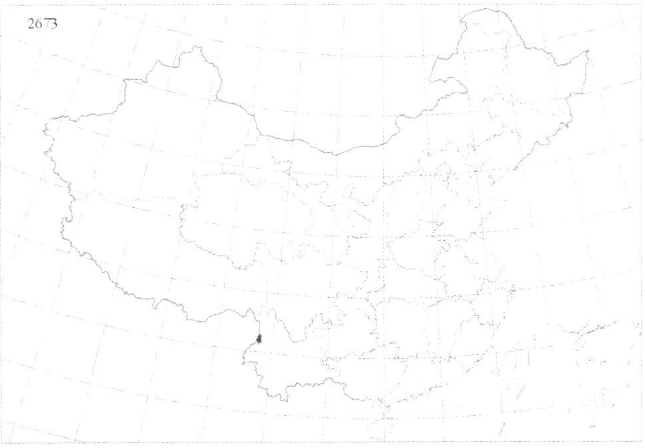

Saxifragaceae 虎耳草科 *Philadelphus* 山梅花属
Philadelphus pekinensis Rupr. 太平花
Shrubs, deciduous

10.4 (2.5~15.8); -4.3 (-15.9~2.9); 23.7 (12.4~28.2); 11.4 (4.8~15.8); 92.4 (25.1~133.1); 26.9 (2.8~80.4); 704 (441~891) ❙ 634 (341~1489); 349 (190~565); 23 (7~196) ❙ -9.6 (-47.7~83.5); 596 (341~891); 188.7 (58.7~334.4)

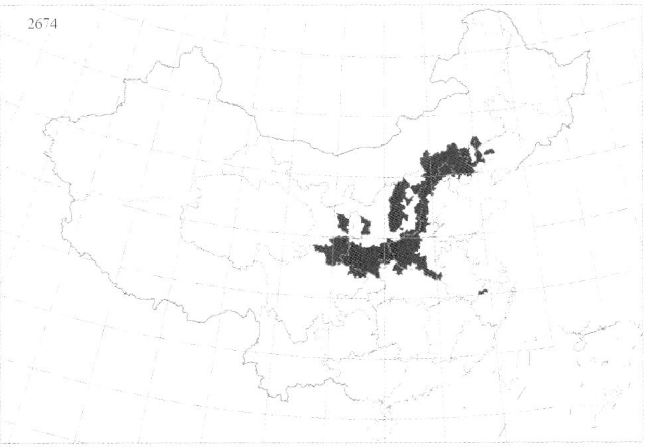

Saxifragaceae 虎耳草科 *Philadelphus* 山梅花属
Philadelphus purpurascens (Koehne) Rehder 紫萼山梅花
Shrubs, deciduous

11.4 (0.3~18.9); 3.8 (-9.9~12.5); 17.6 (9.2~25.3); 11.8 (3.3~19.0); 89.9 (11.9~167.8); 12.4 (0.0~67.5); 661 (391~929) ❙ 959 (579~1414); 549 (362~817); 33 (9~88) ❙ 50.4 (-14.2~128.6); 641 (391~851); 290.4 (171.0~407.1)

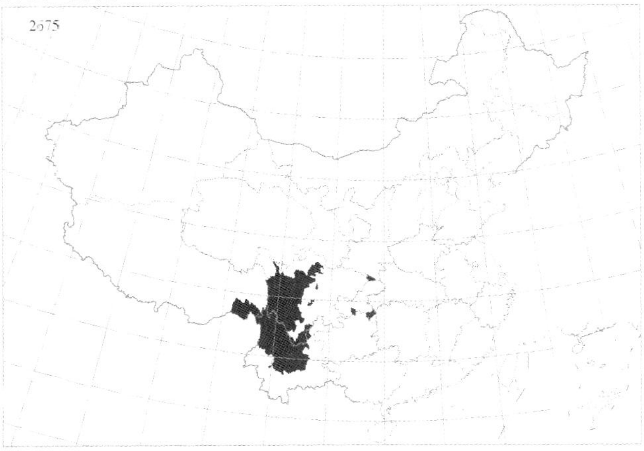

Saxifragaceae 虎耳草科 *Philadelphus* 山梅花属
Philadelphus purpurascens var. venustus (Koehne) S. Y. Hu 美丽山梅花
Shrubs, deciduous

6.0; -1.7; 12.8; 6.3; 35.6; 22.7; 498 ❙ 802; 487; 20 ❙ 60.9; 498; 292.5

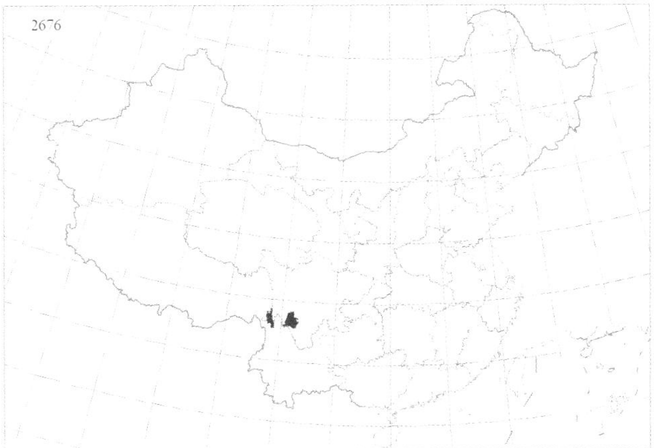

Saxifragaceae 虎耳草科 *Philadelphus* 山梅花属
Philadelphus reevesianus S. Y. Hu 毛药山梅花
Shrubs, deciduous

13.7; 2.3; 24.6; 13.7; 109.4; 4.7; 758 ❙ 1267; 542; 94 ❙ 67.3; 758; 252.6

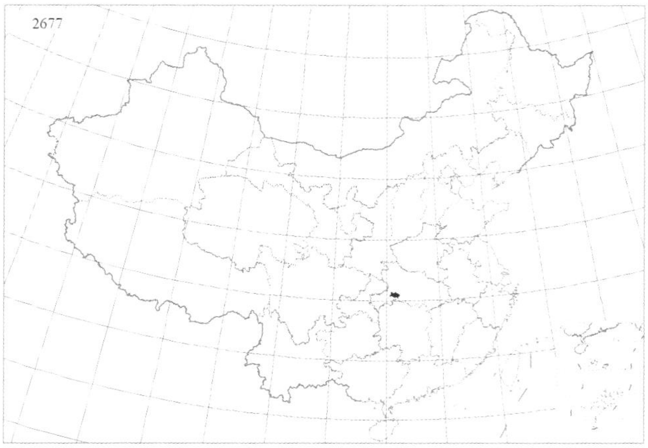

Saxifragaceae 虎耳草科 *Philadelphus* 山梅花属
Philadelphus schrenkii Rupr. 东北山梅花
Shrubs, deciduous

5.2 (-0.9~11.0); -13.9 (-24.7~-5.7); 21.7 (18.5~25.5); 8.7 (6.5~12.0); 69.4 (47.6~101.7); 66.6 (29.5~120.7); 607 (513~757) ❙ 642 (409~1073); 418 (258~691); 19 (8~40) ❙ 6.5 (-24.3~74.1); 572 (409~684); 269.1 (174.5~346.5)

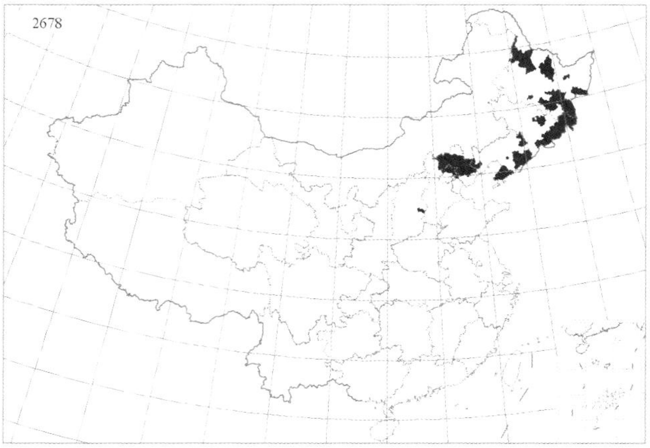

Saxifragaceae 虎耳草科 *Philadelphus* 山梅花属
Philadelphus sericanthus Koehne 绢毛山梅花
Shrubs, deciduous

14.3 (2.2~18.9); 3.5 (-7.8~12.5); 24.1 (9.9~29.3); 14.4 (3.8~19.0); 117.3 (15.5~167.8); 5.3 (0.0~52.8); 785 (413~984) ❙ 1129 (341~1997); 506 (190~816); 85 (8~259) ❙ 42.9 (-40.9~130.7); 769 (341~941); 237.0 (61.2~406.0)

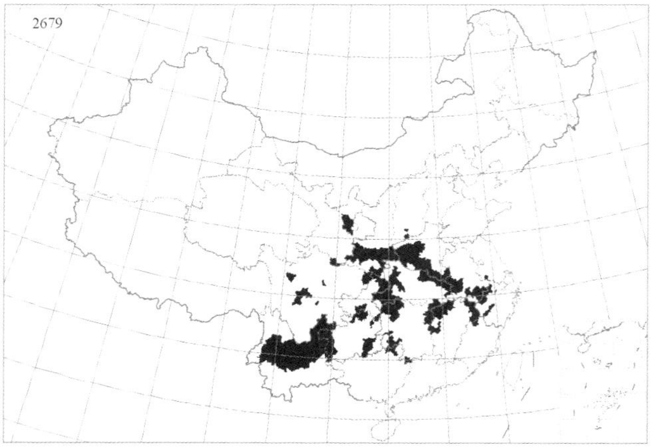

Saxifragaceae 虎耳草科 *Philadelphus* 山梅花属
Philadelphus sericanthus var. kulingensis (Koehne) Hand.-Mazz. 牯岭山梅花
Shrubs, deciduous

16.1 (14.1~18.2); 4.0 (1.6~6.5); 27.8 (26.2~29.3); 16.1 (14.2~18.2); 135.4 (116.7~158.8); 1.7 (0.0~6.5); 882 (796~984) ❙ 1579 (1211~1922); 577 (474~685); 194 (122~259) ❙ 79.5 (28.8~111.3); 879 (796~941); 239.5 (183.9~319.9)

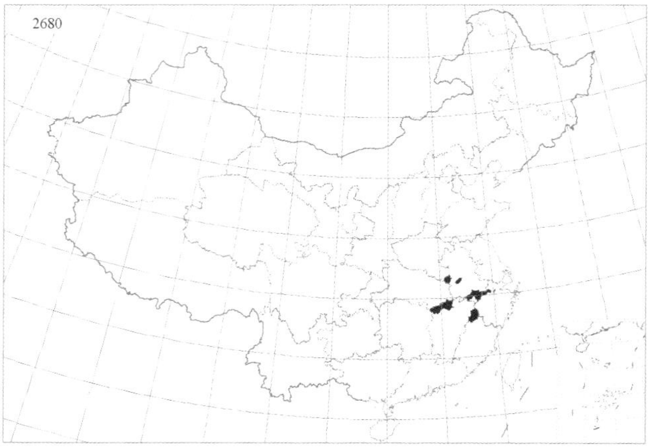

Saxifragaceae 虎耳草科 *Philadelphus* 山梅花属
Philadelphus subcanus Koehne 毛柱山梅花
Shrubs, deciduous

11.8 (-0.3~16.6); 0.2 (-10.9~8.7); 22.6 (9.0~28.8); 12.0 (3.0~16.7); 94.9 (10.4~140.6); 13.2 (0.0~73.4); 713 (378~920) ❙ 969 (460~1571); 471 (263~817); 50 (8~157) ❙ 39.0 (-22.2~128.6); 697 (378~910); 229.1 (139.1~357.7)

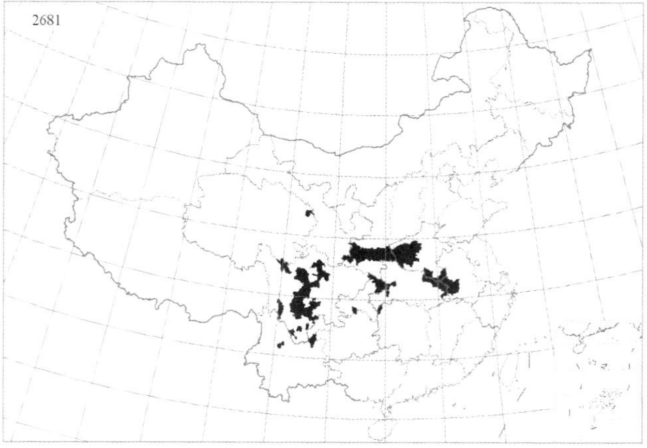

Saxifragaceae 虎耳草科 *Philadelphus* 山梅花属
Philadelphus subcanus var. dubius Koehne 密毛山梅花
Shrubs, deciduous

7.2 (2.2~13.3); -1.9 (-7.1~3.6); 15.2 (9.9~22.1); 8.0 (3.8~13.2); 51.8 (15.5~100.9); 25.1 (1.8~49.1); 546 (413~723) ❙ 978 (800~1081); 541 (427~632); 26 (18~35) ❙ 85.3 (49.1~125.9); 546 (413~723); 257.9 (171.0~341.6)

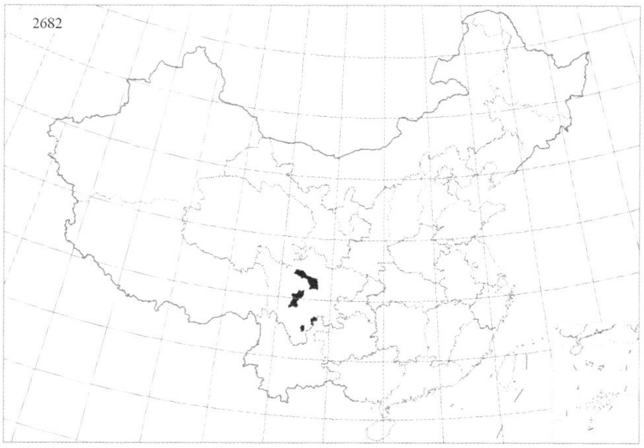

Saxifragaceae 虎耳草科 *Philadelphus* 山梅花属

Philadelphus subcanus var. magdalenae (Koehne) S. Y. Hu 城口山梅花

Shrubs, deciduous

9.0 (2.0~14.9); -0.2 (-7.2~6.1); 17.2 (9.5~24.7); 9.5 (3.8~15.0); 66.6 (15.4~119.6); 18.0 (0.0~51.4); 599 (411~789) ❙ 1046 (729~1410); 567 (460~816); 30 (11~65) ❙ 79.3 (18.6~128.6); 598 (411~789); 255.7 (150.8~331.6)

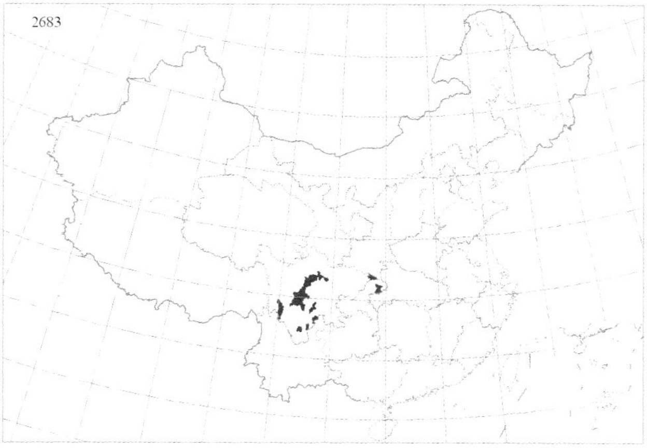

Saxifragaceae 虎耳草科 *Philadelphus* 山梅花属

Philadelphus tenuifolius Rupr. ex Maxim. 薄叶山梅花

Shrubs, deciduous

5.2 (1.7~13.5); -13.6 (-20.6~-1.6); 21.6 (18.5~27.1); 8.7 (6.7~13.7); 68.2 (48.1~117.5); 65.1 (14.6~93.8); 602 (515~820) ❙ 604 (309~1073); 370 (199~691); 20 (7~40) ❙ 0.8 (-42.4~74.1); 521 (309~678); 227.5 (71.4~346.5)

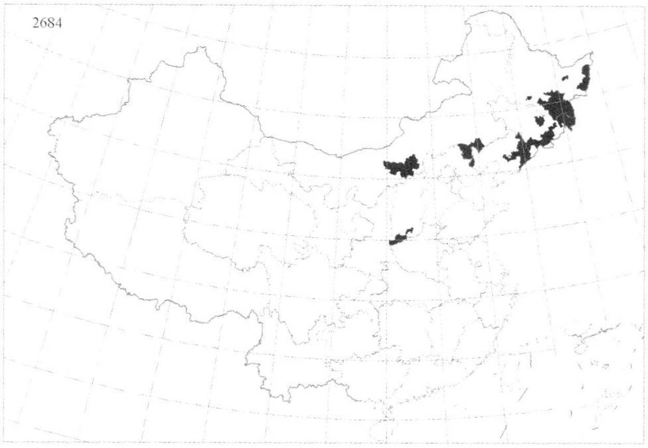

Saxifragaceae 虎耳草科 *Philadelphus* 山梅花属

Philadelphus tetragonus S. M. Hwang 四棱山梅花

Shrubs, deciduous

9.1; -0.7; 18.1; 9.2; 64.1; 14.1; 595 ❙ 822; 420; 19 ❙ 38.2; 595; 275.4

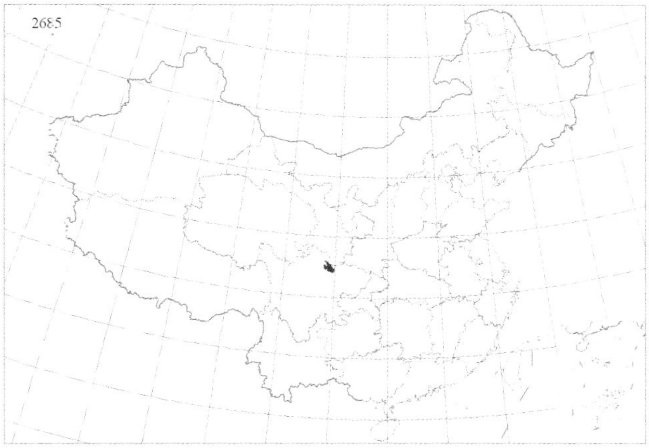

Saxifragaceae 虎耳草科 *Philadelphus* 山梅花属

Philadelphus tomentosus Wall. ex G. Don 绒毛山梅花

Shrubs, deciduous

4.5 (-1.1~13.3); -3.9 (-9.6~5.4); 11.9 (6.2~19.6); 5.9 (2.0~13.4); 34.0 (2.5~100.4); 39.6 (0.0~75.6); 470 (334~706) ❙ 772 (321~1736); 458 (193~1007); 23 (9~51) ❙ 51.9 (-4.0~145.8); 469 (321~706); 235.8 (61.6~447.4)

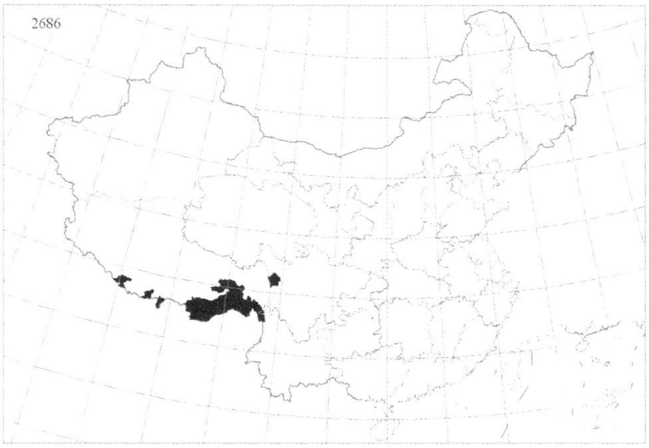

Saxifragaceae 虎耳草科 *Philadelphus* 山梅花属

Philadelphus tsianschanensis Wang et Li 千山山梅花

Shrubs, deciduous

5.8 (1.8~8.2); -13.2 (-18.7~-10.3); 22.1 (19.3~24.3); 8.9 (6.9~10.3); 71.5 (50.5~87.0); 60.9 (47.7~87.9); 609 (524~678) ❙ 907 (735~1073); 574 (457~691); 33 (25~40) ❙ 49.5 (8.4~74.1); 609 (524~678); 303.0 (219.8~335.3)

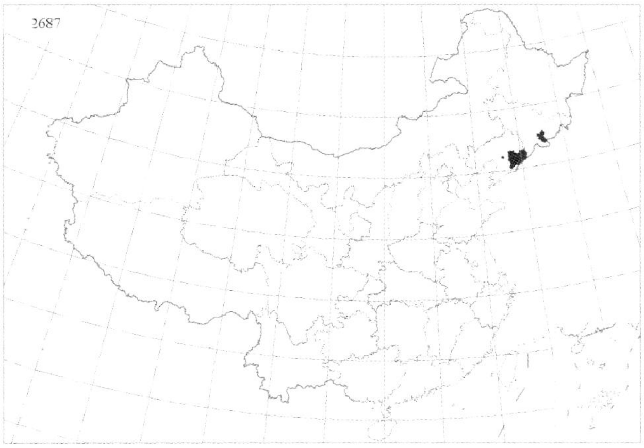

Saxifragaceae 虎耳草科 *Philadelphus* 山梅花属

Philadelphus zhejiangensis S. M. Hwang 浙江山梅花

Shrubs, deciduous

16.0 (13.4~18.2); 4.1 (1.5~8.4); 27.5 (25.2~29.2); 16.0 (13.4~18.3); 133.9 (108.3~159.1); 1.5 (0.0~7.3); 872 (755~984) ❙ 1586 (1117~1997); 567 (427~677); 200 (122~259) ❙ 82.5 (28.5~130.7); 867 (755~938); 242.5 (149.5~319.9)

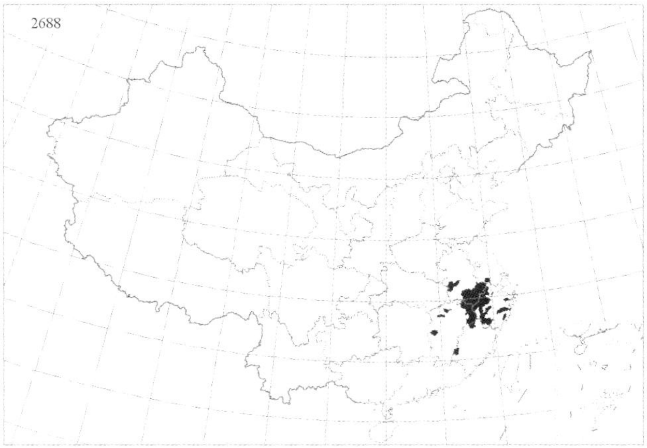

Saxifragaceae 虎耳草科 *Pileostegia* 冠盖藤属
Pileostegia tomentella Hand.-Mazz. 星毛冠盖藤
Shrubs, evergreen

18.3 (14.3~24.0); 8.4 (3.3~19.1); 27.1 (24.2~29.8); 18.4 (14.3~24.0); 160.4 (113.2~228.1); 0.1 (0.0~2.3); 960 (761~1302) ❙ 1645 (1165~2435); 623 (411~1428); 177 (48~257) ❙ 72.3 (19.1~130.7); 956 (761~1221); 242.3 (146.0~620.0)

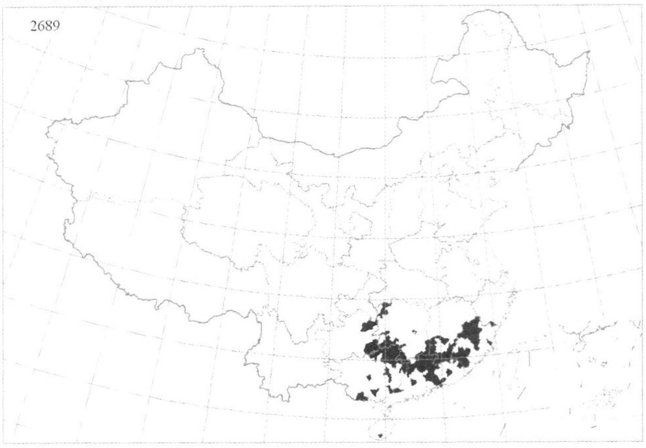

Saxifragaceae 虎耳草科 *Pileostegia* 冠盖藤属
Pileostegia viburnoides Hook. f. et Thomson 冠盖藤
Shrubs, evergreen

18.1 (6.0~24.8); 8.1 (−3.2~20.1); 26.9 (14.4~30.1); 18.1 (6.6~24.8); 157.9 (39.7~237.6); 0.6 (0.0~26.7); 957 (508~1385) ❙ 1570 (711~3495); 622 (357~1847); 159 (15~897) ❙ 66.0 (−6.9~246.5); 950 (508~1273); 231.9 (95.3~620.0)

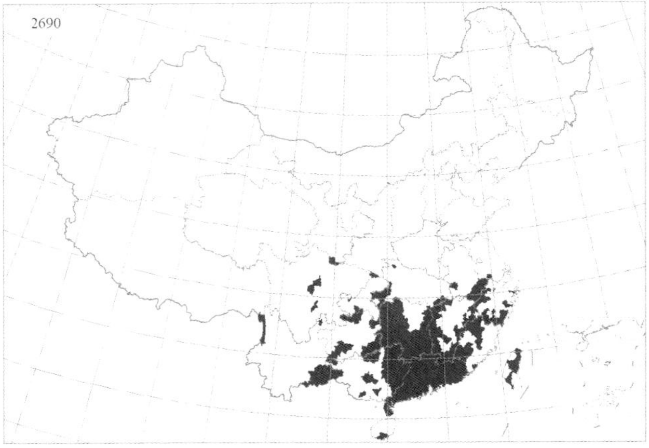

Saxifragaceae 虎耳草科 *Platycrater* 蛛网萼属
Platycrater arguta Siebold et Zucc. 蛛网萼
Shrubs, deciduous

16.8 (15.0~18.2); 5.9 (2.6~7.3); 27.2 (24.7~29.1); 16.9 (15.0~18.2); 142.7 (120.8~158.8); 0.2 (0.0~3.5); 894 (789~984) ❙ 1825 (1572~1997); 615 (530~713); 232 (188~259) ❙ 104.9 (71.7~147.0); 890 (789~961); 256.1 (183.9~303.4)

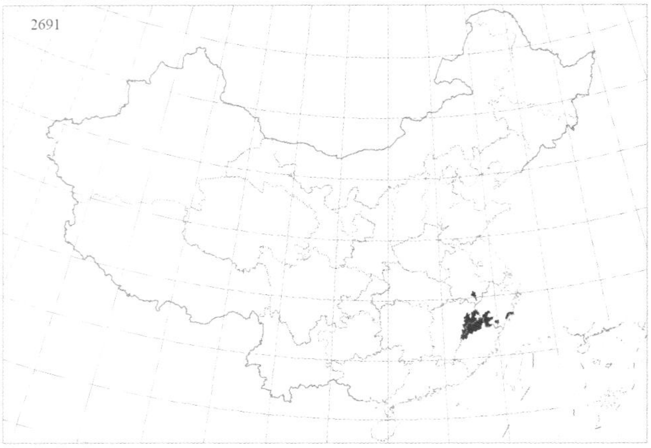

Saxifragaceae 虎耳草科 *Polyosma* 多香木属
Polyosma cambodiana Gagnep. 多香木
Trees

22.8 (20.4~24.8); 16.2 (13.2~20.1); 27.6 (23.6~28.7); 22.9 (20.5~24.8); 214.4 (185.8~237.6); 0.0 (0.0~0.0); 1222 (989~1385) ❙ 1585 (868~2435); 707 (312~1428); 84 (41~130) ❙ 31.5 (−34.3~99.5); 1132 (868~1299); 405.1 (235.1~659.0)

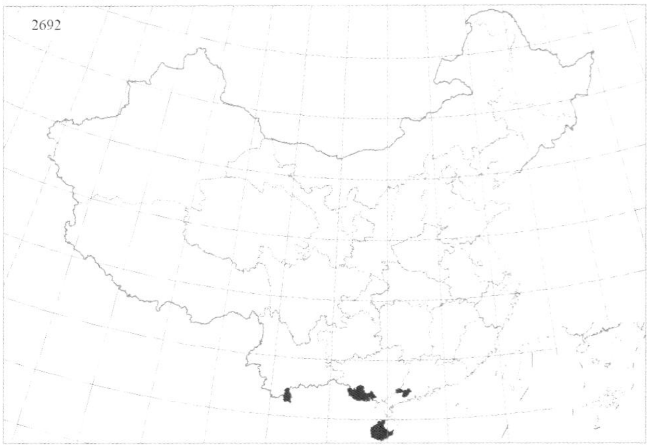

Saxifragaceae 虎耳草科 *Ribes* 茶藨子属
Ribes aciculare Sm. 阿尔泰醋栗
Small shrubs, deciduous

4.7 (−2.8~11.7); −14.4 (−20.9~−7.6); 20.4 (10.9~27.5); 8.4 (3.5~13.0); 65.9 (16.3~113.4); 69.3 (32.4~109.2); 603 (390~820) ❙ 190 (30~450); 88 (21~211); 17 (1~51) ❙ −65.7 (−96.3~3.9); 189 (30~406); 81.4 (2.1~297.3)

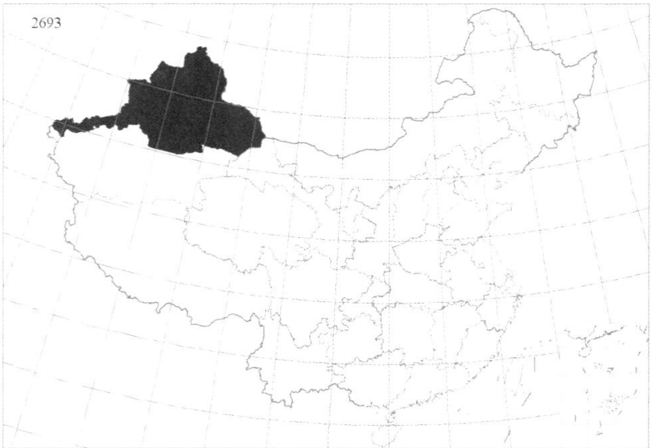

Saxifragaceae 虎耳草科 *Ribes* 茶藨子属
Ribes alpestre Wall. ex Decne. 长刺茶藨子
Shrubs, deciduous

5.8 (−3.3~15.5); −4.7 (−15.4~8.2); 15.3 (7.0~27.8); 7.3 (2.0~15.6); 48.9 (3.7~131.3); 38.3 (0.0~103.2); 528 (332~881) ❙ 724 (281~1478); 396 (175~727); 24 (5~98) ❙ 38.9 (−40.9~125.9); 507 (281~881); 251.5 (32.2~407.1)

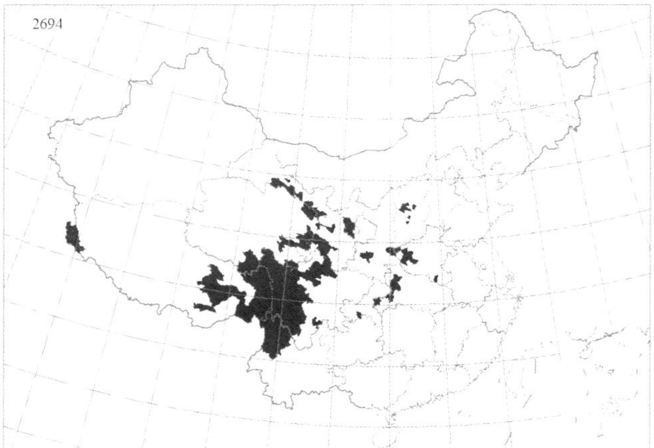

Saxifragaceae 虎耳草科 *Ribes* 茶藨子属

Ribes alpestre var. giganteum Jancz. 大刺茶藨子

Shrubs, deciduous

4.3 (0.3~9.1); -6.9 (-9.9~-0.7); 14.5 (9.3~21.1); 6.0 (3.3~9.2); 37.3 (11.9~71.4); 45.1 (14.1~67.5); 486 (391~607) ❙ 650 (307~950); 350 (173~546); 15 (7~25) ❙ 41.6 (-49.4~128.6); 433 (307~595); 218.0 (42.6~341.6)

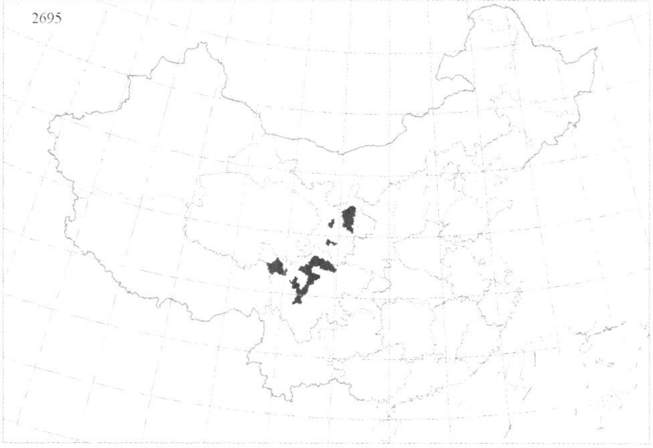

Saxifragaceae 虎耳草科 *Ribes* 茶藨子属

Ribes altissimum Turcz. ex Pojark. 高茶藨子

Shrubs, deciduous

1.9 (-0.6~5.8); -17.1 (-19.5~-14.9); 18.7 (15.9~23.3); 7.0 (5.4~9.7); 52.4 (37.5~80.9); 88.6 (71.2~104.2); 545 (475~668) ❙ 184 (156~251); 77 (56~106); 22 (15~33) ❙ -65.8 (-76.7~-52.0); 184 (156~251); 86.2 (16.6~139.1)

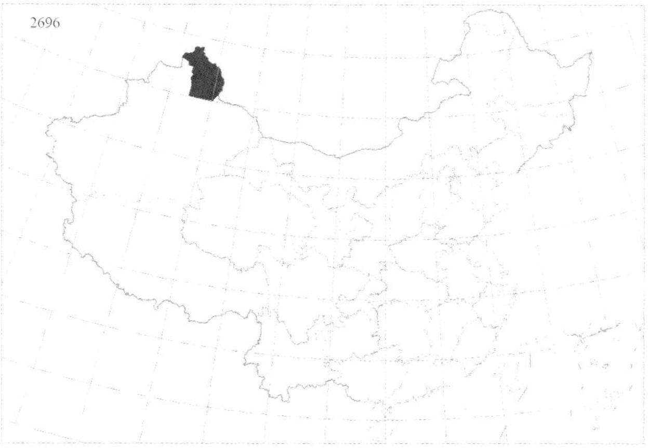

Saxifragaceae 虎耳草科 *Ribes* 茶藨子属

Ribes ambiguum Maxim. 四川蔓茶藨子

Small shrubs, deciduous

15.9 (10.2~18.7); 5.1 (-0.7~8.1); 26.2 (21.0~28.9); 16.0 (10.4~18.8); 133.1 (77.2~166.0); 1.7 (0.0~13.6); 858 (634~1002) ❙ 1150 (899~1521); 488 (420~571); 59 (19~196) ❙ 35.9 (8.0~95.9); 858 (634~1002); 178.8 (108.2~305.6)

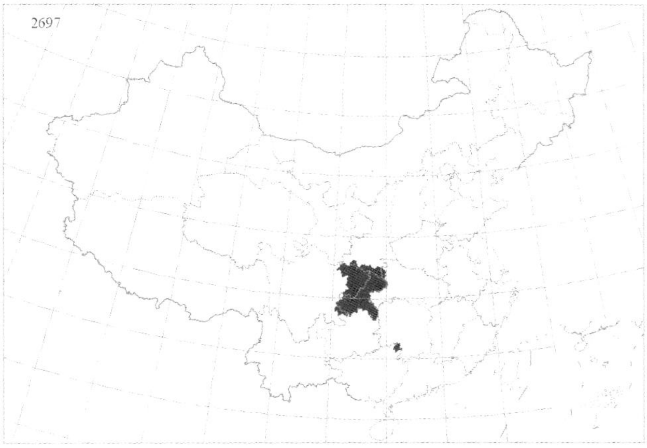

Saxifragaceae 虎耳草科 *Ribes* 茶藨子属

Ribes burejense F. Schmidt 刺果茶藨子

Shrubs, deciduous

8.4 (-0.9~15.1); -8.2 (-25.0~1.8); 22.9 (17.5~27.8); 10.2 (6.5~15.2); 82.9 (45.4~128.0); 41.8 (5.6~120.7); 663 (507~865) ❙ 574 (325~1073); 345 (201~691); 19 (7~53) ❙ -12.5 (-47.7~74.1); 549 (325~817); 184.7 (44.7~346.5)

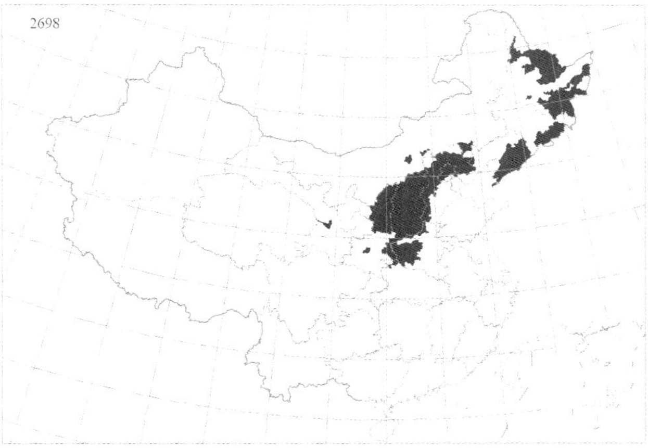

Saxifragaceae 虎耳草科 *Ribes* 茶藨子属

Ribes davidii Franch. 革叶茶藨子

Shrubs, evergreen

12.3 (7.3~15.5); 2.8 (-2.0~5.3); 21.1 (15.0~26.3); 12.4 (7.5~15.6); 93.2 (48.2~128.1); 4.8 (0.0~20.4); 698 (537~830) ❙ 1173 (988~1469); 580 (465~816); 61 (27~192) ❙ 68.9 (35.3~106.2); 698 (537~830); 249.6 (148.9~352.3)

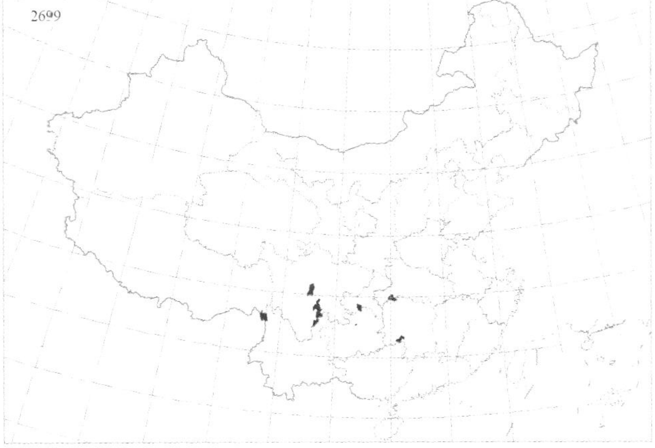

Saxifragaceae 虎耳草科 *Ribes* 茶藨子属

Ribes davidii var. ciliatum L. T. Lu 睫毛茶藨子

Shrubs, evergreen

10.7; 1.9; 18.6; 10.8; 75.5; 6.3; 636 ❙ 1133; 634; 35 ❙ 79.3; 636; 288.2

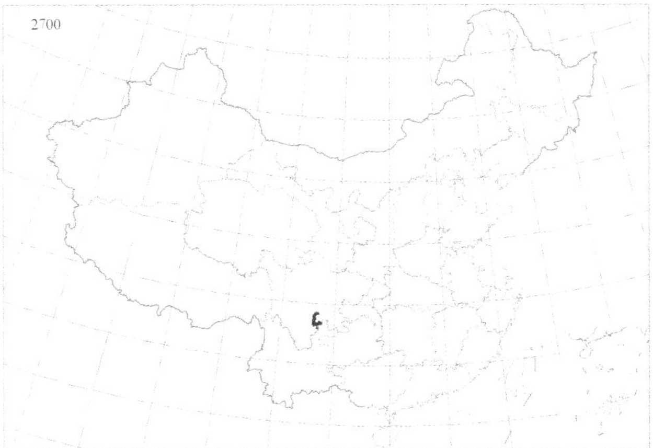

Saxifragaceae 虎耳草科 *Ribes* 茶藨子属
Ribes davidii var. lobatum L. T. Lu 浅裂茶藨子
Shrubs, evergreen

10.6; 1.8; 18.6; 10.7; 75.0; 6.9; 635 | 1050; 576; 26 | 65.4; 635; 282.5

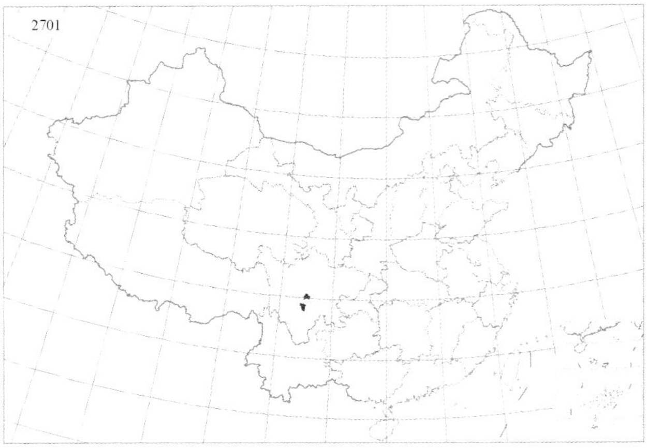

Saxifragaceae 虎耳草科 *Ribes* 茶藨子属
Ribes diacanthum Pall. 双刺茶藨子
Shrubs, deciduous

3.3 (-5.0~12.1); -16.5 (-29.4~-2.5); 20.7 (16.6~25.2); 7.8 (5.0~12.4); 60.9 (34.4~103.1); 81.4 (17.8~154.8); 569 (448~747) | 416 (132~966); 269 (81~616); 12 (2~35) | -25.4 (-80.1~54.7); 390 (132~624); 179.8 (13.6~338.4)

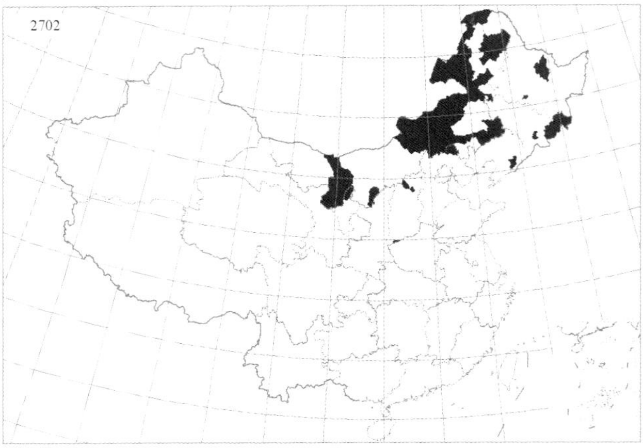

Saxifragaceae 虎耳草科 *Ribes* 茶藨子属
Ribes fargesii Franch. 花茶藨子
Shrubs, deciduous

10.2; -2.5; 21.9; 10.6; 81.7; 19.2; 654 | 935; 437; 34 | 43.8; 633; 223.5

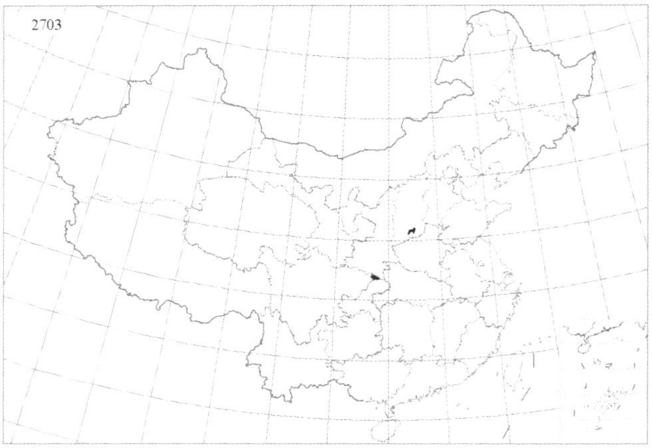

Saxifragaceae 虎耳草科 *Ribes* 茶藨子属
Ribes fasciculatum Siebold et Zucc. 簇花茶藨子
Shrubs, deciduous

12.8 (-1.3~18.6); -0.4 (-12.3~9.3); 25.0 (8.3~29.3); 13.0 (2.5~18.7); 107.3 (7.0~164.0); 13.4 (0.0~82.2); 765 (354~965) | 871 (341~1828); 417 (190~685); 60 (8~253) | 13.2 (-40.9~108.0); 716 (341~965); 196.4 (61.2~334.4)

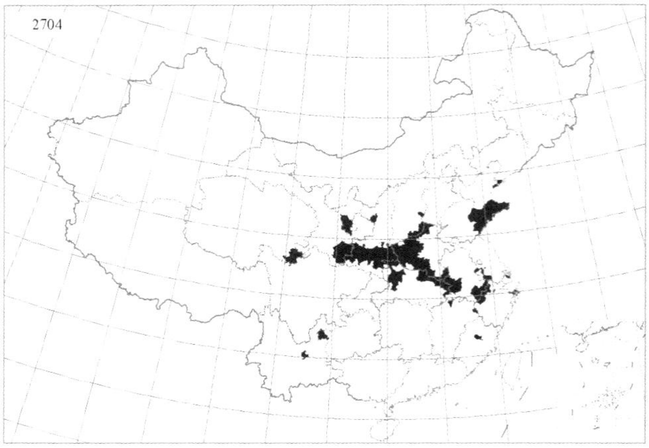

Saxifragaceae 虎耳草科 *Ribes* 茶藨子属
Ribes fasciculatum var. chinense Maxim. 华蔓茶藨子
Shrubs, deciduous

12.1 (-1.4~17.2); -1.4 (-15.9~5.3); 24.7 (8.3~29.3); 12.6 (2.5~17.2); 103.4 (7.0~146.8); 17.5 (0.0~92.1); 750 (354~941) | 850 (282~1828); 409 (182~685); 54 (5~253) | 12.5 (-36.5~108.0); 698 (282~941); 210.8 (101.3~334.4)

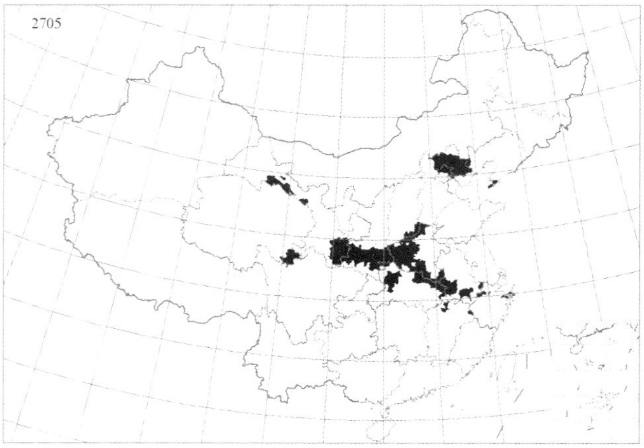

Saxifragaceae 虎耳草科 *Ribes* 茶藨子属
Ribes formosanum Hayata 台湾茶藨子
Shrubs, deciduous

18.3 (16.2~21.7); 12.8 (11.0~15.3); 22.8 (20.1~27.0); 18.3 (16.2~21.7); 159.8 (134.2~200.2); 0.0 (0.0~0.0); 886 (773~1118) | 2471 (1836~2859); 1109 (900~1412); 240 (110~402) | 185.5 (64.2~246.5); 881 (773~1075); 274.8 (202.8~363.7)

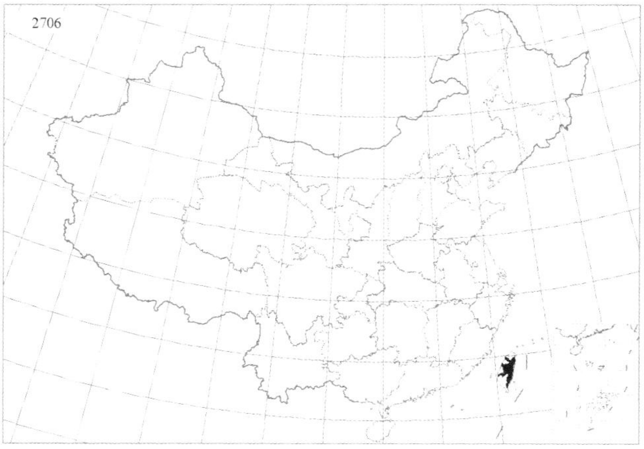

Saxifragaceae 虎耳草科 *Ribes* 茶藨子属
Ribes franchetii Jancz. 鄂西茶藨子
Shrubs, deciduous

11.8 (2.2~15.3); 1.0 (-6.6~5.3); 22.3 (9.9~26.5); 12.1 (3.8~15.4); 93.6 (15.5~125.7); 10.9 (0.0~49.1); 698 (413~832) ❙ 1168 (932~1410); 525 (421~816); 62 (19~113) ❙ 70.5 (31.7~125.9); 698 (413~832); 252.7 (213.4~284.3)

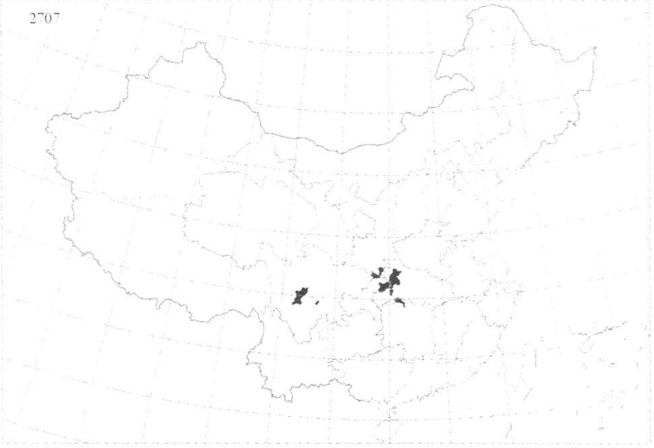

Saxifragaceae 虎耳草科 *Ribes* 茶藨子属
Ribes fuyunense T. C. Ku et F. Konta 富蕴茶藨子
Shrubs, deciduous

2.2 (-0.6~5.8); -17.3 (-19.5~-14.9); 19.1 (15.9~23.3); 7.3 (5.4~9.7); 55.6 (37.5~80.9); 88.3 (71.2~104.2); 559 (475~668) ❙ 169 (156~187); 70 (56~79); 21 (15~25) ❙ -69.2 (-76.7~-65.1); 169 (156~187); 81.7 (19.8~120.8)

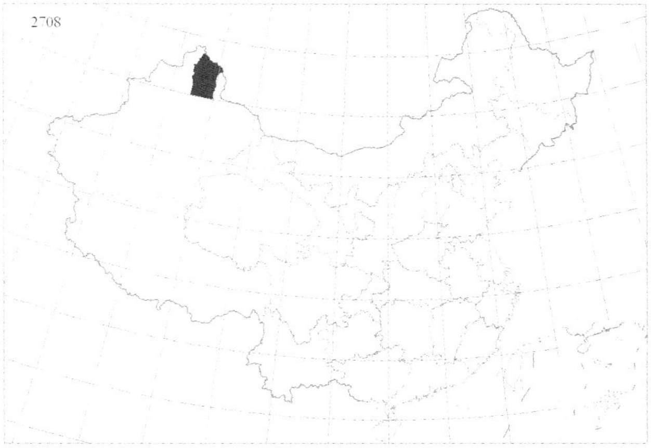

Saxifragaceae 虎耳草科 *Ribes* 茶藨子属
Ribes giraldii Jancz. 陕西茶藨子
Shrubs, deciduous

7.6 (-3.3~15.7); -5.4 (-15.4~3.4); 19.5 (7.2~28.4); 8.9 (2.0~15.8); 66.9 (3.7~131.7); 34.6 (2.1~103.2); 598 (339~880) ❙ 625 (225~1621); 315 (131~616); 26 (4~196) ❙ 4.7 (-58.9~95.5); 552 (225~880); 212.9 (66.4~349.1)

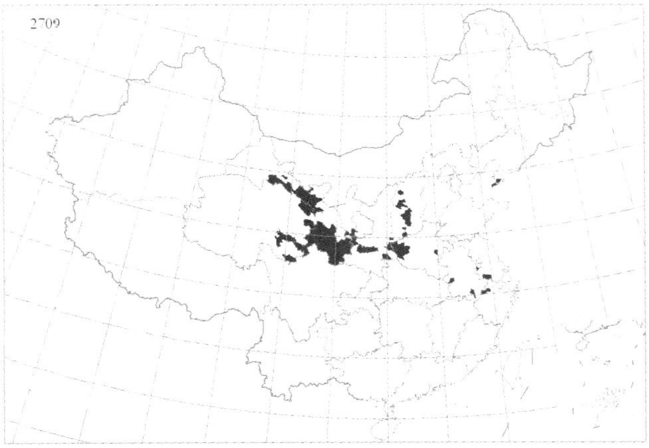

Saxifragaceae 虎耳草科 *Ribes* 茶藨子属
Ribes giraldii var. cuneatum Wang et Li 滨海茶藨子
Shrubs, deciduous

10.5; -4.9; 23.7; 11.3; 93.2; 26.8; 705 ❙ 632; 398; 27 ❙ -10.4; 632; 180.8

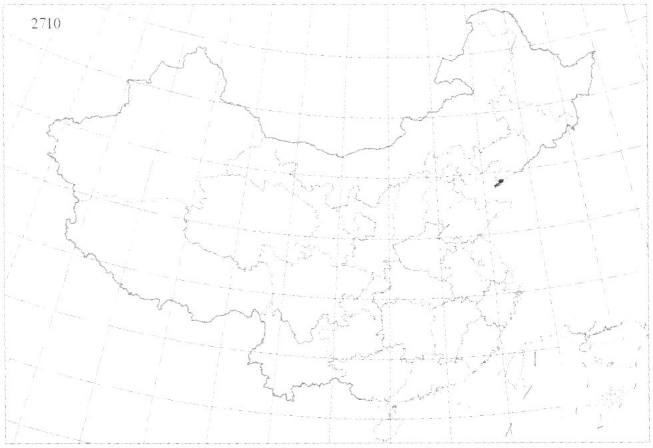

Saxifragaceae 虎耳草科 *Ribes* 茶藨子属
Ribes glabricalycinum L. T. Lu 光萼茶藨子
Small shrubs, deciduous

4.6 (0.9~9.6); -3.8 (-9.5~2.4); 12.0 (9.8~15.7); 5.7 (3.6~9.6); 30.1 (14.0~60.8); 34.0 (5.4~63.1); 473 (403~600) ❙ 838 (628~1041); 494 (407~601); 22 (11~45) ❙ 78.2 (48.6~125.9); 473 (403~600); 297.5 (236.3~352.3)

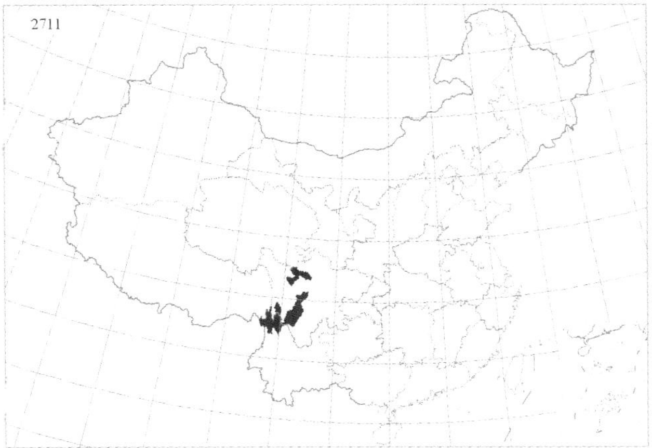

Saxifragaceae 虎耳草科 *Ribes* 茶藨子属
Ribes glabrifolium L. T. Lu 光叶茶藨子
Small shrubs, deciduous

12.4; 0.3; 24.0; 12.5; 99.7; 10.4; 723 ❙ 877; 387; 37 ❙ 21.3; 723; 269.3

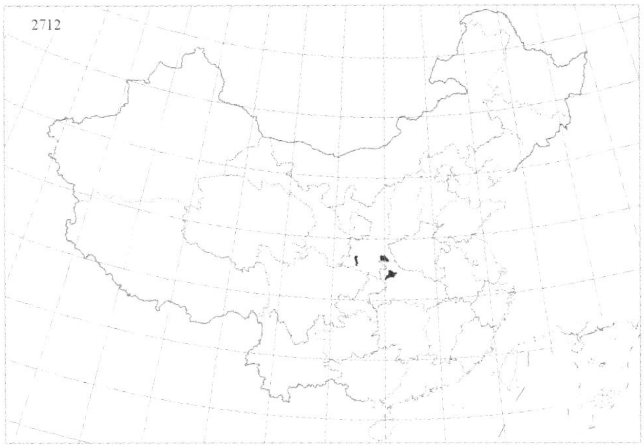

Saxifragaceae 虎耳草科 *Ribes* 茶藨子属
Ribes glaciale Wall. 冰川茶藨子
Shrubs, deciduous

9.7 (−5.5~19.1); −2.4 (−17.3~13.0); 20.5 (5.8~27.8); 10.7 (1.3~19.1); 83.8 (1.2~169.7); 26.2 (0.0~127.1); 655 (299~929) ❙ 745 (63~1689); 402 (45~835); 30 (2~212) ❙ 13.7 (−84.9~128.6); 589 (63~917); 218.4 (8.4~516.7)

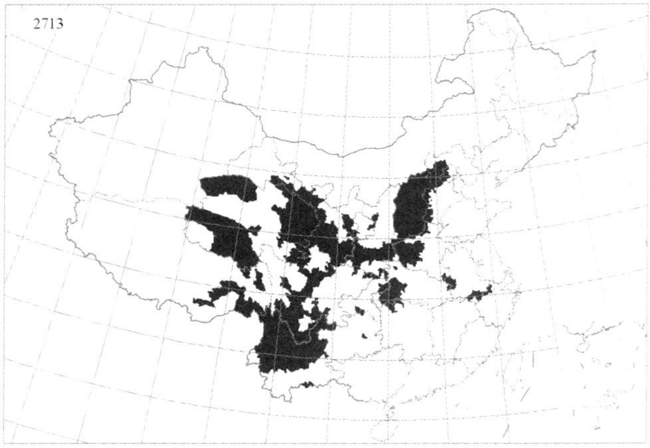

Saxifragaceae 虎耳草科 *Ribes* 茶藨子属
Ribes griffithii Hook. f. et Thomson 曲萼茶藨子
Shrubs, deciduous

7.9 (−1.8~14.4); 0.3 (−10.2~7.6); 14.4 (6.1~19.7); 8.5 (1.8~14.4); 55.9 (1.9~113.1); 20.4 (0.0~82.8); 557 (329~728) ❙ 962 (320~1736); 562 (217~1007); 32 (7~70) ❙ 66.8 (−3.0~145.8); 556 (320~727); 298.7 (72.2~487.2)

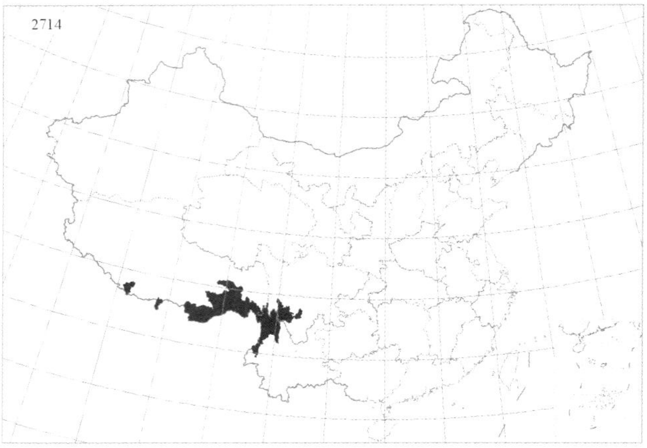

Saxifragaceae 虎耳草科 *Ribes* 茶藨子属
Ribes griffithii var. gongshanense (T. C. Ku) L. T. Lu 贡山茶藨子
Shrubs, deciduous

8.6; 1.3; 15.0; 8.7; 52.6; 8.8; 574 ❙ 1041; 585; 45 ❙ 81.5; 574; 352.3

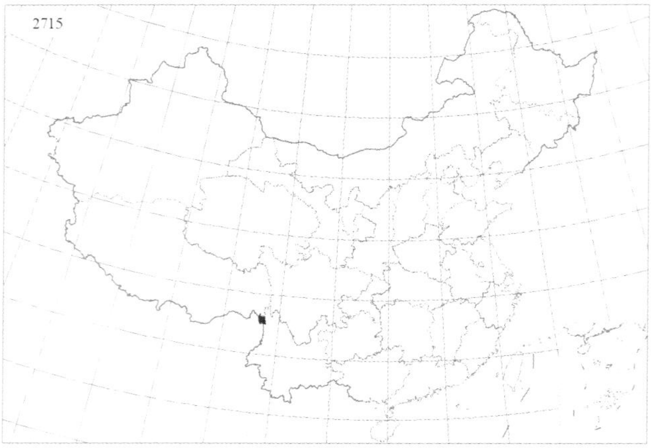

Saxifragaceae 虎耳草科 *Ribes* 茶藨子属
Ribes henryi Franch. 华中茶藨子
Small shrubs, evergreen

13.3 (11.1~15.7); 2.9 (1.8~5.3); 23.0 (19.4~26.5); 13.3 (11.1~15.7); 103.6 (79.8~129.3); 3.8 (0.0~6.9); 735 (648~841) ❙ 1242 (1001~1478); 580 (478~816); 67 (30~94) ❙ 69.4 (46.2~107.2); 735 (648~841); 240.1 (182.2~292.6)

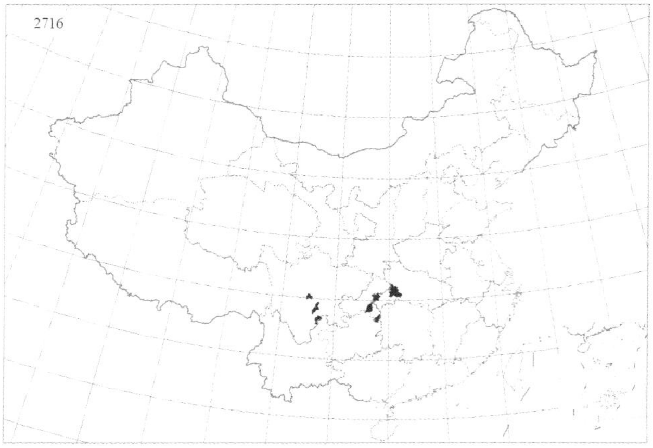

Saxifragaceae 虎耳草科 *Ribes* 茶藨子属
Ribes heterotrichum C. A. Mey. 圆叶茶藨子
Shrubs, deciduous

4.5 (−2.8~10.2); −14.7 (−20.9~−8.5); 20.4 (10.9~27.5); 8.3 (3.5~12.1); 65.7 (16.3~107.6); 70.8 (36.6~109.2); 602 (390~813) ❙ 198 (45~450); 91 (27~211); 18 (3~51) ❙ −64.4 (−94.0~3.9); 197 (45~406); 89.2 (7.6~297.3)

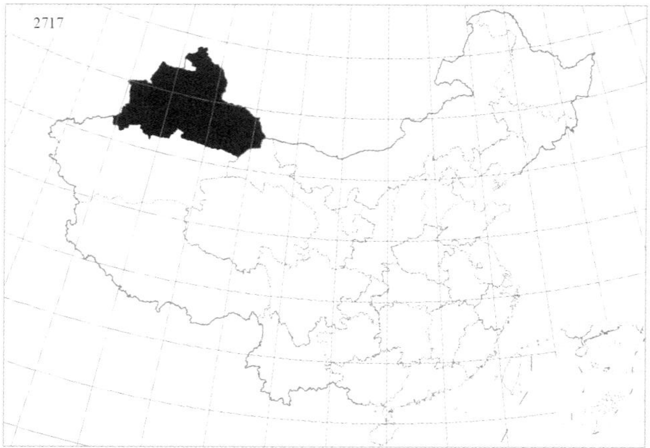

Saxifragaceae 虎耳草科 *Ribes* 茶藨子属
Ribes himalense Royle ex Decne. 糖茶藨子
Small shrubs, deciduous

6.5 (−5.5~14.8); −6.2 (−20.8~7.6); 17.9 (5.8~27.1); 8.2 (1.3~14.9); 59.2 (1.2~119.4); 40.7 (0.0~127.1); 566 (299~820) ❙ 616 (128~1337); 337 (81~727); 18 (2~83) ❙ 11.3 (−80.8~128.6); 491 (128~791); 213.2 (13.6~407.1)

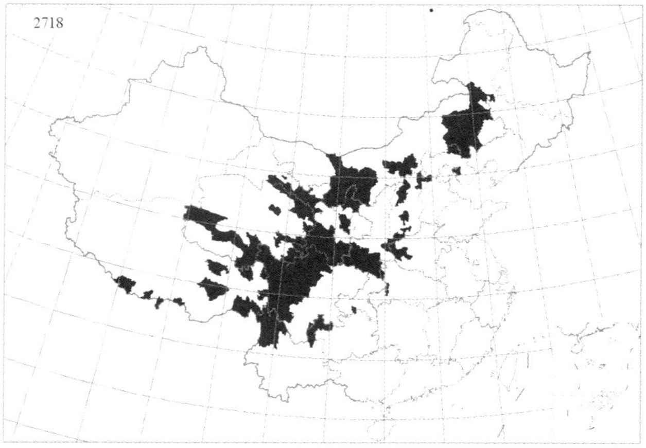

Saxifragaceae 虎耳草科 *Ribes* 茶藨子属
Ribes himalense var. glandulosum Jancz. 疏腺茶藨子
Small shrubs, deciduous

4.7; -4.3; 12.5; 5.6; 31.0; 34.3; 473 ❙ 900; 513; 19 ❙ 91.9; 473; 293.4

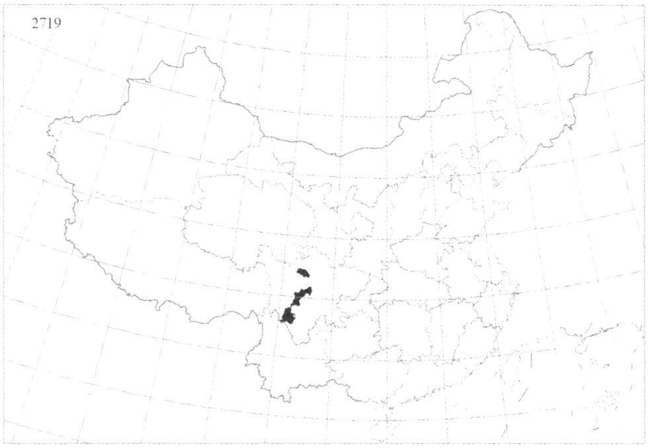

Saxifragaceae 虎耳草科 *Ribes* 茶藨子属
Ribes himalense var. pubicalycinum L. T. Lu et J. T. Pan 毛萼茶藨子
Small shrubs, deciduous

-0.4; -10.9; 8.9; 3.0; 10.2; 74.6; 374 ❙ 611; 354; 14 ❙ 62.2; 374; 284.3

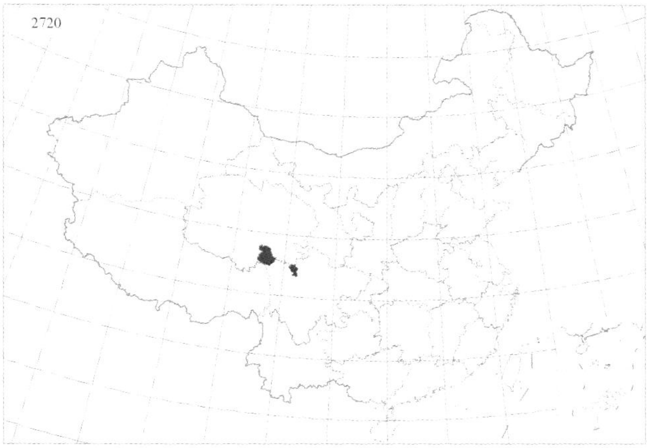

Saxifragaceae 虎耳草科 *Ribes* 茶藨子属
Ribes himalense var. trichophyllum T. C. Ku 异毛茶藨子
Small shrubs, deciduous

7.2 (0.5~13.6); -7.3 (-15.6~0.4); 20.2 (10.7~25.8); 8.8 (3.8~13.6); 67.8 (16.2~113.4); 40.8 (10.0~78.3); 599 (406~786) ❙ 554 (189~856); 311 (114~465); 16 (4~49) ❙ -4.6 (-70.2~86.5); 498 (189~786); 205.8 (33.5~357.7)

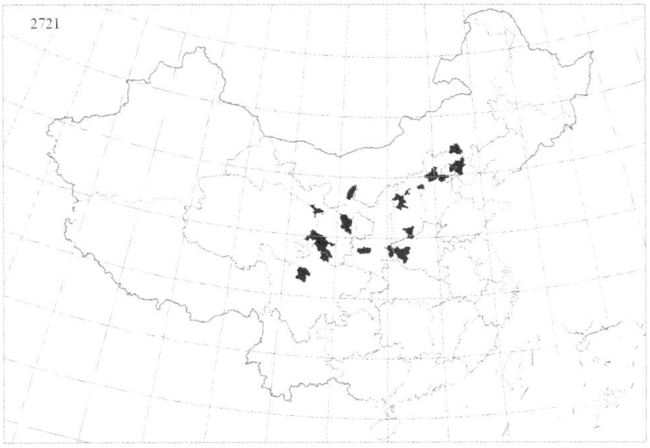

Saxifragaceae 虎耳草科 *Ribes* 茶藨子属
Ribes himalense var. verruculosum (Rehder) L. T. Lu 瘤糖茶藨子
Small shrubs, deciduous

5.7 (1.9~11.7); -9.0 (-12.7~-2.8); 18.6 (11.7~24.6); 7.7 (4.3~12.0); 56.4 (21.1~99.0); 48.0 (18.9~65.0); 552 (427~729) ❙ 529 (374~669); 304 (205~350); 15 (8~23) ❙ -2.4 (-29.3~48.8); 491 (374~624); 195.0 (81.2~289.4)

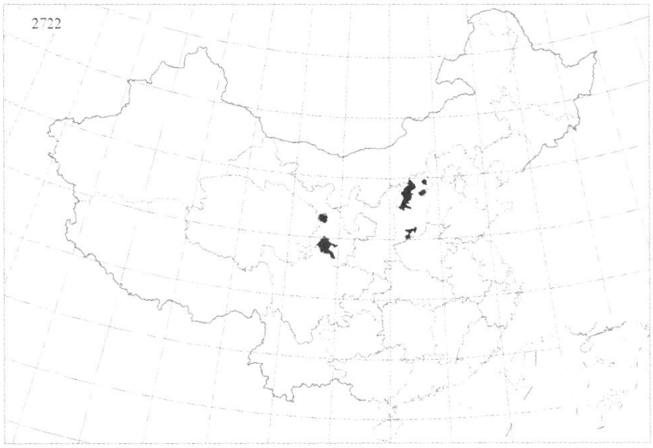

Saxifragaceae 虎耳草科 *Ribes* 茶藨子属
Ribes horridum Rupr. ex Maxim. 密刺茶藨子
Small shrubs, deciduous

3.0 (1.4~4.5); -15.8 (-18.0~-14.1); 19.6 (18.4~21.1); 7.3 (6.4~8.2); 54.2 (45.8~63.5); 77.4 (68.3~88.8); 541 (501~581) ❙ 639 (561~777); 382 (333~473); 22 (16~35) ❙ 18.9 (-2.6~55.1); 537 (501~566); 313.7 (289.3~329.9)

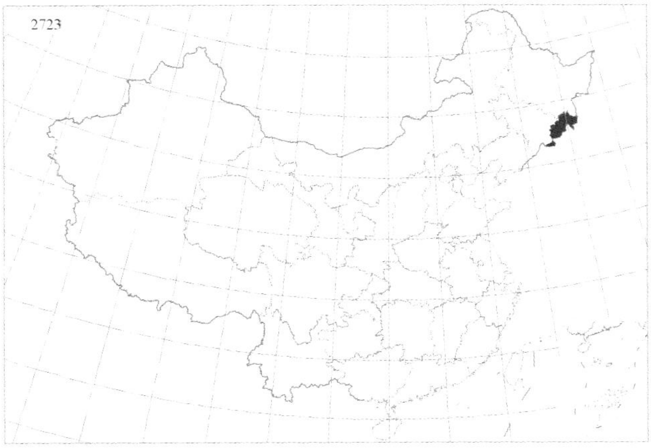

Saxifragaceae 虎耳草科 *Ribes* 茶藨子属
Ribes humile Jancz. 矮醋栗
Small shrubs

5.5 (0.3~14.8); -4.1 (-9.9~3.8); 14.0 (9.3~25.3); 6.6 (3.3~14.8); 40.5 (11.9~119.2); 34.6 (1.4~67.5); 505 (391~795) ❙ 929 (632~1358); 480 (346~606); 28 (12~88) ❙ 86.6 (49.4~125.9); 505 (391~795); 264.4 (171.0~341.6)

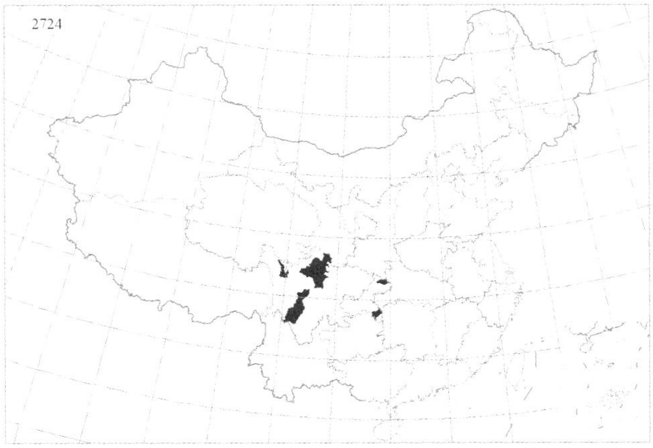

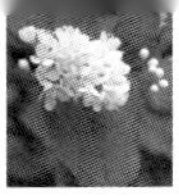

Saxifragaceae 虎耳草科 *Ribes* 茶藨子属

Ribes hunanense C. Y. Yang et C. J. Qi 湖南茶藨子

Vines, semievergreen

15.9 (14.3~17.0); 4.8 (3.3~6.2); 26.2 (24.3~28.1); 16.0 (14.3~17.0); 132.2 (114.1~144.4); 0.6 (0.0~2.3); 846 (765~912) ❙ 1500 (1409~1624); 505 (419~594); 187 (161~207) ❙ 78.1 (55.1~101.8); 840 (765~891); 223.3 (169.7~258.0)

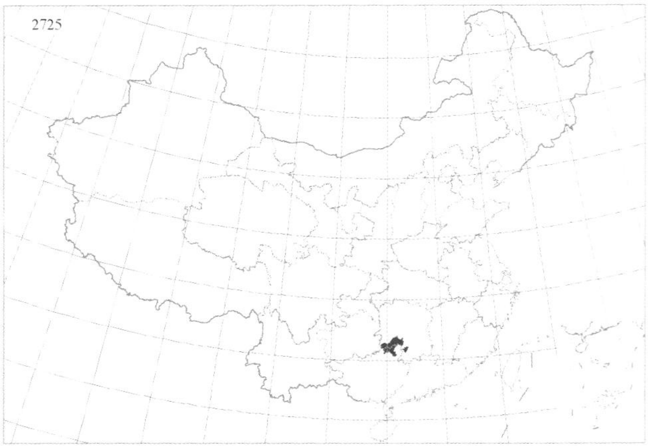

Saxifragaceae 虎耳草科 *Ribes* 茶藨子属

Ribes kialanum Jancz. 康边茶藨子

Shrubs, deciduous

4.6 (0.3~14.8); -4.9 (-9.9~3.8); 13.0 (9.3~25.3); 6.0 (3.3~14.8); 35.1 (11.9~119.2); 39.9 (1.4~67.5); 485 (391~795) ❙ 911 (632~1358); 481 (372~548); 28 (12~88) ❙ 90.6 (61.7~128.6); 485 (391~795); 279.3 (187.9~341.6)

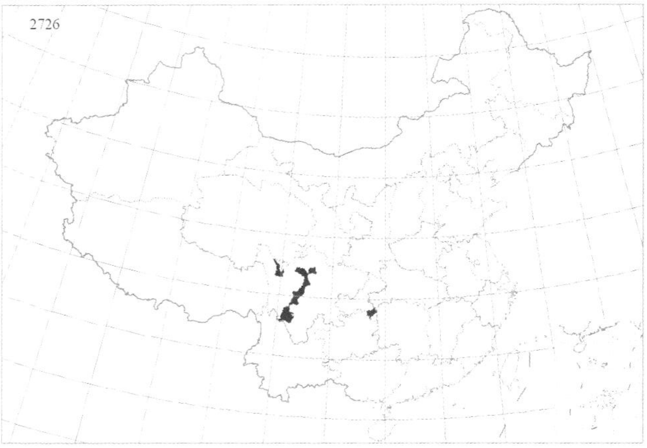

Saxifragaceae 虎耳草科 *Ribes* 茶藨子属

Ribes komarovii Pojark. 长白茶藨子

Shrubs, deciduous

7.5 (0.8~13.3); -9.0 (-20.1~-0.1); 21.9 (10.7~27.2); 9.5 (3.8~13.5); 75.5 (16.2~115.6); 45.4 (11.7~91.9); 632 (411~827) ❙ 651 (373~1073); 378 (201~691); 22 (7~40) ❙ 5.8 (-47.7~74.1); 552 (373~698); 233.2 (88.1~346.5)

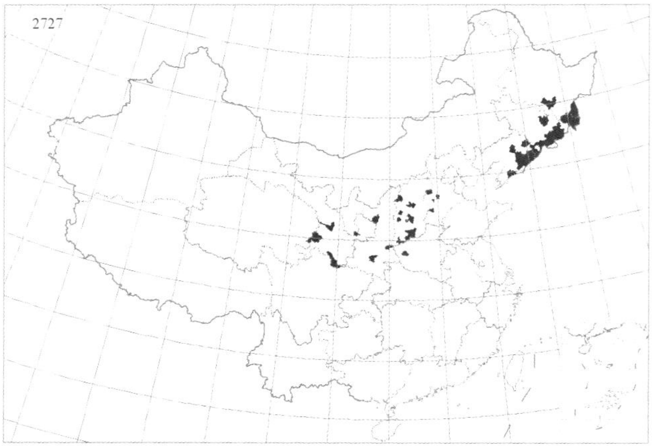

Saxifragaceae 虎耳草科 *Ribes* 茶藨子属

Ribes komarovii var. cuneifolium Liou 楔叶长白茶藨子

Shrubs, deciduous

4.5 (2.2~7.0); -14.4 (-16.9~-11.0); 21.0 (18.5~22.8); 8.2 (6.7~9.3); 63.3 (47.6~76.0); 68.2 (51.4~82.7); 578 (513~624) ❙ 753 (561~1005); 462 (333~635); 26 (16~39) ❙ 29.6 (-2.6~68.8); 575 (513~624); 310.6 (287.1~323.4)

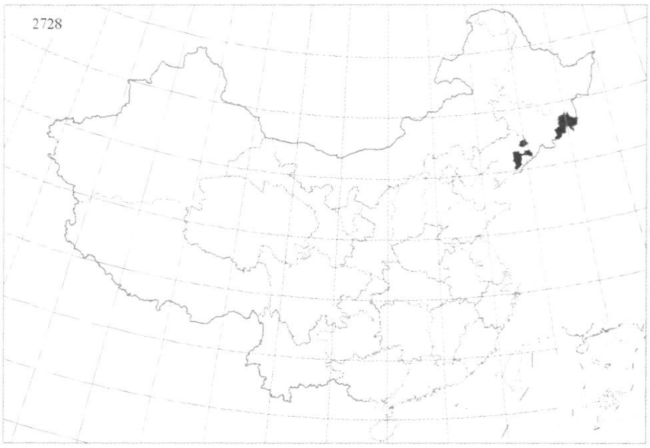

Saxifragaceae 虎耳草科 *Ribes* 茶藨子属

Ribes laciniatum Hook. f. et Thomson 裂叶茶藨子

Shrubs, deciduous

6.1 (-1.8~14.3); -2.2 (-12.3~7.6); 13.3 (6.1~19.7); 7.3 (1.8~14.3); 44.9 (1.9~111.9); 30.9 (0.0~82.8); 515 (329~728) ❙ 770 (320~1431); 463 (217~817); 21 (8~47) ❙ 44.6 (-3.0~113.0); 513 (320~707); 259.5 (61.6~363.7)

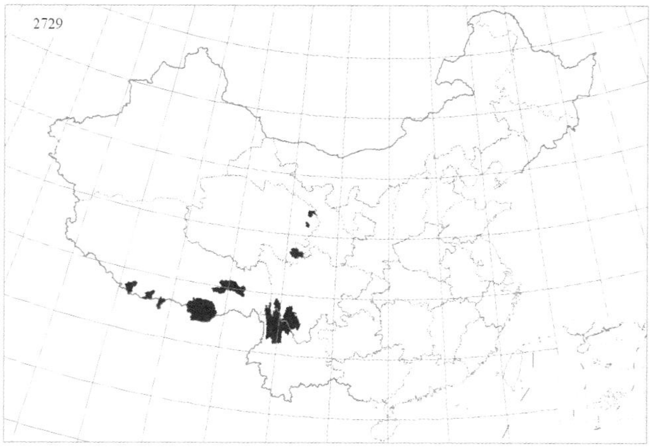

Saxifragaceae 虎耳草科 *Ribes* 茶藨子属

Ribes latifolium Jancz. 阔叶茶藨子

Shrubs, deciduous

3.0 (1.4~4.5); -15.8 (-18.0~-14.1); 19.6 (18.4~21.1); 7.3 (6.4~8.2); 54.2 (45.8~63.5); 77.4 (68.3~88.8); 541 (501~581) ❙ 639 (561~777); 382 (333~473); 22 (16~35) ❙ 18.9 (-2.6~55.1); 537 (501~566); 313.7 (289.3~329.9)

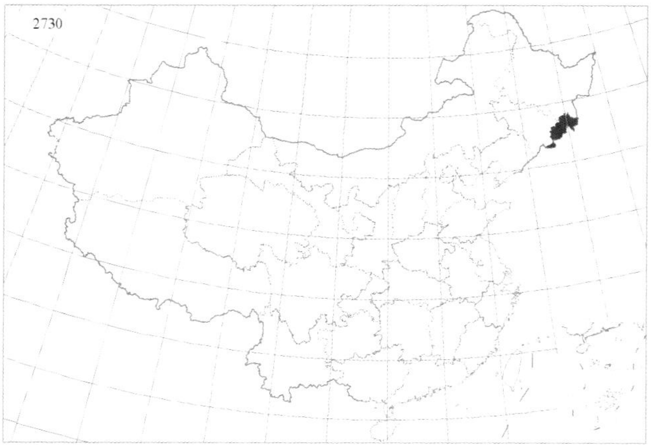

Saxifragaceae 虎耳草科 *Ribes* 茶藨子属
Ribes laurifolium Jancz. 桂叶茶藨子
Shrubs, evergreen

15.2 (8.6~18.9); 7.9 (1.3~12.5); 20.9 (15.0~25.6); 15.3 (8.7~19.0); 124.0 (52.6~167.5); 1.0 (0.0~8.8); 771 (574~914) ❙ 1163 (874~1454); 641 (489~816); 49 (25~71) ❙ 52.3 (6.5~99.9); 757 (574~851); 339.7 (154.4~516.7)

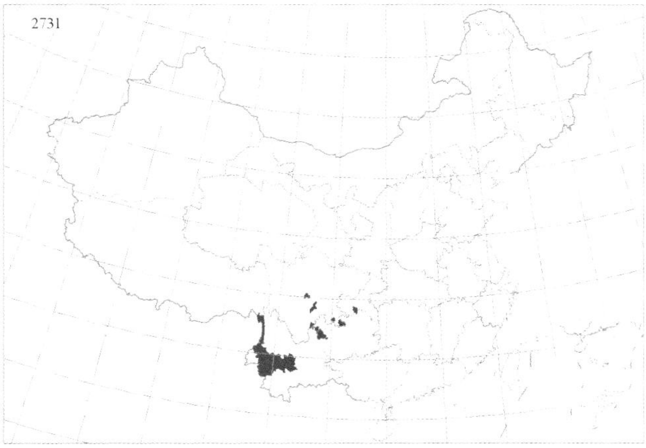

Saxifragaceae 虎耳草科 *Ribes* 茶藨子属
Ribes longiracemosum Franch. 长序茶藨子
Shrubs, deciduous

11.5 (2.2~15.1); 0.1 (-6.6~5.3); 22.1 (9.9~27.5); 11.7 (3.8~15.2); 91.3 (15.5~128.0); 12.5 (0.0~49.1); 699 (413~865) ❙ 881 (626~1414); 446 (306~817); 38 (14~83) ❙ 30.5 (-22.2~125.9); 672 (413~817); 231.1 (139.1~352.3)

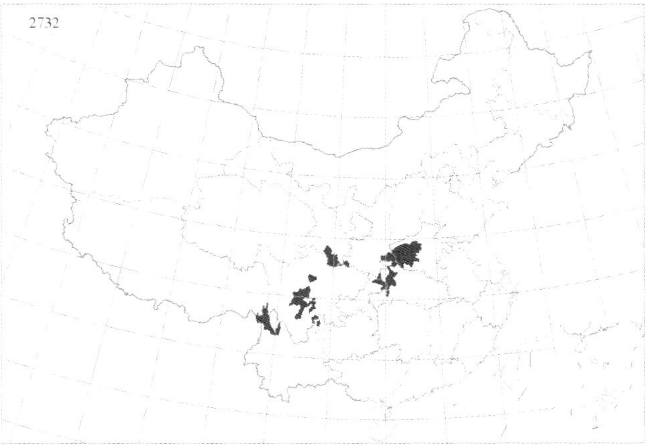

Saxifragaceae 虎耳草科 *Ribes* 茶藨子属
Ribes longiracemosum var. davidii Jancz. 腺毛茶藨子
Shrubs, deciduous

3.8; -4.4; 11.1; 4.8; 23.5; 37.4; 447 ❙ 837; 486; 24 ❙ 90.0; 447; 260.9

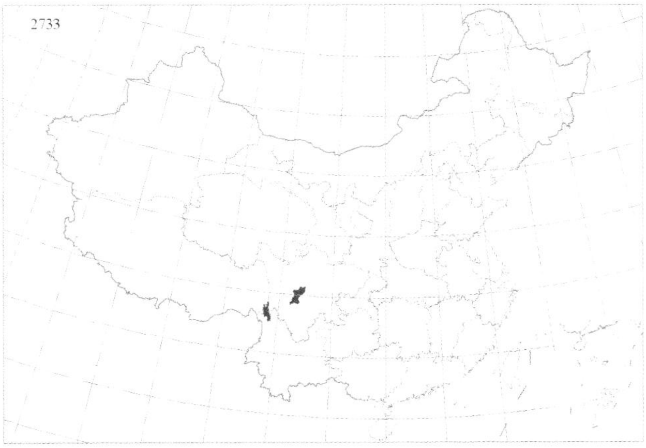

Saxifragaceae 虎耳草科 *Ribes* 茶藨子属
Ribes longiracemosum var. pilosum T. C. Ku 毛长串茶藨子
Shrubs, deciduous

8.6; 1.3; 15.0; 8.7; 52.6; 8.8; 574 ❙ 1041; 585; 45 ❙ 81.5; 574; 352.3

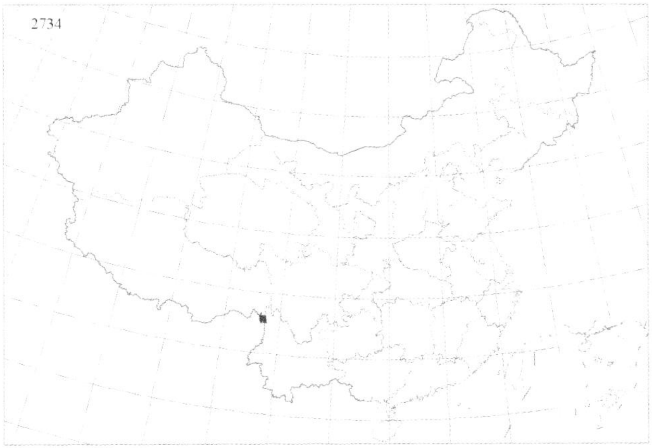

Saxifragaceae 虎耳草科 *Ribes* 茶藨子属
Ribes luridum Hook. f. et Thomson 紫花茶藨子
Shrubs, deciduous

8.4 (0.3~16.7); 0.2 (-9.9~9.7); 15.6 (9.3~24.9); 9.0 (3.3~16.7); 61.4 (11.9~140.9); 19.6 (0.0~67.5); 576 (391~819) ❙ 967 (527~1736); 537 (317~1007); 34 (9~106) ❙ 69.9 (14.4~145.8); 572 (391~791); 305.1 (182.9~447.4)

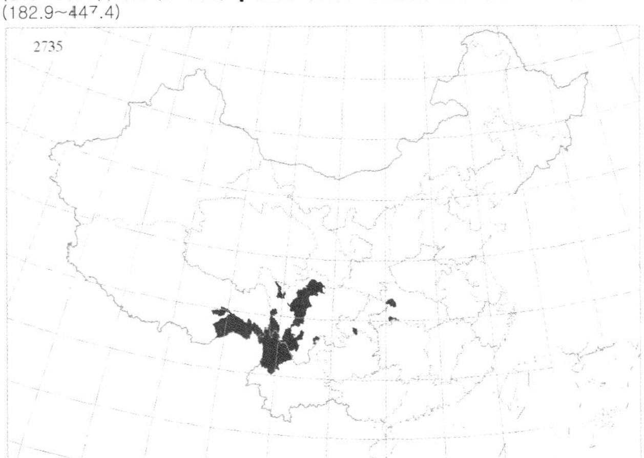

Saxifragaceae 虎耳草科 *Ribes* 茶藨子属
Ribes mandshuricum (Maxim.) Kom. 东北茶藨子
Shrubs, deciduous

8.6 (-4.6~14.4); -7.5 (-28.8~1.8); 22.9 (14.0~28.0); 10.4 (5.0~14.5); 84.1 (32.4~124.7); 40.0 (5.8~149.1); 668 (448~859) ❙ 601 (249~1098); 365 (179~691); 21 (5~65) ❙ -9.4 (-54.1~74.1); 569 (249~812); 190.5 (70.3~346.5)

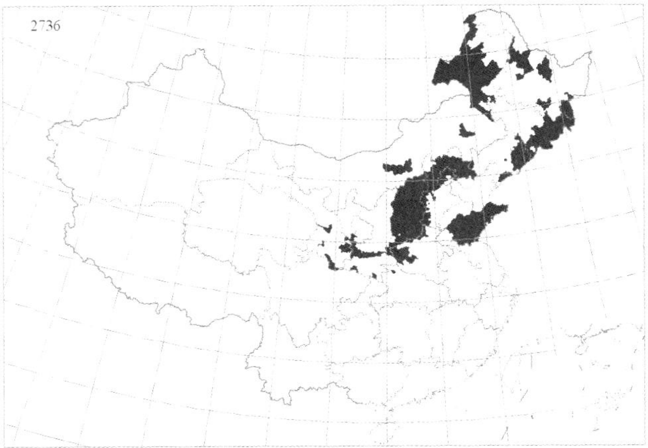

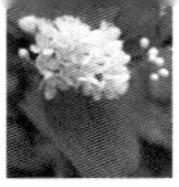

Saxifragaceae 虎耳草科 *Ribes* 茶藨子属
Ribes mandshuricum var. subglabrum Kom. 光耀北茶藨子
Shrubs, deciduous

7.6 (3.3~13.3); -8.1 (-15.3~-0.2); 21.5 (14.0~26.6); 9.4 (5.6~13.4); 73.6 (32.4~114.9); 42.1 (11.7~68.0); 620 (467~804) ❙ 584 (367~1005); 342 (235~635); 18 (10~44) ❙ -5.6 (-37.7~68.8); 546 (367~773); 202.9 (109.4~323.4)

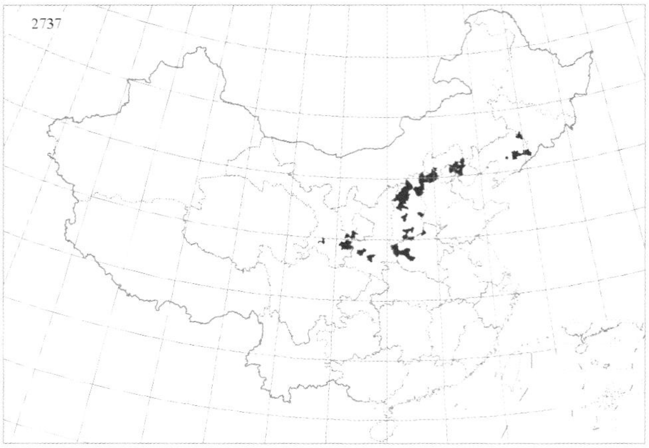

Saxifragaceae 虎耳草科 *Ribes* 茶藨子属
Ribes maximowiczianum Kom. 尖叶茶藨子
Small shrubs, deciduous

3.0 (-0.9~7.0); -17.4 (-24.7~-7.2); 20.6 (17.8~22.8); 7.7 (6.5~9.3); 59.0 (47.6~76.0); 81.8 (38.3~120.7); 565 (513~624) ❙ 663 (495~1073); 405 (256~691); 23 (10~40) ❙ 16.5 (-11.1~74.1); 558 (495~624); 294.5 (183.9~335.3)

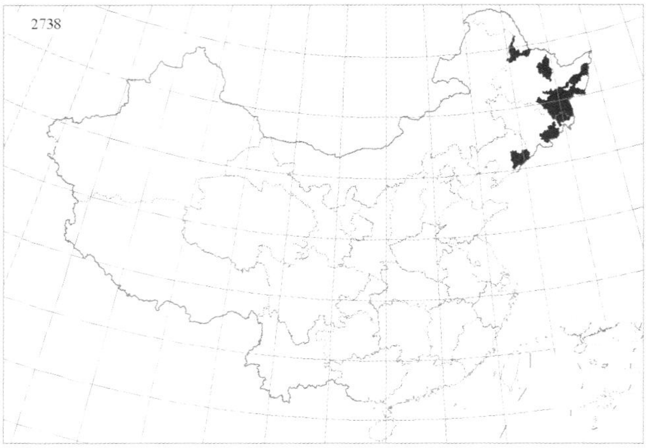

Saxifragaceae 虎耳草科 *Ribes* 茶藨子属
Ribes maximowiczii Batalin 华西茶藨子
Shrubs, deciduous

11.0 (5.7~14.8); 0.7 (-4.1~4.3); 20.5 (14.2~24.9); 11.1 (6.4~14.9); 82.9 (38.1~119.2); 10.7 (0.7~29.9); 659 (504~791) ❙ 1008 (669~1220); 495 (317~688); 38 (15~83) ❙ 54.1 (11.5~101.4); 659 (504~791); 259.8 (147.3~292.6)

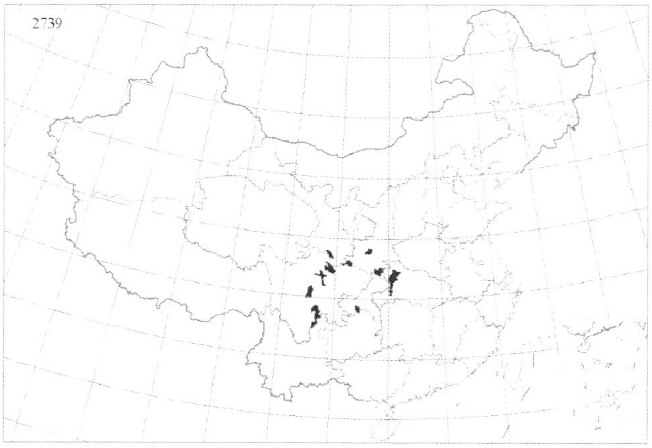

Saxifragaceae 虎耳草科 *Ribes* 茶藨子属
Ribes meyeri Maxim. 天山茶藨子
Shrubs, deciduous

4.0 (-5.7~12.4); -13.4 (-20.9~-0.3); 18.6 (6.3~27.5); 7.6 (1.6~13.2); 58.1 (2.0~114.1); 69.0 (12.1~133.0); 566 (294~813) ❙ 225 (32~677); 112 (19~339); 14 (0~48) ❙ -55.6 (-95.9~55.3); 212 (32~644); 96.0 (1.6~346.5)

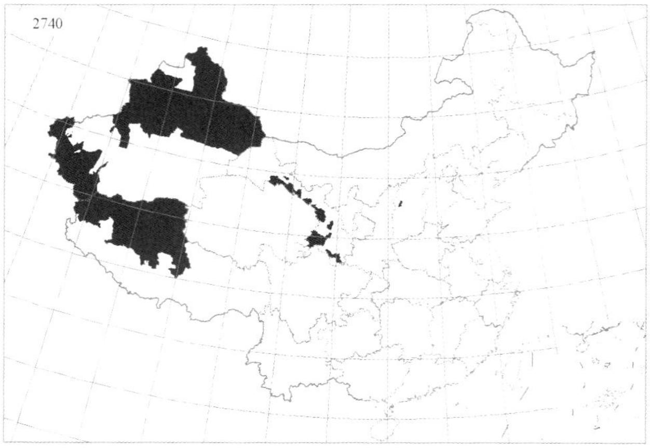

Saxifragaceae 虎耳草科 *Ribes* 茶藨子属
Ribes meyeri var. pubescens L. T. Lu 北疆茶藨子
Shrubs, deciduous

2.6 (-2.8~7.6); -16.5 (-20.3~-13.1); 18.7 (10.9~25.2); 7.3 (3.5~10.7); 54.6 (16.3~93.1); 82.9 (61.0~109.2); 554 (390~721) ❙ 212 (144~357); 96 (56~188); 20 (6~45) ❙ -60.1 (-76.8~-23.3); 212 (144~357); 110.4 (16.6~297.3)

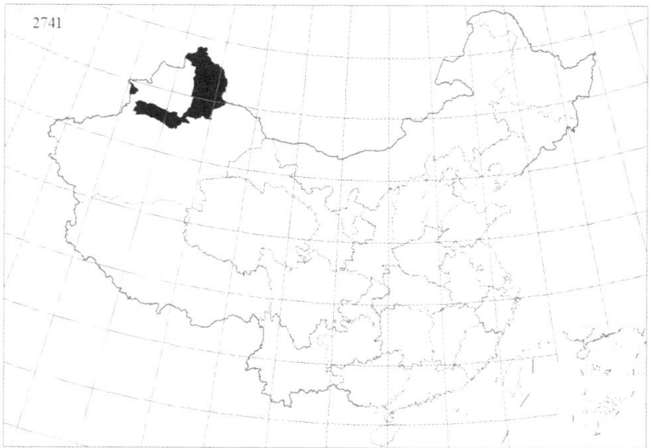

Saxifragaceae 虎耳草科 *Ribes* 茶藨子属
Ribes moupinense Franch. 宝兴茶藨子
Shrubs, deciduous

10.1 (-1.8~17.4); -0.6 (-12.4~11.0); 19.8 (6.1~28.1); 10.6 (1.8~17.4); 80.2 (1.9~149.3); 18.6 (0.0~82.8); 643 (329~924) ❙ 855 (320~1571); 432 (190~816); 33 (6~174) ❙ 33.0 (-40.9~128.6); 627 (320~924); 250.2 (61.2~487.2)

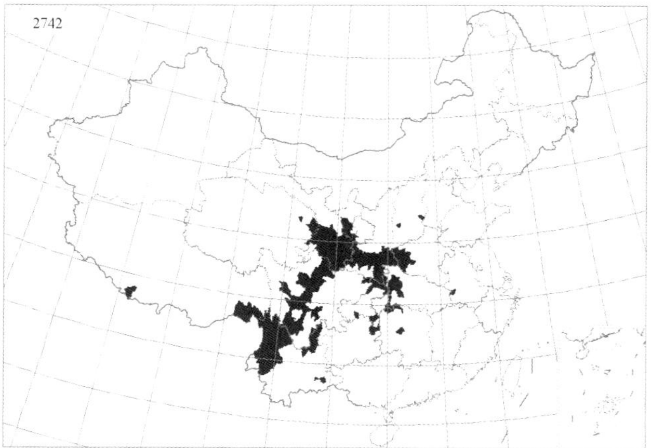

Saxifragaceae 虎耳草科 *Ribes* 茶藨子属

Ribes moupinense var. muliense S. H. Yu et J. M. Xu 木里茶藨子

Shrubs, deciduous

6.6; -1.3; 13.3; 6.8; 39.6; 19.7; 514 ┃ 862; 548; 12 ┃ 67.7; 514; 311.5

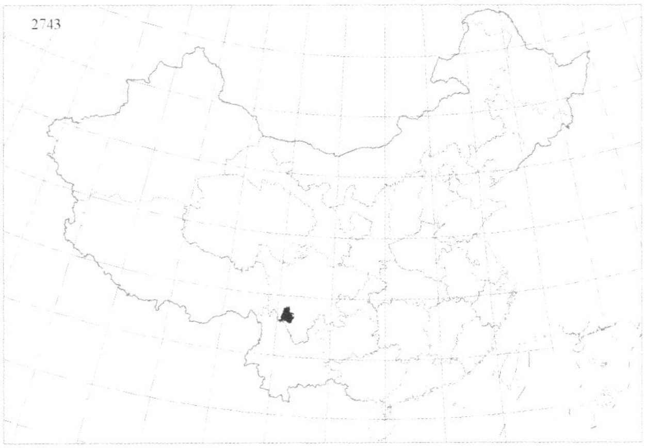

Saxifragaceae 虎耳草科 *Ribes* 茶藨子属

Ribes moupinense var. tripartitum (Batalin) Jancz. 三裂茶藨子

Shrubs, deciduous

10.9 (2.9~14.8); 1.9 (-6.3~7.8); 19.2 (11.2~24.9); 11.1 (4.4~14.9); 81.4 (20.5~119.2); 9.5 (0.0~44.6); 651 (428~791) ┃ 995 (677~1246); 503 (336~651); 38 (14~71) ┃ 54.9 (-0.2~121.7); 650 (428~791); 291.3 (182.9~382.6)

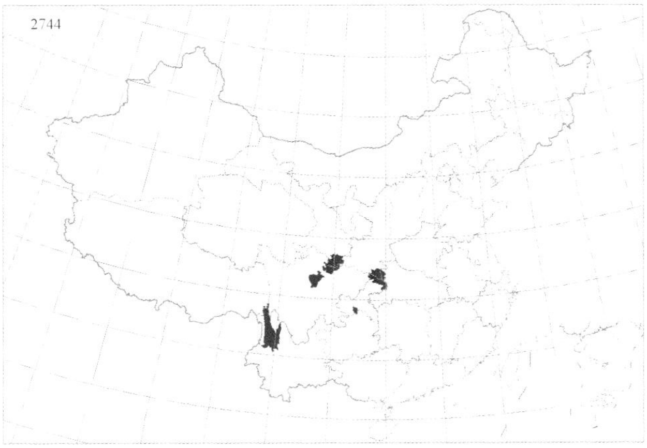

Saxifragaceae 虎耳草科 *Ribes* 茶藨子属

Ribes nigrum L. 黑茶藨子

Shrubs, deciduous

2.2 (-5.8~9.3); -17.9 (-29.8~-9.3); 19.3 (9.6~26.5); 7.3 (2.5~11.8); 55.6 (9.7~104.4); 88.4 (40.7~157.4); 553 (329~786) ┃ 313 (69~777); 173 (24~473); 20 (4~51) ┃ -41.0 (-89.2~55.1); 298 (69~606); 153.9 (14.2~329 9)

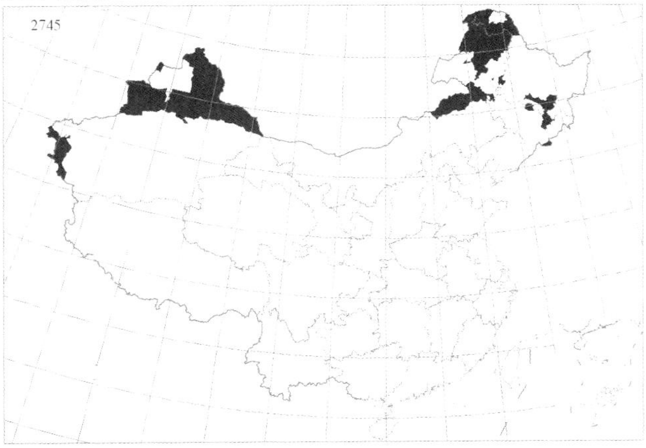

Saxifragaceae 虎耳草科 *Ribes* 茶藨子属

Ribes orientale Desf. 东方茶藨子

Shrubs, deciduous

2.1 (-3.3~13.3); -7.2 (-15.2~5.4); 10.5 (6.1~19.6); 4.4 (1.7~13.4); 21.9 (1.9~100.4); 55.8 (0.0~103.2); 420 (324~706) ┃ 601 (250~1736); 359 (191~1007); 17 (1~103) ┃ 38.0 (-27.8~145.8); 410 (250~706); 222.1 (32.2~447.4)

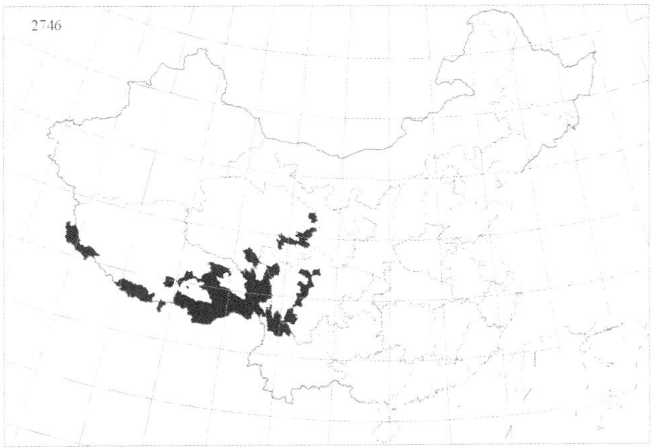

Saxifragaceae 虎耳草科 *Ribes* 茶藨子属

Ribes palczewskii (Jancz.) Pojark. 英吉利茶藨子

Shrubs, deciduous

0.0 (-5.8~7.7); -22.0 (-29.8~-9.1); 19.6 (15.3~22.6); 6.7 (4.4~9.5); 51.2 (28.0~77.8); 110.5 (45.2~157.4); 526 (426~632) ┃ 435 (244~657); 291 (152~415) 1 (5~20) ┃ -16.1 (-60.8~23.8); 421 (244~536); 220.6 (38.8~316.0)

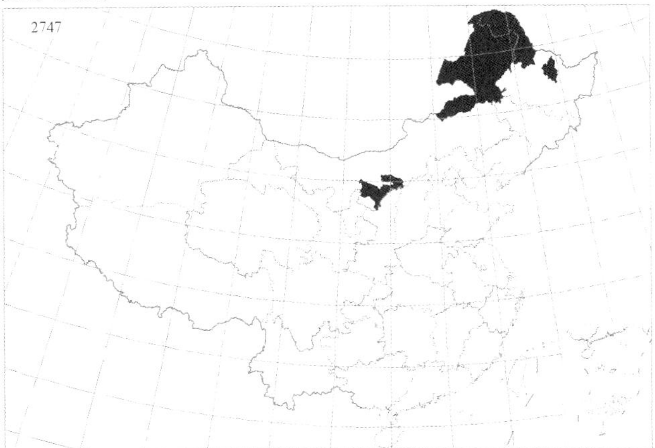

Saxifragaceae 虎耳草科 *Ribes* 茶藨子属

Ribes procumbens Pall. 水葡萄茶藨子

Small shrubs, deciduous

-1.9 (-5.8~3.8); -24.5 (-29.8~-16.8); 18.3 (15.3~22.2); 5.9 (4.4~8.4); 43.6 (28.0~65.9); 125.5 (79.6~157.4); 493 (426~604) ┃ 472 (429~510); 320 (300~339); 13 (6~16) ┃ -3.9 (-28.7~15.9); 453 (426~487); 279.1 (185.5~309.5)

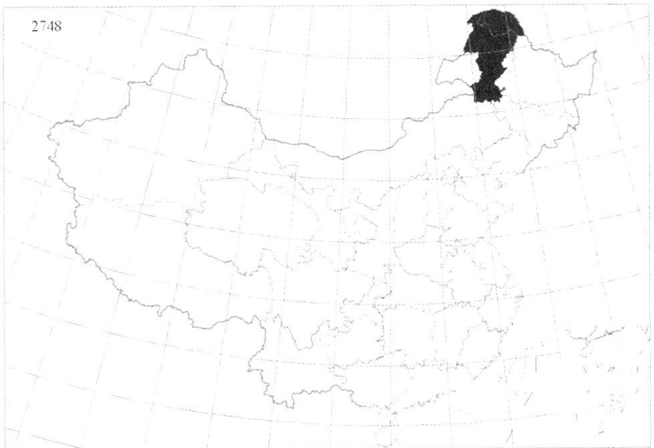

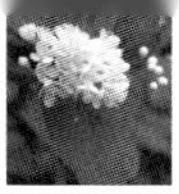

Saxifragaceae 虎耳草科 *Ribes* 茶藨子属

Ribes pseudofasciculatum K. S. Hao 青海茶藨子

Shrubs, deciduous

2.2 (−0.3~6.6); −7.0 (−10.9~−1.3); 10.2 (9.0~13.3); 4.1 (3.0~6.8); 18.0 (10.4~39.6); 51.1 (19.7~73.4); 419 (378~514) ❙ 747 (581~984); 456 (347~601); 14 (11~19) ❙ 77.9 (47.6~125.9); 419 (378~514); 272.0 (208.8~314.5)

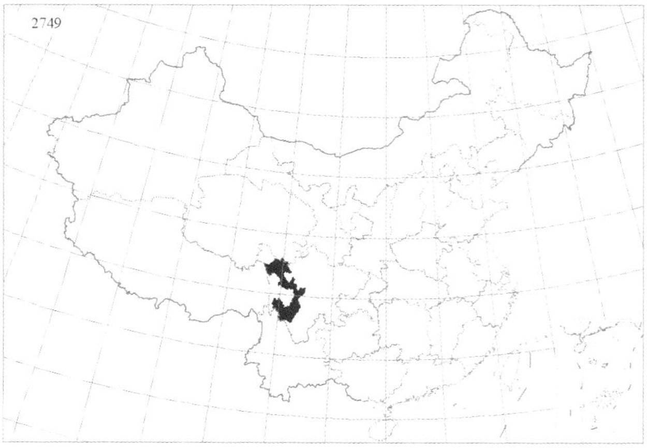

Saxifragaceae 虎耳草科 *Ribes* 茶藨子属

Ribes pulchellum Turcz. 美丽茶藨子

Shrubs, deciduous

6.8 (−1.5~13.5); −9.5 (−25.5~−1.6); 21.2 (9.7~27.1); 9.1 (2.9~13.7); 71.1 (11.3~117.5); 48.8 (14.6~128.7); 613 (359~820) ❙ 458 (128~1023); 282 (81~535); 12 (2~28) ❙ −24.0 (−80.8~128.6); 446 (128~668); 154.9 (13.6~313.2)

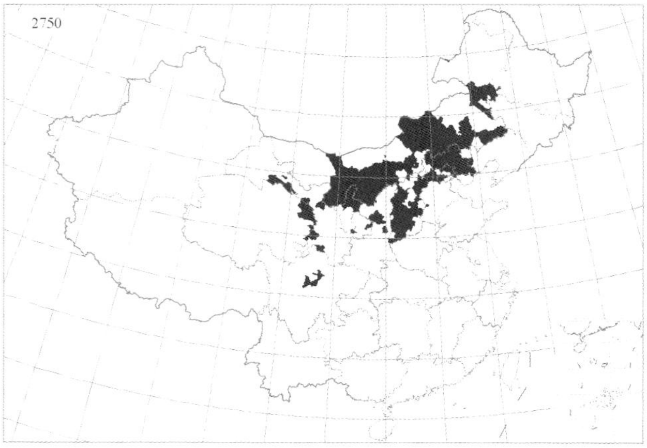

Saxifragaceae 虎耳草科 *Ribes* 茶藨子属

Ribes reclinatum L. 欧洲醋栗

Shrubs, deciduous

3.6 (−1.3~9.3); −15.4 (−20.9~−9.3); 18.3 (13.2~27.1); 7.5 (4.6~12.1); 56.1 (26.7~107.6); 72.8 (40.7~103.9); 564 (440~810) ❙ 293 (140~406); 134 (53~211); 26 (12~48) ❙ −44.4 (−82.8~−7.6); 293 (140~406); 199.8 (50.3~297.3)

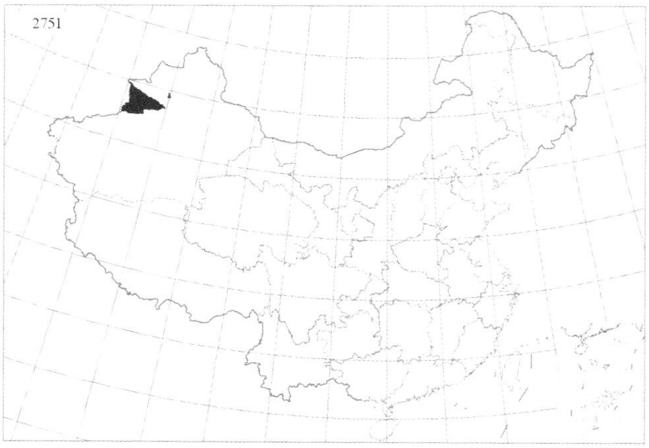

Saxifragaceae 虎耳草科 *Ribes* 茶藨子属

Ribes rubrisepalum L. T. Lu 红萼茶藨子

Shrubs, deciduous

4.8 (1.9~11.3); −5.3 (−9.2~−1.4); 14.0 (9.9~23.7); 6.1 (3.8~11.5); 37.0 (15.4~92.4); 38.3 (15.2~57.8); 490 (413~697) ❙ 829 (621~1005); 429 (318~546); 22 (13~28) ❙ 74.8 (5.8~128.6); 490 (413~697); 291.1 (248.3~357.7)

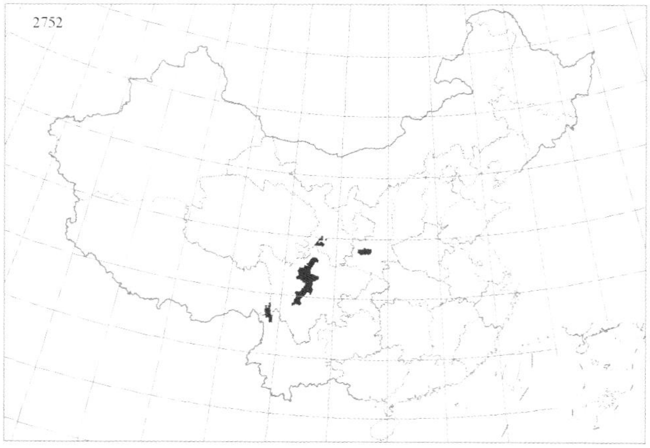

Saxifragaceae 虎耳草科 *Ribes* 茶藨子属

Ribes saxatile Pall. 石生茶藨子

Shrubs, deciduous

4.2 (−2.8~10.2); −14.7 (−20.9~−8.5); 19.9 (10.9~26.1); 8.1 (3.5~11.7); 62.7 (16.3~98.7); 71.2 (36.6~109.2); 587 (390~750) ❙ 201 (45~450); 95 (27~211); 17 (3~51) ❙ −63.1 (−94.0~3.9); 200 (45~406); 94.2 (7.6~297.3)

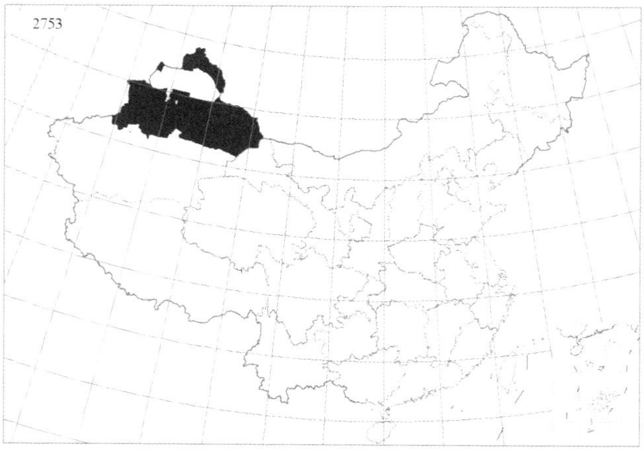

Saxifragaceae 虎耳草科 *Ribes* 茶藨子属

Ribes setchuense Jancz. 四川茶藨子

Shrubs, deciduous

5.8 (0.8~14.9); −4.0 (−10.6~5.3); 14.6 (9.9~23.6); 6.8 (3.8~15.0); 42.7 (15.5~119.6); 32.8 (0.0~66.0); 511 (411~782) ❙ 897 (590~1410); 466 (299~816); 24 (11~49) ❙ 75.8 (26.3~125.9); 511 (411~782); 270.3 (200.8~341.6)

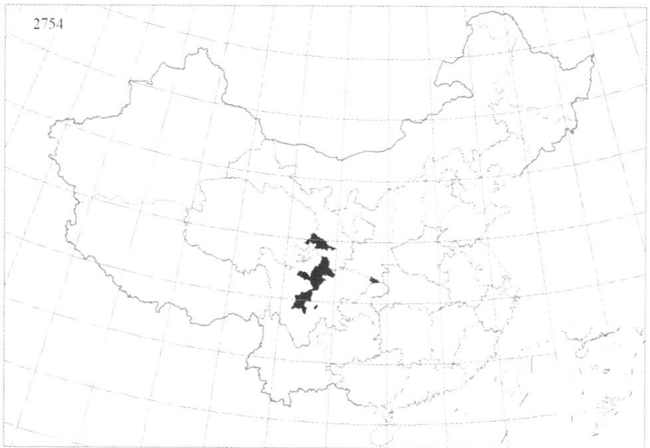

Saxifragaceae 虎耳草科 *Ribes* 茶藨子属
Ribes stenocarpum Maxim. 长果茶藨子
Shrubs, deciduous

3.8 (-3.2~11.3); -8.9 (-15.4~-0.7); 15.0 (7.3~23.7); 6.1 (2.0~11.5); 39.3 (4.0~92.4); 53.0 (14.1~101.5); 487 (335~697) ❙ 490 (148~950); 266 (92~490); 10 (4~25) ❙ 2.5 (-74.8~121.7); 431 (148~697); 198.0 (61.2~357.7)

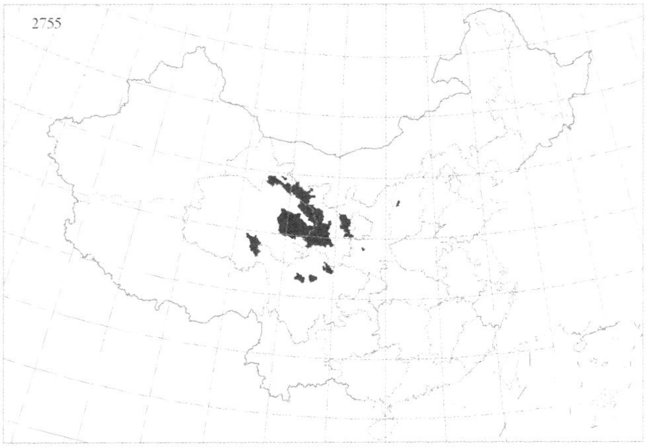

Saxifragaceae 虎耳草科 *Ribes* 茶藨子属
Ribes takare D. Don 渐尖茶藨子
Shrubs, deciduous

10.8 (-1.8~19.2); 0.2 (-12.7~13.3); 20.1 (6.1~27.5); 11.3 (1.8~19.4); 87.8 (1.9~172.6); 17.9 (0.0~82.8); 669 (329~932) ❙ 887 (290~1736); 467 (193~1007); 37 (4~106) ❙ 33.9 (-28.9~145.8); 643 (290~923); 256.8 (61.6~516.7)

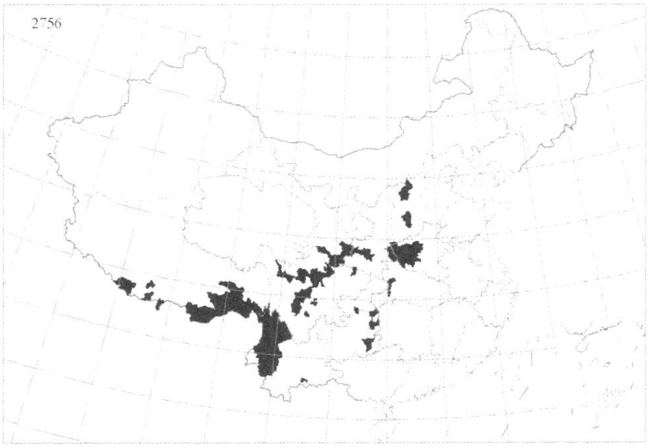

Saxifragaceae 虎耳草科 *Ribes* 茶藨子属
Ribes takare var. desmocarpum (Hook. f. et Thomson) L. T. Lu 束果茶藨子
Shrubs, deciduous

5.2 (-1.8~13.3); -2.9 (-10.2~5.4); 12.3 (6.1~19.6); 6.5 (1.8~13.4); 38.5 (1.9~100.4); 36.0 (0.0~82.8); 488 (329~706) ❙ 794 (290~1736); 469 (193~1007); 28 (4~58) ❙ 48.6 (-26.9~145.8); 478 (290~706); 237.5 (61.6~447.4)

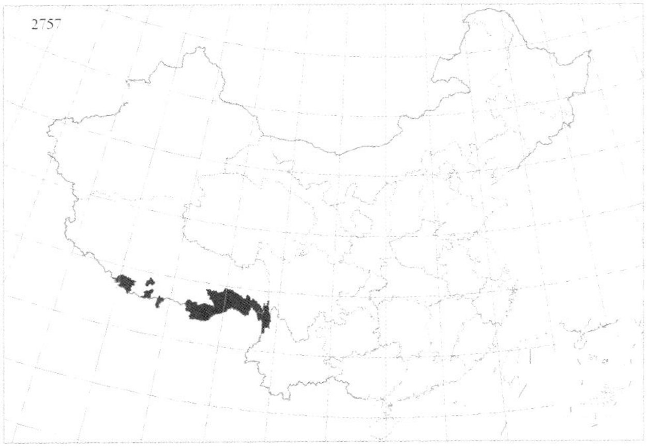

Saxifragaceae 虎耳草科 *Ribes* 茶藨子属
Ribes tenue Jancz. 细枝茶藨子
Shrubs, deciduous

11.0 (-1.4~17.2); 0.4 (-15.4~9.7); 20.9 (8.5~29.3); 11.4 (2.7~17.2); 87.5 (8.3~146.8); 14.4 (0.0~92.1); 680 (363~941) ❙ 1025 (282~1828); 483 (182~817); 60 (5~253) ❙ 51.1 (-40.9~128.6); 669 (282~941); 252.0 (61.2~407.1)

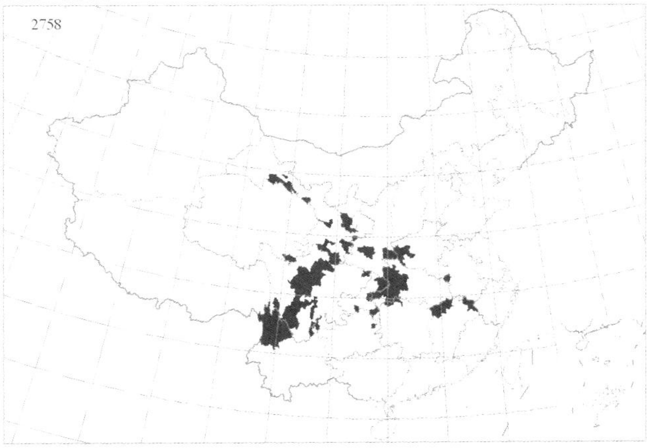

Saxifragaceae 虎耳草科 *Ribes* 茶藨子属
Ribes tianquanense S. H. Yu et J. M. Xu 天全茶藨子
Shrubs, evergreen

13.8; 4.3; 22.5; 13.9; 108.9; 2.3; 748 ❙ 1295; 738; 42 ❙ 72.8; 748; 228.4

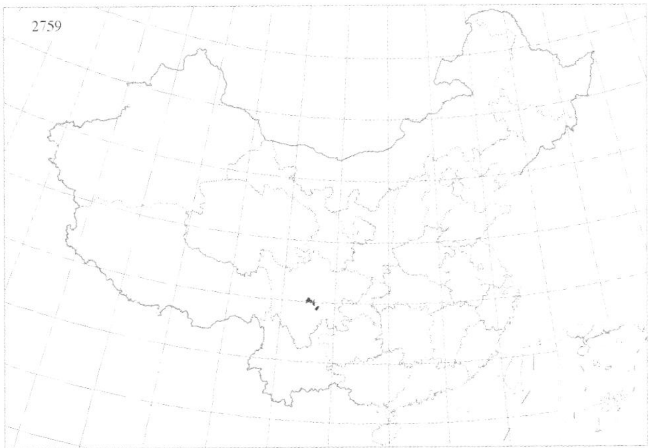

Saxifragaceae 虎耳草科 *Ribes* 茶藨子属
Ribes triste Pall. 矮茶藨子
Shrubs, deciduous

1.1 (-5.0~7.0); -19.9 (-29.4~-11.0); 19.3 (16.5~22.4); 6.9 (5.0~9.3); 51.3 (34.4~76.0); 98.0 (51.4~154.8); 527 (448~624) ❙ 635 (447~966); 397 (300~616); 21 (12~38) ❙ 19.3 (-12.2~64.7); 522 (447~624); 305.9 (276.7~337.1)

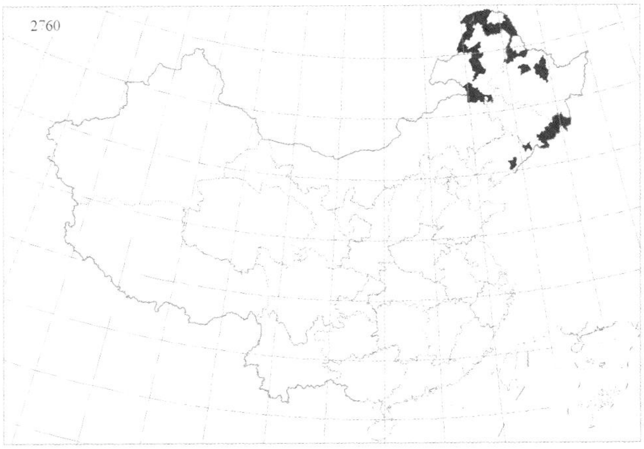

Saxifragaceae 虎耳草科 *Ribes* 茶藨子属
Ribes vilmorinii Jancz. 小果茶藨子
Shrubs, deciduous

7.2 (-1.4~14.1); -4.6 (-15.4~6.5); 17.9 (9.3~24.6); 8.4 (3.3~14.2); 60.0 (11.9~109.8); 32.7 (0.0~92.1); 573 (375~729) ❙ 744 (282~1049); 399 (182~641); 20 (5~35) ❙ 34.4 (-34.2~128.6); 543 (282~717); 247.5 (93.5~344.4)

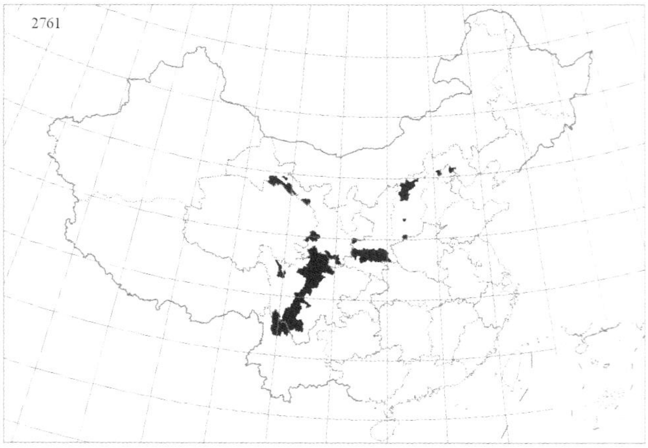

Saxifragaceae 虎耳草科 *Ribes* 茶藨子属
Ribes viridiflorum (Cheng) L. T. Lu et G. Yao 绿花茶藨子
Shrubs, deciduous

15.3 (13.4~18.2); 3.4 (1.5~7.8); 27.0 (25.2~28.7); 15.4 (13.4~18.3); 127.6 (108.3~159.6); 3.3 (0.0~7.3); 841 (755~965) ❙ 1551 (1364~1706); 601 (529~666); 175 (129~220) ❙ 84.8 (55.4~108.0); 841 (755~965); 238.9 (189.8~284.0)

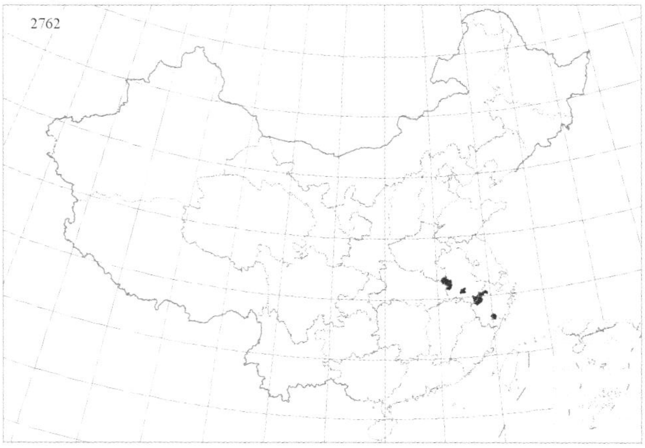

Saxifragaceae 虎耳草科 *Schizophragma* 钻地风属
Schizophragma choufenianum Chun 临桂钻地风
Scandent shrubs, deciduous

17.9 (14.6~21.2); 7.9 (3.8~12.8); 26.5 (24.6~28.6); 17.9 (14.7~21.3); 155.6 (117.7~195.4); 0.2 (0.0~1.5); 934 (779~1115) ❙ 1565 (1260~1740); 642 (544~720); 163 (58~191) ❙ 69.7 (31.6~101.8); 934 (779~1115); 244.5 (185.5~318.5)

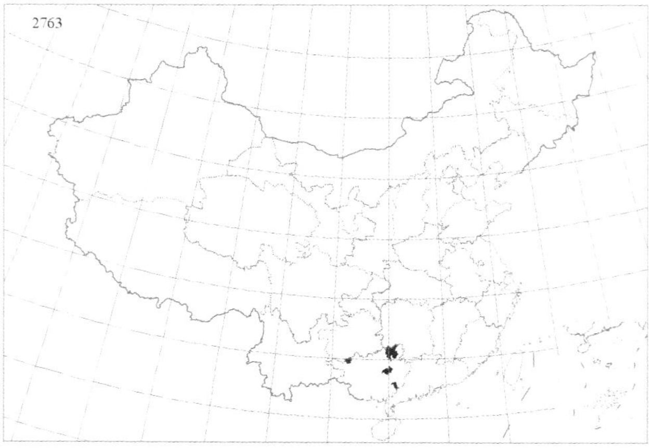

Saxifragaceae 虎耳草科 *Schizophragma* 钻地风属
Schizophragma corylifolium Chun 秦榛钻地风
Scandent shrubs, deciduous

16.5 (14.4~19.0); 4.8 (2.1~9.3); 27.5 (26.3~29.1); 16.5 (14.5~19.1); 139.9 (118.7~168.8); 1.4 (0.0~4.9); 892 (801~986) ❙ 1624 (1274~1922); 572 (454~644); 207 (153~259) ❙ 82.3 (41.9~111.3); 890 (801~986); 253.0 (183.9~319.9)

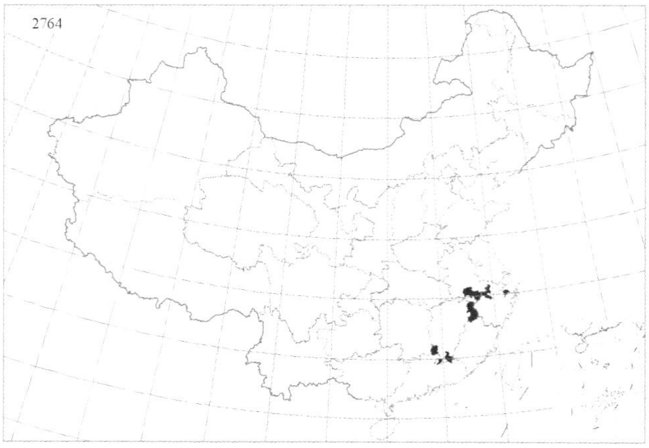

Saxifragaceae 虎耳草科 *Schizophragma* 钻地风属
Schizophragma crassum Hand.-Mazz. 厚叶钻地风
Scandent shrubs, deciduous

10.0 (5.4~18.6); 3.0 (-2.1~12.7); 15.9 (12.3~22.5); 10.1 (5.8~18.7); 69.8 (31.5~163.9); 9.1 (0.0~25.6); 615 (481~888) ❙ 1034 (741~1547); 606 (426~911); 38 (21~63) ❙ 67.2 (52.8~81.5); 615 (481~888); 353.0 (273.5~551.3)

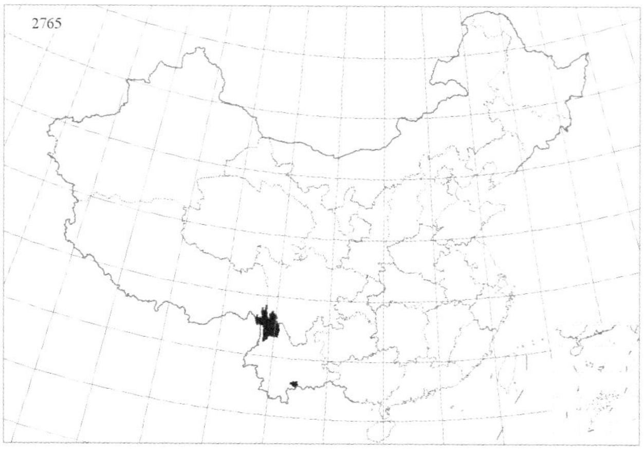

Saxifragaceae 虎耳草科 *Schizophragma* 钻地风属
Schizophragma crassum var. hsitaoanum (Chun) C. F. Wei 维西钻地风
Scandent shrubs, deciduous

9.6; 2.4; 15.7; 9.6; 60.8; 5.4; 600 ❙ 1007; 578; 40 ❙ 67.9; 600; 351.7

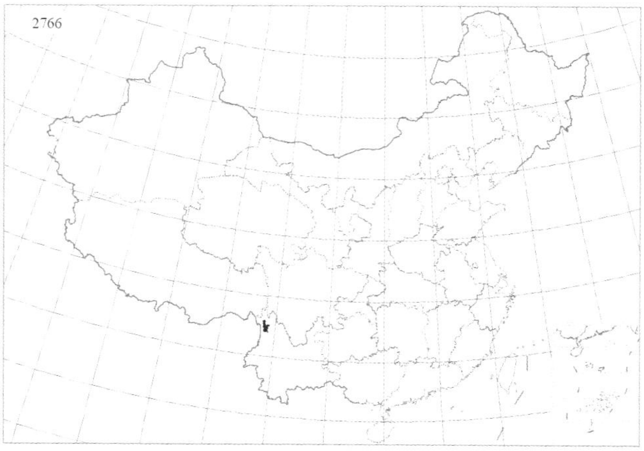

Saxifragaceae 虎耳草科 *Schizophragma* 钻地风属
Schizophragma elliptifolium C. F. Wei 椭圆钻地风
Scandent shrubs, deciduous

13.8 (9.0~16.0); 4.2 (0.6~6.3); 22.7 (16.5~25.8); 13.9 (9.0~16.1); 108.3 (58.7~133.0); 1.7 (0.0~10.3); 746 (584~831) ❙ 1129 (918~1410); 570 (473~816); 54 (32~88) ❙ 52.2 (30.0~101.6); 746 (584~831); 215.1 (170.4~293.1)

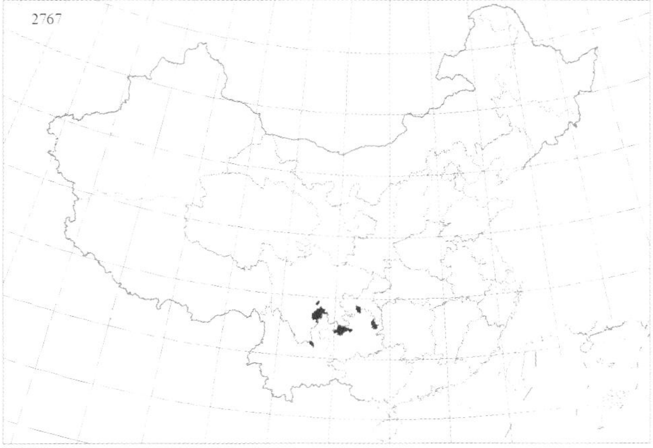

Saxifragaceae 虎耳草科 *Schizophragma* 钻地风属
Schizophragma fauriei Hayata 圆叶钻地风
Scandent shrubs, deciduous

19.2 (15.0~24.1); 12.5 (2.6~18.9); 25.0 (20.3~28.0); 19.2 (15.0~24.1); 170.9 (123.9~229.7); 0.3 (0.0~3.5); 964 (783~1322) ❙ 2511 (1621~3495); 1006 (600~1594); 330 (47~897) ❙ 165.3 (32.5~235.6); 945 (783~1095); 261.4 (154.2~363.7)

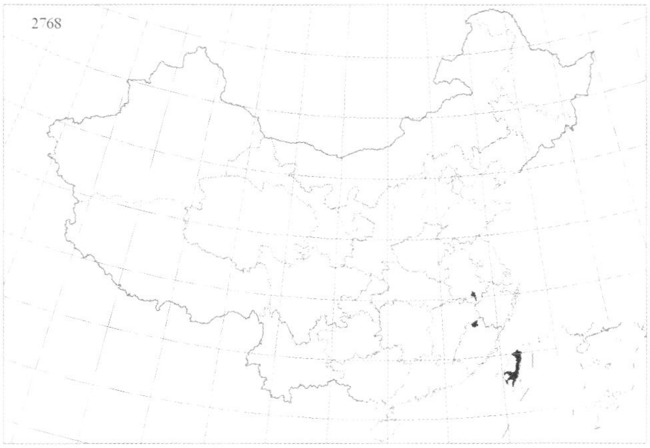

Saxifragaceae 虎耳草科 *Schizophragma* 钻地风属
Schizophragma hypoglaucum Rehder 白背钻地风
Scandent shrubs, deciduous

14.3 (10.1~17.8); 3.2 (-0.7~7.8); 24.9 (17.7~27.4); 14.4 (10.3~17.9); 116.2 (70.1~154.2); 3.5 (0.0~13.6); 788 (621~929) ❙ 1335 (1006~1689); 550 (429~816); 108 (22~221) ❙ 69.5 (18.3~106.4); 788 (621~929); 242.7 (149.6~319.9)

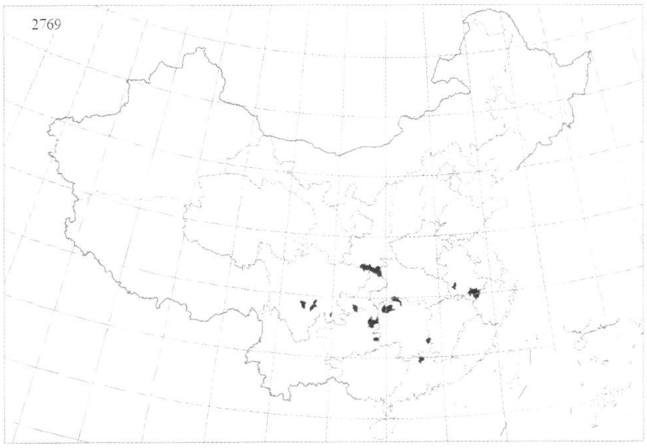

Saxifragaceae 虎耳草科 *Schizophragma* 钻地风属
Schizophragma integrifolium Oliv. 钻地风
Scandent shrubs, deciduous

16.0 (5.7~24.1); 4.8 (-3.5~18.9); 26.5 (14.2~29.8); 16.1 (6.4~24.1); 135.5 (38.1~229.7); 2.4 (0.0~28.1); 870 (504~1322) ❙ 1384 (626~2859); 547 (316~1594); 138 (22~402) ❙ 59.3 (-22.2~235.6); 858 (504~1151); 218.7 (128.7~551.6)

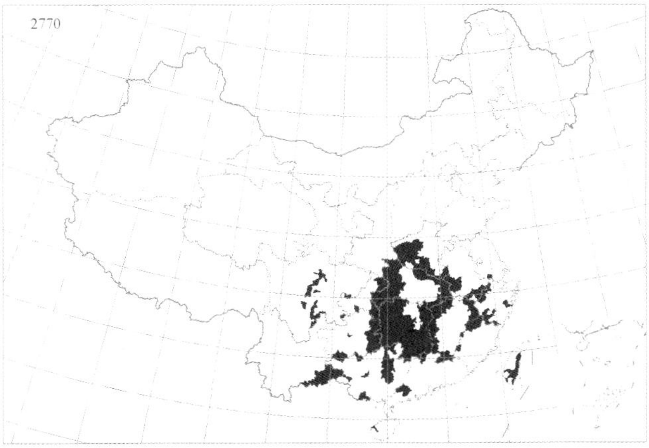

Saxifragaceae 虎耳草科 *Schizophragma* 钻地风属
Schizophragma integrifolium var. glaucescens Rehder 粉绿钻地风
Scandent shrubs, deciduous

16.5 (14.1~18.5); 5.1 (1.6~9.0); 27.2 (24.2~29.4); 16.5 (14.2~18.5); 139.5 (113.2~162.1); 0.9 (0.0~6.5); 884 (761~984) ❙ 1608 (1277~1997); 545 (411~680); 205 (58~259) ❙ 82.5 (39.1~130.7); 877 (761~961); 234.9 (163.7~319.9)

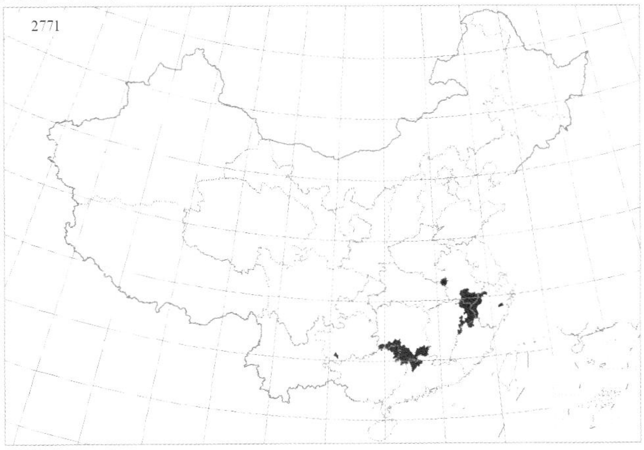

Saxifragaceae 虎耳草科 *Schizophragma* 钻地风属
Schizophragma megalocarpum Chun 大果钻地风
Scandent shrubs, deciduous

11.8; 2.0; 20.7; 11.9; 90.7; 9.1; 696 ❙ 1078; 527; 40 ❙ 60.3; 696; 218.0

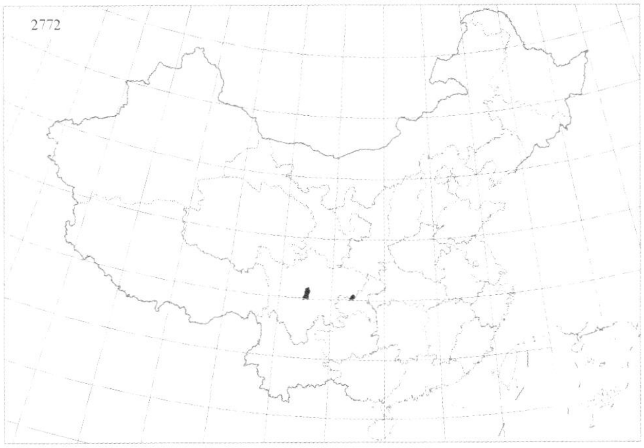

Saxifragaceae 虎耳草科 *Schizophragma* 钻地风属
Schizophragma molle (Rehder) Chun 柔毛钻地风
Scandent shrubs, deciduous

16.9 (7.3~21.2); 6.6 (-2.0~14.2); 26.2 (15.6~29.6); 17.0 (7.5~21.3); 144.4 (48.2~195.2); 0.6 (0.0~20.4); 891 (537~1126) ❙ 1568 (1001~1997); 603 (445~860); 165 (27~259) ❙ 76.8 (24.9~130.7); 889 (537~1126); 255.6 (154.4~551.6)

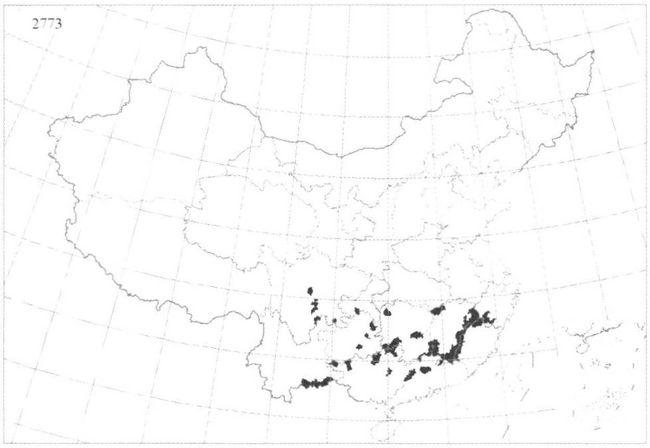

Pittosporaceae 海桐花科 *Pittosporum* 海桐花属
Pittosporum angustilimbum C. Y. Wu 窄叶海桐
Shrubs, evergreen

7.5; -0.4; 14.3; 7.7; 46.6; 16.4; 541 ❙ 950; 555; 25 ❙ 76.2; 541; 294.8

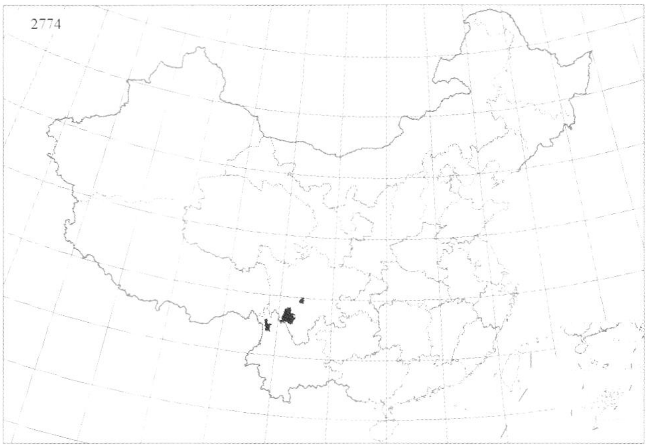

Pittosporaceae 海桐花科 *Pittosporum* 海桐花属
Pittosporum balansae DC. 聚花海桐
Shrubs, evergreen

22.3 (17.6~24.8); 14.8 (7.1~20.1); 28.1 (25.7~29.2); 22.3 (17.7~24.8); 208.2 (152.5~237.6); 0.0 (0.0~0.0); 1196 (906~1385) ❙ 1597 (868~2435); 694 (312~1428); 111 (41~189) ❙ 35.2 (-34.3~99.5); 1156 (868~1360); 301.4 (164.9~620.0)

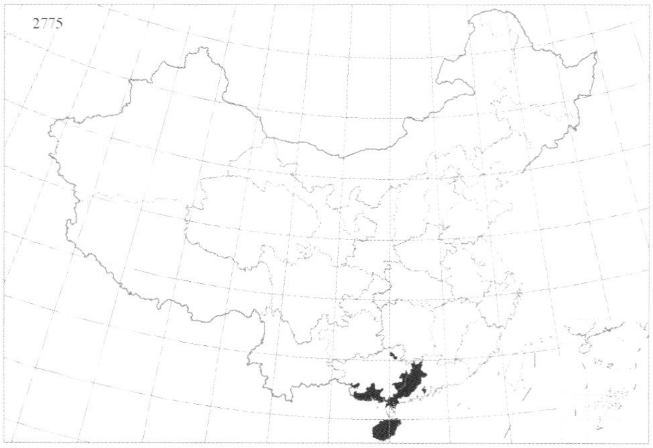

Pittosporaceae 海桐花科 *Pittosporum* 海桐花属
Pittosporum balansae var. angustifolium Gagnep. 窄叶聚花海桐
Shrubs, evergreen

22.3 (17.6~24.8); 14.9 (7.8~20.1); 28.1 (25.7~29.2); 22.4 (17.7~24.8); 208.4 (152.5~237.6); 0.0 (0.0~0.0); 1197 (906~1385) ❙ 1591 (868~2435); 665 (312~1428); 111 (41~177) ❙ 34.7 (-34.3~99.5); 1152 (868~1360); 310.6 (178.7~620.0)

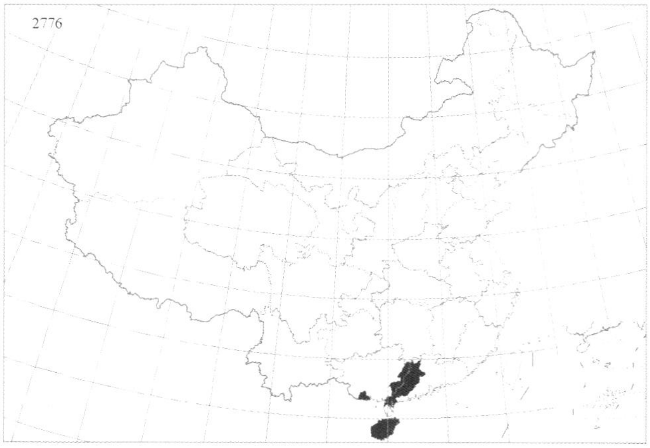

Pittosporaceae 海桐花科 *Pittosporum* 海桐花属
Pittosporum brevicalyx (Oliv.) Gagnep. 短萼海桐
Shrubs or small trees, evergreen

16.3 (5.3~22.4); 6.9 (-2.4~15.1); 24.3 (12.2~30.1); 16.3 (5.7~22.4); 136.5 (30.5~209.1); 0.7 (0.0~26.2); 850 (479~1204) ❙ 1285 (741~1888); 572 (299~1044); 105 (12~250) ❙ 51.0 (-14.2~128.9); 831 (479~1204); 256.3 (142.9~659.0)

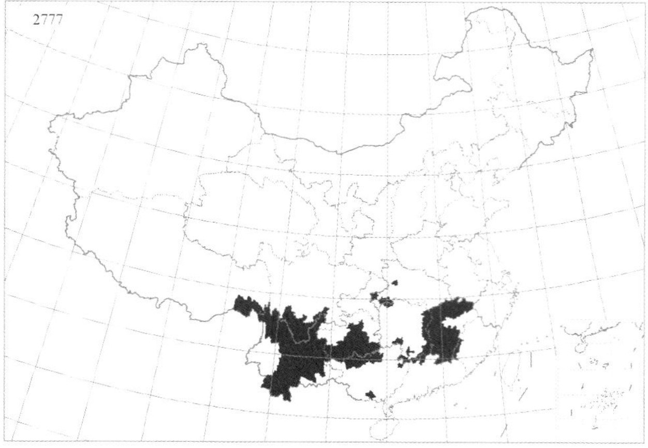

Pittosporaceae 海桐花科 *Pittosporum* 海桐花属
Pittosporum crispulum Gagnep. 皱叶海桐
Shrubs or small trees, evergreen

13.9 (6.0~18.6); 4.2 (-3.2~12.7); 22.8 (14.4~27.3); 14.0 (6.6~18.7); 110.8 (39.7~163.9); 3.0 (0.0~26.7); 758 (508~949) ❙ 1090 (863~1547); 557 (444~911); 52 (22~142) ❙ 45.5 (12.0~101.4); 758 (508~949); 226.3 (130.1~551.3)

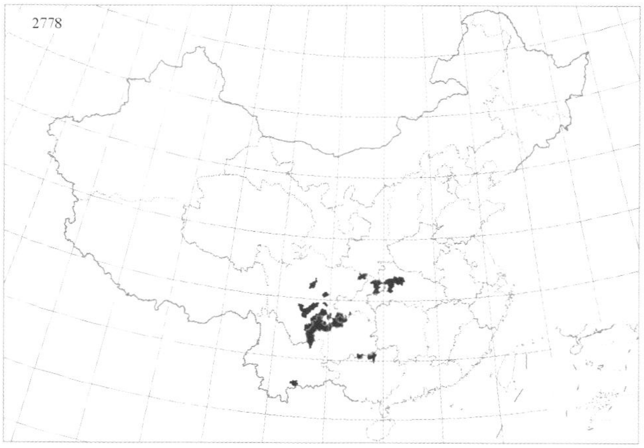

Pittosporaceae 海桐花科 *Pittosporum* 海桐花属
Pittosporum daphniphylloides Hayata 牛耳枫叶海桐
Shrubs or small trees, evergreen

17.9 (16.2~20.8); 12.8 (11.0~15.3); 22.0 (20.1~25.0); 18.0 (16.2~20.8); 155.5 (134.2~189.2); 0.0 (0.0~0.0); 858 (773~1019) ❙ 2639 (2445~2859); 1178 (947~1412); 250 (110~402) ❙ 211.1 (140.0~246.5); 858 (773~1019); 286.0 (207.7~363.7)

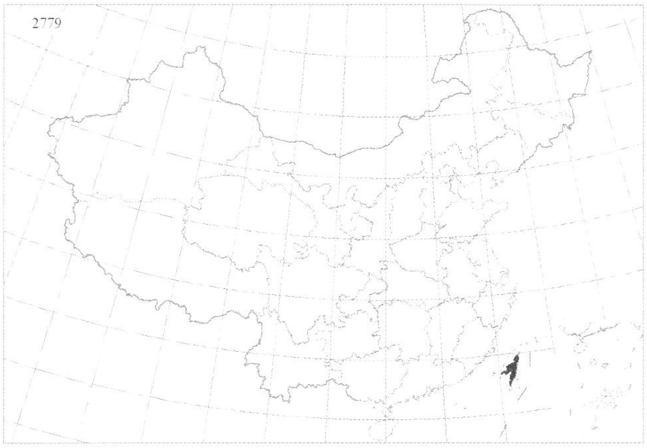

Pittosporaceae 海桐花科 *Pittosporum* 海桐花属
Pittosporum daphniphylloides var. adaphniphylloides (Hu et F. T. Wang) W. T. Wang 大叶海桐
Trees, evergreen

15.0 (7.3~18.5); 4.9 (-2.0~10.9); 24.4 (15.6~27.7); 15.1 (7.5~18.3); 122.1 (48.2~160.0); 1.5 (0.0~20.4); 799 (537~949) ❙ 1213 (956~1492); 567 (465~817); 66 (20~113) ❙ 53.1 (5.5~107.2); 795 (537~949); 214.9 (135.1~491.5)

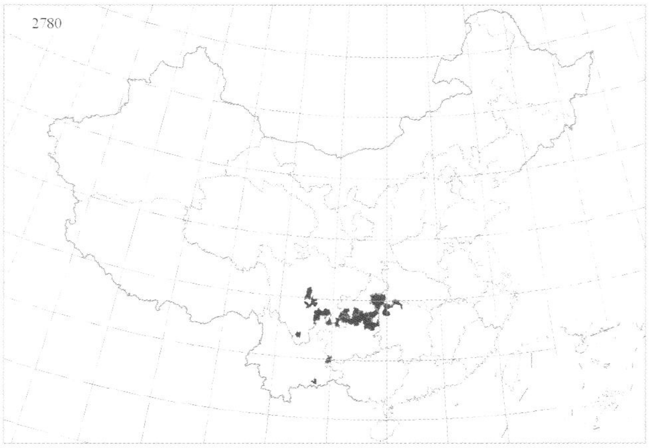

Pittosporaceae 海桐花科 *Pittosporum* 海桐花属
Pittosporum elevaticostatum H. T. Chang et S. Z. Yan 突肋海桐
Shrubs, evergreen

15.5 (10.6~18.9); 4.7 (-0.2~9.5); 25.4 (21.1~28.1); 15.5 (10.6~18.9); 127.7 (79.4~167.3); 1.5 (0.0~12.1); 824 (643~969) ❙ 1277 (1101~1494); 532 (452~670); 96 (46~161) ❙ 55.7 (24.4~94.5); 824 (643~969); 222.7 (148.9~284.3)

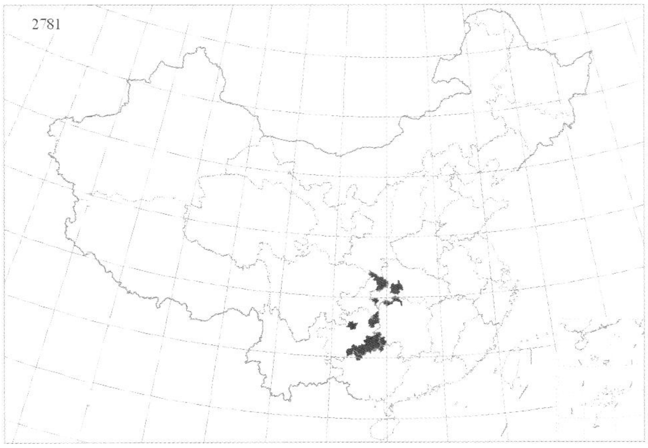

Pittosporaceae 海桐花科 *Pittosporum* 海桐花属
Pittosporum fulvipilosum H. T. Chang et S. Z. Yan 褐毛海桐
Shrubs or small trees, evergreen

19.4 (17.3~22.8); 9.6 (6.5~14.7); 27.7 (26.2~29.8); 19.4 (17.3~22.8); 173.0 (147.9~213.6); 0.0 (0.0~0.0); 1014 (906~1230) ❙ 1626 (1416~1961); 601 (505~893); 181 (106~197) ❙ 61.1 (27.0~95.5); 1011 (906~1230); 237.6 (174.9~275.2)

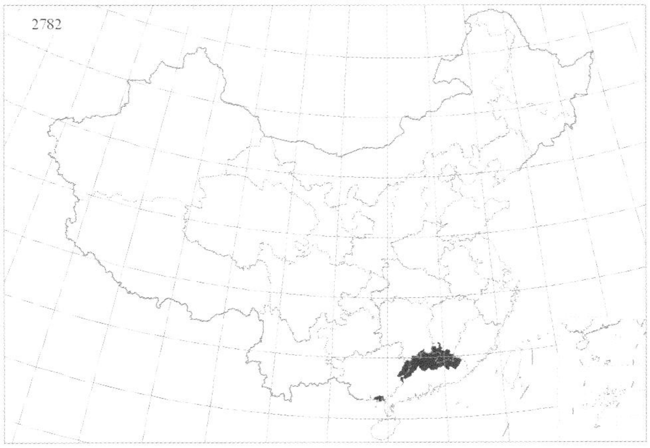

Pittosporaceae 海桐花科 *Pittosporum* 海桐花属
Pittosporum glabratum Lindl. 光叶海桐
Shrubs, evergreen

18.6 (6.0~23.2); 8.8 (-3.2~15.8); 27.1 (14.4~29.9); 18.6 (6.6~23.2); 164.0 (39.7~218.9); 0.5 (0.0~26.7); 983 (508~1257) ❙ 1544 (1001~2435); 641 (420~1428); 147 (28~259) ❙ 58.6 (5.8~130.7); 980 (508~1257); 231.8 (120.3~411.1)

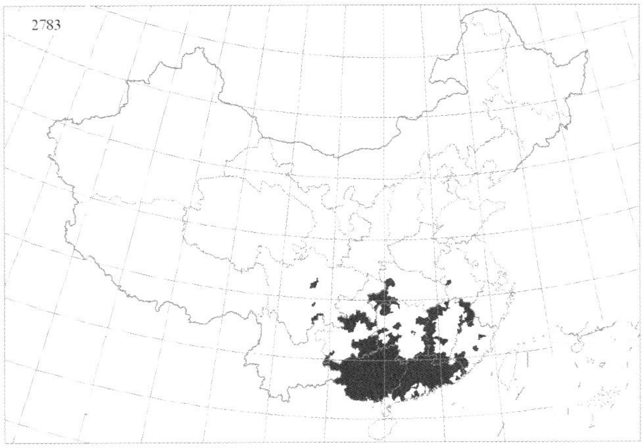

Pittosporaceae 海桐花科 *Pittosporum* 海桐花属
Pittosporum glabratum var. neriifolium Rehder et E. H. Wilson 狭叶海桐
Shrubs, evergreen

17.2 (9.3~22.3); 6.3 (-1.7~14.7); 27.3 (20.1~29.9); 17.3 (9.5~22.4); 148.1 (69.8~208.7); 0.6 (0.0~18.0); 921 (603~1200) ❙ 1482 (931~2037); 549 (323~1026); 157 (51~259) ❙ 61.3 (-1.6~128.9); 909 (603~1200); 209.4 (120.3~318.5)

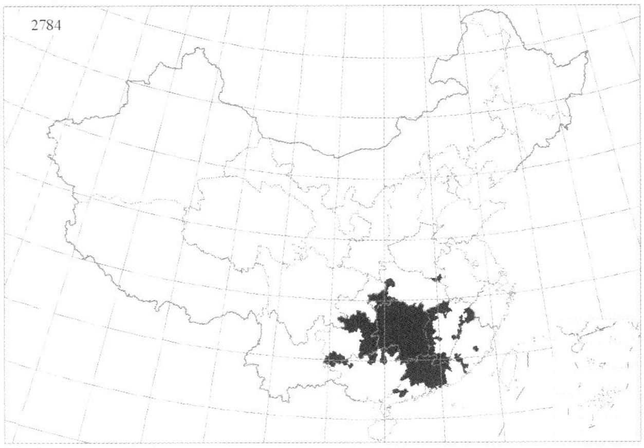

Pittosporaceae 海桐花科 *Pittosporum* 海桐花属

Pittosporum henryi Gowda 小柄果海桐

Shrubs, evergreen

15.2 (14.7~16.2); 4.5 (3.8~6.2); 25.4 (24.9~25.9); 15.2 (14.8~16.3); 123.7 (118.7~135.7); 0.8 (0.0~1.5); 808 (791~845) ❙ 1195 (1055~1280); 493 (473~522); 80 (64~94) ❙ 48.0 (24.9~56.8); 808 (791~845); 182.9 (154.4~211.8)

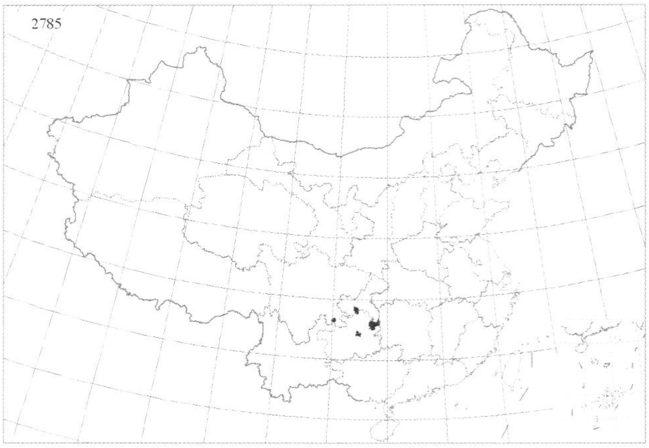

Pittosporaceae 海桐花科 *Pittosporum* 海桐花属

Pittosporum heterophyllum Franch. 异叶海桐

Shrubs, evergreen

11.1 (1.8~18.5); 2.5 (−7.1~10.7); 18.6 (9.5~28.1); 11.4 (3.8~18.3); 84.1 (15.5~160.0); 10.8 (0.0~53.3); 660 (404~958) ❙ 972 (583~1265); 541 (348~663); 31 (11~71) ❙ 52.6 (4.4~125.9); 649 (404~958); 256.4 (128.7~357.7)

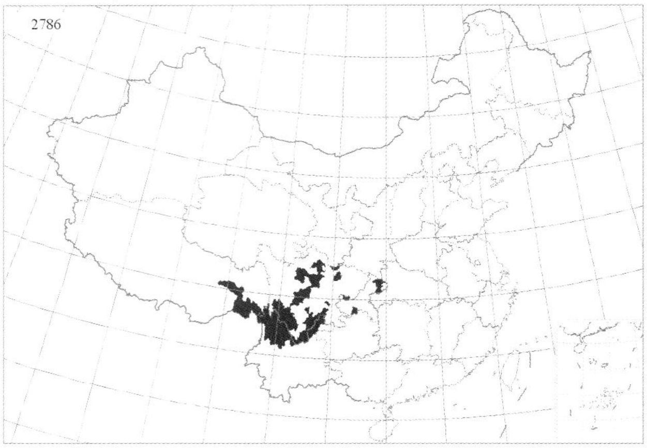

Pittosporaceae 海桐花科 *Pittosporum* 海桐花属

Pittosporum heterophyllum var. ledoides Hand.-Mazz. 带叶海桐

Shrubs, evergreen

15.0; 8.4; 20.2; 15.1; 121.2; 0.0; 753 ❙ 950; 570; 27 ❙ 26.3; 683; 307.7

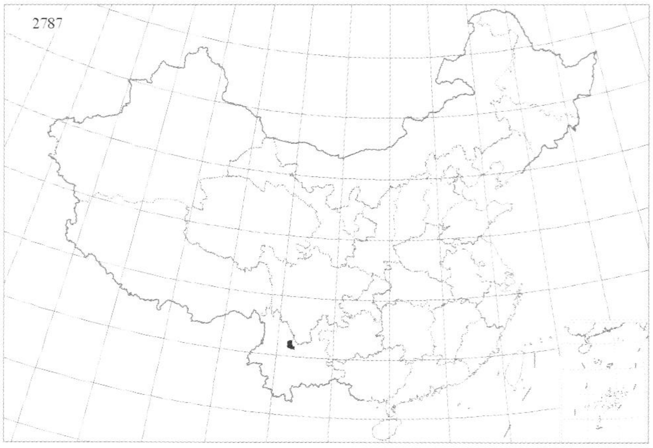

Pittosporaceae 海桐花科 *Pittosporum* 海桐花属

Pittosporum illicioides Makino 海金子

Shrubs, evergreen

16.4 (9.3~21.8); 4.9 (−2.9~15.3); 27.2 (20.1~30.1); 16.4 (9.5~21.8); 138.6 (69.8~202.0); 1.6 (0.0~20.5); 885 (603~1168) ❙ 1410 (626~2859); 522 (316~1412); 156 (26~402) ❙ 59.3 (−22.2~235.6); 871 (603~1134); 209.9 (128.7~332.5)

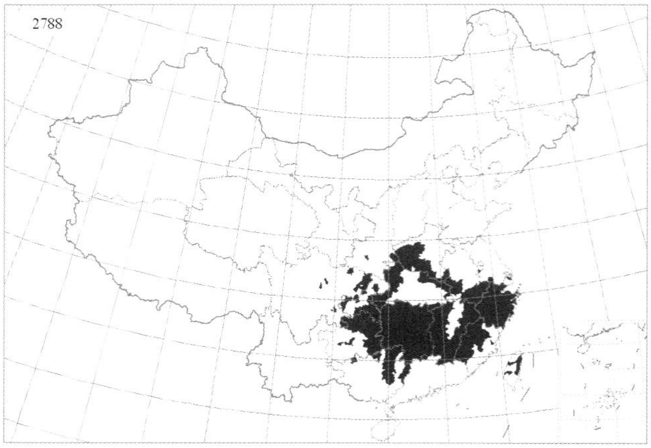

Pittosporaceae 海桐花科 *Pittosporum* 海桐花属

Pittosporum johnstonianum Gowda 滇西海桐

Trees, evergreen

10.2 (5.4~17.6); 2.4 (−2.1~11.0); 16.7 (12.3~26.3); 10.3 (5.8~17.6); 74.4 (31.5~151.6); 11.6 (0.0~25.6); 630 (481~909) ❙ 1051 (741~1496); 584 (426~816); 48 (12~142) ❙ 66.6 (54.0~81.5); 630 (481~909); 318.9 (225.9~434.4)

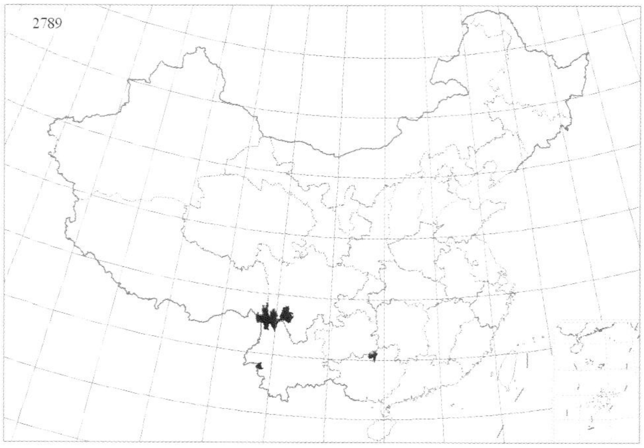

Pittosporaceae 海桐花科 *Pittosporum* 海桐花属

Pittosporum johnstonianum var. glomerulatum C. Y. Wu 密花海桐

Trees, evergreen

17.4; 11.0; 21.4; 17.4; 149.3; 0.0; 834 ❙ 1406; 816; 48 ❙ 68.7; 834; 434.4

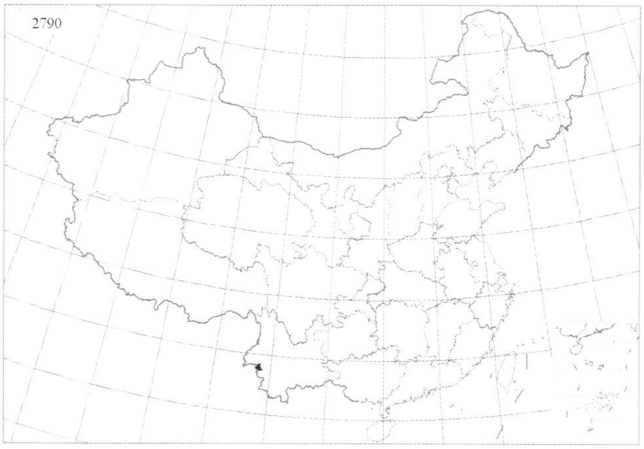

Pittosporaceae 海桐花科 *Pittosporum* 海桐花属
Pittosporum kerrii Craib 羊脆木
Trees, evergreen

16.8 (12.6~21.0); 10.3 (6.0~15.1); 21.3 (17.9~25.6); 16.8 (12.6~21.0); 142.1 (91.6~191.8); 0.0 (0.0~0.0); 821 (678~1053) ❙ 1171 (797~1751); 657 (411~996); 47 (23~77) ❙ 42.3 (-14.2~112.2); 799 (678~1053); 367.5 (207.5~659.0)

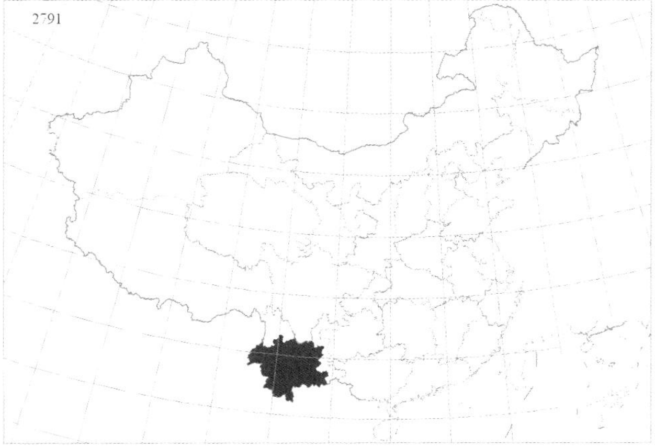

Pittosporaceae 海桐花科 *Pittosporum* 海桐花属
Pittosporum kunmingense H. T. Chang et S. Z. Yan 昆明海桐
Shrubs or trees, evergreen

16.5 (13.6~19.1); 9.4 (6.5~12.5); 21.6 (19.0~26.6); 16.5 (13.6~19.1); 138.0 (103.6~169.2); 0.0 (0.0~0.0); 812 (707~969) ❙ 1060 (907~1348); 601 (521~762); 44 (28~64) ❙ 30.7 (6.5~72.8); 786 (690~969); 288.2 (189.5~395.4)

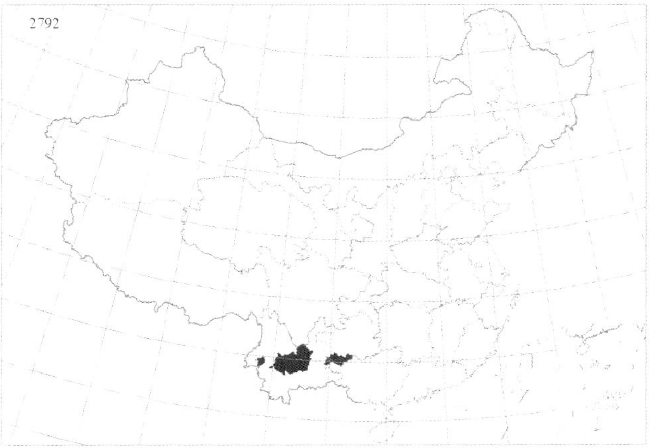

Pittosporaceae 海桐花科 *Pittosporum* 海桐花属
Pittosporum kwangsiense H. T. Chang et S. Z. Yan 广西海桐
Shrubs or trees, evergreen

19.5 (17.8~22.2); 11.0 (8.2~14.2); 25.9 (21.9~28.3); 19.5 (17.9~22.3); 173.8 (154.5~207.5); 0.0 (0.0~0.0); 990 (854~1190) ❙ 1427 (1139~1758); 741 (651~848); 89 (47~189) ❙ 45.0 (15.4~82.1); 986 (816~1190); 332.3 (224.9~551.6)

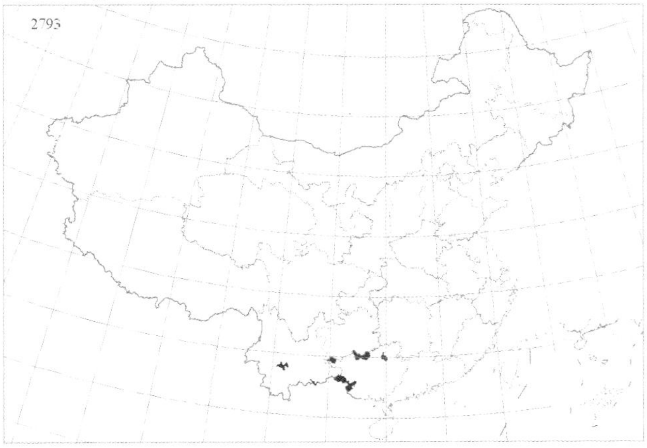

Pittosporaceae 海桐花科 *Pittosporum* 海桐花属
Pittosporum kweichowense Gowda 贵州海桐
Shrubs, evergreen

15.8; 5.9; 24.5; 15.8; 130.4; 0.4; 819 ❙ 1302; 587; 84 ❙ 59.2; 819; 232.2

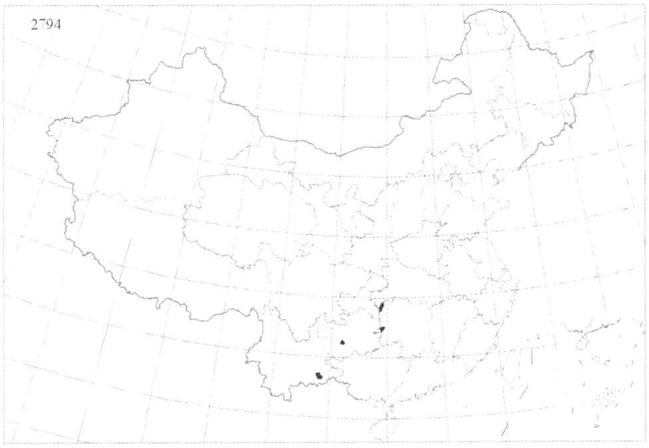

Pittosporaceae 海桐花科 *Pittosporum* 海桐花属
Pittosporum kweichowense var. buxifolium (K. M. Feng ex C. Y. Wu) Z. Y. Zhang et Turland 黄杨叶海桐
Shrubs, evergreen

17.5 (16.6~18.9); 9.7 (9.1~10.6); 23.2 (21.9~25.0); 17.6 (16.7~19.0); 150.9 (140.0~167.5); 0.0 (0.0~0.0); 859 (811~939) ❙ 1241 (1074~1422); 691 (582~813); 49 (47~52) ❙ 44.5 (32.4~65.6); 859 (811~939); 325.9 (277.0~359.7)

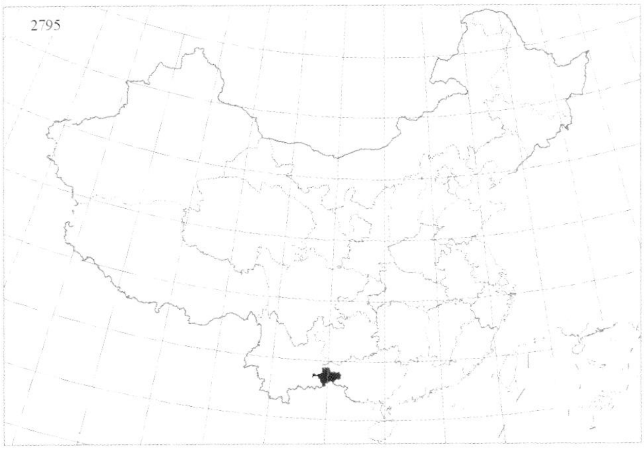

Pittosporaceae 海桐花科 *Pittosporum* 海桐花属
Pittosporum kweichowense var. podocarpifolium (C. Y. Wu) Z. Y. Zhang et Turland 罗汉松叶海桐
Shrubs, evergreen

18.9; 10.6; 25.0; 19.0; 167.5; 0.0; 939 ❙ 1292; 729; 49 ❙ 37.6; 939; 342.5

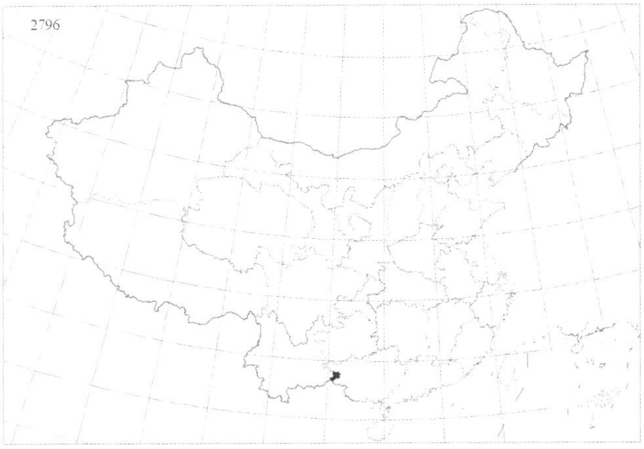

Pittosporaceae 海桐花科 *Pittosporum* 海桐花属
Pittosporum lenticellatum Chun ex H. Peng et Y. F. Deng 卵果海桐
Shrubs, evergreen

18.5 (16.0~21.3); 8.5 (5.3~12.3); 26.8 (24.4~28.7); 18.5 (16.0~21.3); 162.1 (132.2~196.1); 0.0 (0.0~0.0); 962 (826~1130) ❙ 1475 (1236~1776); 637 (523~728); 140 (60~189) ❙ 53.9 (22.7~85.5); 961 (826~1130); 230.3 (185.5~280.9)

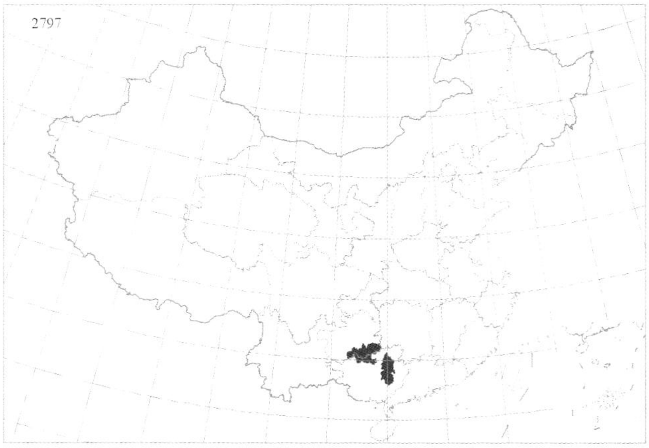

Pittosporaceae 海桐花科 *Pittosporum* 海桐花属
Pittosporum leptosepalum Gowda 薄萼海桐
Shrubs or small trees, evergreen

19.8 (16.7~22.7); 10.3 (5.6~15.0); 27.7 (26.2~29.8); 19.8 (16.8~22.8); 177.7 (141.2~213.1); 0.0 (0.0~0.0); 1037 (885~1221) ❙ 1663 (1302~2435); 679 (497~1428); 169 (86~197) ❙ 61.1 (12.9~99.5); 1035 (885~1221); 249.1 (174.9~327.7)

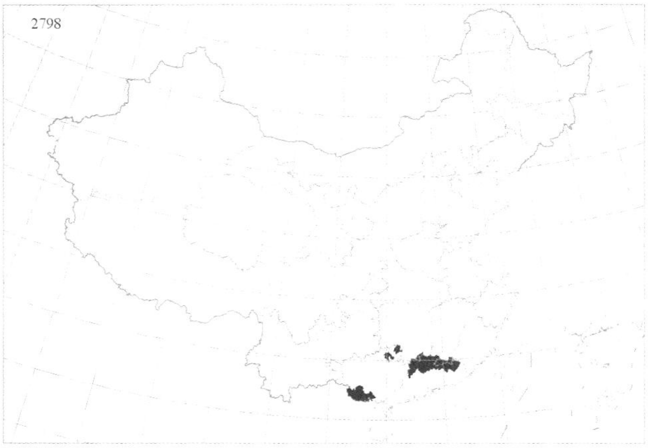

Pittosporaceae 海桐花科 *Pittosporum* 海桐花属
Pittosporum merrillianum Gowda 滇越海桐
Shrubs, evergreen

17.7 (16.3~21.0); 10.3 (9.1~14.2); 22.9 (21.4~25.6); 17.7 (16.4~21.0); 152.5 (136.4~191.8); 0.0 (0.0~0.0); 863 (797~1053) ❙ 1270 (1007~1522); 703 (539~860); 52 (44~68) ❙ 47.4 (19.0~80.6); 862 (797~1053); 378.0 (277.0~551.6)

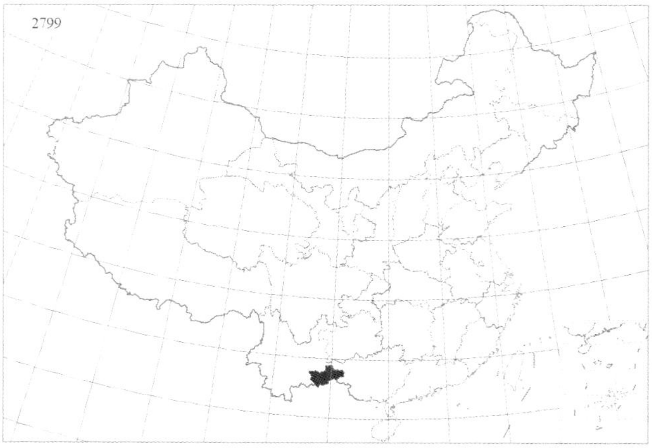

Pittosporaceae 海桐花科 *Pittosporum* 海桐花属
Pittosporum napaulense (DC.) Rehder et E. H. Wilson 滇藏海桐
Shrubs or small trees, evergreen

16.1 (5.3~20.5); 9.5 (-2.4~14.9); 20.5 (12.2~23.7); 16.1 (5.7~20.5); 134.9 (30.5~185.9); 1.5 (0.0~26.2); 802 (479~992) ❙ 1417 (961~1736); 782 (558~1007); 51 (31~70) ❙ 78.8 (47.6~145.8); 791 (479~959); 457.3 (234.2~627.0)

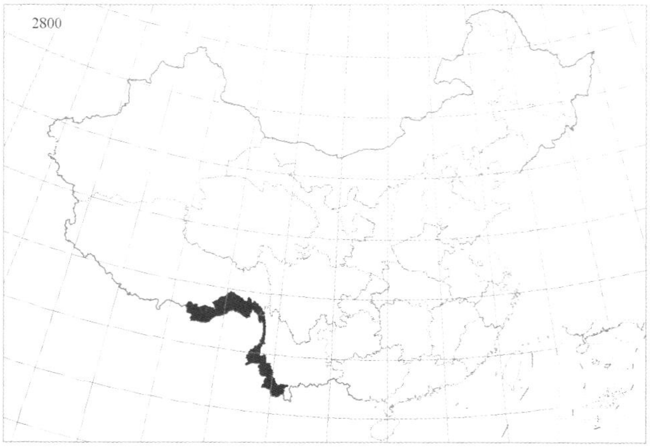

Pittosporaceae 海桐花科 *Pittosporum* 海桐花属
Pittosporum oligophlebium H. T. Chang et S. Z. Yan 贫脉海桐
Shrubs, evergreen

16.7; 9.9; 21.4; 16.8; 141.2; 0.0; 813 ❙ 1153; 647; 50 ❙ 43.0; 802; 353.6

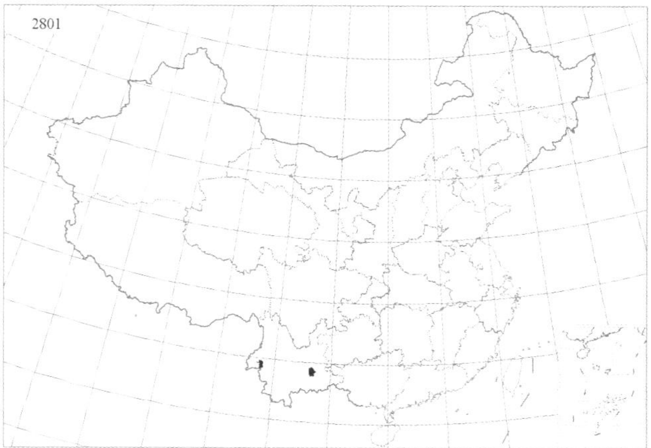

Pittosporaceae 海桐花科 *Pittosporum* 海桐花属
Pittosporum omeiense H. T. Chang et S. Z. Yan 峨眉海桐
Shrubs, evergreen

14.4 (7.3~19.1); 4.6 (-2.0~10.1); 23.2 (15.6~27.1); 14.4 (7.5~19.1); 116.0 (48.2~169.2); 2.9 (0.0~20.4); 773 (537~966) ❙ 1221 (985~1590); 622 (489~946); 59 (27~123) ❙ 60.0 (19.4~106.2); 773 (537~966); 225.6 (130.1~310.8)

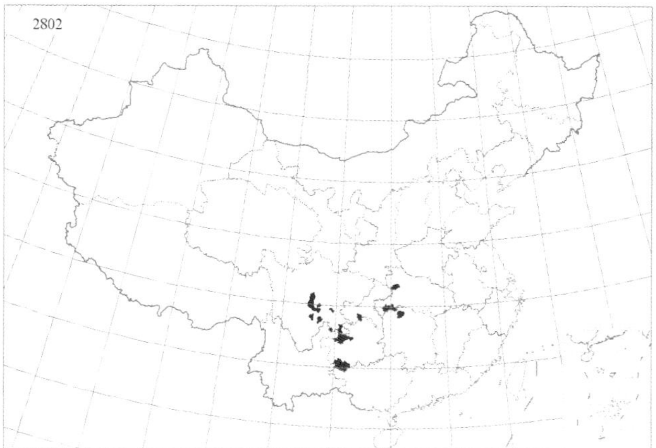

Pittosporaceae 海桐花科 *Pittosporum* 海桐花属
Pittosporum paniculiferum H. T. Chang et S. Z. Yan 圆锥海桐
Trees, evergreen

18.3 (14.9~21.0); 12.0 (5.3~15.1); 22.5 (20.3~25.6); 18.4 (15.0~21.0); 160.7 (119.6~191.8); 0.0 (0.0~0.0); 885 (782~1053) ❙ 1432 (1232~1751); 778 (655~996); 55 (41~77) ❙ 62.1 (36.8~88.8); 873 (780~1053); 459.6 (214.5~659.0)

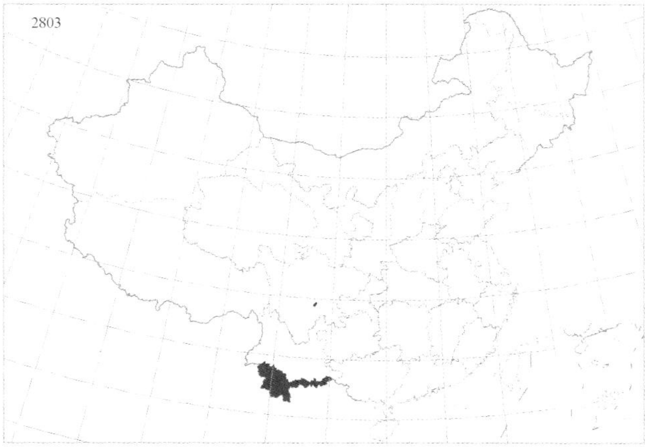

Pittosporaceae 海桐花科 *Pittosporum* 海桐花属
Pittosporum parvicapsulare H. T. Chang et S. Z. Yan 小果海桐
Shrubs, evergreen

17.6 (16.1~22.2); 7.4 (5.3~13.9); 26.4 (25.5~28.3); 17.6 (16.1~22.3); 151.7 (133.1~207.5); 0.0 (0.0~0.0); 916 (838~1190) ❙ 1415 (1268~1559); 574 (476~713); 140 (87~197) ❙ 55.6 (15.4~71.9); 916 (838~1190); 238.1 (180.3~327.7)

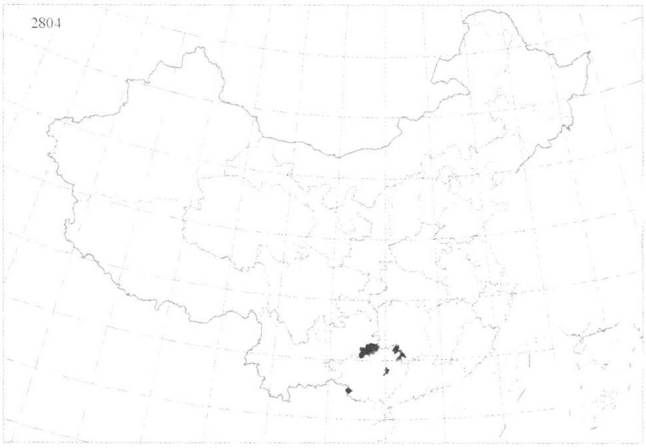

Pittosporaceae 海桐花科 *Pittosporum* 海桐花属
Pittosporum parvilimbum H. T. Chang et S. Z. Yan 小叶海桐
Shrubs, evergreen

20.7; 10.7; 29.0; 20.8; 189.3; 0.0; 1108 ❙ 1428; 639; 135 ❙ 28.9; 1108; 160.7

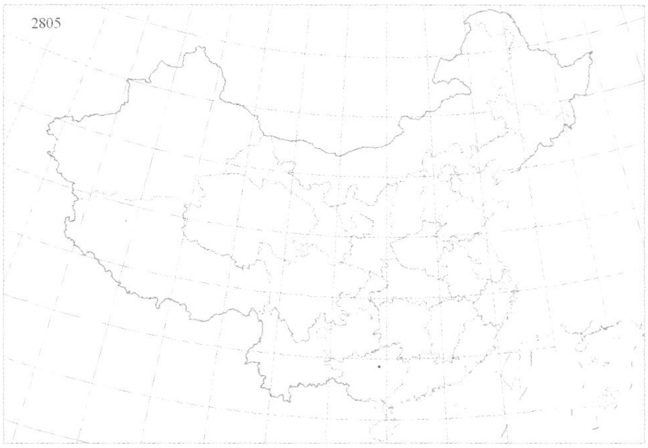

Pittosporaceae 海桐花科 *Pittosporum* 海桐花属
Pittosporum pauciflorum Hook. et Arn. 少花海桐
Shrubs, evergreen

19.4 (14.6~24.8); 9.8 (3.5~20.1); 27.6 (24.6~29.8); 19.4 (14.7~24.8); 172.9 (117.7~237.6); 0.1 (0.0~1.6); 1020 (779~1385) ❙ 1628 (931~2037); 639 (323~1026); 165 (41~225) ❙ 60.7 (-1.6~101.8); 1011 (779~1212); 238.4 (120.3~620.0)

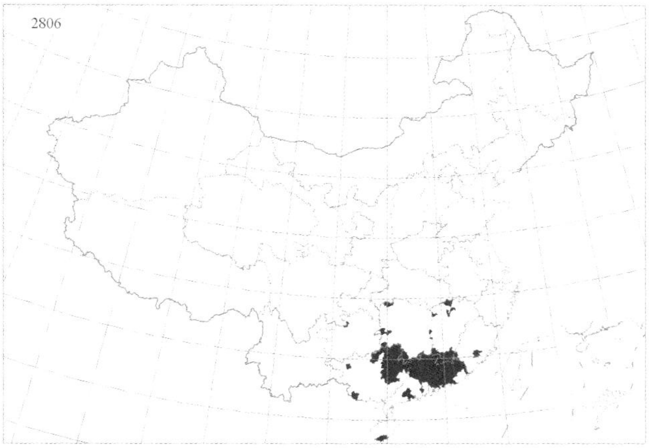

Pittosporaceae 海桐花科 *Pittosporum* 海桐花属
Pittosporum pauciflorum var. oblongum H. T. Chang et S. Z. Yan 长果海桐
Shrubs, evergreen

19.2 (17.3~21.0); 9.3 (6.5~12.2); 27.7 (26.2~29.8); 19.2 (17.3~21.0); 170.5 (147.9~192.6); 0.0 (0.0~0.0); 1001 (906~1113) ❙ 1634 (1486~1852); 590 (505~711); 186 (161~197) ❙ 63.6 (35.6~86.3); 998 (906~1107); 238.8 (174.9~271.3)

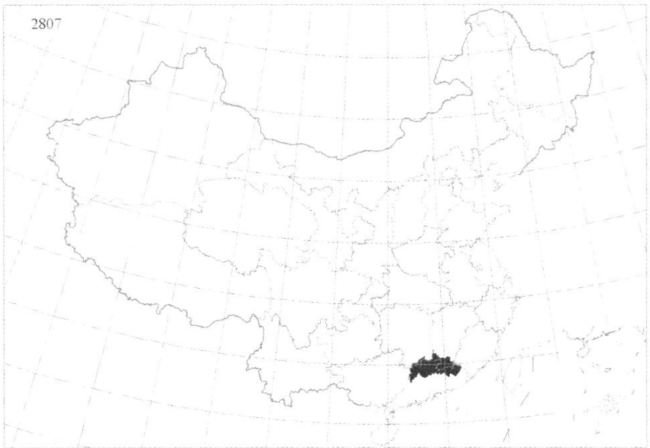

Pittosporaceae 海桐花科 *Pittosporum* 海桐花属
Pittosporum pentandrum var. formosanum (Hayata) Z. Y. Zhang et Turland 台琼海桐
Shrubs or small trees, evergreen

23.3 (18.5~24.8); 17.5 (13.5~20.1); 27.5 (21.7~29.2); 23.3 (18.5~24.8); 220.1 (162.1~237.6); 0.0 (0.0~0.0); 1254 (863~1385) ❙ 1708 (868~2993); 721 (312~1847); 100 (41~547) ❙ 42.4 (-34.3~214.4); 1166 (863~1360); 376.3 (154.2~620.0)

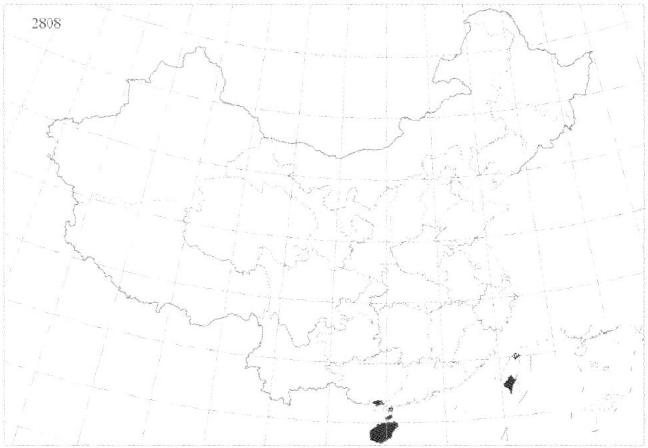

Pittosporaceae 海桐花科 *Pittosporum* 海桐花属
Pittosporum perglabratum H. T. Chang et S. Z. Yan 全秃海桐
Shrubs or small trees, evergreen

14.3 (9.3~16.7); 3.4 (-1.7~6.3); 24.5 (20.1~26.5); 14.3 (9.5~16.7); 114.9 (69.8~140.5); 3.4 (0.0~18.0); 774 (603~863) ❙ 1230 (993~1435); 505 (420~591); 84 (46~106) ❙ 59.9 (36.2~101.4); 774 (603~863); 229.2 (170.4~279.5)

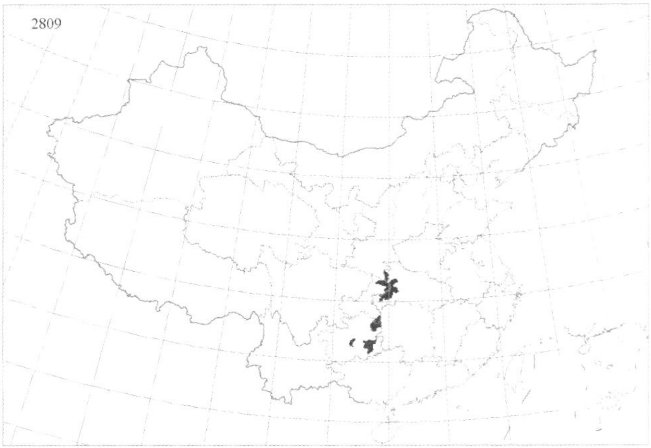

Pittosporaceae 海桐花科 *Pittosporum* 海桐花属
Pittosporum perryanum Gowda 缝线海桐
Shrubs, evergreen

18.4 (14.3~24.8); 9.5 (4.0~20.1); 25.7 (21.4~28.5); 18.4 (14.3~24.8); 161.3 (112.6~237.6); 0.0 (0.0~1.1); 949 (756~1385) ❙ 1388 (958~2005); 622 (392~891); 100 (41~197) ❙ 47.4 (-11.9~85.1); 934 (756~1166); 285.1 (149.6~620.0)

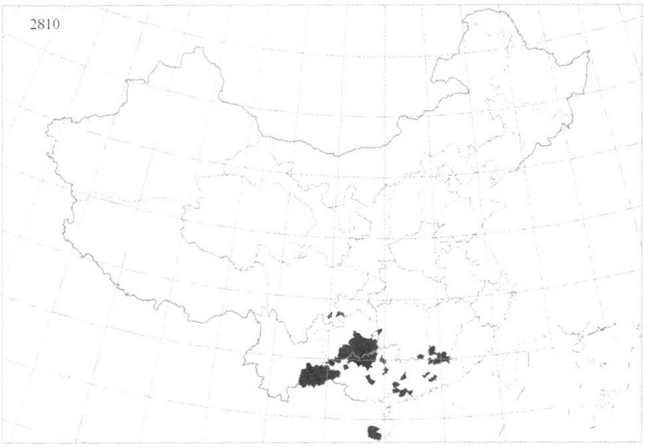

Pittosporaceae 海桐花科 *Pittosporum* 海桐花属
Pittosporum perryanum var. linearifolium H. T. Chang et S. Z. Yan 狭叶缝线海桐
Shrubs, evergreen

16.7; 6.3; 25.7; 16.7; 140.5; 0.0; 863 ❙ 1268; 531; 105 ❙ 47.0; 863; 246.1

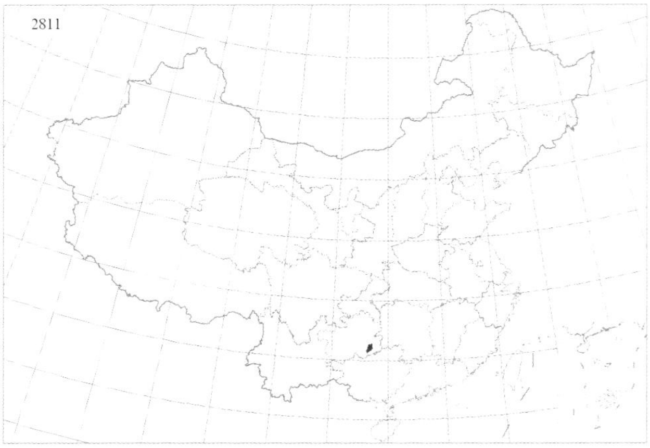

Pittosporaceae 海桐花科 *Pittosporum* 海桐花属
Pittosporum planilobum H. T. Chang et S. Z. Yan 扁片海桐
Trees, evergreen

18.7; 8.4; 27.8; 18.8; 165.3; 0.0; 985 ❙ 1647; 643; 184 ❙ 67.4; 985; 216.9

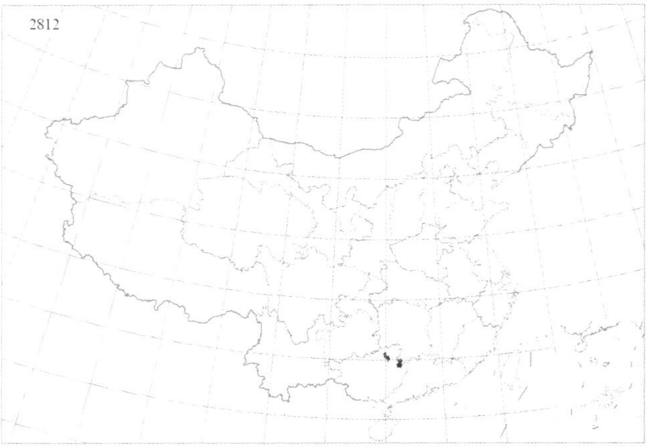

Pittosporaceae 海桐花科 *Pittosporum* 海桐花属
Pittosporum podocarpum Gagnep. 柄果海桐
Shrubs, evergreen

15.2 (5.4~21.3); 6.4 (-4.2~15.1); 22.7 (12.3~29.3); 15.3 (5.8~21.4); 125.5 (31.5~196.3); 2.5 (0.0~29.4); 795 (481~1121) ❙ 1155 (707~1751); 576 (332~996); 60 (15~204) ❙ 45.6 (-14.2~112.2); 786 (481~1121); 286.7 (108.2~659.0)

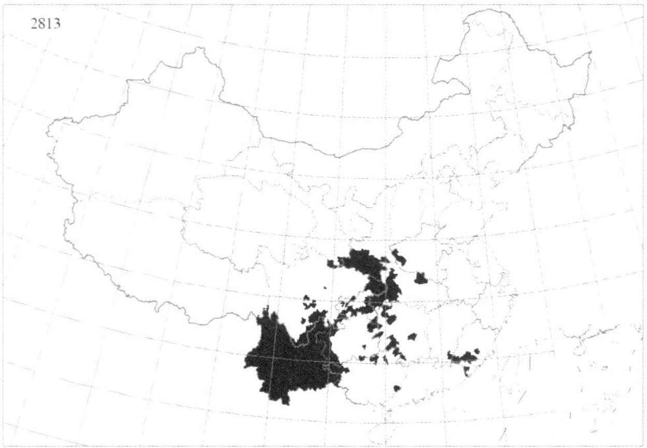

Pittosporaceae 海桐花科 *Pittosporum* 海桐花属
Pittosporum podocarpum var. angustatum Gowda 线叶柄果海桐
Shrubs, evergreen

15.7 (9.4~19.2); 5.6 (-2.5~13.3); 24.9 (19.4~27.7); 15.7 (9.8~19.4); 129.8 (73.9~172.6); 1.4 (0.0~20.6); 824 (619~973) ❙ 1261 (707~1745); 542 (352~817); 89 (15~196) ❙ 53.2 (-0.2~100.5); 823 (619~973); 236.3 (141.4~491.5)

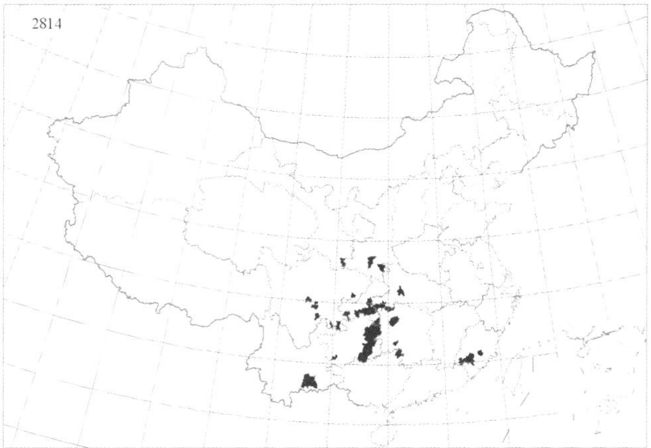

Pittosporaceae 海桐花科 *Pittosporum* 海桐花属
Pittosporum podocarpum var. hejiangense (H. Y. Su) Z. Y. Zhang et Turland 合江海桐 Shrubs, evergreen

17.2; 6.8; 27.0; 17.3; 147.4; 0.0; 905 ❙ 1097; 499; 72 ❙ 21.3; 905; 158.7

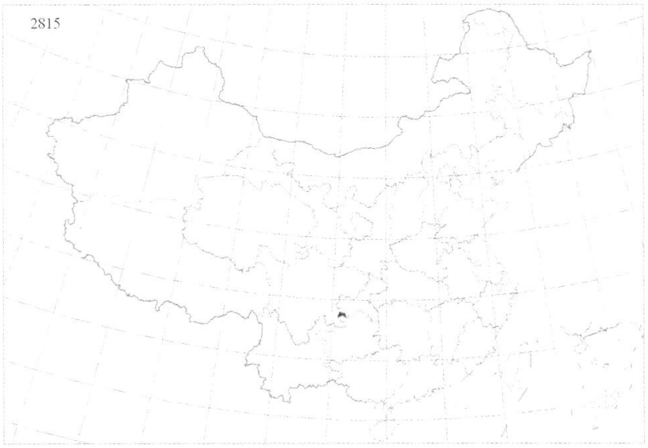

Pittosporaceae 海桐花科 *Pittosporum* 海桐花属
Pittosporum podocarpum var. molle W. D. Han 毛花柄果海桐
Shrubs, evergreen

19.2; 11.0; 26.7; 19.2; 170.8; 0.0; 971 ❙ 1621; 566; 166 ❙ 67.0; 971; 280.5

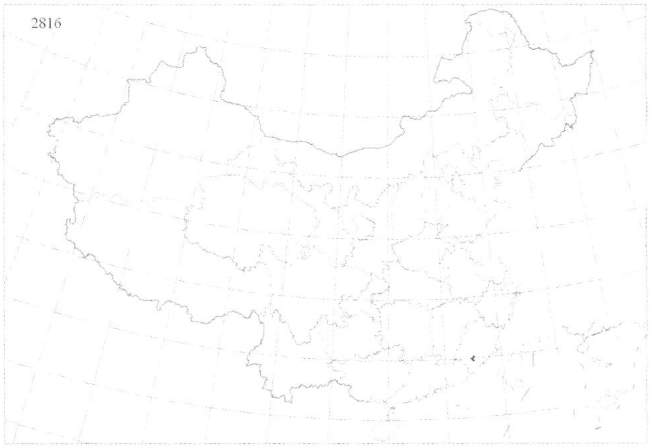

Pittosporaceae 海桐花科 *Pittosporum* 海桐花属
Pittosporum pulchrum Gagnep. 秀丽海桐
Shrubs, evergreen

21.6 (19.7~22.7); 13.2 (10.5~15.0); 27.9 (27.0~28.6); 21.6 (19.8~22.8); 199.2 (177.1~213.1); 0.0 (0.0~0.0); 1140 (1018~1221) ❙ 1686 (1279~2435); 919 (676~1428); 109 (73~138) ❙ 47.4 (14.8~99.5); 1140 (1018~1221); 302.4 (272.3~335.1)

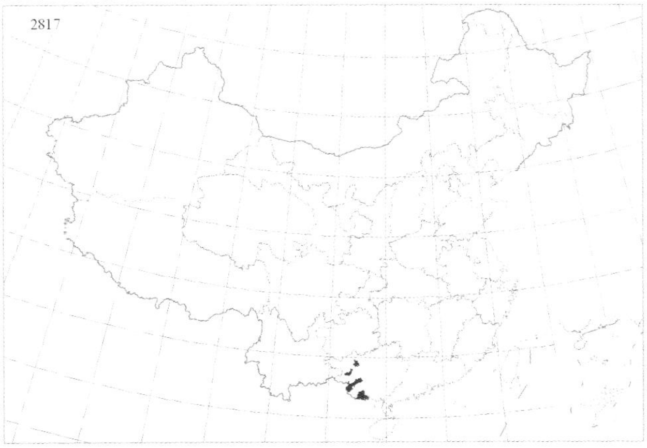

Pittosporaceae 海桐花科 *Pittosporum* 海桐花属
Pittosporum reflexisepalum C. Y. Wu 折萼海桐
Shrubs, evergreen

17.2; 9.1; 23.1; 17.3; 147.5; 0.0; 847 ❙ 1161; 633; 50 ❙ 37.1; 847; 314.4

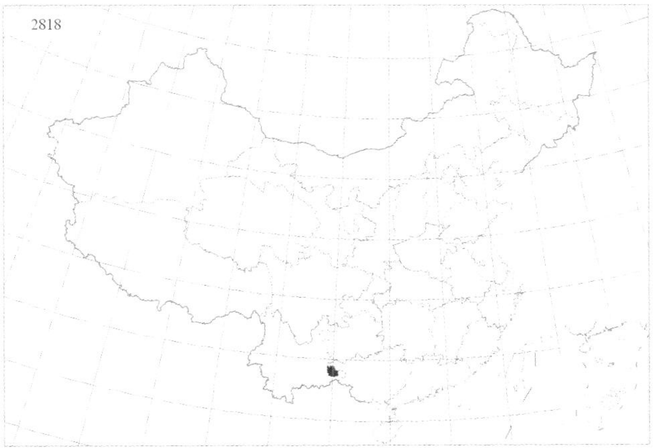

Pittosporaceae 海桐花科 *Pittosporum* 海桐花属
Pittosporum rehderianum Gowda 厚圆果海桐
Shrubs, evergreen

13.0 (9.3~17.2); 1.2 (-2.9~6.8); 24.1 (19.8~28.1); 13.1 (9.5~17.3); 104.6 (69.8~147.4); 8.7 (0.0~20.6); 743 (603~924) ❙ 1039 (622~1492); 460 (298~613); 57 (15~123) ❙ 40.2 (-13.4~107.2); 740 (603~924); 235.9 (133.5~316.6)

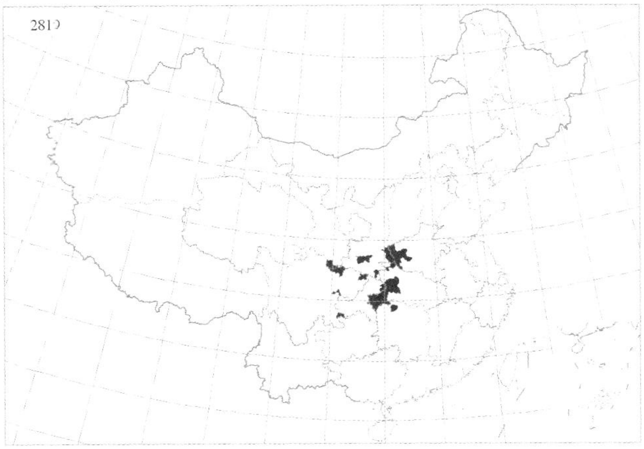

Pittosporaceae 海桐花科 *Pittosporum* 海桐花属
Pittosporum rehderianum var. ternstroemioides (C. Y. Wu) Z. Y. Zhang et Turland 厚皮香海桐 Shrubs, evergreen

16.0; 9.3; 20.4; 16.0; 132.4; 0.0; 780 ❙ 1348; 762; 56 ❙ 72.8; 780; 395.4

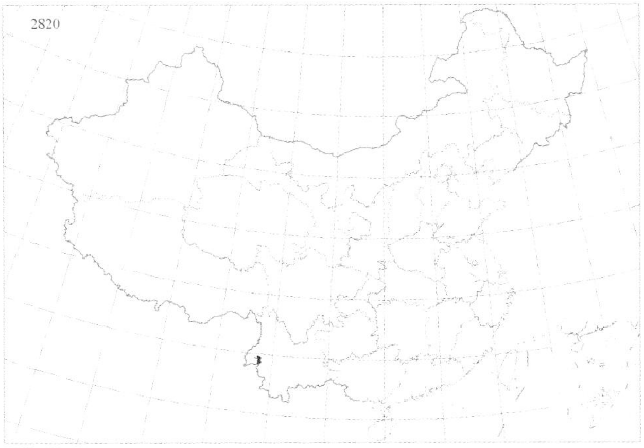

Pittosporaceae 海桐花科 *Pittosporum* 海桐花属

Pittosporum saxicola Rehder et E. H. Wilson 石生海桐

Shrubs, evergreen

4.5; -4.9; 12.8; 5.6; 30.4; 36.0; 470 ❙ 856; 446; 21 ❙ 82.3; 470; 310.1

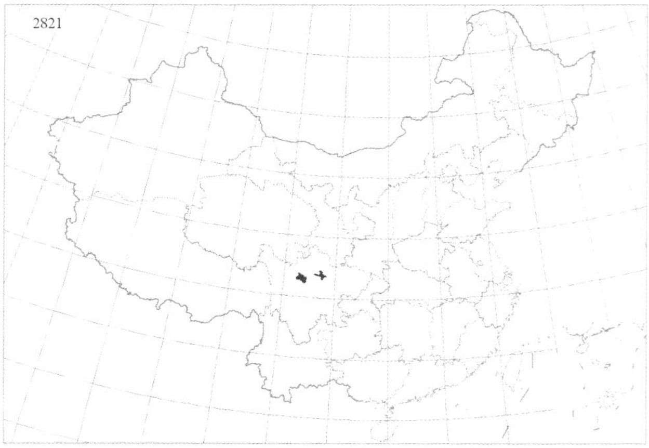

Pittosporaceae 海桐花科 *Pittosporum* 海桐花属

Pittosporum subulisepalum Hu et F. T. Wang 尖萼海桐

Shrubs, evergreen

15.8 (14.4~16.9); 3.4 (2.1~4.6); 27.7 (26.3~29.0); 15.9 (14.5~16.9); 132.3 (118.7~144.3); 2.0 (0.4~4.9); 872 (801~934) ❙ 1513 (1206~1801); 542 (437~622); 189 (153~242) ❙ 74.2 (35.6~112.2); 866 (801~914); 242.4 (149.5~319.9)

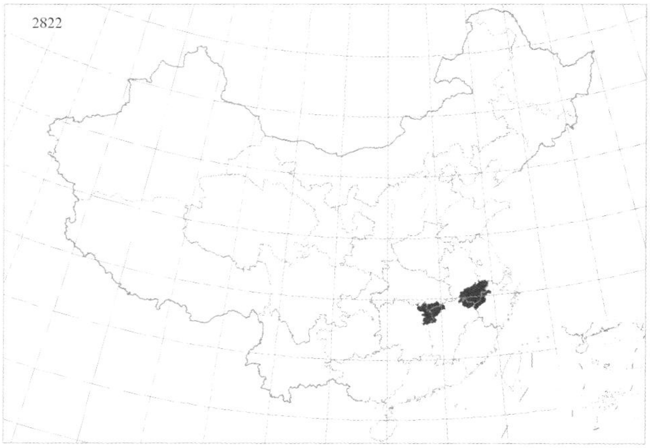

Pittosporaceae 海桐花科 *Pittosporum* 海桐花属

Pittosporum tenuivalvatum H. T. Chang et S. Z. Yan 薄片海桐

Shrubs or small trees, evergreen

19.4; 10.3; 26.4; 19.4; 172.8; 0.0; 988 ❙ 1317; 690; 78 ❙ 33.3; 988; 290.0

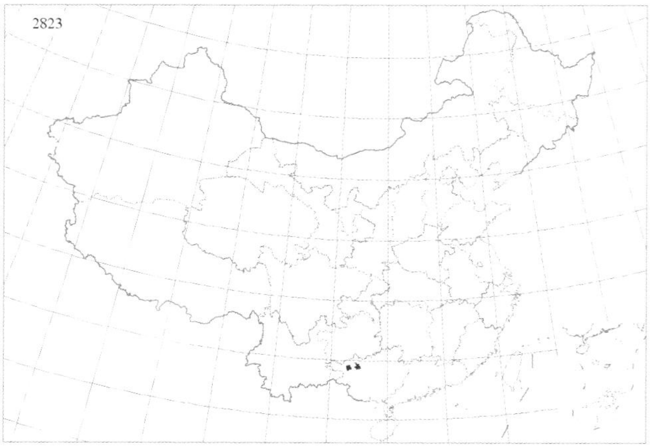

Pittosporaceae 海桐花科 *Pittosporum* 海桐花属

Pittosporum tobira (Thunb.) W. T. Aiton 海桐

Shrubs or small trees, evergreen

17.0 (11.7~24.2); 7.1 (0.9~17.7); 26.2 (18.8~29.4); 17.1 (11.7~24.2); 145.4 (82.6~230.4); 0.5 (0.0~8.5); 895 (658~1331) ❙ 1439 (807~3495); 572 (349~1847); 141 (31~897) ❙ 61.1 (-5.3~224.4); 890 (658~1331); 218.6 (111.9~332.5)

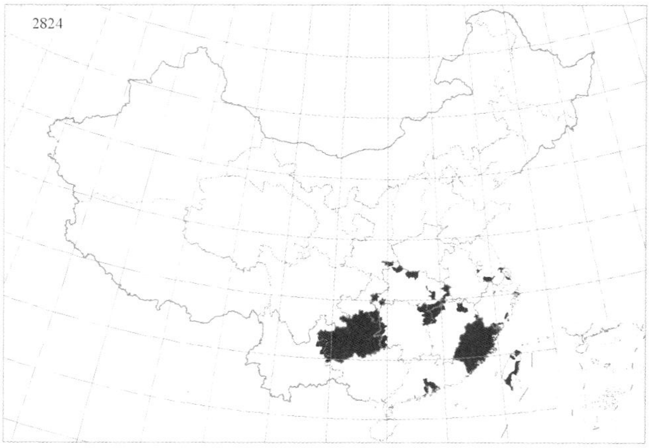

Pittosporaceae 海桐花科 *Pittosporum* 海桐花属

Pittosporum tobira var. calvescens Ohwi 秃序海桐

Shrubs or small trees, evergreen

20.3; 12.0; 28.3; 20.3; 184.2; 0.0; 1056 ❙ 1288; 437; 210 ❙ 22.4; 1027; 206.4

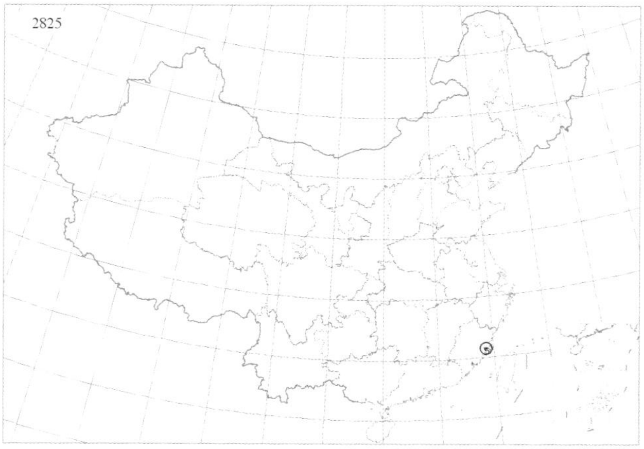

Pittosporaceae 海桐花科 *Pittosporum* 海桐花属

Pittosporum tonkinense Gagnep. 四子海桐

Shrubs or small trees, evergreen

18.2 (14.2~22.2); 9.8 (6.2~14.2); 24.6 (20.1~28.6); 18.3 (14.2~22.3); 159.1 (110.5~207.5); 0.0 (0.0~0.0); 920 (729~1190) ❙ 1309 (958~1758); 685 (531~860); 76 (44~189) ❙ 42.9 (13.3~82.1); 919 (729~1190); 304.2 (189.5~551.6)

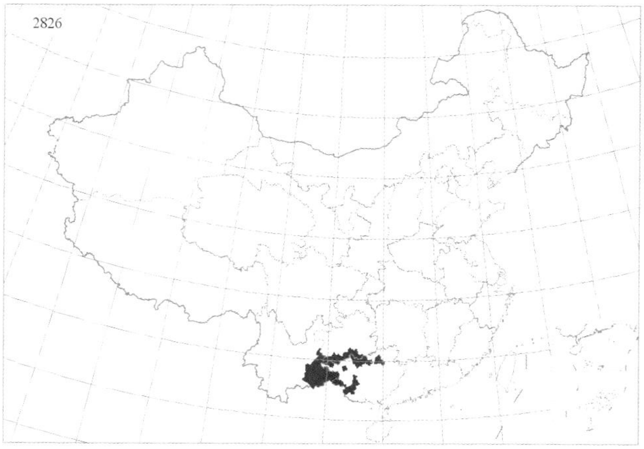

Pittosporaceae 海桐花科 *Pittosporum* 海桐花属
Pittosporum trigonocarpum H. Lév. 棱果海桐
Shrubs, evergreen

15.4 (11.7~19.1); 5.0 (−0.8~10.1); 24.9 (18.8~27.4); 15.5 (11.7~19.1); 126.6 (82.6~169.2); 0.8 (0.0~13.4); 814 (658~969) ❙ 1220 (798~1740); 540 (377~720); 82 (31~187) ❙ 50.1 (−2.0~99.4); 814 (658~969); 202.8 (146.0~282.2)

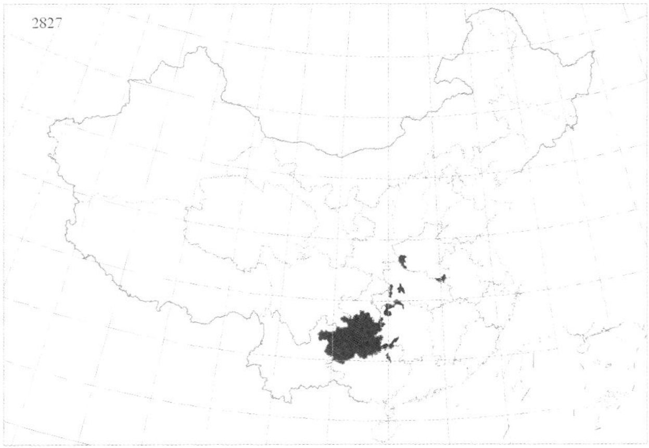

Pittosporaceae 海桐花科 *Pittosporum* 海桐花属
Pittosporum truncatum E. Pritz. 崖花子
Shrubs, evergreen

13.5 (2.6~19.2); 1.8 (−7.0~9.2); 24.5 (11.2~28.4); 13.6 (4.3~19.2); 109.7 (20.0~170.8); 7.3 (0.0~48.2); 767 (427~1002) ❙ 1045 (626~1502); 468 (316~678); 60 (15~161) ❙ 37.1 (−22.2~107.2); 756 (427~1002); 223.4 (139.1~334.4)

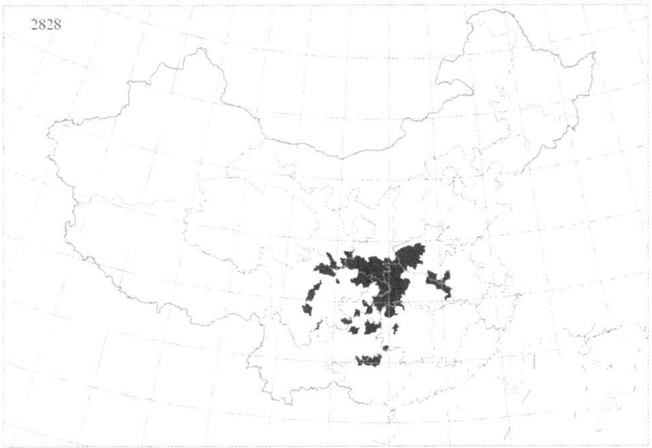

Pittosporaceae 海桐花科 *Pittosporum* 海桐花属
Pittosporum tubiflorum H. T. Chang et S. Z. Yan 管花海桐
Shrubs, evergreen

15.2 (13.5~17.8); 4.5 (2.6~6.0); 25.3 (23.5~29.1); 15.3 (13.6~17.8); 124.0 (106.7~154.1); 0.9 (0.0~4.1); 811 (744~960) ❙ 1278 (1179~1497); 502 (445~613); 101 (70~221) ❙ 58.1 (45.6~100.5); 808 (744~927); 206.7 (182.9~245.8)

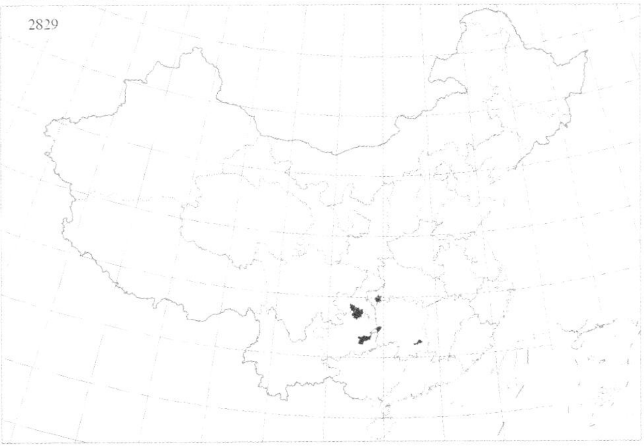

Pittosporaceae 海桐花科 *Pittosporum* 海桐花属
Pittosporum undulatifolium H. T. Chang et S. Z. Yan 波叶海桐
Trees, evergreen

11.7; 2.4; 19.9; 11.8; 89.8; 9.4; 680 ❙ 1183; 602; 49 ❙ 78.4; 680; 225.8

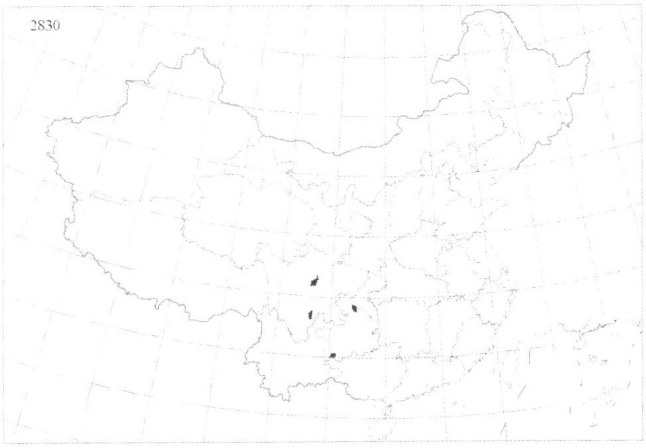

Pittosporaceae 海桐花科 *Pittosporum* 海桐花属
Pittosporum viburnifolium Hayata 荚蒾叶海桐
Shrubs, evergreen

20.3; 15.9; 23.6; 20.4; 184.4; 0.0; 993 ❙ 2853; 1618; 147 ❙ 190.4; 993; 330.3

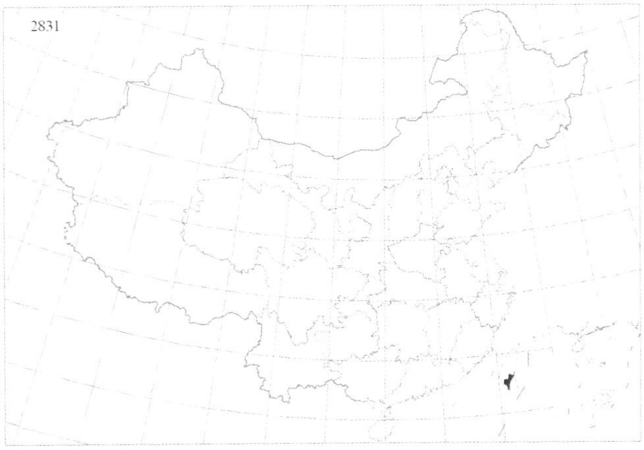

Pittosporaceae 海桐花科 *Pittosporum* 海桐花属
Pittosporum xylocarpum Hu et F. T. Wang 木果海桐
Shrubs, evergreen

16.4 (9.3~21.0); 6.0 (−1.7~14.2); 26.1 (20.0~28.9); 16.5 (9.5~21.0); 138.0 (69.8~191.8); 0.6 (0.0~18.0); 872 (603~1053) ❙ 1131 (862~1590); 550 (376~946); 57 (24~187) ❙ 30.8 (5.6~107.2); 870 (603~1053); 164.3 (108.2~551.6)

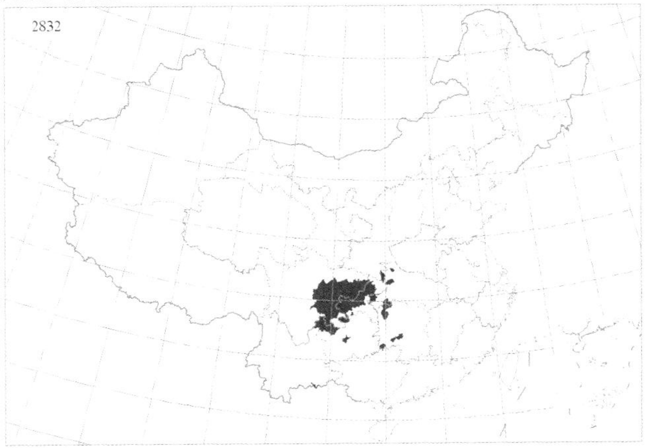

Hamamelidaceae 金缕梅科 *Altingia* 蕈树属
Altingia chinensis (Champ. ex Benth.) Oliv. ex Hance 蕈树
Trees, evergreen

19.0 (14.3~24.8); 9.6 (3.3~20.1); 27.1 (22.4~29.8); 19.1 (14.3~24.8); 168.7 (112.6~237.6); 0.1 (0.0~2.3); 995 (756~1385) ❙ 1609 (1089~2435); 662 (419~1428); 154 (41~259) ❙ 63.1 (1.5~147.0); 991 (756~1221); 250.9 (120.3~620.0)

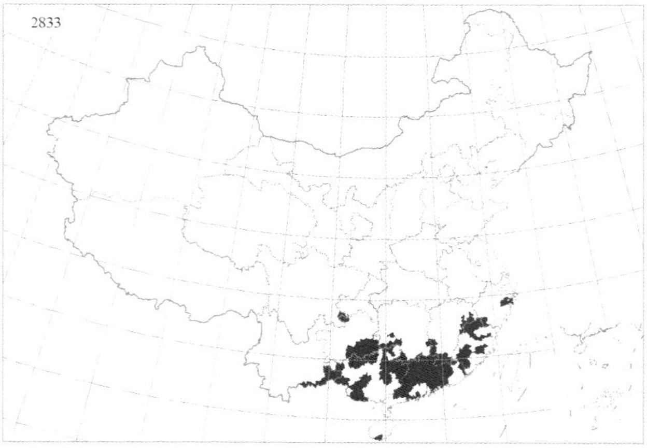

Hamamelidaceae 金缕梅科 *Altingia* 蕈树属
Altingia excelsa Noronha 细青皮
Trees, evergreen

18.1 (13.3~21.0); 11.8 (5.4~15.1); 22.1 (19.1~25.6); 18.1 (13.4~21.0); 157.4 (100.4~191.8); 0.0 (0.0~0.0); 872 (706~1053) ❙ 1419 (1049~1751); 785 (598~1007); 54 (41~77) ❙ 63.7 (21.7~145.8); 858 (706~1053); 467.8 (346.9~659.0)

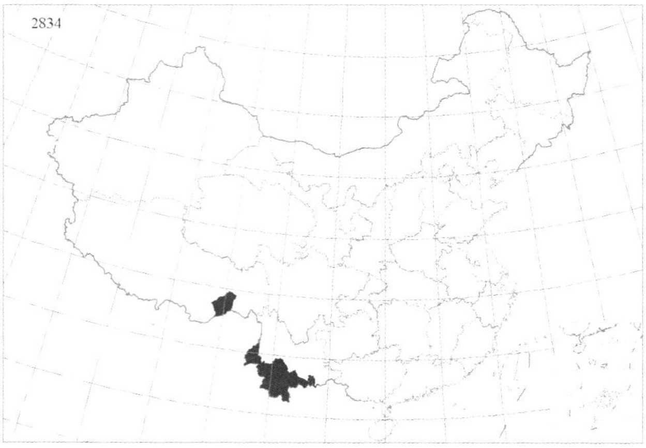

Hamamelidaceae 金缕梅科 *Altingia* 蕈树属
Altingia gracilipes Hemsl. 细柄蕈树
Trees, evergreen

18.9 (14.6~21.7); 9.7 (3.8~15.2); 27.2 (24.3~29.7); 18.9 (14.7~22.7); 167.3 (117.7~212.7); 0.0 (0.0~1.5); 981 (779~1216) ❙ 1628 (1070~1997); 593 (356~994); 185 (73~257) ❙ 68.5 (-5.0~147.0); 973 (779~1216); 251.5 (123.9~332.5)

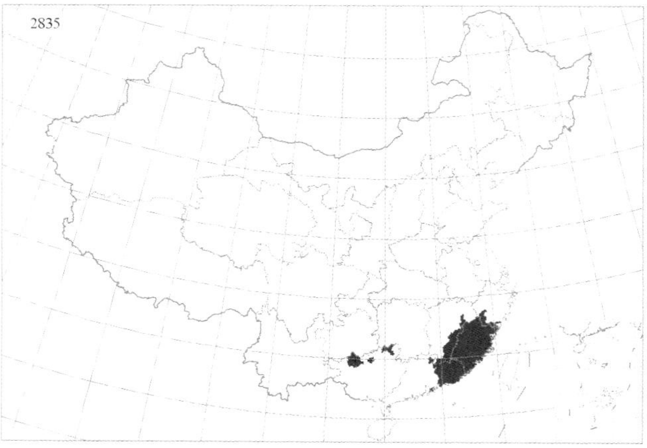

Hamamelidaceae 金缕梅科 *Altingia* 蕈树属
Altingia multinervis Cheng 赤水蕈树
Trees, evergreen

15.9; 5.5; 25.6; 15.9; 131.1; 0.0; 833 ❙ 1089; 503; 69 ❙ 30.9; 833; 181.2

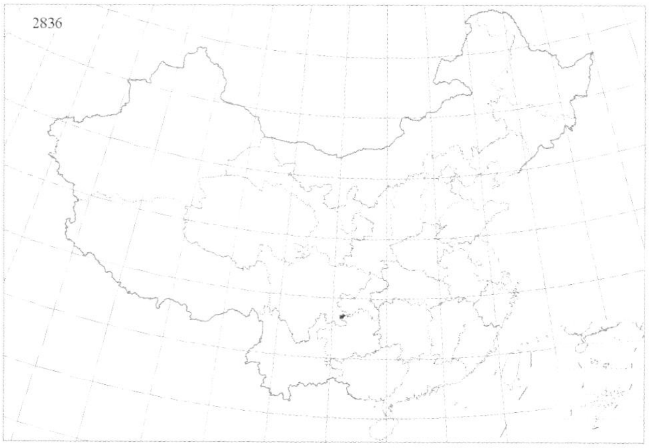

Hamamelidaceae 金缕梅科 *Altingia* 蕈树属
Altingia obovata Merr. et Chun 海南蕈树
Trees, evergreen

23.8 (22.2~24.8); 18.6 (17.2~20.1); 27.5 (25.7~28.4); 23.9 (22.3~24.8); 226.6 (207.2~237.6); 0.0 (0.0~0.0); 1294 (1127~1385) ❙ 1514 (1179~1767); 521 (392~610); 58 (41~92) ❙ 17.6 (-11.9~47.8); 1155 (1055~1299); 519.7 (439.7~620.0)

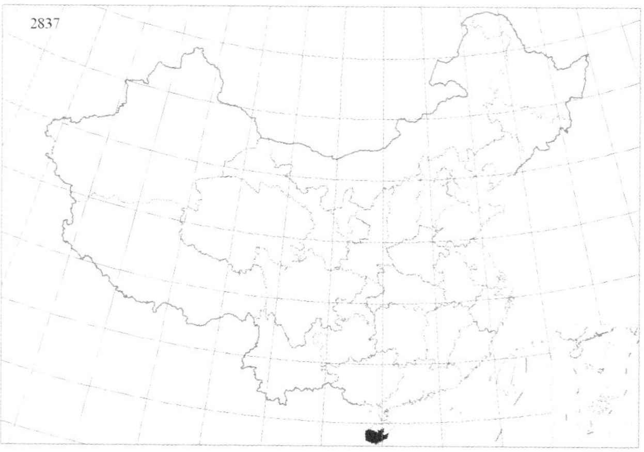

Hamamelidaceae 金缕梅科 *Altingia* 蕈树属
Altingia siamensis Craib 镰尖蕈树
Trees, evergreen

20.4 (18.2~22.4); 12.0 (8.7~14.5); 27.3 (23.1~28.7); 20.4 (18.3~22.4); 185.2 (159.3~209.2); 0.0 (0.0~0.0); 1056 (927~1198) ❙ 1514 (1284~1751); 671 (539~996); 126 (73~195) ❙ 44.0 (20.8~88.8); 1056 (927~1198); 262.8 (164.9~514.1)

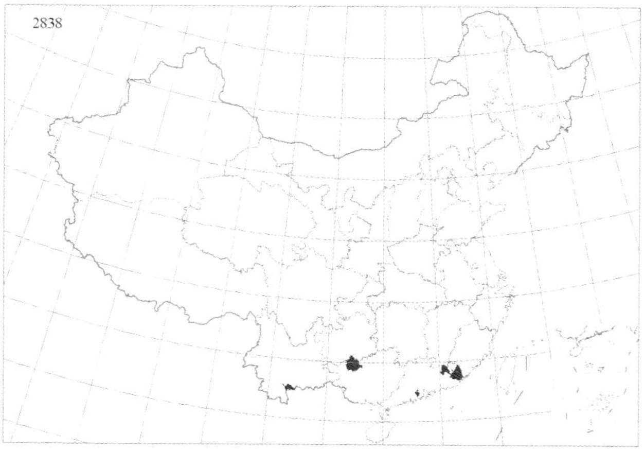

Hamamelidaceae 金缕梅科 *Altingia* 蕈树属
Altingia tenuifolia Chun ex H. T. Chang 薄叶蕈树
Trees, evergreen

17.5 (14.6~18.9); 6.7 (3.8~9.2); 27.2 (24.4~29.1); 17.5 (14.7~19.0); 150.5 (117.7~167.7); 0.1 (0.0~1.5); 924 (779~992) ▎1668 (1342~1847); 571 (531~611); 203 (95~248) ▎81.0 (62.5~101.9); 913 (779~992); 239.6 (215.4~268.9)

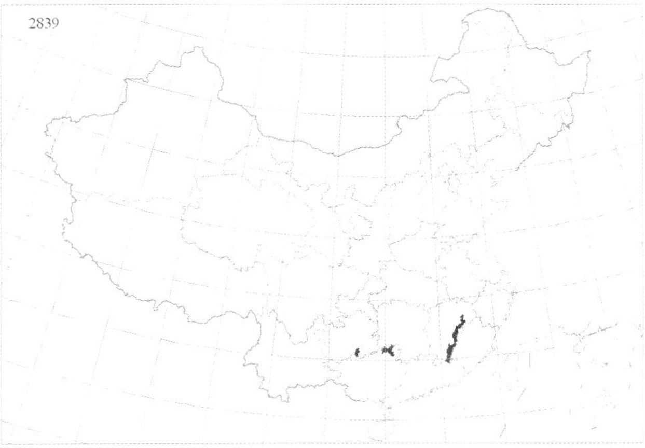

Hamamelidaceae 金缕梅科 *Altingia* 蕈树属
Altingia yunnanensis Rehder et E. H. Wilson 云南蕈树
Trees, evergreen

17.9 (16.3~21.0); 10.6 (7.8~14.2); 23.1 (20.3~26.0); 17.9 (16.4~21.0); 155.2 (136.4~191.8); 0.0 (0.0~0.0); 874 (789~1053) ▎1298 (1049~1547); 726 (582~911); 54 (44~68) ▎48.7 (26.1~80.6); 868 (780~1053); 368.3 (189.5~551.6)

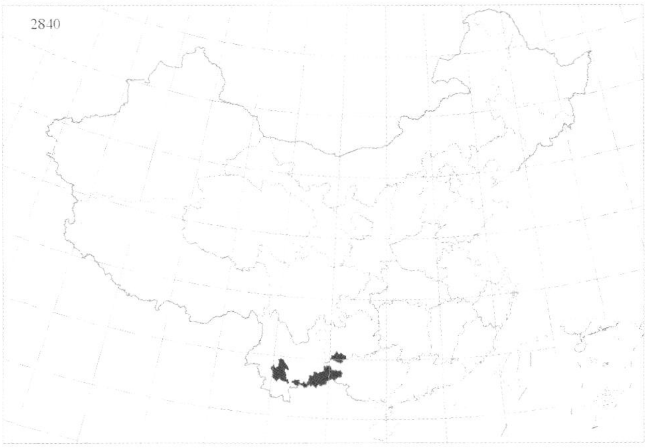

Hamamelidaceae 金缕梅科 *Chunia* 山铜材属
Chunia bucklandioides H. T. Chang 山铜材
Trees, evergreen

24.1 (23.0~24.8); 19.0 (17.7~20.1); 27.6 (26.7~28.4); 24.2 (23.1~24.8); 229.9 (217.0~237.6); 0.0 (0.0~0.0); 1321 (1217~1385) ▎1598 (1405~1725); 555 (473~619); 65 (41~110) ▎21.5 (1.5~41.8); 1197 (1137~1326); 513.6 (352.9~620.0)

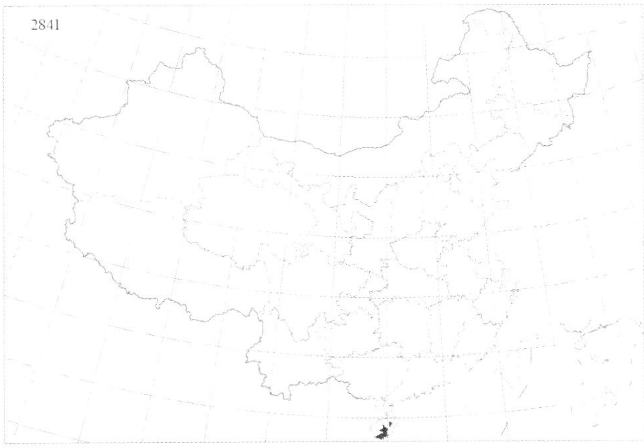

Hamamelidaceae 金缕梅科 *Corylopsis* 蜡瓣花属
Corylopsis alnifolia (H. Lév.) C. K. Schneid. 桤叶蜡瓣花
Shrubs, evergreen

14.5 (13.3~15.6); 4.6 (3.6~5.5); 23.6 (21.6~24.9); 14.6 (13.3~15.6); 115.7 (101.4~127.5); 0.6 (0.0~1.8); 767 (715~808) ▎1129 (999~1206); 537 (501~597); 62 (53~69) ▎47.2 (37.5~59.3); 767 (715~808); 175.6 (153.9~203.8)

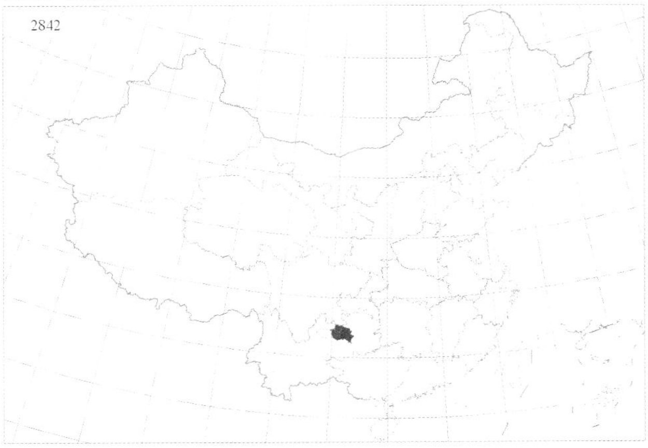

Hamamelidaceae 金缕梅科 *Corylopsis* 蜡瓣花属
Corylopsis brevistyla H. T. Chang 短柱蜡瓣花
Shrubs, deciduous

13.2 (8.6~15.2); 6.1 (1.3~8.7); 18.9 (15.0~21.0); 13.2 (8.7~15.2); 100.0 (52.6~122.5); 1.5 (0.0~8.8); 698 (574~753) ▎1019 (885~1057); 565 (479~610); 47 (34~59) ▎47.1 (26.5~81.5); 697 (574~748); 357.1 (276.9~384.1)

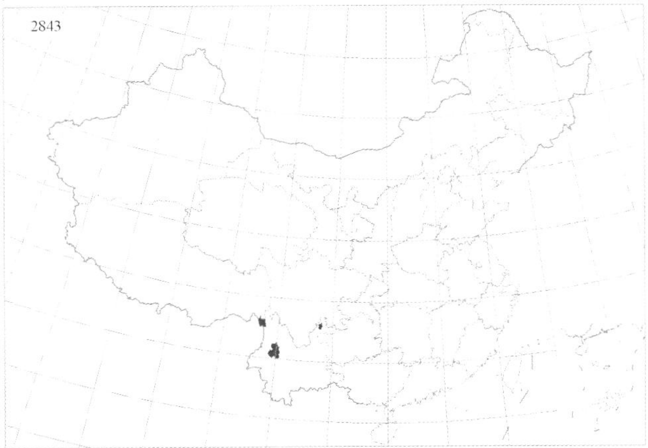

Hamamelidaceae 金缕梅科 *Corylopsis* 蜡瓣花属
Corylopsis glandulifera Hemsl. 腺蜡瓣花
Shrubs, deciduous

16.0 (14.4~17.6); 4.5 (2.1~7.4); 27.4 (24.7~29.3); 16.1 (14.5~17.6); 134.0 (118.7~151.2); 1.0 (0.0~4.9); 868 (789~941) ▎1643 (1296~1950); 590 (463~713); 202 (153~253) ▎90.0 (46.2~147.0); 867 (789~941); 243.2 (197.9~319.9)

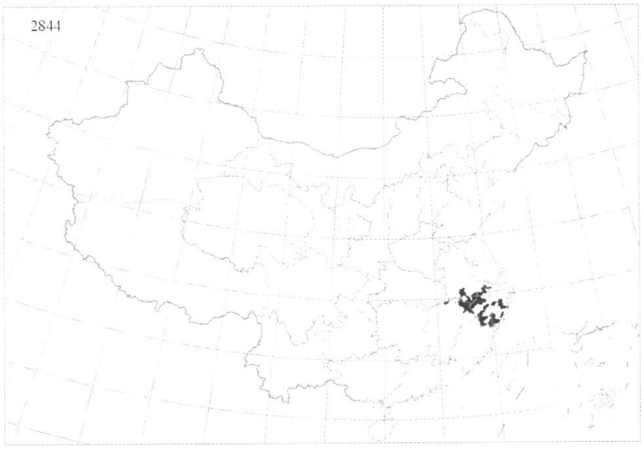

Hamamelidaceae 金缕梅科 *Corylopsis* 蜡瓣花属
Corylopsis glaucescens Hand.-Mazz. 怒江蜡瓣花
Shrubs or small trees, deciduous

12.2 (8.6~16.1); 4.6 (1.3~8.6); 18.5 (15.0~24.7); 12.2 (8.7~16.1); 88.5 (52.6~133.0); 2.2 (0.0~8.8); 678 (574~801) | 1118 (1007~1337); 625 (507~727); 49 (20~69) | 67.3 (27.3~93.0); 667 (574~801); 335.7 (184.8~407.1)

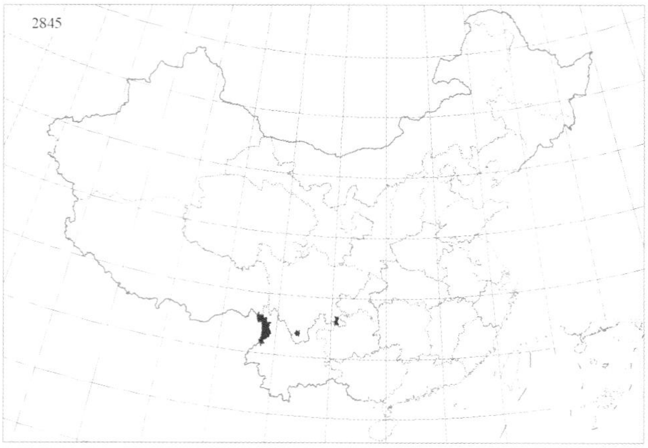

Hamamelidaceae 金缕梅科 *Corylopsis* 蜡瓣花属
Corylopsis henryi Hemsl. 鄂西蜡瓣花
Shrubs, deciduous

14.0 (12.0~16.1); 3.0 (0.6~5.1); 24.5 (23.0~26.6); 14.0 (12.1~16.2); 111.8 (94.7~134.2); 3.8 (0.0~9.7); 767 (697~864) | 1378 (1118~1478); 567 (482~599); 82 (64~92) | 80.4 (52.8~107.2); 767 (697~864); 215.3 (173.7~247.3)

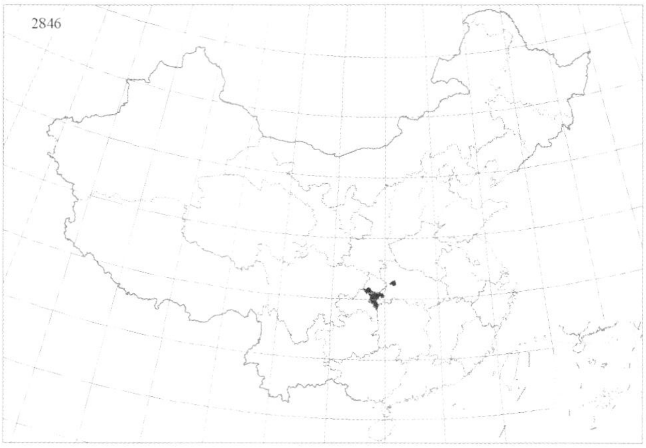

Hamamelidaceae 金缕梅科 *Corylopsis* 蜡瓣花属
Corylopsis microcarpa H. T. Chang 小果蜡瓣花
Shrubs, deciduous

10.9 (2.6~14.0); 1.1 (-7.0~4.5); 19.9 (11.2~24.1); 11.2 (4.3~14.1); 82.8 (20.0~109.4); 11.5 (0.5~48.2); 657 (427~755) | 991 (711~1414); 532 (357~817); 27 (15~47) | 53.1 (11.5~94.4); 657 (427~755); 272.1 (217.6~322.4)

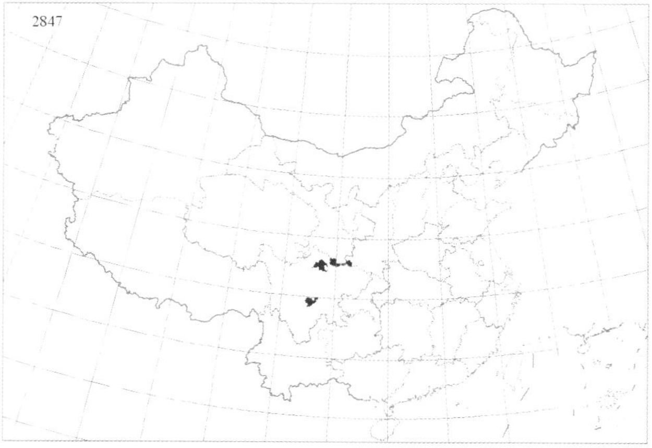

Hamamelidaceae 金缕梅科 *Corylopsis* 蜡瓣花属
Corylopsis multiflora Hance 瑞木
Shrubs or small trees, deciduous or semievergreen

18.5 (13.5~24.8); 8.7 (2.6~20.1); 27.1 (20.1~29.8); 18.6 (13.6~24.8); 162.9 (106.7~237.6); 0.2 (0.0~4.1); 978 (744~1385) | 1494 (868~2680); 603 (312~1428); 143 (41~254) | 54.9 (-34.3~246.5); 966 (744~1360); 243.1 (134.6~620.0)

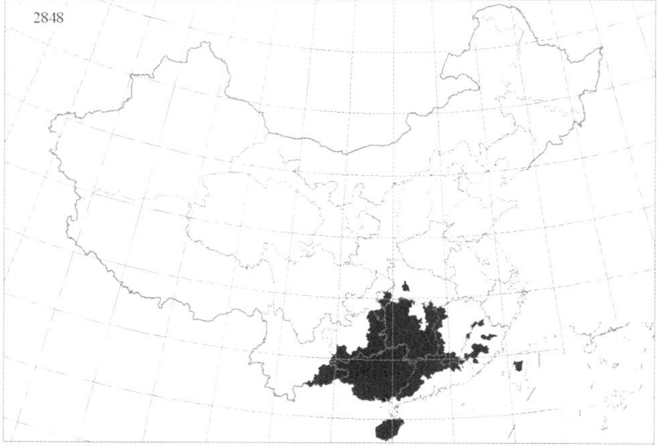

Hamamelidaceae 金缕梅科 *Corylopsis* 蜡瓣花属
Corylopsis multiflora var. nivea H. T. Chang 白背瑞木
Shrubs or small trees, deciduous or semievergreen

17.2 (16.6~18.0); 6.9 (5.9~8.8); 26.7 (26.3~27.2); 17.2 (16.6~18.0); 146.7 (138.9~156.0); 0.0 (0.0~0.0); 895 (866~923) | 1874 (1711~1997); 623 (583~669); 235 (202~254) | 109.6 (86.9~130.7); 895 (866~923); 285.0 (270.5~303.4)

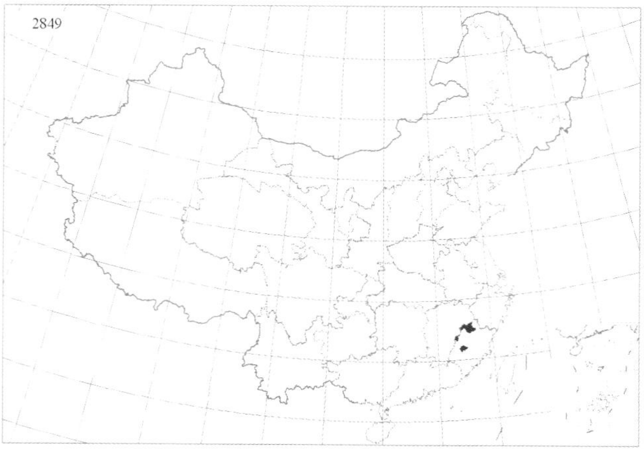

Hamamelidaceae 金缕梅科 *Corylopsis* 蜡瓣花属
Corylopsis obovata H. T. Chang 黔蜡瓣花
Shrubs, deciduous

15.0 (13.9~15.5); 4.1 (3.1~5.0); 25.5 (24.3~26.0); 15.0 (14.0~15.5); 121.3 (110.5~126.6); 1.1 (0.0~2.6); 802 (760~823) | 1277 (1178~1478); 506 (471~599); 87 (81~94) | 59.6 (43.1~94.5); 802 (760~823); 198.4 (170.4~216.8)

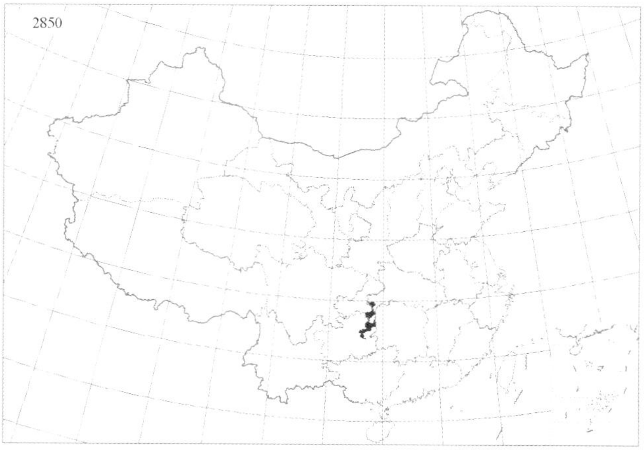

Hamamelidaceae 金缕梅科 *Corylopsis* 蜡瓣花属
Corylopsis omeiensis X. J. Yang 峨眉蜡瓣花
Shrubs, deciduous

13.4 (7.3~19.1); 4.0 (−2.0~10.1); 21.8 (15.6~26.0); 13.5 (7.5~19.1); 105.8 (48.2~169.2); 4.7 (0.0~20.4); 741 (537~966) ❙ 1163 (1001~1410); 620 (519~816); 43 (22~65) ❙ 59.5 (30.0~87.3); 741 (537~966); 255.8 (209.3~318.7)

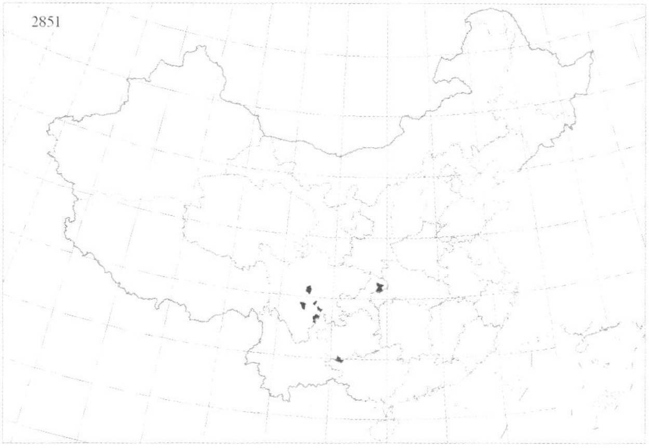

Hamamelidaceae 金缕梅科 *Corylopsis* 蜡瓣花属
Corylopsis pauciflora Siebold et Zucc. 少花瑞木
Shrubs

19.4; 13.3; 24.4; 19.4; 172.6; 0.0; 967 ❙ 2109; 1039; 182 ❙ 128.0; 946; 242.1

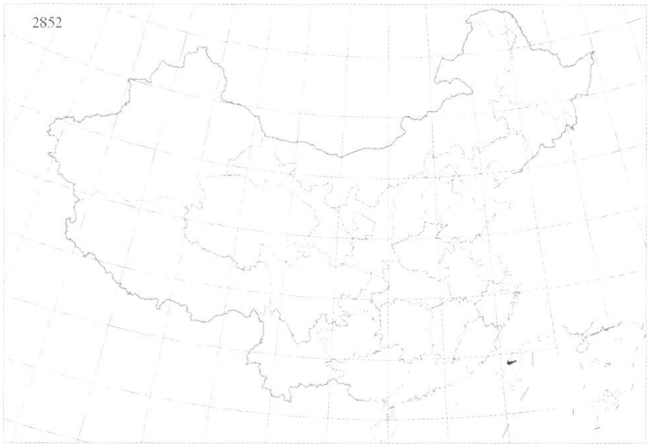

Hamamelidaceae 金缕梅科 *Corylopsis* 蜡瓣花属
Corylopsis platypetala Rehder et E. H. Wilson 阔蜡瓣花
Shrubs, deciduous

15.4 (5.4~18.2); 4.6 (−3.5~10.7); 25.4 (12.3~29.9); 15.4 (5.8~18.3); 126.9 (31.5~159.4); 1.9 (0.0~28.1); 837 (481~986) ❙ 1352 (741~1893); 519 (392~816); 139 (15~227) ❙ 62.4 (4.4~129.9); 818 (481~977); 222.3 (134.6~407.1)

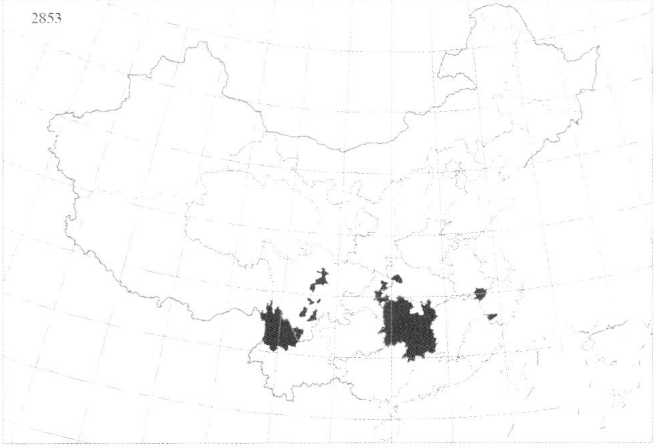

Hamamelidaceae 金缕梅科 *Corylopsis* 蜡瓣花属
Corylopsis rotundifolia H. T. Chang 圆叶蜡瓣花
Shrubs, deciduous

14.7 (10.6~15.6); 3.8 (−0.2~5.0); 25.2 (21.1~26.5); 14.8 (10.6~15.7); 118.9 (79.4~129.2); 1.8 (0.0~12.1); 794 (643~840) ❙ 1281 (1171~1478); 509 (461~599); 86 (46~111) ❙ 62.1 (41.2~94.5); 794 (643~840); 202.1 (170.4~284.3)

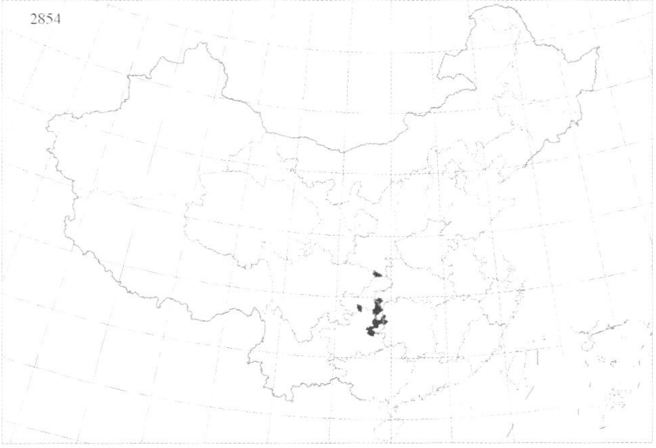

Hamamelidaceae 金缕梅科 *Corylopsis* 蜡瓣花属
Corylopsis sinensis Hemsl. 蜡瓣花
Shrubs, deciduous

16.5 (9.3~22.0); 5.2 (−1.7~14.1); 27.1 (20.1~29.8); 16.6 (9.5~22.1); 139.7 (69.8~205.1); 1.1 (0.0~18.0); 888 (603~1170) ❙ 1474 (862~2014); 532 (376~954); 165 (43~259) ❙ 66.6 (7.2~130.7); 880 (603~1170); 217.9 (134.6~354.5)

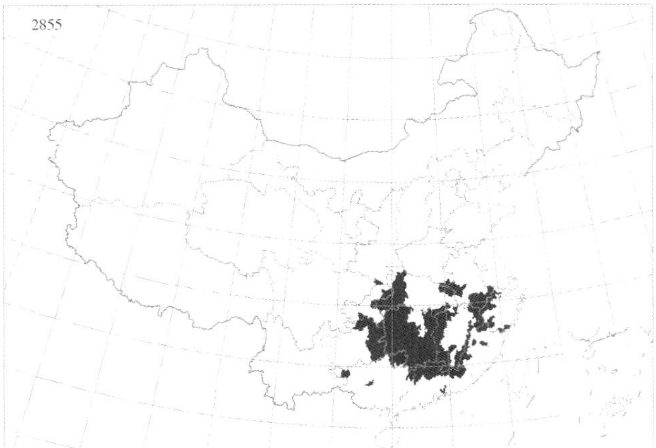

Hamamelidaceae 金缕梅科 *Corylopsis* 蜡瓣花属
Corylopsis sinensis var. calvescens Rehder et E. H. Wilson 秃蜡瓣花
Shrubs, deciduous

15.0 (9.3~18.5); 3.9 (−1.7~8.6); 25.5 (20.1~28.3); 15.1 (9.5~18.5); 123.1 (69.8~162.1); 2.6 (0.0~18.0); 812 (603~953) ❙ 1378 (1118~1689); 537 (465~617); 121 (46~221) ❙ 70.3 (35.7~107.2); 812 (603~953); 226.6 (146.0~319.9)

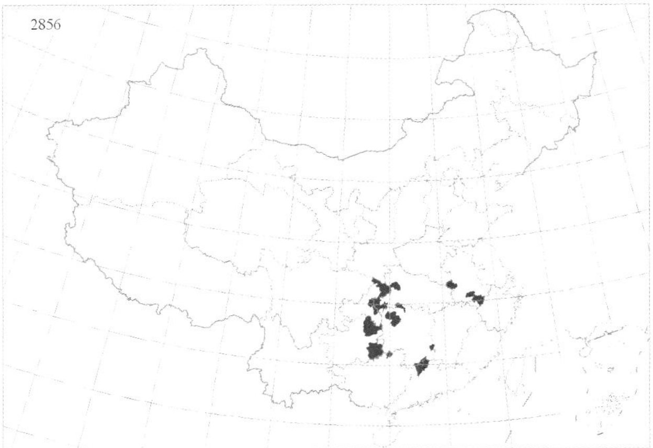

Hamamelidaceae 金缕梅科 *Corylopsis* 蜡瓣花属
Corylopsis stelligera Guillaumin 星毛蜡瓣花
Shrubs or small trees, deciduous

14.5 (11.7~17.0); 4.0 (0.6~6.5); 24.4 (18.8~28.1); 14.6 (11.7~17.0); 116.8 (82.6~144.4); 1.7 (0.0~9.7); 782 (658~912) ❙ 1281 (916~1624); 530 (419~613); 96 (31~198) ❙ 64.0 (35.3~107.2); 780 (658~891); 214.3 (148.9~285.1)

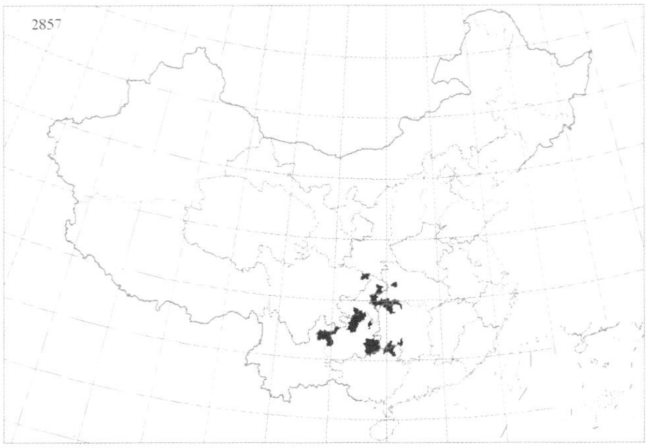

Hamamelidaceae 金缕梅科 *Corylopsis* 蜡瓣花属
Corylopsis trabeculosa He et W. C. Cheng 俅江蜡瓣花
Shrubs or small trees, deciduous

8.6; 1.3; 15.0; 8.7; 52.6; 8.8; 574 ❙ 1041; 585; 45 ❙ 81.5; 574; 352.3

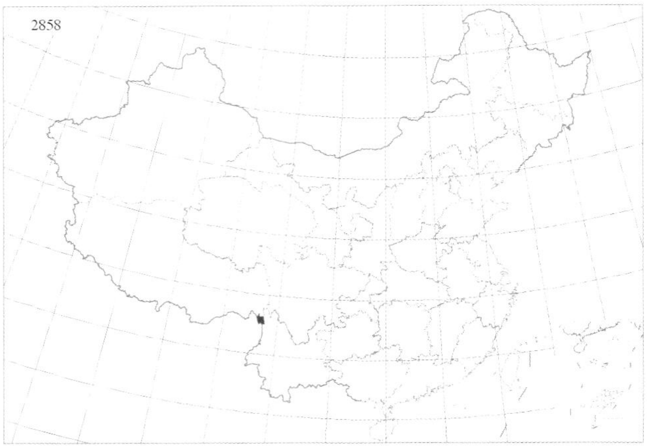

Hamamelidaceae 金缕梅科 *Corylopsis* 蜡瓣花属
Corylopsis veitchiana Bean 红药蜡瓣花
Shrubs, deciduous

14.2 (9.3~16.2); 2.8 (−1.7~5.5); 25.1 (20.1~27.5); 14.2 (9.5~16.2); 114.6 (69.8~135.6); 4.0 (0.0~18.0); 782 (603~877) ❙ 1377 (1132~1770); 549 (473~622); 110 (46~223) ❙ 76.9 (32.5~112.2); 782 (603~877); 232.0 (170.4~299.5)

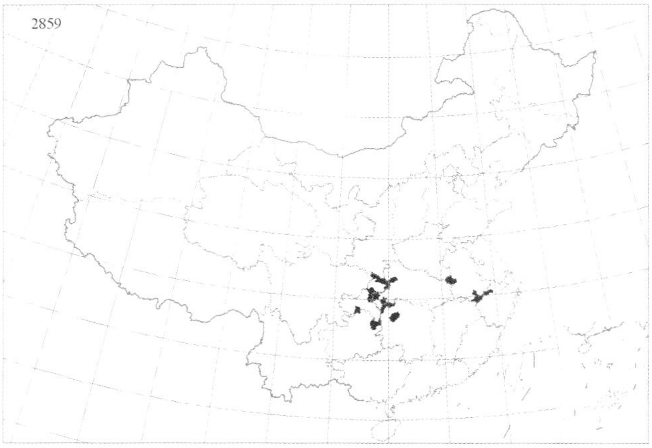

Hamamelidaceae 金缕梅科 *Corylopsis* 蜡瓣花属
Corylopsis velutina Hand.-Mazz. 绒毛蜡瓣花
Shrubs, deciduous

13.9; 6.0; 20.4; 14.0; 108.5; 0.8; 728 ❙ 1020; 609; 23 ❙ 40.7; 697; 311.6

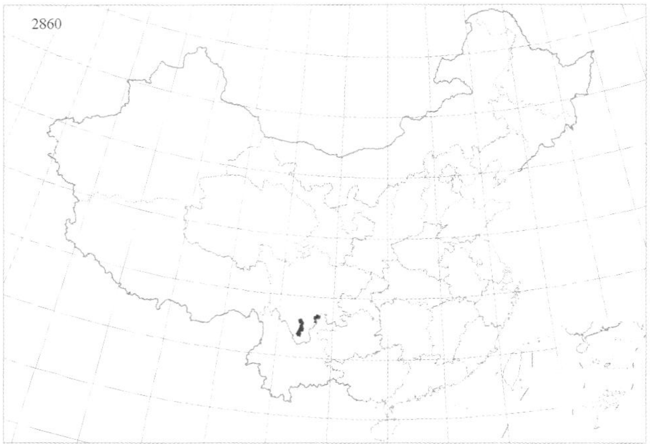

Hamamelidaceae 金缕梅科 *Corylopsis* 蜡瓣花属
Corylopsis willmottiae Rehder et E. H. Wilson 四川蜡瓣花
Shrubs or small trees, deciduous

11.3 (2.0~16.1); 2.0 (−7.2~8.6); 19.7 (9.9~25.6); 11.6 (3.8~16.1); 86.0 (15.4~133.0); 9.3 (0.0~51.4); 668 (413~822) ❙ 1085 (822~1418); 578 (420~817); 36 (18~86) ❙ 66.0 (18.6~128.6); 662 (413~822); 247.4 (147.3~330.0)

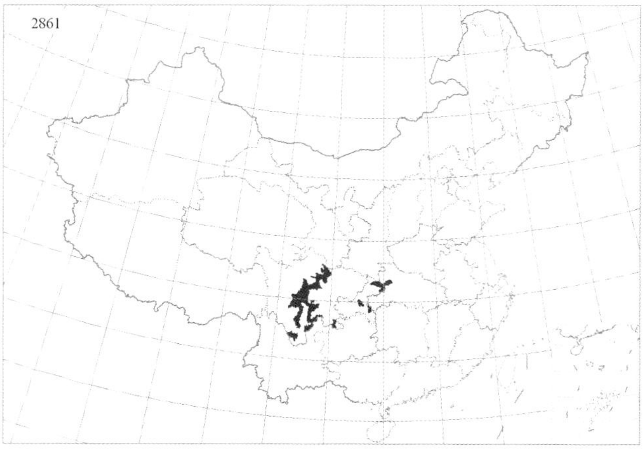

Hamamelidaceae 金缕梅科 *Corylopsis* 蜡瓣花属
Corylopsis yui Hu et W. C. Cheng 长穗蜡瓣花
Shrubs, deciduous

11.1; 4.0; 16.8; 11.1; 76.2; 3.2; 638 ❙ 1207; 669; 57 ❙ 89.0; 638; 381.4

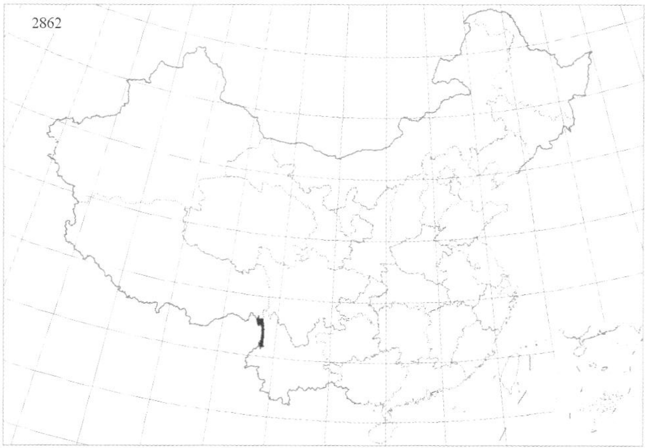

Hamamelidaceae 金缕梅科 *Corylopsis* 蜡瓣花属

Corylopsis yunnanensis Diels 滇蜡瓣花

Shrubs, deciduous

13.2 (8.6~21.0); 6.3 (1.3~14.2); 18.7 (15.0~25.6); 13.2 (8.7~21.0); 99.9 (52.6~191.8); 1.0 (0.0~8.8); 706 (574~1053) ❙ 1099 (937~1522); 631 (553~818); 42 (15~70) ❙ 57.0 (14.4~99.9); 695 (574~1053); 355.2 (265.0~551.6)

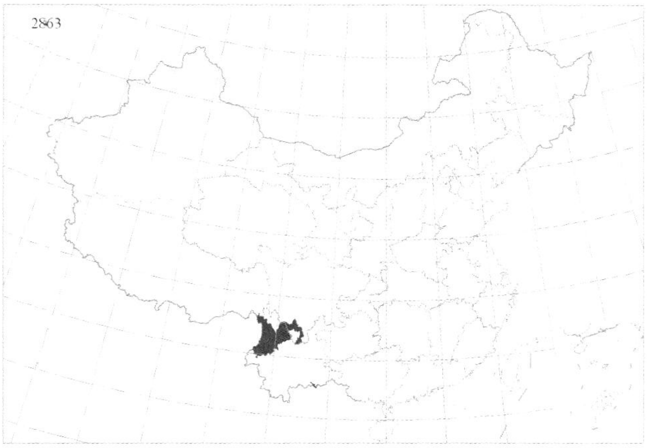

Hamamelidaceae 金缕梅科 *Disanthus* 双花木属

Disanthus cercidifolius subsp. longipes (H. T. Chang) K. Y. Pan 长柄双花木

Shrubs, deciduous

16.9 (14.6~18.8); 5.8 (3.8~8.2); 27.2 (24.6~29.5); 17.0 (14.7~18.8); 143.7 (117.7~165.8); 0.2 (0.0~1.5); 898 (779~991) ❙ 1617 (1418~1922); 539 (400~677); 206 (176~259) ❙ 81.0 (45.4~129.9); 888 (779~991); 225.4 (153.4~279.5)

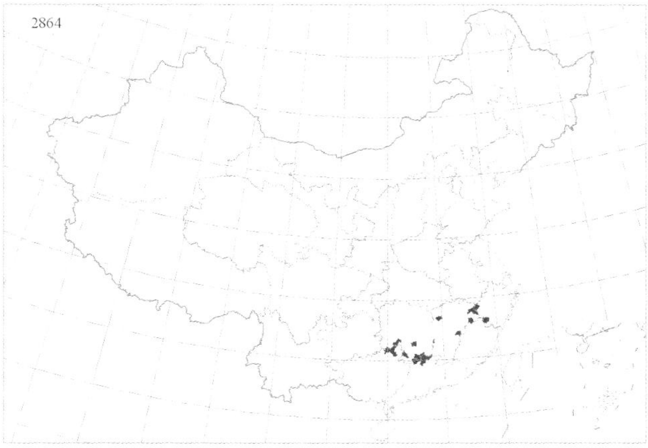

Hamamelidaceae 金缕梅科 *Distyliopsis* 假蚊母树属

Distyliopsis dunnii (Hemsl.) P. K. Endress 尖叶假蚊母树

Shrubs or small trees

18.8 (14.3~22.2); 9.5 (3.7~14.3); 27.1 (22.4~29.8); 18.9 (14.3~22.3); 166.4 (113.2~207.5); 0.0 (0.0~1.5); 977 (761~1190) ❙ 1594 (1070~2024); 602 (356~1056); 174 (48~257) ❙ 64.8 (-5.0~137.5); 971 (761~1190); 248.5 (123.9~413.5)

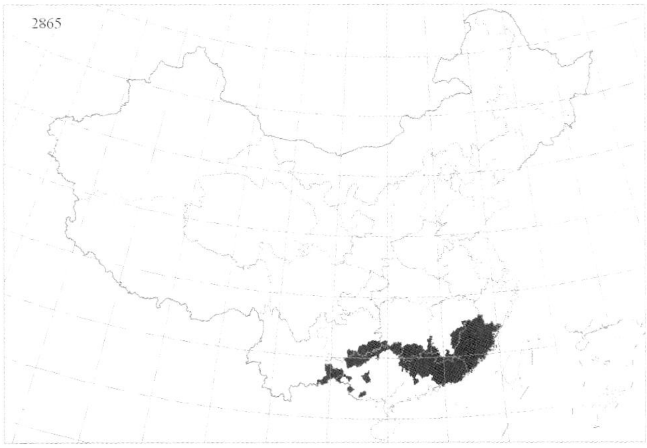

Hamamelidaceae 金缕梅科 *Distyliopsis* 假蚊母树属

Distyliopsis laurifolia (Hemsl.) P. K. Endress 樟叶假蚊母树

Shrubs

17.8 (15.4~21.0); 10.2 (7.4~14.2); 23.2 (20.7~26.6); 17.8 (15.5~21.0); 153.5 (125.7~191.8); 0.0 (0.0~0.0); 871 (767~1053) ❙ 1237 (958~1522); 676 (531~860); 57 (44~107) ❙ 41.9 (13.3~80.6); 868 (767~1053); 354.9 (245.3~551.6)

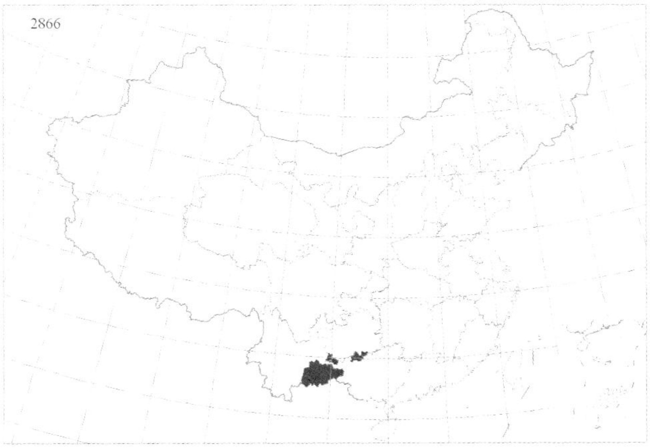

Hamamelidaceae 金缕梅科 *Distyliopsis* 假蚊母树属

Distyliopsis salicifolia (H. L. Li ex E. Walker) P. K. Endress 柳叶假蚊母树

Shrubs, evergreen

23.7 (22.2~24.7); 18.7 (17.2~19.8); 27.2 (25.7~28.2); 23.7 (22.3~24.7); 225.0 (207.2~236.5); 0.0 (0.0~0.0); 1279 (1127~1376) ❙ 1659 (1551~1767); 569 (534~610); 63 (48~88) ❙ 30.6 (15.4~47.8); 1175 (1055~1299); 565.3 (494.7~620.0)

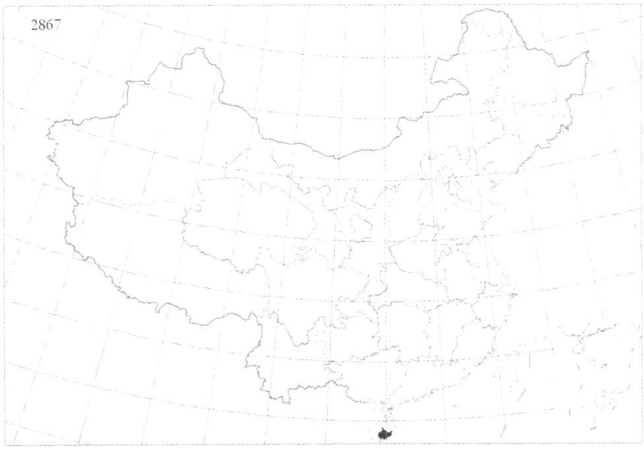

Hamamelidaceae 金缕梅科 *Distyliopsis* 假蚊母树属

Distyliopsis tutcheri (Hemsl.) P. K. Endress 钝叶假蚊母树

Trees

21.6 (17.7~24.8); 13.9 (6.9~20.1); 27.8 (25.7~29.1); 21.7 (17.7~24.8); 200.2 (152.4~237.6); 0.0 (0.0~0.0); 1147 (927~1385) ❙ 1662 (868~2037); 673 (312~1033); 130 (41~197) ❙ 46.7 (-34.3~95.5); 1115 (868~1360); 289.2 (152.9~620.0)

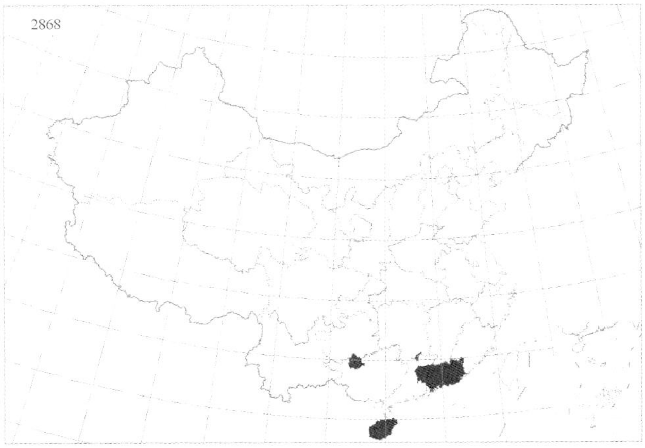

Hamamelidaceae 金缕梅科 *Distyliopsis* 假蚊母树属

Distyliopsis yunnanensis (H. T. Chang) C. Y. Wu 滇假蚊母树

Shrubs or small trees, evergreen

19.4；12.8；23.8；19.4；172.8；0.0；944 ❙ 1488；670；77 ❙ 57.7；928；497.1

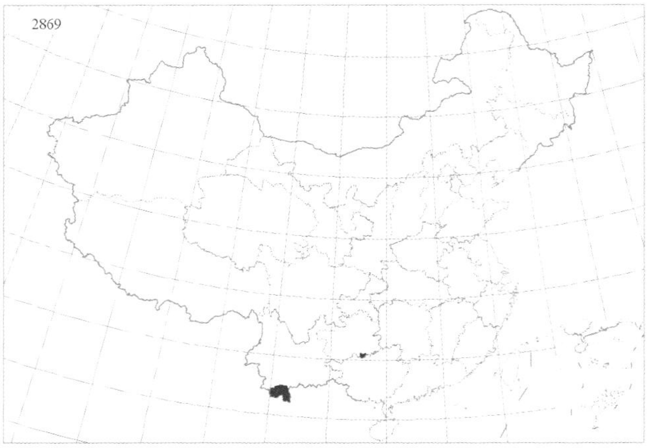

Hamamelidaceae 金缕梅科 *Distylium* 蚊母树属

Distylium buxifolium (Hance) Merr. 小叶蚊母树

Shrubs, evergreen

18.0 (12.6~21.7)；8.1 (1.6~14.3)；26.7 (21.8~29.8)；18.0 (12.7~21.7)；156.3 (98.8~200.5)；0.3 (0.0~6.9)；940 (711~1140) ❙ 1509 (1107~2024)；601 (444~902)；142 (49~254) ❙ 61.3 (21.9~137.5)；939 (711~1140)；235.6 (146.0~411.1)

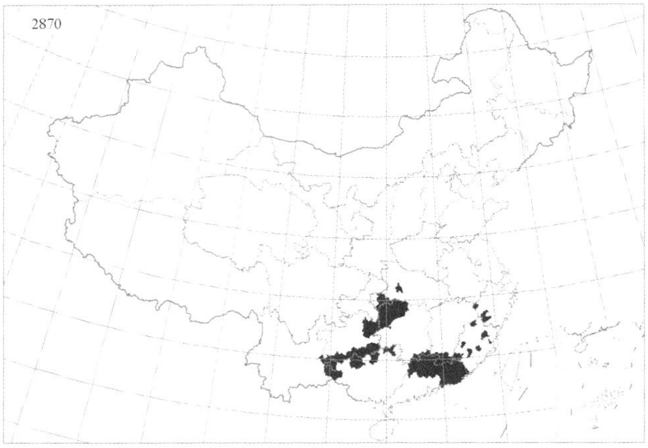

Hamamelidaceae 金缕梅科 *Distylium* 蚊母树属

Distylium chinense (Franch. ex Hemsl.) Diels 中华蚊母树

Shrubs, evergreen

15.1 (10.2~19.1)；4.2 (-0.7~10.1)；25.2 (21.0~27.8)；15.1 (10.4~19.1)；123.5 (77.2~169.2)；2.2 (0.0~13.6)；812 (634~969) ❙ 1228 (863~1466)；526 (375~701)；78 (27~149) ❙ 52.0 (10.5~106.2)；811 (634~969)；214.9 (126.1~305.6)

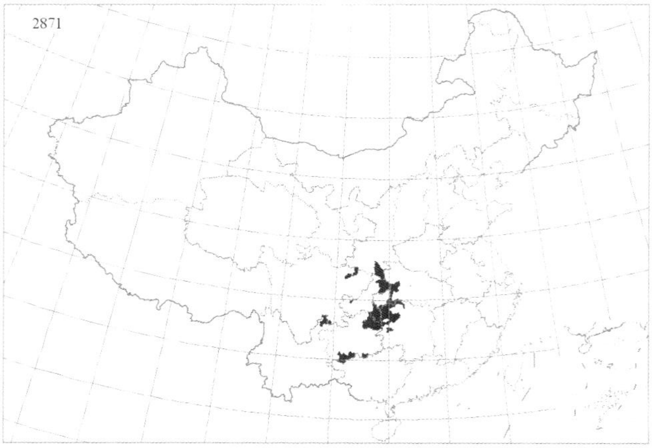

Hamamelidaceae 金缕梅科 *Distylium* 蚊母树属

Distylium chungii (F. P. Metcalf) W. C. Cheng 闽粤蚊母树

Small trees, evergreen

18.4 (14.7~21.7)；9.2 (3.8~14.3)；26.6 (24.3~28.5)；18.5 (14.8~21.7)；161.7 (118.7~200.5)；0.1 (0.0~1.5)；953 (791~1140) ❙ 1537 (1221~1997)；613 (473~723)；150 (60~255) ❙ 63.2 (27.5~137.5)；953 (791~1140)；253.6 (170.4~322.1)

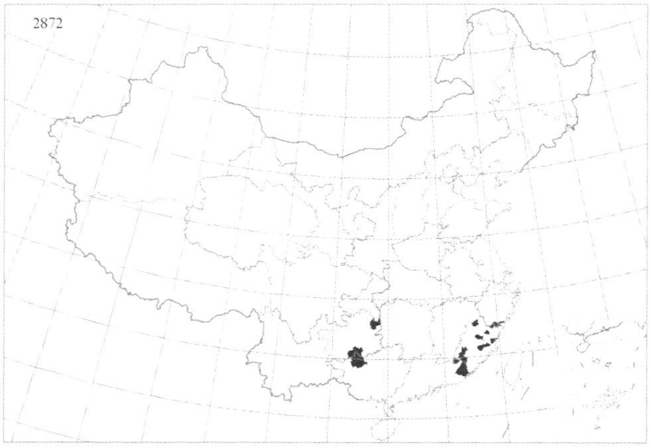

Hamamelidaceae 金缕梅科 *Distylium* 蚊母树属

Distylium cuspidatum H. T. Chang 尖尾蚊母树

Small trees, evergreen

16.1 (13.4~17.7)；6.1 (2.7~9.9)；24.8 (22.4~27.8)；16.1 (13.4~17.7)；134.0 (104.5~152.4)；0.3 (0.0~3.7)；832 (735~937) ❙ 1297 (1105~1504)；572 (458~860)；90 (48~129) ❙ 56.4 (18.0~95.9)；832 (735~937)；238.3 (126.1~413.5)

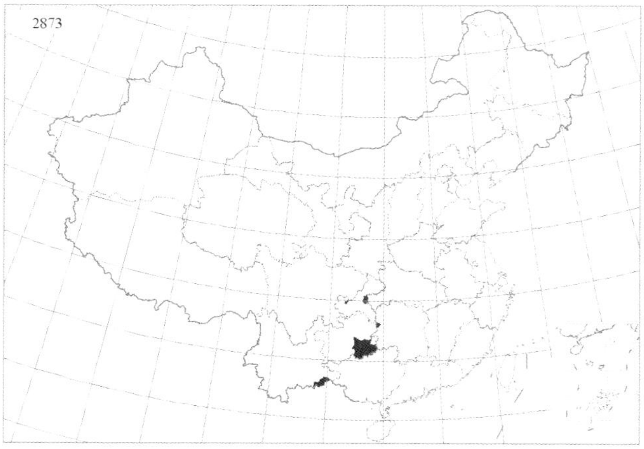

Hamamelidaceae 金缕梅科 *Distylium* 蚊母树属

Distylium dunnianum H. Lév. 窄叶蚊母树

Shrubs or small trees, evergreen

16.1 (11.7~21.3)；6.2 (2.7~11.9)；24.9 (18.8~28.8)；16.2 (11.7~21.3)；134.5 (82.6~196.1)；0.3 (0.0~3.8)；842 (658~1137) ❙ 1241 (916~1662)；565 (444~821)；84 (31~188) ❙ 47.7 (21.9~71.9)；842 (658~1136)；216.2 (146.0~411.1)

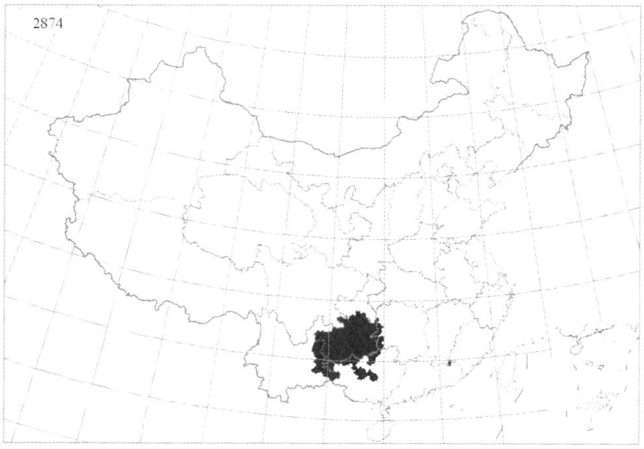

Hamamelidaceae 金缕梅科 *Distylium* 蚊母树属
Distylium elaeagnoides H. T. Chang 鳞毛蚊母树
Shrubs or small trees, evergreen

18.1 (14.6~21.8); 8.0 (3.8~14.0); 26.8 (24.6~28.5); 18.1 (14.7~21.9); 157.3 (117.7~202.2); 0.2 (0.0~1.5); 947 (779~1150) ❙ 1605 (1423~2148); 636 (465~1234); 170 (124~192) ❙ 70.6 (29.9~101.8); 947 (779~1150); 244.8 (210.3~318.8)

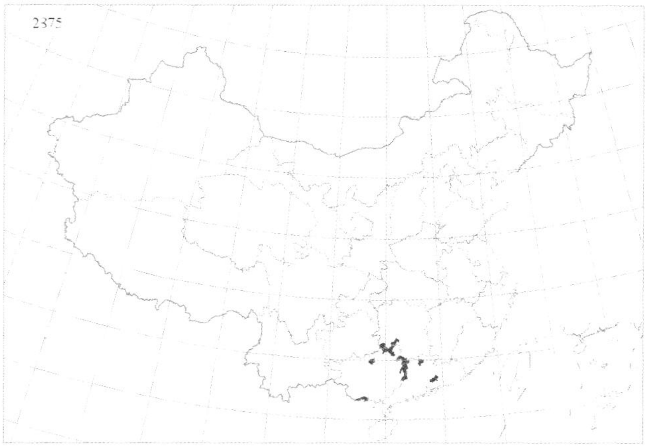

Hamamelidaceae 金缕梅科 *Distylium* 蚊母树属
Distylium gracile Nakai 台湾蚊母树
Small trees, evergreen

17.6; 9.4; 25.4; 17.6; 151.5; 0.0; 894 ❙ 2132; 741; 303 ❙ 139.6; 894; 250.9

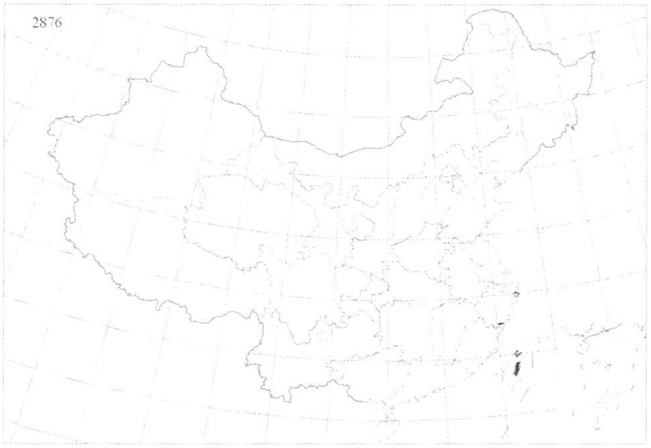

Hamamelidaceae 金缕梅科 *Distylium* 蚊母树属
Distylium macrophyllum H. T. Chang 大叶蚊母树
Shrubs or small trees, evergreen

17.3 (14.3~21.0); 6.7 (3.3~11.9); 26.8 (24.3~28.9); 17.3 (14.3~21.1); 148.3 (114.1~193.0); 0.2 (0.0~2.3); 908 (765~1114) ❙ 1499 (1284~1662); 548 (419~821); 174 (57~211) ❙ 66.2 (30.9~101.8); 901 (765~1114); 226.7 (163.7~371.7)

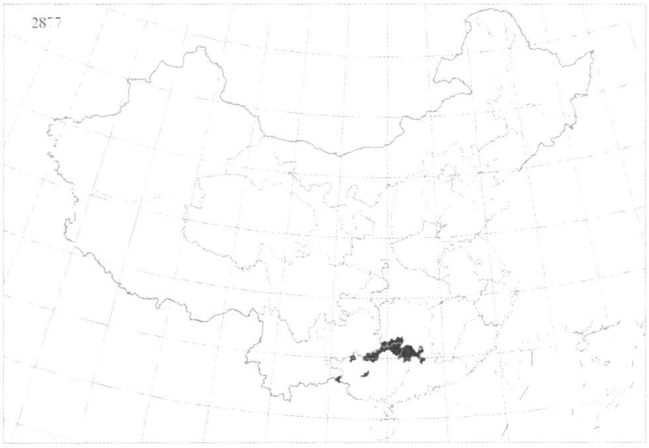

Hamamelidaceae 金缕梅科 *Distylium* 蚊母树属
Distylium myricoides Hemsl. 杨梅蚊母树
Shrubs or small trees, evergreen

17.7 (13.3~24.8); 7.4 (2.1~20.1); 27.0 (21.6~30.1); 17.8 (13.4~24.8); 153.6 (101.4~237.6); 0.3 (0.0~4.9); 940 (715~1385) ❙ 1499 (868~2435); 572 (312~1428); 152 (41~259) ❙ 60.9 (-34.3~147.0); 927 (715~1360); 234.3 (134.6~620.0)

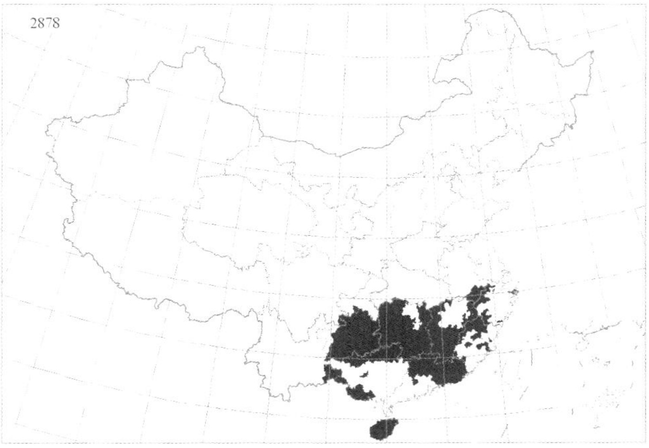

Hamamelidaceae 金缕梅科 *Distylium* 蚊母树属
Distylium pingpienense (Hu) E. Walker 屏边蚊母树
Shrubs or trees, evergreen

16.5 (13.0~21.0); 7.3 (1.9~14.2); 24.5 (22.4~25.9); 16.6 (13.1~21.0); 139.6 (102.9~191.8); 0.8 (0.0~6.1); 846 (730~1053) ❙ 1299 (1141~1522); 631 (473~860); 69 (48~97) ❙ 54.0 (37.1~80.6); 846 (730~1053); 297.2 (170.4~551.6)

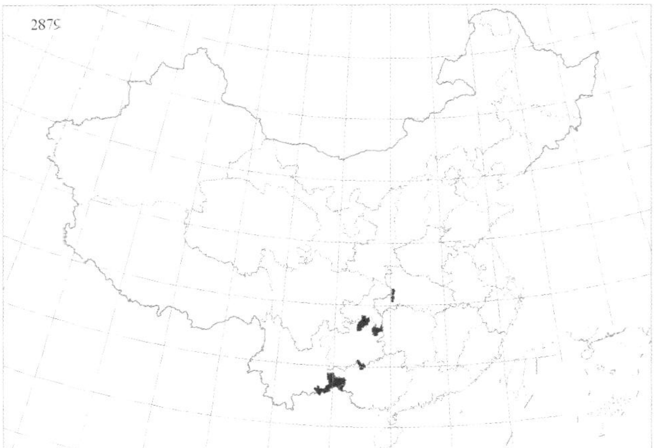

Hamamelidaceae 金缕梅科 *Distylium* 蚊母树属
Distylium racemosum Siebold et Zucc. 蚊母树
Shrubs or trees, evergreen

20.5 (16.5~22.8); 11.8 (5.8~17.7); 27.8 (21.7~29.8); 20.5 (16.5~22.8); 186.4 (138.2~214.1); 0.0 (0.0~0.0); 1078 (861~1225) ❙ 1739 (1447~2992); 735 (481~1847); 154 (75~216) ❙ 62.8 (20.8~214.4); 1076 (861~1225); 220.0 (95.3~363.7)

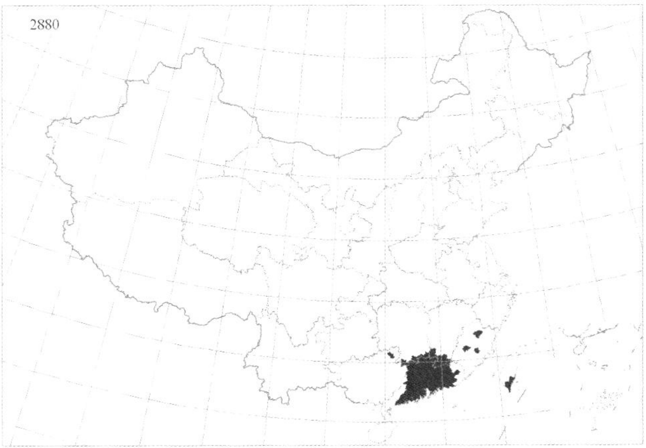

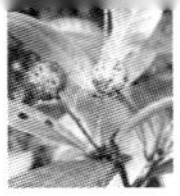

Hamamelidaceae 金缕梅科 *Eustigma* 秀柱花属

Eustigma balansae Oliv. 褐毛秀柱花

Trees, evergreen

20.7 (18.3~21.9); 11.9 (9.4~14.2); 27.6 (25.5~28.7); 20.7 (18.4~22.0); 188.5 (160.9~203.4); 0.0 (0.0~0.0); 1081 (924~1168) ❙ 1531 (1250~1900); 727 (579~1056); 117 (58~161) ❙ 41.2 (27.0~71.0); 1080 (924~1168); 254.0 (185.5~318.5)

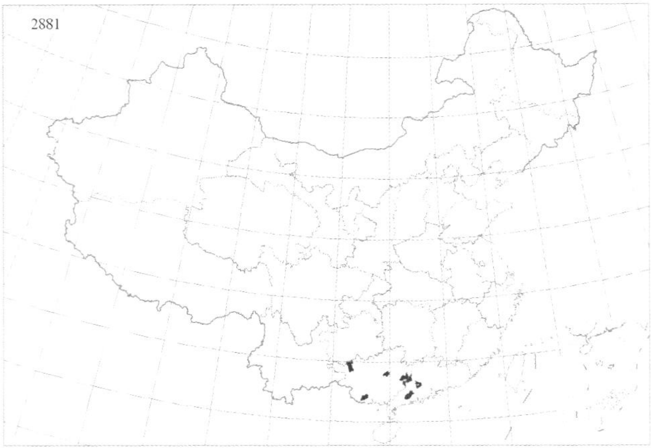

Hamamelidaceae 金缕梅科 *Eustigma* 秀柱花属

Eustigma lenticellatum C. Y. Wu 云南秀柱花

Trees, evergreen

17.8 (16.3~21.0); 10.3 (7.4~14.2); 23.2 (21.4~25.6); 17.8 (16.4~21.0); 153.7 (136.4~191.8); 0.0 (0.0~0.0); 871 (797~1053) ❙ 1315 (1074~1522); 722 (582~860); 58 (47~107) ❙ 51.1 (32.4~80.6); 871 (797~1053); 380.7 (245.3~551.6)

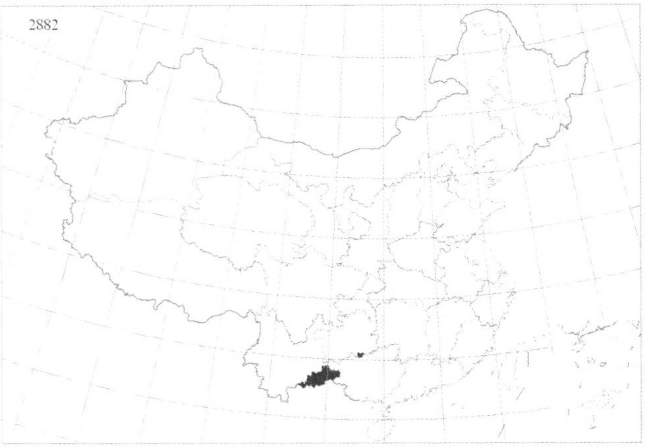

Hamamelidaceae 金缕梅科 *Eustigma* 秀柱花属

Eustigma oblongifolium Gardner et Champ. 秀柱花

Shrubs or small trees, evergreen

20.5 (14.6~24.8); 12.1 (3.8~20.1); 27.5 (20.1~29.1); 20.6 (14.7~24.8); 186.7 (117.7~237.6); 0.0 (0.0~1.5); 1080 (773~1385) ❙ 1660 (868~2680); 664 (312~1186); 149 (41~245) ❙ 56.7 (-34.3~246.5); 1061 (773~1360); 269.6 (152.9~620.0)

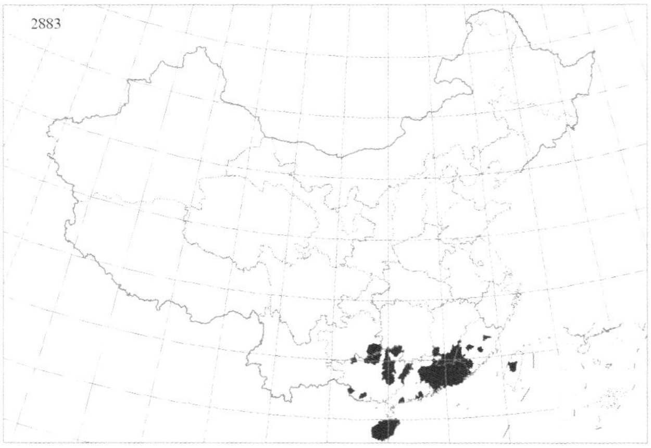

Hamamelidaceae 金缕梅科 *Exbucklandia* 马蹄荷属

Exbucklandia longipetala H. T. Chang 长瓣马蹄荷

Trees, evergreen

16.7 (14.3~19.2); 6.4 (4.0~9.2); 25.7 (23.5~27.5); 16.7 (14.3~19.2); 141.0 (112.6~170.8); 0.1 (0.0~1.1); 867 (756~1002) ❙ 1362 (1218~1564); 565 (476~678); 120 (94~182) ❙ 57.2 (47.0~71.9); 867 (756~1002); 229.8 (197.5~256.4)

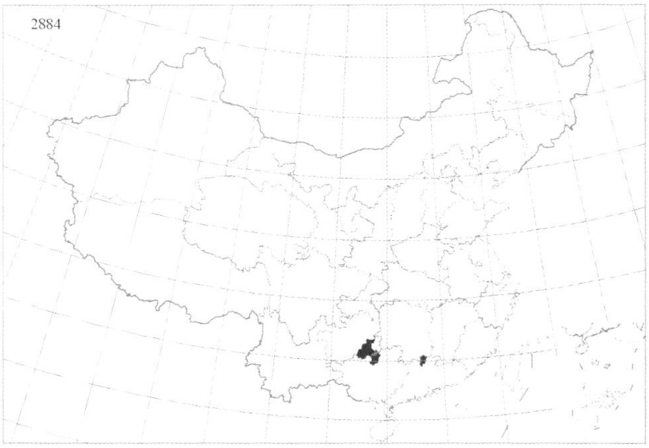

Hamamelidaceae 金缕梅科 *Exbucklandia* 马蹄荷属

Exbucklandia populnea (R. Br. ex Griff.) R. W. Brown 马蹄荷

Trees, evergreen

16.7 (5.3~21.8); 8.6 (-2.4~14.2); 23.1 (12.2~28.5); 16.7 (5.7~21.9); 141.6 (30.5~202.6); 0.7 (0.0~26.2); 848 (479~1150) ❙ 1354 (741~1876); 666 (426~1007); 90 (28~225) ❙ 60.4 (19.0~145.8); 844 (479~1150); 327.3 (189.5~551.6)

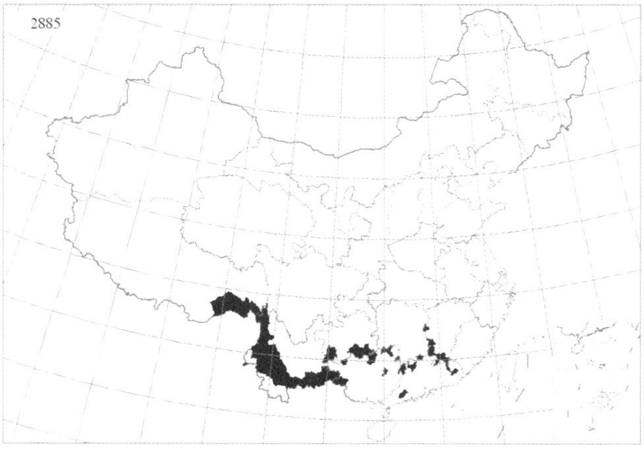

Hamamelidaceae 金缕梅科 *Exbucklandia* 马蹄荷属

Exbucklandia tonkinensis (Lecomte) Steenis 大果马蹄荷

Trees, evergreen

20.0 (14.3~24.8); 11.4 (3.3~20.1); 27.3 (21.4~29.8); 20.1 (14.3~24.8); 181.2 (112.6~237.6); 0.0 (0.0~2.3); 1054 (756~1385) ❙ 1606 (1161~2435); 669 (392~1428); 138 (41~258) ❙ 54.6 (-12.0~130.7); 1043 (756~1360); 257.6 (95.3~620.0)

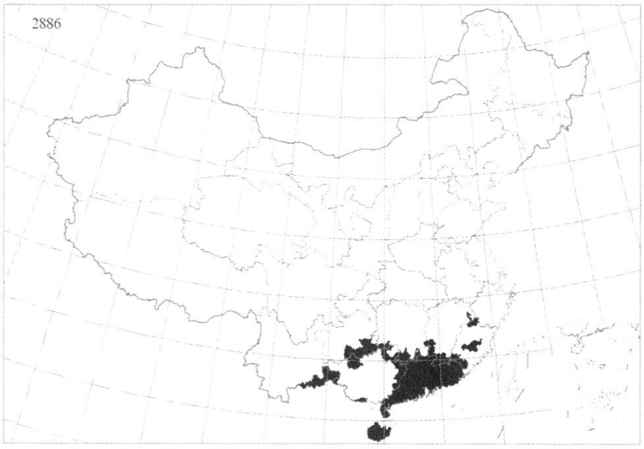

Hamamelidaceae 金缕梅科 *Fortunearia* 牛鼻栓属

Fortunearia sinensis Rehder et E. H. Wilson 牛鼻栓

Shrubs or small trees, deciduous

15.2 (9.3~18.7); 2.8 (-2.9~8.2); 27.0 (20.1~29.5); 15.2 (9.5~18.7); 126.9 (69.8~164.9); 4.4 (0.0~20.5); 848 (603~984) ▎1236 (626~1997); 490 (316~685); 128 (19~259) ▎44.9 (-22.2~130.7); 834 (603~981); 212.1 (131.2~319.9)

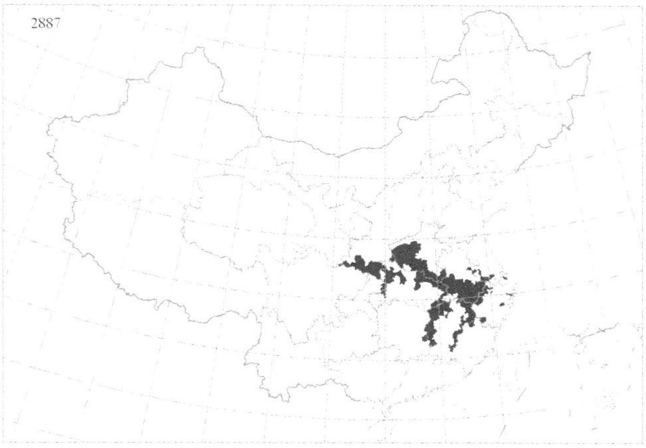

Hamamelidaceae 金缕梅科 *Hamamelis* 金缕梅属

Hamamelis mollis Oliv. 金缕梅

Shrubs or small trees, deciduous

15.4 (9.3~18.8); 3.6 (-1.7~9.2); 26.6 (20.1~29.3); 15.4 (9.5~18.9); 127.3 (69.8~166.3); 2.5 (0.0~18.0); 841 (603~984) ▎1417 (807~1922); 535 (349~685); 145 (37~259) ▎68.6 (-2.0~112.2); 838 (603~970); 225.6 (146.0~319.9)

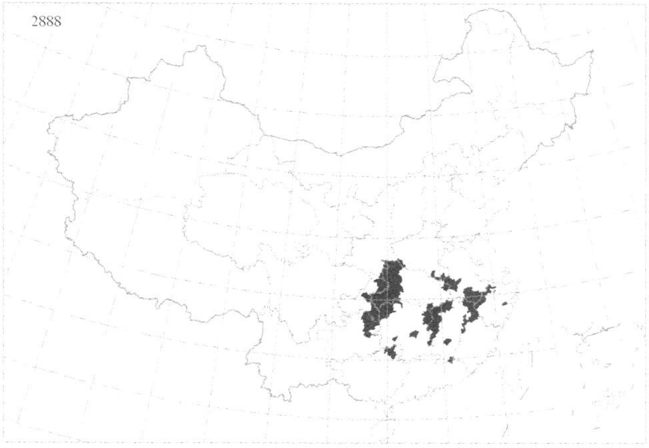

Hamamelidaceae 金缕梅科 *Liquidambar* 枫香属

Liquidambar acalycina H. T. Chang 缺萼枫香树

Trees, deciduous

16.6 (9.3~22.4); 5.6 (-1.7~14.8); 26.9 (20.1~29.3); 16.7 (9.5~22.4); 141.0 (69.8~208.9); 0.8 (0.0~18.0); 887 (603~1191) ▎1560 (862~2030); 565 (376~1033); 175 (47~259) ▎76.4 (7.8~147.0); 884 (603~1191); 232.2 (146.0~411.1)

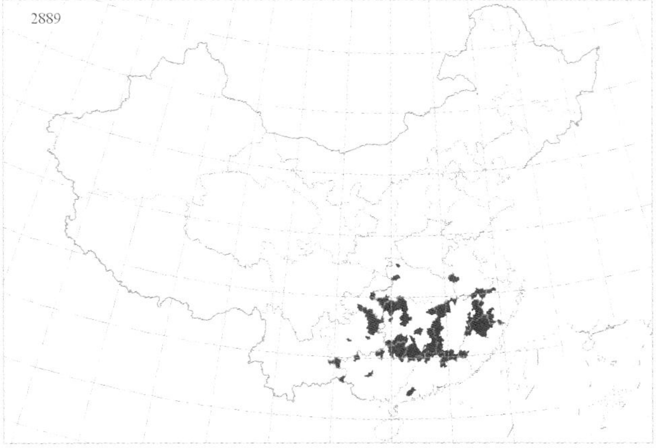

Hamamelidaceae 金缕梅科 *Liquidambar* 枫香属

Liquidambar formosana Hance 枫香树

Trees, deciduous

17.3 (2.2~24.8); 6.7 (-6.6~20.1); 27.0 (9.9~30.1); 17.4 (3.8~24.8); 149.8 (15.5~237.6); 1.5 (0.0~49.1); 930 (413~1385) ▎1392 (626~3495); 564 (299~1594); 138 (16~897) ▎50.4 (-34.3~246.5); 915 (413~1360); 220.8 (95.3~620.0)

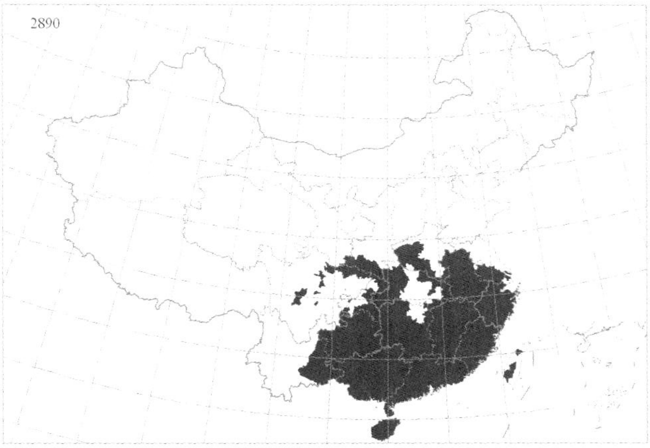

Hamamelidaceae 金缕梅科 *Loropetalum* 檵木属

Loropetalum chinense (R. Br.) Oliv. 檵木

Shrubs or small trees, evergreen

17.0 (3.2~22.5); 6.6 (-6.4~15.2); 26.6 (11.4~30.1); 17.1 (4.7~22.7); 146.2 (22.7~212.7); 1.2 (0.0~43.8); 906 (436~1216) ▎1402 (626~2037); 560 (299~1044); 139 (17~259) ▎55.1 (-22.2~147.0); 895 (436~1216); 228.1 (108.2~659.0)

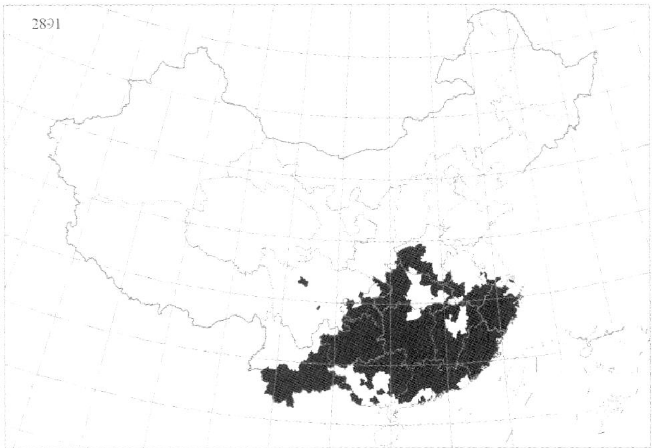

Hamamelidaceae 金缕梅科 *Loropetalum* 檵木属

Loropetalum chinense var. rubrum Yieh 红花檵木

Shrubs or small trees, evergreen

17.1 (16.3~17.7); 4.5 (3.8~5.0); 29.0 (28.1~29.9); 17.1 (16.4~17.7); 146.1 (137.5~153.0); 0.5 (0.0~1.2); 937 (894~976) ▎1495 (1405~1580); 470 (439~499); 204 (196~214) ▎59.9 (43.9~71.2); 896 (874~950); 185.9 (141.7~214.8)

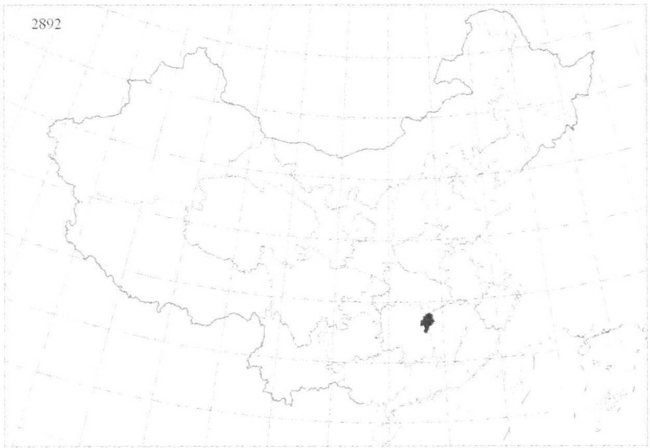

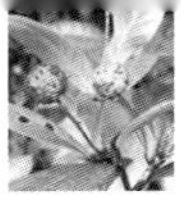

Hamamelidaceae 金缕梅科 *Loropetalum* 檵木属

Loropetalum lanceum Hand.-Mazz. 大果檵木

Trees, evergreen

20.1 (16.1~22.7); 11.2 (5.3~15.0); 27.2 (25.5~28.6); 20.2 (16.1~22.8); 182.2 (133.1~213.1); 0.0 (0.0~0.0); 1062 (838~1221) ❙ 1600 (1268~2435); 819 (523~1428); 113 (86~130) ❙ 51.0 (12.9~99.5); 1062 (838~1221); 269.3 (222.1~327.7)

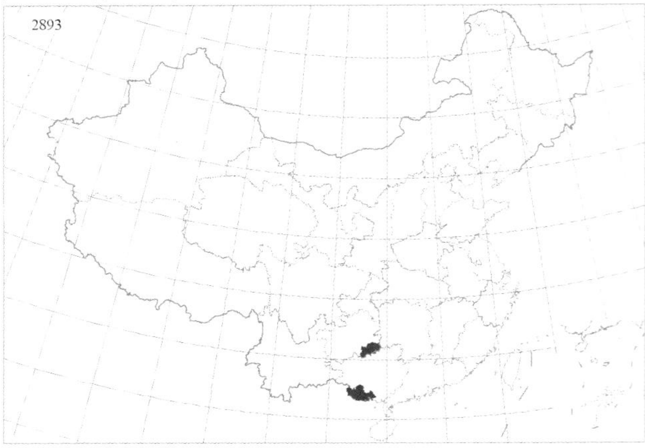

Hamamelidaceae 金缕梅科 *Loropetalum* 檵木属

Loropetalum subcordatum (Benth.) Oliv. 四药门花

Shrubs or small trees, evergreen

22.4 (17.4~24.8); 15.0 (7.4~20.1); 28.2 (25.5~29.2); 22.4 (17.4~24.8); 209.3 (148.9~237.6); 0.0 (0.0~0.0); 1199 (886~1385) ❙ 1662 (868~2037); 712 (312~1033); 115 (41~191) ❙ 40.1 (-34.3~95.5); 1167 (868~1360); 265.6 (95.3~620.0)

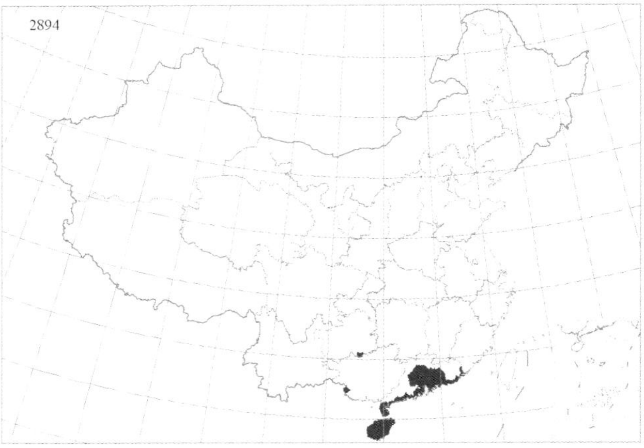

Hamamelidaceae 金缕梅科 *Mytilaria* 壳菜果属

Mytilaria laosensis Lecomte 壳菜果

Trees, evergreen

20.0 (16.3~22.7); 11.9 (7.3~15.0); 26.2 (21.4~28.8); 20.1 (16.4~22.8); 180.9 (136.4~213.1); 0.0 (0.0~0.0); 1033 (797~1221) ❙ 1514 (1018~2435); 779 (554~1428); 95 (47~142) ❙ 46.6 (12.9~99.5); 1033 (797~1221); 296.1 (203.4~551.6)

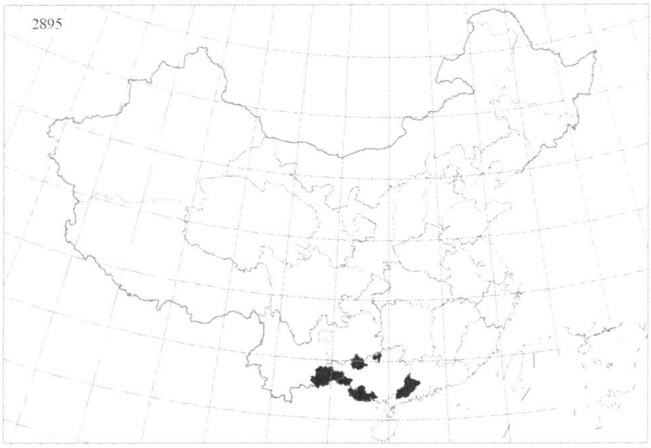

Hamamelidaceae 金缕梅科 *Parrotia* 银缕梅属

Parrotia subaequalis (H. T. Chang) R. M. Hao et H. T. Wei 银缕梅

Small trees, deciduous

14.8 (13.4~15.6); 2.6 (1.5~3.6); 26.9 (25.2~28.1); 14.9 (13.4~15.7); 122.1 (108.3~130.0); 3.8 (1.7~7.3); 825 (755~869) ❙ 1403 (1117~1604); 574 (478~666); 158 (122~200) ❙ 71.2 (28.5~108.0); 825 (755~869); 228.0 (179.2~284.0)

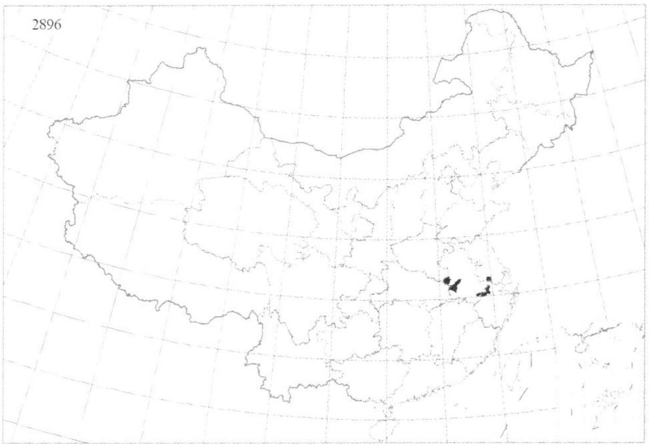

Hamamelidaceae 金缕梅科 *Rhodoleia* 红花荷属

Rhodoleia championii Hook. f. 红花荷

Trees, evergreen

21.9 (17.9~24.8); 14.1 (7.4~20.1); 28.1 (25.7~29.2); 22.0 (18.0~24.8); 203.8 (155.6~237.6); 0.0 (0.0~0.0); 1169 (913~1385) ❙ 1646 (868~2435); 737 (312~1428); 121 (41~194) ❙ 42.0 (-34.3~99.5); 1149 (868~1360); 258.3 (95.3~620.0)

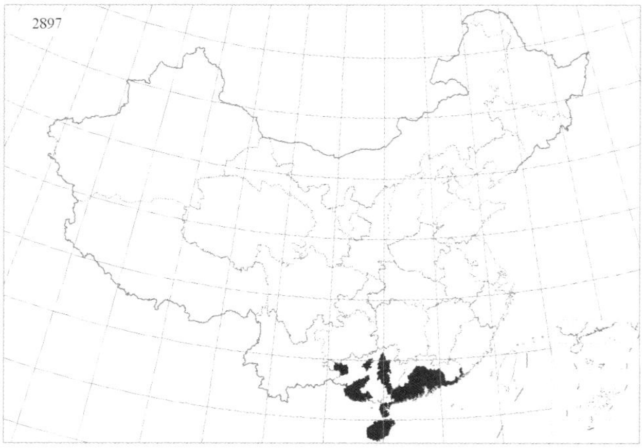

Hamamelidaceae 金缕梅科 *Rhodoleia* 红花荷属

Rhodoleia forrestii Chun ex Exell 绒毛红花荷

Trees, evergreen

13.0 (8.6~16.9); 6.1 (1.3~10.0); 18.2 (15.0~21.2); 13.0 (8.7~16.9); 98.1 (52.6~142.6); 1.6 (0.0~8.8); 695 (574~814) ❙ 1281 (1007~1688); 715 (578~959); 58 (40~70) ❙ 83.9 (53.6~112.2); 695 (574~814); 419.1 (335.1~627.0)

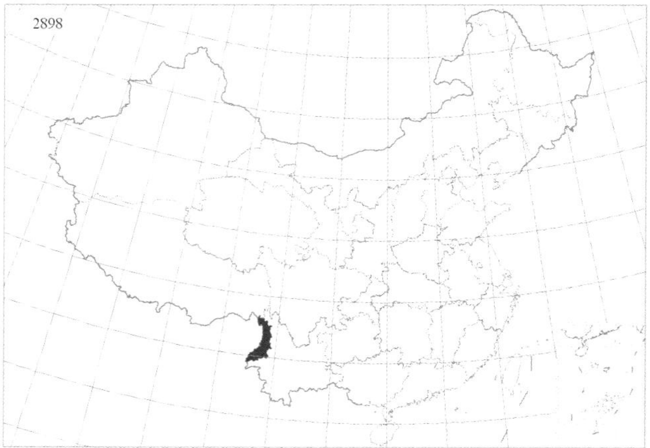

Hamamelidaceae 金缕梅科 *Rhodoleia* 红花荷属
Rhodoleia henryi Tong 小脉红花荷
Trees, evergreen

18.0 (16.3~21.0); 10.8 (7.4~14.2); 23.2 (21.4~25.6); 18.1 (16.4~21.0); 156.8 (136.4~191.8); 0.0 (0.0~0.0); 880 (797~1053) ❙ 1340 (1074~1547); 746 (582~911); 58 (47~107) ❙ 52.5 (32.4~80.6); 879 (797~1053); 401.5 (245.3~551.6)

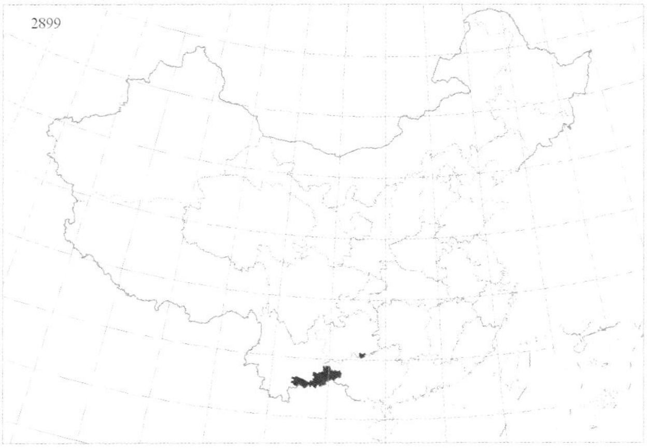

Hamamelidaceae 金缕梅科 *Rhodoleia* 红花荷属
Rhodoleia macrocarpa H. T. Chang 大果红花荷
Trees, evergreen

16.3; 9.1; 21.4; 16.4; 136.4; 0.0; 797 ❙ 1190; 656; 50 ❙ 49.4; 797; 347.3

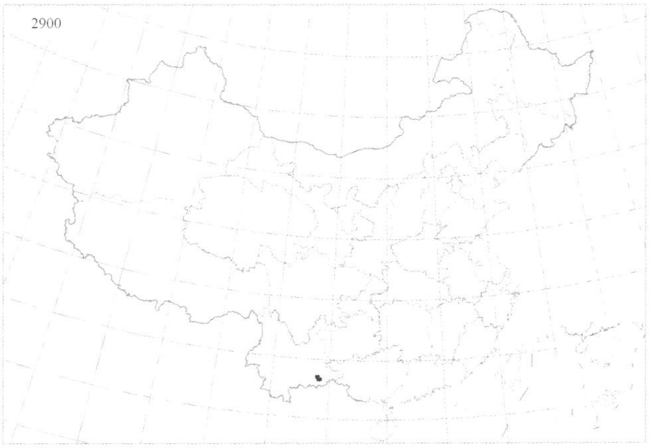

Hamamelidaceae 金缕梅科 *Rhodoleia* 红花荷属
Rhodoleia parvipetala Tong 小花红花荷
Trees, evergreen

20.0 (16.3~23.4); 11.8 (6.3~16.1); 26.5 (21.4~29.2); 20.1 (16.4~23.5); 181.0 (136.4~221.7); 0.0 (0.0~0.0); 1038 (797~1273) ❙ 1511 (1074~2014); 711 (531~954); 111 (43~181) ❙ 46.8 (9.0~83.4); 1036 (797~1273); 277.5 (178.7~551.6)

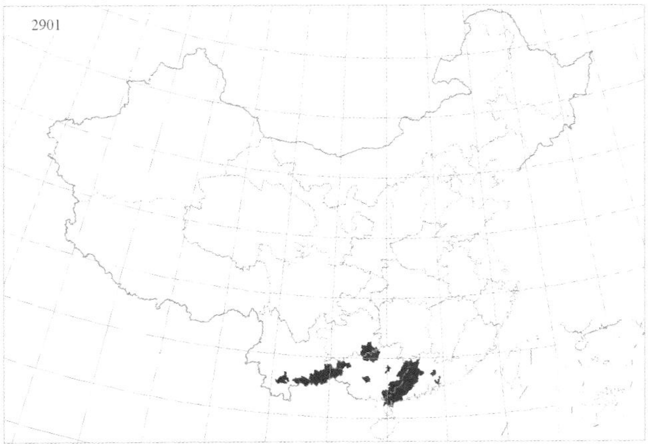

Hamamelidaceae 金缕梅科 *Rhodoleia* 红花荷属
Rhodoleia stenopetala H. T. Chang 窄瓣红花荷
Trees, evergreen

22.8 (16.3~24.8); 16.6 (7.3~20.1); 27.4 (21.4~29.2); 22.9 (16.4~24.8); 214.3 (136.4~237.6); 0.0 (0.0~0.0); 1229 (797~1385) ❙ 1537 (1179~1854); 593 (392~828); 77 (41~142) ❙ 26.9 (-11.9~64.7); 1149 (797~1360); 414.4 (164.9~620.0)

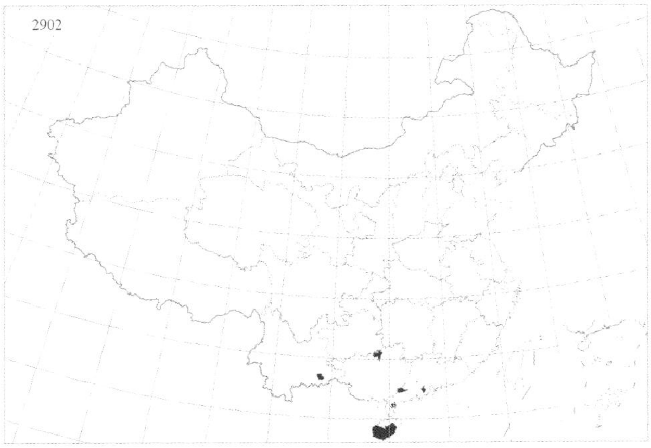

Hamamelidaceae 金缕梅科 *Semiliquidambar* 半枫荷属
Semiliquidambar cathayensis H. T. Chang 半枫荷
Trees, evergreen

18.4 (14.3~24.7); 8.6 (3.3~19.8); 27.0 (23.5~29.8); 18.4 (14.3~24.7); 161.3 (112.6~236.5); 0.1 (0.0~2.3); 965 (756~1376) ❙ 1622 (1089~2435); 606 (445~1428); 172 (48~259) ❙ 69.6 (6.7~147.0); 957 (756~1299); 256.8 (157.7~620.0)

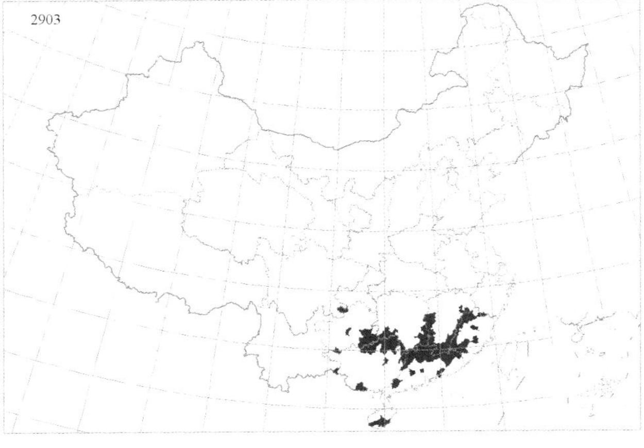

Hamamelidaceae 金缕梅科 *Semiliquidambar* 半枫荷属
Semiliquidambar caudata H. T. Chang 长尾半枫荷
Trees, semievergreen

16.3 (15.0~18.4); 5.9 (4.0~9.0); 26.2 (24.7~27.2); 16.3 (15.1~18.4); 136.2 (120.8~161.2); 0.2 (0.0~1.0); 855 (789~953) ❙ 1804 (1715~1950); 631 (587~713); 213 (195~230) ❙ 112.1 (81.7~147.0); 855 (789~950); 264.4 (245.5~283.9)

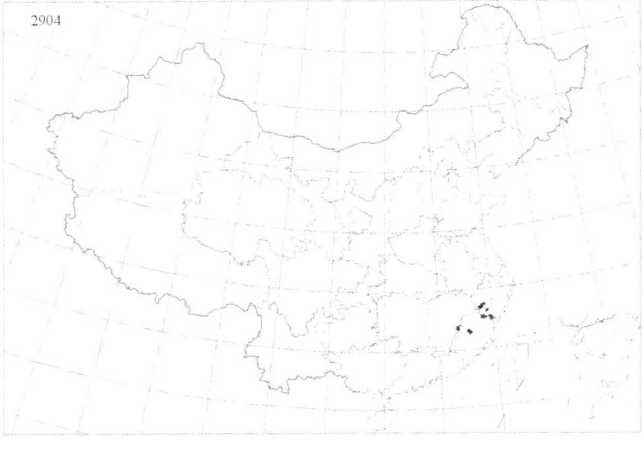

Hamamelidaceae 金缕梅科 *Semiliquidambar* 半枫荷属
Semiliquidambar chingii (F. P. Metcalf) H. T. Chang 细柄半枫荷
Trees, evergreen

18.1 (14.6~20.7); 8.0 (3.8~10.9); 27.1 (24.3~29.8); 18.1 (14.7~20.8); 157.6 (117.7~189.3); 0.1 (0.0~1.5); 946 (779~1113) ▎1692 (1381~1997); 606 (508~695); 196 (107~254) ▎80.5 (35.6~137.5); 942 (779~1072); 252.0 (174.9~303.4)

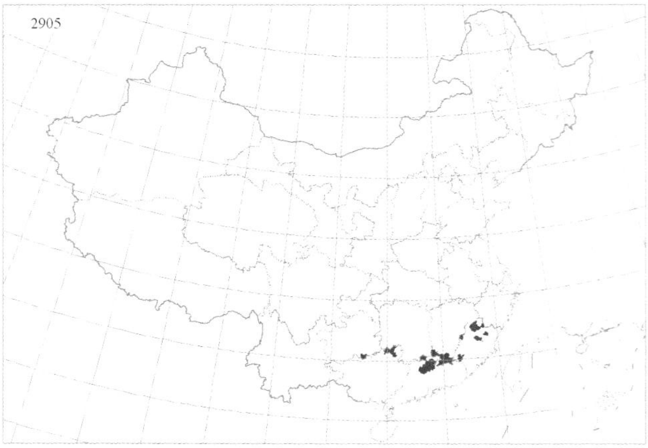

Hamamelidaceae 金缕梅科 *Sinowilsonia* 山白树属
Sinowilsonia henryi Hemsl. 山白树
Shrubs or small trees, deciduous

12.6 (3.5~15.8); -0.4 (-6.5~3.3); 24.7 (12.6~28.0); 12.7 (5.0~15.8); 104.1 (26.3~133.1); 12.7 (2.1~43.4); 751 (446~891) ▎793 (520~1478); 379 (258~628); 36 (14~134) ▎7.1 (-36.5~107.2); 686 (446~891); 210.6 (120.1~334.4)

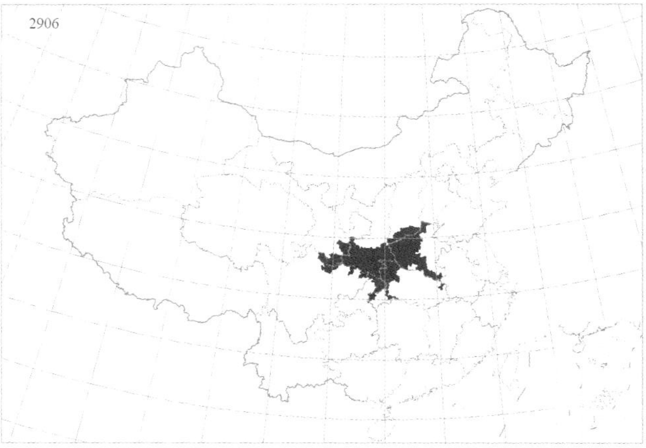

Hamamelidaceae 金缕梅科 *Sinowilsonia* 山白树属
Sinowilsonia henryi var. glabrescens H. T. Chang 秃山白树
Shrubs or small trees, deciduous

11.1; -3.3; 24.0; 11.6; 94.2; 20.7; 708 ▎612; 326; 25 ▎-13.5; 612; 193.0

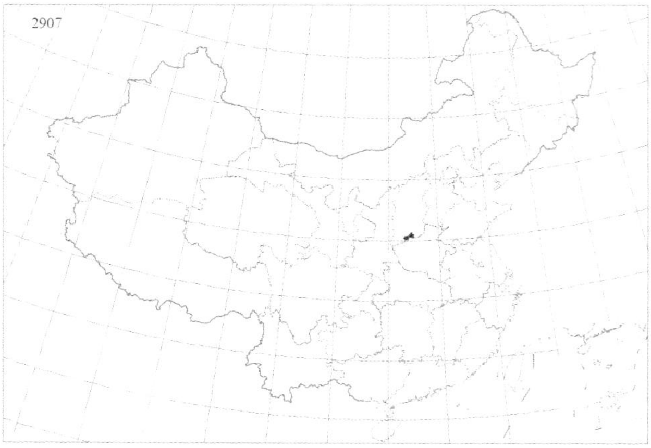

Hamamelidaceae 金缕梅科 *Sycopsis* 水丝梨属
Sycopsis sinensis Oliv. 水丝梨
Trees, evergreen

16.7 (9.3~24.1); 5.7 (-1.7~18.9); 26.8 (20.1~29.9); 16.8 (9.5~24.1); 142.3 (69.8~229.7); 1.3 (0.0~18.0); 895 (603~1322) ▎1444 (798~2759); 540 (349~1594); 152 (29~303) ▎62.0 (-3.3~246.5); 883 (603~1160); 215.5 (134.6~413.5)

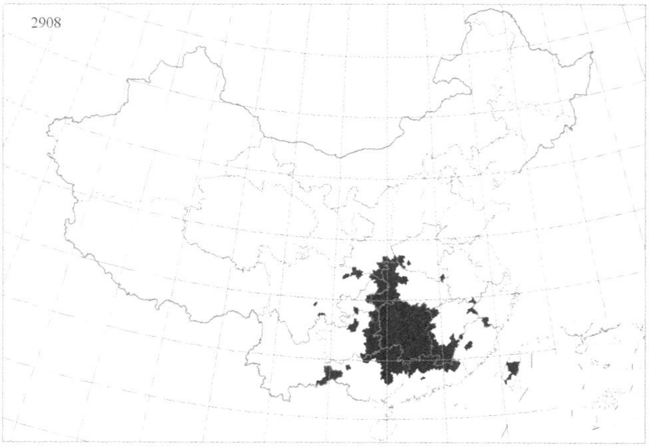

Hamamelidaceae 金缕梅科 *Sycopsis* 水丝梨属
Sycopsis triplinervia H. T. Chang 三脉水丝梨
Shrubs, evergreen

14.4 (12.8~16.6); 4.9 (2.7~7.9); 23.0 (19.3~27.7); 14.5 (12.9~16.6); 114.6 (95.9~139.8); 0.9 (0.0~3.7); 769 (699~895) ▎1127 (879~1444); 537 (472~592); 68 (28~147) ▎45.8 (25.7~95.9); 759 (690~889); 235.5 (185.3~276.9)

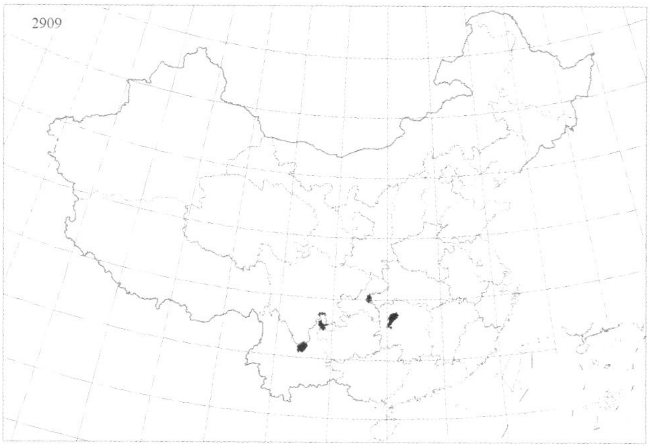

Eucommiaceae 杜仲科 *Eucommia* 杜仲属
Eucommia ulmoides Oliv. 杜仲
Trees, deciduous

15.0 (3.4~22.0); 3.5 (-6.0~13.5); 25.6 (11.7~29.8); 15.0 (4.7~22.0); 124.7 (23.2~203.8); 4.6 (0.0~41.9); 824 (442~1170) ▎1153 (548~1922); 498 (264~816); 86 (15~259) ▎39.8 (-36.5~123.6); 802 (442~1170); 209.4 (111.9~407.1)

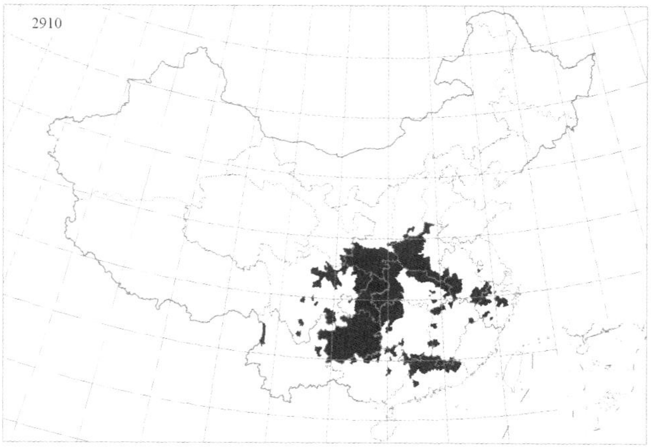

E

Rosaceae 蔷薇科 *Amelanchier* 唐棣属
Amelanchier asiatica (Siebold et Zucc.) Endl. ex Walp. 东亚唐棣
Shrubs or trees, deciduous

15.5 (13.4~17.0); 3.7 (1.5~6.7); 27.1 (25.2~27.9); 15.5 (13.4~17.0); 128.0 (108.3~144.4); 2.0 (0.0~7.3); 844 (755~895) ❙ 1547 (1317~1813); 573 (481~666); 185 (122~226) ❙ 83.7 (53.3~123.6); 844 (755~895); 248.5 (183.4~319.9)

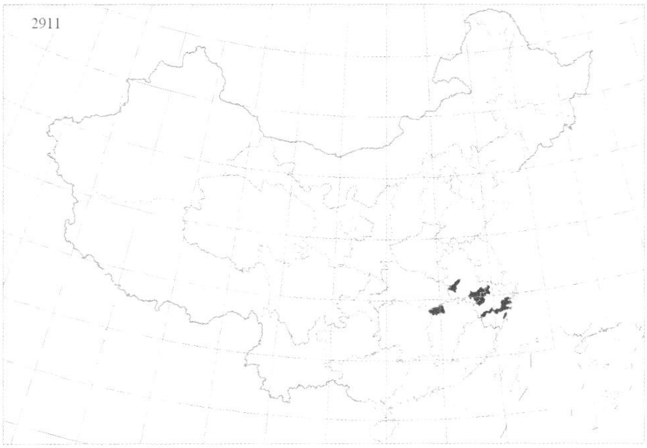

Rosaceae 蔷薇科 *Amelanchier* 唐棣属
Amelanchier sinica (C. K. Schneid.) Chun 唐棣
Trees, deciduous

11.6 (6.2~15.6); -2.1 (-10.6~4.4); 24.1 (15.8~29.5); 12.0 (7.0~15.7); 97.4 (45.2~136.0); 18.1 (0.6~50.1); 722 (522~932) ❙ 746 (358~1663); 345 (200~666); 36 (7~220) ❙ 3.0 (-45.2~108.0); 629 (358~881); 193.5 (44.7~334.4)

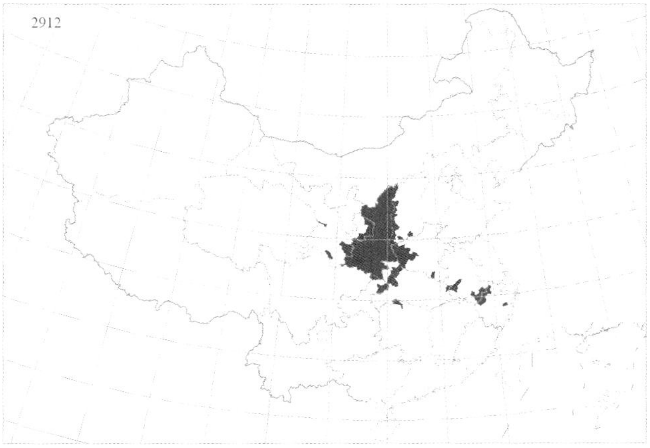

Rosaceae 蔷薇科 *Amygdalus* 桃属
Amygdalus communis L. 扁桃
Shrubs or trees, deciduous

3.6 (-4.1~11.9); -14.9 (-20.9~-5.6); 18.2 (9.8~27.1); 7.5 (2.5~12.7); 56.6 (9.9~107.9); 72.6 (25.2~117.9); 563 (330~810) ❙ 254 (51~406); 116 (23~211); 23 (4~48) ❙ -51.6 (-91.1~-7.6); 254 (51~406); 177.6 (13.7~297.3)

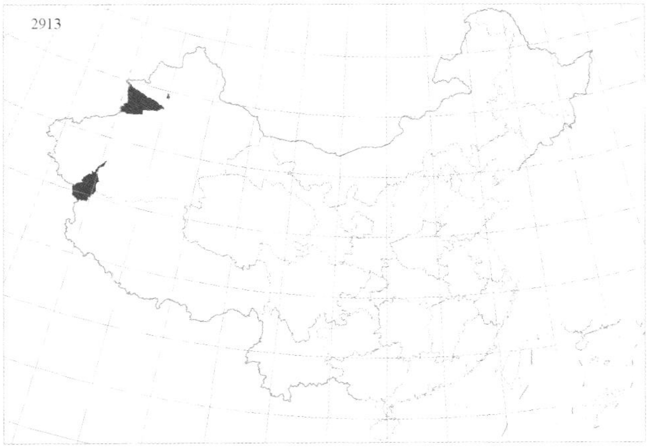

Rosaceae 蔷薇科 *Amygdalus* 桃属
Amygdalus davidiana (Carrière) de Vos ex Henry 山桃
Trees, deciduous

11.4 (-5.7~19.1); -2.9 (-18.7~11.0); 24.2 (7.6~29.5); 12.1 (1.7~19.1); 99.8 (4.8~169.2); 23.0 (0.0~133.0); 732 (294~969) ❙ 709 (74~1621); 381 (46~720); 33 (2~195) ❙ -3.7 (-81.8~125.9); 615 (74~969); 175.7 (13.6~365.3)

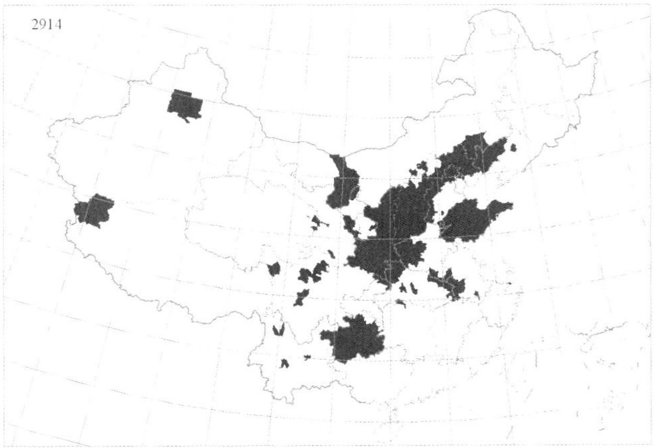

Rosaceae 蔷薇科 *Amygdalus* 桃属
Amygdalus ferganensis (Kostina et Rjabov) T. T. Yu et L. T. Lu 新疆桃
Trees, deciduous

8.2; -9.1; 21.9; 10.1; 81.7; 43.3; 664 ❙ 71; 32; 7 ❙ -88.6; 71; 40.6

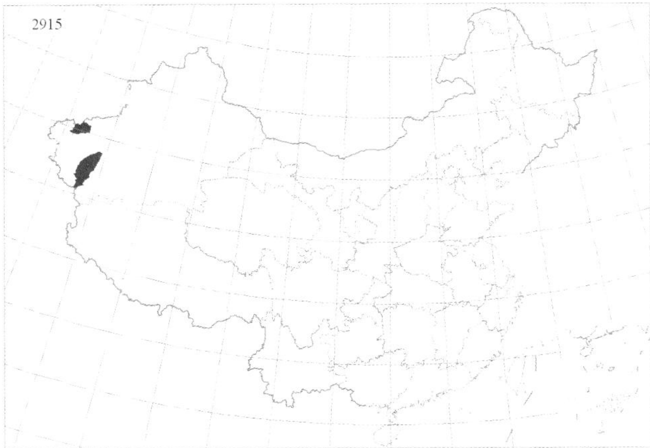

Rosaceae 蔷薇科 *Amygdalus* 桃属
Amygdalus kansuensis (Rehder) Skeels 甘肃桃
Shrubs or trees, deciduous

10.0 (2.2~15.3); -2.3 (-9.7~6.2); 21.3 (9.9~28.2); 10.5 (3.8~15.4); 80.9 (15.5~125.3); 20.7 (0.0~56.4); 651 (413~864) ❙ 761 (373~1439); 364 (201~593); 27 (7~92) ❙ 17.8 (-37.6~125.9); 616 (373~822); 227.8 (46.6~334.4)

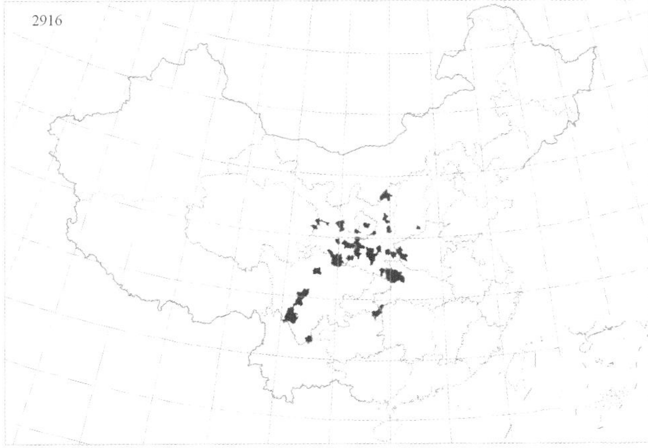

Rosaceae 蔷薇科 *Amygdalus* 桃属

Amygdalus mira (Koehne) Ricker 光核桃

Trees, deciduous

4.6 (-1.8~13.3); -3.9 (-10.6~5.4); 12.0 (6.1~19.6); 6.0 (1.8~13.4); 33.8 (1.9~100.4); 38.2 (0.0~82.8); 473 (329~706) ❙ 714 (246~1736); 430 (192~1007); 19 (0~58) ❙ 43.5 (-33.6~145.8); 455 (246~706); 245.2 (61.6~447.4)

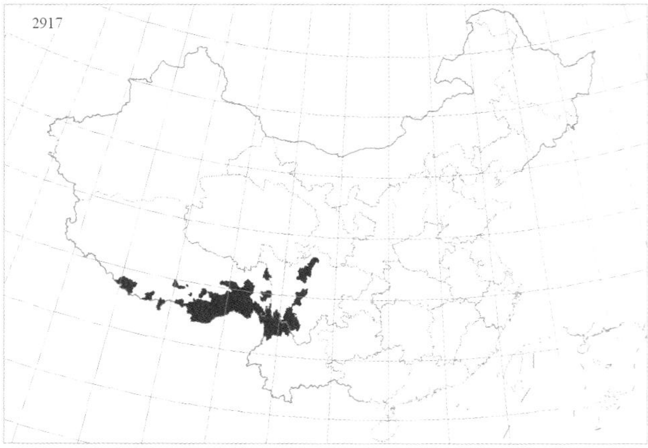

Rosaceae 蔷薇科 *Amygdalus* 桃属

Amygdalus mongolica (Maxim.) Ricker 蒙古扁桃

Shrubs, deciduous

6.5 (-1.4~8.8); -10.6 (-16.5~-7.1); 21.7 (11.5~24.3); 9.0 (3.5~10.3); 71.5 (17.2~85.9); 52.8 (35.3~92.1); 611 (375~679) ❙ 213 (49~416); 132 (34~265); 5 (1~15) ❙ -64.1 (-92.8~-24.8); 213 (49~416); 62.9 (5.1~147.7)

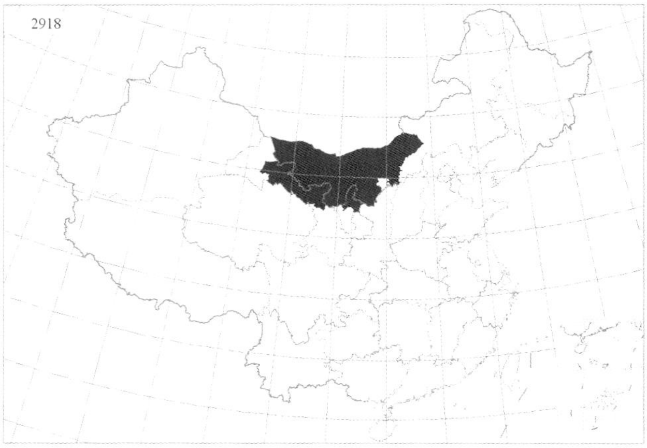

Rosaceae 蔷薇科 *Amygdalus* 桃属

Amygdalus nana L. 矮扁桃

Shrubs, deciduous

2.6 (-1.3~9.0); -16.2 (-20.9~-9.3); 18.3 (13.2~23.3); 7.1 (4.5~10.7); 52.6 (26.7~88.9); 80.7 (40.7~104.2); 547 (433~703) ❙ 257 (152~450); 111 (56~211); 27 (12~51) ❙ -51.3 (-76.8~3.9); 255 (152~406); 139.7 (16.6~297.3)

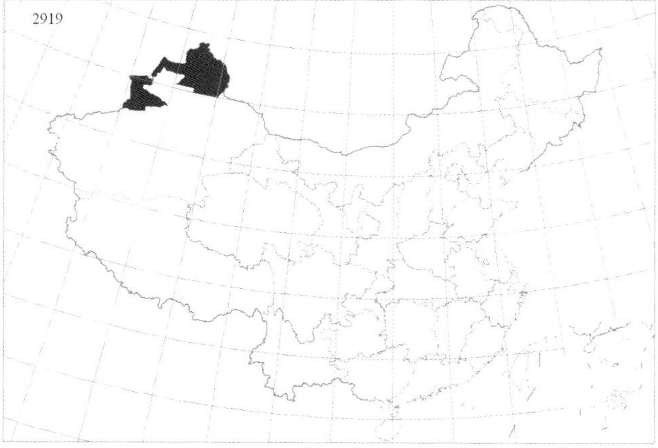

Rosaceae 蔷薇科 *Amygdalus* 桃属

Amygdalus pedunculata Pall. 长梗扁桃

Shrubs or small trees, deciduous

4.4 (0.8~8.6); -14.0 (-20.7~-8.6); 20.7 (18.2~23.6); 8.1 (6.4~10.3); 62.2 (45.9~84.8); 68.7 (41.2~101.4); 574 (506~667) ❙ 311 (137~416); 204 (92~275); 8 (3~15) ❙ -45.0 (-77.5~-20.8); 311 (137~416); 90.8 (25.4~200.5)

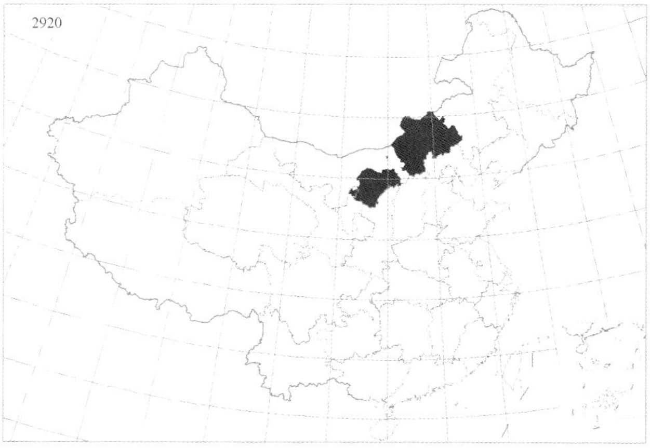

Rosaceae 蔷薇科 *Amygdalus* 桃属

Amygdalus persica L. 桃

Trees, deciduous

15.4 (0.8~24.8); 4.9 (-10.4~20.1); 25.0 (9.9~29.4); 15.6 (3.7~24.8); 130.1 (15.1~237.6); 4.3 (0.0~65.2); 845 (404~1385) ❙ 1264 (471~1997); 535 (247~911); 102 (9~259) ❙ 50.5 (-35.9~132.1); 826 (404~1360); 241.1 (101.3~620.0)

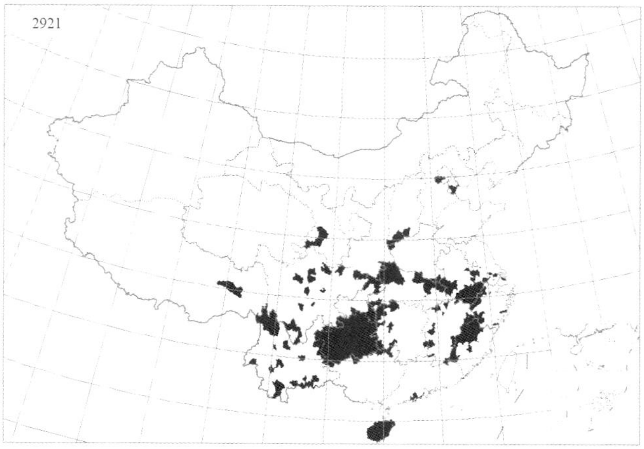

Rosaceae 蔷薇科 *Amygdalus* 桃属

Amygdalus tangutica (Batalin) Korsh. 西康扁桃

Shrubs, deciduous

4.8 (1.9~14.8); -5.6 (-9.2~3.8); 14.2 (11.1~25.3); 6.2 (4.3~14.8); 38.2 (20.5~119.2); 39.7 (1.4~57.8); 493 (427~795) ❙ 792 (590~1358); 384 (299~543); 25 (12~88) ❙ 61.2 (26.3~121.7); 493 (427~795); 276.1 (187.9~341.6)

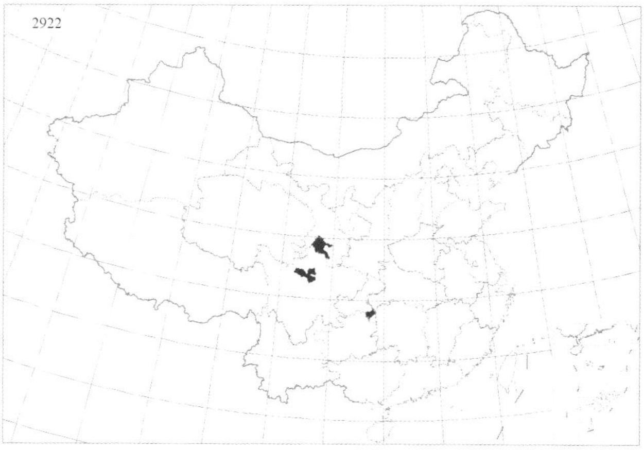

Rosaceae 蔷薇科 *Amygdalus* 桃属
Amygdalus triloba (Lindl.) Ricker 榆叶梅
Shrubs or small trees, deciduous

9.5 (−2.6~17.7); −5.7 (−20.1~5.3); 23.1 (8.7~29.9); 11.0 (2.5~17.7); 89.4 (7.7~153.0); 34.6 (0.0~98.0); 695 (349~976) ❙ 665 (70~1663); 361 (23~648); 40 (5~212) ❙ −5.9 (−91.1~101.3); 582 (70~921); 177.1 (46.6~323.4)

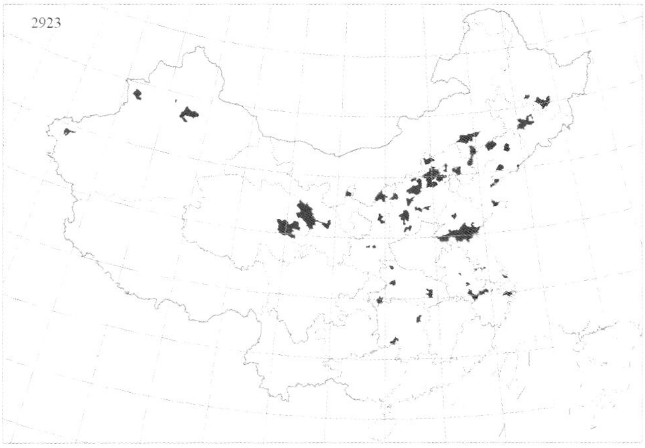

Rosaceae 蔷薇科 *Armeniaca* 杏属
Armeniaca dasycarpa (Ehrb.) Borkh 紫杏
Trees, deciduous

3.6 (−1.3~9.3); −15.4 (−20.9~−9.3); 18.3 (13.2~27.1); 7.5 (4.6~12.1); 56.1 (26.7~107.6); 72.8 (40.7~103.9); 564 (440~810) ❙ 293 (140~406); 134 (53~211); 26 (12~48) ❙ −44.4 (−82.8~−7.6); 293 (140~406); 199.8 (50.3~297.3)

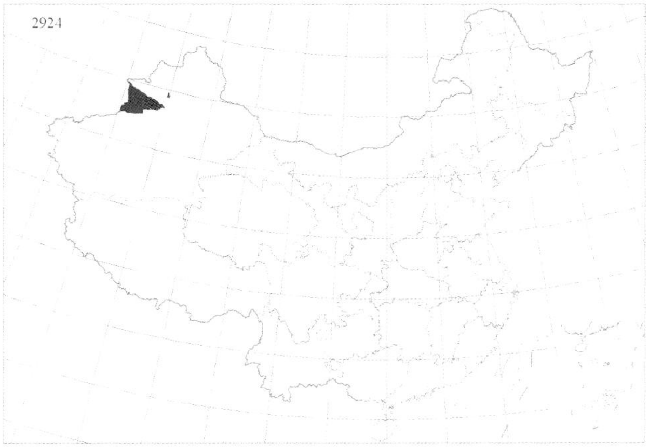

Rosaceae 蔷薇科 *Armeniaca* 杏属
Armeniaca holosericea (Batalin) Kostina 藏杏
Trees, deciduous

6.6 (−1.8~16.2); −4.0 (−12.0~5.5); 16.4 (8.6~28.2); 7.8 (2.4~16.2); 54.7 (7.9~134.6); 34.6 (0.0~88.8); 557 (333~866) ❙ 688 (324~1197); 365 (233~558); 24 (0~98) ❙ 30.4 (−33.2~125.9); 507 (324~854); 206.7 (32.2~337.3)

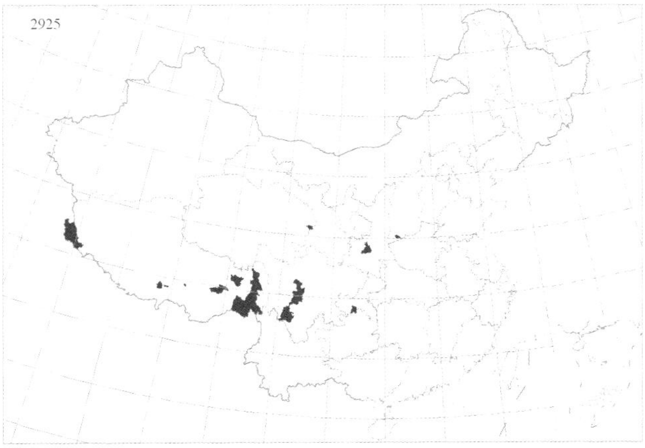

Rosaceae 蔷薇科 *Armeniaca* 杏属
Armeniaca hongpingensis T. T. Yu et C. L. Li 洪平杏
Trees, deciduous

9.3; −1.7; 20.1; 9.5; 69.8; 18.0; 603 ❙ 1214; 516; 71 ❙ 101.4; 603; 279.5

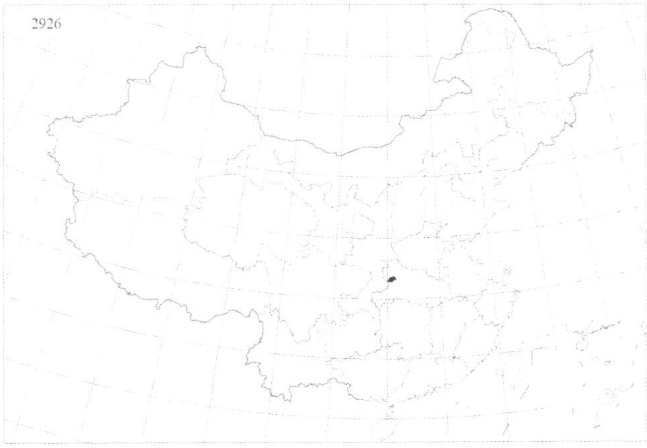

Rosaceae 蔷薇科 *Armeniaca* 杏属
Armeniaca mandshurica (Maxim.) Skvortsov 东北杏
Trees, deciduous

5.0 (1.4~10.5); −13.8 (−20.5~−4.9); 21.1 (18.4~24.8); 8.4 (6.4~11.3); 65.1 (45.8~93.2); 65.1 (26.8~91.9); 586 (501~705) ❙ 623 (367~1073); 394 (235~691); 21 (9~40) ❙ 6.1 (−41.6~74.1); 522 (367~678); 230.3 (91.4~346.5)

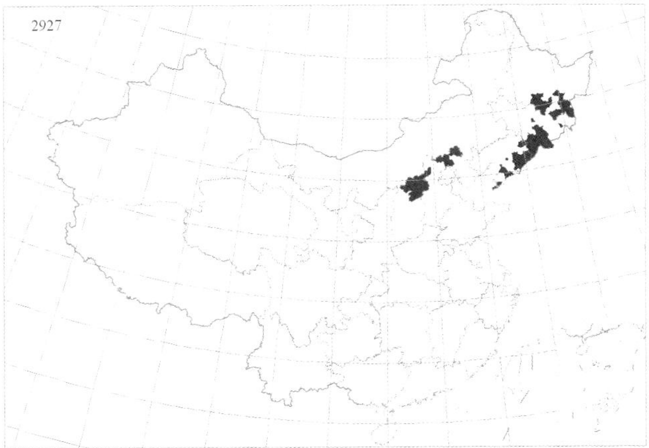

Rosaceae 蔷薇科 *Armeniaca* 杏属
Armeniaca mume Siebold 梅
Shrubs or trees, deciduous

16.0 (0.3~24.1); 7.0 (−9.9~18.9); 23.8 (9.3~29.7); 16.0 (3.3~24.2); 133.8 (11.9~230.1); 1.5 (0.0~67.5); 834 (391~1326) ❙ 1330 (527~3495); 611 (299~1847); 104 (9~897) ❙ 58.7 (−14.2~246.5); 822 (391~1326); 270.6 (111.9~659.0)

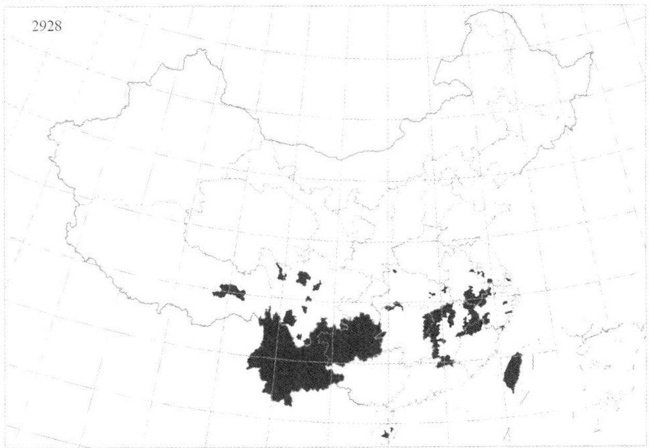

Rosaceae 蔷薇科 *Armeniaca* 杏属
Armeniaca sibirica (L.) Lam. 山杏
Shrubs or small trees, deciduous

7.2 (−4.6~14.2); −10.0 (−28.8~−0.7); 22.3 (10.5~27.8); 9.6 (3.6~14.4); 76.8 (15.5~124.3); 50.3 (12.6~149.1); 638 (406~859) ❙ 494 (63~976); 311 (45~608); 13 (2~40) ❙ −22.1 (−84.9~66.0); 485 (63~698); 165.8 (8.4~330.9)

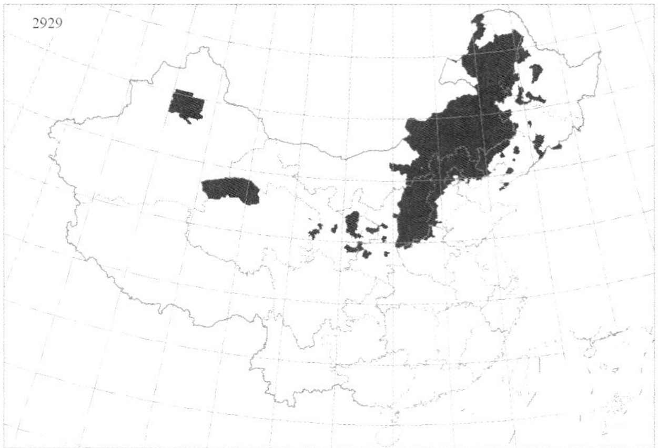

Rosaceae 蔷薇科 *Armeniaca* 杏属
Armeniaca vulgaris Lam. 杏
Trees, deciduous

13.2 (−1.3~21.0); 2.1 (−20.9~15.1); 22.7 (8.5~28.0); 13.6 (2.7~21.0); 111.9 (8.3~191.8); 12.8 (0.0~103.9); 758 (363~1053) ❙ 908 (225~1751); 494 (65~996); 42 (0~129) ❙ 19.5 (−65.1~121.7); 694 (225~1053); 235.1 (60.2~659.0)

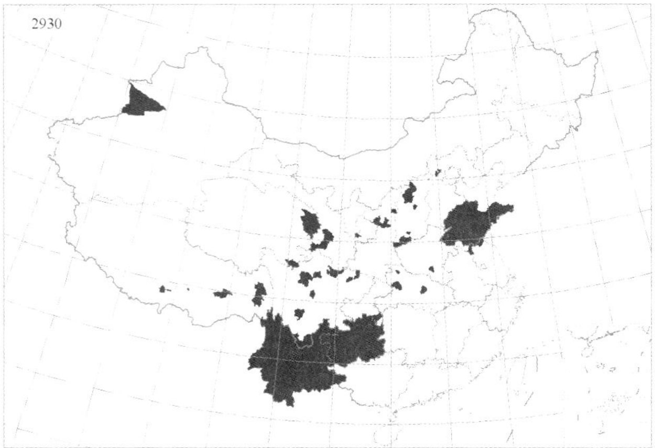

Rosaceae 蔷薇科 *Armeniaca* 杏属
Armeniaca vulgaris var. ansu (Maxim.) T. T. Yu et L. T. Lu 野杏
Trees, deciduous

7.4 (−1.0~13.5); −8.8 (−18.4~−1.6); 21.5 (8.5~27.1); 9.4 (2.7~13.7); 73.8 (8.3~118.0); 45.1 (14.6~90.3); 624 (363~830) ❙ 487 (256~702); 301 (169~499); 13 (5~27) ❙ −21.4 (−57.7~82.4); 483 (256~659); 155.4 (46.8~288.8)

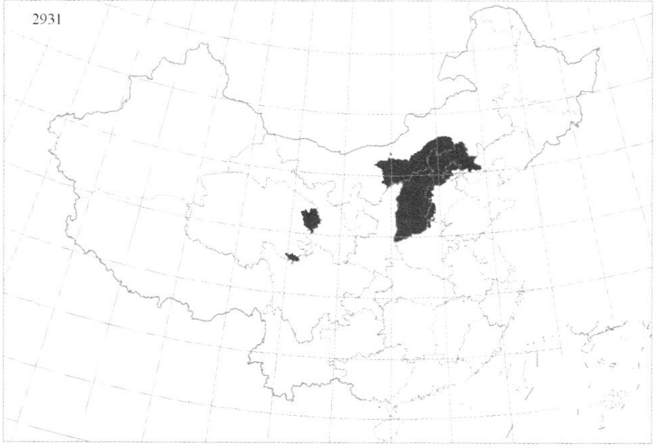

Rosaceae 蔷薇科 *Cerasus* 樱属
Cerasus campanulata (Maxim.) A. N. Vassiljeva 钟花樱桃
Shrubs or trees, deciduous

18.5 (14.3~24.5); 9.3 (2.5~18.9); 27.0 (20.1~29.8); 18.6 (14.3~24.5); 163.1 (113.2~234.4); 0.1 (0.0~3.9); 967 (761~1360) ❙ 1648 (1070~3495); 595 (356~1847); 192 (43~897) ❙ 73.6 (−5.0~246.5); 956 (761~1360); 259.2 (123.9~572.9)

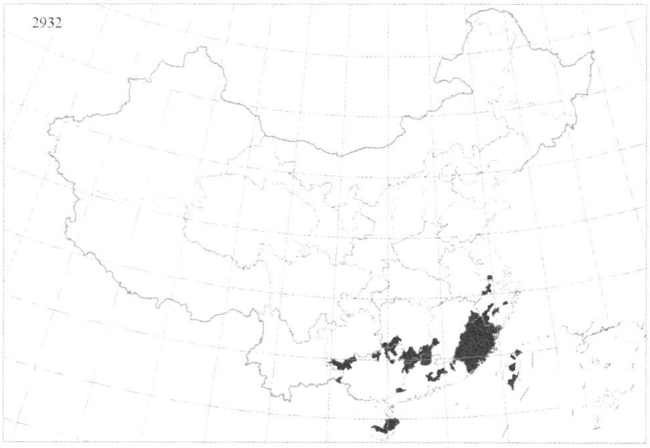

Rosaceae 蔷薇科 *Cerasus* 樱属
Cerasus caudata (Franch.) T. T. Yu et C. L. Li 尖尾樱桃
Trees, deciduous

11.9 (8.6~17.7); 5.1 (1.3~10.9); 17.5 (15.0~22.5); 12.0 (8.7~17.8); 85.3 (52.6~153.4); 1.7 (0.0~8.8); 665 (574~854) ❙ 1137 (968~1368); 646 (578~756); 48 (24~69) ❙ 71.4 (52.8~93.0); 665 (574~854); 371.4 (319.8~491.5)

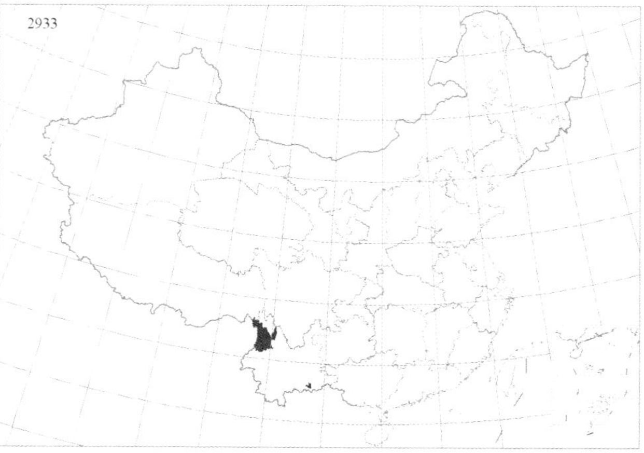

Rosaceae 蔷薇科 *Cerasus* 樱属
Cerasus cerasoides (Buch.-Ham. ex D. Don) S. Y. Sokolov 高盆樱桃
Trees, deciduous

15.1 (−1.1~21.0); 7.9 (−9.6~15.1); 20.5 (6.2~26.6); 15.2 (2.0~21.0); 124.3 (2.5~191.8); 2.5 (0.0~75.6); 771 (334~1053) ❙ 1104 (321~1751); 621 (193~1007); 43 (7~77) ❙ 42.9 (−14.2~145.8); 755 (321~1053); 328.9 (61.6~659.0)

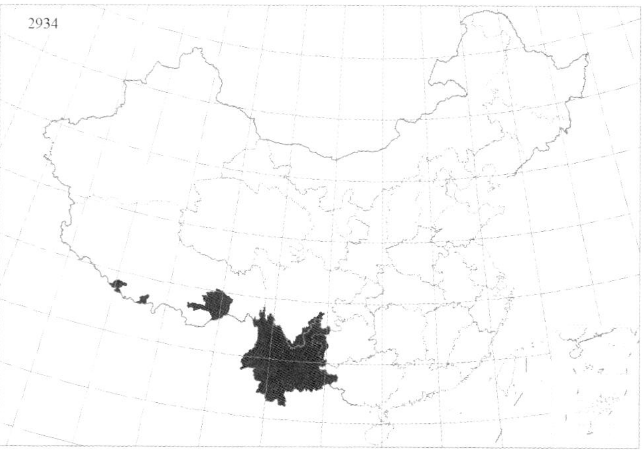

Rosaceae 蔷薇科 *Cerasus* 樱属
Cerasus clarofolia (C. K. Schneid.) T. T. Yu et C. L. Li 微毛樱桃
Shrubs or small trees, deciduous

12.8 (1.8~18.2); 1.5 (-8.2~10.9); 23.4 (9.9~29.9); 13.1 (3.8~18.2); 104.1 (15.5~158.6); 9.8 (0.0~53.3); 745 (404~976) ❙ 1165 (374~1950); 494 (205~816); 95 (8~238) ❙ 55.6 (-35.9~147.0); 721 (374~927); 238.6 (81.2~491.5)

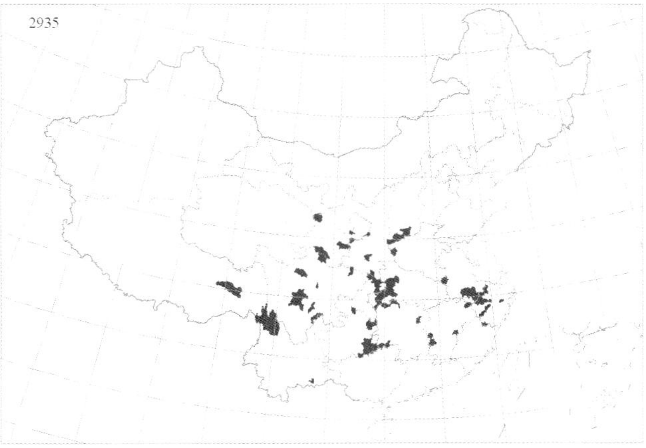

Rosaceae 蔷薇科 *Cerasus* 樱属
Cerasus conadenia (Koehne) T. T. Yu et C. L. Li 锥腺樱桃
Shrubs or trees, deciduous

9.5 (1.8~14.8); -1.7 (-9.7~3.9); 20.0 (9.9~26.1); 10.0 (3.8~14.8); 74.4 (15.4~119.2); 19.7 (1.3~57.8); 631 (404~795) ❙ 838 (516~1358); 410 (283~601); 27 (9~88) ❙ 37.5 (-16.6~128.6); 627 (404~795); 265.8 (169.1~352.3)

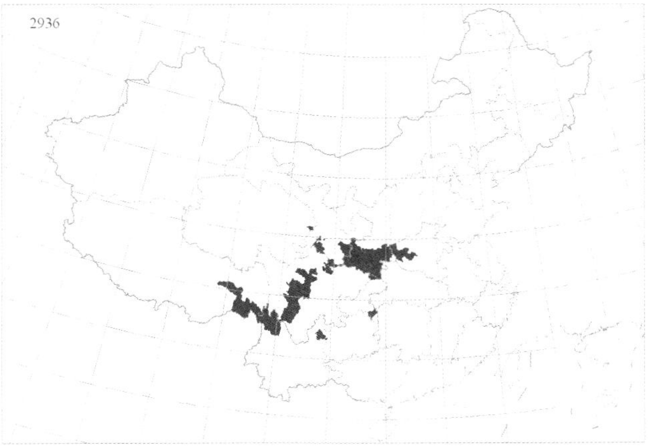

Rosaceae 蔷薇科 *Cerasus* 樱属
Cerasus conradinae (Koehne) T. T. Yu et C. L. Li 华中樱桃
Trees, deciduous

15.3 (5.4~19.0); 5.7 (-2.1~12.0); 23.9 (12.3~29.1); 15.3 (5.8~19.0); 125.4 (31.5~168.4); 1.4 (0.0~25.6); 808 (481~1002) ❙ 1228 (741~1997); 548 (393~817); 94 (15~259) ❙ 52.0 (-14.2~130.7); 799 (481~1002); 241.1 (108.2~352.3)

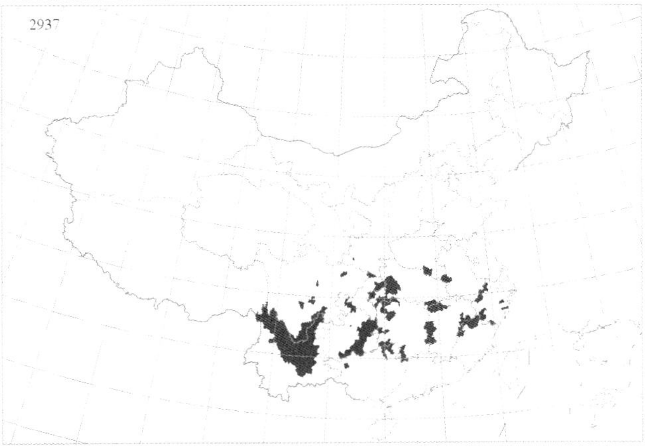

Rosaceae 蔷薇科 *Cerasus* 樱属
Cerasus crataegifolia (Hand.-Mazz.) Hand.-Mazz. 山楂叶樱桃
Shrubs, deciduous

9.5 (5.3~13.3); 1.9 (-2.4~5.4); 15.9 (12.2~19.6); 9.6 (5.7~13.4); 63.0 (30.5~100.4); 9.0 (0.0~26.2); 596 (479~706) ❙ 1077 (741~1736); 634 (426~1007); 31 (15~51) ❙ 79.2 (51.8~145.8); 595 (479~706); 319.5 (234.2~447.4)

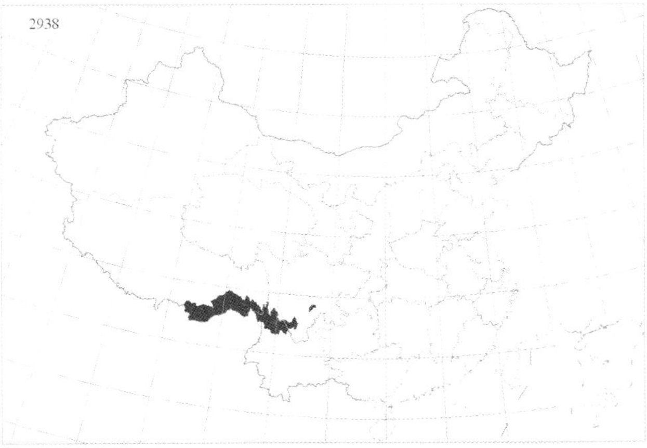

Rosaceae 蔷薇科 *Cerasus* 樱属
Cerasus cyclamina (Koehne) T. T. Yu et C. L. Li 襄阳山樱桃
Trees, deciduous

16.1 (11.1~18.7); 5.2 (1.1~8.1); 26.4 (19.4~28.9); 16.2 (11.3~18.8); 135.1 (80.8~166.0); 1.2 (0.0~8.5); 864 (652~1002) ❙ 1378 (1082~1742); 531 (411~709); 127 (37~221) ❙ 62.4 (8.0~128.9); 859 (652~1002); 199.0 (108.2~292.6)

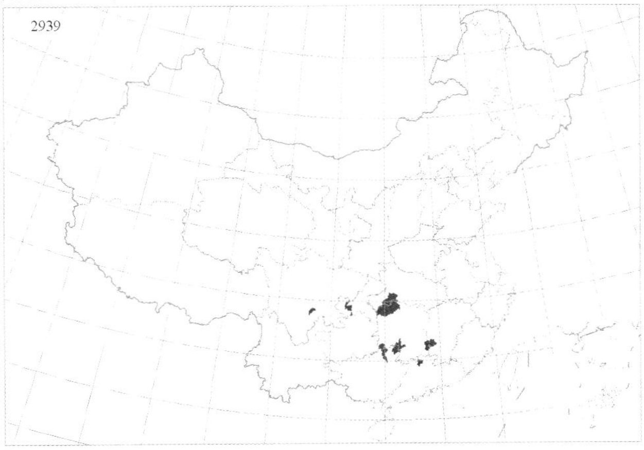

Rosaceae 蔷薇科 *Cerasus* 樱属
Cerasus cyclamina var. biflora (Koehne) T. T. Yu et C. L. Li 双花襄阳山樱桃
Trees, deciduous

15.8; 4.0; 27.0; 15.8; 130.6; 1.0; 854 ❙ 1391; 545; 127 ❙ 62.7; 854; 215.3

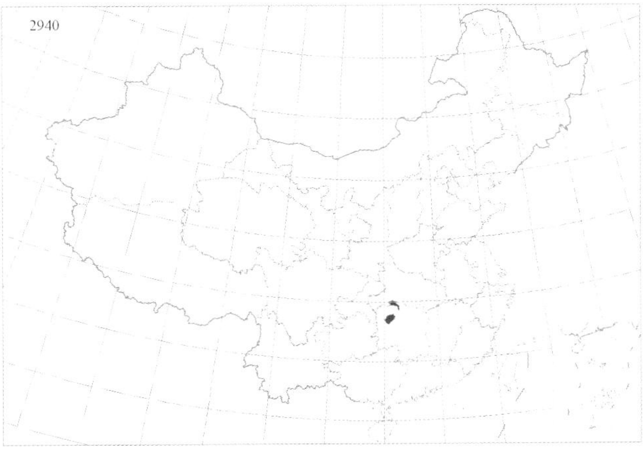

Rosaceae 蔷薇科 *Cerasus* 樱属

Cerasus dictyoneura (Diels) Holub 毛叶欧李

Shrubs, deciduous

10.8 (2.6~15.8); -3.8 (-15.9~2.9); 24.0 (17.5~28.0); 11.6 (6.6~15.8); 94.7 (45.4~133.1); 24.7 (2.8~80.4); 713 (507~891) ❙ 641 (341~1200); 351 (190~536); 24 (8~109) ❙ -9.8 (-43.4~83.5); 605 (341~891); 186.6 (61.2~334.4)

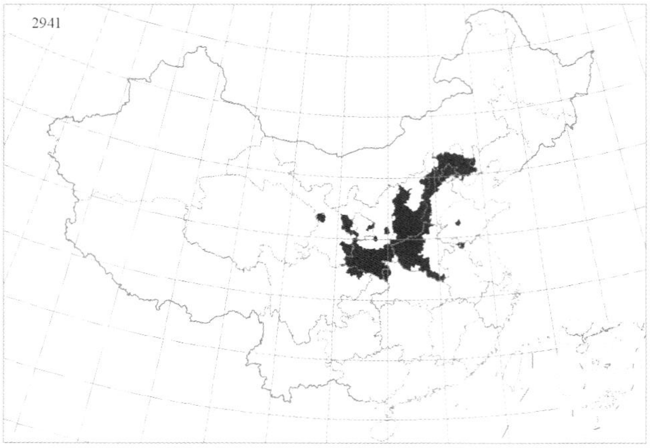

Rosaceae 蔷薇科 *Cerasus* 樱属

Cerasus dielsiana (C. K. Schneid.) T. T. Yu et C. L. Li 尾叶樱桃

Shrubs or trees, deciduous

15.4 (10.2~21.0); 3.5 (-3.8~12.2); 26.7 (19.4~29.4); 15.5 (10.6~21.0); 129.4 (79.8~192.6); 3.8 (0.0~22.5); 850 (648~1107) ❙ 1294 (626~1997); 510 (316~713); 129 (26~254) ❙ 52.0 (-22.2~147.0); 837 (626~1107); 209.2 (108.2~319.9)

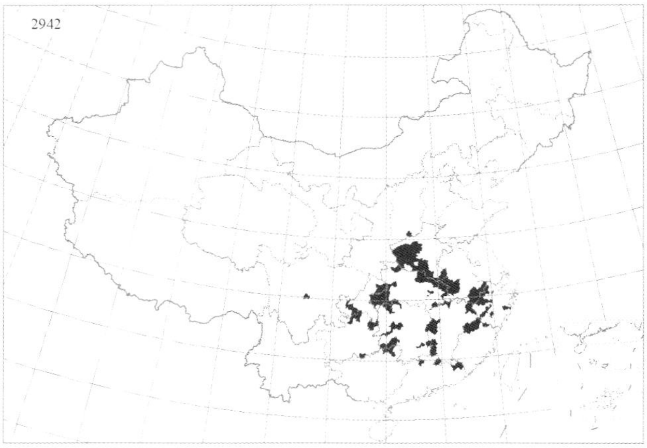

Rosaceae 蔷薇科 *Cerasus* 樱属

Cerasus dielsiana var. abbreviata (Cardot) T. T. Yu et C. L. Li 短梗尾叶樱桃

Shrubs or trees, deciduous

14.8; 4.3; 24.9; 14.9; 119.2; 0.7; 791 ❙ 1197; 489; 71 ❙ 51.2; 791; 182.9

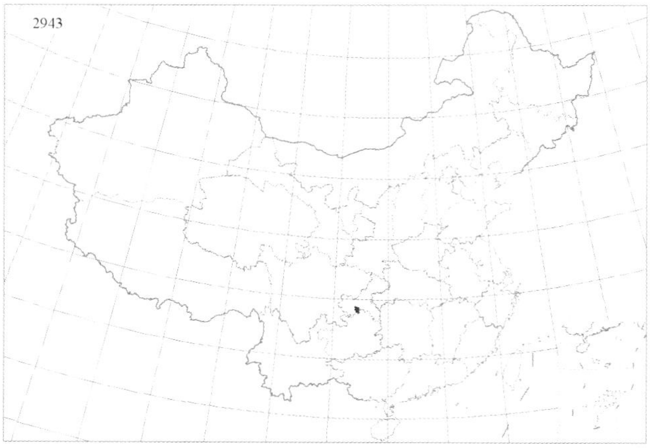

Rosaceae 蔷薇科 *Cerasus* 樱属

Cerasus discoidea T. T. Yu et C. L. Li 迎春樱桃

Trees, deciduous

12.6 (1.8~17.1); 1.8 (-7.1~9.0); 22.6 (9.9~29.3); 12.9 (3.8~17.1); 101.6 (15.5~145.7); 9.4 (0.0~53.3); 737 (404~941) ❙ 1130 (512~1828); 513 (258~816); 81 (12~253) ❙ 54.1 (-39.5~125.9); 711 (404~941); 245.4 (131.6~382.6)

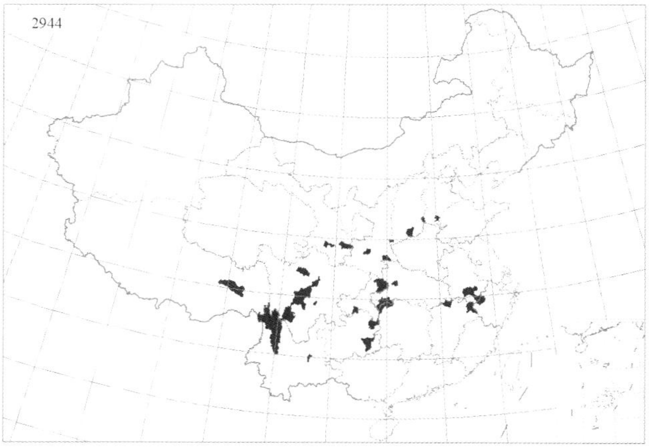

Rosaceae 蔷薇科 *Cerasus* 樱属

Cerasus dolichadenia (Cardot) C. L. Li et S. Y. Jiang 长腺樱桃

Shrubs or small trees, deciduous

10.3 (9.3~11.7); -4.4 (-5.7~-2.8); 23.4 (22.7~24.8); 11.1 (10.4~12.0); 89.0 (82.4~99.3); 24.8 (18.9~30.4); 687 (658~731) ❙ 571 (525~624); 313 (288~350); 20 (15~23) ❙ -16.7 (-23.0~-6.3); 571 (525~624); 173.8 (127.2~205.7)

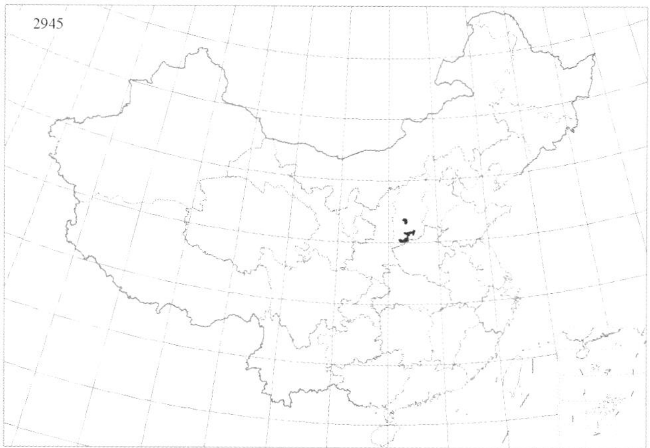

Rosaceae 蔷薇科 *Cerasus* 樱属

Cerasus fruticosa (Pall.) Woronow 草原樱桃

Shrubs, deciduous

4.1 (-1.3~9.3); -15.6 (-20.9~-11.6); 20.3 (13.2~27.5); 8.2 (4.5~12.1); 64.2 (26.7~107.6); 74.5 (51.6~104.2); 597 (433~813) ❙ 231 (107~450); 102 (51~211); 22 (8~51) ❙ -58.6 (-85.3~3.9); 229 (107~406); 108.7 (19.8~297.3)

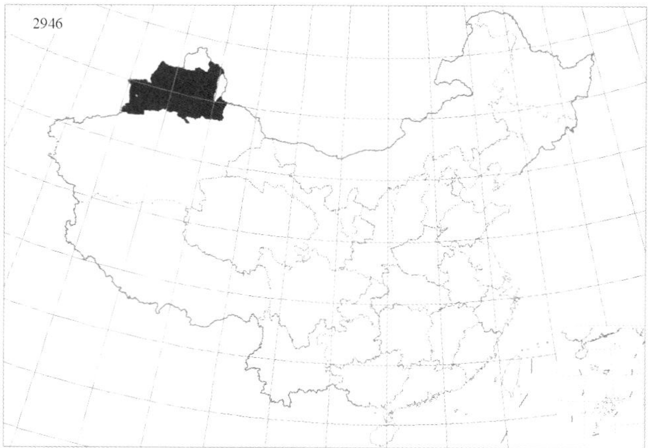

Rosaceae 蔷薇科 *Cerasus* 樱属

Cerasus glandulosa (Thunb.) Sokolov 麦李

Shrubs, deciduous

15.0 (10.2~22.1); 2.5 (-3.7~13.7); 26.9 (16.6~29.4); 15.1 (10.6~22.1); 127.1 (71.3~205.6); 6.0 (0.0~23.0); 848 (633~1182) ▍1103 (551~1922); 491 (245~779); 102 (21~259) ▍29.5 (-37.3~123.6); 824 (551~1182); 185.6 (108.2~319.8)

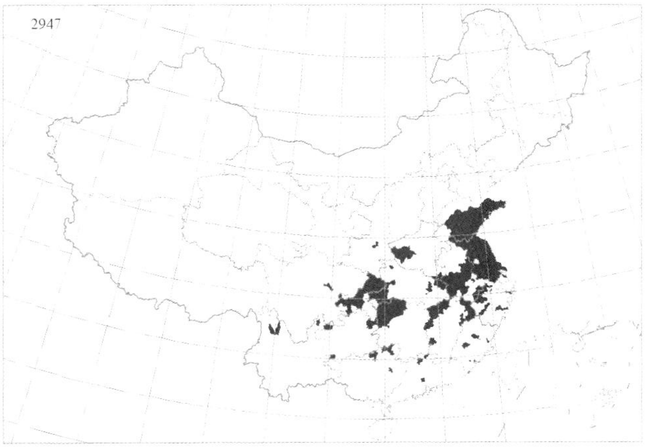

Rosaceae 蔷薇科 *Cerasus* 樱属

Cerasus henryi (C. K. Schneid.) T. T. Yu et C. L. Li 蒙自樱桃

Trees, deciduous

16.2 (9.6~19.3); 9.2 (2.4~13.1); 21.4 (15.7~27.1); 16.2 (9.6~19.4); 135.1 (60.8~173.2); 0.2 (0.0~5.4); 805 (600~969) ▍1064 (939~1740); 593 (524~709); 49 (15~187) ▍32.8 (6.5~85.5); 789 (600~969); 292.7 (172.1~406.7)

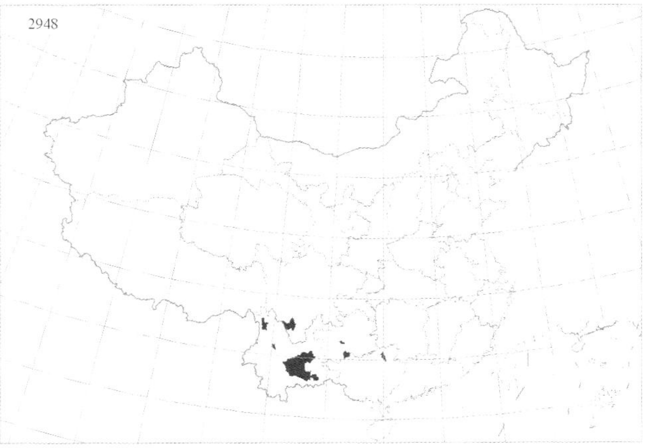

Rosaceae 蔷薇科 *Cerasus* 樱属

Cerasus humilis (Bunge) Sokoloff 欧李

Shrubs, deciduous

9.5 (-0.8~17.0); -6.6 (-25.0~4.4); 23.9 (17.5~29.2); 11.1 (6.4~17.0); 91.4 (45.4~144.9); 36.6 (0.6~118.4); 702 (506~941) ▍615 (132~1571); 362 (81~666); 28 (2~169) ▍-13.4 (-80.1~108.0); 585 (132~926); 165.1 (13.6~333.3)

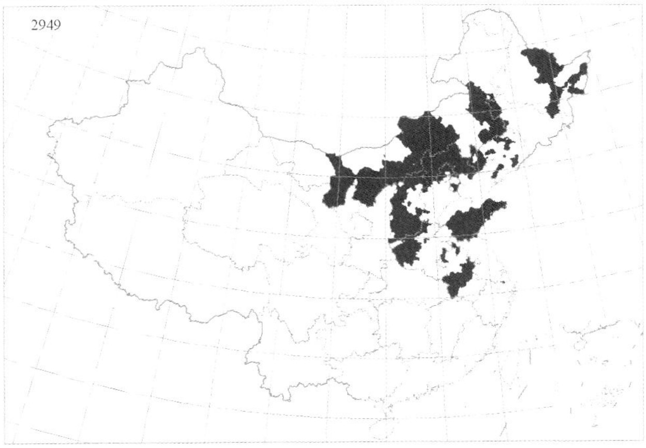

Rosaceae 蔷薇科 *Cerasus* 樱属

Cerasus japonica (Thunb.) Loisel. 郁李

Shrubs, deciduous

13.1 (1.4~19.1); -1.1 (-20.6~10.0); 26.1 (18.4~29.4); 13.7 (6.4~19.1); 114.8 (45.8~169.1); 16.6 (0.0~93.8); 801 (501~973) ▍894 (466~1997); 441 (258~713); 70 (12~254) ▍10.6 (-36.5~147.0); 734 (466~973); 187.9 (110.1~337.1)

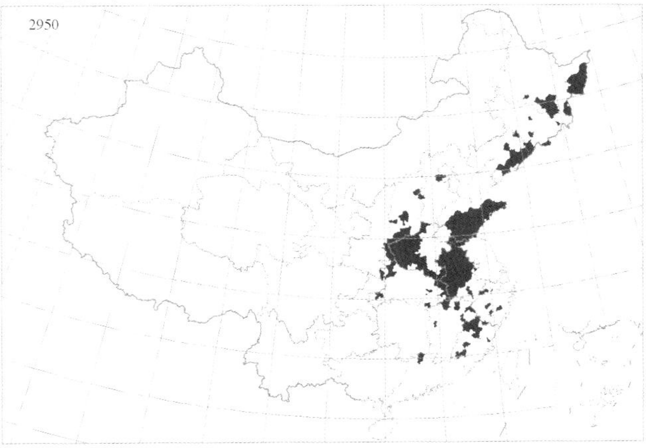

Rosaceae 蔷薇科 *Cerasus* 樱属

Cerasus japonica var. nakaii (H. Lév.) T. T. Yu et C. L. Li 长梗郁李

Shrubs, deciduous

19.2 (18.2~20.1); 9.9 (8.7~10.5); 26.7 (25.9~27.7); 19.3 (18.3~20.1); 171.3 (159.3~181.6); 0.0 (0.0~0.0); 989 (928~1052) ▍1362 (1284~1459); 686 (651~723); 91 (73~111) ▍37.8 (30.9~47.9); 989 (928~1052); 286.0 (248.4~322.1)

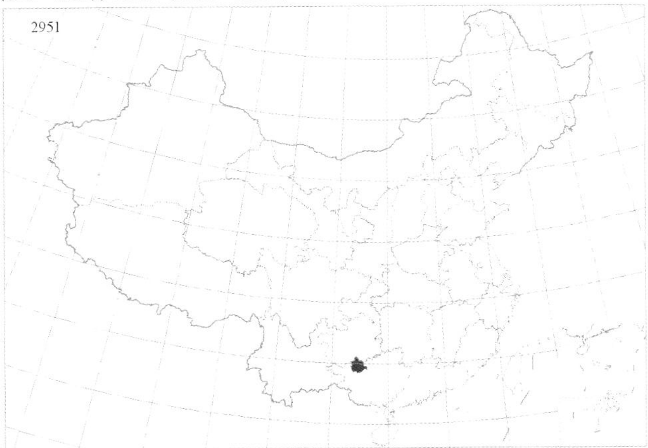

Rosaceae 蔷薇科 *Cerasus* 樱属

Cerasus maximowiczii (Rupr.) Kom. 黑樱桃

Trees, deciduous

2.6 (-0.8~8.2); -18.4 (-25.0~-10.3); 20.6 (18.4~24.3); 7.6 (6.4~10.3); 58.1 (45.8~87.0); 86.2 (47.7~118.4); 562 (501~678) ▍676 (558~1073); 413 (309~691); 24 (15~40) ▍20.2 (-4.8~74.1); 559 (501~678); 298.6 (219.8~346.5)

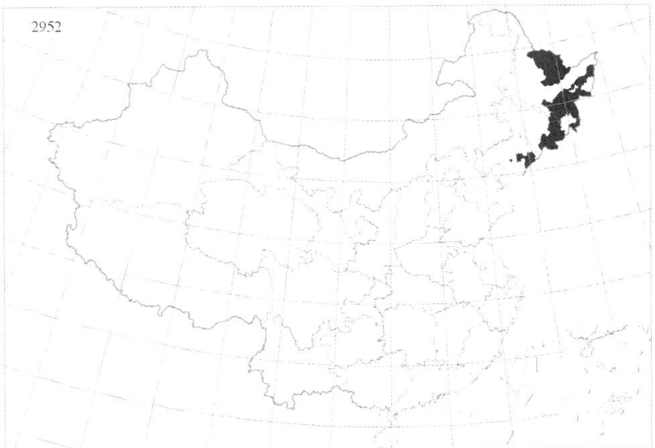

Rosaceae 蔷薇科 *Cerasus* 樱属

Cerasus mugus (Hand.-Mazz.) Hand.-Mazz. 偃樱桃

Shrubs, deciduous

7.8; 0.4; 14.3; 8.0; 47.9; 13.5; 551 ❙ 1003; 574; 39 ❙ 83.4; 551; 312.7

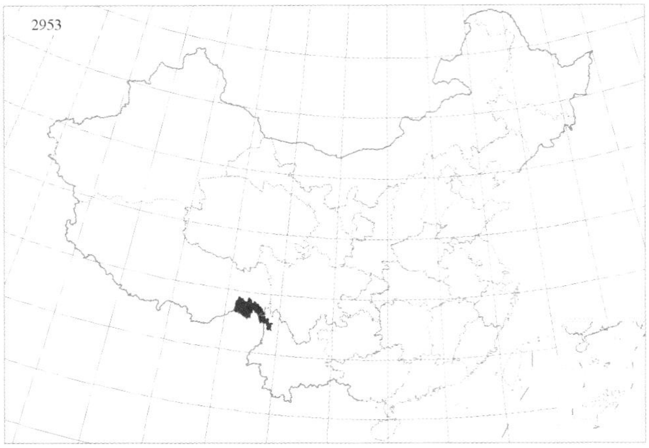

Rosaceae 蔷薇科 *Cerasus* 樱属

Cerasus patentipila (Hand.-Mazz.) T. T. Yu et C. L. Li 散毛樱桃

Shrubs or trees, deciduous

8.8; 1.6; 15.0; 8.9; 55.9; 9.5; 576 ❙ 925; 556; 28 ❙ 60.7; 576; 318.1

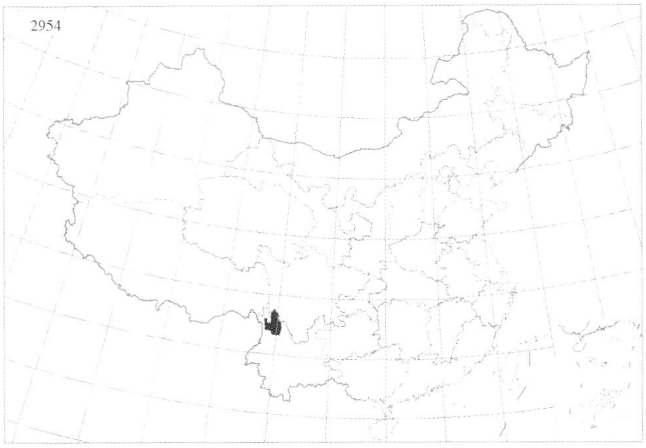

Rosaceae 蔷薇科 *Cerasus* 樱属

Cerasus pleiocerasus (Koehne) T. T. Yu et C. L. Li 雕核樱桃

Trees, deciduous

10.7 (2.2~18.9); 2.1 (-6.6~12.0); 18.1 (9.9~27.6); 11.1 (3.8~19.0); 84.0 (15.5~167.8); 15.2 (0.0~49.1); 649 (413~929) ❙ 909 (607~1358); 496 (273~619); 26 (15~88) ❙ 48.8 (-28.3~125.9); 607 (413~795); 265.1 (174.6~309.7)

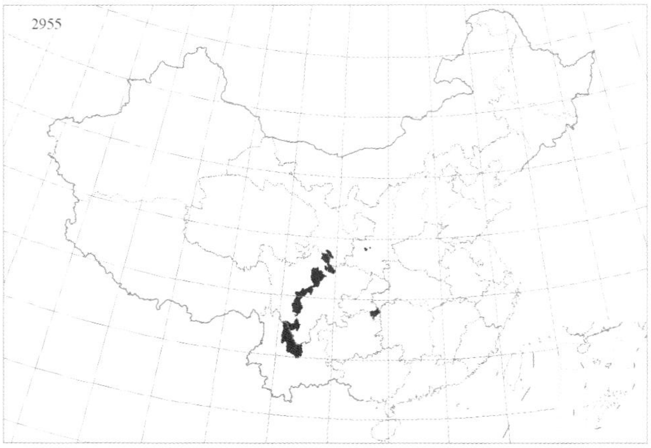

Rosaceae 蔷薇科 *Cerasus* 樱属

Cerasus pogonostyla (Maxim.) T. T. Yu et C. L. Li 毛柱郁李

Shrubs or small trees, deciduous

18.5 (14.1~24.1); 9.6 (1.6~18.9); 26.8 (20.1~29.3); 18.5 (14.2~24.1); 162.6 (116.7~229.7); 0.1 (0.0~6.5); 957 (773~1322) ❙ 1750 (1070~3495); 654 (356~1847); 208 (47~897) ❙ 85.9 (-5.0~246.5); 943 (773~1144); 249.1 (119.4~363.7)

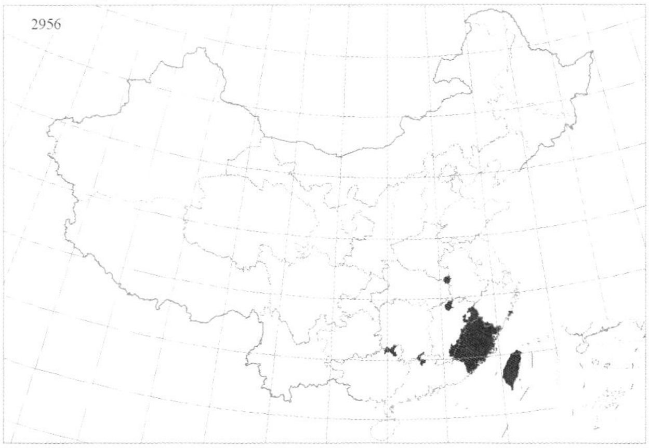

Rosaceae 蔷薇科 *Cerasus* 樱属

Cerasus polytricha (Koehne) T. T. Yu et C. L. Li 多毛樱桃

Shrubs or trees, deciduous

11.1 (2.7~17.4); -1.3 (-7.7~7.4); 22.5 (11.1~28.1); 11.5 (4.4~17.4); 90.6 (20.5~148.9); 16.8 (0.0~49.3); 694 (428~903) ❙ 793 (373~1410); 386 (201~816); 33 (7~107) ❙ 17.2 (-37.6~121.7); 651 (373~903); 228.3 (88.1~357.7)

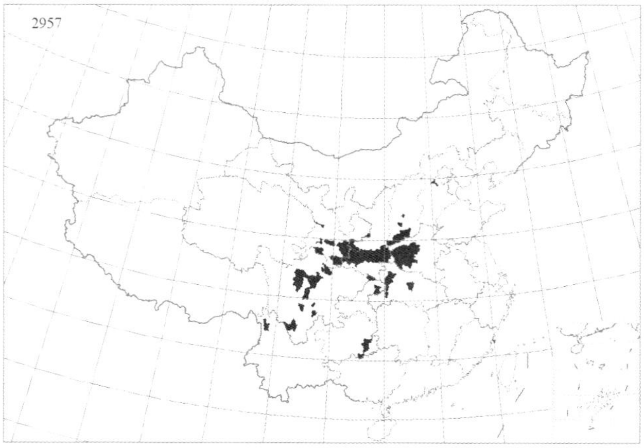

Rosaceae 蔷薇科 *Cerasus* 樱属

Cerasus pseudocerasus (Lindl.) Loudon 樱桃

Trees, deciduous

11.6 (-0.2~19.1); -3.0 (-23.8~10.1); 24.5 (12.2~29.9); 12.9 (5.7~19.1); 106.0 (30.5~169.2); 26.6 (0.0~112.6); 751 (479~991) ❙ 962 (470~1959); 465 (245~1007); 63 (7~259) ❙ 26.8 (-36.5~145.8); 716 (470~991); 217.3 (111.9~447.4)

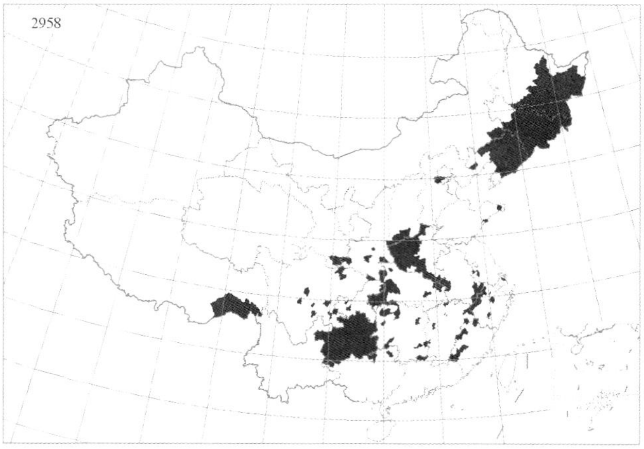

Rosaceae 蔷薇科 *Cerasus* 樱属

Cerasus pusilliflora (Cardot) T. T. Yu et C. L. Li 细花樱桃

Shrubs or trees, deciduous

15.9 (13.6~18.9); 9.2 (6.5~12.5); 20.7 (18.8~23.9); 15.9 (13.6~19.0); 130.7 (103.6~167.8); 0.0 (0.0~0.0); 783 (706~929) ❙ 979 (797~1049); 564 (411~598); 39 (23~54) ❙ 25.6 (−14.2~47.6); 747 (683~851); 289.6 (218.0~382.6)

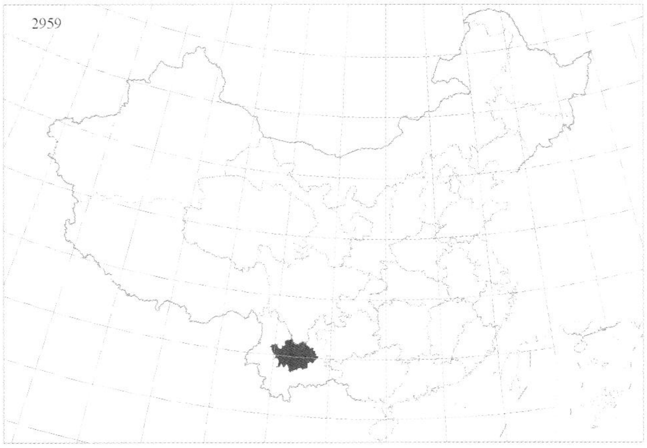

Rosaceae 蔷薇科 *Cerasus* 樱属

Cerasus schneideriana (Koehne) T. T. Yu et C. L. Li 浙闽樱桃

Trees, deciduous

16.8 (13.4~19.2); 6.4 (1.5~10.7); 26.5 (24.3~28.5); 16.9 (13.4~19.2); 142.7 (108.3~170.8); 0.5 (0.0~7.3); 882 (755~973) ❙ 1695 (1172~1997); 614 (479~723); 196 (50~254) ❙ 93.1 (21.9~147.0); 881 (755~973); 269.0 (208.7~411.1)

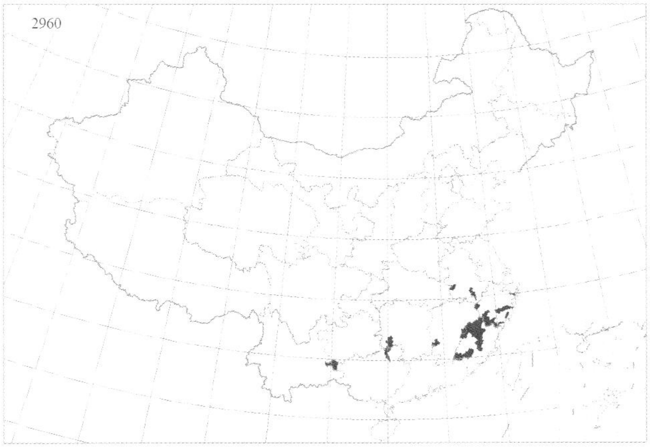

Rosaceae 蔷薇科 *Cerasus* 樱属

Cerasus serrula (Franch.) T. T. Yu et C. L. Li 细齿樱桃

Trees, deciduous

11.4 (−1.0~18.9); 3.4 (−11.8~12.0); 18.0 (8.5~27.5); 11.8 (2.7~19.0); 89.6 (8.3~167.8); 12.7 (0.0~79.5); 663 (363~933) ❙ 944 (510~1736); 532 (311~1007); 35 (8~182) ❙ 46.2 (−14.2~145.8); 647 (363~933); 274.7 (142.9~447.4)

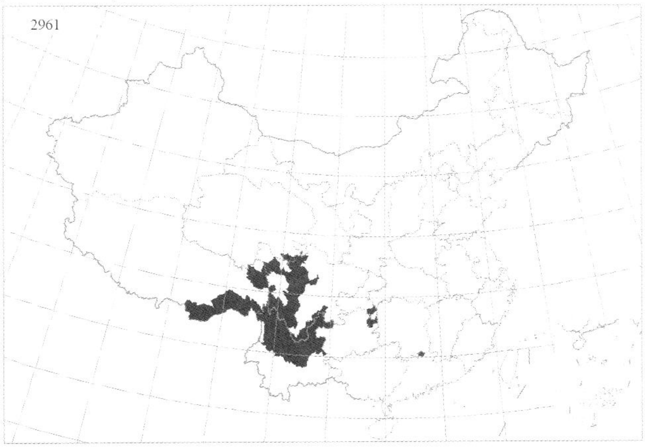

Rosaceae 蔷薇科 *Cerasus* 樱属

Cerasus serrulata (Lindl.) Loudon 山樱桃

Trees, deciduous

14.6 (2.1~21.0); 2.0 (−19.6~11.9); 26.4 (19.0~29.9); 14.9 (7.0~21.1); 124.6 (50.5~193.0); 8.7 (0.0~87.4); 830 (532~1114) ❙ 1264 (517~1950); 498 (258~821); 135 (11~259) ❙ 50.0 (−35.9~147.0); 805 (517~1114); 217.0 (117.3~333.1)

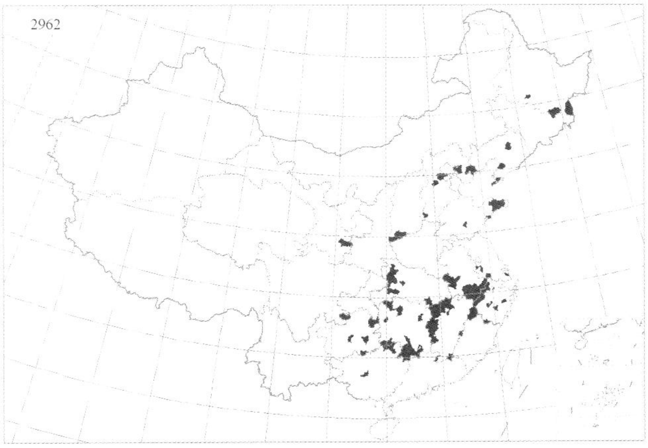

Rosaceae 蔷薇科 *Cerasus* 樱属

Cerasus serrulata var. pubescens (Makino) T. T. Yu et C. L. Li 毛叶山樱桃

Trees, deciduous

14.8 (7.7~18.2); 2.1 (−9.3~7.2); 26.5 (22.6~29.1); 15.1 (9.7~18.2); 127.2 (80.1~158.8); 8.8 (0.0~46.8); 840 (645~984) ❙ 1433 (517~1997); 535 (306~669); 170 (11~259) ❙ 66.3 (−29.6~130.7); 812 (517~961); 239.1 (136.4~319.9)

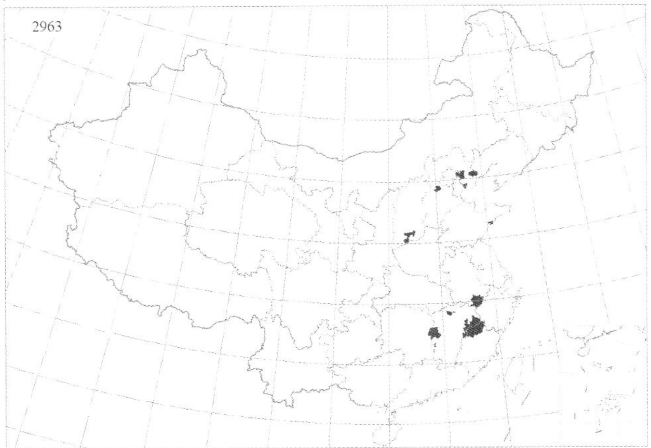

Rosaceae 蔷薇科 *Cerasus* 樱属

Cerasus setulosa (Batalin) T. T. Yu et C. L. Li 刺毛樱桃

Shrubs or small trees, deciduous

10.4 (0.8~15.9); −1.6 (−10.4~5.5); 21.7 (10.4~25.9); 10.9 (3.7~15.9); 84.6 (15.1~131.1); 19.0 (0.0~65.2); 663 (409~833) ❙ 798 (341~1466); 371 (190~602); 32 (7~105) ❙ 20.0 (−40.9~106.2); 645 (341~833); 230.4 (61.2~346.5)

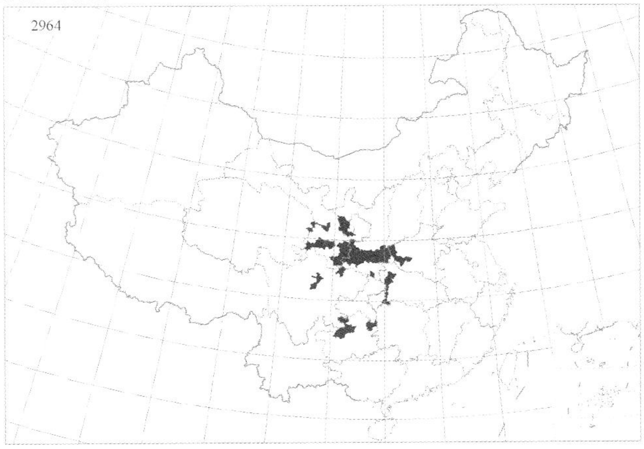

Rosaceae 蔷薇科 *Cerasus* 樱属
Cerasus stipulacea (Maxim.) T. T. Yu et C. L. Li 托叶樱桃
Shrubs or small trees, deciduous

8.7 (−1.3~13.9); −3.3 (−12.4~2.7); 19.8 (8.3~25.6); 9.5 (2.5~13.9); 71.3 (7.0~112.7); 26.4 (3.8~82.2); 615 (354~781) | 732 (373~1164); 350 (201~601); 21 (5~51) | 20.7 (−37.6~128.6); 601 (354~781); 251.5 (81.2~364.1)

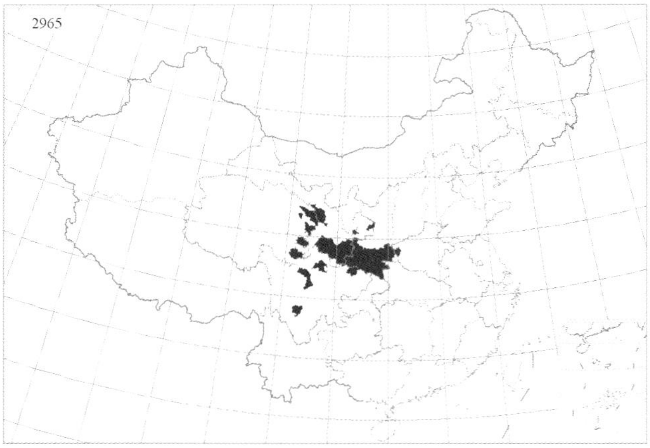

Rosaceae 蔷薇科 *Cerasus* 樱属
Cerasus subhirtella (Miq.) S. Y. Sokolov 大叶早樱
Trees, deciduous

15.6 (13.7~17.1); 3.8 (2.1~11.0); 27.0 (20.1~29.3); 15.6 (13.7~17.1); 129.9 (109.4~145.7); 2.1 (0.0~4.9); 853 (758~941) | 1566 (1267~2680); 614 (529~1186); 174 (94~245) | 85.0 (55.4~246.5); 853 (758~941); 249.0 (193.3~318.0)

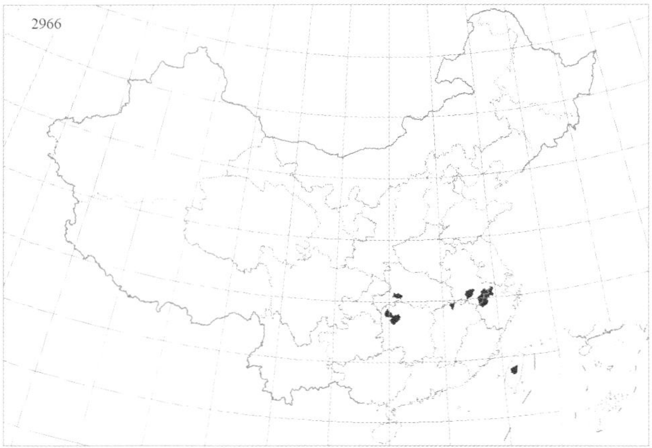

Rosaceae 蔷薇科 *Cerasus* 樱属
Cerasus szechuanica (Batalin) T. T. Yu et C. L. Li 四川樱桃
Shrubs or trees, deciduous

11.1 (2.2~15.3); −1.6 (−7.4~5.3); 23.0 (9.9~27.8); 11.5 (3.8~15.4); 91.9 (15.5~125.7); 17.7 (0.0~49.1); 697 (413~854) | 790 (341~1478); 373 (190~816); 36 (8~113) | 14.8 (−40.9~125.9); 649 (341~832); 213.2 (61.2~323.3)

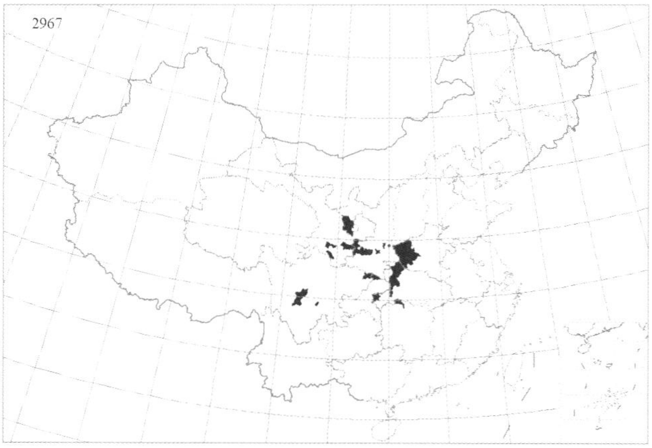

Rosaceae 蔷薇科 *Cerasus* 樱属
Cerasus tatsienensis (Batalin) T. T. Yu et C. L. Li 康定樱桃
Shrubs or small trees, deciduous

10.0 (2.2~17.1); −4.7 (−13.7~5.3); 23.1 (9.9~29.4); 11.0 (3.8~17.1); 88.8 (15.5~146.1); 28.4 (0.0~71.7); 687 (413~952) | 664 (367~1492); 368 (235~817); 27 (8~123) | −3.1 (−40.9~125.9); 566 (367~924); 179.7 (101.9~357.7)

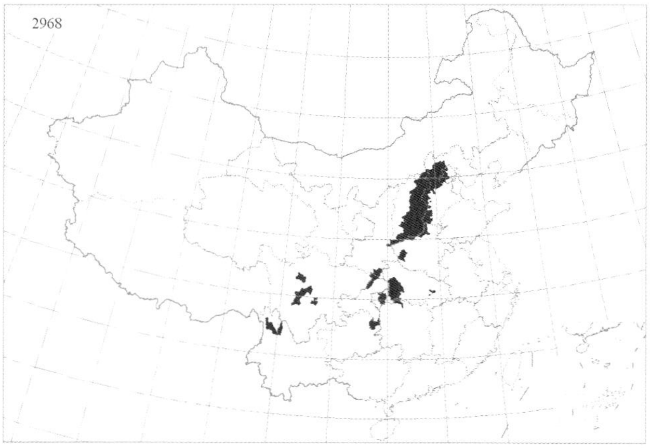

Rosaceae 蔷薇科 *Cerasus* 樱属
Cerasus tianshanica Pojarkov 天山樱桃
Shrubs, deciduous

2.5 (−2.8~9.0); −16.2 (−20.9~−9.3); 17.4 (10.9~23.2); 6.8 (3.5~10.7); 49.3 (16.3~88.9); 78.4 (40.7~109.2); 532 (390~703) | 309 (209~450); 138 (65~211); 29 (6~51) | −39.6 (−65.9~3.9); 306 (209~406); 176.4 (57.8~297.3)

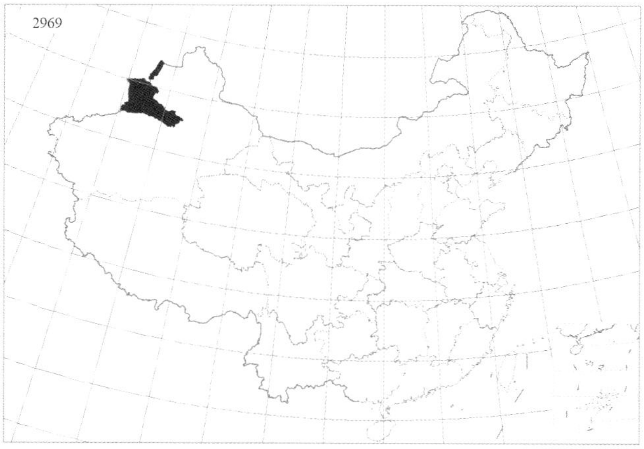

Rosaceae 蔷薇科 *Cerasus* 樱属
Cerasus tomentosa (Thunb) Wall. 毛樱桃
Shrubs or small trees, deciduous

10.8 (−1.3~18.9); −2.4 (−19.2~12.5); 22.6 (9.2~29.3); 11.7 (2.9~19.0); 94.4 (10.1~167.8); 24.0 (0.0~90.7); 704 (359~941) | 788 (132~1950); 423 (81~713); 40 (1~242) | 12.1 (−80.1~147.0); 638 (132~941); 209.6 (13.6~405.0)

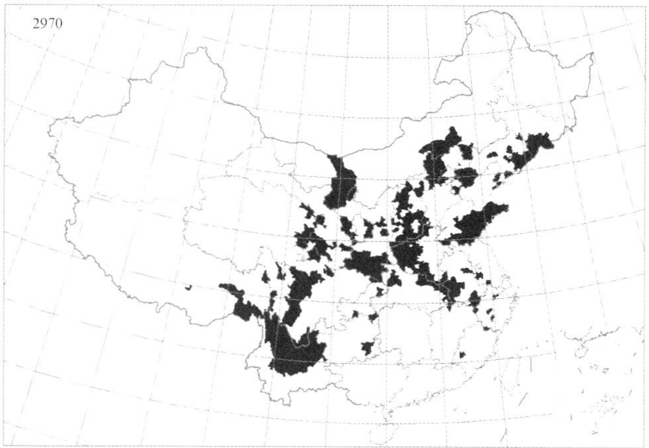

Rosaceae 蔷薇科 *Cerasus* 樱属
Cerasus trichantha (Koehne) C. L. Li et S. Y. Jiang 毛瓣藏樱
Trees, deciduous

4.1; −4.1; 11.3; 5.4; 28.8; 39.3; 455 ❙ 598; 357; 23 ❙ 26.1; 452; 219.8

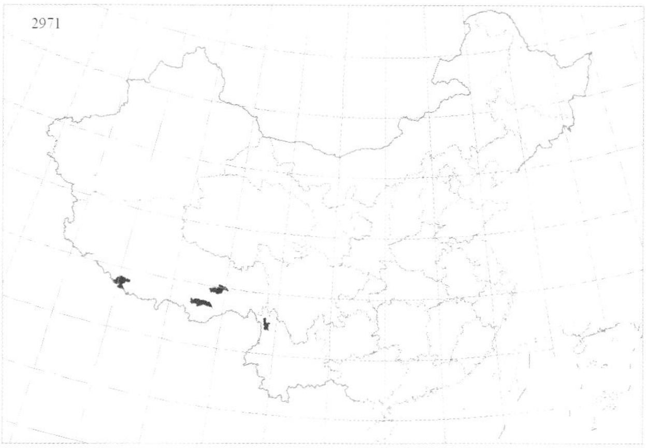

Rosaceae 蔷薇科 *Cerasus* 樱属
Cerasus trichostoma (Koehne) T. T. Yu et C. L. Li 川西樱桃
Trees, deciduous

6.8 (−1.2~15.6); −2.5 (−12.5~5.4); 15.1 (8.5~26.4); 7.8 (2.6~15.6); 51.1 (8.1~128.2); 28.8 (0.0~82.3); 543 (347~839) ❙ 905 (374~1736); 490 (205~1007); 30 (7~112) ❙ 66.2 (−29.3~145.8); 540 (347~839); 263.2 (81.2~447.4)

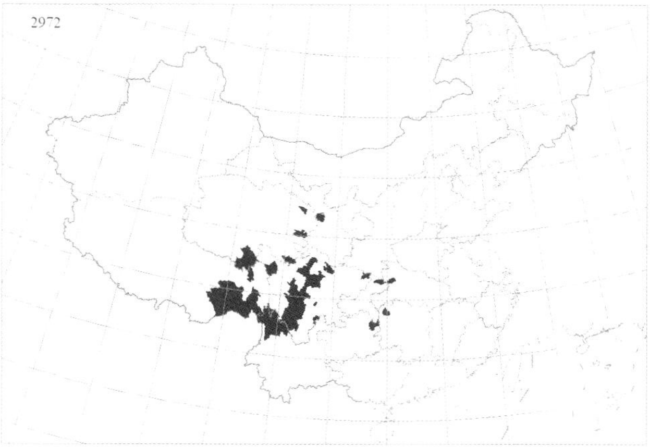

Rosaceae 蔷薇科 *Cerasus* 樱属
Cerasus yedoensis (Matsum.) A. N. Vassiljeva 东京樱花
Trees, deciduous

12.7 (8.4~16.5); −1.6 (−8.9~4.4); 25.8 (23.2~28.9); 13.3 (10.2~16.6); 111.1 (84.0~139.8); 17.8 (0.6~42.8); 784 (661~910) ❙ 918 (590~1377); 471 (398~648); 80 (12~197) ❙ 15.9 (−18.2~54.0); 754 (590~910); 175.9 (131.2~269.1)

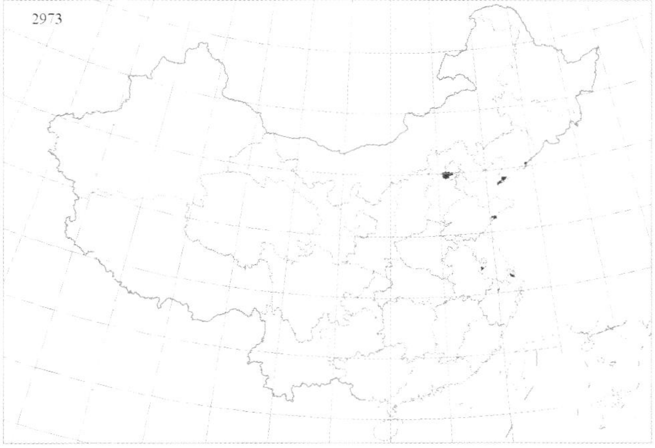

Rosaceae 蔷薇科 *Cerasus* 樱属
Cerasus yunnanensis (Franch.) T. T. Yu et C. L. Li 云南樱桃
Shrubs or trees, deciduous

14.9 (5.4~19.8); 7.4 (−2.1~13.1); 20.7 (12.3~27.1); 14.9 (5.8~19.8); 120.2 (31.5~177.9); 1.3 (0.0~25.6); 762 (481~1010) ❙ 1060 (741~1740); 594 (411~816); 43 (12~187) ❙ 40.3 (−14.2~99.9); 744 (481~1010); 298.7 (142.9~516.7)

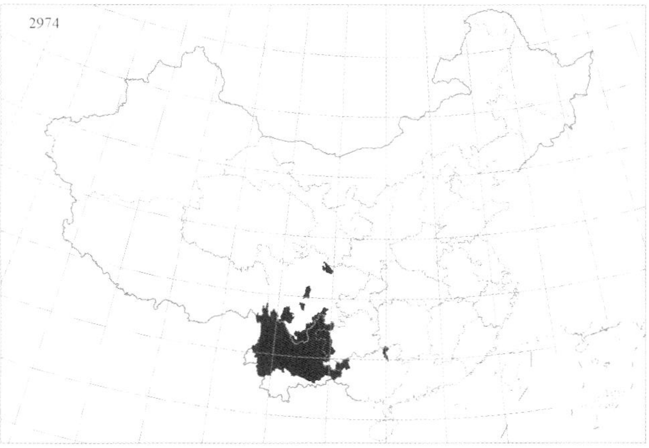

Rosaceae 蔷薇科 *Cerasus* 樱属
Cerasus yunnanensis var. polybotrys (Koehne) T. T. Yu et C. L. Li 多花云南樱桃
Trees, deciduous

10.8; 3.9; 16.6; 10.9; 71.8; 1.3; 633 ❙ 968; 594; 24 ❙ 52.8; 633; 319.8

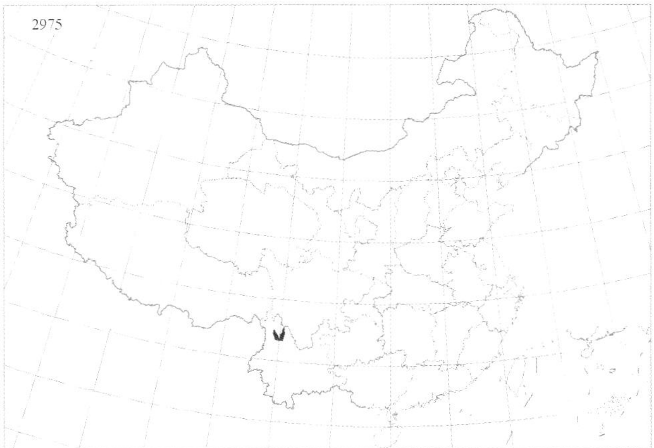

Rosaceae 蔷薇科 *Chaenomeles* 木瓜属
Chaenomeles cathayensis (Hemsl.) C. K. Schneid. 毛叶木瓜
Shrubs or small trees, deciduous

14.0 (2.2~20.7); 2.6 (−6.6~10.5); 24.9 (9.9~30.1); 14.2 (3.8~20.8); 115.9 (15.5~189.3); 6.8 (0.0~49.1); 792 (413~1113) ❙ 1230 (337~1997); 482 (257~685); 110 (1~259) ❙ 53.8 (−26.1~130.7); 783 (337~1032); 244.4 (141.7~351.7)

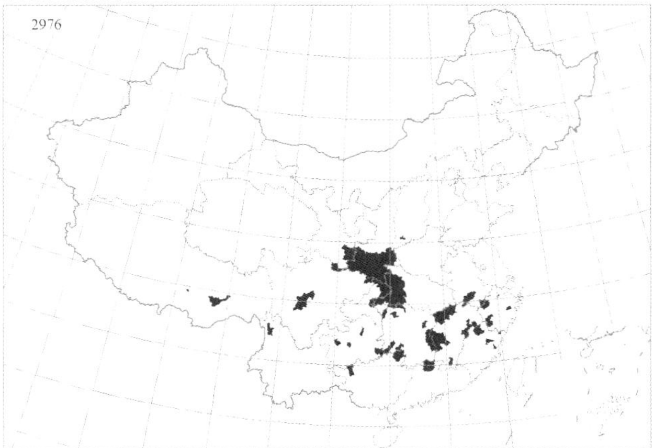

Rosaceae 蔷薇科 *Chaenomeles* 木瓜属
Chaenomeles sinensis (Thouin) Koehne 木瓜
Shrubs or small trees, deciduous

15.1 (7.4~24.6); 2.7 (-9.9~19.5); 26.5 (18.8~29.4); 15.1 (9.5~24.7); 126.7 (69.8~235.9); 5.5 (0.0~50.9); 841 (603~1372) ❙ 1052 (409~1997); 469 (258~779); 86 (10~254) ❙ 25.0 (-36.6~130.7); 778 (409~1299); 194.0 (111.9~548.0)

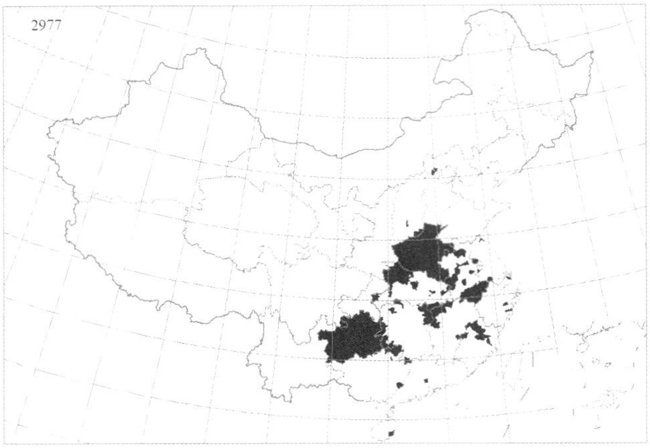

Rosaceae 蔷薇科 *Chaenomeles* 木瓜属
Chaenomeles speciosa (Sweet) Nakai 皱皮木瓜
Shrubs, deciduous

14.3 (1.8~20.7); 3.9 (-7.1~13.1); 23.6 (9.9~29.3); 14.4 (3.8~20.7); 117.6 (15.7~188.9); 5.1 (0.0~53.3); 781 (404~1071) ❙ 1048 (520~1997); 504 (245~816); 63 (12~254) ❙ 34.7 (-36.5~130.7); 749 (404~1071); 235.1 (120.1~516.7)

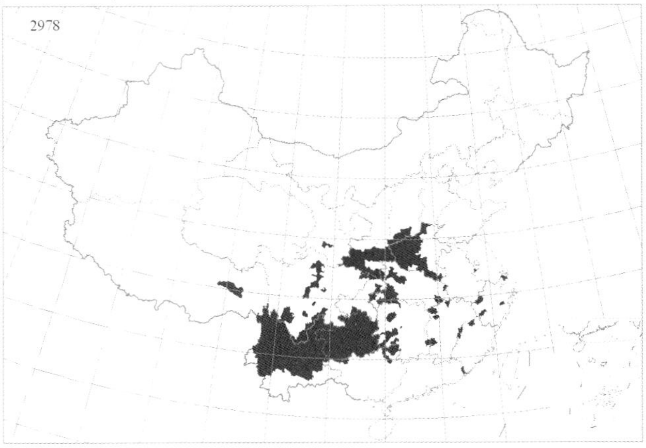

Rosaceae 蔷薇科 *Chaenomeles* 木瓜属
Chaenomeles thibetica T. T. Yu 西藏木瓜
Shrubs or small trees, deciduous

5.2 (1.8~13.3); -3.9 (-7.2~5.4); 13.3 (9.9~20.1); 6.3 (3.8~13.4); 38.1 (15.4~100.4); 34.8 (0.0~53.3); 488 (404~706) ❙ 880 (337~1736); 489 (257~1007); 27 (1~71) ❙ 73.0 (-23.5~145.8); 475 (337~706); 275.9 (169.5~447.4)

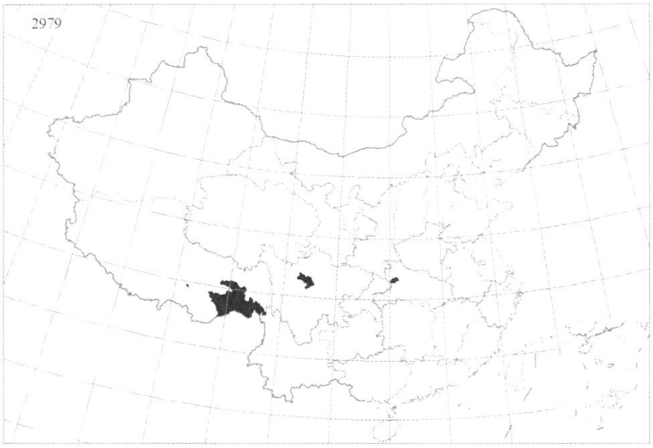

Rosaceae 蔷薇科 *Chamaerhodos* 地蔷薇属
Chamaerhodos altaica (Laxm.) Bunge 阿尔泰地蔷薇
Subshrubs, deciduous

2.4 (-0.6~5.8); -17.0 (-20.7~-12.9); 19.6 (15.9~23.3); 7.3 (5.4~9.7); 54.8 (37.5~80.9); 85.2 (63.1~104.2); 550 (475~668) ❙ 245 (156~401); 144 (56~275); 14 (4~33) ❙ -54.9 (-76.7~-20.8); 245 (156~401); 93.8 (16.6~200.5)

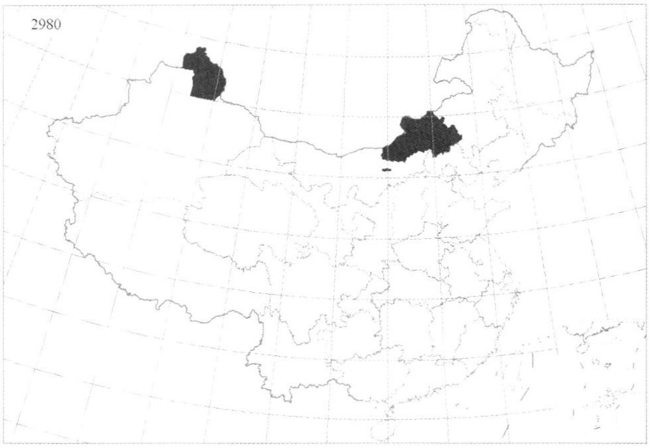

Rosaceae 蔷薇科 *Comarum* 沼委陵菜属
Comarum salesovianum (Stephan) Asch. et Graebn. 西北沼委陵菜
Subshrubs, deciduous

3.0 (-5.9~11.4); -13.0 (-20.9~-6.5); 16.6 (5.4~27.5); 6.6 (1.3~12.7); 47.2 (0.4~109.3); 70.2 (29.5~130.5); 519 (300~813) ❙ 280 (32~611); 153 (21~340); 12 (0~48) ❙ -39.4 (-95.9~54.6); 266 (32~539); 120.4 (1.6~364.1)

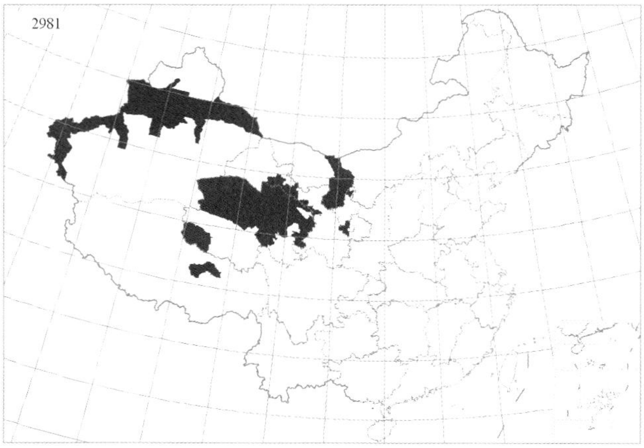

Rosaceae 蔷薇科 *Cotoneaster* 栒子属
Cotoneaster acuminatus Lindl. 尖叶栒子
Shrubs

5.7 (-2.2~16.7); -3.5 (-12.8~9.7); 13.8 (6.1~23.6); 7.1 (1.8~16.7); 45.2 (1.9~140.9); 36.0 (0.0~91.3); 513 (329~819) ❙ 765 (271~1441); 440 (165~817); 23 (4~84) ❙ 46.2 (-44.6~128.6); 499 (271~782); 251.6 (61.6~407.1)

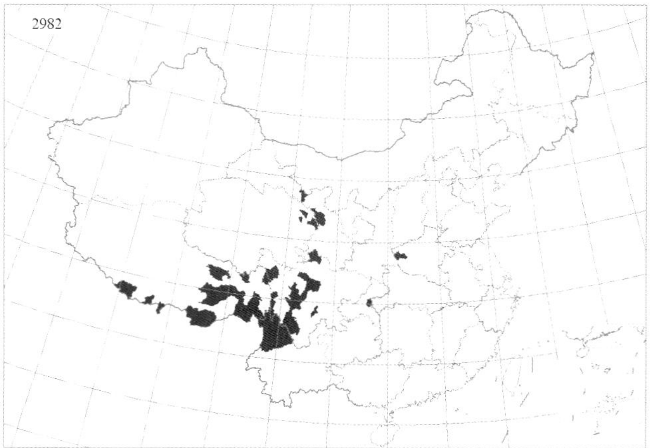

Rosaceae 蔷薇科 *Cotoneaster* 栒子属
Cotoneaster acutifolius Turcz. 灰栒子
Shrubs, deciduous

9.7 (−1.4~17.9); −4.3 (−20.7~11.6); 22.4 (8.5~29.2); 10.8 (2.6~18.0); 86.4 (8.1~155.6); 29.2 (0.0~101.4); 678 (347~936) ❙ 672 (128~2680); 353 (81~1186); 31 (2~308) ❙ −1.1 (−80.8~246.5); 563 (128~930); 184.7 (13.6~407.1)

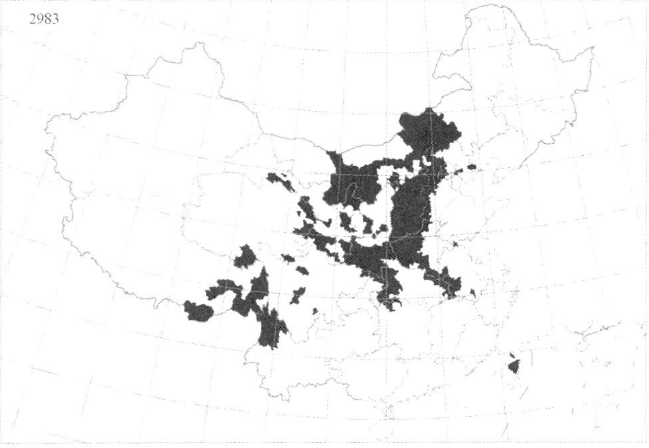

Rosaceae 蔷薇科 *Cotoneaster* 栒子属
Cotoneaster acutifolius var. villosulus Rehder et E. H. Wilson 密毛灰栒子
Shrubs, deciduous

10.6 (−1.8~15.1); −2.5 (−12.7~4.3); 22.6 (6.1~28.0); 11.3 (1.8~15.2); 90.2 (1.9~128.0); 22.2 (0.7~82.8); 695 (329~880) ❙ 755 (320~1437); 383 (190~616); 35 (8~140) ❙ 12.2 (−40.9~125.9); 622 (320~817); 197.9 (61.2~341.6)

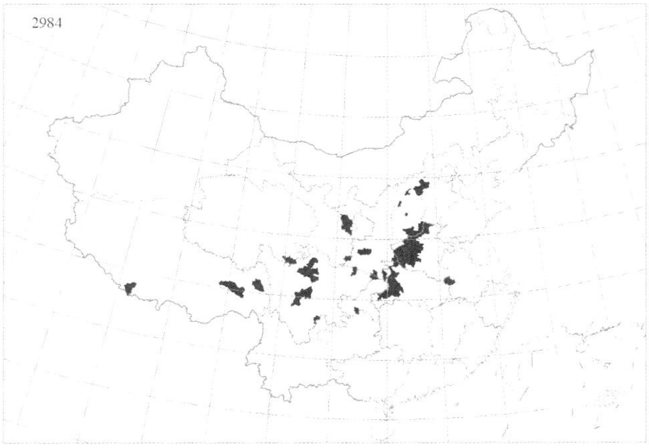

Rosaceae 蔷薇科 *Cotoneaster* 栒子属
Cotoneaster adpressus Bois 匍匐栒子
Shrubs, deciduous

10.3 (−3.3~13.9); 0.9 (−15.2~12.5); 18.6 (6.1~26.8); 11.1 (1.8~19.0); 83.9 (1.9~167.8); 19.5 (0.0~103.2); 647 (329~929) ❙ 899 (246~1478); 472 (191~817); 36 (0~129) ❙ 38.7 (−35.9~128.6); 634 (246~918); 248.3 (72.2~487.2)

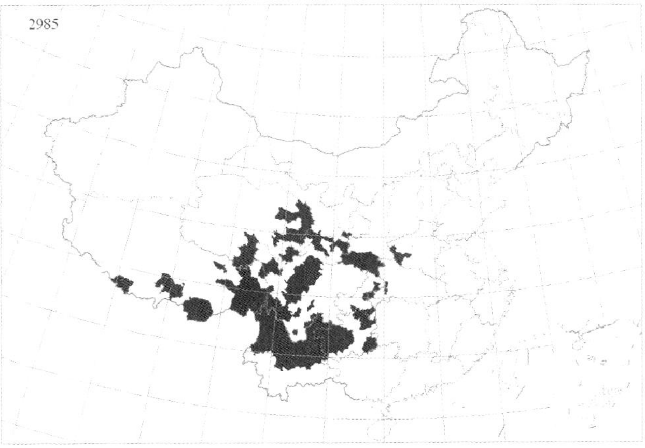

Rosaceae 蔷薇科 *Cotoneaster* 栒子属
Cotoneaster affinis Lindl. 藏边栒子
Shrubs, deciduous

5.1; −2.8; 11.9; 6.8; 41.6; 40.2; 494 ❙ 765; 462; 25 ❙ 40.2; 488; 202.8

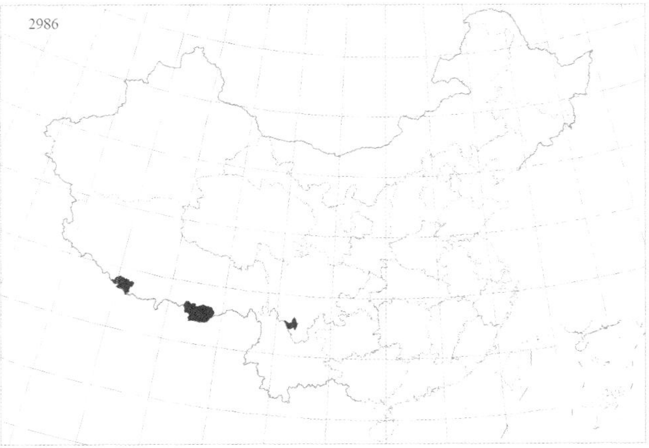

Rosaceae 蔷薇科 *Cotoneaster* 栒子属
Cotoneaster ambiguus Rehder et E. H. Wilson 川康栒子
Shrubs, deciduous

11.7 (−0.8~13.9); 1.1 (−12.5~12.0); 21.4 (9.2~27.7); 12.0 (2.9~19.0); 93.6 (10.1~167.8); 12.7 (0.0~79.5); 691 (374~929) ❙ 897 (196~1478); 443 (117~627); 37 (4~129) ❙ 30.5 (−69.9~125.9); 660 (196~874); 247.2 (47.6~351.7)

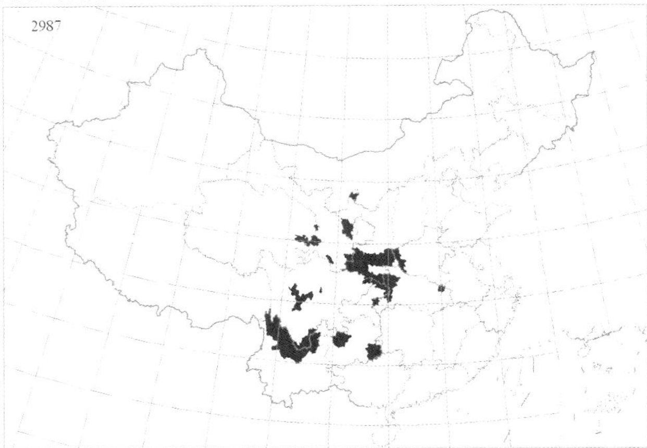

Rosaceae 蔷薇科 *Cotoneaster* 栒子属
Cotoneaster apiculatus Rehder et E. H. Wilson 细尖栒子
Shrubs, deciduous

11.1 (2.2~19.2); 0.5 (−7.1~13.3); 21.0 (9.9~25.4); 11.4 (3.8~19.4); 86.6 (15.5~172.6); 12.9 (0.0~49.1); 672 (413~932) ❙ 977 (373~1492); 467 (201~816); 38 (7~105) ❙ 47.3 (−37.6~125.9); 668 (373~923); 271.4 (88.1~452.0)

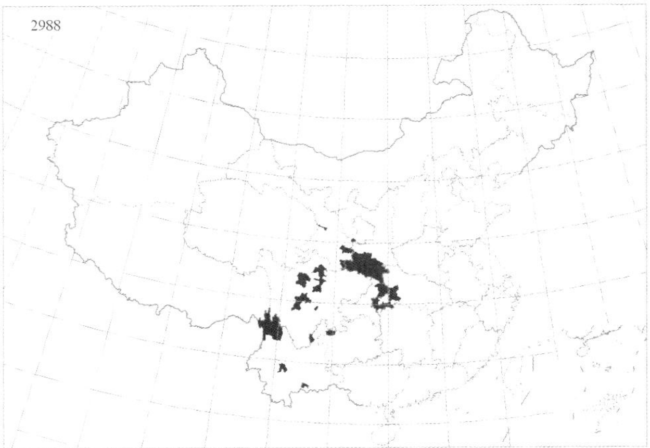

Rosaceae 蔷薇科 *Cotoneaster* 栒子属
Cotoneaster bullatus Bois 泡叶栒子
Shrubs, deciduous

10.1 (2.2~15.2); 0.8 (-6.6~7.2); 18.6 (9.9~25.9); 10.4 (3.8~15.2); 74.6 (15.5~123.9); 13.2 (0.0~49.1); 631 (413~816) ❙ 1118 (741~1736); 555 (425~1007); 48 (12~105) ❙ 78.8 (47.6~145.8); 630 (413~816); 271.5 (170.4~447.4)

2989

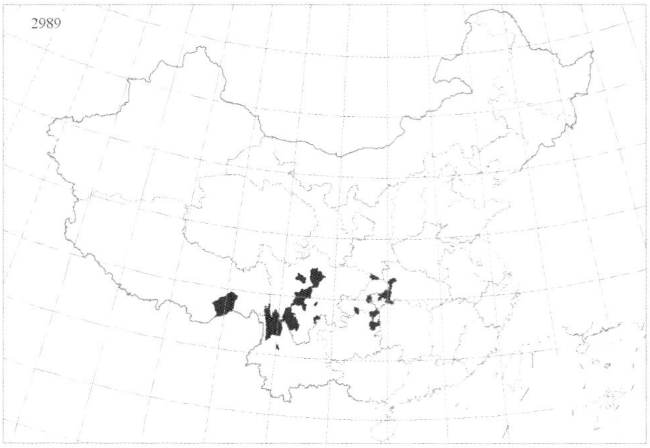

Rosaceae 蔷薇科 *Cotoneaster* 栒子属
Cotoneaster bullatus var. macrophyllus Rehder et E. H. Wilson 大叶泡叶栒子
Shrubs, deciduous

10.4 (2.0~14.9); 1.0 (-7.2~5.3); 18.9 (10.1~23.6); 10.8 (3.8~15.0); 79.1 (15.4~119.6); 13.3 (0.0~51.4); 643 (413~782) ❙ 1134 (943~1410); 634 (507~816); 34 (23~49) ❙ 80.8 (41.9~128.6); 643 (413~782); 234.3 (147.3~292.6)

2990

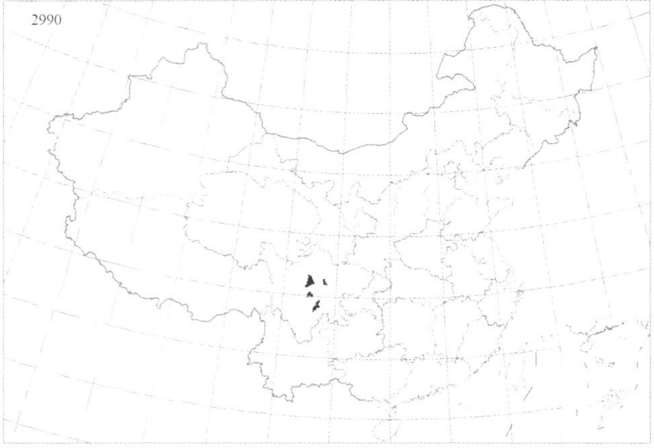

Rosaceae 蔷薇科 *Cotoneaster* 栒子属
Cotoneaster buxifolius Wall. ex Lindl. 黄杨叶栒子
Shrubs, evergreen

13.4 (0.3~19.3); 5.6 (-9.9~13.1); 19.7 (9.3~26.6); 13.6 (3.3~19.4); 106.5 (11.9~173.2); 5.7 (0.0~67.5); 718 (391~939) ❙ 1004 (527~1504); 562 (317~860); 39 (9~88) ❙ 42.6 (-14.2~125.9); 703 (391~912); 277.4 (142.9~491.5)

2991

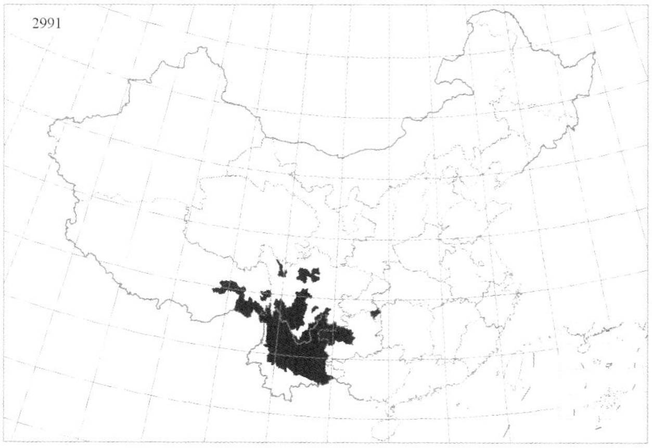

Rosaceae 蔷薇科 *Cotoneaster* 栒子属
Cotoneaster buxifolius var. marginatus Loudon 多花黄杨叶栒子
Shrubs, evergreen or semievergreen

3.5; -5.4; 11.4; 4.8; 24.2; 41.9; 438 ❙ 599; 370; 13 ❙ 35.3; 409; 201.7

2992

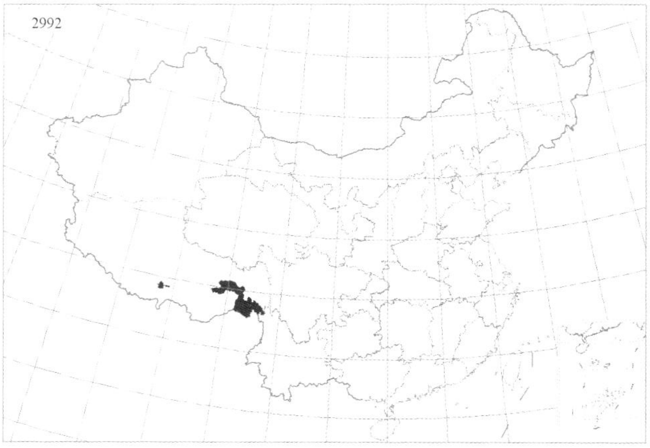

Rosaceae 蔷薇科 *Cotoneaster* 栒子属
Cotoneaster chengkangensis T. T. Yu 镇康栒子
Shrubs, deciduous

13.5; 6.7; 18.5; 13.5; 105.0; 2.7; 717 ❙ 1207; 697; 44 ❙ 68.3; 717; 393.1

2993

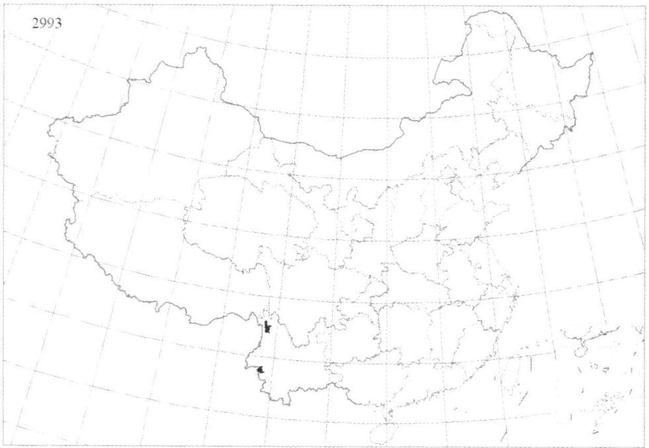

Rosaceae 蔷薇科 *Cotoneaster* 栒子属
Cotoneaster conspicuus Comber ex Marquand 大果栒子
Shrubs, evergreen

3.3; -5.4; 11.1; 4.6; 22.3; 41.8; 427 ❙ 473; 287; 7 ❙ 5.0; 427; 230.7

2994

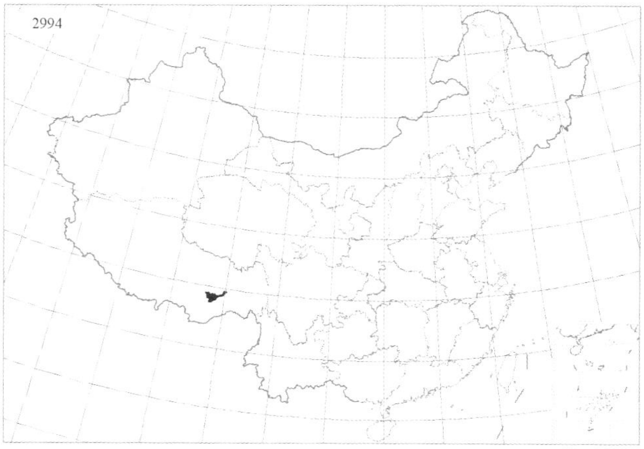

Rosaceae 蔷薇科 *Cotoneaster* 栒子属
Cotoneaster coriaceus Franch. 厚叶栒子
Shrubs, evergreen

14.7 (2.2~21.0); 7.5 (-6.6~14.2); 20.3 (9.9~26.6); 14.8 (3.8~21.0); 119.5 (15.5~191.8); 2.5 (0.0~49.1); 759 (413~1053) ❙ 1048 (741~1522); 599 (411~860); 39 (12~69) ❙ 40.8 (-14.2~125.9); 737 (413~1053); 304.6 (171.0~551.6)

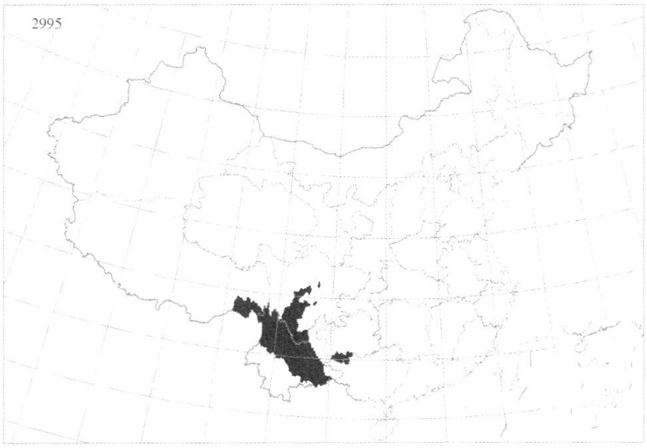

Rosaceae 蔷薇科 *Cotoneaster* 栒子属
Cotoneaster dammeri C. K. Schneid. 矮生栒子
Shrubs, evergreen

12.6 (2.2~19.1); 3.0 (-7.1~13.0); 21.4 (9.9~28.8); 12.7 (3.8~19.1); 97.8 (15.5~169.7); 6.0 (0.0~49.1); 709 (413~994) ❙ 1077 (632~1478); 525 (306~835); 50 (14~113) ❙ 53.5 (-8.1~125.9); 707 (413~991); 241.2 (123.1~506.2)

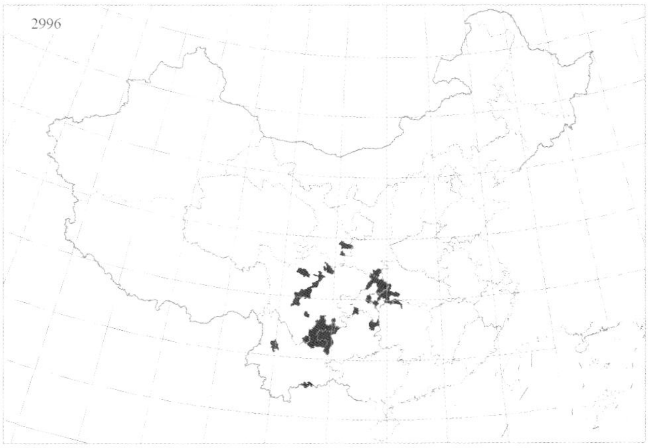

Rosaceae 蔷薇科 *Cotoneaster* 栒子属
Cotoneaster dammeri var. radicans (Dammer ex C. K. Schneid.) C. K. Schneid. 长柄矮生栒子
Shrubs, evergreen

6.8; -2.3; 15.0; 7.5; 47.8; 25.5; 533 ❙ 1005; 554; 26 ❙ 93.1; 533; 255.6

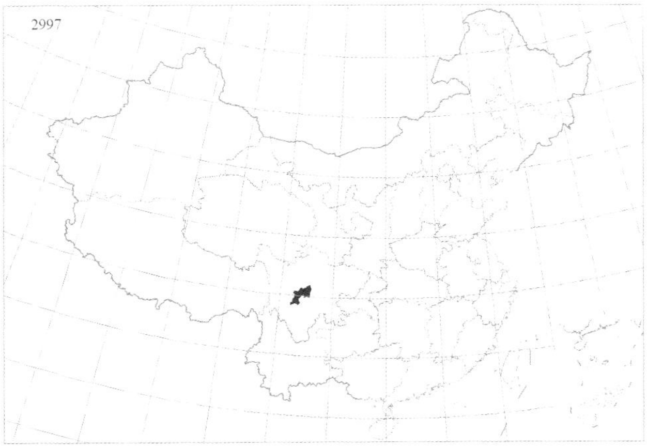

Rosaceae 蔷薇科 *Cotoneaster* 栒子属
Cotoneaster dielsianus E. Pritz. 木帚栒子
Shrubs, evergreen

13.2 (1.8~18.9); 4.6 (-7.2~12.5); 20.7 (9.9~26.8); 13.4 (3.8~19.0); 104.8 (15.4~167.8); 5.4 (0.0~53.3); 722 (404~929) ❙ 1066 (473~1492); 552 (287~816); 48 (7~120) ❙ 49.9 (-14.2~128.6); 712 (404~912); 263.3 (142.9~516.7)

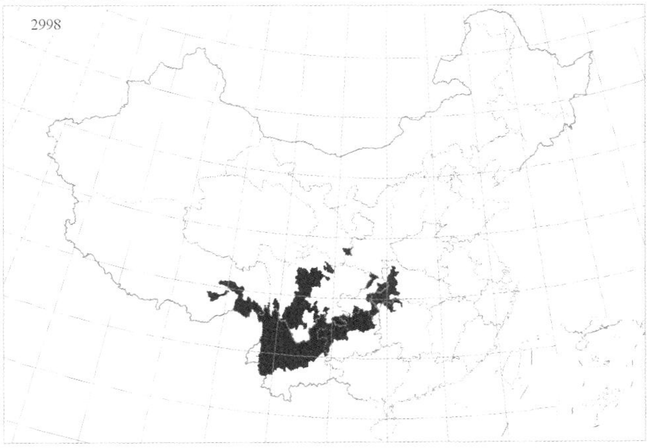

Rosaceae 蔷薇科 *Cotoneaster* 栒子属
Cotoneaster dielsianus var. elegans Rehder et E. H. Wilson 小叶木帚栒子
Shrubs, evergreen

8.2 (2.2~11.1); 0.3 (-6.6~3.9); 15.1 (9.9~19.4); 8.7 (3.8~11.3); 55.1 (15.5~80.8); 15.6 (1.3~49.1); 567 (413~652) ❙ 986 (862~1220); 596 (546~688); 21 (12~37) ❙ 77.1 (51.8~125.9); 565 (413~652); 294.5 (248.3~319.8)

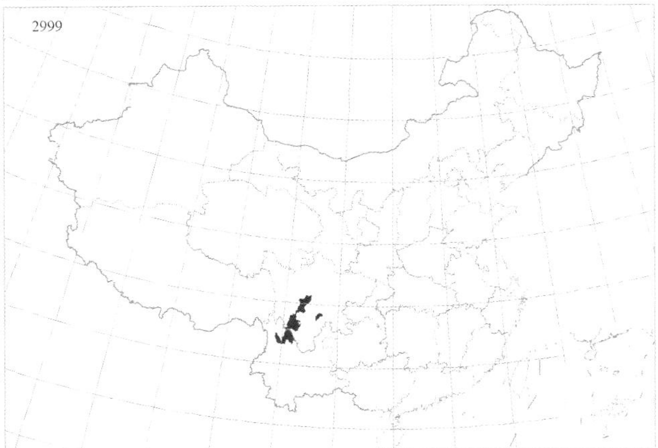

Rosaceae 蔷薇科 *Cotoneaster* 栒子属
Cotoneaster divaricatus Rehder et E. H. Wilson 散生栒子
Shrubs, deciduous

10.1 (-1.2~17.2); -0.4 (-11.7~8.6); 19.9 (8.5~29.3); 10.7 (2.6~17.2); 80.6 (8.1~146.8); 18.4 (0.0~82.3); 652 (347~941) ❙ 960 (341~1590); 460 (190~816); 50 (7~196) ❙ 47.1 (-40.9~128.6); 644 (341~941); 238.1 (61.2~357.7)

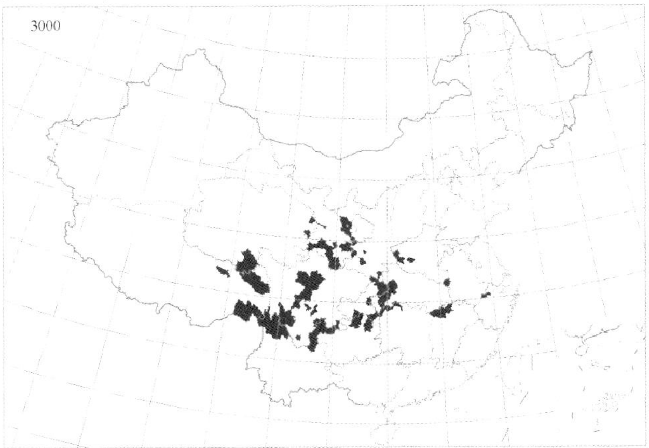

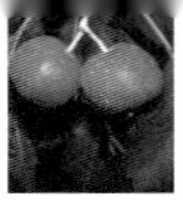

Rosaceae 蔷薇科 *Cotoneaster* 栒子属
Cotoneaster fangianus T. T. Yu 恩施栒子
Shrubs, deciduous

12.8; 2.0; 23.2; 12.8; 99.7; 5.9; 720 ❙ 1356; 557; 71 ❙ 89.1; 720; 243.0

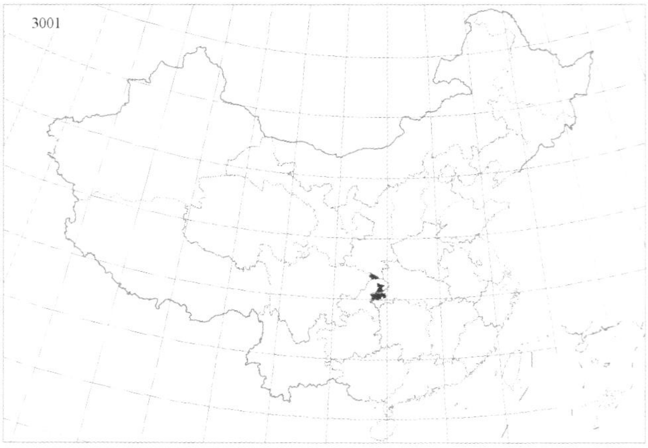

Rosaceae 蔷薇科 *Cotoneaster* 栒子属
Cotoneaster foveolatus Rehder et E. H. Wilson 麻核栒子
Shrubs, deciduous

10.5 (-0.3~18.5); -0.5 (-10.9~10.5); 20.8 (9.0~27.5); 11.0 (3.0~18.3); 83.9 (10.4~160.0); 16.8 (0.0~73.4); 665 (378~906) ❙ 874 (341~1466); 435 (190~816); 35 (7~113) ❙ 35.2 (-40.9~128.6); 638 (341~832); 247.3 (61.2~407.1)

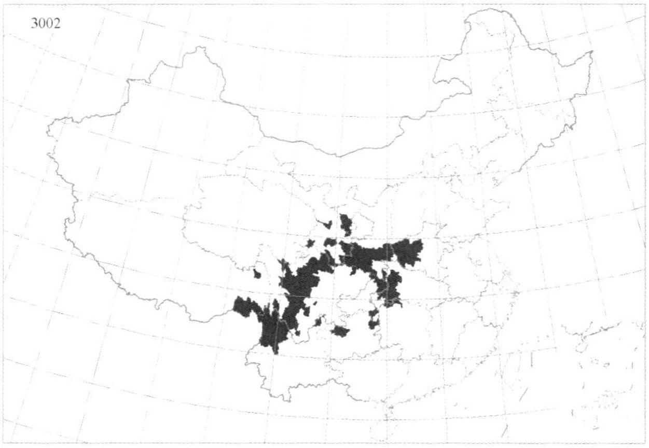

Rosaceae 蔷薇科 *Cotoneaster* 栒子属
Cotoneaster franchetii Bois 西南栒子
Shrubs, semievergreen

14.4 (2.2~18.9); 6.7 (-6.6~12.5); 20.7 (9.9~26.6); 14.5 (3.8~19.0); 115.6 (15.5~167.8); 2.0 (0.0~49.1); 749 (413~929) ❙ 1051 (741~1454); 575 (411~816); 44 (12~94) ❙ 42.1 (-14.2~125.9); 734 (413~912); 282.4 (142.9~516.7)

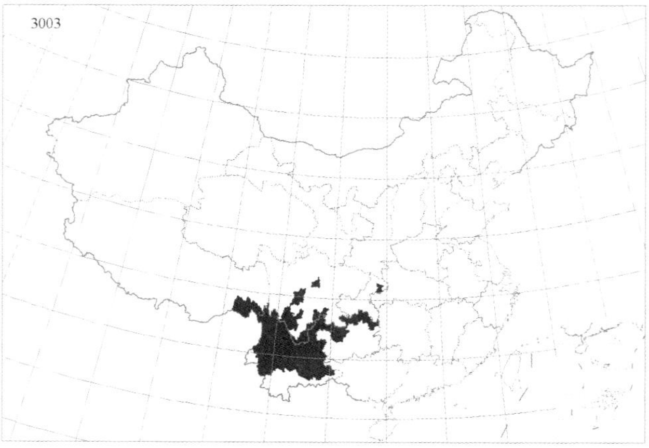

Rosaceae 蔷薇科 *Cotoneaster* 栒子属
Cotoneaster frigidus Wall. ex Lindl. 耐寒栒子
Shrubs or small trees, deciduous

1.8 (-1.8~12.2); -6.6 (-10.2~4.3); 9.3 (6.1~18.5); 4.3 (1.8~12.2); 21.6 (1.9~87.4); 59.7 (0.7~82.8); 408 (329~672) ❙ 566 (320~1431); 347 (193~817); 21 (10~33) ❙ 22.2 (-4.0~113.0); 404 (320~672); 143.8 (61.6~363.7)

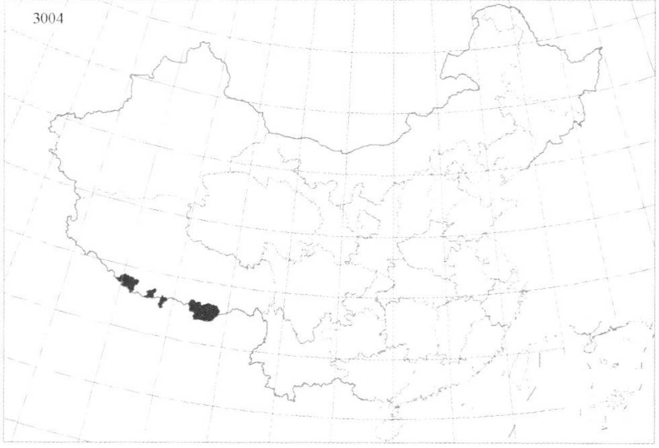

Rosaceae 蔷薇科 *Cotoneaster* 栒子属
Cotoneaster glabratus Rehder et E. H. Wilson 光叶栒子
Shrubs, semievergreen

13.8 (5.4~19.1); 4.5 (-2.1~10.1); 22.0 (12.3~26.8); 13.9 (5.8~19.1); 109.5 (31.5~169.2); 3.5 (0.0~25.6); 750 (481~966) ❙ 1198 (741~1478); 598 (426~816); 59 (21~94) ❙ 61.5 (30.0~101.4); 750 (481~966); 244.2 (146.0~384.7)

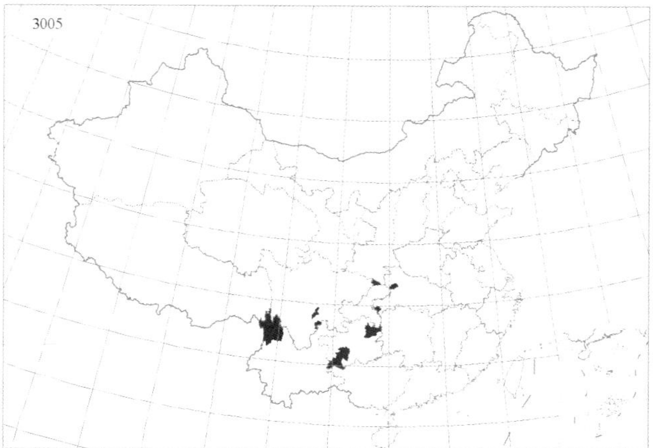

Rosaceae 蔷薇科 *Cotoneaster* 栒子属
Cotoneaster glaucophyllus Franch. 粉叶栒子
Shrubs, semievergreen

15.2 (5.4~21.0); 7.5 (-2.1~14.2); 21.4 (12.3~28.6); 15.3 (5.8~21.0); 124.5 (31.5~191.8); 1.0 (0.0~25.6); 780 (481~1099) ❙ 1103 (741~1688); 609 (411~959); 46 (12~161) ❙ 42.4 (-14.2~112.2); 766 (481~1099); 294.8 (142.9~627.0)

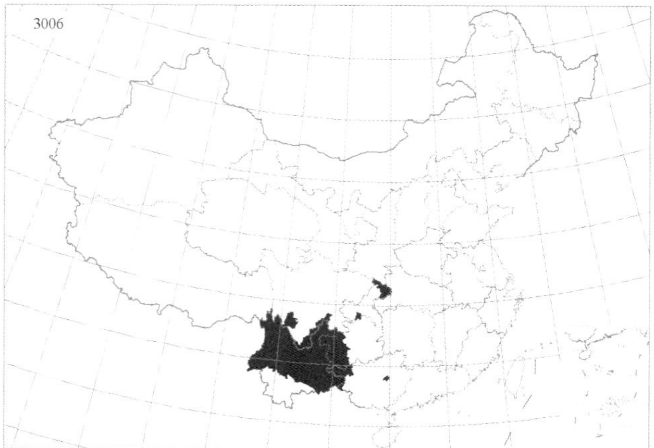

Rosaceae 蔷薇科 *Cotoneaster* 栒子属
Cotoneaster glaucophyllus var. meiophyllus W. W. Sm. 小叶粉叶栒子
Shrubs, semievergreen

14.4 (11.7~15.9); 4.5 (2.7~5.8); 23.3 (18.8~26.0); 14.4 (11.7~15.9); 113.5 (82.6~131.2); 0.7 (0.0~3.8); 763 (658~831) ❙ 1130 (916~1252); 531 (475~686); 63 (31~89) ❙ 48.2 (35.3~70.6); 763 (658~831); 189.4 (148.9~215.5)

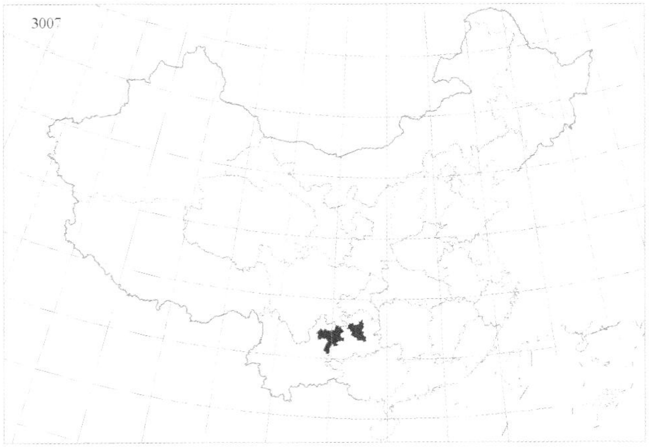

Rosaceae 蔷薇科 *Cotoneaster* 栒子属
Cotoneaster glomerulatus W. W. Sm. 球花栒子
Shrubs, deciduous

9.5 (5.4~14.3); 2.4 (-2.1~7.6); 15.6 (12.3~19.7); 9.7 (5.8~14.3); 64.1 (31.5~111.9); 9.1 (0.0~25.6); 596 (481~728) ❙ 940 (741~1134); 554 (426~651); 33 (21~47) ❙ 58.4 (36.1~78.3); 592 (481~707); 310.7 (273.5~351.7)

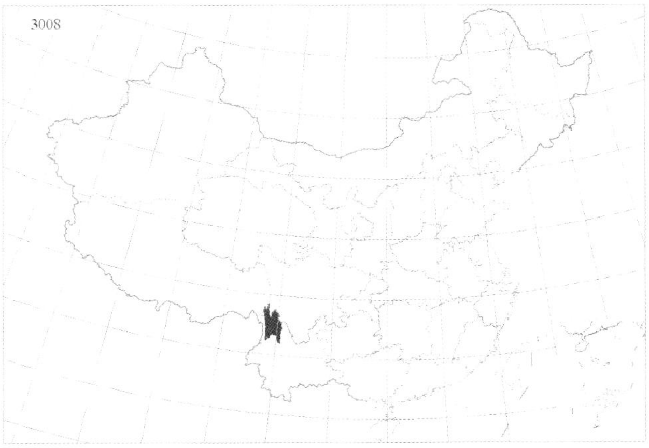

Rosaceae 蔷薇科 *Cotoneaster* 栒子属
Cotoneaster gracilis Rehder et E. H. Wilson 细弱栒子
Shrubs, deciduous

9.3 (0.8~15.5); -3.1 (-11.4~4.1); 20.9 (9.5~27.8); 10.1 (3.8~15.6); 77.3 (15.4~131.3); 24.6 (0.9~66.0); 638 (411~881) ❙ 741 (132~1478); 351 (81~602); 28 (2~105) ❙ 17.8 (-80.1~128.6); 577 (132~881); 223.5 (13.6~357.7)

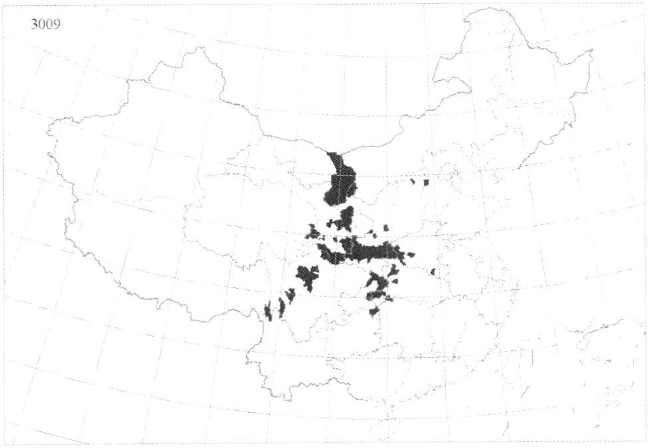

Rosaceae 蔷薇科 *Cotoneaster* 栒子属
Cotoneaster harrovianus E. H. Wilson 蒙自栒子
Shrubs, evergreen

17.5 (16.9~18.1); 10.9 (10.3~11.7); 22.1 (21.4~22.5); 17.6 (16.9~18.0); 150.8 (142.7~156.5); 0.0 (0.0~0.0); 845 (813~863) ❙ 1135 (958~1368); 623 (531~756); 49 (44~59) ❙ 34.4 (13.3~60.1); 838 (813~863); 386.8 (311.9~491.5)

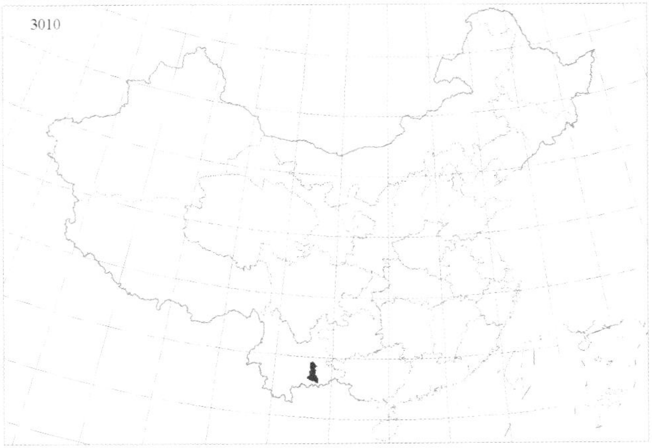

Rosaceae 蔷薇科 *Cotoneaster* 栒子属
Cotoneaster harrysmithii Flinck et B. Hylmo 丹巴栒子
Shrubs, deciduous

5.8; -3.9; 14.6; 7.0; 45.2; 35.1; 519 ❙ 896; 437; 35 ❙ 72.3; 519; 251.3

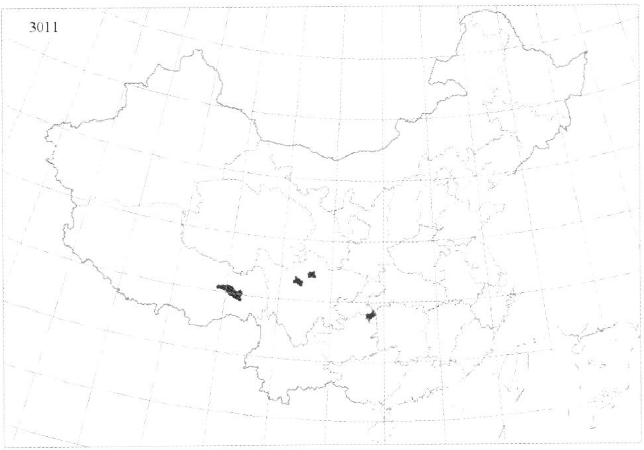

Rosaceae 蔷薇科 *Cotoneaster* 栒子属
Cotoneaster hebephyllus Diels 钝叶栒子
Shrubs, deciduous

6.7 (-1.4~19.6); -2.5 (-15.4~13.9); 14.7 (8.1~25.3); 8.0 (2.6~19.6); 55.3 (8.1~175.1); 34.3 (0.0~92.1); 543 (347~933) ❙ 750 (255~1500); 421 (165~785); 22 (1~88) ❙ 36.4 (-44.6~128.6); 510 (255~882); 256.6 (60.5~530.1)

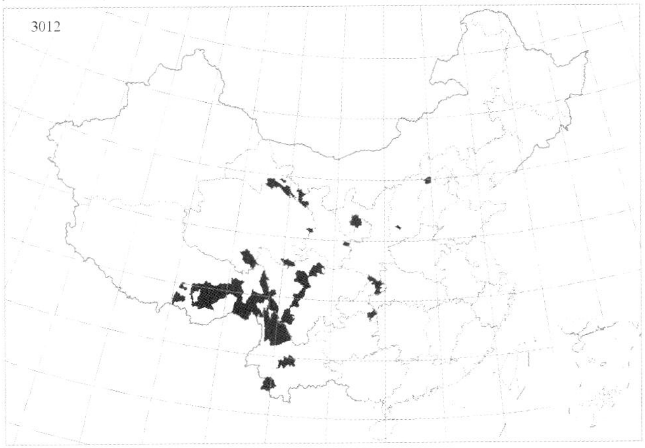

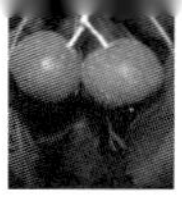

Rosaceae 蔷薇科 *Cotoneaster* 栒子属

Cotoneaster hebephyllus var. incanus W. W. Sm. 灰毛钝叶栒子

Shrubs, deciduous

9.1; 1.9; 15.2; 9.2; 58.3; 8.5; 584 ❙ 905; 564; 20 ❙ 55.4; 582; 301.0

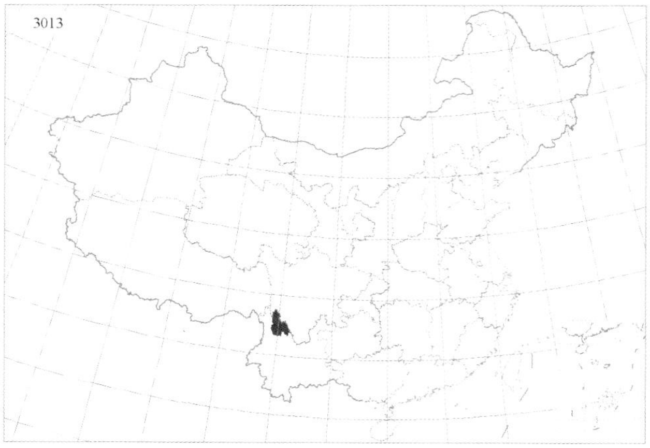

Rosaceae 蔷薇科 *Cotoneaster* 栒子属

Cotoneaster hebephyllus var. majusculus W. W. Sm. 大果钝叶栒子

Shrubs, deciduous

14.1; 6.5; 20.2; 14.2; 109.8; 0.0; 729 ❙ 1022; 615; 18 ❙ 40.1; 707; 306.0

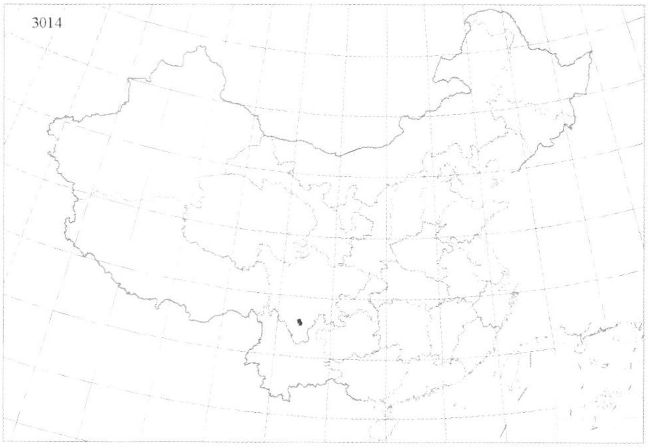

Rosaceae 蔷薇科 *Cotoneaster* 栒子属

Cotoneaster horizontalis Decne. 平枝栒子

Shrubs, deciduous or semievergreen

12.0 (-3.3~17.1); 1.3 (-15.2~6.4); 22.1 (7.2~29.4); 12.4 (2.0~17.1); 96.1 (3.7~146.1); 11.3 (0.0~103.2); 710 (339~952) ❙ 1066 (467~1689); 488 (268~816); 63 (8~220) ❙ 51.3 (-16.9~128.6); 708 (339~941); 235.4 (111.9~407.1)

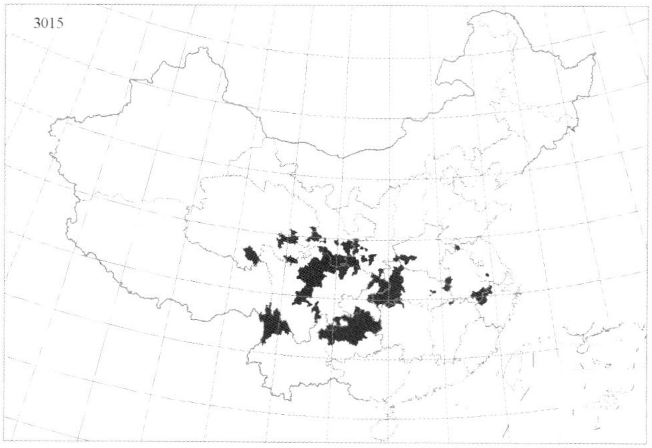

Rosaceae 蔷薇科 *Cotoneaster* 栒子属

Cotoneaster horizontalis var. perpusillus C. K. Schneid. 小叶平枝栒子

Shrubs, deciduous or semievergreen

13.5 (9.5~16.2); 1.5 (-3.3~6.4); 24.6 (18.4~27.5); 13.6 (10.1~16.2); 109.9 (77.6~134.4); 7.8 (0.0~22.7); 765 (637~865) ❙ 911 (626~1337); 431 (308~727); 49 (15~94) ❙ 19.5 (-22.2~92.4); 738 (626~859); 194.0 (139.1~407.1)

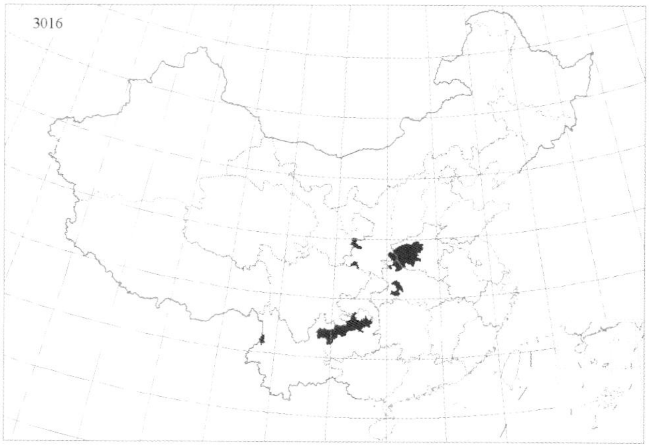

Rosaceae 蔷薇科 *Cotoneaster* 栒子属

Cotoneaster integerrimus Medik. 全缘栒子

Shrubs, deciduous

4.1 (-5.8~11.0); -14.7 (-29.8~-5.7); 20.7 (13.1~25.5); 8.1 (4.4~12.0); 63.2 (26.7~101.7); 73.8 (29.5~157.4); 580 (426~757) ❙ 399 (53~694); 270 (33~499); 11 (2~45) ❙ -30.5 (-92.4~15.9); 394 (53~659); 176.3 (10.3~309.5)

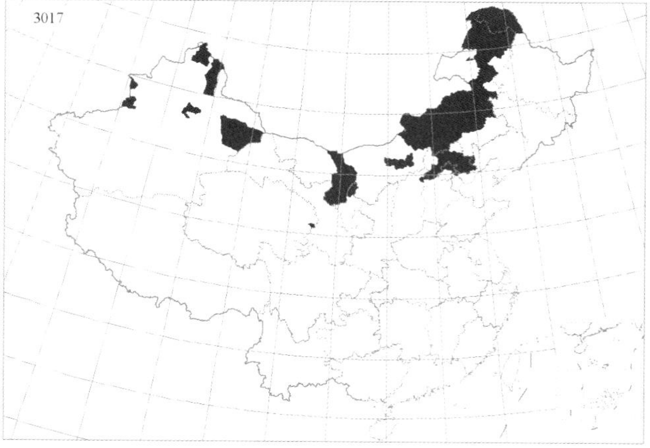

Rosaceae 蔷薇科 *Cotoneaster* 栒子属

Cotoneaster langei G. Klotz 中甸栒子

Shrubs, deciduous or semievergreen

9.8 (2.2~14.6); 2.6 (-6.6~7.8); 15.8 (9.9~20.0); 10.0 (3.8~14.6); 67.2 (15.5~115.2); 9.7 (0.0~49.1); 603 (413~738) ❙ 976 (741~1164); 576 (426~651); 32 (12~58) ❙ 64.0 (30.5~125.9); 598 (413~707); 318.1 (248.3~366.3)

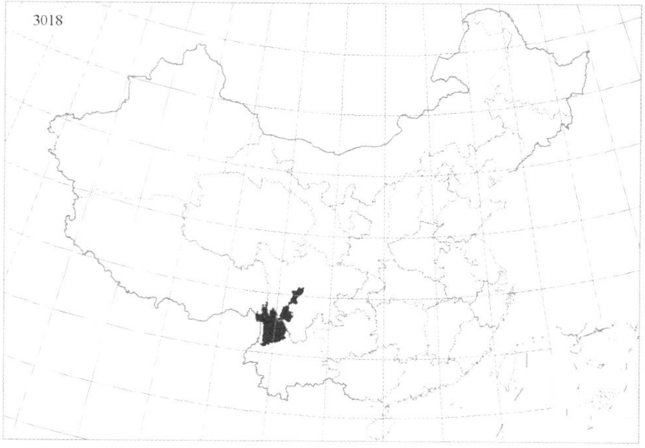

Rosaceae 蔷薇科 *Cotoneaster* 栒子属
Cotoneaster melanocarpus Lodd. 黑果栒子
Shrubs, deciduous

6.0 (-5.8~15.5); -11.9 (-29.8~2.1); 21.7 (13.2~27.8); 9.1 (4.4~15.6); 72.5 (26.7~131.3); 59.5 (4.3~157.4); 621 (426~881) ❙ 428 (53~1077); 267 (33~565); 15 (2~97) ❙ -30.2 (-92.4~57.9); 412 (53~881); 161.7 (10.3~346.5)

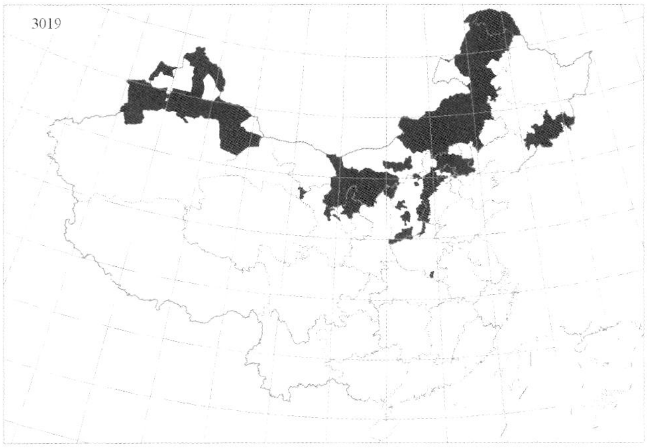

Rosaceae 蔷薇科 *Cotoneaster* 栒子属
Cotoneaster microphyllus Wall. ex Lindl. 小叶栒子
Shrubs, evergreen

12.1 (-1.8~18.9); 4.4 (-10.5~12.5); 18.4 (6.1~26.6); 12.5 (1.8~19.0); 96.8 (1.9~167.8); 10.8 (0.0~82.8); 682 (329~929) ❙ 960 (282~1736); 547 (193~1007); 35 (1~84) ❙ 41.4 (-25.4~145.8); 664 (282~912); 281.8 (61.6~516.7)

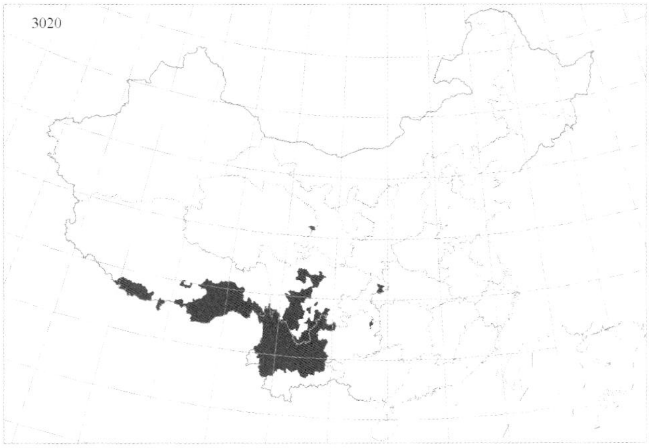

Rosaceae 蔷薇科 *Cotoneaster* 栒子属
Cotoneaster microphyllus var. cochleatus (Franch.) Rehder et E. H. Wilson 白毛小叶栒子
Shrubs, evergreen

8.6; 0.4; 15.5; 9.4; 70.2; 26.7; 590 ❙ 901; 506; 29 ❙ 57.8; 583; 221.9

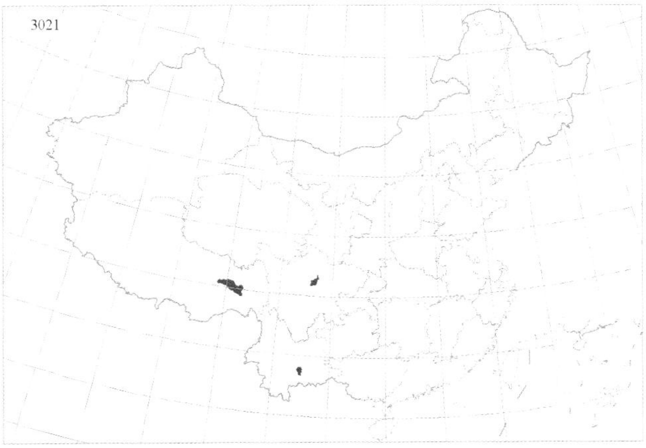

Rosaceae 蔷薇科 *Cotoneaster* 栒子属
Cotoneaster microphyllus var. glacialis Hook. f. ex Wenzig 无毛小叶栒子
Shrubs, evergreen

7.0; -1.4; 14.2; 8.0; 51.6; 27.0; 538 ❙ 1007; 583; 22 ❙ 78.6; 538; 266.6

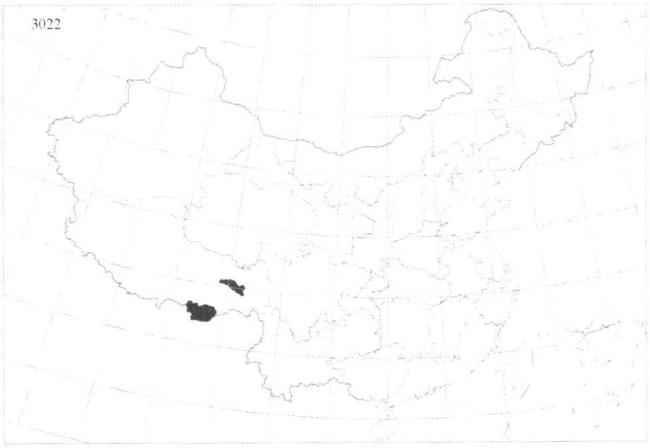

Rosaceae 蔷薇科 *Cotoneaster* 栒子属
Cotoneaster microphyllus var. thymifolius (Baker) Koehne 细叶小叶栒子
Shrubs, evergreen

1.4; -7.2; 9.2; 3.5; 13.8; 56.9; 390 ❙ 477; 286; 18 ❙ 20.9; 385; 176.2

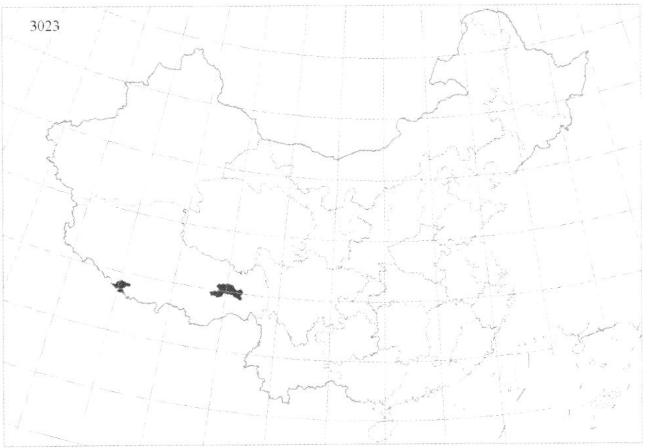

Rosaceae 蔷薇科 *Cotoneaster* 栒子属
Cotoneaster mongolicus Pojark. 蒙古栒子
Shrubs, deciduous

5.3 (0.1~8.8); -12.7 (-21.5~-7.1); 21.2 (15.9~24.3); 8.5 (5.6~10.3); 66.9 (37.7~85.9); 62.6 (35.3~108.6); 592 (465~679) ❙ 257 (49~418); 164 (34~275); 6 (1~12) ❙ -55.3 (-92.8~-20.8); 257 (49~418); 82.5 (5.1~200.5)

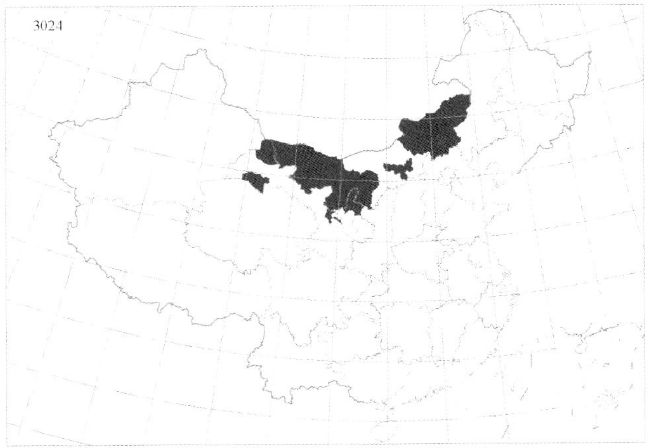

Rosaceae 蔷薇科 *Cotoneaster* 栒子属
Cotoneaster morrisonensis Hayata 台湾栒子
Shrubs, semievergreen

18.2; 12.8; 22.6; 18.3; 159.3; 0.0; 881 ❙ 2661; 1186; 252 ❙ 207.4; 881; 254.2

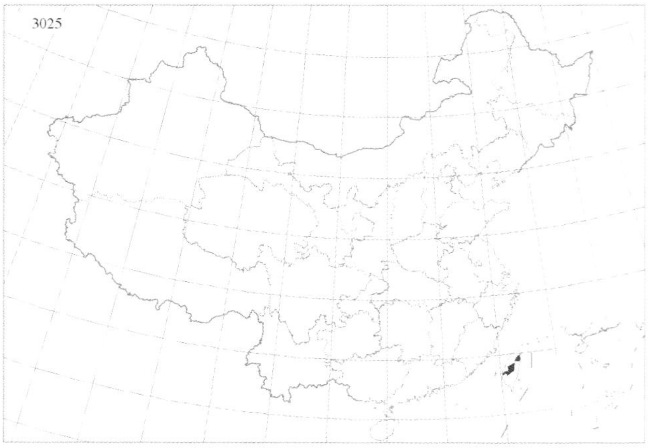

Rosaceae 蔷薇科 *Cotoneaster* 栒子属
Cotoneaster moupinensis Franch. 宝兴栒子
Shrubs, deciduous

11.6 (2.2~16.2); 1.6 (-9.3~7.0); 20.6 (9.9~26.3); 11.8 (3.8~16.2); 88.6 (15.5~134.6); 9.3 (0.0~49.1); 679 (413~854) ❙ 991 (196~1466); 504 (117~817); 41 (4~105) ❙ 47.4 (-69.9~125.9); 662 (196~854); 234.5 (47.6~384.7)

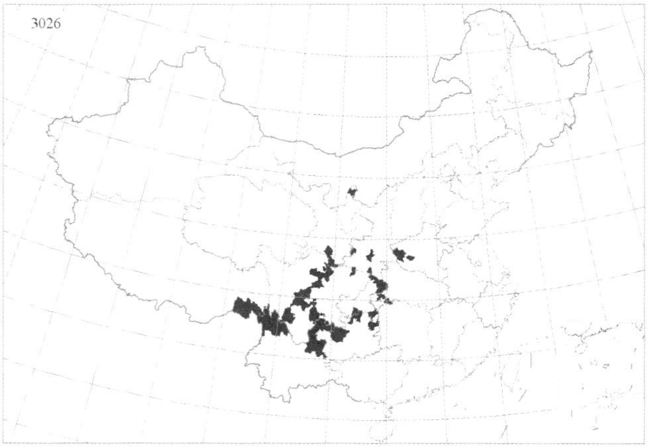

Rosaceae 蔷薇科 *Cotoneaster* 栒子属
Cotoneaster multiflorus Bunge 水栒子
Shrubs, deciduous

8.0 (-5.5~17.5); -6.5 (-21.5~10.9); 20.9 (5.8~28.0); 9.6 (1.3~17.6); 74.9 (1.2~151.2); 38.6 (0.0~127.1); 632 (299~891) ❙ 565 (63~1214); 316 (45~636); 19 (2~109) ❙ -7.9 (-84.9~128.6); 510 (63~891); 181.6 (8.4~364.1)

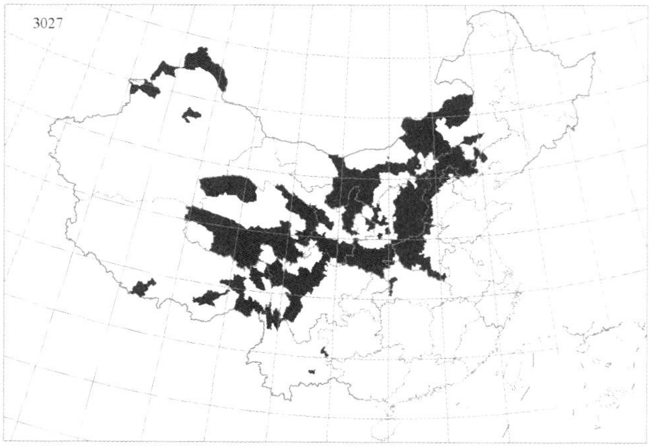

Rosaceae 蔷薇科 *Cotoneaster* 栒子属
Cotoneaster multiflorus var. atropurpureus T. T. Yu 紫果水栒子
Shrubs, deciduous

3.4 (1.8~6.6); -5.5 (-7.2~-1.3); 11.1 (9.9~13.3); 4.7 (3.8~6.8); 22.8 (15.4~39.6); 41.7 (19.7~53.3); 440 (404~514) ❙ 804 (583~943); 465 (348~548); 18 (11~28) ❙ 83.8 (44.2~128.6); 440 (404~514); 284.1 (169.5~357.7)

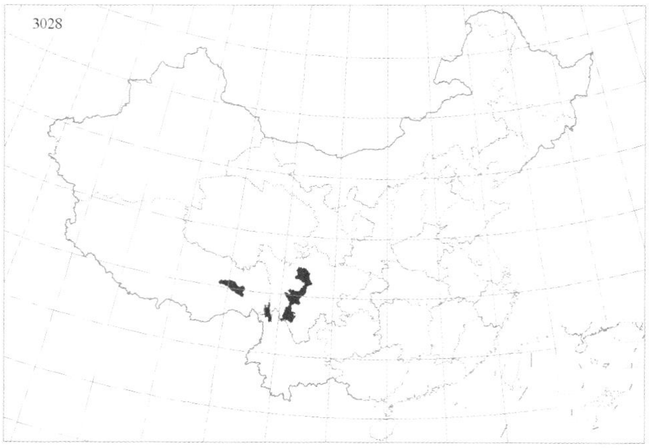

Rosaceae 蔷薇科 *Cotoneaster* 栒子属
Cotoneaster multiflorus var. calocarpus Rehder et E. H. Wilson 大果水栒子
Shrubs, deciduous

8.1 (5.9~12.3); -6.8 (-9.8~-0.8); 21.5 (17.9~24.4); 9.4 (7.3~12.3); 74.2 (51.7~100.7); 36.9 (13.4~47.7); 620 (529~728) ❙ 411 (189~798); 216 (114~381); 13 (4~42) ❙ -33.5 (-70.2~19.6); 391 (189~728); 130.3 (33.5~274.1)

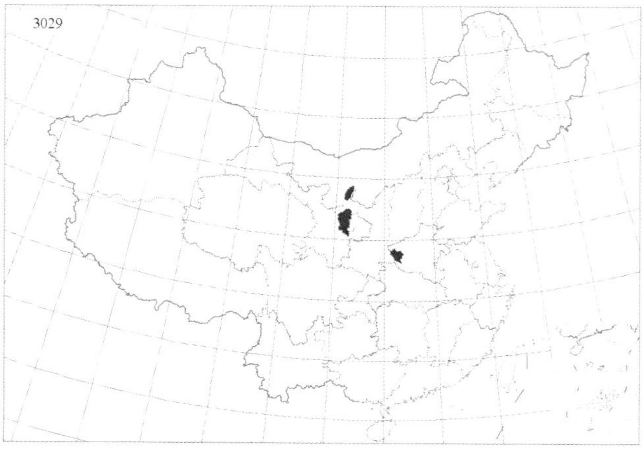

Rosaceae 蔷薇科 *Cotoneaster* 栒子属
Cotoneaster nitens Rehder et E. H. Wilson 光泽栒子
Shrubs, deciduous

5.3 (2.6~14.9); -4.3 (-7.1~5.3); 13.7 (11.1~23.6); 6.5 (4.3~15.0); 40.7 (20.0~119.6); 36.9 (0.0~48.2); 500 (427~782) ❙ 956 (791~1410); 513 (400~816); 26 (17~49) ❙ 92.9 (80.4~121.7); 500 (427~782); 286.7 (214.5~357.7)

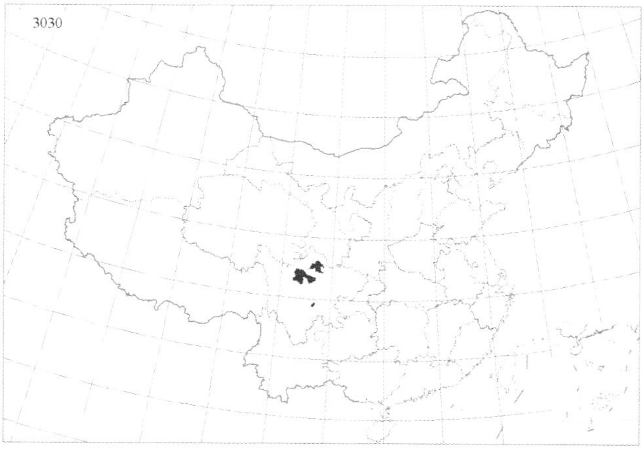

Rosaceae 蔷薇科 *Cotoneaster* 栒子属
Cotoneaster nitidifolius C. Marquand 亮叶栒子
Shrubs, deciduous

3.5; -6.5; 2.6; 5.0; 26.3; 43.4; 446 ▎731; 346; 19 ▎63.8; 446; 284.1

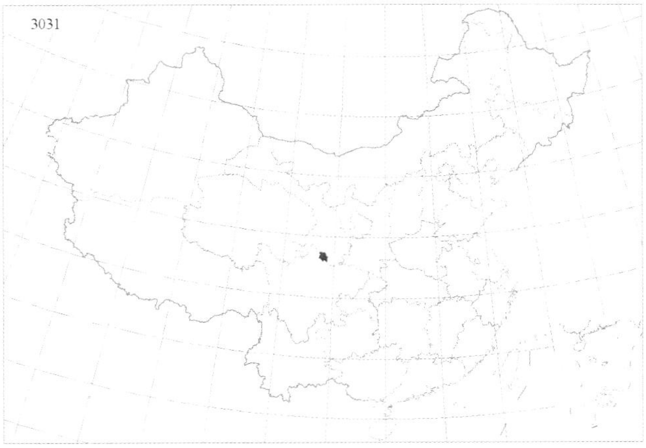

Rosaceae 蔷薇科 *Cotoneaster* 栒子属
Cotoneaster nitidus Jacquem. 两列栒子
Shrubs, deciduous or semievergreen

14.3 (2.2~18.9); 7.3 (-6.6~12.5); 19.7 (9.9~24.6); 14.4 (3.8~19.0); 115.0 (15.5~167.5); 2.5 (0.0~49.1); 739 (413~914) ▎1067 (741~1736); 607 (426~1007); 44 (12~70) ▎46.6 (12.2~145.8); 729 (413~851); 317.0 (189.1~487.2)

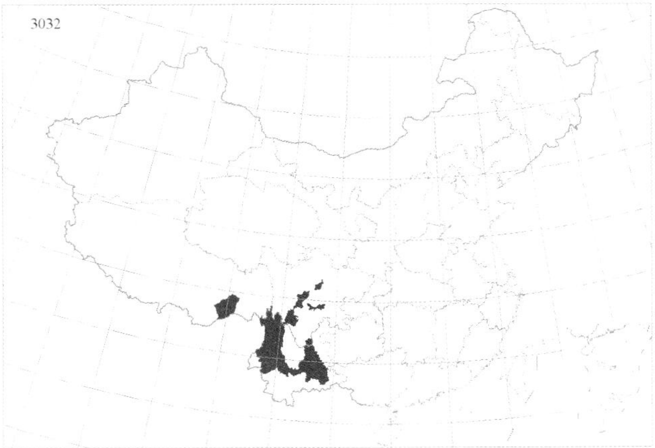

Rosaceae 蔷薇科 *Cotoneaster* 栒子属
Cotoneaster obscurus Rehder et E. H. Wilson 暗红栒子
Shrubs, deciduous

7.2 (0.3~15.3); -2.2 (-9.9~5.4); 15.6 (8.3~26.5); 8.0 (2.9~15.4); 53.5 (9.6~125.7); 27.1 (0.0~67.5); 551 (368~832) ▎959 (389~1736); 504 (246~1007); 34 (7~113) ▎74.2 (0.0~145.8); 551 (368~832); 262.4 (142.5~447.4)

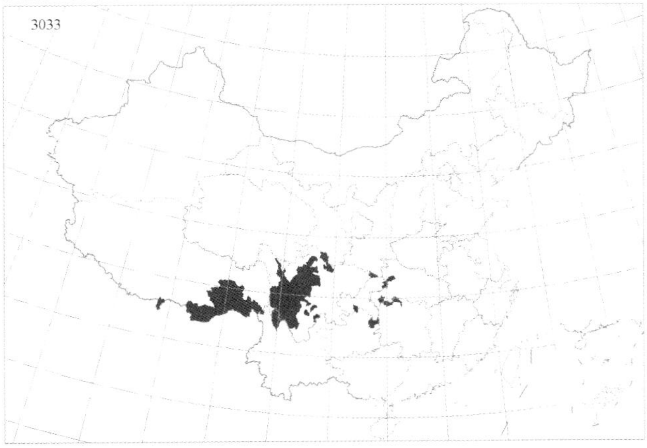

Rosaceae 蔷薇科 *Cotoneaster* 栒子属
Cotoneaster oliganthus Pojark. 少花栒子
Shrubs, deciduous

4.0 (-2.8~9.0); -14.6 (-20.9~-9.3); 19.8 (10.9~24.4); 7.9 (3.5~10.7); 60.4 (16.3~88.9); 71.5 (40.7~109.2); 576 (390~703) ▎225 (53~406); 110 (33~211); 18 (2~51) ▎-58.9 (-92.4~-7.6); 225 (53~406); 117.7 (10.3~297.3)

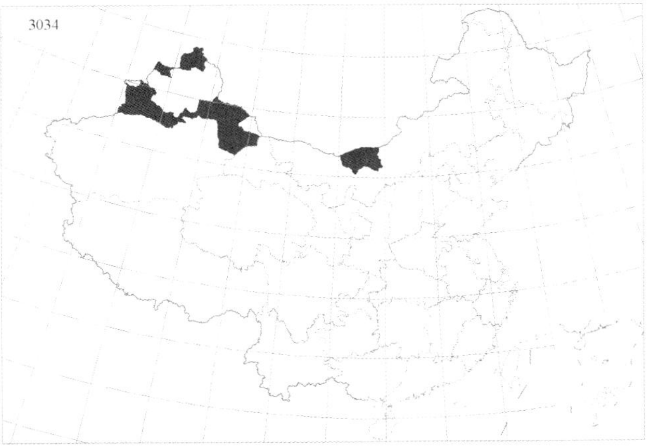

Rosaceae 蔷薇科 *Cotoneaster* 栒子属
Cotoneaster pannosus Franch. 毡毛栒子
Shrubs, semievergreen

13.9 (2.2~18.9); 6.5 (-6.6~12.5); 19.7 (9.5~26.6); 14.0 (3.8~19.0); 110.8 (15.5~167.8); 3.8 (0.0~49.1); 729 (411~929) ▎1011 (628~1454); 576 (346~816); 38 (11~70) ▎41.2 (-14.2~125.9); 710 (411~912); 296.7 (142.9~513.7)

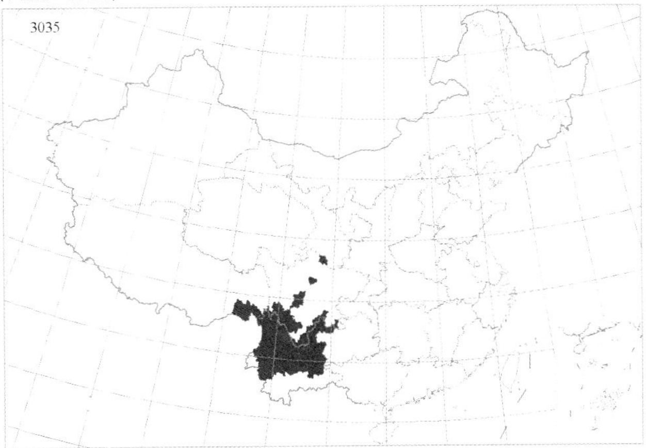

Rosaceae 蔷薇科 *Cotoneaster* 栒子属
Cotoneaster reticulatus Rehder et E. H. Wilson 网脉栒子
Shrubs, deciduous

6.0; -3.2; 14.4; 6.6; 39.7; 26.7; 508 ▎1023; 535; 28 ▎101.2; 508; 200.8

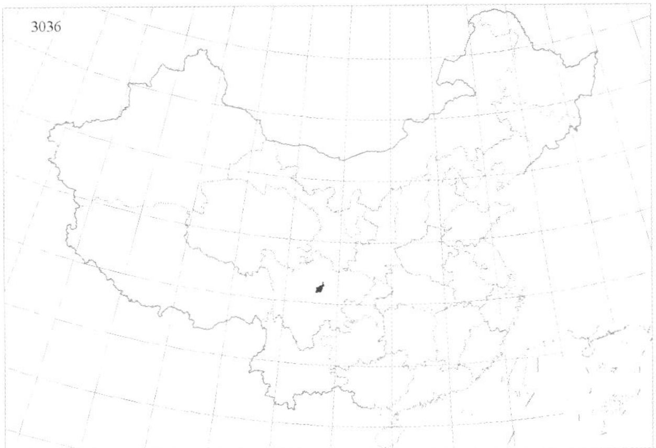

Rosaceae 蔷薇科 *Cotoneaster* 栒子属
Cotoneaster rhytidophyllus Rehder et E. H. Wilson 麻叶栒子
Shrubs, evergreen or semievergreen

13.5 (6.2~16.2); 3.6 (−2.4~8.0); 22.7 (14.1~25.6); 13.6 (6.7~16.3); 106.4 (39.4~135.7); 3.7 (0.0~24.1); 740 (509~845) ❙ 1150 (916~1414); 552 (486~817); 56 (23~88) ❙ 56.9 (24.9~93.8); 740 (509~845); 208.6 (154.4~298.7)

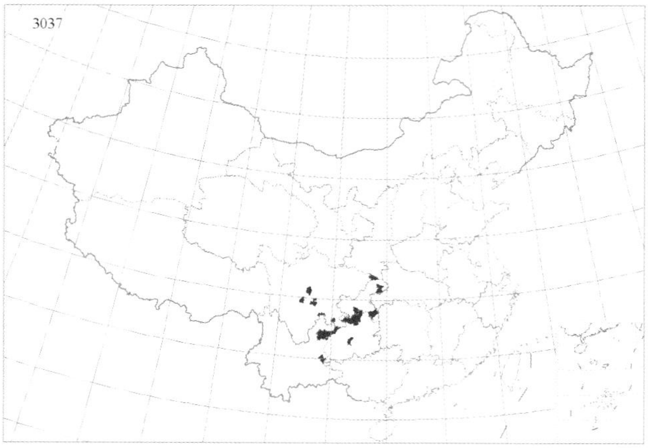

Rosaceae 蔷薇科 *Cotoneaster* 栒子属
Cotoneaster rotundifolius Wall. ex Lindl. 圆叶栒子
Shrubs, evergreen

7.0 (−1.8~17.3); −2.2 (−10.2~11.0); 15.3 (6.1~23.6); 8.1 (1.8~17.3); 55.8 (1.9~148.0); 31.4 (0.0~82.8); 545 (329~832) ❙ 811 (320~1441); 439 (217~816); 27 (0~84) ❙ 41.9 (−26.0~121.7); 522 (320~783); 224.3 (72.2~365.3)

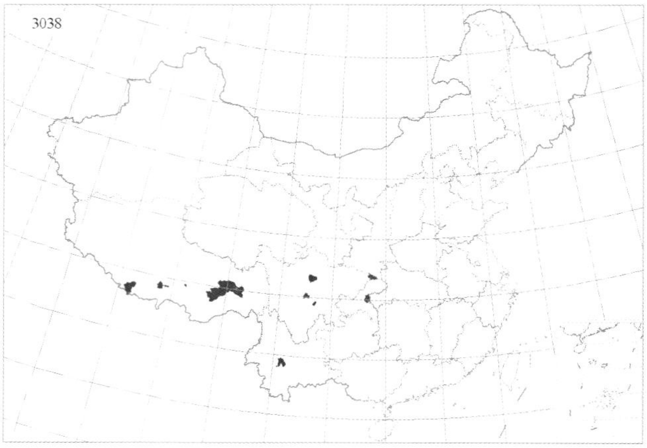

Rosaceae 蔷薇科 *Cotoneaster* 栒子属
Cotoneaster rubens W. W. Sm. 红花栒子
Shrubs, semievergreen

5.6 (−0.4~12.2); −3.7 (−10.0~4.6); 13.9 (8.4~19.9); 6.7 (2.8~12.2); 41.4 (9.4~87.4); 33.5 (0.4~73.3); 503 (360~672) ❙ 685 (264~1431); 405 (190~817); 18 (1~58) ❙ 33.1 (−40.9~125.9); 471 (264~672); 240.2 (61.2~384.7)

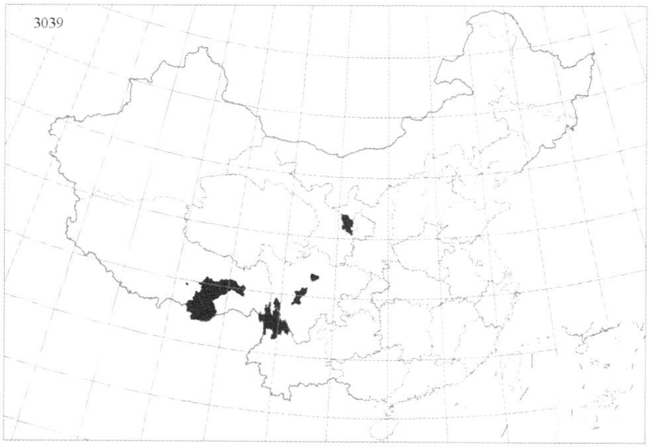

Rosaceae 蔷薇科 *Cotoneaster* 栒子属
Cotoneaster rubens var. minimus T. T. Yu 小叶红花栒子
Shrubs, semievergreen

4.2; −4.9; 12.4; 5.3; 28.4; 37.1; 463 ❙ 957; 508; 24 ❙ 107.9; 463; 244.8

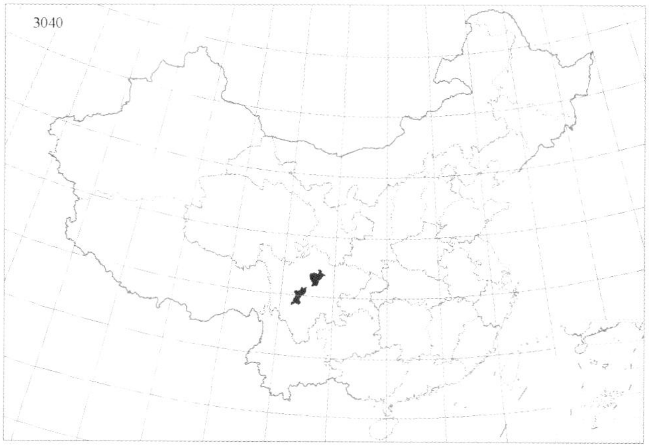

Rosaceae 蔷薇科 *Cotoneaster* 栒子属
Cotoneaster salicifolius Franch. 柳叶栒子
Shrubs, semievergreen or green

12.7 (2.2~16.2); 2.5 (−7.0~7.9); 22.2 (9.9~26.8); 12.8 (3.8~16.3); 99.3 (15.5~135.7); 6.6 (0.0~49.1); 716 (413~859) ❙ 1166 (669~1492); 535 (317~816); 60 (15~113) ❙ 64.5 (24.9~125.9); 716 (413~859); 223.7 (146.0~352.3)

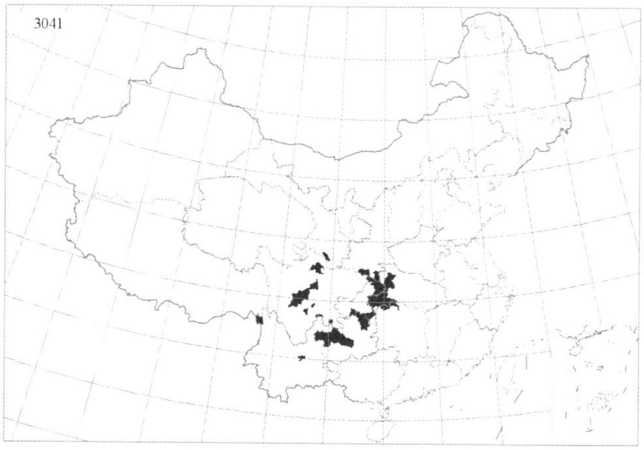

Rosaceae 蔷薇科 *Cotoneaster* 栒子属
Cotoneaster salicifolius var. angustus T. T. Yu 窄叶柳叶栒子
Shrubs, semievergreen or green

11.9; 2.1; 20.7; 12.1; 92.4; 9.1; 693 ❙ 1123; 631; 34 ❙ 66.7; 693; 188.7

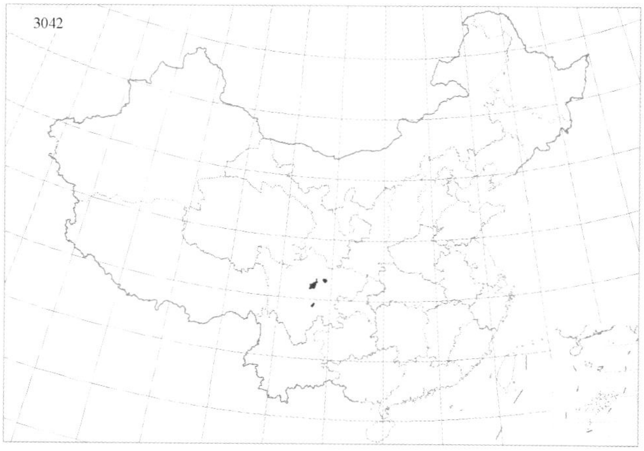

Rosaceae 蔷薇科 *Cotoneaster* 栒子属
Cotoneaster salicifolius var. henryanus (C. K. Schneid.) T. T. Yu 大叶柳叶栒子
Shrubs, semievergreen or green

12.9; 1.7; 23.6; 12.9; 101.4; 6.7; 724 ❙ 1234; 515; 83 ❙ 70.7; 724; 263.2

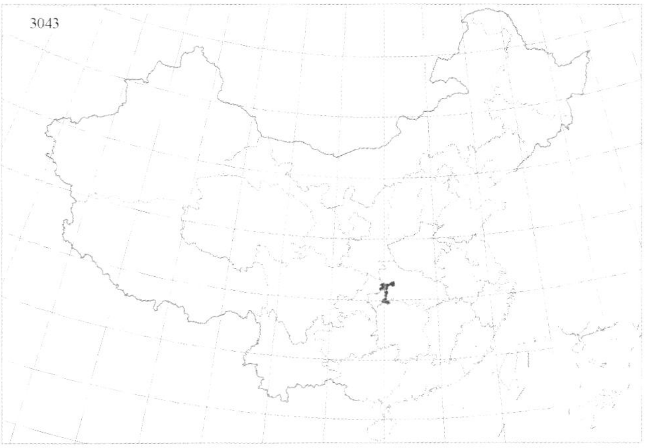

Rosaceae 蔷薇科 *Cotoneaster* 栒子属
Cotoneaster salicifolius var. rugosus (E. Pritz.) Rehder et E. H. Wilson 皱叶柳叶栒子
Shrubs, semievergreen or green

12.2 (7.3~14.9); 1.6 (-2.0~5.3); 22.2 (15.6~25.7); 12.2 (7.5~15.0); 94.2 (48.2~121.2); 7.7 (0.0~20.4); 698 (537~807) ❙ 1274 (964~1492); 583 (421~816); 68 (27~105) ❙ 82.7 (31.7~107.2); 698 (537~807); 248.1 (211.6~285.1)

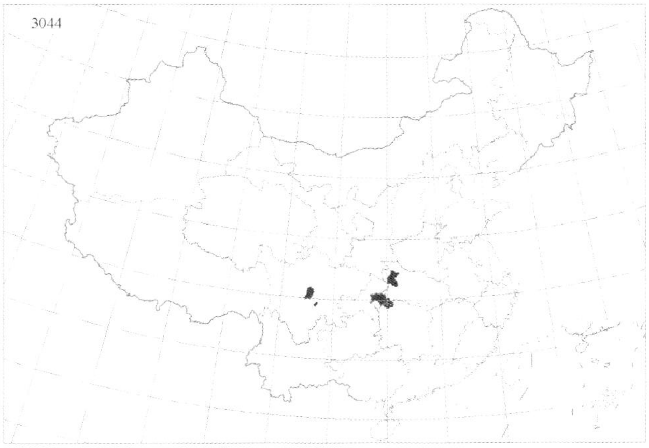

Rosaceae 蔷薇科 *Cotoneaster* 栒子属
Cotoneaster sanguineus T. T. Yu 血色栒子
Shrubs, deciduous

3.3; -5.0; 10.6; 5.2; 28.4; 48.2; 445 ❙ 760; 457; 22 ❙ 60.1; 441; 190.7

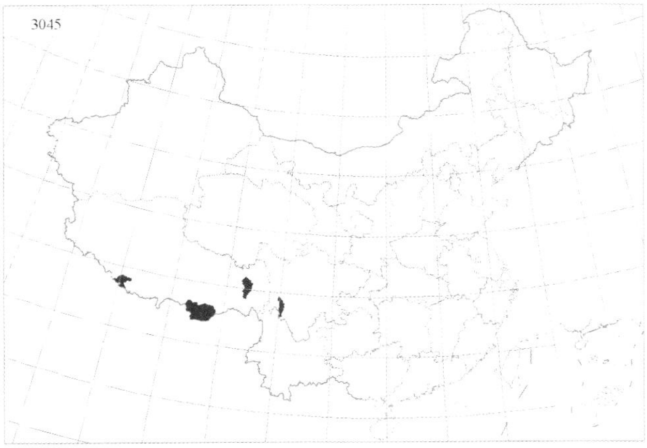

Rosaceae 蔷薇科 *Cotoneaster* 栒子属
Cotoneaster schantungensis G. Klotz 山东栒子
Shrubs, deciduous

13.3; -1.3; 26.4; 13.6; 114.7; 14.9; 811 ❙ 688; 413; 37 ❙ -14.7; 688; 150.1

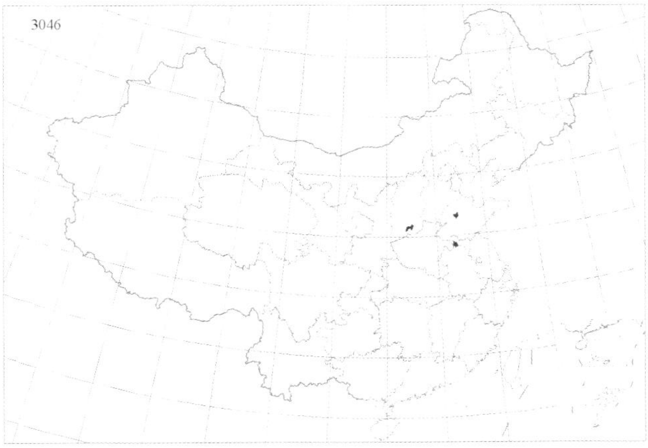

Rosaceae 蔷薇科 *Cotoneaster* 栒子属
Cotoneaster sherriffii G. Klotz 康巴栒子
Shrubs, semievergreen

2.4 (-1.5~6.6); -6.5 (-10.5~-1.3); 10.3 (7.1~13.3); 4.2 (2.2~6.8); 19.8 (4.7~39.6); 50.9 (19.7~82.2); 414 (332~514) ❙ 506 (273~961); 323 (208~558); 11 (1~31) ❙ 18.0 (-30.3~100.7); 380 (273~514); 191.9 (62.8~311.5)

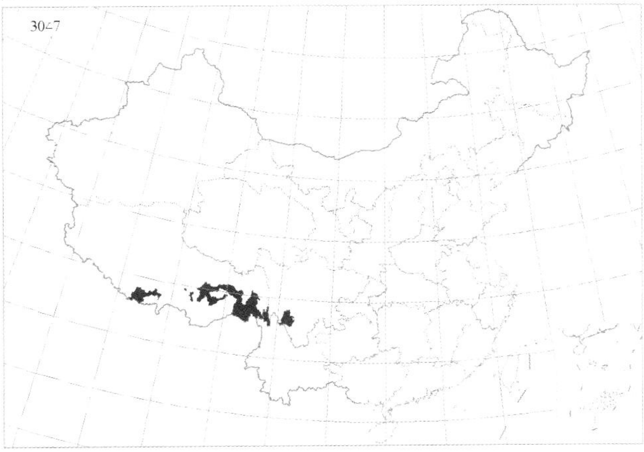

Rosaceae 蔷薇科 *Cotoneaster* 栒子属
Cotoneaster silvestrii Pamp. 华中栒子
Shrubs, deciduous

13.1 (-1.4~17.2); 0.6 (-15.4~5.1); 24.9 (9.3~29.2); 13.4 (3.3~17.2); 110.0 (11.9~146.8); 12.0 (0.0~92.1); 781 (375~941) ❙ 1055 (282~1828); 467 (182~666); 87 (5~253) ❙ 36.4 (-24.8~128.6); 757 (282~936); 203.1 (99.3~357.7)

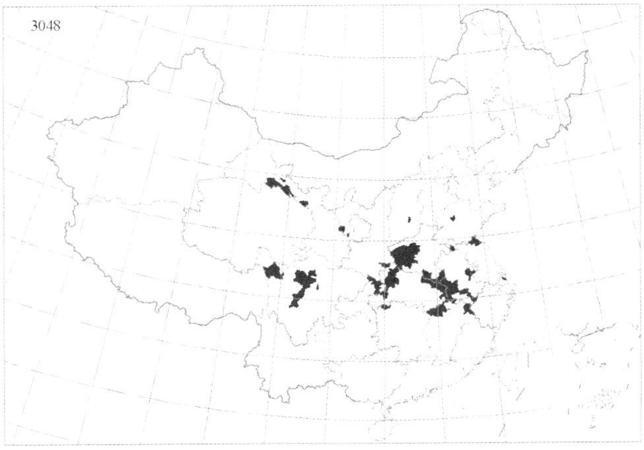

Rosaceae 蔷薇科 *Cotoneaster* 栒子属
Cotoneaster soongoricus (Regel et Herder) Popov 准噶尔栒子
Shrubs, deciduous

4.7 (-4.1~11.4); -10.4 (-20.9~1.3); 17.6 (7.2~26.5); 7.3 (2.0~12.7); 53.3 (3.7~109.3); 56.3 (8.8~114.2); 547 (335~793) ❙ 403 (32~1041); 225 (21~585); 14 (0~51) ❙ -17.0 (-95.9~128.6); 311 (32~577); 151.3 (1.6~357.7)

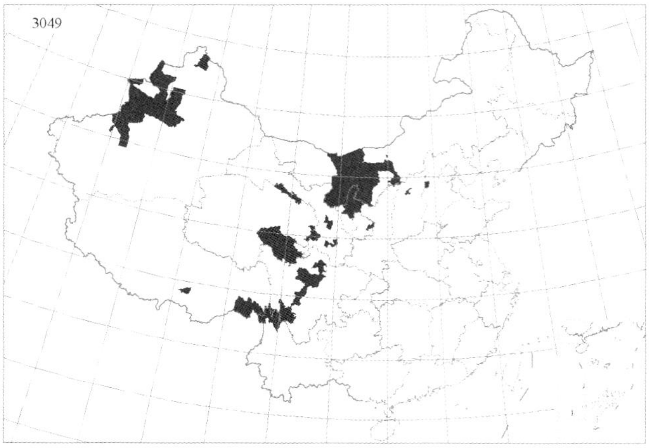

Rosaceae 蔷薇科 *Cotoneaster* 栒子属
Cotoneaster soongoricus var. microcarpus (Rehder et E. H. Wilson) G. Klotz 小果准噶尔栒子 Shrubs, deciduous

2.6; -7.0; 11.2; 4.3; 20.0; 48.2; 427 ❙ 829; 400; 23 ❙ 94.4; 427; 252.2

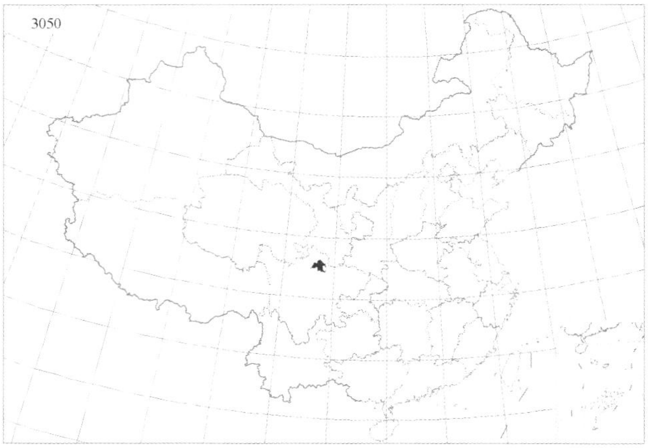

Rosaceae 蔷薇科 *Cotoneaster* 栒子属
Cotoneaster subadpressus T. T. Yu 高山栒子
Shrubs, deciduous or semievergreen

8.3 (5.4~11.2); 1.0 (-2.1~4.6); 14.6 (12.3~16.7); 8.4 (5.8~11.3); 52.4 (31.5~75.8); 12.4 (0.4~25.6); 561 (481~642) ❙ 906 (741~1054); 545 (426~627); 27 (12~40) ❙ 61.3 (52.8~67.9); 561 (481~642); 315.2 (273.5~351.7)

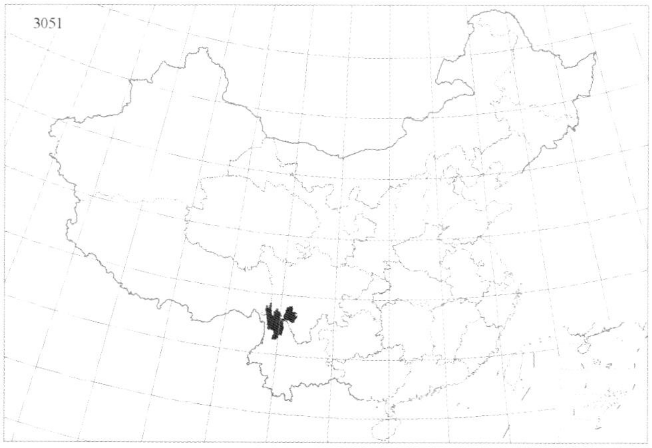

Rosaceae 蔷薇科 *Cotoneaster* 栒子属
Cotoneaster submultiflorus Popov 毛叶水栒子
Shrubs, deciduous

5.5 (-4.1~13.5); -9.0 (-20.9~2.4); 18.2 (7.1~27.1); 7.7 (2.0~13.7); 56.6 (3.7~117.5); 50.2 (5.4~114.2); 556 (332~820) ❙ 460 (128~1246); 263 (62~578); 14 (0~48) ❙ -13.4 (-80.8~128.6); 407 (128~733); 166.7 (13.6~351.7)

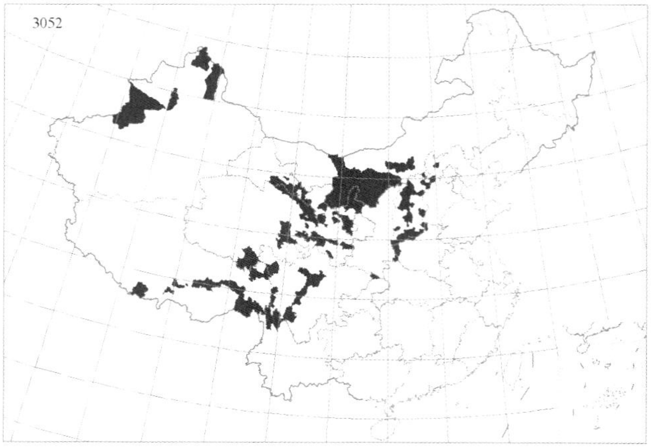

Rosaceae 蔷薇科 *Cotoneaster* 栒子属
Cotoneaster taylorii T. T. Yu 藏南栒子
Shrubs or small trees, deciduous

1.6 (0.0~2.9); -7.4 (-8.7~-5.7); 9.9 (8.1~11.3); 3.7 (2.7~4.4); 15.8 (8.0~21.4); 56.0 (45.2~67.9); 400 (363~423) ❙ 501 (341~560); 308 (226~340); 10 (4~14) ❙ 26.1 (-15.3~54.1); 387 (341~423); 192.4 (111.0~237.6)

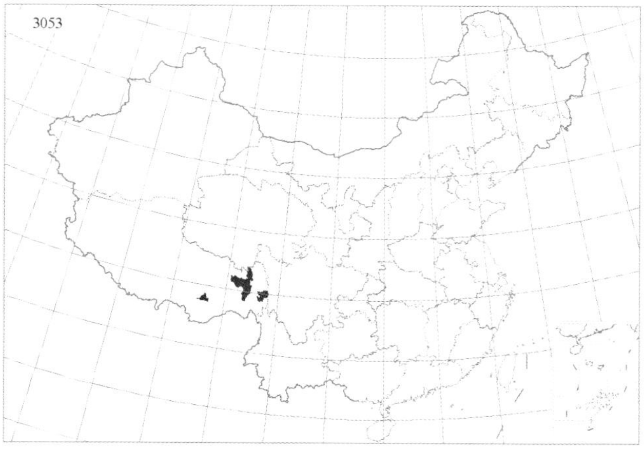

Rosaceae 蔷薇科 *Cotoneaster* 栒子属
Cotoneaster tenuipes Rehder et E. H. Wilson 细枝栒子
Shrubs, deciduous

8.1 (-1.4~15.5); -2.8 (-15.4~8.6); 17.9 (8.5~25.4); 9.1 (2.6~15.6); 65.6 (8.1~127.3); 28.2 (0.0~92.1); 591 (347~817) ❙ 759 (132~1180); 398 (81~651); 22 (2~65) ❙ 32.5 (-80.1~128.6); 536 (132~817); 256.5 (13.6~384.1)

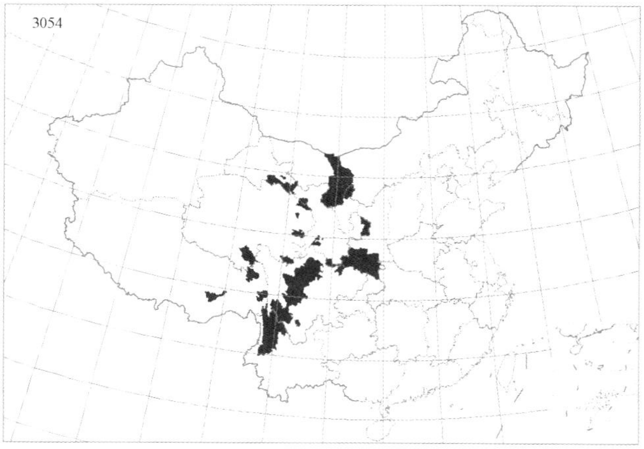

Rosaceae 蔷薇科 *Cotoneaster* 栒子属
Cotoneaster turbinatus Craib 陀螺果栒子
Shrubs, evergreen

14.5 (5.4~13.9); 7.5 (-2.1~12.0); 19.9 (12.3~24.3); 14.5 (5.8~19.0); 115.7 (31.5~167.8); 1.3 (0.0~25.6); 744 (481~929) ❙ 968 (741~1478); 563 (411~627); 35 (15~91) ❙ 31.7 (-14.2~94.5); 710 (481~815); 280.7 (190.3~372.3)

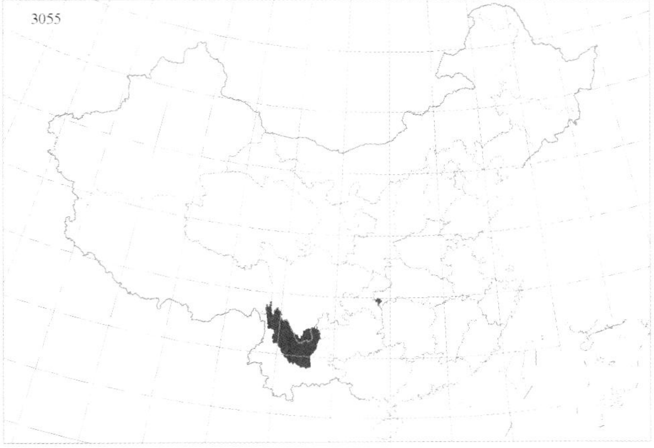

Rosaceae 蔷薇科 *Cotoneaster* 栒子属
Cotoneaster uniflorus Bunge 单花栒子
Shrubs, deciduous

3.4 (-1.3~9.0); -15.3 (-20.9~-8.7); 19.0 (11.0~24.4); 7.5 (3.8~10.7); 56.5 (17.4~88.9); 75.7 (40.7~104.2); 562 (407~703) ❙ 232 (53~450); 105 (33~250); 22 (2~51) ❙ -56.3 (-92.4~3.9); 230 (53~407); 118.0 (10.3~297.3)

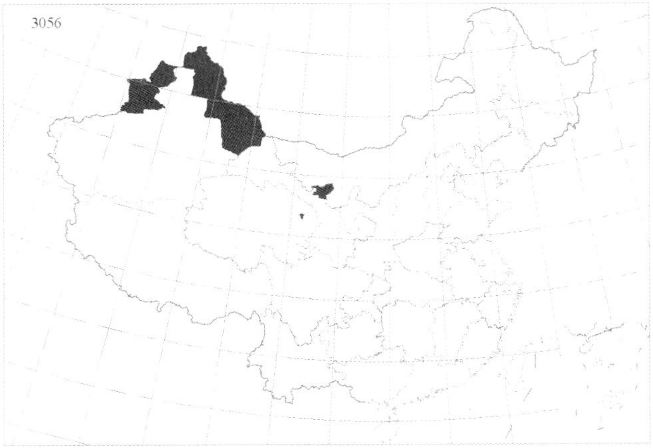

Rosaceae 蔷薇科 *Cotoneaster* 栒子属
Cotoneaster verruculosus Diels 疣枝栒子
Shrubs, deciduous or semievergreen

11.0 (5.3~17.3); 3.9 (-2.4~11.0); 16.8 (12.2~21.6); 11.1 (5.7~17.3); 78.9 (30.5~148.0); 6.3 (0.0~26.2); 638 (479~832) ❙ 1082 (685~1736); 621 (406~1007); 42 (15~70) ❙ 69.8 (26.1~145.8); 634 (479~783); 339.3 (201.9~487.2)

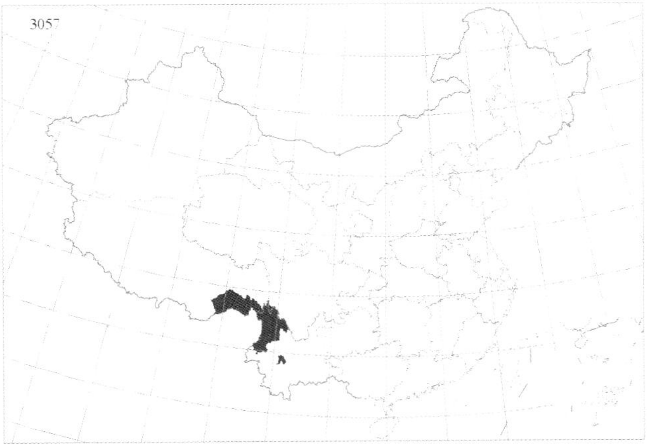

Rosaceae 蔷薇科 *Cotoneaster* 栒子属
Cotoneaster wardii W. W. Sm. 白毛栒子
Shrubs, evergreen

8.5; 1.2; 14.7; 8.6; 53.5; 11.5; 565 ❙ 884; 545; 23 ❙ 57.1; 565; 301.3

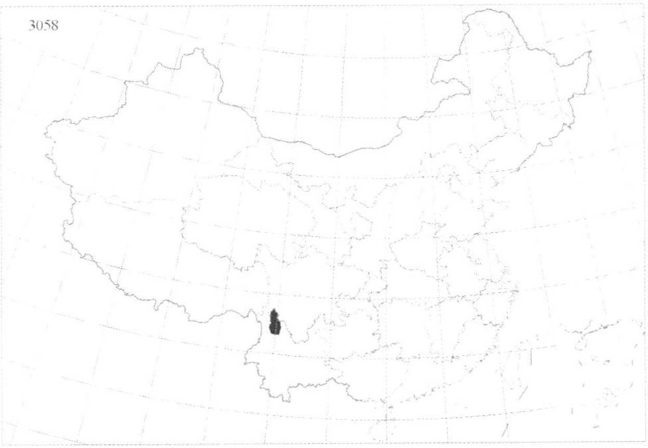

Rosaceae 蔷薇科 *Cotoneaster* 栒子属
Cotoneaster zabelii C. K. Schneid. 西北栒子
Shrubs, deciduous

10.4 (-2.6~16.2); -4.0 (-15.4~4.4); 23.5 (8.5~28.0); 11.3 (2.5~16.2); 92.0 (7.7~135.6); 26.1 (0.6~98.0); 702 (349~881) ❙ 638 (132~1451); 351 (81~600); 25 (2~140) ❙ -8.5 (-80.1~101.4); 586 (132~881); 176.6 (13.6~334.4)

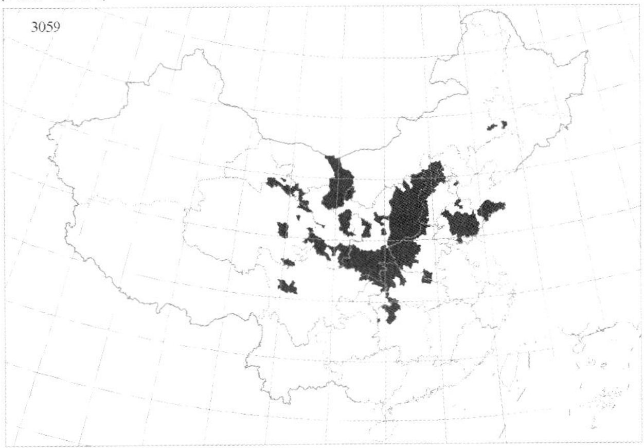

Rosaceae 蔷薇科 *Crataegus* 山楂属
Crataegus altaica (Loudon) Lange 阿尔泰山楂
Trees, deciduous

3.2 (-1.3~9.0); -15.5 (-20.9~-9.3); 18.8 (13.4~23.2); 7.4 (4.7~10.7); 55.4 (28.1~88.9); 75.9 (40.7~104.2); 557 (443~703) ❙ 273 (163~509); 120 (62~304); 27 (12~51) ❙ -50.6 (-71.6~-10.6); 273 (163~509); 154.1 (19.8~297.3)

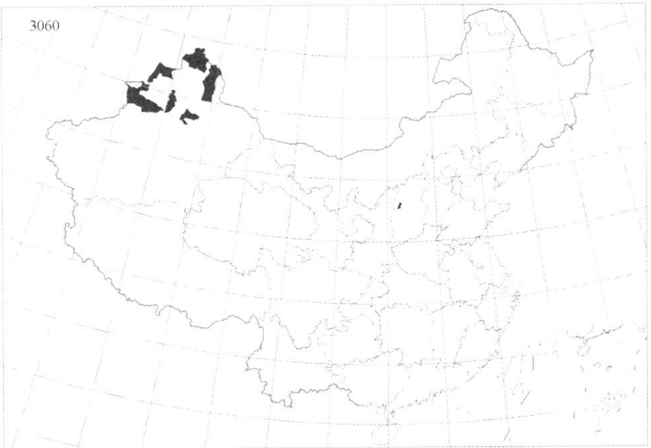

Rosaceae 蔷薇科　*Crataegus* 山楂属

Crataegus aurantia Pojark. 橘红山楂

Shrubs or small trees, deciduous

10.7 (3.3~15.1); -4.4 (-12.7~1.8); 24.1 (17.5~28.0); 11.5 (6.6~15.2); 94.8 (45.4~128.0); 26.3 (5.6~65.0); 715 (507~880) ❙ 577 (397~817); 331 (248~453); 20 (9~53) ❙ -18.7 (-43.4~19.6); 572 (397~817); 164.0 (70.3~274.1)

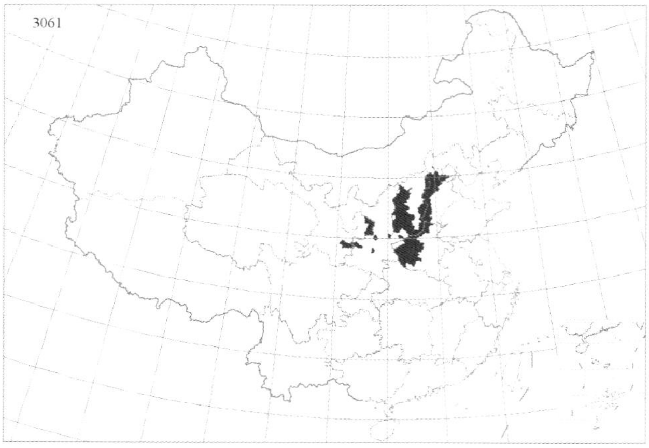

Rosaceae 蔷薇科　*Crataegus* 山楂属

Crataegus chungtienensis W. W. Smith 中甸山楂

Shrubs

6.1; -1.5; 12.7; 6.3; 35.2; 21.8; 496 ❙ 801; 496; 21 ❙ 61.4; 496; 282.9

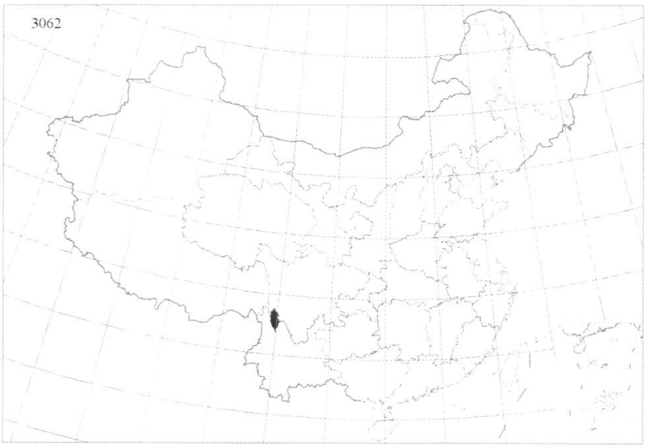

Rosaceae 蔷薇科　*Crataegus* 山楂属

Crataegus cuneata Siebold et Zucc. 野山楂

Shrubs, deciduous

15.6 (5.2~21.8); 3.3 (-10.7~14.2); 27.3 (17.9~30.1); 15.7 (7.7~21.9); 132.4 (57.5~202.6); 4.4 (0.0~54.8); 869 (550~1150) ❙ 1222 (478~1997); 487 (258~891); 132 (14~259) ❙ 39.4 (-36.5~147.0); 835 (478~1150); 199.7 (110.1~382.6)

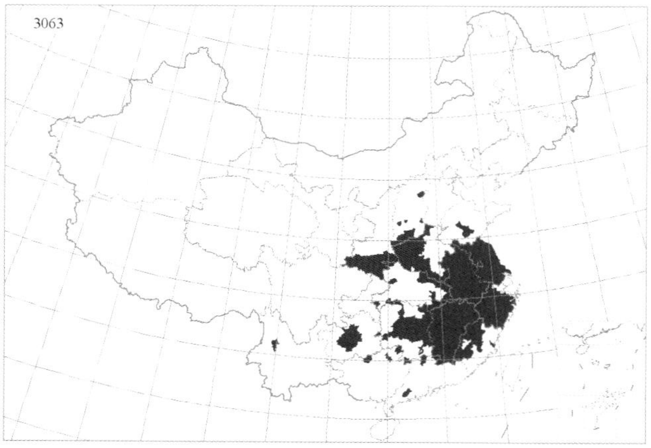

Rosaceae 蔷薇科　*Crataegus* 山楂属

Crataegus dahurica Koehne ex C. K. Schneid. 光叶山楂

Shrubs or small trees, deciduous

-1.3 (-5.8~8.4); -23.3 (-29.8~-6.4); 18.3 (15.3~21.6); 6.0 (4.4~9.7); 44.4 (28.0~74.2); 119.2 (32.8~157.4); 494 (426~624) ❙ 484 (372~702); 320 (258~409); 14 (9~27) ❙ -2.7 (-28.6~15.9); 461 (372~624); 272.0 (183.0~309.5)

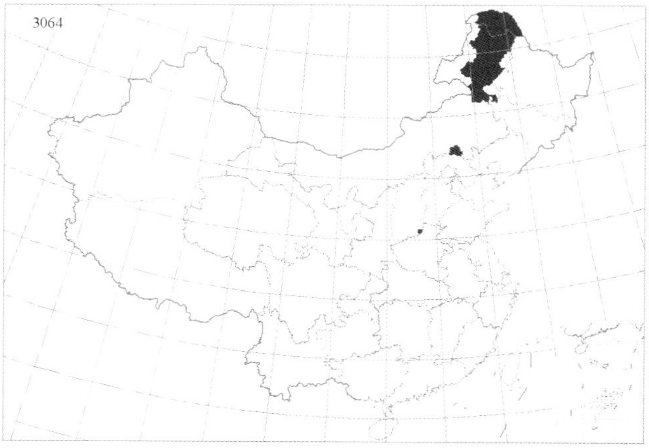

Rosaceae 蔷薇科　*Crataegus* 山楂属

Crataegus dahurica var. laevicalyx (J. X. Huang et al.)T. C. Ku et Spongberg 光萼山楂

Shrubs or small trees, deciduous

2.5; -15.6; 18.6; 6.9; 48.9; 78.3; 523 ❙ 457; 309; 9 ❙ -12.6; 457; 247.3

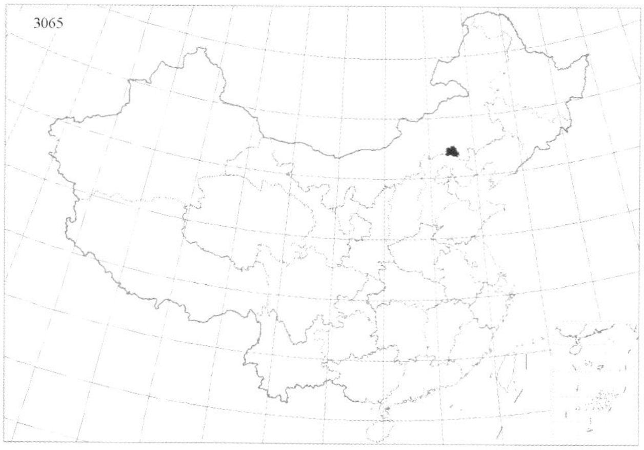

Rosaceae 蔷薇科　*Crataegus* 山楂属

Crataegus hupehensis Sarg. 湖北山楂

Shrubs or trees, deciduous

13.5 (6.0~17.3); 0.6 (-7.6~6.4); 25.6 (14.4~29.3); 13.7 (6.6~17.3); 112.3 (39.7~147.3); 10.0 (0.0~40.5); 786 (508~941) ❙ 1001 (375~1689); 434 (203~685); 78 (8~218) ❙ 26.3 (-36.5~108.0); 748 (375~941); 210.5 (95.7~334.4)

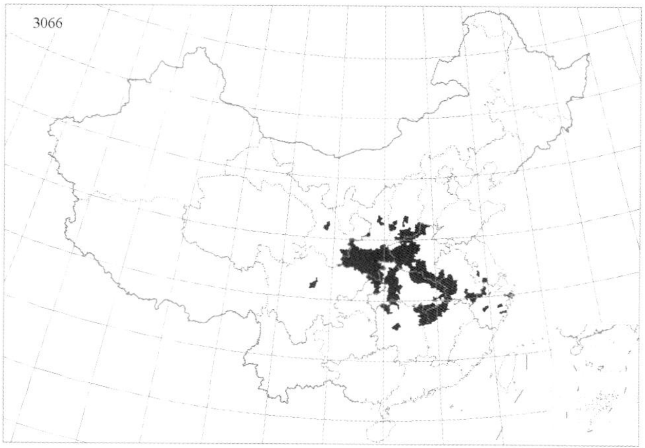

Rosaceae 蔷薇科 *Crataegus* 山楂属
Crataegus kansuensis E. H. Wilson 甘肃山楂
Shrubs or trees, deciduous

9.2 (0.2~15.4); -5.1 (-13.8~4.8); 22.0 (9.9~27.8); 10.3 (3.6~15.5); 81.9 (15.4~128.0); 30.9 (0.2~74.6); 660 (406~865) ❙ 598 (252~1255); 324 (144~546); 20 (5~79) ❙ -7.5 (-59.3~128.6); 548 (252~820); 182.5 (44.5~357.7)

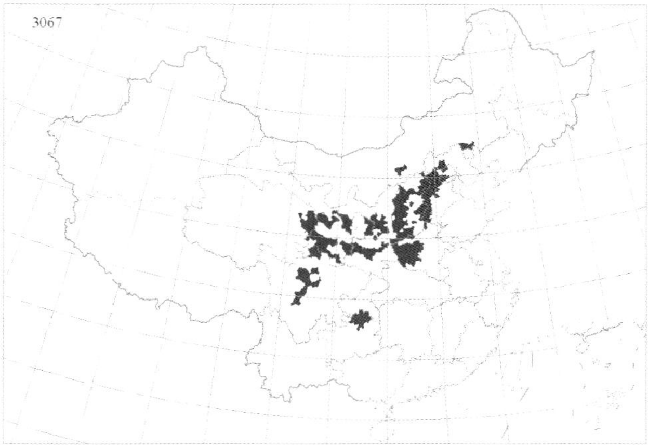

Rosaceae 蔷薇科 *Crataegus* 山楂属
Crataegus maximowiczii C. K. Schneid. 毛山楂
Shrubs or small trees

5.1 (-5.8~13.5); -13.0 (-29.8~-1.2); 21.2 (15.3~27.1); 8.7 (4.4~13.7); 67.7 (28.0~117.5); 65.8 (14.6~157.4); 599 (426~820) ❙ 503 (137~856); 303 (92~447); 15 (3~33) ❙ -15.8 (-77.5~52.6); 487 (137~702); 201.7 (25.4~333.1)

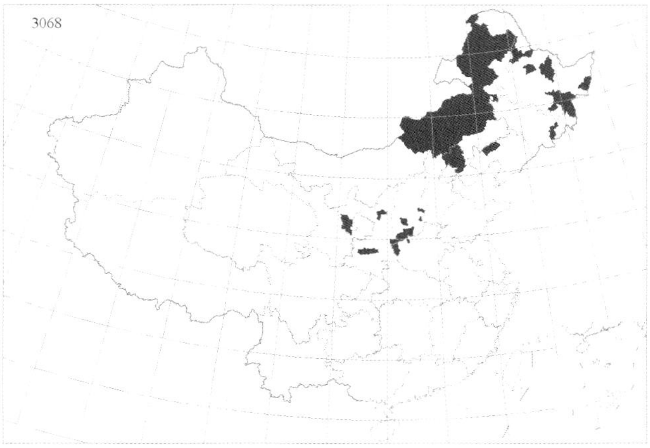

Rosaceae 蔷薇科 *Crataegus* 山楂属
Crataegus oresbia W. W. Smith 滇西山楂
Shrubs

8.7; 1.5; 14.9; 8.8; 54.7; 9.9; 573 ❙ 918; 559; 25 ❙ 60.4; 571; 311.7

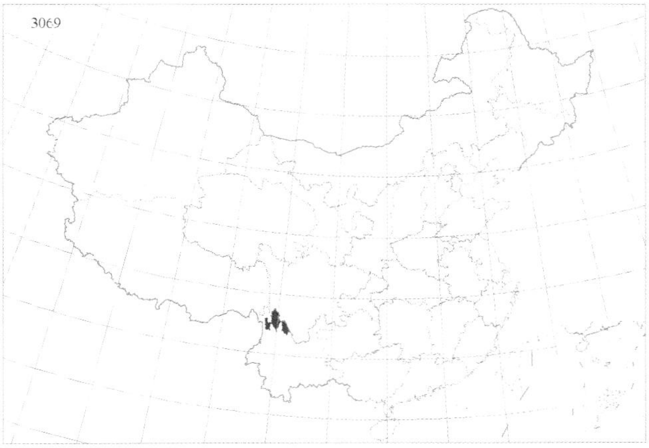

Rosaceae 蔷薇科 *Crataegus* 山楂属
Crataegus pinnatifida Bunge 山楂
Trees, deciduous

10.5 (-5.8~16.2); -5.6 (-29.8~4.3); 24.8 (15.3~28.5); 11.8 (4.4~16.3); 98.6 (28.0~136.5); 31.6 (0.7~157.4); 739 (426~904) ❙ 596 (309~1571); 376 (190~691); 22 (5~152) ❙ -18.1 (-51.1~108.0); 572 (309~904); 176.3 (61.2~346.5)

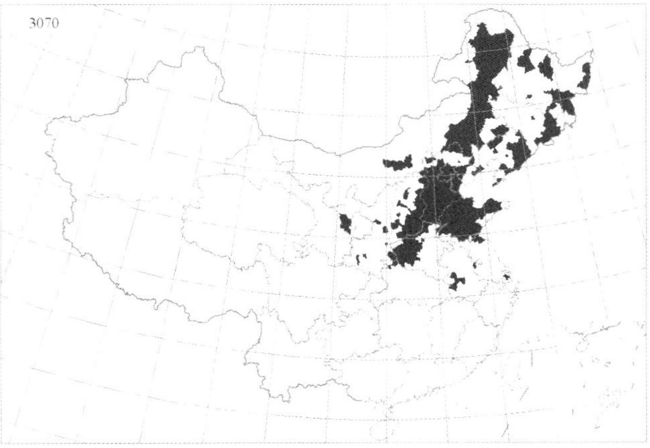

Rosaceae 蔷薇科 *Crataegus* 山楂属
Crataegus remotilobata Raikova ex Popov 裂叶山楂
Trees, deciduous

4.1; -12.9; 18.5; 7.5; 55.4; 65.5; 548 ❙ 368; 217; 14 ❙ -33.6; 368; 152.8

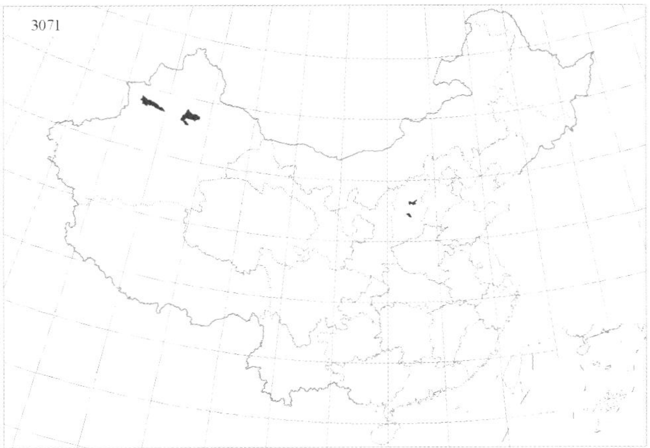

Rosaceae 蔷薇科 *Crataegus* 山楂属
Crataegus sanguinea Pall. 辽宁山楂
Shrubs or small trees, deciduous

5.5 (-5.8~13.5); -12.3 (-29.8~-1.3); 21.3 (15.3~27.1); 8.8 (4.4~13.7); 68.9 (28.0~117.5); 62.1 (14.6~157.4); 604 (426~820) ❙ 446 (137~785); 271 (62~381); 14 (3~51) ❙ -26.2 (-77.5~19.6); 442 (137~679); 162.3 (19.8~309.5)

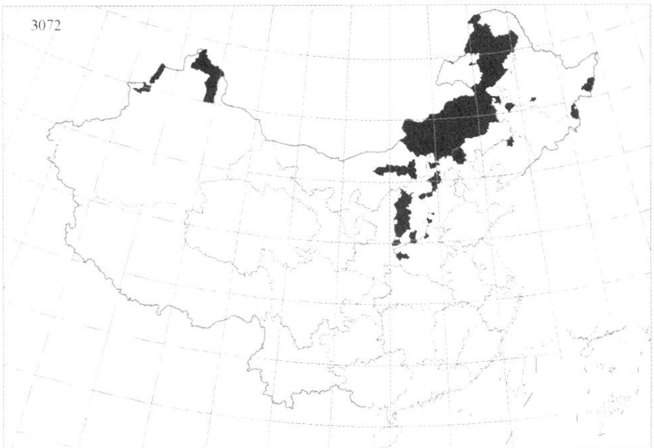

Rosaceae 蔷薇科 *Crataegus* 山楂属
Crataegus scabrifolia (Franch.) Rehder 云南山楂
Trees, deciduous

15.9 (5.4~21.8); 8.5 (-2.1~14.2); 21.4 (12.3~29.0); 15.9 (5.8~21.8); 131.9 (31.5~202.0); 0.6 (0.0~25.6); 802 (481~1168) ❙ 1136 (741~1900); 633 (411~1056); 49 (15~144) ❙ 42.3 (-14.2~99.9); 788 (481~1134); 309.3 (160.8~551.6)

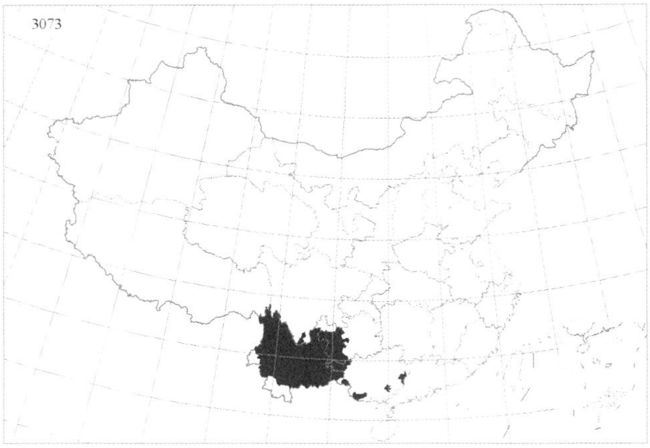

Rosaceae 蔷薇科 *Crataegus* 山楂属
Crataegus shandongensis F. Z. Li 山东山楂
Shrubs, deciduous

13.3; -1.7; 26.8; 13.5; 116.5; 15.9; 817 ❙ 710; 466; 27 ❙ -13.1; 710; 180.5

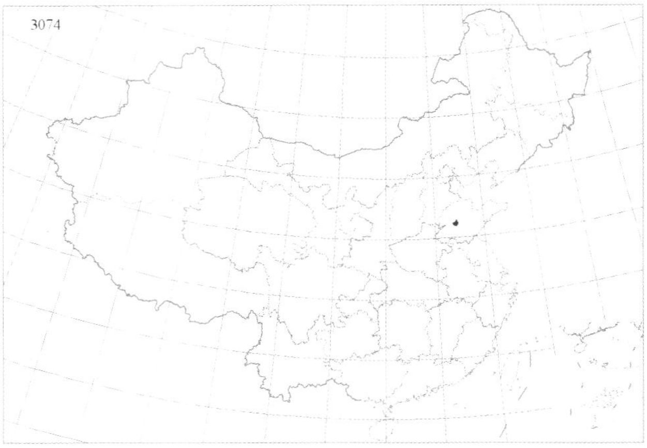

Rosaceae 蔷薇科 *Crataegus* 山楂属
Crataegus shensiensis Pojarkova 陕西山楂
Shrubs

10.7 (7.4~12.2); -2.2 (-4.2~0.2); 23.0 (18.7~24.8); 11.1 (8.2~12.3); 87.8 (58.8~99.3); 18.4 (10.8~29.4); 679 (561~731) ❙ 732 (562~882); 337 (288~391); 24 (15~32) ❙ 9.2 (-23.0~52.6); 647 (561~717); 250.4 (197.4~318.2)

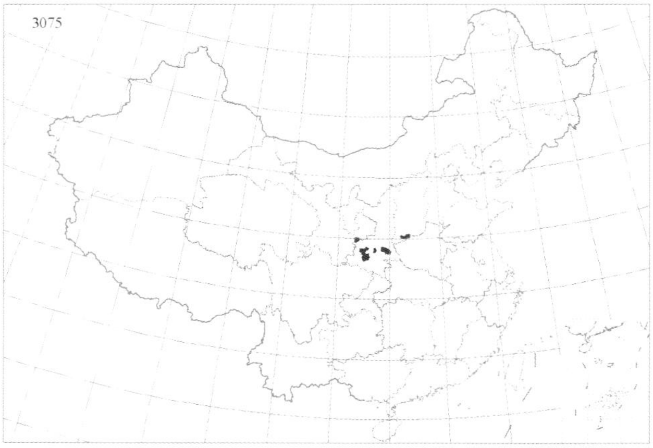

Rosaceae 蔷薇科 *Crataegus* 山楂属
Crataegus songarica K. Koch 准噶尔山楂
Shrubs or trees, deciduous

2.9 (-1.3~9.0); -15.7 (-20.9~-9.3); 17.3 (13.2~23.2); 7.0 (4.6~10.7); 50.4 (26.7~88.9); 74.7 (40.7~103.9); 537 (440~703) ❙ 310 (225~406); 143 (65~211); 27 (12~48) ❙ -40.1 (-65.1~-7.6); 310 (225~406); 216.4 (165.7~297.3)

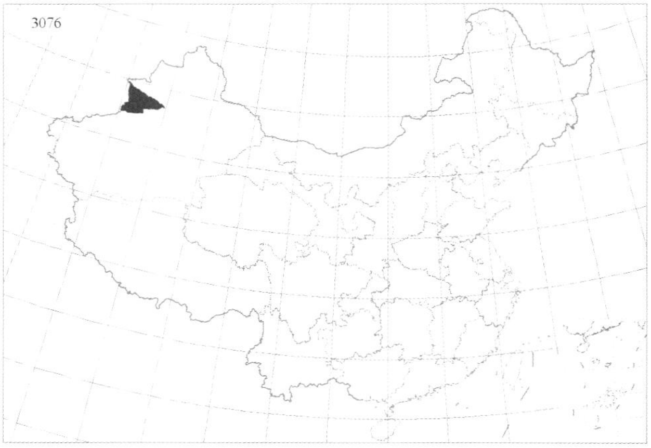

Rosaceae 蔷薇科 *Crataegus* 山楂属
Crataegus wilsonii Sarg. 少毛山楂
Shrubs, deciduous

10.4 (1.9~17.7); -2.4 (-11.8~5.5); 22.3 (11.7~29.5); 11.0 (4.3~17.7); 86.0 (21.1~152.8); 21.1 (0.0~60.2); 671 (427~966) ❙ 816 (367~1663); 379 (235~602); 41 (10~212) ❙ 20.2 (-37.7~106.2); 645 (367~868); 237.1 (124.1~334.4)

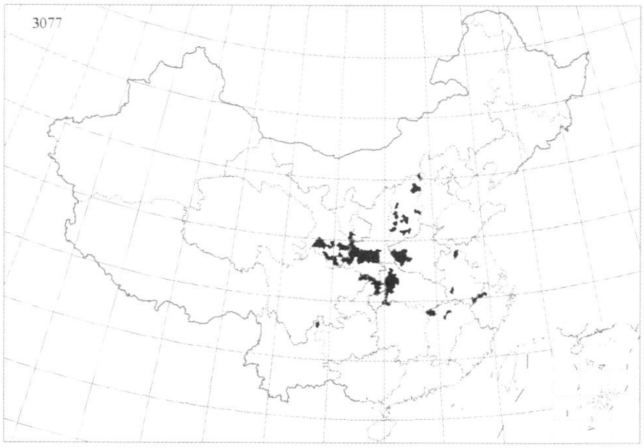

Rosaceae 蔷薇科 *Dichotomanthes* 牛筋条属
Dichotomanthes tristaniicarpa Kurz 牛筋条
Shrubs or small trees, evergreen

15.6 (5.4~21.0); 8.5 (-2.1~15.1); 20.9 (12.3~26.6); 15.7 (5.8~21.0); 128.5 (31.5~191.8); 0.7 (0.0~25.6); 785 (481~1053) ❙ 1123 (741~1751); 632 (411~996); 44 (15~77) ❙ 43.3 (-14.2~112.2); 768 (481~1053); 334.8 (142.9~659.0)

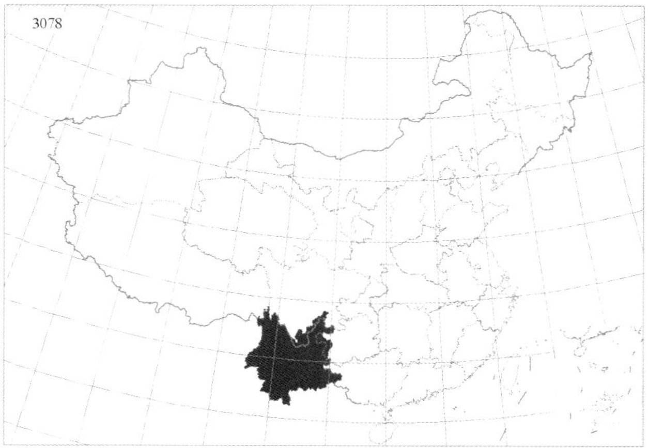

Rosaceae 蔷薇科 *Docynia* 移㭎属

Docynia delavayi (Franch.) C. K. Schneid. 云南移㭎

Trees, evergreen

15.5 (5.4~21.0); 8.4 (-2.1~15.1); 20.9 (12.3~26.6); 15.6 (5.8~21.0); 127.6 (31.5~191.8); 0.9 (0.0~25.6); 783 (481~1053) ▎ 1120 (741~1751); 630 (411~996); 44 (12~77) ▎ 43.5 (-14.2~112.2); 767 (481~1053); 334.0 (142.9~659.0)

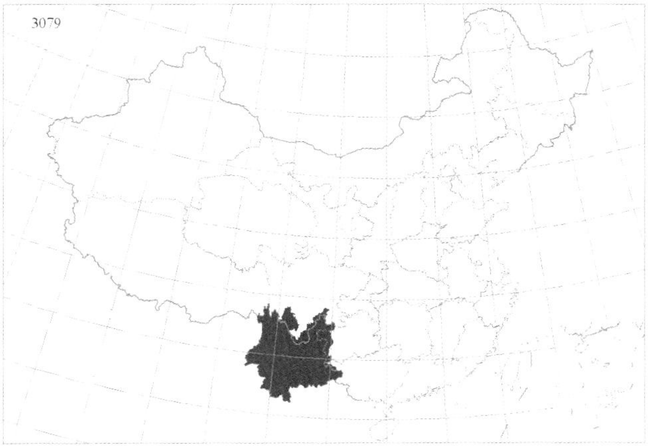

Rosaceae 蔷薇科 *Docynia* 移㭎属

Docynia indica (Wall.) Decne. 移㭎

Trees, deciduous or semievergreen

15.8 (3.0~21.0); 8.9 (-5.6~15.1); 20.8 (10.5~25.6); 15.8 (4.3~21.0); 130.7 (20.1~191.8); 1.0 (0.0~43.9); 789 (423~1053) ▎ 1111 (628~1751); 626 (410~996); 43 (11~77) ▎ 41.0 (-14.2~88.8); 770 (423~1053); 333.7 (190.3~659.0)

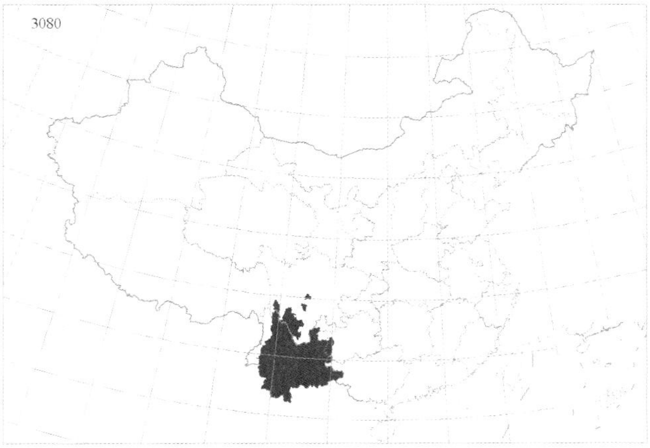

Rosaceae 蔷薇科 *Dryas* 仙女木属

Dryas octopetala var. asiatica (Nakai) Nakai 东亚仙女木

Subshrubs, evergreen

5.2 (-1.1~11.7); -14.4 (-18.7~-7.6); 21.3 (13.2~27.5); 8.8 (4.5~13.0); 70.9 (26.7~113.4); 67.4 (32.4~99.8); 623 (433~820) ▎ 303 (30~1005); 166 (21~635); 18 (1~46) ▎ -44.8 (-96.3~68.8); 250 (30~595); 114.8 (2.1~329.9)

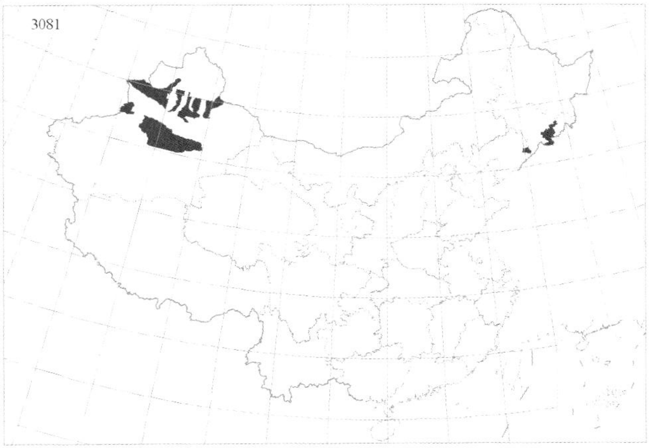

Rosaceae 蔷薇科 *Eriobotrya* 枇杷属

Eriobotrya bengalensis var. angustifolia Cardot 窄叶南亚枇杷

Trees, evergreen

16.6 (13.6~21.0); 9.6 (3.8~14.2); 21.7 (19.0~25.9); 16.6 (13.6~21.0); 139.3 (103.6~191.8); 0.1 (0.0~1.5); 816 (707~1053) ▎ 1067 (797~1547); 595 (411~911); 46 (23~94) ▎ 30.8 (-14.2~74.2); 787 (683~1053); 310.3 (161.9~551.6)

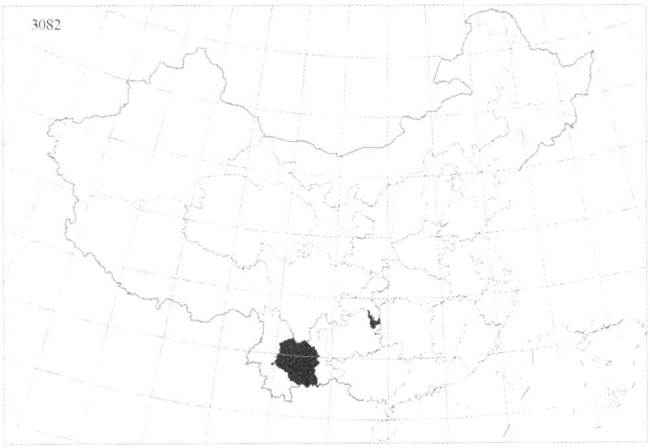

Rosaceae 蔷薇科 *Eriobotrya* 枇杷属

Eriobotrya cavaleriei (H. Lév.) Rehder 大花枇杷

Trees, evergreen

18.3 (7.3~23.2); 8.4 (-2.0~15.8); 26.9 (15.6~29.8); 18.4 (7.5~23.2); 160.8 (48.2~218.9); 0.3 (0.0~20.4); 965 (537~1257) ▎ 1500 (853~2435); 612 (384~ 428); 147 (23~230) ▎ 56.7 (5.8~128.9); 960 (537~1257); 233.9 (126. ~551.6)

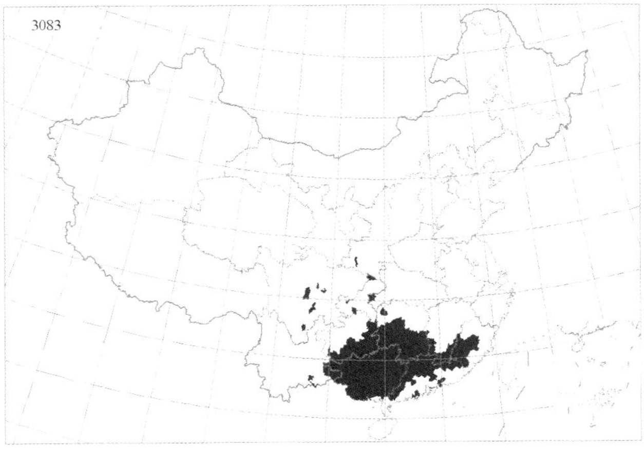

Rosaceae 蔷薇科 *Eriobotrya* 枇杷属

Eriobotrya deflexa (Hemsl.) Nakai 台湾枇杷

Trees, evergreen

21.1 (14.6~24.8); 13.9 (3.8~20.1); 26.9 (20.1~28.9); 21.1 (14.7~24.8); 193.8 (117.7~237.6); 0.0 (0.0~1.5); 1113 (773~1385) ▎ 1858 (868~3495); 803 (312~1847); 159 (41~897) ▎ 74.1 (-34.3~246.5); 1059 (773~1360); 317.0 (154.2~620.0)

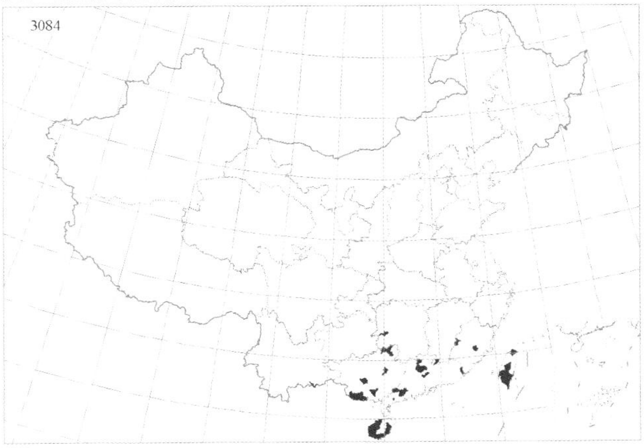

Rosaceae 蔷薇科 *Eriobotrya* 枇杷属

Eriobotrya elliptica Lindl. 椭圆枇杷

Trees, evergreen

15.5; 8.2; 21.1; 15.6; 126.9; 0.0; 780 ❙ 1552; 882; 55 ❙ 103.0; 780; 469.5

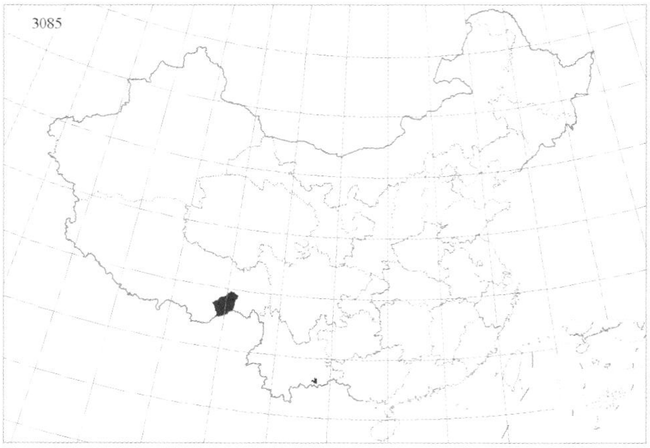

Rosaceae 蔷薇科 *Eriobotrya* 枇杷属

Eriobotrya fragrans Champ. ex Benth. 香花枇杷

Shrubs or small trees, evergreen

19.7 (−1.8~23.6); 11.0 (−10.2~16.7); 26.9 (6.1~29.8); 19.8 (1.8~23.6); 178.2 (1.9~223.5); 0.9 (0.0~82.8); 1037 (329~1281) ❙ 1577 (320~2037); 678 (193~1056); 139 (20~216) ❙ 52.8 (−6.9~145.8); 1036 (320~1273); 235.9 (72.2~551.6)

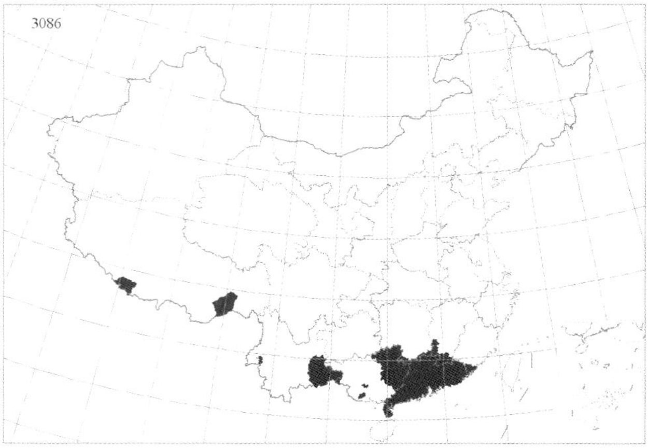

Rosaceae 蔷薇科 *Eriobotrya* 枇杷属

Eriobotrya henryi Nakai 窄叶枇杷

Shrubs or small trees, evergreen

16.9 (13.2~19.3); 10.1 (5.8~13.1); 21.6 (18.4~25.5); 16.9 (13.3~19.4); 142.8 (99.4~173.2); 0.0 (0.0~0.0); 824 (695~939) ❙ 1106 (939~1495); 624 (524~836); 49 (37~107) ❙ 34.4 (6.5~92.4); 802 (695~903); 309.0 (186.7~432.9)

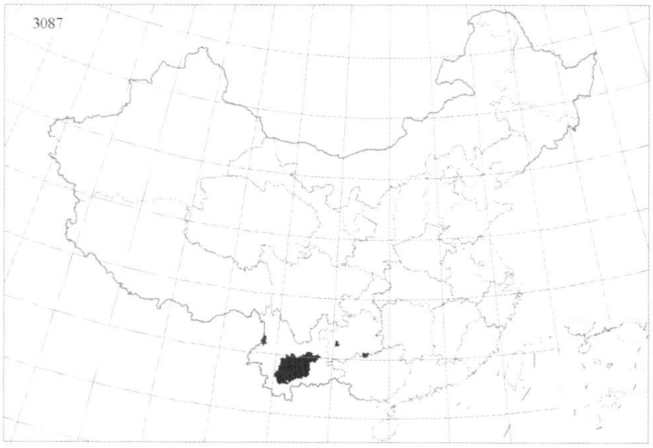

Rosaceae 蔷薇科 *Eriobotrya* 枇杷属

Eriobotrya japonica (Thunb.) Lindl. 枇杷

Trees, evergreen

15.9 (5.4~22.5); 7.1 (−2.1~15.1); 23.4 (12.3~29.9); 16.0 (5.8~22.6); 132.5 (31.5~210.9); 0.9 (0.0~25.6); 824 (481~1206) ❙ 1244 (741~2024); 601 (411~996); 75 (15~223) ❙ 51.2 (−14.2~112.2); 815 (481~1206); 270.3 (120.3~659.0)

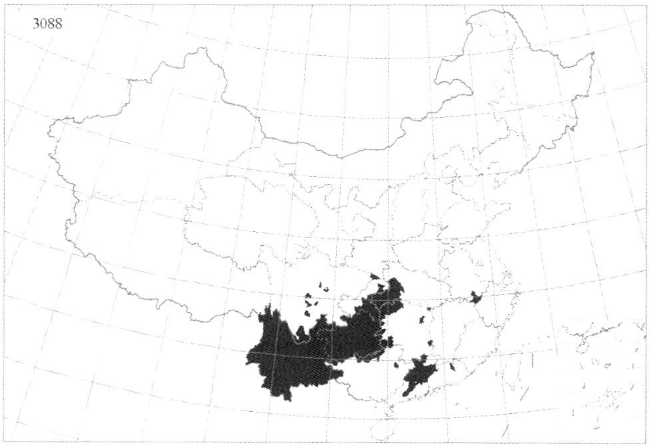

Rosaceae 蔷薇科 *Eriobotrya* 枇杷属

Eriobotrya malipoensis K. C. Kuan 麻栗坡枇杷

Trees, evergreen

18.2; 9.2; 25.5; 18.2; 158.7; 0.0; 919 ❙ 1423; 673; 102 ❙ 55.7; 919; 304.7

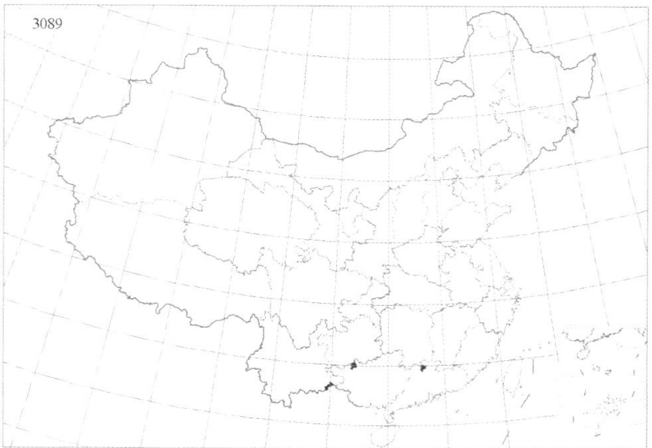

Rosaceae 蔷薇科 *Eriobotrya* 枇杷属

Eriobotrya obovata W. W. Sm. 倒卵叶枇杷

Trees, evergreen

15.8 (11.2~18.9); 9.0 (4.6~12.5); 20.9 (16.7~26.7); 15.9 (11.3~19.0); 130.3 (75.8~167.8); 0.0 (0.0~0.4); 786 (642~929) ❙ 1017 (797~1624); 571 (411~651); 46 (23~191) ❙ 29.9 (−14.2~82.3); 751 (642~916); 298.2 (218.0~394.7)

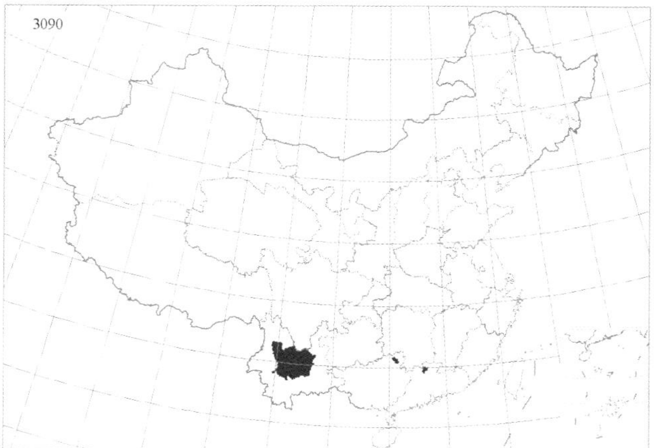

Rosaceae 蔷薇科 *Eriobotrya* 枇杷属

Eriobotrya prinoides Rehder et E. H. Wilson 栎叶枇杷

Trees, evergreen

15.0 (5.4~21.0); 7.7 (-2.1~14.2); 20.7 (12.3~26.6); 15.1 (5.8~21.0); 122.1 (31.5~ 91.8); 1.1 (0.0~25.6); 766 (481~1053) ❙ 1027 (741~1522); 585 (411~860); 39 (12~71) ❙ 35.1 (-14.2~80.6); 745 (481~1053); 299.6 (142.9~551.6)

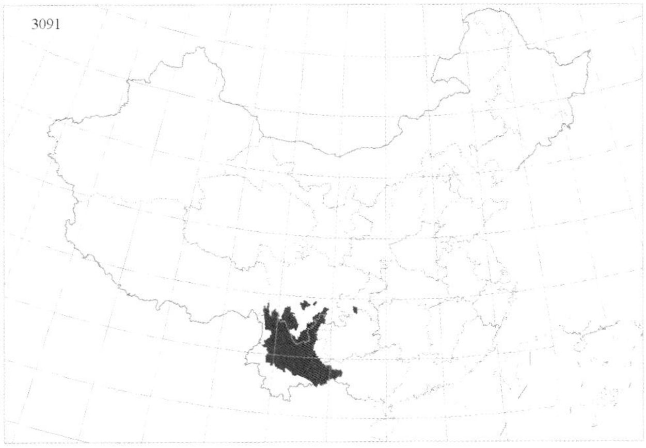

Rosaceae 蔷薇科 *Eriobotrya* 枇杷属

Eriobotrya salwinensis Hand.-Mazz. 怒江枇杷

Trees, evergreen

10.7 (8.6~13.2); 3.7 (1.3~6.4); 16.5 (15.0~18.4); 10.8 (8.7~13.3); 72.5 (52.6~99.4); 3.1 (0.0~8.8); 630 (574~695) ❙ 1153 (1007~1337); 647 (578~727); 52 (40~69) ❙ 82.6 (67.9~93.0); 630 (574~695); 367.1 (339.5~407.1)

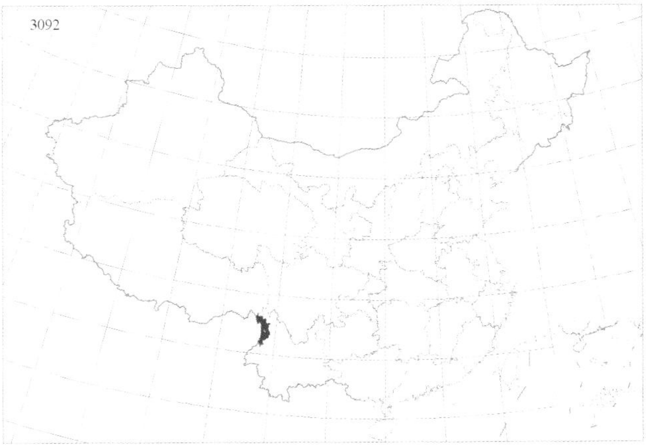

Rosaceae 蔷薇科 *Eriobotrya* 枇杷属

Eriobotrya seguinii (H. Lév.) Cardot ex Guillaumin 小叶枇杷

Shrubs, evergreen

18.0 (15.9~19.8); 9.1 (6.9~11.1); 24.9 (23.0~26.4); 18.0 (16.0~19.8); 156.0 (131.6~177.9); 0.0 (0.0~0.0); 902 (802~1010) ❙ 1258 (1161~1334); 680 (633~729); 56 (43~64) ❙ 40.1 (16.9~64.7); 902 (802~1010); 278.9 (189.5~411.1)

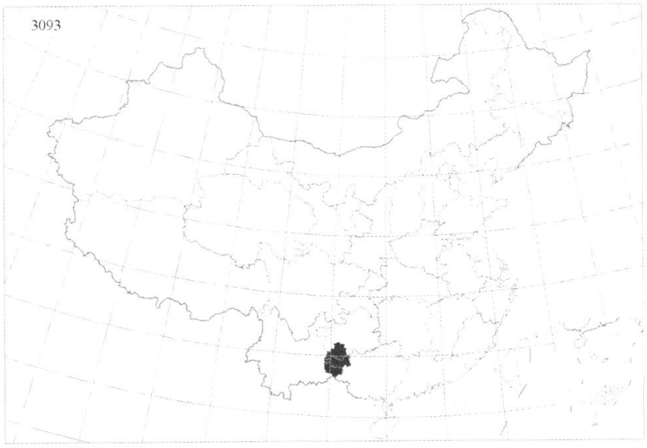

Rosaceae 蔷薇科 *Eriobotrya* 枇杷属

Eriobotrya serrata J. E. Vidal 齿叶枇杷

Trees, evergreen

18.9 (16.4~21.0); 12.6 (9.6~15.1); 22.8 (20.6~25.6); 18.9 (16.4~21.0); 167.1 (137.3~191.8); 0.0 (0.0~0.0); 911 (795~1053) ❙ 1514 (1276~1751); 821 (655~996); 57 (41~77) ❙ 67.1 (36.8~112.2); 899 (795~1053); 511.3 (371.7~659.0)

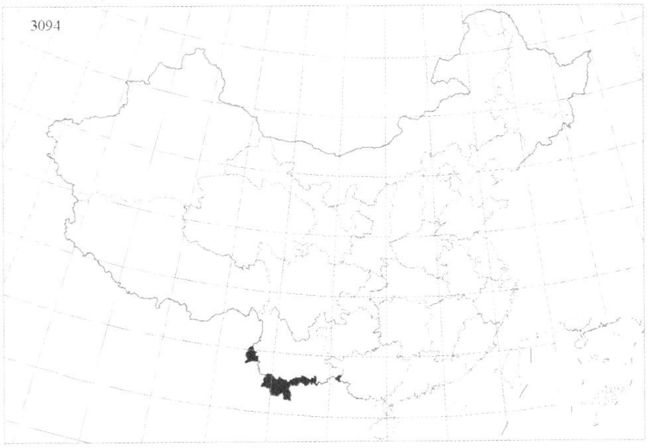

Rosaceae 蔷薇科 *Eriobotrya* 枇杷属

Eriobotrya tengyuehensis W. W. Sm. 腾越枇杷

Trees, evergreen

14.8 (8.6~17.7); 6.8 (1.3~11.0); 21.3 (15.0~26.8); 14.9 (8.7~17.8); 119.2 (52.6~153.4); 1.0 (0.0~8.8); 766 (574~862) ❙ 1262 (1041~1561); 647 (463~897); 67 (44~96) ❙ 66.4 (26.1~99.9); 762 (574~862); 358.1 (187.1~539.7)

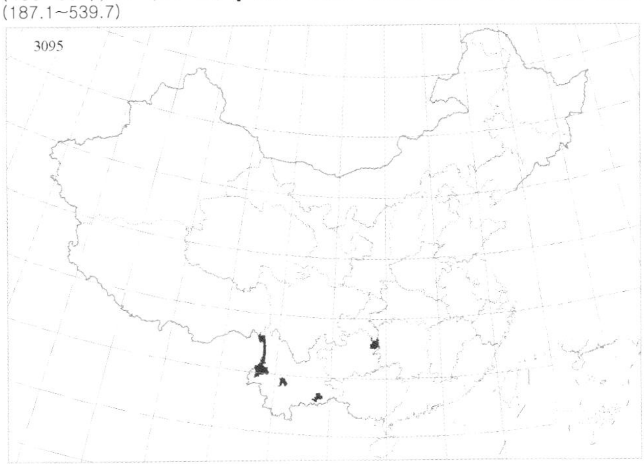

Rosaceae 蔷薇科 *Exochorda* 白鹃梅属

Exochorda giraldii Hesse 红柄白鹃梅

Shrubs, deciduous

11.4 (1.6~16.5); -3.0 (-13.7~4.4); 24.4 (10.9~28.9); 12.1 (4.0~16.6); 99.2 (17.9~139.8); 21.9 (0.6~71.7); 733 (419~910) ❙ 707 (367~1689); 364 (235~617); 39 (8~220) ❙ -4.1 (-43.4~105.7); 624 (367~910); 185.1 (101.9~344.4)

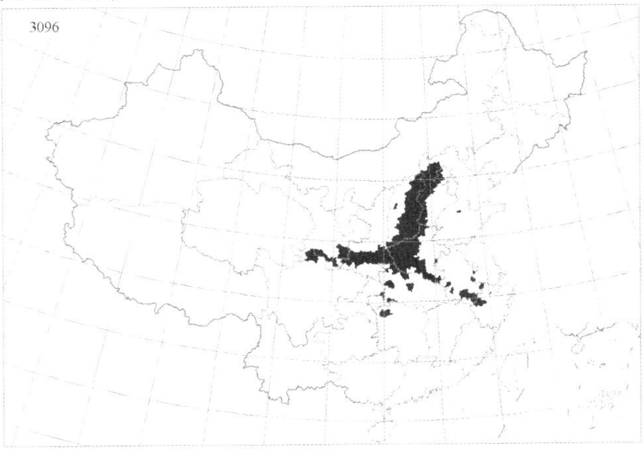

Rosaceae 蔷薇科 *Exochorda* 白鹃梅属
Exochorda giraldii var. wilsonii (Rehder) Rehder 绿柄白鹃梅
Shrubs, deciduous

14.3 (12.0~16.2); 1.8 (-0.8~3.5); 26.2 (22.9~28.3); 14.4 (12.1~16.2); 118.3 (94.7~136.1); 5.8 (1.5~13.4); 808 (696~898) ❙ 1232 (793~1689); 506 (366~617); 116 (42~212) ❙ 52.4 (-2.0~105.7); 807 (696~898); 231.8 (151.3~319.9)

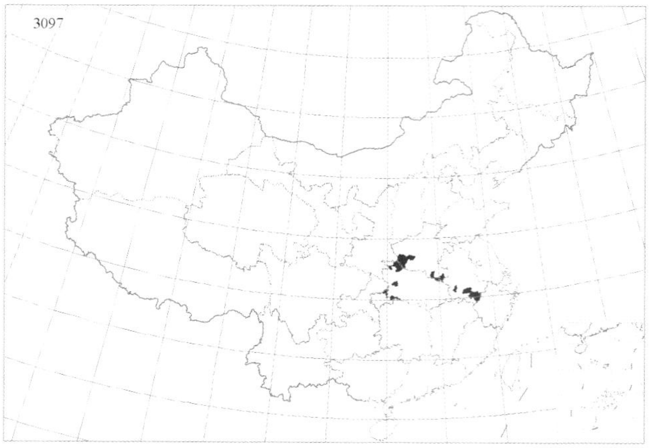

Rosaceae 蔷薇科 *Exochorda* 白鹃梅属
Exochorda racemosa (Lindl.) Rehder 白鹃梅
Shrubs, deciduous

12.8 (3.3~17.3); -1.1 (-12.7~5.9); 25.6 (17.5~29.4); 13.4 (6.6~17.3); 110.6 (45.4~147.3); 16.7 (0.0~65.0); 782 (507~954) ❙ 928 (367~1922); 414 (235~685); 84 (10~259) ❙ 14.6 (-37.7~111.3); 718 (367~941); 174.4 (70.3~318.0)

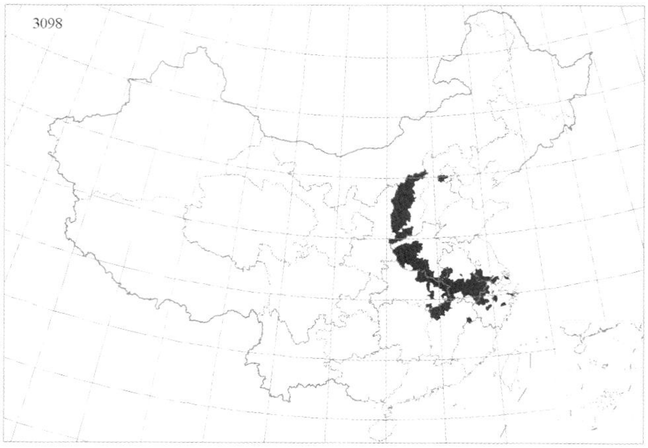

Rosaceae 蔷薇科 *Exochorda* 白鹃梅属
Exochorda serratifolia S. Moore 齿叶白鹃梅
Shrubs, deciduous

8.2 (6.5~10.9); -9.6 (-12.6~-4.4); 23.8 (22.5~25.2); 10.2 (9.2~11.7); 85.7 (75.7~99.1); 46.6 (23.8~56.7); 674 (630~743) ❙ 590 (473~735); 408 (332~465); 13 (7~26) ❙ -12.6 (-28.0~8.4); 583 (473~698); 209.3 (131.6~273.5)

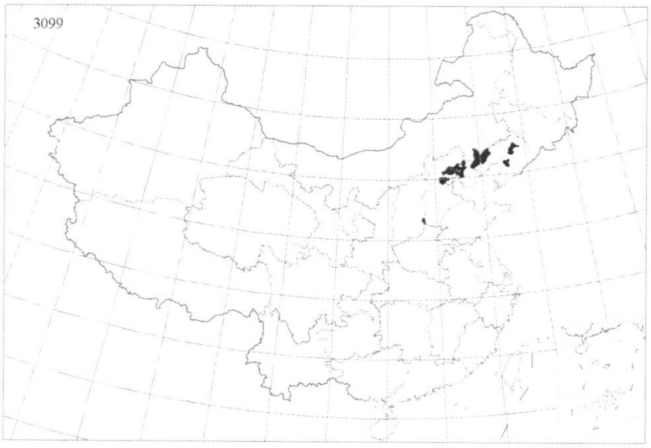

Rosaceae 蔷薇科 *Kerria* 棣棠属
Kerria japonica (L.) DC. 棣棠花
Shrubs, deciduous

15.1 (2.2~20.5); 4.2 (-7.7~13.1); 25.1 (9.9~30.1); 15.2 (3.8~20.5); 125.9 (15.5~186.5); 4.1 (0.0~49.3); 826 (413~1073) ❙ 1198 (341~1997); 500 (190~836); 104 (7~259) ❙ 44.5 (-40.9~147.0); 801 (341~1036); 224.4 (61.2~516.7)

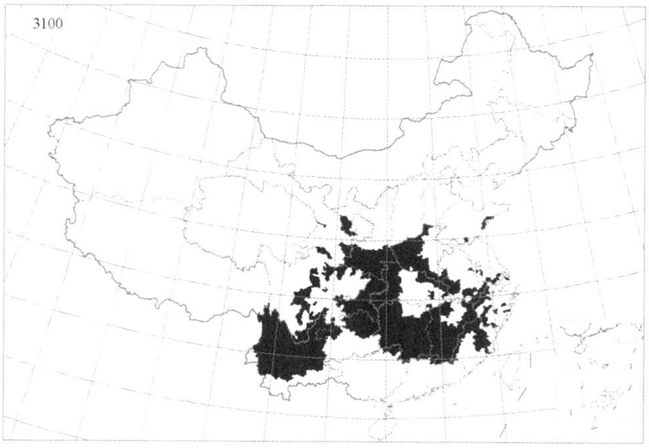

Rosaceae 蔷薇科 *Laurocerasus* 桂樱属
Laurocerasus andersonii (Hook. f.) T. T. Yu et L. T. Lu 云南桂樱
Shrubs or small trees, evergreen

19.0; 13.1; 22.6; 19.1; 168.8; 0.0; 909 ❙ 1495; 775; 62 ❙ 64.5; 891; 427.9

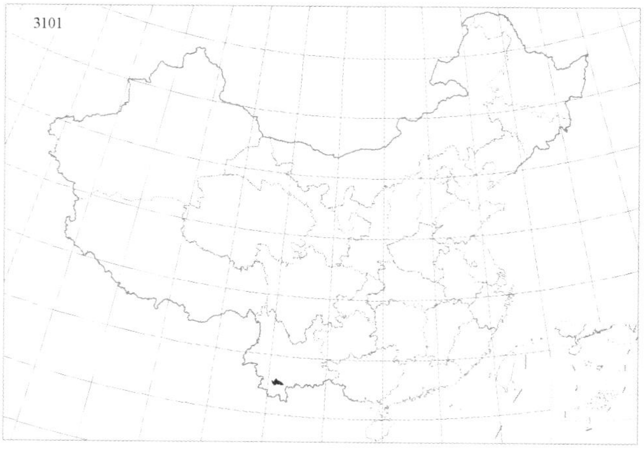

Rosaceae 蔷薇科 *Laurocerasus* 桂樱属
Laurocerasus aquifolioides Chun ex T. T. Yu et L. T. Lu 冬青叶桂樱
Shrubs, evergreen

20.1 (17.7~21.8); 11.0 (7.7~14.2); 27.6 (26.6~28.2); 20.1 (17.8~21.9); 181.3 (153.6~202.6); 0.0 (0.0~0.0); 1048 (916~1150) ❙ 1699 (1564~1876); 667 (556~891); 167 (116~188) ❙ 62.2 (56.1~70.7); 1048 (916~1150); 248.8 (205.8~273.3)

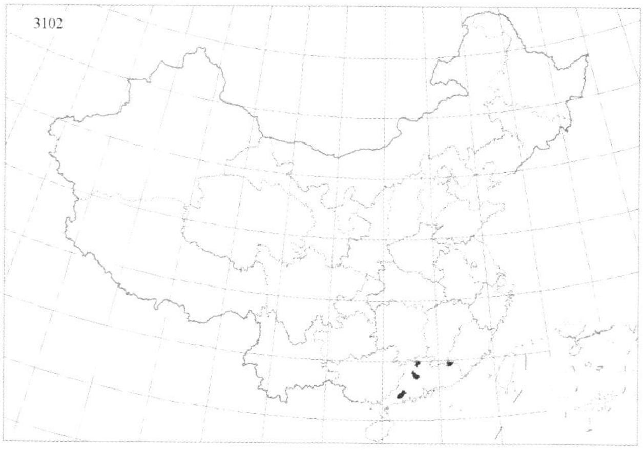

Rosaceae 蔷薇科 *Laurocerasus* 桂樱属
Laurocerasus australis T. T. Yu et L. T. Lu 南方桂樱
Shrubs or small trees, evergreen

19.2 (17.9~21.3); 9.9 (7.4~12.6); 26.6 (25.5~27.9); 19.2 (18.0~21.3); 170.3 (155.6~196.0); 0.0 (0.0~0.0); 984 (918~1123) | 1375 (1182~1740); 678 (561~776); 94 (43~187) | 40.4 (5.8~85.5); 983 (918~1121); 267.5 (189.5~354.5)

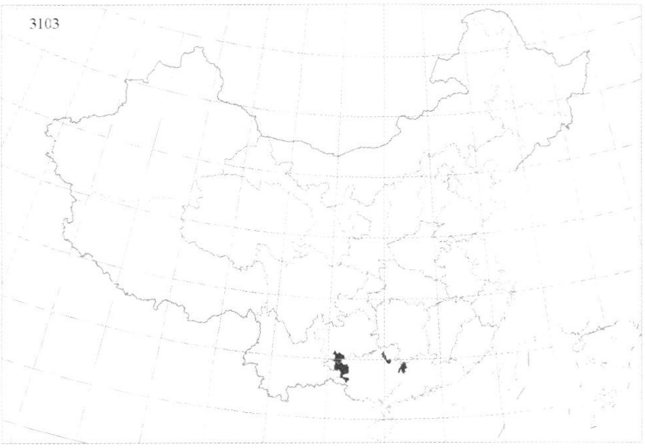

Rosaceae 蔷薇科 *Laurocerasus* 桂樱属
Laurocerasus dolichophylla T. T. Yu et L. T. Lu 长叶桂樱
Trees, evergreen

17.4; 9.7; 22.9; 17.5; 149.7; 0.0; 848 | 1338; 756; 50 | 57.6; 848; 348.0

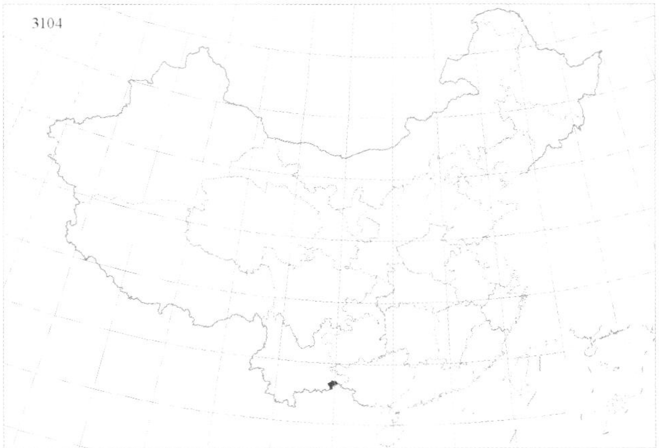

Rosaceae 蔷薇科 *Laurocerasus* 桂樱属
Laurocerasus fordiana (Dunn) T. T. Yu et L. T. Lu 华南桂樱
Shrubs or small trees, evergreen

22.7 (19.1~24.8); 15.6 (10.7~20.1); 28.0 (25.0~29.0); 22.7 (19.1~24.8); 212.6 (169.6~237.6); 0.0 (0.0~0.0); 1219 (945~1385) | 1505 (868~2148); 666 (312~1234); 89 (41~161) | 25.3 (-34.3~86.7); 1124 (868~1264); 347.9 (194.5~620.0)

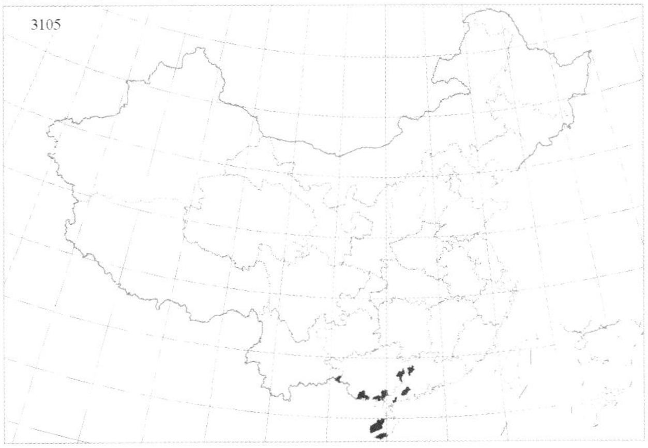

Rosaceae 蔷薇科 *Laurocerasus* 桂樱属
Laurocerasus hypotricha (Rehder) T. T. Yu et L. T. Lu 毛背桂樱
Trees, evergreen

16.7 (2.2~21.8); 7.5 (-6.6~14.2); 24.5 (9.9~28.5); 16.8 (3.8~21.9); 144.5 (15.5~202.6); 3.2 (0.0~49.1); 895 (413~1150) | 1440 (741~1876); 636 (426~891); 124 (19~224) | 62.0 (16.9~125.9); 894 (413~1150); 259.6 (171.0~354.5)

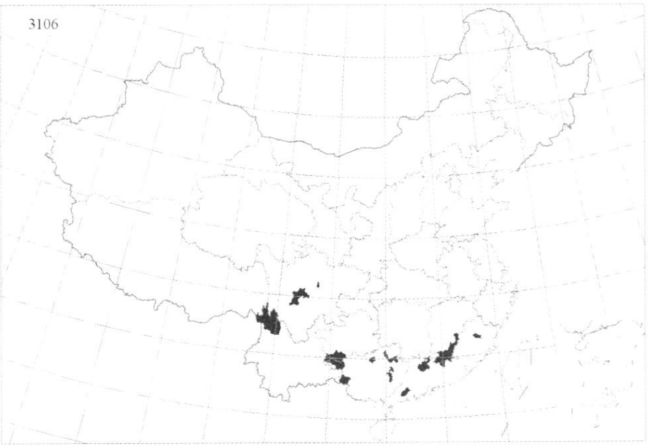

Rosaceae 蔷薇科 *Laurocerasus* 桂樱属
Laurocerasus jenkinsii (Hook. f.) Browicz 坚核桂樱
Trees, evergreen

17.8 (14.4~20.5); 11.1 (7.5~15.1); 22.2 (19.1~23.7); 17.9 (14.4~20.5); 154.3 (113.1~185.9); 0.0 (0.0~0.0); 864 (727~992) | 1464 (1074~1688); 799 (582~959); 56 (42~77) | 70.2 (32.4~112.2); 857 (727~986); 475.8 (277.0~659.0)

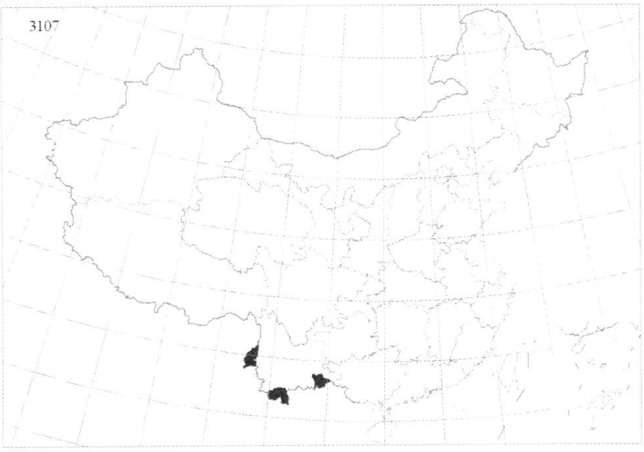

Rosaceae 蔷薇科 *Laurocerasus* 桂樱属
Laurocerasus marginata (Dunn) T. T. Yu et L. T. Lu 全缘桂樱
Shrubs or small trees, evergreen

20.6 (17.7~22.4); 12.0 (7.7~14.8); 27.8 (26.6~28.7); 20.6 (17.8~22.4); 187.6 (153.6~209.2); 0.0 (0.0~0.0); 1077 (916~1198) | 1732 (1447~2030); 711 (539~1033); 158 (82~191) | 61.5 (20.8~95.5); 1077 (916~1198); 234.9 (163.7~275.2)

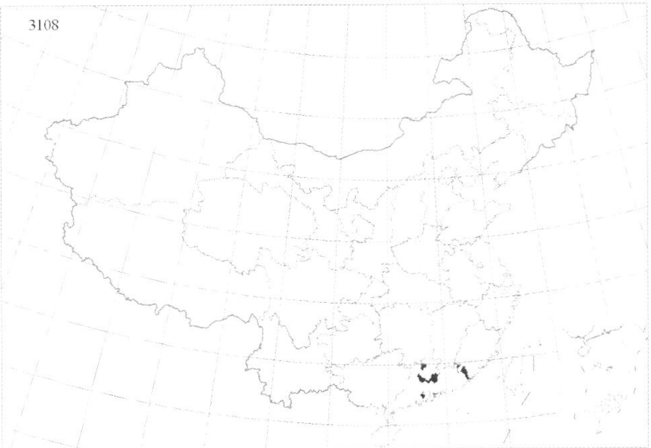

Rosaceae 蔷薇科 *Laurocerasus* 桂樱属
Laurocerasus menghaiensis T. T. Yu et L. T. Lu 勐海桂樱
Trees, evergreen

19.2; 13.8; 22.3; 19.2; 170.5; 0.0; 909 ❙ 1499; 686; 58 ❙ 64.9; 880; 513.9

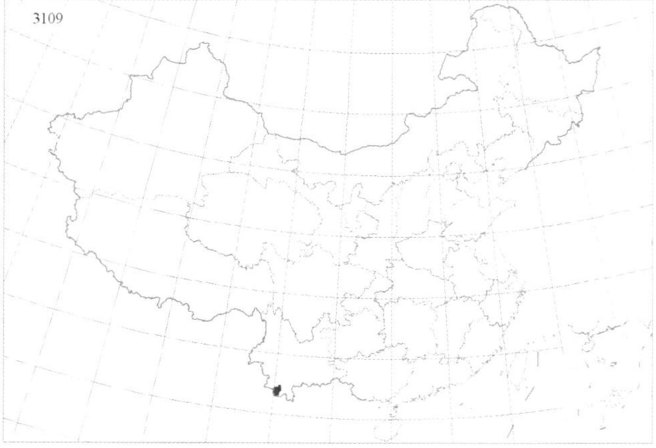

Rosaceae 蔷薇科 *Laurocerasus* 桂樱属
Laurocerasus phaeosticta (Hance) C. K. Schneid. 腺叶桂樱
Shrubs or small trees, evergreen

18.2 (5.3~24.8); 9.0 (−2.4~20.1); 26.2 (12.2~29.8); 18.3 (5.7~24.8); 159.6 (30.5~237.6); 0.3 (0.0~26.2); 955 (479~1385) ❙ 1480 (868~3495); 613 (312~1388); 136 (29~897) ❙ 56.5 (−34.3~246.5); 942 (479~1360); 254.0 (95.3~620.0)

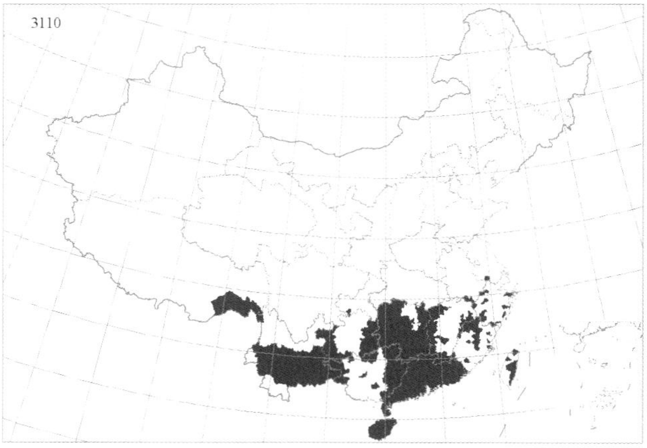

Rosaceae 蔷薇科 *Laurocerasus* 桂樱属
Laurocerasus spinulosa (Siebold et Zucc.) C. K. Schneid. 刺叶桂樱
Trees, evergreen

16.8 (12.0~21.8); 5.7 (0.6~13.5); 27.2 (21.4~29.8); 16.8 (12.0~21.8); 142.5 (92.3~201.9); 0.7 (0.0~9.7); 896 (690~1158) ❙ 1508 (993~2005); 539 (392~827); 171 (38~259) ❙ 68.7 (10.4~147.0); 888 (690~1158); 221.8 (118.7~346.9)

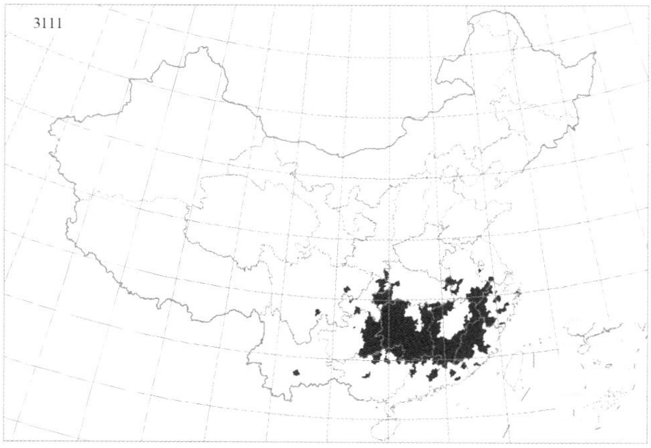

Rosaceae 蔷薇科 *Laurocerasus* 桂樱属
Laurocerasus undulata (Buch.-Ham. ex D. Don) M. Roem. 尖叶桂樱
Trees, evergreen

16.7 (−1.8~22.0); 6.9 (−10.2~15.1); 25.3 (6.1~29.9); 16.8 (1.8~22.1); 142.1 (1.9~204.9); 1.1 (0.0~82.8); 877 (329~1166) ❙ 1357 (320~2005); 569 (217~1007); 121 (11~233) ❙ 54.7 (−14.2~145.8); 862 (320~1166); 252.6 (72.2~659.0)

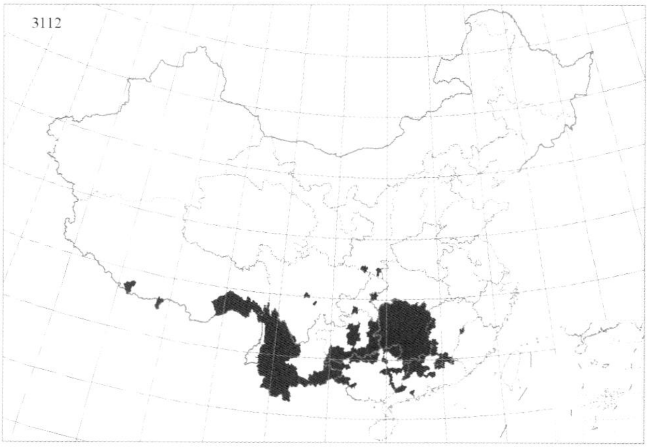

Rosaceae 蔷薇科 *Laurocerasus* 桂樱属
Laurocerasus zippeliana (Miq.) T. T. Yu et L. T. Lu 大叶桂樱
Trees, evergreen

16.5 (5.4~24.2); 7.2 (−2.1~18.3); 24.5 (12.3~29.8); 16.5 (5.8~24.3); 139.1 (31.5~231.2); 0.7 (0.0~25.6); 861 (481~1335) ❙ 1335 (711~3495); 587 (312~1428); 109 (15~897) ❙ 55.0 (−34.3~224.4); 849 (481~1321); 258.2 (95.3~659.0)

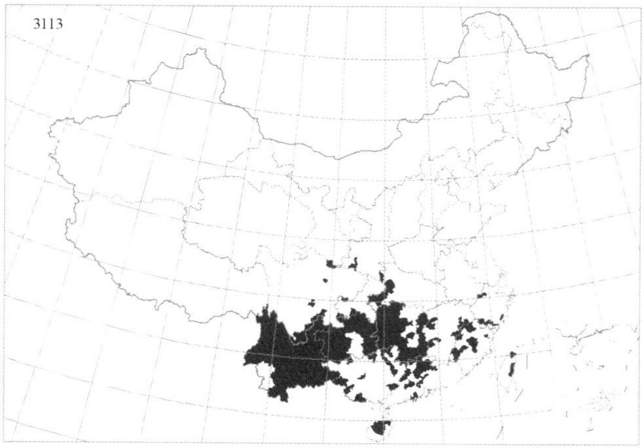

Rosaceae 蔷薇科 *Maddenia* 臭樱属
Maddenia fujianensis Y. T. Chang 福建假稠李
Shrubs, deciduous

16.6; 6.0; 26.3; 16.6; 138.9; 0.0; 866 ❙ 1997; 669; 254 ❙ 130.7; 866; 303.4

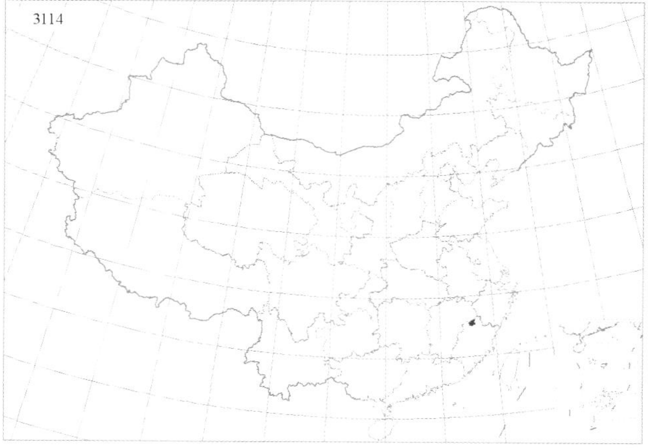

Rosaceae 蔷薇科 *Maddenia* 臭樱属
Maddenia himalaica Hook. f. et Thomson 喜马拉雅臭樱
Trees, deciduous

3.6; -4.8; 11.0; 5.1; 25.9; 42.5; 449 ❙ 671; 393; 23 ❙ 41.9; 449; 221.4

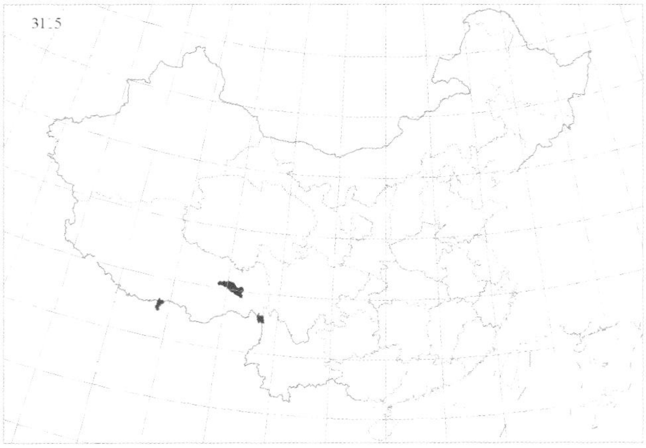

Rosaceae 蔷薇科 *Maddenia* 臭樱属
Maddenia hypoleuca Koehne 臭樱
Shrubs or small trees, deciduous

11.4 (2.2~17.0); -0.4 (-7.4~6.2); 22.5 (9.9~26.8); 11.7 (3.8~17.0); 91.3 (15.5~144.4); 14.3 (0.0~49.1); 692 (413~891) ❙ 925 (341~1624); 415 (190~600); 45 (8~191) ❙ 34.2 (-40.9~125.9); 682 (341~891); 242.7 (61.2~334.4)

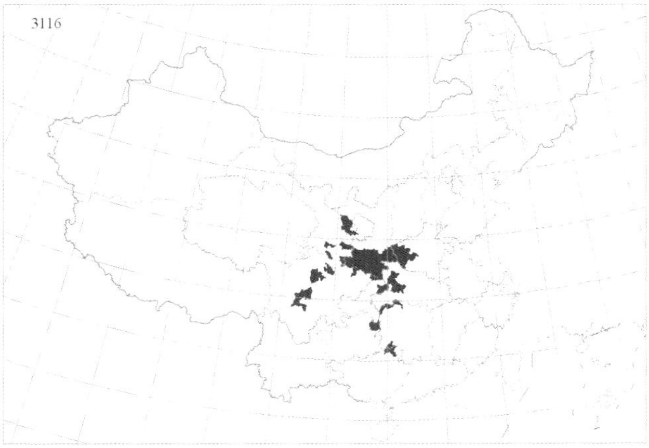

Rosaceae 蔷薇科 *Maddenia* 臭樱属
Maddenia hypoxantha Koehne 四川臭樱
Shrubs, deciduous

5.8; -4.7; 14.8; 7.3; 47.8; 37.4; 522 ❙ 906; 519; 22 ❙ 68.0; 503; 263.9

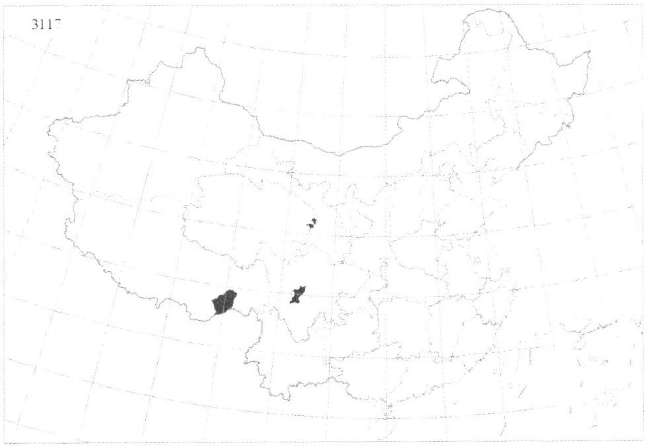

Rosaceae 蔷薇科 *Maddenia* 臭樱属
Maddenia incisoserrata T. T. Yu et T. C. Ku 锐齿臭樱
Shrubs, deciduous

10.7 (0.8~16.2); -1.3 (-10.6~5.3); 22.0 (9.9~27.6); 11.2 (3.8~16.2); 87.0 (15.5~134.4); 18.1 (0.0~66.0); 676 (411~859) ❙ 910 (341~1813); 401 (190~632); 50 (7~226) ❙ 33.4 (-40.9~125.9); 660 (341~859); 238.2 (61.2~334.4)

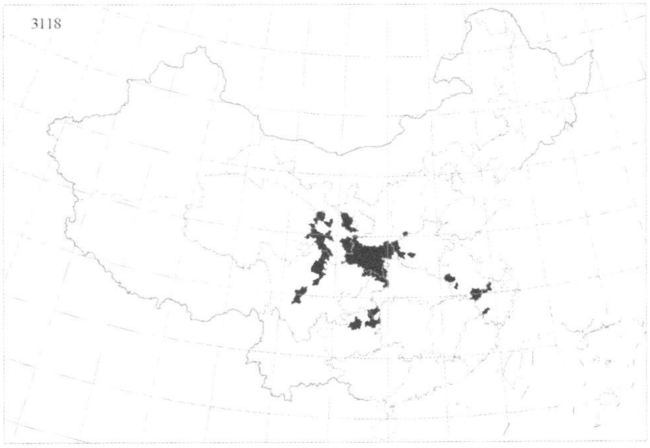

Rosaceae 蔷薇科 *Maddenia* 臭樱属
Maddenia wilsonii Koehne 华西臭樱
Shrubs or small trees, deciduous

10.1 (1.9~15.2); -0.8 (-9.2~5.3); 20.3 (9.9~26.0); 10.5 (3.8~15.2); 79.0 (15.5~123.9); 17.4 (0.0~57.8); 646 (413~816) ❙ 941 (374~1410); 441 (205~816); 37 (8~94) ❙ 48.5 (-29.3~125.9); 643 (374~816); 253.2 (81.2~334.4)

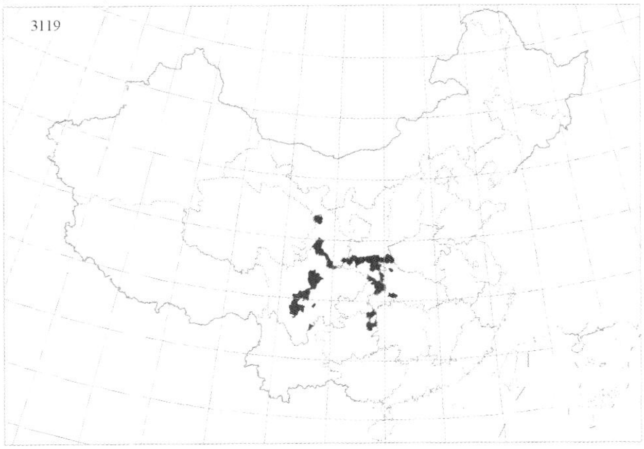

Rosaceae 蔷薇科 *Malus* 苹果属
Malus asiatica Nakai 花红
Trees, deciduous

13.3 (1.9~19.1); 2.4 (-20.1~12.5); 22.9 (9.5~29.3); 13.7 (3.8~19.2); 111.3 (15.5~170.1); 11.4 (0.0~91.9); 752 (411~981) ❙ 999 (225~1637); 489 (65~816); 55 (5~228) ❙ 31.5 (-65.1~125.9); 710 (225~981); 231.4 (33.5~516.7)

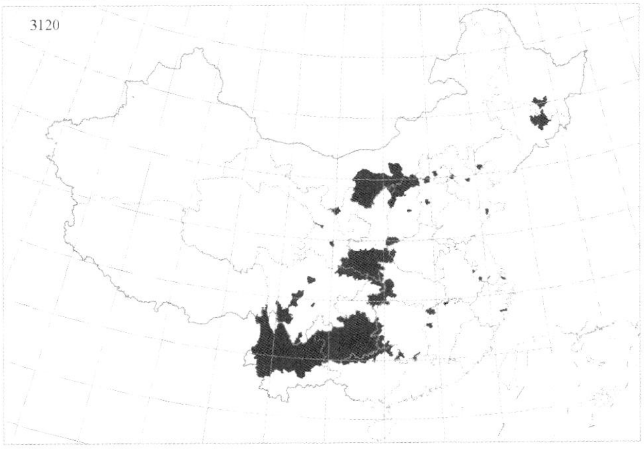

Rosaceae 蔷薇科 *Malus* 苹果属
Malus baccata (L.) Borkh. 山荆子
Trees, deciduous

8.7 (-5.8~17.7); -7.3 (-29.8~7.7); 23.0 (9.7~28.4); 10.6 (2.9~17.8); 85.4 (11.3~153.6); 39.9 (0.0~157.4); 675 (359~916) ❙ 622 (195~1736); 362 (133~1007); 22 (4~182) ❙ -7.4 (-67.0~145.8); 573 (195~916); 194.1 (43.7~447.4)

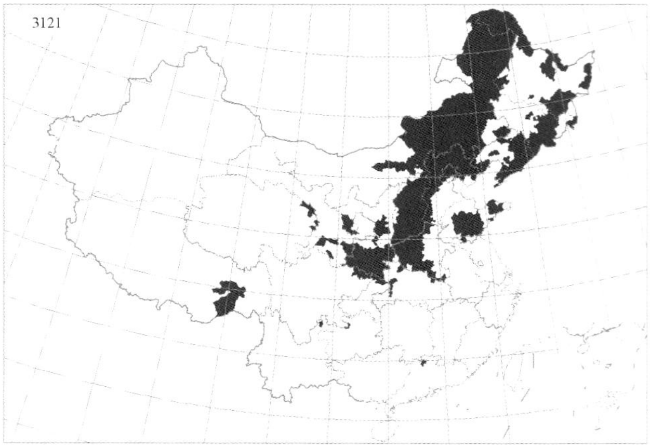

Rosaceae 蔷薇科 *Malus* 苹果属
Malus baccata var. gracilis (Rehder) T. C. Ku 垂枝山荆子
Trees, deciduous

9.8; -4.9; 22.8; 10.6; 84.0; 26.3; 665 ❙ 624; 350; 23 ❙ -6.3; 624; 162.7

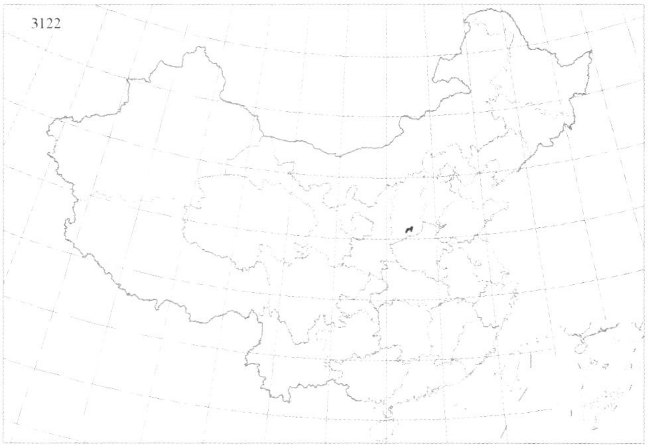

Rosaceae 蔷薇科 *Malus* 苹果属
Malus daochengensis C. L. Li 稻城海棠
Trees, deciduous

7.3 (1.4~10.8); -0.6 (-7.6~3.9); 14.1 (9.2~16.9); 7.8 (3.5~10.9); 47.3 (13.4~71.8); 19.7 (1.3~55.6); 540 (402~633) ❙ 912 (699~1135); 545 (426~644); 23 (10~40) ❙ 69.1 (52.8~89.0); 540 (402~633); 280.2 (208.8~351.7)

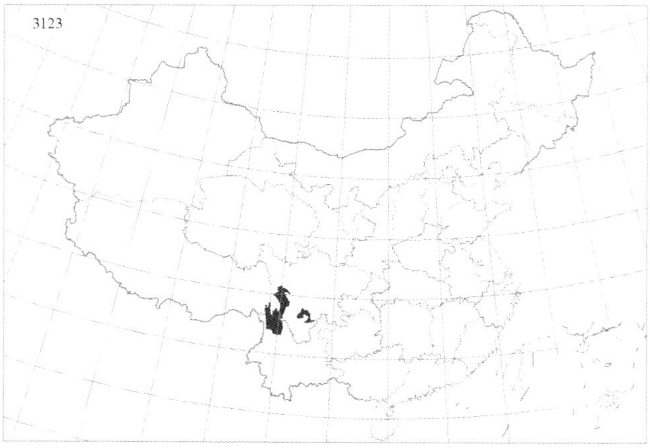

Rosaceae 蔷薇科 *Malus* 苹果属
Malus doumeri (Bois) A. Chev. 台湾海棠
Trees, deciduous

18.0 (10.8~24.1); 7.9 (2.1~18.9); 27.2 (16.6~30.1); 18.1 (10.9~24.1); 157.3 (71.8~229.7); 0.3 (0.0~4.9); 950 (633~1322) ❙ 1661 (968~3495); 626 (403~1594); 192 (24~897) ❙ 76.7 (16.9~246.5); 942 (633~1182); 241.9 (148.9~411.1)

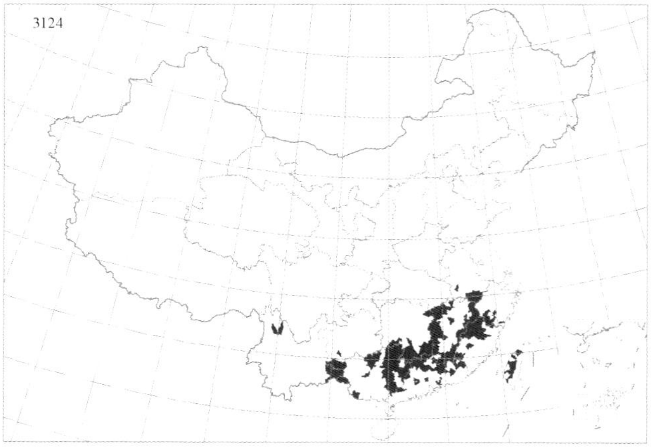

Rosaceae 蔷薇科 *Malus* 苹果属
Malus halliana Koehne 垂丝海棠
Trees, deciduous

14.3 (2.2~18.9); 2.6 (-6.6~12.0); 24.7 (9.9~29.9); 14.4 (3.8~19.0); 119.2 (15.5~167.8); 7.3 (0.0~49.1); 804 (413~976) ❙ 853 (512~1893); 451 (258~816); 49 (11~227) ❙ 8.2 (-36.5~129.9); 712 (413~910); 211.8 (120.1~487.2)

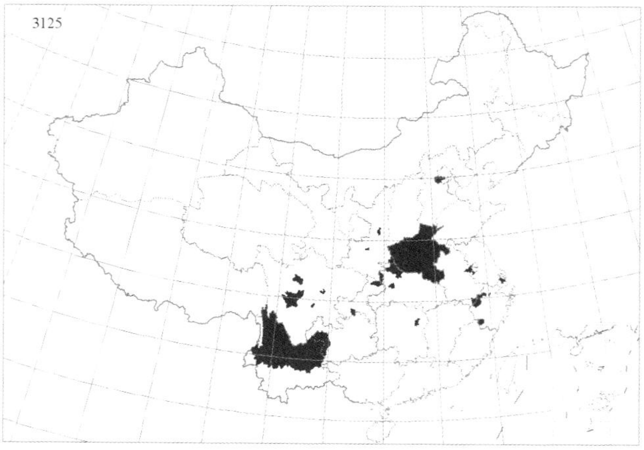

Rosaceae 蔷薇科 *Malus* 苹果属
Malus honanensis Rehder 河南海棠
Shrubs or small trees, deciduous

10.6 (3.4~15.3); -4.2 (-13.7~3.6); 23.8 (11.7~28.2); 11.4 (4.7~15.4); 93.3 (23.2~128.0); 25.7 (1.4~71.7); 708 (442~880) ❙ 605 (367~1435); 336 (201~591); 22 (7~113) ❙ -13.7 (-43.4~92.8); 576 (367~832); 177.9 (88.1~320.6)

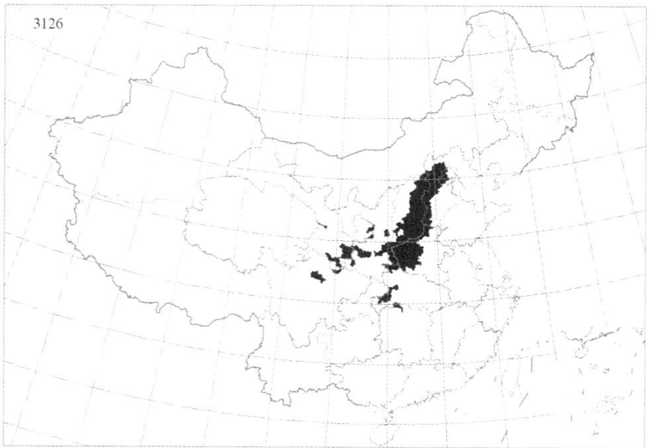

Rosaceae 蔷薇科 *Malus* 苹果属
Malus hupehensis (Pamp.) Rehder 湖北海棠
Trees, deciduous

14.7 (1.9~22.7); 3.2 (-13.7~15.1); 25.2 (9.9~30.1); 15.0 (3.8~22.8); 125.4 (15.4~213.1); 8.1 (0.0~71.7); 825 (413~1221) ❙ 1142 (367~2435); 499 (201~1428); 97 (7~259) ❙ 36.3 (-40.9~147.0); 783 (367~1221); 223.6 (88.1~659.0)

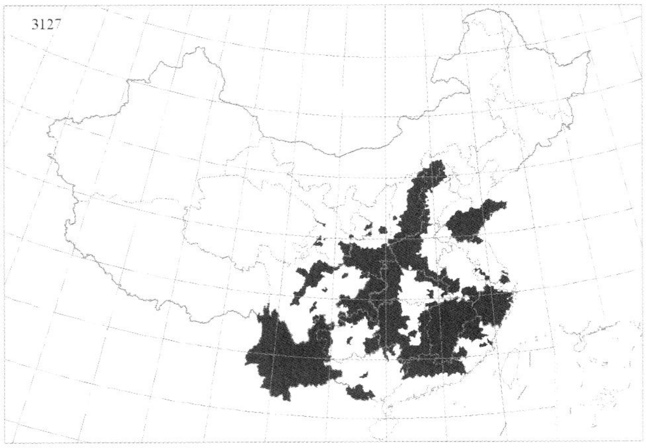

Rosaceae 蔷薇科 *Malus* 苹果属
Malus kansuensis (Batalin) C. K. Schneid. 陇东海棠
Shrubs or small trees, deciduous

9.1 (-4.1~16.8); -3.0 (-17.3~6.5); 20.3 (7.8~27.6); 9.9 (2.1~16.9); 75.9 (5.2~142.7); 25.8 (0.0~114.2); 636 (340~869) ❙ 751 (341~1368); 366 (190~584); 30 (5~129) ❙ 21.5 (-40.9~128.6); 604 (340~869); 221.4 (61.2~357.7)

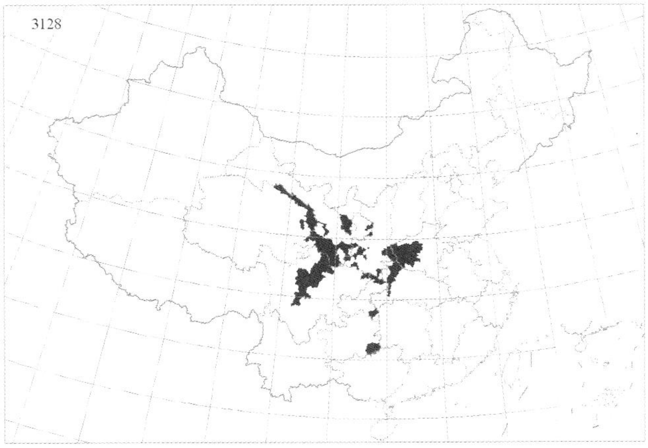

Rosaceae 蔷薇科 *Malus* 苹果属
Malus kansuensis var. calva (Rehder) T. C. Ku et Spongberg 光叶陇东海棠
Shrubs or small trees, deciduous

11.3 (2.2~15.1); -1.4 (-7.4~2.3); 23.3 (9.9~27.5); 11.6 (3.8~15.2); 93.0 (15.5~128.0); 16.8 (4.9~49.1); 702 (413~865) ❙ 770 (341~1214); 366 (190~546); 31 (8~83) ❙ 11.8 (-40.9~125.9); 669 (341~817); 224.9 (61.2~334.4)

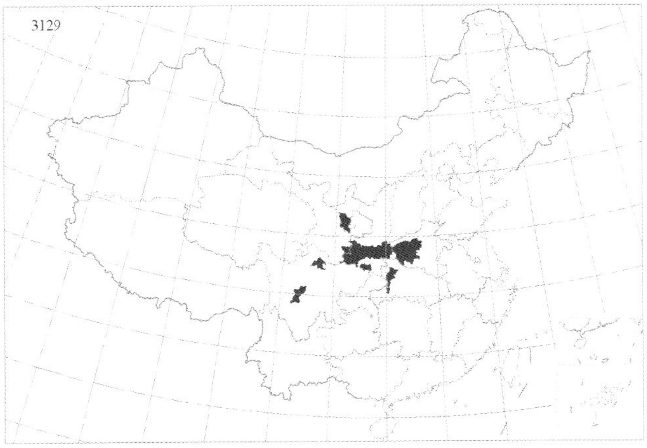

Rosaceae 蔷薇科 *Malus* 苹果属
Malus komarovii (Sarg.) Rehder 山楂海棠
Shrubs or small trees, deciduous

2.6 (1.4~4.1); -16.5 (-18.7~-14.1); 19.2 (18.4~20.9); 7.1 (6.4~8.1); 51.9 (45.8~61.8); 80.6 (70.4~88.8); 530 (501~575) ❙ 682 (561~777); 411 (340~473); 25 (16~35) ❙ 29.4 (-2.5~55.1); 528 (501~561); 318.8 (303.7~329.9)

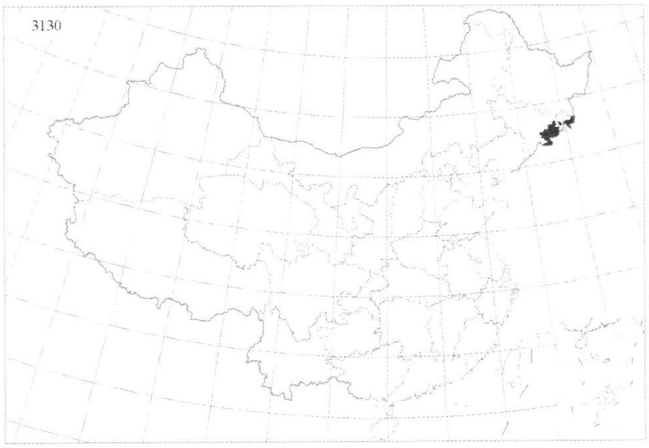

Rosaceae 蔷薇科 *Malus* 苹果属
Malus leiocalyca S. Z. Huang 光萼海棠
Shrubs or small trees, deciduous

17.5 (15.0~24.1); 7.1 (1.4~18.9); 27.0 (20.1~29.3); 17.5 (15.0~24.1); 150.7 (120.8~229.7); 0.6 (0.0~6.0); 923 (773~1322) ❙ 1711 (817~3495); 655 (399~1594); 201 (47~897) ❙ 86.2 (-8.8~246.5); 915 (773~1107); 237.7 (135.7~342.5)

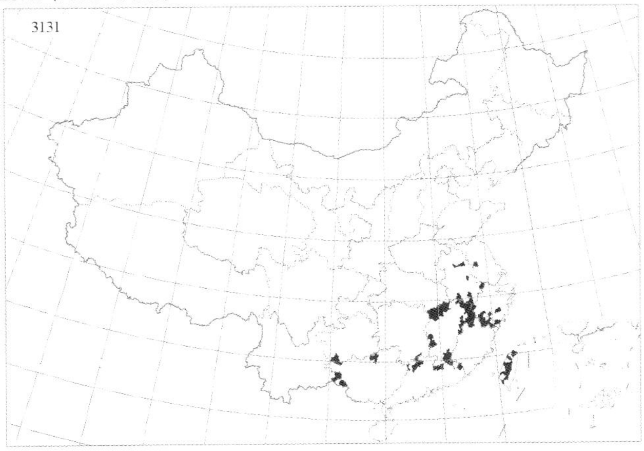

Rosaceae 蔷薇科 *Malus* 苹果属
Malus mandshurica (Maxim.) Kom. ex Juz. 毛山荆子
Trees, deciduous

7.9 (-5.8~15.6); -8.6 (-29.8~5.5); 22.5 (12.4~28.2); 9.9 (4.4~15.6); 79.5 (24.6~127.5); 44.4 (0.0~157.4); 648 (426~864) ❙ 599 (332~1489); 347 (215~691); 20 (5~196) ❙ -6.6 (-40.9~82.7); 549 (332~850); 197.9 (44.7~346.5)

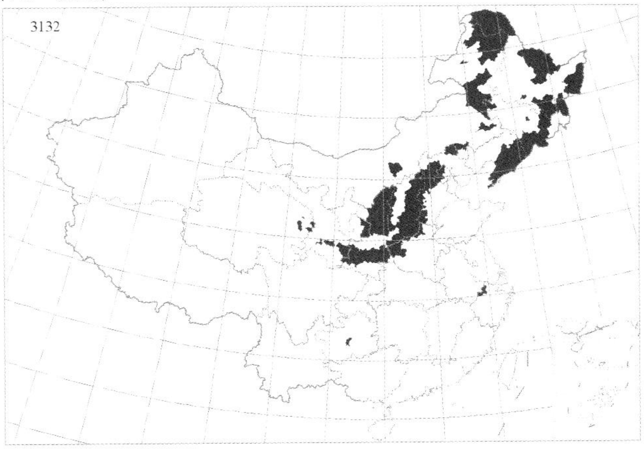

Rosaceae 蔷薇科 *Malus* 苹果属
Malus muliensis T. C. Ku 木里海棠
Trees, deciduous

6.6；-1.3；13.3；6.8；39.6；19.7；514 | 862；548；12 | 67.7；514；311.5

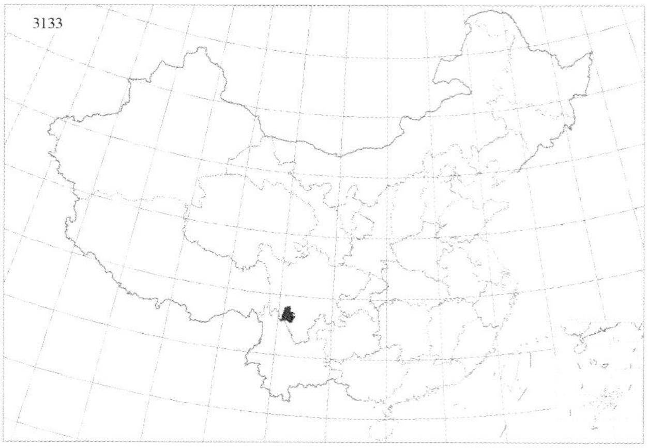

Rosaceae 蔷薇科 *Malus* 苹果属
Malus ombrophila Hand.-Mazz. 沧江海棠
Trees, deciduous

11.1 (5.3~16.7)；3.5 (-2.4~9.7)；17.5 (12.2~23.6)；11.2 (5.7~16.7)；79.2 (30.5~140.9)；5.4 (0.0~26.2)；644 (479~819) | 1005 (741~1410)；586 (426~816)；34 (12~69) | 57.6 (14.4~100.7)；636 (479~782)；301.2 (214.5~407.1)

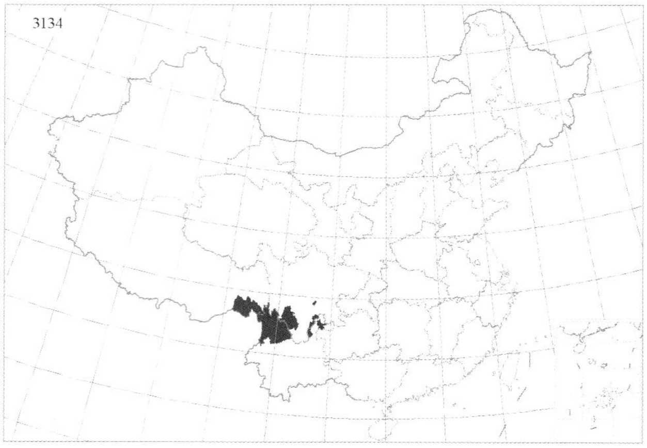

Rosaceae 蔷薇科 *Malus* 苹果属
Malus prattii (Hemsl.) C. K. Schneid. 西蜀海棠
Trees, deciduous

10.7 (2.2~16.2)；2.3 (-6.6~7.6)；18.2 (9.9~26.3)；10.9 (3.8~16.2)；77.6 (15.5~134.6)；8.3 (0.0~49.1)；643 (413~854) | 1069 (741~1414)；595 (426~817)；37 (15~71) | 68.6 (18.6~125.9)；641 (413~854)；273.4 (150.8~407.1)

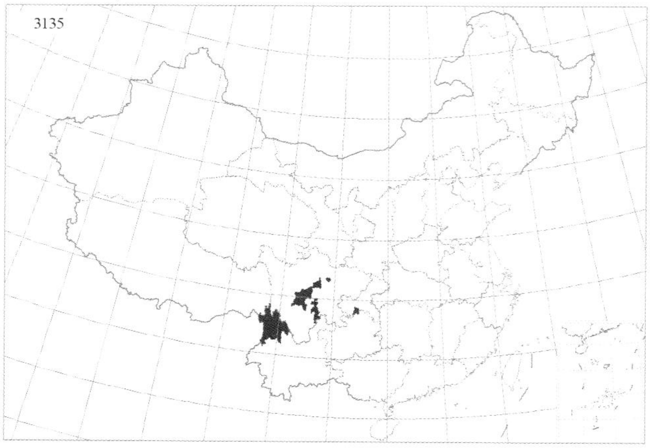

Rosaceae 蔷薇科 *Malus* 苹果属
Malus prunifolia (Willd.) Borkh. 秋子
Trees, deciduous

11.1 (2.2~18.2)；-4.6 (-16.1~11.8)；24.9 (12.3~28.3)；12.0 (5.8~18.2)；100.5 (31.5~158.4)；26.4 (0.0~81.0)；748 (481~874) | 563 (332~1446)；367 (200~688)；19 (7~100) | -23.7 (-51.1~87.2)；552 (332~847)；160.1 (44.7~351.7)

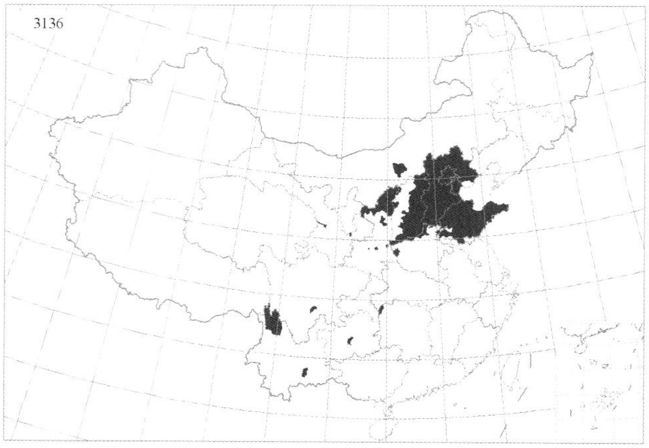

Rosaceae 蔷薇科 *Malus* 苹果属
Malus rockii Rehder 丽江山荆子
Trees, deciduous

8.2 (-1.1~15.0)；0.4 (-9.1~8.2)；14.8 (6.2~20.0)；8.8 (2.0~15.0)；58.5 (2.5~119.9)；20.1 (0.0~75.6)；563 (334~748) | 889 (321~1736)；521 (193~1007)；29 (0~59) | 55.3 (-26.0~145.8)；557 (321~748)；288.5 (79.3~447.4)

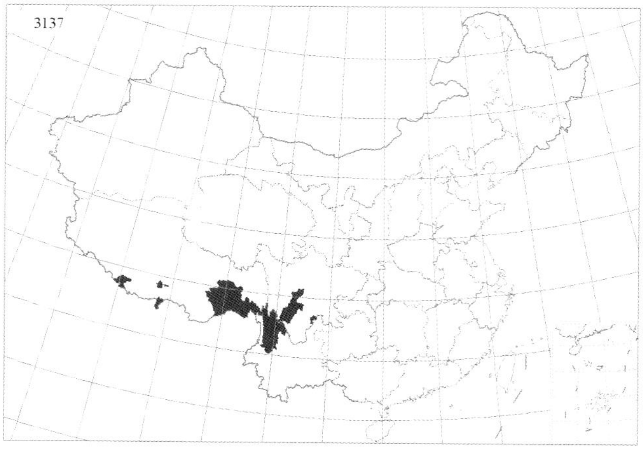

Rosaceae 蔷薇科 *Malus* 苹果属
Malus sieboldii (Regel) Rehder 三叶海棠
Shrubs, deciduous

14.7 (2.5~19.4)；3.2 (-10.4~10.1)；25.3 (12.4~29.3)；14.9 (4.8~19.5)；123.1 (25.1~173.7)；5.8 (0.0~56.4)；812 (441~1038) | 1193 (516~1922)；497 (259~720)；102 (9~259) | 44.8 (-33.2~128.9)；801 (441~1038)；215.4 (134.6~334.4)

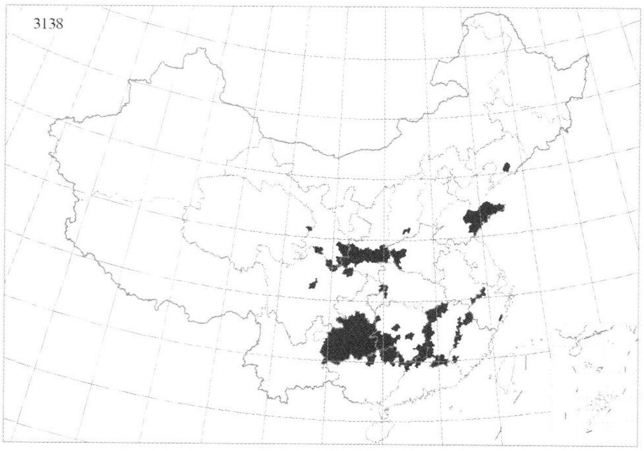

Rosaceae 蔷薇科 *Malus* 苹果属

Malus sieversii (Ledeb.) M. Roem. 新疆野苹果

Trees, deciduous

3.1 (-1.3~9.0); -15.6 (-20.9~-9.3); 17.8 (13.2~23.2); 7.1 (4.6~10.7); 52.0 (26.7~88.9); 74.6 (40.7~103.9); 544 (440~703) ▎310 (225~406); 136 (65~211); 31 (12~51) ▎-41.3 (-65.1~-7.6); 310 (225~406); 205.6 (141.5~297.3)

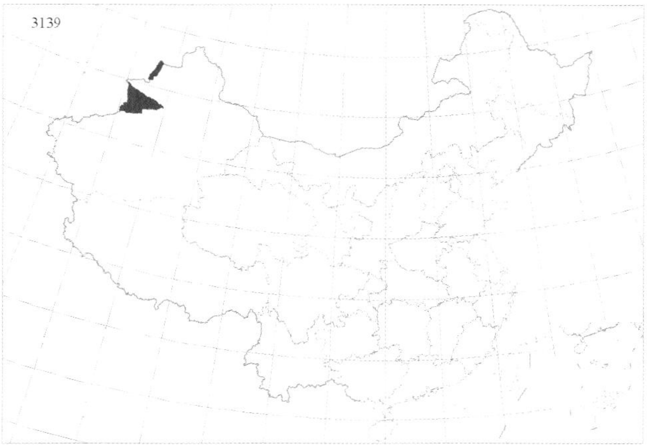

Rosaceae 蔷薇科 *Malus* 苹果属

Malus sikkimensis (Wenz.) Koehne 锡金海棠

Trees, deciduous

6.1 (-0.7~13.3); -1.9 (-9.6~5.4); 13.1 (7.7~19.6); 7.1 (2.5~13.4); 42.9 (6.8~100.4); 28.7 (0.0~74.2); 509 (338~706) ▎860 (370~1736); 511 (246~1007); 24 (7~51) ▎59.1 (0.0~145.8); 509 (338~706); 261.6 (61.6~447.4)

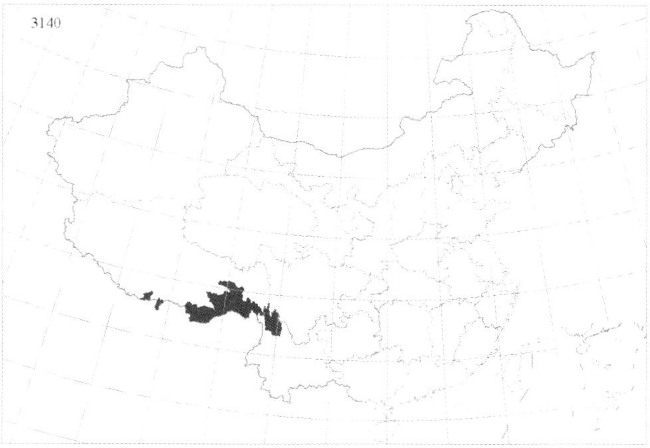

Rosaceae 蔷薇科 *Malus* 苹果属

Malus spectabilis (Aiton) Borkhausen 海棠花

Trees, deciduous

12.7 (0.2~18.9); 3.3 (-12.4~12.0); 20.7 (9.9~28.9); 13.3 (3.8~19.0); 106.6 (15.5~167.8); 13.1 (0.0~74.6); 723 (407~929) ▎842 (386~1489); 465 (219~627); 38 (5~197) ▎16.3 (-39.7~125.9); 665 (386~910); 247.0 (111.8~382.6)

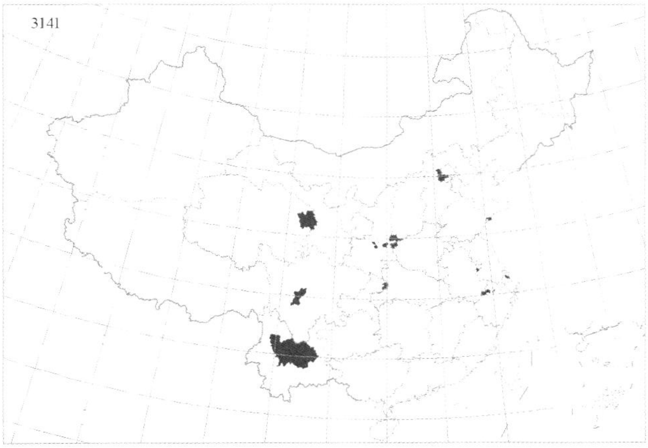

Rosaceae 蔷薇科 *Malus* 苹果属

Malus toringoides (Rehder) Hughes 变叶海棠

Shrubs or small trees, deciduous

3.7 (-1.0~9.1); -7.4 (-12.5~-3.5); 13.6 (8.1~23.0); 5.6 (2.7~10.2); 33.6 (8.0~83.0); 48.3 (27.8~79.5); 471 (363~658) ▎617 (318~950); 345 (223~546); 13 (1~25) ▎37.0 (-31.4~128.6); 428 (318~577); 216.3 (50.7~341.6)

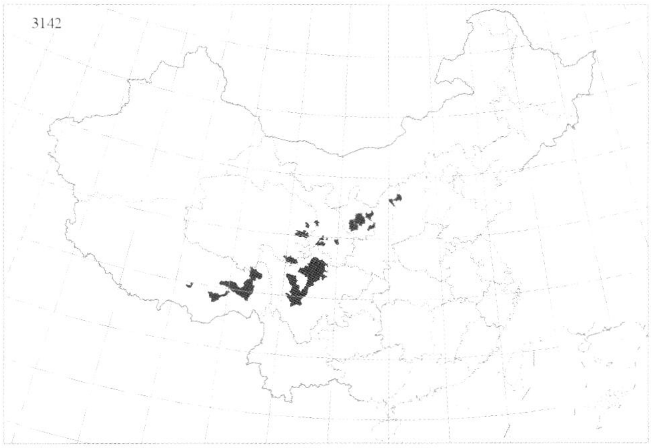

Rosaceae 蔷薇科 *Malus* 苹果属

Malus transitoria (Batalin) C. K. Schneid. 花叶海棠

Shrubs or small trees, deciduous

6.7 (-1.3~13.8); -7.6 (-13.5~6.7); 19.2 (8.5~26.6); 8.4 (2.6~13.8); 63.1 (8.1~113.3); 42.6 (0.0~86.3); 581 (347~800) ▎474 (132~963); 263 (81~592); 11 (2~39) ▎-13.3 (-80.1~125.9); 428 (132~690); 148.8 (13.6~344.4)

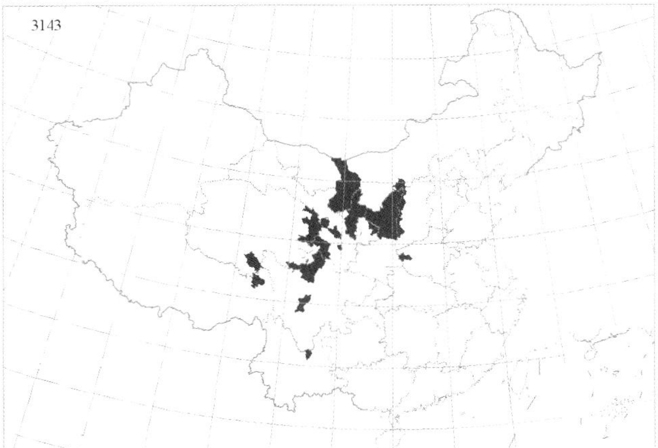

Rosaceae 蔷薇科 *Malus* 苹果属

Malus transitoria var. glabrescens T. T. Yu et T. C. Ku 少毛花叶海棠

Shrubs or small trees, deciduous

2.4; -7.2; 11.3; 4.4; 21.4; 51.6; 423 ▎510; 311; 11 ▎20.7; 423; 237.6

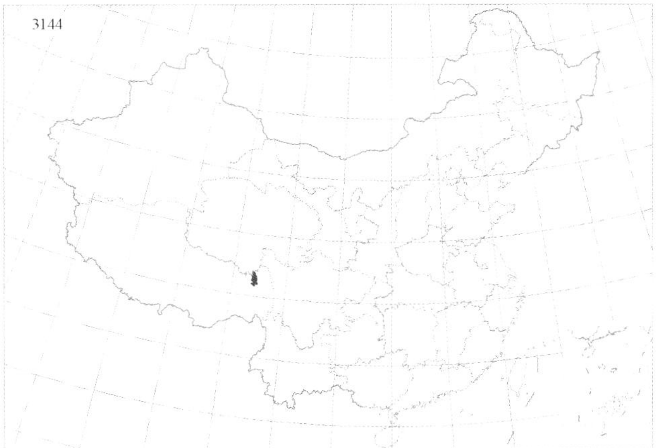

Rosaceae 蔷薇科 *Malus* 苹果属

Malus yunnanensis (Franch.) C. K. Schneid. 滇池海棠

Trees, deciduous

11.5 (2.6~16.2); 1.5 (−7.0~8.7); 20.9 (11.1~26.6); 11.7 (4.3~16.2); 88.6 (20.0~134.6); 9.6 (0.0~48.2); 681 (427~854) | 1018 (473~1466); 489 (287~727); 43 (7~120) | 51.4 (5.0~121.7); 680 (427~854); 275.6 (170.4~407.1)

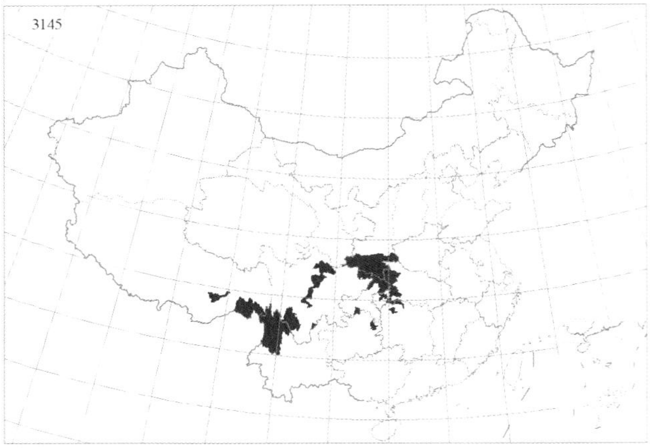

Rosaceae 蔷薇科 *Malus* 苹果属

Malus yunnanensis var. veitchii (Osborn) Rehder 川鄂滇池海棠

Trees, deciduous

11.5 (3.3~15.3); 0.9 (−5.4~5.4); 21.4 (11.1~26.5); 11.7 (4.6~15.4); 89.4 (22.3~125.7); 11.1 (0.0~41.8); 683 (427~832) | 1023 (473~1736); 472 (287~1007); 52 (7~113) | 48.8 (−8.1~145.8); 680 (427~832); 259.2 (170.4~447.4)

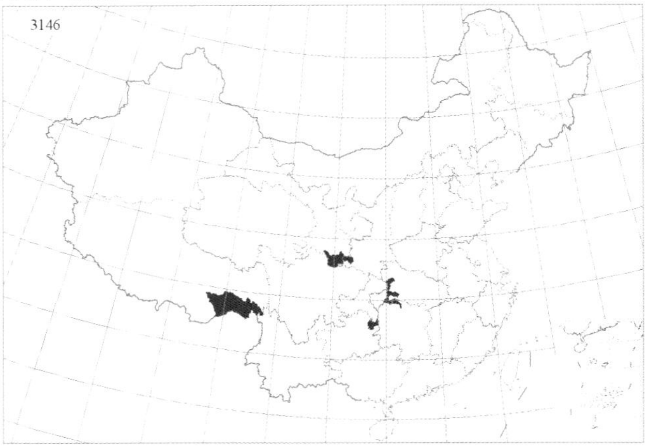

Rosaceae 蔷薇科 *Malus* 苹果属

Malus × micromalus Makino 西府海棠

Trees, deciduous

14.7 (5.0~18.9); 6.6 (−12.3~12.5); 21.1 (16.6~28.2); 15.0 (7.7~19.0); 122.1 (57.5~167.8); 4.6 (0.0~61.5); 767 (550~929) | 951 (245~1454); 543 (65~792); 38 (7~70) | 24.4 (−65.1~99.9); 714 (245~909); 289.8 (88.1~487.2)

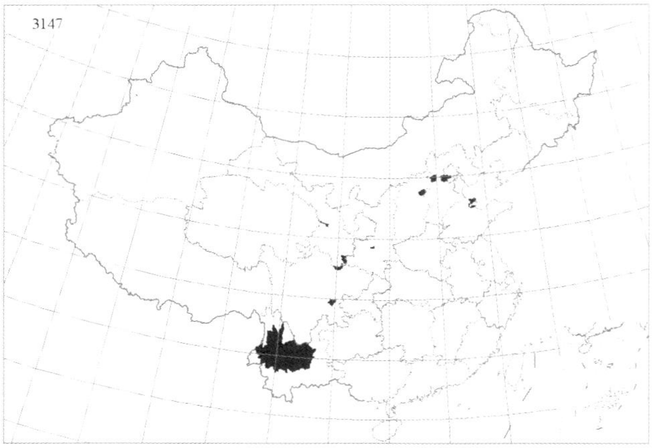

Rosaceae 蔷薇科 *Neillia* 绣线梅属

Neillia affinis Hemsl. 川康绣线梅

Shrubs, deciduous

12.5 (−1.8~21.0); 4.6 (−10.2~14.2); 18.9 (6.1~26.3); 12.8 (1.8~21.0); 100.4 (1.9~191.8); 10.3 (0.0~82.8); 702 (329~1053) | 1119 (320~1547); 627 (217~911); 42 (10~71) | 60.3 (−3.0~125.9); 700 (320~1053); 313.1 (61.6~551.6)

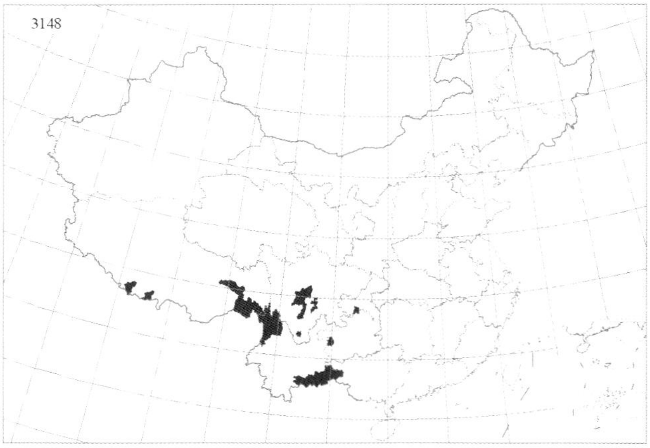

Rosaceae 蔷薇科 *Neillia* 绣线梅属

Neillia affinis var. pauciflora (Rehder) J. E. Vidal 少花川康绣线梅

Shrubs, deciduous

15.1 (8.6~18.1); 7.3 (1.3~11.7); 21.5 (15.0~25.9); 15.2 (8.7~18.0); 123.2 (52.6~156.5); 1.3 (0.0~8.8); 777 (574~863) | 1142 (958~1368); 587 (473~756); 57 (40~94) | 49.0 (13.3~93.0); 775 (574~863); 323.5 (170.4~491.5)

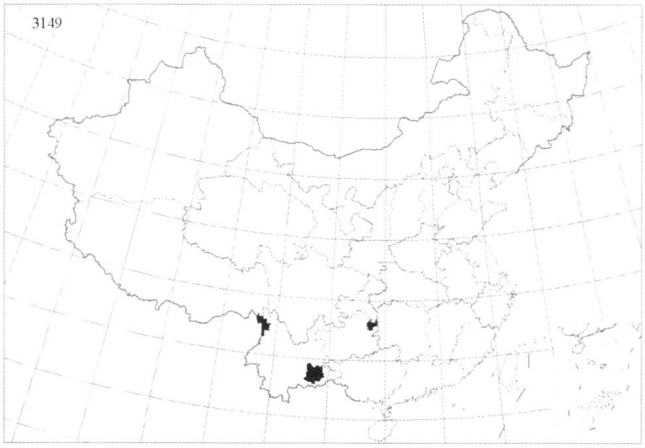

Rosaceae 蔷薇科 *Neillia* 绣线梅属

Neillia breviracemosa T. C. Ku 短序绣线梅

Shrubs, deciduous

13.2; 6.4; 18.4; 13.3; 99.4; 0.0; 695 | 1337; 727; 69 | 92.4; 695; 407.1

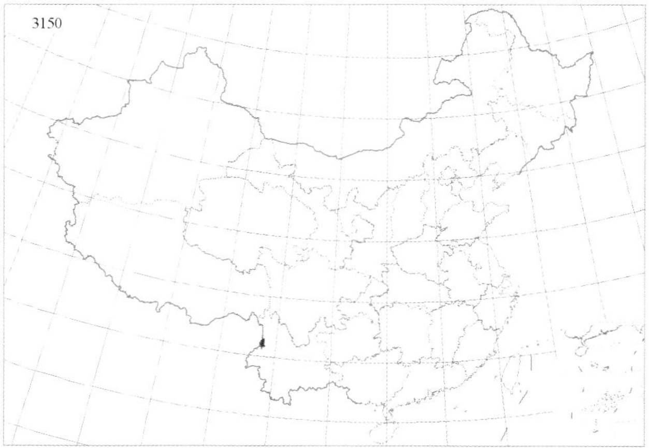

Rosaceae 蔷薇科 *Neillia* 绣线梅属
Neillia densiflora T. T. Yu et T. C. Ku 密花绣线梅
Shrubs, deciduous

0.3；-8.6 8.3；2.9；9.6；65.4；368 | 389；246；11 | 0.0；368；142.5

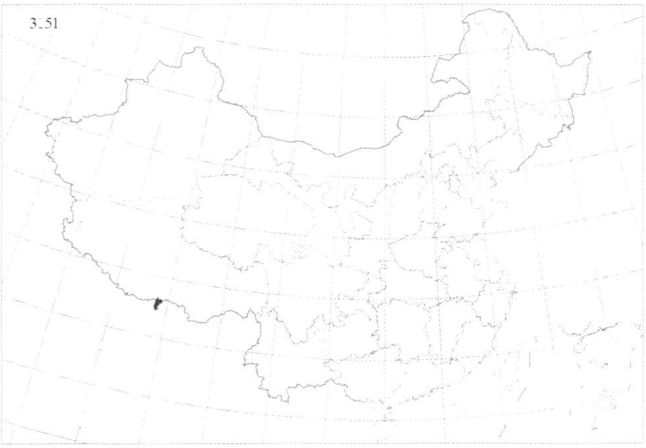

Rosaceae 蔷薇科 *Neillia* 绣线梅属
Neillia fugongensis T. C. Ku 福贡绣线梅
Shrubs, deciduous

11.3；4.3；17.0；11.3；76.6；0.7；644 | 1243；696；58 | 93.0；644；384.7

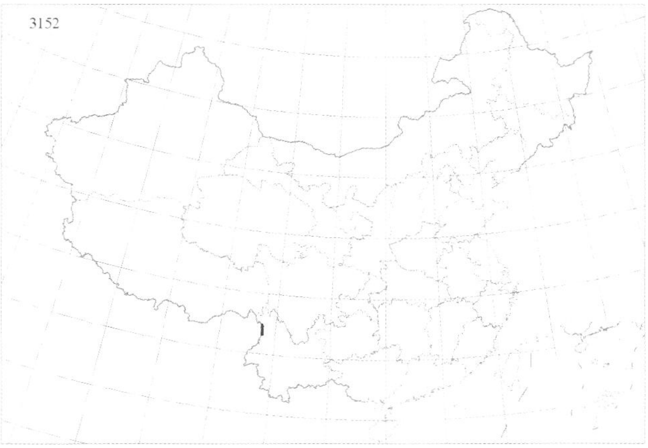

Rosaceae 蔷薇科 *Neillia* 绣线梅属
Neillia gracilis Franch. 矮生绣线梅
Subshrubs, deciduous

13.9 (3.8~18.9)；6.4 (-4.3~12.0)；19.9 (10.8~26.6)；14.0 (4.7~19.0)；109.2 (21.6~167.8)；1.7 (0.0~35.4)；728 (444~929) | 983 (741~1252)；564 (411~636)；36 (15~58) | 36.8 (-14.2~121.7)；708 (444~912)；269.0 (142.9~382.6)

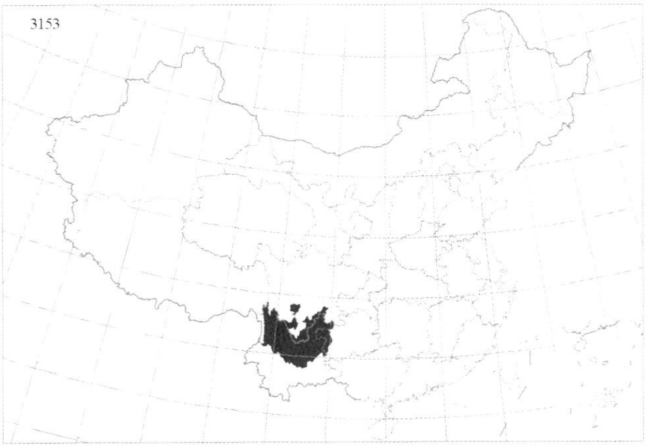

Rosaceae 蔷薇科 *Neillia* 绣线梅属
Neillia grandiflora T. T. Yu et T. C. Ku 大花绣线梅
Shrubs, deciduous

-1.8；-10.2；6.1；1.8；1.9；82.8；329 | 320；217；20 | -3.0；320；72.2

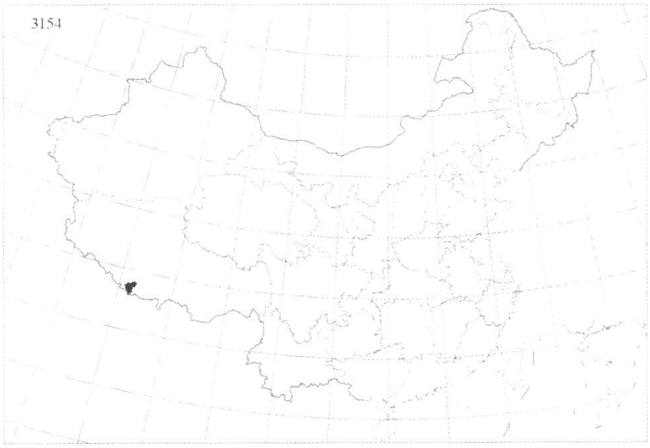

Rosaceae 蔷薇科 *Neillia* 绣线梅属
Neillia jinggangshanensis Z. X. Yu 井冈山绣线梅
Shrubs, deciduous

16.5 (14.3~17.4)；4.8 (3.7~5.7)；27.6 (24.2~29.2)；16.5 (14.3~17.4)；138.8 (113.2~149.4)；0.3 (0.0~1.4)；888 (761~950) | 1609 (1555~1742)；511 (483~581)；216 (200~225) | 82.3 (66.4~128.9)；886 (761~950)；221.8 (186.9~255.1)

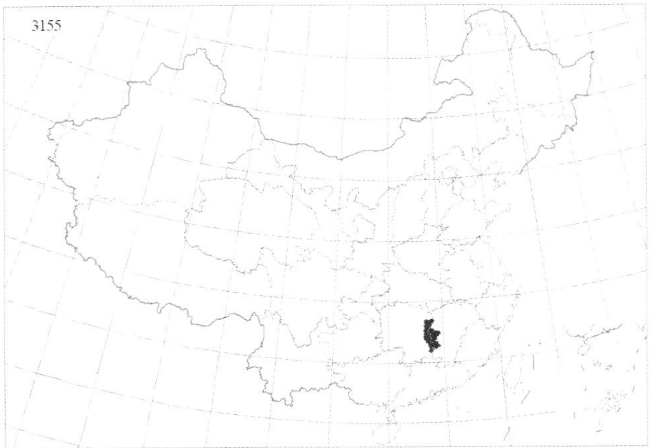

Rosaceae 蔷薇科 *Neillia* 绣线梅属
Neillia ribesioides Rehder 毛叶绣线梅
Shrubs, deciduous

12.9 (2.2~18.9)；3.5 (-7.4~12.5)；21.2 (9.9~27.7)；13.1 (3.8~19.0)；103.1 (15.5~167.8)；7.5 (0.0~49.1)；716 (413~930) | 938 (341~1478)；492 (190~686)；36 (8~118) | 31.9 (-40.9~125.9)；700 (341~930)；263.3 (61.2~406.0)

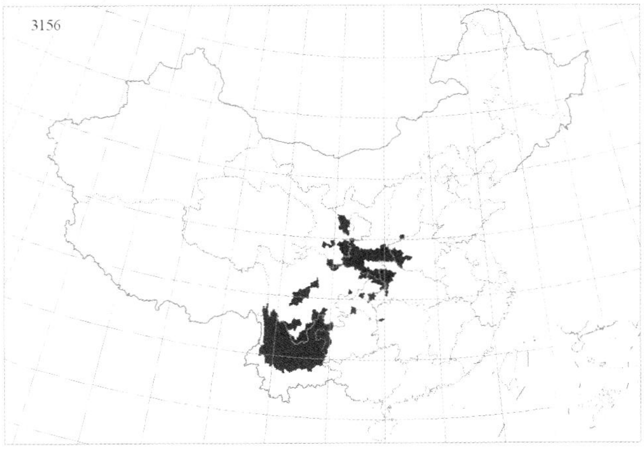

Rosaceae 蔷薇科 *Neillia* 绣线梅属

Neillia rubiflora D. Don 粉花绣线梅

Shrubs, deciduous

9.5 (−1.8~18.2); 2.0 (−10.2~11.9); 15.7 (6.1~22.0); 10.1 (1.8~18.2); 73.5 (1.9~158.7); 18.5 (0.0~82.8); 608 (329~868) | 1004 (320~1435); 574 (217~816); 33 (11~49) | 62.8 (−3.0~101.2); 604 (320~847); 316.2 (72.2~516.7)

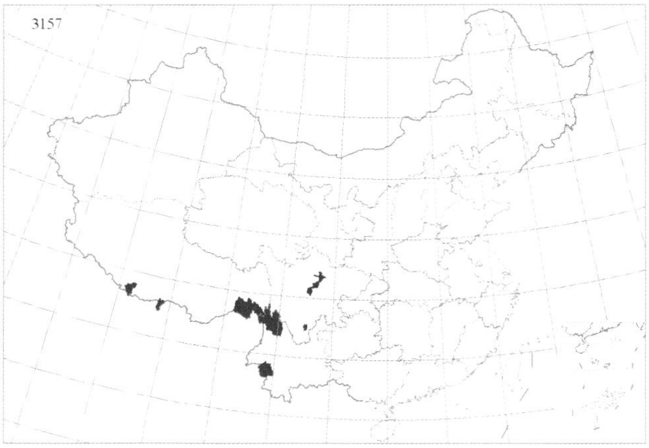

Rosaceae 蔷薇科 *Neillia* 绣线梅属

Neillia serratisepala H. L. Li 云南绣线梅

Shrubs, deciduous

16.3 (1.8~21.0); 9.5 (−7.1~14.2); 21.1 (9.9~26.3); 16.4 (3.8~21.0); 137.7 (15.7~191.8); 1.7 (0.0~53.3); 813 (404~1053) | 1230 (527~1751); 681 (317~996); 54 (9~169) | 51.8 (6.5~99.9); 801 (404~1053); 382.4 (169.5~551.6)

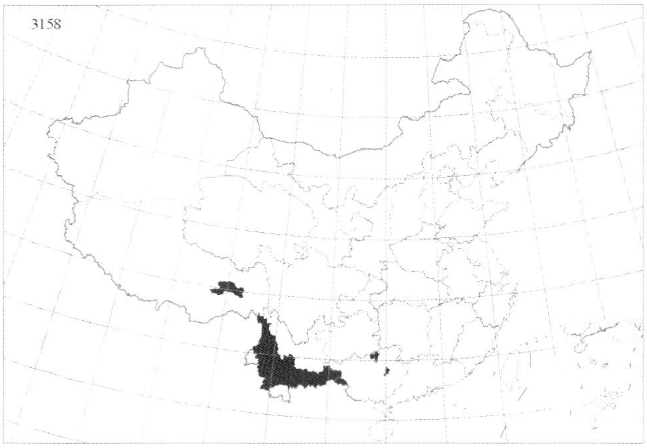

Rosaceae 蔷薇科 *Neillia* 绣线梅属

Neillia sinensis Oliv. 中华绣线梅

Shrubs, deciduous

15.0 (3.0~21.0); 4.5 (−7.7~13.1); 24.5 (12.4~29.8); 15.1 (4.8~21.1); 124.4 (25.1~193.0); 4.0 (0.0~49.3); 810 (441~1114) | 1170 (341~1829); 501 (190~821); 92 (8~230) | 43.4 (−40.9~128.9); 796 (341~1114); 228.6 (61.2~411.1)

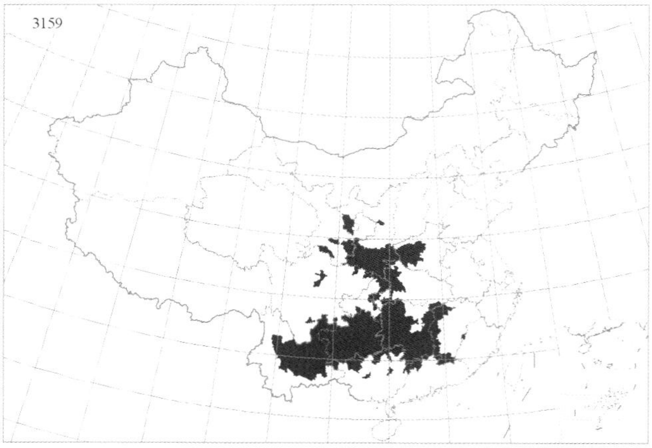

Rosaceae 蔷薇科 *Neillia* 绣线梅属

Neillia sinensis var. caudata Rehder 尾叶中华绣线梅

Shrubs, deciduous

18.8 (16.9~21.0); 12.4 (10.3~14.2); 23.1 (21.4~25.6); 18.8 (16.9~21.0); 165.7 (142.7~191.8); 0.0 (0.0~0.0); 910 (813~1053) | 1386 (1178~1547); 785 (646~911); 58 (51~68) | 52.7 (36.8~74.2); 908 (813~1053); 489.3 (383.4~551.6)

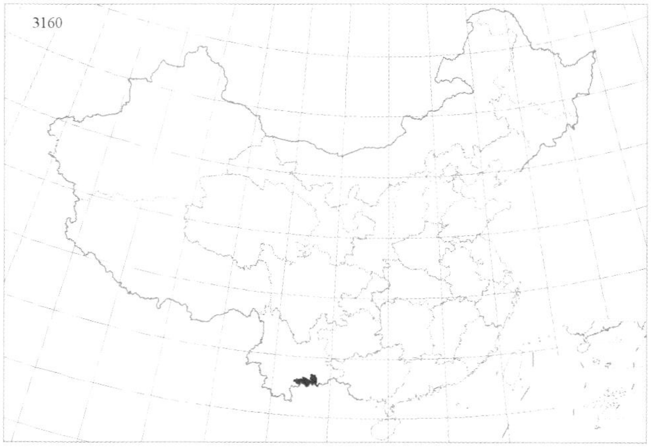

Rosaceae 蔷薇科 *Neillia* 绣线梅属

Neillia sparsiflora Rehder 疏花绣线梅

Shrubs, deciduous

13.0; 6.1; 18.6; 13.1; 97.9; 0.9; 696 | 970; 594; 21 | 40.5; 661; 301.8

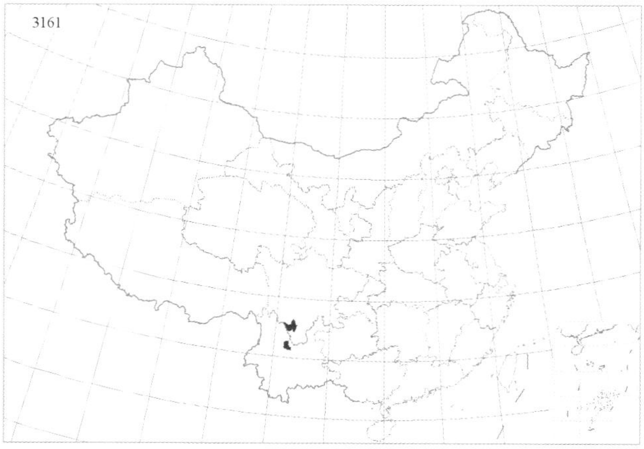

Rosaceae 蔷薇科 *Neillia* 绣线梅属

Neillia thibetica Bureau et Franch. 西康绣线梅

Shrubs, deciduous

10.4 (2.2~16.6); 1.9 (−6.6~6.6); 18.0 (9.9~25.7); 10.6 (3.8~16.6); 74.9 (15.5~139.7); 9.1 (0.0~49.1); 634 (413~866) | 1060 (741~1590); 606 (420~946); 31 (15~59) | 68.6 (38.2~125.9); 632 (413~866); 275.1 (166.3~352.3)

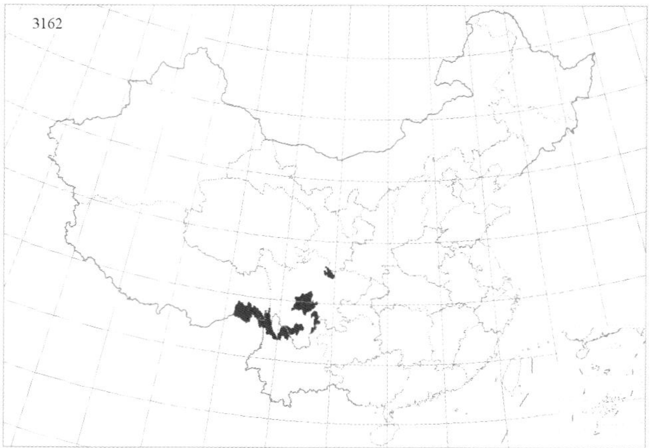

Rosaceae 蔷薇科 *Neillia* 绣线梅属
Neillia thibetica var. lobata (Rehder) T. T. Yu 裂叶西康绣线梅
Shrubs, deciduous

10.1 (6.1~13.7); 2.2 (-2.4~4.7); 16.9 (12.7~21.2); 10.2 (6.3~13.7); 69.4 (35.2~104.8); 7.6 (0.3~24.1); 615 (496~722) ❙ 986 (801~1143); 576 (496~619); 27 (15~50) ❙ 61.2 (51.8~92.9); 614 (496~722); 260.5 (192.4~319.3)

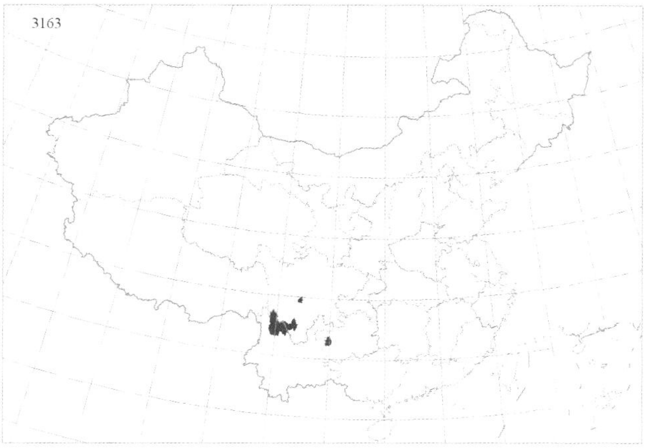

Rosaceae 蔷薇科 *Neillia* 绣线梅属
Neillia thyrsiflora D. Don 绣线梅
Shrubs, deciduous

16.6 (1.8~21.0); 9.8 (-7.1~15.1); 21.3 (9.9~26.3); 16.6 (3.8~21.0); 140.3 (15.7~191.8); 1.2 (0.0~53.3); 820 (404~1053) ❙ 1276 (583~1751); 707 (348~1007); 52 (13~142) ❙ 56.4 (12.5~145.8); 809 (404~1053); 401.6 (169.5~659.0)

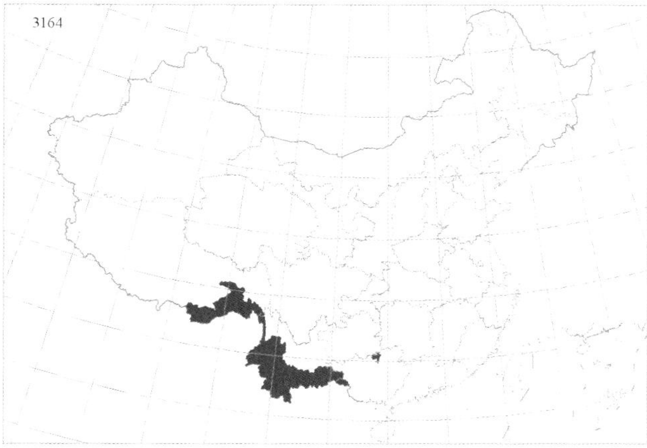

Rosaceae 蔷薇科 *Neillia* 绣线梅属
Neillia thyrsiflora var. tunkinensis (J. E. Vidal) J. E. Vidal 毛果绣线梅
Shrubs, deciduous

15.0 (11.7~19.2); 6.3 (2.7~10.9); 22.2 (19.6~27.5); 15.0 (11.7~19.2); 121.0 (84.6~170.8); 1.1 (0.0~3.8); 777 (661~1002) ❙ 1165 (964~1495); 604 (501~756); 62 (39~135) ❙ 49.9 (39.0~60.1); 777 (661~1002); 269.4 (183.0~491.5)

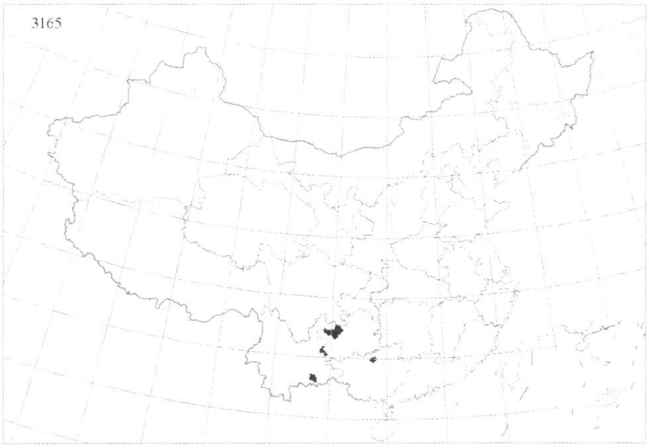

Rosaceae 蔷薇科 *Osteomeles* 小石积属
Osteomeles anthyllidifolia (Sm.) Lindl. 小石积
Shrubs

18.5; 14.1; 21.7; 18.5; 162.1; 0.0; 863 ❙ 2713; 1388; 188 ❙ 214.4; 863; 363.7

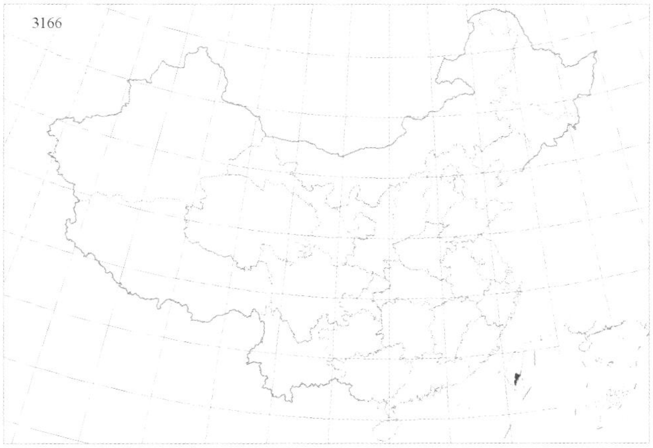

Rosaceae 蔷薇科 *Osteomeles* 小石积属
Osteomeles schwerinae C. K. Schneid. 华西小石积
Shrubs, deciduous or semievergreen

14.5 (2.2~19.3); 7.1 (-7.0~13.5); 20.4 (9.5~26.7); 14.6 (3.9~19.4); 117.7 (15.8~173.2); 3.0 (0.0~48.6); 752 (411~939) ❙ 1054 (669~1751); 589 (317~996); 44 (11~191) ❙ 41.8 (-14.2~121.7); 735 (411~927); 298.7 (142.9~551.3)

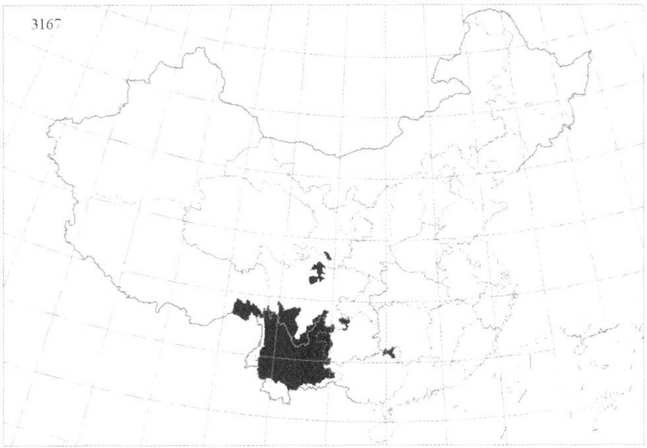

Rosaceae 蔷薇科 *Osteomeles* 小石积属
Osteomeles schwerinae var. microphylla Rehder et E. H. Wilson 小叶华西小石积
Shrubs, deciduous or semievergreen

7.4 (1.9~12.6); -3.4 (-9.2~1.0); 17.3 (11.7~23.5); 8.3 (4.3~12.6); 57.6 (21.1~99.9); 28.2 (8.6~57.8); 565 (427~724) ❙ 682 (621~760); 339 (317~382); 15 (13~17) ❙ 23.6 (2.4~48.8); 565 (427~724); 273.1 (220.9~294.3)

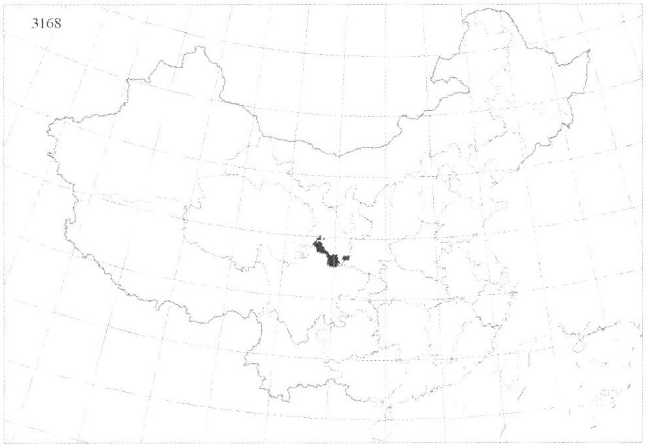

Rosaceae 蔷薇科 *Osteomeles* 小石积属
Osteomeles subrotunda K. Koch 圆叶小石积
Shrubs, evergreen

16.5; 5.8; 26.2; 16.5; 138.8; 0.4; 869 ❙ 1583; 567; 188 ❙ 83.7; 869; 243.8

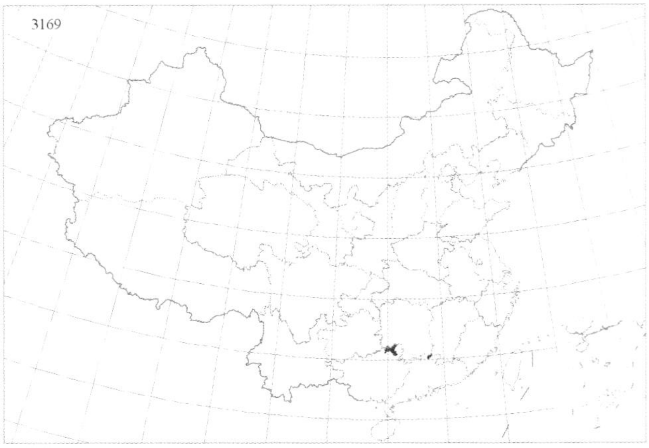

Rosaceae 蔷薇科 *Padus* 稠李属
Padus avium Mill. 稠李
Trees, deciduous

8.1 (−5.8~15.1); −7.6 (−29.8~2.3); 22.2 (11.7~27.5); 10.1 (4.3~15.2); 80.5 (21.1~128.0); 42.6 (4.9~157.4); 652 (426~865) ❙ 583 (190~1164); 322 (80~499); 21 (5~65) ❙ −11.3 (−66.6~83.5); 549 (190~817); 200.1 (54.8~334.4)

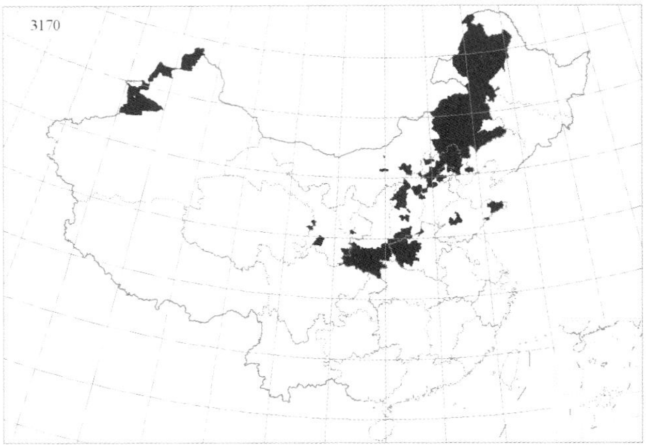

Rosaceae 蔷薇科 *Padus* 稠李属
Padus avium var. pubescens (Regel et Tiling) T. C. Ku et B. M. Barthol. 毛叶稠李
Trees, deciduous

8.7 (3.9~13.5); −6.7 (−15.3~−0.1); 22.5 (14.7~26.6); 10.1 (6.0~13.6); 81.0 (36.3~114.9); 36.4 (11.5~71.1); 652 (482~804) ❙ 584 (132~1073); 335 (81~691); 18 (2~44) ❙ −10.0 (−80.1~74.1); 534 (132~773); 205.0 (13.6~335.3)

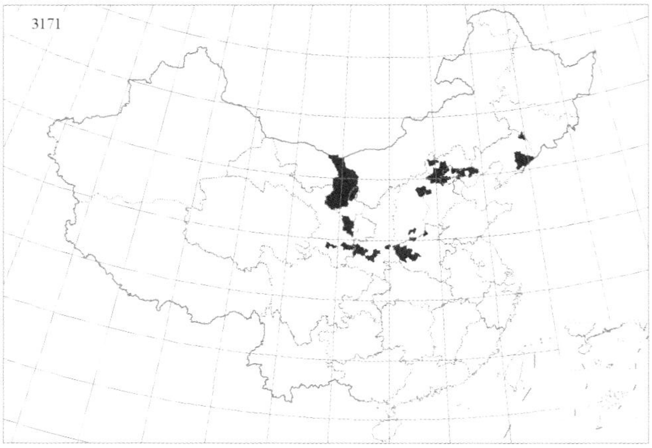

Rosaceae 蔷薇科 *Padus* 稠李属
Padus brachypoda (Batalin) C. K. Schneid. 短梗稠李
Trees, deciduous

12.5 (2.0~18.9); 2.4 (−7.7~12.0); 21.7 (9.9~29.3); 12.7 (3.8~19.0); 99.6 (15.4~167.8); 9.1 (0.0~51.4); 713 (413~964) ❙ 1014 (341~1893); 493 (190~816); 55 (7~227) ❙ 42.7 (−40.9~129.9); 697 (341~955); 259.6 (61.2~406.0)

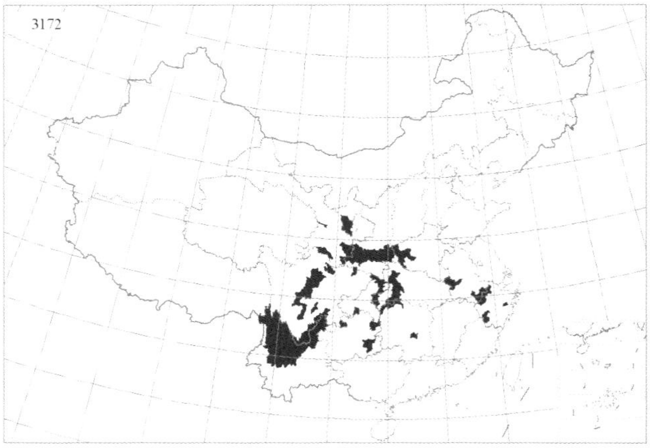

Rosaceae 蔷薇科 *Padus* 稠李属
Padus brunnescens T. T. Yu et T. C. Ku 褐毛稠李
Trees, deciduous

9.6; 1.5; 16.7; 9.6; 63.4; 7.7; 602 ❙ 1107; 634; 26 ❙ 84.4; 602; 295.6

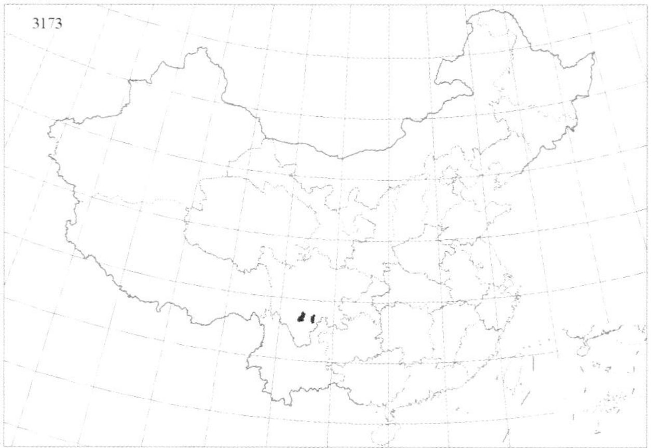

Rosaceae 蔷薇科 *Padus* 稠李属
Padus buergeriana (Miq.) T. T. Yu et T. C. Ku 橉木
Trees, deciduous

15.6 (−1.8~21.0); 5.0 (−10.2~14.2); 25.4 (6.1~29.9); 15.7 (1.8~21.0); 129.9 (1.9~191.8); 2.6 (0.0~82.8); 839 (329~1113) ❙ 1369 (320~1997); 540 (217~909); 138 (11~259) ❙ 62.6 (−23.0~147.0); 829 (320~1053); 244.4 (72.2~551.6)

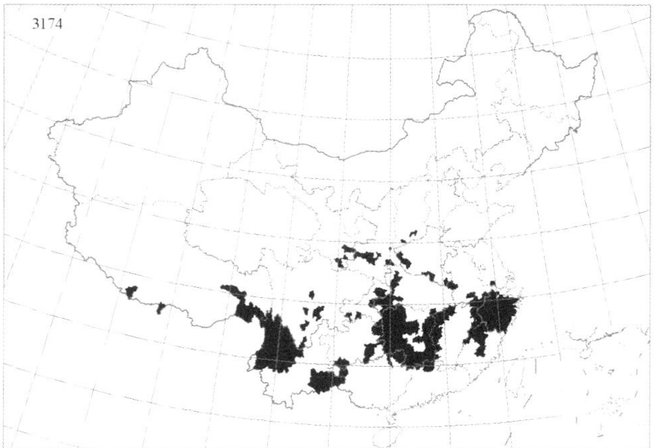

Rosaceae 蔷薇科 *Padus* 稠李属

Padus cornuta (Wall. ex Royle) Carrière 光萼稠李

Trees, deciduous

2.8 (-1.8~12.2); -5.5 (-10.2~4.3); 10.2 (6.1~18.5); 4.7 (1.8~12.2); 24.6 (1.9~87.4); 50.7 (0.7~82.8); 431 (329~672) ❙ 667 (320~1431); 397 (193~817); 23 (11~33) ❙ 41.8 (-4.0~113.0); 427 (320~672); 176.9 (72.2~363.7)

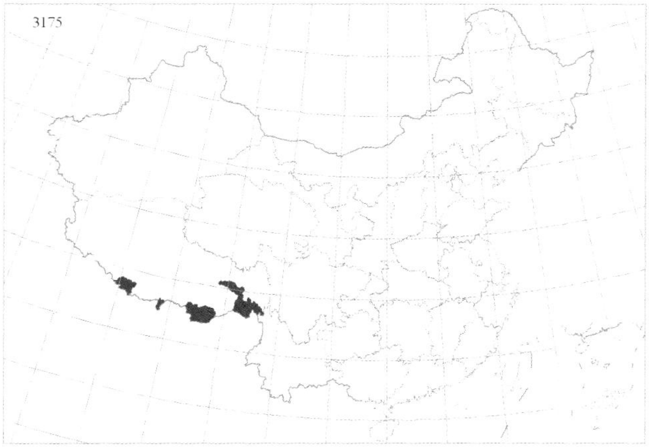

Rosaceae 蔷薇科 *Padus* 稠李属

Padus grayana (Maxim.) C. K. Schneid. 灰叶稠李

Trees, deciduous

15.5 (3.2~19.9); 4.1 (-6.4~11.8); 26.3 (11.4~29.4); 15.6 (4.7~19.9); 129.9 (22.7~179.3); 2.8 (0.0~43.8); 847 (436~1036) ❙ 1354 (626~1997); 512 (316~816); 139 (17~259) ❙ 59.5 (-22.2~130.7); 834 (436~1036); 217.7 (134.6~357.7)

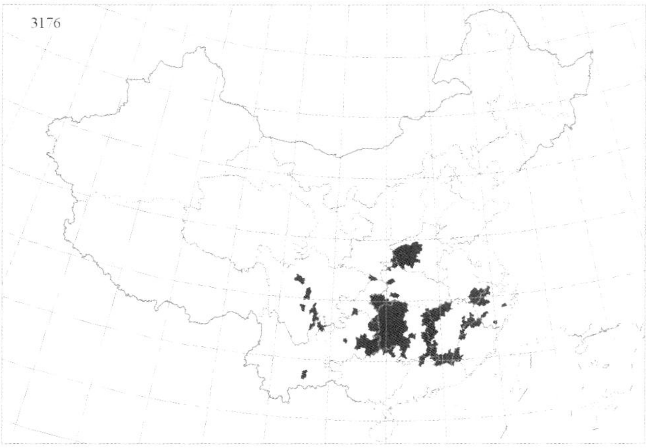

Rosaceae 蔷薇科 *Padus* 稠李属

Padus maackii (Rupr.) Kom. 斑叶稠李

Trees, deciduous

3.7 (-0.2~8.4); -16.8 (-23.8~-8.9); 21.1 (18.4~24.7); 8.1 (6.4~10.2); 62.6 (45.8~87.8); 77.9 (42.8~112.6); 580 (501~686) ❙ 719 (555~1073); 444 (309~691); 26 (13~40) ❙ 23.9 (-5.4~74.1); 576 (501~686); 295.2 (213.9~346.5)

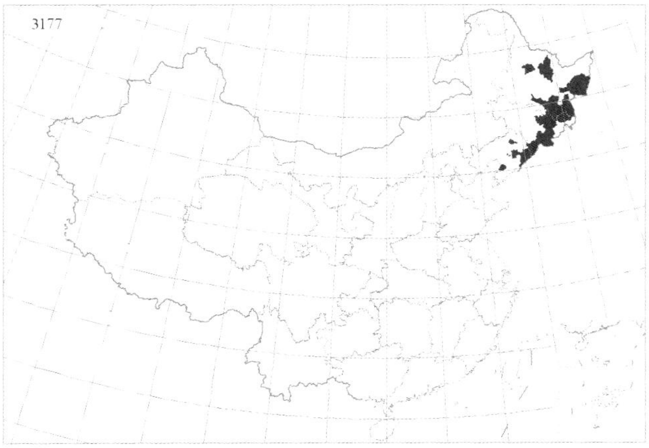

Rosaceae 蔷薇科 *Padus* 稠李属

Padus napaulensis (Ser.) C. K. Schneid. 粗梗稠李

Trees, deciduous

15.8 (1.8~19.3); 7.5 (-7.1~13.5); 22.6 (9.9~28.5); 15.8 (3.8~19.4); 131.5 (15.7~173.2); 1.9 (0.0~53.3); 806 (404~958) ❙ 1212 (583~1793); 603 (348~1007); 77 (13~238) ❙ 50.8 (6.5~145.8); 792 (404~958); 301.7 (153.9~551.3)

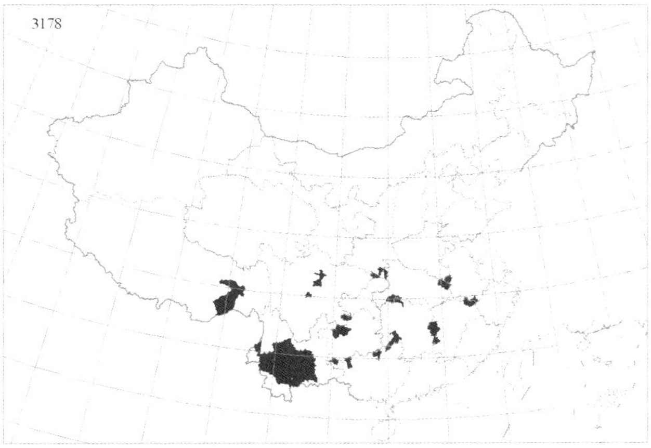

Rosaceae 蔷薇科 *Padus* 稠李属

Padus obtusata (Koehne) T. T. Yu et T. C. Ku 细齿稠李

Trees, deciduous

13.4 (0.3~22.8); 2.9 (-9.9~16.7); 23.0 (9.3~29.3); 13.6 (3.3~22.8); 109.2 (11.9~213.7) 8.0 (0.0~67.5); 758 (391~1209) ❙ 1173 (373~2680); 512 (201~1412); 91 (7~259) ❙ 54.6 (-37.6~246.5); 746 (373~1131); 258.7 (81.2~406.0)

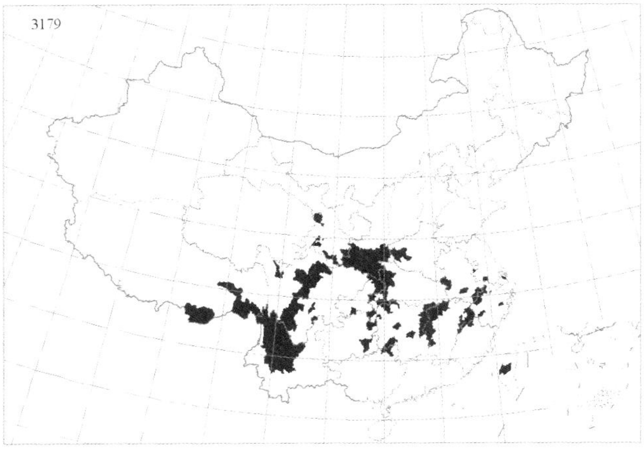

Rosaceae 蔷薇科 *Padus* 稠李属

Padus perulata (Koehne) T. T. Yu et T. C. Ku 宿鳞稠李

Trees, deciduous

12.7 (5.4~19.2); 6.0 (-2.1~13.3); 18.2 (12.3~23.5); 12.8 (5.8~19.4); 95.8 (31.5~172.6); 2.4 (0.0~25.6); 690 (481~932) ❙ 1083 (741~1337); 618 (426~774); 45 (15~69) ❙ 58.0 (30.5~93.0); 685 (481~923); 355.9 (273.5~452.0)

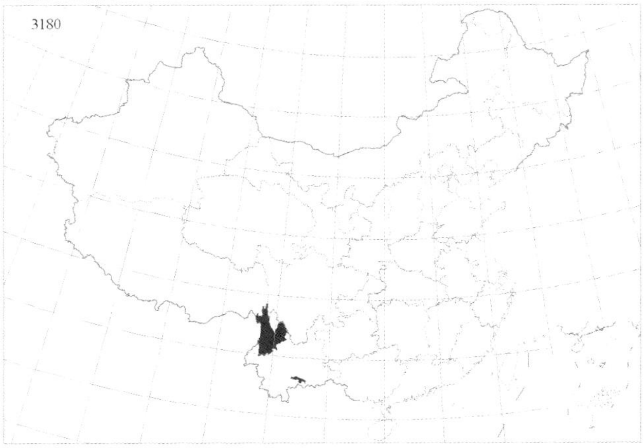

Rosaceae 蔷薇科 *Padus* 稠李属
Padus stellipila (Koehne) T. T. Yu et T. C. Ku 星毛稠李
Trees, deciduous

14.2 (5.7~18.2); 3.0 (-3.5~8.6); 24.8 (14.1~29.1); 14.3 (6.4~18.2); 116.0 (38.1~158.8); 5.6 (0.0~28.1); 793 (504~984) ❙ 1379 (711~1922); 528 (357~677); 140 (15~259) ❙ 72.6 (7.8~129.9); 788 (504~961); 251.4 (170.4~330.0)

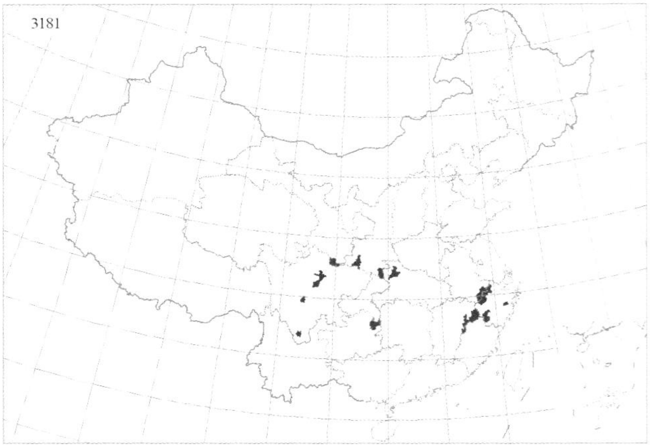

Rosaceae 蔷薇科 *Padus* 稠李属
Padus velutina (Batalin) C. K. Schneid. 毡毛稠李
Trees, deciduous

12.1 (0.3~16.9); 0.3 (-9.9~4.8); 23.3 (9.3~28.4); 12.4 (3.3~16.9); 98.5 (11.9~143.2); 12.9 (0.2~67.5); 724 (391~917) ❙ 1085 (632~1613); 444 (290~613); 77 (14~209) ❙ 50.3 (-16.6~107.2); 715 (391~878); 245.3 (187.9~304.5)

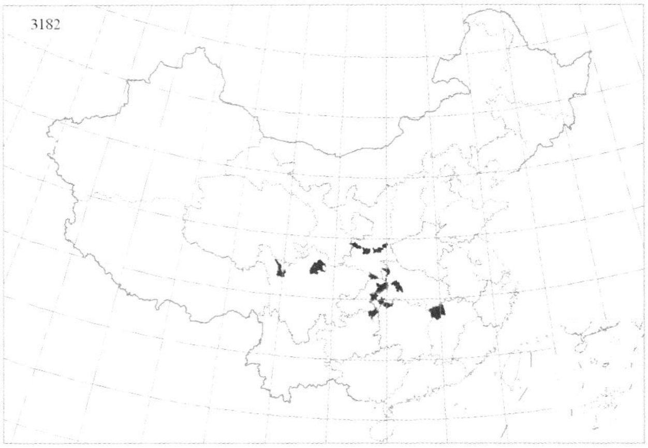

Rosaceae 蔷薇科 *Padus* 稠李属
Padus wilsonii C. K. Schneid. 绢毛稠李
Trees, deciduous

14.8 (5.3~19.8); 3.2 (-3.5~11.8); 25.6 (12.2~29.3); 14.9 (5.7~19.8); 123.0 (30.5~177.9); 5.1 (0.0~28.1); 819 (479~1038) ❙ 1258 (626~1997); 504 (290~816); 118 (15~259) ❙ 52.9 (-22.2~130.7); 805 (479~1038); 229.8 (139.1~434.4)

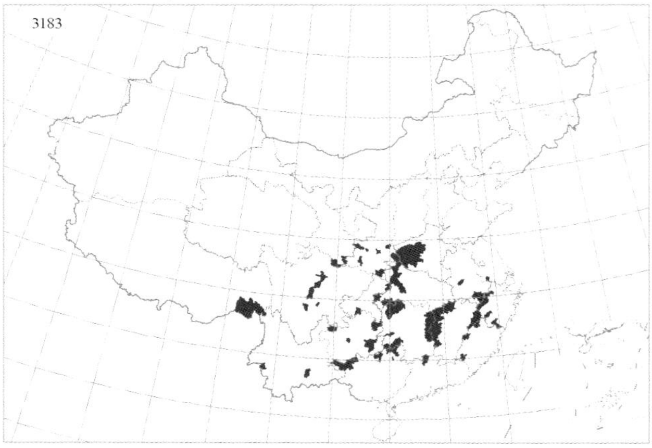

Rosaceae 蔷薇科 *Photinia* 石楠属
Photinia anlungensis T. T. Yu 安龙石楠
Shrubs

16.8; 7.8; 23.6; 16.8; 141.5; 0.0; 836 ❙ 1334; 720; 64 ❙ 59.6; 836; 226.6

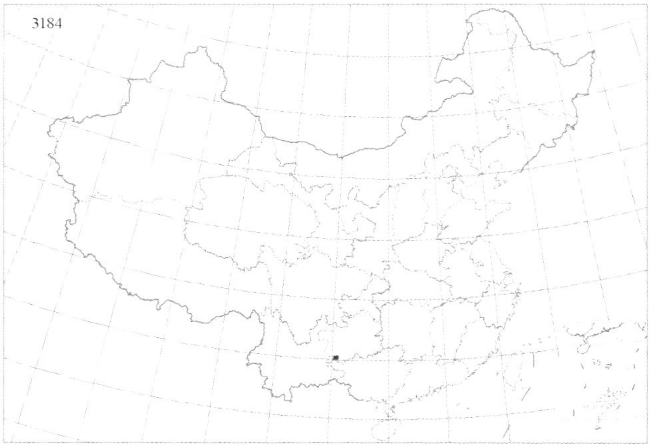

Rosaceae 蔷薇科 *Photinia* 石楠属
Photinia arguta (Decne.) J. E. Vidal 锐齿石楠
Shrubs or trees

15.8 (8.6~20.5); 7.8 (1.3~15.1); 22.4 (15.0~26.6); 15.9 (8.7~20.5); 131.6 (52.6~185.9); 1.0 (0.0~8.8); 821 (574~992) ❙ 1337 (1041~1736); 649 (473~1007); 67 (45~94) ❙ 66.1 (27.5~145.8); 816 (574~986); 371.4 (170.4~659.0)

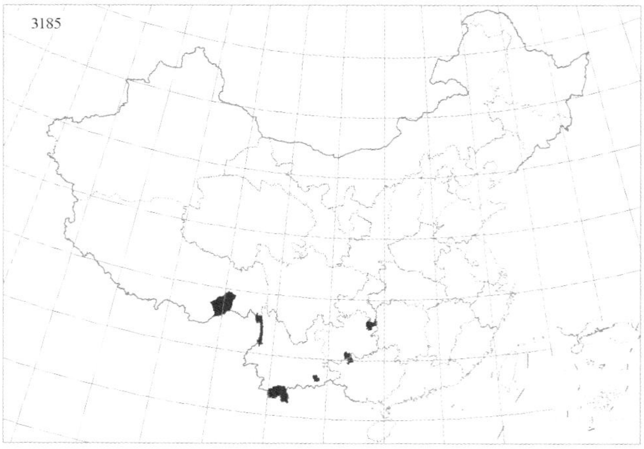

Rosaceae 蔷薇科 *Photinia* 石楠属
Photinia arguta var. hookeri (Decne.) J. E. Vidal 毛果锐齿石楠
Shrubs or trees

17.5; 10.5; 22.3; 17.5; 149.9; 0.0; 846 ❙ 958; 531; 44 ❙ 13.3; 823; 311.9

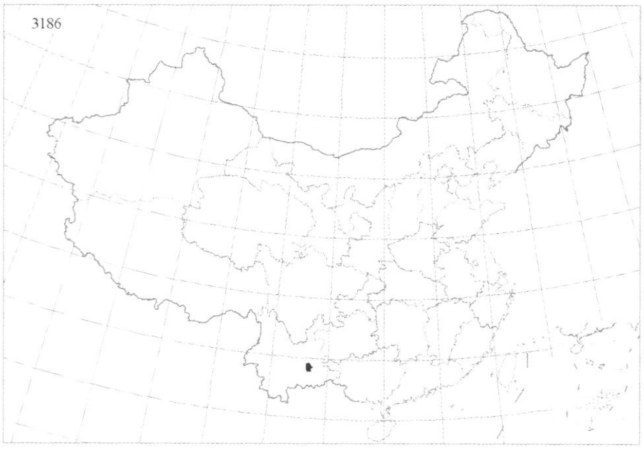

Rosaceae 蔷薇科 *Photinia* 石楠属
Photinia beauverdiana C. K. Schneid. 中华石楠
Shrubs or trees, deciduous

16.1 (7.3~22.3); 5.3 (−2.9~14.2); 26.1 (15.0~30.1); 16.2 (7.5~22.3); 136.0 (48.2~208.1); 1.9 (0.0~20.6); 865 (537~1202) ❙ 1351 (626~1997); 541 (299~996); 130 (15~259) ❙ 55.9 (−22.2~147.0); 852 (537~1202); 233.4 (134.1~551.6)

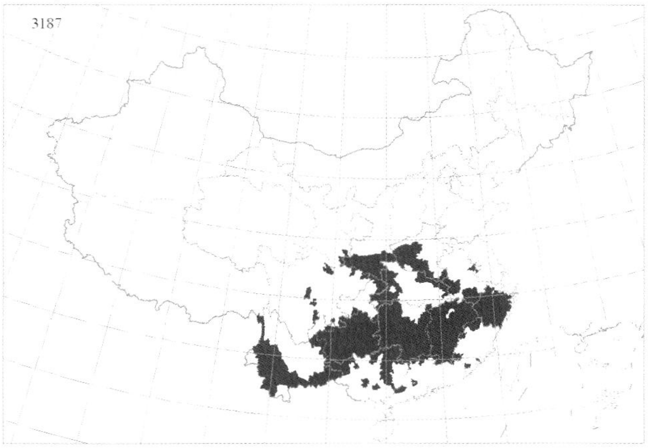

Rosaceae 蔷薇科 *Photinia* 石楠属
Photinia beauverdiana var. brevifolia Cardot 短叶中华石楠
Shrubs or trees, deciduous

14.5 (9.3~17.6); 2.2 (−2.9~7.0); 26.1 (20.1~29.3); 14.5 (9.5~17.6); 119.8 (69.8~151.7); 5.8 (0.0~20.6); 814 (603~941) ❙ 1168 (626~1950); 486 (316~713); 101 (19~242) ❙ 43.3 (−22.2~147.0); 798 (603~941); 213.8 (134.1~311.0)

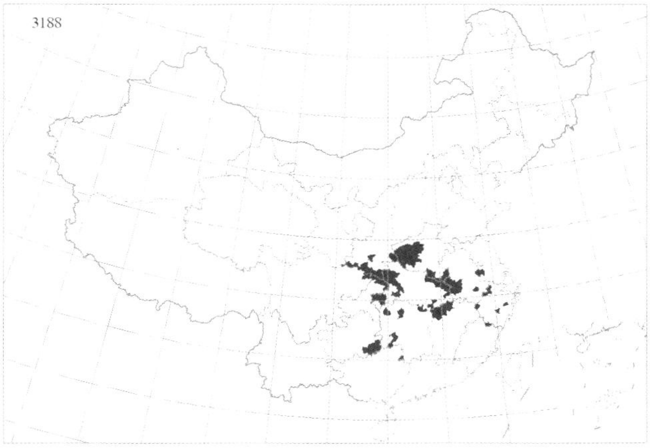

Rosaceae 蔷薇科 *Photinia* 石楠属
Photinia beckii C. K. Schneid. 椭圆叶石楠
Trees

17.3; 10.6; 22.0; 17.3; 148.1; 0.0; 834 ❙ 1273; 701; 55 ❙ 52.5; 834; 437.5

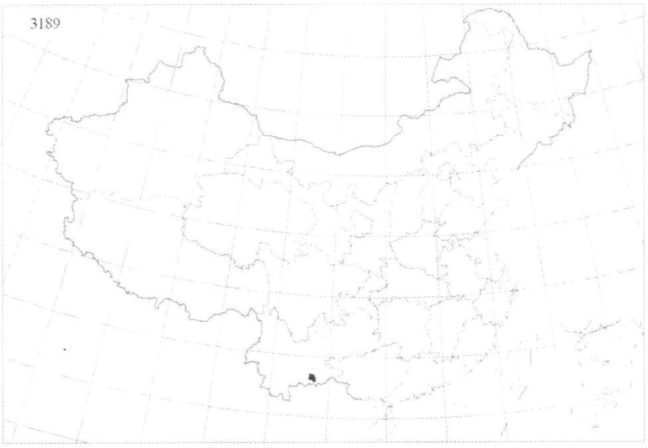

Rosaceae 蔷薇科 *Photinia* 石楠属
Photinia benthamiana Hance 闽粤石楠
Shrubs or trees

20.4 (16.0~24.8); 11.7 (4.0~20.1); 27.9 (21.4~29.8); 20.4 (16.1~24.8); 185.3 (133.8~237.6); 0.0 (0.0~1.0); 1082 (813~1385) ❙ 1634 (1178~2037); 652 (392~1033); 150 (41~257) ❙ 53.1 (−11.9~87.8); 1055 (813~1299); 251.1 (111.9~548.0)

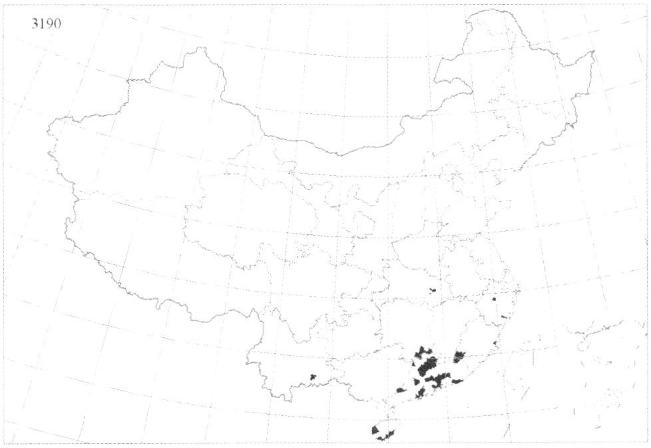

Rosaceae 蔷薇科 *Photinia* 石楠属
Photinia benthamiana var. obovata H. L. Li 倒卵叶闽粤石楠
Shrubs or trees

22.9; 16.3; 28.1; 23.0; 215.6; 0.0; 1237 ❙ 1653; 693; 97 ❙ 34.7; 1182; 278.8

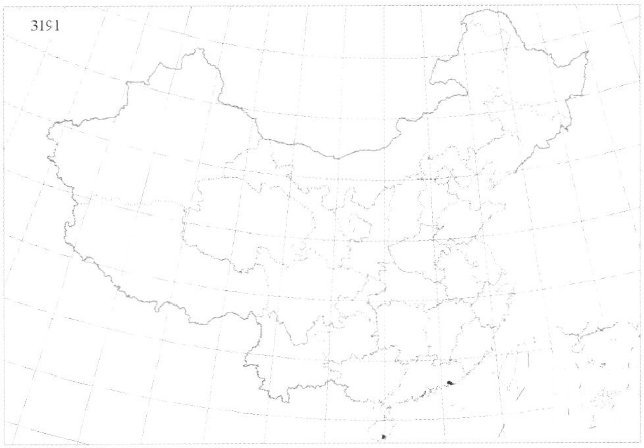

Rosaceae 蔷薇科 *Photinia* 石楠属
Photinia benthamiana var. salicifolia Cardot 柳叶闽粤石楠
Shrubs or trees

22.3 (18.1~24.8); 15.1 (9.5~20.1); 27.7 (24.6~28.6); 22.4 (18.2~24.8); 208.6 (158.1~237.6); 0.0 (0.0~0.0); 1195 (901~1385) ❙ 1565 (1179~2435); 738 (392~1428); 88 (41~145) ❙ 31.8 (−11.9~99.5); 1141 (901~1360); 362.5 (208.5~620.0)

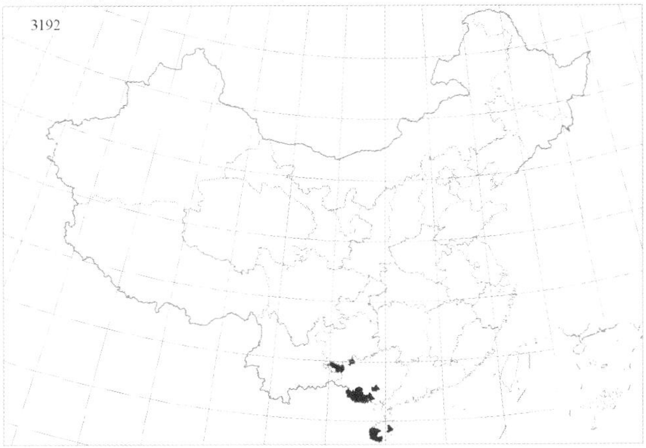

Rosaceae 蔷薇科 *Photinia* 石楠属
Photinia berberidifolia Rehder et E. H. Wilson 小檗叶石楠
Shrubs, evergreen

9.5; 1.2; 16.8; 9.7; 64.9; 10.4; 600 ❙ 1064; 615; 25 ❙ 78.3; 599; 270.0

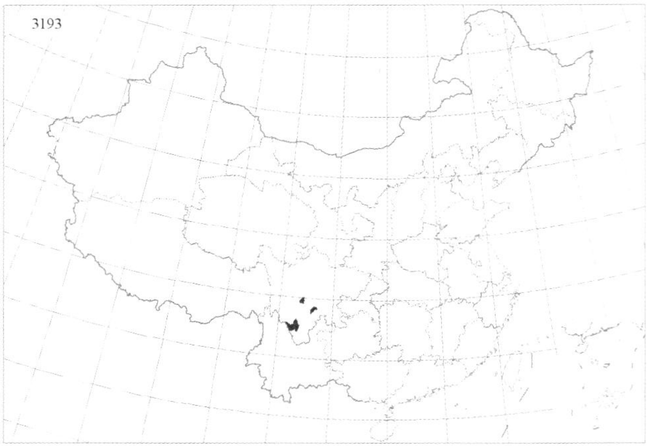

Rosaceae 蔷薇科 *Photinia* 石楠属
Photinia bergerae C. K. Schneid. 湖北石楠
Shrubs

12.7; 2.1; 23.0; 12.8; 99.0; 5.6; 713 ❙ 1478; 592; 85 ❙ 107.2; 713; 234.7

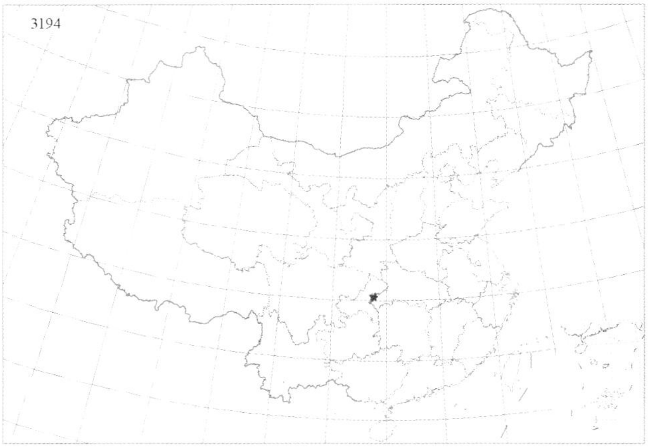

Rosaceae 蔷薇科 *Photinia* 石楠属
Photinia blinii (H. Lév.) Rehder 短叶石楠
Shrubs, deciduous

15.6; 5.5; 24.6; 15.6; 127.5; 0.0; 808 ❙ 1206; 546; 69 ❙ 49.2; 808; 189.4

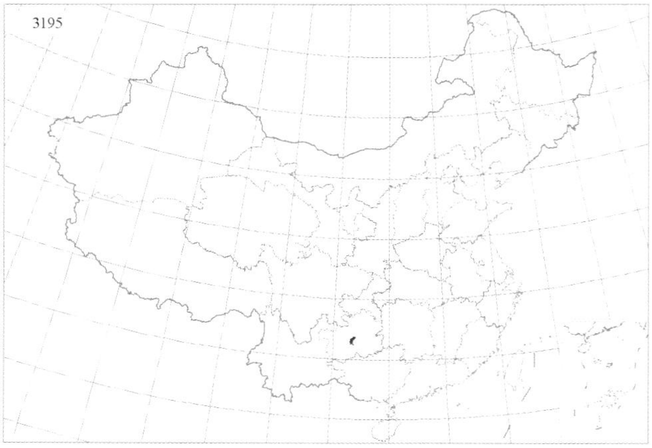

Rosaceae 蔷薇科 *Photinia* 石楠属
Photinia bodinieri H. Lév. 贵州石楠
Trees, evergreen

16.6 (2.2~20.7); 5.4 (-6.6~13.8); 27.0 (9.9~30.1); 16.6 (3.8~20.8); 140.4 (15.5~189.3); 0.9 (0.0~49.1); 890 (413~1113) ❙ 1448 (842~1997); 525 (299~820); 160 (12~259) ❙ 63.0 (-6.7~130.7); 872 (413~1056); 213.5 (111.9~513.9)

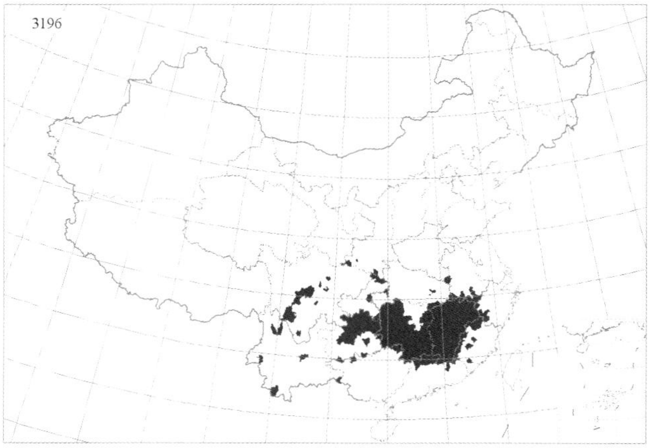

Rosaceae 蔷薇科 *Photinia* 石楠属
Photinia calleryana (Decne.) Cardot 城口石楠
Shrubs or small trees, deciduous

16.2 (14.3~18.9); 6.1 (4.0~9.5); 24.9 (23.5~26.6); 16.2 (14.3~18.9); 134.4 (112.6~167.3); 0.1 (0.0~1.1); 836 (756~969) ❙ 1257 (1160~1381); 563 (486~667); 81 (54~107) ❙ 50.8 (27.5~71.9); 836 (756~969); 225.6 (185.3~262.1)

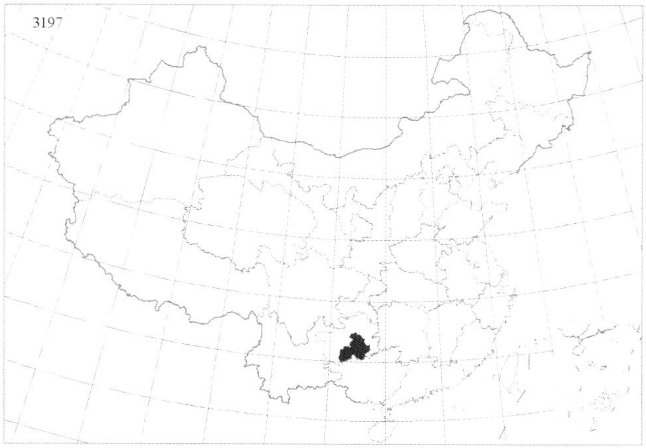

Rosaceae 蔷薇科 *Photinia* 石楠属
Photinia callosa Chun ex K. C. Kuan 厚齿石楠
Shrubs or trees

20.5 (15.5~22.9); 11.6 (4.8~15.2); 27.6 (25.1~28.9); 20.6 (15.6~22.9); 186.7 (127.2~214.8); 0.0 (0.0~0.2); 1078 (815~1232) ❙ 1587 (1172~2435); 766 (562~1428); 134 (43~194) ❙ 48.0 (12.9~99.5); 1078 (815~1232); 259.6 (185.5~411.1)

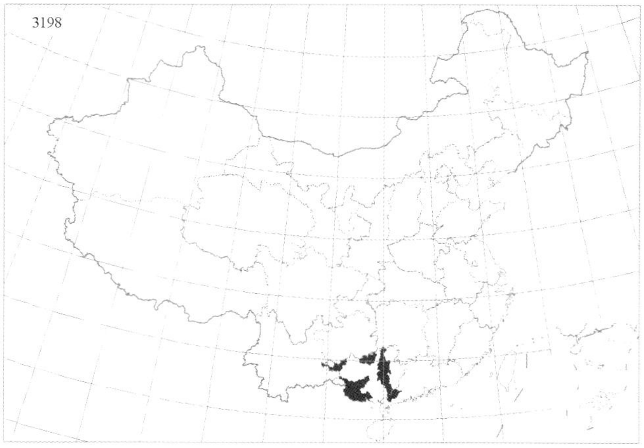

Rosaceae 蔷薇科 *Photinia* 石楠属
Photinia chihsiniana K. C. Kuan 临桂石楠
Trees

18.2 (14.6~22.4); 8.1 (3.8~14.0); 27.1 (24.6~28.7); 18.3 (14.7~22.4); 159.4 (117.7~209.1); 0.2 (0.0~1.5); 962 (779~1204) ▎1692 (1494~2148); 702 (476~1234); 175 (124~197) ▎76.9 (56.8~101.8); 962 (779~1204); 236.6 (180.3~318.8)

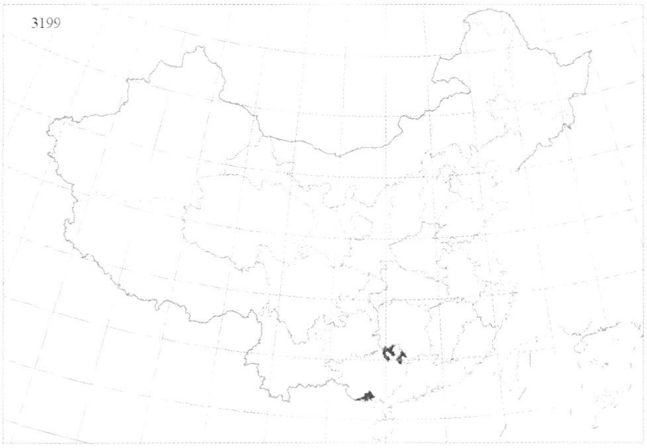

Rosaceae 蔷薇科 *Photinia* 石楠属
Photinia chingiana Hand.-Mazz. 宜山石楠
Shrubs

18.5; 8.6; 26.8; 18.6; 162.9; 0.0; 961 ▎1458; 667; 126 ▎52.3; 961; 233.3

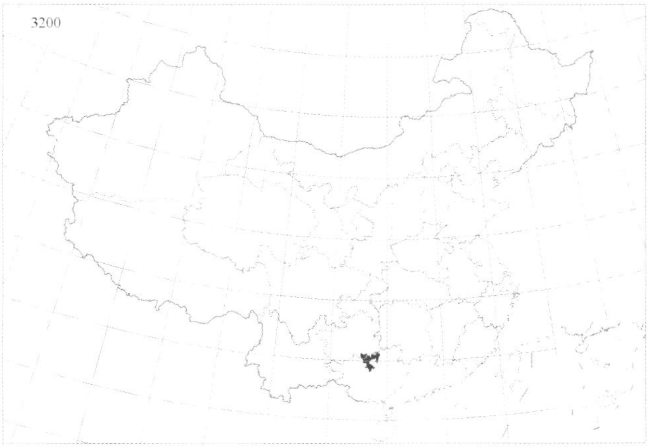

Rosaceae 蔷薇科 *Photinia* 石楠属
Photinia chingshuiensis (T. Shimizu) T. S. Liu et H. J. Su 清水石楠
Shrubs, deciduous

16.4; 11.5; 20.3; 16.5; 137.8; 0.0; 783 ▎2499; 947; 303 ▎219.3; 783; 303.6

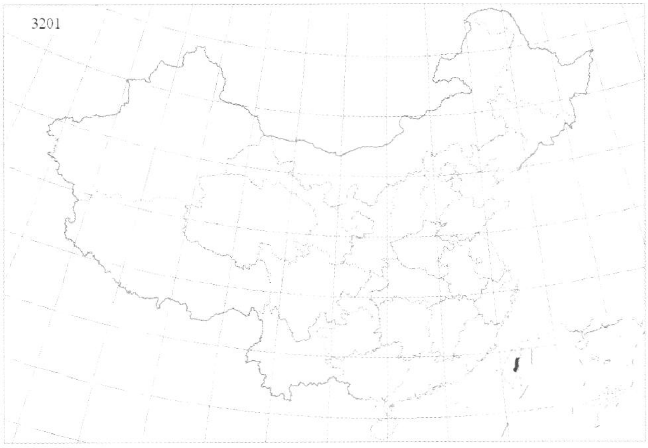

Rosaceae 蔷薇科 *Photinia* 石楠属
Photinia crassifolia H. Lév. 厚叶石楠
Shrubs, evergreen

16.1 (11.7~21.0); 7.5 (2.7~14.2); 23.0 (18.8~25.9); 16.2 (11.7~21.0); 134.3 (82.6~191.8); 0.3 (0.0~3.8); 816 (658~1053) ▎1214 (916~1522); 629 (473~860); 59 (31~95) ▎48.9 (19.0~80.6); 816 (658~1053); 279.3 (160.8~551.6)

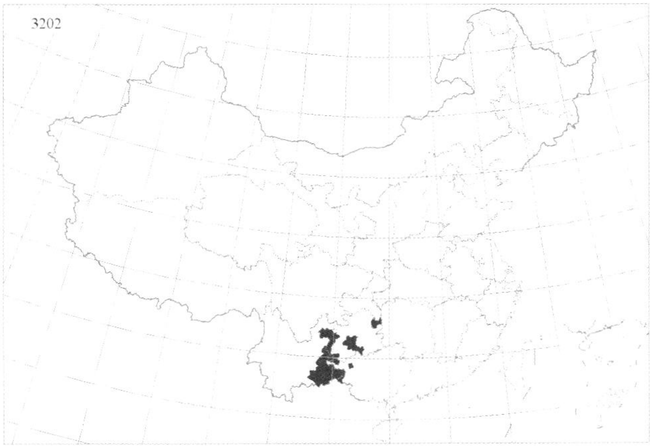

Rosaceae 蔷薇科 *Photinia* 石楠属
Photinia fokienensis (Finet et Franch.) Franch. ex Cardot 福建石楠
Shrubs or trees, deciduous

17.3 (15.8~19.1); 6.5 (3.9~10.0); 27.4 (25.8~29.1); 17.4 (15.9~19.1); 148.4 (131.5~169.1); 0.1 (0.0~1.1); 916 (861~984) ▎1851 (1716~1997); 613 (535~669); 241 (193~259) ▎102.6 (76.3~130.7); 910 (861~973); 260.9 (183.9~303.4)

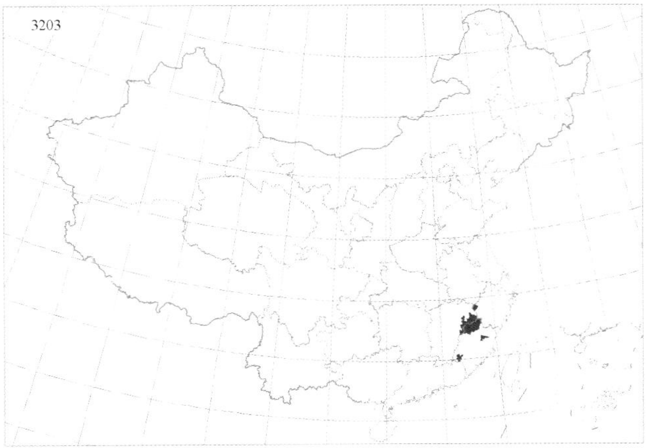

Rosaceae 蔷薇科 *Photinia* 石楠属
Photinia glabra (Thunb.) Maxim. 光叶石楠
Trees, evergreen

16.8 (6.6~22.3); 6.2 (−2.0~15.1); 26.5 (13.3~30.1); 16.8 (6.8~22.3); 142.5 (39.6~208.1); 0.8 (0.0~19.7); 890 (514~1202) ▎1410 (635~1997); 553 (299~1056); 145 (12~259) ▎58.6 (−25.5~147.0); 879 (514~1202); 231.9 (108.2~659.0)

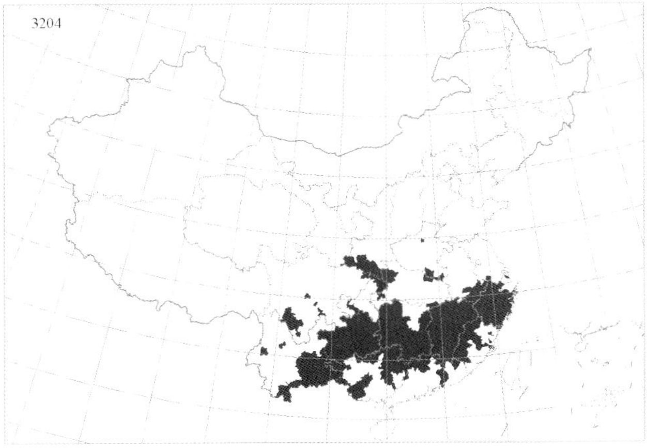

Rosaceae 蔷薇科 *Photinia* 石楠属

Photinia glomerata Rehder et E. H. Wilson 球花石楠

Shrubs or trees, evergreen

15.1 (5.4~19.3); 7.7 (-2.1~13.1); 20.8 (12.3~27.7); 15.1 (5.8~19.4); 122.1 (31.5~173.2); 1.0 (0.0~25.6); 766 (481~939) ❙ 1046 (741~1504); 592 (411~860); 41 (12~85) ❙ 37.4 (-14.2~107.2); 747 (481~939); 294.2 (142.5~491.5)

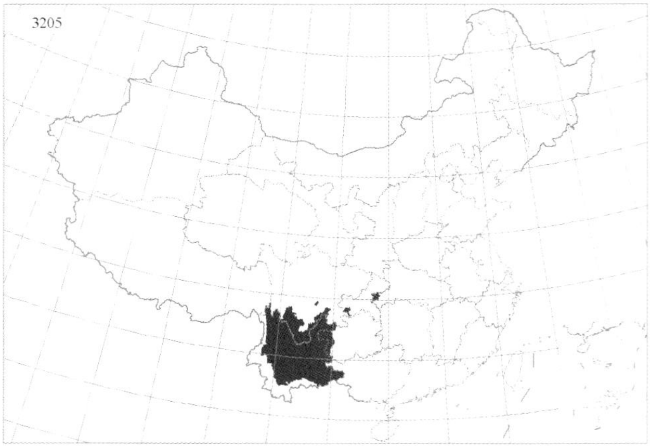

Rosaceae 蔷薇科 *Photinia* 石楠属

Photinia hirsuta Hand.-Mazz. 褐毛石楠

Shrubs or trees, deciduous

17.0 (13.6~19.4); 5.9 (2.1~10.0); 27.4 (24.0~29.9); 17.0 (13.6~19.4); 144.9 (107.0~173.4); 0.5 (0.0~4.9); 904 (746~1029) ❙ 1676 (1265~1997); 571 (437~713); 208 (65~259) ❙ 86.2 (40.5~147.0); 897 (746~1029); 245.1 (141.7~319.9)

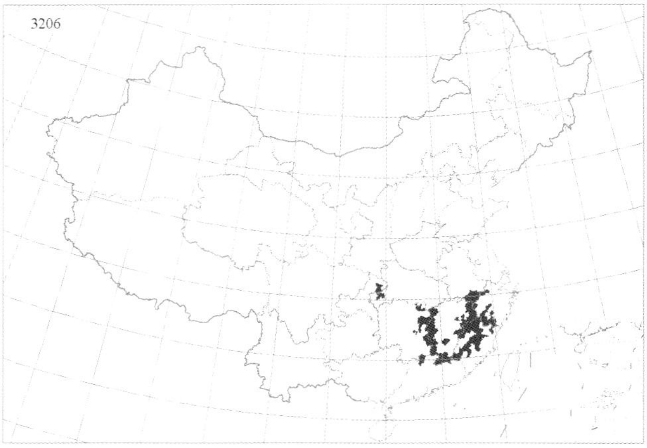

Rosaceae 蔷薇科 *Photinia* 石楠属

Photinia hirsuta var. lobulata T. T. Yu 裂叶褐毛石楠

Shrubs or small trees, deciduous

18.4; 9.3; 26.3; 18.4; 160.8; 0.0; 934 ❙ 1746; 618; 199 ❙ 87.4; 934; 270.5

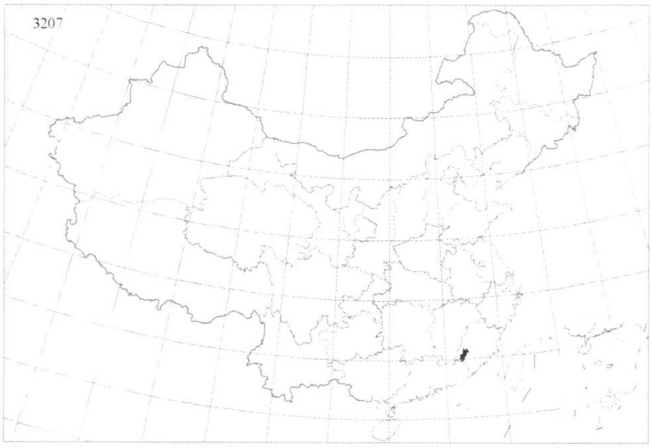

Rosaceae 蔷薇科 *Photinia* 石楠属

Photinia impressivena Hayata 陷脉石楠

Shrubs or trees

19.6 (15.5~24.1); 10.3 (4.8~18.2); 27.4 (22.1~28.9); 19.6 (15.6~24.2); 175.3 (127.2~230.1); 0.0 (0.0~0.2); 1023 (815~1326) ❙ 1618 (1007~2005); 634 (513~827); 168 (44~231) ❙ 59.5 (19.0~95.5); 1020 (815~1326); 247.1 (185.5~538.2)

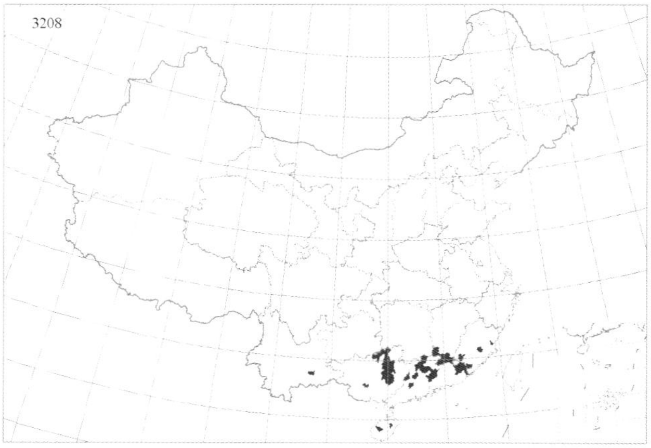

Rosaceae 蔷薇科 *Photinia* 石楠属

Photinia integrifolia Lindl. 全缘石楠

Trees, evergreen

15.5 (1.8~19.8); 8.7 (-7.1~13.5); 20.6 (9.9~26.4); 15.6 (3.8~19.8); 128.6 (15.7~177.9); 1.6 (0.0~53.3); 781 (404~1010) ❙ 1108 (527~1751); 632 (317~1007); 44 (9~77) ❙ 42.7 (-14.2~145.8); 763 (404~1010); 336.2 (169.5~551.3)

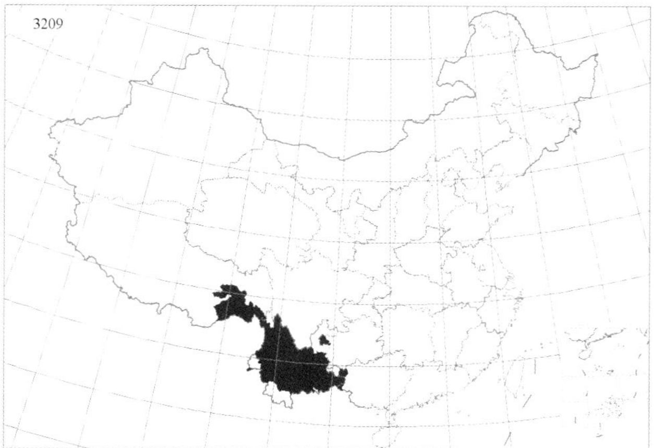

Rosaceae 蔷薇科 *Photinia* 石楠属

Photinia komarovii (H. Lév. et Vaniot) L. T. Lu et C. L. Li 垂丝石楠

Shrubs, deciduous

16.5 (13.5~18.2); 5.2 (2.6~7.2); 27.1 (24.0~29.1); 16.5 (13.6~18.2); 138.7 (106.7~158.8); 0.5 (0.0~4.1); 881 (744~984) ❙ 1668 (1178~1997); 572 (471~669); 198 (81~259) ❙ 88.7 (43.1~130.7); 878 (744~961); 236.1 (170.4~303.4)

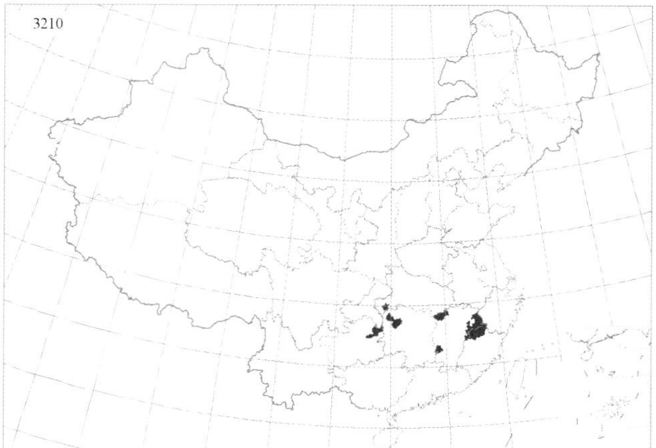

Rosaceae 蔷薇科 *Photinia* 石楠属

Photinia kwangsiensis H. L. Li 广西石楠

Trees

19.4 (17.4~21.8); 9.7 (7.3~12.9); 27.5 (25.5~28.6); 19.5 (17.4~21.8); 173.6 (148.9~201.6); 0.0 (0.0~0.0); 1018 (886~1162) ❙ 1559 (1371~1758); 669 (562~735); 162 (107~189) ❙ 54.4 (22.7~85.5); 1017 (886~1162); 222.5 (185.5~245.3)

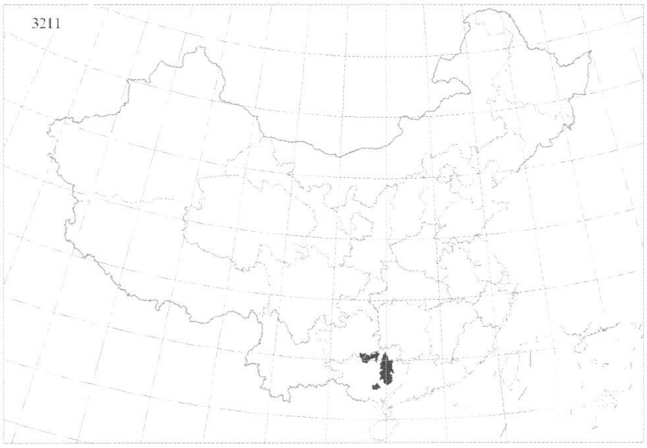

Rosaceae 蔷薇科 *Photinia* 石楠属

Photinia lanuginosa T. T. Yu 绵毛石楠

Shrubs or trees, evergreen

16.1 (14.3~17.5); 4.4 (3.3~6.2); 26.9 (24.3~29.7); 16.1 (14.3~17.5); 134.1 (114.1~150.6); 0.7 (0.0~2.3); 863 (765~964) ❙ 1443 (1316~1624); 488 (442~594); 175 (145~196) ❙ 67.9 (41.2~101.8); 854 (765~900); 202.6 (154.4~258.0)

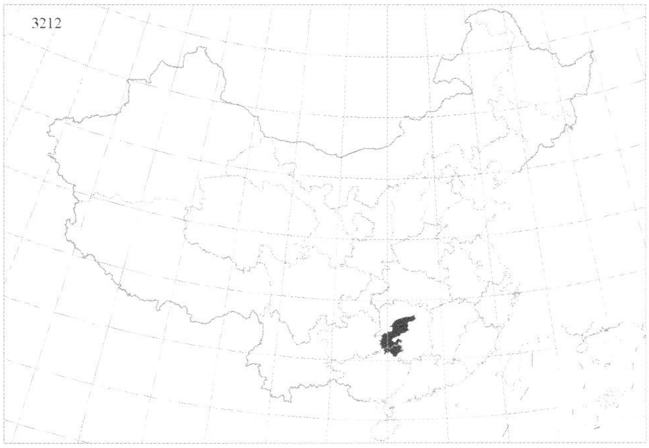

Rosaceae 蔷薇科 *Photinia* 石楠属

Photinia lasiogyna (Franch.) C. K. Schneid. 倒卵叶石楠

Shrubs or small trees

15.8 (5.4~21.0); 7.3 (-2.1~15.1); 22.7 (12.3~30.1); 15.8 (5.8~21.0); 130.4 (31.5~191.8); 0.7 (0.0~25.6); 811 (481~1053) ❙ 1224 (741~1997); 592 (411~996); 84 (12~259) ❙ 50.3 (-14.2~130.7); 794 (481~1053); 288.1 (141.7~659.0)

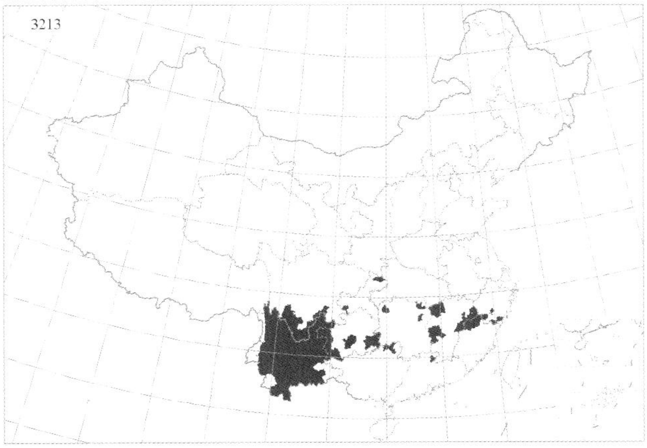

Rosaceae 蔷薇科 *Photinia* 石楠属

Photinia lochengensis T. T. Yu 罗城石楠

Shrubs or trees, evergreen

17.3 (15.9~19.2); 7.1 (5.3~9.2); 26.1 (25.5~27.5); 17.3 (16.0~19.2); 148.1 (131.9~170.8); 0.0 (0.0~0.0); 896 (831~1002) ❙ 1446 (1268~1791); 614 (523~678); 132 (105~195) ❙ 62.2 (47.0~115.7); 896 (831~1002); 238.1 (222.1~256.4)

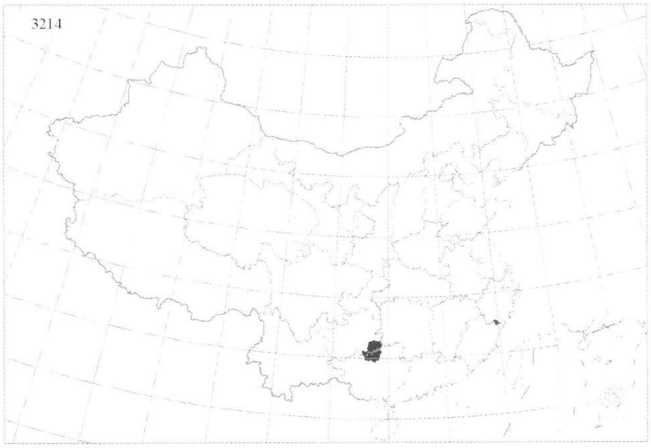

Rosaceae 蔷薇科 *Photinia* 石楠属

Photinia oriformis W. W. Sm. 带叶石楠

Shrubs or trees

14.5 (11.6~16.4); 7.5 (3.1~9.8); 19.8 (18.6~21.2); 14.5 (11.5~16.5); 114.4 (80.9~137.5); 0.3 (0.0~2.8); 736 (654~801) ❙ 969 (911~1009); 574 (534~596); 33 (28~39) ❙ 32.2 (13.8~51.0); 709 (654~740); 276.6 (255.4~295.8)

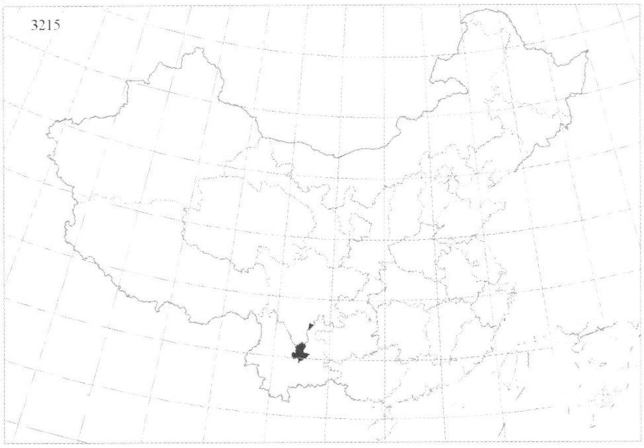

Rosaceae 蔷薇科 *Photinia* 石楠属

Photinia lucida (Decne.) C. K. Schneid. 台湾石楠

Trees, deciduous

19.2; 14.4; 22.8; 19.2; 170.5; 0.0; 948 ❙ 2836; 1517; 176 ❙ 206.5; 948; 293.0

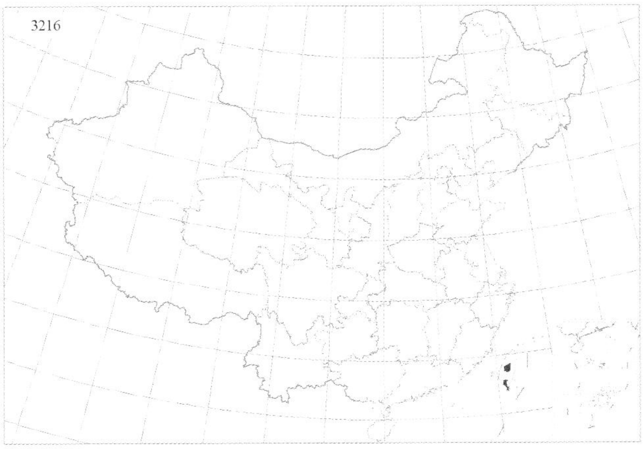

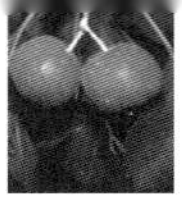

Rosaceae 蔷薇科 *Photinia* 石楠属
Photinia megaphylla T. T. Yu et T. C. Ku 大叶石楠
Shrubs, evergreen

13.3; 5.4; 19.6; 13.4; 100.4; 0.0; 706 ❙ 1736; 1007; 51 ❙ 145.8; 706; 447.4

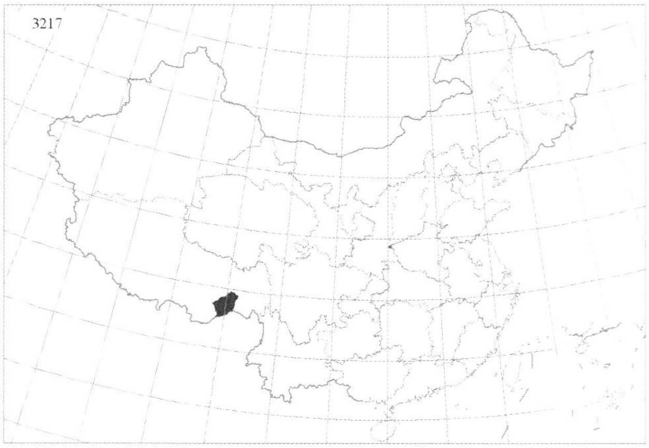

Rosaceae 蔷薇科 *Photinia* 石楠属
Photinia parvifolia (E. Pritz.) C. K. Schneid. 小叶石楠
Shrubs, deciduous

17.9 (9.1~23.6); 7.8 (-1.9~16.7); 27.2 (18.1~29.9); 18.0 (9.2~23.6); 156.4 (64.1~223.5); 0.7 (0.0~18.0); 952 (595~1281) ❙ 1560 (741~3495); 598 (356~1056); 168 (19~897) ❙ 65.8 (-6.9~224.4); 947 (595~1273); 227.7 (95.3~332.5)

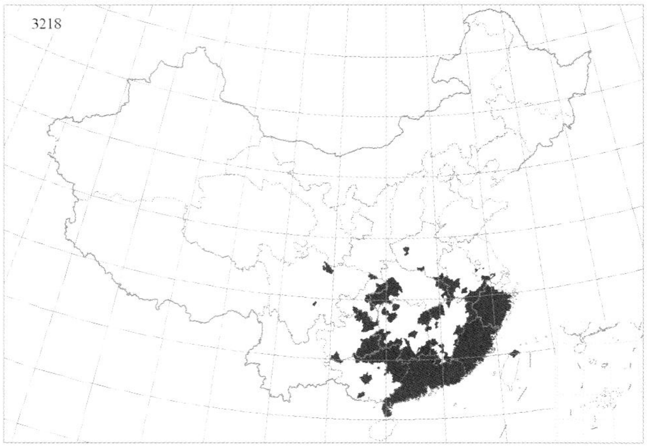

Rosaceae 蔷薇科 *Photinia* 石楠属
Photinia parvifolia var. subparvifolia (Y. K. Li et X. M. Wang) L. T. Lu et C. L. Li 假小叶石楠
Shrubs, deciduous

17.4; 7.4; 25.5; 17.4; 148.9; 0.0; 886 ❙ 1381; 637; 107 ❙ 55.9; 886; 245.3

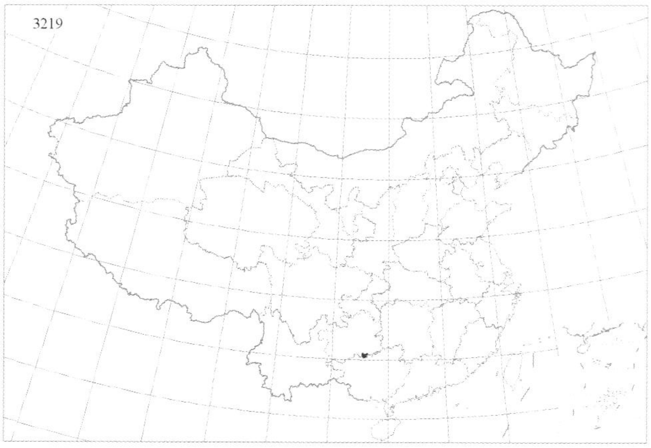

Rosaceae 蔷薇科 *Photinia* 石楠属
Photinia pilosicalyx T. T. Yu 毛果石楠
Shrubs

14.6 (10.6~17.1); 4.3 (-0.2~8.5); 24.2 (21.1~25.6); 14.7 (10.6~17.1); 118.5 (79.4~145.1); 2.6 (0.0~12.1); 782 (643~845) ❙ 1243 (1047~1441); 557 (489~720); 62 (29~88) ❙ 60.0 (24.9~95.9); 782 (643~845); 217.1 (154.4~284.3)

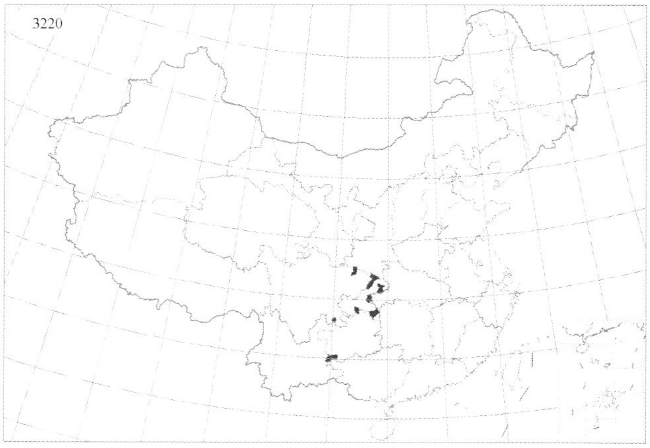

Rosaceae 蔷薇科 *Photinia* 石楠属
Photinia podocarpifolia T. T. Yu 罗汉松叶石楠
Shrubs

18.7 (16.8~21.3); 9.6 (6.8~12.6); 25.8 (23.6~27.9); 18.7 (16.8~21.3); 164.3 (141.5~196.0); 0.0 (0.0~0.0); 950 (836~1123) ❙ 1260 (1182~1334); 665 (566~723); 61 (43~99) ❙ 33.7 (5.8~59.6); 950 (836~1121); 279.9 (189.5~354.5)

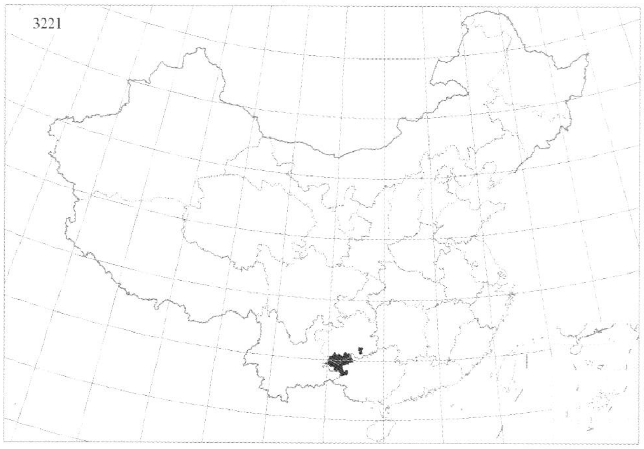

Rosaceae 蔷薇科 *Photinia* 石楠属
Photinia prionophylla (Franch.) C. K. Schneid. 刺叶石楠
Shrubs or trees

13.2 (2.2~18.9); 6.1 (-6.3~12.0); 18.8 (9.5~23.9); 13.3 (3.9~19.0); 103.1 (15.8~167.8); 4.5 (0.0~48.6); 705 (411~929) ❙ 933 (729~1057); 556 (411~627); 28 (11~46) ❙ 35.4 (-14.2~77.4); 665 (411~760); 286.6 (207.5~372.3)

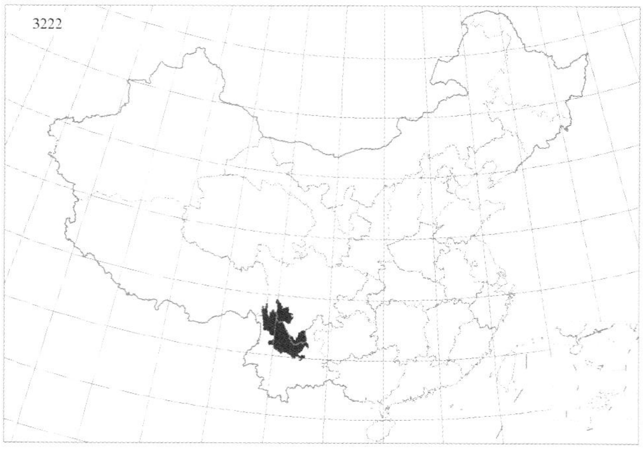

Rosaceae 蔷薇科 *Photinia* 石楠属
Photinia prunifolia (Hook. et Arn.) Lindl. 桃叶石楠
Trees, evergreen

19.1 (13.4~24.3); 9.4 (1.5~19.1); 27.6 (23.1~30.1); 19.1 (13.4~24.3); 169.9 (108.3~231.9); 0.2 (0.0~7.3); 1011 (755~1338) ❙ 1599 (868~2037); 619 (312~1056); 165 (42~259) ❙ 60.5 (-34.3~130.7); 999 (755~1326); 235.4 (95.3~620.0)

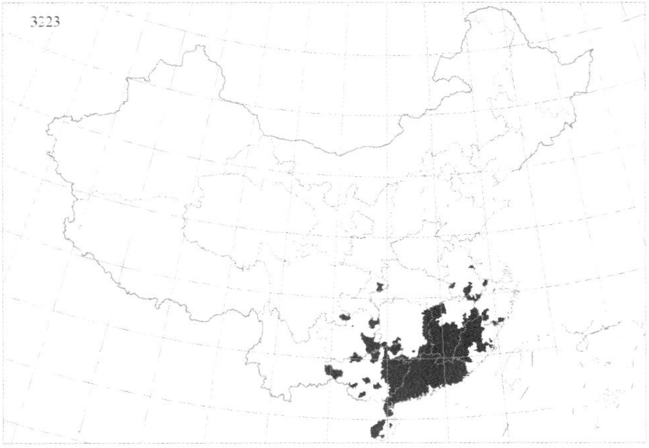

Rosaceae 蔷薇科 *Photinia* 石楠属
Photinia prunifolia var. denticulata T. T. Yu 重齿桃叶石楠
Trees, evergreen

16.9 (15.6~18.2); 6.4 (4.9~7.8); 26.9 (25.8~28.7); 16.9 (15.6~18.3); 142.9 (127.5~159.6); 0.0 (0.0~0.1); 886 (824~965) ❙ 1738 (1624~1893); 636 (594~677); 195 (179~227) ❙ 97.4 (76.7~129.9); 886 (824~965); 232.3 (189.8~269.3)

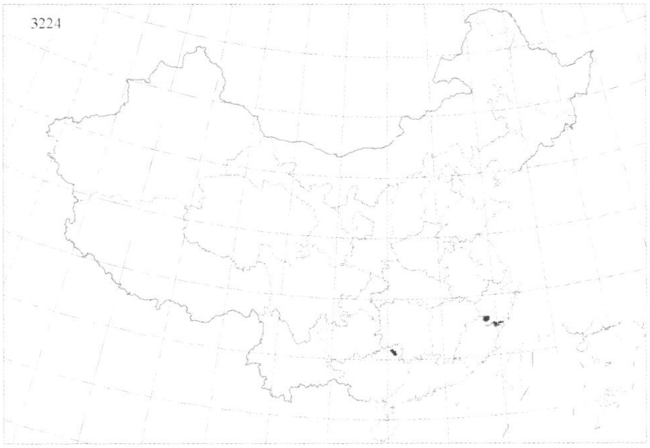

Rosaceae 蔷薇科 *Photinia* 石楠属
Photinia raupingensis K. C. Kuan 饶平石楠
Trees

21.1 (17.3~24.8); 12.6 (6.4~20.1); 28.0 (25.7~29.8); 21.1 (17.3~24.8); 193.3 (147.9~237.6); 0.0 (0.0~0.0); 1112 (906~1385) ❙ 1653 (1187~2435); 714 (455~1428); 139 (41~200) ❙ 49.8 (-6.9~99.5); 1106 (906~1273); 237.5 (95.3~620.0)

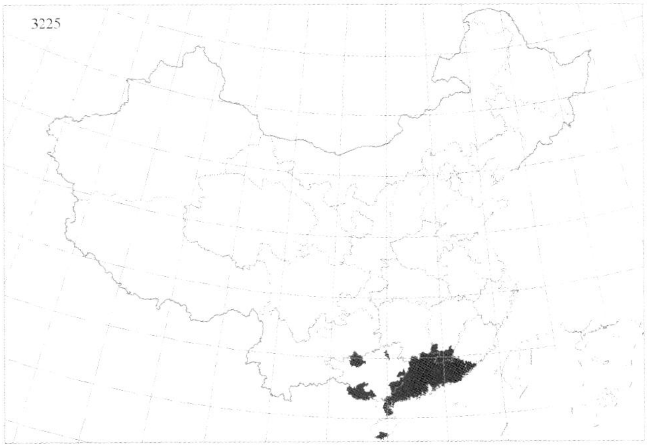

Rosaceae 蔷薇科 *Photinia* 石楠属
Photinia schneideriana Rehder et E. H. Wilson 绒毛石楠
Shrubs or trees

16.8 (10.6~22.1); 5.5 (-0.2~13.7); 27.3 (18.6~29.9); 16.8 (10.6~22.1); 142.3 (79.4~205.6); 0.7 (0.0~12.1); 898 (643~1182) ❙ 1508 (988~2005); 535 (392~858); 171 (28~259) ❙ 68.6 (8.0~130.7); 888 (643~1182); 213.7 (108.2~319.9)

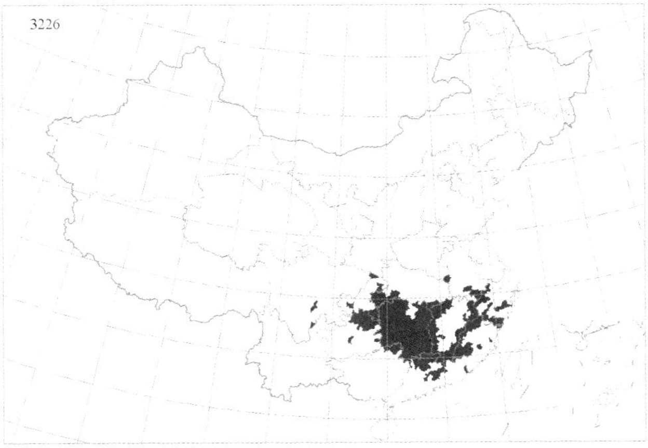

Rosaceae 蔷薇科 *Photinia* 石楠属
Photinia schneideriana var. parviflora (Cardot) L. T. Lu et C. L. Li 小花石楠
Trees, deciduous

17.1 (12.7~22.3); 5.7 (1.1~13.8); 27.6 (23.0~30.1); 17.1 (12.8~22.3); 146.4 (99.0~208.1); 0.8 (0.0~7.9); 919 (713~1202) ❙ 1450 (753~1922); 533 (299~1056); 164 (19~259) ❙ 58.4 (-12.2~128.9); 902 (713~1202); 208.0 (134.1~335.1)

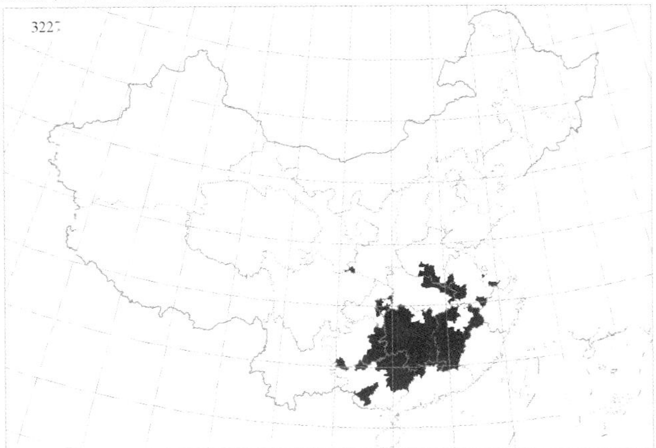

Rosaceae 蔷薇科 *Photinia* 石楠属
Photinia serratifolia (Desf.) Kalkman 石楠
Shrubs or trees, evergreen

16.0 (5.4~21.7); 5.3 (-6.2~15.3); 26.0 (12.3~30.1); 16.1 (5.8~21.7); 134.6 (31.5~200.2); 1.7 (0.0~32.8); 860 (481~1150) ❙ 1315 (508~2713); 539 (312~1412); 131 (12~303) ❙ 52.6 (-37.2~246.5); 843 (481~1148); 224.7 (111.9~551.6)

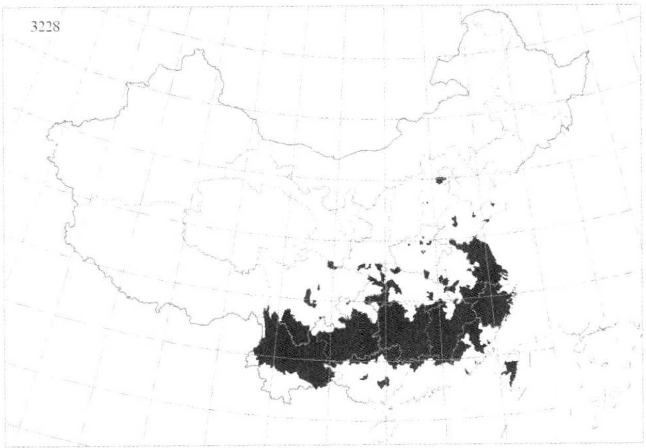

Rosaceae 蔷薇科 *Photinia* 石楠属
Photinia stenophylla Hand.-Mazz. 窄叶石楠
Shrubs

16.9 (16.1~17.6); 6.6 (5.3~7.4); 25.7 (25.5~26.3); 17.0 (16.1~17.6); 143.4 (133.1~151.6); 0.0 (0.0~0.0); 873 (838~909) ❙ 1357 (1268~1496); 579 (523~639); 117 (99~142) ❙ 55.5 (47.0~64.7); 873 (838~909); 233.6 (222.1~246.1)

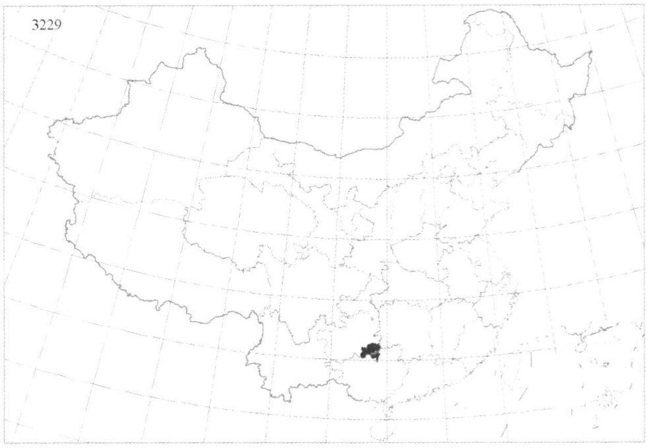

Rosaceae 蔷薇科 *Photinia* 石楠属
Photinia tsaii Rehder 福贡石楠
Shrubs or trees

12.3; 5.5; 17.6; 12.4; 91.4; 3.2; 677 ❙ 1212; 684; 50 ❙ 80.2; 673; 391.9

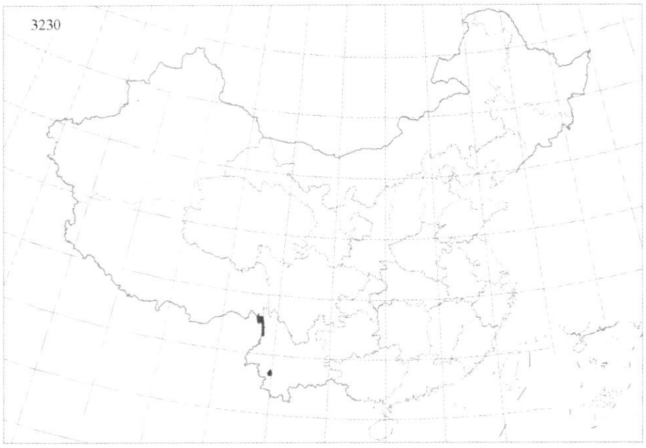

Rosaceae 蔷薇科 *Photinia* 石楠属
Photinia tushanensis T. T. Yu 独山石楠
Shrubs

18.0; 8.2; 25.9; 18.0; 155.9; 0.0; 926 ❙ 1430; 674; 114 ❙ 55.2; 926; 231.1

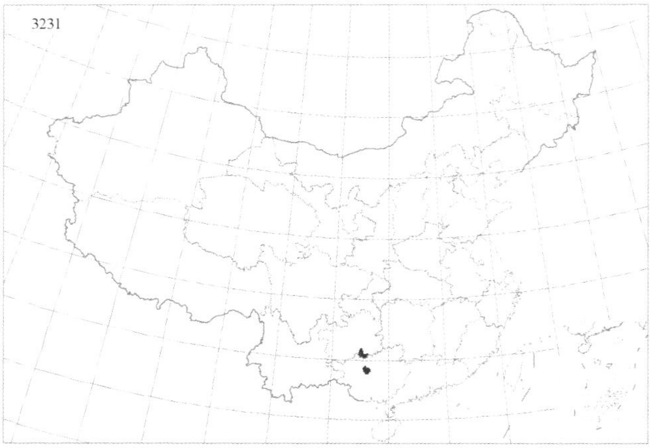

Rosaceae 蔷薇科 *Photinia* 石楠属
Photinia villosa (Thunb.) DC. 毛叶石楠
Shrubs or trees, deciduous

15.7 (9.3~19.9); 3.9 (-3.3~11.0); 26.8 (20.1~30.1); 15.8 (9.5~19.9); 133.4 (69.8~179.3); 3.9 (0.0~22.0); 866 (603~1037) ❙ 1321 (626~1997); 500 (299~820); 142 (17~259) ❙ 51.0 (-22.2~130.7); 846 (603~1036); 220.9 (134.1~371.7)

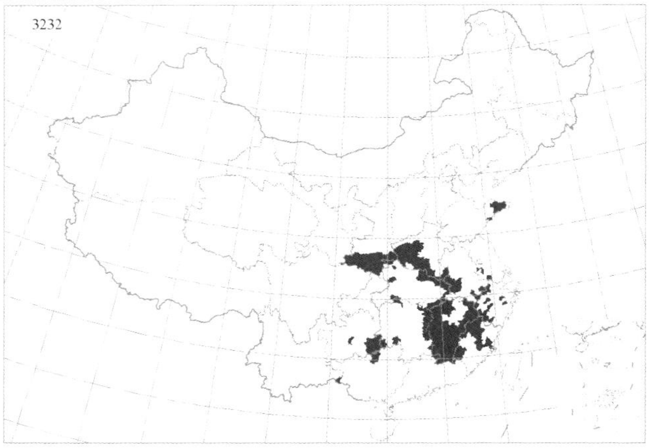

Rosaceae 蔷薇科 *Photinia* 石楠属
Photinia villosa var. sinica Rehder et E. H. Wilson 庐山石楠
Shrubs or trees, deciduous

15.8 (9.3~21.3); 4.2 (-2.9~12.3); 26.6 (18.8~30.1); 15.9 (9.5~21.3); 134.0 (69.8~196.1); 3.7 (0.0~20.6); 867 (603~1130) ❙ 1344 (626~1997); 512 (299~728); 141 (15~259) ❙ 53.8 (-22.2~130.7); 850 (603~1130); 226.8 (139.1~334.4)

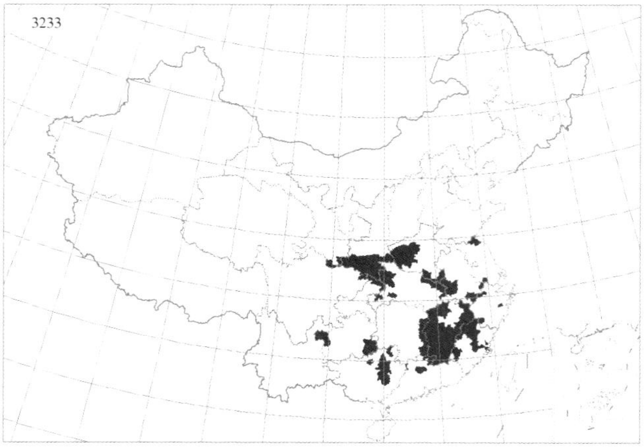

Rosaceae 蔷薇科 *Photinia* 石楠属
Photinia zhejiangensis P. L. Chiu 浙江石楠
Shrubs, evergreen

15.6 (14.6~17.0); 4.8 (3.1~6.2); 25.8 (24.6~26.8); 15.7 (14.7~17.0); 128.7 (117.7~144.4); 0.5 (0.0~2.2); 828 (779~891) ❙ 1723 (1425~1950); 621 (518~713); 202 (161~227) ❙ 108.7 (64.9~147.0); 828 (779~891); 257.7 (239.6~283.9)

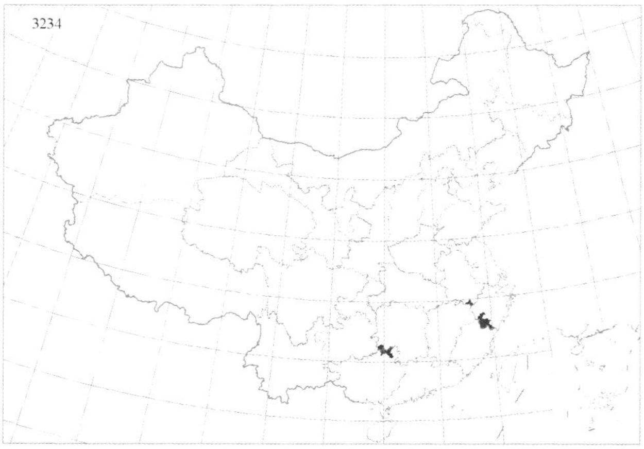

Rosaceae 蔷薇科 *Physocarpus* 风箱果属
Physocarpus amurensis (Maxim.) Maxim. 风箱果
Shrubs

7.7 (2.1~10.5); −10.3 (−20.1~−4.9); 23.2 (21.0~25.0); 9.9 (7.7~11.3); 82.2 (57.8~93.2); 49.8 (26.8~91.9); 658 (564~705) ❙ 611 (535~735); 420 (385~475); 15 (9~27) ❙ −6.8 (−23.1~16.5); 597 (535~698); 240.8 (158.5~323.4)

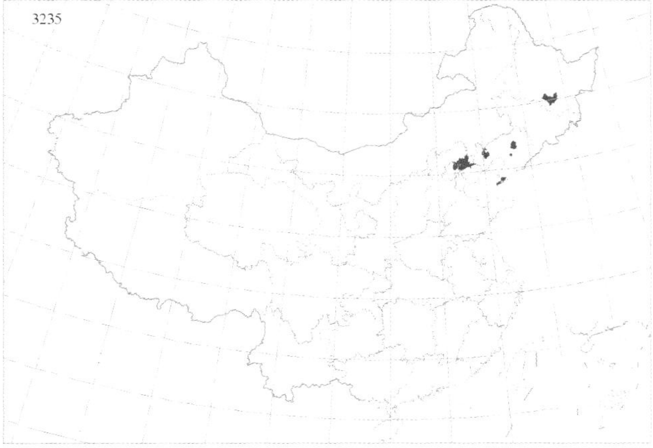

Rosaceae 蔷薇科 *Potaninia* 绵刺属
Potaninia mongolica Maxim. 绵刺
Shrubs

7.4 (6.2~8.1); −9.8 (−12.2~−7.9); 22.7 (20.2~24.0); 9.5 (8.5~10.1); 77.4 (66.0~83.7); 48.0 (42.1~59.5); 634 (585~663) ❙ 174 (84~252); 106 (53~152); 4 (1~6) ❙ −72.4 (−87.1~−57.0); 174 (84~252); 45.6 (7.5~104.3)

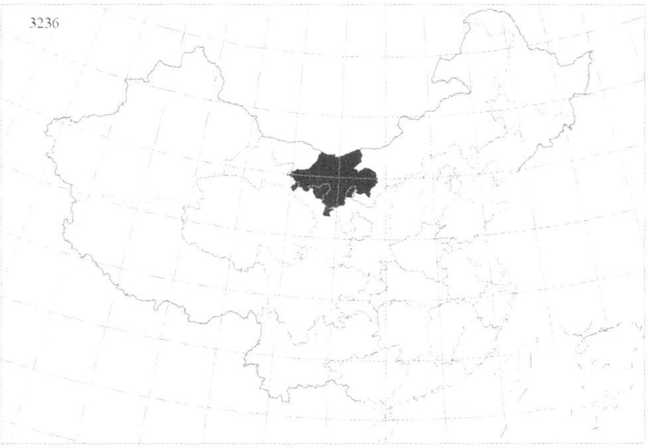

Rosaceae 蔷薇科 *Potentilla* 委陵菜属
Potentilla fruticosa L. 金露梅
Shrubs, deciduous

6.2 (−5.9~18.9); −7.1 (−29.8~12.5); 17.8 (5.4~27.5); 8.7 (1.3~19.0); 65.2 (0.4~167.8); 50.0 (0.0~157.4); 578 (299~929) ❙ 594 (44~1625); 346 (19~817); 21 (1~186) ❙ 2.3 (−94.5~128.6); 468 (44~912); 208.3 (8.4~516.7)

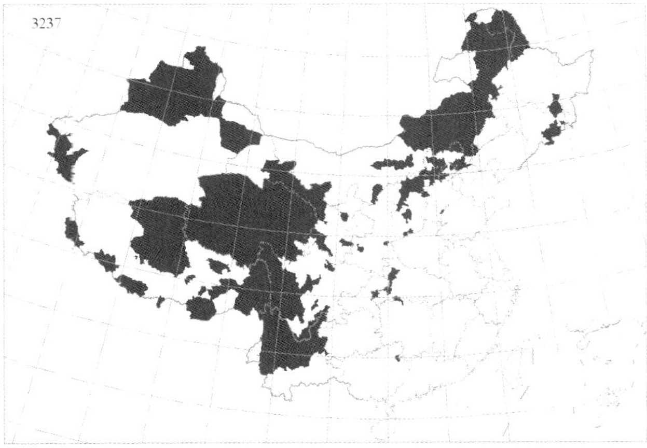

Rosaceae 蔷薇科 *Potentilla* 委陵菜属
Potentilla fruticosa var. albicans Rehder et E. H. Wilson 白毛金露梅
Shrubs, deciduous

3.7 (−2.6~14.1); −6.9 (−17.6~6.5); 12.8 (8.7~20.2); 5.6 (2.5~14.2); 32.3 (7.7~109.8); 47.6 (0.0~98.0); 468 (349~729) ❙ 624 (192~1023); 369 (106~615); 16 (7~31) ❙ 33.9 (−63.2~125.9); 431 (192~707); 213.1 (32.6~319.8)

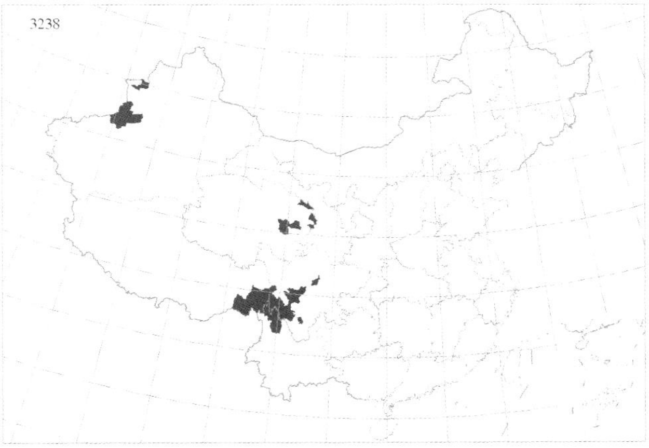

Rosaceae 蔷薇科 *Potentilla* 委陵菜属
Potentilla fruticosa var. arbuscula (D. Don) Maxim. 伏毛金露梅
Shrubs, deciduous

6.7 (0.3~14.9); −1.4 (−9.9~5.3); 13.9 (9.3~23.6); 7.5 (3.3~15.0); 45.8 (11.9~119.6); 24.5 (0.0~67.5); 528 (391~782) ❙ 920 (535~1410); 530 (326~816); 29 (10~58) ❙ 74.0 (27.1~125.9); 528 (391~782); 283.0 (205.0~384.7)

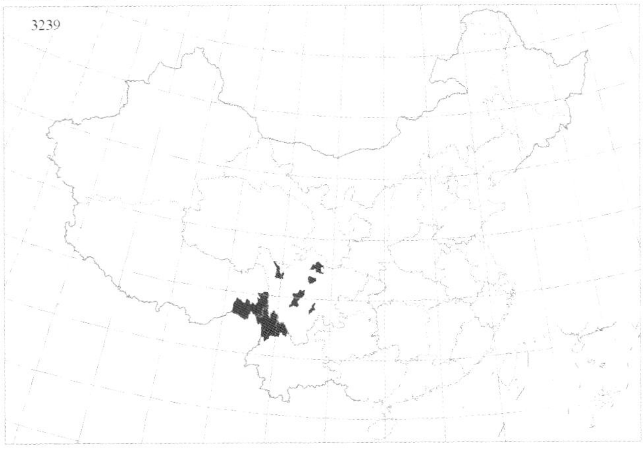

Rosaceae 蔷薇科 *Potentilla* 委陵菜属
Potentilla fruticosa var. pumila Hook. f. 垫状金露梅
Shrubs, deciduous

−1.3 (−5.9~3.3); −13.5 (−17.3~−9.3); 9.9 (5.4~14.3); 3.2 (1.3~5.6); 14.4 (0.4~33.5); 89.0 (53.1~130.5); 376 (300~468) ❙ 367 (105~664); 218 (72~373); 8 (2~15) ❙ −0.1 (−72.6~87.3); 291 (105~450); 151.1 (20.1~307.6)

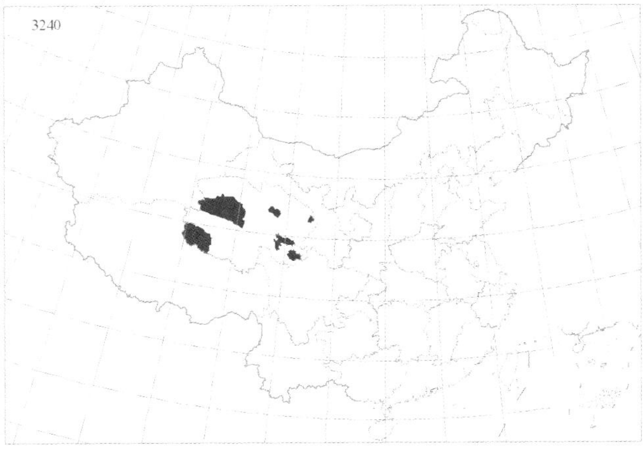

Rosaceae 蔷薇科 *Potentilla* 委陵菜属
Potentilla glabra Lodd. 银露梅
Shrubs, deciduous

5.1 (−5.9~17.2); −8.1 (−29.8~9.1); 17.0 (5.4~27.1); 7.4 (1.3~17.3); 52.2 (0.4~147.5); 50.0 (0.0~157.4); 538 (300~847) ❙ 613 (105~1492); 338 (70~696); 19 (1~113) ❙ 15.3 (−77.3~128.6); 475 (105~847); 212.7 (13.3~384.7)

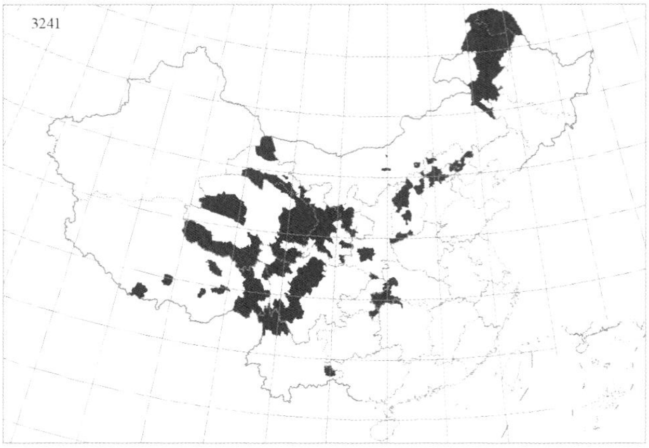

Rosaceae 蔷薇科 *Potentilla* 委陵菜属
Potentilla glabra var. longipetala T. T. Yu et C. L. Li 长瓣银露梅
Shrubs, deciduous

9.1; 1.9; 15.3; 9.1; 56.7; 7.1; 587 ❙ 1024; 581; 42 ❙ 74.7; 587; 352.0

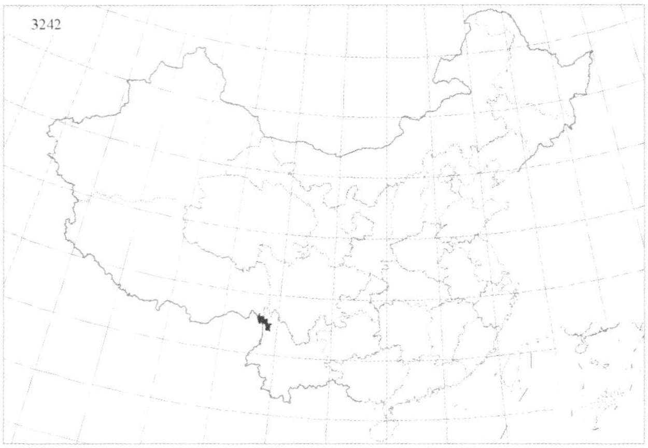

Rosaceae 蔷薇科 *Potentilla* 委陵菜属
Potentilla glabra var. mandshurica (Maxim.) Hand.-Mazz. 白毛银露梅
Shrubs, deciduous

3.4 (−5.9~12.3); −8.8 (−17.3~1.1); 14.2 (5.4~24.7); 5.8 (1.3~12.3); 36.6 (0.4~96.2); 55.4 (8.1~130.5); 478 (300~722) ❙ 552 (105~1023); 309 (72~535); 11 (2~28) ❙ 18.4 (−72.6~128.6); 450 (105~707); 237.2 (20.1~364.1)

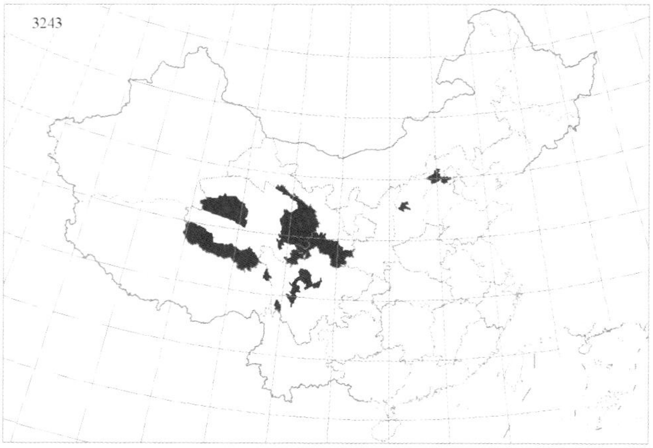

Rosaceae 蔷薇科 *Potentilla* 委陵菜属
Potentilla glabra var. veitchii (E. H. Wilson) Hand.-Mazz. 伏毛银露梅
Shrubs, deciduous

4.6 (−0.3~10.8); −3.7 (−10.9~3.9); 11.9 (9.0~16.6); 5.6 (3.0~10.9); 30.0 (10.4~71.8); 34.6 (1.3~73.4); 470 (378~633) ❙ 802 (535~1007); 473 (326~601); 21 (10~40) ❙ 71.5 (27.1~121.7); 470 (378~633); 265.5 (130.1~351.7)

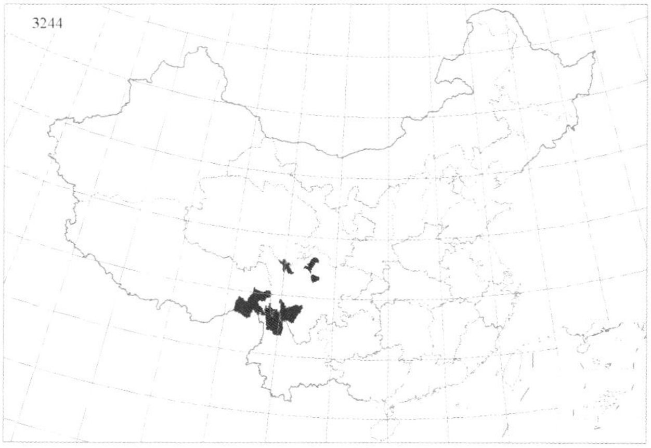

Rosaceae 蔷薇科 *Potentilla* 委陵菜属
Potentilla parvifolia Fisch. ex Lehm. 小叶金露梅
Shrubs, deciduous

3.7 (−5.9~13.0); −10.7 (−29.4~4.6); 16.4 (5.4~26.8); 6.6 (1.3~13.3); 46.7 (0.4~114.1); 62.0 (0.4~154.8); 513 (294~811) ❙ 404 (23~1134); 238 (16~651); 13 (0~103) ❙ −17.2 (−97.2~128.6); 357 (23~684); 148.1 (2.8~364.1)

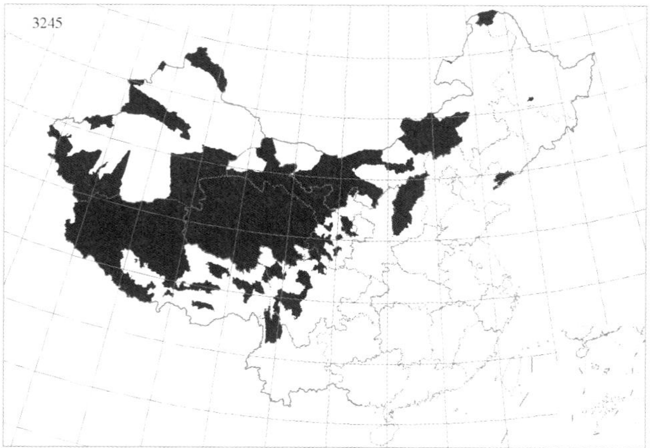

Rosaceae 蔷薇科 *Potentilla* 委陵菜属
Potentilla parvifolia var. hypoleuca Hand.-Mazz. 白毛小叶金露梅
Shrubs, deciduous

0.9 (−3.3~3.3); −10.6 (−15.4~−6.4); 10.9 (7.2~14.3); 3.9 (2.0~5.6); 18.4 (3.7~33.5); 67.6 (43.8~103.2); 409 (339~468) ❙ 552 (63~943); 309 (45~507); 12 (2~23) ❙ 34.7 (−84.9~128.6); 372 (63~450); 234.0 (8.4~357.7)

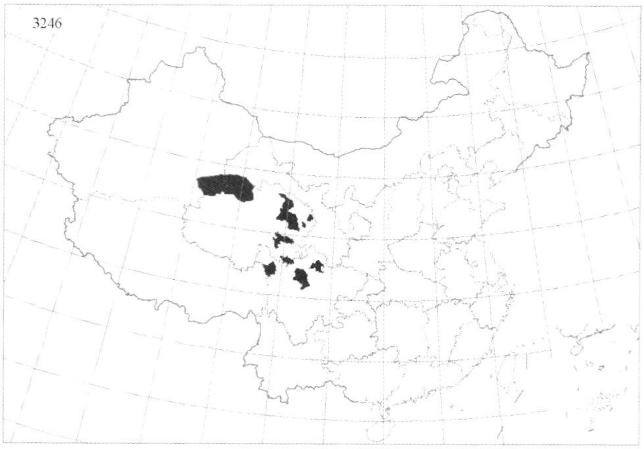

Rosaceae 蔷薇科 *Prinsepia* 扁核木属
Prinsepia scandens Hayata 台湾扁核木
Shrubs, deciduous

18.3 (16.2~21.7); 12.7 (11.0~15.3); 22.8 (20.1~27.0); 18.3 (16.2~21.7); 159.4 (134.2~200.2); 0.0 (0.0~0.0); 887 (773~1118) ❙ 2393 (1836~2713); 1110 (900~1388); 221 (120~308) ❙ 177.9 (64.2~246.5); 878 (773~1075); 284.3 (202.8~363.7)

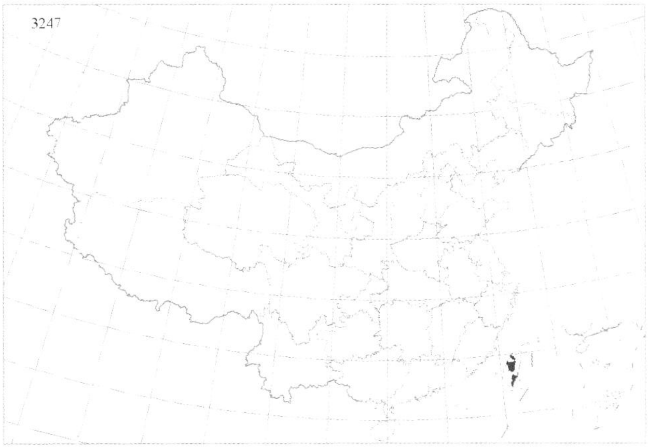

Rosaceae 蔷薇科 *Prinsepia* 扁核木属
Prinsepia sinensis (Oliv.) Oliv. ex Bean 东北蕤核
Shrubs, deciduous

2.9 (-0.9~8.0); -18.2 (-25.0~-9.8); 20.9 (18.4~23.5); 7.8 (6.4~10.0); 59.9 (45.8~83.4); 84.7 (46.4~120.7); 569 (501~658) ❙ 684 (408~1073); 422 (281~691); 24 (7~40) ❙ 19.9 (-31.7~74.1); 560 (408~658); 294.4 (132.8~346.5)

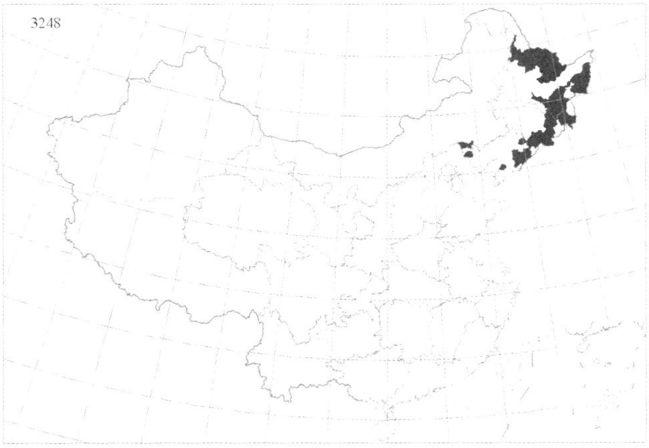

Rosaceae 蔷薇科 *Prinsepia* 扁核木属
Prinsepia uniflora Batalin 蕤核
Shrubs, deciduous

9.4 (1.9~16.2); -5.8 (-13.7~5.5); 22.9 (11.7~27.8); 10.6 (4.3~16.2); 85.3 (21.1~134.6); 32.3 (0.0~71.7); 672 (427~865) ❙ 540 (201~1197); 308 (119~489); 17 (4~71) ❙ -19.3 (-69.9~51.2); 530 (201~854); 152.9 (38.8~289.4)

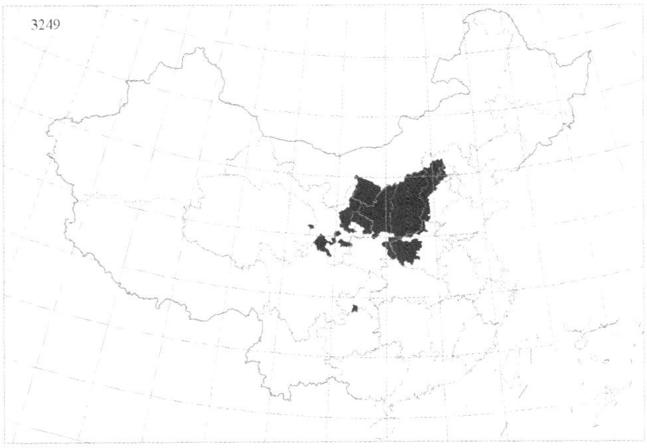

Rosaceae 蔷薇科 *Prinsepia* 扁核木属
Prinsepia uniflora var. serrata Rehder 齿叶蕤核
Shrubs, deciduous

6.6 (2.9~9.9); -5.6 (-7.7~-4.0); 17.7 (11.2~22.7); 7.8 (4.4~10.6); 55.1 (20.5~83.9); 35.2 (24.4~49.3); 554 (428~663) ❙ 625 (466~950); 315 (248~490); 16 (11~25) ❙ 19.6 (-25.0~121.7); 499 (428~588); 221.7 (121.3~285.5)

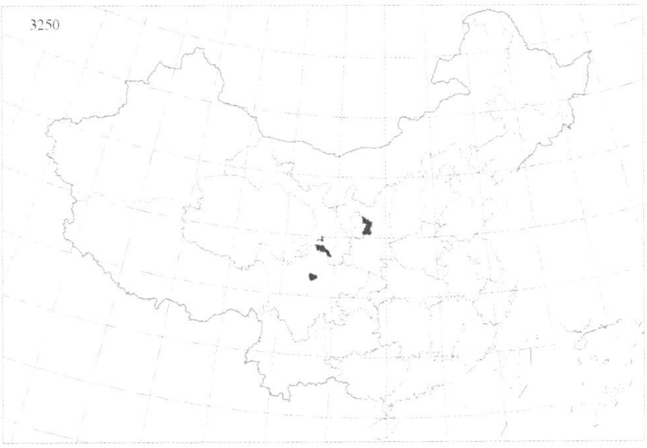

Rosaceae 蔷薇科 *Prinsepia* 扁核木属
Prinsepia utilis Royle 扁核木
Shrubs, deciduous

14.7 (-1.1~19.3); 6.2 (-9.1~13.3); 21.8 (6.2~27.7); 14.8 (2.0~19.4); 120.1 (2.5~173.2); 2.9 (0.0~75.6); 771 (334~981) ❙ 1105 (321~1736); 569 (193~1007); 55 (9~146) ❙ 43.9 (-14.2~145.8); 761 (321~981); 262.1 (79.3~516.7)

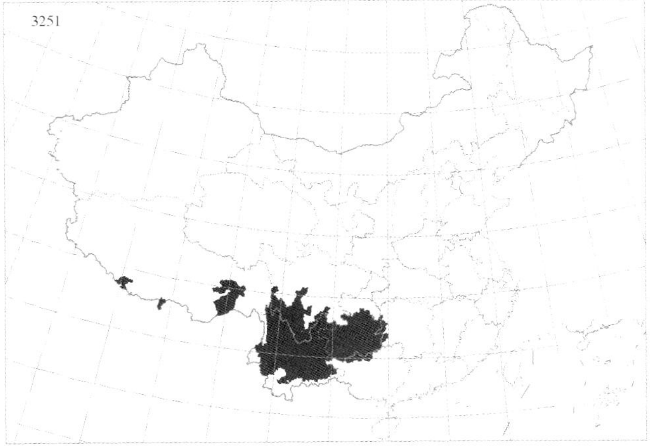

Rosaceae 蔷薇科 *Prunus* 李属
Prunus cerasifera Ehrh. 樱桃李
Shrubs or trees, deciduous

6.5; -11.4; 20.7; 9.0; 70.8; 53.0; 626 ❙ 301; 101; 47 ❙ -50.0; 301; 178.7

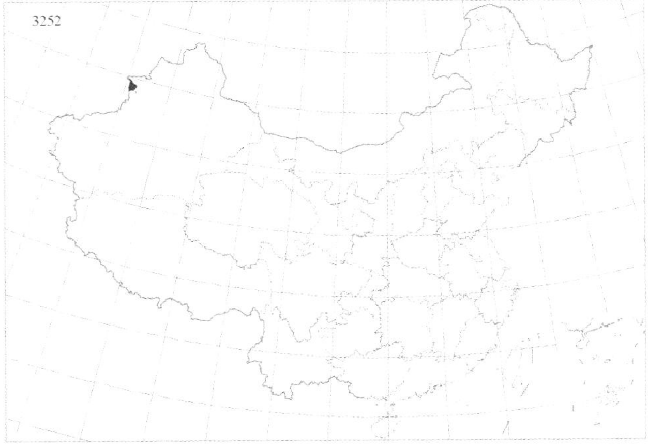

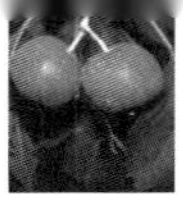

Rosaceae 蔷薇科 *Prunus* 李属

Prunus salicina Lindl. 李

Trees, deciduous

16.2 (1.9~23.6); 5.8 (−11.8~16.7); 25.4 (11.2~29.9); 16.4 (4.3~23.6); 139.7 (20.0~223.5); 4.8 (0.0~60.2); 879 (427~1281) ❙ 1255 (337~2435); 556 (190~1428); 102 (1~254) ❙ 41.4 (−45.2~130.7); 852 (337~1273); 231.1 (44.7~659.0)

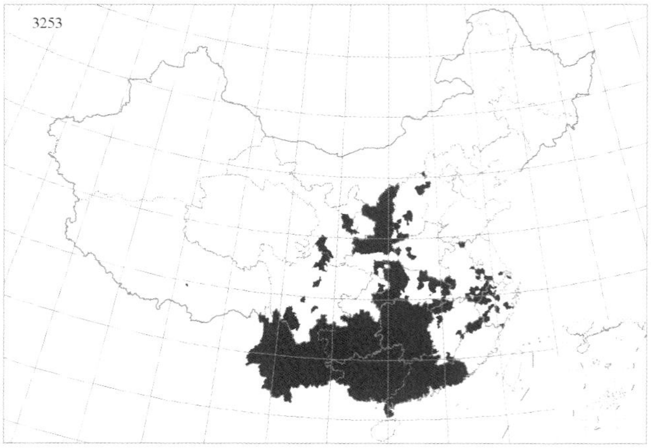

Rosaceae 蔷薇科 *Prunus* 李属

Prunus salicina var. pubipes (Koehne) L. H. Bailey 毛梗李

Trees, deciduous

7.8; 0.4; 14.2; 8.0; 48.9; 14.3; 548 ❙ 877; 546; 19 ❙ 60.7; 548; 304.7

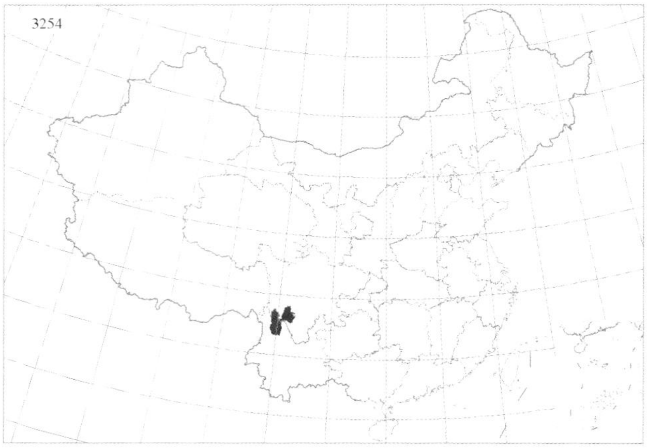

Rosaceae 蔷薇科 *Prunus* 李属

Prunus simonii Carrière 杏李

Trees, deciduous

7.8 (2.6~11.0); −9.4 (−15.9~−5.7); 23.0 (19.1~25.5); 9.9 (7.2~12.0); 81.6 (52.0~101.7); 47.3 (29.5~80.4); 658 (536~757) ❙ 563 (409~694); 409 (258~499); 11 (8~12) ❙ −14.3 (−24.3~5.4); 561 (409~659); 237.4 (174.5~287.6)

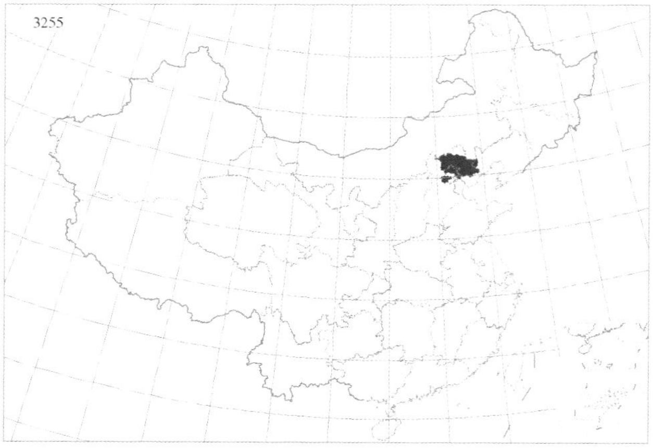

Rosaceae 蔷薇科 *Prunus* 李属

Prunus ussuriensis Kovalev et Kostina 东北李

Trees, deciduous

5.5 (2.5~8.4); −14.6 (−19.6~−11.0); 22.6 (20.0~25.0); 9.0 (7.3~10.6); 72.5 (54.1~90.9); 66.1 (49.4~87.4); 618 (542~700) ❙ 806 (556~966); 509 (367~616); 28 (15~35) ❙ 31.4 (−9.7~57.4); 610 (542~698); 280.1 (177.1~346.5)

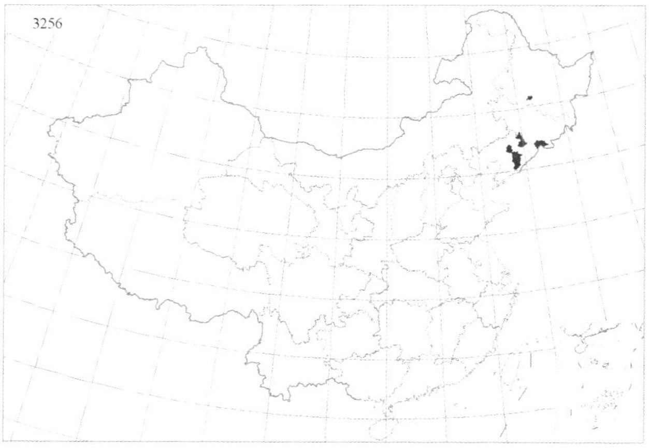

Rosaceae 蔷薇科 *Pygeum* 臀形果属

Pygeum henryi Dunn 云南臀果木

Trees, evergreen

18.2 (8.6~22.2); 11.5 (1.3~15.1); 22.8 (15.0~28.3); 18.2 (8.7~22.3); 158.9 (52.6~207.5); 0.2 (0.0~8.8); 889 (574~1190) ❙ 1381 (1028~1751); 731 (566~996); 65 (41~187) ❙ 56.1 (12.5~93.2); 875 (574~1190); 420.7 (208.7~659.0)

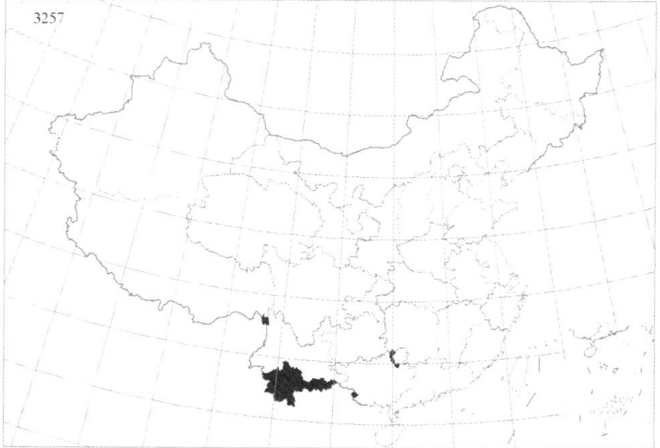

Rosaceae 蔷薇科 *Pygeum* 臀形果属

Pygeum laxiflorum Merr. ex H. L. Li 疏花臀果木

Trees, evergreen

21.9 (21.3~22.7); 13.9 (13.2~15.0); 27.9 (27.3~28.6); 22.0 (21.3~22.8); 203.6 (195.9~213.1); 0.0 (0.0~0.0); 1162 (1111~1221) ❙ 1876 (1302~2435); 1047 (670~1428); 116 (86~130) ❙ 61.1 (12.9~99.5); 1162 (1111~1221); 289.6 (235.1~318.8)

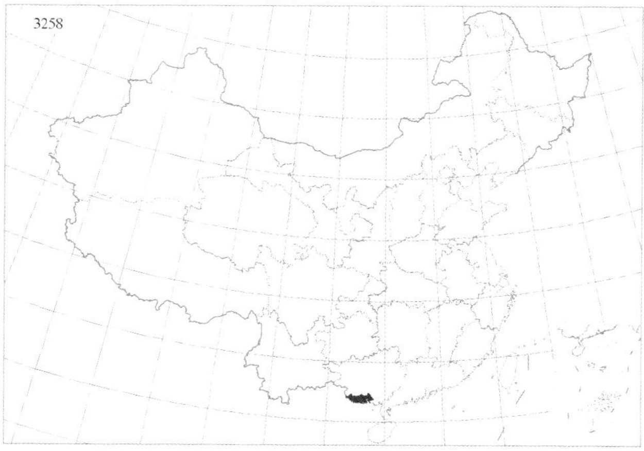

Rosaceae 蔷薇科 *Pygeum* 臀形果属

Pygeum macrocarpum T. T. Yu et L. T. Lu 大果臀果木

Trees, evergreen

19.1 (17.3~21.0); 13.0 (9.9~15.1); 23.2 (21.6~25.6); 19.2 (17.3~21.0); 170.1 (147.1~191.8); 0.0 (0.0~0.0); 924 (833~1053) ❙ 1502 (1276~1751); 800 (655~996); 61 (51~77) ❙ 63.2 (36.8~88.8); 915 (833~1053); 502.8 (359.7~659.0)

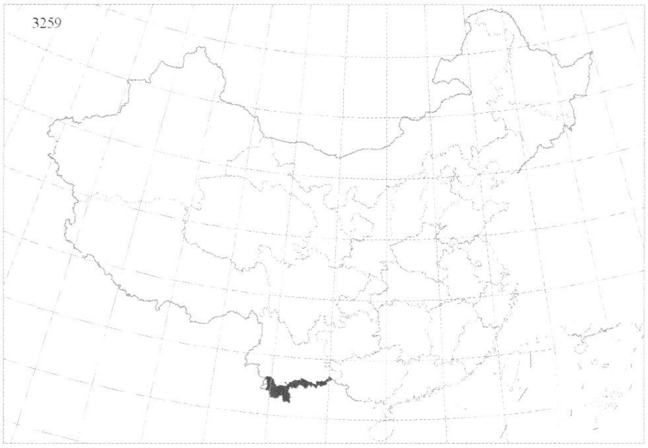

Rosaceae 蔷薇科 *Pygeum* 臀形果属

Pygeum oblongum T. T. Yu et L. T. Lu 长圆臀果木

Trees, evergreen

19.2 (17.7~21.0); 12.9 (10.9~14.2); 23.4 (22.5~25.6); 19.2 (17.8~21.0); 170.7 (153.4~191.8); 0.0 (0.0~0.0); 929 (854~1053) ❙ 1482 (1276~1751); 843 (741~996); 62 (51~77) ❙ 60.0 (36.8~88.8); 927 (854~1053); 511.1 (452.0~551.6)

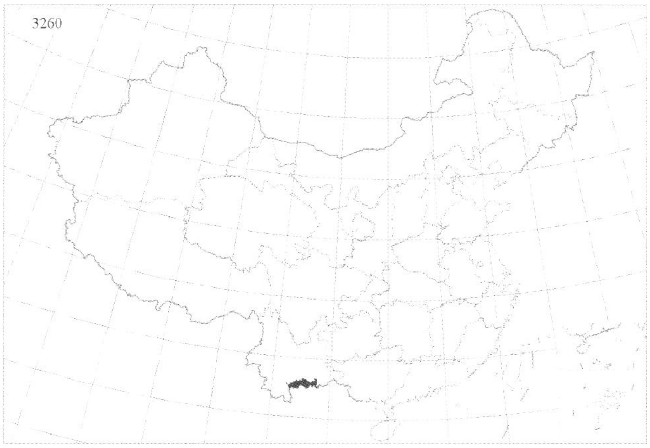

Rosaceae 蔷薇科 *Pygeum* 臀形果属

Pygeum topengii Merr. 臀果木

Trees, evergreen

20.6 (14.3~24.8); 12.4 (3.7~20.1); 27.3 (21.4~29.8); 20.7 (14.3~24.8); 187.9 (113.2~237.6); 0.0 (0.0~1.5); 1083 (761~1385) ❙ 1596 (868~2435); 689 (312~1428); 128 (41~230) ❙ 49.3 (-34.3~128.9); 1069 (761~1360); 271.1 (95.3~659.0)

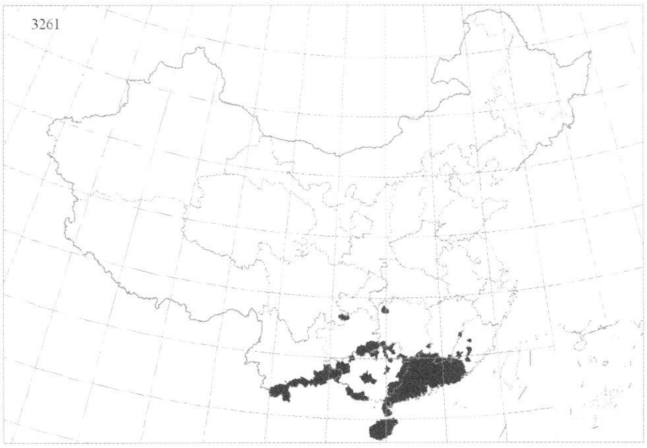

Rosaceae 蔷薇科 *Pyracantha* 火棘属

Pyracantha angustifolia (Franch.) C. K. Schneid. 窄叶火棘

Shrubs or trees, evergreen

13.8 (2.2~18.9); 5.7 (-5.8~12.5); 20.5 (9.4~28.2); 13.9 (3.7~19.0); 109.3 (14.2~167.8); 3.3 (0.0~48.9); 731 (402~929) ❙ 1001 (535~1454); 550 (245~792); 37 (10~88) ❙ 38.5 (-36.2~100.7); 713 (402~912); 279.1 (131.4~487.2)

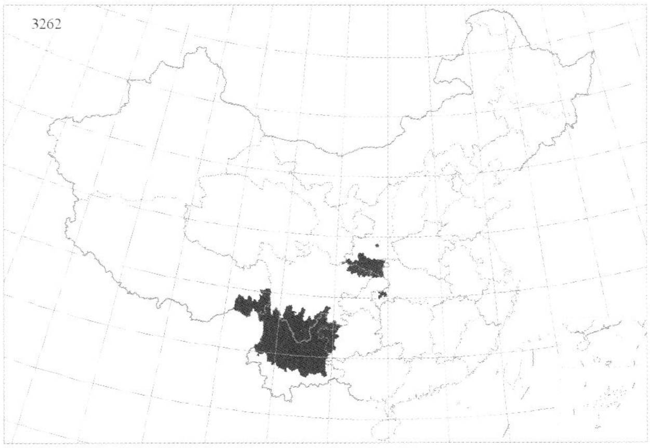

Rosaceae 蔷薇科 *Pyracantha* 火棘属

Pyracantha atalantioides (Hance) Stapf 全缘火棘

Shrubs or trees, evergreen

16.0 (2.2~21.3); 5.5 (-6.6~12.3); 25.6 (9.9~29.7); 16.1 (3.8~21.3); 134.6 (15.5~196.1); 1.5 (0.0~49.1); 855 (413~1130) ❙ 1338 (798~1829); 541 (352~821); 119 (19~230) ❙ 56.8 (-2.0~128.9); 851 (413~1130); 215.3 (141.4~354.5)

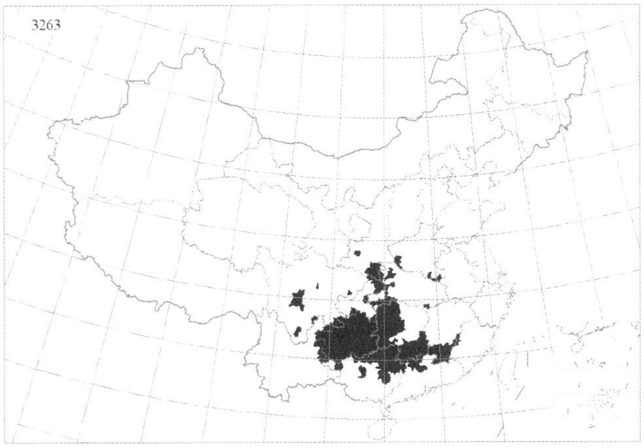

Rosaceae 蔷薇科 *Pyracantha* 火棘属

Pyracantha crenulata (D. Don) M. Roem. 细圆齿火棘

Shrubs or trees, evergreen

14.7 (2.6~20.3); 4.4 (-7.0~12.5); 24.0 (11.2~29.1); 14.8 (4.3~20.4); 120.8 (20.0~184.5); 3.9 (0.0~48.2); 794 (427~1072) ❙ 1126 (512~1777); 513 (258~816); 75 (12~196) ❙ 41.6 (-35.3~107.2); 779 (427~1072); 231.7 (121.6~384.1)

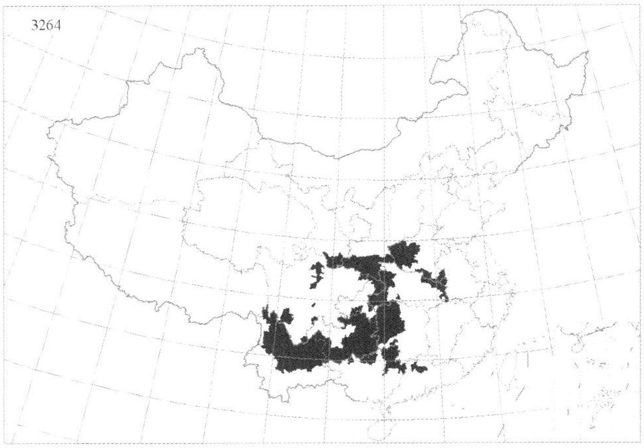

Rosaceae 蔷薇科 *Pyracantha* 火棘属

Pyracantha crenulata var. kansuensis Rehder 细叶细圆齿火棘

Shrubs, evergreen

7.8 (2.6~11.0); -2.0 (-7.0~4.2); 16.7 (11.2~21.3); 8.3 (4.3~11.0); 56.0 (20.0~78.8); 21.9 (0.8~48.2); 564 (427~644) ❙ 838 (669~1134); 426 (317~651); 23 (14~47) ❙ 51.9 (2.4~101.2); 564 (427~644); 268.5 (200.8~339.5)

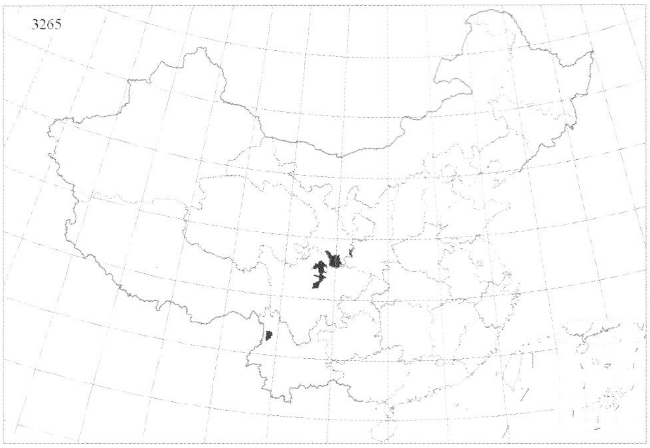

Rosaceae 蔷薇科 *Pyracantha* 火棘属

Pyracantha densiflora T. T. Yu 密花火棘

Shrubs, evergreen

21.5 (17.6~22.7); 13.4 (7.3~15.2); 28.1 (24.6~29.1); 21.6 (17.6~22.7); 199.3 (151.6~212.7); 0.0 (0.0~0.0); 1142 (901~1216) ❙ 1654 (1172~2037); 806 (639~1033); 110 (50~163) ❙ 45.1 (20.8~72.1); 1142 (901~1216); 195.6 (95.3~411.1)

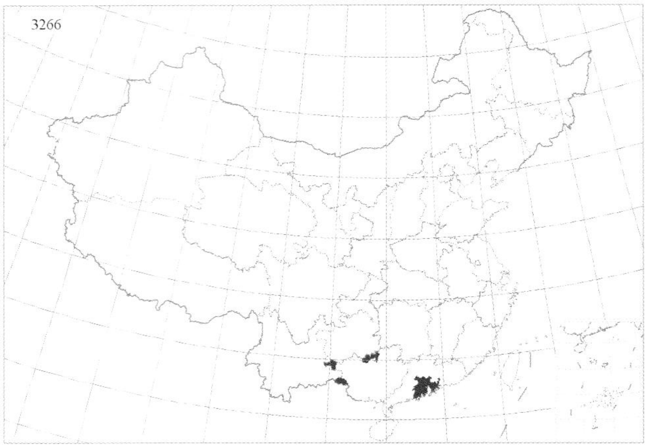

Rosaceae 蔷薇科 *Pyracantha* 火棘属

Pyracantha fortuneana (Maxim.) H. L. Li 火棘

Shrubs, evergreen

15.4 (-1.8~21.1); 5.1 (-10.2~15.1); 24.6 (6.1~30.1); 15.4 (1.8~21.1); 128.1 (1.9~193.7); 3.2 (0.0~82.8); 825 (329~1114) ❙ 1205 (320~1829); 533 (193~996); 93 (11~231) ❙ 45.8 (-35.3~128.9); 811 (320~1114); 240.7 (72.2~659.0)

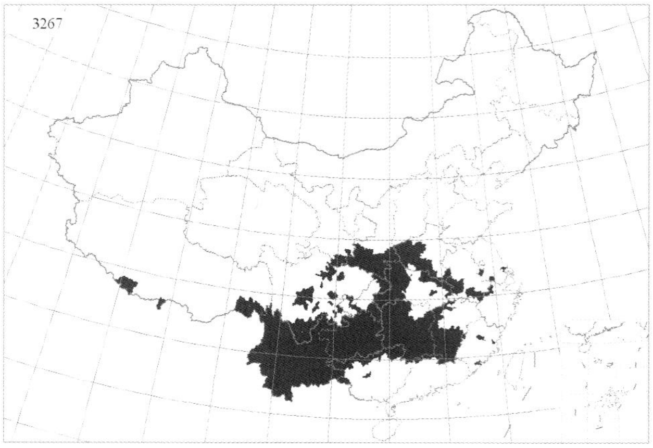

Rosaceae 蔷薇科 *Pyracantha* 火棘属

Pyracantha inermis J. E. Vidal 澜沧火棘

Shrubs, evergreen

18.5 (16.3~20.5); 12.6 (10.2~14.9); 22.0 (20.3~23.7); 18.5 (16.4~20.5); 162.4 (136.6~185.9); 0.0 (0.0~0.0); 884 (789~992) ❙ 1418 (1232~1500); 746 (655~836); 51 (41~66) ❙ 60.7 (47.1~74.2); 859 (780~959); 465.5 (372.4~570.1)

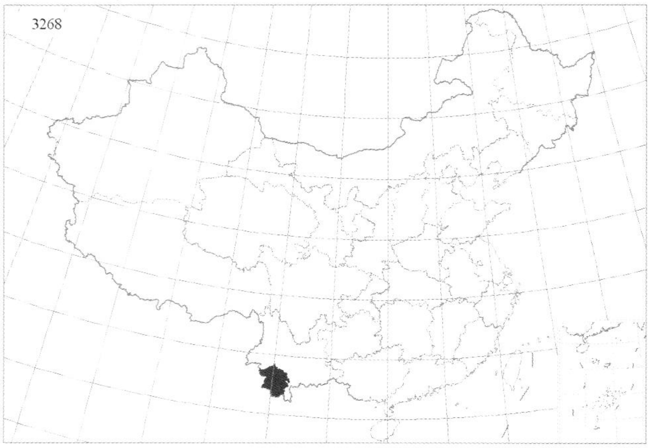

Rosaceae 蔷薇科 *Pyracantha* 火棘属

Pyracantha koidzumii (Hayata) Rehder 台湾火棘

Shrubs, evergreen

19.0; 14.4; 22.5; 19.1; 168.9; 0.0; 923 ❙ 2735; 1394; 199 ❙ 200.0; 923; 321.4

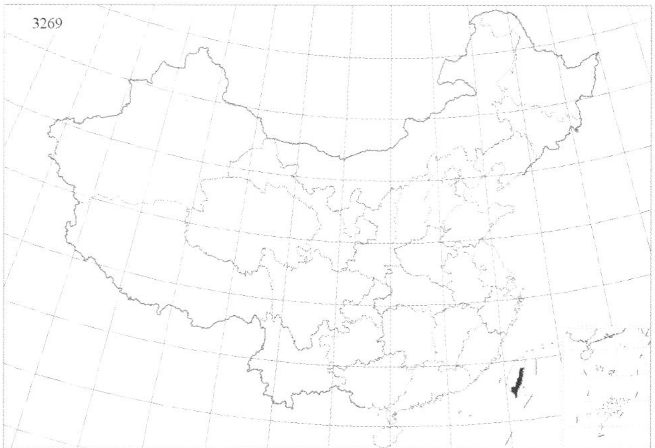

Rosaceae 蔷薇科 *Pyrus* 梨属

Pyrus armeniacifolia T. T. Yu 杏叶梨

Trees, deciduous

4.1; -14.4; 20.4; 8.1; 61.7; 72.2; 589 ❙ 315; 102; 51 ❙ -46.4; 312; 173.1

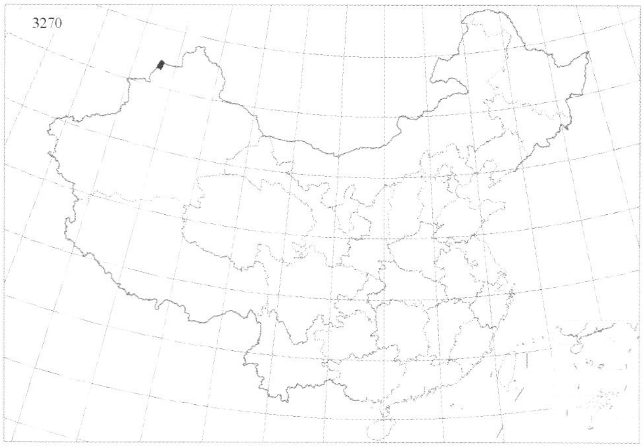

Rosaceae 蔷薇科 *Pyrus* 梨属

Pyrus betulifolia Bunge 杜梨

Trees, deciduous

12.6 (-1.1~19.4); -1.9 (-14.4~9.4); 25.6 (6.2~29.9); 13.1 (2.0~19.4); 109.5 (2.5~173.4); 18.3 (0.0~75.6); 783 (334~1029) ❙ 739 (321~1922); 396 (193~685); 47 (1~259) ❙ -6.3 (-51.1~112.2); 645 (321~1029); 173.2 (44.7~364.1)

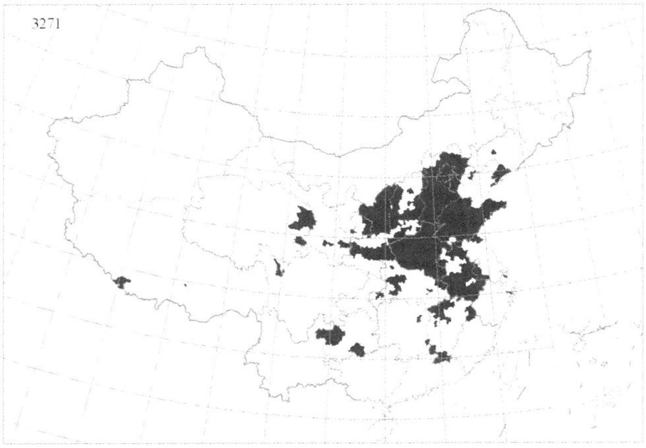

Rosaceae 蔷薇科 *Pyrus* 梨属

Pyrus bretschneideri Rehder 白梨

Trees, deciduous

9.2 (0.4~13.5); -4.8 (-11.3~1.5); 21.8 (10.5~27.1); 10.1 (3.6~13.7); 79.5 (15.5~117.5); 29.0 (7.1~69.4); 648 (406~820) ❙ 600 (373~892); 312 (201~458); 17 (7~35) ❙ -6.6 (-37.6~52.6); 561 (373~760); 185.6 (44.7~334.4)

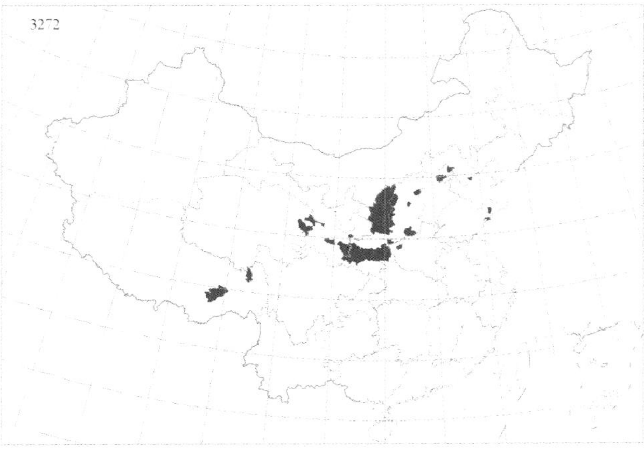

Rosaceae 蔷薇科 *Pyrus* 梨属

Pyrus calleryana Decne. 豆梨

Trees, deciduous

16.7 (0.0~24.8); 5.9 (-11.4~20.1); 26.7 (9.7~30.1); 16.8 (3.3~24.8); 144.5 (12.2~237.6); 3.1 (0.0~72.2); 908 (395~1385) ❙ 1334 (58~3495); 539 (36~1847); 130 (4~897) ❙ 46.3 (-90.8~246.5); 882 (58~1360); 215.2 (11.3~620.0)

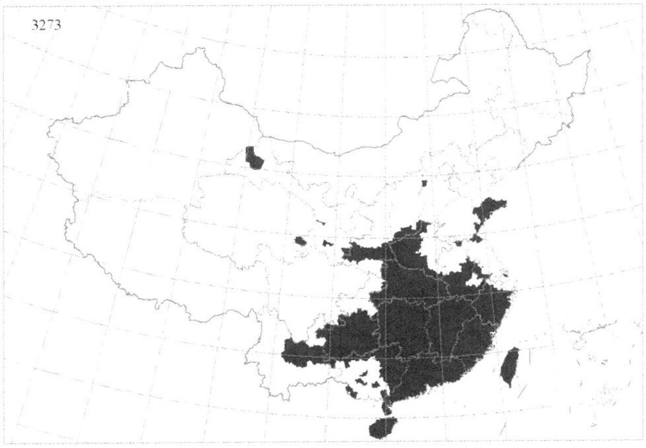

Rosaceae 蔷薇科 *Pyrus* 梨属

Pyrus calleryana var. integrifolia T. T. Yu 全缘叶豆梨

Trees, deciduous

15.8 (14.7~17.3); 3.7 (2.3~5.5); 27.7 (25.9~29.1); 15.8 (14.8~17.3); 132.0 (121.1~147.3); 1.9 (0.0~4.2); 868 (811~938) ❙ 1447 (1060~1813); 531 (463~632); 188 (122~242) ❙ 67.4 (21.1~123.6); 864 (811~903); 223.7 (153.8~284.3)

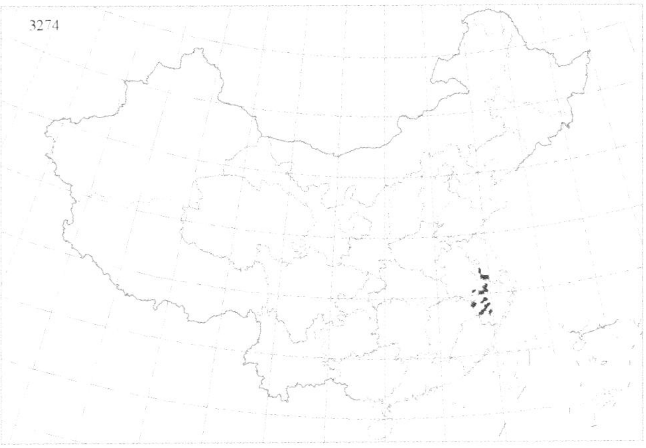

Rosaceae 蔷薇科 *Pyrus* 梨属

Pyrus calleryana var. koehnei (C. K. Schneid.) T. T. Yu 楔叶豆梨

Trees, deciduous

20.0 (14.3~23.6); 10.8 (2.1~16.7); 27.6 (23.1~29.8); 20.0 (14.3~23.6); 180.2 (113.2~223.5); 0.1 (0.0~4.9); 1052 (761~1281) ❙ 1592 (1161~2435); 689 (438~1428); 146 (43~254) ❙ 53.1 (-6.9~130.7); 1050 (761~1273); 235.3 (95.3~411.1)

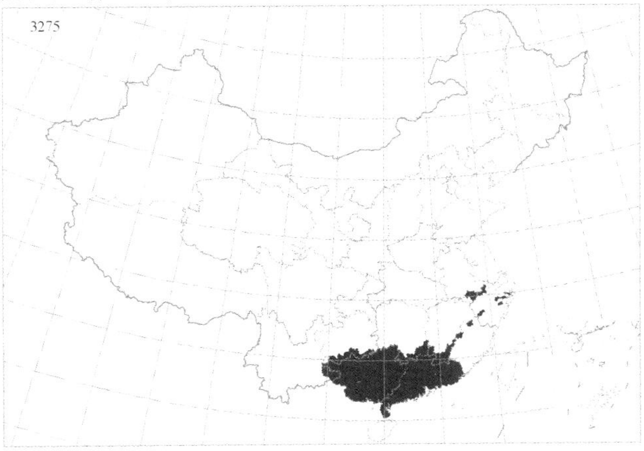

Rosaceae 蔷薇科 *Pyrus* 梨属

Pyrus calleryana var. lanceata Rehder 柳叶豆梨

Trees, deciduous

16.0 (15.0~17.3); 3.7 (1.8~6.0); 27.7 (26.3~29.1); 16.0 (15.0~17.3); 134.3 (123.7~147.3); 1.8 (0.0~5.2); 878 (826~938) ❙ 1576 (960~1997); 562 (460~669); 200 (100~254) ❙ 79.7 (10.2~130.7); 875 (826~921); 241.6 (136.7~303.4)

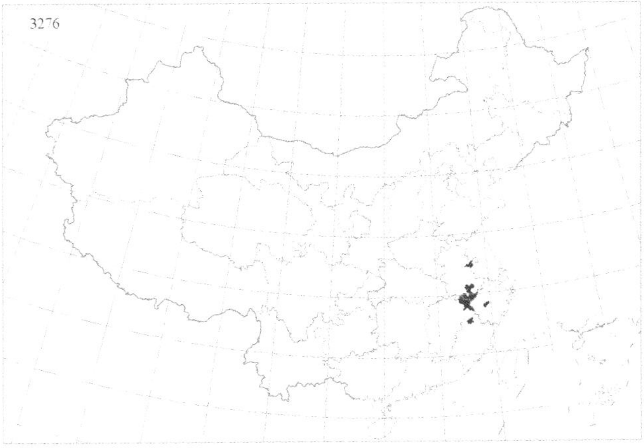

Rosaceae 蔷薇科 *Pyrus* 梨属
Pyrus hopeiensis T. T. Yu 河北梨
Trees, deciduous

7.5 (2.6~11.0); -9.3 (-15.9~-5.6); 22.4 (17.9~25.5); 9.6 (7.2~12.0); 77.8 (51.7~101.7); 46.8 (28.0~80.4); 639 (529~757) ❙ 490 (196~718); 334 (117~499); 10 (4~29) ❙ -23.3 (-69.9~9.1); 486 (196~659); 182.1 (33.5~287.6)

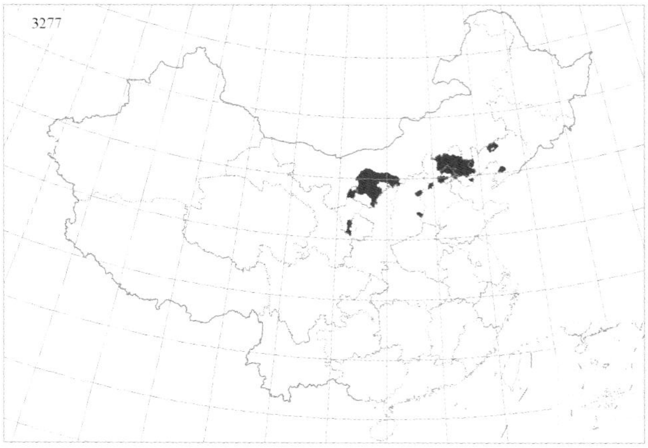

Rosaceae 蔷薇科 *Pyrus* 梨属
Pyrus pashia Buch.-Ham. ex D. Don 川梨
Trees, deciduous

15.1 (-1.1~21.0); 7.3 (-9.1~15.1); 21.4 (6.2~28.9); 15.2 (2.0~21.0); 124.7 (2.5~191.8); 2.6 (0.0~75.6); 782 (334~1053) ❙ 1120 (321~1751); 605 (193~996); 48 (11~129) ❙ 44.1 (-14.2~112.2); 769 (321~1053); 298.3 (79.3~659.0)

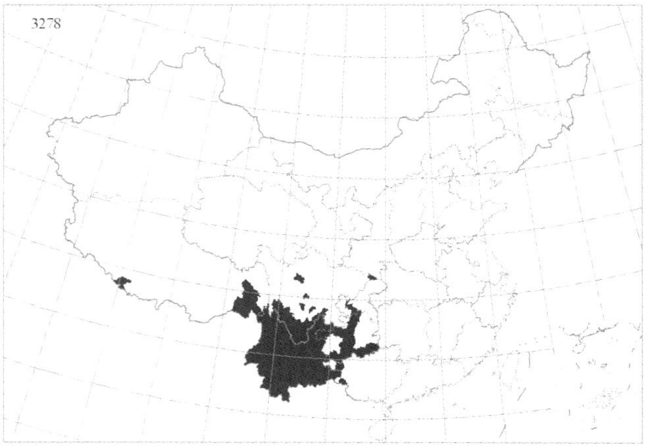

Rosaceae 蔷薇科 *Pyrus* 梨属
Pyrus pashia var. kumaoni Stapf 无毛川梨
Trees, deciduous

16.9; 10.3; 21.4; 16.9; 142.7; 0.0; 813 ❙ 1178; 646; 51 ❙ 44.9; 813; 383.4

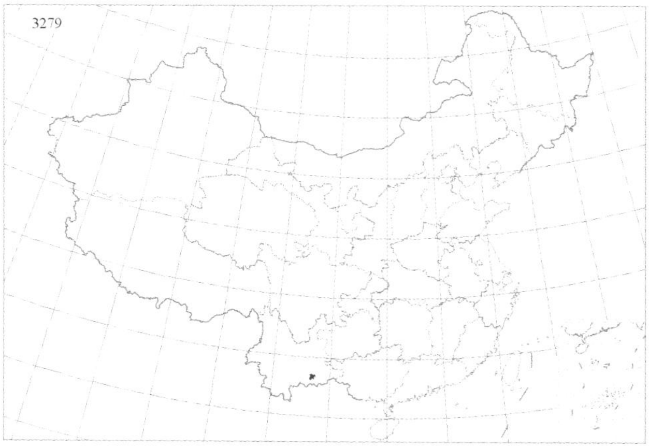

Rosaceae 蔷薇科 *Pyrus* 梨属
Pyrus pashia var. obtusata Cardot 钝叶川梨
Trees, deciduous

14.6; 7.1; 20.3; 14.6; 114.8; 0.0; 741 ❙ 1014; 627; 22 ❙ 37.0; 692; 265.0

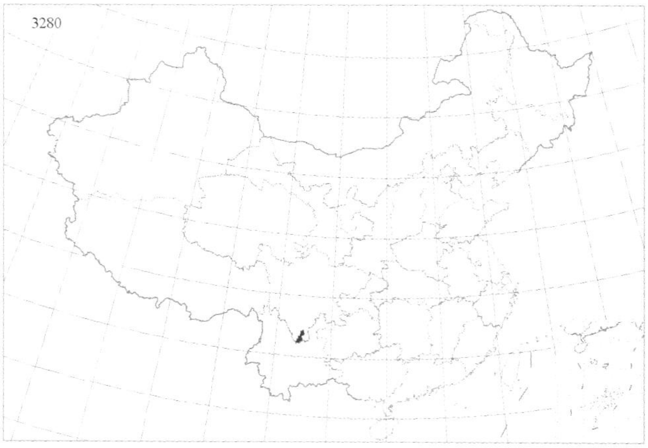

Rosaceae 蔷薇科 *Pyrus* 梨属
Pyrus phaeocarpa Rehder 褐梨
Trees, deciduous

10.9 (3.9~15.1); -4.2 (-13.7~1.8); 24.4 (18.7~28.1); 11.7 (7.3~15.2); 97.3 (52.8~128.0); 25.9 (5.6~71.7); 727 (535~880) ❙ 595 (367~917); 351 (235~560); 23 (8~53) ❙ -17.9 (-40.9~44.0); 587 (367~817); 166.3 (53.8~311.0)

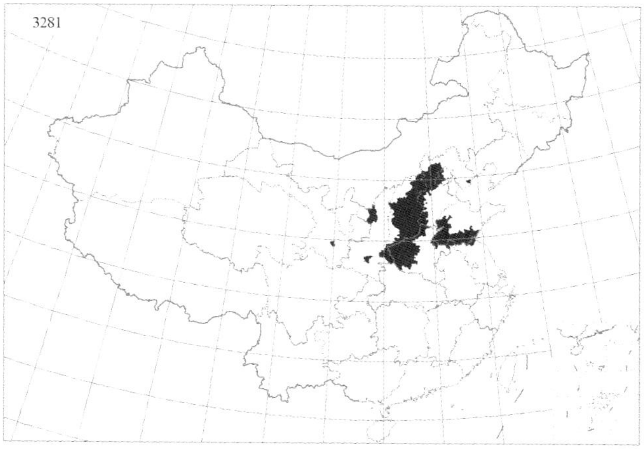

Rosaceae 蔷薇科 *Pyrus* 梨属
Pyrus pseudopashia T. T. Yu 滇梨
Trees, deciduous

11.7 (5.4~16.8); 3.6 (-2.1~7.6); 18.7 (12.3~25.7); 11.8 (5.8~16.9); 86.9 (31.5~142.7); 5.7 (0.0~25.6); 677 (481~869) ❙ 1056 (741~1368); 556 (426~651); 57 (15~129) ❙ 56.6 (36.1~78.3); 674 (481~869); 285.8 (222.1~351.7)

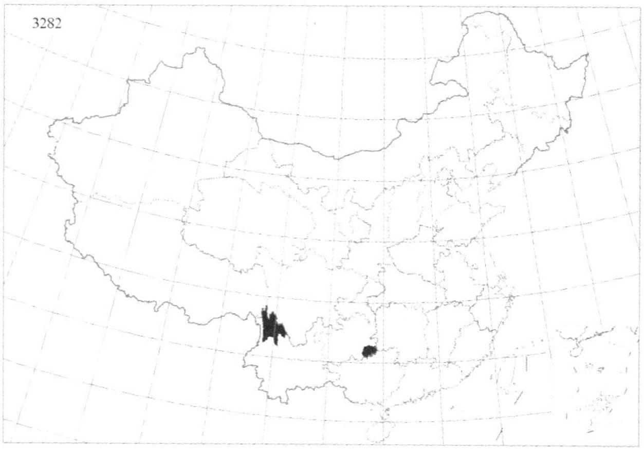

Rosaceae 蔷薇科 *Pyrus* 梨属
Pyrus pyrifolia (Burm. f.) Nakai 沙梨
Trees, deciduous

16.8 (2.2~23.6); 7.0 (-7.3~16.7); 25.4 (9.9~29.9); 16.9 (3.8~23.6); 144.1 (15.5~223.5); 2.1 (0.0~49.1); 894 (413~1281) ❙ 1333 (470~2435); 585 (249~1428); 109 (9~254) ❙ 49.3 (-26.5~130.7); 886 (413~1273); 239.1 (53.3~551.6)

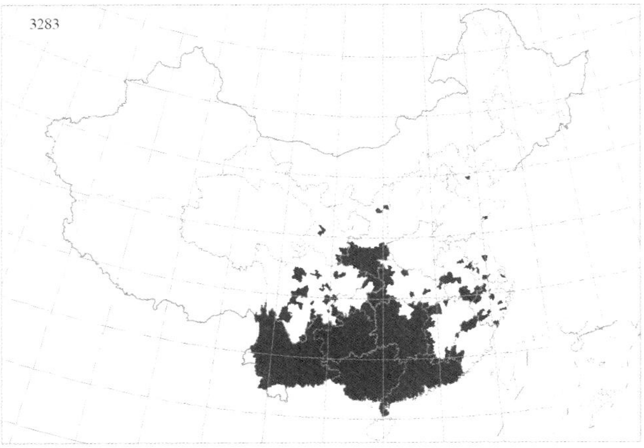

Rosaceae 蔷薇科 *Pyrus* 梨属
Pyrus serrulata Rehder 麻梨
Trees, deciduous

16.1 (4.7~22.4); 4.8 (-12.0~14.8); 26.6 (15.6~30.1); 16.3 (7.5~22.4); 138.6 (48.2~208.9); 4.4 (0.0~59.0); 880 (537~1191) ❙ 1334 (458~2030); 514 (258~1033); 136 (11~259) ❙ 50.3 (-35.3~147.0); 856 (458~1191); 212.3 (70.3~339.5)

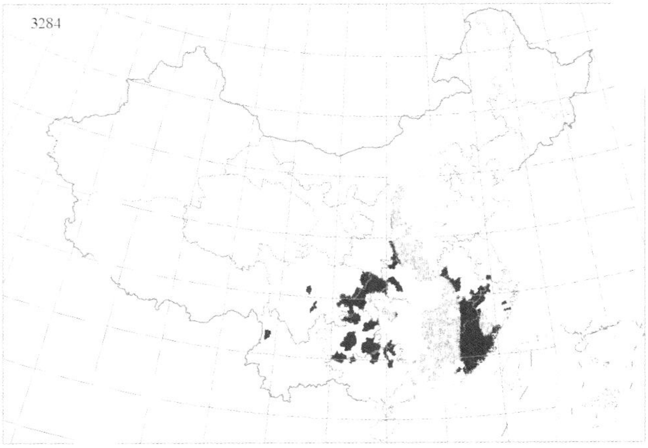

Rosaceae 蔷薇科 *Pyrus* 梨属
Pyrus sinkiangensis T. T. Yu 新疆梨
Trees, deciduous

6.5 (0.4~13.5); -8.4 (-13.9~1.4); 19.1 (10.5~26.1); 8.4 (3.6~13.5); 63.2 (15.5~109.4); 44.7 (7.1~69.4); 589 (406~793) ❙ 364 (32~882); 190 (21~389); 11 (0~48) ❙ -32.4 (-95.9~39.9); 335 (32~767); 141.6 (1.6~316.6)

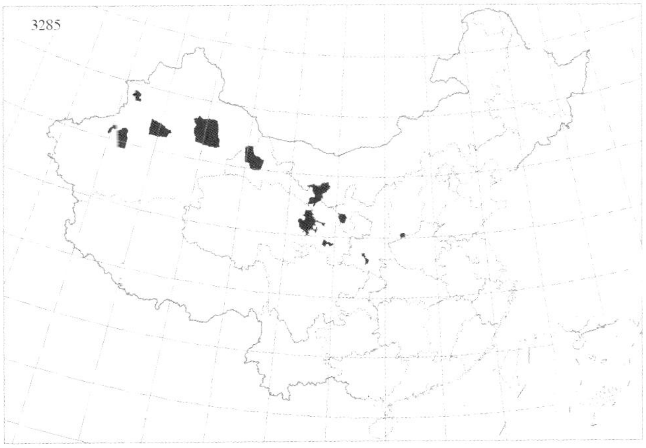

Rosaceae 蔷薇科 *Pyrus* 梨属
Pyrus ussuriensis Maxim. 秋子梨
Trees, deciduous

7.3 (-2.9~15.3); -9.9 (-26.4~3.1); 22.3 (16.3~27.8); 9.7 (5.4~15.4); 77.5 (40.2~127.2); 49.5 (2.7~134.6); 640 (471~859) ❙ 580 (290~1489); 361 (196~691); 19 (5~196) ❙ -9.0 (-48.5~82.7); 544 (290~850); 196.6 (46.6~346.5)

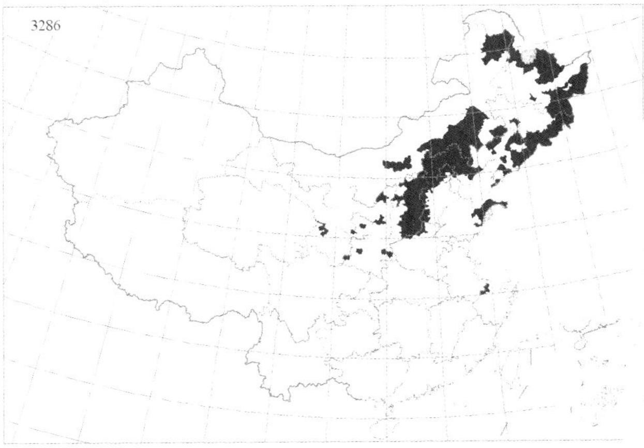

Rosaceae 蔷薇科 *Pyrus* 梨属
Pyrus xerophila T. T. Yu 木梨
Trees, deciduous

9.8 (0.0~15.1); -4.8 (-13.7~1.8); 23.0 (8.1~28.0); 10.8 (2.7~15.2); 87.3 (8.0~128.0); 28.8 (5.6~71.7); 682 (363~880) ❙ 577 (341~882); 323 (190~453); 19 (5~53) ❙ -14.4 (-40.9~54.1); 560 (341~817); 176.4 (61.2~323.3)

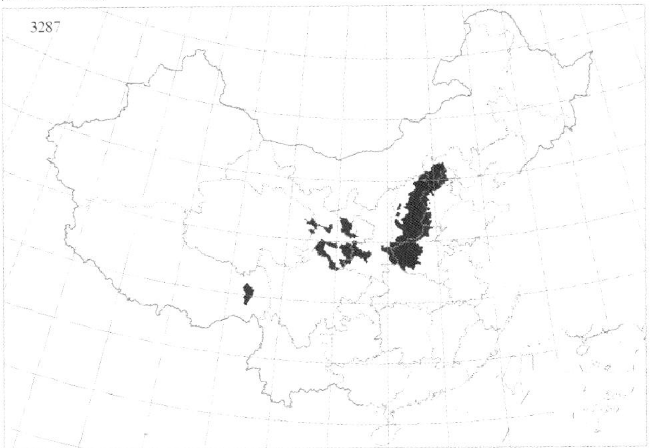

Rosaceae 蔷薇科 *Rhaphiolepis* 石斑木属
Rhaphiolepis ferruginea F. P. Metcalf 锈毛石斑木
Shrubs or trees, evergreen

18.7 (14.6~24.1); 9.4 (3.8~18.0); 27.2 (24.3~29.1); 18.8 (14.7~24.2); 165.[illegible] (117.7~230.2); 0.0 (0.0~1.5); 973 (779~1327) ❙ 1610 (868~1997); 570 ([illegible]~821); 192 (44~259) ❙ 68.8 (-34.3~147.0); 958 (779~1254); 256.4 (123.9~439.7)

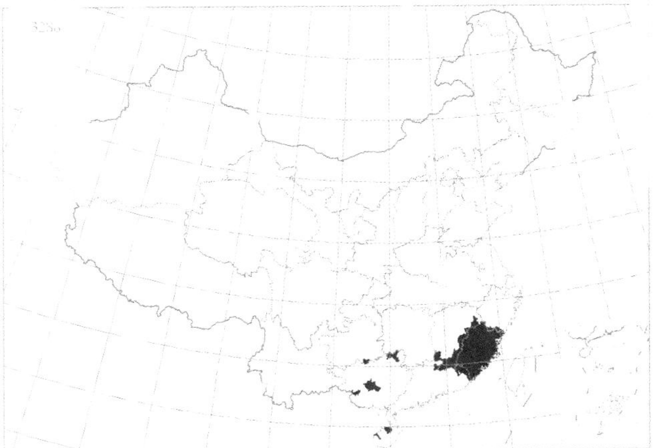

Rosaceae 蔷薇科 *Rhaphiolepis* 石斑木属
Rhaphiolepis ferruginea var. serrata F. P. Metcalf 齿叶绣毛石斑木
Shrubs or trees, evergreen

18.3 (15.0~21.4); 9.1 (4.9~13.6); 27.0 (24.3~29.0); 18.4 (15.1~21.5); 160.6 (120.8~197.7); 0.0 (0.0~0.1); 949 (789~1128) ❙ 1625 (1070~1997); 561 (356~713); 199 (130~257) ❙ 74.3 (-5.0~147.0); 938 (789~1094); 257.0 (123.9~332.5)

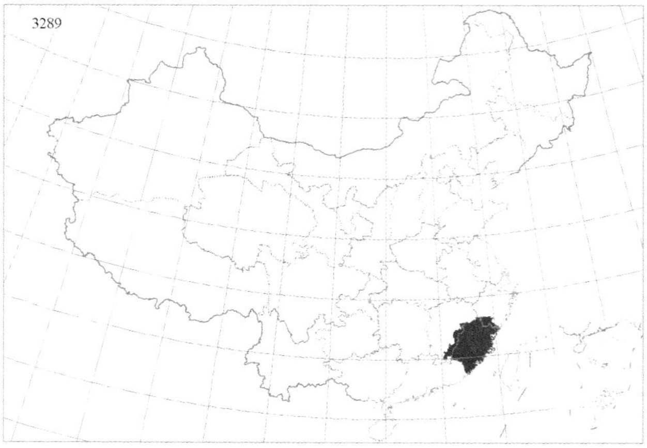

Rosaceae 蔷薇科 *Rhaphiolepis* 石斑木属
Rhaphiolepis indica (L.) Lindl. 石斑木
Shrubs or small trees, evergreen

17.5 (11.7~24.8); 7.1 (1.6~20.1); 27.1 (18.6~30.1); 17.5 (11.7~24.8); 150.6 (82.6~237.6); 0.3 (0.0~6.9); 929 (658~1385) ❙ 1502 (868~2993); 554 (299~1847); 166 (31~547) ❙ 63.4 (-34.3~246.5); 910 (658~1360); 233.3 (123.9~620.0)

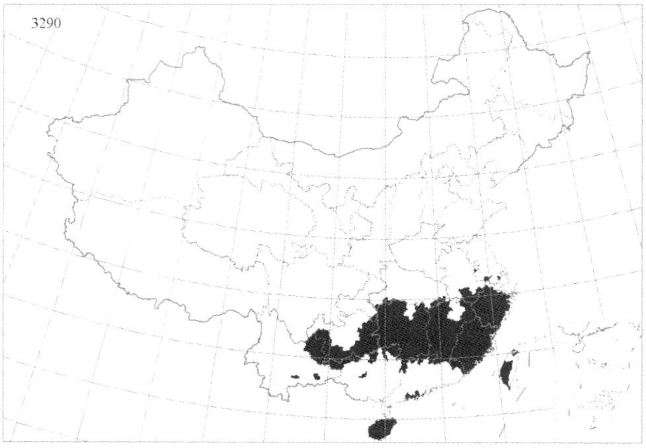

Rosaceae 蔷薇科 *Rhaphiolepis* 石斑木属
Rhaphiolepis integerrima Hook. et Arn. 全缘石斑木
Shrubs or small trees, evergreen

19.2; 13.8; 23.8; 19.2; 170.6; 0.0; 925 ❙ 2853; 1156; 367 ❙ 208.9; 925; 312.6

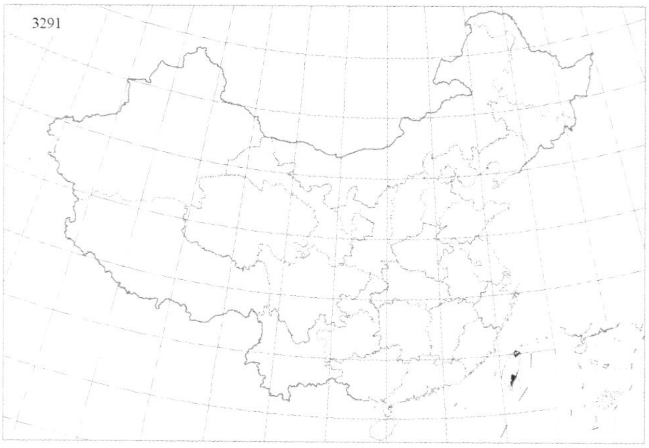

Rosaceae 蔷薇科 *Rhaphiolepis* 石斑木属
Rhaphiolepis lanceolata Hu 细叶石斑木
Shrubs or trees, evergreen

20.2 (14.6~24.8); 11.8 (3.8~20.1); 27.1 (24.6~28.6); 20.3 (14.7~24.8); 183.6 (117.7~237.6); 0.1 (0.0~1.5); 1076 (779~1385) ❙ 1638 (1284~2435); 733 (473~1428); 126 (41~191) ❙ 55.6 (1.5~101.8); 1043 (779~1221); 319.0 (232.5~620.0)

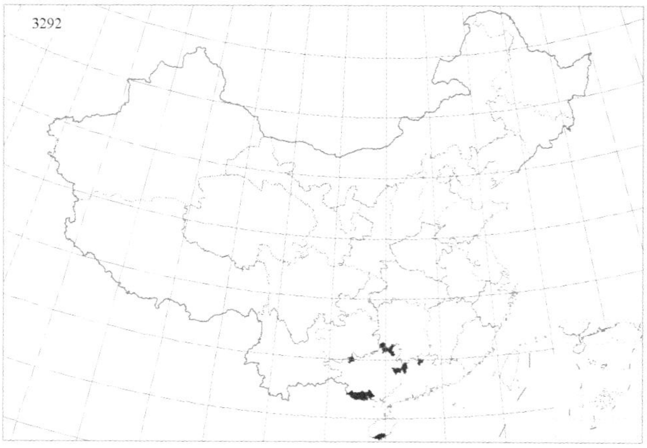

Rosaceae 蔷薇科 *Rhaphiolepis* 石斑木属
Rhaphiolepis major Cardot 大叶石斑木
Shrubs, evergreen

18.2 (15.0~21.4); 8.6 (3.9~13.6); 27.2 (24.3~29.1); 18.3 (15.1~21.5); 159.3 (120.8~197.7); 0.0 (0.0~1.1); 949 (789~1128) ❙ 1639 (1070~1997); 561 (356~713); 204 (130~259) ❙ 75.3 (-5.0~147.0); 940 (789~1107); 253.1 (123.9~332.5)

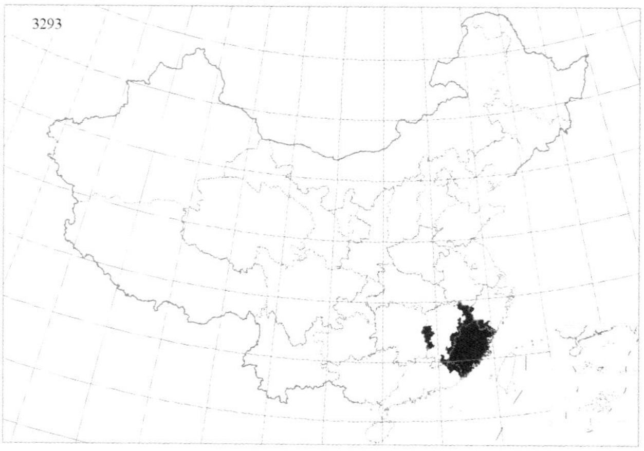

Rosaceae 蔷薇科 *Rhaphiolepis* 石斑木属
Rhaphiolepis salicifolia Lindl. 柳叶石斑木
Shrubs or trees, evergreen

19.9 (14.6~24.8); 11.2 (3.8~20.1); 27.3 (24.6~28.9); 20.0 (14.7~24.8); 179.6 (117.7~237.6); 0.0 (0.0~1.5); 1049 (779~1385) ❙ 1606 (868~2435); 641 (312~1428); 152 (41~254) ❙ 57.9 (-34.3~130.7); 1014 (779~1299); 296.5 (182.7~620.0)

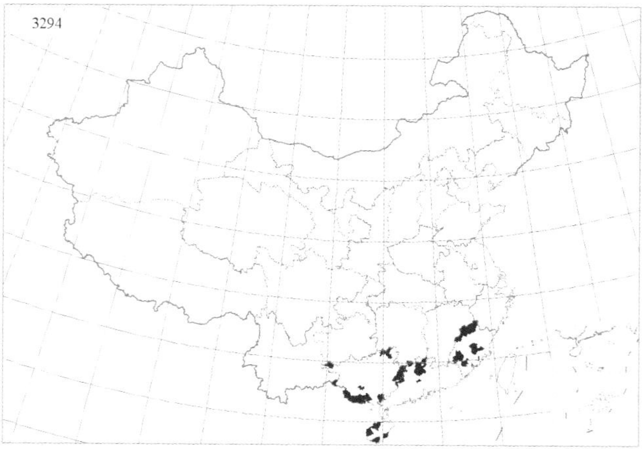

Rosaceae 蔷薇科 *Rhaphiolepis* 石斑木属
Rhaphiolepis umbellata (Thunb.) Makino 厚叶石斑木
Shrubs or trees, evergreen

16.7 (15.6~17.6); 6.0 (4.3~7.5); 27.6 (27.1~27.9); 16.8 (15.7~17.6); 141.4 (128.8~151.2); 0.1 (0.0~0.8); 892 (843~920) ❙ 1503 (1021~1719); 539 (370~644); 179 (139~191) ❙ 68.4 (17.1~87.6); 891 (843~920); 218.7 (176.3~249.2)

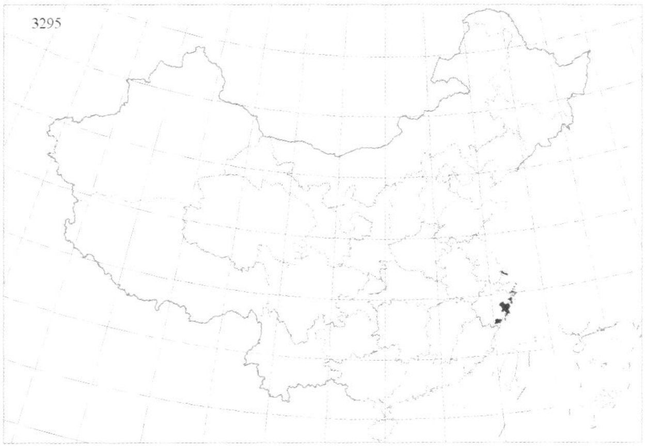

Rosaceae 蔷薇科 *Rhodotypos* 鸡麻属
Rhodotypos scandens (Thunb.) Makino 鸡麻
Shrubs, deciduous

13.6 (7.4~16.9); 0.4 (-7.4~4.4); 26.0 (18.7~29.2); 13.8 (8.2~16.9); 114.1 (58.8~143.8); 10.3 (0.6~36.2); 796 (561~936) ❙ 954 (520~1689); 431 (258~666); 77 (15~220) ❙ 19.0 (-35.9~108.0); 754 (520~930); 194.7 (128.6~323.3)

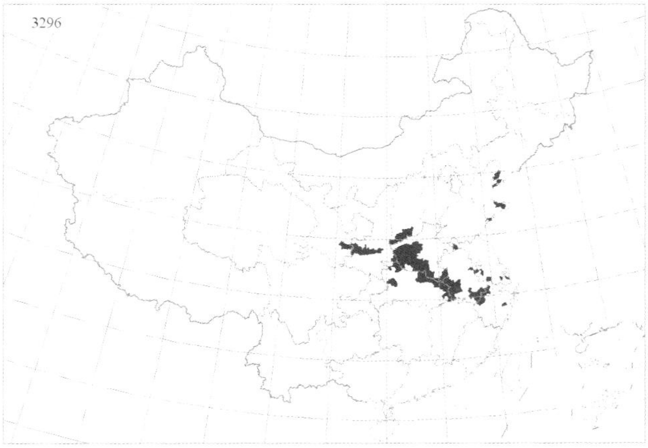

Rosaceae 蔷薇科 *Rosa* 蔷薇属
Rosa acicularis Lindl. 刺蔷薇
Shrubs, evergreen

6.4 (-5.8~14.2); -11.2 (-29.8~-0.7); 21.6 (11.7~27.8); 9.2 (4.3~14.4); 72.7 (21.1~124.3); 55.8 (12.6~157.4); 622 (426~859) ❙ 486 (53~1073); 299 (33~691); 17 (2~51) ❙ -20.5 (-92.4~74.1); 463 (53~690); 177.1 (10.3~346.5)

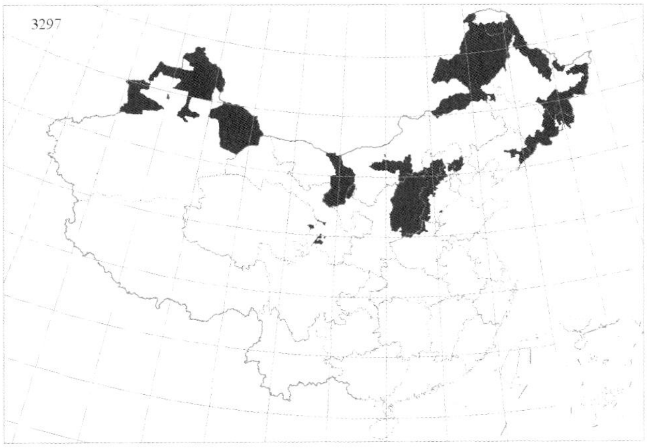

Rosaceae 蔷薇科 *Rosa* 蔷薇属
Rosa albertii Regel 腺齿蔷薇
Shrubs, evergreen

2.8 (-4.1~9.0); -14.2 (-20.9~-7.2); 16.9 (7.8~24.4); 6.6 (2.1~10.7); 47.3 (5.2~88.9); 73.6 (39.3~114.2); 523 (340~703) ❙ 290 (53~545); 151 (33~310); 16 (3~51) ❙ -39.2 (-92.4~45.6); 279 (53~450); 146.3 (10.3~301.5)

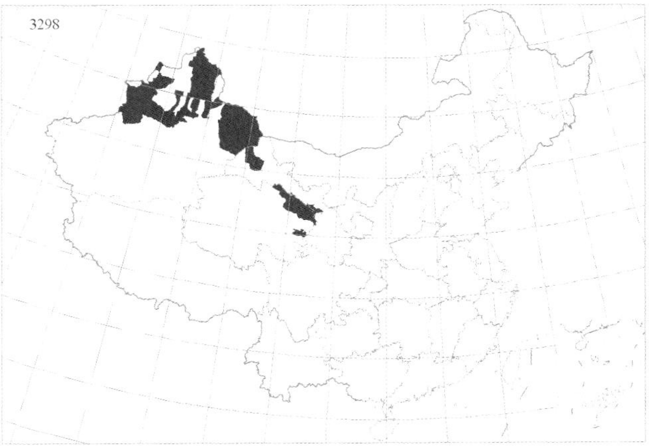

Rosaceae 蔷薇科 *Rosa* 蔷薇属
Rosa anemoniflora Fortune ex Lindl. 银粉蔷薇
Climbing shrubs, evergreen

17.4 (16.6~18.6); 7.4 (6.0~9.3); 26.7 (25.8~27.6); 17.4 (16.6~18.7); 149.3 (138.9~164.0); 0.0 (0.0~0.0); 903 (862~965) ❙ 1818 (1702~1997); 613 (587~669); 222 (198~254) ❙ 101.8 (76.4~130.7); 903 (862~965); 287.7 (264.8~303.4)

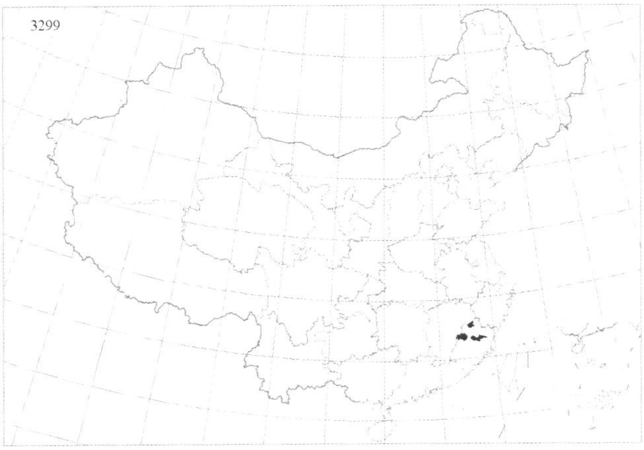

Rosaceae 蔷薇科 *Rosa* 蔷薇属
Rosa baiyushanensis Q. L. Wang 白玉山蔷薇
Shrubs, evergreen

10.5; -4.9; 23.7; 11.3; 93.2; 26.8; 705 ❙ 632; 398; 27 ❙ -10.4; 632; 180.8

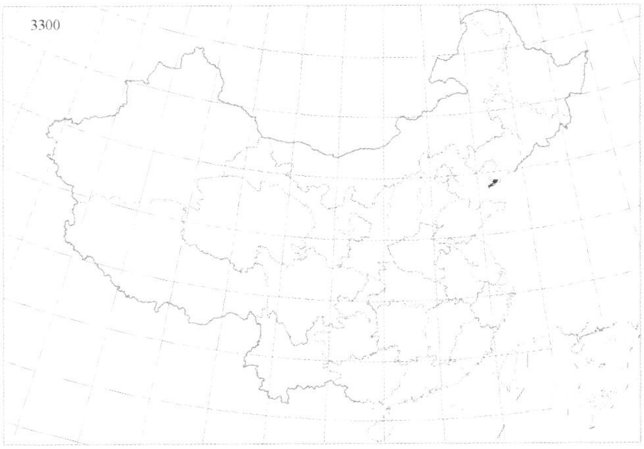

Rosaceae 蔷薇科 *Rosa* 蔷薇属
Rosa banksiae Aiton 木香花
Small shrubs, evergreen

14.5 (5.4~18.9); 4.9 (-2.9~12.5); 22.9 (12.3~29.7); 14.5 (5.8~19.0); 117.6 (31.5~167.8); 3.3 (0.0~25.6); 777 (481~985) ❙ 1044 (512~1702); 528 (258~816); 60 (15~214) ❙ 34.8 (-35.3~108.0); 755 (481~965); 241.9 (121.6~406.0)

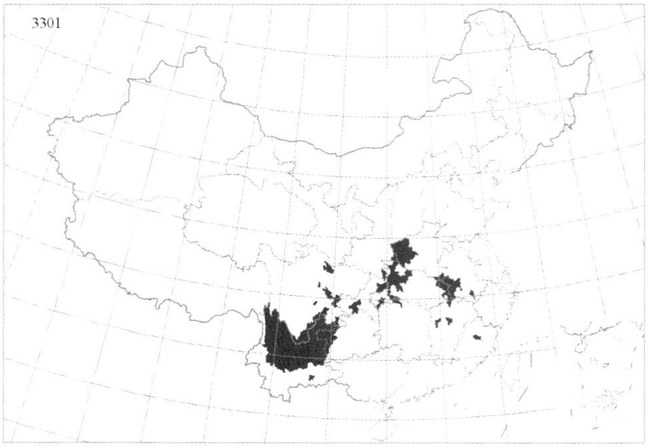

Rosaceae 蔷薇科 *Rosa* 蔷薇属
Rosa banksiae var. normalis Regel 单瓣木香花
Climbing shrubs, evergreen

13.8 (5.7~18.9); 3.5 (-4.1~12.0); 23.2 (14.2~28.8); 13.9 (6.4~19.0); 112.5 (38.1~167.8); 6.1 (0.0~29.9); 762 (504~929) ❙ 992 (626~1571); 493 (316~816); 51 (15~157) ❙ 31.2 (-22.2~108.0); 739 (504~910); 244.1 (133.8~382.6)

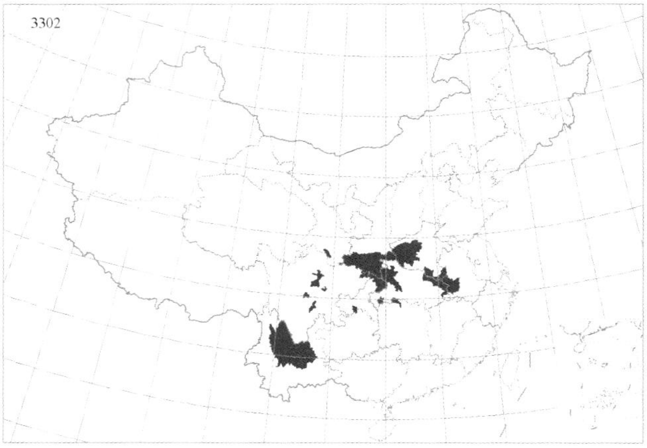

Rosaceae 蔷薇科 *Rosa* 蔷薇属
Rosa banksiopsis Baker 拟木香
Small shrubs

12.4 (6.2~17.2); 0.5 (-4.2~5.5); 23.8 (14.1~29.3); 12.6 (6.7~17.2); 100.7 (39.4~146.8); 10.8 (0.0~29.4); 730 (509~941) ❙ 1017 (622~1663); 440 (290~816); 63 (15~212) ❙ 38.5 (-16.6~101.4); 724 (509~941); 246.6 (170.4~323.3)

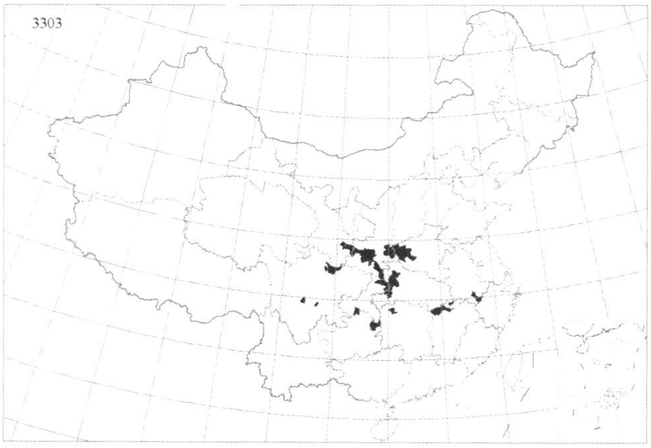

Rosaceae 蔷薇科 *Rosa* 蔷薇属
Rosa beggeriana Schrenk 弯刺蔷薇
Shrubs

5.0 (-1.3~12.3); -13.5 (-20.9~-5.3); 20.2 (13.2~27.1); 8.4 (4.4~13.0); 66.2 (26.2~112.9); 65.5 (24.7~103.9); 602 (413~810) ❙ 189 (26~450); 87 (16~211); 17 (0~51) ❙ -65.4 (-96.7~3.9); 188 (26~406); 92.3 (1.6~297.3)

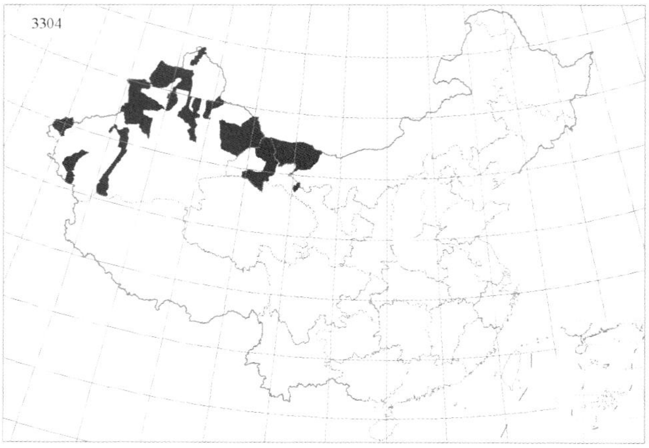

Rosaceae 蔷薇科 *Rosa* 蔷薇属
Rosa bella Rehder et E. H. Wilson 美蔷薇
Shrubs

9.5 (2.2~15.2); -6.0 (-16.0~2.3); 23.2 (14.0~28.0); 10.7 (5.6~15.3); 87.4 (32.4~128.0); 33.2 (4.3~81.1); 682 (467~880) ❙ 548 (202~1198); 321 (134~536); 18 (6~109) ❙ -19.4 (-65.5~39.7); 539 (202~858); 162.3 (50.7~287.6)

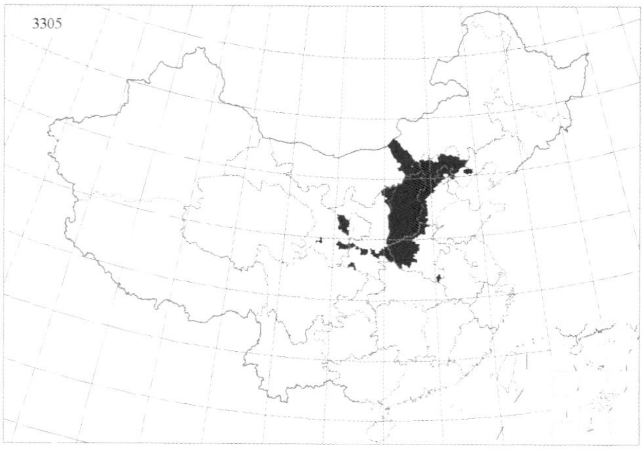

Rosaceae 蔷薇科 *Rosa* 蔷薇属
Rosa berberifolia Pall. 小檗叶蔷薇
Dwarf shrubs

5.0 (-0.6~9.3); -15.0 (-19.5~-12.0); 21.7 (15.9~27.5); 8.9 (5.4~12.1); 71.7 (37.5~107.6); 70.9 (51.6~104.2); 629 (475~813) ❙ 198 (107~357); 84 (51~136); 20 (8~51) ❙ -67.4 (-85.3~-34.9); 198 (107~357); 75.1 (19.8~190.1)

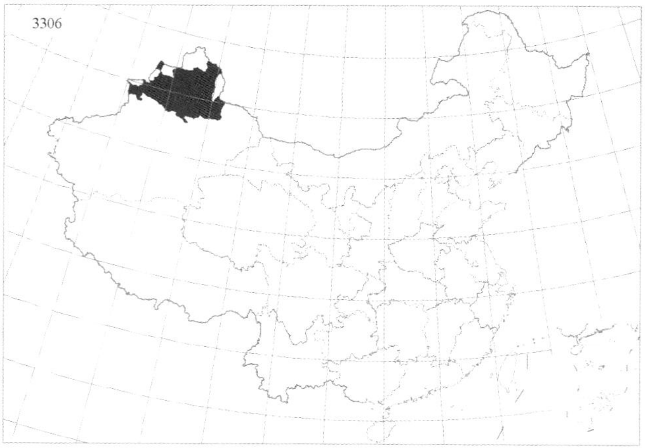

Rosaceae 蔷薇科 *Rosa* 蔷薇属
Rosa bracteata J. C. Wendl. 硕苞蔷薇
Shrubs, evergreen

16.8 (14.3~21.1); 5.9 (2.1~14.6); 27.3 (20.3~29.4); 16.8 (14.3~21.1); 142.5 (114.1~193.7); 0.6 (0.0~4.9); 896 (765~1104) ❙ 1584 (1002~3495); 569 (370~1388); 202 (42~897) ❙ 77.5 (14.0~224.4); 891 (765~1094); 236.1 (131.2~589.2)

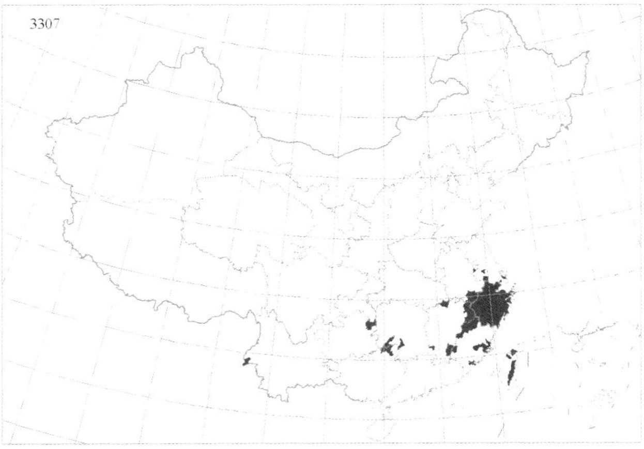

Rosaceae 蔷薇科 *Rosa* 蔷薇属
Rosa brunonii Lindl. 复伞房蔷薇
Scandent shrubs

10.8 (-1.1~14.9); -0.1 (-9.1~5.3); 21.1 (6.2~25.4); 11.0 (2.0~15.0); 83.7 (2.5~119.6); 13.5 (0.0~75.6); 664 (334~782) ❙ 904 (321~1410); 433 (193~816); 29 (14~68) ❙ 37.5 (-4.0~125.9); 664 (321~782); 271.5 (79.3~352.3)

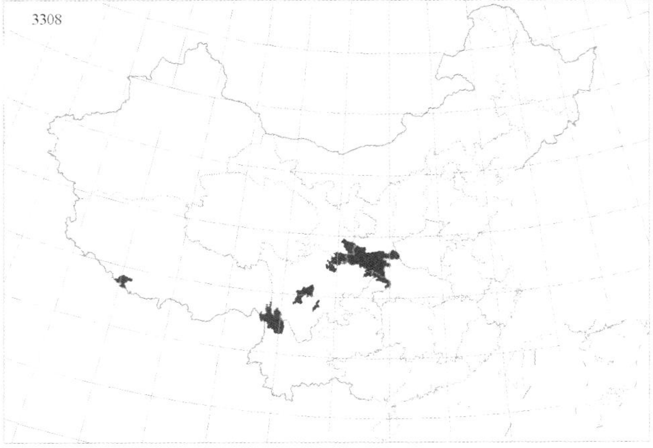

Rosaceae 蔷薇科 *Rosa* 蔷薇属
Rosa calyptopoda Cardot 短角蔷薇
Small shrubs

4.3 (2.2~6.2); -4.8 (-7.0~-2.4); 12.5 (9.9~14.4); 5.4 (3.8~6.7); 28.9 (15.5~39.7); 36.8 (24.1~49.1); 465 (413~509) ❙ 939 (829~1023); 495 (400~546); 24 (19~28) ❙ 103.2 (83.0~125.9); 465 (413~509); 242.1 (200.8~267.5)

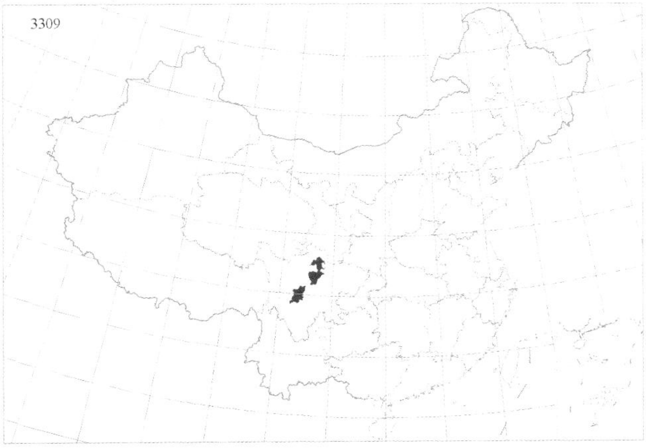

Rosaceae 蔷薇科 *Rosa* 蔷薇属
Rosa caudata Baker 尾萼蔷薇
Shrubs

13.0 (5.7~17.1); 0.5 (-3.5~5.5); 24.7 (14.2~29.3); 13.1 (6.4~17.1); 106.8 (38.1~145.7); 10.5 (0.0~28.1); 759 (504~941) ❙ 966 (626~1621); 442 (290~816); 59 (22~195) ❙ 28.1 (-22.2~106.2); 727 (504~941); 213.6 (139.1~316.6)

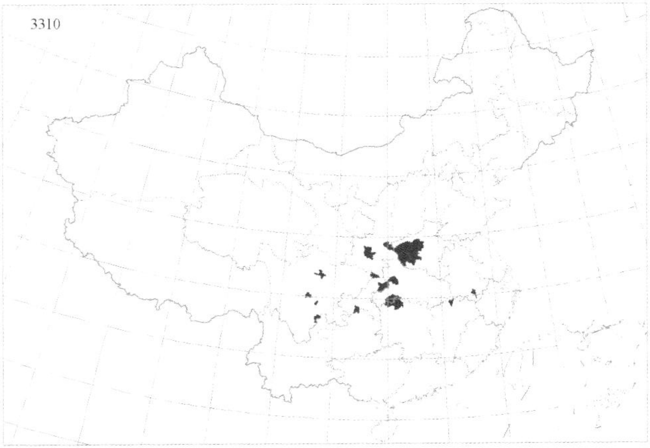

Rosaceae 蔷薇科 *Rosa* 蔷薇属
Rosa caudata var. maxima T. T. Yu et T. C. Ku 大花尾萼蔷薇
Shrubs

11.1; 1.8; 19.4; 11.1; 79.8; 6.9; 648 ❙ 1076; 584; 30 ❙ 66.0; 648; 246.3

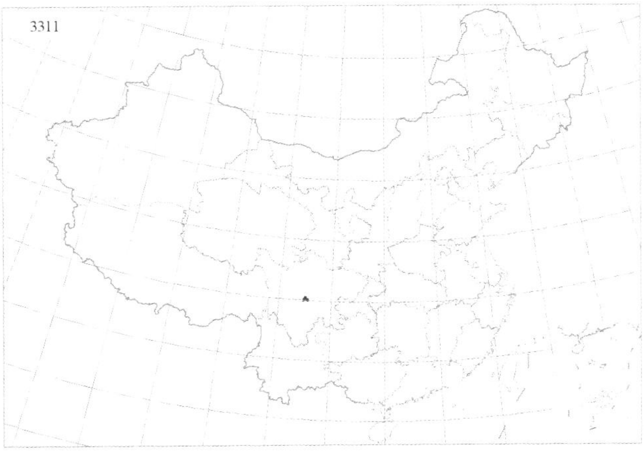

Rosaceae 蔷薇科 *Rosa* 蔷薇属
Rosa chengkouensis T. T. Yu et T. C. Ku 城口蔷薇
Shrubs

11.3; 0.5; 21.8; 11.3; 85.9; 10.3; 667 ❙ 1227; 513; 51 ❙ 84.4; 667; 271.3

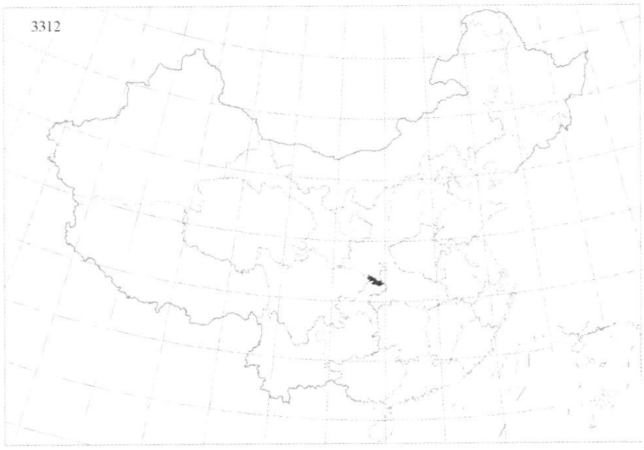

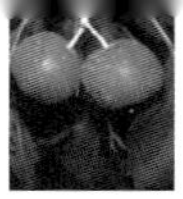

Rosaceae 蔷薇科 *Rosa* 蔷薇属
Rosa corymbulosa Rolfe 伞房蔷薇
Small shrubs

12.4 (7.4~16.2); 0.1 (-4.2~5.5); 24.1 (18.7~27.5); 12.6 (8.2~16.2); 101.1 (58.8~134.6); 11.3 (0.0~29.4); 733 (561~865) ❙ 892 (626~1466); 405 (306~602); 40 (15~113) ❙ 23.0 (-22.2~106.2); 711 (561~854); 234.7 (139.1~334.4)

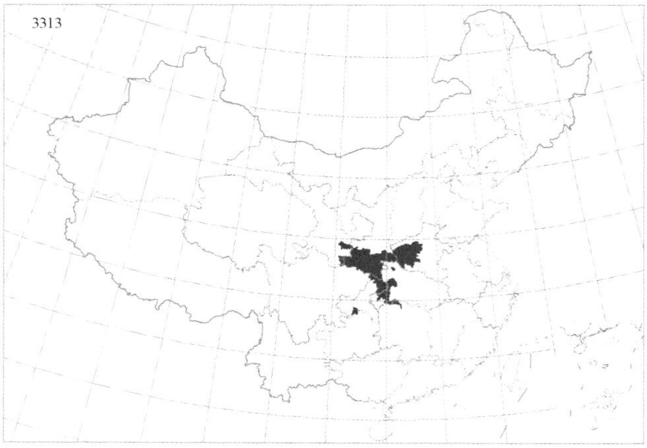

Rosaceae 蔷薇科 *Rosa* 蔷薇属
Rosa cymosa Tratt. 小果蔷薇
Scandent shrubs

16.9 (2.2~23.6); 6.0 (-6.6~16.7); 27.0 (9.9~30.1); 17.0 (3.8~23.6); 145.4 (15.5~223.5); 1.7 (0.0~49.1); 911 (413~1281) ❙ 1378 (626~2680); 549 (299~1412); 142 (14~341) ❙ 51.4 (-22.2~246.5); 899 (413~1273); 211.3 (95.3~359.7)

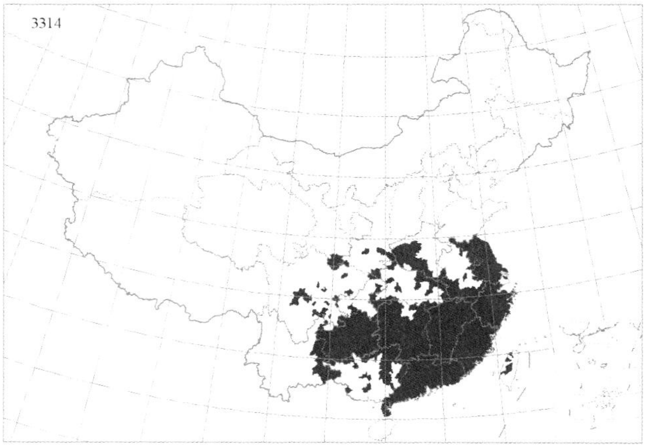

Rosaceae 蔷薇科 *Rosa* 蔷薇属
Rosa cymosa var. puberula T. T. Yu et T. C. Ku 毛叶山木香
Scandent shrubs

12.6 (7.4~18.4); 0.6 (-4.2~9.0); 24.1 (18.7~28.9); 12.7 (8.2~18.4); 102.3 (58.8~161.2); 10.4 (0.0~29.4); 736 (561~950) ❙ 1050 (634~1726); 436 (290~666); 72 (16~213) ❙ 41.1 (-16.6~108.0); 731 (561~950); 264.1 (189.1~334.4)

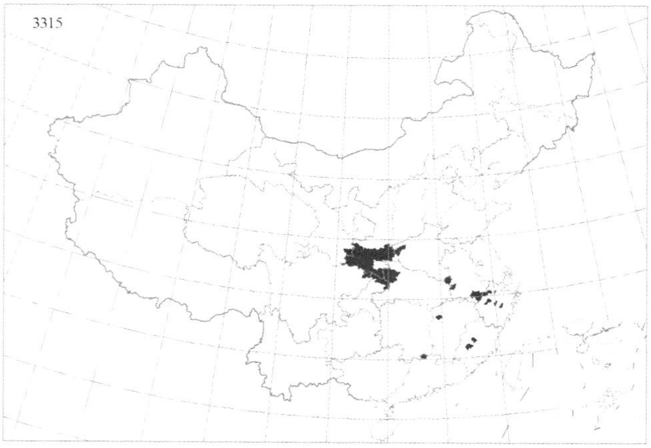

Rosaceae 蔷薇科 *Rosa* 蔷薇属
Rosa daishanensis T. C. Ku 岱山蔷薇
Shrubs

16.5; 5.3; 27.7; 16.5; 138.0; 0.0; 888 ❙ 1363; 481; 183 ❙ 53.5; 888; 222.6

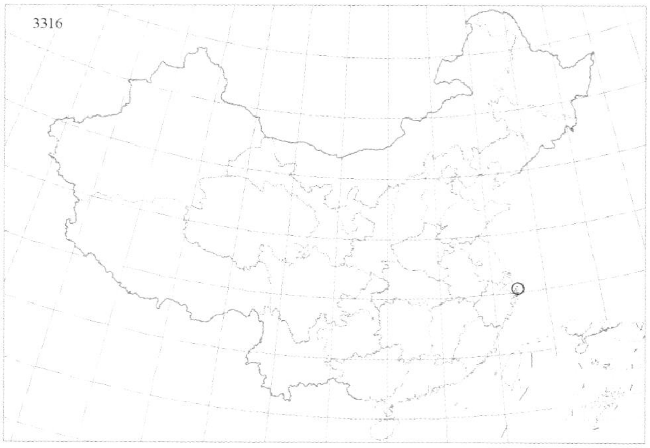

Rosaceae 蔷薇科 *Rosa* 蔷薇属
Rosa davidii Crép. 西北蔷薇
Shrubs

11.1 (0.4~18.9); -0.9 (-12.7~12.0); 21.8 (10.5~28.0); 11.7 (3.6~19.0); 92.2 (15.5~167.8); 18.7 (0.0~69.4); 693 (406~929) ❙ 751 (341~1577); 408 (190~932); 27 (5~71) ❙ 9.5 (-40.9~101.4); 626 (341~912); 210.1 (61.2~351.7)

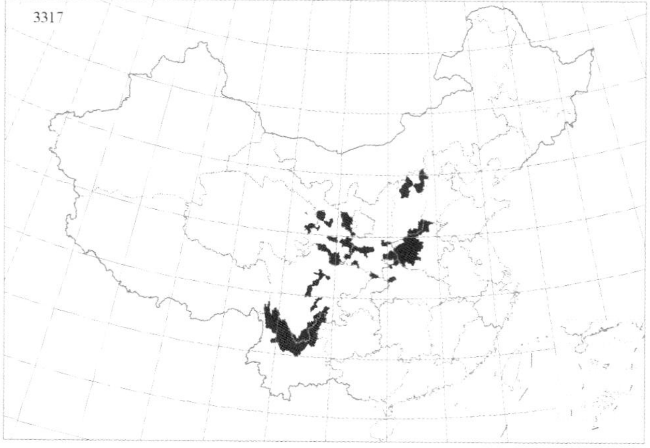

Rosaceae 蔷薇科 *Rosa* 蔷薇属
Rosa davidii var. elongata Rehder et E. H. Wilson 长果西北蔷薇
Shrubs

9.2; 0.1; 17.5; 9.4; 64.5; 12.8; 594 ❙ 1113; 609; 32 ❙ 87.2; 594; 282.5

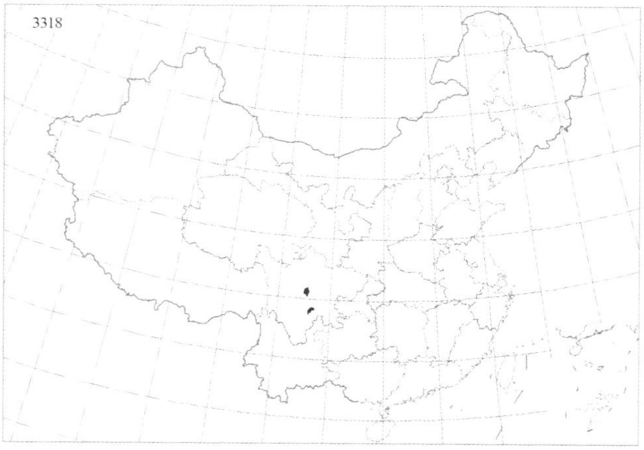

Rosaceae 蔷薇科 *Rosa* 蔷薇属
Rosa davurica Pall. 山刺玫
Shrubs

6.1 (−5.8~16.1); −11.5 (−29.8~5.7); 21.7 (14.4~26.6); 9.0 (4.4~16.1); 71.5 (28.0~133.5); 57.6 (0.0~157.4); 613 (426~847) ❙ 549 (128~1410); 335 (81~816); 16 (2~83) ❙ −10.4 (−80.8~101.2); 497 (128~845); 196.6 (13.6~346.5)

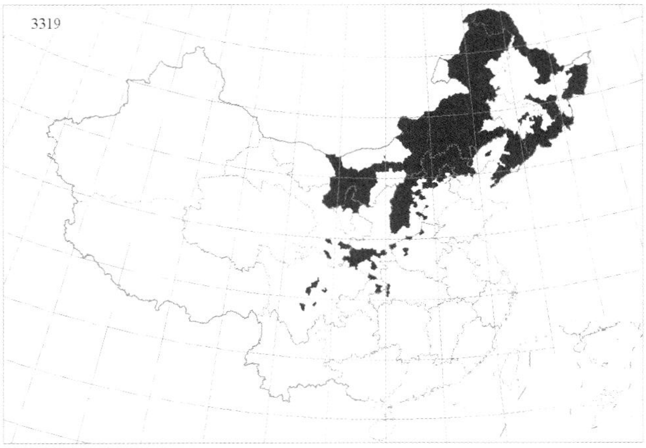

Rosaceae 蔷薇科 *Rosa* 蔷薇属
Rosa davurica var. glabra Liou 光叶山刺玫
Shrubs

10.7; −5.8; 25.3; 11.7; 98.9; 30.6; 743 ❙ 581; 449; 12 ❙ −21.8; 581; 211.2

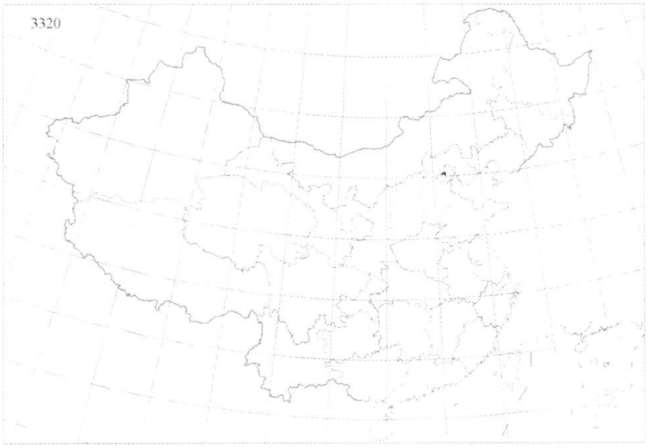

Rosaceae 蔷薇科 *Rosa* 蔷薇属
Rosa davurica var. setacea Liou 多刺山刺玫
Shrubs

10.7; −5.3; 25.3; 11.7; 98.9; 30.6; 743 ❙ 581; 449; 12 ❙ −21.8; 581; 211.2

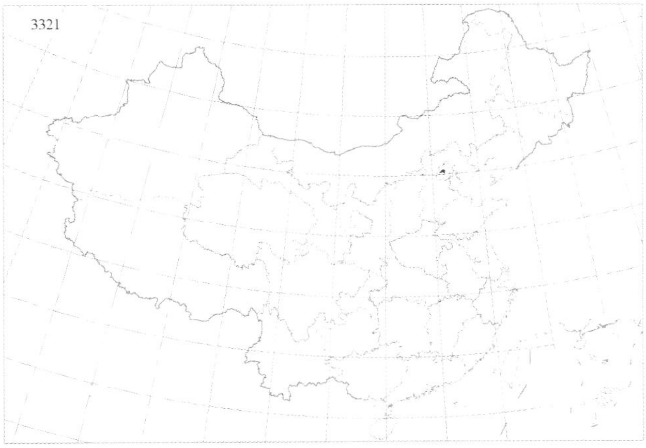

Rosaceae 蔷薇科 *Rosa* 蔷薇属
Rosa dawoensis Pax et Hoffman 道孚蔷薇
Shrubs

2.0; −7.4; 10.1; 4.0; 16.9; 51.9; 416 ❙ 777; 457; 15 ❙ 86.8; 416; 271.2

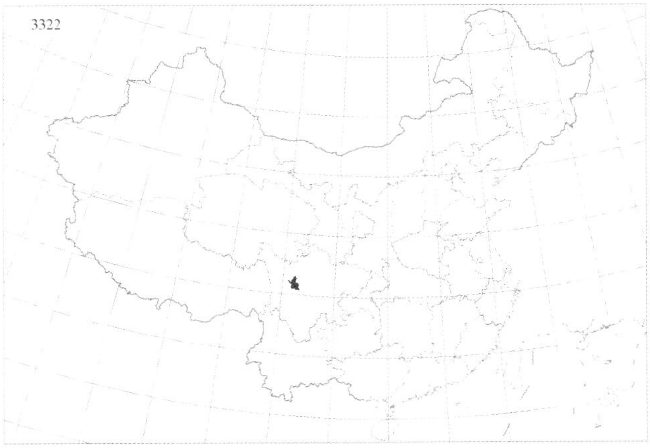

Rosaceae 蔷薇科 *Rosa* 蔷薇属
Rosa deqenensis T. C. Ku 德钦蔷薇
Shrubs

5.4; −2.1 12.3; 5.8; 31.5; 25.6; 481 ❙ 741; 426; 28 ❙ 54.0; 481; 273.5

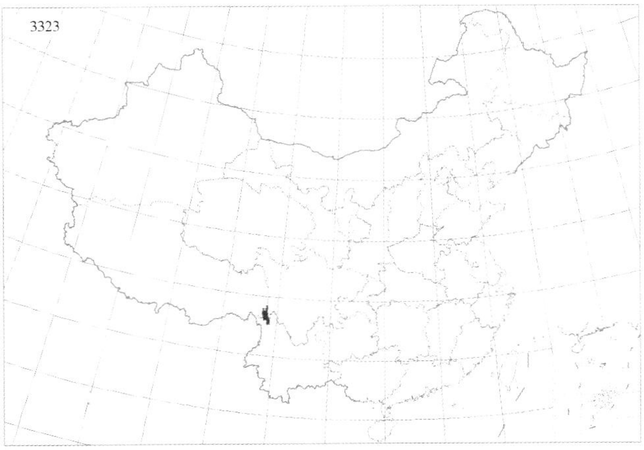

Rosaceae 蔷薇科 *Rosa* 蔷薇属
Rosa derongensis T. C. Ku 得荣蔷薇
Small shrubs

5.6; −2.4; 12.6; 5.9; 32.6; 26.2; 483 ❙ 685; 406; 23 ❙ 41.8; 483; 201.9

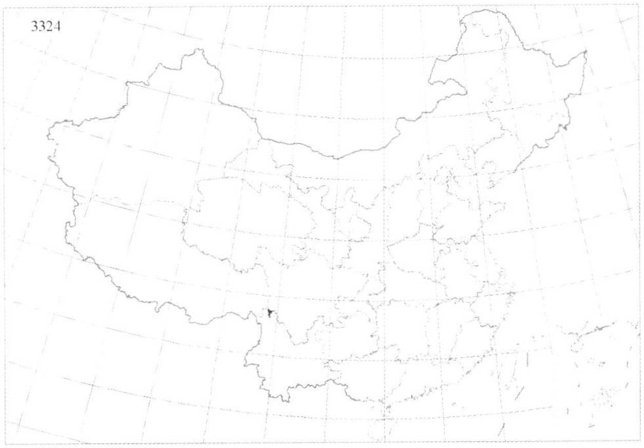

Rosaceae 蔷薇科 *Rosa* 蔷薇属
Rosa duplicata T. T. Yu et T. C. Ku 重齿蔷薇
Small shrubs

4.5; -3.4; 11.7; 5.3; 27.3; 32.3; 460 ❙ 746; 437; 23 ❙ 60.6; 460; 247.5

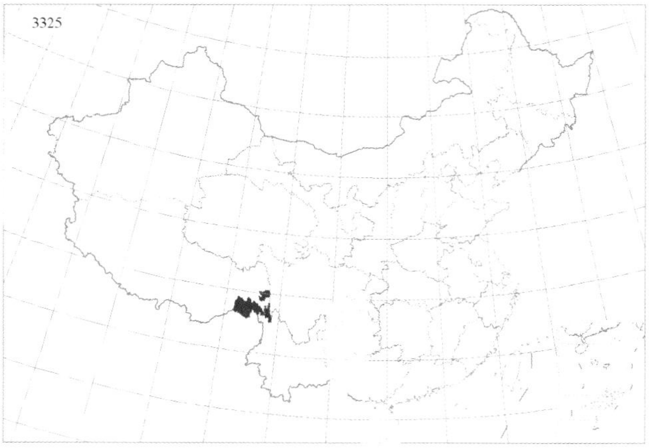

Rosaceae 蔷薇科 *Rosa* 蔷薇属
Rosa fargesiana Boulenger 川东蔷薇
Shrubs, deciduous

10.6; -0.2; 21.1; 10.6; 79.4; 12.1; 643 ❙ 1246; 525; 46 ❙ 93.8; 643; 284.3

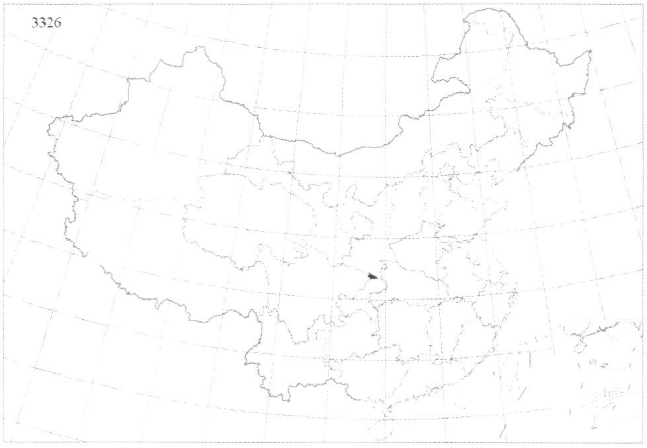

Rosaceae 蔷薇科 *Rosa* 蔷薇属
Rosa farreri Cox 刺毛蔷薇
Small shrubs

11.5 (1.9~16.9); 1.7 (-9.2~7.0); 20.4 (11.7~25.9); 11.9 (4.3~17.0); 91.7 (21.1~143.7); 12.8 (0.0~57.8); 687 (427~881) ❙ 1165 (621~1577); 654 (317~932); 38 (13~60) ❙ 67.5 (28.2~89.0); 687 (427~881); 232.6 (167.5~289.4)

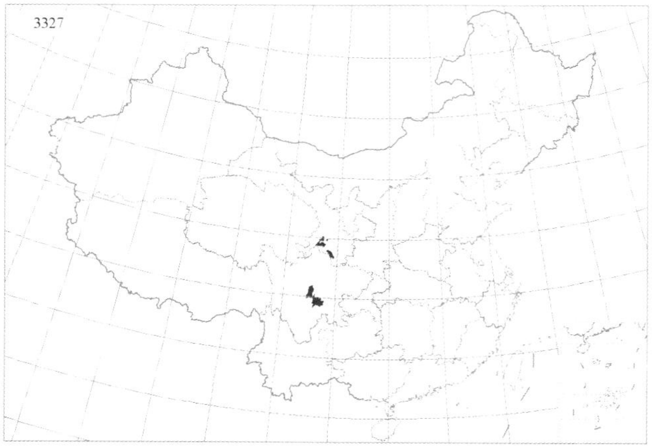

Rosaceae 蔷薇科 *Rosa* 蔷薇属
Rosa fedtschenkoana Regel 腺果蔷薇
Small shrubs

2.7 (-4.3~12.3); -15.5 (-20.9~-5.3); 17.4 (9.6~26.0); 6.9 (2.5~13.0); 50.0 (9.7~112.3); 77.3 (24.7~121.1); 531 (329~801) ❙ 229 (44~406); 104 (19~211); 21 (4~48) ❙ -55.4 (-94.5~-7.6); 229 (44~406); 121.6 (9.3~297.3)

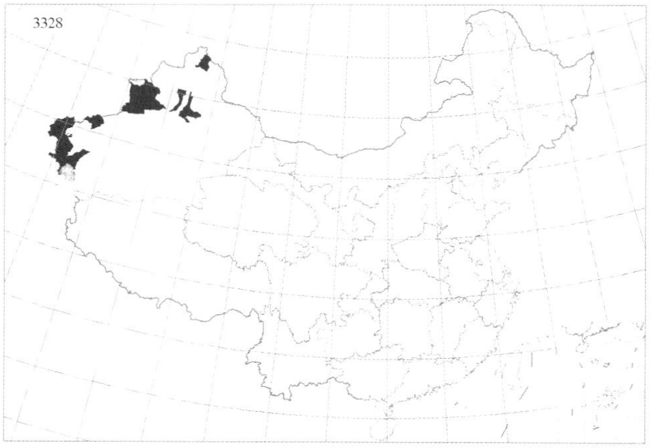

Rosaceae 蔷薇科 *Rosa* 蔷薇属
Rosa filipes Rehder et E. H. Wilson 腺梗蔷薇
Shrubs

10.3 (5.7~14.6); 0.3 (-6.7~7.9); 19.1 (14.2~23.7); 10.7 (6.4~14.6); 78.9 (38.1~115.8); 14.4 (0.0~35.7); 638 (504~736) ❙ 971 (373~1736); 517 (201~1007); 31 (7~51) ❙ 53.7 (-37.6~145.8); 615 (373~732); 246.2 (88.1~447.4)

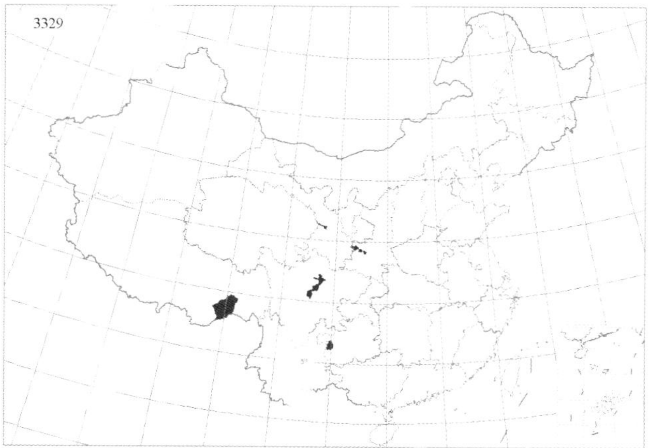

Rosaceae 蔷薇科 *Rosa* 蔷薇属
Rosa foetida var. persiana (Lemaire) Rehder 异味蔷薇
Shrubs

7.0; -9.9; 20.9; 9.4; 74.9; 49.7; 634 ❙ 36; 22; 2 ❙ -93.5; 36; 8.5

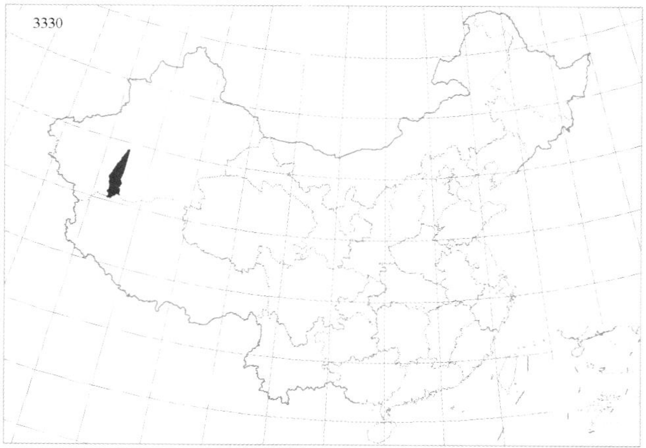

Rosaceae 蔷薇科 *Rosa* 蔷薇属

Rosa forrestiana Boulenger 滇边蔷薇

Small shrubs

6.1 (2.2~10.4); -1.7 (-6.6~3.4); 12.8 (9.5~16.3); 6.7 (3.8~10.5); 38.1 (15.5~68.0); 24.7 (2.6~49.1); 505 (411~623) ❙ 860 (729~1007); 526 (426~603); 21 (11~40) ❙ 72.3 (51.8~125.9); 504 (411~615); 282.4 (208.8~35[illegible].7)

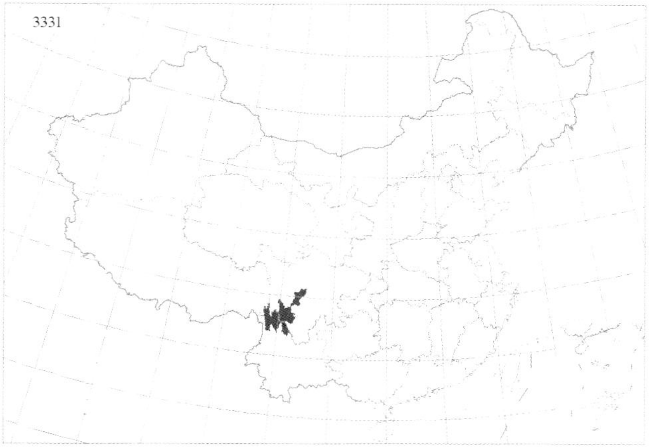

Rosaceae 蔷薇科 *Rosa* 蔷薇属

Rosa giraldii Crép. 陕西蔷薇

Shrubs

8.6 (-4.1~16.2); -4.1 (-17.3~5.5); 20.3 (7.2~27.1); 9.6 (2.0~16.2); 73.8 (3.7~134.6); 29.8 (0.0~114.2); 624 (339~854) ❙ 696 (148~1410); 343 (92~816); 25 (4~83) ❙ 13.4 (-74.8~128.6); 581 (148~854); 226.6 (42.6~357.7)

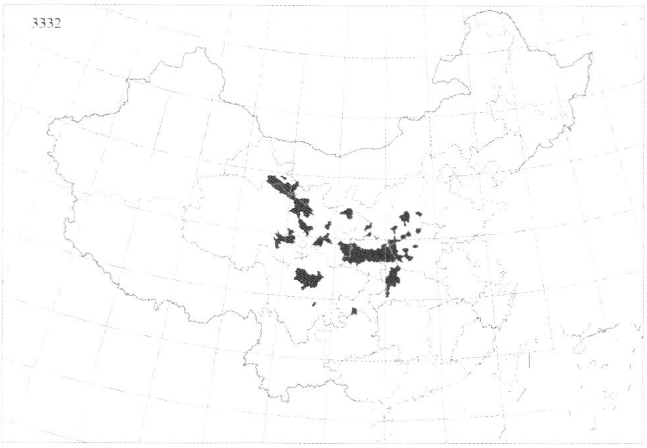

Rosaceae 蔷薇科 *Rosa* 蔷薇属

Rosa giraldii var. bidentata T. T. Yu et T. C. Ku 重齿陕西蔷薇

Shrubs

3.0; -7.7; 12.4; 4.8; 25.1; 49.3; 441 ❙ 656; 323; 15 ❙ 48.8; 441; 282.7

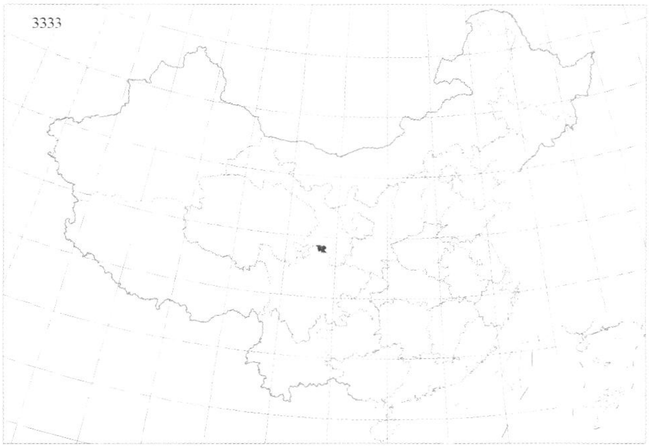

Rosaceae 蔷薇科 *Rosa* 蔷薇属

Rosa giraldii var. venulosa Rehder et E. H. Wilson 毛叶陕西蔷薇

Shrubs

12.5 (10.5~14.5); -0.5 (-2.7~1.8); 24.6 (22.9~26.4); 12.7 (10.8~14.6); 102.7 (85.4~120.7); 12.3 (5.7~19.0); 740 (668~821) ❙ 823 (634~1118); 367 (290~482); 44 (25~75) ❙ 11.7 (-16.6~60.4); 719 (634~821); 221.6 (174.3~247.3)

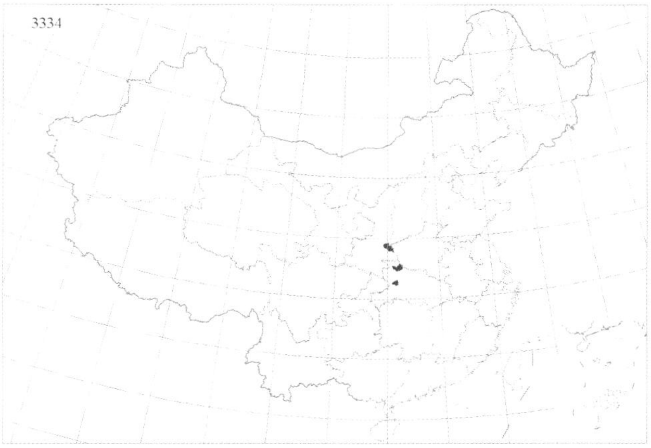

Rosaceae 蔷薇科 *Rosa* 蔷薇属

Rosa glomerata Rehder et E. H. Wilson 绣球蔷薇

Scandent shrubss or vines

13.7 (2.2~18.7); 4.0 (-6.6~8.1); 22.6 (9.9~28.9); 13.8 (3.8~18.8); 109.0 (15.5~166.0); 4.2 (0.0~49.1); 765 (413~1002) ❙ 1105 (741~1435); 531 (426~816); 50 (15~88) ❙ 48.8 (8.0~125.9); 765 (413~1002); 218.1 (108.2~407.1)

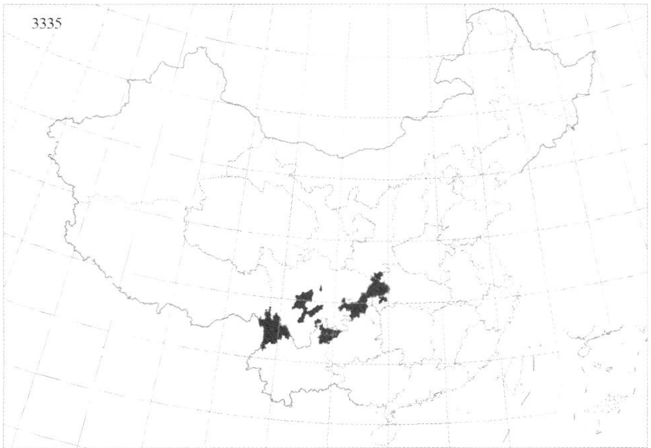

Rosaceae 蔷薇科 *Rosa* 蔷薇属

Rosa graciliflora Rehder et E. H. Wilson 细梗蔷薇

Small shrubs

5.9 (-1.0~14.6); -3.4 (-12.5~7.8); 14.1 (8.5~22.6); 7.2 (2.7~14.6); 45.9 (8.3~115.2); 33.9 (0.0~79.5); 519 (363~738) ❙ 783 (386~1164); 444 (219~651); 21 (5~58) ❙ 52.2 (-29.0~128.6); 511 (363~730); 290.2 (155.3~382.6)

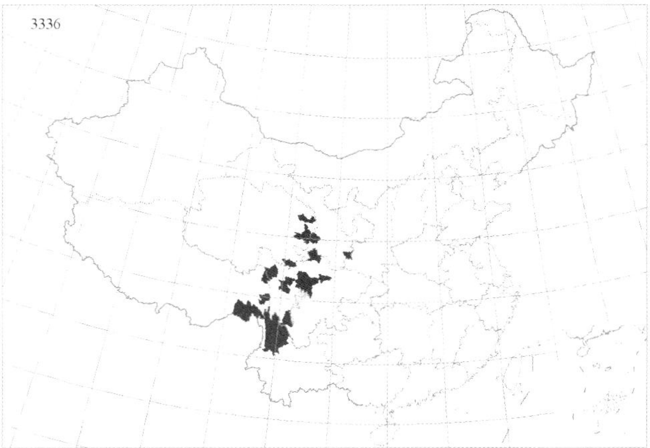

Rosaceae 蔷薇科 *Rosa* 蔷薇属
Rosa helenae Rehder et E. H. Wilson 卵果蔷薇
Shrubs

12.7 (2.2~18.3); 1.7 (-7.0~11.8); 22.9 (9.9~29.3); 12.8 (3.8~18.4); 101.9 (15.5~160.9); 8.9 (0.0~49.1); 733 (413~961) ❙ 975 (626~1679); 474 (290~816); 50 (11~195) ❙ 34.6 (-22.2~125.9); 717 (413~961); 234.9 (139.1~352.3)

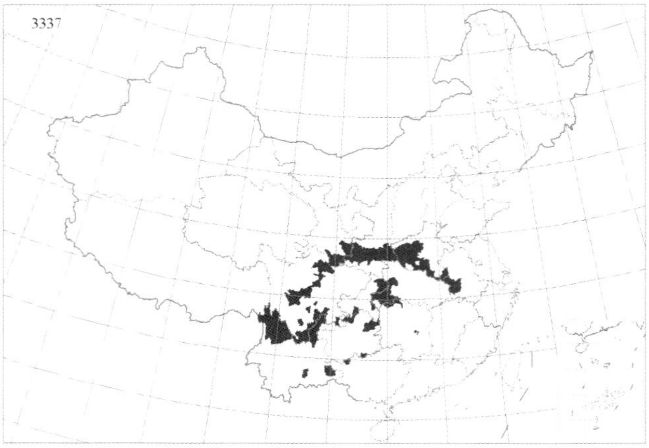

Rosaceae 蔷薇科 *Rosa* 蔷薇属
Rosa henryi Boulenger 软条七蔷薇
Shrubs

16.9 (9.1~23.6); 6.1 (-2.9~16.7); 26.9 (18.1~30.1); 17.0 (9.2~23.6); 145.4 (64.1~223.5); 1.9 (0.0~20.6); 910 (595~1281) ❙ 1404 (512~2037); 547 (258~1033); 142 (15~259) ❙ 54.5 (-35.3~147.0); 897 (512~1273); 217.4 (95.3~371.7)

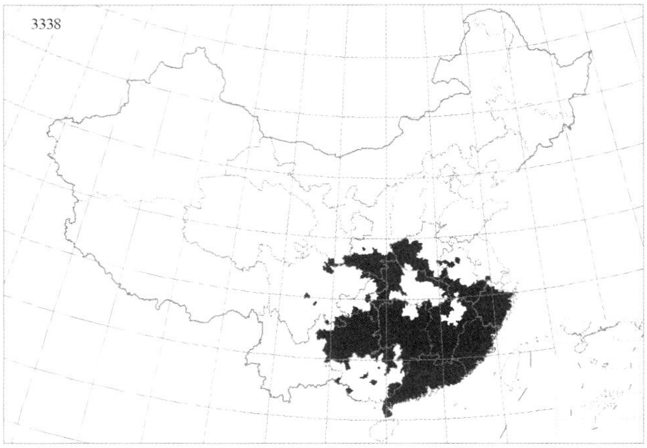

Rosaceae 蔷薇科 *Rosa* 蔷薇属
Rosa hezhangensis T. L. Xu 赫章蔷薇
Shrubs

11.7; 2.7; 19.6; 11.7; 84.6; 3.8; 661 ❙ 964; 516; 39 ❙ 45.7; 661; 199.3

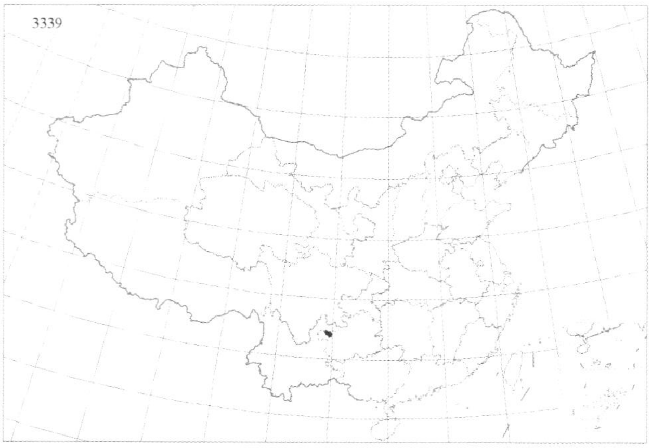

Rosaceae 蔷薇科 *Rosa* 蔷薇属
Rosa hugonis Hemsl. 黄蔷薇
Small shrubs

8.1 (-0.8~14.8); -6.4 (-12.7~3.8); 21.1 (9.2~26.6); 9.5 (2.9~14.8); 73.7 (10.1~119.2); 35.5 (1.4~79.5); 622 (374~800) ❙ 552 (189~1358); 299 (114~546); 16 (4~88) ❙ -9.9 (-70.2~125.9); 511 (189~795); 172.8 (33.5~341.6)

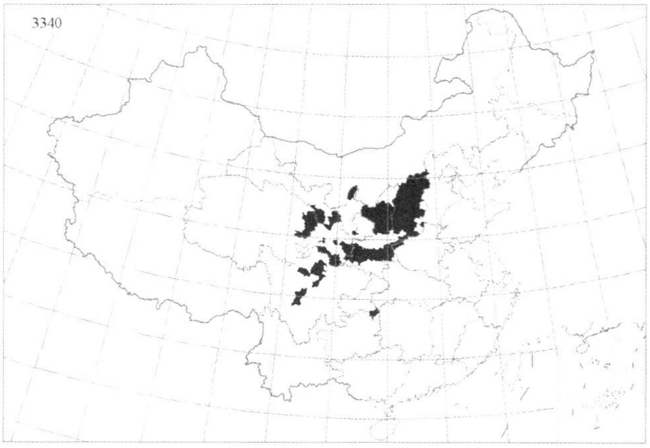

Rosaceae 蔷薇科 *Rosa* 蔷薇属
Rosa kokanica (Regel) Regel ex Juz. 腺叶蔷薇
Small shrubs

1.7; -17.7; 18.5; 6.9; 51.5; 91.0; 541 ❙ 175; 73; 21 ❙ -67.4; 175; 86.7

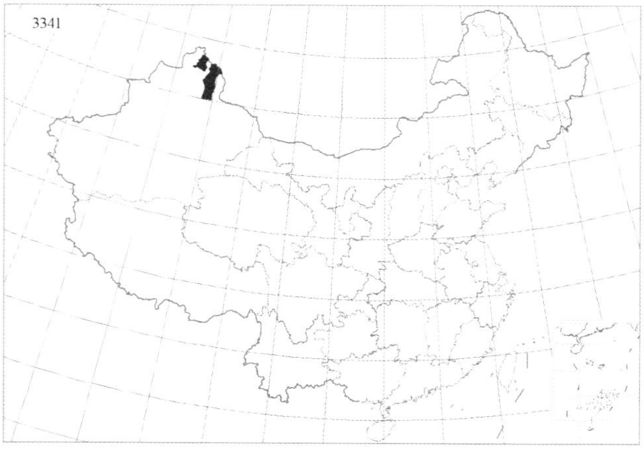

Rosaceae 蔷薇科 *Rosa* 蔷薇属
Rosa koreana Kom. 长白蔷薇
Small shrubs

2.1 (-1.5~4.8); -18.9 (-25.0~-14.1); 20.2 (18.4~23.0); 7.4 (6.2~8.7); 55.5 (45.8~68.9); 89.8 (67.8~124.5); 549 (501~616) ❙ 656 (544~976); 406 (333~608); 21 (13~40) ❙ 19.6 (-9.7~66.0); 543 (501~588); 306.5 (177.1~333.3)

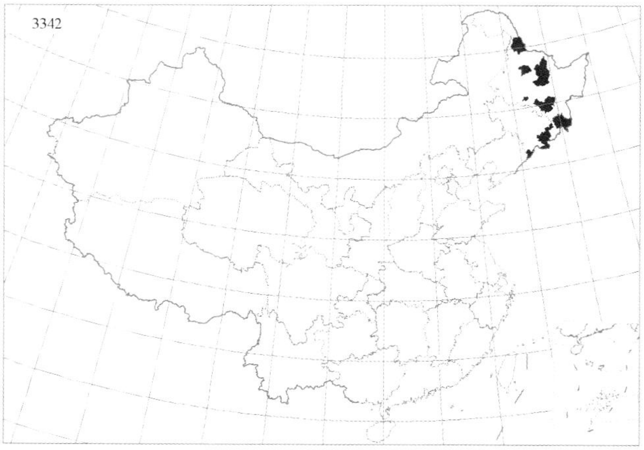

Rosaceae 蔷薇科 *Rosa* 蔷薇属
Rosa kunmingensis T. C. Ku 昆明蔷薇
Shrubs

14.6；7.9 19.6；14.6；115.8；0.0；736 | 1009；596；39 | 37.1；732；278.4

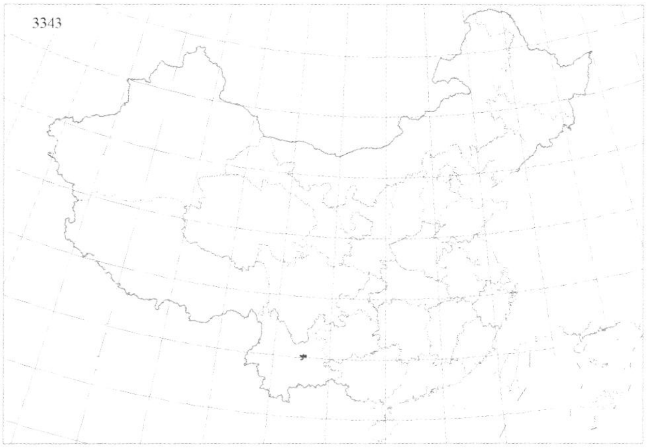

Rosaceae 蔷薇科 *Rosa* 蔷薇属
Rosa kwangtungensis T. T. Yu et Tsai 广东蔷薇
Climbing shrubs

19.3 (14.6~22.5)；9.8 (2.5~14.8)；28.0 (24.6~29.4)；19.4 (14.7~22.6)；173.1 (117.7~210.9)；0.5 (0.0~3.9)；1029 (779~1212) | 1579 (1235~2030)；644 (425~1033)；170 (75~257) | 55.7 (14.0~130.7)；1026 (779~1212)；221.8 (111.9~303.4)

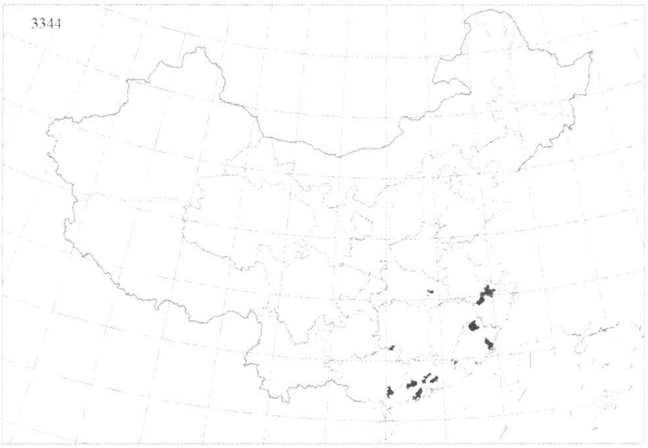

Rosaceae 蔷薇科 *Rosa* 蔷薇属
Rosa kwangtungensis var. mollis F. P. Metcalf 毛叶广东蔷薇
Climbing shrubs

21.3；12.8；28.4；21.3；195.7；0.0；1122 | 1518；607；164 | 35.1；1108；206.9

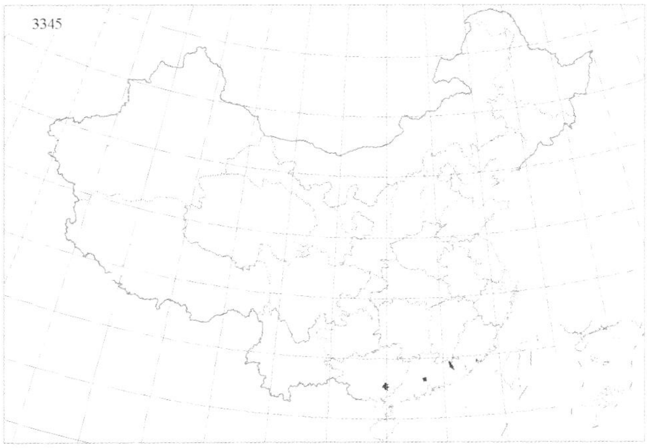

Rosaceae 蔷薇科 *Rosa* 蔷薇属
Rosa kwangtungensis var. plena T. T. Yu et T. C. Ku 重瓣广东蔷薇
Climbing shrubs

21.7；13.1；28.6；21.7；200.2；0.0；1150 | 1674；708；148 | 45.6；1150；201.5

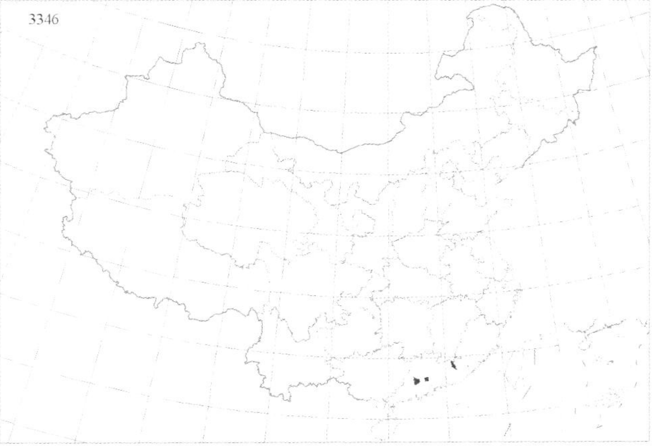

Rosaceae 蔷薇科 *Rosa* 蔷薇属
Rosa kweichowensis T. T. Yu et T. C. Ku 贵州刺梨
Climbing shrubs, evergreen or semievergreen

14.6；4.5；23.9；14.7；116.8；0.8；773 | 1160；528；65 | 50.0；773；180.9

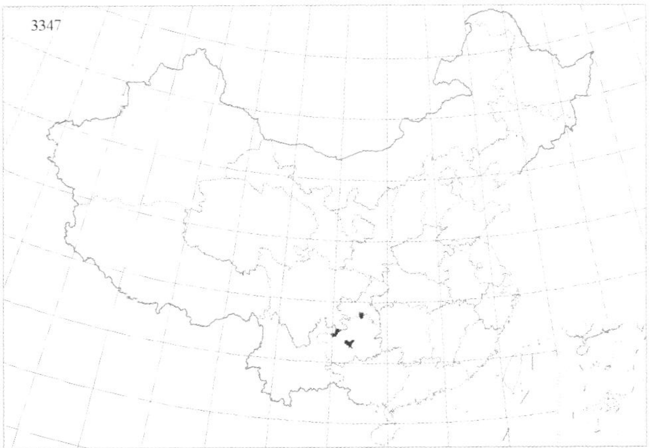

Rosaceae 蔷薇科 *Rosa* 蔷薇属
Rosa laevigata Michx. 金樱子
Climbing shrubs, evergreen

17.3 (3.5~23.6)；7.2 (−6.5~16.7)；26.3 (12.3~30.1)；17.3 (5.0~23.6)；148.8 (26.3~223.5)；1.1 (0.0~43.4)；916 (446~1281) | 1400 (626~3495)；581 (299~1428)；135 (15~897) | 53.2 (−22.2~246.5)；904 (446~1273)；234.8 (95.3~659.0)

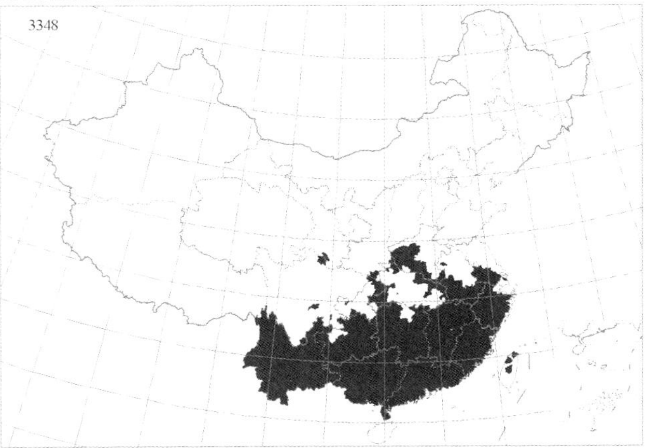

Rosaceae 蔷薇科 *Rosa* 蔷薇属

Rosa langyashanica D. C. Zhang et J. Z. Shao 琅琊山蔷薇

Shrubs, deciduous

15.4; 1.8; 28.0; 15.4; 130.0; 5.0; 875 ❙ 994; 464; 102 ❙ 13.5; 875; 151.6

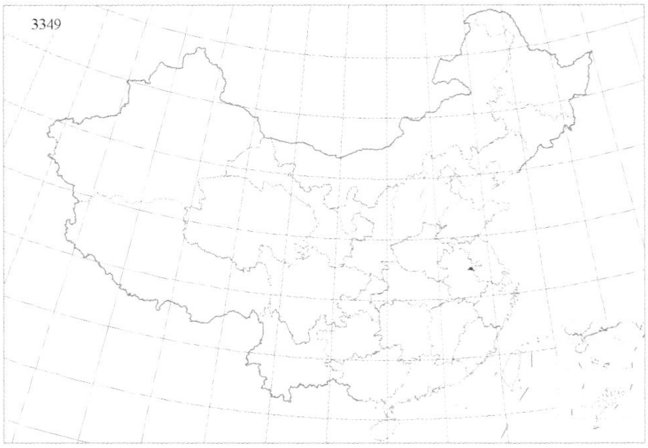

Rosaceae 蔷薇科 *Rosa* 蔷薇属

Rosa lasiosepala F. P. Metcalf 毛萼蔷薇

Scandent shrubs

17.4 (14.3~20.7); 7.2 (3.8~11.1); 26.2 (23.5~29.0); 17.4 (14.3~20.8); 149.3 (112.6~189.3); 0.2 (0.0~1.5); 908 (756~1108) ❙ 1415 (1182~1624); 598 (476~775); 131 (43~196) ❙ 57.4 (16.9~101.8); 908 (756~1108); 240.7 (160.7~354.5)

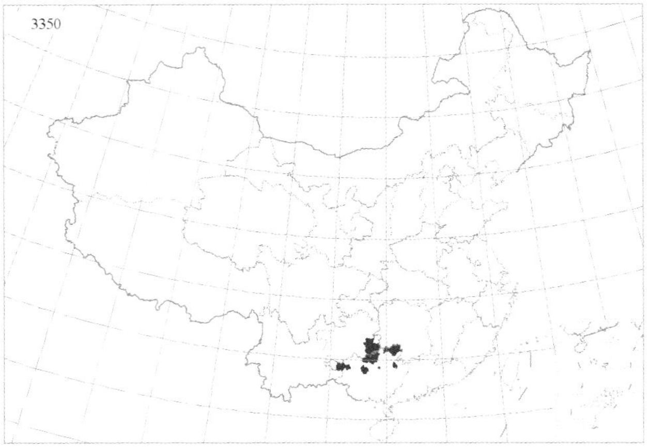

Rosaceae 蔷薇科 *Rosa* 蔷薇属

Rosa laxa Retz. 疏花蔷薇

Shrubs

3.8 (-4.1~12.4); -14.6 (-20.9~-5.4); 19.0 (8.5~27.1); 7.7 (2.5~13.2); 58.9 (8.1~114.1); 73.1 (25.2~117.9); 571 (330~811) ❙ 204 (32~491); 95 (19~301); 18 (0~51) ❙ -60.5 (-95.9~41.8); 200 (32~406); 99.6 (1.6~297.3)

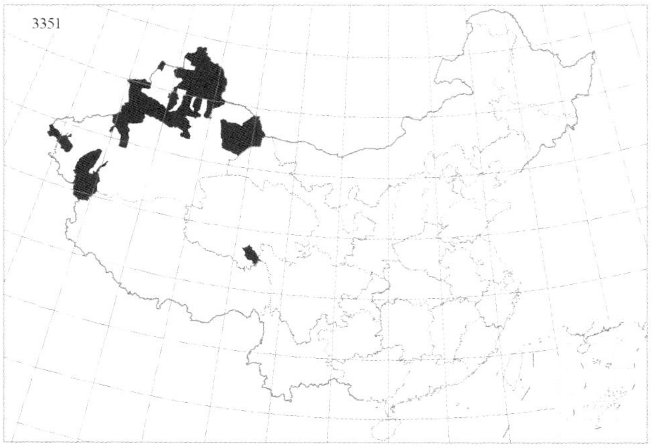

Rosaceae 蔷薇科 *Rosa* 蔷薇属

Rosa lichiangensis T. T. Yu et T. C. Ku 丽江蔷薇

Scandent shrubs

13.9 (10.4~16.7); 7.2 (3.4~9.7); 19.2 (16.3~22.1); 13.9 (10.5~16.7); 107.4 (68.0~140.9); 0.3 (0.0~2.6); 719 (623~819) ❙ 1011 (937~1164); 585 (553~639); 39 (15~59) ❙ 41.7 (14.4~68.1); 698 (615~748); 333.7 (278.6~384.1)

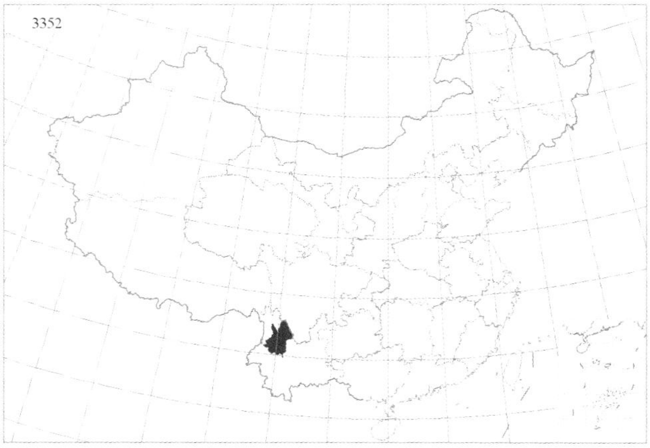

Rosaceae 蔷薇科 *Rosa* 蔷薇属

Rosa longicuspis Bertol. 长尖叶蔷薇

Scandent shrubs

15.3 (2.2~21.0); 8.0 (-6.3~15.1); 20.8 (9.5~26.6); 15.3 (3.9~21.0); 125.1 (15.8~191.8); 1.3 (0.0~48.6); 776 (411~1053) ❙ 1121 (729~1751); 628 (411~996); 44 (11~88) ❙ 45.1 (-14.2~112.2); 759 (411~1053); 329.2 (142.9~659.0)

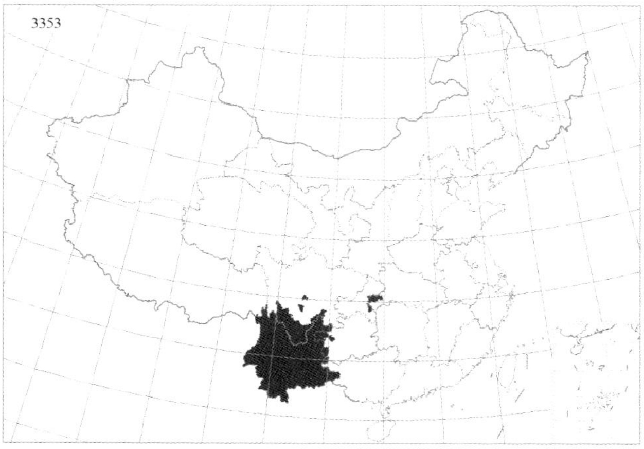

Rosaceae 蔷薇科 *Rosa* 蔷薇属

Rosa longicuspis var. sinowilsonii (Hemsl.) Yü et Ku 多花长尖叶蔷薇

Scandent shrubs

11.1 (5.4~15.3); 2.9 (-2.1~5.3); 18.5 (12.3~25.9); 11.2 (5.8~15.4); 79.5 (31.5~125.0); 5.4 (0.0~25.6); 654 (481~818) ❙ 1123 (741~1410); 614 (426~816); 46 (28~75) ❙ 71.9 (47.4~93.0); 654 (481~818); 295.5 (152.4~384.7)

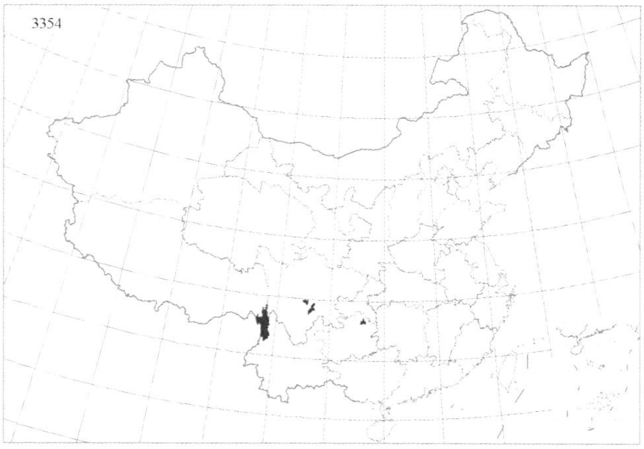

Rosaceae 蔷薇科 *Rosa* 蔷薇属
Rosa luciae Franch. et Roch. ex Crép. 光叶蔷薇
Scandent shrubs

18.7 (15.6~22.4); 8.9 (4.3~15.2); 27.9 (25.9~29.3); 18.9 (15.6~22.7); 166.5 (127.5~212.7); 0.1 (0.0~0.7); 1000 (824~1216) ❙ 1716 (1363~2030); 662 (481~1033); 182 (104~258) ❙ 74.1 (17.4~130.7); 996 (824~1216); 226.6 (163.7~303.4)

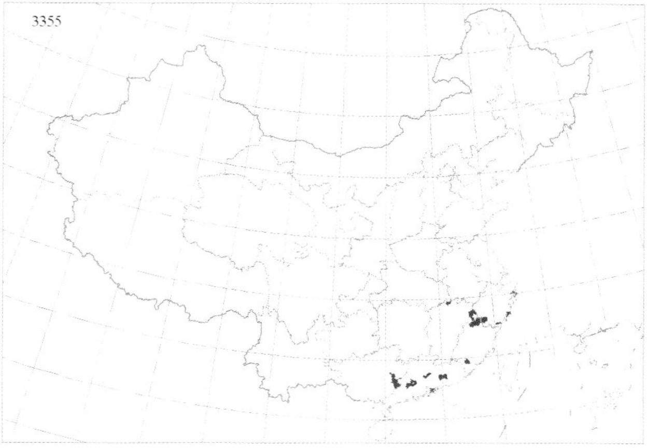

Rosaceae 蔷薇科 *Rosa* 蔷薇属
Rosa luciae var. rosea H. L. Li 粉花光叶蔷薇
Shrubs

16.4; 11.5; 20.3; 16.5; 137.8; 0.0; 783 ❙ 2499; 947; 303 ❙ 219.3; 783; 303.6

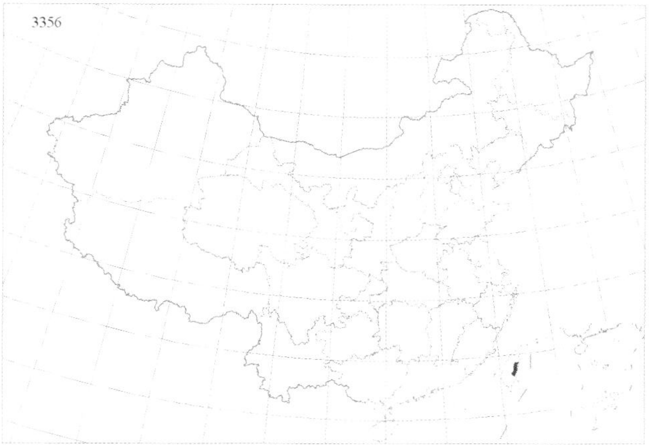

Rosaceae 蔷薇科 *Rosa* 蔷薇属
Rosa lucidissima H. Lév. 亮叶月季
Climbing shrubs, evergreen or semievergreen

15.4 (6.0~21.5); 5.2 (-3.2~13.2); 24.7 (14.4~28.5); 15.4 (6.6~21.5); 126.9 (39.7~198.2); 2.1 (0.0~26.7); 821 (508~1138) ❙ 1213 (890~1776); 558 (443~932); 69 (21~142) ❙ 49.1 (8.5~101.4); 821 (508~1138); 204.0 (118.7~283.9)

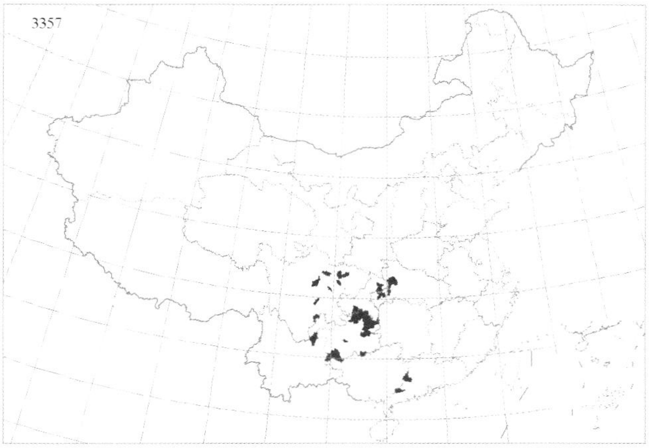

Rosaceae 蔷薇科 *Rosa* 蔷薇属
Rosa ludingensis T. C. Ku 泸定蔷薇
Small shrubs

6.8; -2.2; 14.8; 7.1; 43.8; 22.2; 523 ❙ 993; 535; 25 ❙ 90.1; 523; 246.8

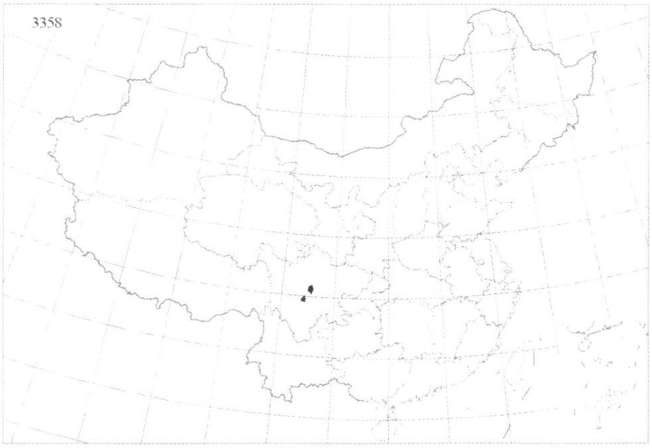

Rosaceae 蔷薇科 *Rosa* 蔷薇属
Rosa macrophylla Lindl. 大叶蔷薇
Shrubs

6.3 (-1.8~14.5); -1.5 (-10.5~7.8); 13.1 (6.1~19.7); 7.3 (1.8~14.5); 45.8 (1.9~113.5); 30.2 (0.0~82.8); 516 (329~730) ❙ 795 (311~1431); 473 (193~817); 26 (5~54) ❙ 49.4 (-6.3~125.9); 513 (311~730); 261.8 (62.8~382.6)

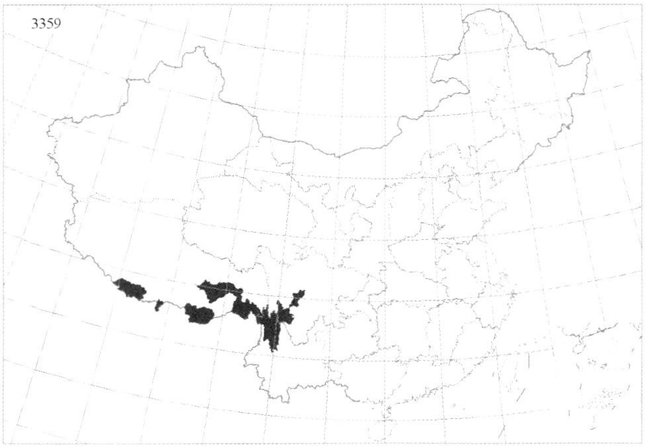

Rosaceae 蔷薇科 *Rosa* 蔷薇属
Rosa macrophylla var. glandulifera T. T. Yu et T. C. Ku 腺果大叶蔷薇
Shrubs

3.0 (-1.1~12.2); -5.7 (-10.0~4.3); 10.6 (6.2~18.5); 4.8 (2.0~12.2); 25.8 (2.5~87.4); 49.8 (0.7~75.6); 432 (334~672) ❙ 585 (321~1431); 351 (193~817); 16 (5~33) ❙ 22.8 (-4.0~113.0); 430 (321~672); 217.0 (79.3~363.7)

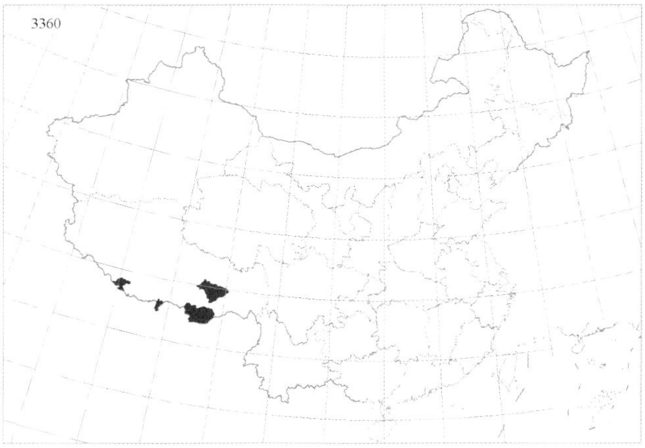

Rosaceae 蔷薇科 *Rosa* 蔷薇属
Rosa mairei H. Lév. 毛叶蔷薇
Dwarf shrubs

12.3 (1.8~18.9); 4.1 (-7.4~12.0); 19.1 (9.5~24.6); 12.5 (3.8~19.0); 95.0 (15.5~167.8); 6.9 (0.0~53.3); 686 (404~929) ❙ 978 (324~1334); 553 (257~720); 36 (0~65) ❙ 44.2 (-26.0~125.9); 669 (324~854); 254.6 (123.3~382.6)

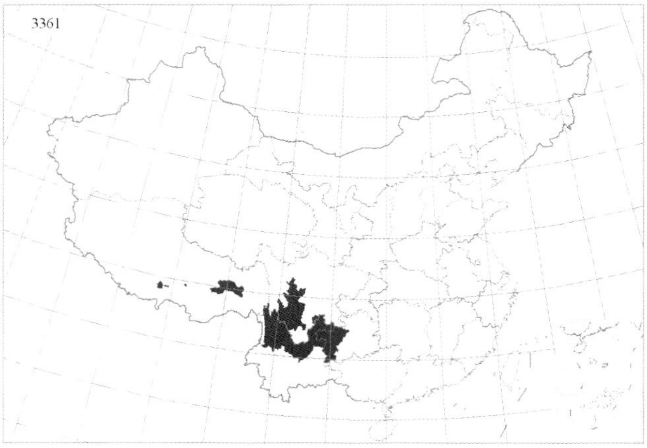

Rosaceae 蔷薇科 *Rosa* 蔷薇属
Rosa maximowicziana Regel 伞花蔷薇
Small shrubs

9.5 (5.2~12.1); -6.6 (-15.6~-1.2); 23.5 (22.3~24.6); 10.8 (9.1~12.4); 89.1 (73.9~103.4); 34.8 (14.2~71.8); 684 (616~746) ❙ 729 (409~1073); 455 (298~691); 32 (6~65) ❙ 7.1 (-36.8~74.1); 648 (409~729); 201.5 (113.4~335.3)

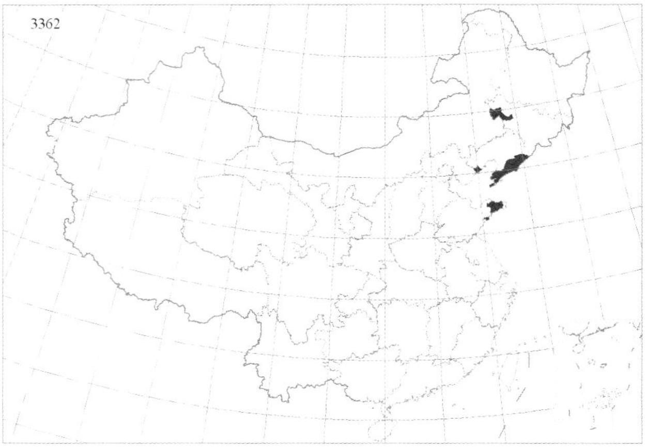

Rosaceae 蔷薇科 *Rosa* 蔷薇属
Rosa miyiensis T. C. Ku 米易蔷薇
Small shrubs

16.1; 8.6; 21.9; 16.1; 133.0; 0.0; 797 ❙ 1014; 630; 20 ❙ 27.3; 719; 330.0

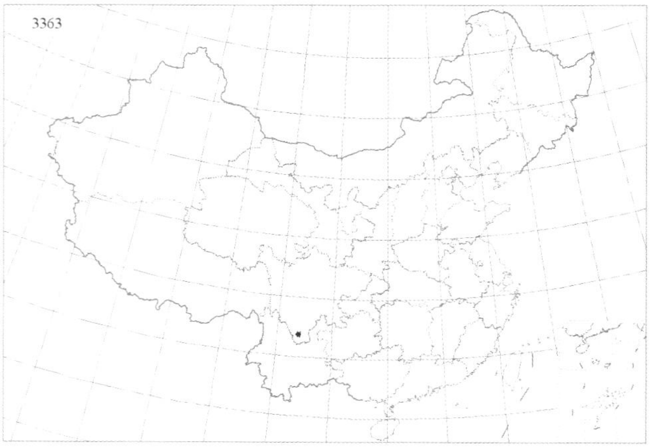

Rosaceae 蔷薇科 *Rosa* 蔷薇属
Rosa morrisonensis Hayata 玉山蔷薇
Small shrubs

20.8; 15.3; 25.0; 20.8; 189.2; 0.0; 1019 ❙ 2445; 1412; 110 ❙ 140.0; 1019; 207.7

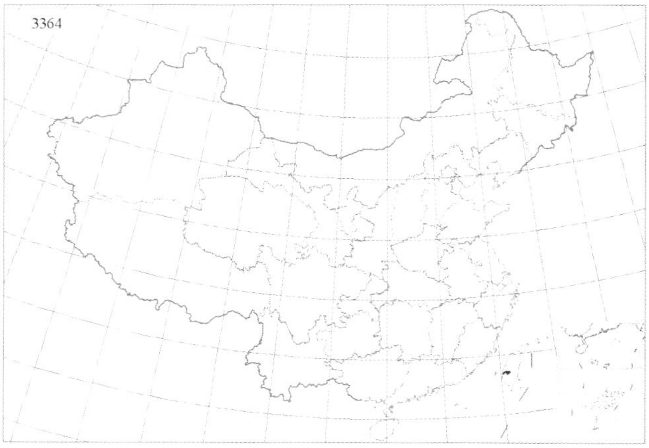

Rosaceae 蔷薇科 *Rosa* 蔷薇属
Rosa moyesii Hemsl. et E. H. Wilson 华西蔷薇
Shrubs

10.6 (-4.1~18.9); -1.8 (-17.3~12.5); 21.5 (7.8~28.0); 11.4 (2.1~19.0); 91.1 (5.2~167.8); 23.3 (0.0~114.2); 685 (340~929) ❙ 716 (341~1358); 401 (190~663); 26 (5~113) ❙ 6.4 (-40.9~128.6); 597 (340~854); 213.4 (61.2~406.0)

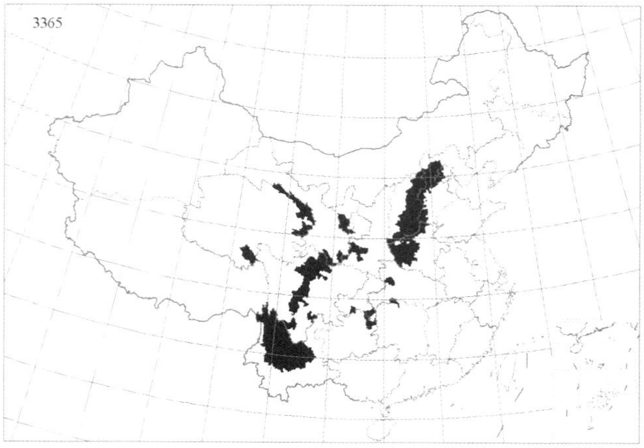

Rosaceae 蔷薇科 *Rosa* 蔷薇属
Rosa moyesii var. pubescens T. T. Yu et H. T. Tsai 毛叶华西蔷薇
Shrubs

3.8; -4.3; 10.8; 4.7; 21.6; 35.4; 444 ❙ 984; 601; 16 ❙ 121.7; 444; 265.2

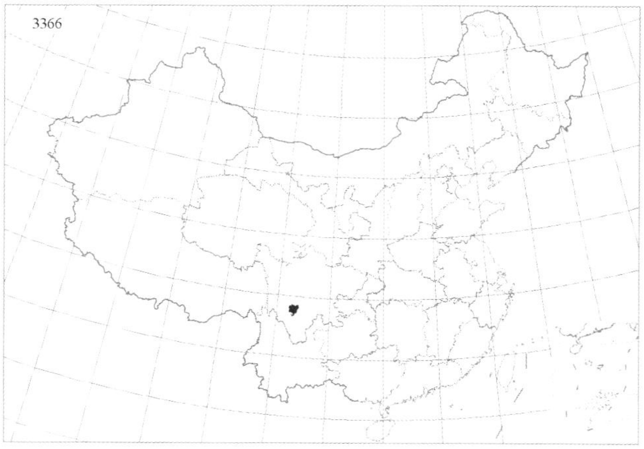

Rosaceae 蔷薇科 *Rosa* 蔷薇属
Rosa multibracteata Hemsl. et E. H. Wilson 多苞蔷薇
Shrubs

7.2 (2.2~19.2); -1.3 (-7.1~13.3); 14.7 (9.9~23.5); 7.9 (3.8~19.4); 51.2 (15.5~172.6); 24.0 (0.0~49.1); 546 (413~932) ❙ 950 (741~1276); 521 (400~741); 27 (15~56) ❙ 78.2 (36.8~125.9); 545 (413~923); 294.3 (200.8~452.0)

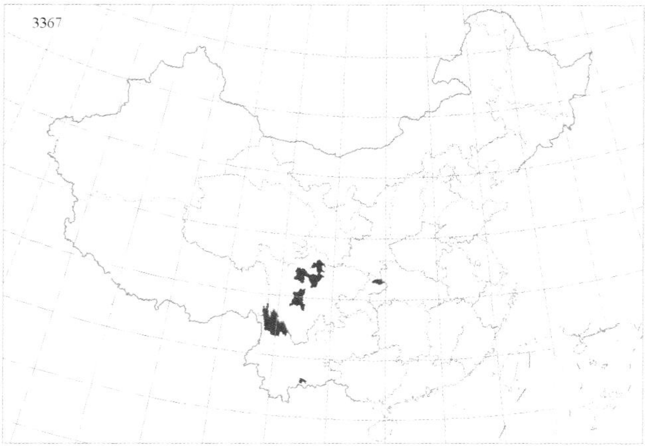

Rosaceae 蔷薇科 *Rosa* 蔷薇属
Rosa multiflora Thunb. 野蔷薇
Scandent shrubs

14.6 (3.3~24.8); 2.6 (-13.7~20.1); 25.7 (12.3~29.9); 14.9 (5.8~24.8); 124.3 (31.5~237.6); 8.5 (0.0~71.7); 829 (481~1385) ❙ 1086 (367~2680); 477 (235~1412); 96 (8~254) ❙ 29.3 (-43.4~246.5); 782 (367~1360); 208.0 (101.9~620.0)

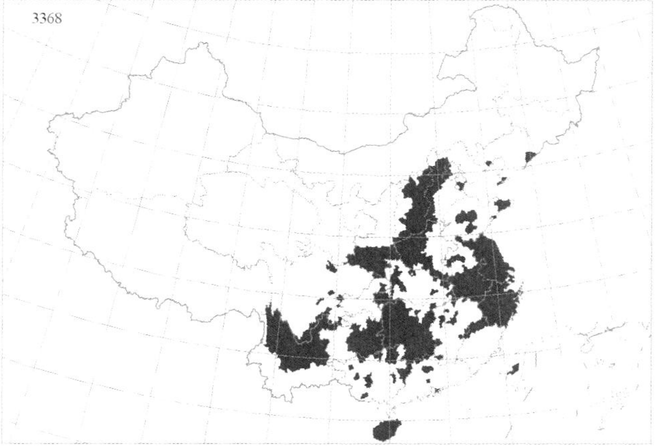

Rosaceae 蔷薇科 *Rosa* 蔷薇属
Rosa multiflora var. cathayensis Rehder et E. H. Wilson 粉团蔷薇
Scandent shrubs

16.1 (2.9~21.9); 4.7 (-6.4~13.6); 26.8 (11.2~30.1); 16.1 (4.4~22.0); 136.1 (20.5~203.4); 2.6 (0.0~44.6); 875 (428~1168) ❙ 1325 (548~1997); 508 (285~816); 141 (16~259) ❙ 51.4 (-26.1~147.0); 856 (428~1168); 212.5 (110.1~382.6)

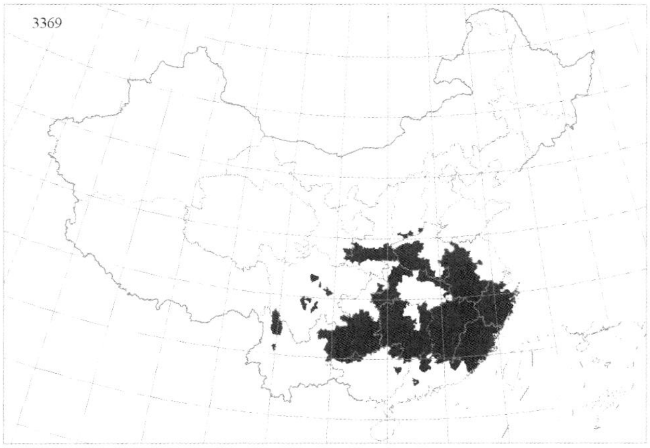

Rosaceae 蔷薇科 *Rosa* 蔷薇属
Rosa murielae Rehder et E. H. Wilson 西南蔷薇
Shrubs

7.8 (2.0~13.6); -1.2 (-7.2~7.2); 16.0 (9.9~23.9); 8.3 (3.8~13.7); 55.5 (15.4~104.1); 20.9 (0.0~51.4); 560 (413~729) ❙ 982 (741~1246); 528 (420~688); 28 (12~46) ❙ 78.3 (36.2~128.6); 560 (413~729); 284.7 (200.8~372.3)

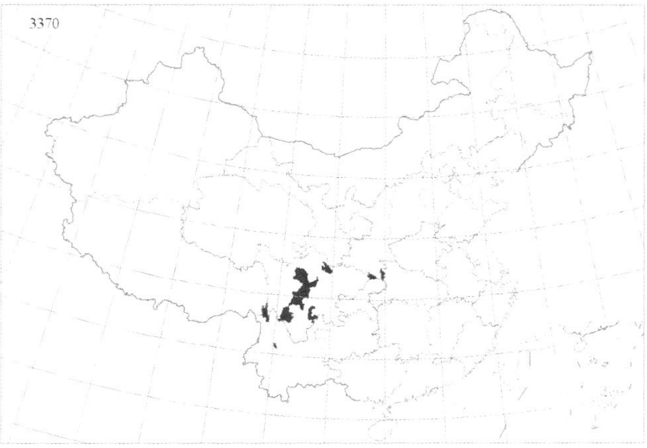

Rosaceae 蔷薇科 *Rosa* 蔷薇属
Rosa odorata (Andrews) Sweet 香水月季
Scandent shrubs, evergreen or semievergreen

14.7 (10.8~21.0); 8.0 (3.9~14.2); 19.8 (16.6~25.6); 14.7 (10.9~21.0); 116.8 (71.8~191.8); 0.1 (0.0~1.3); 752 (633~1053) ❙ 1052 (911~1522); 614 (534~835); 40 (24~68) ❙ 40.1 (13.8~64.0); 737 (633~1053); 332.6 (234.5~551.6)

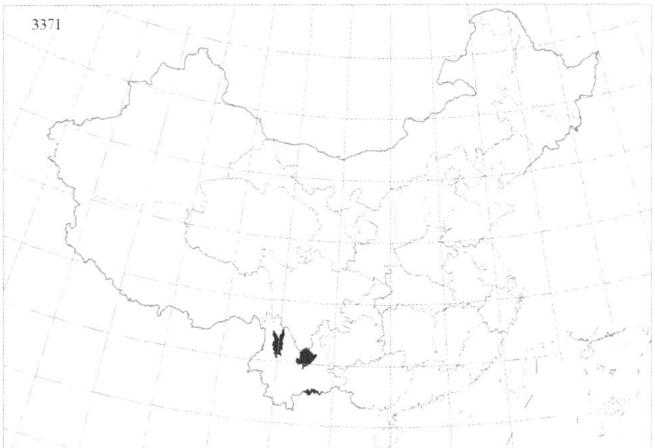

Rosaceae 蔷薇科 *Rosa* 蔷薇属
Rosa odorata var. erubescens (Focke) T. T. Yu et T. C. Ku 粉红香水月季
Scandent shrubs, evergreen or semievergreen

10.8; 3.9; 16.6; 10.9; 71.8; 1.3; 633 ❙ 968; 594; 24 ❙ 52.8; 633; 319.8

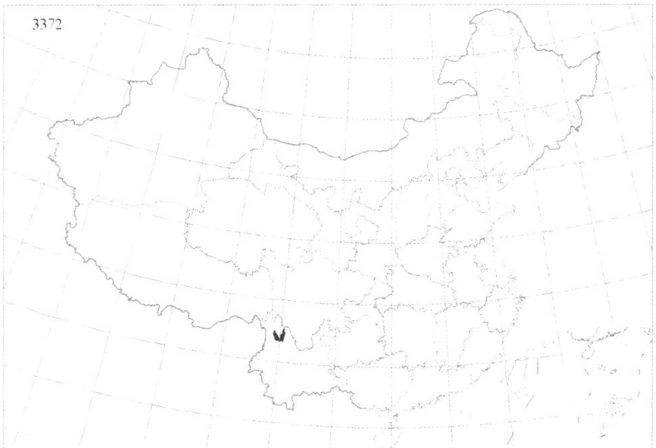

Rosaceae 蔷薇科 *Rosa* 蔷薇属

Rosa odorata var. gigantea (Collett ex Crép.) Rehder et E. H. Wilson 大花香水月季 Scandent shrubs, evergreen or semievergreen

15.0 (5.4~21.0); 8.2 (-2.1~14.2); 20.1 (12.3~25.6); 15.0 (5.8~21.0); 123.2 (31.5~191.8); 3.2 (0.0~25.6); 775 (481~1053) ❙ 1106 (741~1522); 617 (426~818); 46 (24~68) ❙ 44.2 (19.0~67.9); 771 (481~1053); 379.6 (273.5~551.6)

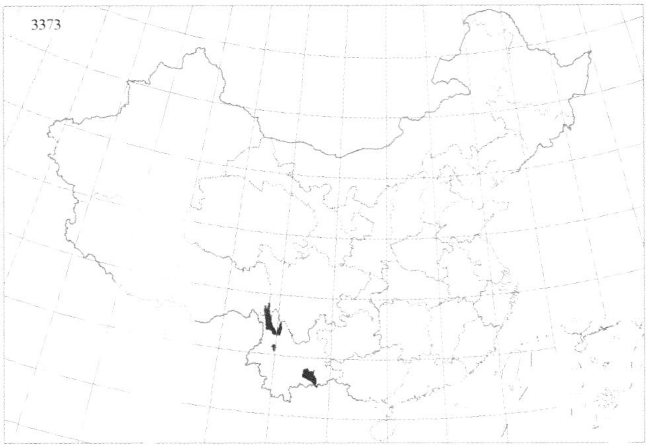

Rosaceae 蔷薇科 *Rosa* 蔷薇属

Rosa odorata var. pseudoindica (Lindl.) Rehder 橘黄香水月季

Scandent shrubs

[illegible] (5.4~16.7); 4.7 (-2.1~9.7); 17.3 (12.3~22.1); 11.7 (5.8~16.7); 84.5 (31.5~140.9); 4.0 (0.0~25.6); 656 (481~819) ❙ 1028 (741~1337); 593 (426~727); 40 (15~69) ❙ 58.0 (14.4~93.0); 646 (481~748); 337.1 (273.5~407.1)

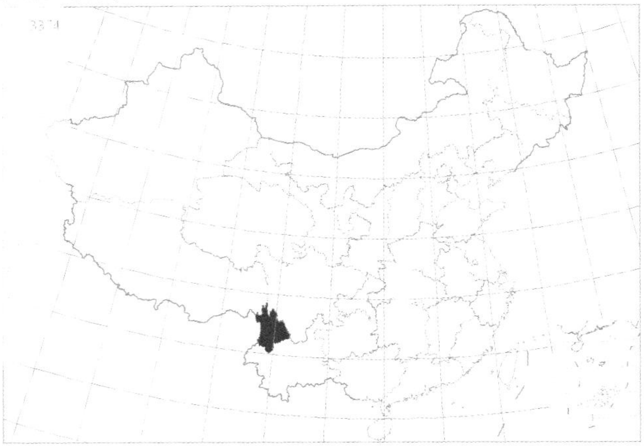

Rosaceae 蔷薇科 *Rosa* 蔷薇属

Rosa omeiensis Rolfe 峨眉蔷薇

Shrubs

9.0 (-4.1~18.9); -2.7 (-17.3~12.5); 19.6 (6.1~29.5); 9.9 (1.8~19.0); 74.8 (1.9~167.5); 26.2 (0.0~114.2); 630 (329~932) ❙ 720 (250~1431); 376 (173~817); 23 (1~83) ❙ 18.0 (-49.4~128.6); 574 (250~851); 222.6 (42.6~407.1)

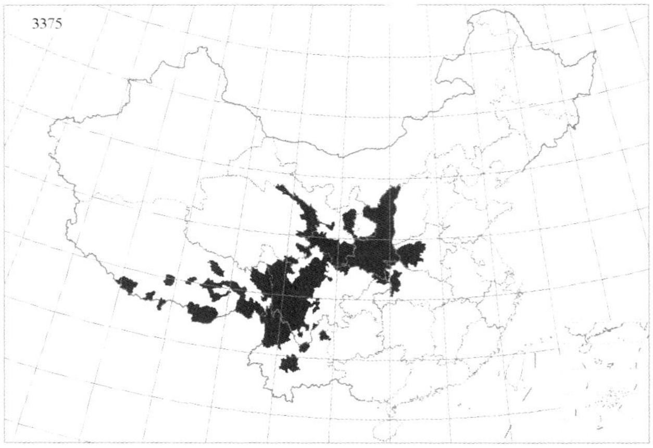

Rosaceae 蔷薇科 *Rosa* 蔷薇属

Rosa oxyacantha Bieb. 尖刺蔷薇

Small shrubs

1.6 (-0.6~5.9); -17.5 (-19.5~-13.8); 18.2 (15.9~23.0); 6.8 (5.4~9.4); 50.2 (37.5~77.9); 90.9 (66.6~104.2); 535 (475~655) ❙ 166 (152~187); 75 (69~79); 18 (15~25) ❙ -68.5 (-76.8~-65.1); 166 (152~187); 92.7 (28.1~120.8)

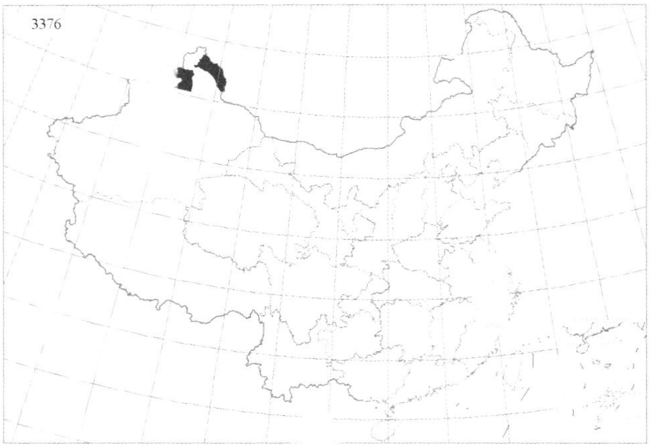

Rosaceae 蔷薇科 *Rosa* 蔷薇属

Rosa persetosa Rolfe 全针蔷薇

Shrubs

6.1; -2.3; 13.4; 7.0; 40.6; 26.7; 514 ❙ 904; 503; 29 ❙ 77.2; 514; 346.7

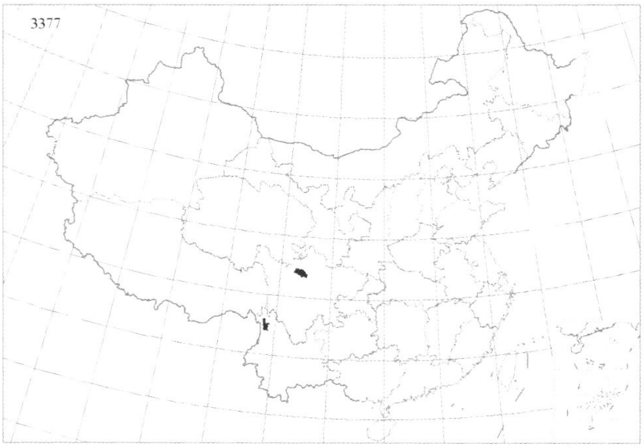

Rosaceae 蔷薇科 *Rosa* 蔷薇属

Rosa pinnatisepala T. C. Ku 羽萼蔷薇

Shrubs

6.6; -1.3; 13.3; 6.8; 39.6; 19.7; 514 ❙ 862; 548; 12 ❙ 67.7; 514; 311.5

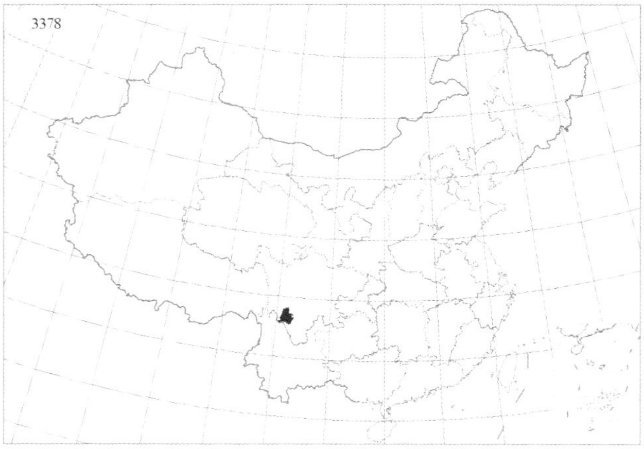

Rosaceae 蔷薇科 *Rosa* 蔷薇属

Rosa platyacantha Schrenk 宽刺蔷薇

Small shrubs

4.4 (-2.8~11.9); -14.7 (-20.9~-5.6); 20.0 (10.9~27.1); 8.2 (3.5~12.7); 64.0 (16.3~107.9); 70.1 (25.2~109.2); 593 (390~810) ❙ 222 (70~450); 102 (23~211); 20 (4~51) ❙ -59.4 (-91.1~3.9); 221 (70~406); 103.2 (10.9~297.3)

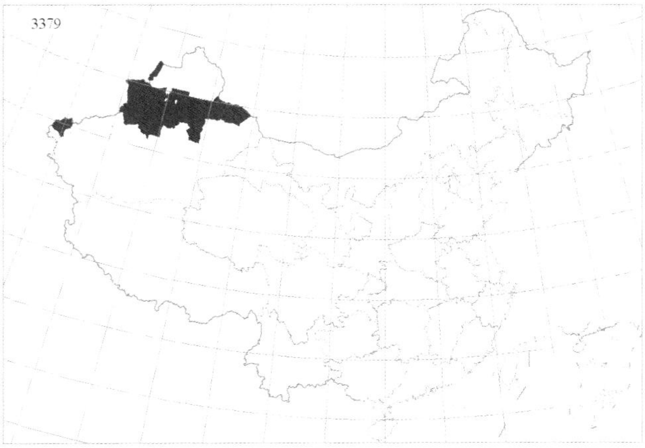

Rosaceae 蔷薇科 *Rosa* 蔷薇属

Rosa praelucens Byhouwer 中甸刺玫

Shrubs

6.1; -1.5; 12.7; 6.3; 35.2; 21.8; 496 ❙ 801; 496; 21 ❙ 61.4; 496; 282.9

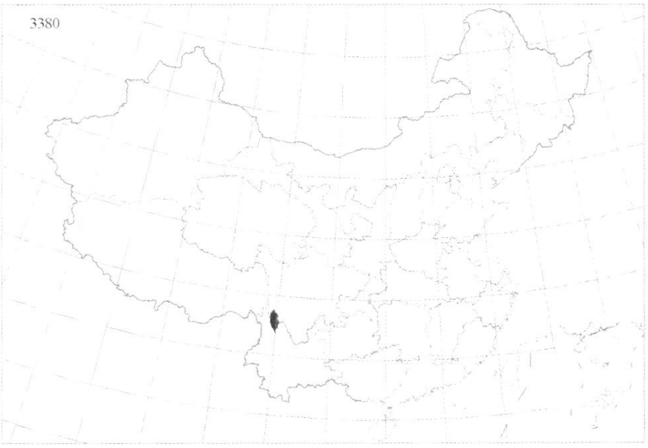

Rosaceae 蔷薇科 *Rosa* 蔷薇属

Rosa prattii Hemsl. 铁杆蔷薇

Shrubs

9.3 (-1.4~18.5); 0.5 (-15.4~10.5); 17.0 (9.9~26.6); 9.8 (3.5~18.3); 69.6 (15.4~160.0); 17.6 (0.0~92.1); 606 (375~906) ❙ 941 (282~1410); 523 (182~816); 29 (5~64) ❙ 58.9 (-24.8~128.6); 600 (282~864); 286.0 (101.3~384.1)

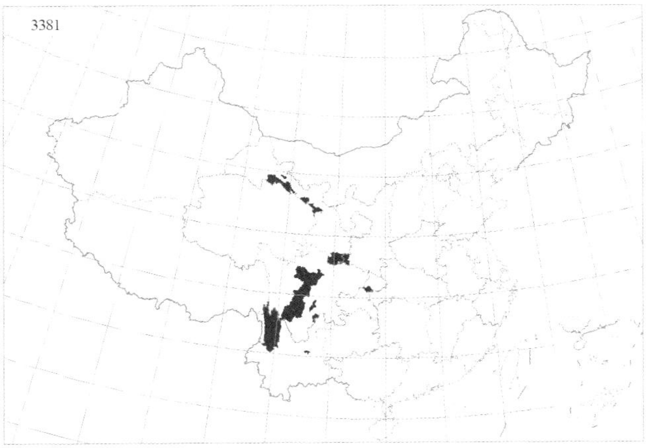

Rosaceae 蔷薇科 *Rosa* 蔷薇属

Rosa pricei Hayata 太鲁阁蔷薇

Shrubs

17.1; 11.8; 21.6; 17.2; 146.1; 0.0; 817 ❙ 2679; 953; 353 ❙ 227.4; 817; 284.7

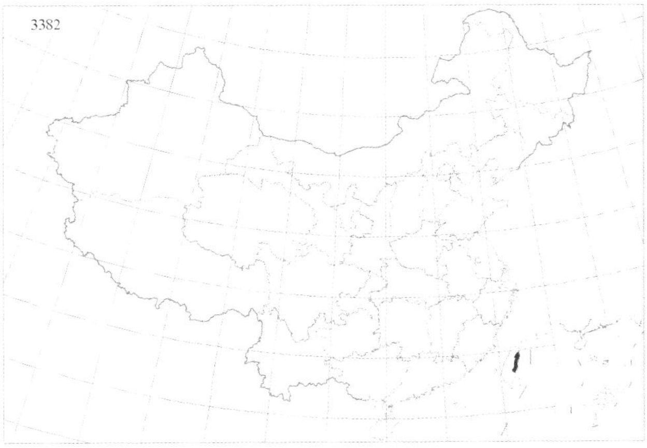

Rosaceae 蔷薇科 *Rosa* 蔷薇属

Rosa primula Boulenger 樱草蔷薇

Small shrubs

9.1 (2.7~14.5); -6.3 (-13.7~2.3); 22.7 (10.8~28.2); 10.4 (4.4~14.6); 83.7 (20.5~126.9); 33.9 (4.9~71.7); 666 (429~874) ❙ 541 (367~981); 318 (235~539); 16 (8~49) ❙ -17.7 (-40.9~92.9); 527 (367~786); 163.1 (70.3~341.6)

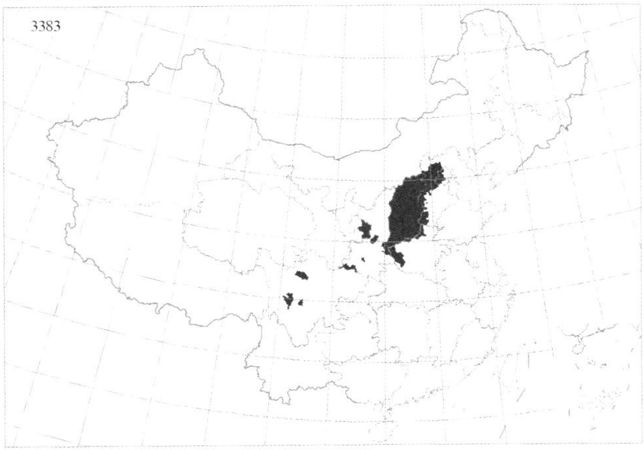

Rosaceae 蔷薇科 *Rosa* 蔷薇属

Rosa pseudobanksiae T. T. Yu et T. C. Ku 粉蕾木香

Scandent shrubs

15.0 (13.6~15.9); 8.5 (7.2~9.5); 20.0 (18.8~20.7); 15.0 (13.7~16.0); 120.6 (104.1~131.6); 0.0 (0.0~0.0); 749 (706~779) ❙ 1021 (986~1044); 570 (556~592); 48 (44~54) ❙ 36.6 (26.7~47.6); 734 (706~751); 351.7 (303.3~382.6)

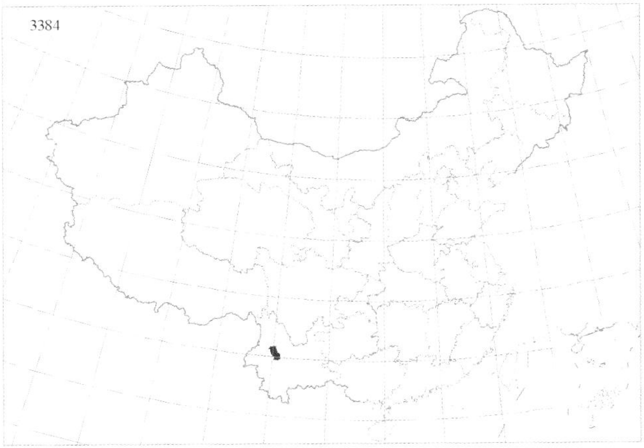

Rosaceae 蔷薇科 *Rosa* 蔷薇属
Rosa roxburghii Tratt. 缫丝花
Shrubs

15.2 (1.8~21.4); 4.7 (−7.1~13.1); 24.7 (9.9~29.4); 15.2 (3.8~21.5); 125.6 (15.7~197.8); 2.9 (0.0~53.3); 816 (404~1140) ❙ 1170 (551~1950); 524 (245~860); 83 (13~259) ❙ 43.7 (−36.2~147.0); 806 (404~1140); 223.2 (108.2~491.5)

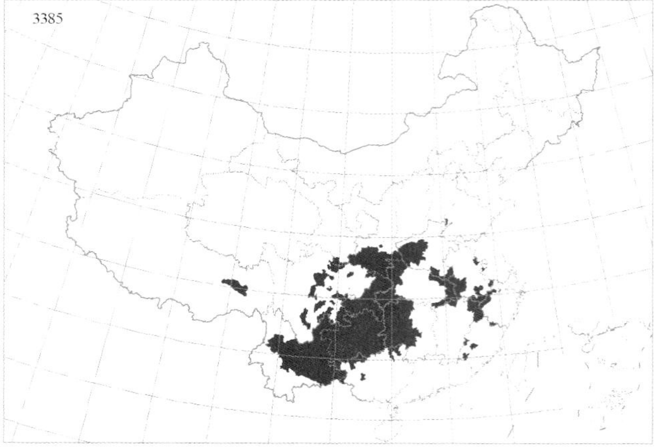

Rosaceae 蔷薇科 *Rosa* 蔷薇属
Rosa rubus H. Lév. et Vaniot 悬钩子蔷薇
Shrubs

15.4 (2.2~22.0); 5.4 (−6.6~14.1); 24.2 (9.9~29.8); 15.4 (3.8~22.1); 127.4 (15.5~205.1); 2.5 (0.0~49.1); 816 (413~1170) ❙ 1189 (622~2014); 541 (298~954); 84 (15~254) ❙ 45.5 (−22.2~130.7); 806 (413~1170); 244.5 (118.7~516.7)

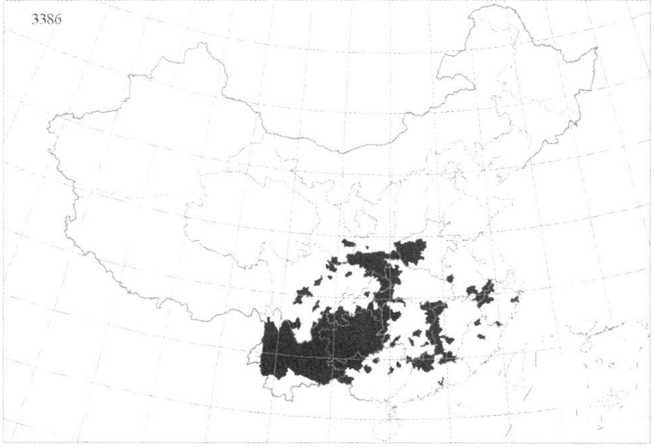

Rosaceae 蔷薇科 *Rosa* 蔷薇属
Rosa rugosa Thunb. 玫瑰
Shrubs

7.0 (−2.9~16.5); −10.9 (−26.4~4.4); 22.5 (11.2~28.9); 9.7 (4.3~16.6); 77.8 (20.0~139.8); 53.4 (0.6~134.6); 643 (427~910) ❙ 593 (409~1377); 349 (235~648); 22 (7~197) ❙ −7.5 (−36.4~94.4); 556 (409~910); 188.1 (44.7~333.3)

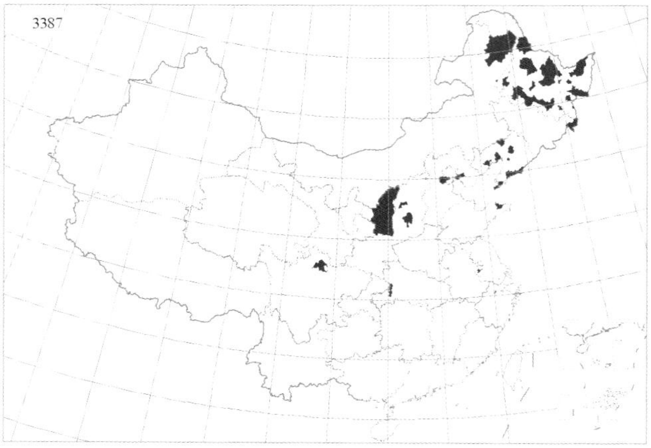

Rosaceae 蔷薇科 *Rosa* 蔷薇属
Rosa sambucina var. pubescens Koidzumi 山蔷薇
Scandent shrubs

19.0 (16.2~22.2); 13.4 (11.0~17.7); 23.6 (20.1~27.0); 19.0 (16.2~22.2); 168.4 (134.2~206.7); 0.0 (0.0~0.0); 939 (773~1123) ❙ 2449 (1836~2992); 1202 (900~1847); 205 (107~308) ❙ 168.3 (64.2~246.5); 930 (773~1123); 270.9 (202.8~296.9)

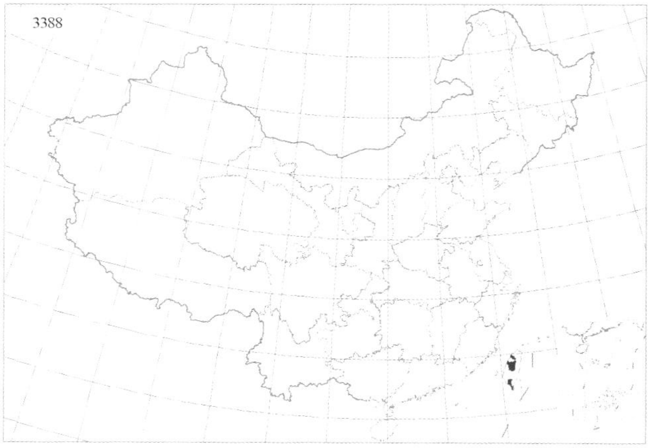

Rosaceae 蔷薇科 *Rosa* 蔷薇属
Rosa saturata Baker 大红蔷薇
Shrubs

12.9 (7.3~16.2); 1.7 (−2.0~5.2); 23.6 (15.6~27.3); 12.9 (7.5~16.2); 102.0 (48.2~134.4); 7.2 (0.0~20.4); 734 (537~866) ❙ 1241 (964~1489); 527 (421~628); 81 (27~196) ❙ 70.9 (29.8~106.2); 734 (537~866); 247.0 (172.9~285.1)

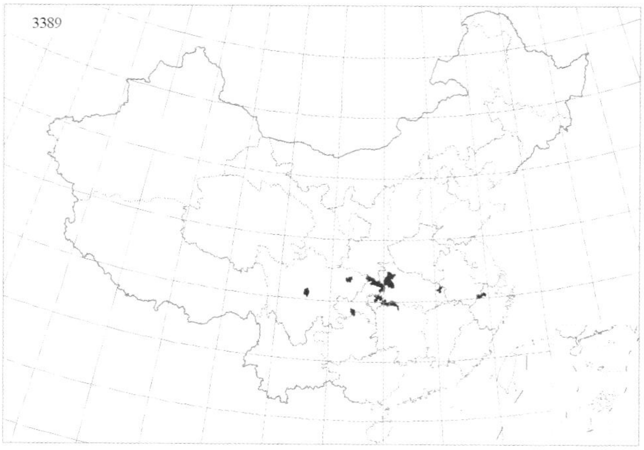

Rosaceae 蔷薇科 *Rosa* 蔷薇属
Rosa sericea Lindl. 绢毛蔷薇
Shrubs

12.3 (−0.7~18.9); 4.5 (−9.6~12.5); 18.6 (7.7~23.9); 12.6 (2.5~19.0); 96.6 (6.8~167.8); 8.9 (0.0~74.2); 683 (338~929) ❙ 959 (256~1348); 542 (200~762); 36 (0~69) ❙ 42.4 (−35.1~125.9); 666 (256~851); 276.3 (61.6~407.1)

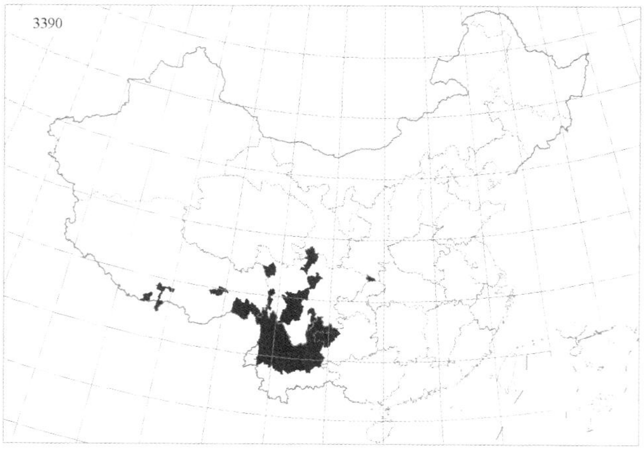

Rosaceae 蔷薇科 *Rosa* 蔷薇属
Rosa sertata Rolfe 钝叶蔷薇
Shrubs

11.2 (-1.4~18.9); -1.1 (-15.4~12.0); 22.4 (9.7~29.3); 11.9 (2.9~19.0); 94.8 (11.3~167.8); 19.4 (0.0~92.1); 702 (359~950) ❙ 843 (282~1997); 408 (174~816); 49 (5~259) ❙ 18.2 (-47.5~130.7); 647 (282~948); 217.5 (44.5~382 6)

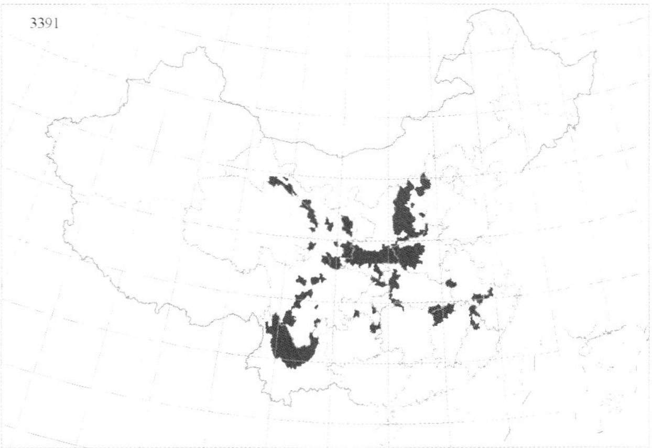

Rosaceae 蔷薇科 *Rosa* 蔷薇属
Rosa setipoda Hemsl. et E. H. Wilson 刺梗蔷薇
Shrubs

8.1 (2.2~13.2); -3.6 (-10.2~1.9); 19.0 (11.2~25.6); 9.0 (4.4~13.2); 65.7 (20.5~110.4); 27.4 (6.1~60.7); 593 (428~773) ❙ 749 (341~1246); 363 (190~531); 26 (7~83) ❙ 26.9 (-40.9~121.7); 575 (341~773); 233.9 (61.2~294.3)

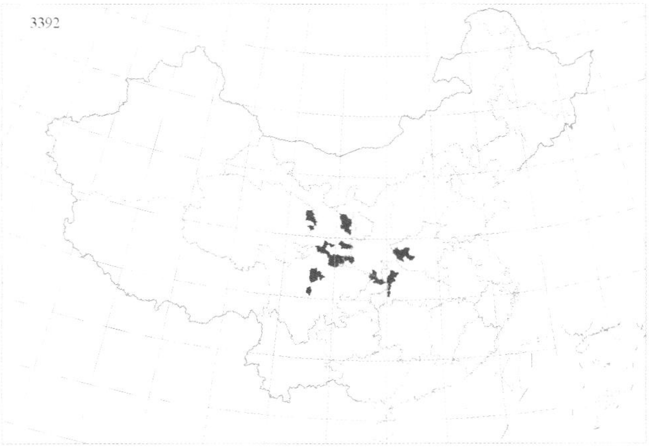

Rosaceae 蔷薇科 *Rosa* 蔷薇属
Rosa shangchengensis T. C. Ku 商城蔷薇
Shrubs

15.2; 2.3; 27.4; 15.3; 127.7; 4.3; 858 ❙ 1198; 536; 109 ❙ 39.7; 858; 181.4

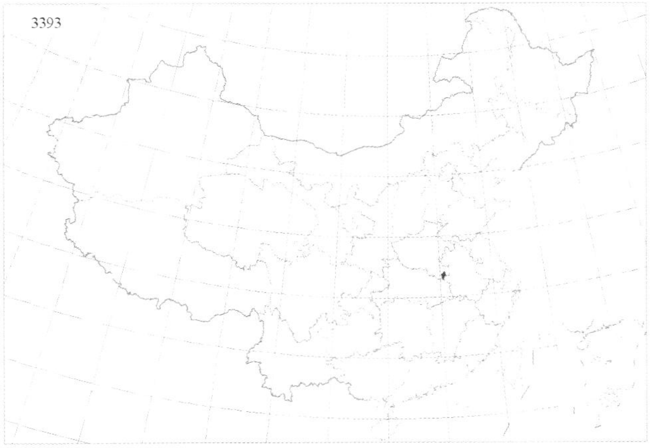

Rosaceae 蔷薇科 *Rosa* 蔷薇属
Rosa sikangensis T. T. Yu et T. C. Ku 川西蔷薇
Small shrubs

3.3 (-2.3~11.7); -5.3 (-12.8~3.9); 10.9 (6.1~18.8); 4.9 (1.7~11.7); 25.5 (1.9~82.6); 45.5 (1.3~91.3); 440 (324~658) ❙ 622 (311~1041); 383 (193~601); 20 (0~103) ❙ 38.5 (-26.0~121.7); 427 (311~658); 203.6 (46.9~352.3)

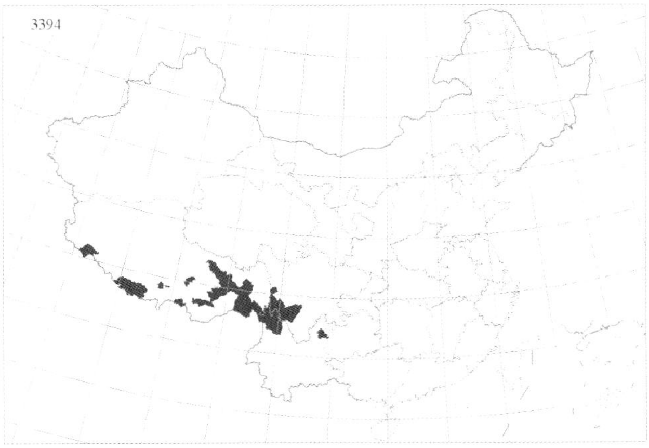

Rosaceae 蔷薇科 *Rosa* 蔷薇属
Rosa sinobiflora T. C. Ku 双花蔷薇
Shrubs

8.6; 1.3; 15.0; 8.7; 52.6; 8.8; 574 ❙ 1041; 585; 45 ❙ 81.5; 574; 352.3

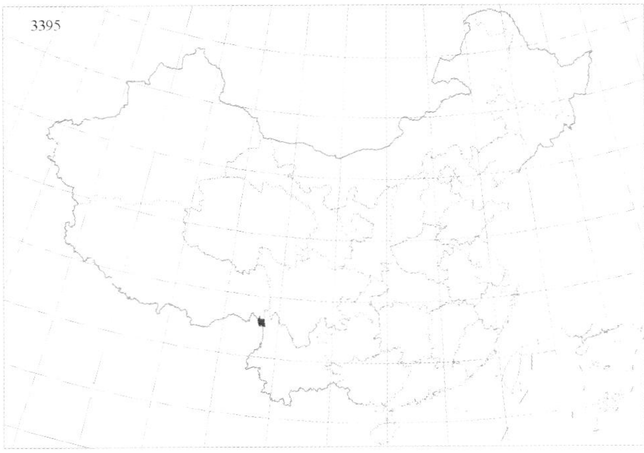

Rosaceae 蔷薇科 *Rosa* 蔷薇属
Rosa soulieana Crép. 川滇蔷薇
Shrubs

5.4 (0.0~14.9); -3.3 (-9.9~7.9); 13.1 (8.1~24.7); 6.5 (2.7~15.0); 39.3 (8.0~119.6); 33.9 (0.0~67.9); 498 (363~782) ❙ 846 (510~1410); 485 (311~816); 22 (10~65) ❙ 71.3 (20.7~128.6); 497 (363~782); 269.2 (111.0~357.7)

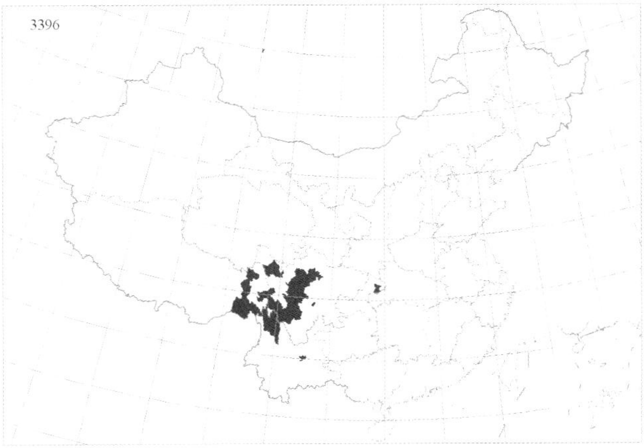

Rosaceae 蔷薇科 *Rosa* 蔷薇属
Rosa spinosissima L. 密刺蔷薇
Dwarf shrubs

3.7 (−4.3~9.3); −15.7 (−20.9~−11.6); 19.9 (9.6~27.5); 7.9 (2.5~12.1); 61.4 (9.7~107.6); 76.7 (51.6~121.1); 583 (329~813) ❙ 216 (69~450); 96 (24~211); 21 (4~51) ❙ −61.0 (−89.2~3.9); 215 (69~406); 99.7 (14.2~297.3)

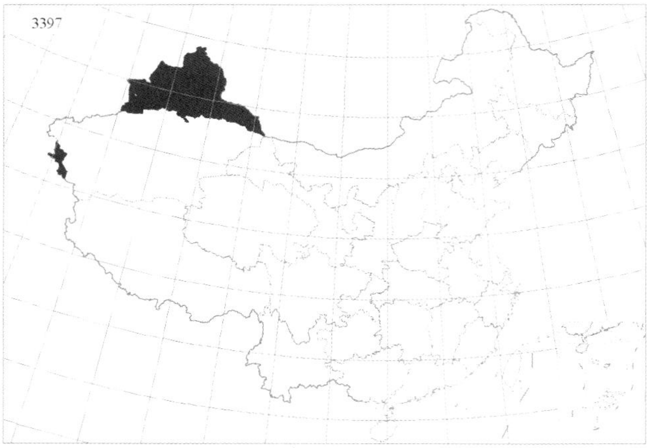

Rosaceae 蔷薇科 *Rosa* 蔷薇属
Rosa sweginzowii Koehne 扁刺蔷薇
Shrubs

9.3 (−1.4~15.5); −3.4 (−15.4~7.8); 20.9 (6.2~29.5); 10.2 (2.0~15.5); 78.7 (2.5~136.0); 26.0 (0.0~92.1); 648 (334~932) ❙ 677 (271~1454); 343 (165~792); 22 (4~83) ❙ 8.1 (−45.2~128.6); 572 (271~817); 203.8 (44.7~487.2)

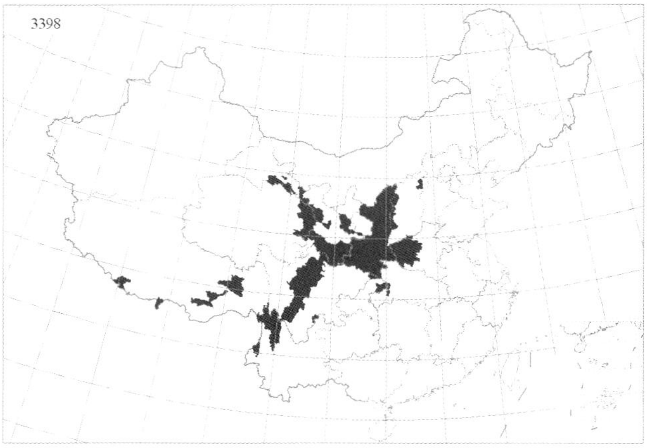

Rosaceae 蔷薇科 *Rosa* 蔷薇属
Rosa taiwanensis Nakai 小金樱子
Shrubs

17.6; 11.6; 22.6; 17.7; 151.8; 0.0; 845 ❙ 2335; 1002; 245 ❙ 176.7; 845; 290.2

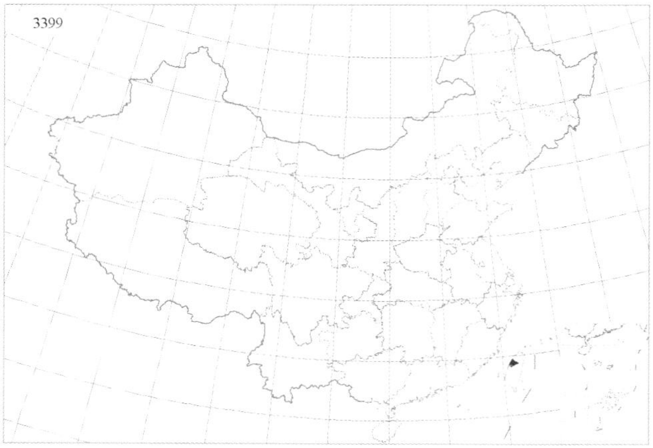

Rosaceae 蔷薇科 *Rosa* 蔷薇属
Rosa taronensis T. T. Yu 俅江蔷薇
Shrubs

8.7; 1.4; 14.9; 8.8; 58.2; 14.0; 570 ❙ 907; 519; 37 ❙ 60.0; 570; 322.8

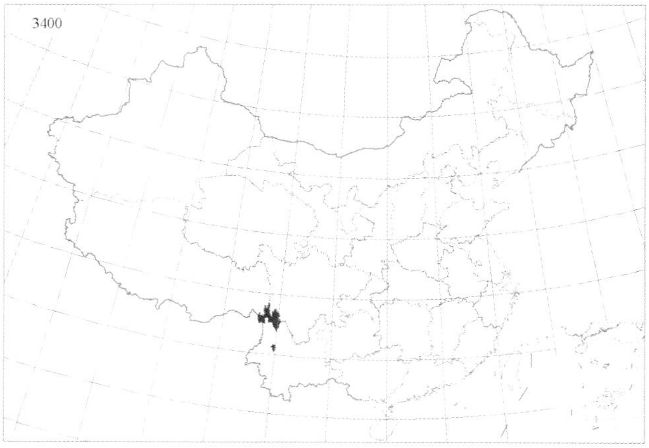

Rosaceae 蔷薇科 *Rosa* 蔷薇属
Rosa transmorrisonensis Hayata 高山蔷薇
Shrubs, evergreen

17.3 (16.2~18.5); 12.0 (11.0~14.1); 21.5 (20.1~23.4); 17.3 (16.2~18.5); 147.6 (134.2~162.1); 0.0 (0.0~0.0); 822 (773~874) ❙ 2512 (2288~2713); 1105 (947~1388); 245 (188~303) ❙ 206.7 (161.7~246.5); 822 (773~874); 307.3 (281.4~363.7)

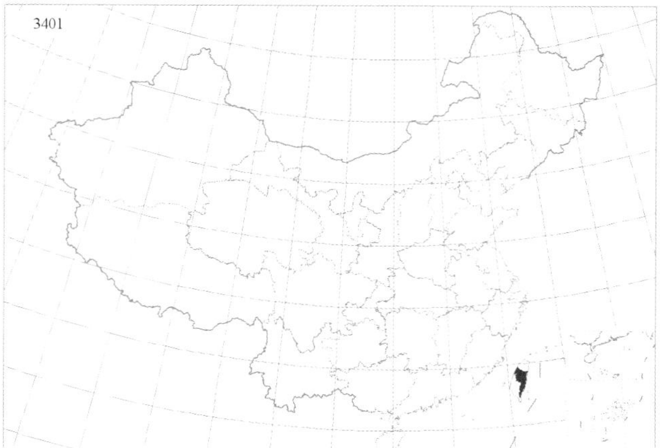

Rosaceae 蔷薇科 *Rosa* 蔷薇属
Rosa tsinglingensis Pax et K. Hoffm. 秦岭蔷薇
Small shrubs

8.4 (4.6~12.5); −4.4 (−8.3~−0.8); 20.5 (14.7~25.0); 9.3 (6.0~12.5); 70.5 (36.3~102.9); 28.5 (13.2~45.6); 609 (482~744) ❙ 627 (341~856); 305 (190~391); 17 (8~25) ❙ 3.1 (−40.9~52.6); 568 (341~697); 215.0 (61.2~323.3)

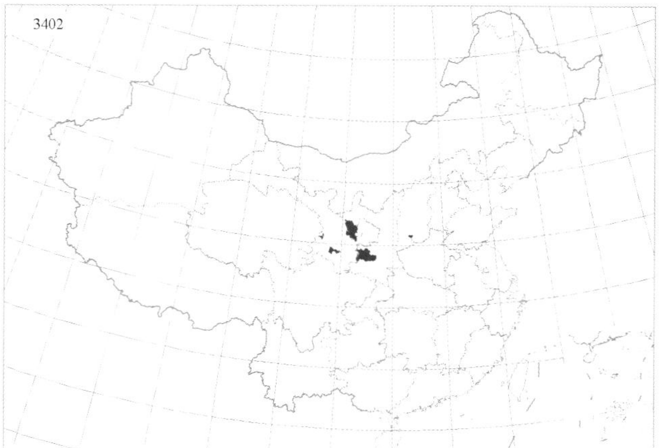

Rosaceae 蔷薇科 *Rosa* 蔷薇属
Rosa uniflorella Buzunova 单花合柱蔷薇
Shrubs

16.5; 5.3; 27.7; 16.5; 138.0; 0.0; 888 | 1363; 481; 183 | 53.5; 888; 222.6

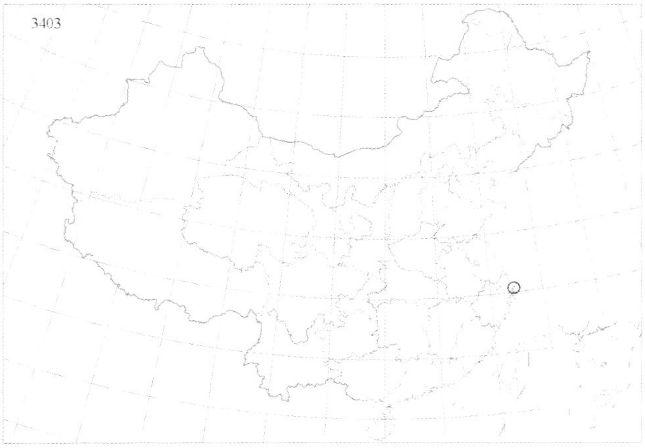

Rosaceae 蔷薇科 *Rosa* 蔷薇属
Rosa webbiana Wallich ex Royle 藏边蔷薇
Shrubs

4.6 (-2.3~12.8); -7.9 (-17.7~4.3); 15.4 (6.2~26.0); 7.2 (1.7~13.0); 49.0 (2.1~112.3); 52.9 (0.7~96.1); 524 (324~801) | 494 (44~1243); 256 (19~696); 38 (4~103) | 0.5 (-94.5~93.0); 316 (44~702); 133.7 (9.3~384.7)

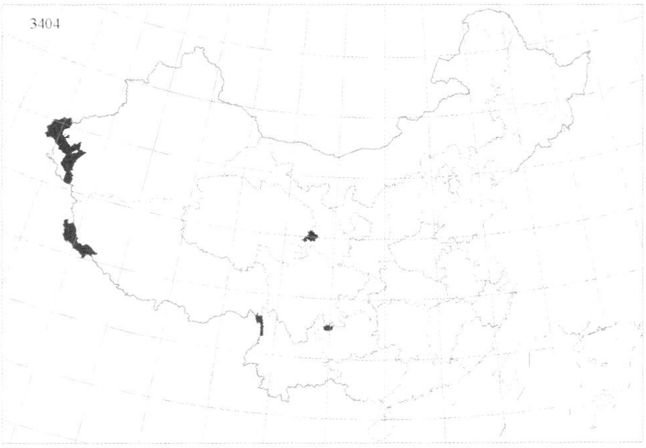

Rosaceae 蔷薇科 *Rosa* 蔷薇属
Rosa weisiensis T. T. Yu et T. C. Ku 维西蔷薇
Shrubs

9.6; 2.4; 15.7; 9.6; 60.8; 5.4; 600 | 1007; 578; 40 | 67.9; 600; 351.7

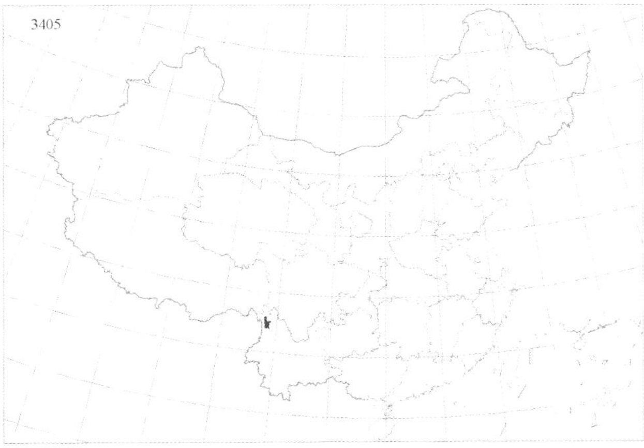

Rosaceae 蔷薇科 *Rosa* 蔷薇属
Rosa willmottiae Hemsl. 小叶蔷薇
Shrubs

3.7 (-4.8~18.0); -7.3 (-18.0~11.6); 13.5 (7.2~23.7); 5.7 (1.9~18.1); 34.9 (3.7~156.9); 50.2 (0.0~121.6); 474 (332~869) | 635 (271~1207); 345 (165~584); 15 (4~56) | 35.0 (-44.6~128.6); 459 (271~815); 247.5 (101.3~341.6)

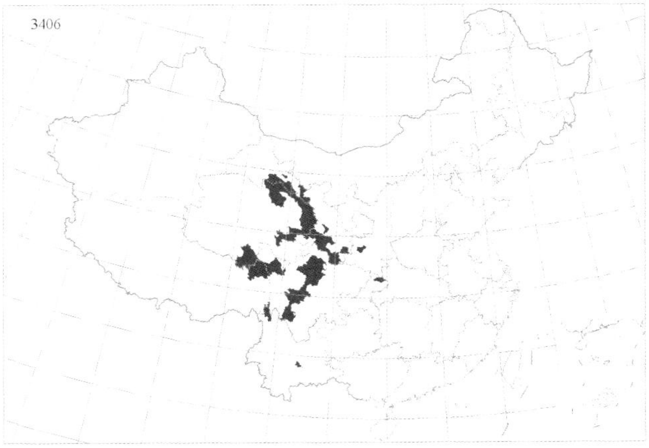

Rosaceae 蔷薇科 *Rosa* 蔷薇属
Rosa xanthina Lindley 黄刺玫
Shrubs

9.0 (4.9~14.5); -6.8 (-13.0~-0.2); 22.9 (17.3~28.2); 10.3 (7.6~14.6); 83.7 (53.6~125.3); 35.5 (10.4~63.6); 664 (544~864) | 498 (84~777); 285 (53~486); 14 (1~34) | -24.9 (-87.1~19.2); 493 (84~729); 148.5 (7.5~260.4)

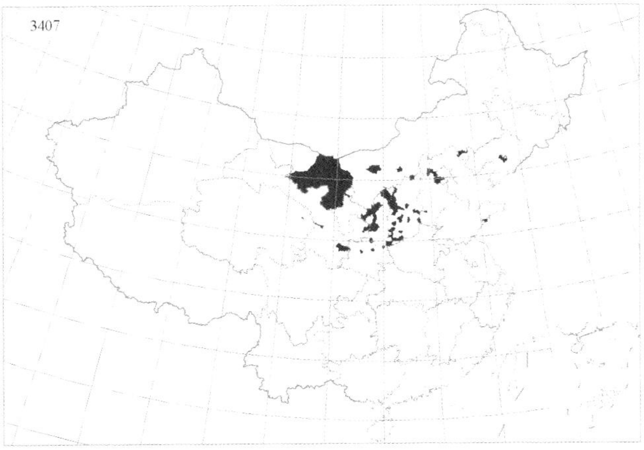

Rosaceae 蔷薇科 *Rosa* 蔷薇属
Rosa zhongdianensis T. C. Ku 中甸蔷薇
Shrubs

6.1; -1.5; 12.7; 6.3; 35.2; 21.8; 496 | 801; 496; 21 | 61.4; 496; 282.9

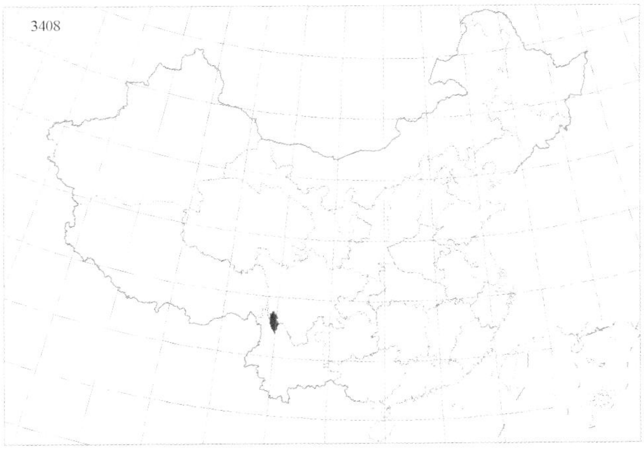

Rosaceae 蔷薇科 *Rubus* 悬钩子属
Rubus acuminatus Sm. 尖叶悬钩子
Shrubs, deciduous

16.4 (13.2~19.1); 8.3 (6.4~10.1); 22.6 (18.4~26.0); 16.5 (13.3~19.1); 137.7 (99.4~169.2); 0.0 (0.0~0.0); 826 (695~966) ❙ 1347 (1232~1561); 735 (666~897); 62 (51~70) ❙ 65.4 (30.0~99.9); 826 (695~966); 327.5 (189.5~539.7)

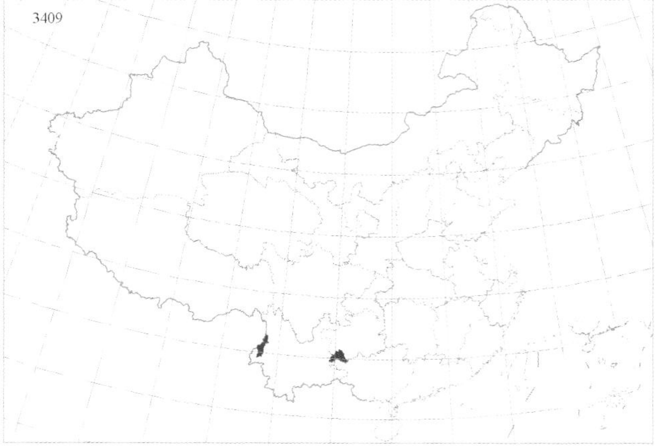

Rosaceae 蔷薇科 *Rubus* 悬钩子属
Rubus acuminatus var. puberulus T. T. Yu et L. T. Lu 柔毛尖叶悬钩子
Shrubs, deciduous

17.4 (15.9~19.1); 8.5 (6.9~10.1); 24.3 (23.0~26.0); 17.4 (16.0~19.1); 149.2 (131.6~169.2); 0.0 (0.0~0.0); 873 (802~966) ❙ 1284 (1232~1334); 693 (666~720); 59 (51~64) ❙ 47.8 (30.0~64.7); 873 (802~966); 237.1 (189.5~283.9)

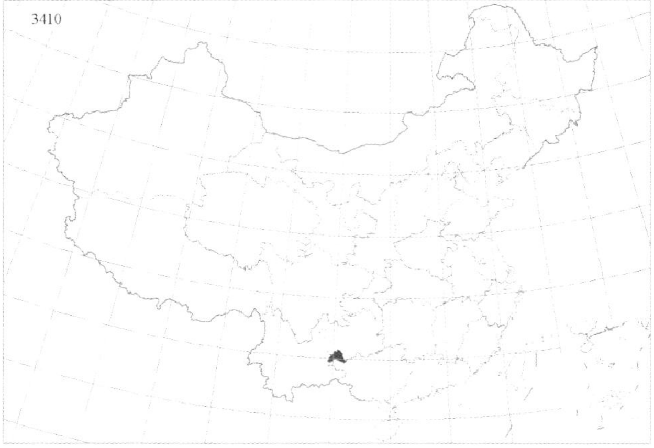

Rosaceae 蔷薇科 *Rubus* 悬钩子属
Rubus adenophorus Rolfe 腺毛莓
Shrubs, deciduous

16.2 (9.3~21.8); 4.4 (-2.9~13.1); 27.2 (20.1~29.9); 16.2 (9.5~21.8); 136.9 (69.8~201.6); 2.3 (0.0~20.5); 879 (603~1160) ❙ 1394 (626~1997); 505 (316~821); 156 (29~259) ❙ 58.1 (-22.2~130.7); 863 (603~1160); 207.8 (134.6~319.9)

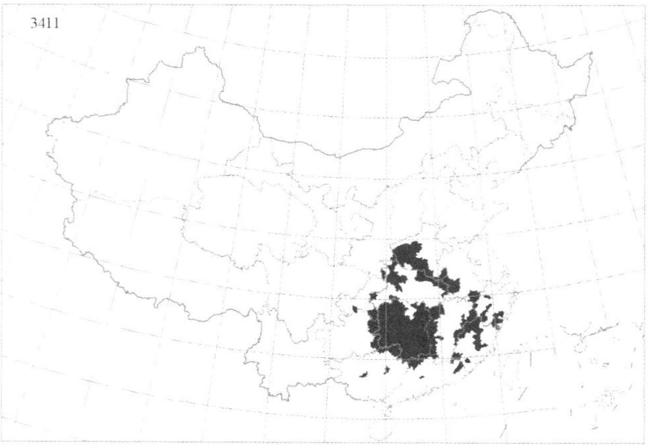

Rosaceae 蔷薇科 *Rubus* 悬钩子属
Rubus alceifolius Poir. 粗叶悬钩子
Scandent shrubs, deciduous

19.2 (14.3~24.8); 9.8 (3.3~20.1); 27.3 (20.1~29.8); 19.2 (14.3~24.8); 171.0 (112.6~237.6); 0.1 (0.0~2.3); 1012 (756~1385) ❙ 1569 (868~3495); 634 (312~1428); 153 (41~897) ❙ 57.2 (-34.3~246.5); 1001 (756~1360); 244.0 (95.3~659.0)

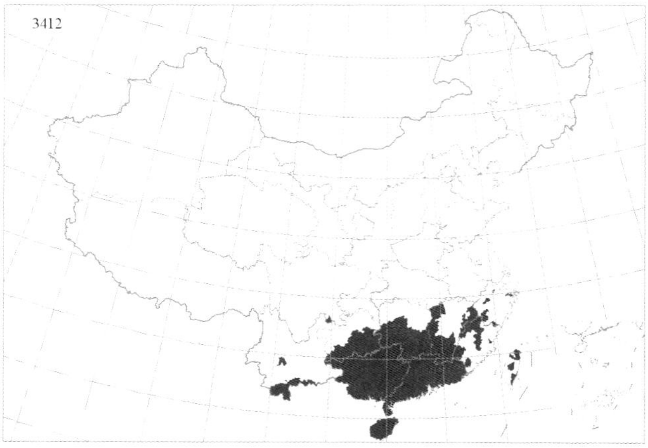

Rosaceae 蔷薇科 *Rubus* 悬钩子属
Rubus alexeterius Focke 刺萼悬钩子
Shrubs, deciduous

13.5 (-0.7~18.9); 6.1 (-9.6~12.0); 19.4 (7.7~26.6); 13.6 (2.5~19.0); 106.8 (6.8~167.8); 4.5 (0.0~74.2); 716 (338~929) ❙ 948 (337~1147); 550 (257~644); 34 (1~57) ❙ 33.5 (-23.5~100.7); 693 (337~912); 271.1 (61.6~382.6)

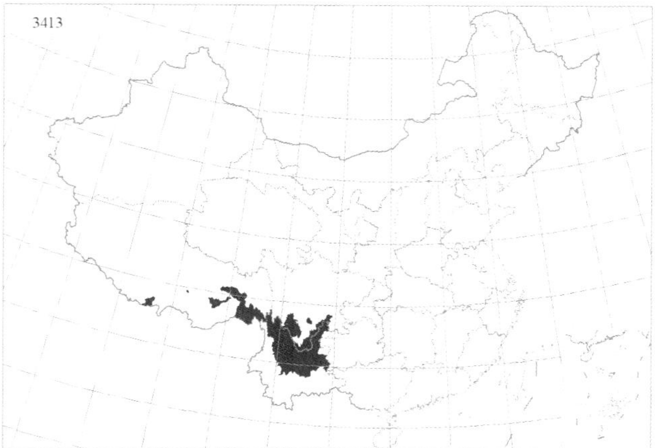

Rosaceae 蔷薇科 *Rubus* 悬钩子属
Rubus alexeterius var. acaenocalyx (H. Hara) T. T. Yu et L. T. Lu 腺毛刺萼悬钩子
Shrubs, deciduous

6.2 (1.8~11.0); -2.0 (-7.1~3.9); 13.3 (9.9~17.0); 7.0 (3.8~11.1); 42.0 (15.7~74.6); 26.9 (1.3~53.3); 510 (404~640) ❙ 702 (337~990); 442 (257~619); 12 (1~24) ❙ 33.5 (-23.5~67.7); 492 (337~638); 257.4 (169.5~319.8)

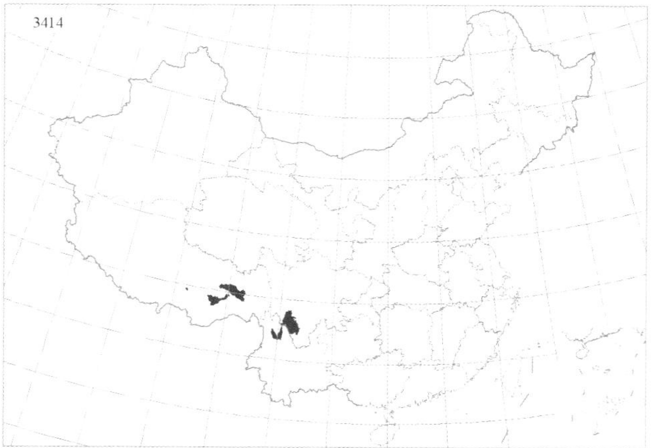

Rosaceae 蔷薇科 *Rubus* 悬钩子属
Rubus alnifoliolatus H. Lév. 桤叶悬钩子
Small shrubs, deciduous

19.3 (16.4~22.2); 14.0 (11.5~17.7); 23.5 (20.3~25.9); 19.3 (16.5~22.2); 171.6 (137.8~206.7); 0.0 (0.0~0.0); 938 (783~1123) ❙ 2750 (2445~2993); 1246 (924~1847); 276 (107~547) ❙ 196.5 (140.0~235.6); 938 (783~1123); 283.2 (207.7~363.7)

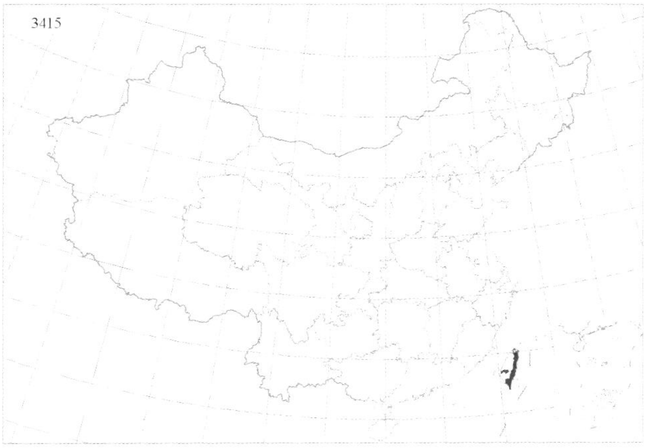

Rosaceae 蔷薇科 *Rubus* 悬钩子属
Rubus amabilis Focke 秀丽莓
Shrubs, deciduous

10.6 (−1.0~18.1); −1.8 (−12.0~6.9); 22.1 (8.5~29.3); 11.2 (2.7~18.1); 88.4 (8.3~157.7); 20.6 (0.0~79.5); 687 (363~965) ❙ 860 (341~1893); 399 (190~816); 51 (7~233) ❙ 26.7 (−40.9~129.9); 637 (341~944); 217.4 (61.2~357.7)

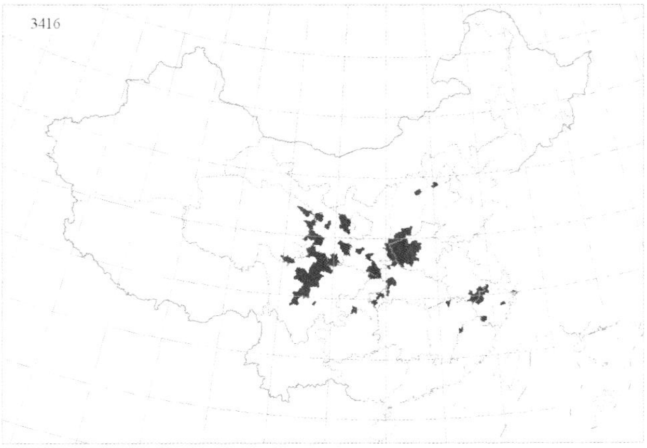

Rosaceae 蔷薇科 *Rubus* 悬钩子属
Rubus amabilis var. aculeatissimus T. T. Yu et L. T. Lu 刺萼秀丽莓
Shrubs, deciduous

12.0; 1.8; 21.5; 12.0; 91.6; 7.4; 693 ❙ 1009; 454; 45 ❙ 44.7; 693; 229.2

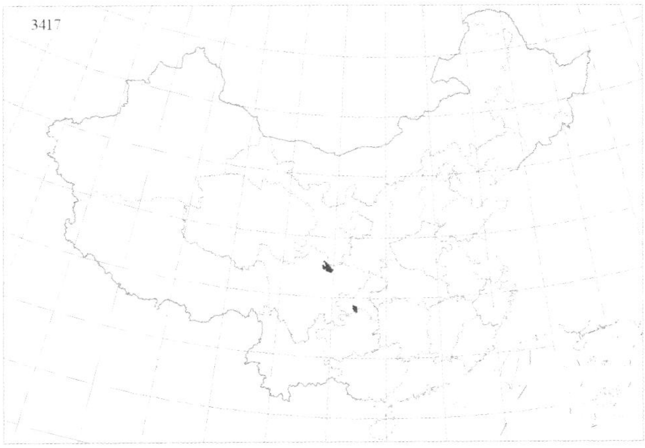

Rosaceae 蔷薇科 *Rubus* 悬钩子属
Rubus amabilis var. microcarpus T. T. Yu et L. T. Lu 小果秀丽莓
Shrubs, deciduous

2.5; −9.7; 13.1; 5.0; 27.9; 56.4; 449 ❙ 516; 283; 9 ❙ 14.7; 449; 218.4

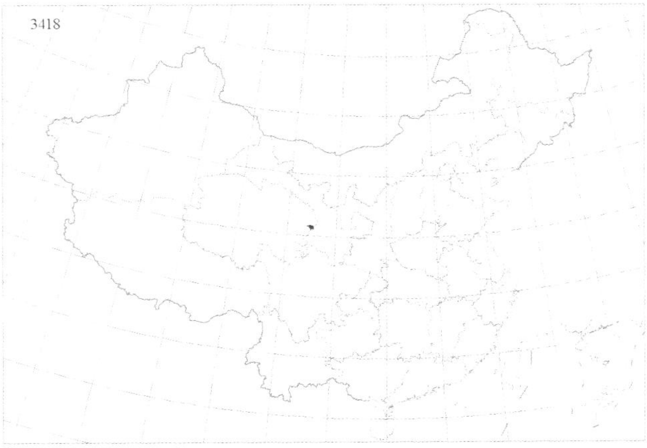

Rosaceae 蔷薇科 *Rubus* 悬钩子属
Rubus amphidasys Focke 周毛悬钩子
Small shrubs, deciduous

16.7 (11.1~21.8); 5.3 (1.6~13.1); 27.4 (19.4~30.1); 16.7 (11.3~21.8); 141.1 (80.8~202.0); 0.5 (0.0~6.9); 895 (652~1168) ❙ 1472 (931~1997); 524 (323~821); 172 (37~259) ❙ 64.8 (−1.6~147.0); 881 (652~1160); 211.2 (134.6~332.5)

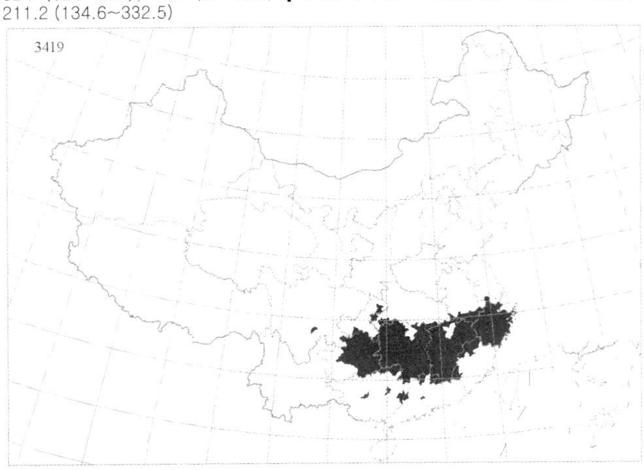

Rosaceae 蔷薇科 *Rubus* 悬钩子属
Rubus angustibracteatus T. T. Yu et L. T. Lu 狭苞悬钩子
Scandent shrubs, deciduous

6.4; −2.9; 14.7; 6.9; 42.0; 25.0; 516 ❙ 983; 509; 27 ❙ 90.5; 516; 245.3

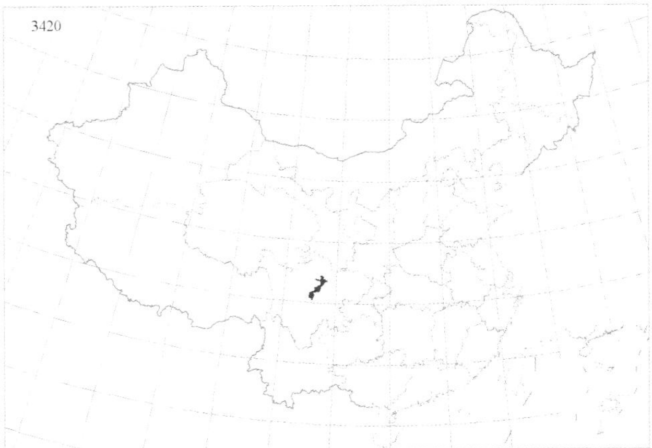

Rosaceae 蔷薇科 *Rubus* 悬钩子属
Rubus arachnoideus Y. C. Liu et F. Y. Lu 灰叶悬钩子
Subshrubs

16.4; 11.5; 20.3; 16.5; 137.8; 0.0; 783 ▎2499; 947; 303 ▎219.3; 783; 303.6

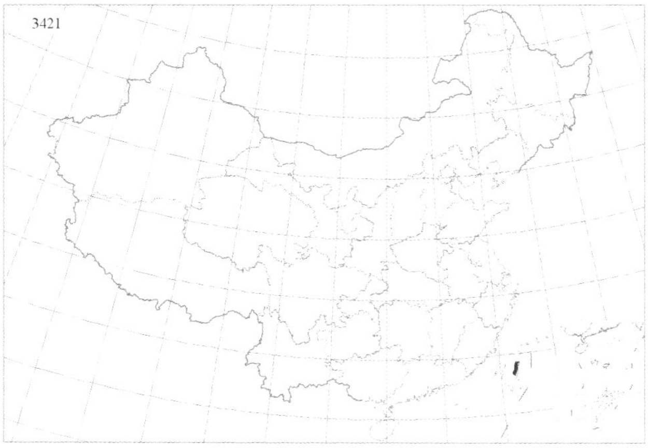

Rosaceae 蔷薇科 *Rubus* 悬钩子属
Rubus assamensis Focke 西南悬钩子
Shrubs, deciduous

15.2 (2.2~19.6); 6.3 (-6.6~13.0); 22.8 (9.9~27.8); 15.3 (3.8~19.6); 124.3 (15.5~175.7); 1.6 (0.0~49.1); 793 (413~965) ▎1195 (932~1736); 630 (475~1007); 54 (19~80) ▎52.5 (6.5~145.8); 788 (413~965); 271.4 (142.9~627.0)

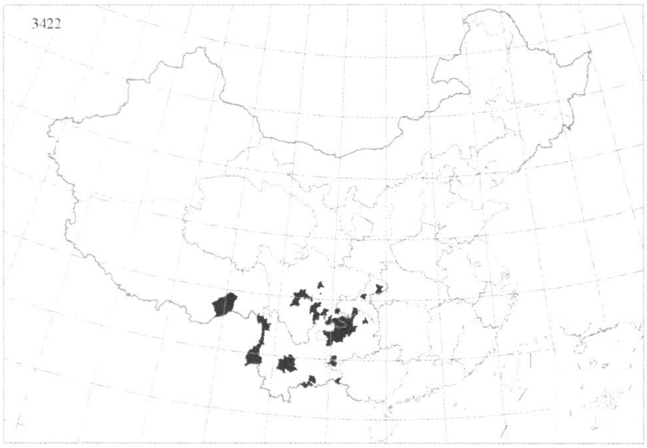

Rosaceae 蔷薇科 *Rubus* 悬钩子属
Rubus aurantiacus Focke 橘红悬钩子
Shrubs, deciduous

14.1 (2.2~18.3); 7.0 (-6.6~11.8); 19.6 (9.9~22.7); 14.3 (3.8~18.3); 113.4 (15.5~159.8); 3.2 (0.0~49.1); 734 (413~882) ▎1040 (911~1431); 597 (524~817); 38 (19~56) ▎45.2 (6.5~125.9); 714 (413~863); 292.0 (192.4~406.7)

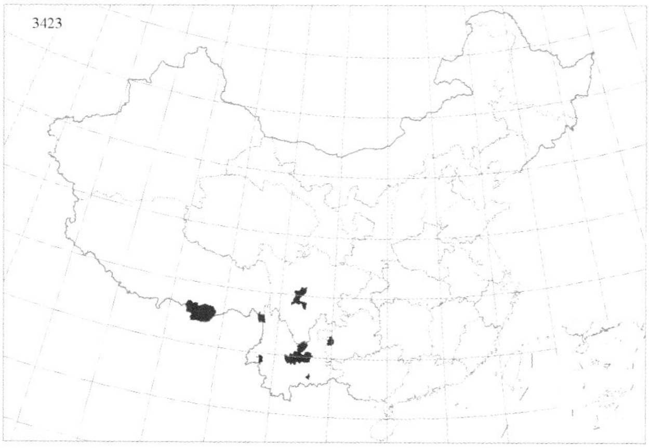

Rosaceae 蔷薇科 *Rubus* 悬钩子属
Rubus aurantiacus var. obtusifolius T. T. Yu et L. T. Lu 钝叶橘红悬钩子
Shrubs, deciduous

15.9 (14.3~18.1); 7.2 (4.6~11.7); 23.0 (21.4~24.4); 15.9 (14.3~18.0); 131.0 (112.3~156.5); 0.0 (0.0~0.4); 801 (751~863) ▎1205 (1107~1298); 617 (561~648); 58 (48~65) ▎50.7 (34.8~62.6); 801 (751~863); 245.8 (160.8~406.7)

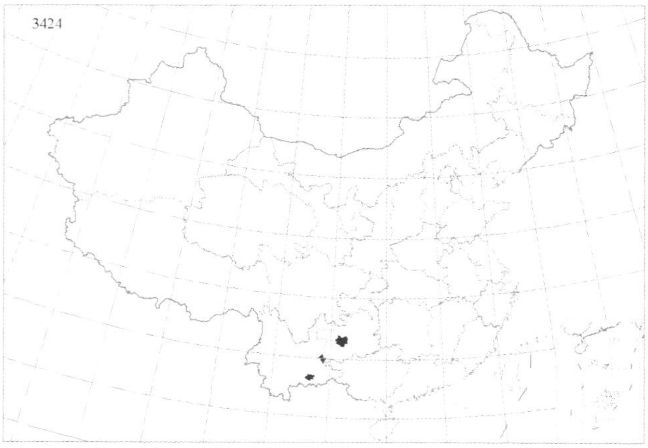

Rosaceae 蔷薇科 *Rubus* 悬钩子属
Rubus austrotibetanus T. T. Yu et L. T. Lu 藏南悬钩子
Shrubs, deciduous

6.7 (-1.8~17.3); -1.2 (-10.2~11.0); 13.5 (6.1~21.6); 7.9 (1.8~17.3); 53.0 (1.9~148.0); 32.1 (0.0~82.8); 531 (329~832) ▎854 (320~1736); 498 (193~1007); 32 (7~69) ▎47.7 (-4.0~145.8); 525 (320~783); 256.2 (61.6~447.4)

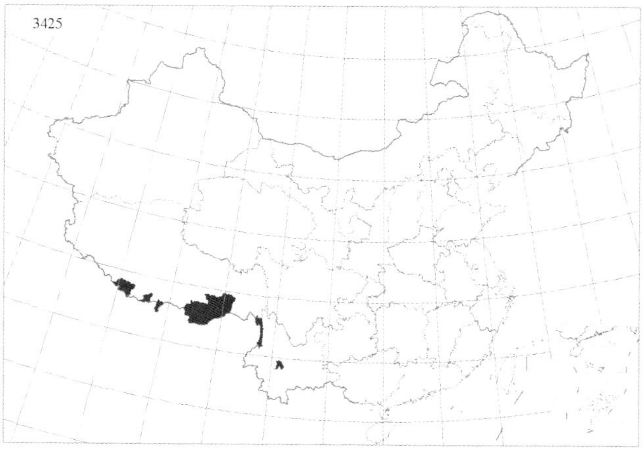

Rosaceae 蔷薇科 *Rubus* 悬钩子属
Rubus bambusarum Focke 竹叶鸡爪茶
Shrubs, evergreen

13.7 (9.3~18.0); 2.8 (-1.7~7.2); 24.1 (20.1~28.1); 13.7 (9.5~18.0); 109.0 (69.8~156.3); 4.4 (0.0~18.0); 755 (603~958) ▎1267 (1070~1492); 529 (453~816); 73 (29~105) ▎69.4 (11.7~107.2); 755 (603~958); 224.9 (128.7~291.4)

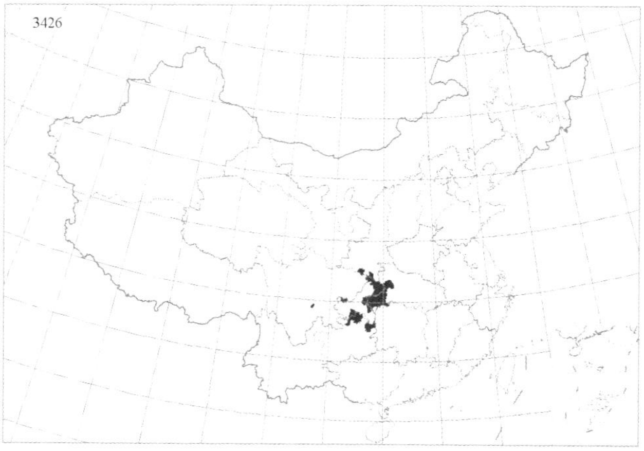

Rosaceae 蔷薇科 *Rubus* 悬钩子属
Rubus biflorus Buch.-Ham. ex Sm. 粉枝莓
Shrubs, deciduous

11.1 (−1.8~18.9); 2.0 (−10.2~12.0); 19.3 (6.1~26.8); 11.5 (1.8~19.0); 87.5 (1.9~167.8); 13.2 (0.0~82.8); 665 (329~929) ❙ 946 (300~1492); 492 (193~651); 39 (1~96) ❙ 44.2 (−37.6~128.6); 646 (300~855); 255.5 (72.2~382.6)

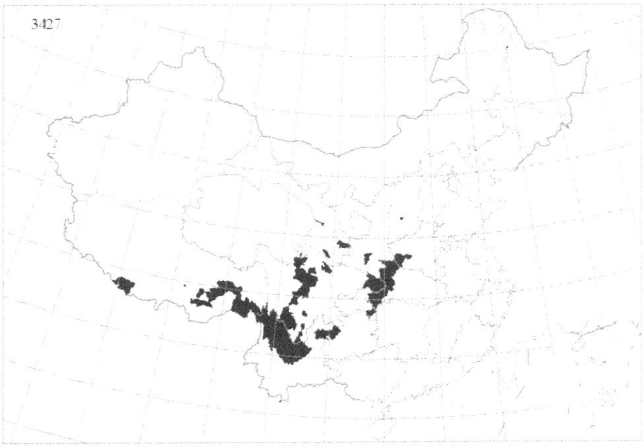

Rosaceae 蔷薇科 *Rubus* 悬钩子属
Rubus biflorus var. adenophorus Franch. 腺毛粉枝莓
Shrubs, deciduous

−1.1; −9.1; 6.2; 2.0; 2.5; 75.6; 334 ❙ 321; 193; 33 ❙ −4.0; 321; 79.3

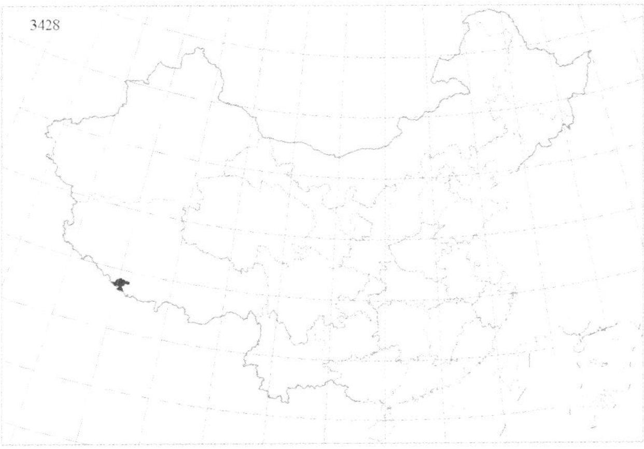

Rosaceae 蔷薇科 *Rubus* 悬钩子属
Rubus biflorus var. pubescens T. T. Yu et L. T. Lu 柔毛粉枝莓
Shrubs, deciduous

11.0; 3.7; 17.0; 11.1; 74.6; 1.8; 640 ❙ 990; 619; 15 ❙ 54.7; 638; 296.0

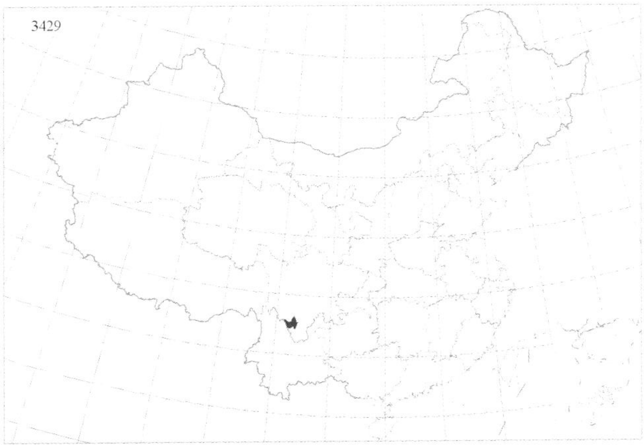

Rosaceae 蔷薇科 *Rubus* 悬钩子属
Rubus bonatianus Focke 滇北悬钩子
Shrubs, deciduous

9.1; 1.1; 16.1; 9.6; 65.5; 15.4; 594 ❙ 995; 575; 30 ❙ 73.7; 588; 278.1

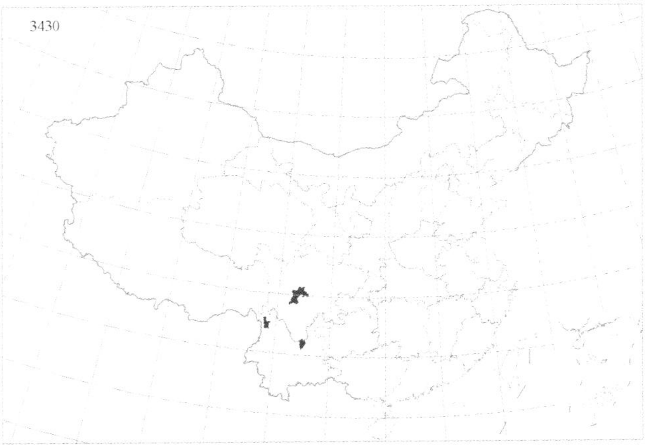

Rosaceae 蔷薇科 *Rubus* 悬钩子属
Rubus brevipetiolatus T. T. Yu et L. T. Lu 短柄悬钩子
Shrubs, deciduous

17.7; 10.9; 22.5; 17.8; 153.4; 0.0; 854 ❙ 1368; 756; 59 ❙ 60.1; 854; 491.5

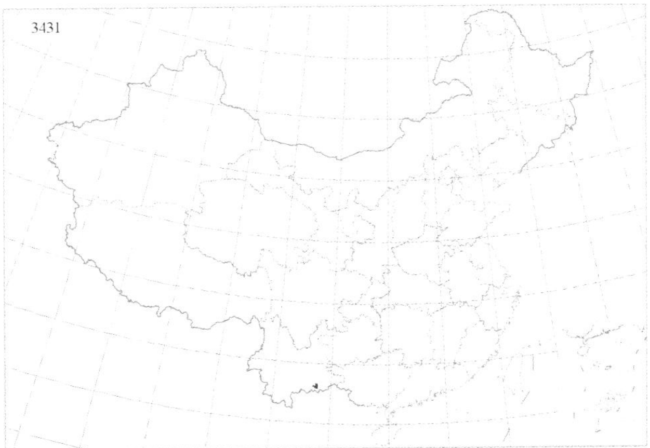

Rosaceae 蔷薇科 *Rubus* 悬钩子属
Rubus buergeri Miq. 寒莓
Small shrubs, deciduous

16.7 (11.1~22.2); 5.8 (1.6~15.3); 26.8 (18.8~30.1); 16.7 (11.1~22.3); 141.1 (79.8~207.5); 0.5 (0.0~6.9); 888 (648~1190) ❙ 1479 (871~3495); 551 (299~1388); 164 (26~897) ❙ 66.6 (−6.7~235.6); 878 (648~1190); 218.4 (126.1~363.7)

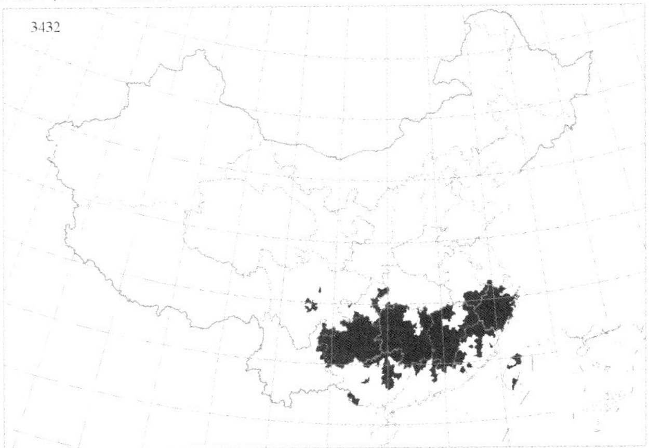

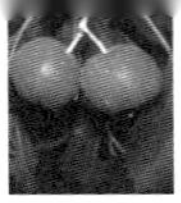

Rosaceae 蔷薇科 *Rubus* 悬钩子属
Rubus caesius L. 欧洲木莓
Shrubs, deciduous

2.5 (-1.3~9.0); -16.1 (-20.9~-9.3); 17.9 (13.2~23.2); 7.0 (4.5~10.7); 50.7 (26.7~88.9); 79.7 (40.7~103.9); 539 (433~703) ❙ 285 (144~450); 124 (63~211); 29 (12~51) ❙ -45.3 (-76.8~3.9); 283 (144~406); 168.1 (41.1~297.3)

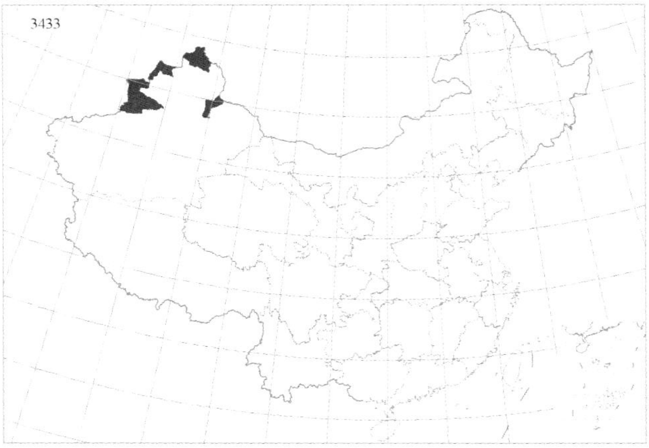

Rosaceae 蔷薇科 *Rubus* 悬钩子属
Rubus calycacanthus H. Lév. 猬莓
Shrubs, deciduous

16.2 (11.7~19.2); 7.5 (2.7~10.7); 23.1 (18.8~26.6); 16.2 (11.7~19.2); 135.3 (82.6~170.8); 0.4 (0.0~3.8); 824 (658~981) ❙ 1197 (879~1574); 641 (472~860); 56 (31~177) ❙ 45.9 (21.9~93.2); 824 (658~981); 267.7 (185.6~413.5)

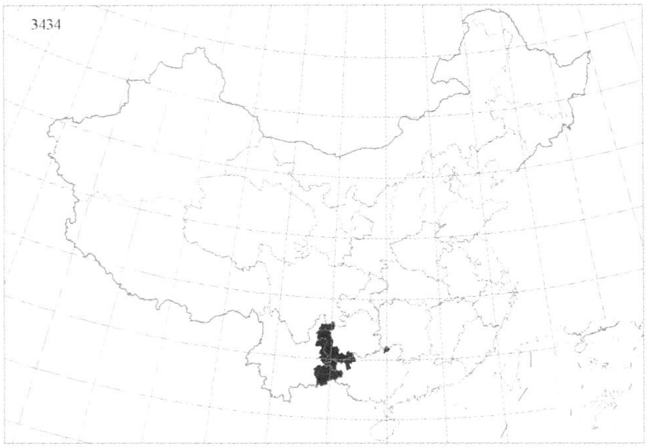

Rosaceae 蔷薇科 *Rubus* 悬钩子属
Rubus caudifolius Wuzhi 尾叶悬钩子
Shrubs, deciduous

16.0 (13.5~18.2); 5.2 (2.6~10.9); 26.0 (22.5~29.1); 16.1 (13.6~18.2); 133.5 (106.7~158.8); 0.7 (0.0~4.1); 849 (744~984) ❙ 1588 (1211~1997); 573 (465~756); 173 (59~259) ❙ 86.5 (45.6~130.7); 847 (744~927); 248.4 (170.4~491.5)

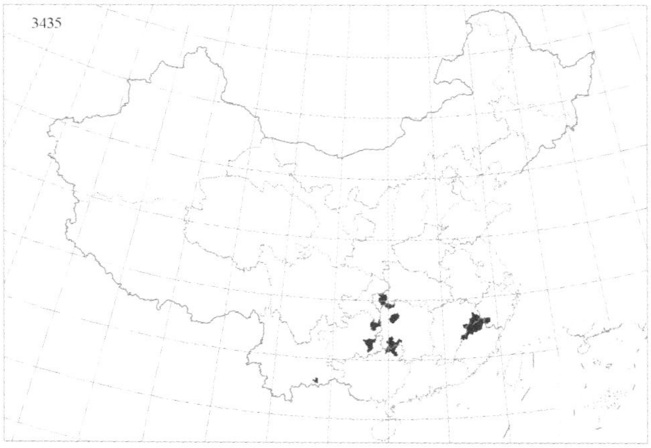

Rosaceae 蔷薇科 *Rubus* 悬钩子属
Rubus chiliadenus Focke 长序莓
Shrubs, deciduous

15.1 (12.0~17.4); 4.6 (0.6~7.9); 24.8 (23.1~26.3); 15.1 (12.1~17.4); 122.8 (94.7~148.7); 1.5 (0.0~9.7); 799 (697~878) ❙ 1209 (1089~1298); 537 (473~661); 71 (58~94) ❙ 51.6 (30.9~64.6); 799 (697~878); 203.0 (170.4~262.4)

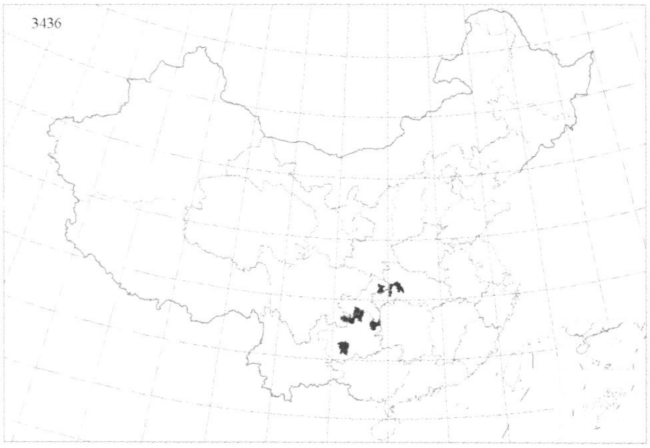

Rosaceae 蔷薇科 *Rubus* 悬钩子属
Rubus chingii Hu 掌叶覆盆子
Shrubs, deciduous

16.6 (13.4~22.1); 5.4 (1.5~13.7); 27.5 (22.1~29.7); 16.7 (13.4~22.1); 141.0 (108.3~205.6); 0.7 (0.0~7.3); 894 (755~1182) ❙ 1565 (1021~1997); 563 (370~779); 193 (55~259) ❙ 75.9 (17.1~147.0); 890 (755~1182); 233.4 (131.2~432.9)

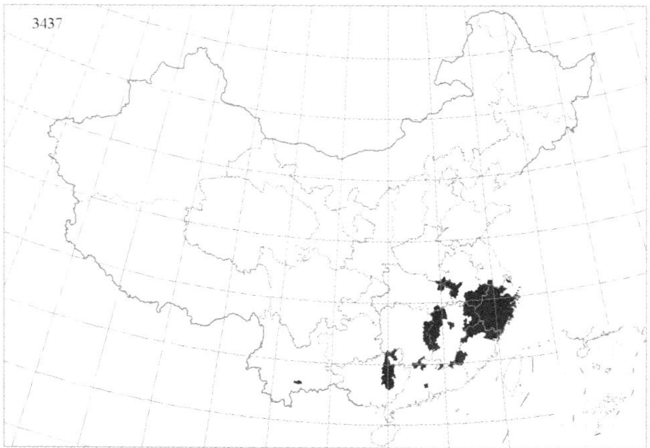

Rosaceae 蔷薇科 *Rubus* 悬钩子属
Rubus chingii var. suavissimus (S. Lee) L. T. Lu 甜茶
Shrubs, deciduous

19.8 (17.9~22.1); 10.3 (7.7~13.7); 27.6 (26.3~28.8); 19.8 (17.9~22.1); 177.5 (155.4~205.6); 0.0 (0.0~0.0); 1035 (918~1182) ❙ 1611 (1512~1784); 624 (507~779); 168 (144~180) ❙ 56.4 (40.2~67.8); 1035 (918~1182); 226.4 (157.7~263.5)

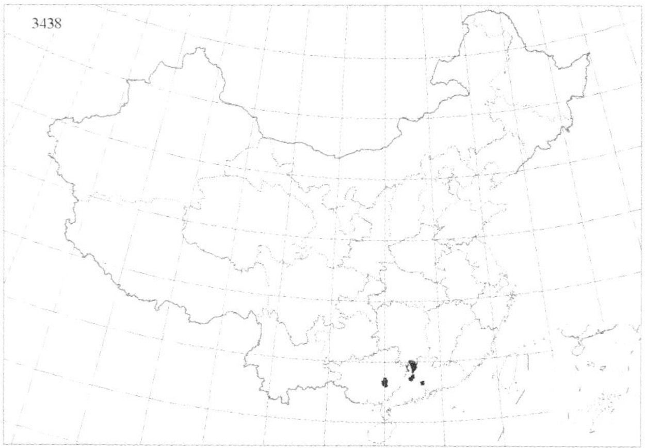

Rosaceae 蔷薇科 *Rubus* 悬钩子属

Rubus chroosepalus Focke 毛萼莓

Shrubs, semievergreen

14.5 (3.8~21.8); 3.8 (-4.3~13.7); 24.5 (10.8~29.1); 14.6 (4.7~21.9); 118.5 (21.6~202.7); 4.0 (0.0~35.4); 792 (444~1159) ❙ 1196 (760~1726); 510 (352~850); 82 (16~228) ❙ 51.2 (5.0~121.7); 792 (444~1159); 232.6 (134.1~334.4)

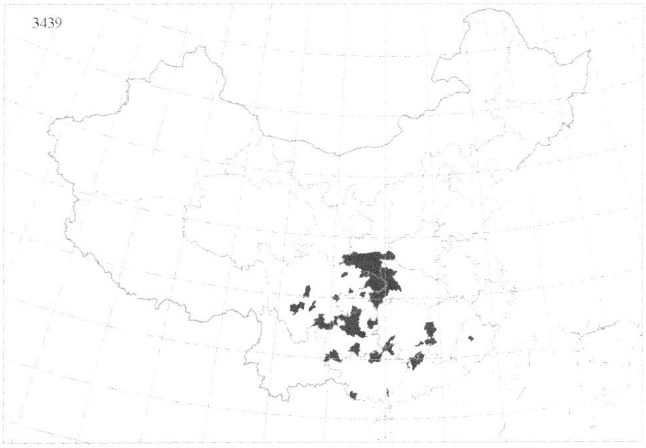

Rosaceae 蔷薇科 *Rubus* 悬钩子属

Rubus chrysobotrys Hand.-Mazz. 黄穗悬钩子

Shrubs, deciduous

15.7 (8.6~19.6); 8.9 (1.3~12.8); 20.3 (15.0~23.5); 15.7 (8.7~19.6); 129.1 (52.6~175.7); 0.5 (0.0~8.8); 782 (574~953) ❙ 1348 (1041~1688); 761 (553~959); 56 (42~70) ❙ 72.9 (34.7~112.2); 780 (574~932); 438.4 (335.1~627.0)

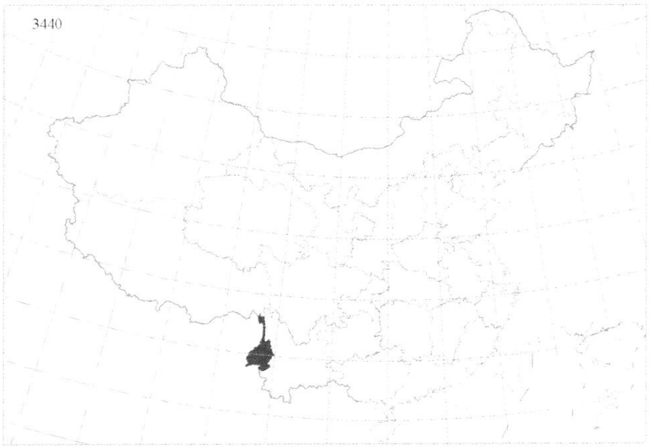

Rosaceae 蔷薇科 *Rubus* 悬钩子属

Rubus cinclidodictyus Cardot 网纹悬钩子

Shrubs, deciduous

15.4 (8.6~19.6); 7.9 (1.3~12.8); 20.8 (15.0~23.6); 15.4 (8.7~19.6); 125.6 (52.6~175.7); 0.9 (0.0~8.8); 780 (574~953) ❙ 1310 (885~1688); 740 (479~959); 51 (34~70) ❙ 68.4 (23.3~112.2); 774 (574~932); 415.9 (219.7~627.0)

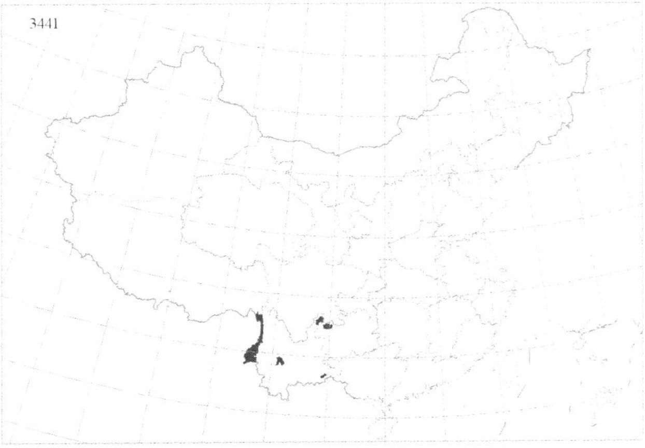

Rosaceae 蔷薇科 *Rubus* 悬钩子属

Rubus cochinchinensis Tratt. 蛇泡筋

Shrubs, deciduous

21.0 (7.3~24.8); 13.2 (-2.0~20.1); 27.0 (15.6~29.0); 21.0 (7.5~24.8); 192.7 (48.2~237.6); 0.3 (0.0~20.4); 1113 (537~1385) ❙ 1543 (868~2435); 686 (312~1428); 113 (27~194) ❙ 41.8 (-34.3~99.5); 1076 (537~1360); 329.3 (185.5~620.0)

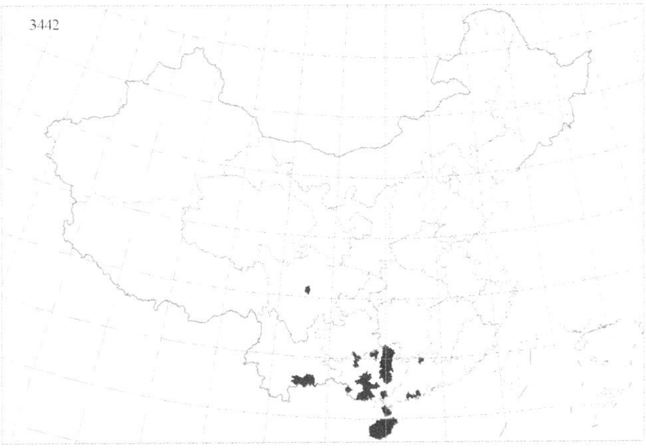

Rosaceae 蔷薇科 *Rubus* 悬钩子属

Rubus cockburnianus Hemsl. 华中悬钩子

Shrubs, deciduous

13.5 (2.9~19.6); 5.2 (-6.3~13.9); 20.5 (11.2~27.8); 13.6 (4.4~19.6); 107.5 (20.5~175.1); 4.6 (0.0~44.6); 726 (428~933) ❙ 972 (527~1500); 538 (292~816); 37 (9~103) ❙ 35.4 (-15.8~121.7); 707 (428~912); 274.8 (142.9~530.1)

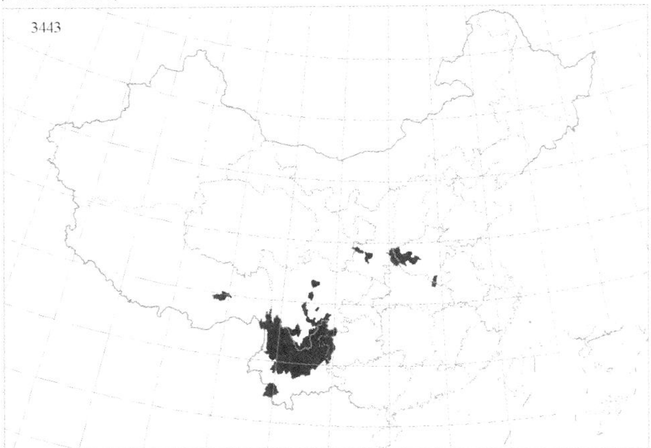

Rosaceae 蔷薇科 *Rubus* 悬钩子属

Rubus columellaris Tutcher 小柱悬钩子

Shrubs, deciduous

17.7 (12.9~22.2); 7.4 (1.9~13.9); 27.0 (21.4~30.1); 17.8 (12.9~22.3); 153.8 (101.2~207.5); 0.3 (0.0~5.9); 933 (723~1190) ❙ 1538 (849~1997); 593 (388~1056); 164 (43~259) ❙ 65.4 (-5.3~130.7); 928 (723~1190); 246.7 (133.5~551.3)

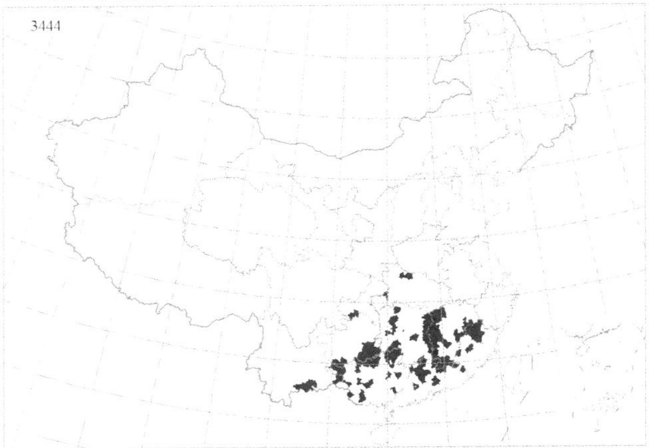

Rosaceae 蔷薇科 *Rubus* 悬钩子属

Rubus columellaris var. villosus T. T. Yu et L. T. Lu 柔毛小柱悬钩子

Shrubs, deciduous

19.4; 10.1; 27.0; 19.4; 172.5; 0.0; 1002 ❙ 1766; 687; 178 ❙ 75.6; 1002; 262.1

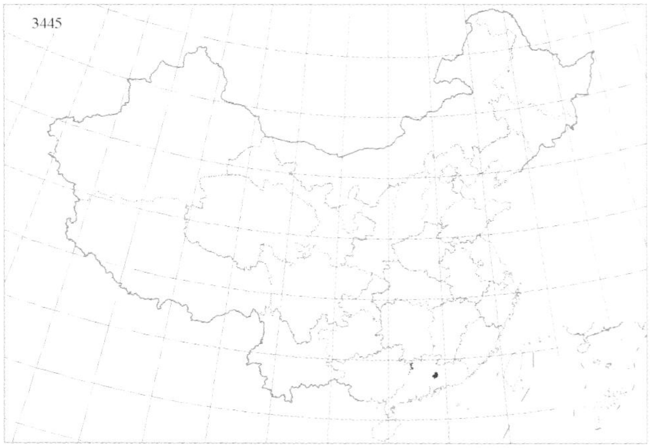

Rosaceae 蔷薇科 *Rubus* 悬钩子属

Rubus corchorifolius L. f. 山莓

Shrubs, deciduous

16.2 (3.3~23.0); 5.2 (-12.7~15.3); 26.2 (15.0~30.1); 16.3 (6.6~23.0); 138.5 (45.4~216.2); 4.0 (0.0~65.0); 878 (507~1247) ❙ 1269 (367~3495); 534 (235~1428); 120 (11~897) ❙ 43.4 (-37.7~246.5); 851 (367~1247); 222.3 (70.3~659.0)

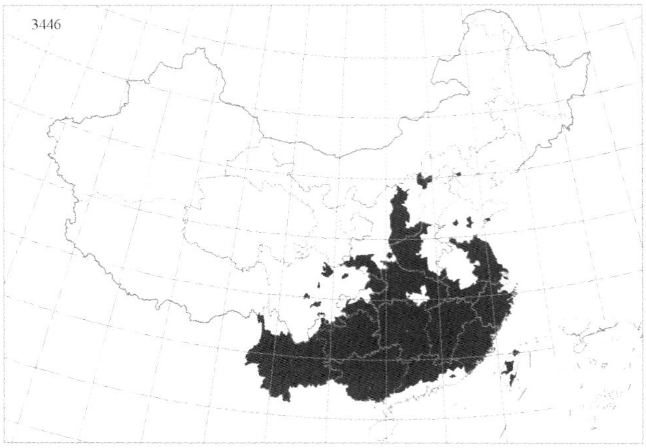

Rosaceae 蔷薇科 *Rubus* 悬钩子属

Rubus coreanus Miq. 插田泡

Shrubs, deciduous

15.7 (3.2~21.4); 4.4 (-7.4~13.6); 26.3 (11.4~29.9); 15.7 (4.7~21.5); 131.5 (22.7~197.7); 3.1 (0.0~43.8); 854 (436~1128) ❙ 1285 (341~1997); 503 (190~821); 128 (8~259) ❙ 49.9 (-40.9~147.0); 841 (341~1114); 213.7 (61.2~411.1)

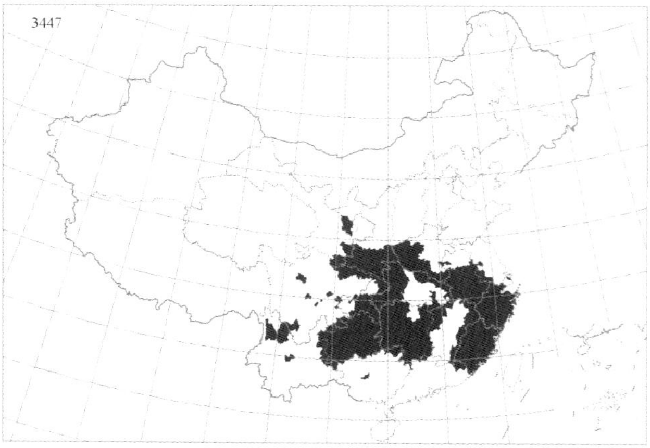

Rosaceae 蔷薇科 *Rubus* 悬钩子属

Rubus coreanus var. tomentosus Cardot 毛叶插田泡

Shrubs, deciduous

13.7 (3.2~19.1); 1.7 (-6.4~9.8); 25.1 (11.4~28.9); 13.8 (4.7~19.2); 112.6 (22.7~170.1); 7.2 (0.0~43.8); 784 (436~981) ❙ 1078 (626~1793); 474 (303~817); 79 (14~238) ❙ 37.5 (-22.2~112.2); 769 (436~981); 220.3 (133.5~357.7)

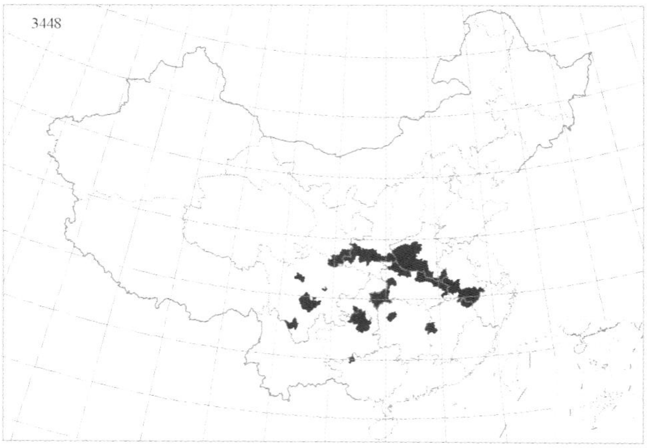

Rosaceae 蔷薇科 *Rubus* 悬钩子属

Rubus crassifolius T. T. Yu et L. T. Lu 厚叶悬钩子

Small shrubs, deciduous

14.1 (11.1~17.7); 0.9 (-3.4~7.7); 26.0 (23.0~28.5); 14.2 (11.6~17.8); 119.6 (95.6~153.6); 10.2 (0.0~22.0); 811 (709~927) ❙ 1055 (636~1630); 478 (392~594); 99 (23~228) ❙ 27.7 (-23.9~101.8); 776 (636~927); 186.2 (136.4~258.0)

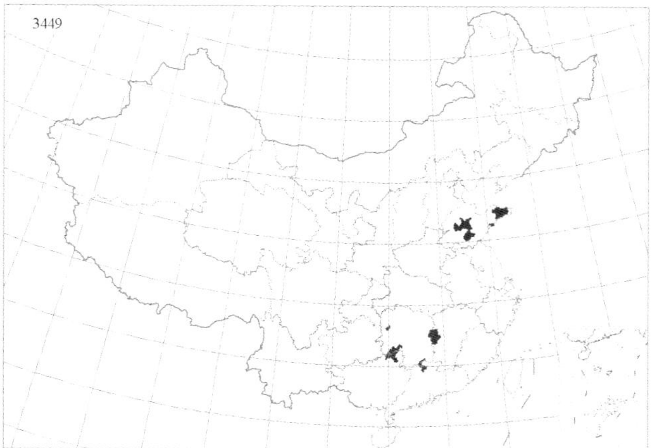

Rosaceae 蔷薇科 *Rubus* 悬钩子属

Rubus crataegifolius Bunge 牛叠肚

Shrubs, deciduous

8.0 (0.1~15.5); -9.1 (-21.2~2.1); 23.0 (17.5~28.5); 10.1 (6.4~15.6); 82.2 (45.4~131.3); 45.3 (4.3~106.1); 662 (501~889) ❙ 606 (397~1077); 382 (254~691); 20 (8~97) ❙ -6.8 (-47.7~74.1); 566 (397~881); 208.9 (110.1~346.5)

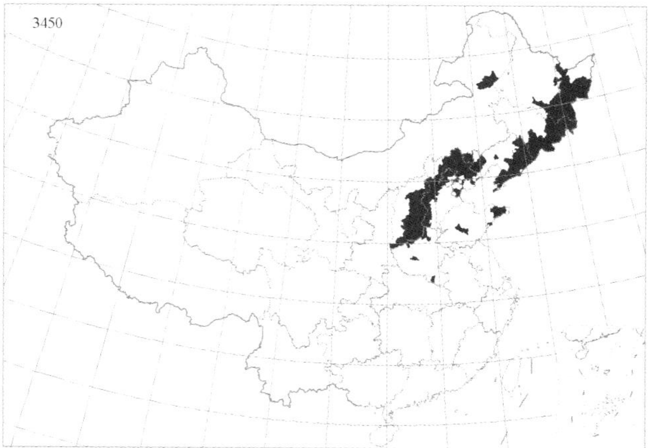

Rosaceae 蔷薇科 *Rubus* 悬钩子属
Rubus croceacanthus H. Lév. 薄瓣悬钩子
Shrubs, deciduous

20.2 (16.2~24.1); 14.5 (11.0~18.9); 25.0 (20.1~28.2); 20.2 (16.2~24.1); 182.9 (134.2~229.7); 0.0 (0.0~0.0); 1019 (773~1322) ❙ 2402 (1498~3495); 1106 (750~1847); 255 (47~897) ❙ 146.2 (25.2~246.5); 983 (773~1144); 245.0 (119.4~363.7)

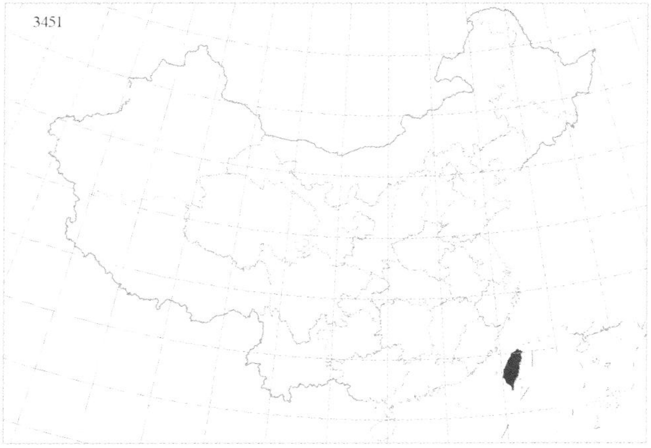

Rosaceae 蔷薇科 *Rubus* 悬钩子属
Rubus delavayi Franch. 三叶悬钩子
Small shrubs, deciduous

15.0 (5.4~18.9); 7.7 (-2.1~12.5); 20.5 (12.3~26.6); 15.0 (5.8~19.0); 120.9 (31.5~167.8); 0.8 (0.0~25.6); 761 (481~939) ❙ 1030 (741~1504); 586 (411~860); 41 (15~69) ❙ 36.5 (-14.2~93.0); 741 (481~939); 295.6 (142.9~413.5)

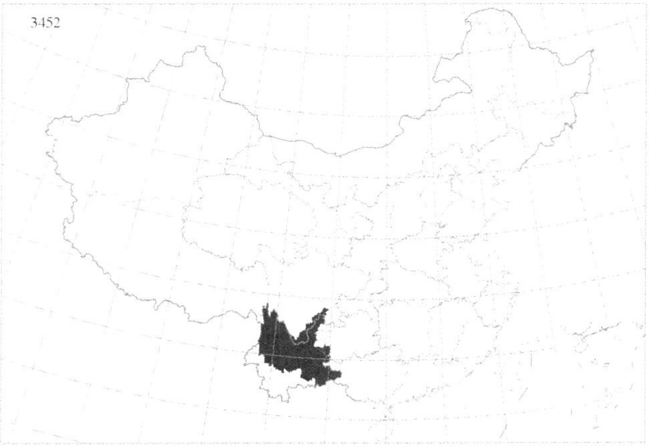

Rosaceae 蔷薇科 *Rubus* 悬钩子属
Rubus dolichophyllus Hand.-Mazz. 长叶悬钩子
Shrubs, deciduous

15.6 (11.7~19.8); 6.6 (2.7~11.1); 22.9 (18.8~26.4); 15.6 (11.7~19.8); 127.8 (82.6~177.9); 0.6 (0.0~3.8); 807 (658~1010) ❙ 1170 (916~1496); 622 (489~720); 56 (31~142) ❙ 45.9 (16.9~70.6); 807 (658~1010); 234.9 (183.0~354.5)

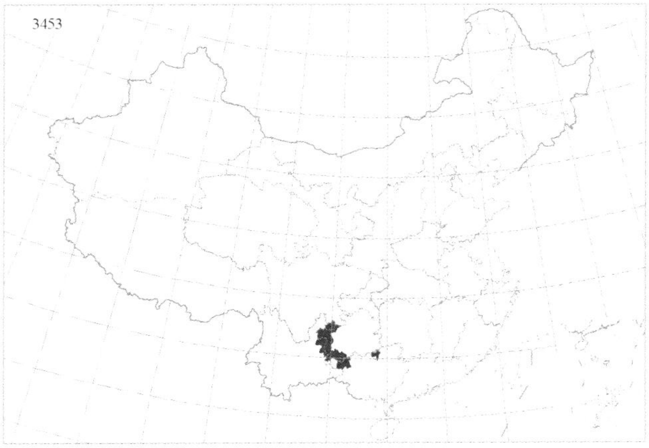

Rosaceae 蔷薇科 *Rubus* 悬钩子属
Rubus dolichophyllus var. pubescens T. T. Yu et L. T. Lu 毛梗长叶悬钩子
Shrubs, deciduous

14.2; 5.6; 20.8; 14.2; 110.9; 0.0; 734 ❙ 1252; 686; 53 ❙ 70.6; 734; 215.5

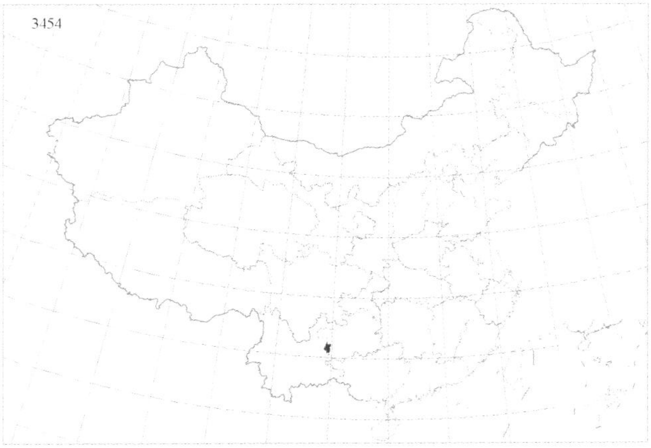

Rosaceae 蔷薇科 *Rubus* 悬钩子属
Rubus doyonensis Hand.-Mazz. 白藨
Scandent shrubs, deciduous

10.0 (2.9~19.2); 2.2 (-6.3~13.3); 16.5 (11.2~23.5); 10.4 (4.4~19.4); 76.9 (20.5~172.6); 16.7 (0.0~44.6); 627 (428~932) ❙ 1119 (853~1428); 610 (425~835); 42 (22~69) ❙ 84.2 (36.8~121.7); 626 (428~923); 347.1 (200.8~506.2)

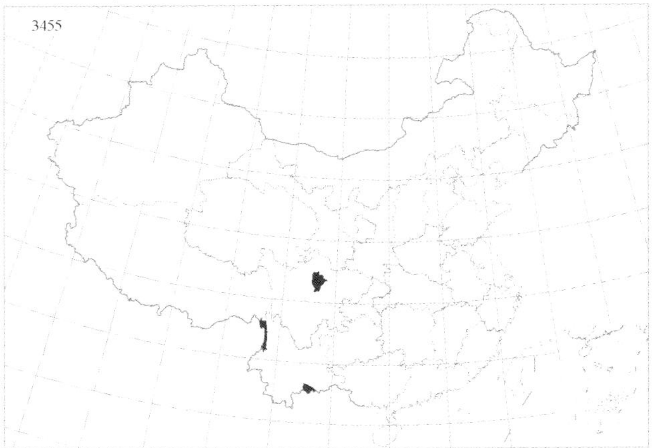

Rosaceae 蔷薇科 *Rubus* 悬钩子属
Rubus dunnii F. P. Metcalf 闽粤悬钩子
Scandent shrubs, deciduous

18.4 (14.6~20.9); 9.2 (3.8~12.7); 26.7 (24.6~28.6); 18.4 (14.7~21.0); 161.4 (117.7~191.5); 0.2 (0.0~1.5); 951 (779~1094) ❙ 1604 (1375~1742); 581 (476~682); 175 (133~199) ❙ 70.7 (31.5~101.8); 951 (779~1094); 261.3 (228.3~305.7)

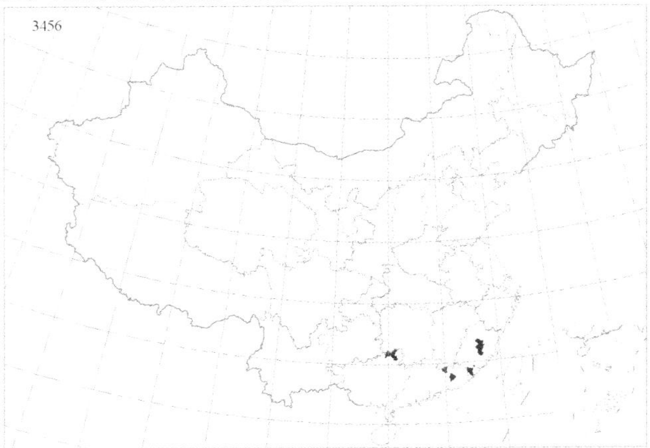

Rosaceae 蔷薇科 *Rubus* 悬钩子属

Rubus dunnii var. glabrescens T. T. Yu et L. T. Lu 光叶闽粤悬钩子

Scandent shrubs, deciduous

20.9; 12.7; 28.6; 21.0; 191.5; 0.0; 1094 ❙ 1515; 521; 149 ❙ 38.5; 1094; 228.3

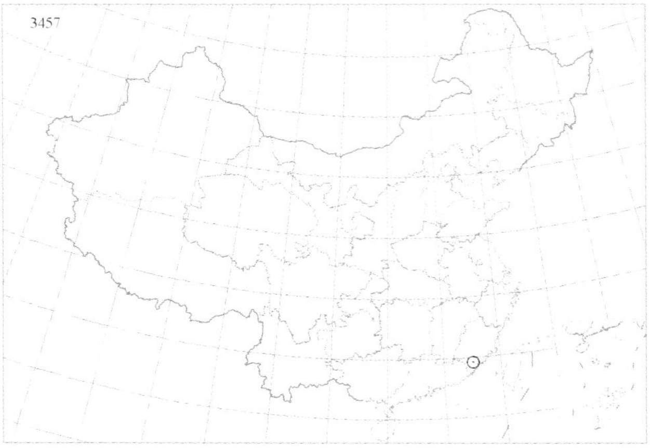

Rosaceae 蔷薇科 *Rubus* 悬钩子属

Rubus ellipticus Sm. 椭圆悬钩子

Shrubs, deciduous

15.6 (-1.8~21.6); 7.8 (-10.2~15.1); 21.7 (6.1~28.5); 15.7 (1.8~21.7); 129.2 (1.9~199.8); 1.8 (0.0~82.8); 798 (329~1153) ❙ 1137 (320~1751); 622 (217~1007); 49 (9~173) ❙ 43.2 (-14.2~145.8); 785 (320~1153); 304.7 (72.2~659.0)

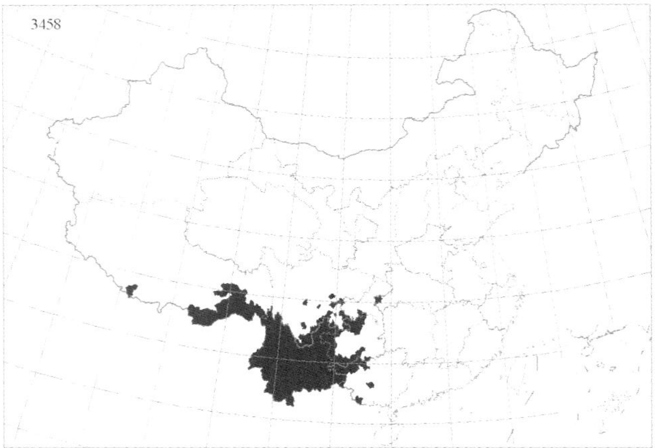

Rosaceae 蔷薇科 *Rubus* 悬钩子属

Rubus ellipticus var. obcordatus Focke 栽秧泡

Shrubs, deciduous

15.2 (1.8~21.0); 6.6 (-7.1~14.2); 22.4 (9.9~26.6); 15.3 (3.8~21.0); 125.4 (15.7~191.8); 1.9 (0.0~53.3); 790 (404~1053) ❙ 1144 (527~1736); 611 (317~1007); 51 (9~97) ❙ 45.4 (12.0~145.8); 787 (404~1053); 264.1 (143.4~551.6)

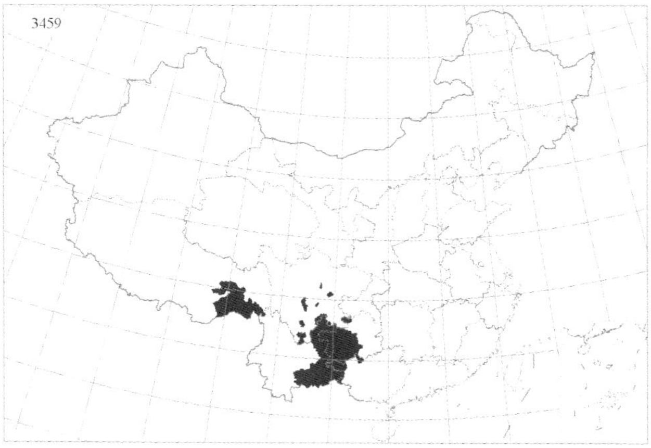

Rosaceae 蔷薇科 *Rubus* 悬钩子属

Rubus erythrocarpus T. T. Yu et L. T. Lu 红果悬钩子

Shrubs, deciduous

9.8; 2.7; 15.9; 9.9; 63.3; 5.0; 606 ❙ 1097; 620; 47 ❙ 80.8; 606; 362.9

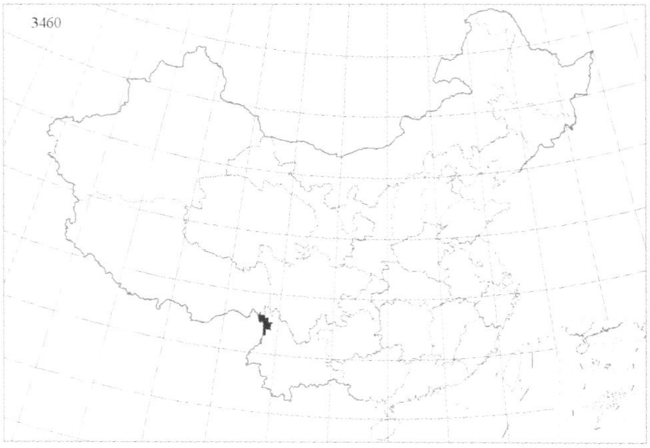

Rosaceae 蔷薇科 *Rubus* 悬钩子属

Rubus erythrocarpus var. weixiensis T. T. Yu et L. T. Lu 腺萼红果悬钩子

Shrubs, deciduous

9.6; 2.4; 15.7; 9.6; 60.8; 5.4; 600 ❙ 1007; 578; 40 ❙ 67.9; 600; 351.7

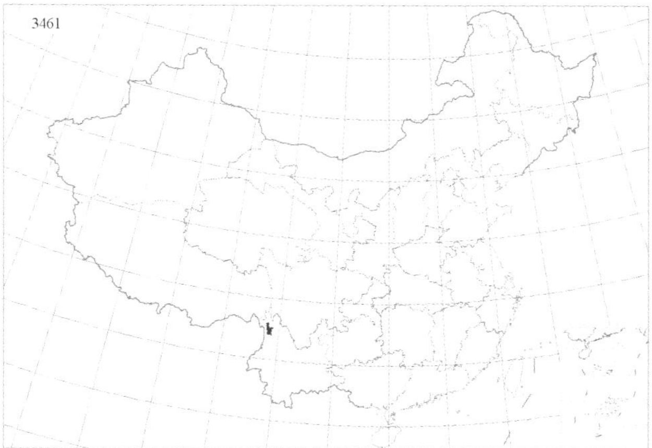

Rosaceae 蔷薇科 *Rubus* 悬钩子属

Rubus eucalyptus Focke 桉叶悬钩子

Shrubs, deciduous

11.9 (2.0~16.2); 1.6 (-7.2~5.5); 21.5 (10.1~26.3); 12.1 (3.8~16.2); 92.8 (15.4~134.6); 9.5 (0.0~51.4); 695 (413~854) ❙ 1137 (711~1478); 526 (357~816); 55 (15~105) ❙ 66.3 (7.8~128.6); 695 (413~854); 232.3 (147.3~292.6)

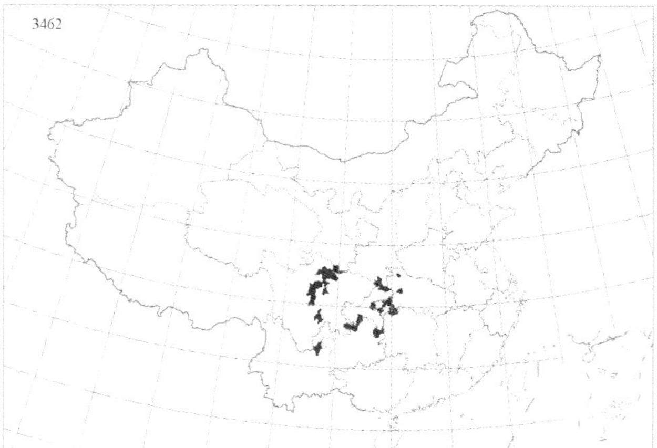

Rosaceae 蔷薇科 *Rubus* 悬钩子属

Rubus eucalyptus var. etomentosus T. T. Yu et L. T. Lu 脱毛桉叶悬钩子

Shrubs, deciduous

2.6; -7.0; 11.2; 4.3; 20.0; 48.2; 427 ❙ 829; 400; 23 ❙ 94.4; 427; 252.2

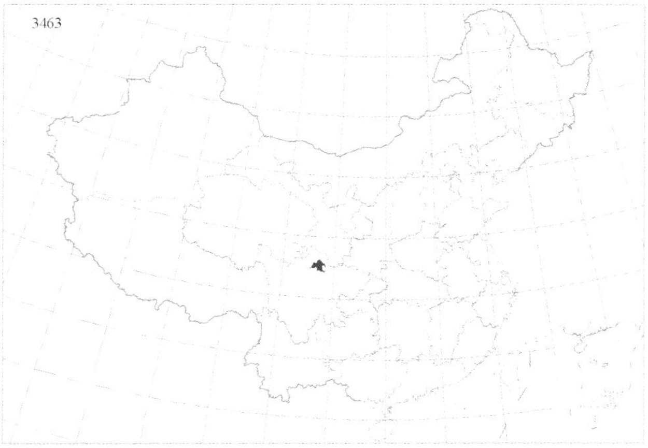

Rosaceae 蔷薇科 *Rubus* 悬钩子属

Rubus eucalyptus var. trullisatus (Focke) T. T. Yu et L. T. Lu 无腺桉叶悬钩子

Shrubs, deciduous

10.9; 1.1; 20.1; 11.0; 82.4; 10.8; 654 ❙ 1221; 624; 41 ❙ 87.2; 654; 257.0

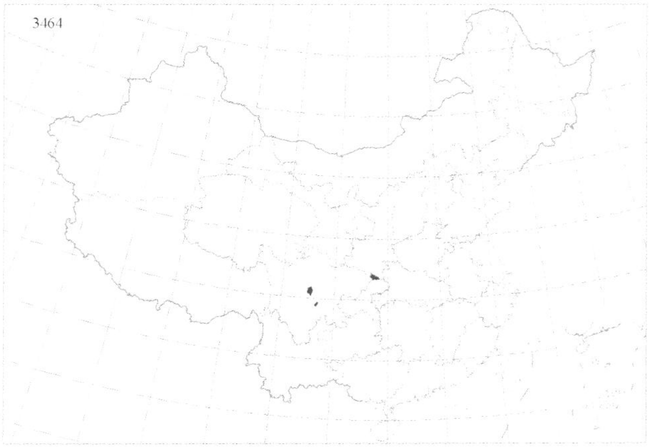

Rosaceae 蔷薇科 *Rubus* 悬钩子属

Rubus eustephanos Focke 大红泡

Shrubs, deciduous

16.0 (10.6~21.6); 5.0 (-1.5~12.6); 26.4 (19.4~29.3); 16.0 (10.6~21.7); 133.9 (79.4~199.8); 1.4 (0.0~15.5); 862 (643~1153) ❙ 1301 (724~1997); 524 (316~816); 107 (24~254) ❙ 51.9 (2.5~130.7); 861 (643~1153); 202.5 (108.2~332.5)

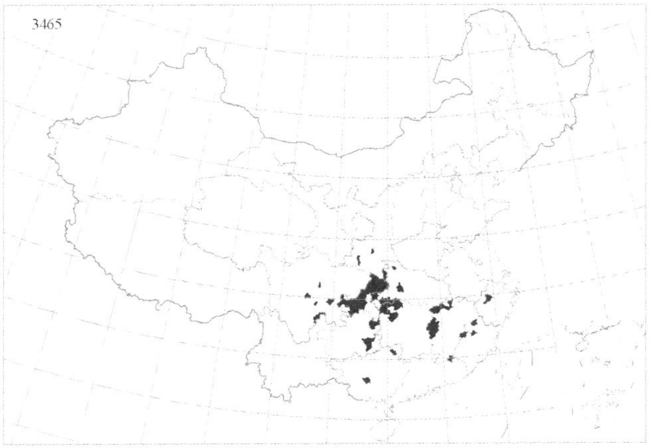

Rosaceae 蔷薇科 *Rubus* 悬钩子属

Rubus eustephanos var. glanduliger T. T. Yu et L. T. Lu 腺毛大红泡

Shrubs, deciduous

15.5 (10.1~18.7); 5.3 (-0.2~8.1); 25.1 (17.7~28.9); 15.6 (10.3~18.8); 128.4 (70.1~166.0); 1.8 (0.0~12.1); 839 (621~1002) ❙ 1133 (1001~1410); 525 (442~816); 54 (22~71) ❙ 38.8 (8.0~93.8); 839 (621~1002); 182.1 (108.2~318.7)

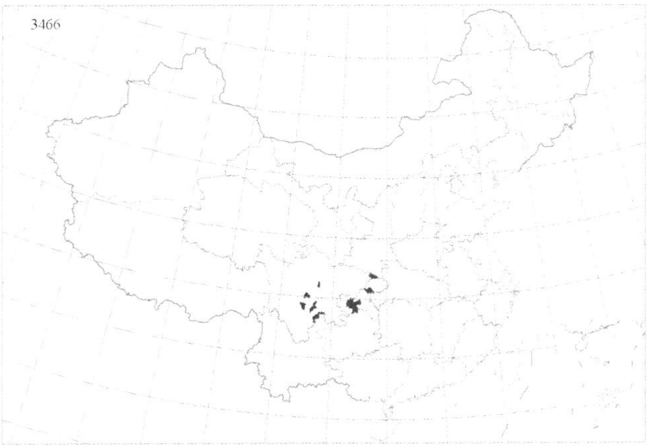

Rosaceae 蔷薇科 *Rubus* 悬钩子属

Rubus faberi Focke 峨眉悬钩子

Shrubs, deciduous

14.9; 5.3; 23.6; 15.0; 119.6; 0.0; 782 ❙ 1410; 816; 49 ❙ 80.4; 782; 214.5

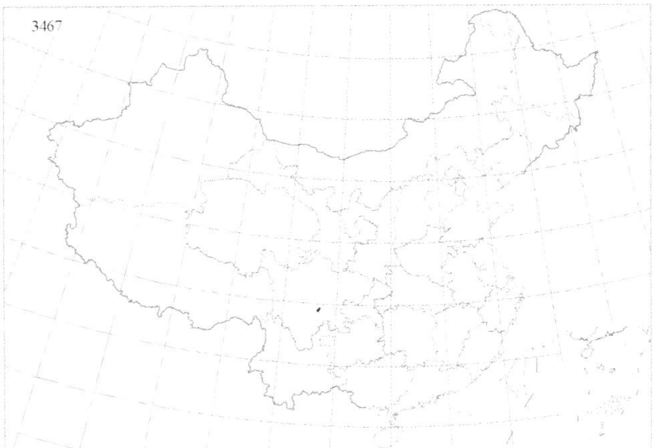

Rosaceae 蔷薇科 *Rubus* 悬钩子属

Rubus fanjingshanensis L. T. Lu ex Boufford et al. 梵净山悬钩子

Shrubs

15.2; 4.1; 25.8; 15.2; 123.4; 0.9; 812 ❙ 1221; 473; 88 ❙ 50.4; 812; 211.8

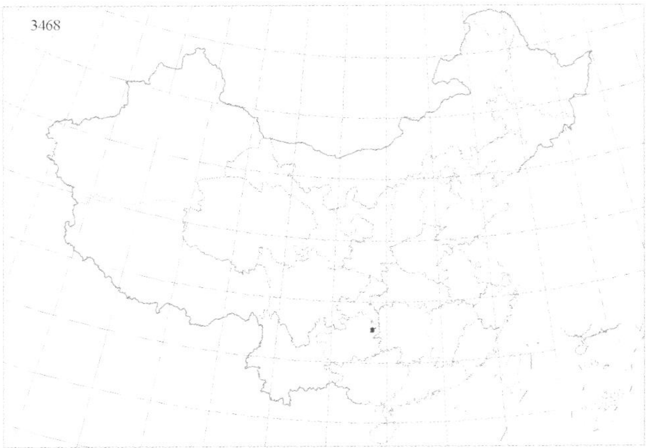

Rosaceae 蔷薇科 *Rubus* 悬钩子属
Rubus feddei H. Lév. et Vaniot 黔桂悬钩子
Shrubs, deciduous

19.1 (15.5~22.2); 10.1 (4.8~13.9); 26.1 (22.7~28.7); 19.1 (15.6~22.3); 169.6 (127.2~207.5); 0.0 (0.0~0.2); 979 (815~1190) ❙ 1395 (1161~1776); 708 (591~820); 95 (43~194) ❙ 43.6 (15.4~93.2); 979 (815~1190); 284.8 (189.5~411.1)

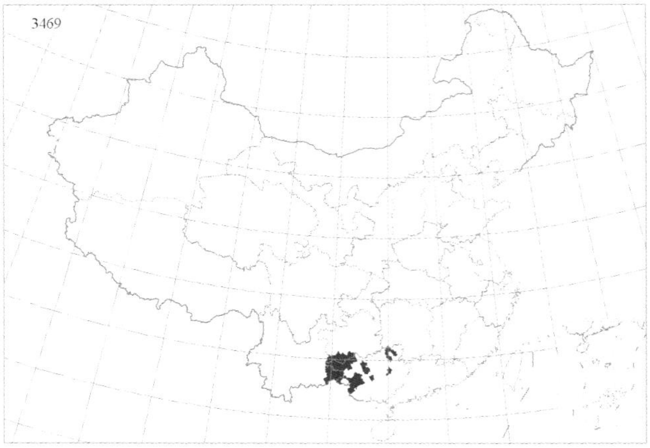

Rosaceae 蔷薇科 *Rubus* 悬钩子属
Rubus flagelliflorus Focke 攀枝莓
Shrubs, deciduous

15.2 (10.2~19.1); 4.7 (-0.7~14.1); 25.0 (21.0~29.5); 15.2 (10.4~19.1); 126.2 (77.2~169.2); 3.3 (0.0~13.6); 818 (634~993) ❙ 1407 (993~2713); 590 (420~1388); 106 (39~258) ❙ 73.1 (27.5~214.4); 815 (634~969); 248.1 (176.4~363.7)

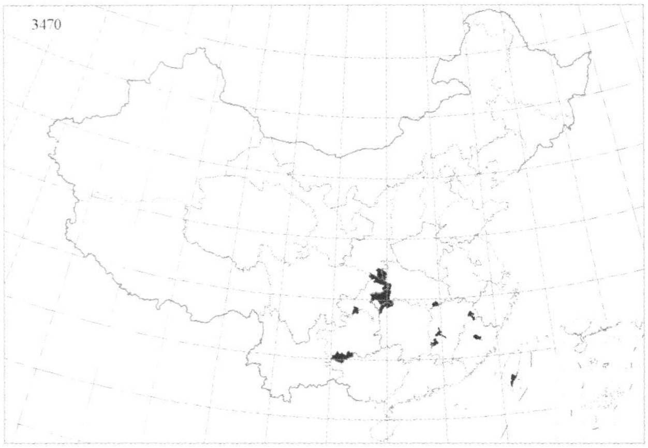

Rosaceae 蔷薇科 *Rubus* 悬钩子属
Rubus flosculosus Focke 弓茎悬钩子
Shrubs, deciduous

11.8 (1.8~16.7); -1.1 (-10.1~6.0); 23.7 (9.9~28.0); 12.1 (3.8~16.8); 97.7 (15.5~141.5); 15.8 (0.0~53.3); 724 (404~896) ❙ 805 (428~1997); 386 (258~1007); 38 (9~254) ❙ 13.7 (-36.5~147.0); 657 (404~896); 213.7 (120.1~447.4)

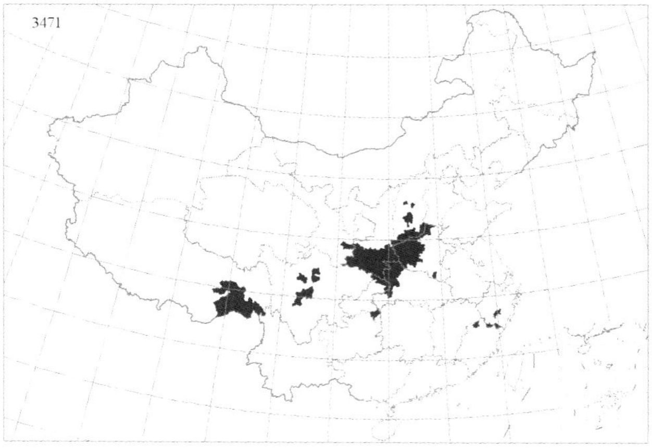

Rosaceae 蔷薇科 *Rubus* 悬钩子属
Rubus flosculosus var. etomentosus T. T. Yu et L. T. Lu 脱毛弓茎悬钩子
Shrubs, deciduous

11.9; 2.0; 20.9; 12.1; 93.6; 10.2; 701 ❙ 1501; 600; 141 ❙ 109.0; 701; 287.9

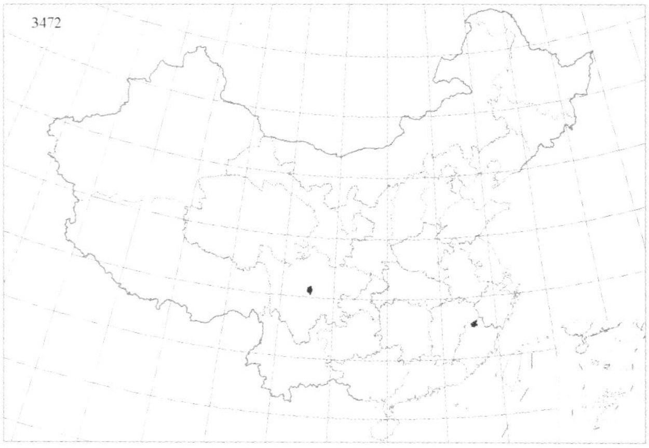

Rosaceae 蔷薇科 *Rubus* 悬钩子属
Rubus foliaceistipulatus T. T. Yu et L. T. Lu 托叶悬钩子
Shrubs, deciduous

14.4; 7.5; 19.1; 14.4; 113.1; 0.0; 727 ❙ 1454; 792; 70 ❙ 99.9; 727; 487.2

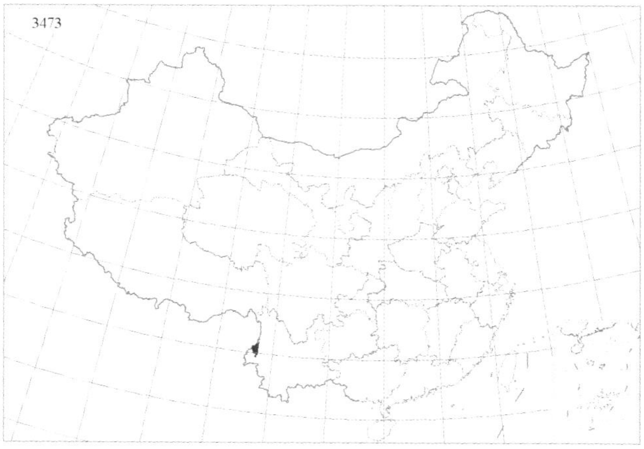

Rosaceae 蔷薇科 *Rubus* 悬钩子属
Rubus formosensis Kuntze 台湾悬钩子
Shrubs

19.3 (15.5~24.1); 12.0 (4.8~18.9); 25.4 (20.1~28.3); 19.3 (15.6~24.1); 171.7 (127.2~229.7); 0.0 (0.0~0.2); 978 (773~1322) ❙ 2177 (1374~3495); 948 (556~1847); 240 (47~897) ❙ 128.4 (15.4~246.5); 960 (773~1190); 249.0 (119.4~363.7)

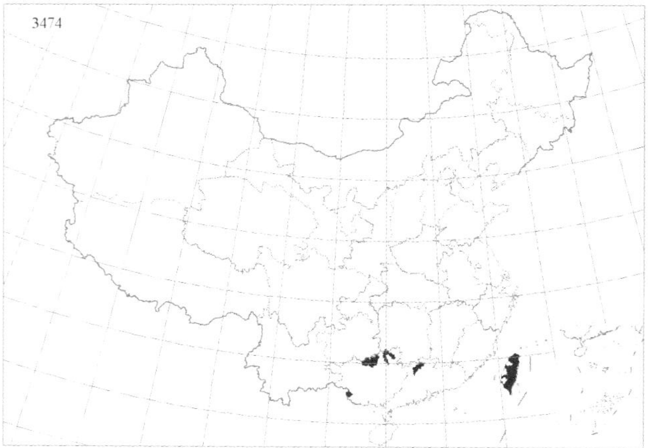

Rosaceae 蔷薇科 *Rubus* 悬钩子属
Rubus forrestianus Handel-Mazzetti 贡山蓬虆
Shrubs

11.5; 4.4; 17.1; 11.5; 82.8; 4.4; 651 ❙ 1248; 688; 57 ❙ 90.7; 651; 419.8

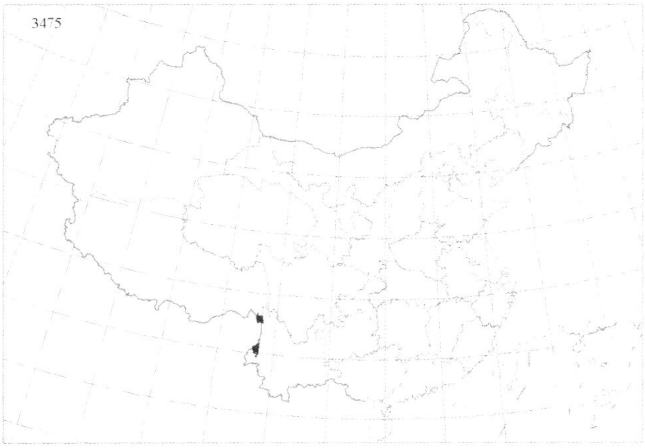

Rosaceae 蔷薇科 *Rubus* 悬钩子属
Rubus fraxinifoliolus Hayata 梣叶悬钩子
Shrubs, deciduous

19.0 (16.2~22.2); 13.4 (11.0~17.7); 23.6 (20.1~27.0); 19.0 (16.2~22.2); 168.5 (134.2~206.7); 0.0 (0.0~0.0); 935 (773~1123) ❙ 2523 (1836~2993); 1162 (900~1847); 248 (107~547) ❙ 175.5 (64.2~246.5); 930 (773~1123); 265.9 (202.8~303.6)

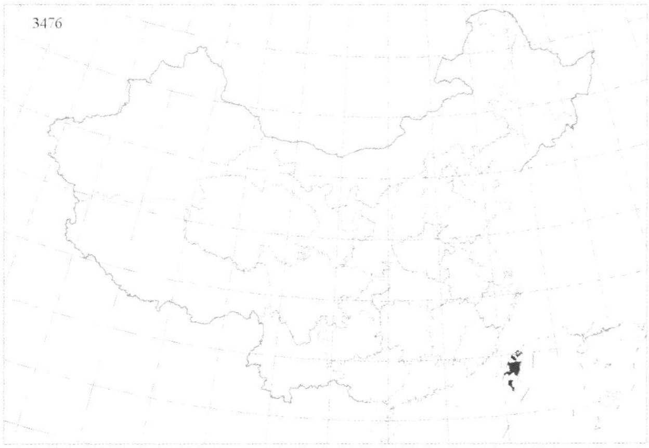

Rosaceae 蔷薇科 *Rubus* 悬钩子属
Rubus fraxinifolius Poir. 兰屿桤叶悬钩子
Shrubs, deciduous

18.5; 14.1; 21.7; 18.5; 162.1; 0.0; 863 ❙ 2713; 1388; 188 ❙ 214.4; 863; 363.7

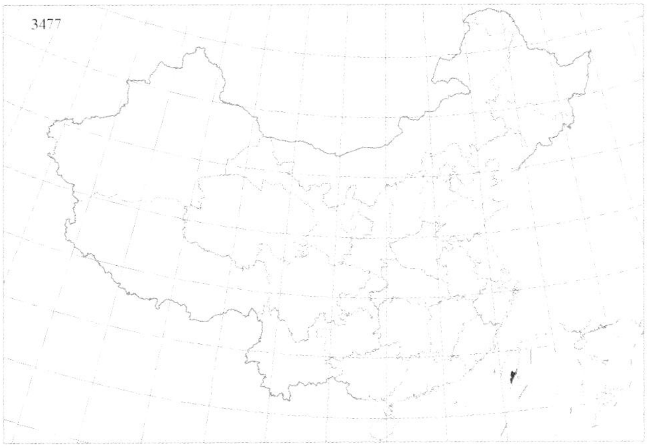

Rosaceae 蔷薇科 *Rubus* 悬钩子属
Rubus fujianensis T. T. Yu et L. T. Lu 福建悬钩子
Shrubs, deciduous

16.8 (15.0~18.2); 5.9 (4.9~7.2); 27.0 (24.7~29.1); 16.8 (15.1~18.2); 142.0 (120.8~158.8); 0.0 (0.0~0.1); 890 (789~984) ❙ 1902 (1759~1997); 634 (535~713); 247 (223~259) ❙ 114.8 (78.7~147.0); 883 (789~927); 261.8 (183.9~303.4)

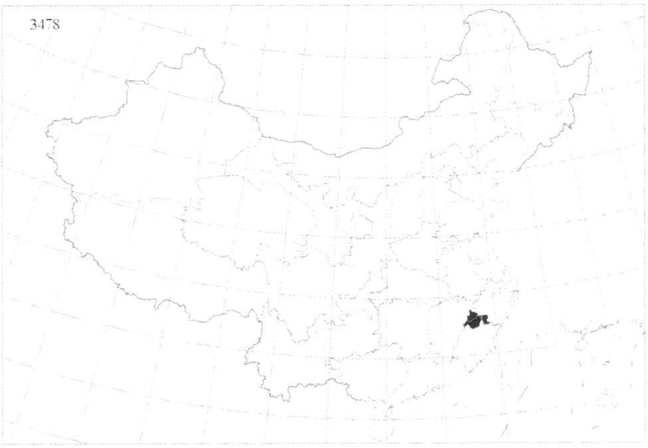

Rosaceae 蔷薇科 *Rubus* 悬钩子属
Rubus fuscifolius T. T. Yu et L. T. Lu 锈叶悬钩子
Shrubs, deciduous

15.6 (8.6~21.0); 8.8 (1.3~14.2); 20.7 (15.0~25.6); 15.6 (8.7~21.0); 128.8 (52.6~191.8); 1.9 (0.0~8.8); 809 (574~1053) ❙ 1321 (1041~1522); 738 (585~835); 57 (45~68) ❙ 67.0 (44.5~93.0); 809 (574~1053); 457.2 (352.3~551.6)

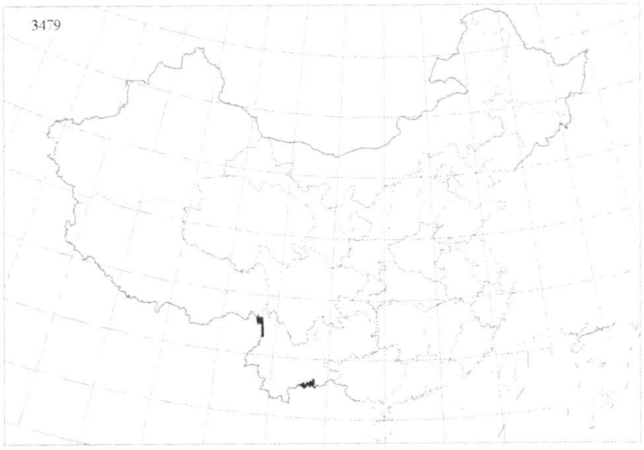

Rosaceae 蔷薇科 *Rubus* 悬钩子属
Rubus fuscorubens Focke 黄毛悬钩子
Shrubs, deciduous

15.3; 4.8; 25.2; 15.4; 126.2; 1.6; 817 ❙ 1147; 542; 71 ❙ 41.6; 817; 207.7

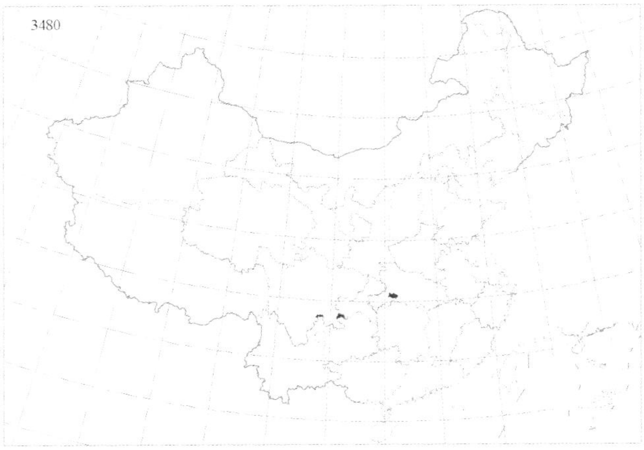

Rosaceae 蔷薇科 *Rubus* 悬钩子属
Rubus glabricarpus W. C. Cheng 光果悬钩子
Shrubs, deciduous

16.5 (14.7~19.1); 5.8 (2.5~10.0); 26.8 (24.7~28.9); 16.5 (14.8~19.1); 139.3 (120.8~169.1); 0.8 (0.0~3.9); 874 (789~973) ❙ 1691 (1377~1997); 601 (493~713); 208 (175~254) ❙ 94.5 (51.2~147.0); 874 (789~973); 271.4 (189.1~305.7)

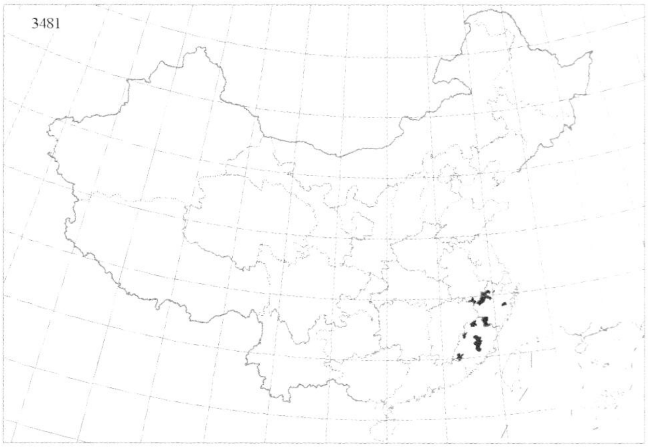

Rosaceae 蔷薇科 *Rubus* 悬钩子属
Rubus glabricarpus var. glabratus C. Z. Zheng et Y. Y. Fang 无毛光果悬钩子
Shrubs, deciduous

17.1 (15.3~18.1); 6.1 (4.2~7.2); 27.4 (26.0~28.7); 17.2 (15.4~18.1); 146.2 (125.8~157.7); 0.1 (0.0~0.8); 908 (818~965) ❙ 1821 (1561~1997); 599 (510~669); 236 (200~259) ❙ 101.1 (66.6~130.7); 904 (818~961); 256.1 (182.2~303.4)

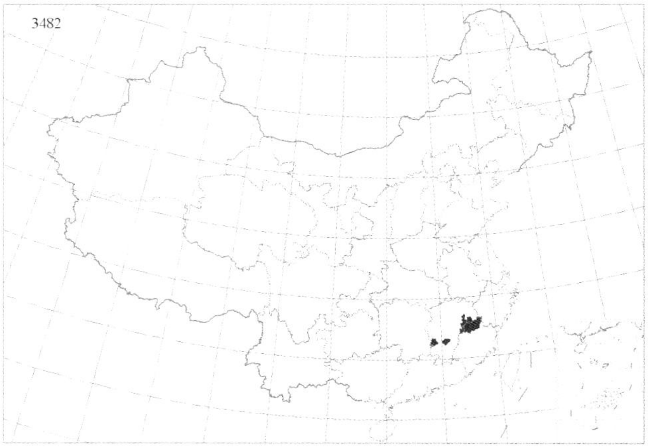

Rosaceae 蔷薇科 *Rubus* 悬钩子属
Rubus glandulosocarpus M. X. Nie 腺果悬钩子
Shrubs, deciduous

16.2; 5.0; 26.8; 16.2; 135.1; 0.4; 860 ❙ 1611; 524; 208 ❙ 88.0; 860; 217.4

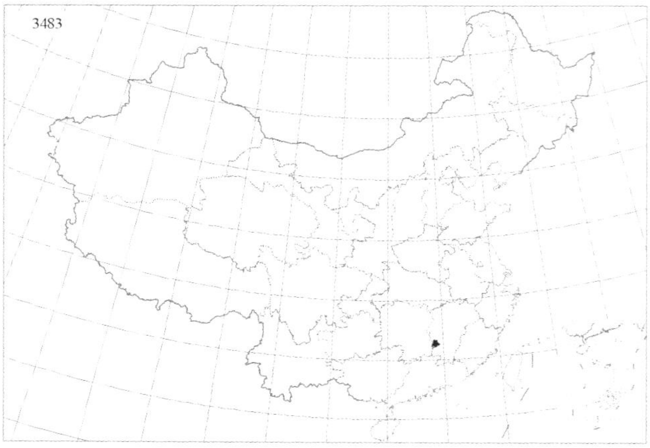

Rosaceae 蔷薇科 *Rubus* 悬钩子属
Rubus gongshanensis T. T. Yu et L. T. Lu 贡山悬钩子
Shrubs, deciduous

8.6; 1.3; 15.0; 8.7; 52.6; 8.8; 574 ❙ 1041; 585; 45 ❙ 81.5; 574; 352.3

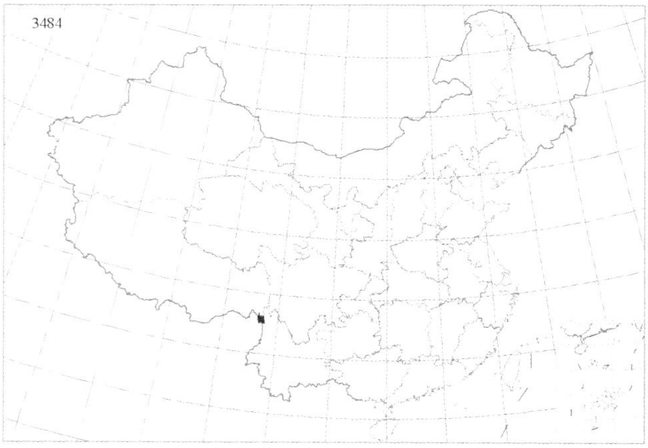

Rosaceae 蔷薇科 *Rubus* 悬钩子属
Rubus gongshanensis var. qiujiangensis T. T. Yu et L. T. Lu 无刺贡山悬钩子
Shrubs, deciduous

8.6; 1.3; 15.0; 8.7; 52.6; 8.8; 574 ❙ 1041; 585; 45 ❙ 81.5; 574; 352.3

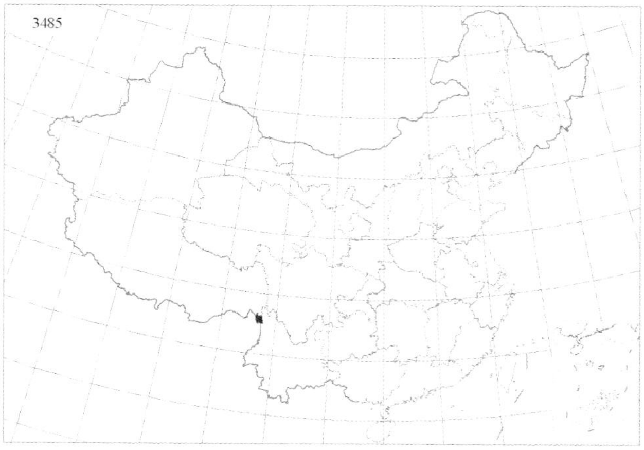

Rosaceae 蔷薇科 *Rubus* 悬钩子属
Rubus grandipaniculatus T. T. Yu et L. T. Lu 大序悬钩子
Shrubs, deciduous

13.3; 2.1; 24.0; 13.3; 105.1; 5.4; 743 ❙ 1135; 474; 52 ❙ 52.4; 743; 246.3

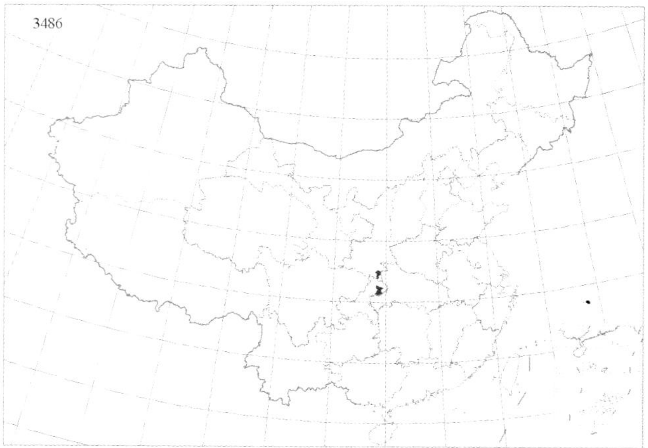

Rosaceae 蔷薇科 *Rubus* 悬钩子属
Rubus grayanus Maxim. 中南悬钩子
Shrubs, deciduous

17.2 (14.3~19.4); 6.2 (3.7~10.0); 27.4 (24.2~29.3); 17.3 (14.3~19.4); 147.2 (113.2~173.4); 0.1 (0.0~1.5); 912 (761~1029) | 1690 (1486~1997); 567 (488~677); 214 (176~259) | 86.1 (57.0~130.7); 910 (761~1029); 236.7 (183.9~303.4)

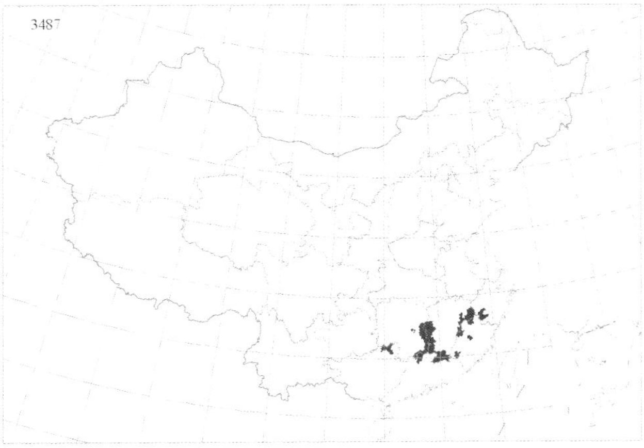

Rosaceae 蔷薇科 *Rubus* 悬钩子属
Rubus grayanus var. trilobatus T. T. Yu et L. T. Lu 三裂中南悬钩子
Shrubs, deciduous

16.7; 6.1; 26.5; 16.7; 140.7; 0.0; 875 | 1887; 631; 235 | 116.2; 875; 277.0

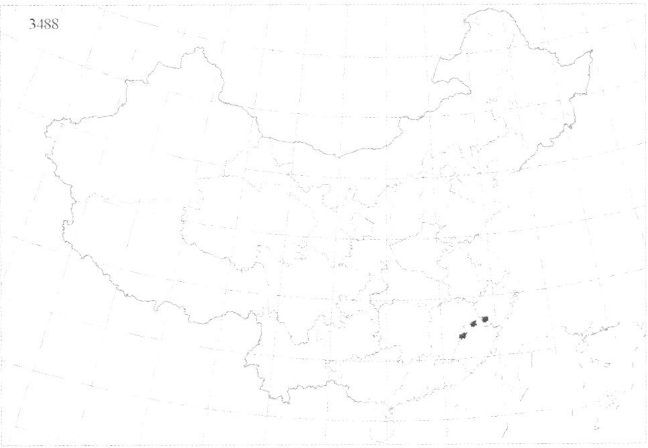

Rosaceae 蔷薇科 *Rubus* 悬钩子属
Rubus gressittii F. P. Metcalf 江西悬钩子
Shrubs, deciduous

19.1 (16.7~22.4); 9.2 (4.9~14.8); 27.7 (26.9~28.5); 19.1 (16.7~22.4); 169.7 (140.6~208.9); 0.0 (0.0~0.1); 1003 (900~1191) | 1734 (1368~2030); 662 (505~1033); 182 (110~259) | 72.9 (52.0~111.3); 1003 (900~1191); 232.7 (163.7~279.5)

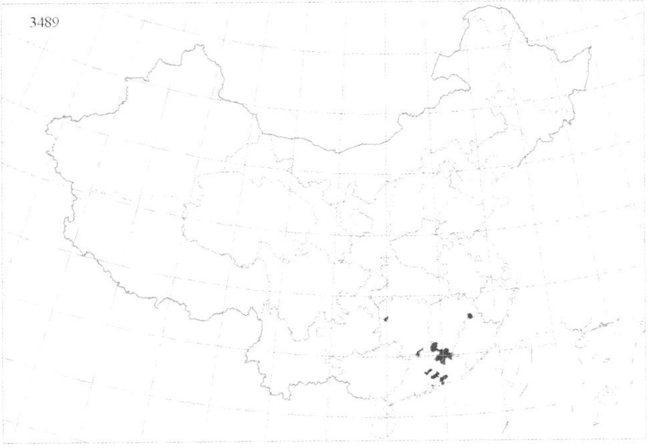

Rosaceae 蔷薇科 *Rubus* 悬钩子属
Rubus gyamdaensis L. T. Lu et Boufford 柔毛悬钩子
Shrubs, deciduous

7.1; -2.0; 15.3; 8.0; 52.1; 26.5; 542 | 904; 507; 23 | 59.8; 542; 272.4

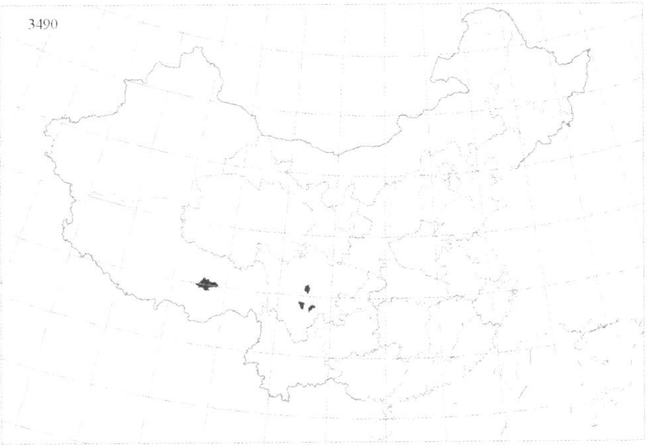

Rosaceae 蔷薇科 *Rubus* 悬钩子属
Rubus hanceanus Kuntze 华南悬钩子
Shrubs, deciduous

18.3 (14.6~22.2); 7.9 (3.8~13.9); 27.4 (24.6~29.0); 18.3 (14.7~22.3); 159.6 (117.7~207.5); 0.1 (0.0~1.5); 960 (779~1190) | 1560 (1348~1829); 558 (403~821); 181 (87~230) | 64.2 (15.4~104.6); 953 (779~1190); 224.9 (151.0~327.7)

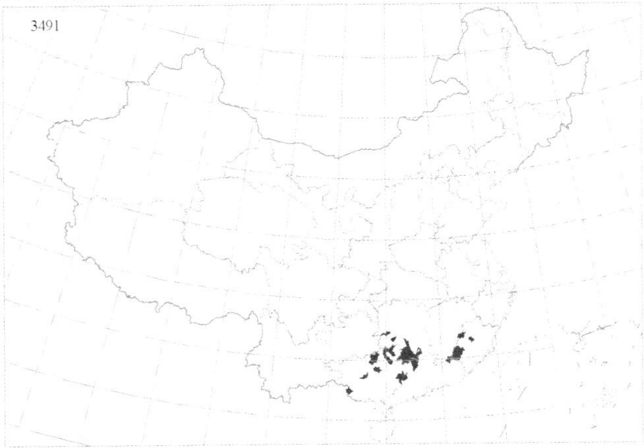

Rosaceae 蔷薇科 *Rubus* 悬钩子属
Rubus hastifolius H. Lév. et Vaniot 戟叶悬钩子
Shrubs, deciduous

16.6 (13.6~22.1); 5.9 (2.6~13.7); 26.3 (23.5~28.8); 16.6 (13.6~22.1); 139.6 (107.0~205.6); 0.5 (0.0~3.9); 874 (746~1182) | 1369 (1168~2005); 540 (427~824); 125 (43~209) | 57.0 (16.9~99.4); 872 (746~1182); 227.0 (157.7~354.5)

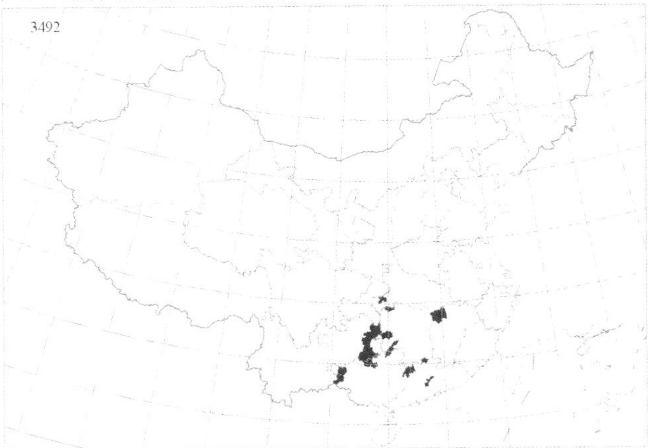

Rosaceae 蔷薇科 *Rubus* 悬钩子属

Rubus hemithyrsus Handel-Mazzetti 半锥莓

Scandent shrubs

11.3; 4.3; 17.0; 11.3; 76.6; 0.7; 644 ❙ 1243; 696; 58 ❙ 93.0; 644; 384.7

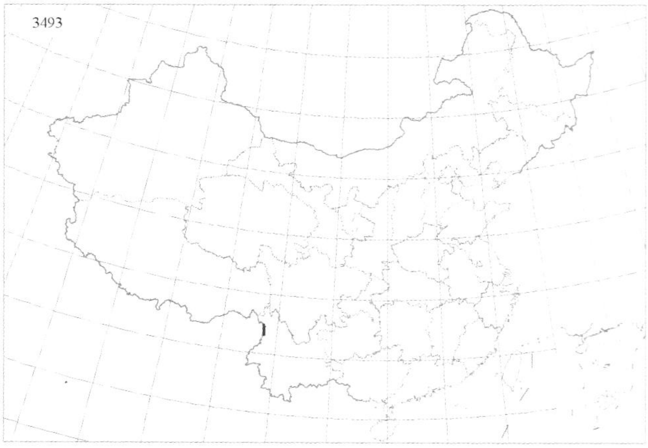

Rosaceae 蔷薇科 *Rubus* 悬钩子属

Rubus henryi Hemsl. et Kuntze 鸡爪茶

Shrubs, evergreen

15.0 (12.0~17.7); 4.4 (0.6~10.9); 24.9 (22.5~27.1); 15.1 (12.1~17.8); 122.4 (94.7~153.4); 1.6 (0.0~9.7); 800 (697~891) ❙ 1297 (1076~1624); 542 (465~756); 94 (48~191) ❙ 62.6 (35.3~107.2); 800 (697~891); 222.5 (146.0~491.5)

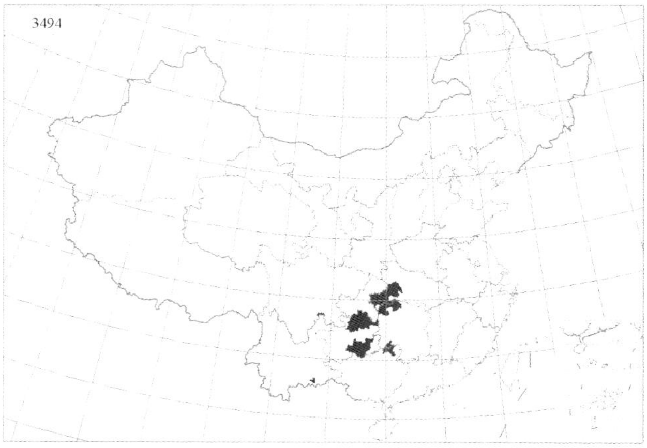

Rosaceae 蔷薇科 *Rubus* 悬钩子属

Rubus henryi var. sozostylus (Focke) T. T. Yu et L. T. Lu 大叶鸡爪茶

Shrubs, evergreen

15.9 (12.6~19.1); 5.8 (1.6~10.1); 24.9 (23.0~27.1); 16.0 (12.7~19.2); 132.6 (98.8~170.1); 0.8 (0.0~6.9); 832 (711~981) ❙ 1307 (1101~1526); 591 (477~816); 83 (49~181) ❙ 58.4 (27.5~107.2); 832 (711~981); 226.6 (148.9~322.1)

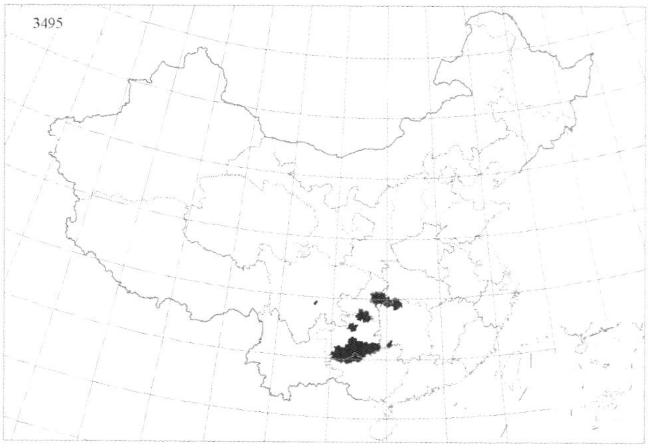

Rosaceae 蔷薇科 *Rubus* 悬钩子属

Rubus hirsutus Thunb. 蓬蘽

Shrubs, deciduous

16.9 (13.4~21.4); 5.7 (1.5~13.6); 27.7 (24.2~29.7); 16.9 (13.4~21.5); 144.0 (108.3~197.7); 1.0 (0.0~7.3); 907 (755~1128) ❙ 1478 (879~1997); 534 (356~713); 181 (68~259) ❙ 64.1 (-5.0~147.0); 899 (755~1094); 223.0 (110.1~332.5)

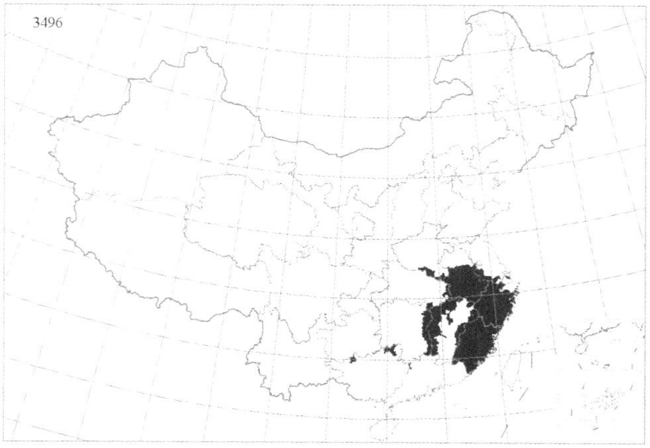

Rosaceae 蔷薇科 *Rubus* 悬钩子属

Rubus hirsutus var. brevipedicellus Z. M. Wu 短梗蓬蘽

Shrubs, deciduous

14.4; 2.1; 26.3; 14.5; 118.7; 4.9; 801 ❙ 1604; 599; 200 ❙ 100.2; 801; 277.3

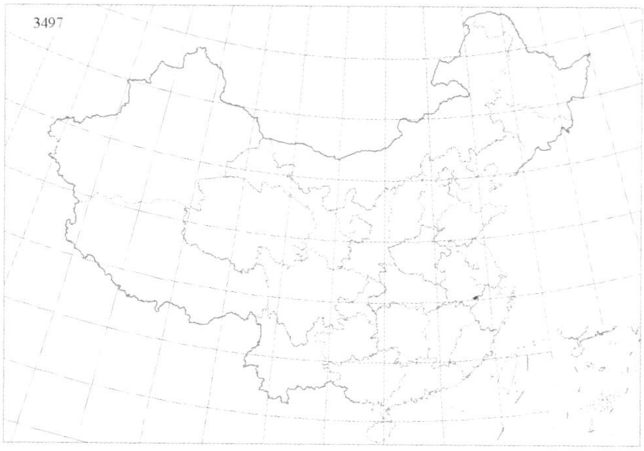

Rosaceae 蔷薇科 *Rubus* 悬钩子属

Rubus howii Merr. et Chun 裂叶悬钩子

Scandent shrubs, deciduous

24.4; 19.5; 27.8; 24.5; 233.5; 0.0; 1350 ❙ 1635; 547; 65 ❙ 21.1; 1219; 554.2

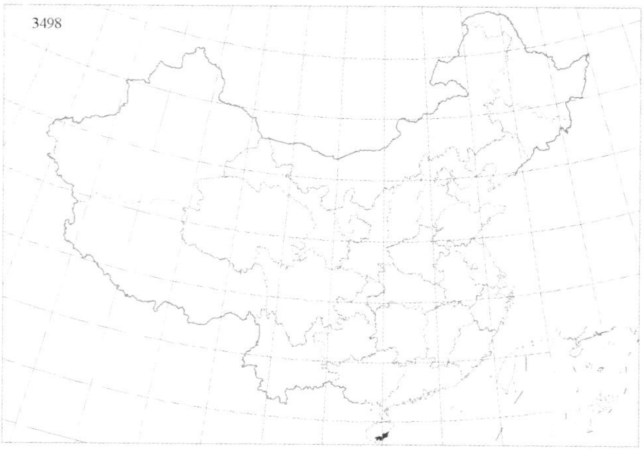

Rosaceae 蔷薇科 *Rubus* 悬钩子属
Rubus huangpingensis T. T. Yu et L. T. Lu 黄平悬钩子
Shrubs, deciduous

15.7; 5.2; 25.6; 15.7; 128.4; 0.0; 824 ❙ 1187; 486; 84 ❙ 44.1; 824; 203.1

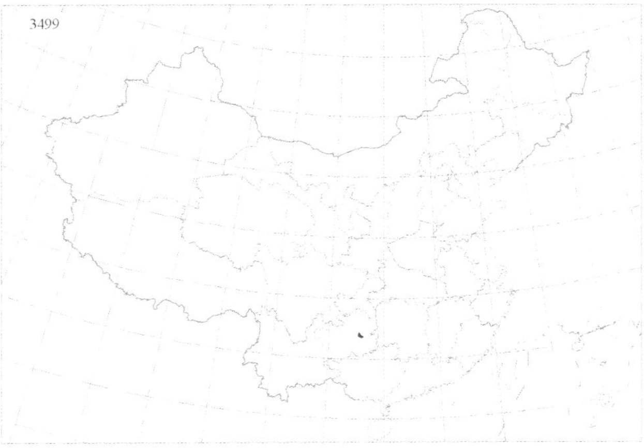

Rosaceae 蔷薇科 *Rubus* 悬钩子属
Rubus humulifolius C. A. Mey. 葎草叶悬钩子
Subshrubs, deciduous

-4.3; -28.5; 17.1; 5.1; 36.5; 147.7; 455 ❙ 478; 312; 15 ❙ 4.9; 455; 288.7

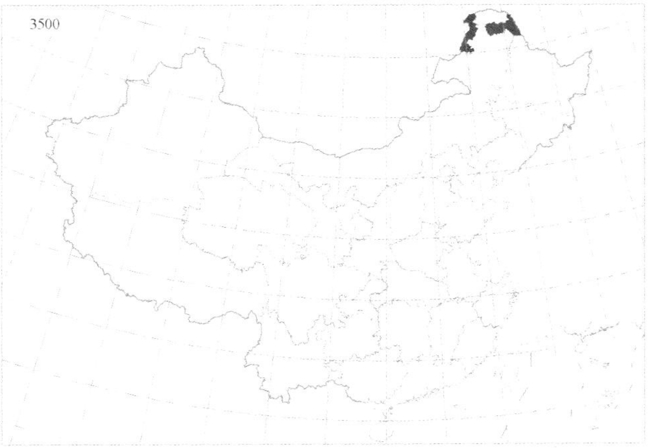

Rosaceae 蔷薇科 *Rubus* 悬钩子属
Rubus hunanensis Hand.-Mazz. 湖南悬钩子
Shrubs, deciduous

16.5 (12.6~21.0); 5.4 (1.6~11.9); 26.9 (20.1~29.3); 16.6 (12.7~21.1); 139.5 (98.8~193.0); 0.6 (0.0~6.9); 881 (711~1114) ❙ 1532 (1132~1997); 553 (403~821); 176 (43~259) ❙ 74.4 (16.9~147.0); 877 (711~1114); 230.1 (151.0~354.5)

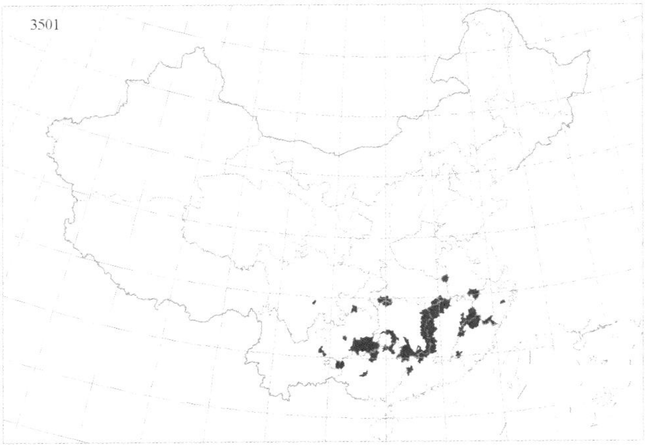

Rosaceae 蔷薇科 *Rubus* 悬钩子属
Rubus hypopitys Focke 滇藏悬钩子
Subshrubs, deciduous

10.9 (4.4~14.4); 3.4 (-4.3~7.5); 16.9 (12.1~19.6); 11.1 (5.3~14.4); 78.4 (28.5~113.1); 7.3 (0.0~34.6); 634 (455~727) ❙ 1225 (538~1736); 691 (337~1007); 50 (9~70) ❙ 88.5 (18.2~145.8); 634 (455~727); 374.5 (168.2~487.2)

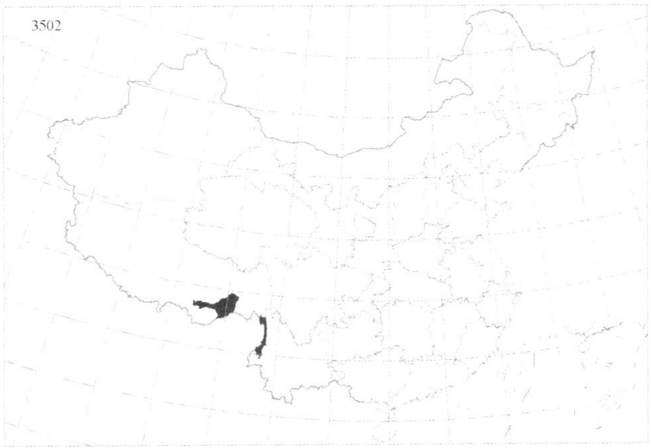

Rosaceae 蔷薇科 *Rubus* 悬钩子属
Rubus ichangensis Hemsl. et Kuntze 宜昌悬钩子
Shrubs, deciduous or semievergreen

15.6 (7.3~21.8); 4.9 (-2.0~14.2); 25.6 (15.6~29.4); 15.7 (7.5~21.9); 129.4 (48.2~202.6); 1.5 (0.0~20.4); 836 (537~1150) ❙ 1246 (711~1876); 518 (357~891); 94 (15~214) ❙ 50.0 (8.0~107.2); 832 (537~1150); 202.6 (108.2~354.5)

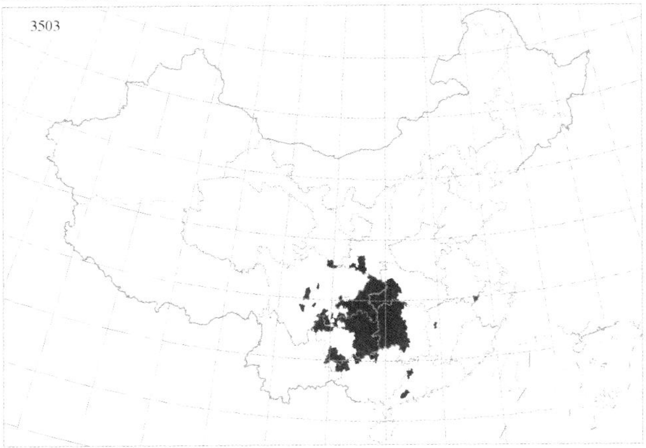

Rosaceae 蔷薇科 *Rubus* 悬钩子属
Rubus idaeopsis Focke 拟覆盆子
Shrubs, deciduous

14.3 (5.3~19.1); 4.2 (-2.4~10.9); 23.3 (12.2~29.1); 14.4 (5.7~19.1); 116.3 (30.5~169.2); 3.9 (0.0~26.2); 780 (479~966) ❙ 1202 (741~1624); 561 (368~816); 88 (28~209) ❙ 54.6 (0.0~101.8); 778 (479~966); 244.4 (179.5~491.5)

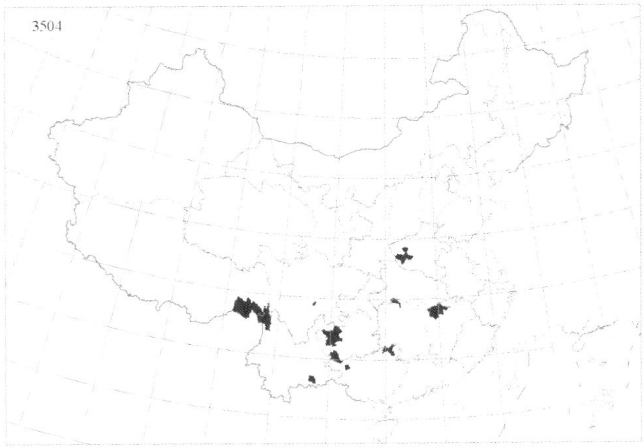

Rosaceae 蔷薇科 *Rubus* 悬钩子属
Rubus idaeus L. 覆盆子
Shrubs, deciduous

8.7 (-1.3~15.1); -7.6 (-23.8~1.8); 22.9 (13.2~27.5); 10.4 (4.5~15.2); 84.5 (26.7~128.0); 39.9 (5.6~112.6); 672 (433~865) ❙ 501 (120~817); 292 (51~500); 19 (8~53) ❙ -25.4 (-85.3~45.0); 496 (120~817); 154.6 (21.1~338.4)

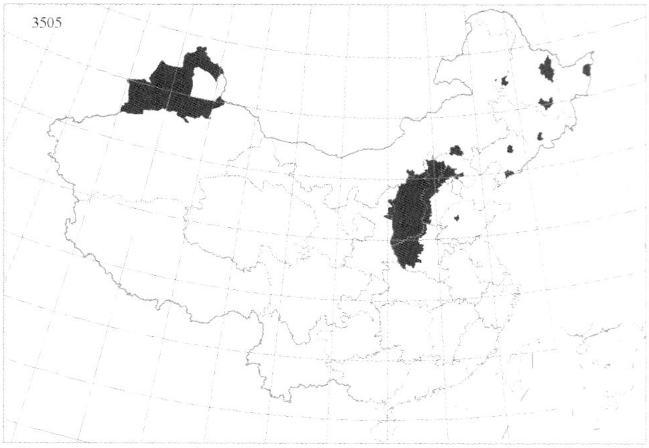

Rosaceae 蔷薇科 *Rubus* 悬钩子属
Rubus idaeus var. borealisinensis T. T. Yu et L. T. Lu 华北覆盆子
Shrubs, deciduous

5.7 (2.2~11.6); -11.4 (-16.0~-4.9); 20.9 (18.5~26.0); 8.5 (6.7~12.4); 66.7 (48.1~105.2); 57.3 (25.9~81.1); 593 (515~779) ❙ 425 (202~646); 284 (134~478); 11 (6~17) ❙ -28.3 (-65.5~-9.7); 425 (202~645); 150.5 (52.7~247.3)

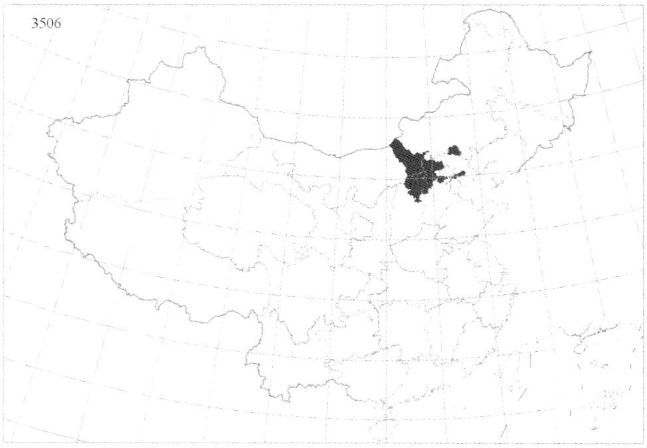

Rosaceae 蔷薇科 *Rubus* 悬钩子属
Rubus impressinervus F. P. Metcalf 陷脉悬钩子
Herbs or subshrubs, deciduous

17.0 (14.6~19.1); 6.6 (3.8~10.0); 26.7 (24.6~29.1); 17.1 (14.7~19.1); 145.0 (117.7~169.1); 0.1 (0.0~1.5); 893 (779~984) ❙ 1734 (1486~1997); 592 (505~713); 214 (175~259) ❙ 95.0 (60.2~147.0); 890 (779~975); 253.9 (183.9~305.7)

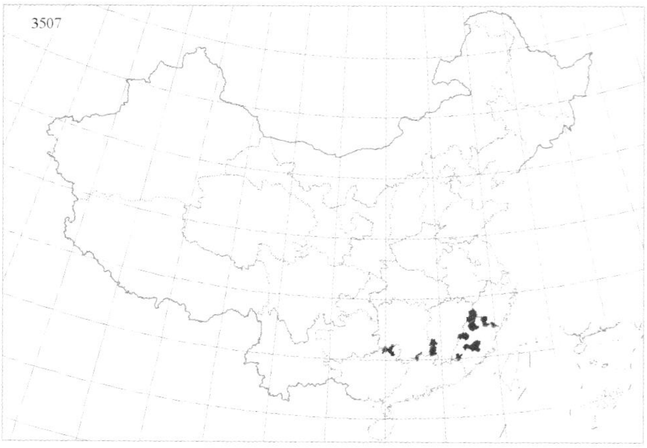

Rosaceae 蔷薇科 *Rubus* 悬钩子属
Rubus innominatus S. Moore 白叶莓
Shrubs, deciduous

15.9 (5.4~22.2); 4.6 (-2.9~13.9); 26.4 (12.3~30.1); 15.9 (5.8~22.3); 133.7 (31.5~207.5); 2.5 (0.0~25.6); 861 (481~1190) ❙ 1313 (512~1997); 515 (257~821); 127 (14~259) ❙ 51.9 (-35.3~147.0); 848 (481~1190); 213.8 (121.6~491.5)

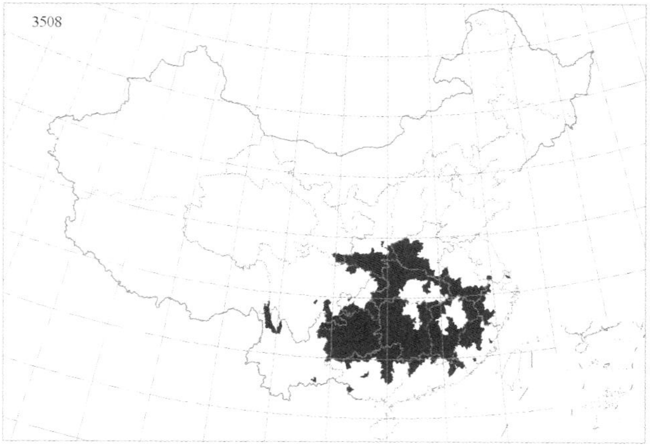

Rosaceae 蔷薇科 *Rubus* 悬钩子属
Rubus innominatus var. aralioides (Hance) T. T. Yu et L. T. Lu 密腺白叶莓
Shrubs, deciduous

18.8 (15.3~21.6); 8.8 (4.2~13.2); 27.6 (25.7~29.3); 18.8 (15.4~21.6); 165.7 (125.8~199.2); 0.0 (0.0~0.8); 981 (818~1139) ❙ 1640 (1268~1953); 588 (473~858); 190 (105~231) ❙ 67.5 (42.4~103.1); 978 (818~1139); 237.5 (182.2~300.0)

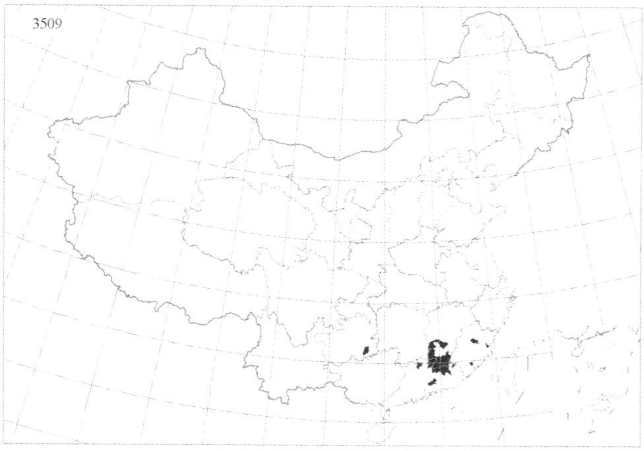

Rosaceae 蔷薇科 *Rubus* 悬钩子属
Rubus innominatus var. kuntzeanus (Hemsl.) L. H. Bailey 无腺白叶莓
Shrubs, deciduous

15.8 (9.3~18.9); 4.3 (-1.7~9.5); 26.7 (20.1~29.8); 15.9 (9.5~18.9); 132.0 (69.8~167.3); 1.7 (0.0~18.0); 856 (603~986) ❙ 1391 (815~1922); 511 (352~709); 147 (31~259) ❙ 62.9 (-2.0~128.9); 847 (603~972); 211.3 (126.1~319.9)

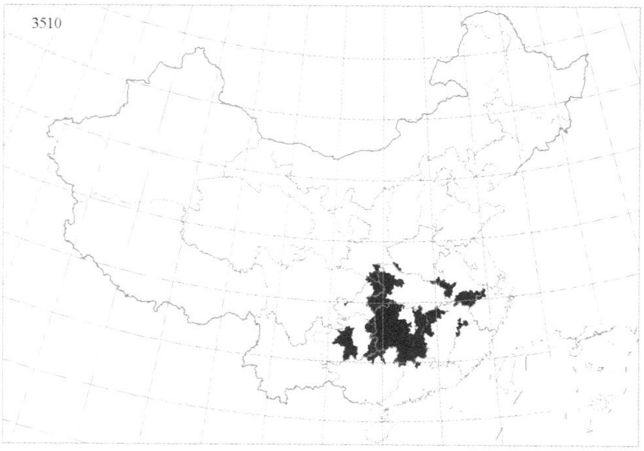

Rosaceae 蔷薇科 *Rubus* 悬钩子属
Rubus innominatus var. macrosepalus F. P. Metcalf 宽萼白叶莓
Shrubs, deciduous

15.0 (13.4~15.9); 3.3 (1.5~5.9); 26.4 (25.2~27.4); 15.0 (13.4~16.0); 123.2 (108.3~131.9); 2.9 (0.0~7.3); 819 (755~851) ❙ 1646 (1436~1791); 623 (582~666); 185 (140~223) ❙ 101.2 (68.8~115.7); 819 (755~851); 233.0 (183.4~260.8)

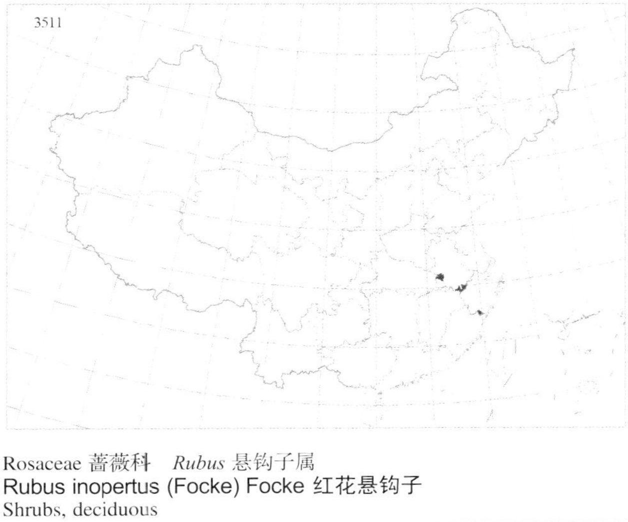

Rosaceae 蔷薇科 *Rubus* 悬钩子属
Rubus innominatus var. quinatus L. H. Bailey 五叶白叶莓
Shrubs, deciduous

16.6 (15.9~17.2); 4.3 (3.7~5.0); 28.6 (27.5~29.3); 16.7 (16.0~17.2); 140.8 (132.7~146.8); 0.7 (0.0~1.3); 914 (868~941) ❙ 1466 (1211~1590); 563 (474~685); 166 (134~193) ❙ 60.7 (28.8~80.0); 913 (868~941); 218.0 (184.7~249.0)

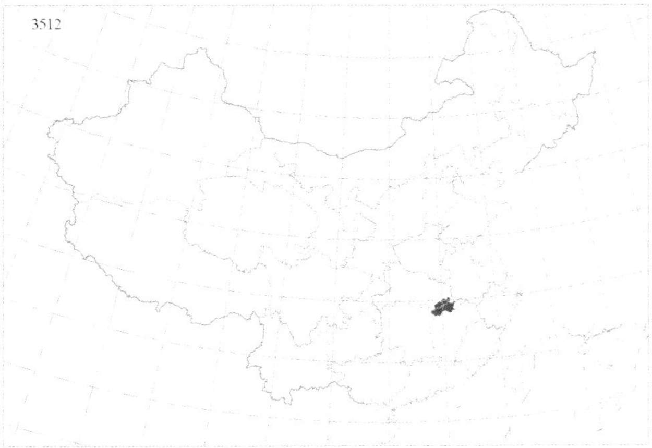

Rosaceae 蔷薇科 *Rubus* 悬钩子属
Rubus inopertus (Focke) Focke 红花悬钩子
Shrubs, deciduous

15.9 (7.3~24.1); 6.7 (-2.0~18.9); 24.1 (15.6~28.2); 15.9 (7.5~24.1); 132.3 (48.2~229.7); 1.5 (0.0~20.4); 835 (537~1322) ❙ 1478 (964~3495); 682 (468~1847); 114 (22~897) ❙ 76.5 (18.5~246.5); 827 (537~1144); 230.6 (119.4~506.2)

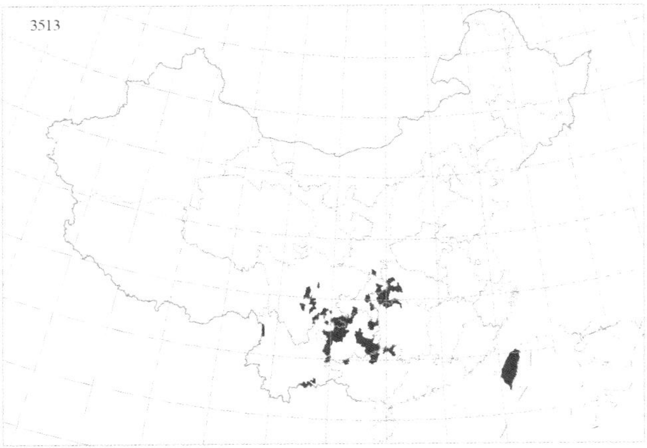

Rosaceae 蔷薇科 *Rubus* 悬钩子属
Rubus irenaeus Focke 灰毛泡
Shrubs, evergreen

16.4 (10.6~22.0); 5.3 (-0.2~14.1); 26.8 (19.4~29.9); 16.5 (10.6~22.1); 138.1 (79.4~205.1); 0.5 (0.0~12.1); 877 (643~1170) ❙ 1418 (964~2014); 525 (392~954); 150 (29~259) ❙ 61.9 (11.7~147.0); 866 (643~1170); 206.3 (128.7~305.7)

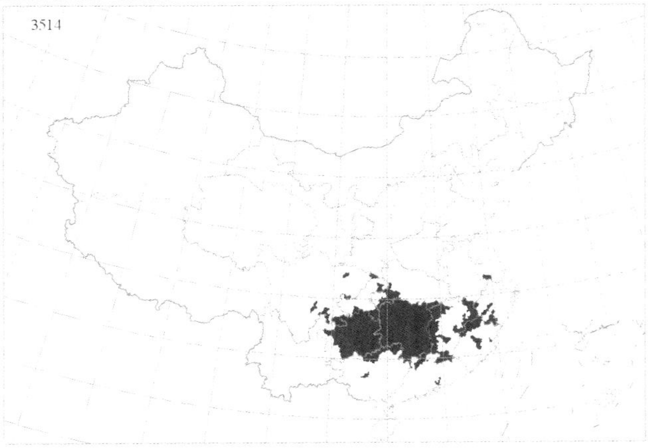

Rosaceae 蔷薇科 *Rubus* 悬钩子属
Rubus irritans Focke 紫色悬钩子
Shrubs, deciduous

0.9 (-4.1~6.1); -11.0 (-17.3~-4.3); 11.5 (6.1~17.5); 4.1 (1.8~7.6); 21.3 (1.9~53.6); 69.9 (35.4~114.2); 415 (329~544) ❙ 469 (271~984); 276 (165~601); 9 (4~20) ❙ 14.1 (-44.6~121.7); 383 (271~463); 208.4 (72.2~301.5)

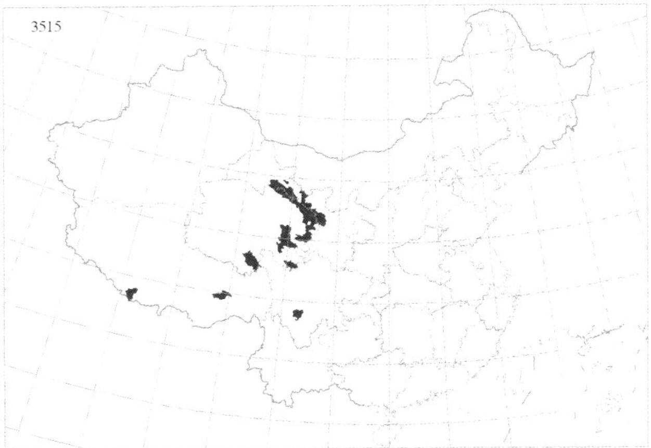

Rosaceae 蔷薇科 *Rubus* 悬钩子属
Rubus jambosoides Hance 蒲桃叶悬钩子
Shrubs, deciduous

17.6 (14.3~21.6); 6.9 (3.7~13.2); 27.5 (24.2~29.7); 17.7 (14.3~21.6); 152.1 (113.2~199.2); 0.1 (0.0~1.5); 931 (761~1139) ❙ 1584 (1410~2005); 534 (400~858); 198 (136~230) ❙ 71.0 (39.6~128.9); 919 (761~1139); 221.1 (151.0~305.7)

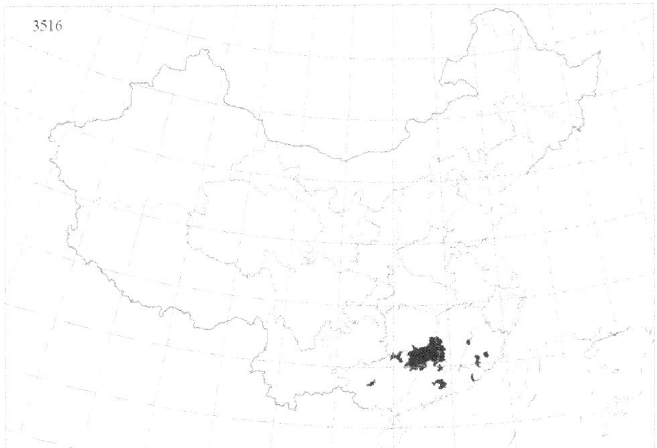

Rosaceae 蔷薇科 *Rubus* 悬钩子属

Rubus jianensis L. T. Lu et Boufford 常绿悬钩子

Shrubs, evergreen

17.4 (15.3~18.8); 6.0 (4.2~7.8); 28.2 (26.0~30.1); 17.5 (15.4~18.9); 149.9 (125.8~166.5); 0.1 (0.0~0.8); 933 (818~1024) ❙ 1565 (1452~1660); 510 (453~575); 205 (189~225) ❙ 68.7 (41.7~103.1); 906 (818~975); 212.5 (148.9~255.1)

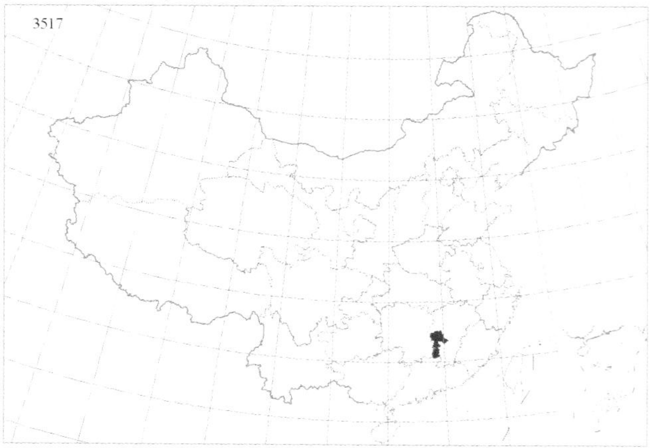

Rosaceae 蔷薇科 *Rubus* 悬钩子属

Rubus jinfoshanensis T. T. Yu et L. T. Lu 金佛山悬钩子

Shrubs, deciduous

17.4; 8.7; 24.8; 17.5; 149.7; 0.2; 888 ❙ 1305; 635; 66 ❙ 47.1; 888; 350.6

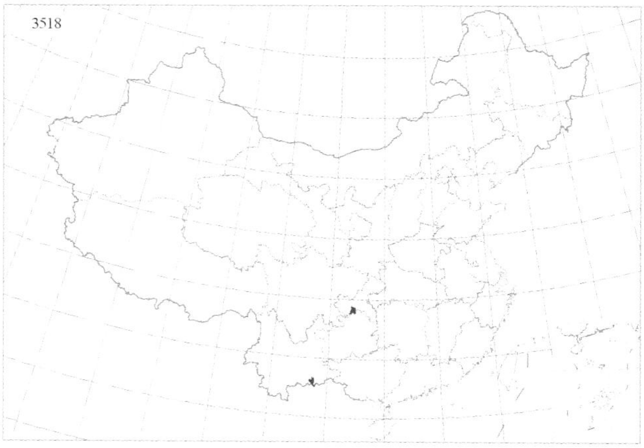

Rosaceae 蔷薇科 *Rubus* 悬钩子属

Rubus kawakamii Hayata 桑叶悬钩子

Shrubs, deciduous

19.4 (16.2~24.1); 14.0 (11.0~18.9); 23.6 (20.1~28.0); 19.4 (16.2~24.1); 172.7 (134.2~229.7); 0.0 (0.0~0.0); 964 (773~1322) ❙ 2407 (1751~2992); 1215 (900~1847); 188 (47~308) ❙ 162.0 (32.5~246.5); 930 (773~1123); 258.1 (154.2~303.6)

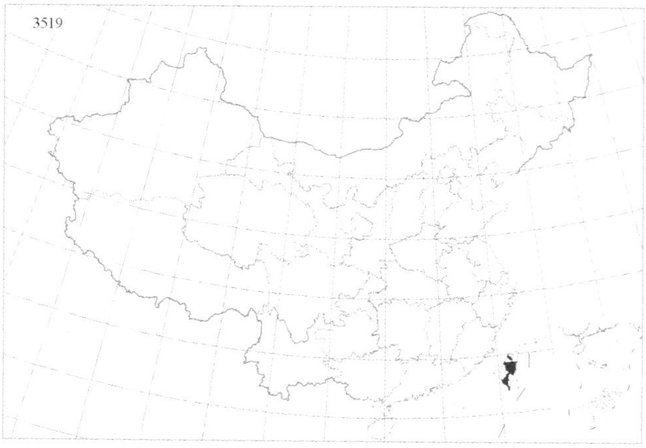

Rosaceae 蔷薇科 *Rubus* 悬钩子属

Rubus komarovii Nakai 绿叶悬钩子

Shrubs, deciduous

1.7 (−5.0~7.7); −18.4 (−29.4~−10.0); 19.5 (17.0~23.0); 7.1 (5.0~9.8); 53.7 (35.2~81.4); 92.7 (48.7~154.8); 535 (450~653) ❙ 572 (457~777); 365 (300~473); 17 (9~35) ❙ 7.6 (−14.5~55.1); 512 (450~569); 291.1 (238.2~329.9)

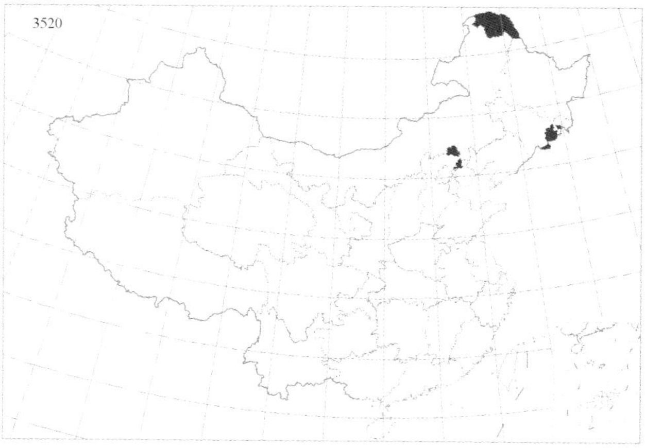

Rosaceae 蔷薇科 *Rubus* 悬钩子属

Rubus kulinganus L. H. Bailey 牯岭悬钩子

Shrubs, deciduous

15.9 (14.7~17.1); 3.8 (2.5~5.0); 27.8 (26.8~29.3); 15.9 (14.8~17.1); 132.4 (121.1~145.7); 1.5 (0.0~3.9); 872 (815~941) ❙ 1511 (1317~1675); 576 (525~685); 174 (122~212) ❙ 73.8 (51.1~101.3); 872 (815~941); 240.6 (183.4~318.0)

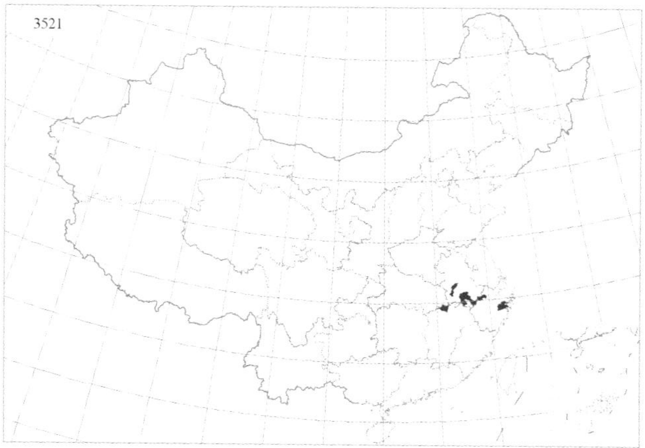

Rosaceae 蔷薇科 *Rubus* 悬钩子属

Rubus kwangsiensis H. L. Li 广西悬钩子

Shrubs, deciduous

16.6 (14.3~19.6); 6.4 (3.3~11.2); 25.6 (24.3~26.7); 16.7 (14.3~19.6); 140.7 (114.1~175.5); 0.8 (0.0~2.3); 866 (765~986) ❙ 1524 (1400~1624); 616 (530~776); 156 (74~191) ❙ 78.0 (42.0~101.8); 866 (765~986); 255.3 (225.9~294.6)

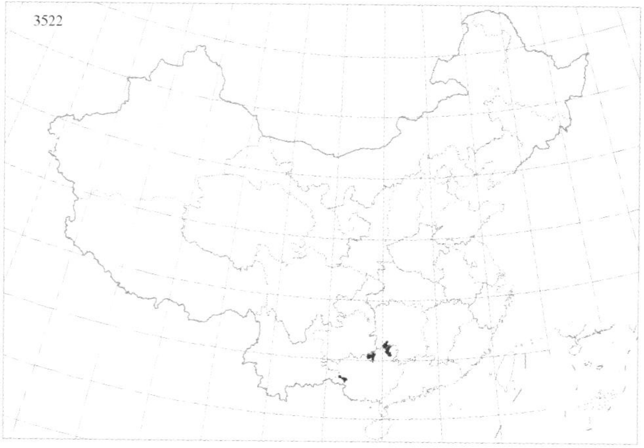

Rosaceae 蔷薇科 *Rubus* 悬钩子属
Rubus lambertianus Ser. 高粱泡
Shrubs, semideciduous

16.5 (6.0~24.8); 5.4 (−3.2~20.1); 26.8 (14.4~30.1); 16.6 (6.6~24.8); 140.9 (39.7~237.6); 2.2 (0.0~26.7); 893 (508~1385) ❙ 1318 (512~2859); 525 (258~1186); 132 (14~402) ❙ 47.7 (−35.3~246.5); 875 (508~1360); 218.1 (108.2~620.0)

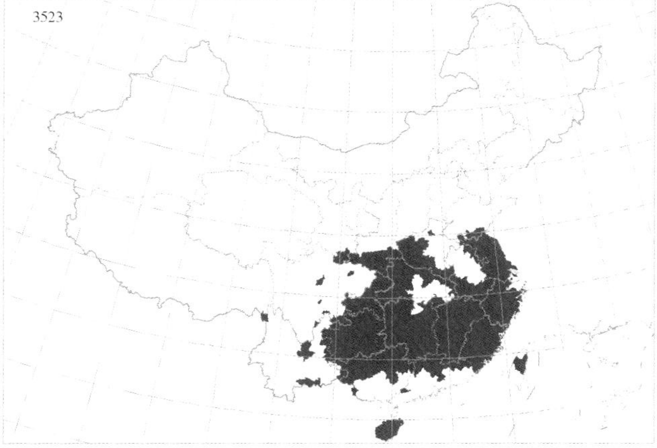

Rosaceae 蔷薇科 *Rubus* 悬钩子属
Rubus lambertianus var. glaber Hemsl. 光滑高粱泡
Shrubs, semideciduous

13.6 (2.2~17.7); 1.4 (−6.6~6.9); 25.1 (9.9~28.8); 13.7 (3.8~17.7); 111.4 (15.5~152.2); 7.9 (0.0~49.1); 780 (413~937) ❙ 1002 (626~1571); 452 (316~688); 64 (14~196) ❙ 29.8 (−22.2~125.9); 762 (413~937); 205.0 (126.1~334.4)

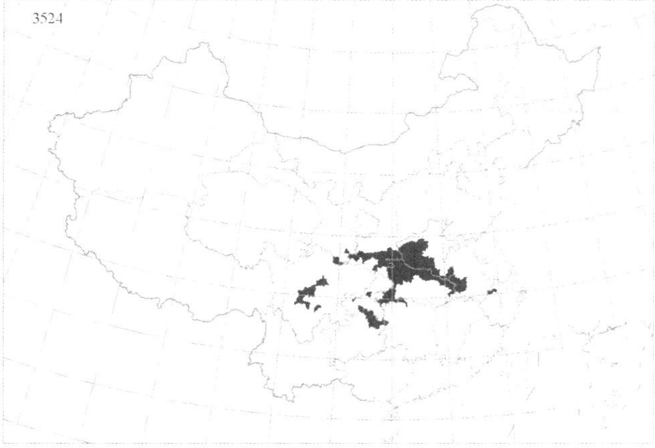

Rosaceae 蔷薇科 *Rubus* 悬钩子属
Rubus lambertianus var. glandulosus Cardot 腺毛高粱泡
Shrubs, semideciduous

15.1 (13.3~18.5); 5.8 (2.6~14.1); 23.4 (21.6~24.6); 15.2 (13.3~18.5); 123.0 (101.4~162.1); 0.6 (0.0~4.1); 784 (715~863) ❙ 1394 (999~2859); 646 (501~1388); 93 (53~402) ❙ 75.9 (37.5~235.6); 784 (715~863); 203.1 (153.9~363.7)

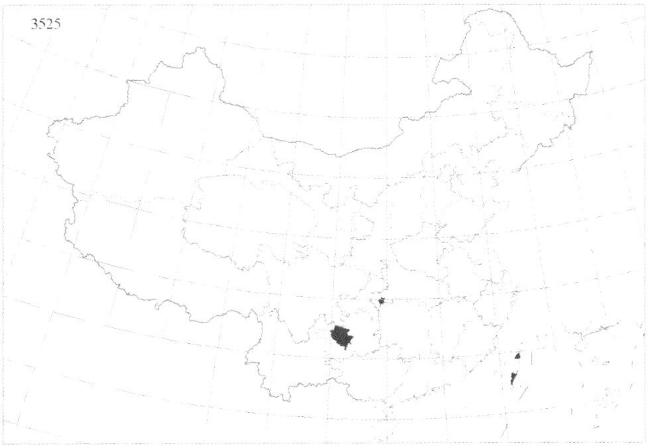

Rosaceae 蔷薇科 *Rubus* 悬钩子属
Rubus lambertianus var. paykouangensis (H. Lév.) Hand.-Mazz. 毛叶高粱泡
Shrubs, semideciduous

16.0 (11.7~19.1); 6.4 (2.7~13.0); 24.4 (18.8~26.8); 16.1 (11.7~19.2); 133.5 (82.6~170.1); 0.4 (0.0~3.8); 829 (658~981) ❙ 1228 (916~1428); 573 (465~835); 74 (31~129) ❙ 48.4 (13.3~71.9); 829 (658~981); 240.7 (146.0~506.2)

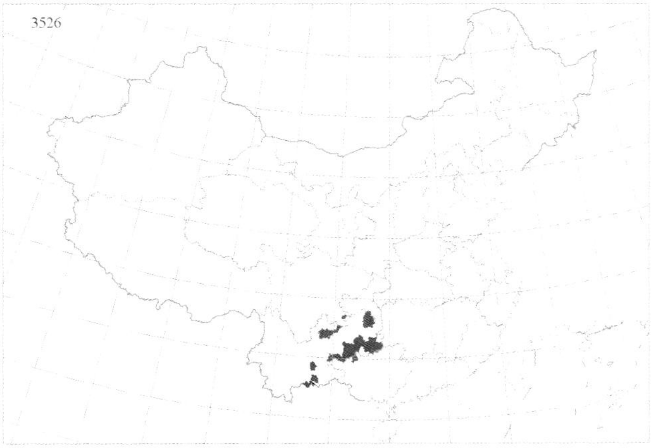

Rosaceae 蔷薇科 *Rubus* 悬钩子属
Rubus lanyuensis Chang 兰屿悬钩子
Shrubs

18.5; 14.1; 21.7; 18.5; 162.1; 0.0; 863 ❙ 2713; 1388; 188 ❙ 214.4; 863; 363.7

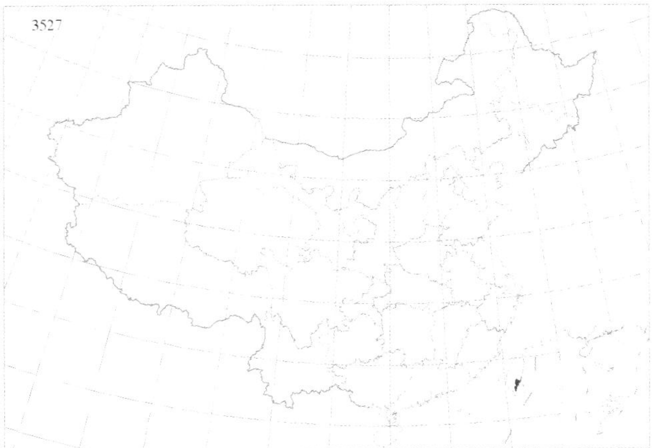

Rosaceae 蔷薇科 *Rubus* 悬钩子属
Rubus lasiostylus Focke 绵果悬钩子
Shrubs, deciduous

11.7 (2.2~14.9); −0.1 (−6.6~3.6); 22.9 (9.9~27.8); 11.9 (3.8~15.0); 93.3 (15.5~123.8); 12.4 (1.6~49.1); 700 (413~854) ❙ 932 (581~1492); 420 (257~613); 40 (12~106) ❙ 35.0 (−32.0~125.9); 693 (413~817); 256.4 (151.3~334.4)

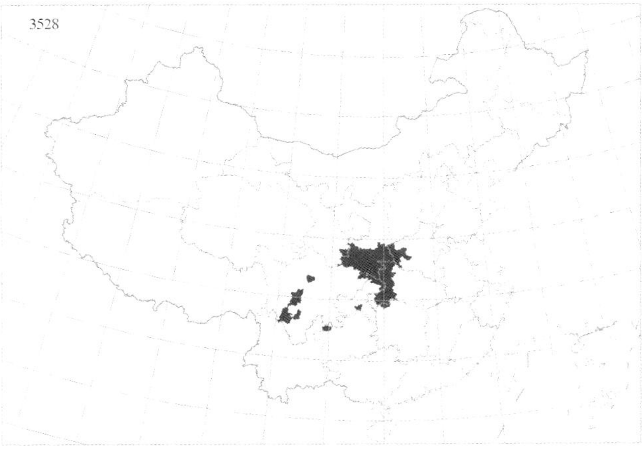

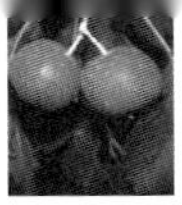

Rosaceae 蔷薇科 *Rubus* 悬钩子属
Rubus lasiostylus var. dizygos Focke 五叶绵果悬钩子
Shrubs, deciduous

12.8; 2.9; 22.0; 12.9; 97.6; 3.2; 711 ❙ 1270; 593; 61 ❙ 77.5; 711; 284.1

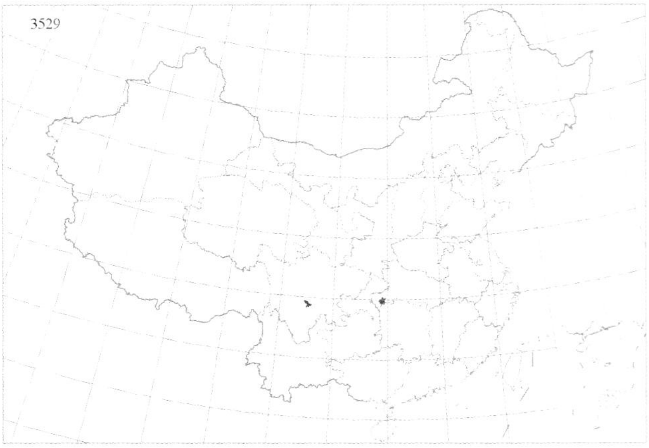

Rosaceae 蔷薇科 *Rubus* 悬钩子属
Rubus lasiotrichos Focke 多毛悬钩子
Shrubs, deciduous

16.6; 8.9; 22.5; 16.6; 139.6; 0.6; 816 ❙ 1280; 633; 84 ❙ 57.6; 804; 374.3

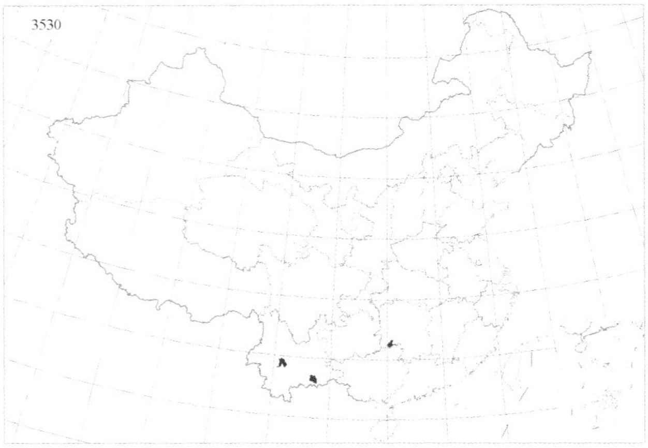

Rosaceae 蔷薇科 *Rubus* 悬钩子属
Rubus latoauriculatus F. P. Metcalf 耳叶悬钩子
Shrubs, deciduous

21.1 (17.6~22.7); 12.5 (7.3~15.0); 27.7 (25.0~28.6); 21.2 (17.6~22.8); 194.0 (151.6~213.1); 0.0 (0.0~0.0); 1114 (909~1221) ❙ 1674 (1302~2435); 870 (639~1428); 132 (57~194) ❙ 50.7 (12.9~99.5); 1114 (909~1221); 269.6 (206.0~371.7)

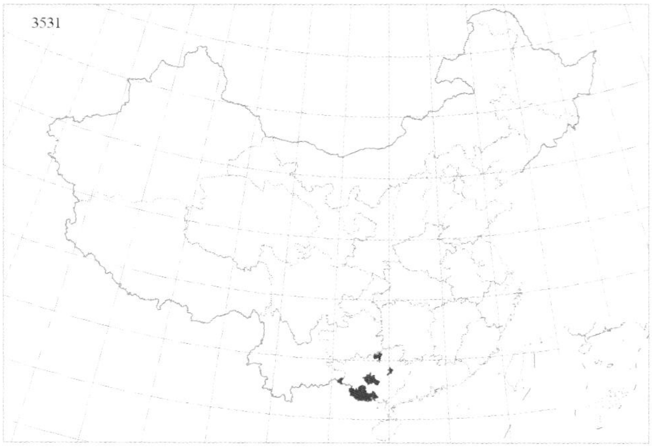

Rosaceae 蔷薇科 *Rubus* 悬钩子属
Rubus laxus Focke 疏松悬钩子
Shrubs, deciduous

18.1 (8.6~21.0); 12.0 (1.3~15.1); 22.1 (15.0~25.6); 18.2 (8.7~21.0); 158.5 (52.6~191.8); 0.4 (0.0~8.8); 878 (574~1053) ❙ 1372 (1041~1751); 739 (585~996); 55 (41~77) ❙ 56.6 (26.1~88.8); 862 (574~1053); 460.1 (352.3~659.0)

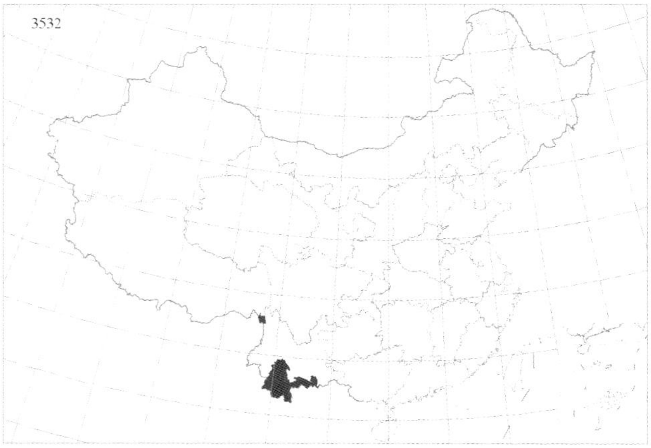

Rosaceae 蔷薇科 *Rubus* 悬钩子属
Rubus leucanthus Hance 白花悬钩子
Shrubs, deciduous

18.6 (9.6~24.6); 9.0 (1.6~19.5); 26.9 (15.7~29.5); 18.7 (9.6~24.7); 164.1 (60.8~235.9); 0.3 (0.0~6.9); 982 (600~1372) ❙ 1534 (899~2030); 623 (403~1056); 144 (19~230) ❙ 58.2 (-12.0~128.9); 974 (600~1360); 254.2 (139.6~620.0)

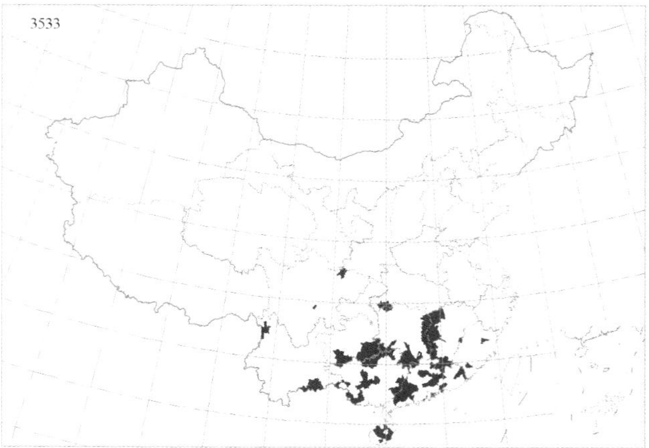

Rosaceae 蔷薇科 *Rubus* 悬钩子属
Rubus lichuanensis T. T. Yu et L. T. Lu 黎川悬钩子
Shrubs, deciduous

18.1; 6.9; 28.5; 18.1; 157.7; 0.0; 965 ❙ 1773; 564; 233 ❙ 83.7; 944; 261.7

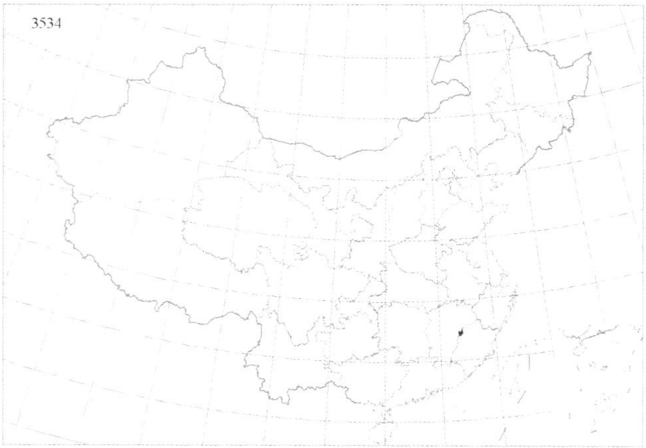

Rosaceae 蔷薇科 *Rubus* 悬钩子属
Rubus lineatus Reinw. 绢毛悬钩子
Shrubs, deciduous

15.3 (2.4~21.0); 8.2 (-7.2~14.2); 20.5 (11.3~25.6); 15.4 (4.4~21.0); 127.3 (21.4~191.8); 3.3 (0.0~51.6); 783 (423~1053) ❙ 1265 (510~1736); 713 (311~1007); 52 (11~70) ❙ 62.8 (20.7~145.8); 780 (423~1053); 405.5 (237.6~55[illegible].6)

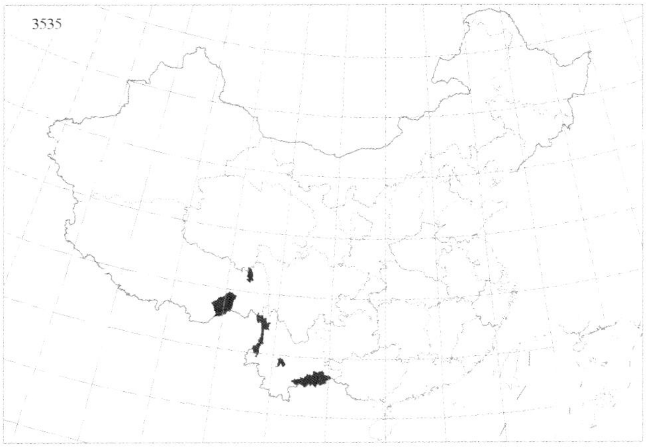

Rosaceae 蔷薇科 *Rubus* 悬钩子属
Rubus lineatus var. angustifolius Hook. f. 狭叶绢毛悬钩子
Shrubs, deciduous

9.1; 1.9; 15.3; 9.1; 56.7; 7.1; 587 ❙ 1024; 581; 42 ❙ 74.7; 587; 352.0

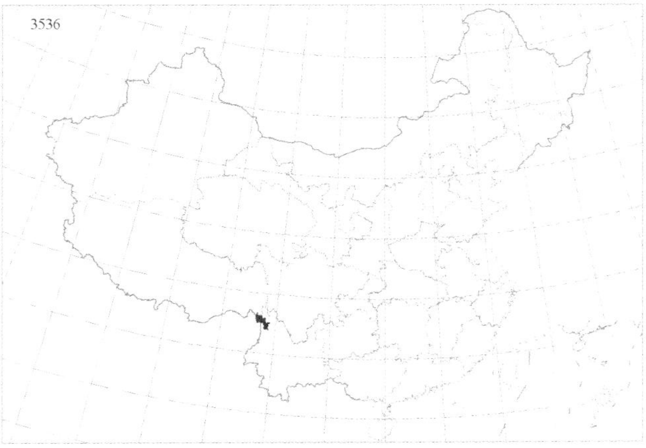

Rosaceae 蔷薇科 *Rubus* 悬钩子属
Rubus lineatus var. glabrescens T. T. Yu et L. T. Lu 光秃绢毛悬钩子
Shrubs, deciduous

18.6 (16.9~21.0); 12.3 (10.3~14.2); 22.9 (21.4~25.6); 18.6 (16.9~21.0); 163.5 (142.7~191.8); 0.0 (0.0~0.0); 898 (813~1053) ❙ 1352 (1163~1547) 765 (641~911); 56 (48~68) ❙ 50.7 (34.8~74.2); 897 (813~1053); 472.0 (383.4~551.6)

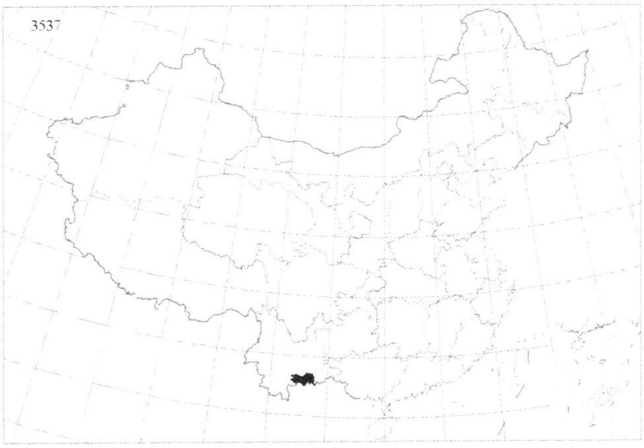

Rosaceae 蔷薇科 *Rubus* 悬钩子属
Rubus lishuiensis T. T. Yu et L. T. Lu 丽水悬钩子
Shrubs, deciduous

16.7; 5.5; 27.9; 16.8; 141.5; 0.0; 896 ❙ 1573; 555; 190 ❙ 75.7; 896; 224.8

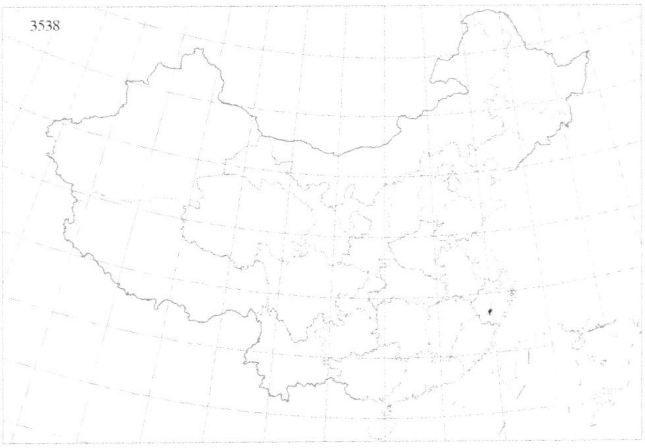

Rosaceae 蔷薇科 *Rubus* 悬钩子属
Rubus lobatus T. T. Yu et L. T. Lu 五裂悬钩子
Shrubs, deciduous

18.5 (14.6~21.8); 8.8 (3.8~14.2); 26.8 (24.6~29.0); 18.5 (14.7~21.9); 162.7 (117.7~202.6); 0.1 (0.0~1.5); 966 (779~1168) ❙ 1562 (1371~1876); 604 (517~891); 167 (116~191) ❙ 64.1 (17.4~101.8); 963 (779~1150); 244.4 (214.2~273.3)

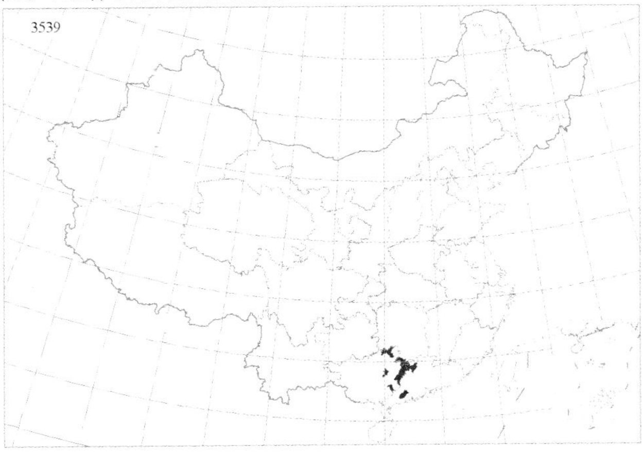

Rosaceae 蔷薇科 *Rubus* 悬钩子属
Rubus lobophyllus Y. K. Shih ex F. P. Metcalf 角裂悬钩子
Shrubs, deciduous

17.2 (14.3~21.8); 7.4 (3.3~14.2); 25.6 (22.2~28.6); 17.2 (14.3~21.9); 146.9 (114.1~202.7); 0.2 (0.0~2.3); 888 (765~1159) ❙ 1425 (1232~1776); 621 (419~828); 121 (58~198) ❙ 61.5 (30.5~101.8); 887 (765~1159); 243.9 (171.2~551.6)

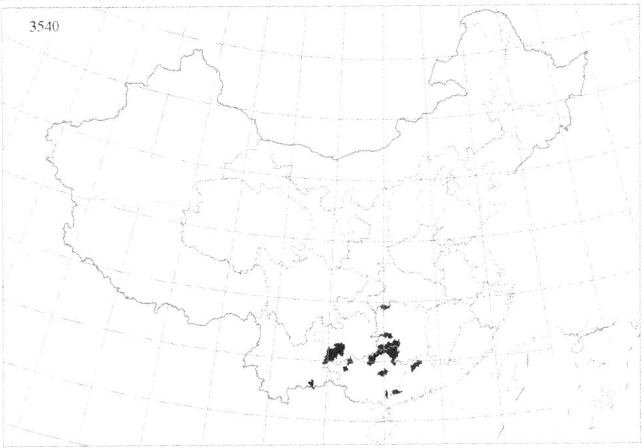

Rosaceae 蔷薇科 *Rubus* 悬钩子属
Rubus lohfauensis F. P. Metcalf 罗浮山悬钩子
Scandent shrubs

21.6；13.2；28.2；21.6；199.2；0.0；1139 ❙ 1953；858；167 ❙ 71.5；1139；240.0

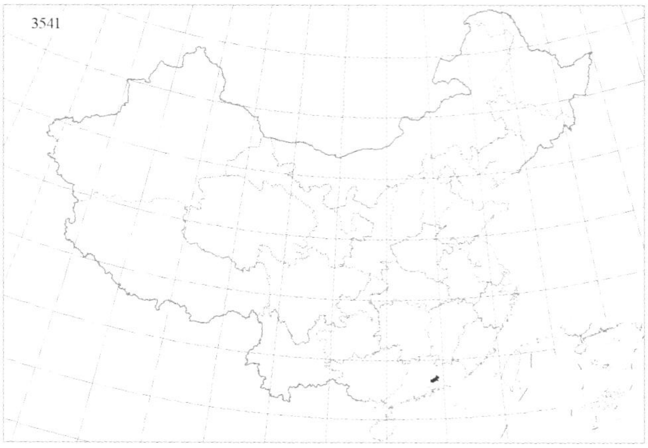

Rosaceae 蔷薇科 *Rubus* 悬钩子属
Rubus lucens Focke 光亮悬钩子
Shrubs, deciduous

19.0 (17.7~21.0)；13.0 (10.9~15.1)；22.9 (21.6~25.6)；19.1 (17.8~21.0)；168.9 (153.4~191.8)；0.0 (0.0~0.0)；916 (854~1053) ❙ 1401 (1028~1751)；758 (566~996)；59 (45~77) ❙ 53.2 (12.5~88.8)；899 (816~1053)；461.8 (339.5~659.0)

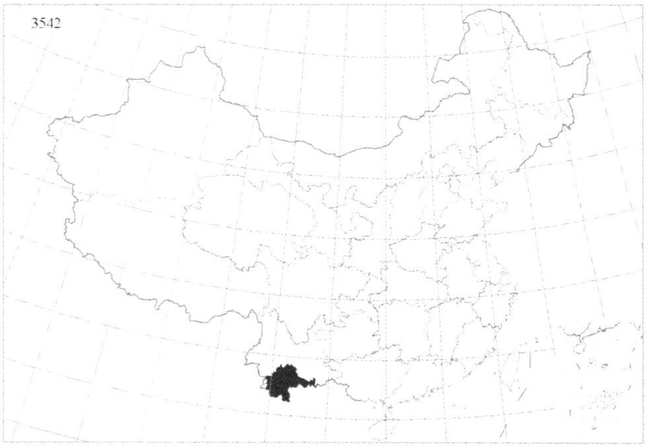

Rosaceae 蔷薇科 *Rubus* 悬钩子属
Rubus luchunensis T. T. Yu et L. T. Lu 绿春悬钩子
Shrubs, deciduous

18.6；12.6；22.6；18.6；163.6；0.0；889 ❙ 1437；840；58 ❙ 61.6；889；496.8

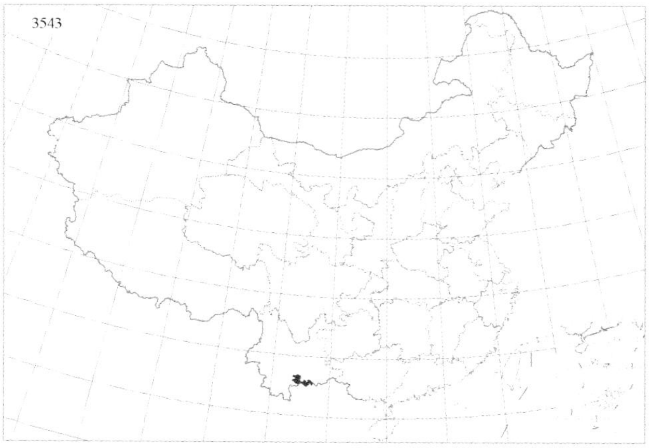

Rosaceae 蔷薇科 *Rubus* 悬钩子属
Rubus luchunensis var. coriaceus T. T. Yu et L. T. Lu 硬叶绿春悬钩子
Shrubs, deciduous

18.4；12.0；22.9；18.5；161.5；0.0；886 ❙ 1398；796；58 ❙ 57.9；886；498.8

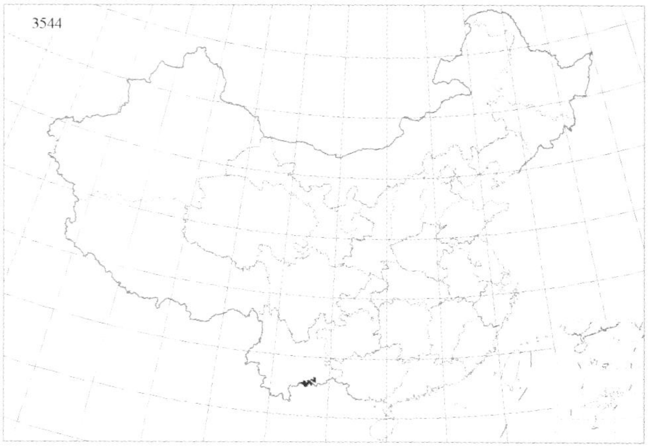

Rosaceae 蔷薇科 *Rubus* 悬钩子属
Rubus lutescens Franch. 黄色悬钩子
Subshrubs, deciduous

7.7 (-1.0~16.7)；-0.4 (-11.8~9.7)；14.5 (8.5~22.1)；8.5 (2.7~16.7)；56.4 (8.3~140.9)；23.9 (0.0~79.5)；556 (363~819) ❙ 871 (520~1243)；521 (317~696)；23 (8~58) ❙ 59.4 (14.4~125.9)；544 (363~721)；287.4 (130.1~384.7)

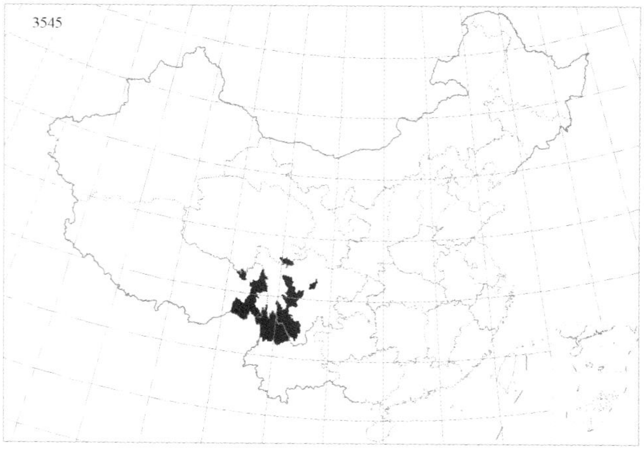

Rosaceae 蔷薇科 *Rubus* 悬钩子属
Rubus macilentus Cambess. 细瘦悬钩子
Shrubs, deciduous

12.9 (-1.8~21.0)；5.3 (-10.2~14.2)；19.1 (6.1~26.6)；13.1 (1.8~21.0)；100.6 (1.9~191.8)；5.2 (0.0~82.8)；702 (329~1053) ❙ 1032 (320~1736)；588 (217~1007)；37 (13~70) ❙ 49.1 (-14.2~145.8)；685 (320~1053)；293.6 (72.2~551.6)

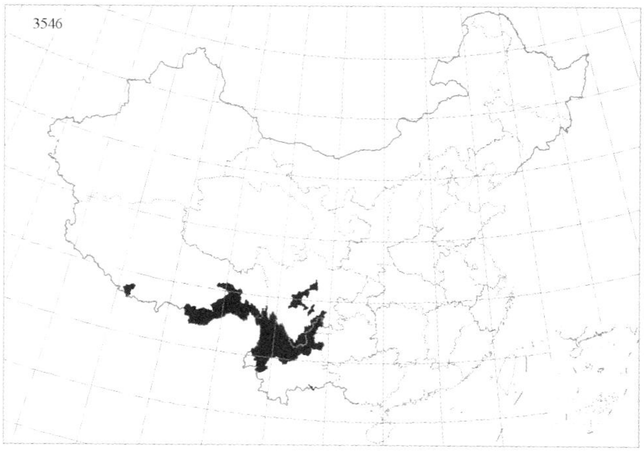

Rosaceae 蔷薇科 *Rubus* 悬钩子属
Rubus macilentus var. angulatus Delavay 棱枝细瘦悬钩子
Shrubs, deciduous

11.9 (8.6~14.5); 5.1 (1.3~7.8); 17.5 (15.0~19.7); 12.0 (8.7~14.5); 85.3 (52.6~113.5); 1.6 (0.0~8.8); 662 (574~730) ❙ 1107 (991~1337); 626 (570~727); 48 (35~69) ❙ 68.0 (36.1~93.0); 660 (574~730); 358.2 (296.9~407.1)

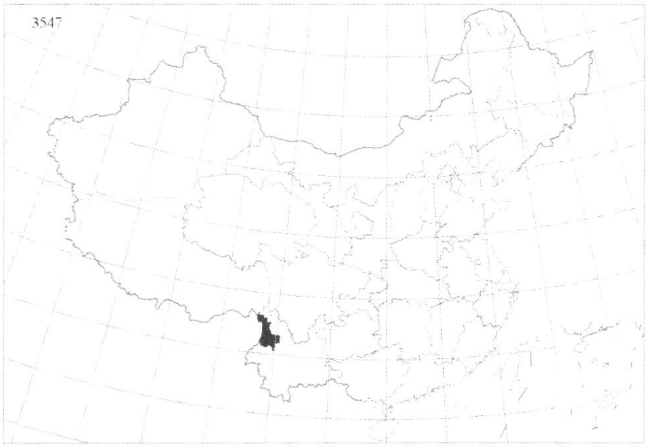

Rosaceae 蔷薇科 *Rubus* 悬钩子属
Rubus malifolius Focke 棠叶悬钩子
Shrubs, deciduous

16.7 (11.1~22.2); 6.1 (1.6~15.2); 26.4 (19.4~29.8); 16.8 (11.1~22.7); 141.7 (79.8~212.7); 0.4 (0.0~6.9); 884 (648~1216) ❙ 1388 (999~2005); 551 (392~994); 131 (30~230) ❙ 57.4 (8.0~128.9); 878 (648~1216); 213.3 (108.2~551.6)

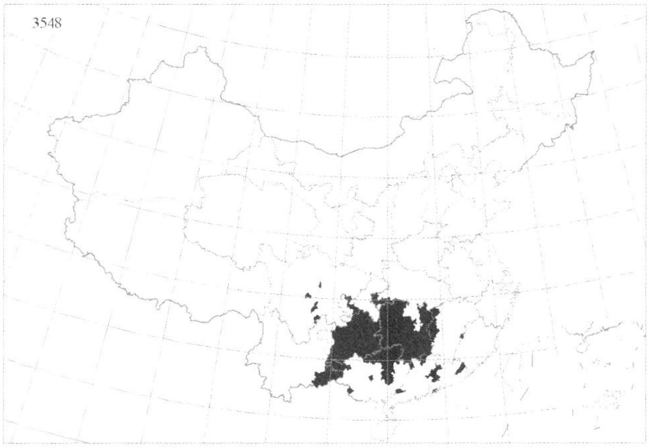

Rosaceae 蔷薇科 *Rubus* 悬钩子属
Rubus malipoensis T. T. Yu et L. T. Lu 麻栗坡悬钩子
Shrubs, deciduous

17.4; 9.7; 22.9; 17.5; 149.7; 0.0; 848 ❙ 1338; 756; 50 ❙ 57.6; 848; 348.0

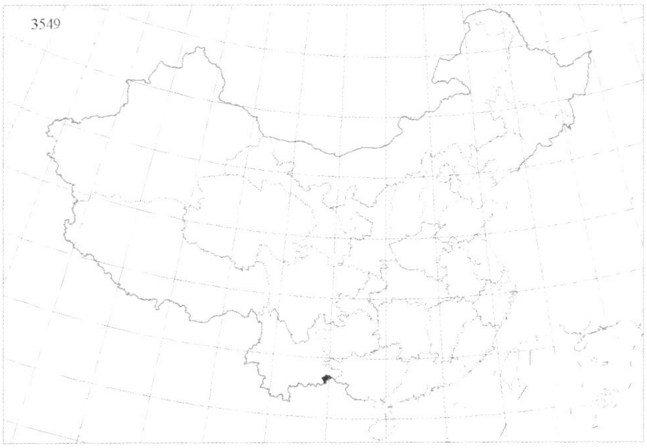

Rosaceae 蔷薇科 *Rubus* 悬钩子属
Rubus mallotifolius C. Y. Wu ex T. T. Yu et L. T. Lu 楸叶悬钩子
Shrubs, deciduous

18.6 (16.9~21.0); 11.9 (9.9~14.2); 23.3 (21.4~25.6); 18.7 (16.9~21.0); 163.8 (142.7~191.8); 0.0 (0.0~0.0); 904 (813~1053) ❙ 1408 (1178~1547); 798 (646~911); 58 (51~68) ❙ 56.1 (36.8~80.6); 903 (813~1053); 465.2 (371.7~551.6)

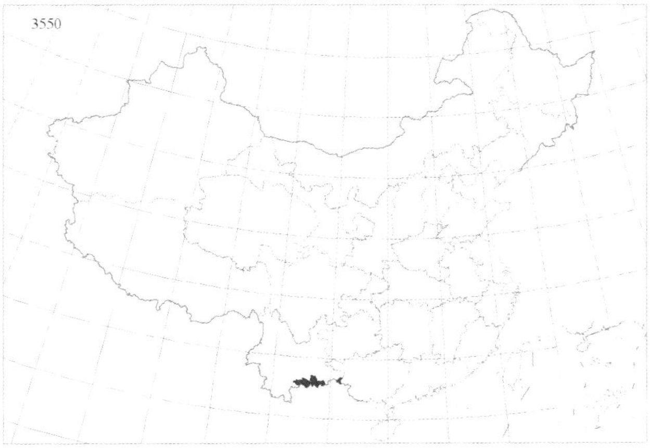

Rosaceae 蔷薇科 *Rubus* 悬钩子属
Rubus menglaensis T. T. Yu et L. T. Lu 勐腊悬钩子
Shrubs, deciduous

20.4; 15.1; 23.6; 20.5; 185.8; 0.0; 989 ❙ 1607; 702; 77 ❙ 62.5; 986; 659.0

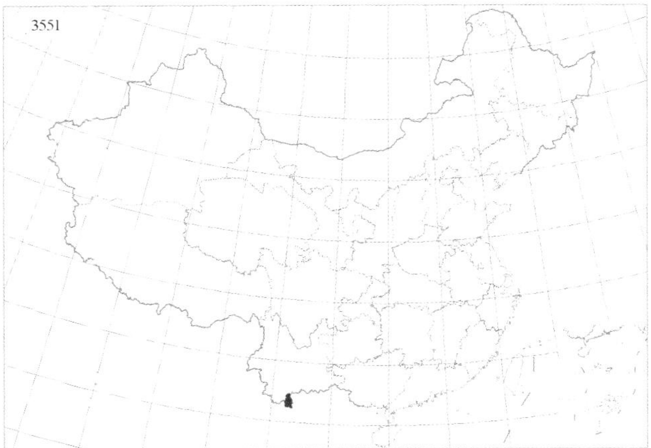

Rosaceae 蔷薇科 *Rubus* 悬钩子属
Rubus mesogaeus Focke 喜阴悬钩子
Shrubs, deciduous

12.3 (1.8~20.8); 1.6 (-7.6~15.3); 22.0 (9.9~28.2); 12.6 (3.8~20.8); 99.2 (15.4~189.2); 10.9 (0.0~53.3); 714 (404~1019) ❙ 914 (341~2680); 464 (190~1412); 38 (7~245) ❙ 29.7 (-40.9~246.5); 677 (341~1019); 236.6 (61.2~407.1)

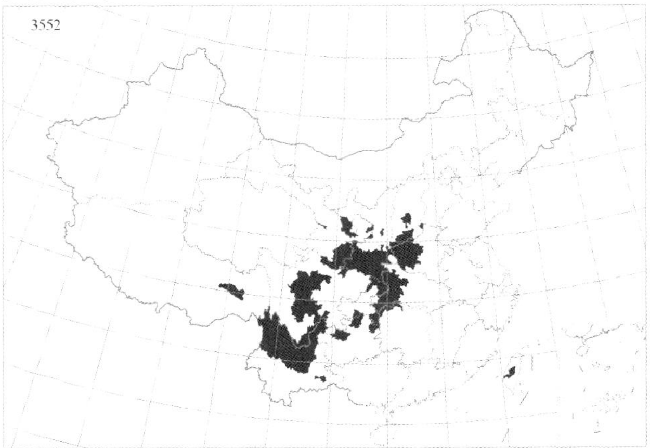

Rosaceae 蔷薇科 *Rubus* 悬钩子属

Rubus mesogaeus var. glabrescens T. T. Yu et L. T. Lu 脱毛喜阴悬钩子

Shrubs, deciduous

13.4; 3.6; 22.4; 13.4; 103.2; 2.1; 731 ❙ 1099; 519; 53 ❙ 50.2; 731; 230.8

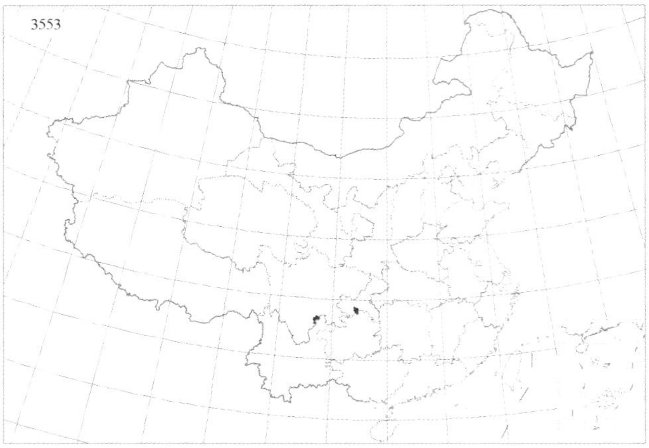

Rosaceae 蔷薇科 *Rubus* 悬钩子属

Rubus mesogaeus var. oxycomus Focke 腺毛喜阴悬钩子

Shrubs, deciduous

4.9 (2.0~10.2); -4.0 (-7.4~2.4); 12.6 (10.1~16.9); 5.9 (4.0~10.3); 32.4 (16.9~68.2); 33.6 (5.1~51.9); 479 (416~620) ❙ 888 (731~1041); 491 (346~606); 21 (15~45) ❙ 86.1 (63.8~121.7); 479 (416~620); 298.0 (221.3~357.7)

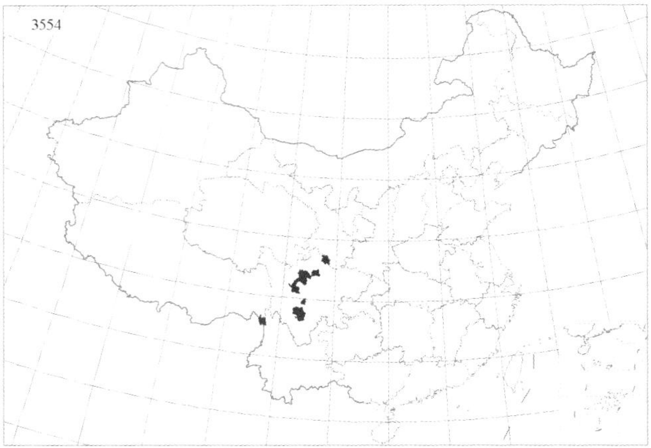

Rosaceae 蔷薇科 *Rubus* 悬钩子属

Rubus metoensis T. T. Yu et L. T. Lu 墨脱悬钩子

Small shrubs, deciduous

13.3; 5.4; 19.6; 13.4; 100.4; 0.0; 706 ❙ 1736; 1007; 51 ❙ 145.8; 706; 447.4

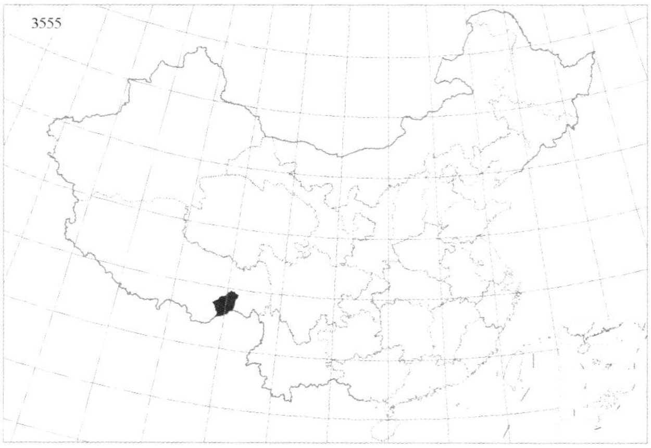

Rosaceae 蔷薇科 *Rubus* 悬钩子属

Rubus multisetosus T. T. Yu et L. T. Lu 刺毛悬钩子

Shrubs, deciduous

14.8; 8.1; 19.9; 14.9; 118.3; 0.0; 751 ❙ 1019; 570; 47 ❙ 37.4; 743; 348.1

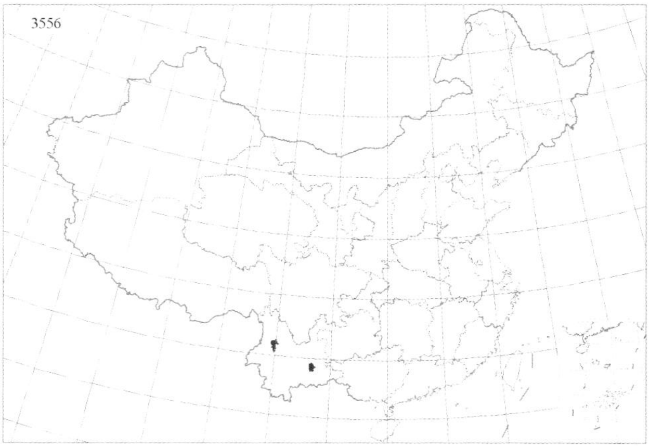

Rosaceae 蔷薇科 *Rubus* 悬钩子属

Rubus nagasawanus Koidz. 高砂悬钩子

Shrubs, deciduous

17.0 (13.7~20.1); 5.8 (3.5~11.5); 27.4 (20.3~30.1); 17.0 (13.7~20.1); 144.4 (104.8~181.6); 0.3 (0.0~1.5); 906 (722~1052) ❙ 1552 (1143~2499); 547 (427~947); 189 (50~303) ❙ 72.4 (28.8~219.3); 889 (722~1052); 220.3 (141.7~303.6)

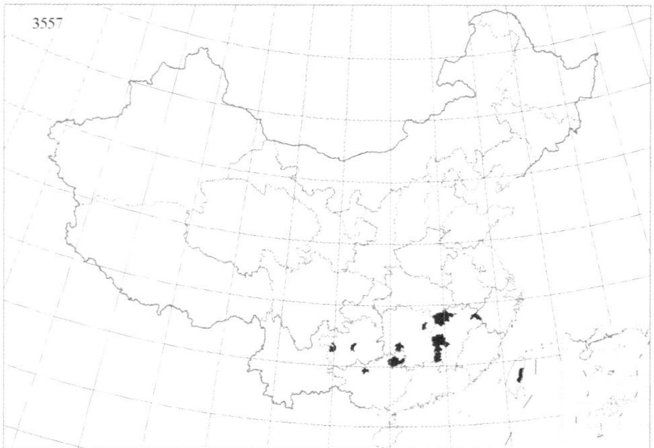

Rosaceae 蔷薇科 *Rubus* 悬钩子属

Rubus neoviburnifolius L. T. Lu et Boufford 荚蒾叶悬钩子

Shrubs, deciduous

16.7 (13.8~19.6); 10.1 (6.7~13.1); 21.3 (19.1~23.9); 16.8 (13.8~19.6); 141.3 (105.9~175.7); 0.0 (0.0~0.0); 816 (714~953) ❙ 1148 (797~1688); 654 (411~959); 46 (23~70) ❙ 40.7 (-14.2~112.2); 792 (690~932); 351.8 (218.0~627.0)

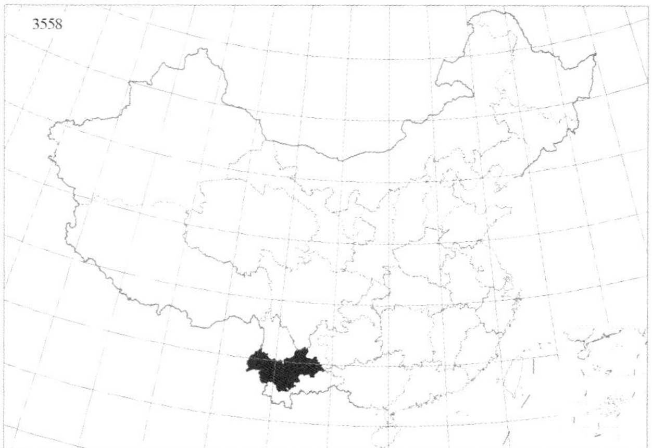

Rosaceae 蔷薇科 *Rubus* 悬钩子属
Rubus niveus Thunb. 红泡刺藤
Shrubs, deciduous

14.7 (-1.8~21.8); 5.7 (-10.2~15.1); 22.4 (6.1~28.4); 14.8 (1.8~21.9); 121.3 (1.9~202.7); 4.3 (0.0~82.8); 784 (329~1162) ❙ 1131 (320~2859); 573 (193~1186); 58 (9~402) ❙ 45.1 (-20.0~246.5); 777 (320~1162); 270.1 (61.6~659.0)

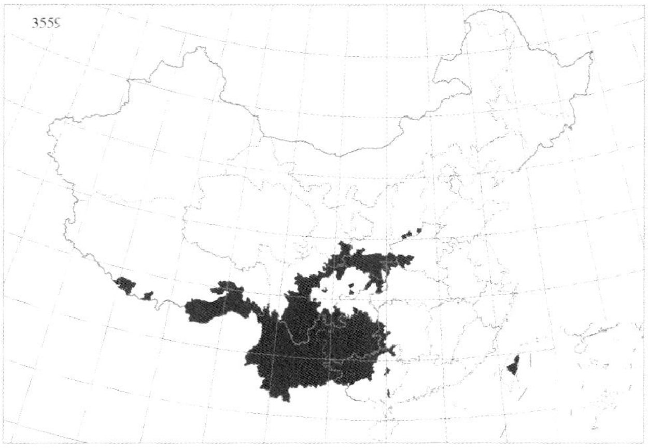

Rosaceae 蔷薇科 *Rubus* 悬钩子属
Rubus oblongus T. T. Yu et L. T. Lu 长圆悬钩子
Shrubs, deciduous

14.6 (11.7~18.5); 5.7 (2.7~10.9); 22.1 (19.6~24.3); 14.6 (11.7~18.3); 116.3 (84.6~160.0); 1.3 (0.0~3.8); 760 (661~906) ❙ 1042 (956~1196); 547 (501~597); 47 (20~60) ❙ 38.8 (5.5~59.3); 741 (661~836); 231.2 (179.5~340.7)

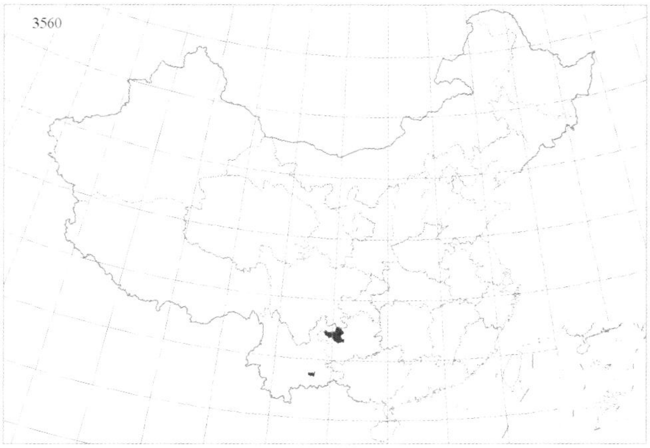

Rosaceae 蔷薇科 *Rubus* 悬钩子属
Rubus ourosepalus Cardot 宝兴悬钩子
Shrubs, deciduous

7.3; -2.0; 15.6; 7.5; 48.2; 20.4; 537 ❙ 1005; 531; 27 ❙ 87.3; 537; 272.3

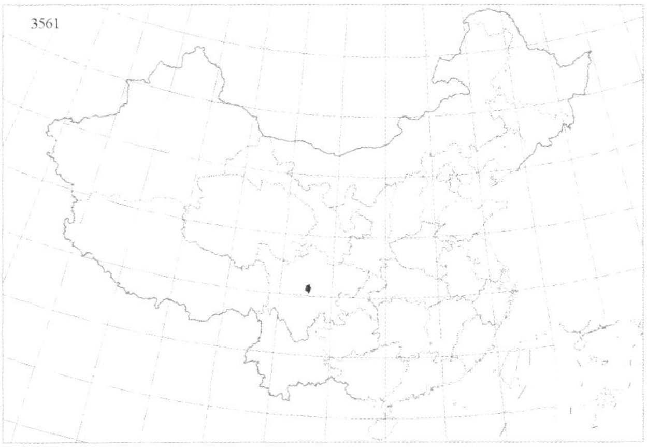

Rosaceae 蔷薇科 *Rubus* 悬钩子属
Rubus pacificus Hance 太平莓
Shrubs, evergreen

16.4 (14.4~22.0); 4.6 (2.1~14.1); 27.6 (24.6~29.8); 16.4 (14.5~22.1); 138.1 (117.7~205.1); 1.0 (0.0~4.9); 888 (779~1170) ❙ 1571 (1002~2014); 560 (419~954); 193 (112~259) ❙ 77.3 (14.0~147.0); 883 (779~1170); 231.3 (134.1~319.9)

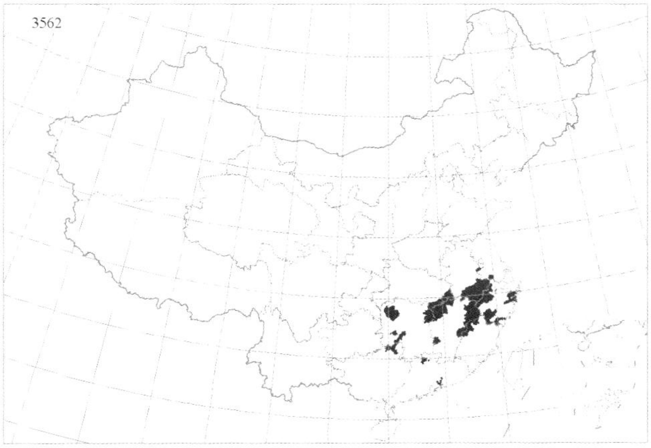

Rosaceae 蔷薇科 *Rubus* 悬钩子属
Rubus panduratus Hand.-Mazz. 琴叶悬钩子
Shrubs, deciduous

17.9 (14.3~21.3); 8.0 (3.3~13.2); 26.4 (22.4~28.6); 18.0 (14.3~21.3); 156.0 (112.6~196.1); 0.2 (0.0~2.3); 936 (756~1130) ❙ 1510 (1221~1900); 631 (476~1056); 149 (55~191) ❙ 62.5 (22.7~101.8); 936 (756~1130); 237.8 (185.5~413.5)

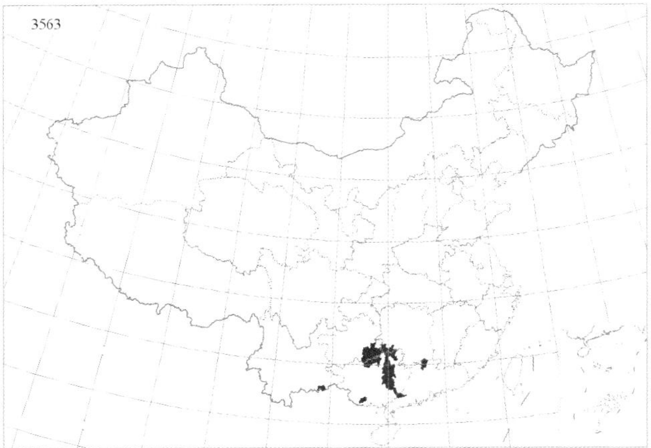

Rosaceae 蔷薇科 *Rubus* 悬钩子属
Rubus panduratus var. etomentosus Hand.-Mazz. 脱毛琴叶悬钩子
Shrubs, deciduous

17.7; 7.5; 26.5; 17.7; 152.6; 0.0; 912 ❙ 1530; 597; 162 ❙ 67.7; 912; 229.2

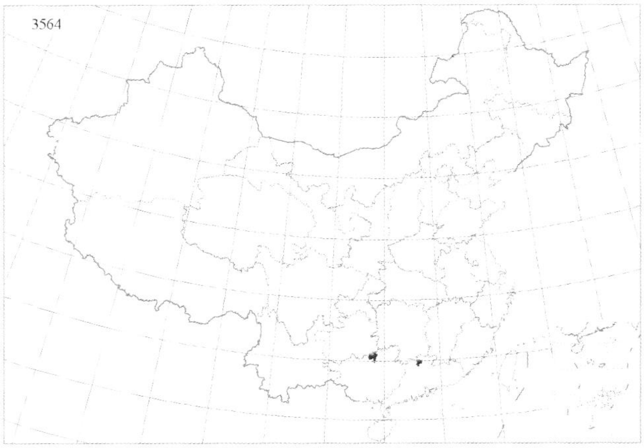

Rosaceae 蔷薇科　*Rubus* 悬钩子属
Rubus paniculatus Sm. 圆锥悬钩子
Shrubs, deciduous

16.0 (−1.8~21.0); 9.4 (−10.2~14.2); 20.7 (6.1~26.4); 16.1 (1.8~21.0); 134.5 (1.9~191.8); 2.2 (0.0~82.8); 797 (329~1053) ❙ 1162 (320~1751); 663 (217~996); 48 (10~77) ❙ 45.5 (−3.0~112.2); 781 (320~1053); 368.3 (61.6~627.0)

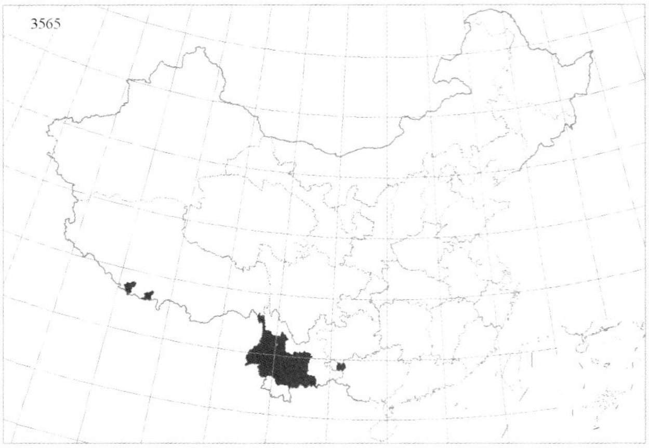

Rosaceae 蔷薇科　*Rubus* 悬钩子属
Rubus paniculatus var. glabrescens T. T. Yu et L. T. Lu 脱毛圆锥悬钩子
Shrubs, deciduous

8.6; 1.3; 15.0; 8.7; 52.6; 8.8; 574 ❙ 1041; 585; 45 ❙ 81.5; 574; 352.3

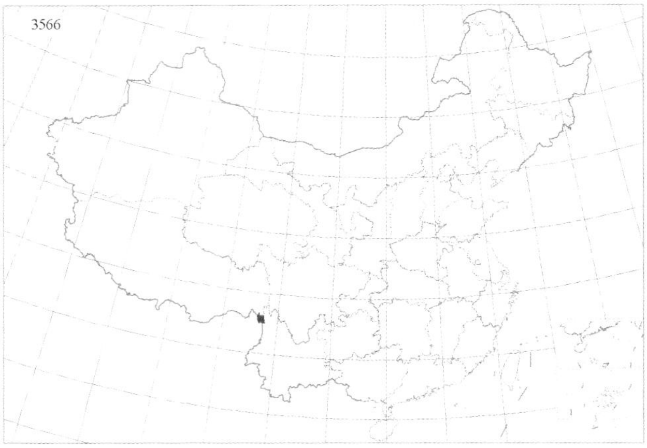

Rosaceae 蔷薇科　*Rubus* 悬钩子属
Rubus pararosifolius F. P. Metcalf 矮空心泡
Shrubs, deciduous

18.6; 9.3; 27.6; 18.7; 164.0; 0.0; 965 ❙ 1702; 591; 199 ❙ 76.4; 965; 300.0

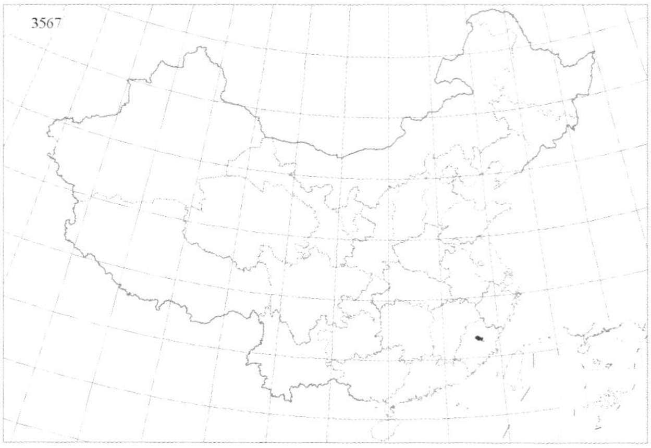

Rosaceae 蔷薇科　*Rubus* 悬钩子属
Rubus parkeri Hance 乌泡子
Shrubs, deciduous

14.6 (9.3~21.0); 3.1 (−2.9~14.2); 25.4 (19.4~28.9); 14.6 (9.5~21.0); 120.1 (69.8~191.8); 4.8 (0.0~20.6); 805 (603~1053) ❙ 1073 (626~1997); 480 (316~818); 66 (15~254) ❙ 33.7 (−22.2~130.7); 795 (603~1053); 205.6 (108.2~551.6)

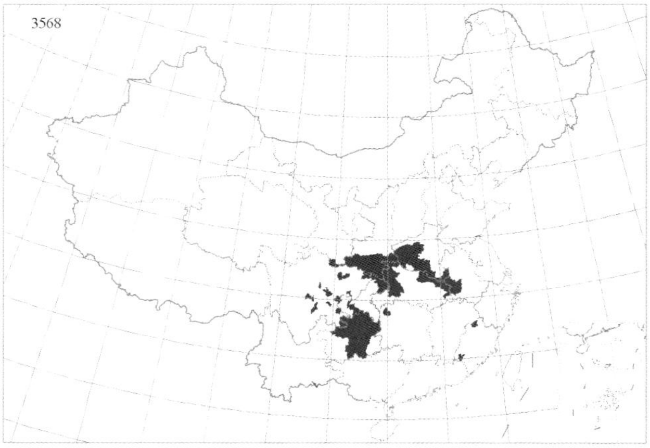

Rosaceae 蔷薇科　*Rubus* 悬钩子属
Rubus parviaraliifolius Hayata 楤叶悬钩子
Shrubs, deciduous

17.6; 12.0; 22.3; 17.6; 151.4; 0.0; 849 ❙ 2758; 1004; 374 ❙ 226.2; 849; 280.0

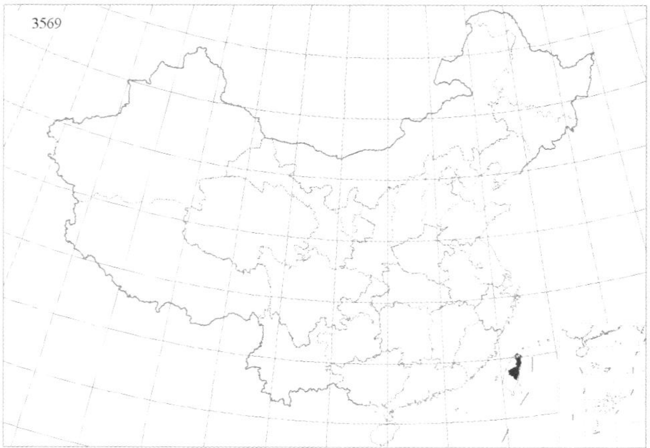

Rosaceae 蔷薇科　*Rubus* 悬钩子属
Rubus parvifolius L. 茅莓
Shrubs, deciduous

15.2 (1.9~24.7); 3.4 (−19.6~19.8); 25.9 (11.1~30.1); 15.5 (4.3~24.7); 132.1 (20.5~236.5); 9.3 (0.0~87.4); 857 (427~1376) ❙ 1155 (341~3495); 505 (190~1847); 100 (6~897) ❙ 32.0 (−45.2~246.5); 814 (341~1360); 204.9 (44.7~572.9)

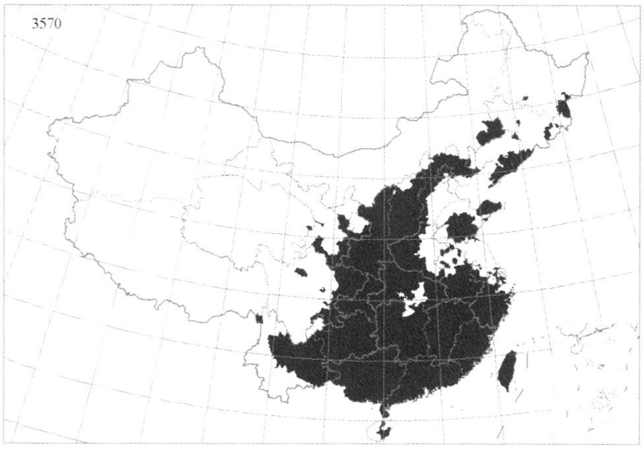

Rosaceae 蔷薇科 *Rubus* 悬钩子属

Rubus parvifolius var. adenochlamys (Focke) Migo 腺花茅莓

Shrubs, deciduous

13.9 (1.9~19.4); 0.7 (-12.7~9.4); 26.2 (11.7~29.5); 14.2 (4.3~19.4); 118.4 (21.1~173.4); 10.8 (0.0~65.0); 816 (427~1029) ❙ 972 (457~1813); 428 (248~666); 84 (9~226) ❙ 17.5 (-36.5~123.6); 757 (427~1029); 187.2 (110.1~323.3)

3571

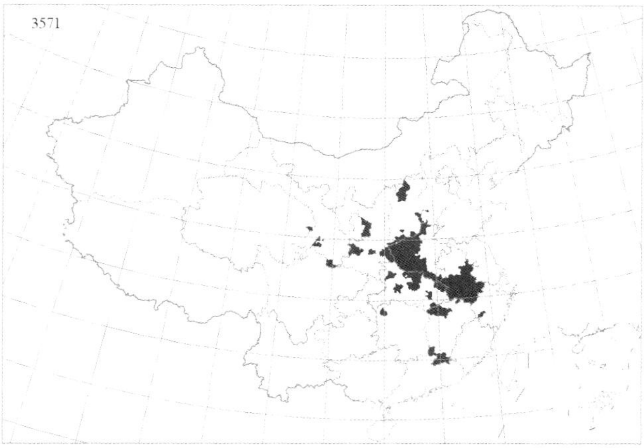

Rosaceae 蔷薇科 *Rubus* 悬钩子属

Rubus paucidentatus T. T. Yu et L. T. Lu 少齿悬钩子

Subshrubs, deciduous

21.0 (17.7~22.7); 12.5 (7.7~15.0); 27.7 (26.6~28.6); 21.1 (17.8~22.8); 193.2 (153.6~213.1); 0.0 (0.0~0.0); 1111 (916~1221) ❙ 1753 (1302~2435); 902 (556~1428); 128 (86~182) ❙ 58.0 (12.9~99.5); 1111 (916~1221); 274.7 (228.7~318.8)

3572

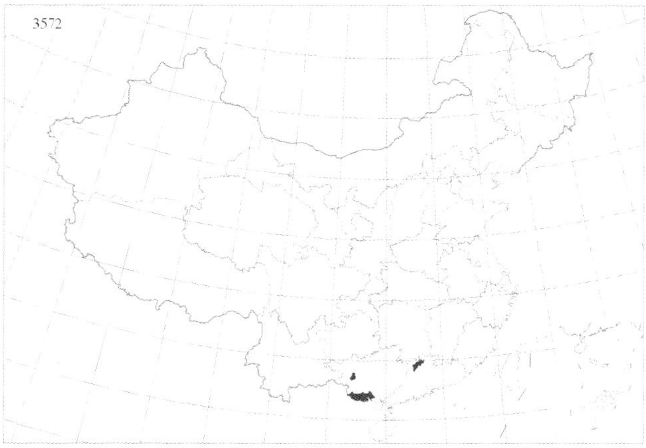

Rosaceae 蔷薇科 *Rubus* 悬钩子属

Rubus paucidentatus var. guangxiensis T. T. Yu et L. T. Lu 广西少齿悬钩子

Subshrubs, deciduous

21.9; 14.1; 27.9; 22.0; 203.7; 0.0; 1161 ❙ 2161; 1240; 126 ❙ 85.7; 1161; 305.5

3573

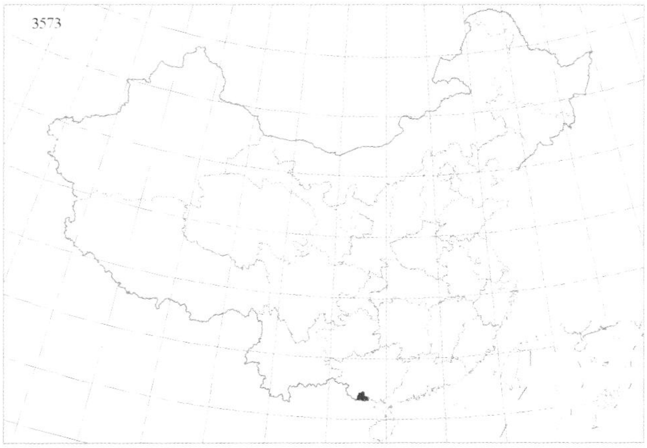

Rosaceae 蔷薇科 *Rubus* 悬钩子属

Rubus pectinarioides H. Hara 匍匐悬钩子

Subshrubs, deciduous

14.9 (8.6~16.7); 4.6 (1.3~6.3); 24.4 (15.0~26.0); 14.9 (8.7~16.7); 119.9 (52.6~140.5); 0.7 (0.0~8.8); 791 (574~863) ❙ 1207 (1041~1736); 525 (473~1007); 79 (45~106) ❙ 53.5 (35.3~145.8); 791 (574~863); 210.9 (148.9~447.4)

3574

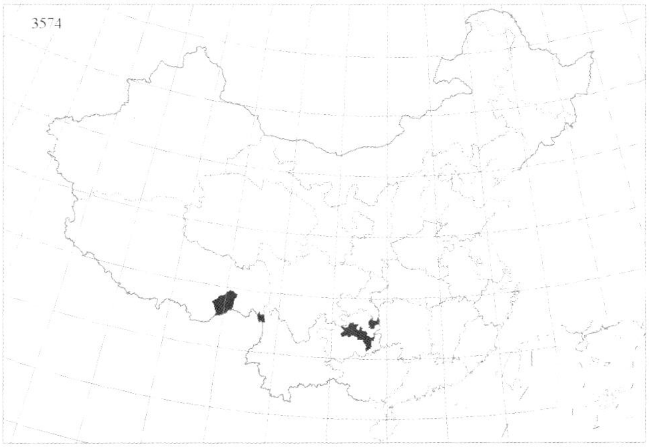

Rosaceae 蔷薇科 *Rubus* 悬钩子属

Rubus pectinaris Focke 梳齿悬钩子

Shrubs

14.9; 5.3; 23.6; 15.0; 119.6; 0.0; 782 ❙ 1410; 816; 49 ❙ 80.4; 782; 214.5

3575

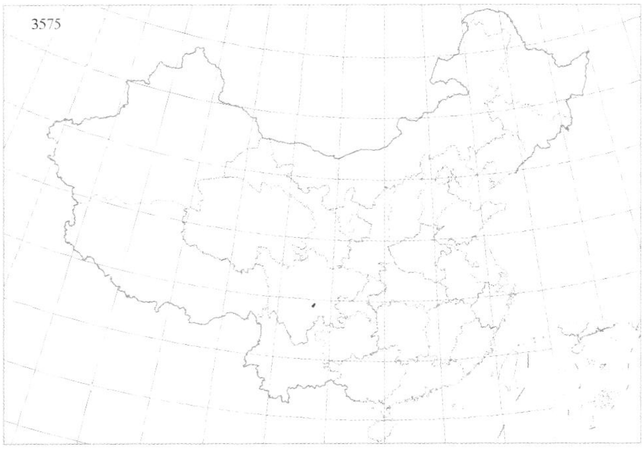

Rosaceae 蔷薇科 *Rubus* 悬钩子属

Rubus pedunculosus D. Don 密毛纤细悬钩子

Shrubs, deciduous

3.4; -5.5; 11.4; 4.7; 23.3; 41.8; 431 ❙ 527; 317; 9 ❙ 22.4; 431; 280.0

3576

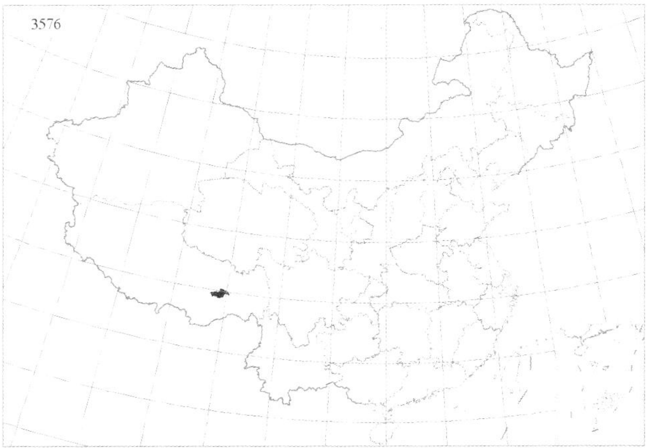

Rosaceae 蔷薇科 *Rubus* 悬钩子属
Rubus peltatus Maxim. 盾叶莓
Shrubs, deciduous

15.7 (12.6~19.1); 4.2 (1.6~10.0); 26.5 (23.0~29.3); 15.7 (12.7~19.1); 129.9 (98.8~169.1); 1.6 (0.0~6.9); 846 (711~984) ❙ 1522 (927~1997); 564 (447~713); 165 (65~259) ❙ 80.0 (9.8~147.0); 845 (711~973); 236.3 (152.4~319.9)

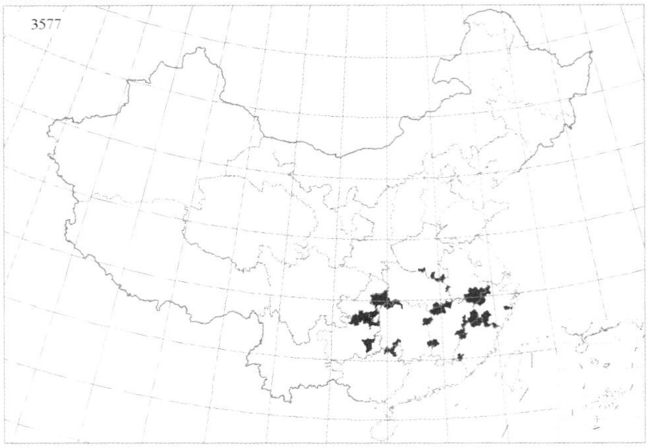

Rosaceae 蔷薇科 *Rubus* 悬钩子属
Rubus penduliflorus C. Y. Wu ex T. T. Yu et L. T. Lu 河口悬钩子
Shrubs, deciduous

21.0; 14.2; 25.6; 21.0; 191.8; 0.0; 1053 ❙ 1522; 818; 68 ❙ 44.5; 1053; 551.6

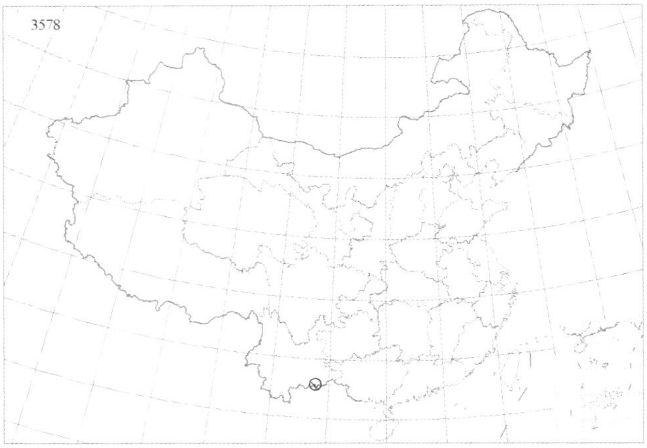

Rosaceae 蔷薇科 *Rubus* 悬钩子属
Rubus pentagonus Wall. ex Focke 掌叶悬钩子
Shrubs, deciduous

15.0 (−1.8~21.0); 7.7 (−10.2~14.2); 20.4 (6.1~25.6); 15.1 (1.8~21.0); 122.6 (1.9~191.8); 2.6 (0.0~82.8); 765 (329~1053) ❙ 1093 (320~1736); 621 (217~1007); 43 (10~70) ❙ 43.1 (−14.2~145.8); 749 (320~1053); 318.1 (61.6~627.0)

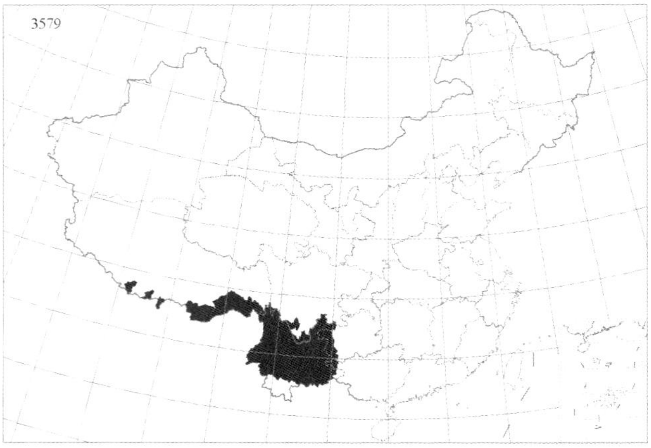

Rosaceae 蔷薇科 *Rubus* 悬钩子属
Rubus pentagonus var. longisepalus T. T. Yu et L. T. Lu 长萼掌叶悬钩子
Shrubs, deciduous

18.5 (16.9~21.0); 11.9 (9.9~14.2); 23.0 (21.4~25.6); 18.5 (16.9~21.0); 162.0 (142.7~191.8); 0.0 (0.0~0.0); 895 (813~1053) ❙ 1348 (1163~1522); 757 (641~860); 55 (48~68) ❙ 51.0 (34.8~80.6); 894 (813~1053); 457.8 (383.4~551.6)

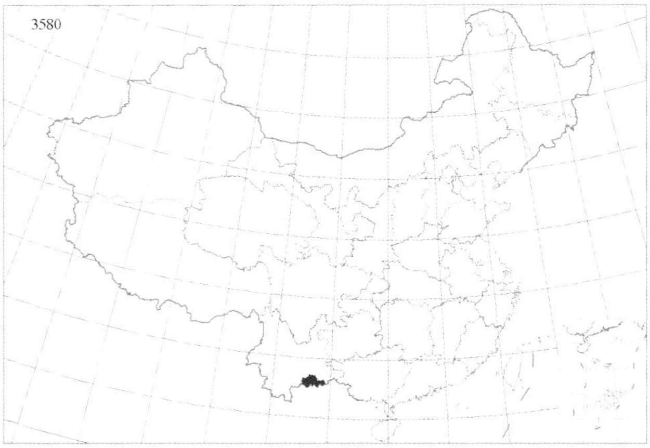

Rosaceae 蔷薇科 *Rubus* 悬钩子属
Rubus pentagonus var. modestus (Focke) T. T. Yu et L. T. Lu 无刺掌叶悬钩子
Shrubs, deciduous

11.2 (2.2~15.2); 2.3 (−6.6~6.4); 19.3 (9.9~25.9); 11.4 (3.8~15.2); 83.1 (15.5~123.9); 7.9 (0.0~49.1); 665 (413~816) ❙ 1158 (932~1448); 587 (473~816); 54 (18~105) ❙ 76.6 (50.4~125.9); 665 (413~816); 274.0 (170.4~407.1)

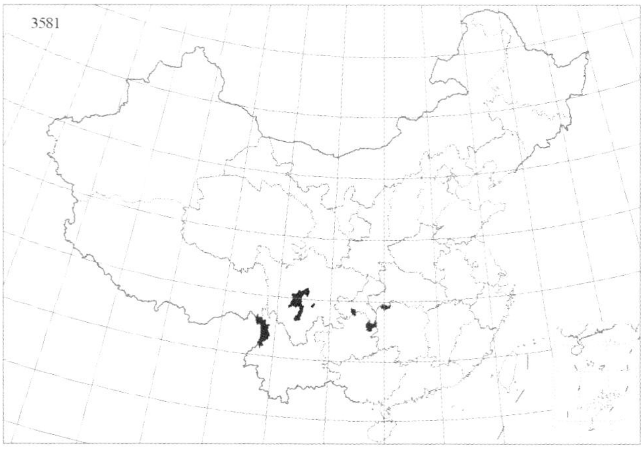

Rosaceae 蔷薇科 *Rubus* 悬钩子属
Rubus phoenicolasius Maxim. 多腺悬钩子
Shrubs, deciduous

11.3 (−2.5~16.6); −2.4 (−13.6~4.7); 23.9 (7.4~28.8); 12.0 (2.2~16.7); 97.5 (4.6~140.6); 20.9 (0.3~93.9); 725 (341~920) ❙ 743 (132~1571); 367 (81~666); 42 (2~157) ❙ 2.2 (−80.1~125.9); 650 (132~910); 186.1 (13.6~323.3)

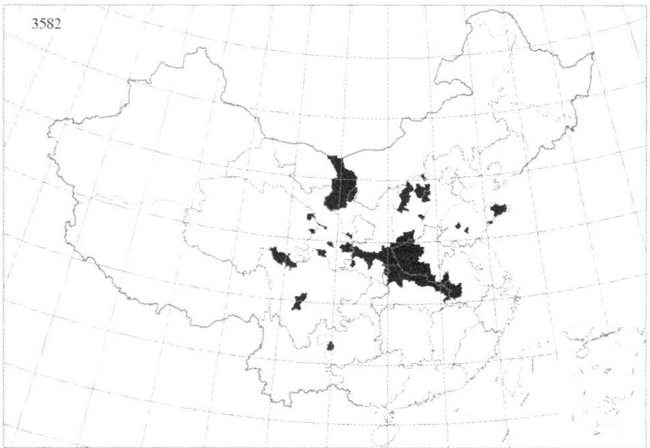

Rosaceae 蔷薇科 *Rubus* 悬钩子属
Rubus pileatus Focke 菰帽悬钩子
Shrubs, deciduous

9.9 (0.8~16.9); -3.5 (-12.7~6.2); 22.1 (10.7~28.2); 10.7 (3.8~16.9); 84.9 (16.2~143.4); 25.2 (0.0~66.0); 670 (411~924) ❙ 675 (341~1410); 344 (190~816); 23 (7~113) ❙ 1.9 (-40.9~121.7); 593 (341~924); 193.9 (58.7~341.6)

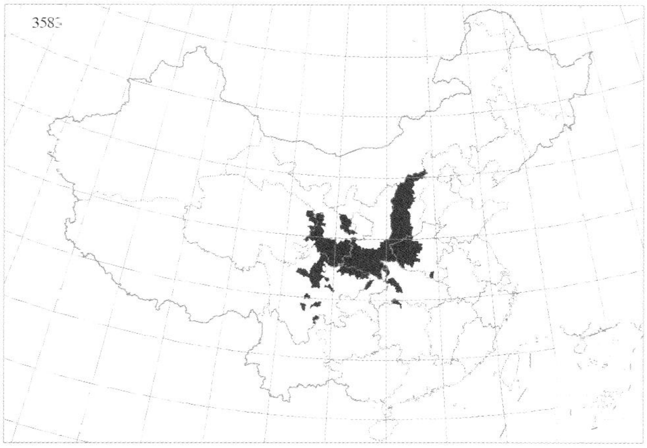

Rosaceae 蔷薇科 *Rubus* 悬钩子属
Rubus piluliferus Focke 陕西悬钩子
Shrubs, deciduous

10.9 (3.4~16.7); -1.1 (-8.7~5.5); 22.3 (11.7~28.9); 11.3 (4.7~16.8); 88.0 (23.2~142.4); 16.4 (0.0~45.4); 679 (442~924) ❙ 843 (537~1492); 390 (257~613); 34 (10~162) ❙ 24.0 (-32.0~107.2); 665 (442~924); 240.2 (132.4~334.4)

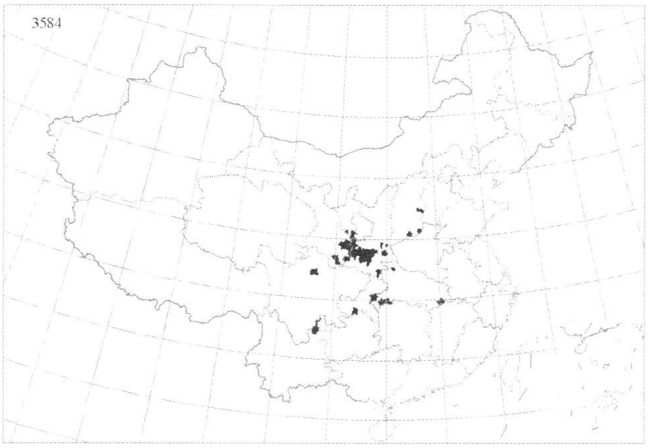

Rosaceae 蔷薇科 *Rubus* 悬钩子属
Rubus pinnatisepalus Hemsl. 羽萼悬钩子
Shrubs, deciduous

14.9 (8.6~19.9); 5.9 (0.9~15.2); 23.5 (15.0~28.7); 15.2 (8.7~22.7); 123.9 (52.6~212.7); 1.2 (0.0~8.8); 802 (574~1216) ❙ 1355 (916~2993); 614 (429~1186); 97 (31~547) ❙ 68.5 (26.1~246.5); 801 (574~1216); 244.8 (148.9~434.4)

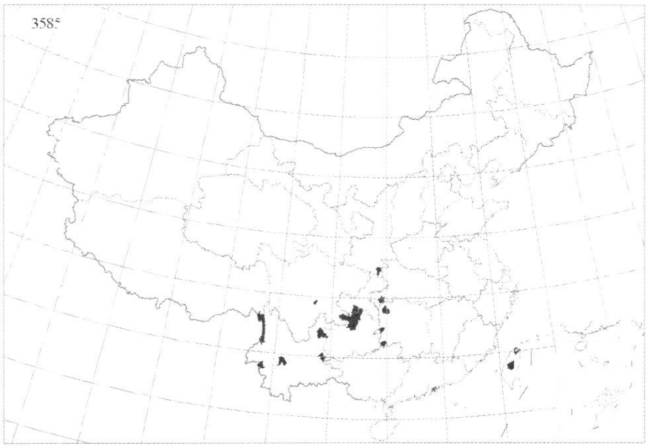

Rosaceae 蔷薇科 *Rubus* 悬钩子属
Rubus pinnatisepalus var. glandulosus T. T. Yu et L. T. Lu 密腺羽萼悬钩子
Shrubs, deciduous

15.5; 4.3; 26.0; 15.5; 126.8; 0.7; 828 ❙ 1285; 501; 99 ❙ 55.2; 828; 193.5

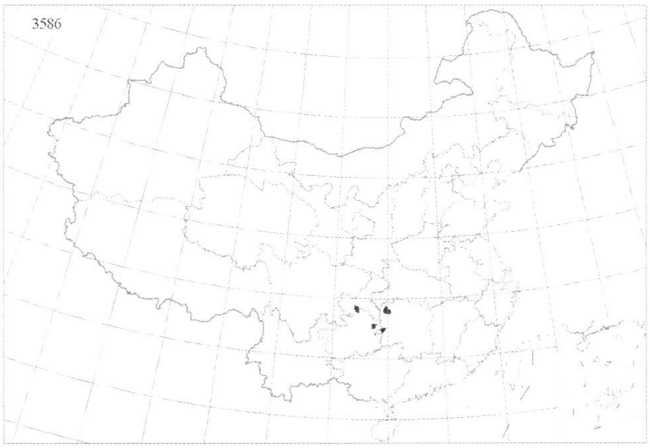

Rosaceae 蔷薇科 *Rubus* 悬钩子属
Rubus pirifolius Sm. 梨叶悬钩子
Shrubs, deciduous

19.1 (7.3~24.8); 10.6 (-2.0~20.1); 26.0 (15.6~28.8); 19.2 (7.5~24.8); 170.2 (48.2~237.6); 0.2 (0.0~20.4); 992 (537~1385) ❙ 1562 (916~3495); 685 (447~1388); 137 (20~897) ❙ 59.4 (-12.0~246.5); 982 (537~1360); 279.2 (164.9~620.0)

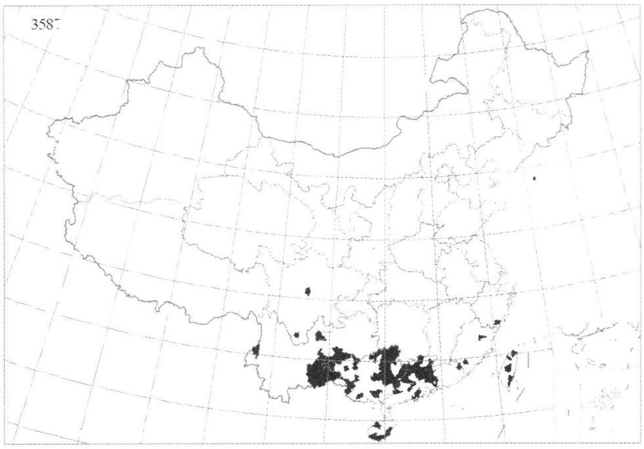

Rosaceae 蔷薇科 *Rubus* 悬钩子属
Rubus pirifolius var. cordatus T. T. Yu et L. T. Lu 心状梨叶悬钩子
Shrubs, deciduous

17.7; 10.9; 22.5; 17.8; 153.4; 0.0; 854 ❙ 1368; 756; 59 ❙ 60.1; 854; 491.5

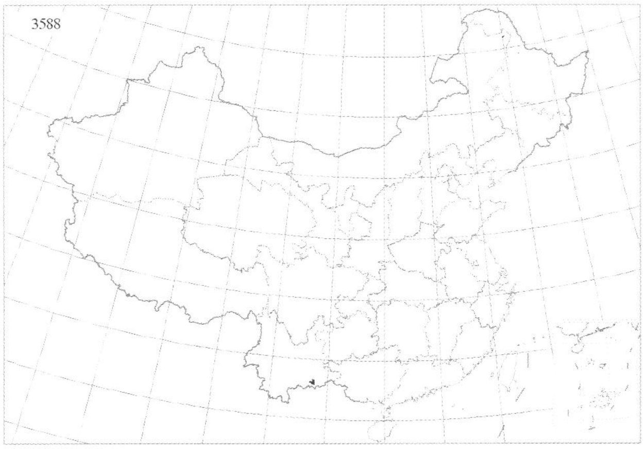

Rosaceae 蔷薇科 *Rubus* 悬钩子属

Rubus pirifolius var. permollis Merr. 柔毛梨叶悬钩子

Shrubs, deciduous

23.5; 17.4; 28.1; 23.6; 222.9; 0.0; 1286 ❙ 1490; 659; 58 ❙ 17.0; 1163; 371.4

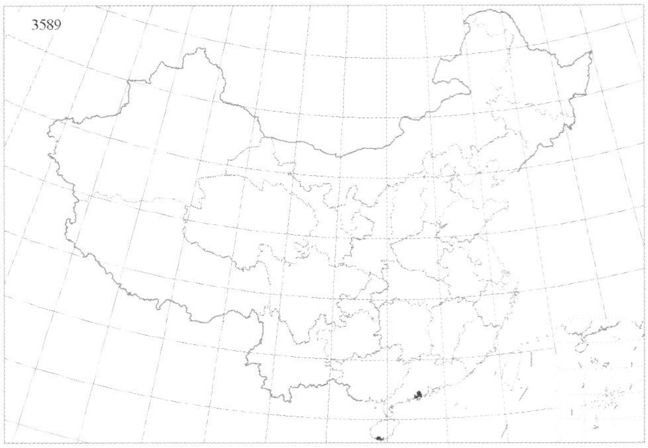

Rosaceae 蔷薇科 *Rubus* 悬钩子属

Rubus pirifolius var. tomentosus Kuntze ex Franch. 绒毛梨叶悬钩子

Shrubs, deciduous

7.3; −2.0; 15.6; 7.5; 48.2; 20.4; 537 ❙ 1005; 531; 27 ❙ 87.3; 537; 272.3

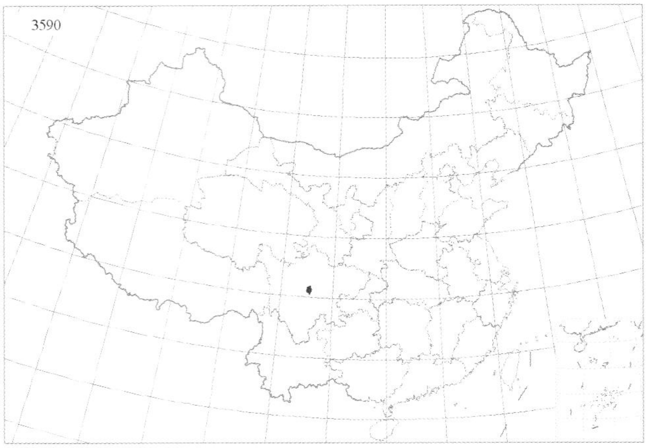

Rosaceae 蔷薇科 *Rubus* 悬钩子属

Rubus platysepalus Handel-Mazzetti 武冈悬钩子

Shrubs

16.4; 4.8; 27.4; 16.5; 138.1; 0.2; 881 ❙ 1409; 445; 184 ❙ 59.8; 878; 169.7

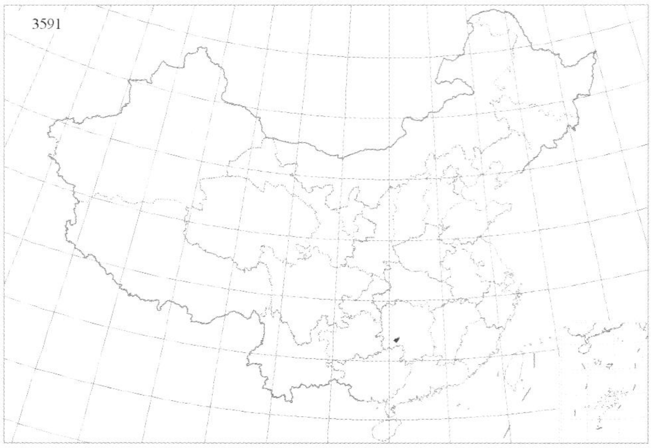

Rosaceae 蔷薇科 *Rubus* 悬钩子属

Rubus playfairianus Hemsl. ex Focke 五叶鸡爪茶

Shrubs, deciduous

13.9 (6.0~17.8); 3.5 (−3.2~7.1); 23.8 (14.4~28.0); 14.0 (6.6~17.8); 111.5 (39.7~154.1); 3.6 (0.0~26.7); 765 (508~944) ❙ 1245 (931~1590); 574 (411~946); 65 (25~113) ❙ 64.5 (12.1~107.2); 765 (508~944); 210.5 (130.1~292.6)

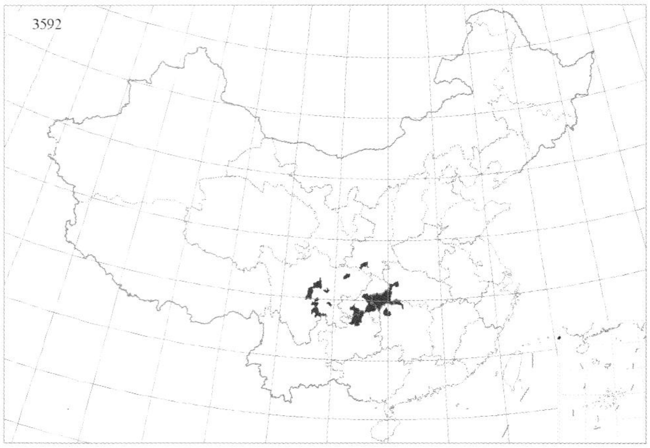

Rosaceae 蔷薇科 *Rubus* 悬钩子属

Rubus pluribracteatus L. T. Lu et Boufford 大乌泡

Shrubs, deciduous

17.3 (10.1~22.7); 8.7 (1.8~15.1); 24.4 (16.6~29.3); 17.4 (10.3~22.8); 148.8 (70.1~213.1); 0.1 (0.0~7.0); 887 (621~1221) ❙ 1346 (797~2435); 639 (411~1428); 100 (22~253) ❙ 52.0 (−14.2~128.9); 878 (621~1221); 292.6 (148.9~659.0)

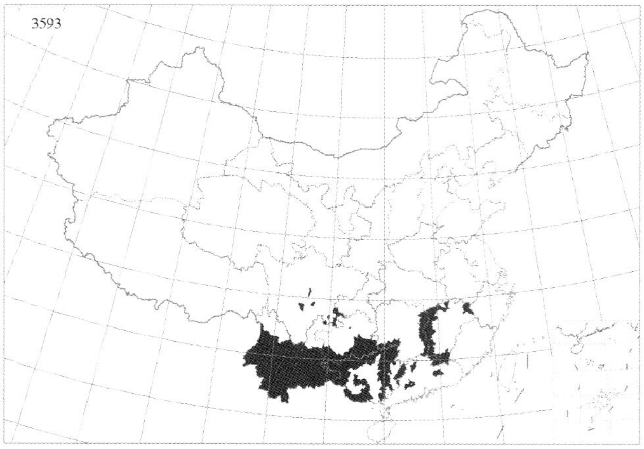

Rosaceae 蔷薇科 *Rubus* 悬钩子属

Rubus poliophyllus Kuntze 毛叶悬钩子

Shrubs, deciduous

17.3 (8.6~20.5); 11.2 (1.3~15.1); 21.4 (15.0~23.7); 17.4 (8.7~20.5); 148.6 (52.6~185.9); 0.4 (0.0~8.8); 846 (574~992) ❙ 1292 (939~1751); 689 (524~996); 53 (37~77) ❙ 53.7 (6.5~93.0); 818 (574~986); 421.9 (301.1~659.0)

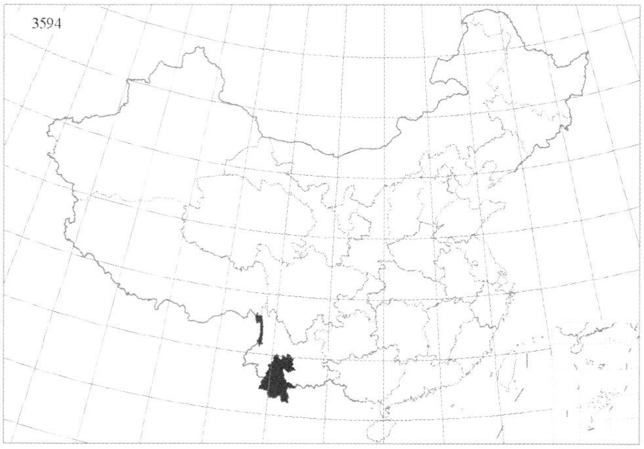

Rosaceae 蔷薇科 *Rubus* 悬钩子属
Rubus polyodontus Hand.-Mazz. 多齿悬钩子
Small shrubs, deciduous

10.0; 2.8; 16.0; 10.0; 64.6; 4.7; 609 ▎1142; 640; 51 ▎87.3; 609; 368.5

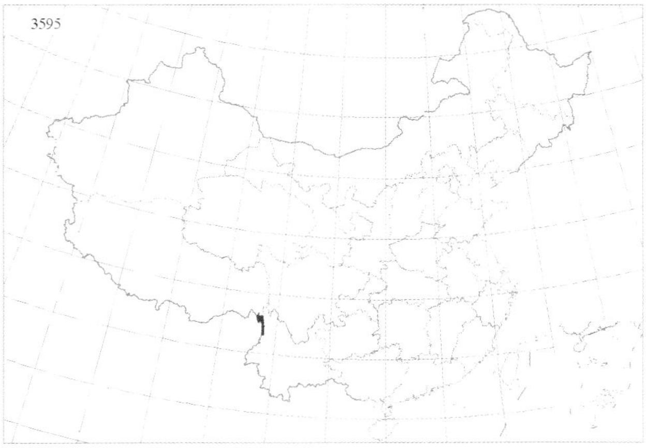

Rosaceae 蔷薇科 *Rubus* 悬钩子属
Rubus preptanthus Focke 早花悬钩子
Shrubs, deciduous

13.6 (5.4~18.9); 6.0 (-2.1~12.0); 19.9 (12.3~26.6); 13.7 (5.8~19.0); 106.2 (31.5~167.8); 2.0 (0.0~25.6); 723 (481~929) ▎963 (741~1178); 560 (411~646); 31 (15~57) ▎35.1 (-14.2~81.5); 693 (481~912); 273.9 (142.9~383.4)

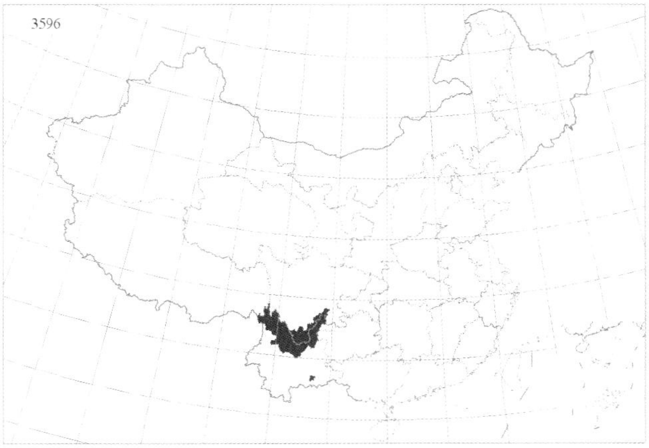

Rosaceae 蔷薇科 *Rubus* 悬钩子属
Rubus pseudopileatus Cardot 假帽莓
Shrubs, deciduous

4.1 (2.0~7.3); -4.9 (-7.2~-2.0); 12.2 (9.9~15.6); 5.3 (3.8~7.5); 27.8 (15.4~48.2); 37.9 (20.4~51.4); 460 (413~537) ▎962 (932~1005); 523 (490~546); 24 (19~27) ▎111.3 (87.3~128.6); 460 (413~537); 256.0 (221.3~272.3)

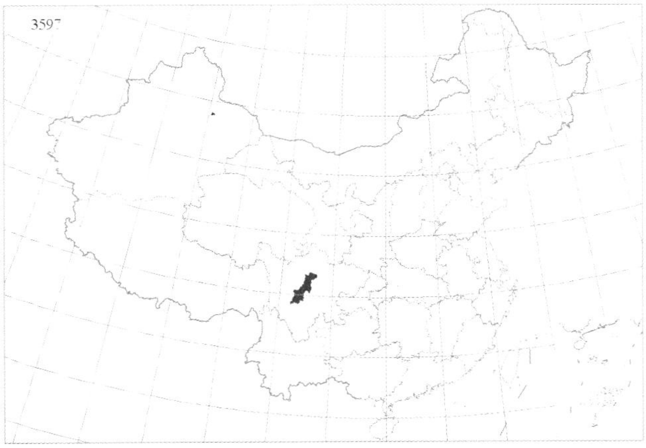

Rosaceae 蔷薇科 *Rubus* 悬钩子属
Rubus pseudopileatus var. glabratus T. T. Yu et L. T. Lu 光梗假帽莓
Shrubs, deciduous

9.3; 0.0; 17.6; 9.7; 68.9; 17.1; 605 ▎1092; 597; 34 ▎84.6; 605; 258.3

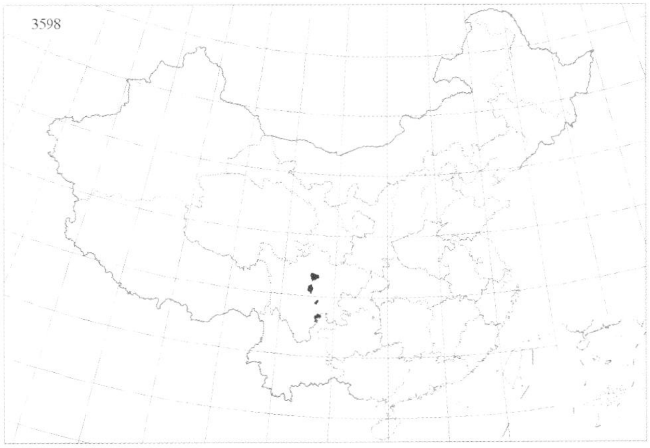

Rosaceae 蔷薇科 *Rubus* 悬钩子属
Rubus pseudopileatus var. kangdingensis T. T. Yu et L. T. Lu 康定假帽莓
Shrubs, deciduous

2.2; -6.6; 9.9; 3.8; 15.5; 49.1; 413 ▎932; 546; 19 ▎125.9; 413; 248.3

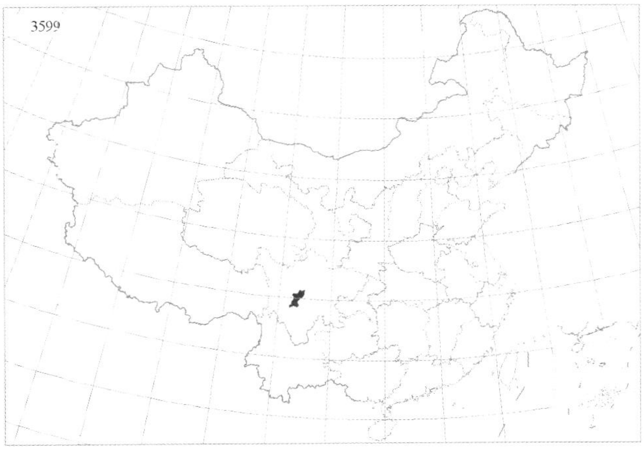

Rosaceae 蔷薇科 *Rubus* 悬钩子属
Rubus ptilocarpus T. T. Yu et L. T. Lu 毛果悬钩子
Shrubs, deciduous

5.4 (-1.0~16.2); -4.6 (-11.8~5.5); 14.5 (8.5~26.3); 7.2 (2.7~16.2); 47.5 (8.3~134.6); 41.7 (0.0~79.5); 524 (363~854) ▎854 (581~1197); 435 (347~520); 30 (12~71) ▎69.6 (32.5~128.6); 524 (363~854); 272.6 (176.4~341.6)

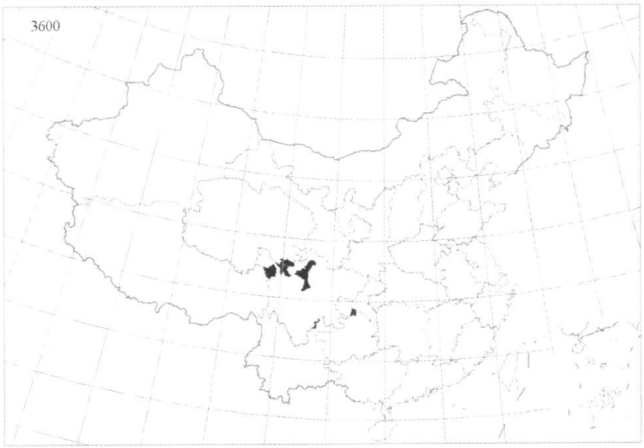

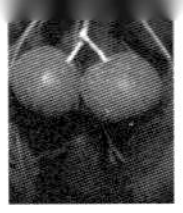

Rosaceae 蔷薇科 *Rubus* 悬钩子属

Rubus ptilocarpus var. degensis T. T. Yu et L. T. Lu 长萼毛果悬钩子

Shrubs, deciduous

0.6; -9.4; 9.5; 3.4; 12.7; 65.4; 394 ❙ 581; 347; 12 ❙ 47.6; 394; 307.8

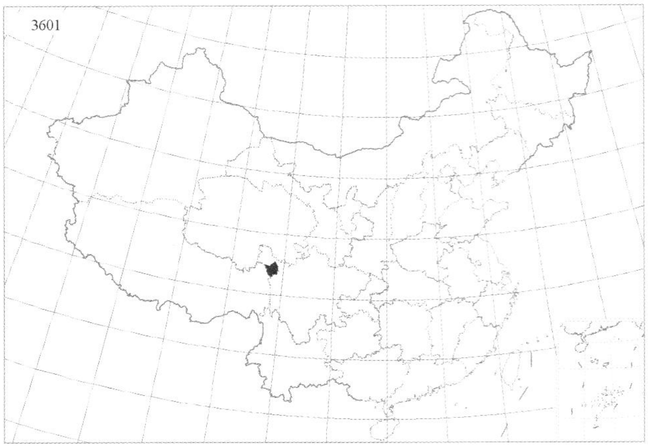

Rosaceae 蔷薇科 *Rubus* 悬钩子属

Rubus pungens Cambess. 针刺悬钩子

Shrubs, deciduous

12.1 (-1.1~20.8); 0.2 (-10.2~15.3); 23.2 (6.2~29.3); 12.5 (2.0~20.8); 99.9 (2.5~189.2); 14.7 (0.0~75.6); 730 (334~1019) ❙ 1065 (321~2859); 460 (190~1412); 81 (7~402) ❙ 43.0 (-40.9~246.5); 701 (321~1019); 222.6 (61.2~351.7)

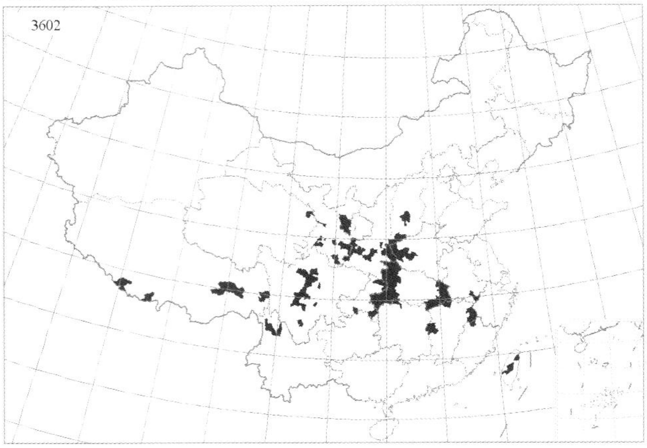

Rosaceae 蔷薇科 *Rubus* 悬钩子属

Rubus pungens var. linearisepalus T. T. Yu et L. T. Lu 线萼针刺悬钩子

Shrubs, deciduous

10.2; 3.2; 16.2; 10.2; 66.3; 3.4; 617 ❙ 988; 586; 32 ❙ 60.4; 617; 335.8

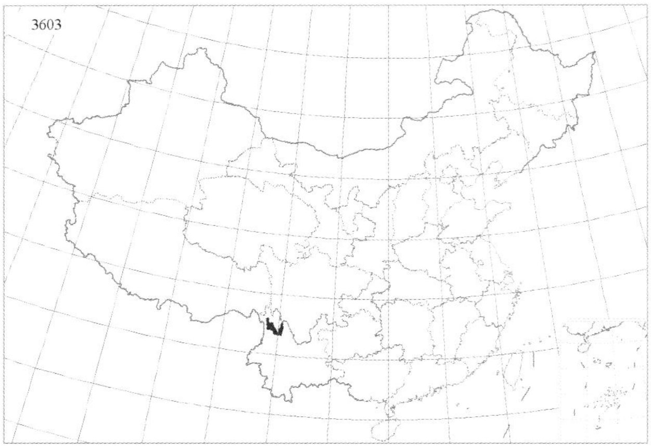

Rosaceae 蔷薇科 *Rubus* 悬钩子属

Rubus pungens var. oldhamii (Miq.) Maxim. 香莓

Shrubs, deciduous

11.2 (1.9~17.1); -0.9 (-12.7~6.5); 22.6 (9.9~27.8); 11.7 (3.8~17.1); 92.5 (15.4~145.5); 17.2 (0.0~65.0); 697 (413~894) ❙ 911 (457~1997); 413 (257~669); 45 (12~254) ❙ 31.2 (-35.0~130.7); 681 (413~894); 236.8 (115.9~351.7)

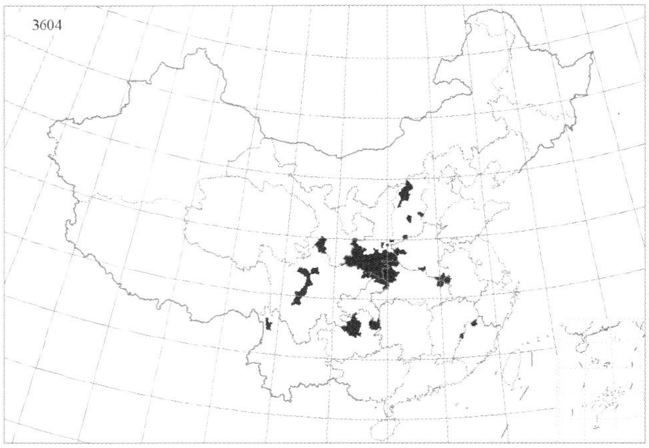

Rosaceae 蔷薇科 *Rubus* 悬钩子属

Rubus quinquefoliolatus T. T. Yu et L. T. Lu 五叶悬钩子

Shrubs, deciduous

16.0 (11.7~21.0); 8.2 (2.7~14.2); 22.0 (18.8~25.6); 16.1 (11.7~21.0); 133.3 (82.6~191.8); 0.3 (0.0~3.8); 804 (658~1053) ❙ 1216 (916~1522); 667 (510~860); 52 (31~68) ❙ 51.4 (19.0~80.6); 803 (658~1053); 306.9 (185.6~551.6)

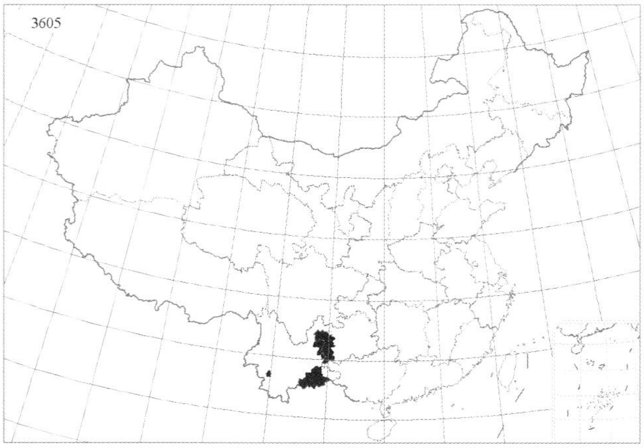

Rosaceae 蔷薇科 *Rubus* 悬钩子属

Rubus raopingensis T. T. Yu et L. T. Lu 饶平悬钩子

Shrubs, deciduous

20.2; 12.1; 27.4; 20.2; 182.7; 0.0; 1036 ❙ 1563; 558; 150 ❙ 51.2; 1036; 266.5

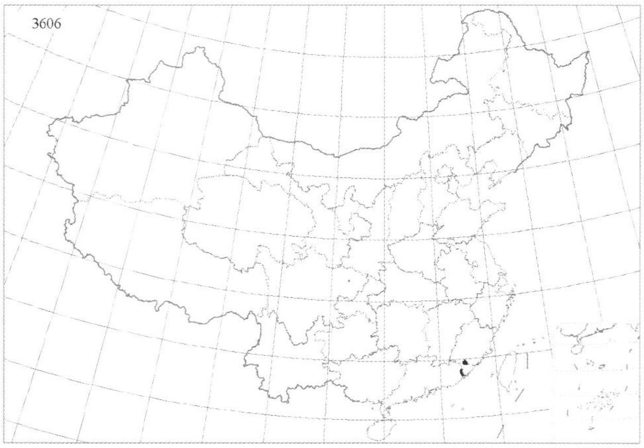

Rosaceae 蔷薇科 *Rubus* 悬钩子属
Rubus reflexus Ker Gawl. 锈毛莓
Shrubs, deciduous

17.6 (12.1~23.4); 7.1 (1.1~16.1); 27.2 (20.1~30.1); 17.7 (12.2~23.5); 152.7 (95.0~221.7); 0.3 (0.0~8.5); 937 (696~1273) ❙ 1524 (871~3495); 574 (299~1033); 166 (37~897) ❙ 63.7 (-6.7~224.4); 924 (696~1273); 224.1 (134.6~491.5)

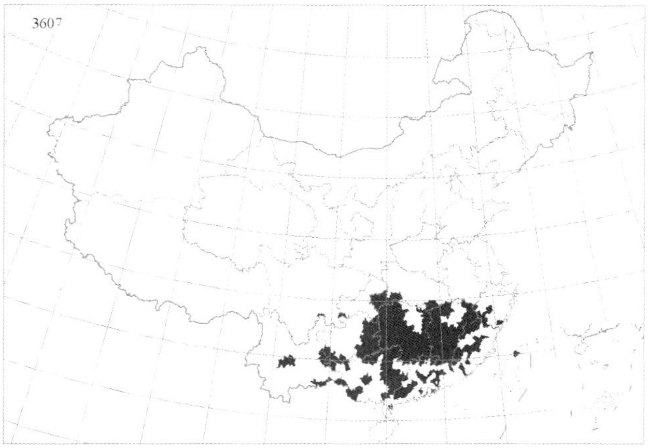

Rosaceae 蔷薇科 *Rubus* 悬钩子属
Rubus reflexus var. hui (Diels ex Hu) F. P. Metcalf 浅裂锈毛莓
Shrubs, deciduous

18.1 (14.3~24.8); 8.2 (3.5~20.1); 27.1 (22.8~29.1); 18.2 (14.3~24.8); 158.4 (112.6~237.6); 0.2 (0.0~1.5); 956 (756~1385) ❙ 1646 (1176~2993); 606 (427~1033); 182 (41~547) ❙ 74.7 (-12.0~235.6); 946 (756~1257); 251.8 (163.7~538.2)

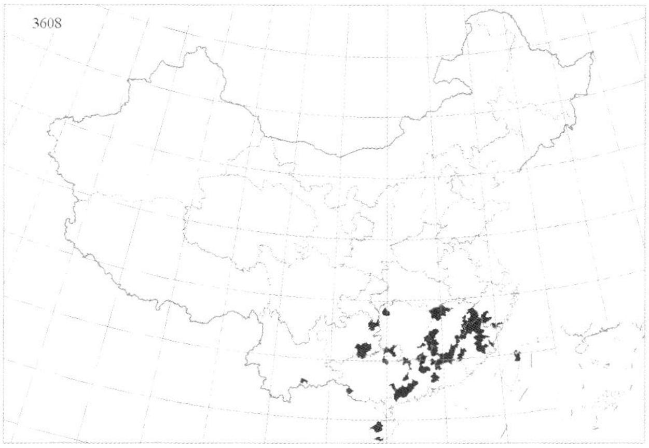

Rosaceae 蔷薇科 *Rubus* 悬钩子属
Rubus reflexus var. lanceolobus F. P. Metcalf 深裂悬钩子
Shrubs, deciduous

19.6 (16.7~23.2); 10.0 (4.8~15.8); 27.7 (25.8~28.9); 19.6 (16.7~23.2); 175.3 (141.0~218.9); 0.0 (0.0~0.2); 1030 (878~1257) ❙ 1664 (1371~2030); 680 (467~1033); 170 (87~259) ❙ 63.4 (15.4~121.9); 1028 (875~1257); 238.2 (163.7~327.7)

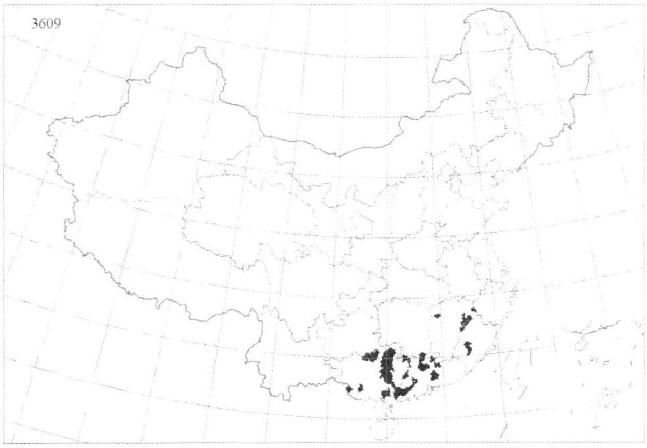

Rosaceae 蔷薇科 *Rubus* 悬钩子属
Rubus reflexus var. macrophyllus T. T. Yu et L. T. Lu 大叶锈毛莓
Shrubs, deciduous

20.1; 12.6; 25.6; 20.1; 181.4; 0.0; 1017 ❙ 1487; 785; 72 ❙ 46.2; 1017; 417.7

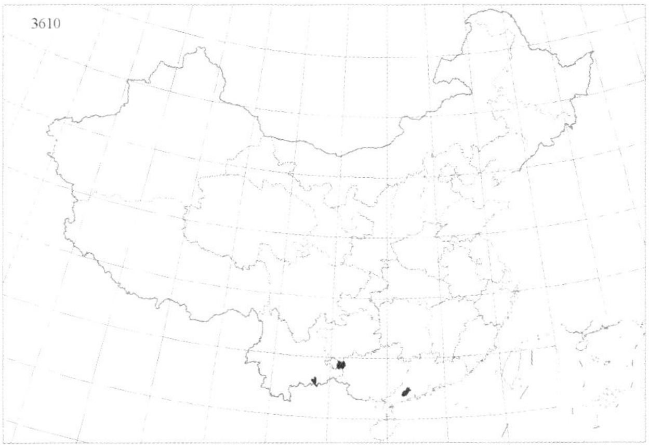

Rosaceae 蔷薇科 *Rubus* 悬钩子属
Rubus reflexus var. orogenes Hand.-Mazz. 长叶锈毛莓
Shrubs, deciduous

15.7 (14.3~17.1); 5.1 (3.8~8.5); 25.5 (23.4~27.6); 15.8 (14.3~17.1); 129.4 (112.6~145.1); 0.3 (0.0~1.5); 827 (756~902) ❙ 1267 (1168~1660); 506 (445~666); 103 (59~216) ❙ 53.3 (38.5~103.1); 827 (756~902); 213.2 (169.7~283.9)

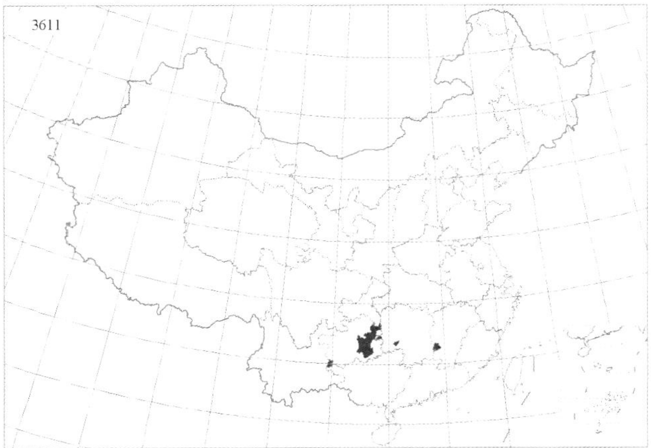

Rosaceae 蔷薇科 *Rubus* 悬钩子属
Rubus refractus H. Lév. 曲萼悬钩子
Shrubs, deciduous

15.5; 5.9; 23.8; 15.6; 127.1; 0.0; 798 ❙ 1298; 648; 63 ❙ 62.6; 798; 194.4

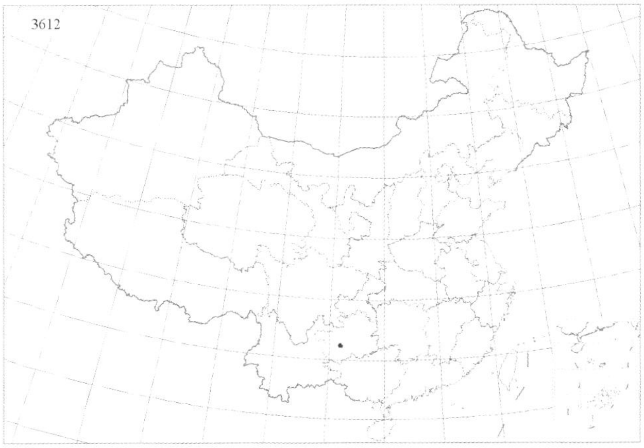

Rosaceae 蔷薇科 *Rubus* 悬钩子属

Rubus reticulatus Wall. ex Hook. f. 网脉悬钩子

Shrubs, deciduous

8.2; 0.1; 15.0; 8.8; 58.5; 20.1; 565 ❙ 1178; 683; 31 ❙ 100.9; 565; 303.7

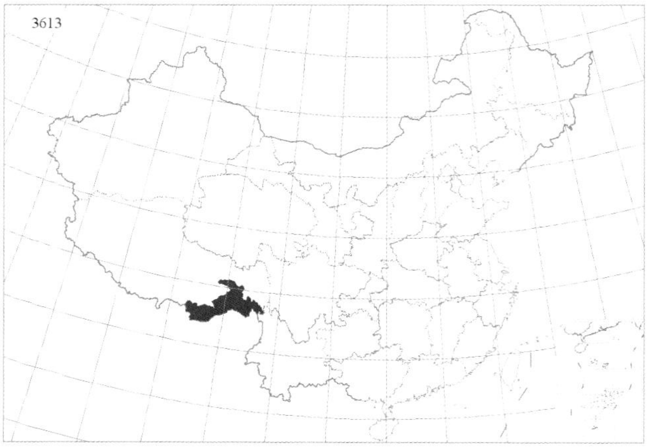

Rosaceae 蔷薇科 *Rubus* 悬钩子属

Rubus rolfei S. Vidal 高山悬钩子

Subshrubs, deciduous

19.3 (16.2~24.1); 13.7 (11.0~18.9); 23.8 (20.1~28.0); 19.3 (16.2~24.1); 171.7 (134.2~229.7); 0.0 (0.0~0.0); 953 (773~1322) ❙ 2421 (1751~2992); 1150 (750~1847); 217 (47~402) ❙ 163.2 (32.5~246.5); 930 (773~1123); 258.1 (154.2~303.6)

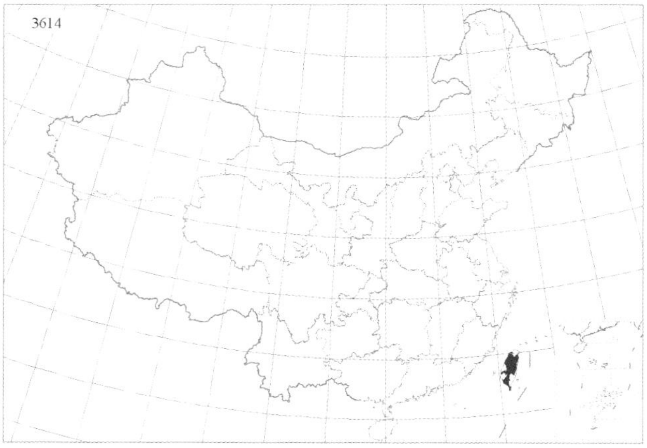

Rosaceae 蔷薇科 *Rubus* 悬钩子属

Rubus rosifolius Sm. 空心泡

Shrubs, deciduous

18.0 (9.3~24.6); 7.9 (-1.7~19.5); 27.1 (19.1~29.7); 18.0 (9.5~24.7); 156.8 (69.8~235.9); 0.5 (0.0~18.0); 949 (603~1372) ❙ 1621 (879~3495); 628 (400~1847); 177 (30~897) ❙ 72.1 (0.7~246.5); 942 (603~1360); 239.9 (123.9~572.9)

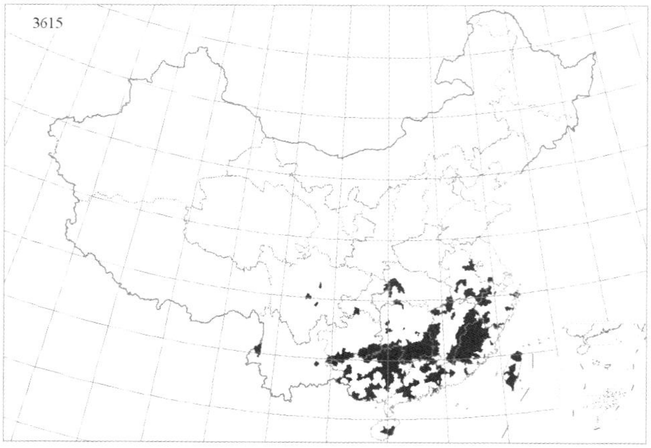

Rosaceae 蔷薇科 *Rubus* 悬钩子属

Rubus rosifolius var. coronarius (Sims) Focke 重瓣空心泡

Shrubs, deciduous

17.6 (16.6~21.1); 6.8 (5.7~13.2); 27.6 (26.3~29.1); 17.6 (16.6~21.1); 151.4 (138.9~193.7); 0.0 (0.0~0.0); 932 (866~1104) ❙ 1833 (1258~1997); 601 (425~669); 245 (185~259) ❙ 98.5 (14.0~130.7); 922 (866~1047); 253.5 (183.9~303.4)

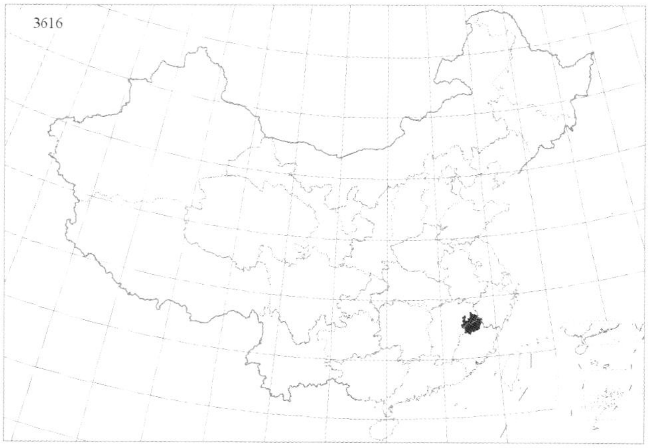

Rosaceae 蔷薇科 *Rubus* 悬钩子属

Rubus rufus Focke 棕红悬钩子

Shrubs, deciduous

16.9 (8.6~21.0); 7.8 (1.3~15.1); 24.6 (15.0~29.1); 16.9 (8.7~21.0); 143.4 (52.6~191.8); 0.3 (0.0~8.8); 865 (574~1081) ❙ 1397 (939~1950); 618 (461~996); 111 (35~259) ❙ 61.9 (6.5~147.0); 858 (574~1081); 296.0 (153.9~659.0)

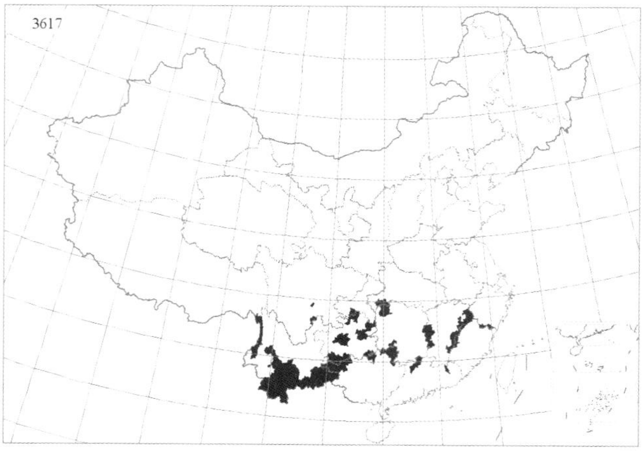

Rosaceae 蔷薇科 *Rubus* 悬钩子属

Rubus rufus var. palmatifidus Cardot 掌裂棕红悬钩子

Shrubs, deciduous

15.8 (14.3~17.4); 5.2 (3.8~7.4); 25.6 (23.5~26.8); 15.8 (14.3~17.4); 130.1 (112.6~148.9); 0.4 (0.0~1.5); 830 (756~886) ❙ 1255 (1165~1381); 511 (461~637); 98 (75~129) ❙ 51.4 (35.7~71.9); 830 (756~886); 214.1 (146.0~246.1)

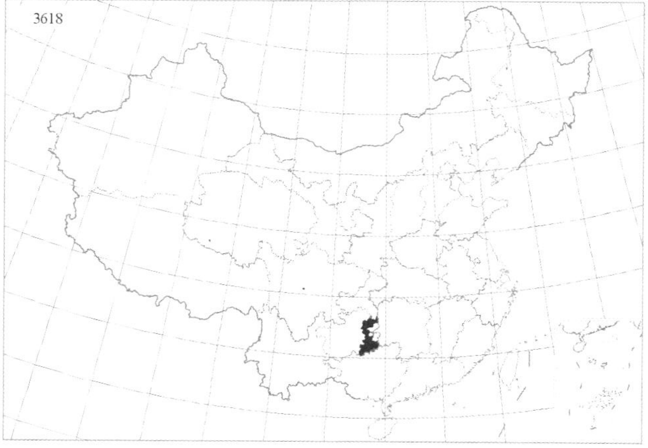

Rosaceae 蔷薇科 *Rubus* 悬钩子属
Rubus sachalinensis H. Lév. 库页悬钩子
Shrubs, deciduous

3.3 (-5.8~12.8); -15.3 (-29.8~-3.9); 19.6 (7.8~27.2); 7.5 (2.1~13.3); 57.1 (5.2~114.7); 76.6 (21.1~157.4); 556 (340~827) ▎396 (63~1073); 247 (45~691); 15 (2~51) ▎-27.4 (-84.9~74.1); 378 (63~658); 172.8 (8.4~338.4)

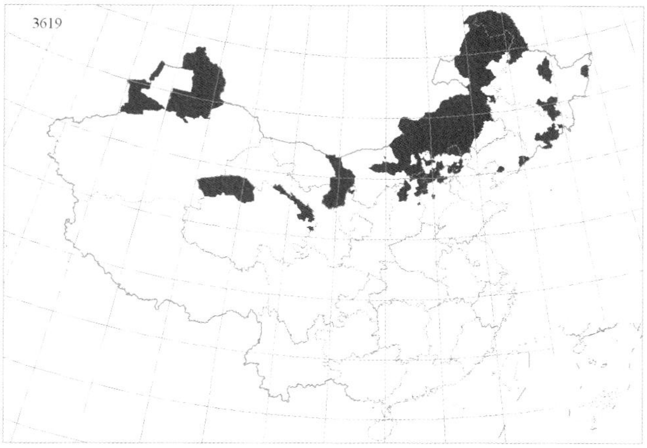

Rosaceae 蔷薇科 *Rubus* 悬钩子属
Rubus salwinensis Hand.-Mazz. 怒江悬钩子
Shrubs, deciduous

21.6; 12.6; 28.5; 21.7; 199.8; 0.0; 1153 ▎1625; 795; 173 ▎40.9; 1153; 237.8

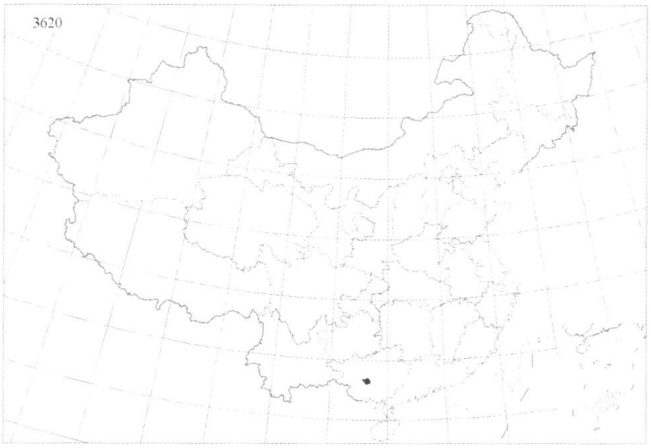

Rosaceae 蔷薇科 *Rubus* 悬钩子属
Rubus septemlobus H. L. Li 裂果悬钩子
Shrubs, deciduous

19.6 (17.9~21.3); 9.8 (7.4~12.3); 27.8 (25.8~28.6); 19.7 (18.0~21.3); 176.1 (155.6~196.1); 0.0 (0.0~0.0); 1032 (913~1130) ▎1596 (1371~1758); 669 (562~728); 175 (161~189) ▎56.0 (22.7~85.5); 1031 (913~1130); 220.0 (185.5~243.9)

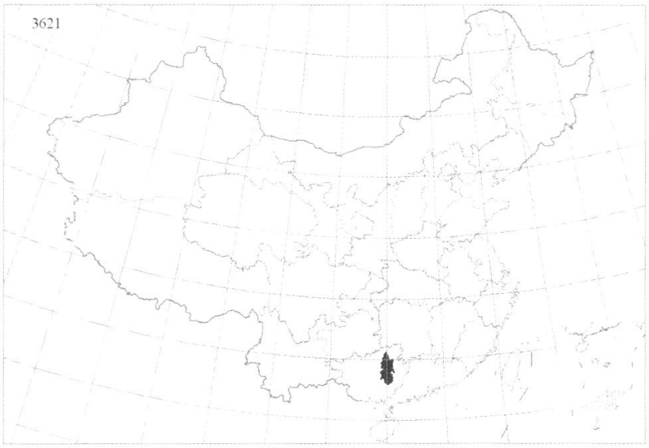

Rosaceae 蔷薇科 *Rubus* 悬钩子属
Rubus setchuenensis Bureau et Franch. 川莓
Shrubs, deciduous

15.3 (1.8~21.0); 6.1 (-7.1~15.1); 23.2 (9.9~29.1); 15.3 (3.8~21.0); 125.4 (15.5~191.8); 1.6 (0.0~53.3); 799 (404~1053) ▎1198 (583~1997); 575 (348~996); 72 (13~254) ▎50.5 (-14.2~130.7); 792 (404~1053); 254.6 (126.1~659.0)

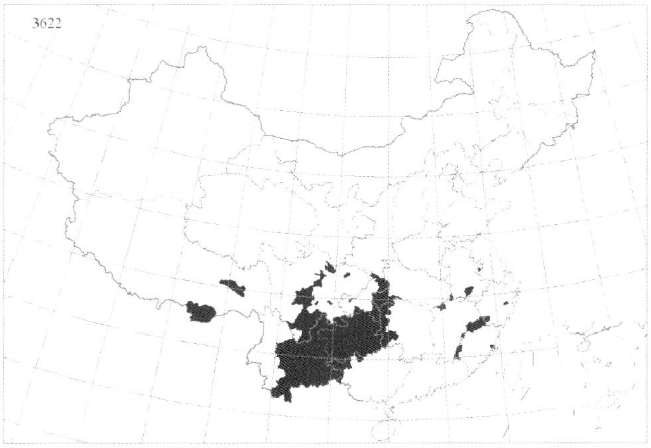

Rosaceae 蔷薇科 *Rubus* 悬钩子属
Rubus shihae F. P. Metcalf 桂滇悬钩子
Shrubs, deciduous

18.1 (14.6~22.2); 7.9 (3.8~13.9); 26.9 (21.9~28.7); 18.1 (14.7~22.3); 157.6 (117.7~207.5); 0.0 (0.0~1.5); 945 (779~1190) ▎1516 (1074~1776); 613 (411~773); 158 (47~207) ▎61.4 (15.4~101.8); 943 (779~1190); 239.1 (170.8~491.5)

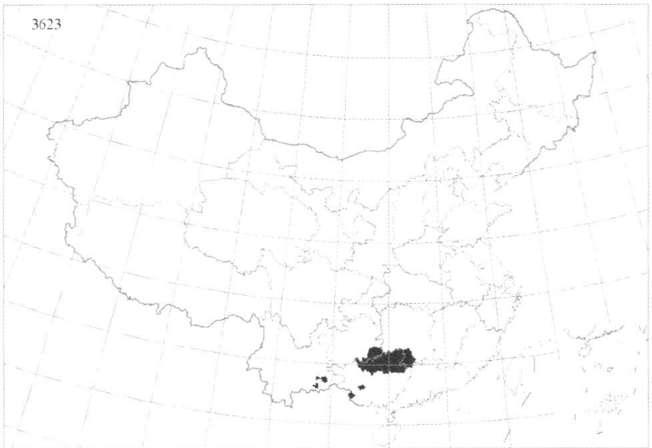

Rosaceae 蔷薇科 *Rubus* 悬钩子属
Rubus sikkimensis Hook. f. 锡金悬钩子
Dwarf shrubs, deciduous

12.2; 4.3; 18.5; 12.2; 87.4; 0.7; 672 ▎1431; 817; 31 ▎113.0; 672; 363.7

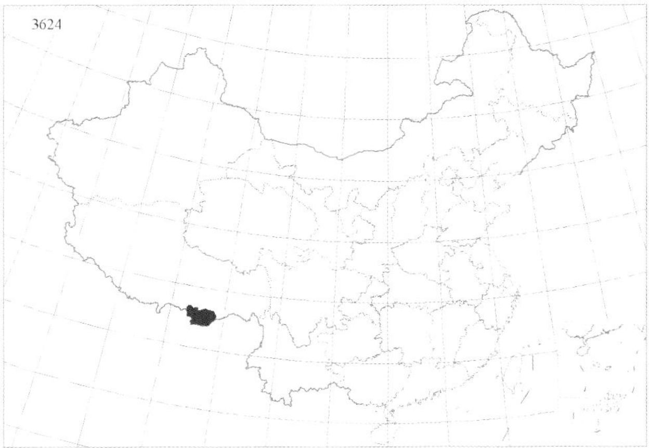

Rosaceae 蔷薇科 *Rubus* 悬钩子属

Rubus simplex Focke 单茎悬钩子

Subshrubs, deciduous

13.1 (2.2~18.9); 3.8 (-6.6~12.0); 21.5 (9.9~26.3); 13.3 (3.8~19.0); 105.4 (15.5~167.8); 7.3 (0.0~49.1); 723 (413~929) ❙ 1007 (632~1448); 508 (306~816); 42 (15~105) ❙ 41.7 (-14.2~125.9); 699 (413~854); 269.5 (176.4~365.3)

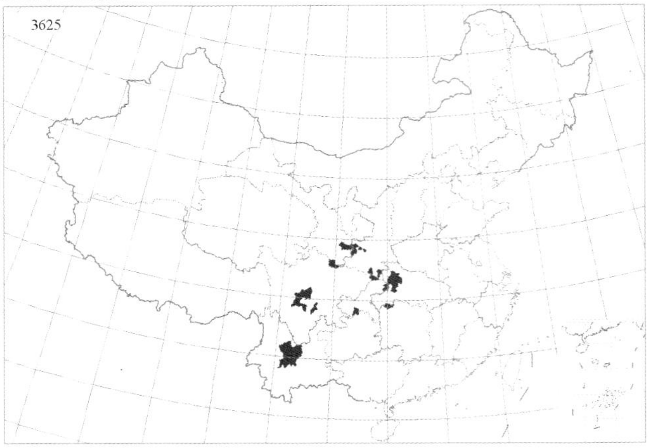

Rosaceae 蔷薇科 *Rubus* 悬钩子属

Rubus spananthus Z. M. Wu et Z. L. Cheng 少花悬钩子

Shrubs, deciduous

14.1; 1.6; 26.2; 14.2; 116.7; 6.5; 796 ❙ 1364; 601; 129 ❙ 71.2; 796; 243.1

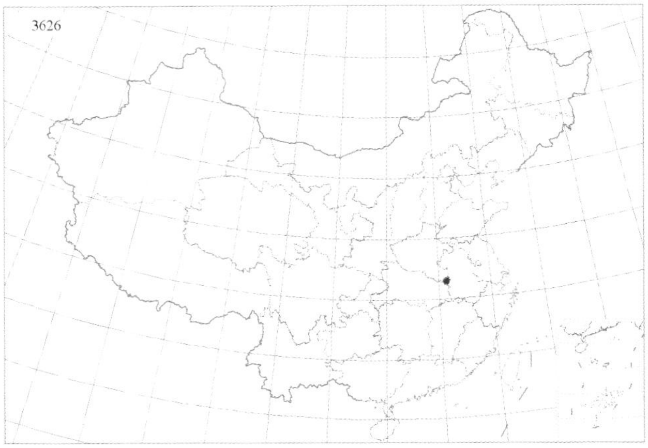

Rosaceae 蔷薇科 *Rubus* 悬钩子属

Rubus spinulosoides F. P. Metcalf 刺毛白叶莓

Shrubs, deciduous

13.5; 1.6; 24.6; 13.6; 109.4; 6.7; 761 ❙ 1145; 516; 67 ❙ 49.5; 761; 201.4

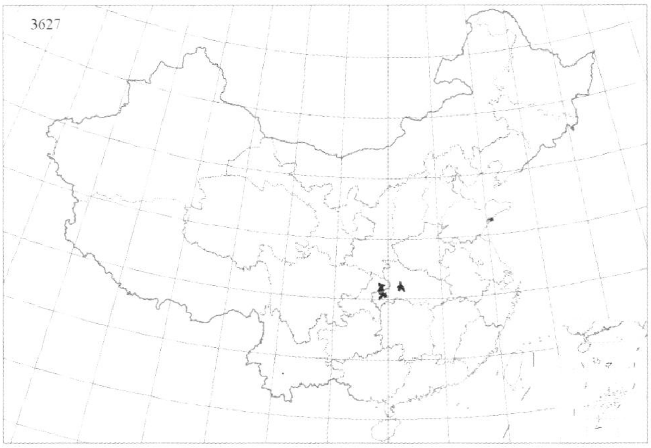

Rosaceae 蔷薇科 *Rubus* 悬钩子属

Rubus stans Focke 直立悬钩子

Shrubs, deciduous

12.7 (1.8~18.9); 5.1 (-10.2~12.0); 18.9 (9.9~26.6); 12.9 (3.8~19.0); 100.3 (15.7~167.8); 7.4 (0.0~60.7); 696 (404~929) ❙ 929 (460~1141); 540 (263~627); 31 (8~57) ❙ 36.5 (-14.2~121.7); 674 (404~912); 266.8 (142.9~351.7)

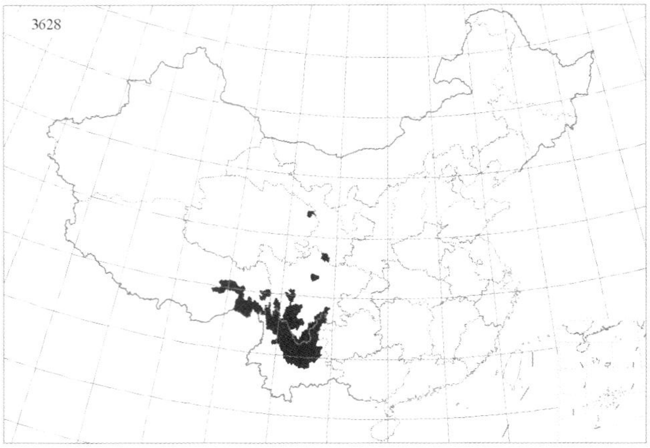

Rosaceae 蔷薇科 *Rubus* 悬钩子属

Rubus stans var. soulieanus (Cardot) T. T. Yu et L. T. Lu 多刺直立悬钩子

Shrubs, deciduous

9.2 (2.0~16.4); 0.2 (-7.4~6.4); 17.2 (9.9~25.6); 9.7 (3.8~16.4); 68.2 (15.4~137.2); 17.3 (0.0~51.9); 606 (413~857) ❙ 1078 (535~1583); 615 (326~938); 29 (10~57) ❙ 80.1 (27.1~128.6); 606 (413~857); 261.3 (171.5~337.3)

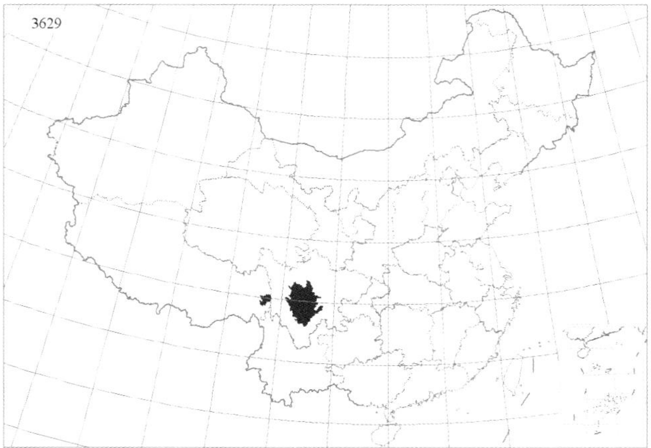

Rosaceae 蔷薇科 *Rubus* 悬钩子属

Rubus stimulans Focke 华西悬钩子

Shrubs, deciduous

11.5 (5.3~15.8); 4.6 (-2.4~9.4); 17.2 (12.2~20.7); 11.6 (5.7~15.9); 84.0 (30.5~131.1); 5.1 (0.0~26.2); 651 (479~778) ❙ 1012 (741~1243); 581 (426~696); 41 (21~59) ❙ 57.4 (26.7~100.7); 645 (479~748); 331.8 (234.2~384.7)

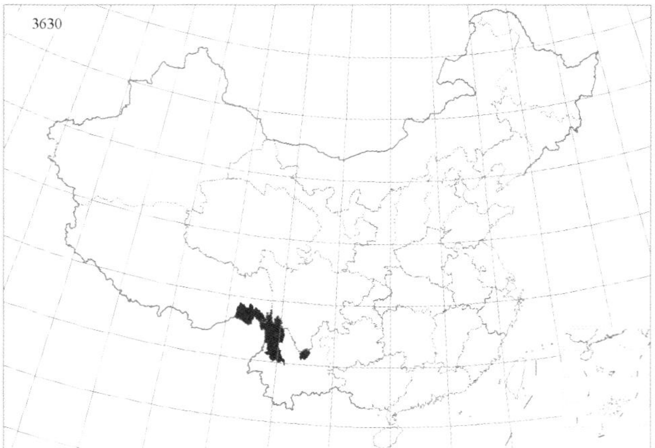

Rosaceae 蔷薇科 *Rubus* 悬钩子属
Rubus stipulosus T. T. Yu et L. T. Lu 巨托悬钩子
Shrubs, deciduous

18.4; 8.6; 26.5; 18.4; 161.1; 0.1; 960 ❙ 1583; 686; 173 ❙ 68.0; 960; 246.2

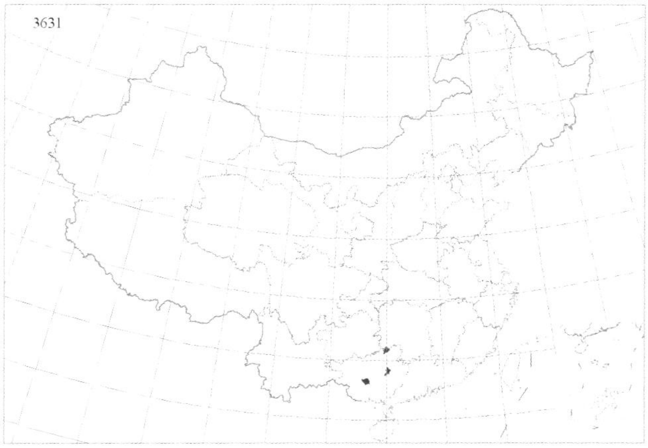

Rosaceae 蔷薇科 *Rubus* 悬钩子属
Rubus subcoreanus T. T. Yu et L. T. Lu 柱序悬钩子
Shrubs, deciduous

11.0 (0.0~14.5); -1.8 (-11.4~0.9); 23.0 (9.7~28.2); 11.4 (3.3~14.6); 91.3 (12.2~125.3); 18.2 (8.7~72.2); 697 (395~864) ❙ 726 (541~1087); 337 (245~477); 24 (11~42) ❙ 6.6 (-36.2~56.2); 638 (395~728); 237.0 (114.4~364.1)

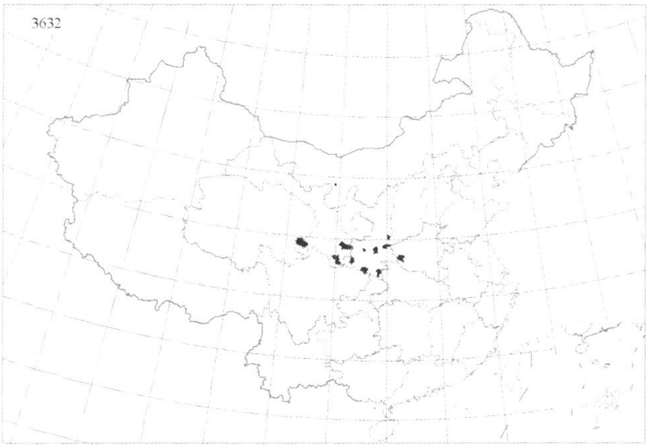

Rosaceae 蔷薇科 *Rubus* 悬钩子属
Rubus subinopertus T. T. Yu et L. T. Lu 紫红悬钩子
Shrubs, deciduous

11.1 (-0.7~16.6); 2.5 (-9.6~6.6); 18.7 (7.7~25.7); 11.4 (2.5~16.6); 82.6 (6.8~139.7); 8.8 (0.0~74.2); 655 (338~866) ❙ 1123 (370~1736); 643 (262~1007); 37 (10~59) ❙ 69.4 (4.7~145.8); 654 (338~866); 270.2 (61.6~447.4)

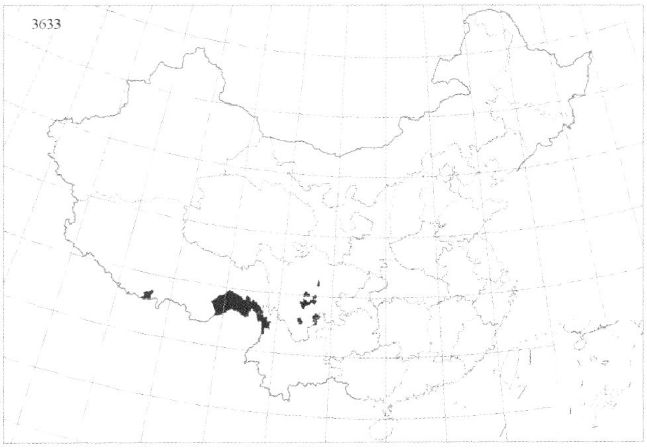

Rosaceae 蔷薇科 *Rubus* 悬钩子属
Rubus subornatus Focke 美饰悬钩子
Shrubs, deciduous

13.0 (0.3~18.9); 5.1 (-8.6~12.0); 19.5 (8.3~27.4); 13.2 (2.9~19.0); 102.9 (9.6~167.8); 6.6 (0.0~65.4); 709 (368~951) ❙ 988 (389~1358); 554 (246~727); 39 (10~88) ❙ 42.2 (-14.2~128.6); 695 (368~951); 264.5 (135.1~407.1)

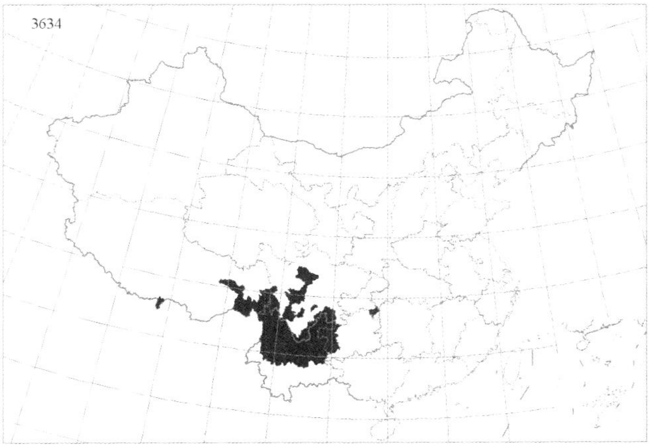

Rosaceae 蔷薇科 *Rubus* 悬钩子属
Rubus subornatus var. melanadenus Focke 黑腺美饰悬钩子
Shrubs, deciduous

9.1 (1.8~16.7); 1.3 (-7.2~9.7); 15.8 (9.9~23.6); 9.6 (3.8~16.7); 66.5 (15.4~140.9); 16.4 (0.0~53.3); 592 (404~819) ❙ 972 (535~1410); 557 (326~816); 33 (10~69) ❙ 67.2 (14.4~128.6); 586 (404~782); 302.3 (169.5~407.1)

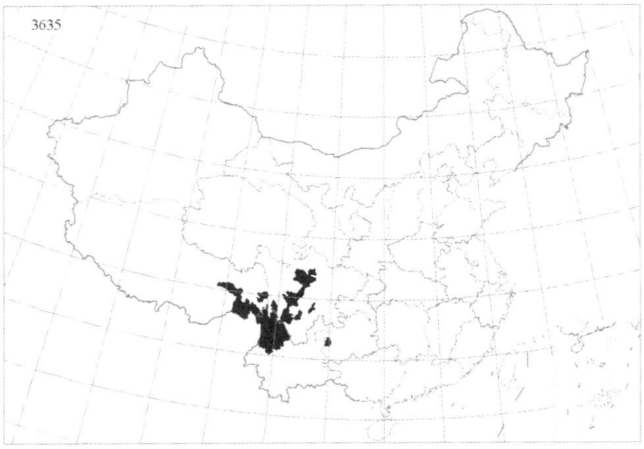

Rosaceae 蔷薇科 *Rubus* 悬钩子属
Rubus subtibetanus Hand.-Mazz. 密刺悬钩子
Shrubs, deciduous

8.5 (2.9~14.1); -3.6 (-8.0~6.5); 19.4 (11.2~23.0); 9.3 (4.4~14.2); 68.8 (20.5~109.8); 26.2 (0.0~44.6); 606 (428~729) ❙ 750 (423~1220); 411 (247~688); 20 (8~37) ❙ 28.5 (-35.2~121.7); 547 (423~707); 210.5 (44.7~306.0)

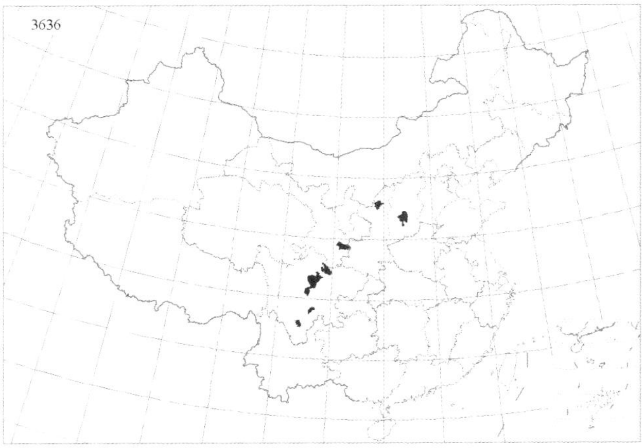

Rosaceae 蔷薇科 *Rubus* 悬钩子属
Rubus sumatranus Miq. 红腺悬钩子
Shrubs, deciduous

18.5 (8.6~24.8); 9.4 (1.1~20.1); 26.4 (15.0~29.8); 18.5 (8.7~24.8); 162.4 (52.6~237.6); 0.4 (0.0~8.8); 963 (574~1385) ❙ 1540 (939~3495); 641 (356~1412); 144 (20~897) ❙ 62.2 (-6.9~246.5); 956 (574~1329); 263.0 (95.3~659.0)

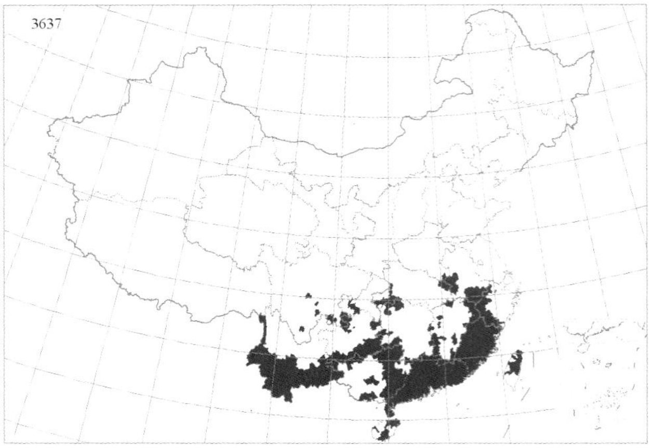

Rosaceae 蔷薇科 *Rubus* 悬钩子属
Rubus swinhoei Hance 木莓
Shrubs, deciduous or semievergreen

16.8 (9.1~22.0); 5.9 (-0.8~14.6); 26.9 (18.1~30.1); 16.8 (9.2~22.1); 142.4 (64.1~205.1); 0.7 (0.0~14.1); 894 (595~1170) ❙ 1487 (798~3495); 550 (299~1186); 162 (19~897) ❙ 66.6 (-6.7~246.5); 882 (595~1170); 218.4 (134.6~332.5)

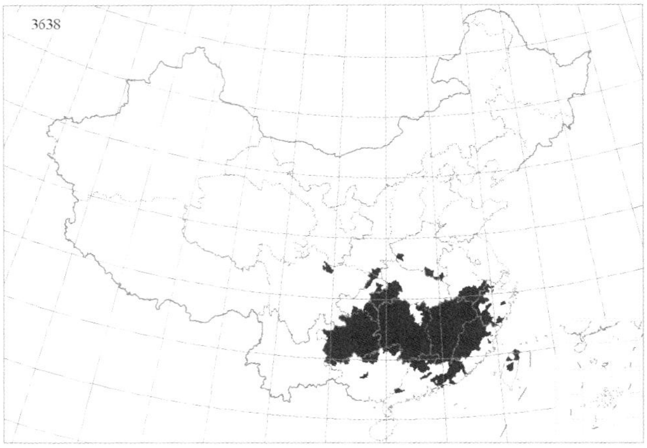

Rosaceae 蔷薇科 *Rubus* 悬钩子属
Rubus taitoensis Hayata 台东刺花悬钩子
Shrubs, deciduous

17.8; 12.5; 22.1; 17.8; 153.9; 0.0; 857 ❙ 2621; 1126; 265 ❙ 210.3; 857; 266.6

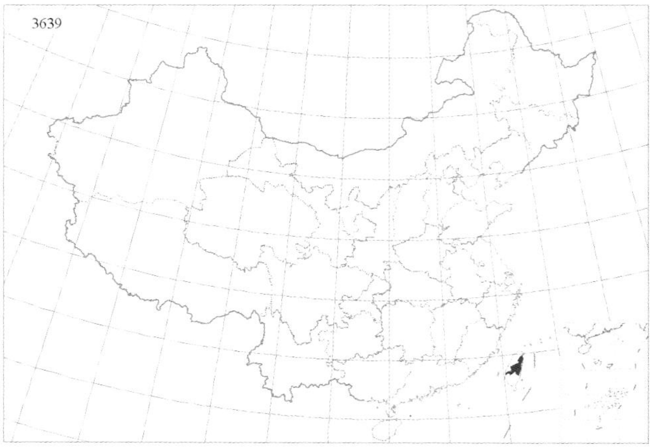

Rosaceae 蔷薇科 *Rubus* 悬钩子属
Rubus taiwanicola Koidz. et Ohwi 小叶悬钩子
Shrubs, deciduous

18.4; 13.2; 22.5; 18.4; 160.9; 0.0; 879 ❙ 2629; 1176; 251 ❙ 202.3; 879; 285.2

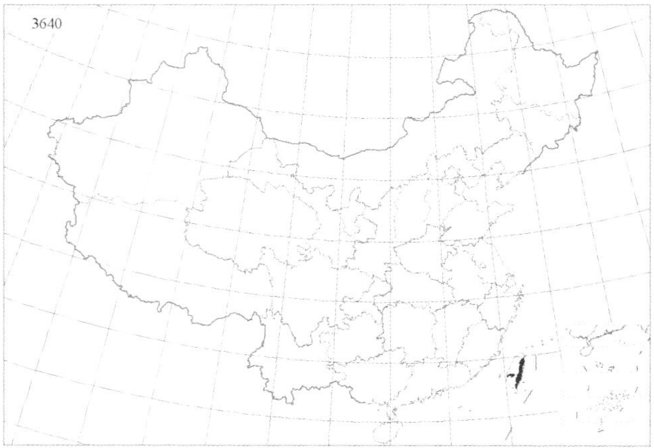

Rosaceae 蔷薇科 *Rubus* 悬钩子属
Rubus taronensis C. Y. Wu ex T. T. Yu et L. T. Lu 独龙悬钩子
Shrubs, deciduous

8.6; 1.3; 15.0; 8.7; 52.6; 8.8; 574 ❙ 1041; 585; 45 ❙ 81.5; 574; 352.3

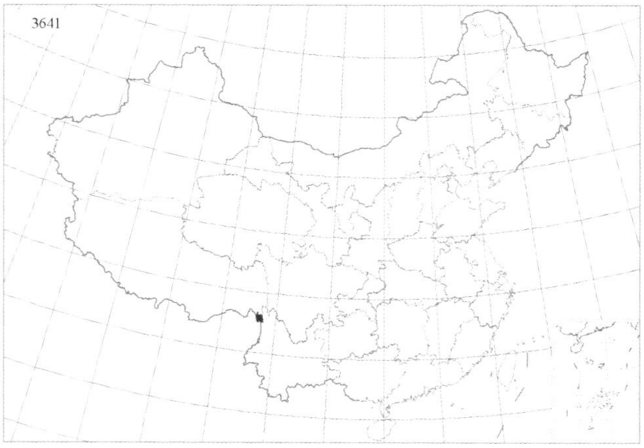

Rosaceae 蔷薇科 *Rubus* 悬钩子属
Rubus tephrodes Hance 灰白毛莓
Shrubs, deciduous

16.7 (9.8~22.2); 5.5 (-2.5~14.4); 27.2 (20.1~29.9); 16.8 (10.2~22.3); 142.1 (79.2~207.5); 0.7 (0.0~20.1); 896 (641~1190) ❙ 1466 (795~2993); 536 (299~1186); 162 (25~551) ❙ 63.9 (-6.7~246.5); 885 (641~1190); 211.2 (111.9~327.7)

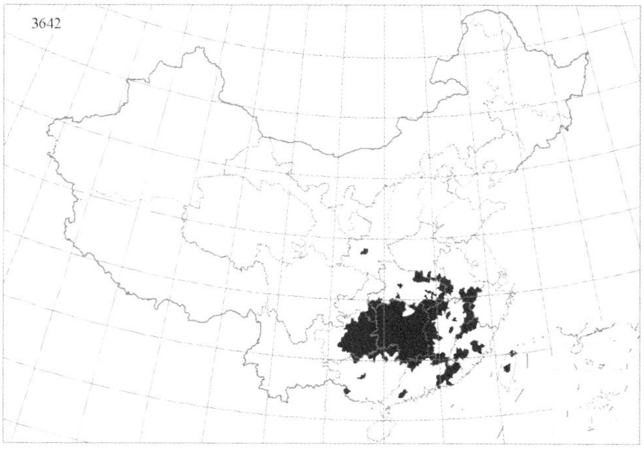

Rosaceae 蔷薇科 *Rubus* 悬钩子属

Rubus tephrodes var. ampliflorus (H. Lév. et Vaniot) Hand.-Mazz. 无腺灰白毛莓

Shrubs, deciduous

16.9 (14.3~19.4); 5.2 (3.3~9.4); 27.9 (24.2~29.9); 16.9 (14.3~19.4); 143.6 (113.2~173.4); 0.4 (0.0~2.3); 910 (761~1029) ❙ 1467 (1171~1865); 498 (392~685); 177 (88~258) ❙ 61.7 (28.8~128.9); 892 (761~1029); 197.1 (134.6~263.5)

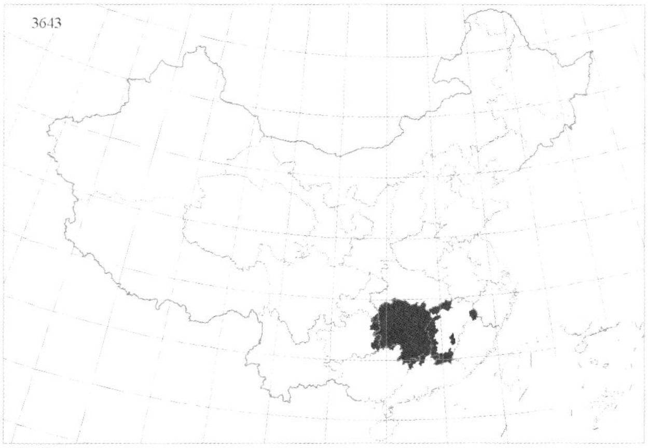

Rosaceae 蔷薇科 *Rubus* 悬钩子属

Rubus thibetanus Franch. 西藏悬钩子

Shrubs, deciduous

9.4 (2.9~13.4); -1.1 (-6.3~3.3); 19.1 (11.2~24.6); 9.7 (4.4~13.4); 70.4 (20.5~108.0); 17.5 (2.4~44.6); 616 (428~760) ❙ 857 (626~1049); 423 (305~574); 22 (13~28) ❙ 44.0 (-2.2~121.7); 615 (428~760); 263.1 (200.8~322.4)

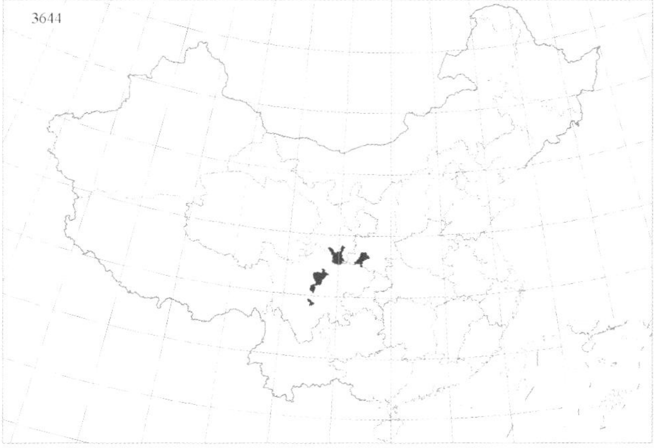

Rosaceae 蔷薇科 *Rubus* 悬钩子属

Rubus tinifolius C. Y. Wu ex T. T. Yu et L. T. Lu 截叶悬钩子

Shrubs, deciduous

18.0 (14.4~21.0); 11.5 (7.5~14.2); 22.4 (19.1~25.6); 18.0 (14.4~21.0); 156.2 (113.1~191.8); 0.0 (0.0~0.0); 874 (727~1053) ❙ 1277 (1049~1522); 715 (598~835); 54 (44~70) ❙ 47.1 (21.3~99.9); 865 (727~1053); 437.6 (294.5~551.6)

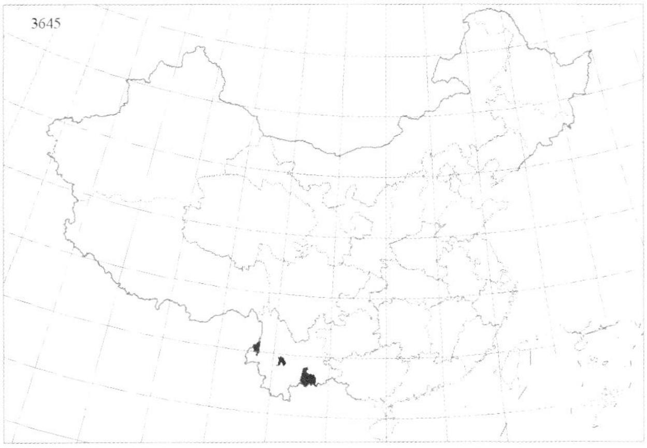

Rosaceae 蔷薇科 *Rubus* 悬钩子属

Rubus treutleri Hook. f. 滇西北悬钩子

Shrubs, deciduous

8.6; 1.3; 15.0; 8.7; 52.6; 8.8; 574 ❙ 1041; 585; 45 ❙ 81.5; 574; 352.3

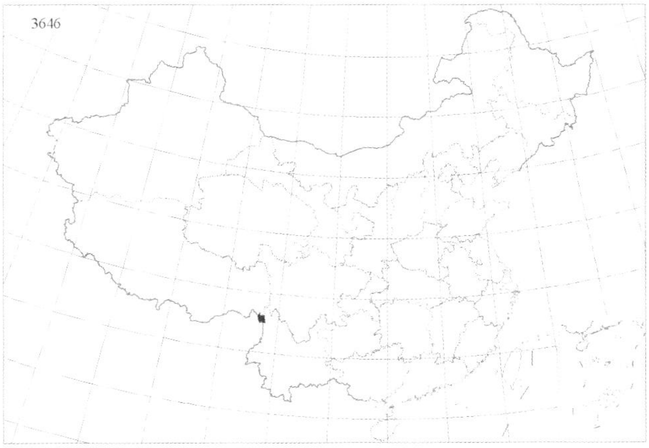

Rosaceae 蔷薇科 *Rubus* 悬钩子属

Rubus trianthus Focke 三花悬钩子

Shrubs, deciduous

16.3 (12.1~21.7); 4.9 (1.1~15.3); 27.2 (20.1~29.9); 16.4 (12.2~21.7); 137.4 (95.0~200.2); 0.9 (0.0~8.5); 879 (696~1118) ❙ 1476 (817~3495); 533 (378~1412); 169 (53~897) ❙ 68.3 (-2.0~246.5); 868 (696~1114); 208.8 (131.2~319.9)

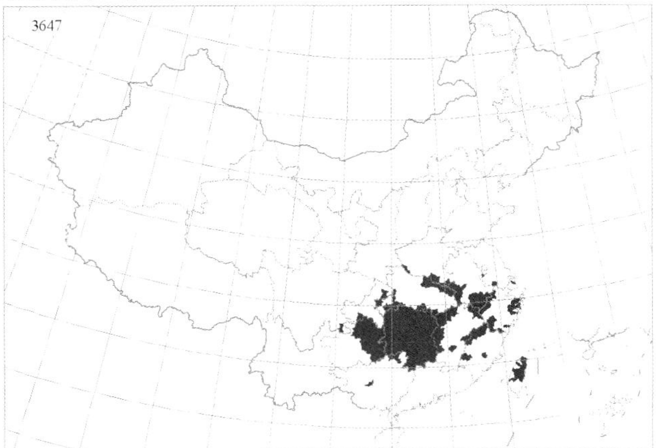

Rosaceae 蔷薇科 *Rubus* 悬钩子属

Rubus tricolor Focke 三色莓

Shrubs, deciduous

13.9 (2.0~18.9); 6.2 (-7.4~12.5); 20.2 (9.9~26.6); 14.0 (3.8~19.0); 110.7 (15.4~167.8); 3.2 (0.0~51.9); 733 (413~929) ❙ 1007 (741~1410); 567 (411~816); 39 (15~71) ❙ 40.0 (-14.2~128.6); 715 (413~912); 272.5 (142.9~406.0)

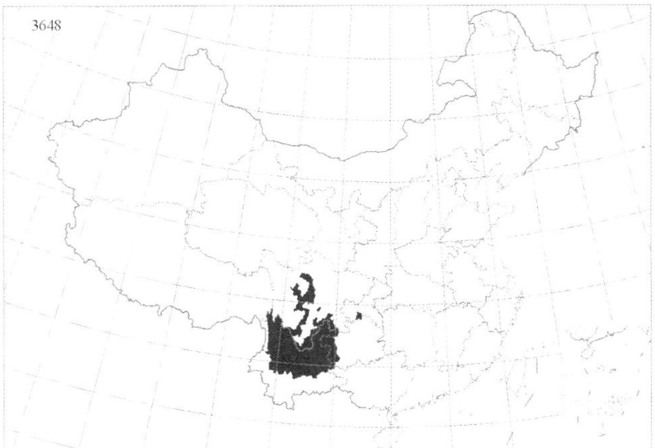

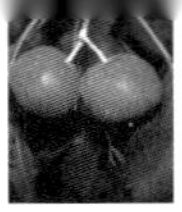

Rosaceae 蔷薇科 *Rubus* 悬钩子属

Rubus trijugus Focke 三对叶悬钩子

Shrubs, deciduous

6.7 (2.0~14.3); -1.5 (-7.4~7.6); 13.8 (9.9~19.7); 7.5 (3.8~14.3); 46.3 (15.4~111.9); 25.3 (0.0~51.9); 528 (413~728) ❙ 886 (543~1220); 525 (330~688); 21 (9~37) ❙ 70.9 (28.5~128.6); 525 (413~707); 291.6 (248.3~341.6)

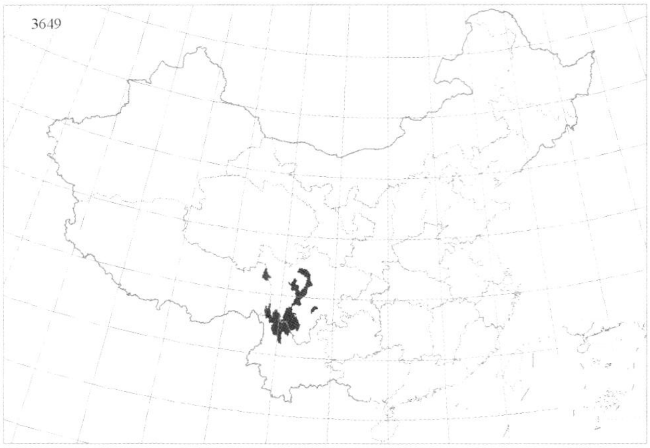

Rosaceae 蔷薇科 *Rubus* 悬钩子属

Rubus tsangii Merr. 光滑悬钩子

Shrubs, deciduous

17.0 (6.6~21.8); 9.8 (-1.3~15.1); 22.3 (13.3~28.5); 17.0 (6.8~21.8); 144.5 (39.6~201.9); 0.3 (0.0~19.7); 840 (514~1158) ❙ 1299 (862~2859); 675 (477~1186); 72 (12~402) ❙ 54.5 (6.5~246.5); 826 (514~1158); 350.4 (189.5~659.0)

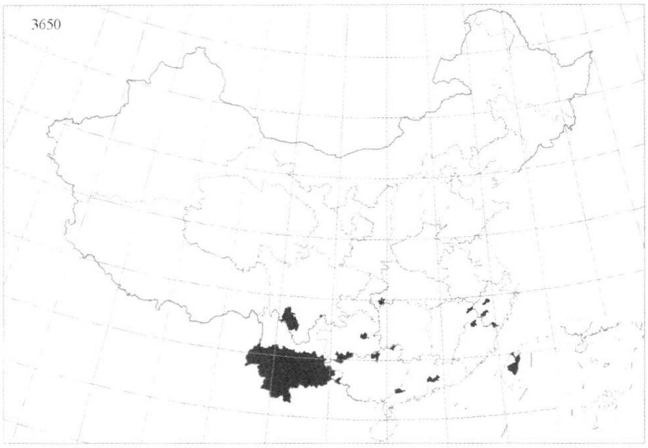

Rosaceae 蔷薇科 *Rubus* 悬钩子属

Rubus tsangii var. yanshanensis (Z. X. Yu et W. T. Ji) L. T. Lu 铅山悬钩子

Shrubs, deciduous

17.4; 5.6; 28.7; 17.5; 149.4; 0.0; 942 ❙ 1302; 442; 167 ❙ 39.3; 834; 210.8

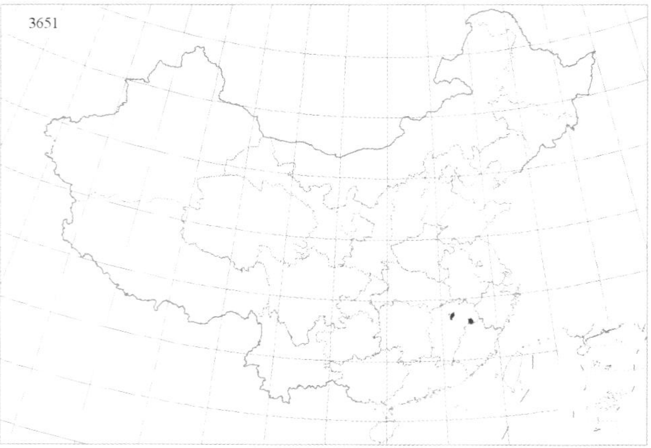

Rosaceae 蔷薇科 *Rubus* 悬钩子属

Rubus tsangorum Hand.-Mazz. 东南悬钩子

Small shrubs, deciduous

17.5 (14.1~21.8); 6.7 (1.6~12.8); 27.4 (24.2~29.3); 17.5 (14.2~21.8); 150.3 (113.2~202.0); 0.3 (0.0~6.5); 924 (761~1168) ❙ 1646 (1211~1997); 571 (419~720); 200 (129~259) ❙ 79.3 (17.4~147.0); 920 (761~1134); 242.4 (169.7~332.5)

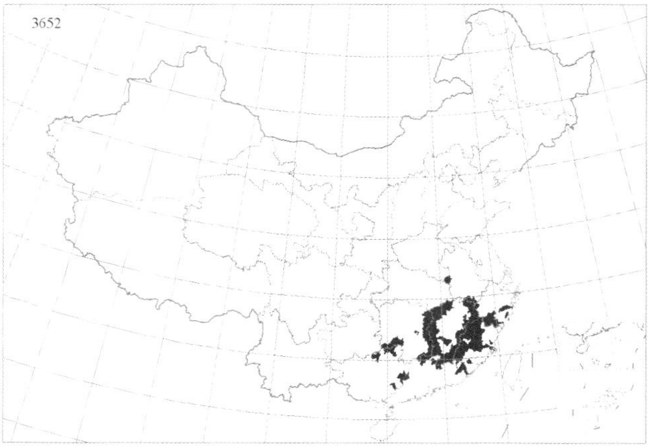

Rosaceae 蔷薇科 *Rubus* 悬钩子属

Rubus wallichianus Wight et Arn. 红毛悬钩子

Shrubs, deciduous

15.8 (7.3~24.1); 7.1 (-2.0~18.9); 23.3 (15.0~28.9); 15.9 (7.5~24.1); 131.6 (48.2~229.7); 1.3 (0.0~20.4); 819 (537~1322) ❙ 1219 (793~2993); 597 (377~1594); 71 (15~547) ❙ 49.8 (-14.2~246.5); 806 (537~1112); 266.5 (108.2~551.6)

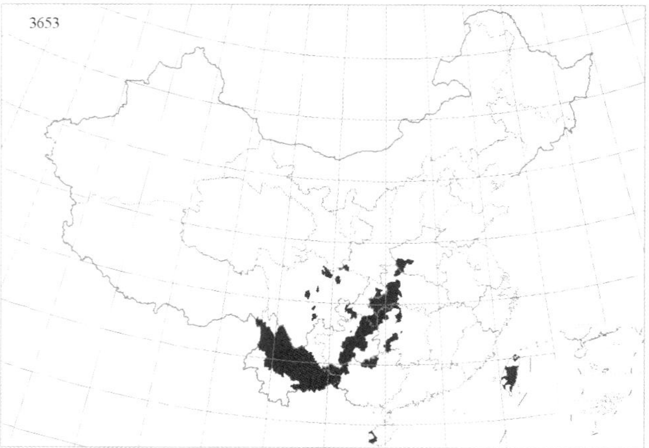

Rosaceae 蔷薇科 *Rubus* 悬钩子属

Rubus wangii F. P. Metcalf 大苞悬钩子

Shrubs, deciduous

20.0 (17.9~21.6); 10.5 (7.4~13.2); 27.8 (25.8~28.6); 20.0 (18.0~21.7); 180.3 (155.6~199.8); 0.0 (0.0~0.0); 1050 (913~1153) ❙ 1613 (1371~1776); 693 (562~828); 169 (125~189) ❙ 54.8 (22.7~85.5); 1049 (913~1153); 225.3 (185.5~254.1)

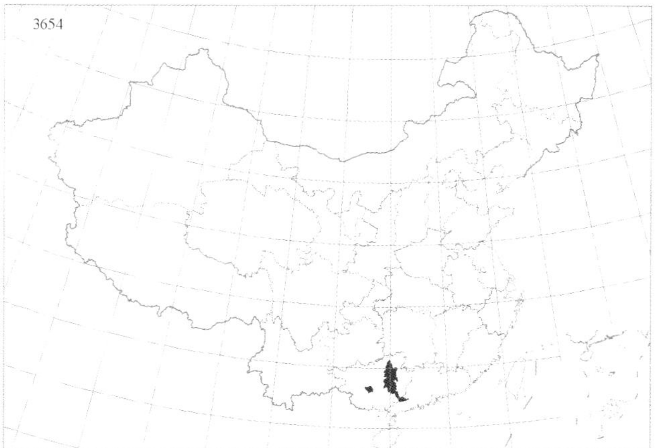

Rosaceae 蔷薇科 *Rubus* 悬钩子属
Rubus wardii Merr. 大花悬钩子
Shrubs or subshrubs, deciduous

11.0；3.4；17.3；11.0；76.5；4.4；640 ▌1389；796；48 ▌113.7；640；399.9

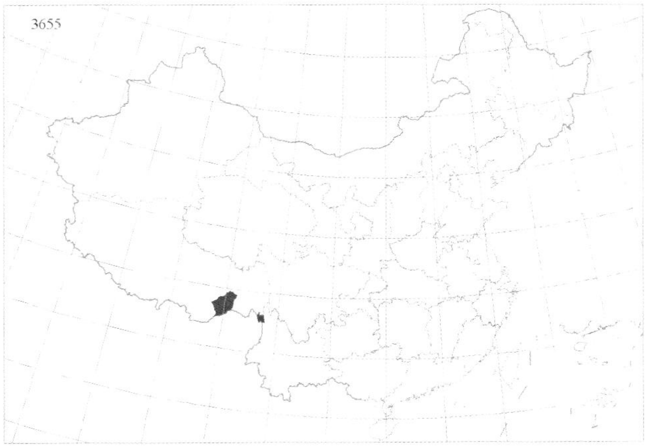

Rosaceae 蔷薇科 *Rubus* 悬钩子属
Rubus wawushanensis T. T. Yu et L. T. Lu 瓦屋山悬钩子
Shrubs, deciduous

10.6；1.3；19.1；10.8；78.8；10.4；642 ▌1210；674；37 ▌88.2；642；244.9

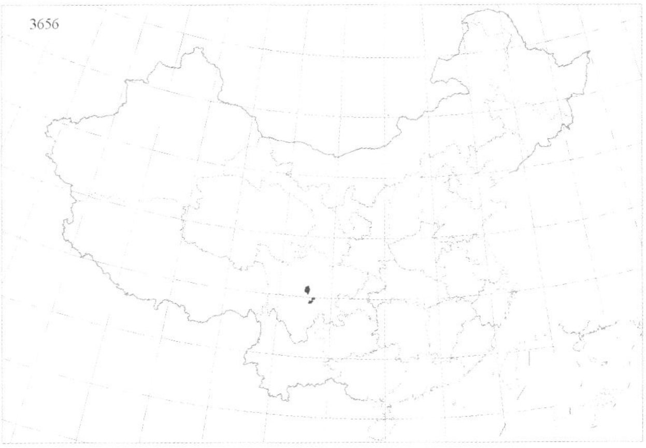

Rosaceae 蔷薇科 *Rubus* 悬钩子属
Rubus wilsonii Duthie 湖北悬钩子
Shrubs, deciduous

13.0；1.9；23.8；13.1；102.9；6.1；730 ▌1202；500；83 ▌64.6；730；262.4

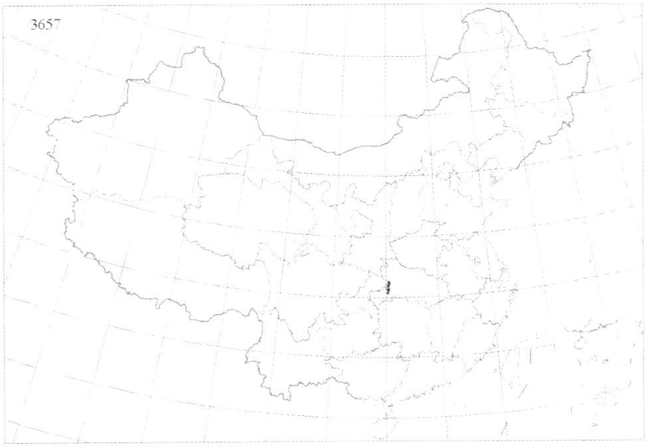

Rosaceae 蔷薇科 *Rubus* 悬钩子属
Rubus wushanensis T. T. Yu et L. T. Lu 巫山悬钩子
Shrubs, deciduous

13.7；2.6；24.4；13.8；109.1；4.0；758 ▌1148；475；68 ▌51.4；758；258.0

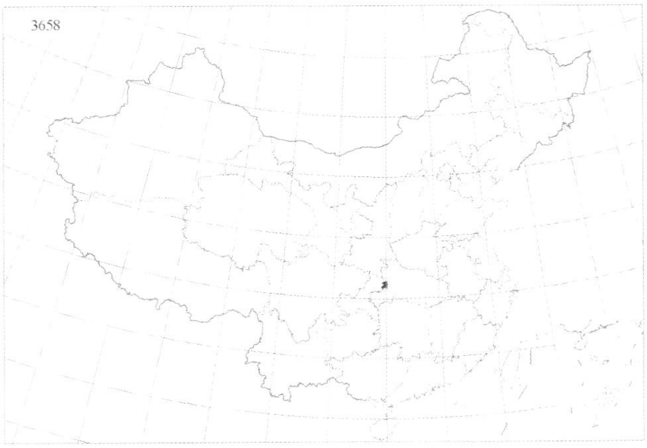

Rosaceae 蔷薇科 *Rubus* 悬钩子属
Rubus wuzhianus L. T. Lu et Boufford 锯叶悬钩子
Shrubs, deciduous

14.2；3.1；24.8；14.3；114.0；2.8；776 ▌1470；607；101 ▌89.9；776；234.4

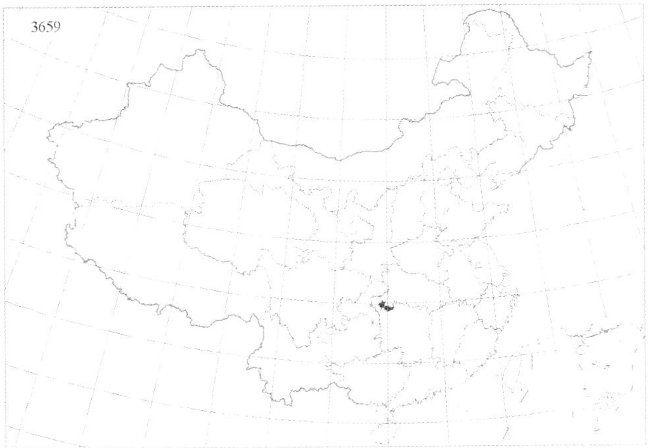

Rosaceae 蔷薇科 *Rubus* 悬钩子属
Rubus xanthocarpus Bureau et Franch. 黄果悬钩子
Subshrubs, deciduous

12.7 (0.4~17.1)；0.6 (−10.7~5.8)；24.1 (9.9~29.4)；13.0 (3.5~17.2)；105.0 (13.3~147.0)；11.7 (0.0~68.1)；760 (402~954) ▌1013 (587~1663)；449 (273~817)；70 (12~212) ▌35.4 (−28.3~128.6)；745 (402~942)；212.8 (111.9~357.7)

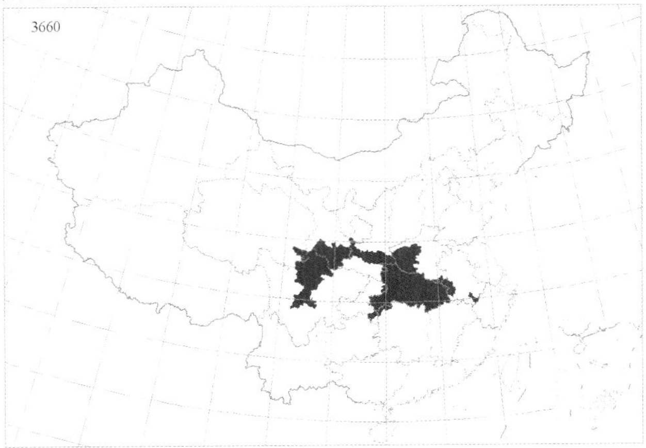

Rosaceae 蔷薇科 *Rubus* 悬钩子属

Rubus xanthoneurus Focke 黄脉莓

Shrubs, deciduous

16.3 (10.6~21.2); 6.1 (−0.2~14.2); 25.5 (21.1~29.0); 16.3 (10.6~21.3); 136.8 (79.4~195.2); 0.7 (0.0~12.1); 854 (643~1126) ❙ 1345 (863~2005); 568 (375~827); 110 (27~254) ❙ 58.0 (10.5~130.7); 852 (643~1126); 227.1 (130.1~551.6)

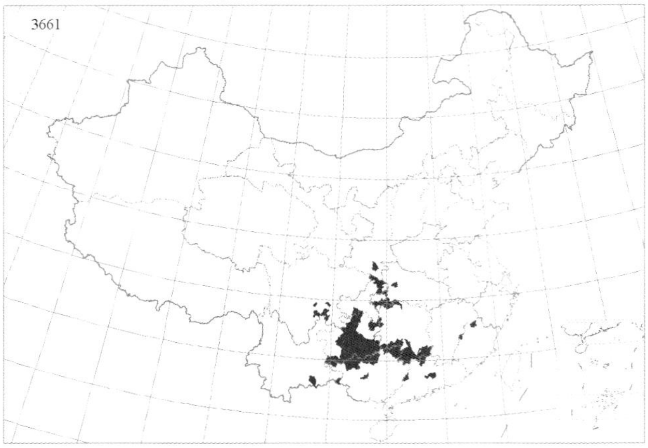

Rosaceae 蔷薇科 *Rubus* 悬钩子属

Rubus xanthoneurus var. glandulosus T. T. Yu et L. T. Lu 腺毛黄脉莓

Shrubs, deciduous

16.7; 6.6; 25.5; 16.8; 141.7; 0.6; 879 ❙ 1397; 608; 118 ❙ 60.5; 879; 241.1

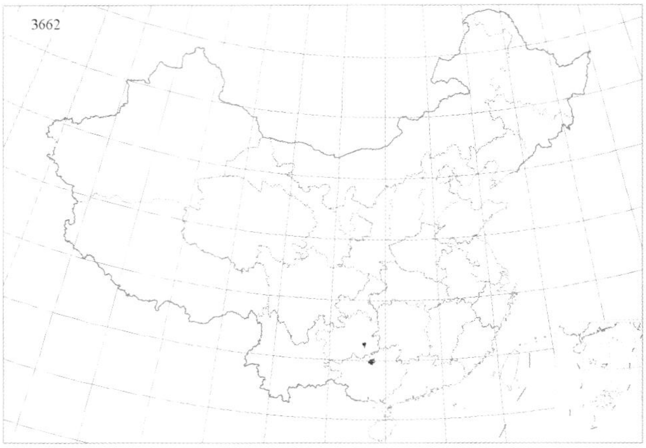

Rosaceae 蔷薇科 *Rubus* 悬钩子属

Rubus xichouensis T. T. Yu et L. T. Lu 西畴悬钩子

Shrubs, deciduous

17.2; 9.5; 22.7; 17.3; 147.0; 0.0; 838 ❙ 1254; 699; 48 ❙ 49.7; 838; 336.3

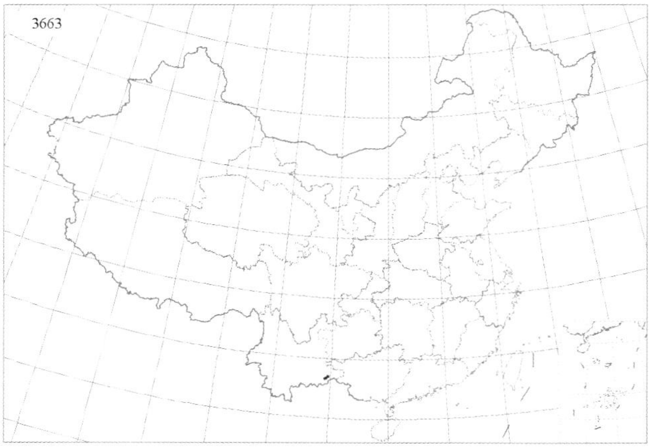

Rosaceae 蔷薇科 *Rubus* 悬钩子属

Rubus yanyunii Y. T. Chang et L. Y. Chen 九仙莓

Shrubs, deciduous

18.2; 9.5; 26.0; 18.2; 158.3; 0.0; 919 ❙ 1658; 608; 184 ❙ 80.7; 919; 286.6

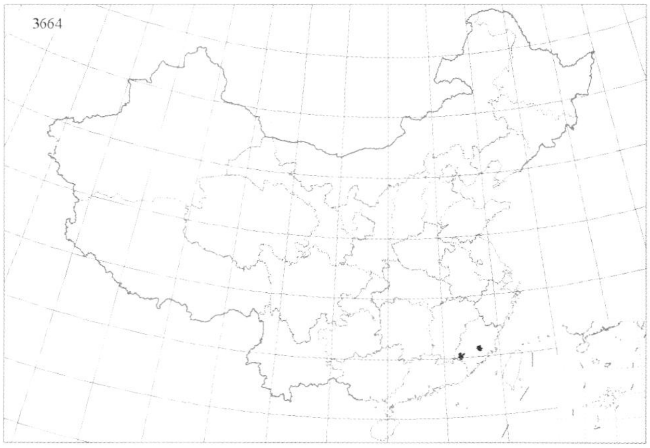

Rosaceae 蔷薇科 *Rubus* 悬钩子属

Rubus yiwuanus W. P. Fang 奕武悬钩子

Shrubs, deciduous

14.9; 5.3; 23.6; 15.0; 119.6; 0.0; 782 ❙ 1410; 816; 49 ❙ 80.4; 782; 214.5

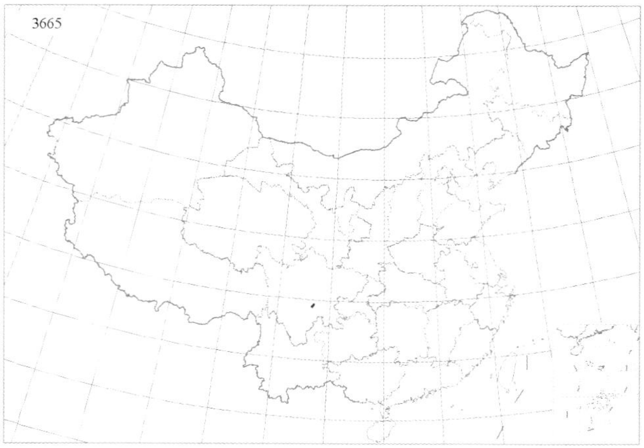

Rosaceae 蔷薇科 *Rubus* 悬钩子属

Rubus yuliensis Y. C. Liu et F. Y. Lu 玉里悬钩子

Shrubs

16.4; 11.5; 20.3; 16.5; 137.8; 0.0; 783 ❙ 2499; 947; 303 ❙ 219.3; 783; 303.6

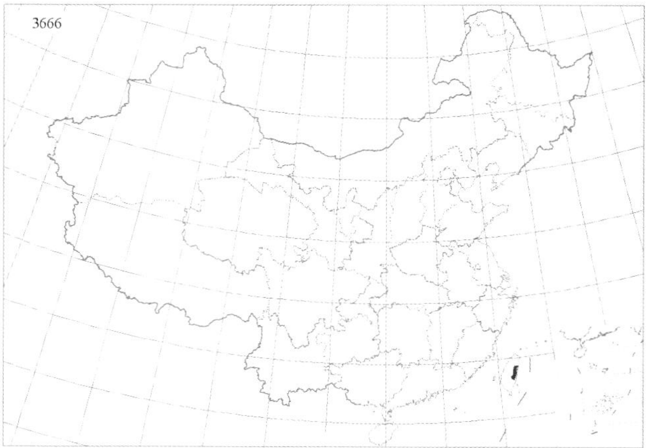

Rosaceae 蔷薇科 *Rubus* 悬钩子属

Rubus yunanicus Kuntze 云南悬钩子

Shrubs

19.8; 4.0; 23.5; 19.8; 177.8; 0.0; 954 | 1446; 745; 61 | 51.7; 938; 538.1

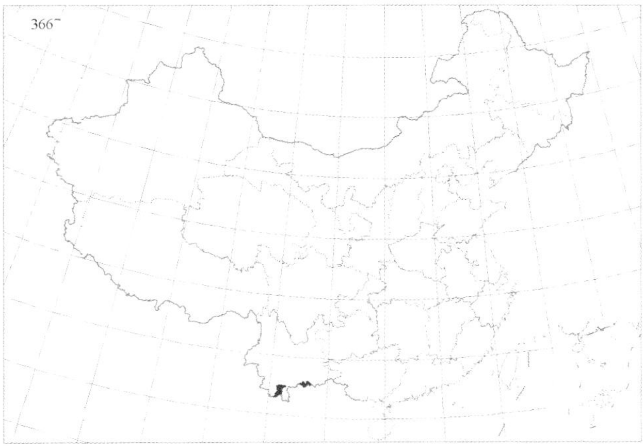

Rosaceae 蔷薇科 *Rubus* 悬钩子属

Rubus zhaogoshanensis T. T. Yu et L. T. Lu 草果山悬钩子

Shrubs, deciduous

17.2; 9.5; 22.7; 17.3; 147.0; 0.0; 838 | 1254; 699; 48 | 49.7; 838; 336.3

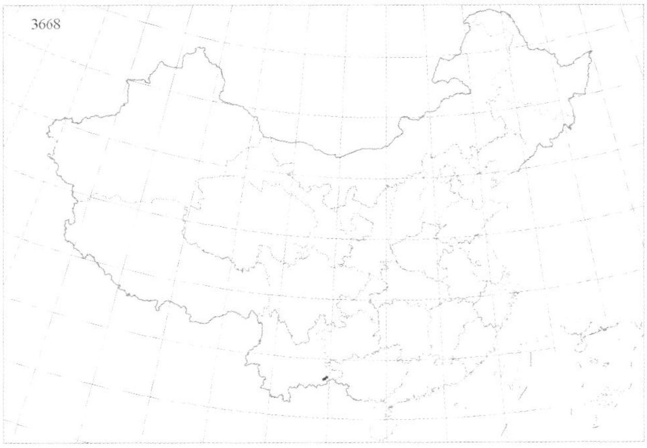

Rosaceae 蔷薇科 *Sibiraea* 鲜卑花属

Sibiraea angustata (Rehder) Hand.-Mazz. 窄叶鲜卑花

Shrubs, deciduous

1.9 (-4.4~10.8); -8.8 (-17.3~3.9); 11.4 (6.9~20.6); 4.5 (1.7~10.9); 22.8 (3.2~7[illegible].8); 59.5 (1.3~115.9); 427 (323~633) | 585 (226~1041); 337 (138~594); 13 (2~45) | 36.4 (-44.9~128.6); 411 (226~633); 234.8 (77.1~364.1)

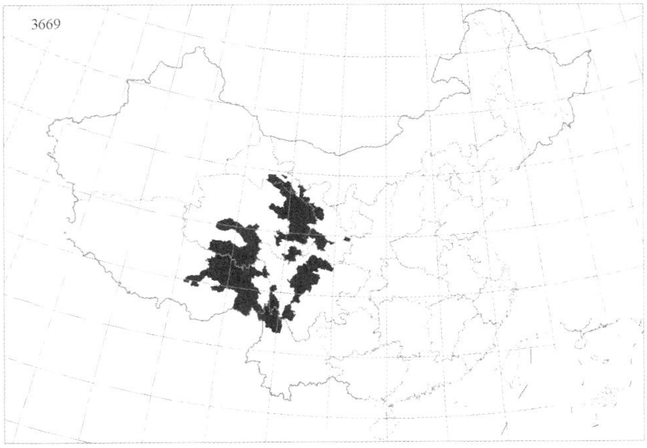

Rosaceae 蔷薇科 *Sibiraea* 鲜卑花属

Sibiraea laevigata (L.) Maxim. 鲜卑花

Shrubs, deciduous

1.0 (-5.5~7.7); -11.3 (-17.3~-4.1); 12.0 (5.8~23.5); 4.4 (1.3~9.8); 24.0 (1.2~81.3); 71.4 (29.9~127.1); 424 (299~654) | 440 (63~731); 250 (45~379); 9 (1~19) | 7.0 (-87.1~87.3); 359 (63~522); 201.6 (7.5~364.1)

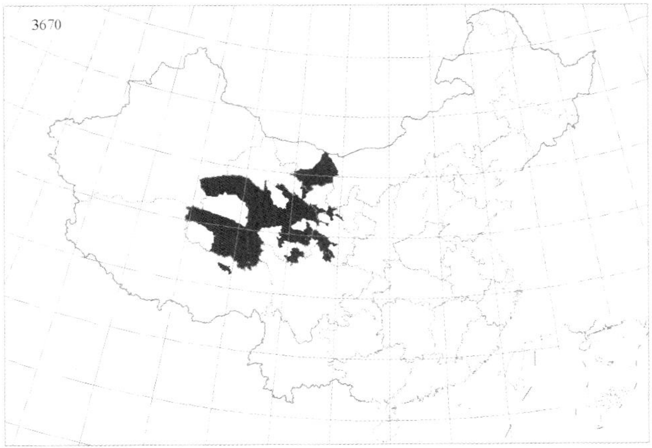

Rosaceae 蔷薇科 *Sibiraea* 鲜卑花属

Sibiraea tomentosa Diels 毛叶鲜卑花

Shrubs, deciduous

8.5; 1.2; 14.7; 8.6; 53.5; 11.5; 565 | 884; 545; 23 | 57.1; 565; 301.3

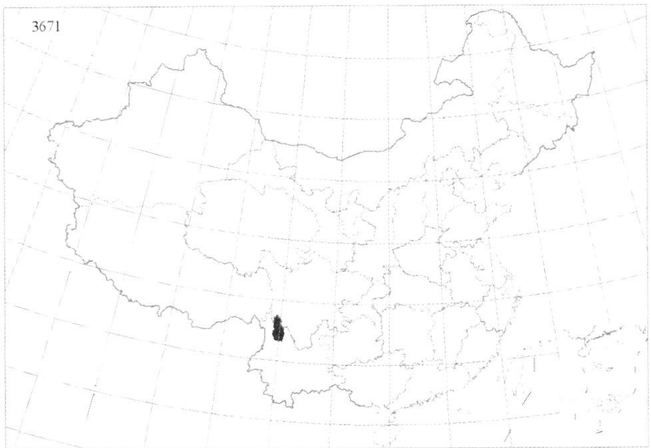

Rosaceae 蔷薇科 *Sorbaria* 珍珠梅属

Sorbaria arborea C. K. Schneid. 高丛珍珠梅

Shrubs, deciduous

11.3 (-0.4~18.9); 0.5 (-10.6~12.0); 21.1 (8.4~29.3); 11.6 (2.8~19.0); 90.1 (9.4~167.8); 14.3 (0.0~73.3); 684 (360~941) | 886 (341~1736); 444 (190~1007); 38 (5~184) | 31.5 (-40.9~145.8); 661 (341~941); 241.1 (61.2~447.4)

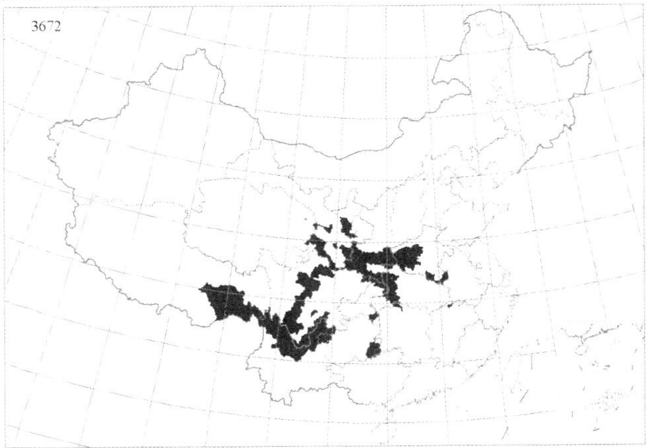

Rosaceae 蔷薇科 *Sorbaria* 珍珠梅属

Sorbaria arborea var. glabrata Rehder 光叶高丛珍珠梅

Shrubs, deciduous

10.6 (2.6~15.1); −1.1 (−7.1~4.1); 21.6 (11.1~26.8); 10.9 (4.3~15.1); 83.6 (20.0~122.7); 16.1 (0.9~48.2); 663 (427~834) ❙ 879 (632~1254); 407 (290~606); 33 (15~83) ❙ 35.2 (−16.6~121.7); 660 (427~817); 265.0 (151.3~357.7)

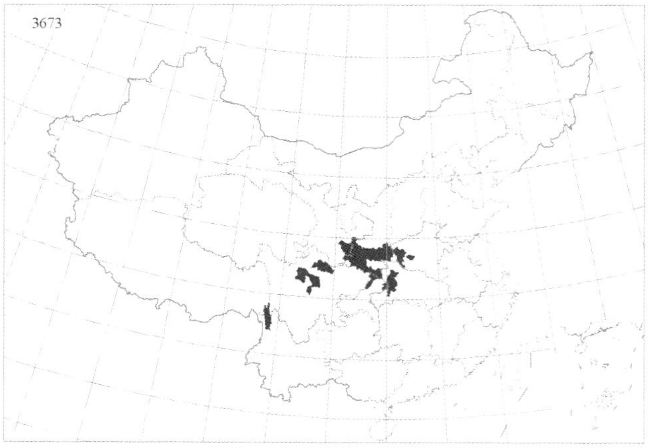

Rosaceae 蔷薇科 *Sorbaria* 珍珠梅属

Sorbaria kirilowii (Regel et Tiling) Maxim. 华北珍珠梅

Shrubs, deciduous

11.4 (0.8~15.7); −3.8 (−14.7~4.4); 24.9 (10.7~28.3); 12.1 (3.8~15.7); 100.8 (16.2~133.0); 24.0 (0.6~74.1); 748 (411~895) ❙ 593 (341~1246); 361 (190~528); 20 (5~103) ❙ −19.4 (−51.1~93.8); 570 (341~854); 172.8 (60.5~334.4)

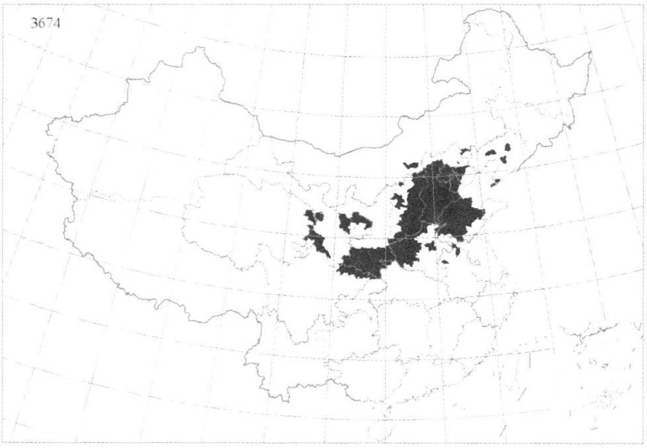

Rosaceae 蔷薇科 *Sorbaria* 珍珠梅属

Sorbaria sorbifolia (L.) A. Braun 珍珠梅

Shrubs, deciduous

3.8 (−5.8~11.6); −16.5 (−29.8~−1.9); 21.2 (15.3~25.1); 8.2 (4.4~11.8); 64.5 (28.0~96.3); 78.6 (16.9~157.4); 585 (426~722) ❙ 693 (137~1073); 435 (92~691); 24 (3~40) ❙ 18.9 (−77.5~74.1); 555 (137~711); 273.5 (25.4~346.5)

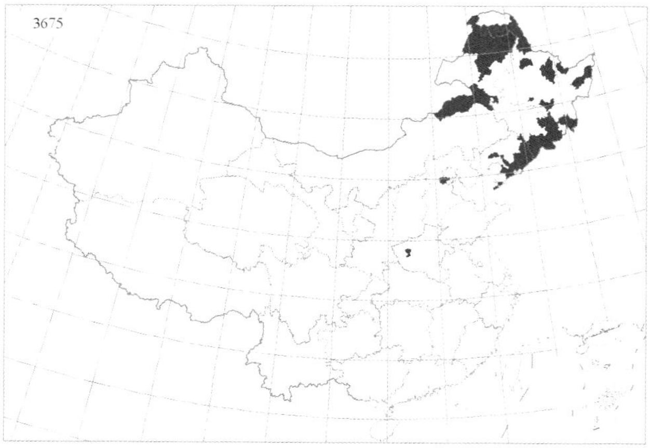

Rosaceae 蔷薇科 *Sorbaria* 珍珠梅属

Sorbaria sorbifolia var. stellipila Maxim. 星毛珍珠梅

Shrubs, deciduous

6.1 (5.1~8.0); −13.3 (−15.5~−9.8); 22.8 (21.9~23.5); 9.2 (8.7~10.0); 74.7 (69.5~83.4); 60.5 (46.4~68.8); 624 (601~658) ❙ 819 (718~942); 515 (446~601); 29 (23~33) ❙ 31.7 (9.1~56.9); 624 (601~658); 290.6 (239.1~323.4)

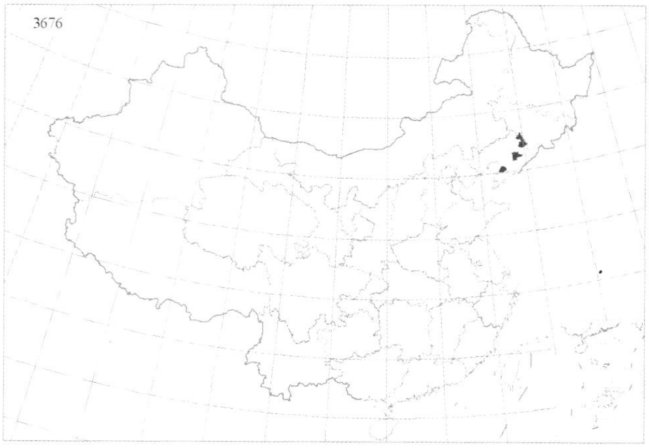

Rosaceae 蔷薇科 *Sorbus* 花楸属

Sorbus albopilosa T. T. Yu et L. T. Lu 白毛花楸

Trees, deciduous

5.9 (−1.8~13.3); −2.3 (−10.2~5.4); 12.9 (6.1~19.6); 7.2 (1.8~13.4); 46.0 (1.9~100.4); 35.0 (0.0~82.8); 511 (329~706) ❙ 967 (320~1736); 569 (217~1007); 29 (11~51) ❙ 71.3 (−3.0~145.8); 509 (320~706); 252.0 (72.2~447.4)

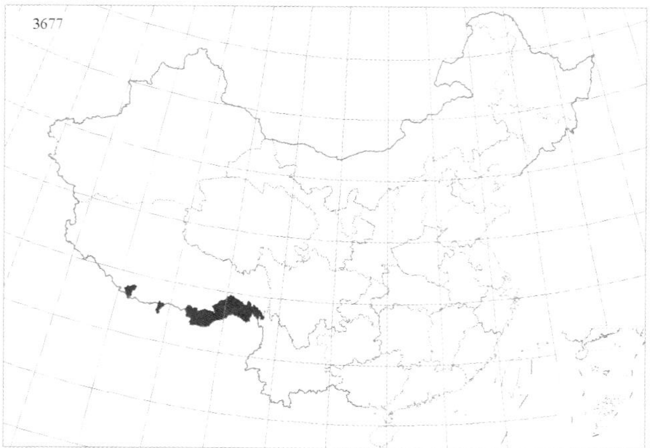

Rosaceae 蔷薇科 *Sorbus* 花楸属

Sorbus alnifolia (Siebold et Zucc.) C. Koch 水榆花楸

Trees, deciduous

11.9 (1.4~18.2); −2.3 (−20.1~11.5); 24.7 (15.8~29.4); 12.6 (6.4~18.3); 104.1 (45.2~159.4); 20.9 (0.0~91.9); 750 (501~984) ❙ 871 (349~2499); 415 (227~947); 63 (8~303) ❙ 13.8 (−44.9~219.3); 676 (349~950); 196.5 (70.3~346.5)

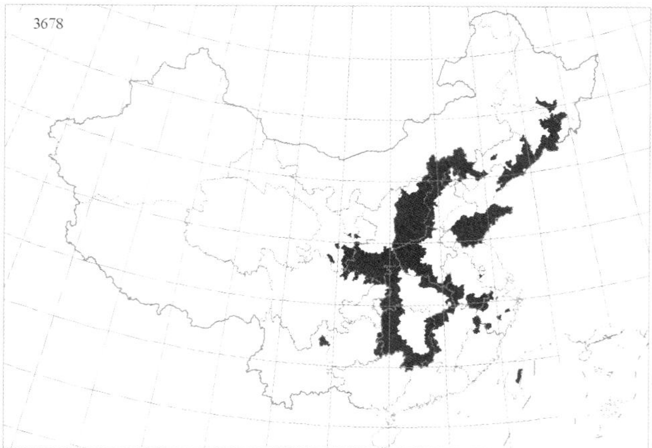

Rosaceae 蔷薇科 *Sorbus* 花楸属

Sorbus alnifolia var. lobulata Rehder 裂叶水榆花楸

Trees, deciduous

11.5 (8.ɛ~13.3)；-3.5 (-7.7~-1.3)；24.9 (22.8~26.5)；12.1 (10.4~13.5)；101.4 (84.0~114.7)；22.5 (14.5~40.0)；742 (665~802) ❙ 722 (624~817)；459 (350~529)；28 (11~37) ❙ -2.8 (-10.4~5.6)；715 (624~800)；166.7 (136.6~205.4)

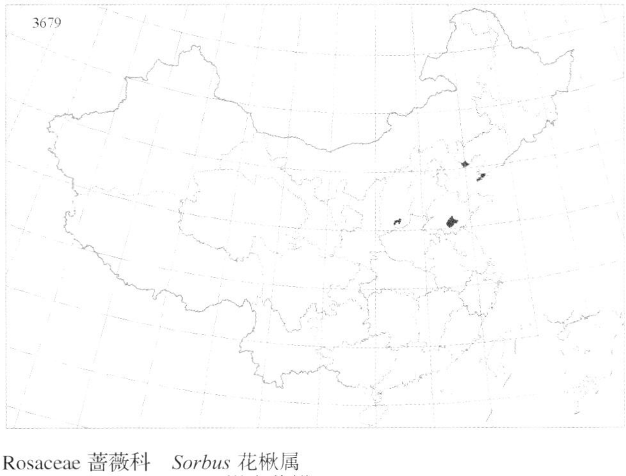

Rosaceae 蔷薇科 *Sorbus* 花楸属

Sorbus amabilis Cheng ex T. T. Yu et K. C. Kuan 黄山花楸

Trees, deciduous

16.0 (13.4~18.2)；4.0 (1.5~6.5)；27.4 (25.2~29.3)；16.0 (13.4~18.2)；133.8 (108.3~158.8)；1.9 (0.0~7.3)；871 (755~984) ❙ 1647 (1317~1997)；594 (492~685)；202 (122~259) ❙ 89.4 (53.3~130.7)；869 (755~941)；244.0 (183.9~319.9)

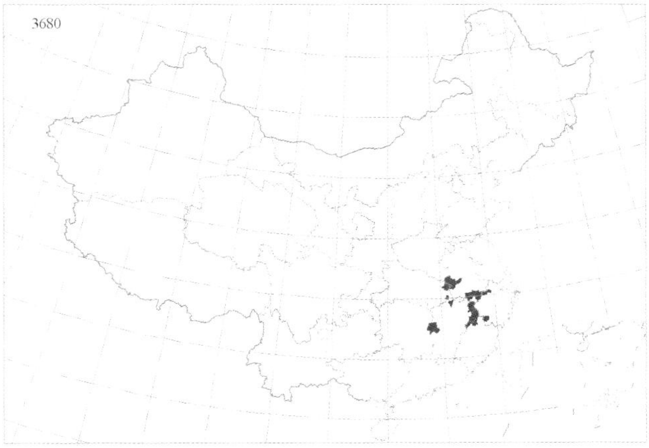

Rosaceae 蔷薇科 *Sorbus* 花楸属

Sorbus arguta T. T. Yu 锐齿花楸

Small trees, deciduous

14.5；6.7；21.1；14.6；115.0；0.0；748 ❙ 997；584；35 ❙ 33.2；723；251.4

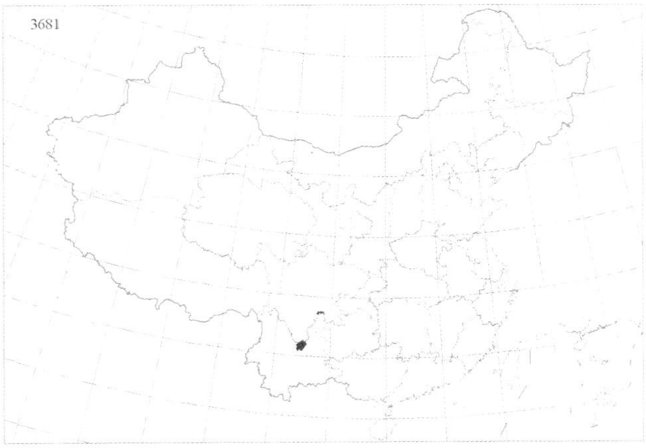

Rosaceae 蔷薇科 *Sorbus* 花楸属

Sorbus aronioides Rehder 毛背花楸

Shrubs or trees, deciduous

15.6 (2.2~21.6)；6.1 (-7.0~14.2)；24.0 (9.9~28.6)；15.7 (3.8~21.7)；129.8 (15.5~199.8)；1.9 (0.0~49.1)；829 (413~1153) ❙ 1286 (801~1758)；594 (400~946)；90 (19~196) ❙ 56.5 (19.4~125.9)；829 (413~1153)；250.0 (146.0~551.6)

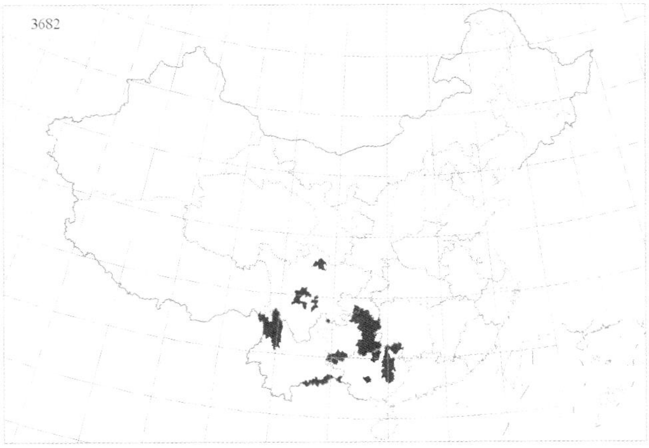

Rosaceae 蔷薇科 *Sorbus* 花楸属

Sorbus astateria (Cardot) Hand.-Mazz. 多变花楸

Shrubs or small trees, deciduous

14.4 (5.4~18.9)；7.2 (-2.1~12.5)；20.1 (12.3~26.6)；14.5 (5.3~19.0)；114.7 (31.5~167.8)；1.2 (0.0~25.6)；743 (481~929) ❙ 989 (741~1431)；569 (411~817)；37 (15~59) ❙ 34.8 (-14.2~113.0)；719 (481~912)；288.8 (142.9~406.0)

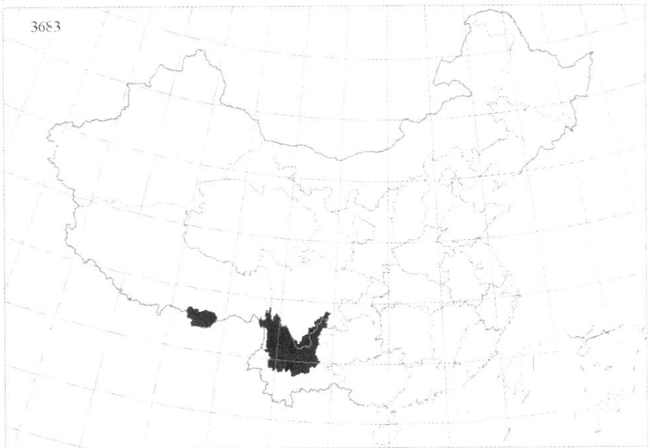

Rosaceae 蔷薇科 *Sorbus* 花楸属

Sorbus caloneura (Stapf) Rehder 美脉花楸

Shrubs or trees, deciduous

16.6 (2.2~22.7)；6.5 (-6.6~15.0)；25.7 (9.9~29.8)；16.7 (3.8~22.8)；141.3 (15.5~213.1)；1.0 (0.0~49.1)；879 (413~1221) ❙ 1376 (760~2435)；582 (377~1428)；117 (17~254) ❙ 57.2 (-2.0~130.7)；875 (413~1221)；229.9 (126.1~627.0)

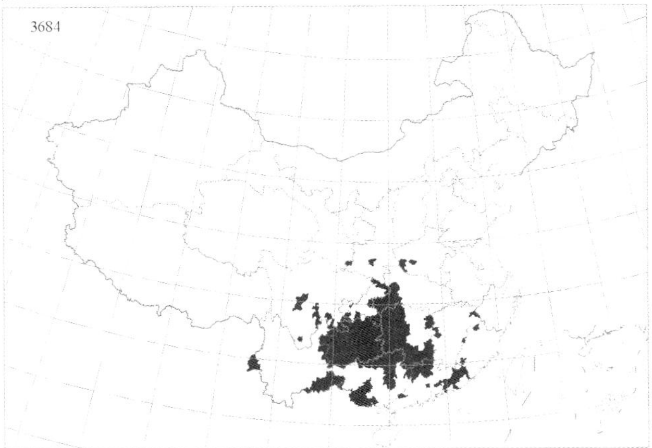

Rosaceae 蔷薇科 *Sorbus* 花楸属
Sorbus caloneura var. kwangtungensis T. T. Yu 广东美脉花楸
Shrubs or trees, deciduous

19.7; 10.2; 27.8; 19.7; 176.5; 0.0; 1031 ❙ 1753; 684; 180 ❙ 69.6; 1031; 234.9

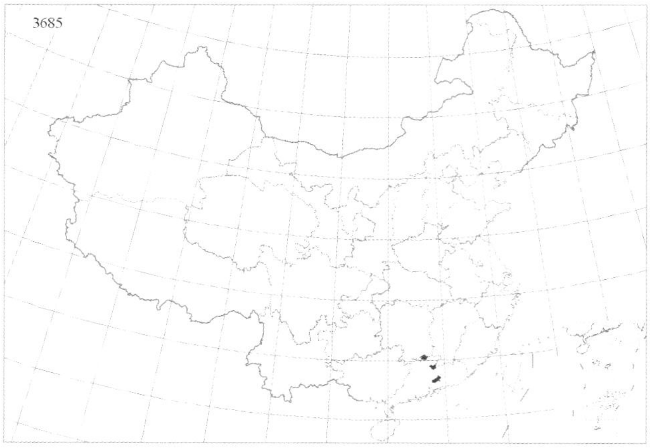

Rosaceae 蔷薇科 *Sorbus* 花楸属
Sorbus coronata (Cardot) T. T. Yu et H. T. Tsai 冠萼花楸
Trees, deciduous

14.2 (5.4~18.9); 6.6 (-2.1~12.5); 20.3 (12.3~26.6); 14.3 (5.8~19.0); 112.2 (31.5~167.8); 1.3 (0.0~25.6); 738 (481~929) ❙ 1019 (741~1736); 569 (411~1007); 42 (12~106) ❙ 39.5 (-14.2~145.8); 720 (481~912); 279.2 (142.9~447.4)

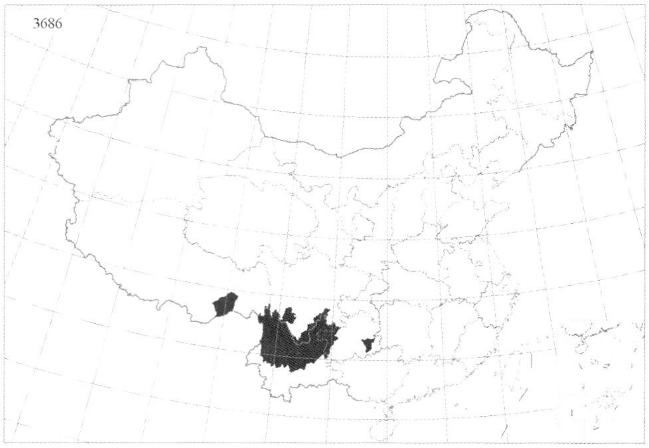

Rosaceae 蔷薇科 *Sorbus* 花楸属
Sorbus coronata var. glabrescens T. T. Yu et L. T. Lu 脱毛冠萼花楸
Trees, deciduous

13.3; 5.4; 19.6; 13.4; 100.4; 0.0; 706 ❙ 1736; 1007; 51 ❙ 145.8; 706; 447.4

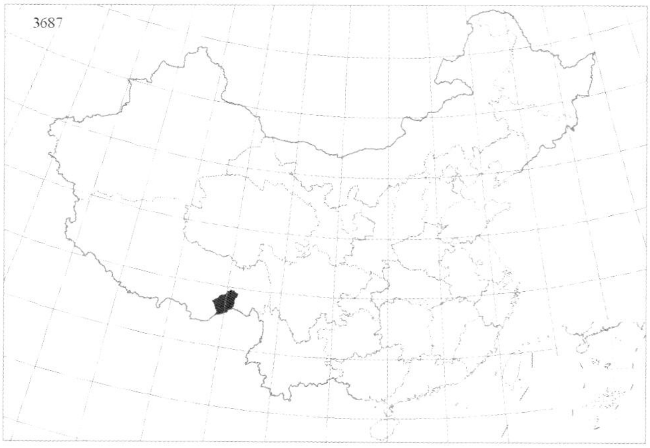

Rosaceae 蔷薇科 *Sorbus* 花楸属
Sorbus corymbifera (Miq.) Khep et Yakovlev 疣果花楸
Trees, deciduous

17.9 (5.4~24.8); 10.1 (-2.1~20.1); 23.9 (12.3~29.8); 18.0 (5.8~24.8); 155.9 (31.5~237.6); 0.4 (0.0~25.6); 909 (481~1385) ❙ 1387 (741~1900); 671 (419~1056); 89 (28~198) ❙ 53.9 (1.5~112.2); 896 (481~1326); 343.7 (174.9~659.0)

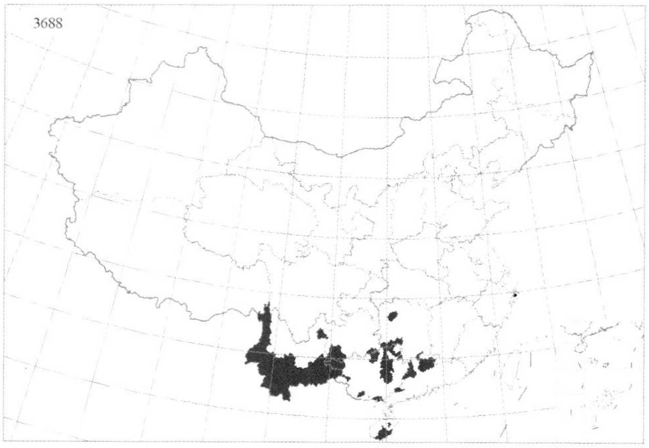

Rosaceae 蔷薇科 *Sorbus* 花楸属
Sorbus cuspidata (Spach) Hedl. 白叶花楸
Trees, deciduous

-1.5; -9.6; 6.1; 1.9; 2.2; 79.2; 332 ❙ 320; 205; 26 ❙ -3.5; 320; 75.7

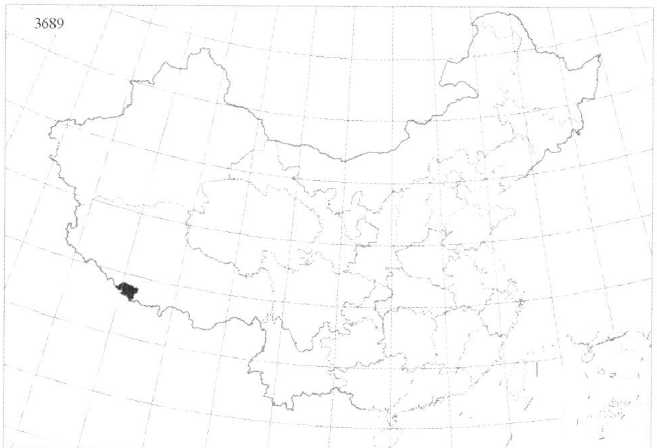

Rosaceae 蔷薇科 *Sorbus* 花楸属
Sorbus discolor (Maxim.) Maxim. 北京花楸
Trees, deciduous

9.9 (-4.6~15.1); -5.4 (-28.8~2.6); 23.6 (11.5~28.2); 11.1 (3.5~15.2); 90.6 (17.2~128.0); 31.4 (3.5~149.1); 697 (375~880) ❙ 592 (252~1621); 354 (144~600); 22 (4~195) ❙ -14.8 (-59.3~95.5); 579 (252~829); 173.5 (50.4~318.2)

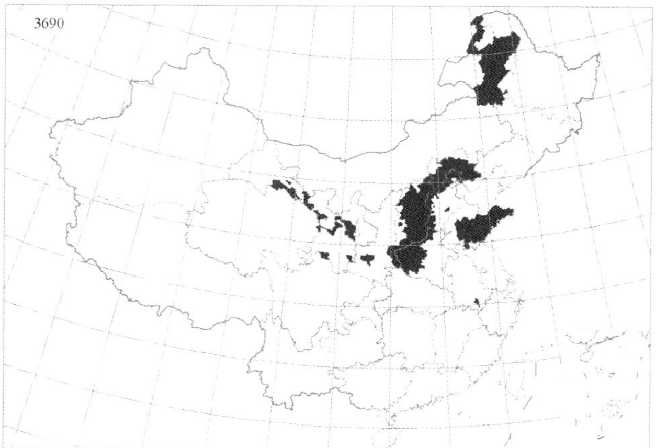

Rosaceae 蔷薇科　*Sorbus* 花楸属

Sorbus dunnii Rehder 棕脉花楸

Small trees, deciduous

16.2 (14.1~19.2); 5.0 (1.6~10.0); 26.7 (23.5~29.1); 16.3 (14.2~19.2); 136.5 (112.6~170.8); 1.0 (0.0~6.5); 868 (756~1002) ❙ 1623 (1218~1997); 588 (465~709); 189 (97~259) ❙ 87.2 (47.0~130.7); 866 (756~1002); 247.2 (183.9~319.9)

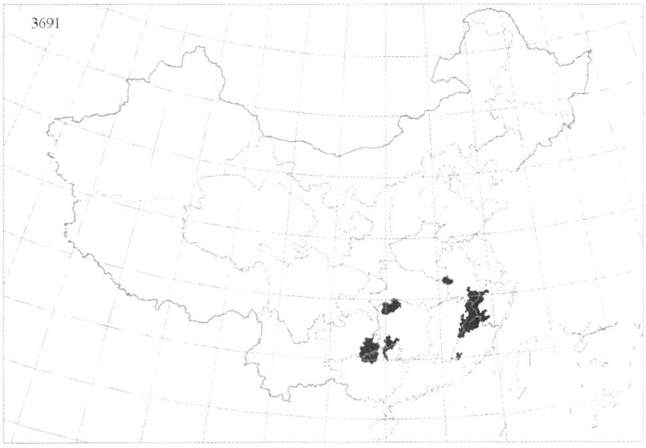

Rosaceae 蔷薇科　*Sorbus* 花楸属

Sorbus epidendron Hand.-Mazz. 附生花楸

Shrubs or trees, deciduous

11.9 (5.4~19.1); 3.9 (-2.1~13.0); 18.9 (12.3~25.4); 12.0 (5.8~19.1); 89.2 (31.5~169.7); 5.9 (0.0~25.6); 681 (481~917) ❙ 1052 (741~1428); 567 (426~835); 46 (21~69) ❙ 56.2 (35.3~93.0); 679 (481~917); 298.7 (148.9~506.2)

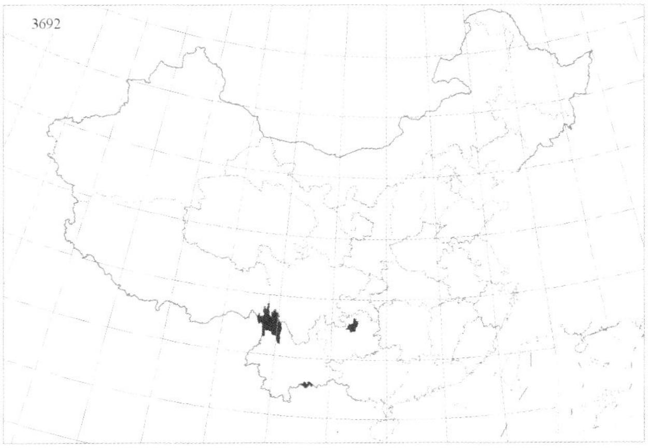

Rosaceae 蔷薇科　*Sorbus* 花楸属

Sorbus esserteauiana Koehne 麻叶花楸

Shrubs or trees, deciduous

8.9 (5.7~13.1); -0.4 (-3.5~3.0); 17.4 (14.2~22.1); 9.2 (6.4~13.1); 62.9 (38.1~100.1); 15.2 (3.0~28.1); 590 (504~720) ❙ 985 (822~1076); 525 (420~584); 26 (19~30) ❙ 68.9 (38.2~101.2); 590 (504~720); 246.2 (147.3~318.7)

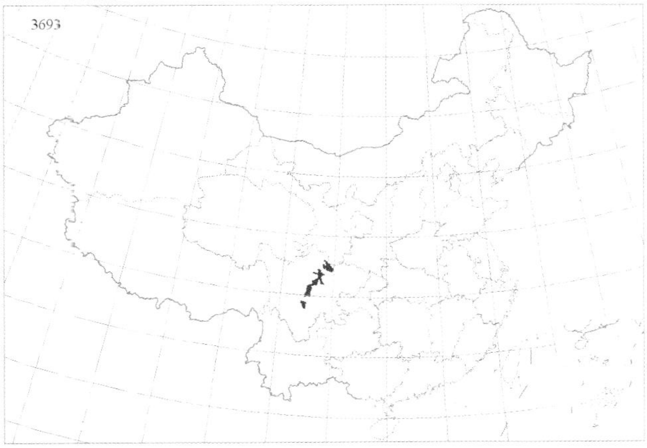

Rosaceae 蔷薇科　*Sorbus* 花楸属

Sorbus ferruginea (Wenz.) Rehder 锈色花楸

Shrubs or trees, deciduous

14.3 (5.4~20.5); 7.6 (-2.1~15.1); 19.3 (12.3~23.7); 14.3 (5.8~20.5); 114.3 (31.5~185.9); 2.4 (0.0~25.6); 746 (481~992) ❙ 1200 (741~1751); 669 (426~996); 49 (15~77) ❙ 61.5 (14.4~99.9); 737 (481~986); 392.0 (248.4~659.0)

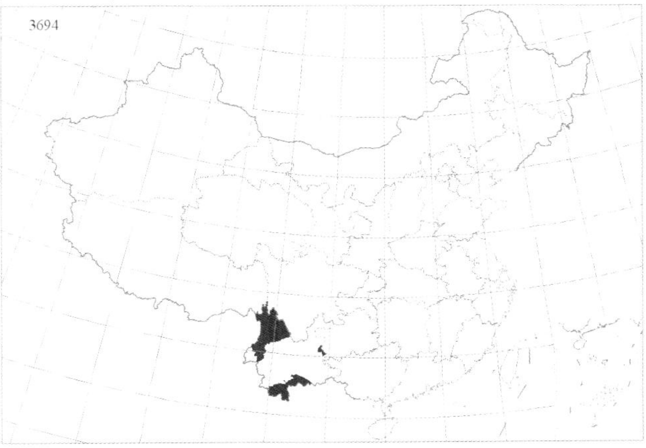

Rosaceae 蔷薇科　*Sorbus* 花楸属

Sorbus filipes Hand.-Mazz. 纤细花楸

Shrubs, deciduous

6.7 (-0.[illegible]~13.2); -1.3 (-10.0~6.4); 13.6 (8.4~18.5); 7.5 (2.8~13.3); 45.5 (9.4~99.4); 24.8 (0.0~73.3); 524 (360~695) ❙ 872 (341~1431); 508 (226~817); 27 (4~69) ❙ 62.6 (-15.3~128.6); 520 (341~695); 293.3 (168.2~407.1)

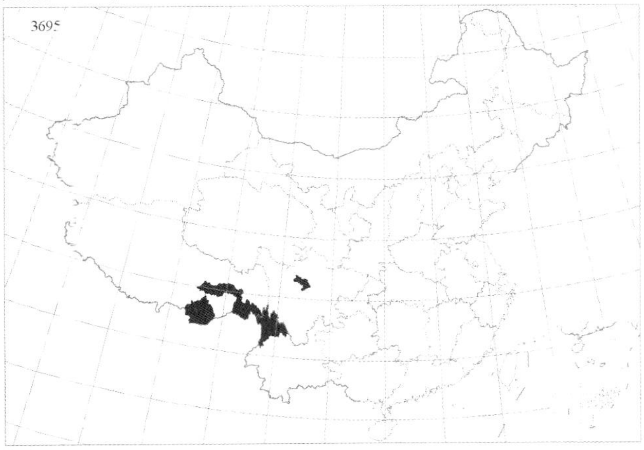

Rosaceae 蔷薇科　*Sorbus* 花楸属

Sorbus folgneri (C. K. Schneid.) Rehder 石灰树

Trees, deciduous

15.6 (2.2~21.6); 4.5 (-6.6~14.1); 25.8 (9.9~30.1); 15.6 (3.8~21.7); 130.7 (15.5~199.8); 3.1 (0.0~49.1); 846 (413~1153) ❙ 1277 (512~1997); 511 (245~902); 119 (14~259) ❙ 50.1 (-36.2~130.7); 828 (413~1153); 218.0 (121.6~334.4)

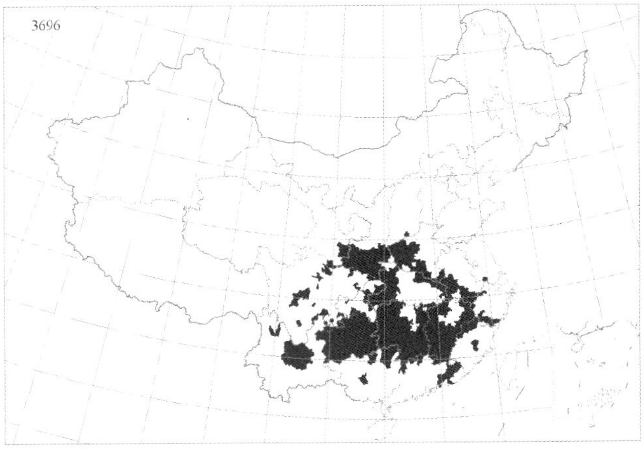

Rosaceae 蔷薇科 *Sorbus* 花楸属

Sorbus folgneri var. duplicatodentata T. T. Yu et L. T. Lu 齿叶石灰树

Trees, deciduous

16.3；4.3；27.8；16.3；136.9；0.7；886 ❙ 1462；524；160 ❙ 65.1；882；203.8

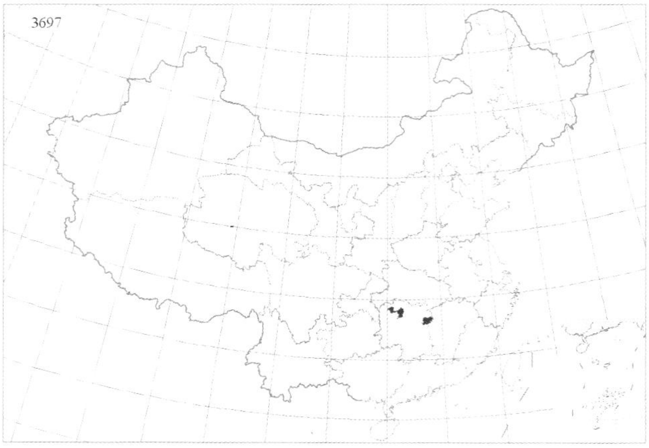

Rosaceae 蔷薇科 *Sorbus* 花楸属

Sorbus foliolosa (Wall.) Spach 尼泊尔花楸

Trees, deciduous

8.9 (0.3~14.3)；1.5 (−8.6~7.6)；15.3 (8.3~19.7)；9.3 (2.9~14.3)；62.4 (9.6~111.9)；14.5 (0.0~65.4)；583 (368~728) ❙ 975 (341~1736)；570 (226~1007)；32 (4~69) ❙ 62.7 (−15.3~145.8)；577 (341~707)；309.8 (142.5~447.4)

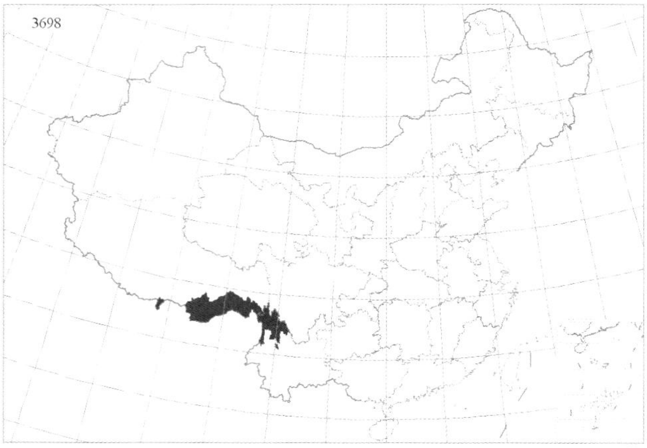

Rosaceae 蔷薇科 *Sorbus* 花楸属

Sorbus globosa T. T. Yu et Tsai 圆果花楸

Trees, deciduous

18.8 (8.6~23.0)；11.0 (1.3~15.1)；24.5 (15.0~29.0)；18.8 (8.7~23.0)；166.1 (52.6~216.2)；0.2 (0.0~8.8)；956 (574~1247) ❙ 1445 (1041~2435)；750 (476~1428)；85 (41~171) ❙ 53.0 (12.9~112.2)；949 (574~1247)；342.4 (189.5~659.0)

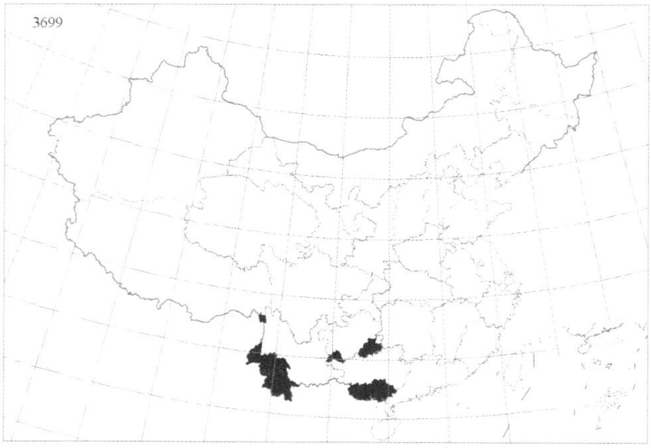

Rosaceae 蔷薇科 *Sorbus* 花楸属

Sorbus glomerulata Koehne 球穗花楸

Shrubs or small trees, deciduous

12.5 (6.1~16.2)；3.4 (−2.4~7.9)；20.7 (12.7~26.5)；12.6 (6.3~16.2)；95.4 (35.2~134.6)；5.4 (0.0~24.1)；703 (496~854) ❙ 1090 (801~1410)；567 (475~816)；49 (21~113) ❙ 56.5 (28.1~92.9)；696 (496~854)；244.2 (176.4~319.8)

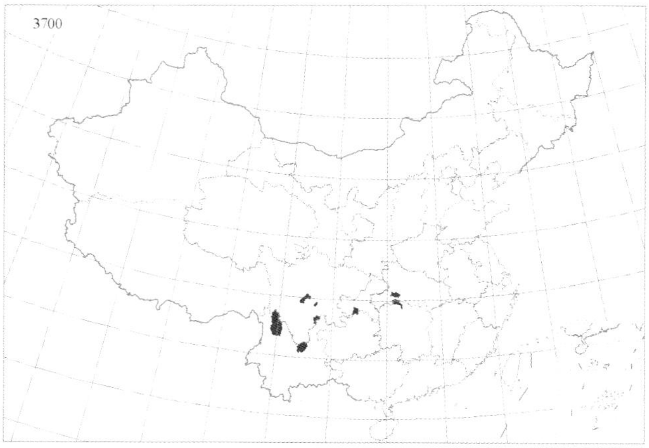

Rosaceae 蔷薇科 *Sorbus* 花楸属

Sorbus guanxianensis T. C. Ku 灌县花楸

Trees, deciduous

13.3；3.2；22.1；13.2；100.9；2.5；723 ❙ 1081；606；32 ❙ 49.4；723；171.0

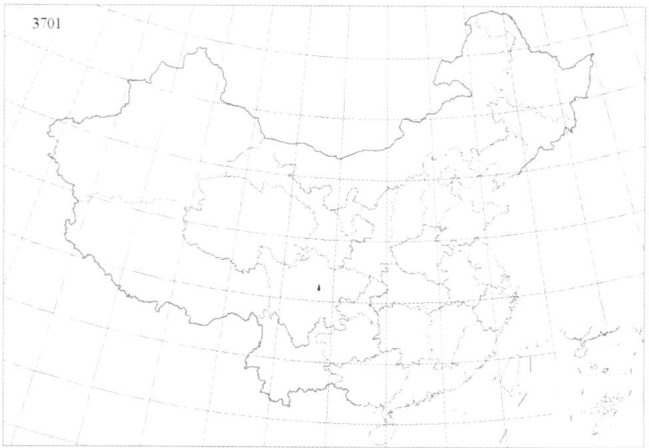

Rosaceae 蔷薇科 *Sorbus* 花楸属

Sorbus helenae Koehne 钝齿花楸

Trees, deciduous

10.1 (3.3~14.9)；0.7 (−6.1~5.3)；18.4 (11.4~24.0)；10.4 (4.6~15.0)；74.6 (22.5~119.6)；13.1 (0.0~42.7)；630 (434~789) ❙ 1047 (812~1410)；584 (452~816)；28 (17~49) ❙ 69.6 (18.6~101.2)；630 (434~789)；236.4 (150.8~337.3)

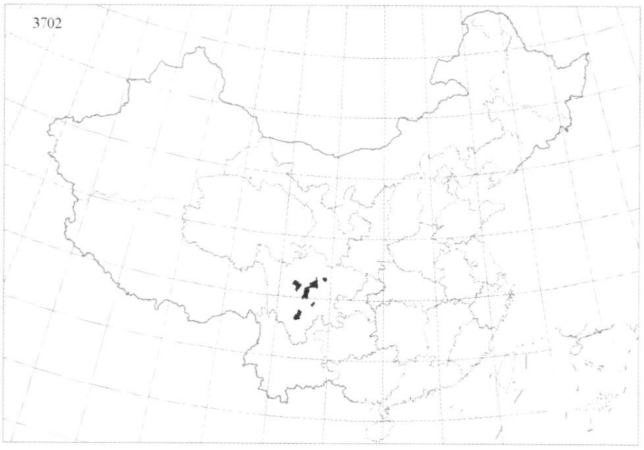

Rosaceae 蔷薇科　*Sorbus* 花楸属
Sorbus helenae var. argutiserrata T. T. Yu 尖齿花楸
Trees, deciduous

3.3；-6.1；11.4；4.6；22.5；42.7；434 ❙ 812；452；17 ❙ 87.0；434；337.3

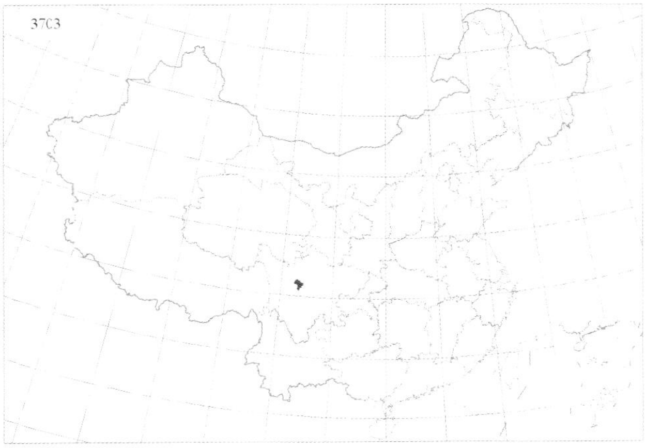

Rosaceae 蔷薇科　*Sorbus* 花楸属
Sorbus hemsleyi (C. K. Schneid.) Rehder 江南花楸
Trees, deciduous

15.3 (2.2~22.3)；4.3 (-7.7~14.1)；25.3 (9.9~29.4)；15.3 (3.8~22.3)；127.3 (15.5~208.1)；3.6 (0.0~49.3)；835 (413~1202) ❙ 1276 (626~1997)；529 (316~902)；112 (12~259) ❙ 52.8 (-22.2~147.0)；826 (413~1202)；225.5 (139.1~384.7)

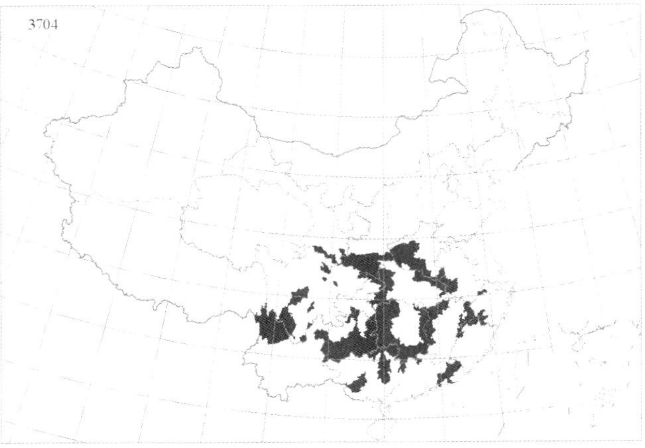

Rosaceae 蔷薇科　*Sorbus* 花楸属
Sorbus hupehensis C. K. Schneid. 湖北花楸
Trees, deciduous

11.1 (-2.5~21.0)；-1.0 (-15.4~14.2)；22.2 (8.3~29.2)；11.7 (2.5~21.0)；92.7 (7.0~191.8)；18.4 (0.0~98.1)；703 (347~1053) ❙ 855 (148~1793)；429 (92~818)；43 (4~238) ❙ 23.6 (-74.8~128.6)；663 (148~1053)；222.6 (61.2~551.6)

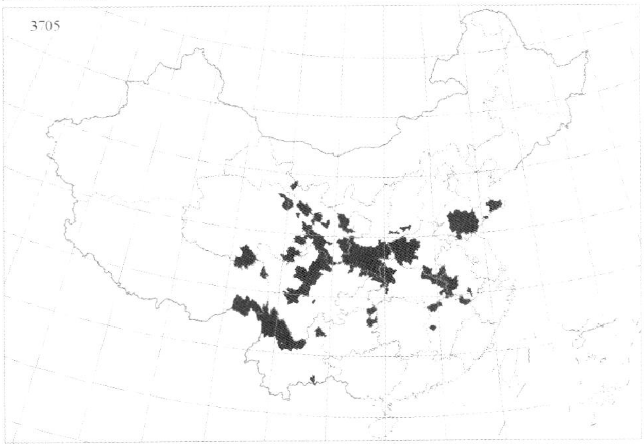

Rosaceae 蔷薇科　*Sorbus* 花楸属
Sorbus insignis (Hook. f.) Hedl. 卷边花楸
Trees, deciduous

12.1 (5.3~19.6)；5.4 (-2.4~13.9)；17.3 (12.2~22.7)；12.2 (5.7~19.6)；94.1 (30.5~175.1)；8.3 (0.0~26.2)；684 (479~933) ❙ 1174 (741~1500)；629 (426~785)；42 (28~58) ❙ 73.4 (53.6~100.7)；670 (479~882)；380.0 (234.2~530.1)

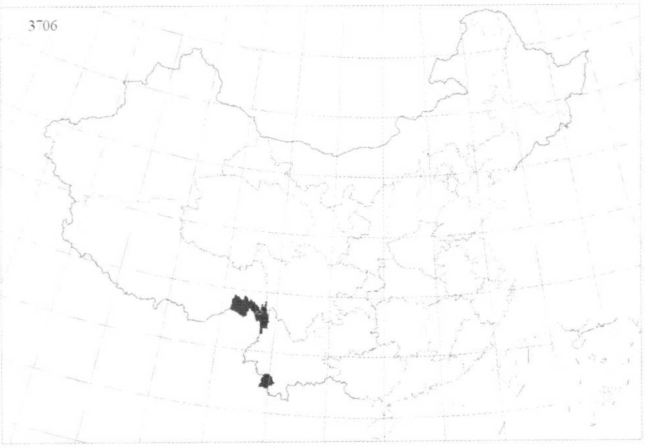

Rosaceae 蔷薇科　*Sorbus* 花楸属
Sorbus keissleri (C. K. Schneid.) Rehder 毛序花楸
Trees, deciduous

16.0 (-0.7~22.3)；5.8 (-9.6~14.2)；25.2 (7.7~29.1)；16.1 (2.5~22.3)；135.0 (6.8~208.1)；2.2 (0.0~74.2)；856 (338~1202) ❙ 1363 (370~1758)；572 (262~839)；119 (10~228) ❙ 60.4 (4.7~107.2)；856 (338~1202)；244.0 (61.6~551.6)

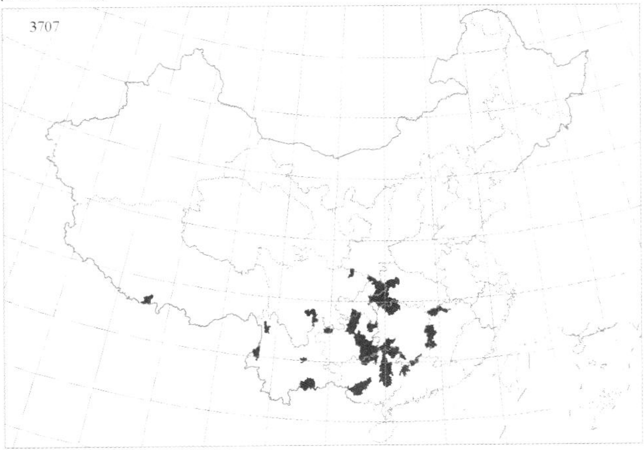

Rosaceae 蔷薇科　*Sorbus* 花楸属
Sorbus kiukiangensis T. T. Yu 俅江花楸
Shrubs or small trees, deciduous

12.1 (3.3~16.7)；5.1 (-5.5~9.8)；17.8 (11.1~22.1)；12.3 (4.6~16.8)；91.4 (22.3~141.1)；5.6 (0.0~41.8)；670 (427~819) ❙ 974 (473~1337)；563 (287~727)；37 (7~69) ❙ 46.2 (5.0~93.0)；649 (427~748)；325.2 (230.7~407.1)

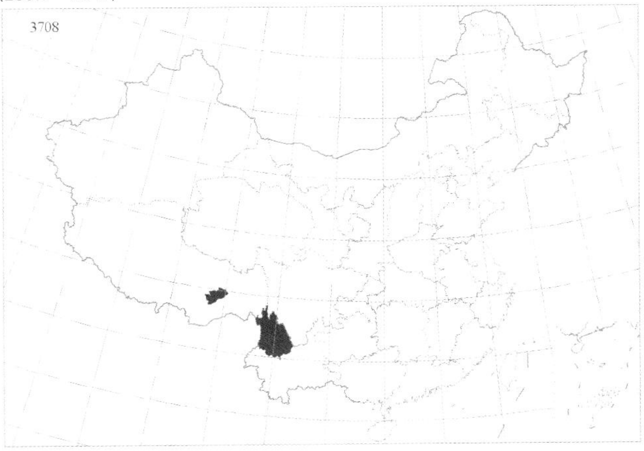

Rosaceae 蔷薇科 *Sorbus* 花楸属
Sorbus kiukiangensis var. glabrescens T. T. Yu 无毛俅江花楸
Shrubs or small trees, deciduous

8.6；1.3；15.0；8.7；52.6；8.8；574 ❙ 1041；585；45 ❙ 81.5；574；352.3

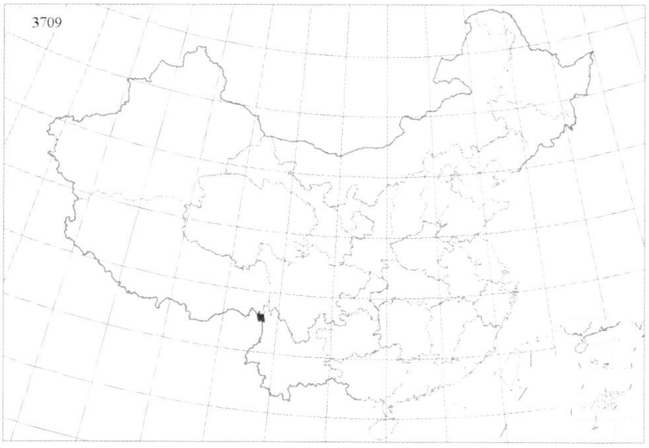

Rosaceae 蔷薇科 *Sorbus* 花楸属
Sorbus koehneana C. K. Schneid. 陕甘花楸
Shrubs or small trees, deciduous

9.3 (−1.3~16.2)；−3.8 (−13.7~8.2)；21.1 (8.5~28.0)；10.3 (2.6~16.2)；80.0 (8.1~134.6)；27.6 (0.0~86.3)；653 (347~880) ❙ 683 (341~1358)；364 (190~651)；24 (5~94) ❙ 8.0 (−40.9~128.6)；574 (341~854)；210.9 (61.2~384.1)

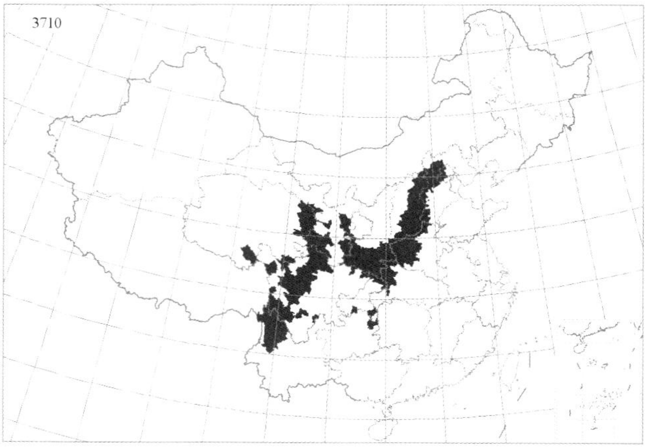

Rosaceae 蔷薇科 *Sorbus* 花楸属
Sorbus lanpingensis L. T. Lu 兰坪花楸
Shrubs, deciduous

10.3；3.3；16.1；10.3；66.9；3.1；618 ❙ 1071；615；44 ❙ 73.1；618；345.6

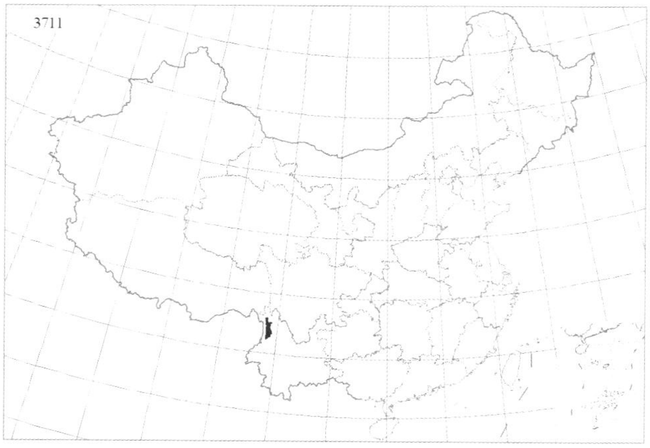

Rosaceae 蔷薇科 *Sorbus* 花楸属
Sorbus megalocarpa Rehder 大果花楸
Shrubs or small trees, deciduous

15.4 (2.0~19.1)；4.9 (−7.2~10.1)；25.0 (10.1~29.4)；15.4 (3.8~19.1)；126.0 (15.4~169.2)；1.1 (0.0~51.4)；818 (413~972) ❙ 1257 (874~1740)；532 (392~817)；98 (23~209) ❙ 54.2 (18.3~128.6)；814 (413~969)；203.7 (134.6~322.4)

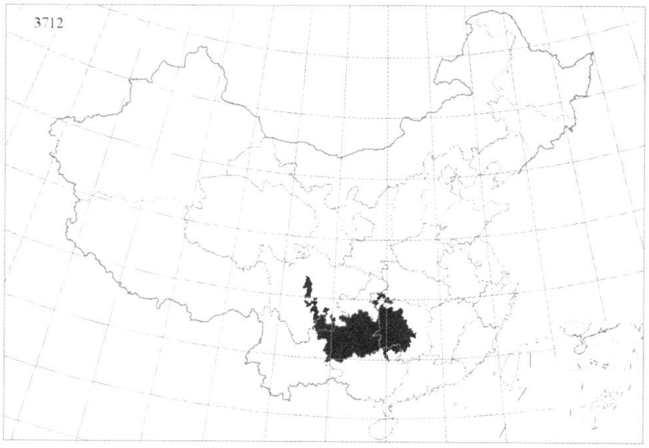

Rosaceae 蔷薇科 *Sorbus* 花楸属
Sorbus megalocarpa var. cuneata Rehder 楔叶大果花楸
Shrubs or small trees, deciduous

15.2 (7.3~17.1)；4.7 (−2.0~8.5)；25.0 (15.6~26.8)；15.2 (7.5~17.1)；123.6 (48.2~145.1)；1.1 (0.0~20.4)；806 (537~871) ❙ 1201 (1005~1299)；498 (448~666)；82 (27~106) ❙ 49.6 (35.7~87.3)；806 (537~871)；204.7 (146.0~283.9)

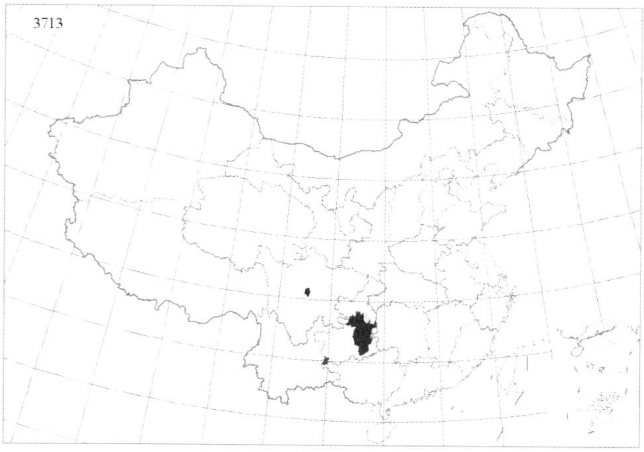

Rosaceae 蔷薇科 *Sorbus* 花楸属
Sorbus meliosmifolia Rehder 泡吹叶花楸
Trees, deciduous

11.0 (2.2~19.1)；1.7 (−7.0~13.0)；19.4 (9.9~26.7)；11.4 (3.8~19.1)；85.9 (15.5~169.7)；13.0 (0.0~49.1)；666 (413~917) ❙ 1148 (822~1624)；573 (400~835)；58 (19~191) ❙ 74.7 (38.2~125.9)；666 (413~917)；279.3 (217.6~506.2)

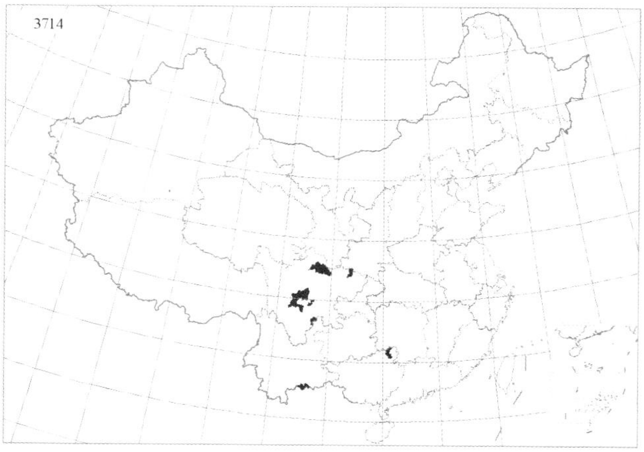

Rosaceae 蔷薇科 *Sorbus* 花楸属
Sorbus monbeigii (Cardot) T. T. Yu 维西花楸
Trees, deciduous

9.6; 2.4; 15.7; 9.6; 60.8; 5.4; 600 ❙ 1007; 578; 40 ❙ 67.9; 600; 351.7

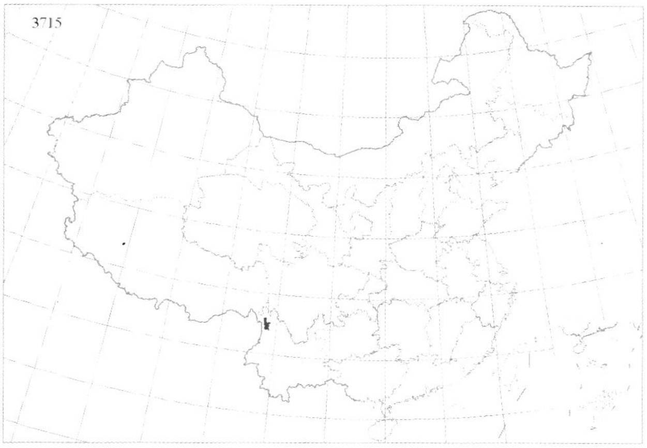

Rosaceae 蔷薇科 *Sorbus* 花楸属
Sorbus multijuga Koehne 多对花楸
Shrubs or small trees, deciduous

6.2 (−0.4~13.6); −2.5 (−10.0~7.2); 14.0 (8.4~19.4); 7.2 (2.8~13.7); 45.1 (9.4~104.1); 29.6 (0.0~73.3); 516 (360~706) ❙ 833 (341~1076); 470 (226~592); 21 (4~46) ❙ 58.9 (−15.3~125.9); 510 (341~706); 247.2 (168.2~372.3)

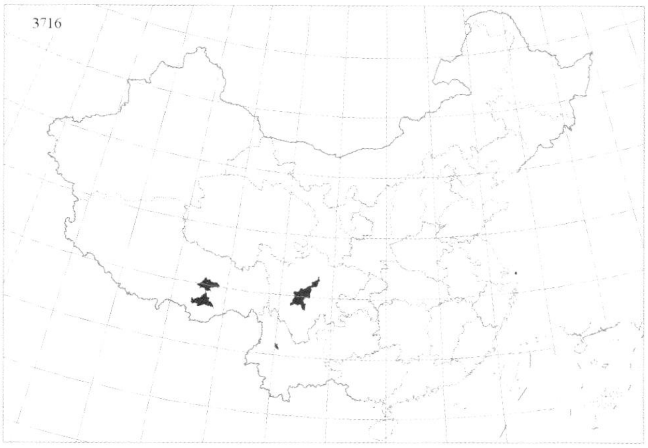

Rosaceae 蔷薇科 *Sorbus* 花楸属
Sorbus obsoletidentata (Cardot) T. T. Yu 宾川花楸
Shrubs or small trees, deciduous

15.3; 8.7; 20.5; 15.4; 124.3; 0.0; 761 ❙ 976; 564; 38 ❙ 28.3; 717; 309.8

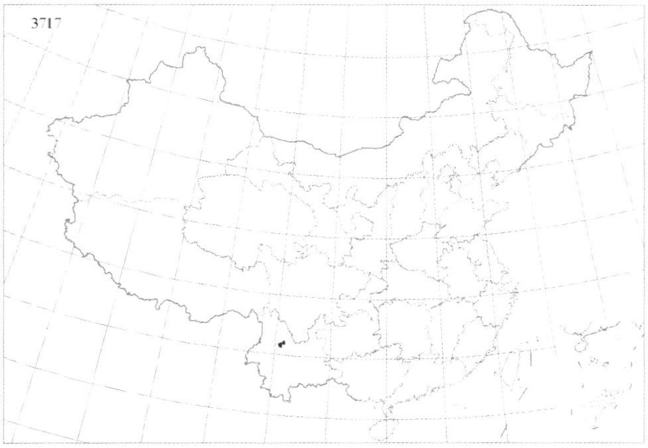

Rosaceae 蔷薇科 *Sorbus* 花楸属
Sorbus ochracea (Hand.-Mazz.) J. E. Vidal 褐毛花楸
Shrubs or trees, deciduous

12.6 (−1.4~21.0); 5.5 (−11.4~15.1); 18.2 (7.9~25.6); 13.1 (2.4~21.0); 105.4 (6.6~191.8); 13.3 (0.0~82.3); 704 (341~1053) ❙ 1059 (510~1607); 590 (311~860); 41 (8~77) ❙ 50.0 (6.5~100.7); 693 (341~1053); 349.3 (111.0~659.0)

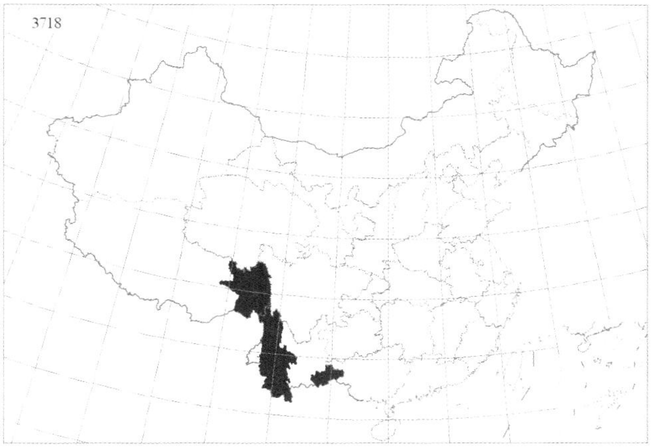

Rosaceae 蔷薇科 *Sorbus* 花楸属
Sorbus oligodonta (Cardot) Hand.-Mazz. 少齿花楸
Trees, deciduous

13.3 (2.2~18.9); 5.9 (−7.0~12.0); 19.3 (9.5~26.6); 13.5 (3.9~19.0); 105.1 (15.8~167.8); 4.5 (0.0~48.6); 713 (411~929) ❙ 971 (473~1431); 558 (287~817); 35 (7~88) ❙ 38.9 (−14.2~121.7); 691 (411~912); 277.5 (142.9~384.1)

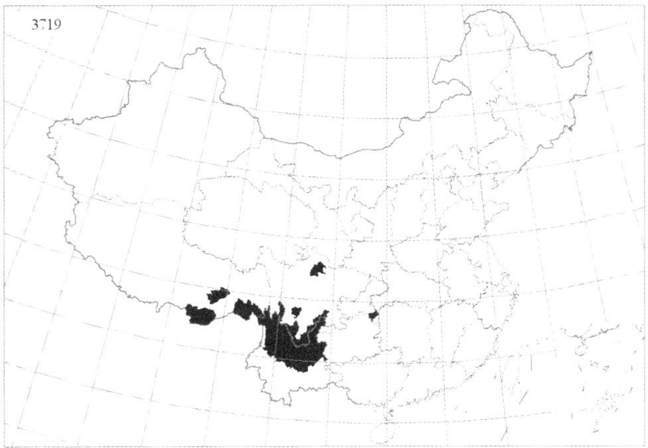

Rosaceae 蔷薇科 *Sorbus* 花楸属
Sorbus pallescens Rehder 灰叶花楸
Trees, deciduous

10.9 (−1.1~18.9); 3.4 (−9.6~12.0); 17.0 (6.2~23.9); 11.3 (2.0~19.0); 83.5 (2.5~167.8); 12.5 (0.0~75.6); 643 (334~929) ❙ 945 (321~1736); 555 (193~1007); 29 (10~58) ❙ 50.9 (−14.2~145.8); 615 (321~799); 282.8 (61.6~447.4)

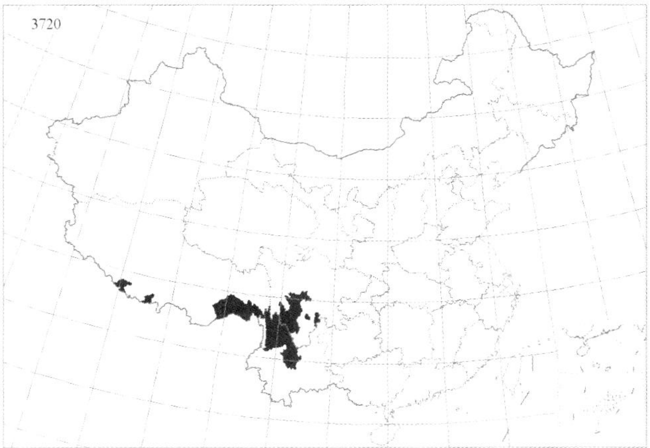

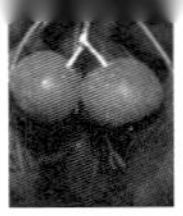

Rosaceae 蔷薇科　*Sorbus* 花楸属
Sorbus pohuashanensis (Hance) Hedl. 花楸树
Trees, deciduous

9.4 (−5.8~15.7); −6.6 (−29.8~2.7); 23.7 (15.3~28.0); 11.0 (4.4~15.7); 90.2 (28.0~132.3); 36.7 (3.2~157.4); 696 (426~889) ❙ 612 (332~1663); 368 (203~691); 24 (5~212) ❙ −11.7 (−43.4~101.3); 582 (332~881); 182.4 (95.7~346.5)

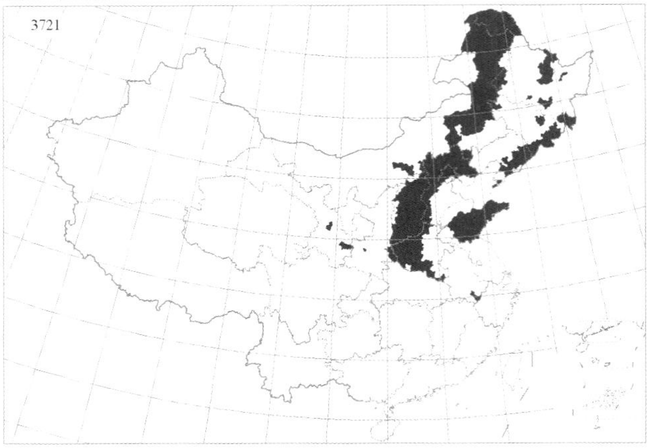

Rosaceae 蔷薇科　*Sorbus* 花楸属
Sorbus poteriifolia Hand.-Mazz. 侏儒花楸
Shrubs, deciduous

7.9; 0.5; 14.3; 8.0; 48.3; 13.3; 552 ❙ 930; 530; 38 ❙ 67.8; 552; 325.8

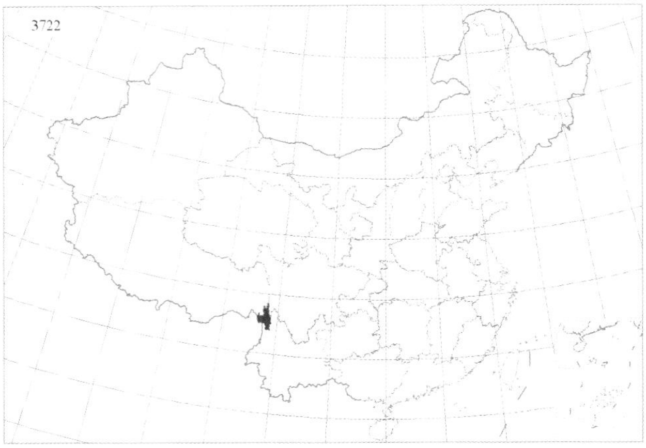

Rosaceae 蔷薇科　*Sorbus* 花楸属
Sorbus prattii Koehne 西康花楸
Shrubs, deciduous

8.0 (−1.8~16.7); −1.2 (−10.6~9.7); 16.2 (6.1~25.9); 8.8 (1.8~16.7); 60.3 (1.9~140.9); 24.1 (0.0~82.8); 573 (329~819) ❙ 880 (320~1431); 477 (203~817); 29 (5~94) ❙ 53.8 (−32.2~128.6); 566 (320~816); 271.1 (72.2~407.1)

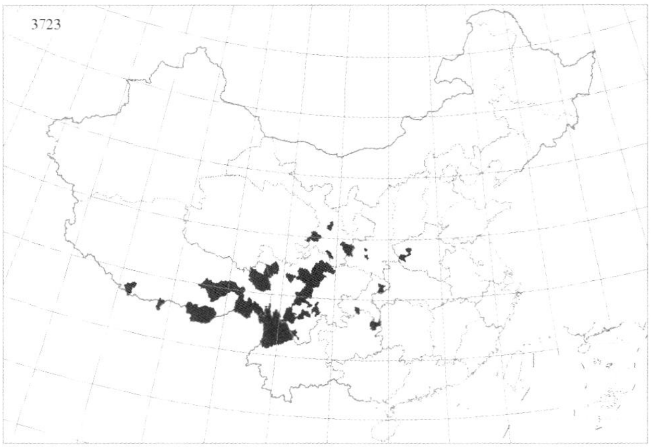

Rosaceae 蔷薇科　*Sorbus* 花楸属
Sorbus prattii var. aestivalis (Koehne) T. T. Yu 多对西康花楸
Shrubs, deciduous

7.7 (0.3~14.9); −1.3 (−8.6~5.3); 15.9 (8.3~23.6); 8.5 (2.9~15.0); 57.6 (9.6~119.6); 24.0 (0.0~65.4); 564 (368~782) ❙ 1023 (389~1414); 575 (246~817); 30 (11~49) ❙ 80.2 (0.0~128.6); 564 (368~782); 245.2 (142.5~292.6)

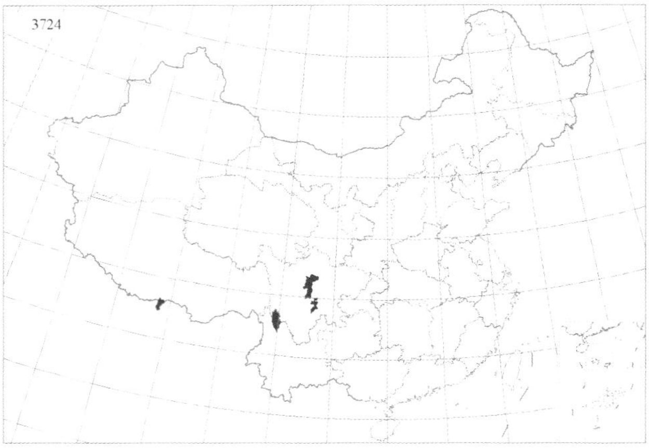

Rosaceae 蔷薇科　*Sorbus* 花楸属
Sorbus pteridophylla Hand.-Mazz. 蕨叶花楸
Shrubs, deciduous

6.9 (−1.8~13.2); −0.7 (−10.2~6.4); 13.5 (6.1~18.4); 7.8 (1.8~13.3); 48.2 (1.9~99.4); 24.4 (0.0~82.8); 533 (329~695) ❙ 857 (320~1337); 510 (217~727); 30 (5~69) ❙ 54.1 (−3.0~93.0); 531 (320~695); 281.8 (61.6~407.1)

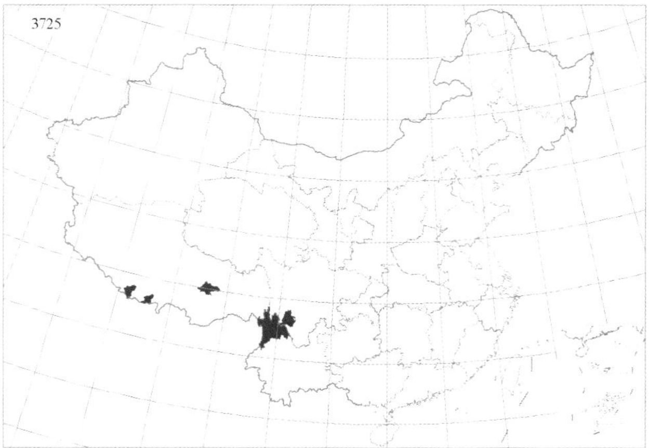

Rosaceae 蔷薇科　*Sorbus* 花楸属
Sorbus pteridophylla var. tephroclada Hand.-Mazz. 灰毛蕨叶花楸
Shrubs, deciduous

8.6; 1.3; 15.0; 8.7; 52.6; 8.8; 574 ❙ 1041; 585; 45 ❙ 81.5; 574; 352.3

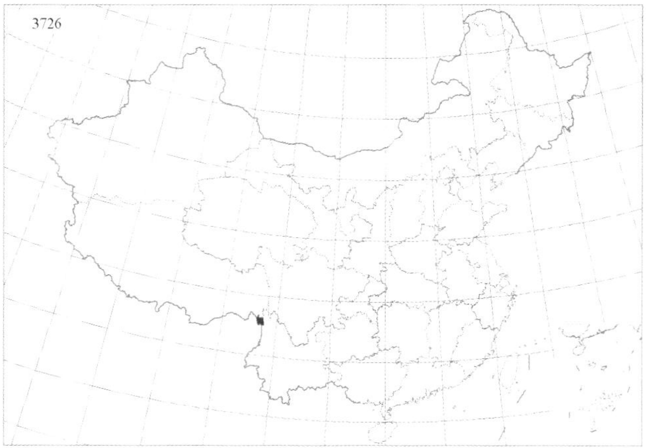

Rosaceae 蔷薇科 *Sorbus* 花楸属
Sorbus randaiensis (Hayata) Koidz. 台湾花楸
Trees, deciduous

18.9; 13.2; 23.5; 18.9; 167.2; 0.0; 932 ❙ 2336; 1169; 180 ❙ 160.6; 921; 245.3

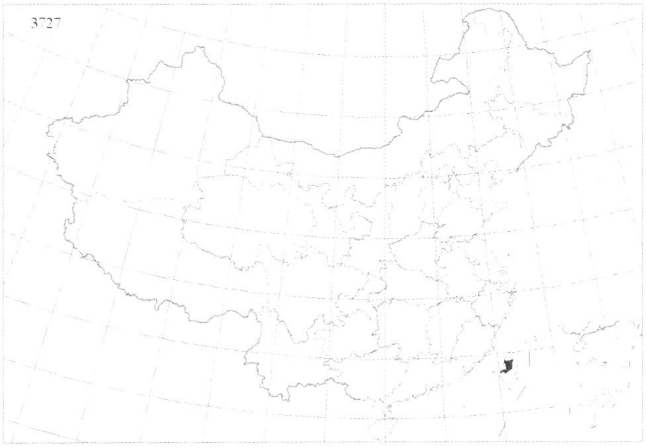

Rosaceae 蔷薇科 *Sorbus* 花楸属
Sorbus reducta Diels 铺地花楸
Shrubs, deciduous

13.3 (2.8~18.9); 5.8 (−7.1~12.0); 19.4 (11.4~26.6); 13.4 (4.6~19.0); 102.5 (23.9~167.8); 2.8 (0.0~50.1); 709 (426~929) ❙ 958 (318~1164); 555 (239~651); 33 (2~59) ❙ 36.5 (−25.4~81.5); 684 (318~912); 279.9 (142.9~384.1)

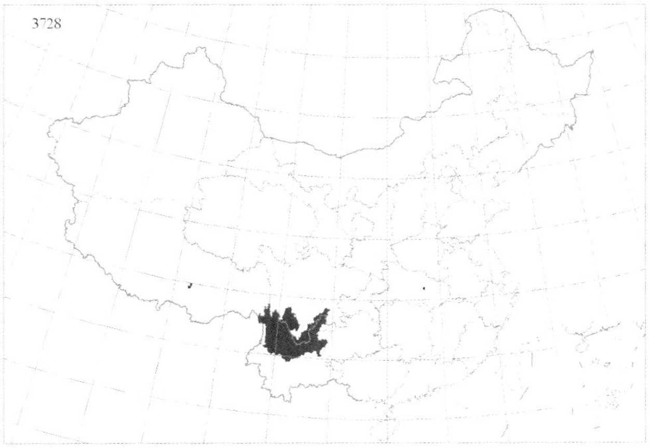

Rosaceae 蔷薇科 *Sorbus* 花楸属
Sorbus reducta var. pubescens L. T. Lu 毛萼铺地花楸
Shrubs, deciduous

8.6; 1.3; 15.0; 8.7; 52.6; 8.8; 574 ❙ 1041; 585; 45 ❙ 81.5; 574; 352.3

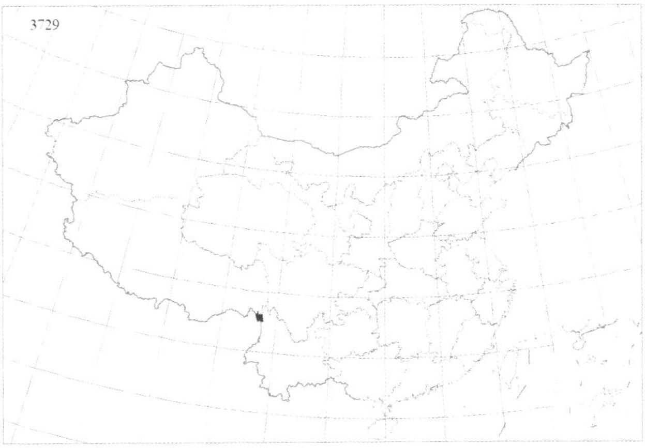

Rosaceae 蔷薇科 *Sorbus* 花楸属
Sorbus rehderiana Koehne 西南花楸
Shrubs or small trees, deciduous

10.1 (−3.3~19.2); 2.0 (−15.4~13.3); 16.8 (6.1~26.6); 10.8 (1.8~19.4); 81.1 (1.9~172.6); 19.8 (0.0~103.2); 628 (329~932) ❙ 897 (148~1561); 512 (92~897); 31 (4~71) ❙ 44.4 (−74.8~128.6); 612 (148~923); 271.3 (61.6~539.7)

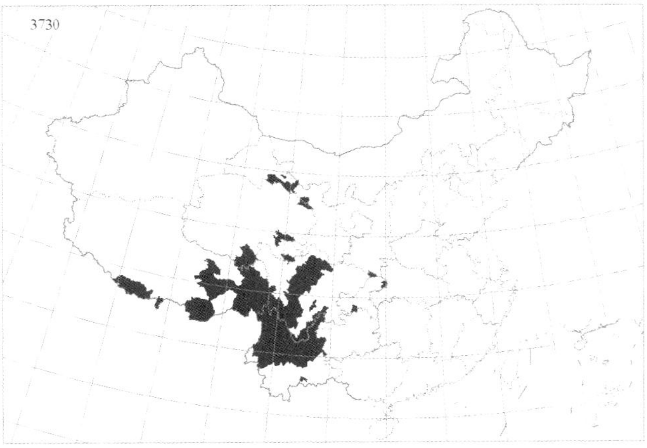

Rosaceae 蔷薇科 *Sorbus* 花楸属
Sorbus rehderiana var. cupreonitens Hand.-Mazz. 锈毛西南花楸
Shrubs or small trees, deciduous

8.9 (−1.5~16.7); 1.6 (−10.5~9.7); 15.1 (6.2~22.1); 9.6 (2.0~16.7); 66.5 (2.5~140.9); 19.6 (0.0~82.2); 585 (332~819) ❙ 863 (311~1164); 509 (193~651); 32 (10~59) ❙ 45.4 (−6.3~100.7); 574 (311~748); 271.3 (61.6~384.1)

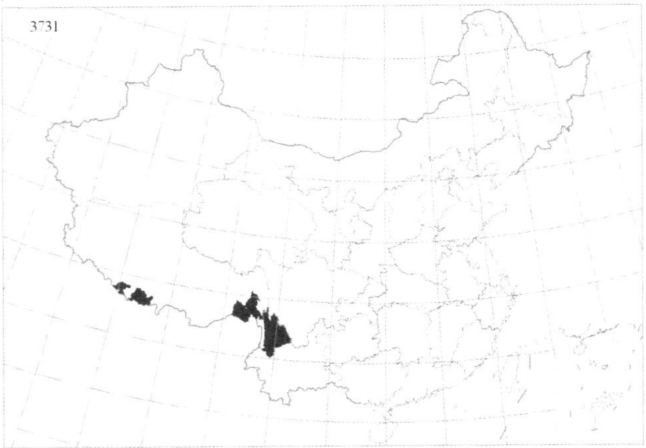

Rosaceae 蔷薇科 *Sorbus* 花楸属
Sorbus rehderiana var. grosseserrata Koehne 巨齿西南花楸
Shrubs or small trees, deciduous

7.2 (2.9~11.3); −0.7 (−6.3~4.3); 14.1 (11.2~17.0); 7.6 (4.4~11.3); 46.1 (20.5~76.6); 19.3 (0.7~44.6); 535 (428~644) ❙ 970 (741~1243); 541 (425~696); 33 (21~58) ❙ 82.4 (52.8~121.7); 535 (428~644); 296.7 (200.8~384.7)

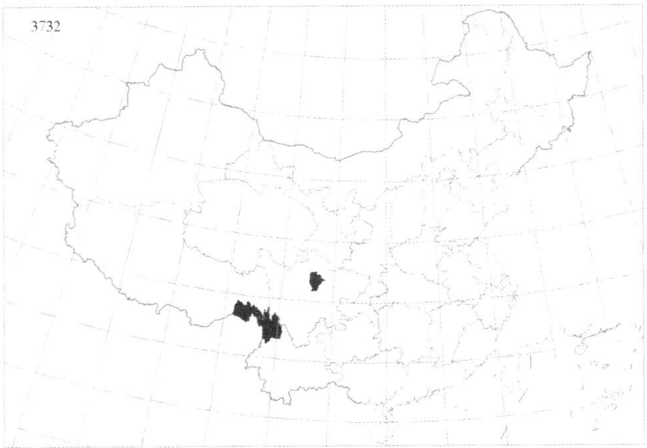

Rosaceae 蔷薇科 *Sorbus* 花楸属
Sorbus rhamnoides (Decne.) Rehder 鼠李叶花楸
Trees, deciduous

15.7 (5.4~21.0); 8.7 (-2.1~14.2); 21.0 (12.3~25.9); 15.8 (5.8~21.0); 130.3 (31.5~191.8); 1.1 (0.0~25.6); 792 (481~1053) ❙ 1182 (741~1751); 651 (426~996); 52 (20~129) ❙ 49.8 (5.5~93.0); 775 (481~1053); 360.5 (170.4~551.6)

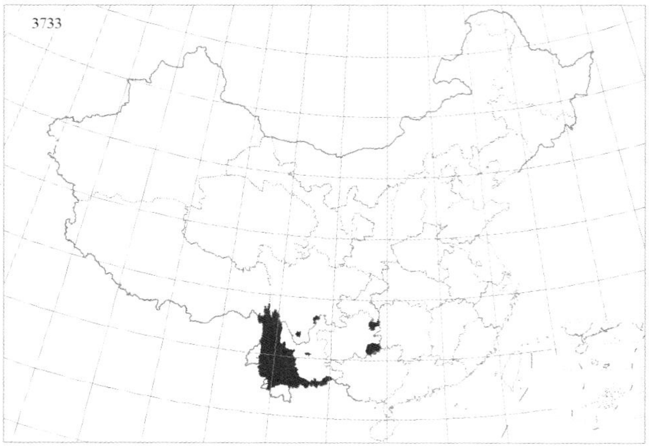

Rosaceae 蔷薇科 *Sorbus* 花楸属
Sorbus rhombifolia C. J. Qi et K. W. Liu 菱叶花楸
Trees, deciduous

16.2 (14.6~17.6); 5.4 (3.8~6.6); 26.1 (24.6~27.7); 16.3 (14.7~17.6); 135.7 (117.7~151.3); 0.3 (0.0~1.5); 855 (779~928) ❙ 1538 (1425~1624); 545 (476~594); 183 (161~197) ❙ 80.7 (61.1~101.8); 854 (779~922); 235.1 (180.3~258.0)

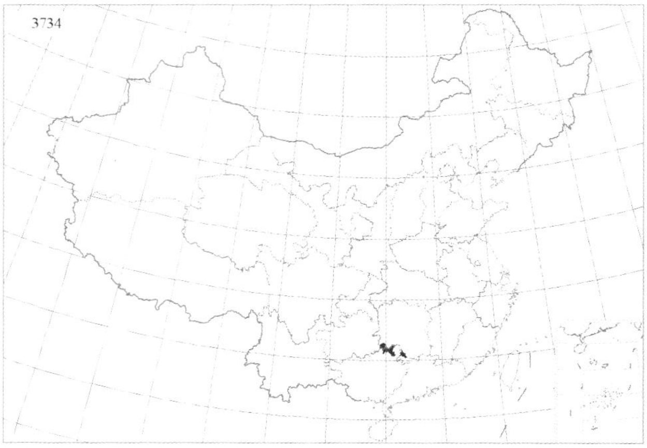

Rosaceae 蔷薇科 *Sorbus* 花楸属
Sorbus rufopilosa C. K. Schneid. 红毛花楸
Shrubs or small trees, deciduous

8.6 (-1.8~15.2); 0.4 (-10.2~7.9); 15.8 (6.1~25.9); 9.2 (1.8~15.2); 63.0 (1.9~123.9); 19.5 (0.0~82.8); 582 (329~816) ❙ 919 (320~1414); 518 (193~817); 31 (3~94) ❙ 57.4 (-10.6~128.6); 577 (320~816); 265.2 (72.2~384.7)

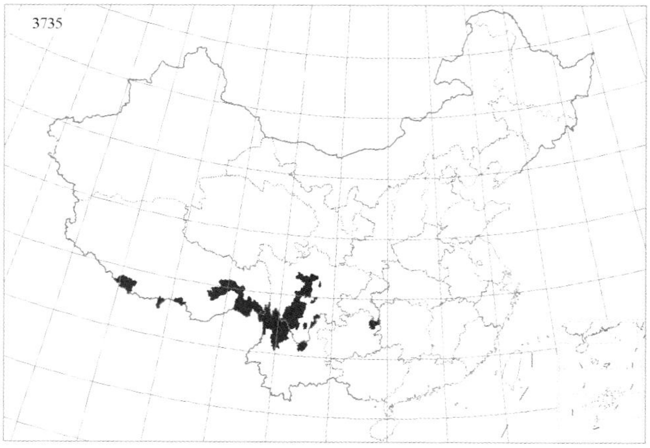

Rosaceae 蔷薇科 *Sorbus* 花楸属
Sorbus salwinensis T. T. Yu et L. T. Lu 怒江花楸
Trees, deciduous

10.0; 2.8; 16.0; 10.0; 64.6; 4.7; 609 ❙ 1142; 640; 51 ❙ 87.3; 609; 368.5

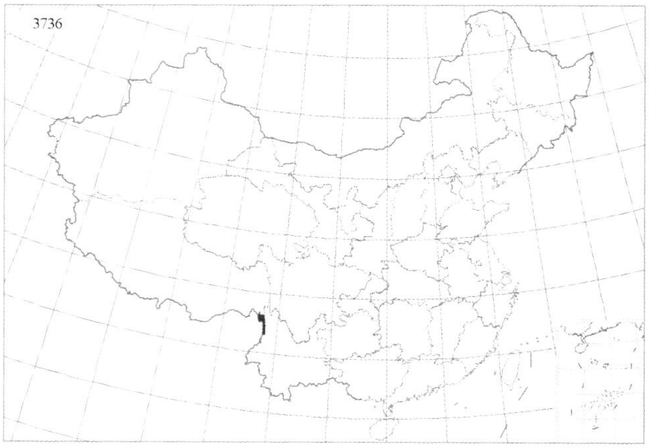

Rosaceae 蔷薇科 *Sorbus* 花楸属
Sorbus sargentiana Koehne 晚绣花楸
Trees, deciduous

12.8 (6.0~15.1); 3.3 (-3.2~5.3); 21.4 (14.4~24.9); 12.9 (6.6~15.1); 97.6 (39.7~121.7); 3.7 (0.0~26.7); 708 (508~791) ❙ 1043 (874~1414); 563 (472~817); 39 (23~71) ❙ 48.7 (18.6~101.2); 708 (508~791); 229.0 (150.8~322.4)

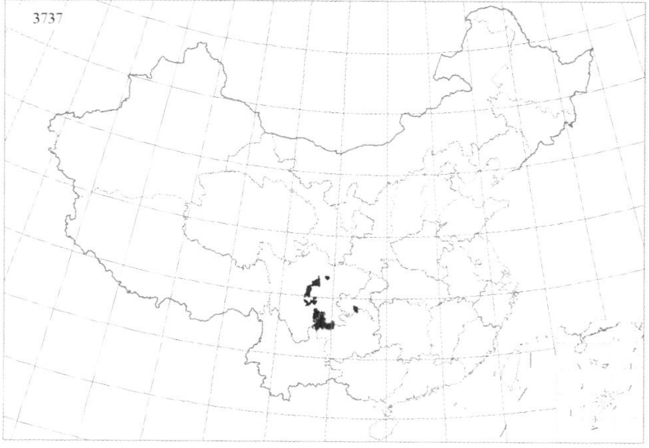

Rosaceae 蔷薇科 *Sorbus* 花楸属
Sorbus scalaris Koehne 梯叶花楸
Shrubs or small trees, deciduous

13.4 (9.6~16.6); 4.2 (2.2~6.6); 21.7 (15.7~25.7); 13.4 (9.6~16.6); 102.9 (60.8~139.7); 1.8 (0.0~5.4); 729 (600~866) ❙ 1208 (1001~1590); 684 (550~946); 42 (26~59) ❙ 65.5 (36.4~89.0); 729 (600~866); 254.4 (166.3~351.7)

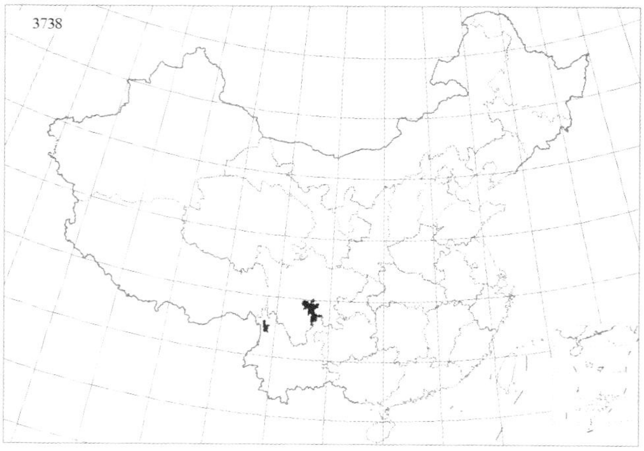

Atlas of Woody Plants in China

Distribution and Climate

Volume II

Edited by

Jingyun Fang

Zhiheng Wang

Zhiyao Tang

高等教育出版社·北京
HIGHER EDUCATION PRESS BEIJING

内容提要

准确翔实的物种分布资料是宏观生态学、生物地理学以及保护生物学等学科的研究基础，也是农业、林业生产实践的重要基础。北美、欧洲和日本在20世纪七八十年代就出版了详细的动、植物物种分布图，为这些区域的生物地理学、宏观生态学研究以及生物多样性保护实践提供了不可缺少的基础资料。中国是全球12个“巨大生物多样性国家”之一，拥有超过30 000种维管植物，其中包括约11 000种木本植物、3200种树木，物种数量远高于同处中高纬度的北美洲和欧洲。但中国目前还没有较为完善的植物分布图集。

本图集在忠实于标本记录的基础上，收集了我国目前已出版的全国、各省区和地方植物志、树木志，以及区域科学考察报告和学术论文中有关的木本植物分布资料，并在此基础上邀请全国各地20多位专家对现有资料进行修订和补充，最终编制了我国已知的全部11 405种木本植物的详尽分布图。图集中的物种名录采用了当前最新的分类系统，与国际标准一致。同时，图集还提供了物种的生活型以及物种分布区的13个气候和初级生产力指标，为生态学、生物地理学和保护生物学的基础研究和实践提供了重要的基础资料。

本图集是目前中国第一部具有完整木本植物名录、包含翔实分布信息和气候特征的专著，可作为生态学、植物学、地理学等相关学科的广大师生以及农林科技工作者的参考书。

图书在版编目（CIP）数据

中国木本植物分布图集 = Atlas of Woody Plants in China: Distribution and Climate: 中、英文 / 方精云，王志恒，唐志尧编著．— 北京：高等教育出版社，2009.9

ISBN 978-7-04-026859-1

I. 中… II. ① 方… ② 王… ③ 唐… III. 木本植物—中国—图集 Ⅳ. S717. 2-64

中国版本图书馆 CIP 数据核字（2009）第 170374 号

策划编辑 林金安 王 莉　**责任编辑** 王 莉 刘思涵　**封面设计** 张志奇
版式设计 张志奇　**责任排版** 高 瑾 张志奇　**责任印制** 陈伟光

出版发行	高等教育出版社	购书热线	010-58581118
社　　址	北京市西城区德外大街4号	咨询电话	400-810-0598
邮政编码	100120	网　　址	http://www.hep.edu.cn http://www.hep.com.cn
经　　销	蓝色畅想图书发行有限公司	网上订购	http://www.landraco.com http://www.landraco.com.cn
印　　刷	涿州市星河印刷有限公司		
开　　本	889×1194 1/16	版　　次	2009年10月第1版
印　　张	41.5	印　　次	2009年10月第1次印刷
字　　数	1 300 000	定　　价	580.00元（全4册）

物 料 号 26859-001

Atlas of Woody Plants in China

Distribution and Climate

Edited by

Jingyun Fang

Zhiheng Wang

Zhiyao Tang

Other Contributors

Xiulian Chi, Zhaodi Guo, Yue Jiang, Lai Jiang, Jing Li, Liping Li, Xin Lin, Chang Liu, Jinglei Liu, Yining Liu, Xiujuan Qiao, Yanlin Ren, Zehao Shen, Kun Tan, Luying Tang, Jie Wang, Shaopeng Wang, Yetao Yan, Xinping Zhang, Yanming Zhang, Rui Zhou, Jiangling Zhu

Review Experts (Focal Areas)

Sizhao Fang (*Guizhou*), Ying Hai (*Xinjiang*), Shushen Lai (*Jiangxi*), Fazeng Li (*Shandong*), Guangzhao Li (*Guangxi*), Zhenyu Li (*Species checklist*), Cunzhu Liang (*Inner Mongolia and Ningxia*), Chudian Liu (*Fujian*), Anwar Mohammat (*Xinjiang*), Yi Ren (*Shaanxi*), Kun Sun (*Gansu*), Shizhou Sun (*North China and Tibet*), Zemin Wu (*Anhui*), Chengdong Xu (*Yunnan*), Qinzhou Yang (*Sichuan and Chongqing*), Huagu Ye (*Guangdong*), Yongzhong Ye (*Henan*), Xunlin Yu (*Hunan*), Zien Zhao (*Hubei*), Xingmin Zhou (*Qinghai*), Zhanjiang Zong (*Northeast China*)

中国木本植物分布图集

方精云　王志恒　唐志尧 编著

参与数据收集和数据库建立的人员（按姓氏拼音排序）

池秀莲、郭兆迪、江玥、姜来、李晶、李利平、林鑫、刘畅、刘晶磊、刘怿宁、乔秀娟、任艳林、沈泽昊、谭琨、汤璐瑛、汪洁、王少鹏、闫业涛、张新平、张延明、周睿、朱江玲

审校专家及负责区域（按姓氏拼音排序）

方嗣昭(贵州)、海鹰(新疆)、赖书绅(江西)、李法曾(山东)、李光照(广西)、李振宇(植物名录)、梁存柱(内蒙古、宁夏)、刘初钿(福建)、安尼瓦尔·买买提(新疆)、任毅(陕西)、孙坤(甘肃)、孙世洲(华北地区和西藏)、吴泽民(安徽)、徐成东(云南)、杨钦周(四川、重庆)、叶华谷(广东)、叶永忠(河南)、喻勋林(湖南)、赵子恩(湖北)、周兴民(青海)、宗占江(东北地区)

Biosketches of Editors | 作者简介

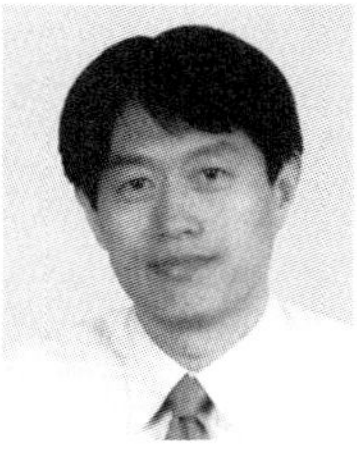

Jingyun Fang | 方精云

Dr. Jingyun Fang is a Cheung Kong Professor at Department of Ecology, Peking University. He obtained his PhD from Osaka City University (Japan) in 1989. His research interests cover vegetation ecology, biogeography of plants, carbon cycle and the application of remote sensing. He is an author of more than 230 publications including five books. He has received several awards including He-Liang-He-Li Science and Technology Progress Award, Cheung Kong Scholars Achievement Award, and National Natural Science Award of the State Council. He is a member of the Chinese Academy of Sciences and the Academy of Science for the Developing World (TWAS), and serves as the editorial board member for several journals including *Frontiers in Ecology and the Environment, Ecosystems, Global Environmental Change, Science China, Ecological Research, BMC Ecology* and others.

北京大学长江特聘教授、中科院院士，第三世界科学院院士。1989 年获日本大阪市立大学生物学博士学位。主要从事植被生态学、生物多样性、生态系统碳循环等方面的科研和教学工作。共发表论文 230 余篇，出版专著 5 部。曾获首届国家杰出青年科学基金 (1994)、教育部自然科学一等奖 (2003)，国家自然科学二等奖 (2004)，长江学者成就奖 (2006)，何梁何利科学技术进步奖 (2007) 等奖项以及全国野外科技工作先进个人称号 (2009)。是国内外多个专业杂志的编委或副主编，包括 *Frontiers in Ecology and the Environment*, *Ecosystems*, *Global Environmental Change*, *Science China*, *Ecological Research*, *BMC Ecology*。

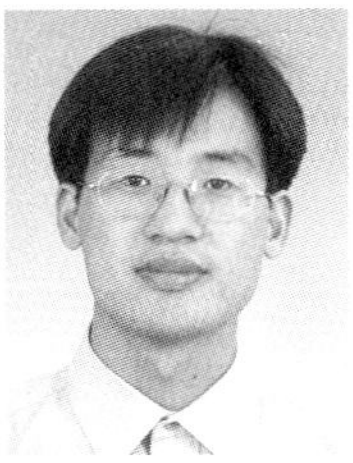

Zhiheng Wang | 王志恒

Dr. Zhiheng Wang is a postdoc researcher in Institute of Environmental Sciences, University of Zurich, Switzerland. He got his PhD at Peking University in 2008, and has research interests in all issues of macroecology. He has been working on the Database of China's Woody Plants for the last six years. His recent focus is on macro-scale patterns of species richness, species range size, species body size, patterns in community structure, and the mechanistic causes of such patterns. His current research involves testing the influence of contemporary climate, evolutionary history, human activities, and spatial scale on species diversity patterns using the synthesis of climatic, phylogenetic, and anthropogenic data. He has published 20 papers, including a recent paper in the Proceedings of the National Academy of Sciences, USA.

目前在瑞士苏黎世大学环境科学研究所从事博士后研究。2008 年毕业于北京大学，获理学博士学位。过去 6 年以来一直致力于“中国木本植物分布数据库”的建设。主要从事宏观生态学和全球变化生态学研究，内容包括物种多样性、物种分布区、个体大小和群落结构的大尺度格局及其进化和气候的形成机制，以及气候变化对物种多样性的影响。发表论文 20 余篇，包括一篇美国科学院院刊论文。

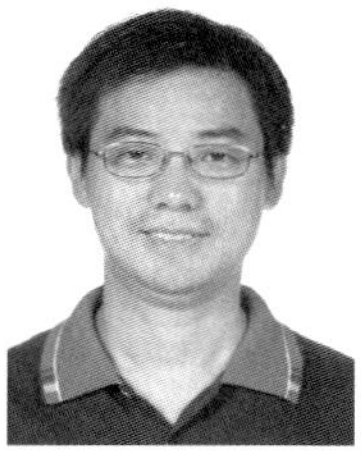

Zhiyao Tang | 唐志尧

Dr. Zhiyao Tang is an associate professor at the Department of Ecology, Peking University. He got his PhD at Peking University in 2003. His research interests include plant community ecology, biodiversity patterns, climate-vegetation interaction, and application of remote sensing in ecology. He has authored more than 40 peer-reviewed articles and two books.

北京大学生态学系副教授。2003 年毕业于北京大学，获理学博士学位。主要从事植被植物群落生态学、生物多样性、植被气候关系以及生态遥感等方面的科研与教学工作。发表论文 40 余篇，专著 2 部。

Preface | 前言

China is one of the earth's 12 mega-biodiversity countries, and the most diverse country in the northern hemisphere. With more than 30,000 species of vascular plants, more than 1/3 of which are woody plants, the number of plant species in China is higher than in Europe and North America combined. However, a comprehensive atlas of the distribution of woody plants in China has not been yet available though atlases of the species distribution in many western countries were compiled several decades ago. For example, an atlas of the United States trees was published in the 1970s. The atlas of vascular plants in Japan with a vertical resolution of 50 m was compiled in the 1970s. A project for mapping distribution of vascular plants in Europe was launched in 1965, and 13 volumes of the atlas have been published to date. These atlases of plants are of both scientific and practical importance, and provide an indispensable database for understanding the patterns and mechanisms of species distribution, utilizing the biological resources, protecting biological diversity, implementing regional development planning, and predicting ecological responses of plants to global environmental changes.

Why has China lacked of such atlas? This might be primarily due to there being too many species and limited information on the distributions of those species. These make some people think that it is not right time to compile the national atlas for China's plants. However, I don't agree with this opinion. Nothing is perfect, especially for the information on species distribution. For example, in 1817, Alexander von Humboldt drew the first isothermal map based on temperature data from only 58 climatic stations globally, which became the basis for climatic division and physiogeographical study for decades. Japan has completed the atlas for all its vascular plants with a vertical resolution of 50 m in the 1970s with limited information on distribution of plants. Compared with these, China has accumulated copious data on plant distribution through the efforts of the several generations of botanists and ecologists over the past half century. These hard-earned data provide an essential base to systematically compile the atlas of China's plants. Therefore, in my opinion, now it is the time to compile the atlas of China's plants, which is an urgent need for the conservation and sustainable use of biological resources.

The idea of compiling the atlas of China's plants initially came into my mind in the mid-1980s when I was a PhD student in Japan. My dissertation was to examine the relationships between climate and vegetation in China and I strongly felt that China was lacking of the species distribution data. I was very impressed when I saw the detailed spatial explicit distribution maps of vascular plants in Japan, and made a copy of the maps of major species and brought them back to China. They are still kept on my bookshelf now. In 1996, I was touched again when I saw the atlas of the United States trees and European species in McGill University, Canada, where I was a visiting scholar. These two shocks made me long hope that we should have an atlas for China's plants in the near future. In 2003 as the first step of compiling an atlas for all the plants of China, we initiated a project to compile the *Atlas for All the China's Woody Plants*, although without specific funds, because distribution information is relatively rich and the checklist is well-documented for the woody plants in China. As the basic data of the atlas, we first started to construct "*Database of China's Woody Plants*". We documented the distribution sites for all the species based on specimen records and all the validated records documented in publications. After six years' efforts, this atlas which documents more than 11,000 woody species was eventually published.

In this atlas, we provide distribution maps for more than 11,000 species at the level of counties. We also documented information on climate, primary production and life forms for each species. This atlas contains information on both species and environmental variables, compared with the previous atlases published in other countries or regions.

As above mentioned, we launched this project without specific financial support. However, we have been lucky to have creative human resources and excellent students, who have been dedicated to the collection and compilation of these basic data, which is difficult to yield articles but is critically significant in scientific research and practical use. Here, I would like especially to mention two major contributors of this atlas, Dr. Zhiheng Wang and Dr. Zhiyao Tang, who took major responsibility to complete the atlas. Zhiheng was my PhD student whose PhD project was to examine patterns of China's woody plants and the mechanisms that shape those patterns. He joined my group in his senior year, and I asked him to postpone his defense for his PhD twice to complete the construction of the database of China's woody plants. For this reason, he stayed in my group as a PhD student for eight years. During this period he led a group of students to overcome many problems that data compilations posed and spent six years and finally completed the database, which makes publishing this atlas possible. Zhiyao donated uncounted hours of his time to collect data, check distribution records, collate species name, design maps, and so on. In addition, many staff and students contributed their time in data collection, data input and species name collation. To all of them, I am grateful.

The atlas was based on many sources of information, including a Chinese version (*Flora Reipublicae Popularis Sinicae*) and English version of "*Flora of China*", and "*Seed Plants of China*" edited by Zhengyi Wu, "*Higher Plants of China*" edited by Liguo Fu, and nu-

merous local flora and species surveying reports. The compilation of this atlas would be impossible without these basic, most essential works. Here, I am greatly indebted to all the authors for their efforts. I would also like to thank the botanists and field biologists who reviewed and corrected the distribution records of species. I want to acknowledge many students in the Department of Ecology, Peking Universiy, who assisted this project by documenting distribution data. Mr. Jin'an Lin and Ms. Li Wang of Higher Education Press (China) spent lots of time editing, collating and publishing this atlas. I will like to express my many thanks for all these people. The compilation of this atlas was partly supported by the National Natural Science Foundation of China (Nos: 40638039, 90711002, 30721140306, 40871030, 40501025, 49971002, 39970044, 39425003).

In spite of our best efforts, errors still remain in this atlas due to our limited knowledge and limited information on species distribution. Please contact us (Email: zytang@urban.pku.edu.cn)and send us your comments so that we can update the atlas in the near future.

Jingyun Fang
Professor of Plant Ecology
September, 2009

中国是世界上物种多样性特别丰富的国家之一，拥有维管植物约30 000余种，其中1/3以上是木本植物，比北美和欧洲的总和还要多。但中国至今尚没有全面记述这些物种地理分布的图集。这应该说是件十分缺憾的事。早在几十年前，西方国家就已经完成了物种分布图的编制工作。比如，美国于20世纪70年代完成了全美木本植物分布图的编制；日本在20世纪70年代出版了所有维管植物的分布图，物种垂直分布范围的精度达50 m；欧洲也于1965年开始了维管植物分布图的编制工作。这些分布图为揭示物种的分布规律、开发和利用生物资源、开展生物多样性保护、进行区域发展规划，以及预测气候变化的生态影响等提供了不可或缺的基本资料，其科学价值和实践意义是不言而喻的。

为什么中国缺乏这类分布图？物种多、物种分布资料不足又散布于各类文献中，造成了编制工作难度大，这可能是最主要的原因。此外，国内对编制物种分布图的保守认识也是一个重要的限制因素。

据我所知，国内一些人认为，我国物种分布资料不足，目前尚不具备编制植物分布图集的条件。但我不同意这种说法。科学技术的进步是无止境的，大自然的很多现象和规律是逐渐被认知的，否则人类文明就不会进步到今天。1817年洪堡依据全球仅有的58个气象站的气温资料，编绘了世界首张全球等温线分布图，是气候分区和自然地理学研究的一个重要里程碑，成为相当长时期内开展相关研究的基石。又如，日本在30多年前就能出版垂直分布范围高达50 m精度的维管植物分布图，不能说他们当时已经掌握了所有物种准确的分布信息。因此，如果等物种分布资料齐全了才编制其分布图，那我们所待何时？事实上，经过广大植物学家、生态学家半个多世纪的努力，我国的物种分布资料已经有了相当深厚的积累，系统编制物种分布图的时机已经成熟，这也是时代发展的迫切要求。当然，随着知识的积累和认识水平的提高，这类分布图的信息需要进一步补充和完善。

编制中国植物分布图是我在学生时代就萌生的愿望。早在20世纪80年代的中后期，我在日本读书，博士期间的研究课题就是关于中国植物分布与气候关系的研究，实感中国物种分布数据之缺乏。当我看到日本详尽的物种分布图时，那种欣喜若狂至今还记忆深刻，当时我就不加思索地把主要物种的分布图全都复印了下来，带回国内，至今还保存着。1996年，我去加拿大McGill大学做访问学者，看到美国和欧洲也有详尽的物种分布图，又一次受到触动。于是，就产生了将来我们也要编制中国植物分布图的想法。2003年，在没有专项经费支持的情况下，我们决定先编制分布资料较多、种名相对稳定的木本植物分布图，于是启动了“中国木本植物分布图的编制”自设课题。次年，作为编制物种分布图的基础工作，课题组开始了“中国木本植物分布数据库”的建设。历经6年的艰苦工作，这本多达2 000余页的图集终于与读者见面了。

本图集的特点是为11 405个物种提供县级分布信息；每个物种的分布信息在忠实于标本记录的基础上，最大可能地吸收了所有物种分布研究的成果；提供物种分布区的气候资料和分布区初级生产力的平均状况。此外，我们也提供了绝大多数物种的主要生活型信息。本图集的这些特点，既克服了西方国家仅提供物种分布信息的局限性，也满足了人们对分布区气候和物种生活型等基础信息的需求，使其更具实用性。

前面讲过，我们是在没有专项经费的条件下开始此项工作的。我们虽然没有专项经费，但有丰厚的人力资源，有非常优秀的同学，他们愿意献身于这类发表成果难、但有重要科学和应用价值的基础性数据的收集和整理工作。这里，我要特别提到本图集的两位主要作者：王志恒和唐志尧，他们是本图集的主要贡献者。王志恒当时是我的直博生，我给他的博士论文题目就是中国木本植物多样性的分布格局及其机制的研究。他自大三就进入我的研究组，直到博士毕业，一共呆了8年。为了完成数据库的建设，我两度让他推延博士毕业。他带领一批同学，克服物种多、

物种名称杂乱、物种分布信息零散不全等困难，历时6年，终于建成了“中国木本植物分布数据库”，使本图集的编制成为可能。另一名主要作者唐志尧，在物种分布数据收集、分布区和种名校核、分布图制作与设计、编辑和出版等方面，做了主要的工作。可以说，他们二位是本图集编制出版的主要贡献者。此外，很多老师和同学在数据收集和录入、种名核校等方面也做了大量繁杂的工作。

本图集凝聚了几代人的心血和汗水，它是在中英文版《中国植物志》、吴征镒先生主编的《中国植物分布数据库》所记载的物种名录、傅立国主编的《中国高等植物》以及众多地方植物志和调查报告中所记载的物种分布信息的基础上编制出版的。因此，我要特别感谢这些作者们，没有他们的工作，就没有本图集的出版。北京大学生态学系的多位同学参与了数据的收集、整理和校对，以及分布图的制作等。数据库建成后邀请各地的有关专家对种名和分布区进行了校验和补充。高等教育出版社的林金安先生、王莉女士为本图集的制作、编辑和出版做了大量艰苦而有成效的工作。在此，对他们表示衷心的感谢。本图集在数据收集、整理和出版过程中，得到国家自然科学基金委的部分资助（编号：40638039, 90711002, 30721140306, 40871030, 40501025, 49971002, 39970044, 39425003）。

由于我们水平有限，以及现有资料的不足，本图集肯定存在一些纰漏，请读者们批评指正(请联系zytang@urban.pku.edu.cn)，以便在修订时改进。

方精云
2009年9月
于北京大学

How to Use This Atlas | 本图集使用说明

1 Arrangement of families, genera and species

The order of families, genera, and species were arranged as follows: the families are organized according to the phylogenetic system of *Flora of China*; within each family, the genera and species are arranged alphabetically. The atlas is published in three volumes, plus a separate volume of species checklist.

2 Arrangement of the climatic variables

The atlas lists mean and range of 12 climatic variables and vegetation net primary productivity. They are: mean annual temperature (MAT, °C), mean temperature of the coldest month (MTCM, °C), mean temperature of the warmest month (MTWM, °C), annual biotemperature (ABT, °C), warmth index (WI, °C), coldness index (CI, °C), potential evapotranspiration (PET, mm), annual precipitation (AP, mm), precipitation of warmest quarter (PWQ, mm), precipitation of coldest quarter (PCQ, mm), moisture index (Im), annual actual evaportranspiration (AET, mm), vegetation net primary productivity (NPP, $g \cdot a^{-1} \cdot m^{-2}$). For those species distributed in less than five counties, only the mean values of the climatic variables are listed. To save space, names of these variables and their units are not showed in the illustration of each species; these variables are listed as the following order, and three categories of variables (thermal index, humid/arid index, and integrative index) are separated by mark " ▌ ".

MAT (°C); MTCM (°C); MTWM (°C); ABT (°C); WI (°C); CI (°C); PET (mm) ▌ AP(mm); PWQ (mm); PCQ (mm) ▌ Im; AET (mm); NPP ($g \cdot a^{-1} \cdot m^{-2}$)

The following illustration explains the content of each map, using *Amentotaxus argotaenia* (Hance) Pilg. as an example. Following the illustration, MAT is 18.0 °C (Range is 9.3~22.9 °C); MTCM 8.0 °C (−1.7~15.2 °C); MTWM 26.8 °C (19.6 ~29.8 °C); ABT 18.0 °C (9.5~22.9 °C); WI 157.1 °C (69.8 ~214.8 °C); CI 0.7 °C (0.0 ~18.0 °C); PET 952 mm (603 ~1232 mm) ▌ AP 1530 mm (711 ~2435 mm); PWQ 605 mm (357 ~1428 mm); PCQ 150 mm (15 ~250mm) ▌ Im 61.9 (21.1 ~145.8); AET 947 mm (603 ~1232 mm); NPP 228.0 $g \cdot a^{-1} \cdot m^{-2}$ (134.6~447.4 $g \cdot a^{-1} \cdot m^{-2}$).

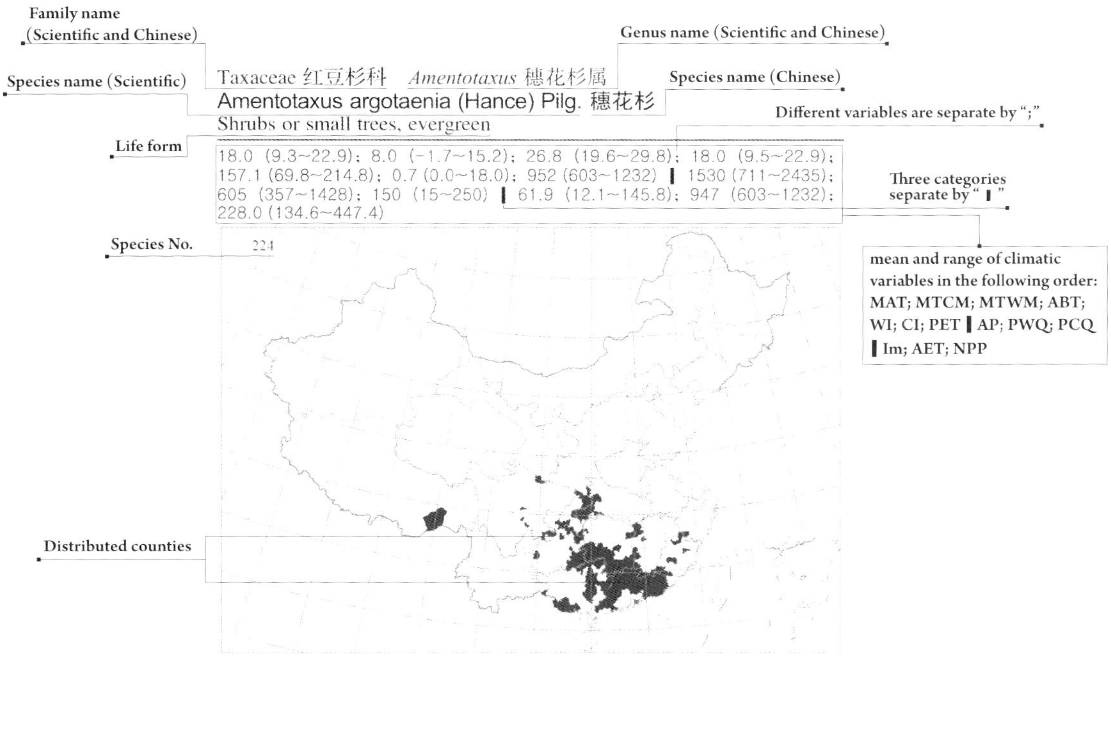

1　科、属、种排列说明

本图集科、属、种排列顺序如下：科序按 *Flora of China* 系统排列；科内（属、种）按字母顺序排列。本图集分三册出版，并附种名索引一册。

2　气候指标表示方法及使用说明

本图集给出每个种分布区的 12 个气候变量和植被净生产力的均值和范围。这些指标的意义和计算见第 xxvi~xxvii 页说明。对于分布区小于 5 个县域的物种，我们仅列出各气候指标的平均值。为节省空间，这些指标按以下顺序给出，并在各类气候指标（温度指标、水分指标和水热综合指标）之间隔以“▌”符号。另外，各指标都省去单位。顺序（含单位）如下：

年均温（℃）；最冷月均温（℃）；最热月均温（℃）；年生物温度（℃）；温暖指数（℃）；寒冷指数（℃）；潜在蒸散量（mm）▌年降水（mm）；最暖季降水（mm）；最冷季降水（mm）▌湿润指数；年实际蒸散量（mm）；植被净第一性生产力（$g \cdot a^{-1} \cdot m^{-2}$）

以穗花杉［*Amentotaxus argotaenia*（Hance）Pilg, 红豆杉科 Taxaceae, 穗花杉属 *Amentotaxus*］为例，用下面的示意图说明分布图及气候指标的含义。穗花杉在其分布区的气候指标（平均值和范围）如下：年均温为 18.0℃（范围为 9.3~22.9℃），最冷月均温 8.0℃（−1.7~15.2℃），最热月均温 26.8℃（19.6~29.8℃），年生物温度 18.0℃（9.5~22.9℃），温暖指数 157.1℃（69.8~214.8℃），寒冷指数 0.7℃（0.0~18.0℃）、潜在蒸散量 952 mm（603~1232 mm）▌年降水 1530 mm（711~2435 mm），最暖季降水 605 mm（357~1428 mm），最冷季降水 150 mm（15~250 mm）▌湿润指数 61.9（21.1~145.8），年实际蒸散量 947 mm（603~1232 mm），植被净第一性生产力 228.0 $g \cdot a^{-1} \cdot m^{-2}$（134.6~447.4 $g \cdot a^{-1} \cdot m^{-2}$）。

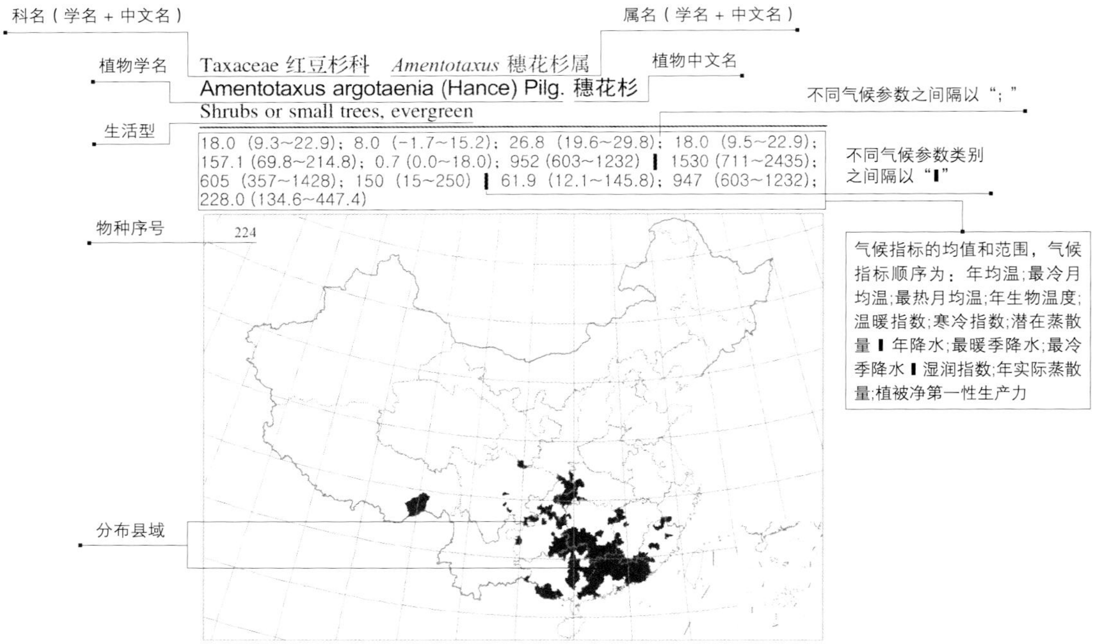

Contents | 目录

Rosaceae 蔷薇科 *Spiraea* 绣线菊属
Spiraea alpina Pall. 高山绣线菊
Shrubs, deciduous

3.7 (-4.3~15.9); -8.5 (-20.1~6.2); 14.6 (6.2~25.7); 6.0 (2.0~15.9); 39.0 (2.5~130.8); 53.9 (0.0~121.1); 488 (329~822) ❙ 568 (71~1318); 314 (24~753); 15 (2~50) ❙ 19.9 (-80.1~128.6); 429 (71~822); 216.4 (13.6~364.1)

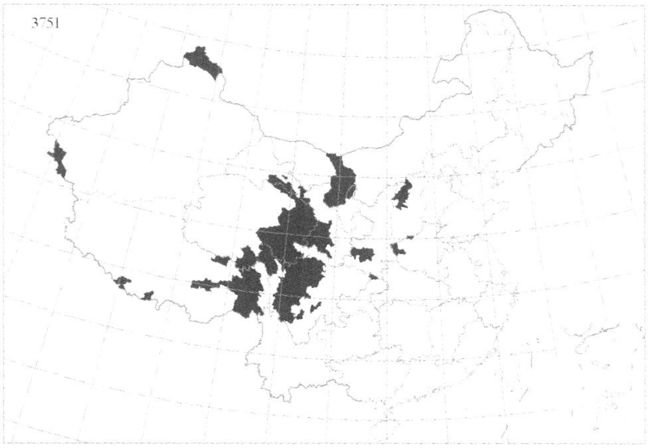

Rosaceae 蔷薇科 *Spiraea* 绣线菊属
Spiraea aquilegiifolia Pall. 耧斗菜叶绣线菊
Dwarf shrubs, deciduous

6.9 (-5.8~15.1); -9.9 (-29.8~1.8); 21.9 (11.5~28.0); 9.4 (3.5~15.2); 75.2 (17.2~128.0); 51.4 (5.6~157.4); 629 (375~880) ❙ 446 (58~817); 267 (36~409); 14 (1~53) ❙ -29.4 (-90.8~19.6); 442 (58~817); 141.6 (7.5~309.5)

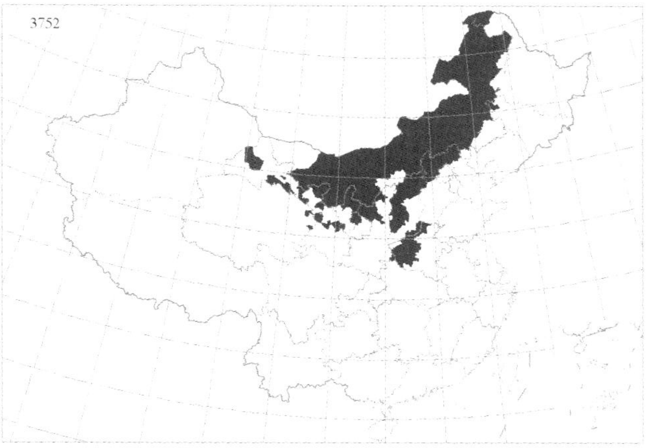

Rosaceae 蔷薇科 *Spiraea* 绣线菊属
Spiraea arcuata Hook. f. 拱枝绣线菊
Small shrubs, deciduous

6.3 (-1.8~15.2); -2.6 (-10.5~7.1); 14.3 (6.1~25.9); 7.7 (1.8~15.2); 52.0 (1.9~123.9); 35.7 (0.0~82.8); 533 (329~816) ❙ 801 (311~1736); 442 (193~1007); 31 (3~94) ❙ 40.9 (-10.6~145.8); 525 (311~816); 200.9 (62.8~447.4)

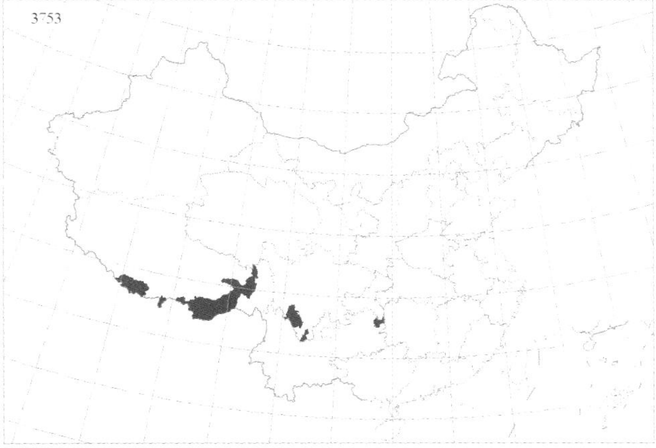

Rosaceae 蔷薇科 *Spiraea* 绣线菊属
Spiraea bella Sims 藏南绣线菊
Shrubs, deciduous

3.1 (-1.8~13.6); -5.1 (-10.2~7.2); 10.5 (6.1~18.8); 4.9 (1.8~13.7); 26.1 (1.9~104.1); 48.2 (0.0~82.8); 433 (329~706) ❙ 541 (320~1042); 328 (193~592); 21 (7~46) ❙ 18.7 (-4.0~54.0); 431 (320~706); 190.4 (61.6~372.3)

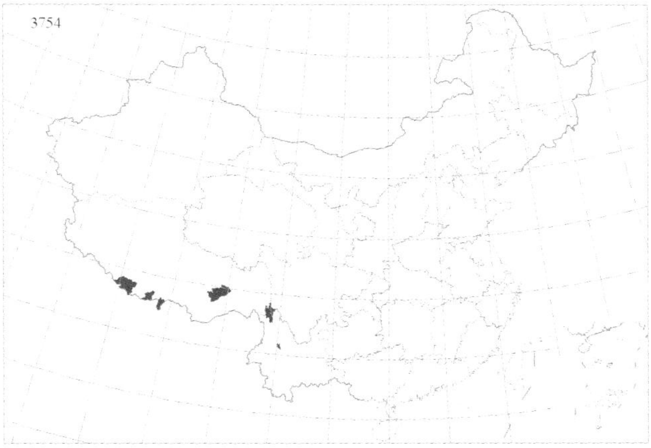

Rosaceae 蔷薇科 *Spiraea* 绣线菊属
Spiraea blumei G. Don 绣球绣线菊
Shrubs, deciduous

11.7 (-4.0~22.0); -2.5 (-28.2~13.5); 24.6 (9.7~29.9); 12.5 (3.3~22.0); 103.0 (12.2~203.8); 22.6 (0.0~146.2); 746 (395~1170) ❙ 840 (101~1997); 392 (61~726); 65 (5~259) ❙ 8.6 (-83.8~130.7); 654 (101~1170); 182.6 (33.5~364.1)

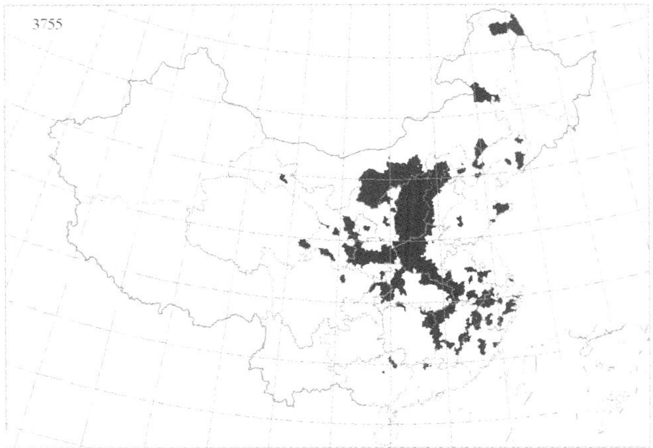

Rosaceae 蔷薇科 *Spiraea* 绣线菊属
Spiraea blumei var. latipetala Hemsl. 宽瓣绣球绣线菊
Shrubs, deciduous

17.7 (15.0~19.9); 6.8 (2.6~10.6); 27.7 (26.6~30.1); 17.8 (15.0~19.9); 153.6 (123.7~179.3); 0.4 (0.0~3.5); 941 (826~1036) ❙ 1675 (1363~1922); 574 (453~644); 210 (169~259) ❙ 78.7 (41.7~111.3); 934 (826~1036); 238.3 (148.9~299.5)

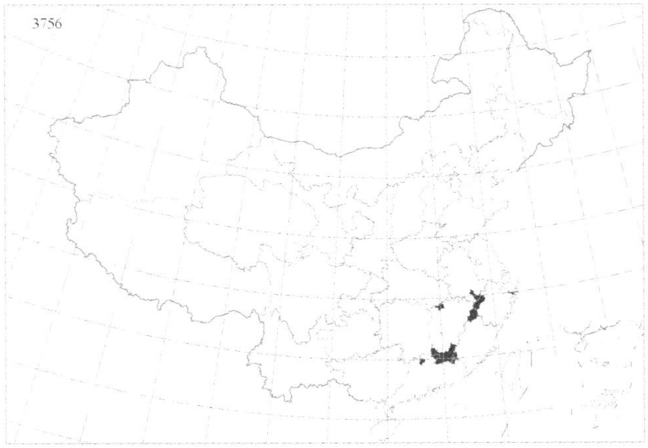

Rosaceae 蔷薇科 *Spiraea* 绣线菊属

Spiraea blumei var. microphylla Rehder 小叶绣球绣线菊

Shrubs, deciduous

9.2 (4.7~12.7); -3.6 (-7.4~-0.1); 21.1 (14.7~25.9); 9.8 (6.0~12.9); 75.4 (36.3~108.3); 24.9 (11.7~41.0); 630 (482~771) ❙ 636 (341~892); 307 (190~391); 19 (8~33) ❙ 1.9 (-40.9~52.6); 576 (341~697); 214.9 (61.2~318.2)

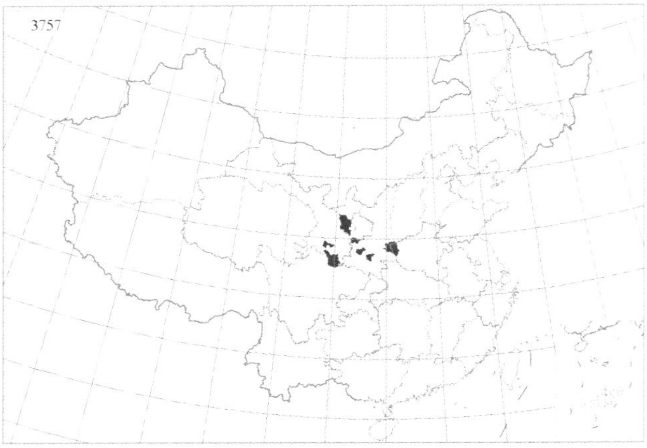

Rosaceae 蔷薇科 *Spiraea* 绣线菊属

Spiraea blumei var. pubicarpa Cheng 毛果绣球绣线菊

Shrubs, deciduous

13.0 (9.8~16.3); 0.0 (-2.7~4.6); 25.3 (21.9~28.2); 13.2 (10.2~16.4); 107.9 (79.2~136.9); 11.6 (0.4~20.1); 764 (641~887) ❙ 986 (581~1801); 399 (257~598); 85 (22~242) ❙ 26.5 (-32.0~109.2); 713 (581~887); 233.4 (182.6~284.0)

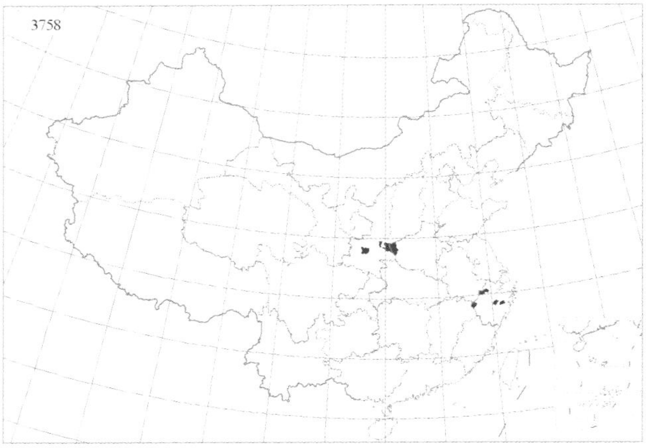

Rosaceae 蔷薇科 *Spiraea* 绣线菊属

Spiraea calcicola W. W. Sm. 石灰岩绣线菊

Dwarf shrubs, deciduous

8.1 (5.4~10.8); 0.8 (-2.1~3.9); 14.5 (12.3~16.6); 8.3 (5.8~10.9); 50.4 (31.5~71.8); 12.6 (1.3~25.6); 557 (481~633) ❙ 912 (741~1041); 536 (426~594); 32 (21~45) ❙ 63.5 (52.8~81.5); 557 (481~633); 316.0 (273.5~352.3)

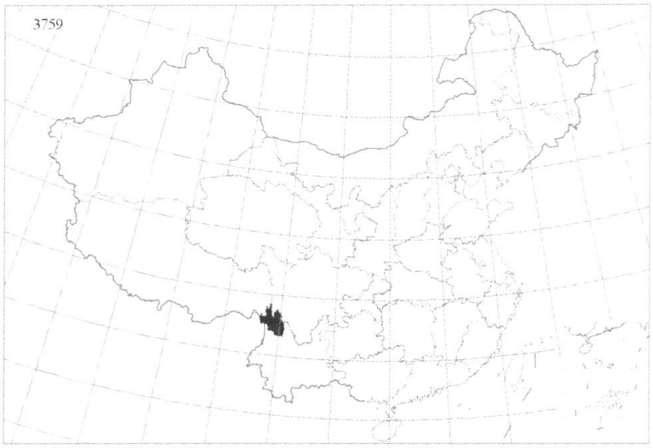

Rosaceae 蔷薇科 *Spiraea* 绣线菊属

Spiraea canescens D. Don 楔叶绣线菊

Shrubs, deciduous

1.7 (-2.2~5.4); -7.5 (-12.8~-2.1); 9.9 (6.1~12.3); 3.8 (1.8~5.8); 16.5 (1.9~31.5); 56.1 (25.6~91.3); 403 (329~481) ❙ 554 (300~961); 338 (208~558); 12 (1~31) ❙ 35.3 (-23.5~128.6); 387 (300~481); 228.8 (72.2~307.8)

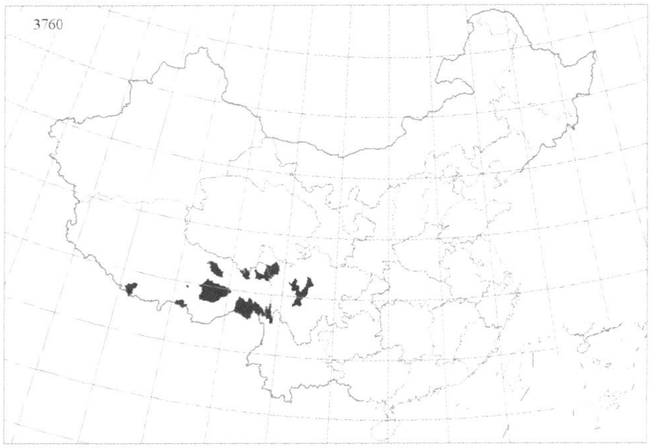

Rosaceae 蔷薇科 *Spiraea* 绣线菊属

Spiraea canescens var. glaucophylla Franch. 粉背楔叶绣线菊

Shrubs, deciduous

7.5 (0.7~15.9); -1.2 (-9.0~6.5); 15.2 (9.2~24.6); 8.2 (3.3~15.9); 54.1 (13.3~130.8); 23.3 (0.0~64.8); 555 (388~822) ❙ 968 (536~1414); 562 (328~817); 23 (10~50) ❙ 75.7 (38.3~128.6); 554 (388~822); 264.9 (189.1~337.3)

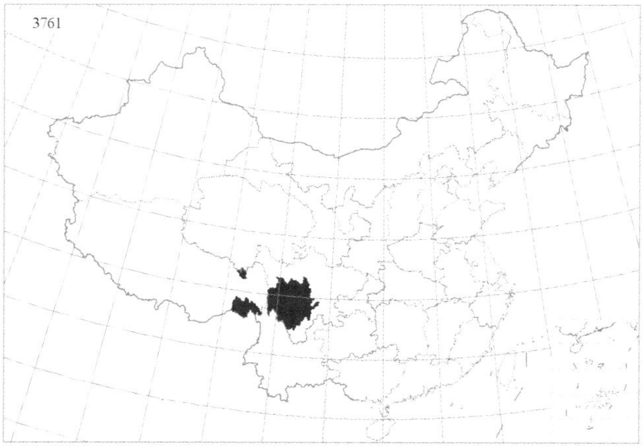

Rosaceae 蔷薇科 *Spiraea* 绣线菊属

Spiraea cantoniensis Lour. 麻叶绣线菊

Shrubs, deciduous

16.7 (13.4~24.0); 5.1 (-0.6~17.7); 27.5 (24.2~29.9); 16.7 (13.4~24.1); 142.3 (108.3~229.0); 1.5 (0.0~12.5); 903 (755~1321) ❙ 1446 (584~2014); 539 (312~954); 164 (21~259) ❙ 60.6 (-34.1~130.7); 887 (584~1321); 214.3 (111.9~332.5)

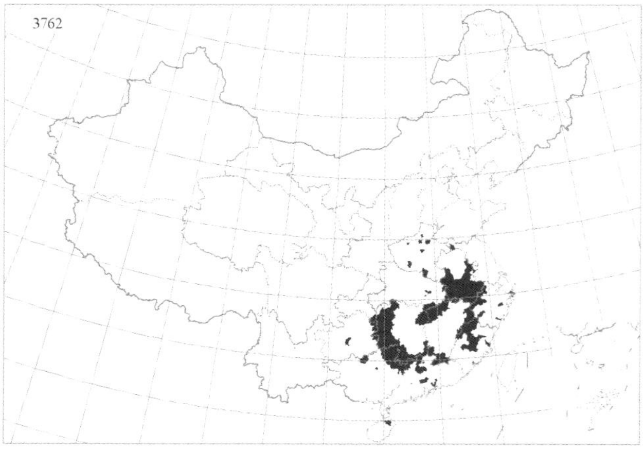

Rosaceae 蔷薇科 *Spiraea* 绣线菊属

Spiraea cantoniensis var. jiangxiensis (Z. X. Yu) L. T. Lu 江西绣线菊

Shrubs, deciduous

16.2; 4.1; 27.8; 16.2; 135.4; 0.9; 882 ❙ 1393; 505; 164 ❙ 58.5; 881; 236.4

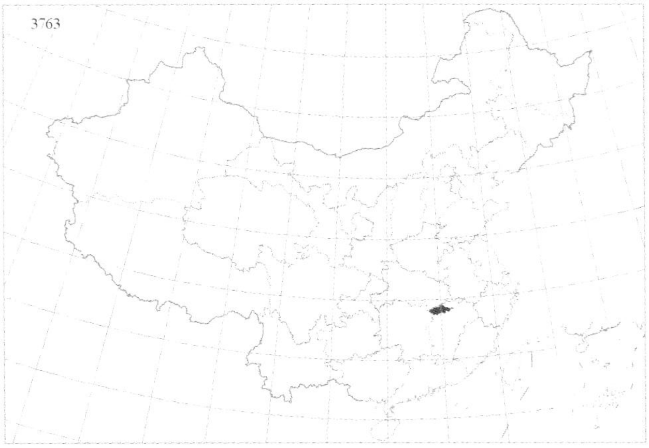

Rosaceae 蔷薇科 *Spiraea* 绣线菊属

Spiraea cantoniensis var. pilosa T. T. Yu 毛萼麻叶绣线菊

Shrubs, deciduous

16.3; 5.2; 26.8; 16.3; 136.9; 0.8; 869 ❙ 1429; 551; 139 ❙ 64.3; 869; 215.8

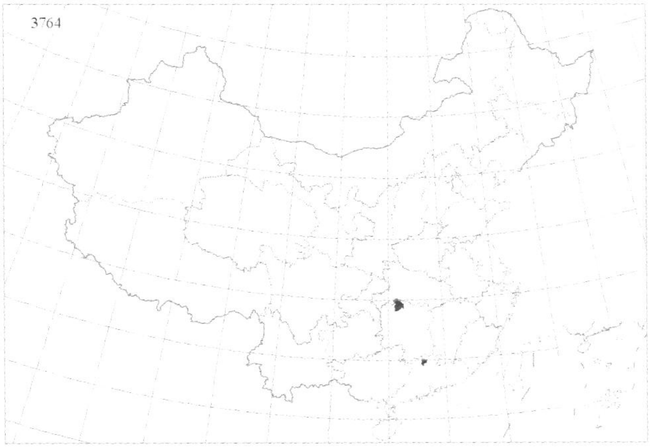

Rosaceae 蔷薇科 *Spiraea* 绣线菊属

Spiraea chamaedryfolia L. 石蚕叶绣线菊

Shrubs, deciduous

7.3 (-1.3~13.6); -10.4 (-23.8~-1.3); 22.6 (13.2~27.5); 9.8 (4.5~13.8); 79.4 (25.7~119.7); 51.0 (14.6~112.6); 652 (433~841) ❙ 535 (120~1073); 335 (51~691); 19 (8~51) ❙ -16.7 (-85.3~74.1); 496 (120~690); 196.0 (21.1~346.5)

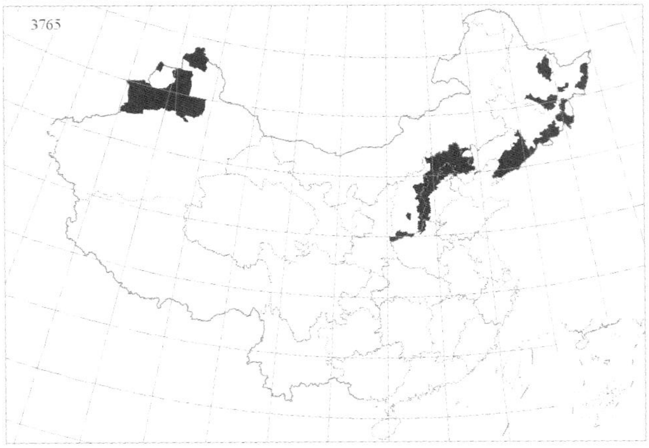

Rosaceae 蔷薇科 *Spiraea* 绣线菊属

Spiraea chinensis Maxim. 中华绣线菊

Shrubs, deciduous

15.5 (3.7~21.8); 3.4 (-18.1~15.2); 26.6 (15.6~30.1); 15.6 (7.3~22.7); 131.6 (48.2~212.7); 5.4 (0.0~84.1); 858 (535~1216) ❙ 1261 (137~2005); 496 (92~994); 128 (3~259) ❙ 45.3 (-77.5~147.0); 819 (137~1216); 203.5 (25.4~335.3)

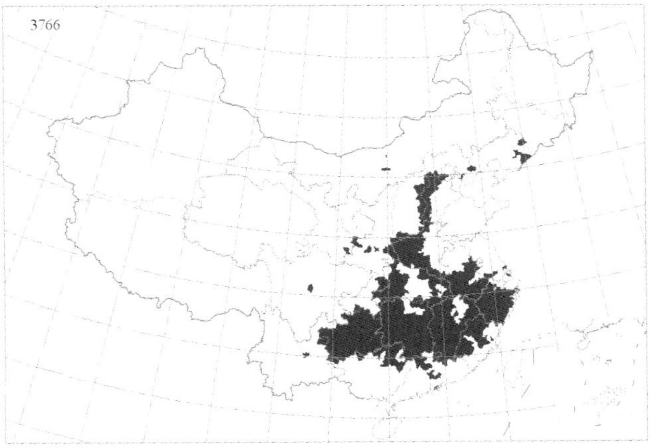

Rosaceae 蔷薇科 *Spiraea* 绣线菊属

Spiraea chinensis var. grandiflora T. T. Yu 大花中华绣线菊

Shrubs, deciduous

12.6; 1.6; 23.2; 12.7; 98.8; 6.9; 711 ❙ 1466; 602; 105 ❙ 106.2; 711; 285.1

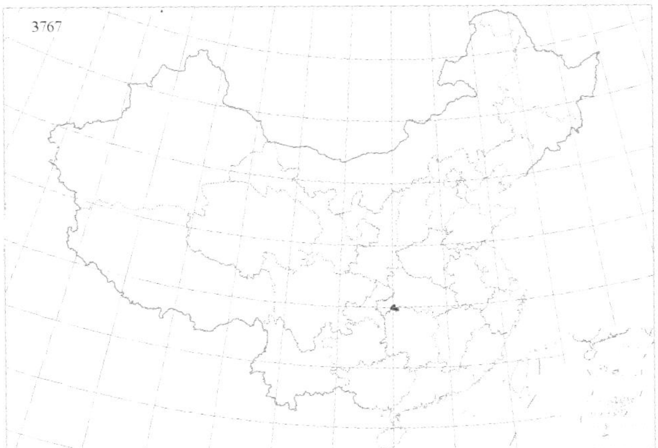

Rosaceae 蔷薇科 *Spiraea* 绣线菊属

Spiraea compsophylla Hand.-Mazz. 粉叶绣线菊

Shrubs, deciduous

10.3 (5.4~17.4); 2.7 (-2.1~7.6); 16.8 (12.3~25.5); 10.4 (5.8~17.4); 73.2 (31.5~148.9); 9.0 (0.0~25.6); 628 (481~886) ❙ 990 (741~1381); 557 (426~637); 43 (21~107) ❙ 58.5 (36.1~81.5); 625 (481~886); 303.2 (245.3~352.3)

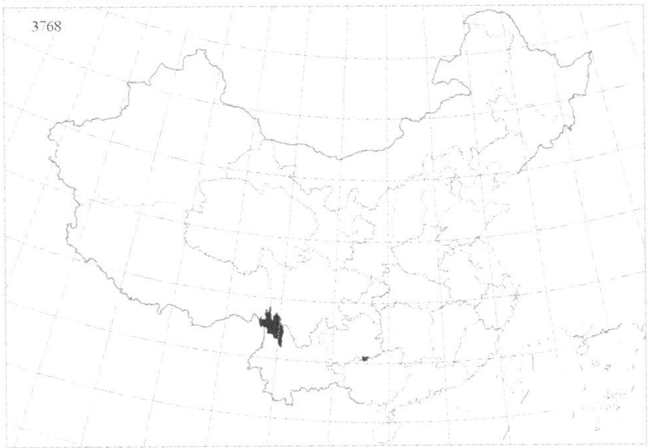

Rosaceae 蔷薇科 *Spiraea* 绣线菊属
Spiraea dahurica (Rupr.) Maxim. 窄叶绣线菊
Shrubs, deciduous

1.8 (-5.8~8.6); -18.9 (-29.8~-9.3); 20.2 (15.3~24.5); 7.4 (4.4~10.5); 57.0 (28.0~89.4); 94.5 (45.6~157.4); 551 (426~692) ❙ 432 (137~866); 295 (92~547); 10 (3~30) ❙ -21.4 (-77.5~38.7); 420 (137~645); 206.7 (25.4~309.5)

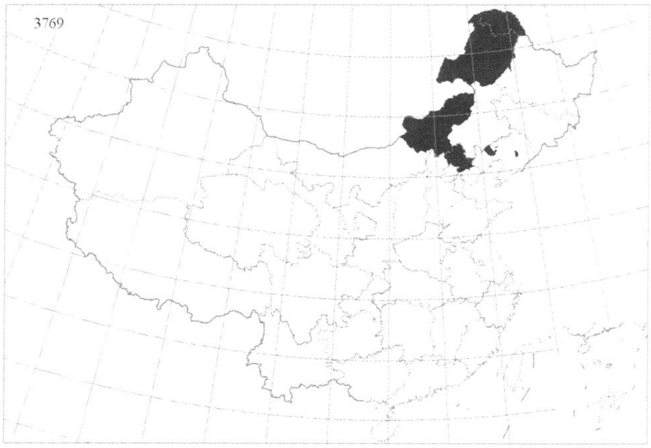

Rosaceae 蔷薇科 *Spiraea* 绣线菊属
Spiraea daochengensis L. T. Lu 稻城绣线菊
Shrubs, deciduous

2.2; -6.3; 9.5; 3.9; 15.8; 48.6; 411 ❙ 729; 482; 11 ❙ 77.4; 411; 208.8

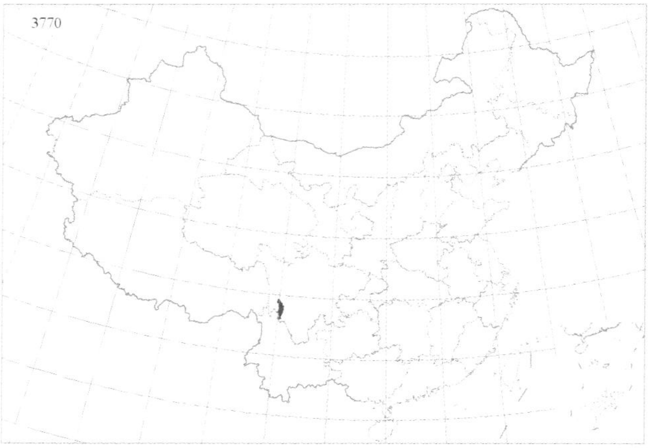

Rosaceae 蔷薇科 *Spiraea* 绣线菊属
Spiraea dasyantha Bunge 毛花绣线菊
Shrubs, deciduous

9.8 (-1.5~18.2); -5.7 (-25.5~6.9); 23.7 (17.5~29.3); 11.2 (6.5~18.2); 91.6 (45.4~158.8); 33.3 (0.0~128.7); 699 (507~984) ❙ 661 (137~1922); 359 (92~644); 38 (3~259) ❙ -8.5 (-77.5~111.3); 576 (137~961); 175.8 (25.4~287.6)

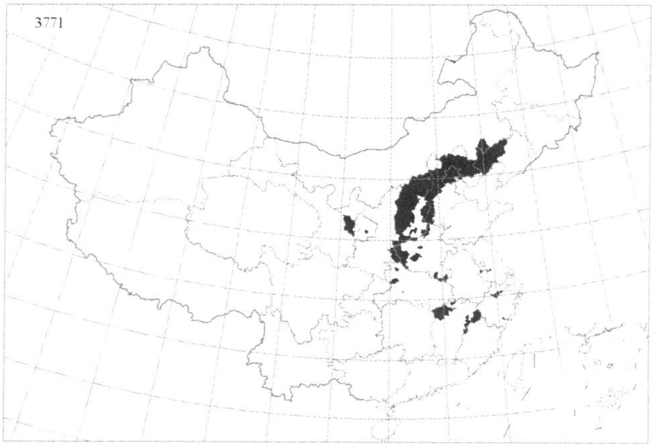

Rosaceae 蔷薇科 *Spiraea* 绣线菊属
Spiraea elegans Pojark. 美丽绣线菊
Shrubs, deciduous

1.7 (-5.8~9.4); -18.8 (-29.8~-7.5); 19.7 (15.3~24.3); 7.2 (4.4~10.9); 54.2 (28.0~91.7); 93.2 (38.2~157.4); 540 (426~704) ❙ 456 (137~976); 305 (92~608); 13 (3~40) ❙ -14.9 (-77.5~66.0); 429 (137~609); 223.3 (25.4~330.9)

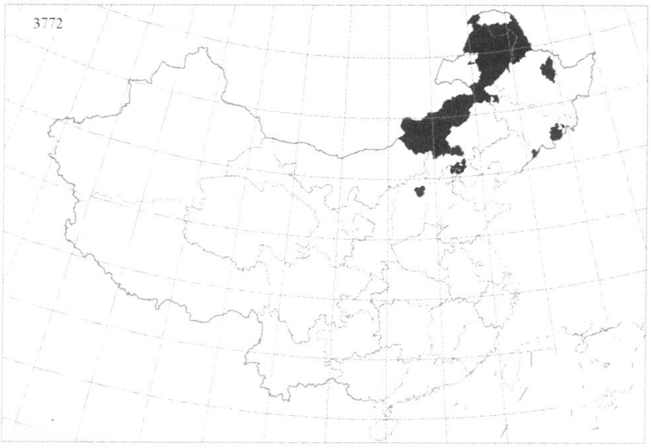

Rosaceae 蔷薇科 *Spiraea* 绣线菊属
Spiraea flexuosa Fisch. ex Cambess. 曲萼绣线菊
Shrubs, deciduous

4.9 (-5.8~11.7); -13.3 (-29.8~-1.6); 20.8 (13.7~27.5); 8.4 (4.4~12.0); 65.1 (27.2~107.3); 65.8 (15.5~157.4); 588 (426~813) ❙ 390 (120~853); 232 (51~527); 14 (2~46) ❙ -31.4 (-85.3~57.4); 369 (120~689); 156.7 (13.6~346.5)

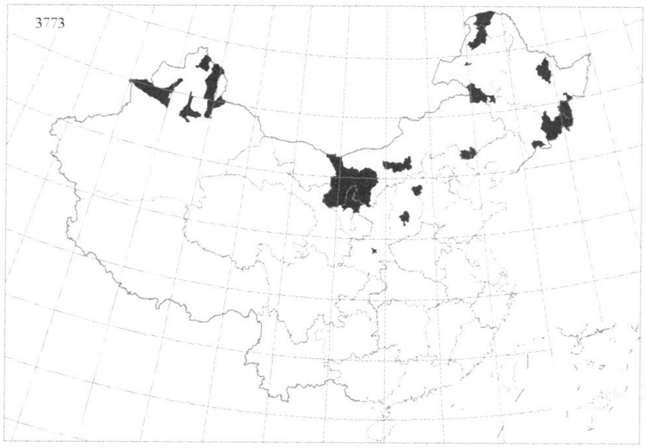

Rosaceae 蔷薇科 *Spiraea* 绣线菊属
Spiraea formosana Hayata 台湾绣线菊
Shrubs, deciduous

19.8; 14.0; 24.6; 19.8; 178.1; 0.0; 984 ❙ 2221; 1163; 158 ❙ 132.0; 970; 230.6

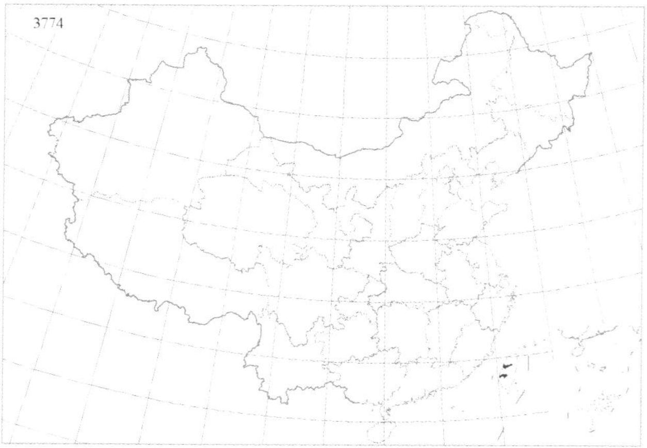

Rosaceae 蔷薇科 *Spiraea* 绣线菊属
Spiraea fritschiana C. K. Schneid. 华北绣线菊
Shrubs, deciduous

11.2 (-0.1~16.6); -3.6 (-23.5~4.4); 24.6 (18.7~28.8); 12.1 (6.9~16.7); 99.5 (52.8~140.6); 24.2 (0.6~113.0); 735 (535~920) ❙ 694 (367~1581); 377 (235~666); 34 (6~185) ❙ -6.2 (-43.4~108.0); 633 (367~910); 180.8 (101.9~334.4)

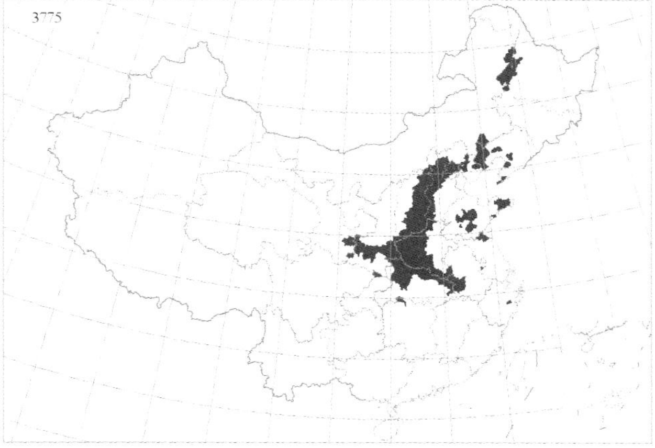

Rosaceae 蔷薇科 *Spiraea* 绣线菊属
Spiraea fritschiana var. angulata (Fritsch ex C. K. Schneid.) Rehder 大叶华北绣线菊
Shrubs, deciduous

12.3 (2.6~15.8); -2.1 (-15.9~3.6); 25.5 (18.7~28.6); 12.8 (7.2~15.8); 107.3 (52.0~133.2); 18.7 (1.4~80.4); 772 (535~893) ❙ 683 (377~1337); 378 (241~545); 35 (8~113) ❙ -11.0 (-37.2~60.7); 660 (377~893); 182.7 (109.4~334.4)

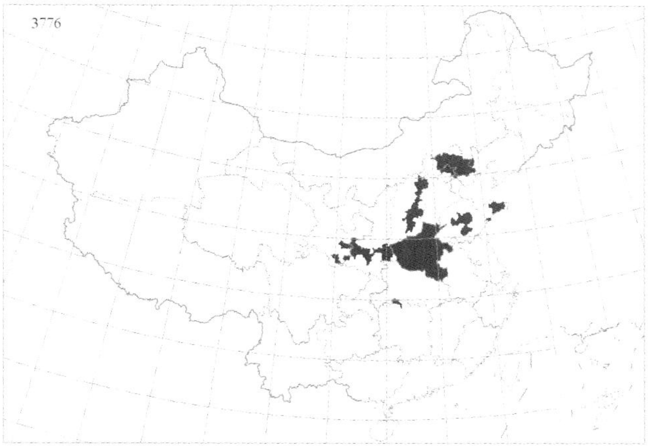

Rosaceae 蔷薇科 *Spiraea* 绣线菊属
Spiraea fritschiana var. parvifolia Liou 小叶华北绣线菊
Shrubs, deciduous

11.3 (7.2~13.8); -4.0 (-10.4~-1.2); 24.9 (22.5~27.3); 12.0 (9.4~14.0); 101.2 (78.0~121.1); 24.6 (14.2~50.9); 742 (636~842) ❙ 666 (526~817); 440 (389~529); 26 (10~65) ❙ -10.3 (-29.5~5.6); 662 (526~800); 184.7 (136.4~273.5)

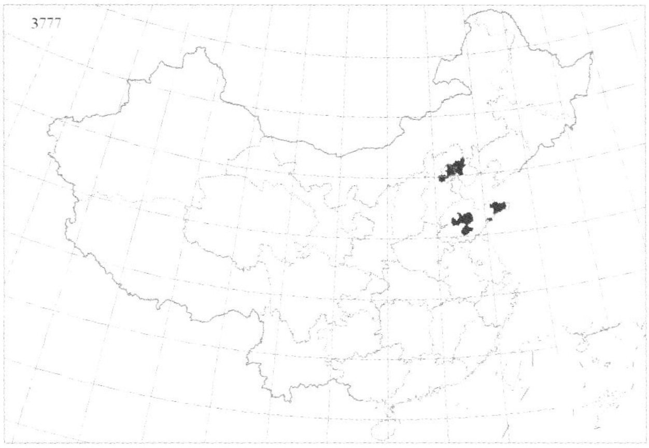

Rosaceae 蔷薇科 *Spiraea* 绣线菊属
Spiraea hailarensis Liou 海拉尔绣线菊
Shrubs, deciduous

-1.3; -25.1; 19.6; 6.3; 48.6; 124.2; 515 ❙ 354; 247; 11 ❙ -31.2; 354; 197.6

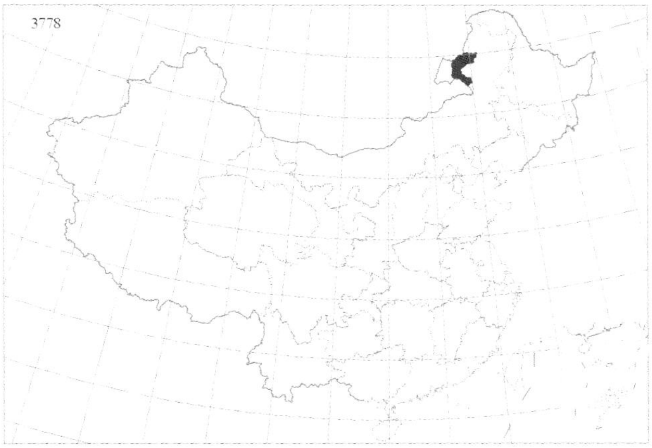

Rosaceae 蔷薇科 *Spiraea* 绣线菊属
Spiraea hayatana H. L. Li 假绣线菊
Shrubs, deciduous

17.0; 11.5; 21.5; 17.0; 144.3; 0.0; 813 ❙ 2770; 1073; 323 ❙ 241.0; 813; 277.5

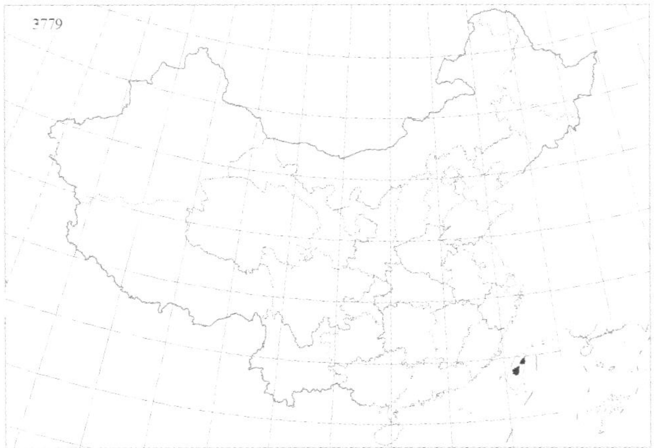

Rosaceae 蔷薇科 *Spiraea* 绣线菊属
Spiraea henryi Hemsl. 翠蓝绣线菊
Shrubs, deciduous

13.5 (2.9~18.2); 2.2 (-11.0~9.0); 24.0 (11.2~28.1); 13.7 (4.4~18.2); 109.9 (20.5~153.8); 6.9 (0.0~49.4); 761 (428~918) ❙ 1060 (626~1492); 489 (316~816); 59 (15~140) ❙ 40.3 (-22.2~121.7); 749 (428~918); 212.2 (139.1~352.3)

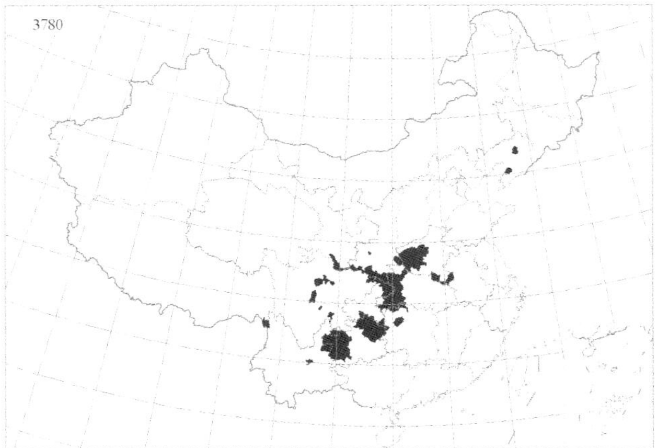

Rosaceae 蔷薇科 *Spiraea* 绣线菊属

Spiraea henryi var. omeiensis T. T. Yu 峨眉翠蓝绣线菊

Shrubs, deciduous

14.7; 3.8; 25.3; 14.8; 118.7; 1.5; 791 ❙ 1240; 486; 84 ❙ 56.8; 791; 170.4

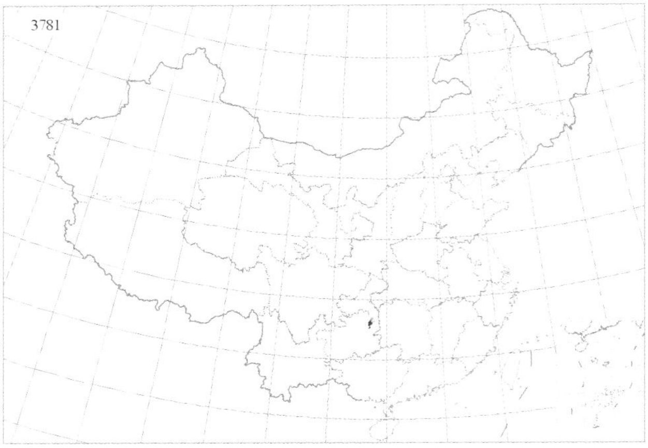

Rosaceae 蔷薇科 *Spiraea* 绣线菊属

Spiraea hingshanensis T. T. Yu et L. T. Lu 兴山绣线菊

Shrubs, deciduous

13.3 (12.0~14.4); 2.1 (0.6~2.6); 24.1 (23.1~25.6); 13.4 (12.1~14.4); 105.8 (94.7~116.7); 5.4 (3.6~9.7); 743 (697~793) ❙ 1262 (1118~1492); 532 (478~613); 84 (75~96) ❙ 70.2 (42.8~100.5); 743 (697~793); 245.9 (216.2~274.9)

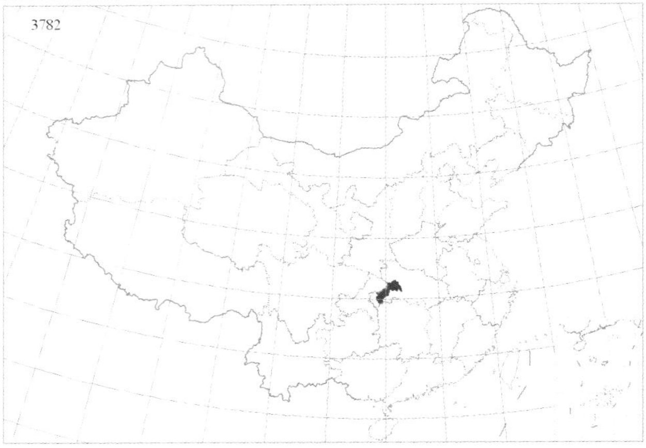

Rosaceae 蔷薇科 *Spiraea* 绣线菊属

Spiraea hirsuta (Hemsl.) C. K. Schneid. 疏毛绣线菊

Shrubs, deciduous

11.3 (2.6~19.1); -2.8 (-13.7~10.4); 24.1 (11.2~29.9); 12.1 (4.3~19.1); 98.7 (20.0~169.1); 22.3 (0.0~71.7); 728 (427~979) ❙ 754 (307~1997); 371 (173~669); 44 (7~254) ❙ 1.6 (-49.4~130.7); 628 (307~977); 182.8 (42.6~334.4)

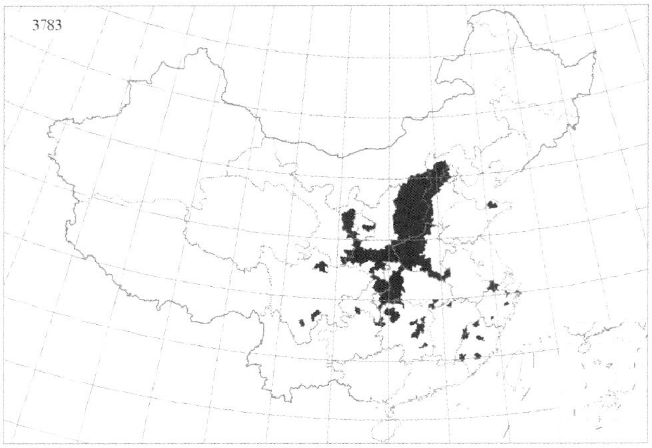

Rosaceae 蔷薇科 *Spiraea* 绣线菊属

Spiraea hirsuta var. rotundifolia (Hemsl.) Rehder 圆叶疏毛绣线菊

Shrubs, deciduous

14.1 (9.8~16.2); 2.4 (-2.5~4.8); 25.2 (21.9~27.5); 14.1 (10.2~16.2); 114.7 (79.2~135.6); 5.1 (0.2~20.1); 783 (641~877) ❙ 1144 (795~1451); 480 (354~591); 81 (25~140) ❙ 46.1 (-2.0~101.3); 782 (641~877); 214.9 (151.3~282.3)

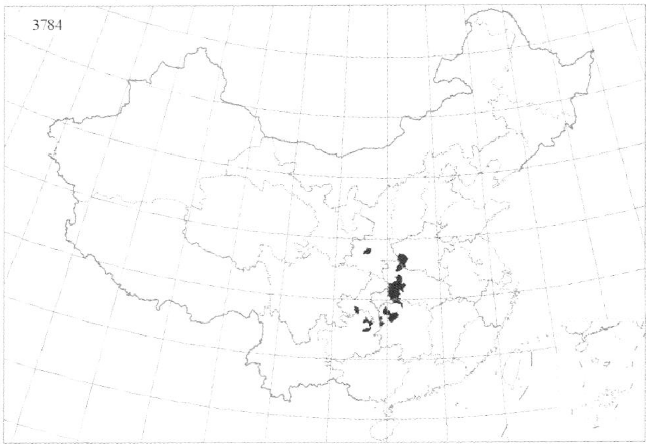

Rosaceae 蔷薇科 *Spiraea* 绣线菊属

Spiraea hypericifolia L. 金丝桃叶绣线菊

Shrubs, deciduous

7.9 (-2.8~15.1); -8.6 (-24.9~1.8); 22.3 (9.7~28.0); 9.9 (2.9~15.2); 80.1 (11.3~128.0); 44.6 (5.6~121.8); 653 (359~880) ❙ 439 (45~857); 243 (27~453); 18 (3~53) ❙ -33.0 (-94.0~35.3); 433 (45~817); 128.6 (7.6~297.3)

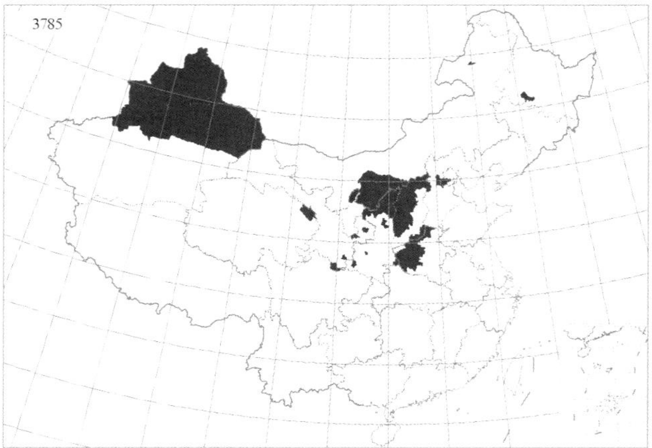

Rosaceae 蔷薇科 *Spiraea* 绣线菊属

Spiraea japonica L. f. 绣线菊

Shrubs, deciduous

14.4 (-1.7~21.0); 3.9 (-12.3~14.2); 23.8 (8.0~29.4); 14.5 (2.4~21.0); 118.8 (6.4~191.8); 6.0 (0.0~86.1); 791 (344~1053) ❙ 1163 (487~1997); 524 (258~1007); 91 (8~254) ❙ 47.5 (-36.5~147.0); 769 (344~1053); 235.1 (111.9~551.6)

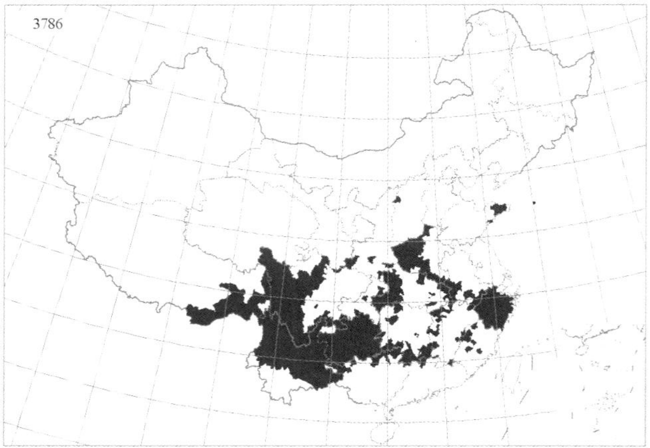

Rosaceae 蔷薇科 *Spiraea* 绣线菊属
Spiraea japonica var. acuminata Franch. 渐尖绣线菊
Shrubs, deciduous

14.8 (1.8~19.8); 3.9 (−7.2~13.3); 24.9 (9.9~29.8); 14.9 (3.8~19.8); 122.2 (15.4~177.9); 4.2 (0.0~53.3); 812 (404~1010) ❙ 1232 (583~1997); 512 (306~1007); 108 (13~254) ❙ 52.0 (−22.2~147.0); 801 (404~1010); 230.1 (111.9~491.5)

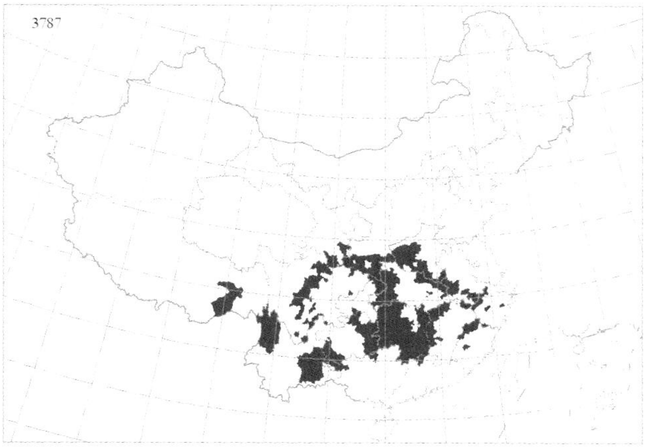

Rosaceae 蔷薇科 *Spiraea* 绣线菊属
Spiraea japonica var. acuta T. T. Yu 急尖绣线菊
Shrubs, deciduous

12.8 (10.8~14.5); 6.2 (3.9~7.8); 18.2 (16.6~19.7); 12.9 (10.9~14.5); 94.8 (71.8~113.5); 0.3 (0.0~1.3); 686 (633~730) ❙ 1026 (968~1057); 596 (570~627); 39 (24~54) ❙ 49.9 (36.1~64.0); 683 (633~730); 345.5 (296.9~382.6)

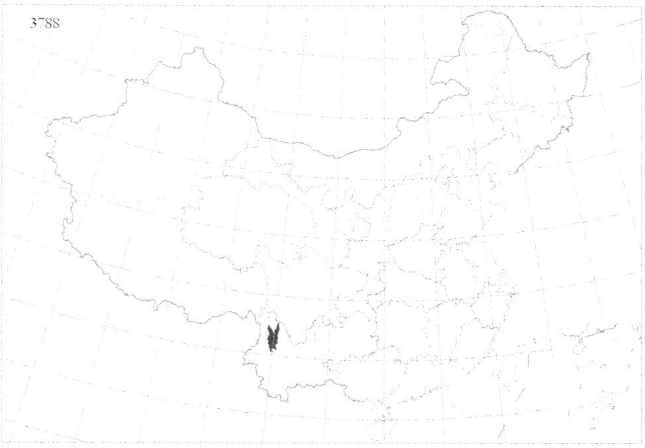

Rosaceae 蔷薇科 *Spiraea* 绣线菊属
Spiraea japonica var. fortunei (Panchon) Rehd. 光叶绣线菊
Shrubs, deciduous

15.1 (2.2~19.8); 4.3 (−6.6~12.5); 24.9 (9.9~29.4); 15.2 (3.8~19.8); 124.9 (15.5~177.9); 3.4 (0.0~49.1); 816 (413~1029) ❙ 1204 (626~1997); 528 (316~720); 97 (14~259) ❙ 46.9 (−22.2~147.0); 803 (413~1029); 225.1 (111.9~406.0)

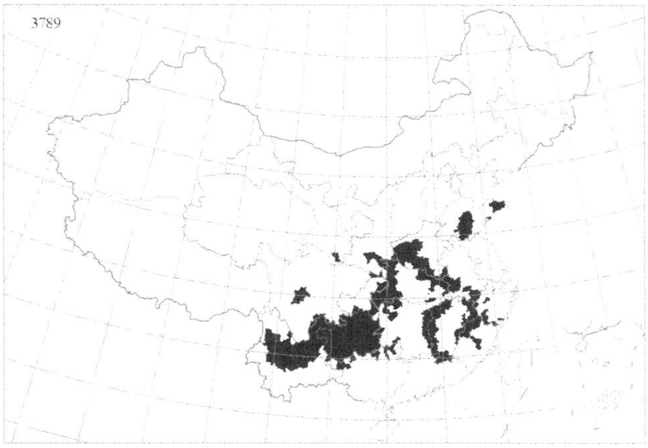

Rosaceae 蔷薇科 *Spiraea* 绣线菊属
Spiraea japonica var. glabra (Regel) Koidz. 无毛绣线菊
Shrubs, deciduous

15.4 (9.3~17.7); 3.5 (−1.7~6.9); 26.8 (20.1~29.2); 15.4 (9.5~17.7); 127.5 (69.8~152.2); 2.3 (0.0~18.0); 844 (603~938) ❙ 1391 (993~1689); 548 (420~666); 134 (46~212) ❙ 65.7 (18.0~108.0); 843 (603~937); 214.0 (126.1~319.9)

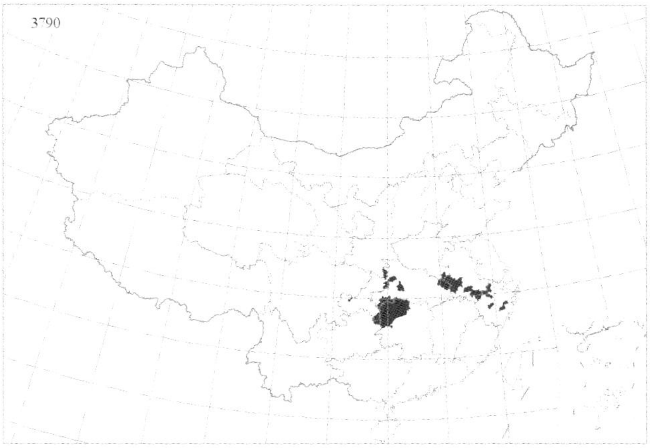

Rosaceae 蔷薇科 *Spiraea* 绣线菊属
Spiraea japonica var. incisa T. T. Yu 锐裂绣线菊
Shrubs, deciduous

10.2 (2.9~14.9); 2.1 (−6.3~7.6); 17.4 (11.2~23.6); 10.5 (4.4~15.0); 73.5 (20.5~119.6); 10.2 (0.0~44.6); 624 (428~782) ❙ 1043 (801~1410); 599 (490~816); 33 (21~49) ❙ 69.6 (36.1~121.7); 621 (428~782); 288.1 (214.5~351.7)

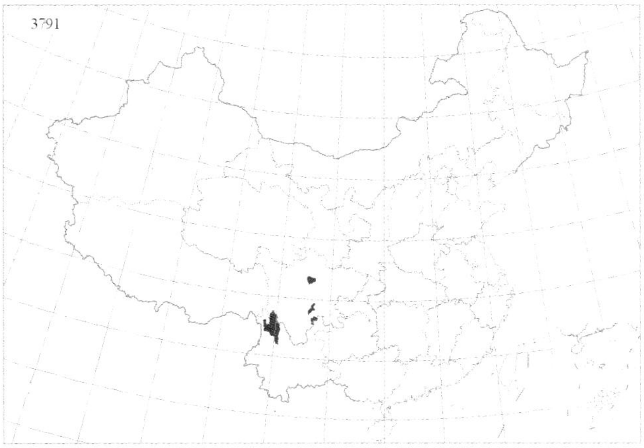

Rosaceae 蔷薇科 *Spiraea* 绣线菊属
Spiraea japonica var. ovalifolia Franch. 椭圆绣线菊
Shrubs, deciduous

14.6 (5.4~18.9); 7.7 (−2.1~12.5); 19.8 (12.3~23.9); 14.6 (5.8~19.0); 117.4 (31.5~167.8); 1.8 (0.0~25.6); 746 (481~929) ❙ 989 (741~1220); 570 (411~688); 38 (15~59) ❙ 34.6 (−14.2~87.3); 716 (481~851); 304.3 (207.5~406.0)

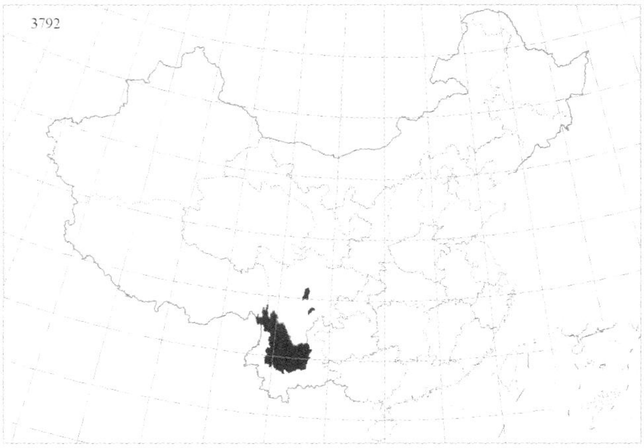

Rosaceae 蔷薇科 *Spiraea* 绣线菊属
Spiraea japonica var. pinnatifida T. T. Yu et L. T. Lu 羽叶绣线菊
Shrubs, deciduous

12.2；4.3；18.5；12.2；87.4；0.7；672 ❙ 1431；817；31 ❙ 113.0；672；363.7

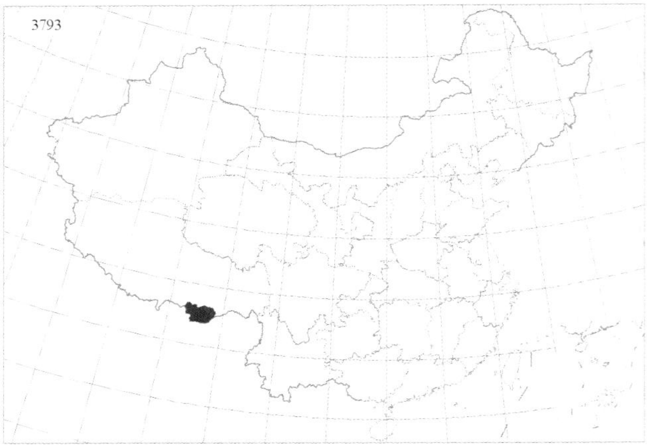

Rosaceae 蔷薇科 *Spiraea* 绣线菊属
Spiraea kwangsiensis T. T. Yu 广西绣线菊
Shrubs, deciduous

18.9 (17.4~22.0)；9.2 (7.1~13.4)；26.9 (25.5~28.7)；18.9 (17.4~22.0)；167.2 (148.9~204.3)；0.0 (0.0~0.0)；980 (886~1176) ❙ 1514 (1182~1776)；694 (637~832)；132 (43~189) ❙ 55.0 (16.9~85.5)；980 (886~1176)；257.0 (203.9~354.5)

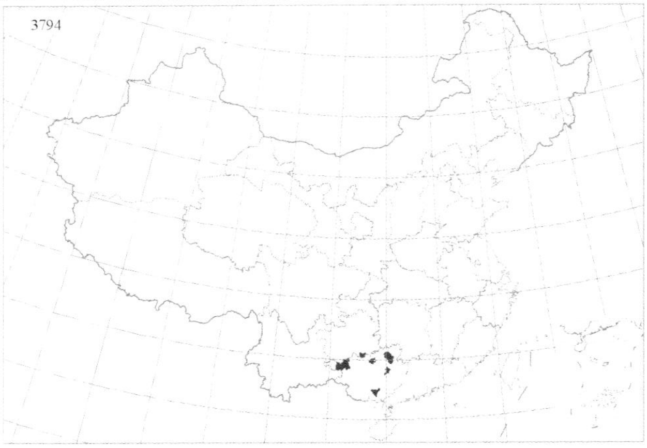

Rosaceae 蔷薇科 *Spiraea* 绣线菊属
Spiraea kweichowensis T. T. Yu et L. T. Lu 贵州绣线菊
Shrubs, deciduous

15.0；4.0；25.7；15.1；122.0；1.1；806 ❙ 1247；484；89 ❙ 54.7；806；189.9

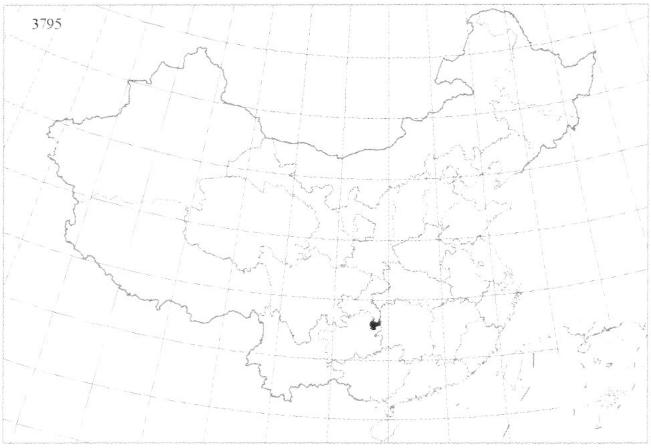

Rosaceae 蔷薇科 *Spiraea* 绣线菊属
Spiraea laeta Rehder 华西绣线菊
Shrubs, deciduous

12.0 (2.2~16.1)；2.3 (-6.6~8.6)；20.8 (9.9~25.8)；12.2 (3.8~16.1)；92.3 (15.5~133.5)；7.5 (0.0~49.1)；689 (413~847) ❙ 996 (669~1265)；514 (317~688)；40 (14~83) ❙ 46.9 (0.0~125.9)；687 (413~845)；231.6 (133.8~357.7)

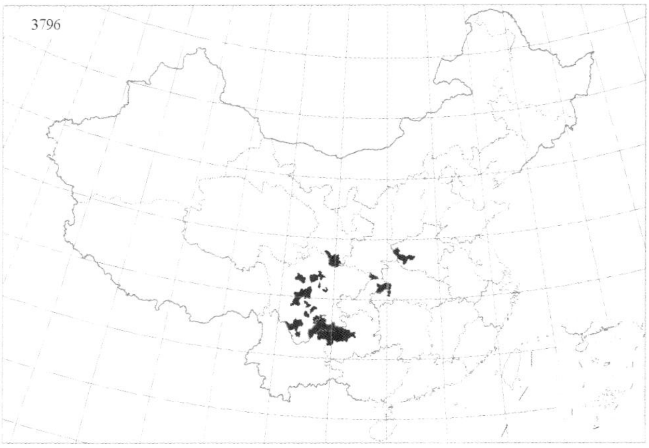

Rosaceae 蔷薇科 *Spiraea* 绣线菊属
Spiraea laeta var. subpubescens Rehder 毛叶华西绣线菊
Shrubs, deciduous

11.0 (10.2~13.0)；-0.7 (-2.9~1.9)；21.9 (19.8~23.8)；11.1 (10.3~13.1)；85.8 (75.8~102.9)；13.6 (6.1~20.5)；669 (634~730) ❙ 823 (677~1202)；385 (336~500)；37 (14~83) ❙ 21.5 (2.4~64.6)；669 (634~730)；256.0 (220.9~279.5)

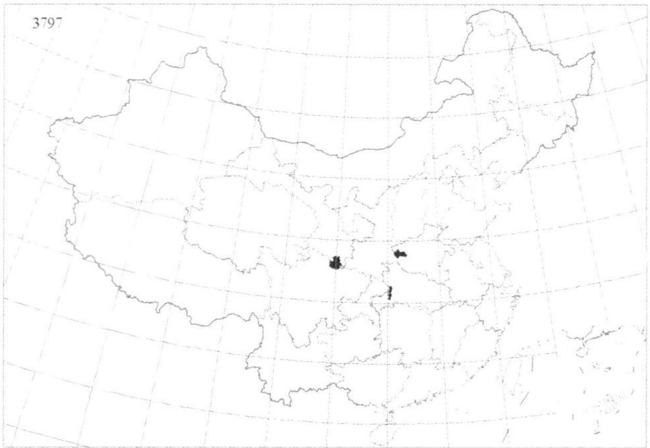

Rosaceae 蔷薇科 *Spiraea* 绣线菊属
Spiraea laeta var. tenuis Rehder 细叶华西绣线菊
Shrubs, deciduous

5.1 (2.2~11.1)；-4.1 (-6.6~1.8)；13.3 (9.9~19.4)；6.1 (3.8~11.1)；35.7 (15.5~79.8)；34.2 (6.9~49.1)；487 (413~648) ❙ 954 (791~1076)；518 (432~584)；24 (17~30) ❙ 99.3 (66.0~125.9)；487 (413~648)；264.1 (200.8~357.7)

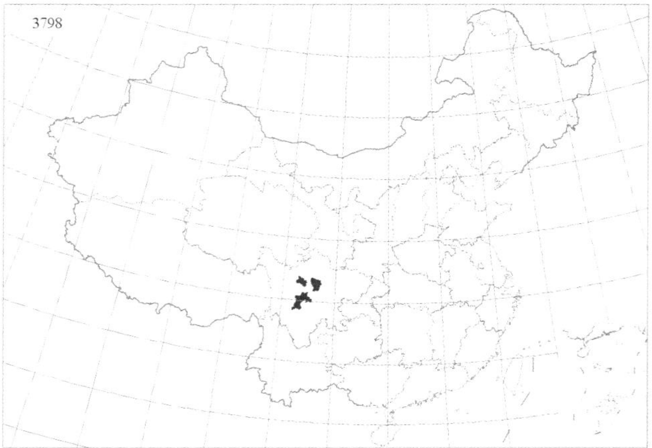

Rosaceae 蔷薇科 *Spiraea* 绣线菊属
Spiraea lichiangensis W. W. Sm. 丽江绣线菊
Shrubs, deciduous

10.8; 3.9; 16.6; 10.9; 71.8; 1.3; 633 ❙ 968; 594; 24 ❙ 52.8; 633; 319.8

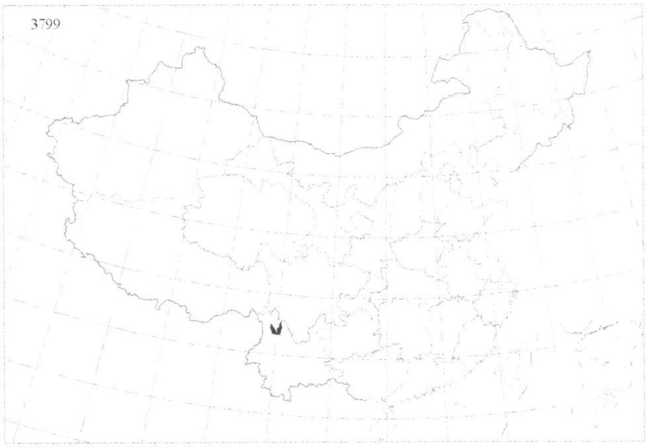

Rosaceae 蔷薇科 *Spiraea* 绣线菊属
Spiraea lobulata T. T. Yu et L. T. Lu 裂叶绣线菊
Shrubs, deciduous

2.6; -6.3; 10.6; 4.2; 19.5; 47.5; 417 ❙ 555; 333; 11 ❙ 33.3; 417; 224.7

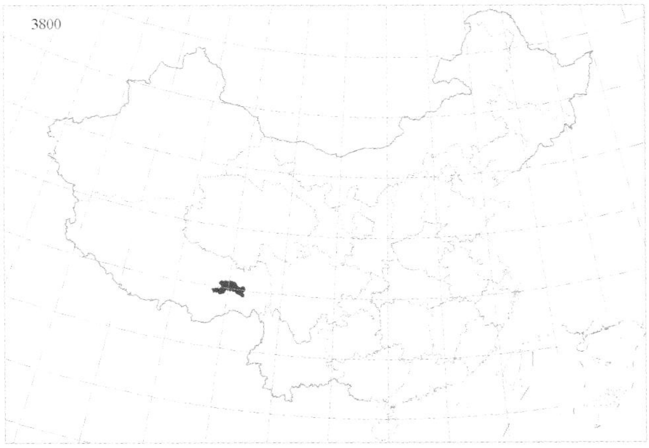

Rosaceae 蔷薇科 *Spiraea* 绣线菊属
Spiraea longigemmis Maxim. 长芽绣线菊
Shrubs, deciduous

10.1 (-1.3~16.1); -2.0 (-13.5~5.3); 21.3 (8.4~28.3); 10.6 (2.8~16.1); 82.3 (9.4~134.7); 20.7 (0.0~86.3); 663 (359~887) ❙ 755 (341~1489); 378 (190~616); 31 (5~196) ❙ 17.8 (-40.9~128.6); 617 (341~887); 226.6 (61.2~352.3)

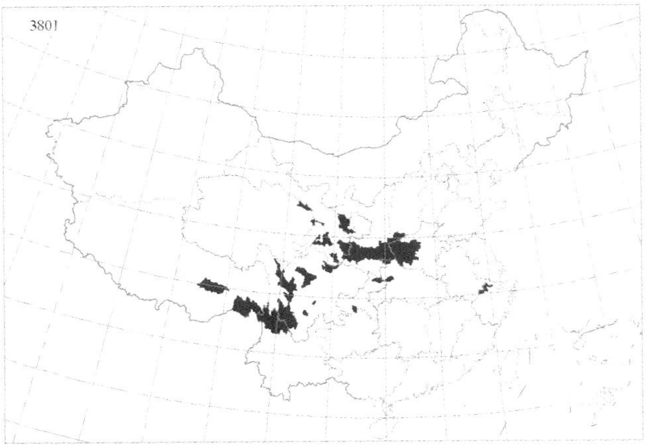

Rosaceae 蔷薇科 *Spiraea* 绣线菊属
Spiraea martini H. Lév. 毛枝绣线菊
Shrubs, deciduous

16.1 (6.1~21.5); 8.5 (-1.5~14.2); 22.0 (12.7~28.7); 16.2 (6.3~21.6); 134.7 (35.2~199.1); 0.5 (0.0~21.8); 815 (496~1131) ❙ 1122 (797~1776); 613 (411~860); 52 (12~189) ❙ 38.0 (-14.2~85.5); 801 (496~1131); 294.2 (160.1~551.6)

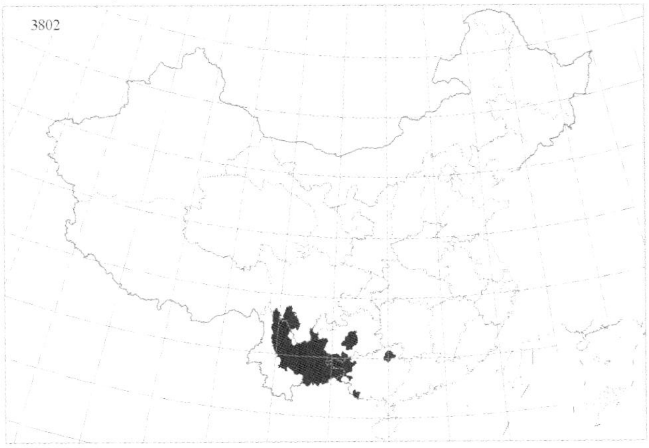

Rosaceae 蔷薇科 *Spiraea* 绣线菊属
Spiraea martini var. tomentosa T. T. Yu 绒毛毛枝绣线菊
Shrubs, deciduous

18.2; 11.8; 22.5; 18.2; 158.4; 0.0; 873 ❙ 1059; 604; 44 ❙ 21.3; 847; 294.5

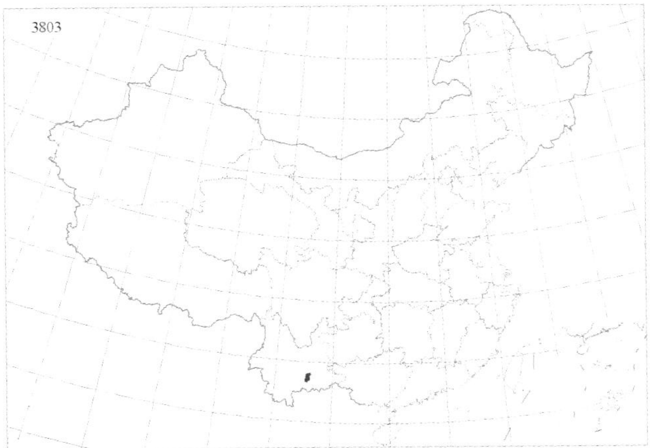

Rosaceae 蔷薇科 *Spiraea* 绣线菊属
Spiraea media Schmidt 欧亚绣线菊
Shrubs, deciduous

7.3 (-5.8~15.1); -9.7 (-29.8~1.8); 22.4 (9.6~28.0); 10.1 (2.5~15.2); 81.3 (9.7~128.0); 53.5 (5.6~157.4); 659 (329~880) ❙ 532 (71~942); 296 (24~601); 25 (3~53) ❙ -19.9 (-78.5~56.9); 508 (71~817); 173.3 (15.7~329.9)

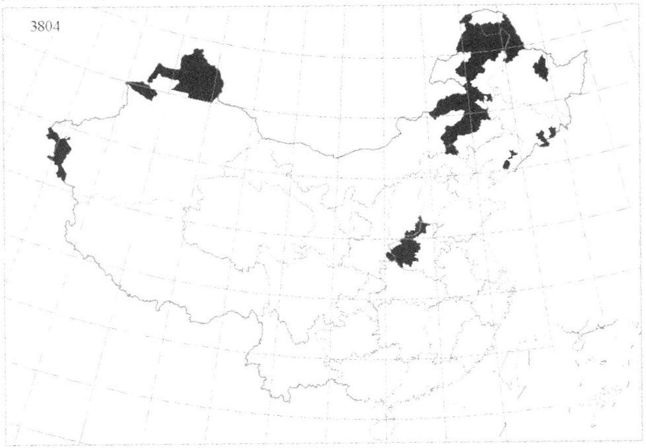

Rosaceae 蔷薇科 *Spiraea* 绣线菊属

Spiraea miyabei var. glabrata Rehder 无毛长蕊绣线菊

Shrubs, deciduous

13.6 (9.3~15.3); 2.0 (-1.7~5.3); 24.7 (20.1~26.9); 13.7 (9.5~15.4); 109.9 (69.8~125.7); 5.8 (0.0~18.0); 763 (603~832) ❙ 1404 (964~1689); 574 (421~816); 127 (49~212) ❙ 83.9 (31.7~105.7); 763 (603~832); 265.6 (213.4~319.9)

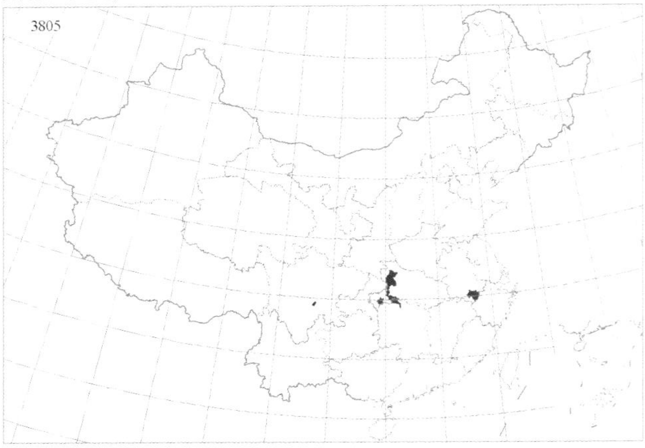

Rosaceae 蔷薇科 *Spiraea* 绣线菊属

Spiraea miyabei var. pilosula Rehder 毛叶长蕊绣线菊

Shrubs, deciduous

13.7; 2.6; 24.4; 13.8; 109.1; 4.0; 758 ❙ 1148; 475; 68 ❙ 51.4; 758; 258.0

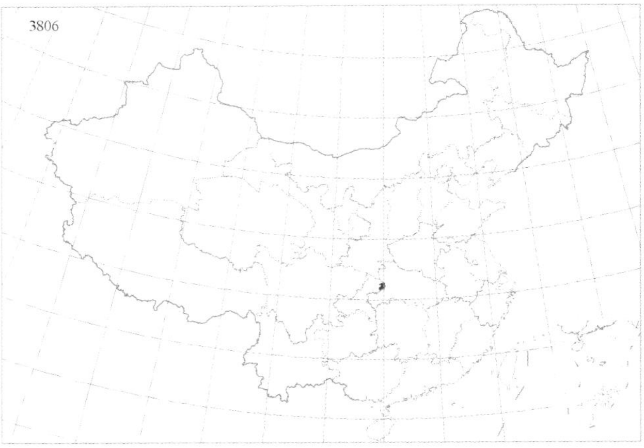

Rosaceae 蔷薇科 *Spiraea* 绣线菊属

Spiraea miyabei var. tenuifolia Rehder 细叶长蕊绣线菊

Shrubs, deciduous

14.9; 5.3; 23.6; 15.0; 119.6; 0.0; 782 ❙ 1410; 816; 49 ❙ 80.4; 782; 214.5

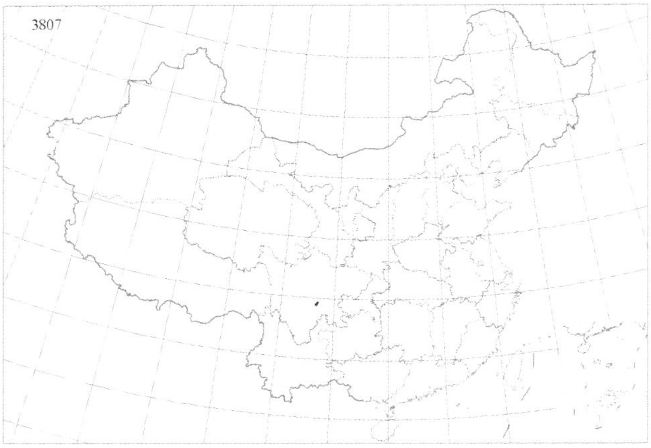

Rosaceae 蔷薇科 *Spiraea* 绣线菊属

Spiraea mollifolia Rehder 毛叶绣线菊

Shrubs, deciduous

7.3 (-2.2~16.8); -2.4 (-12.8~6.6); 15.9 (6.1~26.3); 8.5 (1.8~16.9); 59.8 (1.9~142.7); 32.1 (0.0~91.3); 565 (329~869) ❙ 814 (250~1431); 428 (193~817); 35 (1~129) ❙ 40.7 (-37.6~128.6); 550 (250~869); 218.2 (61.6~363.7)

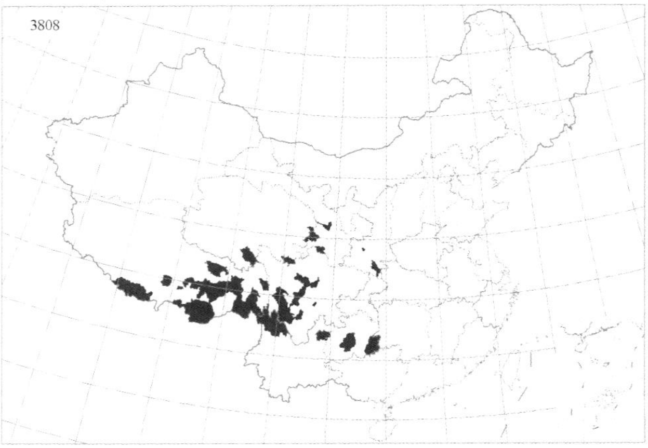

Rosaceae 蔷薇科 *Spiraea* 绣线菊属

Spiraea mollifolia var. glabrata T. T. Yu et L. T. Lu 光秃绣线菊

Shrubs, deciduous

1.7 (-2.2~12.2); -7.0 (-12.8~4.3); 9.6 (6.1~18.5); 4.1 (1.8~12.2); 19.8 (1.9~87.4); 58.7 (0.7~91.3); 406 (329~672) ❙ 587 (311~1431); 360 (193~817); 18 (9~33) ❙ 34.4 (-6.3~113.0); 401 (311~672); 167.6 (61.6~363.7)

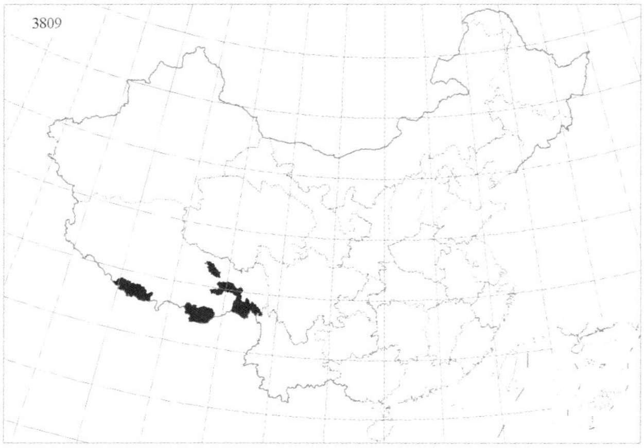

Rosaceae 蔷薇科 *Spiraea* 绣线菊属

Spiraea mongolica Maxim. 蒙古绣线菊

Shrubs, deciduous

8.1 (-5.5~15.1); -6.4 (-17.1~5.4); 21.1 (5.8~28.0); 9.6 (1.3~15.2); 75.6 (1.2~128.0); 37.5 (0.0~127.1); 634 (299~880) ❙ 533 (63~1736); 303 (45~1007); 17 (1~88) ❙ -13.4 (-87.1~145.8); 495 (63~817); 169.9 (7.5~447.4)

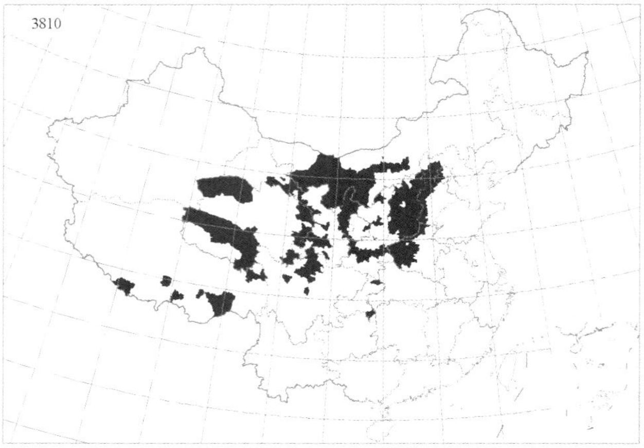

Rosaceae 蔷薇科 *Spiraea* 绣线菊属
Spiraea mongolica var. tomentulosa T. T. Yu 毛枝蒙古绣线菊
Shrubs, deciduous

6.1 (-1.2~8.8); -9.0 (-11.8~-5.4); 19.4 (8.5~24.0); 8.1 (2.6~10.3); 62.3 (8.1~84.8); 48.9 (35.3~82.3); 576 (347~667) ❙ 307 (128~662); 180 (81~379); 6 (2~16) ❙ -37.7 (-80.8~82.4); 265 (128~467); 119.3 (13.6~313.2)

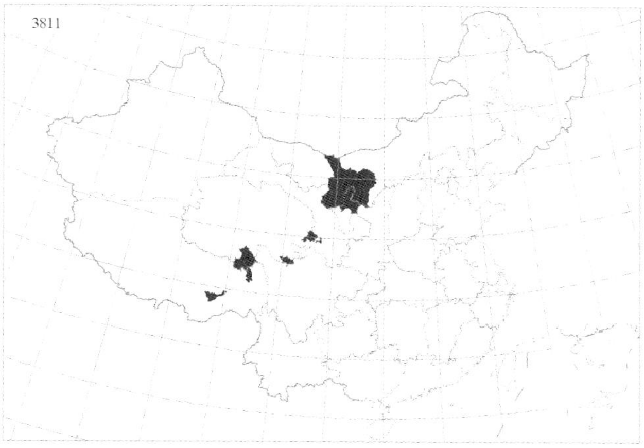

Rosaceae 蔷薇科 *Spiraea* 绣线菊属
Spiraea morrisonicola Hayata 新高山绣线菊
Subshrubs, deciduous

17.0; 11.5; 21.5; 17.0; 144.3; 0.0; 813 ❙ 2770; 1073; 323 ❙ 241.0; 813; 277.5

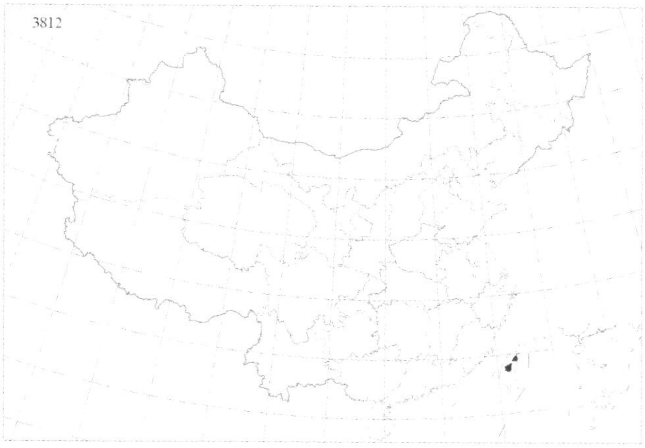

Rosaceae 蔷薇科 *Spiraea* 绣线菊属
Spiraea muliensis T. T. Yu et L. T. Lu 木里绣线菊
Shrubs, deciduous

6.6; -1.3; 13.3; 6.8; 39.6; 19.7; 514 ❙ 862; 548; 12 ❙ 67.7; 514; 311.5

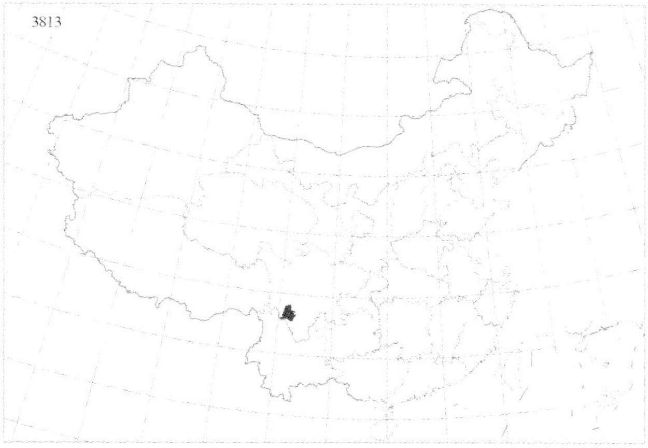

Rosaceae 蔷薇科 *Spiraea* 绣线菊属
Spiraea myrtilloides Rehder 细枝绣线菊
Shrubs, deciduous

7.2 (-3.3~16.1); -3.6 (-15.3~8.2); 17.0 (7.2~27.8); 8.4 (2.0~16.1); 59.2 (3.7~134.4); 32.3 (0.0~103.2); 569 (339~878) ❙ 773 (317~1444); 400 (196~816); 24 (5~169) ❙ 38.7 (-29.0~128.6); 564 (317~878); 269.0 (138.5~384.1)

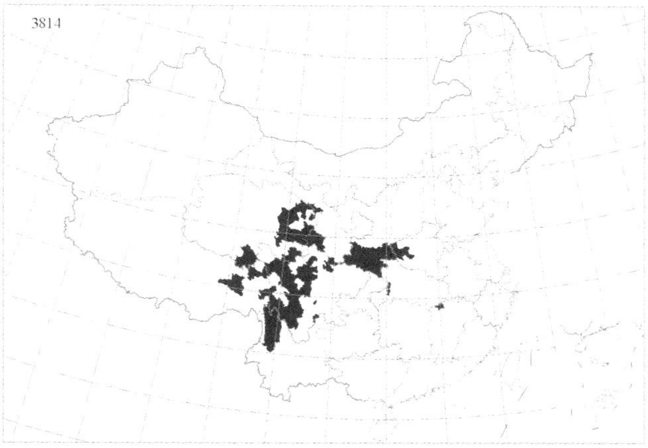

Rosaceae 蔷薇科 *Spiraea* 绣线菊属
Spiraea ningshiaensis T. T. Yu et L. T. Lu 宁夏绣线菊
Shrubs, deciduous

8.0 (7.5~8.7); -9.2 (-9.8~-8.1); 23.0 (22.6~23.8); 9.8 (9.5~10.3); 80.5 (77.7~83.8); 44.5 (39.2~47.7); 647 (633~663) ❙ 189 (132~205); 112 (81~120); 4 (2~5) ❙ -70.8 (-80.1~-68.0); 189 (132~205); 69.8 (13.6~116.9)

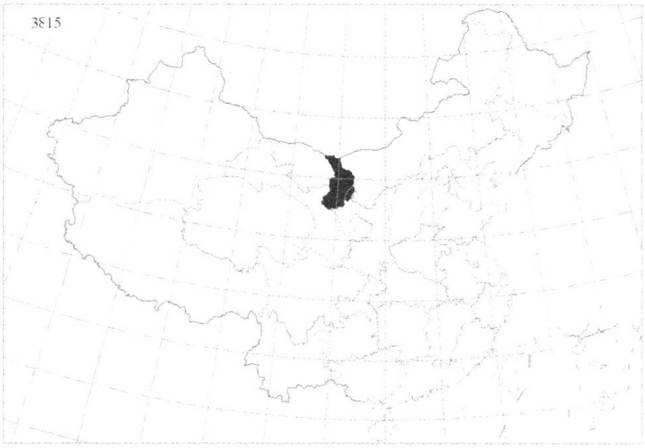

Rosaceae 蔷薇科 *Spiraea* 绣线菊属
Spiraea nishimurae Kitag. 金州绣线菊
Shrubs, deciduous

8.1 (6.0~10.5); -7.4 (-10.1~-4.9); 21.8 (20.0~23.7); 9.6 (8.2~11.3); 75.7 (62.8~93.2); 38.0 (26.8~50.4); 627 (570~705) ❙ 558 (485~718); 333 (299~446); 18 (13~29) ❙ -11.0 (-22.3~9.1); 550 (485~658); 180.4 (127.2~239.1)

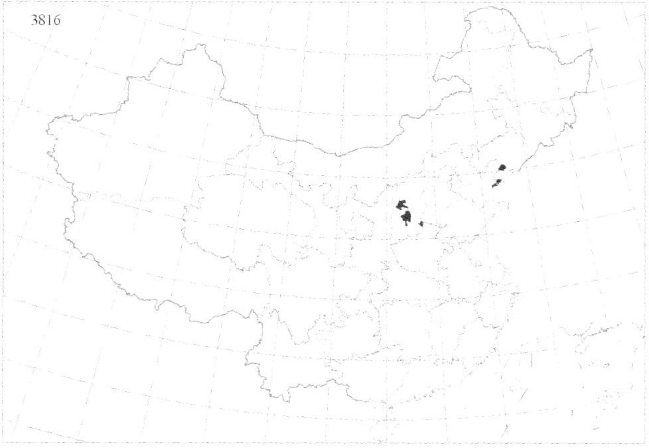

Rosaceae 蔷薇科 *Spiraea* 绣线菊属
Spiraea ovalis Rehder 广椭绣线菊
Shrubs, deciduous

10.2 (0.6~15.2); -1.8 (-9.4~4.3); 21.4 (9.0~27.8); 10.7 (3.3~15.2); 84.1 (11.9~128.0); 20.6 (0.7~65.4); 672 (389~865) ▎787 (512~1280); 387 (257~644); 34 (10~94) ▎23.6 (-35.3~128.6); 627 (389~817); 217.4 (121.6~341.6)

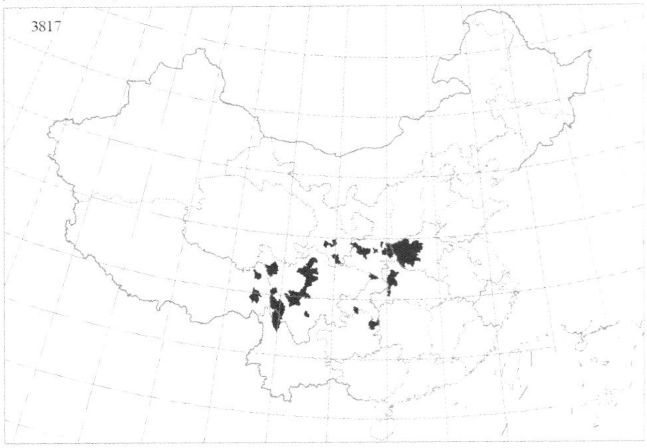

Rosaceae 蔷薇科 *Spiraea* 绣线菊属
Spiraea papillosa Rehder 乳突绣线菊
Shrubs, deciduous

7.7; -0.1; 14.3; 8.3; 55.7; 23.0; 552 ▎928; 558; 22 ▎76.6; 545; 285.6

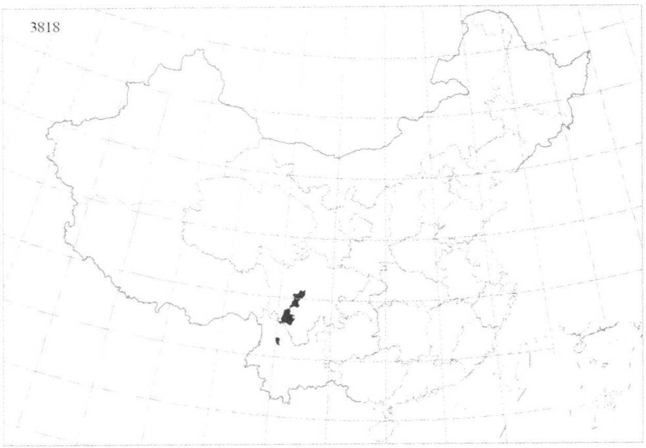

Rosaceae 蔷薇科 *Spiraea* 绣线菊属
Spiraea papillosa var. yunnanensis T. T. Yu 云南乳突绣线菊
Shrubs, deciduous

10.4; 3.1; 16.5; 10.6; 75.8; 9.9; 621 ▎926; 564; 23 ▎51.9; 611; 304.2

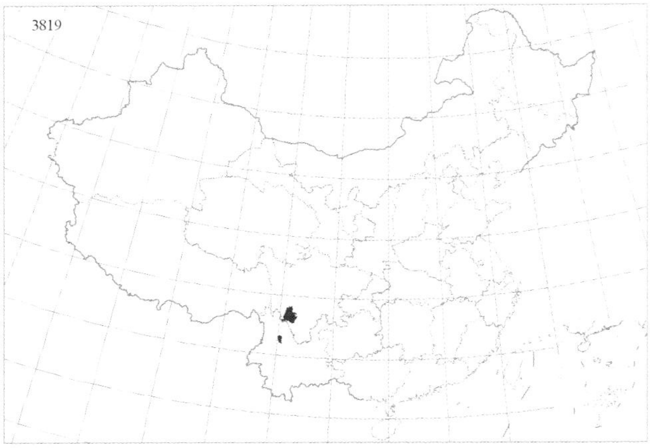

Rosaceae 蔷薇科 *Spiraea* 绣线菊属
Spiraea prostrata Maxim. 平卧绣线菊
Dwarf shrubs, deciduous

12.1 (9.3~14.5); -0.4 (-4.0~1.8); 23.9 (19.8~26.4); 12.2 (9.5~14.6); 98.2 (69.8~120.7); 12.6 (5.7~23.0); 722 (603~821) ▎800 (585~1214); 362 (292~516); 34 (15~71) ▎11.9 (-18.5~101.4); 687 (585~821); 230.5 (174.3~279.5)

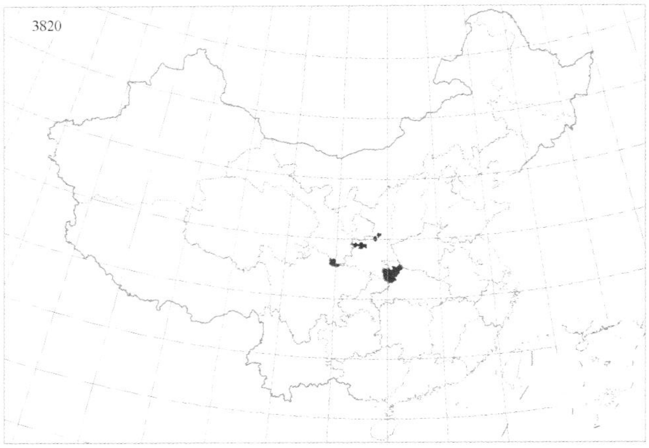

Rosaceae 蔷薇科 *Spiraea* 绣线菊属
Spiraea prunifolia Siebold et Zucc. 李叶绣线菊
Shrubs, deciduous

15.3 (2.3~22.1); 2.8 (-7.4~15.3); 27.1 (10.8~29.9); 15.4 (4.3~22.7); 129.2 (21.0~212.7); 5.0 (0.0~53.3); 859 (415~1216) ▎1149 (306~2680); 486 (234~1186); 120 (1~341) ▎33.1 (-36.2~246.5); 831 (306~1216); 193.4 (110.1~319.9)

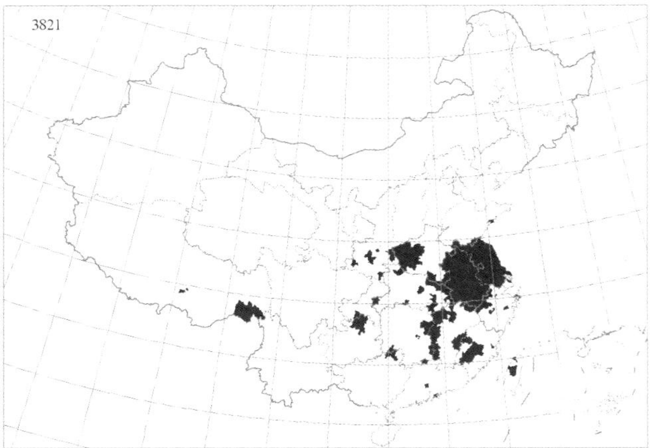

Rosaceae 蔷薇科 *Spiraea* 绣线菊属
Spiraea prunifolia var. simpliciflora (Nakai) Nakai 单瓣李叶绣线菊
Shrubs, deciduous

16.9 (15.1~18.5); 4.8 (2.1~7.8); 28.4 (25.9~29.9); 17.0 (15.1~18.5); 144.0 (125.8~162.3); 0.5 (0.0~4.6); 919 (818~986) ▎1543 (1039~1922); 516 (381~685); 198 (107~259) ▎68.3 (7.5~129.9); 901 (818~977); 204.1 (134.6~279.5)

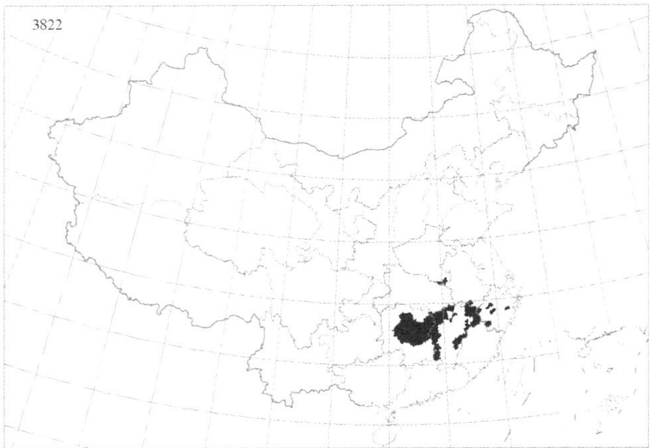

Rosaceae 蔷薇科 *Spiraea* 绣线菊属
Spiraea pubescens Turcz. 土庄绣线菊
Shrubs, deciduous

10.1 (-5.3~17.3); -5.8 (-29.8~6.6); 24.2 (10.7~28.8); 11.4 (3.8~17.3); 94.4 (16.2~147.7); 33.0 (0.0~157.4); 718 (411~932) | 584 (195~1621); 355 (133~691); 21 (4~195) | -17.9 (-67.0~107.2); 557 (195~917); 171.7 (43.7~345.5)

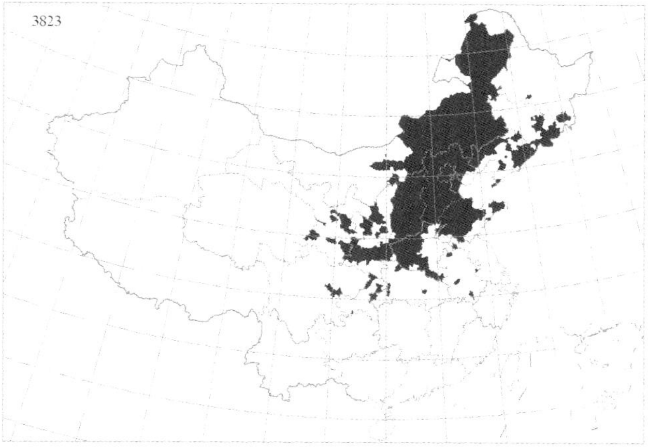

Rosaceae 蔷薇科 *Spiraea* 绣线菊属
Spiraea pubescens var. lasiocarpa Nakai 毛果土庄绣线菊
Shrubs, deciduous

11.0 (7.4~13.4); -1.6 (-4.2~1.2); 23.0 (18.7~25.6); 11.2 (8.2~13.4); 89.2 (58.8~110.4); 16.6 (7.6~29.4); 684 (561~773) | 774 (689~856); 357 (303~398); 29 (18~44) | 13.9 (-2.2~52.6); 681 (561~773); 262.0 (189.3~334.4)

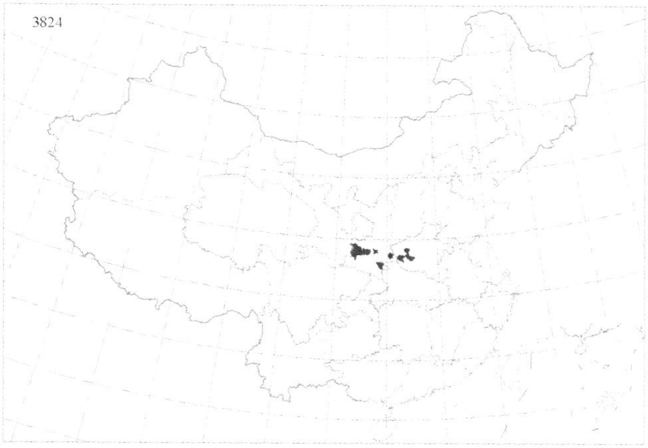

Rosaceae 蔷薇科 *Spiraea* 绣线菊属
Spiraea purpurea Hand.-Mazz. 紫花绣线菊
Shrubs, deciduous

8.9 (1.8~17.4); 0.8 (-7.1~11.0); 15.7 (9.9~21.4); 9.5 (3.8~17.4); 67.4 (15.7~149.3); 20.3 (0.0~53.3); 589 (404~834) | 1095 (583~1736); 618 (346~1007); 33 (13~51) | 82.3 (44.2~145.8); 589 (404~834); 316.0 (169.5~447.4)

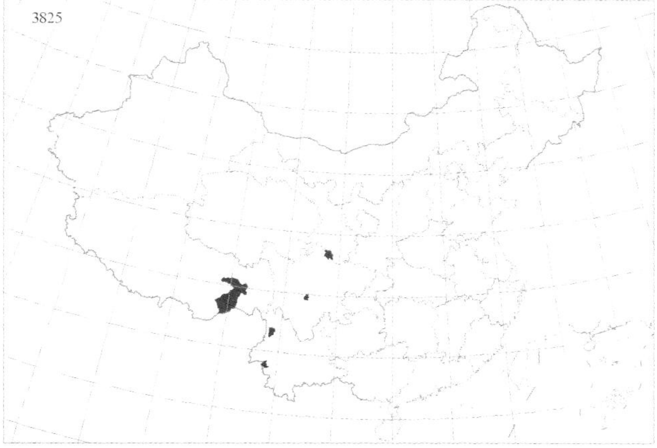

Rosaceae 蔷薇科 *Spiraea* 绣线菊属
Spiraea rosthornii E. Pritz. ex Diels 南川绣线菊
Shrubs, deciduous

10.6 (-4.1~16.6); -1.4 (-17.3~7.8); 21.7 (7.8~28.8); 11.2 (2.1~16.7); 87.4 (5.2~140.6); 19.7 (0.0~114.2); 684 (340~920) | 894 (341~1770); 420 (190~816); 52 (5~223) | 32.1 (-40.9~128.6); 655 (340~914); 233.0 (61.2~382.6)

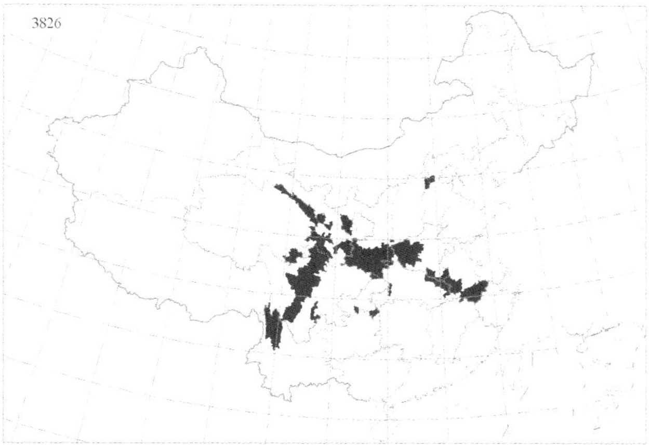

Rosaceae 蔷薇科 *Spiraea* 绣线菊属
Spiraea salicifolia L. 柳叶绣线菊
Shrubs, deciduous

6.3 (-5.8~14.2); -11.5 (-29.8~-0.7); 22.0 (15.3~27.8); 9.3 (4.4~14.4); 74.1 (28.0~124.3); 57.4 (12.6~157.4); 626 (426~859) | 547 (297~1073); 351 (203~691); 16 (5~40) | -11.5 (-44.3~74.1); 519 (297~689); 205.6 (101.9~346.5)

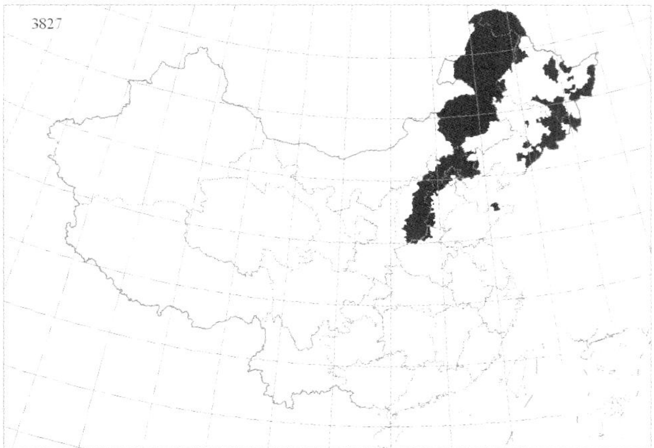

Rosaceae 蔷薇科 *Spiraea* 绣线菊属
Spiraea sargentiana Rehder 茂汶绣线菊
Shrubs, deciduous

12.1 (3.4~16.7); 0.8 (-6.0~7.9); 22.6 (11.7~27.9); 12.3 (4.7~16.8); 97.3 (23.2~141.5); 11.7 (0.0~41.9); 714 (442~896) | 1051 (720~1893); 478 (350~677); 65 (22~227) | 49.5 (-13.5~129.9); 707 (442~896); 234.9 (141.2~290.3)

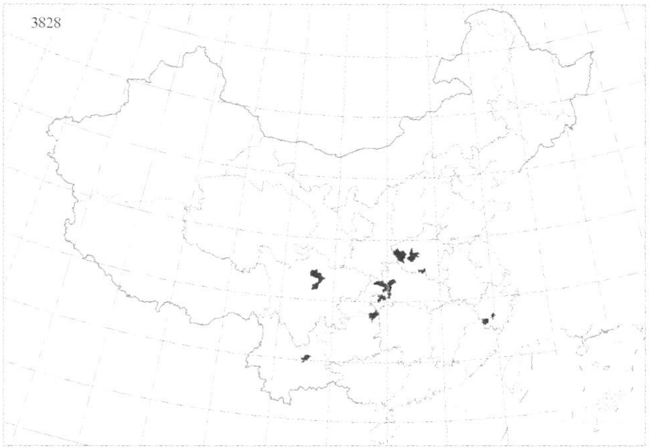

Rosaceae 蔷薇科 *Spiraea* 绣线菊属
Spiraea schneideriana Rehder 川滇绣线菊
Shrubs, deciduous

8.6 (0.0~17.6); -0.6 (-9.4~10.3); 16.9 (8.1~28.0); 9.3 (2.7~17.7); 66.1 (8.0~151.9); 21.9 (0.0~67.9); 595 (358~927) ❙ 973 (320~1997); 492 (225~1007); 49 (1~256) ❙ 61.4 (-23.5~145.8); 591 (320~927); 274.6 (111.0~447.4)

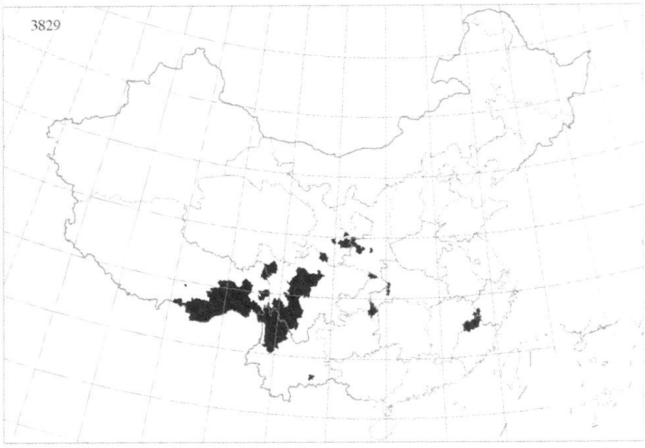

Rosaceae 蔷薇科 *Spiraea* 绣线菊属
Spiraea schneideriana var. amphidoxa Rehder 无毛川滇绣线菊
Shrubs, deciduous

6.6 (0.6~14.9); -3.0 (-9.4~5.4); 15.2 (9.5~24.6); 7.5 (3.4~15.0); 48.6 (12.7~119.6); 28.5 (0.0~65.4); 536 (394~782) ❙ 860 (535~1736); 478 (306~1007); 22 (9~51) ❙ 64.4 (-20.0~145.8); 531 (394~782); 282.0 (162.7~447.4)

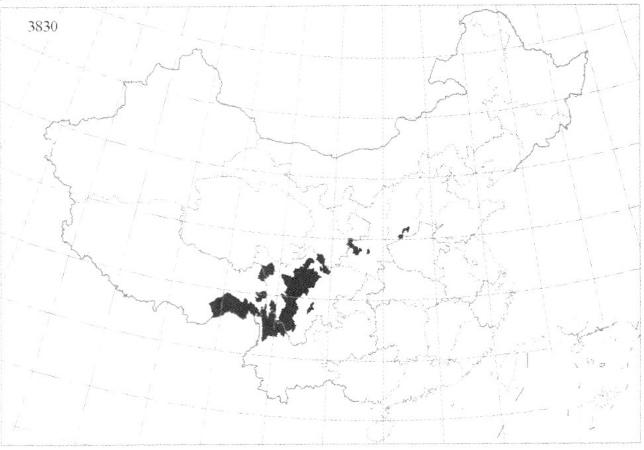

Rosaceae 蔷薇科 *Spiraea* 绣线菊属
Spiraea schochiana Rehder 滇中绣线菊
Shrubs, deciduous

15.6 (13.2~18.9); 8.7 (5.1~12.0); 20.7 (19.0~23.9); 15.6 (13.2~19.0); 127.6 (98.5~167.8); 0.0 (0.0~0.0); 775 (700~929) ❙ 982 (797~1252); 567 (411~686); 38 (23~53) ❙ 27.4 (-14.2~70.6); 741 (683~815); 270.7 (200.4~316.6)

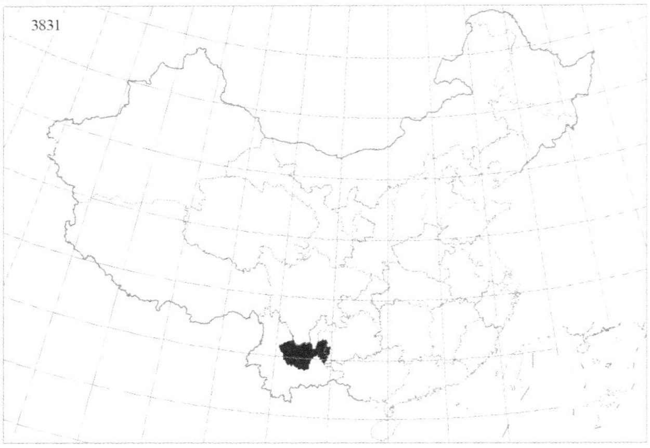

Rosaceae 蔷薇科 *Spiraea* 绣线菊属
Spiraea sericea Turcz. 绢毛绣线菊
Shrubs, deciduous

8.1 (-5.8~16.2); -8.3 (-29.8~5.6); 22.6 (9.7~28.0); 10.2 (3.3~16.3); 81.9 (12.2~135.4); 44.6 (0.0~157.4); 658 (395~880) ❙ 601 (137~1073); 344 (92~691); 21 (3~53) ❙ -7.6 (-77.5~121.7); 557 (137~865); 195.0 (25.4~364.1)

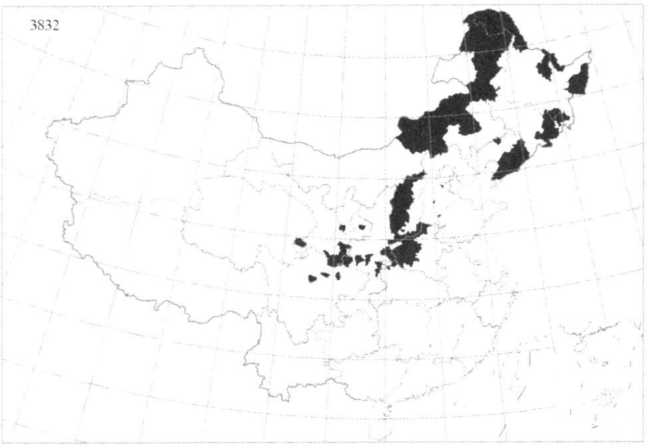

Rosaceae 蔷薇科 *Spiraea* 绣线菊属
Spiraea siccanea (W. W. Sm.) Rehder 干地绣线菊
Shrubs, deciduous

12.5 (9.6~14.5); 5.8 (2.4~7.8); 17.9 (15.7~19.7); 12.5 (9.6~14.5); 91.1 (60.8~113.5); 0.8 (0.0~5.4); 676 (600~730) ❙ 1061 (991~1164); 606 (570~651); 45 (35~58) ❙ 57.6 (36.1~78.3); 674 (600~730); 351.3 (296.9~382.6)

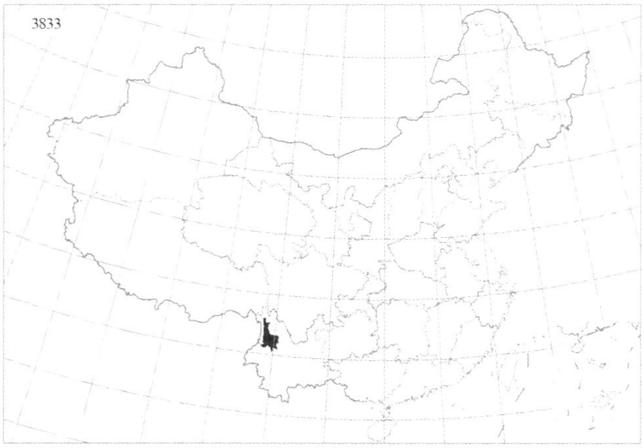

Rosaceae 蔷薇科 *Spiraea* 绣线菊属
Spiraea sublobata Hand.-Mazz. 浅裂绣线菊
Shrubs, deciduous

13.0 (3.8~18.5); 4.6 (-4.3~10.5); 20.3 (10.8~24.7); 13.1 (4.7~18.3); 100.9 (21.6~160.0); 4.2 (0.0~35.4); 711 (444~906) ❙ 1082 (952~1410); 619 (519~816); 31 (16~65) ❙ 55.0 (5.5~121.7); 690 (444~822); 273.8 (189.1~331.6)

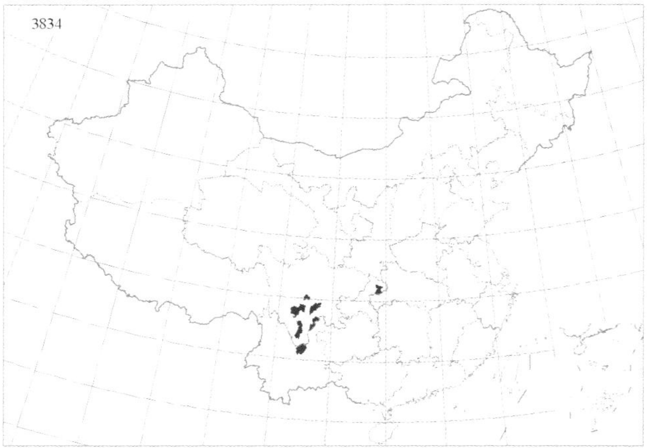

Rosaceae 蔷薇科 *Spiraea* 绣线菊属
Spiraea tarokoensis Hayata 太鲁阁绣线菊
Dwarf shrubs, deciduous

16.4；11.5；20.3；16.5；137.8；0.0；783 | 2499；947；303 | 219.3；783；303.6

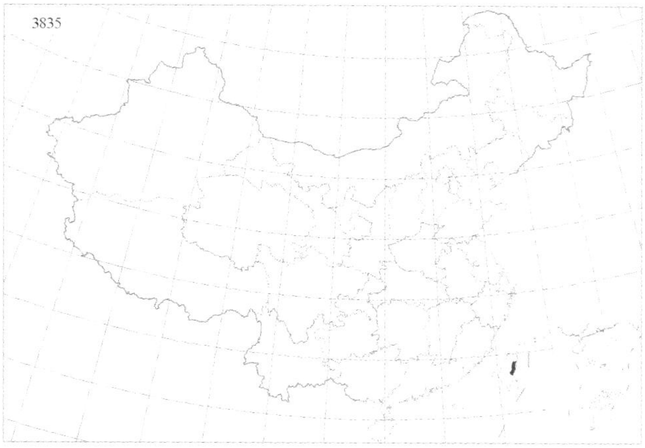

Rosaceae 蔷薇科 *Spiraea* 绣线菊属
Spiraea teniana Rehder 伏毛绣线菊
Shrubs, deciduous

15.0；8.4；20.1；15.0；120.5；0.0；750 | 955；572；29 | 27.4；686；306.0

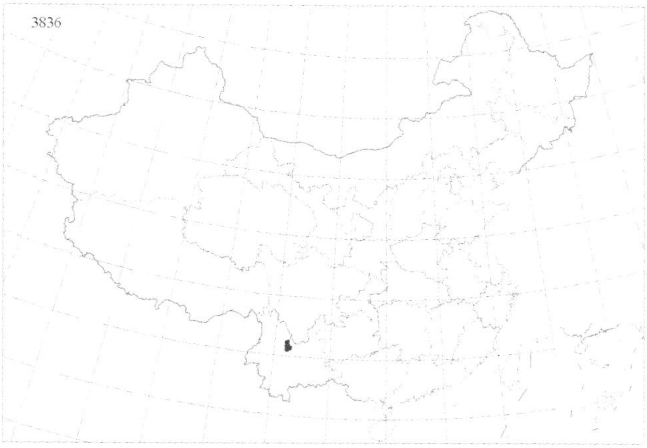

Rosaceae 蔷薇科 *Spiraea* 绣线菊属
Spiraea thunbergii Siebold ex Blume 珍珠绣线菊
Shrubs, deciduous

14.5 (12.3~17.6)；1.0 (-1.2~5.6)；27.2 (24.4~29.0)；14.6 (12.3~17.6)；123.0 (100.7~151.4)；8.7 (0.0~15.0)；839 (728~953) | 873 (607~1516)；386 (273~469)；77 (19~214) | 3.0 (-29.3~59.0)；738 (607~896)；179.3 (134.1~260.3)

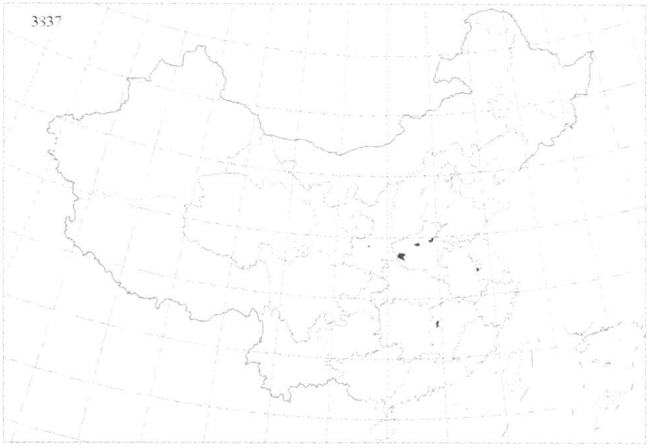

Rosaceae 蔷薇科 *Spiraea* 绣线菊属
Spiraea trichocarpa Nakai 毛果绣线菊
Shrubs, deciduous

7.3 (-4.6~13.5)；-11.0 (-28.8~-1.6)；23.4 (16.6~27.1)；9.9 (5.0~13.7)；81.5 (34.4~117.5)；53.4 (14.6~149.1)；656 (448~820) | 641 (314~1073)；404 (230~691)；21 (5~40) | -1.5 (-40.5~74.1)；555 (314~690)；205.8 (97.2~335.3)

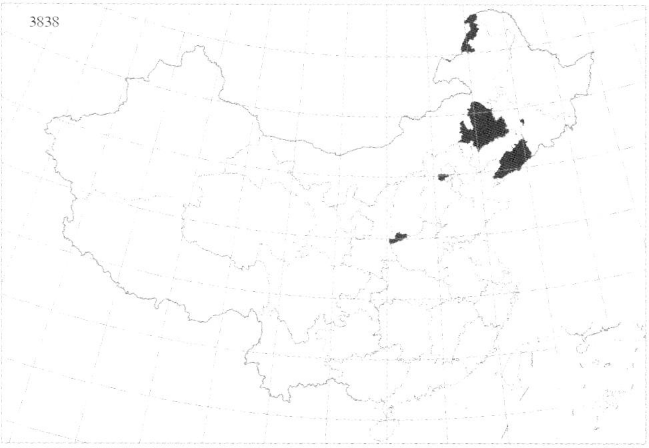

Rosaceae 蔷薇科 *Spiraea* 绣线菊属
Spiraea trilobata L. 三裂绣线菊
Shrubs, deciduous

8.9 (-2.9~17.0)；-7.1 (-26.4~5.1)；23.1 (10.7~28.5)；10.5 (3.8~17.0)；85.2 (16.2~144.4)；38.2 (0.0~134.6)；673 (411~921) | 573 (128~1828)；332 (81~648)；23 (2~253) | -15.5 (-80.8~105.7)；535 (128~921)；169.6 (13.6~346.5)

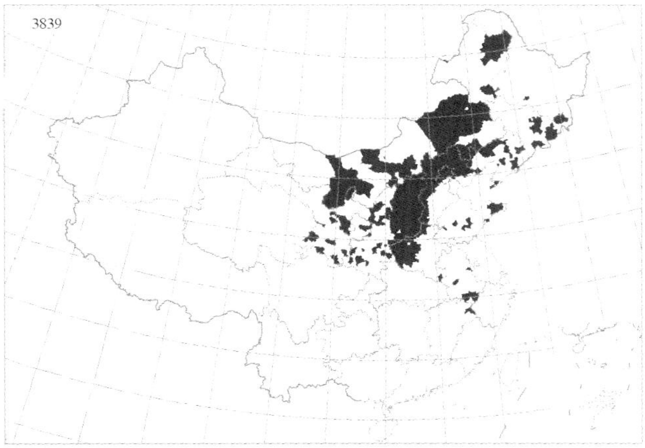

Rosaceae 蔷薇科 *Spiraea* 绣线菊属
Spiraea trilobata var. pubescens T. T. Yu 毛叶三裂绣线菊
Shrubs, deciduous

6.1 (2.9~11.2)；-10.8 (-14.7~-3.9)；21.1 (18.5~24.5)；8.7 (6.9~11.8)；67.7 (49.0~96.6)；53.9 (22.2~74.1)；594 (521~722) | 388 (183~524)；244 (114~295)；12 (4~16) | -34.2 (-71.0~-17.3)；388 (183~524)；119.2 (26.5~185.7)

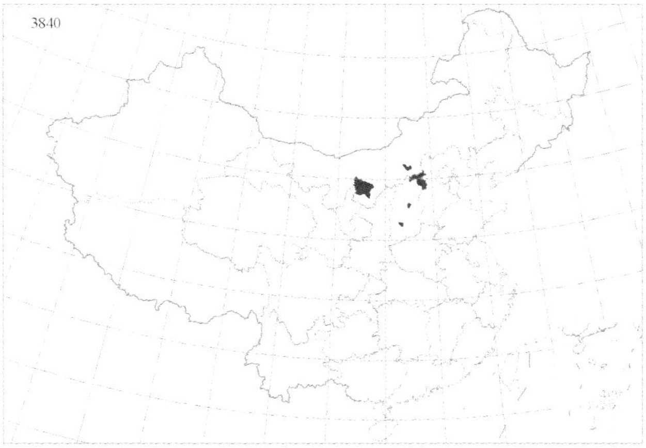

Rosaceae 蔷薇科 *Spiraea* 绣线菊属
Spiraea uratensis Franch. 乌拉绣线菊
Shrubs, deciduous

9.7 (3.0~13.4)； -3.6 (-14.9~1.5)； 22.2 (12.4~25.6)； 10.5 (4.8~13.4)； 82.2 (25.1~110.4)； 25.0 (7.1~74.1)； 656 (441~773) ❙ 686 (187~892)； 327 (127~408)； 23 (5~44) ❙ 4.5 (-67.8~52.6)； 609 (187~773)； 230.8 (35.7~334.4)

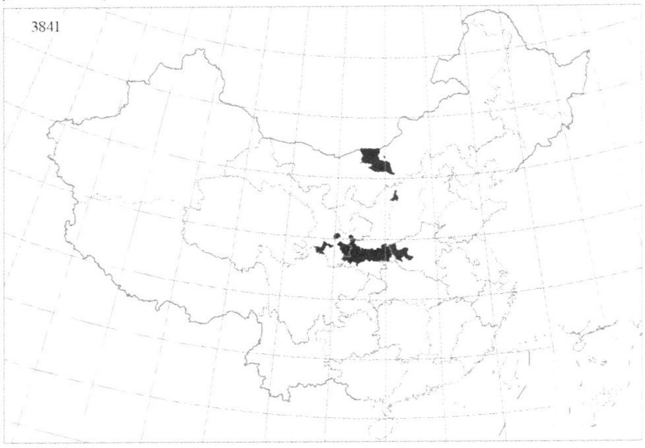

Rosaceae 蔷薇科 *Spiraea* 绣线菊属
Spiraea veitchii Hemsl. 鄂西绣线菊
Shrubs, deciduous

12.7 (2.0~18.9)； 2.7 (-7.2~12.0)； 21.6 (9.9~27.5)； 12.8 (3.8~19.0)； 101.3 (15.4~167.8)； 8.6 (0.0~51.4)； 720 (413~929) ❙ 931 (626~1414)； 486 (308~817)； 39 (12~113) ❙ 33.4 (-22.2~128.6)； 691 (413~863)； 251.4 (136.8~384.1)

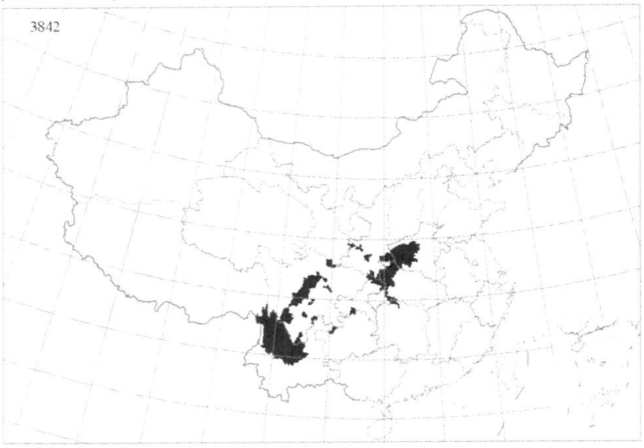

Rosaceae 蔷薇科 *Spiraea* 绣线菊属
Spiraea velutina Franch. 绒毛绣线菊
Shrubs, deciduous

11.1 (-1.8~18.5)； 3.9 (-10.2~10.5)； 17.1 (6.1~24.3)； 11.3 (1.8~18.3)； 81.9 (1.9~160.0)； 8.0 (0.0~82.8)； 646 (329~906) ❙ 1020 (320~1431)； 587 (217~817)； 38 (15~69) ❙ 58.6 (-3.0~113.0)； 639 (320~813)； 318.8 (72.2~407.1)

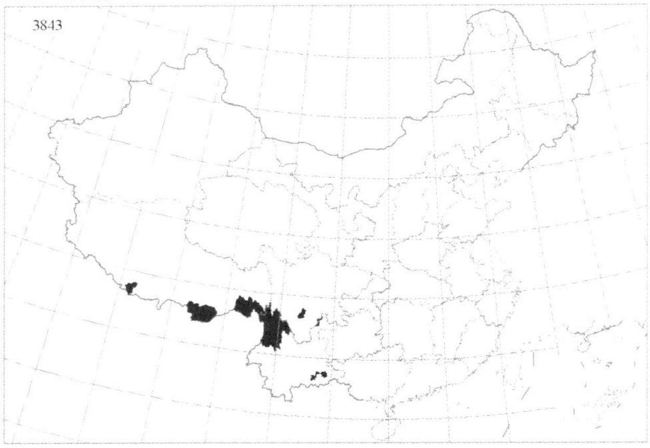

Rosaceae 蔷薇科 *Spiraea* 绣线菊属
Spiraea velutina var. glabrescens T. T. Yu et L. T. Lu 脱毛绣线菊
Shrubs, deciduous

6.4； -1.5； 13.2； 7.6； 47.3； 30.8； 525 ❙ 931； 540； 32 ❙ 63.9； 522； 262.7

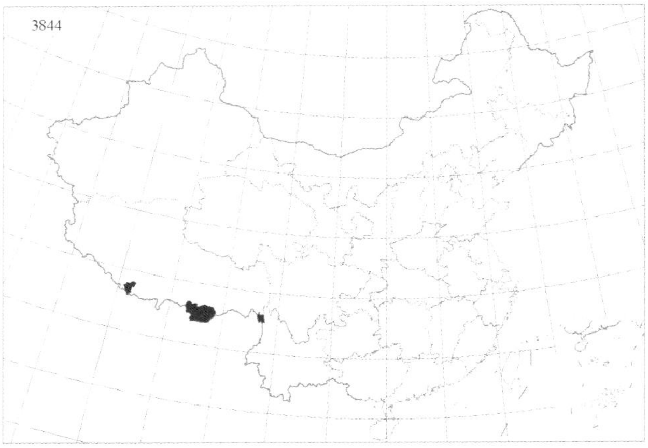

Rosaceae 蔷薇科 *Spiraea* 绣线菊属
Spiraea wilsonii Duthie 陕西绣线菊
Shrubs, deciduous

11.3 (2.0~16.3)； -0.1 (-7.2~8.2)； 21.8 (9.9~28.2)； 11.5 (3.8~16.4)； 90.2 (15.4~136.6)； 14.3 (0.0~51.4)； 691 (413~890) ❙ 861 (551~1492)； 428 (245~613)； 37 (11~109) ❙ 28.8 (-36.2~128.6)； 668 (413~890)； 240.1 (131.4~357.7)

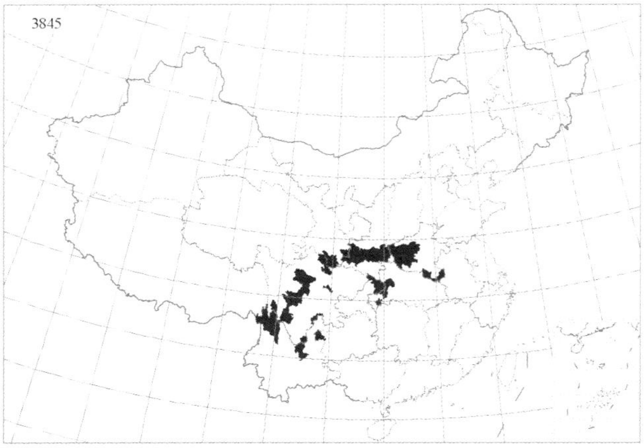

Rosaceae 蔷薇科 *Spiraea* 绣线菊属
Spiraea xizangensis L. T. Lu 西藏绣线菊
Shrubs, deciduous

-1.3 (-3.3~0.8)； -11.9 (-15.2~-9.3)； 8.4 (7.2~9.9)； 2.7 (2.0~3.6)； 8.8 (3.7~15.4)； 83.4 (65.4~103.2)； 355 (335~388) ❙ 406 (250~662)； 261 (198~379)； 7 (1~16) ❙ 14.7 (-27.5~82.4)； 316 (250~363)； 222.0 (131.4~288.8)

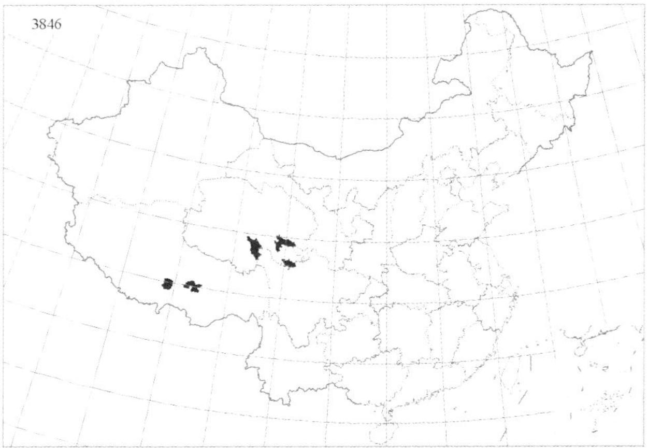

Rosaceae 蔷薇科 *Spiraea* 绣线菊属
Spiraea yunnanensis Franch. 云南绣线菊
Shrubs, deciduous

5.5 (−3.2~14.8); −3.1 (−14.4~4.3); 13.0 (7.3~25.3); 6.6 (2.0~14.8); 38.9 (4.0~119.2); 32.9 (0.7~101.5); 499 (335~795) ❙ 822 (250~1358); 474 (198~696); 25 (1~88) ❙ 62.0 (−27.5~125.9); 489 (250~795); 258.5 (130.1~384.7)

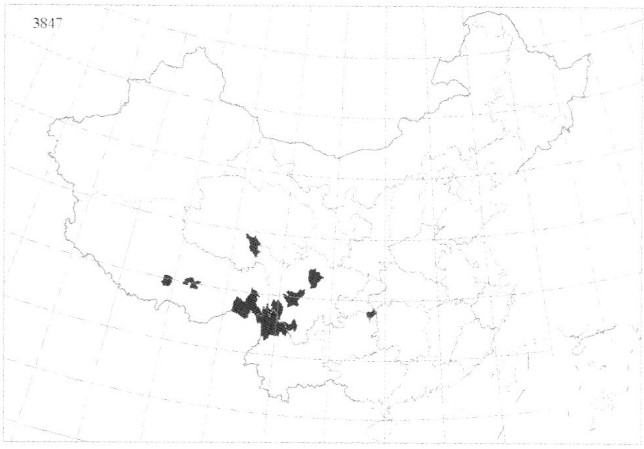

Rosaceae 蔷薇科 *Spiraea* 绣线菊属
Spiraea × vanhouttei (Briot) Carrière 菱叶绣线菊
Shrubs, deciduous

16.8; 5.3; 27.5; 16.8; 142.9; 1.1; 899 ❙ 1581; 540; 203 ❙ 77.0; 880; 231.5

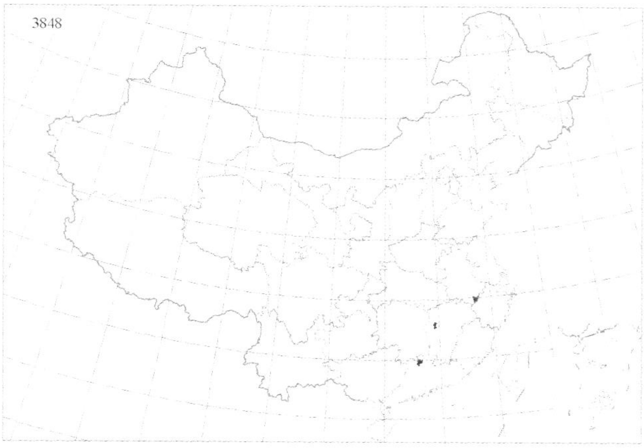

Rosaceae 蔷薇科 *Stephanandra* 野珠兰属
Stephanandra chinensis Hance 野珠兰
Shrubs, deciduous

16.2 (5.2~21.0); 4.4 (−14.5~11.9); 27.4 (20.3~30.1); 16.3 (8.6~21.1); 137.1 (68.0~193.0); 2.3 (0.0~65.0); 882 (595~1114) ❙ 1419 (626~2499); 517 (299~947); 164 (29~303) ❙ 60.7 (−22.2~219.3); 866 (595~1114); 213.3 (134.1~332.5)

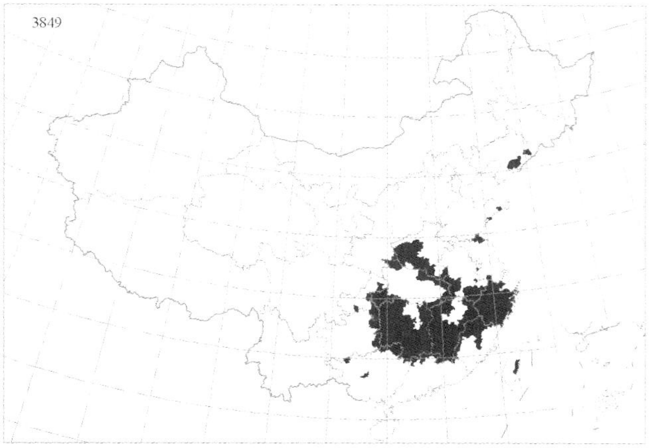

Rosaceae 蔷薇科 *Stephanandra* 野珠兰属
Stephanandra incisa (Thunb.) Zabel 深裂野珠兰
Shrubs, deciduous

12.1 (5.2~16.4); −2.3 (−14.5~11.5); 24.9 (20.3~27.9); 12.6 (8.6~16.5); 105.0 (68.0~137.8); 19.8 (0.0~65.0); 750 (595~876) ❙ 965 (670~2499); 515 (392~947); 71 (27~303) ❙ 28.2 (−13.7~219.3); 732 (595~829); 208.3 (136.4~335.3)

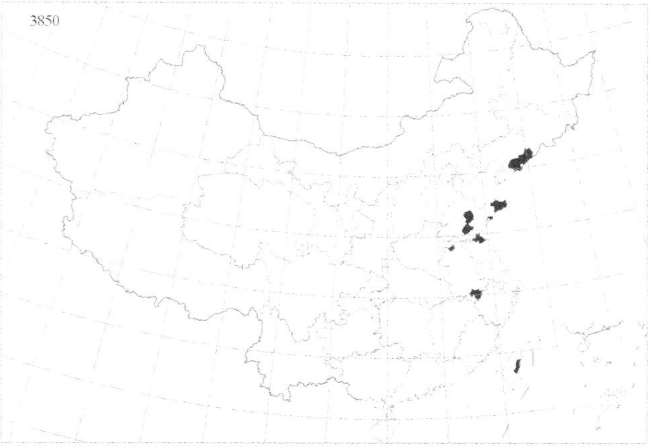

Rosaceae 蔷薇科 *Stranvaesia* 红果树属
Stranvaesia amphidoxa C. K. Schneid. 毛萼红果树
Shrubs or small trees, evergreen

15.7 (10.6~21.3); 5.4 (−0.2~12.3); 24.9 (18.6~29.3); 15.7 (10.6~21.3); 129.3 (79.4~196.1); 0.8 (0.0~12.1); 828 (643~1130) ❙ 1272 (863~1950); 553 (444~728); 96 (25~223) ❙ 53.9 (11.7~147.0); 828 (643~1130); 214.0 (126.1~322.1)

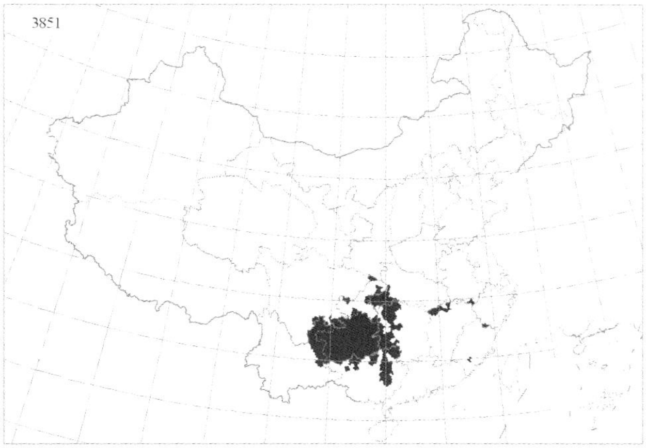

Rosaceae 蔷薇科 *Stranvaesia* 红果树属
Stranvaesia amphidoxa var. amphileia (Hand.-Mazz.) T. T. Yu 湖南红果树
Shrubs or small trees, evergreen

16.1 (14.3~19.2); 5.5 (3.3~10.1); 25.7 (22.2~27.5); 16.1 (14.3~19.2); 134.1 (114.1~170.8); 0.5 (0.0~2.3); 846 (765~1002) ❙ 1360 (1165~1740); 559 (444~720); 115 (51~195) ❙ 61.1 (30.0~101.8); 846 (765~1002); 214.0 (146.0~283.9)

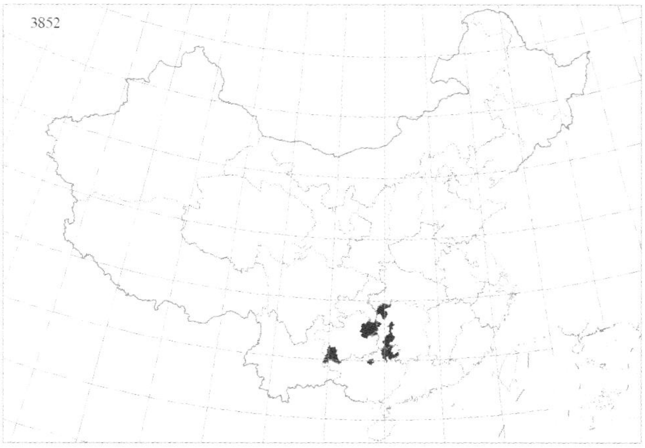

Rosaceae 蔷薇科 *Stranvaesia* 红果树属
Stranvaesia davidiana Decne. 红果树
Shrubs or small trees, evergreen

15.1 (2.2~21.7); 5.0 (-7.0~15.3); 24.3 (9.9~29.3); 15.2 (3.8~21.7); 124.1 (15.5~200.2); 2.4 (0.0~49.1); 810 (413~1118) ❙ 1311 (707~2859); 561 (357~1412); 106 (12~402) ❙ 61.9 (-2.0~246.5); 807 (413~1114); 236.3 (146.0~506.2)

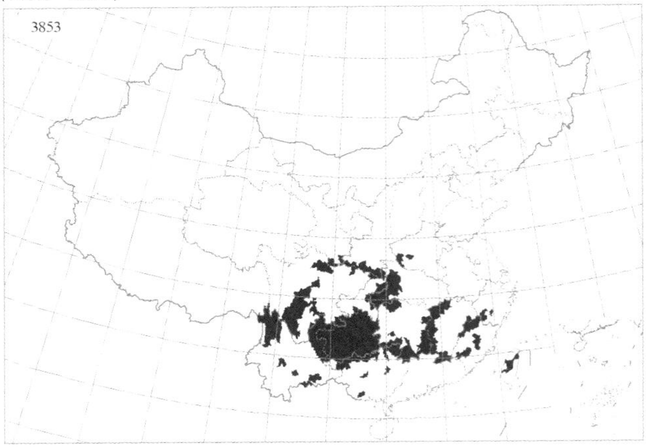

Rosaceae 蔷薇科 *Stranvaesia* 红果树属
Stranvaesia davidiana var. undulata (Decne.) Rehder et E. H. Wilson 波叶红果树
Shrubs or small trees, evergreen

15.7 (2.2~21.3); 4.9 (-6.6~13.2); 25.6 (9.9~29.7); 15.7 (3.8~21.3); 130.8 (15.5~196.1); 2.3 (0.0~49.1); 845 (413~1130) ❙ 1387 (793~1950); 540 (377~1056); 135 (15~259) ❙ 65.0 (-2.0~147.0); 841 (413~1130); 221.2 (126.1~318.5)

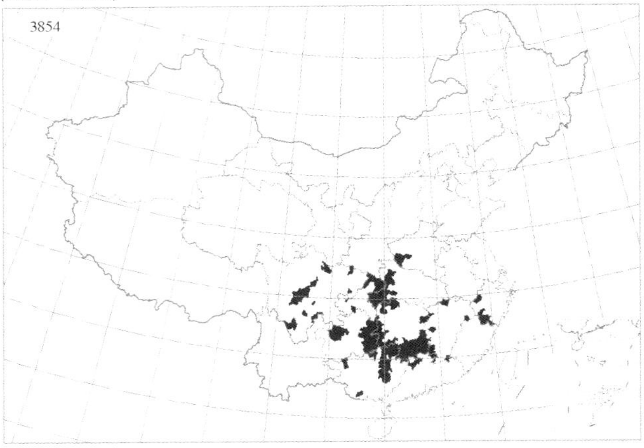

Rosaceae 蔷薇科 *Stranvaesia* 红果树属
Stranvaesia nussia (Buch.-Ham. ex D. Don) Decne. 印缅红果树
Trees, evergreen

5.3; -2.4; 12.2; 5.7; 30.5; 26.2; 479 ❙ 961; 558; 31 ❙ 100.7; 479; 234.2

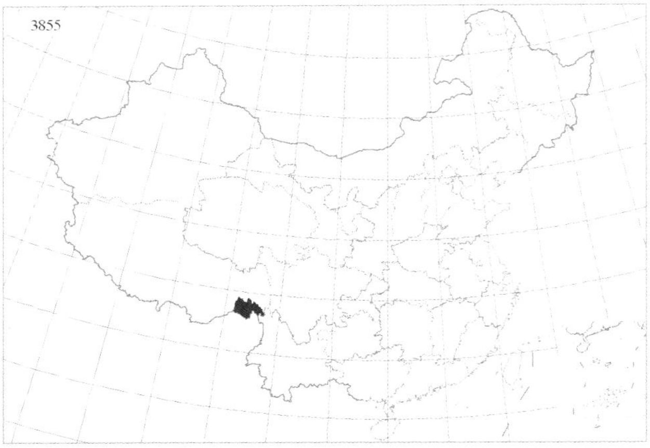

Rosaceae 蔷薇科 *Stranvaesia* 红果树属
Stranvaesia oblanceolata (Rehder et E. H. Wilson) Stapf 滇南红果树
Shrubs, evergreen

18.3 (11.1~21.3); 10.7 (1.8~15.1); 23.7 (19.1~28.7); 18.3 (11.1~21.3); 159.9 (79.8~196.0); 0.1 (0.0~6.9); 906 (648~1123) ❙ 1356 (958~1751); 722 (531~996); 68 (30~166) ❙ 50.6 (5.8~112.2); 900 (648~1121); 375.0 (190.9~659.0)

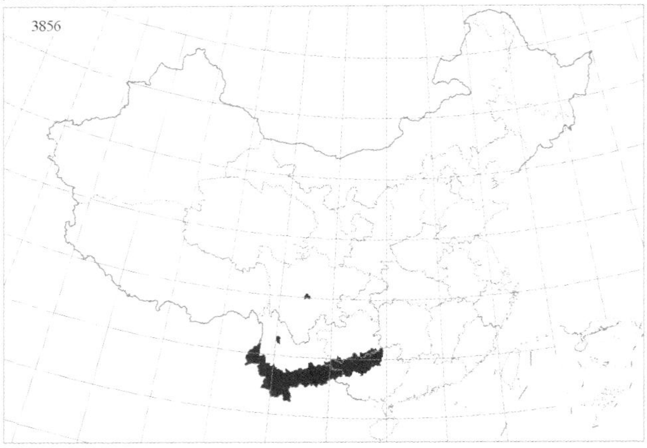

Rosaceae 蔷薇科 *Stranvaesia* 红果树属
Stranvaesia tomentosa T. T. Yu et T. C. Ku 绒毛红果树
Shrubs or small trees, evergreen

15.5; 4.9; 25.6; 15.5; 127.6; 1.1; 829 ❙ 1219; 497; 71 ❙ 49.4; 829; 176.5

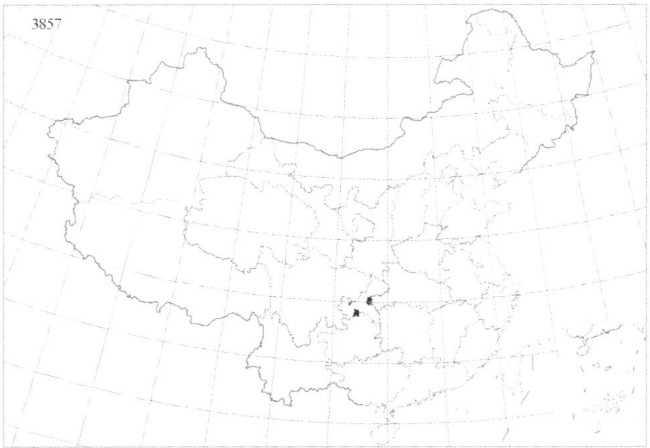

Connaraceae 牛栓藤科 *Agelaea* 栗豆藤属
Agelaea trinervis (Llanos) Merr. 栗豆藤
Vines or shrubs

23.9 (22.2~24.8); 18.8 (17.2~20.1); 27.5 (25.7~28.5); 24.0 (22.3~24.8); 227.7 (207.2~237.6); 0.0 (0.0~0.0); 1302 (1127~1385) ❙ 1573 (1405~1666); 556 (473~637); 62 (41~109) ❙ 21.8 (1.5~47.8); 1175 (1055~1321); 503.3 (283.8~620.0)

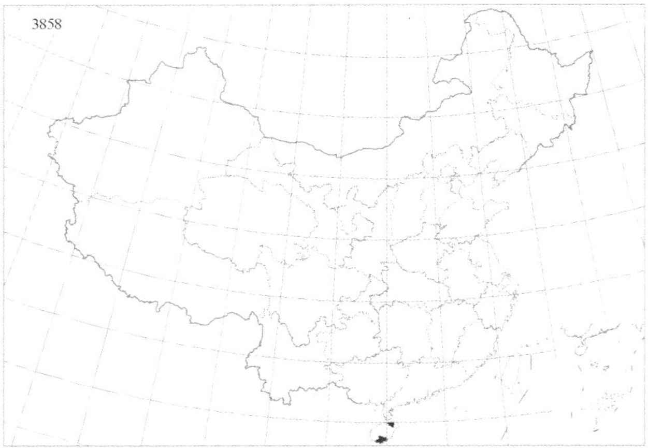

Connaraceae 牛栓藤科 *Cnestis* 螯毛果属
Cnestis palala (Lour.) Merr. 螯毛果
Vines or shrubs

24.6；19.2；28.4；24.6；235.2；0.0；1366 ❙ 1810；594；102 ❙ 32.6；1329；488.3

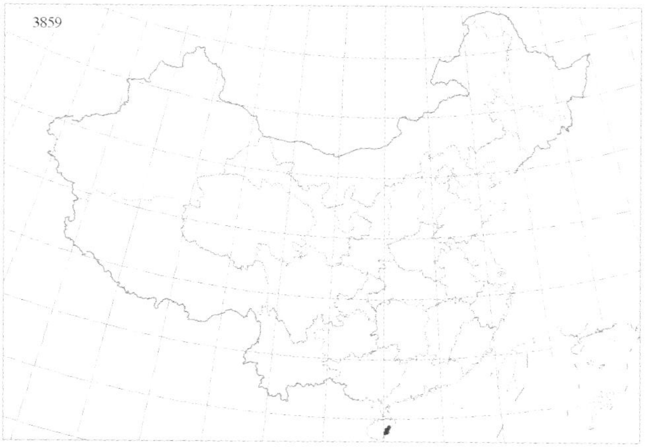

Connaraceae 牛栓藤科 *Connarus* 牛栓藤属
Connarus paniculatus Roxb. 牛栓藤
Vines or shrubs

16.8 (11.4~24.8)；9.6 (2.7~20.1)；22.4 (18.6~28.3)；16.9 (11.4~24.8)；142.7 (80.7~237.6)；0.3 (0.0~3.7)；867 (652~1385) ❙ 1140 (797~1767)；605 (411~911)；43 (21~102) ❙ 32.3 (-14.2~80.6)；825 (652~1299)；337.7 (190.3~620.0)

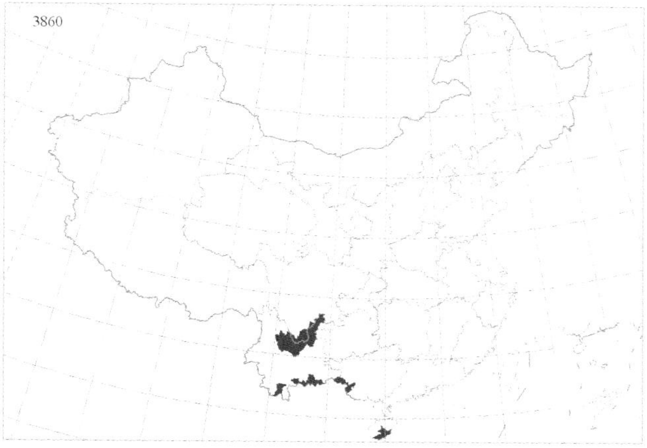

Connaraceae 牛栓藤科 *Connarus* 牛栓藤属
Connarus yunnanensis G. Schellenb. 云南牛栓藤
Scandent shrubs

19.5 (16.9~22.2)；12.2 (9.9~14.9)；24.5 (21.4~28.3)；19.5 (16.9~22.3)；173.8 (142.7~207.5)；0.0 (0.0~0.0)；969 (813~1190) ❙ 1421 (1178~1522)；772 (646~860)；67 (51~102) ❙ 48.4 (15.4~80.6)；966 (813~1190)；426.4 (313.7~570.1)

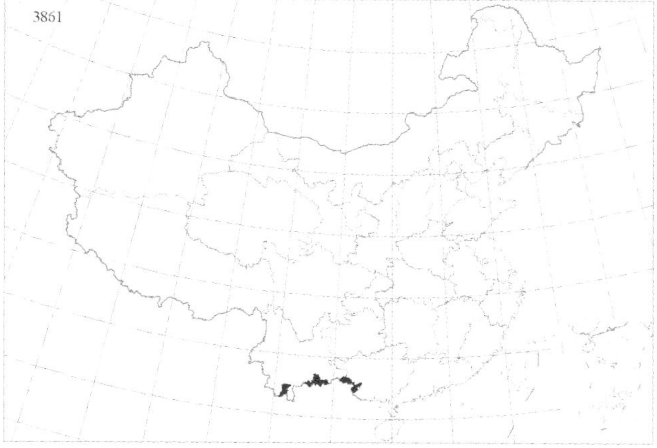

Connaraceae 牛栓藤科 *Ellipanthus* 单叶豆属
Ellipanthus glabrifolius Merr. 单叶豆
Shrubs or trees

24.0 (22.2~24.8)；18.4 (17.2~20.1)；28.0 (25.7~28.8)；24.0 (22.3~24.8)；228.3 (207.2~237.6)；0.0 (0.0~0.0)；1310 (1127~1385) ❙ 1501 (868~1854)；534 (312~712)；74 (41~115) ❙ 15.0 (-34.3~47.8)；1191 (868~1360)；440.3 (248.2~620.0)

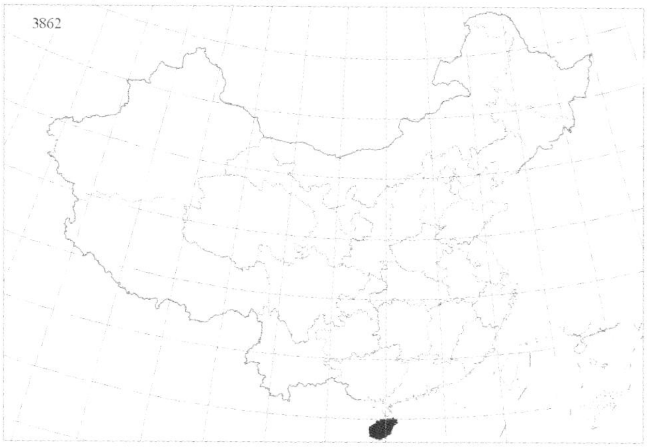

Connaraceae 牛栓藤科 *Rourea* 红叶藤属
Rourea caudata Planch. 长尾红叶藤
Vines or shrubs

18.5 (8.6~21.8)；11.3 (1.3~15.1)；23.7 (15.0~28.4)；18.6 (8.7~21.8)；163.2 (52.6~202.2)；0.5 (0.0~8.8)；934 (574~1154) ❙ 1439 (1041~1736)；715 (574~1007)；82 (44~176) ❙ 57.3 (12.9~145.8)；925 (574~1154)；394.1 (225.6~659.0)

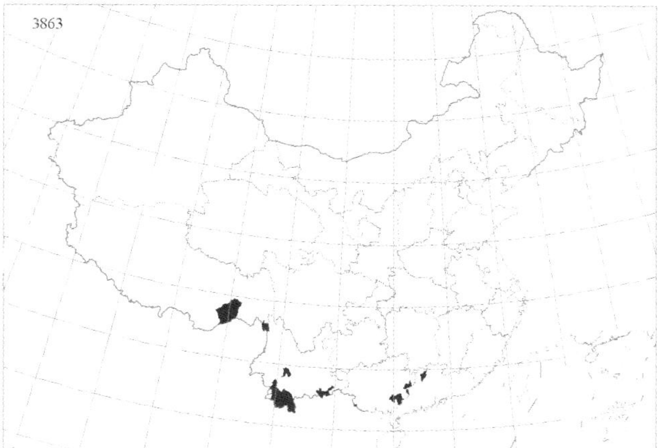

Connaraceae 牛栓藤科 *Rourea* 红叶藤属
Rourea microphylla (Hook. et Arn.) Planch. 小叶红叶藤
Scandent shrubs

20.9 (14.4~24.8)；12.8 (7.1~20.1)；27.5 (19.1~29.8)；20.9 (14.4~24.8)；191.2 (113.1~237.6)；0.0 (0.0~0.0)；1097 (727~1385) ❙ 1607 (868~2435)；685 (312~1428)；133 (41~217) ❙ 48.5 (-34.3~112.2)；1083 (727~1360)；270.9 (95.3~659.0)

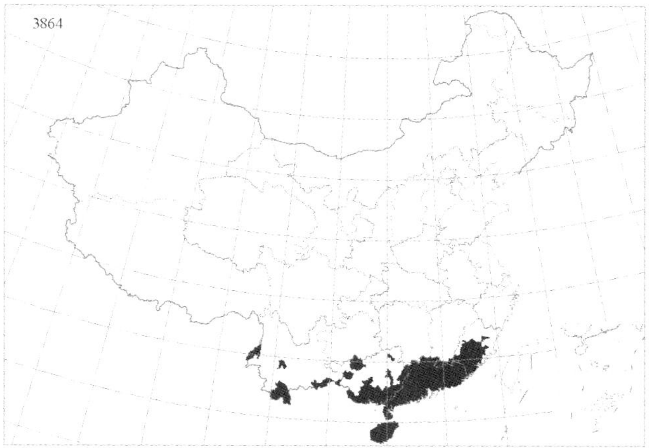

Connaraceae 牛栓藤科 *Rourea* 红叶藤属
Rourea minor (Gaertn.) Leenh. 红叶藤
Vines or shrubs

21.2 (8.6~24.8); 13.3 (1.3~20.1); 27.5 (15.0~29.8); 21.2 (8.7~24.8); 194.5 (52.6~237.6); 0.1 (0.0~8.8); 1115 (574~1385) ❙ 1619 (868~2713); 698 (312~1388); 127 (41~198) ❙ 47.4 (-34.3~214.4); 1098 (574~1360); 273.1 (95.3~659.0)

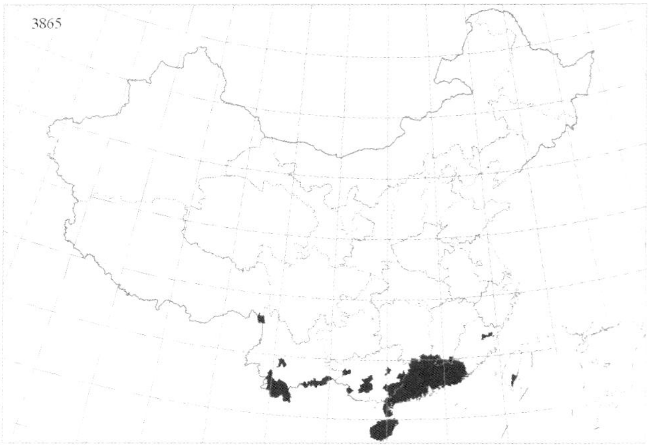

Connaraceae 牛栓藤科 *Roureopsis* 朱果藤属
Roureopsis emarginata (Jack) Merr. 朱果藤
Vines or shrubs

19.1 (17.3~21.0); 12.0 (9.9~15.1); 24.1 (22.4~25.6); 19.1 (17.3~21.0); 169.7 (147.1~191.8); 0.0 (0.0~0.0); 938 (833~1053) ❙ 1487 (1368~1607); 801 (702~860); 65 (55~77) ❙ 59.3 (44.5~80.6); 937 (833~1053); 466.8 (313.7~659.0)

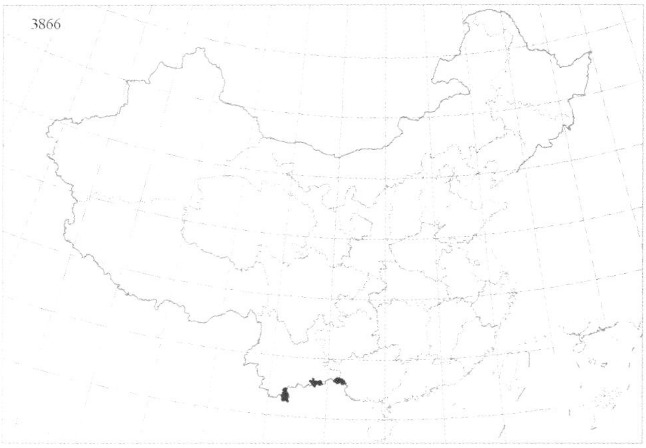

Fabaceae 豆科 *Abrus* 相思子属
Abrus precatorius L. 相思子
Vines

19.5 (8.6~24.8); 12.5 (1.3~20.1); 24.5 (15.0~29.0); 19.5 (8.7~24.8); 174.5 (52.6~237.6); 0.4 (0.0~8.8); 1012 (574~1385) ❙ 1486 (1041~2114); 772 (392~1357); 81 (41~167) ❙ 51.6 (-11.9~112.2); 985 (574~1247); 368.5 (199.6~659.0)

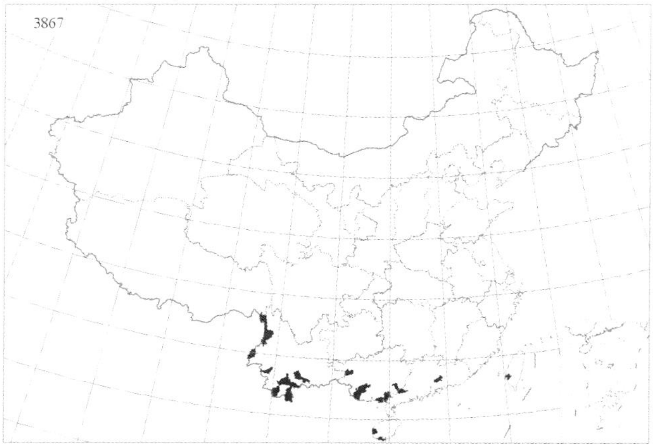

Fabaceae 豆科 *Abrus* 相思子属
Abrus pulchellus Wallich ex Thwaites 美丽相思子
Vines

20.1; 14.3; 23.5; 20.2; 181.9; 0.0; 974 ❙ 1543; 755; 62 ❙ 58.5; 956; 570.7

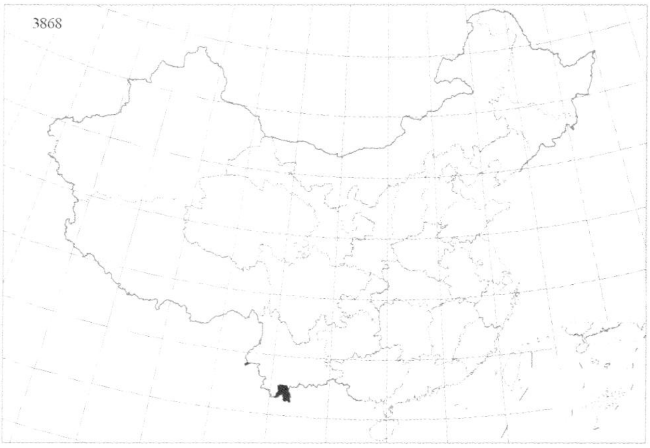

Fabaceae 豆科 *Abrus* 相思子属
Abrus pulchellus subsp. mollis (Hance) Verdcourt 毛相思子
Vines

22.3 (19.4~24.8); 15.0 (10.6~20.1); 28.1 (25.7~29.0); 22.4 (19.5~24.8); 208.6 (173.5~237.6); 0.0 (0.0~0.0); 1193 (980~1385) ❙ 1613 (868~2148); 680 (312~1234); 114 (41~191) ❙ 36.5 (-34.3~95.5); 1155 (868~1360); 296.0 (139.6~620.0)

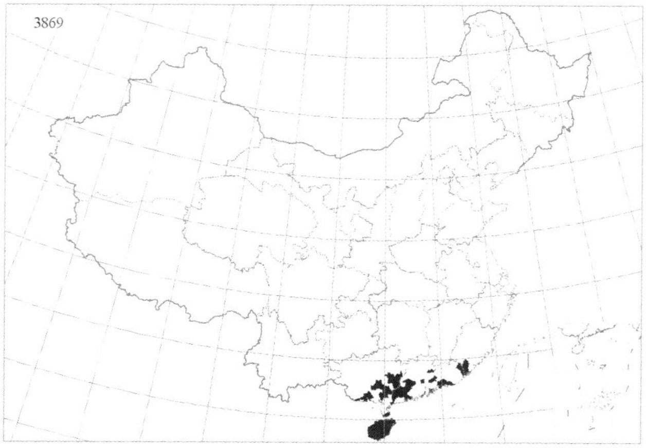

Fabaceae 豆科 *Acacia* 金合欢属
Acacia caesia (L.) Willd. 尖叶相思
Vines

18.6 (6.2~24.8); 10.2 (-2.4~20.1); 25.3 (14.1~28.7); 18.7 (6.7~24.8); 164.8 (39.4~237.6); 0.8 (0.0~24.1); 960 (509~1385) ❙ 1529 (981~2993); 769 (473~1847); 103 (23~547) ❙ 62.2 (1.5~246.5); 948 (509~1137); 280.1 (154.2~527.5)

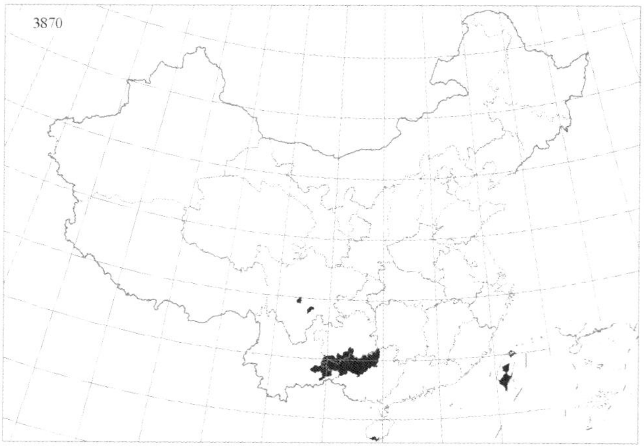

Fabaceae 豆科 *Acacia* 金合欢属
Acacia catechu (L. f.) Willd. 儿茶
Small trees, deciduous

22.8 (16.3~24.8); 16.9 (7.4~20.1); 27.0 (20.3~28.8); 22.8 (16.4~24.8); 213.9 (136.6~237.6); 0.0 (0.0~0.0); 1218 (789~1385) ❙ 1566 (868~2759); 631 (312~1594); 79 (41~179) ❙ 32.7 (-34.3~202.5); 1124 (780~1360); 417.1 (154.2~659.0)

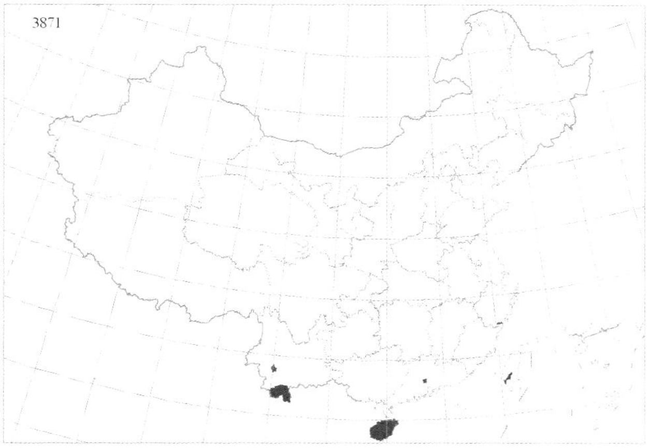

Fabaceae 豆科 *Acacia* 金合欢属
Acacia concinna (Willd.) DC. 藤金合欢
Vines

18.0 (2.2~24.8); 9.4 (-6.6~20.1); 25.3 (9.9~29.3); 18.1 (3.8~24.8); 157.9 (15.5~237.6); 0.7 (0.0~49.1); 940 (413~1385) ❙ 1437 (797~2030); 614 (312~1033); 117 (18~257) ❙ 55.1 (-34.3~125.9); 919 (413~1360); 294.8 (126.1~659.0)

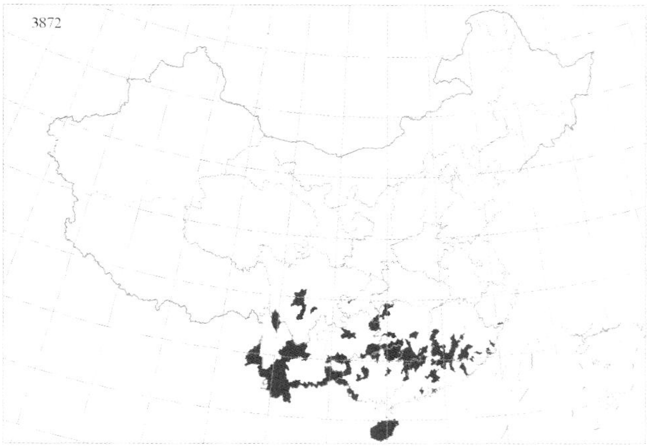

Fabaceae 豆科 *Acacia* 金合欢属
Acacia confusa Merr. 台湾相思
Trees, evergreen

19.1 (11.2~24.8); 11.4 (4.6~20.1); 25.3 (16.7~29.2); 19.1 (11.3~24.8); 169.3 (75.8~237.6); 0.0 (0.0~0.4); 980 (642~1385) ❙ 1465 (797~2993); 665 (312~1847); 115 (23~547) ❙ 51.4 (-34.3~246.5); 960 (642~1360); 301.1 (119.4~659.0)

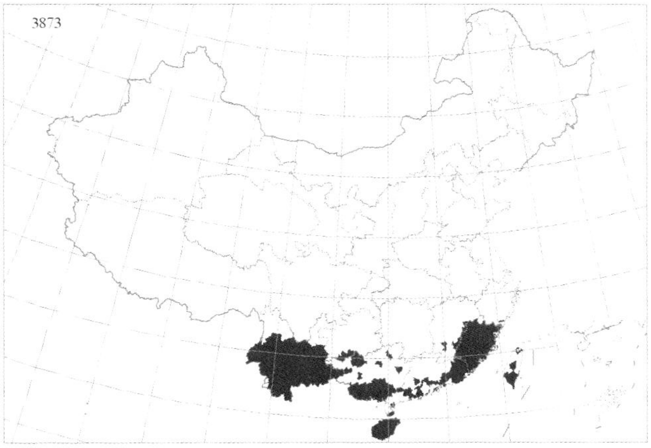

Fabaceae 豆科 *Acacia* 金合欢属
Acacia delavayi Franch. 光叶金合欢
Vines

13.6 (2.2~19.1); 5.9 (-6.6~12.0); 19.8 (9.5~26.6); 13.7 (3.8~19.1); 107.8 (15.5~169.2); 4.4 (0.0~49.1); 727 (411~969) ❙ 986 (628~1334); 570 (406~723); 33 (11~64) ❙ 33.3 (-14.2~125.9); 702 (411~969); 272.8 (142.9~383.4)

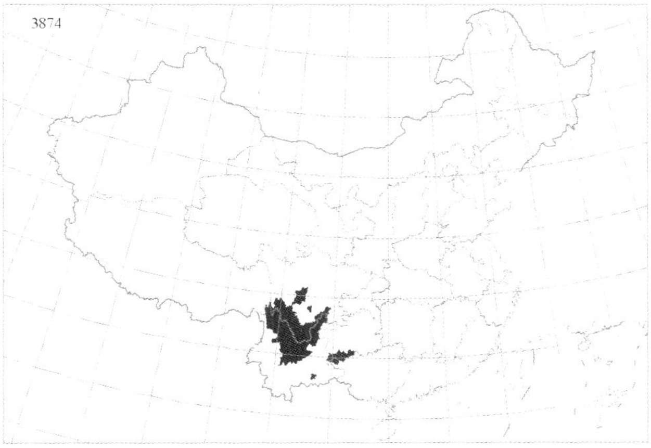

Fabaceae 豆科 *Acacia* 金合欢属
Acacia delavayi var. kunmingensis C. Chen et H. Sun 昆明金合欢
Vines

16.9; 10.3; 21.4; 16.9; 142.7; 0.0; 813 ❙ 1178; 646; 51 ❙ 44.9; 813; 383.4

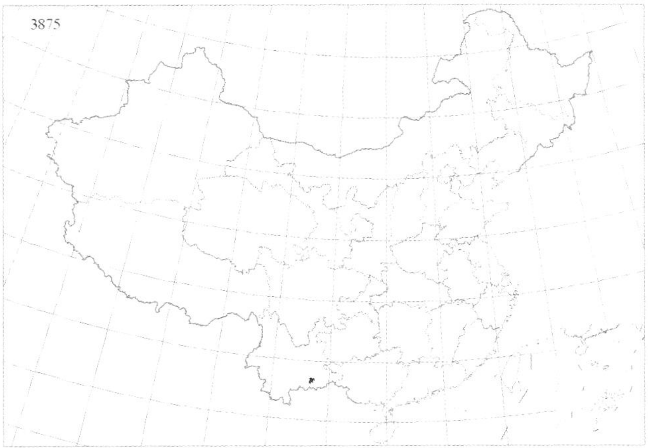

Fabaceae 豆科 *Acacia* 金合欢属
Acacia farnesiana (L.) Willd. 金合欢
Shrubs or small trees

16.6 (5.4~24.8); 9.6 (-2.4~20.1); 22.1 (12.3~28.8); 16.7 (5.8~24.8); 141.7 (31.5~237.6); 0.9 (0.0~25.6); 852 (481~1385) ❙ 1250 (741~2992); 690 (411~1847); 52 (15~191) ❙ 47.1 (-14.2~202.5); 825 (481~1331); 336.2 (119.4~659.0)

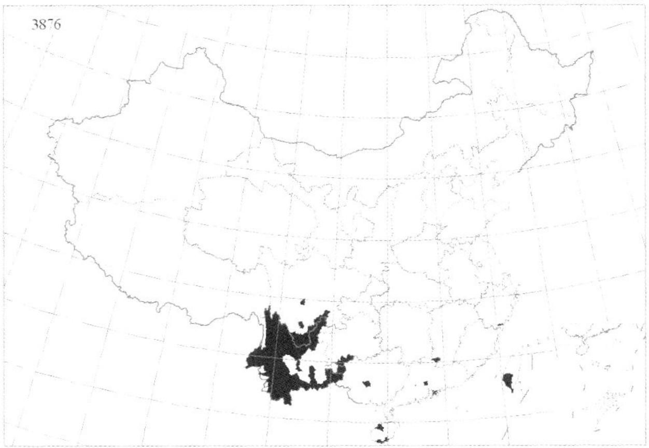

Fabaceae 豆科 *Acacia* 金合欢属

Acacia megaladena Desv. 钝叶金合欢

Vines

17.4 (13.2~22.2); 10.7 (5.4~15.1); 22.0 (18.4~28.3); 17.5 (13.3~22.3); 149.7 (99.4~207.5); 0.0 (0.0~0.0); 849 (695~1190) ❙ 1253 (939~1751); 696 (524~1007); 52 (38~87) ❙ 47.9 (6.5~145.8); 837 (695~1190); 389.8 (234.5~659.0)

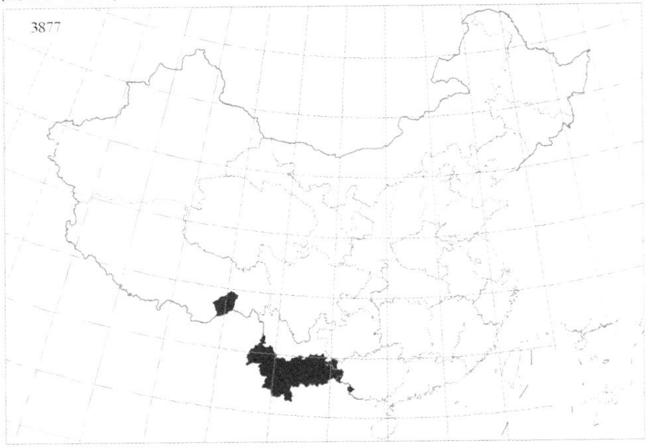

Fabaceae 豆科 *Acacia* 金合欢属

Acacia pennata (L.) Willd. 羽叶金合欢

Vines

17.5 (2.2~24.8); 9.9 (-6.6~20.1); 23.7 (9.9~29.0); 17.6 (3.8~24.8); 152.3 (15.5~237.6); 1.6 (0.0~49.1); 925 (413~1385) ❙ 1287 (741~2035); 624 (312~994); 74 (12~208) ❙ 42.2 (-34.3~125.9); 897 (413~1360); 310.8 (133.6~659.0)

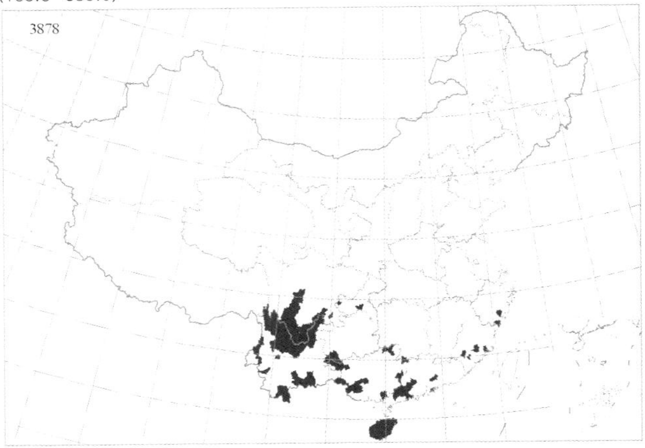

Fabaceae 豆科 *Acacia* 金合欢属

Acacia pruinescens Kurz 粉被金合欢

Vines or shrubs

18.0 (13.2~20.4); 10.9 (6.4~15.1); 22.9 (18.4~25.5); 18.0 (13.3~20.5); 156.2 (99.4~185.8); 0.0 (0.0~0.0); 877 (695~989) ❙ 1375 (1049~1751); 753 (598~996); 56 (42~77) ❙ 57.4 (21.9~92.4); 870 (695~986); 426.6 (314.4~659.0)

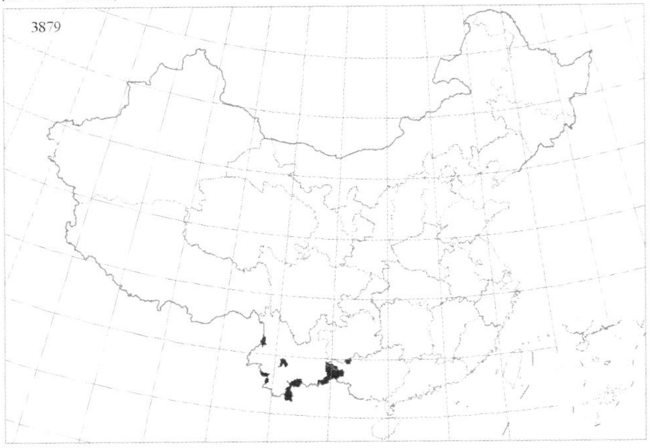

Fabaceae 豆科 *Acacia* 金合欢属

Acacia teniana Harms 无刺金合欢

Shrubs or small trees

10.6 (2.2~15.0); 2.3 (-6.6~8.4); 17.8 (9.9~21.0); 10.9 (3.8~15.1); 78.1 (15.5~121.2); 10.3 (0.0~49.1); 635 (413~753) ❙ 969 (885~1056); 554 (479~615); 28 (19~35) ❙ 58.3 (26.3~125.9); 623 (413~699); 274.4 (248.3~307.7)

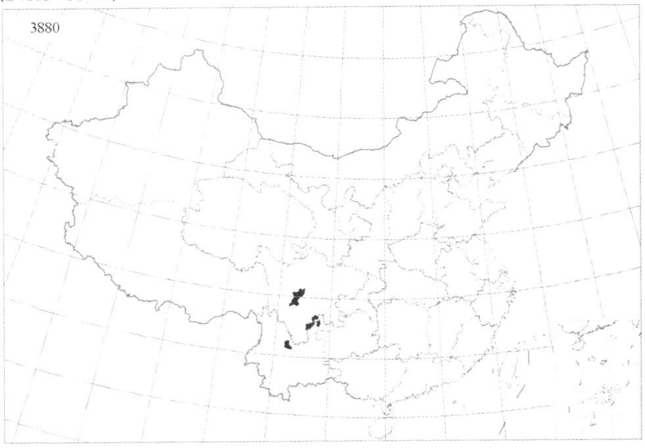

Fabaceae 豆科 *Acacia* 金合欢属

Acacia tonkinensis I. C. Nielsen 滇南金合欢

Scandent shrubs

20.4; 15.1; 23.6; 20.5; 185.8; 0.0; 989 ❙ 1607; 702; 77 ❙ 62.5; 986; 659.0

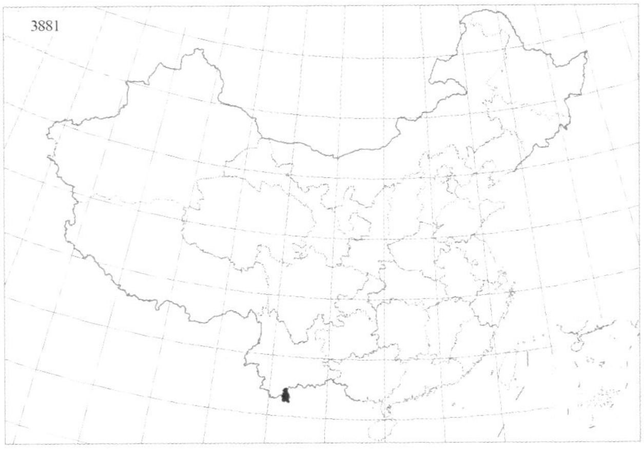

Fabaceae 豆科 *Acacia* 金合欢属

Acacia vietnamensis I. C. Nielsen 越南金合欢

Scandent shrubs

16.7 (14.3~17.7); 5.9 (3.3~7.8); 26.9 (24.3~29.3); 16.8 (14.3~17.7); 141.8 (114.1~152.4); 0.4 (0.0~2.3); 885 (765~961) ❙ 1552 (1425~1679); 558 (517~611); 172 (112~195) ❙ 75.7 (64.9~99.4); 885 (765~961); 223.9 (193.3~257.1)

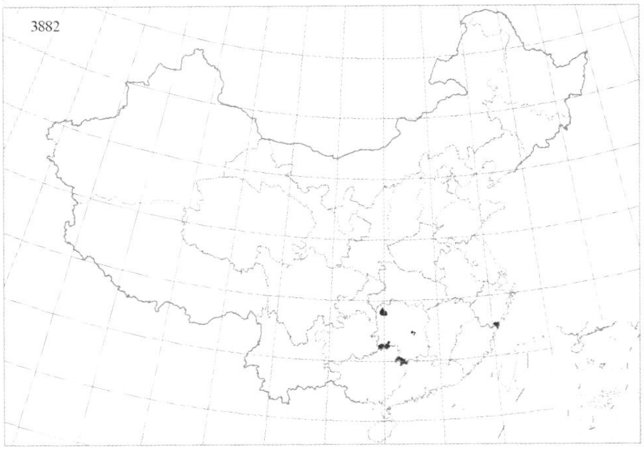

Fabaceae 豆科 *Acacia* 金合欢属
Acacia yunnanensis Franch. 云南相思树
Shrubs

12.9 (2.2~21.6); 5.1 (-6.6~12.6); 19.3 (9.5~28.5); 13.0 (3.8~21.7); 99.5 (15.5~199.8); 4.7 (0.0~49.1); 704 (411~1153) ❙ 971 (729~1625); 567 (411~795); 31 (11~173) ❙ 41.4 (-14.2~125.9); 679 (411~1153); 269.6 (142.9~351.7)

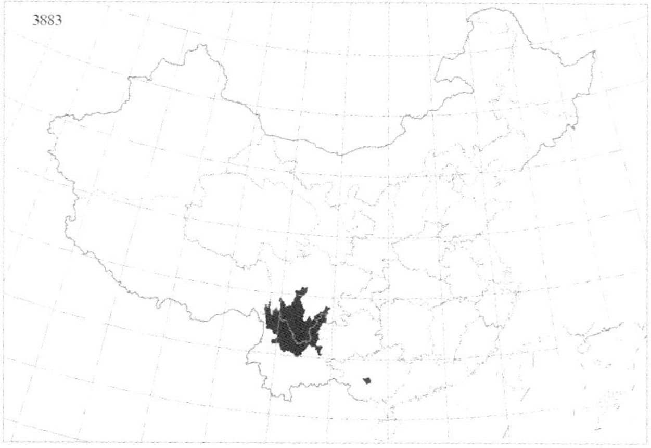

Fabaceae 豆科 *Acrocarpus* 顶果树属
Acrocarpus fraxinifolius Wight ex Arn. 顶果树
Trees

19.5 (14.4~22.2); 11.8 (7.5~15.1); 25.1 (19.1~28.4); 19.5 (14.4~22.3); 174.4 (113.1~207.5); 0.0 (0.0~0.0); 986 (727~1190) ❙ 1378 (1049~1607); 730 (598~911); 76 (43~141) ❙ 41.4 (12.9~99.9); 980 (727~1190); 360.7 (231.5~659.0)

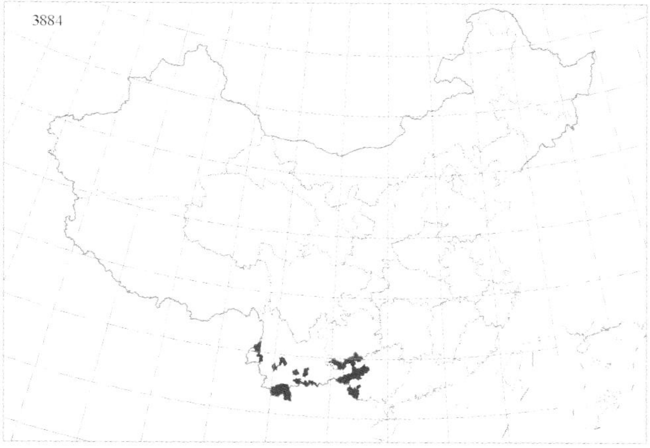

Fabaceae 豆科 *Adenanthera* 海红豆属
Adenanthera microsperma Teijsm. et Binn. 海红豆
Shrubs

20.7 (15.5~24.8); 13.0 (4.8~20.1); 26.7 (20.1~29.0); 20.7 (15.6~24.8); 188.6 (127.2~237.6); 0.0 (0.0~0.2); 1074 (773~1385) ❙ 1577 (868~2993); 740 (312~1847); 111 (41~547) ❙ 49.8 (-34.3~246.5); 1047 (773~1269); 301.5 (139.6~659.0)

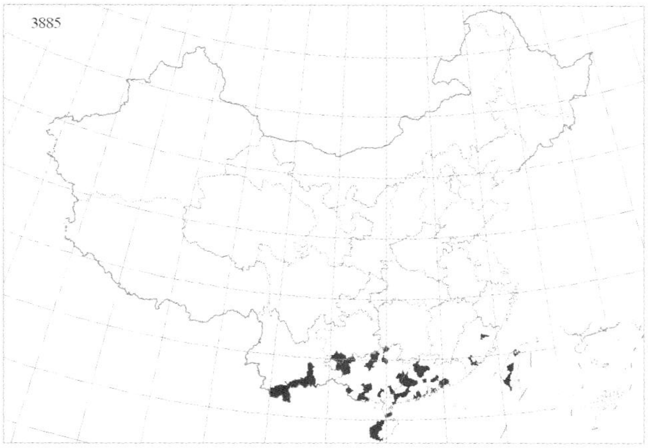

Fabaceae 豆科 *Afgekia* 猪腰豆属
Afgekia filipes (Dunn) R. Geesink 猪腰豆
Scandent shrubs

19.2 (15.9~22.2); 12.2 (9.5~15.1); 24.1 (20.3~28.3); 19.2 (16.0~22.3); 170.9 (131.6~207.5); 0.0 (0.0~0.0); 953 (779~1190) ❙ 1388 (1014~1771); 743 (568~996); 70 (44~173) ❙ 46.5 (5.8~88.8); 941 (751~1190); 378.5 (220.1~659.0)

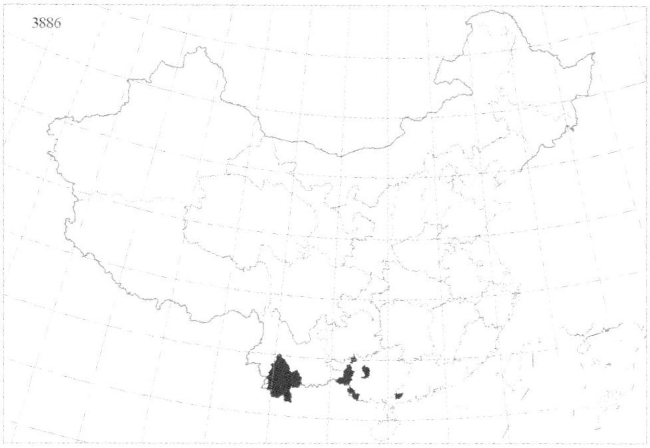

Fabaceae 豆科 *Aganope* 双束鱼腾属
Aganope dinghuensis (P. Y. Chen) T. C. Chen et Pedley 鼎湖鱼藤
Vines, evergreen

22.0; 13.5; 28.8; 22.0; 203.6; 0.0; 1169 ❙ 1641; 724; 132 ❙ 40.5; 1169; 205.0

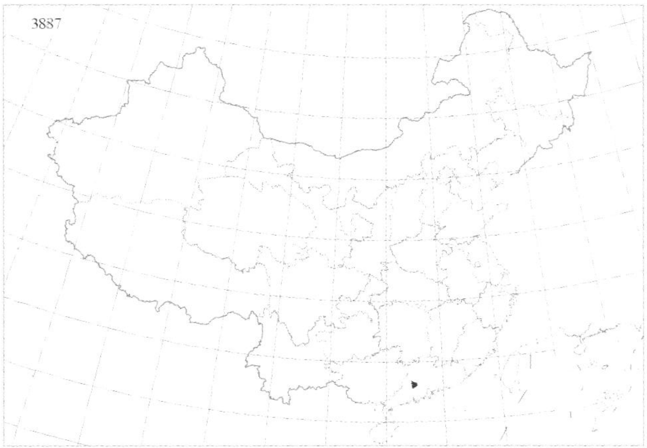

Fabaceae 豆科 *Aganope* 双束鱼腾属
Aganope latifolia (Prain) T. C. Chen et Pedley 大叶鱼藤
Trees, evergreen

21.2 (17.7~22.3); 12.8 (10.9~14.2); 27.5 (22.5~28.9); 21.3 (17.8~22.3); 195.2 (153.4~208.1); 0.0 (0.0~0.0); 1116 (854~1202) ❙ 1523 (1368~1703); 775 (708~839); 123 (57~194) ❙ 37.3 (15.4~60.1); 1116 (854~1202); 300.3 (201.4~551.6)

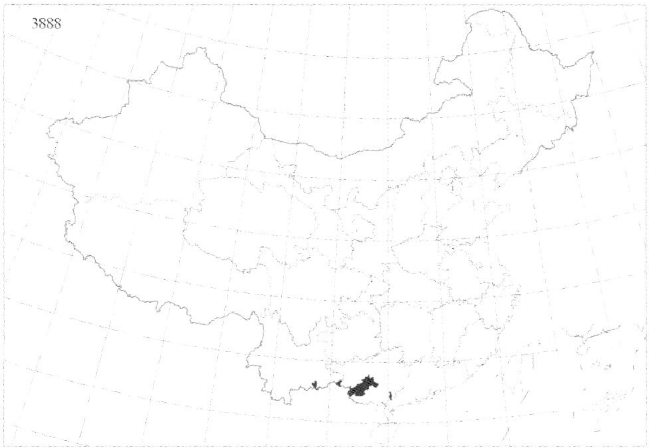

Fabaceae 豆科 *Aganope* 双束鱼腾属
Aganope thyrsiflora (Bentham) Polhill 密锥花鱼藤
Scandent shrubss or vines, evergreen

22.2 (19.1~24.8); 14.9 (13.0~20.1); 27.7 (23.3~28.8); 22.3 (19.1~24.8); 207.1 (169.7~237.6); 0.0 (0.0~0.0); 1181 (917~1385) ❙ 1675 (1302~2435); 802 (455~1428); 101 (41~132) ❙ 42.8 (1.5~99.5); 1151 (917~1299); 336.9 (203.4~572.9)

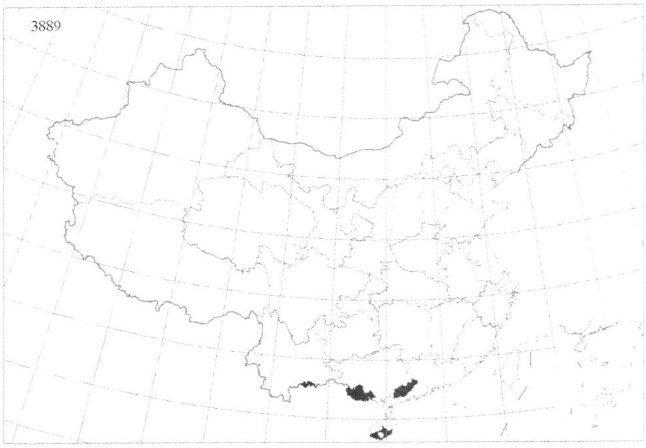

Fabaceae 豆科 *Albizia* 合欢属
Albizia attopeuensis (Pierre) T. L. Wu 海南合欢
Trees, deciduous

24.8; 20.1; 28.0; 24.8; 237.6; 0.0; 1385 ❙ 1405; 473; 41 ❙ 1.5; 1137; 527.5

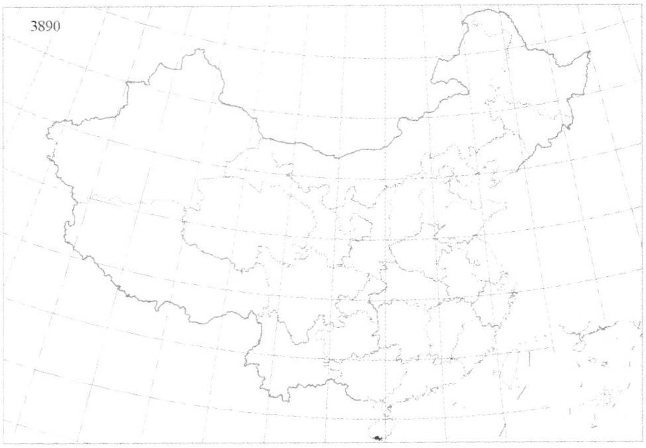

Fabaceae 豆科 *Albizia* 合欢属
Albizia chinensis (Osbeck) Merr. 楹树
Trees, deciduous

18.2 (6.2~24.8); 10.0 (-2.4~20.1); 24.7 (14.1~29.4); 18.2 (6.7~24.8); 159.2 (39.4~237.6); 0.3 (0.0~24.1); 943 (509~1385) ❙ 1355 (797~2435); 650 (312~1428); 93 (22~254) ❙ 44.9 (-34.3~130.7); 927 (509~1360); 288.2 (95.3~659.0)

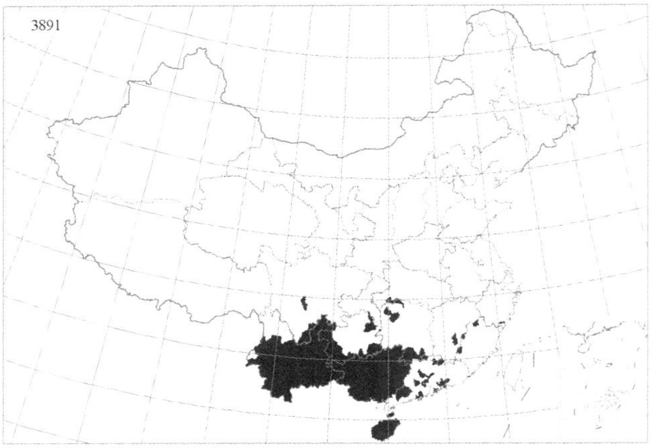

Fabaceae 豆科 *Albizia* 合欢属
Albizia corniculata (Lour.) Merr. 天香藤
Scandent shrubss or vines, deciduous

21.1 (16.9~24.8); 12.9 (7.1~20.1); 27.9 (25.2~29.1); 21.2 (17.0~24.8); 194.2 (143.6~237.6); 0.0 (0.0~0.0); 1118 (862~1385) ❙ 1658 (868~2435); 702 (312~1428); 139 (41~198) ❙ 50.1 (-34.3~100.8); 1095 (862~1360); 268.2 (139.6~620.0)

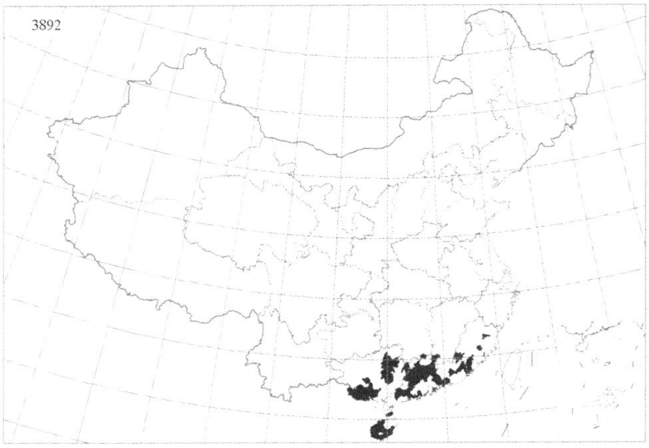

Fabaceae 豆科 *Albizia* 合欢属
Albizia crassiramea Lace 白花合欢
Trees, deciduous

18.9 (17.2~21.0); 11.9 (9.1~15.1); 23.7 (22.3~25.8); 18.9 (17.3~21.0); 166.9 (147.0~191.8); 0.0 (0.0~0.0); 920 (833~1053) ❙ 1431 (1161~1751); 768 (633~996); 60 (48~77) ❙ 55.8 (30.5~88.8); 915 (833~1053); 440.1 (291.2~659.0)

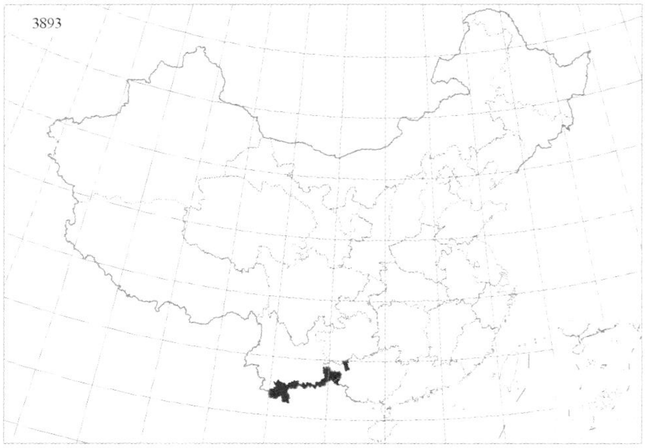

Fabaceae 豆科 *Albizia* 合欢属
Albizia garrettii I. C. Nielsen 黄毛合欢
Trees, deciduous

17.4 (14.4~19.6); 9.3 (4.8~12.8); 23.6 (19.1~28.7); 17.4 (14.4~19.6); 148.9 (113.1~175.7); 0.0 (0.0~0.2); 868 (727~1038) ❙ 1346 (939~1776); 701 (468~959); 75 (39~191) ❙ 55.7 (6.5~112.2); 855 (727~1038); 353.6 (120.2~627.0)

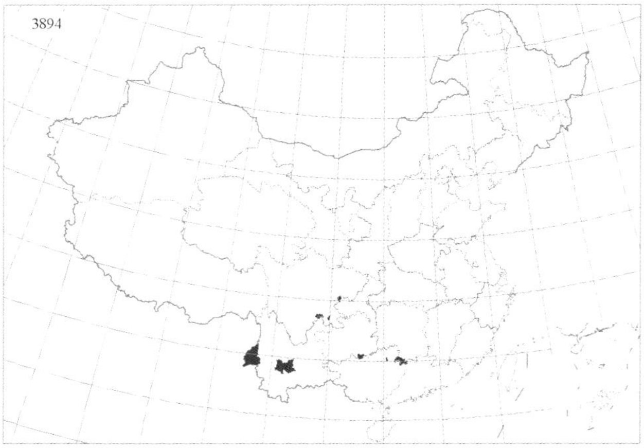

Fabaceae 豆科 *Albizia* 合欢属
Albizia julibrissin Durazz. 合欢
Trees, deciduous

14.8 (-1.8~24.2); 2.7 (-12.7~18.1); 25.9 (6.1~29.9); 15.0 (1.8~24.3); 126.4 (1.9~231.2); 7.6 (0.0~82.8); 836 (329~1335) ❙ 1075 (320~2993); 491 (217~1388); 91 (8~547) ❙ 27.5 (-51.1~246.5); 764 (320~1329); 211.3 (72.2~659.0)

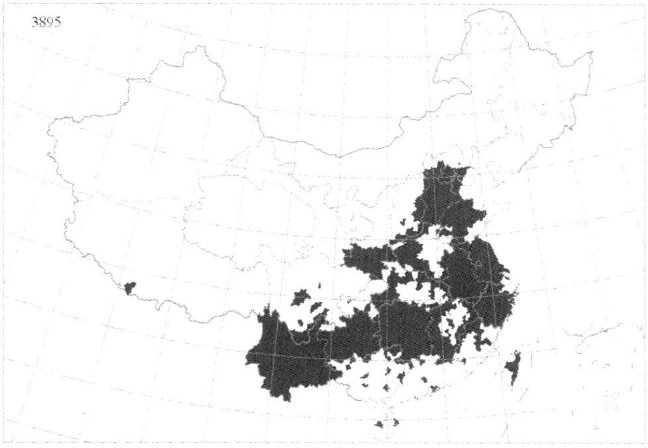

Fabaceae 豆科 *Albizia* 合欢属
Albizia kalkora (Roxb.) Prain 山槐
Shrubs or small trees, deciduous

15.6 (5.4~24.3); 4.4 (-7.5~18.7); 25.9 (12.3~29.9); 15.7 (5.8~24.3); 132.1 (31.5~231.9); 4.2 (0.0~38.3); 853 (481~1338) ❙ 1194 (491~2435); 525 (258~1428); 105 (10~341) ❙ 39.4 (-40.3~172.5); 822 (481~1221); 222.0 (110.1~659.0)

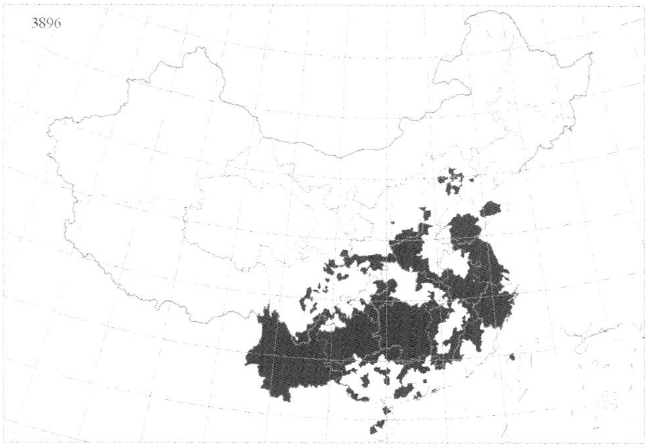

Fabaceae 豆科 *Albizia* 合欢属
Albizia lucidior (Steud.) I. C. Nielsen 光叶合欢
Trees, deciduous

17.8 (10.1~22.3); 10.4 (2.3~15.1); 23.1 (16.9~28.9); 17.8 (10.2~22.3); 153.7 (63.1~208.1); 0.1 (0.0~5.2); 885 (618~1202) ❙ 1291 (952~1751); 700 (531~996); 65 (15~194) ❙ 46.8 (5.8~112.2); 877 (618~1202); 357.7 (188.6~659.0)

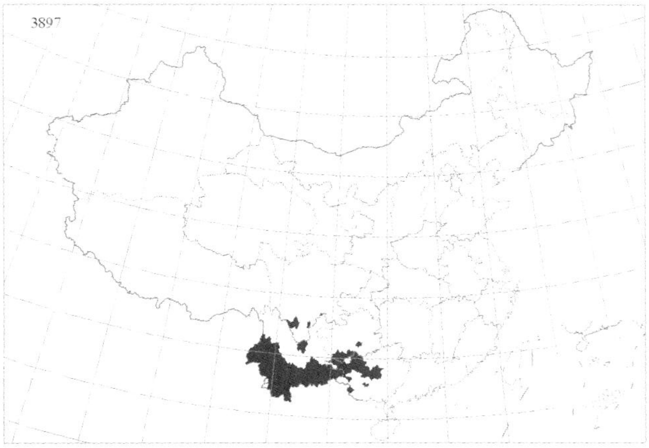

Fabaceae 豆科 *Albizia* 合欢属
Albizia mollis (Wall.) Boivin 毛叶合欢
Trees, deciduous

14.3 (-1.8~21.0); 6.9 (-10.2~14.2); 20.1 (6.1~26.6); 14.4 (1.8~21.0); 115.5 (1.9~191.8); 3.5 (0.0~82.8); 747 (329~1053) ❙ 1031 (320~1530); 591 (193~909); 38 (11~69) ❙ 39.2 (-14.2~121.7); 727 (320~1053); 294.0 (72.2~551.6)

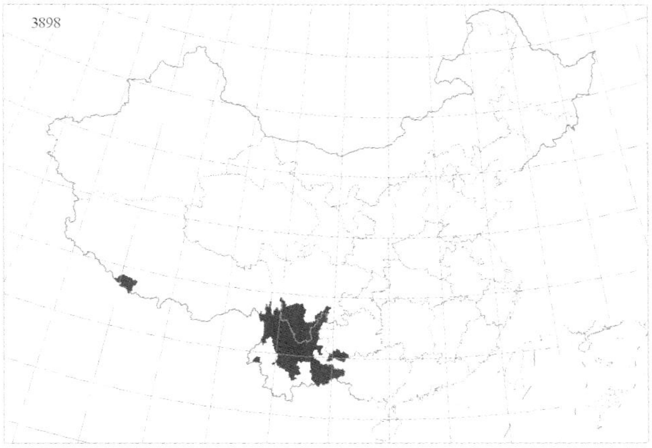

Fabaceae 豆科 *Albizia* 合欢属
Albizia odoratissima (L. f.) Benth. 香合欢
Trees, evergreen

18.6 (12.6~24.8); 11.3 (3.8~20.1); 24.2 (17.9~28.9); 18.7 (12.6~24.8); 164.2 (91.6~237.6); 0.0 (0.0~1.5); 946 (678~1385) ❙ 1359 (868~2035); 678 (312~1007); 75 (20~197) ❙ 46.4 (-34.3~145.8); 919 (678~1299); 355.7 (153.3~659.0)

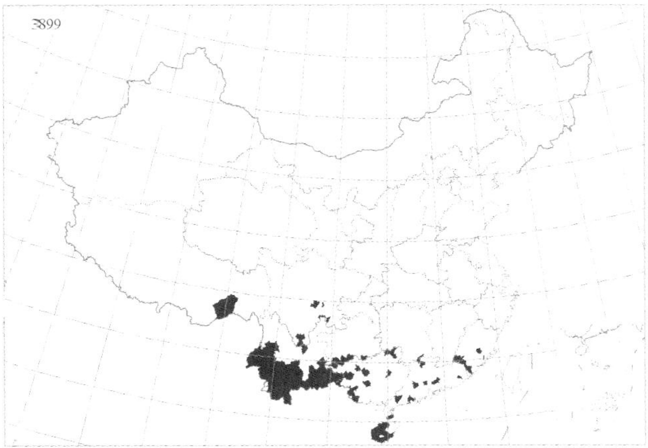

Fabaceae 豆科 *Albizia* 合欢属
Albizia procera (Roxb.) Benth. 黄豆树
Trees, deciduous

21.8 (16.3~24.8); 15.9 (10.2~20.1); 25.9 (20.3~28.8); 21.8 (16.4~24.8); 201.8 (136.6~237.6); 0.0 (0.0~0.0); 1140 (789~1385) ❙ 1595 (868~2759); 720 (312~1594); 79 (41~308) ❙ 45.2 (-34.3~202.5); 1066 (780~1360); 400.8 (119.4~659.0)

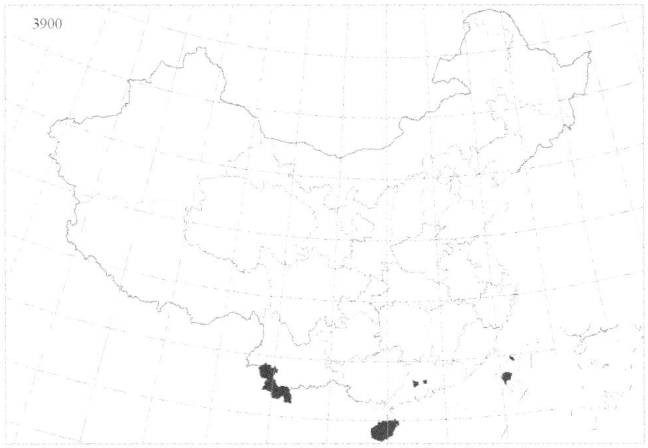

Fabaceae 豆科 *Albizia* 合欢属
Albizia sherriffii Baker 藏合欢
Trees, deciduous

9.1; 1.4; 15.6; 9.3; 61.2; 11.7; 586 ❙ 1246; 717; 42 ❙ 109.3; 586; 344.6

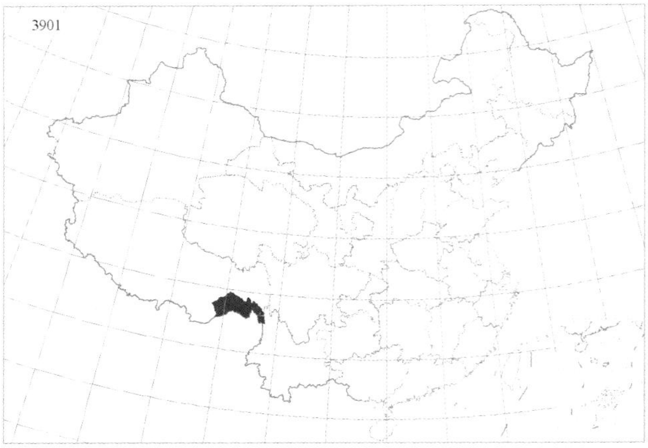

Fabaceae 豆科 *Alhagi* 骆驼刺属
Alhagi sparsifolia Shafarenko ex Keller et Shaparenko 骆驼刺
Subshrubs

5.3; −11.5; 20.5; 8.2; 64.8; 60.8; 584 ❙ 66; 44; 1 ❙ −88.3; 66; 7.0

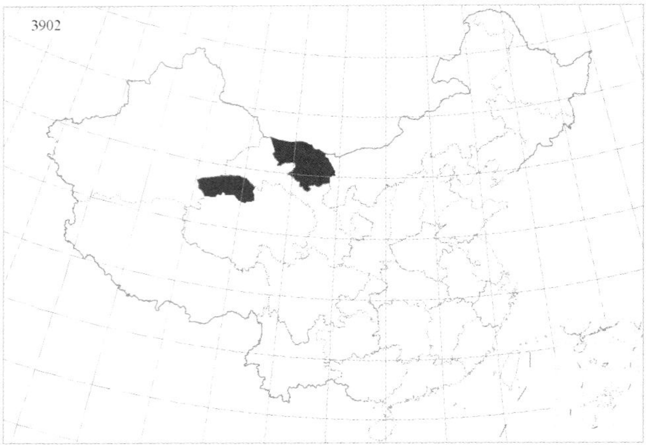

Fabaceae 豆科 *Ammodendron* 银砂槐属
Ammodendron bifolium (Pall.) Yakovlev 银砂槐
Shrubs

4.0; −13.5; 18.3; 7.3; 52.8; 65.2; 549 ❙ 357; 136; 45 ❙ −34.9; 357; 190.1

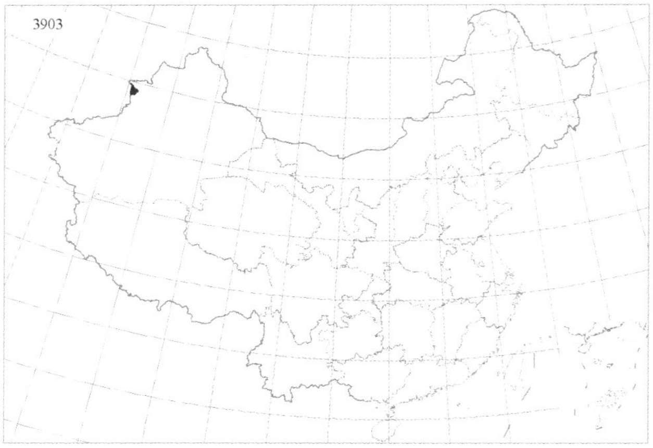

Fabaceae 豆科 *Ammopiptanthus* 沙冬青属
Ammopiptanthus mongolicus (Maxim. ex Kom.) S. H. Cheng 沙冬青
Shrubs, evergreen

7.1 (0.1~11.9); −9.6 (−16.8~−5.6); 21.9 (13.9~25.5); 9.2 (4.8~12.7); 73.8 (28.5~107.9); 47.9 (25.2~86.7); 621 (440~783) ❙ 254 (49~580); 148 (23~298); 7 (1~26) ❙ −57.4 (−92.8~7.5); 252 (49~539); 63.1 (5.1~229.4)

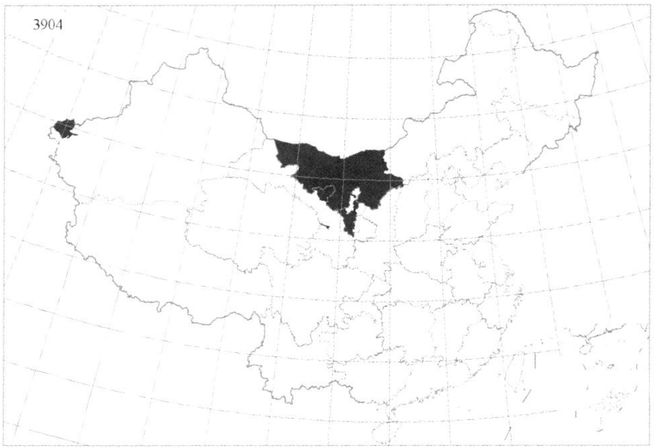

Fabaceae 豆科 *Amorpha* 紫穗槐属
Amorpha fruticosa L. 紫穗槐
Shrubs, deciduous

16.9 (2.2~21.0); 6.2 (−6.6~12.2); 26.7 (9.9~29.9); 17.0 (3.8~21.0); 145.2 (15.5~192.6); 1.8 (0.0~49.1); 909 (413~1113) ❙ 1418 (520~1865); 545 (260~816); 148 (18~258) ❙ 56.0 (−35.9~125.9); 890 (413~1107); 210.7 (126.1~331.6)

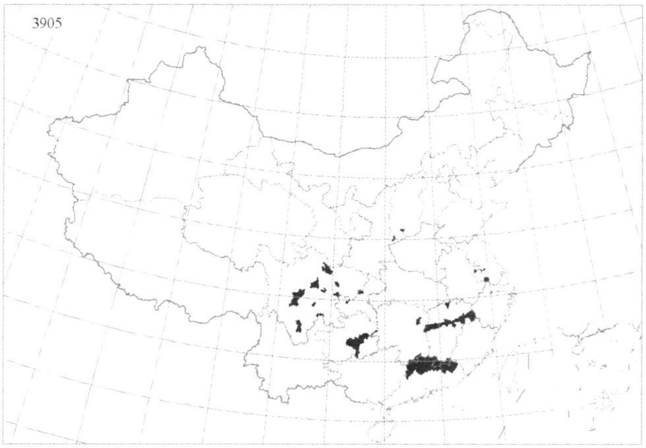

Fabaceae 豆科 *Antheroporum* 肿荚豆属
Antheroporum glaucum Z. Wei 粉叶肿荚豆
Trees, evergreen

18.2 (14.4~20.5); 11.6 (7.5~15.1); 22.4 (19.1~25.0); 18.2 (14.4~20.5); 158.9 (113.1~185.9); 0.0 (0.0~0.0); 882 (727~992) ❙ 1497 (1161~1688); 809 (633~959); 57 (42~77) ❙ 71.1 (37.1~112.2); 873 (727~986); 499.9 (314.4~659.0)

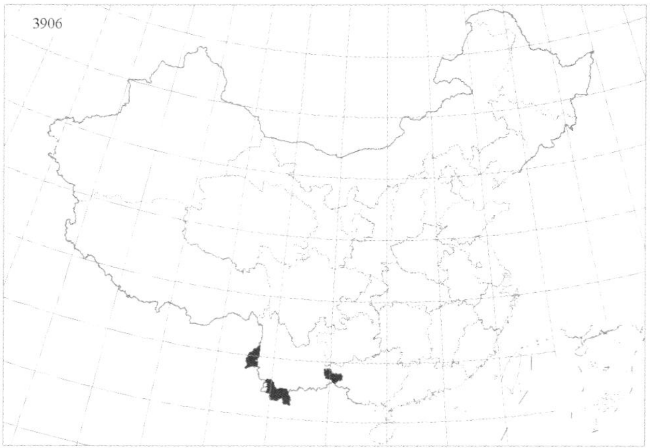

Fabaceae 豆科 *Antheroporum* 肿荚豆属
Antheroporum harmandii Gagnep. 肿荚豆
Trees, evergreen

18.5 (16.8~20.5); 10.7 (7.8~15.1); 24.1 (21.6~26.4); 18.5 (16.8~20.5); 162.6 (141.5~185.9); 0.0 (0.0~0.0); 914 (836~1010) ❙ 1325 (1161~1607); 699 (633~813); 56 (43~77) ❙ 45.5 (16.9~74.2); 908 (836~1010); 366.3 (189.5~659.0)

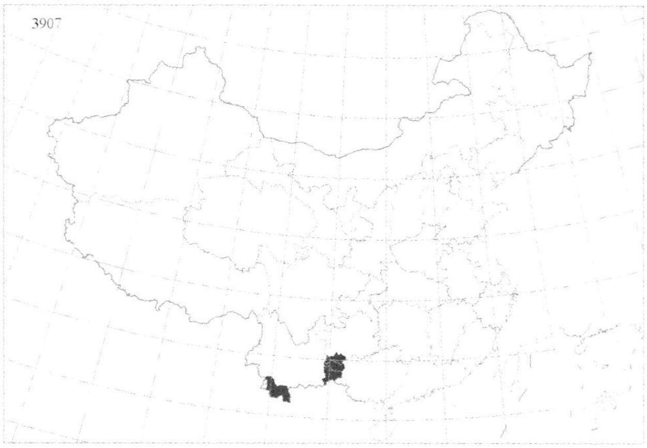

Fabaceae 豆科 *Archidendron* 猴耳环属
Archidendron alternifoliolatum (T. L. Wu) I. C. Nielsen 长叶棋子豆
Trees, deciduous

19.5 (17.7~21.0); 13.5 (10.9~15.1); 23.3 (22.3~25.6); 19.5 (17.8~21.0); 174.2 (153.4~191.8); 0.0 (0.0~0.0); 941 (854~1053) ❙ 1523 (1368~1751); 795 (655~996); 66 (56~77) ❙ 62.3 (44.5~88.8); 933 (854~1053); 544.7 (491.5~659.0)

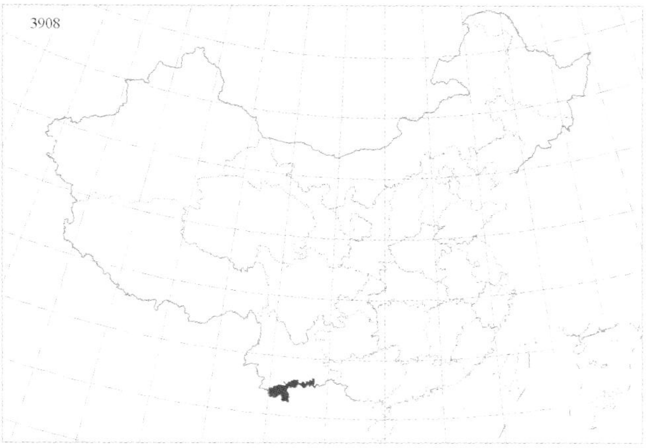

Fabaceae 豆科 *Archidendron* 猴耳环属
Archidendron balansae (Oliv.) I. C. Nielsen 锈毛棋子豆
Trees, deciduous

18.3 (17.2~21.0); 10.9 (9.1~14.2); 23.5 (22.4~25.6); 18.3 (17.3~21.0); 159.6 (147.0~191.8); 0.0 (0.0~0.0); 892 (833~1053) ❙ 1369 (1161~1522); 768 (633~860); 55 (48~68) ❙ 53.9 (37.1~80.6); 892 (833~1053); 414.4 (314.4~551.6)

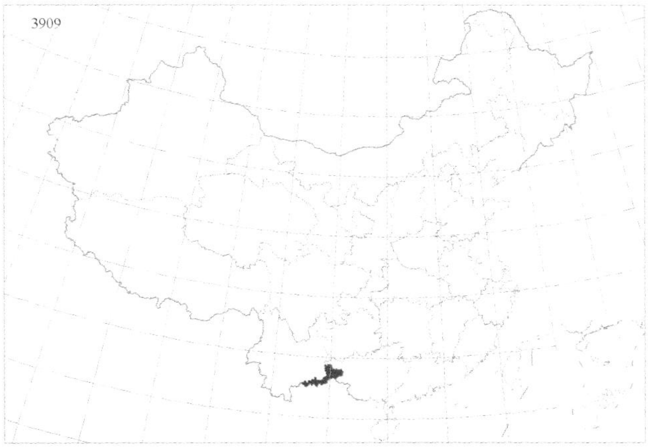

Fabaceae 豆科 *Archidendron* 猴耳环属
Archidendron chevalieri (Kosterm.) I. C. Nielsen 坛腺棋子豆
Small trees, deciduous

21.4 (17.7~22.7); 13.6 (10.9~15.0); 27.1 (22.5~28.6); 21.4 (17.8~22.8); 197.1 (153.4~213.1); 0.0 (0.0~0.0); 1119 (854~1221) ❙ 1724 (1302~2435); 952 (670~1428); 101 (59~130) ❙ 54.1 (12.9~99.5); 1119 (854~1221); 345.3 (235.1~551.6)

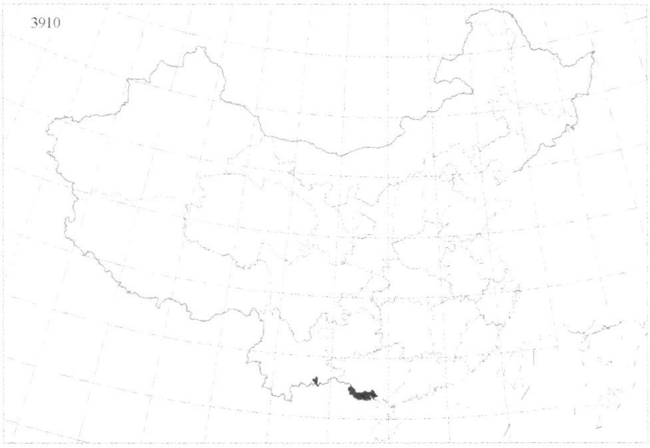

Fabaceae 豆科 *Archidendron* 猴耳环属
Archidendron clypearia (Jack) I. C. Nielsen 猴耳环
Trees, evergreen

19.8 (3.6~24.8); 11.7 (1.3~20.1); 26.2 (15.0~29.0); 19.9 (8.7~24.8); 178.3 (52.6~237.6); 0.1 (0.0~8.8); 1028 (574~1385) ❙ 1505 (868~2435); 699 (312~1428); 112 (41~231) ❙ 49.1 (-34.3~145.8); 1013 (574~1360); 310.1 (139.6~659.0)

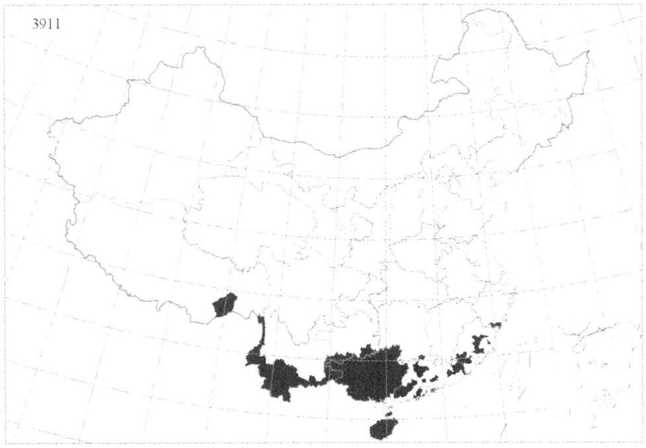

Fabaceae 豆科 *Archidendron* 猴耳环属
Archidendron cordifolium (T. L. Wu) I. C. Nielsen 心叶猴耳环
Trees, evergreen

19.4; 12.6; 24.1; 19.4; 172.6; 0.0; 954 ❙ 1445; 787; 64 ❙ 52.3; 954; 521.5

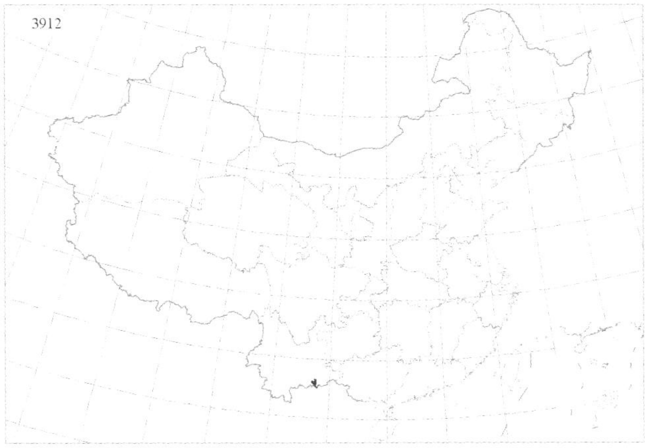

Fabaceae 豆科 *Archidendron* 猴耳环属

Archidendron eberhardtii I. C. Nielsen 大棋子豆

Trees, deciduous

21.2; 12.8; 27.4; 21.3; 195.0; 0.0; 1112 ❙ 1419; 748; 93 ❙ 29.3; 1112; 300.3

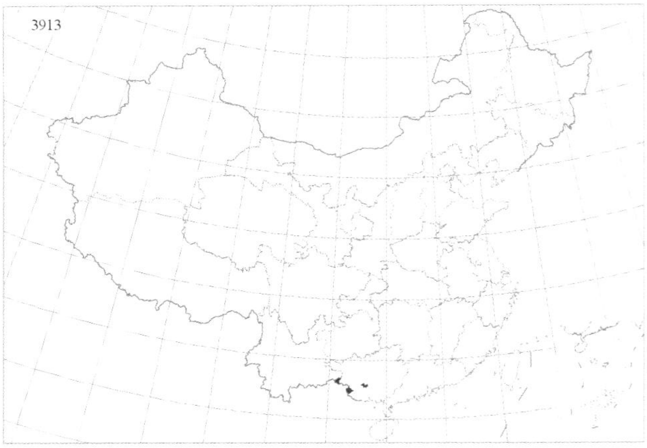

Fabaceae 豆科 *Archidendron* 猴耳环属

Archidendron ellipticum (Blume) I. C. Nielsen 椭圆叶猴耳环

Trees, evergreen

18.0 (16.4~21.0); 11.2 (9.1~14.2); 22.7 (20.6~25.6); 18.1 (16.4~21.0); 156.9 (137.3~191.8); 0.0 (0.0~0.0); 874 (795~1053) ❙ 1269 (939~1688); 710 (524~959); 51 (37~68) ❙ 45.5 (6.5~112.2); 850 (744~1053); 390.7 (293.0~627.0)

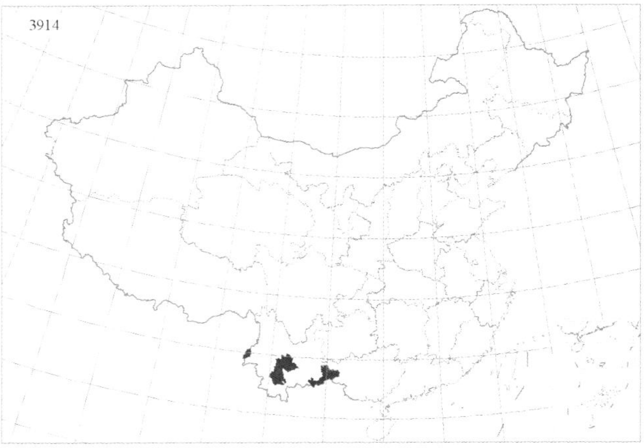

Fabaceae 豆科 *Archidendron* 猴耳环属

Archidendron kerrii (Gagnep.) I. C. Nielsen 碟腺棋子豆

Small trees, deciduous

19.3 (16.3~22.7); 12.2 (9.1~15.1); 24.3 (21.4~28.6); 19.3 (16.4~22.8); 172.0 (136.4~213.1); 0.0 (0.0~0.0); 960 (797~1221) ❙ 1505 (1074~2435); 820 (582~1428); 71 (43~130) ❙ 56.6 (12.9~99.5); 956 (797~1221); 402.4 (235.1~659.0)

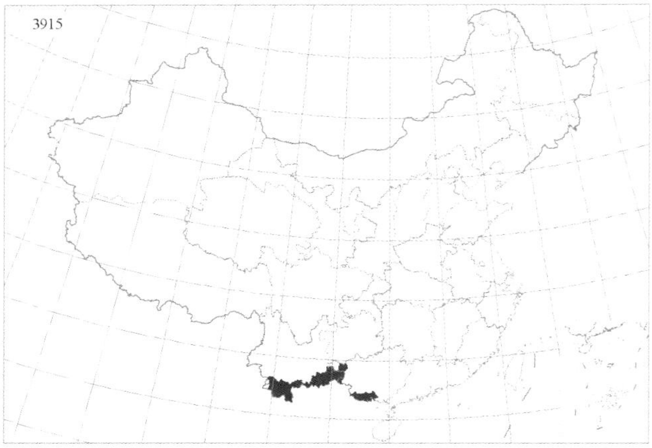

Fabaceae 豆科 *Archidendron* 猴耳环属

Archidendron laoticum (Gagnep.) I. C. Nielsen 老挝棋子豆

Trees, deciduous

20.0; 14.6; 23.2; 20.1; 180.8; 0.0; 963 ❙ 1523; 681; 67 ❙ 58.4; 942; 581.0

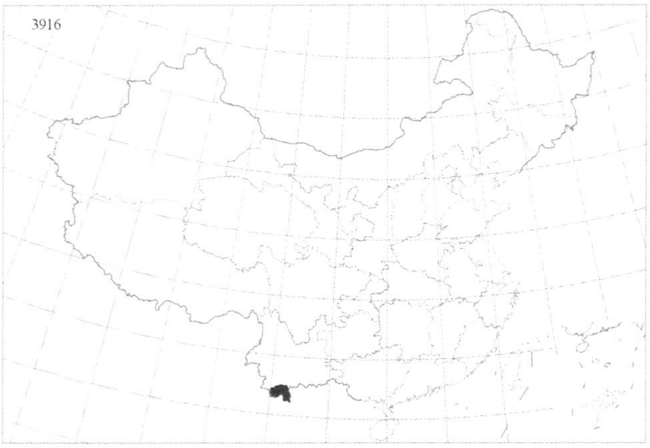

Fabaceae 豆科 *Archidendron* 猴耳环属

Archidendron lucidum (Benth.) I. C. Nielsen 亮叶猴耳环

Trees, evergreen

18.5 (10.6~24.8); 9.6 (-0.2~20.1); 26.4 (19.4~28.9); 18.6 (10.6~24.8); 163.0 (79.4~237.6); 0.3 (0.0~12.1); 964 (643~1385) ❙ 1505 (985~2993); 654 (442~1847); 118 (37~547) ❙ 58.5 (-12.0~246.5); 955 (643~1216); 257.0 (108.2~659.0)

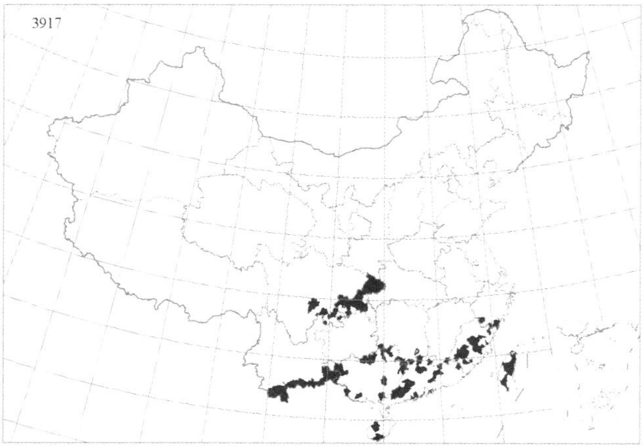

Fabaceae 豆科 *Archidendron* 猴耳环属

Archidendron robinsonii (Gagnep.) I. C. Nielsen 棋子豆

Trees, deciduous

19.9 (16.9~22.7); 12.4 (9.5~15.0); 25.3 (21.4~28.6); 19.9 (16.9~22.8); 178.9 (142.7~213.1); 0.0 (0.0~0.0); 1011 (813~1221) ❙ 1610 (1178~2435); 901 (646~1428); 83 (48~130) ❙ 59.1 (12.9~99.5); 1011 (813~1221); 367.7 (235.1~551.6)

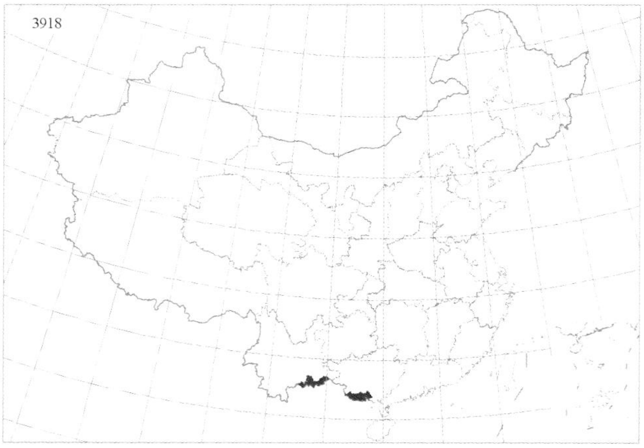

Fabaceae 豆科 *Archidendron* 猴耳环属
Archidendron tonkinense I. C. Nielsen 绢毛棋子豆
Small trees, deciduous

21.4; 13.4; 27.3; 21.5; 197.5; 0.0; 1121 ❙ 1740; 953; 116 ❙ 55.4; 1121; 289.5

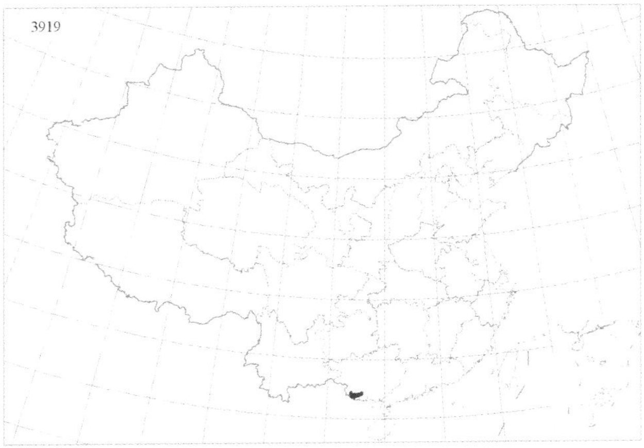

Fabaceae 豆科 *Archidendron* 猴耳环属
Archidendron turgidum (Merr.) I. C. Nielsen 大叶合欢
Small trees, deciduous

20.3 (17.2~22.8); 12.0 (7.1~15.5); 26.7 (22.4~28.8); 20.3 (17.3~22.8); 184.0 (147.0~214.1); 0.0 (0.0~0.0); 1050 (833~1225) ❙ 1553 (1161~2435); 777 (513~1428); 110 (48~182) ❙ 48.5 (5.8~99.5); 1050 (833~1225); 295.6 (188.6~551.3)

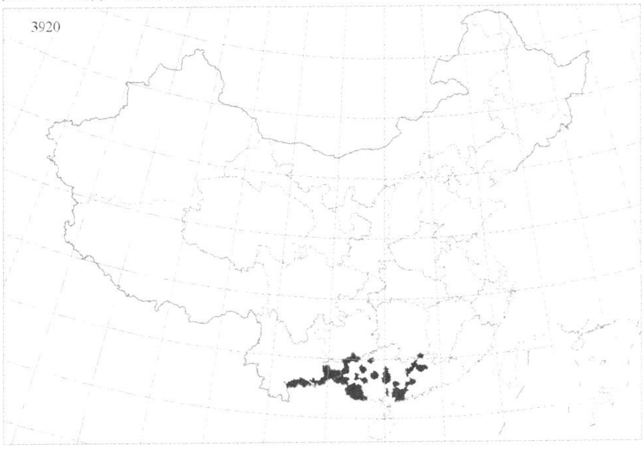

Fabaceae 豆科 *Archidendron* 猴耳环属
Archidendron utile (Chun et F. C. How) I. C. Nielsen 薄叶猴耳环
Shrubs, evergreen

19.9 (14.6~24.0); 11.8 (3.8~19.1); 26.3 (22.3~28.6); 19.9 (14.7~24.0); 179.4 (117.7~228.1); 0.1 (0.0~1.5); 1030 (779~1302) ❙ 1663 (1302~2435); 767 (539~1428); 124 (48~194) ❙ 63.5 (12.9~101.8); 1017 (779~1221); 346.7 (204.3~659.0)

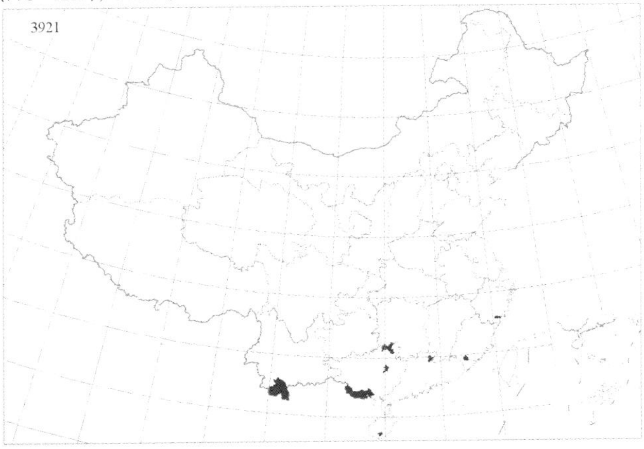

Fabaceae 豆科 *Archidendron* 猴耳环属
Archidendron xichouensis (C. Chen et H. Sun) T. L. Wu 巨腺棋子豆
Trees, deciduous

17.4; 9.8; 22.8; 17.4; 148.9; 0.0; 843 ❙ 1393; 791; 52 ❙ 65.3; 843; 369.8

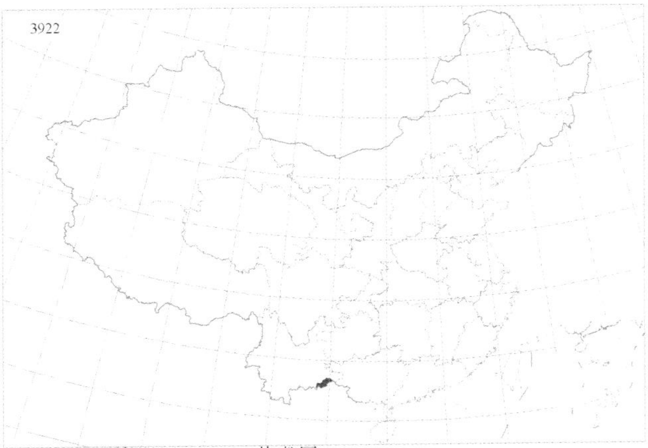

Fabaceae 豆科 *Astragalus* 黄耆属
Astragalus arbuscula Pall. 木黄耆
Shrubs, deciduous

4.1 (-1.3~9.3); -15.5 (-20.9~-9.3); 20.6 (13.4~27.5); 8.2 (4.5~12.1); 65.1 (27.2~107.6); 75.1 (40.7~104.2); 601 (433~813) ❙ 217 (120~450); 91 (51~185); 23 (12~51) ❙ -62.0 (-85.3~3.9); 216 (120~404); 91.0 (16.6~213.5)

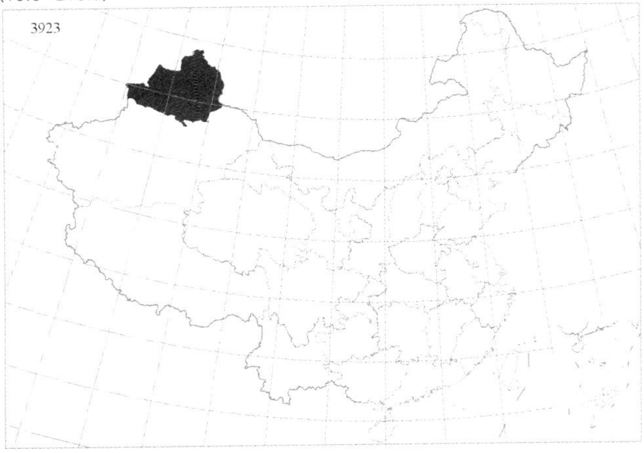

Fabaceae 豆科 *Astragalus* 黄耆属
Astragalus dendroides Kar. et Kir. 树黄耆
Shrubs, deciduous

2.9 (-1.3~9.0); -15.7 (-20.9~-9.3); 17.3 (13.2~23.2); 7.0 (4.6~10.7); 50.4 (26.7~88.9); 74.7 (40.7~103.9); 537 (440~703) ❙ 310 (225~406); 143 (65~211); 27 (12~48) ❙ -40.1 (-65.1~-7.6); 310 (225~406); 216.4 (165.7~297.3)

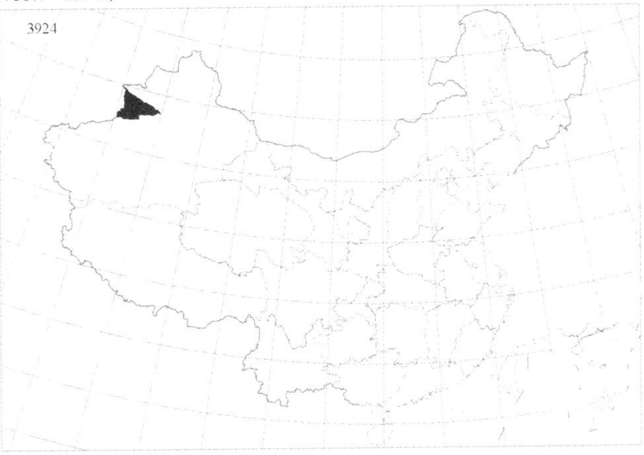

Fabaceae 豆科 *Astragalus* 黄耆属
Astragalus henryi Oliv. 秦岭黄耆
Small shrubs, deciduous

11.1 (7.4~13.0); -0.6 (-4.2~1.9); 22.3 (18.7~24.2); 11.3 (8.2~13.1); 87.5 (58.8~102.9); 13.8 (5.9~29.4); 674 (561~732) ❙ 948 (702~1335); 418 (315~550); 46 (15~86) ❙ 41.3 (-0.2~101.4); 673 (561~732); 274.9 (243.3~318.2)

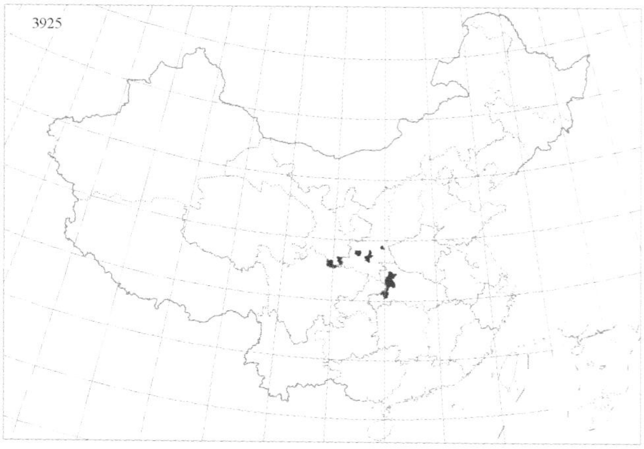

Fabaceae 豆科 *Astragalus* 黄耆属
Astragalus oplites Benth. ex baker 刺叶柄黄耆
Small shrubs, deciduous

6.1 (-4.3~12.5); -8.8 (-20.1~-5.1); 18.5 (6.1~26.5); 8.5 (1.7~13.2); 67.1 (1.9~114.0); 53.7 (23.1~121.1); 603 (324~817) ❙ 191 (53~620); 93 (20~306); 29 (7~103) ❙ -48.7 (-93.3~91.4); 144 (53~333); 51.0 (15.7~120.2)

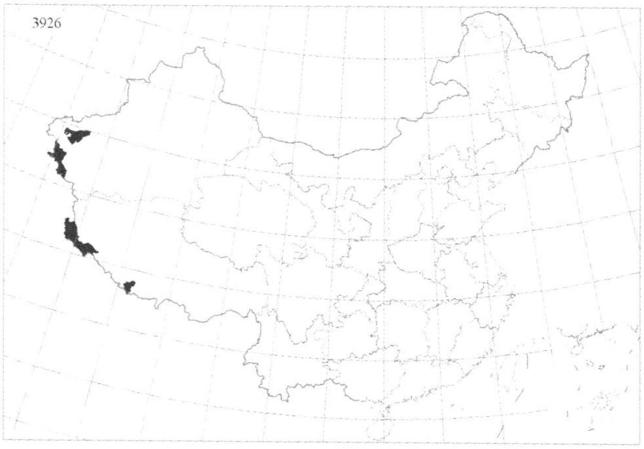

Fabaceae 豆科 *Bauhinia* 羊蹄甲属
Bauhinia acuminata L. 白花羊蹄甲
Shrubs or small trees, deciduous

19.2 (13.8~24.2); 11.4 (6.7~17.6); 25.3 (19.1~29.0); 19.2 (13.8~24.2); 170.7 (105.9~230.4); 0.0 (0.0~0.0); 982 (714~1331) ❙ 1483 (963~1841); 680 (425~835); 111 (29~189) ❙ 52.2 (10.8~99.9); 976 (690~1331); 318.0 (120.3~659.0)

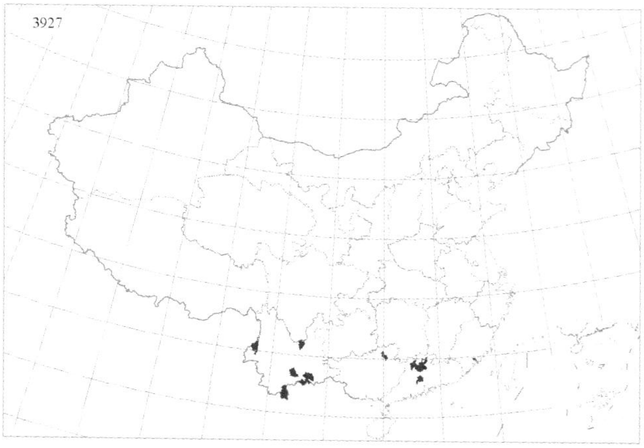

Fabaceae 豆科 *Bauhinia* 羊蹄甲属
Bauhinia apertilobata Merr. et F. P. Metcalf 阔裂叶羊蹄甲
Vines

18.8 (14.6~21.9); 9.1 (3.8~13.5); 27.3 (24.3~28.8); 18.8 (14.7~22.0); 165.7 (117.7~203.4); 0.0 (0.0~1.5); 979 (779~1168) ❙ 1698 (1459~2005); 623 (507~902); 188 (132~259) ❙ 74.9 (26.8~137.5); 978 (779~1168); 249.9 (182.7~302.2)

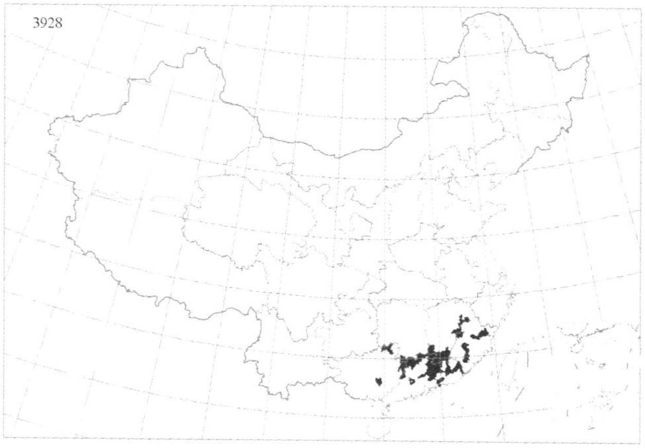

Fabaceae 豆科 *Bauhinia* 羊蹄甲属
Bauhinia aurea H. Lév. 火索藤
Vines

18.2 (10.1~22.6); 9.6 (1.8~15.3); 25.2 (16.6~29.0); 18.3 (10.3~22.6); 159.9 (70.1~211.6); 0.4 (0.0~7.0); 949 (621~1206) ❙ 1326 (879~2024); 679 (472~957); 80 (22~194) ❙ 40.9 (5.8~87.2); 947 (621~1206); 279.4 (142.9~659.0)

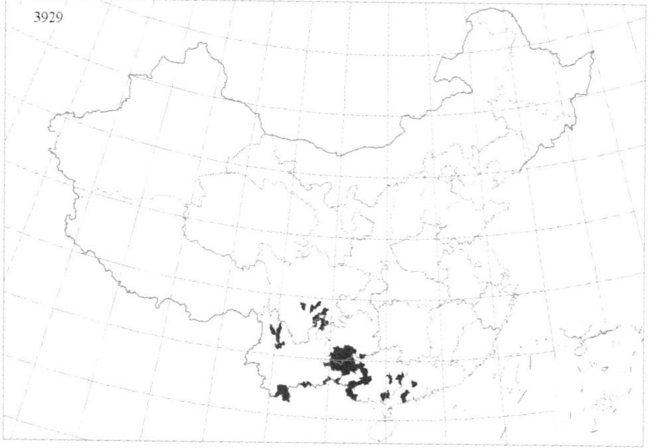

Fabaceae 豆科 *Bauhinia* 羊蹄甲属
Bauhinia bohniana L. Chen 丽江羊蹄甲
Shrubs

11.8; 5.0; 17.5; 11.9; 83.9; 1.3; 661 ❙ 968; 592; 25 ❙ 46.9; 652; 305.7

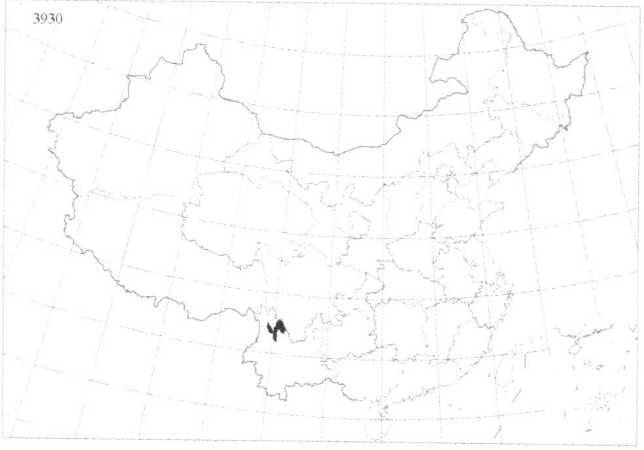

Fabaceae 豆科 *Bauhinia* 羊蹄甲属
Bauhinia brachycarpa Wall. ex Benth. 鞍叶羊蹄甲
Shrubs or small trees

13.4 (1.2~22.2)；4.5 (-7.6~13.9)；21.0 (9.0~29.0)；13.6 (3.3~22.2)；109.6 (11.9~206.5)；8.3 (0.0~56.5)；746 (390~1188) ❙ 1068 (520~1625)；565 (317~848)；47 (10~173) ❙ 46.4 (-14.2~128.6)；737 (390~1188)；266.4 (127.1~530.1)

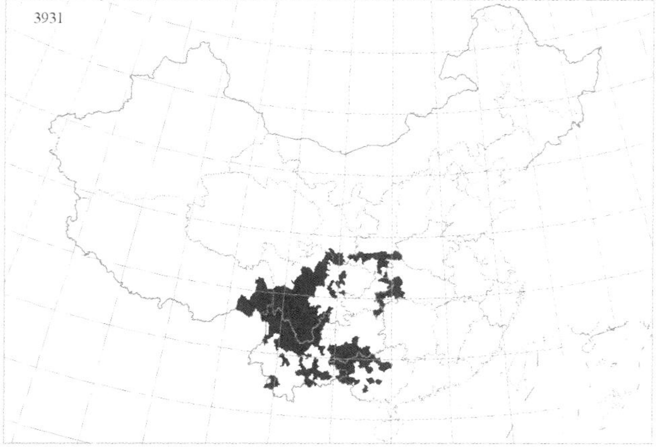

Fabaceae 豆科 *Bauhinia* 羊蹄甲属
Bauhinia calciphila D. X. Zhang et T. C. Chen 石山羊蹄甲
Vines

15.2 (10.1~19.3)；8.2 (2.3~13.1)；20.5 (16.9~23.9)；15.2 (10.2~19.4)；123.8 (68.1~173.2)；1.0 (0.0~5.2)；776 (618~939) ❙ 1014 (797~1178)；580 (411~646)；34 (15~51) ❙ 33.5 (-14.2~70.8)；745 (618~903)；299.0 (218.0~383.4)

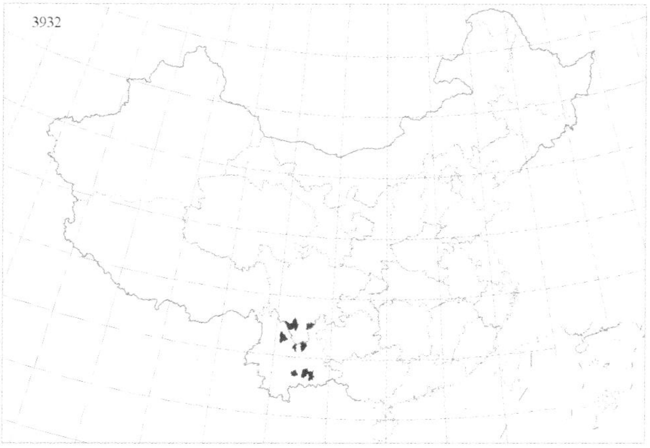

Fabaceae 豆科 *Bauhinia* 羊蹄甲属
Bauhinia carcinophylla Merrill 蟹钳叶羊蹄甲
Vines

15.2；7.3；21.3；15.3；130.5；7.3；846 ❙ 1314；746；64 ❙ 56.0；846；284.8

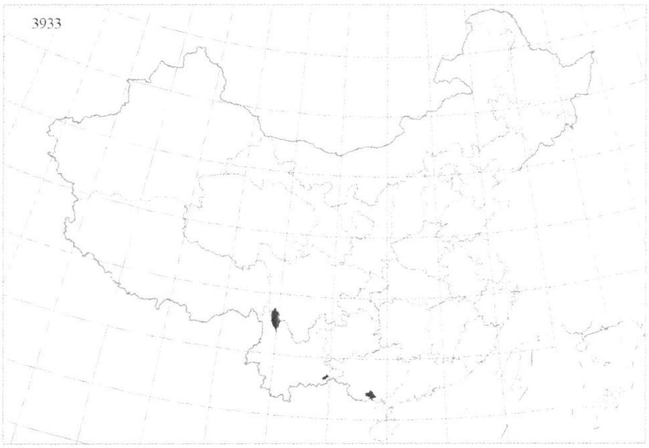

Fabaceae 豆科 *Bauhinia* 羊蹄甲属
Bauhinia cercidifolia D. X. Zhang 紫荆叶羊蹄甲
Vines

21.9；13.2；28.6；22.0；203.8；0.0；1176 ❙ 1518；773；138 ❙ 29.1；1176；272.3

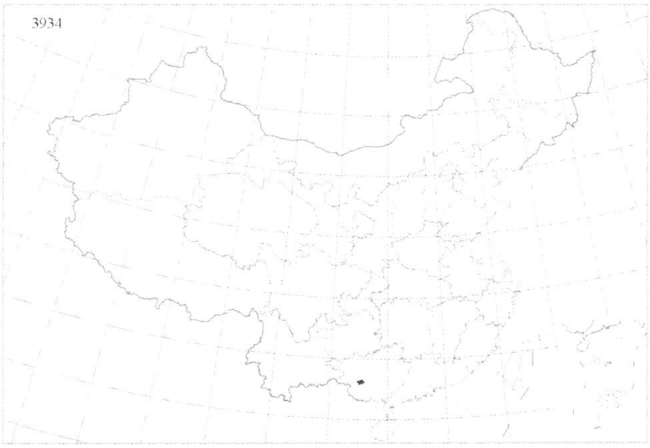

Fabaceae 豆科 *Bauhinia* 羊蹄甲属
Bauhinia chalcophylla L. Chen 多花羊蹄甲
Vines

15.5；7.9；21.5；15.6；128.2；1.4；793 ❙ 1144；646；46 ❙ 43.7；785；302.2

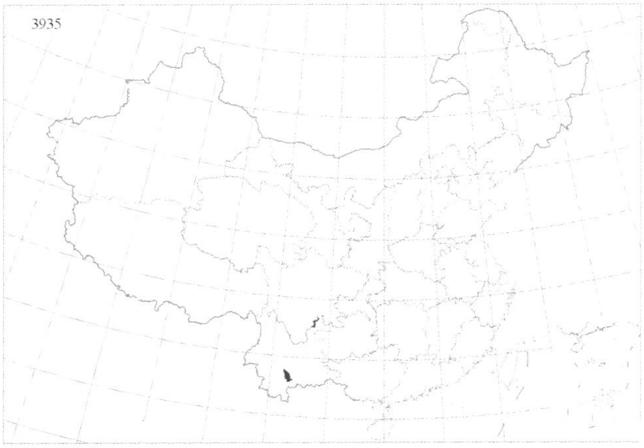

Fabaceae 豆科 *Bauhinia* 羊蹄甲属
Bauhinia championii (Benth.) Benth. 龙须藤
Vines, deciduous

18.0 (13.4~24.6)；7.9 (2.6~19.5)；27.0 (19.1~29.8)；18.0 (13.4~24.7)；156.4 (102.7~235.9)；0.2 (0.0~3.6)；948 (721~1372) ❙ 1504 (958~2993)；604 (392~1428)；149 (43~547) ❙ 60.0 (-12.0~219.3)；942 (721~1360)；226.7 (134.6~548.0)

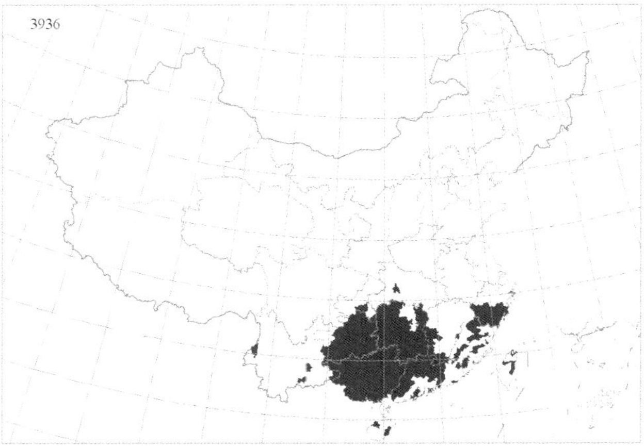

Fabaceae 豆科 *Bauhinia* 羊蹄甲属
Bauhinia coccinea subsp. tonkinensis (Gagnep.) K. Larsen et S. S. Larsen 绯红羊蹄甲
Shrubs

18.6; 11.8; 23.2; 18.6; 162.6; 0.0; 907 ❙ 1356; 740; 59 ❙ 49.8; 907; 475.5

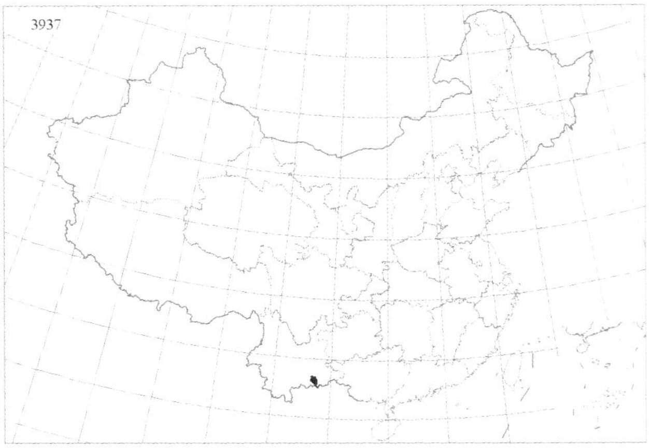

Fabaceae 豆科 *Bauhinia* 羊蹄甲属
Bauhinia comosa Craib 石山羊蹄甲
Vines

14.1 (5.4~19.3); 6.5 (-2.1~13.1); 20.2 (12.3~26.6); 14.1 (5.8~19.4); 111.0 (31.5~173.2); 1.7 (0.0~25.6); 737 (481~939) ❙ 971 (741~1178); 563 (411~646); 32 (15~57) ❙ 33.5 (-14.2~70.8); 707 (481~912); 276.9 (142.9~383.4)

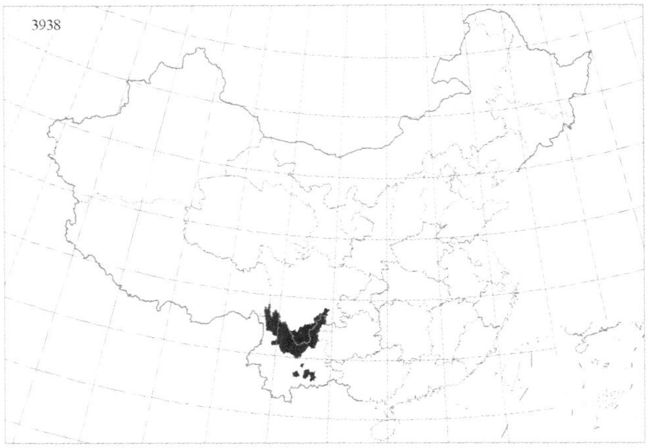

Fabaceae 豆科 *Bauhinia* 羊蹄甲属
Bauhinia corymbosa Roxb. 首冠藤
Vines

20.4 (14.6~24.8); 12.0 (3.8~20.1); 27.4 (24.6~28.7); 20.5 (14.7~24.8); 186.2 (117.7~237.6); 0.0 (0.0~1.5); 1082 (779~1385) ❙ 1639 (1179~2435); 705 (392~1428); 136 (41~216) ❙ 54.1 (-11.9~101.8); 1062 (779~1360); 289.0 (163.7~620.0)

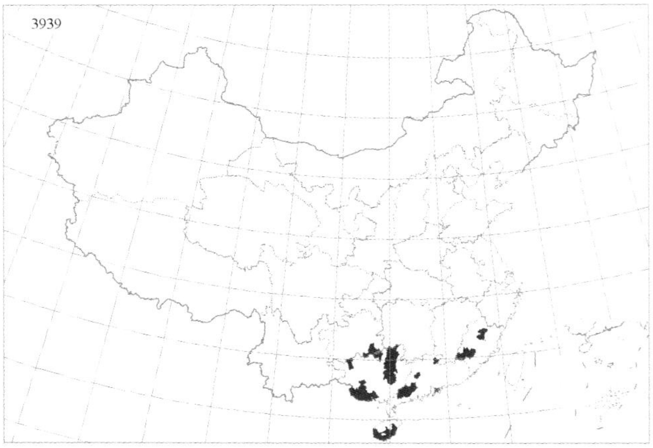

Fabaceae 豆科 *Bauhinia* 羊蹄甲属
Bauhinia corymbosa var. longipes Hosok. 长序首冠藤
Vines

24.7; 19.8; 27.9; 24.7; 236.5; 0.0; 1376 ❙ 1588; 534; 58 ❙ 15.4; 1212; 494.7

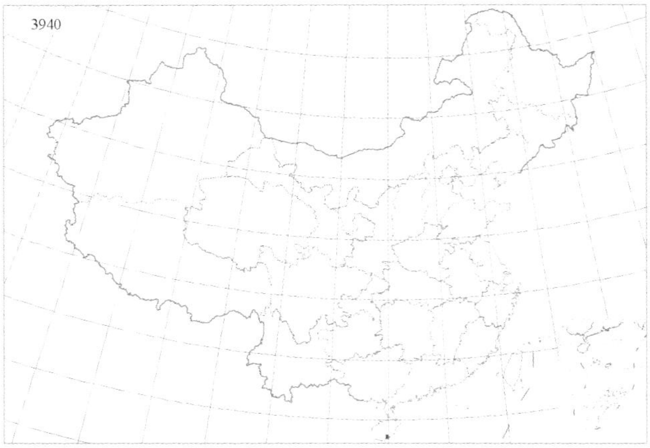

Fabaceae 豆科 *Bauhinia* 羊蹄甲属
Bauhinia damiaoshanensis T. C. Chen 大苗山羊蹄甲
Vines

18.3; 8.7; 26.1; 18.3; 160.0; 0.0; 933 ❙ 1373; 657; 101 ❙ 47.6; 933; 258.6

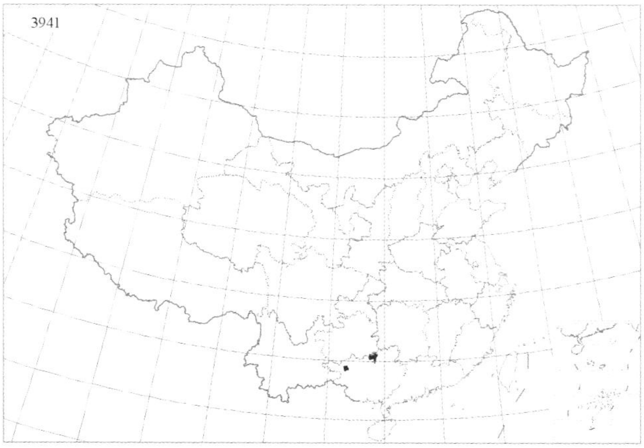

Fabaceae 豆科 *Bauhinia* 羊蹄甲属
Bauhinia delavayi Franch. 薄荚羊蹄甲
Vines

13.4 (5.4~18.9); 5.8 (-2.1~12.0); 19.7 (12.3~26.6); 13.5 (5.8~19.0); 103.7 (31.5~167.8); 2.3 (0.0~25.6); 716 (481~929) ❙ 964 (741~1337); 560 (411~727); 31 (12~69) ❙ 36.7 (-14.2~92.4); 688 (481~912); 275.1 (142.9~407.1)

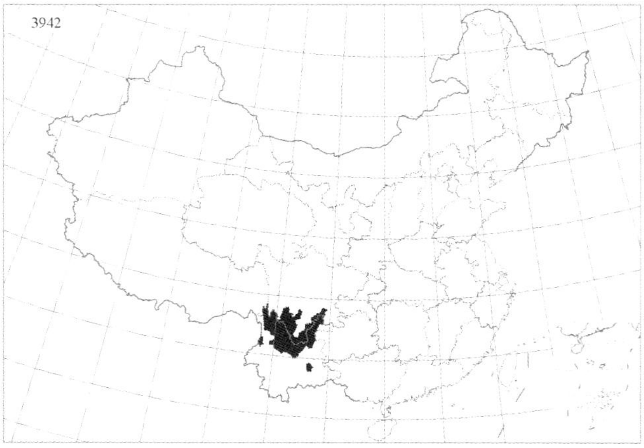

Fabaceae 豆科 *Bauhinia* 羊蹄甲属
Bauhinia didyma L. Chen 孪叶羊蹄甲
Vines

21.4 (17.6~22.7); 13.3 (7.3~15.0); 27.6 (26.3~28.6); 21.5 (17.6~22.8); 197.6 (151.6~213.1); 0.0 (0.0~0.0); 1129 (909~1221) ❙ 1823 (1302~2435); 940 (639~1428); 121 (86~142) ❙ 61.3 (12.9~99.5); 1129 (909~1221); 270.4 (225.9~318.8)

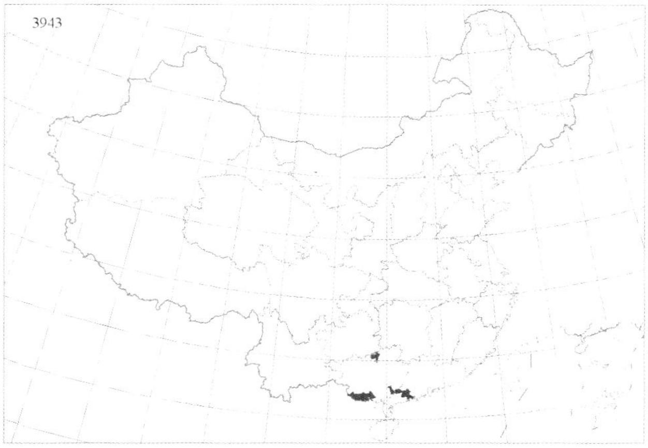

Fabaceae 豆科 *Bauhinia* 羊蹄甲属
Bauhinia erythropoda Hayata 锈荚藤
Vines

22.2 (17.7~24.8); 15.9 (10.9~20.1); 26.6 (22.5~28.8); 22.2 (17.8~24.8); 206.5 (153.4~237.6); 0.0 (0.0~0.0); 1166 (854~1385) ❙ 1636 (1368~2148); 713 (473~1234); 84 (41~132) ❙ 42.3 (1.5~86.7); 1124 (854~1360); 453.1 (203.4~659.0)

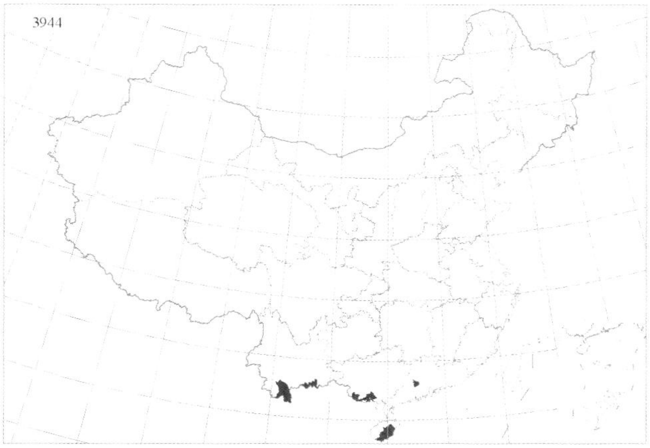

Fabaceae 豆科 *Bauhinia* 羊蹄甲属
Bauhinia esquirolii Gagnepain 元江羊蹄甲
Vines

15.4 (10.8~19.3); 8.0 (3.9~13.1); 21.3 (16.6~24.1); 15.4 (10.9~19.4); 125.5 (71.8~173.2); 0.2 (0.0~1.3); 786 (633~939) ❙ 1026 (797~1244); 560 (411~637); 43 (23~78) ❙ 32.6 (-14.2~58.8); 750 (633~903); 283.7 (218.0~349.8)

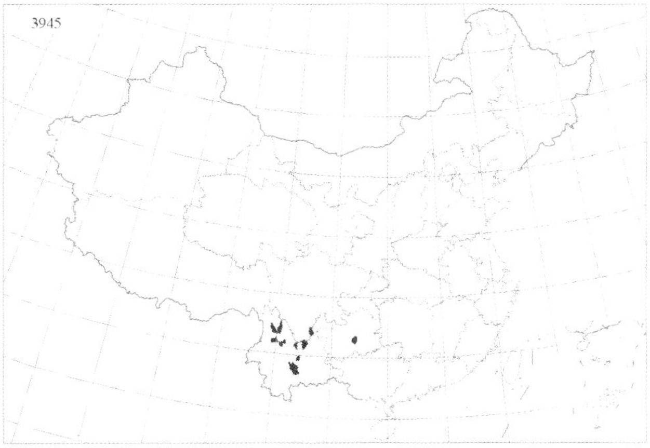

Fabaceae 豆科 *Bauhinia* 羊蹄甲属
Bauhinia glauca (Wall. ex Benth.) Wall. 粉叶羊蹄甲
Vines

16.9 (9.3~22.7); 6.5 (-2.5~15.2); 26.4 (19.1~29.9); 17.0 (9.5~22.8); 144.6 (69.8~213.1); 1.1 (0.0~20.6); 897 (603~1221) ❙ 1443 (760~2435); 567 (352~1428); 139 (17~259) ❙ 60.9 (5.0~147.0); 890 (603~1221); 230.1 (122.7~659.0)

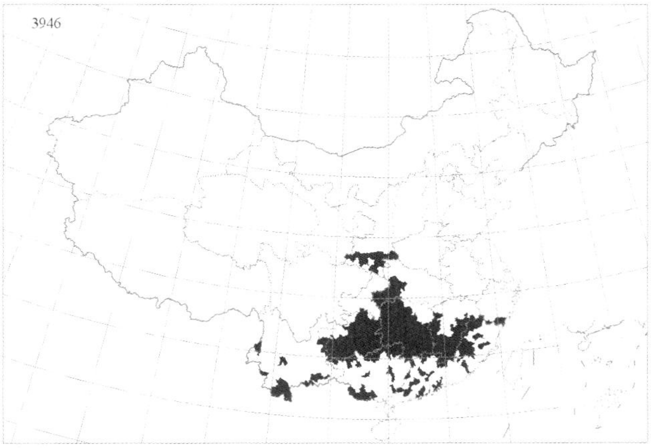

Fabaceae 豆科 *Bauhinia* 羊蹄甲属
Bauhinia glauca subsp. tenuiflora (Watt ex C. B. Clarke) K. Larsen et S. S. Larsen 薄叶羊蹄甲 Vines

16.6 (10.1~21.7); 6.2 (-0.2~15.1); 26.0 (17.7~29.3); 16.6 (10.3~21.7); 140.5 (70.1~200.7); 1.1 (0.0~12.1); 880 (621~1147) ❙ 1406 (711~1953); 572 (357~932); 124 (15~228) ❙ 60.3 (9.3~128.9); 877 (621~1147); 243.2 (123.1~659.0)

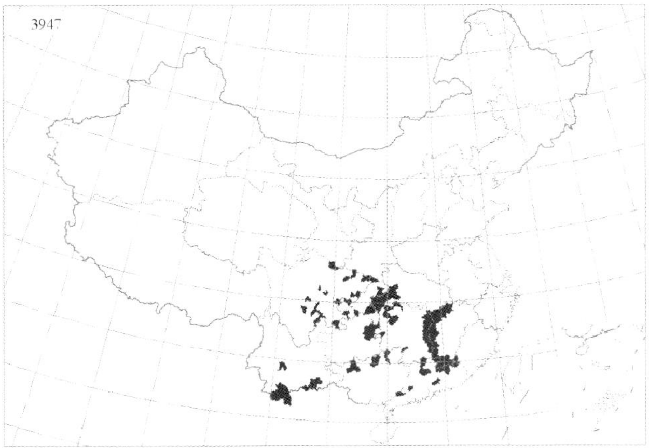

Fabaceae 豆科 *Bauhinia* 羊蹄甲属
Bauhinia hainanensis Merr. et Chun ex L. Chen 海南羊蹄甲
Vines

24.1; 18.8; 27.9; 24.2; 229.9; 0.0; 1322 ❙ 1302; 441; 44 ❙ -1.5; 1099; 469.4

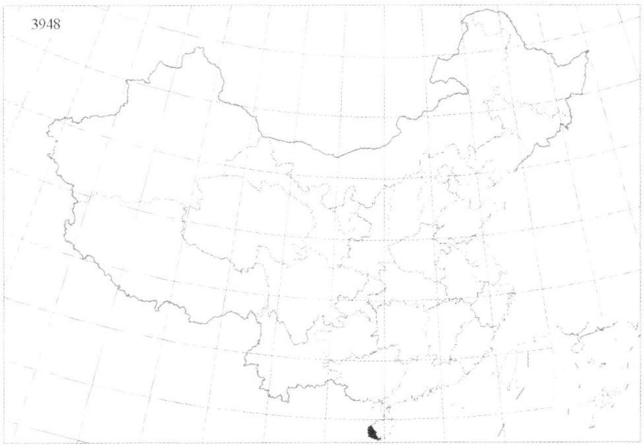

Fabaceae 豆科 *Bauhinia* 羊蹄甲属
Bauhinia hirsuta Weinm. 粗毛羊蹄甲
Shrubs

19.0; 12.5; 23.4; 19.0; 168.1; 0.0; 928 ❙ 1376; 767; 58 ❙ 48.4; 928; 480.4

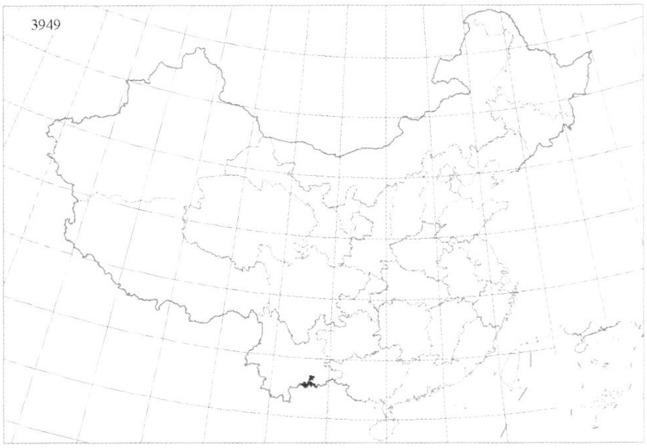

Fabaceae 豆科 *Bauhinia* 羊蹄甲属
Bauhinia hypochrysa T. C. Chen 绸缎藤
Vines

22.2; 13.9; 28.3; 22.3; 207.5; 0.0; 1190 ❙ 1374; 713; 87 ❙ 15.4; 1190; 327.7

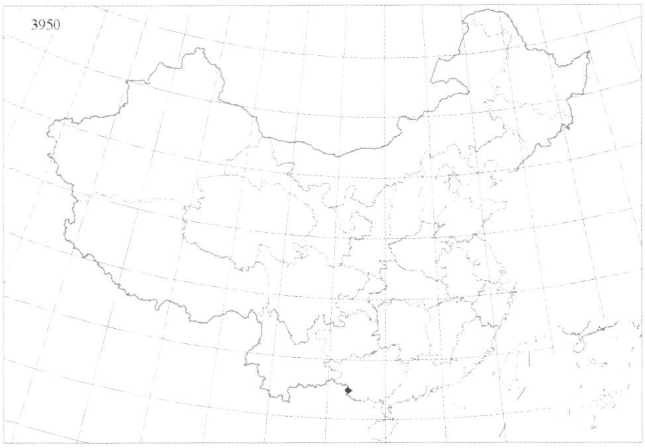

Fabaceae 豆科 *Bauhinia* 羊蹄甲属
Bauhinia hypoglauca T. Tang et F. T. Wang ex T. Chen 滇南羊蹄甲
Vines

18.7; 10.7; 24.4; 18.8; 165.0; 0.0; 949 ❙ 1203; 642; 61 ❙ 28.3; 949; 306.3

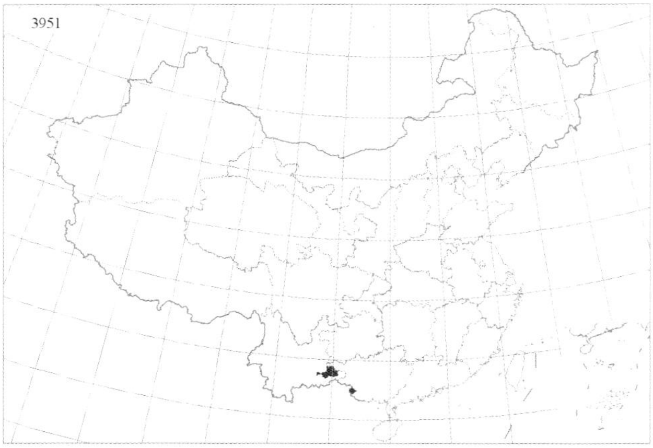

Fabaceae 豆科 *Bauhinia* 羊蹄甲属
Bauhinia japonica Maxim. 日本羊蹄甲
Vines

19.3 (14.6~24.8); 9.8 (3.8~20.1); 27.6 (24.6~29.0); 19.4 (14.7~24.8); 172.4 (117.7~237.6); 0.1 (0.0~1.5); 1030 (779~1385) ❙ 1603 (1269~2435); 630 (403~1428); 160 (41~211) ❙ 58.7 (-1.0~101.8); 1003 (779~1331); 241.4 (151.0~527.5)

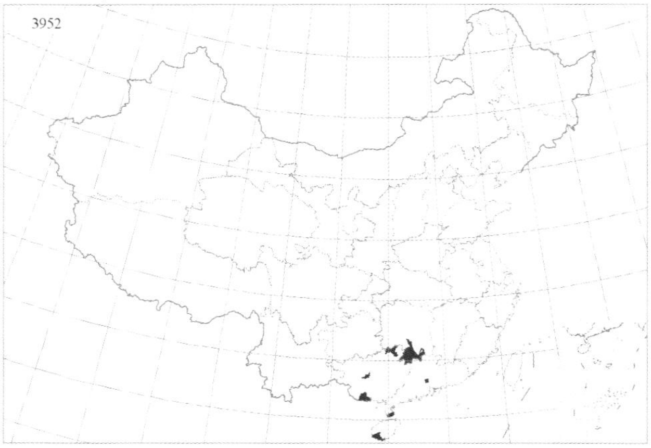

Fabaceae 豆科 *Bauhinia* 羊蹄甲属
Bauhinia khasiana Baker 牛蹄麻
Vines

23.7 (21.0~24.8); 18.2 (14.2~20.1); 27.5 (25.6~28.4); 23.7 (21.0~24.8); 224.7 (191.8~237.6); 0.0 (0.0~0.0); 1286 (1053~1385) ❙ 1386 (1179~1588); 519 (392~818); 50 (41~68) ❙ 9.0 (-11.9~44.5); 1110 (1053~1212); 487.3 (439.7~551.6)

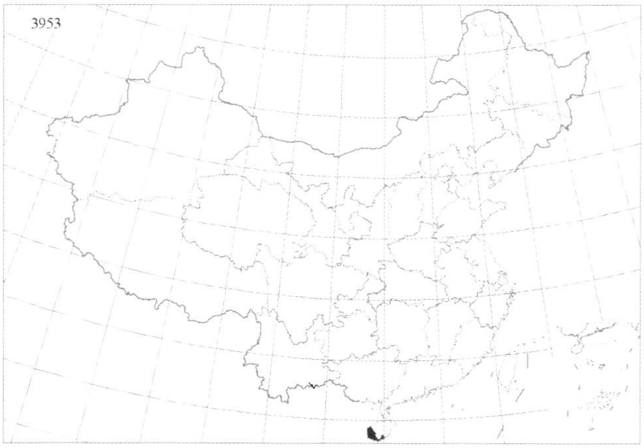

Fabaceae 豆科 *Bauhinia* 羊蹄甲属
Bauhinia khasiana var. tomentella T. C. Chen 毛叶牛蹄麻
Vines

21.0; 14.2; 25.6; 21.0; 191.8; 0.0; 1053 ❙ 1522; 818; 68 ❙ 44.5; 1053; 551.6

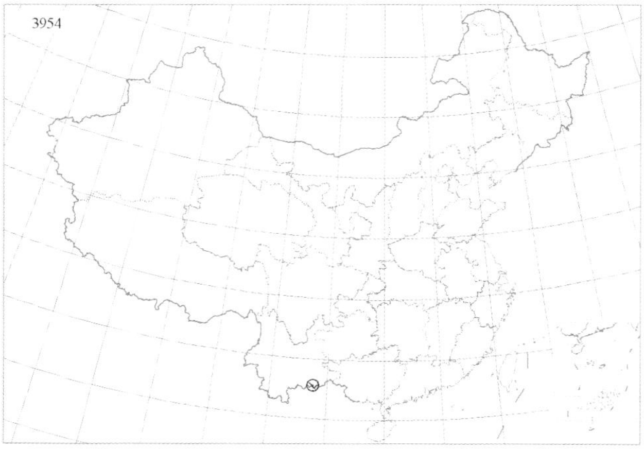

Fabaceae 豆科 *Bauhinia* 羊蹄甲属
Bauhinia lingyuenensis T. C. Chen 凌云羊蹄甲
Vines

19.0; 10.2; 25.8; 19.0; 168.4; 0.0; 958 ▎1250; 675; 60 ▎30.5; 958; 291.2

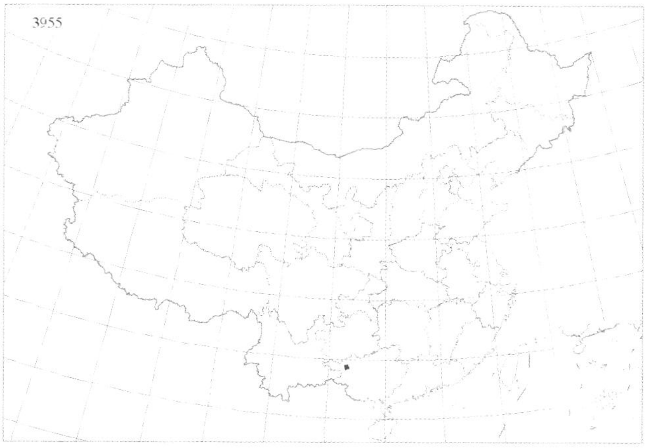

Fabaceae 豆科 *Bauhinia* 羊蹄甲属
Bauhinia longistipes T. C. Chen 长柄羊蹄甲
Vines

14.6; 7.8; 20.0; 14.6; 115.2; 0.0; 738 ▎963; 581; 26 ▎30.5; 689; 309.7

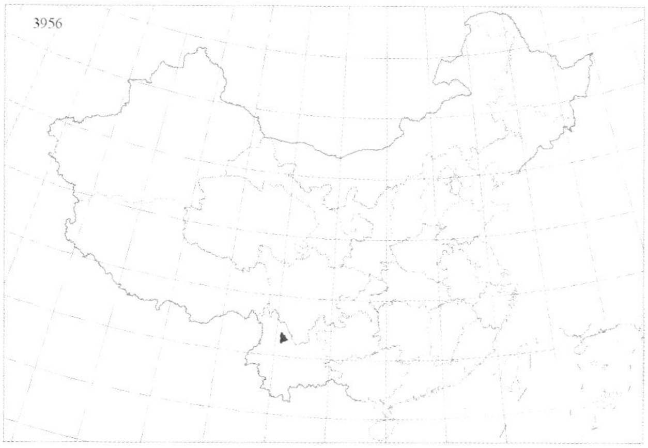

Fabaceae 豆科 *Bauhinia* 羊蹄甲属
Bauhinia nervosa (Wall. ex Benth.) Baker 棒花羊蹄甲
Shrubs

20.5; 15.0; 23.7; 20.5; 185.9; 0.0; 990 ▎1536; 678; 72 ▎55.1; 973; 614.6

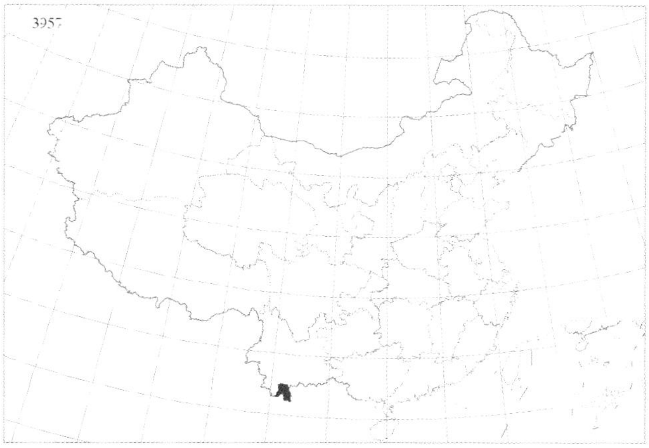

Fabaceae 豆科 *Bauhinia* 羊蹄甲属
Bauhinia ornata Kurz 缅甸羊蹄甲
Vines

23.2 (19.1~24.8); 16.8 (8.9~20.1); 27.8 (25.0~28.8); 23.2 (19.1~24.8); 218.4 (169.6~237.6); 0.0 (0.0~0.0); 1252 (945~1385) ▎1495 (868~1854); 585 (312~820); 80 (41~188) ▎20.8 (-34.3~71.1); 1162 (868~1360); 414.8 (203.9~620.0)

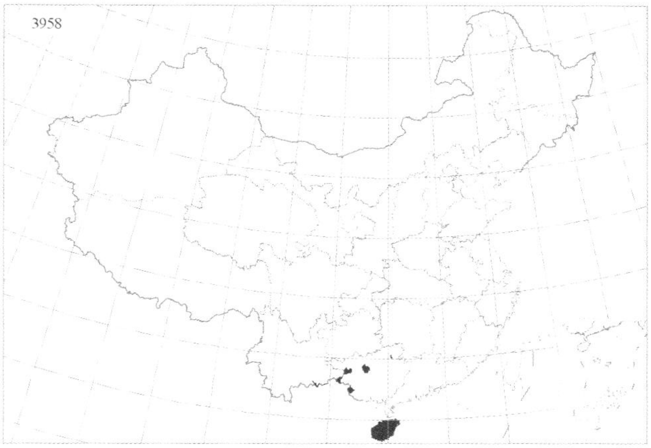

Fabaceae 豆科 *Bauhinia* 羊蹄甲属
Bauhinia ornata var. kerrii (Gagnep.) K. Larsen et S. S. Larsen 褐毛羊蹄甲
Vines

21.5 (14.8~24.8); 14.9 (5.0~20.1); 26.3 (22.3~28.9); 21.5 (14.8~24.8); 198.1 (118.0~237.6); 0.0 (0.0~0.0); 1124 (774~1385) ▎1516 (955~2035); 672 (455~969); 76 (41~141) ▎36.8 (-1.4~71.8); 1082 (774~1360); 416.0 (194.5~659.0)

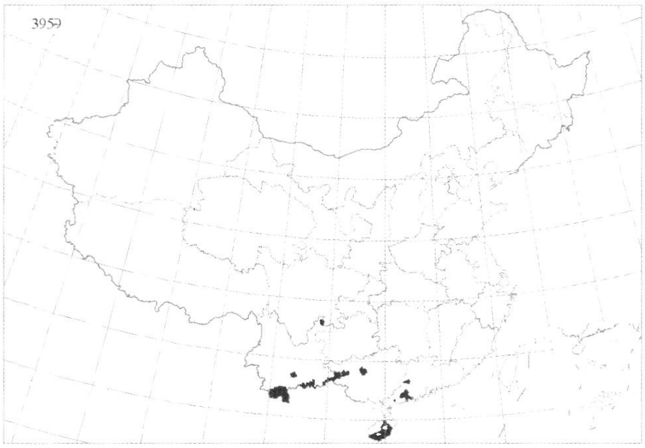

Fabaceae 豆科 *Bauhinia* 羊蹄甲属
Bauhinia ovatifolia T. C. Chen 卵叶羊蹄甲
Vines

20.6; 11.8; 27.2; 20.6; 187.7; 0.0; 1071 ▎1447; 758; 107 ▎35.3; 1071; 264.5

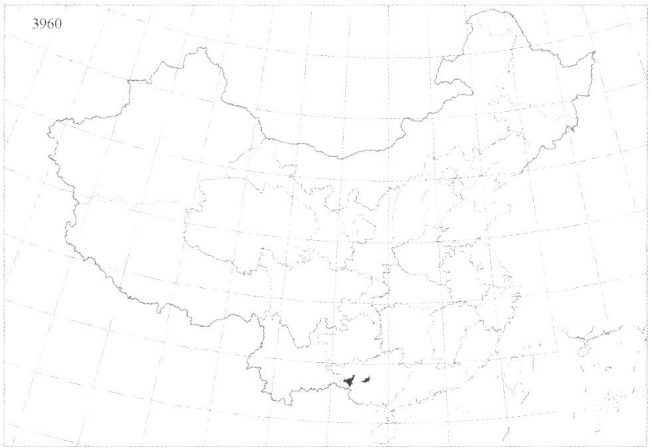

Fabaceae 豆科 *Bauhinia* 羊蹄甲属
Bauhinia paucinervata T. C. Chen 少脉羊蹄甲
Vines

21.5; 13.1; 27.8; 21.6; 199.0; 0.0; 1138 ❙ 1425; 752; 96 ❙ 25.6; 1138; 312.9

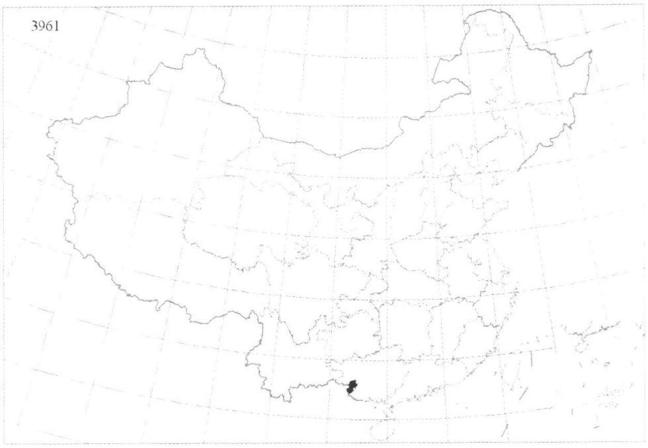

Fabaceae 豆科 *Bauhinia* 羊蹄甲属
Bauhinia purpurea L. 羊蹄甲
Shrubs or trees, evergreen

21.0 (17.3~24.7); 12.9 (6.5~19.8); 27.7 (23.3~29.0); 21.1 (17.3~24.7); 193.4 (147.9~236.5); 0.0 (0.0~0.0); 1116 (907~1376) ❙ 1728 (1269~3495); 700 (480~1033); 171 (48~897) ❙ 57.6 (-1.4~224.4); 1103 (907~1331); 271.4 (157.7~620.0)

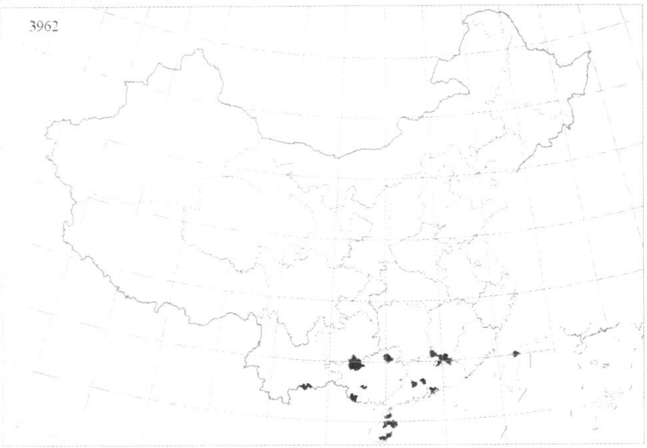

Fabaceae 豆科 *Bauhinia* 羊蹄甲属
Bauhinia pyrrhoclada Drake 红毛羊蹄甲
Vines

20.5 (18.3~24.6); 12.0 (9.3~19.5); 26.9 (23.3~28.6); 20.5 (18.4~24.7); 186.3 (160.9~235.9); 0.0 (0.0~0.0); 1065 (917~1372) ❙ 1463 (1182~2148); 737 (561~1234); 97 (43~173) ❙ 37.7 (15.4~86.7); 1061 (917~1299); 321.6 (226.9~548.0)

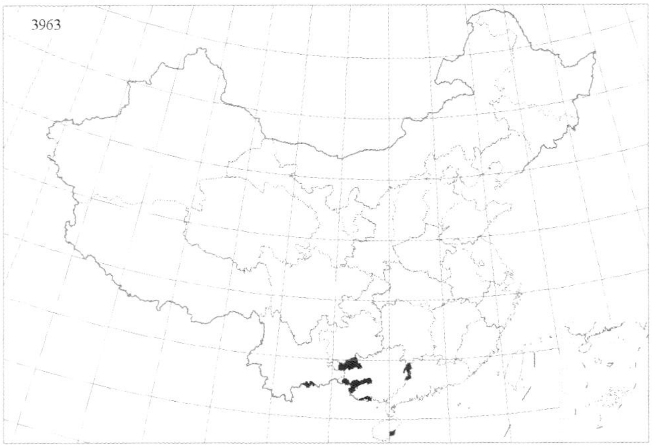

Fabaceae 豆科 *Bauhinia* 羊蹄甲属
Bauhinia quinnanensis T. C. Chen 黔南羊蹄甲
Vines

18.0 (14.8~21.9); 8.9 (3.8~13.2); 25.2 (23.0~28.6); 18.0 (14.8~22.0); 156.5 (119.2~203.8); 0.1 (0.0~1.4); 917 (795~1176) ❙ 1298 (1172~1518); 678 (543~773); 68 (50~138) ❙ 43.0 (21.9~70.8); 917 (795~1176); 263.8 (187.9~411.1)

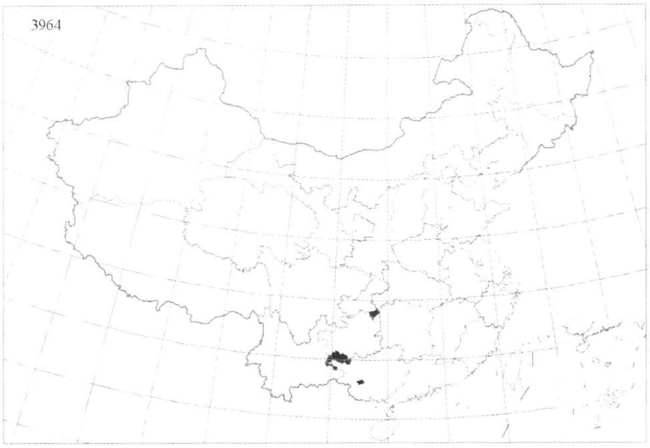

Fabaceae 豆科 *Bauhinia* 羊蹄甲属
Bauhinia racemosa Lam. 总状花羊蹄甲
Small trees, deciduous

19.3; 13.1; 23.6; 19.4; 173.2; 0.0; 939 ❙ 1143; 637; 50 ❙ 21.7; 903; 346.9

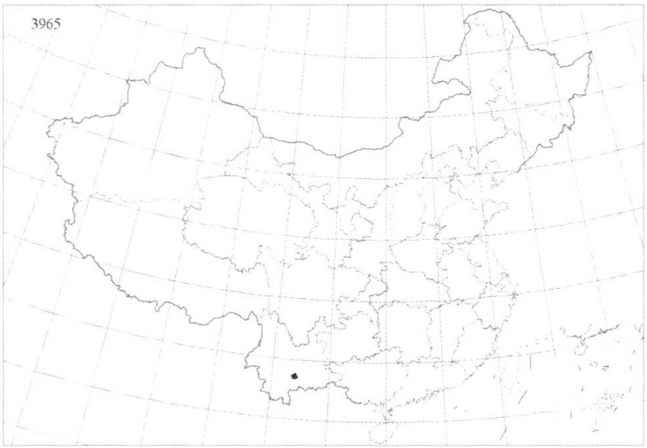

Fabaceae 豆科 *Bauhinia* 羊蹄甲属
Bauhinia rubrovillosa K. Larsen et S. S. Larsen 红背叶羊蹄甲
Vines

19.9; 11.9; 25.8; 20.0; 179.9; 0.0; 1025 ❙ 1398; 763; 70 ❙ 40.5; 1025; 343.7

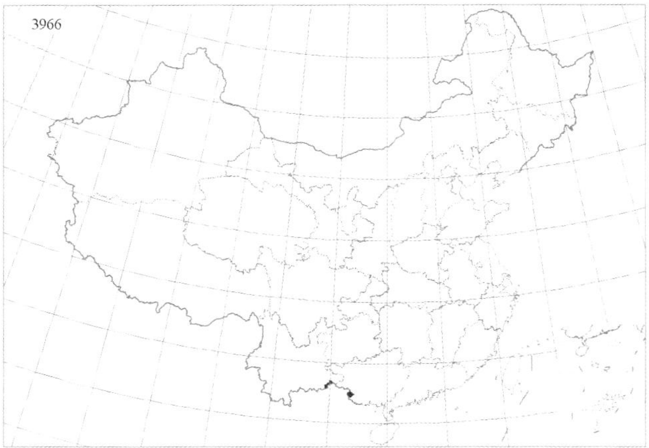

Fabaceae 豆科 *Bauhinia* 羊蹄甲属
Bauhinia scandens L. 攀缘羊蹄甲
Vines

23.5 (19.0~24.8); 17.7 (10.2~20.1); 27.7 (25.8~28.5); 23.6 (19.0~24.8); 222.8 (168.4~237.6); 0.0 (0.0~0.0); 1284 (958~1385) ❙ 1464 (1179~1354); 530 (392~675); 63 (41~115) ❙ 14.6 (-11.9~36.4); 1150 (958~1360); 472.9 (291.2~620.0)

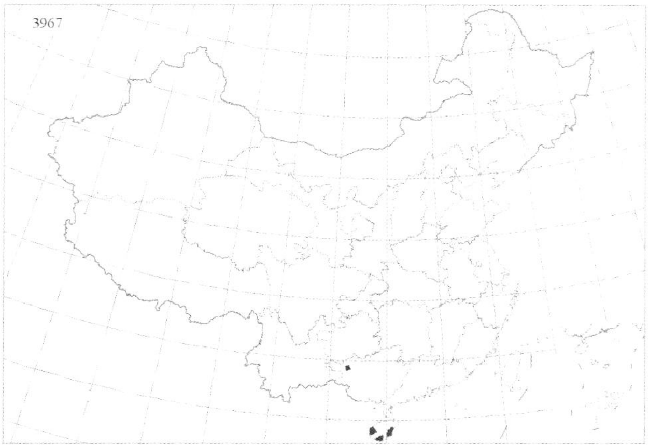

Fabaceae 豆科 *Bauhinia* 羊蹄甲属
Bauhinia tianlinensis T. C. Chen et D. X. Zhang 田林羊蹄甲
Vines

19.8; 11.1; 26.4; 19.8; 177.9; 0.0; 1010 ❙ 1182; 673; 43 ❙ 16.9; 1010; 354.5

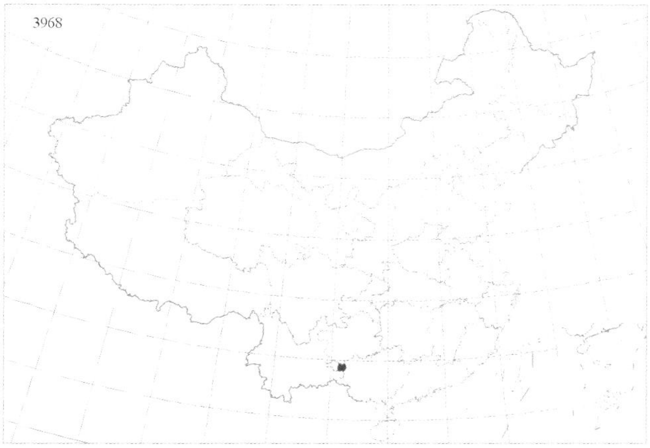

Fabaceae 豆科 *Bauhinia* 羊蹄甲属
Bauhinia tomentosa L. 黄花羊蹄甲
Shrubs

22.0; 14.0; 28.8; 22.2; 206.4; 0.0; 1184 ❙ 1758; 805; 128 ❙ 48.4; 1184; 193.0

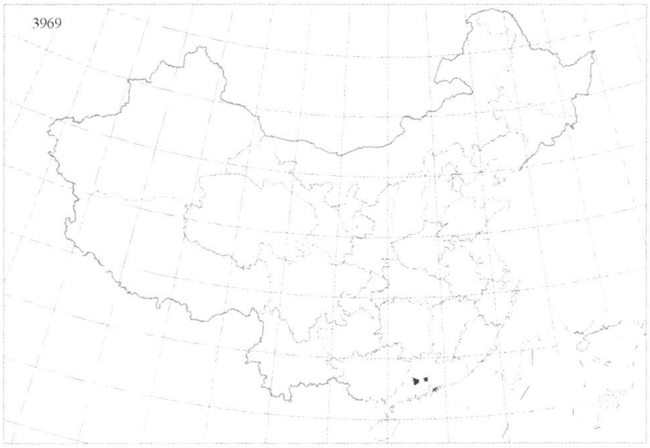

Fabaceae 豆科 *Bauhinia* 羊蹄甲属
Bauhinia touranensis Gagnep. 囊托羊蹄甲
Vines

18.6 (14.3~22.2); 10.1 (4.0~15.1); 25.2 (21.4~28.6); 18.6 (14.3~22.3); 163.7 (112.6~207.5); 0.1 (0.0~1.1); 942 (756~1190) ❙ 1361 (1178~1625); 682 (476~911); 82 (43~173) ❙ 45.5 (15.4~74.2); 940 (756~1190); 338.0 (189.5~659.0)

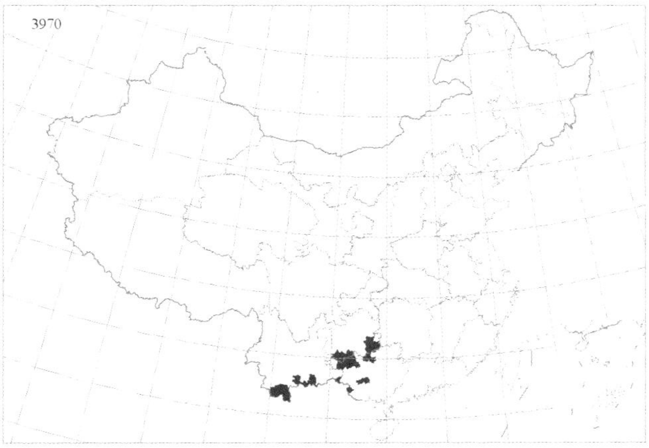

Fabaceae 豆科 *Bauhinia* 羊蹄甲属
Bauhinia variegata L. 洋紫荆
Trees, deciduous

20.1 (13.2~24.8); 12.8 (4.5~20.1); 25.8 (18.4~29.0); 20.2 (13.3~24.8); 182.2 (99.4~237.6); 0.0 (0.0~0.5); 1049 (695~1385) ❙ 1422 (868~1965); 634 (312~994); 91 (41~231) ❙ 38.6 (-34.3~99.9); 1012 (695~1360); 331.1 (120.3~659.0)

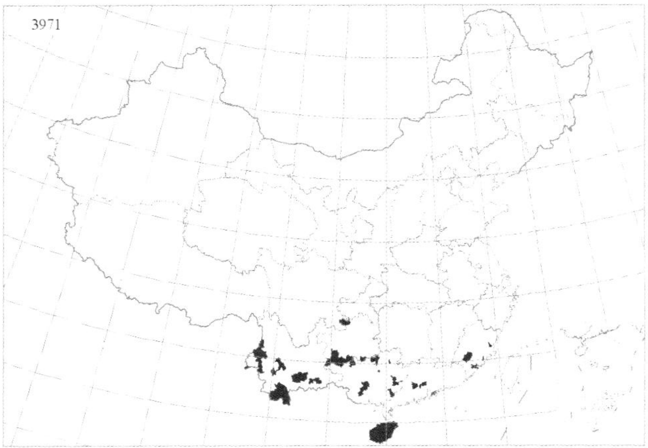

Fabaceae 豆科 *Bauhinia* 羊蹄甲属
Bauhinia variegata var. candida (Roxb.) Voigt 白花羊紫荆
Trees, deciduous

18.0 (14.2~23.8); 11.4 (6.2~18.1); 22.4 (19.1~28.2); 18.0 (14.2~23.8); 156.5 (110.5~225.8); 0.0 (0.0~0.0); 878 (727~1297) ❙ 1413 (958~2993); 777 (531~1847); 64 (41~547) ❙ 60.3 (13.3~203.4); 865 (727~1144); 410.2 (119.4~659.0)

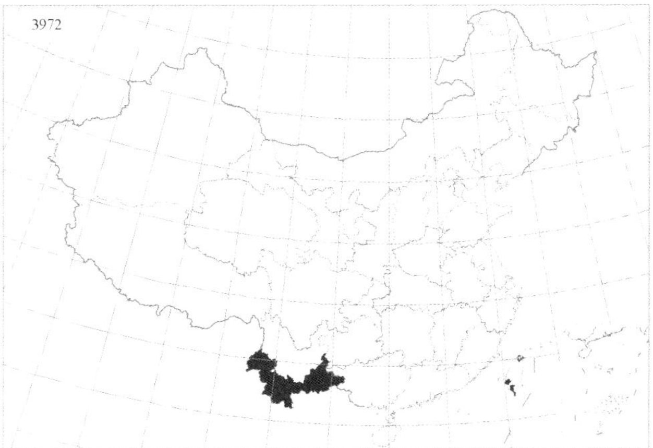

Fabaceae 豆科 *Bauhinia* 羊蹄甲属

Bauhinia venustula T. C. Chen 小巧羊蹄甲

Vines

21.8; 12.9; 28.6; 21.8; 201.9; 0.0; 1165 ❙ 1498; 712; 148 ❙ 28.6; 1163; 205.2

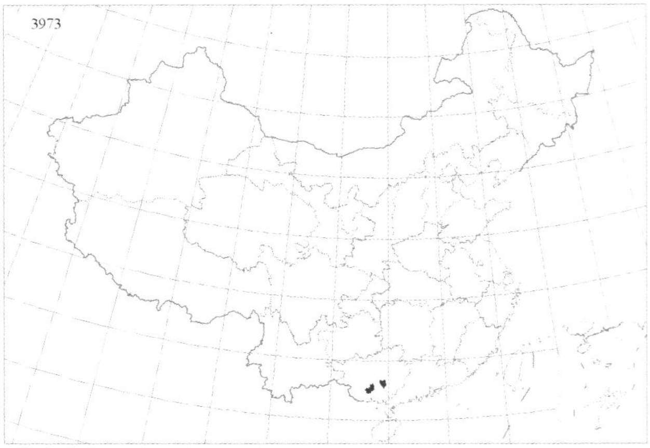

Fabaceae 豆科 *Bauhinia* 羊蹄甲属

Bauhinia viridescens Desv. 绿花羊蹄甲

Shrubs or small trees

22.0 (19.1~24.8); 16.3 (11.0~20.1); 25.8 (22.3~28.4); 22.0 (19.1~24.8); 203.9 (169.6~237.6); 0.0 (0.0~0.0); 1140 (909~1385) ❙ 1402 (1179~1607); 578 (392~820); 54 (41~77) ❙ 27.7 (-11.9~64.9); 1021 (880~1137); 499.0 (371.7~659.0)

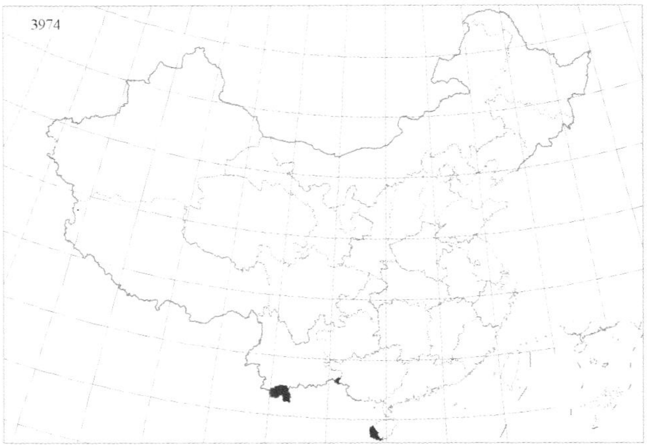

Fabaceae 豆科 *Bauhinia* 羊蹄甲属

Bauhinia wallichii J. F. Macbride 圆叶羊蹄甲

Vines

20.4; 15.1; 23.6; 20.5; 185.8; 0.0; 989 ❙ 1607; 702; 77 ❙ 62.5; 986; 659.0

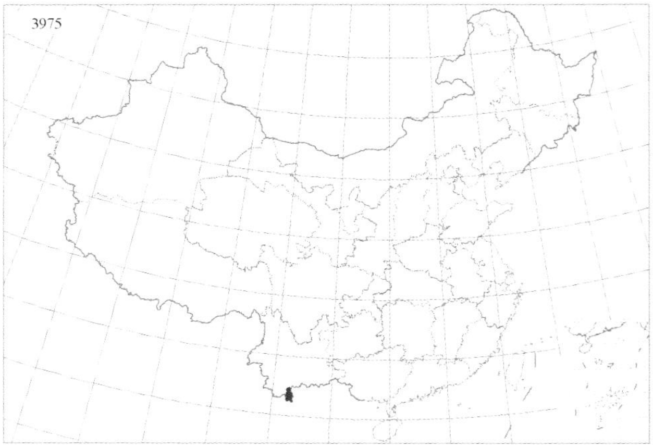

Fabaceae 豆科 *Bauhinia* 羊蹄甲属

Bauhinia wuzhengyii S. S. Larsen 征镒羊蹄甲

Vines

17.6; 9.9; 23.2; 17.7; 152.4; 0.0; 859 ❙ 1422; 813; 52 ❙ 65.6; 859; 359.7

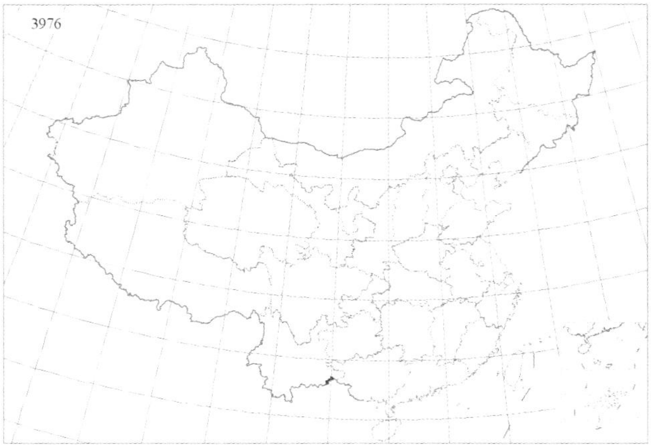

Fabaceae 豆科 *Bauhinia* 羊蹄甲属

Bauhinia yunnanensis Franch. 云南羊蹄甲

Vines

14.2 (5.4~19.3); 6.3 (-2.1~13.1); 20.8 (12.3~26.6); 14.3 (5.8~19.4); 113.1 (31.5~173.2); 1.8 (0.0~25.6); 750 (481~969) ❙ 1018 (741~1410); 580 (411~816); 35 (15~107) ❙ 37.3 (-14.2~87.2); 725 (481~969); 267.6 (142.9~351.7)

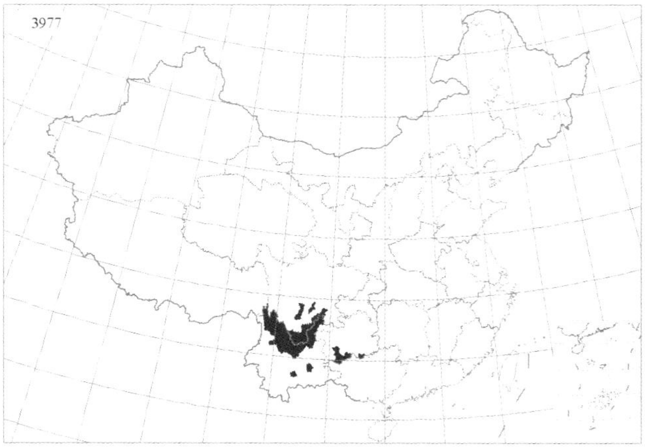

Fabaceae 豆科 *Bauhinia* 羊蹄甲属

Bauhinia × blakeana Dunn. 红花羊蹄甲

Trees, evergreen

20.2 (17.6~24.2); 11.9 (7.3~17.6); 27.5 (25.5~28.8); 20.4 (17.6~24.2); 185.1 (151.6~230.4); 0.0 (0.0~0.0); 1066 (909~1331) ❙ 1620 (1258~2037); 703 (425~1033); 134 (58~193) ❙ 53.0 (10.8~89.4); 1063 (909~1331); 239.2 (157.7~322.1)

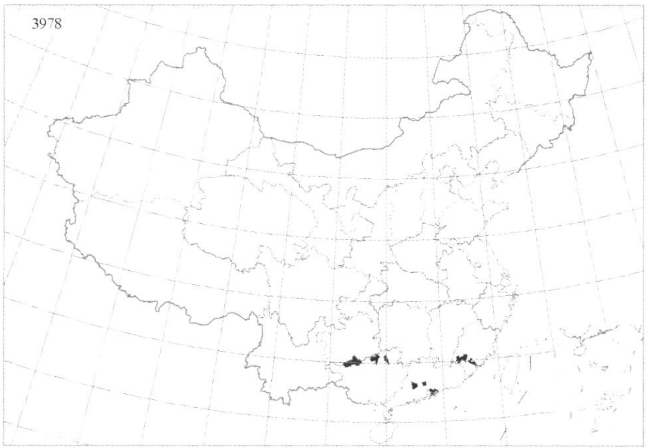

Fabaceae 豆科 *Bowringia* 藤槐属
Bowringia callicarpa Champ. ex Benth. 藤槐
Vines

21.3 (14.6~24.8); 13.2 (3.8~20.1); 27.8 (24.6~29.0); 21.3 (14.7~24.8); 196.2 (117.7~237.6); 0.0 (0.0~1.5); 1132 (779~1385) ▍1638 (868~2435); 697 (312~1428); 134 (41~198) ▍47.1 (-34.3~101.8); 1106 (779~1360); 283.9 (157.7~620.0)

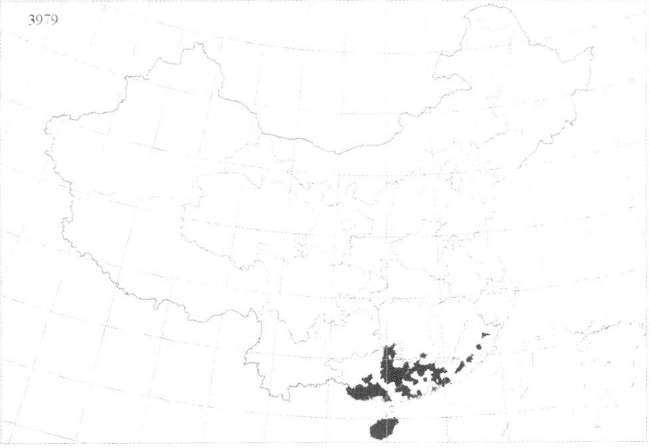

Fabaceae 豆科 *Butea* 紫矿属
Butea monosperma (Lam.) Taub. 紫矿
Trees

17.9 (15.6~20.5); 11.6 (9.2~15.1); 21.9 (20.3~23.9); 18.0 (15.6~20.5); 155.4 (127.6~185.9); 0.0 (0.0~0.0); 862 (767~992) ▍1339 (797~1688); 740 (411~959); 49 (23~77) ▍55.3 (-14.2~112.2); 834 (709~986); 440.9 (218.0~659.0)

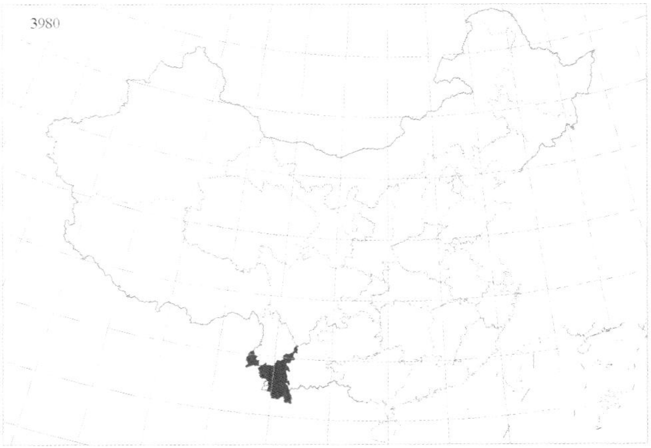

Fabaceae 豆科 *Caesalpinia* 云实属
Caesalpinia bonduc Wight et Arn. 刺果苏木
Vines

18.5 (11.1~24.8); 9.1 (0.6~20.1); 26.8 (18.4~29.2); 18.6 (11.3~24.8); 163.2 (80.8~237.6); 0.3 (0.0~9.7); 980 (652~1385) ▍1451 (868~2993); 634 (312~1847); 113 (27~547) ▍49.8 (-34.3~214.4); 968 (652~1360); 231.3 (108.2~620.0)

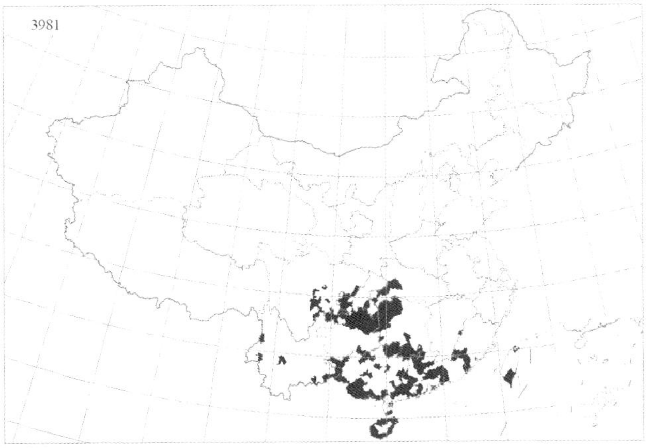

Fabaceae 豆科 *Caesalpinia* 云实属
Caesalpinia caesia Hand.-Mazz. 粉叶苏木
Vines

21.9 (21.3~22.7); 13.9 (13.2~15.0); 27.9 (27.3~28.6); 22.0 (21.3~22.8); 203.6 (195.9~213.1); 0.0 (0.0~0.0); 1162 (1111~1221) ▍1876 (1302~2435); 1047 (670~1428); 116 (86~130) ▍61.1 (12.9~99.5); 1162 (1111~1221); 289.6 (235.1~318.8)

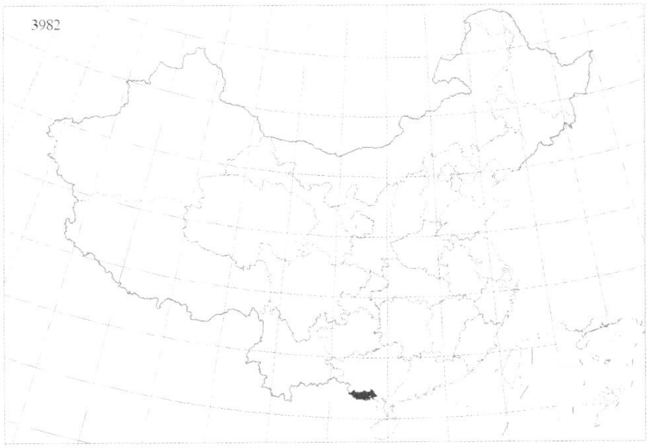

Fabaceae 豆科 *Caesalpinia* 云实属
Caesalpinia cucullata Roxb. 见血飞
Vines

17.5 (8.6~21.0); 11.2 (1.3~15.1); 21.8 (15.0~25.6); 17.5 (8.7~21.0); 150.7 (52.5~191.8); 0.6 (0.0~8.8); 858 (574~1053) ▍1316 (976~1607); 702 (564~911); 55 (35~77) ▍53.5 (26.1~81.5); 845 (574~1053); 454.5 (296.9~659.0)

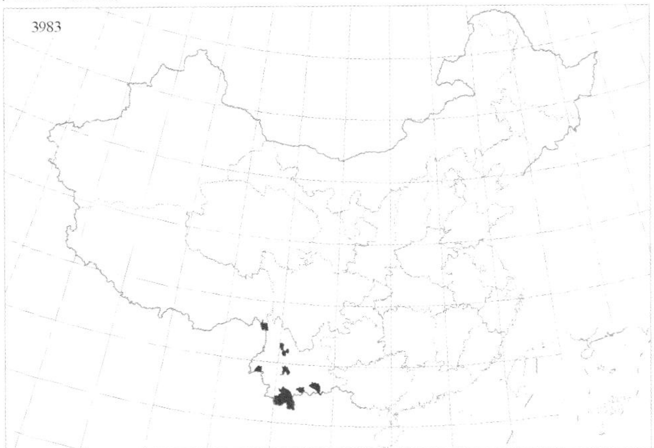

Fabaceae 豆科 *Caesalpinia* 云实属
Caesalpinia decapetala (Roth) Merr. 云实
Vines

16.5 (3.8~24.8); 5.9 (-4.3~20.1); 26.2 (10.8~30.1); 16.6 (4.7~24.8); 140.5 (21.6~237.6); 1.7 (0.0~35.4); 887 (444~1385) ▍1347 (596~2992); 555 (305~1847); 126 (15~259) ▍52.2 (-34.3~202.5); 874 (444~1269); 227.1 (108.2~659.0)

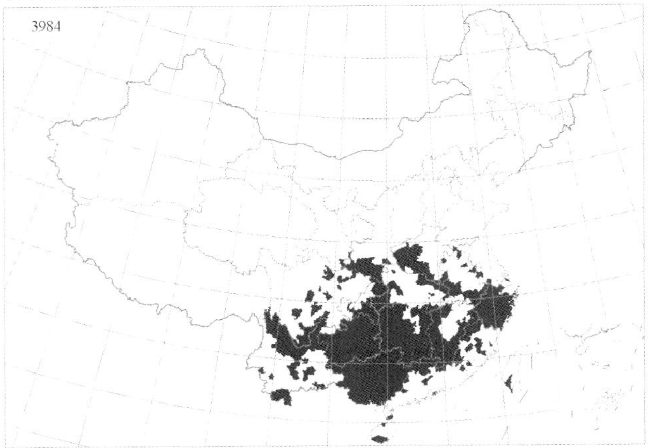

Fabaceae 豆科 *Caesalpinia* 云实属
Caesalpinia digyna Rottler 肉荚云实
Large climbers

24.8; 20.1; 28.0; 24.8; 237.6; 0.0; 1385 ❙ 1405; 473; 41 ❙ 1.5; 1137; 527.5

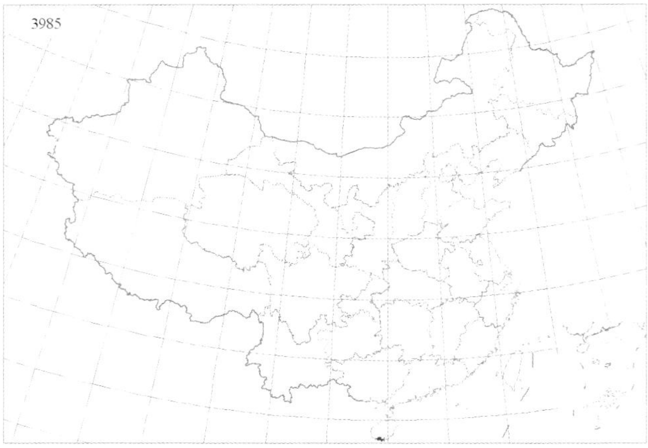

Fabaceae 豆科 *Caesalpinia* 云实属
Caesalpinia enneaphylla Roxb. 九羽见血飞
Large climbers

19.4 (14.3~22.2); 12.1 (7.6~15.1); 24.5 (19.7~28.3); 19.4 (14.3~22.3); 173.2 (111.9~207.5); 0.0 (0.0~0.0); 977 (728~1190) ❙ 1430 (991~2148); 768 (580~1234); 72 (35~125) ❙ 47.6 (5.8~86.7); 969 (707~1190); 383.0 (275.8~659.0)

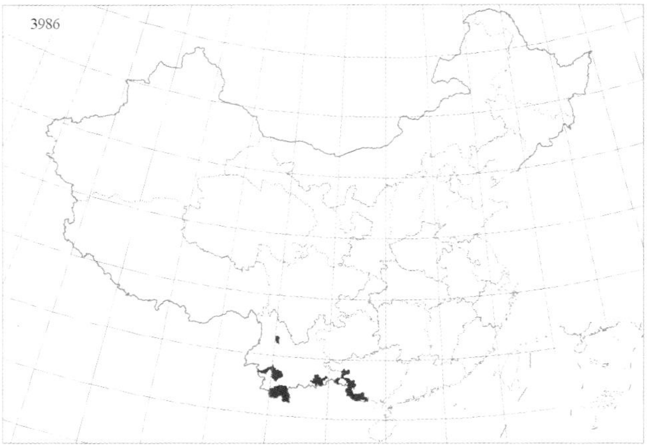

Fabaceae 豆科 *Caesalpinia* 云实属
Caesalpinia hymenocarpa (Prain) Hattink 膜荚见血飞
Vines

21.3; 12.6; 27.9; 21.3; 196.0; 0.0; 1123 ❙ 1188; 634; 56 ❙ 5.8; 1121; 315.6

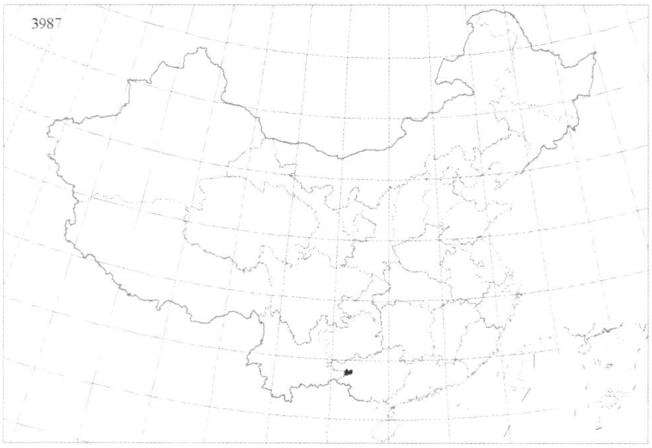

Fabaceae 豆科 *Caesalpinia* 云实属
Caesalpinia magnifoliolata F. P. Metcalf 大叶云实
Vines

19.7 (15.3~22.7); 10.8 (5.0~15.0); 26.6 (20.3~28.8); 19.7 (15.4~22.8); 176.3 (124.7~213.1); 0.0 (0.0~0.0); 1017 (789~1221) ❙ 1498 (1136~2435); 744 (506~1428); 110 (41~194) ❙ 47.4 (12.9~99.5); 1016 (780~1221); 268.1 (162.1~551.6)

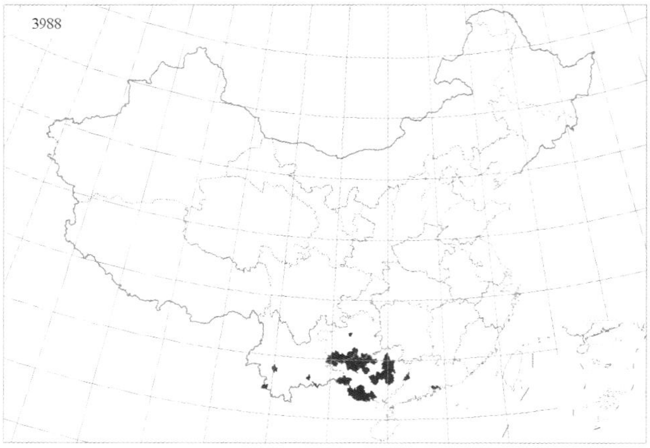

Fabaceae 豆科 *Caesalpinia* 云实属
Caesalpinia millettii Hook. et Arn. 小叶云实
Vines

18.6 (14.6~22.7); 8.5 (3.5~15.1); 27.4 (23.6~29.2); 18.6 (14.7~22.8); 163.6 (117.7~213.1); 0.2 (0.0~1.6); 979 (779~1221) ❙ 1609 (1172~2435); 647 (427~1428); 167 (50~228) ❙ 65.5 (12.9~106.4); 977 (779~1221); 252.6 (186.9~659.0)

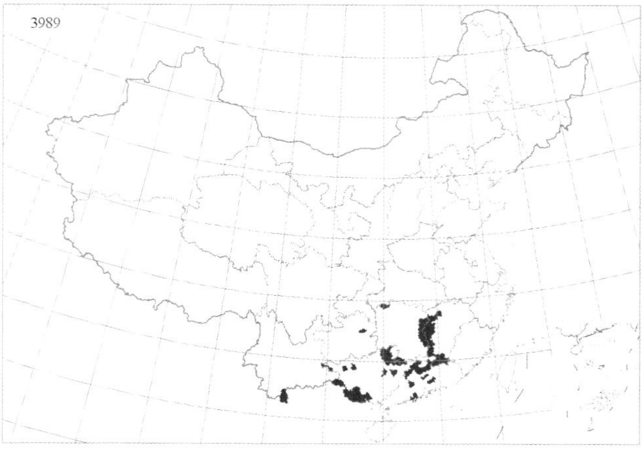

Fabaceae 豆科 *Caesalpinia* 云实属
Caesalpinia mimosoides Lam. 含羞云实
Vines

18.1 (13.1~21.0); 11.5 (6.4~15.1); 22.6 (18.4~26.9); 18.1 (13.2~21.0); 157.3 (98.1~191.8); 0.0 (0.0~0.0); 884 (692~1053) ❙ 1394 (1049~1852); 708 (598~818); 69 (44~194) ❙ 58.2 (21.3~99.9); 872 (692~1053); 444.4 (271.3~659.0)

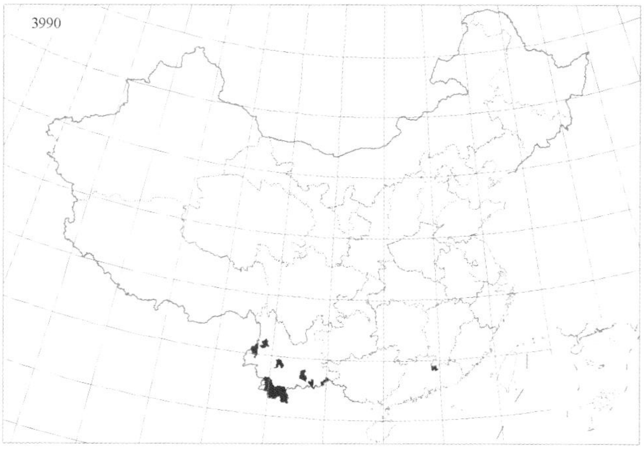

Fabaceae 豆科 *Caesalpinia* 云实属
Caesalpinia minax Hance 喙荚云实
Vines

18.6 (5.4~24.1); 10.0 (−2.1~18.9); 25.5 (12.3~29.0); 18.6 (5.8~24.1); 163.8 (31.5~229.7); 0.4 (0.0~25.6); 972 (481~1322) ❙ 1411 (741~2759); 677 (411~1594); 108 (15~231) ❙ 45.7 (−14.2~202.5); 963 (481~1247); 260.2 (120.3~659.0)

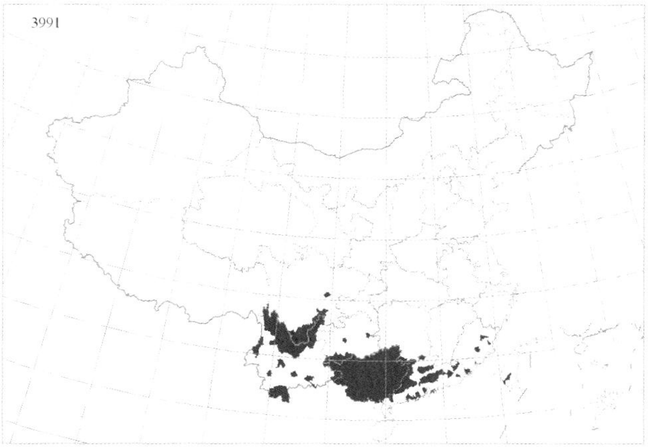

Fabaceae 豆科 *Caesalpinia* 云实属
Caesalpinia pulcherrima (L.) Sw. 金凤花
Shrubs or small trees

23.4 (21.1~24.8); 16.8 (13.5~20.1); 28.3 (27.5~28.8); 23.4 (21.1~24.8); 220.5 (193.2~237.6); 0.0 (0.0~0.0); 1269 (1094~1385) ❙ 1527 (868~1965); 599 (312~994); 90 (41~133) ❙ 21.7 (−34.3~61.6); 1145 (868~1329); 337.5 (203.4~548.0)

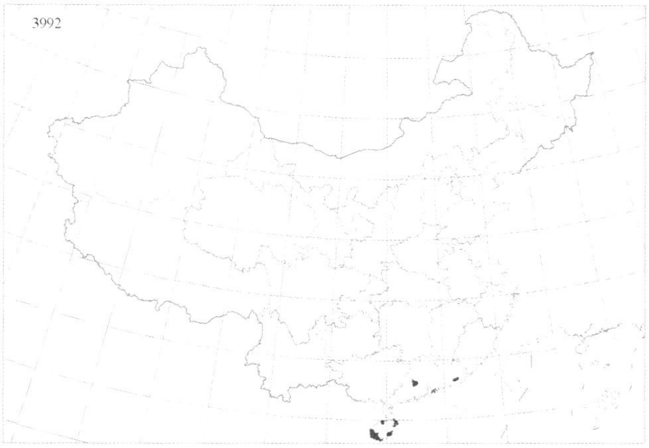

Fabaceae 豆科 *Caesalpinia* 云实属
Caesalp nia sappan L. 苏木
Small trees

17.7 (5.4~24.8); 10.2 (−2.1~20.1); 23.5 (12.3~29.0); 17.7 (5.8~24.8); 153.7 (31.5~237.6); 0.7 (0.0~25.6); 918 (481~1385) ❙ 1284 (741~2993); 669 (441~1847); 69 (15~547) ❙ 40.7 (−14.2~203.4); 895 (481~1299); 309.8 (139.6~659.0)

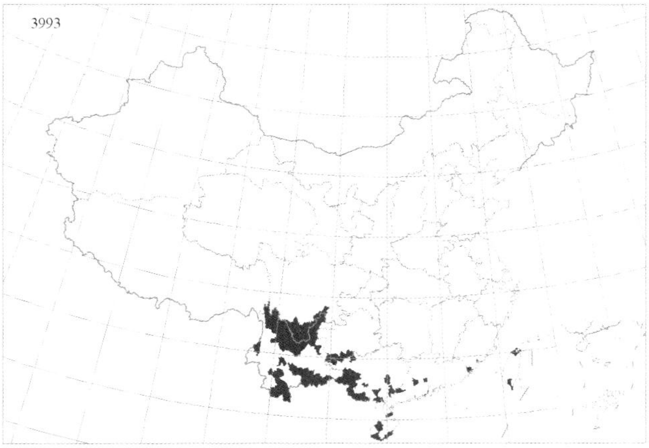

Fabaceae 豆科 *Caesalpinia* 云实属
Caesalpinia sinensis (Hemsl.) J. E. Vidal 鸡嘴簕
Vines

19.3 (12.0~23.4); 10.2 (0.6~16.1); 26.6 (22.0~29.2); 19.3 (12.1~23.5); 172.1 (94.7~221.7); 0.4 (0.0~9.7); 1006 (697~1273) ❙ 1377 (1064~2030); 688 (456~1033); 91 (42~194) ❙ 37.8 (5.8~72.1); 1006 (697~1273); 258.0 (137.9~516.7)

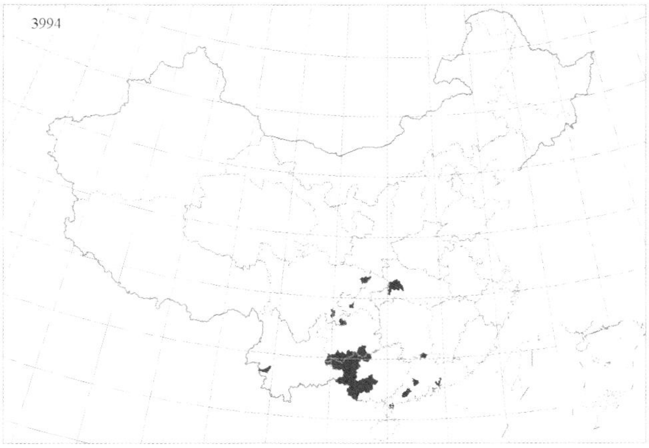

Fabaceae 豆科 *Caesalpinia* 云实属
Caesalpinia tortuosa Roxburgh 扭果苏木
Vines

20.0; 14.6; 23.2; 20.1; 180.8; 0.0; 963 ❙ 1523; 681; 67 ❙ 58.4; 942; 581.0

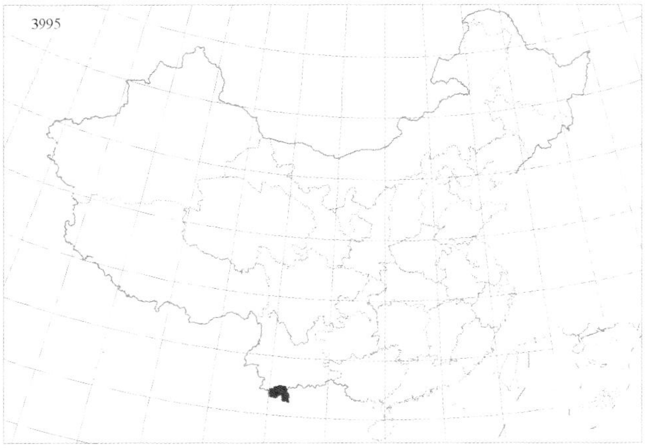

Fabaceae 豆科 *Caesalpinia* 云实属
Caesalpinia vernalis Champ. ex Benth. 春云实
Vines

18.4 (14.7~22.4); 8.9 (2.5~15.2); 27.5 (25.7~28.7); 18.6 (14.8~22.7); 163.2 (120.2~212.7); 0.3 (0.0~3.9); 979 (803~1216) ❙ 1685 (1363~2037); 646 (476~1033); 175 (104~242) ❙ 73.8 (31.5~129.9); 979 (803~1216); 240.3 (163.7~292.1)

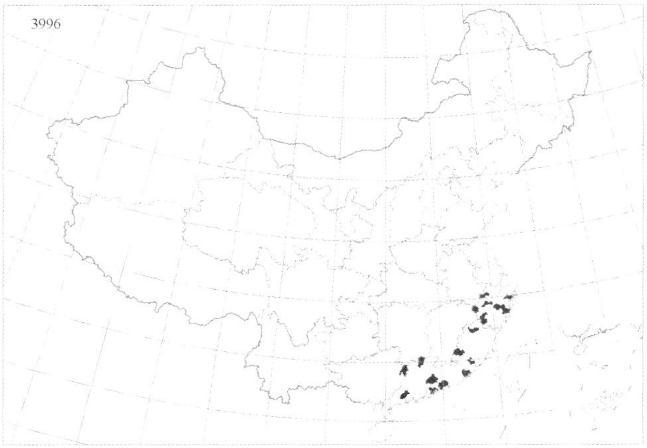

Fabaceae 豆科 *Cajanus* 木豆属
Cajanus cajan (L.) Merr. 木豆
Shrubs

17.3 (5.4~24.8); 9.4 (−2.1~20.1); 24.0 (12.3~29.3); 17.4 (5.8~24.8); 149.8 (31.5~237.6); 0.6 (0.0~25.6); 908 (481~1385) ❙ 1369 (741~2993); 643 (392~1594); 100 (15~547) ❙ 51.5 (−14.2~246.5); 884 (481~1326); 272.0 (119.4~548.0)

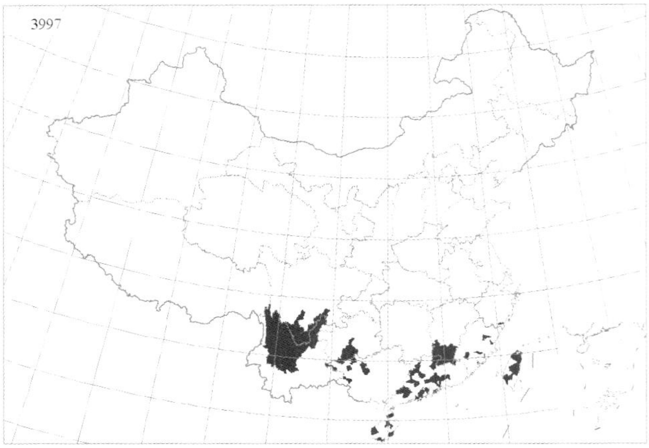

Fabaceae 豆科 *Cajanus* 木豆属
Cajanus crassus (Prain ex King) Maesen 虫豆
Vines

19.8 (16.3~23.4); 12.9 (9.1~17.5); 24.8 (21.4~29.2); 19.9 (16.4~23.5); 178.6 (136.4~221.7); 0.0 (0.0~0.0); 1004 (797~1273) ❙ 1540 (1163~2435); 838 (544~1428); 73 (48~130) ❙ 53.6 (11.9~99.5); 996 (797~1273); 398.9 (178.7~551.6)

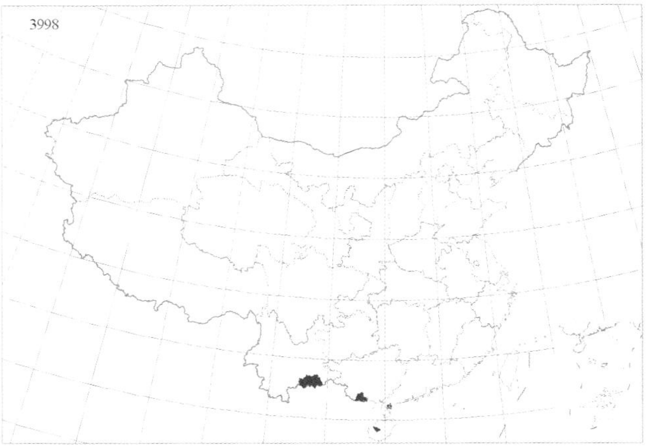

Fabaceae 豆科 *Cajanus* 木豆属
Cajanus grandiflorus (Benth. ex Baker) Maesen 大花虫豆
Vines

17.7 (16.9~18.2); 11.1 (10.3~11.8); 22.1 (21.4~22.5); 17.7 (16.9~18.2); 152.3 (142.7~158.4); 0.0 (0.0~0.0); 850 (813~873) ❙ 1077 (958~1178); 599 (531~646); 46 (44~51) ❙ 26.9 (13.3~44.9); 836 (813~863); 338.8 (294.5~406.7)

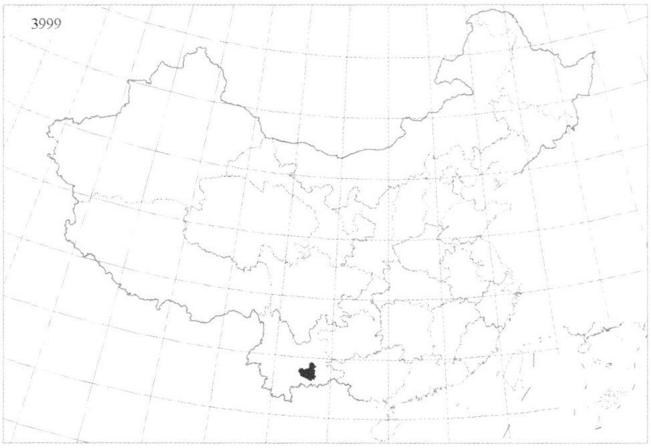

Fabaceae 豆科 *Callerya* 崖豆藤属
Callerya bonatiana (Pamp.) P. K. Lôc 滇桂崖豆藤
Vines

16.4 (13.8~18.9); 9.8 (6.7~12.5); 21.1 (19.3~23.9); 16.4 (13.8~19.0); 137.0 (105.9~167.8); 0.0 (0.0~0.0); 804 (714~929) ❙ 967 (797~1139); 560 (411~651); 36 (23~47) ❙ 21.0 (−14.2~37.1); 747 (683~851); 293.9 (218.0~365.3)

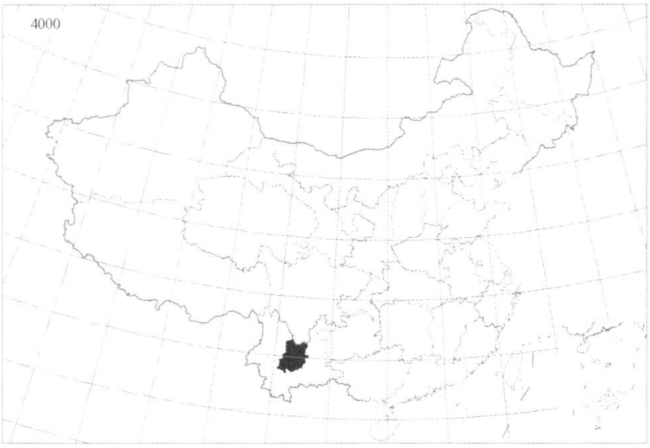

Fabaceae 豆科 *Callerya* 崖豆藤属
Callerya championii (Benth.) Z. Y. Zhu 绿花崖豆藤
Vines

20.1 (16.7~22.4); 11.4 (5.6~15.2); 27.7 (26.5~28.8); 20.3 (16.8~22.7); 183.6 (141.2~212.7); 0.0 (0.0~0.0); 1064 (885~1216) ❙ 1667 (1284~2030); 713 (476~1033); 152 (73~240) ❙ 57.9 (25.0~104.6); 1064 (885~1216); 235.4 (163.7~322.1)

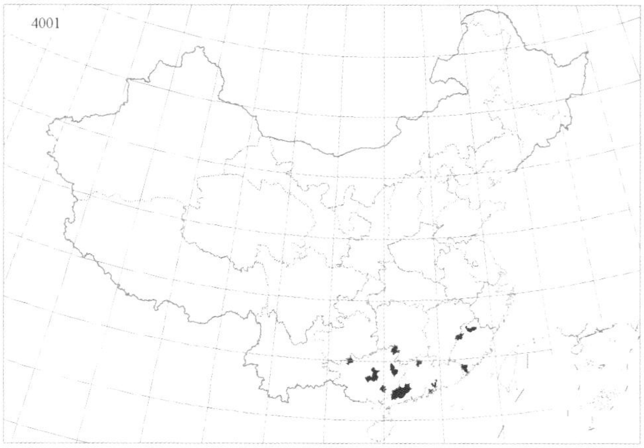

Fabaceae 豆科 *Callerya* 崖豆藤属
Callerya cinerea (Benth.) Schot 灰毛崖豆藤
Scandent shrubs

15.1 (3.0~20.5); 8.5 (−5.6~15.1); 19.9 (10.5~23.7); 15.3 (4.3~20.5); 127.9 (20.1~185.9); 6.6 (0.0~43.9); 789 (423~992) ❙ 1330 (628~1751); 725 (410~1007); 50 (11~77) ❙ 70.6 (36.8~145.8); 781 (423~986); 445.2 (234.2~659.0)

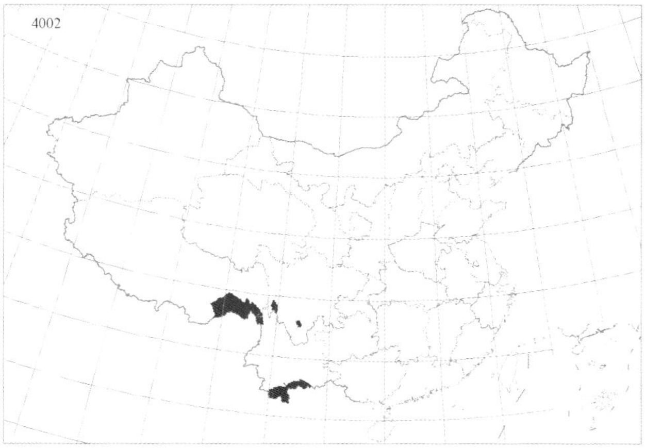

Fabaceae 豆科 *Callerya* 崖豆藤属

Callerya congestiflora (T. C. Chen) Z. Wei et Pedley 密花崖豆藤

Vines

16.8 (11.1~22.2); 5.6 (1.8~13.9); 27.2 (19.4~29.3); 16.8 (11.1~22.3); 142.6 (79.8~207.5); 0.6 (0.0~6.9); 899 (648~1190) ❙ 1532 (985~1922); 552 (427~812); 179 (30~259) ❙ 70.8 (15.4~128.9); 895 (648~1190); 227.5 (149.6~327.7)

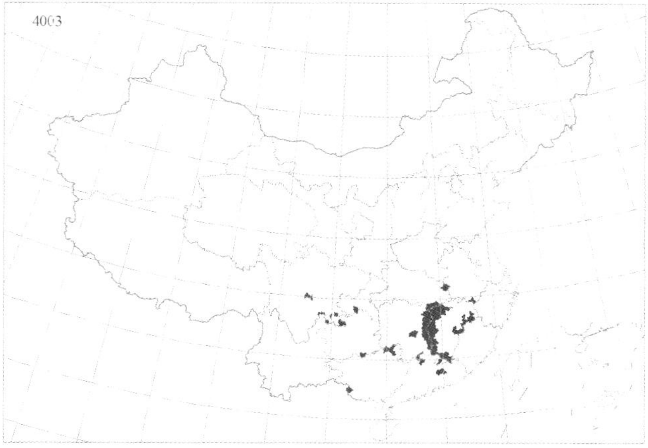

Fabaceae 豆科 *Callerya* 崖豆藤属

Callerya dielsiana (T. C. Chen ex Z. Wei) X. Y. Zhu 香花崖豆藤

Scandent shrubs

17.0 (3.5~24.8); 7.3 (−6.5~20.1); 25.5 (10.8~30.1); 17.0 (4.7~24.8); 145.0 (21.6~237.6); 0.9 (0.0~43.4); 895 (444~1385) ❙ 1393 (677~2435); 596 (312~1428); 121 (14~259) ❙ 56.5 (−34.3~147.0); 884 (444~1360); 249.8 (108.2~659.0)

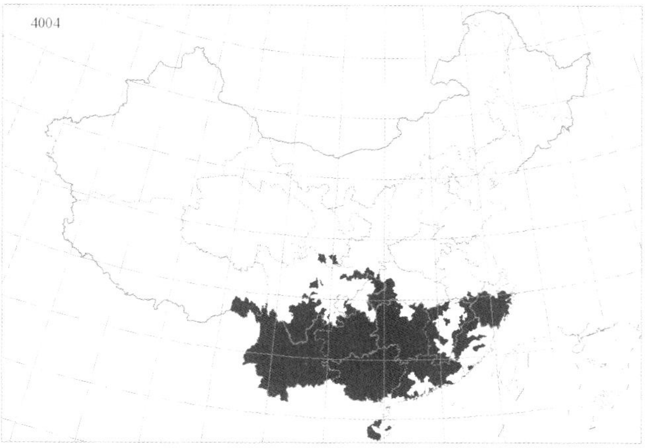

Fabaceae 豆科 *Callerya* 崖豆藤属

Callerya dielsiana var. herterocarpa (Chun ex T. C. Chen) X. Y. Zhu 异果崖豆藤

Scandent shrubs

18.4 (13.9~22.0); 8.4 (3.1~13.5); 27.3 (24.3~28.8); 18.5 (14.0~22.0); 161.9 (110.5~203.8); 0.1 (0.0~2.6); 969 (760~1170) ❙ 1662 (1178~2005); 615 (445~902); 179 (43~259) ❙ 72.5 (16.9~130.7); 967 (760~1170); 243.7 (169.7~354.5)

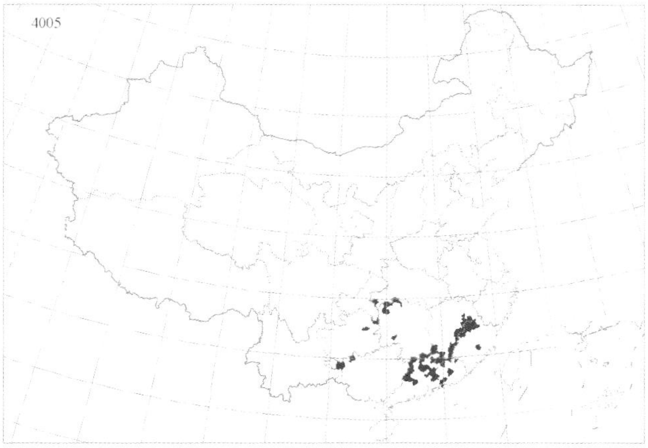

Fabaceae 豆科 *Callerya* 崖豆藤属

Callerya dielsiana var. solida (T. C. Chen ex Z. Wei) X. Y. Zhu 雪峰山崖豆藤

Scandent shrubs

16.3; 4.7; 27.2; 16.4; 136.6; 0.3; 874 ❙ 1378; 447; 169 ❙ 57.7; 872; 188.0

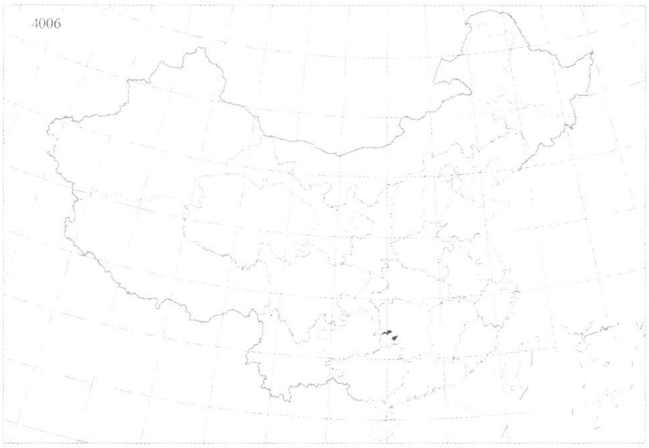

Fabaceae 豆科 *Callerya* 崖豆藤属

Callerya dorwardii (Collett et Hemsley) Z. Wei et Pedley 滇缅崖豆藤

Vines

17.0 (8.6~21.0); 10.5 (1.3~15.1); 21.5 (15.0~25.6); 17.1 (8.7~21.0); 144.8 (52.6~191.8); 0.2 (0.0~8.8); 832 (574~1053) ❙ 1264 (939~1751); 702 (524~996); 51 (15~77) ❙ 52.3 (6.5~112.2); 817 (574~1053); 408.7 (277.0~659.0)

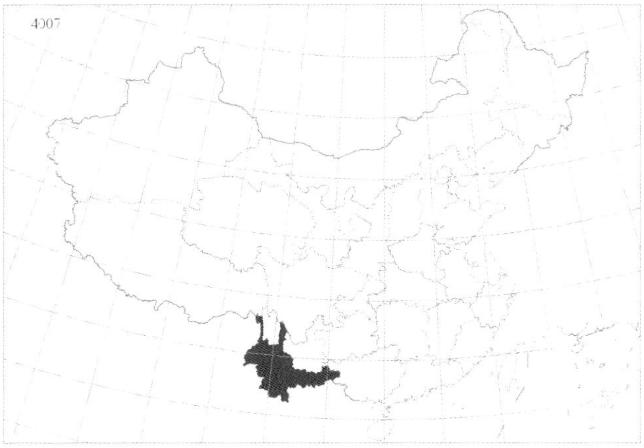

Fabaceae 豆科 *Callerya* 崖豆藤属

Callerya eurybotrya (Drake) Schot 宽序崖豆藤

Scandent shrubs

18.9 (14.6~22.3); 9.6 (3.8~15.1); 26.7 (22.4~29.4); 19.0 (14.7~22.3); 167.9 (117.7~208.1); 0.0 (0.0~1.5); 982 (779~1202) ❙ 1467 (1161~1776); 654 (403~996); 132 (43~217) ❙ 51.0 (5.8~101.8); 978 (779~1202); 275.7 (151.0~659.0)

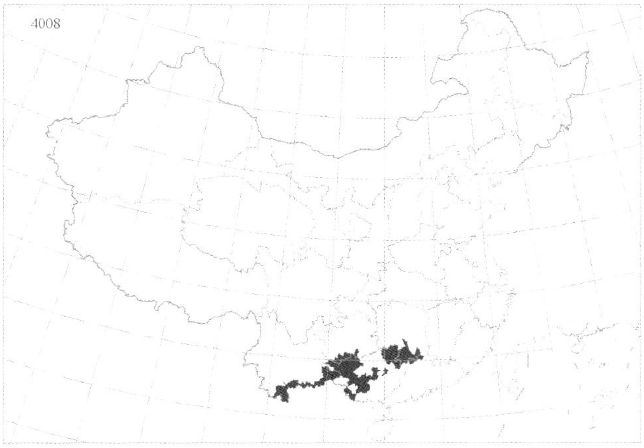

Fabaceae 豆科 *Callerya* 崖豆藤属
Callerya fordii (Dunn) Schot 广东崖豆藤
Vines

20.9 (18.0~22.0); 12.0 (7.7~13.7); 28.1 (25.8~29.0); 20.9 (18.0~22.0); 191.2 (156.4~203.8); 0.0 (0.0~0.0); 1103 (913~1170) ❙ 1544 (1367~1675); 669 (507~768); 142 (97~180) ❙ 40.9 (17.4~69.8); 1100 (913~1170); 228.9 (203.4~273.6)

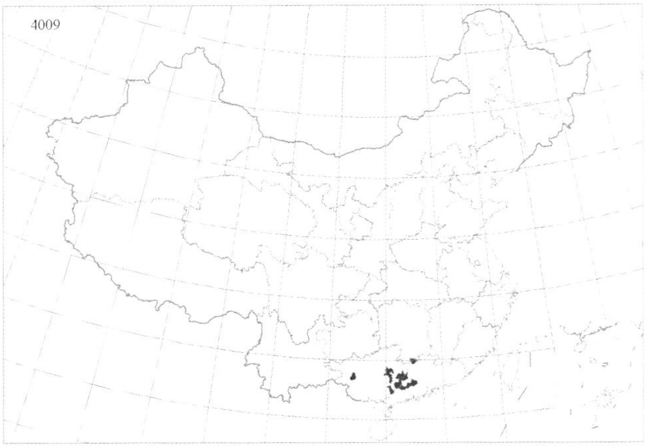

Fabaceae 豆科 *Callerya* 崖豆藤属
Callerya kiangsiensis (Z. Wei) Z. Wei et Pedley 江西崖豆藤
Vines

16.2 (14.3~18.2); 4.2 (2.5~6.1); 28.0 (24.2~29.6); 16.3 (14.3~18.2); 136.5 (113.2~158.7); 1.0 (0.0~3.9); 889 (761~989) ❙ 1460 (931~1997); 517 (323~669); 191 (113~254) ❙ 64.9 (-1.6~130.7); 877 (761~941); 226.1 (180.9~303.4)

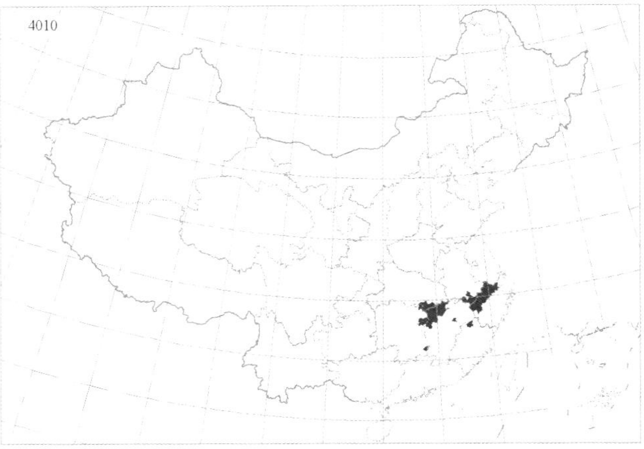

Fabaceae 豆科 *Callerya* 崖豆藤属
Callerya longipedunculata (Z. Wei) X. Y. Zhu 长梗崖豆藤
Vines

18.0 (13.4~22.3); 9.3 (6.1~13.5); 24.7 (19.0~28.9); 18.0 (13.5~22.3); 156.2 (102.1~208.1); 0.0 (0.0~0.0); 914 (706~1202) ❙ 1229 (918~1446); 665 (546~723); 59 (32~136) ❙ 35.2 (16.9~59.6); 914 (706~1202); 272.1 (189.5~354.5)

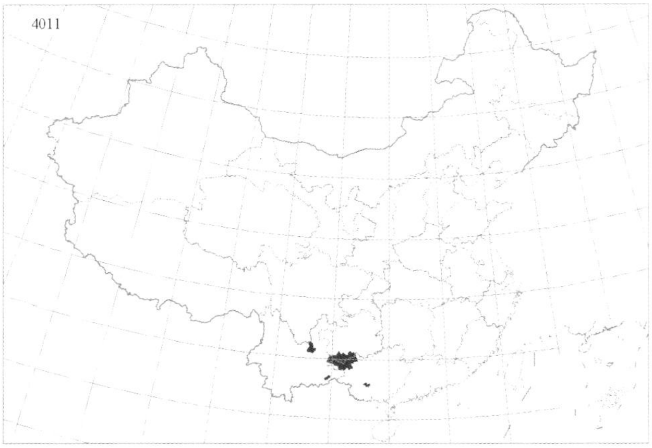

Fabaceae 豆科 *Callerya* 崖豆藤属
Callerya nitida (Benth.) R. Geesink 亮叶崖豆藤
Scandent shrubs

18.3 (14.3~24.8); 8.6 (3.3~20.1); 26.7 (20.1~29.8); 18.3 (14.3~24.8); 159.7 (112.6~237.6); 0.1 (0.0~2.3); 958 (756~1385) ❙ 1549 (868~2680); 625 (312~1428); 148 (41~259) ❙ 64.1 (-34.3~246.5); 947 (756~1254); 249.6 (139.6~620.0)

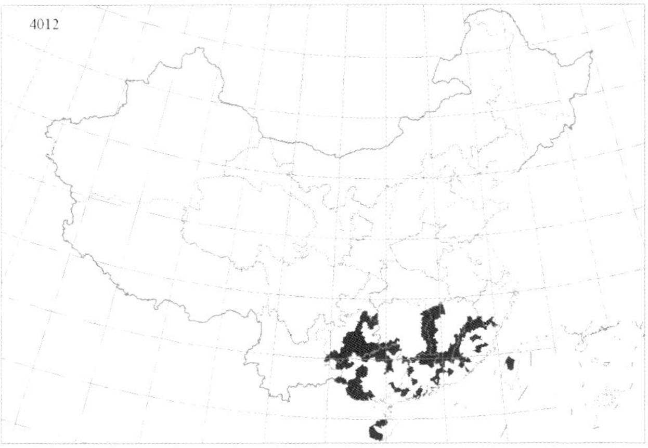

Fabaceae 豆科 *Callerya* 崖豆藤属
Callerya nitida var. hirsutissima (Z. Wei) X. Y. Zhu 丰城崖豆藤
Scandent shrubs

18.2 (16.4~22.4); 8.4 (4.8~14.8); 27.0 (25.2~28.4); 18.2 (16.5~22.4); 158.8 (138.1~208.9); 0.0 (0.0~0.2); 946 (864~1191) ❙ 1675 (1409~2030); 611 (445~1033); 188 (110~254) ❙ 77.5 (59.8~130.7); 946 (864~1191); 242.0 (163.7~305.7)

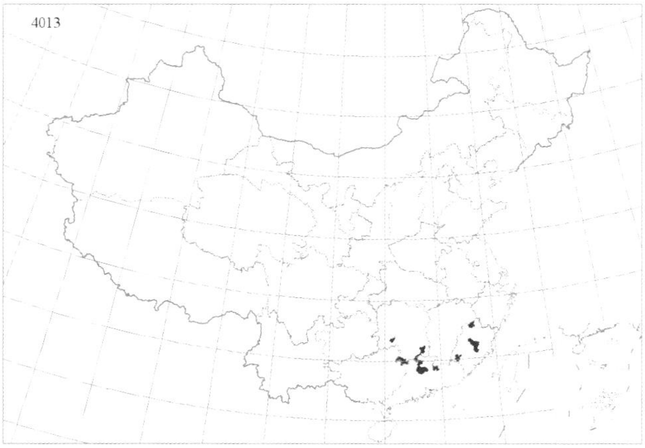

Fabaceae 豆科 *Callerya* 崖豆藤属
Callerya nitida var. minor Z. Wei ex H. Sun 峨眉崖豆藤
Scandent shrubs

17.9 (14.1~21.0); 7.4 (3.5~12.2); 27.3 (20.2~29.8); 17.9 (14.2~21.0); 155.4 (109.8~192.6); 0.2 (0.0~1.5); 944 (729~1113) ❙ 1576 (1014~1852); 562 (427~816); 184 (18~225) ❙ 67.7 (27.3~128.9); 938 (707~1107); 236.6 (174.9~330.0)

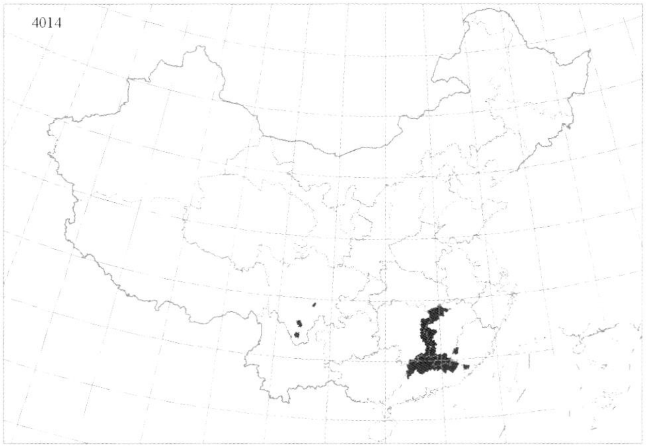

Fabaceae 豆科 *Callerya* 崖豆藤属

Callerya oosperma (Dunn) Z. Wei et Pedley 皱果崖豆藤

Vines

19.6 (5.4~24.8); 11.5 (-2.1~20.1); 25.9 (12.3~29.0); 19.7 (5.8~24.8); 176.1 (31.5~237.6); 0.3 (0.0~25.6); 1016 (481~1385) ❙ 1438 (741~2435); 694 (392~1428); 90 (28~194) ❙ 43.2 (-12.0~99.5); 999 (481~1329); 313.4 (160.8~659.0)

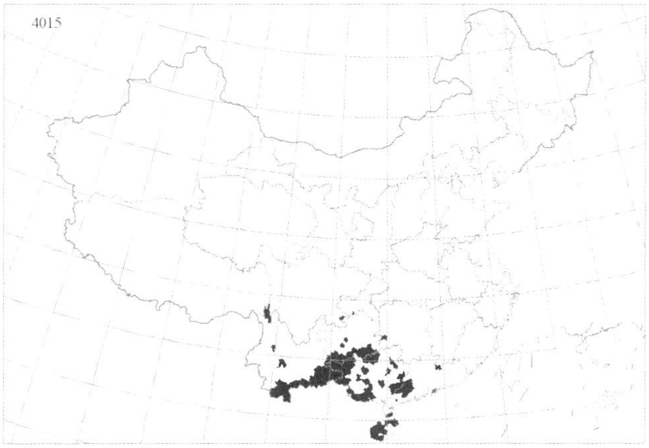

Fabaceae 豆科 *Callerya* 崖豆藤属

Callerya reticulata (Benth.) Schot 网络崖豆藤

Vines, evergreen

17.5 (10.6~24.8); 6.9 (-0.2~20.1); 27.2 (18.8~30.1); 17.5 (10.6~24.8); 151.2 (79.4~237.6); 0.6 (0.0~12.1); 935 (643~1385) ❙ 1457 (827~2993); 573 (299~1847); 153 (28~551) ❙ 57.0 (-34.3~235.6); 921 (643~1331); 221.0 (111.9~620.0)

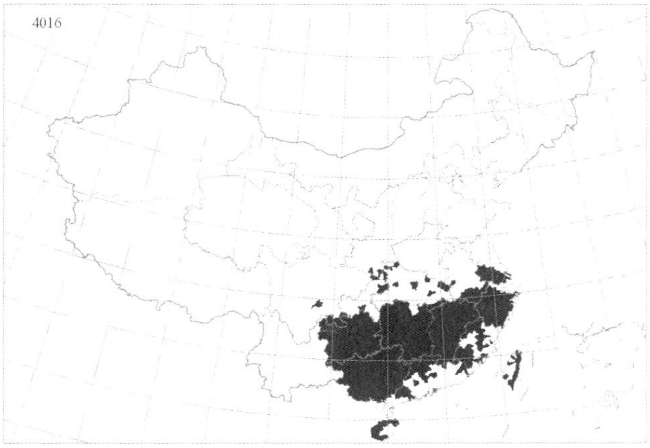

Fabaceae 豆科 *Callerya* 崖豆藤属

Callerya reticulata var. stenophylla (Merr. et Chun) X. Y. Zhu 线叶崖豆藤

Vines, evergreen

22.3 (18.9~24.8); 15.2 (8.5~20.1); 28.0 (27.2~28.7); 22.4 (18.9~24.8); 208.4 (166.9~237.6); 0.0 (0.0~0.0); 1205 (992~1385) ❙ 1523 (1405~1620); 601 (473~777); 100 (41~166) ❙ 28.6 (1.5~63.3); 1124 (992~1212); 361.9 (164.9~620.0)

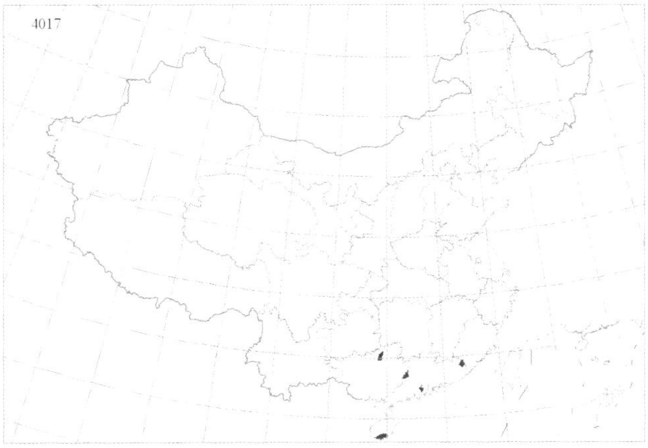

Fabaceae 豆科 *Callerya* 崖豆藤属

Callerya sericosema (Hance) Z. Wei et Pedley 锈毛崖豆藤

Scandent shrubs

15.5 (10.6~19.2); 5.2 (-0.2~10.7); 25.0 (18.8~28.9); 15.6 (10.6~19.2); 127.6 (79.4~170.8); 0.8 (0.0~12.1); 820 (643~1002) ❙ 1213 (916~1492); 540 (444~817); 77 (29~174) ❙ 48.8 (8.0~100.5); 820 (643~1002); 200.6 (108.2~411.1)

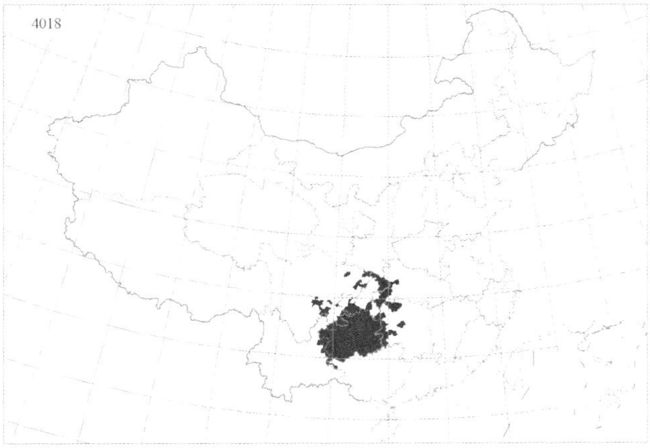

Fabaceae 豆科 *Callerya* 崖豆藤属

Callerya speciosa (Champ. ex Benth.) Schot 美丽崖豆藤

Vines

21.6 (14.6~24.8); 13.8 (3.8~20.1); 27.7 (23.2~29.0); 21.6 (14.7~24.8); 199.5 (117.7~237.6); 0.0 (0.0~1.5); 1151 (779~1385) ❙ 1570 (868~2148); 692 (312~1234); 114 (41~250) ❙ 39.4 (-34.3~121.8); 1117 (779~1360); 299.8 (139.6~620.0)

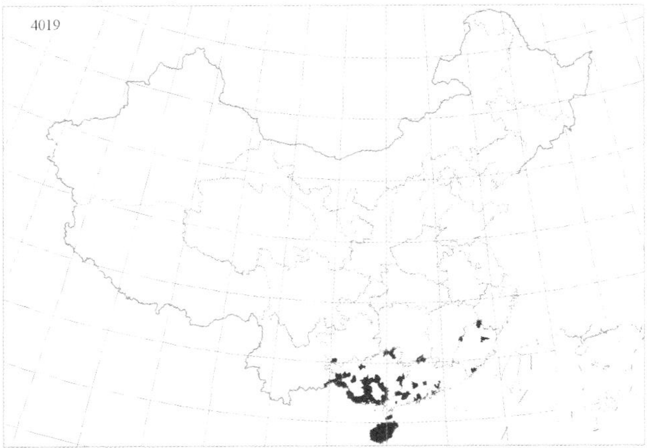

Fabaceae 豆科 *Callerya* 崖豆藤属

Callerya sphaerosperma (Z. Wei) Z. Wei et Pedley 球子崖豆藤

Scandent shrubs

18.6 (16.8~22.2); 9.8 (7.8~13.9); 25.4 (23.4~28.3); 18.6 (16.8~22.3); 163.5 (141.5~207.5); 0.0 (0.0~0.0); 943 (836~1190) ❙ 1288 (1232~1374); 696 (666~723); 62 (51~87) ❙ 37.9 (15.4~59.6); 943 (836~1190); 269.7 (189.5~327.7)

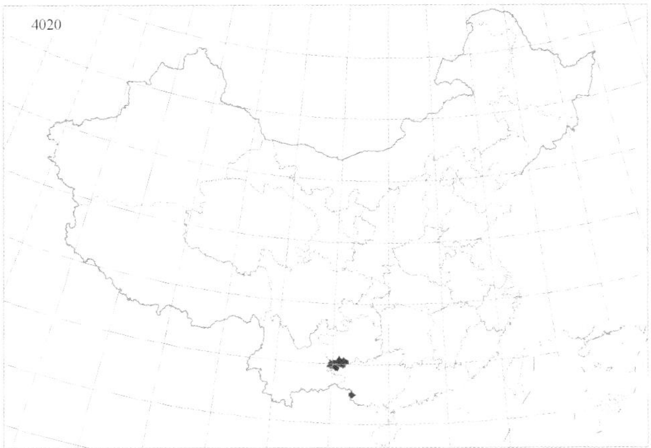

Fabaceae 豆科 *Callerya* 崖豆藤属

Callerya tsui (F. P. Metcalf) Z. Wei et Pedley 喙果崖豆藤

Vines

20.0 (14.6~24.8); 11.3 (3.8~20.1); 26.9 (21.4~29.8); 20.0 (14.7~24.8); 180.3 (117.7~237.6); 0.0 (0.0~1.5); 1045 (779~1385) ❙ 1519 (868~2435); 684 (312~1428); 120 (41~194) ❙ 47.3 (-34.3~101.8); 1025 (779~1264); 282.1 (174.9~620.0)

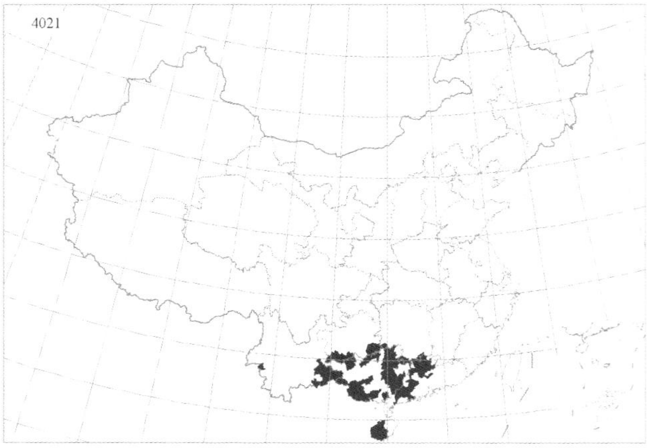

Fabaceae 豆科 *Calliandra* 朱缨花属

Calliandra umbrosa (Wall.) Benth. 云南朱缨花

Shrubs

16.4; 9.6; 20.6; 16.4; 137.3; 0.0; 795 ❙ 1688; 959; 62 ❙ 112.2; 795; 627.0

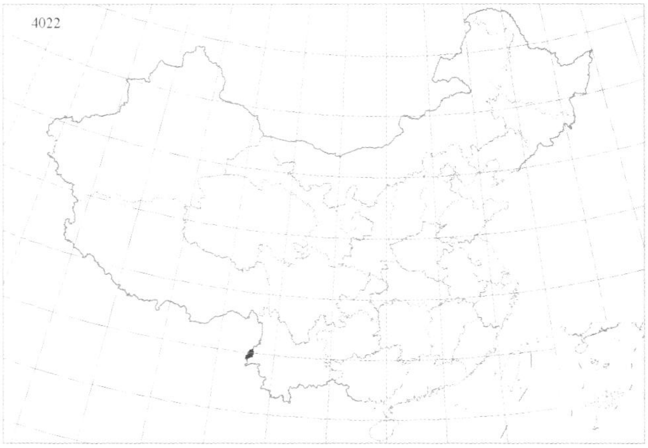

Fabaceae 豆科 *Calophaca* 丽豆属

Calophaca chinensis Boriss. 华丽豆

Small shrubs

3.6; -16.3; 20.1; 7.9; 60.4; 76.9; 581 ❙ 265; 107; 30 ❙ -54.1; 265; 99.9

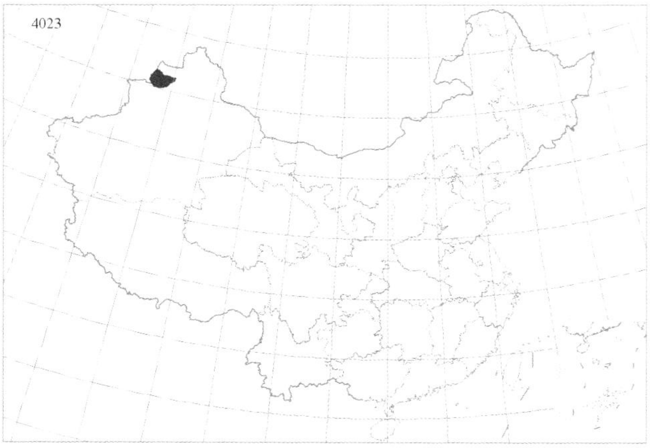

Fabaceae 豆科 *Calophaca* 丽豆属

Calophaca sinica Rehder 丽豆

Shrubs

6.2 (4.1~9.0); -10.4 (-12.9~-6.7); 20.7 (18.9~23.3); 8.6 (7.4~10.3); 66.4 (53.4~84.1); 52.1 (34.7~64.8); 588 (538~664) ❙ 437 (367~515); 271 (235~304); 12 (8~16) ❙ -25.3 (-37.7~-10.6); 437 (367~515); 139.4 (53.8~225.2)

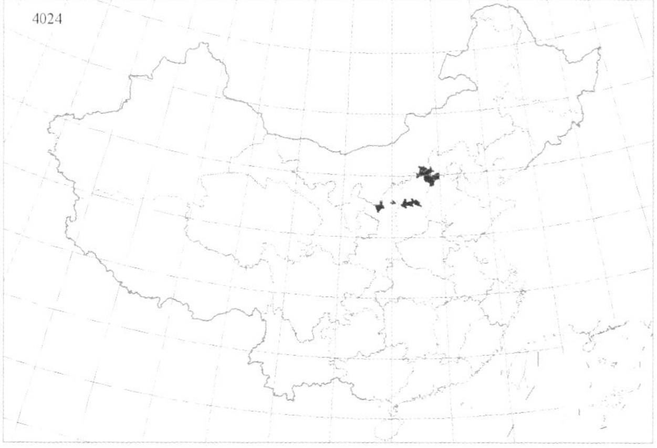

Fabaceae 豆科 *Calophaca* 丽豆属

Calophaca soongorica Kar. et Kir. 新疆丽豆

Shrubs

3.4; -15.5; 19.8; 7.7; 58.2; 76.5; 573 ❙ 296; 109; 40 ❙ -48.3; 295; 157.7

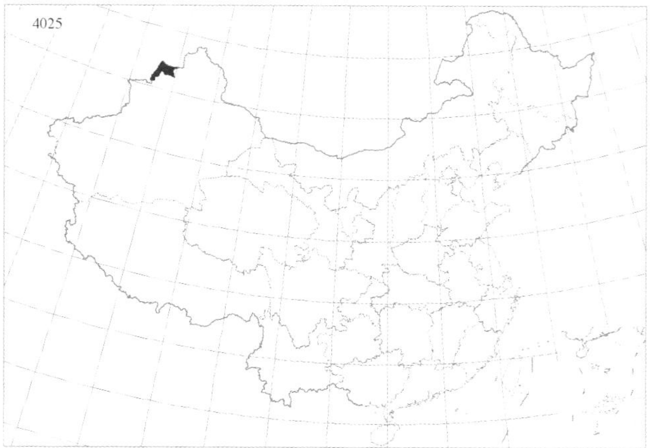

Fabaceae 豆科 *Campylotropis* 杭子梢属

Campylotropis argentea Schindl. 银叶杭子梢

Small shrubs, deciduous

18.0 (16.9~19.3); 11.6 (10.3~13.1); 22.3 (21.4~23.6); 18.0 (16.9~19.4); 156.2 (142.7~173.2); 0.0 (0.0~0.0); 865 (813~939) ❙ 1108 (1007~1178); 617 (539~646); 47 (44~51) ❙ 28.3 (19.0~44.9); 850 (813~903); 344.6 (294.5~406.7)

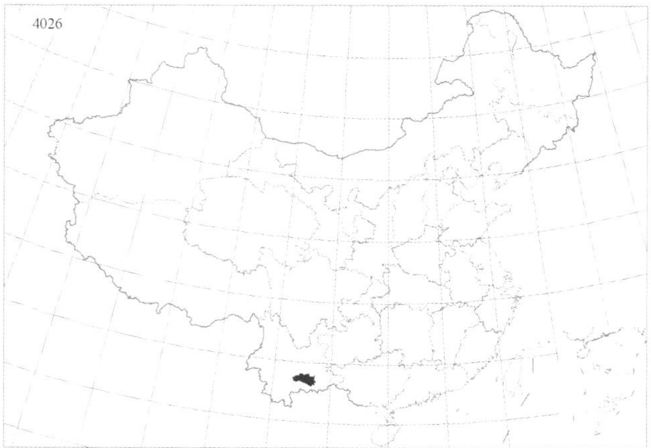

Fabaceae 豆科 *Campylotropis* 杭子梢属
Campylotropis bonii Schindl. 密脉杭子梢
Shrubs, deciduous

20.7; 11.9; 27.5; 20.8; 189.1; 0.0; 1084 ❙ 1429; 733; 98 ❙ 33.4; 1084; 267.4

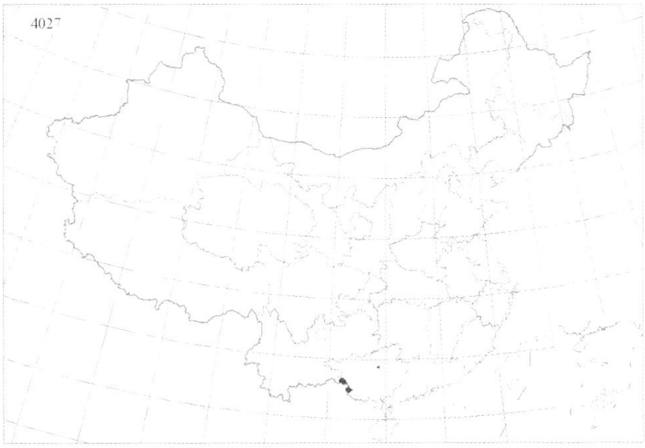

Fabaceae 豆科 *Campylotropis* 杭子梢属
Campylotropis brevifolia Ricker 短序杭子梢
Shrubs, deciduous

8.8; 1.2; 15.2; 8.9; 57.1; 10.8; 577 ❙ 926; 583; 13 ❙ 61.2; 576; 303.7

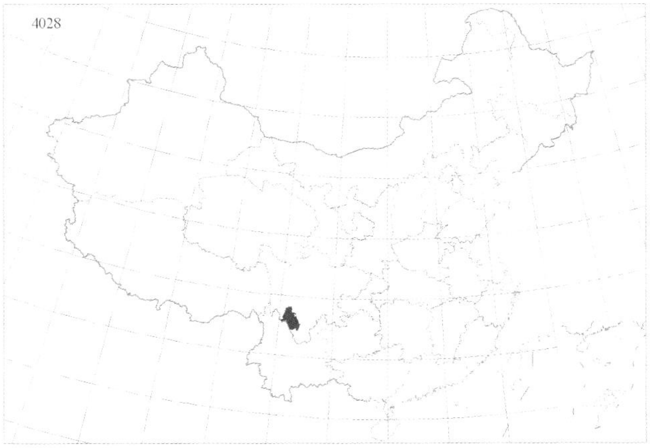

Fabaceae 豆科 *Campylotropis* 杭子梢属
Campylotropis capillipes (Franch.) Schindl. 细花梗杭子梢
Shrubs, deciduous

13.9 (5.4~21.3); 6.4 (-2.1~12.6); 20.1 (12.3~27.9); 14.0 (5.8~21.3); 109.5 (31.5~196.0); 1.7 (0.0~25.6); 737 (481~1123) ❙ 979 (741~1188); 567 (411~673); 33 (15~58) ❙ 35.1 (-14.2~78.3); 711 (481~1121); 279.0 (142.9~372.3)

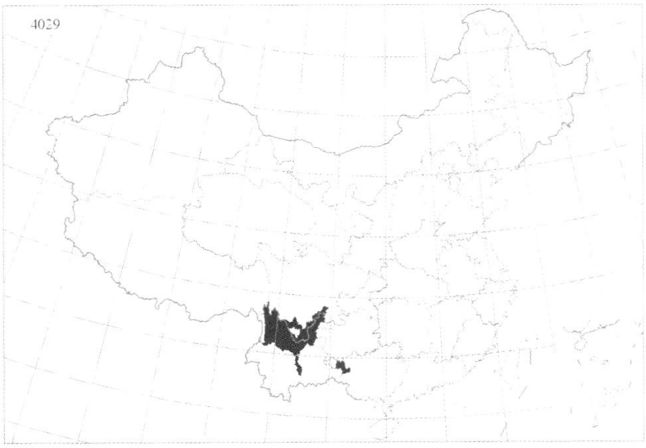

Fabaceae 豆科 *Campylotropis* 杭子梢属
Campylotropis capillipes subsp. prainii (Collett et Hemsl.) Iokawa & H. Ohashi 草山杭子梢
Shrubs, deciduous

16.5 (6.6~22.3); 9.6 (-1.3~15.1); 21.6 (13.3~28.9); 16.6 (6.8~22.3); 139.6 (39.6~208.1); 0.8 (0.0~19.7); 826 (514~1202) ❙ 1188 (862~1751); 672 (524~996); 44 (12~136) ❙ 44.6 (4.4~88.8); 804 (514~1202); 362.4 (201.4~659.0)

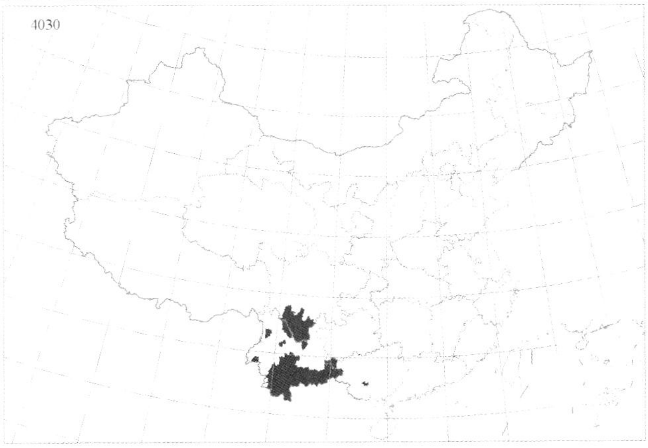

Fabaceae 豆科 *Campylotropis* 杭子梢属
Campylotropis cytisoides f. parviflora (Kutz) Iokawa et H. Ohashi 小花杭子梢
Shrubs, deciduous

17.9 (14.4~20.5); 11.7 (7.5~15.1); 21.9 (19.1~23.7); 18.0 (14.4~20.5); 155.6 (113.1~185.9); 0.0 (0.0~0.0); 863 (727~992) ❙ 1350 (1028~1751); 745 (566~996); 53 (41~77) ❙ 56.7 (12.5~112.2); 848 (727~986); 442.5 (294.5~659.0)

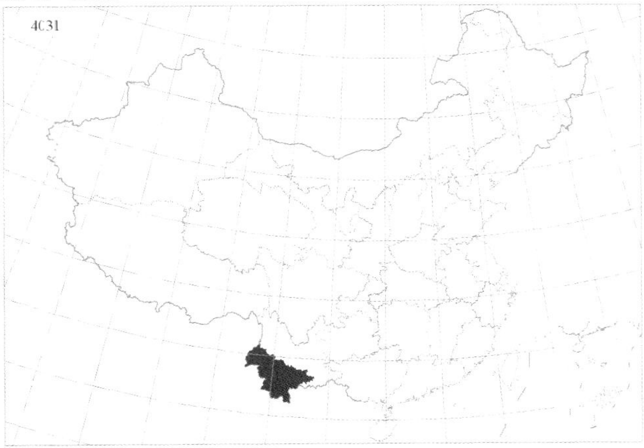

Fabaceae 豆科 *Campylotropis* 杭子梢属
Campylotropis delavayi (Franch.) Schindl. 西南杭子梢
Shrubs, deciduous

15.0 (5.4~22.2); 7.4 (-2.1~13.9); 21.0 (12.3~28.5); 15.1 (5.8~22.3); 121.8 (31.5~207.5); 1.2 (0.0~25.6); 773 (481~1190) ❙ 1103 (741~1625); 610 (411~836); 47 (12~173) ❙ 43.9 (-14.2~107.2); 753 (481~1190); 299.2 (142.9~516.7)

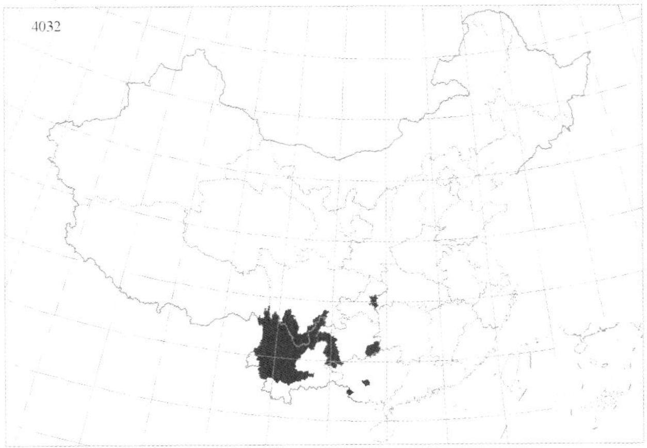

Fabaceae 豆科　*Campylotropis* 荒子梢属
Campylotropis diversifolia (Hemsl.) Schindl. 异叶荒子梢
Shrubs or small trees, deciduous

17.5 (16.3~19.3); 10.7 (9.0~13.1); 22.1 (21.4~23.6); 17.5 (16.4~19.4); 150.3 (136.4~173.2); 0.0 (0.0~0.0); 846 (797~939) ❙ 1089 (958~1190); 602 (531~656); 47 (44~51) ❙ 29.0 (13.3~49.4); 834 (797~903); 331.5 (277.0~406.7)

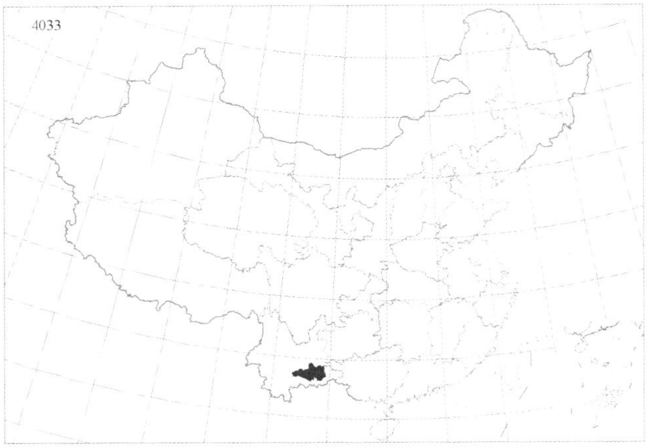

Fabaceae 豆科　*Campylotropis* 荒子梢属
Campylotropis grandifolia Schindl. 大叶荒子梢
Shrubs, deciduous

17.5; 10.5; 22.3; 17.5; 149.9; 0.0; 846 ❙ 958; 531; 44 ❙ 13.3; 823; 311.9

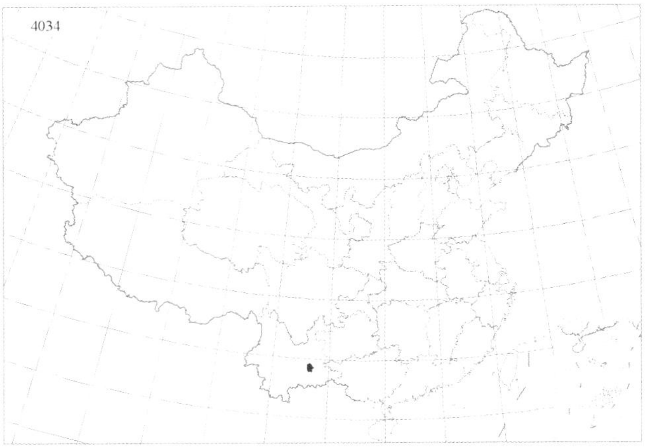

Fabaceae 豆科　*Campylotropis* 荒子梢属
Campylotropis harmsii Schindl. 思茅荒子梢
Shrubs, deciduous

18.8 (16.3~22.2); 11.1 (6.5~15.1); 24.6 (20.3~28.7); 18.9 (16.4~22.3); 166.7 (136.6~207.5); 0.0 (0.0~0.0); 941 (789~1190) ❙ 1375 (1007~1751); 715 (539~996); 77 (42~166) ❙ 47.0 (5.8~88.8); 936 (780~1190); 361.9 (190.9~659.0)

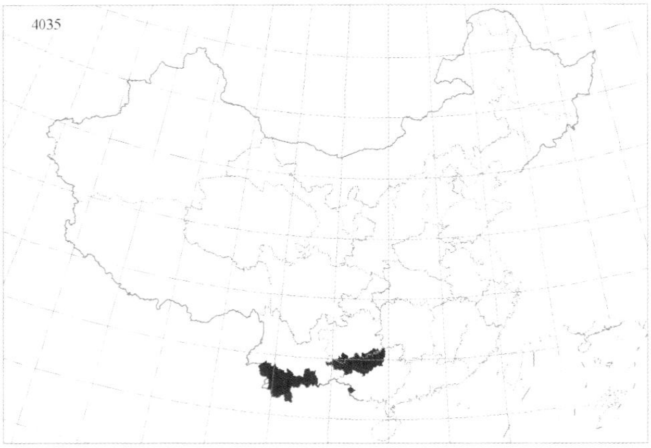

Fabaceae 豆科　*Campylotropis* 荒子梢属
Campylotropis henryi (Schindl.) Schindl. 元江荒子梢
Shrubs, deciduous

15.2 (5.4~21.0); 7.8 (-2.1~14.2); 21.0 (12.3~27.0); 15.3 (5.8~21.0); 124.1 (31.5~191.8); 0.8 (0.0~25.6); 775 (481~1053) ❙ 1074 (741~1751); 604 (411~996); 44 (15~153) ❙ 39.3 (-14.2~93.0); 759 (481~1053); 298.0 (142.9~551.6)

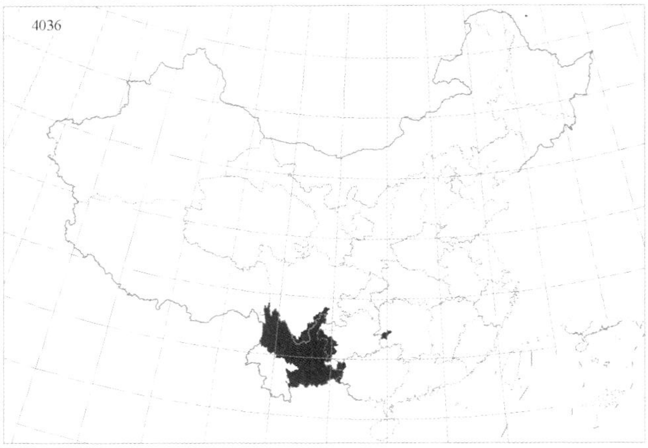

Fabaceae 豆科　*Campylotropis* 荒子梢属
Campylotropis hirtella (Franch.) Schindl. 毛荒子梢
Shrubs, deciduous

14.0 (2.2~19.3); 6.7 (-6.3~13.1); 19.8 (9.5~26.6); 14.1 (3.9~19.4); 112.0 (15.8~173.2); 3.2 (0.0~48.6); 735 (411~939) ❙ 1013 (628~1495); 581 (406~836); 37 (11~69) ❙ 40.1 (-14.2~100.7); 715 (411~912); 290.0 (142.9~432.9)

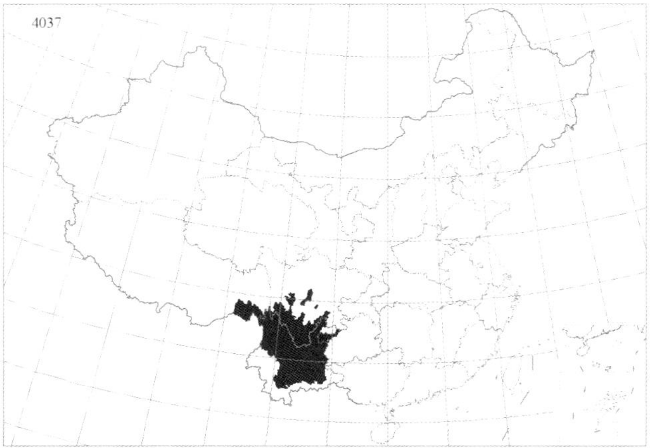

Fabaceae 豆科　*Campylotropis* 荒子梢属
Campylotropis latifolia (Dunn) Schindl. 阔叶荒子梢
Shrubs, deciduous

17.5 (12.4~19.3); 11.0 (4.7~13.3); 22.0 (18.6~23.6); 17.5 (12.4~19.4); 150.6 (89.1~173.2); 0.0 (0.0~0.3); 851 (675~939) ❙ 1111 (896~1336); 627 (528~774); 46 (31~55) ❙ 30.7 (13.3~54.8); 838 (675~923); 347.0 (205.2~452.0)

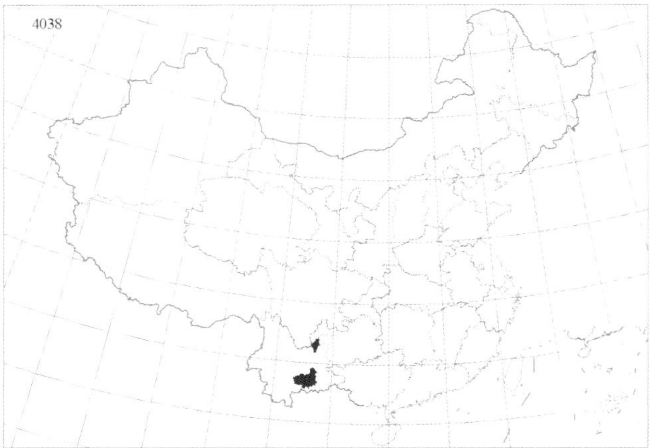

Fabaceae 豆科 *Campylotropis* 杭子梢属

Campylotropis macrocarpa (Bunge) Rehder 杭子梢

Shrubs, deciduous

14.3 (2.2~24.1); 2.6 (−16.1~18.1); 24.9 (9.9~29.9); 14.6 (3.8~24.2); 120.9 (15.5~230.2); 9.0 (0.0~77.4); 807 (413~1329) ❙ 1077 (377~2499); 493 (241~1412); 84 (6~308) ❙ 31.9 (−40.9~219.3); 770 (377~1329); 217.0 (70.3~627.0)

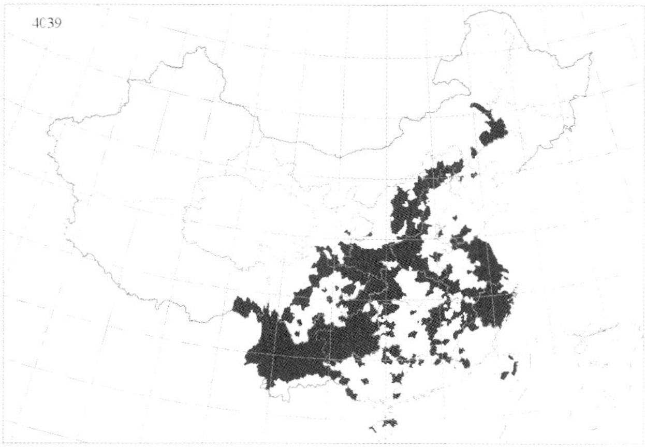

Fabaceae 豆科 *Campylotropis* 杭子梢属

Campylotropis macrocarpa var. hupehensis (Pamp.) Iokawa et H. Ohashi 太白山杭子梢

Shrubs, deciduous

8.6 (3.0~13.0); −5.9 (−12.7~−0.1); 21.6 (12.4~26.6); 9.8 (4.8~13.3); 76.7 (25.1~113.3); 33.0 (11.7~65.0); 634 (441~800) ❙ 564 (341~856); 303 (190~391); 17 (8~39) ❙ −10.1 (−40.9~52.6); 538 (341~711); 177.3 (61.2~318.2)

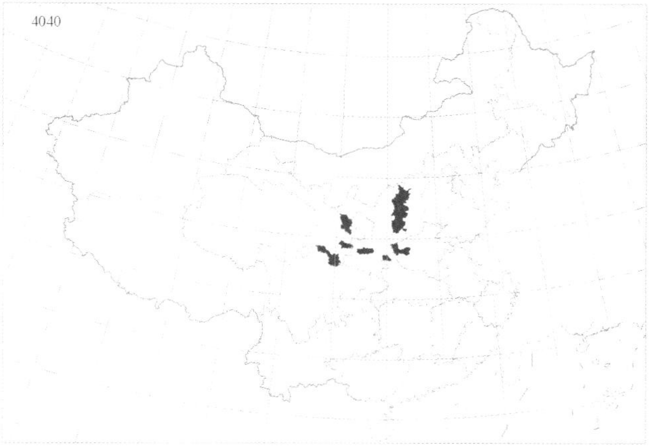

Fabaceae 豆科 *Campylotropis* 杭子梢属

Campylotropis pauciflora C. J. Chen 少花杭子梢

Shrubs, deciduous

14.3; 7.6; 19.7; 14.3; 111.9; 0.0; 728 ❙ 991; 580; 35 ❙ 36.1; 707; 296.9

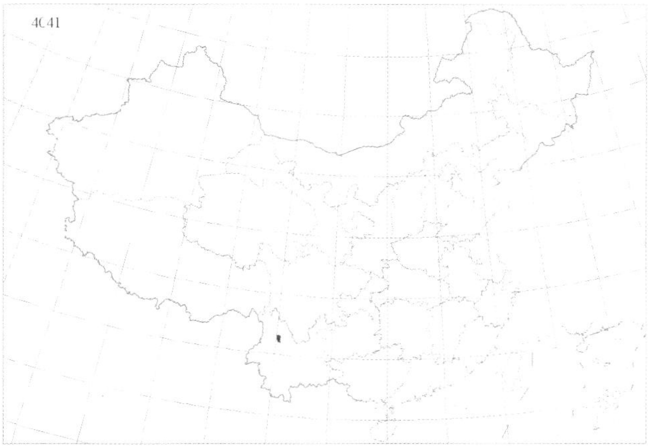

Fabaceae 豆科 *Campylotropis* 杭子梢属

Campylotropis pinetorum (Kurz) Schindl. 松林杭子梢

Shrubs, deciduous

17.4 (13.6~21.0); 10.5 (6.5~15.1); 22.2 (19.0~26.6); 17.4 (13.6~21.0); 149.2 (103.6~191.8); 0.0 (0.0~0.0); 850 (707~1053) ❙ 1222 (911~1751); 681 (524~996); 50 (28~77) ❙ 43.7 (6.5~112.2); 837 (690~1053); 367.0 (189.5~659.0)

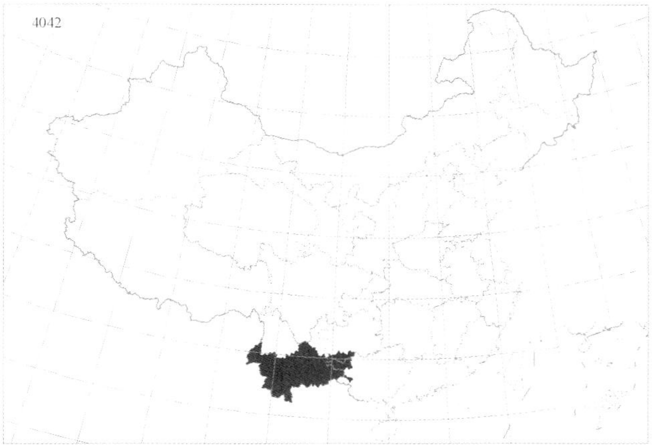

Fabaceae 豆科 *Campylotropis* 杭子梢属

Campylotropis pinetorum subsp. velutina (Dunn) H. Ohashi 绒毛叶杭子梢

Shrubs, deciduous

17.9 (15.4~21.0); 10.1 (7.4~14.2); 23.6 (20.7~26.6); 17.9 (15.5~21.0); 155.2 (125.7~191.8); 0.0 (0.0~0.0); 883 (767~1053) ❙ 1245 (958~1547); 686 (531~911); 54 (43~74) ❙ 41.2 (13.3~80.6); 881 (767~1053); 338.2 (189.5~551.6)

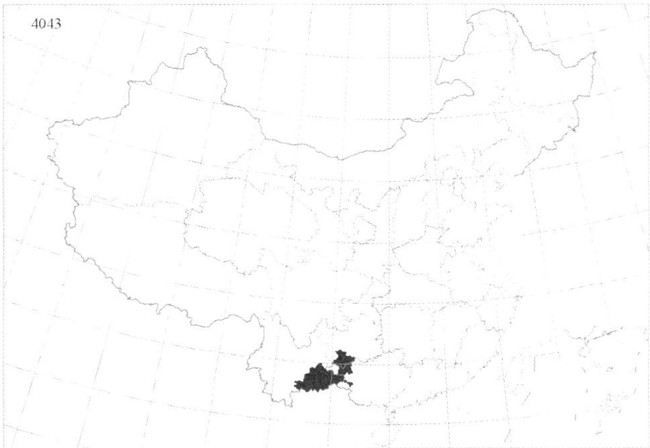

Fabaceae 豆科 *Campylotropis* 杭子梢属

Campylotropis polyantha (Franch.) Schindl. 小雀花

Shrubs, deciduous

13.0 (−1.8~19.3); 5.1 (−10.2~13.1); 19.3 (6.1~27.7); 13.2 (1.8~19.4); 103.6 (1.9~173.2); 7.7 (0.0~82.8); 709 (329~966) ❙ 1007 (320~1736); 565 (217~1007); 37 (9~69) ❙ 45.0 (−14.2~145.8); 694 (320~966); 278.2 (72.2~447.4)

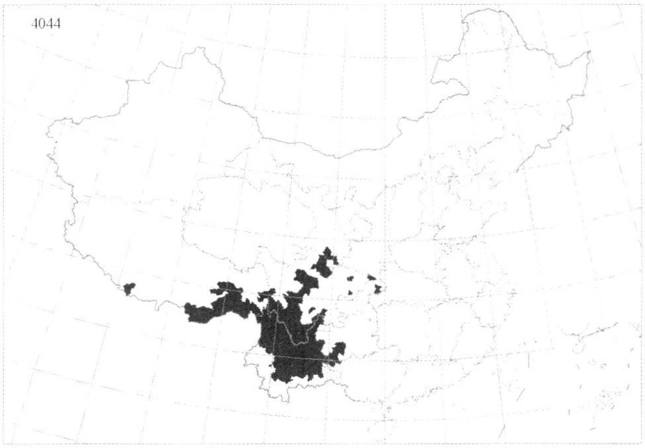

Fabaceae 豆科 *Campylotropis* 荒子梢属
Campylotropis polyantha f. leiocarpa (Pamp.) Iokawa et H. Ohashi 光果小雀花 Shrubs, deciduous

8.4 (2.0~14.8); 0.2 (-7.2~7.8); 15.5 (9.9~24.9); 8.9 (3.8~14.9); 59.4 (15.4~119.2); 18.5 (0.0~51.4); 572 (413~791) ❙ 950 (685~1197); 527 (346~651); 30 (17~71) ❙ 69.8 (30.5~128.6); 568 (413~791); 293.3 (182.9~366.3)

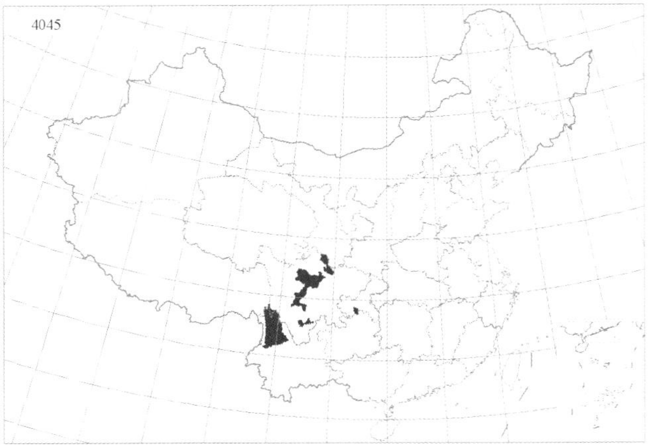

Fabaceae 豆科 *Campylotropis* 荒子梢属
Campylotropis polyantha var. neglecta (Schindl.) Iokawa et H. Ohashi 蒙自荒子梢 Shrubs, deciduous

16.9; 10.3; 21.4; 16.9; 142.7; 0.0; 813 ❙ 1178; 646; 51 ❙ 44.9; 813; 383.4

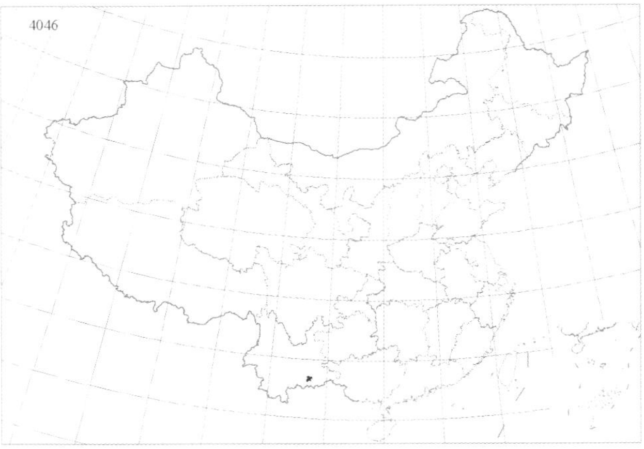

Fabaceae 豆科 *Campylotropis* 荒子梢属
Campylotropis sulcata Schindl. 槽茎荒子梢
Shrubs, deciduous

15.6 (2.9~20.5); 9.3 (-5.7~15.1); 20.1 (10.6~23.7); 15.9 (4.3~20.5); 135.8 (19.9~185.9); 7.5 (0.0~45.2); 806 (421~992) ❙ 1250 (535~1607); 614 (326~762); 54 (10~77) ❙ 52.5 (27.1~72.8); 796 (421~986); 459.6 (234.9~659.0)

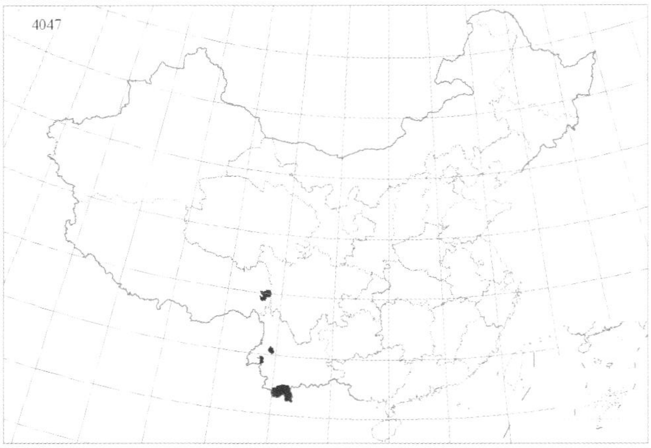

Fabaceae 豆科 *Campylotropis* 荒子梢属
Campylotropis tenuiramea P. Y. Fu 细枝荒子梢
Shrubs, deciduous

13.8; 6.7; 19.3; 13.8; 105.9; 0.0; 714 ❙ 963; 592; 29 ❙ 35.0; 690; 266.2

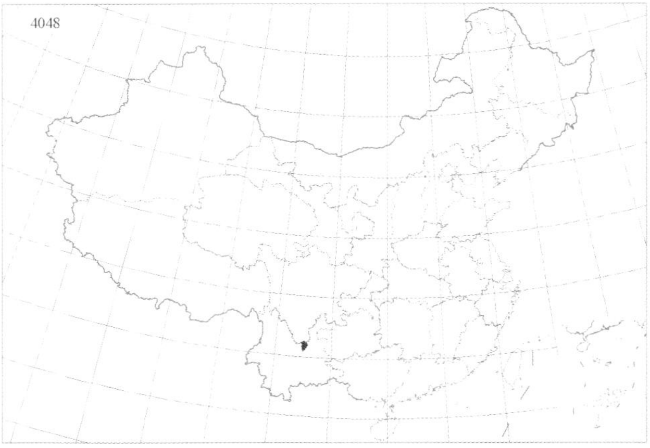

Fabaceae 豆科 *Campylotropis* 荒子梢属
Campylotropis teretiracemosa P. C. Li et C. J. Chen 柱序荒子梢
Shrubs, deciduous

16.1; 8.6; 21.9; 16.1; 133.0; 0.0; 797 ❙ 1014; 630; 20 ❙ 27.3; 719; 330.0

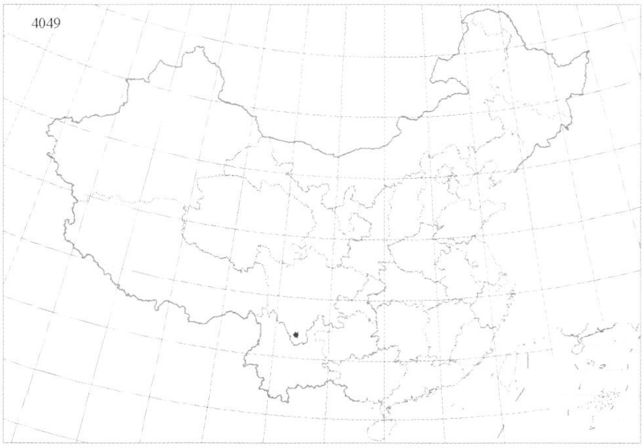

Fabaceae 豆科 *Campylotropis* 荒子梢属
Campylotropis trigonoclada (Franch.) Schindl. 三棱枝荒子梢
Shrubs, deciduous

15.4 (2.2~19.8); 7.5 (-6.6~13.1); 21.8 (9.9~26.9); 15.5 (3.8~19.8); 126.7 (15.5~177.9); 1.0 (0.0~49.1); 788 (413~1010) ❙ 1108 (741~1500); 593 (411~836); 50 (15~107) ❙ 41.4 (-14.2~125.9); 774 (413~1010); 281.7 (142.9~516.7)

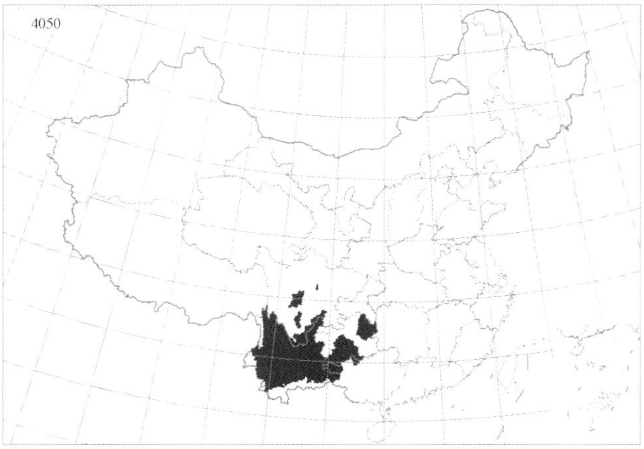

Fabaceae 豆科 *Campylotropis* 杭子梢属
Campylotropis trigonoclada var. bonatiana (Pamp.) Iokawa et H. Ohashi 马尿藤 Shrubs or subshrubs, deciduous

16.0 (11.2~19.3); 9.1 (4.6~13.3); 20.9 (16.7~23.9); 16.0 (11.3~19.4); 132.1 (75.8~173.2); 0.0 (0.0~0.4); 790 (642~939) ❙ 1037 (797~1456); 592 (411~836); 42 (23~62) ❙ 31.6 (-14.2~70.6); 766 (642~923); 301.4 (200.4~452.0)

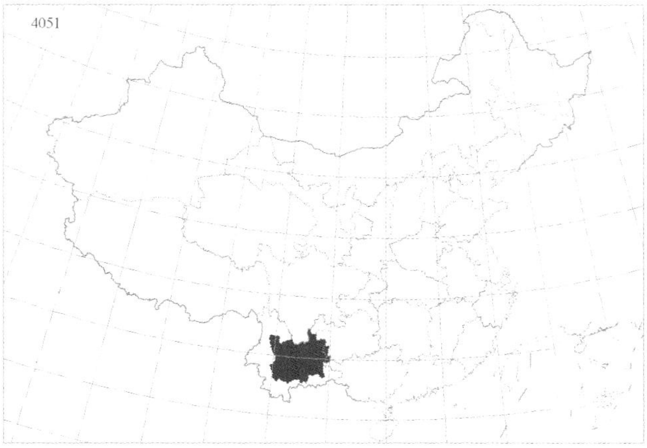

Fabaceae 豆科 *Campylotropis* 杭子梢属
Campylotropis wenshanica P. Y. Fu 秋杭子梢
Shrubs, deciduous

16.8; 9.3; 22.1; 16.8; 141.7; 0.0; 817 ❙ 1222; 678; 49 ❙ 49.6; 817; 341.8

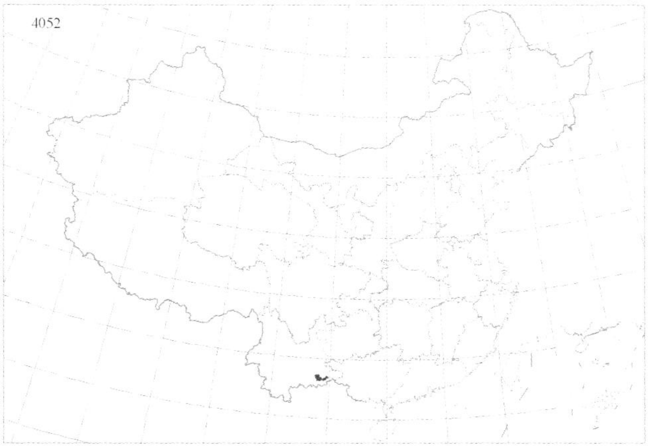

Fabaceae 豆科 *Campylotropis* 杭子梢属
Campylotropis wilsonii Schindl. 小叶杭子梢
Small shrubs, deciduous

4.0 (1.2~6.1); -4.7 (-7.1~-1.5); 11.7 (9.0~14.4); 5.1 (3.3~6.6); 26.0 (11.9~39.7); 38.1 (21.8~56.5); 452 (390~508) ❙ 806 (578~1023); 459 (345~546); 20 (10~28) ❙ 78.1 (38.4~125.9); 452 (390~508); 236.5 (130.1~299.8)

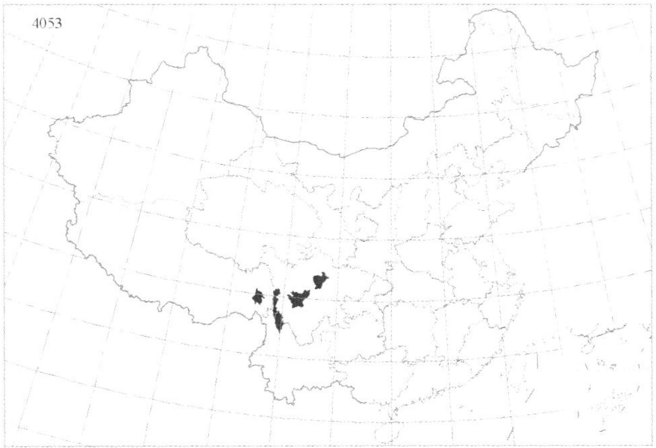

Fabaceae 豆科 *Campylotropis* 杭子梢属
Campylotropis yunnanensis (Franch.) Schindl. 滇杭子梢
Shrubs, deciduous

13.5 (2.9~18.9); 6.0 (-5.7~12.0); 19.6 (10.6~26.6); 13.6 (4.3~19.0); 105.5 (19.9~167.8); 2.7 (0.0~45.2); 718 (421~929) ❙ 973 (535~1180); 558 (326~651); 35 (10~65) ❙ 37.1 (-14.2~78.3); 696 (421~912); 288.9 (142.9~406.0)

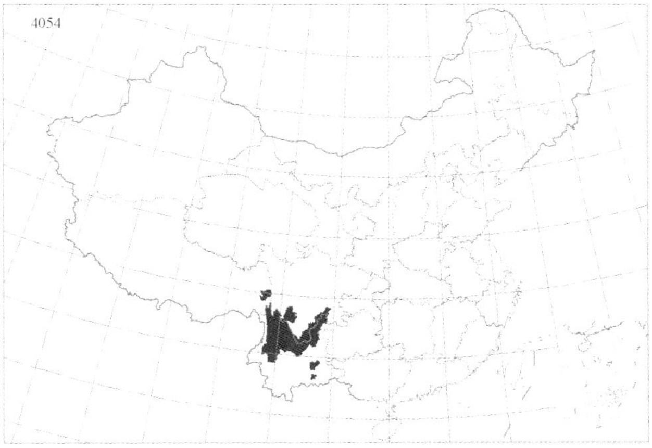

Fabaceae 豆科 *Campylotropis* 杭子梢属
Campylotropis yunnanensis subsp. filipes (Ricker) Iokawa et H. Ohashi 丝梗杭子梢 Shrubs, deciduous

5.2; -2.8; 12.1; 5.7; 30.6; 27.6; 479 ❙ 923; 574; 14 ❙ 94.7; 479; 288.4

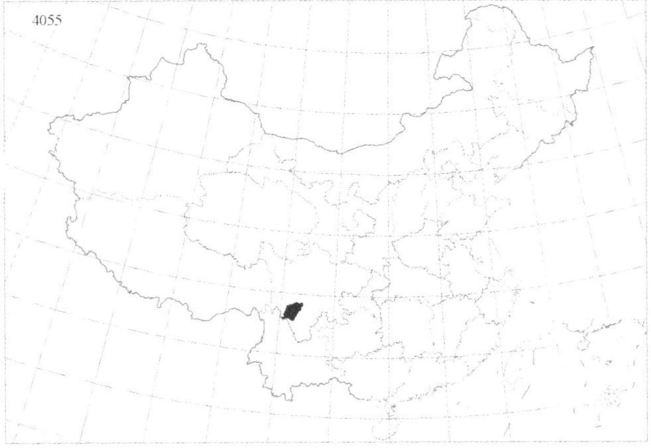

Fabaceae 豆科 *Caragana* 锦鸡儿属
Caragana acanthophylla Kom. 刺叶锦鸡儿
Shrubs, deciduous

5.1 (-2.8~10.8); -14.1 (-20.9~-8.0); 20.7 (10.9~27.5); 8.6 (3.5~12.2); 68.7 (16.3~107.6); 66.7 (34.1~109.2); 615 (390~813) ❙ 197 (70~406); 94 (39~211); 16 (3~48) ❙ -65.1 (-90.6~-7.6); 197 (70~406); 95.7 (10.9~297.3)

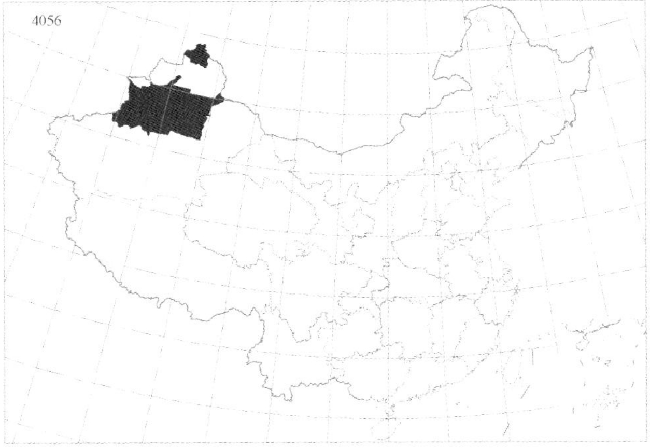

Fabaceae 豆科 *Caragana* 锦鸡儿属

Caragana aegacanthoides (R. Parker) L. B. Chaudhary et S. K. Srivastava 萨迦锦鸡儿

Shrubs

-1.1; -9.1; 6.2; 2.0; 2.5; 75.6; 334 ❙ 321; 193; 33 ❙ -4.0; 321; 79.3

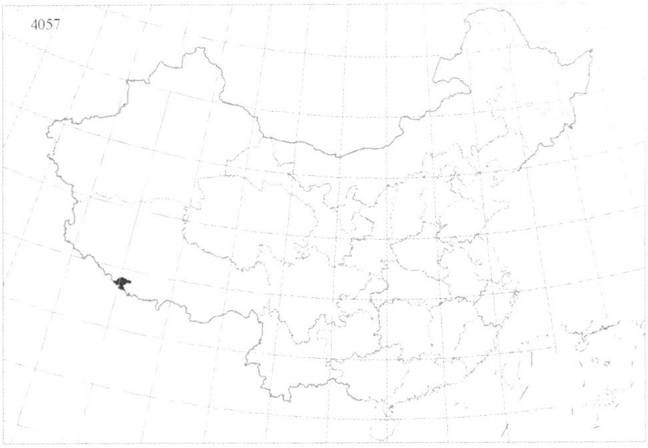

Fabaceae 豆科 *Caragana* 锦鸡儿属

Caragana altaica (Kom.) Pojark. 阿尔泰锦鸡儿

Shrubs, deciduous

1.9 (-0.6~5.8); -17.1 (-19.5~-14.9); 18.7 (15.9~23.3); 7.0 (5.4~9.7); 52.4 (37.5~80.9); 88.6 (71.2~104.2); 545 (475~668) ❙ 184 (156~251); 77 (56~106); 22 (15~33) ❙ -65.8 (-76.7~-52.0); 184 (156~251); 86.2 (16.6~139.1)

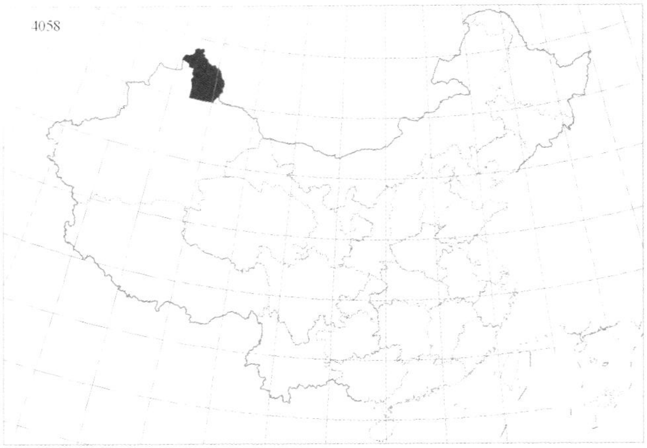

Fabaceae 豆科 *Caragana* 锦鸡儿属

Caragana arborescens Lam. 树锦鸡儿

Shrubs or small trees, deciduous

8.5 (-5.8~15.1); -7.5 (-29.8~1.8); 22.7 (12.4~28.0); 10.3 (4.4~15.2); 83.0 (25.1~128.0); 40.6 (5.6~157.4); 663 (426~880) ❙ 554 (75~1018); 311 (39~648); 19 (4~53) ❙ -16.2 (-89.7~56.9); 535 (75~817); 175.0 (14.6~334.4)

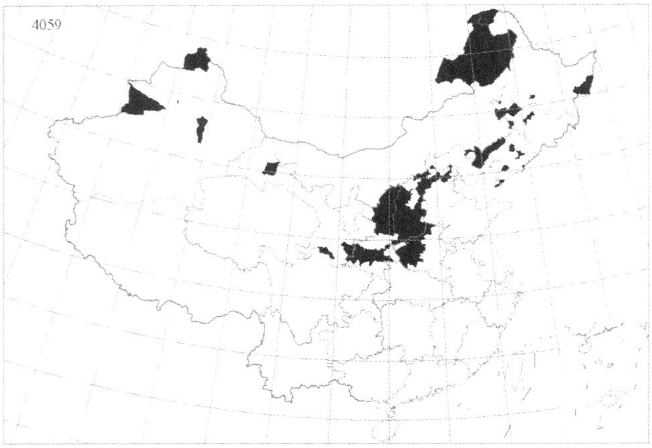

Fabaceae 豆科 *Caragana* 锦鸡儿属

Caragana aurantiaca Koehne 镰叶锦鸡儿

Shrubs, deciduous

5.5 (-2.8~11.4); -13.4 (-20.9~-6.7); 20.8 (10.9~27.5); 8.8 (3.5~12.7); 70.2 (16.3~109.3); 63.4 (29.5~109.2); 621 (390~813) ❙ 180 (32~450); 87 (21~211); 14 (0~48) ❙ -67.5 (-95.9~3.9); 179 (32~406); 76.9 (1.6~297.3)

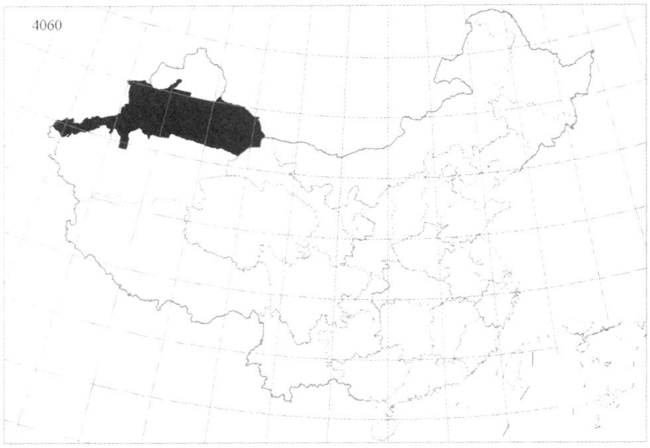

Fabaceae 豆科 *Caragana* 锦鸡儿属

Caragana bicolor Kom. 二色锦鸡儿

Shrubs, deciduous

2.7 (-1.4~14.3); -6.4 (-11.4~7.6); 10.7 (7.9~19.7); 4.5 (2.4~14.3); 21.9 (6.6~111.9); 49.1 (0.0~82.3); 427 (341~728) ❙ 621 (255~991); 373 (190~619); 13 (1~35) ❙ 44.5 (-33.6~128.6); 408 (255~707); 239.2 (101.7~357.7)

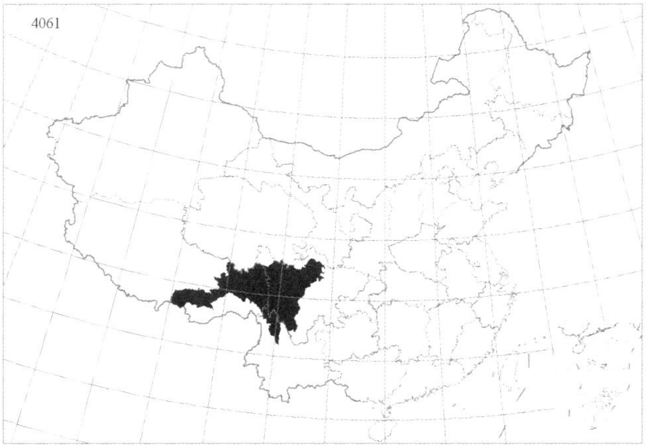

Fabaceae 豆科 *Caragana* 锦鸡儿属

Caragana boisii C. K. Schneid. 扁刺锦鸡儿

Shrubs, deciduous

6.9 (2.0~14.8); -3.1 (-8.4~3.8); 15.9 (9.5~25.3); 7.7 (3.8~14.8); 51.7 (15.4~119.2); 28.4 (1.4~51.4); 546 (411~795) ❙ 828 (505~1358); 442 (277~619); 21 (11~88) ❙ 58.2 (-26.0~128.6); 537 (411~795); 253.9 (60.8~341.6)

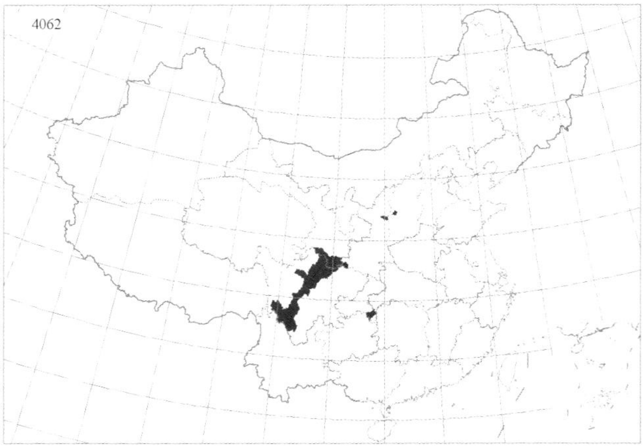

Fabaceae 豆科 *Caragana* 锦鸡儿属
Caragana brachypoda Pojark. 矮脚锦鸡儿
Dwarf shrubs, deciduous

5.6 (0.1~8.8); -12.2 (-21.5~-7.0); 21.3 (16.0~24.3); 8.6 (6.2~10.3); 67.4 (40.3~85.3); 60.0 (35.3~108.6); 595 (489~679) ❙ 261 (49~416); 166 (34~270); 6 (1~15) ❙ -55.0 (-92.8~-23.3); 261 (49~416); 75.8 (5.1~182.4)

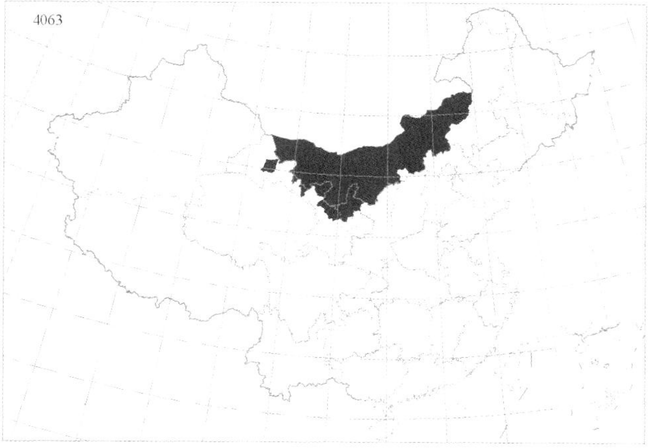

Fabaceae 豆科 *Caragana* 锦鸡儿属
Caragana brevifolia Kom. 短叶锦鸡儿
Shrubs, deciduous

1.8 (-5.5~15.3); -10.0 (-18.0~5.4); 12.3 (5.8~24.7); 4.8 (1.3~15.3); 27.1 (1.2~124.2); 64.4 (0.0~127.1); 440 (299~801) ❙ 517 (63~1736); 293 (45~1007); 11 (2~64) ❙ 20.3 (-84.9~145.8); 386 (63~801); 221.9 (8.4~447.4)

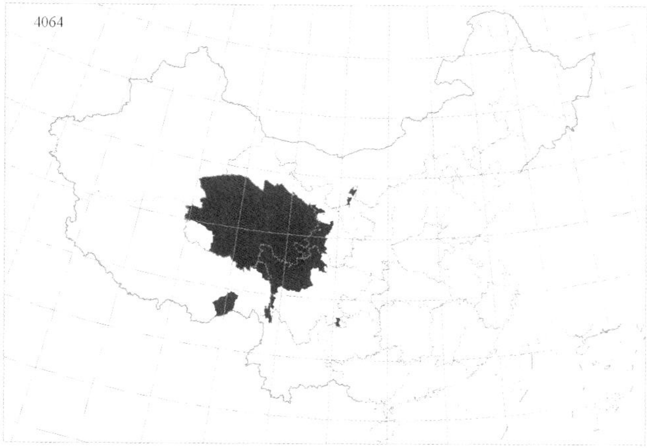

Fabaceae 豆科 *Caragana* 锦鸡儿属
Caragana camilli-schneideri Kom. 北疆锦鸡儿
Shrubs, deciduous

3.7 (-2.8~10.2); -14.9 (-20.9~-8.5); 18.8 (10.9~27.1); 7.7 (3.5~12.1); 58.0 (16.3~107.6); 72.6 (36.6~109.2); 571 (390~810) ❙ 241 (70~406); 112 (44~211); 22 (3~51) ❙ -54.7 (-90.6~-7.6); 241 (70~406); 141.2 (25.3~297.3)

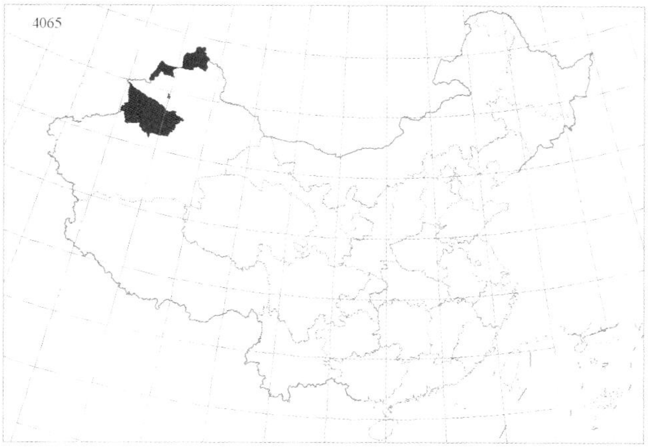

Fabaceae 豆科 *Caragana* 锦鸡儿属
Caragana changduensis Y. X. Liou 昌都锦鸡儿
Shrubs, deciduous

1.6 (-1.2~5.3); -7.6 (-11.7~-2.4); 9.9 (8.1~12.2); 3.8 (2.6~5.7); 16.0 (8.0~30.5); 56.2 (26.2~82.3); 404 (347~479) ❙ 589 (311~961); 359 (232~558); 12 (2~31) ❙ 45.5 (-20.0~100.7); 401 (311~479); 241.8 (111.0~307.8)

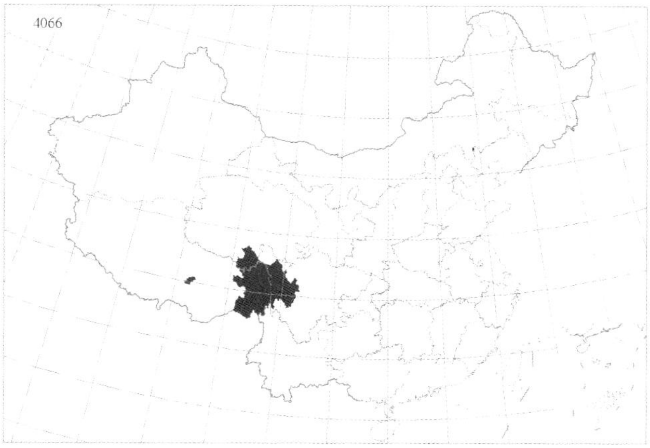

Fabaceae 豆科 *Caragana* 锦鸡儿属
Caragana chinghaiensis Y. X. Liou 青海锦鸡儿
Shrubs, deciduous

-0.9 (-3.3~0.9); -12.4 (-15.3~-9.9); 9.3 (7.2~11.7); 3.0 (2.0~4.2); 11.1 (3.7~21.4); 80.8 (67.5~103.2); 377 (339~424) ❙ 517 (380~656); 302 (225~382); 11 (7~15) ❙ 36.6 (-2.8~73.6); 375 (339~412); 273.4 (189.8~364.1)

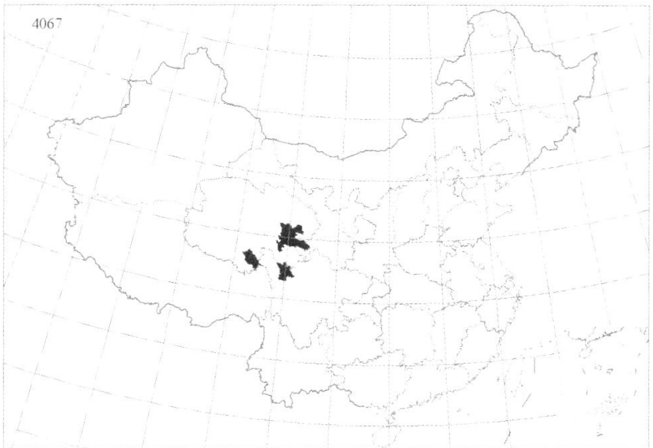

Fabaceae 豆科 *Caragana* 锦鸡儿属
Caragana chinghaiensis var. minima Y. X. Liou 小青海锦鸡儿
Shrubs, deciduous

-0.2; -10.5; 9.1; 3.1; 10.8; 72.1; 377 ❙ 590; 350; 13 ❙ 56.2; 377; 304.2

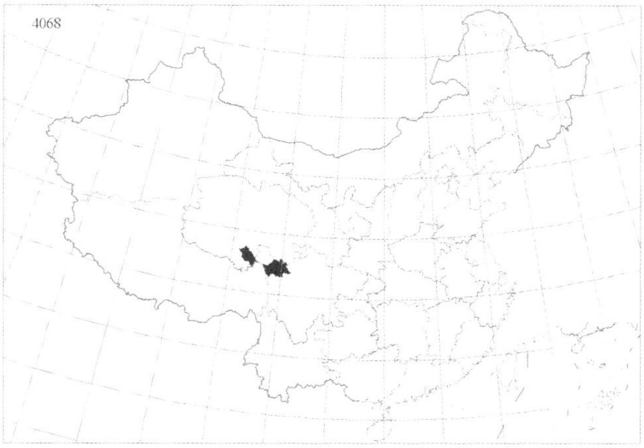

Fabaceae 豆科 *Caragana* 锦鸡儿属
Caragana chumbica Prain 高山锦鸡
Shrubs, deciduous

0.4; -9.1; 8.9; 3.2; 12.2; 66.5; 374 ❙ 254; 194; 1 ❙ -31.8; 254; 104.2

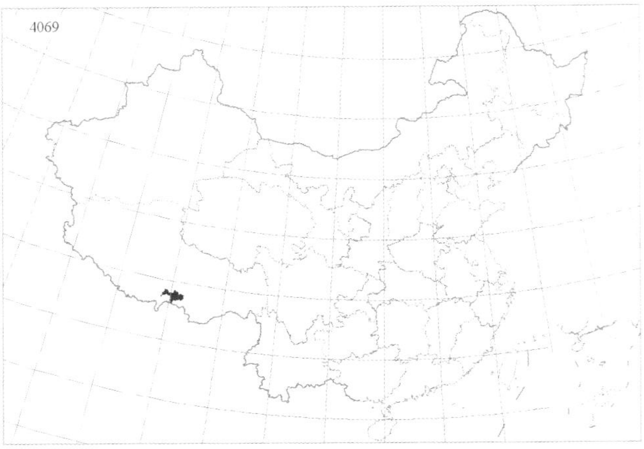

Fabaceae 豆科 *Caragana* 锦鸡儿属
Caragana crassispina C. Marquand 粗刺锦鸡儿
Shrubs, deciduous

3.4; -5.4; 11.2; 4.6; 22.8; 41.8; 429 ❙ 500; 302; 8 ❙ 13.7; 429; 255.3

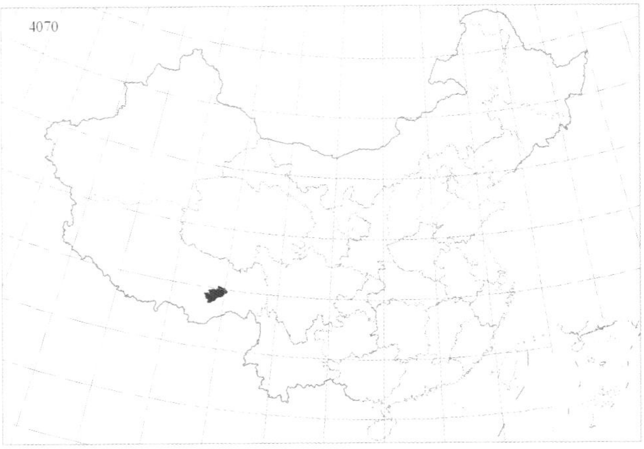

Fabaceae 豆科 *Caragana* 锦鸡儿属
Caragana cuneatoalata Y. X. Liou 楔翼锦鸡儿
Shrubs, deciduous

4.4; -4.3; 12.1; 5.3; 28.5; 34.6; 455 ❙ 538; 337; 9 ❙ 18.2; 455; 168.2

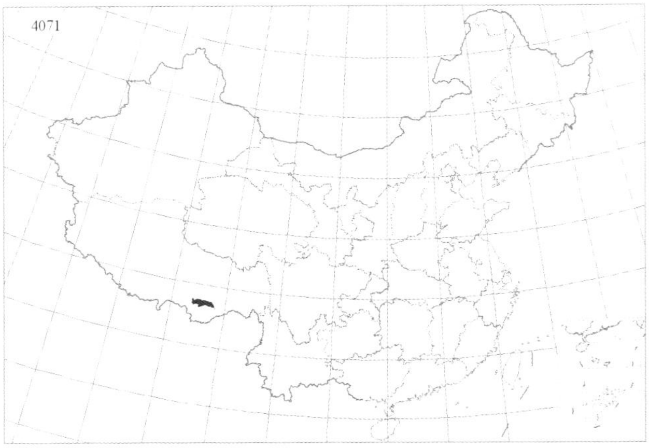

Fabaceae 豆科 *Caragana* 锦鸡儿属
Caragana dasyphylla Pojark. 粗毛锦鸡儿
Dwarf shrubs, deciduous

6.4 (-4.3~12.5); -12.0 (-20.9~-5.1); 21.3 (9.6~26.5); 9.3 (2.5~13.2); 75.1 (9.7~114.0); 57.6 (23.1~121.1); 640 (329~822) ❙ 151 (28~450); 72 (17~211); 12 (0~51) ❙ -72.9 (-96.6~3.9); 150 (28~406); 66.5 (1.6~297.3)

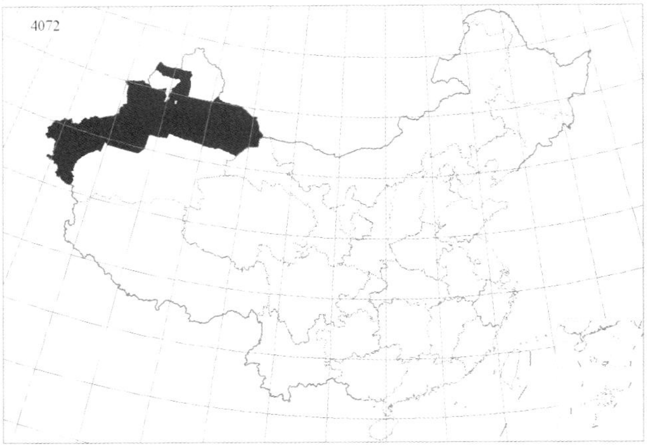

Fabaceae 豆科 *Caragana* 锦鸡儿属
Caragana davazamcii Sanchir 沙地锦鸡儿
Shrubs, deciduous

6.4 (-1.5~13.5); -10.3 (-25.5~1.1); 21.2 (10.1~27.1); 8.9 (3.3~13.7); 70.1 (13.4~117.5); 52.5 (8.1~128.7); 608 (390~820) ❙ 397 (111~707); 234 (74~362); 10 (2~22) ❙ -34.2 (-82.3~19.7); 396 (111~707); 113.4 (13.6~284.0)

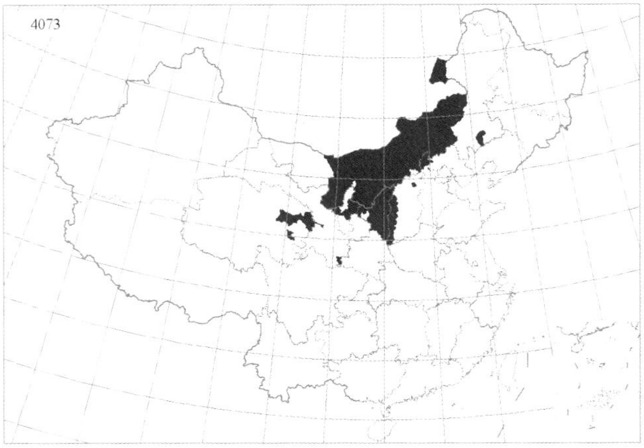

Fabaceae 豆科 *Caragana* 锦鸡儿属
Caragana densa Kom. 密叶锦鸡儿
Shrubs, deciduous

3.5 (-2.8~14.8); -10.7 (-20.9~3.8); 15.4 (8.3~27.5); 6.3 (2.5~14.8); 42.2 (7.0~119.2); 59.4 (1.4~109.2); 508 (347~820) ❙ 474 (30~1358); 254 (21~546); 15 (1~88) ❙ 4.1 (-96.3~128.6); 351 (30~795); 197.3 (2.1~364.1)

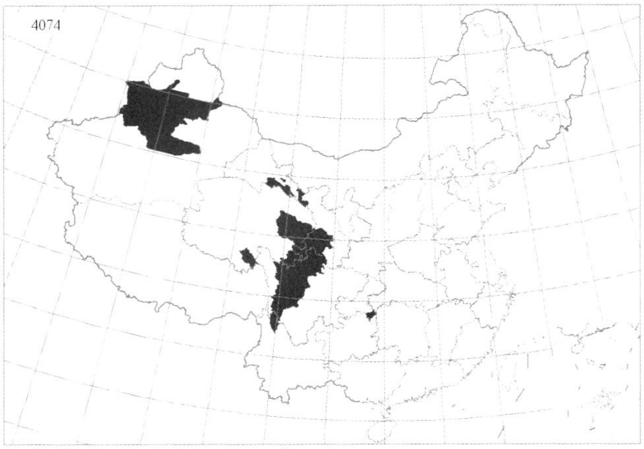

Fabaceae 豆科 *Caragana* 锦鸡儿属
Caragana erinacea Kom. 川西锦鸡儿
Shrubs, deciduous

1.5 (-4.1~6.6); -9.4 (-17.3~-1.5); 11.2 (7.0~18.9); 4.2 (2.0~8.0); 21.1 (3.7~57.9); 62.1 (21.8~114.2); 417 (327~557) ❙ 525 (212~950); 308 (156~546); 10 (0~28) ❙ 26.2 (-40.5~128.6); 390 (212~529); 227.9 (62.8~364.1)

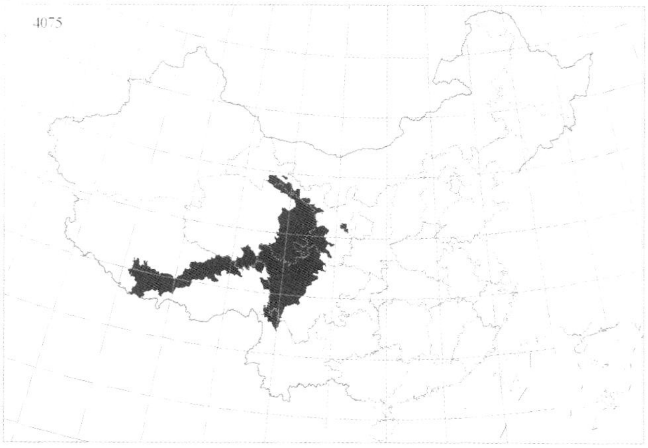

Fabaceae 豆科 *Caragana* 锦鸡儿属
Caragana franchetiana Kom. 云南锦鸡儿
Shrubs, deciduous

7.5 (-2.2~18.9); -0.9 (-12.8~12.0); 14.7 (6.1~26.6); 8.5 (1.8~19.0); 59.1 (1.9~167.8); 28.5 (0.0~91.3); 557 (329~929) ❙ 752 (252~1243); 449 (191~696); 21 (0~58) ❙ 34.9 (-35.9~128.6); 525 (252~912); 243.6 (62.8~384.7)

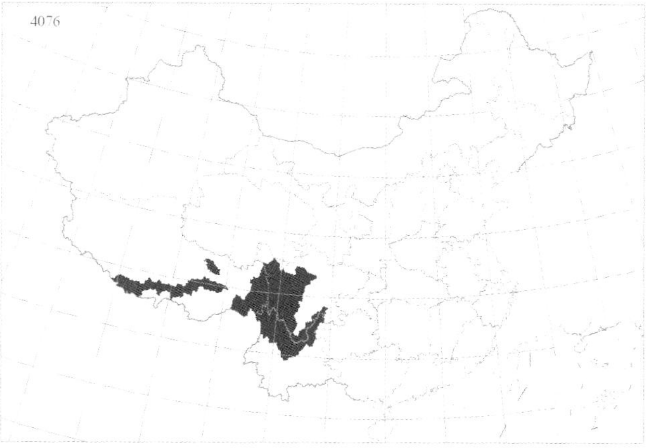

Fabaceae 豆科 *Caragana* 锦鸡儿属
Caragana franchetiana var. gyirongensis (C. C. Ni) Y. X. Liou 吉隆锦鸡儿
Shrubs

-1.1; -9.[illegible]; 6.2; 2.0; 2.5; 75.6; 334 ❙ 321; 193; 33 ❙ -4.0; 321; 79.3

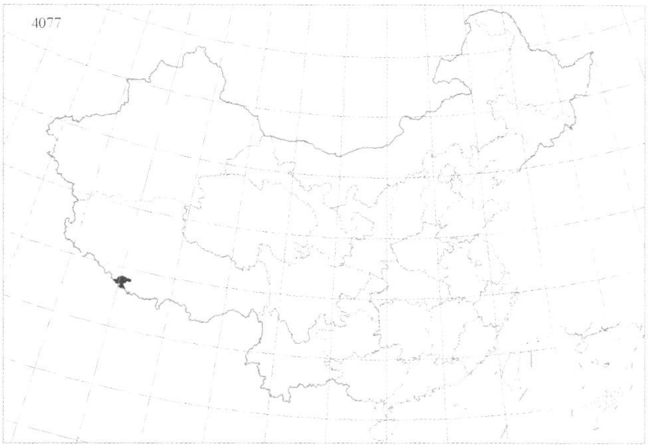

Fabaceae 豆科 *Caragana* 锦鸡儿属
Caragana frutex (L.) K. Koch 黄刺条锦鸡儿
Shrubs, deciduous

8.9 (-2.8~15.1); -7.2 (-20.9~1.8); 22.8 (10.5~27.8); 10.5 (3.4~15.2); 85.3 (14.4~128.0); 38.5 (5.6~109.2); 676 (390~865) ❙ 507 (70~843); 303 (44~533); 20 (2~53) ❙ -25.8 (-90.6~19.6); 503 (70~817); 156.4 (13.6~306.1)

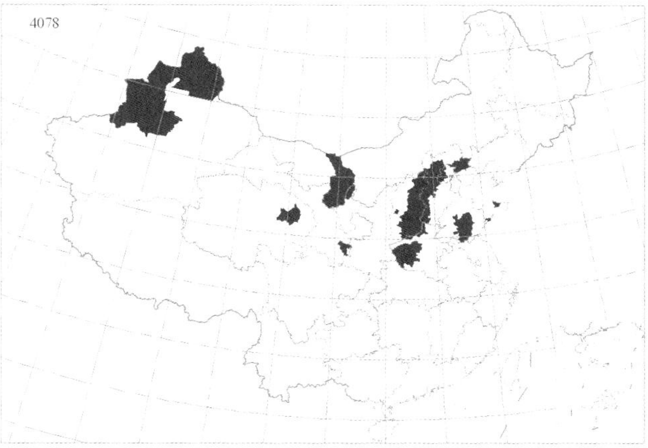

Fabaceae 豆科 *Caragana* 锦鸡儿属
Caragana fruticosa (Pall.) Besser 极东锦鸡儿
Shrubs, deciduous

1.0 (-5.0~3.9); -22.0 (-29.4~-16.9); 20.7 (17.0~22.1); 7.4 (5.0~8.4); 56.4 (35.2~65.3); 104.0 (77.6~154.8); 559 (450~603) ❙ 575 (468~657); 345 (300~416); 20 (11~35) ❙ 2.6 (-11.0~18.6); 546 (450~591); 260.7 (196.0~323.4)

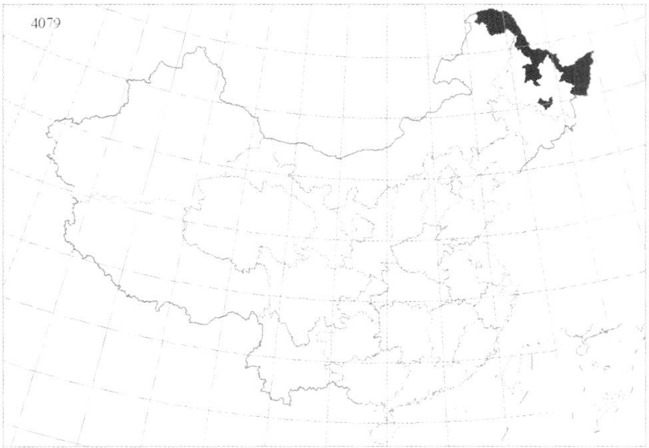

Fabaceae 豆科 *Caragana* 锦鸡儿属
Caragana gerardiana Royle ex Benth. 印度锦鸡儿
Shrubs, deciduous

-1.5 (-2.3~0.1); -10.5 (-12.0~-9.1); 7.1 (5.4~9.5); 2.2 (1.6~3.2); 4.8 (0.4~12.3); 81.8 (71.1~89.7); 337 (324~377) ❙ 421 (282~620); 243 (158~328); 42 (10~103) ❙ 24.5 (-14.2~91.4); 328 (282~377); 105.4 (32.2~290.1)

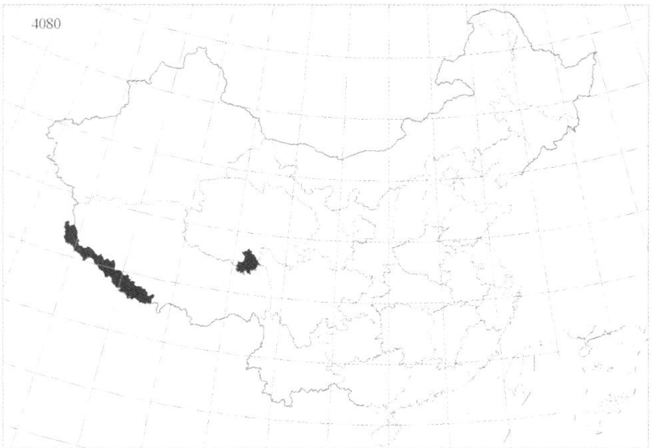

Fabaceae 豆科 *Caragana* 锦鸡儿属
Caragana jubata (Pall.) Poir. 鬼箭锦鸡儿
Shrubs, deciduous

5.5 (-5.7~14.2); -8.5 (-20.9~6.4); 17.8 (5.4~27.8); 7.7 (1.3~14.4); 56.4 (0.4~124.3); 49.4 (0.0~133.0); 556 (294~859) ❙ 450 (30~1337); 263 (19~727); 13 (0~103) ❙ -14.1 (-96.3~128.6); 397 (30~744); 154.6 (2.1~407.1)

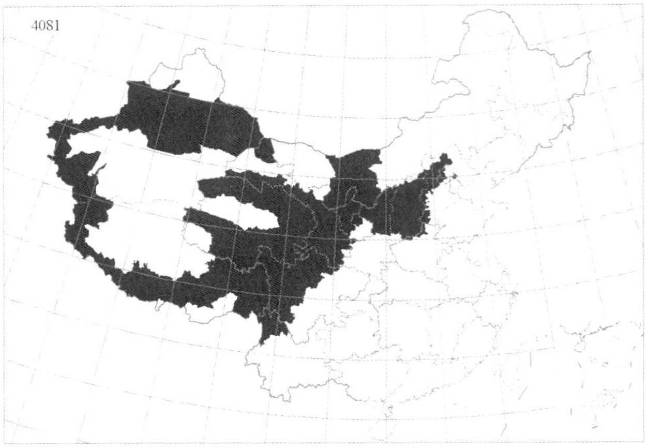

Fabaceae 豆科 *Caragana* 锦鸡儿属
Caragana jubata var. biaurita Y. X. Liou 两耳鬼箭
Shrubs, deciduous

-4.0; -15.4; 7.0; 1.8; 3.4; 110.5; 325 ❙ 382; 244; 9 ❙ 17.4; 320; 150.3

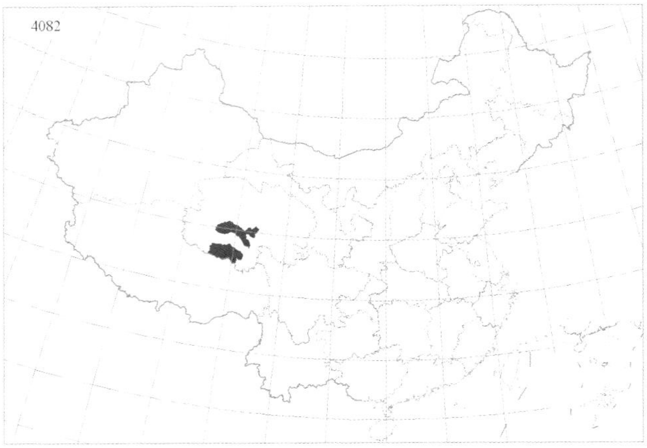

Fabaceae 豆科 *Caragana* 锦鸡儿属
Caragana jubata var. recurva Y. X. Liou 弯耳鬼箭
Shrubs, deciduous

1.8 (-1.2~5.5); -8.1 (-11.7~-5.8); 10.5 (8.5~17.5); 4.0 (2.6~7.3); 18.2 (8.1~51.3); 55.7 (42.6~82.3); 415 (347~530) ❙ 713 (374~943); 409 (205~546); 15 (8~23) ❙ 73.2 (-29.3~128.6); 404 (347~436); 270.0 (81.2~357.7)

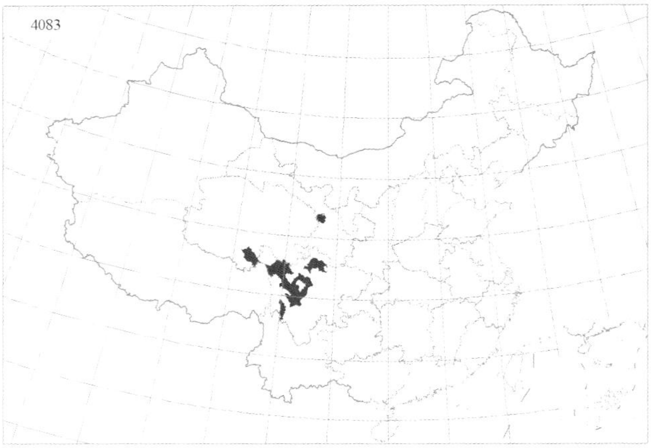

Fabaceae 豆科 *Caragana* 锦鸡儿属
Caragana junatovii Gorbunova 通天河锦鸡儿
Shrubs, deciduous

-2.9 (-5.5~0.1); -13.9 (-17.1~-10.1); 7.6 (5.8~9.5); 2.2 (1.3~3.2); 5.8 (1.2~12.3); 99.6 (71.1~127.1); 336 (299~377) ❙ 406 (270~533); 254 (183~328); 9 (5~12) ❙ 19.4 (-9.9~41.8); 328 (270~377); 201.3 (88.8~290.1)

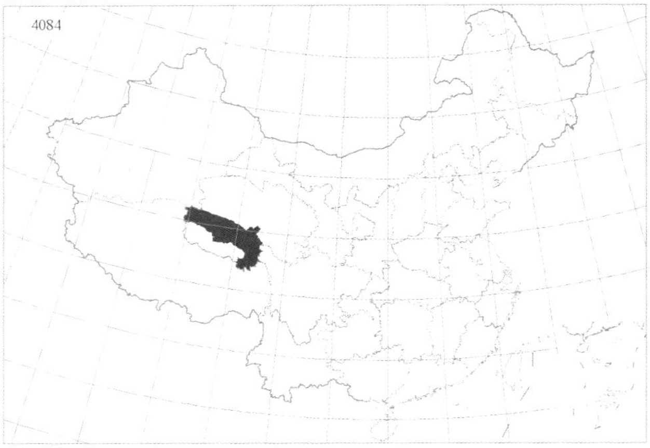

Fabaceae 豆科 *Caragana* 锦鸡儿属
Caragana kansuensis Pojark. 甘肃锦鸡儿
Dwarf shrubs, deciduous

6.3 (2.2~10.1); -10.2 (-16.1~-2.8); 20.9 (17.5~23.4); 8.7 (6.6~11.0); 67.4 (45.4~88.2); 50.8 (21.0~81.1); 594 (507~684) ❙ 429 (115~632); 258 (67~319); 11 (2~15) ❙ -27.4 (-82.2~7.5); 429 (115~628); 114.7 (21.5~241.9)

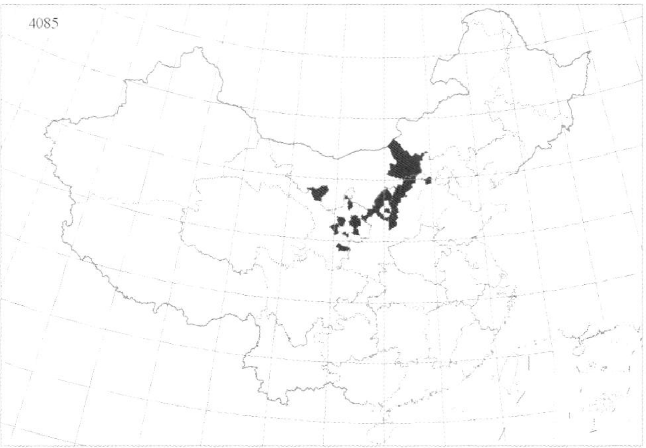

Fabaceae 豆科 *Caragana* 锦鸡儿属
Caragana kirghisorum Pojark. 囊萼锦鸡儿
Shrubs, deciduous

2.9 (-2.8~9.3); -15.4 (-20.9~-9.3); 17.1 (10.9~27.1); 6.8 (3.5~12.1); 49.3 (16.3~107.6); 74.6 (40.7~109.2); 532 (390~810) ❙ 255 (131~406); 123 (53~211); 20 (6~48) ❙ -49.7 (-82.8~-7.6); 255 (131~406); 136.1 (27.0~297.3)

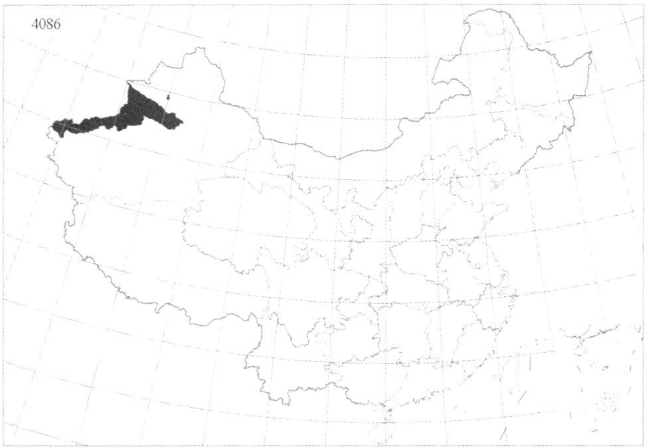

Fabaceae 豆科 *Caragana* 锦鸡儿属
Caragana korshinskii Kom. 柠条锦鸡儿
Shrubs or small trees, deciduous

6.7 (-1.4~13.0); -9.4 (-16.0~-2.0); 21.0 (9.7~26.6); 8.9 (2.9~13.3); 69.3 (11.3~113.3); 48.1 (16.7~92.1); 602 (359~800) ❙ 346 (35~773); 205 (23~345); 9 (1~32) ❙ -41.3 (-94.9~35.3); 343 (35~684); 95.4 (5.1~268.4)

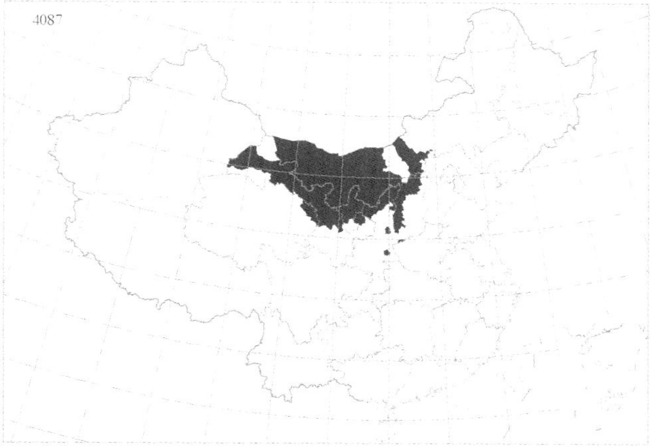

Fabaceae 豆科 *Caragana* 锦鸡儿属
Caragana kozlowii Kom. 沧江锦鸡儿
Shrubs, deciduous

0.9 (-3.2~3.3); -9.0 (-14.4~-6.8); 9.9 (7.3~14.3); 3.6 (2.0~5.6); 15.0 (4.0~33.5); 63.2 (50.8~101.5); 397 (335~468) ❙ 567 (425~777); 340 (251~457); 12 (8~15) ❙ 43.1 (-3.9~86.8); 396 (335~450); 261.0 (130.1~315.4)

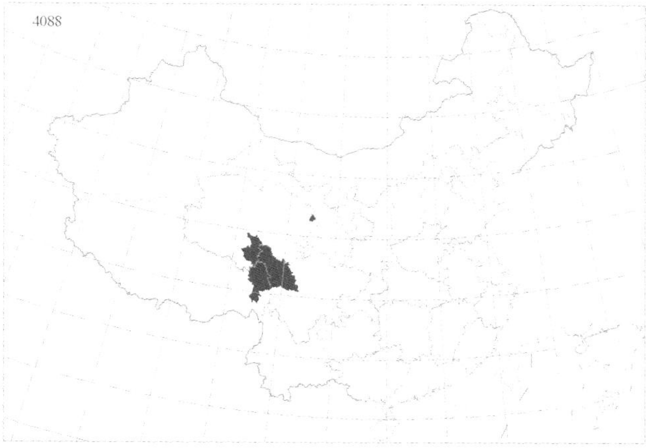

Fabaceae 豆科 *Caragana* 锦鸡儿属
Caragana laeta Kom. 阿拉套锦鸡儿
Shrubs, deciduous

4.1 (-4.3~12.3); -13.9 (-20.9~-5.3); 18.3 (9.6~27.1); 7.7 (2.5~13.0); 57.9 (9.7~112.3); 68.6 (24.7~121.1); 566 (329~810) ❙ 201 (32~450); 95 (19~211); 17 (0~48) ❙ -60.4 (-95.9~3.9); 199 (32~406); 97.3 (1.6~297.3)

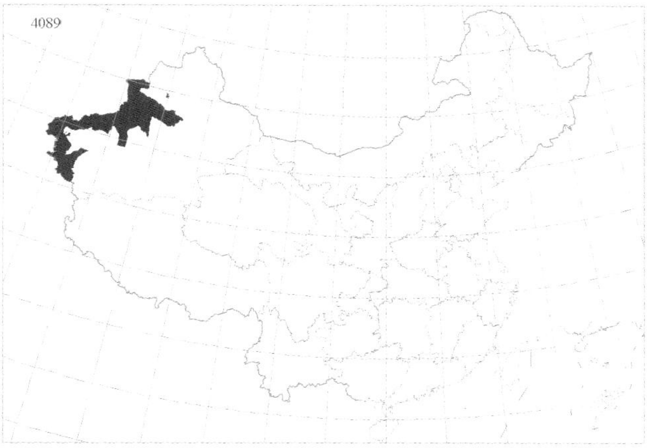

Fabaceae 豆科 *Caragana* 锦鸡儿属
Caragana leucophloea Pojark. 白皮锦鸡儿
Shrubs, deciduous

4.5 (-2.1~11.7); -14.3 (-21.5~-7.6); 20.4 (10.7~27.5); 8.2 (3.2~13.0); 64.3 (14.3~113.4); 70.1 (32.4~108.6); 592 (364~820) ❙ 182 (23~450); 93 (16~245); 14 (1~51) ❙ -67.1 (-97.2~3.9); 181 (23~406); 71.1 (2.1~297.3)

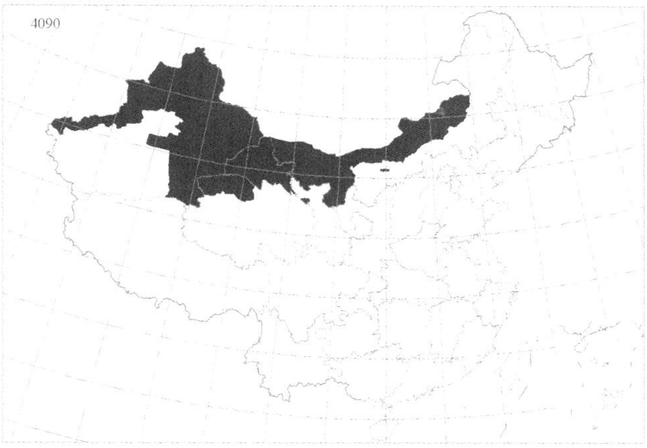

Fabaceae 豆科 *Caragana* 锦鸡儿属
Caragana leucospina Kom. 白刺锦鸡儿
Shrubs, deciduous

5.4 (-2.8~10.8); -13.6 (-20.9~-6.7); 20.7 (10.9~27.5); 8.7 (3.5~12.2); 69.0 (16.3~107.6); 64.2 (29.5~109.2); 615 (390~813) ❙ 184 (45~450); 89 (27~211); 14 (3~48) ❙ -66.8 (-94.0~3.9); 183 (45~406); 78.1 (7.6~297.3)

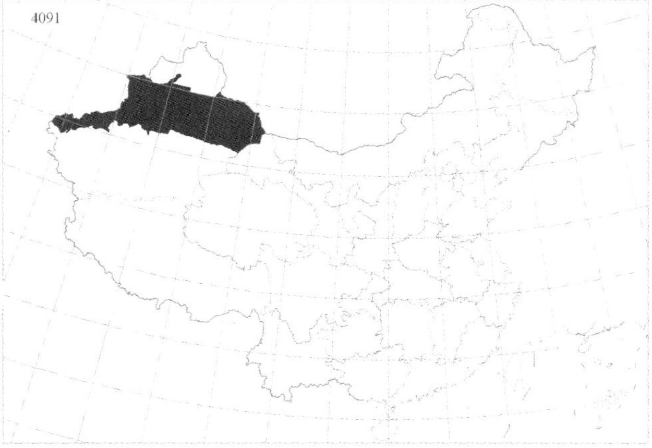

Fabaceae 豆科 *Caragana* 锦鸡儿属
Caragana leveillei Kom. 毛掌叶锦鸡儿
Shrubs, deciduous

10.6 (2.6~15.8); -4.4 (-15.9~2.5); 24.0 (17.5~28.2); 11.5 (6.6~15.8); 94.7 (45.4~133.1); 27.2 (3.4~80.4); 714 (507~891) ❙ 600 (349~1200); 335 (227~536); 24 (8~109) ❙ -16.1 (-43.4~56.2); 582 (349~891); 166.1 (58.7~297.4)

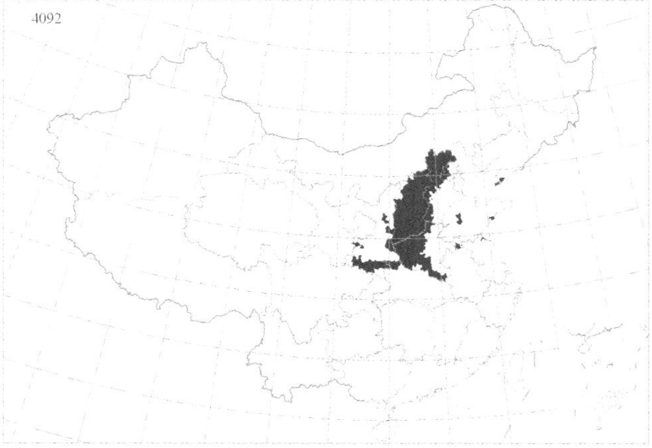

Fabaceae 豆科 *Caragana* 锦鸡儿属
Caragana licentiana Hand.-Mazz. 白毛锦鸡儿
Shrubs, deciduous

7.7 (2.2~13.0); -7.7 (-12.0~-2.4); 21.0 (13.1~26.6); 9.3 (4.9~13.3); 72.4 (27.4~113.3); 39.9 (17.1~60.7); 616 (445~800) ❙ 491 (373~551); 279 (201~309); 12 (7~17) ❙ -19.6 (-37.6~0.0); 491 (373~551); 144.6 (70.3~225.2)

4093
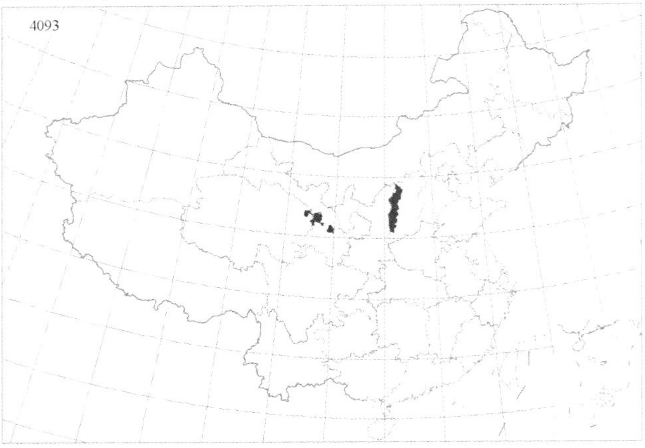

Fabaceae 豆科 *Caragana* 锦鸡儿属
Caragana limprichtii Harms 绢毛锦鸡儿
Shrubs, deciduous

1.3; -8.4; 9.7; 3.7; 14.7; 58.6; 405 ❙ 699; 416; 14 ❙ 72.3; 405; 275.8

4094
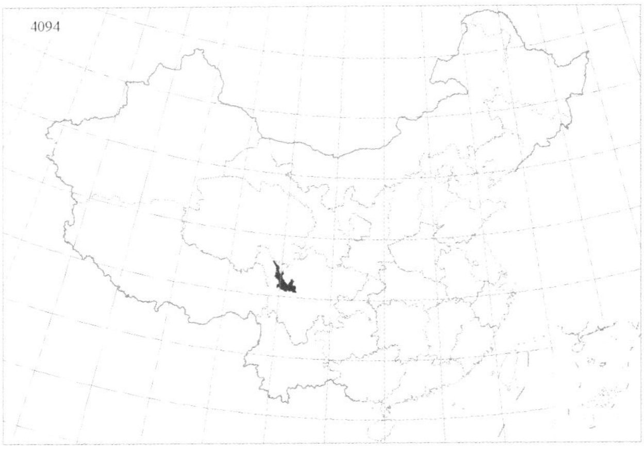

Fabaceae 豆科 *Caragana* 锦鸡儿属
Caragana liouana Zhao Y. Chang et Yakovlev 中间锦鸡儿
Shrubs

4.8 (-0.6~9.0); -12.9 (-21.5~-6.7); 20.5 (10.1~24.2); 8.1 (3.3~10.5); 62.6 (13.4~85.4); 64.5 (35.7~108.6); 576 (390~671) ❙ 332 (111~516); 212 (74~319); 8 (2~14) ❙ -40.9 (-82.3~19.7); 330 (111~515); 99.8 (13.9~284.0)

4095
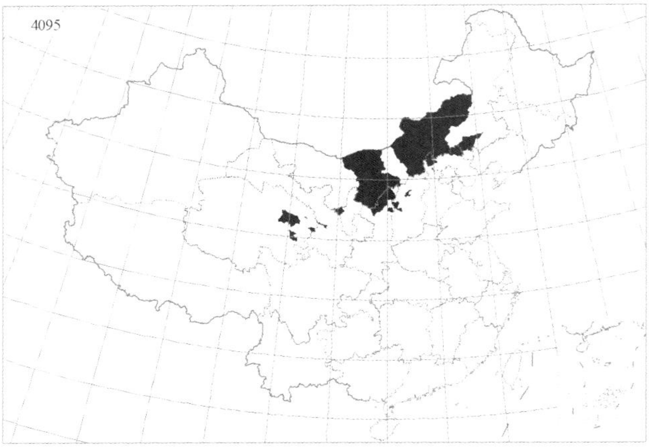

Fabaceae 豆科 *Caragana* 锦鸡儿属
Caragana litwinowii Kom. 金州锦鸡儿
Shrubs, deciduous

10.0 (8.0~12.5); -6.5 (-10.0~-1.2); 24.5 (23.0~26.3); 11.3 (10.0~12.9); 94.5 (83.4~110.3); 33.5 (14.2~46.7); 712 (658~793) ❙ 632 (553~724); 400 (371~448); 24 (9~65) ❙ -10.9 (-27.3~9.1); 626 (553~689); 176.6 (91.1~239.1)

4096
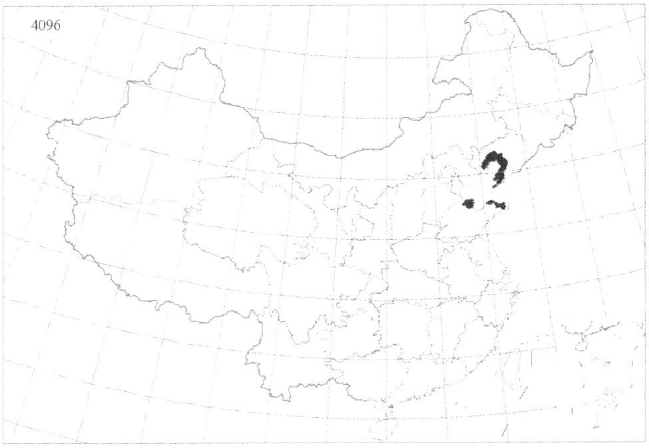

Fabaceae 豆科 *Caragana* 锦鸡儿属
Caragana manshurica Kom. 东北锦鸡儿
Shrubs, deciduous

6.8; -12.9; 23.6; 9.6; 80.1; 58.1; 648 ❙ 816; 513; 27 ❙ 26.0; 648; 278.9

4097
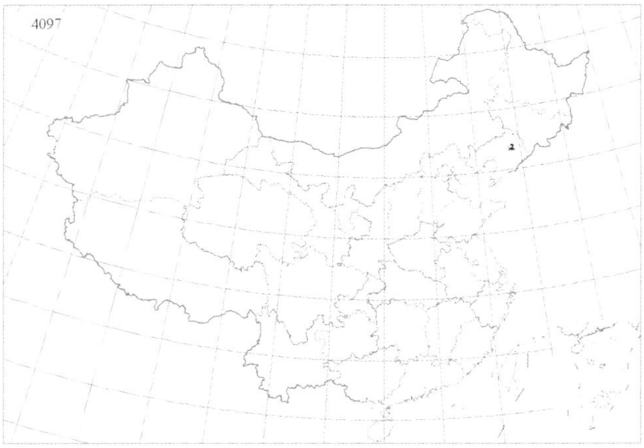

Fabaceae 豆科 *Caragana* 锦鸡儿属
Caragana microphylla Lam. 小叶锦鸡儿
Shrubs, deciduous

8.6 (-1.9~15.0); -7.7 (-25.9~1.5); 23.0 (11.0~28.0); 10.3 (3.8~15.1); 84.1 (17.4~129.0); 40.7 (6.5~128.7); 669 (407~880) ❙ 512 (111~917); 314 (74~560); 17 (2~53) ❙ -24.0 (-82.3~19.6); 507 (111~817); 148.5 (13.6~287.6)

4098
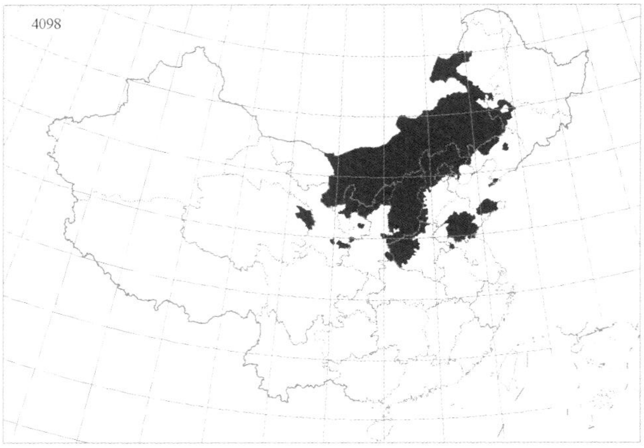

Fabaceae 豆科 *Caragana* 锦鸡儿属

Caragana opulens Kom. 甘蒙锦鸡儿

Shrubs, deciduous

5.2 (-5.5~13.0); -8.9 (-18.1~-2.4); 17.7 (5.8~26.6); 7.4 (1.3~13.3); 53.5 (1.2~113.3); 50.5 (17.1~127.1); 543 (299~800) ❙ 460 (63~950); 265 (45~546); 11 (1~25) ❙ -9.2 (-84.9~128.6); 403 (63~579); 155.7 (8.4~364.1)

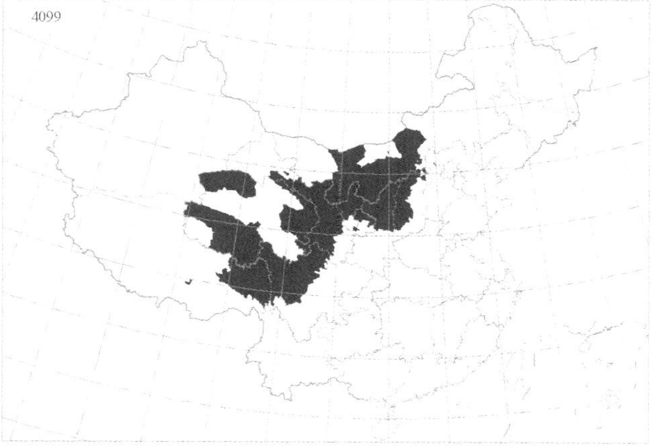

Fabaceae 豆科 *Caragana* 锦鸡儿属

Caragana oreophila W. W. Sm. 滇西锦鸡儿

Shrubs, deciduous

8.5; 1.2; 14.7; 8.6; 53.5; 11.5; 565 ❙ 884; 545; 23 ❙ 57.1; 565; 301.3

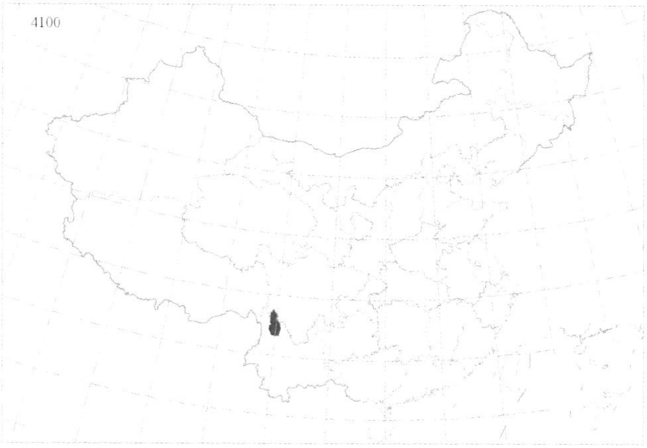

Fabaceae 豆科 *Caragana* 锦鸡儿属

Caragana pekinensis Kom. 北京锦鸡儿

Shrubs, deciduous

9.5 (3.9~14.5); -6.3 (-13.9~0.1); 23.4 (18.7~28.0); 10.8 (7.3~14.6); 88.1 (52.8~127.2); 33.6 (10.7~71.7); 687 (535~880) ❙ 523 (358~702); 326 (234~453); 16 (8~28) ❙ -23.3 (-43.4~12.7); 521 (358~649); 160.4 (101.9~250.3)

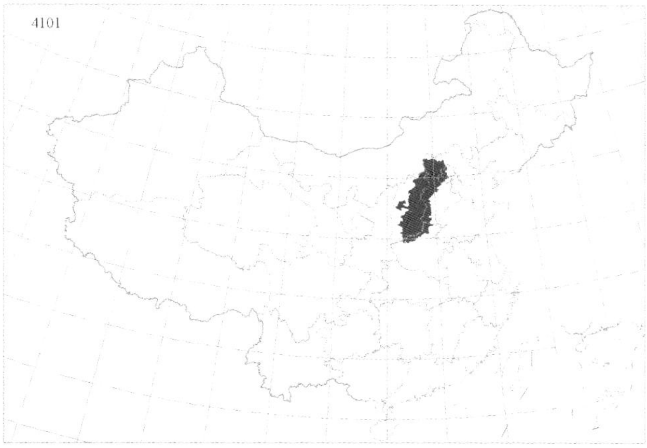

Fabaceae 豆科 *Caragana* 锦鸡儿属

Caragana pleiophylla (Regel) Pojark. 多叶锦鸡儿

Shrubs, deciduous

4.7 (-2.8~11.9); -13.4 (-20.9~-5.6); 18.7 (10.9~25.7); 8.0 (3.5~12.7); 60.9 (16.3~109.3); 64.3 (25.2~109.2); 580 (390~793) ❙ 206 (32~406); 102 (21~211); 15 (0~48) ❙ -60.3 (-95.9~-7.6); 206 (32~406); 109.3 (1.6~297.3)

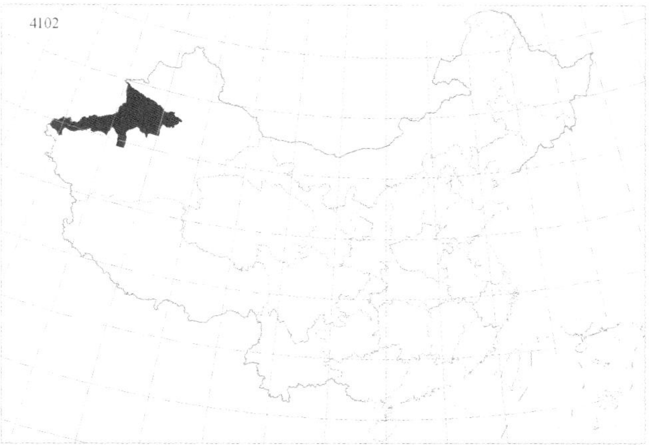

Fabaceae 豆科 *Caragana* 锦鸡儿属

Caragana polourensis Franch. 昆仑锦鸡儿

Small shrubs, deciduous

7.1 (-4.3~12.4); -10.0 (-20.1~-5.2); 20.9 (9.6~26.8); 9.5 (2.5~13.2); 76.9 (9.7~114.1); 51.2 (24.3~121.1); 645 (329~822) ❙ 75 (21~457); 42 (15~259); 4 (0~26) ❙ -85.3 (-97.3~12.5); 74 (21~406); 22.8 (1.6~195.4)

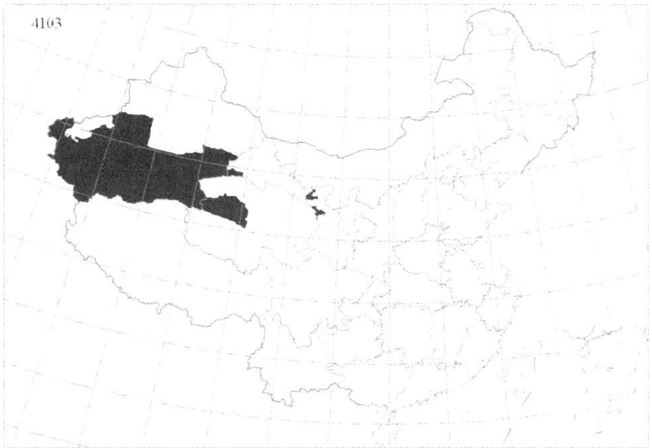

Fabaceae 豆科 *Caragana* 锦鸡儿属

Caragana potaninii Kom. 五台锦鸡儿

Small shrubs, deciduous

4.1 (0.1~9.0); -14.6 (-21.5~-7.2); 20.8 (18.7~23.3); 8.0 (6.6~10.3); 61.8 (49.3~84.1); 71.5 (35.8~108.6); 572 (519~664) ❙ 344 (137~509); 224 (92~312); 9 (3~17) ❙ -39.3 (-77.5~-9.8); 344 (137~509); 126.5 (25.4~228.3)

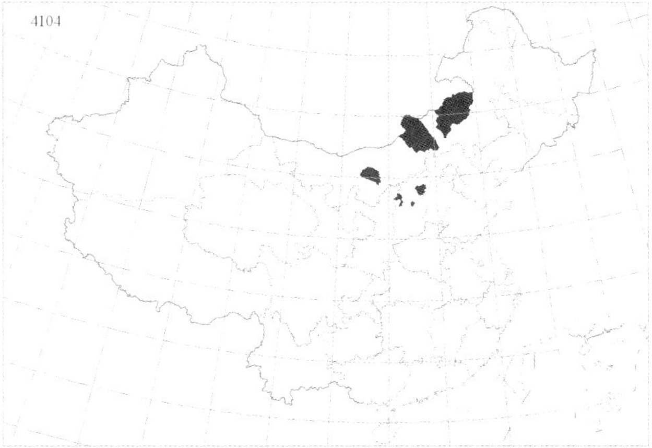

Fabaceae 豆科 *Caragana* 锦鸡儿属
Caragana pruinosa Kom. 粉刺锦鸡儿
Shrubs, deciduous

4.4 (-4.3~12.5); -12.9 (-20.1~-5.1); 18.3 (9.6~26.5); 7.7 (2.5~13.2); 58.5 (9.7~114.0); 65.2 (23.1~121.1); 565 (329~817) ❙ 160 (53~406); 79 (20~211); 12 (1~26) ❙ -67.3 (-93.3~-7.6); 160 (53~406); 66.0 (7.5~261.4)

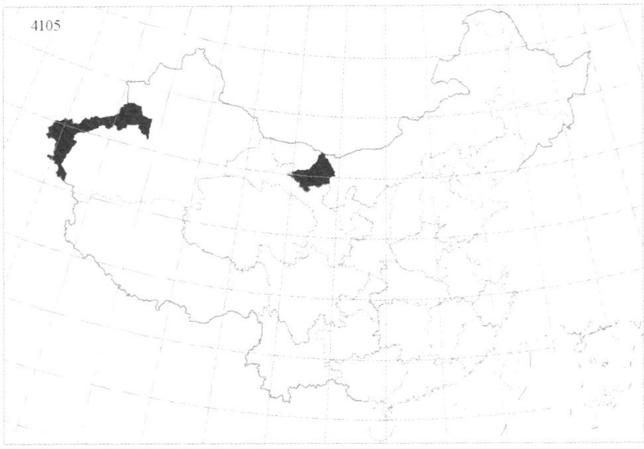

Fabaceae 豆科 *Caragana* 锦鸡儿属
Caragana pumila Pojark. 草原锦鸡儿
Shrubs, deciduous

3.8 (-2.8~9.3); -15.5 (-20.9~-9.3); 20.0 (10.9~27.5); 8.0 (3.5~12.1); 62.2 (16.3~107.6); 75.6 (40.7~109.2); 588 (390~813) ❙ 221 (69~450); 98 (44~211); 21 (4~51) ❙ -60.0 (-89.2~3.9); 220 (69~406); 103.4 (14.2~297.3)

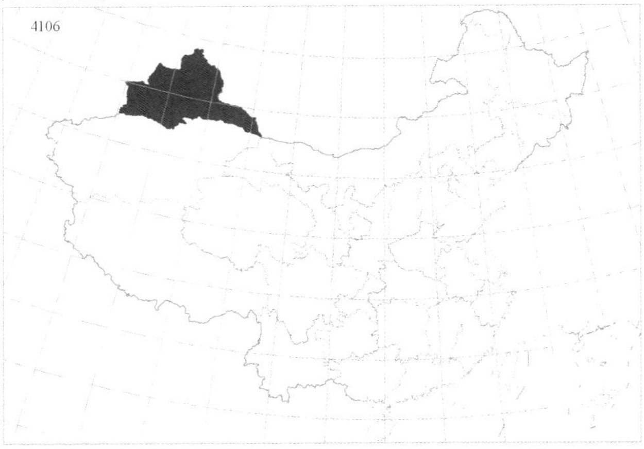

Fabaceae 豆科 *Caragana* 锦鸡儿属
Caragana purdomii Rehder 秦晋锦鸡儿
Shrubs, deciduous

8.4 (0.8~13.5); -7.0 (-12.7~-1.6); 22.0 (10.7~27.1); 9.8 (3.8~13.7); 77.8 (16.2~117.5); 36.8 (14.6~66.0); 639 (411~820) ❙ 512 (377~640); 294 (229~377); 15 (8~26) ❙ -19.0 (-37.2~44.8); 510 (377~640); 154.0 (51.3~313.2)

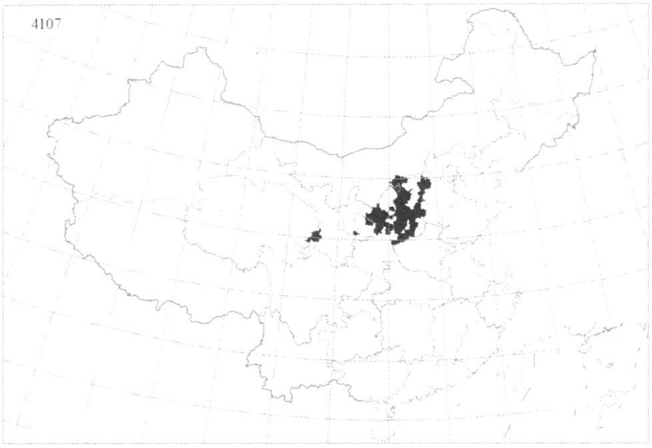

Fabaceae 豆科 *Caragana* 锦鸡儿属
Caragana pygmaea (L.) DC. 矮锦鸡儿
Shrubs, deciduous

5.6 (0.1~13.0); -12.0 (-21.5~-2.4); 21.1 (18.4~26.6); 8.6 (6.6~13.3); 67.2 (47.5~113.3); 59.8 (17.1~108.6); 594 (515~800) ❙ 398 (111~551); 248 (74~321); 11 (3~17) ❙ -33.1 (-82.3~-9.4); 398 (111~551); 127.4 (13.9~235.8)

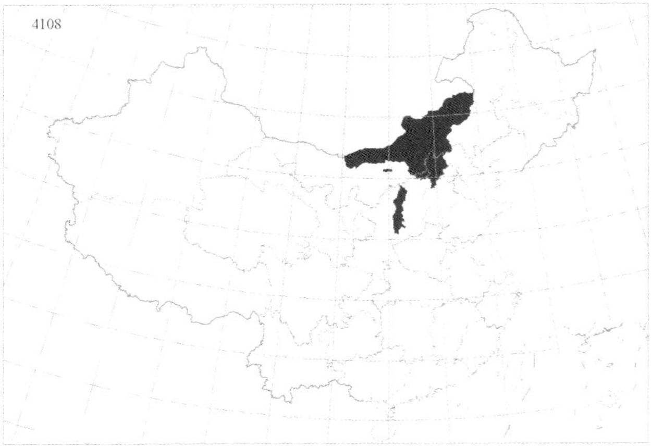

Fabaceae 豆科 *Caragana* 锦鸡儿属
Caragana pygmaea var. angustissima C. K. Schneid. 窄叶矮锦鸡儿
Shrubs, deciduous

2.9 (0.1~4.7); -16.3 (-21.5~-12.9); 20.0 (18.4~23.0); 7.4 (6.6~8.7); 55.4 (47.5~69.0); 79.9 (64.5~108.6); 548 (515~611) ❙ 305 (137~407); 205 (92~270); 8 (3~11) ❙ -43.8 (-77.5~-23.3); 305 (137~407); 101.3 (25.4~182.4)

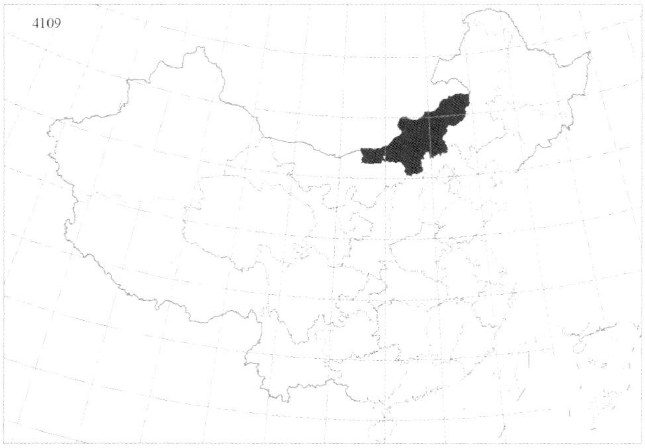

Fabaceae 豆科 *Caragana* 锦鸡儿属
Caragana pygmaea var. parviflora H. C. Fu 小花矮锦鸡儿
Shrubs, deciduous

3.4 (0.8~4.7); -16.9 (-20.7~-13.8); 21.5 (19.7~23.0); 8.0 (6.8~8.7); 62.1 (51.2~69.0); 81.0 (67.3~101.4); 578 (526~611) ❙ 201 (137~257); 136 (92~178); 5 (3~7) ❙ -64.9 (-77.5~-51.1); 201 (137~257); 48.7 (25.4~90.6)

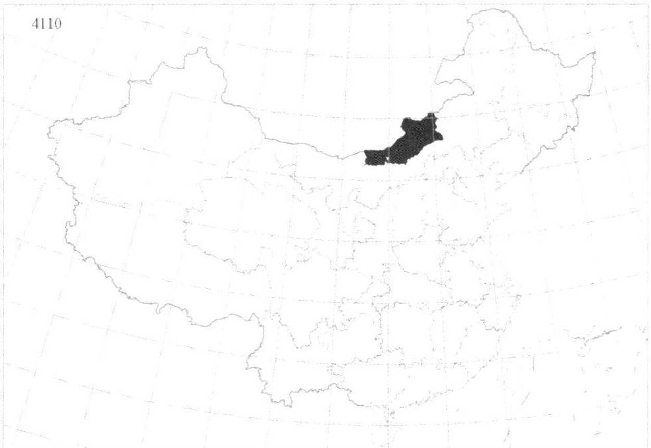

Fabaceae 豆科 *Caragana* 锦鸡儿属
Caragana qingheensis Z. Y. Chang et al. 青河锦鸡儿
Shrubs

1.9; -17.2; 18.6; 6.9; 50.9; 88.1; 538 ❙ 160; 70; 17 ❙ -70.2; 160; 45.6

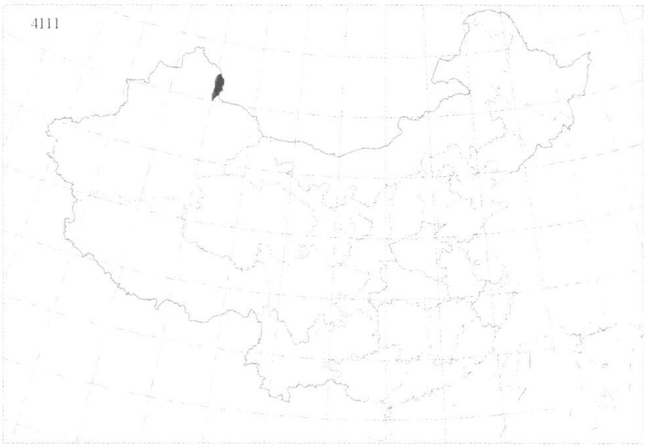

Fabaceae 豆科 *Caragana* 锦鸡儿属
Caragana roborovskyi Kom. 荒漠锦鸡儿
Shrubs, deciduous

4.8 (-5.5~9.7); -11.0 (-20.9~-6.4); 18.5 (5.8~27.5); 7.6 (1.3~12.1); 56.7 (1.2~107.6); 58.3 (35.3~127.1); 557 (299~813) ❙ 299 (35~657); 166 (23~357); 9 (1~48) ❙ -39.9 (-94.9~64.6); 280 (35~515); 117.4 (7.4~364.1)

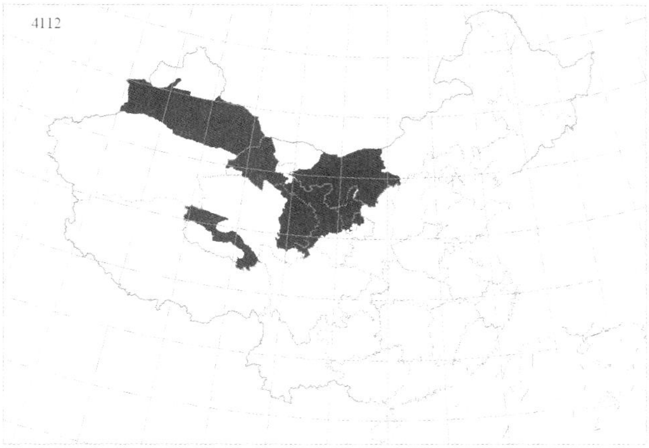

Fabaceae 豆科 *Caragana* 锦鸡儿属
Caragana rosea Turcz. ex Maxim. 红花锦鸡儿
Shrubs, deciduous

10.2 (0.4~18.1); -5.1 (-20.6~6.3); 23.9 (9.9~29.5); 11.3 (3.5~18.1); 92.6 (13.3~157.2); 29.6 (0.0~93.8); 708 (402~973) ❙ 628 (229~1621); 356 (148~691); 25 (5~206) ❙ -10.8 (-63.6~108.0); 580 (229~930); 179.1 (33.5~349.1)

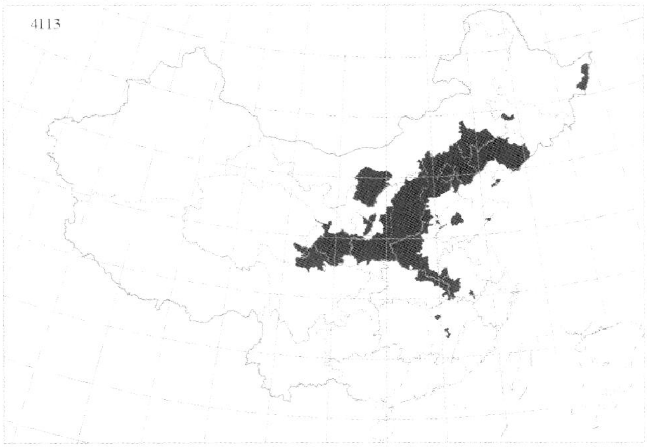

Fabaceae 豆科 *Caragana* 锦鸡儿属
Caragana shensiensis C. W. Chang 秦岭锦鸡儿
Shrubs, deciduous

12.6; 1.0; 23.5; 12.6; 99.9; 8.6; 724 ❙ 760; 382; 17 ❙ 5.0; 724; 294.3

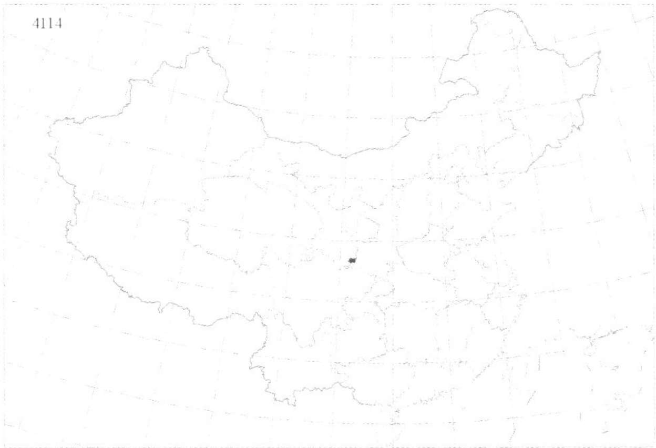

Fabaceae 豆科 *Caragana* 锦鸡儿属
Caragana sinica (Buc'hoz) Rehder 锦鸡儿
Shrubs, deciduous

14.2 (3.9~21.7); 1.7 (-13.7~12.9); 25.9 (14.4~29.9); 14.6 (6.6~21.8); 121.3 (39.7~201.3); 10.0 (0.0~71.7); 817 (508~1162) ❙ 1104 (367~1997); 470 (235~820); 101 (8~259) ❙ 32.4 (-47.7~147.0); 768 (367~1162); 197.4 (101.9~371.7)

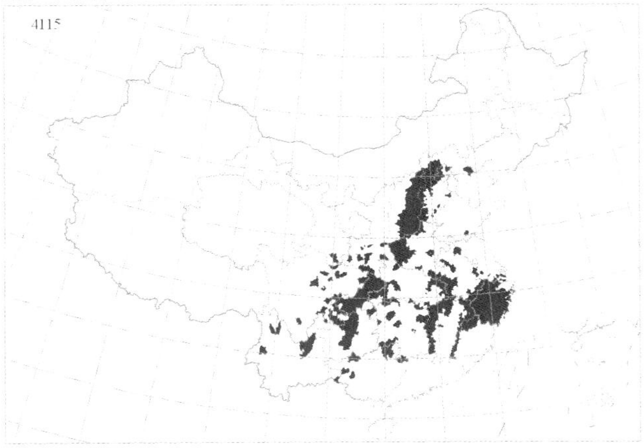

Fabaceae 豆科 *Caragana* 锦鸡儿属
Caragana soongorica Grubov 准噶尔锦鸡儿
Shrubs, deciduous

4.7 (-2.8~10.2); -14.5 (-20.9~-8.1); 20.5 (10.9~27.5); 8.4 (3.5~12.1); 66.5 (16.3~107.6); 69.8 (36.6~109.2); 605 (390~813) ❙ 190 (45~450); 89 (27~211); 16 (3~48) ❙ -65.9 (-94.0~3.9); 189 (45~406); 84.2 (7.6~297.3)

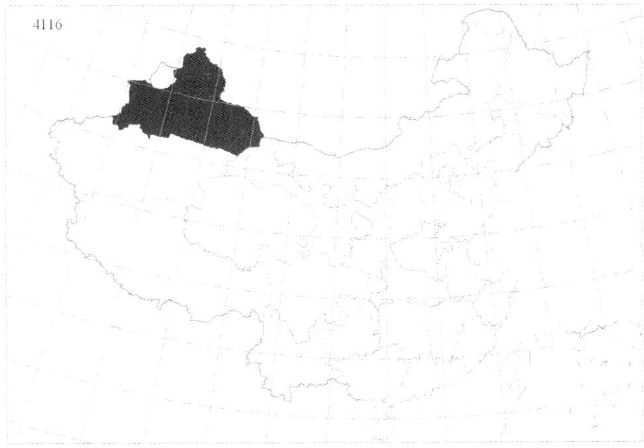

Fabaceae 豆科 *Caragana* 锦鸡儿属

Caragana spinosa (L.) Vahl ex Hornem. 多刺锦鸡儿

Dwarf shrubs, deciduous

4.4 (-0.6~10.8); -14.6 (-19.5~-8.0); 20.2 (15.9~25.0); 8.3 (5.4~12.2); 64.7 (37.5~104.2); 71.5 (34.1~104.2); 597 (475~771) ❙ 167 (70~315); 77 (41~129); 16 (3~51) ❙ -70.7 (-90.6~-46.4); 167 (70~312); 61.4 (16.6~173.1)

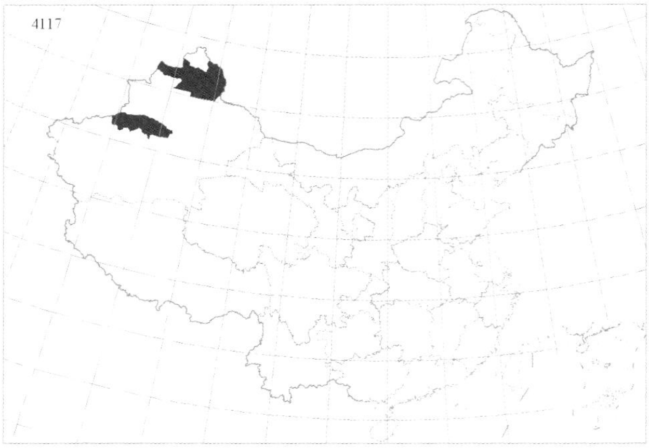

Fabaceae 豆科 *Caragana* 锦鸡儿属

Caragana stenophylla Pojark. 狭叶锦鸡儿

Shrubs, deciduous

5.1 (-5.8~11.7); -13.1 (-29.8~-3.9); 21.0 (11.5~27.5); 8.5 (4.0~12.3); 66.4 (19.5~107.6); 64.9 (21.4~157.4); 593 (414~813) ❙ 328 (49~596); 200 (34~373); 11 (1~51) ❙ -43.4 (-92.8~15.9); 327 (49~596); 107.5 (5.1~309.5)

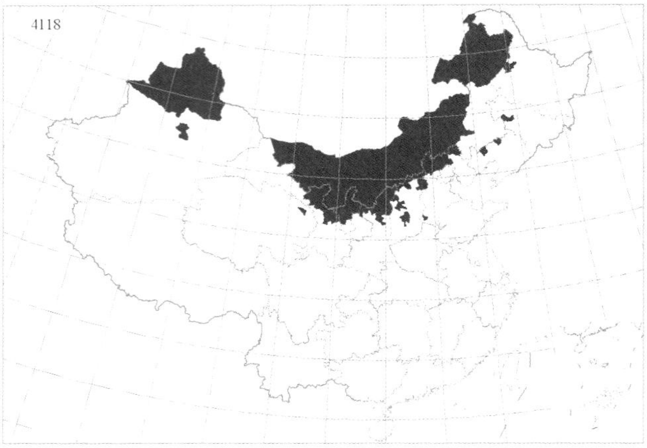

Fabaceae 豆科 *Caragana* 锦鸡儿属

Caragana stipitata Kom. 柄荚锦鸡儿

Shrubs, deciduous

10.3 (3.3~15.1); -4.9 (-13.7~1.8); 23.8 (17.5~28.5); 11.3 (6.6~15.2); 92.4 (45.4~128.7); 28.2 (5.6~71.7); 704 (507~886) ❙ 563 (367~1246); 318 (223~525); 19 (8~53) ❙ -19.5 (-47.7~93.8); 553 (367~817); 162.0 (50.7~311.0)

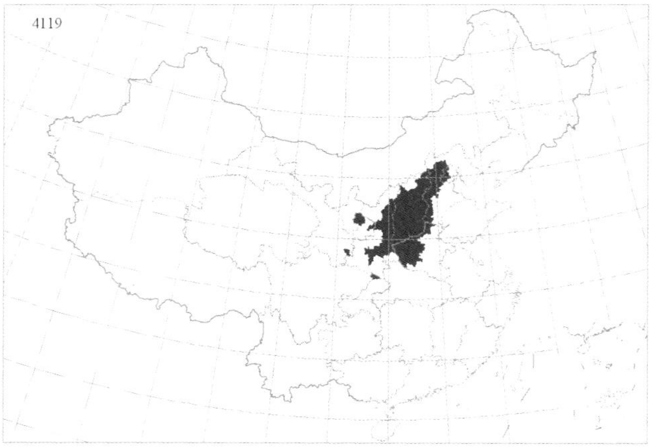

Fabaceae 豆科 *Caragana* 锦鸡儿属

Caragana tangutica Maxim. ex Kom. 青甘锦鸡儿

Shrubs, deciduous

3.1 (-4.1~9.1); -8.8 (-17.3~-3.8); 13.7 (7.2~21.4); 5.4 (2.0~9.7); 33.0 (3.7~74.3); 55.4 (24.9~114.2); 464 (335~624) ❙ 540 (63~950); 296 (45~546); 11 (2~25) ❙ 21.3 (-84.9~128.6); 412 (63~577); 215.1 (8.4~364.1)

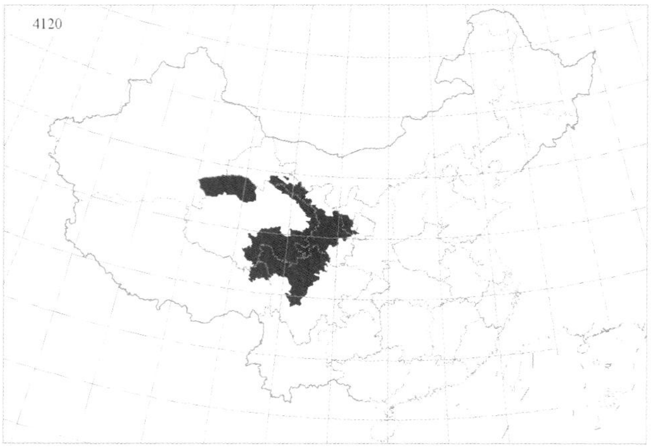

Fabaceae 豆科 *Caragana* 锦鸡儿属

Caragana tekesiensis Y. Z. Zhao et D. W. Zhou 特克斯锦鸡儿

Shrubs

-0.5; -18.2; 13.3; 4.7; 27.6; 93.3; 444 ❙ 375; 199; 17 ❙ -15.5; 375; 245.2

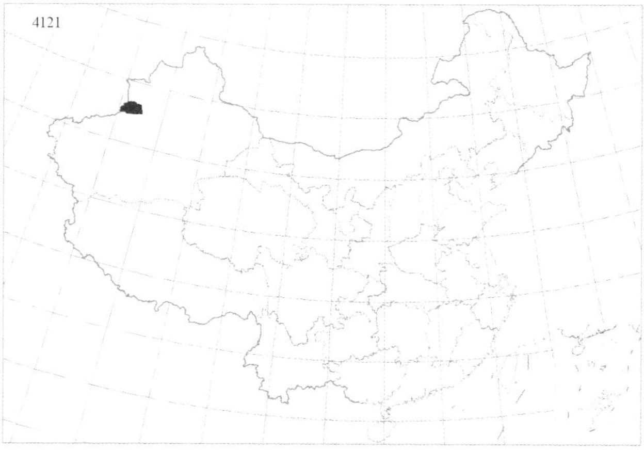

Fabaceae 豆科 *Caragana* 锦鸡儿属

Caragana tibetica Kom. 毛刺锦鸡儿

Shrubs, deciduous

4.0 (-5.9~18.5); -8.5 (-17.3~10.5); 15.0 (5.4~24.3); 6.3 (1.3~18.3); 42.2 (0.4~160.0); 53.9 (0.0~130.5); 498 (299~906) ❙ 489 (63~1042); 283 (45~641); 10 (0~28) ❙ 5.3 (-84.9~128.6); 371 (63~786); 177.6 (8.4~364.1)

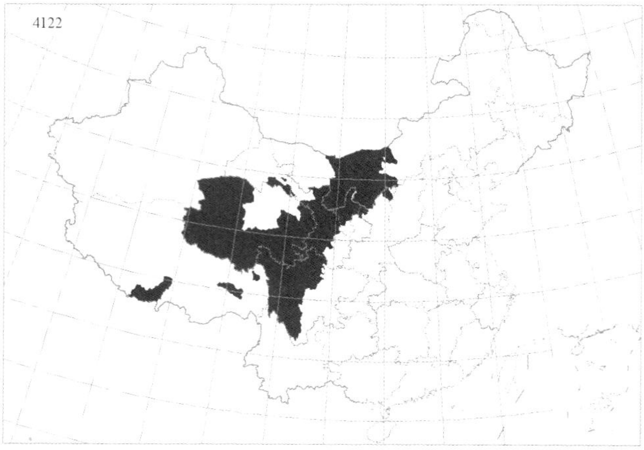

Fabaceae 豆科 *Caragana* 锦鸡儿属

Caragana tragacanthoides (Pall.) Poir. 中亚锦鸡儿

Shrubs, deciduous

3.0 (-0.6~9.0); -16.2 (-19.5~-13.8); 19.9 (15.9~27.5); 7.7 (5.4~12.0); 59.4 (37.5~107.3); 82.4 (59.0~104.2); 578 (475~813) ❙ 193 (120~315); 79 (51~115); 23 (13~51) ❙ -65.6 (-85.3~-46.4); 193 (120~312); 86.5 (16.6~173.1)

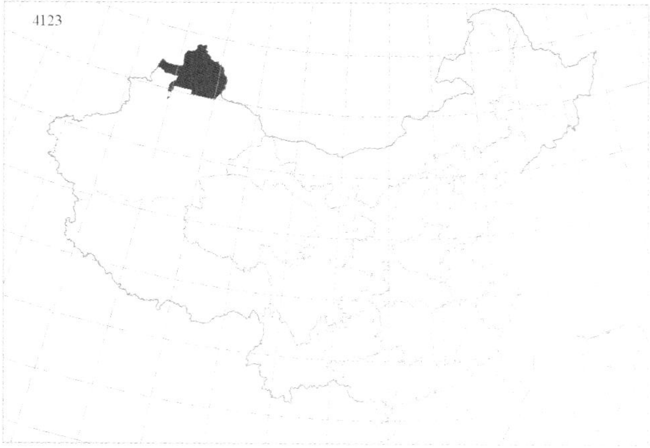

Fabaceae 豆科 *Caragana* 锦鸡儿属

Caragana turfanensis (Krassn.) Kom. 吐鲁番锦鸡儿

Shrubs, deciduous

4.3 (-0.5~11.9); -13.0 (-17.6~-5.6); 18.1 (12.9~25.5); 7.5 (4.5~12.7); 55.6 (25.5~107.9); 64.2 (25.2~92.0); 557 (433~783) ❙ 255 (70~406); 117 (23~211); 22 (9~45) ❙ -51.3 (-91.1~-7.6); 255 (70~406); 128.9 (32.6~261.4)

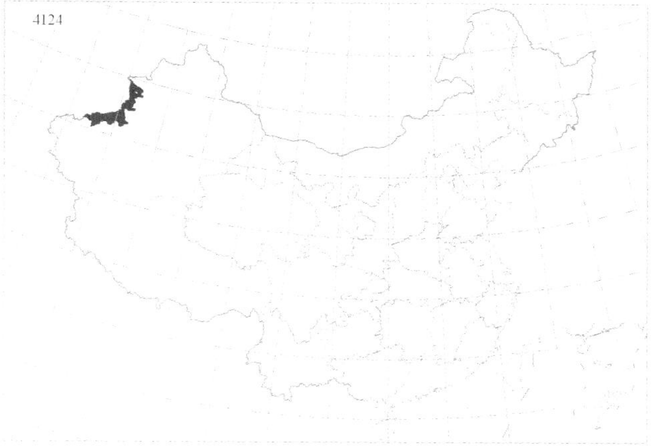

Fabaceae 豆科 *Caragana* 锦鸡儿属

Caragana turkestanica Kom. 新疆锦鸡儿

Shrubs, deciduous

4.8 (-2.8~10.2); -14.2 (-20.9~-8.1); 20.4 (10.9~27.1); 8.4 (3.5~12.1); 66.3 (16.3~107.6); 68.2 (36.6~109.2); 604 (390~810) ❙ 202 (45~450); 94 (27~211); 18 (3~51) ❙ -63.6 (-94.0~3.9); 201 (45~406); 94.1 (7.6~297.3)

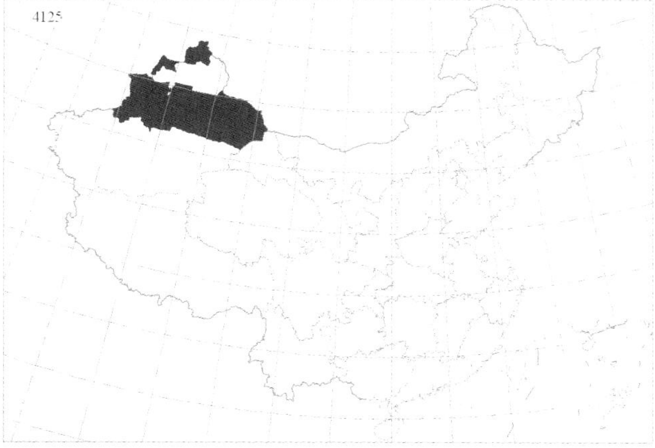

Fabaceae 豆科 *Caragana* 锦鸡儿属

Caragana ussuriensis (Regel) Pojark. 乌苏里锦鸡儿

Shrubs, deciduous

2.8 (1.9~3.9); -18.2 (-21.5~-14.1); 20.7 (19.0~21.9); 7.8 (7.0~8.3); 58.7 (50.5~64.6); 85.1 (70.4~97.6); 570 (532~597) ❙ 596 (558~658); 337 (309~362); 24 (18~35) ❙ 4.6 (-5.9~20.9); 561 (532~591); 268.8 (199.1~333.1)

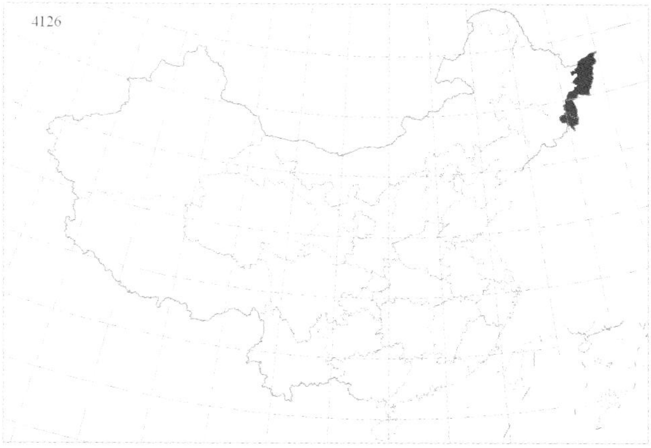

Fabaceae 豆科 *Caragana* 锦鸡儿属

Caragana versicolor Benth. 变色锦鸡儿

Dwarf shrubs, deciduous

1.9 (-5.7~18.5); -7.7 (-18.7~10.5); 10.5 (5.4~24.3); 4.6 (1.3~18.3); 24.5 (0.4~160.0); 61.9 (0.0~133.0); 424 (294~906) ❙ 640 (74~1042); 372 (46~641); 21 (5~103) ❙ 48.1 (-75.0~128.6); 407 (74~786); 220.5 (16.7~357.7)

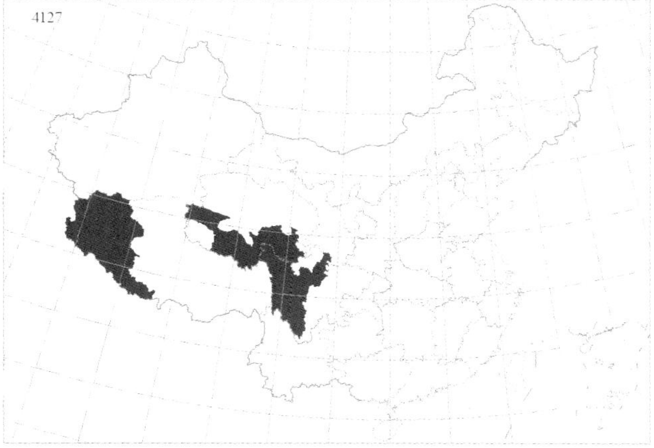

Fabaceae 豆科 *Caragana* 锦鸡儿属

Caragana zahlbruckneri C. K. Schneid. 南口锦鸡儿

Shrubs

6.7 (2.1~11.6); -10.5 (-20.1~-4.8); 21.8 (18.5~26.1); 9.1 (6.7~12.5); 73.2 (48.1~105.6); 52.7 (25.4~91.9); 621 (515~782) ❙ 487 (325~694); 334 (216~499); 11 (8~21) ❙ -21.6 (-41.6~16.5); 485 (325~659); 189.3 (91.4~323.4)

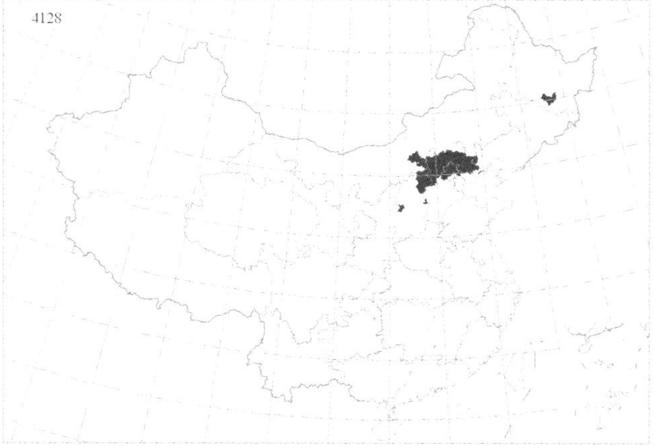

Fabaceae 豆科 *Cassia* 决明属
Cassia javanica subsp. agnes (de Wit) K. Larsen 神黄豆
Trees

19.2 (17.3~20.5); 13.4 (11.0~15.1); 22.8 (21.6~23.7); 19.3 (17.3~20.5); 171.2 (148.0~185.9); 0.0 (0.0~0.0); 923 (832~992) ❙ 1410 (1049~1607); 698 (598~775); 61 (44~77) ❙ 52.2 (26.1~64.9); 899 (783~986); 481.8 (354.7~659.0)

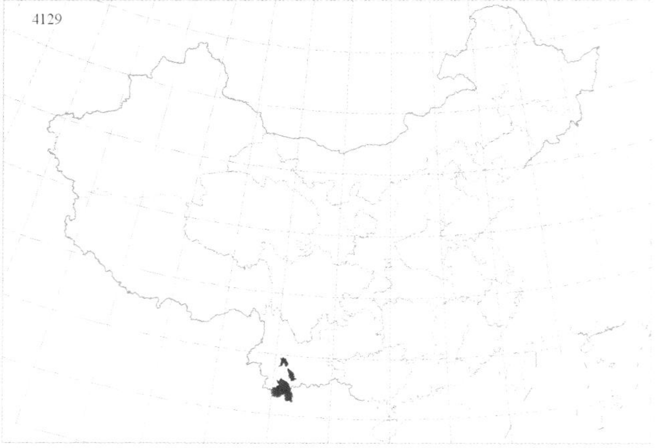

Fabaceae 豆科 *Cercis* 紫荆属
Cercis chinensis Bunge 紫荆
Shrubs, deciduous

15.2 (7.4~22.1); 3.6 (−4.2~15.2); 26.0 (18.7~29.9); 15.3 (8.2~22.7); 126.6 (58.8~212.7); 3.7 (0.0~29.4); 836 (561~1216) ❙ 1193 (548~1997); 504 (264~994); 98 (17~259) ❙ 42.5 (−28.9~130.7); 824 (548~1216); 206.1 (111.9~354.5)

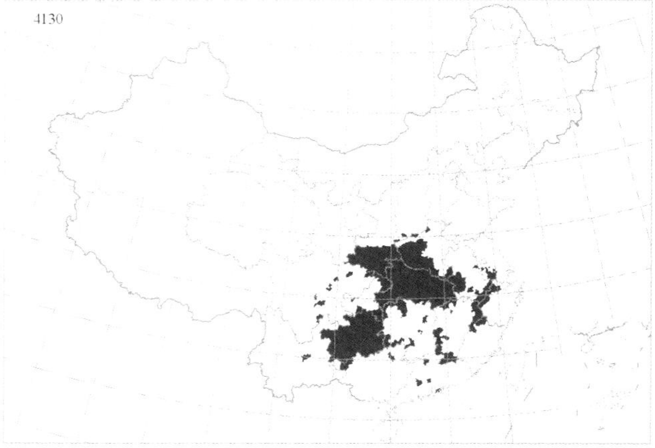

Fabaceae 豆科 *Cercis* 紫荆属
Cercis chingii Chun 黄山紫荆
Shrubs, deciduous

17.0 (14.4~19.4); 5.7 (2.1~9.4); 27.6 (26.0~29.4); 17.0 (14.5~19.4); 145.2 (118.7~173.4); 0.7 (0.0~4.9); 911 (801~1029) ❙ 1630 (1235~1922); 562 (473~644); 206 (123~259) ❙ 79.7 (29.7~111.3); 905 (801~1029); 237.1 (111.9~319.9)

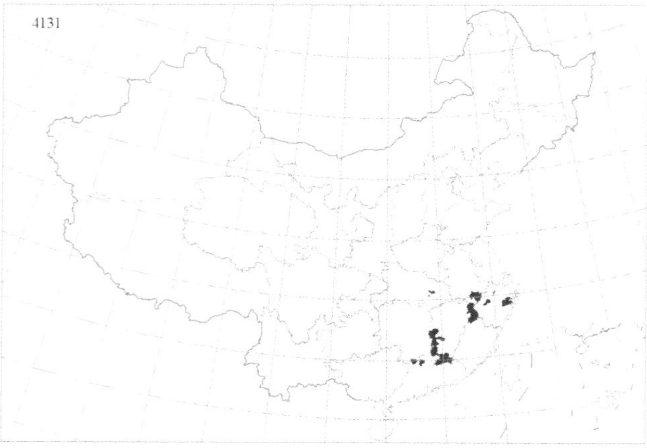

Fabaceae 豆科 *Cercis* 紫荆属
Cercis chuniana F. P. Metcalf 广西紫荆
Trees, deciduous

17.4 (14.3~21.0); 6.7 (3.3~11.9); 27.1 (24.2~29.0); 17.4 (14.3~21.1); 149.0 (113.2~193.0); 0.1 (0.0~2.3); 914 (761~1114) ❙ 1602 (1268~1997); 551 (403~821); 196 (105~258) ❙ 76.2 (47.0~130.7); 909 (761~1114); 230.5 (151.0~303.4)

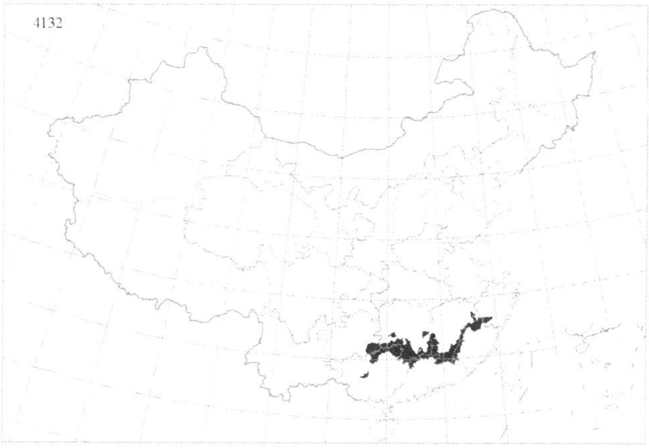

Fabaceae 豆科 *Cercis* 紫荆属
Cercis glabra Pamp. 湖北紫荆
Trees, deciduous

14.3 (5.4~19.8); 3.5 (−4.2~12.0); 24.1 (12.3~30.1); 14.4 (5.8~19.8); 117.0 (31.5~177.9); 5.2 (0.0~29.4); 784 (481~1029) ❙ 1079 (512~1745); 495 (258~723); 68 (15~220) ❙ 37.2 (−35.3~106.2); 764 (481~1029); 231.3 (121.6~411.1)

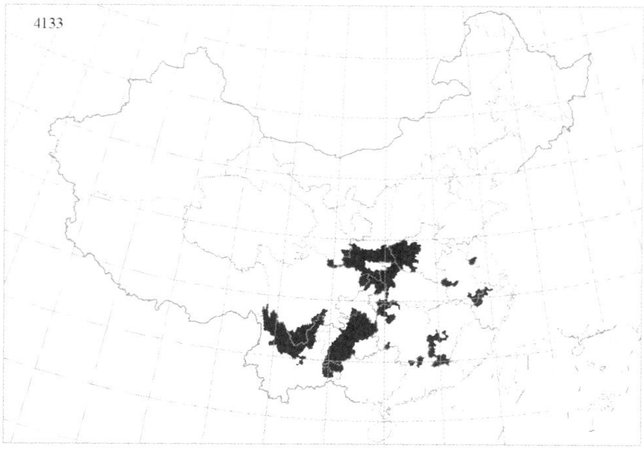

Fabaceae 豆科 *Cercis* 紫荆属
Cercis racemosa Oliv. 垂丝紫荆
Trees, deciduous

14.6 (9.3~19.1); 3.9 (−2.5~10.1); 24.4 (20.1~27.5); 14.6 (9.5~19.1); 118.8 (69.8~169.2); 3.7 (0.0~20.6); 788 (603~969) ❙ 1199 (845~1624); 535 (375~816); 74 (24~191) ❙ 52.6 (17.9~107.2); 788 (603~969); 234.2 (162.8~311.0)

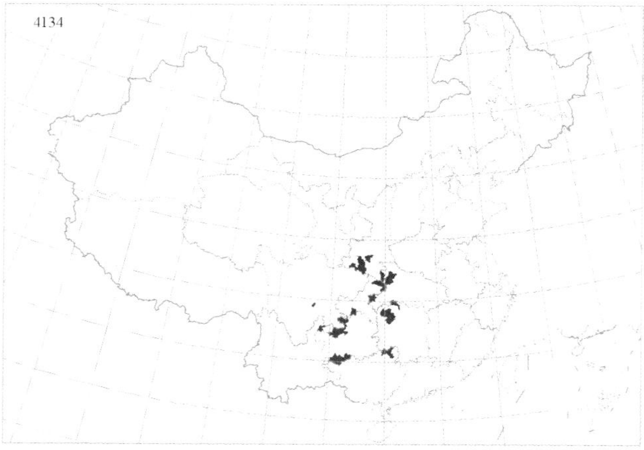

Fabaceae 豆科 *Chesneya* 雀儿豆属
Chesneya acaulis (Baker) Kitam. 无茎雀儿豆
Shrubs, deciduous

-1.8; -12.0; 8.6; 2.4; 7.9; 88.8; 333 ❙ 509; 233; 98 ❙ 52.9; 333; 32.2

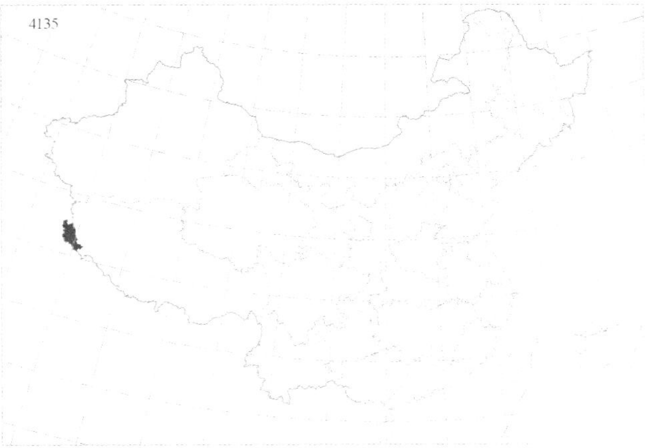

Fabaceae 豆科 *Chesneya* 雀儿豆属
Chesneya crassipes Boriss. 长梗雀儿豆
Herbs or subshrubs, deciduous

-1.8; -12.0; 8.6; 2.4; 7.9; 88.8; 333 ❙ 509; 233; 98 ❙ 52.9; 333; 32.2

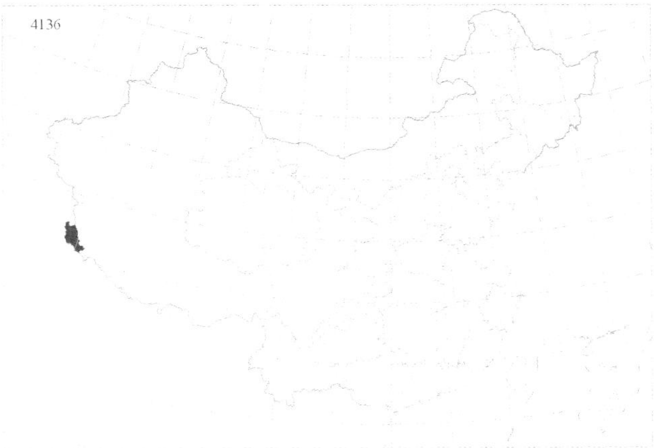

Fabaceae 豆科 *Chesneya* 雀儿豆属
Chesneya nubigena (D. Don) Ali 云雾雀儿豆
Subshrubs, deciduous

5.7 (-2.3~14.3); -2.2 (-10.9~7.6); 12.7 (6.1~19.7); 7.0 (1.7~14.3); 43.3 (1.9~1[illegible].9); 34.1 (0.0~89.7); 503 (324~728) ❙ 778 (311~1736); 462 (193~1007); 30 (4~103) ❙ 46.1 (-15.3~145.8); 497 (311~707); 234.6 (46.9~447.4)

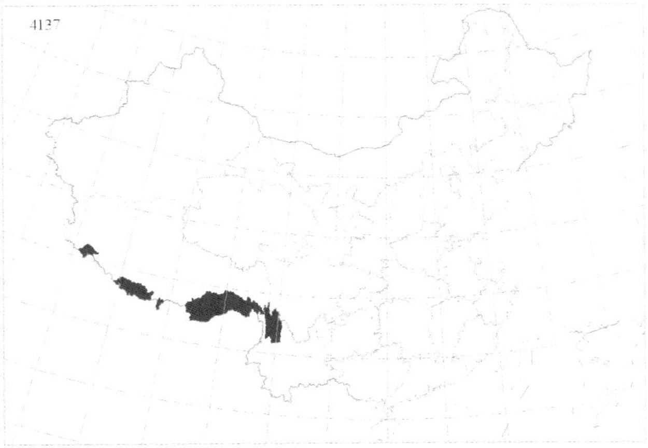

Fabaceae 豆科 *Chesneya* 雀儿豆属
Chesneya polystichoides (Hand.-Mazz.) S. H. Cheng 川滇雀儿豆
Herbs or subshrubs, deciduous

9.6 (2.2~16.1); 2.2 (-6.3~8.6); 15.7 (9.5~21.9); 9.8 (3.9~16.1); 67.4 (15.8~133.0); 12.3 (0.0~48.6); 600 (411~797) ❙ 975 (729~1337); 582 (426~727); 30 (11~69) ❙ 65.8 (27.3~121.7); 587 (411~719); 305.9 (208.8~407.1)

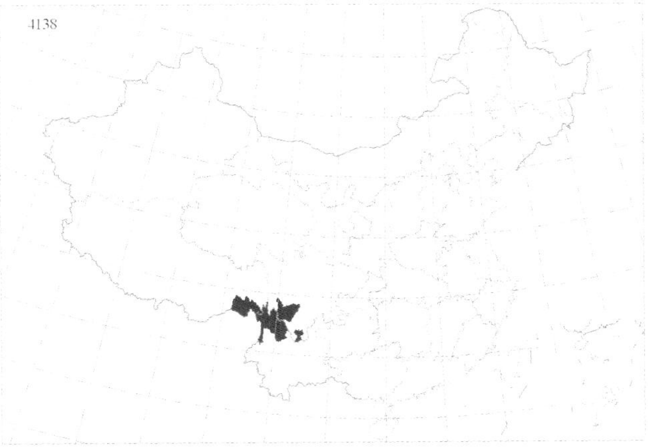

Fabaceae 豆科 *Chesneya* 雀儿豆属
Chesneya spinosa P. C. Li 刺柄雀儿豆
Herbs or subshrubs, deciduous

1.3 (-2.1~4.4); -7.9 (-10.3~-4.3); 9.6 (5.7~12.1); 3.6 (1.7~5.3); 16.0 (1.0~28.5); 59.6 (34.6~85.3); 392 (328~455) ❙ 326 (252~538); 226 (158~337); 8 (0~38) ❙ -17.6 (-35.9~18.2); 314 (252~455); 125.2 (72.1~217.5)

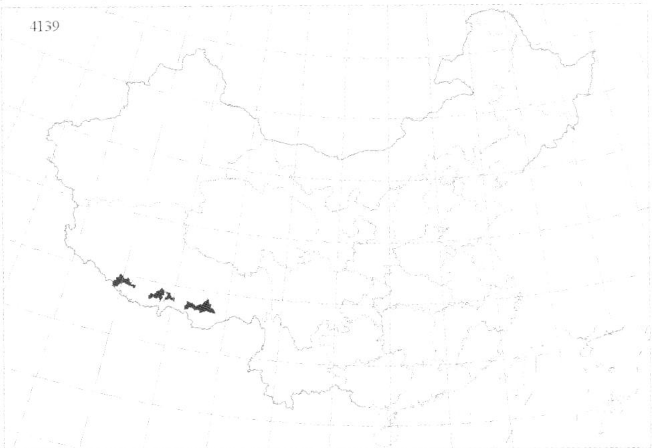

Fabaceae 豆科 *Chesniella* 旱雀豆属
Chesniella ferganensis (Korsh.) Boriss. 甘肃旱雀豆
Herbs or subshrubs, deciduous

5.4 (-2.8~10.2); -13.6 (-20.9~-8.1); 21.0 (10.9~27.5); 8.8 (3.5~12.1); 70.1 (16.3~107.6); 64.4 (36.6~109.2); 619 (390~813) ❙ 185 (45~450); 90 (27~211); 14 (1~48) ❙ -67.0 (-94.0~3.9); 184 (45~406); 81.9 (5.1~297.3)

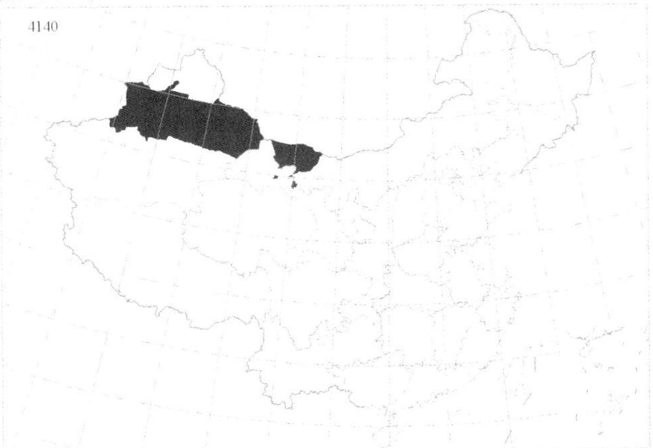

Fabaceae 豆科 *Chesniella* 旱雀豆属
Chesniella mongolica (Maxim.) Boriss. 蒙古旱雀豆
Herbs or subshrubs, deciduous

7.6 (6.2~8.7); -9.9 (-12.2~-8.1); 23.2 (22.4~24.2); 9.7 (9.1~10.3); 79.8 (74.4~84.9); 48.0 (39.2~59.5); 645 (622~671) ❙ 177 (111~244); 109 (74~152); 4 (2~5) ❙ -72.4 (-82.3~-60.8); 177 (111~244); 58.6 (13.6~116.9)

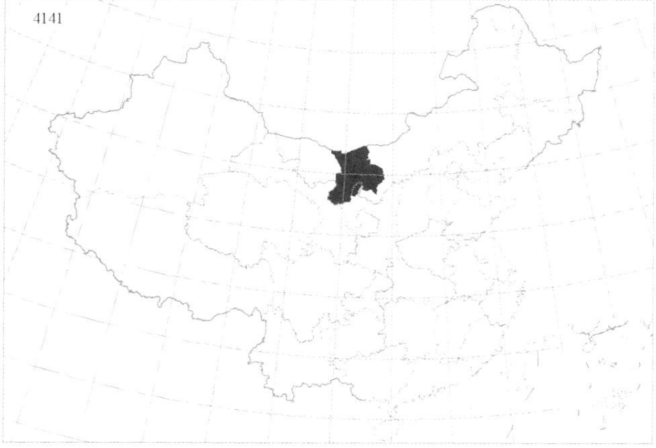

Fabaceae 豆科 *Cladrastis* 香槐属
Cladrastis delavayi (Franch.) Prain 小花香槐
Trees, deciduous

14.2 (2.2~19.3); 4.5 (-6.7~13.3); 22.6 (9.5~28.4); 14.3 (3.8~19.4); 115.7 (15.5~173.2); 5.2 (0.0~49.1); 765 (411~939) ❙ 1043 (373~1997); 525 (201~836); 57 (7~254) ❙ 36.6 (-37.6~130.7); 747 (373~933); 261.2 (88.1~516.7)

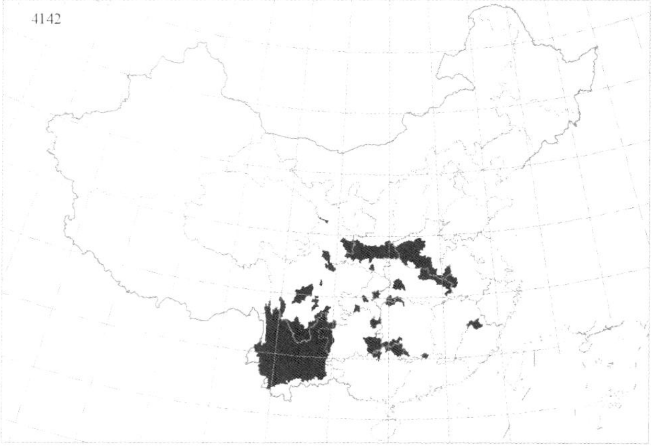

Fabaceae 豆科 *Cladrastis* 香槐属
Cladrastis parvifolia C. Y. Ma 小叶香槐
Trees, deciduous

17.9; 7.4; 27.1; 18.0; 155.6; 0.0; 938 ❙ 1740; 709; 187 ❙ 85.5; 938; 208.7

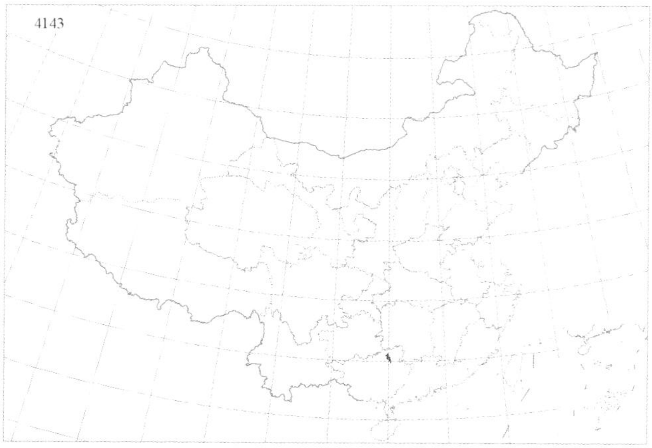

Fabaceae 豆科 *Cladrastis* 香槐属
Cladrastis platycarpa (Maxim.) Makino 翅荚香槐
Trees, deciduous

16.6 (11.7~23.0); 5.8 (1.6~15.1); 26.6 (19.6~29.9); 16.7 (11.7~23.0); 140.7 (84.6~216.2); 0.5 (0.0~6.5); 884 (661~1247) ❙ 1393 (964~1829); 541 (381~997); 140 (39~230) ❙ 58.0 (7.5~128.9); 874 (661~1247); 209.6 (134.6~359.7)

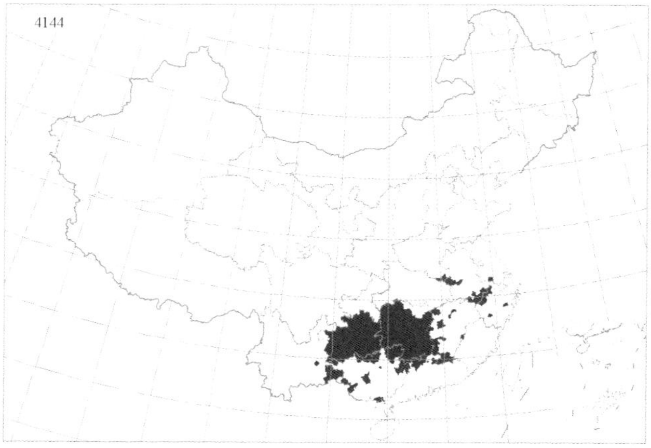

Fabaceae 豆科 *Cladrastis* 香槐属
Cladrastis scandens C. Y. Ma 藤香槐
Vines or shrubs, deciduous

18.3 (16.8~19.1); 9.2 (7.8~10.1); 25.5 (23.6~26.6); 18.4 (16.8~19.1); 160.2 (141.5~169.2); 0.0 (0.0~0.0); 927 (836~969) ❙ 1271 (1236~1334); 677 (617~720); 57 (51~64) ❙ 37.7 (27.5~59.6); 927 (836~969); 242.9 (189.5~282.2)

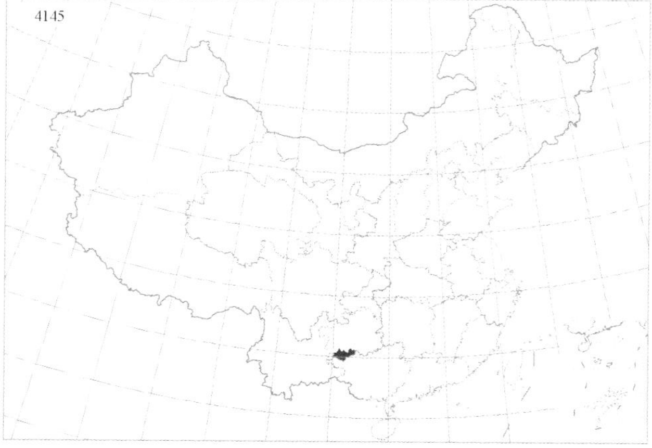

Fabaceae 豆科 *Cladrastis* 香槐属
Cladrastis wilsonii Takeda 香槐
Trees, deciduous

15.0 (5.4~19.4); 3.9 (-4.2~12.5); 25.5 (12.3~29.9); 15.1 (5.8~19.5); 124.4 (31.5~173.7); 3.8 (0.0~29.4); 823 (481~1038) ❙ 1259 (626~1997); 511 (290~792); 122 (15~259) ❙ 51.9 (-22.2~147.0); 809 (481~1038); 239.0 (134.6~491.5)

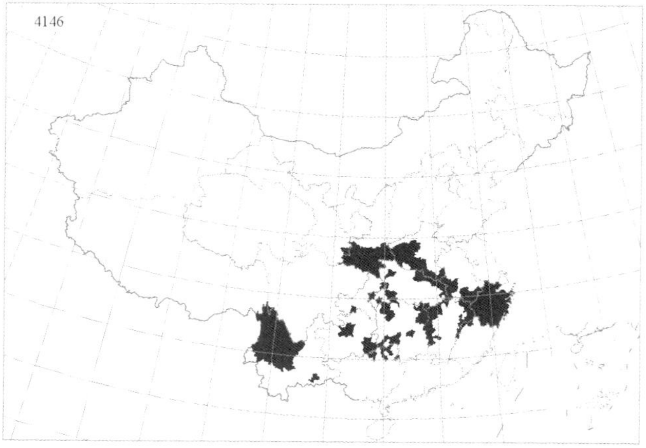

Fabaceae 豆科 *Clitoria* 蝶豆属
Clitoria hanceana Hemsl. 广东蝶豆
Subshrubs, deciduous

21.1 (17.8~22.7); 12.4 (7.1~15.2); 28.3 (26.9~29.8); 21.2 (17.8~22.8); 194.9 (154.0~213.1); 0.0 (0.0~0.0); 1125 (933~1221) ❙ 1694 (1446~2435); 792 (508~1428); 144 (80~189) ❙ 51.0 (20.3~99.5); 1119 (933~1221); 225.9 (157.7~318.8)

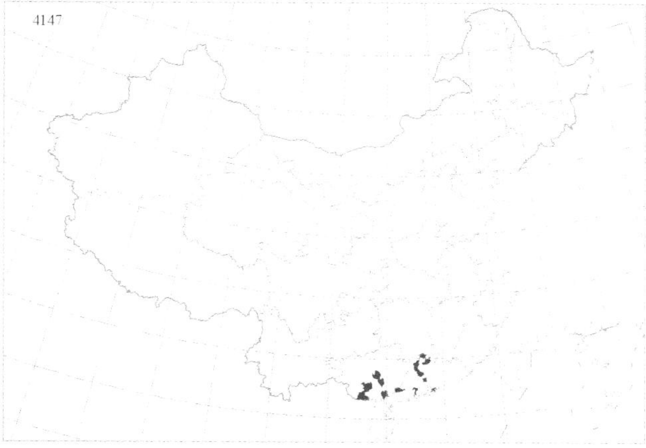

Fabaceae 豆科 *Clitoria* 蝶豆属
Clitoria mariana L. 三叶蝶豆
Subshrubs, deciduous

17.2 (13.6~21.0); 10.2 (4.8~15.1); 22.2 (18.8~26.0); 17.3 (13.7~21.0); 147.1 (104.1~191.8); 0.0 (0.0~0.2); 842 (706~1053) ❙ 1250 (958~1751); 681 (501~996); 54 (41~77) ❙ 48.4 (12.2~112.2); 832 (706~1053); 370.9 (160.1~659.0)

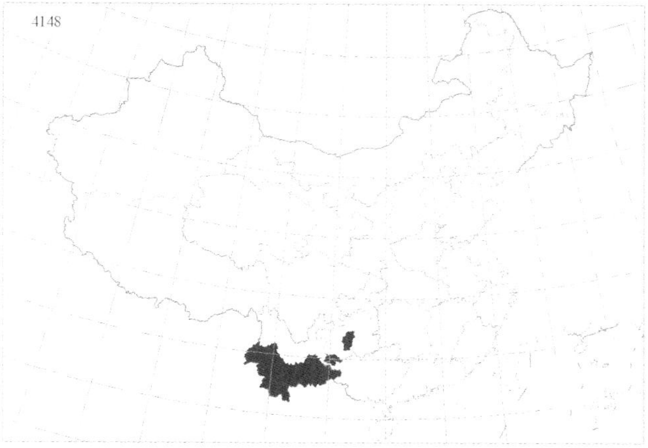

Fabaceae 豆科 *Codariocalyx* 舞草属
Codariocalyx gyroides Roxb. ex Hassk. 圆叶舞草
Shrubs, deciduous

19.3 (14.0~24.8); 11.3 (3.8~20.1); 25.4 (20.3~28.8); 19.3 (14.1~24.8); 171.8 (109.4~237.6); 0.0 (0.0~1.5); 982 (748~1385) ❙ 1463 (1028~1953); 702 (473~1056); 96 (41~209) ❙ 50.7 (-1.4~101.8); 968 (748~1254); 331.5 (130.1~659.0)

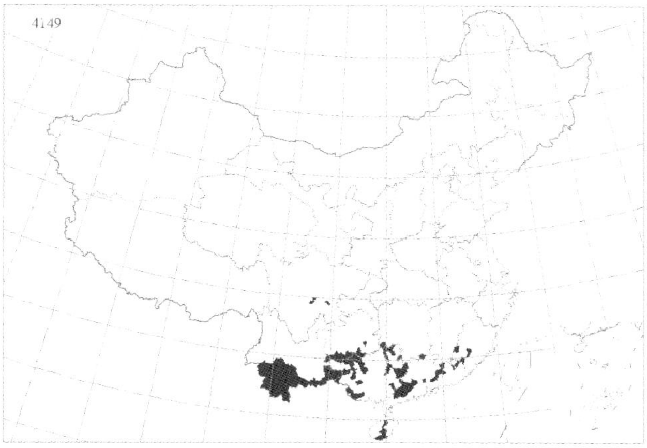

Fabaceae 豆科 *Codariocalyx* 舞草属
Codariocalyx motorius (Houtt.) H. Ohashi 舞草
Small shrubs, deciduous

16.9 (5.4~24.1); 9.1 (-2.1~18.9); 23.0 (12.3~29.0); 16.9 (5.8~24.1); 143.6 (31.5~229.7); 0.4 (0.0~25.6); 859 (481~1322) ❙ 1229 (741~2992); 651 (411~1847); 67 (15~303) ❙ 43.5 (-14.2~219.3); 844 (481~1202); 295.0 (133.8~659.0)

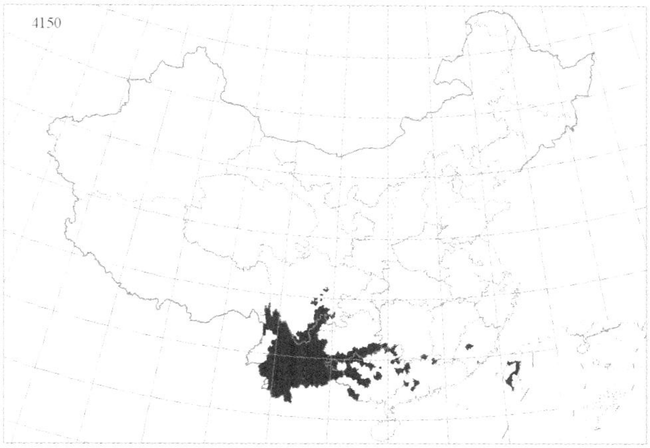

Fabaceae 豆科 *Colutea* 鱼鳔槐属
Colutea arborescens L. 鱼鳔槐
Shrubs, deciduous

9.3; -7.0; 24.0; 10.9; 90.6; 35.3; 695 ❙ 647; 432; 23 ❙ -6.5; 627; 203.5

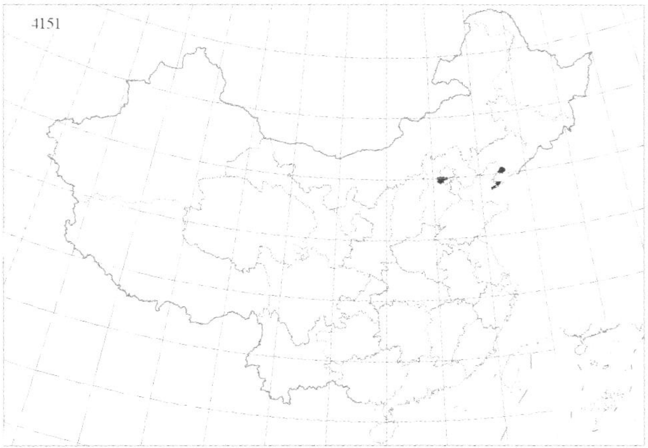

Fabaceae 豆科 *Colutea* 鱼鳔槐属
Colutea delavayi Franch. 膀胱豆
Shrubs, deciduous

11.9 (3.8~16.7); 4.8 (-4.3~9.8); 17.7 (10.8~22.1); 12.0 (4.7~16.8); 88.4 (21.6~141.1); 5.6 (0.0~35.4); 665 (444~819) ❙ 954 (801~1057); 575 (496~627); 25 (12~42) ❙ 47.5 (11.7~121.7); 638 (444~721); 298.2 (232.7~351.6)

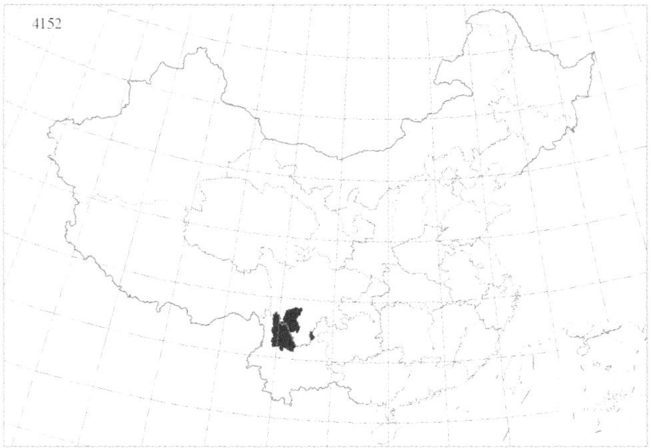

Fabaceae 豆科 *Colutea* 鱼鳔槐属

Colutea nepalensis Sims 尼泊尔鱼鳔槐

Shrubs, deciduous

-2.2 (-5.9~0.2); -14.8 (-18.0~-12.0); 9.7 (5.4~12.3); 2.9 (1.3~4.1); 12.4 (0.4~22.0); 98.1 (79.3~130.5); 360 (300~410) ❙ 266 (105~509); 158 (72~233); 20 (2~98) ❙ -23.8 (-72.6~52.9); 241 (105~333); 55.9 (20.1~112.4)

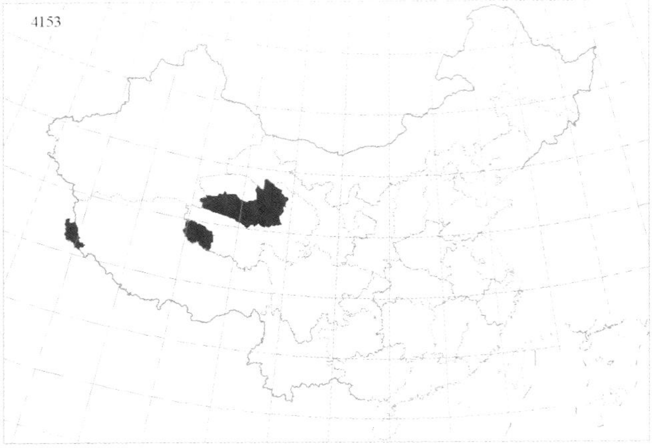

Fabaceae 豆科 *Corethrodendron* 山竹子属

Corethrodendron fruticosum (Pallas) B.H. Choi et H. Ohashi 山竹子

Subshrubs, deciduous

3.6 (-4.6~8.3); -15.9 (-28.8~-7.9); 20.9 (16.6~24.1); 7.9 (5.0~10.1); 61.6 (34.4~85.4); 78.3 (39.5~149.1); 574 (448~671) ❙ 358 (137~528); 244 (92~350); 8 (3~15) ❙ -37.2 (-77.5~-0.4); 358 (137~528); 130.1 (25.4~289.8)

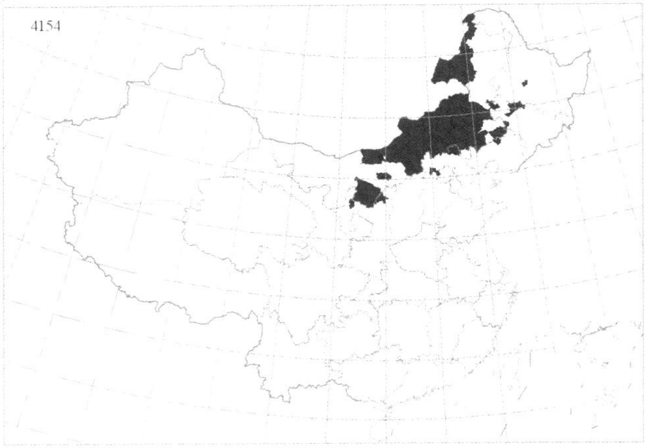

Fabaceae 豆科 *Corethrodendron* 山竹子属

Corethrodendron fruticosum var. mongolicum (Turczaninow) Turczaninow ex Kitagawa 蒙古山竹子

Subshrubs, deciduous

7.9 (7.1~8.5); -8.9 (-10.6~-7.5); 22.7 (21.9~23.1); 9.8 (9.4~10.2); 79.7 (76.2~83.0); 43.9 (39.0~50.1); 643 (627~658) ❙ 407 (225~465); 249 (130~312); 9 (5~11) ❙ -36.8 (-64.1~-29.0); 407 (225~465); 68.5 (44.7~186.8)

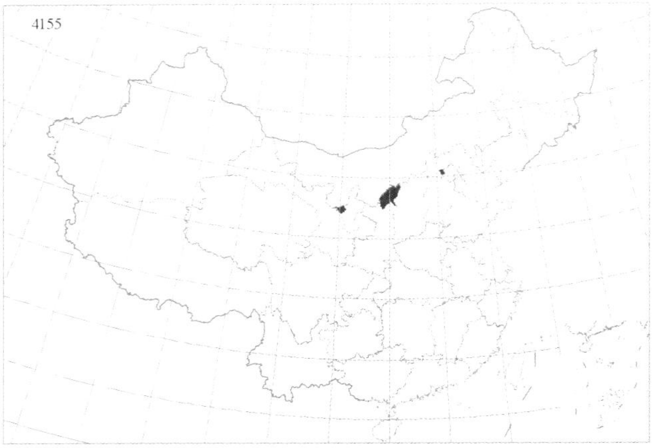

Fabaceae 豆科 *Corethrodendron* 山竹子属

Corethrodendron lignosum (Trautvetter) L. R. Xu et B. H. Choi 木山竹子

Subshrubs, deciduous

2.0 (-4.6~6.4); -18.5 (-28.8~-11.7); 20.3 (16.6~23.8); 7.4 (5.0~9.6); 56.7 (34.4~80.4); 92.3 (58.5~149.1); 552 (448~656) ❙ 376 (290~470); 262 (196~315); 8 (5~15) ❙ -31.4 (-48.5~-0.4); 376 (290~470); 161.6 (79.1~289.8)

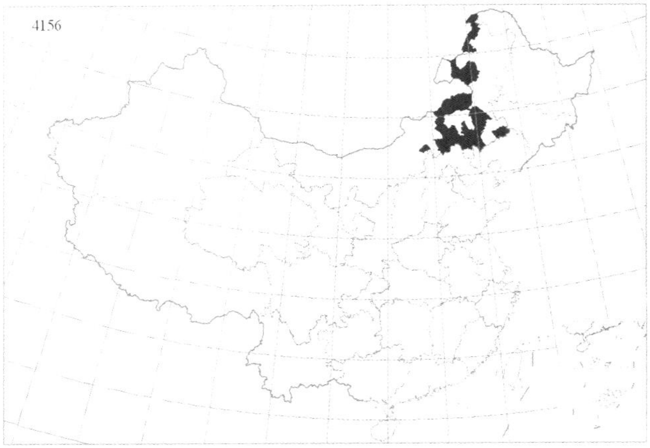

Fabaceae 豆科 *Corethrodendron* 山竹子属

Corethrodendron multijugum (Maxim.) B. H. Choi et H. Ohashi 红花山竹子

Herbs or subshrubs, deciduous

5.9 (-5.9~15.6); -8.2 (-20.9~2.5); 18.4 (5.4~29.5); 8.1 (1.3~15.7); 61.1 (0.4~136.0); 50.1 (3.4~133.0); 575 (294~932) ❙ 447 (22~1478); 234 (15~592); 16 (1~103) ❙ -19.6 (-97.2~107.2); 405 (22~881); 156.4 (2.8~364.1)

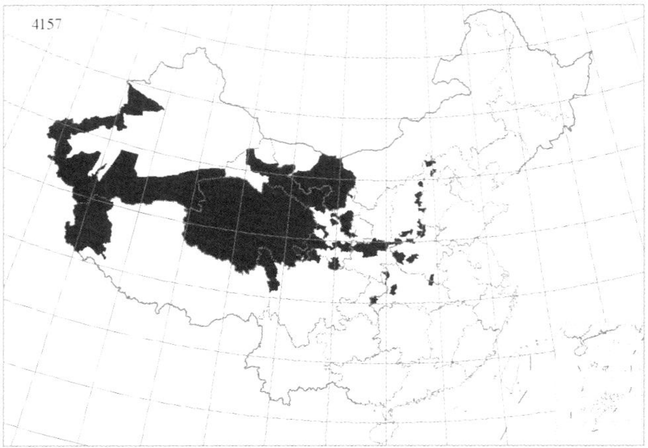

Fabaceae 豆科 *Corethrodendron* 山竹子属

Corethrodendron scoparium Fisch. et Basiner 细枝山竹子

Subshrubs, deciduous

5.2 (-2.6~8.6); -11.7 (-19.5~-7.5); 20.3 (8.7~24.3); 8.2 (2.5~10.3); 64.0 (7.7~85.9); 60.4 (39.0~104.2); 583 (349~679) ❙ 204 (49~467); 119 (34~268); 8 (1~33) ❙ -62.8 (-92.8~19.7); 202 (49~437); 64.7 (5.1~284.0)

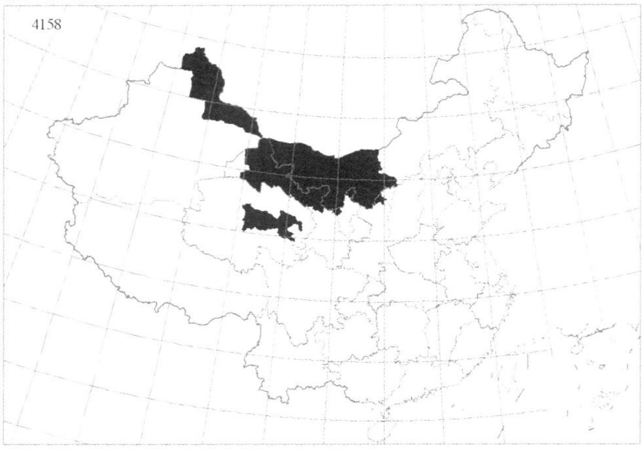

Fabaceae 豆科 *Craspedolobium* 巴豆藤属
Craspedolobium unijugum (Gagnepain) Z. Wei et Pedley 巴豆藤
Scandent shrubs

16.0 (10.[illegible]~21.0); 8.9 (1.8~15.1); 21.4 (16.3~26.6); 16.1 (10.3~21.0); 133.2 (68.0~191.8); 0.3 (0.0~7.0); 802 (620~1053) ❙ 1113 (797~1751); 627 (411~996); 43 (15~77) ❙ 39.0 (-14.2~88.8); 781 (615~1053); 317.4 (185.6~659.0)

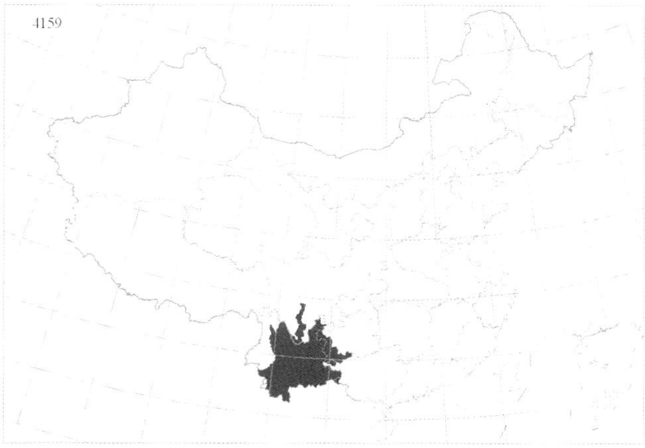

Fabaceae 豆科 *Cytisus* 金雀花属
Cytisus scoparius (L.) Link 金雀儿
Shrubs

10.6; -0.2; 21.1; 10.6; 79.4; 12.1; 643 ❙ 1246; 525; 46 ❙ 93.8; 643; 284.3

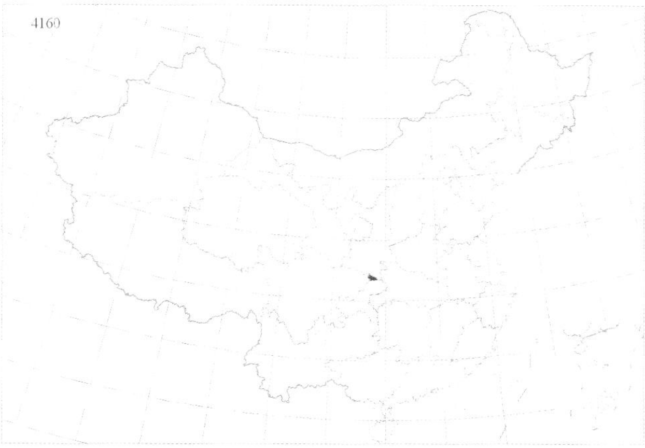

Fabaceae 豆科 *Dalbergia* 黄檀属
Dalbergia assamica Benth. 秧青
Trees, deciduous

17.7 (13.2~21.0); 11.0 (5.4~15.1); 22.1 (18.4~25.6); 17.7 (13.3~21.0); 152.3 (99.4~191.8); 0.0 (0.0~0.0); 856 (695~1053) ❙ 1293 (939~1751); 716 (503~1007); 52 (37~77) ❙ 51.5 (6.5~145.8); 841 (695~1053); 407.7 (181.2~659.0)

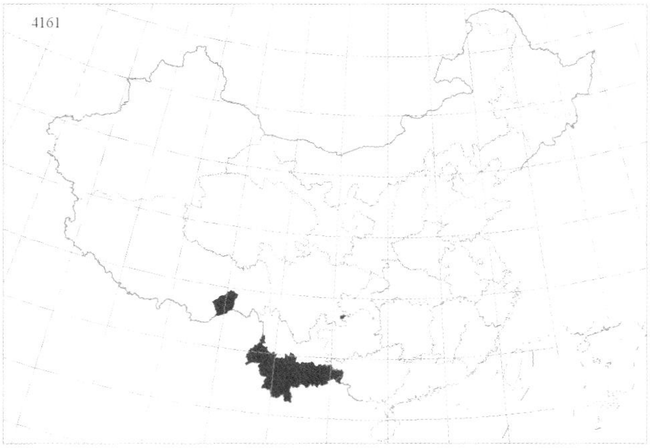

Fabaceae 豆科 *Dalbergia* 黄檀属
Dalbergia balansae Prain 南岭黄檀
Trees, deciduous

18.4 (10.2~24.7); 8.4 (2.4~19.8); 27.3 (16.9~29.8); 18.5 (10.3~24.7); 162.2 (68.2~236.5); 0.2 (0.0~5.1); 975 (620~1376) ❙ 1497 (1036~2435); 604 (392~1428); 148 (18~253) ❙ 55.1 (-11.9~129.9); 966 (620~1326); 236.5 (111.7~572.9)

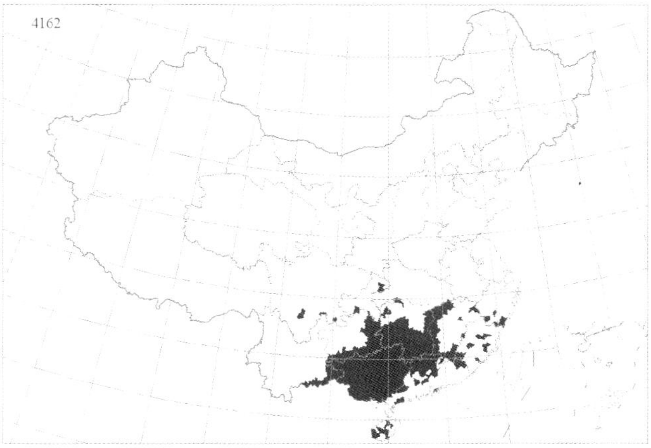

Fabaceae 豆科 *Dalbergia* 黄檀属
Dalbergia benthami Prain 两粤黄檀
Climbing shrubs, deciduous

20.7 (14.6~24.8); 12.9 (3.8~20.1); 27.1 (20.3~29.2); 20.7 (14.7~24.8); 188.9 (117.7~237.6); 0.1 (0.0~1.5); 1095 (779~1385) ❙ 1683 (868~3495); 702 (312~1428); 148 (41~897) ❙ 59.0 (-34.3~224.4); 1056 (779~1326); 294.3 (95.3~620.0)

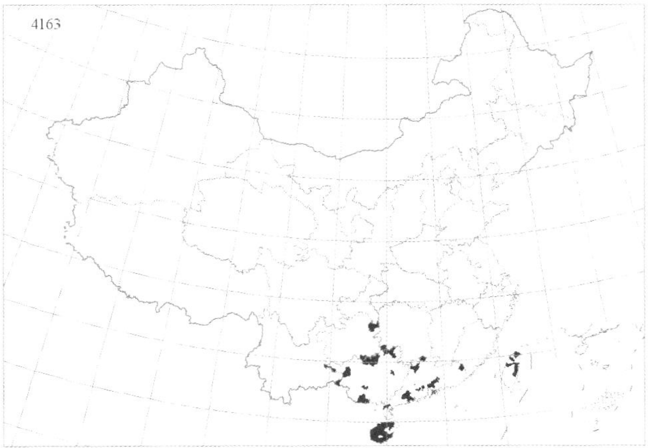

Fabaceae 豆科 *Dalbergia* 黄檀属
Dalbergia burmanica Prain 缅甸黄檀
Trees, deciduous

18.1 (14.7~21.0); 11.4 (3.8~14.9); 22.9 (20.3~25.9); 18.2 (14.8~21.0); 158.4 (118.7~191.8); 0.2 (0.0~1.5); 884 (789~1053) ❙ 1337 (1028~1522); 693 (473~818); 57 (41~94) ❙ 51.7 (12.5~74.2); 863 (780~1053); 414.8 (170.4~570.1)

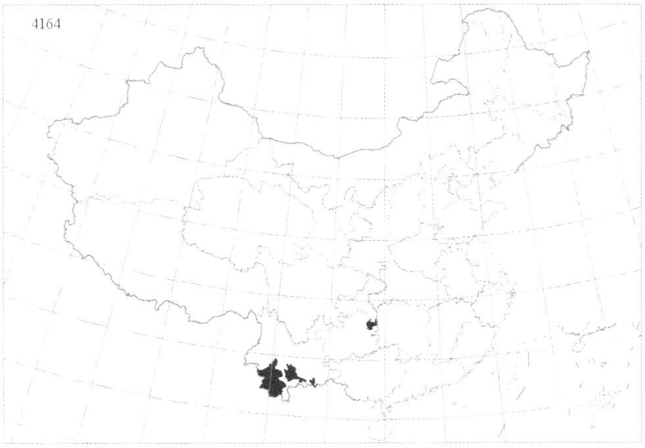

Fabaceae 豆科 *Dalbergia* 黄檀属

Dalbergia candenatensis (Dennst.) Prain 弯枝黄檀

Vines, deciduous

21.3 (15.5~24.8); 13.6 (4.8~20.1); 27.8 (25.1~29.2); 21.5 (15.6~24.8); 197.9 (127.2~237.6); 0.0 (0.0~0.2); 1146 (815~1385) ❙ 1664 (1405~2030); 699 (473~1033); 124 (41~177) ❙ 48.1 (1.5~93.2); 1126 (815~1329); 255.9 (163.7~527.5)

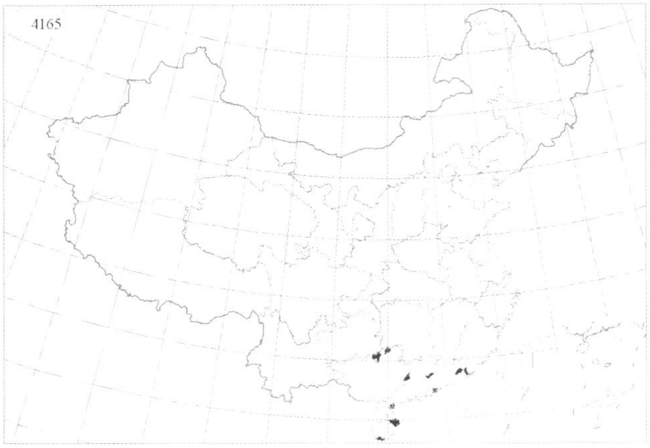

Fabaceae 豆科 *Dalbergia* 黄檀属

Dalbergia cultrata Graham ex Bentham 黑黄檀

Trees

18.7 (15.9~21.0); 12.7 (9.5~15.1); 22.5 (20.6~25.6); 18.8 (16.0~21.0); 165.3 (131.6~191.8); 0.0 (0.0~0.0); 900 (779~1053) ❙ 1431 (1014~1751); 770 (568~996); 56 (41~77) ❙ 59.1 (30.2~88.8); 884 (751~1053); 478.5 (318.7~659.0)

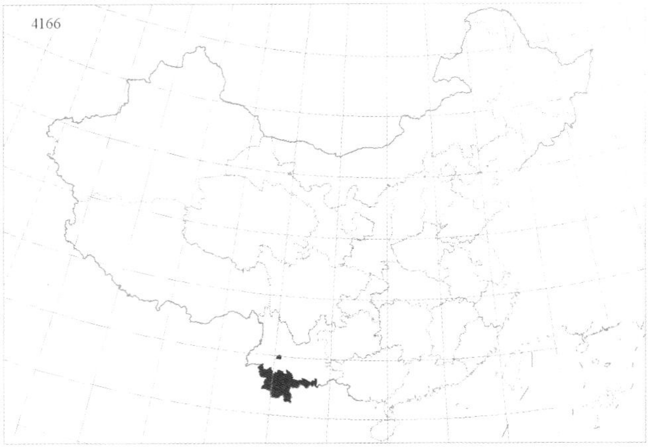

Fabaceae 豆科 *Dalbergia* 黄檀属

Dalbergia dyeriana Prain ex Harms 大金刚藤

Large climbers, deciduous

14.8 (7.3~22.2); 4.0 (-2.5~13.9); 24.9 (15.6~28.7); 14.9 (7.5~22.3); 121.5 (48.2~207.5); 3.3 (0.0~20.6); 802 (537~1190) ❙ 1250 (711~1893); 526 (349~911); 93 (15~227) ❙ 55.8 (-2.0~129.9); 802 (537~1190); 233.7 (126.1~551.3)

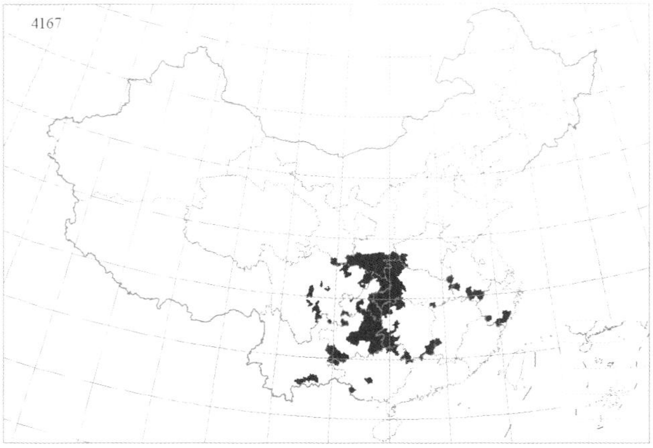

Fabaceae 豆科 *Dalbergia* 黄檀属

Dalbergia hainanensis Merr. et Chun 海南黄檀

Trees, deciduous

24.0 (23.1~24.8); 18.8 (17.5~20.1); 27.7 (27.2~28.4); 24.1 (23.2~24.8); 228.9 (218.4~237.6); 0.0 (0.0~0.0); 1314 (1232~1385) ❙ 1405 (1179~1588); 483 (392~544); 48 (41~58) ❙ 7.0 (-11.9~20.7); 1128 (1056~1212); 504.4 (439.7~620.0)

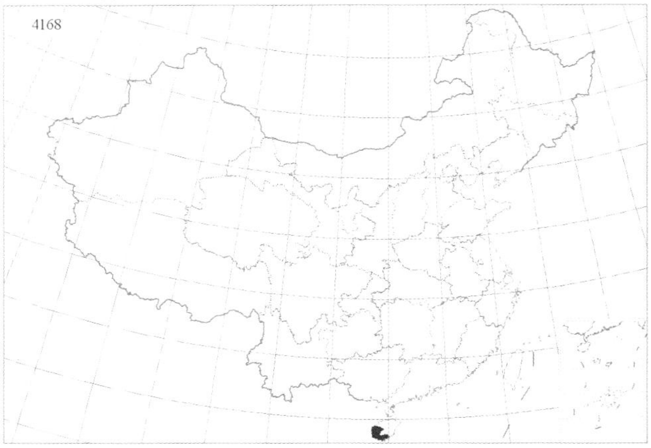

Fabaceae 豆科 *Dalbergia* 黄檀属

Dalbergia hancei Benth. 藤黄檀

Vines, deciduous

17.7 (10.6~24.7); 7.6 (-0.2~19.8); 26.9 (18.8~29.8); 17.8 (10.6~24.7); 154.0 (79.4~236.5); 0.3 (0.0~12.1); 940 (643~1376) ❙ 1490 (871~2435); 597 (299~1428); 145 (29~259) ❙ 59.8 (-6.7~147.0); 934 (643~1326); 226.8 (128.7~620.0)

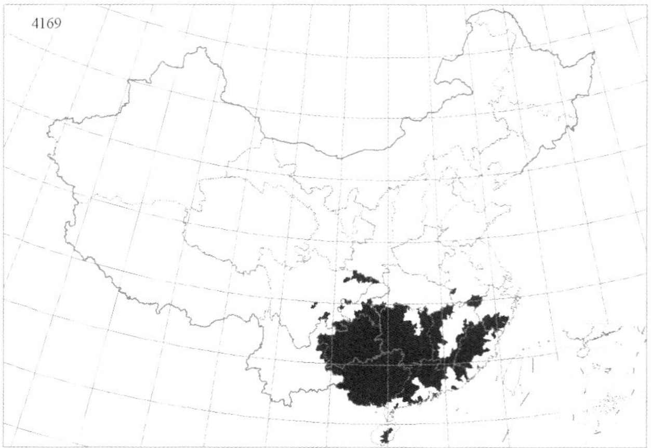

Fabaceae 豆科 *Dalbergia* 黄檀属

Dalbergia henryana Prain 蒙自黄檀

Large climbers, deciduous

17.0 (14.3~21.0); 8.7 (4.0~14.2); 23.6 (21.4~25.7); 17.1 (14.3~21.0); 145.0 (112.6~191.8); 0.2 (0.0~1.1); 851 (756~1053) ❙ 1311 (1178~1522); 668 (476~860); 70 (48~106) ❙ 54.5 (37.6~80.6); 851 (756~1053); 345.9 (225.8~551.6)

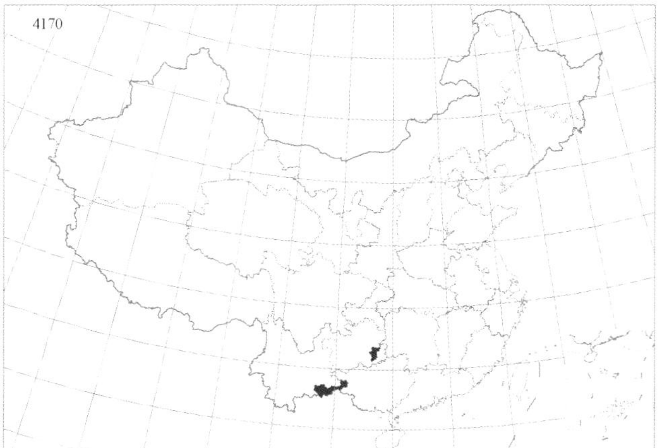

Fabaceae 豆科 *Dalbergia* 黄檀属
Dalbergia hupeana Hance 黄檀
Trees, deciduous

15.6 (7.4~22.7); 3.7 (-4.2~15.0); 26.7 (18.1~29.9); 15.7 (8.2~22.8); 132.2 (58.8~213.1); 4.1 (0.0~29.4); 862 (561~1221) | 1235 (604~2435); 511 (290~1428); 122 (14~259) | 42.3 (-24.1~147.0); 845 (561~1221); 208.1 (103.2~551.6)

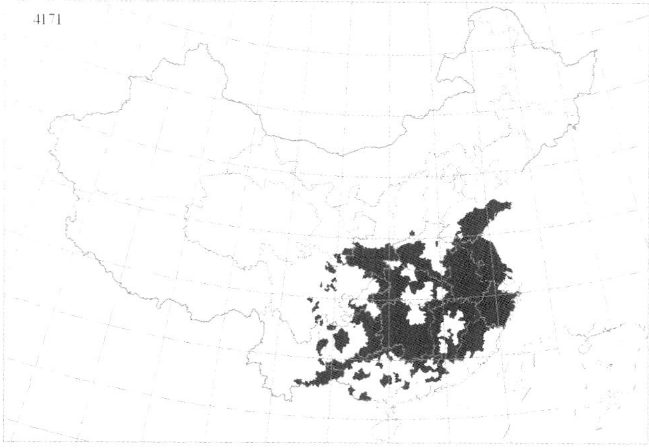

Fabaceae 豆科 *Dalbergia* 黄檀属
Dalbergia jingxiensis S. Y. Liu 靖西黄檀
Shrubs

19.2; 11.0; 25.2; 19.2; 170.5; 0.0; 952 | 1485; 848; 72 | 55.9; 952; 313.7

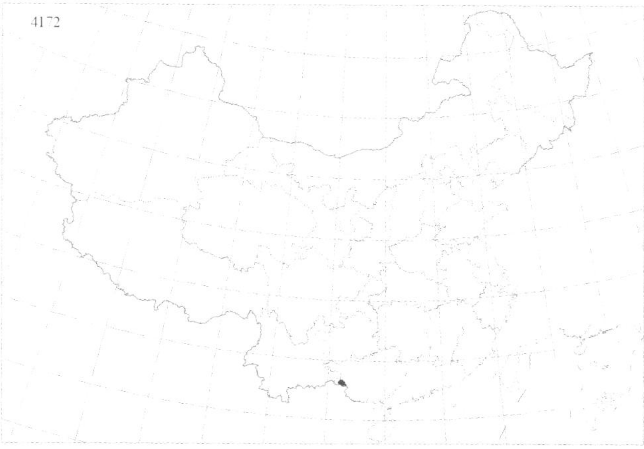

Fabaceae 豆科 *Dalbergia* 黄檀属
Dalbergia kingiana Prain 滇南黄檀
Shrubs, deciduous

18.6; 12.8; 22.1; 18.6; 163.7; 0.0; 885 | 1497; 780; 57 | 69.3; 865; 433.7

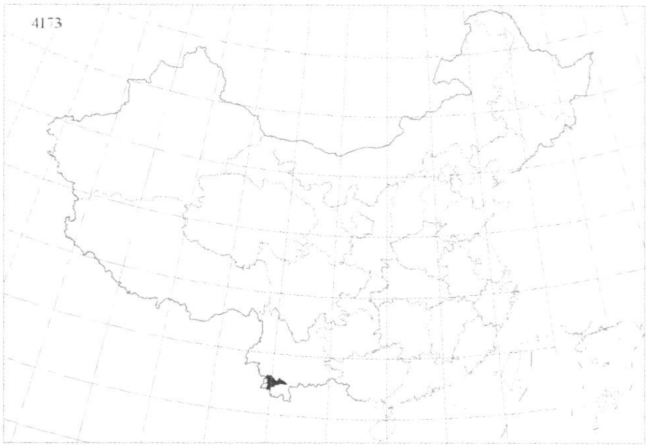

Fabaceae 豆科 *Dalbergia* 黄檀属
Dalbergia mengsuoensis Y. Y. Qian 勐梭黄檀
Shrubs

18.9; 13.1; 22.3; 19.0; 167.5; 0.0; 900 | 1462; 738; 43 | 62.4; 862; 473.8

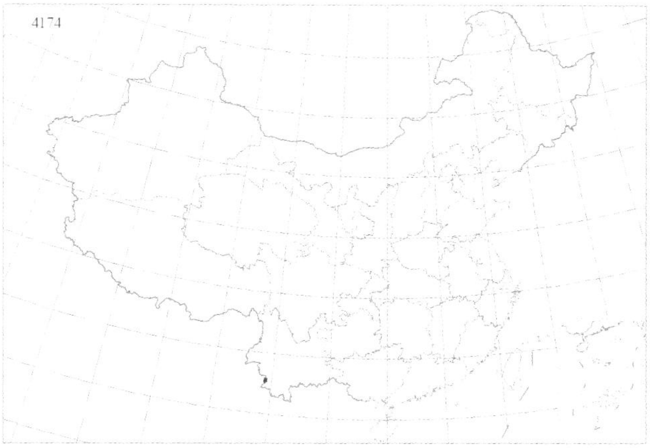

Fabaceae 豆科 *Dalbergia* 黄檀属
Dalbergia millettii Benth. 香港黄檀
Vines, deciduous

17.3 (14.3~22.4); 6.7 (3.3~15.2); 27.3 (24.2~29.3); 17.4 (14.3~22.7); 148.5 (113.2~212.7); 0.2 (0.0~2.3); 918 (761~1216) | 1609 (1315~2030); 585 (403~1033); 184 (104~230) | 76.3 (45.5~147.0); 914 (761~1216); 225.7 (151.0~283.9)

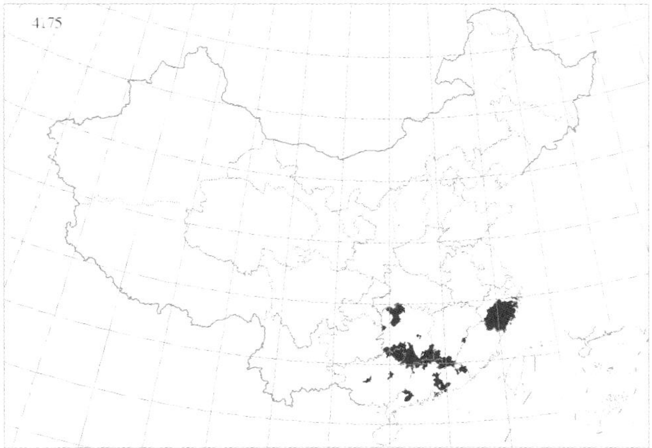

Fabaceae 豆科 *Dalbergia* 黄檀属
Dalbergia mimosoides Franch. 象鼻藤
Shrubs, deciduous

15.7 (1.8~22.2); 6.3 (-7.1~15.1); 24.0 (9.9~29.8); 15.8 (3.8~22.2); 131.1 (15.5~206.9); 1.9 (0.0~53.3); 830 (404~1177) | 1289 (583~1997); 572 (336~1056); 103 (12~259) | 55.6 (-14.2~130.7); 818 (404~1177); 264.9 (108.2~659.0)

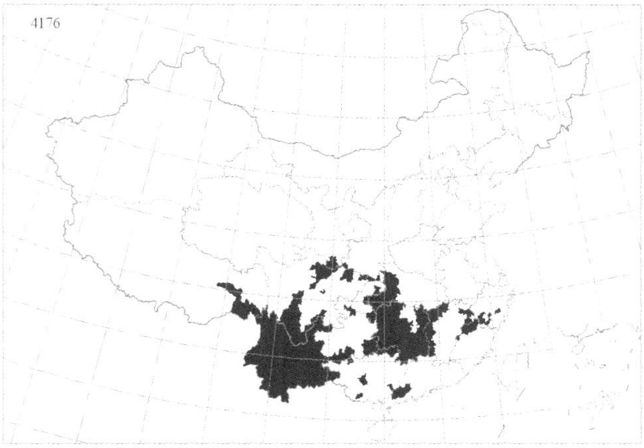

Fabaceae 豆科 *Dalbergia* 黄檀属
Dalbergia obtusifolia (Baker) Prain 钝叶黄檀
Trees, deciduous

18.1 (16.8~19.1); 10.3 (7.6~12.8); 23.9 (21.6~26.6); 18.2 (16.8~19.2); 157.8 (141.5~170.1); 0.0 (0.0~0.0); 893 (832~981) ❙ 1277 (1049~1456); 700 (598~836); 57 (44~73) ❙ 43.3 (26.1~66.9); 884 (783~981); 321.9 (189.5~491.5)

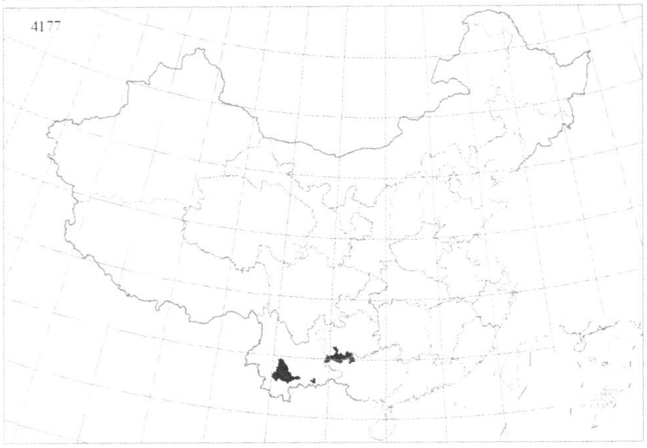

Fabaceae 豆科 *Dalbergia* 黄檀属
Dalbergia odorifera T. C. Chen ex Chun et al. 降香黄檀
Trees, deciduous

22.4 (18.2~24.8); 15.2 (8.7~20.1); 27.9 (25.9~28.8); 22.4 (18.3~24.8); 209.2 (159.3~237.6); 0.0 (0.0~0.0); 1202 (928~1385) ❙ 1450 (868~2030); 617 (312~1033); 84 (41~144) ❙ 22.4 (-34.3~70.4); 1113 (868~1329); 320.7 (139.6~538.2)

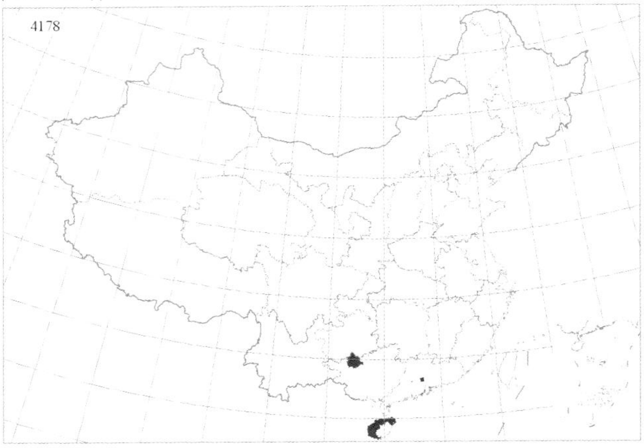

Fabaceae 豆科 *Dalbergia* 黄檀属
Dalbergia peishaensis Chun et T. Chen 白沙黄檀
Vines, deciduous

20.9; 13.0; 27.5; 20.9; 190.9; 0.0; 1111 ❙ 1272; 513; 64 ❙ 14.8; 1043; 324.4

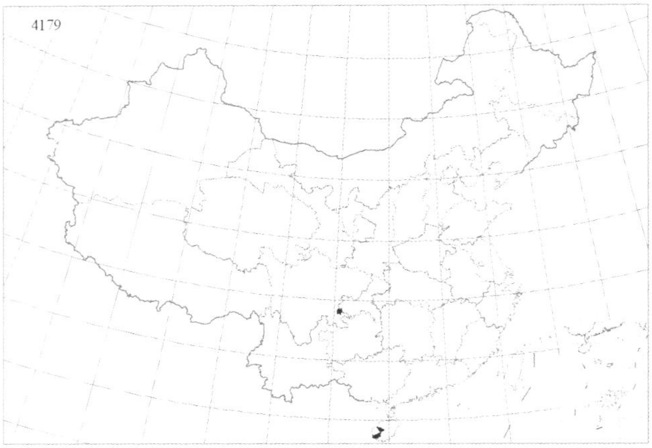

Fabaceae 豆科 *Dalbergia* 黄檀属
Dalbergia pinnata (Lour.) Prain 斜叶黄檀
Trees, deciduous

19.6 (1.8~24.8); 12.4 (-7.1~20.1); 24.9 (9.9~28.8); 19.7 (3.8~24.8); 177.2 (15.7~237.6); 1.0 (0.0~53.3); 1014 (404~1385) ❙ 1448 (583~2435); 729 (312~1428); 78 (13~173) ❙ 45.6 (-34.3~145.8); 987 (404~1360); 355.4 (157.7~659.0)

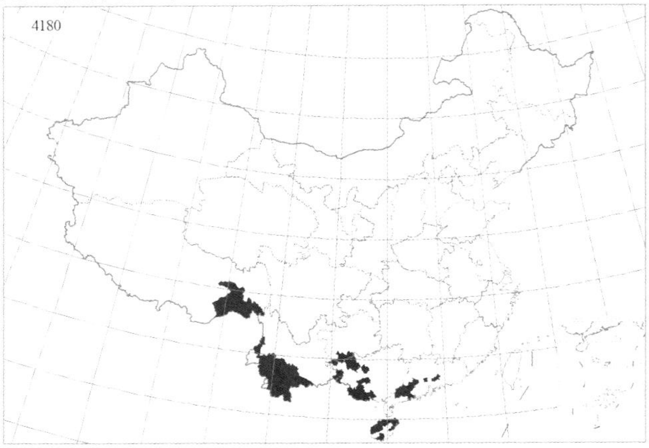

Fabaceae 豆科 *Dalbergia* 黄檀属
Dalbergia polyadelpha Prain 多体蕊黄檀
Trees, deciduous

18.5 (16.3~21.3); 11.8 (8.8~14.2); 23.0 (20.3~27.9); 18.5 (16.4~21.3); 162.4 (136.6~196.0); 0.0 (0.0~0.0); 899 (789~1123) ❙ 1314 (976~1751); 737 (556~996); 55 (42~117) ❙ 46.7 (5.8~88.8); 884 (780~1121); 401.7 (238.3~551.6)

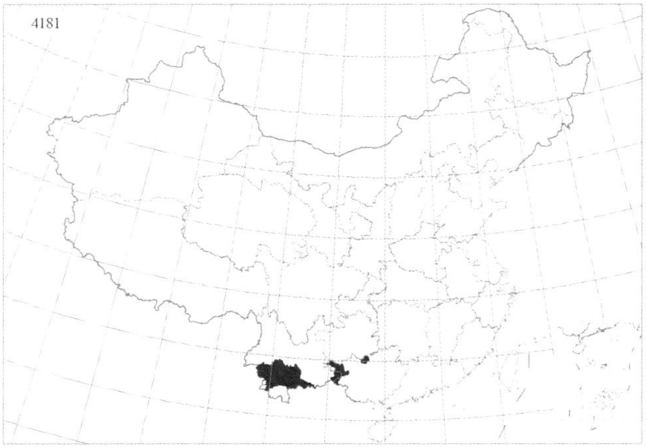

Fabaceae 豆科 *Dalbergia* 黄檀属
Dalbergia rimosa Roxb. 多裂黄檀
Vines, deciduous

19.6 (14.6~22.2); 11.9 (3.8~15.1); 25.4 (21.4~29.8); 19.7 (14.7~22.3); 176.2 (117.7~207.5); 0.1 (0.0~1.5); 996 (779~1190) ❙ 1495 (1163~1900); 741 (508~1056); 100 (48~194) ❙ 52.0 (5.8~101.8); 989 (779~1190); 365.0 (174.9~659.0)

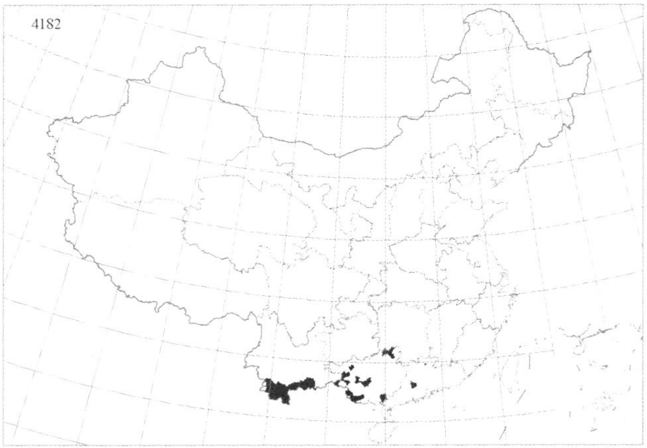

Fabaceae 豆科 *Dalbergia* 黄檀属
Dalbergia sericea Spreng. 毛叶黄檀
Trees, deciduous

12.0; 5.1; 17.3; 12.2; 97.2; 13.1; 683 ❙ 1254; 735; 47 ❙ 87.5; 683; 392.8

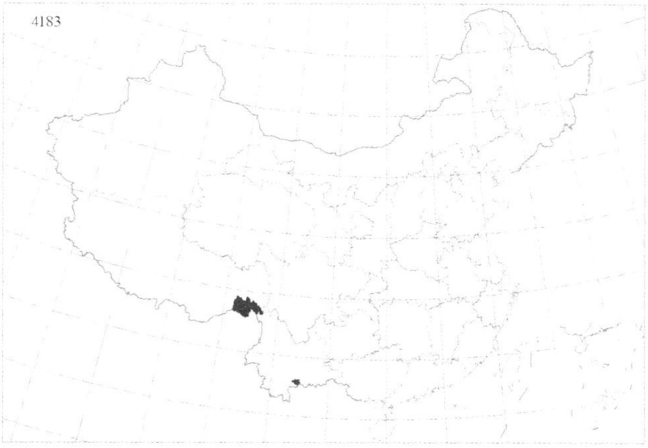

Fabaceae 豆科 *Dalbergia* 黄檀属
Dalbergia stenophylla Prain 狭叶黄檀
Vines, deciduous

13.8 (2.2~20.7); 4.1 (-6.6~12.8); 22.6 (9.9~27.7); 14.0 (3.8~20.7); 113.8 (15.5~189.0); 7.4 (0.0~49.1); 769 (413~1083) ❙ 1291 (822~2005); 584 (420~824); 74 (19~177) ❙ 71.2 (30.0~125.9); 769 (413~1083); 237.2 (182.9~284.3)

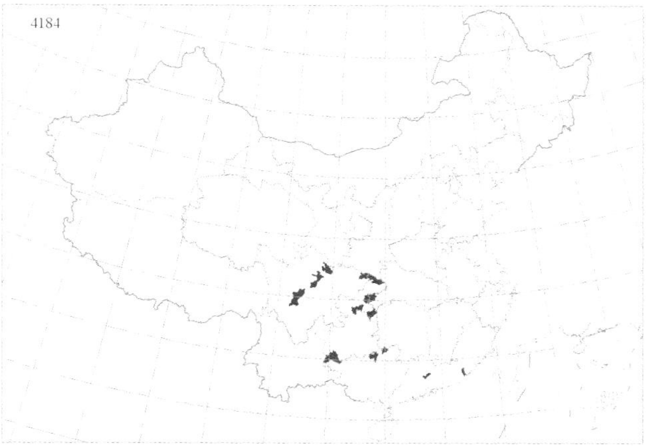

Fabaceae 豆科 *Dalbergia* 黄檀属
Dalbergia stipulacea Wall. 托叶黄檀
Large climbers, deciduous

18.8 (17.1~21.0); 12.7 (9.5~15.1); 22.7 (20.8~25.6); 18.9 (17.1~21.0); 166.4 (145.0~191.8); 0.0 (0.0~0.0); 905 (815~1053) ❙ 1426 (1143~1751); 771 (637~996); 56 (41~77) ❙ 57.9 (21.7~88.8); 890 (802~1053); 470.6 (316.1~659.0)

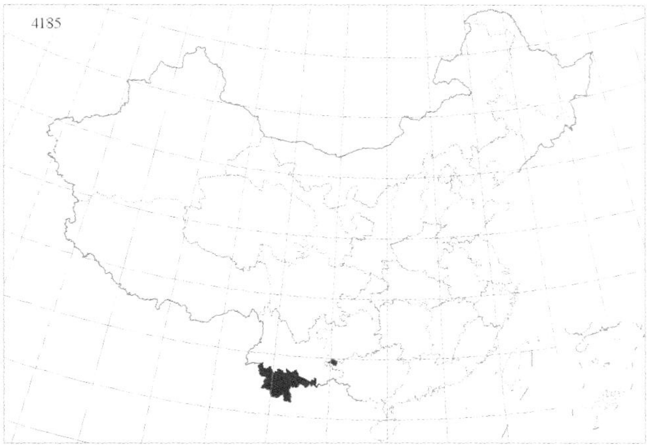

Fabaceae 豆科 *Dalbergia* 黄檀属
Dalbergia tonkinensis Prain 越南黄檀
Trees, deciduous

21.2 (19.1~22.7); 13.0 (11.0~15.0); 27.4 (25.0~28.6); 21.3 (19.1~22.8); 195.2 (169.6~213.1); 0.0 (0.0~0.0); 1112 (945~1221) ❙ 1571 (1188~2435); 859 (634~1428); 96 (56~130) ❙ 41.4 (5.8~99.5); 1112 (945~1221); 303.0 (256.6~371.7)

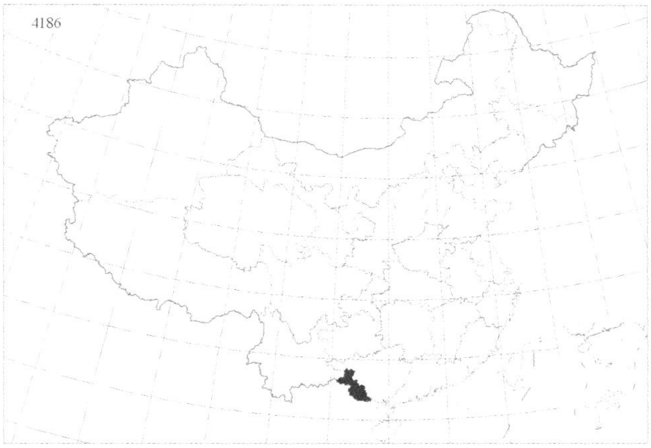

Fabaceae 豆科 *Dalbergia* 黄檀属
Dalbergia tsoi Merr. et Chun 红果黄檀
Vines, deciduous

24.1 (23.0~24.8); 18.7 (17.7~20.1); 27.8 (26.7~28.7); 24.1 (23.1~24.8); 229.2 (217.0~237.6); 0.0 (0.0~0.0); 1316 (1217~1385) ❙ 1441 (1176~1725); 502 (392~619); 58 (41~110) ❙ 9.8 (-12.0~41.8); 1154 (1056~1326); 481.1 (352.9~620.0)

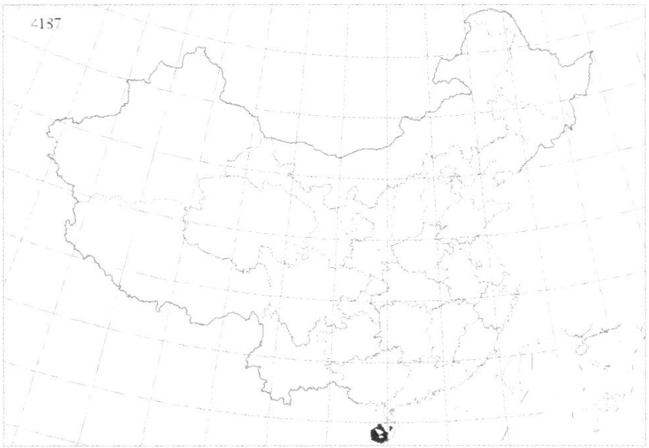

Fabaceae 豆科 *Dalbergia* 黄檀属
Dalbergia yunnanensis Franch. 滇黔黄檀
Large climbers, deciduous

15.6 (5.4~21.0); 7.9 (-2.1~14.2); 21.7 (12.3~28.1); 15.7 (5.8~21.0); 128.9 (31.5~191.8); 0.8 (0.0~25.6); 797 (481~1053) ❙ 1088 (741~1624); 597 (411~911); 49 (15~191) ❙ 37.2 (-14.2~101.8); 780 (481~1053); 292.5 (131.4~551.6)

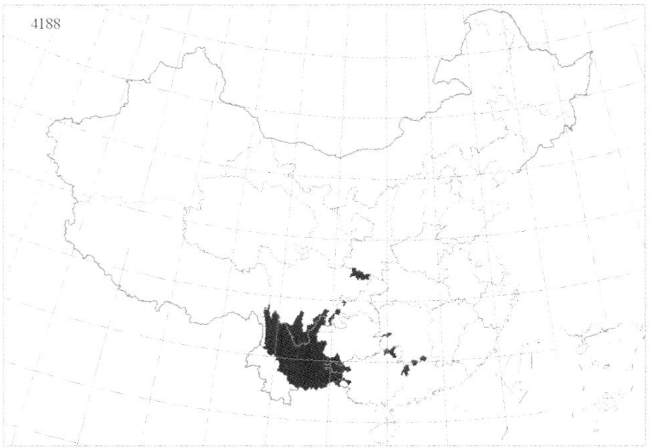

Fabaceae 豆科 *Delonix* 凤凰木属
Delonix regia (Bojer ex Hook.) Raf. 凤凰木
Trees

21.9 (15.2~24.8); 14.8 (4.4~20.1); 27.5 (23.4~29.0); 22.0 (15.2~24.8); 203.9 (123.1~237.6); 0.0 (0.0~0.6); 1177 (802~1385) ❙ 1475 (868~2114); 645 (312~1357); 82 (41~136) ❙ 27.0 (-34.3~71.8); 1114 (802~1360); 333.7 (119.4~620.0)

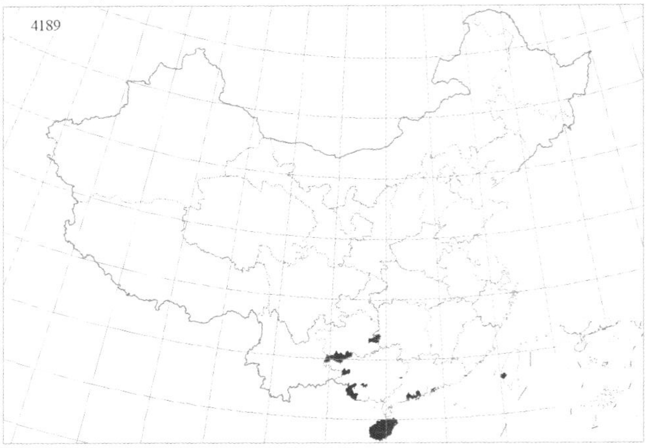

Fabaceae 豆科 *Dendrolobium* 假木豆属
Dendrolobium dispermum (Hayata) Schindl. 两节假木豆
Shrubs

21.9; 17.0; 25.4; 21.9; 202.8; 0.0; 1119 ❙ 2501; 1537; 94 ❙ 133.8; 1043; 234.5

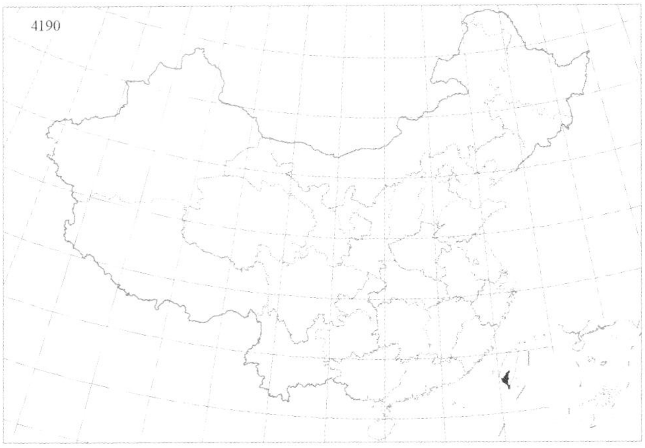

Fabaceae 豆科 *Dendrolobium* 假木豆属
Dendrolobium lanceolatum (Dunn) Schindl. 单节假木豆
Shrubs

22.7 (16.7~24.8); 16.3 (5.9~20.1); 27.8 (26.7~28.7); 22.8 (16.7~24.8); 213.4 (141.0~237.6); 0.0 (0.0~0.0); 1241 (878~1385) ❙ 1468 (1176~1950); 508 (392~666); 91 (41~249) ❙ 24.7 (-12.0~121.9); 1093 (878~1254); 431.8 (273.8~620.0)

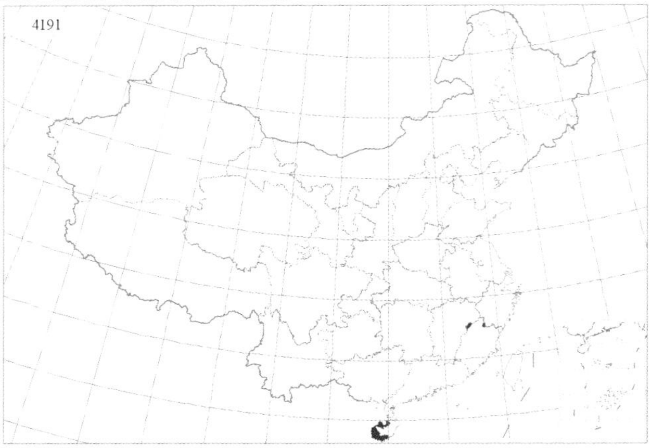

Fabaceae 豆科 *Dendrolobium* 假木豆属
Dendrolobium triangulare (Retz.) Schindl. 假木豆
Shrubs

18.8 (13.2~24.8); 11.2 (4.6~20.1); 24.4 (18.4~29.0); 18.8 (13.3~24.8); 165.6 (99.4~237.6); 0.0 (0.0~0.4); 955 (695~1385) ❙ 1360 (868~2992); 697 (312~1847); 74 (37~188) ❙ 44.5 (-34.3~214.4); 933 (695~1331); 324.6 (119.4~659.0)

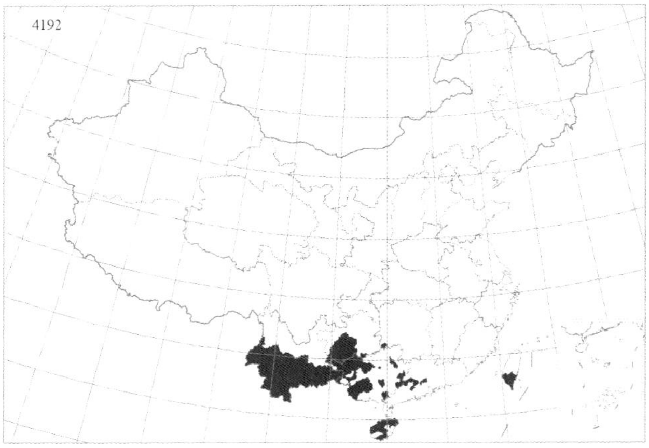

Fabaceae 豆科 *Dendrolobium* 假木豆属
Dendrolobium umbellatum (L.) Benth. 伞花假木豆
Shrubs or small trees

21.0; 16.3; 24.5; 21.1; 192.6; 0.0; 1055 ❙ 2554; 1499; 117 ❙ 154.0; 998; 266.8

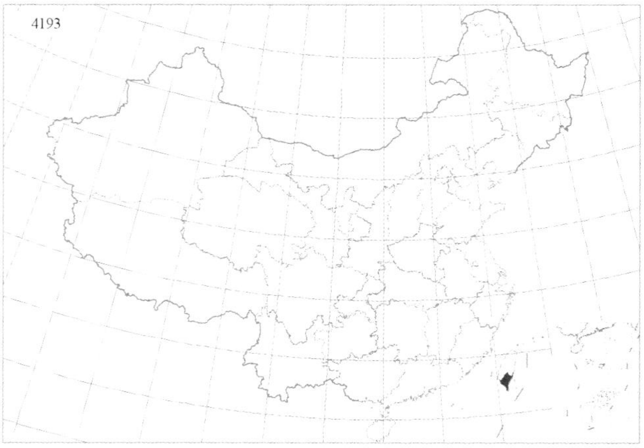

Fabaceae 豆科 *Derris* 鱼藤属
Derris alborubra Hemsl. 白花鱼藤
Vines, evergreen

21.5 (17.3~24.8); 13.5 (6.5~20.1); 28.0 (24.6~29.0); 21.6 (17.3~24.8); 199.3 (147.9~237.6); 0.0 (0.0~0.0); 1154 (901~1385) ❙ 1684 (1293~2435); 704 (455~1428); 131 (41~224) ❙ 48.0 (1.5~99.5); 1128 (901~1360); 292.6 (120.3~620.0)

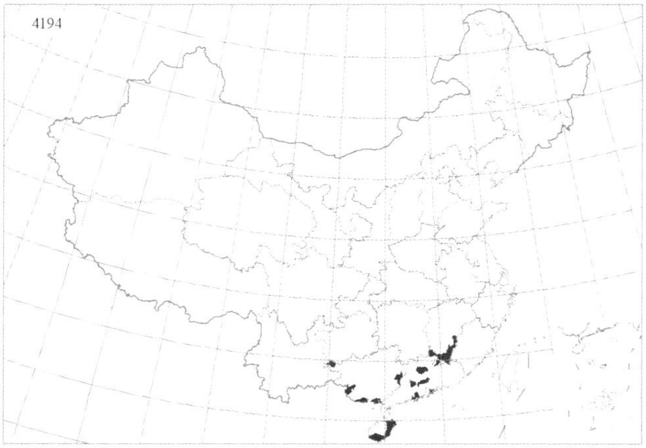

Fabaceae 豆科 *Derris* 鱼藤属
Derris breviramosa F. C. How 短枝鱼藤
Vines, evergreen

24.1；18.6；27.9；24.1；229.7；0.0；1319 ❙ 1363；493；53 ❙ 3.6；1143；515.9

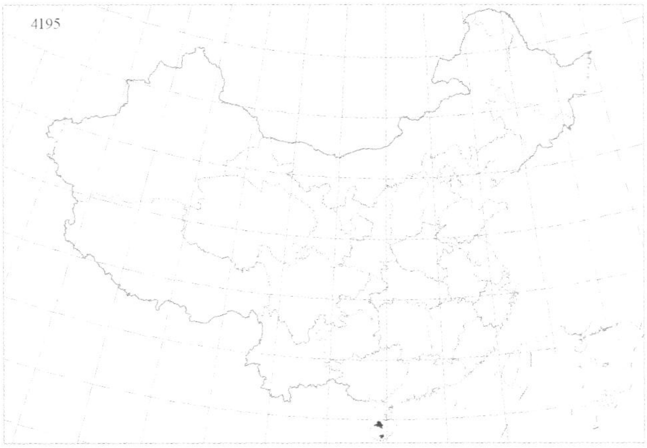

Fabaceae 豆科 *Derris* 鱼藤属
Derris caudatilimba F. C. How 尾叶鱼藤
Vines, evergreen

20.4 (17.7~23.2)；13.2 (10.7~15.8)；25.8 (22.5~29.0)；20.5 (17.8~23.2)；185.6 (153.4~218.9)；0.0 (0.0~0.0)；1035 (854~1257) ❙ 1595 (1368~1953)；776 (592~996)；101 (56~183) ❙ 55.1 (28.9~88.8)；1033 (854~1257)；389.0 (160.7~659.0)

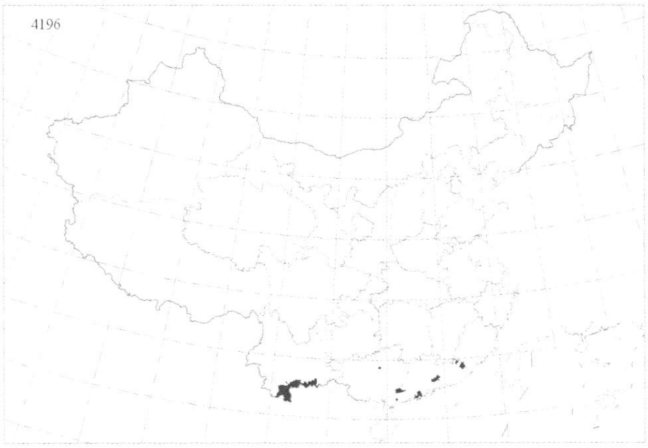

Fabaceae 豆科 *Derris* 鱼藤属
Derris cavaleriei Gagnep. 黔桂鱼藤
Vines, evergreen

18.3 (15.9~21.3)；9.4 (6.9~12.6)；25.2 (23.0~27.9)；18.4 (16.0~21.3)；160.6 (131.6~196.0)；0.0 (0.0~0.0)；928 (802~1123) ❙ 1261 (1172~1334)；679 (634~723)；59 (50~73) ❙ 37.2 (5.8~64.7)；928 (802~1121)；284.7 (189.5~411.1)

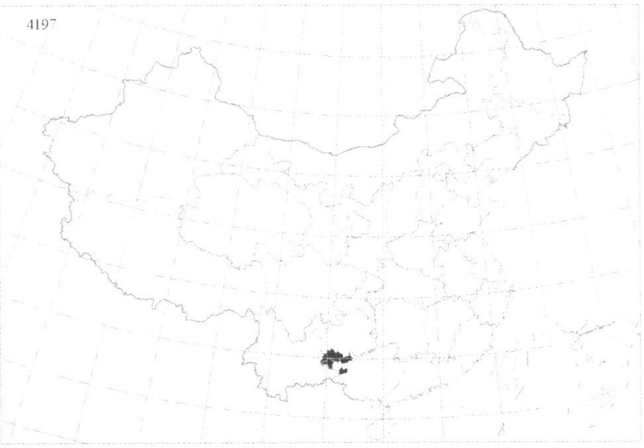

Fabaceae 豆科 *Derris* 鱼藤属
Derris eriocarpa F. C. How 毛果鱼藤
Vines, evergreen

19.2 (16.3~22.3)；11.2 (7.8~15.1)；25.1 (21.4~28.9)；19.2 (16.4~22.3)；170.7 (136.4~208.1)；0.0 (0.0~0.0)；966 (797~1202) ❙ 1375 (1074~1751)；727 (582~996)；75 (43~194) ❙ 43.5 (5.8~88.8)；964 (797~1202)；347.1 (189.5~659.0)

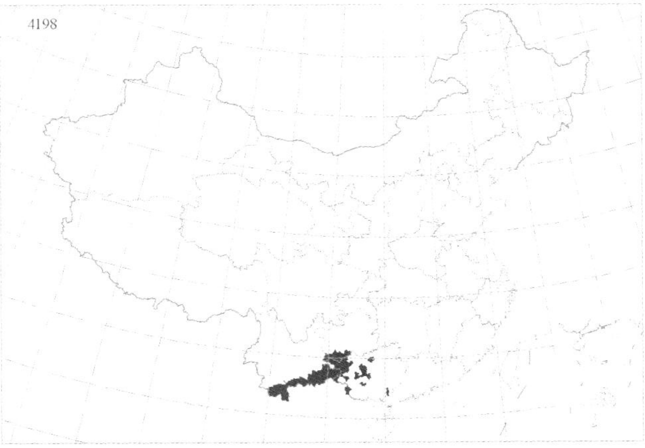

Fabaceae 豆科 *Derris* 鱼藤属
Derris ferruginea Benth. 锈毛鱼藤
Vines, evergreen

21.6 (16.8~24.8)；14.5 (7.8~20.1)；26.8 (21.4~28.8)；21.6 (16.8~24.8)；199.6 (141.5~237.6)；0.0 (0.0~0.0)；1136 (813~1385) ❙ 1544 (868~2435)；713 (312~1428)；88 (41~146) ❙ 38.0 (-34.3~99.5)；1096 (813~1360)；360.4 (157.7~659.0)

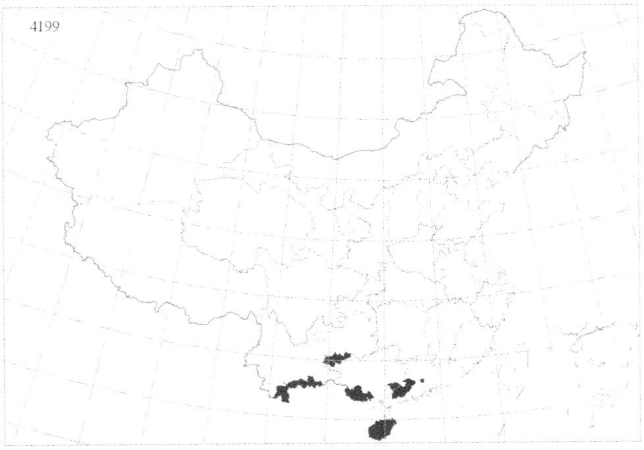

Fabaceae 豆科 *Derris* 鱼藤属
Derris fordii Oliv. 中南鱼藤
Vines, evergreen

17.8 (12.1~22.2)；7.7 (1.1~15.2)；26.9 (20.2~30.1)；17.9 (12.2~22.7)；155.1 (95.0~212.7)；0.3 (0.0~8.5)；937 (696~1216) ❙ 1542 (946~1997)；599 (383~994)；162 (38~259) ❙ 65.6 (5.8~130.7)；929 (696~1216)；247.7 (148.9~551.6)

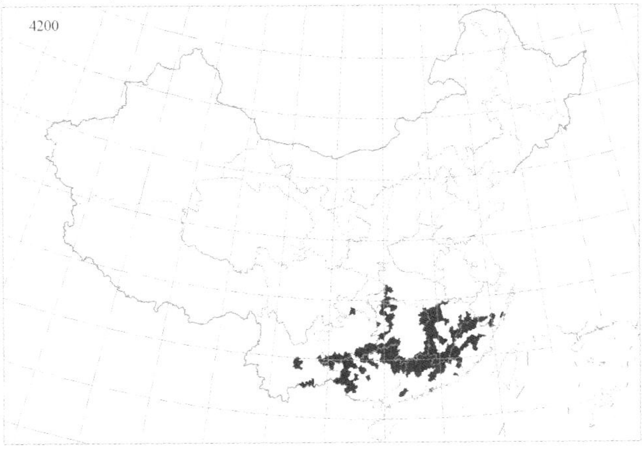

Fabaceae 豆科 *Derris* 鱼藤属
Derris fordii var. lucida F. C. How 亮叶中南鱼藤
Vines, evergreen

19.6 (15.4~22.4); 10.5 (4.9~14.9); 27.0 (22.3~28.7); 19.6 (15.4~22.4); 175.5 (124.9~208.9); 0.0 (0.0~0.1); 1019 (810~1191) ❙ 1539 (958~2035); 720 (476~1033); 126 (44~204) ❙ 51.5 (13.3~124.8); 1018 (810~1191); 254.6 (163.7~452.0)

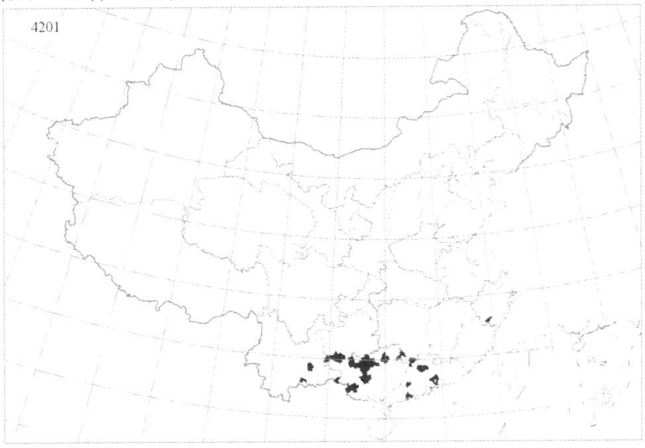

Fabaceae 豆科 *Derris* 鱼藤属
Derris harrowiana (Diels) Z. Wei 大理鱼藤
Vines, evergreen

14.5; 7.8; 19.6; 14.5; 113.5; 0.0; 730 ❙ 1044; 570; 54 ❙ 43.0; 730; 382.6

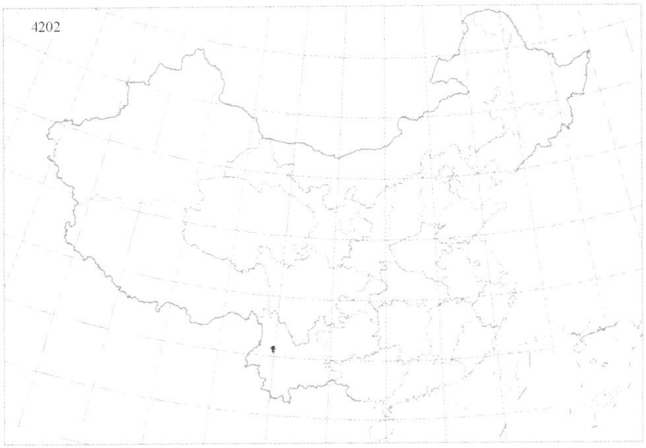

Fabaceae 豆科 *Derris* 鱼藤属
Derris laxiflora Benth. 疏花鱼藤
Vines, evergreen

20.1; 14.4; 24.9; 20.1; 181.5; 0.0; 995 ❙ 2758; 1244; 300 ❙ 177.2; 995; 285.7

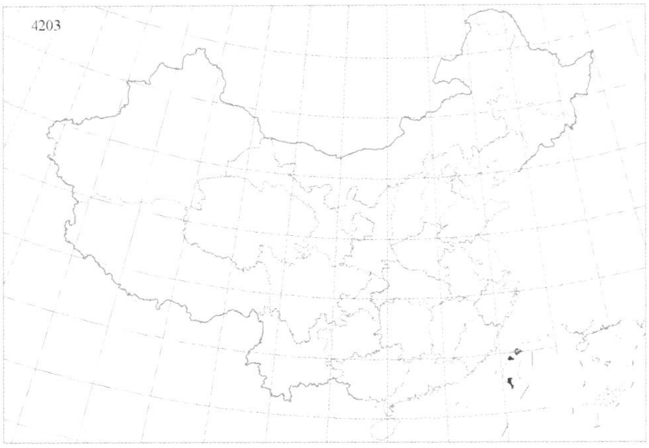

Fabaceae 豆科 *Derris* 鱼藤属
Derris marginata (Roxb.) Benth. 边荚鱼藤
Vines, evergreen

16.0 (5.4~22.1); 7.4 (−2.1~15.1); 23.0 (12.3~28.8); 16.0 (5.8~22.1); 133.0 (31.5~205.6); 0.9 (0.0~25.6); 828 (481~1182) ❙ 1244 (741~1953); 617 (411~996); 76 (15~242) ❙ 49.8 (−14.2~101.9); 815 (481~1182); 277.9 (142.9~659.0)

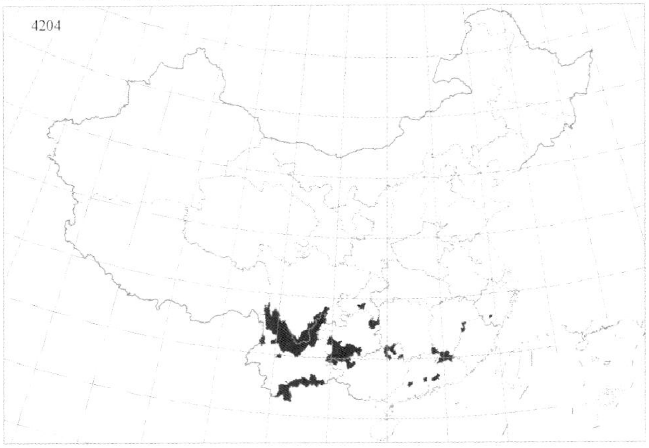

Fabaceae 豆科 *Derris* 鱼藤属
Derris palmifolia Chun et F. C. How 掌叶鱼藤
Vines, evergreen

17.1; 10.9; 21.7; 17.2; 146.1; 0.0; 835 ❙ 1223; 699; 53 ❙ 45.5; 831; 443.8

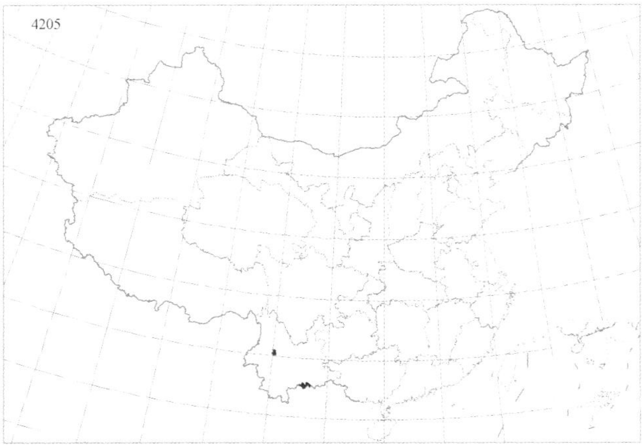

Fabaceae 豆科 *Derris* 鱼藤属
Derris robusta (Roxb. ex DC.) Benth. 大鱼藤树
Trees, deciduous

18.6 (16.9~21.0); 12.5 (10.3~15.1); 22.7 (21.4~25.6); 18.7 (16.9~21.0); 164.1 (142.7~191.8); 0.0 (0.0~0.0); 897 (813~1053) ❙ 1310 (958~1751); 714 (531~996); 55 (44~77) ❙ 45.8 (12.5~88.8); 879 (783~1053); 428.4 (294.5~659.0)

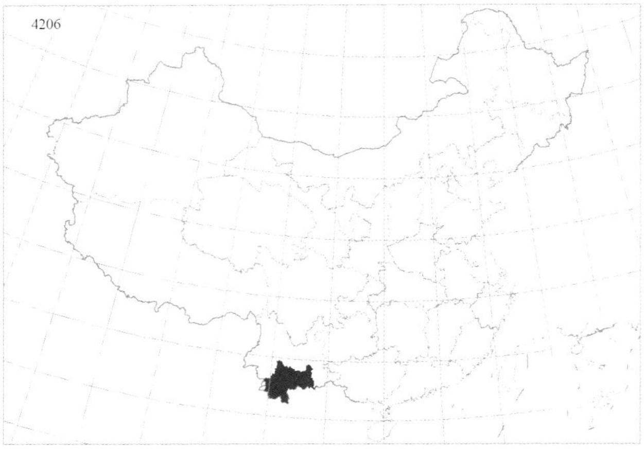

Fabaceae 豆科 *Derris* 鱼藤属
Derris scabricaulis (Franch.) Gagnep. 粗茎鱼藤
Vines, evergreen

15.2 (1.8~23.6); 8.2 (-7.1~16.7); 20.6 (9.9~28.6); 15.3 (3.8~23.6); 125.3 (15.7~223.5); 2.1 (0.0~53.3); 778 (404~1281) ▎1108 (527~1751); 622 (317~1007); 43 (9~85) ▎43.5 (-14.2~145.8); 758 (404~1269); 339.6 (142.9~659.0)

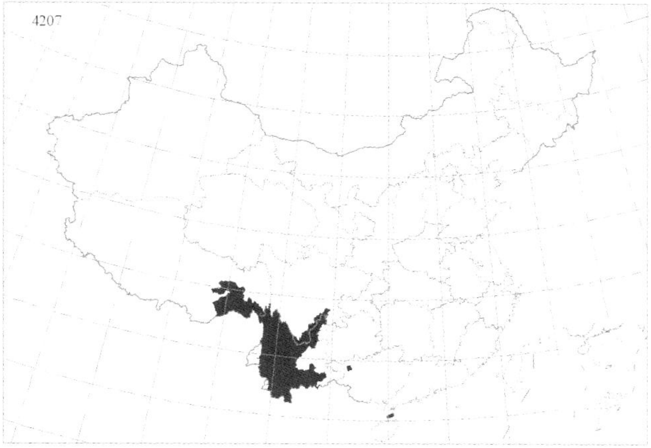

Fabaceae 豆科 *Derris* 鱼藤属
Derris tonkinensis Gagnep. 东京鱼藤
Shrubs or trees, evergreen

20.4 (17.8~22.7); 11.6 (7.1~15.0); 27.2 (25.0~28.7); 20.4 (17.8~22.8); 185.0 (153.8~213.1); 0.0 (0.0~0.0); 1064 (918~1221) ▎1534 (1188~2435); 796 (617~1428); 103 (51~189) ▎44.7 (5.8~99.5); 1063 (918~1221); 277.4 (189.5~371.7)

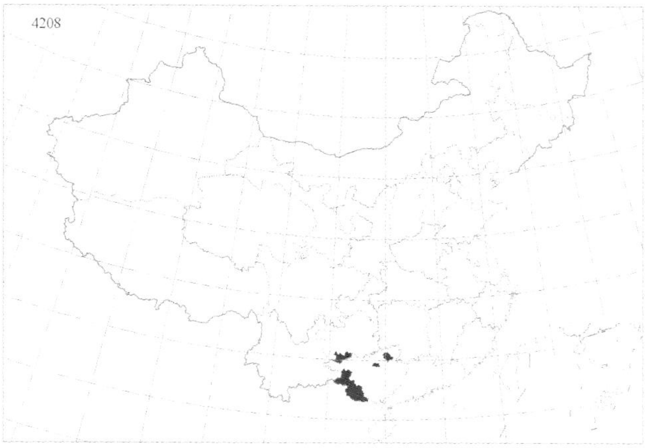

Fabaceae 豆科 *Derris* 鱼藤属
Derris tonkinensis var. compacta Gagnep. 大叶东京鱼藤
Shrubs or trees, evergreen

22.1; 14.0; 28.3; 22.1; 205.5; 0.0; 1175 ▎1522; 766; 95 ▎29.7; 1175; 250.7

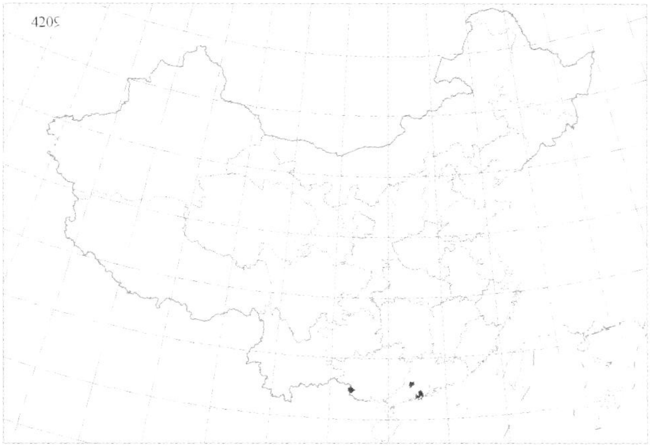

Fabaceae 豆科 *Derris* 鱼藤属
Derris trifoliata Lour. 鱼藤
Vines, evergreen

22.0 (17.4~24.8); 14.9 (7.4~20.1); 27.5 (21.7~29.0); 22.1 (17.4~24.8); 204.7 (148.9~237.6); 0.0 (0.0~0.0); 1169 (863~1385) ▎1803 (1269~2992); 862 (473~1847); 112 (41~188) ▎58.3 (-1.4~214.4); 1142 (863~1331); 278.3 (120.3~620.0)

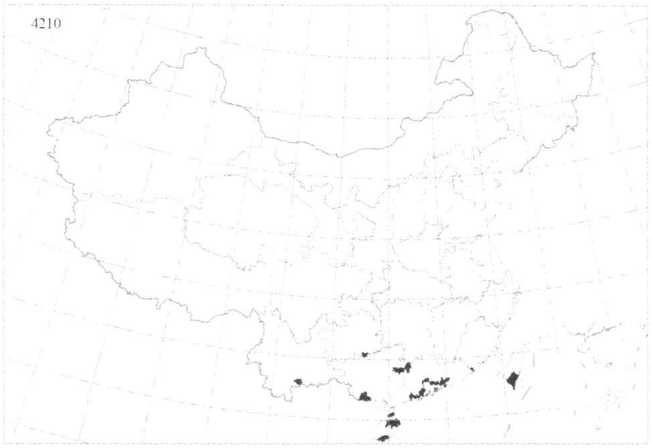

Fabaceae 豆科 *Desmodium* 山蚂蝗属
Desmodium amethystinum Dunn 紫水晶山蚂蝗
Shrubs

19.0; 13.1; 22.6; 19.1; 168.8; 0.0; 909 ▎1495; 775; 62 ▎64.5; 891; 427.9

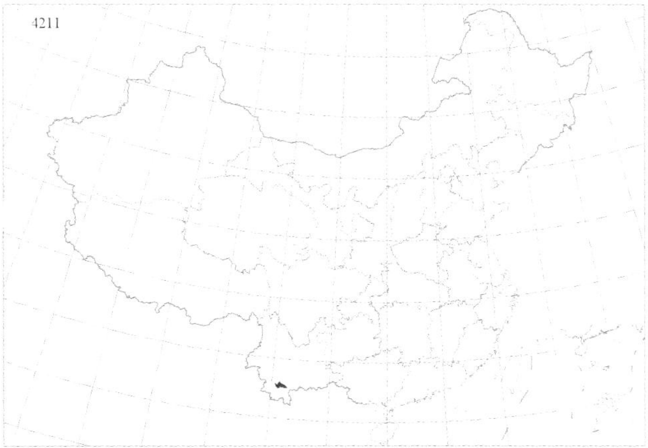

Fabaceae 豆科 *Desmodium* 山蚂蝗属
Desmodium callianthum Franch. 美花山蚂蝗
Shrubs

5.8 (1.8~17.7); -2.6 (-7.2~10.9); 13.0 (9.9~22.5); 6.9 (3.8~17.8); 45.2 (15.7~153.4); 35.0 (0.0~53.6); 505 (403~854) ▎692 (341~1368); 414 (226~756); 17 (4~59) ▎30.7 (-15.3~67.7); 495 (341~854); 283.4 (169.5~491.5)

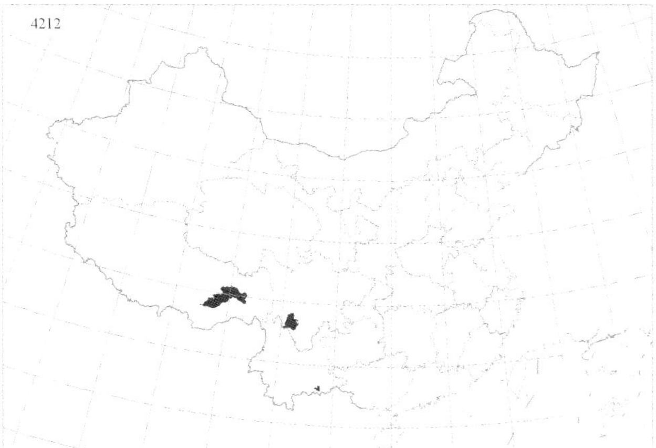

Fabaceae 豆科 *Desmodium* 山蚂蝗属
Desmodium concinnum DC. 凹叶山蚂蝗
Shrubs

18.0 (14.4~21.0); 11.4 (2.3~15.1); 22.4 (19.1~28.5); 18.1 (14.4~21.0); 157.2 (113.1~191.8); 0.1 (0.0~3.4); 874 (727~1053) ❙ 1387 (1028~1751); 770 (469~996); 57 (42~126) ❙ 59.3 (12.5~112.2); 862 (727~1053); 440.5 (134.8~659.0)

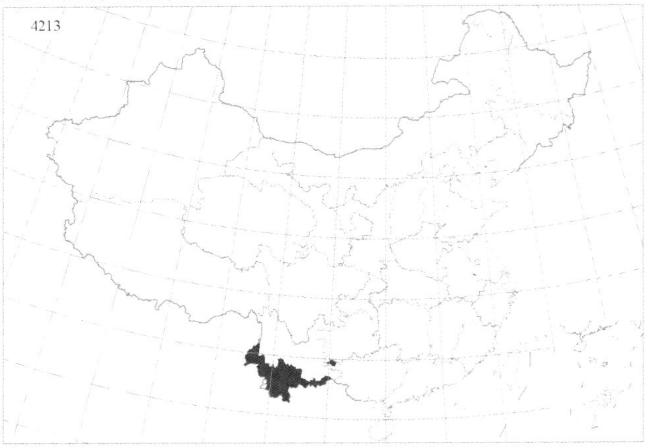

Fabaceae 豆科 *Desmodium* 山蚂蝗属
Desmodium diffusum DC. 单序山蚂蝗
Shrubs

17.5 (11.3~21.0); 10.9 (3.3~15.1); 22.1 (17.0~25.6); 17.6 (11.3~21.0); 150.9 (76.6~191.8); 0.1 (0.0~2.2); 856 (644~1053) ❙ 1368 (942~1751); 743 (489~996); 56 (42~77) ❙ 60.1 (23.3~93.0); 848 (644~1053); 440.1 (219.7~659.0)

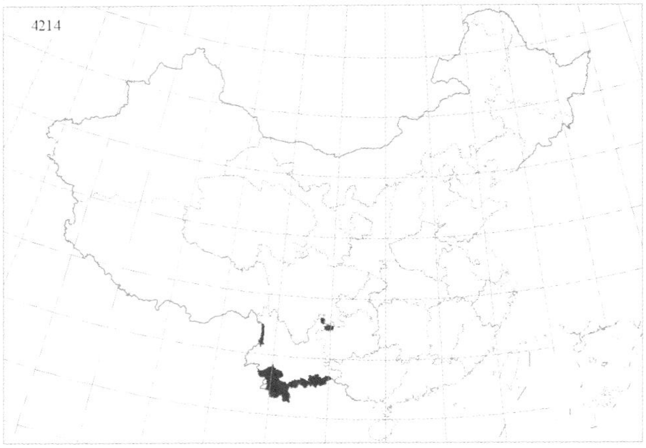

Fabaceae 豆科 *Desmodium* 山蚂蝗属
Desmodium elegans DC. 圆锥山蚂蝗
Shrubs

10.9 (-1.8~20.5); 2.9 (-11.4~15.1); 17.4 (6.1~26.6); 11.5 (1.8~20.5); 87.7 (1.9~185.9); 16.7 (0.0~82.8); 652 (329~992) ❙ 939 (255~1751); 532 (191~1007); 32 (1~88) ❙ 44.1 (-31.1~145.8); 634 (255~986); 281.9 (60.4~659.0)

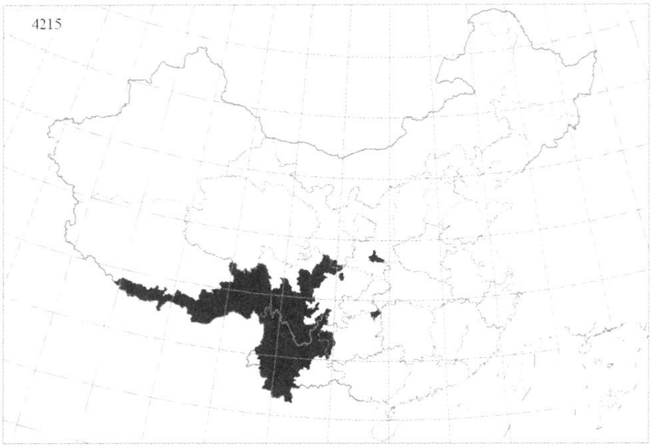

Fabaceae 豆科 *Desmodium* 山蚂蝗属
Desmodium elegans var. handelii (Schindl.) H. Ohashi 盐源山蚂蝗
Shrubs

-0.5; -9.9; 7.9; 2.6; 8.1; 73.5; 343 ❙ 258; 191; 2 ❙ -24.7; 258; 60.4

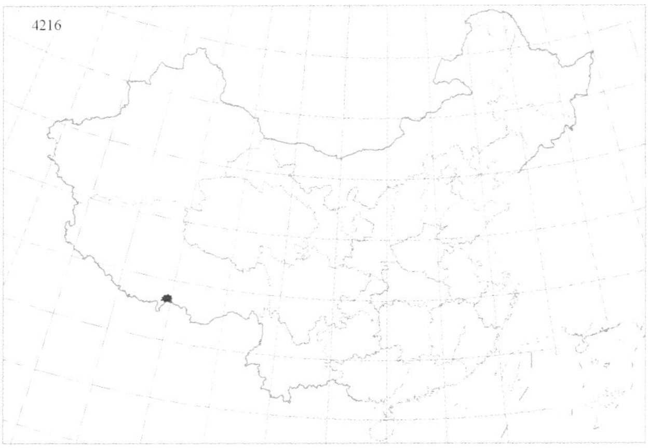

Fabaceae 豆科 *Desmodium* 山蚂蝗属
Desmodium gangeticum (L.) DC. 大叶山蚂蝗
Shrubs

19.5 (11.6~24.8); 11.8 (3.1~20.1); 25.5 (18.4~29.0); 19.6 (11.5~24.8); 174.9 (80.9~237.6); 0.1 (0.0~2.8); 1005 (654~1385) ❙ 1545 (963~3495); 748 (392~1847); 109 (18~897) ❙ 56.0 (-12.0~235.6); 985 (654~1269); 308.4 (119.4~659.0)

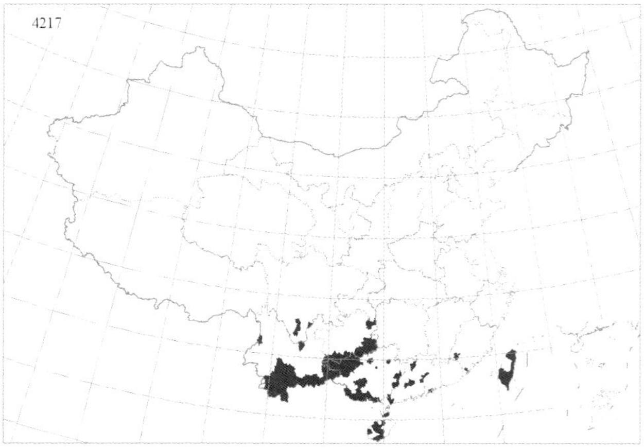

Fabaceae 豆科 *Desmodium* 山蚂蝗属
Desmodium gracillimum Hemsl. 细叶山蚂蝗
Subshrubs

21.9; 17.0; 25.4; 21.9; 202.8; 0.0; 1119 ❙ 2501; 1537; 94 ❙ 133.8; 1043; 234.5

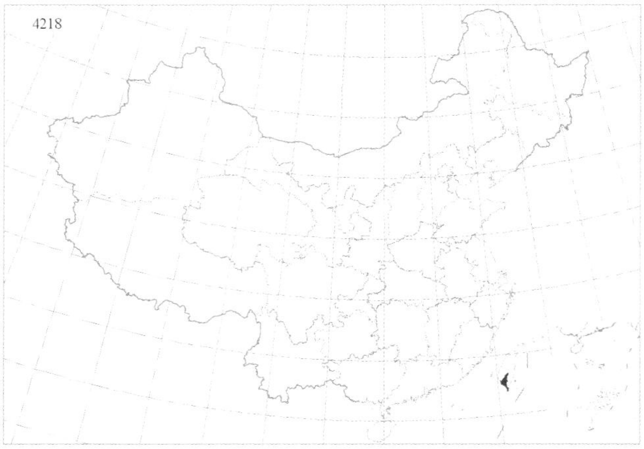

Fabaceae 豆科 *Desmodium* 山蚂蝗属
Desmodium griffithianum Benth. 疏果山蚂蝗
Subshrubs

15.1 (5.4~21.0); 7.9 (−2.1~14.2); 20.7 (12.3~26.6); 15.2 (5.8~21.0); 123.2 (31.5~191.8); 1.0 (0.0~25.6); 772 (481~1053) | 1070 (741~1547); 610 (411~911); 41 (15~70) | 39.5 (−14.2~99.9); 751 (481~1053); 315.2 (142.9~551.6)

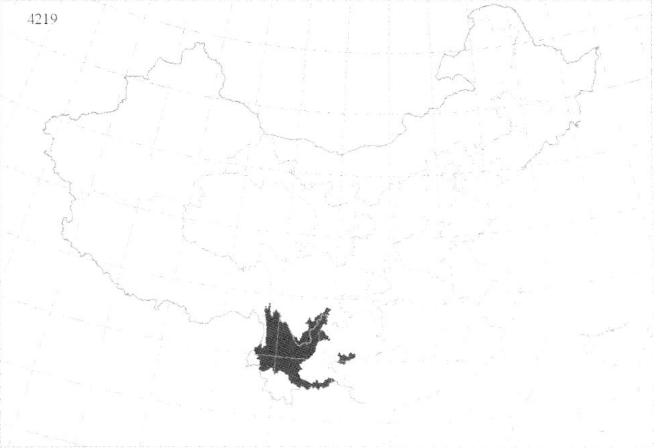

Fabaceae 豆科 *Desmodium* 山蚂蝗属
Desmodium heterocarpon (L.) DC. 假地豆
Shrubs or subshrubs

17.3 (5.4~24.8); 8.3 (−2.1~20.1); 25.1 (12.3~30.1); 17.4 (5.8~24.8); 148.8 (31.5~237.6); 0.4 (0.0~25.6); 905 (481~1385) | 1401 (741~3495); 616 (312~1847); 116 (15~897) | 55.8 (−34.3~246.5); 890 (481~1360); 264.6 (133.8~659.0)

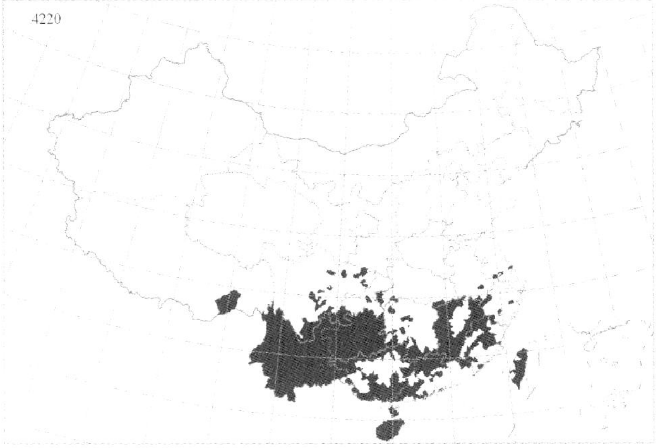

Fabaceae 豆科 *Desmodium* 山蚂蝗属
Desmodium hispidum Franch. 粗硬毛山蚂蝗
Shrubs or subshrubs, deciduous

14.0 (5.4~18.9); 7.1 (−2.1~12.0); 19.4 (12.3~23.9); 14.0 (5.8~19.0); 110.6 (31.5~167.8); 2.6 (0.0~25.6); 729 (481~929) | 947 (741~1178); 556 (411~646); 29 (15~51) | 32.4 (−14.2~67.9); 680 (481~813); 293.4 (218.0~383.4)

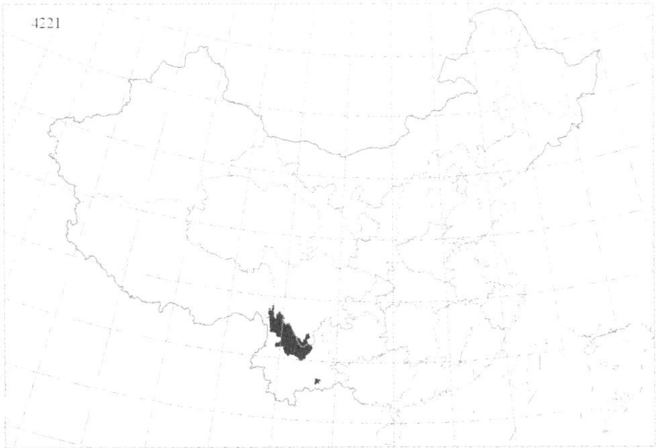

Fabaceae 豆科 *Desmodium* 山蚂蝗属
Desmodium laxiflorum DC. 大叶拿身草
Shrubs or subshrubs

17.4 (1.8~24.1); 9.0 (−7.1~18.9); 24.1 (9.9~29.0); 17.4 (3.8~24.1); 149.4 (15.7~229.7); 0.6 (0.0~53.3); 892 (404~1322) | 1429 (583~3495); 686 (348~1847); 102 (13~897) | 60.8 (5.8~246.5); 885 (404~1204); 292.8 (119.4~659.0)

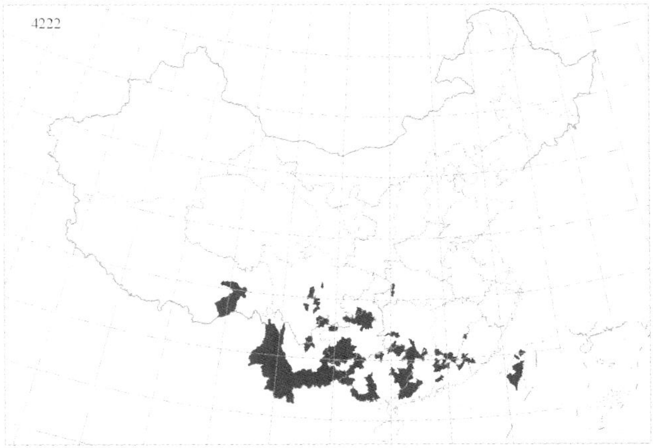

Fabaceae 豆科 *Desmodium* 山蚂蝗属
Desmodium megaphyllum Zoll. et Moritzi 滇南山蚂蝗
Shrubs

17.7 (13.2~20.5); 11.4 (6.4~15.1); 21.8 (18.4~23.7); 17.8 (13.3~20.5); 153.2 (99.4~185.9); 0.0 (0.0~0.0); 856 (695~992) | 1347 (939~1688); 745 (524~959); 53 (37~77) | 57.9 (6.5~112.2); 836 (695~986); 437.9 (234.5~659.0)

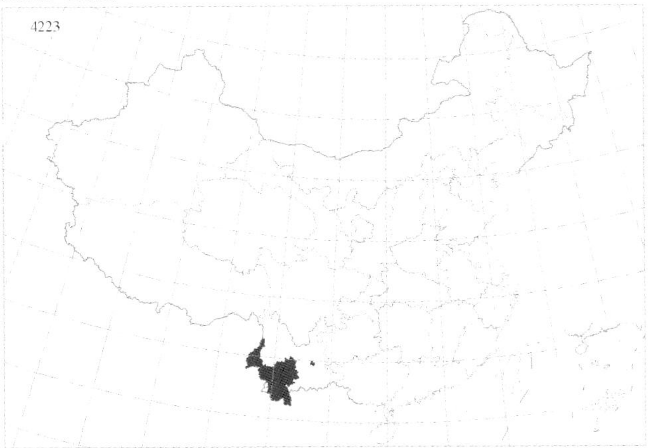

Fabaceae 豆科 *Desmodium* 山蚂蝗属
Desmodium multiflorum DC. 饿蚂蝗
Shrubs

16.1 (−1.8~21.7); 6.6 (−10.2~15.3); 24.5 (6.1~29.9); 16.2 (1.8~21.7); 134.8 (1.9~200.2); 0.9 (0.0~82.8); 842 (329~1118) | 1338 (320~2680); 577 (193~1186); 113 (15~259) | 58.5 (−14.2~246.5); 831 (320~1114); 254.4 (72.2~659.0)

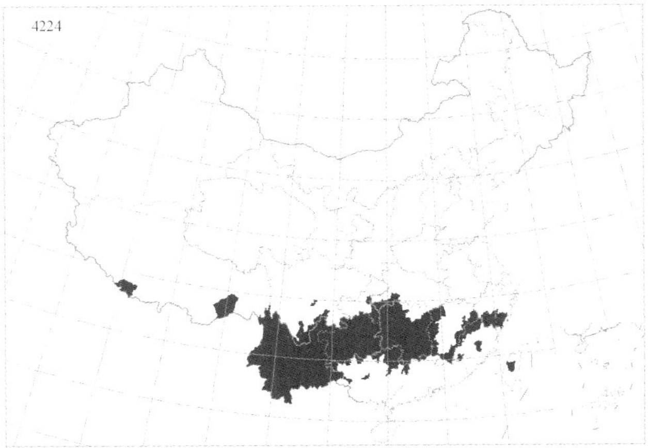

Fabaceae 豆科 *Desmodium* 山蚂蝗属
Desmodium oblongum Wall. ex Benth. 长圆叶山蚂蝗
Shrubs

18.0 (15.9~21.3); 11.6 (9.3~15.1); 22.1 (20.3~27.3); 18.1 (16.0~21.3); 156.9 (131.6~195.9); 0.0 (0.0~0.0); 870 (779~1111) ❙ 1395 (1014~1900); 771 (568~1056); 55 (41~125) ❙ 60.3 (21.7~112.2); 856 (751~1111); 444.0 (297.2~659.0)

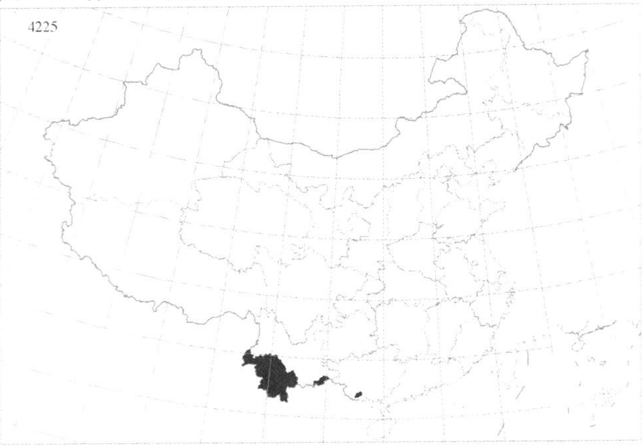

Fabaceae 豆科 *Desmodium* 山蚂蝗属
Desmodium renifolium (L.) Schindl. 肾叶山蚂蝗
Subshrubs

19.1 (16.0~24.3); 12.8 (9.3~18.7); 23.3 (20.3~28.7); 19.2 (16.0~24.3); 170.2 (132.4~231.9); 0.0 (0.0~0.0); 945 (780~1338) ❙ 1430 (868~2992); 784 (312~1847); 56 (41~107) ❙ 54.1 (-34.3~166.5); 903 (780~1144); 431.7 (119.4~659.0)

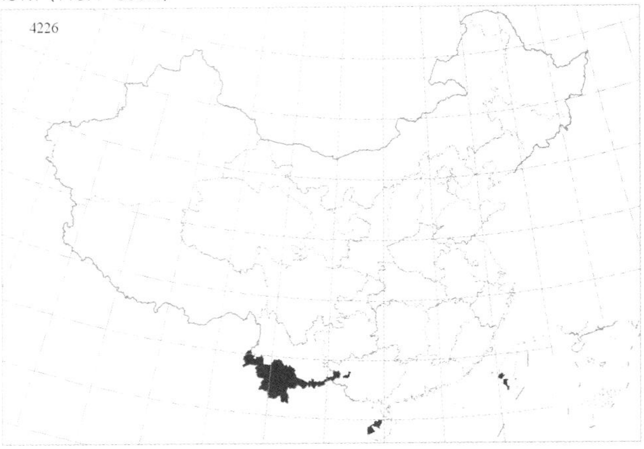

Fabaceae 豆科 *Desmodium* 山蚂蝗属
Desmodium reticulatum Champ. ex Benth. 显脉山绿豆
Subshrubs

22.0 (20.2~23.6); 14.1 (11.9~17.7); 28.3 (26.7~29.0); 22.1 (20.2~23.6); 204.9 (183.0~223.5); 0.0 (0.0~0.0); 1172 (1036~1281) ❙ 1687 (1269~2435); 785 (551~1428); 123 (68~167) ❙ 44.2 (-1.0~99.5); 1166 (1036~1269); 249.8 (157.7~572.9)

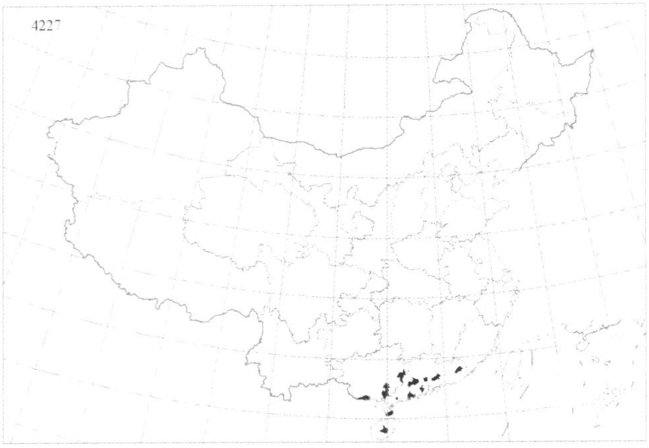

Fabaceae 豆科 *Desmodium* 山蚂蝗属
Desmodium rubrum (Lour.) DC. 赤山蚂蝗
Subshrubs

23.0 (17.5~24.6); 16.2 (10.5~19.5); 28.1 (22.3~29.2); 23.1 (17.5~24.7); 216.6 (149.9~235.9); 0.0 (0.0~0.0); 1242 (846~1372) ❙ 1530 (868~2435); 704 (312~1428); 87 (42~130) ❙ 24.1 (-34.3~99.5); 1170 (823~1321); 313.5 (178.7~548.0)

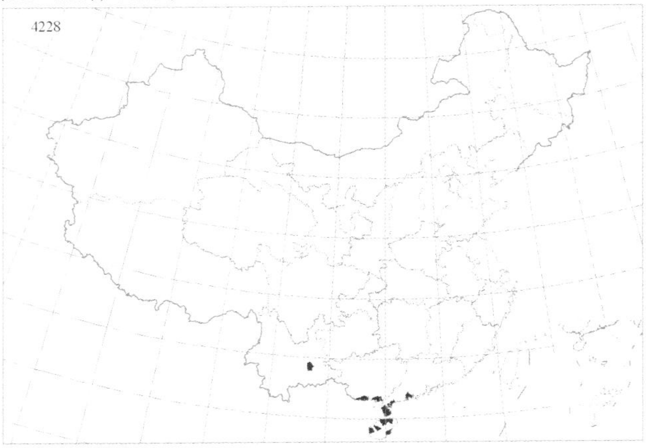

Fabaceae 豆科 *Desmodium* 山蚂蝗属
Desmodium sequax Wall. 长波叶山蚂蝗
Shrubs

15.9 (1.8~24.1); 6.1 (-7.2~18.9); 24.4 (9.9~29.9); 15.9 (3.8~24.1); 132.8 (15.4~229.7); 1.9 (0.0~53.3); 838 (404~1322) ❙ 1271 (583~3495); 574 (316~1594); 99 (13~897) ❙ 51.6 (-22.2~246.5); 823 (404~1153); 244.6 (119.4~659.0)

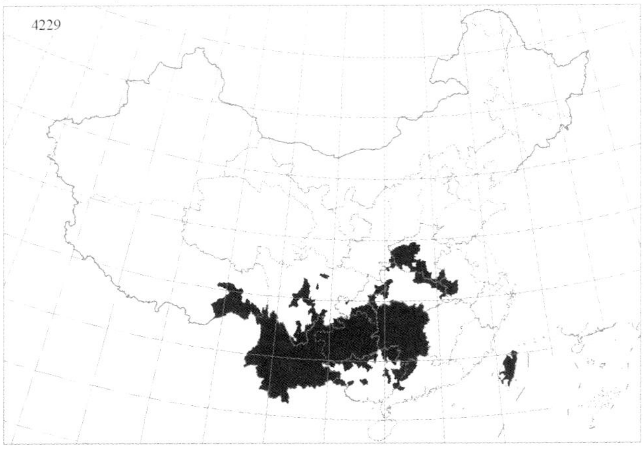

Fabaceae 豆科 *Desmodium* 山蚂蝗属
Desmodium stenophyllum Pamp. 狭叶山蚂蝗
Shrubs

6.6; -1.3; 13.3; 6.8; 39.6; 19.7; 514 ❙ 862; 548; 12 ❙ 67.7; 514; 311.5

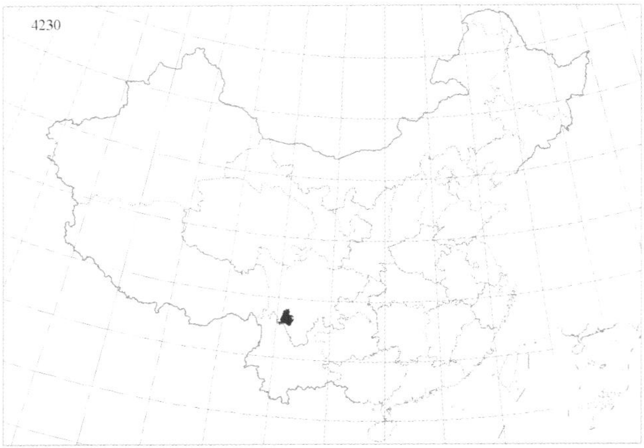

Fabaceae 豆科 *Desmodium* 山蚂蝗属

Desmodium styracifolium (Osbeck) Merr. 广东金钱草

Subshrubs

22.3; 13.5; 28.9; 22.3; 208.1; 0.0; 1202 ❙ 1446; 710; 136 ❙ 20.3; 1202; 20 .4

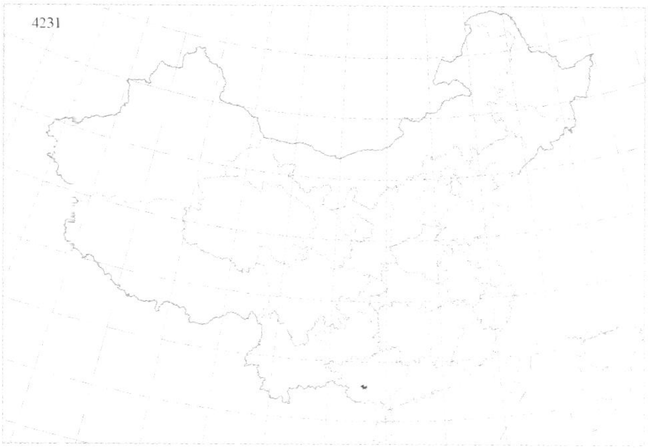

Fabaceae 豆科 *Desmodium* 山蚂蝗属

Desmodium velutinum (Willd.) DC. 绒毛山蚂蝗

Shrubs or subshrubs

18.8 (8.6~24.8); 11.7 (1.3~20.1); 24.1 (15.0~28.9); 18.8 (8.7~24.8); 166.0 (52.6~237.6); 0.1 (0.0~8.8); 952 (574~1385) ❙ 1514 (868~3495); 755 (312~1847); 94 (41~897) ❙ 62.2 (-34.3~246.5); 927 (574~1329); 341.0 (119.4~659.0)

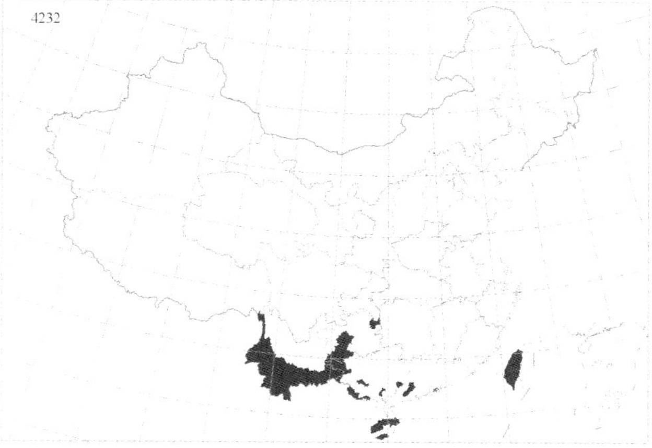

Fabaceae 豆科 *Desmodium* 山蚂蝗属

Desmodium yunnanense Franch. 云南山蚂蝗

Shrubs

14.5 (5.4~18.9); 7.7 (-2.1~12.5); 19.8 (12.3~24.3); 14.6 (5.8~19.0); 116.3 (31.5~167.8); 1.6 (0.0~25.6); 745 (481~929) ❙ 1016 (741~1454); 586 (411~792); 38 (12~70) ❙ 38.6 (-14.2~99.9); 709 (481~851); 323.5 (207.5~487.2)

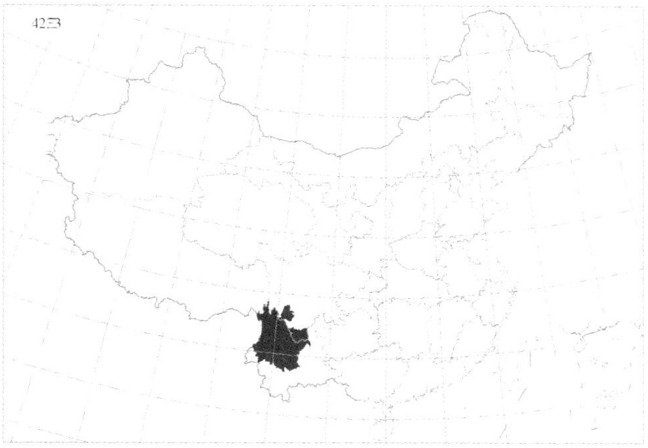

Fabaceae 豆科 *Desmodium* 山蚂蝗属

Desmodium zonatum Miq. 单叶拿身草

Small shrubs

19.8 (14.1~24.8); 13.0 (3.5~20.1); 24.9 (19.1~28.7); 19.8 (14.2~24.8); 178.2 (111.5~237.6); 0.0 (0.0~1.7); 1009 (727~1385) ❙ 1650 (1014~3495); 719 (447~1388); 134 (41~897) ❙ 68.2 (-12.0~224.4); 972 (727~1212); 384.5 (189.5~659.0)

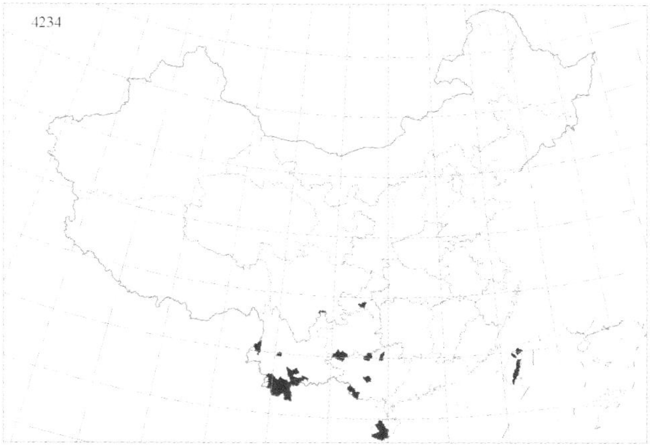

Fabaceae 豆科 *Dicerma* 两节豆属

Dicerma biarticulatum Gagnep. 两节豆

Subshrubs

24.0 (22.2~24.8); 18.9 (17.2~20.1); 27.5 (25.7~28.4); 24.0 (22.3~24.8); 228.0 (207.2~237.6); 0.0 (0.0~0.0); 1304 (1127~1385) ❙ 1462 (1179~1666); 498 (392~597); 48 (41~58) ❙ 13.1 (-11.9~47.8); 1120 (1055~1212); 523.9 (455.2~620.0)

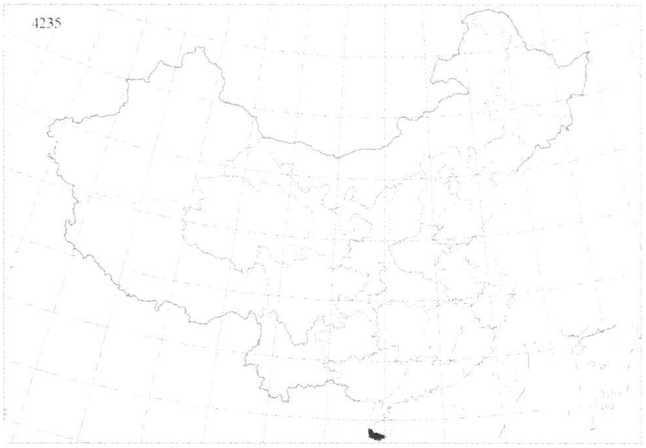

Fabaceae 豆科 *Dunbaria* 野扁豆属

Dunbaria circinalis (Benth.) Baker 卷圈野扁豆

Vines

19.0; 13.1; 22.8; 19.0; 168.5; 0.0; 908 ❙ 1598; 894; 67 ❙ 75.9; 902; 497.8

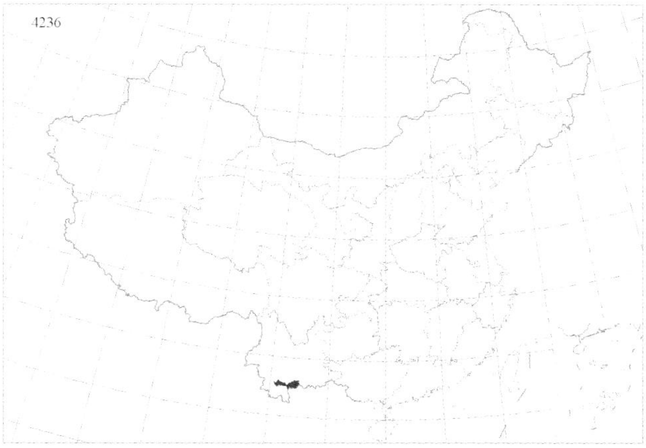

Fabaceae 豆科 *Dysolobium* 镰瓣豆属
Dysolobium grande (Wall. ex Benth.) Prain 镰瓣豆
Vines

17.0 (14.9~19.1); 8.7 (7.6~10.1); 23.3 (20.1~26.6); 17.1 (14.9~19.1); 144.6 (119.1~169.2); 0.0 (0.0~0.0); 852 (748~969) ❙ 1158 (989~1334); 632 (571~720); 53 (43~64) ❙ 36.0 (25.4~59.6); 851 (748~969); 262.2 (189.5~321.8)

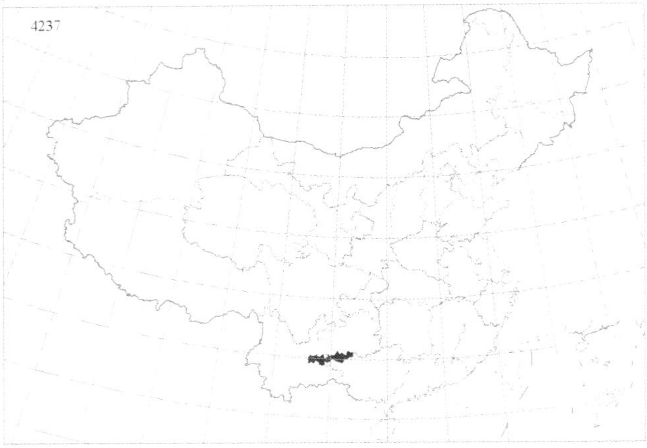

Fabaceae 豆科 *Entada* 榼藤子属
Entada parvifolia Merr. 小叶榼藤
Scandent shrubs

22.2; 17.7; 25.5; 22.2; 206.7; 0.0; 1123 ❙ 2992; 1847; 107 ❙ 166.5; 1123; 296.9

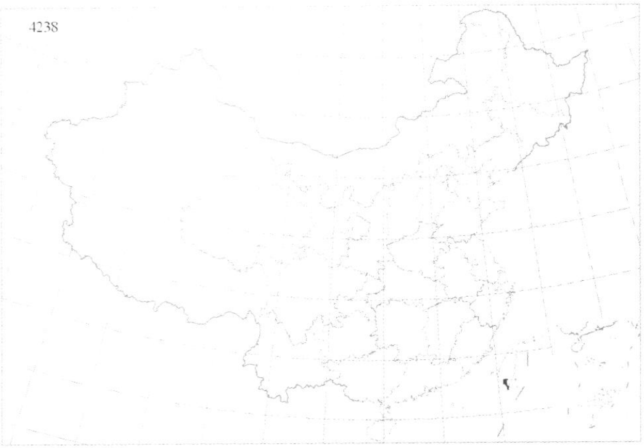

Fabaceae 豆科 *Entada* 榼藤子属
Entada phaseoloides (L.) Merr. 榼藤
Large climbers, evergreen

20.1 (5.3~24.8); 12.7 (-2.4~20.1); 25.8 (12.2~29.2); 20.2 (5.7~24.8); 182.6 (30.5~237.6); 0.3 (0.0~26.2); 1049 (479~1385) ❙ 1571 (868~2993); 746 (312~1847); 103 (31~547) ❙ 54.1 (-34.3~246.5); 1026 (479~1360); 345.0 (122.7~659.0)

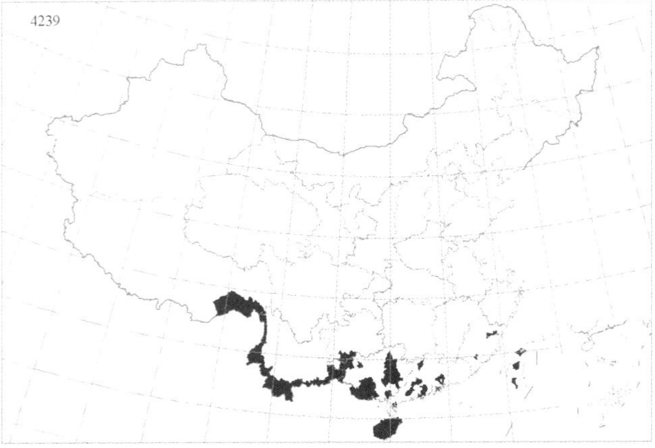

Fabaceae 豆科 *Entada* 榼藤子属
Entada rheedii Sprengel 眼镜豆
Large climbers, evergreen

18.2 (13.3~20.8); 12.1 (5.4~15.3); 22.3 (19.1~25.0); 18.3 (13.4~20.8); 159.4 (100.4~189.2); 0.0 (0.0~0.0); 881 (706~1019) ❙ 1496 (939~2713); 808 (524~1412); 66 (39~188) ❙ 70.9 (6.5~214.4); 859 (706~1019); 422.3 (207.7~659.0)

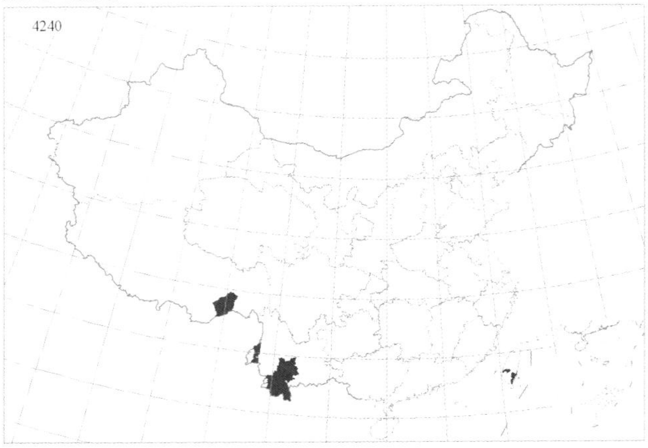

Fabaceae 豆科 *Eremosparton* 无叶豆属
Eremosparton songoricum (Litv.) Vassilcz. 准噶尔无叶豆
Shrubs, deciduous

6.1 (4.9~7.6); -14.0 (-15.0~-13.1); 23.1 (21.9~25.2); 9.6 (8.8~10.7); 80.0 (71.1~93.1); 66.4 (61.0~71.6); 658 (620~721) ❙ 178 (144~199); 73 (63~90); 17 (14~21) ❙ -72.9 (-76.8~-68.6); 178 (144~199); 53.7 (41.1~69.4)

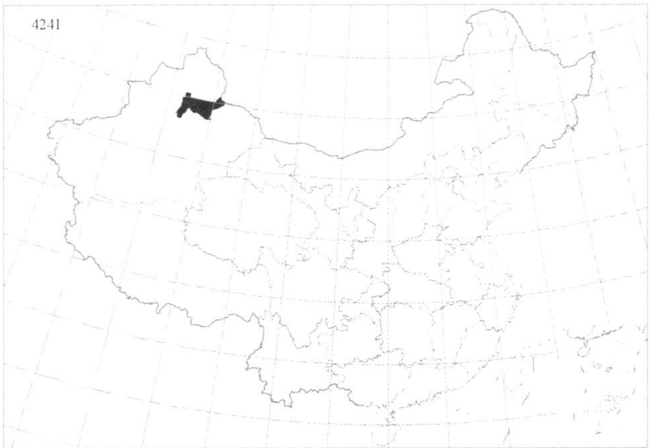

Fabaceae 豆科 *Erythrina* 刺桐属
Erythrina arborescens Roxb. 鹦哥花
Trees

16.4 (-1.1~24.8); 8.9 (-9.1~20.1); 22.3 (6.2~28.8); 16.5 (2.0~24.8); 138.7 (2.5~237.6); 1.1 (0.0~75.6); 847 (334~1385) ❙ 1192 (321~1876); 622 (193~1007); 53 (15~173) ❙ 42.7 (-34.3~145.8); 823 (321~1360); 318.8 (79.3~659.0)

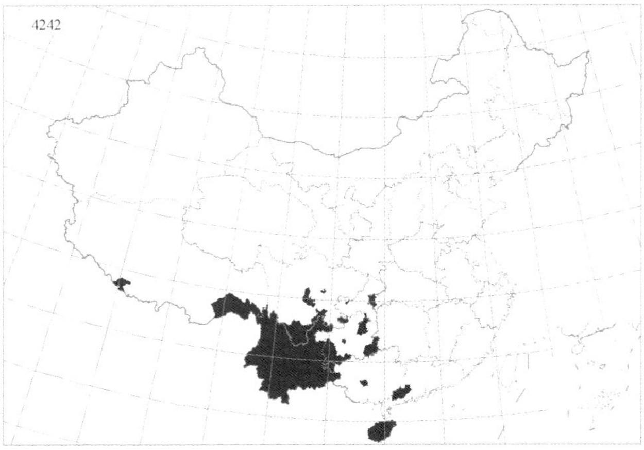

Fabaceae 豆科 *Erythrina* 刺桐属
Erythrina stricta Roxb. 劲直刺桐
Trees

18.1 (5.3~22.7); 11.4 (-2.4~15.1); 22.8 (12.2~28.6); 18.1 (5.7~22.8); 158.2 (30.5~213.1); 0.6 (0.0~26.2); 895 (479~1221) ❙ 1298 (797~2435); 719 (411~1428); 57 (23~130) ❙ 45.2 (-14.2~145.8); 874 (479~1221); 368.5 (218.0~659.0)

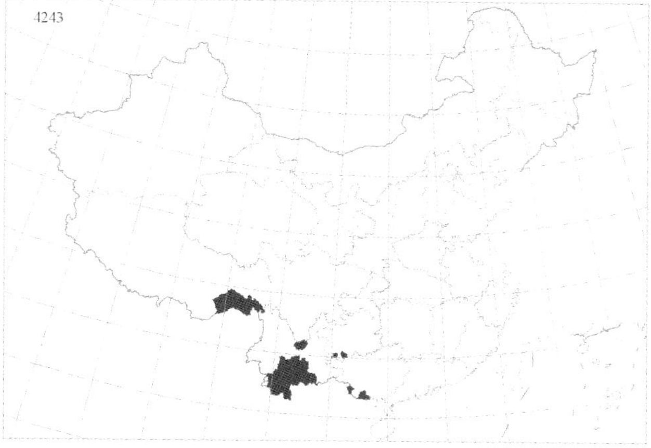

Fabaceae 豆科 *Erythrina* 刺桐属
Erythrina subumbrans (Hassik.) Merr. 翅果刺桐
Trees

19.5 (18.2~21.0); 13.7 (12.4~15.1); 23.1 (21.6~25.6); 19.5 (18.2~21.0); 174.2 (158.6~191.8); 0.0 (0.0~0.0); 938 (861~1053) ❙ 1535 (1428~1751); 796 (655~996); 64 (52~77) ❙ 64.1 (44.5~88.8); 927 (839~1053); 526.0 (427.9~659.0)

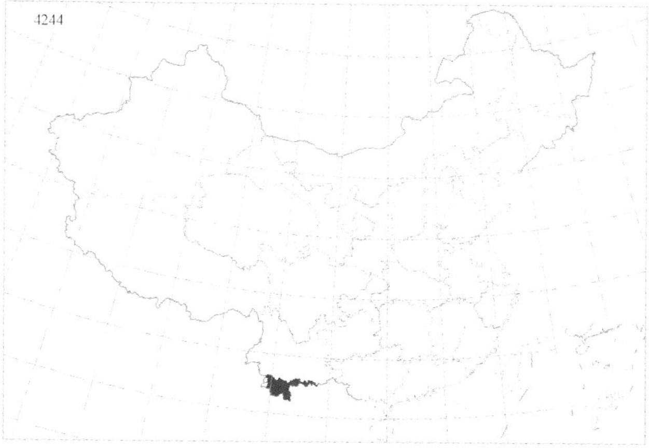

Fabaceae 豆科 *Erythrina* 刺桐属
Erythrina variegata L. 刺桐
Trees

21.0 (14.9~24.8); 13.5 (4.8~20.1); 27.1 (23.9~28.9); 21.1 (15.0~24.8); 192.9 (119.9~237.6); 0.0 (0.0~0.2); 1117 (781~1385) ❙ 1499 (868~2993); 642 (312~1847); 102 (41~547) ❙ 36.8 (-34.3~203.4); 1073 (781~1360); 309.2 (157.7~620.0)

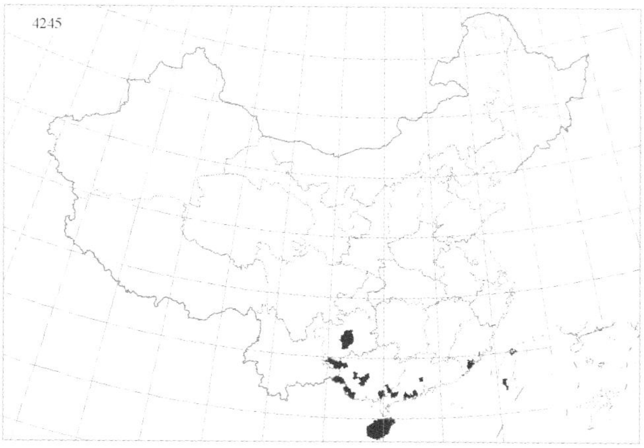

Fabaceae 豆科 *Erythrina* 刺桐属
Erythrina yunnanensis Tsai et Yu ex S. Lee 云南刺桐
Trees

15.5; 8.2; 21.1; 15.6; 126.9; 0.0; 780 ❙ 1552; 882; 55 ❙ 103.0; 780; 469.5

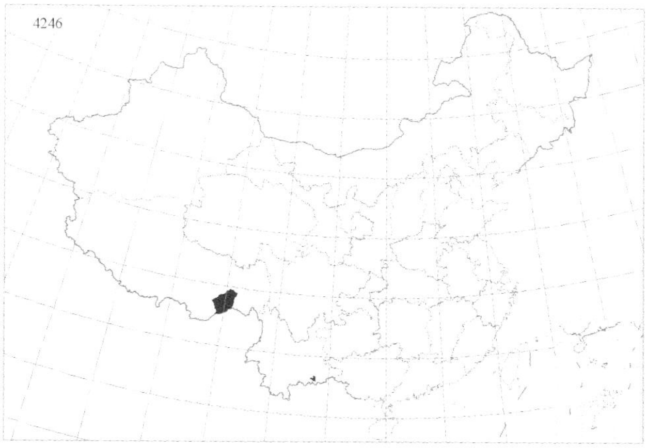

Fabaceae 豆科 *Erythrophleum* 格木属
Erythrophleum fordii Oliv. 格木
Trees

21.1 (15.9~24.8); 13.7 (4.9~20.1); 27.1 (20.1~29.0); 21.2 (16.0~24.8); 193.9 (131.7~237.6); 0.0 (0.0~0.1); 1110 (773~1385) ❙ 1743 (858~3495); 767 (312~1847); 146 (41~897) ❙ 62.4 (-34.3~246.5); 1078 (773~1360); 281.6 (119.4~620.0)

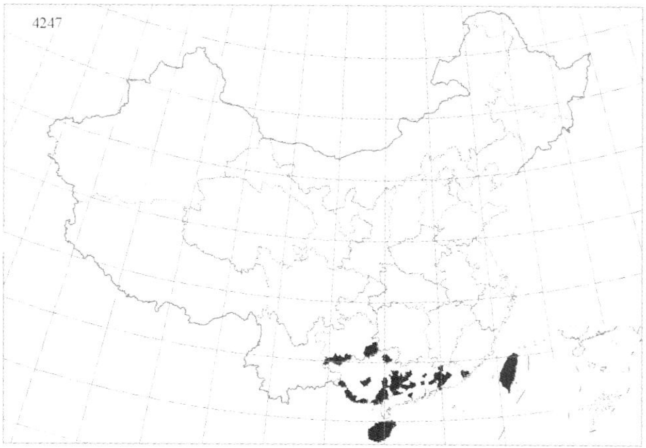

Fabaceae 豆科 *Euchresta* 山豆根属
Euchresta formosana (Hayata) Ohwi 台湾山豆根
Shrubs

19.5 (16.2~24.1); 13.9 (11.0~18.9); 24.0 (20.1~28.0); 19.5 (16.2~24.1); 174.1 (134.2~229.7); 0.0 (0.0~0.0); 972 (773~1322) ❙ 2508 (1751~2993); 1155 (900~1594); 255 (47~547) ❙ 171.5 (32.5~246.5); 926 (773~1095); 248.3 (154.2~289.1)

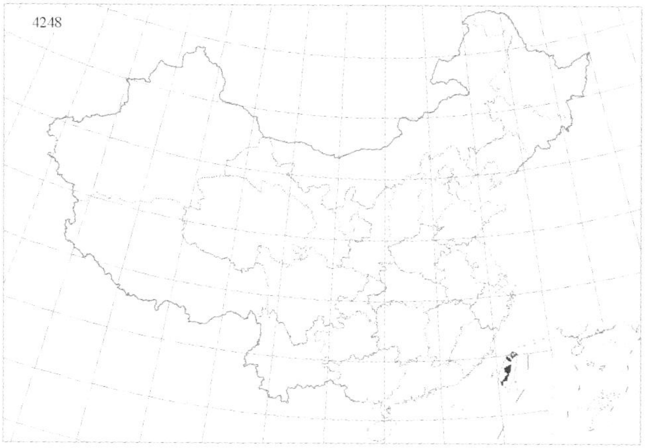

Fabaceae 豆科 *Euchresta* 山豆根属
Euchresta horsfieldii (Lesch.) Benn. 伏毛山豆根
Shrubs

18.4 (13.3~21.0); 11.8 (5.4~15.1); 22.9 (19.6~25.6); 18.4 (13.4~21.0); 160.9 (100.4~191.8); 0.0 (0.0~0.0); 899 (706~1053) ❙ 1515 (1368~1736); 787 (655~1007); 61 (51~77) ❙ 71.5 (44.5~145.8); 891 (706~1053); 500.8 (359.7~659.0)

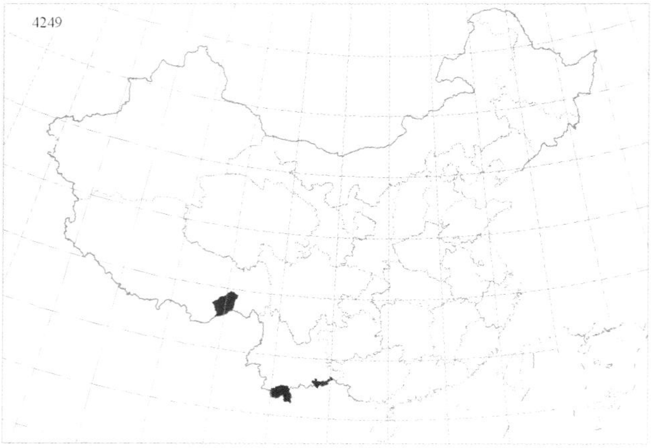

Fabaceae 豆科 *Euchresta* 山豆根属
Euchresta japonica Hook. f. ex Regel 山豆根
Shrubs

17.1 (12.9~24.1); 6.3 (1.9~18.0); 27.1 (23.4~29.3); 17.2 (12.9~24.2); 146.4 (101.2~230.2); 0.4 (0.0~5.9); 910 (723~1329) ❙ 1572 (1236~1820); 558 (427~712); 181 (51~246) ❙ 74.2 (27.5~128.9); 906 (723~1329); 231.4 (186.9~282.2)

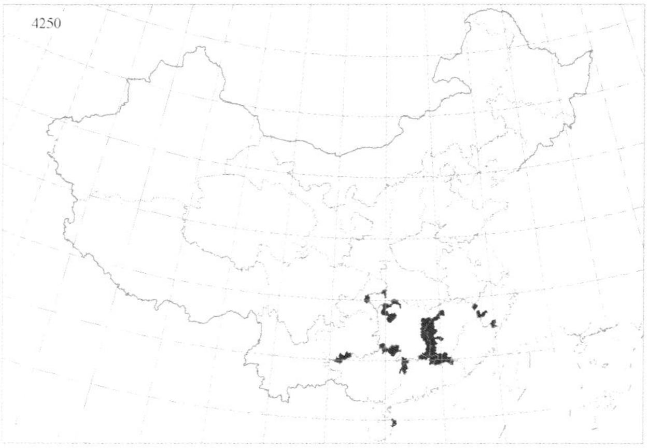

Fabaceae 豆科 *Euchresta* 山豆根属
Euchresta tubulosa Dunn 管萼山豆根
Shrubs

15.5 (12.1~20.4); 4.9 (1.1~15.1); 25.4 (22.9~28.4); 15.6 (12.2~20.5); 128.5 (95.0~185.8); 1.8 (0.0~8.5); 828 (696~989) ❙ 1330 (985~1607); 592 (465~932); 91 (47~190) ❙ 62.1 (18.3~107.2); 827 (696~986); 228.3 (135.3~659.0)

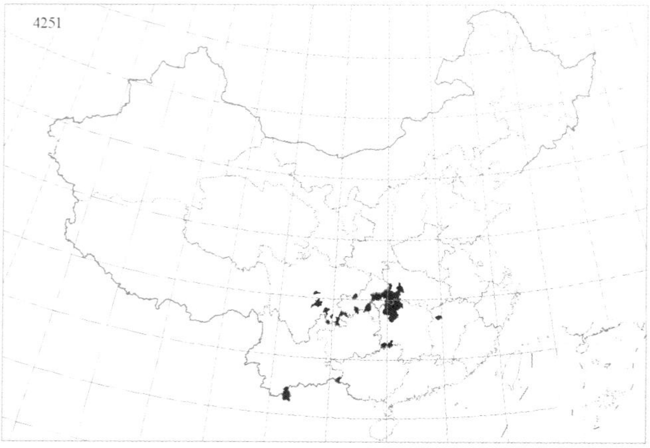

Fabaceae 豆科 *Euchresta* 山豆根属
Euchresta tubulosa var. brevituba C. Chen 短萼山豆根
Shrubs

20.4; 15.1; 23.6; 20.5; 185.8; 0.0; 989 ❙ 1607; 702; 77 ❙ 62.5; 986; 659.0

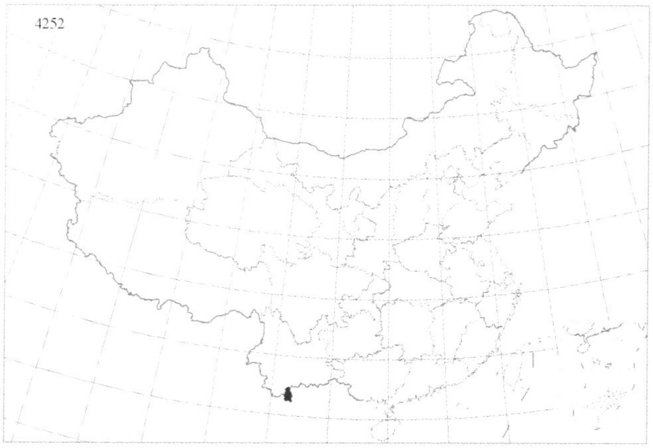

Fabaceae 豆科 *Euchresta* 山豆根属
Euchresta tubulosa var. longiracemosa (S. K. Lee et H. Q. Wen) C. Chen 长序山豆根
Shrubs

19.1; 11.0; 25.0; 19.1; 169.6; 0.0; 945 ❙ 1438; 820; 57 ❙ 52.2; 945; 371.7

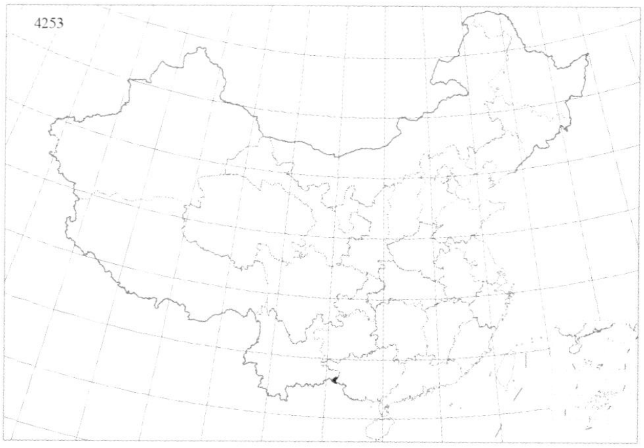

Fabaceae 豆科 *Flemingia* 千斤拔属
Flemingia chappar Buch.-Ham. ex Wall. 墨江千斤拔
Shrubs

19.7; 13.8; 23.3; 19.7; 177.0; 0.0; 951 ❙ 1404; 713; 62 ❙ 47.6; 927; 462.4

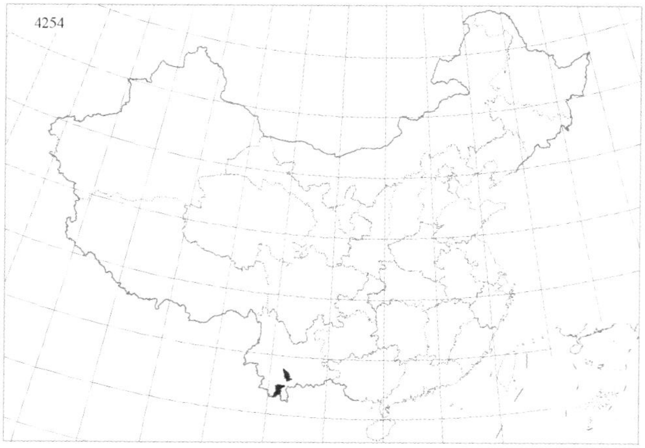

Fabaceae 豆科 *Flemingia* 千斤拔属
Flemingia fluminalis C. B. Clarke ex Prain 河边千斤拔
Small shrubs

15.3 (5.4~22.7); 7.8 (-2.1~15.1); 21.4 (12.3~28.6); 15.4 (5.8~22.8); 125.9 (31.5~213.1); 1.3 (0.0~25.6); 796 (481~1221) ▎1122 (741~2435); 632 (411~1428); 42 (15~130) ▎40.8 (-14.2~99.5); 773 (481~1221); 315.8 (137.9~659.0)

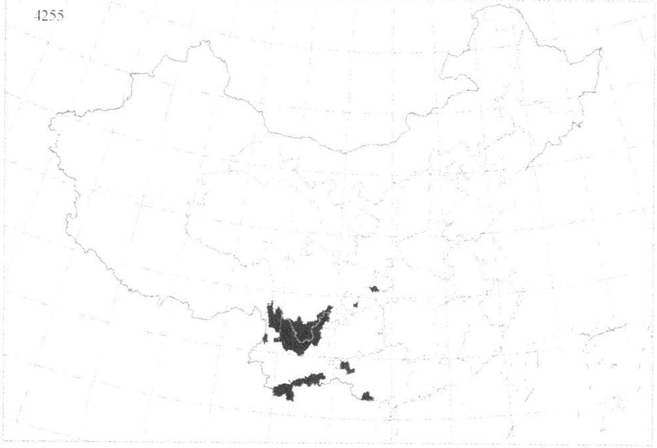

Fabaceae 豆科 *Flemingia* 千斤拔属
Flemingia glutinosa (Prain) Y. T. Wei et S. K. Lee 腺毛千斤拔
Subshrubs

19.0 (16.9~22.2); 12.3 (9.5~15.1); 23.6 (21.4~28.3); 19.0 (16.9~22.3); 168.2 (142.7~207.5); 0.0 (0.0~0.0); 928 (813~1190) ▎1433 (1178~1751); 776 (646~996); 63 (48~87) ▎55.5 (15.4~88.8); 924 (813~1190); 453.8 (322.1~659.0)

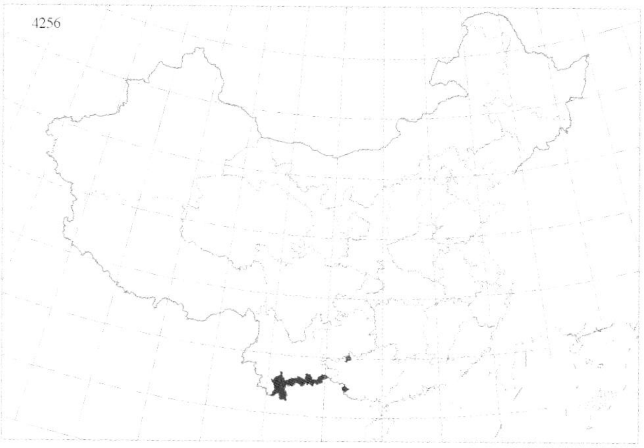

Fabaceae 豆科 *Flemingia* 千斤拔属
Flemingia grahamiana Wight et Arn. 绒毛千斤拔
Shrubs

17.4 (12.6~21.0); 10.9 (6.0~15.1); 21.9 (17.9~25.6); 17.5 (12.6~21.0); 149.5 (91.6~191.8); 0.0 (0.0~0.0); 847 (678~1053) ▎1248 (797~1751); 701 (411~996); 45 (19~77) ▎47.0 (-14.2~112.2); 815 (678~1053); 399.0 (218.0~659.0)

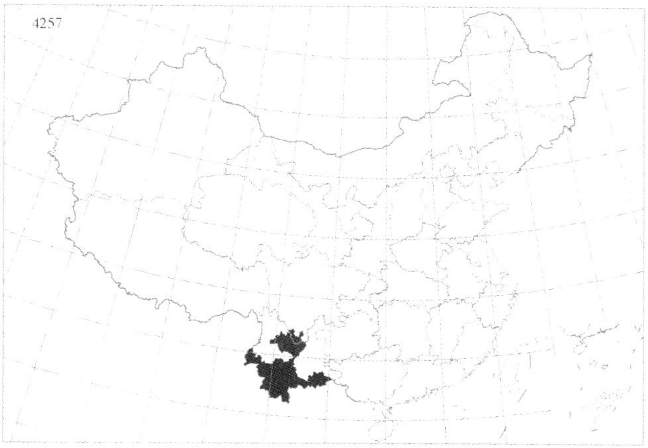

Fabaceae 豆科 *Flemingia* 千斤拔属
Flemingia involucrata Benth. 总苞千斤拔
Shrubs

17.9; 11.4; 21.9; 18.0; 155.7; 0.0; 858 ▎1491; 864; 48 ▎73.9; 848; 552.9

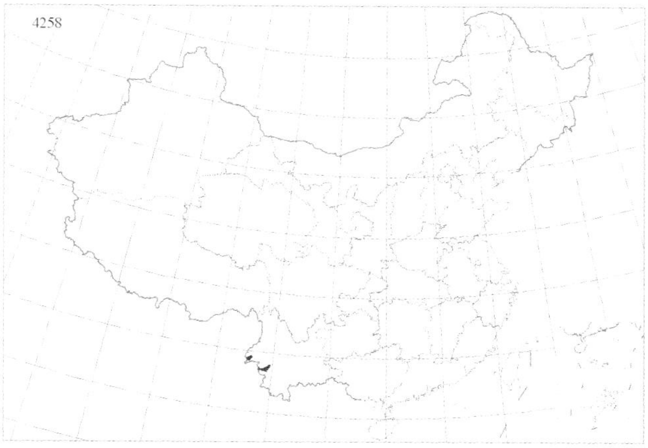

Fabaceae 豆科 *Flemingia* 千斤拔属
Flemingia kweichowensis T. Tang & F. T. Wang ex Y. T. Wei et S. K. Lee 贵州千斤拔
Shrubs

14.7 (11.4~17.7); 5.9 (2.7~10.9); 21.9 (18.8~25.7); 14.7 (11.4~17.8); 117.5 (80.7~153.4); 0.9 (0.0~3.7); 766 (652~869) ▎1099 (863~1368); 560 (479~756); 60 (25~129) ▎42.5 (26.5~60.1); 766 (652~869); 269.7 (190.3~491.5)

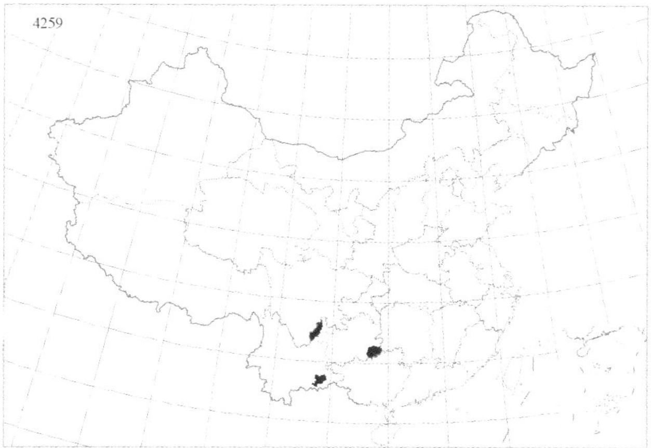

Fabaceae 豆科 *Flemingia* 千斤拔属
Flemingia latifolia Benth. 宽叶千斤拔
Shrubs

15.7 (5.4~22.2); 8.5 (-2.1~15.1); 21.1 (12.3~28.3); 15.7 (5.8~22.3); 129.4 (31.5~207.5); 0.7 (0.0~25.6); 791 (481~1190) ▎1129 (741~1900); 634 (411~1056); 45 (15~125) ▎43.1 (-14.2~112.2); 774 (481~1190); 329.9 (142.9~659.0)

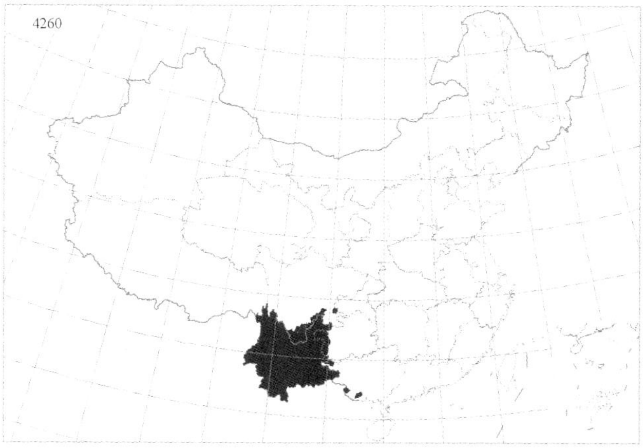

Fabaceae 豆科 *Flemingia* 千斤拔属

Flemingia latifolia var. hainanensis Y. T. Wei et S. K. Lee 海南千斤拔

Shrubs

23.9 (23.0~24.7); 18.4 (17.5~19.8); 27.8 (26.7~28.7); 24.0 (23.1~24.7); 227.5 (217.0~236.5); 0.0 (0.0~0.0); 1304 (1217~1376) ❙ 1530 (1176~1767); 538 (444~619); 69 (44~110) ❙ 17.5 (-12.0~41.8); 1196 (1091~1326); 479.8 (352.9~572.9)

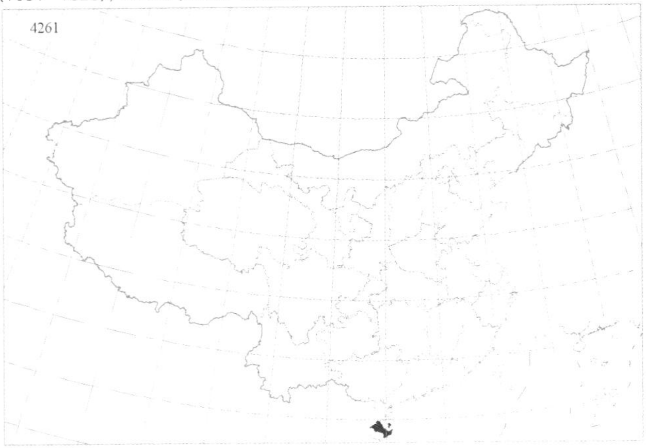

Fabaceae 豆科 *Flemingia* 千斤拔属

Flemingia lineata (L.) Roxb. ex W. T. Aiton 细叶千斤拔

Small shrubs

20.5 (18.6~24.1); 15.1 (12.7~18.9); 24.1 (22.5~28.0); 20.5 (18.7~24.1); 185.8 (163.9~229.7); 0.0 (0.0~0.0); 1011 (888~1322) ❙ 1976 (1428~2992); 1150 (702~1847); 79 (47~127) ❙ 97.6 (32.5~202.5); 978 (888~1123); 419.2 (154.2~659.0)

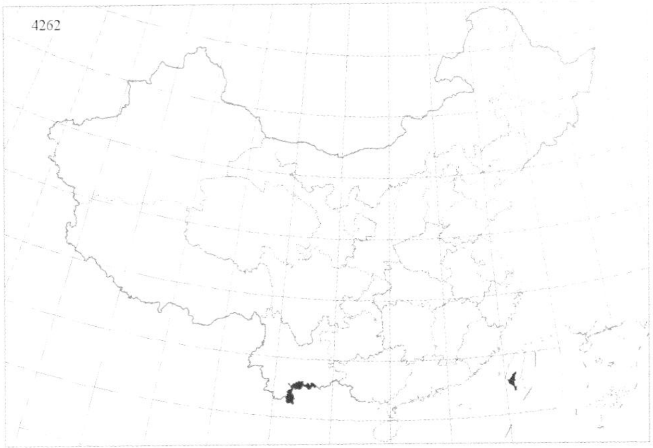

Fabaceae 豆科 *Flemingia* 千斤拔属

Flemingia macrophylla (Willd.) Kuntze ex Prain 大叶千斤拔

Shrubs

17.6 (5.4~24.7); 9.4 (-2.1~19.8); 24.3 (12.3~29.5); 17.6 (5.8~24.7); 151.9 (31.5~236.5); 0.3 (0.0~25.6); 907 (481~1376) ❙ 1366 (741~2993); 653 (392~1847); 94 (15~547) ❙ 50.9 (-14.2~246.5); 893 (481~1299); 286.8 (120.3~659.0)

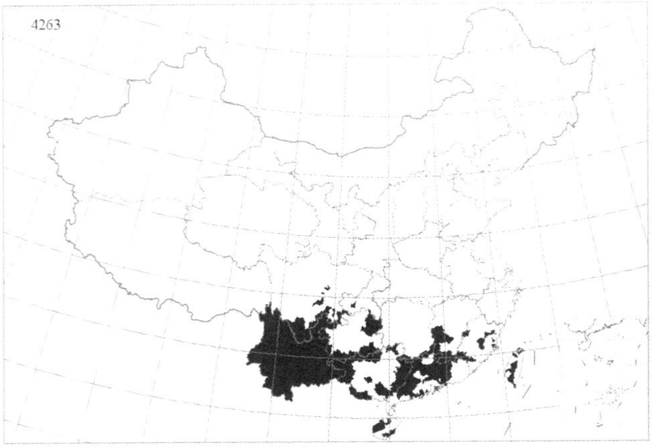

Fabaceae 豆科 *Flemingia* 千斤拔属

Flemingia mengpengensis Y. T. Wei et S. K. Lee 勐板千斤拔

Shrubs

18.9; 13.0; 22.5; 19.0; 167.5; 0.0; 911 ❙ 1507; 759; 62 ❙ 65.6; 910; 546.7

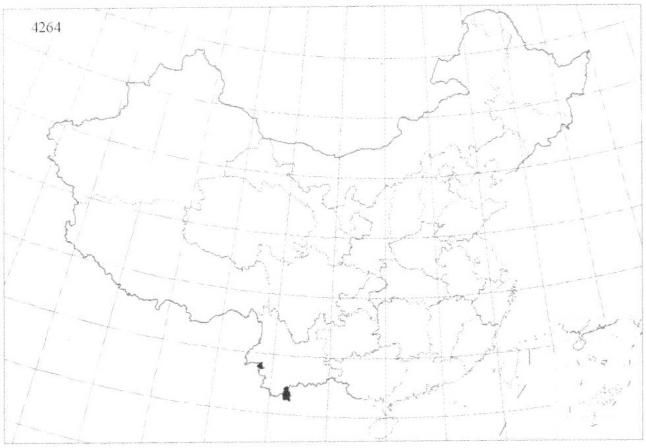

Fabaceae 豆科 *Flemingia* 千斤拔属

Flemingia paniculata Wall. 锥序千斤拔

Shrubs

19.3; 13.7; 22.8; 19.4; 172.6; 0.0; 930 ❙ 1405; 660; 61 ❙ 50.3; 902; 527.1

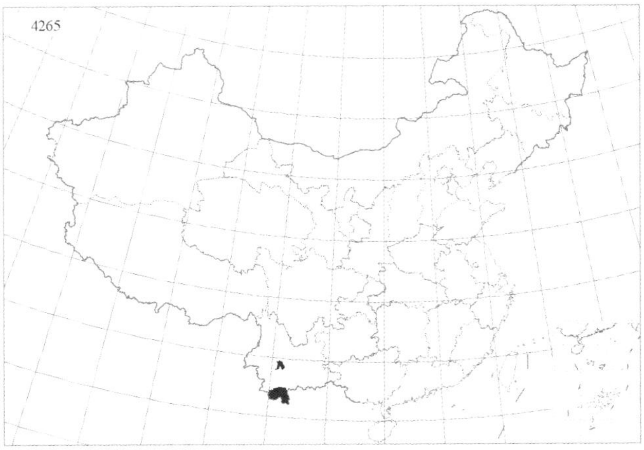

Fabaceae 豆科 *Flemingia* 千斤拔属

Flemingia philippinensis Merr. et Rolfe 千斤拔

Shrubs

18.7 (10.6~24.6); 9.4 (-0.2~19.5); 26.7 (19.1~29.8); 18.7 (10.6~24.7); 164.8 (79.4~235.9); 0.3 (0.0~12.1); 980 (643~1372) ❙ 1547 (868~3495); 630 (312~1847); 151 (18~897) ❙ 59.6 (-34.3~224.4); 963 (643~1331); 247.2 (135.6~548.0)

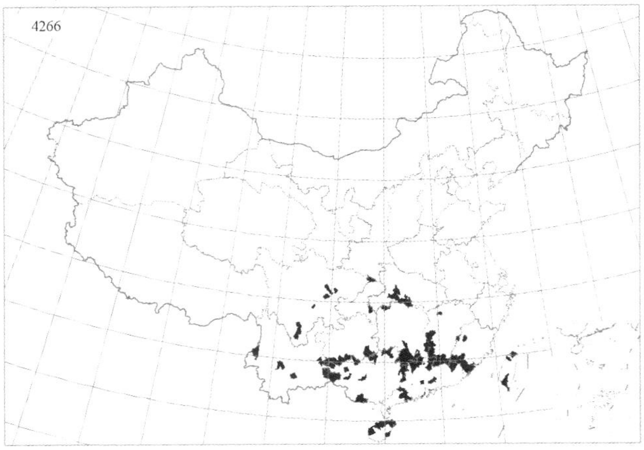

Fabaceae 豆科 *Flemingia* 千斤拔属

Flemingia procumbens Roxb. 矮千斤拔

Shrubs

12.6；6.0；17.9；12.6；91.6；0.0；678 ❙ 1057；610；42 ❙ 56.0；678；349.8

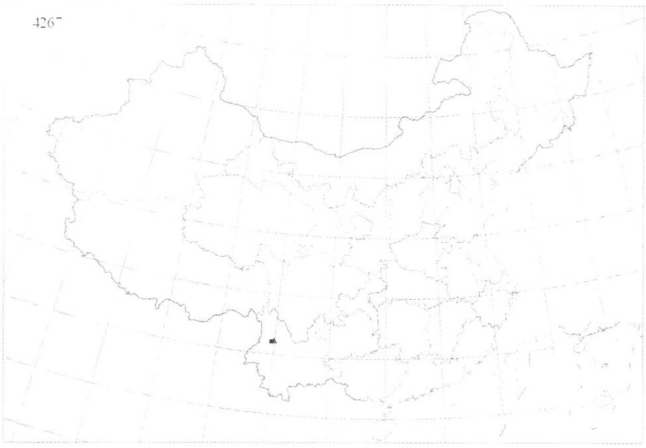

Fabaceae 豆科 *Flemingia* 千斤拔属

Flemingia stricta Roxb. ex W. T. Aiton 长叶千斤拔

Shrubs

18.2；11.9；22.0；18.2；158.7；0.0；868 ❙ 1366；783；45 ❙ 57.4；847；516.7

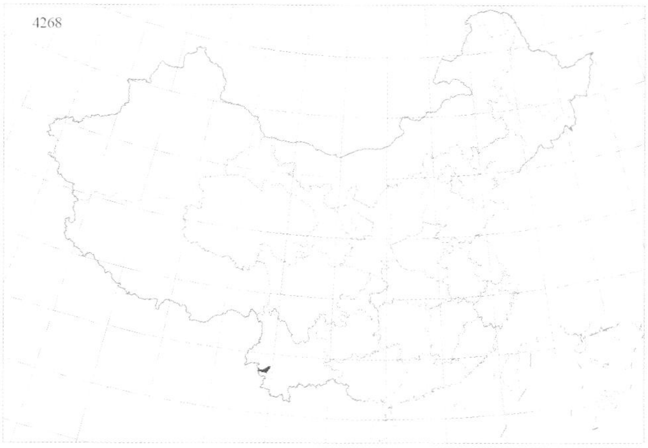

Fabaceae 豆科 *Flemingia* 千斤拔属

Flemingia strobilifera (L.) R. Br. 球穗千斤拔

Shrubs

17.6 (5.4~24.7)；10.1 (-2.1~19.8)；23.3 (12.3~29.0)；17.7 (5.8~24.7)；152.6 (3[illegible].5~236.5)；0.4 (0.0~25.6)；896 (481~1376) ❙ 1314 (741~2992)；686 (41[illegible]~1847)；76 (15~202) ❙ 47.0 (-14.2~202.5)；879 (481~1221)；322.6 (119.4~659.0)

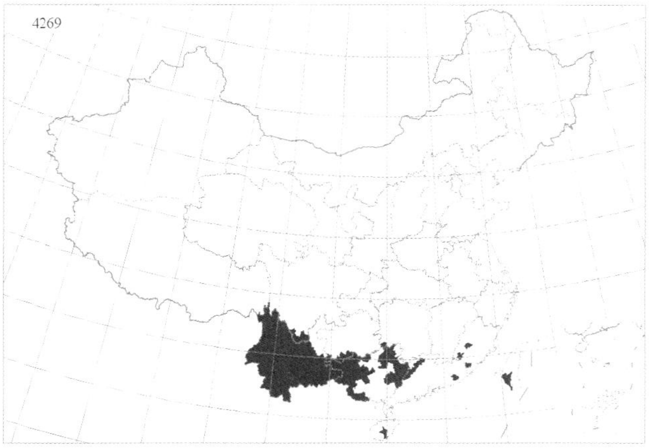

Fabaceae 豆科 *Flemingia* 千斤拔属

Flemingia wallichii Wight et Arn. 云南千斤拔

Shrubs

14.7 (10.1~18.9)；7.4 (2.3~12.0)；20.5 (16.9~24.3)；14.8 (10.2~19.0)；117.5 (68.1~167.8)；0.4 (0.0~5.2)；753 (618~929) ❙ 952 (797~1056)；561 (411~627)；28 (18~46) ❙ 27.8 (-14.2~70.8)；706 (618~786)；269.8 (190.3~322.5)

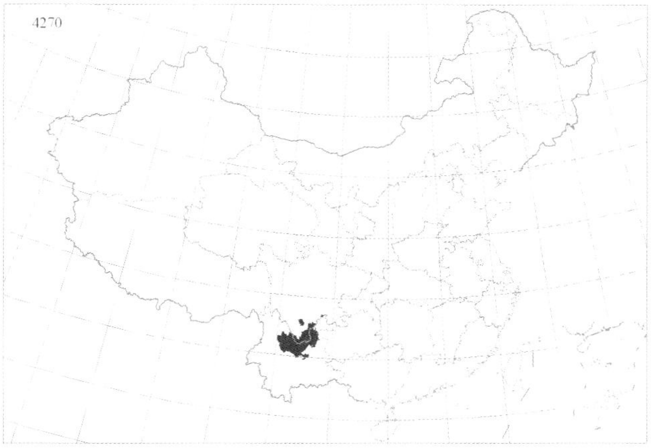

Fabaceae 豆科 *Fordia* 干花豆属

Fordia cauliflora Hemsl. 干花豆

Shrubs

20.3 (15.9~22.6)；12.0 (6.2~15.3)；26.8 (22.3~29.0)；20.4 (16.0~22.6)；184.5 (131.6~211.6)；0.0 (0.0~0.0)；1054 (802~1212) ❙ 1478 (1182~2035)；729 (513~996)；96 (43~191) ❙ 41.2 (5.8~88.8)；1052 (802~1212)；283.5 (122.7~659.0)

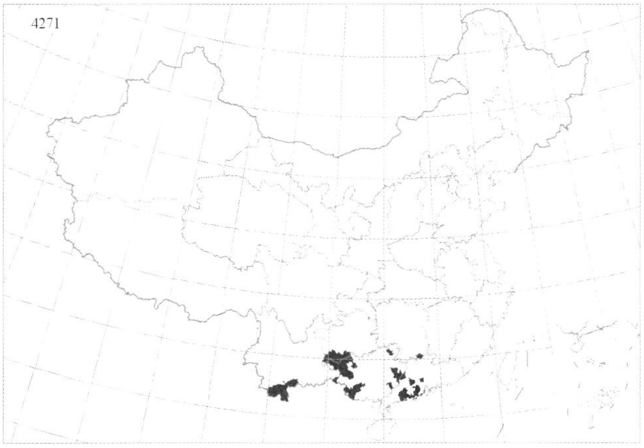

Fabaceae 豆科 *Fordia* 干花豆属

Fordia microphylla Dunn ex Z. Wei 小叶干花豆

Shrubs

18.2 (16.0~21.0)；11.6 (9.1~15.1)；22.6 (20.3~28.2)；18.2 (16.0~21.1)；158.8 (132.4~193.0)；0.0 (0.0~0.0)；881 (780~1114) ❙ 1352 (1007~1751)；751 (539~996)；55 (42~173) ❙ 53.8 (12.5~112.2)；869 (780~1114)；425.8 (220.1~659.0)

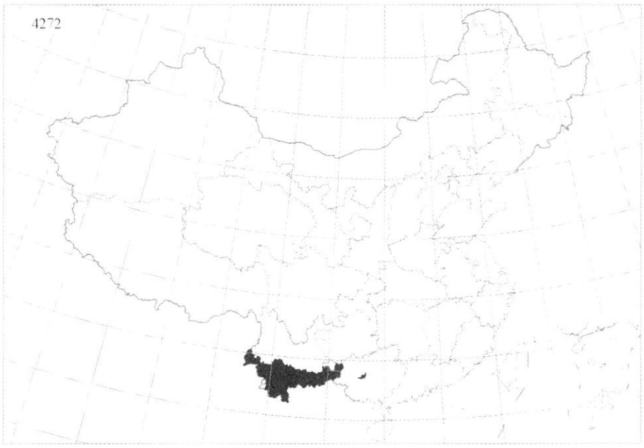

Fabaceae 豆科 *Gleditsia* 皂荚属

Gleditsia australis Hemsl. 小果皂荚

Trees, deciduous

20.9 (14.6~24.8); 12.6 (3.8~20.1); 27.7 (24.6~28.9); 21.0 (14.7~24.8); 191.7 (117.7~237.6); 0.0 (0.0~1.5); 1117 (779~1385) | 1577 (1179~2030); 645 (392~1033); 131 (41~196) | 44.8 (-11.9~101.8); 1082 (779~1329); 288.0 (157.7~620.0)

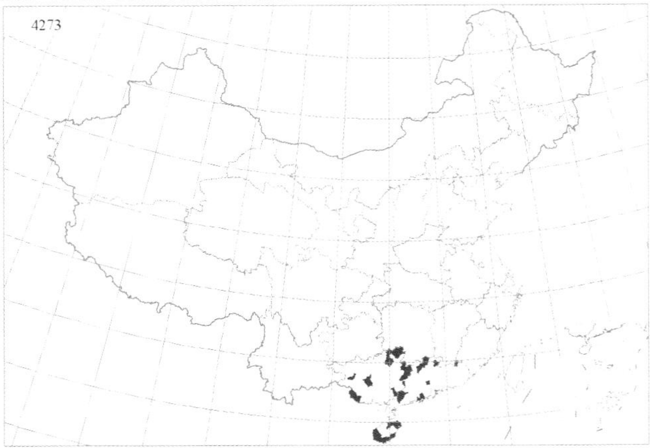

Fabaceae 豆科 *Gleditsia* 皂荚属

Gleditsia fera (Lour.) Merr. 华南皂荚

Trees, deciduous

20.8 (15.6~24.8); 12.6 (4.1~20.1); 27.5 (22.3~28.9); 20.9 (15.6~24.8); 190.9 (128.2~237.6); 0.0 (0.0~0.9); 1104 (839~1385) | 1576 (868~2030); 669 (312~1044); 126 (41~197) | 45.6 (-34.3~85.5); 1073 (839~1329); 294.8 (122.7~659.0)

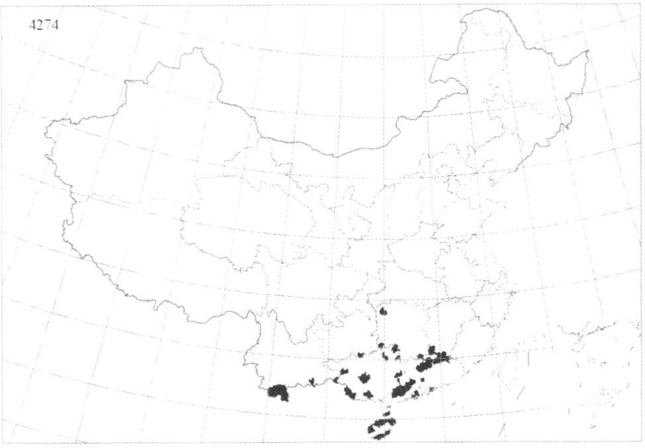

Fabaceae 豆科 *Gleditsia* 皂荚属

Gleditsia japonica Miquel 山皂荚

Trees, deciduous

12.3 (3.1~19.4); -1.9 (-18.8~12.0); 25.1 (18.7~29.9); 13.0 (7.3~19.5); 108.3 (52.8~173.7); 20.0 (0.0~84.9); 768 (535~1038) | 838 (367~1997); 425 (235~718); 59 (8~254) | 7.0 (-40.9~130.7); 682 (367~1038); 188.6 (101.9~383.4)

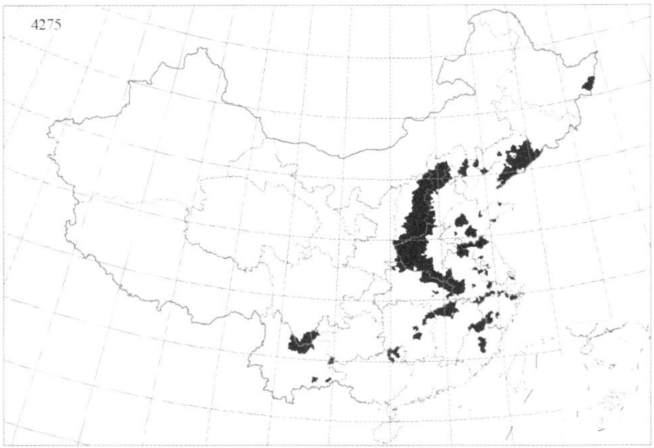

Fabaceae 豆科 *Gleditsia* 皂荚属

Gleditsia japonica var. delavayi (Franch.) L. C. Li 滇皂荚

Trees, deciduous

17.6 (16.1~19.1); 10.2 (4.8~13.0); 23.0 (21.4~26.8); 17.6 (16.2~19.1); 151.3 (134.2~169.7); 0.0 (0.0~0.2); 859 (813~917) | 1222 (958~1428); 659 (463~835); 58 (44~96) | 42.2 (13.3~60.1); 853 (813~917); 361.4 (187.1~506.2)

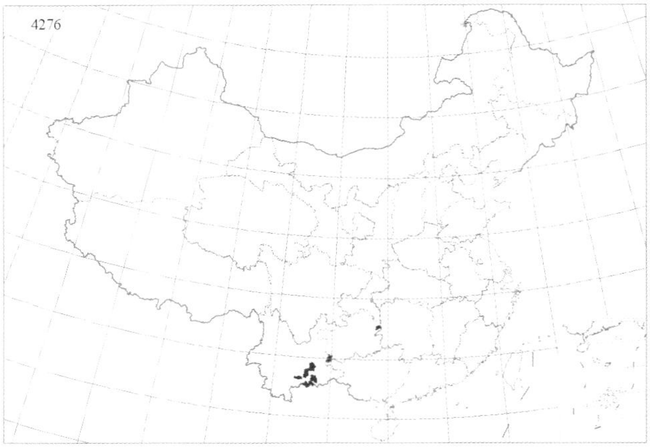

Fabaceae 豆科 *Gleditsia* 皂荚属

Gleditsia japonica var. velutina L. C. Li 绒毛皂荚

Trees, deciduous

17.5; 5.5; 28.7; 17.5; 150.0; 0.0; 944 | 1559; 509; 200 | 65.0; 936; 180.2

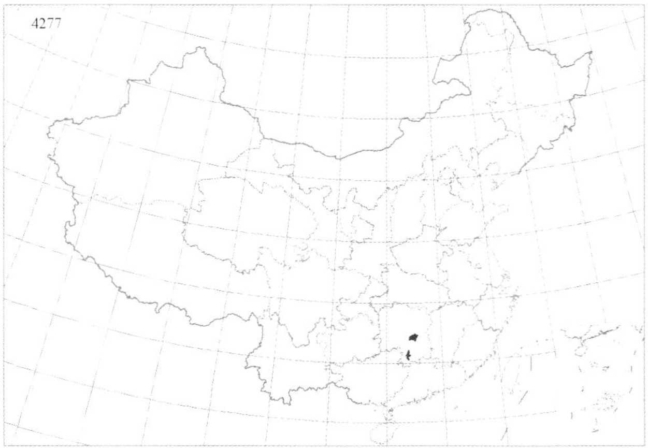

Fabaceae 豆科 *Gleditsia* 皂荚属

Gleditsia microphylla D. A. Gordon ex Isely 野皂荚

Shrubs or small trees, deciduous

10.8 (3.3~17.4); -3.9 (-13.7~6.1); 24.1 (17.5~29.2); 11.7 (6.6~17.4); 96.1 (45.4~149.4); 25.4 (0.0~71.7); 719 (507~950) | 648 (367~1621); 351 (235~616); 30 (8~213) | -10.5 (-40.9~95.5); 604 (367~950); 170.5 (70.3~334.4)

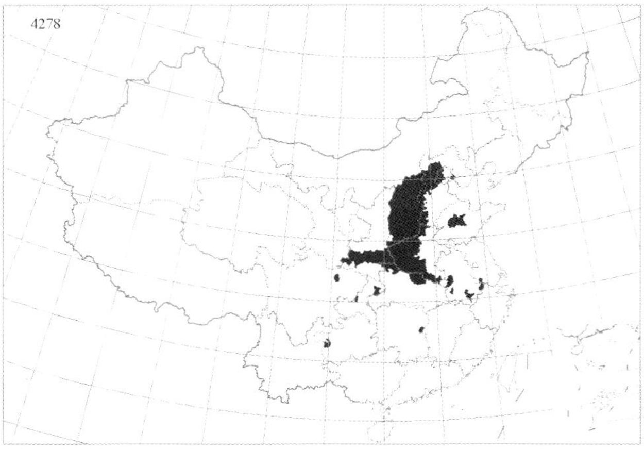

Fabaceae 豆科 *Gleditsia* 皂荚属

Gleditsia sinensis Lam. 皂荚

Trees, deciduous

14.0 (2.0~22.2); 1.0 (-15.9~14.0); 25.9 (10.1~30.1); 14.3 (3.8~22.3); 120.2 (15.4~207.5); 11.1 (0.0~80.4); 817 (413~1190) ❙ 986 (377~2148); 454 (241~1234); 82 (8~259) ❙ 19.2 (-48.6~130.7); 743 (377~1190); 192.1 (70.3~371.7)

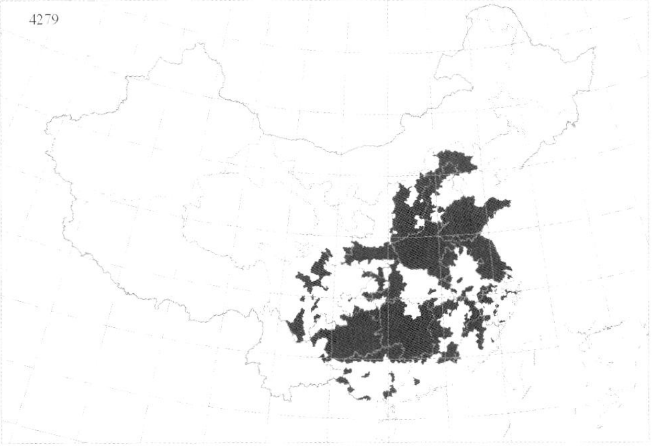

Fabaceae 豆科 *Gleditsia* 皂荚属

Gleditsia triacanthos Mill. 美国皂荚

Trees, deciduous

4.4 (-1.3~9.3); -14.9 (-20.9~-9.3); 19.7 (13.2~27.1); 8.1 (4.6~12.1); 63.5 (26.7~107.6); 70.1 (40.7~103.9); 594 (440~810) ❙ 254 (140~406); 117 (53~211); 21 (12~48) ❙ -53.9 (-82.8~-7.6); 254 (140~406); 152.4 (50.3~297.3)

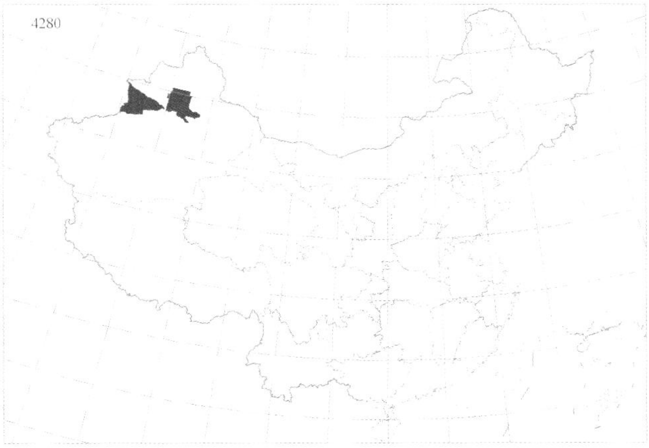

Fabaceae 豆科 *Gymnocladus* 肥皂荚属

Gymnocladus chinensis Baill. 肥皂荚

Trees, deciduous

16.6 (12.0~22.2); 5.4 (0.6~14.2); 27.0 (20.1~30.1); 16.6 (12.1~22.3); 140.3 (92.0~207.5); 0.9 (0.0~9.7); 889 (685~1190) ❙ 1397 (807~2014); 535 (299~954); 145 (37~259) ❙ 57.5 (-6.7~130.7); 877 (685~1190); 207.5 (110.1~551.6)

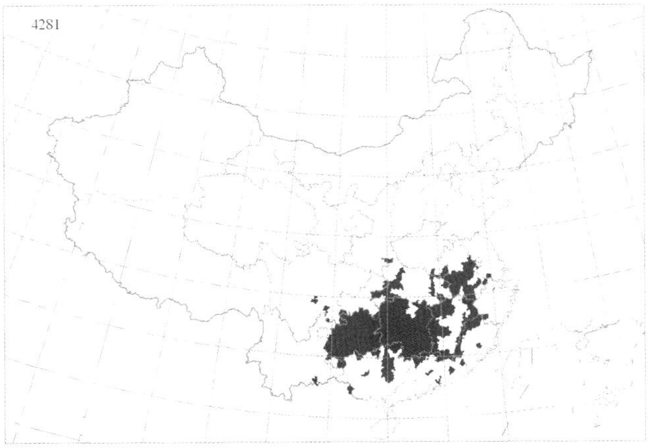

Fabaceae 豆科 *Halimodendron* 铃铛刺属

Halimodendron halodendron (Pall.) Voss 铃铛刺

Shrubs, deciduous

6.5 (-4.1~12.5); -11.9 (-20.9~-5.1); 21.7 (9.8~27.5); 9.5 (2.5~13.2); 76.8 (9.9~114.1); 57.9 (23.1~117.9); 648 (330~822) ❙ 130 (21~450); 63 (15~185); 11 (0~51) ❙ -77.4 (-97.3~3.9); 129 (21~404); 51.6 (1.6~297.3)

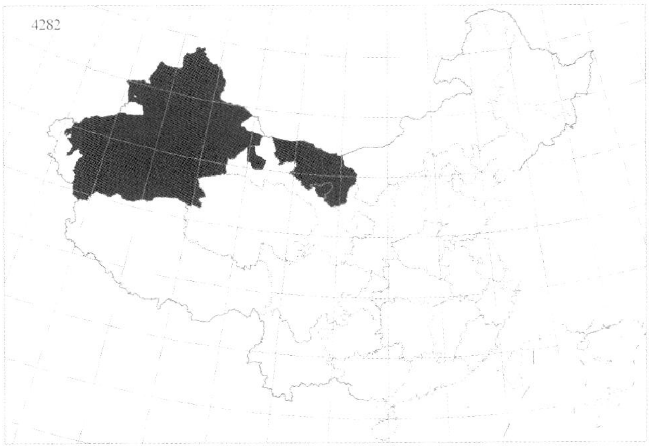

Fabaceae 豆科 *Hylodesmum* 长柄山蚂蝗属

Hylodesmum leptopus (A. Gray ex Benth.) H. Ohashi et R. R. Mill 细长柄山蚂蝗

Shrubs

18.0 (13.3~24.8); 8.4 (3.3~20.1); 26.7 (20.1~29.5); 18.1 (13.3~24.8); 157.2 (101.9~237.6); 0.2 (0.0~2.1); 946 (718~1385) ❙ 1687 (999~3495); 662 (452~1847); 180 (41~897) ❙ 80.6 (1.5~246.5); 935 (718~1221); 249.3 (113.4~659.0)

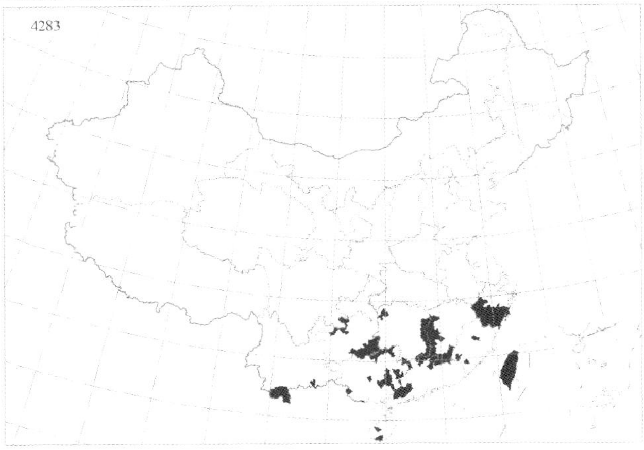

Fabaceae 豆科 *Hylodesmum* 长柄山蚂蝗属

Hylodesmum podocarpum subsp. oxyphyllum (DC.) H. Ohashi et R. R. Mill 尖叶长柄山蚂蝗

Subshrubs

6.1 (1.4~9.5); -13.1 (-19.2~-7.2); 22.4 (18.4~24.8); 9.1 (6.4~10.8); 73.6 (45.8~91.5); 60.1 (35.9~88.8); 621 (501~701) ❙ 825 (651~1073); 515 (413~691); 31 (24~40) ❙ 34.4 (-5.7~74.1); 618 (501~690); 279.5 (164.3~346.5)

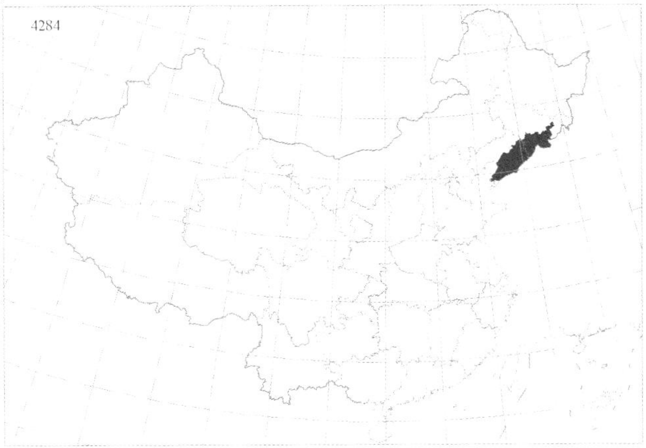

Fabaceae 豆科 *Indigofera* 木蓝属

Indigofera cassioides Rottler ex DC. 椭圆叶木蓝

Shrubs, deciduous

19.2 (17.5~21.3); 10.9 (8.7~13.3); 25.5 (22.1~27.9); 19.2 (17.6~21.3); 170.4 (151.2~196.0); 0.0 (0.0~0.0); 967 (846~1123) ❙ 1271 (1007~1567); 681 (539~776); 67 (44~141) ❙ 31.8 (5.8~54.8); 964 (836~1121); 307.4 (189.5~452.0)

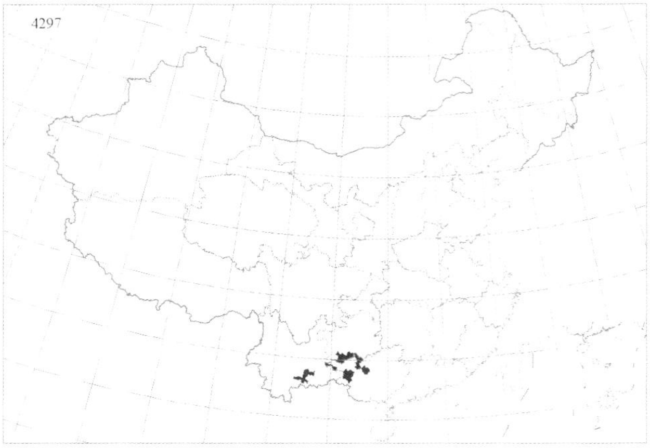

Fabaceae 豆科 *Indigofera* 木蓝属

Indigofera caudata Dunn 尾叶木蓝

Shrubs, deciduous

18.7 (16.8~20.5); 11.5 (7.8~15.1); 23.7 (21.6~26.6); 18.7 (16.8~20.5); 164.5 (141.5~185.9); 0.0 (0.0~0.0); 913 (832~992) ❙ 1339 (1049~1751); 720 (598~996); 58 (44~77) ❙ 46.8 (21.7~88.8); 901 (783~986); 382.5 (189.5~659.0)

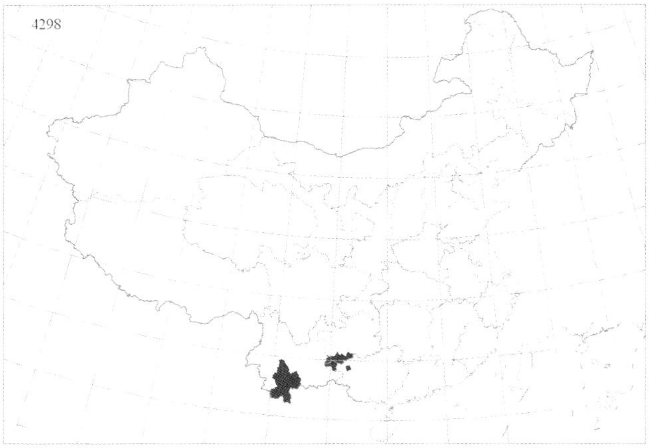

Fabaceae 豆科 *Indigofera* 木蓝属

Indigofera chenii S. S. Chien 南京木蓝

Shrubs, deciduous

15.5; 2.3; 28.1; 15.6; 130.7; 3.9; 879 ❙ 1002; 448; 114 ❙ 14.0; 879; 134.1

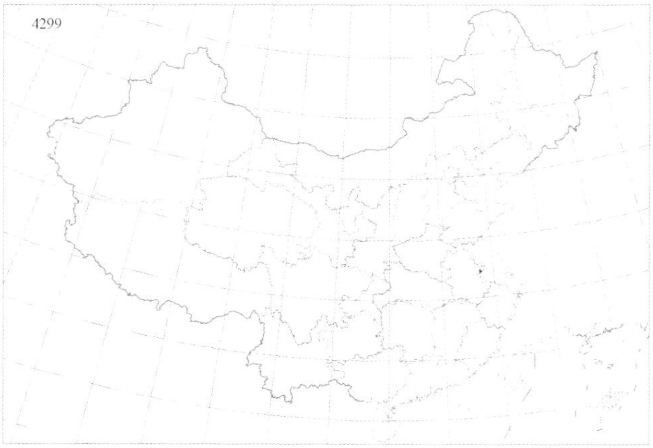

Fabaceae 豆科 *Indigofera* 木蓝属

Indigofera colutea (Burm. f.) Merr. 疏花木蓝

Herbs or subshrubs, deciduous

23.8 (23.0~24.8); 17.7 (15.1~20.1); 28.3 (27.5~29.0); 23.9 (23.0~24.8); 226.5 (216.2~237.6); 0.0 (0.0~0.0); 1303 (1247~1385) ❙ 1343 (868~1745); 553 (312~997); 63 (41~102) ❙ 3.6 (-34.3~39.9); 1111 (868~1264); 366.4 (198.9~527.5)

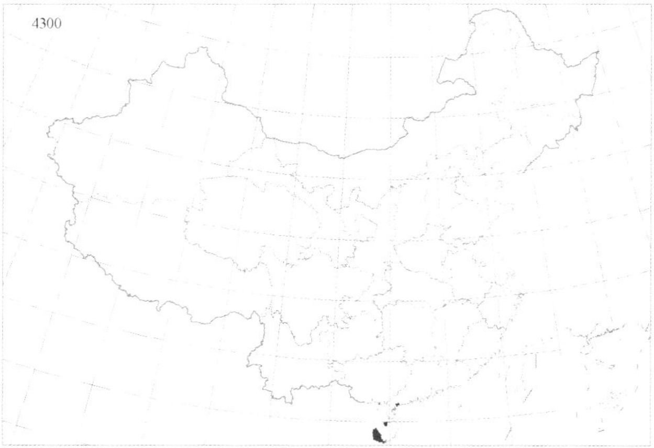

Fabaceae 豆科 *Indigofera* 木蓝属

Indigofera cylindracea Wall. ex Baker 筒果木蓝

Small shrubs, deciduous

-1.6; -10.3; 6.6; 2.0; 3.3; 82.5; 330 ❙ 315; 222; 15 ❙ -4.6; 315; 67.5

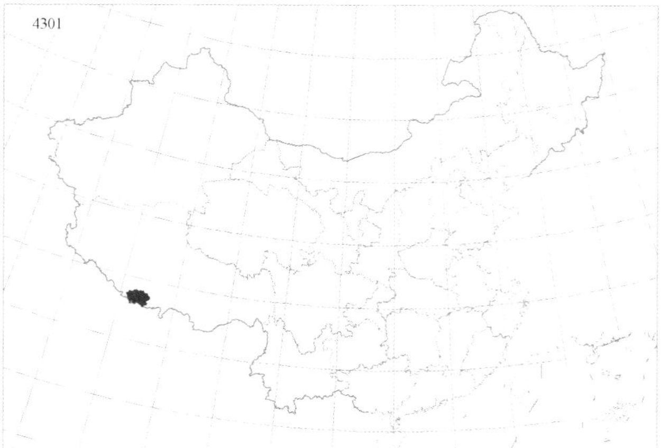

Fabaceae 豆科 *Indigofera* 木蓝属

Indigofera decora Lindl. 庭藤

Shrubs, deciduous

16.6 (9.3~22.4); 5.4 (-2.9~15.2); 27.0 (20.1~29.8); 16.7 (9.5~22.7); 142.4 (69.8~212.7); 2.2 (0.0~20.5); 897 (603~1216) ❙ 1455 (512~2030); 555 (258~1056); 156 (20~259) ❙ 62.0 (-35.3~130.7); 887 (512~1216); 228.4 (121.6~330.0)

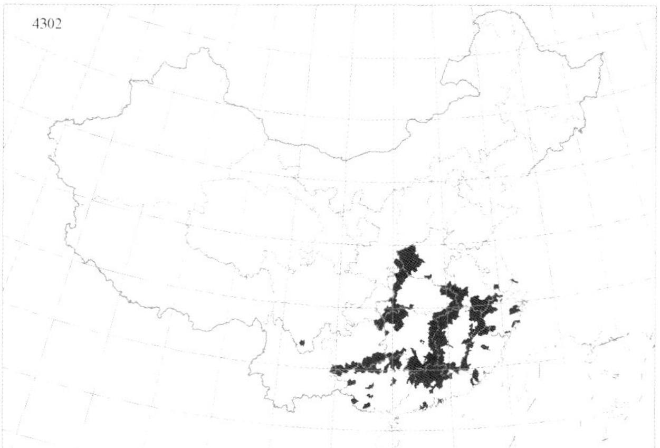

Fabaceae 豆科 *Indigofera* 木蓝属

Indigofera decora var. chalara (Craib) Y. Y. Fang et C. Z. Zheng 兴山木蓝 Shrubs

12.0；0.6 23.1；12.1；94.7；9.7；697 ❙ 1118；482；75 ❙ 60.4；697；247.3

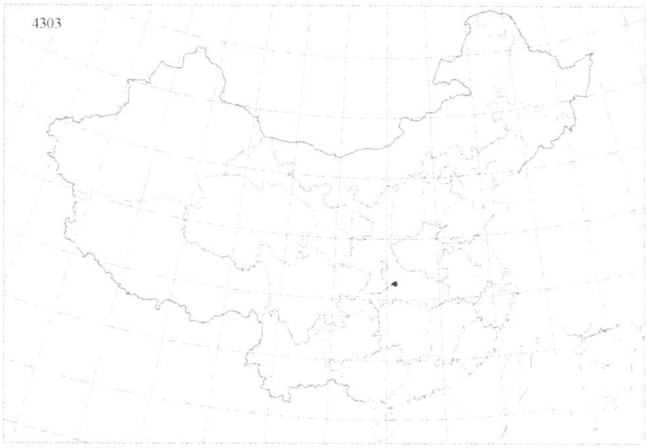

Fabaceae 豆科 *Indigofera* 木蓝属

Indigofera decora var. cooperi (Craib) Y. Y. Fang et C. Z. Zheng 宁波木蓝 Shrubs, deciduous

16.4 (14.4~19.4)；5.2 (2.1~9.3)；27.1 (24.7~29.1)；16.4 (14.5~19.4)；137.8 (118.7~173.4)；0.9 (0.0~4.9)；876 (789~1029) ❙ 1660 (1363~1950)；593 (463~713)；201 (181~242) ❙ 90.8 (53.5~147.0)；875 (789~1029)；247.0 (197.9~319.9)

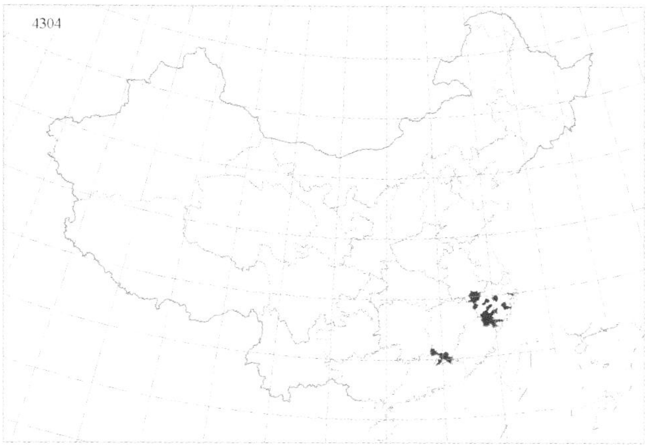

Fabaceae 豆科 *Indigofera* 木蓝属

Indigofera decora var. ichangensis (Craib) Y. Y. Fang et C. Z. Zheng 宜昌木蓝 Shrubs, deciduous

15.3 (10.2~20.2)；3.3 (-2.9~11.1)；26.6 (20.1~29.3)；15.4 (10.6~20.3)；128.5 (83.1~183.3)；4.1 (0.0~20.5)；846 (656~1056) ❙ 1211 (626~1791)；504 (316~685)；108 (20~221) ❙ 42.6 (-22.2~115.7)；830 (626~1056)；206.1 (139.1~330.0)

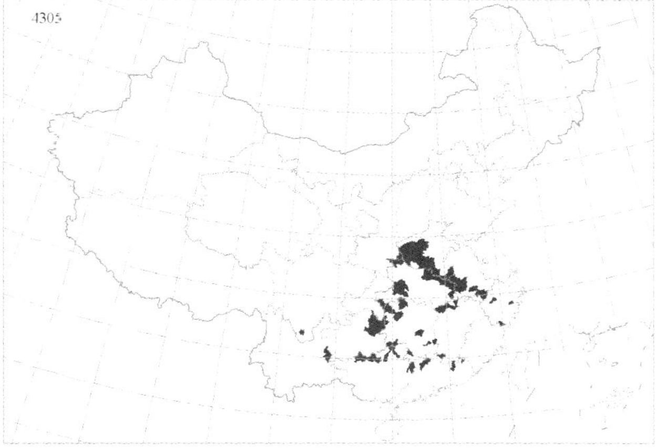

Fabaceae 豆科 *Indigofera* 木蓝属

Indigofera delavayi Franch. 滇木蓝
Shrubs, deciduous

13.0 (2.2~18.9)；5.9 (-6.3~12.0)；18.7 (9.5~23.9)；13.1 (3.9~19.0)；101.5 (15.8~167.8)；4.7 (0.0~48.6)；702 (411~929) ❙ 928 (729~1057)；556 (411~627)；26 (11~42) ❙ 35.6 (-14.2~77.4)；660 (411~760)；284.1 (207.5~351.7)

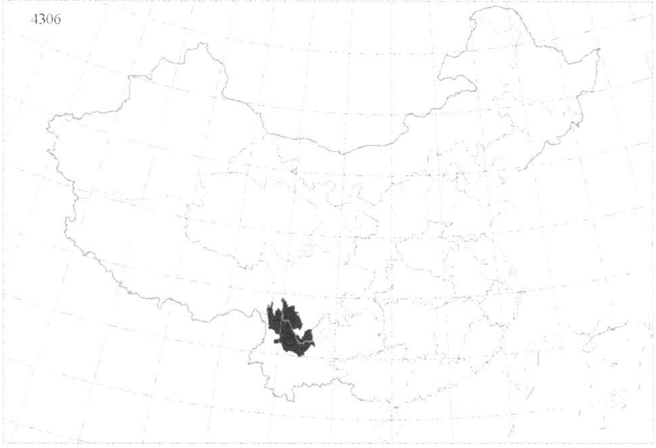

Fabaceae 豆科 *Indigofera* 木蓝属

Indigofera densifructa Y. Y. Fang et C. Z. Zheng 密果木蓝
Shrubs, deciduous

16.6 (14.6~17.8)；4.9 (3.7~7.1)；27.5 (24.6~29.7)；16.6 (14.7~17.8)；139.7 (117.7~154.1)；0.5 (0.0~1.5)；889 (779~977) ❙ 1460 (1211~1679)；486 (419~589)；180 (89~221) ❙ 64.6 (44.3~101.8)；879 (779~977)；188.5 (134.6~258.0)

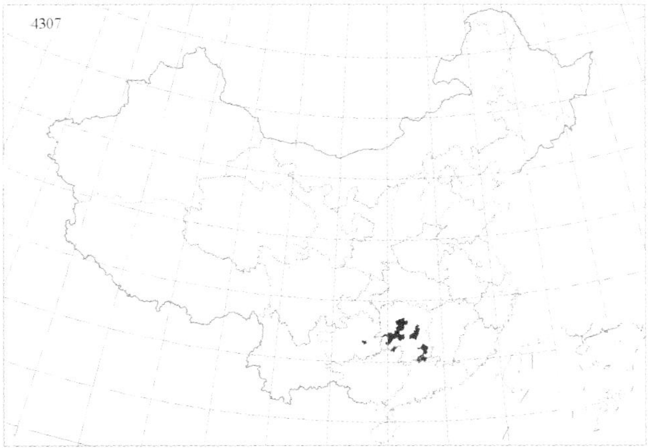

Fabaceae 豆科 *Indigofera* 木蓝属

Indigofera dichroa Craib 川西木蓝
Shrubs, deciduous

15.0 (11.1~18.9)；7.8 (2.2~12.0)；20.7 (19.3~23.9)；15.0 (11.3~19.0)；121.0 (80.8~167.8)；0.7 (0.0~5.3)；762 (652~929) ❙ 974 (797~1220)；567 (411~688)；28 (18~38) ❙ 30.3 (-14.2~87.2)；701 (652~760)；279.6 (218.0~309.8)

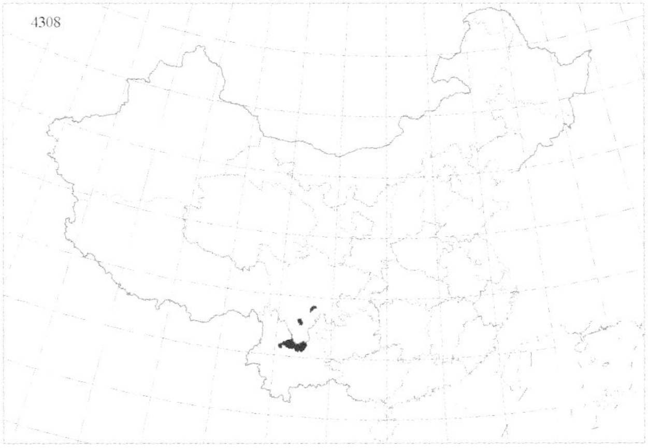

Fabaceae 豆科 *Indigofera* 木蓝属
Indigofera dolichochaete Craib 长齿木蓝
Shrubs, deciduous

12.4 (6.6~18.6); 5.1 (-1.3~12.7); 18.3 (13.3~22.5); 12.5 (6.8~18.7); 94.2 (39.6~163.9); 4.8 (0.0~19.7); 686 (514~888) ❙ 1076 (862~1547); 648 (548~911); 30 (12~63) ❙ 56.6 (37.0~74.2); 676 (514~888); 340.6 (255.4~551.3)

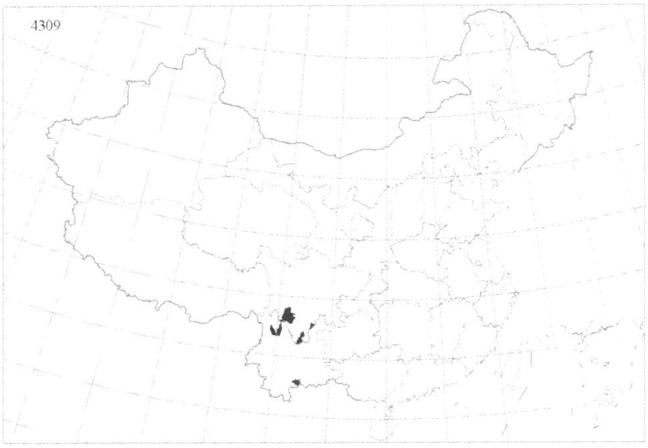

Fabaceae 豆科 *Indigofera* 木蓝属
Indigofera dumetorum Craib 黄花木蓝
Shrubs, deciduous

10.7 (3.8~18.6); 3.5 (-4.3~12.7); 16.6 (10.8~22.5); 10.8 (4.7~18.7); 77.9 (21.6~163.9); 9.5 (0.0~35.4); 634 (444~888) ❙ 978 (741~1547); 592 (426~911); 26 (12~63) ❙ 57.3 (14.4~121.7); 619 (444~888); 319.8 (265.2~551.3)

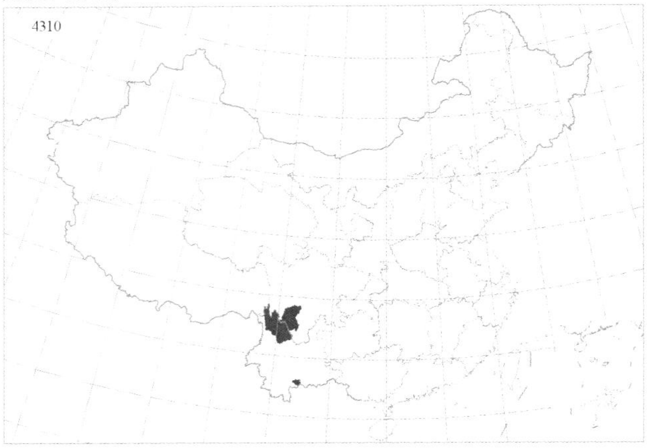

Fabaceae 豆科 *Indigofera* 木蓝属
Indigofera esquirolii H. Lév. 黔南木蓝
Shrubs, deciduous

16.8 (13.2~22.3); 7.3 (4.0~13.5); 24.8 (19.6~28.9); 16.9 (13.2~22.3); 142.4 (98.5~208.1); 0.1 (0.0~1.1); 864 (700~1202) ❙ 1258 (911~1446); 595 (476~723); 83 (32~136) ❙ 46.2 (13.8~71.9); 861 (700~1202); 243.2 (188.8~322.1)

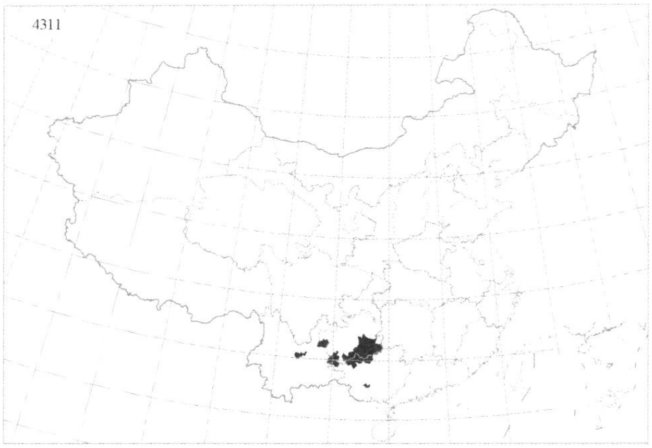

Fabaceae 豆科 *Indigofera* 木蓝属
Indigofera fortunei Craib 华东木蓝
Shrubs, deciduous

15.1 (7.4~17.2); 2.1 (-4.2~6.7); 27.5 (18.7~29.4); 15.1 (8.2~17.2); 126.6 (58.8~146.8); 5.3 (0.0~29.4); 854 (561~952) ❙ 1078 (626~1793); 475 (316~666); 111 (25~238) ❙ 25.8 (-22.2~112.2); 837 (561~942); 182.9 (110.1~319.9)

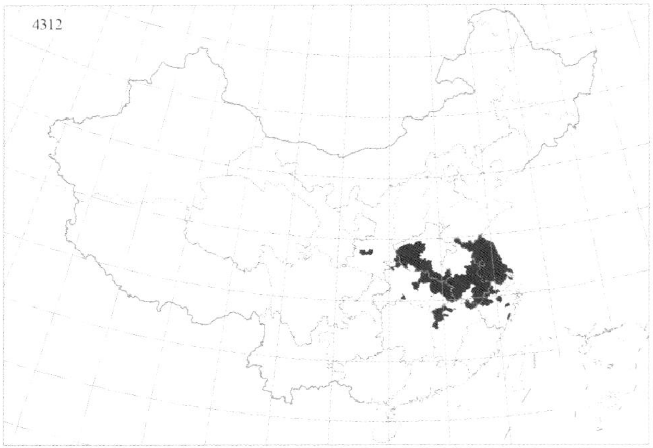

Fabaceae 豆科 *Indigofera* 木蓝属
Indigofera galegoides DC. 假大青蓝
Shrubs or subshrubs, deciduous

22.5 (17.4~24.8); 16.1 (7.4~20.1); 27.1 (22.3~28.7); 22.5 (17.4~24.8); 210.0 (148.9~237.6); 0.0 (0.0~0.0); 1195 (886~1385) ❙ 1650 (1176~2992); 749 (447~1847); 85 (41~138) ❙ 41.7 (-12.0~202.5); 1127 (880~1360); 387.6 (154.2~620.0)

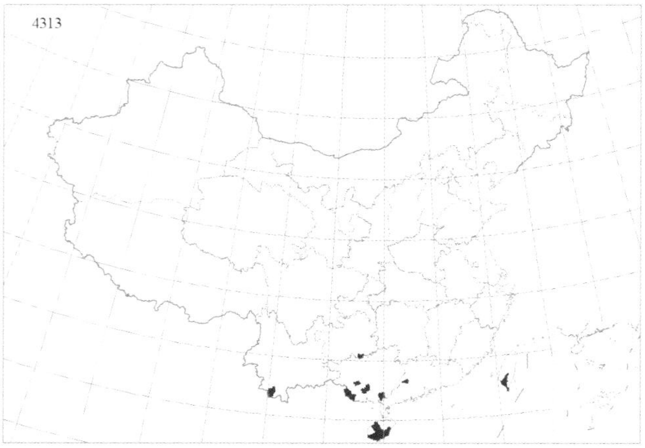

Fabaceae 豆科 *Indigofera* 木蓝属
Indigofera hamiltonii Graham ex Duthie et Prain 腾冲木蓝
Shrubs

14.4; 7.5; 19.1; 14.4; 113.1; 0.0; 727 ❙ 1454; 792; 70 ❙ 99.9; 727; 487.2

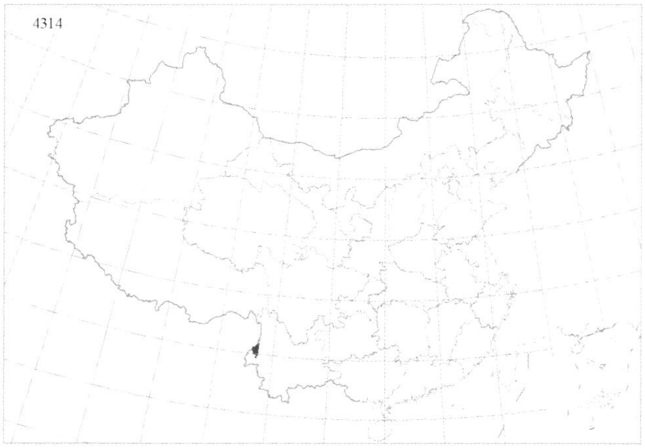

Fabaceae 豆科 *Indigofera* 木蓝属

Indigofera hancockii Craib 苍山木蓝

Shrubs, deciduous

12.9 (2.2~18.9); 5.3 (-7.1~12.0); 19.1 (9.5~26.6); 13.1 (3.8~19.0); 100.9 (15.5~167.8); 5.7 (0.0~49.1); 702 (411~929) ❙ 976 (579~1368); 566 (365~774); 32 (10~59) ❙ 41.9 (-14.2~125.9); 680 (411~912); 288.3 (142.9~491.5)

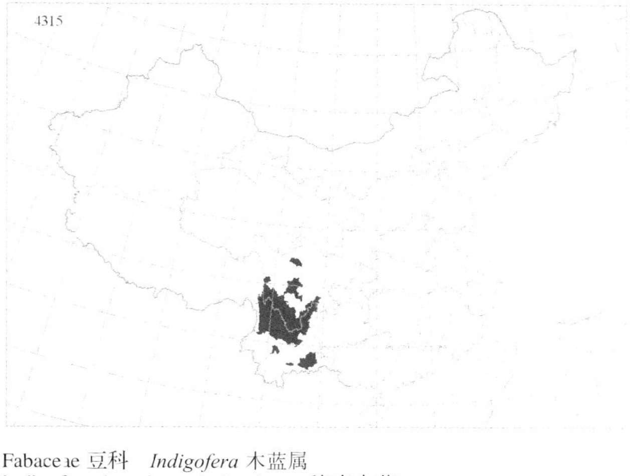

Fabaceae 豆科 *Indigofera* 木蓝属

Indigofera hebepetala var. glabra Ali 光叶毛瓣木蓝

Shrubs, deciduous

0.4 (-1.8~4.4); -8.4 (-10.2~-4.3); 8.5 (6.1~12.1); 3.1 (1.8~5.3); 11.6 (1.9~28.5); 65.7 (34.6~82.8); 371 (329~455) ❙ 368 (290~538); 248 (193~337); 15 (4~33) ❙ -2.2 (-26.9~18.2); 345 (290~455); 95.0 (61.6~168.2)

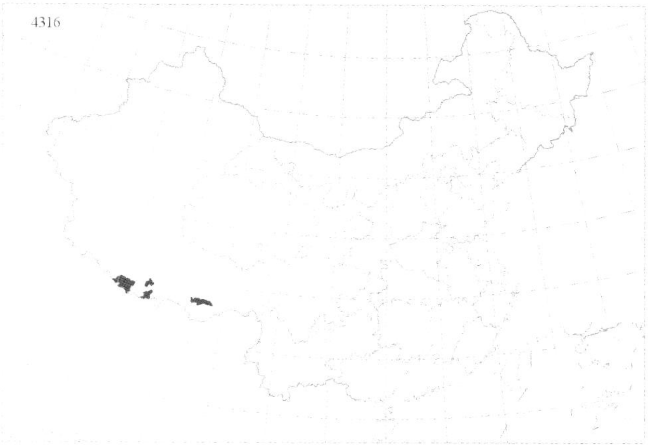

Fabaceae 豆科 *Indigofera* 木蓝属

Indigofera hendecaphylla Jacq. 穗序木蓝

Shrubs, deciduous

15.5 (5.4~22.0); 8.8 (-2.1~15.2); 20.8 (12.3~28.8); 15.7 (5.8~22.7); 129.4 (31.5~212.7); 1.1 (0.0~25.6); 795 (481~1216) ❙ 1086 (741~1965); 613 (411~994); 44 (15~132) ❙ 37.6 (-14.2~99.9); 763 (481~1216); 313.1 (203.4~487.2)

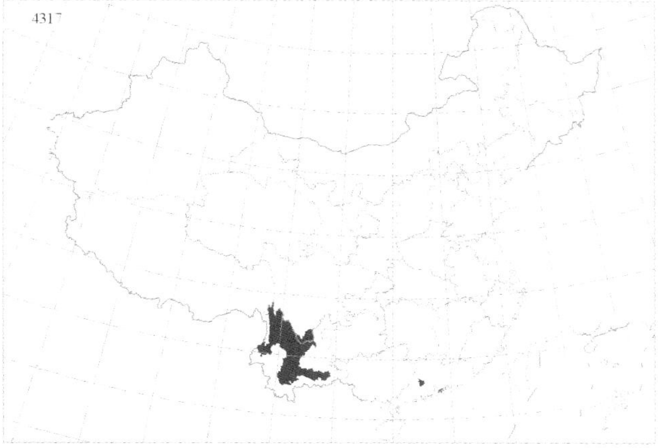

Fabaceae 豆科 *Indigofera* 木蓝属

Indigofera henryi Craib 长梗木蓝

Shrubs, deciduous

11.2 (1.8~17.7); 3.4 (-7.2~10.9); 17.5 (9.9~22.5); 11.6 (3.8~17.8); 87.6 (15.4~153.4); 12.8 (0.0~53.3); 652 (404~854) ❙ 995 (527~1368); 565 (317~756); 33 (9~59) ❙ 56.8 (19.0~128.6); 648 (404~854); 282.5 (169.5~491.5)

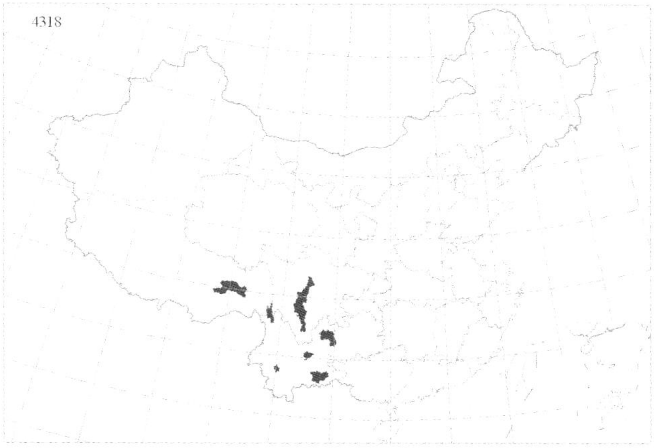

Fabaceae 豆科 *Indigofera* 木蓝属

Indigofera heterantha Wall. 异花木蓝

Shrubs, deciduous

5.8; -1.8; 12.4; 7.1; 44.4; 34.2; 507 ❙ 746; 458; 25 ❙ 40.3; 504; 244.5

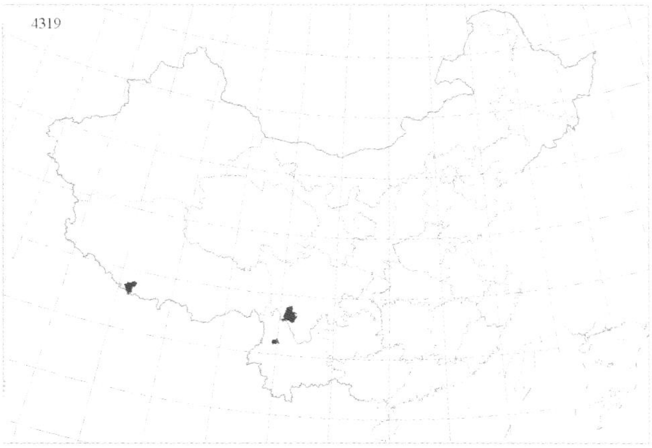

Fabaceae 豆科 *Indigofera* 木蓝属

Indigofera hirsuta L. 硬毛木蓝

Shrubs, deciduous

20.7 (14.6~24.8); 12.9 (3.8~20.1); 27.2 (20.1~29.2); 20.7 (14.7~24.8); 188.8 (117.7~237.6); 0.0 (0.0~1.5); 1090 (773~1385) ❙ 1621 (868~2992); 706 (312~1847); 128 (41~341) ❙ 52.9 (-34.3~246.5); 1055 (773~1360); 281.9 (119.4~620.0)

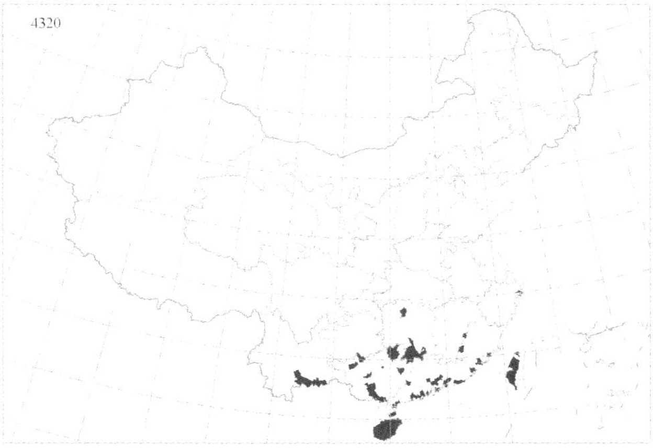

Fabaceae 豆科 *Indigofera* 木蓝属
Indigofera howellii Craib et W. W. Sm. 长序木蓝
Shrubs, deciduous

7.1 (2.2~14.4); -0.5 (-6.6~7.5); 13.6 (9.5~19.1); 7.6 (3.8~14.4); 46.3 (15.5~113.1); 20.4 (0.0~49.1); 533 (411~727) ❙ 952 (729~1454); 567 (426~792); 28 (11~70) ❙ 80.2 (51.8~125.9); 532 (411~727); 303.0 (208.8~487.2)

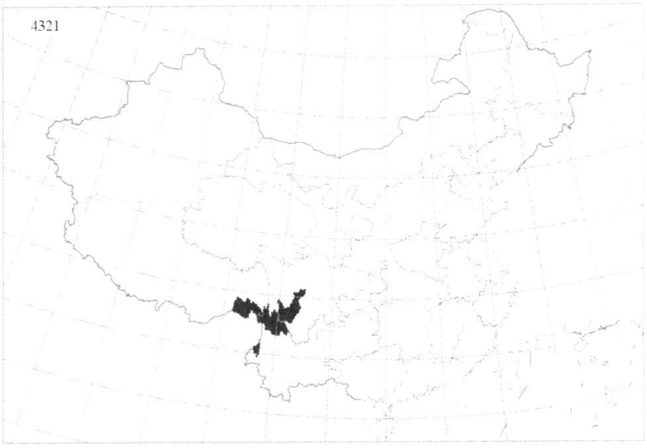

Fabaceae 豆科 *Indigofera* 木蓝属
Indigofera jikongensis Y. Y. Fang et C. Z. Zheng 鸡公木蓝
Shrubs, deciduous

15.8; 2.6; 28.0; 15.8; 133.0; 3.4; 888 ❙ 1121; 491; 100 ❙ 26.2; 886; 162.8

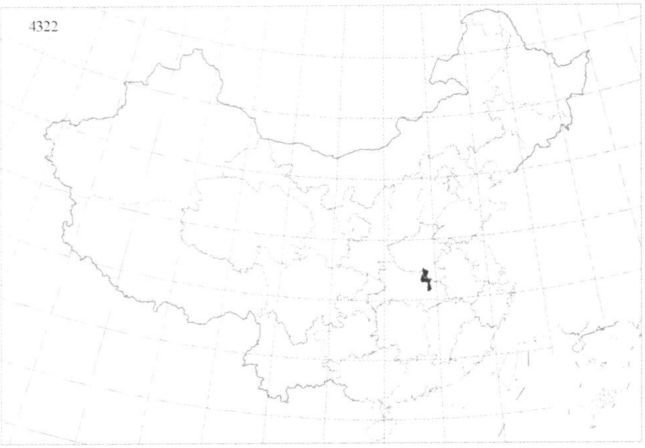

Fabaceae 豆科 *Indigofera* 木蓝属
Indigofera jindongensis Y. Y. Fang et C. Z. Zheng 景东木蓝
Shrubs, deciduous

17.3; 11.0; 21.6; 17.3; 148.0; 0.0; 832 ❙ 1049; 598; 44 ❙ 26.1; 783; 365.3

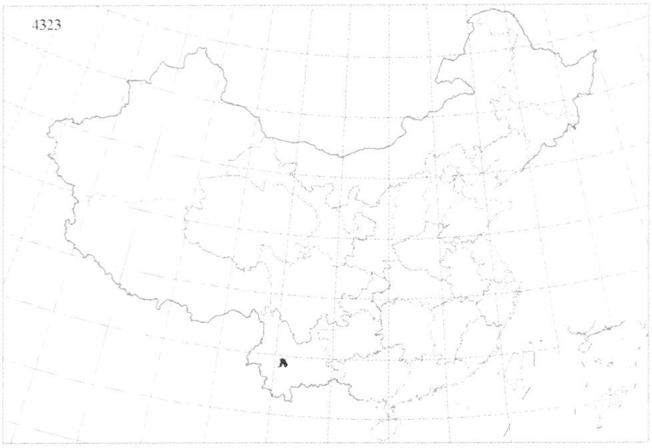

Fabaceae 豆科 *Indigofera* 木蓝属
Indigofera kirilowii Maxim. ex Palibin 花木蓝
Small shrubs, deciduous

10.9 (-2.9~20.1); -4.3 (-26.4~11.0); 24.6 (17.7~29.3); 11.9 (5.4~20.1); 98.7 (40.2~181.7); 27.5 (0.0~134.6); 731 (471~1045) ❙ 743 (377~1689); 405 (241~685); 39 (4~218) ❙ 0.8 (-40.5~108.0); 651 (377~1045); 196.2 (97.2~334.4)

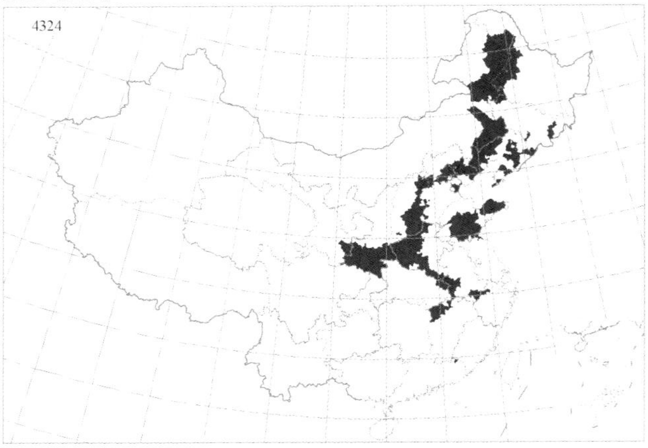

Fabaceae 豆科 *Indigofera* 木蓝属
Indigofera lacei Craib 思茅木蓝
Subshrubs, deciduous

13.8 (6.6~20.5); 7.3 (-1.3~14.9); 18.8 (13.3~23.7); 13.9 (6.8~20.5); 108.8 (39.6~185.9); 2.4 (0.0~19.7); 733 (514~992) ❙ 1138 (862~1499); 627 (548~775); 40 (12~66) ❙ 55.3 (36.1~67.7); 723 (514~959); 381.4 (296.9~570.1)

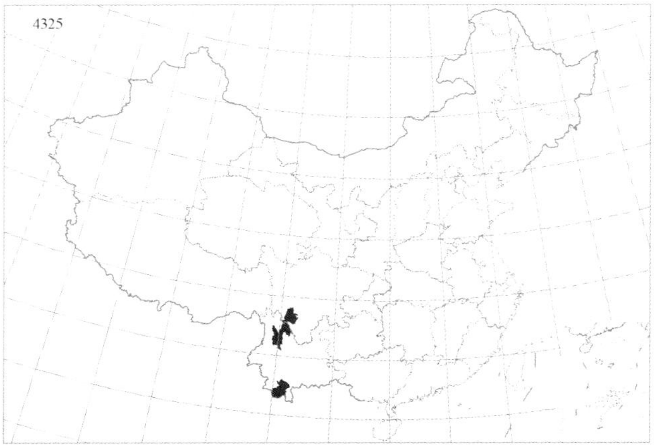

Fabaceae 豆科 *Indigofera* 木蓝属
Indigofera lenticellata Craib 岷谷木蓝
Shrubs, deciduous

6.5 (0.0~16.7); -1.9 (-8.7~9.7); 13.8 (8.1~22.1); 7.3 (2.7~16.7); 46.2 (8.0~140.9); 28.2 (0.0~67.9); 522 (363~819) ❙ 856 (535~1081); 499 (326~619); 20 (10~32) ❙ 68.5 (14.4~128.6); 513 (363~723); 264.8 (111.0~357.7)

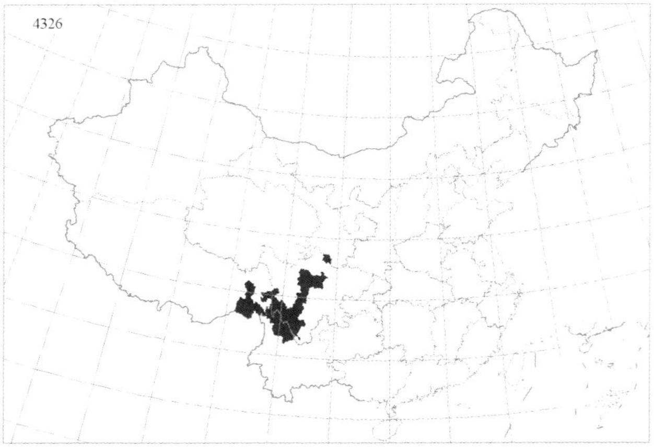

Fabaceae 豆科 *Indigofera* 木蓝属

Indigofera linifolia (Linnaeus f.) Retzius 单叶木蓝

Shrubs

16.8 (11.3~22.2); 9.5 (3.1~17.7); 22.6 (18.6~27.8); 16.8 (11.5~22.2); 141.9 (80.9~206.7); 0.2 (0.0~2.8); 854 (654~1128) | 1292 (797~2992); 722 (411~1847); 52 (20~137) | 48.5 (-14.2~202.5); 823 (654~1128); 258.5 (113.7~363.2)

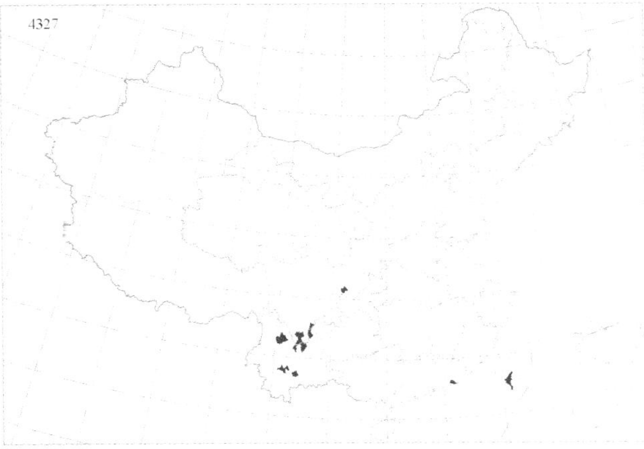

Fabaceae 豆科 *Indigofera* 木蓝属

Indigofera linnaei Ali 九叶木蓝

Shrubs

22.1; 16.6; 25.8; 22.1; 205.4; 0.0; 1162 | 1274; 555; 46 | 11.6; 1020; 437.2

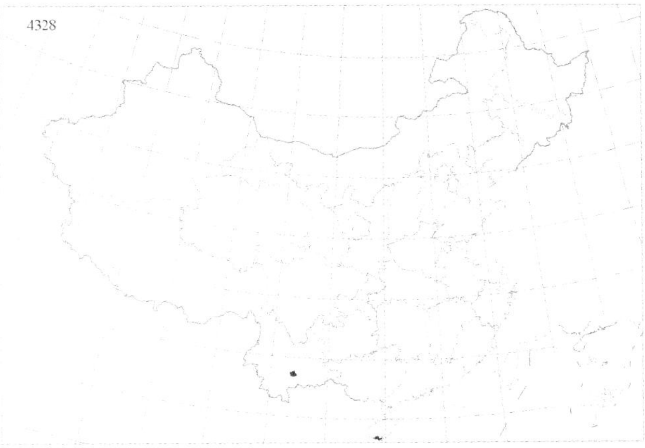

Fabaceae 豆科 *Indigofera* 木蓝属

Indigofera mairei Pamp. 西南木蓝

Shrubs, deciduous

13.2 (3.2~18.9); 5.2 (-5.8~12.0); 19.9 (10.8~26.2); 13.2 (4.5~19.0); 102.2 (21.4~167.8); 3.9 (0.0~42.6); 712 (430~929) | 1010 (707~1337); 563 (357~727); 37 (11~69) | 44.6 (-14.2~121.7); 692 (430~878); 272.1 (148.9~407.1)

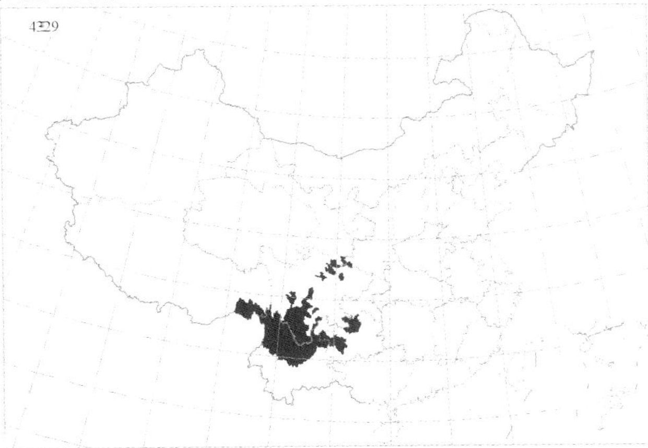

Fabaceae 豆科 *Indigofera* 木蓝属

Indigofera mengtzeana Craib 蒙自木蓝

Shrubs, deciduous

12.4 (6.6~16.9); 4.8 (-1.3~10.3); 18.5 (13.3~21.4); 12.5 (6.8~16.9); 93.9 (39.6~142.7); 4.7 (0.0~19.7); 680 (514~813) | 1027 (862~1178); 605 (548~663); 30 (12~51) | 53.4 (34.8~101.6); 669 (514~813); 294.0 (255.4~383.4)

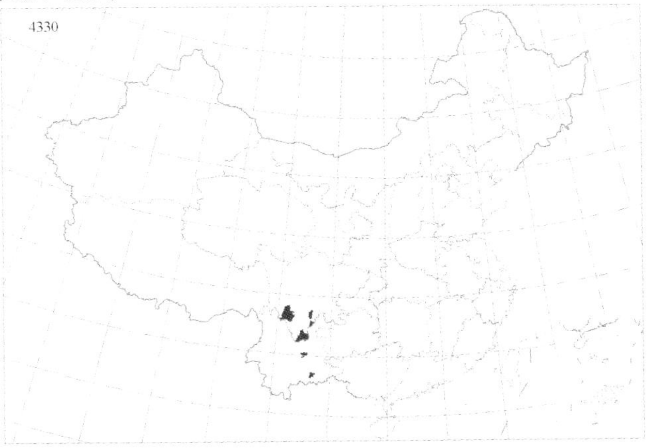

Fabaceae 豆科 *Indigofera* 木蓝属

Indigofera muliensis Y. Y. Fang et C. Z. Zheng 木里木蓝

Shrubs, deciduous

9.3 (5.4~18.5); 2.4 (-2.1~10.5); 16.0 (12.3~24.3); 9.9 (5.8~18.3); 67.5 (31.5~160.0); 9.7 (0.0~25.6); 611 (481~906) | 927 (741~1134); 560 (426~651); 26 (12~47) | 54.9 (5.5~78.3); 595 (481~786); 312.7 (273.5~351.7)

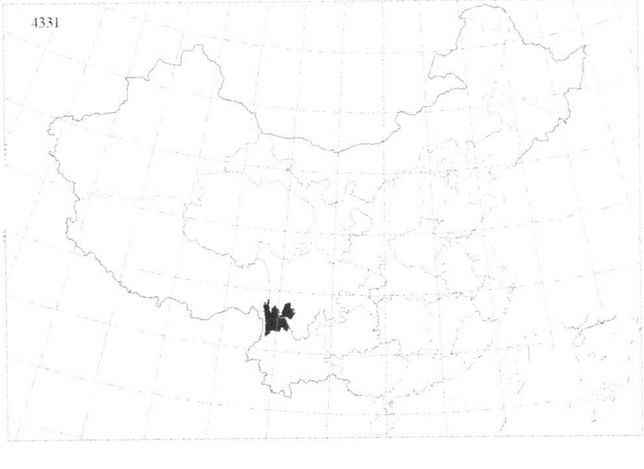

Fabaceae 豆科 *Indigofera* 木蓝属

Indigofera myosurus Craib 华西木蓝

Shrubs, deciduous

11.2 (9.1~14.9); 2.2 (-0.7~5.3); 19.3 (16.9~23.6); 11.3 (9.2~15.0); 81.5 (64.1~119.6); 6.1 (0.0~14.1); 656 (595~782) | 1061 (822~1410); 591 (420~816); 29 (18~49) | 60.9 (38.2~80.4); 656 (595~782); 272.0 (214.5~318.7)

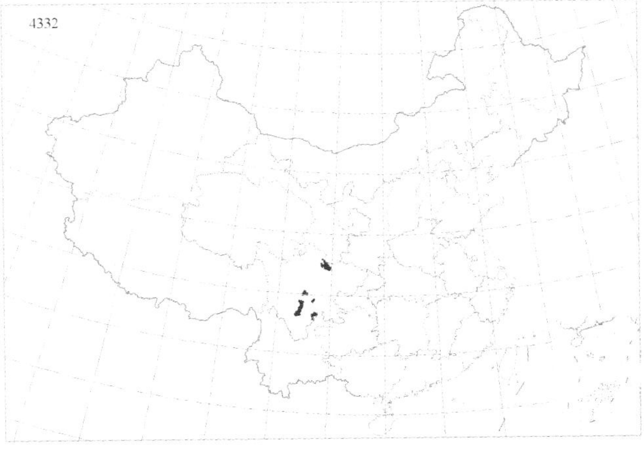

Fabaceae 豆科 *Indigofera* 木蓝属
Indigofera nigrescens Kurz ex King et Prain 黑叶木蓝
Shrubs, deciduous

16.5 (5.3~24.1); 7.5 (-2.7~18.9); 24.2 (12.2~29.7); 16.6 (5.7~24.1); 139.6 (30.5~229.7); 0.7 (0.0~26.2); 855 (479~1322) ❙ 1363 (702~2992); 607 (299~1847); 115 (15~303) ❙ 59.5 (-14.2~219.3); 840 (479~1123); 285.6 (134.6~659.0)

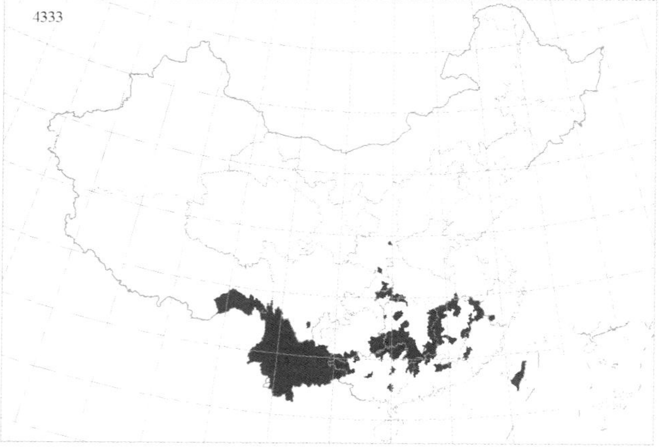

Fabaceae 豆科 *Indigofera* 木蓝属
Indigofera pampaniniana Craib 昆明木蓝
Shrubs, deciduous

15.2 (12.4~18.3); 8.4 (4.7~11.8); 20.3 (18.6~22.7); 15.3 (12.4~18.3); 123.4 (89.1~159.8); 0.0 (0.0~0.3); 763 (675~882) ❙ 959 (896~1049); 563 (524~598); 36 (29~44) ❙ 26.3 (6.5~37.1); 731 (675~799); 276.5 (205.2~365.3)

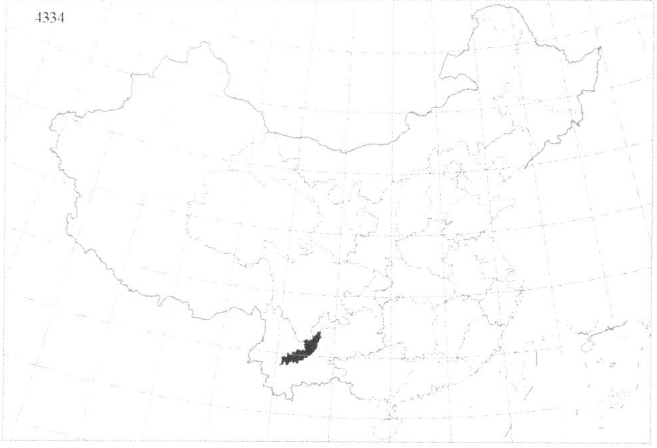

Fabaceae 豆科 *Indigofera* 木蓝属
Indigofera parkesii Craib 浙江木蓝
Small shrubs, deciduous

16.6 (14.4~21.3); 5.3 (2.1~12.6); 27.6 (25.8~29.2); 16.7 (14.5~21.3); 140.9 (118.7~196.0); 0.6 (0.0~4.9); 894 (801~1123) ❙ 1636 (1188~1950); 564 (463~677); 207 (56~259) ❙ 83.5 (5.8~129.9); 889 (801~1121); 242.7 (183.5~318.0)

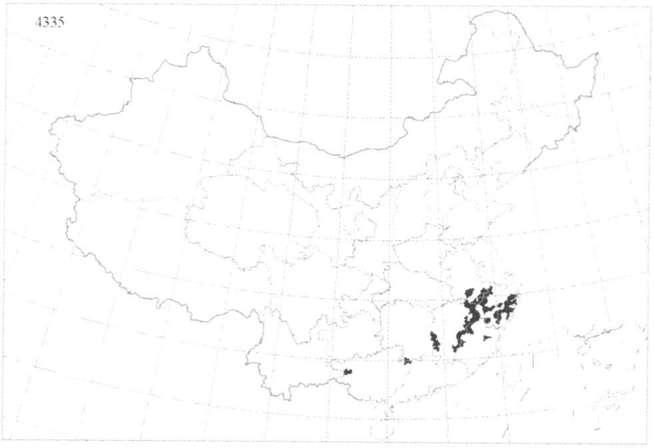

Fabaceae 豆科 *Indigofera* 木蓝属
Indigofera parkesii var. longipedunculata (Y. Y. Fang et C. Z. Zheng) X. F. Gao & B. Schrire 长总梗木蓝
Subshrubs, deciduous

17.0 (15.9~18.2); 5.4 (4.2~6.9); 28.0 (27.5~29.1); 17.1 (15.9~18.2); 145.2 (131.9~158.8); 0.2 (0.0~0.8); 919 (866~984) ❙ 1750 (1360~1922); 576 (498~623); 237 (185~259) ❙ 90.4 (57.0~111.3); 912 (866~961); 243.7 (183.9~279.5)

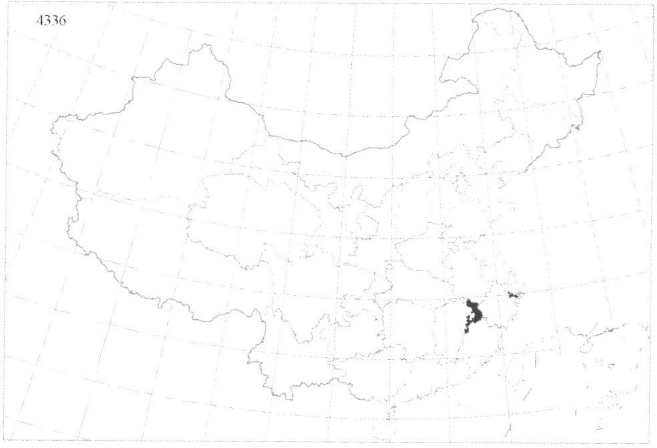

Fabaceae 豆科 *Indigofera* 木蓝属
Indigofera parkesii var. polyphylla Y. Y. Fang et C. Z. Zheng 多叶浙江木蓝
Small shrubs, deciduous

16.1; 3.9; 27.6; 16.1; 134.6; 1.2; 879 ❙ 1757; 609; 227 ❙ 100.4; 879; 261.5

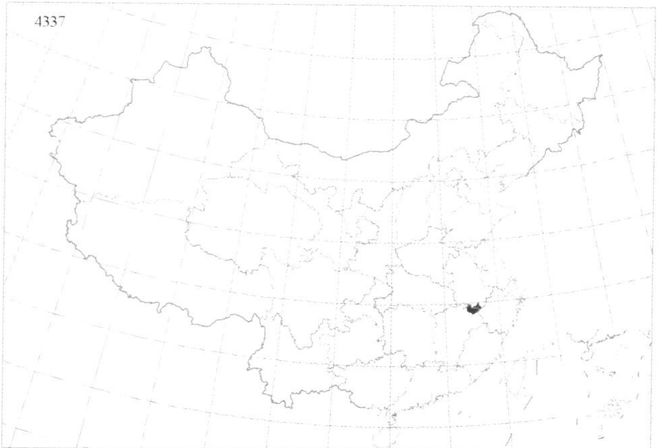

Fabaceae 豆科 *Indigofera* 木蓝属
Indigofera pedicellata Wight et Arn. 长梗木蓝
Herbs or subshrubs, deciduous

20.2 (16.2~24.1); 14.5 (11.0~18.9); 25.0 (20.1~28.2); 20.2 (16.2~24.1); 182.9 (134.2~229.7); 0.0 (0.0~0.0); 1019 (773~1322) ❙ 2402 (1498~3495); 1106 (750~1847); 255 (47~897) ❙ 146.2 (25.2~246.5); 983 (773~1144); 245.0 (119.4~363.7)

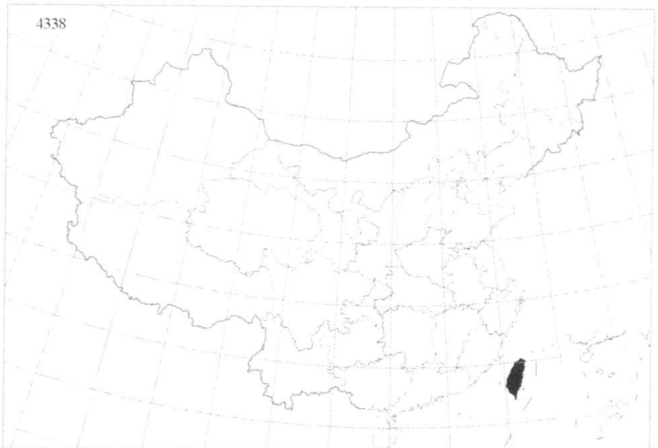

Fabaceae 豆科 *Indigofera* 木蓝属

Indigofera pendula Franch. 垂序木蓝

Shrubs, deciduous

13.3 (2.2~18.9); 6.1 (-6.6~12.0); 19.0 (9.9~24.3); 13.4 (3.8~19.0); 103.7 (15.5~167.8); 3.8 (0.0~49.1); 710 (413~929) ❙ 981 (741~1190); 575 (411~656); 31 (12~58) ❙ 41.9 (-14.2~125.9); 680 (413~836); 300.4 (205.2~333.4)

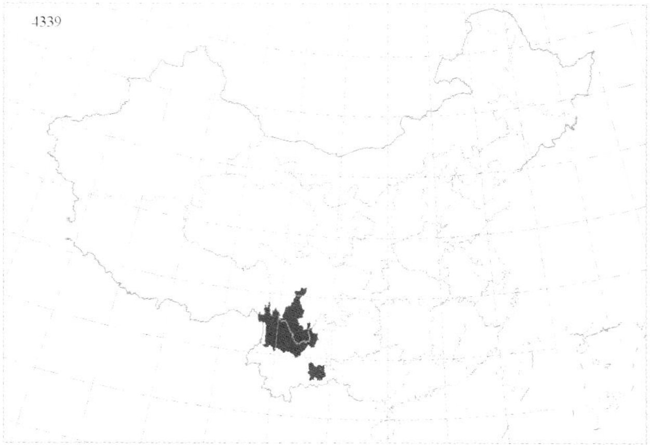

Fabaceae 豆科 *Indigofera* 木蓝属

Indigofera penduloides Y. Y. Fang et C. Z. Zheng 拟垂序木蓝

Shrubs, deciduous

15.2; 8.7; 20.1; 15.2; 122.5; 0.0; 753 ❙ 1019; 562; 50 ❙ 35.2; 746; 381.5

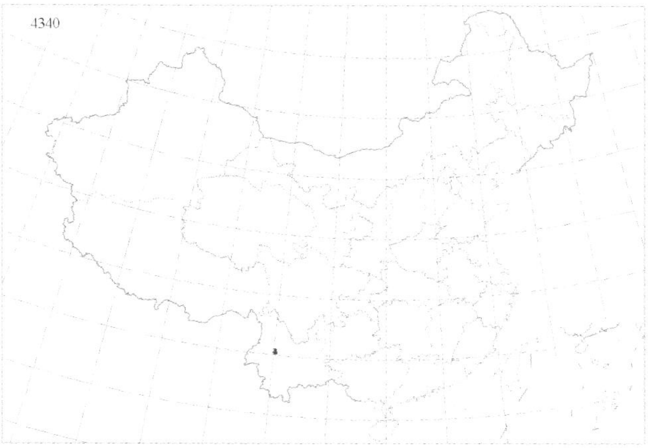

Fabaceae 豆科 *Indigofera* 木蓝属

Indigofera ramulosissima Hosok. 多枝木蓝

Dwarf shrubs, deciduous

16.4; 11.5; 20.3; 16.5; 137.8; 0.0; 783 ❙ 2499; 947; 303 ❙ 219.3; 783; 303.6

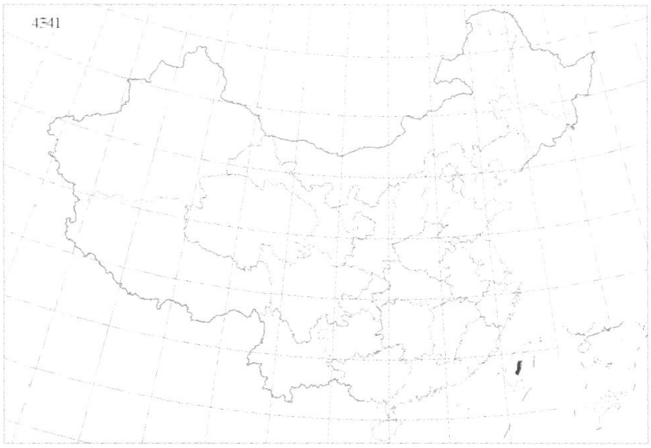

Fabaceae 豆科 *Indigofera* 木蓝属

Indigofera reticulata Franch. 网叶木蓝

Dwarf subshrubs, deciduous

13.5 (2.2~18.9); 5.9 (-6.3~12.7); 19.7 (9.5~26.6); 13.6 (3.9~19.0); 105.7 (15.8~167.8); 2.9 (0.0~48.6); 721 (411~929) ❙ 990 (729~1431); 573 (411~817); 32 (11~57) ❙ 39.8 (-14.2~113.0); 693 (411~912); 278.6 (142.9~403.0)

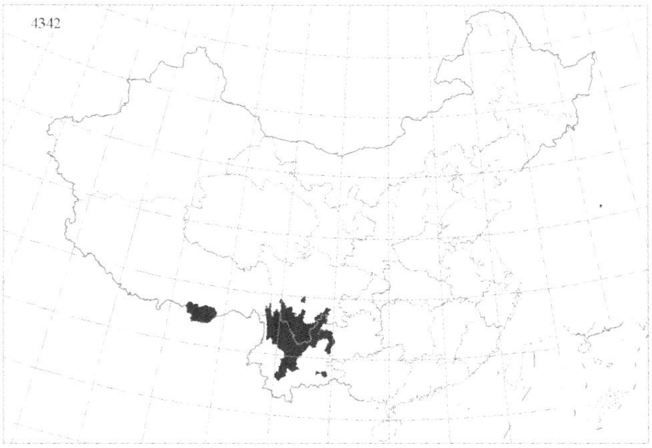

Fabaceae 豆科 *Indigofera* 木蓝属

Indigofera rigioclada Craib 硬叶木蓝

Shrubs, deciduous

9.0 (1.8~18.5); 1.2 (-7.1~10.5); 15.5 (9.9~24.3); 9.4 (3.8~18.3); 64.7 (15.5~160.0); 16.9 (0.0~53.3); 589 (404~906) ❙ 893 (527~1076); 535 (317~630); 23 (9~42) ❙ 55.3 (5.5~125.9); 575 (404~786); 295.9 (169.5~351.7)

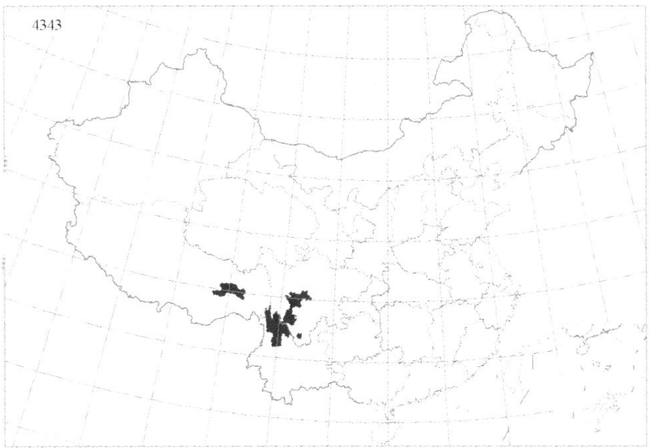

Fabaceae 豆科 *Indigofera* 木蓝属

Indigofera scabrida Dunn 腺毛木蓝

Shrubs, deciduous

14.2 (5.4~19.1); 6.7 (-2.1~13.1); 20.2 (12.3~26.9); 14.3 (5.8~19.1); 112.5 (31.5~169.1); 1.5 (0.0~25.6); 740 (481~973) ❙ 1049 (741~1716); 590 (411~816); 42 (12~197) ❙ 42.5 (-14.2~87.2); 723 (481~973); 300.1 (142.9~491.5)

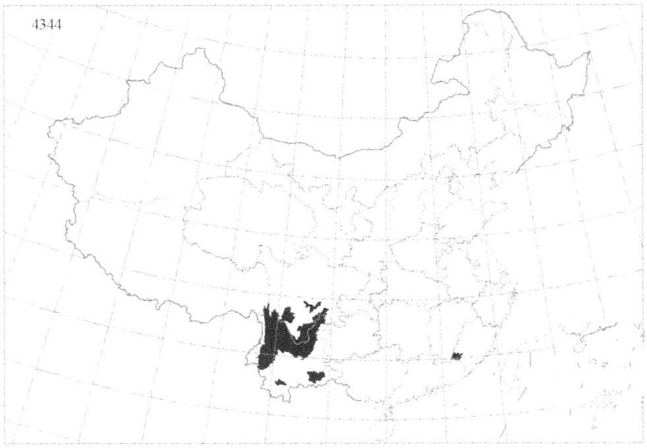

Fabaceae 豆科 *Indigofera* 木蓝属
Indigofera sensitiva Franch. 敏感木蓝
Dwarf shrubs, deciduous

14.0 (7.3~16.7); 6.8 (-2.0~9.8); 19.7 (15.6~22.1); 14.0 (7.5~16.8); 110.6 (48.2~141.1); 2.5 (0.0~20.4); 726 (537~819) ❙ 979 (913~1057); 574 (522~627); 28 (22~42) ❙ 37.4 (11.7~87.3); 677 (537~721); 292.5 (260.2~349.8)

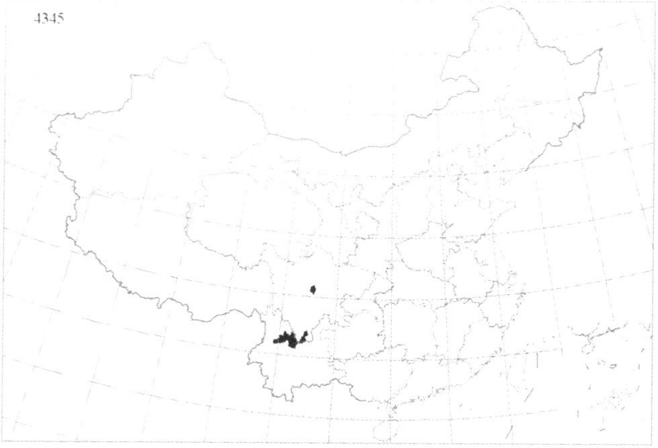

Fabaceae 豆科 *Indigofera* 木蓝属
Indigofera shipingensis X. F. Gao 石屏木蓝
Shrubs

17.9; 11.6; 22.1; 17.9; 155.1; 0.0; 858 ❙ 1098; 634; 46 ❙ 27.9; 836; 295.5

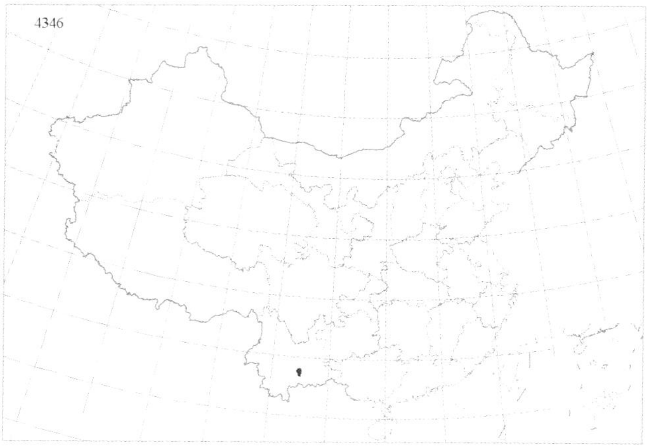

Fabaceae 豆科 *Indigofera* 木蓝属
Indigofera silvestrii Pamp. 刺序木蓝
Shrubs, deciduous

10.3 (1.8~16.9); 1.0 (-7.2~8.4); 18.7 (9.9~27.6); 10.7 (3.8~17.0); 78.6 (15.4~143.7); 14.5 (0.0~53.3); 646 (404~881) ❙ 1018 (527~1577); 523 (317~932); 42 (9~94) ❙ 61.5 (-4.0~128.6); 645 (404~881); 244.7 (134.4~407.1)

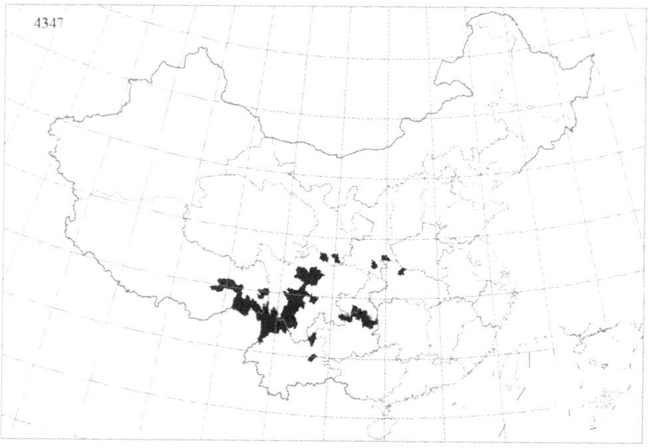

Fabaceae 豆科 *Indigofera* 木蓝属
Indigofera stachyodes Lindl. 茸毛木蓝
Shrubs, deciduous

16.1 (-1.8~21.0); 8.7 (-10.2~15.1); 21.8 (6.1~27.7); 16.2 (1.8~21.0); 135.6 (1.9~191.8); 1.4 (0.0~82.8); 812 (329~1065) ❙ 1145 (320~1751); 625 (193~996); 52 (20~141) ❙ 41.0 (-14.2~100.5); 798 (320~1065); 314.6 (72.2~659.0)

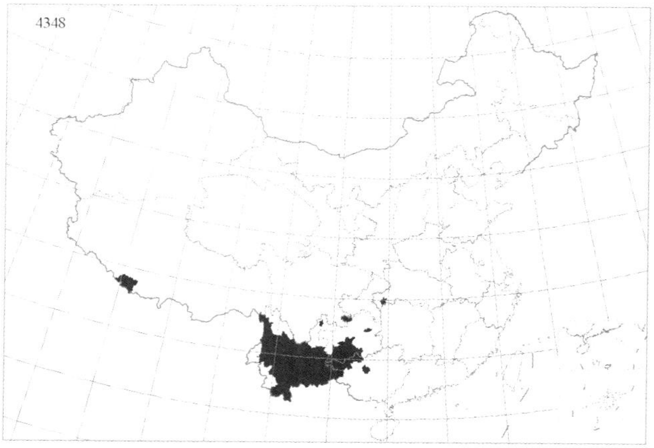

Fabaceae 豆科 *Indigofera* 木蓝属
Indigofera sticta Craib 矮木蓝
Shrubs or subshrubs, deciduous

16.9; 10.3; 21.5; 16.9; 143.1; 0.0; 816 ❙ 1062; 597; 44 ❙ 30.2; 788; 333.4

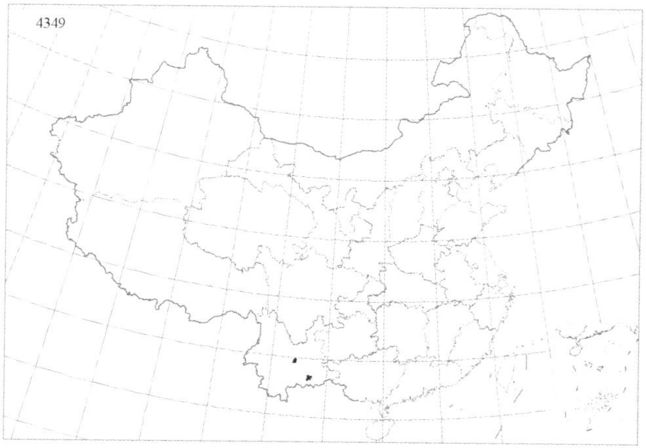

Fabaceae 豆科 *Indigofera* 木蓝属
Indigofera suffruticosa Mill. 野青树
Shrubs or subshrubs, deciduous

18.4 (11.7~24.8); 9.7 (2.7~20.1); 25.8 (18.8~29.0); 18.4 (11.7~24.8); 161.3 (82.6~237.6); 0.2 (0.0~3.8); 962 (658~1385) ❙ 1433 (868~2993); 642 (312~1847); 99 (31~547) ❙ 51.0 (-34.3~214.4); 945 (658~1360); 263.7 (119.4~659.0)

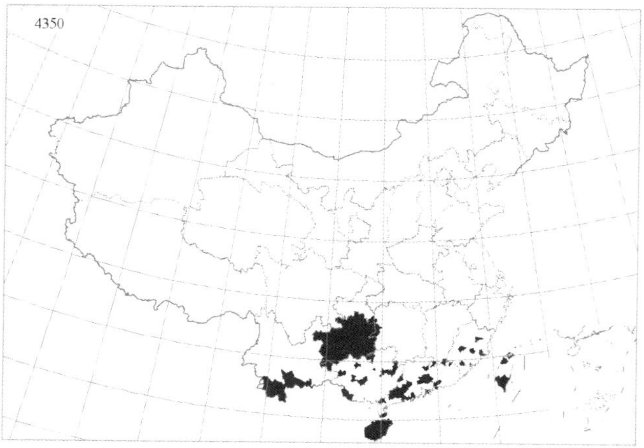

Fabaceae 豆科 *Indigofera* 木蓝属
Indigofera szechuensis Craib 四川木蓝
Shrubs, deciduous

6.6 (0.3~17.5 ; -2.0 (-9.9~10.5); 14.0 (9.3~22.3); 7.5 (3.3~17.5); 48.5 (11.9~149.9); 29.2 (0.0~67.5); 527 (389~846) ❙ 845 (527~1246); 484 (317~615); 22 (9~46) ❙ 64.9 (13.3~128.6); 522 (389~823); 268.2 (161.1~357.7)

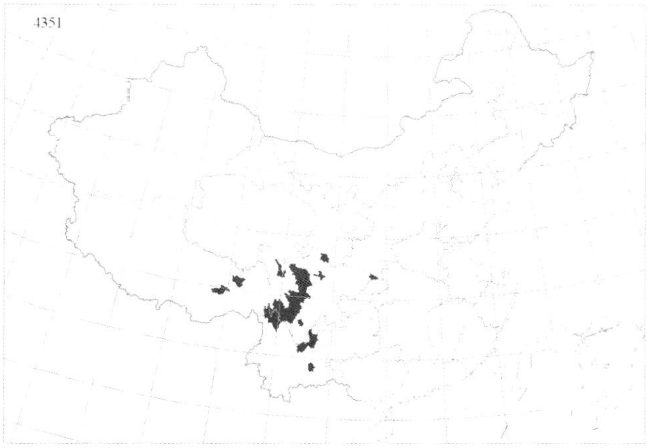

Fabaceae 豆科 *Indigofera* 木蓝属
Indigofera tinctoria L. 木蓝
Subshrubs, deciduous

17.3 (11.7~24.8); 7.4 (1.5~20.1); 26.3 (18.8~29.0); 17.3 (11.7~24.8); 149.7 (82.6~237.6); 1.8 (0.0~7.3); 938 (658~1385) ❙ 1522 (916~2759); 662 (455~1594); 132 (31~245) ❙ 66.2 (1.5~246.5); 916 (658~1360); 258.2 (119.4~572.9)

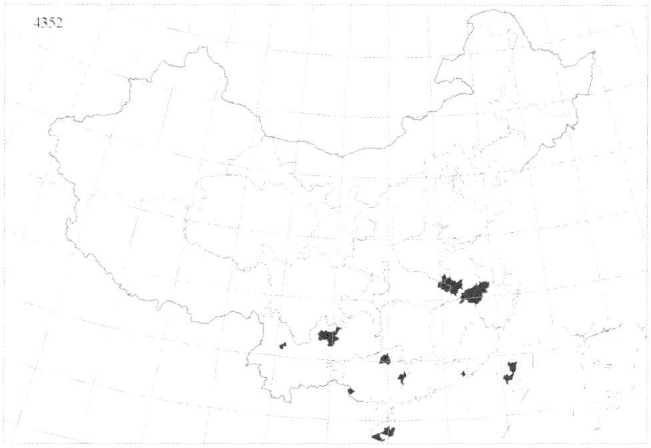

Fabaceae 豆科 *Indigofera* 木蓝属
Indigofera trifoliata L. 三叶木蓝
Herbs or subshrubs, deciduous

22.2; 17.7; 25.5; 22.2; 206.7; 0.0; 1123 ❙ 2992; 1847; 107 ❙ 166.5; 1123; 296.9

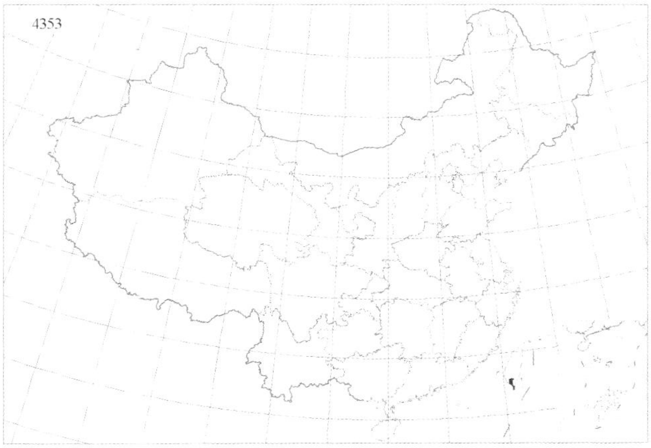

Fabaceae 豆科 *Indigofera* 木蓝属
Indigofera trifoliata var. zhengkangensis H. Sun 镇康三叶木蓝
Subshrubs

17.3; 11.0; 21.6; 17.3; 148.0; 0.0; 832 ❙ 1049; 598; 44 ❙ 26.1; 783; 365.3

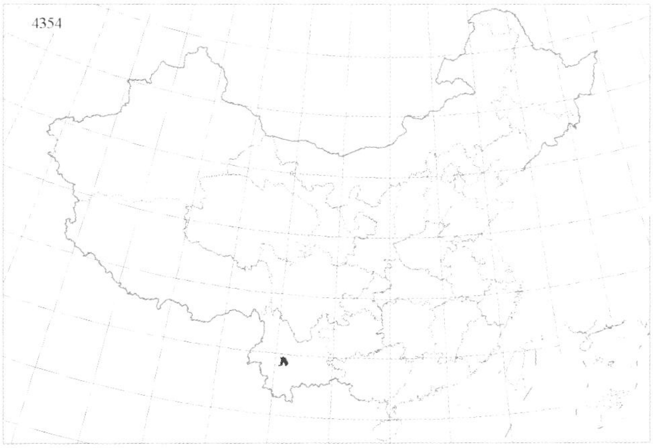

Fabaceae 豆科 *Indigofera* 木蓝属
Indigofera venulosa Champion ex Bentham 脉叶木蓝
Shrubs, deciduous

16.0; 8.5; 27.6; 18.3; 159.8; 0.1; 975 ❙ 1774; 749; 160 ❙ 84.7; 975; 232.1

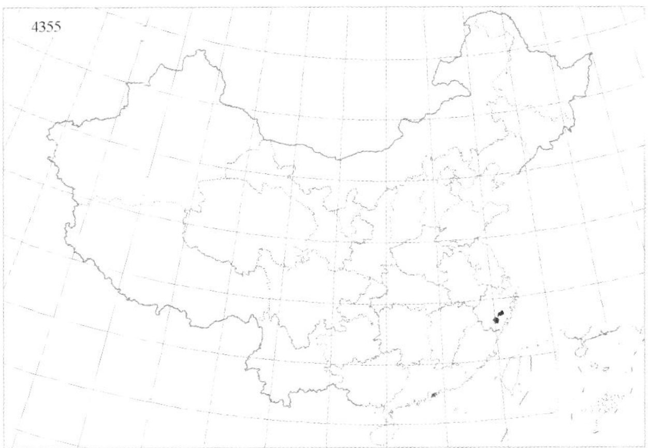

Fabaceae 豆科 *Indigofera* 木蓝属
Indigofera wightii Graham 海南木蓝
Small shrubs, deciduous

13.1 (2.2~24.8); 6.0 (-6.3~20.1); 18.7 (9.5~28.0); 13.3 (3.9~24.8); 106.1 (15.8~237.6); 8.5 (0.0~48.6); 730 (411~1385) ❙ 951 (628~1405); 529 (406~644); 29 (11~48) ❙ 36.8 (-14.2~92.9); 677 (411~1137); 292.5 (201.9~527.5)

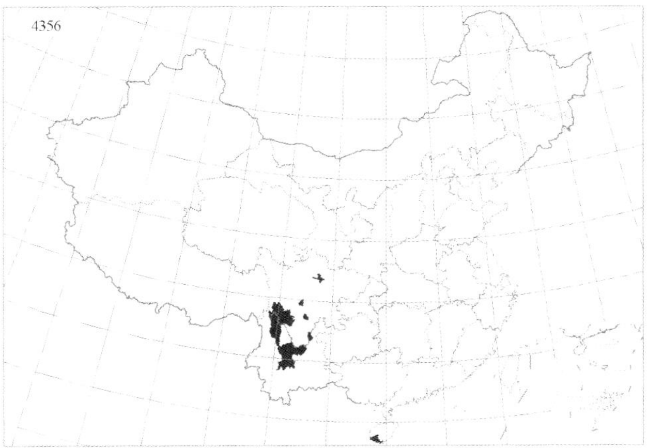

Fabaceae 豆科 *Lespedeza* 胡枝子属

Lespedeza bicolor Turcz. 胡枝子

Shrubs, deciduous

12.8 (−4.0~22.3)；−0.8 (−28.2~14.7)；25.3 (13.1~29.9)；13.7 (5.0~22.4)；114.1 (27.9~208.3)；19.4 (0.0~146.2)；788 (448~1188) ❙ 1016 (137~2993)；449 (92~954)；89 (3~547) ❙ 24.9 (−77.5~203.4)；732 (137~1188)；199.0 (25.4~346.5)

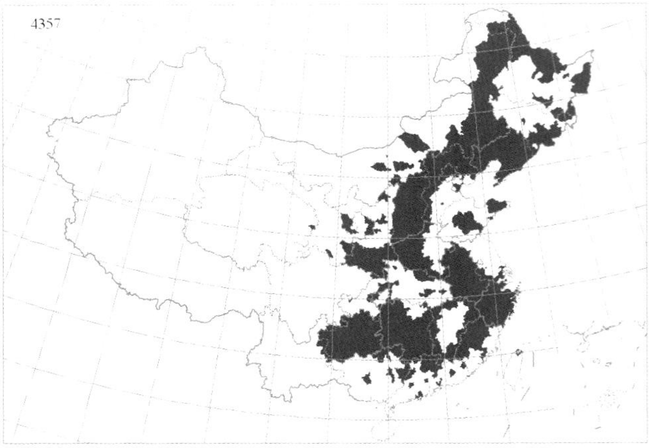

Fabaceae 豆科 *Lespedeza* 胡枝子属

Lespedeza buergeri Miq. 绿叶胡枝子

Shrubs, deciduous

14.6 (3.5~19.4)；2.0 (−6.5~9.3)；26.5 (12.6~29.7)；14.6 (5.0~19.4)；121.7 (26.3~173.4)；6.5 (0.0~43.4)；825 (446~1029) ❙ 1111 (512~1922)；466 (258~685)；108 (14~259) ❙ 33.4 (−35.9~115.7)；801 (446~1029)；207.8 (116.6~334.4)

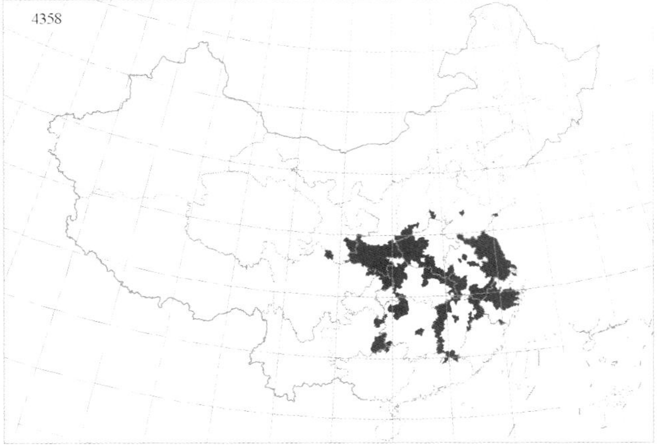

Fabaceae 豆科 *Lespedeza* 胡枝子属

Lespedeza caraganae Bunge 长叶胡枝子

Shrubs, deciduous

10.9 (2.2~15.8)；−4.3 (−16.0~2.5)；24.5 (18.5~28.0)；11.8 (6.7~15.8)；97.7 (48.1~133.1)；26.4 (3.4~81.1)；729 (515~891) ❙ 600 (325~1200)；349 (216~536)；25 (7~109) ❙ −17.9 (−47.7~41.4)；590 (325~891)；162.5 (53.8~274.1)

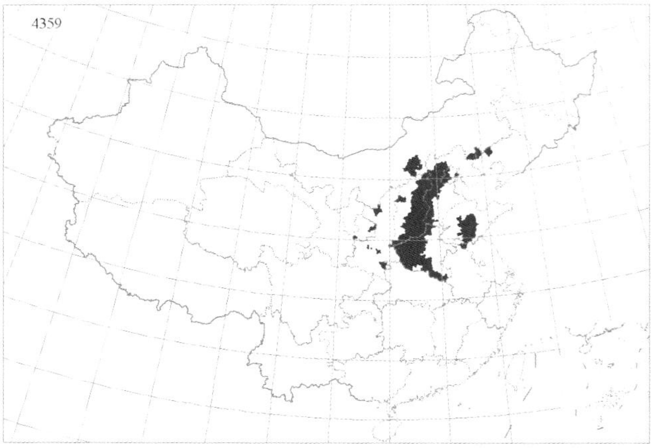

Fabaceae 豆科 *Lespedeza* 胡枝子属

Lespedeza chinensis G. Don 中华胡枝子

Small shrubs, deciduous

17.6 (6.2~23.6)；7.1 (−2.4~16.7)；27.4 (14.1~29.8)；17.7 (6.7~23.6)；153.3 (39.4~223.5)；1.0 (0.0~24.1)；946 (509~1281) ❙ 1523 (730~2993)；593 (370~1388)；163 (21~547) ❙ 62.5 (−16.9~246.5)；940 (509~1273)；219.0 (95.3~363.7)

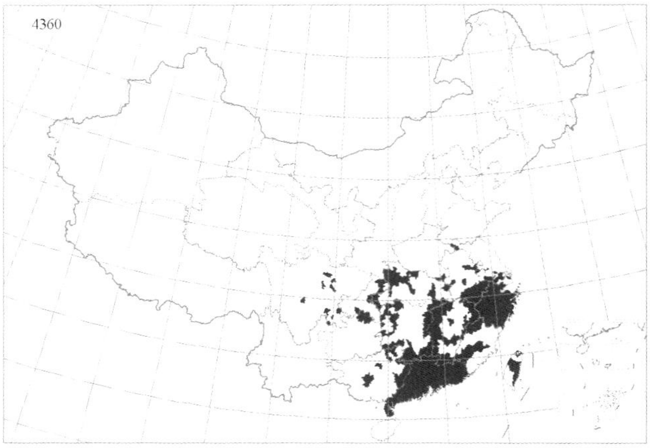

Fabaceae 豆科 *Lespedeza* 胡枝子属

Lespedeza cuneata (Dum. Cours.) G. Don 截叶铁扫帚

Herbs or subshrubs, deciduous

14.9 (−1.8~24.1)；3.8 (−12.7~18.9)；25.0 (6.1~29.9)；15.1 (1.8~24.1)；125.3 (1.9~229.7)；6.1 (0.0~82.8)；823 (329~1322) ❙ 1175 (320~3495)；516 (193~1594)；100 (9~897) ❙ 41.5 (−36.5~246.5)；794 (320~1202)；223.3 (70.3~516.7)

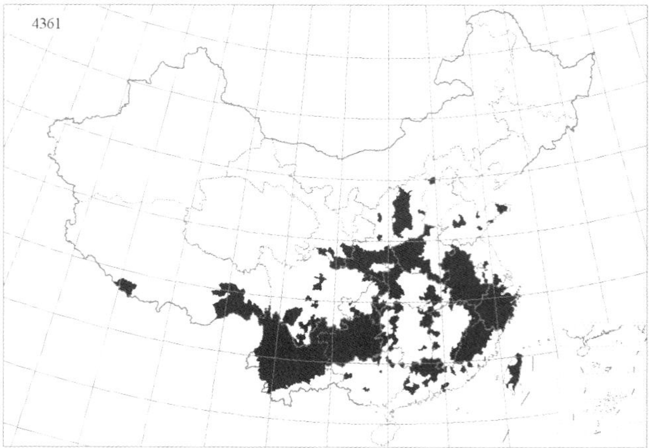

Fabaceae 豆科 *Lespedeza* 胡枝子属

Lespedeza cyrtobotrya Miq. 短梗胡枝子

Shrubs, deciduous

10.9 (1.4~18.2)；−4.0 (−20.1~7.5)；24.3 (17.5~29.3)；12.0 (6.4~18.2)；98.3 (45.4~158.8)；27.0 (0.0~93.6)；727 (501~984) ❙ 808 (377~1922)；406 (241~691)；51 (8~259) ❙ 9.3 (−40.9~112.2)；657 (377~961)；203.0 (109.4~337.1)

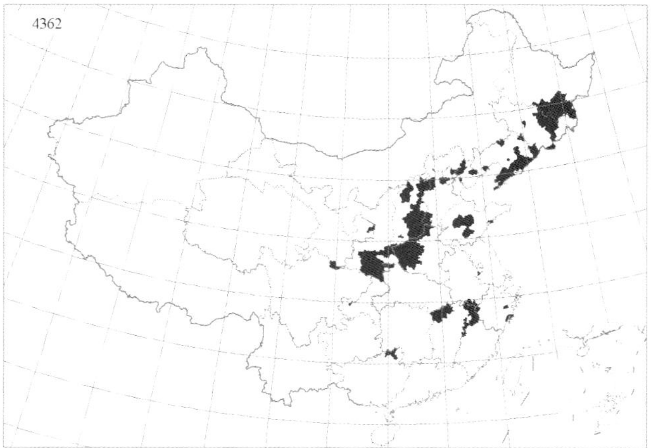

Fabaceae 豆科 *Lespedeza* 胡枝子属
Lespedeza daurica (Laxm.) Schindl. 兴安胡枝子
Herbs or subshrubs, deciduous

10.6 (-4.6~22.0)；-4.0 (-28.8~15.3)；23.9 (8.7~29.3)；11.8 (2.5~22.0)；96.7 (7.7~203.8)；28.4 (0.0~149.1)；723 (349~1170) ❙ 743 (49~2859)；380 (34~1045)；51 (1~402) ❙ 0.9 (-92.8~235.6)；622 (49~1170)；179.7 (5.1~357.7)

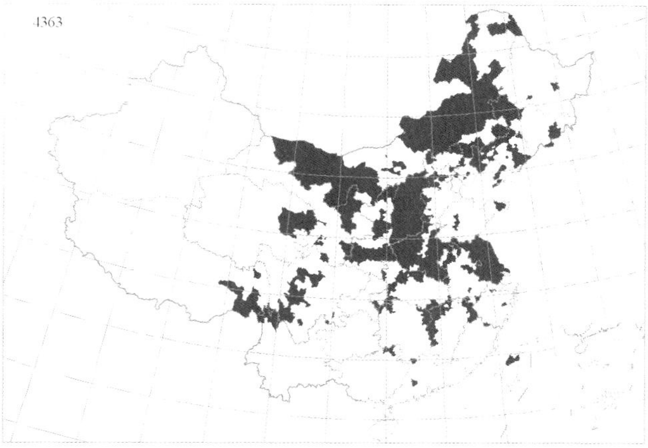

Fabaceae 豆科 *Lespedeza* 胡枝子属
Lespedeza davidii Franch. 大叶胡枝子
Shrubs, deciduous

16.8 (2.2~22.1)；5.3 (-6.6~13.7)；27.7 (9.9~30.1)；16.9 (3.8~22.1)；143.0 (15.5~205.6)；0.7 (0.0~49.1)；907 (413~1182) ❙ 1478 (871~2005)；519 (299~824)；178 (19~259) ❙ 63.5 (-6.7~128.9)；888 (413~1182)；210.4 (111.9~319.9)

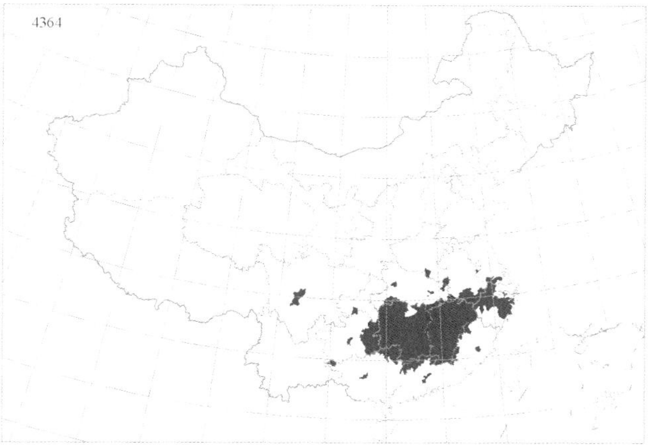

Fabaceae 豆科 *Lespedeza* 胡枝子属
Lespedeza dunnii Schindl. 春花胡枝子
Shrubs, deciduous

16.3 (10.6~19.)；5.4 (-0.2~10.0)；27.1 (21.1~29.2)；16.4 (10.6~19.1)；137.8 (79.4~163.1)；1.1 (0.0~12.1)；874 (643~973) ❙ 1617 (730~1997)；583 (424~713)；189 (46~254) ❙ 85.6 (-16.9~147.0)；867 (643~973)；233.3 (139.2~305.7)

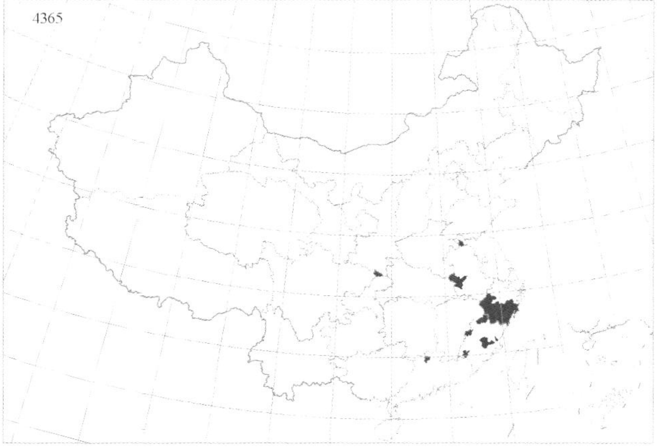

Fabaceae 豆科 *Lespedeza* 胡枝子属
Lespedeza floribunda Bunge 多花胡枝子
Small shrubs, deciduous

12.1 (1.9~22.5)；-1.8 (-16.1~15.0)；24.7 (9.5~29.9)；12.7 (3.8~22.6)；105.3 (15.4~210.9)；20.1 (0.0~81.0)；759 (411~1206) ❙ 812 (205~1997)；403 (119~862)；58 (5~259) ❙ 5.2 (-68.3~130.7)；674 (205~1206)；180.2 (33.5~357.7)

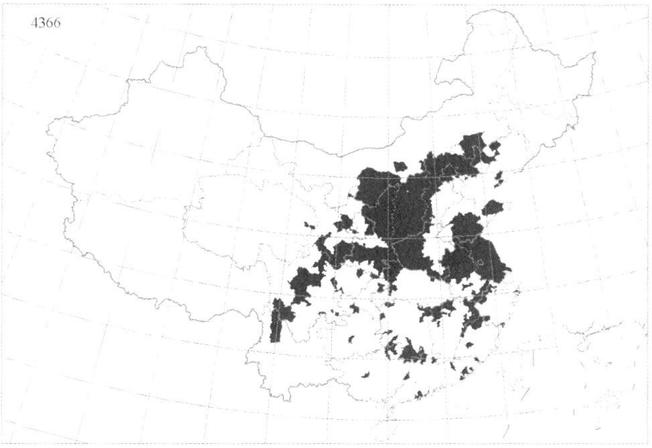

Fabaceae 豆科 *Lespedeza* 胡枝子属
Lespedeza fordii Schindl. 广东胡枝子
Shrubs, deciduous

16.9 (13.0~21.8)；5.6 (1.7~13.1)；27.5 (20.8~29.4)；17.0 (13.1~21.8)；144.4 (102.9~202.0)；0.7 (0.0~6.1)；908 (730~1168) ❙ 1519 (969~1950)；542 (419~770)；184 (53~253) ❙ 67.9 (13.4~129.9)；902 (730~1160)；222.4 (134.6~319.9)

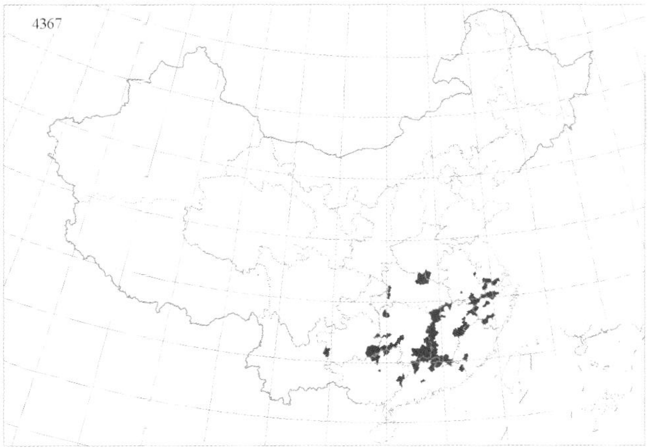

Fabaceae 豆科 *Lespedeza* 胡枝子属
Lespedeza forrestii Schindl. 矮生胡枝子
Shrubs or subshrubs, deciduous

13.3 (5.4~18.9)；5.6 (-2.1~12.0)；19.6 (12.3~26.6)；13.3 (5.8~19.0)；102.3 (31.5~167.8)；2.3 (0.0~25.6)；713 (481~929) ❙ 968 (741~1141)；567 (411~651)；29 (12~57) ❙ 38.0 (-14.2~78.3)；684 (481~912)；276.9 (142.9~351.7)

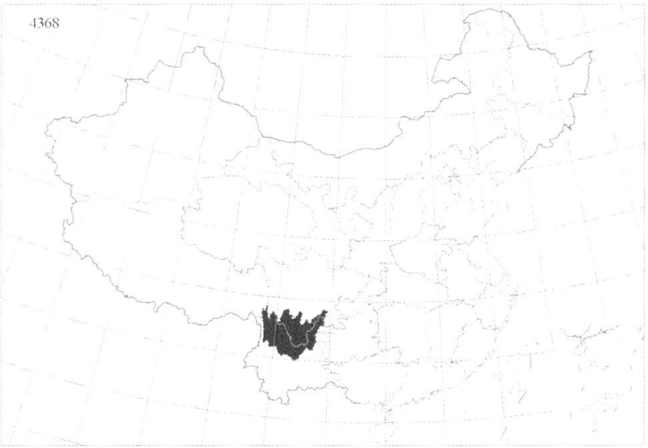

Fabaceae 豆科 *Lespedeza* 胡枝子属

Lespedeza inschanica (Maxim.) Schindl. 阴山胡枝子

Shrubs, deciduous

11.1 (-4.6~16.6); -3.7 (-28.8~6.5); 24.5 (12.4~28.8); 12.1 (4.8~16.7); 99.7 (25.1~140.6); 25.9 (0.0~149.1); 735 (441~920) ❙ 768 (249~1571); 410 (179~666); 39 (6~157) ❙ 4.8 (-54.1~108.0); 666 (249~912); 201.9 (91.4~346.5)

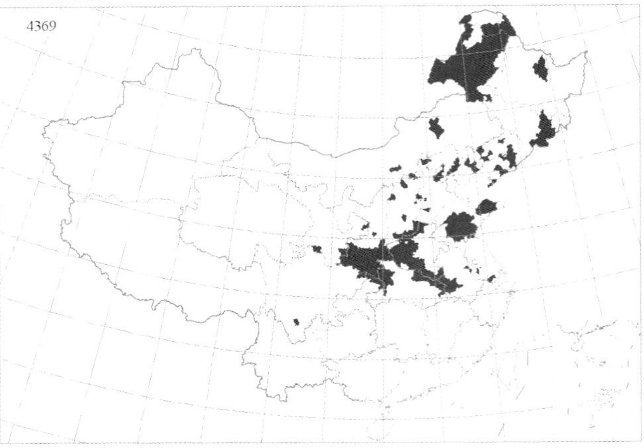

Fabaceae 豆科 *Lespedeza* 胡枝子属

Lespedeza juncea (L. f.) Pers. 尖叶铁扫帚

Small shrubs, deciduous

13.1 (-5.8~23.8); 0.9 (-29.8~18.1); 23.9 (6.1~29.9); 14.0 (1.8~23.8); 116.7 (1.9~225.8); 18.5 (0.0~157.4); 784 (329~1297) ❙ 1046 (137~2993); 503 (92~1847); 77 (3~547) ❙ 29.2 (-77.5~246.5); 734 (137~1221); 226.2 (25.4~659.0)

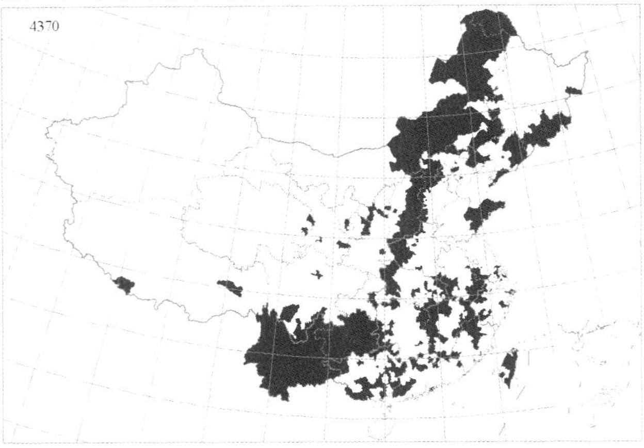

Fabaceae 豆科 *Lespedeza* 胡枝子属

Lespedeza maximowiczii C. K. Schneid. 宽叶胡枝子

Shrubs, deciduous

14.6 (10.2~17.2); 1.9 (-2.9~5.0); 26.8 (22.4~29.3); 14.7 (10.6~17.2); 121.6 (83.1~146.8); 6.1 (0.0~20.5); 826 (656~941) ❙ 1213 (720~1663); 499 (350~685); 126 (33~220) ❙ 45.4 (-13.5~101.3); 821 (656~941); 228.3 (134.1~318.0)

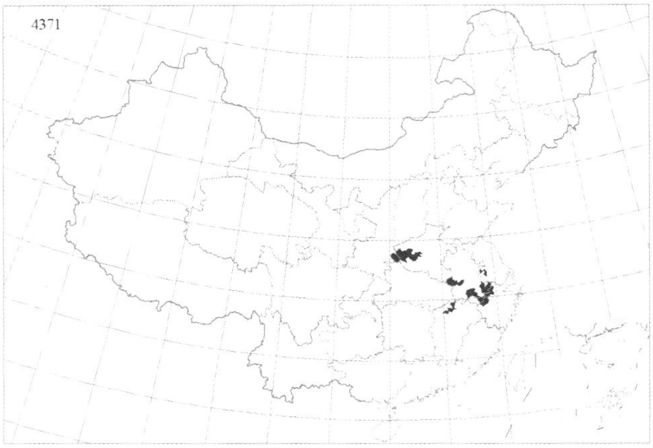

Fabaceae 豆科 *Lespedeza* 胡枝子属

Lespedeza mucronata Ricker 短叶胡枝子

Subshrubs, deciduous

19.0 (15.9~21.7); 9.8 (5.5~14.3); 27.5 (25.8~28.6); 19.0 (16.0~21.7); 168.3 (131.9~200.5); 0.0 (0.0~0.0); 992 (831~1140) ❙ 1651 (1485~1997); 616 (511~680); 186 (123~256) ❙ 69.8 (30.2~130.7); 992 (831~1140); 237.1 (188.9~303.4)

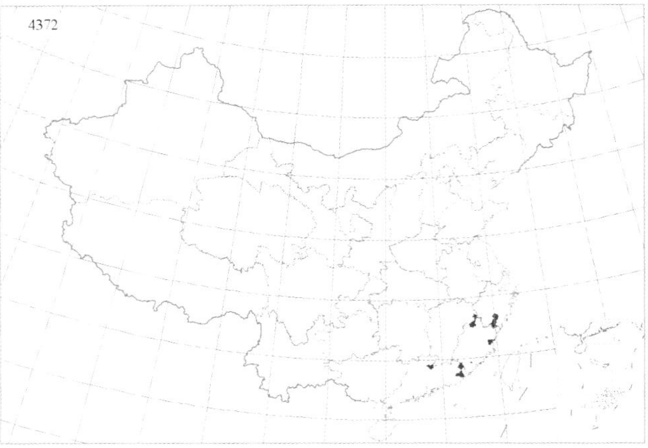

Fabaceae 豆科 *Lespedeza* 胡枝子属

Lespedeza potaninii V. N. Vassiljev 牛枝子

Subshrubs

6.7 (0.4~13.3); -8.6 (-16.0~2.4); 20.3 (9.9~26.8); 8.7 (3.6~13.5); 66.6 (15.5~115.5); 45.8 (5.4~81.1); 596 (404~811) ❙ 454 (111~1076); 267 (74~584); 11 (2~40) ❙ -21.4 (-82.3~121.7); 415 (111~736); 143.1 (13.9~357.7)

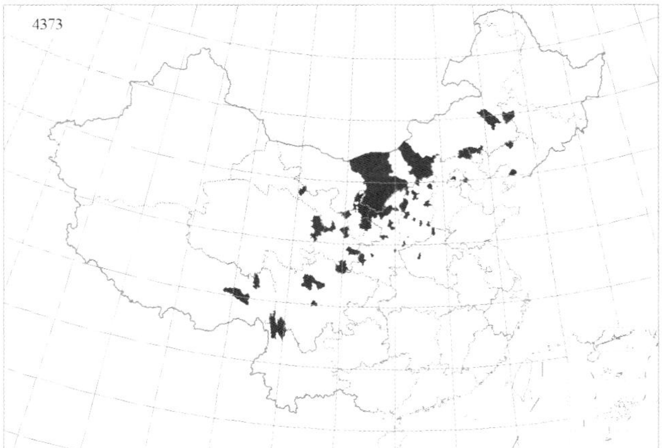

Fabaceae 豆科 *Lespedeza* 胡枝子属

Lespedeza thunbergii (A. DC.) Nakai 日本胡枝子

Shrubs, deciduous

15.3 (3.0~24.1); 3.5 (-12.5~18.9); 26.4 (11.7~30.1); 15.5 (4.7~24.1); 130.4 (23.2~229.7); 5.9 (0.0~62.6); 855 (441~1322) ❙ 1207 (377~2993); 499 (241~1594); 117 (8~547) ❙ 39.7 (-39.5~246.5); 820 (377~1216); 205.2 (90.7~487.2)

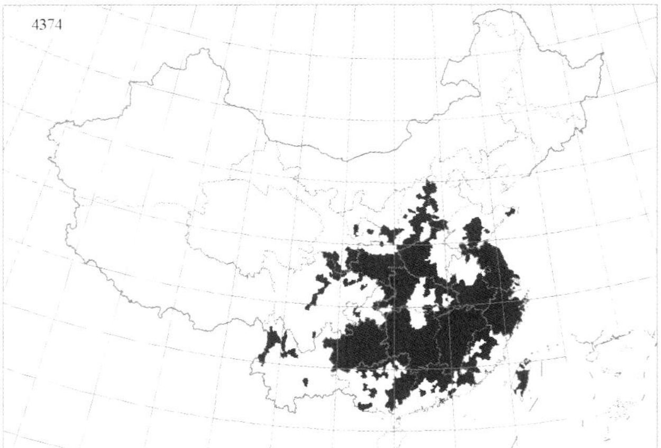

Fabaceae 豆科 *Lespedeza* 胡枝子属

Lespedeza tomentosa (Thunb.) Siebold ex Maxim. 绒毛胡枝子

Shrubs, deciduous

12.9 (−4.6~21.1); −0.8 (−28.8~13.2); 25.4 (8.3~29.4); 13.6 (2.5~21.1); 112.7 (7.0~193.7); 17.3 (0.0~149.1); 786 (354~1114) ❙ 932 (249~1997); 442 (179~821); 74 (5~259) ❙ 16.6 (−54.1~147.0); 708 (249~1114); 198.2 (70.3~491.5)

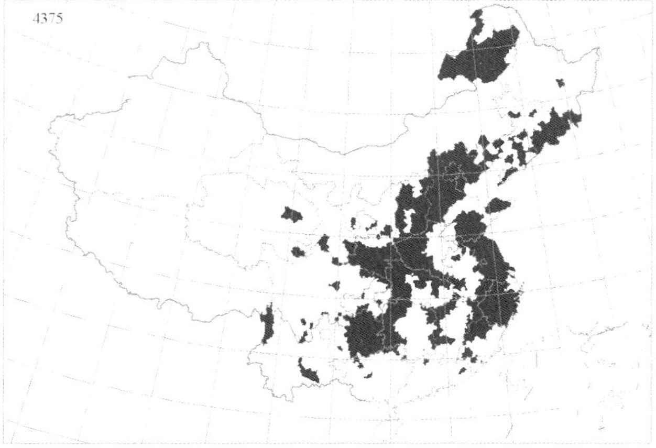

Fabaceae 豆科 *Lespedeza* 胡枝子属

Lespedeza virgata (Thunb.) DC. 细梗胡枝子

Small shrubs, deciduous

13.7 (2.5~22.0); 0.7 (−15.9~15.3); 25.8 (17.9~29.8); 14.1 (6.9~22.0); 117.7 (48.9~204.3); 12.4 (0.0~80.4); 805 (523~1176) ❙ 1036 (307~2993); 465 (173~1388); 91 (7~547) ❙ 26.1 (−49.4~246.5); 756 (307~1176); 199.8 (42.6~363.7)

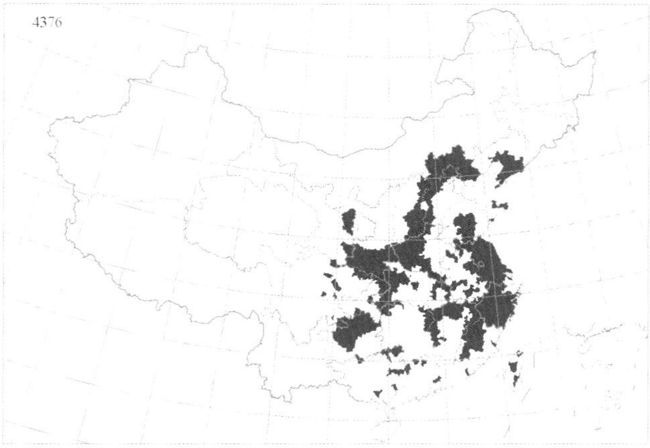

Fabaceae 豆科 *Lespedeza* 胡枝子属

Lespedeza virgata var. macrovirgata (Kitag.) Kitag. 大细梗胡枝子

Small shrubs, deciduous

8.6 (7.0~10.5); −9.2 (−11.0~−4.9); 24.1 (22.4~25.1); 10.4 (9.3~11.3); 87.6 (76.0~93.2); 43.7 (26.8~51.4); 681 (624~710) ❙ 719 (576~1018); 451 (374~648); 25 (11~36) ❙ 6.3 (−18.0~54.7); 645 (576~684); 221.0 (164.3~311.2)

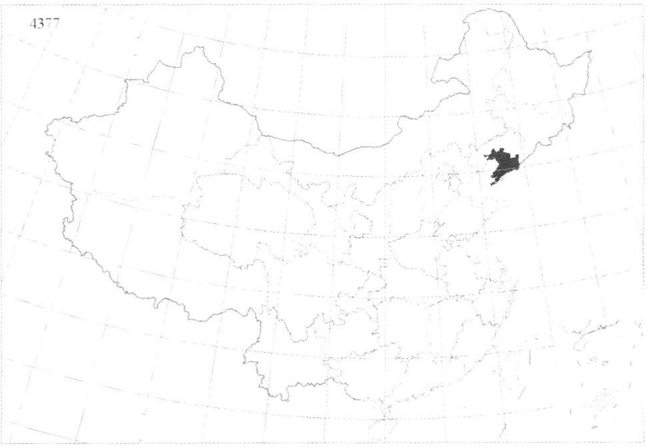

Fabaceae 豆科 *Leucaena* 银合欢属

Leucaena leucocephala (Lam.) de Wit 银合欢

Shrubs or small trees, evergreen

20.5 (13.3~24.8); 12.8 (2.4~20.1); 26.8 (20.2~29.0); 20.6 (13.2~24.8); 187.4 (100.9~237.6); 0.1 (0.0~4.4); 1084 (723~1385) ❙ 1521 (868~2993); 674 (312~1847); 109 (18~547) ❙ 43.2 (−34.3~214.4); 1046 (707~1360); 289.7 (119.4~620.0)

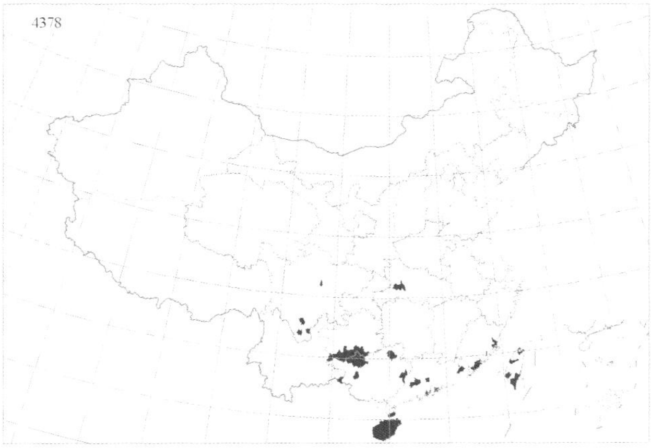

Fabaceae 豆科 *Lysidice* 仪花属

Lysidice rhodostegia Hance 仪花

Shrubs or small trees

20.2 (14.6~23.4); 11.8 (5.7~16.1); 26.9 (19.6~29.2); 20.3 (14.6~23.5); 183.5 (115.8~221.7); 0.0 (0.0~0.0); 1054 (736~1273) ❙ 1495 (1009~2037); 750 (548~1033); 104 (39~197) ❙ 42.5 (5.8~80.6); 1054 (732~1273); 262.3 (157.7~551.6)

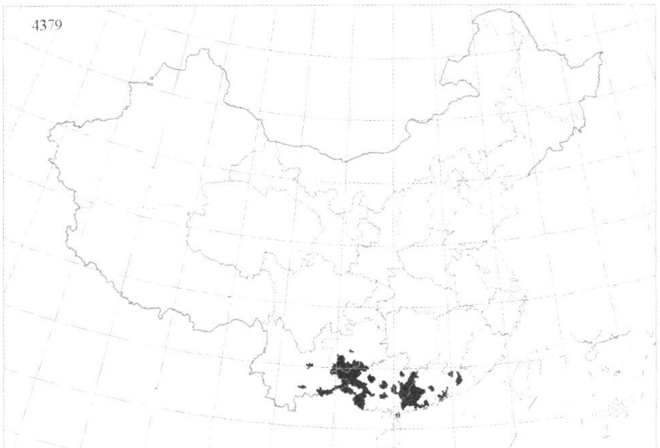

Fabaceae 豆科 *Maackia* 马鞍树属

Maackia amurensis Rupr. et Maxim. 朝鲜槐

Trees, deciduous

7.5 (−0.8~14.1); −10.3 (−25.0~−0.4); 23.0 (18.4~27.1); 10.0 (6.4~14.1); 81.5 (45.8~121.7); 50.8 (12.5~118.4); 657 (501~839) ❙ 705 (513~1073); 429 (290~691); 27 (9~65) ❙ 8.9 (−27.3~74.1); 628 (501~750); 231.1 (135.6~346.5)

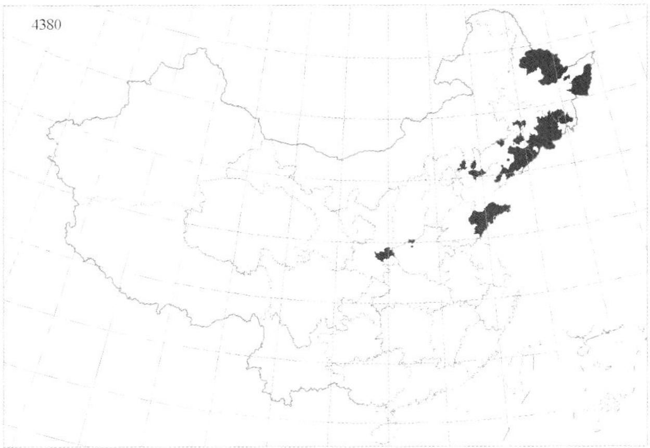

Fabaceae 豆科 *Maackia* 马鞍树属
Maackia australis (Dunn) Takeda 华南马鞍树
Shrubs, deciduous

22.2 (21.9~22.5); 14.3 (13.5~15.2); 28.6 (28.3~28.8); 22.3 (22.0~22.7); 207.3 (203.4~212.7); 0.0 (0.0~0.0); 1186 (1168~1216) ❙ 1766 (1575~2014); 850 (722~994); 114 (75~142) ❙ 48.8 (32.5~72.1); 1186 (1168~1216); 197.8 (139.6~217.4)

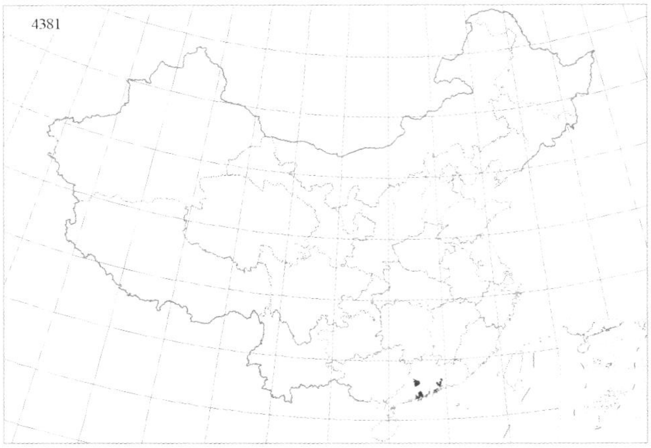

Fabaceae 豆科 *Maackia* 马鞍树属
Maackia chekiangensis S. S. Chien 浙江马鞍树
Shrubs, deciduous

16.3 (15.0~17.7); 4.2 (2.7~5.5); 28.2 (26.8~29.5); 16.3 (15.0~17.7); 137.1 (123.7~152.8); 1.1 (0.0~3.2); 894 (826~966) ❙ 1310 (871~1663); 478 (299~595); 163 (107~212) ❙ 48.4 (-6.7~101.3); 844 (755~900); 205.3 (159.4~256.1)

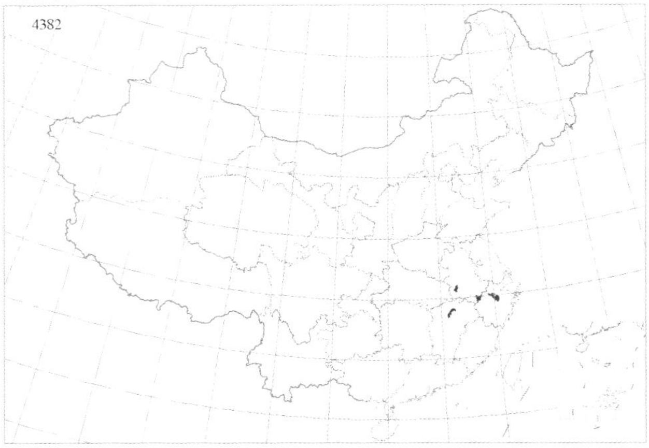

Fabaceae 豆科 *Maackia* 马鞍树属
Maackia floribunda (Miq.) Takeda 多花马鞍树
Shrubs or small trees, deciduous

19.9; 13.5; 25.9; 19.9; 179.1; 0.0; 987 ❙ 2993; 924; 547 ❙ 203.4; 987; 261.4

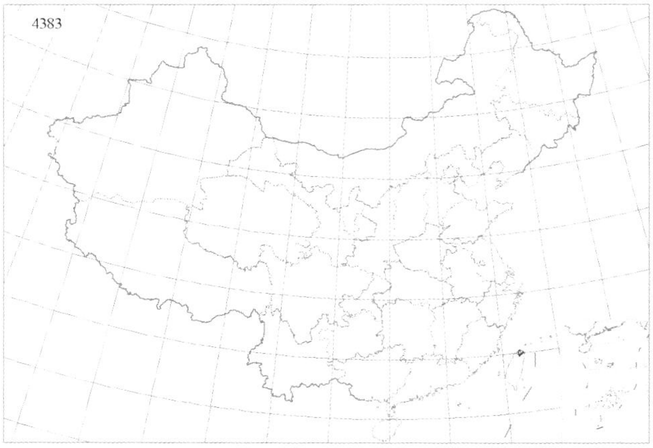

Fabaceae 豆科 *Maackia* 马鞍树属
Maackia hupehensis Takeda 马鞍树
Trees, deciduous

14.9 (7.4~18.2); 2.7 (-4.2~7.8); 26.6 (18.7~29.3); 15.0 (8.2~18.3); 124.2 (58.8~159.6); 5.1 (0.0~29.4); 832 (561~984) ❙ 1242 (512~1950); 484 (258~713); 129 (16~259) ❙ 47.7 (-35.3~147.0); 817 (512~965); 220.9 (121.6~334.4)

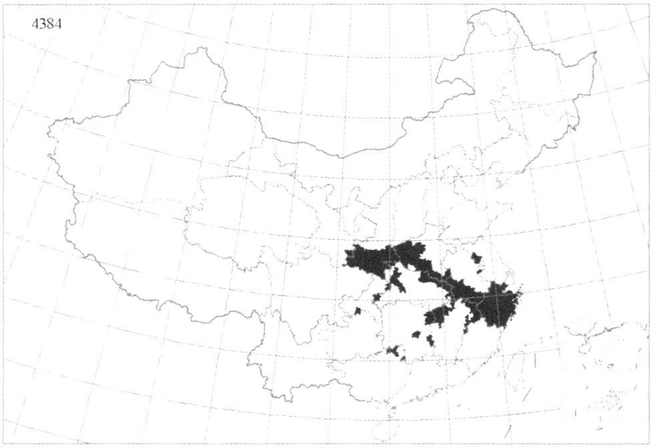

Fabaceae 豆科 *Maackia* 马鞍树属
Maackia tenuifolia (Hemsl.) Hand.-Mazz. 光叶马鞍树
Shrubs or small trees, deciduous

14.9 (7.4~17.1); 2.2 (-4.2~5.3); 27.2 (18.7~29.3); 15.0 (8.2~17.1); 125.2 (58.8~145.7); 5.4 (0.0~29.4); 846 (561~941) ❙ 1084 (626~1662); 467 (316~666); 110 (25~218) ❙ 27.6 (-22.2~108.0); 828 (561~941); 183.1 (131.1~318.2)

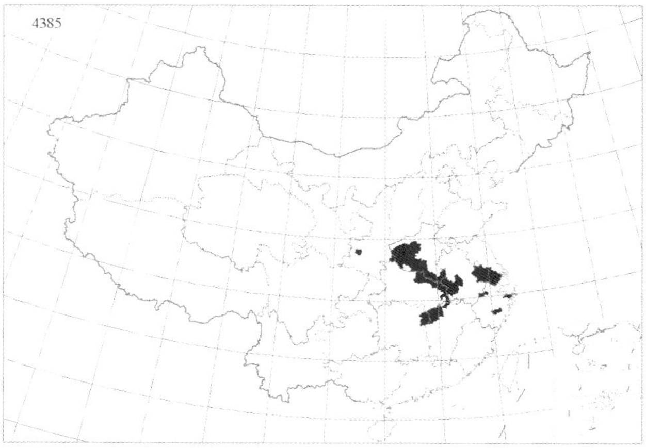

Fabaceae 豆科 *Millettia* 鸡血藤属
Millettia cubittii Dunn 红河鸡血藤
Trees

19.1 (17.3~21.0); 13.1 (11.0~15.1); 23.1 (21.6~25.6); 19.2 (17.3~21.0); 170.3 (148.0~191.8); 0.0 (0.0~0.0); 924 (832~1053) ❙ 1393 (1049~1751); 752 (598~996); 59 (44~77) ❙ 50.7 (21.7~88.8); 907 (783~1053); 475.5 (346.9~659.0)

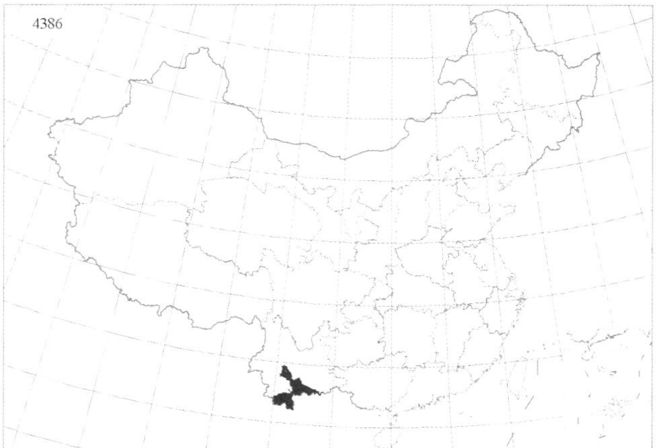

Fabaceae 豆科 *Millettia* 鸡血藤属
Millettia entadoides Z. Wei 榼藤子鸡血藤
Vines

17.4；11.0；21.4；17.4；149.3；0.0；834 ❙ 1406；816；48 ❙ 68.7；834；434.4

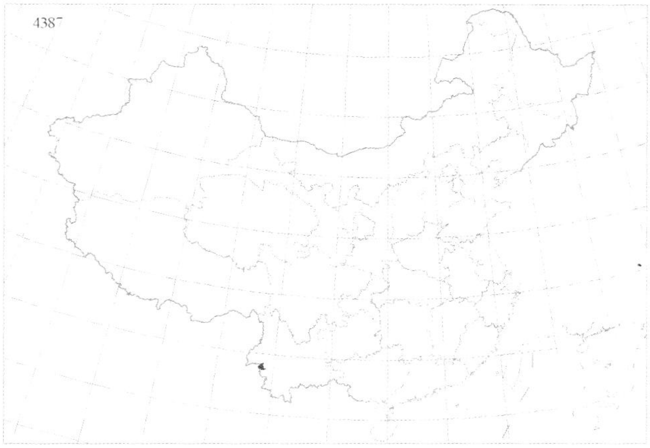

Fabaceae 豆科 *Millettia* 鸡血藤属
Millettia erythrocalyx Gagnep. 红萼鸡血藤
Trees

20.5；15.0；23.7；20.5；185.9；0.0；990 ❙ 1536；678；72 ❙ 55.1；973；614.6

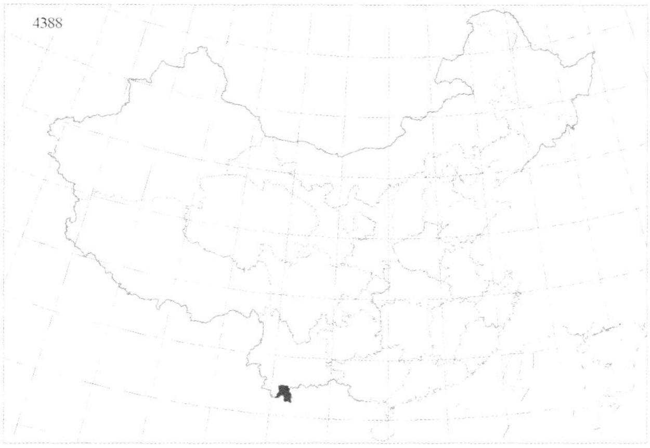

Fabaceae 豆科 *Millettia* 鸡血藤属
Millettia griffithii Dunn 孟连鸡血藤
Shrubs

18.3 (14.4~20.5)；12.1 (7.5~15.1)；22.1 (19.1~23.7)；18.3 (14.4~20.5)；159.8 (113.1~185.9)；0.0 (0.0~0.0)；879 (727~992) ❙ 1510 (1348~1751)；823 (655~996)；55 (41~77) ❙ 72.5 (47.6~112.2)；866 (727~986)；513.5 (395.4~659.0)

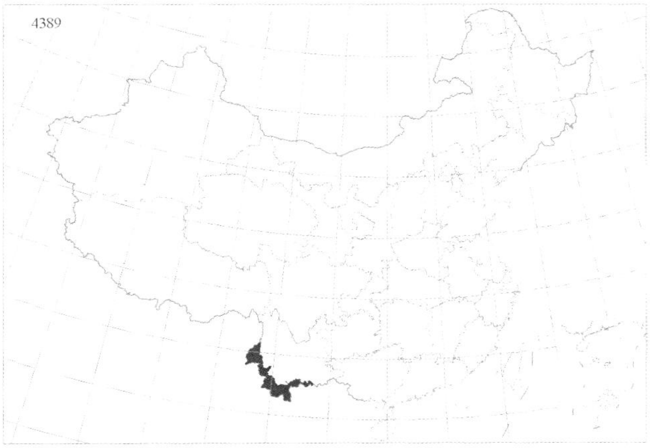

Fabaceae 豆科 *Millettia* 鸡血藤属
Millettia ichthyochtona Drake 闹鱼鸡血藤
Trees

18.6 (16.9~21.5)；11.6 (9.5~14.2)；23.6 (21.4~27.4)；18.7 (16.9~21.6)；164.0 (142.7~199.1)；0.0 (0.0~0.0)；914 (813~1131) ❙ 1393 (1178~1581)；780 (646~860)；61 (48~106) ❙ 53.1 (36.8~80.6)；913 (813~1131)；419.5 (281.7~551.6)

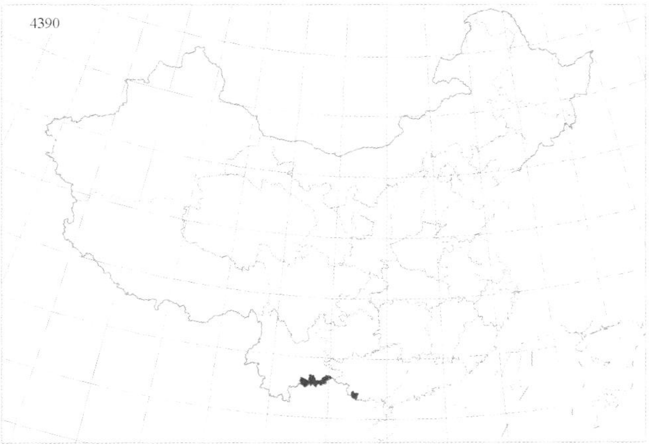

Fabaceae 豆科 *Millettia* 鸡血藤属
Millettia lantsangensis Z. Wei 澜沧鸡血藤
Vines

19.0；13.3；22.2；19.0；167.9；0.0；901 ❙ 1474；718；49 ❙ 63.8；866；489.3

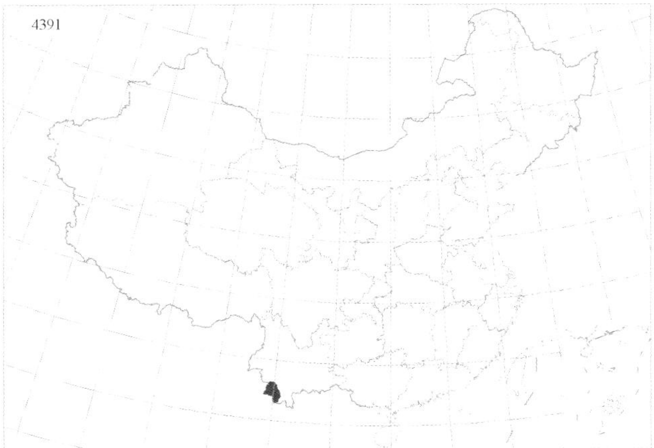

Fabaceae 豆科 *Millettia* 鸡血藤属
Millettia leptobotrya Dunn 思茅鸡血藤
Trees

18.9 (16.3~20.5)；12.7 (6.3~15.1)；23.1 (21.6~25.4)；19.0 (16.3~20.5)；167.8 (136.0~185.9)；0.0 (0.0~0.0)；915 (844~992) ❙ 1487 (1035~1751)；755 (528~996)；64 (52~77) ❙ 62.4 (22.6~88.8)；902 (839~986)；481.9 (179.0~659.0)

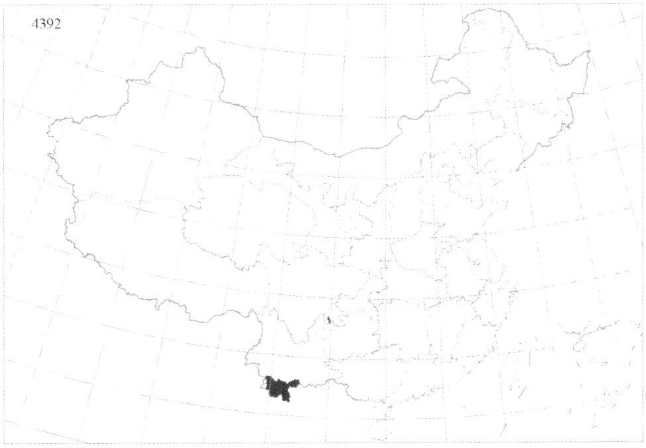

Fabaceae 豆科 *Millettia* 鸡血藤属

Millettia macrostachya Collett et Hemsl. 大穗鸡血藤

Trees

18.6; 12.7; 22.5; 18.7; 163.9; 0.0; 888 ❙ 1547; 911; 63 ❙ 74.2; 888; 551.3

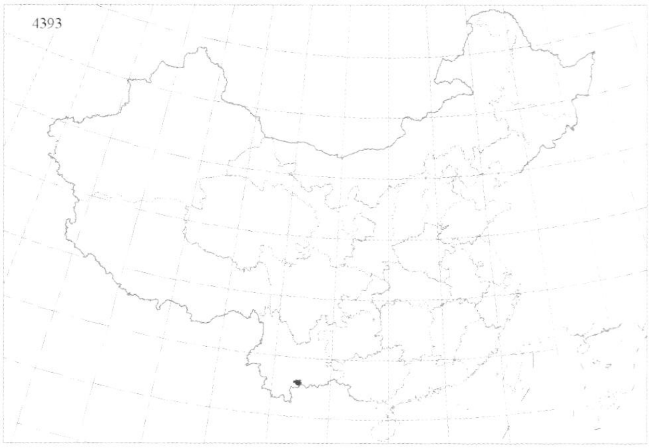

Fabaceae 豆科 *Millettia* 鸡血藤属

Millettia oraria (Hance) Dunn 香港鸡血藤

Shrubs or small trees

21.5 (18.2~22.7); 13.4 (7.7~15.0); 27.9 (27.3~28.6); 21.5 (18.2~22.8); 198.4 (158.5~213.1); 0.0 (0.0~0.0); 1137 (949~1221) ❙ 1991 (1512~2435); 1025 (507~1428); 129 (110~180) ❙ 74.6 (59.3~99.5); 1137 (949~1221); 255.2 (163.7~318.8)

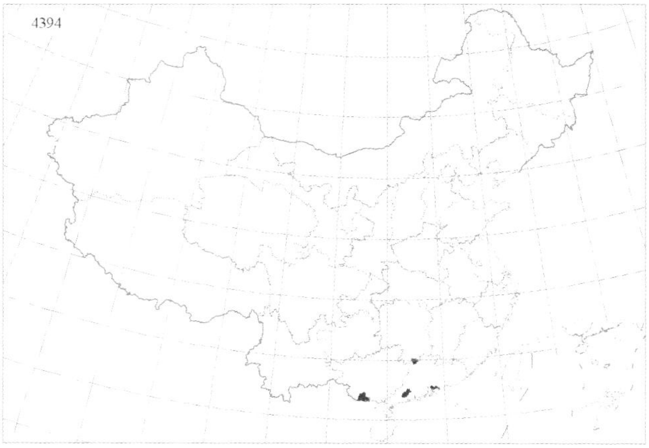

Fabaceae 豆科 *Millettia* 鸡血藤属

Millettia pachycarpa Benth. 厚果鸡血藤

Vines

17.6 (5.3~23.2); 8.6 (-3.5~15.8); 25.1 (12.2~29.8); 17.6 (5.7~23.2); 151.8 (30.5~218.9); 0.4 (0.0~28.1); 910 (479~1257) ❙ 1389 (879~2993); 640 (423~1428); 105 (20~551) ❙ 53.5 (5.5~246.5); 905 (479~1257); 261.9 (108.2~659.0)

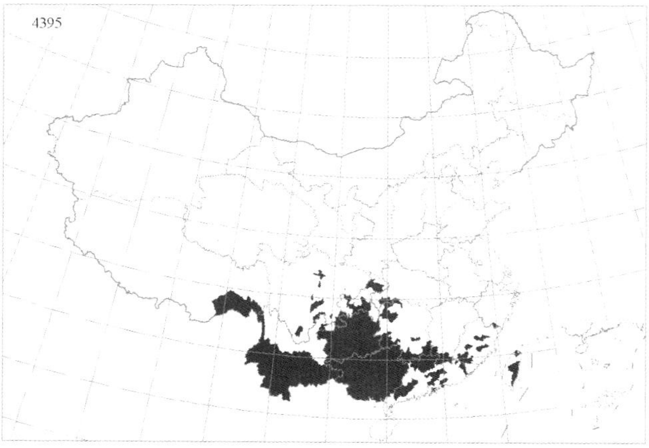

Fabaceae 豆科 *Millettia* 鸡血藤属

Millettia pachyloba Drake 海南鸡血藤

Vines

20.5 (14.6~24.8); 12.6 (5.4~20.1); 26.5 (19.6~28.9); 20.5 (14.6~24.8); 186.3 (115.8~237.6); 0.0 (0.0~0.0); 1063 (736~1385) ❙ 1475 (868~2435); 710 (312~1428); 95 (39~194) ❙ 40.8 (-34.3~99.5); 1037 (732~1326); 331.9 (157.7~659.0)

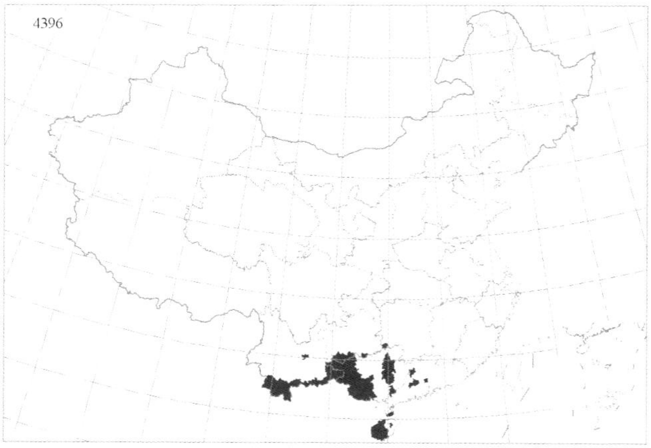

Fabaceae 豆科 *Millettia* 鸡血藤属

Millettia pulchra (Benth.) Kurz 印度鸡血藤

Shrubs or small trees

19.2 (11.3~24.8); 11.7 (3.8~20.1); 25.0 (17.0~29.0); 19.2 (11.3~24.8); 170.6 (76.6~237.6); 0.0 (0.0~1.5); 979 (644~1385) ❙ 1554 (958~3495); 733 (452~1847); 118 (39~897) ❙ 60.5 (1.5~246.5); 962 (644~1329); 300.9 (119.4~659.0)

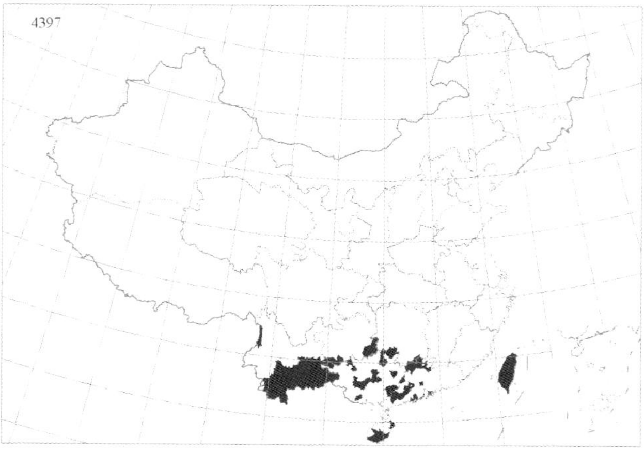

Fabaceae 豆科 *Millettia* 鸡血藤属

Millettia pulchra var. chinensis Dunn 华南小叶鸡血藤

Shrubs or small trees

18.5 (17.4~19.2); 10.5 (9.4~11.8); 24.3 (21.4~26.6); 18.5 (17.4~19.2); 161.9 (149.3~170.8); 0.0 (0.0~0.0); 915 (834~981) ❙ 1236 (1059~1406); 680 (604~816); 54 (44~73) ❙ 35.8 (21.3~68.7); 909 (834~981); 356.1 (294.5~434.4)

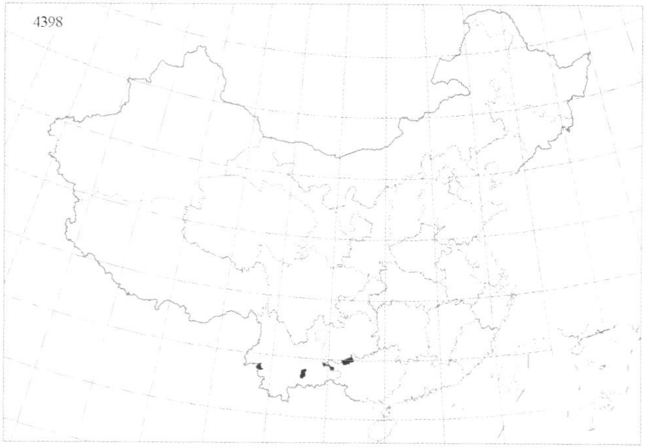

Fabaceae 豆科 *Millettia* 鸡血藤属

Millettia pulchra var. laxior (Dunn) Z. Wei 疏叶鸡血藤

Shrubs or small trees

19.0 (11.3~24.8); 10.1 (4.3~20.1); 26.4 (17.0~29.0); 19.0 (11.3~24.8); 168.0 (76.6~237.6); 0.0 (0.0~0.7); 998 (644~1385) | 1487 (1107~1840); 627 (425~889); 134 (41~202) | 51.9 (1.5~93.0); 984 (644~1232); 260.2 (157.7~527.5)

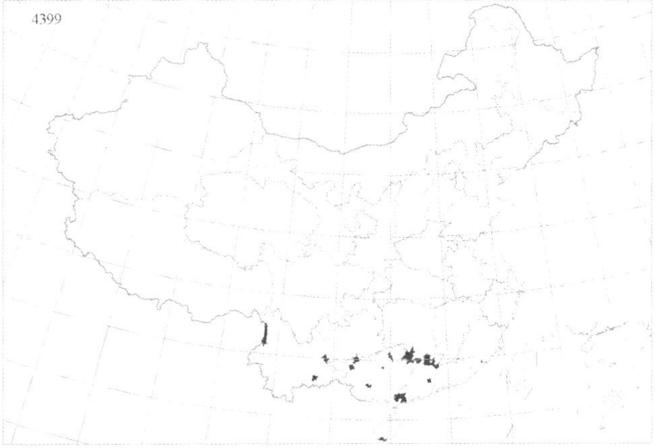

Fabaceae 豆科 *Millettia* 鸡血藤属

Millettia pulchra var. tomentosa Prain 绒叶印度鸡血藤

Shrubs or small trees

23.2; 15.8; 28.9; 23.2; 218.9; 0.0; 1257 | 1686; 824; 102 | 34.1; 1257; 194.5

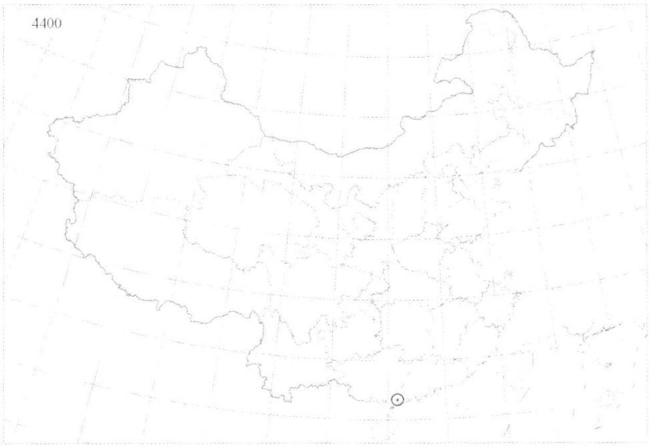

Fabaceae 豆科 *Millettia* 鸡血藤属

Millettia sapindifolia T. C. Chen 无患子叶鸡血藤

Vines

18.9 (18.2~19.2); 9.7 (8.7~11.0); 26.1 (25.2~27.5); 18.9 (18.3~19.2); 167.1 (159.3~170.8); 0.0 (0.0~0.0); 960 (928~1002) | 1335 (1236~1495); 694 (617~848); 76 (51~135) | 39.0 (27.5~55.9); 960 (928~1002); 277.3 (254.3~313.7)

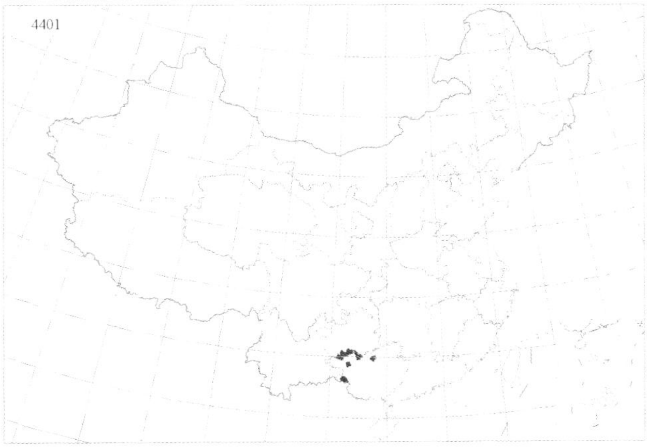

Fabaceae 豆科 *Millettia* 鸡血藤属

Millettia tetraptera Kurz 四翅鸡血藤

Trees

20.4; 15.1; 23.6; 20.5; 185.8; 0.0; 989 | 1607; 702; 77 | 62.5; 986; 659.0

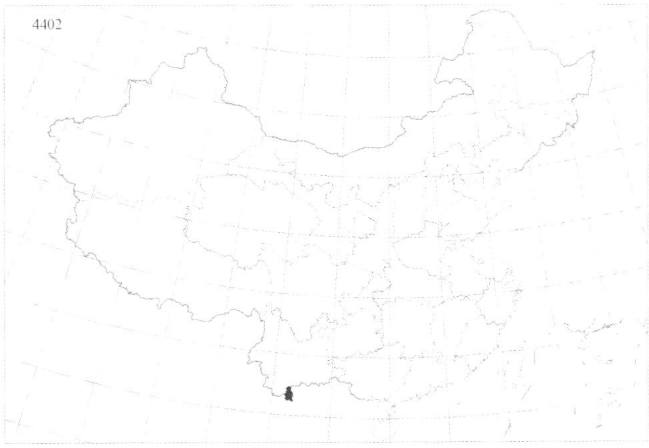

Fabaceae 豆科 *Millettia* 鸡血藤属

Millettia velutina Dunn 绒毛鸡血藤

Trees

17.0 (10.2~21.2); 9.8 (2.4~15.1); 22.2 (16.9~28.5); 17.0 (10.3~21.3); 144.6 (68.2~195.2); 0.1 (0.0~5.1); 840 (620~1126) | 1203 (797~1777); 655 (411~996); 55 (18~191) | 43.0 (-14.2~112.2); 819 (620~1126); 341.1 (176.4~559.0)

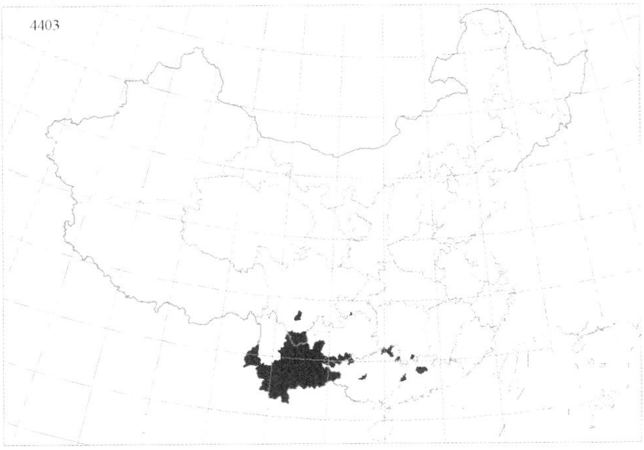

Fabaceae 豆科 *Mimosa* 含羞草属

Mimosa bimucronata (DC.) Kuntze 光荚含羞草

Shrubs, deciduous

23.5 (19.1~24.8); 17.6 (9.8~20.1); 27.8 (25.7~28.8); 23.6 (19.2~24.8); 223.1 (170.1~237.6); 0.0 (0.0~0.0); 1280 (981~1385) | 1476 (868~1854); 546 (312~712); 72 (41~115) | 15.8 (-34.3~47.8); 1173 (868~1360); 430.6 (248.2~620.0)

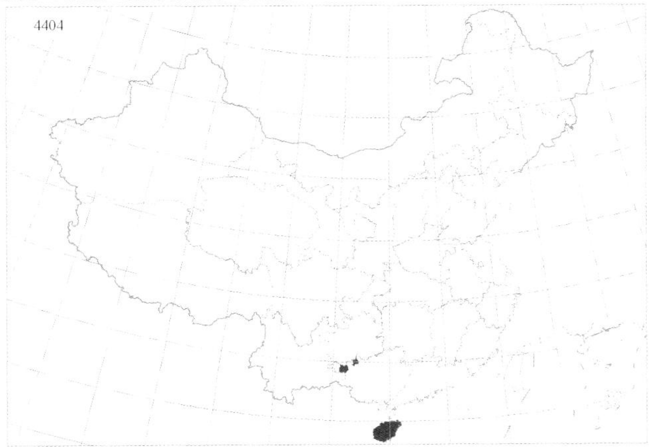

Fabaceae 豆科 *Mucuna* 黧豆属
Mucuna birdwoodiana Tutcher 白花油麻藤
Large climbers, evergreen

18.2 (11.7~23.4); 8.8 (2.7~16.1); 26.4 (18.6~29.8); 18.3 (11.7~23.5); 159.8 (82.6~221.7); 0.2 (0.0~3.8); 958 (658~1273) ❙ 1437 (896~2435); 642 (444~1428); 117 (31~254) ❙ 50.6 (5.8~130.7); 958 (658~1273); 227.8 (133.6~411.1)

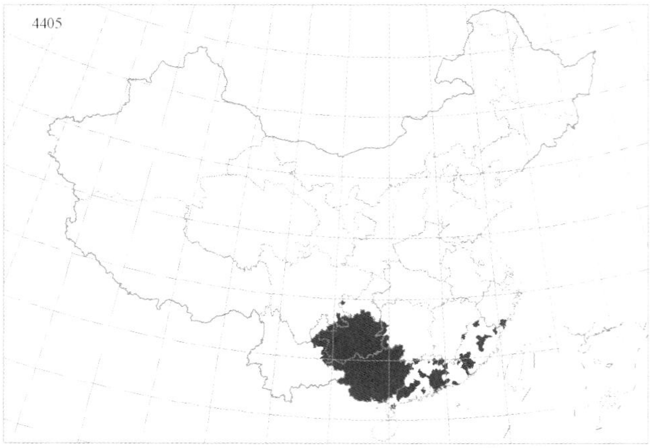

Fabaceae 豆科 *Mucuna* 黧豆属
Mucuna bracteata DC. 黄毛黧豆
Vines, evergreen

19.5; 11.6; 25.3; 19.5; 174.0; 0.0; 988 ❙ 1140; 635; 48 ❙ 16.3; 972; 345.1

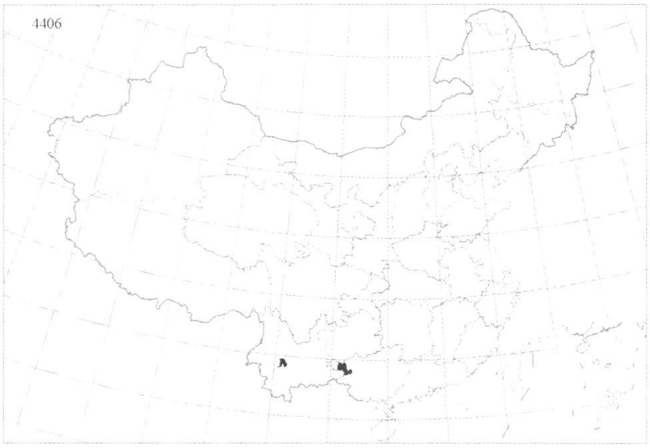

Fabaceae 豆科 *Mucuna* 黧豆属
Mucuna calophylla W. W. Sm. 美叶油麻藤
Vines, evergreen

14.6 (10.1~18.9); 7.6 (2.3~12.0); 20.1 (16.3~23.9); 14.6 (10.2~19.0); 116.1 (68.0~167.8); 0.4 (0.0~5.2); 747 (618~929) ❙ 955 (797~1057); 568 (411~627); 27 (15~42) ❙ 29.6 (-14.2~70.8); 696 (615~760); 281.8 (207.5~349.8)

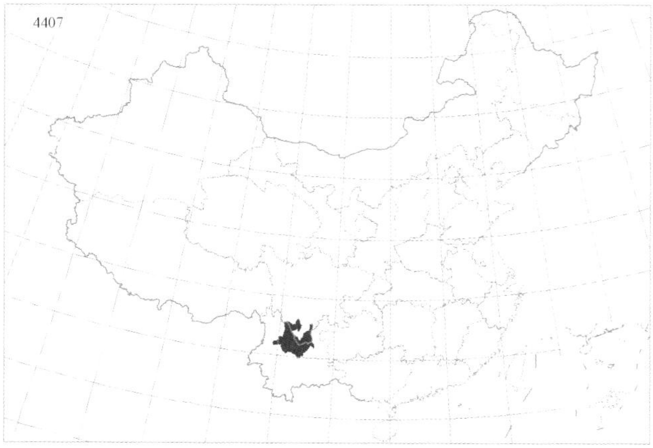

Fabaceae 豆科 *Mucuna* 黧豆属
Mucuna championii Benth. 港油麻藤
Vines, evergreen

19.2; 11.1; 27.7; 20.1; 181.3; 0.0; 1060 ❙ 1813; 813; 147 ❙ 71.7; 1060; 197.8

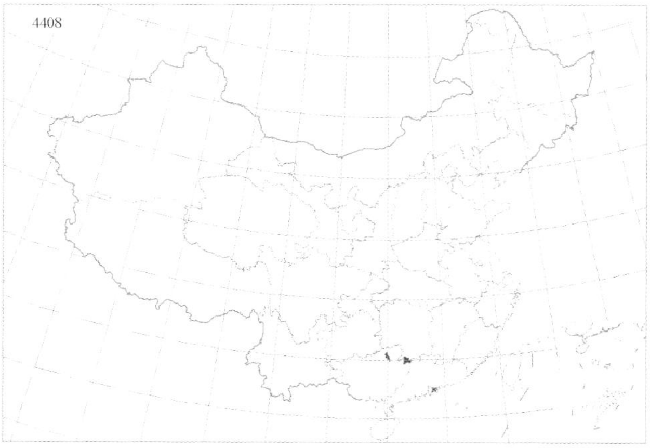

Fabaceae 豆科 *Mucuna* 黧豆属
Mucuna cyclocarpa F. P. Metcalf 闽黧豆
Vines, evergreen

16.9 (14.6~18.2); 5.6 (3.8~7.5); 27.4 (24.6~29.3); 16.9 (14.7~18.2); 143.2 (117.7~158.3); 0.2 (0.0~1.5); 900 (779~961) ❙ 1654 (931~1997); 584 (323~685); 201 (113~254) ❙ 84.6 (-1.6~130.7); 888 (779~961); 239.9 (198.4~303.4)

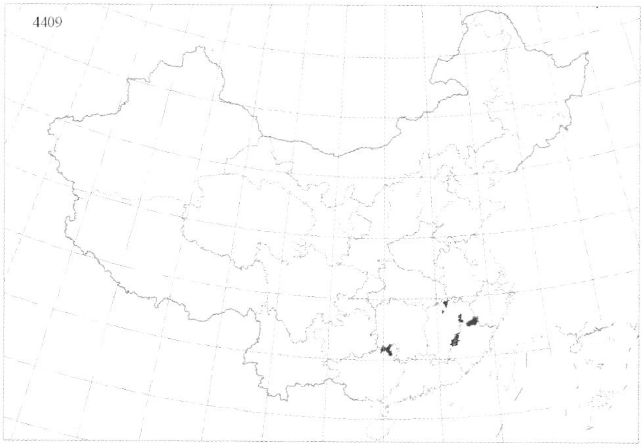

Fabaceae 豆科 *Mucuna* 黧豆属
Mucuna gigantea (Willd.) DC. 巨黧豆
Vines, evergreen

23.0 (18.5~24.5); 17.4 (14.1~18.9); 27.1 (21.7~28.8); 23.1 (18.5~24.5); 216.7 (162.1~234.4); 0.0 (0.0~0.0); 1230 (863~1360) ❙ 1901 (868~2992); 906 (312~1847); 107 (47~188) ❙ 65.4 (-34.3~214.4); 1162 (863~1360); 300.4 (154.2~428.6)

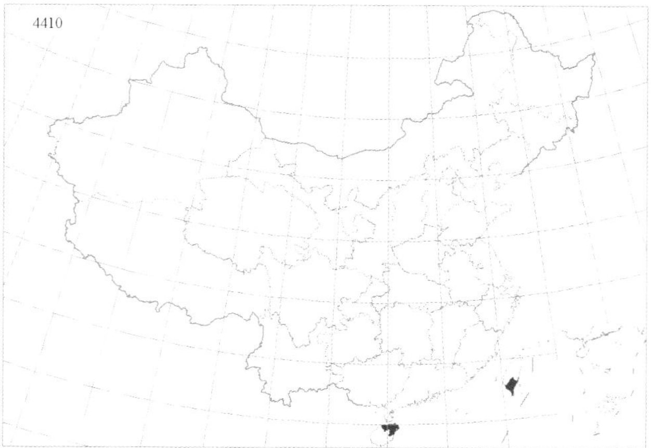

Fabaceae 豆科 *Mucuna* 黧豆属

Mucuna hainanensis Hayata 海南黧豆

Scandent shrubs, evergreen

22.0 (18.5~24.8); 15.0 (9.1~20.1); 27.3 (21.7~29.0); 22.1 (18.5~24.8); 205.0 (162.1~237.6); 0.0 (0.0~0.0); 1166 (863~1385) ❙ 1584 (868~2713); 703 (312~1388); 106 (41~188) ❙ 39.0 (-34.3~214.4); 1119 (863~1299); 356.8 (157.7~659.0)

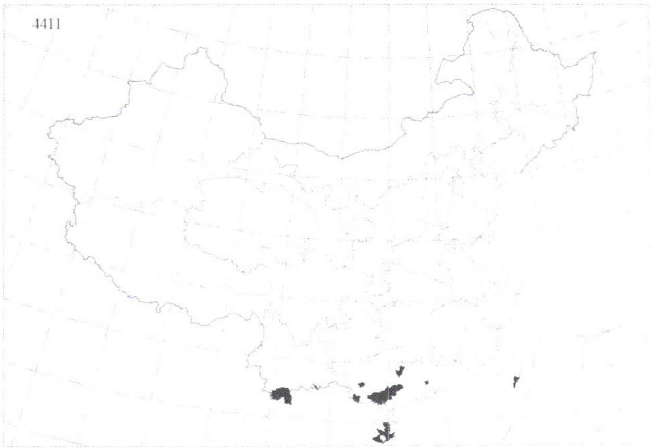

Fabaceae 豆科 *Mucuna* 黧豆属

Mucuna interrupta Gagnep. 间序油麻藤

Vines, evergreen

20.0; 14.6; 23.2; 20.1; 180.8; 0.0; 963 ❙ 1523; 681; 67 ❙ 58.4; 942; 581.0

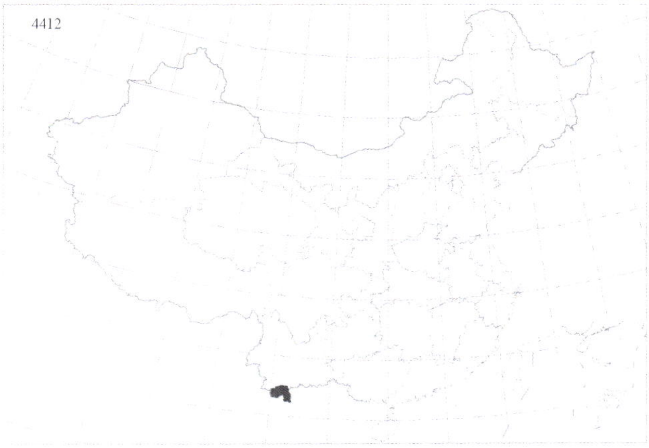

Fabaceae 豆科 *Mucuna* 黧豆属

Mucuna lamellata Wilmot-Dear 褶皮黧豆

Vines, evergreen

17.0 (14.3~22.0); 5.6 (2.2~13.4); 27.8 (24.3~29.3); 17.1 (14.3~22.0); 145.8 (114.1~204.3); 1.0 (0.0~4.0); 919 (765~1176) ❙ 1498 (931~1953); 561 (323~858); 178 (113~259) ❙ 63.4 (-1.6~112.2); 914 (765~1176); 221.6 (134.1~284.0)

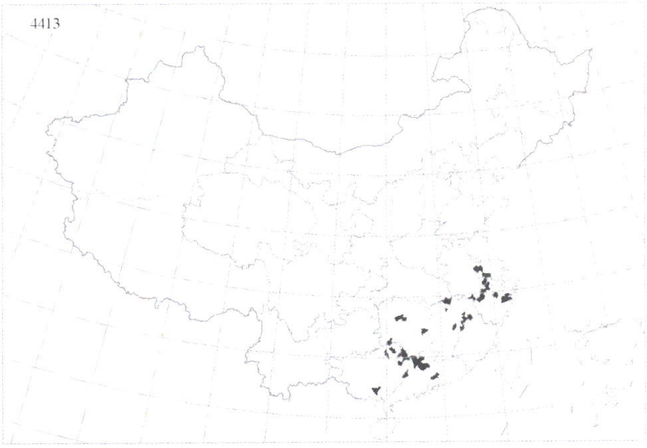

Fabaceae 豆科 *Mucuna* 黧豆属

Mucuna macrobotrys Hance 大球油麻藤

Vines, evergreen

18.1 (10.8~22.9); 11.8 (3.9~15.2); 22.8 (16.6~29.0); 18.3 (10.9~23.0); 159.8 (71.8~215.7); 0.0 (0.0~1.3); 905 (633~1239) ❙ 1323 (911~2035); 696 (524~994); 60 (24~177) ❙ 45.8 (6.5~74.2); 879 (633~1239); 371.0 (203.1~659.0)

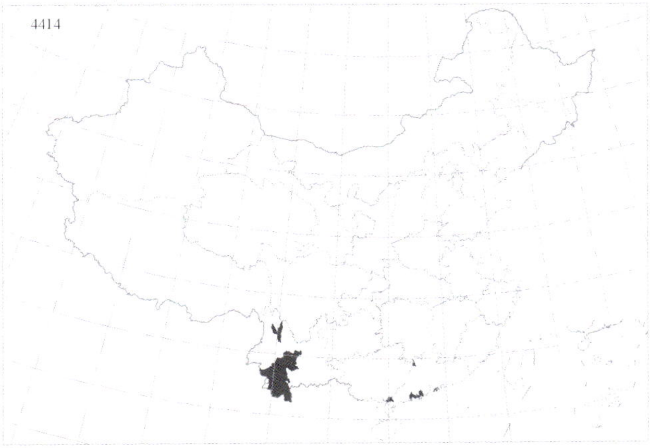

Fabaceae 豆科 *Mucuna* 黧豆属

Mucuna macrocarpa Wall. 大果油麻藤

Large climbers, evergreen

19.6 (8.6~24.1); 12.8 (1.3~18.9); 24.9 (15.0~29.0); 19.6 (8.7~24.1); 175.8 (52.6~229.7); 0.3 (0.0~8.8); 1003 (574~1322) ❙ 1824 (1007~3495); 876 (499~1847); 138 (40~897) ❙ 85.9 (19.2~246.5); 985 (574~1294); 320.3 (95.3~659.0)

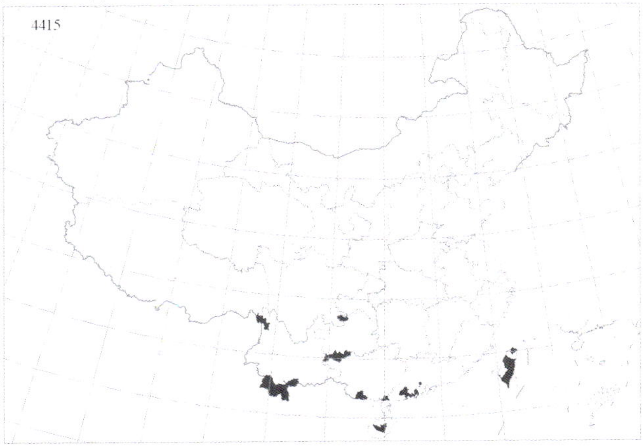

Fabaceae 豆科 *Mucuna* 黧豆属

Mucuna sempervirens Hemsl. 常春油麻藤

Vines, evergreen

15.8 (6.2~22.2); 5.5 (-2.5~15.1); 25.3 (14.1~29.5); 15.9 (6.7~22.3); 131.6 (39.4~207.5); 1.5 (0.0~24.1); 840 (509~1190) ❙ 1338 (760~1997); 554 (299~996); 118 (17~259) ❙ 59.0 (-6.7~147.0); 834 (509~1190); 248.2 (126.1~659.0)

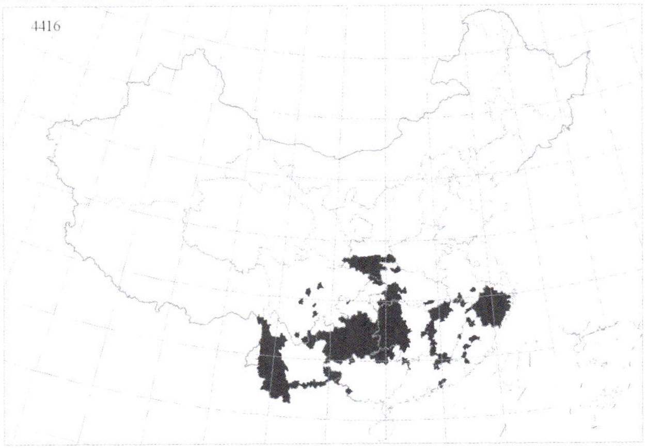

Fabaceae 豆科 *Ohwia* 小槐花属
Ohwia caudata (Thunb.) Ohashi 小槐花
Shrubs or subshrubs

17.0 (9.3~23.0); 6.0 (-2.9~15.1); 27.2 (18.4~30.1); 17.1 (9.5~23.0); 146.4 (69.8~216.2); 1.6 (0.0~20.5); 915 (603~1247) ❙ 1392 (596~3495); 559 (305~1428); 146 (27~897) ❙ 52.2 (-24.7~224.4); 903 (596~1247); 220.4 (108.2~659.0)

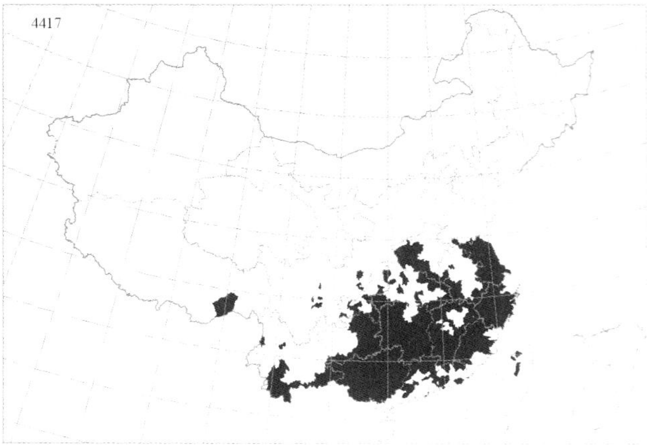

Fabaceae 豆科 *Ormocarpum* 链荚木属
Ormocarpum cochinchinense (Lour.) Merr. 链荚木
Shrubs, evergreen

18.5; 14.1; 21.7; 18.5; 162.1; 0.0; 863 ❙ 2713; 1388; 188 ❙ 214.4; 863; 363.7

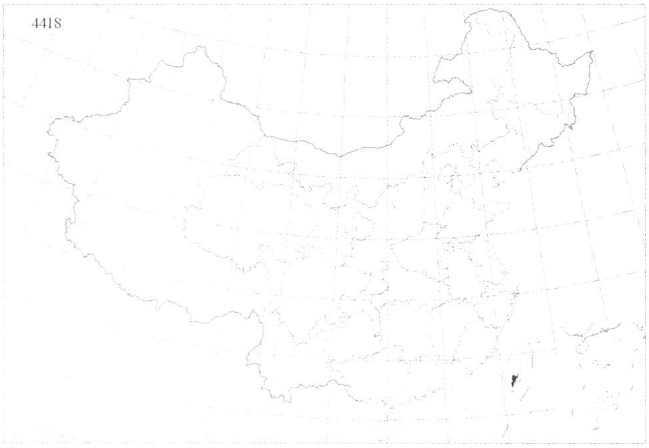

Fabaceae 豆科 *Ormosia* 红豆属
Ormosia apiculata L. Chen 喙顶红豆
Trees, evergreen

18.5; 9.8; 25.2; 18.6; 163.2; 0.0; 930 ❙ 1272; 699; 59 ❙ 37.0; 930; 303.7

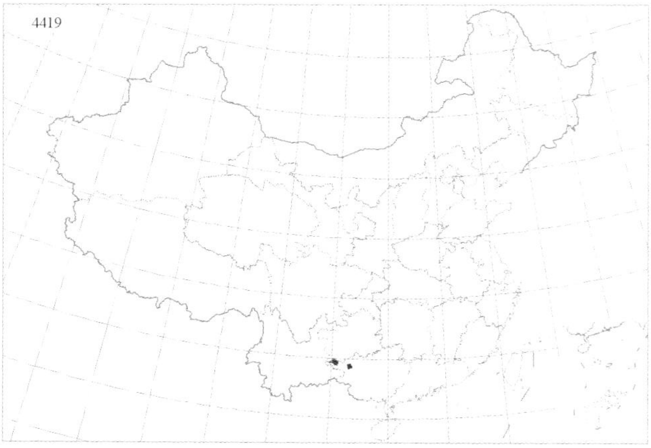

Fabaceae 豆科 *Ormosia* 红豆属
Ormosia balansae Drake 长脐红豆
Trees, evergreen

21.4 (14.6~24.8); 13.6 (3.8~20.1); 27.7 (24.6~28.7); 21.5 (14.7~24.8); 197.7 (117.7~237.6); 0.0 (0.0~1.5); 1149 (779~1385) ❙ 1601 (868~2435); 667 (312~1428); 122 (41~197) ❙ 43.5 (-34.3~101.8); 1094 (779~1360); 337.3 (208.1~620.0)

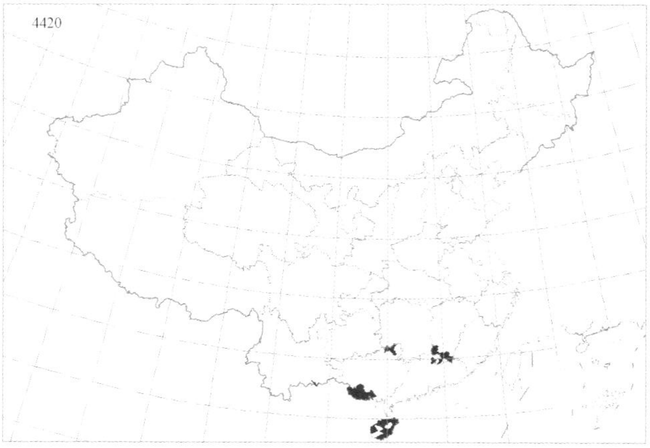

Fabaceae 豆科 *Ormosia* 红豆属
Ormosia elliptica Q. W. Yao et R. H. Chang 厚荚红豆
Trees, evergreen

18.4 (16.8~20.7); 9.0 (6.0~12.8); 26.7 (24.6~27.7); 18.5 (16.8~20.7); 161.8 (141.5~188.9); 0.0 (0.0~0.0); 950 (864~1071) ❙ 1664 (1293~1959); 636 (539~723); 177 (58~250) ❙ 76.1 (39.6~121.8); 950 (864~1071); 264.1 (208.7~316.1)

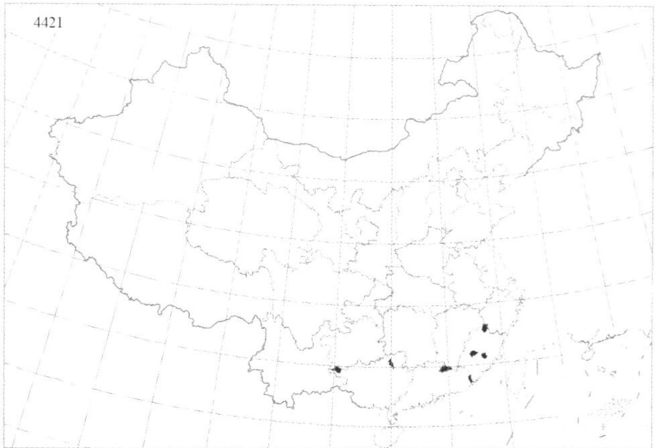

Fabaceae 豆科 *Ormosia* 红豆属
Ormosia emarginata (Hook. et Arn.) Benth. 凹叶红豆
Small trees, evergreen

23.0 (21.8~24.8); 16.5 (14.0~20.1); 27.9 (25.7~28.7); 23.0 (21.9~24.8); 216.6 (202.2~237.6); 0.0 (0.0~0.0); 1236 (1127~1385) ❙ 1772 (1380~2435); 800 (455~1428); 85 (41~130) ❙ 44.6 (1.5~99.5); 1177 (1055~1299); 364.2 (139.6~620.0)

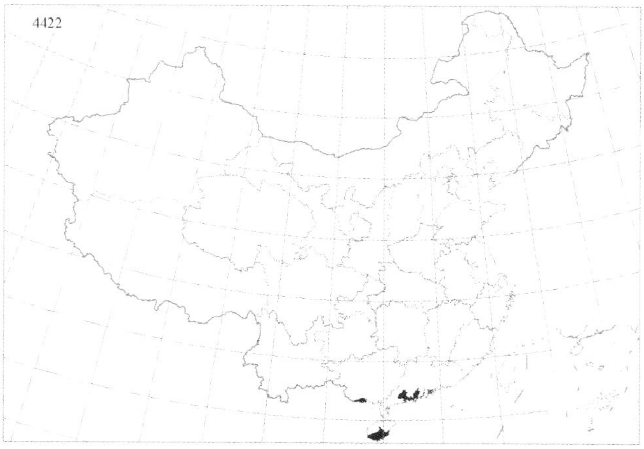

Fabaceae 豆科 *Ormosia* 红豆属

Ormosia eugeniifolia Tsiang ex R. H. Chang 蒲桃叶红豆

Trees, evergreen

21.3; 13.2; 27.3; 21.3; 195.9; 0.0; 1111 | 1900; 1056; 125 | 71.0; 1111; 297.2

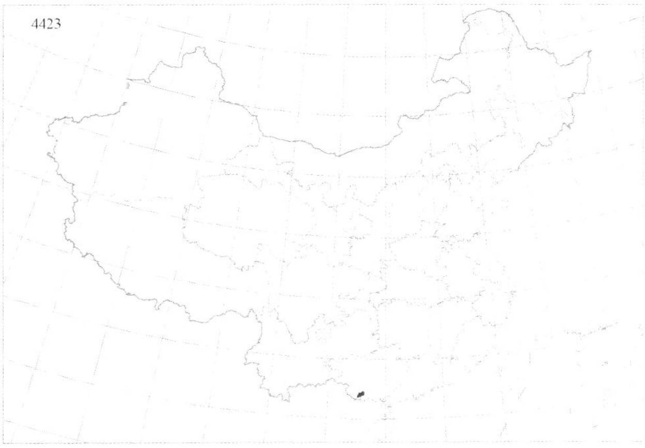

Fabaceae 豆科 *Ormosia* 红豆属

Ormosia ferruginea R. H. Chang 锈枝红豆

Small trees, evergreen

21.8; 14.2; 27.8; 21.9; 202.6; 0.0; 1150 | 1876; 891; 116 | 63.2; 1150; 273.3

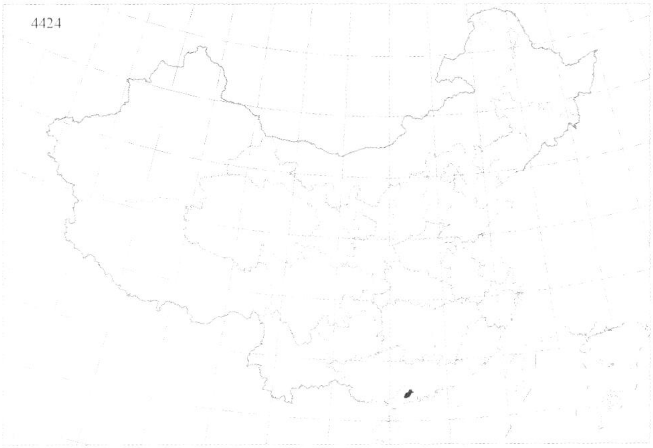

Fabaceae 豆科 *Ormosia* 红豆属

Ormosia fordiana Oliv. 肥荚红豆

Trees, evergreen

20.5 (14.3~24.8); 12.3 (3.7~20.1); 26.9 (22.1~29.2); 20.6 (14.3~24.8); 186.6 (113.2~237.6); 0.0 (0.0~1.4); 1072 (761~1385) | 1546 (1098~2435); 715 (392~1428); 115 (41~218) | 46.4 (-12.0~128.9); 1055 (761~1329); 304.4 (178.7~659.0)

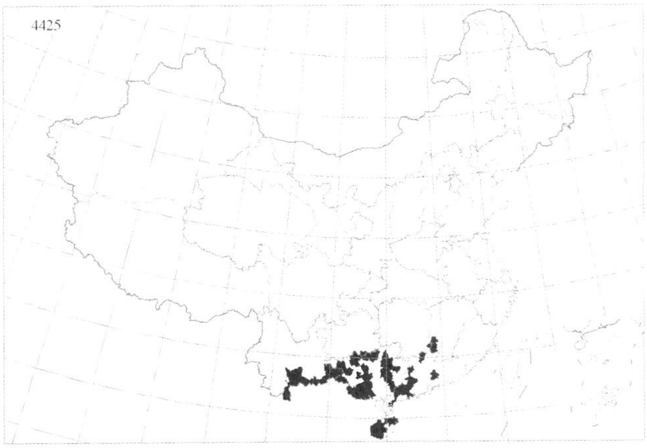

Fabaceae 豆科 *Ormosia* 红豆属

Ormosia formosana Kaneh. 台湾红豆

Trees, evergreen

19.3; 13.8; 23.6; 19.3; 171.5; 0.0; 958 | 2472; 1278; 179 | 167.2; 947; 267.6

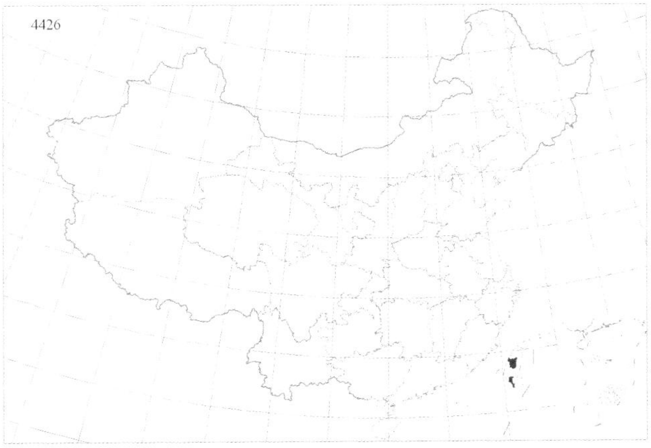

Fabaceae 豆科 *Ormosia* 红豆属

Ormosia glaberrima Y. C. Wu 光叶红豆

Trees, evergreen

19.7 (14.3~24.5); 10.3 (3.7~18.9); 27.6 (24.2~29.0); 19.7 (14.3~24.5); 176.8 (113.2~234.4); 0.1 (0.0~1.5); 1040 (761~1360) | 1633 (1302~2435); 679 (500~1428); 161 (68~218) | 59.2 (12.9~128.9); 1038 (761~1360); 245.0 (157.7~572.9)

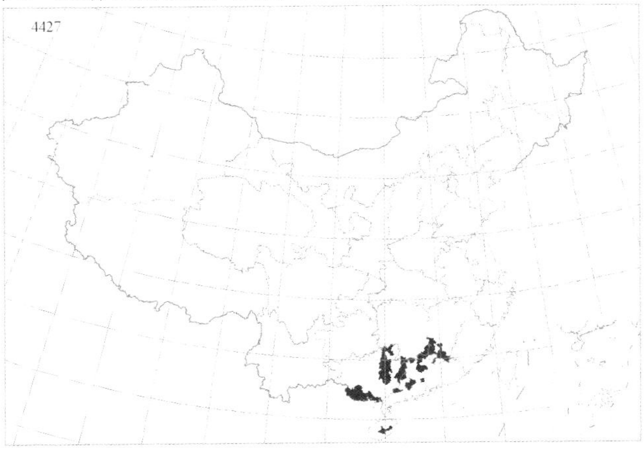

Fabaceae 豆科 *Ormosia* 红豆属

Ormosia hekouensis R. H. Chang 河口红豆

Trees, evergreen

20.7; 14.7; 24.6; 20.7; 188.8; 0.0; 1021 | 1564; 760; 73 | 53.5; 1020; 605.3

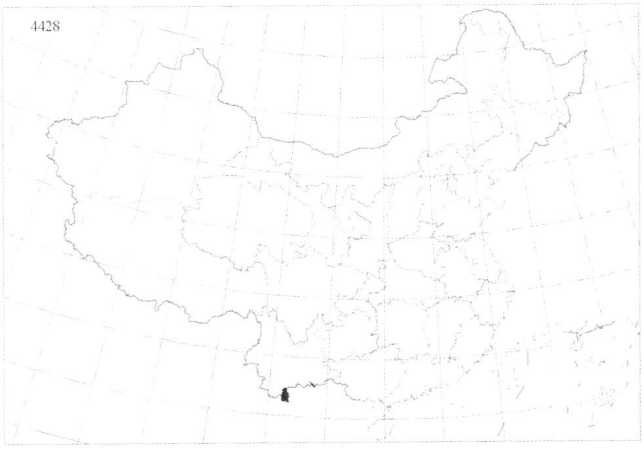

Fabaceae 豆科 *Ormosia* 红豆属

Ormosia henryi Prain 花榈木

Trees, evergreen

17.1 (12.1~23.9); 6.4 (1.1~18.7); 27.1 (20.6~30.1); 17.2 (12.2~23.9); 146.8 (95.0~226.8); 0.4 (0.0~8.5); 910 (696~1294) ❙ 1515 (931~2435); 571 (323~1428); 163 (32~259) ❙ 67.1 (-1.6~147.0); 901 (696~1221); 234.4 (111.7~659.0)

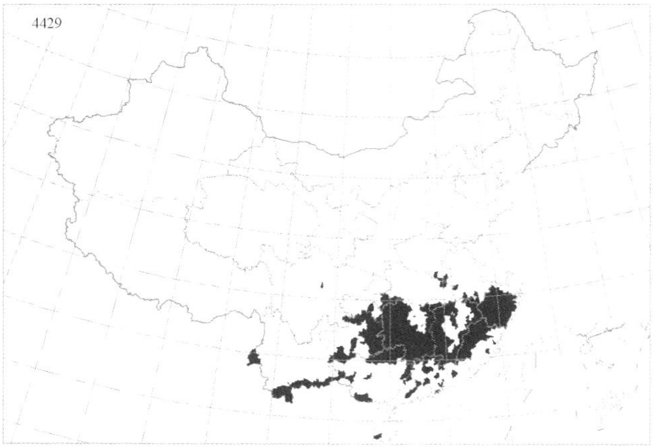

Fabaceae 豆科 *Ormosia* 红豆属

Ormosia hosiei Hemsl. et Wilson 红豆树

Trees, evergreen or deciduous

15.1 (9.1~21.2); 3.7 (-2.9~12.5); 25.9 (18.1~28.9); 15.2 (9.2~21.2); 126.5 (64.1~194.6); 4.2 (0.0~20.6); 832 (595~1114) ❙ 1179 (626~1959); 499 (316~816); 93 (15~250) ❙ 41.6 (-22.2~147.0); 820 (595~1114); 209.7 (108.2~322.1)

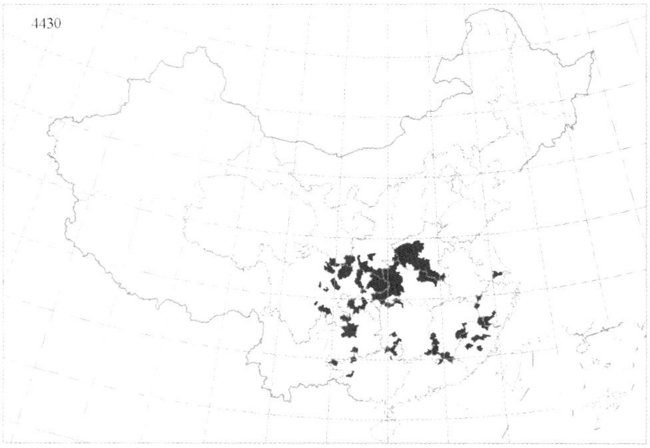

Fabaceae 豆科 *Ormosia* 红豆属

Ormosia howii Merr. et Chun et Merr. ex L. Chen 缘毛红豆

Trees, evergreen

23.7 (22.2~24.8); 18.8 (17.2~20.1); 27.1 (25.7~28.0); 23.8 (22.3~24.8); 225.3 (207.2~237.6); 0.0 (0.0~0.0); 1281 (1127~1385) ❙ 1587 (1405~1725); 551 (473~610); 53 (41~68) ❙ 25.1 (1.5~47.8); 1143 (1055~1212); 561.2 (494.7~620.0)

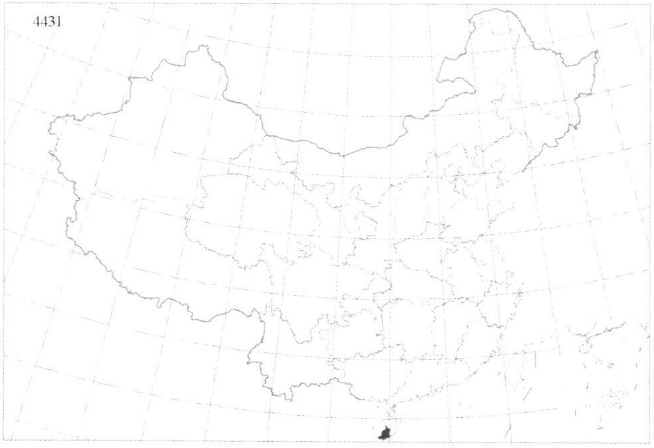

Fabaceae 豆科 *Ormosia* 红豆属

Ormosia indurata L. Chen 韧荚红豆

Trees, evergreen

21.4 (18.4~22.7); 13.3 (9.0~15.2); 27.9 (26.7~28.7); 21.5 (18.4~22.8); 197.8 (161.0~213.1); 0.0 (0.0~0.0); 1132 (948~1221) ❙ 1805 (1302~2435); 883 (566~1428); 135 (86~204) ❙ 59.9 (12.9~99.5); 1132 (948~1221); 267.0 (187.2~327.7)

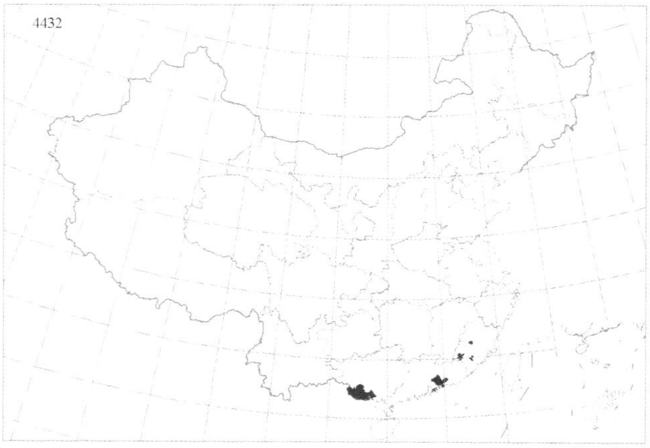

Fabaceae 豆科 *Ormosia* 红豆属

Ormosia inflata Merr. et Chun 胀荚红豆

Trees, evergreen

24.0 (22.2~24.8); 18.4 (17.2~20.1); 28.0 (25.7~28.8); 24.0 (22.3~24.8); 228.3 (207.2~237.6); 0.0 (0.0~0.0); 1310 (1127~1385) ❙ 1501 (868~1854); 534 (312~712); 74 (41~115) ❙ 15.0 (-34.3~47.8); 1191 (868~1360); 440.3 (248.2~620.0)

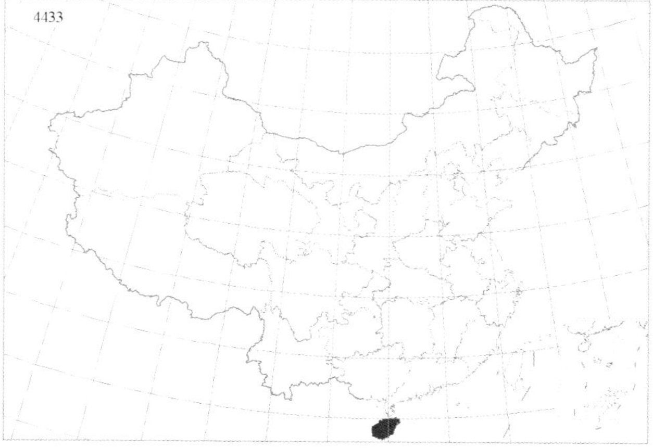

Fabaceae 豆科 *Ormosia* 红豆属

Ormosia longipes L. Chen 纤柄红豆

Trees, evergreen

18.7 (16.9~21.0); 12.2 (10.3~14.2); 23.1 (21.4~25.6); 18.7 (16.9~21.0); 164.3 (142.7~191.8); 0.0 (0.0~0.0); 905 (813~1053) ❙ 1409 (1178~1547); 793 (646~911); 59 (51~68) ❙ 55.9 (44.5~74.2); 905 (813~1053); 496.8 (383.4~551.6)

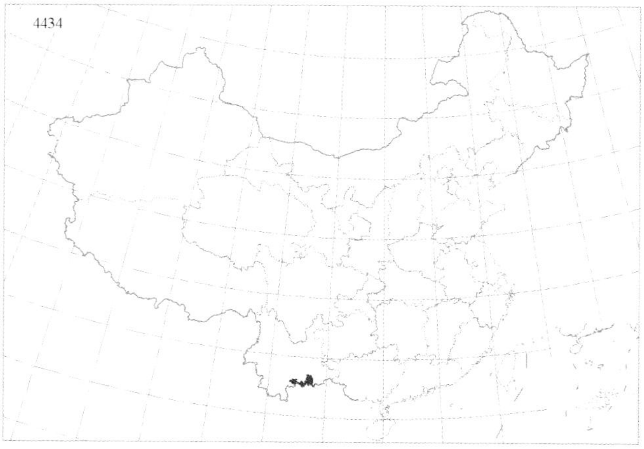

Fabaceae 豆科 *Ormosia* 红豆属
Ormosia merrilliana L. Chen 云开红豆
Trees, evergreen

21.5 (18.9~22.9); 13.1 (10.2~15.2); 28.0 (25.0~29.0); 21.5 (19.0~23.0); 198.4 (167.5~215.7); 0.0 (0.0~0.0); 1136 (939~1239) ❙ 1653 (1250~2435); 824 (665~1428); 118 (49~181) ❙ 45.5 (12.9~99.5); 1136 (939~1239); 256.7 (157.7~342.5)

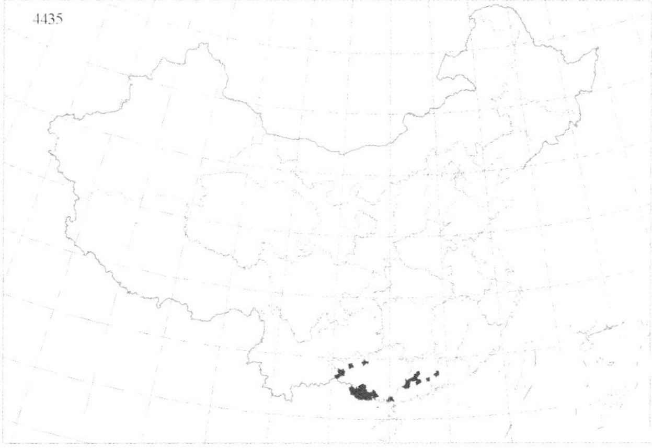

Fabaceae 豆科 *Ormosia* 红豆属
Ormosia microphylla Merr. et L. Chen 小叶红豆
Shrubs or trees, evergreen

18.6 (13.4~22.7); 9.2 (3.6~15.0); 26.5 (22.3~28.6); 18.6 (13.4~22.8); 163.4 (102.7~213.1); 0.1 (0.0~1.8); 970 (721~1221) ❙ 1493 (1072~2435); 660 (452~1428); 127 (43~202) ❙ 54.6 (5.8~101.8); 970 (721~1221); 250.2 (162.8~354.5)

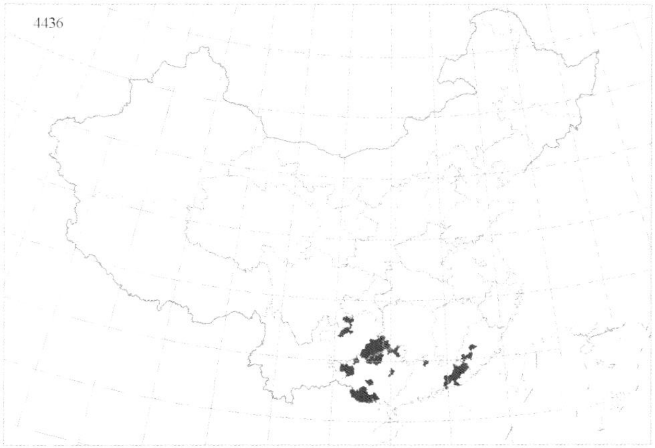

Fabaceae 豆科 *Ormosia* 红豆属
Ormosia microphylla var. tomentosa R. H. Chang 绒毛小叶红豆
Shrubs or trees, evergreen

20.4; 12.0; 27.7; 20.5; 185.5; 0.0; 1057 ❙ 1591; 607; 162 ❙ 50.6; 1057; 266.9

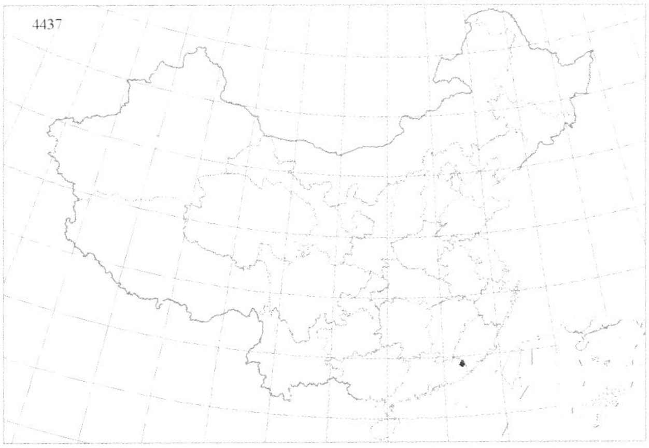

Fabaceae 豆科 *Ormosia* 红豆属
Ormosia nanningensis L. Chen 南宁红豆
Trees, evergreen

21.9 (21.3~22.7); 13.9 (13.2~15.0); 27.9 (27.3~28.6); 22.0 (21.3~22.8); 203.6 (195.9~213.1); 0.0 (0.0~0.0); 1162 (1111~1221) ❙ 1876 (1302~2435); 1047 (670~1428); 116 (86~130) ❙ 61.1 (12.9~99.5); 1162 (1111~1221); 289.6 (235.1~318.8)

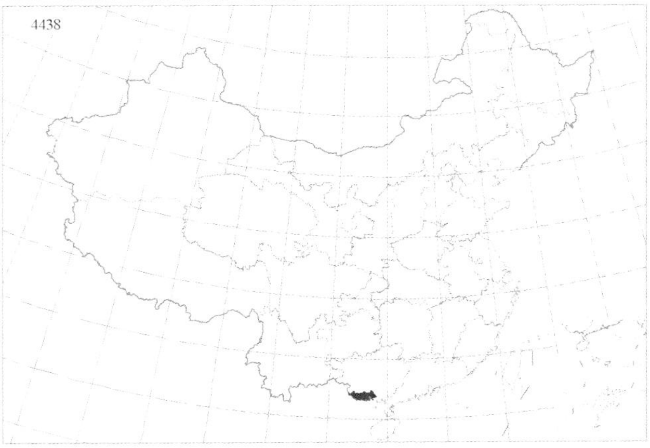

Fabaceae 豆科 *Ormosia* 红豆属
Ormosia napoensis Z. Wei et R. H. Chang 那坡红豆
Small trees, evergreen

19.1; 11.0; 25.0; 19.1; 169.6; 0.0; 945 ❙ 1438; 820; 57 ❙ 52.2; 945; 371.7

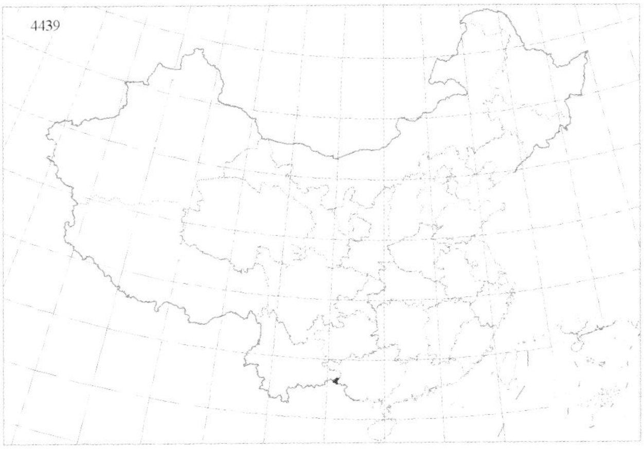

Fabaceae 豆科 *Ormosia* 红豆属
Ormosia nuda (K. C. How) R. H. Chang et Q. W. Yao 秃叶红豆
Trees, evergreen

16.4 (12.7~19.1); 6.4 (2.1~11.0); 25.3 (21.6~27.8); 16.4 (12.8~19.1); 137.7 (99.0~169.2); 0.4 (0.0~5.6); 852 (713~969) ❙ 1292 (1049~1624); 577 (472~812); 90 (44~191) ❙ 52.8 (15.8~107.2); 851 (713~969); 218.3 (152.6~365.3)

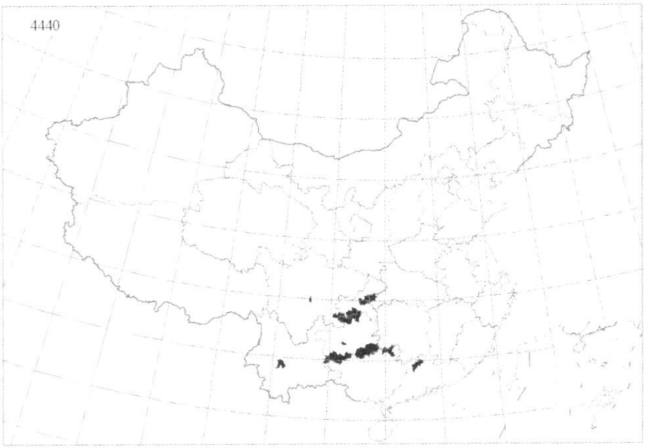

Fabaceae 豆科 *Ormosia* 红豆属

Ormosia olivacea L. Chen 榄绿红豆

Trees, evergreen

18.5 (17.1~21.0); 11.9 (9.4~15.1); 22.9 (20.8~25.6); 18.5 (17.1~21.0); 162.2 (145.0~191.8); 0.0 (0.0~0.0); 894 (815~1053) ❙ 1377 (1049~1607); 743 (598~860); 54 (42~77) ❙ 54.3 (26.1~80.6); 879 (783~1053); 449.7 (318.5~659.0)

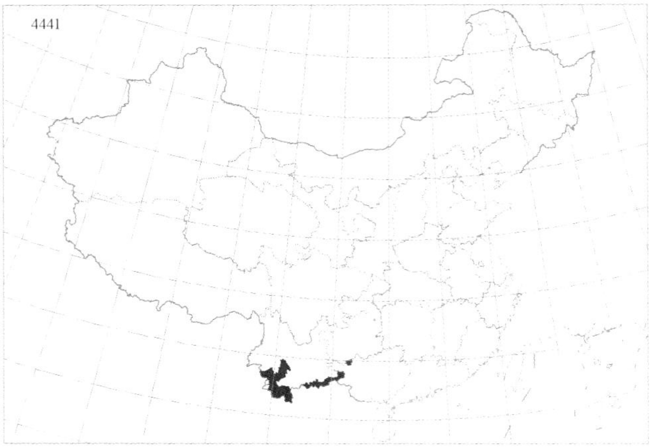

Fabaceae 豆科 *Ormosia* 红豆属

Ormosia pachycarpa Champ. ex Benth. 茸荚红豆

Trees, evergreen

21.1 (19.1~22.1); 12.6 (9.8~15.2); 28.2 (26.6~28.8); 21.3 (19.2~22.7); 195.2 (170.1~212.7); 0.0 (0.0~0.0); 1122 (981~1216) ❙ 1587 (1284~1965); 717 (599~994); 129 (73~164) ❙ 41.1 (26.8~61.6); 1122 (981~1216); 226.2 (157.7~322.1)

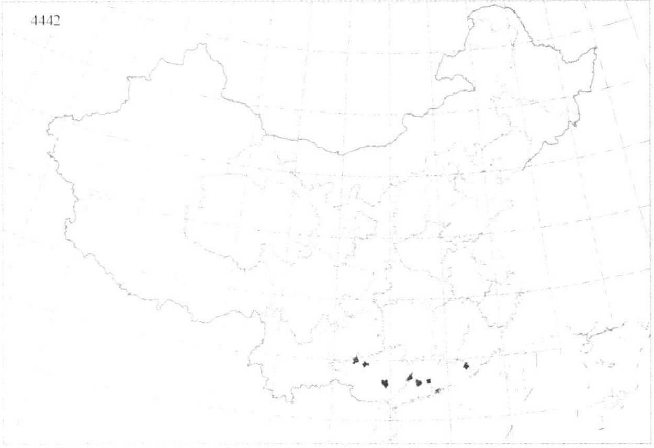

Fabaceae 豆科 *Ormosia* 红豆属

Ormosia pachycarpa var. tenuis Chun ex R. H. Chang 薄毛茸荚红豆

Trees, evergreen

22.0; 13.5; 28.8; 22.0; 203.8; 0.0; 1170 ❙ 1647; 726; 132 ❙ 41.0; 1170; 203.4

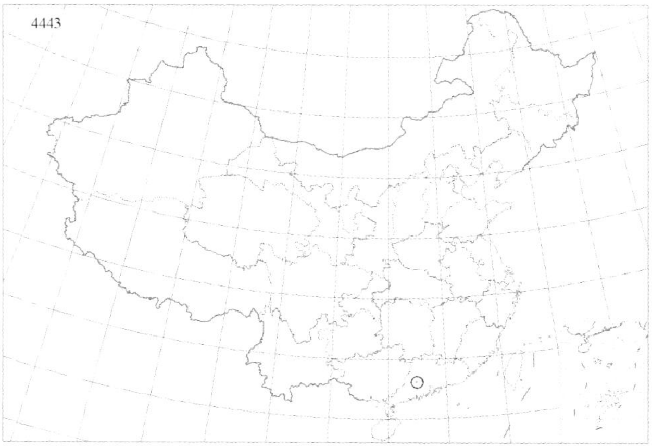

Fabaceae 豆科 *Ormosia* 红豆属

Ormosia pachyptera L. Chen 菱荚红豆

Trees, evergreen

19.7; 11.9; 25.3; 19.7; 176.8; 0.0; 997 ❙ 1393; 763; 68 ❙ 42.6; 997; 397.0

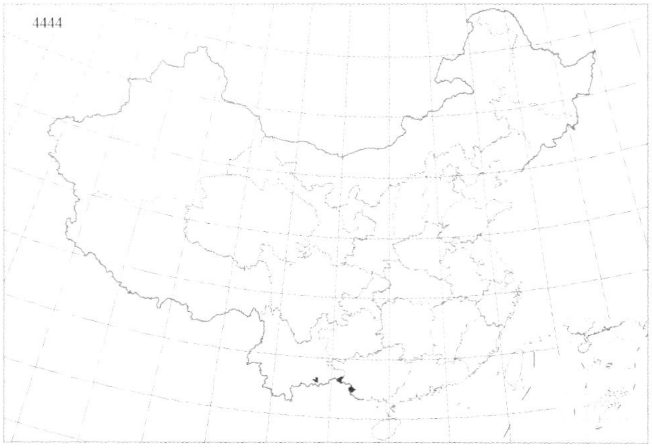

Fabaceae 豆科 *Ormosia* 红豆属

Ormosia pingbianensis W. C. Cheng et R. H. Chang 屏边红豆

Trees, evergreen

18.1 (16.3~21.5); 10.7 (8.5~14.2); 23.4 (21.4~27.4); 18.1 (16.4~21.6); 157.7 (136.4~199.1); 0.0 (0.0~0.0); 891 (797~1131) ❙ 1337 (1074~1581); 743 (582~860); 59 (47~106) ❙ 50.5 (32.4~80.6); 891 (797~1131); 381.2 (277.0~551.6)

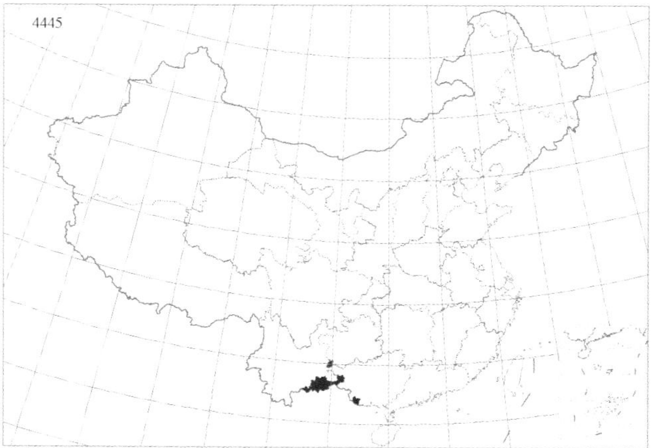

Fabaceae 豆科 *Ormosia* 红豆属

Ormosia pinnata (Lour.) Merr. 海南红豆

Shrubs or trees, evergreen

22.7 (17.2~24.8); 15.7 (6.3~20.1); 28.0 (23.3~29.0); 22.7 (17.2~24.8); 212.8 (146.7~237.6); 0.0 (0.0~0.0); 1223 (902~1385) ❙ 1568 (868~2435); 658 (312~1428); 99 (41~183) ❙ 30.0 (-34.3~99.5); 1163 (868~1360); 338.4 (120.3~620.0)

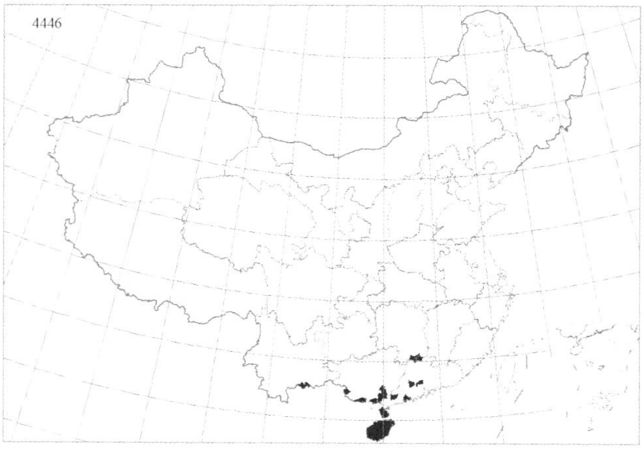

Fabaceae 豆科 *Ormosia* 红豆属

Ormosia pubescens R. H. Chang 柔毛红豆

Trees, evergreen

22.0; 14.1; 23.0; 22.1; 205.1; 0.0; 1171 ❙ 2093; 1191; 126 ❙ 78.5; 1171; 287.9

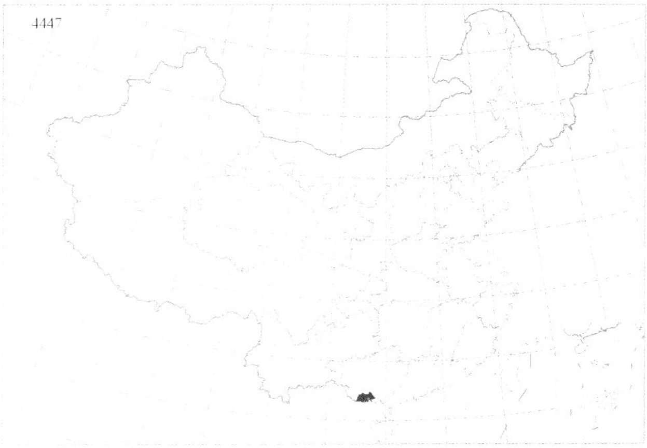

Fabaceae 豆科 *Ormosia* 红豆属

Ormosia purpureiflora L. Chen 紫花红豆

Shrubs or small trees, evergreen

21.3; 12.9; 27.9; 21.3; 195.7; 0.0; 1118 ❙ 1832; 797; 153 ❙ 64.5; 1118; 241.1

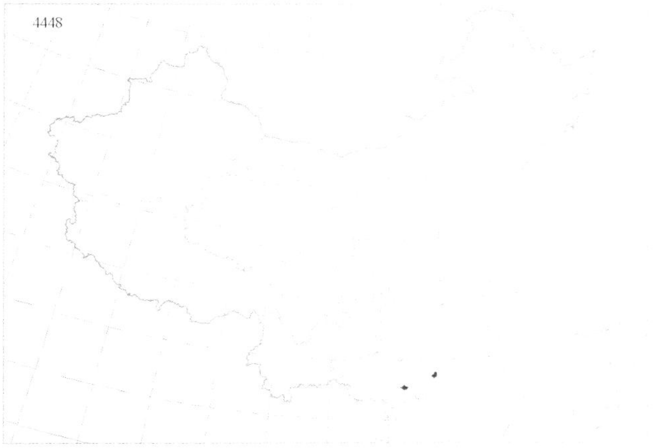

Fabaceae 豆科 *Ormosia* 红豆属

Ormosia saxatilis K. M. Lan 岩生红豆

Trees, evergreen

15.6 (14.3~18.9) 5.6 (4.6~9.5); 24.4 (22.8~26.6); 15.6 (14.3~18.9); 127.5 (112.3~167.3); 0.1 (0.0~0.4); 811 (751~969) ❙ 1209 (1128~1335); 561 (506~638); 72 (60~129) ❙ 49.7 (27.5~59.3); 811 (751~969); 188.6 (153.9~254.3)

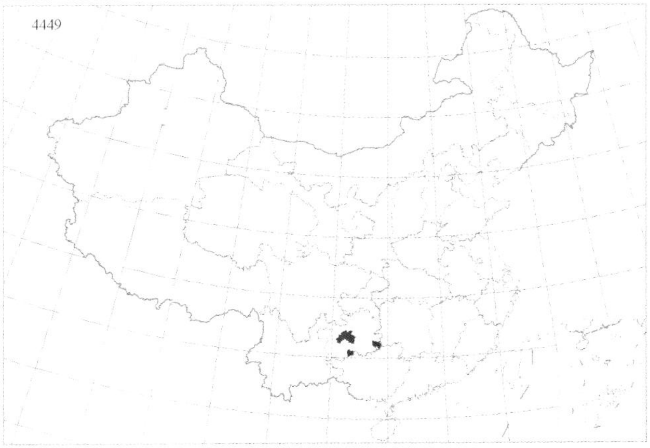

Fabaceae 豆科 *Ormosia* 红豆属

Ormosia semicastrata Hance 软荚红豆

Trees, evergreen

19.4 (14.6~24.8); 10.1 (3.8~20.1); 27.5 (24.6~29.8); 19.5 (14.7~24.8); 173.6 (117.7~237.6); 0.0 (0.0~1.5); 1022 (779~1385) ❙ 1620 (1179~2435); 626 (392~1428); 163 (41~230) ❙ 60.2 (-11.9~104.6); 1009 (779~1360); 254.0 (139.6~620.0)

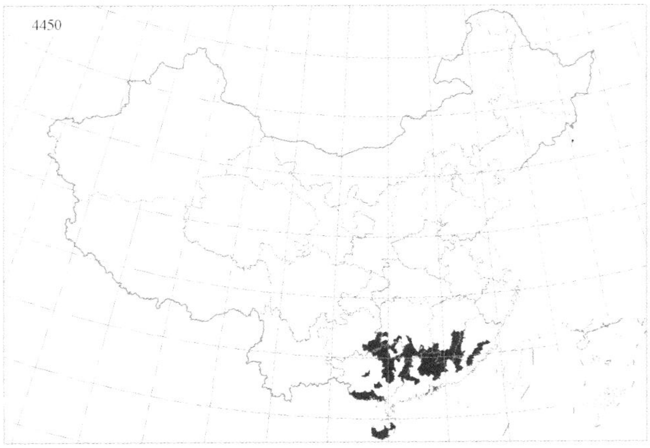

Fabaceae 豆科 *Ormosia* 红豆属

Ormosia semicastrata f. litchiifolia F. C. How 荔枝叶红豆

Trees, evergreen

18.2; 7.8; 27.4; 18.2; 158.4; 0.0; 947 ❙ 1535; 533; 183 ❙ 62.1; 947; 226.8

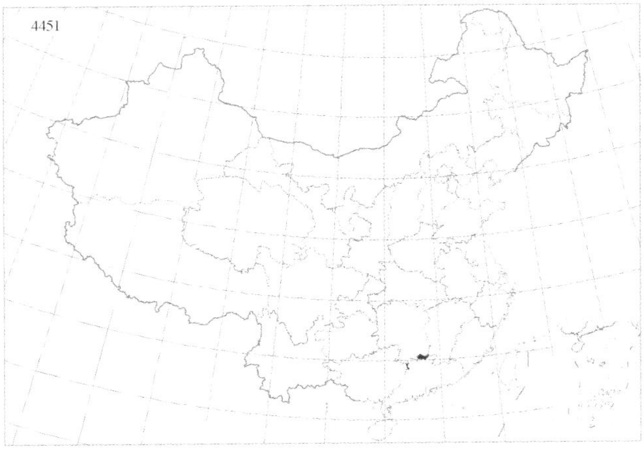

Fabaceae 豆科 *Ormosia* 红豆属

Ormosia semicastrata f. pallida F. C. How 苍叶红豆

Trees, evergreen

19.2 (14.3~22.7); 9.6 (4.0~15.0); 27.2 (23.5~28.8); 19.2 (14.3~22.8); 170.6 (112.6~213.1); 0.0 (0.0~1.1); 1009 (756~1221) ❙ 1580 (1218~2435); 679 (452~1428); 150 (86~203) ❙ 57.5 (12.9~99.5); 1009 (756~1221); 236.0 (185.5~327.7)

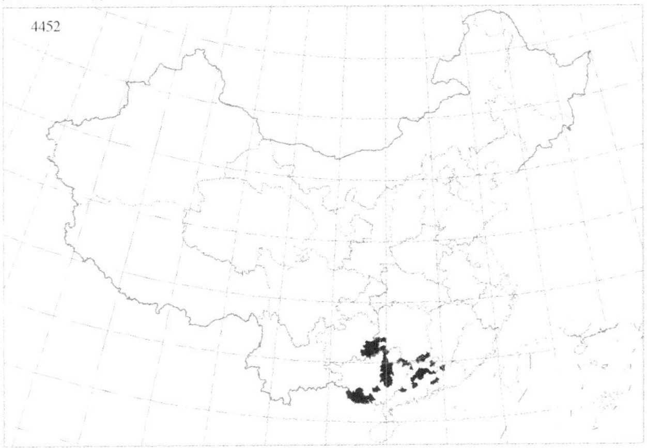

Fabaceae 豆科 *Ormosia* 红豆属
Ormosia sericeolucida L. Chen 亮毛红豆
Trees, evergreen

20.5 (14.6~22.7); 11.9 (3.8~15.0); 27.4 (24.6~28.6); 20.6 (14.7~22.8); 187.2 (117.7~213.1); 0.1 (0.0~1.5); 1090 (779~1221) ❙ 1723 (1302~2435); 866 (544~1428); 131 (86~191) ❙ 60.5 (12.9~101.8); 1090 (779~1221); 278.7 (235.1~327.7)

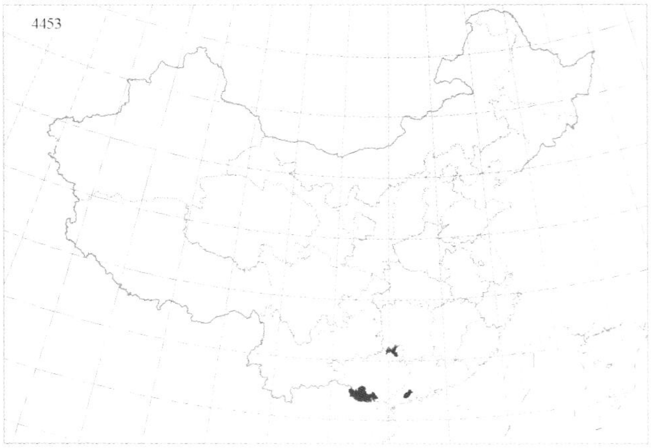

Fabaceae 豆科 *Ormosia* 红豆属
Ormosia simplicifolia Merr. et Chun 单叶红豆
Shrubs or small trees

21.8 (19.1~24.7); 14.3 (11.0~19.8); 27.4 (25.0~28.6); 21.9 (19.1~24.7); 202.4 (169.6~236.5); 0.0 (0.0~0.0); 1155 (945~1376) ❙ 1589 (1188~2435); 797 (534~1428); 89 (48~130) ❙ 38.2 (5.8~99.5); 1131 (945~1299); 356.9 (256.6~620.0)

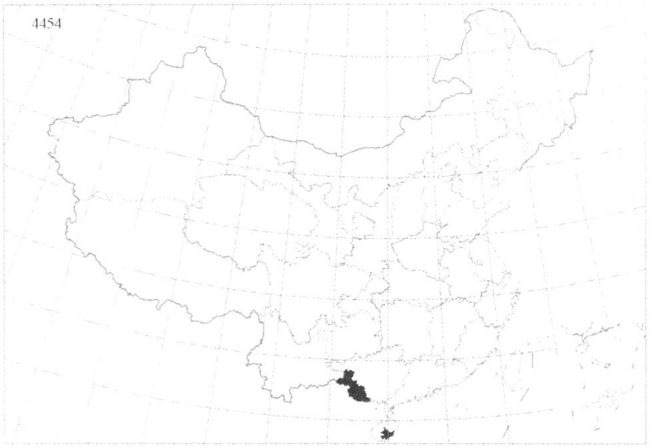

Fabaceae 豆科 *Ormosia* 红豆属
Ormosia striata Dunn 槽纹红豆
Trees, evergreen

19.0 (16.9~21.0); 12.8 (9.9~15.1); 23.0 (21.4~25.6); 19.0 (16.9~21.0); 168.3 (142.7~191.8); 0.0 (0.0~0.0); 919 (813~1053) ❙ 1455 (1178~1607); 737 (646~860); 62 (51~77) ❙ 58.7 (44.5~80.6); 909 (813~1053); 501.4 (383.4~659.0)

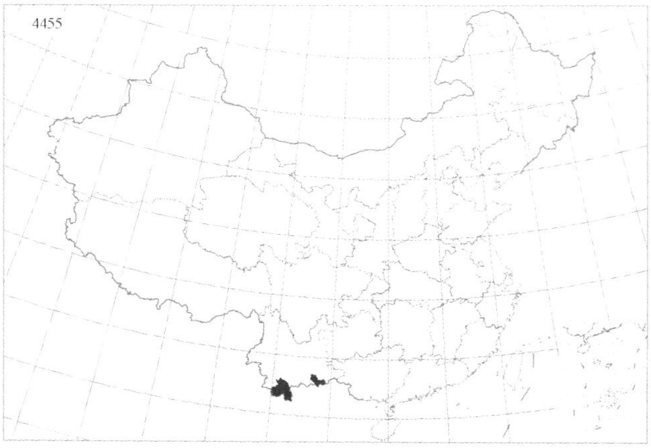

Fabaceae 豆科 *Ormosia* 红豆属
Ormosia xylocarpa Chun ex Merr. et L. Chen 木荚红豆
Trees, evergreen

18.7 (14.3~24.8); 8.8 (3.3~20.1); 27.4 (24.2~29.8); 18.7 (14.3~24.8); 164.8 (113.2~237.6); 0.1 (0.0~2.3); 989 (761~1385) ❙ 1563 (868~2435); 574 (312~1428); 166 (41~254) ❙ 60.5 (-34.3~130.7); 970 (761~1360); 251.9 (134.6~620.0)

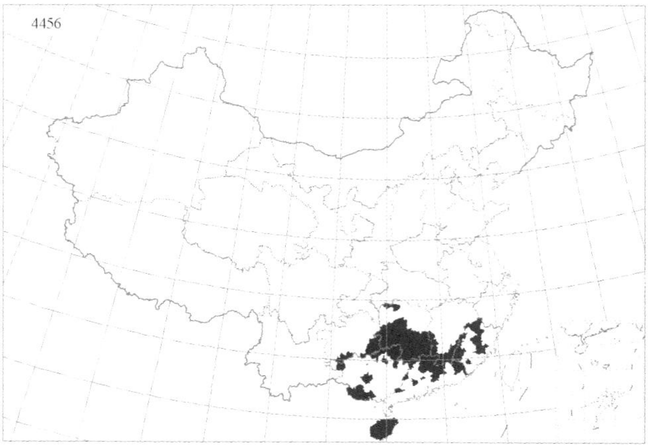

Fabaceae 豆科 *Ormosia* 红豆属
Ormosia yunnanensis Prain 云南红豆
Trees, evergreen

18.4 (16.3~21.0); 12.0 (9.1~15.1); 22.6 (20.3~25.6); 18.4 (16.4~21.0); 161.2 (136.6~191.8); 0.0 (0.0~0.0); 886 (789~1053) ❙ 1379 (1049~1751); 758 (598~996); 55 (42~77) ❙ 55.6 (26.1~88.8); 873 (780~1053); 443.9 (314.4~659.0)

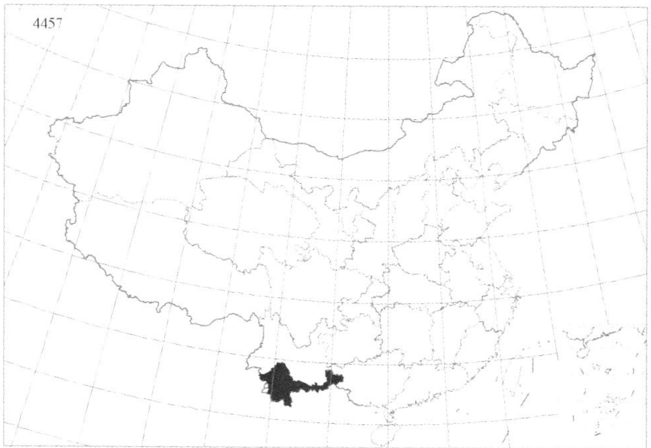

Fabaceae 豆科 *Oxytropis* 棘豆属
Oxytropis aciphylla Boriss. 猫头刺
Dwarf subshrubs, deciduous

5.1 (-2.6~11.2); -11.9 (-21.5~-5.7); 20.2 (8.7~26.8); 8.2 (2.5~12.6); 63.2 (7.7~109.2); 61.2 (30.3~108.6); 579 (349~800) ❙ 246 (21~520); 152 (16~310); 7 (0~33) ❙ -56.0 (-97.3~14.7); 245 (21~520); 75.0 (3.0~276.4)

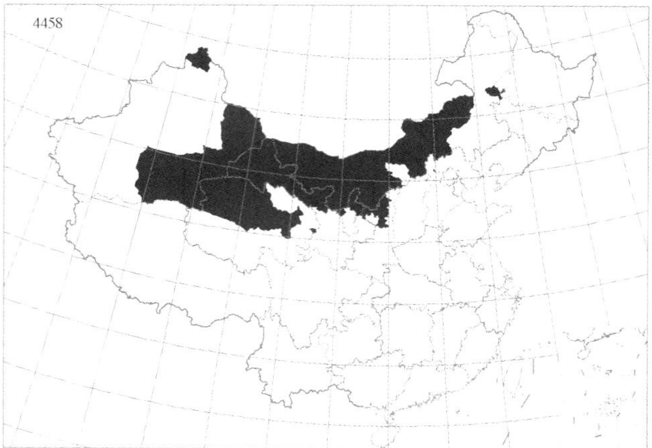

Fabaceae 豆科 *Oxytropis* 棘豆属
Oxytropis densiflora var. multiramosa (P. C. Li) X. Y. Zhu et H. Ohashi 多枝棘豆
Subshrubs, deciduous

2.4; -7.2; 11.3; 4.4; 21.4; 51.6; 423 ▎510; 311; 11 ▎20.7; 423; 237.6

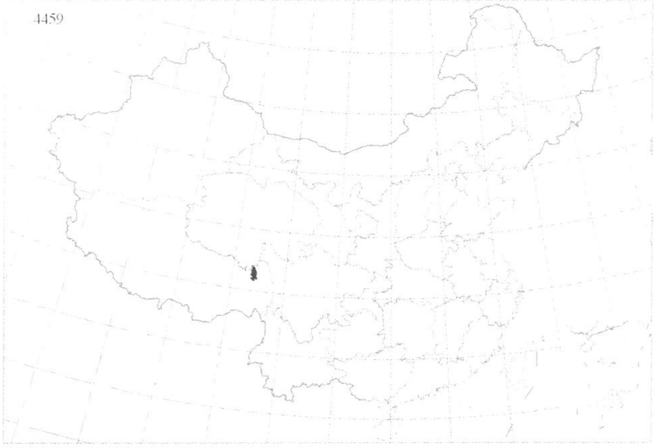

Fabaceae 豆科 *Oxytropis* 棘豆属
Oxytropis hystrix Schrenk 猬刺棘豆
Dwarf shrubs, deciduous

2.7 (-1.1~4.1); -16.3 (-18.6~-14.4); 18.8 (13.7~20.8); 7.2 (4.5~8.3); 53.5 (27.2~64.4); 80.2 (72.2~99.8); 551 (433~597) ▎313 (229~450); 124 (102~185); 36 (20~51) ▎-41.2 (-61.7~3.9); 304 (229~404); 124.4 (58.3~173.1)

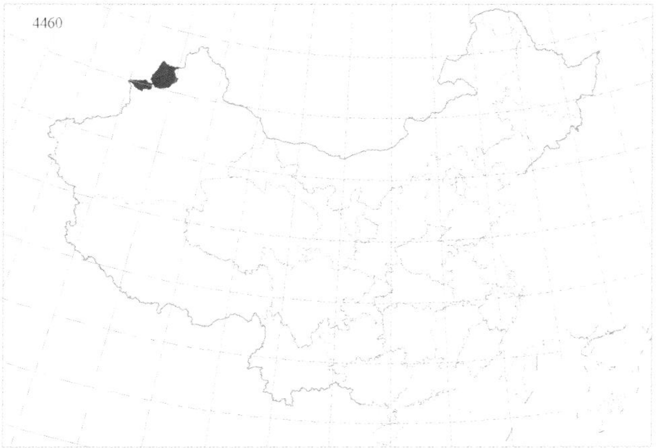

Fabaceae 豆科 *Oxytropis* 棘豆属
Oxytropis tragacanthoides Fisch. ex DC. 胶黄耆状棘豆
Dwarf shrubs, deciduous

4.1 (-4.1~11.2); -13.3 (-19.5~-6.7); 19.3 (7.3~27.5); 7.7 (2.0~12.6); 58.7 (4.1~109.2); 69.6 (34.4~114.2); 566 (335~813) ▎176 (23~373); 93 (16~235); 12 (1~51) ▎-66.1 (-97.2~0.0); 175 (23~373); 64.6 (3.0~208.7)

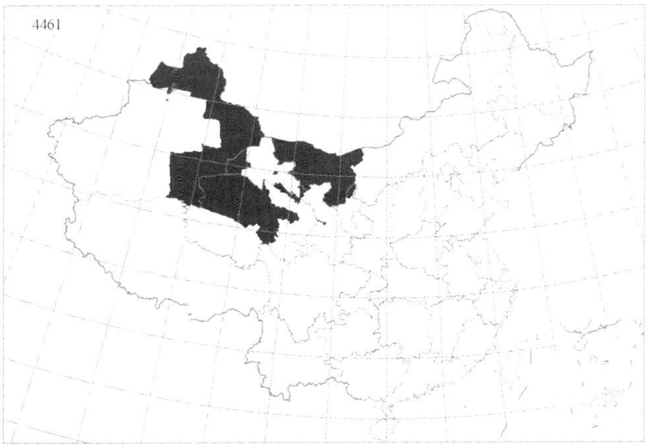

Fabaceae 豆科 *Paraderris* 拟鱼藤属
Paraderris canarensis (Dalzell) Adema 兰屿鱼藤
Vines, evergreen

18.5; 14.1; 21.7; 18.5; 162.1; 0.0; 863 ▎2713; 1388; 188 ▎214.4; 863; 363.7

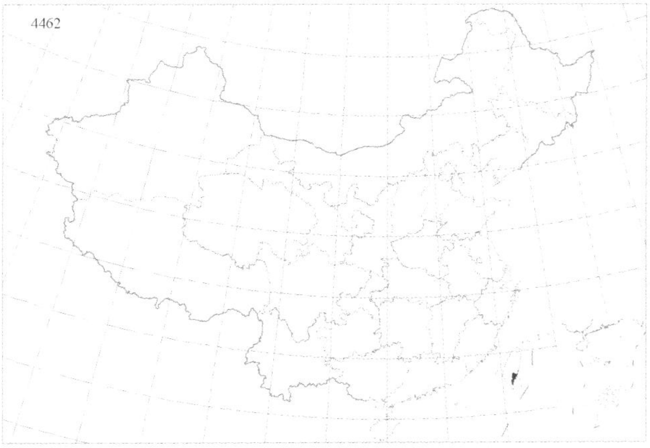

Fabaceae 豆科 *Paraderris* 拟鱼藤属
Paraderris elliptica (Wall.) Adema 毛鱼藤
Vines, evergreen

23.0 (17.3~24.8); 16.4 (6.5~20.1); 27.9 (25.7~29.0); 23.0 (17.3~24.8); 216.1 (147.9~237.6); 0.0 (0.0~0.0); 1244 (907~1385) ▎1507 (868~1854); 562 (312~776); 92 (41~197) ▎23.1 (-34.3~78.8); 1154 (868~1360); 390.2 (160.7~620.0)

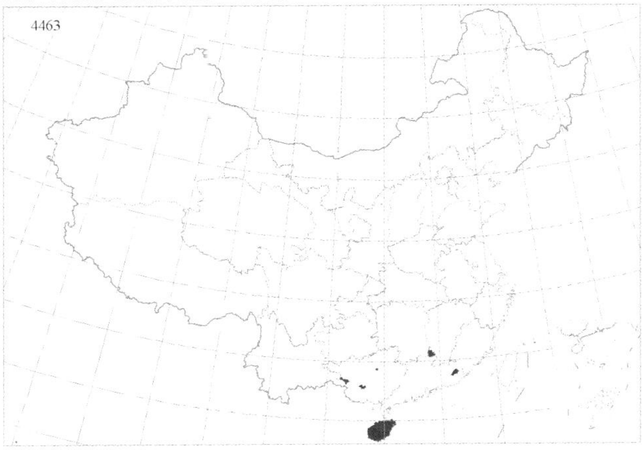

Fabaceae 豆科 *Paraderris* 拟鱼藤属
Paraderris glauca (Merrill et Chun) T. C. Chen et Pedley 粉叶鱼藤
Vines, evergreen

23.2 (21.3~24.8); 16.5 (13.2~20.1); 28.1 (27.2~28.8); 23.3 (21.3~24.8); 219.3 (195.9~237.6); 0.0 (0.0~0.0); 1259 (1111~1385) ▎1540 (868~2435); 690 (312~1428); 86 (41~130) ▎23.6 (-34.3~99.5); 1167 (868~1331); 363.3 (235.1~620.0)

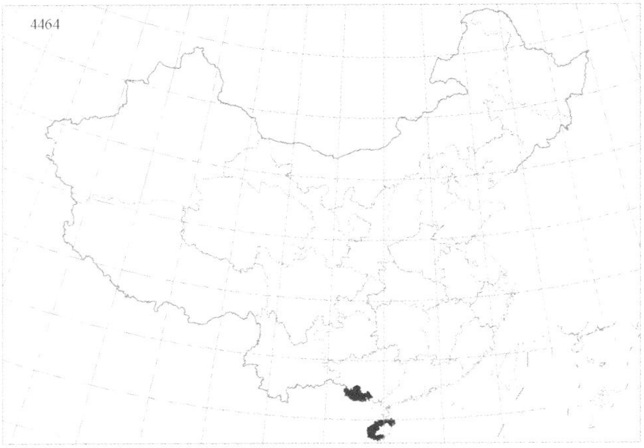

Fabaceae 豆科 *Paraderris* 拟鱼藤属

Paraderris hainanensis (Hayata) Adema 海南鱼藤

Vines, evergreen

24.0 (22.2~24.8); 18.4 (17.2~20.1); 28.0 (25.7~28.8); 24.0 (22.3~24.8); 228.3 (207.2~237.6); 0.0 (0.0~0.0); 1310 (1127~1385) ❙ 1501 (868~1854); 534 (312~712); 74 (41~115) ❙ 15.0 (-34.3~47.8); 1191 (868~1360); 440.3 (248.2~620.0)

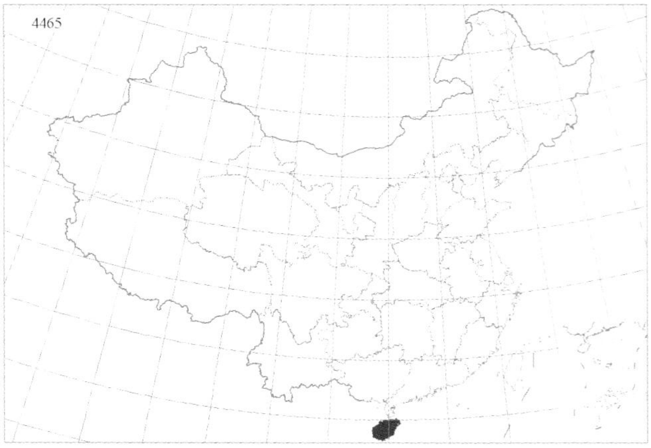

Fabaceae 豆科 *Paraderris* 拟鱼藤属

Paraderris hancei (Hemsley) T. C. Chen et Pedley 粤东鱼藤

Vines, evergreen

22.0 (21.1~22.9); 13.5 (11.9~15.2); 28.7 (28.4~29.0); 22.0 (21.1~23.0); 204.3 (193.2~215.7); 0.0 (0.0~0.0); 1175 (1115~1239) ❙ 1549 (1371~1784); 704 (565~839); 129 (97~144) ❙ 31.9 (17.4~50.9); 1170 (1107~1239); 217.8 (157.7~264.4)

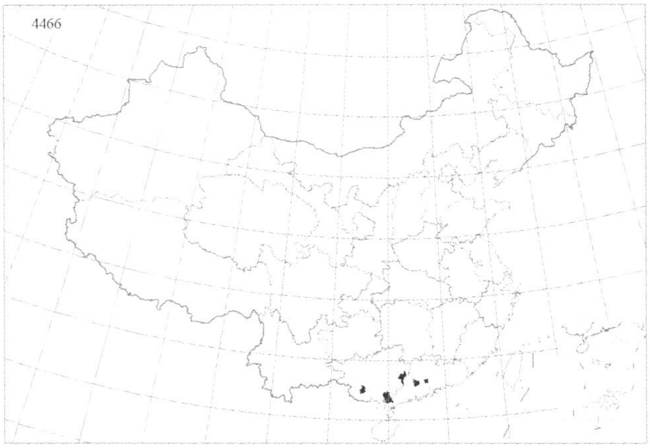

Fabaceae 豆科 *Paraderris* 拟鱼藤属

Paraderris malaccensis (Bentham) Adema 异翅鱼藤

Vines, evergreen

23.9 (22.1~24.8); 18.1 (13.7~20.1); 28.0 (25.7~28.8); 23.9 (22.1~24.8); 227.1 (205.6~237.6); 0.0 (0.0~0.0); 1303 (1127~1385) ❙ 1515 (868~1854); 546 (312~779); 77 (41~144) ❙ 16.8 (-34.3~50.9); 1191 (868~1360); 426.2 (157.7~620.0)

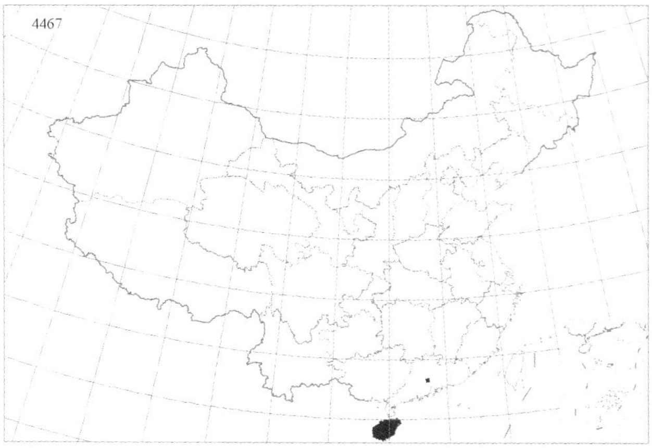

Fabaceae 豆科 *Peltophorum* 盾柱木属

Peltophorum tonkinense (Pierre) Gagnep. 银珠

Trees, deciduous

23.9 (21.5~24.8); 18.3 (13.6~20.1); 27.9 (27.2~28.7); 23.9 (21.6~24.8); 227.1 (199.1~237.6); 0.0 (0.0~0.0); 1305 (1131~1385) ❙ 1384 (868~1767); 505 (312~850); 60 (41~106) ❙ 6.5 (-34.3~39.9); 1121 (868~1299); 460.0 (281.7~620.0)

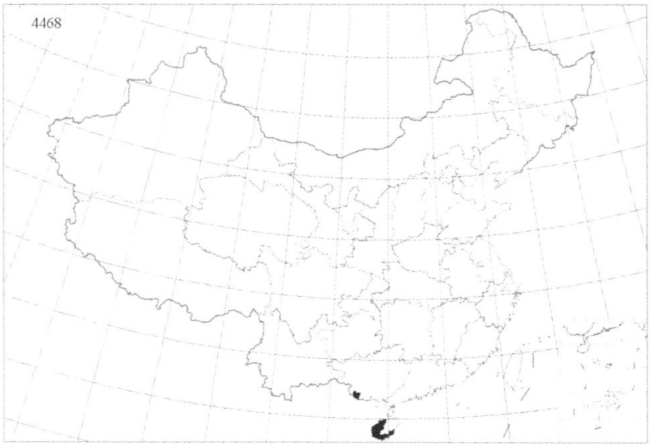

Fabaceae 豆科 *Phyllodium* 排钱草属

Phyllodium elegans (Lour.) Desv. 毛排钱树

Shrubs

21.1 (16.5~24.8); 12.8 (5.8~20.1); 27.9 (25.2~29.8); 21.2 (16.5~24.8); 193.8 (138.2~237.6); 0.0 (0.0~0.0); 1115 (861~1385) ❙ 1648 (868~2037); 692 (312~1033); 141 (41~205) ❙ 49.7 (-34.3~100.8); 1102 (861~1299); 241.5 (95.3~620.0)

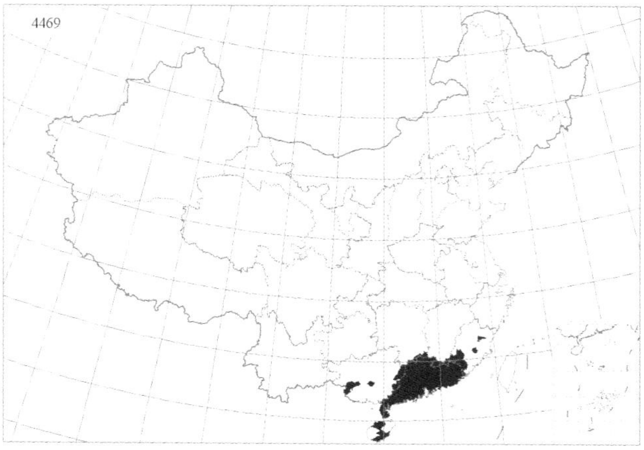

Fabaceae 豆科 *Phyllodium* 排钱草属

Phyllodium kurzianum (Kuntze) H. Ohashi 长柱排钱树

Shrubs

20.8; 14.3; 25.3; 20.8; 190.0; 0.0; 1040 ❙ 1541; 674; 93 ❙ 48.9; 1028; 481.9

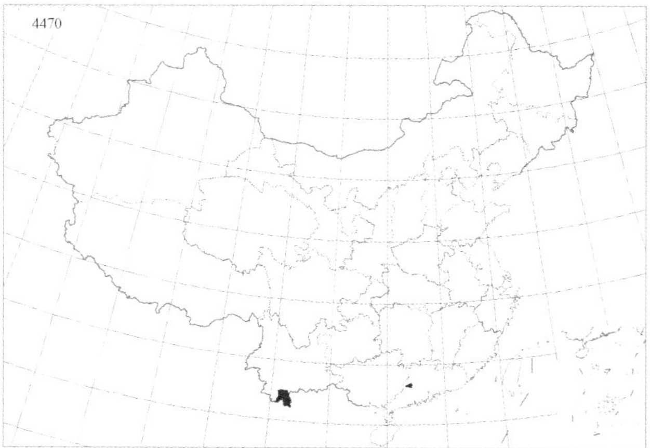

Fabaceae 豆科 *Phyllodium* 排钱草属
Phyllodium longipes (Craib) Schindl. 长叶排钱树
Shrubs

21.2 (17.7~23.2)；13.4 (9.5~15.8)；26.9 (22.3~29.0)；21.2 (17.8~23.2)；194.5 (153.4~218.9)；0.0 (0.0~0.0)；1101 (854~1257) ❙ 1587 (1188~2435)；785 (565~1428)；105 (56~164) ❙ 44.7 (5.8~99.5)；1095 (854~1257)；323.1 (194.5~659.0)

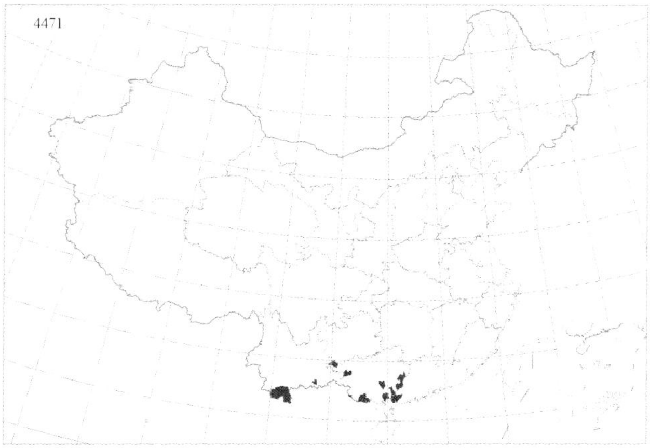

Fabaceae 豆科 *Phyllodium* 排钱草属
Phyllodium pulchellum (L.) Desv. 排钱树
Shrubs

19.7 (14.4~24.8)；12.3 (6.4~20.1)；25.5 (19.1~29.0)；19.8 (14.4~24.8)；177.0 (113.1~237.6)；0.0 (0.0~0.0)；1006 (727~1385) ❙ 1596 (868~3495)；733 (312~1847)；125 (41~897) ❙ 61.6 (-34.3~246.5)；984 (727~1329)；312.0 (119.4~659.0)

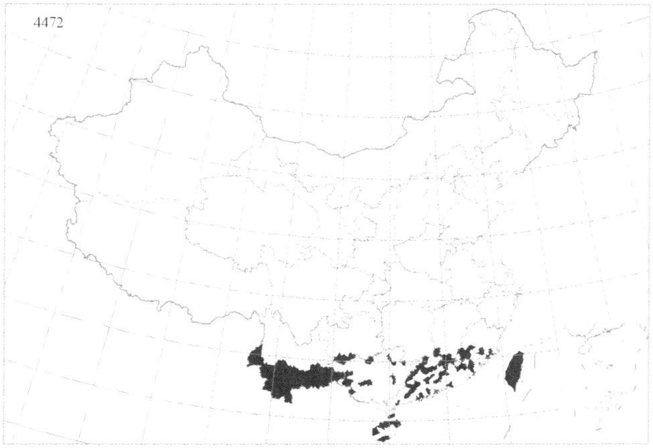

Fabaceae 豆科 *Piptanthus* 黄花木属
Piptanthus nepalensis (Hook.) D. Don ex Sweet 尼泊尔黄花木
Shrubs, deciduous

7.1 (-1.8~16.1)；-1.5 (-10.5~9.0)；14.6 (6.1~23.1)；8.0 (1.8~16.1)；53.0 (1.9~133.0)；27.[illegible] (0.0~82.8)；541 (329~797) ❙ 810 (311~1736)；466 (193~1007)；25 (4~69) ❙ 47.9 (-15.3~145.8)；536 (311~772)；269.0 (61.6~447.4)

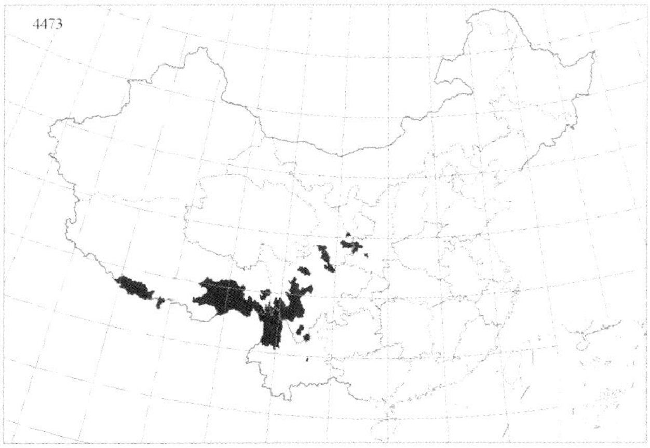

Fabaceae 豆科 *Piptanthus* 黄花木属
Piptanthus tomentosus Franch. 绒叶黄花木
Shrubs, deciduous

13.0 (2.0~18.9)；5.8 (-7.4~12.0)；18.8 (10.1~23.9)；13.2 (4.0~19.0)；102.0 (16.9~167.8)；5.4 (0.0~51.9)；703 (416~929) ❙ 954 (741~1410)；567 (411~816)；28 (11~49) ❙ 39.6 (-14.2~100.7)；663 (416~799)；289.8 (207.5~372.3)

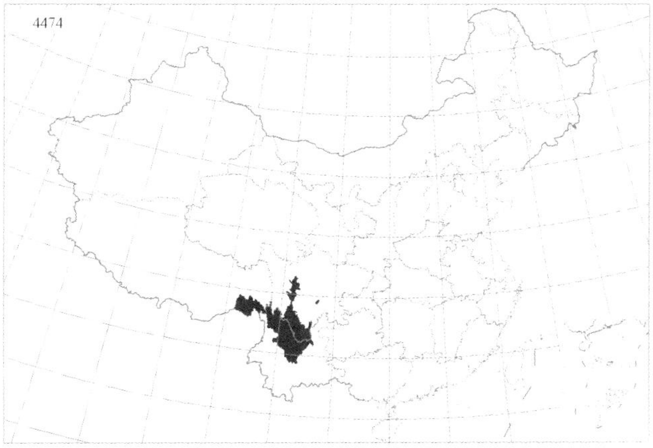

Fabaceae 豆科 *Pithecellobium* 牛蹄豆属
Pithecellobium dulce (Roxb.) Benth. 牛蹄豆
Trees, evergreen

18.9 (16.0~24.1)；10.7 (5.3~18.9)；26.0 (22.8~28.7)；19.2 (16.0~24.2)；169.8 (132.2~230.2)；0.0 (0.0~0.0)；989 (826~1329) ❙ 1577 (1236~2993)；747 (523~1594)；115 (47~547) ❙ 61.7 (-1.0~203.4)；976 (826~1329)；252.8 (154.2~551.6)

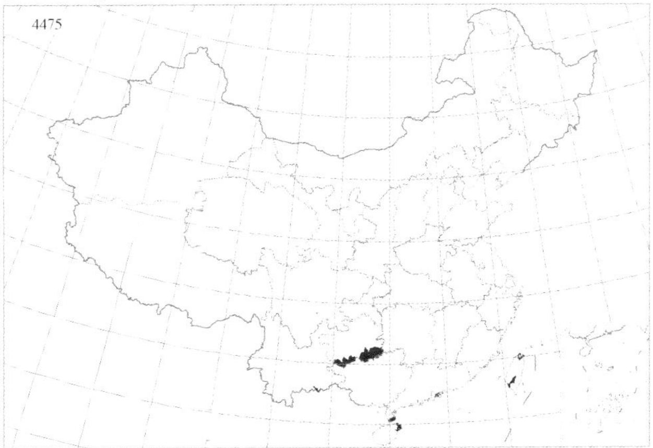

Fabaceae 豆科 *Pongamia* 水黄皮属
Pongamia pinnata (L.) Pierre 水黄皮
Trees

22.1 (17.8~24.8)；15.0 (12.1~20.1)；27.7 (21.7~29.0)；22.1 (17.9~24.8)；205.8 (154.4~237.6)；0.0 (0.0~0.0)；1172 (852~1385) ❙ 1921 (1258~3495)；875 (425~1847)；161 (41~897) ❙ 69.0 (-1.0~235.6)；1152 (852~1360)；253.3 (122.7~548.0)

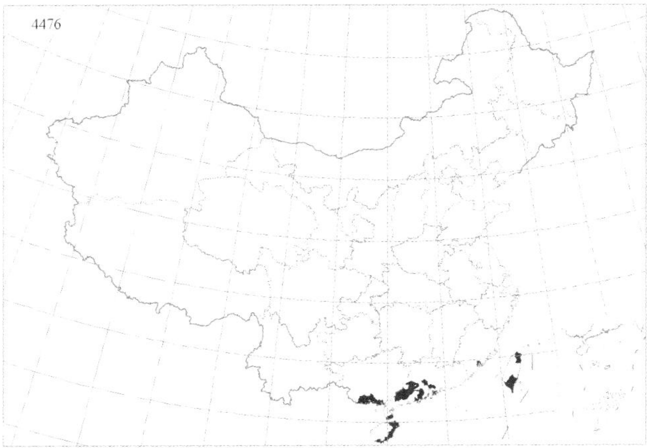

Fabaceae 豆科 *Pterocarpus* 紫檀属

Pterocarpus indicus Willd. 紫檀

Trees

21.3 (16.3~24.6); 14.4 (9.8~19.5); 26.4 (20.3~29.2); 21.3 (16.4~24.7); 195.6 (136.6~235.9); 0.0 (0.0~0.0); 1104 (789~1372) ❙ 1791 (1176~3495); 765 (447~1026); 165 (47~897) ❙ 65.6 (-12.0~224.4); 1088 (780~1331); 335.0 (139.6~659.0)

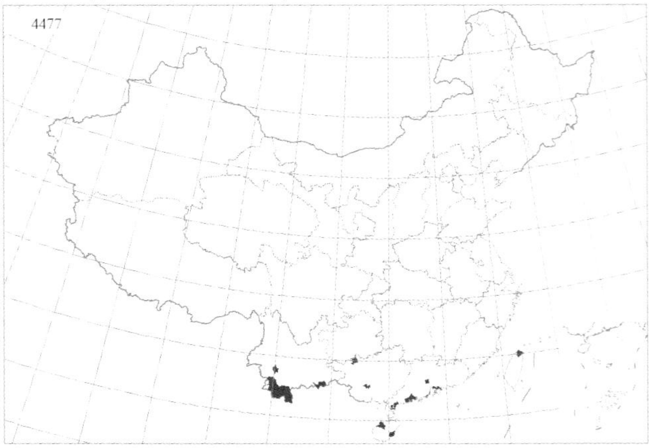

Fabaceae 豆科 *Pterolobium* 老虎刺属

Pterolobium macropterum Kurz 大翅老虎刺

Large climbers

19.3 (17.5~24.8); 12.8 (10.5~20.1); 23.7 (22.1~28.0); 19.3 (17.5~24.8); 172.0 (149.9~237.6); 0.0 (0.0~0.0); 961 (846~1385) ❙ 1162 (939~1405); 610 (473~774); 46 (39~55) ❙ 22.4 (1.5~54.8); 902 (799~1137); 367.5 (294.5~527.5)

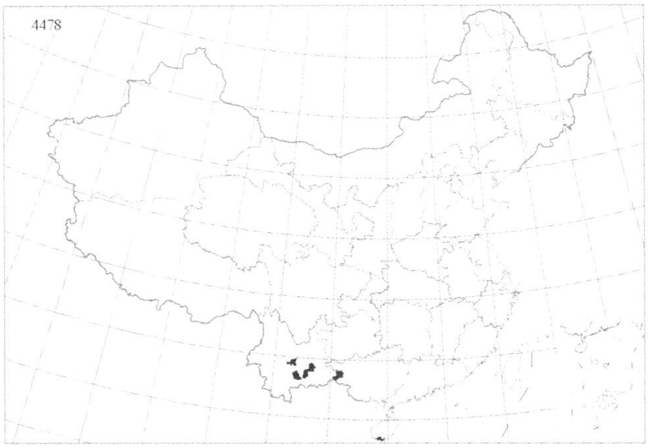

Fabaceae 豆科 *Pterolobium* 老虎刺属

Pterolobium punctatum Hemsl. 老虎刺

Vines or shrubs

17.0 (9.6~24.8); 7.0 (0.6~20.1); 25.8 (15.7~29.2); 17.0 (9.6~24.8); 144.7 (60.8~237.6); 0.5 (0.0~9.7); 893 (600~1385) ❙ 1368 (916~2148); 578 (403~1234); 117 (26~230) ❙ 54.2 (1.5~128.9); 887 (600~1202); 233.0 (146.0~620.0)

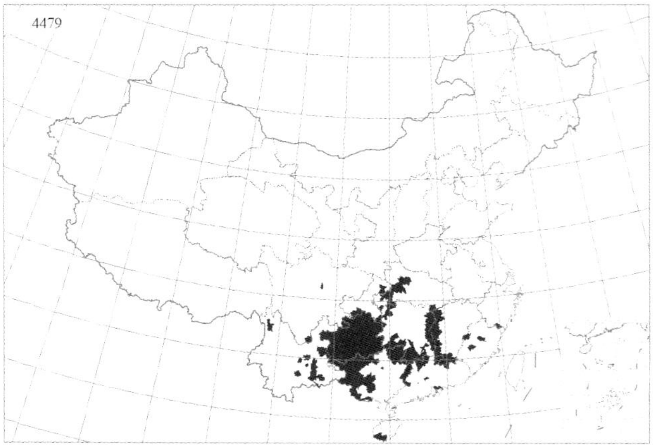

Fabaceae 豆科 *Pueraria* 葛属

Pueraria alopecuroides Craib 密花葛

Scandent shrubs, deciduous

19.0 (16.9~21.0); 13.0 (10.3~15.1); 22.9 (21.4~25.6); 19.1 (16.9~21.0); 168.7 (142.7~191.8); 0.0 (0.0~0.0); 918 (813~1053) ❙ 1394 (1143~1607); 710 (637~818); 59 (48~77) ❙ 52.0 (21.7~74.2); 904 (813~1053); 479.1 (346.9~659.0)

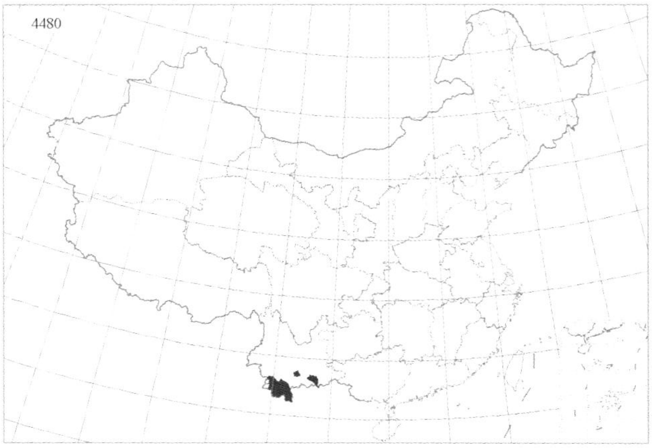

Fabaceae 豆科 *Pueraria* 葛属

Pueraria calycina Franch. 黄毛萼葛

Vines, deciduous

13.6; 6.8; 19.0; 13.6; 103.8; 0.4; 710 ❙ 969; 580; 30 ❙ 37.2; 680; 313.1

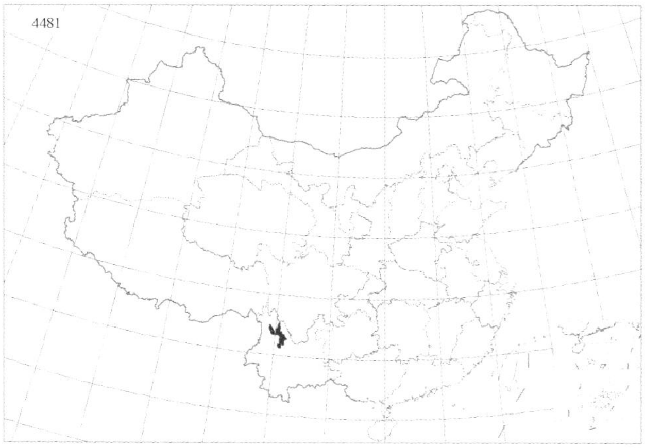

Fabaceae 豆科 *Pueraria* 葛属

Pueraria stricta Kurz 小花野葛

Shrubs, deciduous

19.8; 14.2; 23.3; 19.9; 178.9; 0.0; 957 ❙ 1428; 670; 63 ❙ 49.2; 932; 522.5

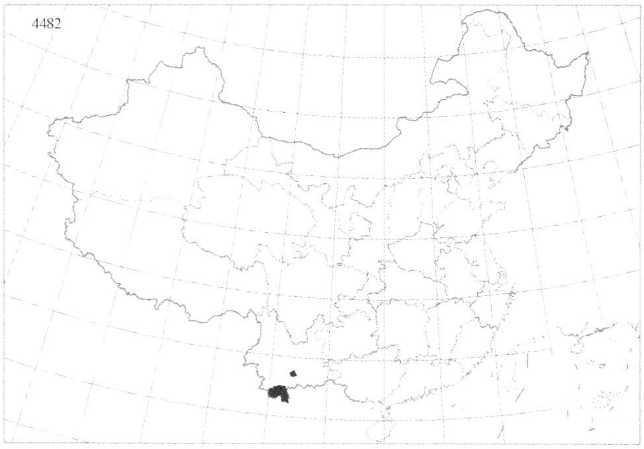

Fabaceae 豆科 *Rhynchosia* 鹿藿属
Rhynchosia rothii Benth. ex Ait. 绒叶鹿藿
Vines

19.4；13.3；24.4；19.4；172.6；0.0；967 ❙ 2109；1039；182 ❙ 128.0；946；242.1

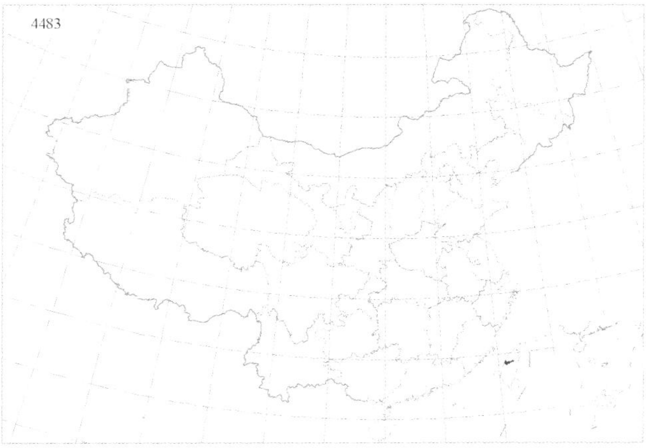

Fabaceae 豆科 *Rhynchosia* 鹿藿属
Rhynchosia rufescens (Willd.) DC. 淡红鹿藿
Shrubs

18.8 (16.9~22.2)；12.1 (9.1~15.1)；23.5 (21.4~28.3)；18.9 (16.9~22.3)；166.4 (142.7~207.5)；0.0 (0.0~0.0)；918 (813~1190) ❙ 1399 (1161~1751)；759 (633~996)；59 (48~87) ❙ 53.0 (15.4~88.8)；915 (813~1190)；447.3 (314.4~659.0)

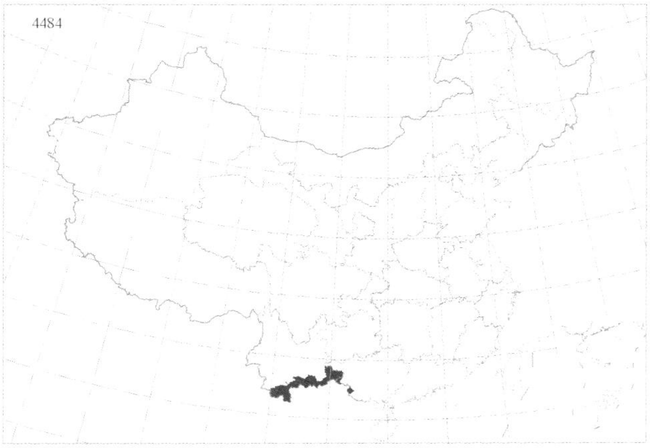

Fabaceae 豆科 *Robinia* 刺槐属
Robinia pseudoacacia L. 刺槐
Trees, deciduous

12.6 (-2.8~19.4)；0.2 (-20.9~10.1)；23.6 (9.2~29.9)；13.4 (2.9~19.5)；109.0 (10.1~173.7)；17.5 (0.0~109.2)；763 (374~1038) ❙ 1024 (45~1776)；437 (23~720)；85 (3~221) ❙ 30.7 (-94.0~125.9)；686 (45~1038)；187.1 (7.6~364.1)

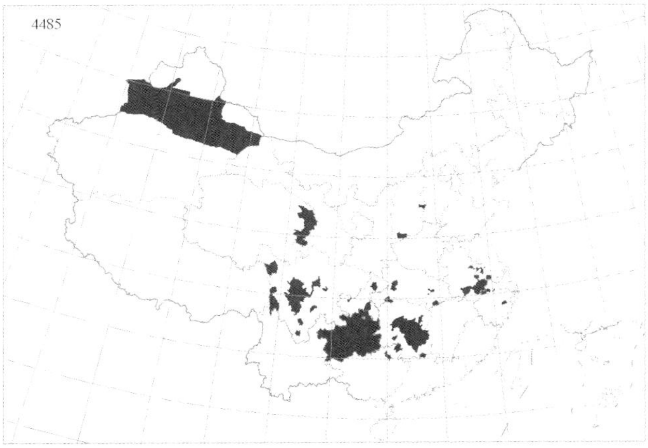

Fabaceae 豆科 *Salweenia* 冬麻豆属
Salweenia wardii Baker f. 冬麻豆
Shrubs, evergreen

1.5；-7.8；9.8；3.7；15.4；56.9；402 ❙ 633；381；13 ❙ 57.5；402；217.9

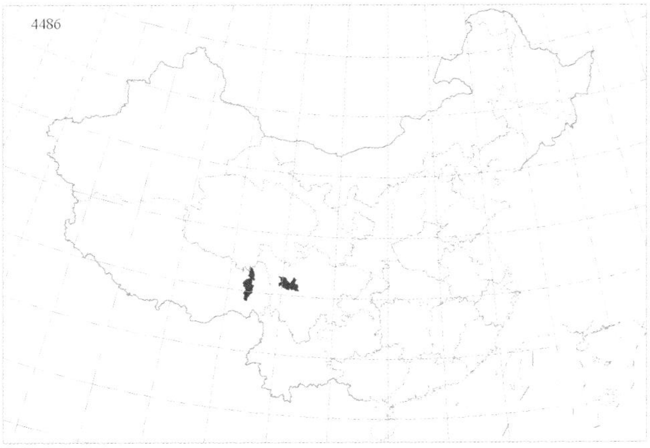

Fabaceae 豆科 *Saraca* 无忧花属
Saraca dives Pierre 中国无忧花
Trees

19.9 (15.8~22.7)；12.2 (9.0~15.0)；25.6 (20.6~28.6)；19.9 (15.8~22.8)；179.4 (129.7~213.1)；0.0 (0.0~0.0)；1014 (776~1221) ❙ 1460 (1011~2435)；809 (586~1428)；78 (44~138) ❙ 44.7 (5.8~99.5)；1013 (772~1221)；354.9 (256.6~551.6)

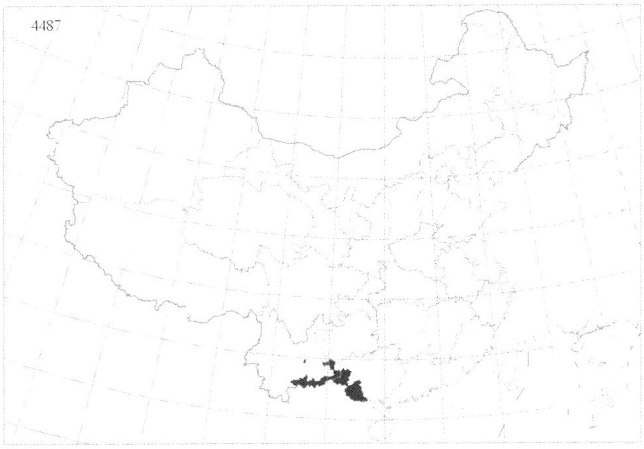

Fabaceae 豆科 *Saraca* 无忧花属
Saraca griffithiana Prain 云南无忧花
Trees

17.5；11.2；21.6；17.5；150.6；0.0；841 ❙ 1617；935；63 ❙ 93.2；841；589.2

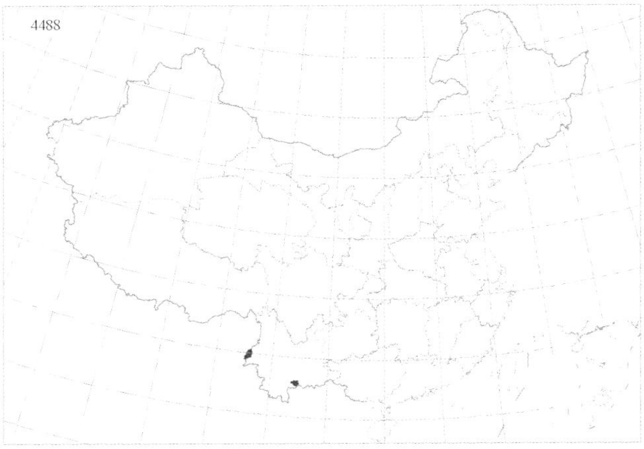

Fabaceae 豆科 *Sarcodum* 耀花豆属
Sarcodum scandens Lour. 耀花豆
Vines

24.0 (22.2~24.8); 18.4 (17.2~20.1); 28.0 (25.7~28.8); 24.0 (22.3~24.8); 228.3 (207.2~237.6); 0.0 (0.0~0.0); 1310 (1127~1385) ❙ 1501 (868~1854); 534 (312~712); 74 (41~115) ❙ 15.0 (-34.3~47.8); 1191 (868~1360); 440.3 (248.2~620.0)

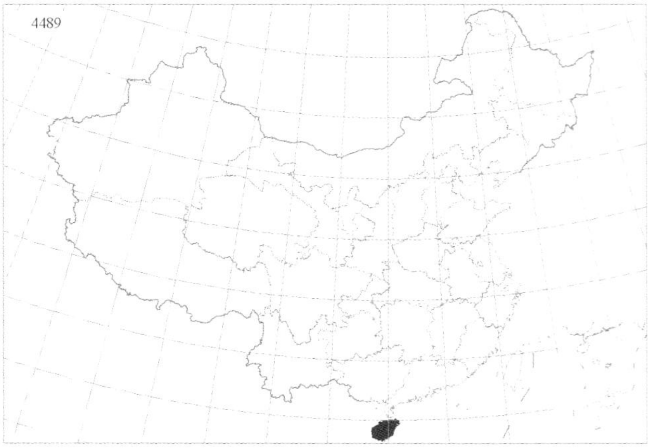

Fabaceae 豆科 *Senna* 番泻决明属
Senna alata (L.) Roxb. 翅柄决明
Shrubs

22.0 (20.4~24.7); 14.4 (11.9~19.8); 28.1 (25.5~29.1); 22.1 (20.4~24.7); 204.7 (184.9~236.5); 0.0 (0.0~0.0); 1167 (1039~1376) ❙ 1762 (1447~2992); 812 (534~1847); 121 (48~188) ❙ 51.5 (15.4~166.5); 1160 (1039~1225); 225.9 (95.3~620.0)

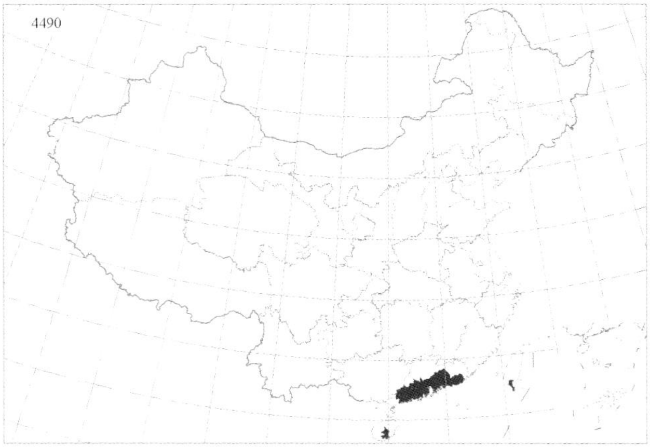

Fabaceae 豆科 *Senna* 番泻决明属
Senna hirsuta (L.) H. S. Irwin et Barneby 毛荚决明
Shrubs

23.0 (19.2~24.8); 17.2 (12.7~20.1); 27.0 (22.3~28.8); 23.0 (19.2~24.8); 215.9 (170.5~237.6); 0.0 (0.0~0.0); 1223 (909~1385) ❙ 1660 (868~2993); 652 (312~1847); 102 (41~547) ❙ 40.1 (-34.3~203.4); 1137 (868~1360); 430.2 (204.4~659.0)

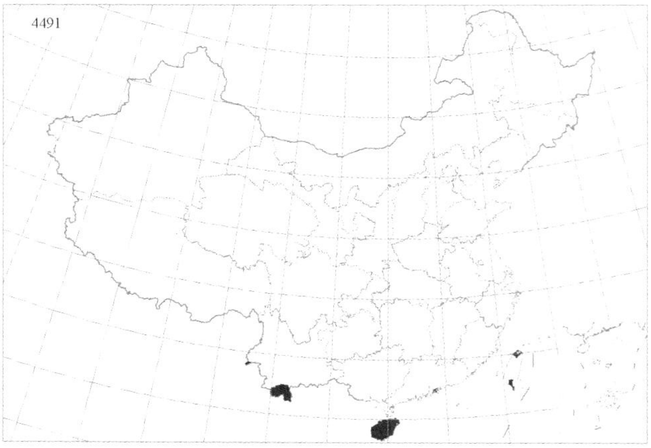

Fabaceae 豆科 *Senna* 番泻决明属
Senna occidentalis (L.) Link 望江南
Shrubs or subshrubs

18.6 (10.2~24.8); 9.4 (-0.7~20.1); 26.4 (17.0~29.4); 18.6 (10.4~24.8); 164.2 (74.6~237.6); 0.5 (0.0~13.6); 976 (634~1385) ❙ 1447 (797~2993); 650 (312~1847); 117 (15~547) ❙ 50.4 (-34.3~246.5); 966 (634~1331); 256.7 (108.2~659.0)

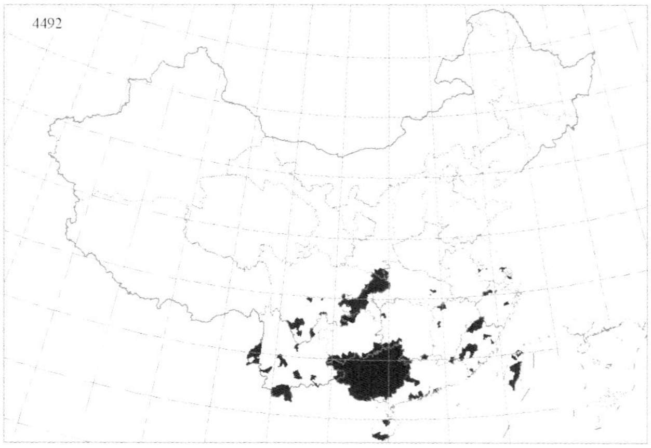

Fabaceae 豆科 *Senna* 番泻决明属
Senna siamea (Lam.) H. S. Irwin et Barneby 铁刀木
Trees

21.8 (15.4~24.8); 15.6 (8.2~20.1); 26.4 (20.1~28.8); 21.8 (15.4~24.8); 201.9 (125.4~237.6); 0.0 (0.0~0.0); 1150 (770~1385) ❙ 1597 (868~2993); 694 (312~1847); 91 (19~547) ❙ 44.3 (-34.3~246.5); 1070 (696~1360); 386.6 (154.2~659.0)

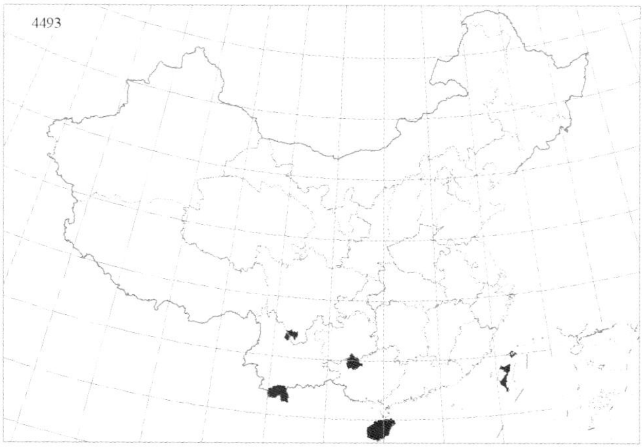

Fabaceae 豆科 *Senna* 番泻决明属
Senna sophera (L.) Roxb. 槐叶决明
Shrubs

17.1 (5.4~24.8); 7.8 (-2.1~20.1); 25.2 (12.3~29.7); 17.1 (5.8~24.8); 146.4 (31.5~237.6); 1.0 (0.0~25.6); 907 (481~1385) ❙ 1266 (741~1965); 569 (312~994); 87 (15~230) ❙ 42.1 (-34.3~128.9); 886 (481~1360); 248.7 (108.2~659.0)

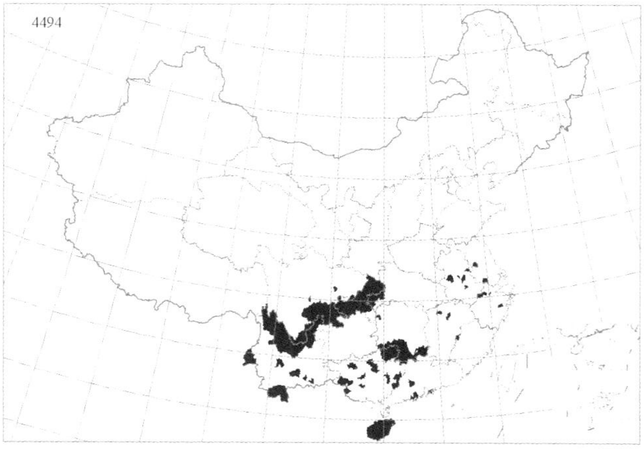

Fabaceae 豆科 *Senna* 番泻决明属

Senna sulfurea (Collad.) H. S. Irwin et Barneby 粉叶决明

Shrubs or small trees

24.0 (22.2~24.8)；18.4 (17.2~20.1)；28.0 (25.7~28.8)；24.0 (22.3~24.8)；228.3 (207.2~237.6)；0.0 (0.0~0.0)；1310 (1127~1385) ▌1501 (868~1854)；534 (312~712)；74 (41~115) ▌15.0 (-34.3~47.8)；1191 (868~1360)；440.3 (248.2~620.0)

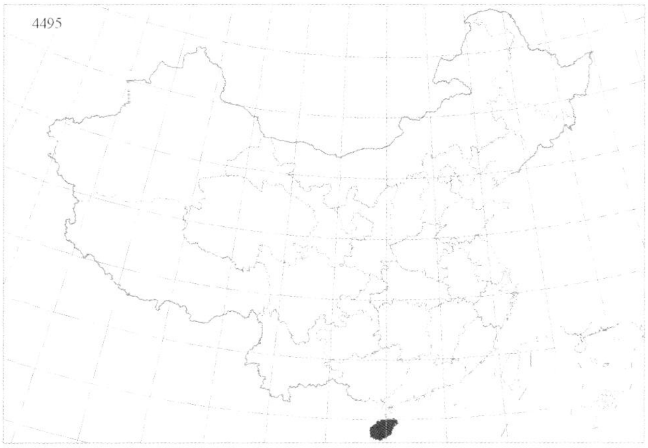

Fabaceae 豆科 *Sesbania* 田菁属

Sesbania grandiflora (L.) Pers. 大花田菁

Small trees

23.6 (22.0~24.8)；17.3 (13.5~20.1)；28.3 (27.9~28.8)；23.7 (22.0~24.8)；223.9 (203.8~237.6)；0.0 (0.0~0.0)；1297 (1170~1385) ▌1638 (1405~1784)；616 (473~779)；93 (41~144) ▌27.5 (1.5~50.9)；1200 (1137~1299)；386.3 (157.7~548.0)

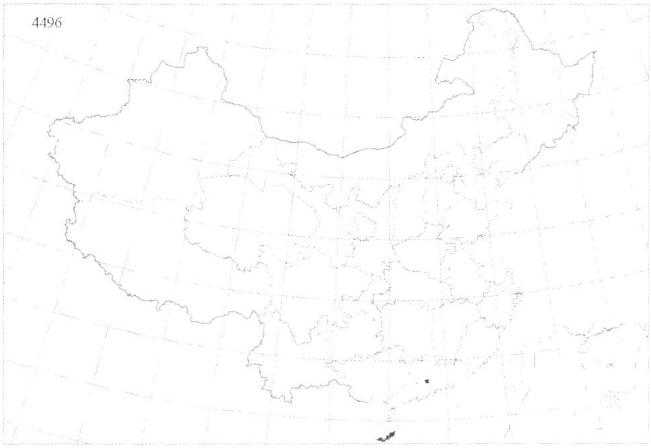

Fabaceae 豆科 *Sindora* 油楠属

Sindora glabra Merr. ex de Wit. 油楠

Trees

24.0 (23.1~24.8)；18.4 (17.5~20.1)；28.0 (27.2~28.7)；24.0 (23.2~24.8)；228.4 (218.4~237.6)；0.0 (0.0~0.0)；1311 (1232~1385) ▌1248 (868~1487)；438 (312~544)；51 (41~71) ▌-4.5 (-34.3~20.7)；1078 (868~1142)；445.1 (288.3~538.2)

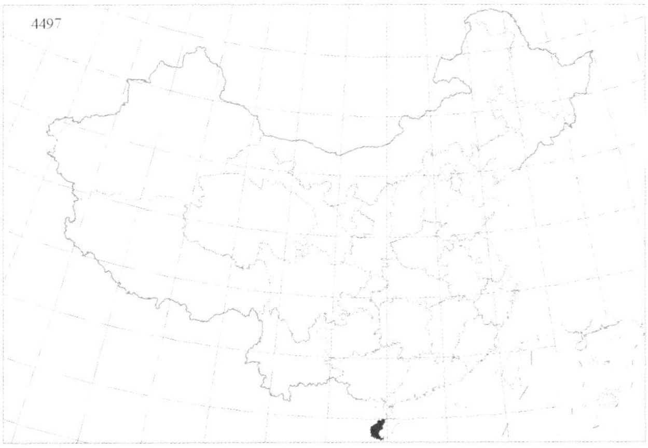

Fabaceae 豆科 *Smithia* 坡油甘属

Smithia blanda Hohen. ex Baker 黄花合叶豆

Subshrubs

15.4 (12.6~18.9)；7.6 (3.6~12.5)；21.8 (17.9~24.9)；15.5 (12.6~19.0)；125.9 (91.6~167.8)；0.1 (0.0~1.8)；781 (678~929) ▌1042 (797~1298)；568 (411~651)；43 (20~78) ▌34.1 (-14.2~62.6)；746 (678~851)；259.4 (153.9~372.3)

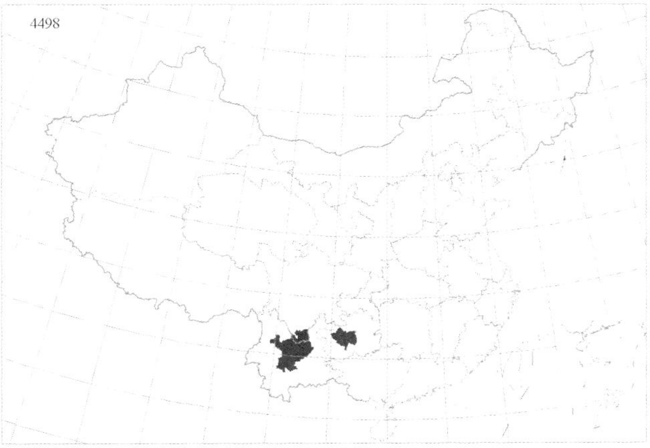

Fabaceae 豆科 *Sophora* 苦参属

Sophora albescens J. St. -Hil. 白花槐

Shrubs or subshrubs, deciduous

12.8 (2.2~19.1)；4.0 (-6.6~10.1)；20.3 (9.9~26.6)；13.0 (3.8~19.1)；102.2 (15.5~169.2)；8.[illegible] (0.0~49.1)；725 (413~969) ▌1109 (791~1381)；606 (432~720)；39 (17~107) ▌58.3 (27.5~125.9)；721 (413~969)；260.3 (172.9~357.7)

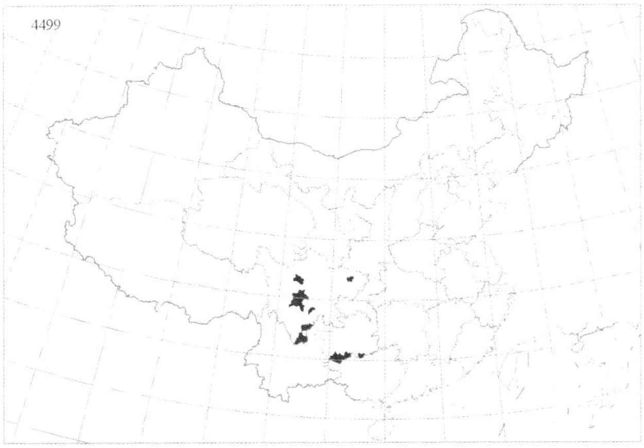

Fabaceae 豆科 *Sophora* 苦参属

Sophora alopecuroides L. 苦豆子

Subshrubs, deciduous

8.6 (-4.8~15.5)；-7.3 (-20.9~5.4)；22.5 (7.4~29.5)；10.4 (1.9~15.5)；84.0 (4.3~136.0)；39.8 (0.0~121.6)；673 (329~932) ▌389 (21~1736)；206 (15~1007)；16 (0~58) ▌-42.3 (-97.3~145.8)；378 (21~821)；113.3 (1.6~447.4)

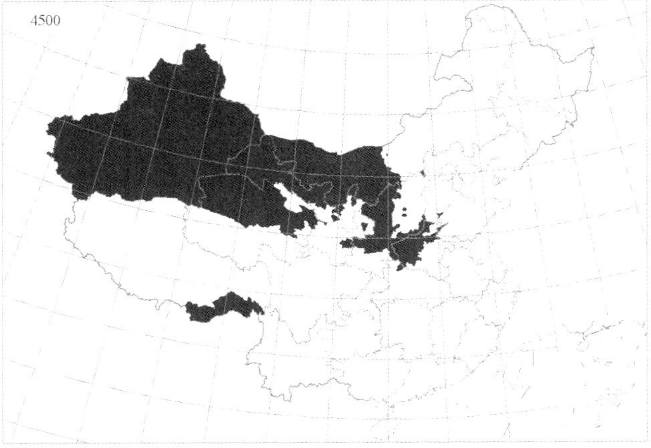

Fabaceae 豆科 *Sophora* 苦参属

Sophora benthamii Steenis 尾叶槐

Shrubs, deciduous

16.1 (13.3~17.3); 9.4 (5.4~11.0); 20.9 (19.6~21.6); 16.2 (13.4~17.3); 133.8 (100.4~148.0); 0.0 (0.0~0.0); 791 (706~832) ❙ 1212 (1049~1736); 684 (594~1007); 49 (44~55) ❙ 55.7 (26.1~145.8); 774 (706~809); 403.7 (365.3~447.4)

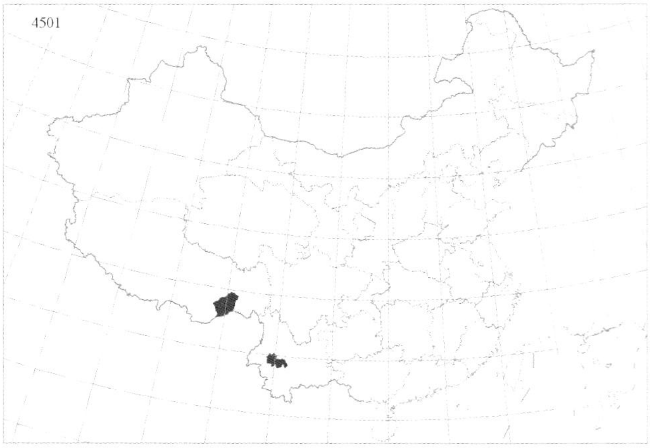

Fabaceae 豆科 *Sophora* 苦参属

Sophora brachygyna C. Y. Ma 短蕊槐

Trees, deciduous

16.1 (14.1~18.2); 4.3 (1.6~6.5); 27.3 (24.3~29.3); 16.1 (14.2~18.2); 134.7 (114.1~158.8); 1.3 (0.0~6.5); 872 (765~984) ❙ 1575 (969~1922); 560 (419~685); 201 (122~259) ❙ 81.0 (13.4~124.8); 869 (765~941); 232.5 (144.7~284.0)

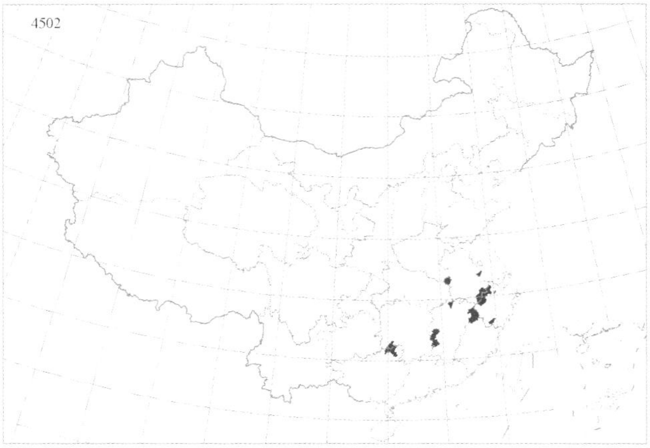

Fabaceae 豆科 *Sophora* 苦参属

Sophora davidii Kom. ex Pavol. 白刺槐

Shrubs or small trees, deciduous

12.1 (-0.4~19.3); 0.1 (-12.7~13.1); 22.7 (8.1~29.5); 12.6 (2.7~19.4); 101.7 (8.0~173.2); 16.2 (0.0~73.3); 729 (360~993) ❙ 833 (252~1950); 434 (190~932); 40 (0~259) ❙ 14.8 (-50.1~147.0); 656 (252~969); 204.1 (44.7~432.9)

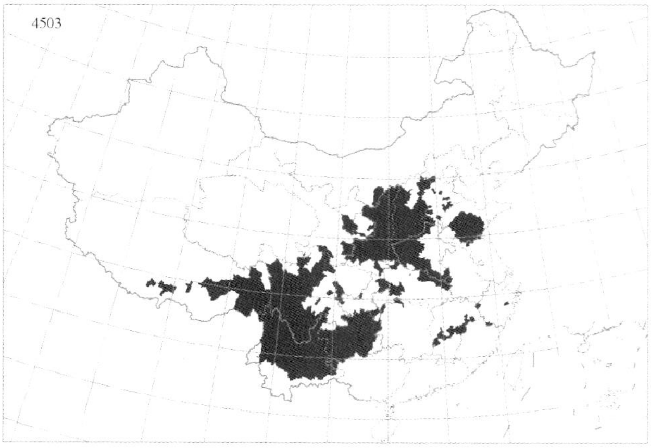

Fabaceae 豆科 *Sophora* 苦参属

Sophora davidii var. chuansiensis (C. Y. Ma) C. Y. Ma 川西白刺槐

Shrubs or small trees, deciduous

10.9 (2.7~14.6); 2.8 (-7.1~7.1); 17.7 (11.1~20.7); 11.0 (4.4~14.6); 77.7 (20.5~114.8); 6.7 (0.0~48.0); 639 (429~741) ❙ 1023 (800~1178); 600 (427~663); 23 (12~35) ❙ 62.0 (29.7~101.6); 631 (429~717); 294.1 (255.4~341.6)

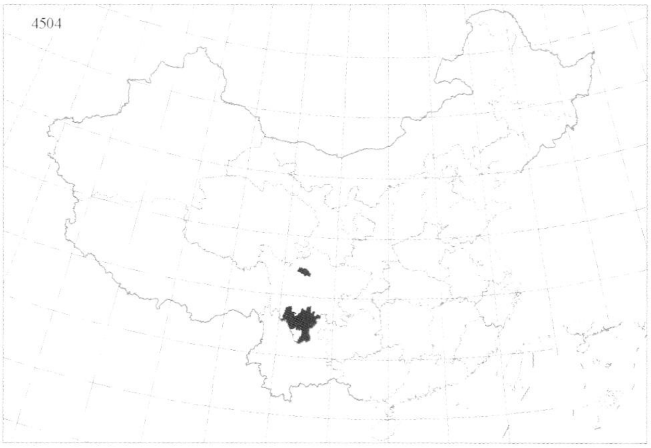

Fabaceae 豆科 *Sophora* 苦参属

Sophora dunnii Prain 柳叶槐

Small shrubs, deciduous

16.8; 10.2; 21.4; 16.8; 142.0; 0.0; 814 ❙ 1138; 666; 37 ❙ 39.8; 768; 378.0

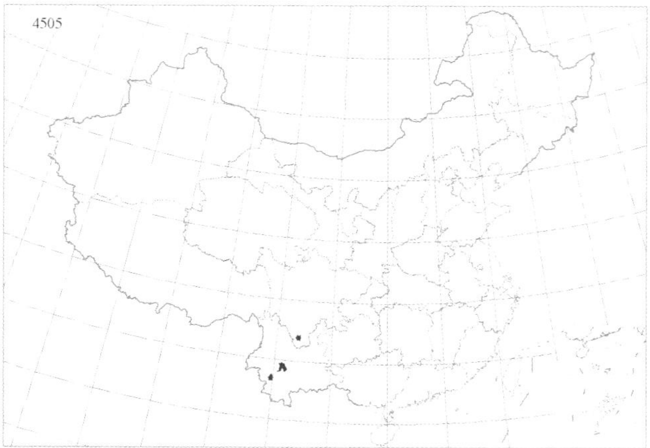

Fabaceae 豆科 *Sophora* 苦参属

Sophora flavescens Aiton 苦参

Herbs or subshrubs, deciduous

13.1 (-5.8~22.3); -0.1 (-29.8~17.7); 25.0 (10.8~30.1); 13.9 (4.4~22.3); 115.4 (21.6~208.1); 17.2 (0.0~157.4); 790 (426~1202) ❙ 973 (195~2993); 469 (133~1847); 73 (4~547) ❙ 21.1 (-67.0~246.5); 716 (195~1202); 212.5 (43.7~659.0)

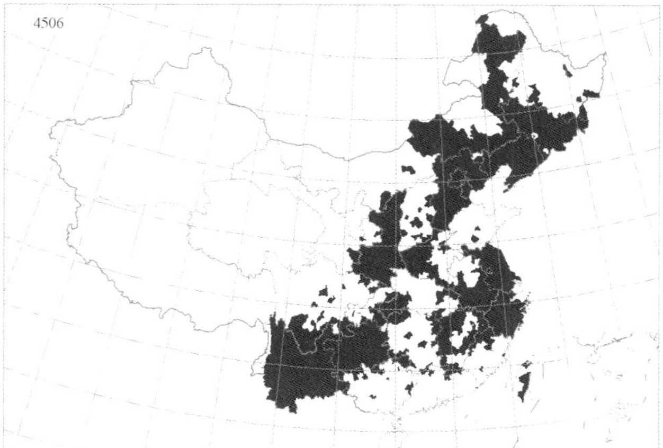

Fabaceae 豆科 *Sophora* 苦参属
Sophora franchetiana Dunn 闽槐
Shrubs or small trees, deciduous

18.2 (14.6~21.0); 8.1 (3.8~12.2); 27.2 (24.6~29.1); 18.2 (14.7~21.0); 159.1 (117.7~192.6); 0.1 (0.0~1.5); 952 (779~1107) ▎1645 (1485~1852); 585 (445~711); 196 (170~253) ▎73.9 (41.1~102.5); 950 (779~1107); 243.9 (202.2~305.7)

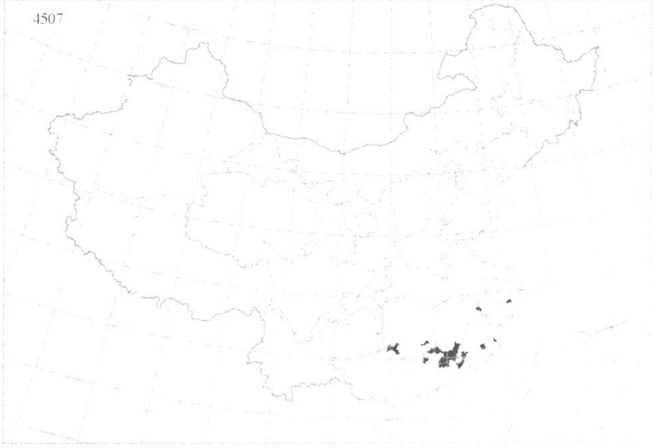

Fabaceae 豆科 *Sophora* 苦参属
Sophora japonica L. 槐
Trees, deciduous

13.9 (-4.3~22.4); 2.1 (-24.4~15.2); 24.2 (9.6~29.9); 14.4 (2.5~22.7); 119.3 (9.7~212.7); 12.4 (0.0~122.1); 796 (329~1216) ▎1049 (21~2014); 466 (15~996); 84 (0~259) ▎28.3 (-97.3~128.9); 709 (21~1216); 213.5 (1.6~659.0)

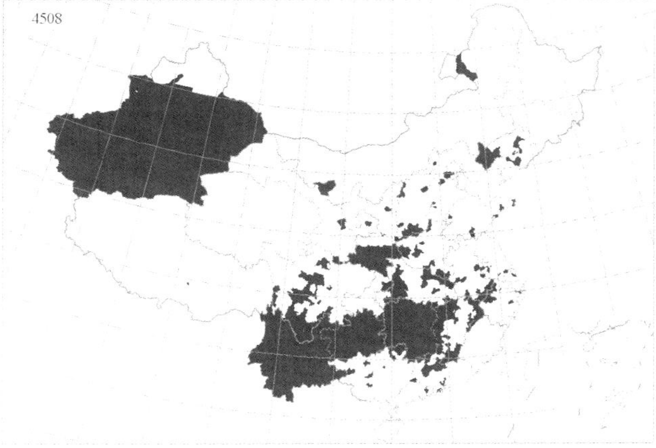

Fabaceae 豆科 *Sophora* 苦参属
Sophora japonica var. pubescens (Tausch) Bosse 毛叶槐
Trees, deciduous

7.5 (2.0~14.6); -0.7 (-7.2~7.8); 14.7 (10.1~20.0); 8.4 (3.8~14.6); 55.8 (15.4~115.2); 24.8 (0.0~51.4); 552 (413~738) ▎949 (829~1023); 528 (400~603); 25 (15~35) ▎80.5 (30.5~128.6); 541 (413~707); 271.2 (200.8~309.7)

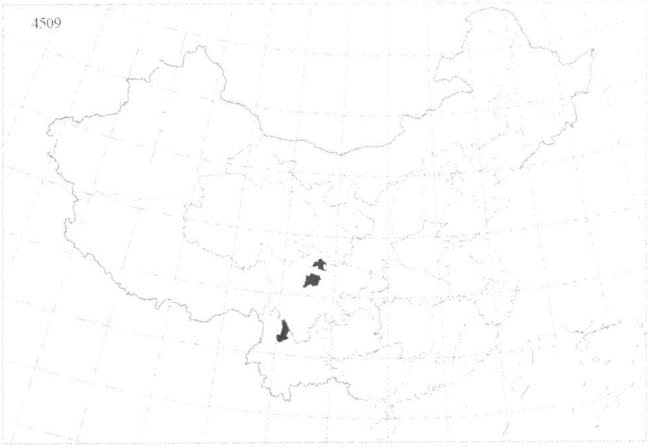

Fabaceae 豆科 *Sophora* 苦参属
Sophora microcarpa C. Y. Ma 细果槐
Shrubs, deciduous

18.9 (16.8~20.5); 11.0 (7.8~15.1); 24.7 (22.3~26.6); 18.9 (16.8~20.5); 166.8 (141.5~185.9); 0.0 (0.0~0.0); 936 (836~992) ▎1357 (1236~1607); 683 (617~723); 61 (51~77) ▎45.2 (27.5~64.9); 929 (836~986); 363.8 (189.5~659.0)

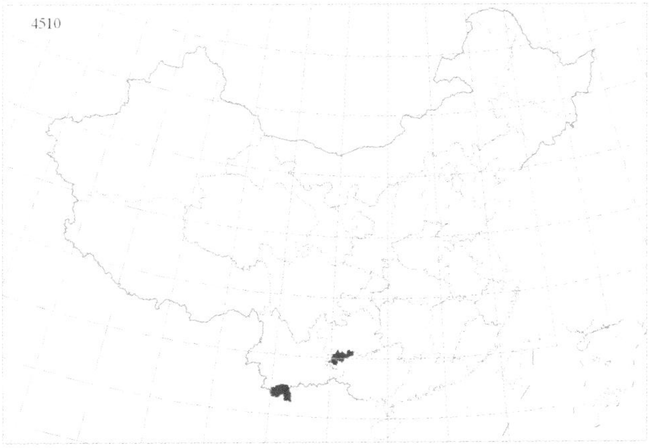

Fabaceae 豆科 *Sophora* 苦参属
Sophora praetorulosa Chun et T. C. Chen ex P. T. Li 疏节槐
Subshrubs, deciduous

23.9; 18.4; 27.9; 23.9; 227.3; 0.0; 1301 ▎1267; 430; 45 ▎-2.5; 1086; 450.1

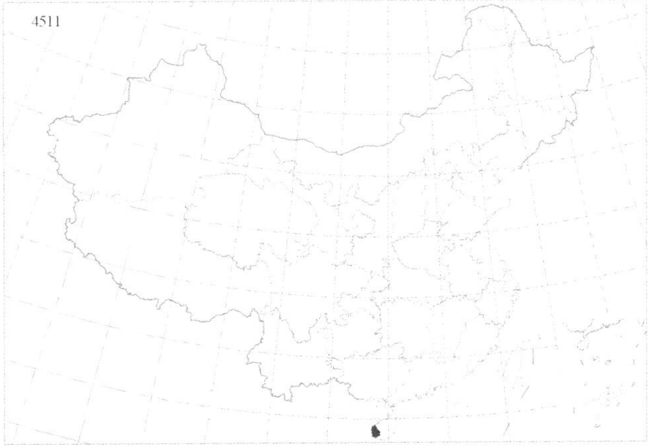

Fabaceae 豆科 *Sophora* 苦参属
Sophora prazeri Prain 锈毛槐
Shrubs

15.7 (2.0~21.0); 7.6 (-7.2~15.1); 22.3 (9.9~28.7); 15.8 (3.8~21.0); 130.7 (15.4~191.8); 1.3 (0.0~51.4); 808 (413~1081) ▎1176 (711~1776); 602 (357~996); 63 (15~200) ▎45.7 (-14.2~128.6); 796 (413~1081); 291.8 (118.7~659.0)

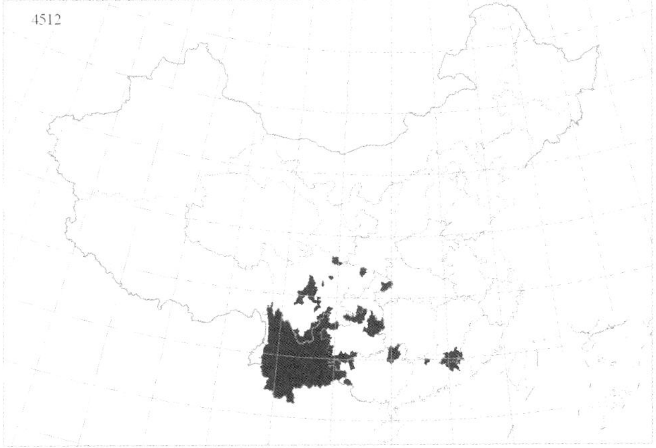

Fabaceae 豆科 *Sophora* 苦参属
Sophora tomentosa L. 绒毛槐
Shrubs or small trees, deciduous

22.4 (18.5~24.8); 16.7 (13.5~20.1); 26.7 (21.7~28.7); 22.4 (18.5~24.8); 209.4 (162.1~237.6); 0.0 (0.0~0.0); 1185 (863~1385) ❙ 2094 (1405~2992); 1042 (473~1847); 104 (41~188) ❙ 87.5 (1.5~214.4); 1113 (863~1299); 331.8 (154.2~548.0)

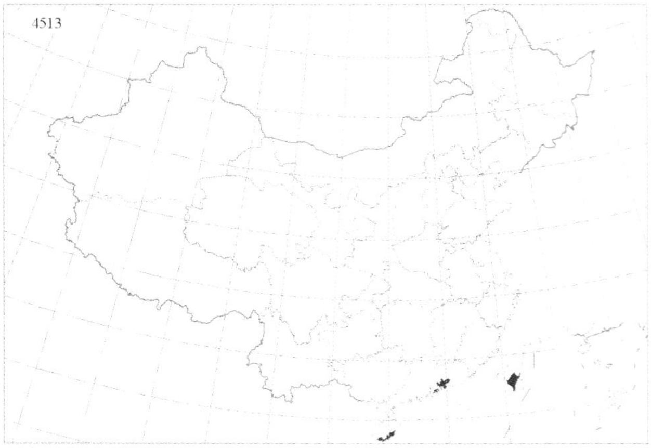

Fabaceae 豆科 *Sophora* 苦参属
Sophora tonkinensis Gagnep. 越南槐
Shrubs, deciduous

18.2 (15.5~21.9); 9.5 (5.3~14.2); 25.0 (20.4~28.6); 18.2 (15.6~22.0); 158.7 (127.1~203.8); 0.0 (0.0~0.0); 922 (780~1176) ❙ 1326 (1007~1662); 681 (465~860); 79 (43~191) ❙ 44.6 (5.8~80.6); 922 (780~1176); 289.4 (146.0~551.6)

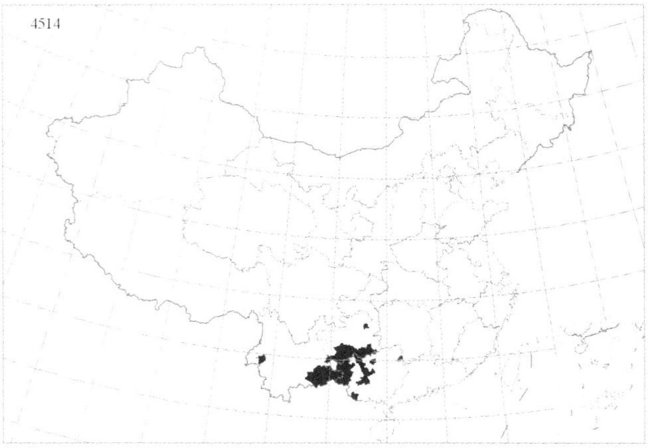

Fabaceae 豆科 *Sophora* 苦参属
Sophora tonkinensis var. purpurescens C. Y. Ma 紫花越南槐
Shrubs

16.8; 7.8; 23.6; 16.8; 141.5; 0.0; 836 ❙ 1334; 720; 64 ❙ 59.6; 836; 226.6

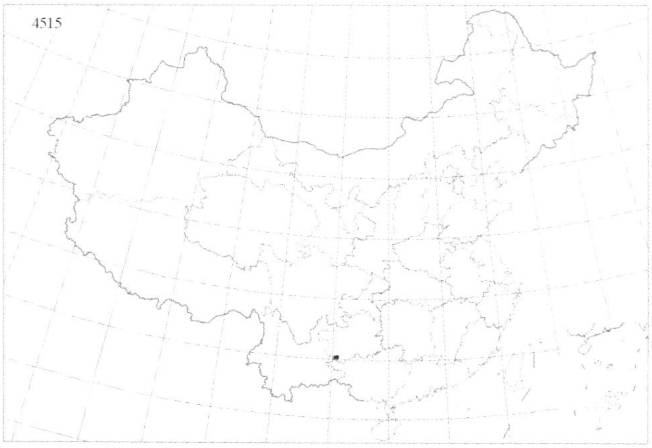

Fabaceae 豆科 *Sophora* 苦参属
Sophora velutina Lindl. 短绒槐
Shrubs, deciduous

15.1 (2.2~21.0); 7.6 (-6.6~15.1); 21.0 (9.9~26.6); 15.2 (3.8~21.0); 123.3 (15.5~191.8); 1.8 (0.0~49.1); 774 (413~1053) ❙ 1069 (741~1751); 597 (411~996); 42 (12~77) ❙ 39.3 (-14.2~125.9); 756 (413~1053); 303.3 (142.9~659.0)

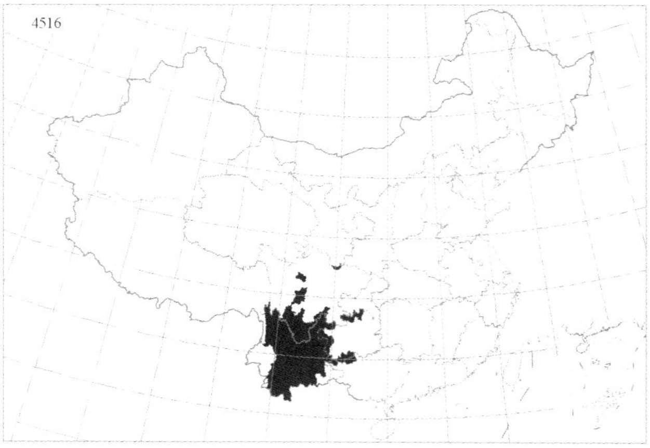

Fabaceae 豆科 *Sophora* 苦参属
Sophora velutina var. cavaleriei (H. Lév.) P. C. Tsoong 光叶短绒槐
Shrubs, deciduous

18.1; 12.0; 22.1; 18.1; 157.3; 0.0; 863 ❙ 1336; 774; 55 ❙ 54.8; 863; 432.9

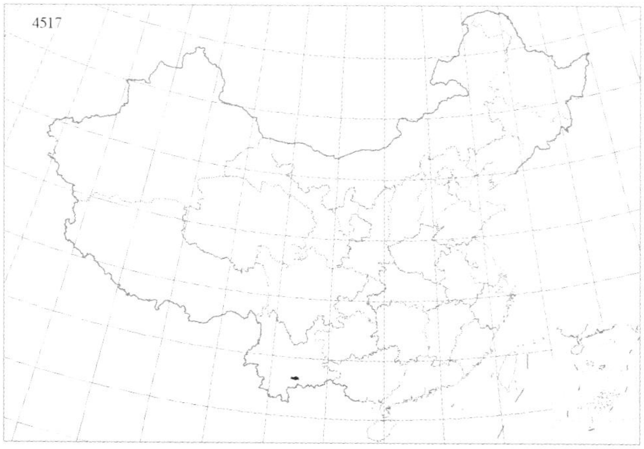

Fabaceae 豆科 *Sophora* 苦参属
Sophora velutina var. dolichopoda C. Y. Ma 长颈槐
Shrubs, deciduous

18.7; 9.5; 26.0; 18.7; 164.9; 0.0; 950 ❙ 1255; 666; 56 ❙ 32.2; 950; 247.0

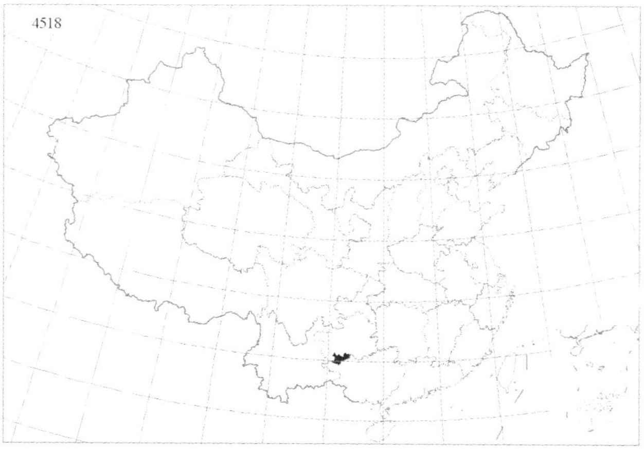

Fabaceae 豆科 *Sophora* 苦参属
Sophora velutina var. scandens C. Y. Ma 攀援槐
Shrubs, deciduous

10.2; 2.4; 16.9; 10.3; 68.2; 5.1; 620 ❙ 1036; 606; 18 ❙ 67.1; 620; 298.0

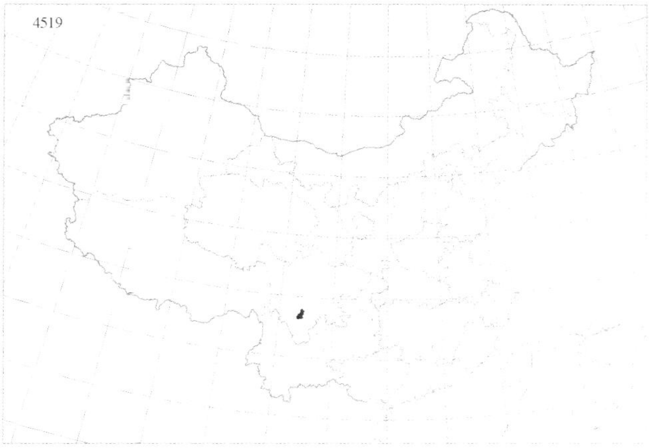

Fabaceae 豆科 *Sophora* 苦参属
Sophora xanthantha C. Y. Ma 黄花槐
Herbs or subshrubs, deciduous

19.3; 13.1; 23.6; 19.4; 173.2; 0.0; 939 ❙ 1143; 637; 50 ❙ 21.7; 903; 346.9

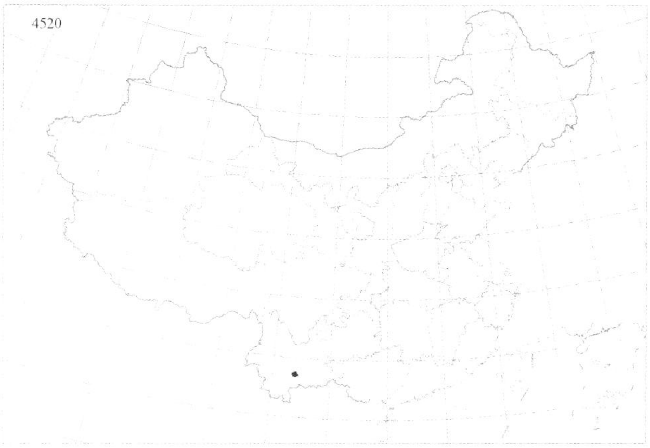

Fabaceae 豆科 *Sophora* 苦参属
Sophora yunnanensis C. Y. Ma 云南槐
Shrubs or small trees, deciduous

18.6; 12.3; 22.8; 18.7; 164.1; 0.0; 899 ❙ 1121; 635; 48 ❙ 24.8; 870; 321.2

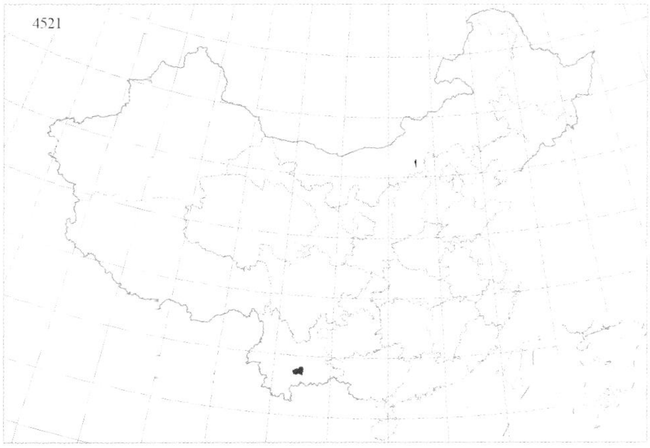

Fabaceae 豆科 *Spatholobus* 密花豆属
Spatholobus biauritus C. F. Wei 双耳密花豆
Vines

17.3; 11.0; 21.6; 17.3; 148.0; 0.0; 832 ❙ 1049; 598; 44 ❙ 26.1; 783; 365.3

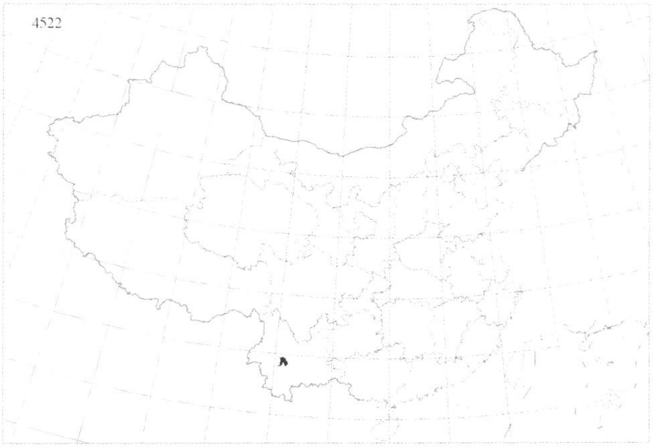

Fabaceae 豆科 *Spatholobus* 密花豆属
Spatholobus discolor C. F. Wei 变色密花豆
Vines

18.6; 12.7; 22.5; 18.7; 163.9; 0.0; 888 ❙ 1547; 911; 63 ❙ 74.2; 888; 551.3

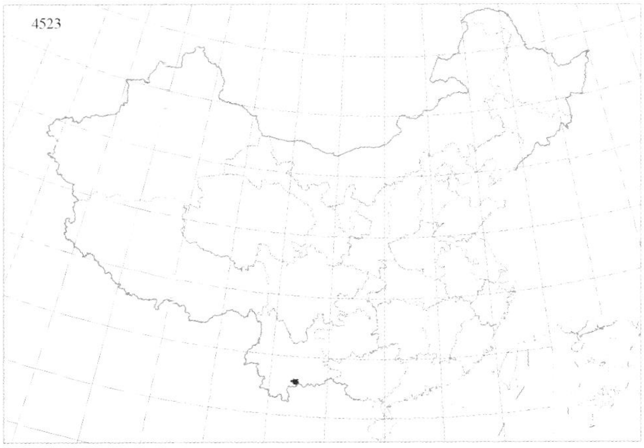

Fabaceae 豆科 *Spatholobus* 密花豆属
Spatholobus gengmaensis C. F. Wei 耿马密花豆
Vines

18.2; 11.9; 22.0; 18.2; 158.7; 0.0; 868 ❙ 1366; 783; 45 ❙ 57.4; 847; 516.7

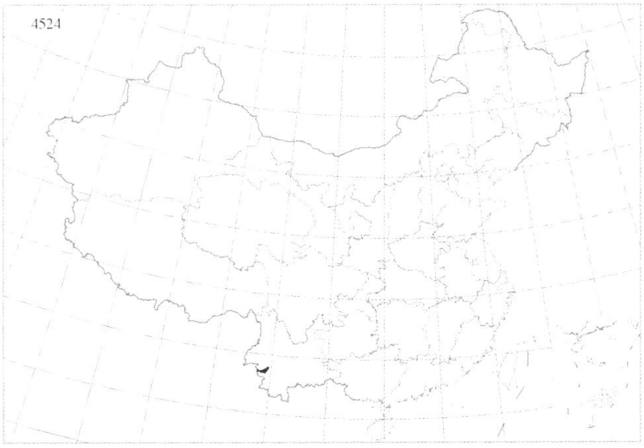

Fabaceae 豆科 *Spatholobus* 密花豆属
Spatholobus harmandii Gagnep. 光叶密花豆
Vines

23.5 (21.3~24.8); 17.5 (13.2~20.1); 27.7 (27.2~28.7); 23.5 (21.3~24.8); 222.0 (195.9~237.6); 0.0 (0.0~0.0); 1271 (1111~1385) ❙ 1470 (1176~1900); 595 (447~1056); 65 (41~125) ❙ 17.6 (-12.0~71.0); 1128 (1111~1142); 446.0 (297.2~538.2)

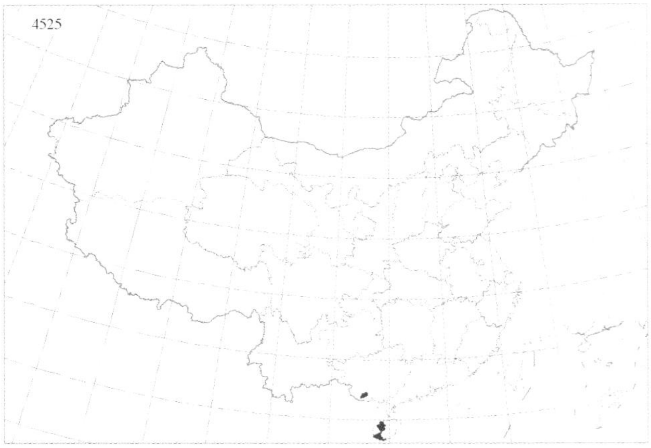

Fabaceae 豆科 *Spatholobus* 密花豆属
Spatholobus pulcher Dunn 美丽密花豆
Vines

19.3 (18.2~20.5); 13.7 (12.4~15.1); 22.8 (21.6~23.7); 19.4 (18.2~20.5); 172.3 (158.6~185.9); 0.0 (0.0~0.0); 925 (861~992) ❙ 1552 (1464~1751); 787 (655~996); 65 (52~77) ❙ 68.1 (47.6~88.8); 910 (839~986); 525.1 (427.9~659.0)

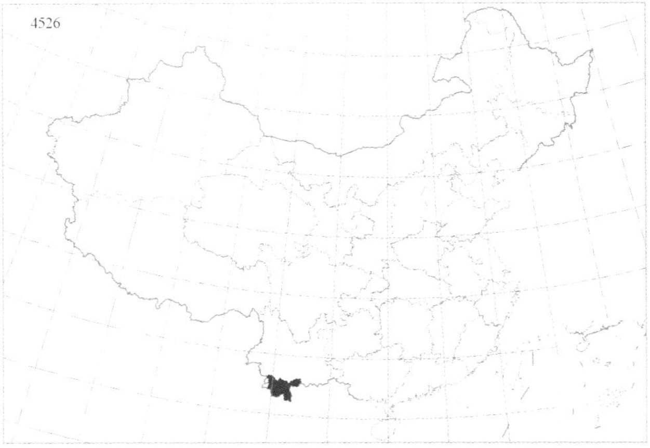

Fabaceae 豆科 *Spatholobus* 密花豆属
Spatholobus sinensis Chun et T. C. Chen 红血藤
Vines

22.7 (21.3~24.8); 15.7 (13.2~20.1); 27.9 (25.7~28.6); 22.8 (21.3~24.8); 213.2 (195.9~237.6); 0.0 (0.0~0.0); 1220 (1111~1385) ❙ 1707 (1302~2435); 805 (473~1428); 97 (41~130) ❙ 41.0 (-1.4~99.5); 1177 (1055~1299); 357.1 (209.2~620.0)

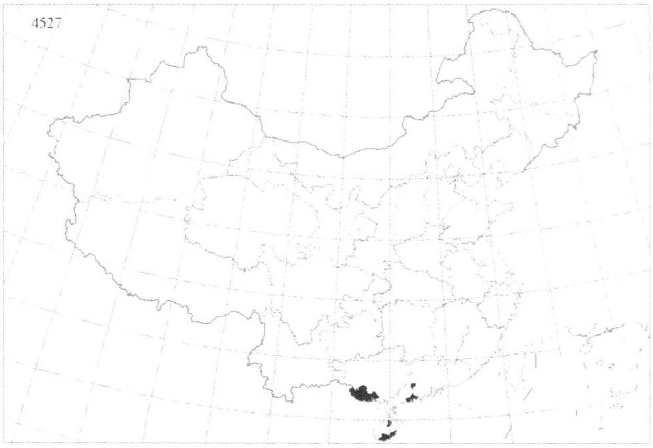

Fabaceae 豆科 *Spatholobus* 密花豆属
Spatholobus suberectus Dunn 密花豆
Vines

19.0 (8.6~22.7); 11.3 (1.3~15.1); 24.9 (15.0~29.0); 19.0 (8.7~22.8); 168.3 (52.6~213.1); 0.1 (0.0~8.8); 960 (574~1221) ❙ 1429 (1028~2435); 717 (425~1428); 94 (43~235) ❙ 49.8 (5.8~100.8); 953 (574~1221); 332.8 (160.7~659.0)

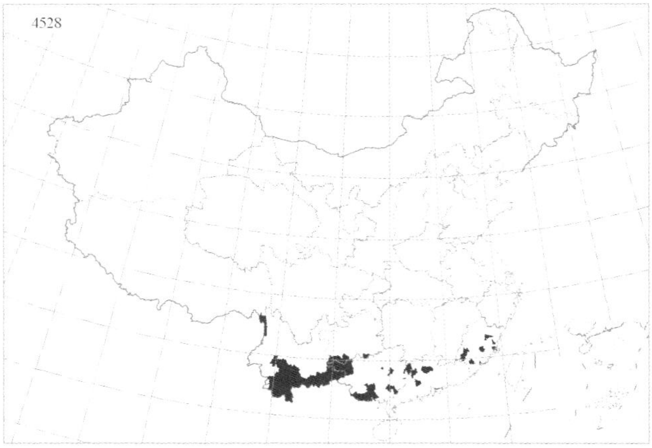

Fabaceae 豆科 *Spatholobus* 密花豆属
Spatholobus uniauritus C. F. Wei 单耳密花豆
Vines

20.5; 14.9; 23.7; 20.5; 185.9; 0.0; 992 ❙ 1464; 655; 66 ❙ 47.6; 959; 570.1

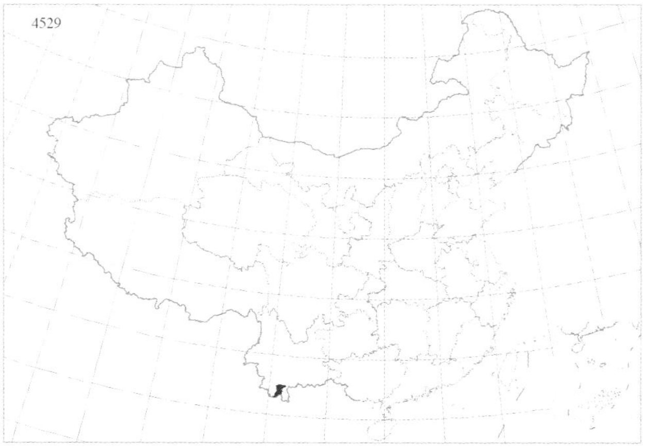

Fabaceae 豆科 *Spatholobus* 密花豆属
Spatholobus varians Dunn 云南密花豆
Vines

19.4 (18.2~20.5); 13.9 (12.4~15.1); 22.8 (21.6~23.7); 19.5 (18.2~20.5); 173.9 (158.6~185.9); 0.0 (0.0~0.0); 932 (861~992) ❙ 1513 (1464~1607); 721 (655~785); 63 (52~77) ❙ 62.8 (47.6~74.2); 911 (839~986); 522.1 (427.9~659.0)

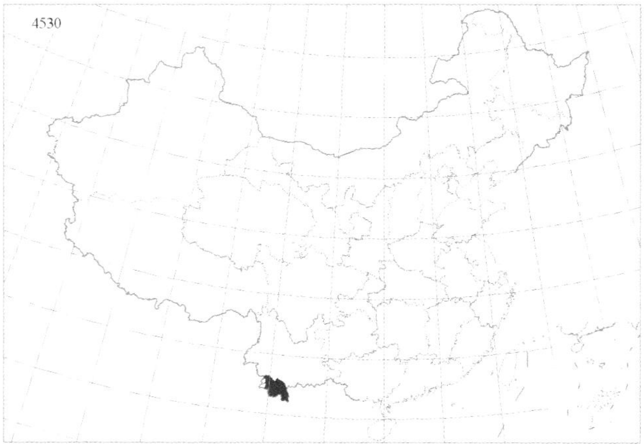

Fabaceae 豆科 *Sphaerophysa* 苦马豆属
Sphaerophysa salsula (Pall.) DC. 苦马豆
Herbs or subshrubs, deciduous

6.6 (−4.8~15.5); −10.3 (−20.9~0.5); 21.0 (7.4~29.5); 9.2 (1.9~15.5); 72.9 (4.3~136.0); 53.4 (9.3~121.6); 629 (329~932) ❙ 298 (21~795); 163 (15~453); 12 (0~51) ❙ −49.7 (−97.3~54.6); 291 (21~702); 106.9 (1.6~364.1)

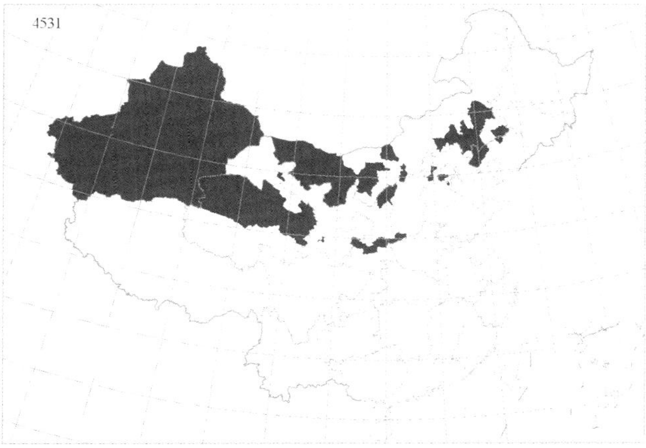

Fabaceae 豆科 *Tadehagi* 葫芦茶属
Tadehagi pseudotriquetrum (DC.) Yang et P. H. Huang 蔓茎葫芦茶
Subshrubs

18.6 (16.4~19.7); 9.4 (5.4~11.5); 26.8 (26.3~27.5); 18.6 (16.5~19.7); 163.4 (137.7~176.5); 0.0 (0.0~0.0); 952 (864~1001) ❙ 1579 (1425~1716); 565 (517~642); 180 (151~231) ❙ 65.9 (54.8~76.3); 952 (864~1001); 254.4 (214.2~290.3)

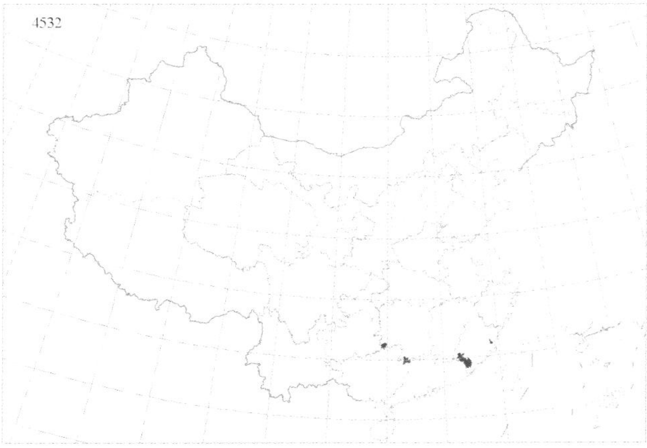

Fabaceae 豆科 *Tadehagi* 葫芦茶属
Tadehagi triquetrum (L.) H. Ohashi 葫芦茶
Shrubs or subshrubs

19.8 (14.3~24.8); 11.5 (3.8~20.1); 26.5 (19.1~29.8); 19.8 (14.3~24.8); 178.1 (112.6~237.6); 0.0 (0.0~1.5); 1028 (727~1385) ❙ 1552 (868~3495); 683 (312~1428); 127 (41~897) ❙ 52.8 (−34.3~224.4); 1017 (727~1360); 281.7 (95.3~659.0)

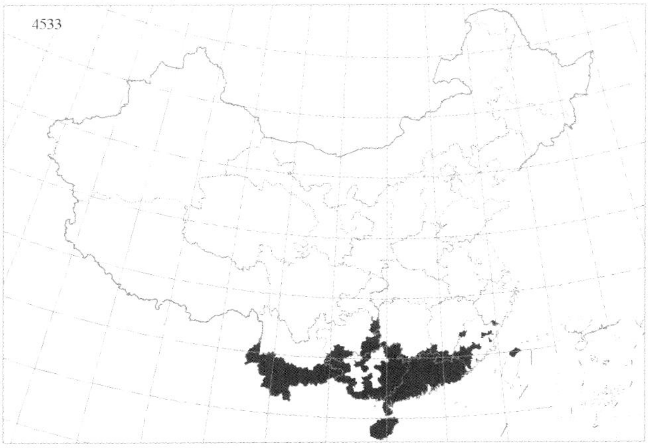

Fabaceae 豆科 *Trifidacanthus* 三叉刺属
Trifidacanthus unifoliolatus Merr. 三叉刺
Small shrubs

24.0 (23.0~24.8); 18.3 (17.7~20.1); 28.1 (26.7~28.7); 24.1 (23.1~24.8); 228.7 (217.0~237.6); 0.0 (0.0~0.0); 1314 (1217~1385) ❙ 1375 (868~1813); 496 (312~712); 68 (41~113) ❙ 5.0 (−34.3~41.8); 1147 (868~1329); 403.4 (248.2~572.9)

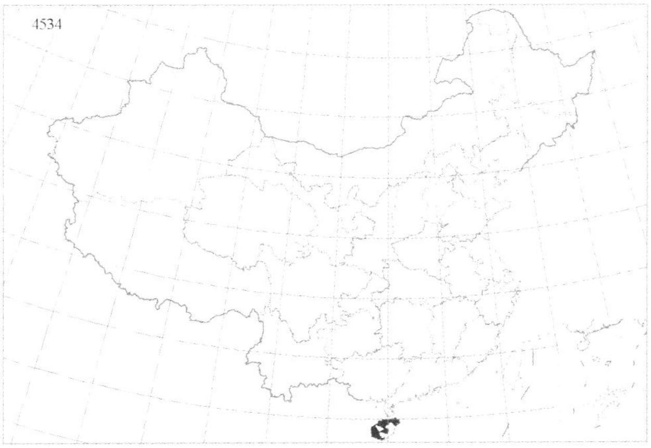

Fabaceae 豆科 *Ulex* 荆豆属
Ulex europaeus L. 荆豆
Shrubs

10.6; −0.2; 21.1; 10.6; 79.4; 12.1; 643 ❙ 1246; 525; 46 ❙ 93.8; 643; 284.3

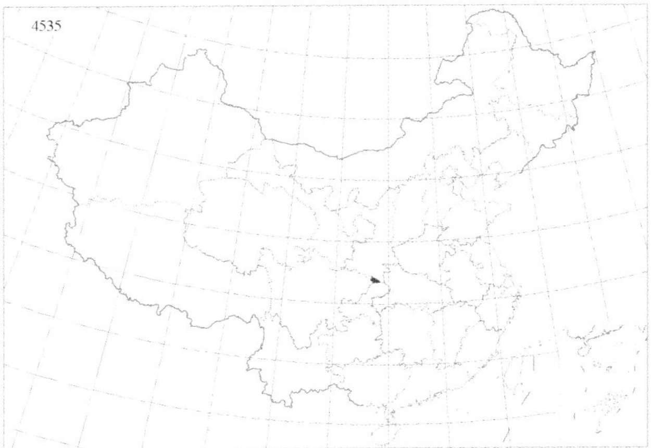

Fabaceae 豆科 *Uraria* 狸尾豆属
Uraria cochinchinensis Schindler 短序算珠豆
Subshrubs

19.8; 12.4; 25.1; 19.8; 177.6; 0.0; 990 ❙ 1580; 824; 82 ❙ 59.2; 990; 418.5

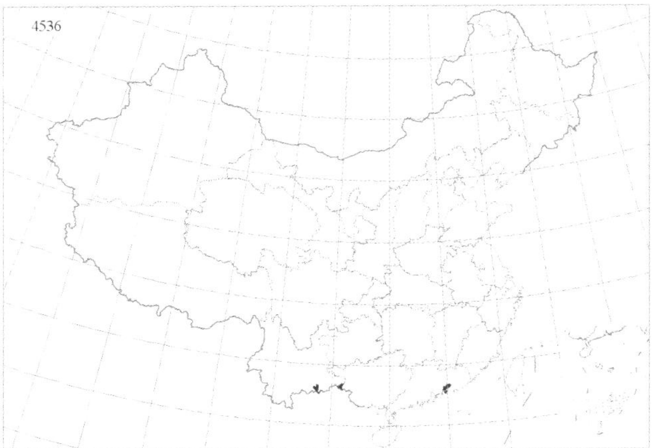

Fabaceae 豆科 *Uraria* 狸尾豆属
Uraria cordifolia Wallich 算珠豆
Shrubs

19.1 (16.8~21.3); 11.6 (7.8~15.1); 24.4 (21.6~27.9); 19.1 (16.8~21.3); 169.3 (141.5~196.0); 0.0 (0.0~0.0); 942 (836~1123) ❙ 1368 (1143~1751); 723 (617~996); 60 (43~77) ❙ 46.1 (5.8~88.8); 934 (836~1121); 374.5 (189.5~659.0)

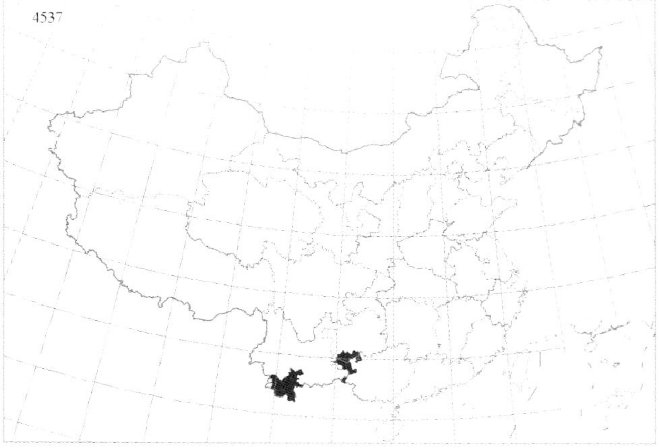

Fabaceae 豆科 *Uraria* 狸尾豆属
Uraria crinita (L.) Desv. 猫尾草
Subshrubs, deciduous

20.1 (14.3~24.8); 11.8 (3.7~20.1); 27.0 (20.1~29.0); 20.2 (14.3~24.8); 182.2 (113.2~237.6); 0.0 (0.0~1.4); 1048 (761~1385) ❙ 1659 (868~2993); 729 (312~1847); 141 (41~547) ❙ 61.5 (-34.3~246.5); 1031 (761~1329); 279.0 (119.4~659.0)

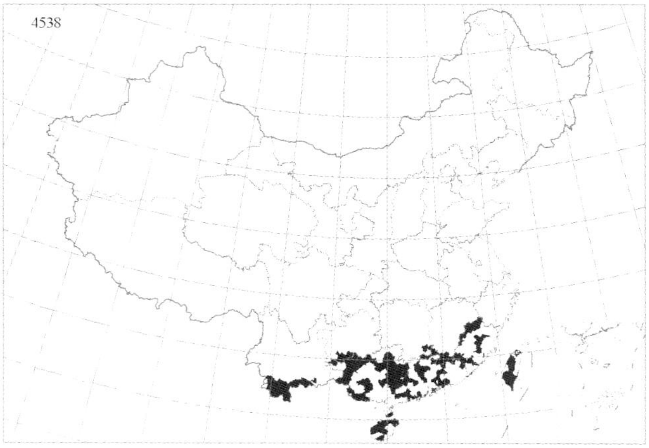

Fabaceae 豆科 *Uraria* 狸尾豆属
Uraria lacei Craib 滇南狸尾豆
Shrubs, deciduous

15.9 (5.4~21.3); 8.3 (-2.1~15.1); 21.9 (12.3~28.3); 16.0 (5.8~21.3); 133.1 (31.5~196.0); 1.5 (0.0~25.6); 819 (481~1123) ❙ 1174 (741~1777); 611 (426~911); 65 (15~187) ❙ 43.8 (5.8~85.5); 801 (481~1121); 320.2 (189.5~659.0)

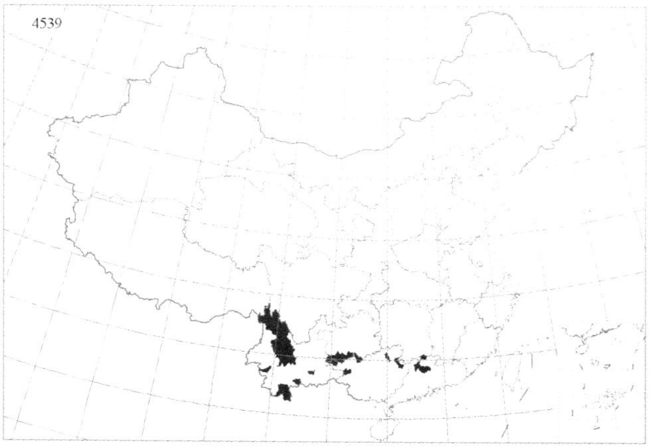

Fabaceae 豆科 *Uraria* 狸尾豆属
Uraria lagopodioides (L.) Desv. ex DC. 狸尾豆
Herbs or subshrubs, deciduous

20.2 (16.2~24.1); 14.5 (11.0~18.9); 25.2 (20.1~28.7); 20.4 (16.2~24.1); 184.4 (134.2~229.7); 0.0 (0.0~0.0); 1029 (773~1322) ❙ 2381 (1498~3495); 1100 (750~1847); 248 (47~897) ❙ 142.2 (25.2~246.5); 994 (773~1216); 243.0 (119.4~363.7)

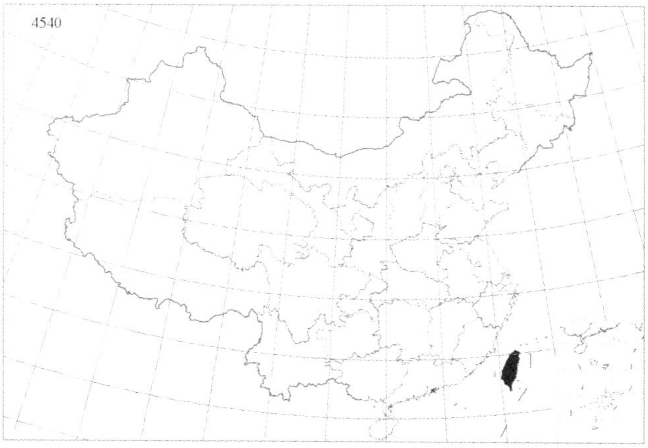

Fabaceae 豆科 *Uraria* 狸尾豆属
Uraria neglecta Prain 长苞狸尾豆
Herbs or subshrubs, deciduous

18.0 (14.6~20.1); 8.1 (3.8~12.7); 26.8 (22.5~28.5); 18.0 (14.7~20.1); 156.3 (117.7~181.6); 0.1 (0.0~1.5); 932 (779~1052) ❙ 1667 (1459~1959); 611 (481~911); 188 (63~250) ❙ 79.7 (38.7~121.8); 932 (779~1052); 264.9 (197.2~551.3)

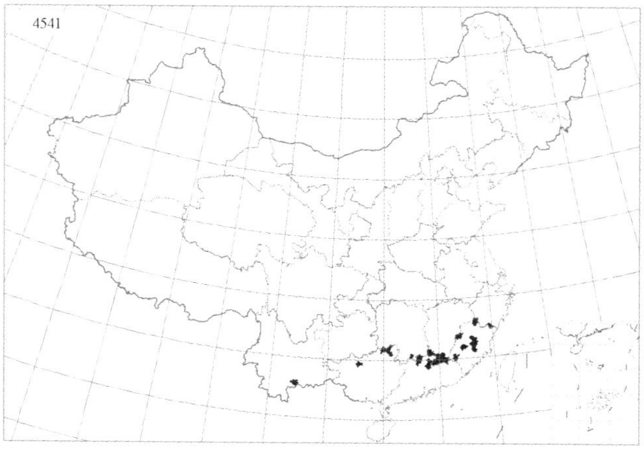

Fabaceae 豆科 *Uraria* 狸尾豆属
Uraria picta (Jacq.) Desv. ex DC. 美花狸尾豆
Shrubs or subshrubs, deciduous

17.2 (10.4~24.7); 10.0 (3.4~19.8); 22.7 (16.3~28.9); 17.3 (10.5~24.7); 147.5 (68.0~236.5); 0.1 (0.0~2.6); 875 (623~1376) ❙ 1146 (797~2759); 640 (411~1594); 51 (15~173) ❙ 31.2 (-14.2~202.5); 828 (615~1212); 287.5 (154.2~494.7)

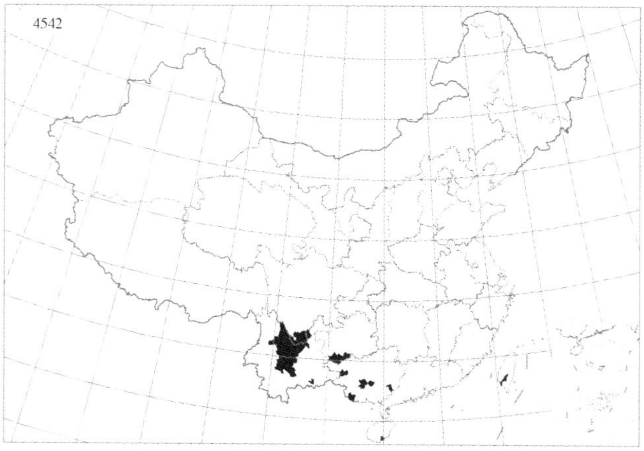

Fabaceae 豆科 *Uraria* 狸尾豆属

Uraria rufescens (DC.) Schindl. 钩柄狸尾豆

Subshrubs, deciduous

21.1 (5.3~24.8); 14.0 (-2.4~20.1); 26.6 (12.2~28.7); 21.2 (5.7~24.8); 195.4 (30.5~237.6); 1.5 (0.0~26.2); 1138 (479~1385) ❙ 1410 (868~1991); 588 (312~902); 88 (31~198) ❙ 29.8 (-34.3~100.8); 1054 (479~1329); 349.9 (234.2~572.9)

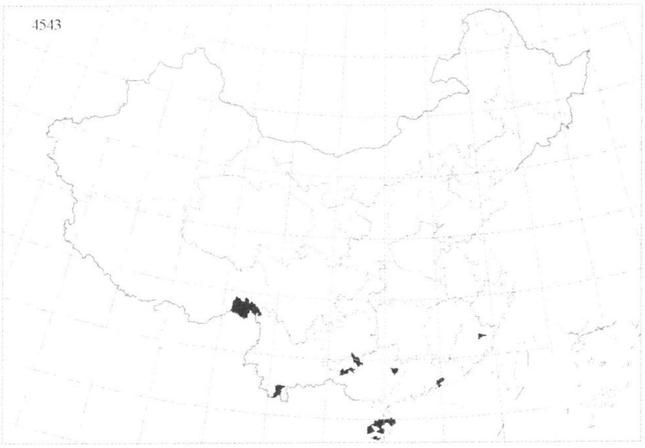

Fabaceae 豆科 *Uraria* 狸尾豆属

Uraria sinensis (Hemsley) Franchet 中华狸尾豆

Subshrubs

14.7 (5.3~20.1); 6.1 (-2.4~13.8); 22.1 (12.2~28.9); 14.8 (5.7~20.1); 118.9 (30.5~181.6); 1.7 (0.0~26.2); 775 (479~1052) ❙ 1057 (741~1547); 567 (408~911); 44 (12~111) ❙ 38.0 (-14.2~100.7); 761 (479~1052); 247.5 (108.2~551.3)

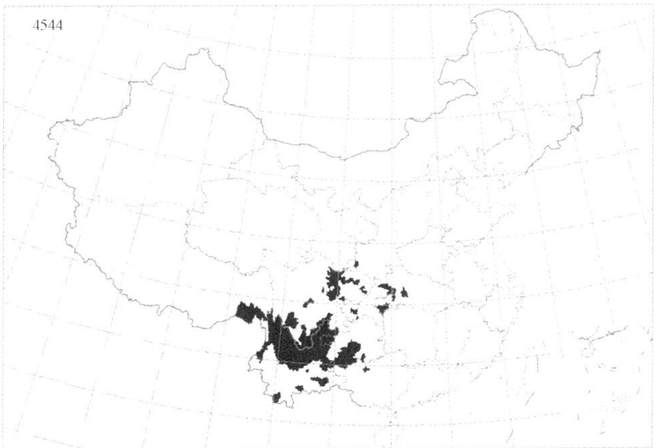

Fabaceae 豆科 *Wisteria* 紫藤属

Wisteria brevidentata Rehder 短梗紫藤

Vines, deciduous

16.5 (15.9~17.2); 4.1 (3.6~5.0); 28.4 (27.5~29.3); 16.5 (16.0~17.2); 139.1 (132.7~146.8); 0.9 (0.0~1.4); 904 (868~941) ❙ 1471 (1211~1590); 536 (474~615); 175 (134~199) ❙ 63.0 (28.8~80.0); 901 (868~941); 217.1 (198.6~243.5)

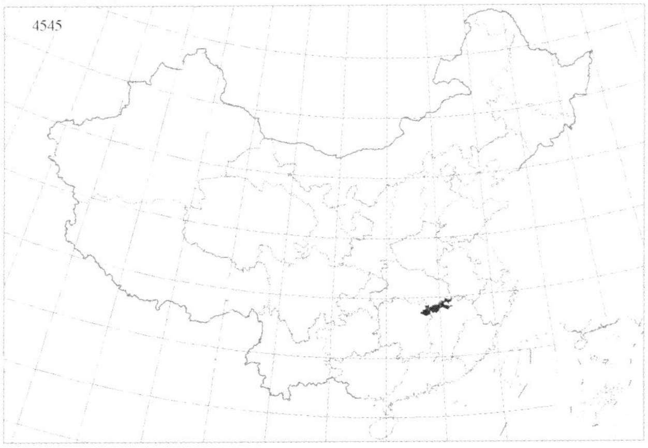

Fabaceae 豆科 *Wisteria* 紫藤属

Wisteria sinensis (Sims) Sweet 紫藤

Vines, deciduous

15.5 (7.4~24.2); 3.3 (-6.2~17.6); 27.0 (18.7~30.1); 15.6 (8.2~24.2); 131.1 (58.8~230.4); 4.8 (0.0~32.8); 862 (561~1331) ❙ 1197 (479~1950); 478 (245~821); 119 (12~259) ❙ 38.1 (-45.2~147.0); 822 (479~1331); 202.7 (108.2~334.4)

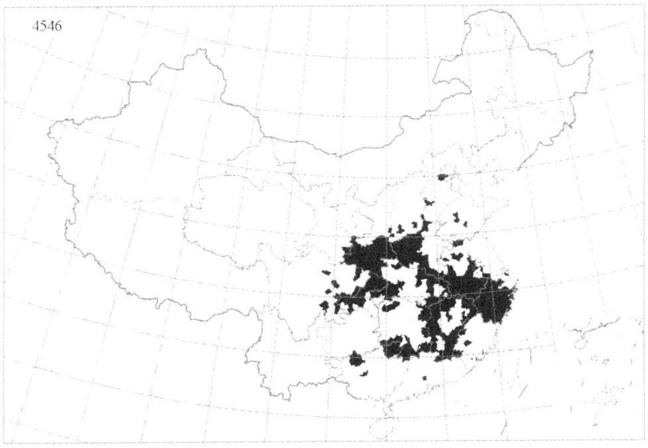

Fabaceae 豆科 *Wisteria* 紫藤属

Wisteria villosa Rehder 藤萝

Vines

13.8; -0.3; 26.7; 14.2; 119.0; 13.5; 822 ❙ 981; 438; 92 ❙ 15.5; 739; 183.5

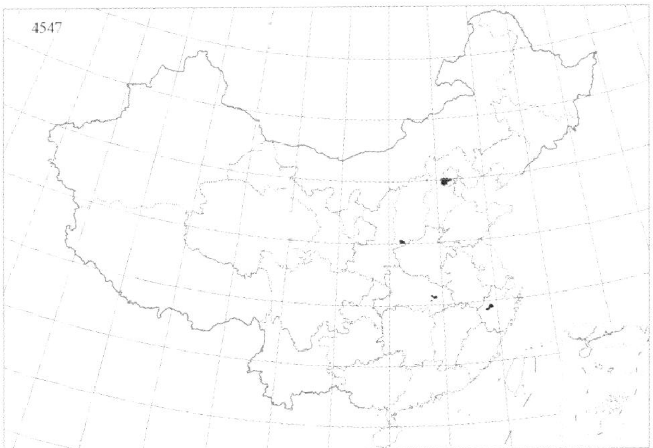

Fabaceae 豆科 *Zenia* 翅荚木属

Zenia insignis Chun 任豆

Trees, deciduous

18.4 (14.3~22.2); 8.7 (3.3~14.2); 26.8 (22.3~29.0); 18.5 (14.3~22.3); 162.0 (114.1~207.5); 0.1 (0.0~2.3); 963 (765~1190) ❙ 1468 (1163~1876); 625 (403~911); 138 (43~211) ❙ 54.0 (5.8~101.8); 959 (765~1190); 249.9 (151.0~551.6)

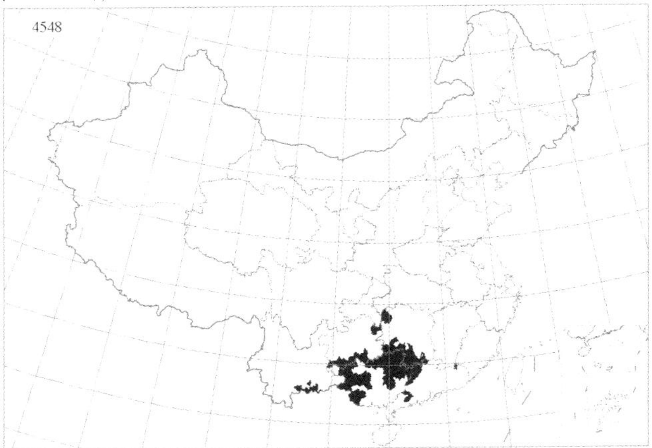

Linaceae 亚麻科 *Reinwardtia* 石海椒属

Reinwardtia indica Dumort. 石海椒

Small shrubs, evergreen

15.4 (10.1~21.0); 7.7 (1.8~14.2); 21.5 (16.6~28.1); 15.4 (10.3~21.0); 125.8 (70.1~191.8); 0.5 (0.0~7.0); 783 (621~1053) ❙ 1042 (797~1522); 577 (411~835); 44 (18~83) ❙ 33.8 (-14.2~80.4); 761 (621~1053); 279.2 (126.1~551.6)

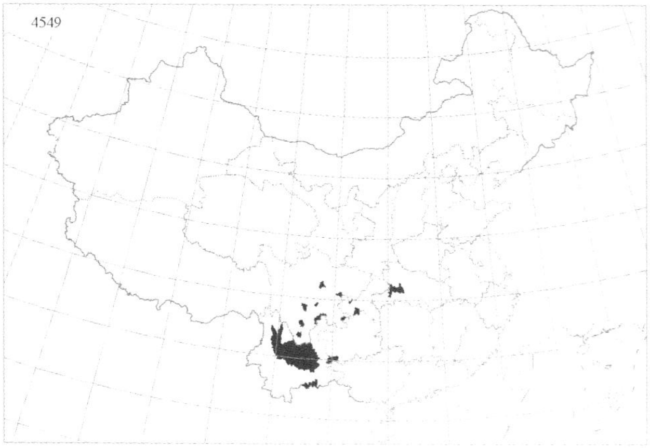

Linaceae 亚麻科 *Tirpitzia* 青篱柴属

Tirpitzia ovoidea Chun et How ex W. L. Sha 米念芭

Shrubs, evergreen

20.8 (18.7~22.3); 11.8 (8.8~13.9); 27.8 (25.2~29.0); 20.9 (18.8~22.3); 190.2 (165.2~208.1); 0.0 (0.0~0.0); 1096 (952~1202) ❙ 1468 (1188~1703); 725 (565~850); 124 (56~194) ❙ 34.4 (5.8~55.9); 1094 (952~1202); 254.0 (160.7~335.1)

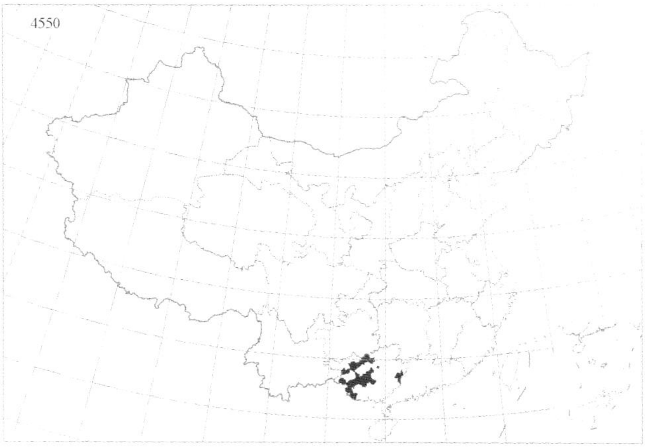

Linaceae 亚麻科 *Tirpitzia* 青篱柴属

Tirpitzia sinensis (Hemsl.) H. Hallier 青篱柴

Shrubs or small trees, evergreen

17.2 (14.2~22.3); 7.6 (3.6~14.2); 25.4 (21.4~29.0); 17.3 (14.2~22.3); 147.5 (112.3~208.1); 0.1 (0.0~1.7); 890 (751~1202) ❙ 1328 (1074~1776); 612 (461~860); 94 (43~195) ❙ 50.0 (5.8~101.8); 889 (751~1202); 243.7 (146.0~551.6)

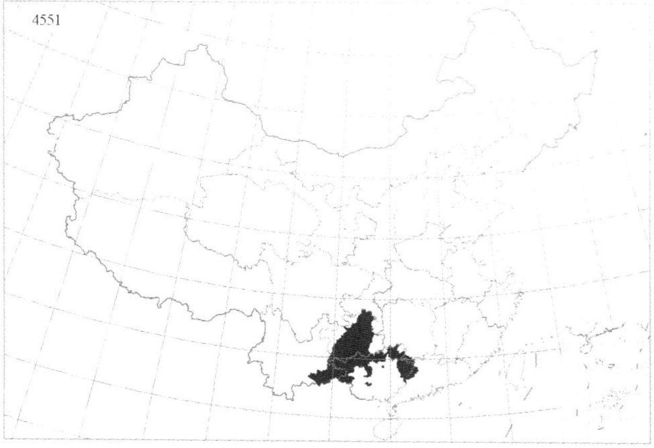

Erythroxylaceae 古柯科 *Erythroxylum* 古柯属

Erythroxylum sinense C. Y. Wu 东方古柯

Shrubs or small trees, evergreen

18.0 (8.6~24.1); 7.9 (1.3~19.1); 27.0 (15.0~29.8); 18.1 (8.7~24.2); 157.1 (52.6~230.2); 0.2 (0.0~8.8); 951 (574~1329) ❙ 1541 (1041~2435); 597 (392~1428); 164 (43~254) ❙ 63.9 (-1.4~147.0); 942 (574~1329); 241.0 (134.6~620.0)

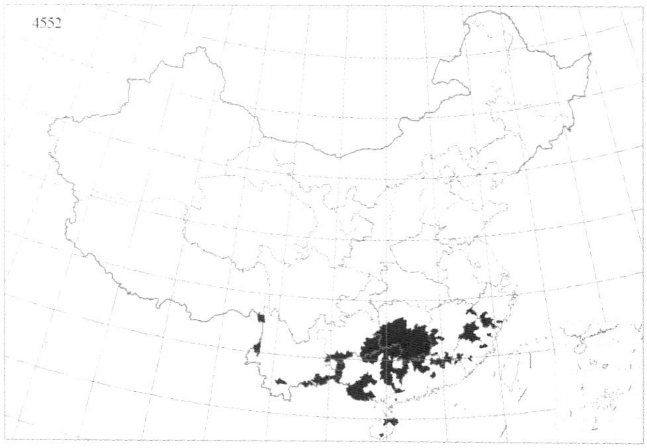

Erythroxylaceae 古柯科 *Ixonanthes* 黏木属

Ixonanthes reticulata Jack 黏木

Trees, evergreen

20.5 (14.1~24.8); 12.3 (3.5~20.1); 27.1 (21.4~29.2); 20.5 (14.2~24.8); 186.6 (111.5~237.6); 0.0 (0.0~1.7); 1073 (762~1385) ❙ 1591 (1074~2435); 719 (444~1428); 121 (41~231) ❙ 49.9 (-6.9~101.8); 1064 (762~1273); 267.2 (95.3~659.0)

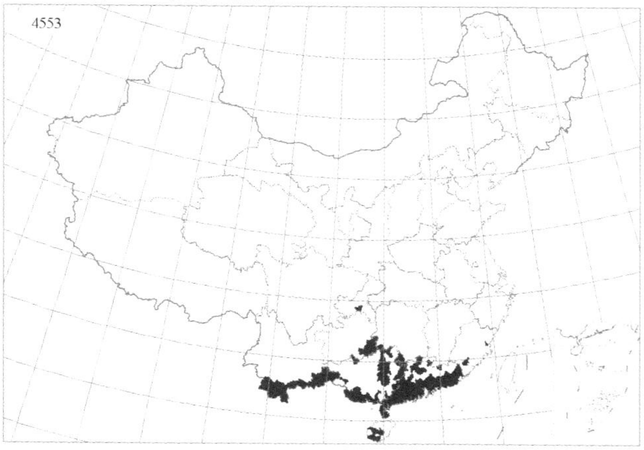

Zygophyllaceae 蒺藜科 *Tetraena* 四合木属

Tetraena mongolica Maxim. 四合木

Shrubs

7.9 (7.1~8.8); -9.2 (-10.4~-7.6); 23.0 (22.4~24.0); 9.8 (9.3~10.3); 80.3 (75.2~84.8); 44.8 (37.3~49.8); 646 (622~667) ❙ 198 (128~282); 119 (81~166); 4 (2~6) ❙ -69.3 (-80.8~-55.2); 198 (128~282); 60.0 (13.6~116.9)

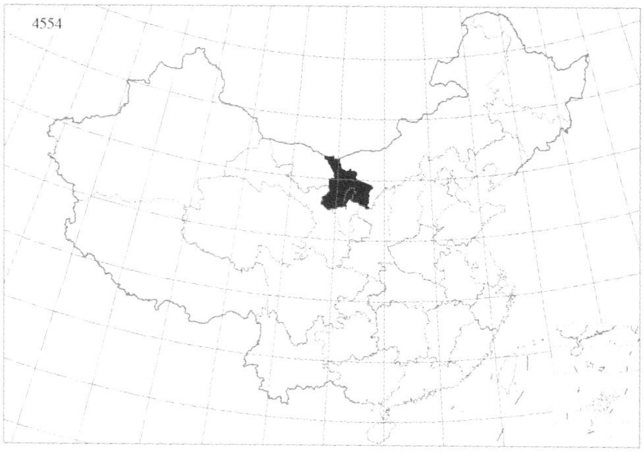

Zygophyllaceae 蒺藜科 *Zygophyllum* 霸王属
Zygophyllum kaschgaricum Boriss 喀什霸王
Shrubs

8.3 (-4.3~12.5); -9.0 (-20.1~-5.1); 22.2 (9.6~26.5); 10.4 (2.5~13.2); 85.5 (9.7~114.0); 44.9 (23.1~121.1); 682 (329~817) ❙ 75 (41~215); 31 (17~80); 9 (2~26) ❙ -86.8 (-95.0~-51.1); 75 (41~215); 36.5 (2.2~120.2)

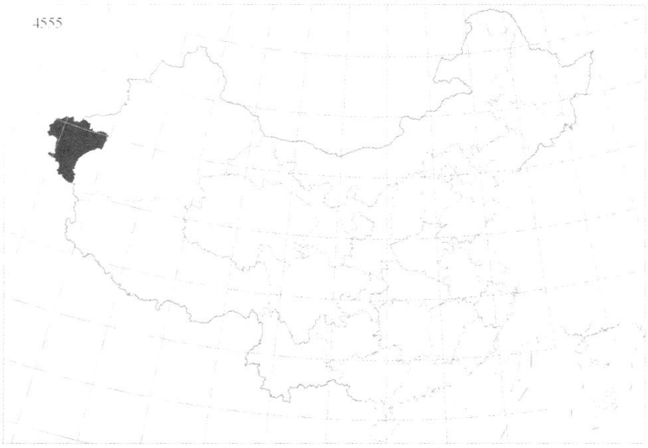

Zygophyllaceae 蒺藜科 *Zygophyllum* 霸王属
Zygophyllum xanthoxylon (Bunge) Maxim. 霸王
Shrubs

6.1 (-4.3~12.5); -11.5 (-20.9~-5.1); 20.8 (9.6~27.5); 9.0 (2.5~13.2); 71.3 (9.7~114.1); 57.0 (23.1~121.1); 621 (329~822) ❙ 155 (21~550); 84 (15~307); 8 (0~45) ❙ -71.5 (-97.3~35.3); 153 (21~463); 56.1 (1.6~297.3)

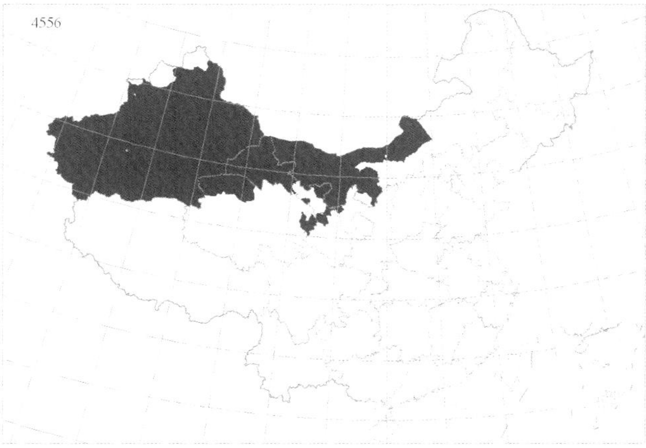

Rutaceae 芸香科 *Acronychia* 山油柑属
Acronychia pedunculata (L.) Miq. 山油柑
Shrubs or small trees, evergreen

21.0 (14.4~24.8); 14.0 (7.5~20.1); 26.2 (19.1~29.1); 21.0 (14.4~24.8); 192.5 (113.1~237.6); 0.0 (0.0~0.0); 1090 (727~1385) ❙ 1696 (868~3495); 807 (312~1847); 117 (41~897) ❙ 60.1 (-34.3~246.5); 1063 (727~1360); 318.3 (95.3~659.0)

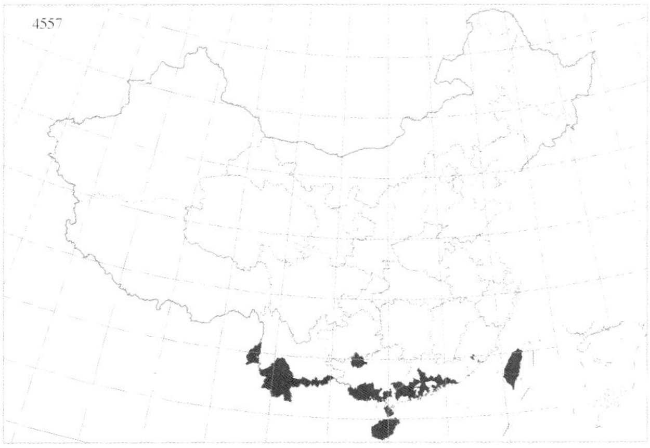

Rutaceae 芸香科 *Aegle* 木橘属
Aegle marmelos (L.) Corrêa 木橘
Trees, deciduous

18.9; 12.9; 22.5; 18.9; 166.9; 0.0; 911 ❙ 1531; 775; 63 ❙ 69.9; 897; 593.2

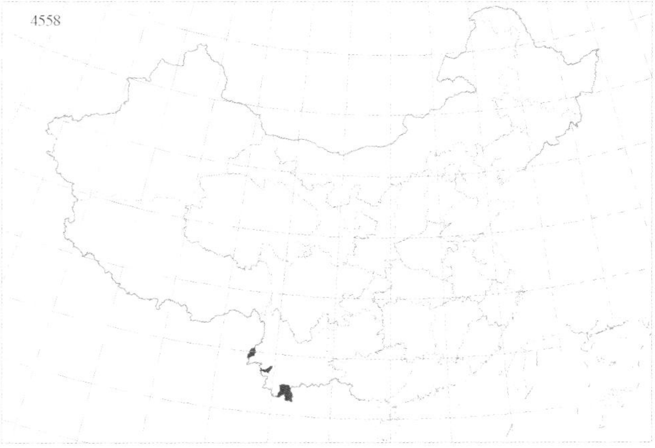

Rutaceae 芸香科 *Atalantia* 酒饼簕属
Atalantia acuminata C. C. Huang 尖叶酒饼簕
Trees, evergreen

19.4 (12.8~21.9); 10.7 (2.7~13.2); 26.1 (21.9~28.6); 19.4 (12.8~22.0); 173.6 (97.9~203.8); 0.2 (0.0~3.7); 996 (712~1176) ❙ 1302 (968~1518); 705 (543~848); 69 (26~138) ❙ 31.4 (5.8~55.9); 996 (712~1176); 303.5 (144.6~411.1)

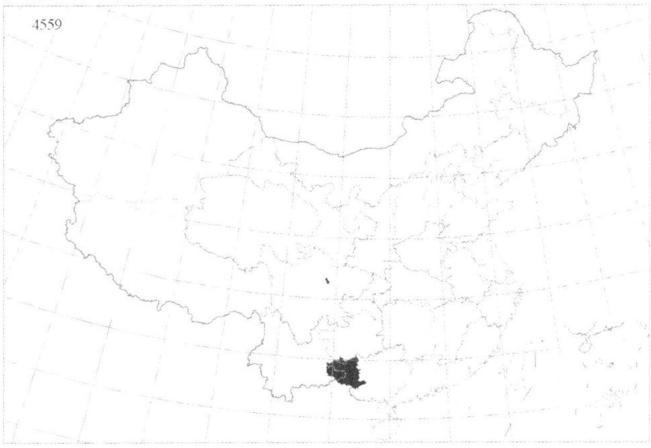

Rutaceae 芸香科 *Atalantia* 酒饼簕属
Atalantia buxifolia (Poir.) Oliv. 酒饼簕
Shrubs, evergreen

22.0 (14.7~24.8); 14.1 (4.4~20.1); 28.2 (24.4~29.2); 22.0 (14.8~24.8); 204.1 (117.8~237.6); 0.0 (0.0~0.6); 1170 (779~1385) ❙ 1644 (929~2435); 753 (392~1428); 119 (25~188) ❙ 41.2 (-11.9~99.5); 1156 (779~1360); 240.9 (95.3~620.0)

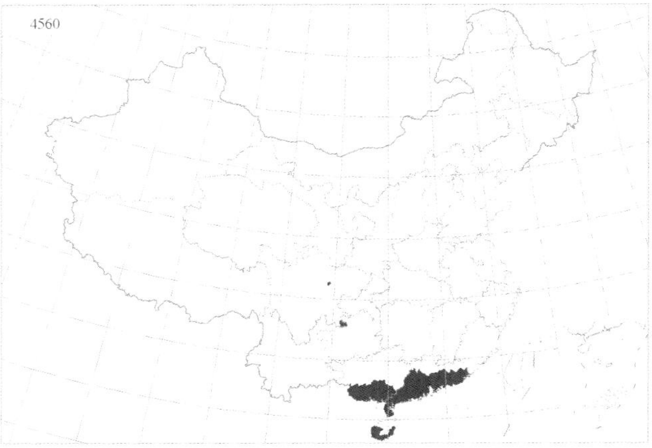

Rutaceae 芸香科 *Atalantia* 酒饼簕属
Atalantia dasycarpa C. C. Huang 厚皮酒饼簕
Trees, evergreen

20.4 (17.2~22.7)；12.4 (9.5~15.0)；26.3 (22.4~28.6)；20.4 (17.3~22.8)；185.1 (147.0~213.1)；0.0 (0.0~0.0)；1050 (833~1221) ❙ 1520 (1188~2435)；836 (634~1428)；84 (48~130) ❙ 45.7 (5.8~99.5)；1050 (833~1221)；335.2 (256.6~551.6)

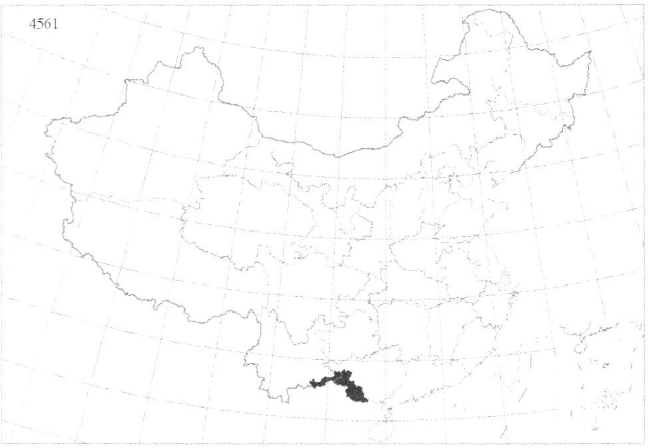

Rutaceae 芸香科 *Atalantia* 酒饼簕属
Atalantia fongkaica C. C. Huang 封开酒饼簕
Shrubs

21.2；12.2；28.5；21.3；195.2；0.0；1126 ❙ 1462；599；142 ❙ 29.9；1126；226.9

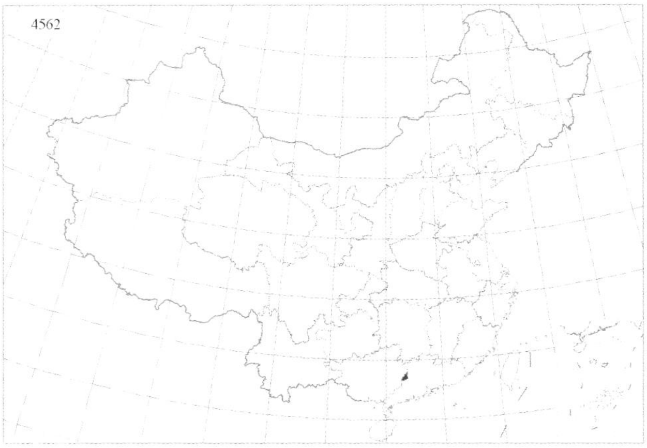

Rutaceae 芸香科 *Atalantia* 酒饼簕属
Atalantia guillauminii Swingle 大果酒饼簕
Trees, evergreen

20.0；13.0；25.1；20.0；180.1；0.0；1002 ❙ 1525；797；82 ❙ 52.8；1002；426.7

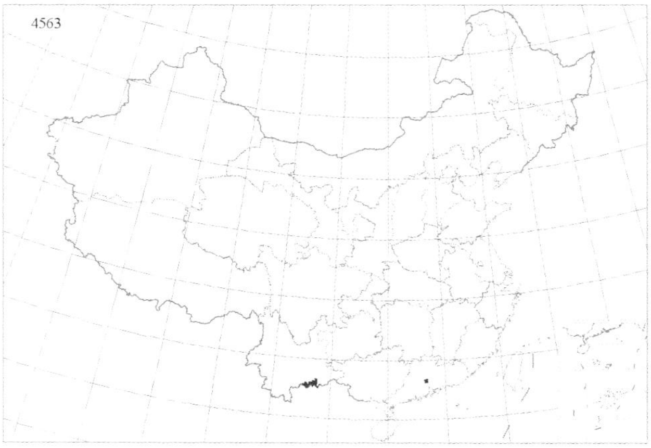

Rutaceae 芸香科 *Atalantia* 酒饼簕属
Atalantia henryi (Swingle) C. C. Huang 薄皮酒饼簕
Trees, evergreen

19.2 (17.2~22.2)；12.6 (9.5~15.1)；23.7 (21.6~28.3)；19.3 (17.3~22.3)；171.3 (147.0~207.5)；0.0 (0.0~0.0)；941 (833~1190) ❙ 1466 (1254~1751)；776 (655~996)；63 (41~102) ❙ 57.1 (15.4~88.8)；931 (833~1190)；455.7 (313.7~659.0)

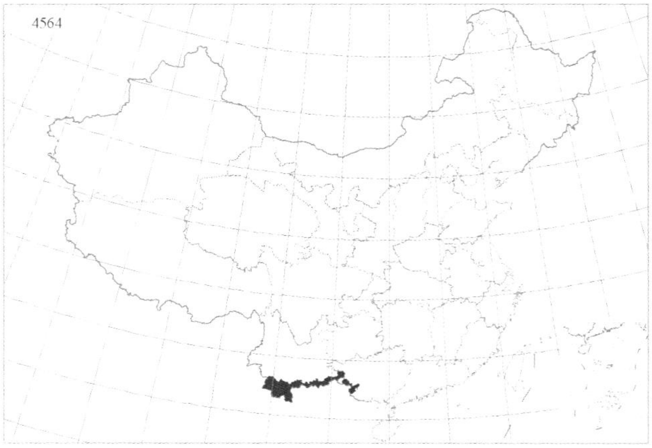

Rutaceae 芸香科 *Atalantia* 酒饼簕属
Atalantia kwangtungensis Merr. 广东酒饼簕
Shrubs, evergreen

23.3 (21.3~24.8)；16.7 (13.2~20.1)；28.1 (25.7~29.2)；23.3 (21.3~24.8)；220.1 (195.9~237.6)；0.0 (0.0~0.0)；1261 (1111~1385) ❙ 1566 (868~2435)；676 (312~1428)；88 (41~130) ❙ 25.2 (-34.3~99.5)；1192 (868~1360)；362.7 (178.7~620.0)

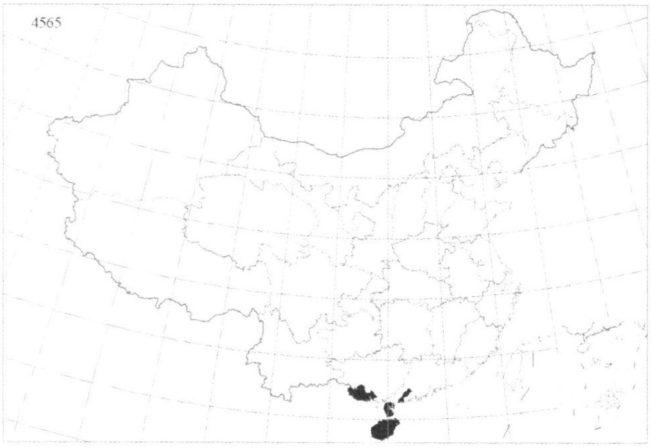

Rutaceae 芸香科 *Citrus* 柑橘属
Citrus cavaleriei H. Lév. ex Cavalerie 宜昌橙
Shrubs or trees, evergreen

15.8 (5.7~21.7)；5.4 (-3.5~12.9)；25.3 (14.2~29.9)；15.8 (6.4~21.8)；131.1 (38.1~201.3)；1.2 (0.0~28.1)；839 (504~1162) ❙ 1244 (711~1758)；543 (357~816)；86 (15~212) ❙ 49.6 (8.0~106.2)；837 (504~1162)；202.4 (108.2~432.9)

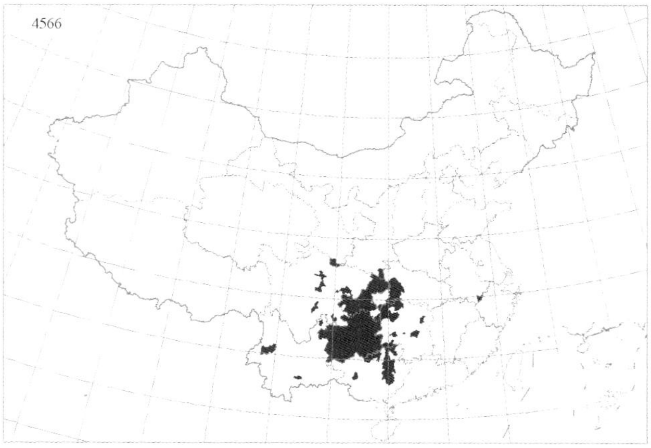

Rutaceae 芸香科 *Citrus* 柑橘属
Citrus hystrix DC. 箭叶橙
Trees, evergreen

19.0 (13.5~24.8); 11.6 (2.9~20.1); 24.8 (22.3~28.0); 19.1 (13.5~24.8); 169.4 (105.3~237.6); 0.2 (0.0~3.0); 967 (744~1385) | 1337 (806~1751); 658 (444~996); 59 (16~142) | 40.0 (-2.2~88.8); 934 (744~1137); 401.4 (127.1~659.0)

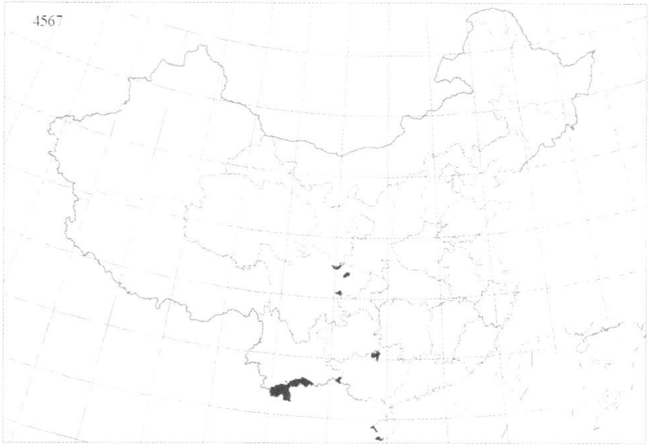

Rutaceae 芸香科 *Citrus* 柑橘属
Citrus japonica Thunberg 金柑
Trees, evergreen

19.6 (15.0~24.8); 10.8 (1.8~20.1); 27.4 (20.1~29.8); 19.7 (15.1~24.8); 176.4 (120.8~237.6); 0.1 (0.0~5.0); 1032 (773~1385) | 1667 (868~3495); 663 (312~1847); 168 (27~897) | 65.2 (-34.3~246.5); 1017 (773~1331); 243.0 (95.3~620.0)

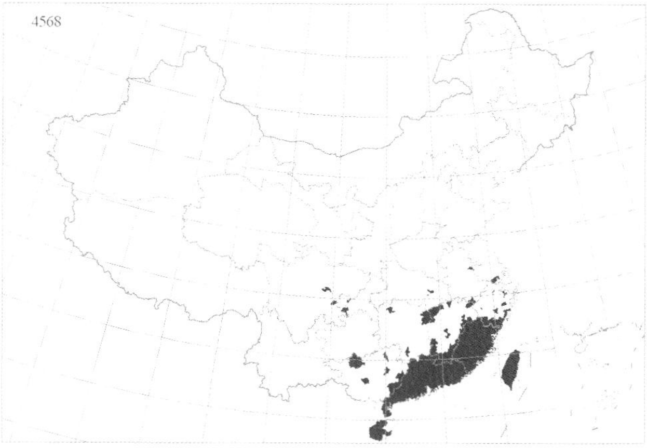

Rutaceae 芸香科 *Citrus* 柑橘属
Citrus maxima (Burm.) Merr. 柚
Trees, evergreen

17.5 (8.6~24.8); 8.3 (-1.7~20.1); 25.5 (15.0~29.8); 17.5 (8.7~24.8); 150.7 (52.6~237.6); 0.6 (0.0~18.0); 914 (574~1385) | 1395 (862~3495); 615 (299~1847); 99 (19~897) | 54.2 (-34.3~246.5); 900 (574~1360); 253.5 (108.2~659.0)

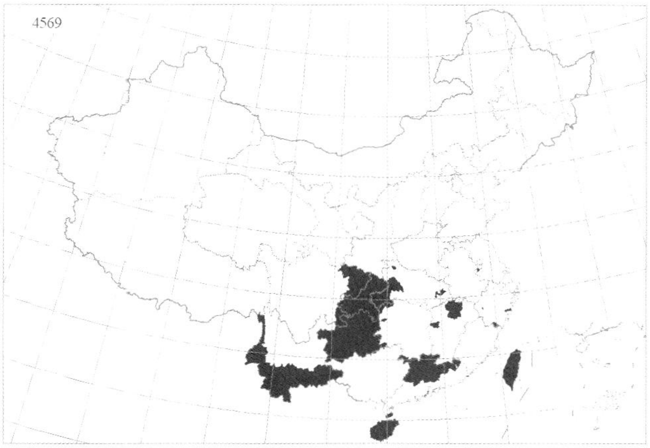

Rutaceae 芸香科 *Citrus* 柑橘属
Citrus medica L. 香橼
Shrubs or trees, evergreen

17.4 (6.6~24.8); 9.4 (-1.3~20.1); 24.0 (13.3~29.0); 17.5 (6.8~24.8); 149.8 (39.6~237.6); 0.4 (0.0~19.7); 895 (514~1385) | 1318 (797~3495); 636 (312~1847); 79 (12~897) | 48.6 (-34.3~246.5); 875 (514~1360); 286.8 (108.2~659.0)

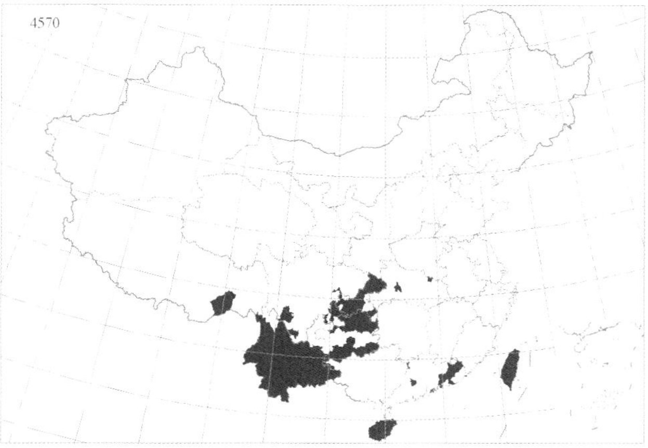

Rutaceae 芸香科 *Citrus* 柑橘属
Citrus reticulata Blanco 柑橘
Trees, evergreen

18.3; 12.9; 22.7; 18.4; 160.4; 0.0; 893 | 2403; 1163; 199 | 179.2; 882; 284.3

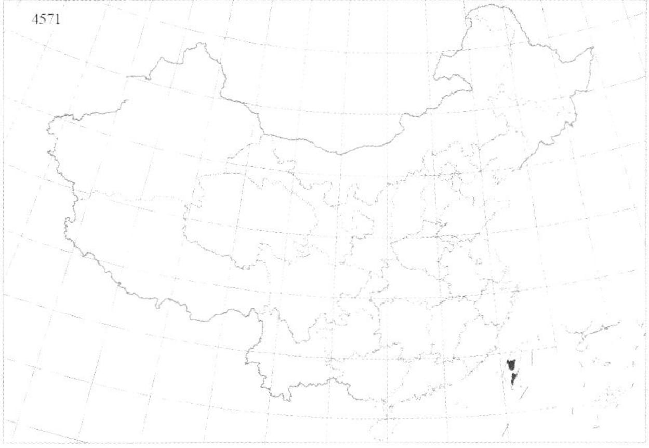

Rutaceae 芸香科 *Citrus* 柑橘属
Citrus trifoliata L. 枳
Trees, deciduous

16.0 (7.4~22.1); 4.2 (-6.2~13.7); 27.0 (18.7~30.1); 16.1 (8.2~22.1); 135.2 (58.8~205.6); 2.7 (0.0~32.8); 875 (561~1182) | 1275 (590~1997); 507 (299~821); 134 (12~259) | 45.1 (-23.1~147.0); 857 (561~1182); 201.4 (110.1~372.3)

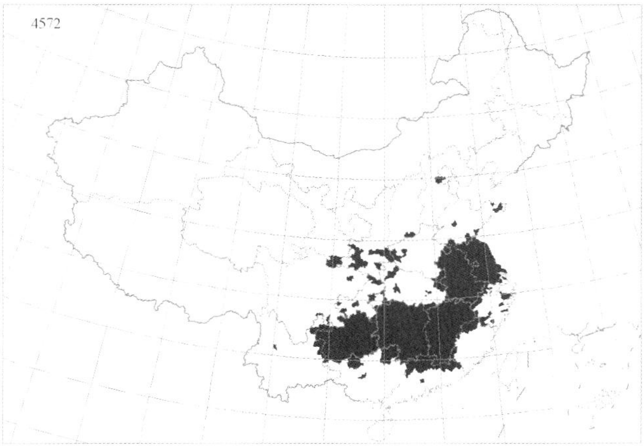

Rutaceae 芸香科 *Citrus* 柑橘属
Citrus × aurantium L. 酸橙
Small trees, evergreen

17.0 (9.0~24.8); 6.6 (-7.2~20.1); 26.5 (19.6~29.7); 17.0 (10.3~24.8); 146.4 (75.8~237.6); 2.5 (0.0~35.8); 916 (634~1385) ❙ 1235 (428~1990); 519 (266~1007); 90 (12~253) ❙ 34.7 (-35.6~145.8); 894 (428~1212); 221.7 (108.2~551.6)

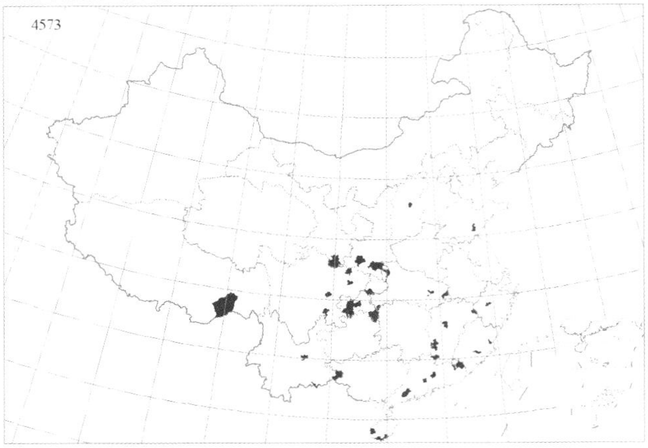

Rutaceae 芸香科 *Citrus* 柑橘属
Citrus × junos Siebold ex Tanaka 香橙
Trees, evergreen

15.6 (9.3~19.1); 4.9 (-1.7~11.7); 25.6 (16.6~29.3); 15.6 (9.5~19.1); 129.7 (69.8~169.2); 2.0 (0.0~18.0); 839 (603~1006) ❙ 1243 (711~1770); 540 (349~816); 95 (15~234) ❙ 49.3 (5.6~112.2); 837 (603~1002); 217.8 (108.2~406.7)

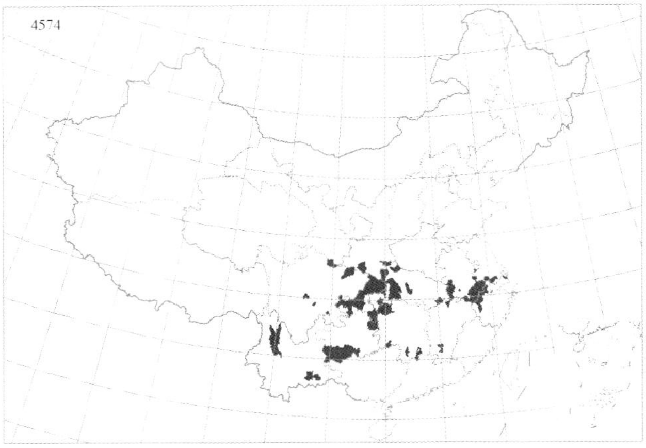

Rutaceae 芸香科 *Citrus* 柑橘属
Citrus × limon (L.) Burm. f. 柠檬
Trees, evergreen

20.2 (14.6~24.7); 11.9 (3.8~19.8); 26.9 (20.1~29.8); 20.3 (14.7~24.7); 183.2 (117.7~236.5); 0.0 (0.0~1.5); 1052 (773~1376) ❙ 1615 (1028~3495); 740 (447~1847); 132 (42~897) ❙ 55.6 (-6.9~246.5); 1045 (773~1360); 263.6 (95.3~659.0)

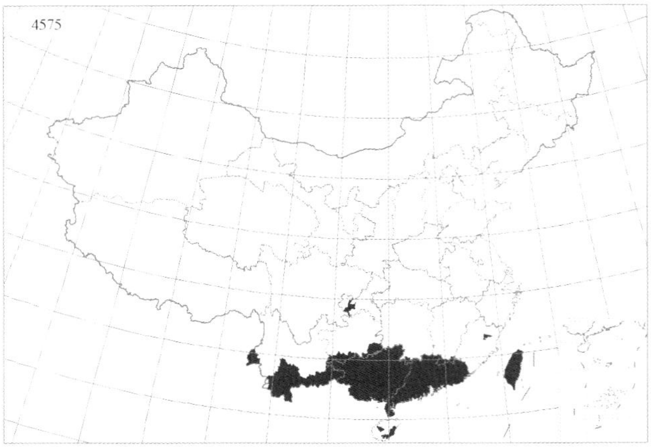

Rutaceae 芸香科 *Citrus* 柑橘属
Citrus × polytrifolia Govaerts 富民枳
Trees, evergreen

14.9; 8.2; 20.1; 15.0; 119.8; 0.0; 748 ❙ 956; 573; 34 ❙ 27.9; 716; 272.5

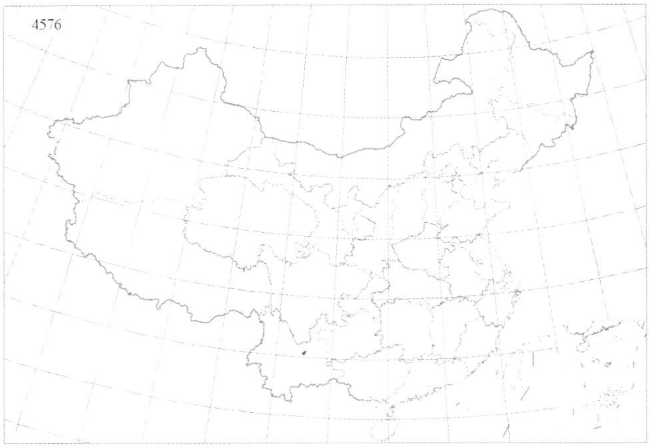

Rutaceae 芸香科 *Clausena* 黄皮属
Clausena anisum-olens (Blanco) Merr. 细叶黄皮
Trees

19.2 (15.5~22.4); 10.8 (4.8~14.5); 26.1 (21.4~28.7); 19.3 (15.5~22.4); 171.3 (126.5~209.2); 0.0 (0.0~0.2); 993 (813~1198) ❙ 1402 (1159~2713); 711 (451~1388); 88 (48~188) ❙ 43.1 (5.8~214.4); 992 (813~1198); 296.9 (146.0~551.6)

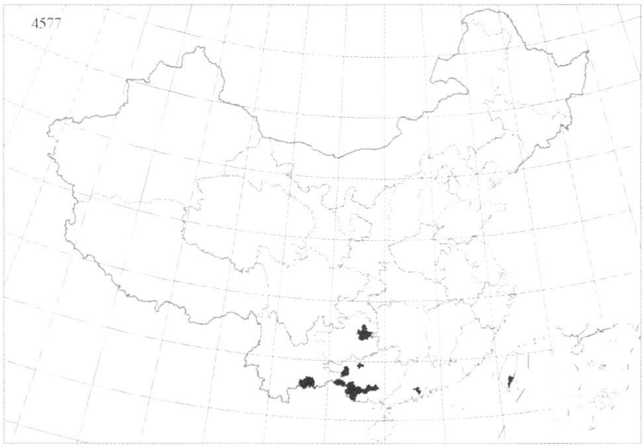

Rutaceae 芸香科 *Clausena* 黄皮属
Clausena dunniana H. Lév. 齿叶黄皮
Trees, deciduous

17.2 (12.0~22.3); 7.8 (0.6~15.1); 25.3 (19.1~29.8); 17.3 (12.1~22.4); 147.6 (92.0~208.2); 0.2 (0.0~9.7); 890 (685~1198) ❙ 1364 (985~2024); 615 (403~996); 99 (41~230) ❙ 53.6 (15.4~112.2); 886 (685~1198); 259.6 (111.7~659.0)

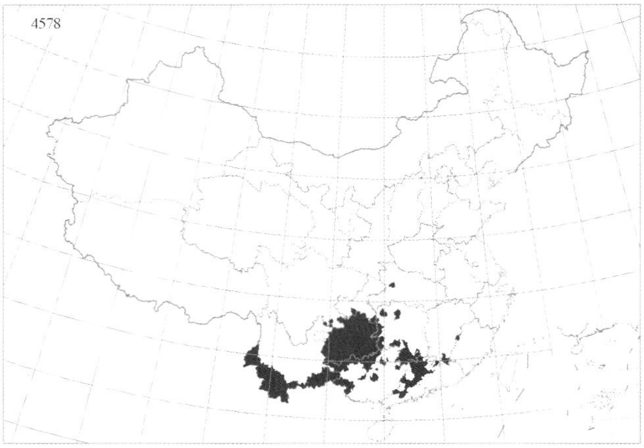

Rutaceae 芸香科 *Clausena* 黄皮属

Clausena dunniana var. robusta C. C. Huang 毛齿叶黄皮

Trees, deciduous

17.1 (15.3~19.1); 6.9 (3.6~10.3); 25.9 (21.4~28.4); 17.1 (15.4~19.1); 145.5 (125.7~169.2); 0.2 (0.0~1.4); 884 (802~969) ▎1365 (1178~1559); 585 (419~720); 115 (51~198) ▎54.8 (27.5~76.9); 882 (802~969); 227.7 (177.4~383.4)

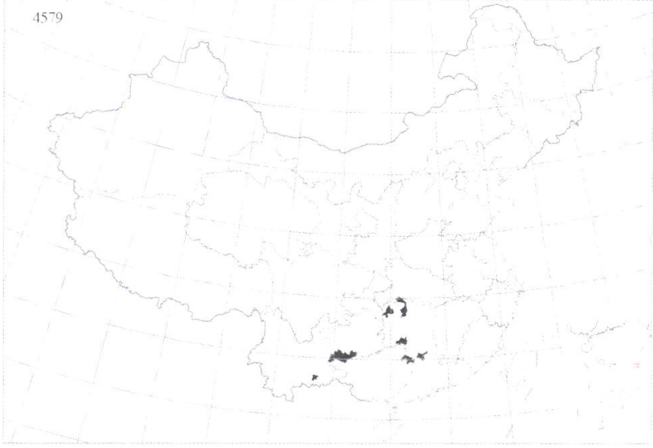

Rutaceae 芸香科 *Clausena* 黄皮属

Clausena emarginata C. C. Huang 小黄皮

Trees

18.3 (14.4~22.2); 11.4 (7.5~15.1); 23.0 (19.1~28.6); 18.4 (14.4~22.3); 160.2 (113.1~207.5); 0.0 (0.0~0.0); 897 (727~1190) ▎1315 (939~1751); 722 (524~996); 58 (37~138) ▎47.6 (5.8~112.2); 884 (727~1190); 400.0 (248.2~659.0)

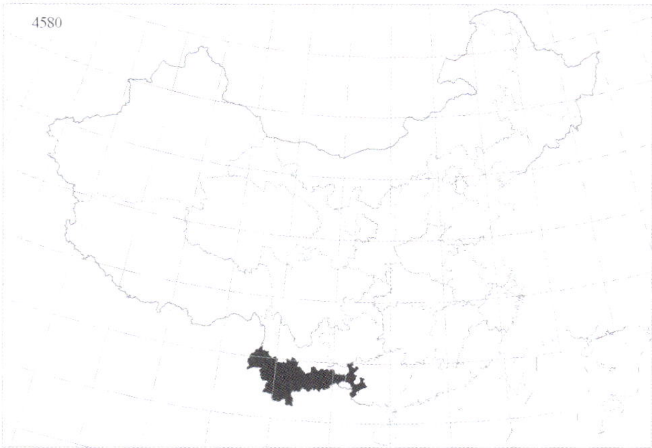

Rutaceae 芸香科 *Clausena* 黄皮属

Clausena excavata N. L. Burman 假黄皮

Shrubs

19.0 (12.8~24.3); 12.1 (3.3~20.1); 24.2 (19.1~29.0); 19.1 (12.9~24.8); 168.7 (97.1~237.6); 0.0 (0.0~2.2); 961 (702~1385) ▎1483 (868~3495); 731 (312~1847); 100 (32~897) ▎56.3 (-34.3~246.5); 935 (702~1299); 333.7 (119.4~659.0)

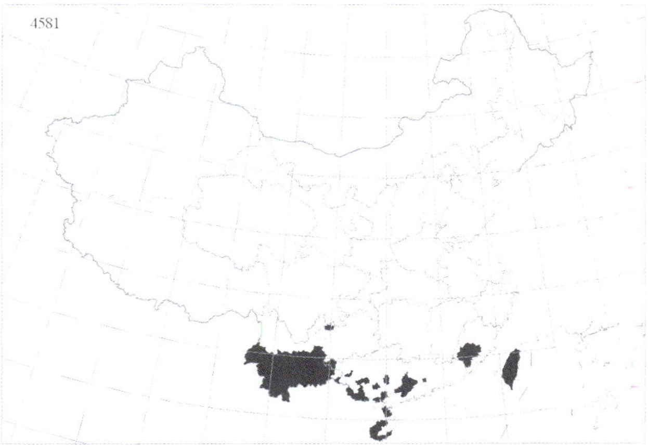

Rutaceae 芸香科 *Clausena* 黄皮属

Clausena hainanensis C. C. Huang et F. W. Xing 海南黄皮

Shrubs or trees

23.6; 17.8; 27.8; 23.6; 223.3; 0.0; 1271 ▎1242; 444; 44 ▎-2.2; 1091; 439.7

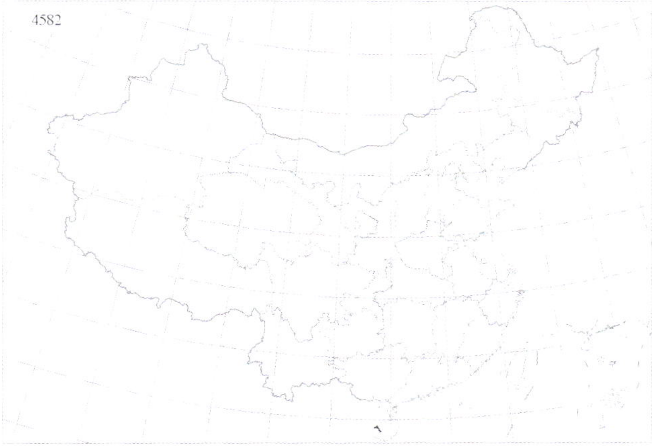

Rutaceae 芸香科 *Clausena* 黄皮属

Clausena lansium (Lour.) Skeels 黄皮

Trees

20.2 (12.4~24.8); 11.8 (3.8~20.1); 27.0 (18.6~29.8); 20.2 (12.4~24.8); 182.6 (89.1~237.6); 0.0 (0.0~1.5); 1056 (675~1385) ▎1588 (868~3495); 705 (312~1847); 136 (31~897) ▎52.6 (-34.3~246.5); 1045 (675~1360); 253.2 (95.3~620.0)

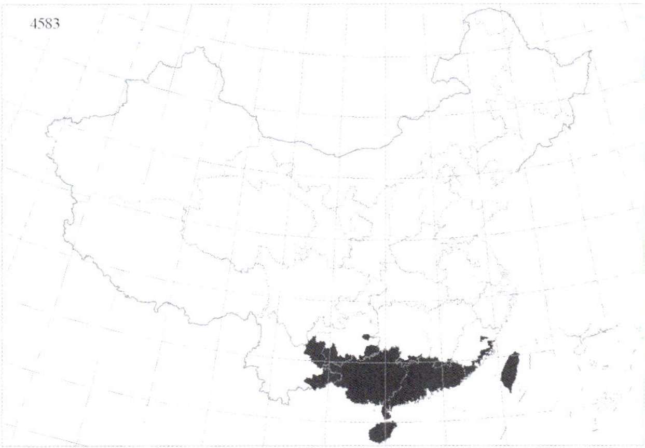

Rutaceae 芸香科 *Clausena* 黄皮属

Clausena lenis Drake 光滑黄皮

Trees

19.9 (13.2~24.8); 13.8 (6.4~20.1); 24.0 (18.4~28.8); 19.9 (13.3~24.8); 179.4 (99.4~237.6); 0.0 (0.0~0.0); 1010 (695~1385) ▎1414 (868~1854); 687 (312~996); 60 (37~115) ▎44.6 (-34.3~112.2); 960 (695~1360); 446.2 (248.2~659.0)

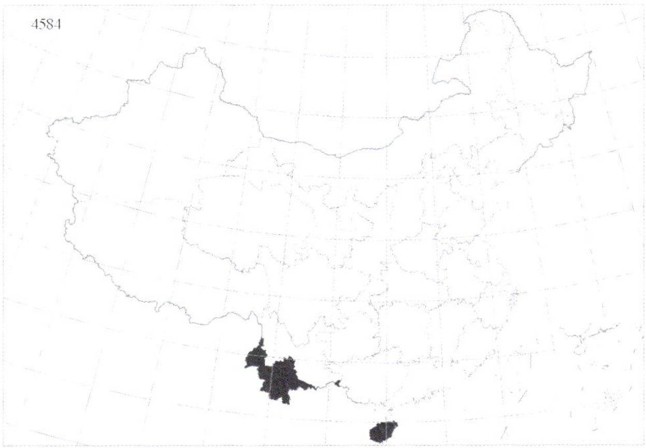

Rutaceae 芸香科 *Clausena* 黄皮属
Clausena odorata C. C. Huang 香花黄皮
Trees

19.0; 12.8; 23.0; 19.0; 168.1; 0.0; 911 ❙ 1345; 771; 58 ❙ 47.6; 895; 354.7

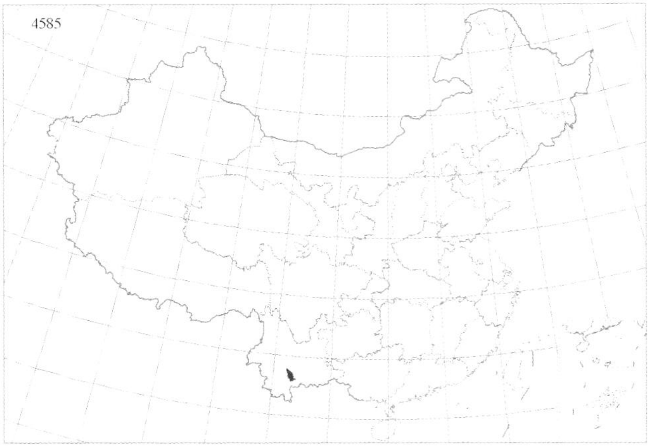

Rutaceae 芸香科 *Clausena* 黄皮属
Clausena vestita D. D. Tao 毛叶黄皮
Trees

11.8 (5.4~16.2); 4.9 (-2.1~9.4); 17.4 (12.3~20.8); 11.9 (5.8~16.3); 86.6 (31.5~135.1); 4.0 (0.0~25.6); 660 (481~790) ❙ 1107 (741~1454); 628 (426~792); 47 (21~70) ❙ 67.7 (36.1~99.9); 659 (481~790); 352.5 (273.5~487.2)

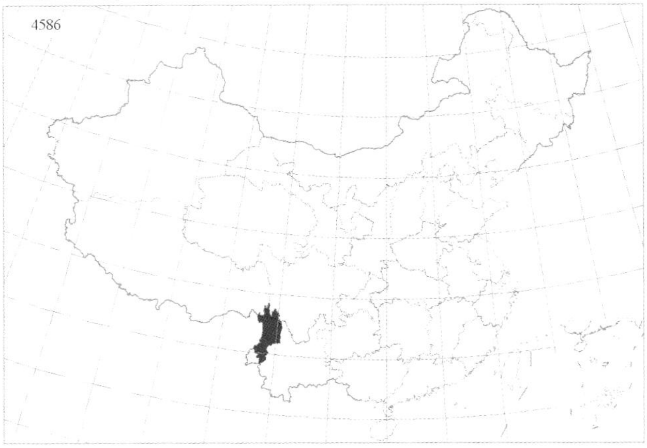

Rutaceae 芸香科 *Clausena* 黄皮属
Clausena yunnanensis C. C. Huang 云南黄皮
Trees

19.8 (17.2~22.2); 12.2 (9.5~14.2); 25.3 (22.5~28.3); 19.8 (17.3~22.3); 178.1 (147.0~207.5); 0.0 (0.0~0.0); 1002 (838~1190) ❙ 1423 (1254~1581); 776 (699~850); 71 (48~106) ❙ 43.6 (15.4~60.1); 1002 (838~1190); 393.4 (281.7~551.6)

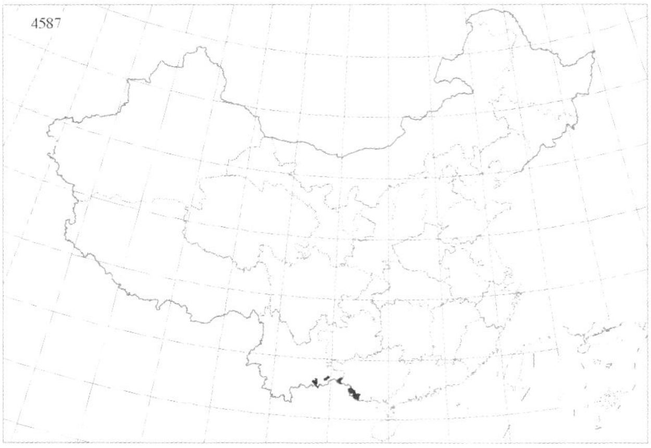

Rutaceae 芸香科 *Clausena* 黄皮属
Clausena yunnanensis var. longgangensis C. F. Liang et Y. X. Lu 弄岗黄皮
Trees

22.2; 13.9; 28.3; 22.3; 207.5; 0.0; 1190 ❙ 1374; 713; 87 ❙ 15.4; 1190; 327.7

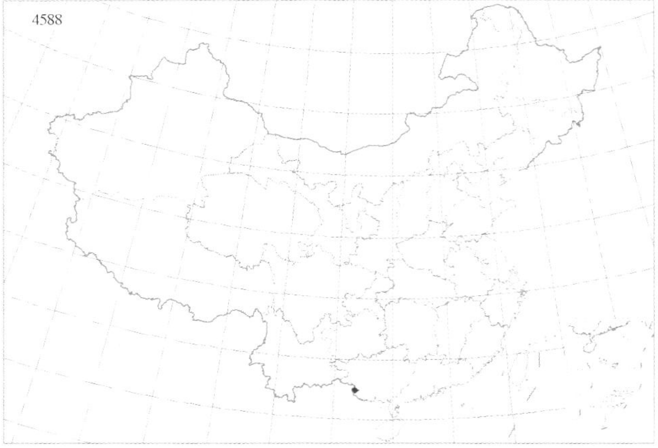

Rutaceae 芸香科 *Glycosmis* 山小橘属
Glycosmis cochinchinensis (Lour.) Pierre 山橘树
Shrubs or trees

21.6 (17.2~24.8); 15.4 (9.1~20.1); 25.9 (22.1~28.8); 21.6 (17.3~24.8); 199.8 (147.0~237.6); 0.0 (0.0~0.0); 1127 (833~1385) ❙ 1500 (868~2992); 673 (312~1847); 71 (41~144) ❙ 37.0 (-34.3~166.5); 1066 (833~1360); 431.9 (225.6~659.0)

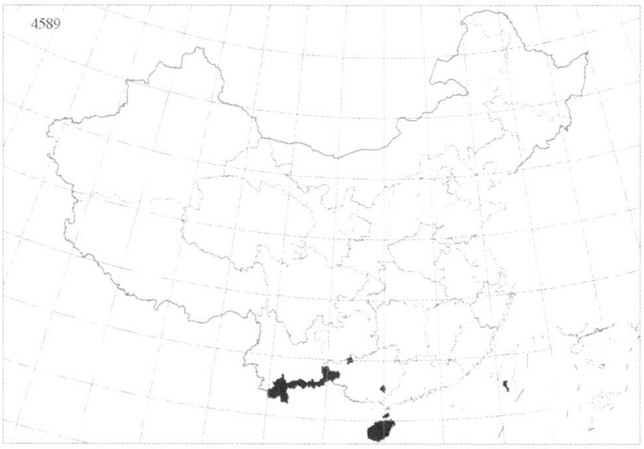

Rutaceae 芸香科 *Glycosmis* 山小橘属
Glycosmis craibii Tanaka 毛山小橘
Trees, evergreen

18.2 (13.2~23.0); 11.5 (6.4~17.7); 22.8 (18.4~28.7); 18.3 (13.3~23.1); 159.3 (99.4~217.0); 0.0 (0.0~0.0); 893 (695~1217) ❙ 1379 (1049~1751); 740 (594~996); 61 (41~142) ❙ 55.4 (15.4~99.9); 882 (695~1190); 423.5 (190.9~659.0)

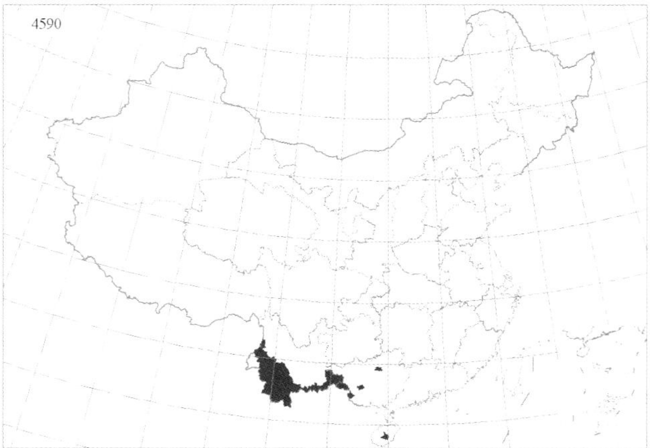

Rutaceae 芸香科 *Glycosmis* 山小橘属
Glycosmis craibii var. glabra (Craib) Tanaka 光叶山小橘
Trees, evergreen

24.1 (23.0~24.8); 19.0 (17.7~20.1); 27.6 (26.7~28.4); 24.1 (23.1~24.8); 229.7 (217.0~237.6); 0.0 (0.0~0.0); 1319 (1217~1385) ❙ 1471 (1179~1725) 500 (392~610); 51 (41~68) ❙ 12.1 (-11.9~41.8); 1138 (1056~1212); 520.9 (455.2~620.0)

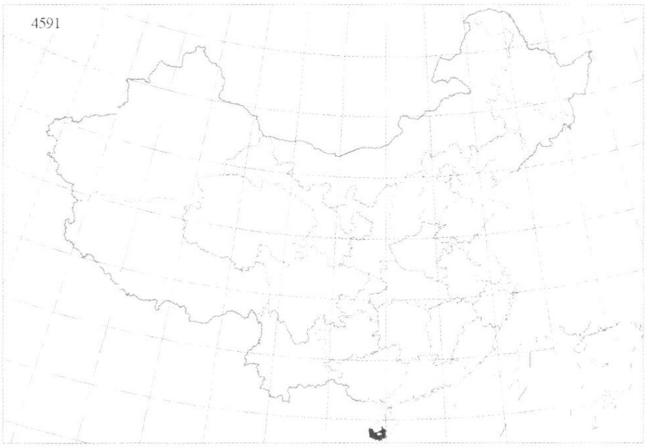

Rutaceae 芸香科 *Glycosmis* 山小橘属
Glycosmis esquirolii (H. Lév.) Tanaka 锈毛山小橘
Trees, evergreen

18.5 (15.1~22.2); 10.7 (5.0~15.1); 24.2 (22.3~28.3); 18.5 (15.2~22.3); 162.1 (121.9~207.5); 0.0 (0.0~0.0); 916 (788~1190) ❙ 1372 (1161~1751); 728 (545~996); 62 (48~87) ❙ 50.6 (15.4~88.8); 914 (788~1190); 375.5 (189.5~659.0)

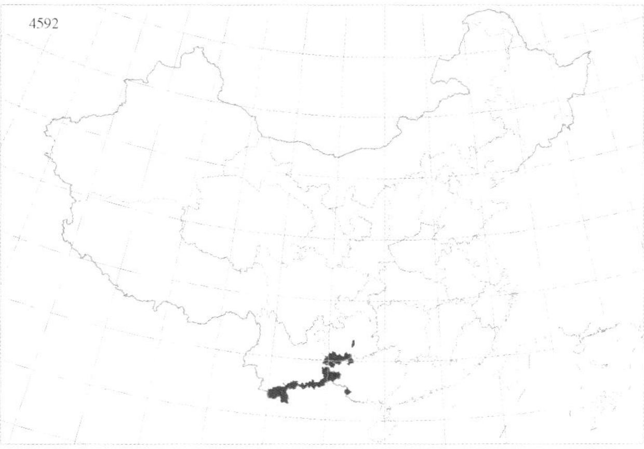

Rutaceae 芸香科 *Glycosmis* 山小橘属
Glycosmis longifolia (Oliver) Tanaka 长叶山小橘
Trees

15.3; 8.7; 19.9; 15.4; 124.3; 0.0; 764 ❙ 1372; 771; 58 ❙ 80.5; 764; 420.8

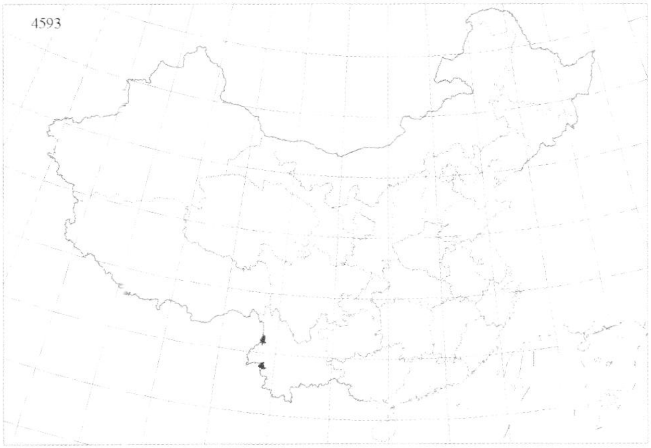

Rutaceae 芸香科 *Glycosmis* 山小橘属
Glycosmis lucida Wall. ex C. C. Huang 亮叶山小橘
Trees, evergreen

18.4 (16.3~20.5); 12.4 (10.2~15.1); 22.2 (20.3~23.7); 18.5 (16.4~20.5); 161.6 (136.6~185.9); 0.0 (0.0~0.0); 883 (789~992) ❙ 1364 (1028~1751); 738 (566~996); 53 (41~77) ❙ 54.6 (12.5~88.8); 861 (780~986); 448.4 (339.5~659.0)

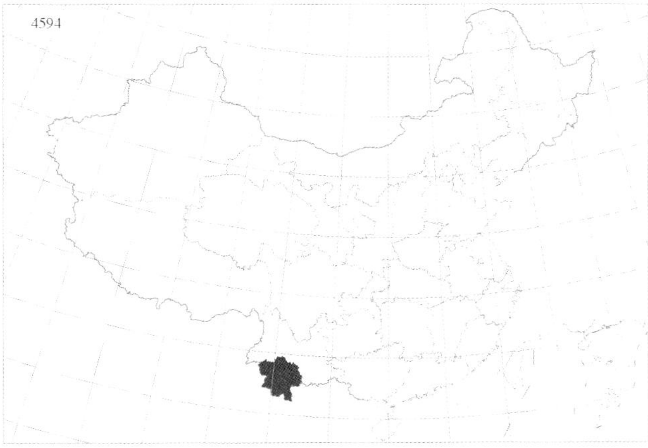

Rutaceae 芸香科 *Glycosmis* 山小橘属
Glycosmis montana Pierre 海南山小橘
Shrubs or trees, evergreen

23.7 (18.9~24.8); 18.0 (10.6~20.1); 27.8 (25.0~28.8); 23.8 (19.0~24.8); 225.2 (167.5~237.6); 0.0 (0.0~0.0); 1291 (939~1385) ❙ 1491 (868~1854); 543 (312~729); 72 (41~115) ❙ 16.1 (-34.3~47.8); 1179 (868~1360); 435.4 (248.2~620.0)

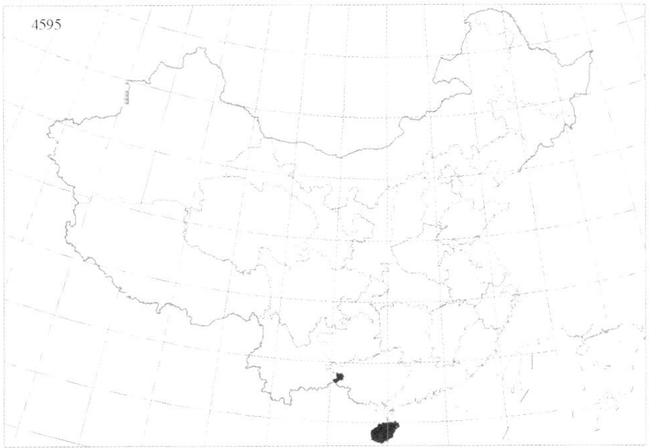

Rutaceae 芸香科 *Glycosmis* 山小橘属
Glycosmis oligantha C. C. Huang 少花山小橘
Shrubs or trees, evergreen

22.2; 13.7; 28.5; 22.2; 206.8; 0.0; 1190 ❙ 1593; 843; 116 ❙ 33.8; 1190; 278.3

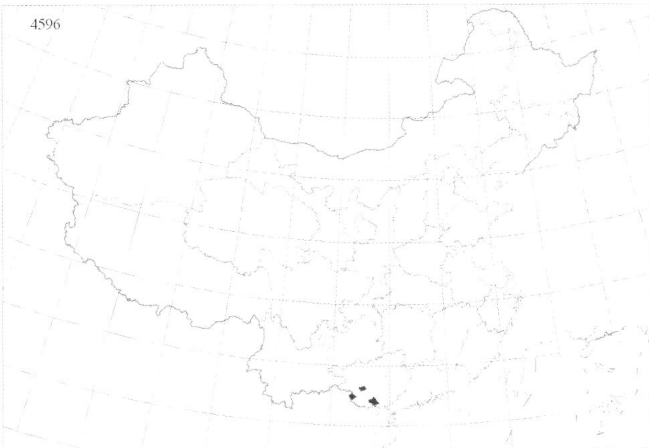

Rutaceae 芸香科 *Glycosmis* 山小橘属
Glycosmis parviflora (Sims) Kurz 小花山小橘
Shrubs or trees, evergreen

19.9 (14.4~24.5); 12.1 (3.8~18.9); 26.2 (19.1~29.2); 20.0 (14.4~24.5); 179.9 (113.1~234.4); 0.0 (0.0~1.5); 1025 (727~1360) ❙ 1657 (1161~3495); 766 (419~1847); 133 (41~897) ❙ 64.2 (5.8~246.5); 1018 (727~1360); 291.5 (95.3~659.0)

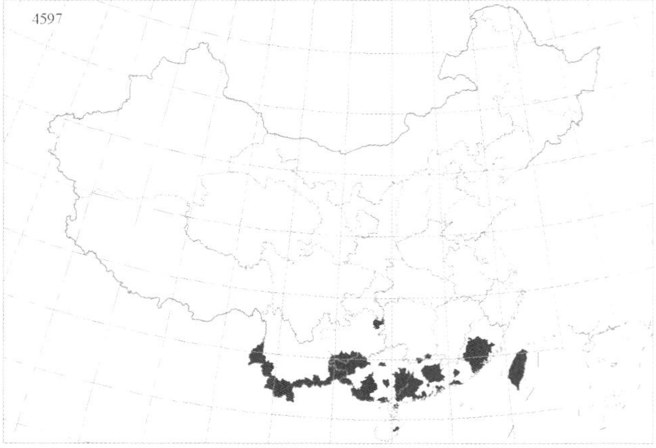

Rutaceae 芸香科 *Glycosmis* 山小橘属
Glycosmis pentaphylla (Retzius) Candolle 山小橘
Trees, evergreen

18.6 (16.8~20.5); 12.7 (10.4~15.1); 22.2 (20.8~23.7); 18.7 (16.9~20.5); 163.8 (142.6~185.9); 0.0 (0.0~0.0); 890 (811~992) ❙ 1432 (1275~1607); 743 (655~816); 53 (41~77) ❙ 61.1 (47.1~74.2); 869 (802~986); 487.0 (403.0~659.0)

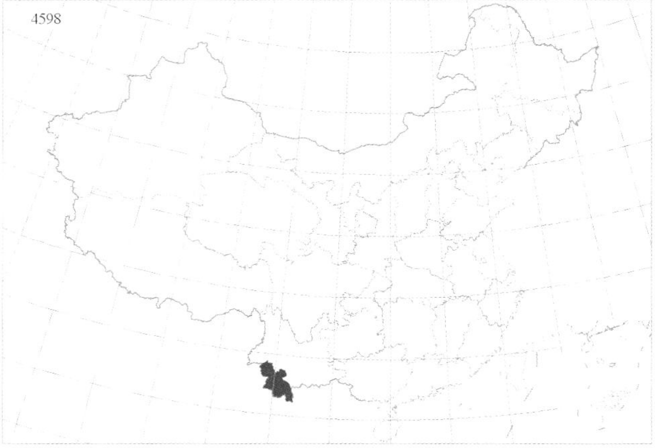

Rutaceae 芸香科 *Glycosmis* 山小橘属
Glycosmis pseudoracemosa (Guillaumin) Swingle 华山小橘
Shrubs or trees, evergreen

18.4 (16.3~22.2); 10.8 (9.1~14.2); 23.8 (21.4~28.3); 18.4 (16.4~22.3); 160.7 (136.4~207.5); 0.0 (0.0~0.0); 908 (797~1190) ❙ 1320 (1074~1522); 731 (582~860); 57 (47~87) ❙ 46.6 (15.4~80.6); 908 (797~1190); 379.9 (277.0~551.6)

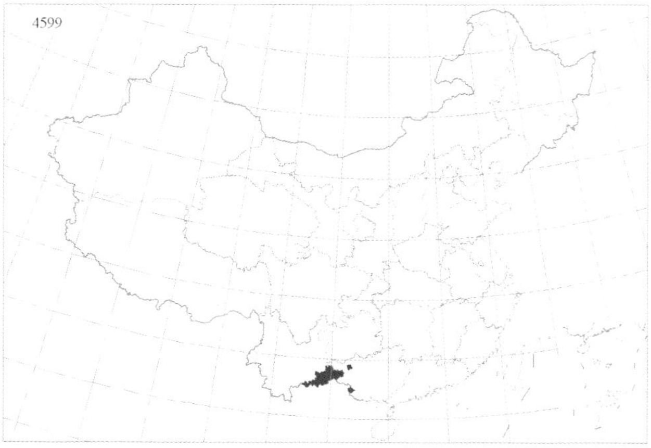

Rutaceae 芸香科 *Glycosmis* 山小橘属
Glycosmis xizangensis (C. Y. Wu et H. Li) D. D. Tao 西藏山小橘
Shrubs or trees, evergreen

13.3; 5.4; 19.6; 13.4; 100.4; 0.0; 706 ❙ 1736; 1007; 51 ❙ 145.8; 706; 447.4

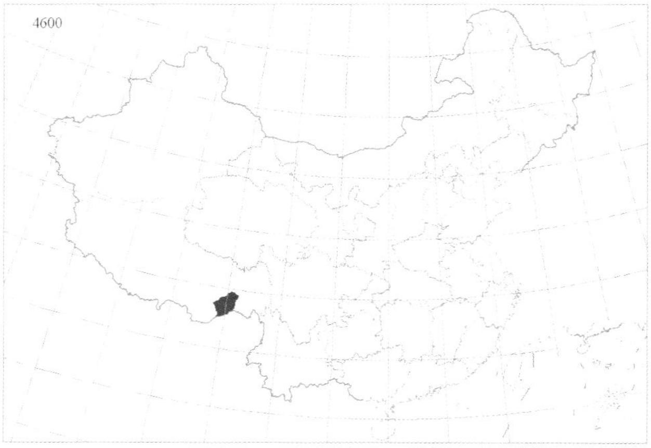

Rutaceae 芸香科 *Luvunga* 三叶藤橘属
Luvunga scandens (Roxb.) Buch.-Ham. ex Wight et Arn. 三叶藤
Vines, evergreen

20.7 (17.2~24.8); 14.3 (9.1~20.1); 25.2 (21.4~28.4); 20.7 (17.3~24.8); 188.7 (147.0~237.6); 0.0 (0.0~0.0); 1074 (833~1385) ❙ 1478 (1161~1767); 676 (473~860); 60 (41~110) ❙ 41.4 (1.5~80.6); 1024 (833~1326); 454.3 (314.4~620.0)

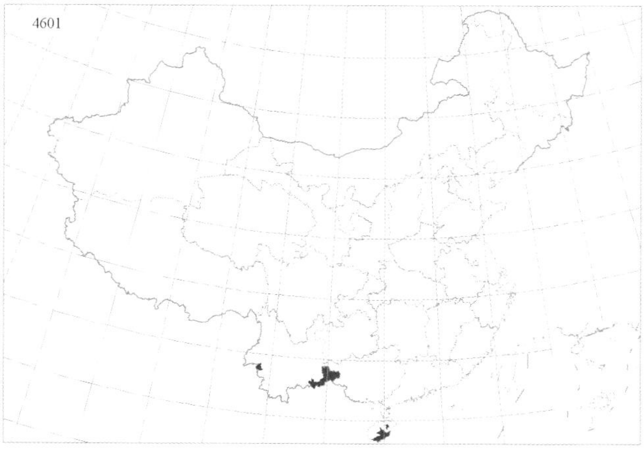

Rutaceae 芸香科 *Maclurodendron* 贡甲属
Maclurodendron oligophlebium (Merr.) T. G. Hartley 贡甲
Trees, evergreen

23.8 (21.8~24.8); 18.0 (14.2~20.1); 28.0 (25.7~28.8); 23.8 (21.9~24.8); 226.1 (202.6~237.6); 0.0 (0.0~0.0); 1297 (1127~1385) ❙ 1516 (868~1876); 562 (312~891); 76 (41~116) ❙ 17.5 (-34.3~63.2); 1190 (868~1360); 419.3 (164.9~620.0)

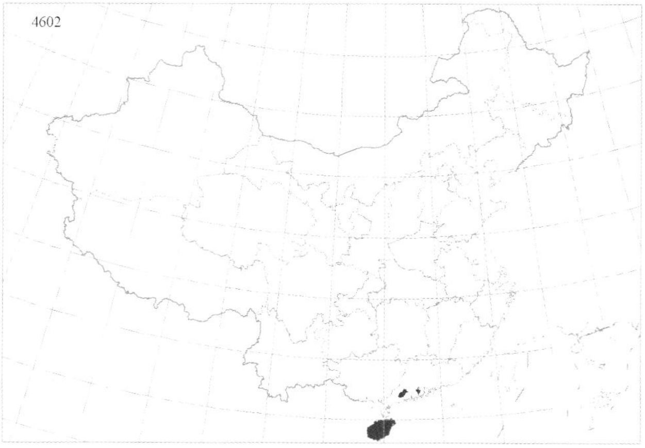

Rutaceae 芸香科 *Melicope* 蜜茱萸属
Melicope chunii (Merr.) T. G. Hartley 海南蜜茱萸
Shrubs or trees

24.1 (22.9~2[illegible].8); 18.5 (15.2~20.1); 28.3 (27.2~29.0); 24.2 (23.0~24.8); 230.1 (215.7~237.6); 0.0 (0.0~0.0); 1328 (1239~1385) ❙ 1426 (1176~1767); 522 (392~714); 63 (41~97) ❙ 7.4 (-12.0~28.8); 1170 (1056~1299); 460.9 (203.1~620.0)

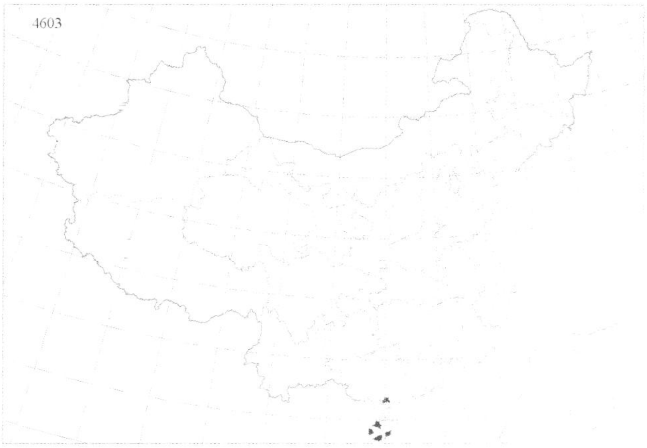

Rutaceae 芸香科 *Melicope* 蜜茱萸属
Melicope glomerata (Craib) T. G. Hartley 密果蜜茱萸
Shrubs or trees

20.5; 15.0; 23.7; 20.5; 185.9; 0.0; 990 ❙ 1536; 678; 72 ❙ 55.1; 973; 614.6

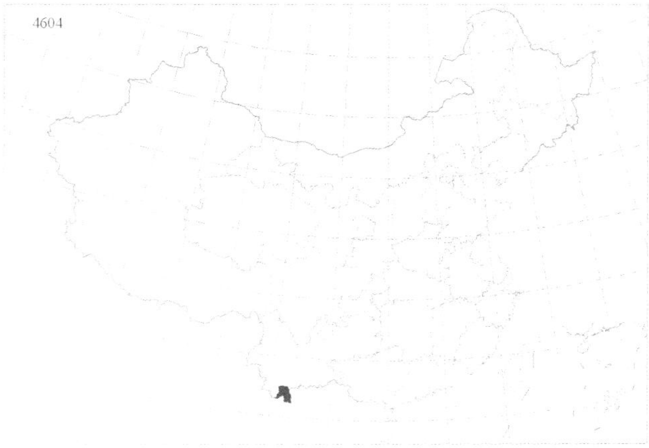

Rutaceae 芸香科 *Melicope* 蜜茱萸属
Melicope lunu-ankenda (Gaertn.) Hartley 三刈叶蜜茱萸
Trees, evergreen

13.3; 5.4; 19.6; 13.4; 100.4; 0.0; 706 ❙ 1736; 1007; 51 ❙ 145.8; 706; 447.4

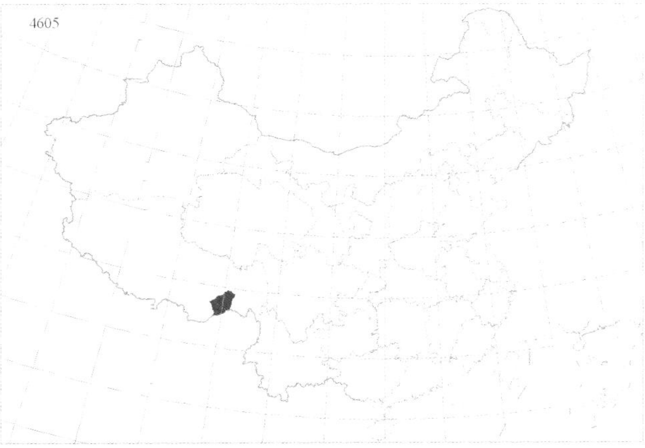

Rutaceae 芸香科 *Melicope* 蜜茱萸属
Melicope patulinervia (Merr. et Chun) C. C. Huang 蜜茱萸
Shrubs, evergreen

23.9 (22.2~24.8); 18.8 (17.2~20.1); 27.5 (25.7~28.4); 24.0 (22.3~24.8); 227.4 (207.2~237.6); 0.0 (0.0~0.0); 1300 (1127~1385) ❙ 1601 (1380~1767); 549 (455~619); 64 (41~110) ❙ 23.9 (1.5~47.8); 1181 (1055~1326); 520.3 (352.9~620.0)

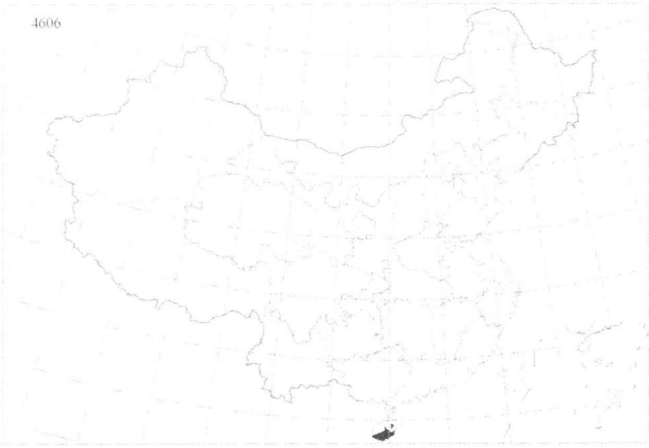

Rutaceae 芸香科 *Melicope* 蜜茱萸属
Melicope pteleifolia (Champ. ex Benth.) Hartley 三椏苦
Shrubs or trees, evergreen

19.4 (8.6~24.8); [illegible]1.3 (1.3~20.1); 25.9 (15.0~29.8); 19.4 (8.7~24.8); 173.4 (52.6~237.6); 0.0 (0.0~8.8); 1004 (574~1385) ❙ 1498 (868~3495); 678 (312~1428); [illegible]20 (28~897) ❙ 50.5 (-34.3~246.5); 993 (574~1360); 281.5 (95.3~659.0)

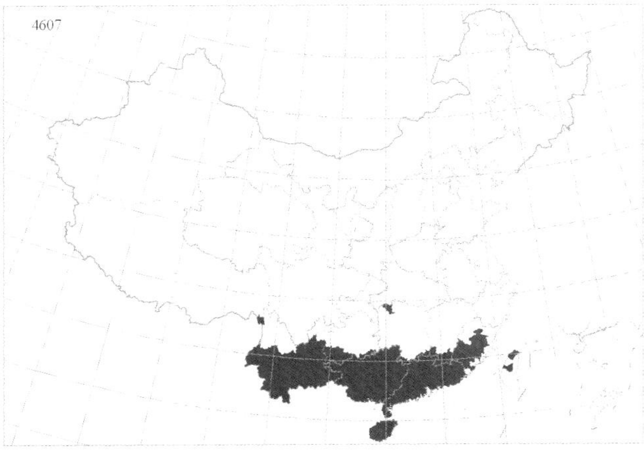

Rutaceae 芸香科 *Melicope* 蜜茱萸属
Melicope semecarpifolia (Merr.) T. G. Hartley 台湾蜜茱萸
Small trees

19.9 (17.8~22.2); 14.3 (12.1~17.7); 24.7 (21.7~27.4); 19.9 (17.9~22.2); 179.0 (154.4~206.7); 0.0 (0.0~0.0); 980 (852~1123) ❙ 2871 (2713~2993); 1192 (844~1847); 359 (107~551) ❙ 195.9 (159.6~235.6); 980 (852~1123); 285.3 (238.5~363.7)

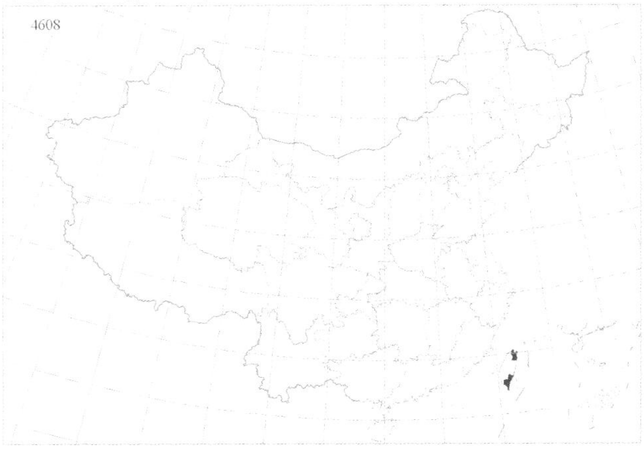

Rutaceae 芸香科 *Melicope* 蜜茱萸属
Melicope triphylla (Lam.) Merr. 三叶蜜茱萸
Shrubs or trees, evergreen

20.3; 15.9; 23.6; 20.4; 184.4; 0.0; 993 ❙ 2853; 1618; 147 ❙ 190.4; 993; 330.3

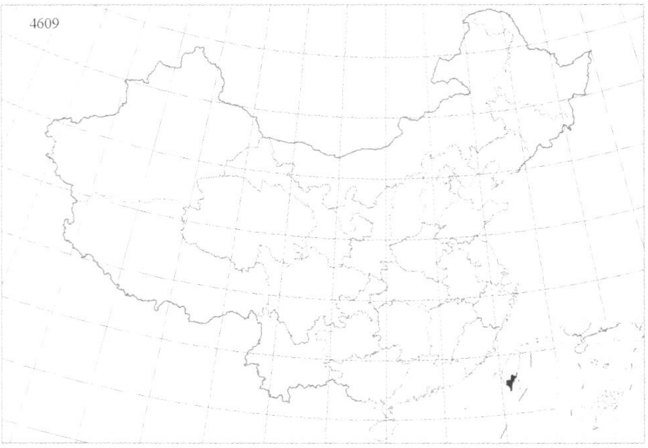

Rutaceae 芸香科 *Melicope* 蜜茱萸属
Melicope viticina (Wallich ex Kurz) T. G. Hartley 单叶蜜茱萸
Shrubs or trees, evergreen

19.4 (18.2~20.5); 13.8 (12.4~15.1); 22.7 (21.6~23.7); 19.4 (18.2~20.5); 173.2 (158.6~185.9); 0.0 (0.0~0.0); 928 (861~992) ❙ 1494 (1434~1607); 715 (655~785); 57 (41~77) ❙ 61.4 (47.6~74.2); 900 (839~986); 516.3 (427.9~659.0)

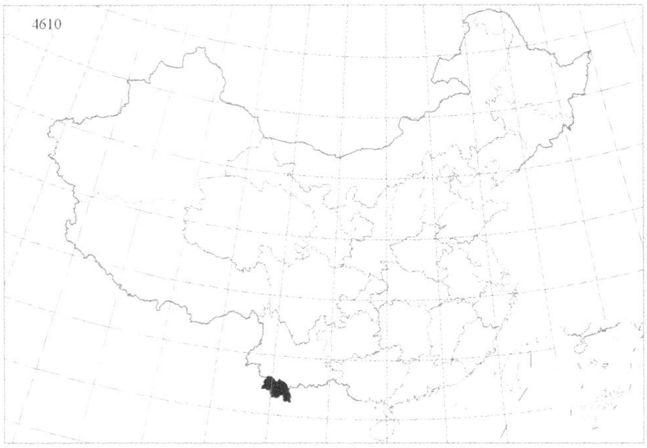

Rutaceae 芸香科 *Micromelum* 小芸木属
Micromelum falcatum (Lour.) Tanaka 大管
Trees

19.8 (14.4~24.8); 12.9 (7.5~20.1); 24.6 (19.1~29.2); 19.8 (14.4~24.8); 177.8 (113.1~237.6); 0.0 (0.0~0.0); 1004 (727~1385) ❙ 1430 (868~2435); 723 (312~1428); 71 (41~189) ❙ 45.3 (-34.3~112.2); 973 (727~1360); 394.7 (160.7~659.0)

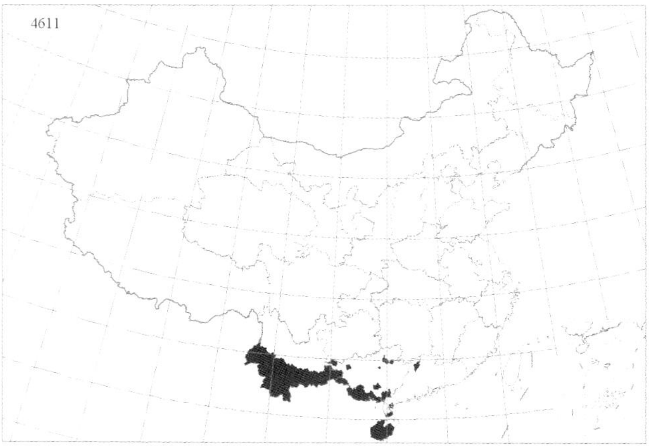

Rutaceae 芸香科 *Micromelum* 小芸木属
Micromelum integerrimum (Buch.-Ham. ex Colebr.) M. Roem. 小芸木
Trees

19.2 (2.9~24.8); 11.7 (-5.7~20.1); 24.8 (10.6~29.8); 19.3 (4.3~24.8); 171.6 (19.9~237.6); 0.3 (0.0~45.2); 984 (421~1385) ❙ 1395 (535~2435); 684 (312~1428); 83 (10~194) ❙ 43.4 (-34.3~145.8); 962 (421~1360); 340.2 (111.7~659.0)

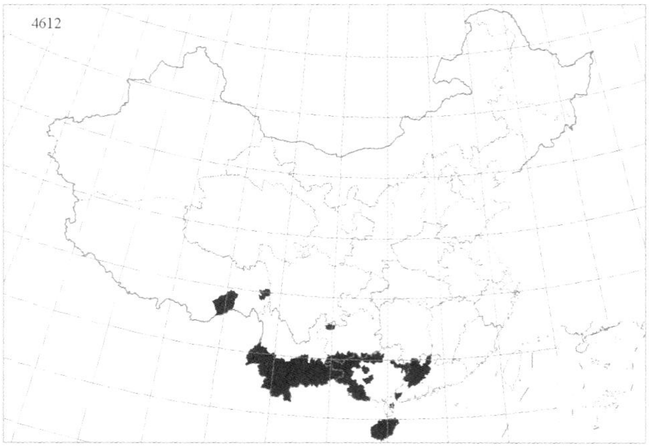

Rutaceae 芸香科 *Murraya* 九里香属
Murraya alata Drake 翼叶九里香
Shrubs, evergreen

23.7 (22.2~24.8); 17.7 (14.7~20.1); 28.1 (25.7~29.2); 23.8 (22.3~24.8); 225.2 (207.2~237.6); 0.0 (0.0~0.0); 1291 (1127~1385) ❙ 1460 (868~1854); 571 (312~997); 71 (41~115) ❙ 13.5 (-34.3~47.8); 1177 (868~1360); 406.7 (178.7~620.0)

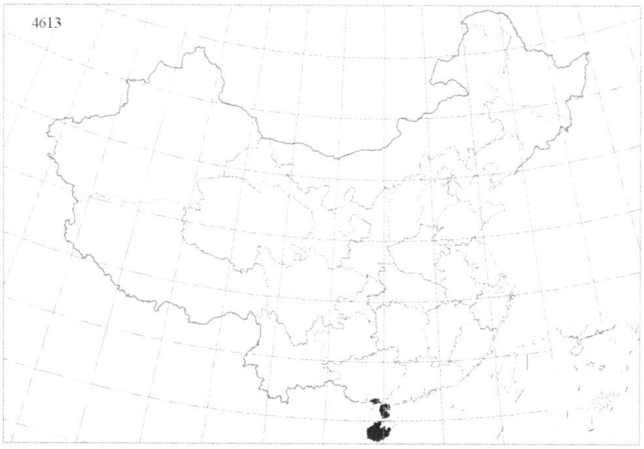

Rutaceae 芸香科 *Murraya* 九里香属
Murraya crenulata (Turcz.) Oliv. 兰屿九里香
Trees, evergreen

18.5; 14.1; 21.7; 18.5; 162.1; 0.0; 863 ❙ 2713; 1388; 188 ❙ 214.4; 863; 363.7

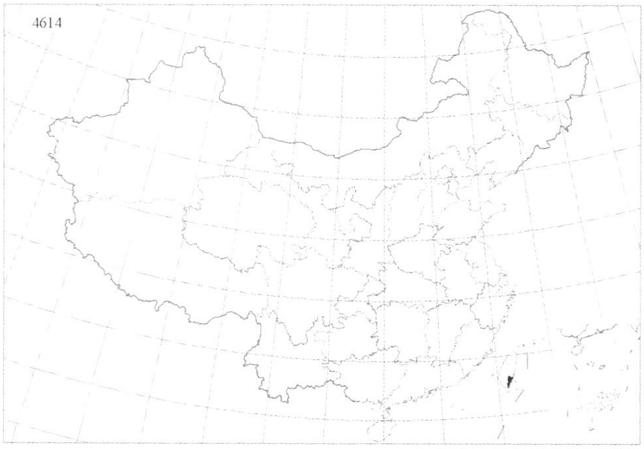

Rutaceae 芸香科 *Murraya* 九里香属
Murraya euchrestifolia Hayata 豆叶九里香
Trees, evergreen

19.0 (9.6~24.3); 12.0 (2.4~18.9); 24.0 (15.7~28.7); 19.0 (9.6~24.3); 167.9 (60.8~231.9); 0.1 (0.0~5.4); 950 (600~1338) ❙ 1545 (868~3495); 789 (312~1847); 100 (39~897) ❙ 64.9 (-34.3~246.5); 927 (600~1221); 346.0 (119.4~659.0)

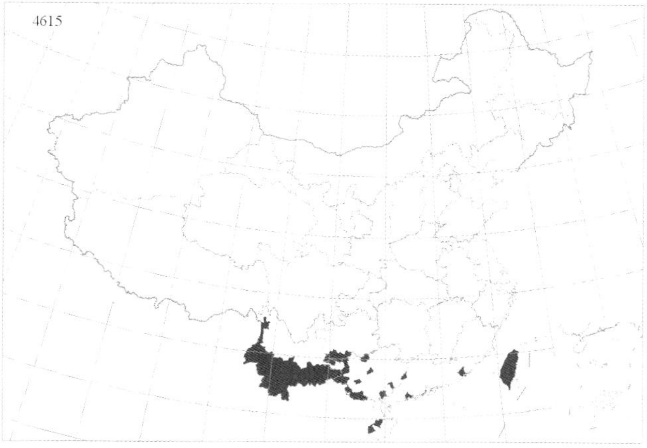

Rutaceae 芸香科 *Murraya* 九里香属
Murraya exotica L. 九里香
Trees, evergreen

19.6 (13.2~24.8); 11.5 (3.8~20.1); 26.0 (18.4~29.8); 19.6 (13.3~24.8); 175.3 (99.4~237.6); 0.0 (0.0~1.5); 1007 (695~1385) ❙ 1547 (868~3495); 701 (312~1847); 126 (41~897) ❙ 55.8 (-34.3~246.5); 993 (695~1360); 295.1 (119.4~659.0)

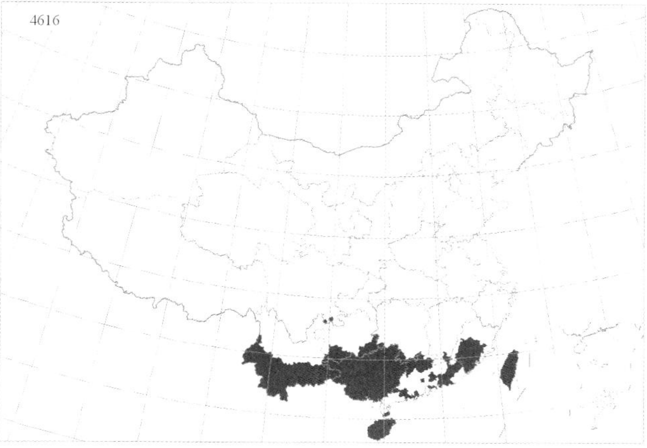

Rutaceae 芸香科 *Murraya* 九里香属
Murraya koenigii (L.) Spreng. 调料九里香
Shrubs or trees, evergreen

17.9 (11.1~24.8); 10.7 (2.2~20.1); 23.3 (19.4~28.8); 18.0 (11.3~24.8); 156.0 (80.8~237.6); 0.2 (0.0~5.3); 894 (652~1385) ❙ 1234 (806~1751); 645 (312~996); 52 (16~120) ❙ 40.1 (-34.3~88.8); 861 (652~1139); 366.8 (120.7~659.0)

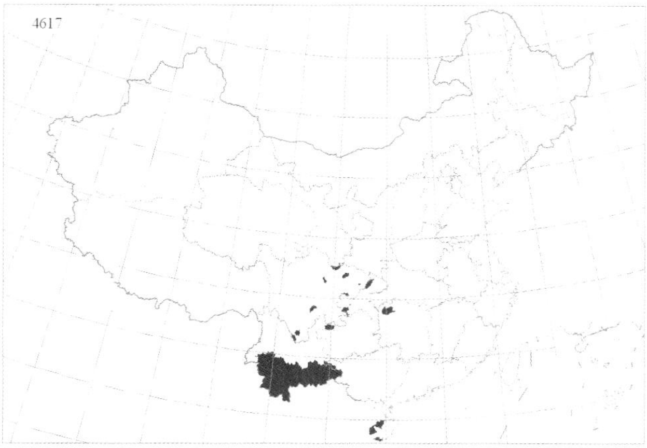

Rutaceae 芸香科 *Murraya* 九里香属
Murraya kwangsiensis (C. C. Huang) C. C. Huang 广西九里香
Trees, evergreen

20.0 (16.3~22.3); 11.6 (9.1~13.9); 26.3 (21.4~28.9); 20.1 (16.4~22.3); 180.8 (136.4~208.1); 0.0 (0.0~0.0); 1036 (797~1202) ❙ 1407 (1074~1662); 745 (582~860); 94 (47~173) ❙ 37.3 (5.8~80.6); 1036 (797~1202); 294.1 (201.4~413.5)

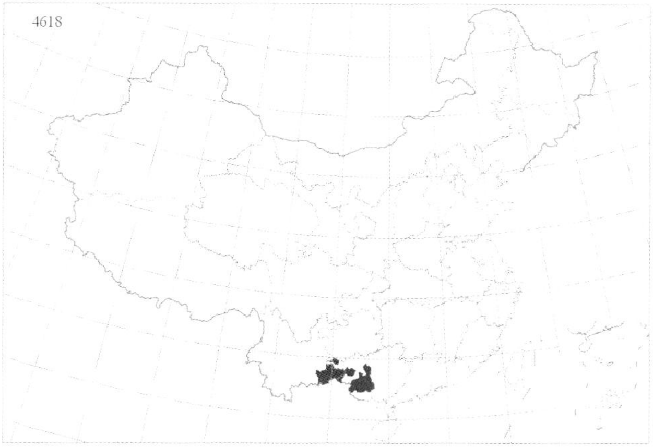

Rutaceae 芸香科 *Murraya* 九里香属
Murraya microphylla (Merr. et Chun) Swingle 小叶九里香
Shrubs or trees, evergreen

22.0; 15.4; 27.0; 22.0; 204.4; 0.0; 1188 ❙ 1389; 522; 66 ❙ 21.4; 1045; 409.3

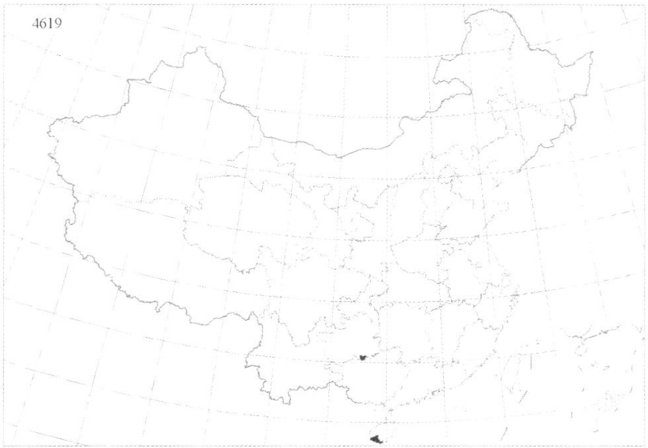

Rutaceae 芸香科 *Murraya* 九里香属
Murraya tetramera C. C. Huang 四数九里香
Trees, evergreen

22.2; 13.9; 28.3; 22.3; 207.5; 0.0; 1190 ❙ 1374; 713; 87 ❙ 15.4; 1190; 327.7

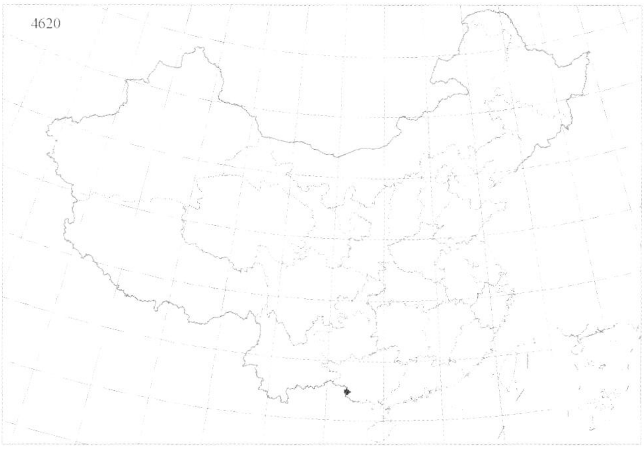

Rutaceae 芸香科 *Orixa* 臭常山属

Orixa japonica Thunb. 臭常山

Shrubs or trees, deciduous

15.4 (9.3~19.1); 4.1 (-2.9~10.1); 26.0 (16.6~29.8); 15.4 (9.5~19.1); 127.7 (69.8~169.2); 2.6 (0.0~20.5); 836 (603~1002) ❙ 1193 (512~1893); 512 (258~816); 97 (21~230) ❙ 43.1 (-35.3~129.9); 828 (512~1002); 198.9 (108.2~319.9)

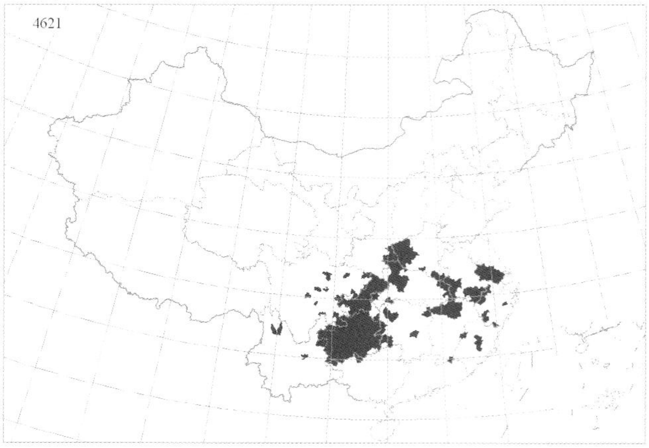

Rutaceae 芸香科 *Paramignya* 单叶藤橘属

Paramignya confertifolia Swingle 单叶藤橘

Vines

21.5 (14.7~24.8); 14.5 (4.4~20.1); 26.6 (22.4~29.0); 21.5 (14.8~24.8); 198.2 (117.8~237.6); 0.0 (0.0~0.6); 1128 (779~1385) ❙ 1491 (868~2435); 708 (312~1428); 77 (25~130) ❙ 34.6 (-34.3~99.5); 1083 (779~1360); 386.8 (135.9~659.0)

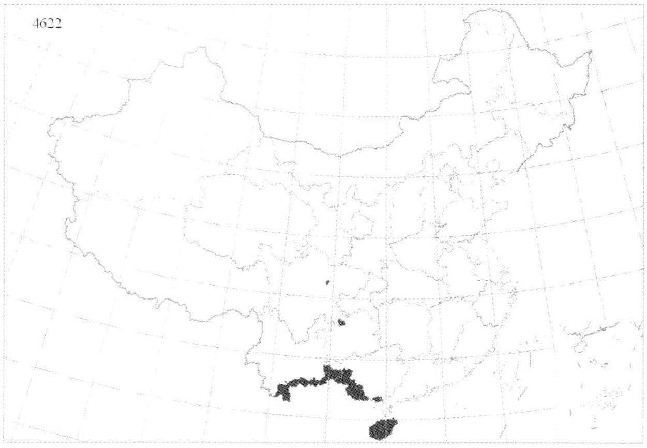

Rutaceae 芸香科 *Phellodendron* 黄檗属

Phellodendron amurense Rupr. 黄檗

Trees, deciduous

7.9 (-2.9~21.1); -9.6 (-26.4~13.2); 23.3 (17.7~28.9); 10.4 (5.4~21.1); 85.1 (40.2~193.7); 49.3 (0.0~134.6); 673 (471~1104) ❙ 731 (392~1770); 431 (260~691); 37 (6~223) ❙ 8.4 (-43.4~112.2); 600 (392~1050); 240.3 (97.2~346.5)

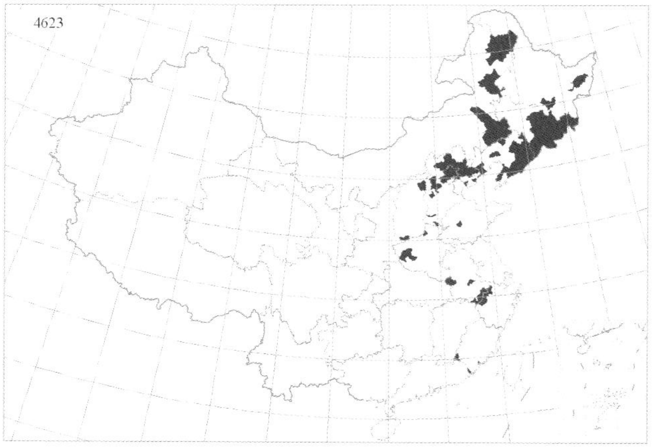

Rutaceae 芸香科 *Phellodendron* 黄檗属

Phellodendron chinense C. K. Schneid. 川黄檗

Trees, deciduous

15.3 (6.2~21.3); 4.5 (-4.2~12.3); 25.2 (13.3~29.8); 15.4 (6.7~21.3); 128.4 (39.4~196.1); 3.8 (0.0~29.4); 835 (509~1130) ❙ 1251 (512~1997); 532 (258~860); 102 (12~259) ❙ 49.5 (-35.3~130.7); 827 (509~1130); 233.6 (111.9~491.5)

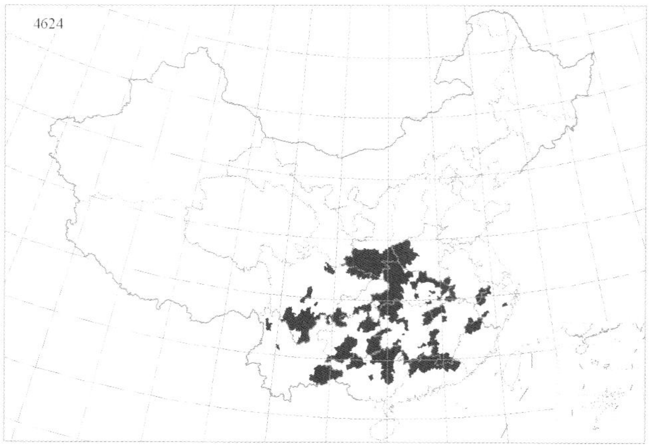

Rutaceae 芸香科 *Phellodendron* 黄檗属

Phellodendron chinense var. glabriusculum C. K. Schneid. 秃叶黄檗

Trees

14.6 (2.9~21.1); 4.5 (-6.3~12.0); 23.9 (11.2~28.4); 14.7 (4.4~21.2); 119.4 (20.5~194.2); 3.8 (0.0~44.6); 794 (428~1121) ❙ 1231 (822~1997); 563 (375~817); 86 (19~254) ❙ 56.9 (15.8~130.7); 793 (428~1121); 234.5 (149.6~354.5)

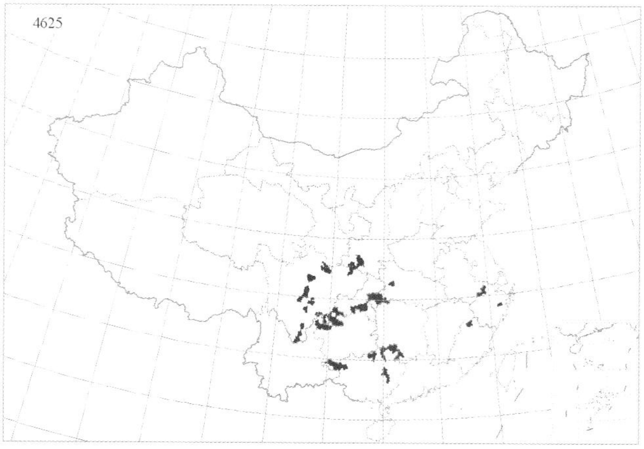

Rutaceae 芸香科 *Skimmia* 茵芋属

Skimmia arborescens T. Anderson ex Gamble 乔木茵芋

Trees, evergreen

20.7 (17.6~22.7); 11.6 (7.3~15.0); 27.9 (25.8~28.9); 20.7 (17.6~22.8); 189.0 (151.6~213.1); 0.0 (0.0~0.0); 1095 (909~1221) ❙ 1629 (1302~2435); 783 (562~1428); 145 (86~189) ❙ 49.7 (12.9~99.5); 1094 (909~1221); 251.7 (185.5~335.1)

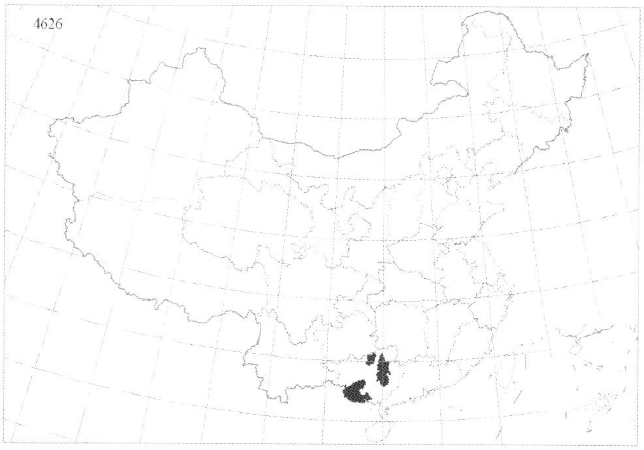

Rutaceae 芸香科 *Skimmia* 茵芋属

Skimmia laureola (DC.) Siebold et Zucc. ex Walp. 月桂茵芋

Shrubs, evergreen

16.0 (5.3~22.0); 8.2 (-2.4~15.1); 22.2 (12.2~28.8); 16.0 (5.7~22.0); 133.4 (30.5~203.8); 1.2 (0.0~26.2); 819 (479~1170) ❙ 1229 (741~2005); 646 (411~1056); 61 (15~194) ❙ 50.6 (-14.2~145.8); 806 (479~1170); 318.9 (142.9~659.0)

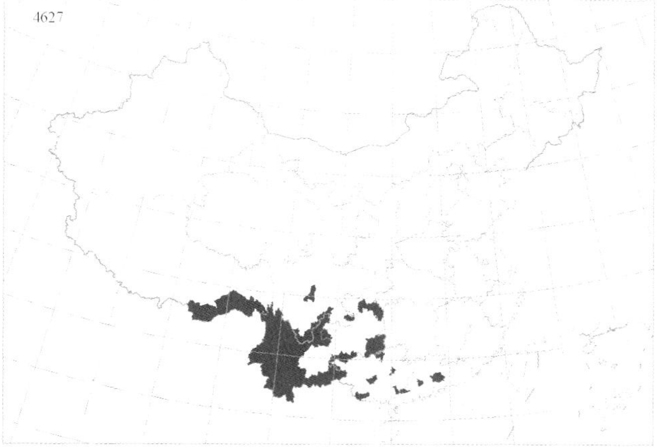

Rutaceae 芸香科 *Skimmia* 茵芋属

Skimmia melanocarpa Rehder et E. H. Wilson 黑果茵芋

Shrubs, evergreen

12.8 (5.3~21.0); 3.3 (-3.5~14.2); 21.5 (12.2~27.7); 12.9 (5.7~21.0); 101.6 (30.5~191.8); 7.0 (0.0~28.1); 720 (479~1053) ❙ 1124 (677~1751); 576 (336~1007); 41 (14~88) ❙ 57.6 (2.4~145.8); 720 (479~1053); 296.4 (141.4~551.6)

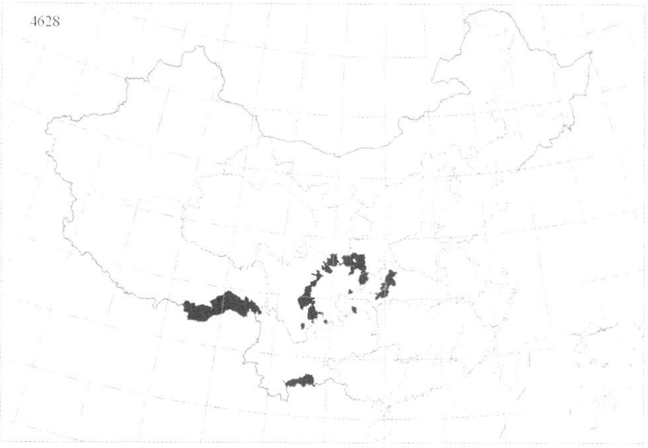

Rutaceae 芸香科 *Skimmia* 茵芋属

Skimmia reevesiana (Fortune) Fortune 茵芋

Shrubs, evergreen

16.8 (2.6~24.7); 6.5 (-7.0~19.8); 26.3 (11.2~29.9); 16.8 (4.3~24.7); 142.8 (20.0~236.5); 0.8 (0.0~48.2); 889 (427~1376) ❙ 1491 (731~3495); 563 (346~1594); 157 (19~897) ❙ 68.7 (-5.0~246.5); 879 (427~1212); 227.3 (123.9~572.9)

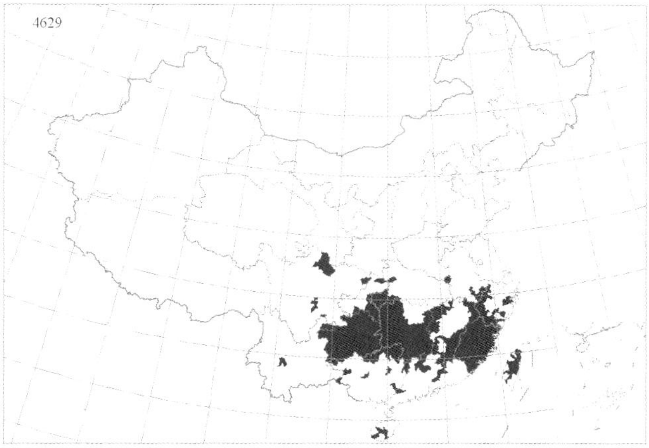

Rutaceae 芸香科 *Tetradium* 四数花属

Tetradium austrosinense (Hand.-Mazz.) Hartley 华南吴茱萸

Trees

19.0 (14.9~22.2); 10.3 (3.6~14.9); 26.1 (21.4~29.8); 19.1 (15.0~22.3); 168.8 (121.2~207.5); 0.0 (0.0~1.6); 974 (797~1190) ❙ 1549 (1059~2005); 670 (505~996); 137 (41~197) ❙ 59.7 (15.4~93.2); 969 (797~1190); 301.2 (174.9~570.1)

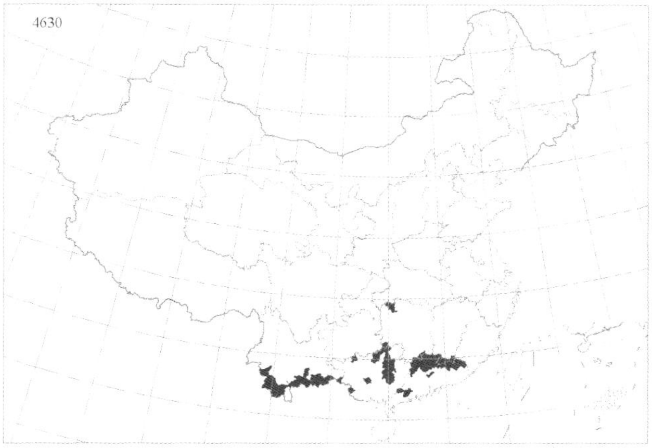

Rutaceae 芸香科 *Tetradium* 四数花属

Tetradium calcicola (Chun ex Huang) Hartley 石山吴茱萸

Trees

17.7 (16.1~20.3); 8.5 (5.3~11.0); 25.0 (22.4~27.7); 17.8 (16.1~20.4); 153.3 (133.1~184.4); 0.0 (0.0~0.0); 896 (826~1065) ❙ 1377 (1161~1567); 683 (523~860); 91 (48~142) ❙ 54.0 (37.1~80.6); 896 (826~1065) 283.5 (215.4~413.5)

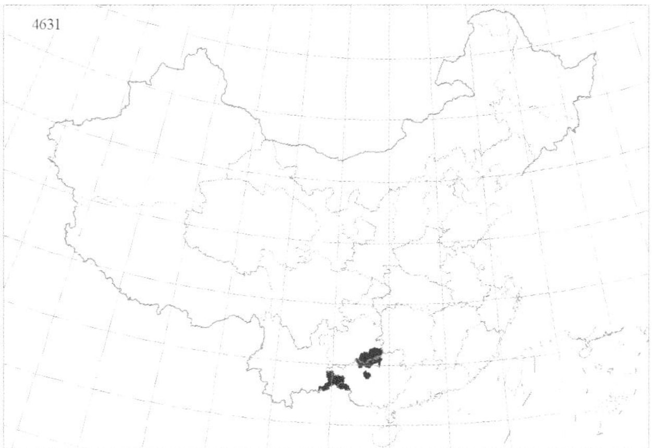

Rutaceae 芸香科 *Tetradium* 四数花属

Tetradium daniellii (Benn.) Hemsl. 臭檀吴茱萸

Trees

12.2 (0.9~17.6); -1.3 (-12.3~10.5); 24.4 (9.8~29.1); 12.5 (3.6~17.7); 102.8 (14.0~152.4); 16.2 (0.0~63.1); 749 (403~941) ❙ 806 (341~1793); 424 (190~816); 41 (7~238) ❙ 9.1 (-47.7~128.6); 677 (341~936); 197.6 (61.2~359.7)

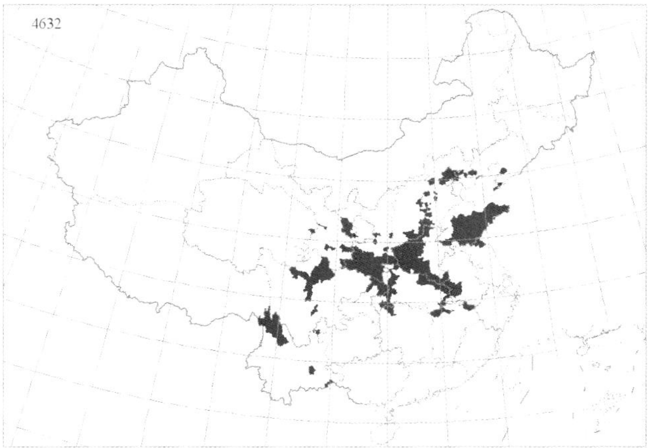

Rutaceae 芸香科 *Tetradium* 四数花属
Tetradium fraxinifolium (Hook.) Hartley 无腺吴萸
Trees

16.2 (8.6~19.6); 9.5 (1.3~13.9); 20.7 (15.0~23.3); 16.2 (8.7~19.6); 134.9 (52.6~175.1); 0.6 (0.0~8.8); 802 (574~942) ❙ 1347 (1028~1736); 747 (566~1007); 52 (41~70) ❙ 70.2 (12.5~145.8); 785 (574~921); 451.8 (339.5~627.0)

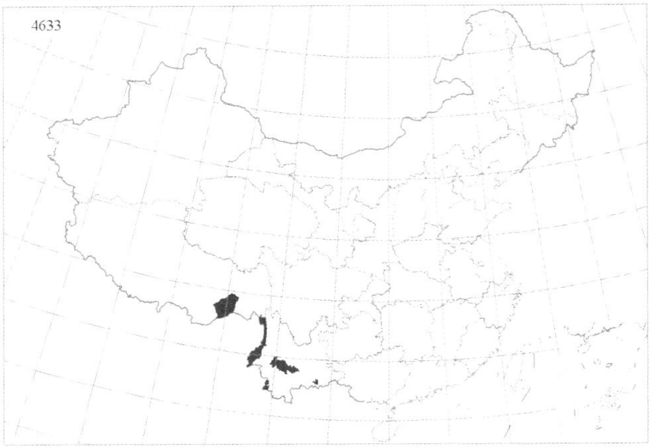

Rutaceae 芸香科 *Tetradium* 四数花属
Tetradium glabrifolium (Champ. ex Benth.) Hartley 楝叶吴萸
Trees

17.2 (8.6~24.8); 7.2 (−1.7~20.1); 26.1 (15.0~30.1); 17.2 (8.7~24.8); 147.5 (52.6~237.6); 0.6 (0.0~18.0); 908 (574~1385) ❙ 1430 (862~3495); 586 (376~1428); 132 (20~897) ❙ 58.3 (−12.0~224.4); 896 (574~1326); 242.5 (111.9~659.0)

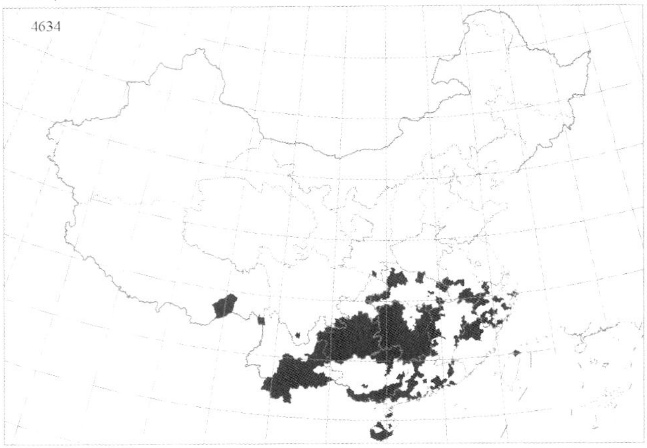

Rutaceae 芸香科 *Tetradium* 四数花属
Tetradium ruticarpum (A. Juss.) Hartley 吴茱萸
Trees

17.1 (2.6~23.6); 7.4 (−7.0~16.7); 25.8 (11.2~30.1); 17.2 (4.3~23.6); 147.2 (20.0~223.5); 1.0 (0.0~48.2); 903 (427~1281) ❙ 1398 (711~2859); 583 (299~1186); 126 (15~402) ❙ 55.4 (−14.2~246.5); 894 (427~1273); 244.3 (95.3~659.0)

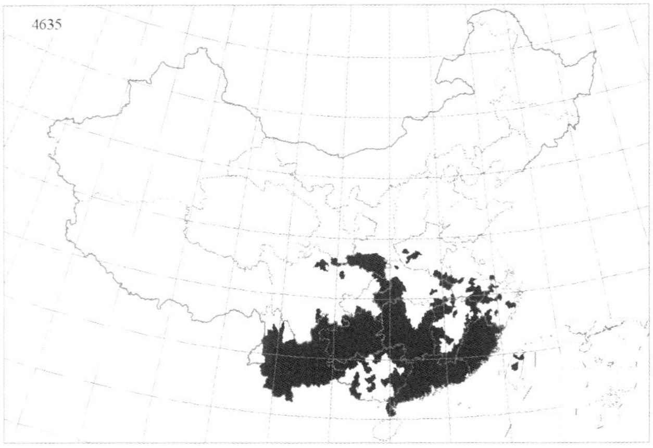

Rutaceae 芸香科 *Tetradium* 四数花属
Tetradium trichotomum Lour. 牛科吴萸
Shrubs or trees

17.6 (7.3~24.5); 9.4 (−2.0~19.1); 24.1 (15.0~28.9); 17.7 (7.5~24.5); 153.1 (48.2~234.4); 1.0 (0.0~20.4); 905 (537~1360) ❙ 1354 (842~1990); 679 (375~1056); 77 (19~189) ❙ 50.7 (5.8~112.2); 898 (537~1360); 331.0 (150.8~659.0)

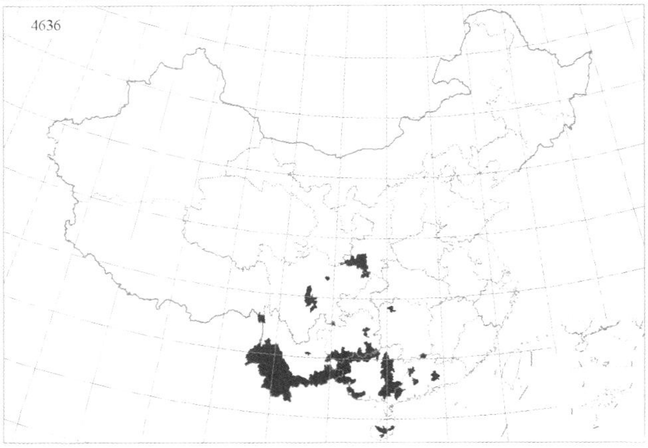

Rutaceae 芸香科 *Toddalia* 飞龙掌血属
Toddalia asiatica (L.) Lam. 飞龙掌血
Vines

17.5 (2.6~24.8); 8.5 (−7.0~20.1); 25.2 (11.2~29.8); 17.5 (4.3~24.8); 151.4 (20.0~237.6); 1.1 (0.0~48.2); 919 (427~1385) ❙ 1397 (707~2992); 618 (312~1847); 110 (15~303) ❙ 53.7 (−34.3~246.5); 909 (427~1360); 258.9 (95.3~659.0)

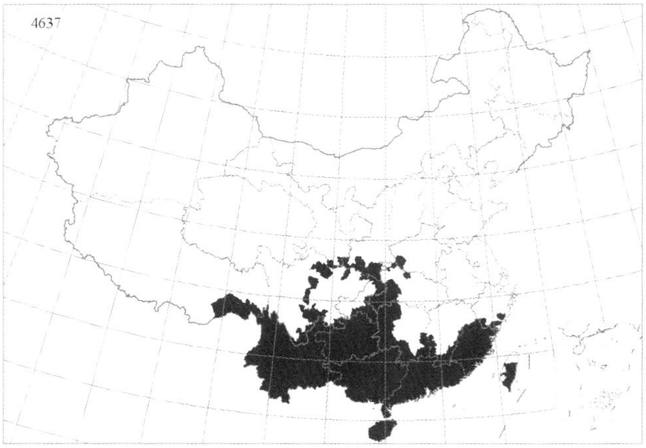

Rutaceae 芸香科 *Zanthoxylum* 花椒属
Zanthoxylum acanthopodium DC. 刺花椒
Shrubs, trees or vines, evergreen

15.6 (−1.8~21.0); 8.0 (−10.2~15.1); 21.5 (6.1~26.6); 15.7 (1.8~21.0); 129.0 (1.9~191.8); 1.5 (0.0~82.8); 793 (329~1053) ❙ 1136 (320~1751); 622 (217~996); 49 (12~129) ❙ 43.7 (−14.2~112.2); 779 (320~1053); 313.1 (72.2~659.0)

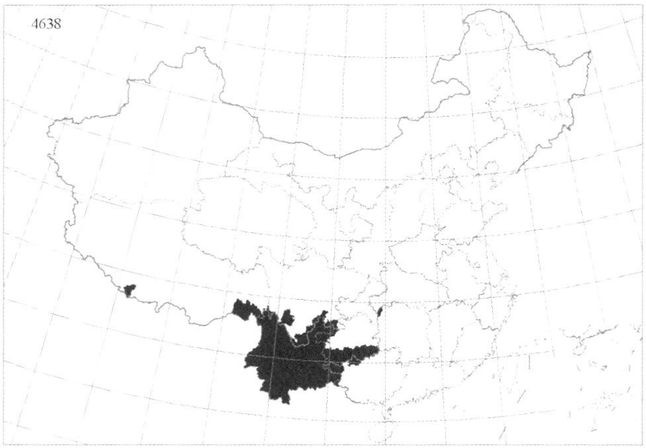

Rutaceae 芸香科 *Zanthoxylum* 花椒属
Zanthoxylum ailanthoides Siebold et Zucc. 椿叶花椒
Trees, evergreen

18.3 (14.3~22.4); 8.6 (2.3~17.7); 27.0 (20.1~29.8); 18.3 (14.3~22.4); 159.7 (114.1~208.9); 0.1 (0.0~4.3); 951 (765~1191) | 1641 (1070~3495); 601 (356~1847); 192 (43~897) | 74.3 (-5.0~246.5); 944 (765~1191); 250.8 (123.9~551.6)

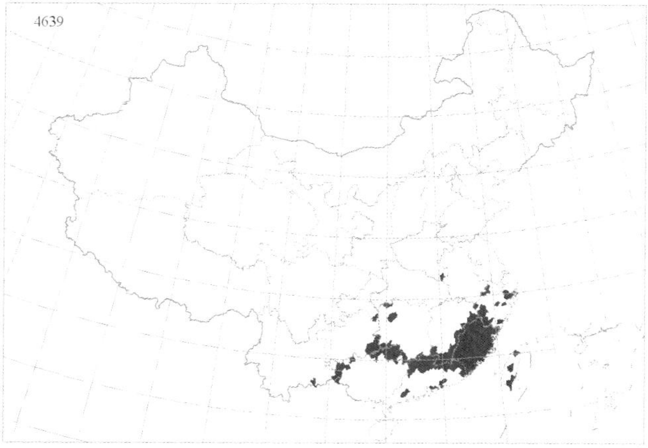

Rutaceae 芸香科 *Zanthoxylum* 花椒属
Zanthoxylum armatum DC. 竹叶花椒
Trees, evergreen

15.6 (-1.8~23.4); 4.6 (-10.2~16.1); 25.6 (6.1~30.1); 15.7 (1.8~23.5); 132.0 (1.9~221.7); 4.4 (0.0~82.8); 851 (329~1273) | 1191 (320~3495); 523 (193~1007); 103 (12~897) | 39.2 (-45.2~224.4); 823 (320~1273); 223.8 (72.2~659.0)

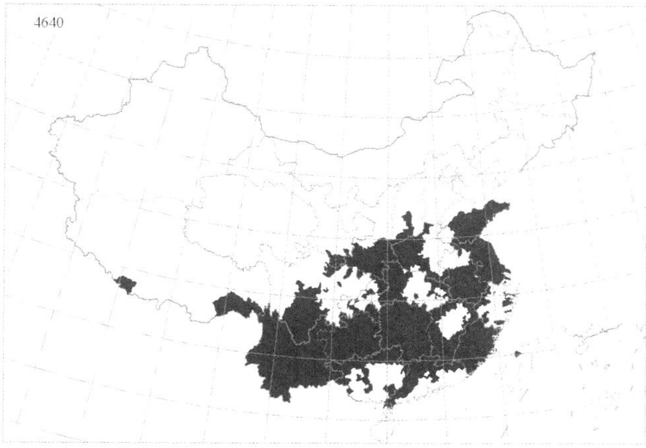

Rutaceae 芸香科 *Zanthoxylum* 花椒属
Zanthoxylum armatum var. ferrugineum (Rehder et E. H. Wilson) C. C. Huang 毛竹叶花椒
Trees, evergreen

17.1 (7.4~21.0); 6.7 (-4.2~12.2); 26.4 (18.7~29.8); 17.2 (8.2~21.0); 148.2 (58.8~192.6); 2.1 (0.0~29.4); 911 (561~1113) | 1439 (711~1852); 564 (332~775); 140 (15~204) | 57.4 (5.0~99.4); 909 (561~1107); 244.5 (169.7~354.5)

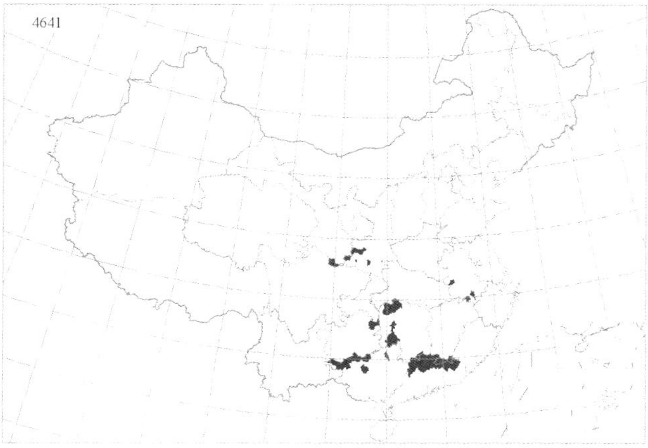

Rutaceae 芸香科 *Zanthoxylum* 花椒属
Zanthoxylum austrosinense C. C. Huang 岭南花椒
Shrubs or trees, evergreen

16.9 (12.1~19.4); 6.0 (1.1~9.6); 27.1 (22.9~30.1); 17.0 (12.2~19.5); 144.1 (95.0~173.7); 0.7 (0.0~8.5); 898 (696~1038) | 1615 (1202~1997); 554 (403~718); 195 (83~259) | 80.7 (41.7~147.0); 889 (696~1038); 231.6 (148.9~305.7)

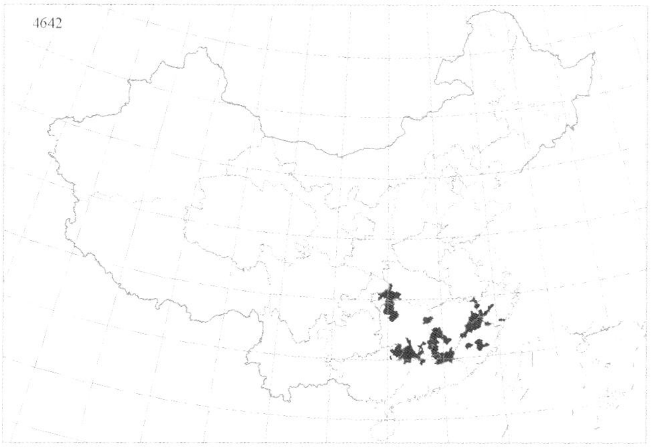

Rutaceae 芸香科 *Zanthoxylum* 花椒属
Zanthoxylum austrosinense var. pubescens C. C. Huang 毛叶岭南花椒
Shrubs or trees, evergreen

15.4; 3.9; 26.2; 15.4; 125.8; 1.2; 828 | 1450; 592; 112 | 75.0; 828; 208.1

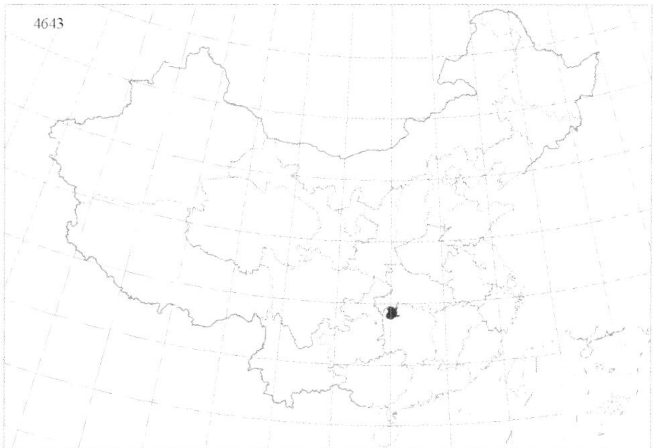

Rutaceae 芸香科 *Zanthoxylum* 花椒属
Zanthoxylum avicennae (Lam.) DC. 簕欓花椒
Trees, deciduous

20.9 (16.2~24.8); 13.1 (7.1~20.1); 27.3 (20.1~29.8); 21.0 (16.2~24.8); 191.8 (134.2~237.6); 0.0 (0.0~0.0); 1095 (773~1385) | 1710 (1143~3495); 739 (392~1847); 148 (41~897) | 59.3 (-11.9~246.5); 1080 (773~1360); 264.0 (95.3~659.0)

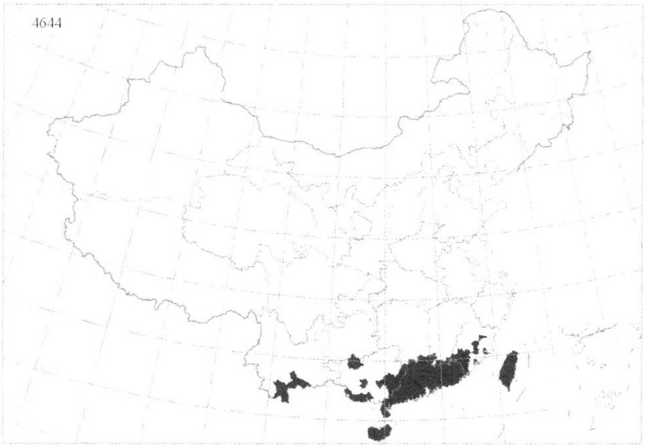

Rutaceae 芸香科 *Zanthoxylum* 花椒属

Zanthoxylum bungeanum Maxim. 花椒

Trees, deciduous

13.4 (−1.8~22.2); 1.2 (−13.7~15.1); 24.3 (6.1~29.3); 13.8 (1.8~22.3); 113.8 (1.9~207.5); 12.4 (0.0~82.8); 783 (329~1190) ❙ 896 (70~1997); 465 (23~1007); 49 (5~254) ❙ 14.9 (−91.1~145.8); 695 (70~1190); 207.3 (44.9~659.0)

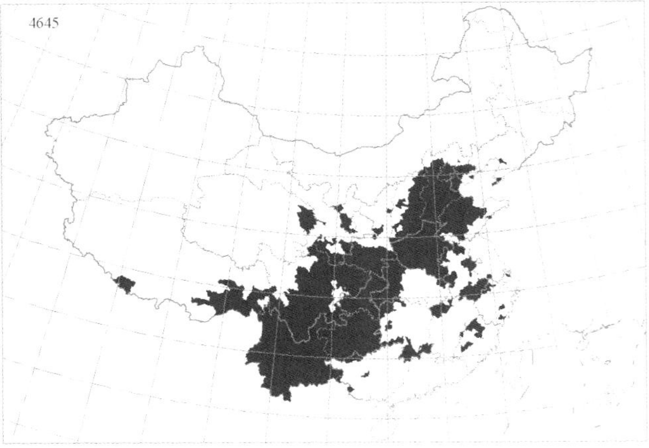

Rutaceae 芸香科 *Zanthoxylum* 花椒属

Zanthoxylum bungeanum var. pubescens C. C. Huang 毛叶花椒

Trees, deciduous

11.0 (2.0~15.5); −1.2 (−7.7~2.4); 22.4 (9.9~27.6); 11.4 (3.8~15.6); 90.3 (15.4~130.8); 17.4 (4.0~51.4); 698 (413~879) ❙ 800 (626~1098); 390 (316~584); 35 (14~71) ❙ 21.5 (−22.2~128.6); 667 (413~837); 215.5 (134.4~357.7)

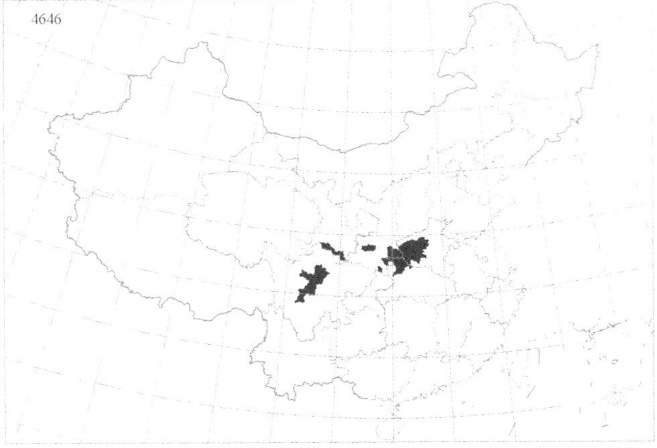

Rutaceae 芸香科 *Zanthoxylum* 花椒属

Zanthoxylum bungeanum var. punctatum C. C. Huang 油叶花椒

Trees, deciduous

11.6 (7.3~14.9); 2.2 (−2.0~5.3); 20.3 (15.6~25.3); 11.7 (7.5~15.0); 87.4 (48.2~119.6); 7.1 (0.0~20.4); 677 (537~795) ❙ 1175 (1005~1410); 609 (531~816); 44 (22~88) ❙ 73.8 (64.7~87.3); 677 (537~795); 247.9 (187.9~318.7)

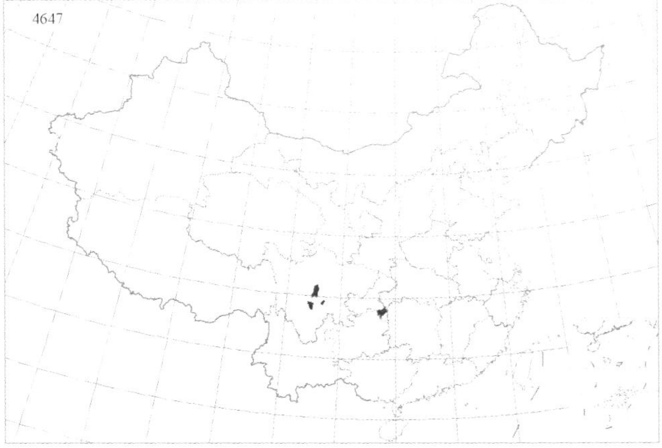

Rutaceae 芸香科 *Zanthoxylum* 花椒属

Zanthoxylum calcicola C. C. Huang 石山花椒

Vines or shrubs, evergreen

18.1 (11.9~22.2); 9.9 (2.9~14.2); 24.5 (20.0~28.4); 18.2 (12.0~22.3); 158.2 (87.2~207.5); 0.2 (0.0~3.4); 915 (671~1190) ❙ 1271 (1001~1522); 675 (473~860); 63 (35~107) ❙ 40.3 (12.9~80.6); 914 (671~1190); 325.0 (170.4~551.6)

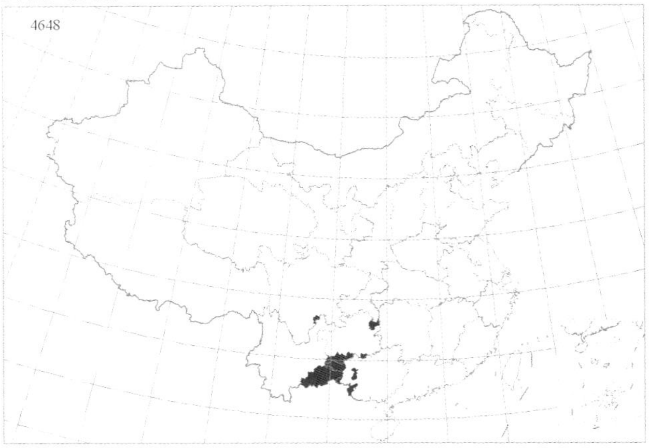

Rutaceae 芸香科 *Zanthoxylum* 花椒属

Zanthoxylum collinsiae Craib 糙叶花椒

Vines, evergreen

18.6 (16.8~20.5); 10.2 (7.8~14.9); 25.0 (23.4~26.4); 18.6 (16.8~20.5); 163.5 (141.5~185.9); 0.0 (0.0~0.0); 931 (836~1010) ❙ 1283 (1182~1464); 680 (655~720); 57 (43~66) ❙ 38.3 (16.9~59.6); 927 (836~1010); 328.3 (226.6~570.1)

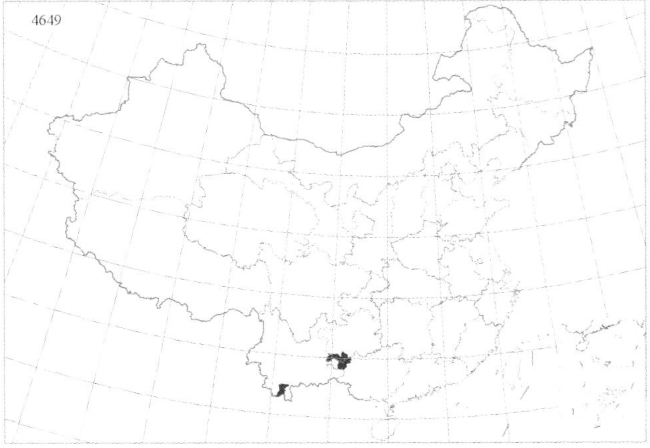

Rutaceae 芸香科 *Zanthoxylum* 花椒属

Zanthoxylum dimorphophyllum Hemsl. 异叶花椒

Trees, deciduous

17.0 (0.3~23.6); 7.5 (−9.9~17.8); 25.5 (9.3~29.8); 17.1 (3.3~23.6); 146.7 (11.9~223.5); 1.6 (0.0~67.5); 903 (391~1281) ❙ 1343 (632~2992); 594 (345~1847); 100 (14~259) ❙ 49.4 (−6.9~166.5); 899 (391~1273); 230.0 (95.3~551.6)

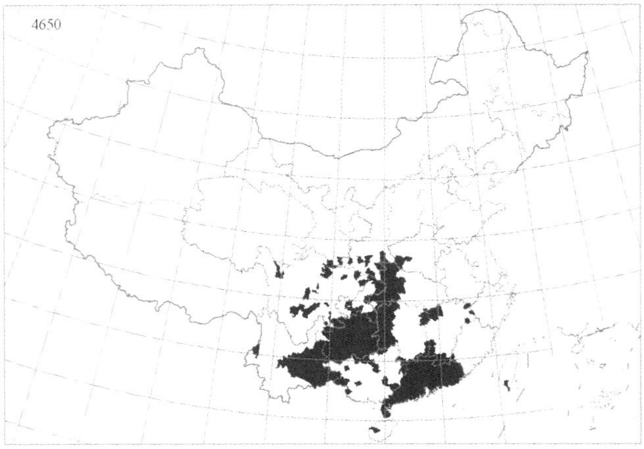

Rutaceae 芸香科 *Zanthoxylum* 花椒属

Zanthoxylum dimorphophyllum var. spinifolium Rehder et E. H. Wilson 刺异叶花椒 Trees, evergreen

15.1 (10.3~18.2); 4.7 (-0.1~9.0); 24.5 (19.6~28.0); 15.1 (10.3~18.2); 122.5 (75.8~158.8); 1.3 (0.0~11.7); 798 (634~918) ❙ 1208 (707~1492); 538 (349~700); 74 (15~120) ❙ 51.5 (-0.2~107.2); 798 (634~918); 205.8 (146.0~285.1)

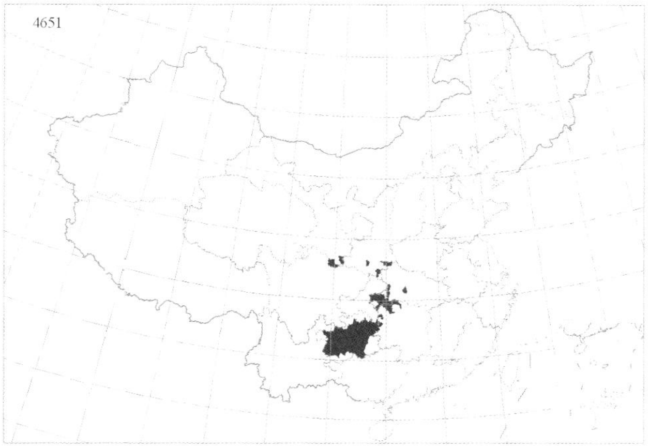

Rutaceae 芸香科 *Zanthoxylum* 花椒属

Zanthoxylum dissitum Hemsl. 蚬壳花椒

Vines, evergreen

15.6 (2.6~24.8); 5.5 (-7.0~20.1); 24.8 (11.2~29.4); 15.7 (4.3~24.8); 130.2 (20.0~237.6); 2.1 (0.0~48.2); 833 (427~1385) ❙ 1243 (711~1776); 544 (349~996); 89 (15~225) ❙ 50.4 (-11.9~128.9); 826 (427~1273); 235.2 (108.2~659.0)

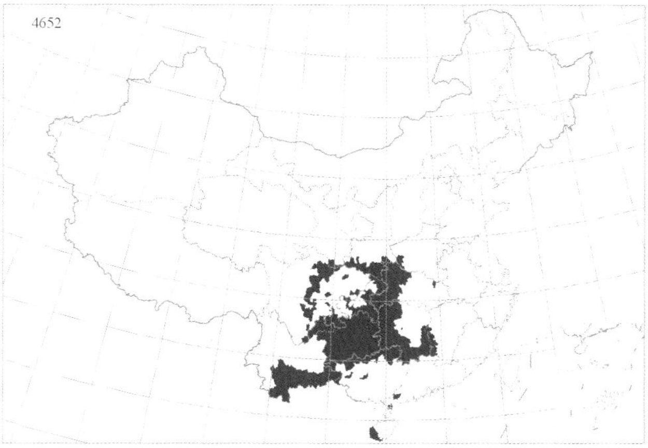

Rutaceae 芸香科 *Zanthoxylum* 花椒属

Zanthoxylum dissitum var. acutiserratum C. C. Huang 针边蚬壳花椒 Vines, evergreen

12.3; 3.3; 20.5; 12.4; 92.1; 3.5; 691 ❙ 1251; 711; 38 ❙ 81.1; 691; 272.6

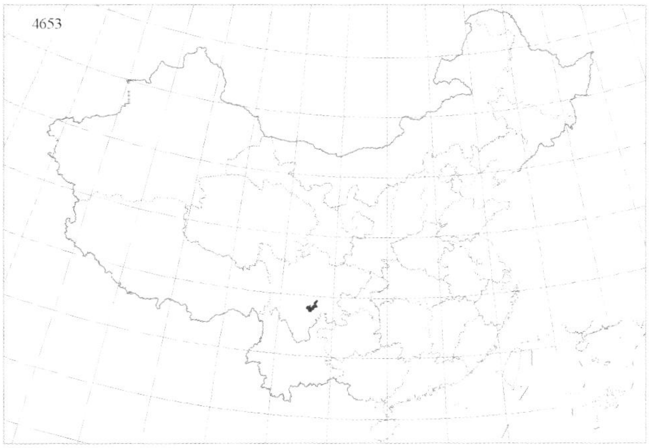

Rutaceae 芸香科 *Zanthoxylum* 花椒属

Zanthoxylum dissitum var. hispidum (Reeder et S. Y. Cheo) C. C. Huang 刺蚬壳花椒 Vines, evergreen

11.4 (7.3~14.9); 2.1 (-2.0~5.3); 19.7 (15.6~23.6); 11.5 (7.5~15.0); 83.7 (48.2~119.6); 6.5 (0.0~20.4); 662 (537~782) ❙ 1125 (1001~1410); 624 (531~816); 34 (22~49) ❙ 70.4 (49.1~87.3); 662 (537~782); 250.7 (171.0~318.7)

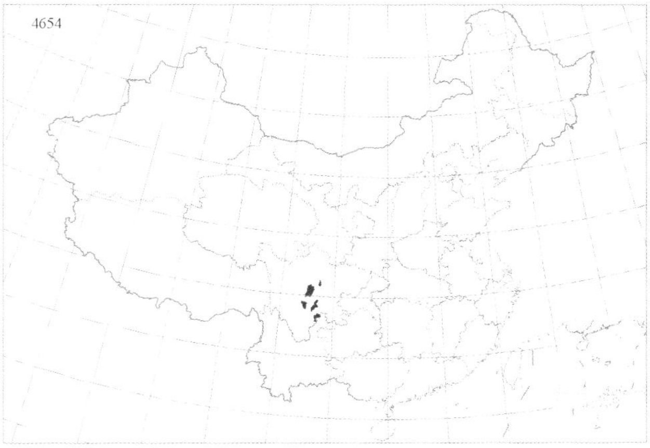

Rutaceae 芸香科 *Zanthoxylum* 花椒属

Zanthoxylum dissitum var. lanciforme C. C. Huang 长叶蚬壳花椒

Vines, evergreen

13.3; 3.6; 22.2; 13.3; 101.9; 1.7; 718 ❙ 999; 501; 53 ❙ 39.0; 718; 183.0

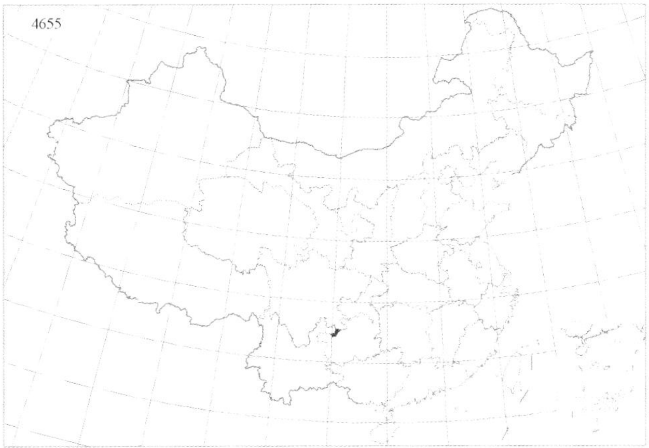

Rutaceae 芸香科 *Zanthoxylum* 花椒属

Zanthoxylum echinocarpum Hemsl. 刺壳花椒

Vines, evergreen

16.9 (2.9~22.0); 6.6 (-6.3~13.4); 26.2 (11.2~29.8); 16.9 (4.4~22.0); 144.1 (20.5~204.3); 1.1 (0.0~44.6); 892 (428~1176) ❙ 1342 (936~1852); 566 (419~860); 104 (23~221) ❙ 51.8 (8.0~121.7); 890 (428~1176); 213.6 (108.2~413.5)

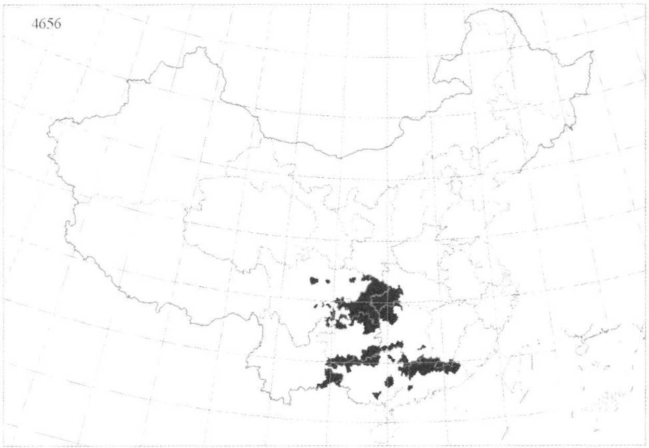

Rutaceae 芸香科 *Zanthoxylum* 花椒属

Zanthoxylum echinocarpum var. tomentosum C. C. Huang 毛刺壳花椒

Vines, evergreen

18.2 (16.8~19.1); 9.5 (7.8~10.3); 25.0 (21.4~26.6); 18.3 (16.8~19.2); 159.0 (141.5~170.1); 0.0 (0.0~0.0); 916 (813~981) ❙ 1262 (1178~1334); 675 (617~723); 59 (51~73) ❙ 38.3 (27.5~59.6); 916 (813~981); 292.4 (226.6~383.4)

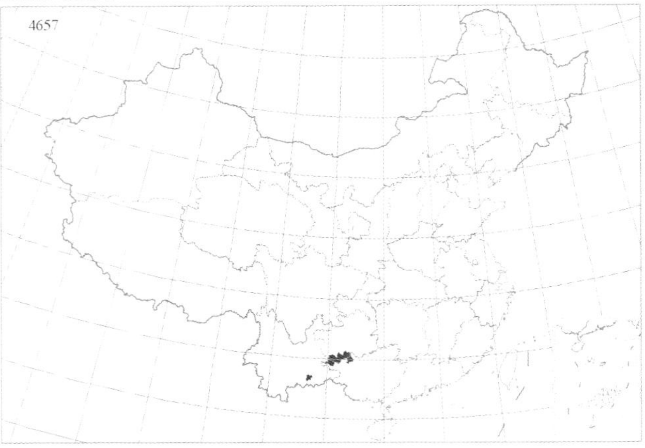

Rutaceae 芸香科 *Zanthoxylum* 花椒属

Zanthoxylum esquirolii H. Lév. 贵州花椒

Shrubs or trees, evergreen

14.7 (2.2~21.0); 5.9 (-6.6~14.2); 22.4 (9.9~28.8); 14.8 (3.8~21.0); 119.3 (15.5~191.8); 1.9 (0.0~49.1); 776 (413~1053) ❙ 1106 (731~1522); 563 (346~860); 54 (12~124) ❙ 44.2 (-14.2~125.9); 766 (413~1053); 243.9 (112.1~551.6)

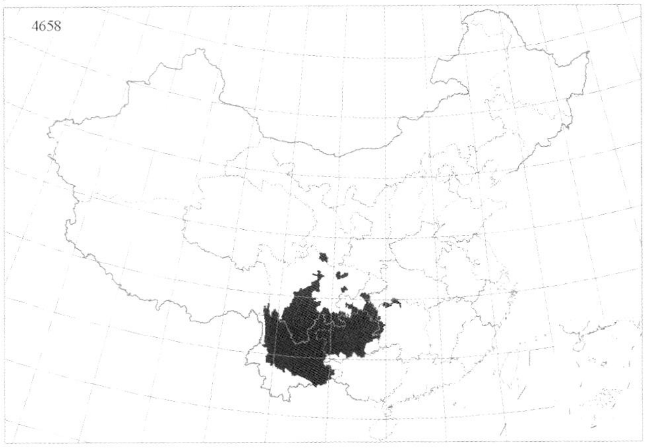

Rutaceae 芸香科 *Zanthoxylum* 花椒属

Zanthoxylum glomeratum C. C. Huang 密果花椒

Shrubs, evergreen

17.9 (16.1~19.1); 8.3 (5.3~11.0); 25.7 (23.6~26.9); 17.9 (16.1~19.1); 155.0 (133.1~169.6); 0.0 (0.0~0.0); 911 (836~969) ❙ 1325 (1236~1496); 647 (523~820); 86 (51~142) ❙ 45.9 (27.5~64.7); 911 (836~969); 255.7 (189.5~371.7)

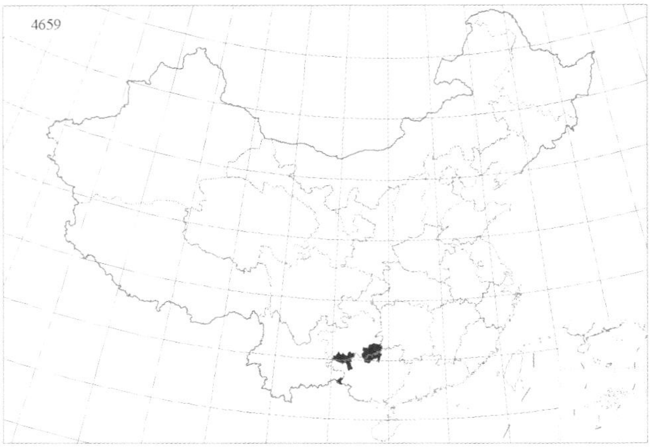

Rutaceae 芸香科 *Zanthoxylum* 花椒属

Zanthoxylum integrifolium (Merr.) Merr. 兰屿花椒

Trees, evergreen

18.5; 14.1; 21.7; 18.5; 162.1; 0.0; 863 ❙ 2713; 1388; 188 ❙ 214.4; 863; 363.7

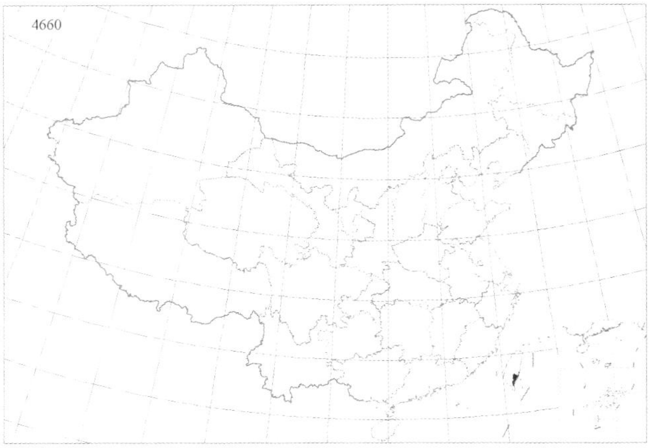

Rutaceae 芸香科 *Zanthoxylum* 花椒属

Zanthoxylum khasianum Hook. f. 云南花椒

Trees, evergreen

17.3 (14.4~19.6); 10.5 (7.5~12.8); 21.5 (19.1~23.5); 17.3 (14.4~19.6); 147.7 (113.1~175.7); 0.0 (0.0~0.0); 839 (727~953) ❙ 1482 (1049~1688); 854 (598~959); 54 (42~70) ❙ 77.6 (26.1~112.2); 829 (727~932); 491.2 (365.3~627.0)

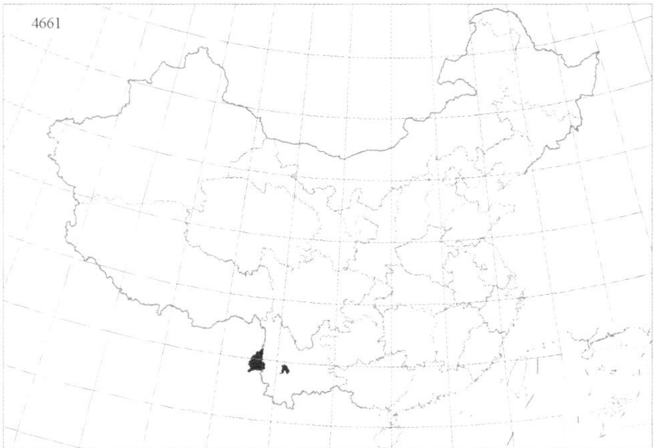

Rutaceae 芸香科 *Zanthoxylum* 花椒属

Zanthoxylum kwangsiense (Hand.-Mazz.) Chun ex C. C. Huang 广西花椒

Vines, evergreen

17.8 (16.0~19.2); 7.6 (4.7~9.5); 26.4 (25.5~27.5); 17.8 (16.0~19.2); 154.2 (132.4~170.8); 0.1 (0.0~0.3); 923 (847~1002) ❙ 1368 (1234~1496); 604 (452~678); 113 (60~142) ❙ 48.6 (27.5~64.7); 923 (847~1002); 237.6 (206.2~256.4)

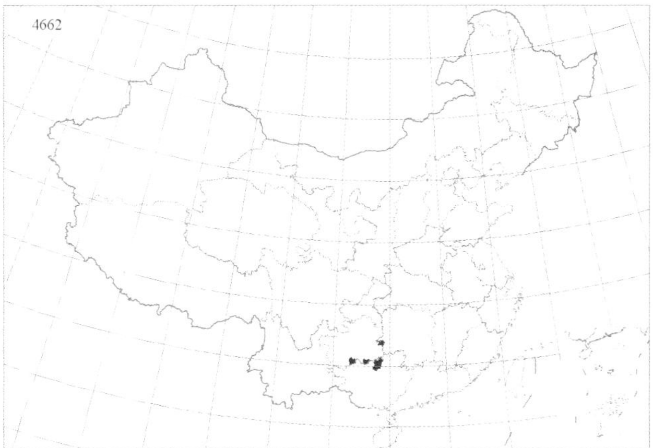

Rutaceae 芸香科 *Zanthoxylum* 花椒属
Zanthoxylum laetum Drake 拟蚬壳花椒
Vines, evergreen

20.3 (14.9~24.8); 12.3 (3.6~20.1); 26.3 (22.3~29.2); 20.3 (15.0~24.8); 184.2 (121.2~237.6); 0.0 (0.0~1.6); 1050 (807~1385) ❙ 1452 (1161~2435) 740 (455~1428); 80 (41~194) ❙ 39.5 (1.5~99.5); 1035 (807~1273); 336.9 (178.7~659.0)

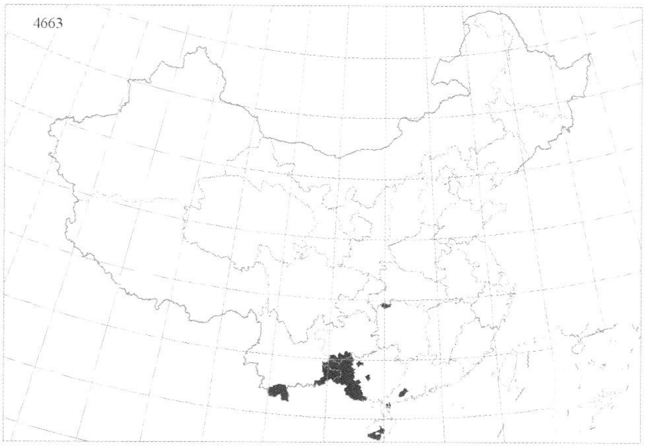

Rutaceae 芸香科 *Zanthoxylum* 花椒属
Zanthoxylum leiboicum C. C. Huang 雷波花椒
Shrubs, evergreen

12.9; 3.8; 20.9; 12.9; 96.6; 2.0; 705 ❙ 1022; 566; 37 ❙ 45.5; 705; 248.7

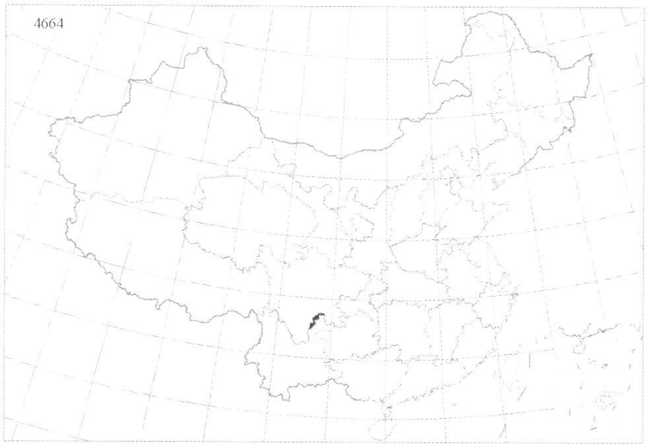

Rutaceae 芸香科 *Zanthoxylum* 花椒属
Zanthoxylum liboense C. C. Huang 荔波花椒
Vines or shrubs, evergreen

17.4; 7.4; 25.5; 17.4; 148.9; 0.0; 886 ❙ 1381; 637; 107 ❙ 55.9; 886; 245.3

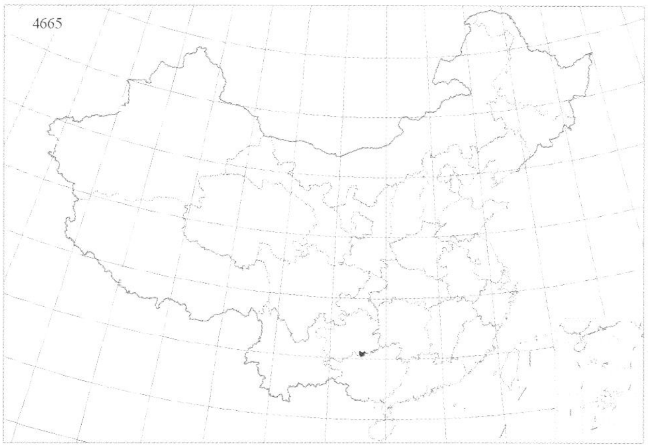

Rutaceae 芸香科 *Zanthoxylum* 花椒属
Zanthoxylum macranthum (Hand.-Mazz.) C. C. Huang 大花花椒
Vines, evergreen

15.8 (5.4~21.0); 8.2 (-2.1~15.1); 21.8 (12.3~29.3); 15.8 (5.8~21.0); 131.1 (31.5~191.8); 1.1 (0.0~25.6); 802 (481~1053) ❙ 1162 (741~1751); 607 (378~1007); 60 (23~216) ❙ 45.5 (-14.2~145.8); 784 (481~1053); 315.2 (150.8~659.0)

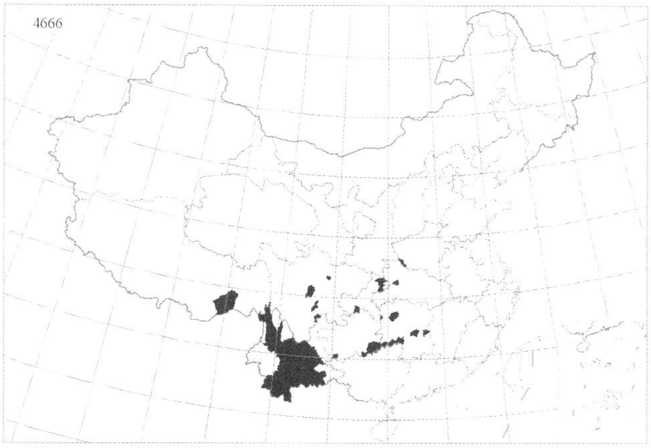

Rutaceae 芸香科 *Zanthoxylum* 花椒属
Zanthoxylum micranthum Hemsl. 小花花椒
Trees, deciduous

15.1 (10.6~19.1); 4.0 (-0.8~10.1); 25.4 (21.1~27.6); 15.1 (10.6~19.1); 124.6 (79.4~169.2); 3.0 (0.0~13.4); 815 (643~969) ❙ 1192 (793~1571); 520 (361~720); 80 (23~192) ❙ 47.0 (-4.0~108.0); 814 (643~969); 207.8 (130.1~284.3)

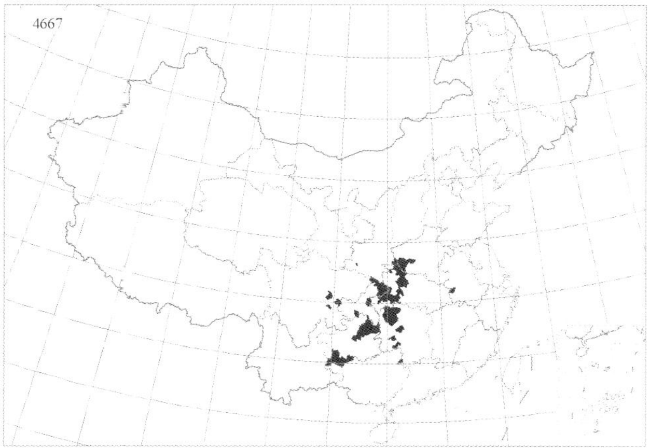

Rutaceae 芸香科 *Zanthoxylum* 花椒属
Zanthoxylum molle Rehder 朵花椒
Trees, deciduous

15.9 (12.3~18.2); 4.0 (-0.8~7.4); 27.3 (23.5~29.8); 15.9 (12.3~18.2); 133.0 (100.7~158.5); 1.7 (0.0~13.4); 866 (728~992) ❙ 1427 (793~1950); 538 (299~713); 161 (42~253) ❙ 65.2 (-6.7~147.0); 856 (728~964); 225.4 (111.9~319.9)

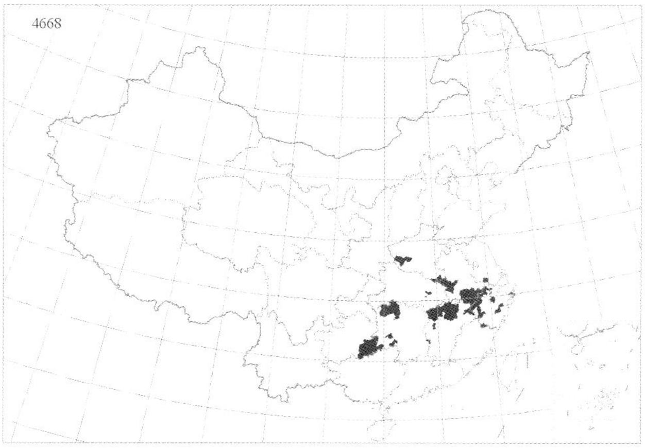

Rutaceae 芸香科 *Zanthoxylum* 花椒属

Zanthoxylum motuoense C. C. Huang 墨脱花椒

Trees, deciduous

9.3；1.5；15.9；9.6；65.5；13.1；592 ❙ 1348；783；41 ❙ 123.3；592；340.8

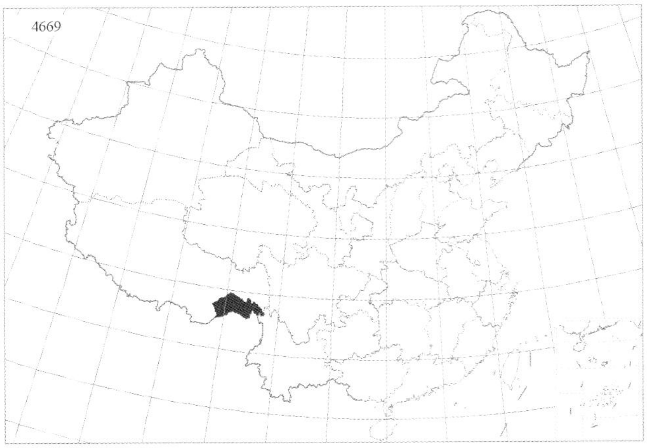

Rutaceae 芸香科 *Zanthoxylum* 花椒属

Zanthoxylum multijugum Franch. 多叶花椒

Vines, evergreen

16.0 (5.4~21.7)；9.2 (−2.1~14.2)；21.0 (12.3~28.4)；16.1 (5.8~21.8)；133.6 (31.5~201.3)；0.7 (0.0~25.6)；801 (481~1162) ❙ 1086 (741~1547)；619 (411~911)；44 (15~102) ❙ 36.4 (−14.2~80.6)；777 (481~1162)；331.3 (218.0~551.6)

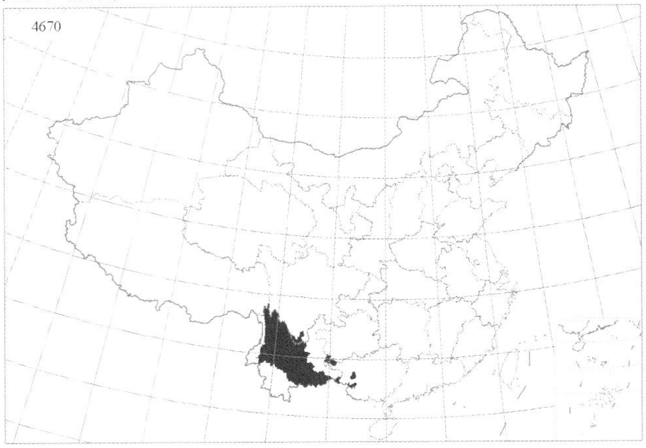

Rutaceae 芸香科 *Zanthoxylum* 花椒属

Zanthoxylum myriacanthum (C. C. Huang) C. C. Huang 大叶臭花椒

Trees, deciduous

19.6 (14.3~24.8)；11.0 (3.8~20.1)；27.1 (20.1~29.8)；19.7 (14.3~24.8)；176.3 (112.6~237.6)；0.0 (0.0~1.5)；1024 (756~1385) ❙ 1664 (868~3495)；676 (312~1847)；163 (41~897) ❙ 65.8 (−34.3~246.5)；1011 (756~1326)；253.8 (95.3~659.0)

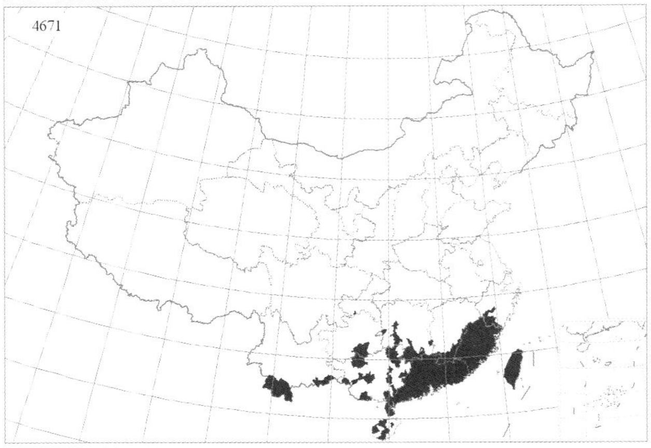

Rutaceae 芸香科 *Zanthoxylum* 花椒属

Zanthoxylum myriacanthum var. pubescens (C. C. Huang) C. C. Huang 毛大叶臭花椒

Trees, evergreen

20.0；14.6；23.2；20.1；180.8；0.0；963 ❙ 1523；681；67 ❙ 58.4；942；581.0

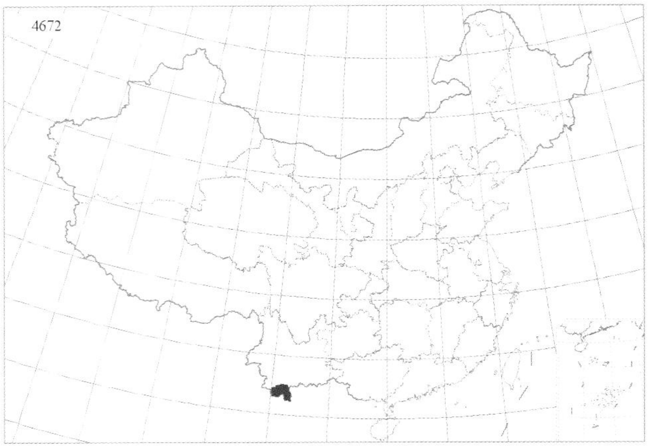

Rutaceae 芸香科 *Zanthoxylum* 花椒属

Zanthoxylum nitidum (Roxb.) DC. 两面针

Shrubs, evergreen

20.4 (8.6~24.8)；12.3 (1.3~20.1)；27.1 (15.0~29.8)；20.5 (8.7~24.8)；185.5 (52.6~237.6)；0.1 (0.0~8.8)；1069 (574~1385) ❙ 1647 (868~3495)；717 (312~1847)；141 (40~897) ❙ 57.4 (−34.3~246.5)；1054 (574~1360)；268.2 (95.3~659.0)

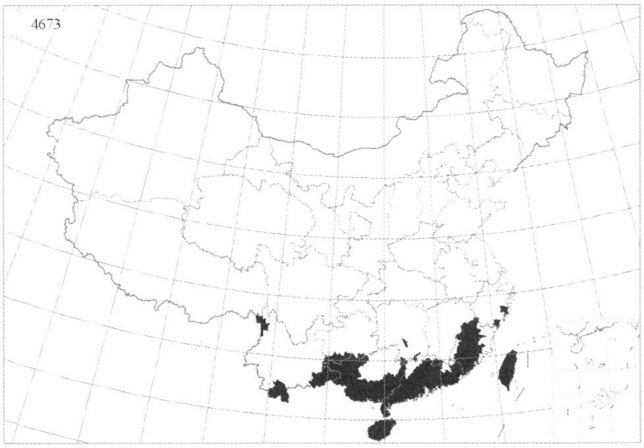

Rutaceae 芸香科 *Zanthoxylum* 花椒属

Zanthoxylum oxyphyllum Edgew. 尖叶花椒

Shrubs or small trees, evergreen

14.6 (−1.8~18.9)；7.6 (−10.2~12.5)；19.7 (6.1~25.5)；14.7 (1.8~19.0)；119.3 (1.9~167.5)；4.2 (0.0~82.8)；748 (329~914) ❙ 1080 (320~1736)；609 (217~1007)；45 (9~107) ❙ 45.7 (−3.0~145.8)；728 (320~886)；329.0 (72.2~516.7)

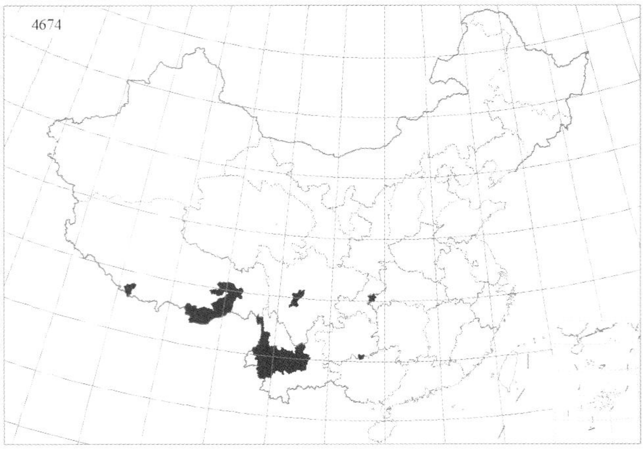

Rutaceae 芸香科 *Zanthoxylum* 花椒属
Zanthoxylum piasezkii Maxim. 川陕花椒
Shrubs or trees, evergreen

8.4 (2.0~14.6); -2.4 (-7.7~7.1); 18.3 (9.9~25.6); 9.0 (3.8~14.6); 65.3 (15.4~114.8); 24.3 (0.0~51.4); 593 (413~773) ❙ 815 (622~1164); 409 (298~627); 25 (14~51) ❙ 44.4 (-13.4~128.6); 588 (413~773); 267.7 (189.3~357.7)

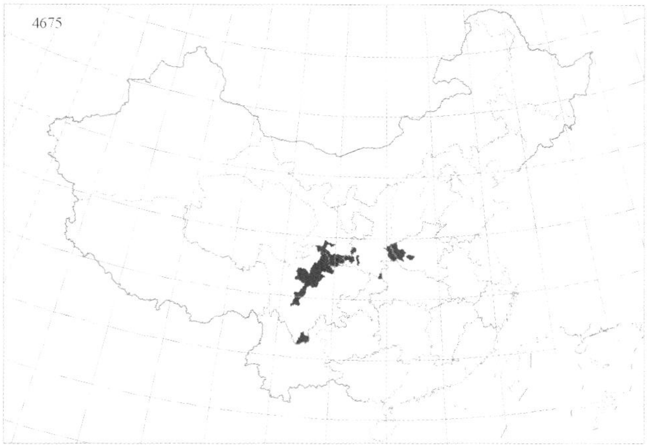

Rutaceae 芸香科 *Zanthoxylum* 花椒属
Zanthoxylum pteracanthum Rehder et E. H. Wilson 翼刺花椒
Shrubs or trees, evergreen

12.3; 0.6; 23.5; 12.3; 97.7; 9.6; 710 ❙ 1077; 472; 71 ❙ 51.9; 710; 246.5

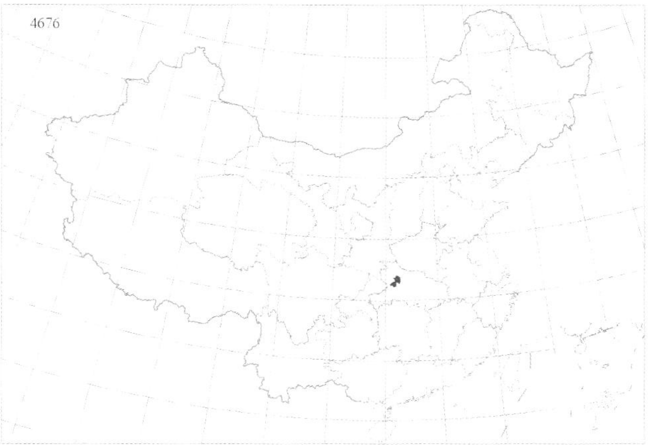

Rutaceae 芸香科 *Zanthoxylum* 花椒属
Zanthoxylum rhombifoliolatum C. C. Huang 菱叶花椒
Shrubs, evergreen

15.3 (14.2~17.0); 4.6 (3.6~6.1); 25.6 (24.4~27.4); 15.4 (14.2~17.1); 125.1 (112.5~144.9); 0.6 (0.0~1.7); 819 (765~909) ❙ 1224 (1133~1299); 497 (478~515); 73 (64~92) ❙ 49.7 (31.7~62.4); 819 (765~909); 178.7 (150.2~191.6)

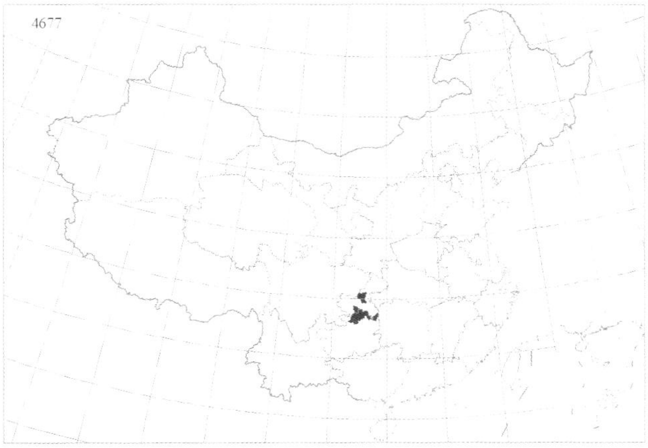

Rutaceae 芸香科 *Zanthoxylum* 花椒属
Zanthoxylum scandens Blume 花椒簕
Shrubs, evergreen

17.7 (5.4~24.6); 8.3 (-2.1~19.5); 25.9 (12.3~29.8); 17.7 (5.8~24.7); 152.9 (31.5~235.9); 0.3 (0.0~25.6); 923 (481~1372) ❙ 1467 (741~3495); 603 (356~1847); 138 (23~897) ❙ 59.8 (-14.2~246.5); 914 (481~1360); 246.2 (95.3~659.0)

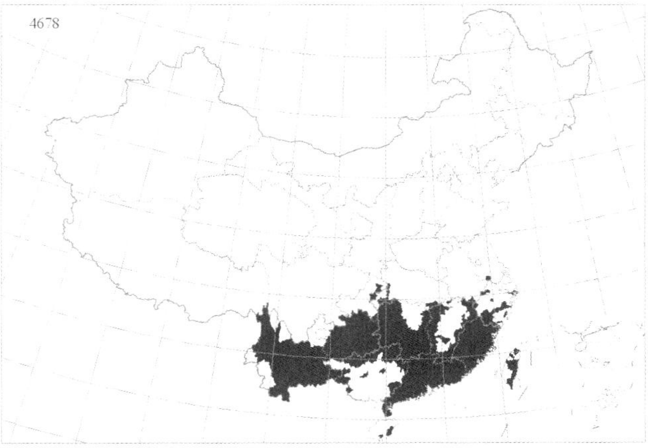

Rutaceae 芸香科 *Zanthoxylum* 花椒属
Zanthoxylum schinifolium Siebold et Zucc. 青花椒
Shrubs, evergreen

13.9 (2.6~22.2); 0.8 (-15.9~17.7); 25.8 (14.1~30.1); 14.3 (6.6~22.2); 119.7 (39.4~206.7); 12.7 (0.0~80.4); 812 (507~1139) ❙ 1034 (349~2992); 464 (227~1847); 91 (8~402) ❙ 24.2 (-43.4~235.6); 755 (349~1139); 188.3 (101.0~335.3)

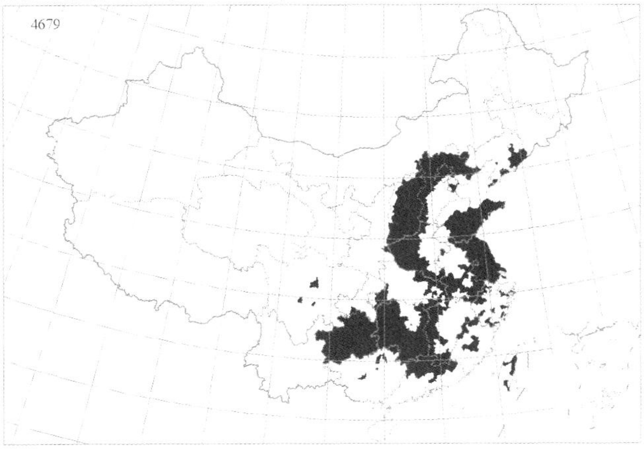

Rutaceae 芸香科 *Zanthoxylum* 花椒属
Zanthoxylum simulans Hance 野花椒
Shrubs or small trees, evergreen

15.1 (4.2~21.7); 2.8 (-12.5~15.3); 26.5 (18.7~30.1); 15.3 (7.3~21.7); 128.3 (52.8~200.2); 7.0 (0.0~62.6); 845 (535~1128) ❙ 1149 (377~2382); 483 (241~1045); 112 (8~308) ❙ 34.2 (-37.2~191.8); 809 (377~1094); 195.5 (55.7~332.5)

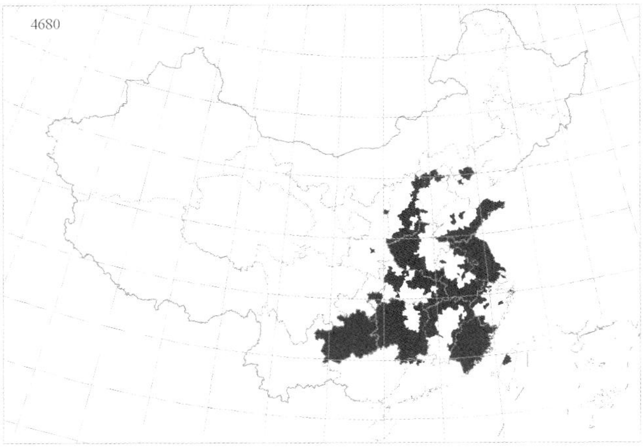

Rutaceae 芸香科 *Zanthoxylum* 花椒属

Zanthoxylum stenophyllum Hemsl. 狭叶花椒

Shrubs, evergreen

12.5 (2.0~21.0); 2.4 (-7.2~14.2); 21.9 (9.9~28.4); 12.7 (3.8~21.0); 100.3 (15.4~191.8); 9.2 (0.0~51.4); 720 (413~1053) ❙ 1084 (535~1526); 523 (317~860); 50 (10~181) ❙ 53.0 (-8.1~128.6); 718 (413~1053); 261.2 (128.7~551.6)

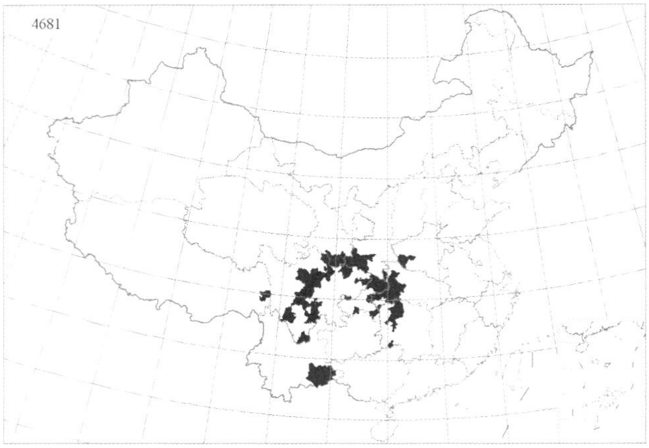

Rutaceae 芸香科 *Zanthoxylum* 花椒属

Zanthoxylum stipitatum C. C. Huang 梗花椒

Shrubs or trees, evergreen

16.9 (14.3~19.4); 5.4 (3.3~9.3); 27.6 (24.2~29.9); 16.9 (14.3~19.4); 143.4 (113.2~173.4); 0.3 (0.0~2.3); 905 (761~1029) ❙ 1503 (1215~1922); 502 (392~644); 185 (100~259) ❙ 66.7 (41.2~128.9); 890 (761~1029); 202.2 (134.6~279.5)

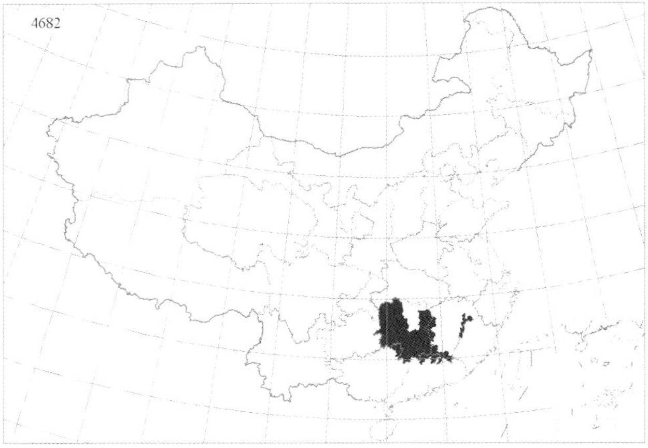

Rutaceae 芸香科 *Zanthoxylum* 花椒属

Zanthoxylum tomentellum Hook. f. 毡毛花椒

Vines, evergreen

13.1 (8.6~18.9); 6.0 (1.3~9.5); 18.7 (15.0~26.6); 13.1 (8.7~18.9); 98.6 (52.6~167.3); 1.4 (0.0~8.8); 704 (574~969) ❙ 1091 (1007~1236); 604 (553~651); 50 (35~65) ❙ 57.1 (27.5~81.5); 704 (574~969); 349.1 (254.3~384.1)

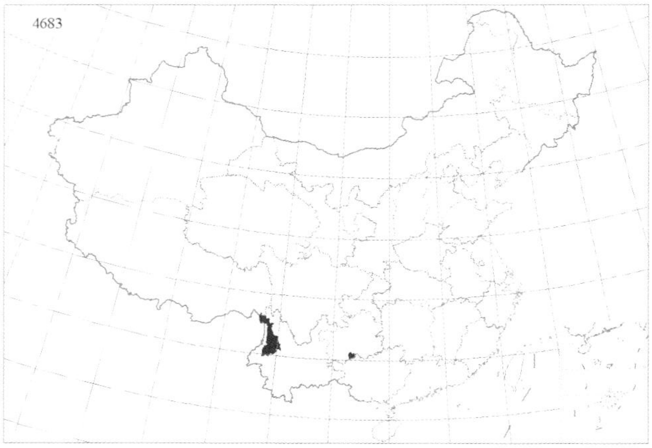

Rutaceae 芸香科 *Zanthoxylum* 花椒属

Zanthoxylum undulatifolium Hemsl. 浪叶花椒

Trees, evergreen

12.4 (7.4~15.3); 0.5 (-4.2~3.6); 23.8 (18.7~26.8); 12.5 (8.2~15.4); 99.8 (58.8~125.7); 10.1 (1.4~29.4); 725 (561~834) ❙ 987 (622~1448); 431 (298~600); 49 (20~113) ❙ 36.7 (-13.4~101.4); 722 (561~832); 254.3 (151.3~334.4)

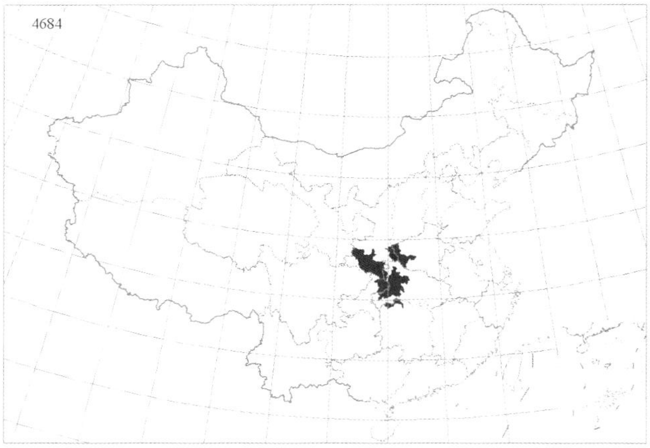

Rutaceae 芸香科 *Zanthoxylum* 花椒属

Zanthoxylum wutaiense Chen 屏东花椒

Shrubs, evergreen

22.2; 17.7; 25.5; 22.2; 206.7; 0.0; 1123 ❙ 2992; 1847; 107 ❙ 166.5; 1123; 296.9

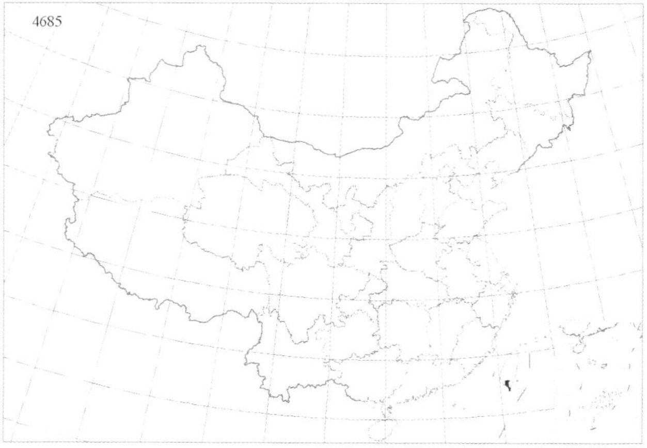

Rutaceae 芸香科 *Zanthoxylum* 花椒属

Zanthoxylum xichouense C. C. Huang 西畴花椒

Scandent shrubs, evergreen

17.2; 9.5; 22.7; 17.3; 147.0; 0.0; 838 ❙ 1254; 699; 48 ❙ 49.7; 838; 336.3

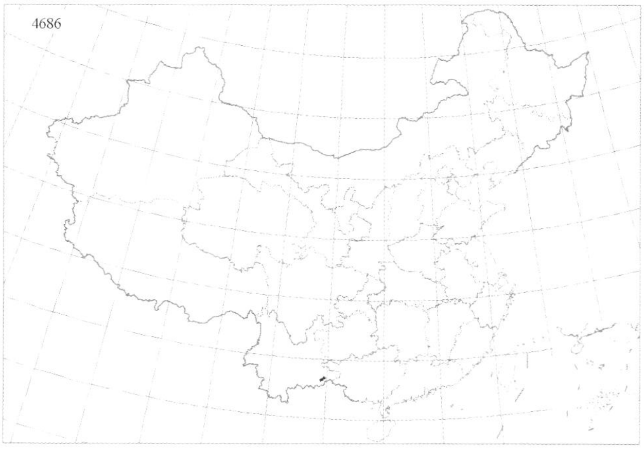

Rutaceae 芸香科 *Zanthoxylum* 花椒属
Zanthoxylum yuanjiangense C. C. Huang 元江花椒
Vines, evergreen

18.9 (18.1~19.3); 12.9 (12.0~13.5); 23.0 (22.1~23.6); 19.0 (18.1~19.4); 168.2 (157.3~173.2); 0.0 (0.0~0.0); 911 (863~939) ❙ 1413 (1143~1751); 815 (637~996); 59 (50~77) ❙ 55.4 (21.7~88.8); 904 (863~927); 467.2 (346.9~551.3)

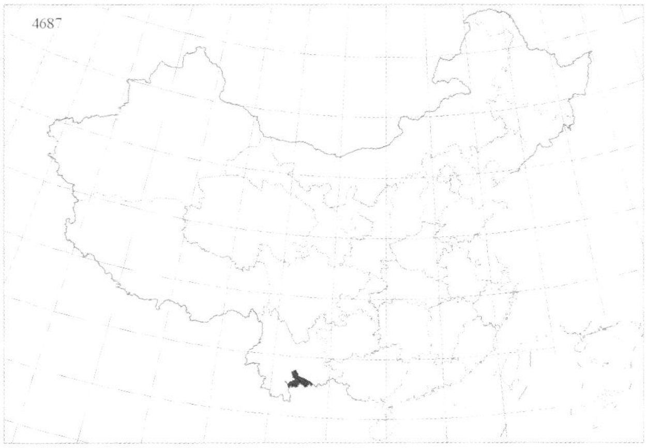

Simaroubaceae 苦木科 *Ailanthus* 臭椿属
Ailanthus altissima (Miller) Swingle 臭椿
Trees, deciduous

13.8 (0.4~22.1); 1.0 (−15.9~15.1); 25.3 (9.5~29.9); 14.2 (3.6~22.1); 118.8 (15.4~205.6); 12.8 (0.0~80.4); 806 (406~1182) ❙ 968 (203~2859); 454 (118~1007); 78 (4~402) ❙ 18.1 (−69.4~235.6); 722 (203~1182); 201.3 (43.1~659.0)

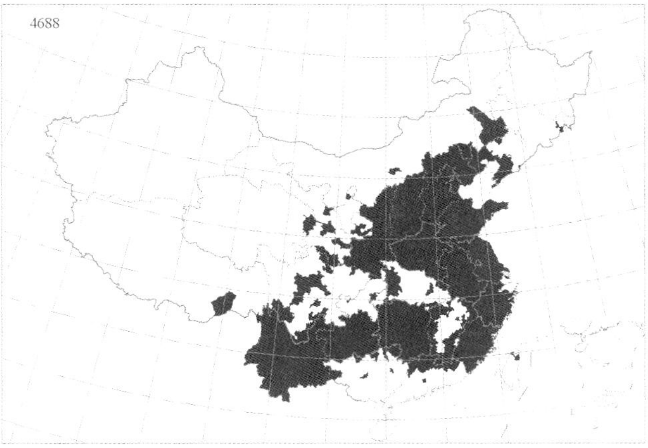

Simaroubaceae 苦木科 *Ailanthus* 臭椿属
Ailanthus altissima var. sutchuenensis (Dode) Rehder et E. H. Wilson 大果臭椿
Trees, deciduous

14.7 (5.4~18.9); 6.9 (−2.1~12.0); 21.0 (12.3~27.8); 14.7 (5.8~19.0); 118.1 (31.5~167.8); 1.6 (0.0~25.6); 758 (481~929) ❙ 1038 (741~1562); 556 (411~636); 52 (12~193) ❙ 38.0 (−14.2~106.2); 735 (481~878); 276.7 (200.4~382.6)

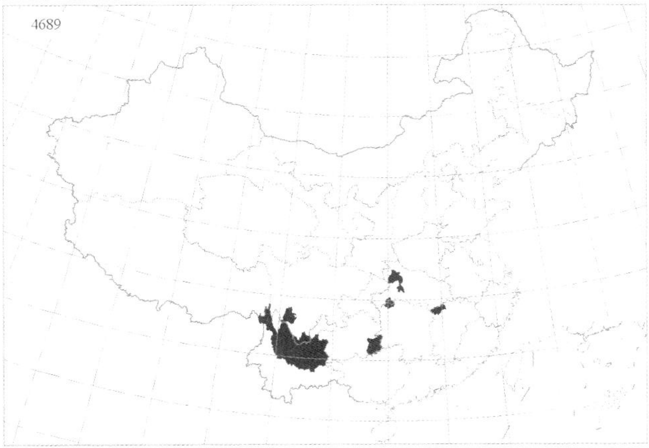

Simaroubaceae 苦木科 *Ailanthus* 臭椿属
Ailanthus fordii Noot. 常绿臭椿
Trees, evergreen

21.8 (19.2~22.8); 14.5 (12.6~15.5); 27.7 (22.3~28.9); 21.9 (19.2~22.8); 202.6 (170.5~214.1); 0.0 (0.0~0.0); 1148 (909~1225) ❙ 1720 (1258~2037); 802 (425~1033); 114 (58~185) ❙ 49.9 (14.0~79.9); 1142 (880~1225); 238.8 (116.6~659.0)

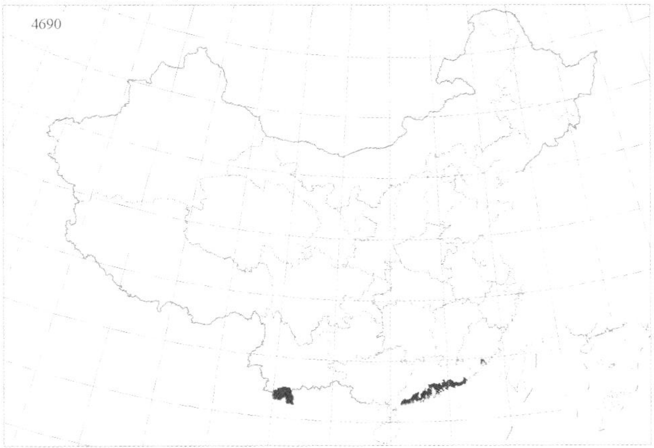

Simaroubaceae 苦木科 *Ailanthus* 臭椿属
Ailanthus giraldii Dode 毛臭椿
Trees, deciduous

14.4 (12.0~17.5); 3.0 (−0.8~6.6); 25.3 (22.8~27.8); 14.5 (12.0~17.5); 118.2 (93.2~149.9); 4.5 (0.0~13.4); 801 (696~934) ❙ 1049 (793~1265); 465 (377~541); 39 (19~58) ❙ 31.2 (−2.0~62.1); 800 (696~934); 216.0 (142.9~288.4)

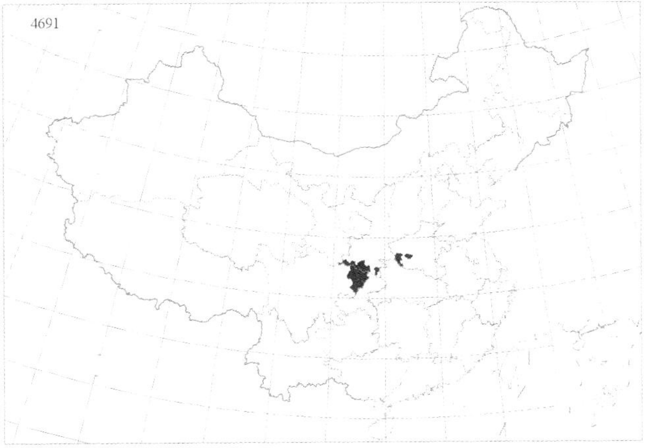

Simaroubaceae 苦木科 *Ailanthus* 臭椿属
Ailanthus guangxiensis S. L. Mo 广西臭椿
Trees

21.9; 13.8; 27.9; 21.9; 203.3; 0.0; 1161 ❙ 1478; 781; 97 ❙ 27.6; 1161; 304.7

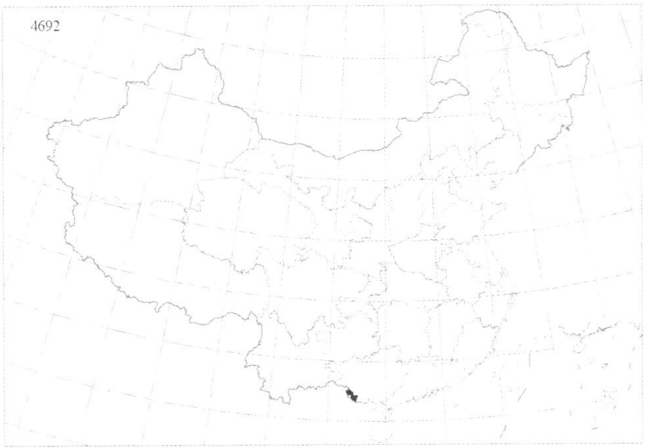

Simaroubaceae 苦木科 *Ailanthus* 臭椿属
Ailanthus triphysa (Dennst.) Alston 岭南臭椿
Trees, evergreen

21.5 (19.2~23.6); 14.2 (10.7~16.7); 27.0 (22.3~29.0); 21.6 (19.2~23.6); 198.7 (170.5~223.5); 0.0 (0.0~0.0); 1119 (909~1281) ❙ 1584 (1269~1776); 750 (551~997); 102 (58~161) ❙ 42.9 (-1.0~64.9); 1113 (880~1269); 327.9 (139.6~659.0)

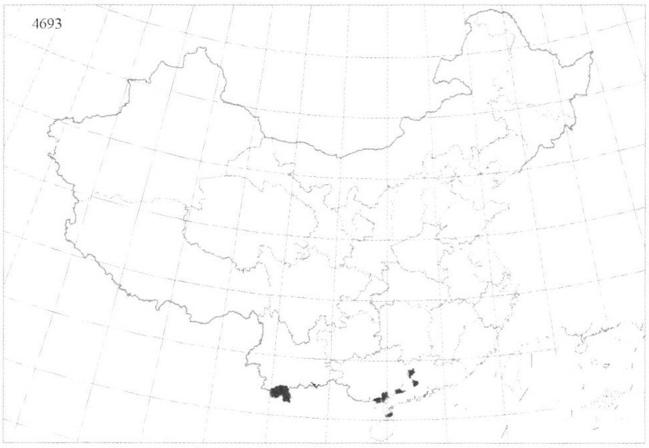

Simaroubaceae 苦木科 *Ailanthus* 臭椿属
Ailanthus vilmoriniana Dode 刺臭椿
Trees

14.0 (5.4~20.4); 3.6 (-3.7~15.1); 23.2 (12.3~29.4); 14.1 (5.8~20.5); 115.1 (31.5~185.8); 6.2 (0.0~25.6); 772 (481~1010) ❙ 979 (604~1607); 516 (350~820); 49 (12~183) ❙ 27.2 (-23.0~101.4); 743 (481~1010); 239.0 (111.9~659.0)

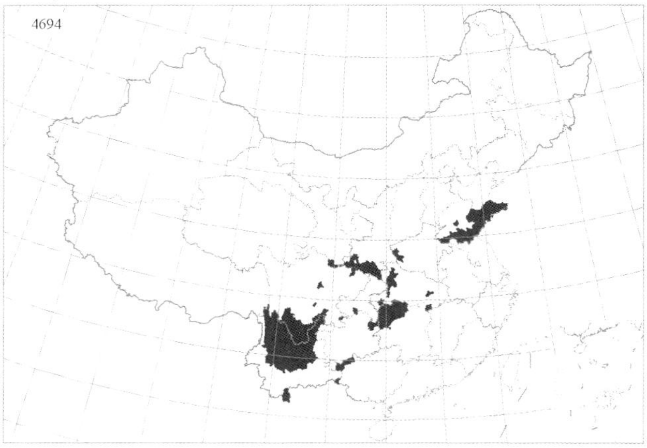

Simaroubaceae 苦木科 *Brucea* 鸦胆子属
Brucea javanica (L.) Merr. 鸦胆子
Shrubs or small trees

20.8 (15.9~24.8); 12.8 (4.7~20.1); 27.5 (20.1~29.8); 20.9 (16.0~24.8); 190.7 (131.7~237.6); 0.0 (0.0~0.3); 1094 (773~1385) ❙ 1677 (868~3495); 718 (312~1847); 146 (41~897) ❙ 56.4 (-34.3~246.5); 1078 (773~1360); 256.7 (95.3~659.0)

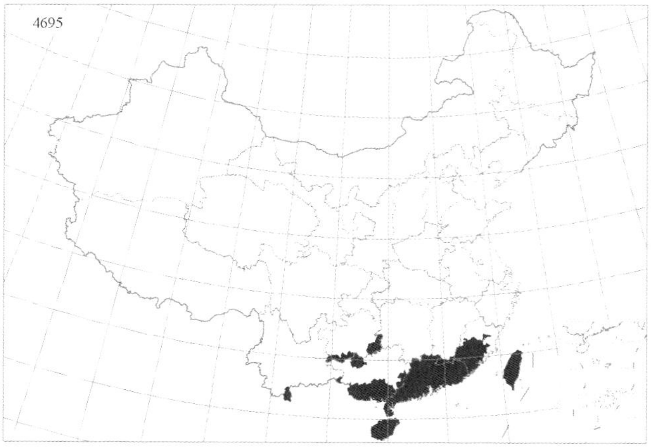

Simaroubaceae 苦木科 *Brucea* 鸦胆子属
Brucea mollis Wall. ex Kurz 柔毛鸦胆子
Shrubs or small trees

19.3 (13.3~24.8); 12.7 (5.4~20.1); 23.9 (19.6~28.4); 19.3 (13.4~24.8); 172.0 (100.4~237.6); 0.0 (0.0~0.0); 961 (706~1385) ❙ 1358 (1007~1751); 707 (392~1007); 57 (41~106) ❙ 44.1 (-11.9~145.8); 929 (706~1212); 413.7 (275.8~659.0)

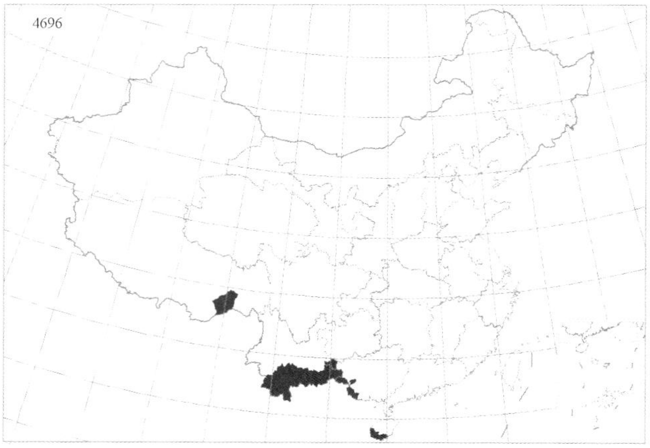

Simaroubaceae 苦木科 *Harrisonia* 牛筋果属
Harrisonia perforata (Blanco) Merr. 牛筋果
Shrubs

23.9 (22.2~24.8); 18.0 (15.9~20.1); 28.1 (25.7~29.2); 23.9 (22.3~24.8); 227.2 (207.2~237.6); 0.0 (0.0~0.0); 1303 (1127~1385) ❙ 1469 (868~1854); 546 (312~712); 75 (41~115) ❙ 12.9 (-34.3~47.8); 1201 (868~1360); 401.1 (178.7~620.0)

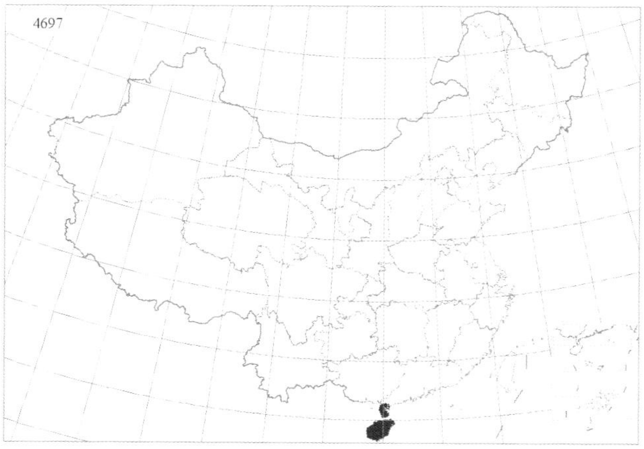

Simaroubaceae 苦木科 *Picrasma* 苦木属
Picrasma chinensis P. Y. Chen 中国苦树
Trees, deciduous

18.5 (13.3~22.2); 10.9 (5.4~15.1); 24.0 (19.6~28.3); 18.5 (13.4~22.3); 162.4 (100.4~207.5); 0.0 (0.0~0.0); 917 (706~1190) ❙ 1366 (1049~1751); 736 (582~1007); 60 (43~96) ❙ 50.5 (15.4~145.8); 913 (706~1190); 386.7 (189.5~659.0)

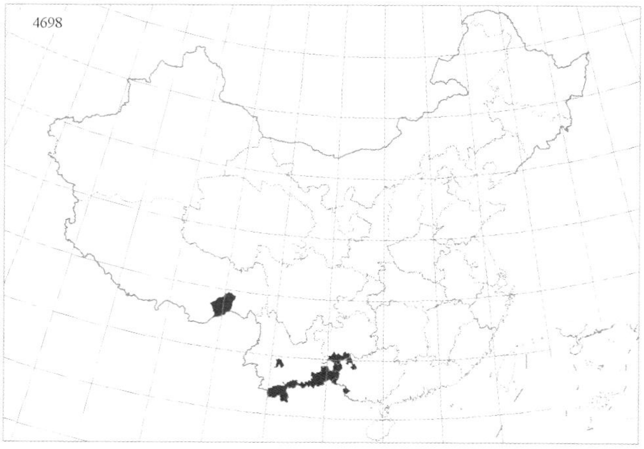

Simaroubaceae 苦木科 *Picrasma* 苦木属
Picrasma quassioides (D. Don) Benn. 苦树
Trees, deciduous

15.0 (−1.1~24.8); 3.7 (−14.5~20.1); 25.2 (6.2~29.8); 15.2 (2.0~24.8); 128.2 (2.5~237.6); 7.9 (0.0~75.6); 836 (334~1385) ❙ 1134 (321~2680); 510 (193~1186); 92 (7~259) ❙ 34.2 (−43.4~246.5); 798 (321~1360); 227.0 (79.3~659.0)

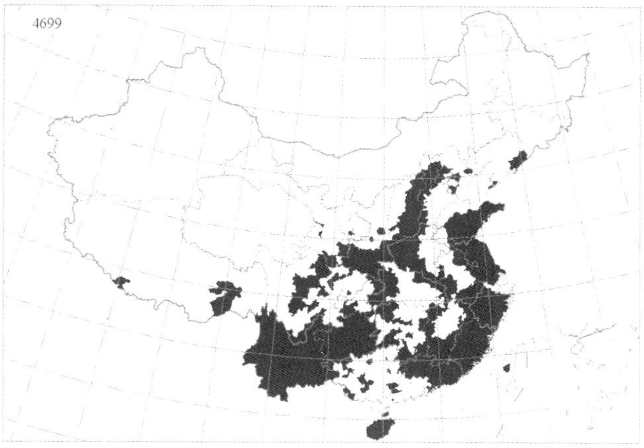

Simaroubaceae 苦木科 *Suriana* 海人树属
Suriana maritima L. 海人树
Shrubs or small trees

20.0 (11.6~24.8); 14.0 (−1.2~20.1); 25.0 (20.1~28.2); 20.1 (11.8~24.8); 181.4 (93.8~237.6); 0.6 (0.0~13.8); 1021 (697~1385) ❙ 2280 (703~3495); 1045 (410~1847); 236 (41~897) ❙ 133.0 (0.0~246.5); 977 (697~1144); 255.3 (119.4~527.5)

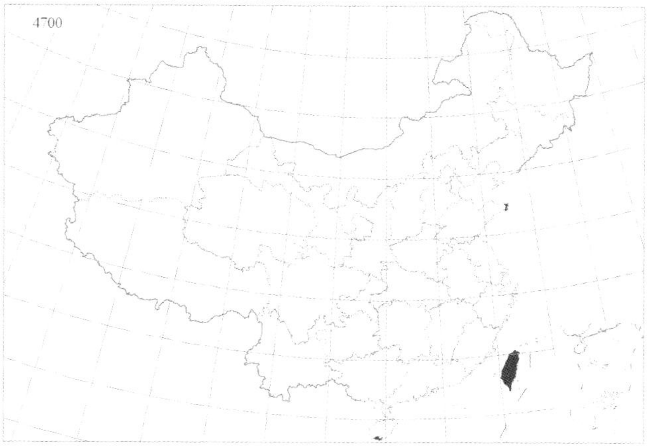

Burseraceae 橄榄科 *Canarium* 橄榄属
Canarium bengalense Roxb. 方榄
Trees, evergreen

20.0 (17.2~22.2); 12.2 (9.5~14.2); 25.6 (22.5~28.3); 20.0 (17.3~22.3); 180.3 (147.0~207.5); 0.0 (0.0~0.0); 1018 (838~1190) ❙ 1412 (1254~1581); 767 (670~850); 74 (48~106) ❙ 40.8 (12.9~65.6); 1018 (838~1190); 373.2 (281.7~551.6)

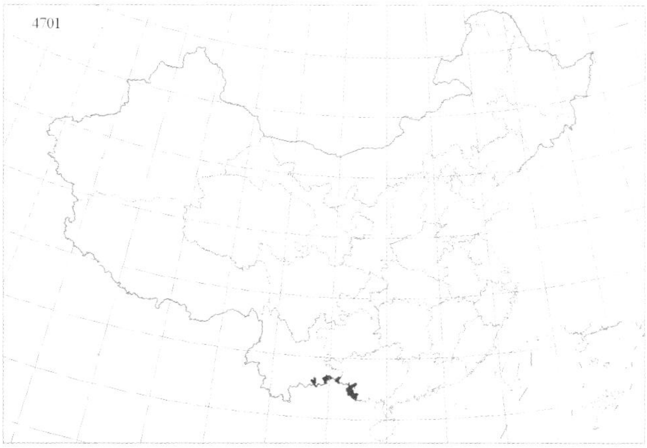

Burseraceae 橄榄科 *Canarium* 橄榄属
Canarium parvum Leenh. 小叶榄
Shrubs or small trees, evergreen

19.4; 12.6; 24.1; 19.4; 172.6; 0.0; 954 ❙ 1445; 787; 64 ❙ 52.3; 954; 521.5

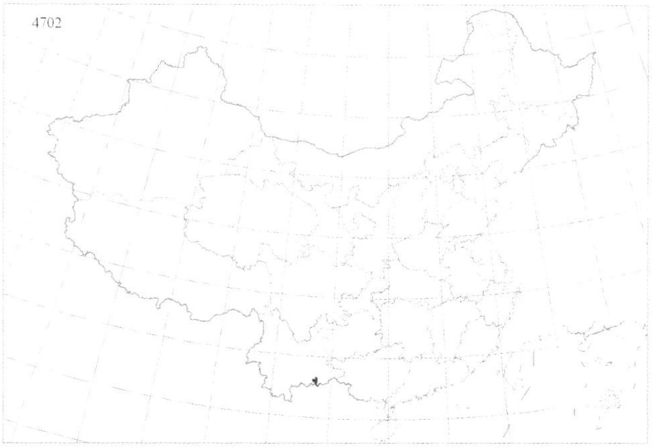

Burseraceae 橄榄科 *Canarium* 橄榄属
Canarium pimela Leenh. 乌榄
Trees, evergreen

21.2 (16.3~24.8); 13.9 (7.3~20.1); 26.7 (20.3~29.2); 21.2 (16.4~24.8); 194.9 (136.6~237.6); 0.0 (0.0~0.0); 1107 (789~1385) ❙ 1552 (1179~2435); 720 (392~1428); 92 (41~169) ❙ 42.1 (−11.9~99.5); 1080 (780~1360); 345.5 (139.6~659.0)

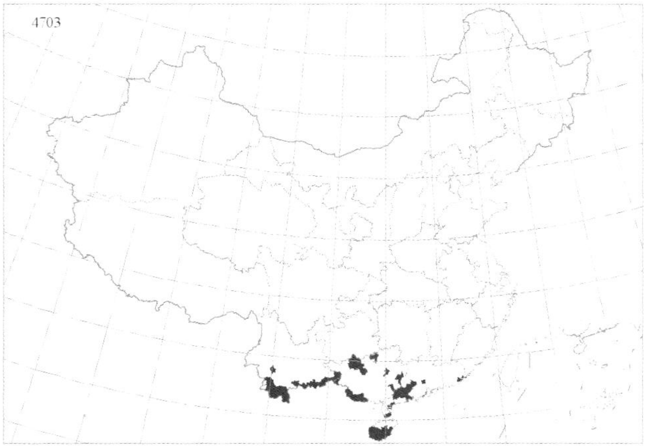

Burseraceae 橄榄科 *Canarium* 橄榄属
Canarium strictum Roxb. 滇榄
Trees, evergreen

20.6; 14.4; 24.5; 20.6; 187.4; 0.0; 1020 ❙ 1486; 689; 72 ❙ 47.6; 1004; 517.7

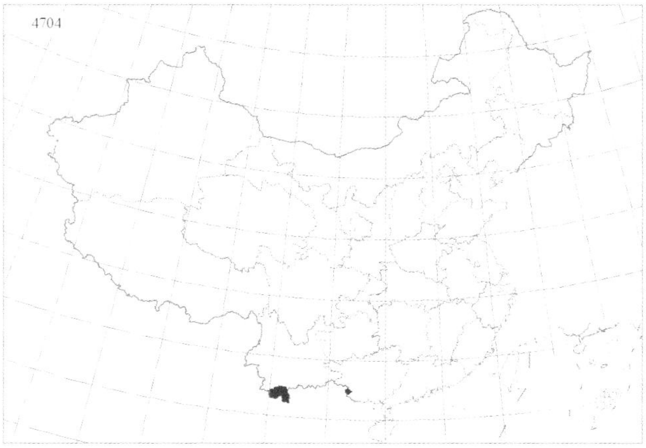

Burseraceae 橄榄科 *Canarium* 橄榄属
Canarium subulatum Guillaumin 毛叶榄
Trees, evergreen

20.6 (13.3~24.8); 12.8 (3.6~20.1); 26.7 (20.1~29.8); 20.6 (13.3~24.8); 187.7 (101.9~237.6); 0.0 (0.0~1.7); 1075 (718~1385) ❙ 1537 (868~2435); 712 (312~1428); 107 (41~244) ❙ 44.6 (-34.3~191.8); 1058 (718~1360); 285.8 (95.3~659.0)

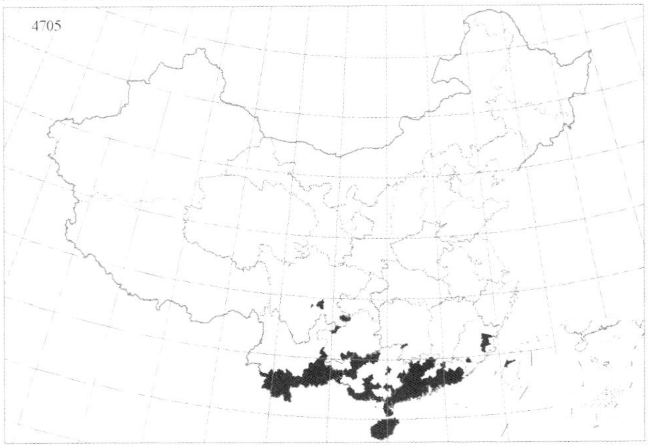

Burseraceae 橄榄科 *Canarium* 橄榄属
Canarium tonkinense Engl. 越榄
Trees, evergreen

19.4; 12.6; 24.1; 19.4; 172.6; 0.0; 954 ❙ 1445; 787; 64 ❙ 52.3; 954; 521.5

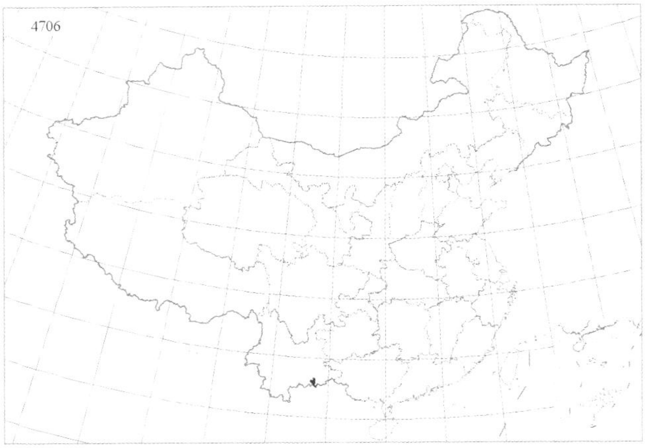

Burseraceae 橄榄科 *Garuga* 白头树属
Garuga floribunda Decne. var. gamblei (King ex Smith) Kalkman 多花白头树
Trees, deciduous

21.9 (18.9~24.8); 14.9 (10.5~20.1); 27.0 (22.3~28.9); 21.9 (19.0~24.8); 203.2 (167.5~237.6); 0.0 (0.0~0.0); 1154 (909~1385) ❙ 1469 (1176~1767); 656 (444~850); 85 (41~178) ❙ 29.5 (-12.0~64.9); 1099 (880~1299); 395.9 (188.6~659.0)

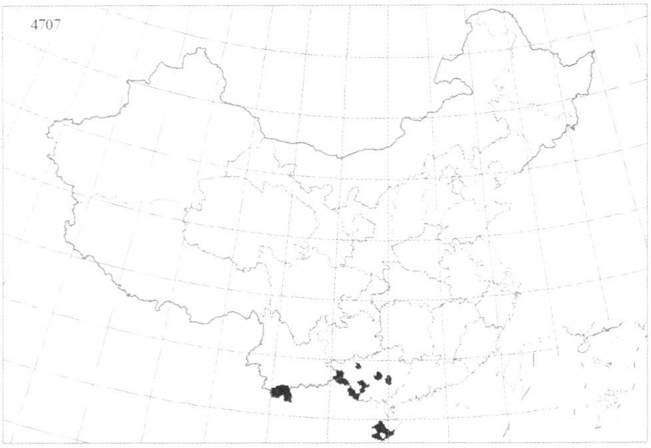

Burseraceae 橄榄科 *Garuga* 白头树属
Garuga forrestii W. W. Sm. 白头树
Trees, deciduous

15.0 (5.4~22.2); 7.7 (-2.1~14.2); 20.7 (12.3~28.3); 15.0 (5.8~22.3); 121.7 (31.5~207.5); 1.2 (0.0~25.6); 775 (481~1190) ❙ 1056 (741~1581); 602 (411~850); 41 (15~106) ❙ 38.1 (-14.2~93.0); 752 (481~1190); 308.1 (142.9~551.6)

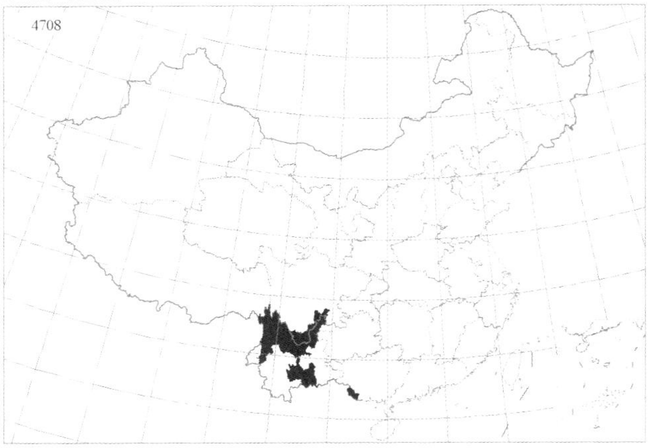

Burseraceae 橄榄科 *Garuga* 白头树属
Garuga pierrei Guillaumin 光叶白头树
Trees, deciduous

20.0; 14.6; 23.2; 20.1; 180.8; 0.0; 963 ❙ 1523; 681; 67 ❙ 58.4; 942; 581.0

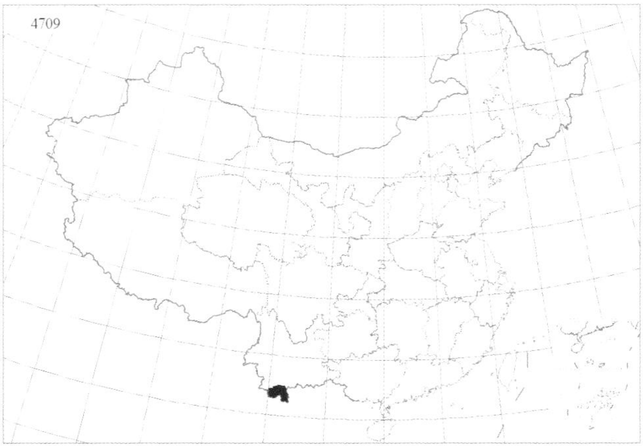

Burseraceae 橄榄科 *Garuga* 白头树属
Garuga pinnata Roxb. 羽叶白头树
Trees, deciduous

21.0 (11.9~24.8); 14.2 (2.9~20.1); 25.9 (20.0~28.8); 21.1 (12.0~24.8); 193.0 (87.2~237.6); 0.2 (0.0~3.4); 1096 (671~1385) ❙ 1443 (1001~1900); 664 (455~1056); 74 (35~171) ❙ 33.9 (1.5~71.0); 1049 (671~1294); 417.5 (188.6~659.0)

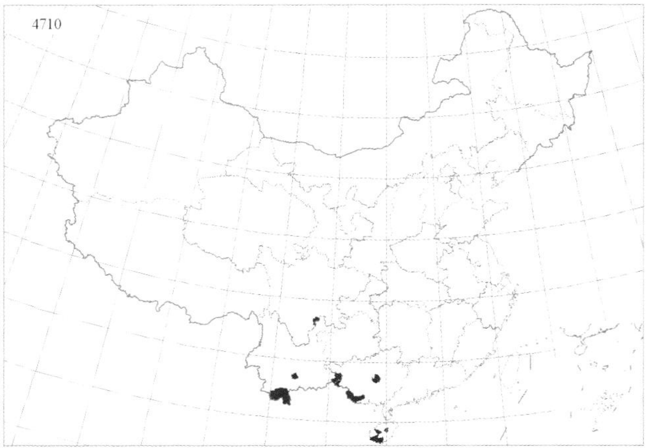

Burseraceae 橄榄科 *Protium* 马蹄果属
Protium serratum (Wall. ex Colebr.) Engl. 马蹄果
Trees, deciduous

17.4 (14.4~20.4); 11.0 (7.5~15.1); 21.4 (19.1~23.6); 17.4 (14.4~20.5); 148.8 (113.1~185.8); 0.0 (0.0~0.0); 844 (727~989) ❙ 1461 (1348~1607); 790 (702~909); 60 (45~77) ❙ 74.3 (57.4~99.9); 839 (727~986); 508.2 (395.4~659.0)

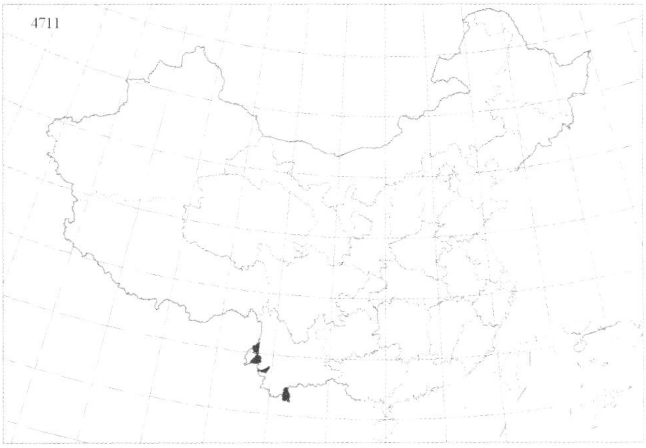

Meliaceae 楝科 *Aglaia* 米仔兰属
Aglaia edulis (Roxburgh) Wallich 马肾果
Trees, evergreen

17.6; 9.9; 23.2; 17.7; 152.4; 0.0; 859 ❙ 1422; 813; 52 ❙ 65.6; 859; 359.7

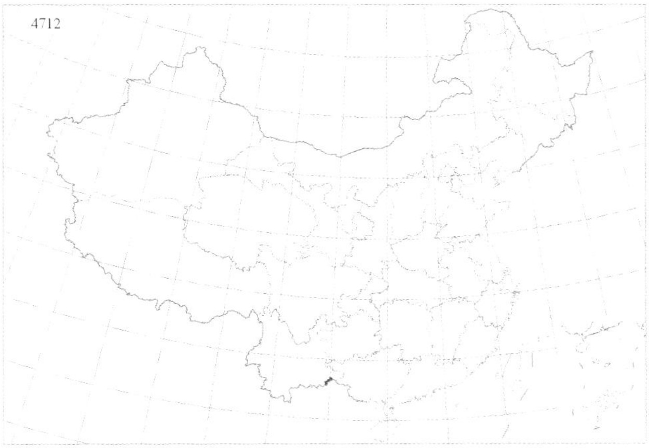

Meliaceae 楝科 *Aglaia* 米仔兰属
Aglaia elaeagnoidea (A. Jussieu) Bentham 山楝
Trees, evergreen

20.9 (17.2~24.8); 13.8 (9.0~20.1); 26.0 (21.7~29.0); 20.9 (17.3~24.8); 190.8 (147.0~237.6); 0.0 (0.0~0.0); 1079 (833~1385) ❙ 1540 (1161~2992); 764 (392~1847); 77 (41~188) ❙ 46.3 (-11.9~214.4); 1044 (833~1329); 374.2 (154.2~659.0)

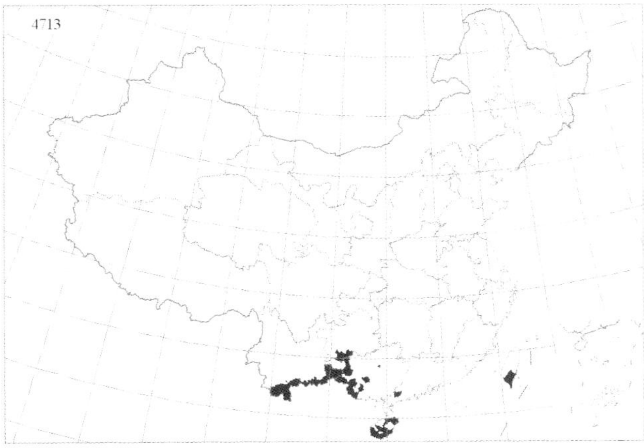

Meliaceae 楝科 *Aglaia* 米仔兰属
Aglaia lawii (Wight) C. J. Saldanha et Ramamorthy 望谟崖摩
Shrubs or trees, evergreen

19.4 (13.3~24.8); 11.9 (5.4~20.1); 24.9 (19.6~29.0); 19.4 (13.4~24.8); 173.0 (100.4~237.6); 0.0 (0.0~0.0); 985 (706~1385) ❙ 1381 (868~1953); 679 (312~1007); 79 (41~192) ❙ 42.5 (-34.3~145.8); 958 (706~1326); 352.5 (189.5~659.0)

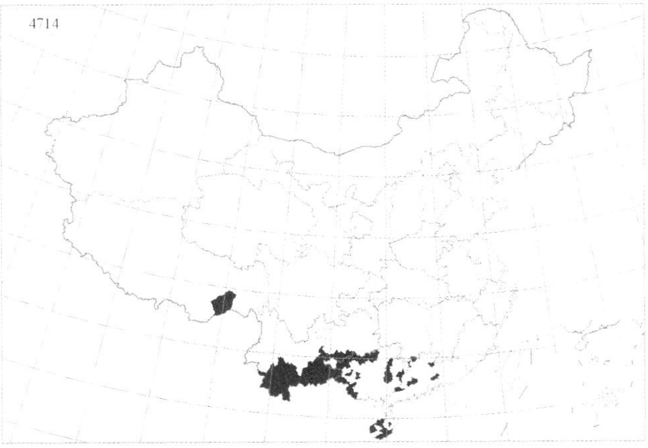

Meliaceae 楝科 *Aglaia* 米仔兰属
Aglaia odorata Lour. 米仔兰
Shrubs or small trees, evergreen

21.8 (16.4~24.8); 14.1 (5.4~20.1); 28.0 (25.0~29.1); 21.9 (16.5~24.8); 202.7 (137.7~237.6); 0.0 (0.0~0.0); 1160 (864~1385) ❙ 1625 (868~2435); 747 (312~1428); 112 (41~194) ❙ 40.9 (-34.3~99.5); 1135 (864~1326); 264.5 (95.3~620.0)

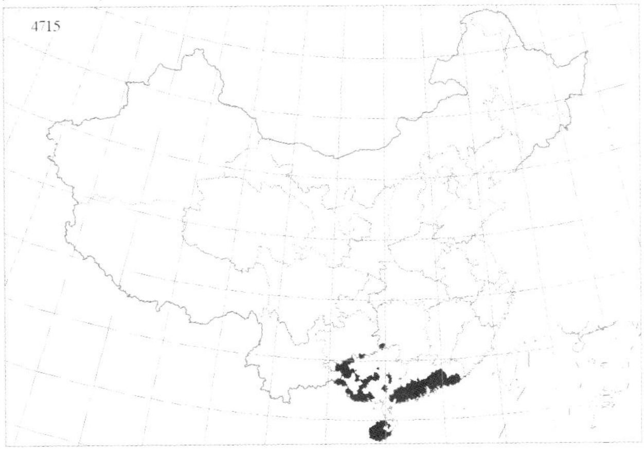

Meliaceae 楝科 *Aglaia* 米仔兰属
Aglaia perviridis Hiern 碧绿米仔兰
Trees, evergreen

18.4 (16.3~21.0); 12.1 (9.1~15.1); 22.6 (20.3~25.6); 18.5 (16.4~21.0); 161.8 (136.4~191.8); 0.0 (0.0~0.0); 889 (789~1053) ❙ 1440 (1190~1751); 779 (655~996); 58 (47~77) ❙ 62.1 (44.5~88.8); 880 (780~1053); 460.6 (336.3~659.0)

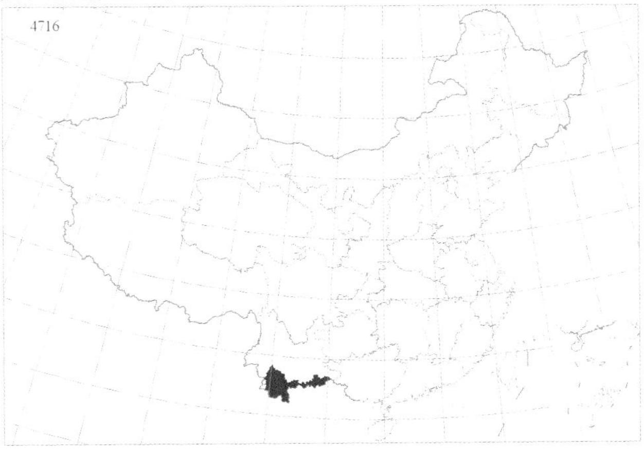

Meliaceae 楝科 *Aglaia* 米仔兰属

Aglaia rimosa (Blanco) Merr. 椭圆叶米仔兰

Shrubs or small trees, evergreen

20.8 (18.5~24.1); 15.7 (13.5~18.9); 24.8 (21.7~28.0); 20.8 (18.5~24.1); 189.9 (162.1~229.7); 0.0 (0.0~0.0); 1041 (863~1322) ❙ 2642 (1751~2993); 1384 (924~1847); 203 (47~547) ❙ 163.8 (32.5~214.4); 996 (863~1123); 265.7 (154.2~363.7)

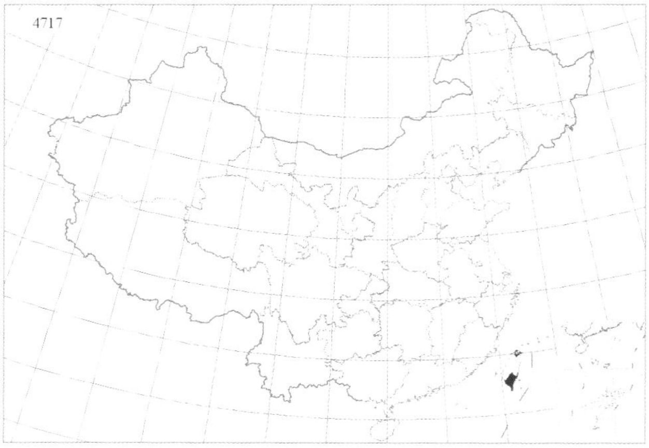

Meliaceae 楝科 *Aglaia* 米仔兰属

Aglaia spectabilis (Miquel) S. S. Jain et Bennet 曲梗崖摩

Trees

18.7 (17.2~20.5); 12.2 (9.5~15.1); 23.0 (22.4~23.7); 18.8 (17.3~20.5); 165.0 (147.0~185.9); 0.0 (0.0~0.0); 904 (833~992) ❙ 1507 (1254~1751); 805 (655~996); 63 (48~77) ❙ 67.0 (47.6~88.8); 899 (833~986); 486.3 (336.3~659.0)

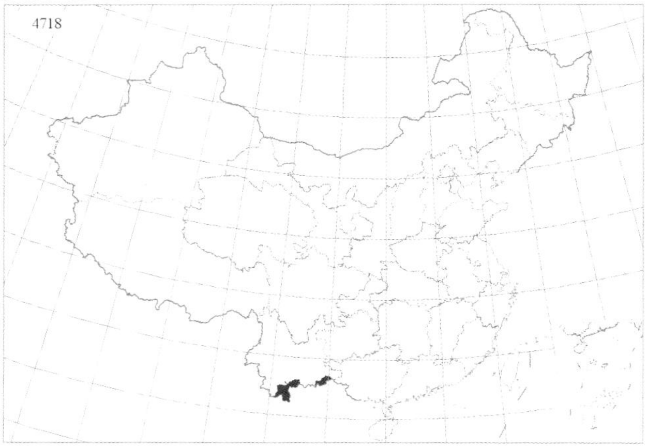

Meliaceae 楝科 *Aglaia* 米仔兰属

Aglaia teysmanniana (Miquel) Miquel 星毛崖摩

Small trees

20.0; 13.6; 24.4; 20.1; 180.7; 0.0; 985 ❙ 1475; 827; 62 ❙ 50.1; 985; 528.9

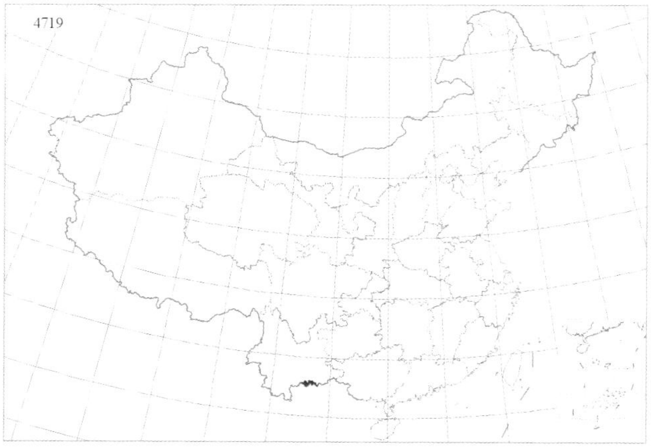

Meliaceae 楝科 *Aphanamixis* 山楝属

Aphanamixis polystachya (Wall.) R. Parker 山楝

Trees

21.4 (16.3~24.8); 14.6 (7.4~20.1); 26.5 (20.3~29.2); 21.5 (16.4~24.8); 197.9 (136.6~237.6); 0.0 (0.0~0.0); 1124 (789~1385) ❙ 1565 (868~2435); 734 (312~1428); 87 (41~186) ❙ 42.0 (-34.3~99.5); 1088 (780~1360); 357.6 (139.6~659.0)

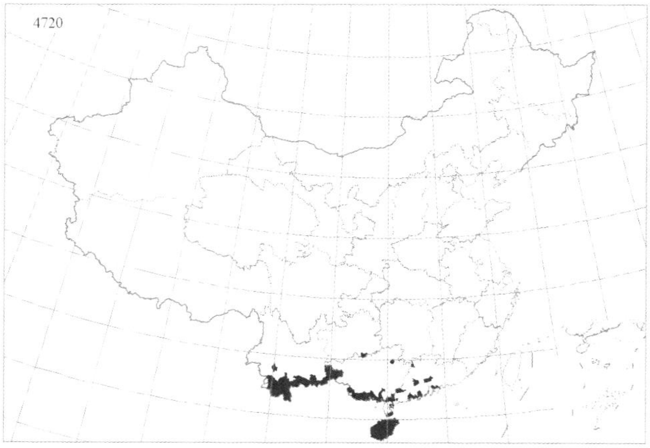

Meliaceae 楝科 *Chisocheton* 溪桫属

Chisocheton cumingianus subsp. balansae (C. Candolle) Mabberley 溪桫

Trees

19.9 (17.7~22.7); 13.5 (10.9~15.1); 24.2 (22.1~28.6); 19.9 (17.8~22.8); 179.0 (153.4~213.1); 0.0 (0.0~0.0); 980 (854~1221) ❙ 1635 (1276~2435); 888 (655~1428); 77 (51~130) ❙ 66.0 (36.8~99.5); 973 (854~1221); 467.8 (297.2~659.0)

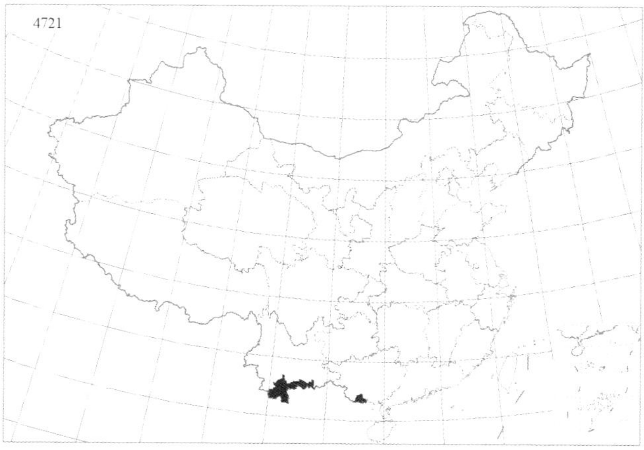

Meliaceae 楝科 *Chukrasia* 麻楝属

Chukrasia tabularis A. Juss. 麻楝

Trees

19.1 (13.3~24.8); 11.6 (5.4~20.1); 24.8 (19.1~28.9); 19.2 (13.4~24.8); 170.0 (100.4~237.6); 0.0 (0.0~0.0); 968 (706~1385) ❙ 1420 (868~2037); 705 (312~1033); 82 (41~188) ❙ 48.5 (-34.3~145.8); 948 (706~1269); 345.3 (157.7~659.0)

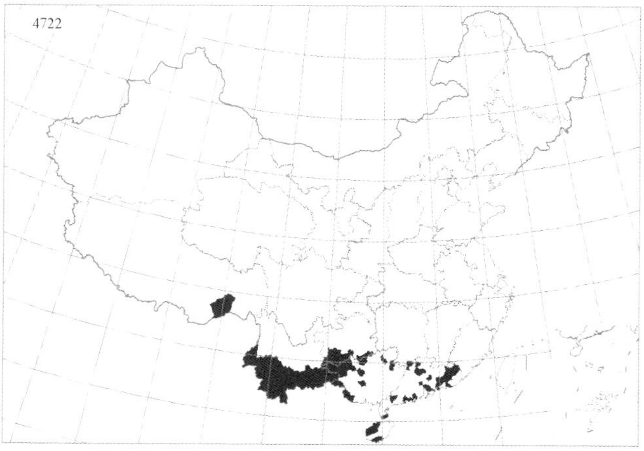

Meliaceae 楝科 *Cipadessa* 浆果楝属
Cipadessa baccifera (Roth) Miq. 浆果楝
Shrubs or small trees

17.1 (7.3~23.0); 8.3 (-2.0~15.1); 24.4 (15.6~29.0); 17.1 (7.5~23.0); 145.7 (48.2~216.2); 0.3 (0.0~20.4); 882 (537~1247) ▎1266 (797~2435); 621 (411~1428); 80 (18~196) ▎44.1 (-14.2~112.2); 875 (537~1247); 262.1 (108.2~659.0)

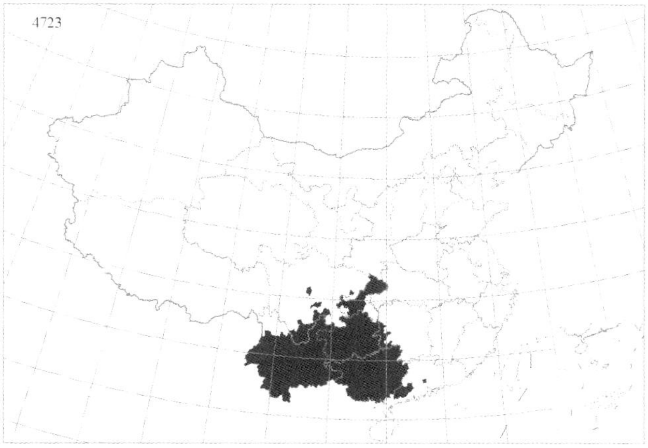

Meliaceae 楝科 *Dysoxylum* 桎木属
Dysoxylum arborescens (Blume) Miquel 兰屿桎木
Trees

18.5; 14.1; 21.7; 18.5; 162.1; 0.0; 863 ▎2713; 1388; 188 ▎214.4; 863; 363.7

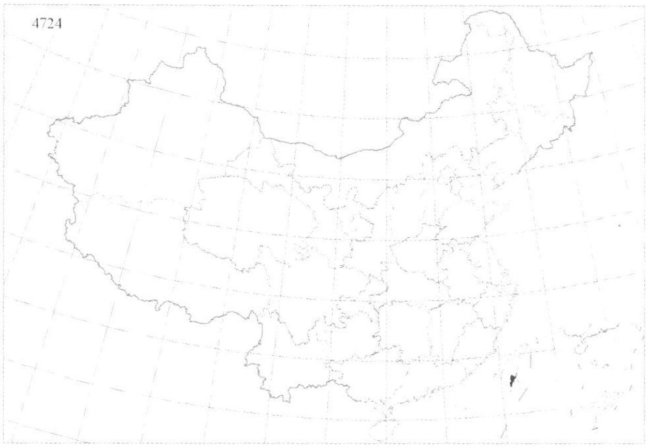

Meliaceae 楝科 *Dysoxylum* 桎木属
Dysoxylum cumingianum C. DC. 肯氏桎木
Small trees

18.5; 14.1; 21.7; 18.5; 162.1; 0.0; 863 ▎2713; 1388; 188 ▎214.4; 863; 363.7

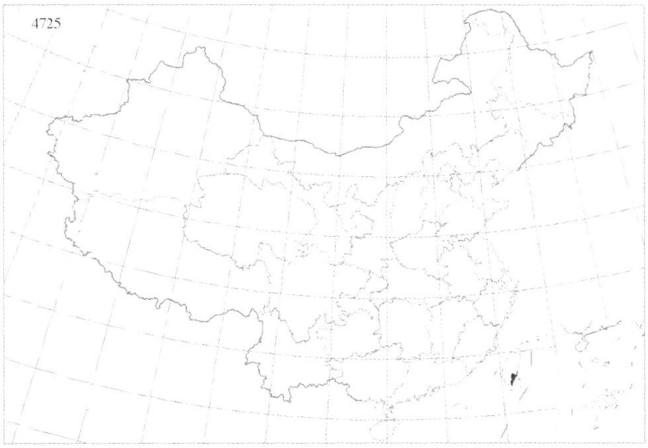

Meliaceae 楝科 *Dysoxylum* 桎木属
Dysoxylum densiflorum (Blume) Miquel 密花桎木
Trees, evergreen

20.5; 15.0; 23.7; 20.5; 185.9; 0.0; 990 ▎1536; 678; 72 ▎55.1; 973; 614.6

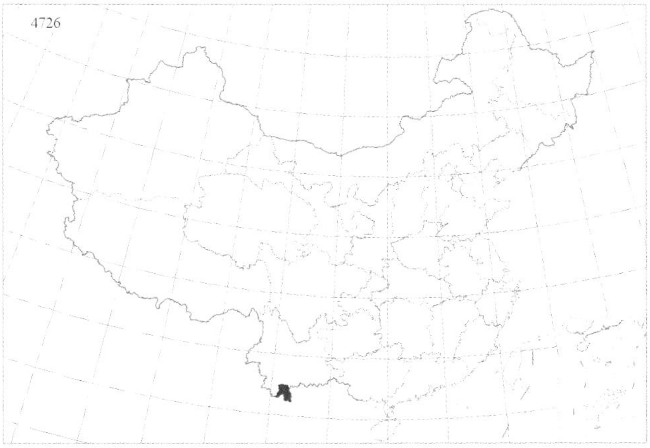

Meliaceae 楝科 *Dysoxylum* 桎木属
Dysoxylum excelsum Blume 桎木
Trees

18.4 (13.0~21.0); 11.6 (5.4~15.1); 23.2 (19.6~25.6); 18.5 (13.4~21.0); 161.7 (100.4~191.8); 0.0 (0.0~0.0); 900 (706~1053) ▎1466 (1161~1751); 795 (633~1007); 59 (48~77) ▎64.5 (37.1~145.8); 895 (706~1053); 462.9 (314.4~659.0)

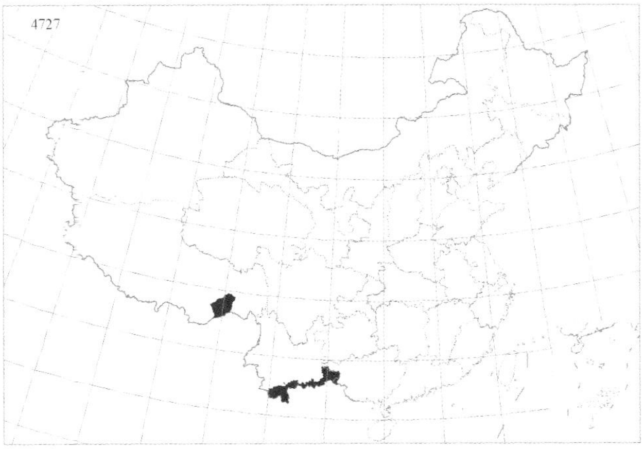

Meliaceae 楝科 *Dysoxylum* 桎木属
Dysoxylum gotadhora (Buchanan-Hamilton) Mabberley 红果桎木
Trees

21.4 (16.0~24.8); 15.3 (8.5~20.1); 25.7 (20.4~28.8); 21.5 (16.0~24.8); 197.6 (132.4~237.6); 0.0 (0.0~0.0); 1113 (780~1385) ▎1459 (868~1854); 650 (312~996); 73 (41~169) ▎34.7 (-34.3~88.8); 1052 (780~1360); 430.3 (203.4~659.0)

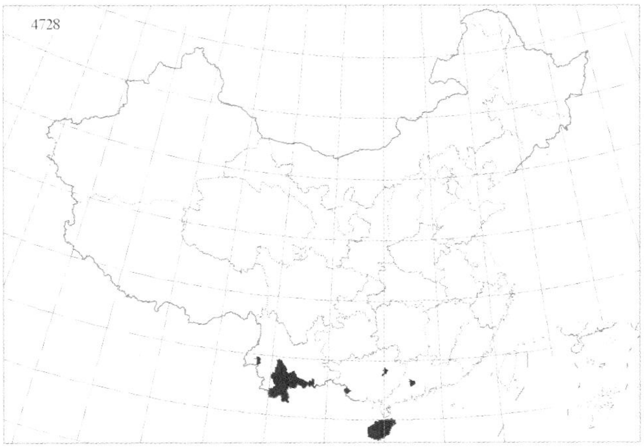

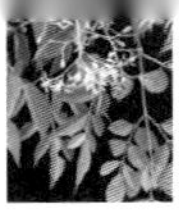

Meliaceae 楝科 *Dysoxylum* 樫木属
Dysoxylum grande Hiern 多脉樫木
Trees

22.5 (19.2~24.8); 15.6 (11.8~20.1); 27.5 (22.3~28.6); 22.5 (19.2~24.8); 210.3 (170.5~237.6); 0.0 (0.0~0.0); 1197 (909~1385) ❙ 1635 (1179~2435); 738 (392~1428); 92 (41~194) ❙ 37.9 (-11.9~99.5); 1141 (880~1360); 388.6 (213.5~620.0)

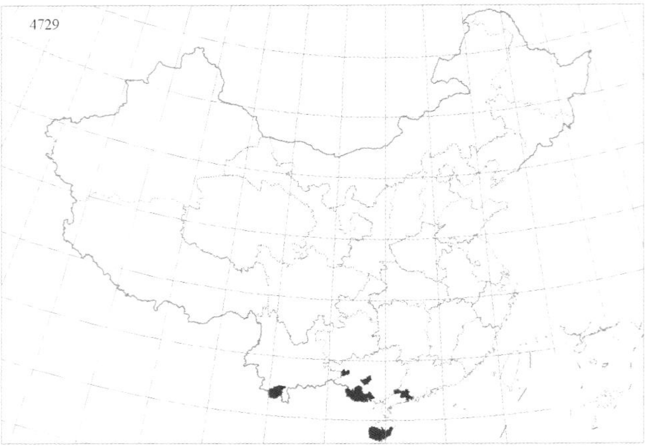

Meliaceae 楝科 *Dysoxylum* 樫木属
Dysoxylum hongkongense (Tutcher) Merr. 香港樫木
Trees

21.2 (14.4~24.8); 13.5 (7.3~20.1); 27.1 (19.1~28.9); 21.2 (14.4~24.8); 194.7 (113.1~237.6); 0.0 (0.0~0.0); 1115 (727~1385) ❙ 1680 (868~2992); 778 (312~1847); 115 (41~189) ❙ 53.9 (-34.3~214.4); 1081 (727~1299); 312.7 (139.6~659.0)

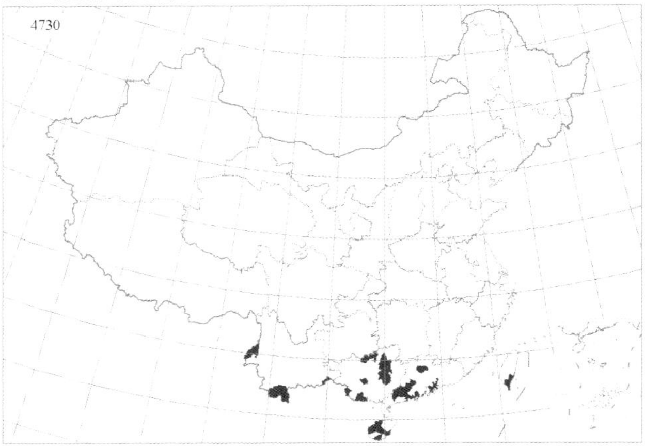

Meliaceae 楝科 *Dysoxylum* 樫木属
Dysoxylum laxiracemosum C. Y. Wu et H. Li 总序樫木
Trees

19.6 (18.6~20.5); 13.9 (12.7~15.1); 23.2 (22.5~23.7); 19.6 (18.7~20.5); 175.6 (163.9~185.9); 0.0 (0.0~0.0); 943 (888~992) ❙ 1559 (1428~1751); 820 (655~996); 68 (56~77) ❙ 65.8 (47.6~88.8); 935 (888~986); 560.1 (506.2~659.0)

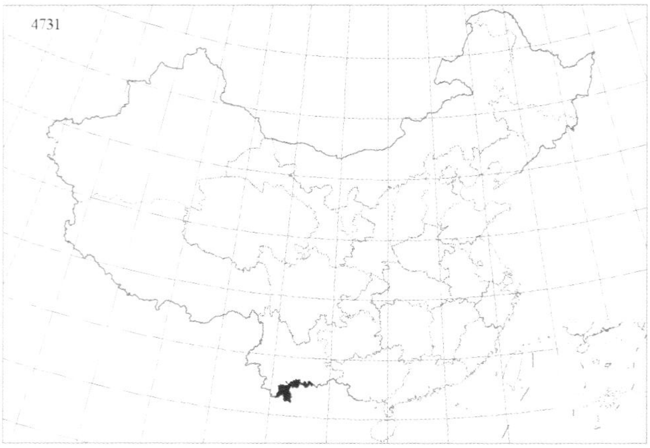

Meliaceae 楝科 *Dysoxylum* 樫木属
Dysoxylum lenticellatum C. Y. Wu et H. Li 皮孔樫木
Trees, deciduous

18.7 (17.2~21.0); 12.0 (9.1~15.1); 23.2 (21.6~25.6); 18.7 (17.3~21.0); 164.5 (147.0~191.8); 0.0 (0.0~0.0); 905 (832~1053) ❙ 1419 (1049~1751); 764 (598~996); 59 (44~77) ❙ 56.8 (26.1~88.8); 897 (783~1053); 463.5 (314.4~659.0)

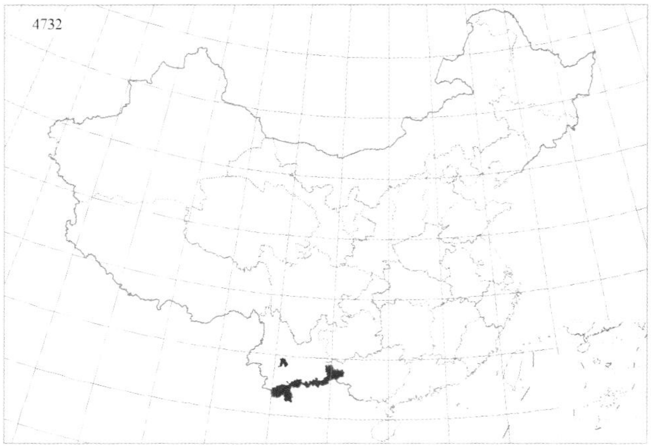

Meliaceae 楝科 *Dysoxylum* 樫木属
Dysoxylum mollissimum Blume 海南樫木
Trees

20.1 (17.2~24.2); 13.5 (9.1~19.1); 24.6 (22.3~28.9); 20.1 (17.3~24.3); 181.2 (147.0~231.2); 0.0 (0.0~0.0); 1010 (833~1335) ❙ 1433 (1161~1751); 723 (447~996); 63 (48~102) ❙ 45.0 (-12.0~88.8); 985 (833~1257); 446.9 (194.5~659.0)

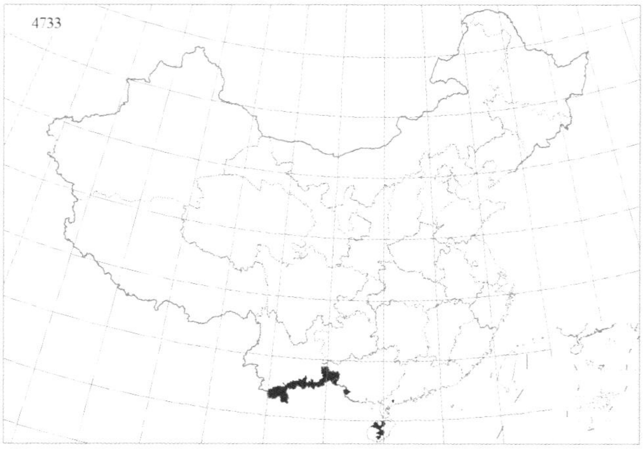

Meliaceae 楝科 *Dysoxylum* 樫木属
Dysoxylum pallens Hiern 少花樫木
Trees

19.5 (17.7~21.0); 13.5 (10.9~15.1); 23.3 (22.3~25.6); 19.5 (17.8~21.0); 174.2 (153.4~191.8); 0.0 (0.0~0.0); 941 (854~1053) ❙ 1523 (1368~1751); 795 (655~996); 66 (56~77) ❙ 62.3 (44.5~88.8); 933 (854~1053); 544.7 (491.5~659.0)

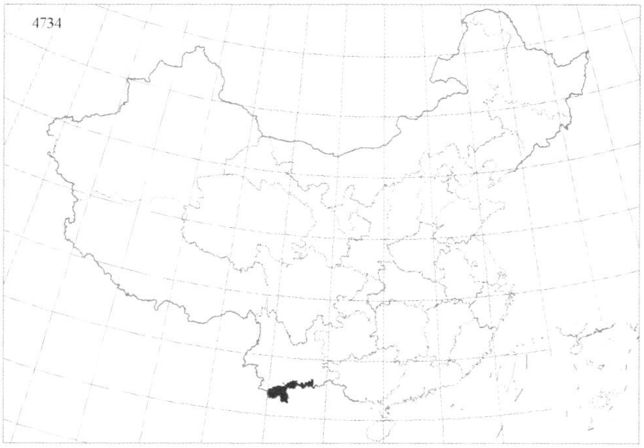

Meliaceae 楝科 *Dysoxylum* 樫木属
Dysoxylum parasiticum (Osbeck) Koestermans 大花樫木
Trees

18.5；14.1；21.7；18.5；162.1；0.0；863 ❙ 2713；1388；188 ❙ 214.4；863；363.7

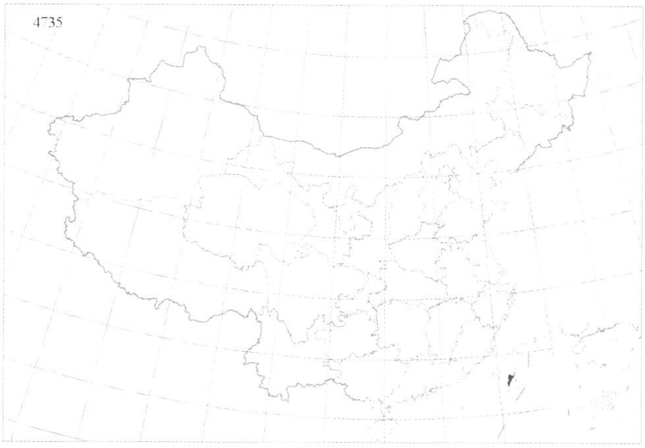

Meliaceae 楝科 *Heynea* 鹧鸪花属
Heynea trijuga Roxb. 鹧鸪花
Trees

18.1 (13.1~24.8)；11.2 (5.3~20.1)；23.2 (18.4~28.8)；18.2 (13.2~24.8)；158.2 (98.1~237.6)；0.0 (0.0~0.0)；914 (692~1385) ❙ 1277 (797~2435)；658 (312~1428)；61 (23~169) ❙ 41.0 (-34.3~112.2)；885 (683~1360)；356.4 (189.5~659.0)

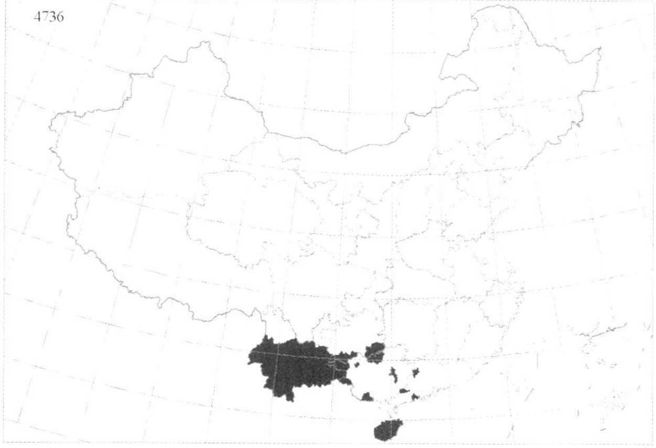

Meliaceae 楝科 *Heynea* 鹧鸪花属
Heynea velutina F. C. How et T. C. Chen 茸果鹧鸪花
Shrubs

22.2 (15.2~24.8)；15.1 (4.1~20.1)；27.5 (22.5~28.6)；22.2 (15.2~24.8)；206.9 (123.4~237.6)；0.0 (0.0~0.9)；1192 (812~1385) ❙ 1647 (1221~2435)；762 (444~1428)；93 (41~131) ❙ 39.7 (-2.2~99.5)；1143 (812~1360)；381.4 (211.1~620.0)

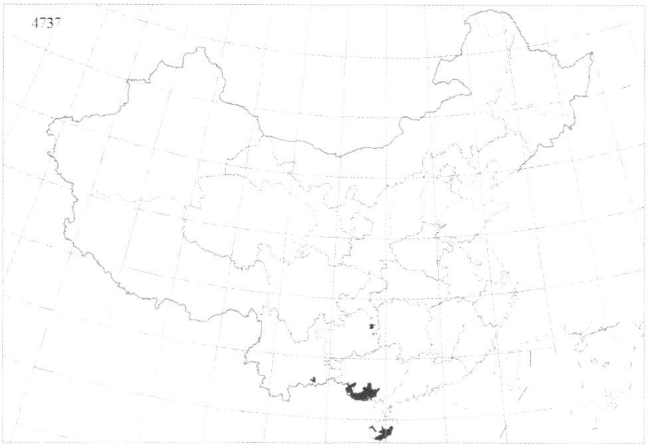

Meliaceae 楝科 *Melia* 楝属
Melia azedarach L. 楝
Trees, deciduous

16.3 (2.9~24.8)；5.3 (-8.6~20.1)；26.3 (10.6~30.1)；16.4 (4.3~24.8)；140.4 (19.9~237.6)；3.7 (0.0~45.2)；892 (421~1385) ❙ 1214 (465~2993)；531 (251~1428)；106 (10~547) ❙ 35.6 (-43.7~203.4)；850 (421~1360)；217.9 (50.7~659.0)

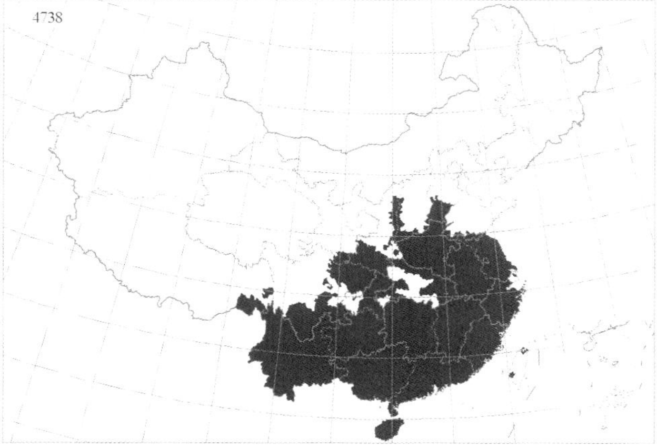

Meliaceae 楝科 *Munronia* 地黄连属
Munronia pinnata (Wallich) W. Theobald 羽状地黄连
Dwarf subshrubs

16.5 (5.4~24.6)；9.1 (-2.1~19.5)；22.1 (12.3~29.0)；16.6 (5.8~24.7)；139.3 (31.5~235.9)；0.5 (0.0~25.6)；834 (481~1372) ❙ 1156 (741~1876)；629 (411~996)；50 (15~173) ❙ 39.2 (-14.2~88.8)；816 (481~1299)；316.8 (142.9~659.0)

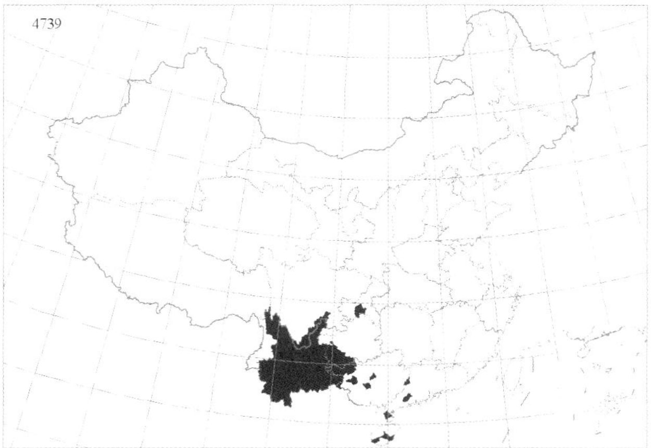

Meliaceae 楝科 *Munronia* 地黄连属
Munronia unifoliolata Oliv. 单叶地黄连
Dwarf subshrubs

14.8 (9.3~24.8)；4.3 (-1.7~20.1)；24.5 (18.6~28.3)；14.8 (9.5~24.8)；119.9 (69.8~237.6)；1.9 (0.0~18.0)；799 (603~1385) ❙ 1136 (863~1478)；507 (390~630)；72 (20~180) ❙ 43.1 (1.5~107.2)；792 (603~1137)；212.9 (135.6~527.5)

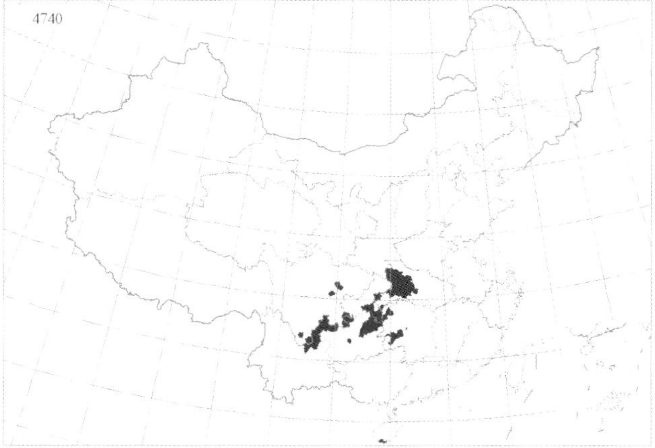

Meliaceae 楝科 *Reinwardtiodendron* 雷楝属

Reinwardtiodendron humile (Hasskarl) Mabberley 雷楝

Shrubs or small trees

23.1；17.5；27.2；23.2；218.4；0.0；1232 ❙ 1487；544；53 ❙ 20.7；1142；538.2

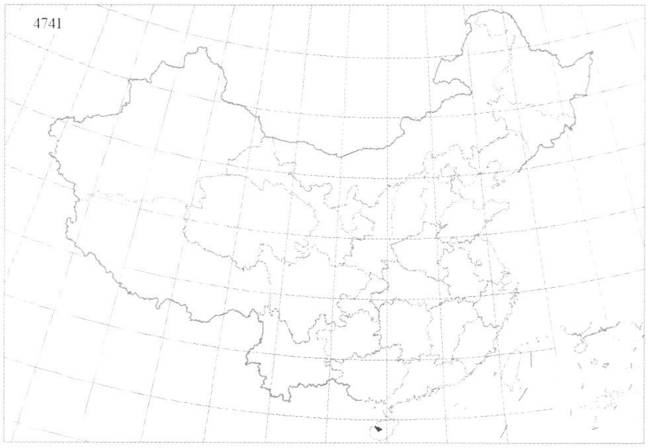

Meliaceae 楝科 *Toona* 香椿属

Toona ciliata M. Roem. 红椿

Trees, deciduous

16.3 (5.3~24.8)；7.1 (−2.9~20.1)；24.3 (12.2~29.9)；16.4 (5.7~24.8)；138.2 (30.5~237.6)；1.5 (0.0~26.2)；857 (479~1385) ❙ 1260 (626~2037)；583 (316~1033)；86 (15~259) ❙ 47.7 (−22.2~130.7)；842 (479~1326)；266.3 (111.7~659.0)

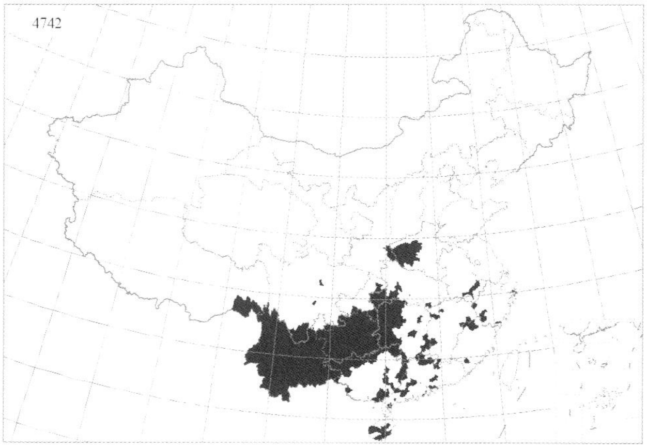

Meliaceae 楝科 *Toona* 香椿属

Toona fargesii A. Chevalier 红花香椿

Trees, evergreen

18.2；8.7；26.9；18.2；158.8；0.0；940 ❙ 1790；612；206 ❙ 92.4；940；282.0

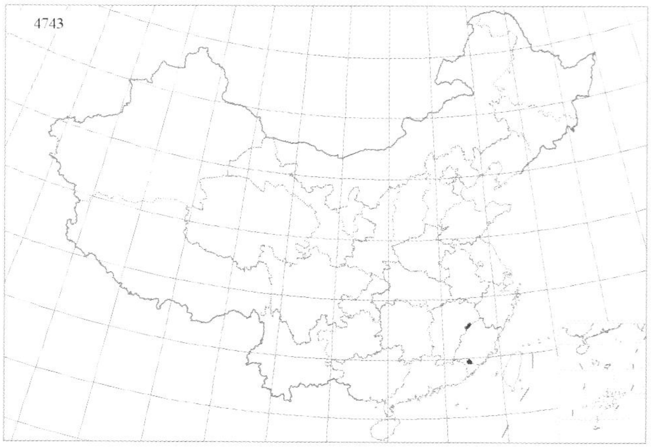

Meliaceae 楝科 *Toona* 香椿属

Toona sinensis (Juss.) Roem. 香椿

Trees, deciduous

16.2 (0.2~24.8)；6.0 (−12.4~20.1)；25.5 (9.9~30.1)；16.4 (3.8~24.8)；138.8 (15.4~237.6)；3.3 (0.0~74.6)；875 (404~1385) ❙ 1290 (386~2435)；546 (219~1428)；113 (5~259) ❙ 47.5 (−38.3~147.0)；856 (386~1360)；242.1 (111.8~659.0)

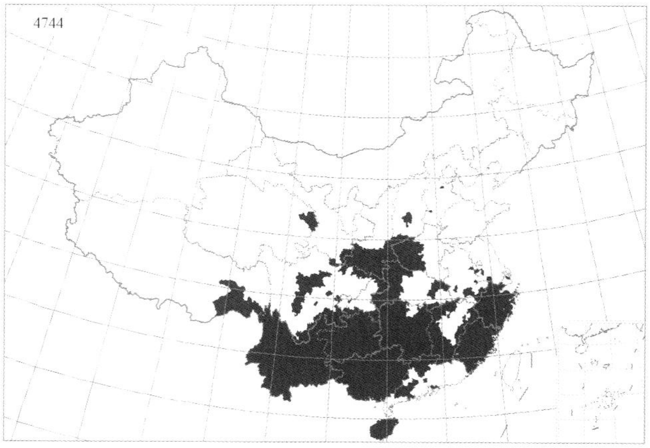

Meliaceae 楝科 *Toona* 香椿属

Toona sureni (Blume) Merr. 紫椿

Trees, deciduous

19.0；13.1；22.6；19.1；168.8；0.0；909 ❙ 1495；775；62 ❙ 64.5；891；427.9

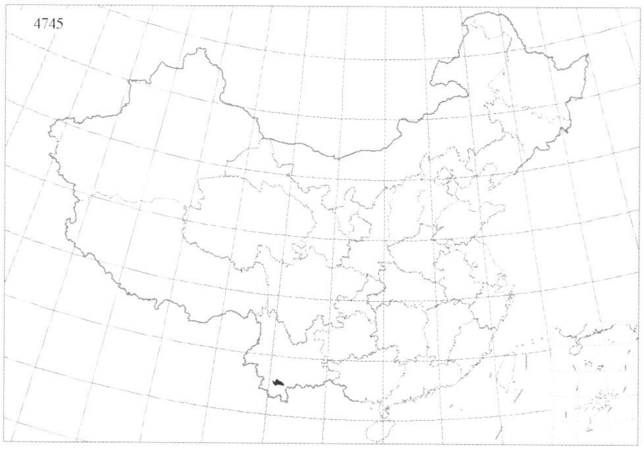

Meliaceae 楝科 *Turraea* 杜楝属

Turraea pubescens Hell. 杜楝

Shrubs

23.6 (22.8~24.8)；17.2 (14.7~20.1)；28.5 (27.2~29.2)；23.7 (22.8~24.8)；224.3 (213.6~237.6)；0.0 (0.0~0.0)；1288 (1230~1385) ❙ 1338 (868~1745)；568 (312~997)；69 (41~106) ❙ 4.3 (−34.3~39.9)；1155 (868~1273)；338.7 (178.7~538.2)

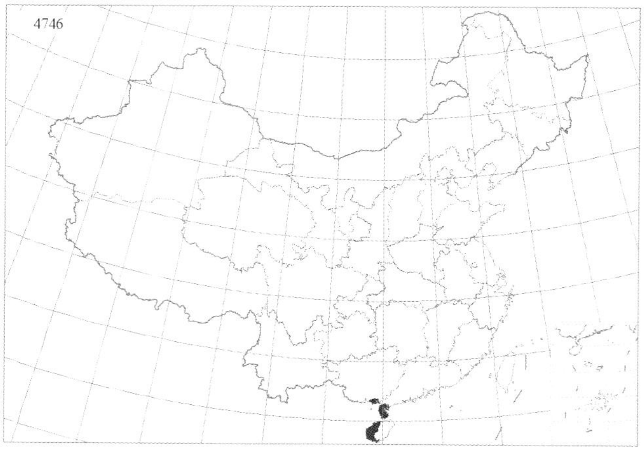

Meliaceae 楝科 *Walsura* 割舌树属
Walsura pinnata Hasskarl 越南割舌树
Trees

21.8 (18.2~24.6); 16.1 (12.4~20.1); 25.6 (21.6~28.6); 21.8 (18.2~24.8); 202.1 (158.6~237.6); 0.0 (0.0~0.0); 1129 (861~1385) ❙ 1616 (1143~2435); 727 (455~1428); 67 (41~130) ❙ 45.6 (1.5~99.5); 1050 (839~1221); 486.4 (300.5~659.0)

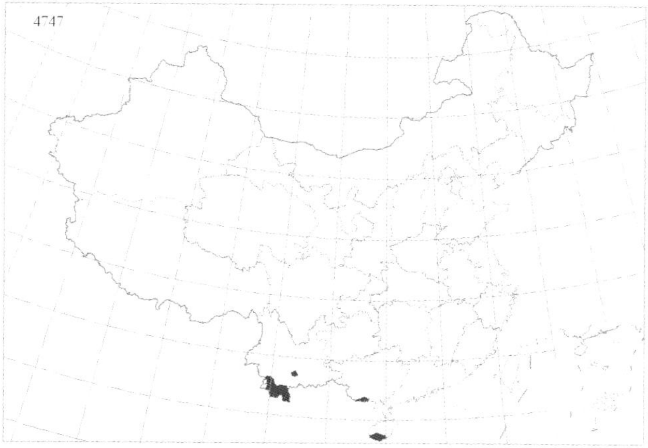

Meliaceae 楝科 *Walsura* 割舌树属
Walsura robusta Roxb. 割舌树
Trees

22.5 (18.9~24.8); 15.9 (10.6~20.1); 27.2 (22.3~28.7); 22.5 (19.0~24.8); 210.1 (167.5~237.6); 0.0 (0.0~0.0); 1194 (909~1385) ❙ 1399 (868~1625); 608 (312~850); 78 (41~173) ❙ 19.8 (-34.3~64.9); 1092 (868~1254); 409.2 (237.8~659.0)

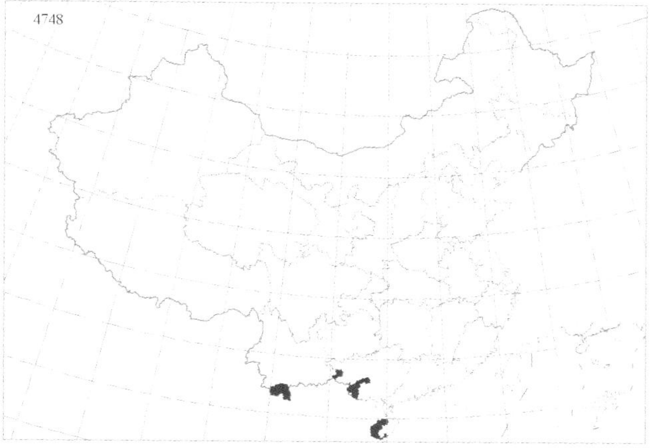

Meliaceae 楝科 *Xylocarpus* 木果楝属
Xylocarpus granatum K. D. Koenig 木果楝
Shrubs or trees

24.2 (24.0~24.8); 18.3 (17.6~20.1); 28.5 (28.0~28.8); 24.3 (24.1~24.8); 231.5 (229.0~237.6); 0.0 (0.0~0.0); 1338 (1321~1385) ❙ 1531 (1309~1813); 577 (473~712); 90 (41~113) ❙ 14.5 (-1.4~36.4); 1274 (1137~1331); 339.0 (248.2~527.5)

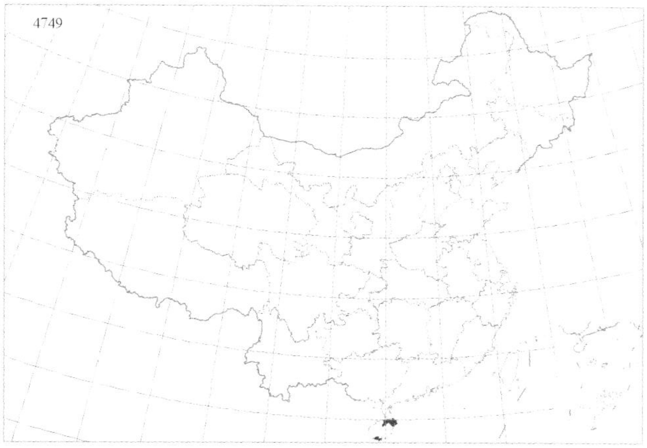

Malpighiaceae 金虎尾科 *Aspidopterys* 盾翅藤属
Aspidopterys esquirolii H. Lév. 花江盾翅藤
Vines, evergreen

18.7 (15.2~21.3); 9.3 (5.8~12.6); 26.1 (23.0~28.7); 18.7 (15.2~21.3); 164.3 (122.5~196.0); 0.0 (0.0~0.0); 957 (779~1123) ❙ 1347 (1172~1620); 679 (588~775); 87 (43~166) ❙ 41.4 (5.8~64.7); 957 (779~1121); 258.2 (186.7~411.1)

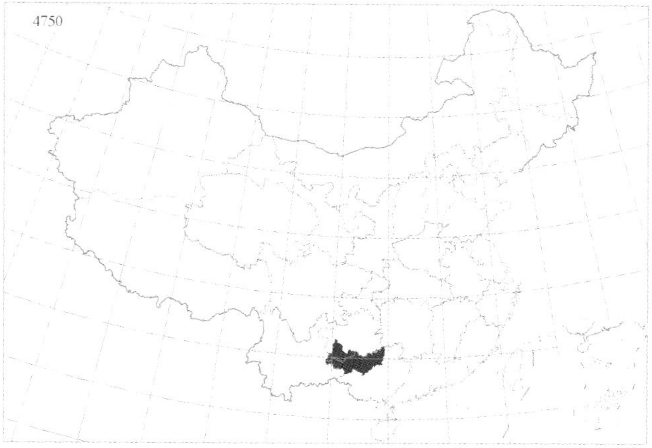

Malpighiaceae 金虎尾科 *Aspidopterys* 盾翅藤属
Aspidopterys floribunda Hutch. 多花盾翅藤
Scandent shrubs, evergreen

18.2 (8.6~21.0); 12.1 (1.3~15.1); 22.4 (15.0~25.6); 18.3 (8.7~21.0); 160.2 (52.6~191.8); 0.8 (0.0~8.8); 895 (574~1053) ❙ 1426 (1041~1751); 757 (585~996); 61 (45~77) ❙ 60.2 (36.8~88.8); 889 (574~1053); 504.1 (352.3~659.0)

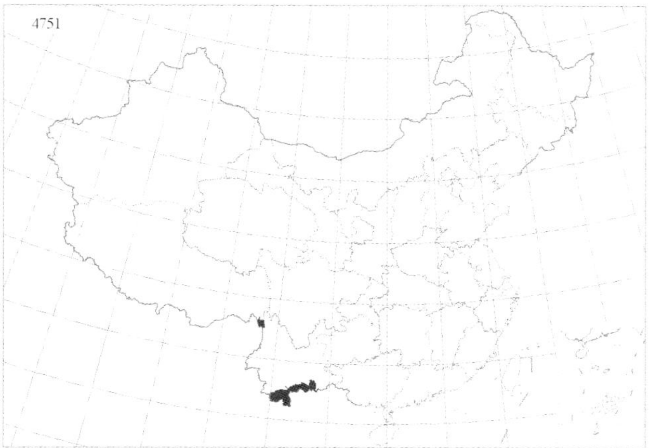

Malpighiaceae 金虎尾科 *Aspidopterys* 盾翅藤属
Aspidopterys glabriuscula (Wall.) A. Juss. 盾翅藤
Vines, evergreen

19.4 (8.6~24.8); 12.9 (1.3~20.1); 23.9 (15.0~28.3); 19.4 (8.7~24.8); 173.6 (52.6~237.6); 0.3 (0.0~8.8); 980 (574~1385) ❙ 1385 (939~1900); 713 (455~1056); 61 (37~125) ❙ 44.7 (1.5~145.8); 943 (574~1212); 425.9 (281.7~659.0)

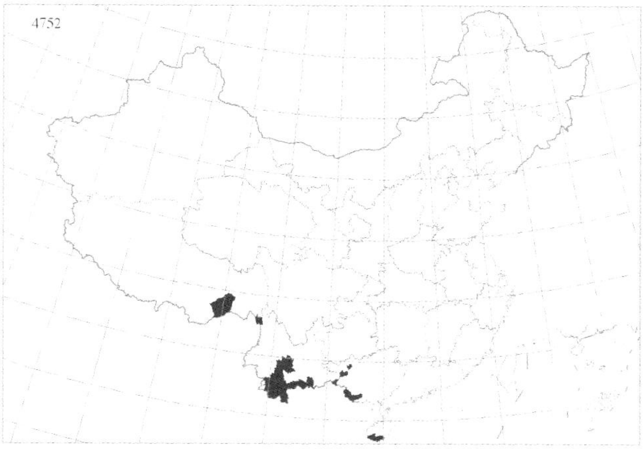

Malpighiaceae 金虎尾科 *Aspidopterys* 盾翅藤属

Aspidopterys henryi Hutch. 蒙自盾翅藤

Vines, evergreen

17.6 (15.4~21.0); 10.5 (8.1~14.2); 22.6 (20.7~25.6); 17.7 (15.5~21.0); 151.8 (125.7~191.8); 0.0 (0.0~0.0); 859 (767~1053) ❙ 1319 (1007~1522); 738 (561~860); 54 (48~68) ❙ 53.5 (31.4~80.6); 859 (767~1053); 407.2 (274.9~551.6)

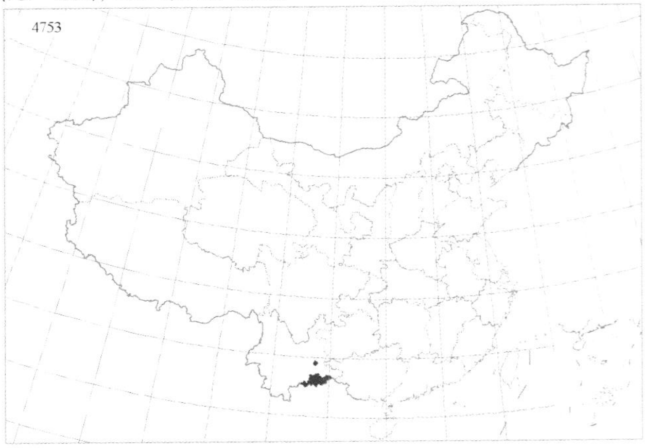

Malpighiaceae 金虎尾科 *Aspidopterys* 盾翅藤属

Aspidopterys nutans Hook. f. 毛叶盾翅藤

Vines, evergreen

18.8 (16.0~22.2); 11.7 (7.3~15.1); 23.6 (20.3~29.0); 18.8 (16.0~22.3); 165.9 (132.4~207.5); 0.0 (0.0~0.0); 933 (780~1190) ❙ 1472 (1192~1703); 787 (639~959); 75 (42~194) ❙ 60.1 (15.4~112.2); 926 (780~1190); 417.4 (160.7~659.0)

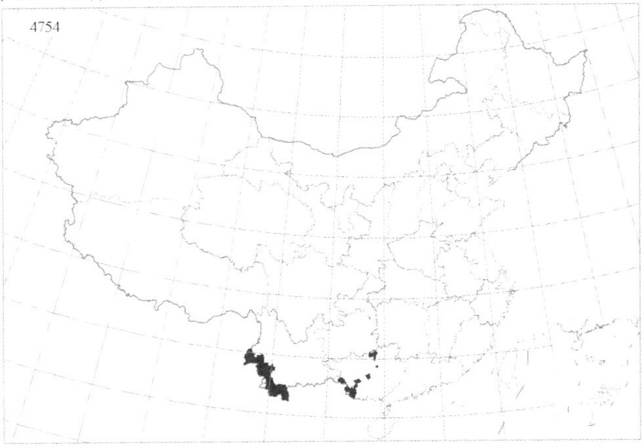

Malpighiaceae 金虎尾科 *Aspidopterys* 盾翅藤属

Aspidopterys obcordata Hemsl. 倒心盾翅藤

Vines, evergreen

20.3 (18.6~24.3); 14.6 (12.7~18.7); 24.0 (22.3~28.4); 20.3 (18.7~24.3); 184.0 (163.9~231.9); 0.0 (0.0~0.0); 1003 (888~1338) ❙ 1473 (1179~1751); 724 (392~996); 60 (42~77) ❙ 50.4 (-11.9~88.8); 957 (880~1142); 518.8 (427.9~659.0)

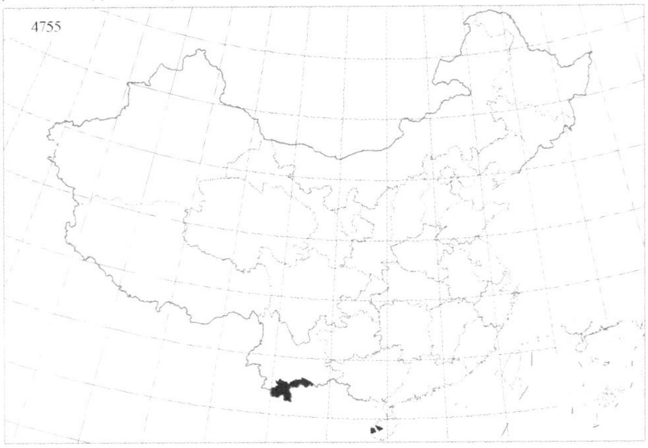

Malpighiaceae 金虎尾科 *Aspidopterys* 盾翅藤属

Aspidopterys obcordata var. hainanensis Arènes 海南盾翅藤

Vines, evergreen

19.4 (13.2~22.3); 11.1 (6.4~14.3); 25.9 (18.4~28.8); 19.5 (13.3~22.4); 173.8 (99.4~208.7); 0.0 (0.0~0.0); 1005 (695~1197) ❙ 1495 (1232~1953); 764 (617~1056); 92 (51~167) ❙ 49.6 (27.5~92.4); 1005 (695~1197); 269.8 (161.4~506.2)

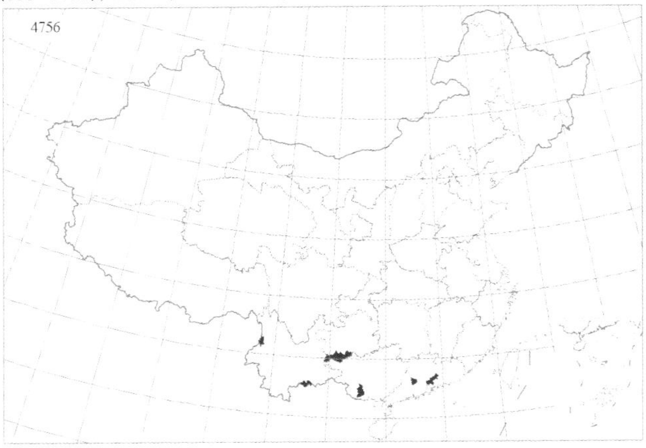

Malpighiaceae 金虎尾科 *Hiptage* 风筝果属

Hiptage acuminata Wall. ex A. Juss. 尖叶风筝果

Scandent shrubs, evergreen

15.5 (11.3~18.9); 8.9 (4.3~12.5); 20.2 (17.0~23.2); 15.5 (11.3~19.0); 126.5 (76.6~167.5); 0.1 (0.0~0.7); 778 (644~914) ❙ 1208 (1028~1454); 672 (566~792); 55 (44~70) ❙ 59.6 (12.5~99.9); 753 (644~851); 391.2 (339.5~487.2)

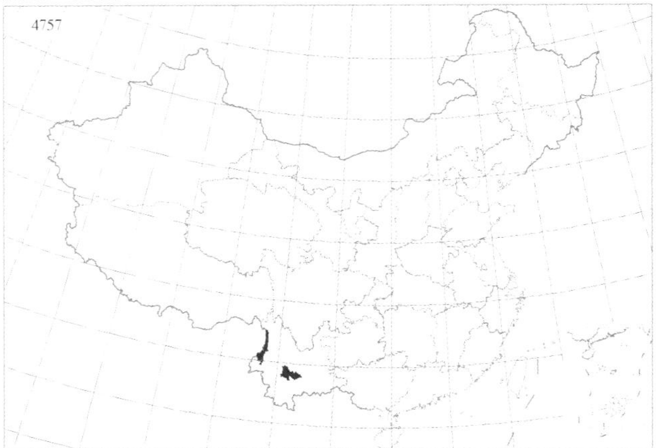

Malpighiaceae 金虎尾科 *Hiptage* 风筝果属

Hiptage benghalensis (L.) Kurz 风筝果

Scandent shrubss or vines, evergreen

20.1 (14.4~24.8); 12.7 (7.1~20.1); 25.8 (19.1~29.2); 20.1 (14.4~24.8); 181.7 (113.1~237.6); 0.0 (0.0~0.0); 1030 (727~1385) ❙ 1619 (939~3495); 778 (425~1847); 115 (37~897) ❙ 58.7 (1.5~246.5); 1020 (727~1326); 297.5 (95.3~659.0)

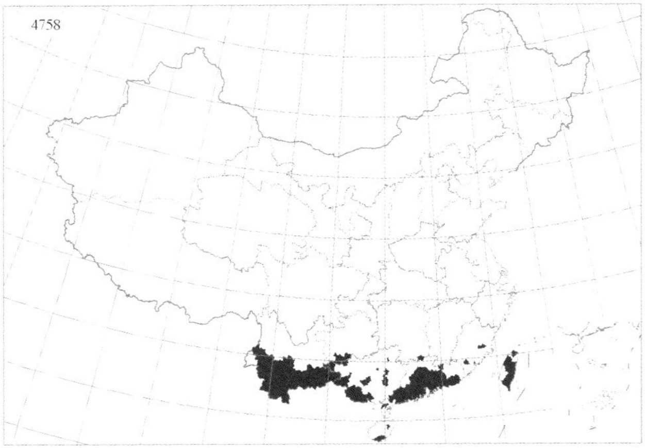

Malpighiaceae 金虎尾科 *Hiptage* 风筝果属
Hiptage benghalensis var. tonkinensis (Dop) S. K. Chen 越南风筝果
Scandent shrubss or vines, evergreen

18.7 (16.3~20.5); 12.7 (9.1~15.1); 22.2 (20.8~23.7); 18.7 (16.4~20.5); 164.4 (136.4~185.9); 0.0 (0.0~0.0); 893 (797~992) ❙ 1437 (1190~1607); 729 (655~803); 53 (41~77) ❙ 61.0 (47.6~74.2); 871 (797~986); 492.9 (347.3~659.0)

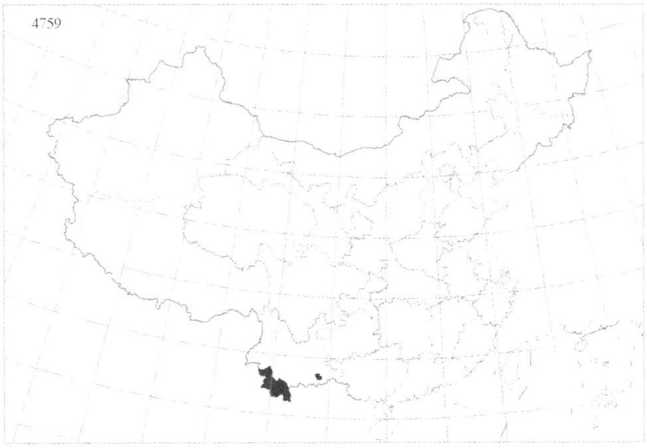

Malpighiaceae 金虎尾科 *Hiptage* 风筝果属
Hiptage candicans Hook. f. 白花风筝果
Shrubs, small trees or vines, evergreen

18.9 (17.1~21.0); 13.0 (11.0~15.1); 22.5 (20.8~25.6); 19.0 (17.1~21.0); 167.7 (145.0~191.8); 0.0 (0.0~0.0); 910 (815~1053) ❙ 1462 (1352~1607); 749 (655~818); 54 (41~77) ❙ 61.2 (44.5~74.2); 889 (802~1053); 505.1 (427.9~659.0)

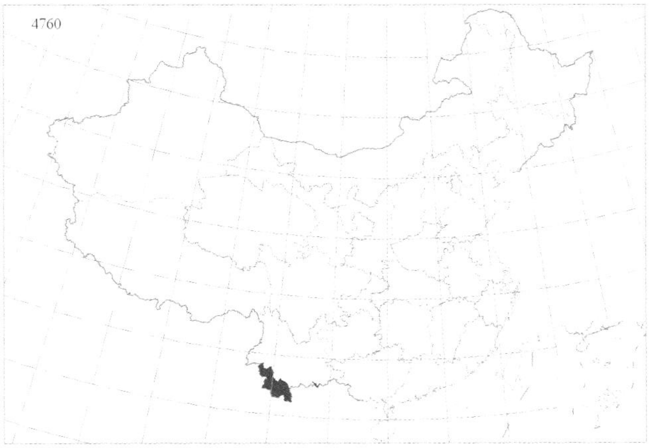

Malpighiaceae 金虎尾科 *Hiptage* 风筝果属
Hiptage candicans var. harmandiana (Pierre) Dop 越南白花风筝果
Shrubs, small trees or vines, evergreen

17.4; 11.0; 21.4; 17.4; 149.3; 0.0; 834 ❙ 1406; 816; 48 ❙ 68.7; 834; 434.4

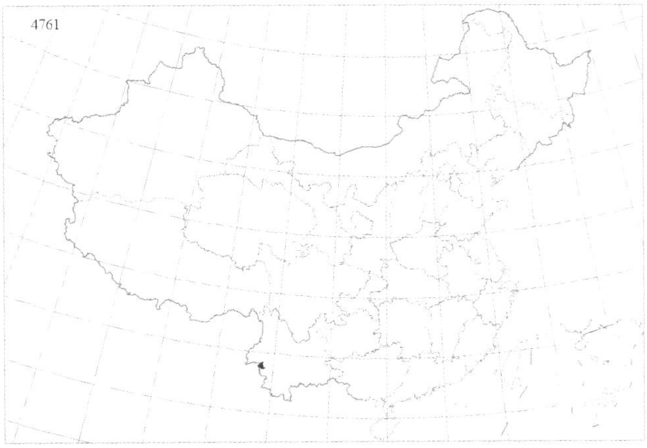

Malpighiaceae 金虎尾科 *Hiptage* 风筝果属
Hiptage fraxinifolia F. N. Wei 白蜡风筝果
Scandent shrubs, evergreen

21.8; 12.9; 28.5; 21.8; 201.6; 0.0; 1162 ❙ 1519; 735; 145 ❙ 30.7; 1162; 208.5

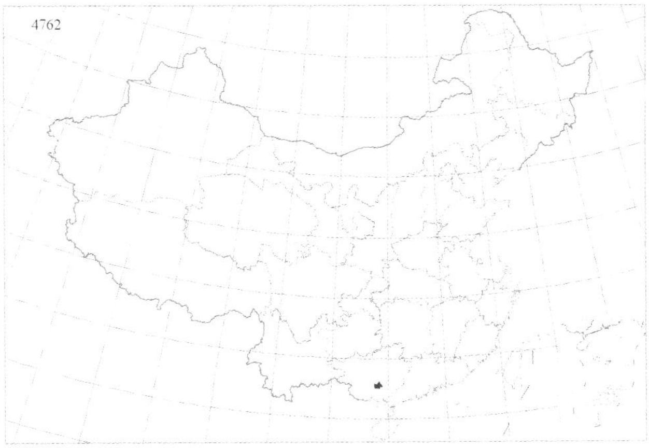

Malpighiaceae 金虎尾科 *Hiptage* 风筝果属
Hiptage minor Dunn 小花风筝果
Shrubs, evergreen

17.5 (15.4~19.3); 9.9 (7.8~13.1); 23.0 (20.6~26.6); 17.5 (15.5~19.4); 150.2 (125.7~173.2); 0.0 (0.0~0.0); 859 (767~969) ❙ 1153 (958~1368); 633 (531~756); 51 (44~64) ❙ 34.4 (13.3~60.1); 854 (767~969); 311.2 (189.5~491.5)

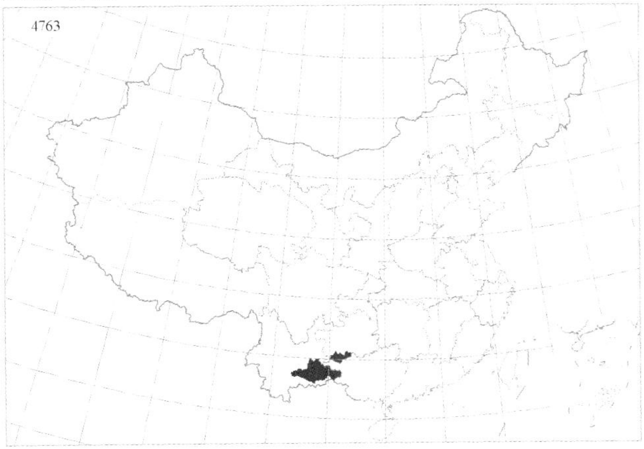

Malpighiaceae 金虎尾科 *Hiptage* 风筝果属
Hiptage multiflora F. N. Wei 多花风筝果
Scandent shrubs, evergreen

22.2; 13.9; 28.3; 22.3; 207.5; 0.0; 1190 ❙ 1374; 713; 87 ❙ 15.4; 1190; 327.7

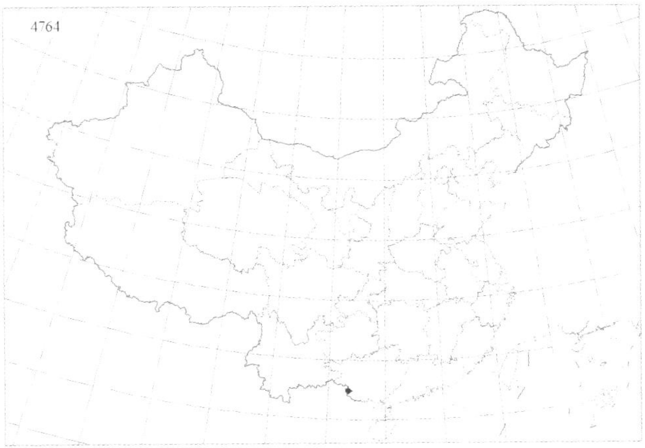

Malpighiaceae 金虎尾科 *Hiptage* 风筝果属
Hiptage tianyangensis F. N. Wei 田阳风筝果
Scandent shrubs, evergreen

19.4；10.4；26.6；19.5；173.8；0.0；999 ❙ 1255；665；59 ❙ 26.1；999；269.3

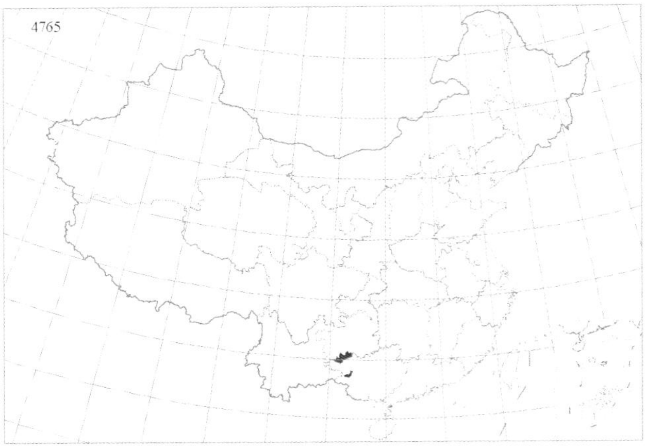

Malpighiaceae 金虎尾科 *Ryssopterys* 翅实藤属
Ryssopterys timoriensis (DC.) Blume ex A. Juss. 翅实藤
Vines

20.0；15.4；23.3；20.0；180.2；0.0；966 ❙ 2821；1610；140 ❙ 194.5；966；304.3

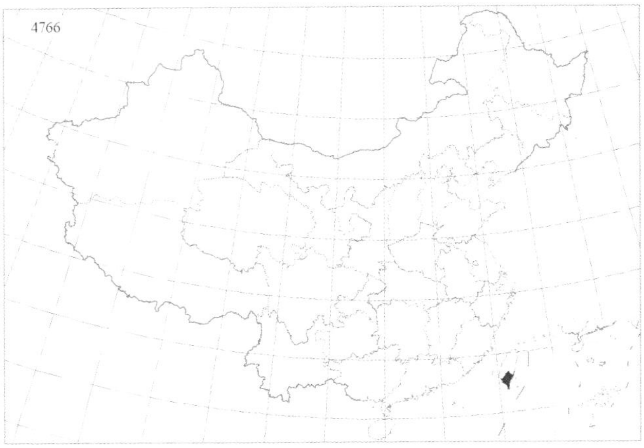

Malpighiaceae 金虎尾科 *Tristellateia* 三星果属
Tristellateia australasiae A. Rich. 三星果
Vines

20.0 (17.9~22.2)；14.1 (11.6~17.7)；25.3 (21.7~27.4)；20.1 (18.0~22.2)；181.0 (155.6~206.7)；0.0 (0.0~0.0)；996 (863~1123) ❙ 2793 (2208~3495)；1068 (750~1847)；420 (107~897) ❙ 180.9 (125.7~224.4)；996 (863~1123)；276.0 (226.8~363.7)

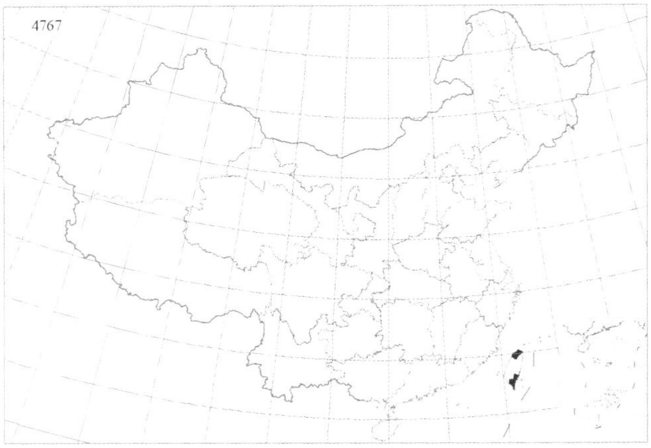

Polygalaceae 远志科 *Polygala* 远志属
Polygala arcuata Hayata 台湾远志
Subshrubs

19.0；14.3；22.4；19.0；167.7；0.0；920 ❙ 2795；1474；180 ❙ 209.1；920；316.6

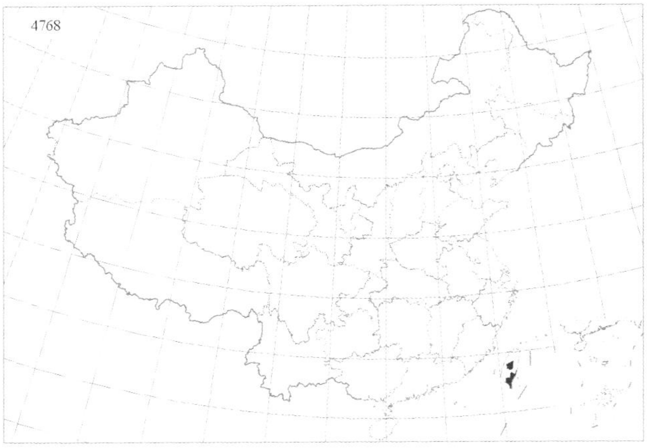

Polygalaceae 远志科 *Polygala* 远志属
Polygala arillata Buchanan-Hamilton ex D. Don 荷包山桂花
Shrubs or small trees

15.3 (-1.8~21.6)；5.6 (-10.2~15.1)；23.8 (6.1~29.3)；15.3 (1.8~21.7)；126.6 (1.9~199.8)；2.9 (0.0~82.8)；812 (329~1153) ❙ 1199 (320~1997)；562 (217~1007)；82 (9~259) ❙ 47.7 (-22.2~145.8)；800 (320~1153)；257.9 (72.2~659.0)

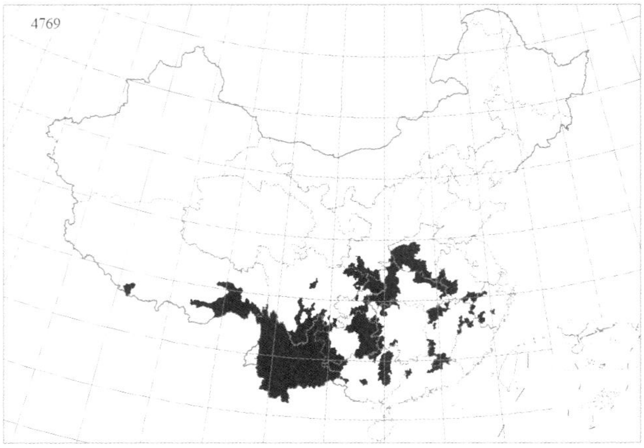

Polygalaceae 远志科 *Polygala* 远志属
Polygala barbellata S. K. Chen 髯毛远志
Shrubs

19.1；12.7；23.5；19.1；169.7；0.0；928 ❙ 1466；830；62 ❙ 58.6；928；525.1

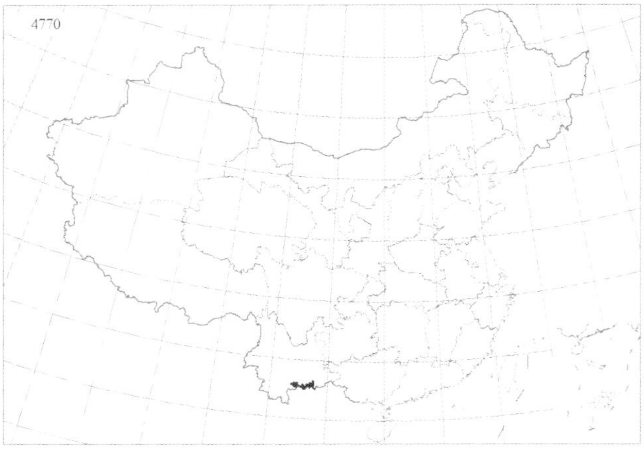

Polygalaceae 远志科 *Polygala* 远志属
Polygala bawanglingensis F. W. Xing et Z. X. Li 坝王远志
Shrubs

23.6；17.8；27.8；23.6；223.3；0.0；1271 ▎1242；444；44 ▎-2.2；1091；439.7

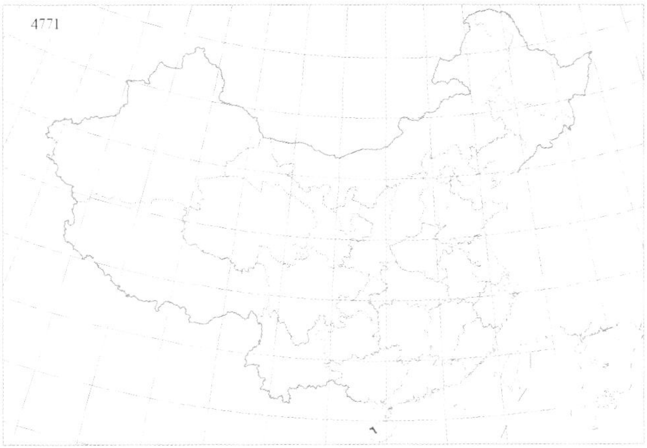

Polygalaceae 远志科 *Polygala* 远志属
Polygala caudata Rehder et E. H. Wilson 尾叶远志
Shrubs

16.0 (11.4~22.0)；6.0 (1.1~13.4)；24.8 (18.8~28.7)；16.0 (11.4~22.0)；132.4 (80.7~204.3)；0.5 (0.0~8.5)；833 (652~1176) ▎1265 (874~2005)；563 (444~860)；88 (25~221) ▎51.9 (13.3~106.4)；833 (652~1176)；222.8 (146.0~491.5)

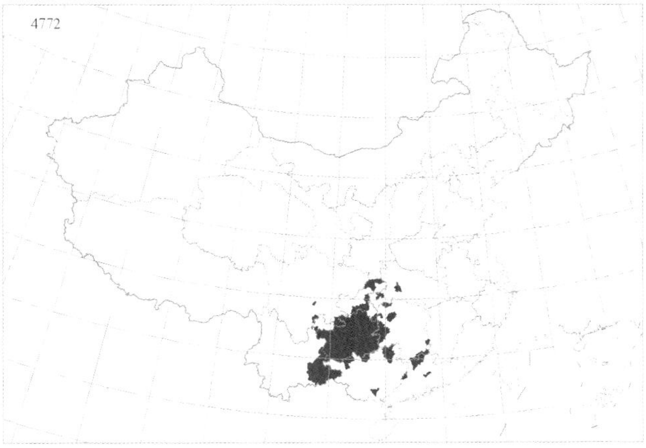

Polygalaceae 远志科 *Polygala* 远志属
Polygala didyma C. Y. Wu 肾果远志
Shrubs or small trees

8.6；1.3；15.0；8.7；52.6；8.8；574 ▎1041；585；45 ▎81.5；574；352.3

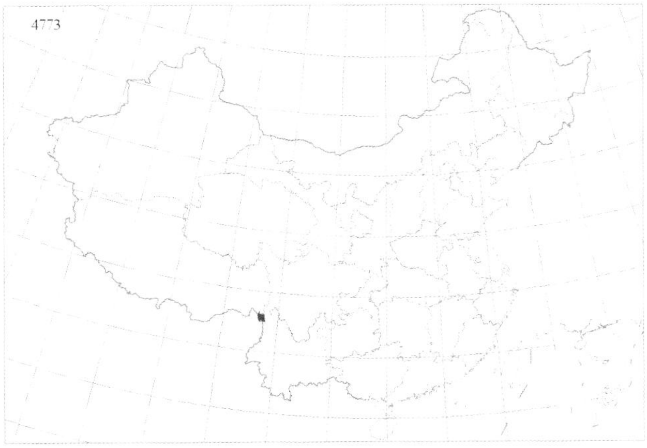

Polygalaceae 远志科 *Polygala* 远志属
Polygala fallax Hemsl. 黄花倒水莲
Shrubs or small trees

18.4 (11.7~23.6)；8.9 (2.7~16.7)；26.8 (18.8~29.8)；18.5 (11.7~23.6)；161.7 (82.6~223.5)；0.1 (0.0~3.8)；965 (658~1281) ▎1517 (916~2435)；615 (356~1428)；145 (31~259) ▎58.5 (-6.9~137.5)；960 (658~1273)；234.0 (95.3~659.0)

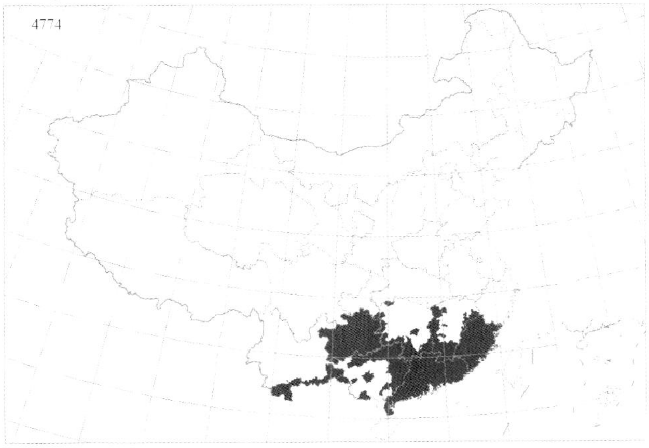

Polygalaceae 远志科 *Polygala* 远志属
Polygala globulifera Dunn 球冠远志
Shrubs

17.6 (8.6~20.5)；11.4 (1.3~15.1)；21.7 (15.0~23.7)；17.6 (8.7~20.5)；152.1 (52.6~185.9)；0.4 (0.0~8.8)；855 (574~992) ▎1333 (1028~1736)；725 (566~1007)；52 (42~77) ▎57.3 (12.5~145.8)；834 (574~986)；431.4 (339.5~659.0)

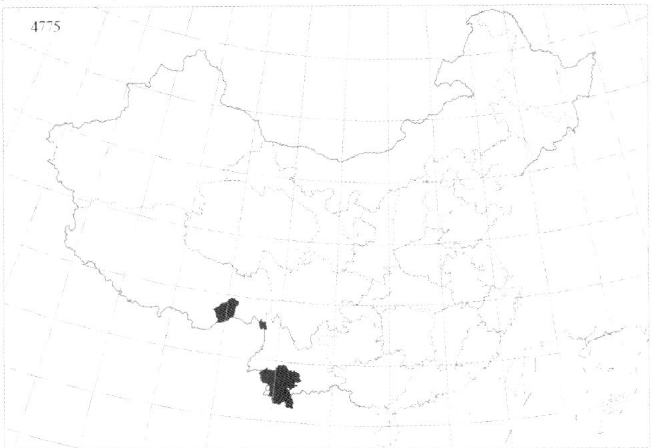

Polygalaceae 远志科 *Polygala* 远志属
Polygala globulifera var. longiracemosa S. K. Chen 长序球冠远志
Shrubs

11.0；3.4；17.3；11.0；76.5；4.4；640 ▎1389；796；48 ▎113.7；640；399.9

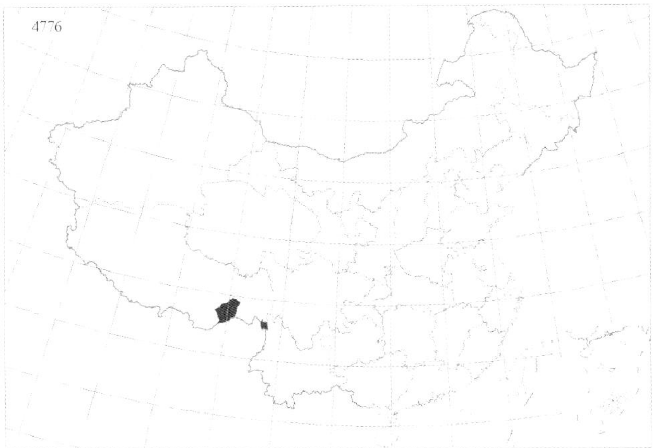

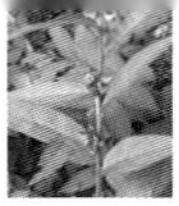

Polygalaceae 远志科 *Polygala* 远志属

Polygala hainanensis Chun et F. C. How 海南远志

Subshrubs

23.1 (19.0~24.8); 16.6 (9.1~20.1); 28.0 (25.7~29.8); 23.2 (19.0~24.8); 218.0 (168.2~237.6); 0.0 (0.0~0.0); 1255 (989~1385) ❙ 1538 (868~1854); 549 (312~712); 98 (41~194) ❙ 24.5 (-34.3~76.3); 1157 (868~1360); 395.8 (174.9~620.0)

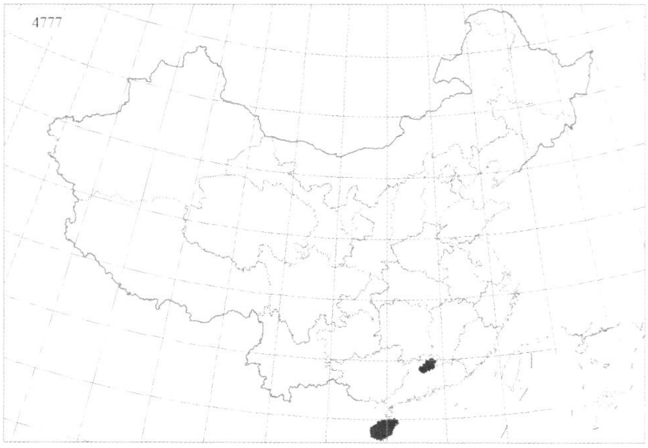

Polygalaceae 远志科 *Polygala* 远志属

Polygala hainanensis var. strigosa Chun et F. C. How 粗毛海南远志

Subshrubs

24.0; 19.1; 27.2; 24.0; 228.1; 0.0; 1302 ❙ 1551; 539; 48 ❙ 19.1; 1147; 620.0

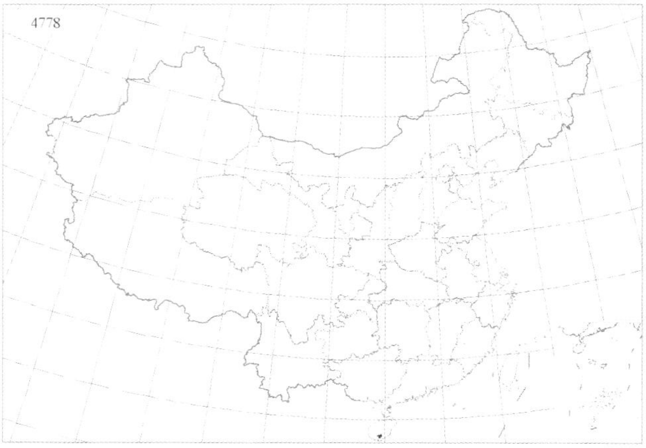

Polygalaceae 远志科 *Polygala* 远志属

Polygala karensium Kurz 密花远志

Shrubs, evergreen

18.0 (4.4~22.4); 11.1 (-4.3~15.1); 22.8 (12.1~28.6); 18.0 (5.3~22.4); 156.8 (28.5~209.1); 0.6 (0.0~34.6); 883 (455~1204) ❙ 1340 (538~2148); 732 (337~1234); 60 (9~142) ❙ 51.5 (12.5~145.8); 872 (455~1204); 389.8 (168.2~659.0)

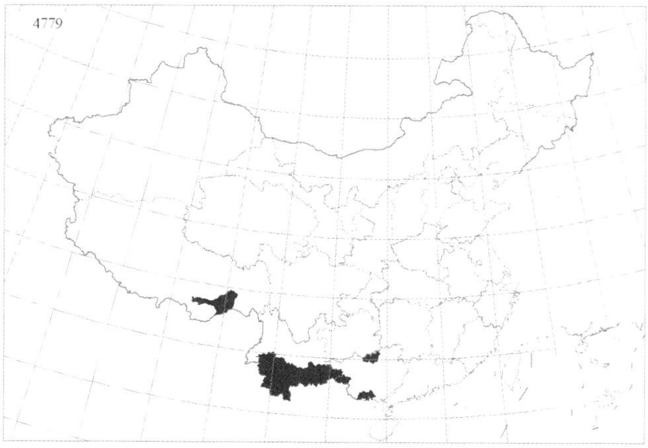

Polygalaceae 远志科 *Polygala* 远志属

Polygala karensium var. obcordata (C. Y. Wu et S. K. Chen) S. K. Chen et J. Parnell 小叶密花远志

Shrubs, evergreen

18.0 (14.2~21.0); 11.4 (6.2~15.1); 22.4 (20.1~25.6); 18.0 (14.2~21.0); 156.3 (110.5~191.8); 0.0 (0.0~0.0); 871 (729~1053) ❙ 1311 (958~1751); 712 (531~996); 53 (41~77) ❙ 50.3 (13.3~88.8); 861 (729~1053); 410.9 (248.4~659.0)

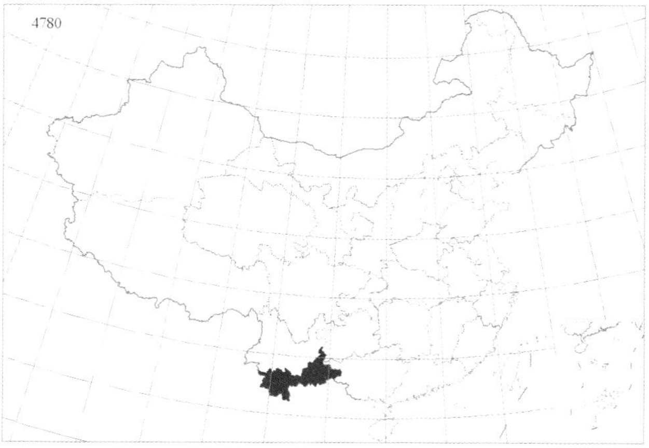

Polygalaceae 远志科 *Polygala* 远志属

Polygala koi Merr. 曲江远志

Subshrubs, evergreen

18.3 (14.2~24.0); 8.5 (3.3~19.1); 27.0 (20.1~29.8); 18.5 (14.2~24.0); 161.9 (110.5~228.1); 0.1 (0.0~2.3); 968 (729~1302) ❙ 1627 (1147~1965); 627 (419~1056); 175 (48~218) ❙ 69.6 (19.1~128.9); 960 (729~1216); 241.4 (163.7~620.0)

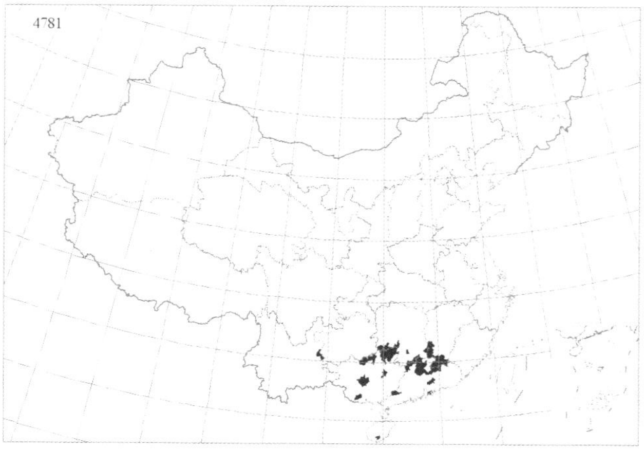

Polygalaceae 远志科 *Polygala* 远志属

Polygala latouchei Franch. 大叶金牛

Subshrubs, evergreen

19.7 (15.0~22.8); 10.6 (4.2~15.5); 27.4 (24.3~28.9); 19.7 (15.1~22.8); 176.9 (120.8~214.1); 0.0 (0.0~0.8); 1030 (789~1225) ❙ 1696 (1302~2435); 699 (512~1428); 167 (57~254) ❙ 66.8 (12.9~147.0); 1030 (789~1225); 255.6 (185.5~371.7)

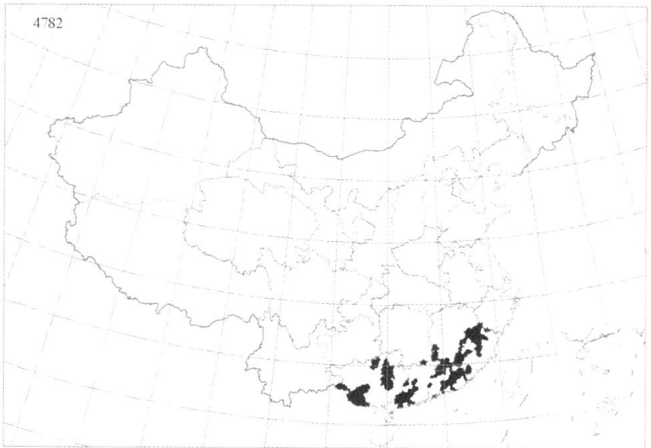

Polygalaceae 远志科 *Polygala* 远志属
Polygala oligosperma C. Y. Wu 少籽远志
Shrubs, evergreen

19.2; 13.8; 22.3; 19.2; 170.5; 0.0; 909 ❙ 1499; 686; 58 ❙ 64.9; 880; 513.9

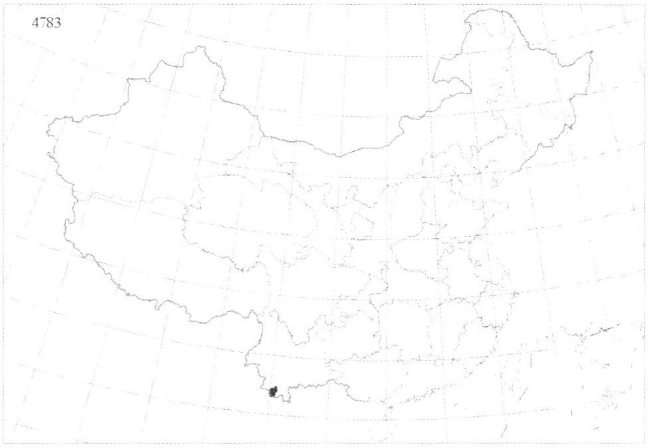

Polygalaceae 远志科 *Polygala* 远志属
Polygala resinosa S. K. Chen 斑果远志
Shrubs, evergreen

22.2; 13.9; 28.3; 22.3; 207.5; 0.0; 1190 ❙ 1374; 713; 87 ❙ 15.4; 1190; 327.7

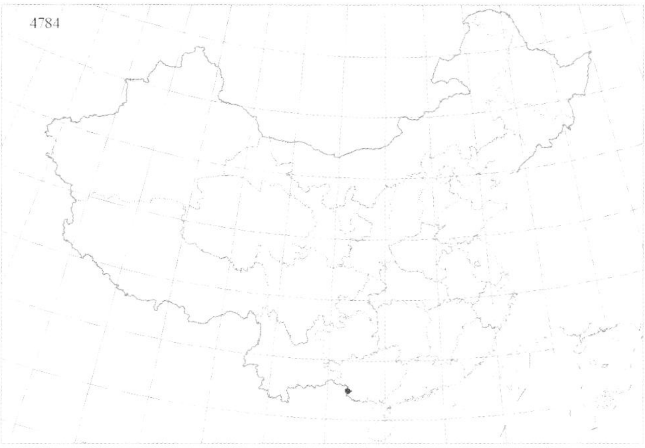

Polygalaceae 远志科 *Polygala* 远志属
Polygala tricholopha Chodat 红花远志
Shrubs, evergreen

21.0 (16.4~24.8); 15.0 (9.6~20.1); 25.0 (20.6~28.3); 21.1 (16.4~24.8); 192.8 (137.3~237.6); 0.0 (0.0~0.0); 1091 (795~1385) ❙ 1582 (1374~1767); 736 (473~959); 58 (41~88) ❙ 51.5 (1.5~112.2); 1029 (795~1299); 527.2 (327.7~627.0)

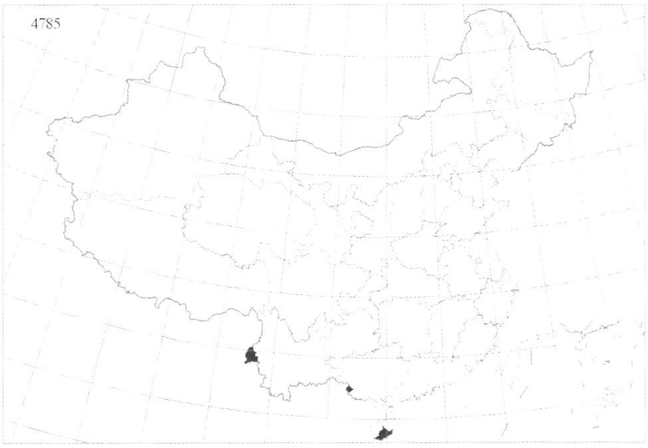

Polygalaceae 远志科 *Polygala* 远志属
Polygala wattersii Hance 长毛籽远志
Shrubs or small trees, evergreen

15.7 (1.8~22.2); 5.6 (-7.1~14.2); 24.8 (9.9~29.8); 15.8 (3.8~22.3); 131.4 (15.7~207.5); 2.7 (0.0~53.3); 837 (404~1190) ❙ 1277 (527~1803); 579 (317~1007); 87 (9~216) ❙ 53.9 (0.0~145.8); 836 (404~1190); 231.4 (108.2~551.6)

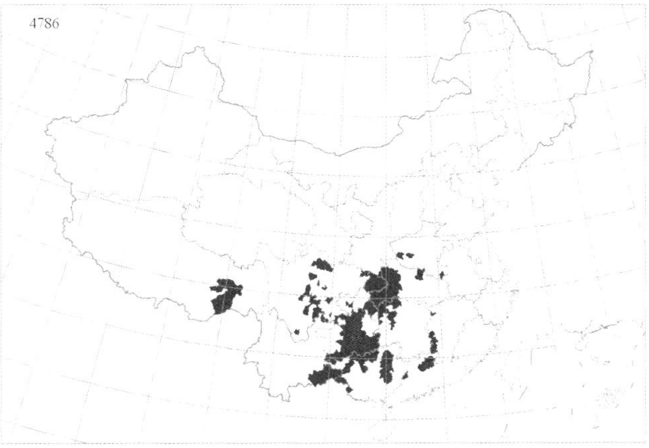

Polygalaceae 远志科 *Securidaca* 蝉翼藤属
Securidaca inappendiculata Hassk. 蝉翼藤
Scandent shrubs

20.7 (17.1~24.8); 12.9 (7.4~20.1); 26.5 (20.8~29.0); 20.7 (17.1~24.8); 188.3 (145.0~237.6); 0.0 (0.0~0.0); 1069 (815~1385) ❙ 1552 (1188~2435); 756 (455~1428); 108 (41~194) ❙ 46.8 (1.5~99.5); 1053 (802~1257); 332.3 (163.7~659.0)

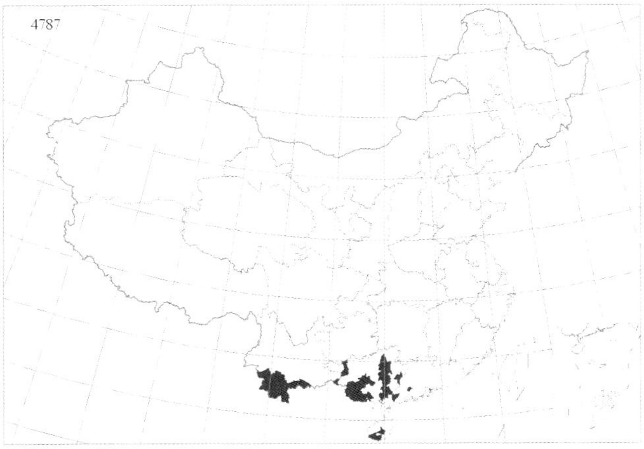

Polygalaceae 远志科 *Securidaca* 蝉翼藤属
Securidaca yaoshanensis Hao 瑶山蝉翼藤
Scandent shrubs

20.6 (17.2~22.7); 12.5 (8.5~15.0); 26.7 (22.4~28.8); 20.6 (17.3~22.8); 187.4 (147.0~213.1); 0.0 (0.0~0.0); 1067 (833~1221) ❙ 1668 (1254~2435); 868 (599~1428); 108 (48~169) ❙ 56.8 (12.9~99.5); 1067 (833~1221); 302.2 (203.4~551.6)

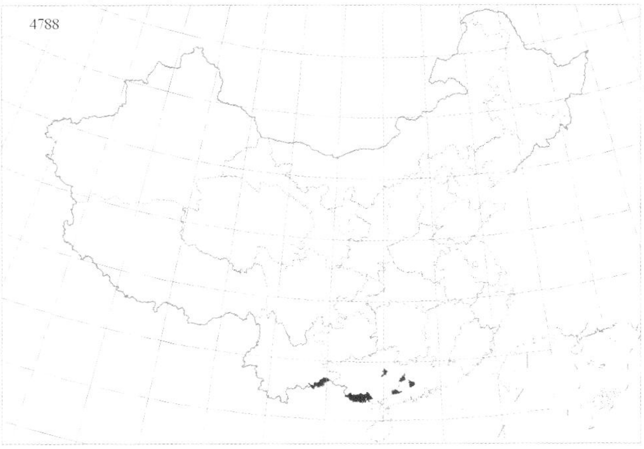

Polygalaceae 远志科 *Xanthophyllum* 黄叶树属

Xanthophyllum flavescens Roxburgh 泰国黄叶树

Trees

19.0 (16.9~21.0); 12.9 (9.9~15.1); 23.1 (21.4~25.6); 19.1 (16.9~21.0); 168.8 (142.7~191.8); 0.0 (0.0~0.0); 919 (813~1053) ❙ 1468 (1178~1751); 782 (646~996); 62 (51~77) ❙ 60.1 (36.8~88.8); 912 (813~1053); 509.7 (383.4~659.0)

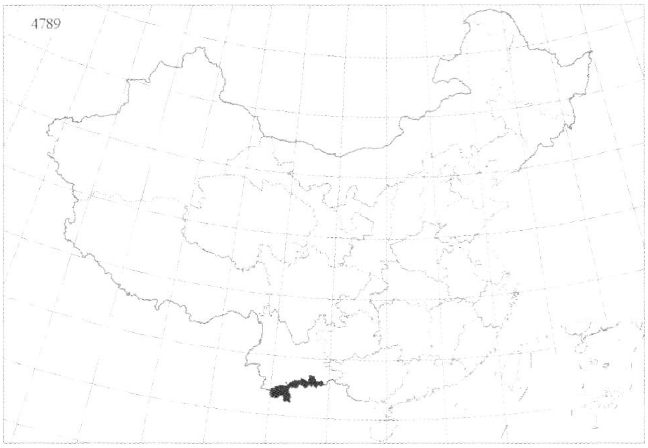

Polygalaceae 远志科 *Xanthophyllum* 黄叶树属

Xanthophyllum hainanense Hu 黄叶树

Trees

21.8 (17.9~24.8); 13.8 (7.4~20.1); 28.1 (25.7~29.8); 21.8 (17.9~24.8); 202.0 (155.4~237.6); 0.0 (0.0~0.0); 1163 (913~1385) ❙ 1616 (868~2435); 693 (312~1428); 128 (41~194) ❙ 40.7 (-34.3~99.5); 1130 (868~1360); 289.9 (174.9~620.0)

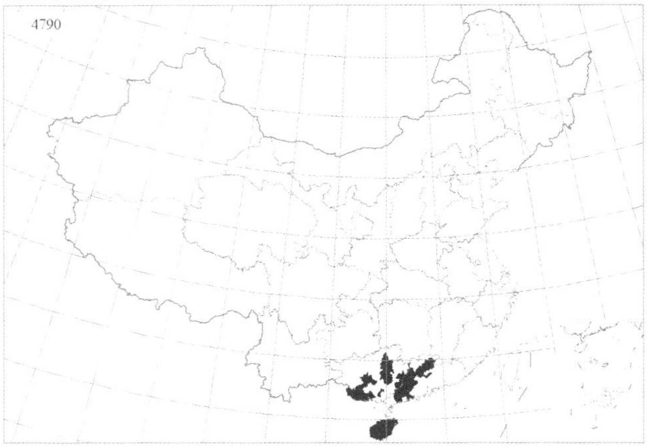

Polygalaceae 远志科 *Xanthophyllum* 黄叶树属

Xanthophyllum oliganthum C. Y. Wu 少花黄叶树

Shrubs or small trees

19.4; 12.6; 24.1; 19.4; 172.6; 0.0; 954 ❙ 1445; 787; 64 ❙ 52.3; 954; 521.5

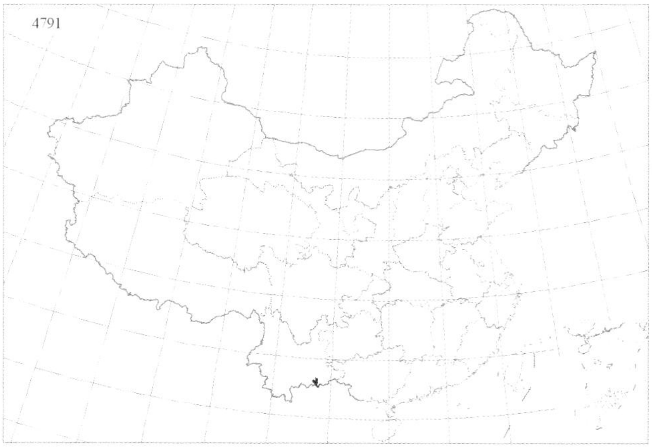

Polygalaceae 远志科 *Xanthophyllum* 黄叶树属

Xanthophyllum yunnanense C. Y. Wu 云南黄叶树

Trees

19.8; 14.3; 23.0; 19.9; 178.2; 0.0; 950 ❙ 1482; 671; 62 ❙ 56.3; 920; 542.0

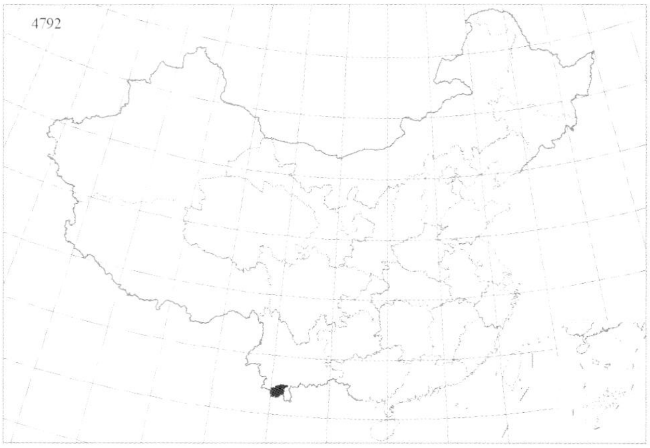

D

Dichapetalaceae 毒鼠子科 *Dichapetalum* 毒鼠子属

Dichapetalum gelonioides (Roxb.) Engl. 毒鼠子

Shrubs or small trees

21.3 (17.2~24.8); 15.7 (9.5~20.1); 25.1 (22.6~28.0); 21.4 (17.3~24.8); 196.6 (147.0~237.6); 0.0 (0.0~0.0); 1099 (838~1385) ❙ 1518 (1254~1725); 640 (473~813); 57 (41~77) ❙ 41.5 (1.5~65.6); 1025 (838~1212); 515.9 (336.3~659.0)

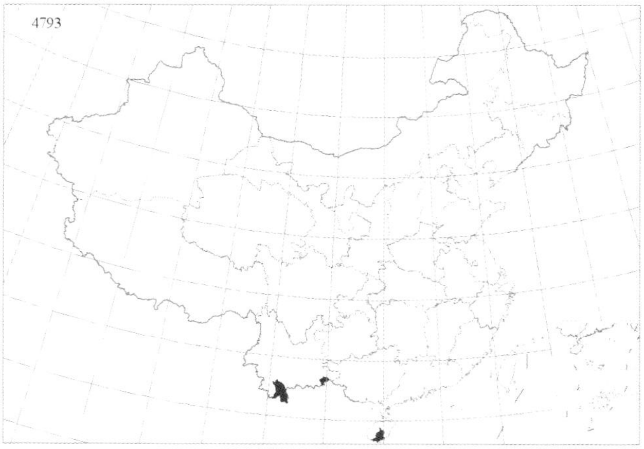

Dichapetalaceae 毒鼠子科 *Dichapetalum* 毒鼠子属

Dichapetalum longipetalum (Turcz.) Engl. 海南毒鼠子

Scandent shrubs

23.6 (21.3~24.8); 17.6 (13.2~20.1); 28.0 (25.7~28.8); 23.6 (21.3~24.8); 223.8 (195.9~237.6); 0.0 (0.0~0.0); 1282 (1111~1385) ❙ 1577 (868~2435); 636 (312~1428); 82 (41~130) ❙ 24.2 (-34.3~99.5); 1188 (868~1360); 408.3 (235.0~620.0)

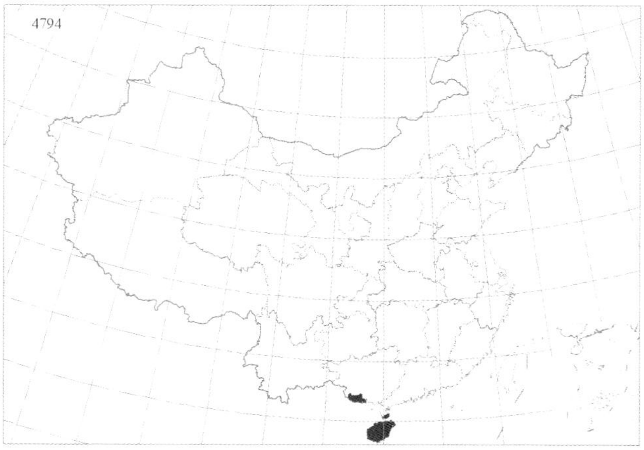

Euphorbiaceae 大戟科 *Acalypha* 铁苋菜属
Acalypha akoensis Hayata 屏东铁苋菜
Shrubs

19.6; 14.6; 23.5; 19.7; 175.8; 0.0; 953 ❙ 2664; 1398; 180 ❙ 180.8; 953; 319.8

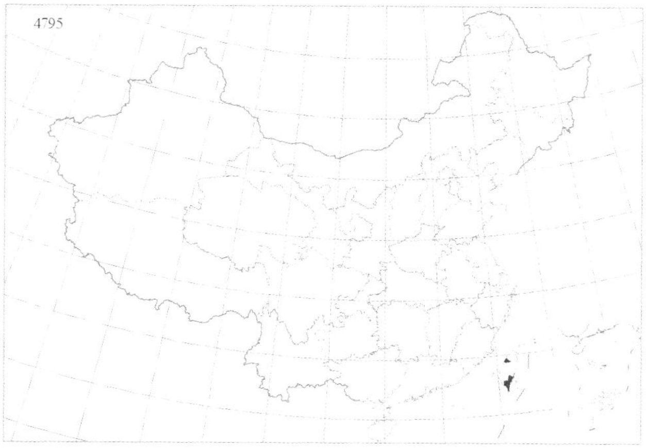

Euphorbiaceae 大戟科 *Acalypha* 铁苋菜属
Acalypha angatensis Blanco 台湾铁苋菜
Shrubs

18.3; 12.5; 23.0; 18.3; 159.8; 0.0; 903 ❙ 2299; 1088; 203 ❙ 167.5; 888; 257.8

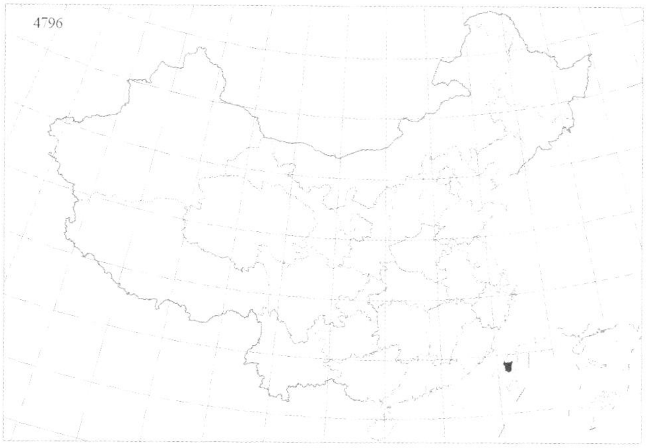

Euphorbiaceae 大戟科 *Acalypha* 铁苋菜属
Acalypha caturus Blume 尖尾铁苋菜
Small trees

20.3; 15.9; 23.6; 20.4; 184.4; 0.0; 993 ❙ 2853; 1618; 147 ❙ 190.4; 993; 330.3

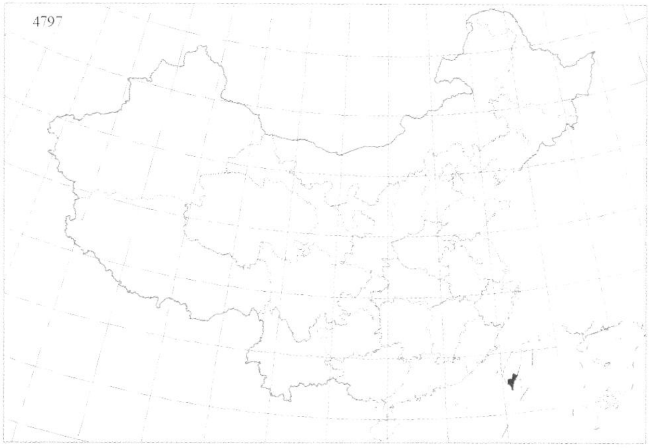

Euphorbiaceae 大戟科 *Acalypha* 铁苋菜属
Acalypha hainanensis Merr. et Chun 海南铁苋菜
Small shrubs

24.8; 20.1; 28.0; 24.8; 237.6; 0.0; 1385 ❙ 1405; 473; 41 ❙ 1.5; 1137; 527.5

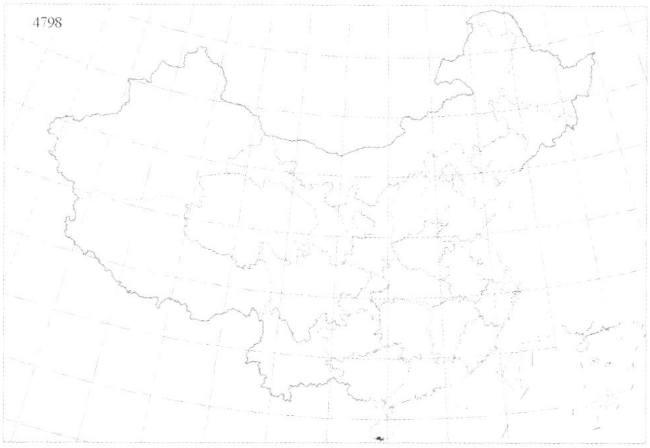

Euphorbiaceae 大戟科 *Acalypha* 铁苋菜属
Acalypha hispida Burm. f. 红穗铁苋菜
Shrubs, evergreen

23.8 (21.7~24.8); 18.0 (14.0~20.1); 28.0 (25.7~28.8); 23.8 (21.7~24.8); 226.1 (200.5~237.6); 0.0 (0.0~0.0); 1297 (1127~1385) ❙ 1513 (868~1854); 553 (312~862); 77 (41~125) ❙ 17.2 (-34.3~47.8); 1190 (868~1360); 414.5 (139.6~620.0)

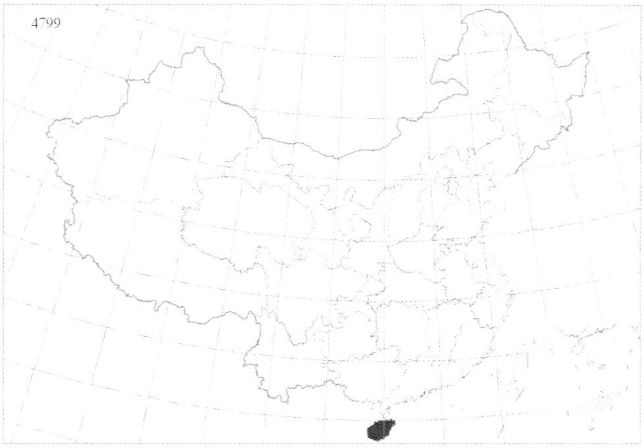

Euphorbiaceae 大戟科 *Acalypha* 铁苋菜属
Acalypha kerrii Craib 卵叶铁苋菜
Shrubs

20.7 (19.1~22.2); 12.5 (10.7~13.9); 27.0 (23.3~29.0); 20.8 (19.1~22.3); 189.6 (169.6~207.5); 0.0 (0.0~0.0); 1077 (917~1190) ❙ 1456 (1374~1529); 774 (639~848); 95 (56~138) ❙ 36.6 (15.4~55.9); 1077 (917~1190); 320.0 (160.7~506.2)

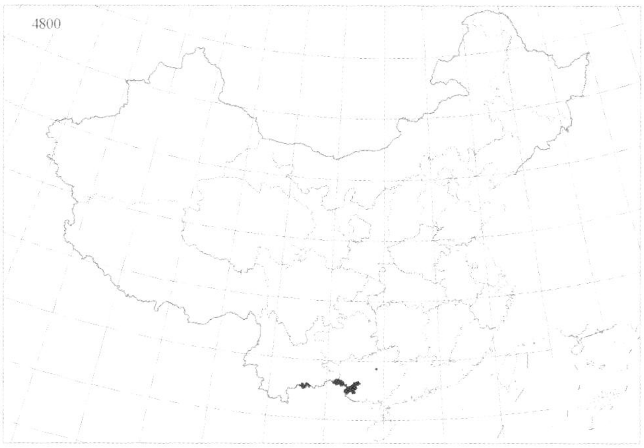

Euphorbiaceae 大戟科 *Acalypha* 铁苋菜属
Acalypha mairei (H. Lév.) C. K. Schneid. 毛叶铁苋菜
Shrubs, deciduous

13.5 (2.2~19.3); 5.8 (-6.3~13.1); 19.8 (9.5~26.6); 13.6 (3.9~19.4); 106.7 (15.8~173.2); 3.7 (0.0~48.6); 725 (411~981) ❙ 984 (729~1438); 571 (411~820); 32 (11~73) ❙ 38.6 (-14.2~121.7); 699 (411~981); 277.1 (142.9~383.4)

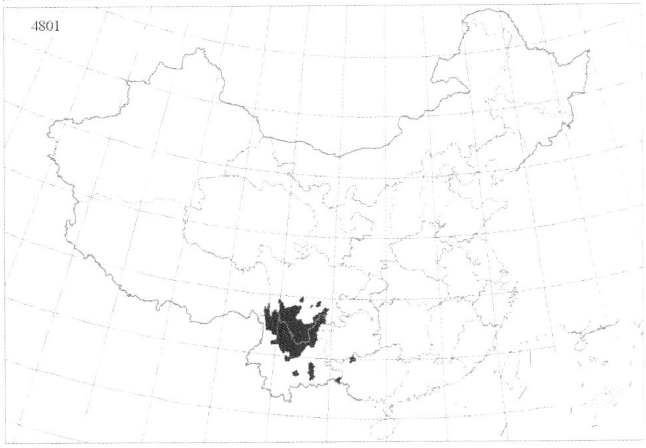

Euphorbiaceae 大戟科 *Acalypha* 铁苋菜属
Acalypha matsudae Hayata 恒春铁苋菜
Shrubs

21.9; 17.0; 25.4; 21.9; 202.8; 0.0; 1119 ❙ 2501; 1537; 94 ❙ 133.8; 1043; 234.5

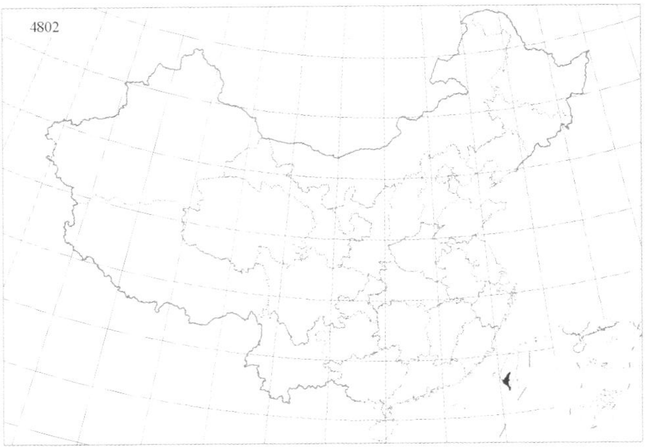

Euphorbiaceae 大戟科 *Acalypha* 铁苋菜属
Acalypha schneideriana Pax et Hoffm. 丽江铁苋菜
Shrubs, deciduous

13.3 (2.2~19.1); 4.3 (-6.7~12.0); 21.1 (9.5~28.1); 13.4 (3.9~19.1); 104.9 (15.8~169.2); 5.3 (0.0~48.6); 728 (411~969) ❙ 986 (373~1577); 535 (201~932); 35 (7~94) ❙ 36.8 (-37.6~101.4); 710 (373~969); 261.4 (88.1~407.1)

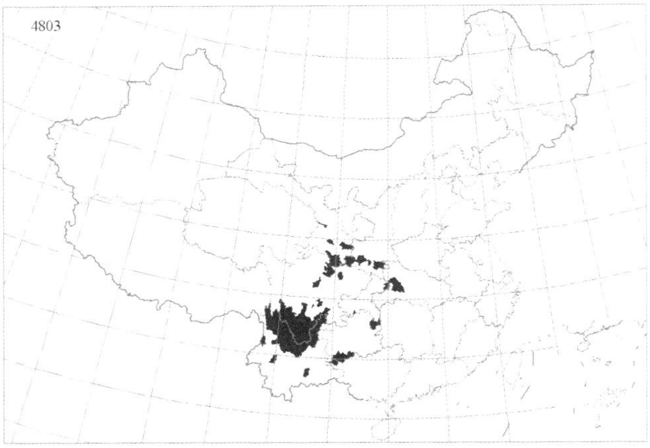

Euphorbiaceae 大戟科 *Actephila* 喜光花属
Actephila excelsa (Dalzell) Müll. Arg. 毛喜光花
Shrubs

20.5 (17.9~22.2); 13.6 (11.9~15.1); 25.1 (21.6~28.4); 20.5 (18.0~22.3); 186.6 (155.8~207.5); 0.0 (0.0~0.0); 1034 (854~1190) ❙ 1476 (1366~1607); 755 (655~850); 75 (41~116) ❙ 44.5 (15.4~68.0); 1021 (833~1190); 455.0 (273.1~659.0)

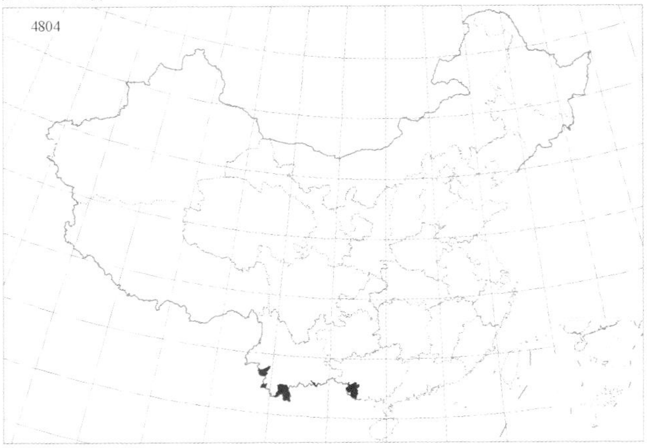

Euphorbiaceae 大戟科 *Actephila* 喜光花属
Actephila merrilliana Chun 喜光花
Shrubs

23.8 (22.1~24.8); 17.9 (13.7~20.1); 28.0 (25.7~28.8); 23.9 (22.1~24.8); 226.2 (205.6~237.6); 0.0 (0.0~0.0); 1298 (1127~1385) ❙ 1509 (868~1854); 554 (312~779); 78 (41~144) ❙ 16.7 (-34.3~50.9); 1191 (868~1360); 421.5 (157.7~620.0)

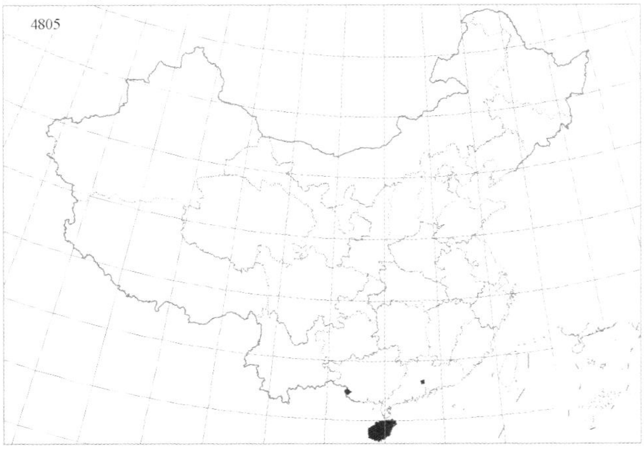

Euphorbiaceae 大戟科 *Alchornea* 山麻杆属
Alchornea davidii Franch. 山麻杆
Shrubs, deciduous

15.6 (2.2~21.9); 3.7 (-6.6~14.2); 26.7 (9.9~30.1); 15.6 (3.8~22.0); 130.7 (15.5~203.8); 3.5 (0.0~49.1); 856 (413~1176) ❙ 1199 (512~1922); 500 (258~835); 114 (17~259) ❙ 39.8 (-35.3~129.9); 836 (413~1176); 202.1 (110.1~551.6)

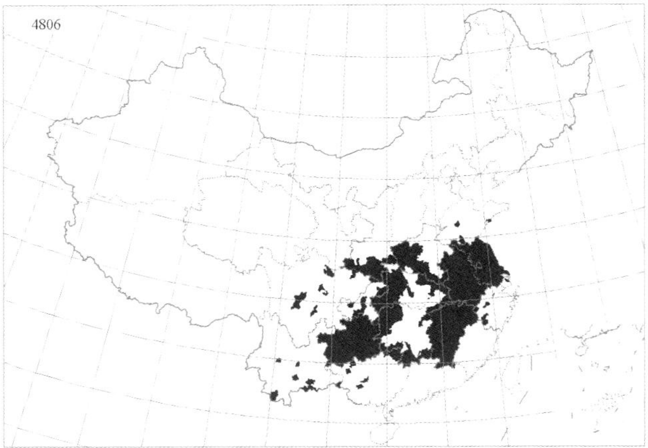

Euphorbiaceae 大戟科 *Alchornea* 山麻杆属
Alchornea hunanensis H. S. Kiu 湖南山麻杆
Shrubs

16.0 (14.3~17.7); 4.3 (3.3~5.4); 27.0 (24.3~29.9); 16.0 (14.3~17.7); 132.9 (114.1~153.0); 0.8 (0.0~2.3); 862 (765~976) ❙ 1422 (1337~1526); 521 (439~600); 148 (105~198) ❙ 65.7 (43.9~99.4); 849 (765~900); 204.9 (141.7~257.1)

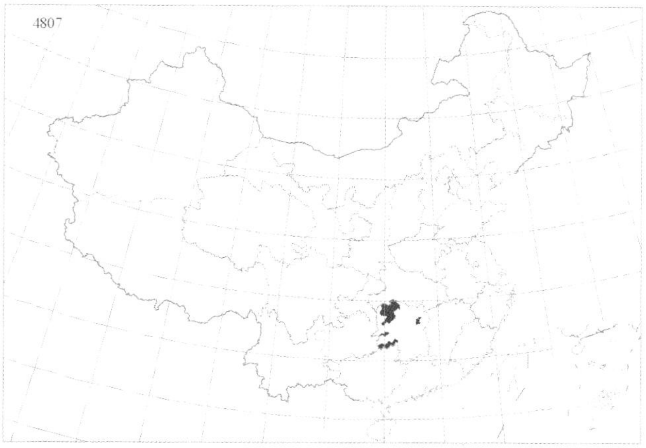

Euphorbiaceae 大戟科 *Alchornea* 山麻杆属
Alchornea kelungensis Hayata 厚柱山麻杆
Shrubs

20.5; 14.0; 26.6; 20.5; 186.1; 0.0; 1032 ❙ 3244; 873; 722 ❙ 213.9; 1032; 244.1

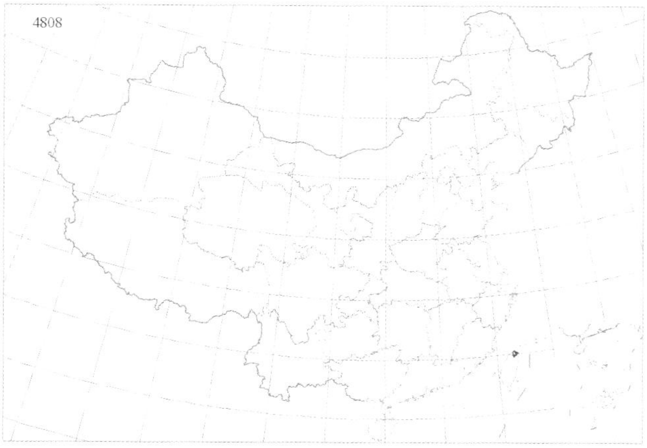

Euphorbiaceae 大戟科 *Alchornea* 山麻杆属
Alchornea rugosa (Lour.) Müll. Arg. 羽脉山麻杆
Shrubs or trees

20.8 (14.6~24.8); 12.2 (3.8~20.1); 27.5 (23.1~29.2); 20.8 (14.7~24.8); 189.6 (117.7~237.6); 0.0 (0.0~1.5); 1097 (779~1385) ❙ 1513 (868~2435); 688 (312~1428); 120 (41~196) ❙ 39.9 (-34.3~101.8); 1078 (779~1360); 280.0 (157.7~620.0)

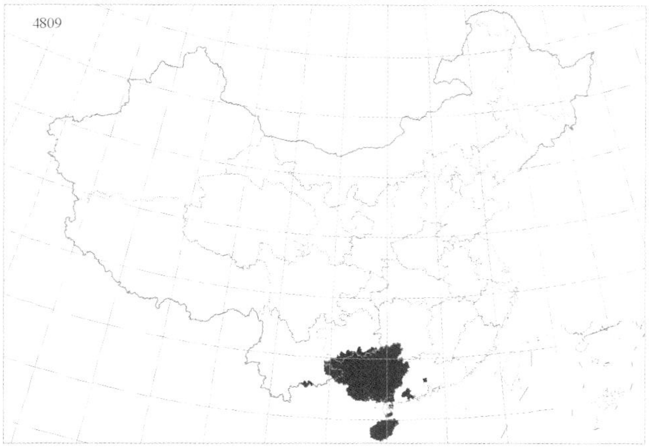

Euphorbiaceae 大戟科 *Alchornea* 山麻杆属
Alchornea rugosa var. pubescens (Pax et K. Hoffm.) H. S. Kiu 海南山麻杆
Shrubs or trees

23.2 (22.0~24.8); 16.1 (13.3~20.1); 28.4 (27.8~28.9); 23.2 (22.0~24.8); 218.5 (204.5~237.6); 0.0 (0.0~0.0); 1260 (1179~1385) ❙ 1401 (1176~1600); 615 (444~839); 90 (41~136) ❙ 12.1 (-12.0~34.7); 1156 (1091~1202); 341.4 (201.4~527.5)

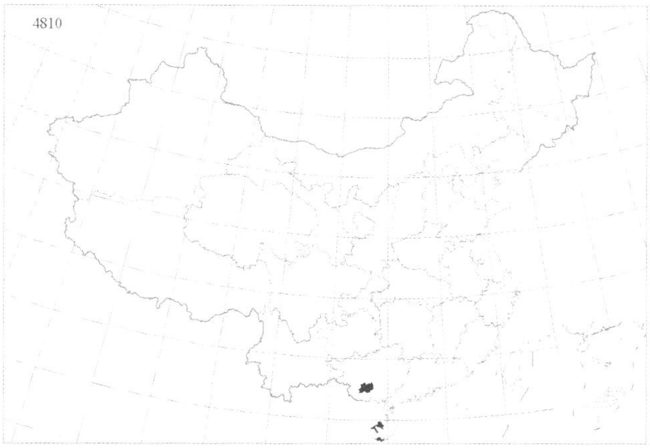

Euphorbiaceae 大戟科 *Alchornea* 山麻杆属
Alchornea tiliifolia (Benth.) Müll. Arg. 椴叶山麻杆
Shrubs or trees

19.9 (14.6~23.2); 11.2 (3.8~15.8); 26.7 (22.3~29.0); 19.9 (14.7~23.2); 178.9 (117.7~218.9); 0.0 (0.0~1.5); 1031 (779~1257) ❙ 1556 (1143~2435); 748 (507~1428); 119 (50~191) ❙ 52.0 (12.9~101.8); 1029 (779~1257); 286.8 (160.8~659.0)

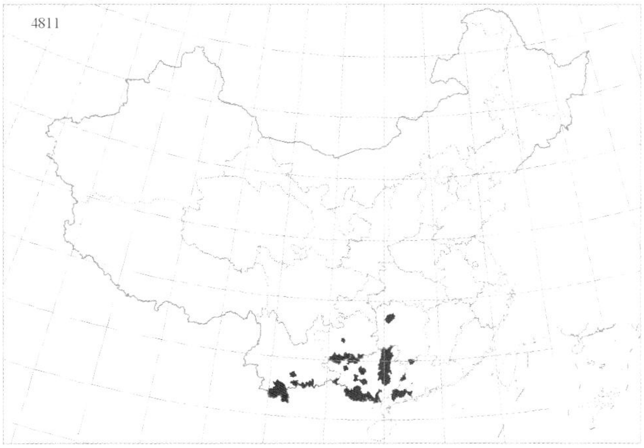

Euphorbiaceae 大戟科 *Alchornea* 山麻杆属
Alchornea trewioides (Benth.) Müll. Arg. 红背山麻杆
Shrubs

19.5 (11.3~24.8); 10.4 (3.3~20.1); 27.2 (17.0~29.9); 19.5 (11.3~24.8); 174.6 (76.6~237.6); 0.1 (0.0~2.3); 1028 (644~1385) ❙ 1539 (868~2435); 655 (312~1428); 135 (41~258) ❙ 51.8 (-34.3~130.7); 1018 (644~1360); 245.5 (95.3~620.0)

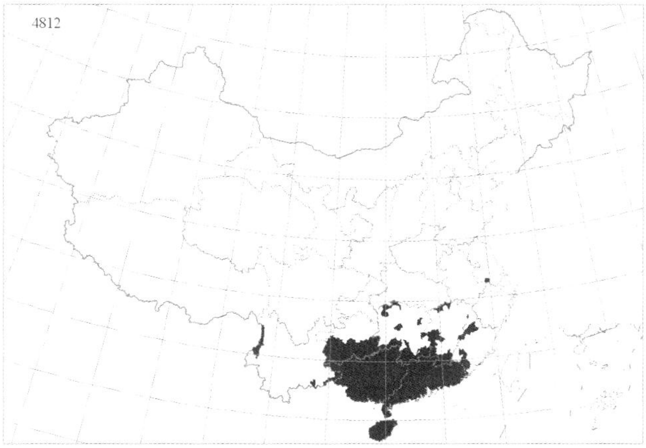

Euphorbiaceae 大戟科 *Alchornea* 山麻杆属

Alchornea trewioides var. sinica (Benth.) Müll. Arg. 绿背山麻杆

Shrubs

19.1 (17.1~21.4); 11.0 (8.5~13.0); 25.1 (23.3~28.2); 19.2 (17.1~21.5); 170.1 (145.1~197.8); 0.0 (0.0~0.0); 961 (845~1140) ❙ 1369 (1232~1498); 752 (666~835); 73 (56~130) ❙ 43.1 (30.5~55.7); 961 (845~1140); 340.2 (248.2~506.2)

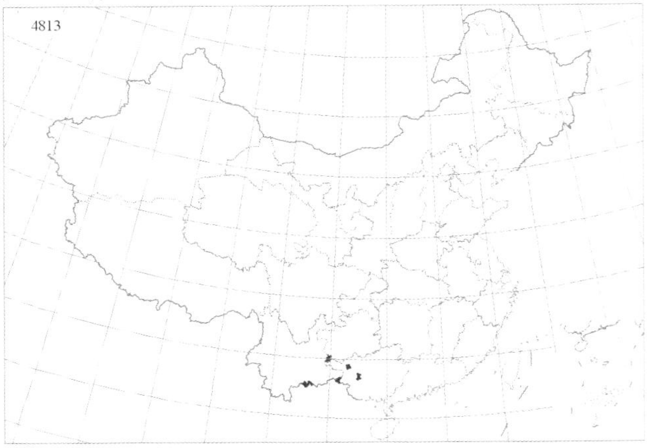

Euphorbiaceae 大戟科 *Aleurites* 石栗属

Aleurites moluccana (L.) Willd. 石栗

Trees, evergreen

21.2 (13.4~24.8); 14.2 (2.7~20.1); 26.4 (20.6~28.9); 21.2 (13.4~24.8); 194.9 (104.5~237.6); 0.1 (0.0~3.7); 1110 (735~1385) ❙ 1451 (868~1900); 646 (312~1056); 83 (41~178) ❙ 33.5 (-34.3~95.9); 1066 (735~1360); 372.3 (157.7~659.0)

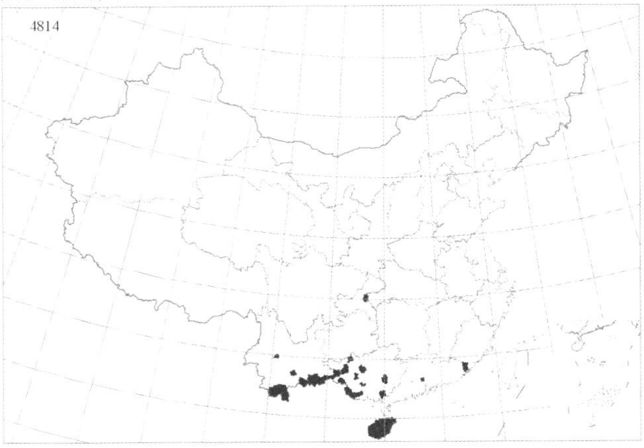

Euphorbiaceae 大戟科 *Antidesma* 五月茶属

Antidesma acidum Retz. 西南五月茶

Shrubs or small trees, evergreen

18.8 (15.5~21.6); 12.2 (5.9~15.1); 23.4 (22.1~28.5); 18.8 (15.6~21.7); 165.9 (127.1~199.8); 0.0 (0.0~0.0); 917 (798~1153) ❙ 1394 (1059~1625); 737 (604~911); 67 (44~173) ❙ 52.3 (21.3~80.6); 909 (798~1153); 437.1 (194.4~659.0)

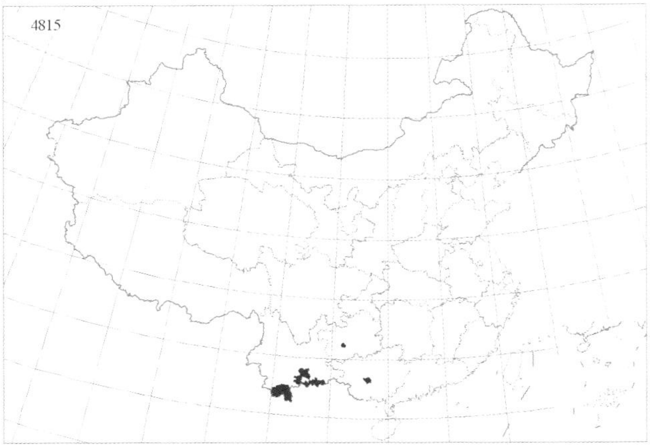

Euphorbiaceae 大戟科 *Antidesma* 五月茶属

Antidesma bunius (L.) Spreng. 五月茶

Trees, evergreen

19.9 (13.2~24.8); 11.0 (3.5~20.1); 27.3 (18.4~30.1); 19.9 (13.3~24.8); 179.2 (99.4~237.6); 0.1 (0.0~1.7); 1049 (695~1385) ❙ 1571 (868~2435); 669 (299~1428); 133 (41~254) ❙ 52.0 (-34.3~145.8); 1030 (695~1360); 259.5 (95.3~659.0)

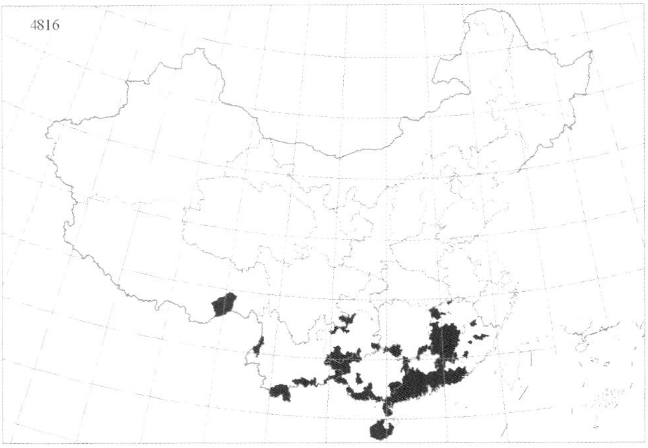

Euphorbiaceae 大戟科 *Antidesma* 五月茶属

Antidesma fordii Hemsl. 黄毛五月茶

Small trees, evergreen

20.6 (16.9~24.8); 12.3 (7.1~20.1); 27.2 (20.8~29.0); 20.6 (16.9~24.8); 187.6 (142.7~237.6); 0.0 (0.0~0.0); 1073 (813~1385) ❙ 1558 (1143~2435); 714 (425~1428); 126 (41~257) ❙ 46.2 (-2.2~99.5); 1066 (802~1269); 285.2 (157.7~659.0)

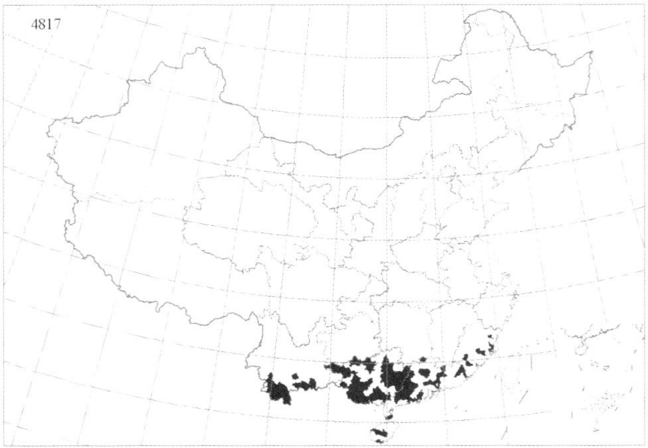

Euphorbiaceae 大戟科 *Antidesma* 五月茶属

Antidesma ghaesembilla Gaertn. 方叶五月茶

Trees, evergreen

22.0 (18.9~24.8); 14.4 (10.6~20.1); 28.0 (22.6~29.2); 22.1 (19.0~24.8); 204.9 (167.5~237.6); 0.0 (0.0~0.0); 1170 (909~1385) ❙ 1631 (868~2037); 720 (312~1056); 115 (41~197) ❙ 40.6 (-34.3~95.5); 1149 (868~1360); 266.3 (95.3~659.0)

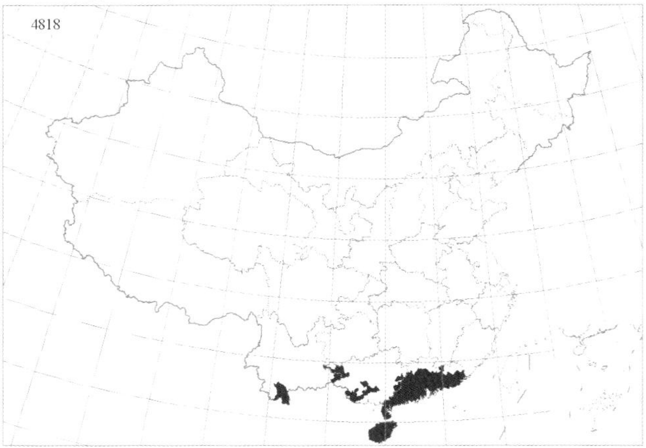

Euphorbiaceae 大戟科 *Antidesma* 五月茶属

Antidesma hainanense Merr. 海南五月茶

Shrubs, evergreen

22.8 (19.2~24.8); 16.8 (13.8~20.1); 27.0 (22.3~29.2); 22.8 (19.2~24.8); 214.2 (170.5~237.6); 0.0 (0.0~0.0); 1213 (909~1385) ❙ 1560 (1405~1725); 656 (473~893); 72 (41~110) ❙ 30.4 (1.5~64.9); 1148 (880~1326); 445.4 (178.7~620.0)

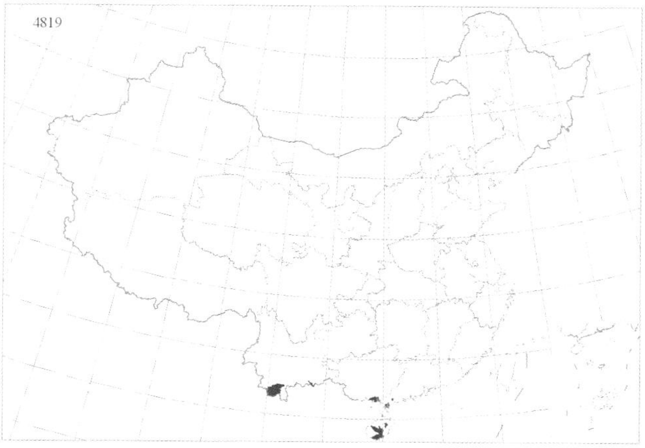

Euphorbiaceae 大戟科 *Antidesma* 五月茶属

Antidesma hontaushanense C. E. Chang 河头山五月茶

Trees

18.5; 14.1; 21.7; 18.5; 162.1; 0.0; 863 ❙ 2713; 1388; 188 ❙ 214.4; 863; 363.7

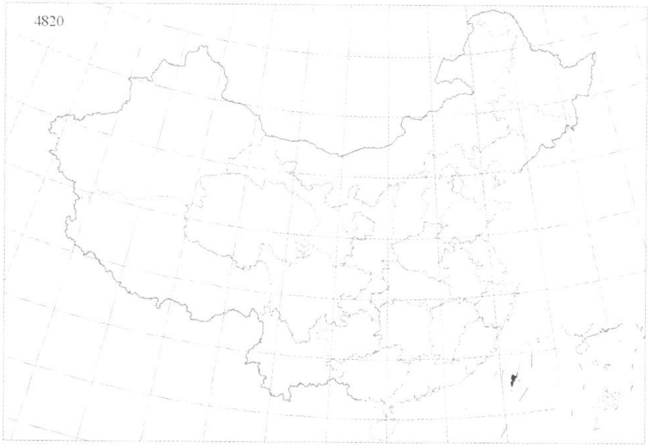

Euphorbiaceae 大戟科 *Antidesma* 五月茶属

Antidesma japonicum Siebold et Zucc. 酸味子

Shrubs, evergreen

18.3 (6.1~24.8); 8.3 (-2.0~20.1); 27.1 (12.7~30.1); 18.3 (6.3~24.8); 160.2 (35.2~237.6); 0.5 (0.0~21.8); 969 (496~1385) ❙ 1494 (750~3495); 601 (299~1847); 145 (20~897) ❙ 55.8 (-34.3~246.5); 955 (496~1360); 233.1 (95.3~659.0)

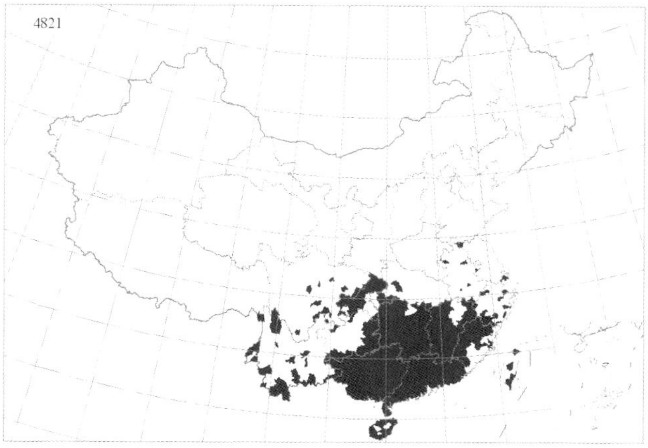

Euphorbiaceae 大戟科 *Antidesma* 五月茶属

Antidesma maclurei Merr. 多花五月茶

Trees, evergreen

24.1 (23.0~24.8); 19.0 (17.7~20.1); 27.6 (26.7~28.4); 24.1 (23.1~24.8); 229.7 (217.0~237.6); 0.0 (0.0~0.0); 1319 (1217~1385) ❙ 1471 (1179~1725); 500 (392~610); 51 (41~68) ❙ 12.1 (-11.9~41.8); 1138 (1056~1212); 520.9 (455.2~620.0)

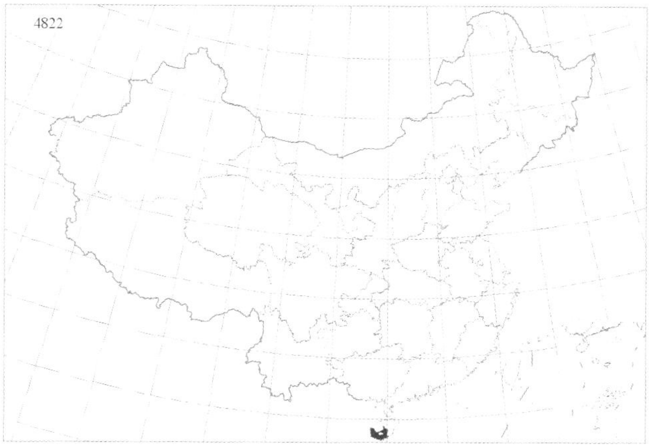

Euphorbiaceae 大戟科 *Antidesma* 五月茶属

Antidesma montanum Blume 山地五月茶

Trees, evergreen

20.5 (13.3~24.8); 13.2 (4.3~20.1); 26.0 (19.6~28.8); 20.5 (13.4~24.8); 186.1 (100.4~237.6); 0.0 (0.0~0.7); 1061 (706~1385) ❙ 1509 (868~2992); 701 (312~1847); 86 (41~303) ❙ 46.3 (-34.3~219.3); 1027 (706~1360); 365.3 (182.9~659.0)

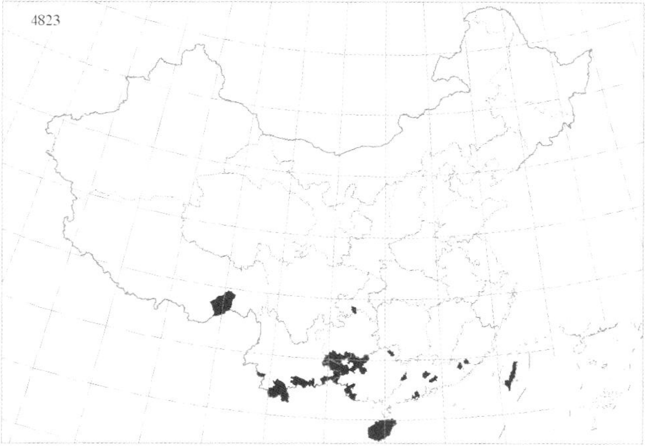

Euphorbiaceae 大戟科 *Antidesma* 五月茶属

Antidesma montanum var. microphyllum Petra ex Hoffmam. 小叶五月茶

Shrubs, evergreen

18.8 (12.8~24.8); 9.9 (2.6~20.1); 26.1 (19.1~28.9); 18.8 (12.9~24.8); 165.7 (96.0~237.6); 0.1 (0.0~4.0); 971 (699~1385) ❙ 1471 (868~2030); 639 (312~1033); 118 (36~228) ❙ 53.2 (-34.3~128.9); 957 (699~1326); 285.9 (143.8~659.0)

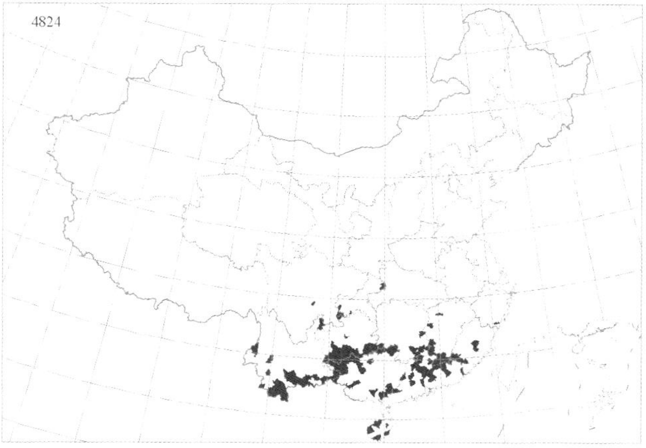

Euphorbiaceae 大戟科 *Antidesma* 五月茶属
Antidesma nienkui Merr. et Chun 大果五月茶
Trees, evergreen

23.9; 18.9; 27.3; 23.9; 227.3; 0.0; 1301 ❙ 1565; 542; 55 ❙ 21.6; 1150; 550.2

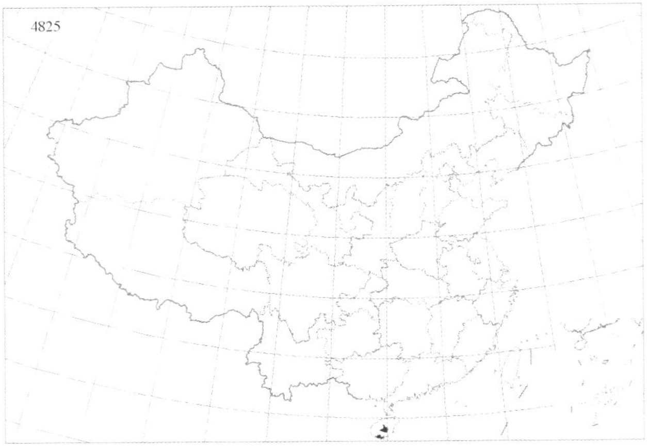

Euphorbiaceae 大戟科 *Antidesma* 五月茶属
Antidesma sootepense Craib 泰北五月茶
Trees, evergreen

19.8; 13.0; 24.3; 19.8; 177.7; 0.0; 968 ❙ 1451; 738; 62 ❙ 49.9; 952; 470.9

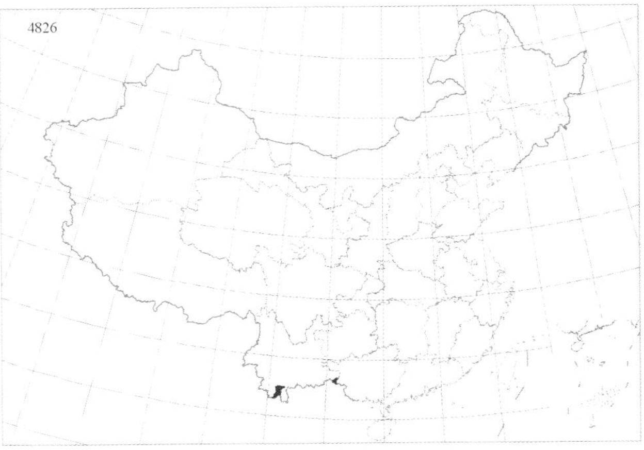

Euphorbiaceae 大戟科 *Aporosa* 银柴属
Aporosa dioica (Roxb.) Müll. Arg. 银柴
Trees

21.9 (17.3~24.8); 14.7 (9.5~20.1); 27.4 (21.6~29.2); 22.0 (17.3~24.8); 203.6 (148.0~237.6); 0.0 (0.0~0.0); 1159 (832~1385) ❙ 1578 (868~2435); 729 (312~1428); 97 (41~177) ❙ 37.5 (-34.3~99.5); 1126 (783~1360); 325.6 (120.3~659.0)

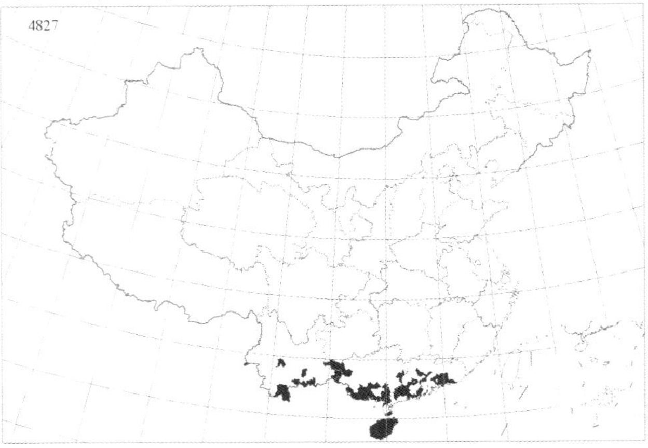

Euphorbiaceae 大戟科 *Aporosa* 银柴属
Aporosa planchoniana Baill. ex Müll. Arg. 全缘叶银柴
Shrubs

18.9; 12.9; 22.9; 18.9; 166.8; 0.0; 902 ❙ 1487; 873; 59 ❙ 65.0; 902; 528.7

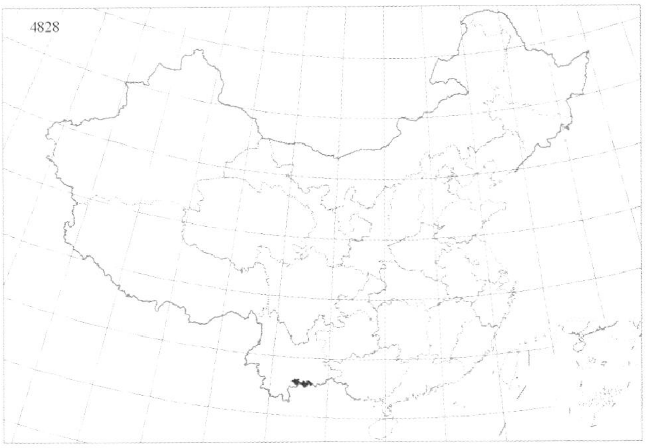

Euphorbiaceae 大戟科 *Aporosa* 银柴属
Aporosa villosa (Lindl.) Baill. 毛银柴
Shrubs or small trees

19.9 (14.4~24.7); 12.9 (7.5~19.8); 24.7 (19.1~28.9); 19.9 (14.4~24.7); 178.7 (113.1~236.5); 0.0 (0.0~0.0); 1004 (727~1376) ❙ 1439 (1049~2005); 727 (455~911); 73 (42~188) ❙ 45.8 (5.8~99.9); 980 (727~1232); 404.4 (203.9~659.0)

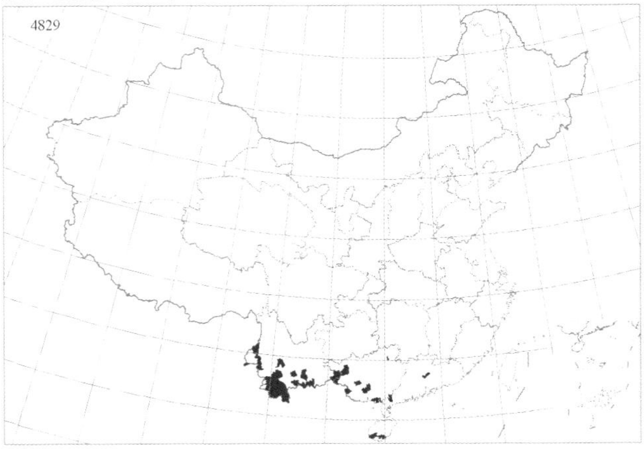

Euphorbiaceae 大戟科 *Aporosa* 银柴属
Aporosa yunnanensis (Pax et K. Hoffm.) F. P. Metcalf 云南银柴
Small trees, evergreen

18.9 (11.7~24.8); 10.6 (2.7~20.1); 25.6 (18.8~29.0); 19.0 (11.7~24.8); 167.9 (82.6~237.6); 0.1 (0.0~3.8); 981 (658~1385) ❙ 1450 (916~2435); 706 (455~1428); 90 (31~197) ❙ 49.1 (-1.4~99.5); 971 (658~1257); 280.2 (122.7~659.0)

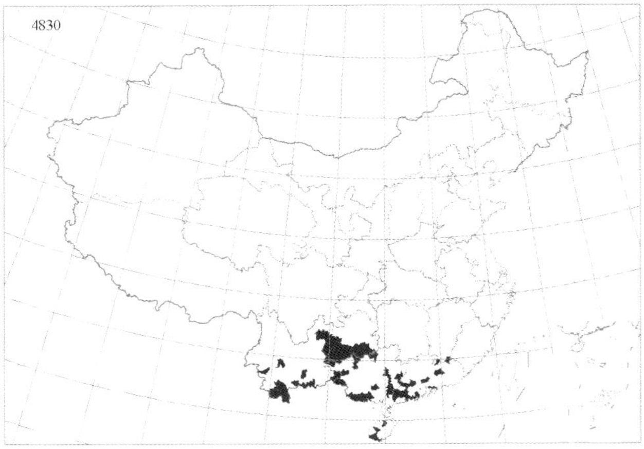

Euphorbiaceae 大戟科 *Baccaurea* 木奶果属
Baccaurea ramiflora Lour. 木奶果
Trees, evergreen

22.1 (16.9~24.8); 15.5 (10.3~20.1); 26.8 (21.4~29.2); 22.1 (16.9~24.8); 205.6 (142.7~237.6); 0.0 (0.0~0.0); 1166 (813~1385) ❙ 1531 (868~2435); 698 (312~1428); 81 (41~130) ❙ 34.0 (-34.3~99.5); 1114 (813~1360); 401.3 (178.7~659.0)

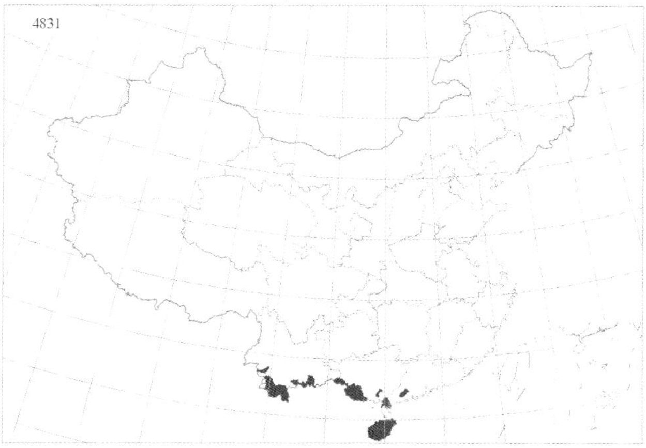

Euphorbiaceae 大戟科 *Balakata* 浆果乌桕属
Balakata baccata (Roxburgh) Esser 浆果乌桕
Trees

18.6 (13.3~20.5); 12.6 (5.4~15.1); 22.5 (19.6~23.7); 18.6 (13.4~20.5); 163.6 (100.4~185.9); 0.0 (0.0~0.0); 901 (706~992) ❙ 1539 (1428~1736); 796 (655~1007); 62 (51~77) ❙ 73.6 (47.6~145.8); 890 (706~986); 525.1 (427.9~659.0)

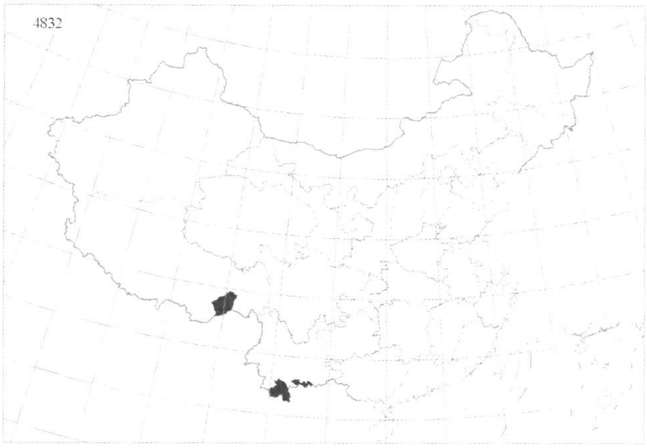

Euphorbiaceae 大戟科 *Baliospermum* 斑籽属
Baliospermum bilobatum T. L. Chin 西藏斑籽木
Shrubs or small trees

13.3; 5.4; 19.6; 13.4; 100.4; 0.0; 706 ❙ 1736; 1007; 51 ❙ 145.8; 706; 447.4

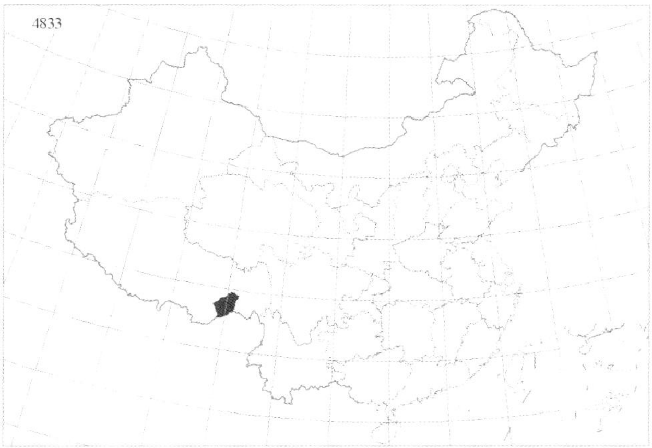

Euphorbiaceae 大戟科 *Baliospermum* 斑籽属
Baliospermum calycinum Müll. Arg. 云南斑籽木
Shrubs

17.0 (5.3~20.5); 10.6 (-2.4~15.1); 21.2 (12.2~23.7); 17.0 (5.7~20.5); 145.5 (30.5~185.9); 1.2 (0.0~26.2); 831 (479~992) ❙ 1403 (961~1736); 769 (558~1007); 54 (31~77) ❙ 70.7 (21.7~145.8); 819 (479~986); 461.5 (234.2~659.0)

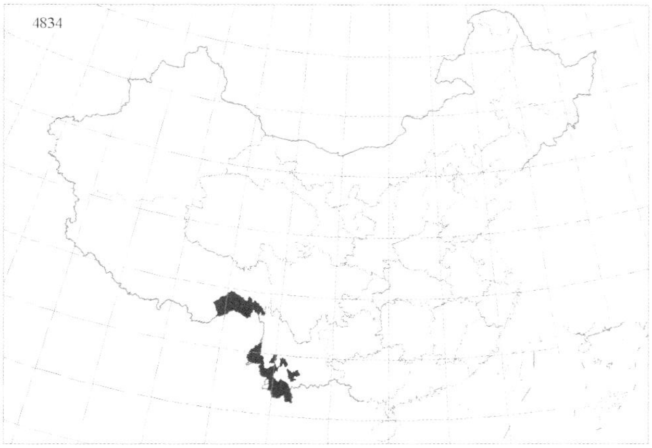

Euphorbiaceae 大戟科 *Bischofia* 秋枫属
Bischofia javanica Blume 秋枫
Trees, evergreen or semievergreen

18.3 (9.3~24.8); 9.1 (-1.7~20.1); 26.3 (18.8~30.1); 18.4 (9.5~24.8); 161.2 (69.8~237.6); 0.6 (0.0~18.0); 964 (603~1385) ❙ 1458 (558~3495); 625 (299~1847); 123 (19~897) ❙ 52.6 (-36.5~235.6); 948 (558~1360); 247.9 (95.3~659.0)

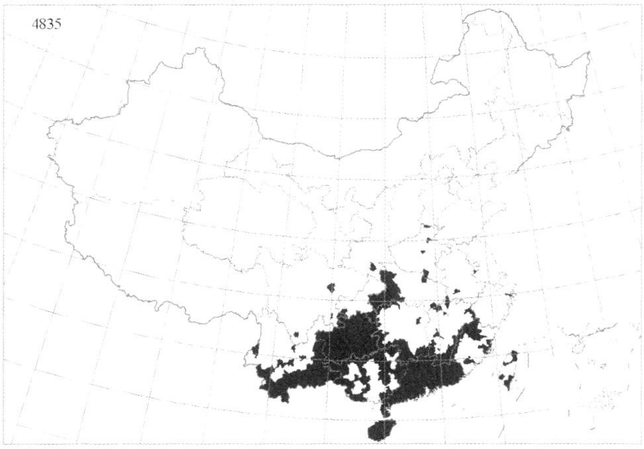

Euphorbiaceae 大戟科 *Bischofia* 秋枫属
Bischofia polycarpa (H. Lév.) Airy Shaw 重阳木
Trees, deciduous

15.8 (9.3~24.1); 3.8 (-2.9~18.9); 27.0 (20.1~29.9); 15.9 (9.5~24.1); 134.4 (69.8~229.7); 3.8 (0.0~20.5); 874 (603~1322) ❙ 1166 (512~2114); 488 (258~1357); 110 (19~257) ❙ 33.1 (-36.5~147.0); 825 (512~1269); 192.1 (110.1~332.5)

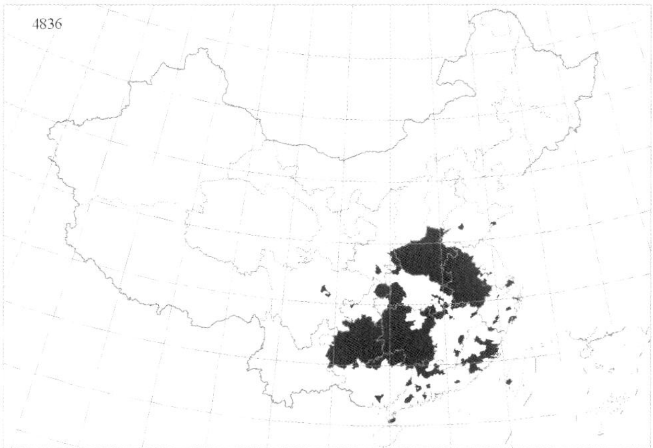

Euphorbiaceae 大戟科 *Blachia* 留萼木属
Blachia andamanica (Kurz) Hook. f. 大果留萼木
Shrubs

22.2; 13.9; 28.3; 22.3; 207.5; 0.0; 1190 ❙ 1374; 713; 87 ❙ 15.4; 1190; 327.7

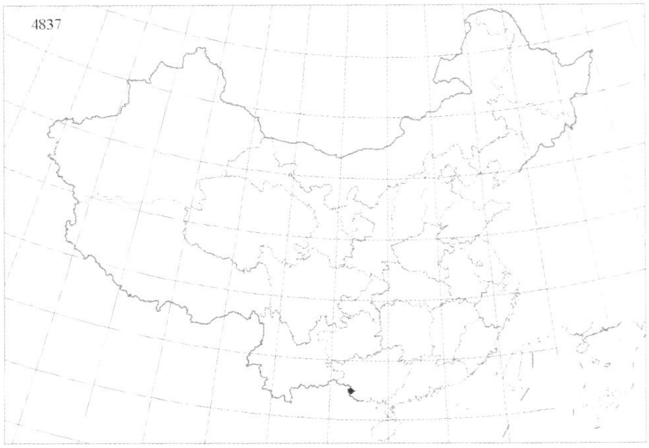

Euphorbiaceae 大戟科 *Blachia* 留萼木属
Blachia pentzii (Müll. Arg.) Benth. 留萼木
Shrubs

23.7 (20.5~24.8); 17.9 (14.2~20.1); 27.7 (23.7~28.8); 23.7 (20.5~24.8); 224.5 (185.9~237.6); 0.0 (0.0~0.0); 1282 (992~1385) ❙ 1490 (868~1854); 553 (312~818); 74 (41~115) ❙ 17.1 (-34.3~47.8); 1178 (868~1360); 442.0 (235.0~620.0)

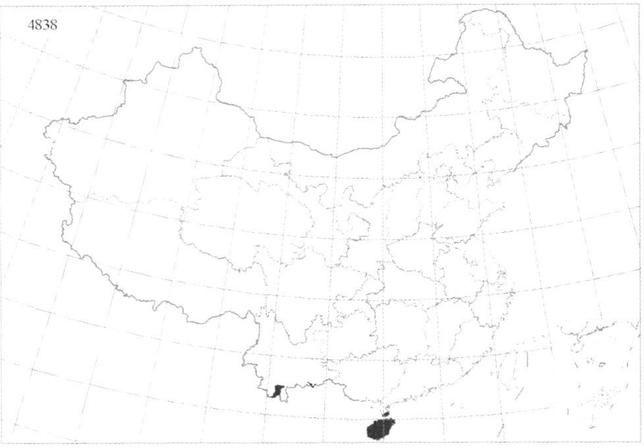

Euphorbiaceae 大戟科 *Blachia* 留萼木属
Blachia siamensis Gagnep. 海南留萼木
Shrubs

24.8; 20.1; 28.0; 24.8; 237.6; 0.0; 1385 ❙ 1405; 473; 41 ❙ 1.5; 1137; 527.5

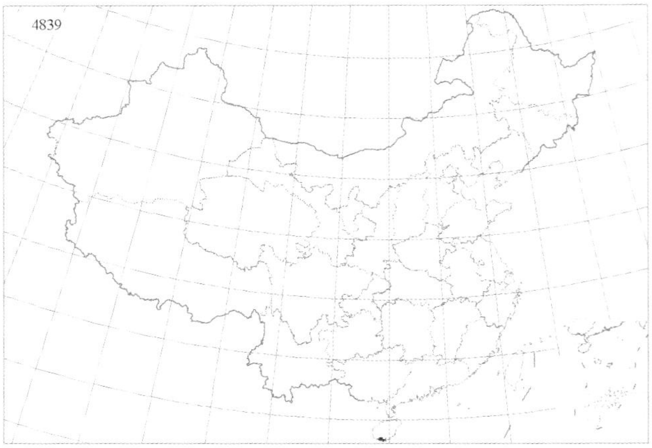

Euphorbiaceae 大戟科 *Breynia* 黑面神属
Breynia fruticosa (L.) Hook. f. 黑面神
Shrubs

19.8 (11.7~24.8); 11.1 (1.9~20.1); 27.0 (18.8~30.1); 19.8 (11.7~24.8); 178.0 (82.6~237.6); 0.1 (0.0~6.1); 1038 (658~1385) ❙ 1523 (868~2435); 662 (312~1428); 129 (19~226) ❙ 48.2 (-34.3~101.8); 1024 (658~1360); 255.4 (95.3~659.0)

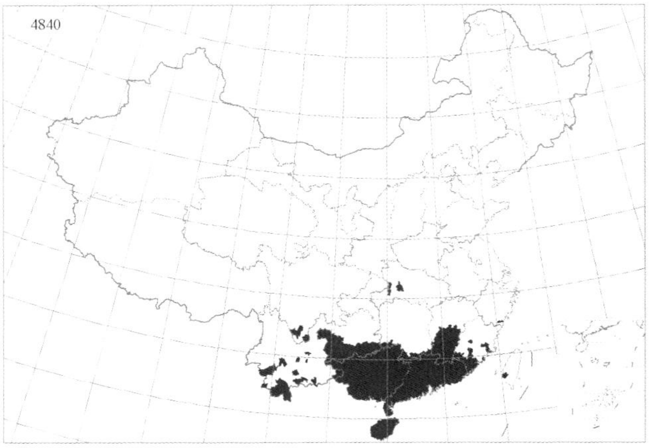

Euphorbiaceae 大戟科 *Breynia* 黑面神属
Breynia retusa (Dennst.) Alston 钝叶黑面神
Shrubs, deciduous

18.8 (13.3~24.8); 10.3 (3.8~20.1); 25.6 (19.4~29.0); 18.9 (13.4~24.8); 166.3 (100.4~237.6); 0.0 (0.0~1.5); 965 (706~1385) ❙ 1404 (868~1803); 660 (312~1007); 103 (38~194) ❙ 47.6 (-34.3~145.8); 949 (706~1202); 301.4 (160.7~659.0)

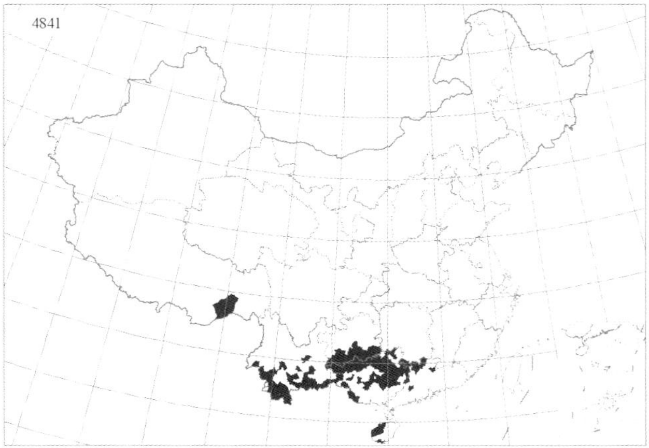

Euphorbiaceae 大戟科 *Breynia* 黑面神属
Breynia rostrata Merr. 喙果黑面神
Shrubs or trees, evergreen

20.6 (15.9~24.8); 12.4 (5.9~20.1); 27.3 (21.6~29.8); 20.7 (16.0~24.8); 187.8 (131.9~237.6); 0.0 (0.0~0.0); 1077 (831~1385) ❙ 1588 (868~2435); 672 (312~1428); 137 (41~257) ❙ 49.4 (-34.3~115.7); 1058 (815~1360); 288.4 (164.9~659.0)

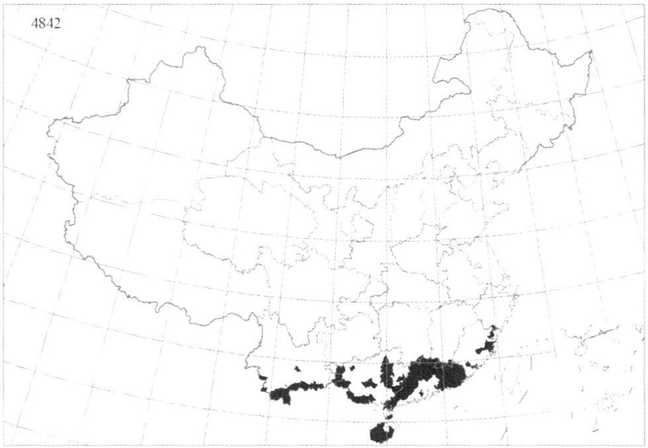

Euphorbiaceae 大戟科 *Breynia* 黑面神属
Breynia vitis-idaea (Burm.) C. E. C. Fisch. 小叶黑面神
Shrubs

19.8 (15.5~24.1); 11.4 (4.8~18.9); 26.6 (20.3~29.0); 19.8 (15.6~24.1); 177.6 (126.7~229.7); 0.0 (0.0~0.2); 1021 (783~1322) ❙ 1590 (1049~3495); 742 (411~1847); 142 (42~897) ❙ 57.7 (7.3~224.4); 1010 (783~1257); 254.6 (119.4~659.0)

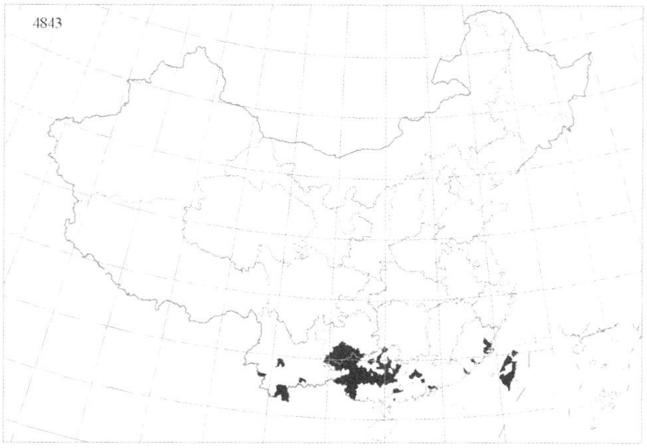

Euphorbiaceae 大戟科 *Bridelia* 土密树属
Bridelia affinis Craib 硬叶土蜜树
Shrubs or small trees

16.5; 9.8; 21.3; 16.5; 138.4; 0.0; 804 ❙ 1052; 580; 48 ❙ 31.1; 794; 358.9

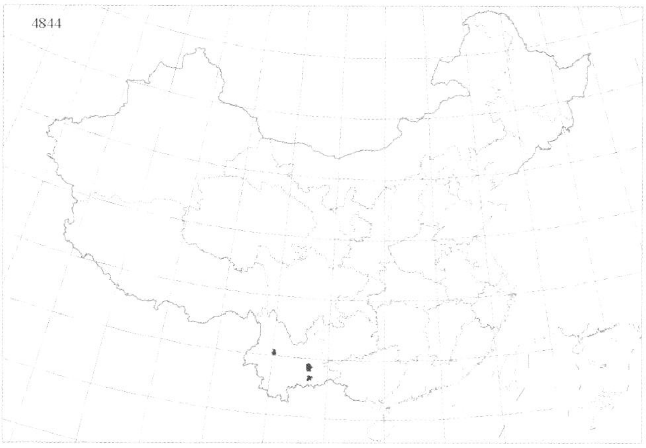

Euphorbiaceae 大戟科 *Bridelia* 土密树属
Bridelia balansae Tutcher 禾串树
Trees

19.8 (14.6~24.8); 11.9 (3.8~20.1); 26.3 (20.3~28.9); 19.9 (14.7~24.8); 178.6 (117.7~237.6); 0.0 (0.0~1.5); 1031 (779~1385) ❙ 1539 (868~3495); 700 (312~1594); 113 (41~897) ❙ 51.7 (-34.3~235.6); 1009 (779~1360); 297.7 (108.2~659.0)

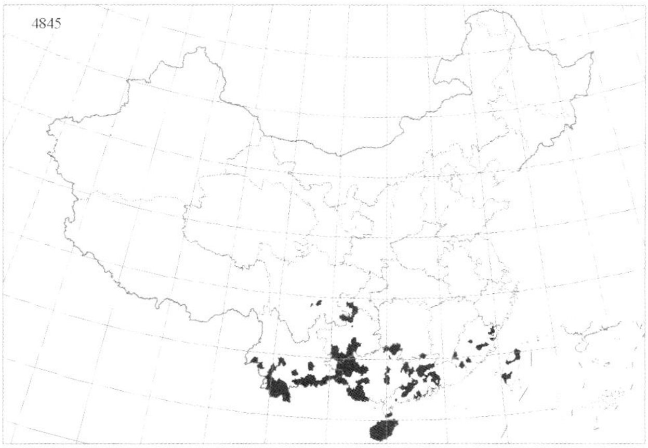

Euphorbiaceae 大戟科 *Bridelia* 土密树属
Bridelia glauca Blume 膜叶土蜜树
Trees

20.8 (18.1~23.6); 13.2 (10.9~16.7); 26.7 (22.1~28.7); 20.9 (18.1~23.6); 190.5 (157.3~223.5); 0.0 (0.0~0.0); 1076 (863~1281) ❙ 1611 (1269~2005); 709 (551~824); 121 (55~182) ❙ 51.1 (-1.0~85.1); 1073 (863~1269); 308.2 (206.6~659.0)

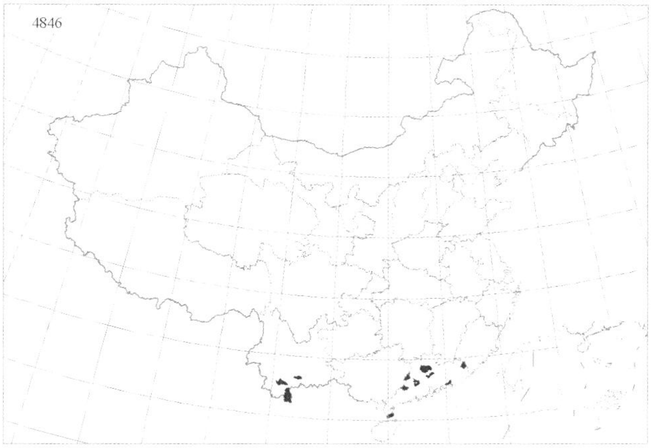

Euphorbiaceae 大戟科 *Bridelia* 土密树属
Bridelia parvifolia Kuntze 圆叶土蜜树
Shrubs

24.2; 18.4; 28.5; 24.3; 231.6; 0.0; 1337 ❙ 1177; 420; 50 ❙ -11.9; 1098; 433.5

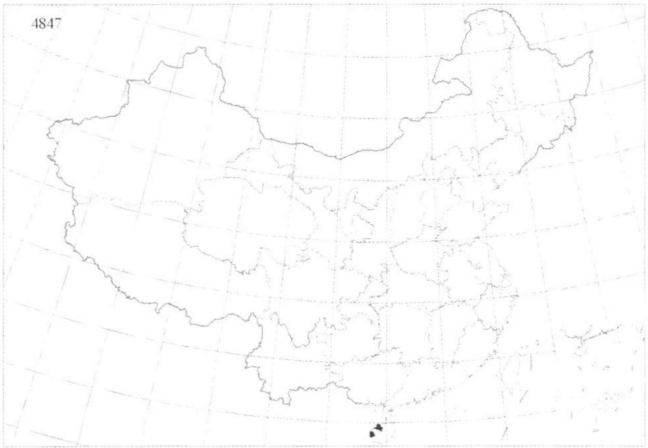

Euphorbiaceae 大戟科 *Bridelia* 土密树属
Bridelia stipularis (L.) Blume 土蜜藤
Scandent shrubss or vines

20.2 (14.4~24.8); 13.6 (7.4~20.1); 24.9 (19.1~29.2); 20.3 (14.4~24.8); 183.0 (113.1~237.6); 0.0 (0.0~0.0); 1036 (727~1385) ❙ 1420 (868~2992); 708 (312~1847); 64 (41~175) ❙ 40.5 (-34.3~166.5); 988 (727~1360); 403.8 (119.4~659.0)

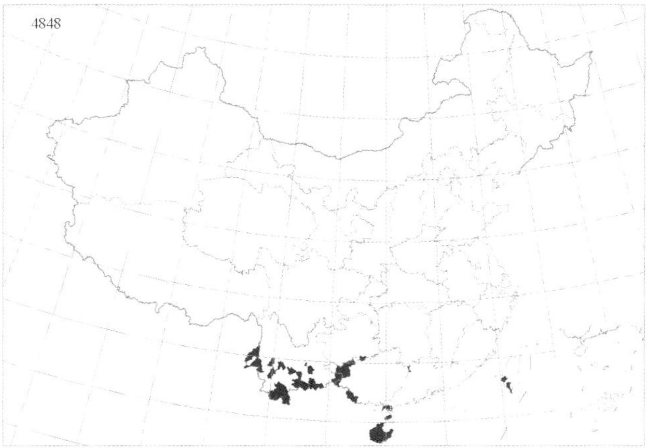

Euphorbiaceae 大戟科 *Bridelia* 土密树属
Bridelia tomentosa Blume 土蜜树
Shrubs or small trees, evergreen

21.5 (15.9~24.8); 14.0 (5.5~20.1); 27.5 (20.1~29.2); 21.6 (15.9~24.8); 199.1 (131.2~237.6); 0.0 (0.0~0.0); 1136 (773~1385) ❙ 1611 (868~2992); 732 (312~1847); 115 (41~246) ❙ 43.5 (-34.3~246.5); 1115 (751~1360); 269.9 (95.3~659.0)

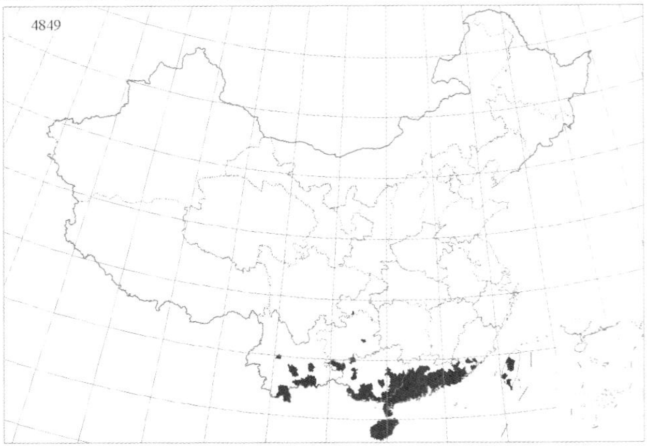

Euphorbiaceae 大戟科 *Cephalomappa* 肥牛木属
Cephalomappa sinensis (Chun et F. C. How) Kosterm. 肥牛树
Trees, evergreen

20.8 (19.1~22.2); 12.7 (9.8~14.2); 26.8 (23.6~28.6); 20.8 (19.2~22.3); 190.1 (170.1~207.5); 0.0 (0.0~0.0); 1077 (939~1190) ❙ 1433 (1143~1900); 769 (637~1056); 88 (43~138) ❙ 33.3 (12.9~71.0); 1074 (903~1190); 331.9 (272.3~551.6)

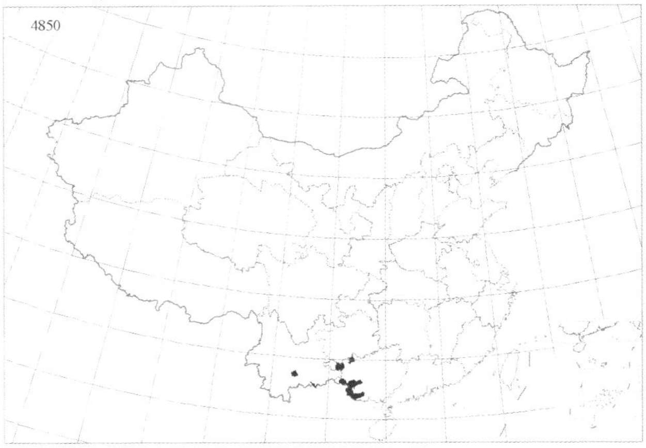

Euphorbiaceae 大戟科 *Chaetocarpus* 刺果树属
Chaetocarpus castanocarpus (Roxb.) Thwaites 刺果树
Trees

20.5; 15.0; 23.7; 20.5; 185.9; 0.0; 990 ❙ 1536; 678; 72 ❙ 55.1; 973; 614.6

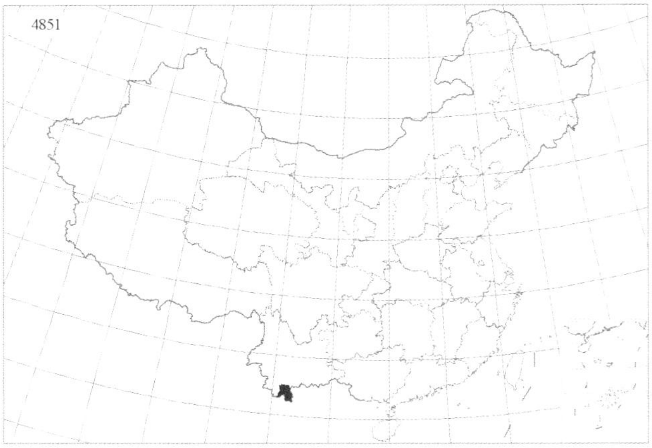

Euphorbiaceae 大戟科 *Cladogynos* 白大凤属
Cladogynos orientalis Zipp. ex Span. 白大凤
Shrubs

21.3 (19.1~22.2); 12.8 (11.0~13.9); 27.6 (25.0~28.6); 21.3 (19.1~22.3); 196.0 (169.6~207.5); 0.0 (0.0~0.0); 1120 (945~1190) ❙ 1425 (1182~1625); 752 (670~850); 98 (43~173) ❙ 27.5 (12.9~52.2); 1120 (945~1190); 298.8 (237.8~371.7)

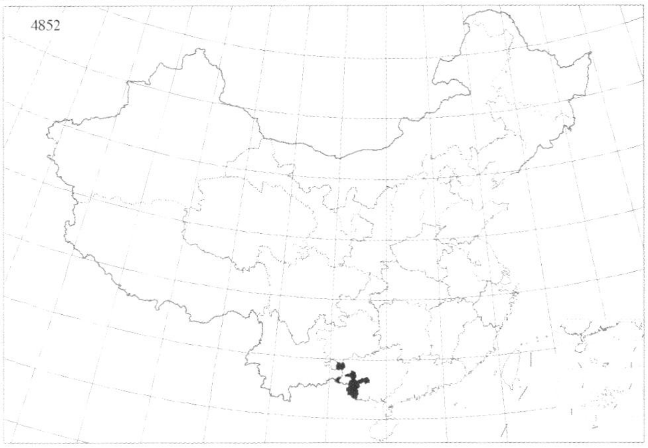

Euphorbiaceae 大戟科 *Claoxylon* 白桐树属
Claoxylon brachyandrum Pax et Hoffm. 台湾白桐树
Shrubs or small trees

18.5; 14.1; 21.7; 18.5; 162.1; 0.0; 863 ❙ 2713; 1388; 188 ❙ 214.4; 863; 363.7

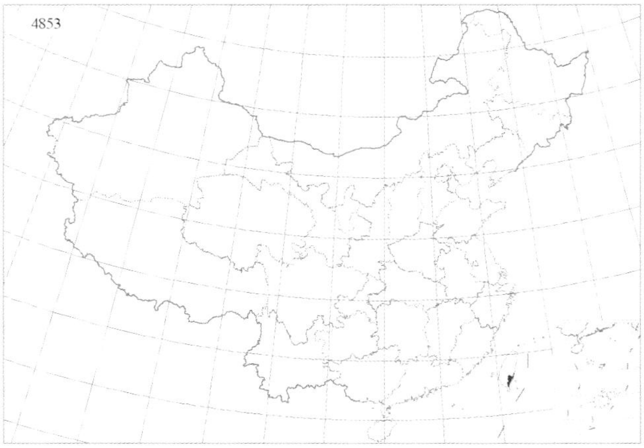

Euphorbiaceae 大戟科 *Claoxylon* 白桐树属
Claoxylon hainanense Pax et K. Hoffm. 海南白桐树
Shrubs or small trees

24.0 (22.2~24.8); 18.3 (16.7~20.1); 28.0 (25.7~28.8); 24.0 (22.3~24.8); 228.0 (207.2~237.6); 0.0 (0.0~0.0); 1308 (1127~1385) ❙ 1490 (868~1854); 534 (312~712); 74 (41~115) ❙ 14.2 (-34.3~47.8); 1195 (868~1360); 430.1 (235.0~620.0)

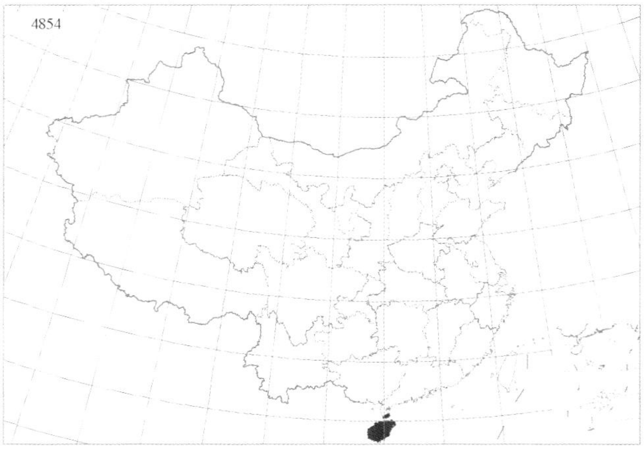

Euphorbiaceae 大戟科 *Claoxylon* 白桐树属
Claoxylon indicum (Reinw. ex Blume) Hassk. 白桐树
Shrubs or small trees

21.9 (17.3~24.8); 14.4 (9.8~20.1); 27.8 (22.4~29.2); 22.0 (17.3~24.8); 203.9 (147.1~237.6); 0.0 (0.0~0.0); 1163 (833~1385) ❙ 1611 (868~2037); 724 (312~1044); 111 (41~188) ❙ 39.6 (-34.3~85.1); 1143 (833~1360); 272.8 (95.3~659.0)

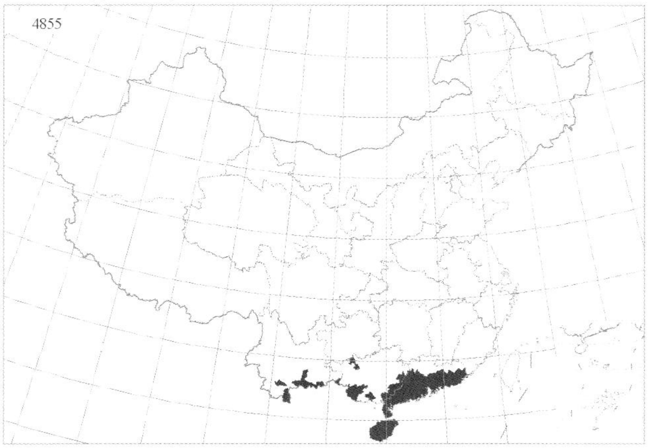

Euphorbiaceae 大戟科 *Claoxylon* 白桐树属
Claoxylon khasianum Hook. f. 膜叶白桐树
Shrubs or small trees

18.8 (17.3~21.0); 12.2 (9.9~14.9); 23.3 (21.6~25.6); 18.8 (17.3~21.0); 165.9 (147.1~191.8); 0.0 (0.0~0.0); 913 (832~1053) ❙ 1397 (1049~1547); 770 (598~911); 58 (44~68) ❙ 53.3 (26.1~80.6); 903 (783~1053); 474.0 (342.5~570.1)

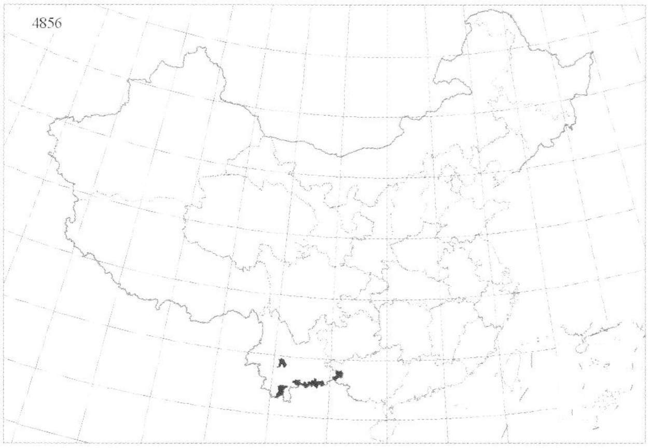

Euphorbiaceae 大戟科 *Claoxylon* 白桐树属
Claoxylon longifolium (Blume) Endl. et Hassk. 长叶白桐树
Shrubs or small trees

18.9 (17.3~21.0); 12.8 (10.9~14.9); 22.9 (21.6~25.6); 18.9 (17.3~21.0); 167.2 (148.0~191.8); 0.0 (0.0~0.0); 912 (832~1053) ❙ 1375 (1049~1522); 741 (598~835); 56 (42~68) ❙ 51.0 (26.1~68.0); 896 (783~1053); 487.6 (365.3~570.1)

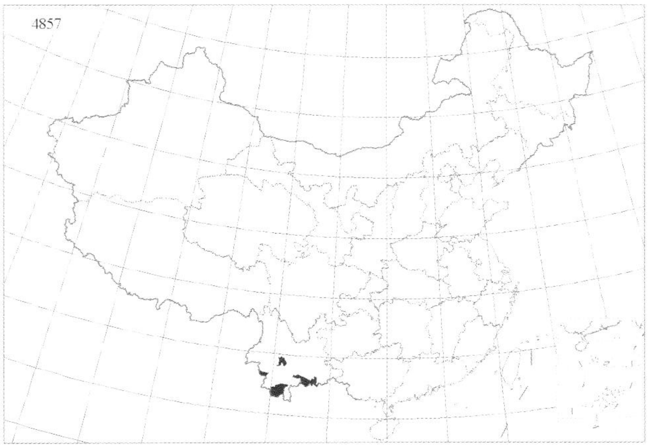

Euphorbiaceae 大戟科 *Cleidiocarpon* 蝴蝶果属
Cleidiocarpon cavaleriei (Lév.) Airy Shaw 蝴蝶果
Trees, evergreen

19.8 (15.9~24.1); 11.3 (6.9~18.3); 26.4 (21.6~28.8); 19.9 (16.0~24.1); 178.6 (131.6~229.4); 0.0 (0.0~0.0); 1024 (802~1321) ❙ 1404 (1049~2148); 731 (572~1234); 86 (43~173) ❙ 37.9 (5.8~86.7); 1021 (783~1294); 289.1 (157.7~474.7)

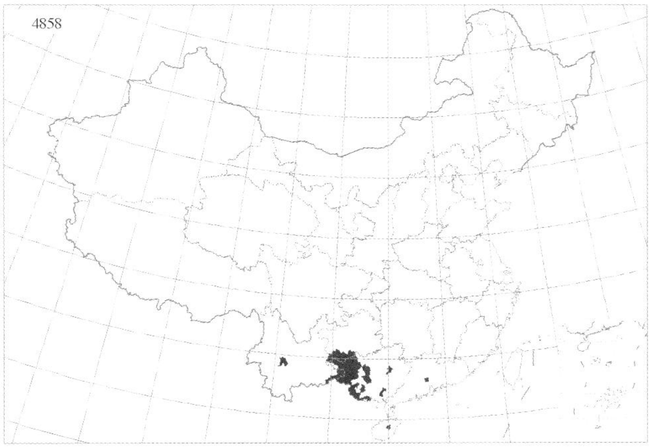

Euphorbiaceae 大戟科 *Cleidion* 棒柄花属
Cleidion bracteosum Gagnep. 灰岩棒柄花
Small trees

19.6 (16.8~22.2); 11.3 (7.4~15.1); 25.9 (22.3~28.3); 19.7 (16.8~22.3); 175.9 (141.5~207.5); 0.0 (0.0~0.0); 999 (836~1190) ❙ 1399 (1182~1710); 719 (617~911); 81 (43~186) ❙ 41.2 (5.8~74.2); 996 (836~1190); 336.3 (226.6~659.0)

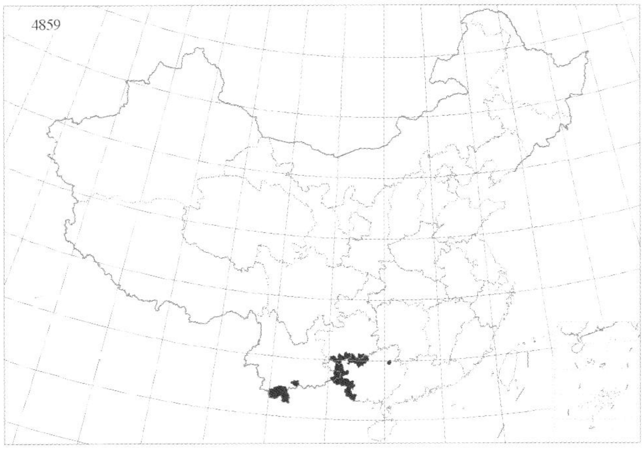

Euphorbiaceae 大戟科 *Cleidion* 棒柄花属
Cleidion brevipetiolatum Pax et K. Hoffm. 棒柄花
Small trees

19.6 (13.2~24.8); 11.9 (5.4~20.1); 25.5 (18.4~28.8); 19.7 (13.3~24.8); 175.9 (99.4~237.6); 0.0 (0.0~0.0); 1006 (695~1385) ❙ 1405 (868~2030); 690 (312~1033); 82 (41~186) ❙ 42.6 (-34.3~145.8); 977 (695~1212); 343.0 (163.7~659.0)

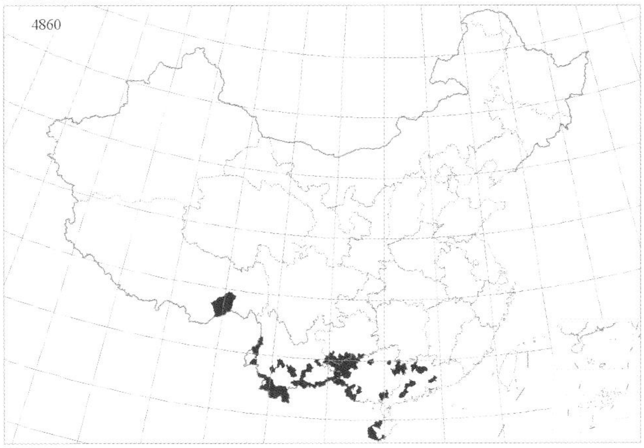

Euphorbiaceae 大戟科 *Cleidion* 棒柄花属
Cleidion spiciflorum (Burm. f.) Merr. 长棒柄花
Small trees

19.1 (17.3~20.5); 13.1 (10.9~15.1); 22.9 (21.6~23.7); 19.1 (17.3~20.5); 169.5 (148.0~185.9); 0.0 (0.0~0.0); 919 (832~992) ❙ 1355 (1049~1607); 672 (598~756); 59 (44~77) ❙ 47.2 (21.7~64.9); 894 (783~986); 491.1 (346.9~659.0)

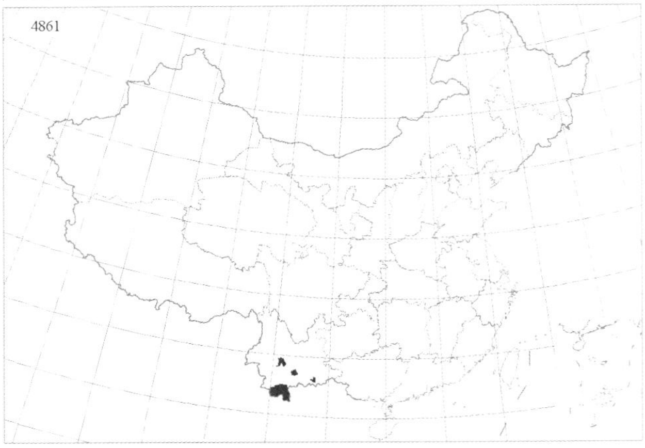

Euphorbiaceae 大戟科 *Cleistanthus* 闭花木属
Cleistanthus concinnus Croizat 东方闭花木
Shrubs

23.7; 18.3; 27.6; 23.8; 225.1; 0.0; 1282 ❙ 1311; 450; 46 ❙ 2.2; 1101; 447.5

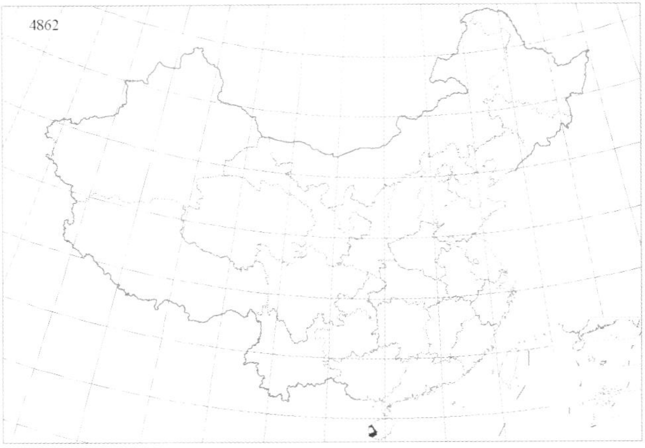

Euphorbiaceae 大戟科 *Cleistanthus* 闭花木属
Cleistanthus macrophyllus Hook. f. 大叶闭花木
Trees

20.0; 13.6; 24.4; 20.1; 180.7; 0.0; 985 ❙ 1475; 827; 62 ❙ 50.1; 985; 528.9

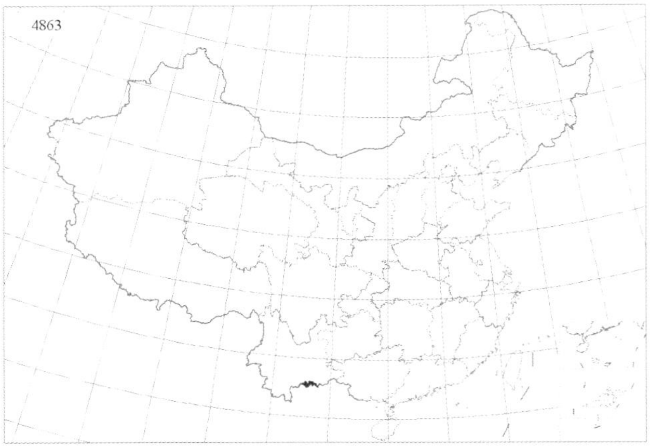

Euphorbiaceae 大戟科 *Cleistanthus* 闭花木属
Cleistanthus pedicellatus Hook. f. 米咀闭花木
Shrubs or small trees

22.2; 13.9; 28.3; 22.3; 207.5; 0.0; 1190 ❙ 1374; 713; 87 ❙ 15.4; 1190; 327.7

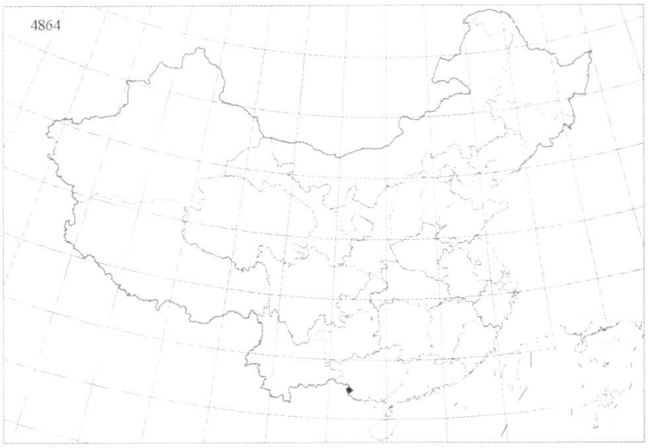

Euphorbiaceae 大戟科 *Cleistanthus* 闭花木属
Cleistanthus petelotii Merr. ex Croizat 假肥牛树
Trees

21.9 (21.5~22.2); 13.6 (13.1~13.9); 28.1 (27.4~28.5); 22.0 (21.6~22.3); 203.9 (199.1~207.5); 0.0 (0.0~0.0); 1168 (1131~1190) ❙ 1507 (1374~1600); 793 (713~850); 108 (87~128) ❙ 29.2 (15.4~39.9); 1168 (1131~1190); 294.8 (256.6~335.1)

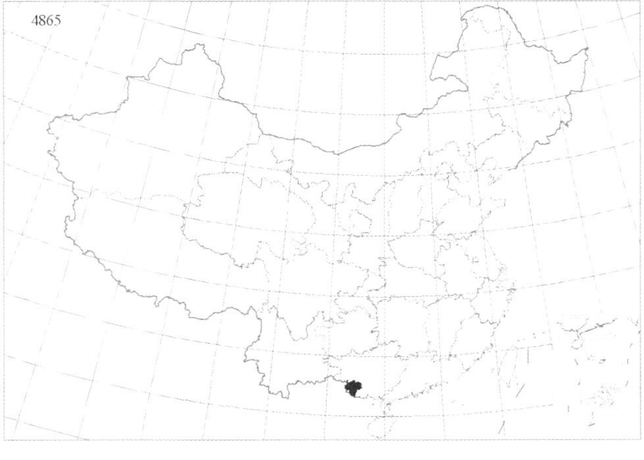

Euphorbiaceae 大戟科 *Cleistanthus* 闭花木属
Cleistanthus sumatranus (Miq.) Müll. Arg. 闭花木
Trees, evergreen

22.4 (19.0~24.8); 15.4 (10.2~20.1); 27.6 (23.3~29.0); 22.5 (19.0~24.8); 209.6 (168.4~237.6); 0.0 (0.0~0.0); 1196 (917~1385) ❙ 1534 (868~2435); 707 (312~1428); 88 (41~173) ❙ 29.6 (-34.3~99.5); 1131 (868~1360); 381.0 (203.1~659.0)

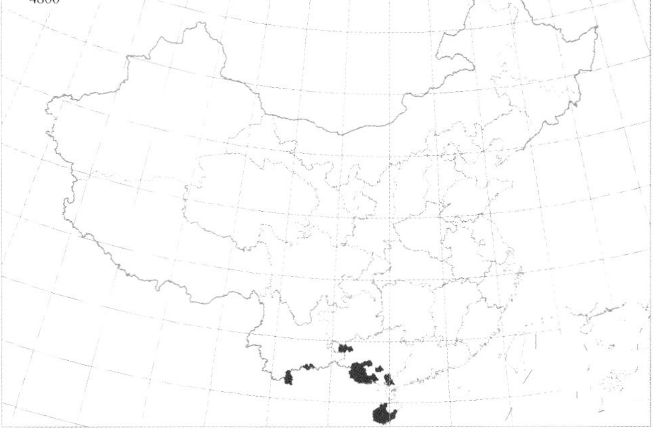

Euphorbiaceae 大戟科 *Cleistanthus* 闭花木属
Cleistanthus tomentosus Hance 锈毛闭花木
Small trees

24.1；18.5；28.1；24.1；229.1；0.0；1320 | 1460；564；57 | 10.8；1186；442.[illegible]

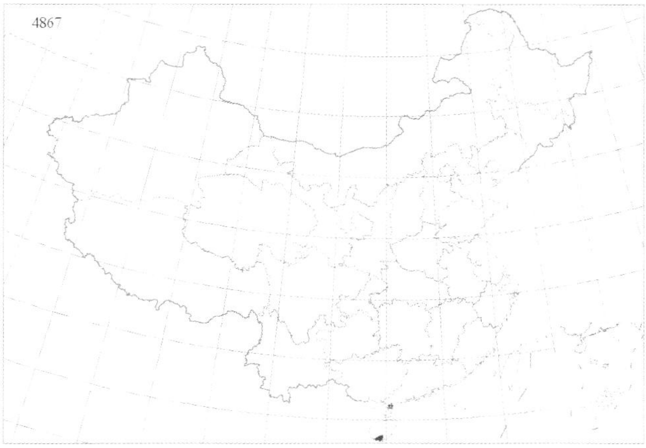

Euphorbiaceae 大戟科 *Cleistanthus* 闭花木属
Cleistanthus tonkinensis Jabl. 馒头果
Shrubs or small trees

21.1 (17.6~23.2)；13.8 (9.9~15.8)；26.3 (22.3~28.9)；21.1 (17.7~23.2)；193.2 (152.4~218.9)；0.0 (0.0~0.0)；1085 (859~1257) | 1705 (1163~2435)；902 (641~1428)；94 (48~130) | 56.7 (27.0~99.5)；1082 (859~1257)；372.7 (194.5~659.0)

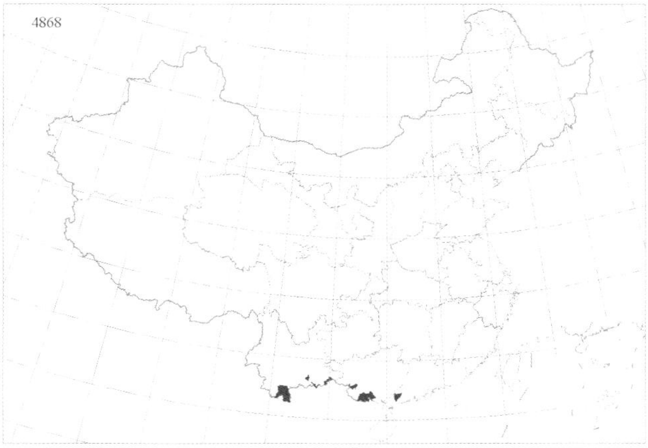

Euphorbiaceae 大戟科 *Cnesmone* 粗毛藤属
Cnesmone hainanensis (Merr. et Chun) Croizat 海南粗毛藤
Vines or shrubs

23.9 (22.8~24.7)；17.4 (14.7~19.8)；28.5 (27.9~29.0)；23.9 (22.8~24.7)；226.9 (213.6~236.5)；0.0 (0.0~0.0)；1310 (1230~1376) | 1599 (1269~1813)；671 (480~997)；94 (58~113) | 22.2 (-1.4~39.9)；1270 (1212~1329)；323.8 (199.6~548.0)

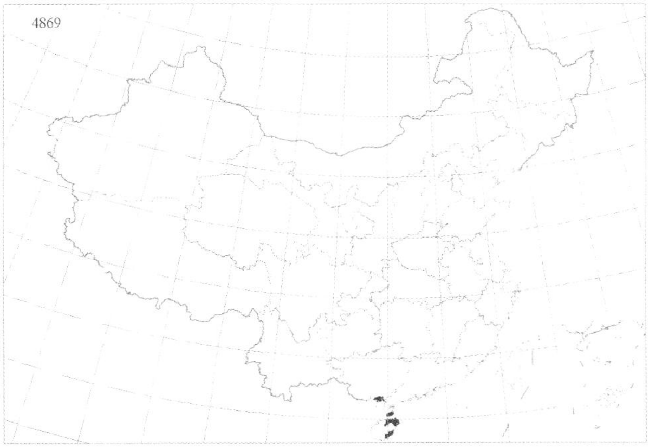

Euphorbiaceae 大戟科 *Cnesmone* 粗毛藤属
Cnesmone mairei (H. Lév.) Croizat 粗毛藤
Vines

15.8 (13.8~18.9)；9.0 (6.7~12.0)；21.0 (19.3~23.9)；15.8 (13.8~19.0)；129.8 (105.9~167.8)；0.0 (0.0~0.0)；784 (714~929) | 937 (797~991)；550 (411~592)；29 (21~38) | 20.5 (-14.2~36.1)；710 (683~760)；280.6 (218.0~309.8)

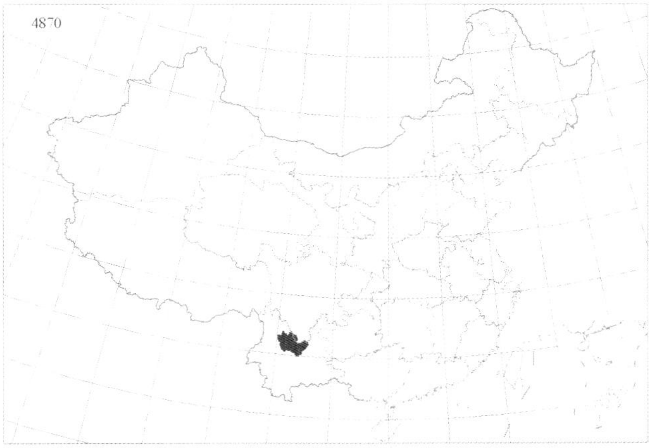

Euphorbiaceae 大戟科 *Cnesmone* 粗毛藤属
Cnesmone tonkinensis (Gagnep.) Croizat 灰岩粗毛藤
Vines

23.6 (21.9~24.8)；17.4 (13.2~20.1)；28.0 (27.2~28.6)；23.6 (22.0~24.8)；223.3 (203.8~237.6)；0.0 (0.0~0.0)；1286 (1176~1385) | 1499 (1374~1767)；587 (455~773)；75 (41~138) | 16.8 (1.5~29.1)；1177 (1112~1299)；458.4 (272.3~620.0)

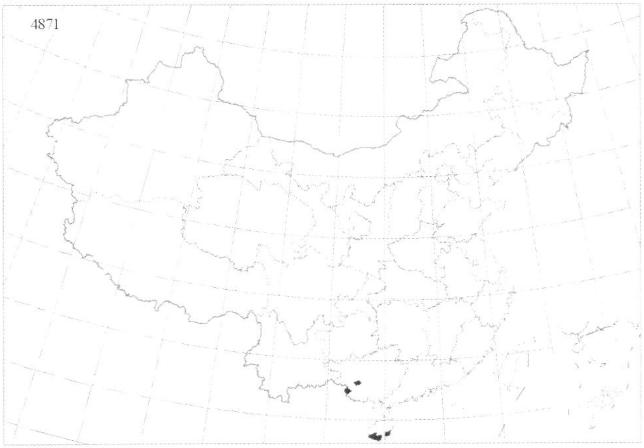

Euphorbiaceae 大戟科 *Croton* 巴豆属
Croton cascarilloides Raeusch. 银叶巴豆
Shrubs

21.9 (16.2~24.8)；15.8 (11.0~20.1)；26.6 (20.1~28.8)；21.9 (16.2~24.8)；203.0 (134.2~237.6)；0.0 (0.0~0.0)；1148 (773~1385) | 1863 (868~3495)；804 (312~1847)；155 (41~897) | 72.4 (-34.3~246.5)；1086 (773~1360)；340.6 (119.4~659.0)

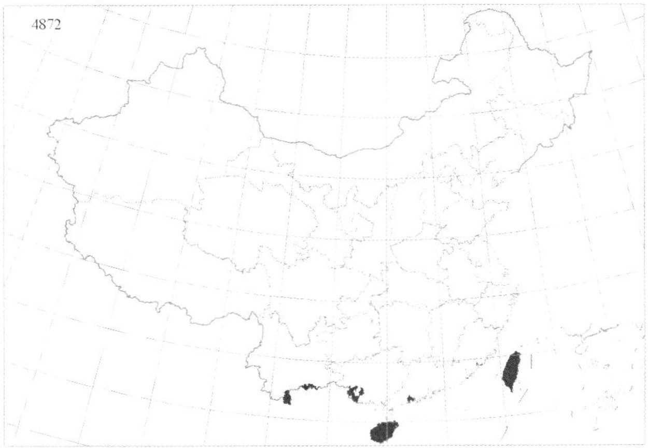

Euphorbiaceae 大戟科 *Croton* 巴豆属

Croton caudatus Geiseler 卵叶巴豆

Shrubs

19.6 (19.0~20.5); 13.9 (12.8~15.1); 23.1 (22.3~23.7); 19.7 (19.0~20.5); 175.9 (168.1~185.9); 0.0 (0.0~0.0); 942 (909~992) ❙ 1463 (1345~1607); 719 (655~835); 59 (41~77) ❙ 55.3 (47.6~64.9); 920 (880~986); 522.3 (354.7~659.0)

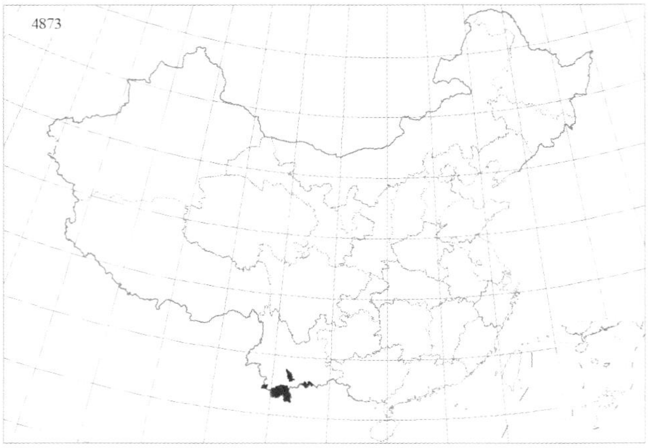

Euphorbiaceae 大戟科 *Croton* 巴豆属

Croton chunianus Croizat 光果巴豆

Shrubs

23.9; 18.9; 27.4; 24.0; 227.5; 0.0; 1298 ❙ 1466; 497; 48 ❙ 12.9; 1130; 537.6

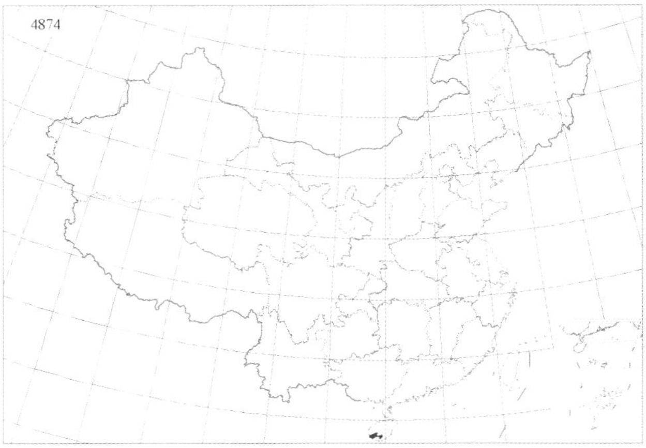

Euphorbiaceae 大戟科 *Croton* 巴豆属

Croton cnidophyllus Radcl.-Sm. et Govaerts 荨麻叶巴豆

Shrubs

18.2 (15.9~19.6); 9.5 (6.9~13.9); 24.9 (22.7~26.6); 18.2 (16.0~19.6); 158.4 (131.6~175.1); 0.0 (0.0~0.0); 912 (802~981) ❙ 1288 (1232~1434); 674 (617~720); 58 (41~73) ❙ 42.0 (27.5~64.7); 907 (802~981); 287.3 (189.5~530.1)

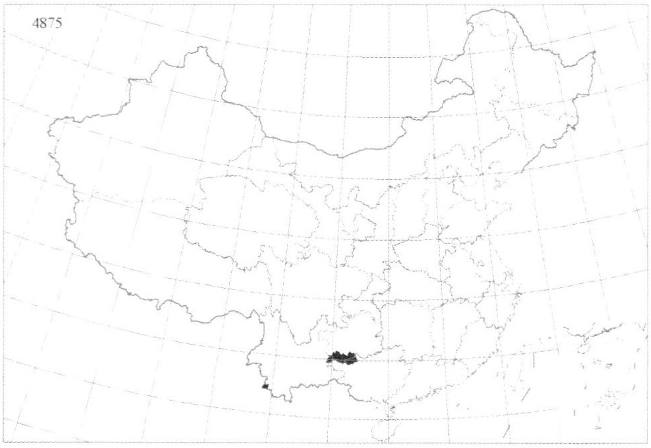

Euphorbiaceae 大戟科 *Croton* 巴豆属

Croton crassifolius Geiseler 鸡骨香

Shrubs

19.9 (2.2~24.8); 10.8 (-6.6~20.1); 27.7 (9.9~30.1); 20.0 (3.8~24.8); 179.9 (15.5~237.6); 0.7 (0.0~49.1); 1058 (413~1385) ❙ 1587 (932~2035); 612 (425~1033); 157 (19~230) ❙ 52.8 (-6.9~125.9); 1035 (413~1329); 239.6 (139.6~570.1)

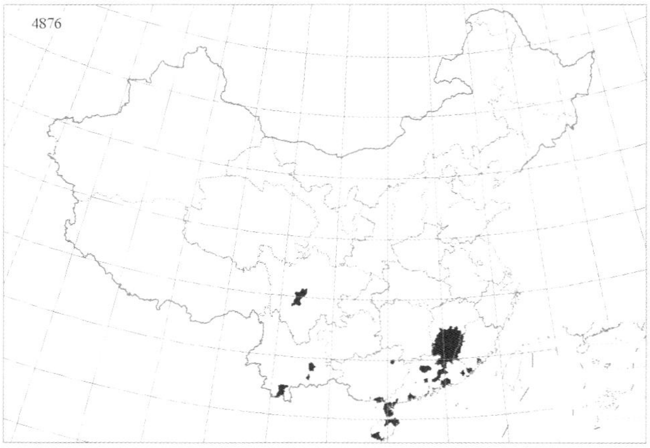

Euphorbiaceae 大戟科 *Croton* 巴豆属

Croton damayeshu Y. T. Chang 大麻叶巴豆

Small trees

18.8; 12.7; 23.1; 18.9; 166.9; 0.0; 907 ❙ 1254; 727; 50 ❙ 37.9; 896; 417.6

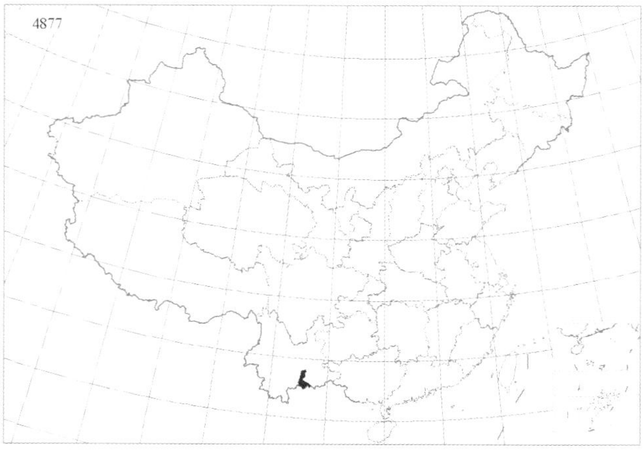

Euphorbiaceae 大戟科 *Croton* 巴豆属

Croton euryphyllus W. W. Sm. 石山巴豆

Shrubs

15.6 (5.4~22.2); 7.4 (-2.1~13.9); 22.3 (12.3~29.0); 15.6 (5.8~22.3); 129.1 (31.5~207.5); 1.4 (0.0~25.6); 817 (481~1190) ❙ 1142 (741~1776); 601 (411~909); 63 (12~188) ❙ 40.3 (-14.2~92.4); 800 (481~1190); 271.6 (142.9~506.2)

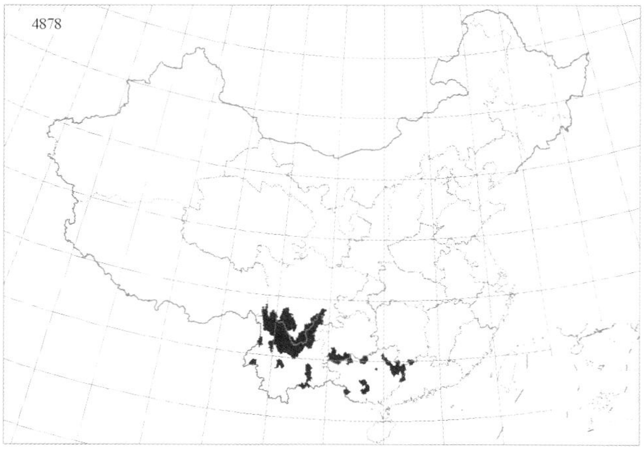

Euphorbiaceae 大戟科 *Croton* 巴豆属
Croton hancei Benth. 香港巴豆
Shrubs or small trees

22.4 (22.3~22.6); 14.7 (14.3~15.2); 28.7 (28.4~28.9); 22.5 (22.3~22.7); 210.1 (207.9~212.7); 0.0 (0.0~0.0); 1203 (1185~1216) ❙ 1797 (1599~2037); 888 (771~1033); 111 (85~138) ❙ 49.5 (31.8~71.8); 1203 (1185~1216); 156.8 (95.3~204.4)

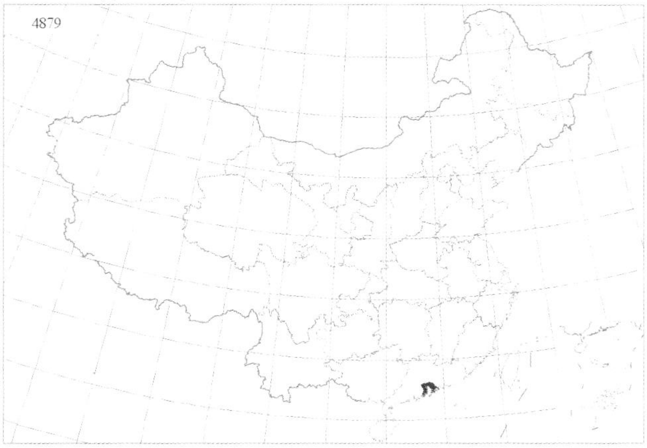

Euphorbiaceae 大戟科 *Croton* 巴豆属
Croton howii Merr. et Chun ex Y. T. Chang 宽昭巴豆
Shrubs

24.2 (23.0~24.8); 19.3 (17.7~20.1); 27.6 (26.7~28.2); 24.3 (23.1~24.8); 231.0 (217.0~237.6); 0.0 (0.0~0.0); 1330 (1217~1385) ❙ 1607 (1405~1767); 545 (473~610); 61 (41~88) ❙ 21.3 (1.5~41.8); 1192 (1137~1299); 552.6 (494.7~620.0)

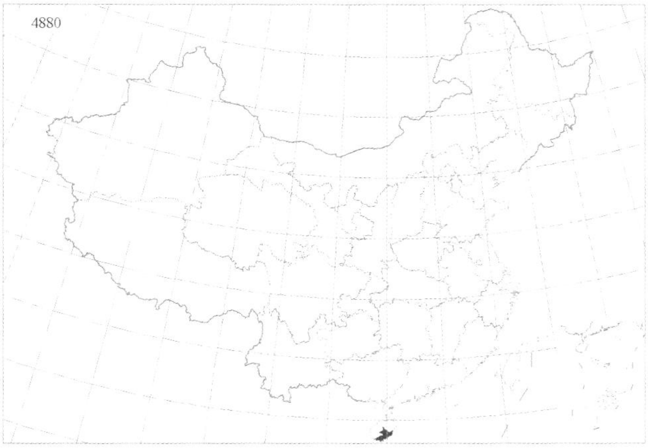

Euphorbiaceae 大戟科 *Croton* 巴豆属
Croton joufra Roxburgh 长果巴豆
Trees

19.1; 13.0; 23.3; 19.1; 169.7; 0.0; 917 ❙ 1428; 835; 56 ❙ 55.7; 917; 506.2

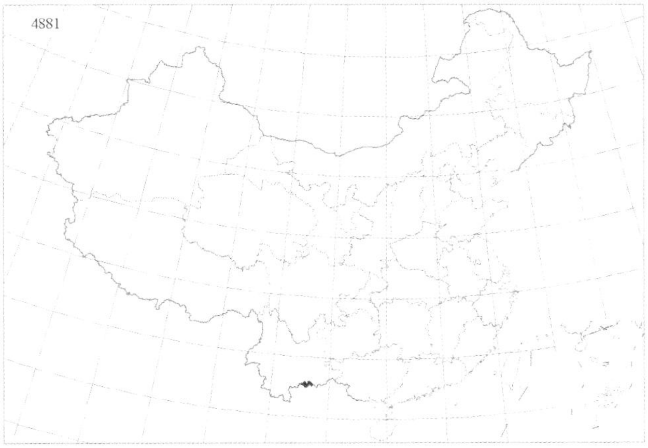

Euphorbiaceae 大戟科 *Croton* 巴豆属
Croton kongensis Gagnep. 越南巴豆
Shrubs

23.2 (20.4~24.8); 17.8 (14.9~20.1); 26.9 (23.6~28.7); 23.2 (20.5~24.8); 218.9 (185.8~237.6); 0.0 (0.0~0.0); 1241 (989~1385) ❙ 1380 (1176~1607); 521 (392~702); 55 (41~77) ❙ 14.4 (-12.0~62.5); 1083 (959~1212); 508.3 (411.8~659.0)

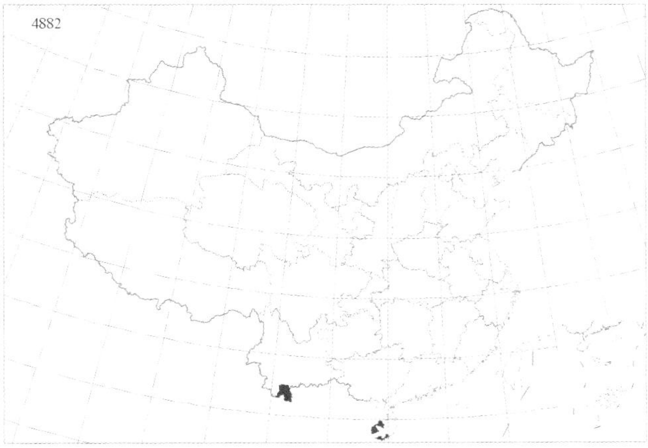

Euphorbiaceae 大戟科 *Croton* 巴豆属
Croton lachnocarpus Benth. 毛果巴豆
Shrubs

19.3 (14.3~23.3); 9.8 (3.7~16.0); 27.5 (23.0~30.1); 19.4 (14.3~23.3); 172.5 (112.6~220.1); 0.0 (0.0~1.5); 1017 (756~1264) ❙ 1606 (1175~2435); 654 (419~1428); 155 (51~259) ❙ 59.0 (15.4~128.9); 1012 (756~1264); 224.1 (95.3~327.7)

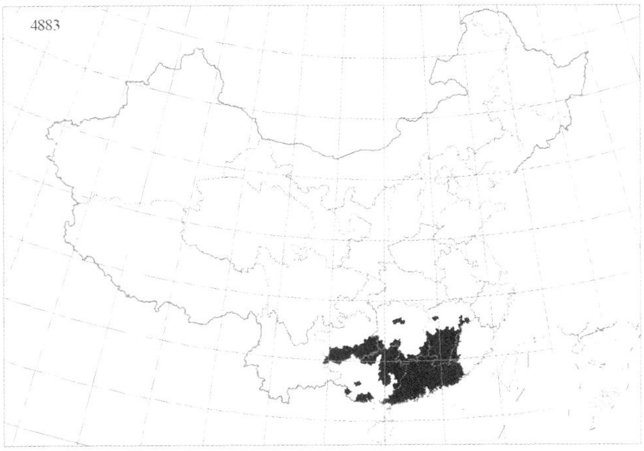

Euphorbiaceae 大戟科 *Croton* 巴豆属
Croton laevigatus Vahl 光叶巴豆
Shrubs or small trees

22.8 (14.4~24.8); 16.9 (7.5~20.1); 26.9 (19.1~28.8); 22.8 (14.4~24.8); 213.7 (113.1~237.6); 0.0 (0.0~0.0); 1225 (727~1385) ❙ 1480 (868~1854); 573 (312~795); 75 (41~173) ❙ 23.7 (-34.3~99.9); 1129 (727~1360); 441.0 (237.8~620.0)

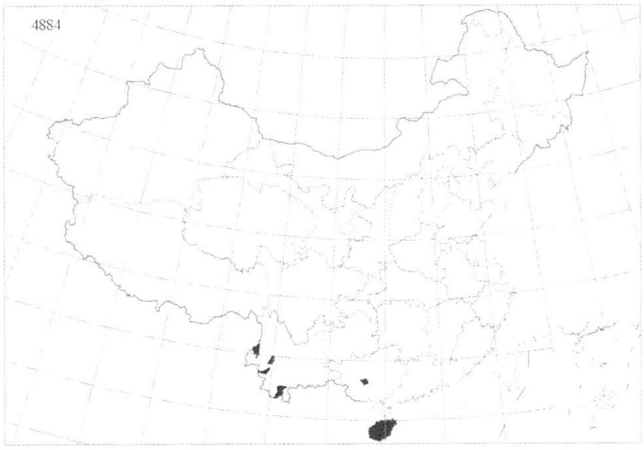

Euphorbiaceae 大戟科 *Croton* 巴豆属
Croton laniflorus Geiseler 疏齿巴豆
Shrubs

24.4；19.6；27.8；24.5；233.9；0.0；1353 ❙ 1574；526；59 ❙ 16.4；1194；565.2

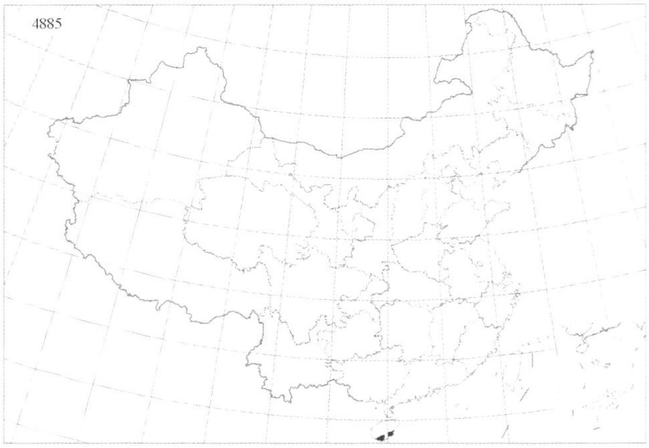

Euphorbiaceae 大戟科 *Croton* 巴豆属
Croton laui Merr. et F. P. Metcalf 海南巴豆
Shrubs, deciduous or semideciduous

24.2 (23.6~24.8)；19.0 (17.8~20.1)；27.9 (27.5~28.4)；24.3 (23.6~24.8)；231.2 (223.3~237.6)；0.0 (0.0~0.0)；1333 (1271~1385) ❙ 1359 (1179~1588)；460 (392~534)；47 (41~58) ❙ 1.9 (-11.9~15.4)；1122 (1056~1212)；474.5 (439.7~527.5)

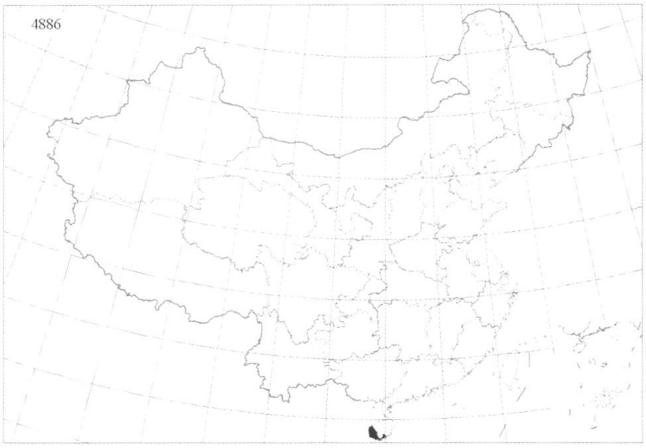

Euphorbiaceae 大戟科 *Croton* 巴豆属
Croton lauioides Radcliffe-Smith et Govaerts 榄绿巴豆
Shrubs or small trees

23.3；17.2；27.9；23.3；220.1；0.0；1267 ❙ 1640；682；79 ❙ 32.3；1143；400.4

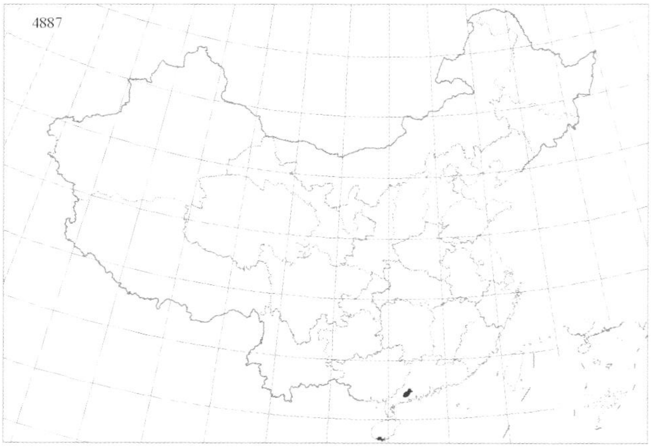

Euphorbiaceae 大戟科 *Croton* 巴豆属
Croton mangelong Y. T. Chang 曼哥龙巴豆
Trees

17.1；10.8；21.3；17.2；146.3；0.0；827 ❙ 1225；691；48 ❙ 47.7；813；460.9

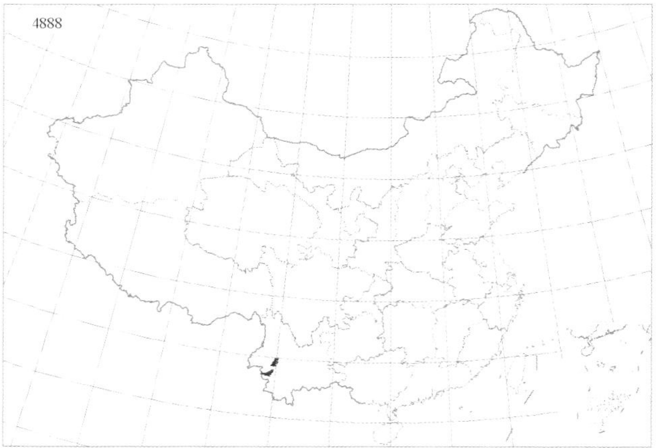

Euphorbiaceae 大戟科 *Croton* 巴豆属
Croton merrillianus Croizat 厚叶巴豆
Shrubs

22.8 (21.3~24.7)；16.0 (13.2~19.8)；27.7 (26.7~28.6)；22.8 (21.3~24.7)；213.9 (195.9~236.5)；0.0 (0.0~0.0)；1224 (1111~1376) ❙ 1788 (1302~2435)；853 (534~1428)；96 (48~130) ❙ 47.2 (12.9~99.5)；1179 (1111~1299)；397.3 (235.1~620.0)

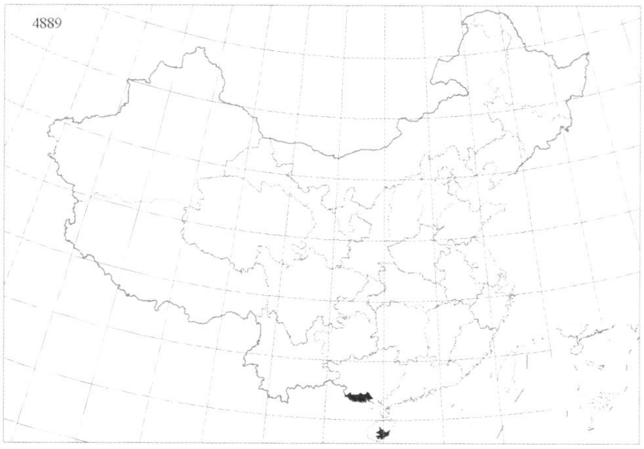

Euphorbiaceae 大戟科 *Croton* 巴豆属
Croton purpurascens Y. T. Chang 淡紫毛巴豆
Shrubs

18.8；9.1；27.3；18.9；166.7；0.0；980 ❙ 1644；603；180 ❙ 67.8；980；229.9

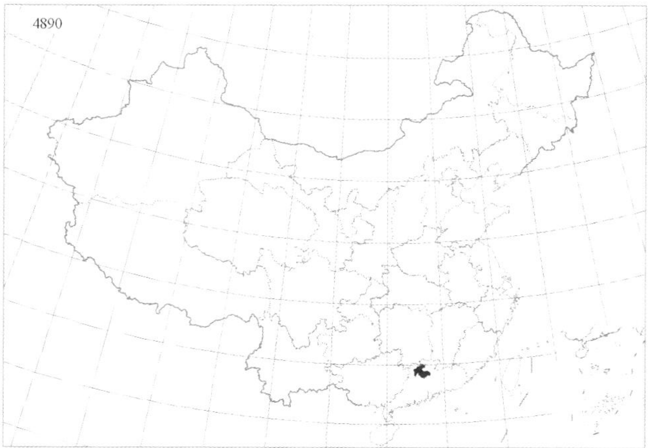

Euphorbiaceae 大戟科 *Croton* 巴豆属
Croton tiglium L. 巴豆
Shrubs or small trees, evergreen

19.4 (3.8~24.8); 10.3 (-4.3~20.1); 27.1 (10.8~30.1); 19.4 (4.7~24.8); 173.3 (2[illegible].6~237.6); 0.3 (0.0~35.4); 1020 (444~1385) ❙ 1514 (868~2680); 644 (312~1428); 132 (16~259) ❙ 50.4 (-34.3~246.5); 1008 (444~1360); 244.5 (95.3~620.0)

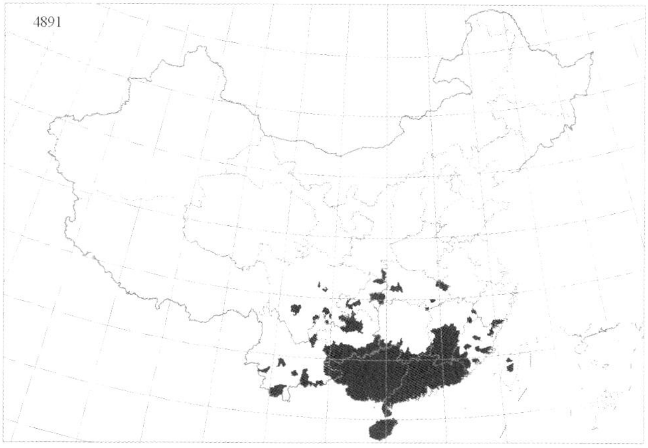

Euphorbiaceae 大戟科 *Croton* 巴豆属
Croton yanhuii Y. T. Chang 延辉巴豆
Shrubs

20.0; 14.6; 23.2; 20.1; 180.8; 0.0; 963 ❙ 1523; 681; 67 ❙ 58.4; 942; 581.0

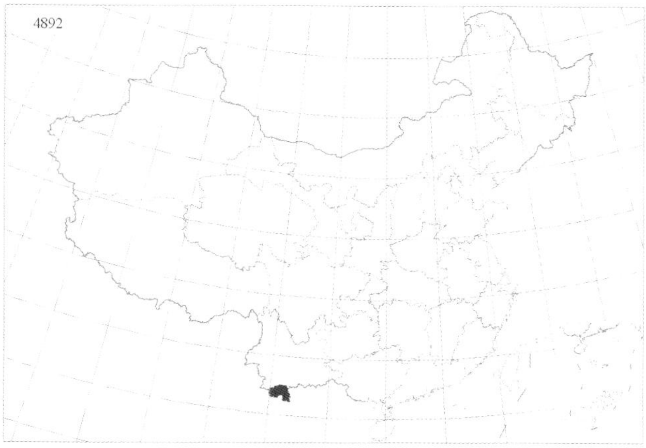

Euphorbiaceae 大戟科 *Croton* 巴豆属
Croton yunnanensis W. W. Sm. 云南巴豆
Shrubs

13.1 (2.2~18.9); 5.4 (-6.6~12.0); 19.5 (9.9~26.6); 13.2 (3.8~19.0); 101.7 (15.5~167.8); 3.6 (0.0~49.1); 708 (413~929) ❙ 956 (741~1141); 554 (411~627); 31 (15~57) ❙ 38.0 (-14.2~125.9); 681 (413~912); 268.1 (142.9~372.3)

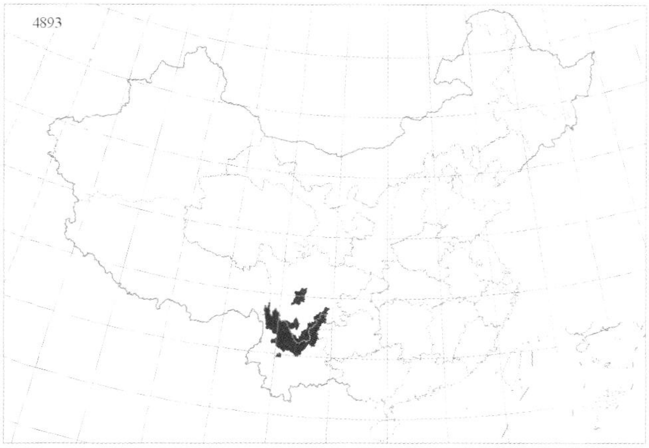

Euphorbiaceae 大戟科 *Dalechampia* 黄蓉花属
Dalechampia bidentata Blume 黄蓉花
Shrubs

18.5 (14.4~20.5); 12.6 (7.5~15.1); 22.2 (19.1~23.7); 18.5 (14.4~20.5); 162.0 (113.1~185.9); 0.0 (0.0~0.0); 893 (727~992) ❙ 1428 (1049~1607); 701 (598~792); 63 (44~77) ❙ 60.9 (26.1~99.9); 871 (727~986); 503.9 (365.3~659.0)

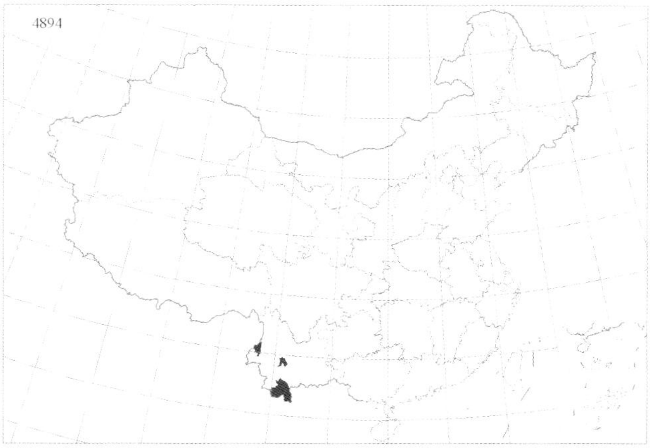

Euphorbiaceae 大戟科 *Deutzianthus* 东京桐属
Deutzianthus tonkinensis Gagnep. 东京桐
Trees, deciduous

20.8 (17.3~22.2); 13.1 (9.9~14.2); 26.4 (22.4~28.5); 20.9 (17.3~22.3); 190.3 (147.1~207.5); 0.0 (0.0~0.0); 1077 (833~1190) ❙ 1514 (1374~1600); 820 (713~911); 91 (55~128) ❙ 43.1 (15.4~80.6); 1077 (833~1190); 373.8 (256.6~551.6)

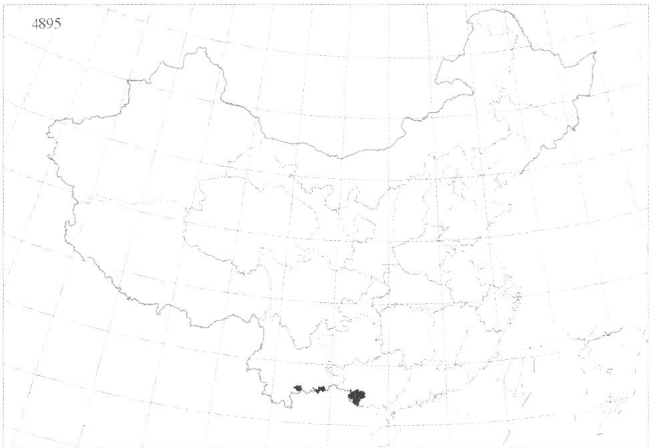

Euphorbiaceae 大戟科 *Dimorphocalyx* 异萼木属
Dimorphocalyx poilanei Gagnep. 异萼木
Shrubs or small trees

24.7; 20.0; 27.9; 24.8; 237.1; 0.0; 1380 ❙ 1496; 503; 50 ❙ 8.4; 1175; 511.1

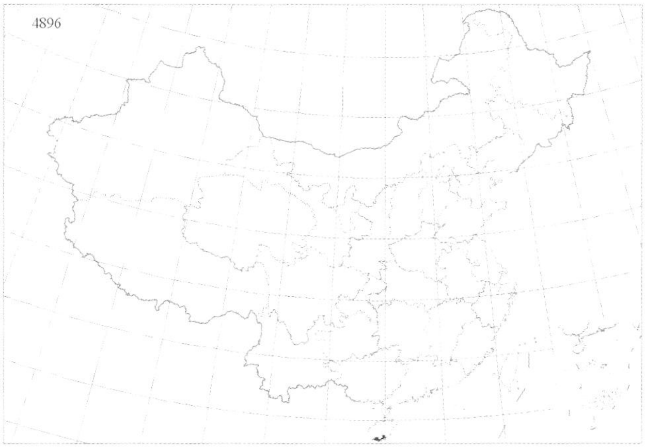

Euphorbiaceae 大戟科 *Discocleidion* 丹麻杆属

Discocleidion rufescens (Franch.) Pax et K. Hoffm. 毛丹麻杆

Shrubs or small trees, deciduous

14.1 (2.6~21.3); 2.6 (−7.0~13.3); 24.9 (11.2~29.5); 14.2 (4.3~21.3); 116.5 (20.0~196.1); 6.6 (0.0~48.2); 792 (427~1137) ❙ 1096 (597~1776); 463 (290~741); 76 (14~191) ❙ 38.4 (−28.0~106.2); 786 (427~1136); 224.1 (108.2~452.0)

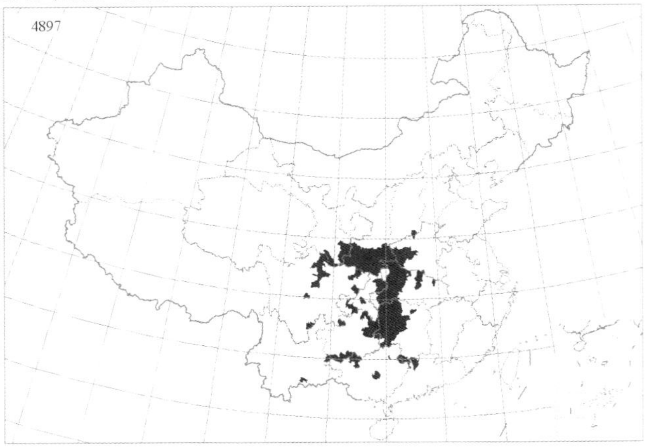

Euphorbiaceae 大戟科 *Discocleidion* 丹麻杆属

Discocleidion ulmifolium (Müll. Arg.) Pax et K. Hoffm. 丹麻杆

Shrubs, deciduous

16.4; 4.8; 27.4; 16.5; 138.0; 0.5; 884 ❙ 1853; 618; 246 ❙ 109.8; 884; 264.3

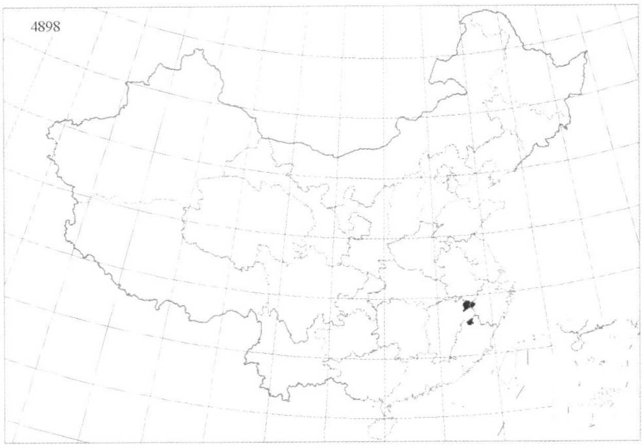

Euphorbiaceae 大戟科 *Drypetes* 核果木属

Drypetes arcuatinervia Merr. et Chun 拱网核果木

Shrubs

22.6 (20.4~24.8); 15.5 (11.9~20.1); 28.0 (23.6~29.2); 22.7 (20.5~24.8); 212.1 (185.8~237.6); 0.0 (0.0~0.0); 1213 (989~1385) ❙ 1582 (868~2435); 726 (312~1428); 98 (41~146) ❙ 32.1 (−34.3~99.5); 1147 (868~1273); 328.1 (178.7~659.0)

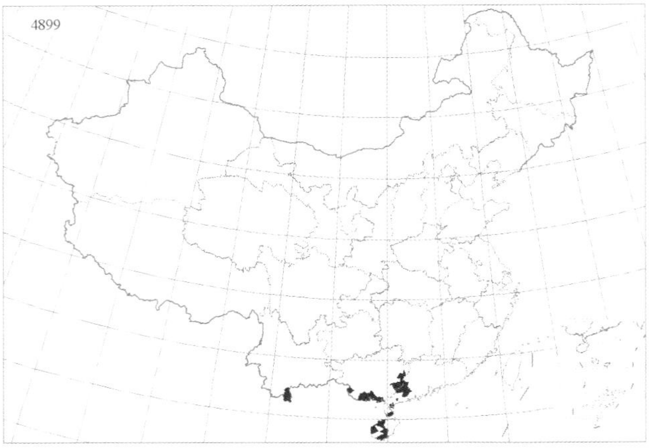

Euphorbiaceae 大戟科 *Drypetes* 核果木属

Drypetes congestiflora Chun et T. C. Chen 密花核果木

Trees

22.6 (20.3~24.8); 15.4 (10.9~20.1); 27.9 (27.5~28.3); 22.6 (20.4~24.8); 211.2 (184.4~237.6); 0.0 (0.0~0.0); 1215 (1065~1385) ❙ 1436 (1374~1567); 636 (455~775); 84 (41~141) ❙ 19.6 (1.5~47.2); 1129 (1065~1190); 375.6 (232.6~527.5)

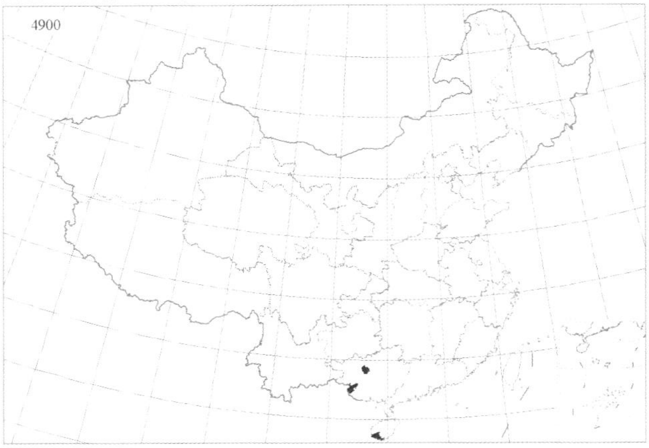

Euphorbiaceae 大戟科 *Drypetes* 核果木属

Drypetes cumingii (Baill.) Pax et K. Hoffm. 青枣核果木

Trees

22.3 (17.2~24.8); 15.9 (9.5~20.1); 27.0 (22.3~29.0); 22.4 (17.3~24.8); 208.7 (147.0~237.6); 0.0 (0.0~0.0); 1184 (838~1385) ❙ 1451 (868~1840); 626 (312~889); 73 (41~125) ❙ 25.2 (−34.3~64.9); 1093 (838~1269); 403.3 (198.9~659.0)

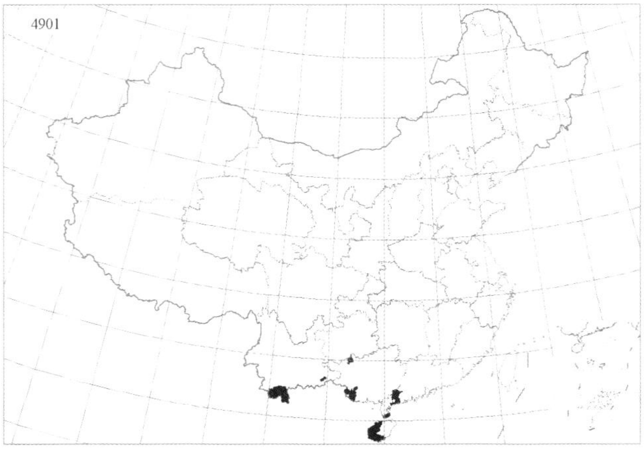

Euphorbiaceae 大戟科 *Drypetes* 核果木属

Drypetes hainanensis Merr. 海南核果木

Trees

23.7 (22.2~24.8); 18.3 (13.9~20.1); 27.6 (25.7~28.5); 23.8 (22.3~24.8); 225.4 (207.2~237.6); 0.0 (0.0~0.0); 1289 (1127~1385) ❙ 1518 (1179~1854); 541 (392~713); 62 (41~115) ❙ 18.3 (−11.9~47.8); 1164 (1055~1360); 499.9 (327.7~620.0)

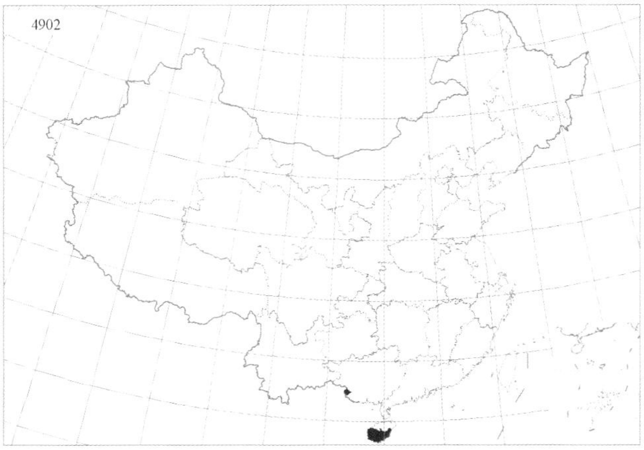

Euphorbiaceae 大戟科 *Drypetes* 核果木属
Drypetes hoaensis Gagnep. 勐腊核果木
Trees

20.4; 15.1; 23.6; 20.5; 185.8; 0.0; 989 ❙ 1607; 702; 77 ❙ 62.5; 986; 659.0

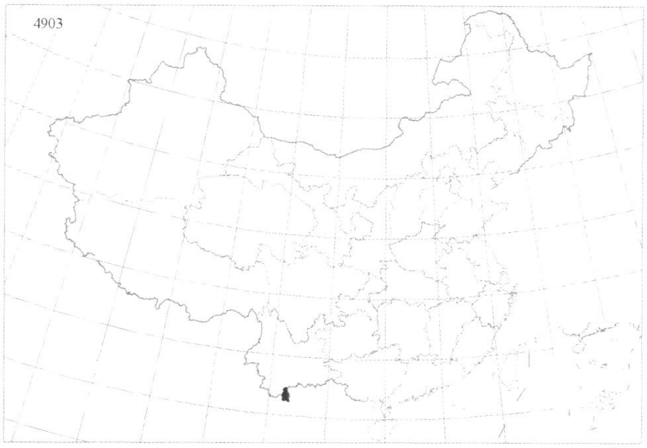

Euphorbiaceae 大戟科 *Drypetes* 核果木属
Drypetes indica (Müll. Arg.) Pax et K. Hoffm. 核果木
Trees

20.1 (14.9~24.8); 12.7 (4.7~20.1); 25.9 (21.4~28.9); 20.2 (15.0~24.8); 181.9 (119.9~237.6); 0.0 (0.0~0.3); 1048 (781~1385) ❙ 1531 (1179~2992); 730 (392~1847); 79 (41~194) ❙ 49.2 (-11.9~202.5); 1007 (781~1360); 339.9 (119.4~659.0)

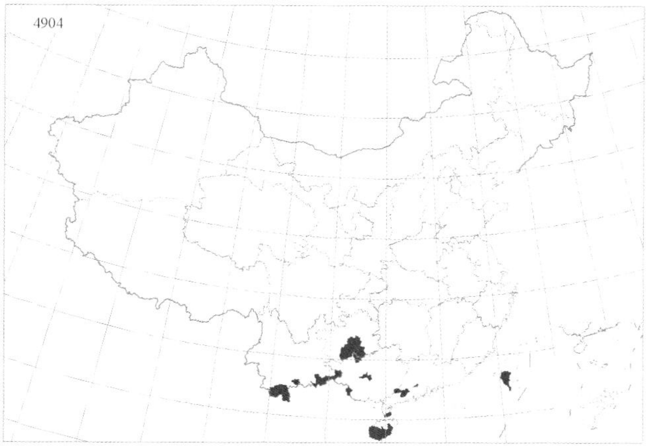

Euphorbiaceae 大戟科 *Drypetes* 核果木属
Drypetes ntegrifolia Merr. et Chun 全缘叶核果木
Shrubs

22.7; 15.3; 28.2; 22.7; 212.8; 0.0; 1230 ❙ 1398; 588; 91 ❙ 14.6; 1145; 362.6

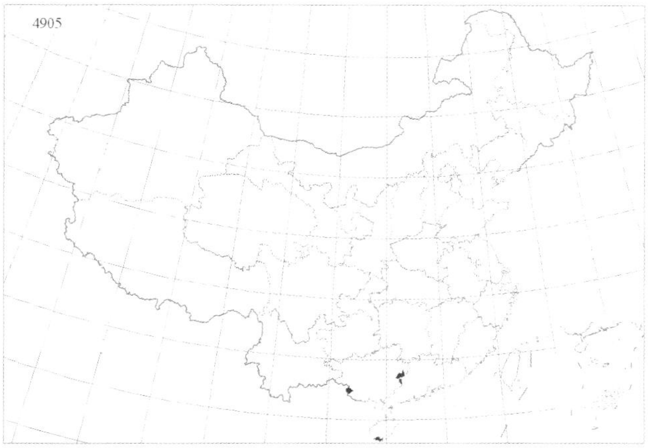

Euphorbiaceae 大戟科 *Drypetes* 核果木属
Drypetes littoralis (C. B. Rob.) Merr. 滨海核果木
Small trees, evergreen

20.3; 15.9; 23.6; 20.4; 184.4; 0.0; 993 ❙ 2853; 1618; 147 ❙ 190.4; 993; 330.3

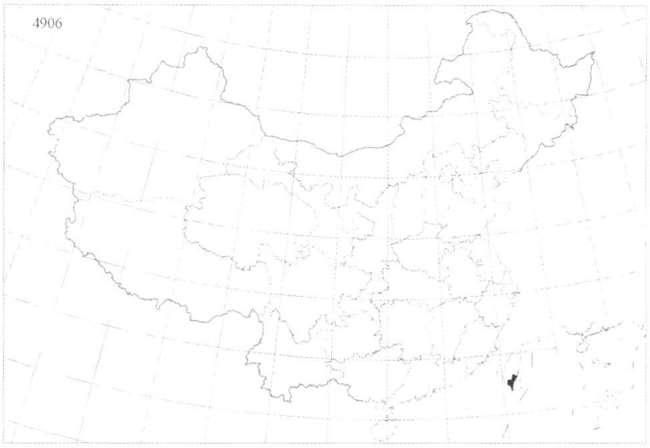

Euphorbiaceae 大戟科 *Drypetes* 核果木属
Drypetes longistipitata P. T. Li 细柄核果木
Trees

24.4; 19.6; 27.6; 24.4; 232.9; 0.0; 1343 ❙ 1478; 506; 45 ❙ 10.3; 1142; 573.8

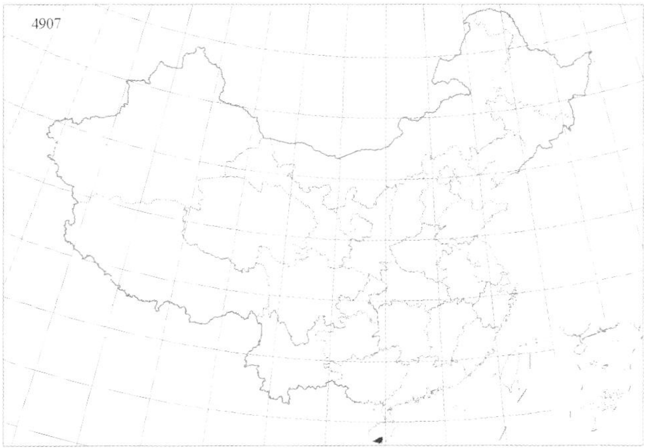

Euphorbiaceae 大戟科 *Drypetes* 核果木属
Drypetes obtusa Merr. et Chun 钝叶核果木
Small trees

23.2; 17.0; 27.8; 23.2; 218.5; 0.0; 1247 ❙ 1790; 786; 79 ❙ 44.7; 1169; 391.9

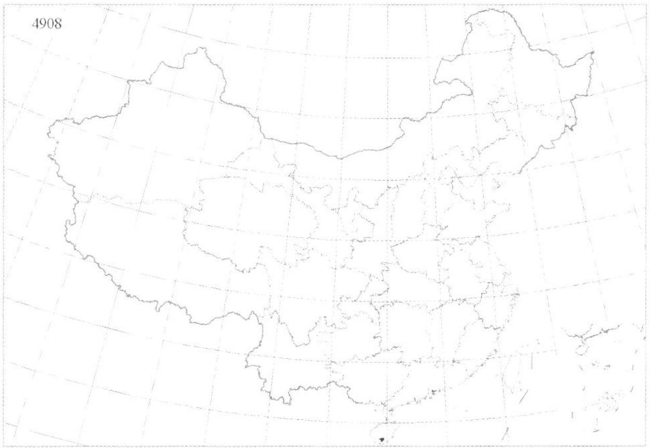

Euphorbiaceae 大戟科 *Drypetes* 核果木属

Drypetes perreticulata Gagnep. 网脉核果木

Trees

22.4 (16.8~24.8); 16.0 (7.8~20.1); 27.0 (22.5~28.7); 22.4 (16.8~24.8); 209.2 (141.5~237.6); 0.0 (0.0~0.0); 1190 (836~1385) ❙ 1499 (868~2035); 630 (312~969); 75 (41~128) ❙ 28.6 (-34.3~74.2); 1099 (836~1360); 430.8 (226.6~659.0)

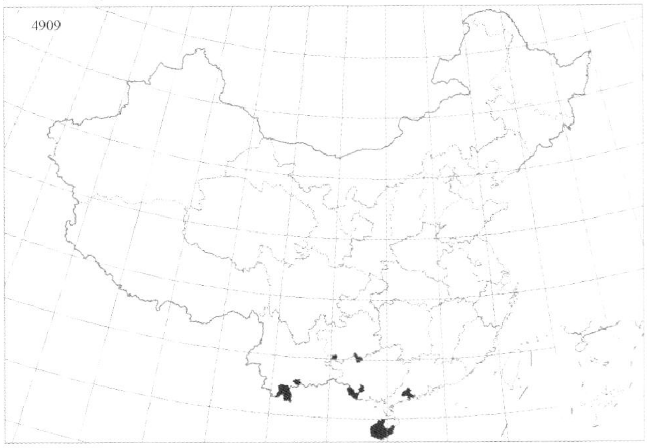

Euphorbiaceae 大戟科 *Endospermum* 黄桐属

Endospermum chinense Benth. 黄桐

Trees, evergreen

21.7 (17.7~24.8); 13.7 (7.4~20.1); 28.1 (22.5~29.2); 21.8 (17.8~24.8); 201.4 (153.4~237.6); 0.0 (0.0~0.0); 1154 (854~1385) ❙ 1633 (868~2435); 745 (312~1428); 124 (41~189) ❙ 42.7 (-34.3~99.5); 1138 (854~1273); 240.8 (95.3~620.0)

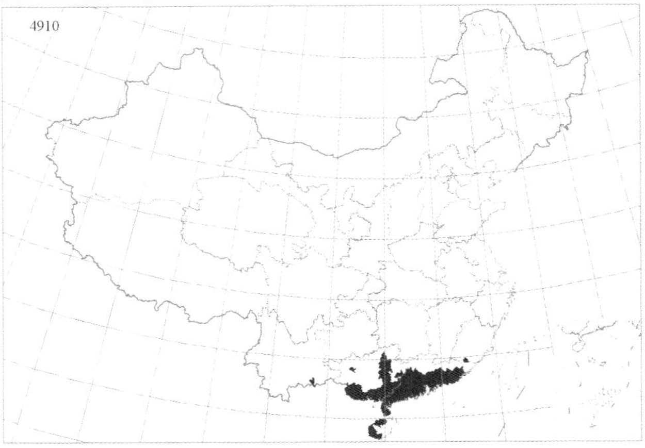

Euphorbiaceae 大戟科 *Epiprinus* 风轮桐属

Epiprinus siletianus (Baill.) Croizat 风轮桐

Shrubs or small trees

22.9 (18.6~24.8); 17.5 (12.7~20.1); 26.6 (22.3~28.7); 23.0 (18.7~24.8); 215.5 (163.9~237.6); 0.0 (0.0~0.0); 1218 (888~1385) ❙ 1471 (868~1854); 558 (312~911); 62 (41~115) ❙ 23.9 (-34.3~74.2); 1088 (868~1360); 509.7 (288.3~659.0)

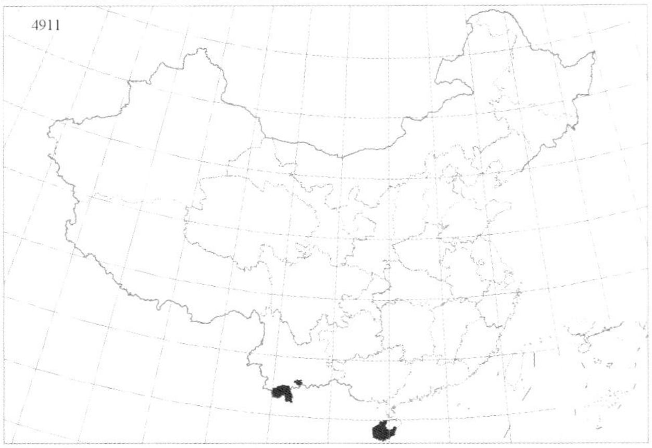

Euphorbiaceae 大戟科 *Erismanthus* 轴花属

Erismanthus sinensis Oliv. 轴花木

Shrubs or trees

23.9 (23.1~24.3); 18.1 (17.5~18.7); 28.0 (27.2~28.7); 23.9 (23.2~24.3); 226.8 (218.4~231.9); 0.0 (0.0~0.0); 1298 (1232~1338) ❙ 1222 (868~1487); 432 (312~544); 53 (42~71) ❙ -5.5 (-34.3~20.7); 1068 (868~1142); 431.4 (288.3~538.2)

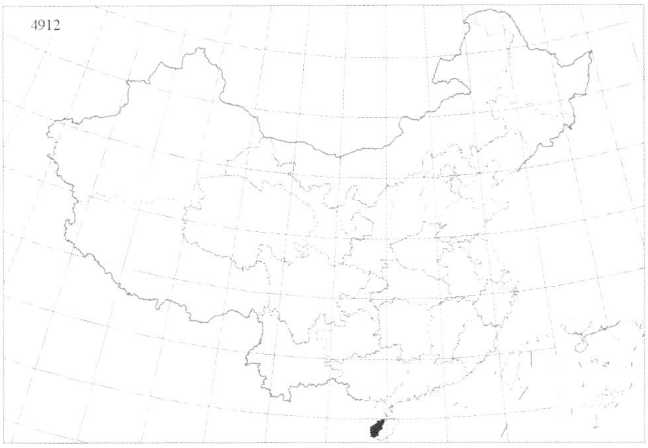

Euphorbiaceae 大戟科 *Euphorbia* 大戟属

Euphorbia hainanensis Croizat 海南大戟

Shrubs, evergreen

23.9; 18.4; 27.9; 23.9; 227.3; 0.0; 1301 ❙ 1267; 430; 45 ❙ -2.5; 1086; 450.1

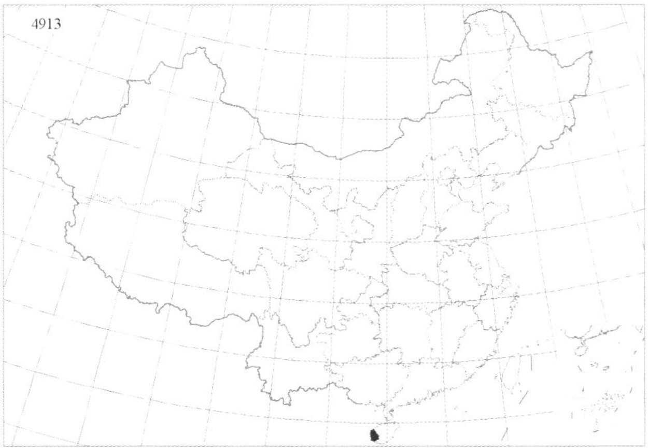

Euphorbiaceae 大戟科 *Euphorbia* 大戟属

Euphorbia royleana Boiss. 霸王鞭

Shrubs, evergreen

14.0 (5.4~19.3); 6.3 (-2.1~13.1); 20.3 (12.3~27.3); 14.0 (5.8~19.4); 110.4 (31.5~173.2); 2.0 (0.0~25.6); 738 (481~949) ❙ 972 (741~1438); 559 (411~820); 33 (15~65) ❙ 33.4 (-14.2~70.8); 710 (481~949); 269.2 (135.1~371.7)

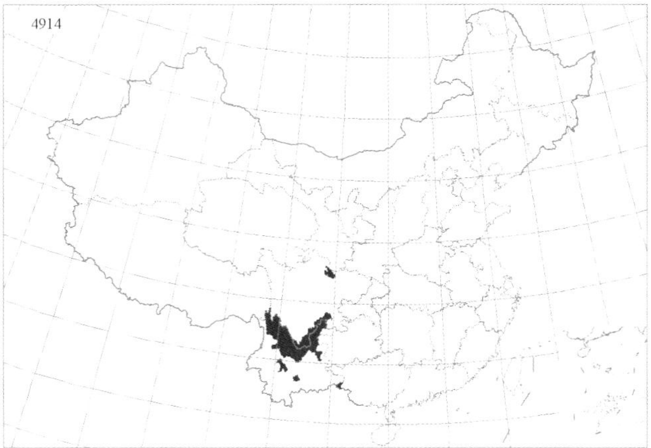

Euphorbiaceae 大戟科 *Excoecaria* 海漆属
Excoecaria acerifolia Didr. 云南土沉香
Shrubs, evergreen

13.5 (1.8~ 9.3); 4.1 (-7.1~13.8); 21.8 (9.5~28.4); 13.6 (3.8~19.4); 107.8 (15.5~173.2); 5.7 (0.0~53.3); 744 (404~939) ❙ 1110 (527~1736); 530 (317~1007); 62 (9~149) ❙ 50.9 (-14.2~145.8); 734 (404~912); 243.2 (142.9~513.9)

4915

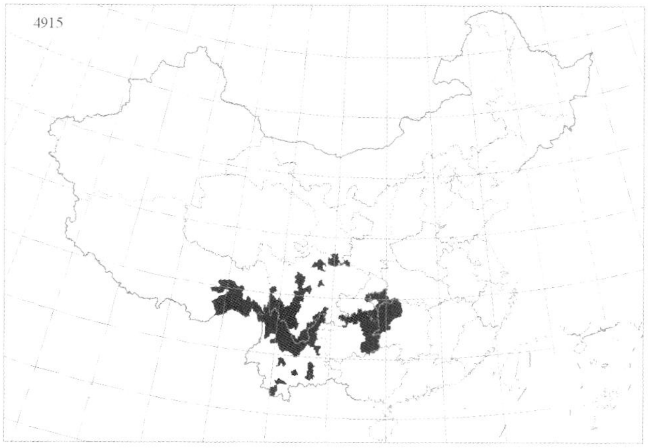

Euphorbiaceae 大戟科 *Excoecaria* 海漆属
Excoecaria agallocha L. 海漆
Trees, evergreen

21.2 (15.7~24.8); 14.3 (4.2~20.1); 26.8 (20.1~29.1); 21.2 (15.7~24.8); 194.3 (129.7~237.6); 0.0 (0.0~0.8); 1103 (773~1385) ❙ 2006 (1190~3495); 929 (418~1847); 176 (41~897) ❙ 90.0 (1.5~246.5); 1072 (773~1247); 261.9 (119.4~620.0)

4916

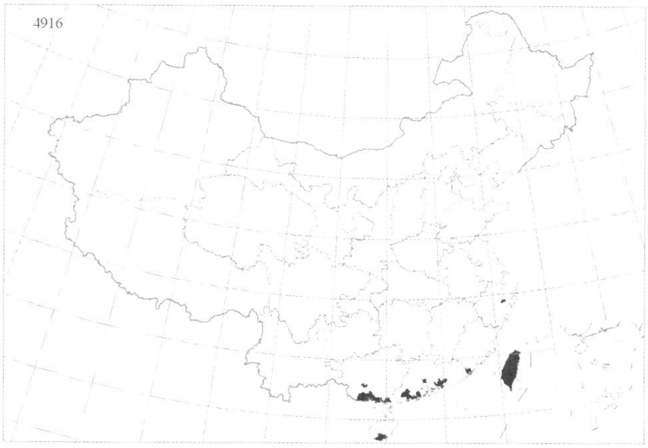

Euphorbiaceae 大戟科 *Excoecaria* 海漆属
Excoecaria cochinchinensis Lour. 红背桂
Shrubs, evergreen

22.6 (16.0~24.8); 16.2 (4.3~20.1); 27.4 (22.3~28.8); 22.6 (16.1~24.8); 211.3 (133.5~237.6); 0.0 (0.0~0.8); 1214 (847~1385) ❙ 1445 (868~1854); 562 (312~862); 74 (27~139) ❙ 20.2 (-34.3~64.9); 1128 (845~1360); 397.9 (126.1~659.0)

4917

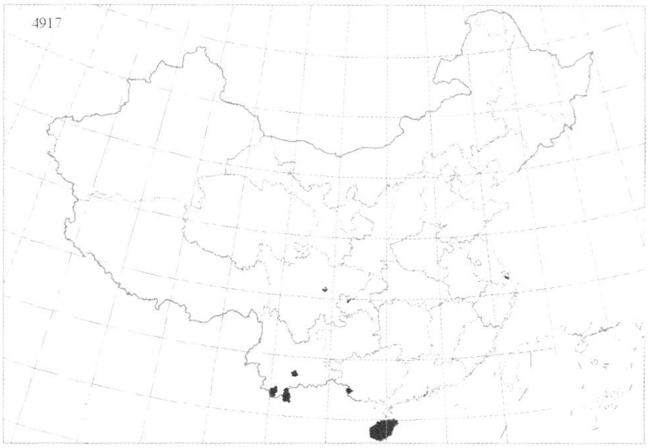

Euphorbiaceae 大戟科 *Excoecaria* 海漆属
Excoecaria cochinchinensis var. viridis (Pax et Hoffm.) Merr. 绿背桂花
Shrubs

23.1 (19.3~24.8); 17.3 (12.6~20.1); 27.4 (22.8~28.7); 23.2 (19.3~24.8); 218.2 (171.9~237.6); 0.0 (0.0~0.0); 1241 (912~1385) ❙ 1630 (868~2992); 711 (312~1847); 73 (41~127) ❙ 35.4 (-34.3~202.5); 1123 (868~1360); 412.2 (154.2~620.0)

4918

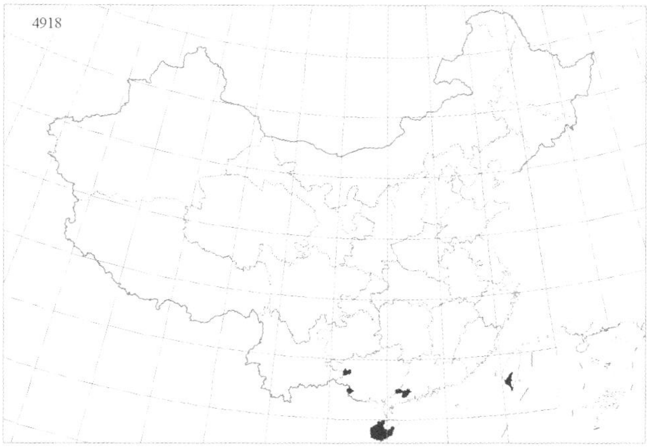

Euphorbiaceae 大戟科 *Excoecaria* 海漆属
Excoecaria kawakamii Hayata 兰屿土沉香
Shrubs

18.5; 14.1; 21.7; 18.5; 162.1; 0.0; 863 ❙ 2713; 1388; 188 ❙ 214.4; 863; 363.7

4919

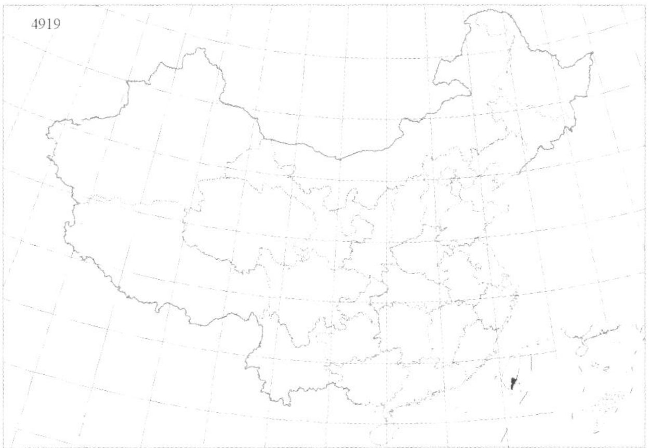

Euphorbiaceae 大戟科 *Excoecaria* 海漆属
Excoecaria venenata S. K. Lee et F. N. Wei 鸡尾木
Shrubs

21.5 (19.1~22.3); 13.1 (11.0~13.9); 27.7 (25.0~28.9); 21.5 (19.1~22.3); 198.1 (169.6~208.1); 0.0 (0.0~0.0); 1132 (945~1202) ❙ 1478 (1374~1581); 776 (710~850); 105 (57~136) ❙ 31.3 (15.4~52.2); 1132 (945~1202); 284.0 (201.4~371.7)

4920

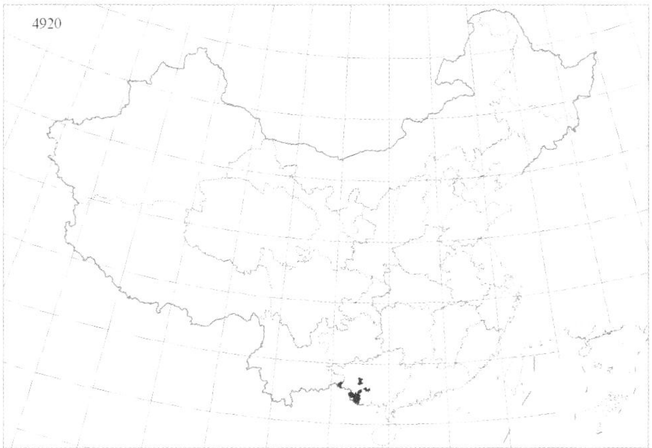

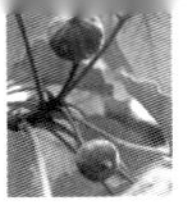

Euphorbiaceae 大戟科 *Falconeria* 异序乌桕属
Falconeria insignis Royle 异序乌桕
Trees, deciduous

21.6 (17.5~24.8)；15.9 (10.9~20.1)；25.5 (22.1~28.4)；21.6 (17.6~24.8)；199.2 (151.2~237.6)；0.0 (0.0~0.0)；1124 (846~1385) | 1350 (1007~1607)；574 (392~835)；52 (41~77) | 23.9 (-11.9~62.5)；1019 (836~1212)；483.2 (340.7~659.0)

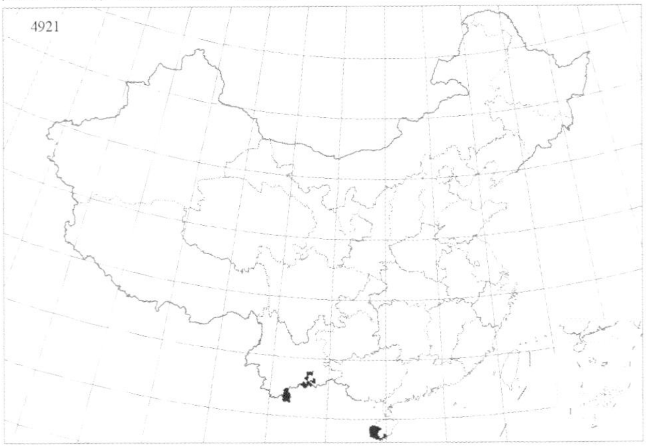

Euphorbiaceae 大戟科 *Flueggea* 白饭树属
Flueggea acicularis (Croizat) G. L. Webster 毛白饭树
Shrubs

14.8 (11.1~20.4)；6.1 (1.9~15.1)；22.3 (19.4~23.8)；14.9 (11.3~20.5)；120.3 (80.8~185.8)；2.0 (0.0~6.1)；779 (652~989) | 1149 (885~1607)；591 (479~741)；52 (34~83) | 48.0 (23.3~87.2)；777 (652~986)；333.6 (210.2~659.0)

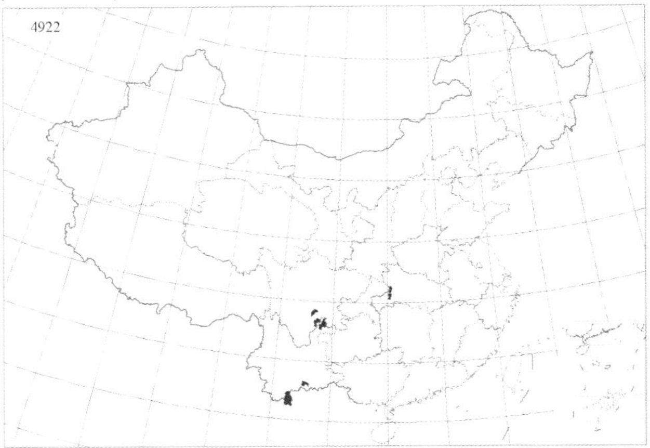

Euphorbiaceae 大戟科 *Flueggea* 白饭树属
Flueggea leucopyrus Willd. 聚花白饭树
Shrubs

13.3 (5.4~18.9)；5.5 (-2.4~12.0)；19.7 (12.3~26.6)；13.3 (5.8~19.0)；102.7 (31.5~167.8)；2.9 (0.0~26.2)；713 (481~929) | 956 (685~1202)；554 (406~641)；31 (12~83) | 36.0 (-14.2~70.8)；684 (481~912)；269.2 (142.9~351.7)

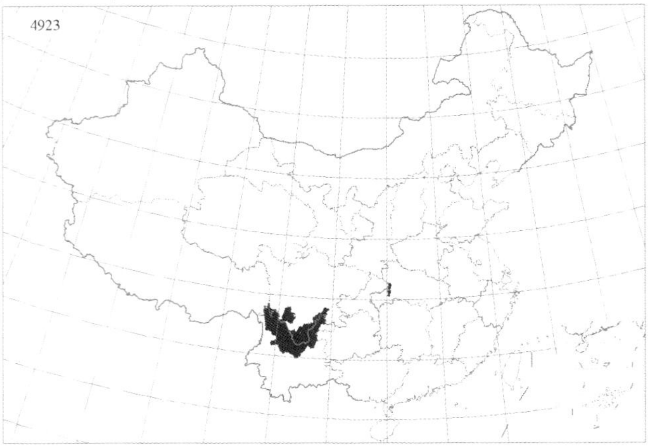

Euphorbiaceae 大戟科 *Flueggea* 白饭树属
Flueggea suffruticosa (Pall.) Baill. 一叶萩
Shrubs, deciduous

12.2 (-4.6~24.7)；-1.7 (-28.8~19.8)；24.6 (10.1~29.9)；13.2 (3.8~24.7)；109.4 (15.4~236.5)；22.6 (0.0~149.1)；768 (413~1376) | 919 (132~2992)；437 (81~1847)；68 (2~303) | 16.0 (-80.1~219.3)；690 (132~1326)；195.5 (13.6~551.3)

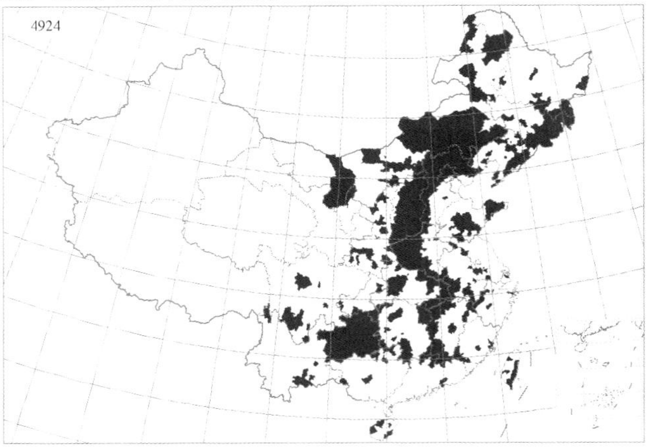

Euphorbiaceae 大戟科 *Flueggea* 白饭树属
Flueggea virosa (Roxb. ex Willd.) Voigt 白饭树
Shrubs, deciduous

19.1 (5.4~24.8)；10.9 (-3.4~20.1)；25.9 (12.3~29.9)；19.2 (5.8~24.8)；170.6 (31.5~237.6)；0.8 (0.0~25.6)；1006 (481~1385) | 1501 (670~3495)；682 (312~1847)；117 (15~897) | 50.7 (-34.3~246.5)；986 (481~1360)；258.3 (95.3~659.0)

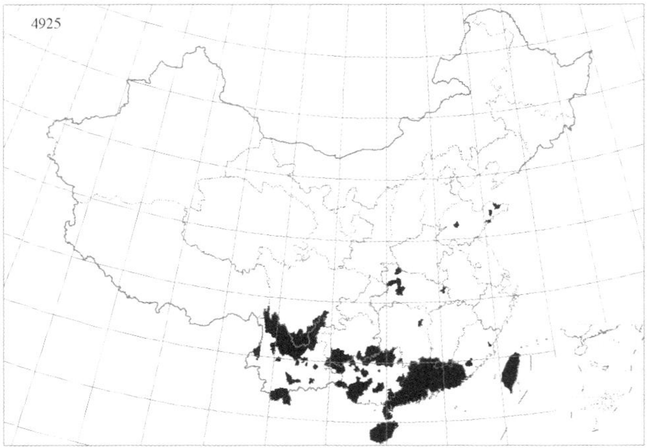

Euphorbiaceae 大戟科 *Glochidion* 算盘子属
Glochidion arborescens Blume 白毛算盘子
Trees

18.0 (13.2~20.5)；11.8 (6.4~15.1)；22.1 (18.4~23.7)；18.1 (13.3~20.5)；157.0 (99.4~185.9)；0.0 (0.0~0.0)；873 (695~992) | 1334 (939~1616)；711 (524~946)；56 (39~77) | 53.8 (6.5~92.4)；850 (695~986)；453.3 (316.6~659.0)

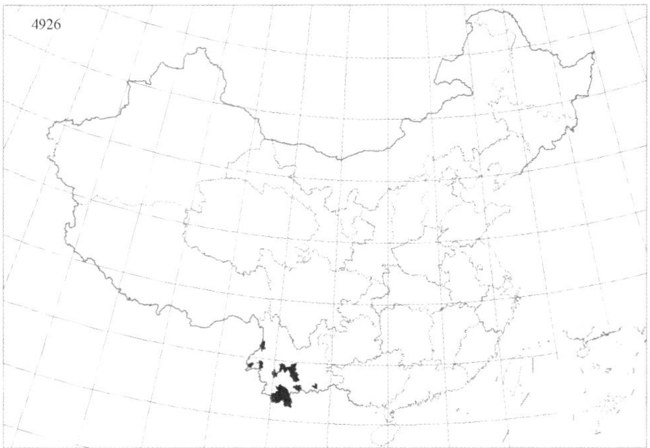

Euphorbiaceae 大戟科 *Glochidion* 算盘子属
Glochidion chademenosocarpum Hayata 线药算盘子
Shrubs

20.2 (16.2~24.1); 14.5 (11.0~18.9); 25.0 (20.1~28.2); 20.2 (16.2~24.1); 182.9 (134.2~229.7); 0.0 (0.0~0.0); 1019 (773~1322) ❙ 2402 (1498~3495); 1106 (750~1847); 255 (47~897) ❙ 146.2 (25.2~246.5); 983 (773~1144); 245.0 (119.4~363.7)

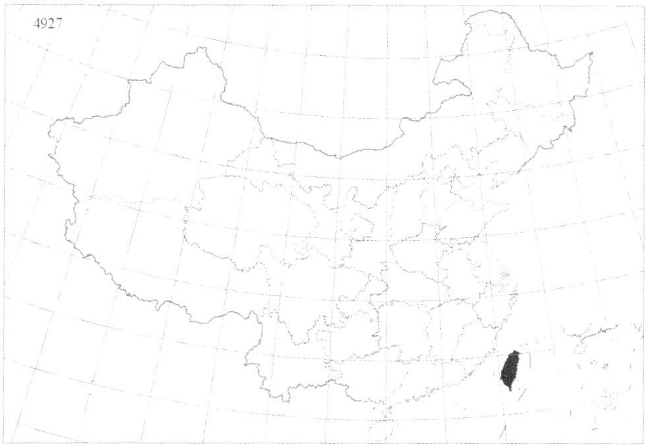

Euphorbiaceae 大戟科 *Glochidion* 算盘子属
Glochidion coccineum (Buch.-Ham.) Müll. Arg. 红算盘子
Shrubs or trees, evergreen

20.7 (15.9~24.8); 12.6 (6.9~20.1); 27.0 (21.4~28.7); 20.7 (16.0~24.8); 188.4 (131.6~237.6); 0.0 (0.0~0.0); 1088 (797~1385) ❙ 1444 (868~1854); 635 (312~862); 106 (41~189) ❙ 35.6 (-34.3~85.5); 1048 (797~1360); 291.5 (139.6~620.0)

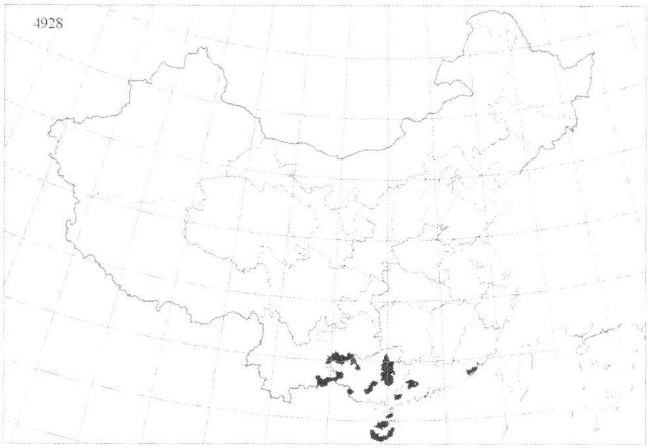

Euphorbiaceae 大戟科 *Glochidion* 算盘子属
Glochidion daltonii (Müll. Arg.) Kurz 革叶算盘子
Shrubs or trees, evergreen

16.1 (5.4~21.7); 6.8 (-2.1~13.2); 24.0 (12.3~30.1); 16.1 (5.8~21.7); 134.2 (31.5~200.8); 1.0 (0.0~25.6); 842 (481~1153) ❙ 1240 (741~1745); 571 (411~909); 88 (15~226) ❙ 47.4 (-14.2~107.2); 823 (481~1153); 253.8 (118.7~506.2)

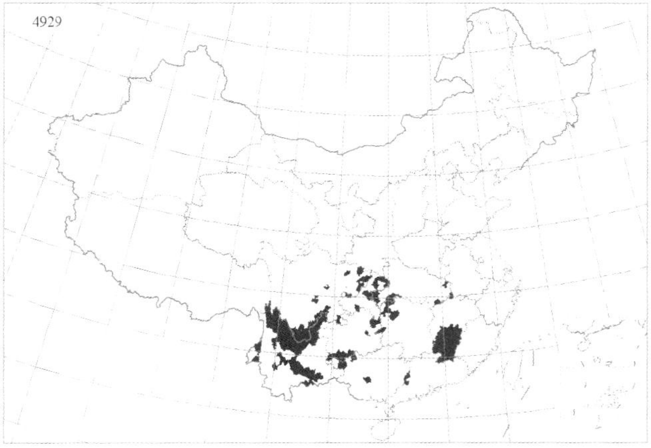

Euphorbiaceae 大戟科 *Glochidion* 算盘子属
Glochidion ellipticum Wight 四裂算盘子
Trees, evergreen

20.6 (13.2~24.8); 13.5 (4.8~20.1); 26.0 (18.4~29.0); 20.7 (13.3~24.8); 188.2 (99.4~237.6); 0.0 (0.0~0.2); 1069 (695~1385) ❙ 1542 (868~3495); 696 (312~1033); 110 (41~897) ❙ 47.0 (-34.3~224.4); 1031 (695~1360); 361.9 (139.6~659.0)

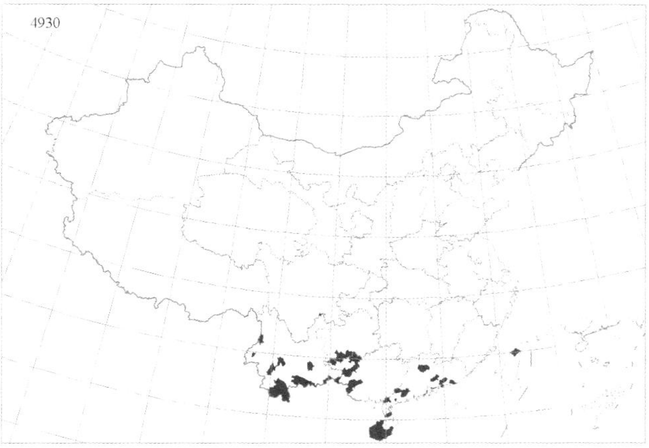

Euphorbiaceae 大戟科 *Glochidion* 算盘子属
Glochidion eriocarpum Champ. ex Benth. 毛果算盘子
Shrubs

19.9 (13.2~24.8); 11.7 (3.8~20.1); 26.6 (18.4~29.8); 19.9 (13.3~24.8); 179.4 (99.4~237.6); 0.0 (0.0~1.5); 1034 (695~1385) ❙ 1618 (868~3495); 711 (312~1847); 137 (39~897) ❙ 59.2 (-34.3~246.5); 1020 (695~1329); 272.3 (95.3~659.0)

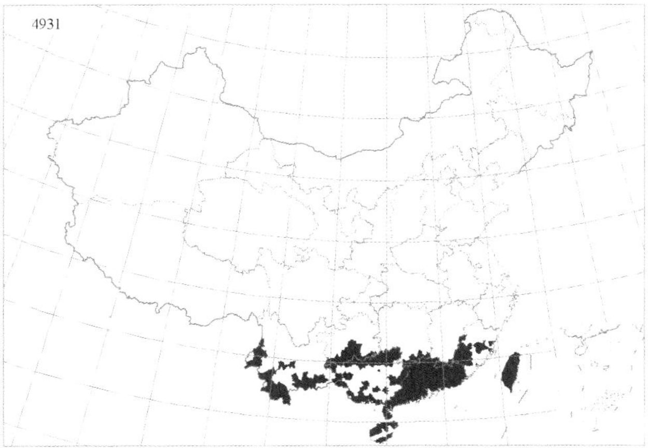

Euphorbiaceae 大戟科 *Glochidion* 算盘子属
Glochidion heyneanum (Wight et Arn.) Wight 绒毛算盘子
Trees

17.6 (16.3~20.4); 11.2 (9.1~15.1); 22.0 (20.3~23.6); 17.7 (16.4~20.5); 151.8 (136.4~185.8); 0.0 (0.0~0.0); 851 (789~989) ❙ 1159 (911~1607); 629 (524~756); 49 (32~77) ❙ 35.9 (6.5~62.5); 818 (725~986); 392.7 (293.0~659.0)

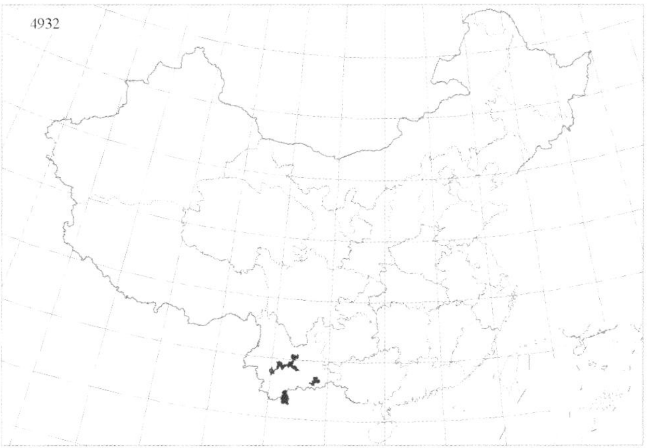

Euphorbiaceae 大戟科 *Glochidion* 算盘子属

Glochidion hirsutum (Roxb.) Voigt 厚叶算盘子

Shrubs or small trees

21.0 (13.3~24.8); 13.6 (5.4~20.1); 26.9 (19.1~29.2); 21.1 (13.4~24.8); 193.1 (100.4~237.6); 0.0 (0.0~0.0); 1099 (706~1385) ❙ 1663 (868~3495); 764 (312~1847); 129 (38~897) ❙ 54.6 (-34.3~246.5); 1080 (706~1360); 278.0 (95.3~659.0)

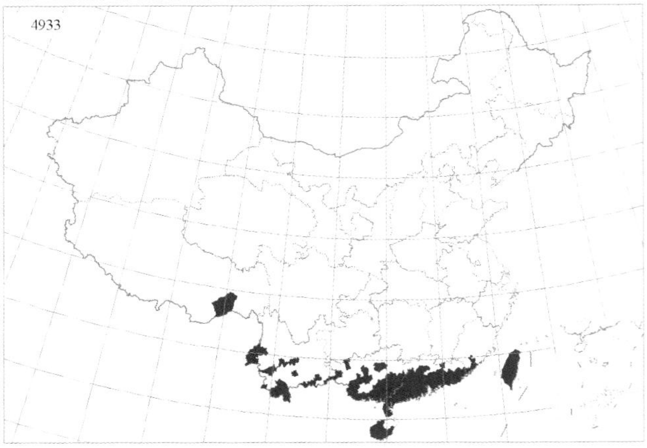

Euphorbiaceae 大戟科 *Glochidion* 算盘子属

Glochidion khasicum (Müll. Arg.) Hook. f. 长柱算盘子

Shrubs or small trees

18.3 (12.7~21.6); 9.8 (4.7~15.1); 25.0 (19.2~28.5); 18.4 (12.7~21.7); 160.6 (93.0~199.8); 0.0 (0.0~0.3); 938 (687~1153) ❙ 1310 (922~1703); 680 (506~848); 79 (26~194) ❙ 38.9 (16.9~62.5); 936 (687~1153); 321.6 (210.2~659.0)

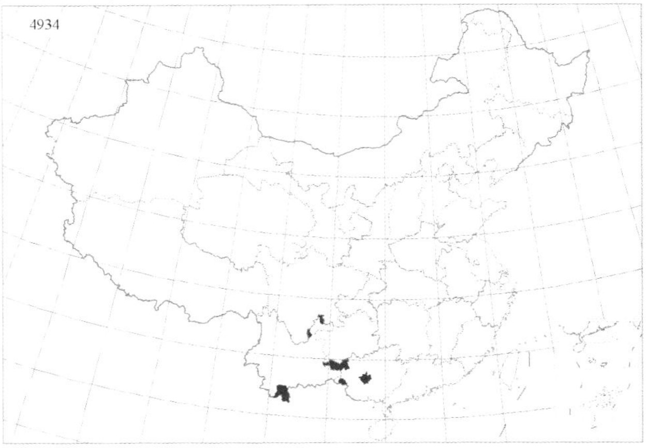

Euphorbiaceae 大戟科 *Glochidion* 算盘子属

Glochidion kusukusense Hayata 台湾算盘子

Shrubs

20.2 (16.2~24.1); 14.5 (11.0~18.9); 25.0 (20.1~28.2); 20.2 (16.2~24.1); 182.9 (134.2~229.7); 0.0 (0.0~0.0); 1019 (773~1322) ❙ 2402 (1498~3495); 1106 (750~1847); 255 (47~897) ❙ 146.2 (25.2~246.5); 983 (773~1144); 245.0 (119.4~363.7)

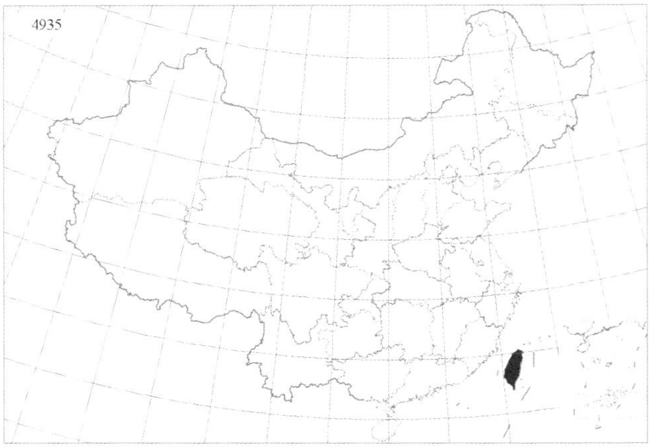

Euphorbiaceae 大戟科 *Glochidion* 算盘子属

Glochidion lanceolarium (Roxb.) Voigt 艾胶算盘子

Shrubs, evergreen

21.9 (15.5~24.8); 14.5 (8.6~20.1); 27.8 (20.4~29.2); 22.0 (15.6~24.8); 203.9 (127.3~237.6); 0.0 (0.0~0.0); 1164 (768~1385) ❙ 1620 (868~2435); 739 (312~1428); 108 (41~197) ❙ 40.3 (-34.3~99.5); 1138 (768~1360); 279.4 (95.3~659.0)

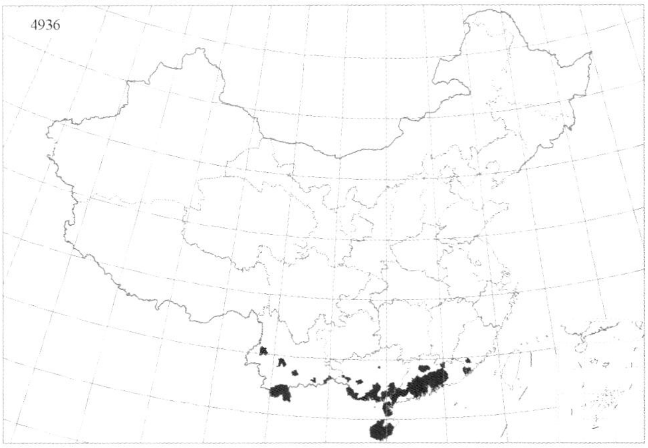

Euphorbiaceae 大戟科 *Glochidion* 算盘子属

Glochidion lanceolatum Hayata 披针叶算盘子

Trees

20.6 (16.2~23.8); 15.3 (11.0~18.1); 24.8 (20.1~28.2); 20.7 (16.2~23.8); 188.0 (134.2~225.8); 0.0 (0.0~0.0); 1048 (773~1297) ❙ 2542 (1757~2993); 1312 (924~1847); 203 (58~547) ❙ 155.9 (35.5~246.5); 1002 (773~1144); 256.5 (119.4~363.7)

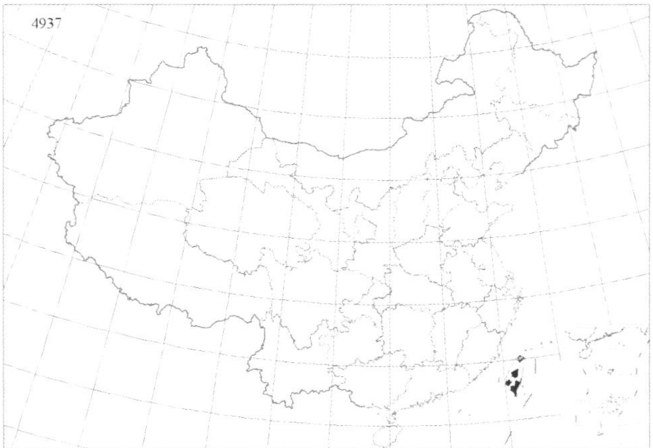

Euphorbiaceae 大戟科 *Glochidion* 算盘子属

Glochidion lutescens Blume 山漆茎

Trees

21.7; 15.2; 26.4; 21.7; 200.5; 0.0; 1159 ❙ 1355; 550; 51 ❙ 22.3; 1013; 434.0

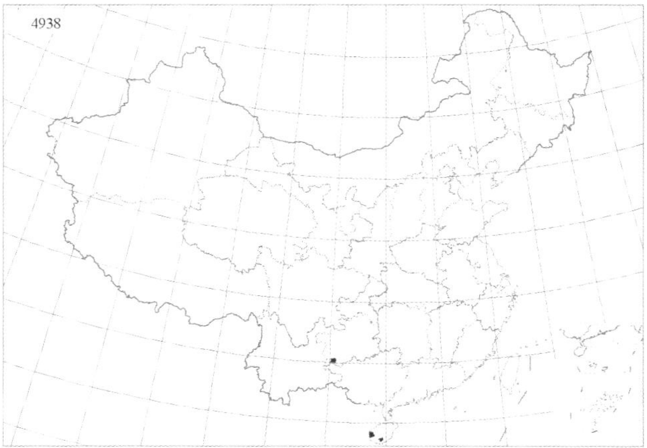

Euphorbiaceae 大戟科 *Glochidion* 算盘子属
Glochidion medogense T. L. Chin 墨脱算盘子
Shrubs or trees

13.3; 5.4; 19.6; 13.4; 100.4; 0.0; 706 ❙ 1736; 1007; 51 ❙ 145.8; 706; 447.4

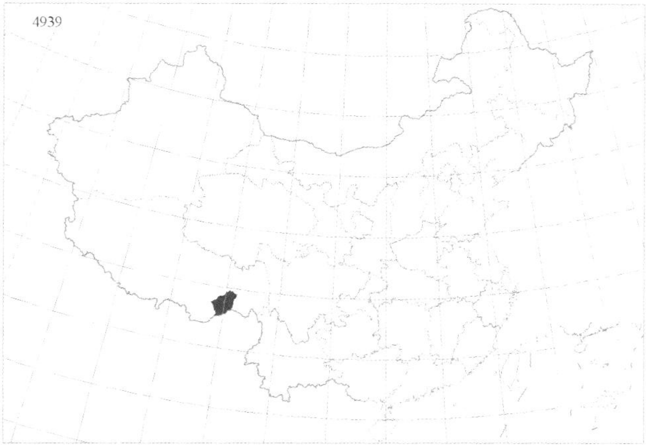

Euphorbiaceae 大戟科 *Glochidion* 算盘子属
Glochidion nubigenum Hook. f. 云雾算盘子
Trees

13.3; 5.4; 19.6; 13.4; 100.4; 0.0; 706 ❙ 1736; 1007; 51 ❙ 145.8; 706; 447.4

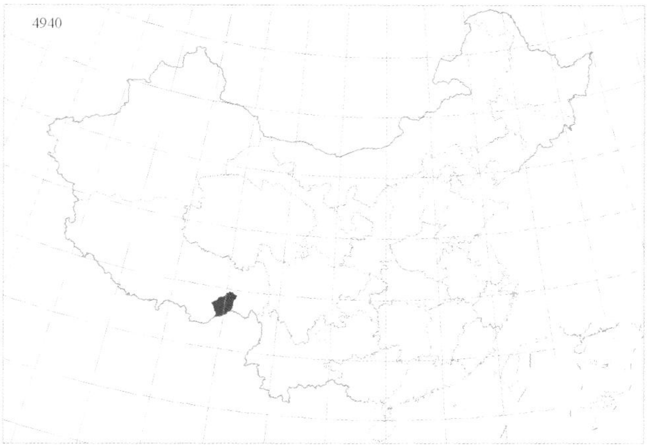

Euphorbiaceae 大戟科 *Glochidion* 算盘子属
Glochidion oblatum Hook. f. 宽果算盘子
Shrubs

19.8 (16.9~22.0); 12.7 (10.3~14.2); 24.9 (21.4~28.8); 19.8 (16.9~22.0); 177.6 (142.7~203.8); 0.0 (0.0~0.0); 995 (813~1170) ❙ 1475 (1178~1647); 726 (646~818); 83 (51~132) ❙ 49.2 (40.1~64.9); 990 (813~1170); 391.7 (203.4~551.6)

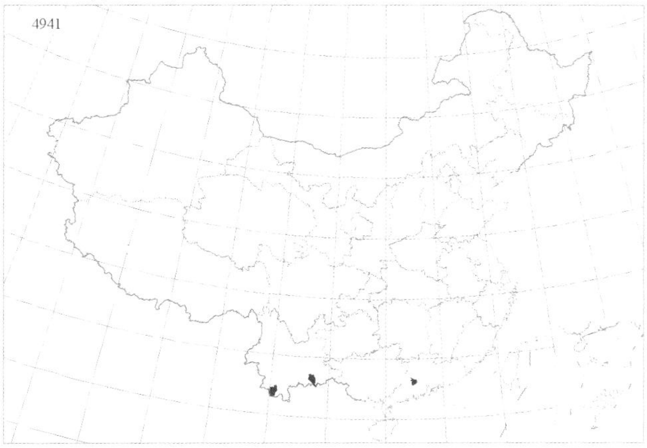

Euphorbiaceae 大戟科 *Glochidion* 算盘子属
Glochidion philippicum (Cav.) C. B. Rob. 甜叶算盘子
Trees

19.7 (13.7~24.8); 12.2 (3.8~20.1); 25.6 (19.7~29.0); 19.8 (13.7~24.8); 177.4 (104.7~237.6); 0.0 (0.0~1.5); 1016 (714~1385) ❙ 1618 (868~3495); 779 (312~1847); 123 (19~897) ❙ 62.8 (-34.3~246.5); 984 (692~1269); 296.3 (119.4~659.0)

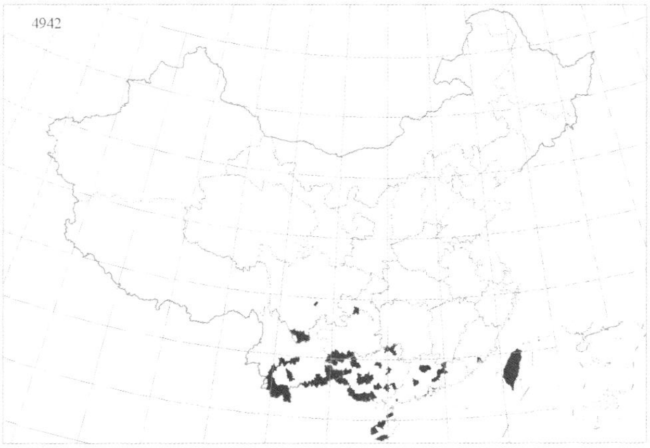

Euphorbiaceae 大戟科 *Glochidion* 算盘子属
Glochidion puberum (L.) Hutch. 算盘子
Shrubs or small trees

17.0 (6.2~24.1); 6.1 (-4.1~18.9); 27.1 (15.8~30.1); 17.1 (7.0~24.2); 146.9 (45.2~230.1); 1.8 (0.0~29.9); 920 (522~1326) ❙ 1366 (512~3495); 558 (258~1594); 134 (15~897) ❙ 48.5 (-35.3~246.5); 904 (512~1326); 211.9 (95.3~659.0)

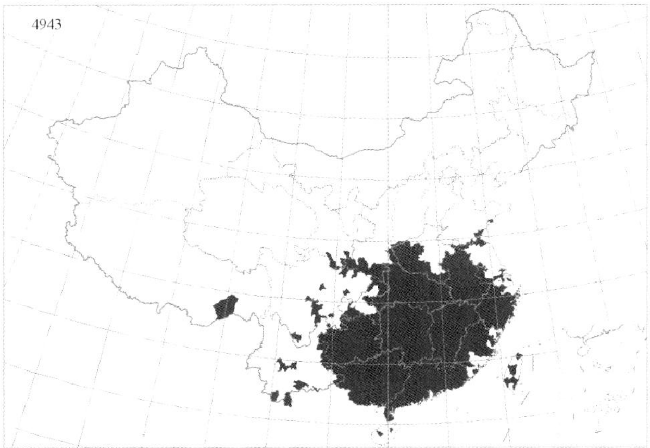

Euphorbiaceae 大戟科 *Glochidion* 算盘子属
Glochidion rubrum Blume 台闽算盘子
Shrubs or trees

19.7 (16.1~24.1); 12.5 (3.9~18.9); 26.4 (20.1~28.9); 19.8 (16.1~24.1); 177.4 (134.2~229.7); 0.1 (0.0~1.1); 1012 (773~1322) ❙ 1934 (1244~3495); 824 (411~1847); 221 (47~897) ❙ 96.8 (14.0~246.5); 990 (773~1144); 236.0 (119.4~363.7)

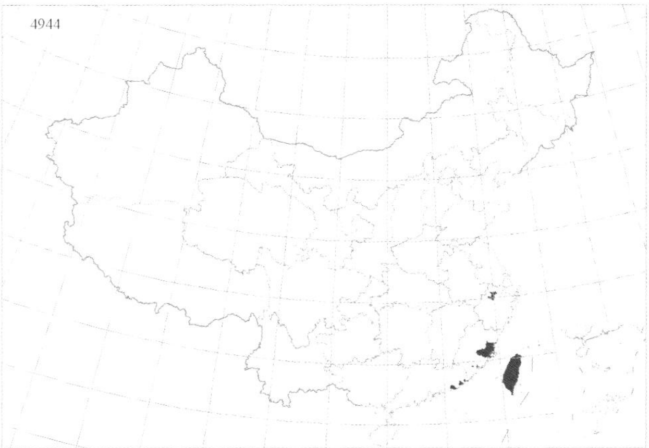

Euphorbiaceae 大戟科 *Glochidion* 算盘子属
Glochidion sphaerogynum (Müll. Arg.) Kurz 圆果算盘子
Shrubs or trees

20.7 (16.2~24.8); 13.8 (6.3~20.1); 25.6 (20.3~29.2); 20.7 (16.2~24.8); 188.8 (134.5~237.6); 0.0 (0.0~0.0); 1069 (789~1385) ❙ 1448 (868~2435); 700 (312~1428); 73 (41~194) ❙ 38.5 (-34.3~112.2); 1027 (780~1360); 389.8 (178.7~659.0)

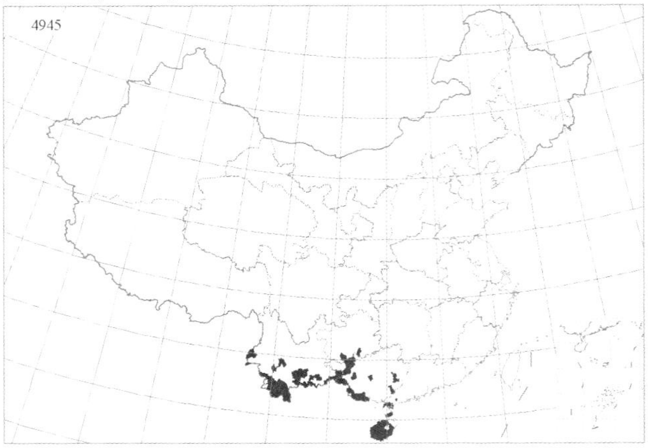

Euphorbiaceae 大戟科 *Glochidion* 算盘子属
Glochidion suishaense Hayata 水社算盘子
Shrubs

20.2 (16.2~24.1); 14.5 (11.0~18.9); 25.0 (20.1~28.2); 20.2 (16.2~24.1); 182.9 (134.2~229.7); 0.0 (0.0~0.0); 1019 (773~1322) ❙ 2402 (1498~3495); 1106 (750~1847); 255 (47~897) ❙ 146.2 (25.2~246.5); 983 (773~1144); 245.0 (119.4~363.7)

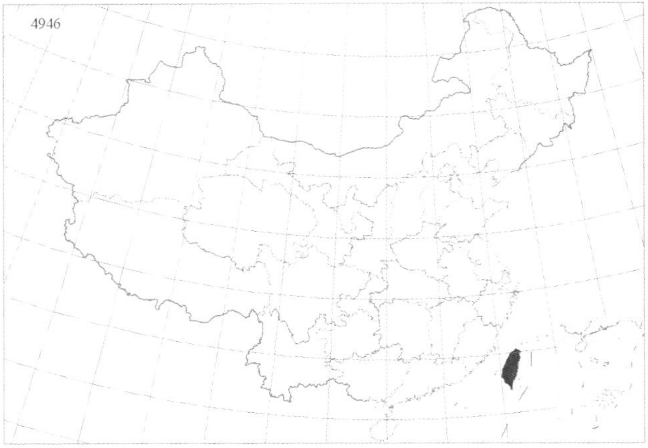

Euphorbiaceae 大戟科 *Glochidion* 算盘子属
Glochidion thomsonii (Müll. Arg.) Hook. f. 青背叶算盘子
Shrubs

13.3; 5.4; 19.6; 13.4; 100.4; 0.0; 706 ❙ 1736; 1007; 51 ❙ 145.8; 706; 447.4

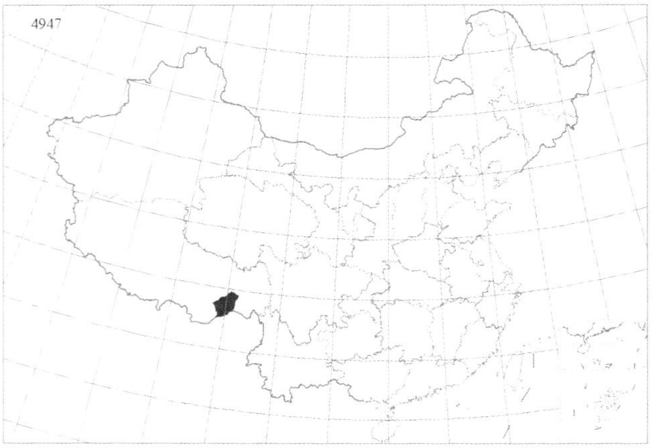

Euphorbiaceae 大戟科 *Glochidion* 算盘子属
Glochidion triandrum (Blanco) C. B. Rob. 里白算盘子
Shrubs or small trees

19.3 (11.3~24.3); 11.8 (3.6~19.1); 25.4 (17.0~28.8); 19.3 (11.3~24.3); 172.2 (76.6~231.9); 0.1 (0.0~1.4); 990 (644~1338) ❙ 1630 (868~3495); 721 (312~1847); 134 (19~897) ❙ 68.3 (-34.3~246.5); 960 (644~1326); 315.9 (119.4~659.0)

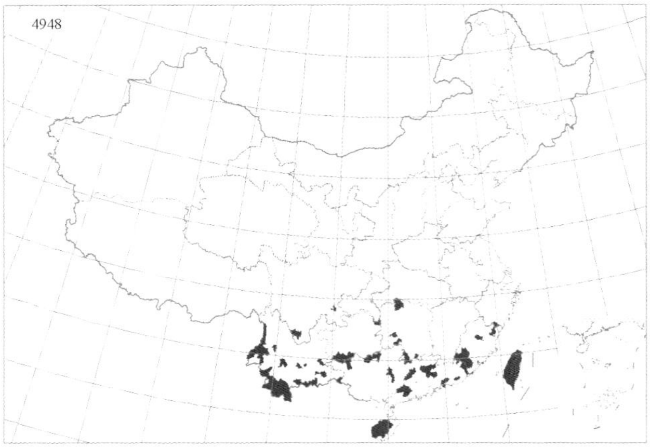

Euphorbiaceae 大戟科 *Glochidion* 算盘子属
Glochidion wilsonii Hutch. 湖北算盘子
Shrubs

16.5 (9.3~21.6); 5.0 (-1.7~12.6); 27.3 (20.1~30.1); 16.6 (9.5~21.7); 140.1 (69.8~199.8); 1.4 (0.0~18.0); 892 (603~1153) ❙ 1428 (793~1997); 528 (299~795); 157 (36~259) ❙ 60.2 (-6.7~130.7); 879 (603~1153); 216.4 (108.2~319.9)

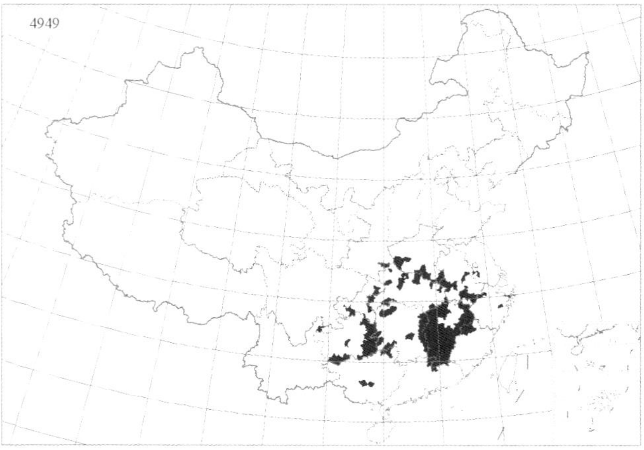

Euphorbiaceae 大戟科 *Glochidion* 算盘子属
Glochidion wrightii Benth. 白背算盘子
Shrubs or trees

20.6 (14.3~24.8); 12.4 (4.0~20.1); 27.1 (19.6~29.2); 20.6 (14.3~24.8); 187.5 (112.6~237.6); 0.0 (0.0~1.1); 1081 (730~1385) ❙ 1534 (868~2037); 689 (312~1033); 112 (38~230) ❙ 43.4 (-34.3~104.6); 1064 (717~1360); 267.3 (95.3~659.0)

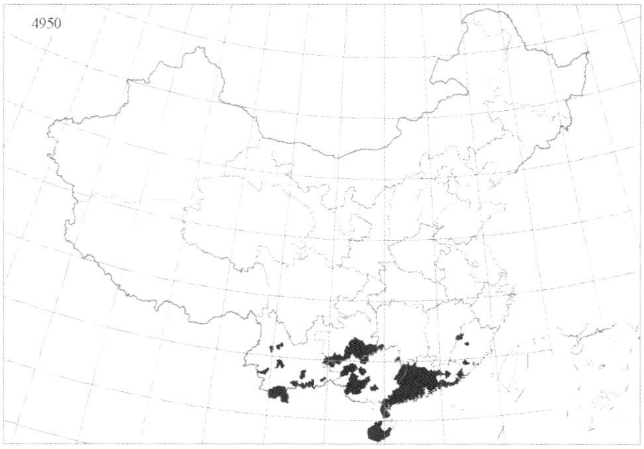

Euphorbiaceae 大戟科 *Glochidion* 算盘子属
Glochidion zeylanicum (Gaertn.) A.Juss. 香港算盘子
Shrubs or trees

21.4 (16.2~24.8); 13.8 (7.7~20.1); 27.4 (20.1~29.1); 21.4 (16.2~24.8); 196.9 (134.2~237.6); 0.0 (0.0~0.0); 1122 (773~1385) ❙ 1680 (868~3495); 729 (312~1388); 137 (41~897) ❙ 52.6 (-34.3~246.5); 1102 (773~1360); 277.0 (95.3~659.0)

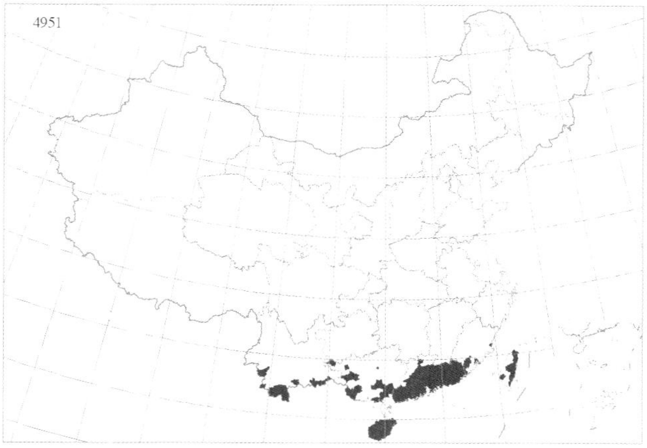

Euphorbiaceae 大戟科 *Hancea* 粗毛野桐属
Hancea hookeriana Seem. 粗毛野桐
Shrubs or small trees

22.8 (20.7~24.8); 15.8 (11.7~20.1); 28.1 (25.7~28.9); 22.8 (20.8~24.8); 213.9 (189.5~237.6); 0.0 (0.0~0.0); 1224 (1081~1385) ❙ 1648 (868~2435); 731 (312~1428); 99 (41~168) ❙ 36.0 (-34.3~99.5); 1172 (868~1360); 326.7 (122.7~620.0)

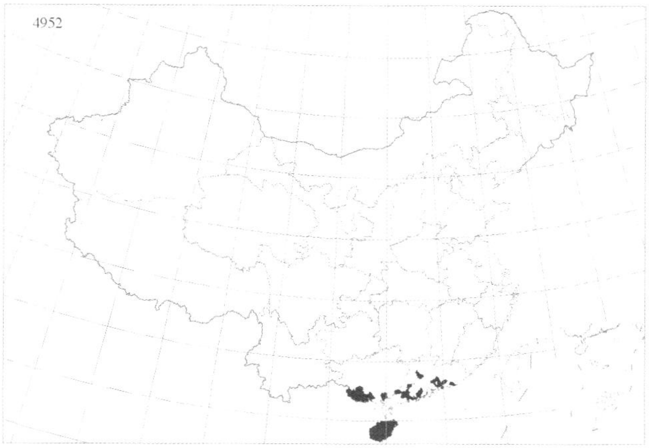

Euphorbiaceae 大戟科 *Homalanthus* 澳杨属
Homalanthus fastuosus (Lind.) F. Vill. 圆叶澳杨
Shrubs or trees

23.7 (18.5~24.8); 18.1 (14.1~20.1); 27.7 (21.7~28.8); 23.7 (18.5~24.8); 225.0 (162.1~237.6); 0.0 (0.0~0.0); 1287 (863~1385) ❙ 1562 (868~2713); 576 (312~1388); 79 (41~188) ❙ 24.9 (-34.3~214.4); 1175 (863~1360); 436.5 (248.2~620.0)

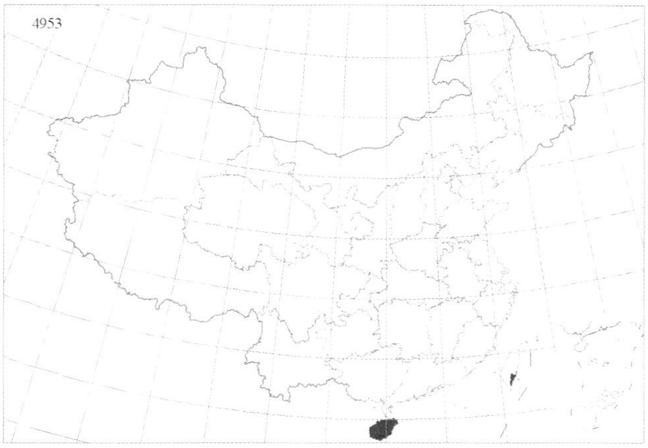

Euphorbiaceae 大戟科 *Homonoia* 水杨柳属
Homonoia riparia Lour. 水柳
Shrubs

20.1 (11.6~24.8); 12.8 (3.1~20.1); 25.6 (18.4~29.0); 20.2 (11.5~24.8); 182.2 (80.9~237.6); 0.1 (0.0~2.8); 1049 (654~1385) ❙ 1456 (868~2992); 723 (312~1847); 77 (19~178) ❙ 41.6 (-34.3~202.5); 1017 (654~1360); 341.7 (154.2~659.0)

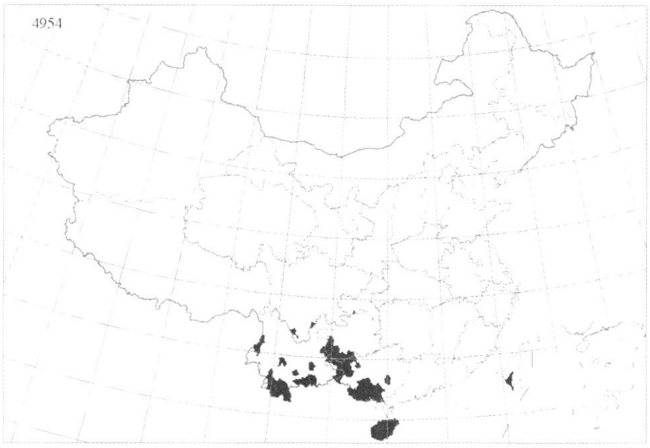

Euphorbiaceae 大戟科 *Jatropha* 麻风树属
Jatropha curcas L. 麻风树
Shrubs or small trees

18.4 (5.4~24.8); 11.3 (-2.1~20.1); 24.0 (12.3~28.9); 18.5 (5.8~24.8); 162.5 (31.5~237.6); 0.7 (0.0~25.6); 961 (481~1385) ❙ 1426 (741~3495); 703 (312~1847); 89 (12~897) ❙ 50.6 (-34.3~246.5); 928 (481~1360); 306.4 (119.4~659.0)

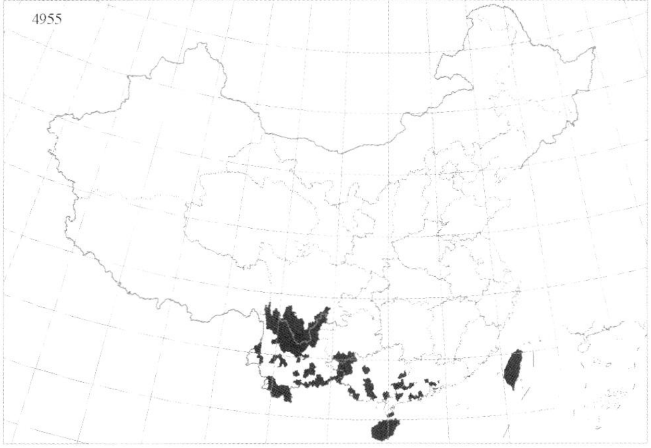

Euphorbiaceae 大戟科 *Koilodepas* 白茶树属
Koilodepas hainanense (Merr.) Croizat 白茶树
Shrubs or trees

24.0 (22.2~24.8); 18.4 (17.2~20.1); 28.0 (25.7~28.8); 24.0 (22.3~24.8); 228.3 (207.2~237.6); 0.0 (0.0~0.0); 1310 (1127~1385) ❙ 1501 (868~1854); 534 (312~712); 74 (41~115) ❙ 15.0 (-34.3~47.8); 1191 (868~1360); 440.3 (248.2~620.0)

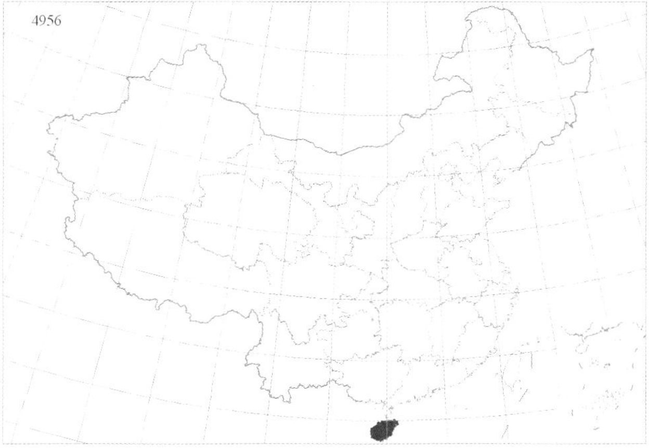

Euphorbiaceae 大戟科 *Lasiococca* 轮叶戟属
Lasiococca comberi var. pseudoverticillata (Merr.) H. S. Kiu 轮叶戟
Trees

22.7 (18.6~24.8); 17.3 (12.7~20.1); 26.4 (22.3~28.7); 22.8 (18.7~24.8); 213.0 (163.9~237.6); 0.0 (0.0~0.0); 1202 (888~1385) ❙ 1384 (868~1607); 546 (312~911); 56 (41~77) ❙ 19.4 (-34.3~74.2); 1048 (868~1212); 501.9 (288.3~659.0)

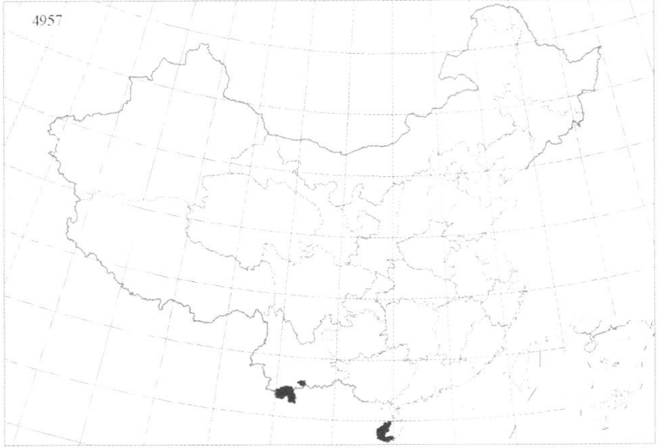

Euphorbiaceae 大戟科 *Leptopus* 雀舌木属
Leptopus australis (Zoll. et Moritzi) Pojark. 薄叶雀舌木
Small shrubs

23.3 (19.1~24.8); 17.8 (11.0~20.1); 27.1 (25.0~28.0); 23.3 (19.1~24.8); 219.7 (169.6~237.6); 0.0 (0.0~0.0); 1260 (945~1385) ❙ 1472 (1380~1588); 564 (455~820); 51 (41~58) ❙ 19.0 (1.5~52.2); 1111 (945~1212); 493.8 (371.7~620.0)

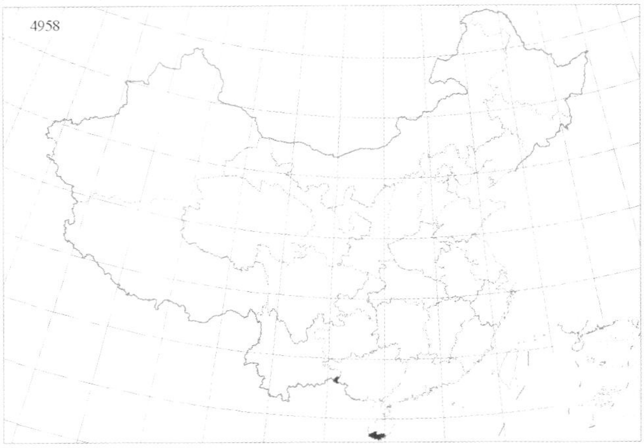

Euphorbiaceae 大戟科 *Leptopus* 雀舌木属
Leptopus chinensis (Bunge) Pojark. 雀儿舌头
Shrubs

12.4 (-1.1~20.5); -0.9 (-15.9~14.9); 24.3 (6.2~29.3); 13.0 (2.0~20.5); 107.0 (2.5~185.9); 17.2 (0.0~81.2); 760 (334~1040) ❙ 807 (63~1961); 431 (23~1007); 40 (2~228) ❙ 6.4 (-91.1~145.8); 651 (63~1040); 195.3 (8.4~570.1)

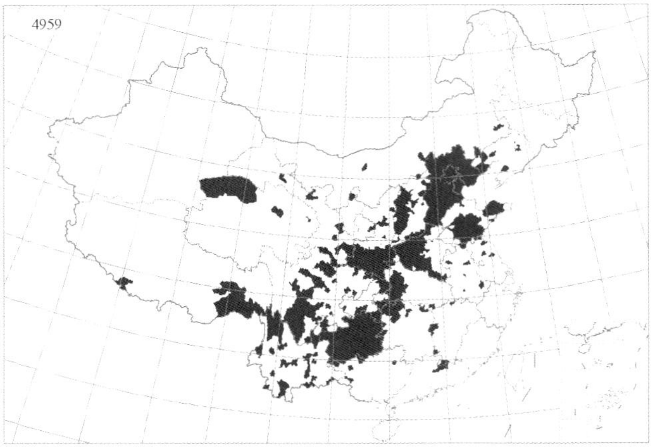

Euphorbiaceae 大戟科 *Leptopus* 雀舌木属
Leptopus clarkei (Hook. f.) Pojark. 缘腺雀舌木
Shrubs

15.2 (3.8~21.6); 6.1 (-4.3~14.9); 23.0 (10.8~28.5); 15.2 (4.7~21.7); 123.7 (21.6~199.8); 1.1 (0.0~35.4); 794 (444~1153) ❙ 1148 (741~1625); 566 (411~835); 63 (15~173) ❙ 45.4 (-14.2~121.7); 784 (444~1153); 244.6 (142.9~570.1)

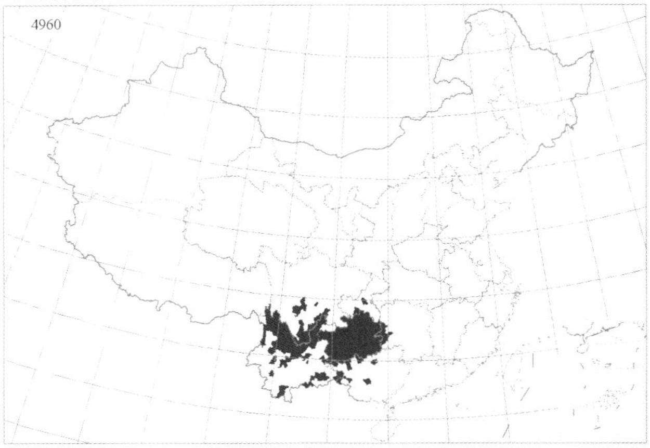

Euphorbiaceae 大戟科 *Leptopus* 雀舌木属
Leptopus fangdingianus (P. T. Li) Uorontsovals Petra Hoffmam. 方鼎木
Shrubs

19.1; 11.0; 25.1; 19.2; 170.0; 0.0; 949 ❙ 1461; 834; 65 ❙ 54.0; 949; 342.7

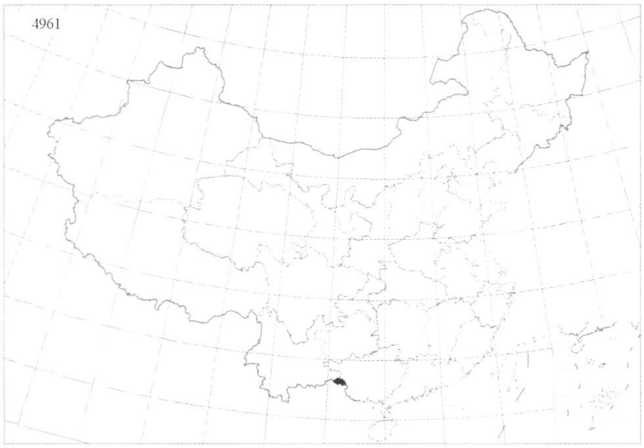

Euphorbiaceae 大戟科 *Leptopus* 雀舌木属
Leptopus hainanensis (Merr. et Chun) Pojark. 海南雀舌木
Shrubs

23.9 (22.2~24.8); 18.6 (17.2~20.1); 27.6 (25.7~28.5); 23.9 (22.3~24.8); 227.0 (207.2~237.6); 0.0 (0.0~0.0); 1298 (1127~1385) ❙ 1531 (1179~1854); 525 (392~622); 60 (41~115) ❙ 18.5 (-11.9~47.8); 1161 (1055~1360); 515.5 (428.6~620.0)

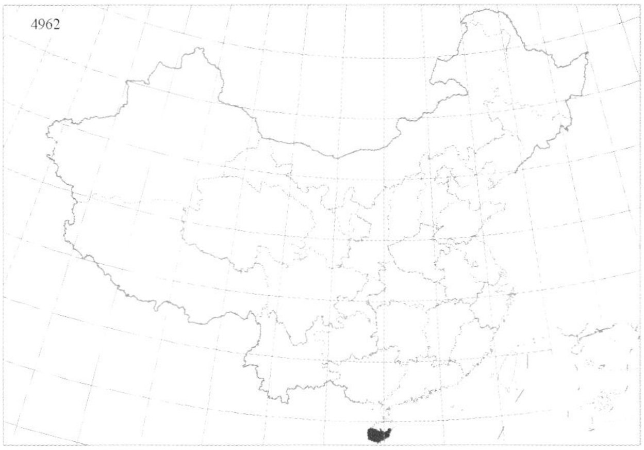

Euphorbiaceae 大戟科 *Macaranga* 血桐属
Macaranga andamanica Kurz 轮苞血桐
Shrubs or trees

20.7 (14.8~24.3); 12.4 (4.3~19.1); 27.3 (22.6~29.0); 20.7 (14.9~24.3); 188.8 (119.2~231.9); 0.0 (0.0~0.7); 1090 (791~1338) ❙ 1536 (1132~2435); 736 (392~1428); 105 (42~194) ❙ 42.1 (-12.0~99.5); 1073 (791~1257); 283.2 (176.4~620.0)

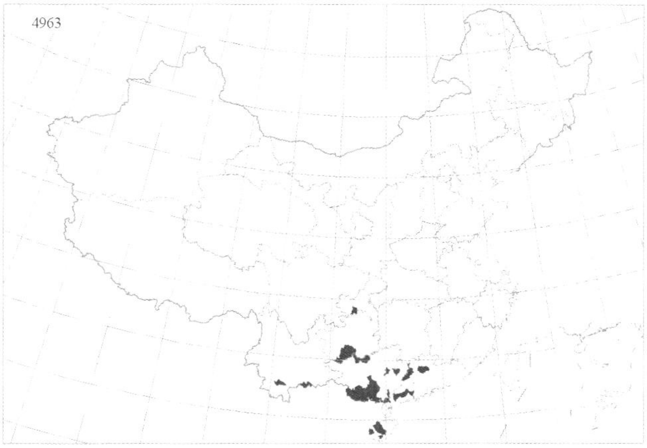

Euphorbiaceae 大戟科 *Macaranga* 血桐属
Macaranga denticulata (Blume) Müll. Arg. 中平树
Trees

19.3 (8.6~24.8); 10.7 (1.3~20.1); 26.3 (15.0~30.1); 19.4 (8.7~24.8); 172.3 (52.6~237.6); 0.1 (0.0~8.8); 1004 (574~1385) ❙ 1494 (868~2435); 654 (312~1428); 120 (41~230) ❙ 51.5 (-34.3~145.8); 977 (574~1360); 315.5 (148.9~659.0)

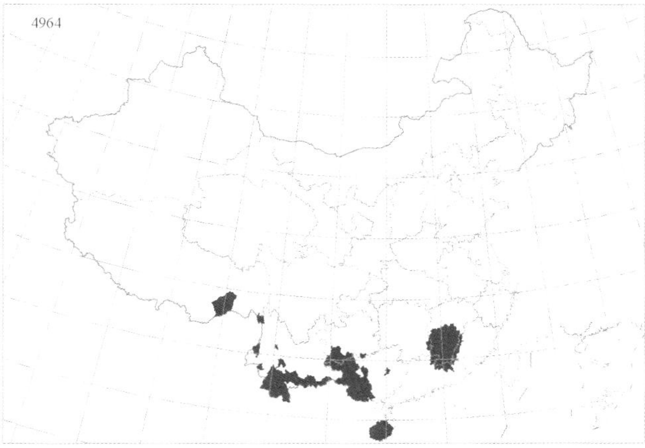

Euphorbiaceae 大戟科 *Macaranga* 血桐属
Macaranga henryi (Pax et K. Hoffm.) Rehder 草鞋木
Shrubs or trees

19.0 (13.7~22.7); 10.4 (4.7~15.1); 25.8 (20.6~28.9); 19.1 (13.7~22.8); 168.9 (104.8~213.1); 0.0 (0.0~0.3); 976 (722~1221) ❙ 1451 (1011~2435); 726 (523~1428); 101 (44~194) ❙ 49.3 (12.9~99.5); 975 (722~1221); 288.7 (185.5~659.0)

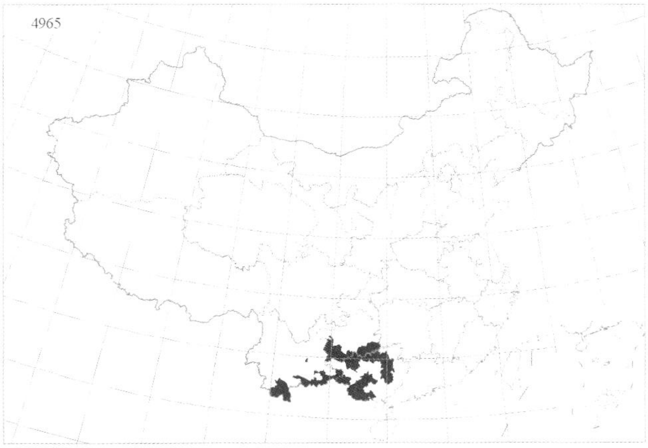

Euphorbiaceae 大戟科 *Macaranga* 血桐属
Macaranga indica Wight 印度血桐
Trees

19.0 (8.6~23.2); 10.8 (1.3~15.8); 25.3 (15.0~28.9); 19.1 (8.7~23.2); 169.0 (52.6~218.9); 0.2 (0.0~8.8); 978 (574~1257) ❙ 1452 (929~2435); 747 (499~1428); 85 (25~177) ❙ 49.6 (5.8~145.8); 976 (574~1257); 312.4 (135.9~659.0)

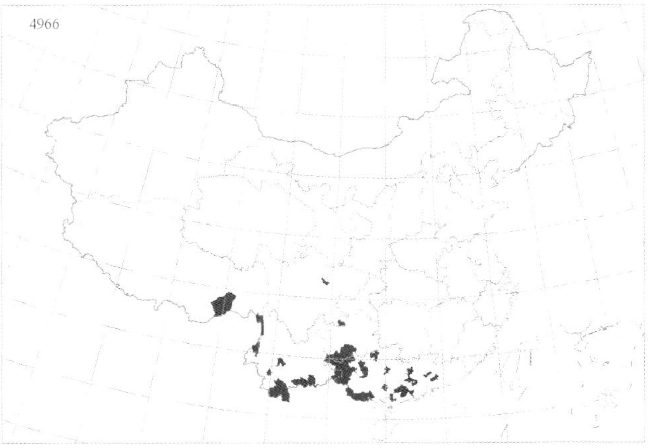

Euphorbiaceae 大戟科 *Macaranga* 血桐属
Macaranga kurzii (Kuntze) Pax et Hoffm. 尾叶血桐
Shrubs or trees

19.4 (14.4~22.2); 11.4 (7.4~15.1); 25.5 (19.1~28.6); 19.5 (14.4~22.3); 173.5 (113.1~207.5); 0.0 (0.0~0.0); 985 (727~1190) ❙ 1495 (1098~1900); 736 (562~1056); 107 (42~189) ❙ 53.2 (5.8~99.9); 981 (727~1190); 350.8 (185.5~659.0)

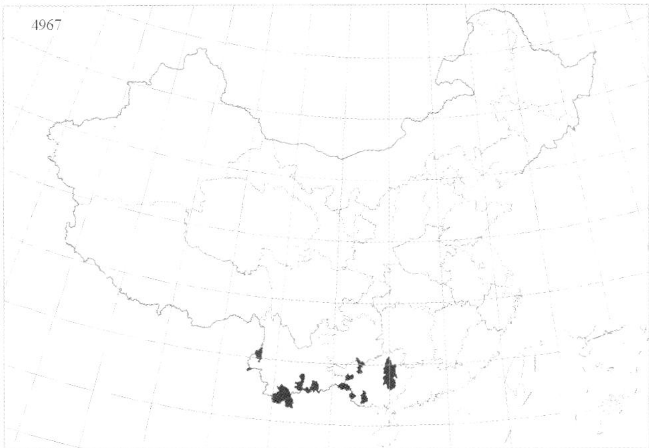

Euphorbiaceae 大戟科 *Macaranga* 血桐属
Macaranga lowii King ex Hook. f. 刺果血桐
Trees

23.2 (20.2~24.7); 16.8 (12.0~19.8); 28.0 (25.7~28.8); 23.2 (20.2~24.7); 218.7 (183.0~236.5); 0.0 (0.0~0.0); 1251 (1036~1376) ❙ 1621 (1176~2037); 632 (392~1033); 94 (42~167) ❙ 30.8 (-12.0~71.8); 1190 (1036~1360); 370.4 (139.6~620.0)

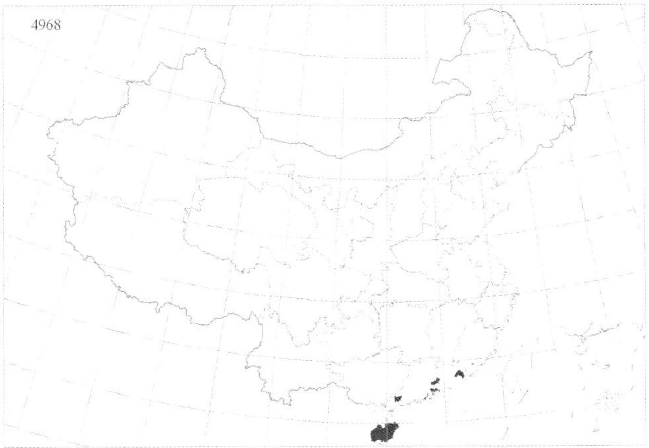

Euphorbiaceae 大戟科 *Macaranga* 血桐属
Macaranga pustulata King ex Hook. f. 泡腺血桐
Shrubs or trees

15.9 (5.3~21.0); 9.2 (-2.4~14.2); 20.6 (12.2~25.6); 15.9 (5.7~21.0); 132.9 (30.5~191.8); 2.1 (0.0~26.2); 799 (479~1053) ❙ 1305 (939~1736); 733 (524~1007); 53 (31~70) ❙ 66.0 (6.5~145.8); 787 (479~1053); 430.6 (234.2~589.2)

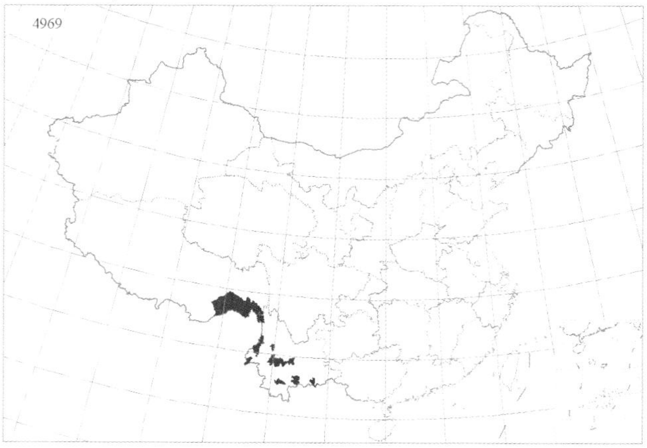

Euphorbiaceae 大戟科 *Macaranga* 血桐属
Macaranga sampsonii Hance 鼎湖血桐
Shrubs or small trees

21.0 (14.6~24.8); 12.9 (3.8~20.1); 27.6 (23.3~29.0); 21.1 (14.7~24.8); 192.7 (117.7~237.6); 0.0 (0.0~1.5); 1110 (779~1385) ❙ 1625 (868~2435); 687 (312~1428); 131 (41~198) ❙ 48.8 (-34.3~101.8); 1082 (779~1360); 296.9 (163.7~620.0)

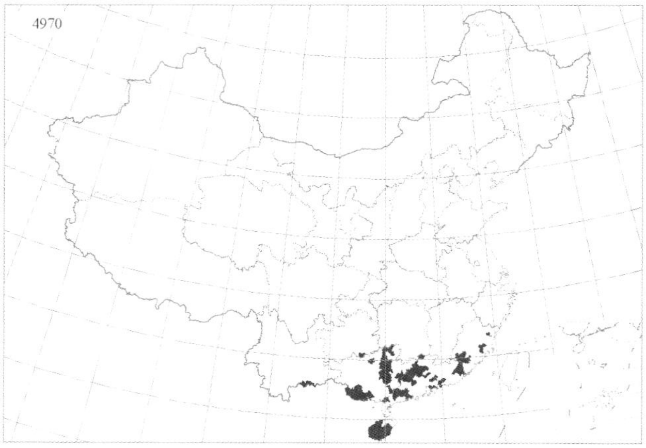

Euphorbiaceae 大戟科 *Macaranga* 血桐属
Macaranga sinensis (Baill.) Müll. Arg. 台湾血桐
Shrubs or small trees

18.5; 14.1; 21.7; 18.5; 162.1; 0.0; 863 ❙ 2713; 1388; 188 ❙ 214.4; 863; 363.7

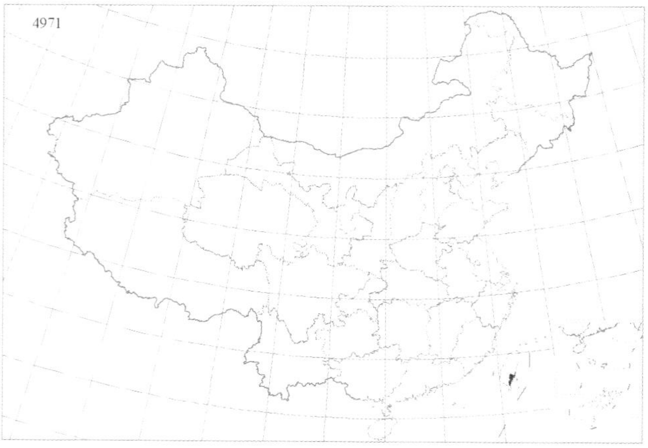

Euphorbiaceae 大戟科 *Macaranga* 血桐属
Macaranga tanarius (L.) Müll. Arg. 血桐
Trees

20.7 (16.2~24.1); 14.5 (11.0~18.9); 25.8 (20.1~28.8); 20.7 (16.2~24.1); 188.4 (134.2~229.7); 0.0 (0.0~0.0); 1057 (773~1322) ❙ 2263 (1498~3495); 1052 (750~1847); 222 (47~897) ❙ 124.7 (25.2~246.5); 1028 (773~1207); 230.1 (119.4~363.7)

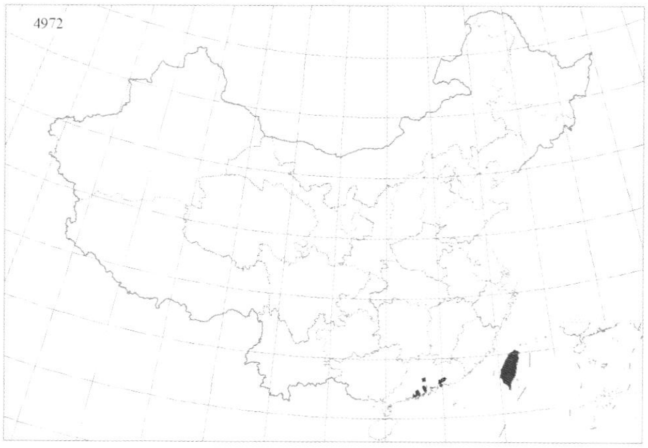

Euphorbiaceae 大戟科 *Mallotus* 野桐属
Mallotus anomalus Merr. et Chun 锈毛野桐
Shrubs

23.4 (18.0~24.8); 17.4 (8.5~20.1); 27.7 (25.8~28.7); 23.4 (18.0~24.8); 220.9 (156.4~237.6); 0.0 (0.0~0.0); 1271 (913~1385) ❙ 1473 (868~2148); 566 (312~1234); 74 (41~169) ❙ 18.4 (-34.3~86.7); 1107 (868~1299); 447.3 (243.9~620.0)

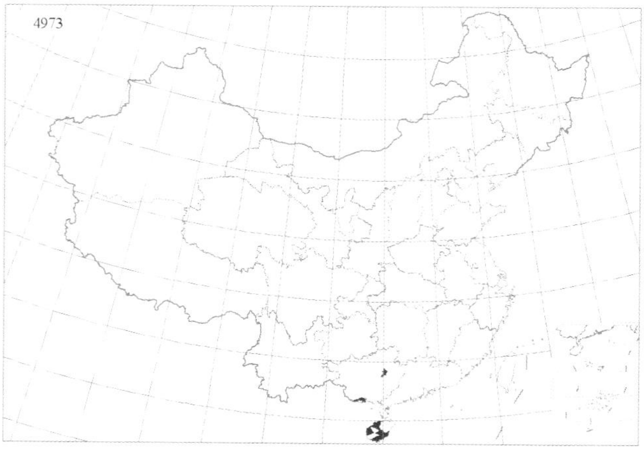

Euphorbiaceae 大戟科 *Mallotus* 野桐属
Mallotus apelta (Lour.) Müll. Arg. 白背叶
Shrubs

17.8 (9.3~24.8); 7.3 (-2.9~20.1); 27.3 (20.1~30.1); 17.8 (9.5~24.8); 155.1 (69.8~237.6); 1.5 (0.0~20.5); 953 (603~1385) ❙ 1422 (626~2435); 569 (299~1428); 141 (19~259) ❙ 50.1 (-34.3~147.0); 938 (603~1360); 227.4 (95.3~620.0)

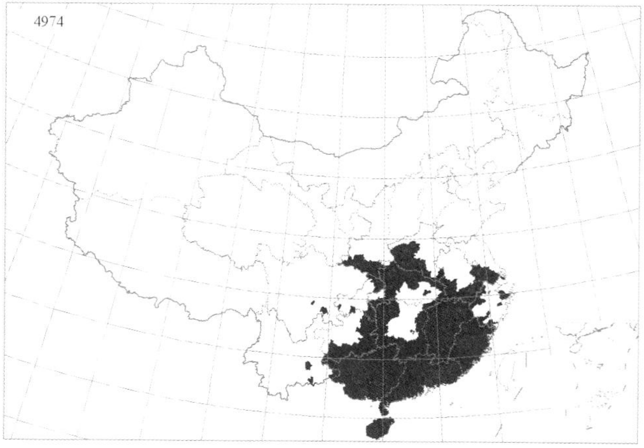

Euphorbiaceae 大戟科 *Mallotus* 野桐属
Mallotus apelta var. kwangsiensis F. P. Metcalf 广西白背叶
Small trees

19.5 (17.7~22.2); 10.1 (7.7~13.9); 27.5 (25.2~28.5); 19.6 (17.8~22.3); 175.0 (153.6~207.5); 0.0 (0.0~0.0); 1019 (916~1190) ▎1571 (1374~1777); 636 (507~848); 156 (72~194) ▎55.2 (15.4~71.6); 1019 (916~1190); 255.8 (226.5~327.7)

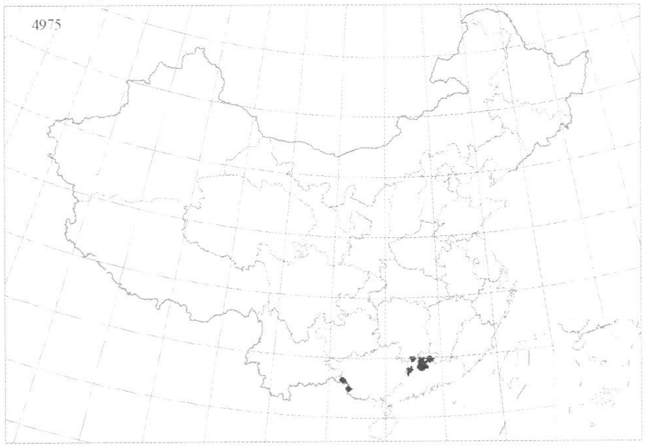

Euphorbiaceae 大戟科 *Mallotus* 野桐属
Mallotus barbatus (Wall.) Müll. Arg. 毛桐
Shrubs or small trees

17.4 (6.0~23.2); 7.6 (-3.2~15.8); 26.0 (14.4~29.3); 17.5 (6.6~23.2); 150.2 (39.7~218.9); 0.5 (0.0~26.7); 917 (508~1257) ▎1383 (916~2435); 612 (444~1428); 111 (28~216) ▎52.1 (5.8~107.2); 916 (508~1257); 233.2 (126.1~659.0)

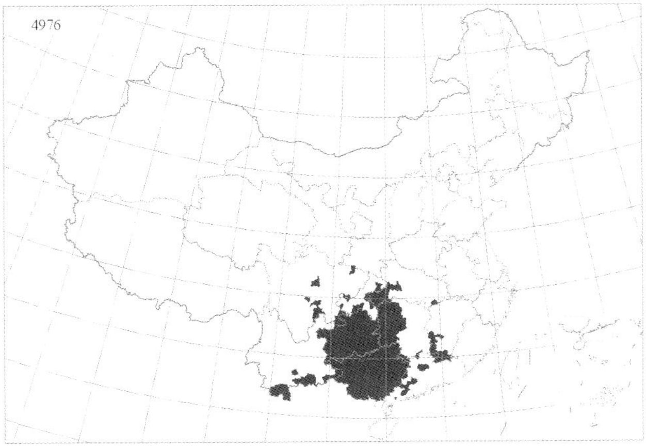

Euphorbiaceae 大戟科 *Mallotus* 野桐属
Mallotus barbatus var. croizatianus (F. P. Metcalf) S. M. Hwang 石山毛桐
Shrubs

20.4; 11.4; 27.4; 20.4; 184.8; 0.0; 1062 ▎1505; 764; 116 ▎41.4; 1061; 274.2

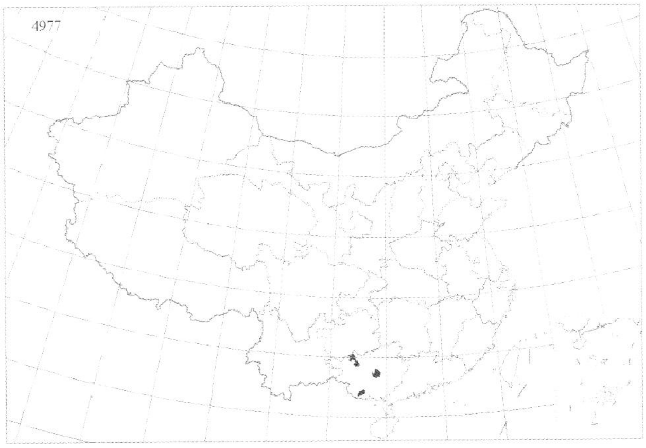

Euphorbiaceae 大戟科 *Mallotus* 野桐属
Mallotus barbatus var. pedicellaris Croizat 长梗毛桐
Small trees

14.4; 3.4; 24.9; 14.5; 115.9; 2.4; 783 ▎1465; 603; 94 ▎87.8; 783; 215.6

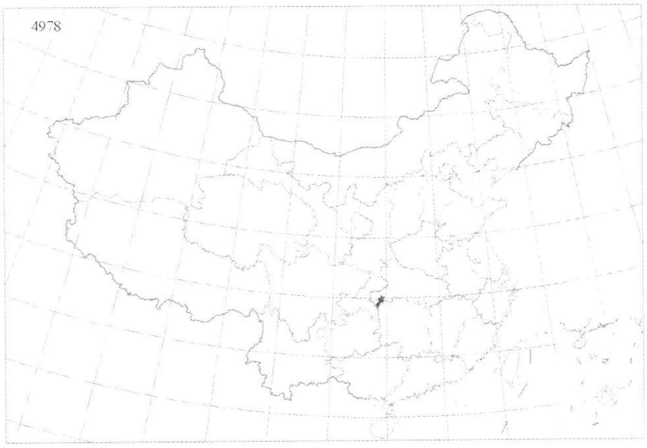

Euphorbiaceae 大戟科 *Mallotus* 野桐属
Mallotus conspurcatus Croizat 桂野桐
Shrubs

15.7 (9.3~21.2); 5.7 (-1.7~12.5); 24.6 (20.1~28.4); 15.8 (9.5~21.2); 134.0 (69.8~194.6); 5.0 (0.0~18.0); 854 (603~1121) ▎1380 (1202~1578); 628 (500~848); 94 (71~168) ▎68.0 (14.8~106.2); 854 (603~1121); 268.0 (218.0~313.7)

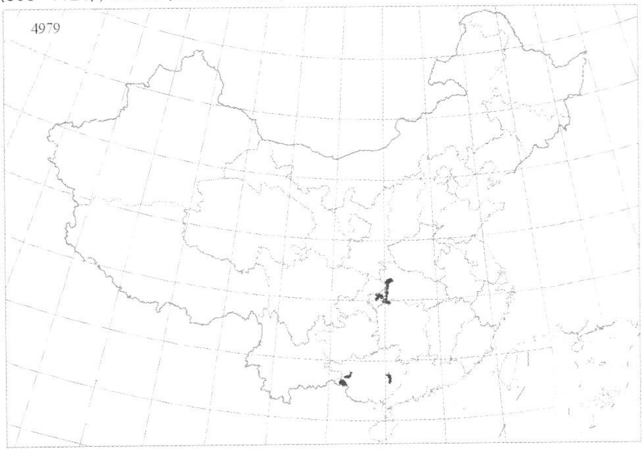

Euphorbiaceae 大戟科 *Mallotus* 野桐属
Mallotus decipiens Müll. Arg. 短柄野桐
Shrubs

19.3 (16.9~21.0); 13.2 (10.3~14.9); 23.3 (21.4~25.6); 19.3 (16.9~21.0); 172.1 (142.7~191.8); 0.0 (0.0~0.0); 937 (813~1053) ▎1418 (1178~1522); 728 (646~835); 60 (51~68) ▎51.5 (44.5~64.9); 925 (813~1053); 505.0 (383.4~570.1)

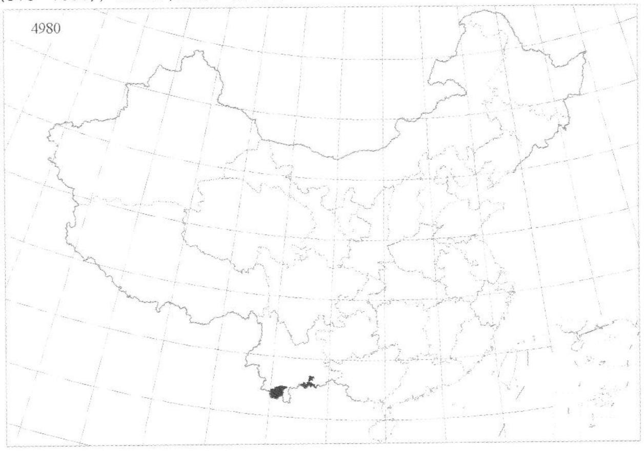

Euphorbiaceae 大戟科 *Mallotus* 野桐属

Mallotus dunnii F. P. Metcalf 南平野桐

Shrubs

18.9 (14.6~22.7); 9.2 (3.8~15.0); 27.1 (24.6~28.6); 18.9 (14.7~22.8); 166.8 (117.7~213.1); 0.2 (0.0~1.5); 991 (779~1221) ❙ 1705 (1462~2435); 723 (505~1428); 165 (124~191) ❙ 72.9 (29.9~101.8); 991 (779~1221); 255.2 (188.8~318.8)

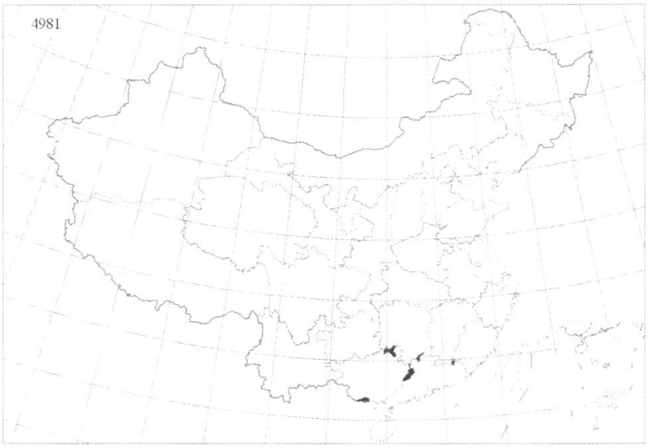

Euphorbiaceae 大戟科 *Mallotus* 野桐属

Mallotus esquirolii H. Lév. 长叶野桐

Shrubs or small trees

18.2 (15.9~24.0); 9.8 (4.7~19.1); 24.6 (20.3~28.9); 18.2 (16.0~24.0); 158.6 (131.6~228.1); 0.0 (0.0~0.3); 921 (785~1302) ❙ 1307 (1049~1647); 665 (452~893); 69 (43~129) ❙ 43.7 (5.8~74.2); 911 (779~1230); 314.8 (189.5~620.0)

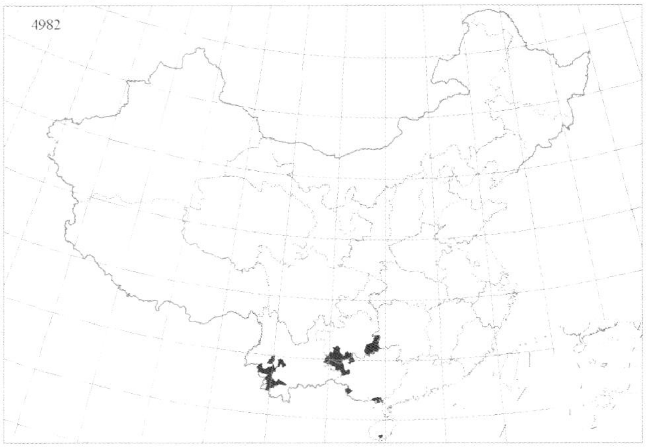

Euphorbiaceae 大戟科 *Mallotus* 野桐属

Mallotus garrettii Airy Shaw 粉绿野桐

Shrubs

19.5; 13.1; 23.7; 19.6; 174.7; 0.0; 947 ❙ 1418; 690; 58 ❙ 50.1; 926; 475.5

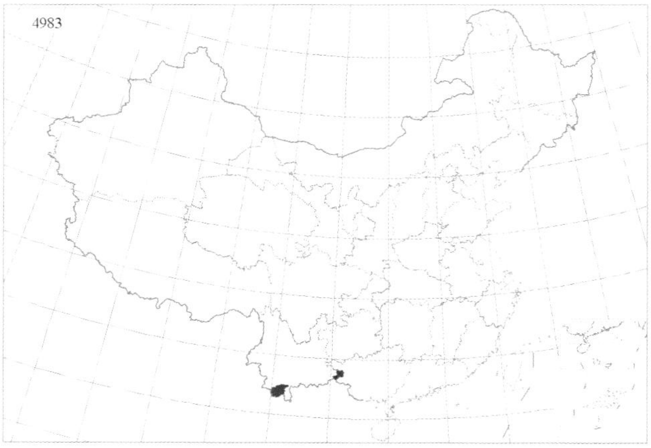

Euphorbiaceae 大戟科 *Mallotus* 野桐属

Mallotus japonicus (L. f.) Müll. Arg. 野梧桐

Shrubs or small trees

16.7 (11.7~22.0); 5.6 (-0.9~15.3); 27.0 (20.0~29.3); 16.7 (11.8~22.0); 141.8 (87.2~204.5); 1.3 (0.0~13.5); 895 (671~1172) ❙ 1494 (800~3495); 584 (352~1388); 164 (26~897) ❙ 67.5 (4.2~246.5); 892 (671~1172); 229.9 (110.1~551.6)

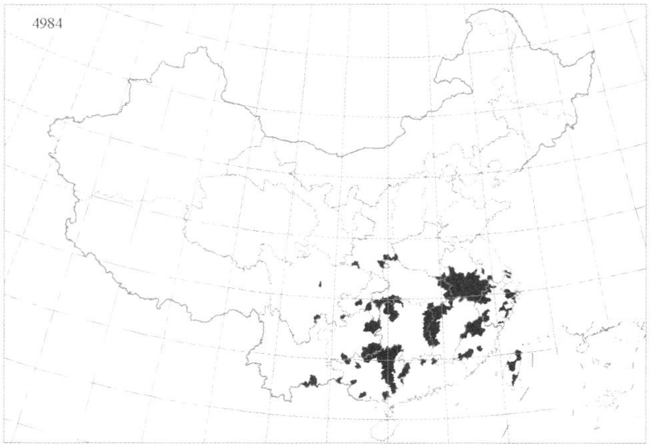

Euphorbiaceae 大戟科 *Mallotus* 野桐属

Mallotus lianus Croizat 东南野桐

Shrubs or small trees

18.4 (12.1~22.4); 8.7 (1.1~14.2); 26.9 (22.5~29.8); 18.5 (12.2~22.4); 161.8 (95.0~209.1); 0.2 (0.0~8.5); 962 (696~1204) ❙ 1618 (1232~2148); 632 (448~1234); 169 (48~259) ❙ 69.8 (17.4~130.7); 960 (696~1204); 252.0 (170.8~551.6)

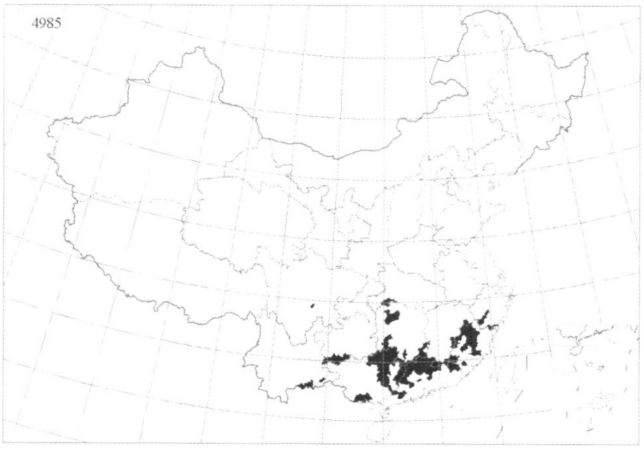

Euphorbiaceae 大戟科 *Mallotus* 野桐属

Mallotus lotingensis F. P. Metcalf 罗定野桐

Shrubs

21.8; 13.7; 28.1; 21.8; 201.8; 0.0; 1151 ❙ 1675; 768; 124 ❙ 45.5; 1151; 221.6

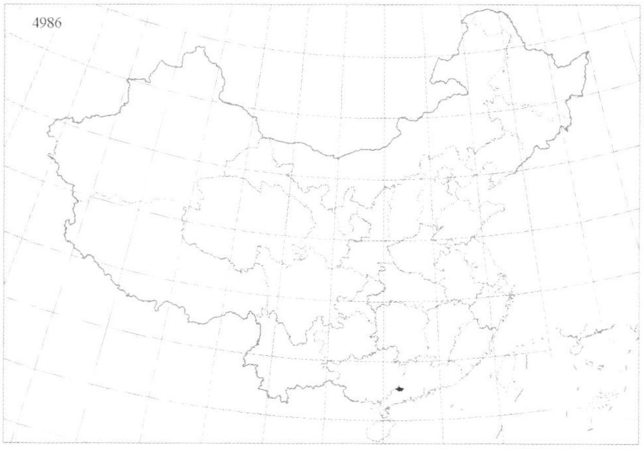

Euphorbiaceae 大戟科 *Mallotus* 野桐属
Mallotus metcalfianus Croizat 褐毛野桐
Small trees

21.1 (16.9~22.7); 13.4 (10.3~15.0); 26.8 (21.4~28.6); 21.2 (16.9~22.8); 194.2 (142.7~213.1); 0.0 (0.0~0.0); 1105 (813~1221) ❙ 1652 (1178~2435); 905 (646~1428); 100 (51~130) ❙ 49.6 (12.9~99.5); 1105 (813~1221); 335.1 (235.1~551.6)

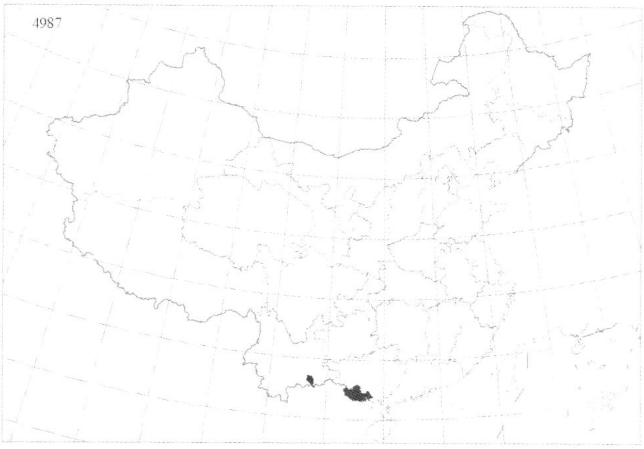

Euphorbiaceae 大戟科 *Mallotus* 野桐属
Mallotus microcarpus Pax et Hoffm. 小果野桐
Shrubs

17.8 (12.1~22.2); 7.7 (1.1~13.9); 26.7 (22.9~29.1); 17.9 (12.2~22.3); 155.0 (95.0~207.5); 0.4 (0.0~8.5); 936 (696~1190) ❙ 1490 (1178~1841); 591 (419~850); 150 (51~198) ❙ 60.7 (15.4~106.2); 934 (696~1190); 235.2 (169.7~327.7)

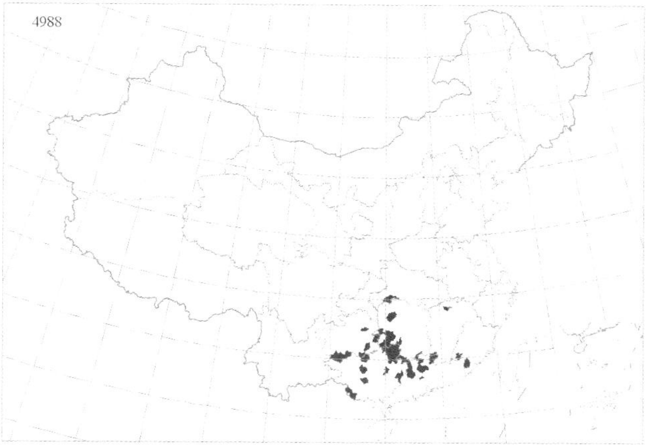

Euphorbiaceae 大戟科 *Mallotus* 野桐属
Mallotus millietii H. Lév. 贵州野桐
Scandent shrubs, deciduous

15.4 (7.3~23.6); 4.5 (-2.9~17.8); 25.2 (15.6~29.2); 15.5 (7.5~23.6); 129.2 (48.2~223.3); 3.9 (0.0~20.5); 833 (537~1271) ❙ 1155 (512~1742); 540 (258~820); 73 (15~221) ❙ 39.2 (-35.3~108.0); 815 (512~1190); 226.8 (121.6~659.0)

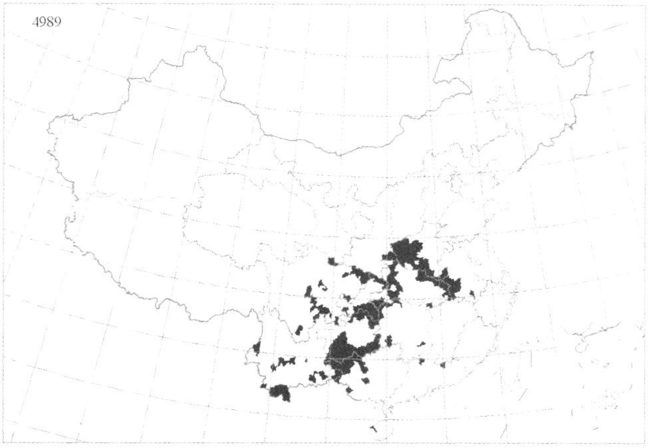

Euphorbiaceae 大戟科 *Mallotus* 野桐属
Mallotus nepalensis Müll. Arg. 尼泊尔野桐
Shrubs or small trees

15.9 (-1.8~22.2); 5.4 (-10.2~14.2); 25.4 (6.1~29.8); 16.0 (1.8~22.3); 133.0 (1.9~207.5); 1.9 (0.0~82.8); 848 (329~1190) ❙ 1351 (320~1997); 554 (193~1007); 124 (13~254) ❙ 59.8 (-4.0~145.8); 842 (320~1190); 232.7 (72.2~551.6)

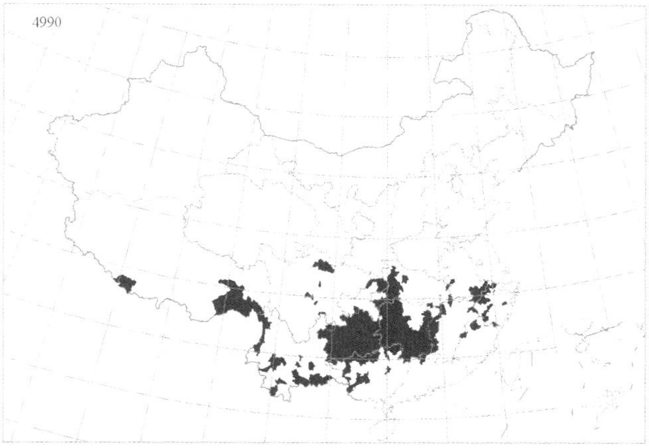

Euphorbiaceae 大戟科 *Mallotus* 野桐属
Mallotus oreophilus Müll. Arg. 山地野桐
Trees

16.5; 6.0; 25.6; 16.6; 138.8; 0.0; 857 ❙ 1324; 545; 119 ❙ 54.5; 857; 231.1

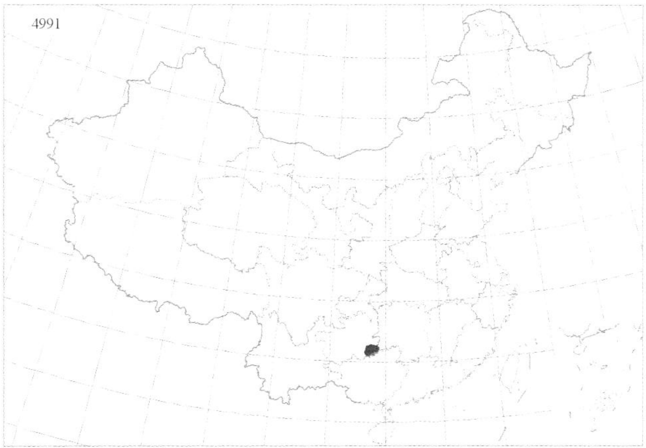

Euphorbiaceae 大戟科 *Mallotus* 野桐属
Mallotus paniculatus (Lam.) Müll. Arg. 白楸
Shrubs or trees

20.6 (14.6~24.8); 12.9 (3.8~20.1); 26.8 (20.1~29.0); 20.7 (14.7~24.8); 188.3 (117.7~237.6); 0.0 (0.0~1.5); 1078 (773~1385) ❙ 1663 (868~3495); 727 (312~1847); 138 (41~897) ❙ 58.4 (-34.3~246.5); 1056 (773~1360); 310.2 (122.7~659.0)

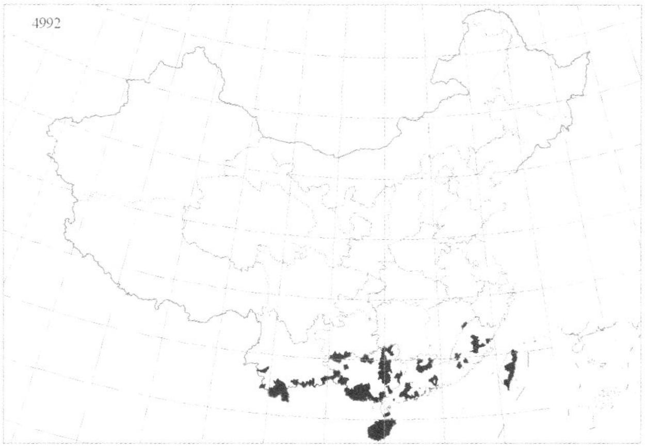

Euphorbiaceae 大戟科 *Mallotus* 野桐属

Mallotus peltatus (Geiseler) Müll. Arg. 山苦茶

Shrubs or small trees

23.2 (18.1~24.8); 16.9 (9.5~20.1); 27.8 (24.6~28.9); 23.2 (18.2~24.8); 218.5 (158.1~237.6); 0.0 (0.0~0.0); 1251 (901~1385) ❙ 1548 (868~2035); 620 (312~969); 80 (41~125) ❙ 25.0 (-34.3~71.8); 1168 (868~1360); 384.9 (194.5~620.0)

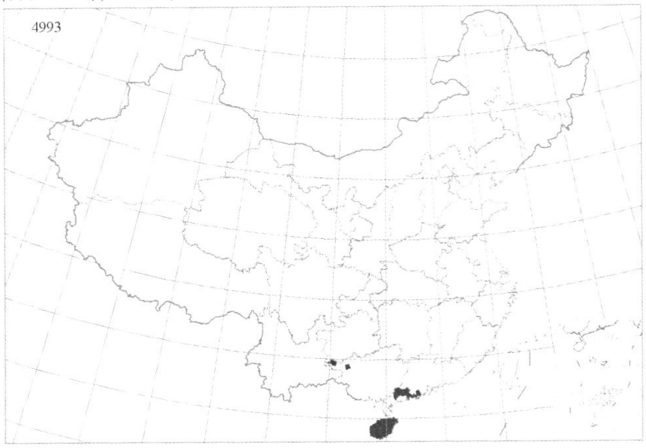

Euphorbiaceae 大戟科 *Mallotus* 野桐属

Mallotus philippensis (Lam.) Müll. Arg. 粗糠柴

Trees, evergreen

17.7 (2.2~24.8); 7.9 (-6.6~20.1); 26.4 (9.9~30.1); 17.7 (3.8~24.8); 153.4 (15.5~237.6); 0.8 (0.0~49.1); 938 (413~1385) ❙ 1438 (711~3495); 599 (312~1847); 129 (15~897) ❙ 54.6 (-34.3~246.5); 924 (413~1360); 235.9 (95.3~659.0)

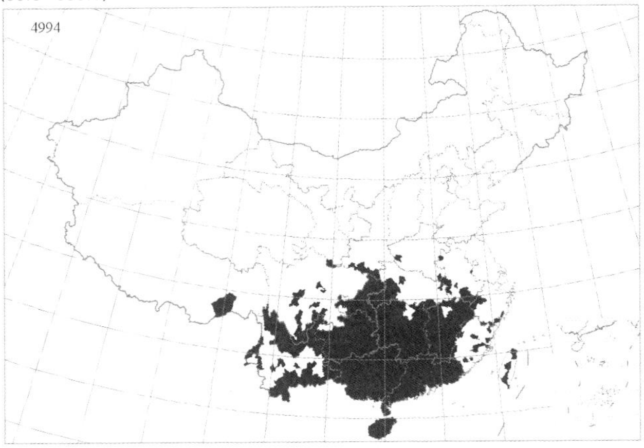

Euphorbiaceae 大戟科 *Mallotus* 野桐属

Mallotus philippensis var. reticulatus (Dunn) F. P. Metcalf 网脉粗糠柴

Shrubs

19.1 (17.5~20.2); 10.2 (7.2~12.1); 27.5 (27.1~28.0); 19.1 (17.6~20.2); 169.8 (150.9~182.5); 0.0 (0.0~0.0); 985 (916~1037) ❙ 1591 (1317~1828); 541 (448~593); 195 (151~235) ❙ 62.4 (30.7~99.7); 985 (916~1037); 264.9 (206.7~300.0)

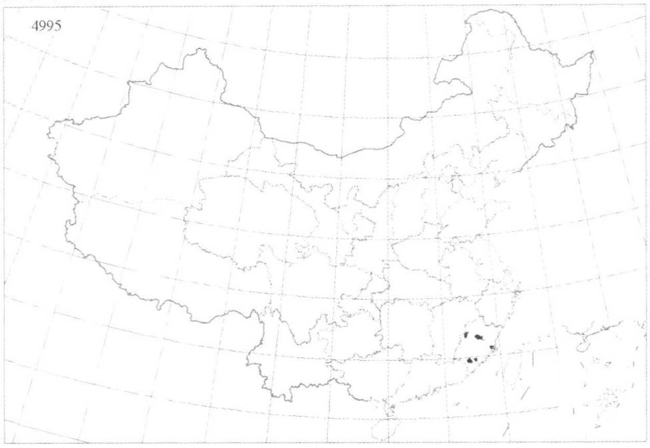

Euphorbiaceae 大戟科 *Mallotus* 野桐属

Mallotus repandus (Willd.) Müll. Arg. 石岩枫

Scandent shrubs, evergreen

16.6 (-1.1~24.8); 5.7 (-9.1~20.1); 26.7 (6.2~30.1); 16.7 (2.0~24.8); 141.7 (2.5~237.6); 2.0 (0.0~75.6); 896 (334~1385) ❙ 1384 (321~3495); 542 (193~1847); 140 (17~897) ❙ 54.8 (-35.3~224.4); 881 (321~1329); 221.1 (79.3~659.0)

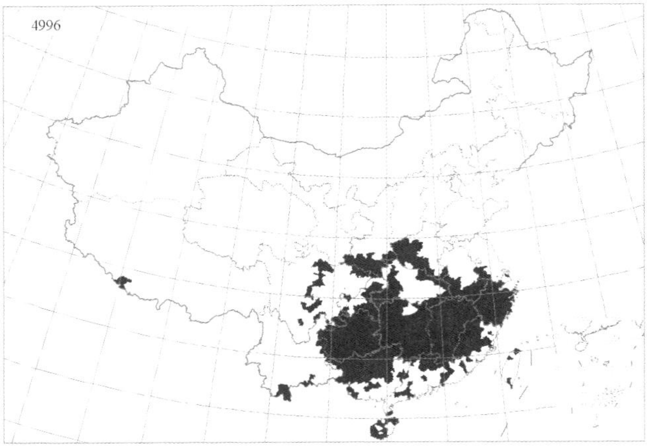

Euphorbiaceae 大戟科 *Mallotus* 野桐属

Mallotus tenuifolius Pax 野桐

Shrubs or small trees

14.8 (10.3~18.5); 3.7 (-0.3~7.8); 25.3 (19.8~29.4); 14.9 (10.3~18.5); 121.9 (75.8~162.3); 3.3 (0.0~12.1); 809 (634~975) ❙ 1287 (711~1997); 520 (357~816); 107 (15~254) ❙ 58.7 (5.0~130.7); 805 (634~975); 232.5 (123.3~303.4)

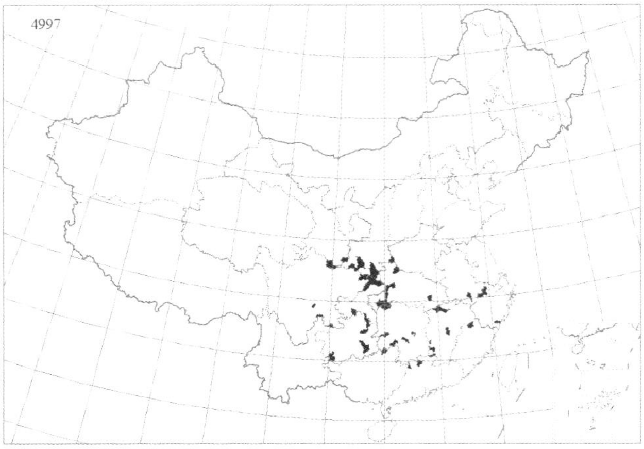

Euphorbiaceae 大戟科 *Mallotus* 野桐属

Mallotus tenuifolius var. castanopsis (F. P. Metcalf) H. S. Kiu 乐昌野桐

Shrubs or small trees

17.8; 7.1; 27.5; 17.8; 154.0; 0.0; 933 ❙ 1503; 513; 182 ❙ 61.1; 933; 197.5

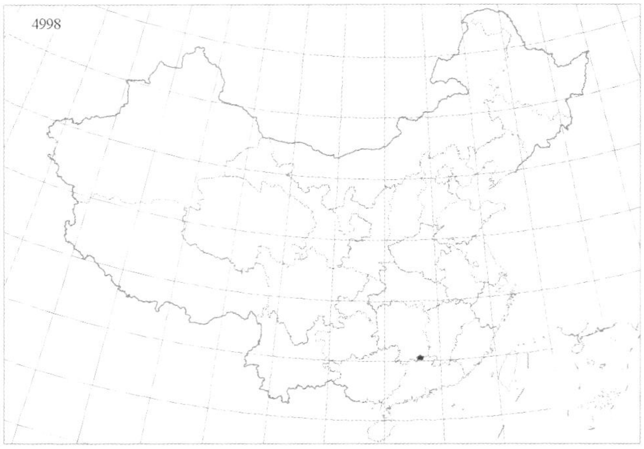

Euphorbiaceae 大戟科 *Mallotus* 野桐属
Mallotus tenuifolius var. paxii (Pampanini) H. S. Kiu 红叶野桐
Shrubs

16.3 (9.3~21.0); 4.8 (-1.7~11.9); 27.2 (20.1~29.9); 16.4 (9.5~21.1); 137.6 (69.8~193.0); 1.2 (0.0~18.0); 882 (603~1114) ❙ 1440 (800~1997); 512 (361~321); 165 (25~259) ❙ 63.8 (9.1~147.0); 870 (603~1114); 209.6 (111.9~305.7)

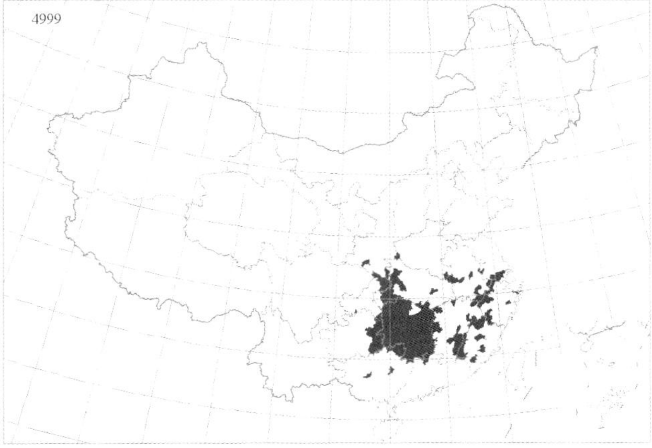

Euphorbiaceae 大戟科 *Mallotus* 野桐属
Mallotus tetracoccus (Roxb.) Kurz 四果野桐
Small trees

18.2 (13.3~21.0); 11.8 (5.4~15.1); 22.3 (19.6~25.6); 18.2 (13.4~21.0); 158.5 (100.4~191.8); 0.0 (0.0~0.0); 877 (706~1053) ❙ 1401 (1049~1736); 754 (598~1007); 54 (41~77) ❙ 60.5 (26.1~145.8); 861 (706~1053); 446.3 (336.3~659.0)

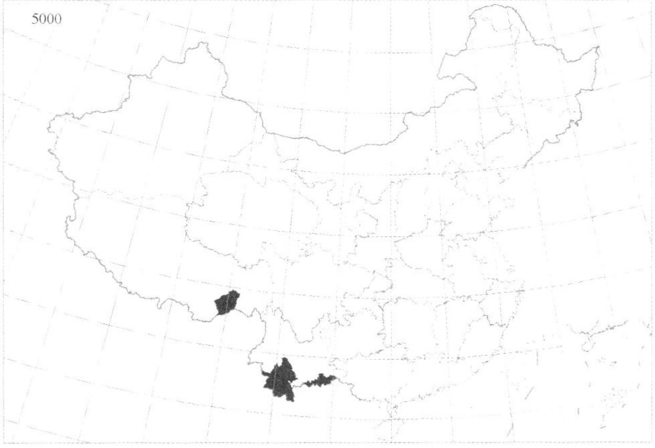

Euphorbiaceae 大戟科 *Mallotus* 野桐属
Mallotus tiliifolius (Blume) Müll. Arg. 椴叶野桐
Small trees

22.0 (16.2~24.8); 16.4 (11.0~20.1); 26.4 (20.1~28.8); 22.1 (16.2~24.8); 205.0 (134.2~237.6); 0.0 (0.0~0.0); 1161 (773~1385) ❙ 1963 (868~3495); 827 (312~1847); 167 (41~897) ❙ 82.3 (-34.3~246.5); 1084 (773~1360); 340.1 (119.4~620.0)

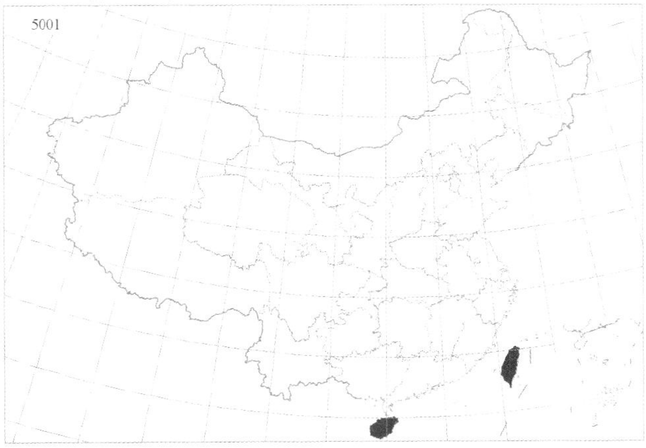

Euphorbiaceae 大戟科 *Mallotus* 野桐属
Mallotus yunnanensis Pax. et K. Hoffmam 云南野桐
Shrubs

21.0 (13.2~24.8); 14.0 (6.4~20.1); 26.1 (18.4~28.8); 21.0 (13.3~24.8); 192.5 (99.4~237.6); 0.0 (0.0~0.0); 1100 (695~1385) ❙ 1412 (958~1854); 645 (392~848); 71 (41~128) ❙ 30.6 (-11.9~92.4); 1060 (695~1360); 385.0 (226.6~659.0)

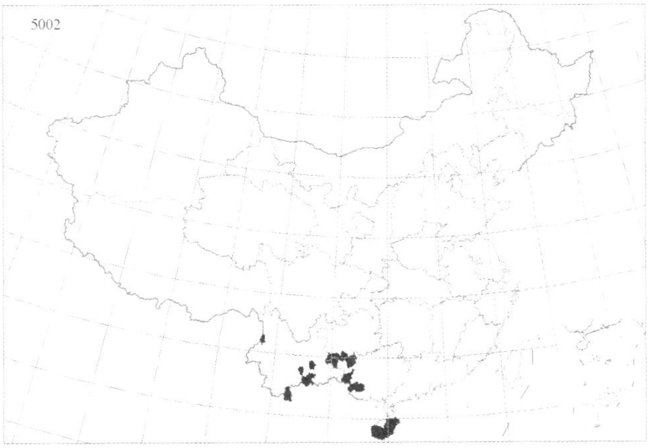

Euphorbiaceae 大戟科 *Margaritaria* 蓝子木属
Margaritaria indica (Dalzell) Airy Shaw 蓝子木
Trees

21.6 (17.3~24.1); 16.0 (12.1~18.9); 25.8 (22.8~28.3); 21.6 (17.9~24.1); 199.5 (154.4~229.7); 0.0 (0.0~0.0); 1115 (852~1322) ❙ 2186 (1374~2992); 1123 (455~1847); 136 (47~402) ❙ 109.9 (6.7~235.6); 1047 (852~1190); 292.0 (154.2~455.2)

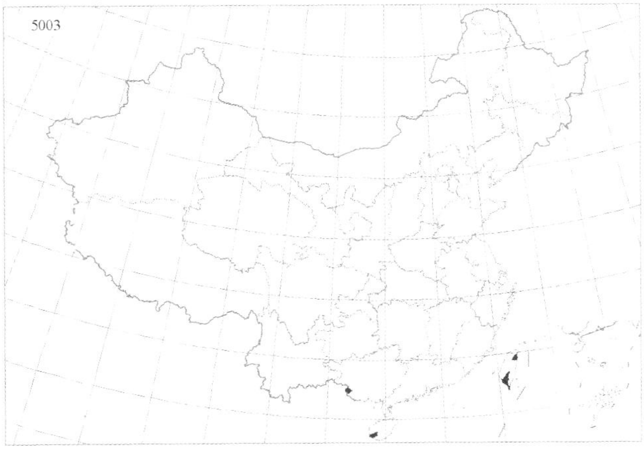

Euphorbiaceae 大戟科 *Megistostigma* 大柱藤属
Megistostigma yunnanense Croizat 云南大柱藤
Vines

20.0; 14.6; 23.2; 20.1; 180.8; 0.0; 963 ❙ 1523; 681; 67 ❙ 58.4; 942; 581.0

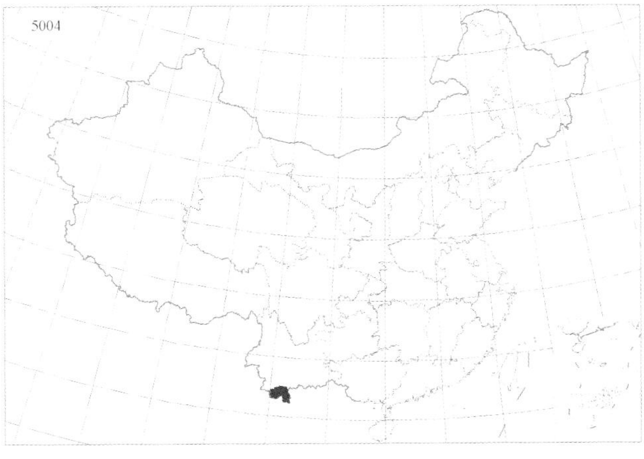

Euphorbiaceae 大戟科 *Melanolepis* 黑鳞木属
Melanolepis multiglandulosa (Reinw. ex Blume) Reichb. f. et Zoll. 墨鳞
Small trees

20.0 (16.4~22.2); 14.4 (11.5~17.7); 24.7 (20.3~27.4); 20.0 (16.5~22.2); 180.1 (137.8~206.7); 0.0 (0.0~0.0); 990 (783~1123) ❙ 2848 (2445~3495); 1169 (821~1847); 386 (107~897) ❙ 189.7 (140.0~224.4); 990 (783~1123); 271.2 (207.7~363.7)

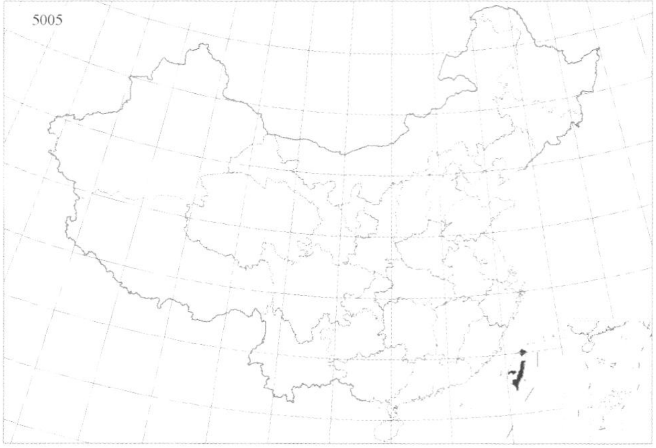

Euphorbiaceae 大戟科 *Microdesmis* 小盘木属
Microdesmis caseariifolia Planch. 小盘木
Shrubs or trees

22.4 (19.0~24.8); 15.0 (9.8~20.1); 28.0 (25.0~29.0); 22.5 (19.0~24.8); 209.4 (168.4~237.6); 0.0 (0.0~0.0); 1199 (945~1385) ❙ 1601 (868~2435); 732 (312~1428); 99 (41~173) ❙ 34.5 (-34.3~99.5); 1161 (868~1360); 312.7 (139.6~620.0)

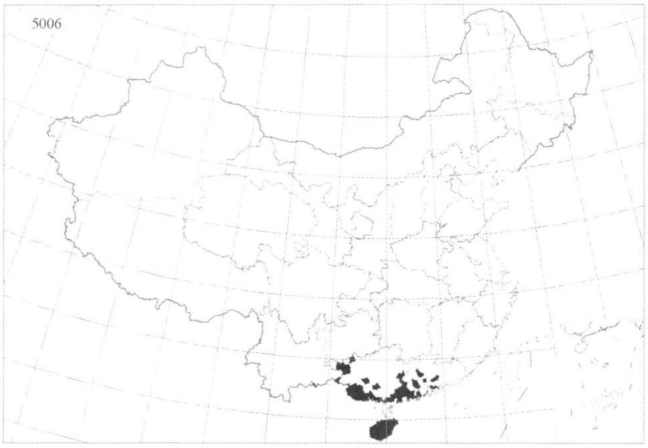

Euphorbiaceae 大戟科 *Neoshirakia* 白木乌桕属
Neoshirakia atrobadiomaculata (F. P. Metcalf) Esser et P. T. Li 斑子乌桕
Shrubs

17.6 (14.3~21.8); 6.5 (2.7~13.4); 27.9 (24.2~30.1); 17.7 (14.3~21.8); 152.3 (113.2~201.9); 0.2 (0.0~3.2); 940 (761~1158) ❙ 1571 (871~2005); 544 (299~827); 191 (107~254) ❙ 67.9 (-6.7~130.7); 916 (755~1158); 225.6 (148.9~303.4)

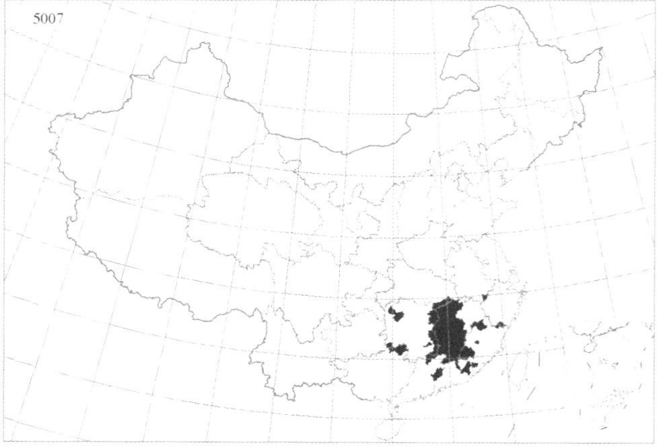

Euphorbiaceae 大戟科 *Neoshirakia* 白木乌桕属
Neoshirakia japonica (Siebold et Zuccarini) Esser 白木乌桕
Shrubs or trees, deciduous

16.2 (9.3~21.0); 4.5 (-2.9~12.0); 27.1 (20.1~30.1); 16.3 (9.5~21.1); 137.4 (69.8~193.0); 2.6 (0.0~20.6); 882 (603~1114) ❙ 1358 (512~2005); 514 (258~824); 146 (21~259) ❙ 53.3 (-35.3~130.7); 861 (512~1114); 209.4 (111.9~322.1)

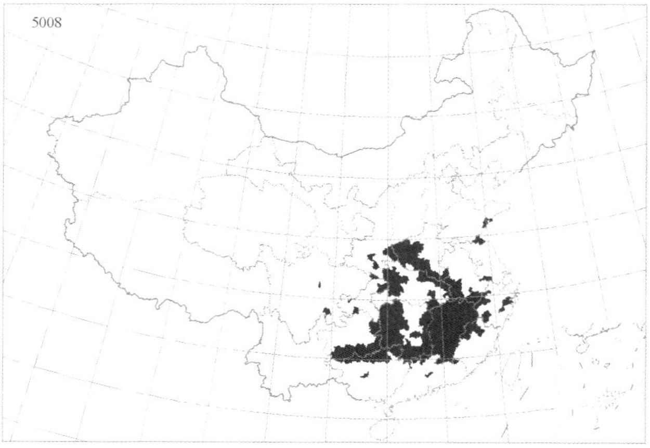

Euphorbiaceae 大戟科 *Ostodes* 叶轮木属
Ostodes katharinae Pax et Hoffm. 云南叶轮木
Trees

16.6 (5.3~20.5); 10.3 (-2.4~15.1); 20.9 (12.2~23.7); 16.7 (5.7~20.5); 141.5 (30.5~185.9); 1.2 (0.0~26.2); 820 (479~992) ❙ 1317 (939~1736); 716 (524~1007); 52 (31~77) ❙ 62.7 (6.5~145.8); 802 (479~986); 428.8 (234.2~659.0)

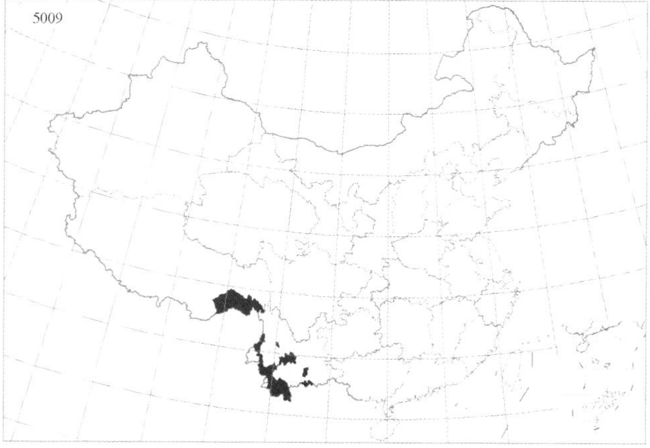

Euphorbiaceae 大戟科 *Ostodes* 叶轮木属
Ostodes paniculata Blume 叶轮木
Trees

21.8 (13.3~24.8); 16.0 (5.4~20.1); 25.8 (19.6~28.8); 21.8 (13.4~24.8); 202.2 (100.4~237.6); 0.0 (0.0~0.0); 1151 (706~1385) ❙ 1464 (868~1854); 601 (312~1007); 63 (41~144) ❙ 32.8 (-34.3~145.8); 1042 (706~1360); 474.6 (157.7~659.0)

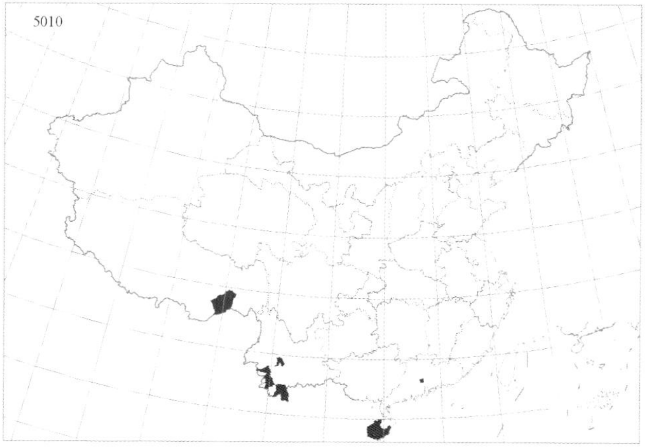

Euphorbiaceae 大戟科 *Phyllanthodendron* 珠子木属
Phyllanthodendron anthopotamicum (Hand.-Mazz.) Croizat 珠子木
Shrubs

20.0 (15.0~24.8); 12.4 (4.4~20.1); 25.6 (20.4~29.2); 20.0 (15.0~24.8); 180.3 (121.0~237.6); 0.0 (0.0~0.6); 1026 (780~1385) ▎1397 (868~1900); 702 (312~1056); 72 (41~141) ▎39.1 (-34.3~88.8); 992 (780~1269); 346.8 (189.5~659.0)

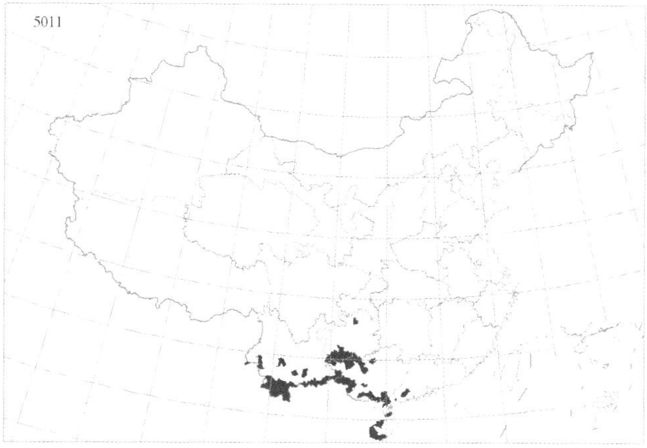

Euphorbiaceae 大戟科 *Phyllanthodendron* 珠子木属
Phyllanthodendron breynioides P. T. Li 龙州珠子木
Shrubs

21.4; 13.2; 27.5; 21.4; 196.8; 0.0; 1121 ▎1499; 773; 83 ▎33.7; 1121; 294.0

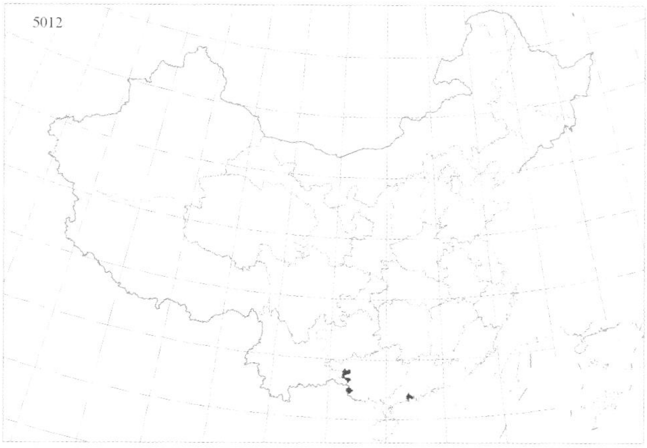

Euphorbiaceae 大戟科 *Phyllanthodendron* 珠子木属
Phyllanthodendron caudatifolium P. T. Li 尾叶珠子木
Shrubs

17.1; 8.5; 23.4; 17.1; 145.1; 0.0; 845 ▎1232; 666; 59 ▎45.8; 845; 283.9

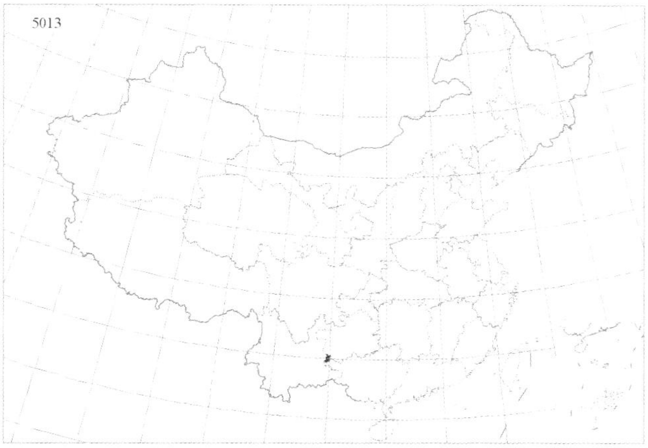

Euphorbiaceae 大戟科 *Phyllanthodendron* 珠子木属
Phyllanthodendron dunnianum H. Lév. 枝翅珠子木
Shrubs or small trees

20.6 (15.1~24.8); 12.9 (5.0~20.1); 26.5 (21.4~29.0); 20.6 (15.2~24.8); 187.7 (121.9~237.6); 0.0 (0.0~0.0); 1072 (788~1385) ▎1483 (868~2435); 693 (312~1428); 93 (41~197) ▎40.0 (-34.3~99.5); 1046 (783~1360); 328.4 (164.9~659.0)

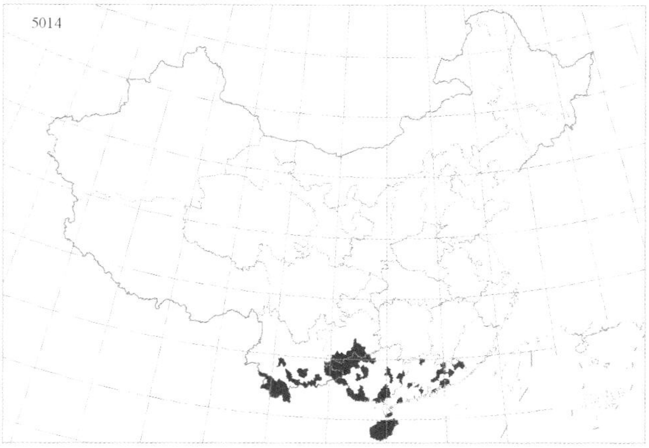

Euphorbiaceae 大戟科 *Phyllanthodendron* 珠子木属
Phyllanthodendron lativenium Croizat 宽脉珠子木
Shrubs

18.1 (16.8~19.1); 9.1 (7.8~10.1); 25.1 (23.4~26.6); 18.1 (16.8~19.1); 157.7 (141.5~169.2); 0.0 (0.0~0.0); 912 (836~969) ▎1268 (1232~1334); 682 (617~723); 58 (51~64) ▎39.7 (27.5~59.6); 912 (836~969); 259.3 (189.5~316.1)

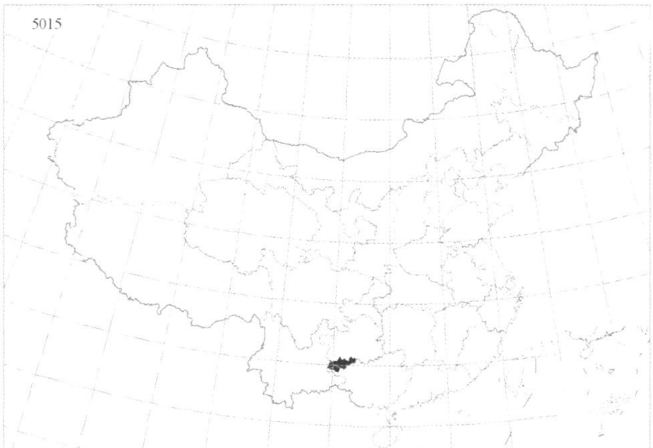

Euphorbiaceae 大戟科 *Phyllanthodendron* 珠子木属
Phyllanthodendron moi (P. T. Li) P. T. Li 弄岗珠子木
Shrubs

21.9; 13.8; 27.9; 21.9; 203.3; 0.0; 1161 ▎1478; 781; 97 ▎27.6; 1161; 304.7

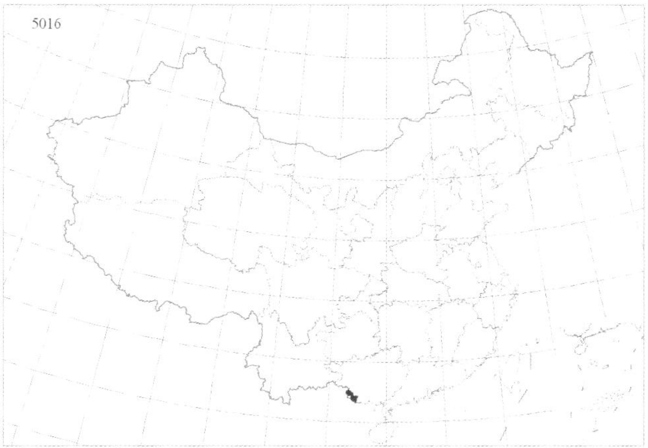

Euphorbiaceae 大戟科 *Phyllanthodendron* 珠子木属
Phyllanthodendron orbicularifolium P. T. Li 圆叶珠子木
Shrubs

20.7；12.5；26.8；20.7；189.0；0.0；1071 ▎1429；780；79 ▎35.6；1071；320.7

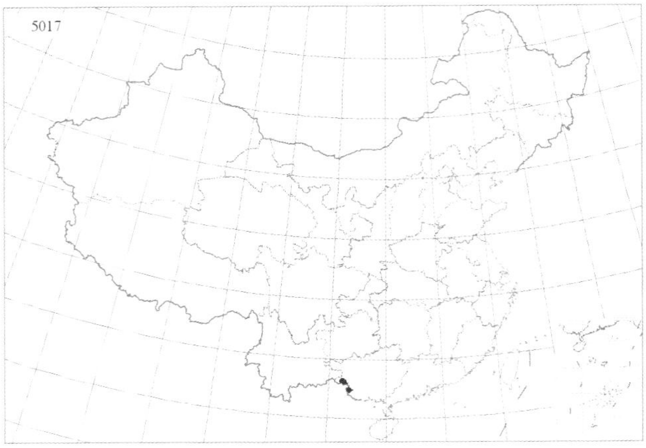

Euphorbiaceae 大戟科 *Phyllanthodendron* 珠子木属
Phyllanthodendron petraeum P. T. Li 岩生珠子木
Shrubs

22.2；13.9；28.3；22.3；207.5；0.0；1190 ▎1374；713；87 ▎15.4；1190；327.7

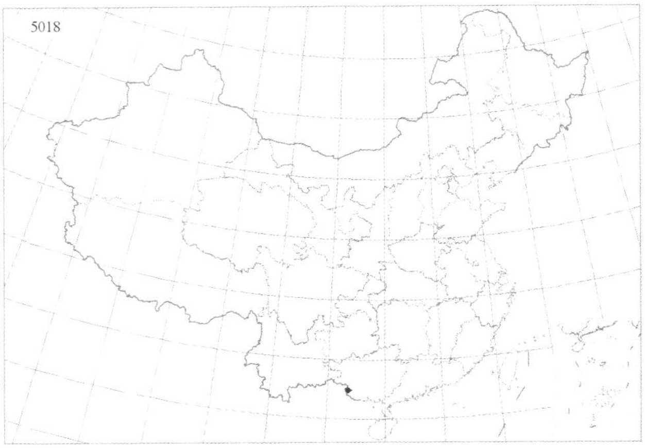

Euphorbiaceae 大戟科 *Phyllanthodendron* 珠子木属
Phyllanthodendron roseum Craib et Hutch. 玫花珠子木
Shrubs or small trees

19.2；13.8；22.3；19.2；170.5；0.0；909 ▎1499；686；58 ▎64.9；880；513.9

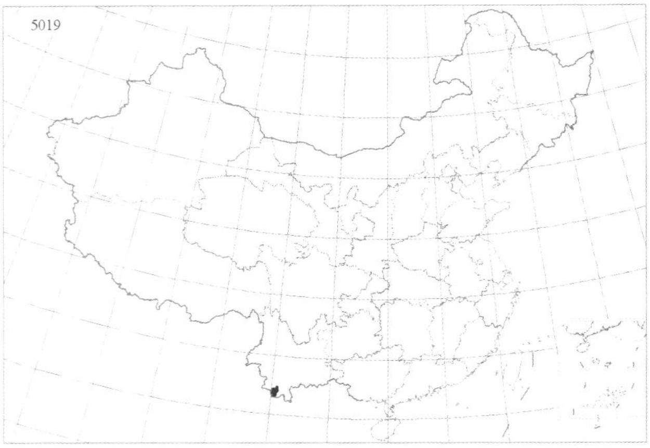

Euphorbiaceae 大戟科 *Phyllanthodendron* 珠子木属
Phyllanthodendron yunnanense Croizat 云南珠子木
Shrubs or small trees

19.0；13.2；22.4；19.0；168.4；0.0；904 ▎1411；720；55 ▎56.0；874；458.4

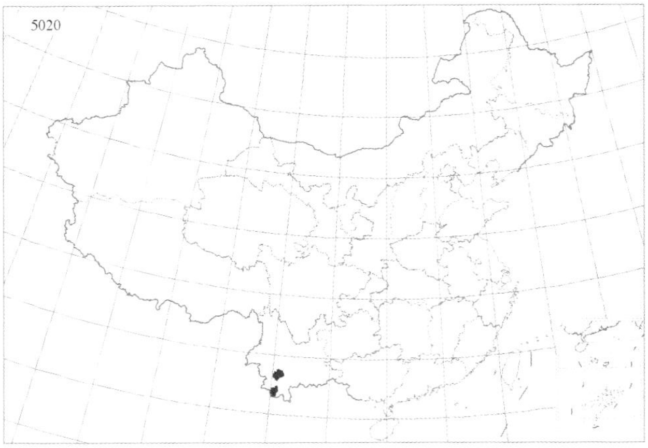

Euphorbiaceae 大戟科 *Phyllanthus* 叶下珠属
Phyllanthus bodinieri (H. Lév.) Rehder 贵州叶下珠
Shrubs

17.4；8.2；25.1；17.4；149.4；0.0；897 ▎1360；669；80 ▎52.6；897；208.9

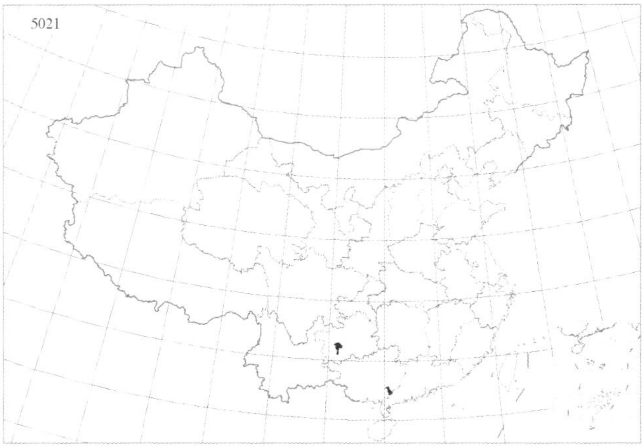

Euphorbiaceae 大戟科 *Phyllanthus* 叶下珠属
Phyllanthus chekiangensis Croizat et F. P. Metcalf 浙江叶下珠
Shrubs

17.1 (13.0~22.7)；6.9 (1.9~15.2)；26.4 (19.1~29.0)；17.2 (13.1~22.7)；147.1 (102.9~212.7)；0.5 (0.0~6.1)；904 (727~1216) ▎1564 (963~1997)；599 (425~994)；167 (26~259) ▎73.7 (14.0~130.7)；898 (689~1216)；264.4 (116.6~659.0)

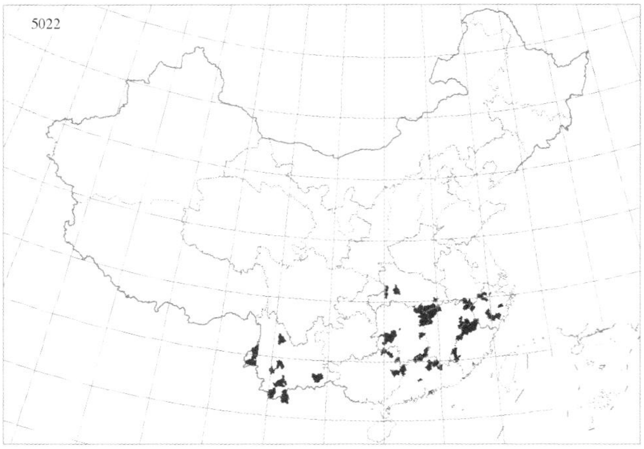

Euphorbiaceae 大戟科 *Phyllanthus* 叶下珠属
Phyllanthus clarkei Hook. f. 滇藏叶下珠
Shrubs

16.1 (−1.8~22.5); 8.5 (−10.2~15.0); 21.8 (6.1~28.7); 16.3 (1.8~22.6); 139.4 (1.9~210.9); 5.7 (0.0~82.8); 840 (329~1206) ❙ 1243 (320~1758); 689 (193~ 007); 53 (12~106) ❙ 48.1 (−4.0~145.8); 834 (320~1206); 333.4 (72.2~627.0)

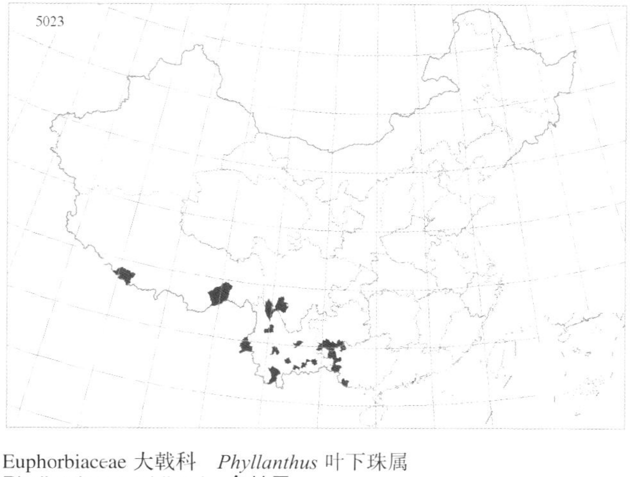

Euphorbiaceae 大戟科 *Phyllanthus* 叶下珠属
Phyllanthus cochinchinensis Müll. Arg. 越南叶下珠
Shrubs

19.3 (−1.1~24.8); 11.7 (−9.1~20.1); 25.4 (6.2~29.2); 19.4 (2.0~24.8); 173.4 (2.5~237.6); 1.0 (0.0~75.6); 1019 (334~1385) ❙ 1419 (321~2435); 672 (193~1428); 89 (12~204) ❙ 40.4 (−34.3~145.8); 996 (321~1360); 284.4 (79.3~659.0)

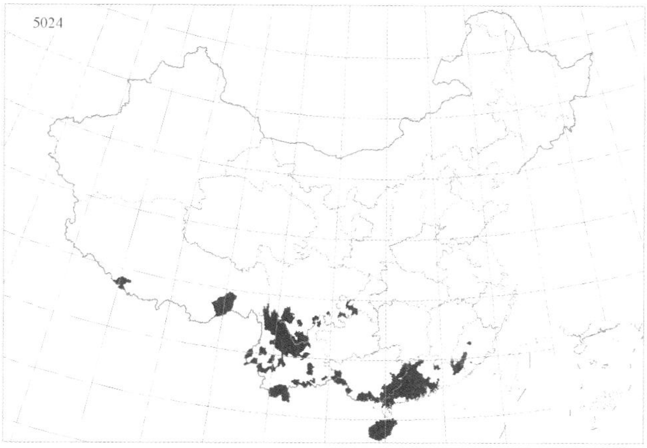

Euphorbiaceae 大戟科 *Phyllanthus* 叶下珠属
Phyllanthus emblica L. 余甘子
Trees

18.7 (2.9~24.8); 10.8 (−6.3~20.1); 25.0 (11.2~30.1); 18.7 (4.4~24.8); 165.0 (20.5~237.6); 0.4 (0.0~44.6); 969 (428~1385) ❙ 1438 (741~3495); 677 (299~1847); 106 (12~897) ❙ 49.7 (−34.3~246.5); 951 (428~1360); 278.5 (95.3~659.0)

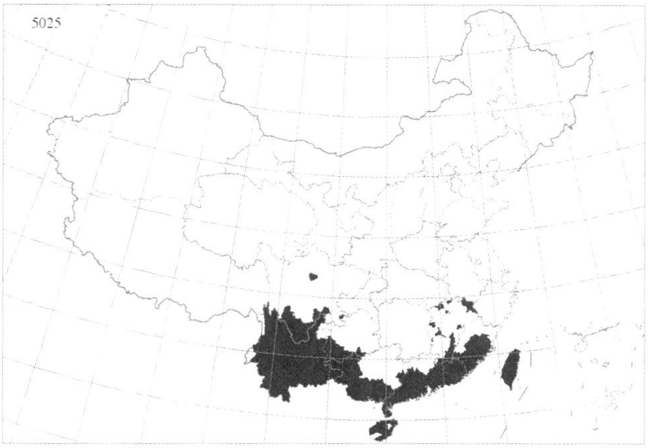

Euphorbiaceae 大戟科 *Phyllanthus* 叶下珠属
Phyllanthus fangchengensis P. T. Li 尖叶下珠
Shrubs

21.9 (21.3~22.7); 13.9 (13.2~15.0); 27.9 (27.3~28.6); 22.0 (21.3~22.8); 203.6 (195.9~213.1); 0.0 (0.0~0.0); 1162 (1111~1221) ❙ 1876 (1302~2435); 1047 (670~1428); 116 (86~130) ❙ 61.1 (12.9~99.5); 1162 (1111~1221); 289.6 (235.1~318.8)

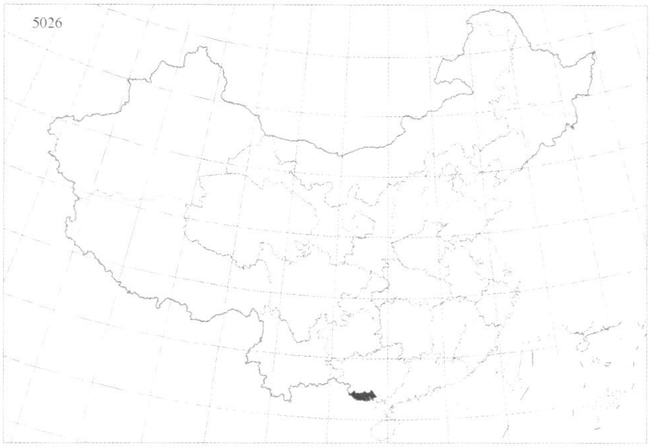

Euphorbiaceae 大戟科 *Phyllanthus* 叶下珠属
Phyllanthus fimbricalyx P. T. Li 穗萼叶下珠
Shrubs

17.9; 11.9; 21.6; 18.0; 155.8; 0.0; 854 ❙ 1435; 803; 42 ❙ 68.0; 833; 505.1

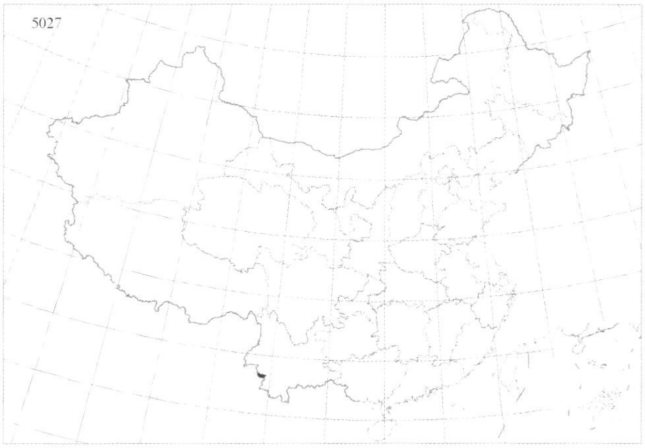

Euphorbiaceae 大戟科 *Phyllanthus* 叶下珠属
Phyllanthus flexuosus (Siebold et Zucc.) Müll. Arg. 落萼叶下珠
Shrubs

16.6 (−1.8~22.3); 6.2 (−10.2~15.1); 26.0 (6.1~30.1); 16.7 (1.8~22.3); 141.3 (1.9~208.1); 1.6 (0.0~82.8); 882 (329~1202) ❙ 1341 (320~1997); 558 (193~996); 126 (15~259) ❙ 51.9 (−35.3~147.0); 866 (320~1202); 235.5 (72.2~659.0)

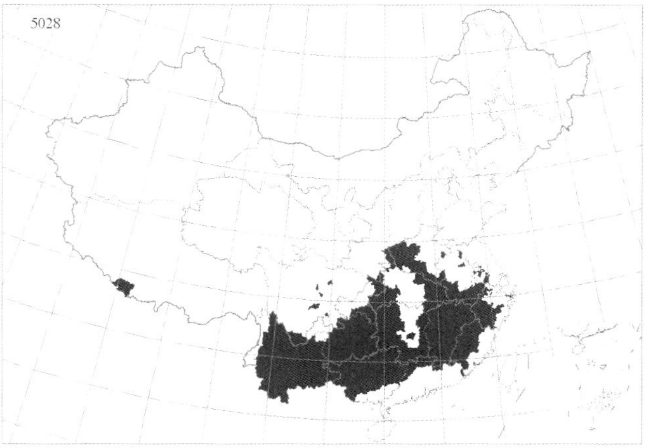

Euphorbiaceae 大戟科 *Phyllanthus* 叶下珠属
Phyllanthus forrestii W. W. Sm. 刺果叶下珠
Shrubs

15.5 (6.1~19.2); 6.8 (−1.5~13.8); 22.8 (12.7~29.5); 15.5 (6.3~19.2); 127.5 (35.2~170.5); 1.5 (0.0~21.8); 810 (496~968) ❙ 1171 (801~1606); 569 (496~686); 77 (19~216) ❙ 44.7 (4.4~71.6); 779 (496~968); 283.4 (172.9~513.9)

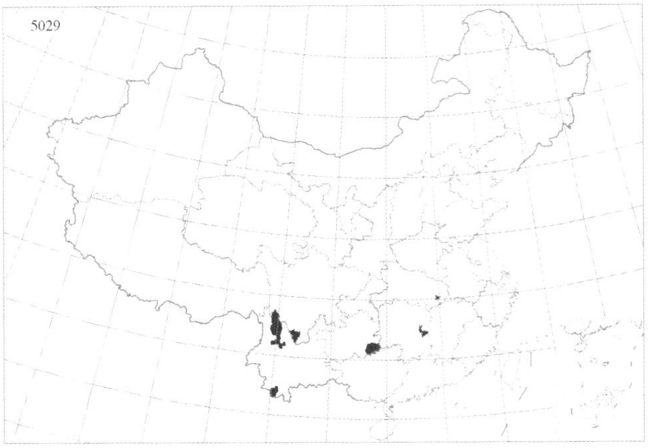

Euphorbiaceae 大戟科 *Phyllanthus* 叶下珠属
Phyllanthus franchetianus H. Lév. 云贵叶下珠
Shrubs

15.4 (11.6~19.3); 7.5 (2.9~13.3); 21.7 (18.6~23.6); 15.5 (11.5~19.4); 126.6 (80.9~173.2); 1.1 (0.0~3.4); 791 (654~939) ❙ 1088 (885~1366); 608 (479~783); 41 (28~51) ❙ 38.0 (21.7~57.4); 782 (654~923); 337.6 (236.5~516.7)

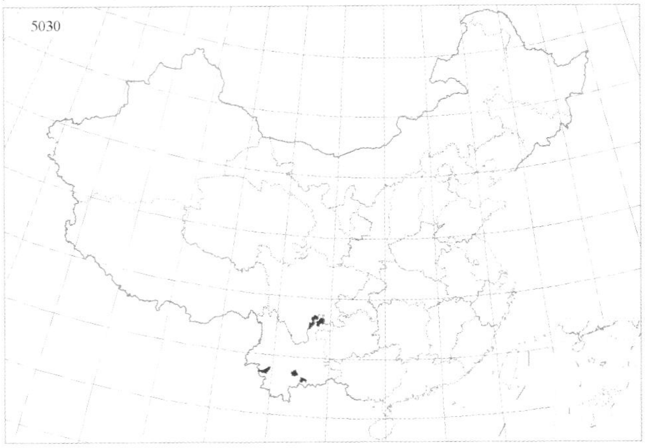

Euphorbiaceae 大戟科 *Phyllanthus* 叶下珠属
Phyllanthus glaucus Wall. ex Müll. Arg. 青灰叶下珠
Shrubs

16.6 (−1.8~22.8); 5.6 (−10.2~14.7); 26.9 (6.1~30.1); 16.7 (1.8~22.8); 141.1 (1.9~213.6); 1.5 (0.0~82.8); 892 (329~1230) ❙ 1421 (320~2024); 543 (193~1056); 157 (12~259) ❙ 59.4 (−12.6~147.0); 882 (320~1230); 221.0 (72.2~530.1)

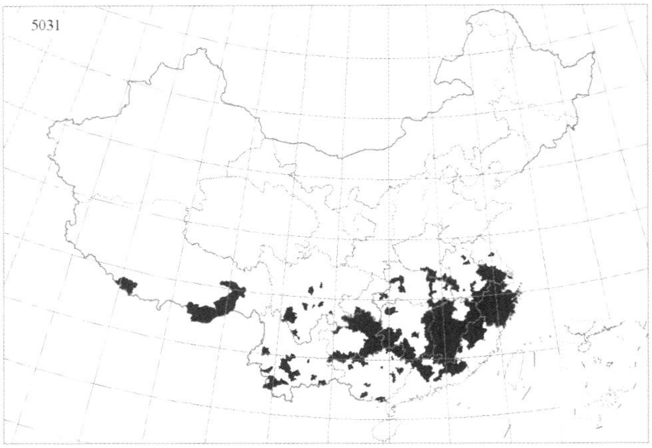

Euphorbiaceae 大戟科 *Phyllanthus* 叶下珠属
Phyllanthus gracilipes (Miq.) Müll. Arg. 毛果叶下珠
Shrubs

19.8; 11.1; 26.4; 19.8; 177.9; 0.0; 1010 ❙ 1182; 673; 43 ❙ 16.9; 1010; 354.5

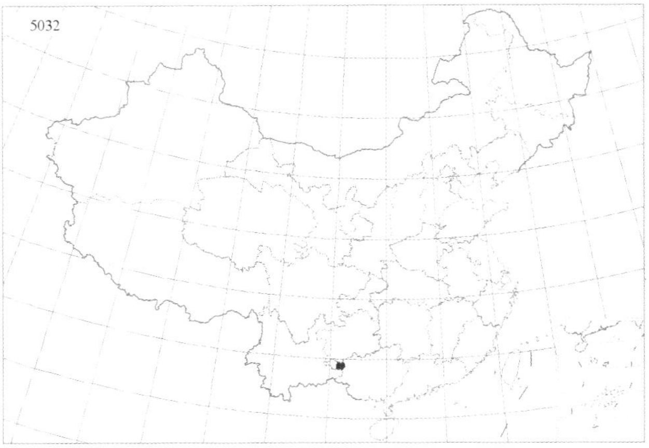

Euphorbiaceae 大戟科 *Phyllanthus* 叶下珠属
Phyllanthus guangdongensis P. T. Li 广东叶下珠
Shrubs

20.2; 11.2; 27.7; 20.3; 183.1; 0.0; 1060 ❙ 1616; 662; 150 ❙ 53.2; 1060; 242.0

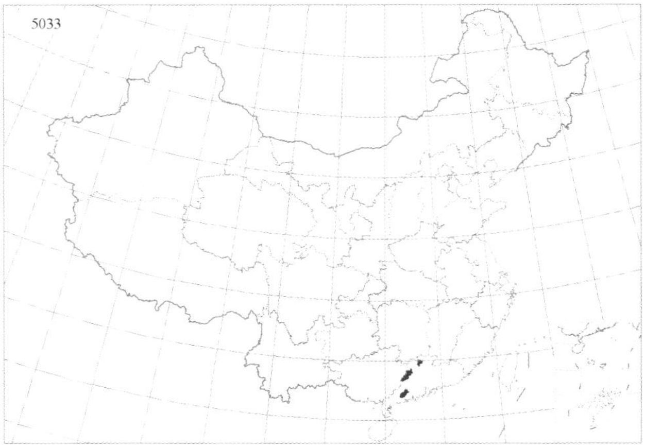

Euphorbiaceae 大戟科 *Phyllanthus* 叶下珠属
Phyllanthus hainanensis Merr. 海南叶下珠
Shrubs

22.9 (18.1~24.8); 17.0 (9.5~20.1); 27.3 (24.6~28.4); 23.0 (18.2~24.8); 215.5 (158.1~237.6); 0.0 (0.0~0.0); 1238 (901~1385) ❙ 1300 (1179~1405); 497 (392~723); 47 (41~58) ❙ 7.5 (−11.9~43.5); 1060 (901~1137); 438.8 (316.1~527.5)

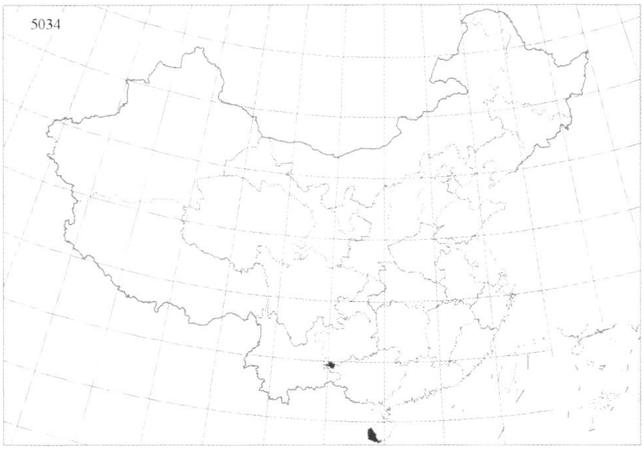

Euphorbiaceae 大戟科 *Phyllanthus* 叶下珠属
Phyllanthus maderaspatensis L. 麻德拉斯叶下珠
Shrubs

15.2; 28.7. 22.7; 212.7; 0.0; 1216 ▎1965; 994; 104 ▎61.6; 1216; 204.4

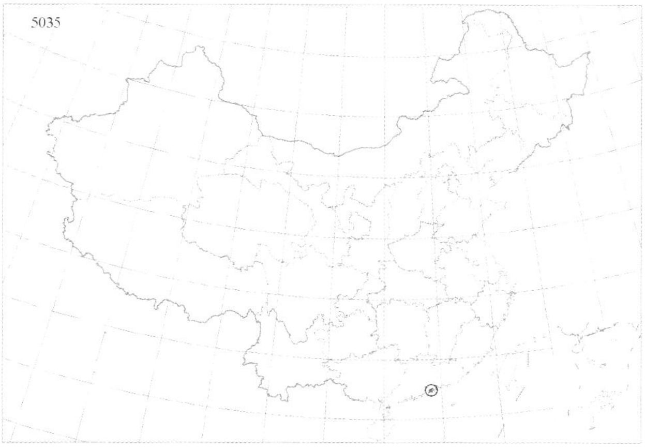

Euphorbiaceae 大戟科 *Phyllanthus* 叶下珠属
Phyllanthus nanellus P. T. Li 单花水油甘
Shrubs

24.0 (22.2~24.8); 18.4 (17.2~20.1); 28.0 (25.7~28.8); 24.0 (22.3~24.8); 228.3 (207.2~237.6); 0.0 (0.0~0.0); 1310 (1127~1385) ▎1501 (868~1854); 534 (312~712); 74 (41~115) ▎15.0 (-34.3~47.8); 1191 (868~1360); 440.3 (248.2~620.0)

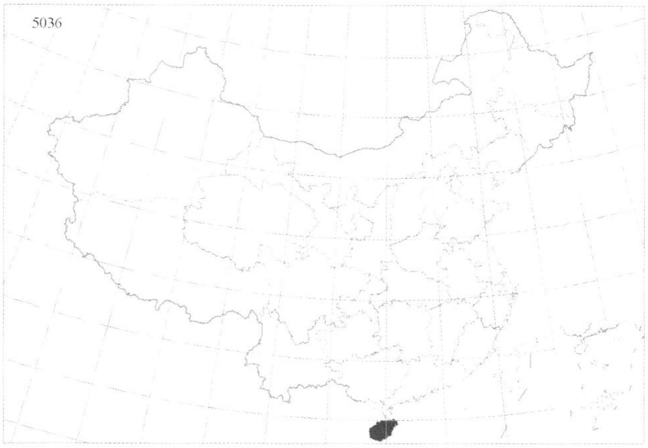

Euphorbiaceae 大戟科 *Phyllanthus* 叶下珠属
Phyllanthus oligospermus Hayata 少子叶下珠
Shrubs, deciduous

17.9; 11.6; 23.4; 18.0; 155.6; 0.0; 864 ▎2355; 900; 308 ▎172.5; 864; 284.4

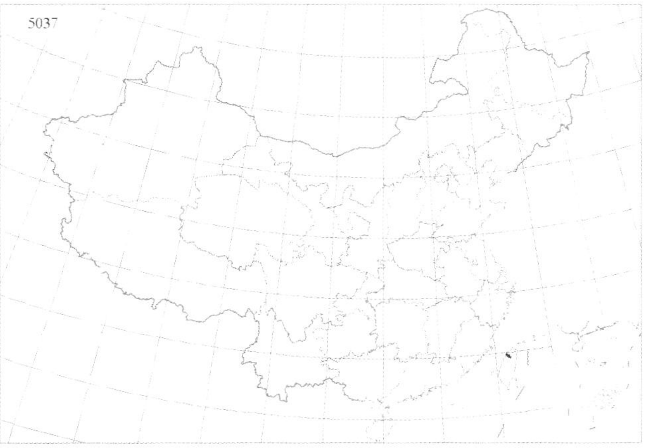

Euphorbiaceae 大戟科 *Phyllanthus* 叶下珠属
Phyllanthus pachyphyllus Müll. Arg. 崖县叶下珠
Shrubs

24.8; 20.1; 28.0; 24.8; 237.6; 0.0; 1385 ▎1405; 473; 41 ▎1.5; 1137; 527.5

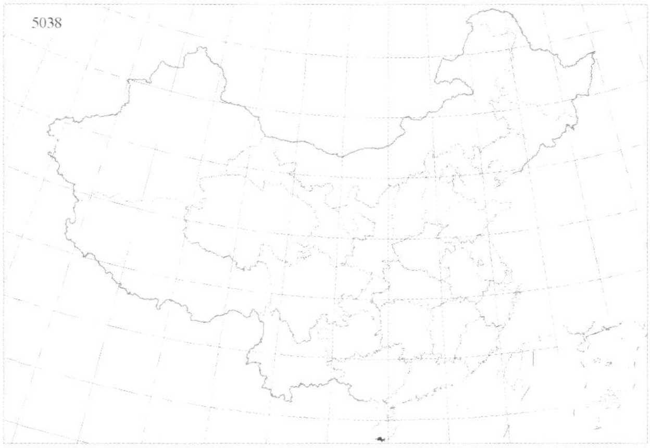

Euphorbiaceae 大戟科 *Phyllanthus* 叶下珠属
Phyllanthus pulcher Wall. ex Müll. Arg. 云桂叶下珠
Shrubs

19.3 (18.1~20.5); 12.8 (9.4~15.1); 23.7 (22.3~25.5); 19.4 (18.2~20.5); 172.3 (158.1~185.9); 0.0 (0.0~0.0); 938 (901~992) ▎1426 (1260~1607); 706 (655~835); 59 (41~77) ▎52.0 (36.3~64.9); 921 (880~986); 487.7 (316.1~659.0)

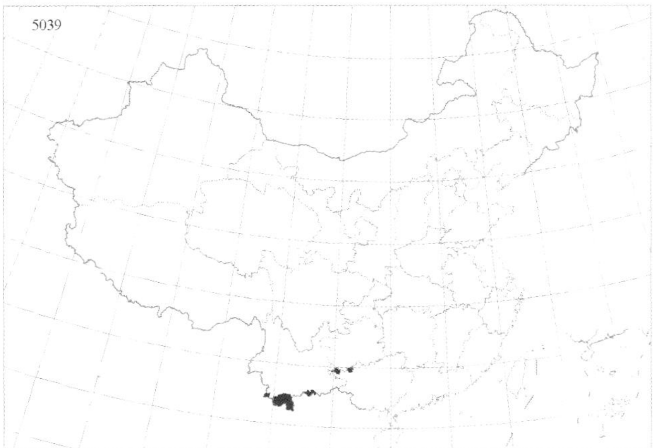

Euphorbiaceae 大戟科 *Phyllanthus* 叶下珠属
Phyllanthus reticulatus Poir. 小果叶下珠
Shrubs

19.3 (6.6~24.8); 10.3 (-2.5~20.1); 26.9 (13.3~30.1); 19.3 (6.8~24.8); 172.1 (39.6~237.6); 0.3 (0.0~20.4); 1013 (514~1385) ▎1513 (795~2992); 664 (299~1847); 126 (12~257) ▎51.0 (-34.3~246.5); 1001 (514~1360); 253.5 (95.3~659.0)

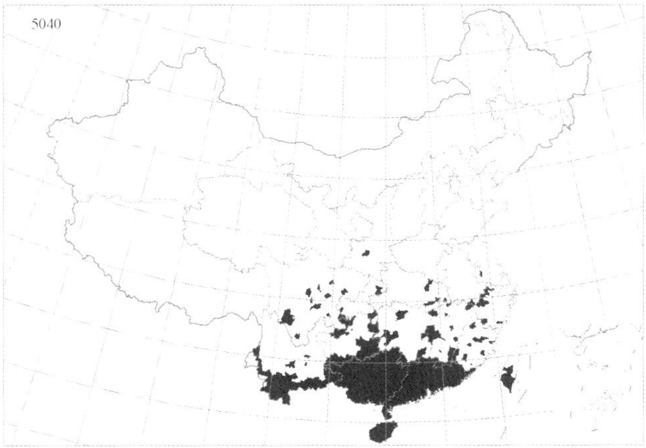

Euphorbiaceae 大戟科 *Phyllanthus* 叶下珠属
Phyllanthus rheophyticus M. G. Gilbert et P. T. Li 水油甘
Shrubs

23.7 (21.1~24.8); 17.7 (11.9~20.1); 28.1 (25.7~29.2); 23.8 (21.1~24.8); 225.1 (193.2~237.6); 0.0 (0.0~0.0); 1292 (1115~1385) ❙ 1478 (868~1854); 544 (312~712); 80 (41~144) ❙ 14.8 (-34.3~47.8); 1192 (868~1360); 401.3 (178.7~620.0)

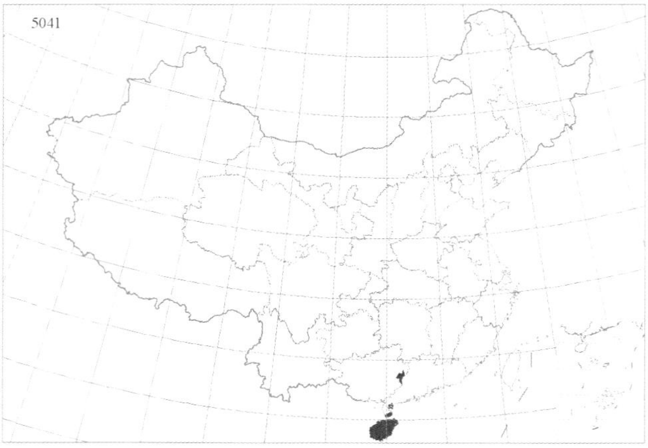

Euphorbiaceae 大戟科 *Phyllanthus* 叶下珠属
Phyllanthus sootepensis Craib 云泰叶下珠
Shrubs

19.1; 13.4; 22.5; 19.1; 169.7; 0.0; 909 ❙ 1497; 730; 60 ❙ 64.7; 886; 470.9

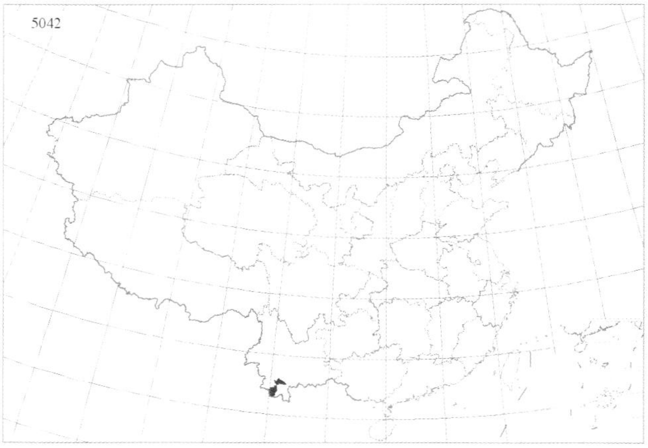

Euphorbiaceae 大戟科 *Phyllanthus* 叶下珠属
Phyllanthus taxodiifolius Beille 落羽杉叶下珠
Shrubs

18.6; 10.5; 24.3; 18.6; 163.4; 0.0; 922 ❙ 1335; 768; 51 ❙ 46.4; 922; 384.5

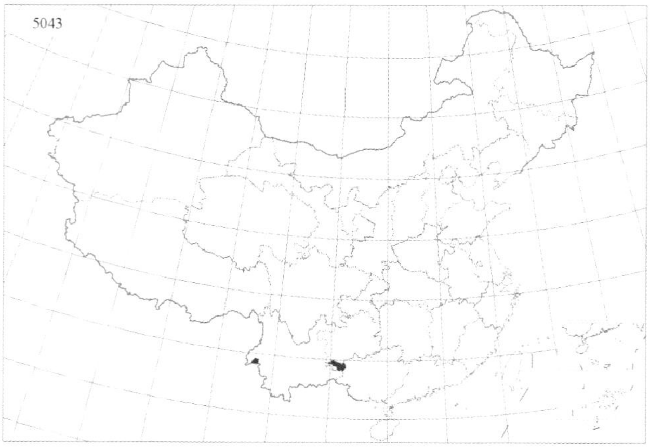

Euphorbiaceae 大戟科 *Phyllanthus* 叶下珠属
Phyllanthus tsarongensis W. W. Sm. 西南叶下珠
Shrubs

10.8 (0.7~18.3); 3.3 (-8.8~11.8); 16.8 (9.5~22.7); 11.2 (3.3~18.3); 82.9 (12.7~159.8); 13.0 (0.0~63.9); 641 (389~882) ❙ 965 (514~1348); 571 (314~762); 30 (10~56) ❙ 55.6 (6.5~125.9); 624 (389~799); 303.9 (234.2~395.4)

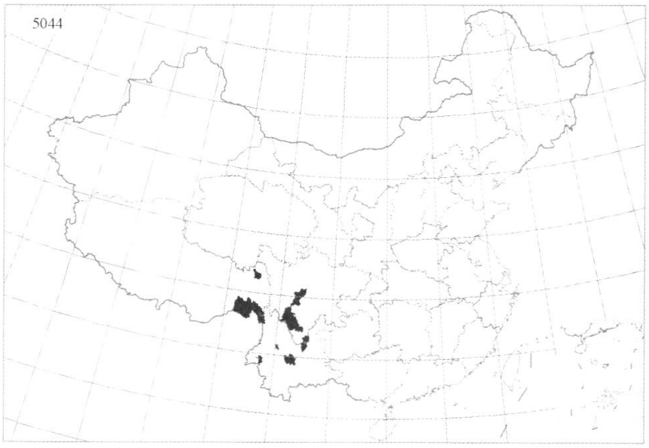

Euphorbiaceae 大戟科 *Phyllanthus* 叶下珠属
Phyllanthus tsiangii P. T. Li 红叶下珠
Shrubs or small trees

24.0 (22.2~24.8); 18.3 (16.7~20.1); 28.0 (25.7~28.8); 24.0 (22.3~24.8); 228.0 (207.2~237.6); 0.0 (0.0~0.0); 1308 (1127~1385) ❙ 1490 (868~1854); 534 (312~712); 74 (41~115) ❙ 14.2 (-34.3~47.8); 1195 (868~1360); 430.1 (235.0~620.0)

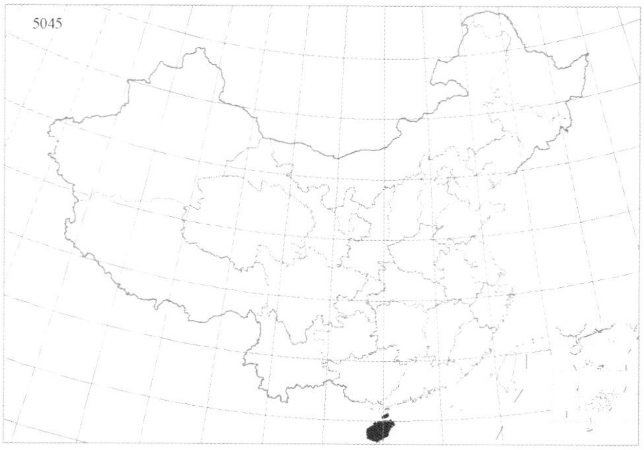

Euphorbiaceae 大戟科 *Putranjiva* 假黄杨属
Putranjiva formosana Kaneh. et Sasaki ex Shimada 台湾假黄杨
Small trees

19.6; 13.7; 24.4; 19.6; 175.8; 0.0; 974 ❙ 2444; 1069; 229 ❙ 165.2; 974; 256.4

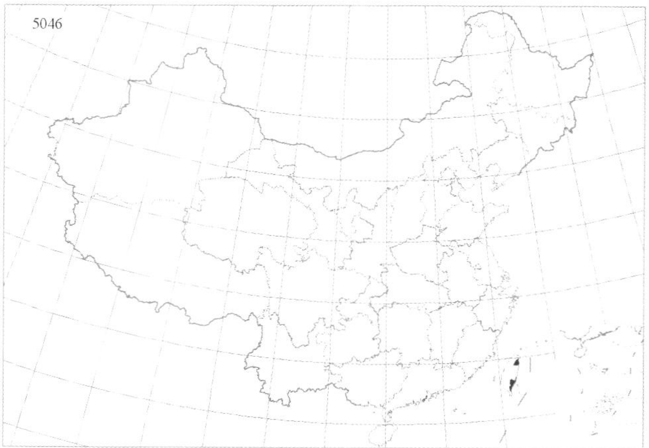

Euphorbiaceae 大戟科 *Richeriella* 龙胆木属
Richeriella gracilis (Merr.) Pax et K. Hoffm. 龙胆木
Trees

23.9 (22.2~24.8)；18.5 (17.2~20.1)；27.7 (25.7~28.7)；24.0 (22.3~24.8)；227.5 (207.2~237.6)；0.0 (0.0~0.0)；1302 (1127~1385) ❙ 1453 (868~1854)．503 (312~622)；61 (41~115) ❙ 12.1 (-34.3~47.8)；1137 (868~1360)；490.1 (288.3~620.0)

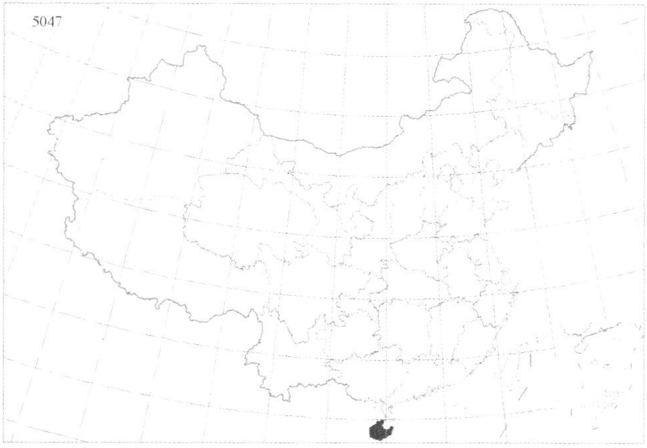

Euphorbiaceae 大戟科 *Ricinus* 蓖麻属
Ricinus communis L. 蓖麻
Herbs or shrubs

17.5 (2.2~24.8)；8.8 (-7.0~20.1)；24.8 (9.5~29.5)；17.5 (3.8~24.8)；151.5 (15.5~237.6)；1.1 (0.0~49.1)；913 (411~1385) ❙ 1372 (685~2993)；620 (312~1847)；111 (11~547) ❙ 51.9 (-34.3~214.4)；900 (411~1360)；282.3 (126.1~659.0)

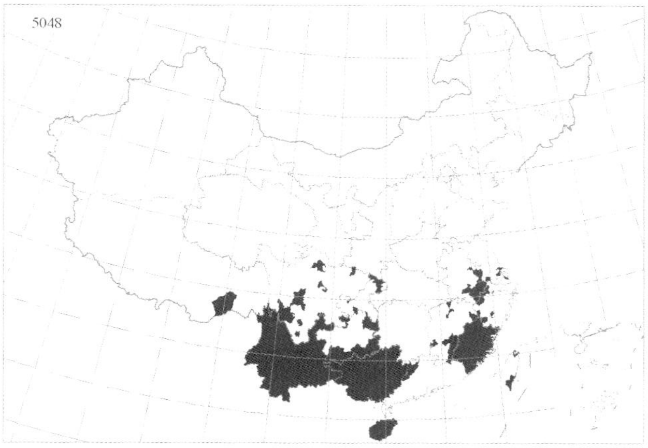

Euphorbiaceae 大戟科 *Sauropus* 守宫木属
Sauropus bonii Beille 茎花守宫木
Shrubs

22.0；14.4；27.7；22.0；204.5；0.0；1181 ❙ 1475；644；83 ❙ 25.0；1157；399.2

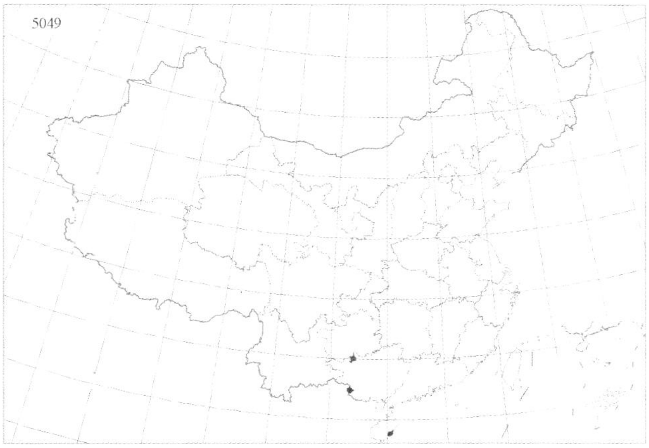

Euphorbiaceae 大戟科 *Sauropus* 守宫木属
Sauropus delavayi Croizat 石山守宫木
Shrubs

22.2；13.9；28.3；22.3；207.5；0.0；1190 ❙ 1374；713；87 ❙ 15.4；1190；327.7

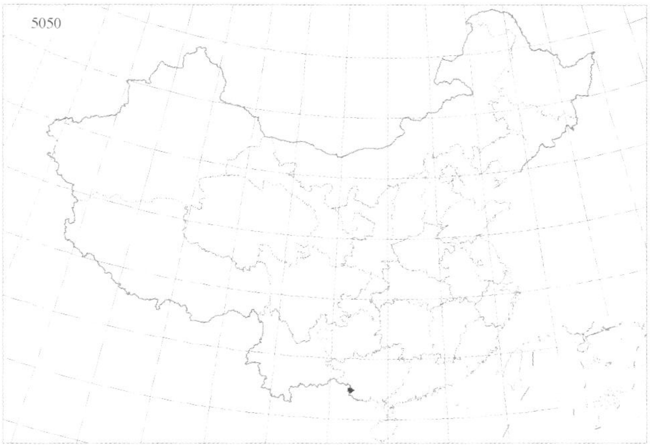

Euphorbiaceae 大戟科 *Sauropus* 守宫木属
Sauropus garrettii Craib 苍叶守宫木
Shrubs

17.7 (7.3~24.8)；8.8 (-2.0~20.1)；25.2 (15.6~29.7)；17.8 (7.5~24.8)；153.8 (48.2~237.6)；0.8 (0.0~20.4)；923 (537~1385) ❙ 1401 (936~2148)；647 (400~1234)；94 (23~221) ❙ 54.0 (1.5~112.2)；910 (537~1360)；287.1 (150.8~659.0)

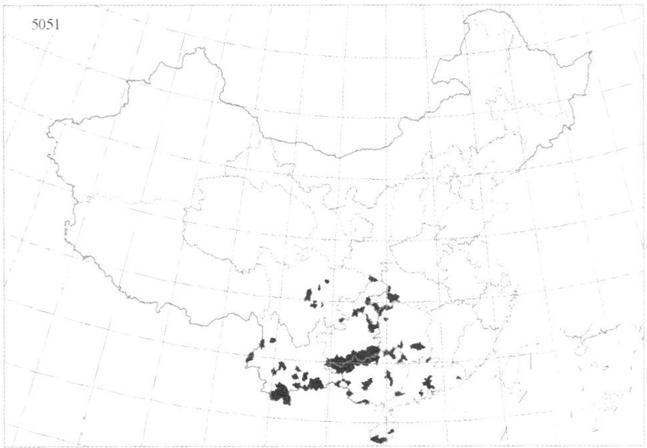

Euphorbiaceae 大戟科 *Sauropus* 守宫木属
Sauropus macranthus Hassk. 长梗守宫木
Shrubs

21.7 (18.1~24.8)；15.4 (9.2~20.1)；26.2 (22.3~28.7)；21.8 (18.2~24.8)；201.5 (158.1~237.6)；0.0 (0.0~0.0)；1137 (888~1385) ❙ 1459 (868~1854)；637 (312~911)；68 (41~135) ❙ 31.8 (-34.3~74.2)；1054 (868~1360)；457.1 (256.4~659.0)

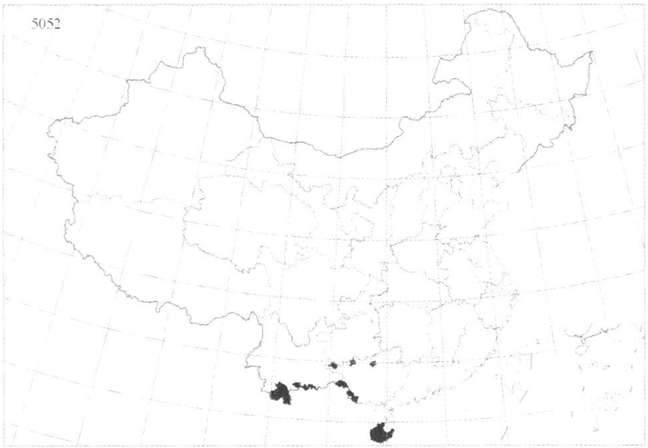

Euphorbiaceae 大戟科 *Sauropus* 守宫木属
Sauropus pierrei (Beille) Croizat 盈江守宫木
Shrubs

16.4；9.6；20.6；16.4；137.3；0.0；795 ❙ 1688；959；62 ❙ 112.2；795；627.0

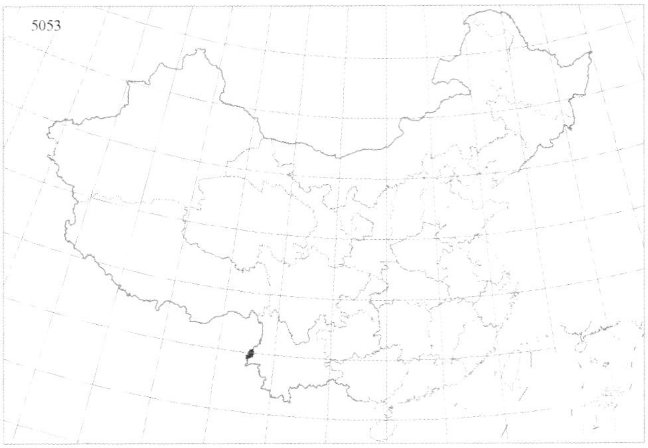

Euphorbiaceae 大戟科 *Sauropus* 守宫木属
Sauropus quadrangularis (Willd.) Müll. Arg. 方枝守宫木
Shrubs

19.8 (17.3~24.3)；11.8 (8.7~18.7)；25.9 (22.4~28.6)；19.9 (17.3~24.3)；178.6 (147.1~231.9)；0.0 (0.0~0.0)；1016 (833~1338) ❙ 1470 (1179~1710)；743 (392~860)；95 (42~194) ❙ 47.3 (-11.9~80.6)；996 (833~1188)；361.0 (236.8~551.6)

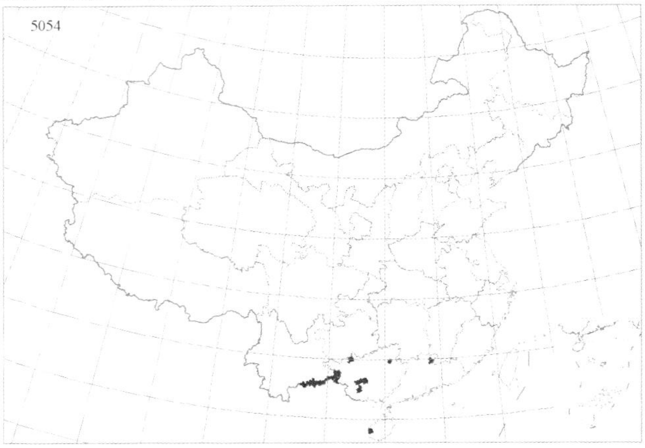

Euphorbiaceae 大戟科 *Sauropus* 守宫木属
Sauropus repandus Müll. Arg. 波萼守宫木
Shrubs

19.2；13.8；22.3；19.2；170.5；0.0；909 ❙ 1499；686；58 ❙ 64.9；880；513.9

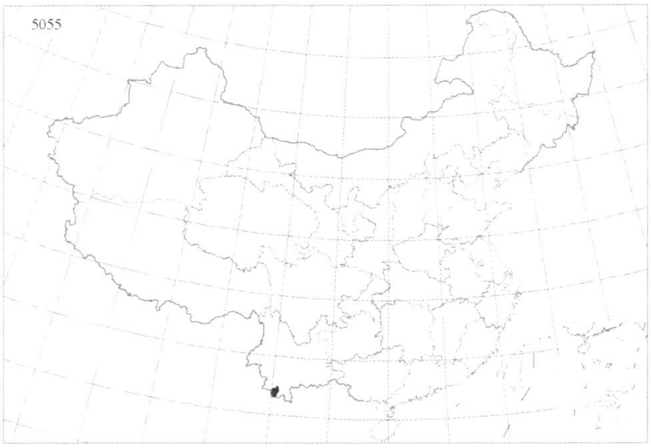

Euphorbiaceae 大戟科 *Sauropus* 守宫木属
Sauropus reticulatus S. L. Mo ex P. T. Li 网脉守宫木
Shrubs

16.2；8.7；21.8；16.2；134.9；0.0；824 ❙ 1411；787；71 ❙ 74.1；824；360.4

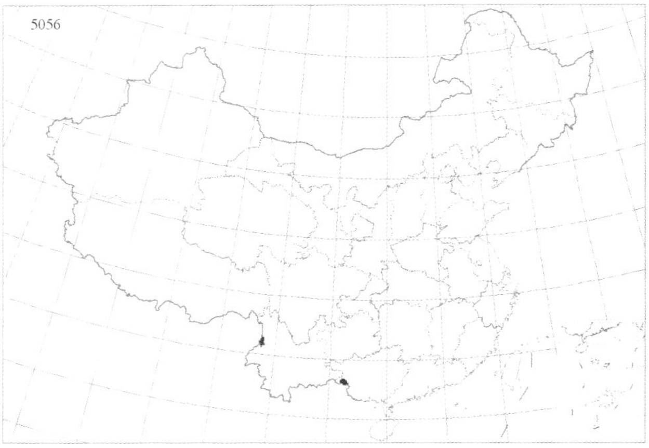

Euphorbiaceae 大戟科 *Strophioblachia* 宿萼木属
Strophioblachia fimbricalyx Boerl. 宿萼木
Shrubs

22.6 (17.7~24.8)；16.5 (10.9~20.1)；26.9 (22.1~28.8)；22.6 (17.8~24.8)；211.5 (153.4~237.6)；0.0 (0.0~0.0)；1204 (854~1385) ❙ 1487 (1059~1854)；611 (392~820)；72 (41~115) ❙ 25.1 (-11.9~60.1)；1137 (847~1360)；432.7 (248.2~620.0)

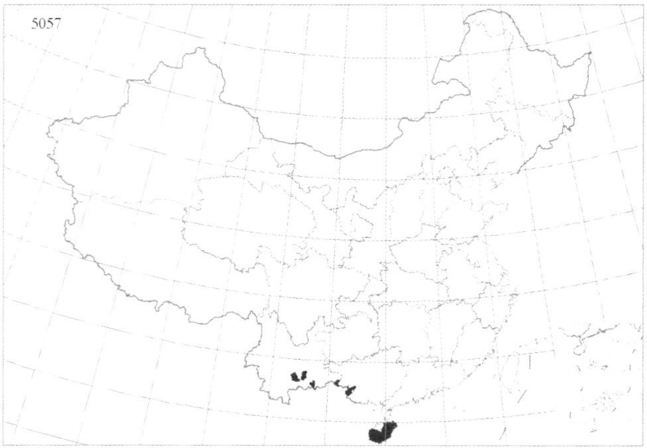

Euphorbiaceae 大戟科 *Sumbaviopsis* 白叶桐属
Sumbaviopsis albicans (Blume) J. J. Sm. 白叶桐
Trees

18.6 (14.9~21.0)；11.9 (5.3~15.1)；23.4 (21.6~25.6)；18.6 (15.0~21.0)；163.6 (119.6~191.8)；0.0 (0.0~0.0)；911 (782~1053) ❙ 1461 (1318~1607)；768 (655~911)；59 (42~77) ❙ 61.1 (44.5~80.4)；900 (782~1053)；456.6 (189.1~659.0)

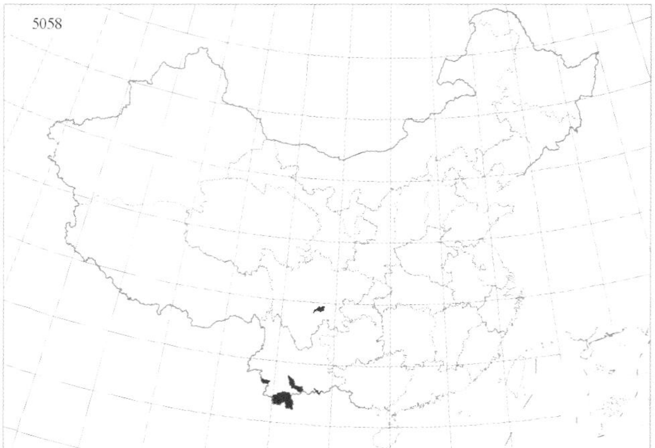

Euphorbiaceae 大戟科 *Suregada* 白树属
Suregada aequorea (Hance) Seem. 台湾白树
Small trees

21.0; 16.3; 24.5; 21.1; 192.6; 0.0; 1055 ❙ 2554; 1499; 117 ❙ 154.0; 998; 266.8

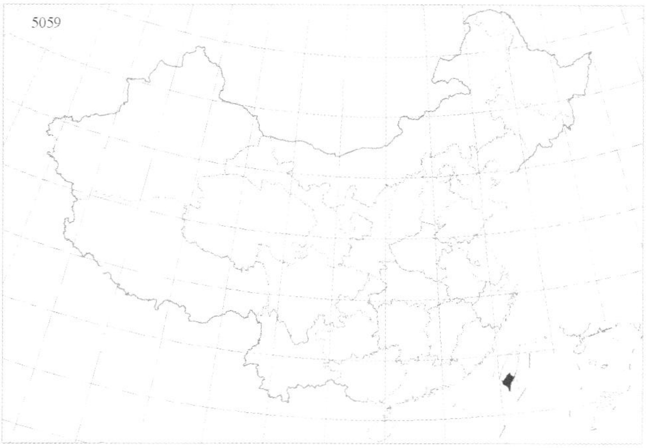

Euphorbiaceae 大戟科 *Suregada* 白树属
Suregada multiflora (Jussieu) Baillon 白树
Shrubs or trees

23.0 (18.2~24.8); 16.6 (11.9~20.1); 27.7 (22.0~29.2); 23.1 (18.2~24.8); 216.6 (158.7~237.6); 0.0 (0.0~0.0); 1235 (868~1385) ❙ 1586 (868~2435); 691 (312~1428); 85 (41~144) ❙ 30.2 (-34.3~99.5); 1173 (847~1360); 367.1 (157.7~659.0)

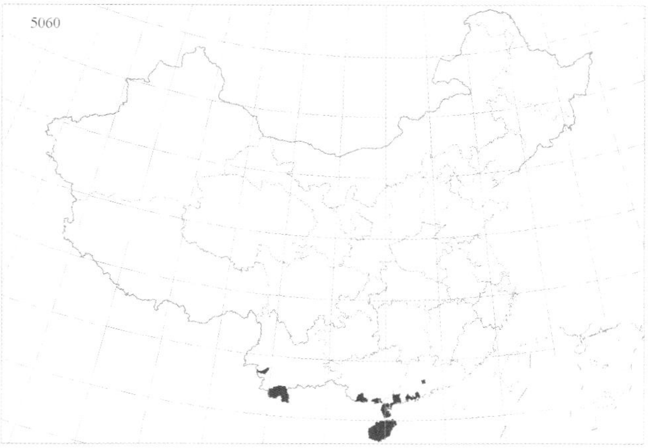

Euphorbiaceae 大戟科 *Trevia* 滑桃树属
Trevia nudiflora L. 滑桃树
Trees

22.1 (18.2~24.8); 16.2 (12.4~20.1); 26.0 (21.6~28.9); 22.1 (18.2~24.8); 205.1 (153.6~237.6); 0.0 (0.0~0.0); 1154 (861~1385) ❙ 1498 (1242~1854); 639 (444~911); 68 (41~136) ❙ 33.4 (-2.2~74.2); 1080 (839~1360) 487.9 (201.4~659.0)

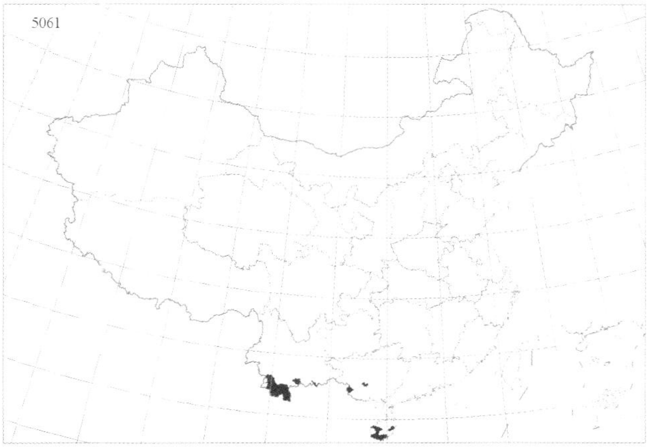

Euphorbiaceae 大戟科 *Triadica* 乌桕属
Triadica cochinchinensis Lour. 山乌桕
Shrubs or trees, deciduous

18.1 (9.3~24.8); 8.2 (-1.7~20.1); 26.7 (19.8~30.1); 18.1 (9.5~24.8); 157.9 (69.8~237.6); 0.6 (0.0~18.0); 952 (603~1385) ❙ 1486 (711~3495); 601 (299~1594); 144 (15~897) ❙ 57.6 (-34.3~246.5); 937 (603~1360); 256.9 (123.1~659.0)

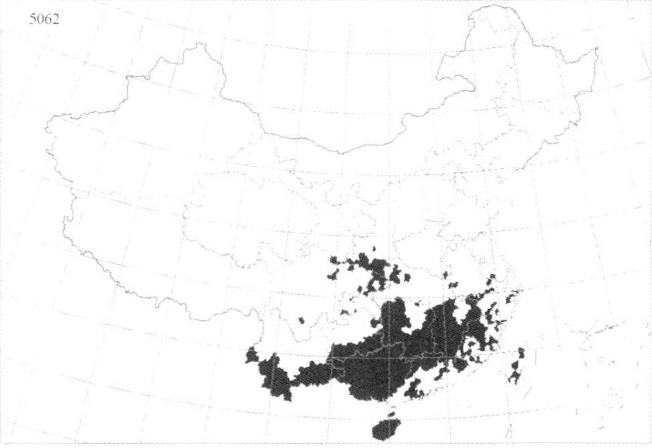

Euphorbiaceae 大戟科 *Triadica* 乌桕属
Triadica rotundifolia (Hemsley) Esser 圆叶乌桕
Shrubs or trees, deciduous

18.4 (13.3~22.3); 8.9 (3.2~14.9); 26.3 (21.4~29.0); 18.4 (13.2~22.3); 161.0 (100.9~208.2); 0.1 (0.0~2.5); 951 (723~1190) ❙ 1465 (1074~2035); 645 (419~969); 126 (32~202) ❙ 55.0 (5.8~101.8); 950 (723~1190); 248.9 (160.7~506.2)

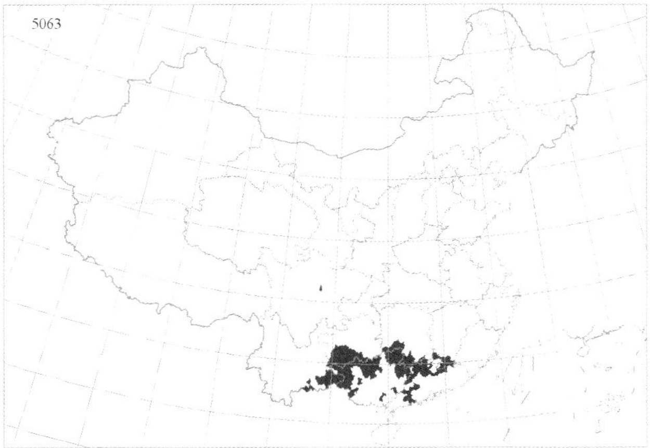

Euphorbiaceae 大戟科 *Triadica* 乌桕属
Triadica sebifera (L.) Small 乌桕
Trees

16.8 (2.2~24.5); 6.1 (-6.6~19.1); 26.5 (9.9~30.1); 16.9 (3.8~24.5); 144.7 (15.5~234.4); 2.5 (0.0~49.1); 908 (413~1360) ❙ 1330 (427~3495); 555 (258~1428); 124 (12~897) ❙ 46.3 (-48.6~235.6); 889 (413~1360); 217.3 (95.3~620.0)

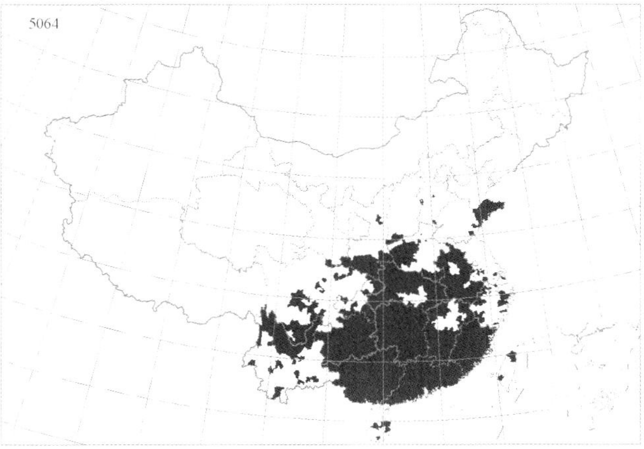

Euphorbiaceae 大戟科 *Trigonostemon* 三宝木属

Trigonostemon bonianus Gagnep. 勐仑三宝木

Shrubs or small trees, evergreen

20.3 (17.2~22.2); 13.4 (9.5~15.1); 25.1 (22.3~28.3); 20.4 (17.3~22.3); 184.6 (147.0~207.5); 0.0 (0.0~0.0); 1024 (838~1190) ❙ 1424 (1188~1607); 706 (634~850); 71 (48~106) ❙ 40.8 (5.8~64.9); 1015 (838~1190); 429.2 (281.7~659.0)

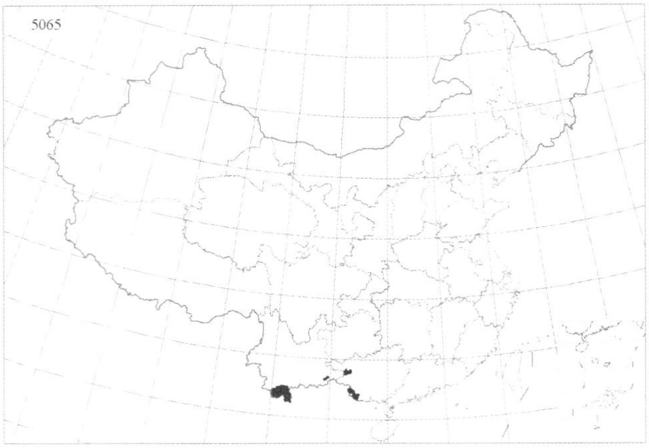

Euphorbiaceae 大戟科 *Trigonostemon* 三宝木属

Trigonostemon chinensis Merr. 三宝木

Shrubs, evergreen

22.0 (18.1~24.8); 14.9 (9.5~20.1); 27.3 (22.3~28.7); 22.0 (18.0~24.8); 204.4 (156.5~237.6); 0.0 (0.0~0.0); 1168 (863~1385) ❙ 1537 (1163~2435); 732 (444~1428); 77 (41~130) ❙ 32.7 (-12.0~99.5); 1109 (863~1221); 363.8 (163.7~572.9)

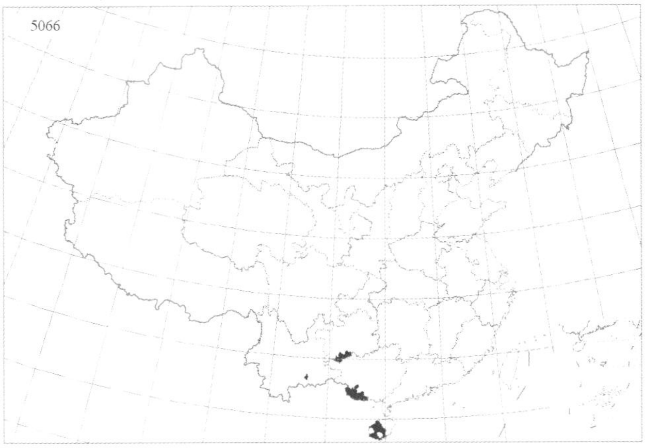

Euphorbiaceae 大戟科 *Trigonostemon* 三宝木属

Trigonostemon flavidus Gagnep. 异叶三宝木

Shrubs, evergreen

23.9 (22.2~24.8); 18.6 (17.2~20.1); 27.6 (25.7~28.7); 23.9 (22.3~24.8); 227.4 (207.2~237.6); 0.0 (0.0~0.0); 1301 (1127~1385) ❙ 1502 (1176~1854); 519 (392~622); 60 (41~115) ❙ 16.0 (-12.0~47.8); 1160 (1055~1360); 506.9 (411.8~620.0)

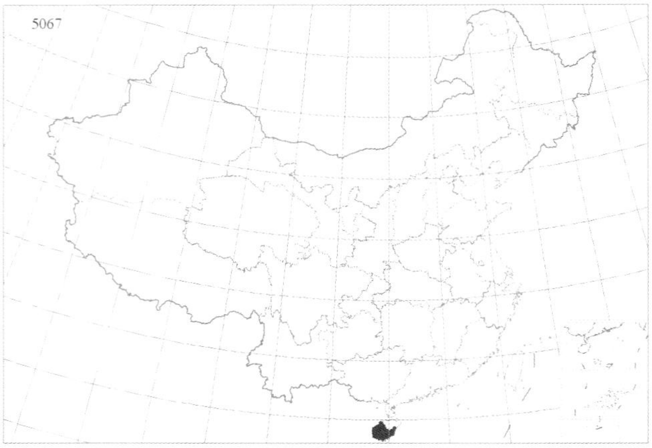

Euphorbiaceae 大戟科 *Trigonostemon* 三宝木属

Trigonostemon fragilis (Gagnep.) Airy Shaw 黄花三宝木

Shrubs, evergreen

23.6 (21.5~24.8); 17.5 (12.9~20.1); 28.0 (25.7~28.8); 23.7 (21.6~24.8); 224.0 (199.1~237.6); 0.0 (0.0~0.0); 1285 (1127~1385) ❙ 1495 (868~1854); 574 (312~850); 79 (41~116) ❙ 16.8 (-34.3~47.8); 1187 (868~1360); 414.0 (248.2~620.0)

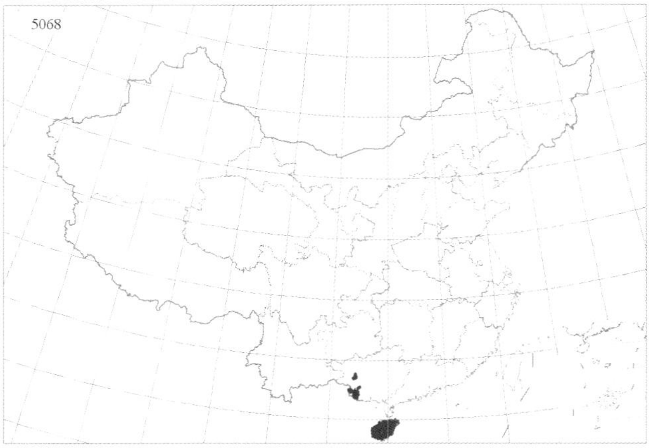

Euphorbiaceae 大戟科 *Trigonostemon* 三宝木属

Trigonostemon howii Merr. et Chun 长序三宝木

Shrubs, evergreen

24.8; 20.1; 28.0; 24.8; 237.6; 0.0; 1385 ❙ 1405; 473; 41 ❙ 1.5; 1137; 527.5

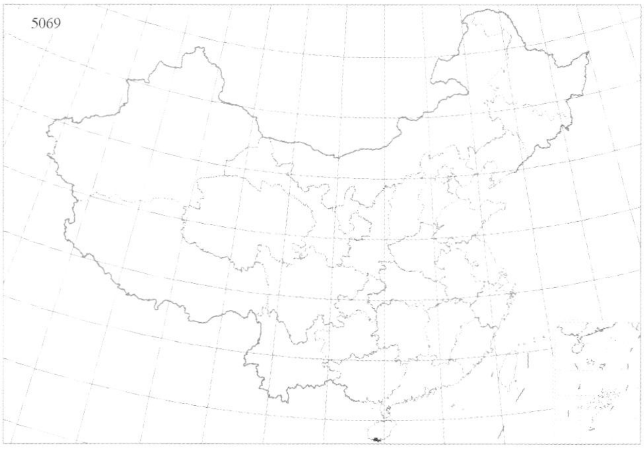

Euphorbiaceae 大戟科 *Trigonostemon* 三宝木属

Trigonostemon thyrsoideus Stapf 长梗三宝木

Shrubs or small trees, evergreen

20.2 (17.1~22.7); 12.7 (8.5~15.1); 25.6 (22.3~28.6); 20.2 (17.1~22.8); 182.7 (145.1~213.1); 0.0 (0.0~0.0); 1024 (845~1221) ❙ 1532 (1163~2435); 815 (634~1428); 83 (48~173) ❙ 49.7 (5.8~99.5); 1020 (845~1221); 384.2 (237.8~659.0)

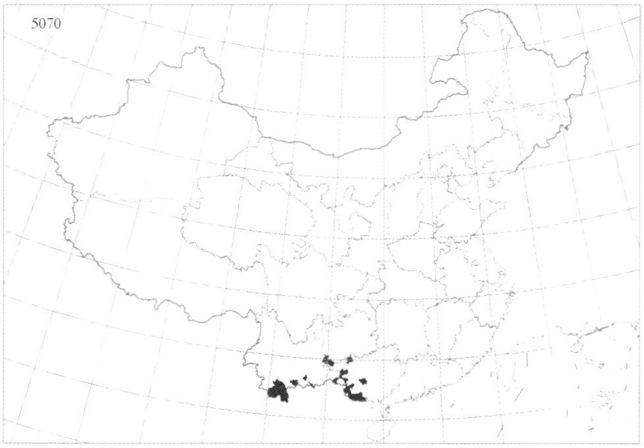

Euphorbiaceae 大戟科 *Trigonostemon* 三宝木属
Trigonostemon xyphophylloides (Croizat) L. K. Dai et T. L. Wu 剑叶三宝木
Shrubs, evergreen

24.7; 20.0 27.9; 24.8; 237.1; 0.0; 1380 ❙ 1496; 503; 50 ❙ 8.4; 1175; 511.1

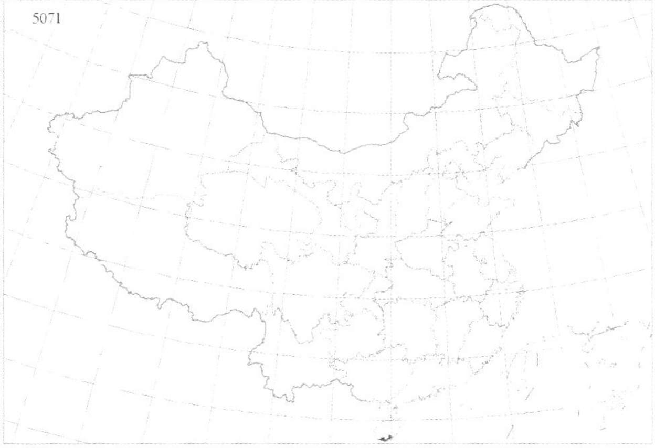

Euphorbiaceae 大戟科 *Vernicia* 油桐属
Vernicia fordii (Hemsl.) Airy Shaw 油桐
Trees, deciduous

16.5 (2.2~24.8); 6.3 (-6.6~20.1); 25.7 (9.9~29.9); 16.6 (3.8~24.8); 140.9 (15.5~237.6); 2.1 (0.0~49.1); 885 (413~1385) ❙ 1299 (512~2037); 558 (258~1033); 110 (15~259) ❙ 47.2 (-35.3~130.7); 870 (413~1360); 239.1 (95.3~659.0)

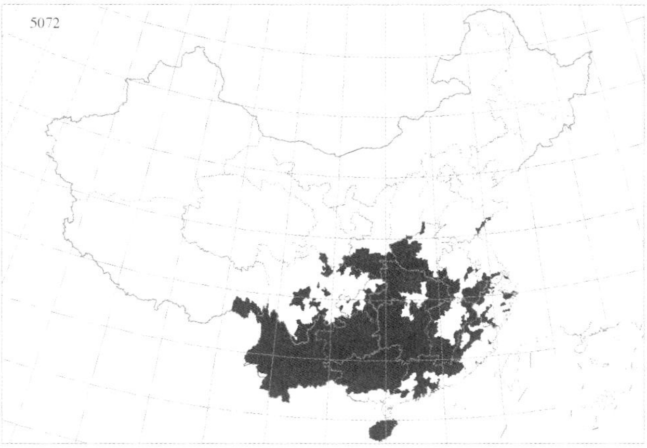

Euphorbiaceae 大戟科 *Vernicia* 油桐属
Vernicia montana Lour. 木油桐
Trees, deciduous

18.6 (10.6~24.8); 8.7 (-0.2~20.1); 27.3 (16.6~30.1); 18.6 (10.6~24.8); 163.6 (71.8~237.6); 0.2 (0.0~12.1); 983 (633~1385) ❙ 1521 (868~2148); 607 (299~1234); 149 (24~259) ❙ 56.6 (-34.3~130.7); 968 (633~1360); 238.4 (95.3~659.0)

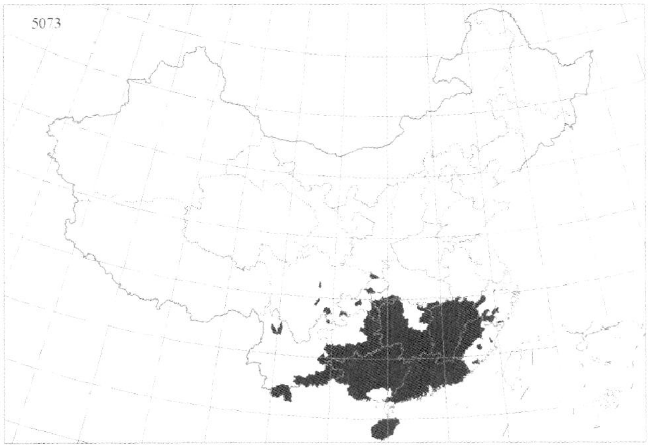

Daphniphyllaceae 交让木科 *Daphniphyllum* 虎皮楠属
Daphniphyllum angustifolium Hutch. 狭叶虎皮楠
Shrubs or small trees, evergreen

16.3 (10.2~19.4); 5.3 (-0.7~9.4); 26.5 (21.0~29.3); 16.4 (10.4~19.4); 137.9 (77.2~173.4); 1.8 (0.0~13.6); 876 (634~1037) ❙ 1547 (853~1922); 557 (384~816); 169 (24~259) ❙ 77.0 (16.4~128.9); 872 (634~1029); 236.4 (167.5~305.6)

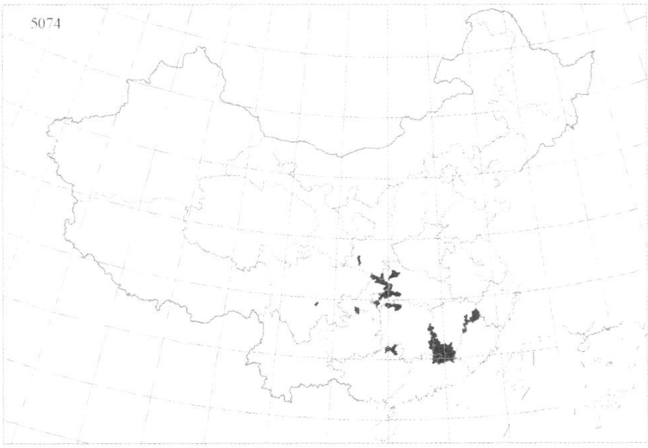

Daphniphyllaceae 交让木科 *Daphniphyllum* 虎皮楠属
Daphniphyllum calycinum Benth. 牛耳枫
Shrubs, evergreen

20.1 (14.3~24.8); 10.9 (3.3~20.1); 27.8 (24.3~30.1); 20.1 (14.3~24.8); 181.5 (114.1~237.6); 0.0 (0.0~2.3); 1062 (765~1385) ❙ 1598 (868~2435); 654 (312~1428); 152 (41~259) ❙ 52.6 (-34.3~130.7); 1048 (765~1360); 245.3 (95.3~620.0)

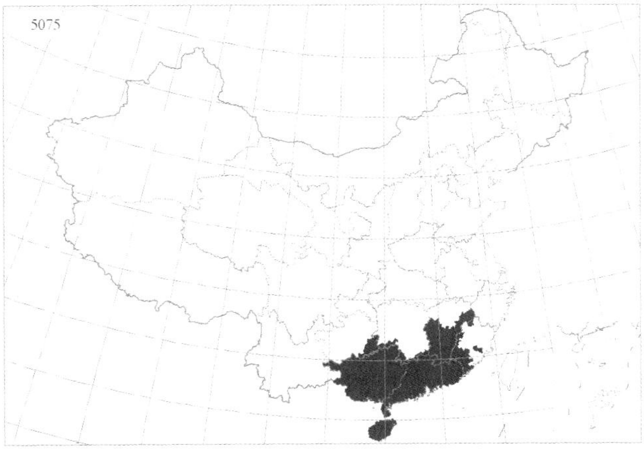

Daphniphyllaceae 交让木科 *Daphniphyllum* 虎皮楠属
Daphniphyllum chartaceum K. Rosenthal 纸叶虎皮楠
Trees, evergreen

17.9; 11.2; 22.4; 18.0; 155.5; 0.0; 860 ❙ 1461; 799; 52 ❙ 69.9; 849; 399.6

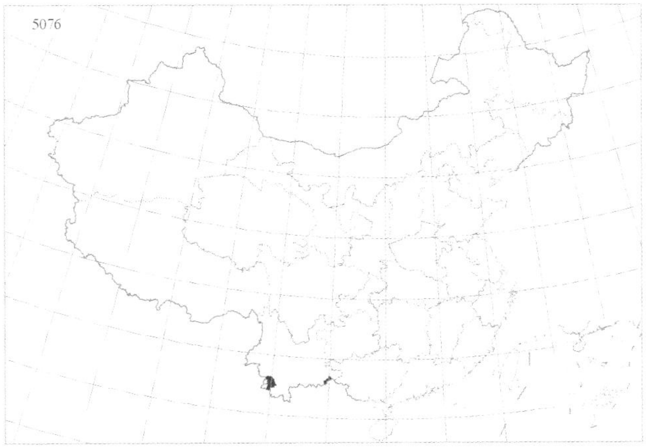

Daphniphyllaceae 交让木科 *Daphniphyllum* 虎皮楠属
Daphniphyllum himalense (Benth.) Müll. Arg. 西藏虎皮楠
Trees, evergreen

15.8 (8.6~19.1); 8.4 (1.3~13.0); 21.4 (15.0~23.6); 15.9 (8.7~19.1); 131.7 (52.6~169.7); 1.1 (0.0~8.8); 798 (574~917) ❙ 1347 (1041~1736); 759 (585~1007); 53 (45~64) ❙ 70.9 (49.4~145.8); 798 (574~917); 383.4 (226.6~506.2)

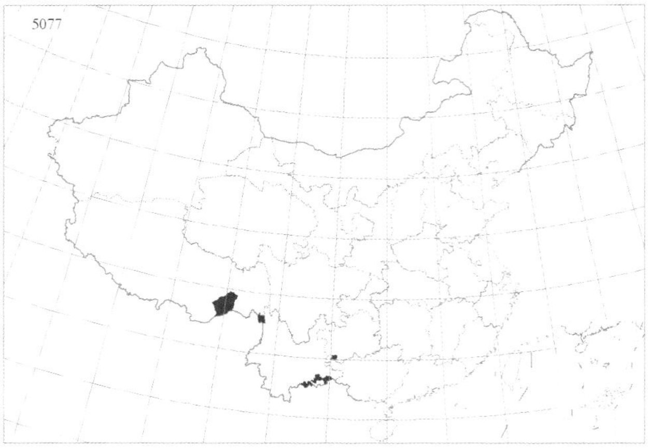

Daphniphyllaceae 交让木科 *Daphniphyllum* 虎皮楠属
Daphniphyllum longeracemosum K. Rosenthal 长序虎皮楠
Trees, evergreen

19.7 (16.9~22.3); 11.0 (7.3~14.2); 26.4 (21.4~28.9); 19.7 (16.9~22.3); 176.5 (142.7~208.1); 0.0 (0.0~0.0); 1014 (813~1202) ❙ 1504 (1178~1758); 734 (562~911); 125 (51~194) ❙ 49.7 (20.3~85.5); 1013 (813~1202); 298.9 (185.5~551.6)

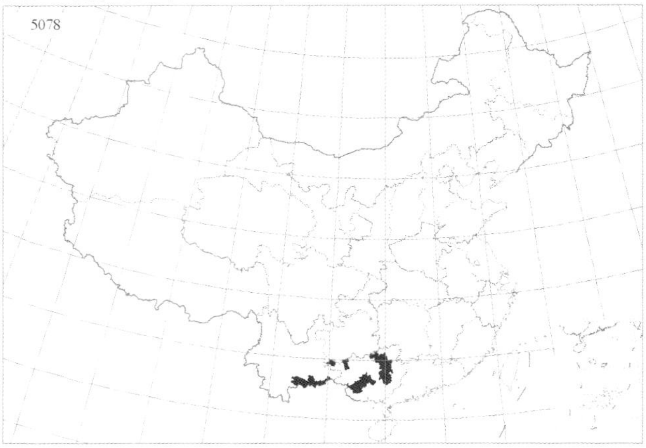

Daphniphyllaceae 交让木科 *Daphniphyllum* 虎皮楠属
Daphniphyllum macropodum Miq. 交让木
Shrubs or small trees, evergreen

16.6 (7.3~24.1); 5.4 (-2.0~18.9); 27.1 (15.0~30.1); 16.7 (7.5~24.1); 141.2 (48.2~229.7); 1.0 (0.0~20.4); 896 (537~1322) ❙ 1423 (807~3495); 539 (299~1594); 151 (18~897) ❙ 59.2 (-6.7~235.6); 882 (537~1206); 208.5 (108.2~551.3)

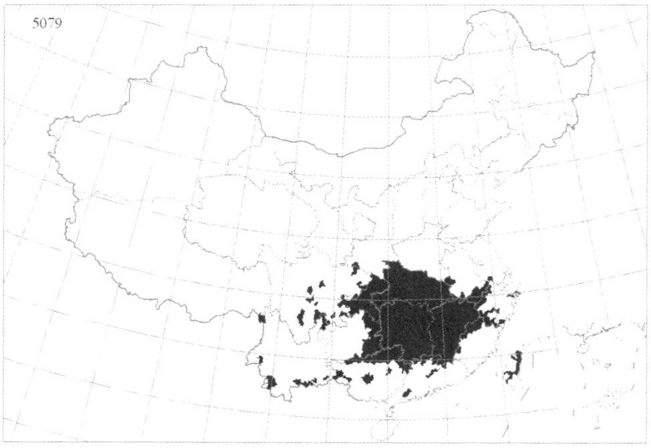

Daphniphyllaceae 交让木科 *Daphniphyllum* 虎皮楠属
Daphniphyllum majus Müll. Arg. 大叶虎皮楠
Shrubs or small trees, evergreen

18.3 (17.3~19.2); 11.9 (9.9~13.8); 22.5 (22.1~23.5); 18.3 (17.3~19.4); 159.9 (147.1~172.6); 0.0 (0.0~0.0); 878 (833~932) ❙ 1348 (958~1504); 732 (531~860); 55 (44~62) ❙ 53.6 (13.3~80.6); 867 (823~923); 434.8 (311.9~513.9)

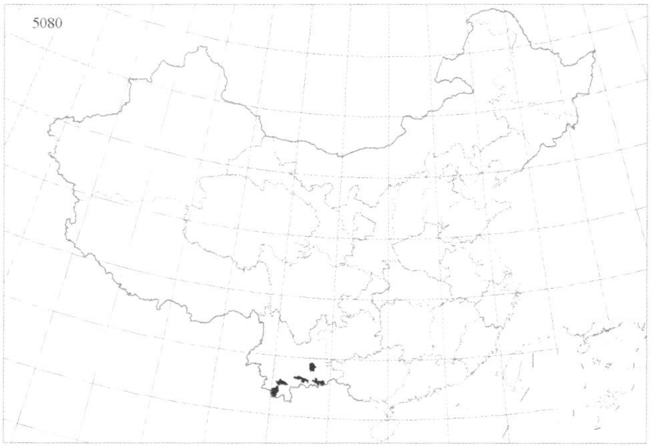

Daphniphyllaceae 交让木科 *Daphniphyllum* 虎皮楠属
Daphniphyllum oldhamii (Hemsl.) Rosenthal 虎皮楠
Trees, evergreen

17.9 (9.3~24.1); 7.8 (-1.7~18.9); 27.0 (19.1~30.1); 17.9 (9.5~24.1); 155.4 (69.8~229.7); 0.4 (0.0~18.0); 945 (603~1322) ❙ 1536 (871~3495); 605 (299~1847); 155 (18~897) ❙ 63.9 (-6.9~246.5); 937 (603~1273); 223.0 (95.3~487.2)

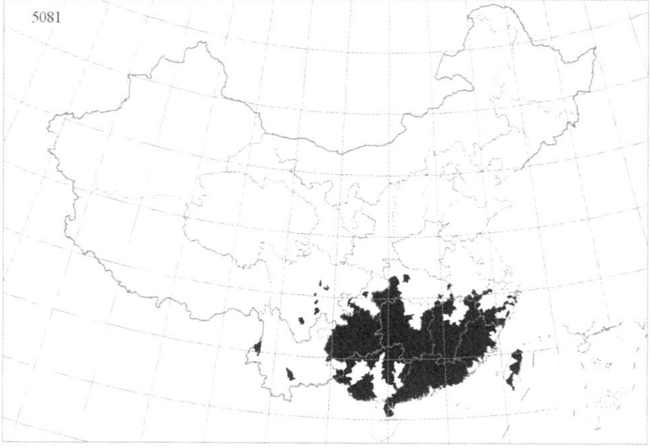

Daphniphyllaceae 交让木科 *Daphniphyllum* 虎皮楠属
Daphniphyllum paxianum K. Rosenthal 显脉虎皮楠
Shrubs or small trees, evergreen

19.6 (3.3~24.8); 11.9 (-6.1~20.1); 25.6 (11.4~28.9); 19.7 (4.6~24.8); 176.6 (22.5~237.6); 0.8 (0.0~42.7); 1031 (434~1385) ❙ 1462 (812~2435); 675 (312~1428); 87 (17~194) ❙ 46.4 (-34.3~145.8); 998 (434~1360); 347.3 (176.4~620.0)

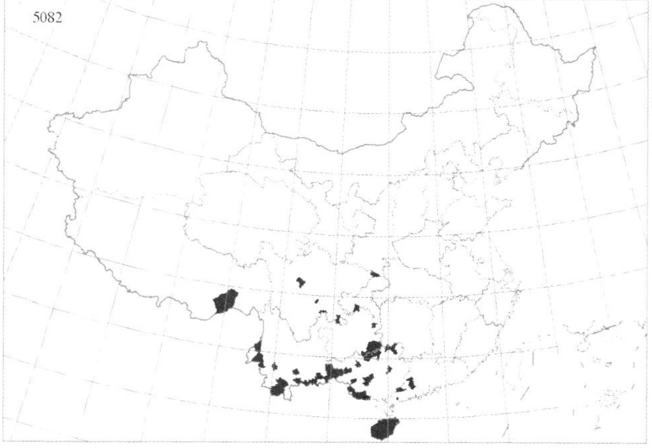

Daphniphyllaceae 交让木科 *Daphniphyllum* 虎皮楠属
Daphniphyllum subverticillatum Merr. 假轮叶虎皮楠
Shrubs, evergreen

20.4 (18.3~21.8); 12.1 (9.8~14.2); 27.4 (25.9~28.5); 20.4 (18.4~21.9); 185.2 (160.5~202.6); 0.0 (0.0~0.0); 1057 (921~1158) ❙ 1802 (1494~2005); 728 (539~891); 164 (116~191) ❙ 71.1 (39.6~95.5); 1057 (921~1158); 261.3 (213.4~290.3)

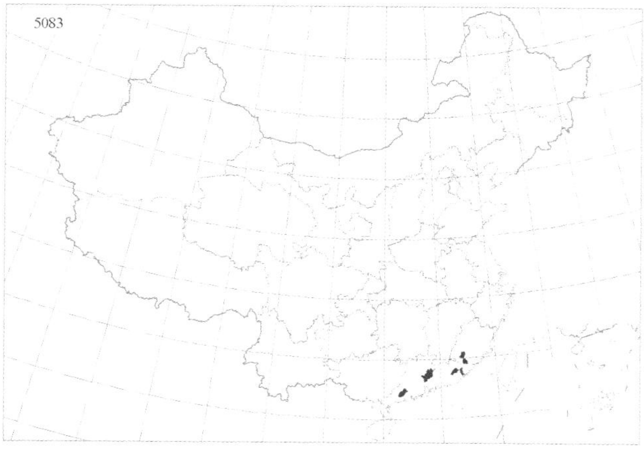

Buxaceae 黄杨科 *Buxus* 黄杨属
Buxus austroyunnanensis Hatus. 滇南黄杨
Small shrubs, evergreen

18.7 (16.9~20.5); 12.8 (10.4~15.1); 22.5 (20.8~23.7); 18.8 (17.0~20.5); 165.3 (143.4~185.9); 0.0 (0.0~0.0); 902 (815~992) ❙ 1335 (946~1607); 683 (548~785); 55 (38~77) ❙ 47.9 (15.5~74.2); 875 (763~986); 456.3 (283.5~659.0)

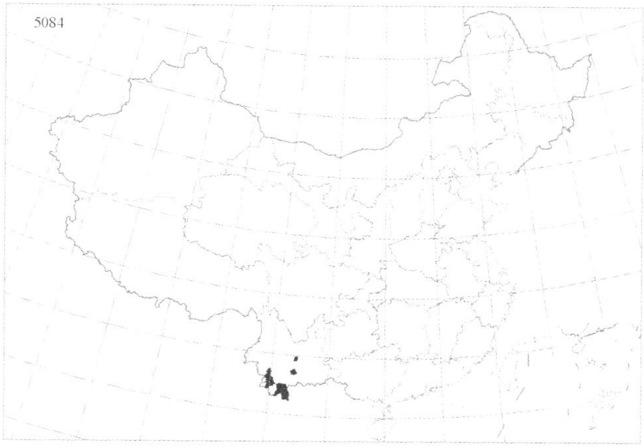

Buxaceae 黄杨科 *Buxus* 黄杨属
Buxus hainanensis Merr. 海南黄杨
Shrubs, evergreen

23.6; 17.5; 27.9; 23.6; 223.2; 0.0; 1288 ❙ 1491; 595; 76 ❙ 16.5; 1153; 473.3

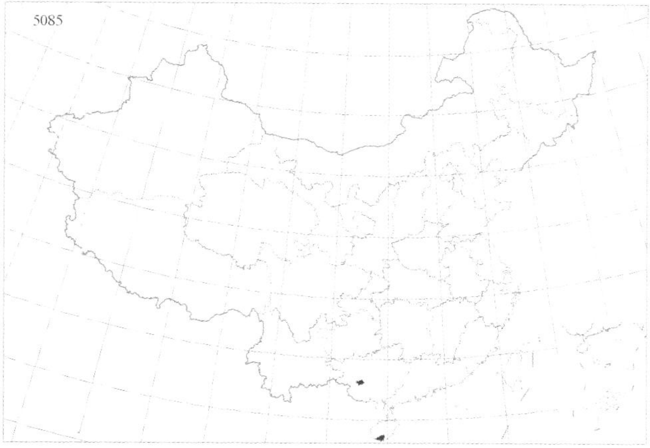

Buxaceae 黄杨科 *Buxus* 黄杨属
Buxus harlandii non Hance: Rehd et Wils. 匙叶黄杨
Small shrubs, evergreen

17.0 (6.2~24.6); 6.8 (-4.2~19.5); 26.2 (13.3~29.8); 17.1 (6.7~24.7); 146.1 (39.4~235.9); 1.3 (0.0~29.4); 905 (509~1372) ❙ 1438 (720~2435); 588 (332~1428); 134 (12~259) ❙ 59.2 (-15.5~147.0); 901 (509~1360); 231.8 (111.9~620.0)

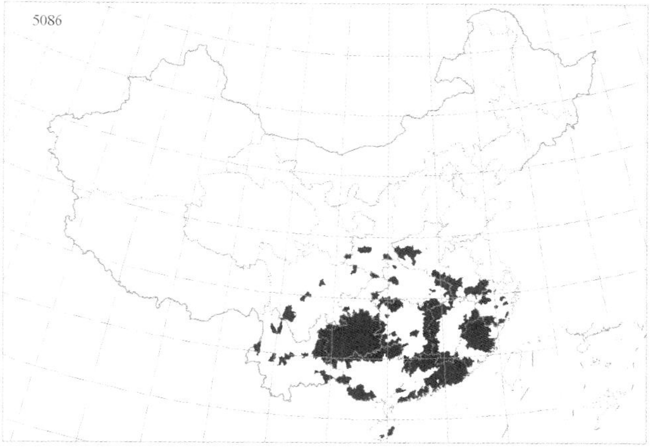

Buxaceae 黄杨科 *Buxus* 黄杨属
Buxus hebecarpa Hatus. 毛果黄杨
Shrubs, evergreen

13.0; 3.3; 21.5; 13.0; 99.7; 3.4; 715 ❙ 1243; 700; 40 ❙ 73.2; 715; 230.4

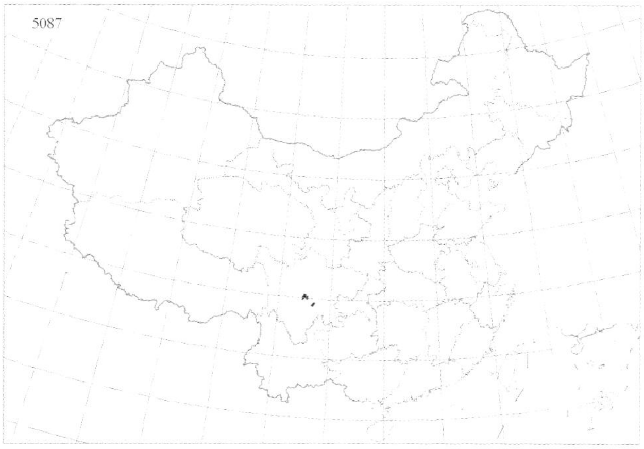

Buxaceae 黄杨科 *Buxus* 黄杨属
Buxus henryi Mayr 大花黄杨
Shrubs, evergreen

15.1 (7.3~20.1); 4.4 (-2.0~10.5); 25.0 (15.6~28.7); 15.1 (7.5~20.1); 124.3 (48.2~181.6); 2.7 (0.0~20.4); 813 (537~1052) ❙ 1287 (807~1624); 542 (349~723); 94 (27~204) ❙ 60.0 (5.6~107.2); 811 (537~1052); 227.5 (150.2~322.1)

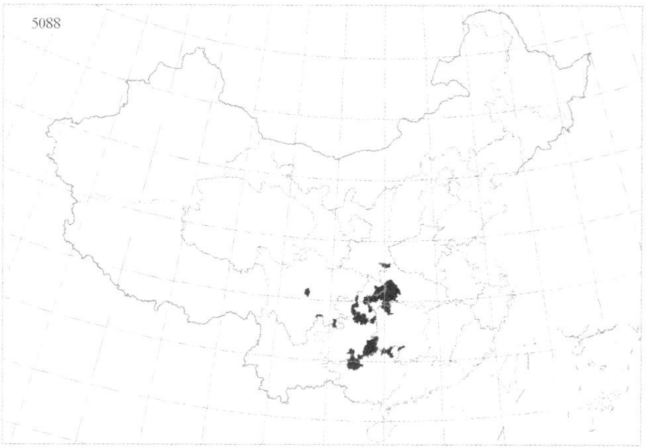

Buxaceae 黄杨科 *Buxus* 黄杨属

Buxus ichangensis Hatus. 宜昌黄杨

Shrubs, evergreen

14.7 (13.0~16.9); 3.3 (1.9~5.5); 25.5 (23.8~28.1); 14.8 (13.1~16.9); 119.9 (102.9~143.4); 2.8 (0.0~6.1); 805 (730~924) ▎1190 (1121~1448); 510 (475~600); 81 (66~105) ▎48.8 (24.4~79.3); 805 (730~924); 219.7 (166.0~262.4)

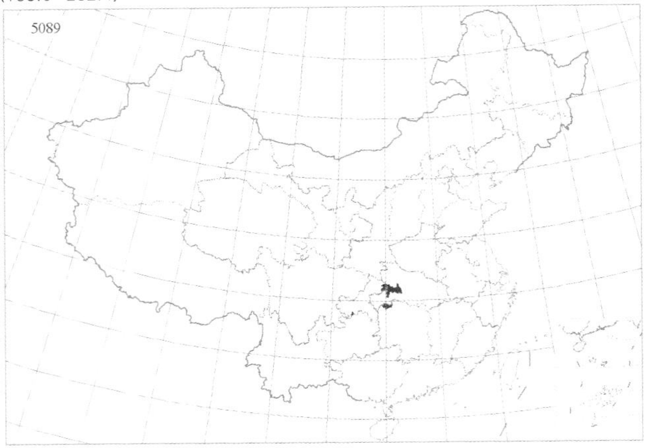

Buxaceae 黄杨科 *Buxus* 黄杨属

Buxus latistyla Gagnep. 阔柱黄杨

Shrubs, evergreen

19.6 (18.1~21.6); 10.6 (8.7~13.1); 26.5 (22.3~28.5); 19.6 (18.0~21.7); 175.1 (156.5~199.8); 0.0 (0.0~0.0); 1005 (863~1153) ▎1394 (1163~1703); 709 (641~809); 100 (48~194) ▎38.6 (27.0~53.1); 1005 (863~1153); 288.4 (213.5~406.7)

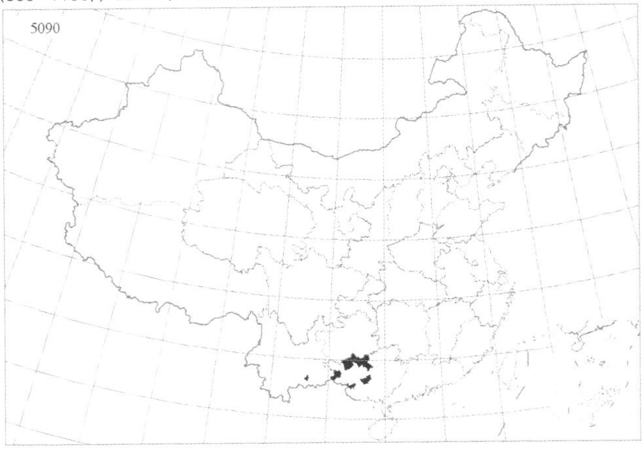

Buxaceae 黄杨科 *Buxus* 黄杨属

Buxus linearifolia M. Cheng 线叶黄杨

Small shrubs, evergreen

22.0; 14.1; 28.0; 22.1; 205.1; 0.0; 1171 ▎2093; 1191; 126 ▎78.5; 1171; 287.9

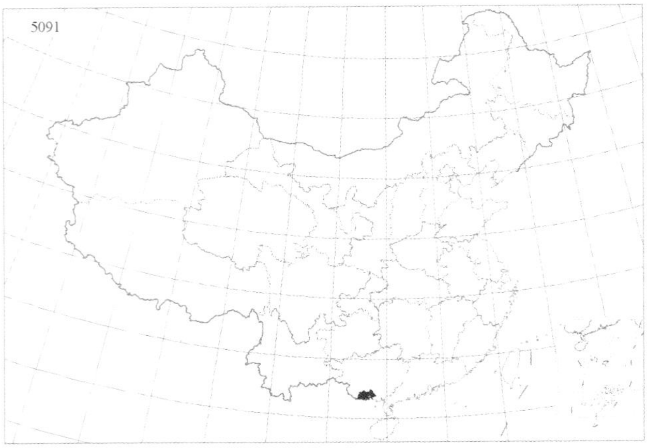

Buxaceae 黄杨科 *Buxus* 黄杨属

Buxus megistophylla H. Lév. 大叶黄杨

Shrubs or small trees, evergreen

19.1 (13.8~23.4); 9.4 (2.4~16.1); 27.4 (23.8~29.8); 19.1 (13.9~23.5); 169.5 (111.2~221.7); 0.1 (0.0~4.4); 1003 (767~1273) ▎1552 (1121~2005); 622 (411~902); 150 (51~204) ▎56.1 (9.0~101.8); 1000 (767~1273); 227.9 (166.9~282.2)

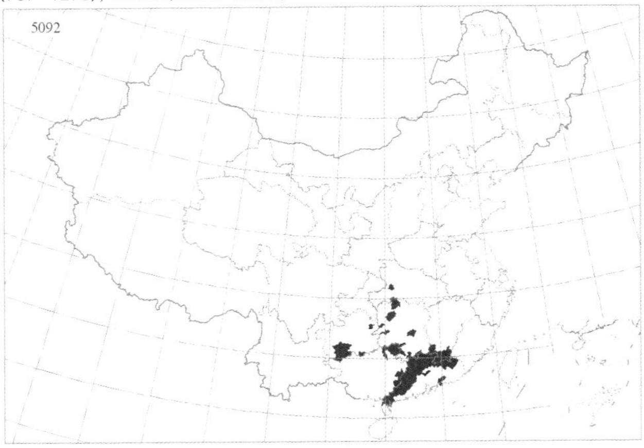

Buxaceae 黄杨科 *Buxus* 黄杨属

Buxus mollicula W. W. Sm. 软毛黄杨

Shrubs, evergreen

11.5 (6.6~15.3); 4.5 (-1.3~8.7); 17.3 (13.3~20.5); 11.6 (6.8~15.4); 83.8 (39.6~124.3); 4.7 (0.0~19.7); 654 (514~761) ▎943 (862~976); 578 (548~603); 23 (12~38) ▎46.2 (28.3~67.7); 634 (514~717); 310.2 (300.3~319.8)

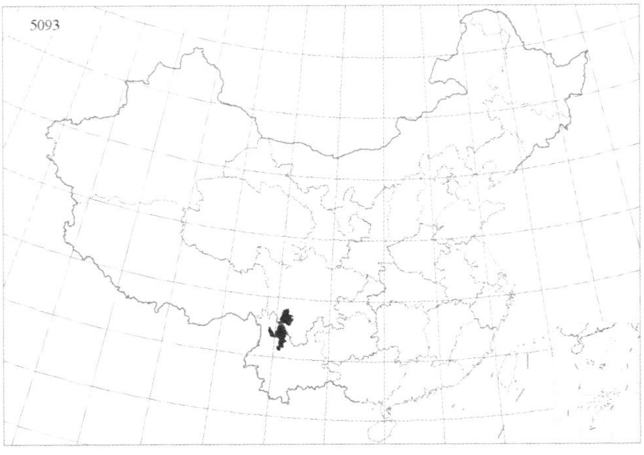

Buxaceae 黄杨科 *Buxus* 黄杨属

Buxus mollicula var. glabra Hand.-Mazz. 变光软毛黄杨

Shrubs, evergreen

8.7; 1.3; 15.0; 8.8; 55.7; 10.5; 574 ▎915; 571; 18 ▎60.3; 574; 315.6

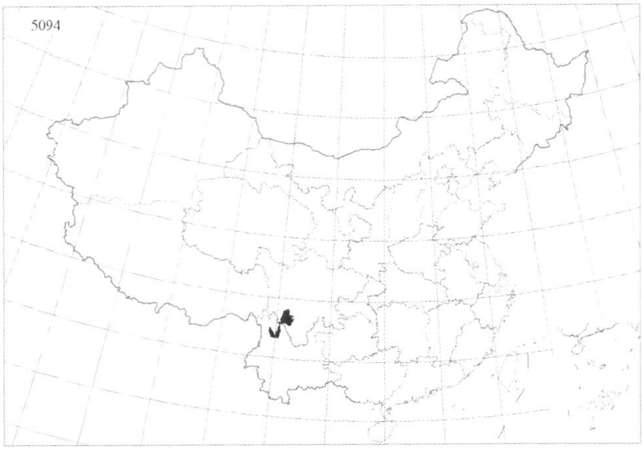

Buxaceae 黄杨科 *Buxus* 黄杨属
Buxus myrica H. Levl. 杨梅黄杨
Shrubs, evergreen

17.5 (8.6~24.8); 7.6 (-1.7~20.1); 26.3 (15.0~29.7); 17.6 (8.7~24.8); 151.5 (52.6~237.6); 0.6 (0.0~18.0); 930 (574~1385) ❙ 1436 (868~2005); 558 (312~891); 135 (31~254) ❙ 57.3 (-34.3~130.7); 909 (574~1360); 245.2 (145.3~620.0)

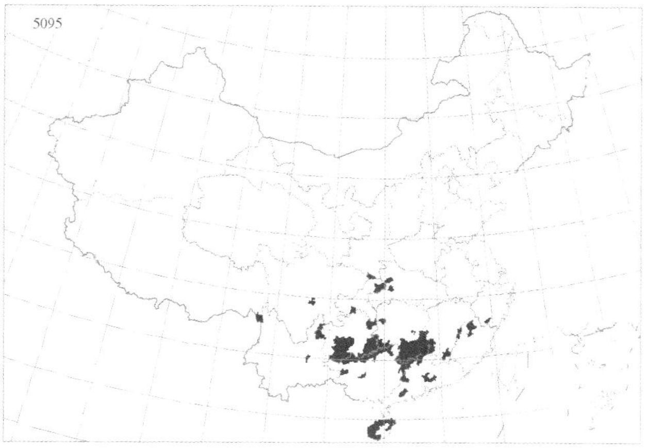

Buxaceae 黄杨科 *Buxus* 黄杨属
Buxus myrica var. angustifolia Gagnep. 狭叶杨梅黄杨
Shrubs, evergreen

19.0 (15.4~22.7); 10.0 (5.4~15.0); 26.3 (23.6~28.6); 19.1 (15.4~22.8); 169.0 (124.8~213.1); 0.0 (0.0~0.0); 989 (790~1221) ❙ 1444 (1194~2435); 750 (497~1428); 86 (51~196) ❙ 46.2 (12.9~99.5); 989 (790~1221); 254.3 (160.8~327.7)

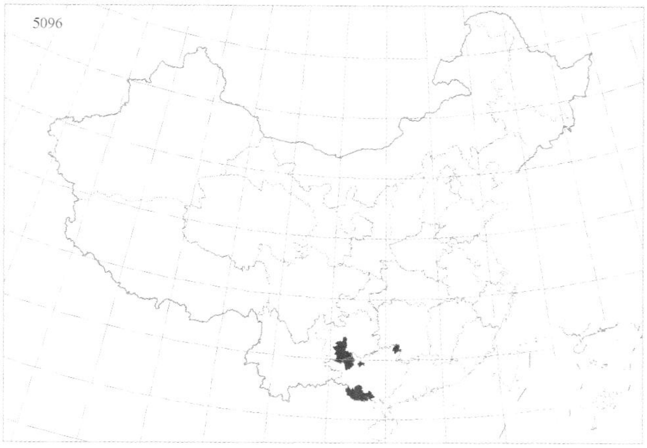

Buxaceae 黄杨科 *Buxus* 黄杨属
Buxus pubiramea Merr. et Chun 毛枝黄杨
Shrubs, evergreen

24.0; 19.[illegible]; 27.2; 24.0; 228.1; 0.0; 1302 ❙ 1551; 539; 48 ❙ 19.1; 1147; 620.0

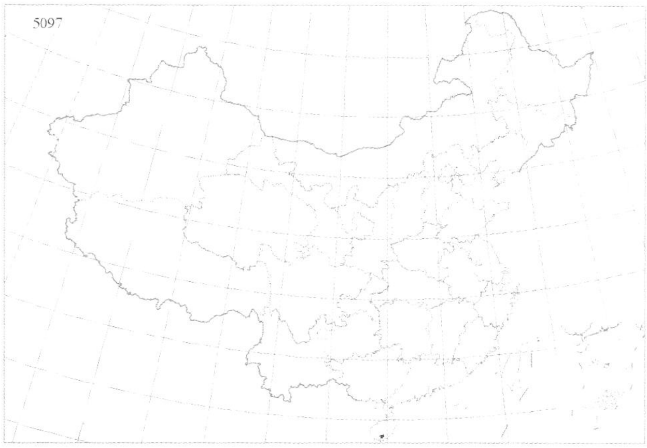

Buxaceae 黄杨科 *Buxus* 黄杨属
Buxus rugulosa Hatus. 皱叶黄杨
Shrubs, evergreen

10.9 (0.6~22.2); 1.3 (-9.4~14.7); 19.6 (9.1~29.3); 11.4 (3.2~22.2); 87.8 (12.4~206.9); 16.7 (0.0~65.4); 680 (371~1177) ❙ 1083 (300~1997); 510 (208~842); 71 (3~259) ❙ 59.3 (-19.2~130.7); 677 (300~1177); 254.1 (176.4~383.4)

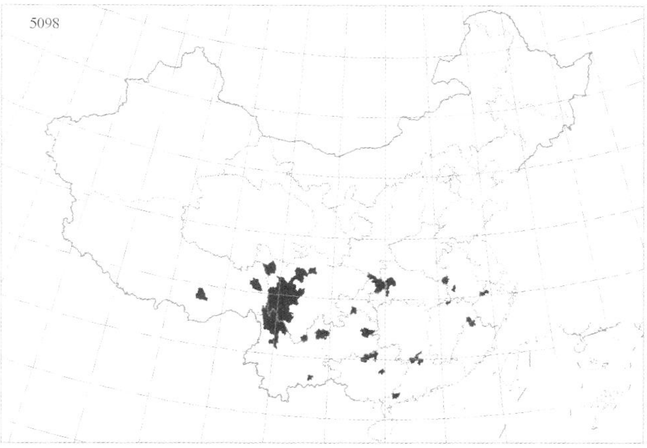

Buxaceae 黄杨科 *Buxus* 黄杨属
Buxus rugulosa var. prostrata (W. W. Sm.) M. Cheng 平卧皱叶黄杨
Shrubs, evergreen

11.0 (5.3~16.0); 3.5 (-2.4~8.0); 17.1 (12.2~21.8); 11.1 (5.7~16.0); 78.7 (30.5~132.2); 6.6 (0.0~26.2); 638 (479~793) ❙ 1124 (801~1736); 659 (496~1007); 34 (15~52) ❙ 76.6 (39.7~145.8); 638 (479~793); 327.4 (234.2~447.4)

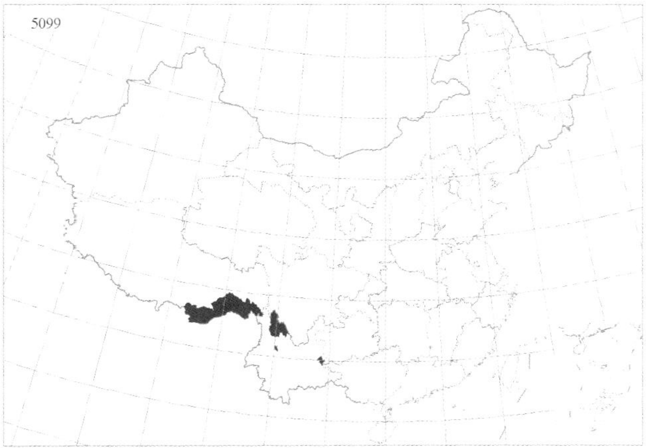

Buxaceae 黄杨科 *Buxus* 黄杨属
Buxus rugulosa var. rupicola (W. W. Sm.) Hatus. 岩生黄杨
Shrubs, evergreen

7.3 (2.0~12.6); -0.5 (-7.2~6.0); 14.1 (10.3~17.9); 7.9 (3.9~12.6); 49.7 (17.3~91.6); 22.1 (0.0~52.8); 540 (410~678) ❙ 811 (510~1057); 484 (311~610); 25 (10~42) ❙ 47.6 (20.7~67.9); 540 (410~678); 292.0 (210.1~351.7)

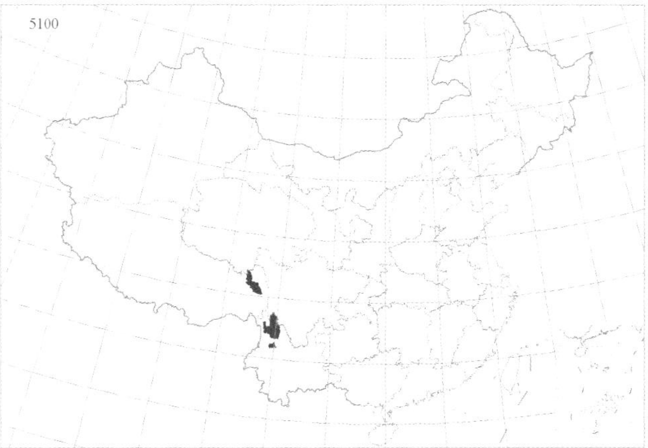

Buxaceae 黄杨科 *Buxus* 黄杨属

Buxus sinica (Rehder et E. H. Wilson) M. Cheng 黄杨

Shrubs or small trees, evergreen

16.9 (2.2~24.1); 6.6 (-7.1~19.1); 26.3 (9.9~29.9); 16.9 (3.8~24.1); 144.9 (15.5~229.7); 1.9 (0.0~49.1); 906 (413~1322) ❙ 1434 (551~2993); 587 (245~1847); 132 (14~551) ❙ 59.0 (-36.2~219.3); 901 (413~1273); 222.8 (95.3~620.0)

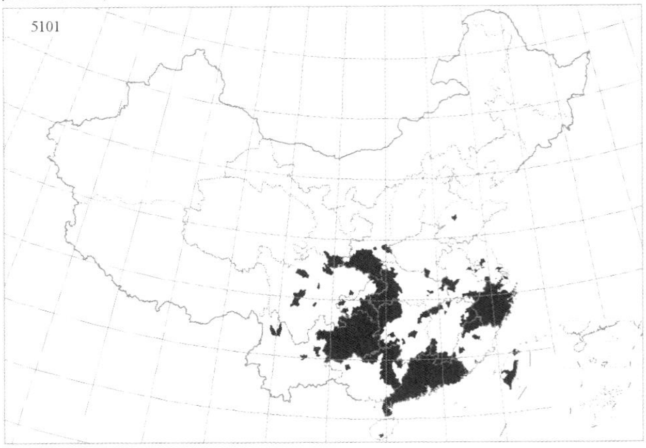

Buxaceae 黄杨科 *Buxus* 黄杨属

Buxus sinica var. aemulans (Rehder et E. H. Wilson) M. Cheng 尖叶黄杨

Shrubs or small trees, evergreen

18.3 (6.2~22.5); 8.3 (-4.1~15.0); 27.2 (15.8~29.8); 18.4 (7.0~22.6); 161.2 (45.2~210.9); 0.7 (0.0~29.9); 966 (522~1206) ❙ 1671 (669~2005); 607 (317~902); 187 (15~259) ❙ 73.8 (28.2~130.7); 964 (522~1206); 245.8 (139.6~303.4)

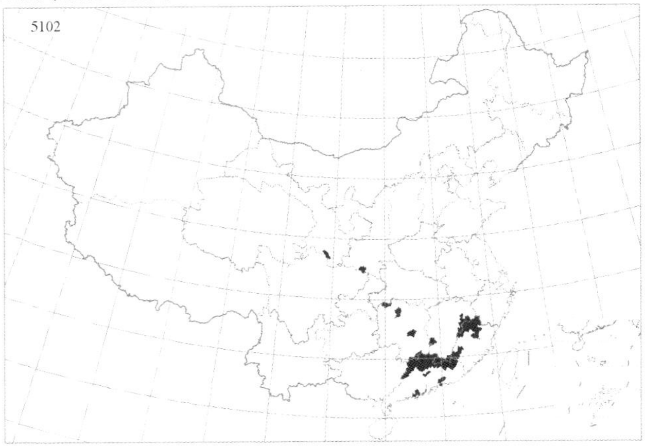

Buxaceae 黄杨科 *Buxus* 黄杨属

Buxus sinica var. intermedia (Kaneh.) M. Cheng 中间黄杨

Shrubs or small trees, evergreen

15.1; 3.6; 26.1; 15.2; 123.5; 1.5; 820 ❙ 1392; 573; 109 ❙ 70.0; 820; 218.2

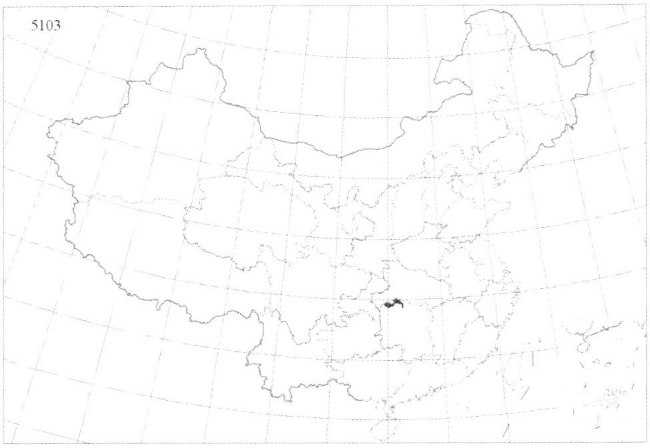

Buxaceae 黄杨科 *Buxus* 黄杨属

Buxus sinica var. parvifolia M. Cheng 小叶黄杨

Shrubs or small trees, evergreen

14.9 (13.4~16.6); 2.8 (1.5~6.0); 26.6 (25.2~27.5); 14.9 (13.4~16.6); 122.9 (108.3~138.9); 3.6 (0.0~7.3); 822 (755~866) ❙ 1572 (1317~1997); 603 (552~669); 181 (122~254) ❙ 91.3 (53.3~130.7); 822 (755~866); 259.6 (183.4~319.9)

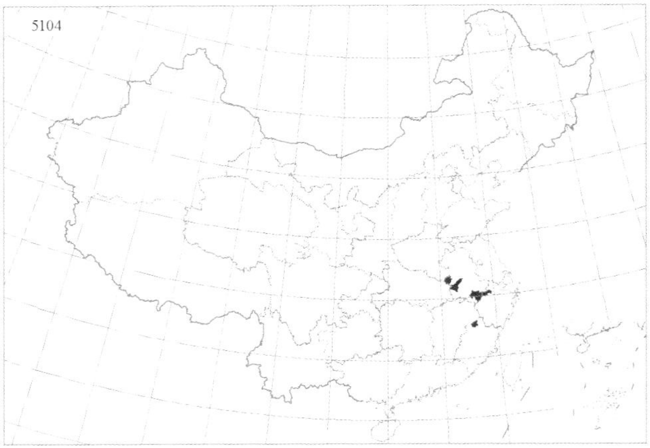

Buxaceae 黄杨科 *Buxus* 黄杨属

Buxus sinica var. vacciniifolia M. Cheng 越橘叶黄杨

Shrubs or small trees, evergreen

17.7; 7.3; 27.1; 17.7; 153.0; 0.0; 922 ❙ 1525; 530; 179 ❙ 65.5; 922; 210.6

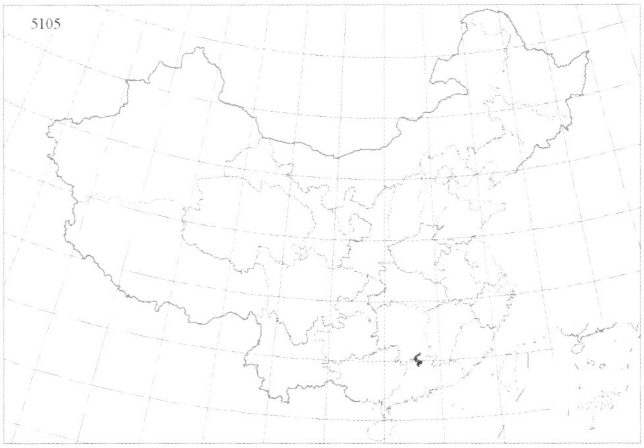

Buxaceae 黄杨科 *Buxus* 黄杨属

Buxus stenophylla Hance 狭叶黄杨

Shrubs, evergreen

16.9 (14.2~22.3); 6.5 (3.6~14.3); 26.5 (23.5~30.1); 16.9 (14.2~22.4); 143.4 (112.5~208.7); 0.2 (0.0~1.7); 892 (756~1197) ❙ 1408 (931~1922); 534 (323~878); 136 (64~259) ❙ 57.7 (-1.6~111.3); 884 (756~1197); 211.9 (146.0~285.2)

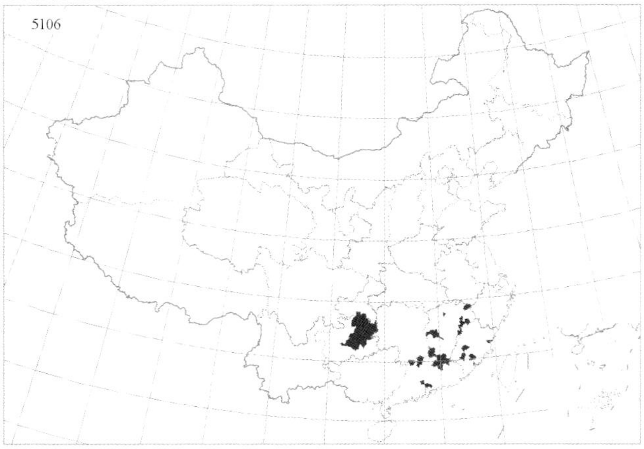

Buxaceae 黄杨科 *Pachysandra* 板凳果属
Pachysandra axillaris Franch. 板凳果
Subshrubs, evergreen

15.9 (5.4~21.0); 6.4 (−2.1~15.3); 24.3 (12.3~30.1); 16.0 (5.8~21.1); 132.3 (31.5~193.0); 0.9 (0.0~25.6); 837 (481~1114) ❙ 1303 (741~2445); 574 (403~1412); 101 (15~259) ❙ 55.4 (−14.2~191.8); 828 (481~1114); 236.4 (128.7~395.4)

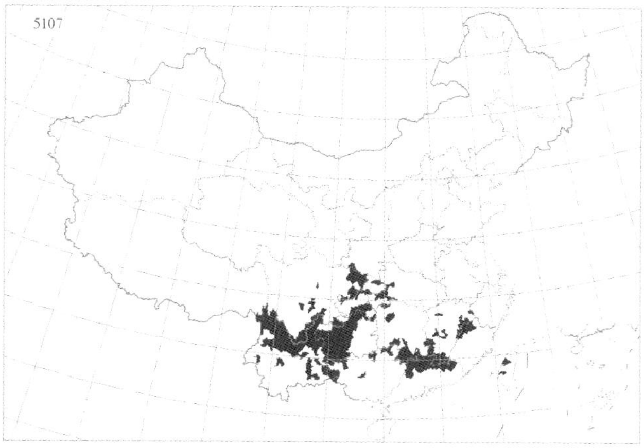

Buxaceae 黄杨科 *Pachysandra* 板凳果属
Pachysandra terminalis Siebold et Zucc. 顶花板凳果
Subshrubs, evergreen

14.6 (3.0~21.0); 3.0 (−7.7~11.9); 25.4 (12.4~29.3); 14.7 (4.8~21.1); 120.9 (25.1~193.0); 5.7 (0.0~49.3); 809 (441~1114) ❙ 1218 (373~1663); 500 (201~821); 106 (7~220) ❙ 50.0 (−37.6~107.2); 803 (373~1114); 227.0 (88.1~334.4)

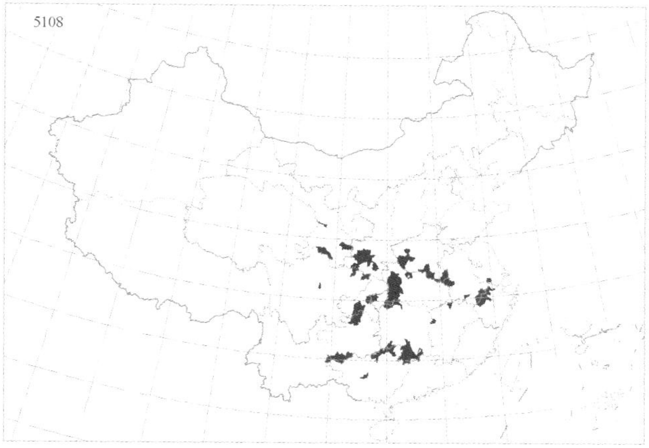

Buxaceae 黄杨科 *Sarcococca* 野扇花属
Sarcococca hookeriana Baill. 羽脉野扇花
Shrubs or small trees, evergreen

12.6 (−1.8~21.0); 3.4 (−11.4~12.0); 20.7 (6.1~29.0); 12.9 (1.8~21.1); 101.2 (1.9~193.0); 9.4 (0.0~82.8); 717 (329~1114) ❙ 1068 (320~1662); 522 (193~821); 63 (8~211) ❙ 49.0 (−14.2~121.7); 707 (320~1114); 246.3 (72.2~434.4)

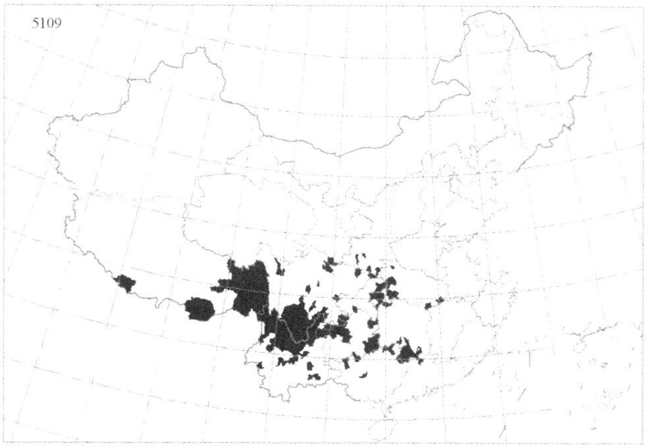

Buxaceae 黄杨科 *Sarcococca* 野扇花属
Sarcococca longifolia M. Cheng et K. F. Wu 长叶野扇花
Shrubs, evergreen

21.2; 12.8; 27.8; 21.3; 195.4; 0.0; 1115 ❙ 1621; 720; 146 ❙ 45.4; 1115; 243.1

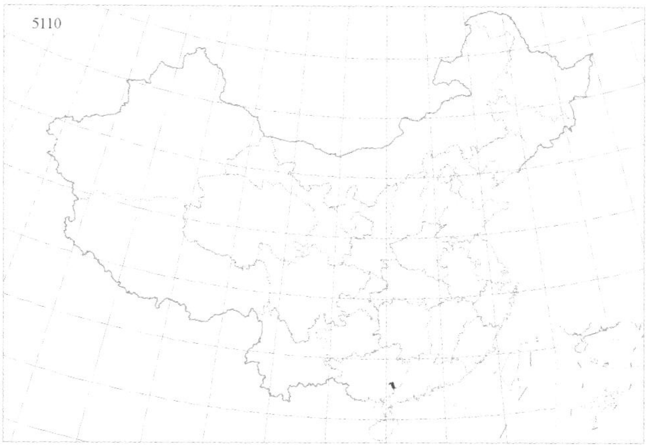

Buxaceae 黄杨科 *Sarcococca* 野扇花属
Sarcococca longipetiolata M. Cheng 长叶柄野扇花
Shrubs, evergreen

17.7 (12.0~23.6); 6.9 (0.6~16.7); 27.6 (23.1~29.8); 17.7 (12.1~23.6); 153.0 (94.7~223.5); 0.3 (0.0~9.7); 939 (697~1281) ❙ 1580 (931~1997); 556 (323~778); 188 (73~259) ❙ 69.6 (−1.6~130.7); 929 (697~1273); 232.2 (111.9~322.1)

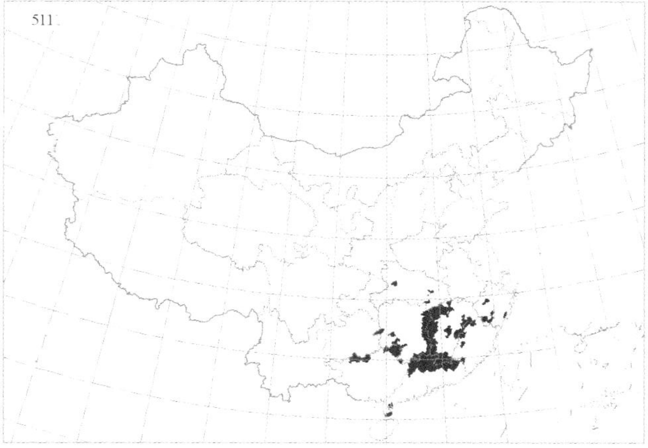

Buxaceae 黄杨科 *Sarcococca* 野扇花属
Sarcococca ruscifolia Stapf 野扇花
Shrubs, evergreen

15.0 (2.2~21.0); 4.7 (−6.6~15.2); 24.3 (9.9~29.9); 15.1 (3.8~22.7); 123.4 (15.5~212.7); 2.9 (0.0~49.1); 808 (413~1216) ❙ 1184 (622~1965); 527 (298~994); 88 (12~228) ❙ 46.9 (−22.2~128.9); 794 (413~1216); 232.8 (133.8~551.6)

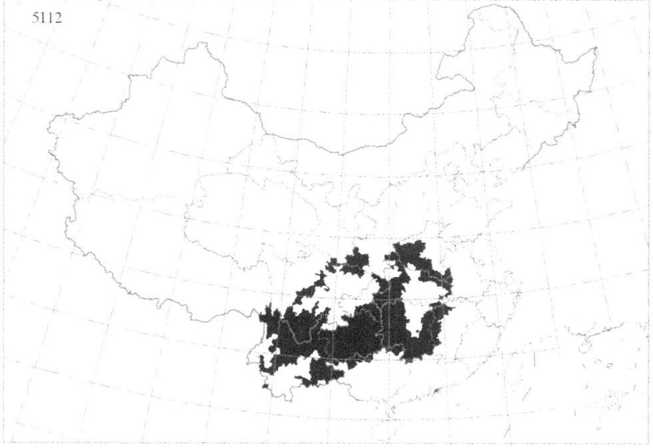

Buxaceae 黄杨科 *Sarcococca* 野扇花属
Sarcococca saligna (D. Don) Müll. Arg. 柳叶野扇花
Shrubs, evergreen

-1.8; -10.2; 6.1; 1.8; 1.9; 82.8; 329 ❙ 320; 217; 20 ❙ -3.0; 320; 72.2

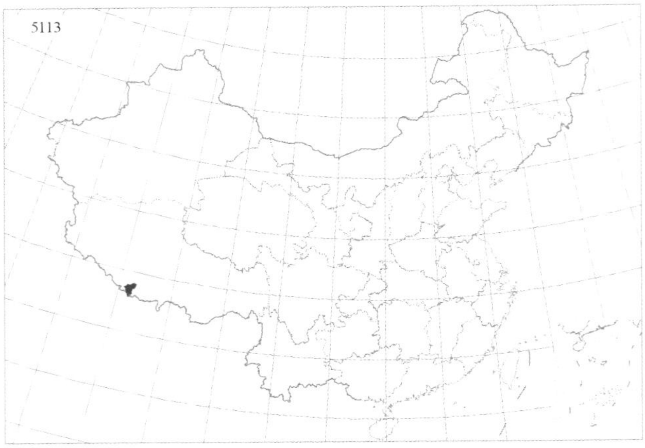

Buxaceae 黄杨科 *Sarcococca* 野扇花属
Sarcococca vagans Stapf 海南野扇花
Shrubs, evergreen

22.8 (17.1~24.8); 16.6 (8.8~20.1); 27.1 (20.8~28.8); 22.8 (17.1~24.8); 213.6 (145.0~237.6); 0.0 (0.0~0.0); 1217 (815~1385) ❙ 1497 (868~1854); 587 (312~771); 78 (41~137) ❙ 25.3 (-34.3~65.9); 1131 (802~1360); 430.2 (238.3~659.0)

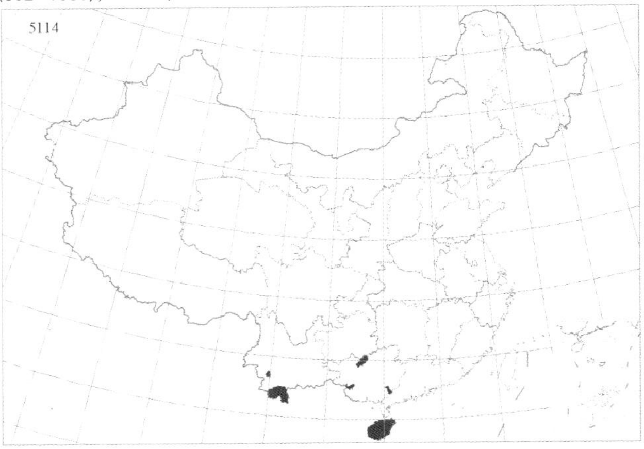

C

Buxaceae 黄杨科 *Sarcococca* 野扇花属
Sarcococca wallichii Stapf 云南野扇花
Shrubs, evergreen

11.4 (-1.8~17.4); 4.4 (-10.2~11.0); 16.9 (6.1~21.6); 12.0 (1.8~17.4); 91.5 (1.9~149.3); 14.0 (0.0~82.8); 657 (329~834) ❙ 1011 (320~1454); 572 (193~816); 46 (20~70) ❙ 49.8 (-4.0~99.9); 651 (320~834); 330.3 (72.2~487.2)

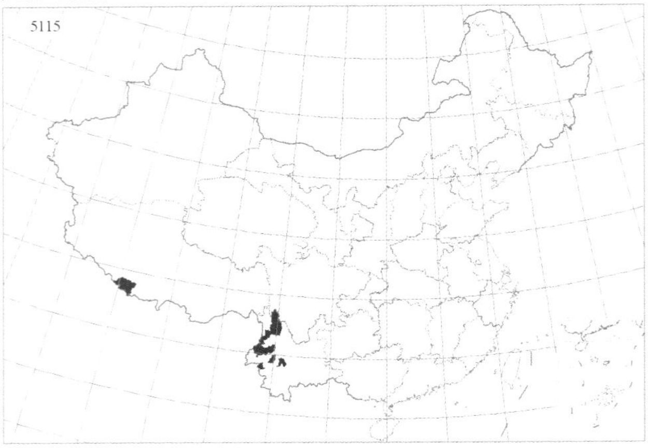

Coriariaceae 马桑科 *Coriaria* 马桑属
Coriaria intermedia Matsum. 台湾马桑
Shrubs, deciduous

19.4 (16.2~24.1); 14.1 (11.0~18.9); 23.7 (20.1~28.0); 19.5 (16.2~24.1); 173.5 (134.2~229.7); 0.0 (0.0~0.0); 960 (773~1322) ❙ 2466 (1751~2992); 1212 (750~1847); 203 (47~402) ❙ 167.2 (32.5~246.5); 935 (773~1123); 261.6 (154.2~363.7)

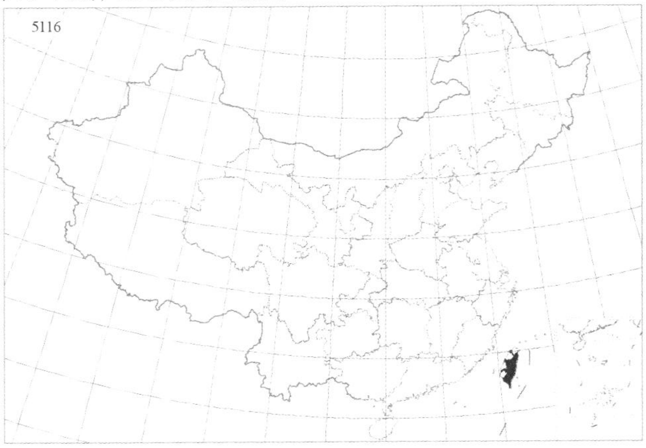

Coriariaceae 马桑科 *Coriaria* 马桑属
Coriaria nepalensis Wall. 马桑
Shrubs, deciduous

13.9 (-1.1~22.7); 3.3 (-9.1~15.0); 23.5 (6.2~29.4); 14.1 (2.0~22.8); 114.2 (2.5~213.1); 6.6 (0.0~75.6); 776 (334~1221) ❙ 1080 (321~2435); 513 (193~1428); 66 (7~195) ❙ 39.9 (-37.6~145.8); 758 (321~1221); 235.9 (79.3~551.6)

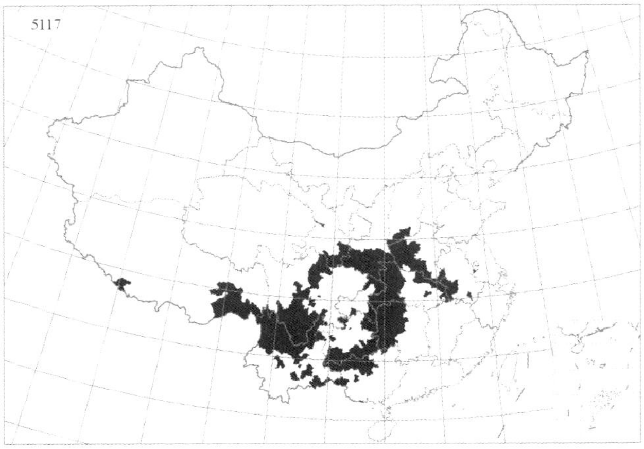

Coriariaceae 马桑科 *Coriaria* 马桑属
Coriaria terminalis Hayata 草马桑
Subshrubs, deciduous

5.6 (-0.7~13.3); -3.1 (-9.6~5.4); 13.3 (7.7~19.6); 6.6 (2.5~13.4); 39.7 (6.8~100.4); 32.6 (0.0~74.2); 498 (338~706) ❙ 913 (370~1736); 515 (246~1007); 26 (10~51) ❙ 76.4 (0.0~145.8); 498 (338~706); 239.8 (61.6~447.4)

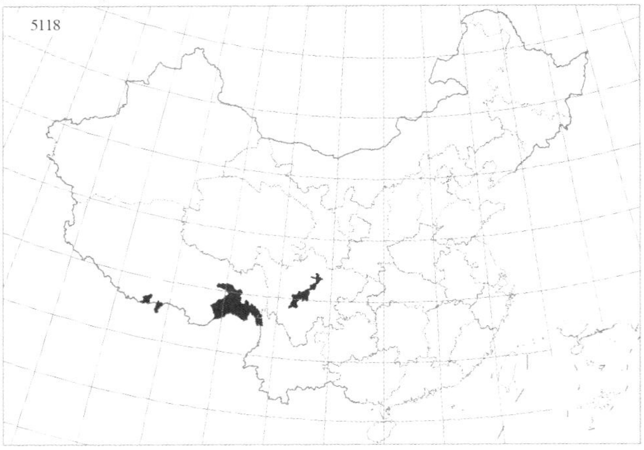

Anacardiaceae 漆树科 *Buchanania* 山様子属
Buchanania arborescens (Blume) Blume 山様子
Trees, evergreen

21.1 (16.2~24.1); 16.0 (11.0~18.9); 24.9 (20.1~28.2); 21.1 (16.2~24.1); 193.7 (134.2~229.7); 0.0 (0.0~0.0); 1085 (773~1322) ❙ 2388 (1751~2992); 1393 (1168~1847); 117 (47~245) ❙ 136.7 (32.5~246.5); 1004 (773~1123); 222.4 (119.4~296.9)

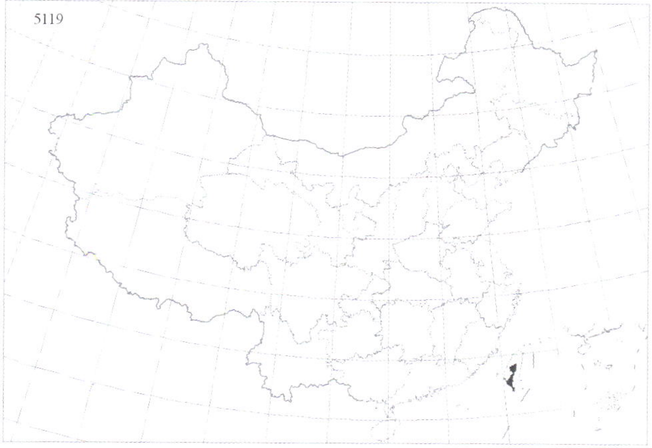

Anacardiaceae 漆树科 *Buchanania* 山様子属
Buchanania latifolia Roxb. 豆腐果
Trees, deciduous

21.1 (16.7~24.8); 15.2 (10.3~20.1); 25.1 (21.3~28.7); 21.1 (16.8~24.8); 193.4 (141.5~237.6); 0.0 (0.0~0.0); 1088 (811~1385) ❙ 1341 (868~1854); 608 (312~996); 58 (37~115) ❙ 26.1 (-34.3~88.8); 1002 (744~1360); 433.3 (288.3~620.0)

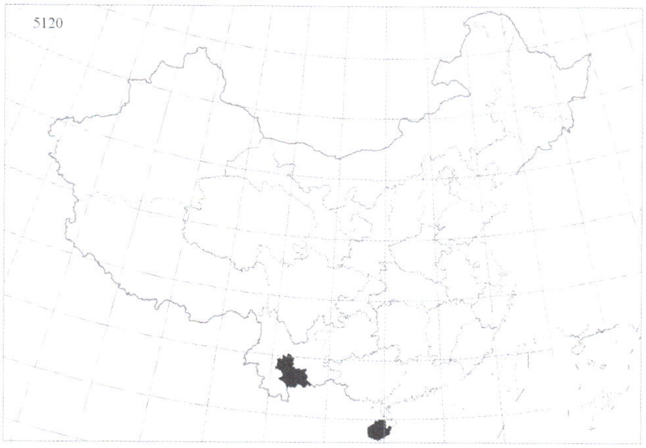

Anacardiaceae 漆树科 *Buchanania* 山様子属
Buchanania microphylla Engl. 小叶山様子
Trees

23.3 (19.2~24.8); 17.2 (10.7~20.1); 27.6 (25.4~28.7); 23.3 (19.2~24.8); 219.6 (170.8~237.6); 0.0 (0.0~0.0); 1256 (961~1385) ❙ 1430 (868~1854); 551 (312~850); 66 (41~115) ❙ 14.5 (-34.3~47.8); 1138 (868~1360); 451.2 (281.7~620.0)

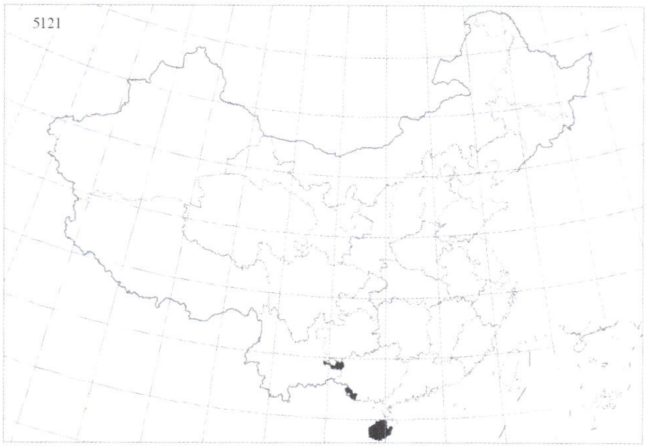

Anacardiaceae 漆树科 *Buchanania* 山様子属
Buchanania yunnanensis C. Y. Wu 云南山様子
Trees, deciduous

20.5; 14.9; 23.7; 20.5; 185.9; 0.0; 992 ❙ 1464; 655; 66 ❙ 47.6; 959; 570.1

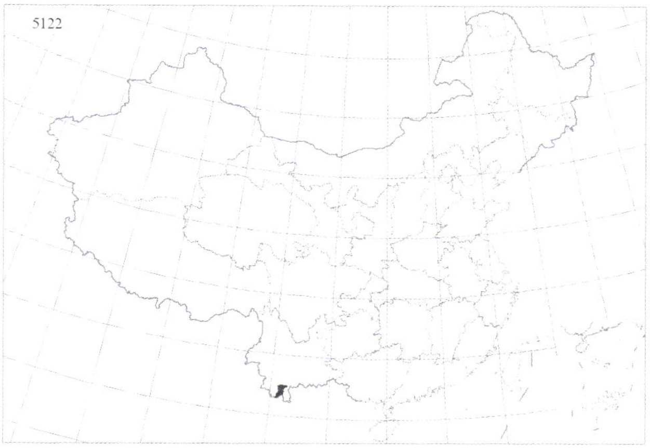

Anacardiaceae 漆树科 *Choerospondias* 南酸枣属
Choerospondias axillaris (Roxb.) B. L. Burtt et A. W. Hill 南酸枣
Trees, deciduous

17.9 (5.7~24.7); 8.1 (-4.1~19.8); 26.6 (14.2~29.9); 18.0 (6.4~24.7); 156.0 (38.1~236.5); 0.6 (0.0~29.9); 945 (504~1376) ❙ 1451 (669~2435); 598 (312~1428); 134 (14~259) ❙ 55.0 (-34.3~147.0); 935 (504~1299); 237.4 (95.3~659.0)

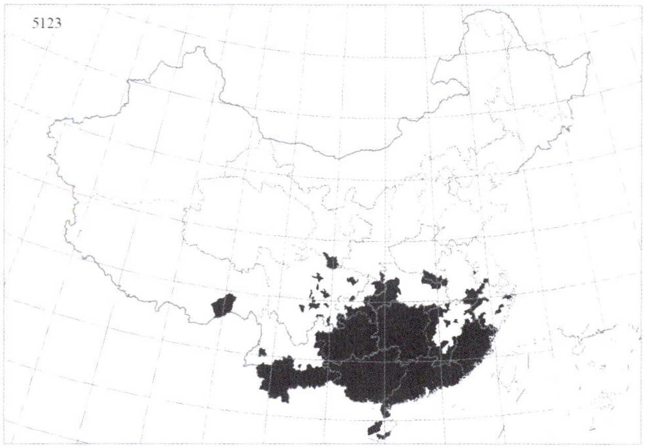

Anacardiaceae 漆树科 *Choerospondias* 南酸枣属
Choerospondias axillaris var. pubinervis (Rehder et E. H. Wilson) B. L. Burtt et A. W. Hill 毛脉南酸枣
Trees, deciduous

16.3 (10.3~21.3); 4.8 (-0.1~12.6); 27.1 (19.8~29.9); 16.3 (10.3~21.3); 136.9 (75.8~196.0); 0.7 (0.0~11.7); 877 (634~1123) ❙ 1385 (711~1742); 495 (357~817); 151 (15~225) ❙ 58.2 (5.8~128.9); 863 (634~1121); 196.6 (126.1~315.6)

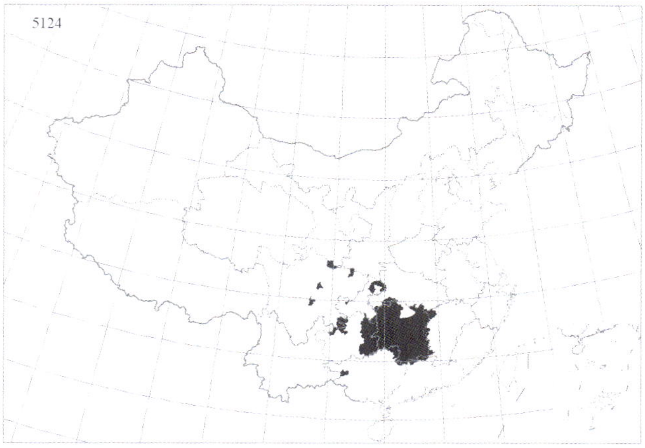

Anacardiaceae 漆树科 *Cotinus* 黄栌属
Cotinus coggygria Scopoli 黄栌
Shrubs, deciduous

13.8 (5.4~19.1); 3.3 (−11.0~12.0); 23.2 (12.3~28.7); 14.1 (5.8~19.1); 113.5 (31.5~169.2); 6.9 (0.0~51.8); 764 (481~969) ❙ 1004 (436~1581); 504 (258~720); 54 (8~205) ❙ 31.2 (−41.2~107.2); 727 (436~969); 224.1 (117.3~384.1)

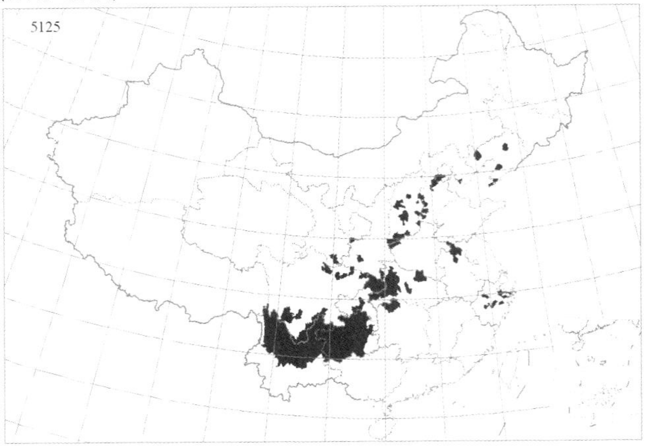

Anacardiaceae 漆树科 *Cotinus* 黄栌属
Cotinus coggygria var. cinerea Engl. 灰毛黄栌
Shrubs, deciduous

12.1 (4.2~17.1); −2.2 (−12.5~4.0); 25.1 (18.7~29.4); 12.6 (7.3~17.1); 104.9 (52.8~146.1); 18.6 (1.0~62.6); 759 (535~952) ❙ 713 (377~1235); 397 (241~560); 34 (8~123) ❙ −5.7 (−37.2~101.4); 668 (377~921); 174.9 (109.4~311.0)

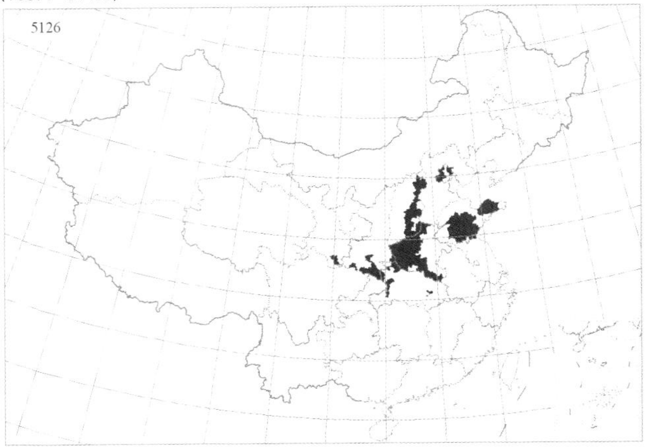

Anacardiaceae 漆树科 *Cotinus* 黄栌属
Cotinus coggygria var. glaucophylla C. Y. Wu 粉背黄栌
Shrubs, deciduous

11.3 (2.5~16.9); −2.8 (−15.9~6.7); 24.0 (12.4~28.4); 12.0 (4.8~16.9); 97.9 (25.1~143.4); 21.7 (0.0~80.4); 726 (441~924) ❙ 724 (349~1631); 390 (227~638); 32 (8~213) ❙ −0.2 (−43.4~107.2); 634 (349~924); 187.5 (70.3~334.4)

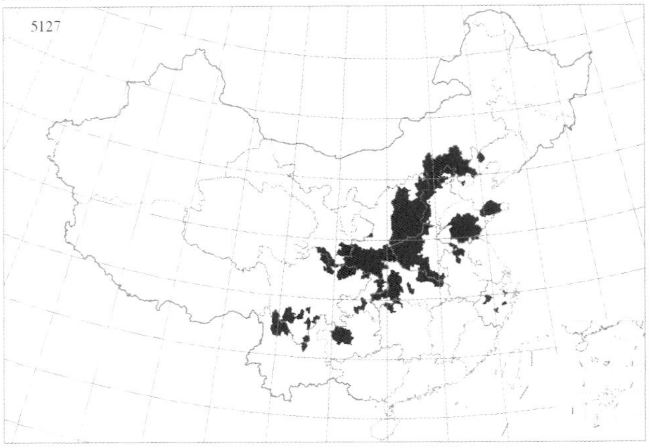

Anacardiaceae 漆树科 *Cotinus* 黄栌属
Cotinus nana W. W. Sm. 矮黄栌
Dwarf shrubs, deciduous

9.8 (6.1~12.8); 1.6 (−1.5~3.9); 16.9 (12.7~21.1); 9.9 (6.3~12.9); 67.7 (35.2~96.0); 9.2 (1.3~21.8); 608 (496~699) ❙ 879 (801~968); 518 (472~594); 25 (12~36) ❙ 46.8 (25.7~67.7); 608 (496~699); 284.1 (229.2~319.8)

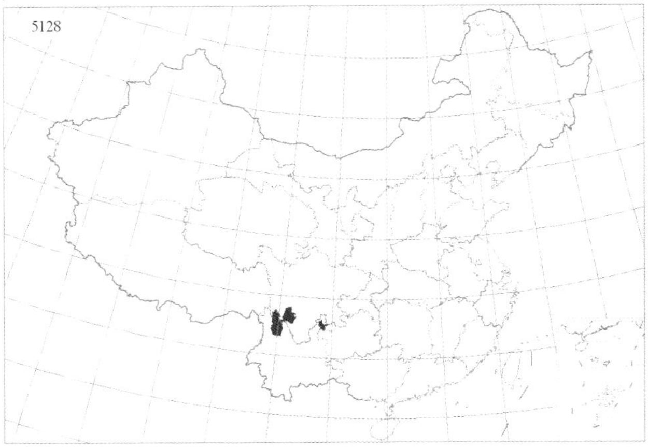

Anacardiaceae 漆树科 *Cotinus* 黄栌属
Cotinus szechuanensis Pénzes 四川黄栌
Shrubs, deciduous

6.6 (2.9~14.8); −3.0 (−6.3~3.8); 15.4 (11.2~25.3); 7.4 (4.4~14.8); 48.1 (20.5~119.2); 28.5 (1.4~44.6); 536 (428~795) ❙ 1021 (853~1358); 491 (425~543); 38 (22~88) ❙ 93.9 (70.8~121.7); 536 (428~795); 241.8 (187.9~290.3)

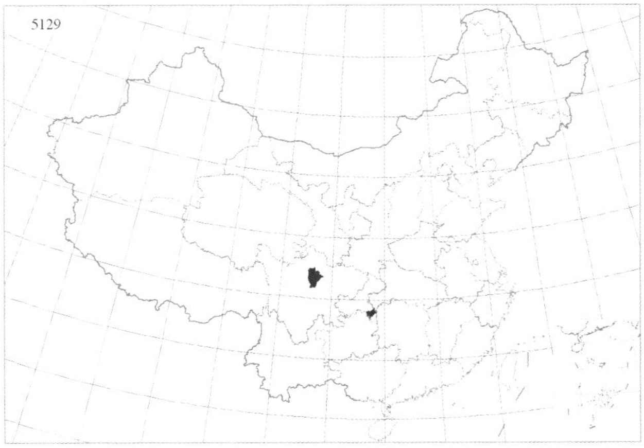

Anacardiaceae 漆树科 *Dobinea* 九子母属
Dobinea delavayi (Baill.) Baill. 羊角天麻
Herbs or subshrubs

14.1 (2.2~18.9); 7.1 (−6.3~12.0); 19.6 (9.5~24.3); 14.2 (3.9~19.0); 111.8 (15.8~167.8); 2.2 (0.0~48.6); 732 (411~929) ❙ 964 (729~1058); 572 (411~641); 30 (11~54) ❙ 34.0 (−14.2~77.4); 693 (411~799); 295.2 (207.5~382.6)

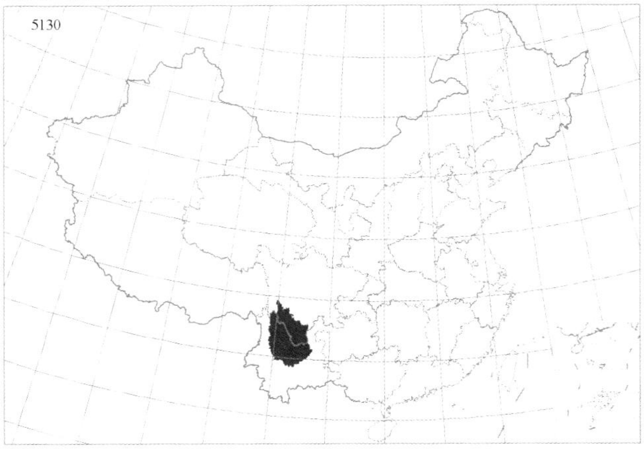

Anacardiaceae 漆树科 *Dobinea* 九子母属
Dobinea vulgaris Buch.-Ham. ex D. Don 九子母
Shrubs

7.2 (3.3~13.3); -0.6 (-5.4~5.4); 14.0 (11.1~19.6); 7.6 (4.6~13.4); 47.5 (22.3~100.4); 20.5 (0.0~41.8); 533 (427~706) | 990 (473~1736); 573 (287~1007); 32 (7~51) | 77.4 (5.0~145.8); 533 (427~706); 307.6 (230.7~447.4)

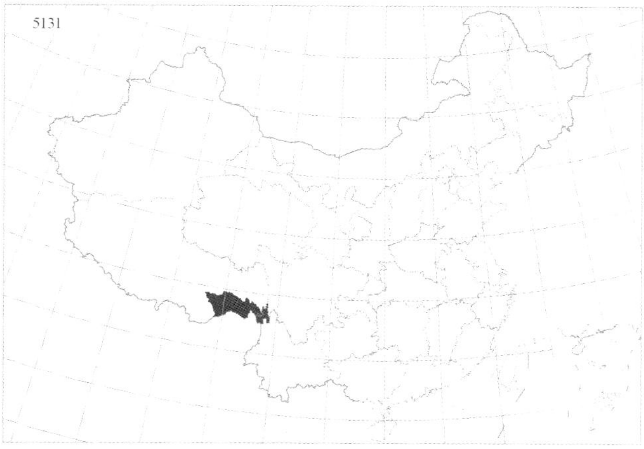

Anacardiaceae 漆树科 *Dracontomelon* 人面子属
Dracontomelon duperreanum Pierre 人面子
Trees, evergreen

21.5 (16.9~24.8); 14.5 (9.1~20.1); 26.7 (21.4~29.0); 21.6 (16.9~24.8); 199.0 (142.7~237.6); 0.0 (0.0~0.0); 1131 (813~1385) | 1502 (868~2037); 679 (312~1056); 89 (41~175) | 35.2 (-34.3~88.8); 1090 (813~1360); 368.4 (163.7~659.0)

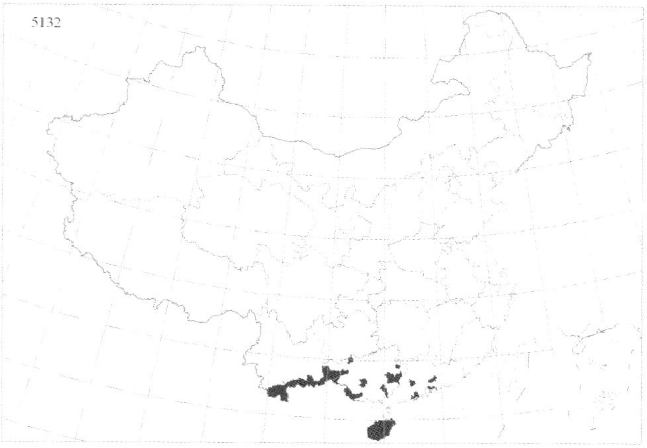

Anacardiaceae 漆树科 *Dracontomelon* 人面子属
Dracontomelon macrocarpum H. L. Li 大果人面子
Trees

20.4; 15.1; 23.6; 20.5; 185.8; 0.0; 989 | 1607; 702; 77 | 62.5; 986; 659.0

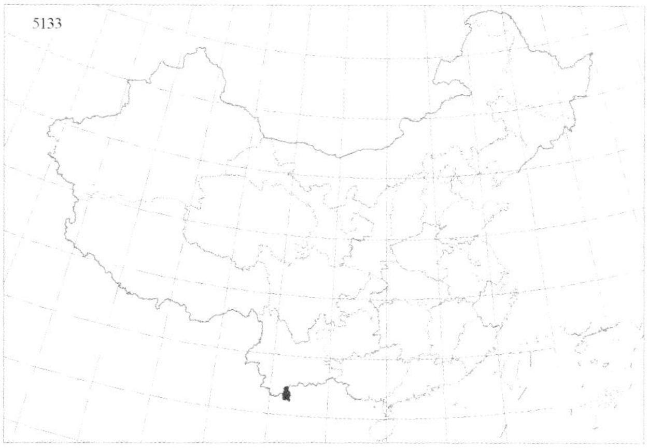

Anacardiaceae 漆树科 *Drimycarpus* 辛果漆属
Drimycarpus anacardiifolius C. Y. Wu et T. L. Ming 大果辛果漆
Trees

17.9; 11.9; 21.6; 18.0; 155.8; 0.0; 854 | 1435; 803; 42 | 68.0; 833; 505.1

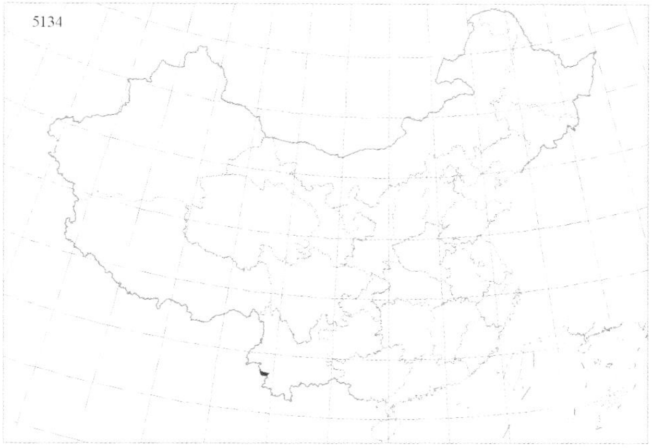

Anacardiaceae 漆树科 *Drimycarpus* 辛果漆属
Drimycarpus racemosus (Roxb.) Hook. f. 辛果漆
Trees

19.0 (17.2~21.8); 12.2 (9.1~15.1); 23.6 (22.3~27.7); 19.0 (17.3~21.9); 168.0 (147.0~202.2); 0.0 (0.0~0.0); 926 (833~1150) | 1461 (1161~2148); 796 (633~1234); 62 (48~124) | 57.6 (34.8~88.8); 921 (833~1150); 451.4 (314.4~659.0)

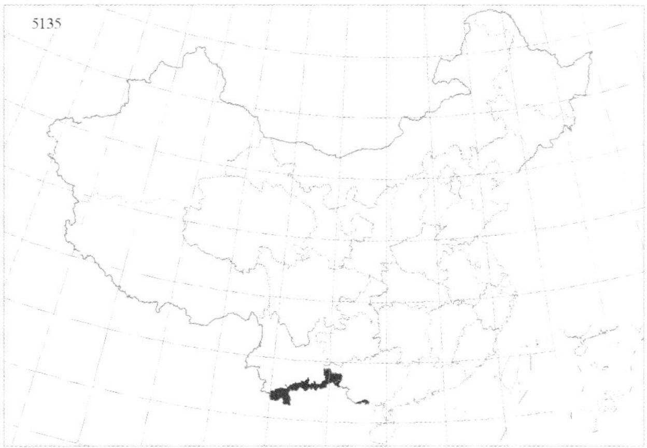

Anacardiaceae 漆树科 *Lannea* 厚皮树属
Lannea coromandelica (Houtt.) Merr. 厚皮树
Trees, deciduous

19.2 (8.6~24.8); 12.5 (1.3~20.1); 23.8 (15.0~29.2); 19.2 (8.7~24.8); 170.8 (52.6~237.6); 0.1 (0.0~8.8); 972 (574~1385) | 1384 (868~2435); 715 (312~1428); 65 (37~173) | 45.3 (-34.3~112.2); 941 (574~1299); 387.0 (111.7~659.0)

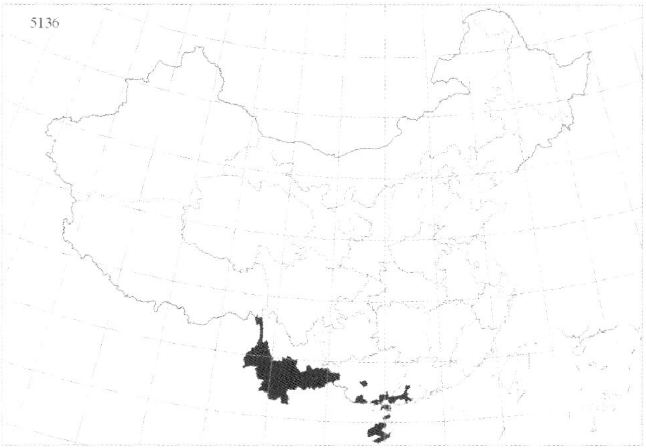

Anacardiaceae 漆树科 *Mangifera* 杧果属

Mangifera laurina Blume 长梗杧果

Trees, evergreen

18.1 (16.3~21.0); 11.5 (9.1~14.2); 22.5 (20.6~25.6); 18.1 (16.4~21.0); 157.4 (136.4~191.8); 0.0 (0.0~0.0); 874 (795~1053) ❙ 1352 (1007~1688); 770 (539~959); 52 (42~68) ❙ 55.1 (19.0~112.2); 867 (795~1053); 441.1 (277.0~627.0)

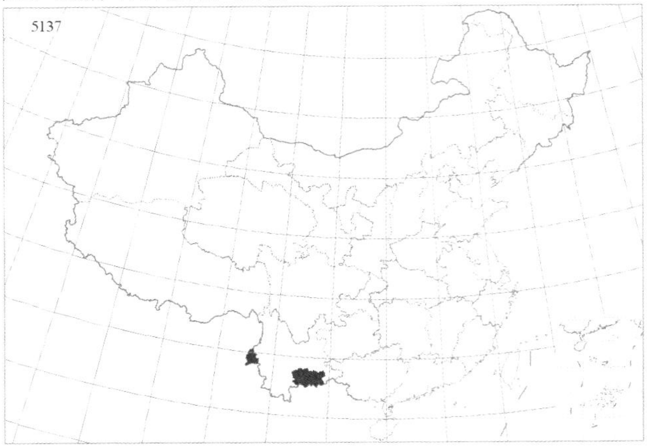

Anacardiaceae 漆树科 *Mangifera* 杧果属

Mangifera persiciforma C. Y. Wu et T. L. Ming 扁桃

Shrubs or trees, evergreen

19.6 (14.9~22.7); 11.3 (8.2~15.0); 25.7 (20.1~28.9); 19.6 (15.0~22.8); 175.3 (119.8~213.1); 0.0 (0.0~0.0); 1004 (748~1221) ❙ 1433 (956~2435); 779 (573~1428); 82 (34~138) ❙ 43.1 (5.8~99.5); 1003 (716~1221); 304.0 (201.4~413.5)

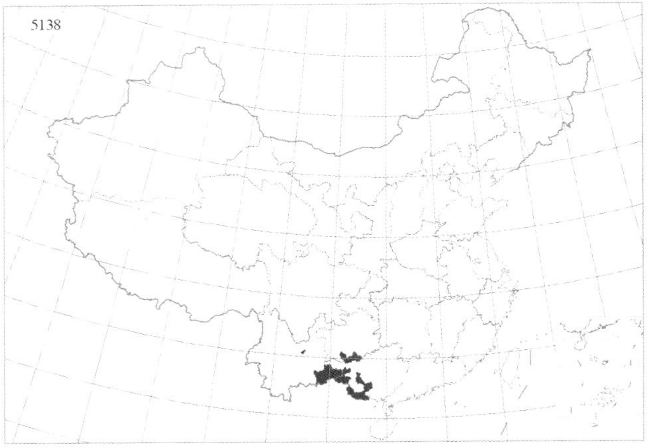

Anacardiaceae 漆树科 *Mangifera* 杧果属

Mangifera sylvatica Roxb. 林生杧果

Trees, evergreen

18.7 (16.3~22.2); 12.5 (10.2~15.1); 22.7 (20.3~28.3); 18.7 (16.4~22.3); 164.5 (136.6~207.5); 0.0 (0.0~0.0); 901 (789~1190) ❙ 1338 (939~1751); 727 (524~996); 54 (37~87) ❙ 48.9 (6.5~88.8); 876 (744~1190); 435.1 (301.1~659.0)

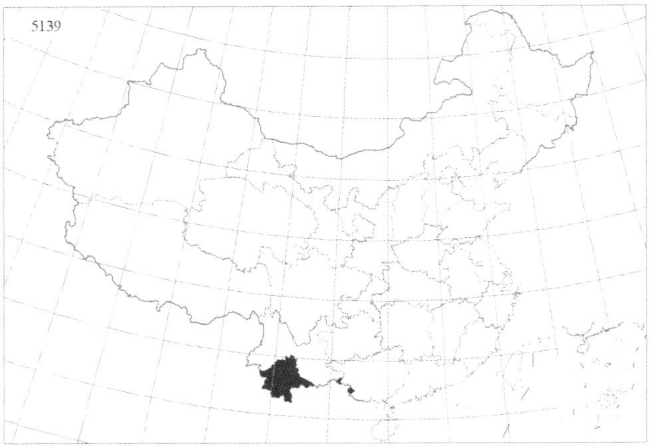

Anacardiaceae 漆树科 *Pegia* 藤漆属

Pegia nitida Colebr. 藤漆

Vines

17.2 (11.2~22.2); 9.8 (3.8~15.1); 22.9 (16.7~28.5); 17.3 (11.3~22.2); 147.5 (75.8~206.5); 0.1 (0.0~1.5); 857 (642~1188) ❙ 1275 (939~1751); 676 (461~1007); 60 (35~177) ❙ 49.4 (6.5~145.8); 847 (642~1188); 361.4 (146.0~659.0)

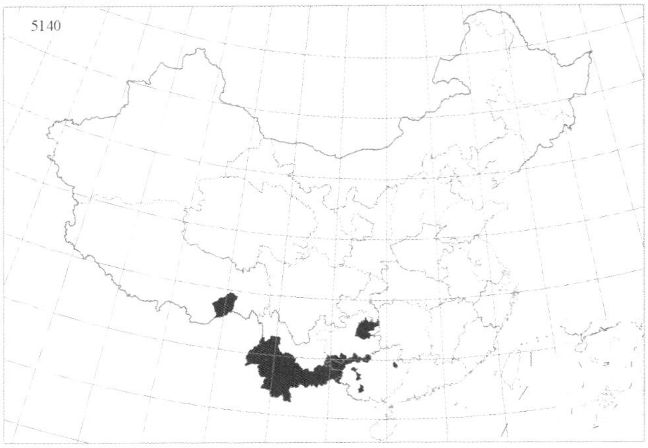

Anacardiaceae 漆树科 *Pegia* 藤漆属

Pegia sarmentosa (Lecomte) Hand.-Mazz. 利黄藤

Vines

20.3 (17.2~23.4); 12.0 (7.4~16.1); 26.7 (22.1~29.2); 20.3 (17.3~23.5); 184.2 (147.0~221.7); 0.0 (0.0~0.0); 1050 (833~1273) ❙ 1490 (1143~1876); 726 (547~996); 107 (43~194) ❙ 43.1 (9.0~88.8); 1048 (833~1273); 288.9 (178.7~551.6)

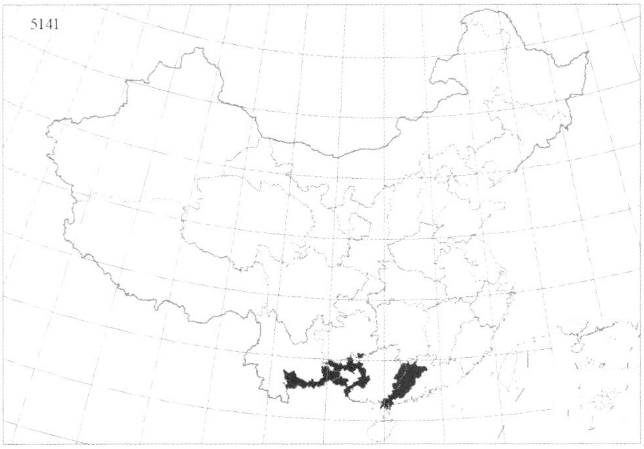

Anacardiaceae 漆树科 *Pistacia* 黄连木属

Pistacia chinensis Bunge 黄连木

Trees, deciduous

15.4 (2.0~24.8); 4.1 (-13.7~20.1); 25.6 (9.9~30.1); 15.6 (3.8~24.8); 132.8 (15.4~237.6); 7.0 (0.0~71.7); 858 (413~1385) ❙ 1153 (367~3495); 522 (235~1847); 96 (8~897) ❙ 32.9 (-42.0~246.5); 814 (367~1273); 219.6 (95.3~659.0)

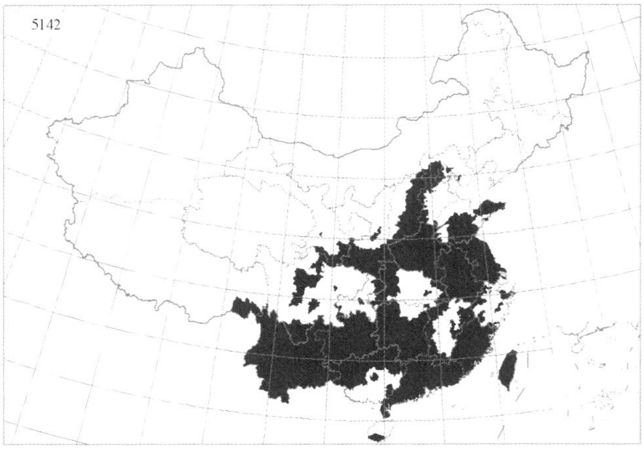

Anacardiaceae 漆树科 *Pistacia* 黄连木属
Pistacia weinmanniifolia J. Poiss. ex Franch. 清香木
Shrubs or small trees, evergreen

15.6 (2.2~22.2); 8.0 (-6.6~15.1); 21.5 (9.5~28.6); 15.7 (3.8~22.3); 129.9 (15.5~207.5); 2.0 (0.0~49.1); 803 (411~1190) ❙ 1156 (535~1751); 635 (326~996); 51 (10~173) ❙ 45.3 (-14.2~125.9); 790 (411~1190); 312.6 (171.2~659.0)

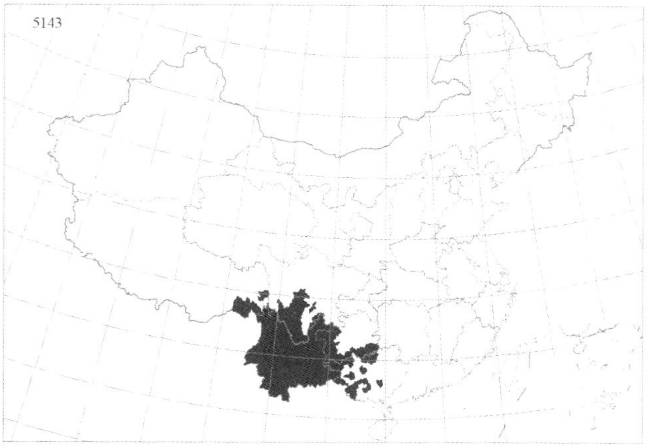

Anacardiaceae 漆树科 *Rhus* 盐肤木属
Rhus chinensis Mill. 盐麸木
Shrubs or small trees, deciduous

15.4 (2.2~24.8); 4.0 (-17.4~20.1); 25.6 (9.5~30.1); 15.7 (3.8~24.8); 133.3 (15.5~237.6); 8.1 (0.0~80.4); 861 (411~1385) ❙ 1169 (341~3495); 526 (190~1847); 98 (8~897) ❙ 33.8 (-43.4~246.5); 822 (341~1360); 219.7 (61.2~659.0)

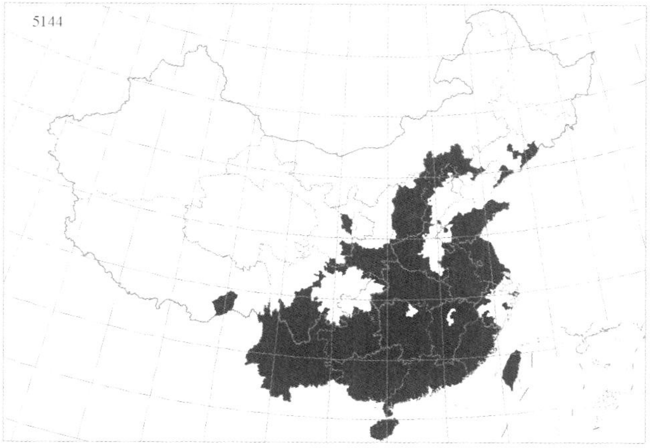

Anacardiaceae 漆树科 *Rhus* 盐肤木属
Rhus chinensis var. roxburghii (DC.) Rehder 滨盐麸木
Shrubs or small trees, deciduous

18.1 (5.4~24.2); 9.8 (-2.1~19.1); 24.9 (12.3~29.0); 18.2 (5.8~24.3); 159.2 (31.5~231.2); 1.0 (0.0~25.6); 959 (481~1335) ❙ 1443 (741~2435); 684 (426~1428); 110 (12~192) ❙ 52.5 (-12.0~101.8); 950 (481~1257); 287.8 (157.7~620.0)

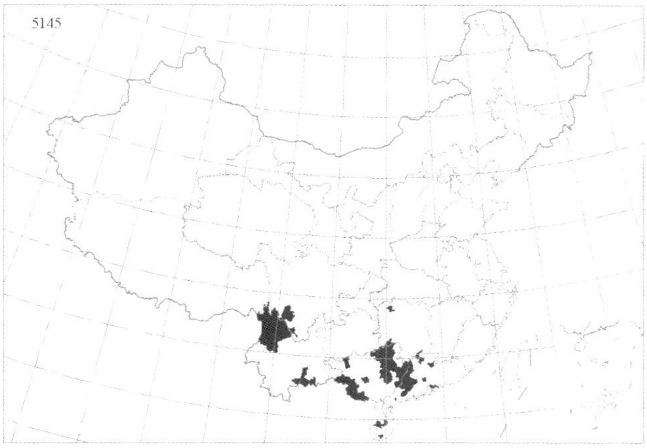

Anacardiaceae 漆树科 *Rhus* 盐肤木属
Rhus hypoleuca Champ. ex Benth. 白背麸杨
Shrubs or small trees, deciduous

19.0 (15.0~24.1); 10.4 (3.9~18.9); 26.7 (20.1~29.1); 19.1 (15.1~24.1); 168.8 (120.8~229.7); 0.0 (0.0~1.1); 988 (773~1322) ❙ 1957 (1425~3495); 783 (505~1847); 217 (47~897) ❙ 102.8 (25.2~246.5); 974 (773~1216); 244.9 (119.4~363.7)

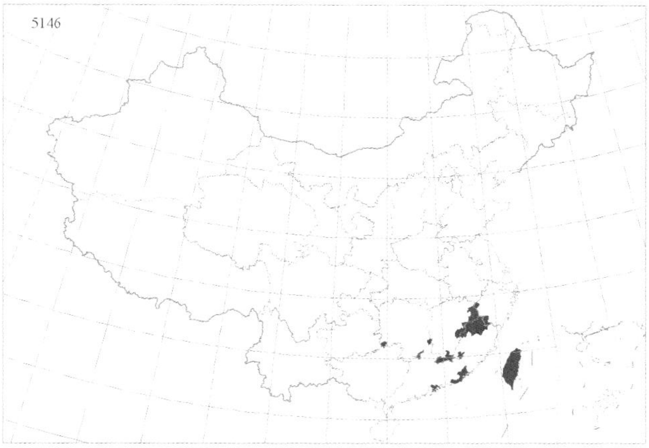

Anacardiaceae 漆树科 *Rhus* 盐肤木属
Rhus potaninii Maxim. 青麸杨
Trees, deciduous

11.7 (2.0~18.9); -0.5 (-13.7~12.5); 22.7 (10.1~28.8); 12.3 (3.8~19.0); 99.0 (15.4~167.8); 17.7 (0.0~71.7); 717 (413~949) ❙ 809 (341~1893); 425 (190~816); 36 (8~253) ❙ 12.8 (-43.3~129.9); 646 (341~949); 211.9 (61.2~491.5)

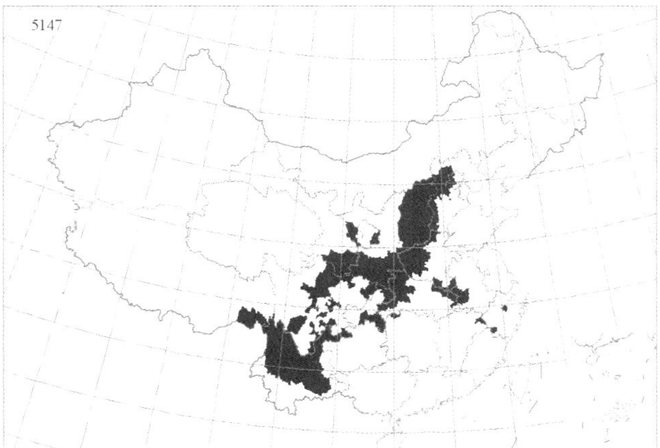

Anacardiaceae 漆树科 *Rhus* 盐肤木属
Rhus punjabensis var. pilosa Engl. 毛麸杨
Trees, deciduous

13.8 (-1.1~18.9); 4.5 (-9.1~12.5); 21.8 (6.2~29.4); 13.9 (2.0~19.0); 111.0 (2.5~167.8); 5.1 (0.0~75.6); 747 (334~952) ❙ 1033 (321~1736); 538 (193~1007); 53 (12~195) ❙ 39.4 (-35.3~145.8); 723 (321~930); 262.6 (79.3~516.7)

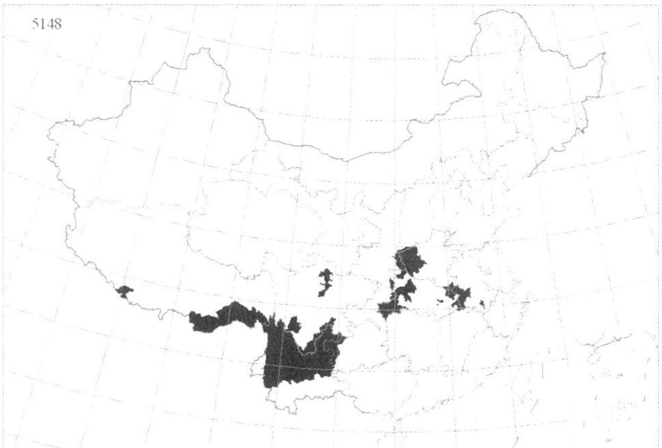

Anacardiaceae 漆树科 *Rhus* 盐肤木属
Rhus punjabensis var. sinica (Diels) Rehder et E. H. Wilson 红麸杨
Trees, deciduous

14.2 (-1.1~19.1); 3.9 (-9.1~11.8); 23.5 (6.2~29.4); 14.3 (2.0~19.1); 115.4 (2.5~169.2); 4.6 (0.0~75.6); 773 (334~969) ❙ 1093 (321~1590); 525 (193~946); 62 (12~157) ❙ 42.1 (-26.1~121.7); 763 (321~969); 225.4 (79.3~491.5)

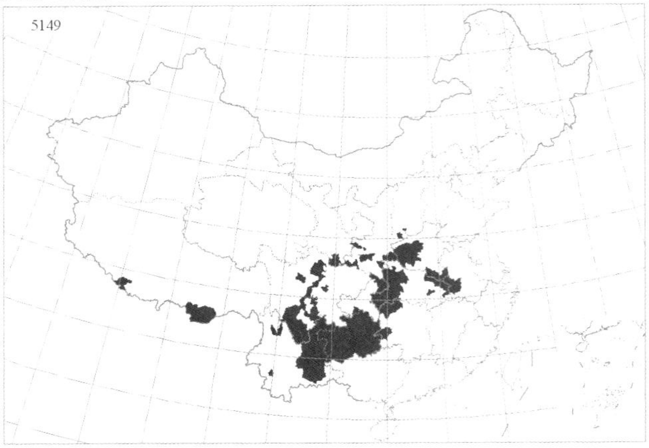

Anacardiaceae 漆树科 *Rhus* 盐肤木属
Rhus teniana Hand.-Mazz. 滇麸杨
Shrubs, deciduous

15.2; 8.6; 20.3; 15.2; 122.7; 0.0; 757 ❙ 963; 567; 32 ❙ 27.3; 700; 308.8

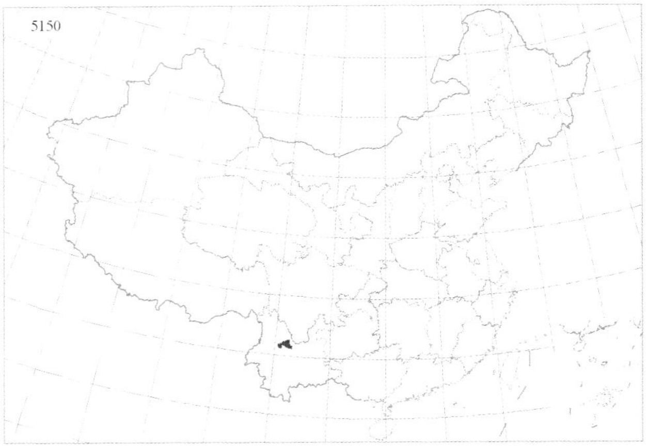

Anacardiaceae 漆树科 *Rhus* 盐肤木属
Rhus wilsonii Hemsl. 川麸杨
Shrubs, deciduous

14.7 (6.2~18.9); 6.7 (-2.4~12.5); 21.4 (14.1~27.6); 14.8 (6.7~19.0); 118.7 (39.4~167.8); 1.4 (0.0~24.1); 767 (509~959) ❙ 995 (797~1318); 561 (411~753); 37 (20~71) ❙ 31.8 (-14.2~101.6); 738 (509~959); 254.9 (135.1~365.3)

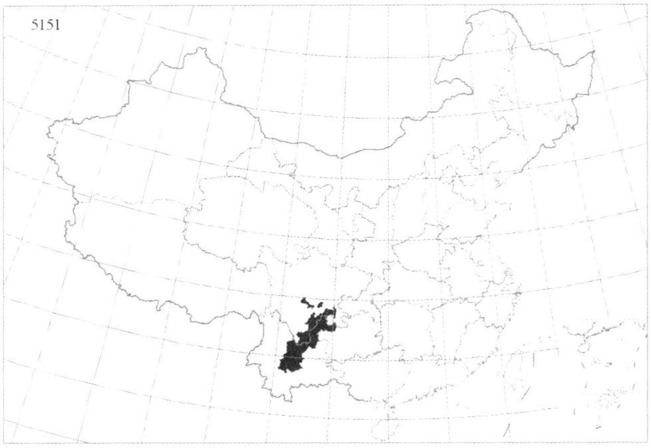

Anacardiaceae 漆树科 *Semecarpus* 肉托果属
Semecarpus cuneiformis Blanco 钝叶肉托果
Trees

18.5; 14.1; 21.7; 18.5; 162.1; 0.0; 863 ❙ 2713; 1388; 188 ❙ 214.4; 863; 363.7

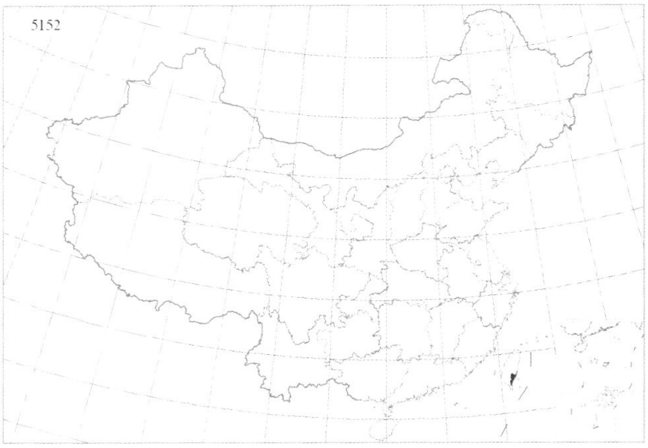

Anacardiaceae 漆树科 *Semecarpus* 肉托果属
Semecarpus longifolius Blume 大叶肉托果
Trees, evergreen

19.2; 14.2; 23.3; 19.3; 171.4; 0.0; 939 ❙ 2799; 1277; 286 ❙ 200.9; 939; 306.4

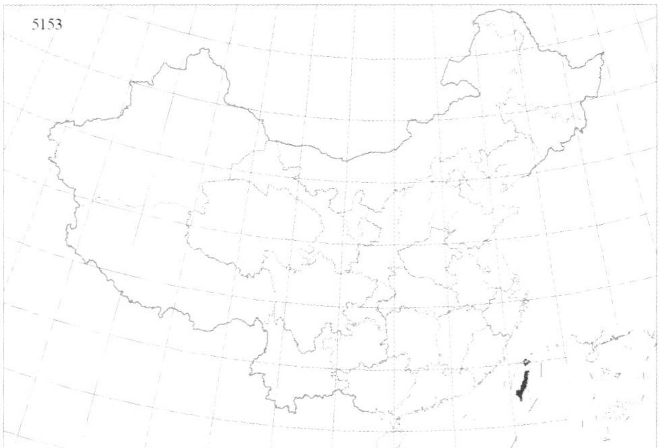

Anacardiaceae 漆树科 *Semecarpus* 肉托果属
Semecarpus microcarpus Wall. ex Hook. f. 小果肉托果
Trees, deciduous

18.1 (14.4~20.5); 11.8 (7.5~15.1); 21.9 (19.1~23.7); 18.1 (14.4~20.5); 157.3 (113.1~185.9); 0.0 (0.0~0.0); 869 (727~992) ❙ 1479 (1275~1688); 799 (655~959); 54 (41~77) ❙ 71.0 (47.6~112.2); 855 (727~986); 499.2 (395.4~659.0)

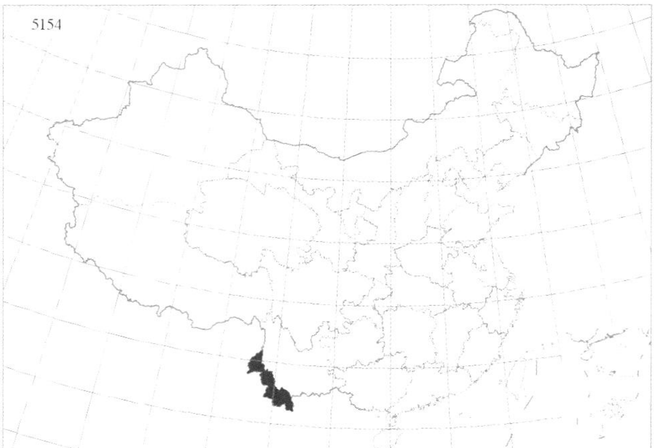

Anacardiaceae 漆树科 *Semecarpus* 肉托果属

Semecarpus reticulatus Lecomte 网脉肉托果

Trees

19.5 (18.6~20.5); 13.9 (12.7~15.1); 23.0 (22.3~23.7); 19.6 (18.7~20.5); 174.6 (163.9~185.9); 0.0 (0.0~0.0); 935 (888~992) ❙ 1560 (1464~1751); 787 (655~996); 67 (58~77) ❙ 67.1 (47.6~88.8); 922 (880~986); 539.4 (427.9~659.0)

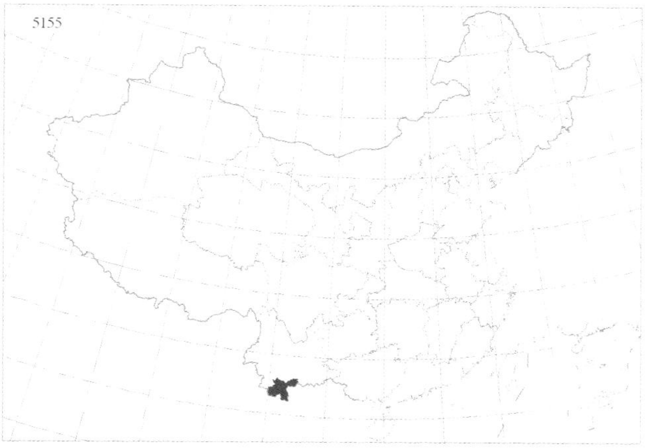

Anacardiaceae 漆树科 *Spondias* 槟榔青属

Spondias haplophylla Airy Shaw et Forman 单叶槟榔青

Trees

19.0; 13.1; 22.6; 19.1; 168.8; 0.0; 909 ❙ 1495; 775; 62 ❙ 64.5; 891; 427.9

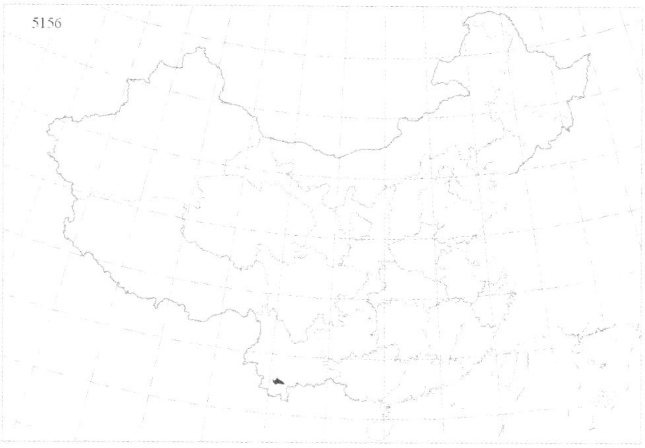

Anacardiaceae 漆树科 *Spondias* 槟榔青属

Spondias lakonensis Pierre 岭南酸枣

Trees, deciduous

20.7 (16.3~24.8); 13.0 (5.4~20.1); 26.8 (21.4~29.8); 20.8 (16.4~24.8); 189.4 (136.4~237.6); 0.0 (0.0~0.0); 1078 (797~1385) ❙ 1596 (1074~2148); 742 (455~1234); 111 (41~255) ❙ 49.0 (1.5~97.5); 1067 (797~1225); 289.1 (95.3~659.0)

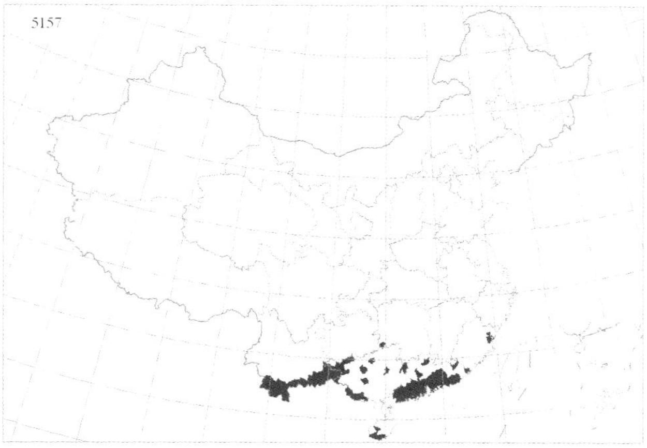

Anacardiaceae 漆树科 *Spondias* 槟榔青属

Spondias lakonensis var. hirsuta C. Y. Wu et T. L. Ming 毛叶岭南酸枣

Trees, deciduous

19.5 (17.7~21.0); 13.5 (10.9~15.1); 23.3 (22.3~25.6); 19.5 (17.8~21.0); 174.0 (153.4~191.8); 0.0 (0.0~0.0); 940 (854~1053) ❙ 1496 (1276~1751); 789 (655~996); 64 (51~77) ❙ 59.5 (36.8~88.8); 932 (854~1053); 534.4 (452.0~659.0)

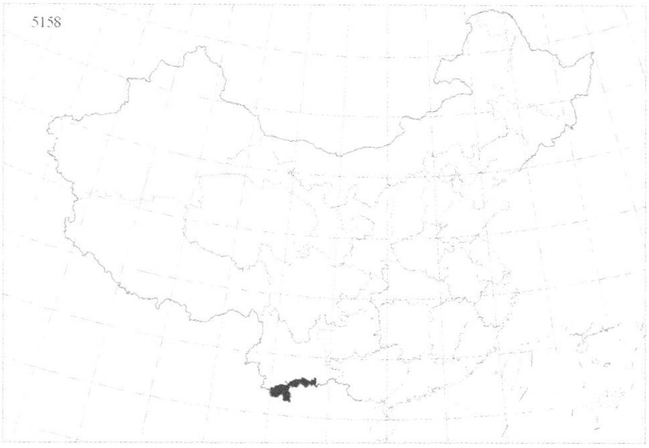

Anacardiaceae 漆树科 *Spondias* 槟榔青属

Spondias pinnata (L. f.) Kurz 槟榔青

Trees, deciduous

20.0 (14.6~24.8); 13.7 (3.8~20.1); 24.2 (20.3~28.7); 20.0 (14.7~24.8); 180.2 (117.7~237.6); 0.0 (0.0~1.5); 1010 (779~1385) ❙ 1449 (868~2435); 712 (312~1428); 65 (38~191) ❙ 47.2 (-34.3~112.2); 957 (763~1360); 444.3 (240.7~659.0)

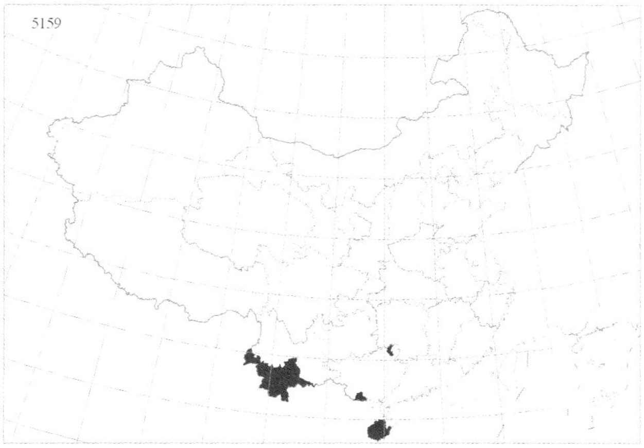

Anacardiaceae 漆树科 *Terminthia* 三叶漆属

Terminthia paniculata (Wall. ex G. Don) C. Y. Wu et T. L. Ming 三叶漆

Shrubs or small trees, deciduous

18.2 (16.0~20.5); 12.0 (9.3~15.1); 22.2 (20.3~23.7); 18.3 (16.0~20.5); 159.3 (132.4~185.9); 0.0 (0.0~0.0); 876 (780~992) ❙ 1371 (939~1751); 759 (524~996); 52 (37~77) ❙ 56.8 (6.5~112.2); 856 (744~986); 448.7 (295.5~659.0)

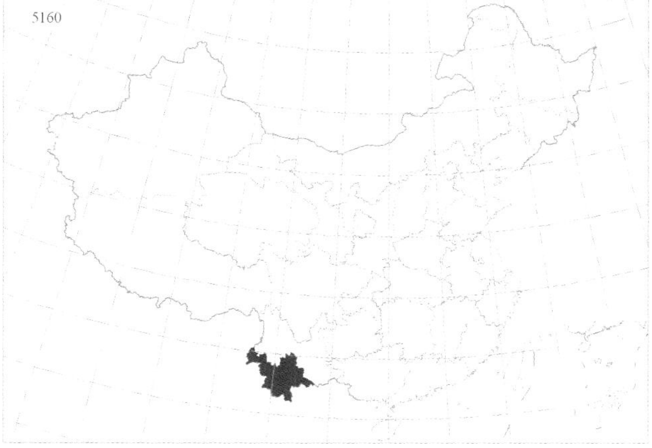

Anacardiaceae 漆树科 *Toxicodendron* 漆树属
Toxicodendron acuminatum (DC.) C. Y. Wu et T. L. Ming 尖叶漆
Trees, deciduous

17.1 (-1.8~21.0); 10.4 (-10.2~15.1); 21.6 (6.1~25.8); 17.3 (1.8~21.0); 149.8 (1.9~191.8); 4.3 (0.0~82.8); 848 (329~1053) ❙ 1297 (320~1751); 721 (193~996); 52 (10~77) ❙ 51.3 (-4.0~112.2); 836 (320~1053); 405.5 (61.6~659.0)

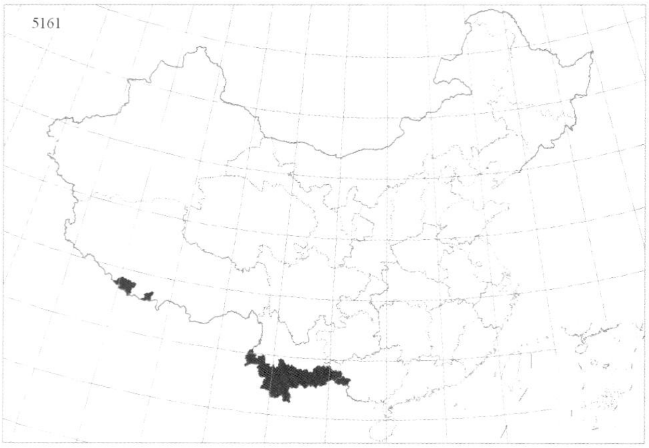

Anacardiaceae 漆树科 *Toxicodendron* 漆树属
Toxicodendron calcicola C. Y. Wu ex T. L. Ming 石山漆
Shrubs or small trees, deciduous

17.9 (16.3~21.0); 10.6 (7.4~14.2); 23.0 (21.4~25.6); 17.9 (16.4~21.0); 154.8 (136.4~191.8); 0.0 (0.0~0.0); 872 (797~1053) ❙ 1247 (958~1522); 689 (531~860); 54 (44~107) ❙ 43.0 (13.3~80.6); 868 (797~1053); 371.4 (245.3~551.6)

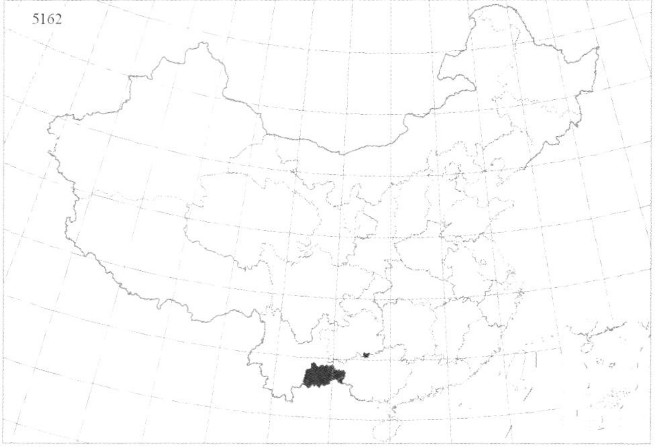

Anacardiaceae 漆树科 *Toxicodendron* 漆树属
Toxicodendron delavayi (Franch.) F. A. Barkley 小漆树
Small shrubs, deciduous

15.5 (3.8~21.0); 8.0 (-4.3~15.1); 21.2 (10.8~26.6); 15.5 (4.7~21.0); 127.3 (21.6~191.8); 1.1 (0.0~35.4); 786 (444~1053) ❙ 1118 (741~1751); 626 (411~996); 44 (15~129) ❙ 43.1 (-14.2~121.7); 770 (444~1053); 317.5 (130.0~659.0)

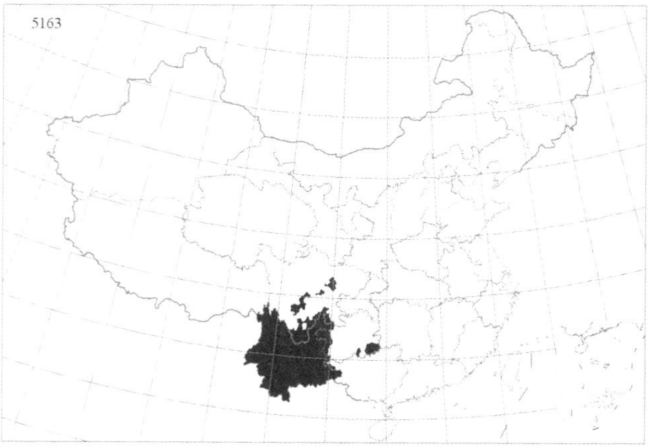

Anacardiaceae 漆树科 *Toxicodendron* 漆树属
Toxicodendron delavayi var. angustifolium C. Y. Wu 狭叶小漆树
Small shrubs, deciduous

11.6; 3.1; 18.6; 11.5; 80.9; 2.8; 654 ❙ 988; 561; 28 ❙ 51.0; 654; 255.4

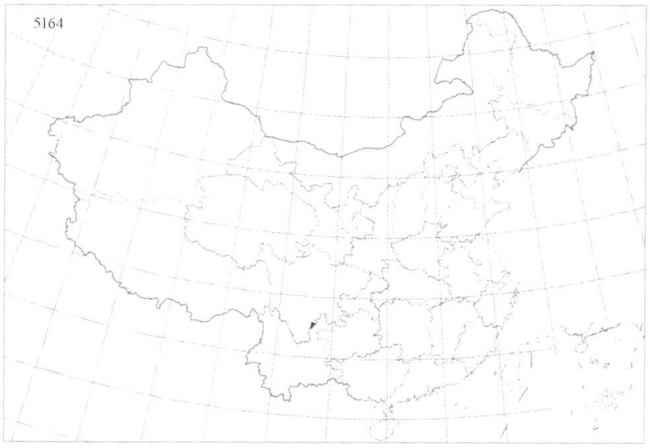

Anacardiaceae 漆树科 *Toxicodendron* 漆树属
Toxicodendron delavayi var. quinquejugum (Rehder et E. H. Wilson) C. Y. Wu et T. L. Ming 多叶小漆树 Small shrubs, deciduous

9.1 (2.2~14.3); 1.3 (-6.6~7.6); 16.0 (9.9~19.7); 9.4 (3.8~14.3); 62.3 (15.5~111.9); 11.9 (0.0~49.1); 590 (413~728) ❙ 1004 (801~1135); 585 (496~644); 24 (15~35) ❙ 72.7 (36.1~125.9); 588 (413~707); 284.3 (221.3~319.8)

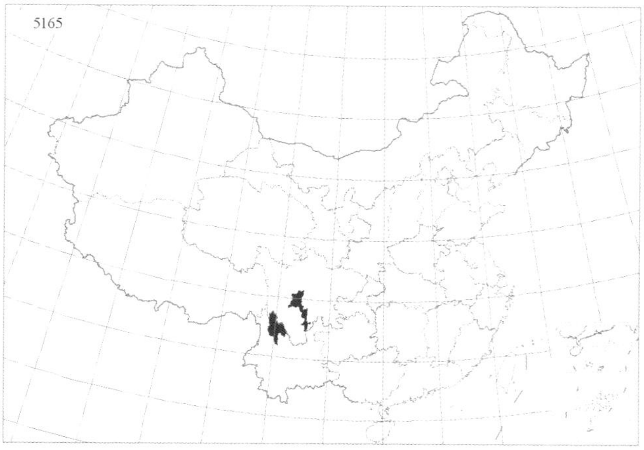

Anacardiaceae 漆树科 *Toxicodendron* 漆树属
Toxicodendron fulvum (Craib) C. Y. Wu et T. L. Ming 黄毛漆
Trees, deciduous

18.4 (14.4~20.5); 12.2 (7.5~15.1); 22.2 (19.1~23.7); 18.5 (14.4~20.5); 161.5 (113.1~185.9); 0.0 (0.0~0.0); 888 (727~992) ❙ 1529 (1434~1688); 809 (655~959); 56 (41~77) ❙ 73.5 (47.6~112.2); 871 (727~986); 519.6 (427.9~659.0)

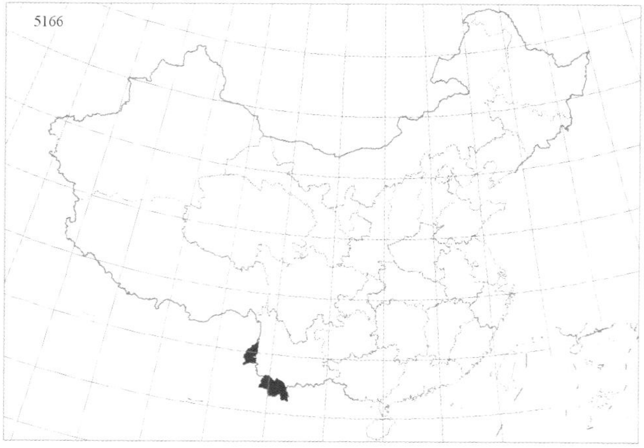

Aquifoliaceae 冬青科 *Ilex* 冬青属
Ilex atrata W. W. Sm. 黑果冬青
Trees, evergreen

15.1 (8.6~18.9); 8.5 (1.3~12.5); 20.1 (15.0~23.9); 15.2 (8.7~19.0); 122.5 (52.6~167.8); 0.3 (0.0~8.8); 761 (574~929) ❙ 1075 (797~1736); 608 (411~1007); 45 (23~70) ❙ 43.2 (-14.2~145.8); 730 (574~851); 344.0 (218.0~487.2)

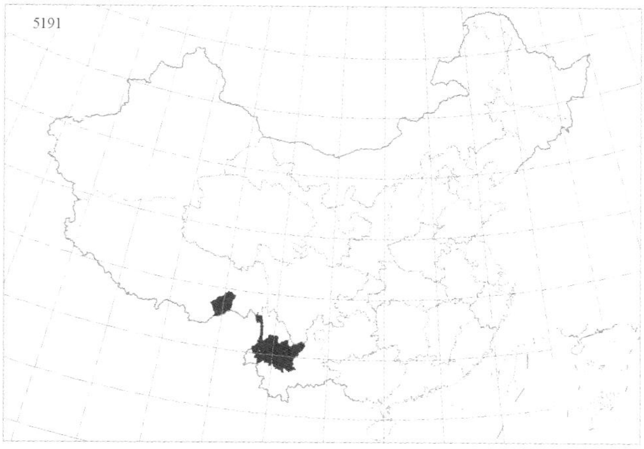

Aquifoliaceae 冬青科 *Ilex* 冬青属
Ilex atrata var. wangii S. Y. Hu 长梗黑果冬青
Trees, evergreen

14.9; 8.1; 20.1; 15.0; 121.6; 2.2; 771 ❙ 1438; 834; 54 ❙ 89.3; 771; 464.3

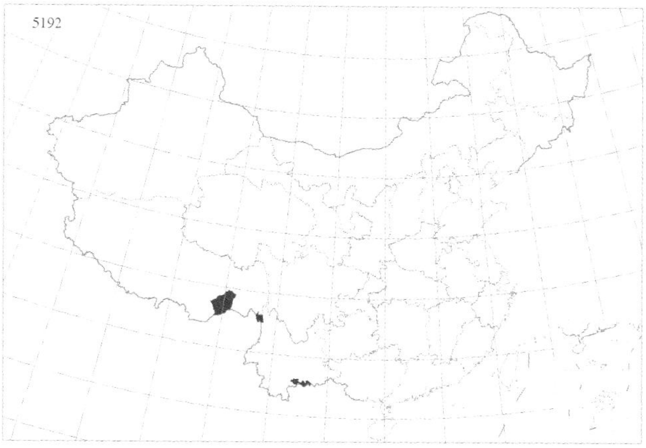

Aquifoliaceae 冬青科 *Ilex* 冬青属
Ilex austrosinensis C. J. Tseng 两广冬青
Shrubs or trees, evergreen

20.5 (17.7~23.1); 11.6 (7.7~17.5); 27.9 (26.6~28.8); 20.5 (17.8~23.2); 186.3 (153.6~218.4); 0.0 (0.0~0.0); 1084 (916~1232) ❙ 1485 (1104~1703); 650 (526~809); 127 (53~194) ❙ 37.3 (15.8~70.7); 1072 (916~1170); 242.0 (151.7~538.2)

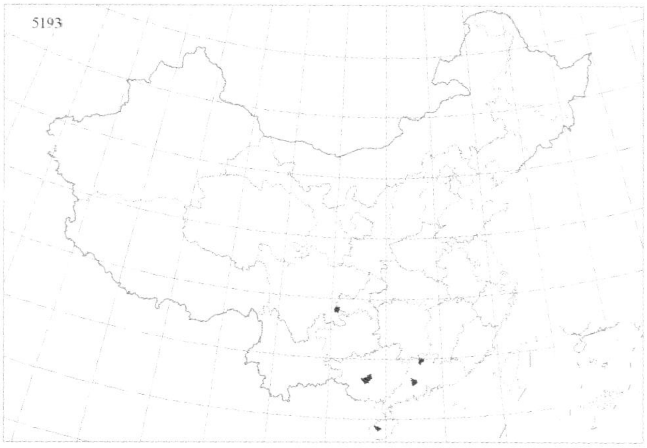

Aquifoliaceae 冬青科 *Ilex* 冬青属
Ilex bidens C. Y. Wu ex Y. R. Li 双齿冬青
Trees, evergreen

19.1; 13.0; 23.3; 19.1; 169.7; 0.0; 917 ❙ 1428; 835; 56 ❙ 55.7; 917; 506.2

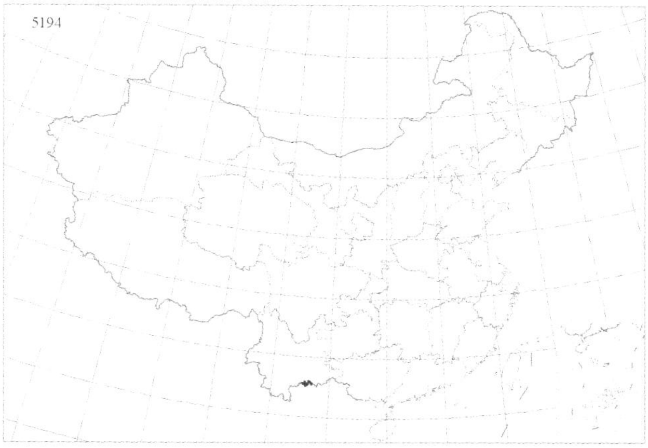

Aquifoliaceae 冬青科 *Ilex* 冬青属
Ilex bioritsensis Hayata 刺叶冬青
Shrubs or small trees, evergreen

15.2 (5.4~24.1); 6.6 (-2.1~18.9); 22.6 (12.3~28.9); 15.2 (5.8~24.1); 124.1 (31.5~229.7); 1.2 (0.0~25.6); 797 (481~1322) ❙ 1255 (741~3495); 617 (411~1847); 82 (12~897) ❙ 56.9 (-14.2~246.5); 784 (481~1148); 249.8 (119.4~487.2)

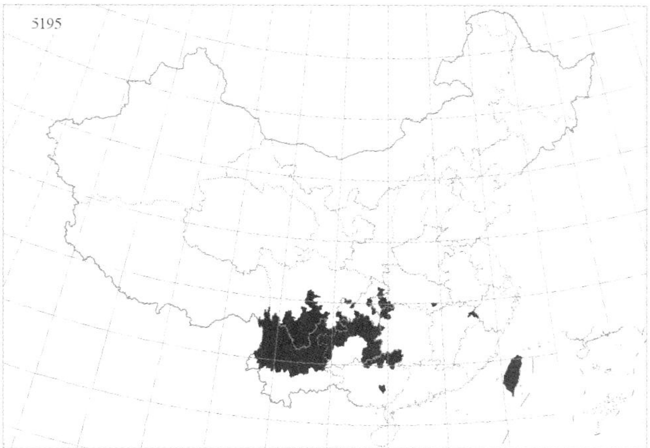

Aquifoliaceae 冬青科 *Ilex* 冬青属
Ilex brachyphylla (Hand.-Mazz.) S. Y. Hu 短叶冬青
Small trees, evergreen

16.1 (14.3~17.6); 4.9 (3.3~6.9); 26.6 (24.3~27.7); 16.2 (14.3~17.6); 135.0 (114.1~151.3); 0.7 (0.0~2.3); 860 (765~928) ❙ 1460 (1337~1526); 502 (445~591); 171 (112~202) ❙ 70.4 (59.8~99.4); 857 (765~922); 201.1 (169.7~257.1)

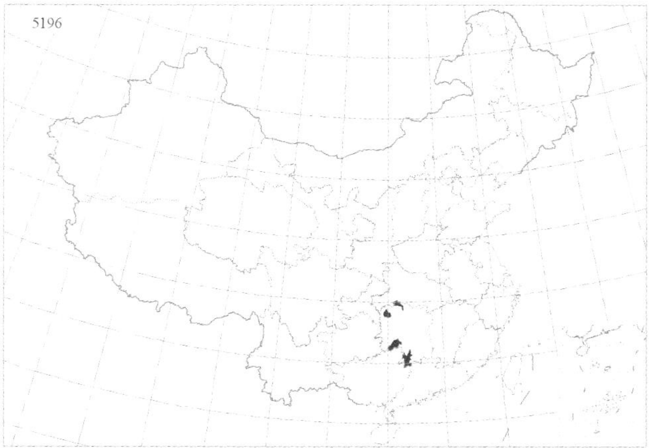

Aquifoliaceae 冬青科 *Ilex* 冬青属
Ilex buergeri Miq. 短梗冬青
Shrubs or trees, evergreen

16.9 (14.4~22.2); 5.6 (2.1~14.7); 27.7 (24.7~30.1); 16.9 (14.5~22.3); 143.9 (118.7~207.5); 0.6 (0.0~4.9); 909 (789~1190) ❙ 1602 (1048~1997); 575 (418~891); 194 (87~259) ❙ 77.4 (15.4~147.0); 900 (789~1190); 232.6 (130.9~327.7)

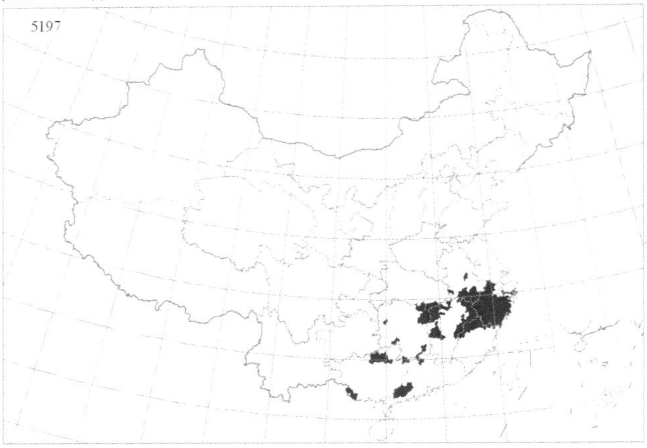

Aquifoliaceae 冬青科 *Ilex* 冬青属
Ilex buxoides S. Y. Hu 黄杨冬青
Trees, evergreen

20.3 (15.3~22.7); 11.6 (4.2~15.0); 27.5 (25.2~28.8); 20.4 (15.4~22.8); 184.3 (125.8~213.1); 0.0 (0.0~0.8); 1066 (818~1221) ❙ 1698 (1302~2435); 772 (495~1428); 147 (86~216) ❙ 60.9 (12.9~103.1); 1066 (818~1221); 261.5 (185.5~327.7)

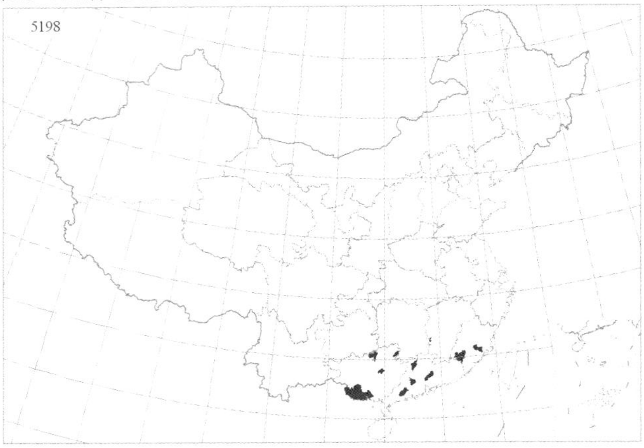

Aquifoliaceae 冬青科 *Ilex* 冬青属
Ilex cauliflora H. W. Li ex Y. R. Li 茎花冬青
Shrubs, evergreen

17.6; 9.9; 23.2; 17.7; 152.4; 0.0; 859 ❙ 1422; 813; 52 ❙ 65.6; 859; 359.7

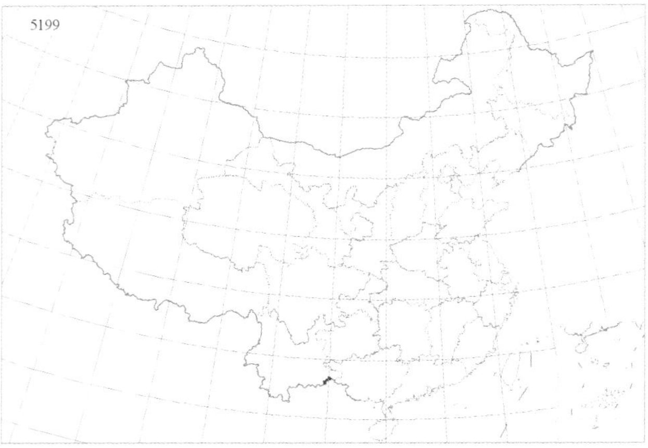

Aquifoliaceae 冬青科 *Ilex* 冬青属
Ilex centrochinensis S. Y. Hu 华中枸骨
Shrubs, evergreen

15.2 (9.3~19.3); 5.9 (-1.7~13.1); 23.3 (16.3~29.3); 15.2 (9.5~19.4); 123.9 (68.0~173.2); 1.4 (0.0~18.0); 796 (603~964) ❙ 1150 (797~1679); 540 (384~774); 76 (15~204) ❙ 44.6 (-14.2~107.2); 781 (603~961); 249.6 (164.1~432.9)

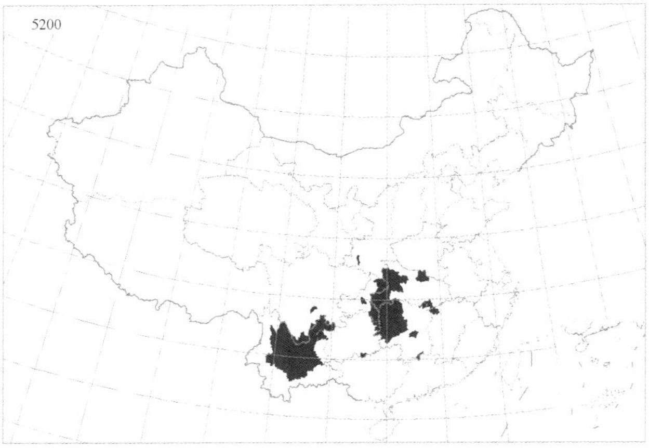

Aquifoliaceae 冬青科 *Ilex* 冬青属
Ilex chamaebuxus C. Y. Wu et Y. R. Li 矮杨梅冬青
Small trees, evergreen

17.4; 9.7; 22.9; 17.5; 149.7; 0.0; 848 ❙ 1338; 756; 50 ❙ 57.6; 848; 348.0

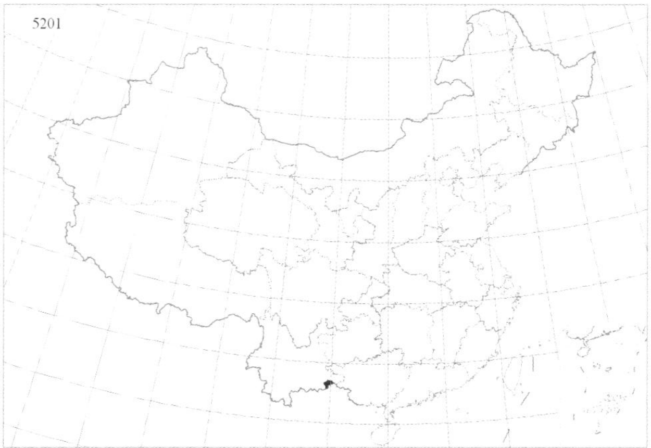

Aquifoliaceae 冬青科 *Ilex* 冬青属
Ilex championii Loes. 凹叶冬青
Shrubs or trees, evergreen

18.1 (14.3~22.4); 7.8 (3.3~15.2); 27.3 (24.2~29.8); 18.1 (14.3~22.7); 157.9 (113.2~212.7); 0.2 (0.0~2.3); 957 (761~1216) ❙ 1564 (1215~2037); 590 (392~1033); 168 (84~230) ❙ 64.4 (22.7~128.9); 952 (761~1216); 219.4 (145.3~305.7)

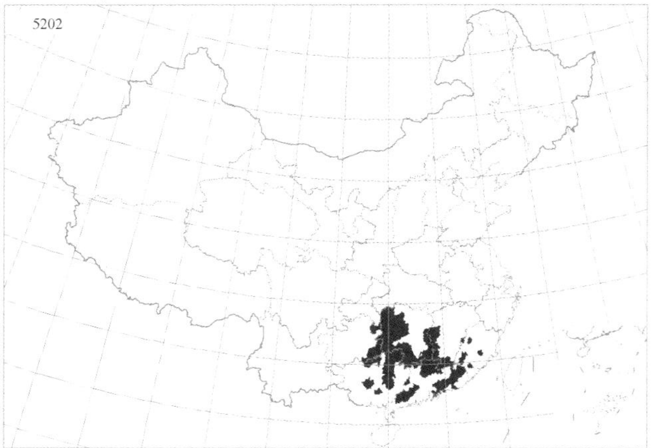

Aquifoliaceae 冬青科 *Ilex* 冬青属
Ilex chapaensis Merr. 沙坝冬青
Trees, deciduous

18.6 (13.6~24.8); 11.2 (6.2~20.1); 24.2 (18.8~29.0); 18.7 (13.6~24.8); 163.9 (103.6~237.6); 0.0 (0.0~0.0); 945 (706~1385) ❙ 1369 (797~2435); 685 (411~1428); 84 (23~205) ❙ 45.0 (-14.2~112.2); 929 (683~1360); 323.5 (122.7~659.0)

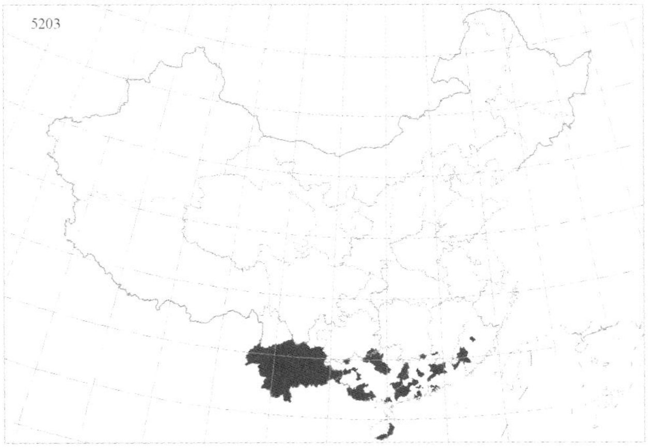

Aquifoliaceae 冬青科 *Ilex* 冬青属
Ilex chartaceifolia C. Y. Wu ex Y. R. Li 纸叶冬青
Trees, evergreen

11.6; 4.6; 17.3; 11.6; 83.0; 4.4; 652 ❙ 1043; 577; 49 ❙ 62.3; 652; 367.4

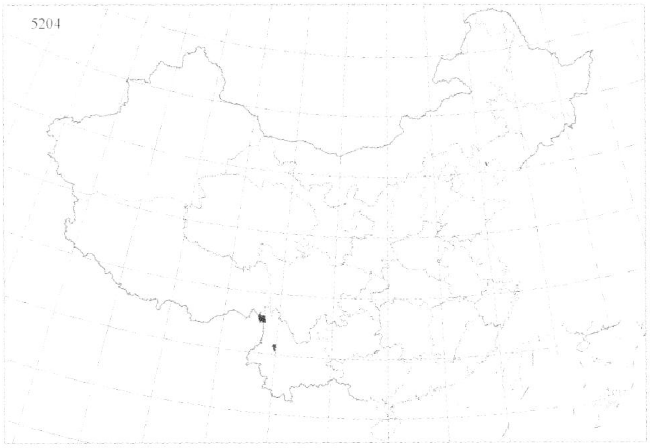

Aquifoliaceae 冬青科 *Ilex* 冬青属
Ilex chengkouensis C. J. Tseng 城口冬青
Trees, evergreen

12.0; [illegible].2; 22.6; 12.0; 92.4; 8.2; 694 ❙ 1227; 526; 39 ❙ 78.0; 694; 270.1

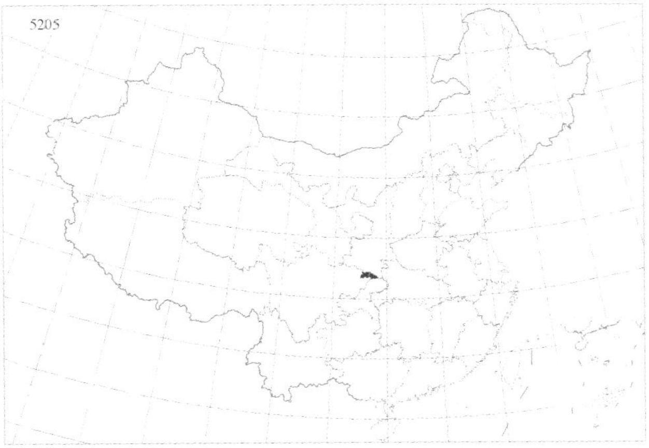

Aquifoliaceae 冬青科 *Ilex* 冬青属
Ilex cheniana T. R. Dudley 龙陵冬青
Small trees, evergreen

13.7; 6.9; 18.7; 13.8; 105.8; 0.4; 714 ❙ 1256; 716; 50 ❙ 75.1; 714; 400.8

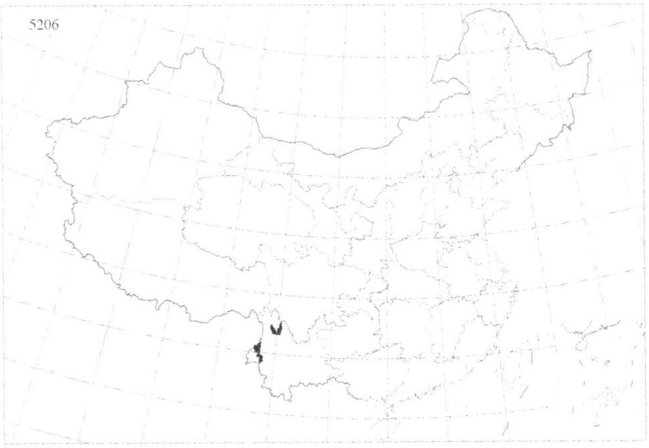

Aquifoliaceae 冬青科 *Ilex* 冬青属
Ilex chinensis Sims 冬青
Trees, evergreen

16.4 (2.2~23.9); 6.2 (-6.6~18.7); 25.7 (9.9~29.9); 16.5 (3.8~23.9); 138.8 (15.5~226.8); 1.0 (0.0~49.1); 873 (413~1294) ❙ 1347 (741~2435); 563 (299~1428); 125 (12~259) ❙ 54.6 (-14.2~147.0); 863 (413~1221); 236.2 (103.2~659.0)

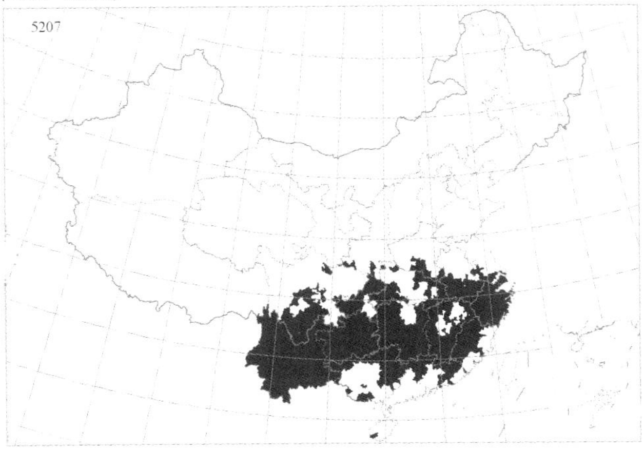

Aquifoliaceae 冬青科 *Ilex* 冬青属
Ilex chingiana Hu et T. Tang 苗山冬青
Trees, evergreen

18.6 (14.3~22.2); 8.8 (3.3~13.9); 26.9 (24.3~28.8); 18.7 (14.3~22.3); 164.3 (114.1~207.5); 0.1 (0.0~2.3); 976 (765~1190) ❙ 1509 (1234~1776); 642 (452~850); 148 (58~196) ❙ 56.2 (15.4~101.8); 976 (765~1190); 241.6 (185.5~335.1)

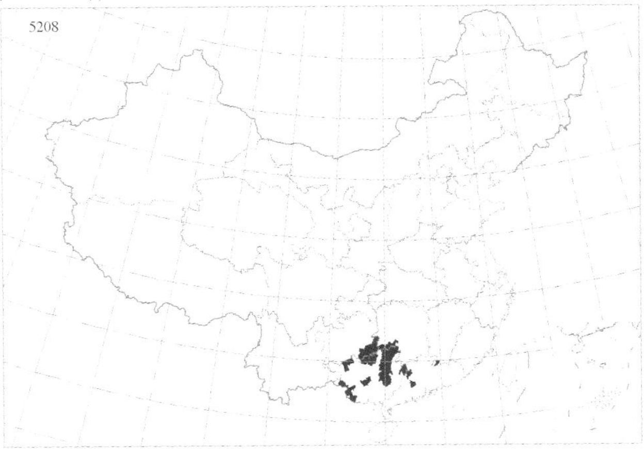

Aquifoliaceae 冬青科 *Ilex* 冬青属
Ilex chuniana S. Y. Hu 铁仔冬青
Shrubs or small trees, evergreen

23.8 (22.2~24.8); 18.0 (14.7~20.1); 28.0 (25.7~28.8); 23.9 (22.3~24.8); 226.5 (207.2~237.6); 0.0 (0.0~0.0); 1299 (1127~1385) ❙ 1517 (868~1854); 564 (312~862); 75 (41~115) ❙ 17.3 (-34.3~47.8); 1192 (868~1360); 415.3 (139.6~620.0)

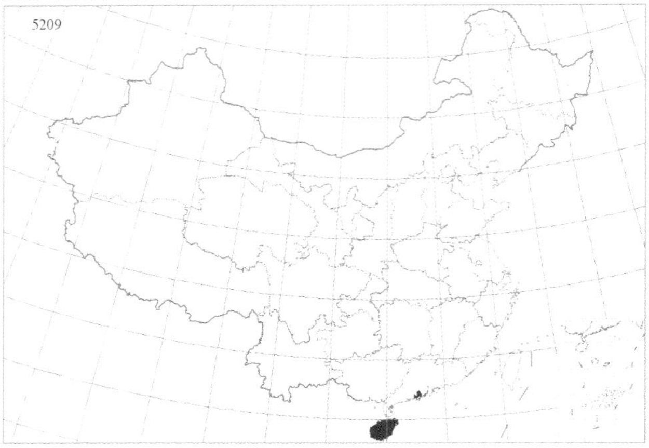

Aquifoliaceae 冬青科 *Ilex* 冬青属
Ilex ciliospinosa Loes. 纤齿枸骨
Shrubs or small trees, evergreen

14.0 (5.3~18.3); 4.0 (-2.4~10.9); 23.2 (12.2~28.4); 14.1 (5.7~18.4); 111.5 (30.5~160.7); 2.5 (0.0~26.2); 768 (479~965) ❙ 1205 (741~1624); 538 (426~817); 86 (12~191) ❙ 57.3 (15.8~107.2); 766 (479~965); 226.3 (151.7~383.4)

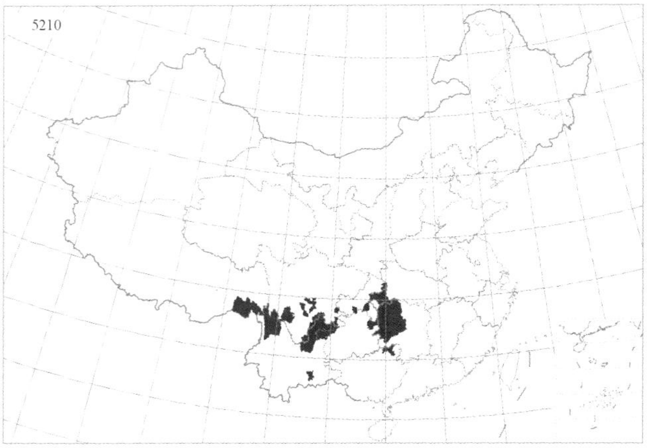

Aquifoliaceae 冬青科 *Ilex* 冬青属
Ilex cinerea Champ. ex Benth. 灰冬青
Shrubs or small trees, evergreen

23.8 (22.2~24.8); 17.9 (14.7~20.1); 28.0 (25.7~28.8); 23.8 (22.3~24.8); 225.9 (207.2~237.6); 0.0 (0.0~0.0); 1295 (1127~1385) ❙ 1537 (868~1965); 584 (312~994); 76 (41~115) ❙ 19.3 (-34.3~61.6); 1193 (868~1360); 405.7 (139.6~620.0)

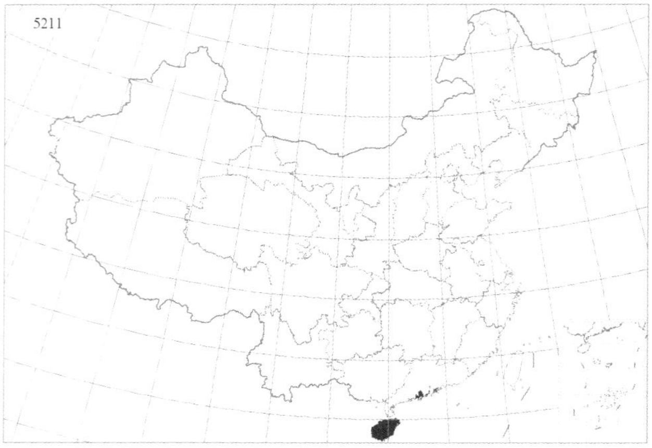

Aquifoliaceae 冬青科 *Ilex* 冬青属
Ilex cochinchinensis (Lour.) Loes. 越南冬青
Trees, evergreen

21.4 (16.2~24.7); 14.6 (6.2~19.8); 26.8 (20.1~29.2); 21.4 (16.2~24.7); 197.3 (134.2~236.5); 0.0 (0.0~0.0); 1121 (773~1376) ❙ 1902 (868~3495); 828 (312~1847); 171 (42~897) ❙ 79.2 (-34.3~246.5); 1075 (773~1331); 285.4 (119.4~620.0)

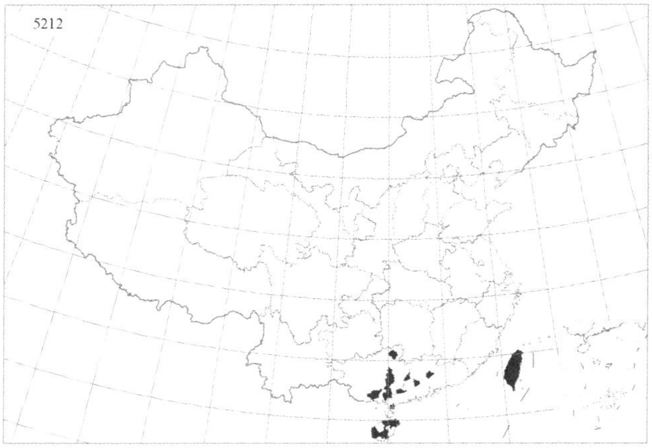

Aquifoliaceae 冬青科 *Ilex* 冬青属
Ilex confertiflora Merr. 密花冬青
Shrubs or small trees, evergreen

20.1 (12.7~24.8); 11.1 (-4.2~20.1); 27.4 (25.0~28.8); 20.1 (13.3~24.8); 182.0 (114.5~237.6); 0.5 (0.0~22.0); 1067 (802~1385) ❙ 1609 (481~2435); 712 (373~1428); 134 (11~197) ❙ 52.1 (-42.0~99.5); 1046 (481~1360); 273.3 (158.4~620.0)

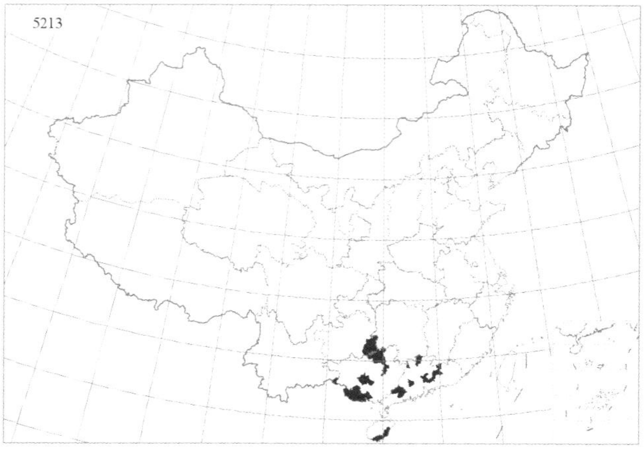

Aquifoliaceae 冬青科 *Ilex* 冬青属
Ilex confertiflora var. kwangsiensis S. Y. Hu 广西密花冬青
Shrubs or small trees, evergreen

20.4 (15.5~22.7); 11.2 (4.8~15.0); 27.7 (25.1~28.6); 20.4 (15.6~22.8); 184.8 (127.2~213.1); 0.0 (0.0~0.2); 1074 (815~1221) ❙ 1631 (1188~2435); 769 (562~1428); 145 (56~189) ❙ 53.1 (5.8~99.5); 1074 (815~1221); 252.0 (185.5~327.7)

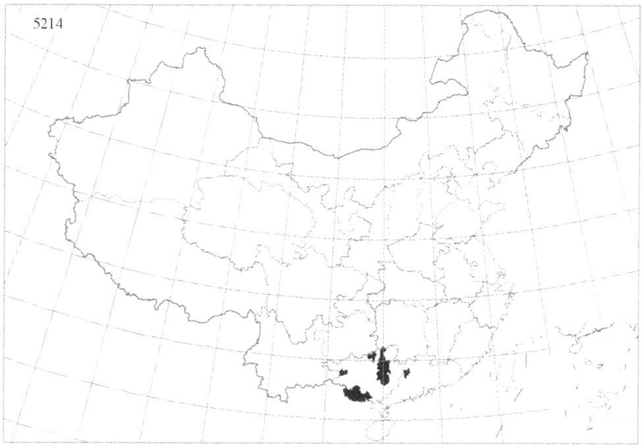

Aquifoliaceae 冬青科 *Ilex* 冬青属

Ilex corallina Franch. 珊瑚冬青

Shrubs, evergreen

15.1 (2.2~21.0); 6.3 (-6.6~15.1); 22.6 (9.5~28.1); 15.2 (3.8~21.0); 123.7 (15.5~191.3); 1.9 (0.0~49.1); 790 (411~1080) ❙ 1159 (711~1751); 589 (346~996); 57 (11~202) ❙ 48.0 (-14.2~125.9); 782 (411~1080); 264.6 (120.2~659.0)

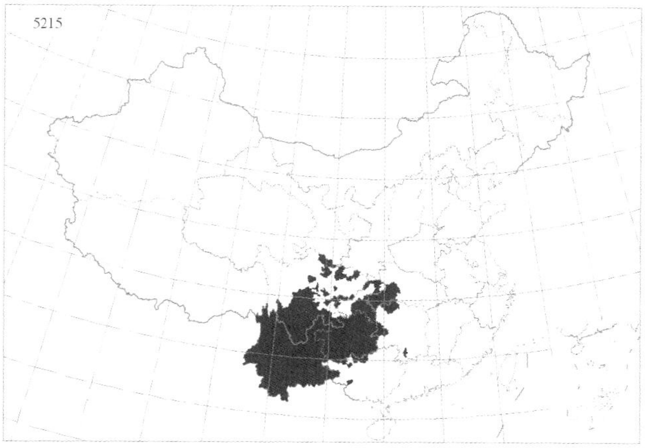

Aquifoliaceae 冬青科 *Ilex* 冬青属

Ilex corallina var. loeseneri H. Lév. 刺叶珊瑚冬青

Shrubs or trees

13.3 (9.6~17.5); 5.5 (2.4~10.9); 19.9 (15.7~25.4); 13.4 (9.6~17.6); 102.1 (60.8~151.2); 1.5 (0.0~5.4); 722 (600~846) ❙ 985 (899~1042); 573 (512~641); 29 (19~44) ❙ 39.2 (10.0~67.9); 716 (600~836); 308.3 (197.9~351.7)

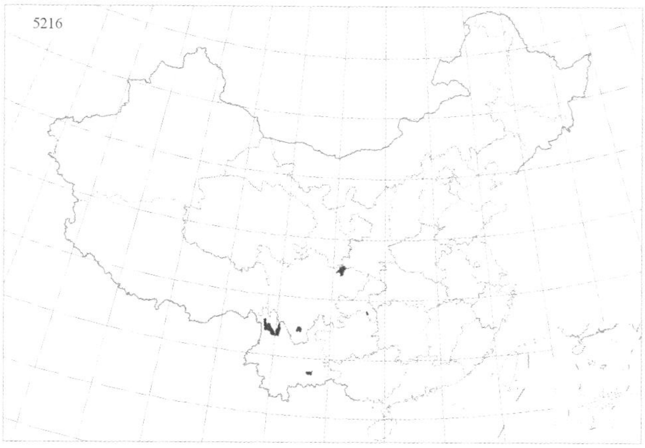

Aquifoliaceae 冬青科 *Ilex* 冬青属

Ilex cornuta Lindl. et Paxton 枸骨

Shrubs or small trees, evergreen

16.6 (9.3~24.8); 5.6 (-2.0~20.1); 26.9 (16.7~30.1); 16.7 (9.5~24.8); 141.0 (62.7~237.6); 1.2 (0.0~18.0); 897 (600~1385) ❙ 1340 (590~1950); 522 (299~836); 140 (23~259) ❙ 50.4 (-34.3~147.0); 876 (590~1360); 225.1 (111.9~620.0)

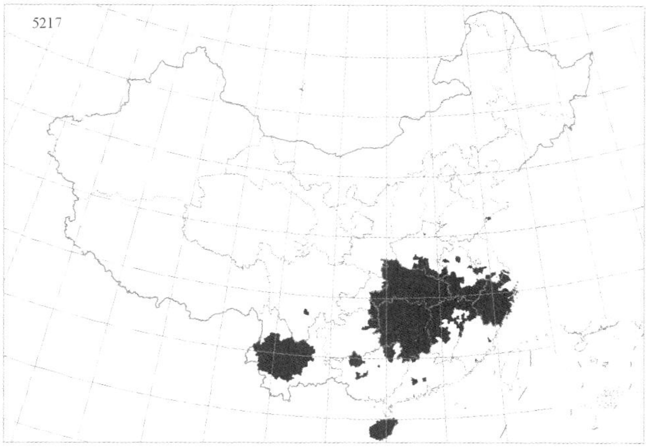

Aquifoliaceae 冬青科 *Ilex* 冬青属

Ilex crenata Thunb. 齿叶冬青

Shrubs, evergreen

18.0 (13.0~24.3); 7.9 (1.9~18.7); 27.4 (20.1~29.3); 18.1 (13.1~24.3); 157.7 (102.9~231.9); 0.2 (0.0~6.1); 960 (730~1338) ❙ 1716 (1179~3495); 614 (392~1186); 205 (42~897) ❙ 81.6 (-11.9~246.5); 949 (730~1326); 248.0 (139.6~572.9)

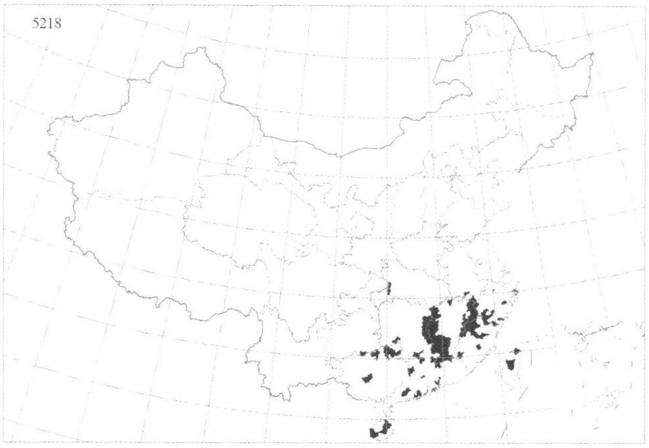

Aquifoliaceae 冬青科 *Ilex* 冬青属

Ilex cupreonitens C. Y. Wu ex Y. R. Li 铜光冬青

Trees, evergreen

18.3 (16.3~21.0); 11.4 (9.1~14.2); 23.1 (21.4~25.6); 18.3 (16.4~21.0); 159.7 (136.4~191.8); 0.0 (0.0~0.0); 891 (797~1053) ❙ 1402 (1190~1522); 785 (656~860); 58 (50~68) ❙ 58.1 (44.5~80.6); 891 (797~1053); 462.0 (347.3~551.6)

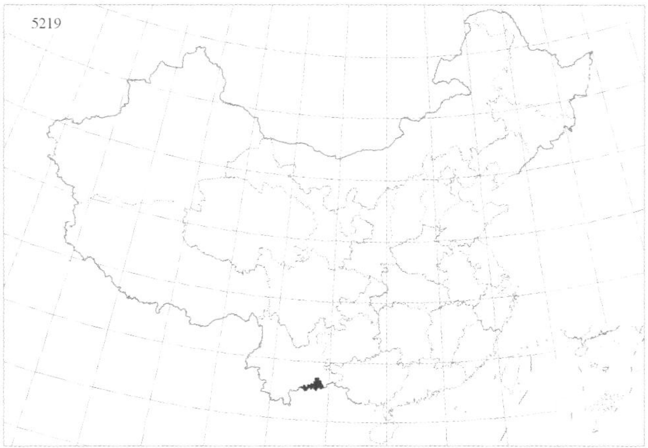

Aquifoliaceae 冬青科 *Ilex* 冬青属

Ilex cyrtura Merr. 弯尾冬青

Trees, evergreen

17.0 (8.6~21.3); 6.9 (1.3~12.8); 25.8 (15.0~28.7); 17.0 (8.7~21.3); 144.6 (52.6~196.1); 0.5 (0.0~8.8); 893 (574~1130) ❙ 1411 (1041~2005); 597 (452~1007); 124 (43~196) ❙ 59.1 (16.9~145.8); 893 (574~1130); 232.9 (162.1~447.4)

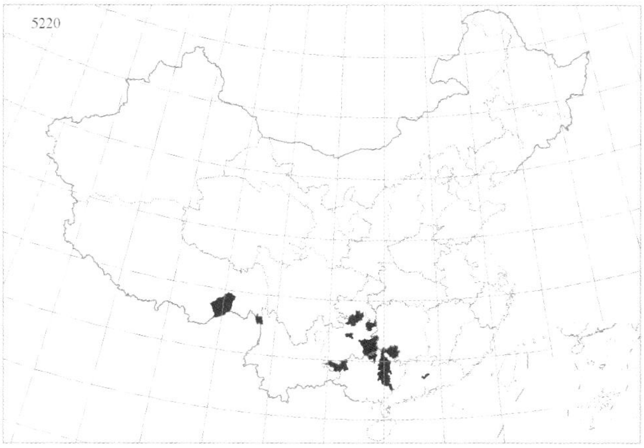

Aquifoliaceae 冬青科 *Ilex* 冬青属
Ilex dabieshanensis K. Yao et M. B. Deng 大别山冬青
Small trees, evergreen

14.3；2.0；26.3；14.3；117.5；5.4；801 ❙ 1437；616；140 ❙ 79.5；801；235.9

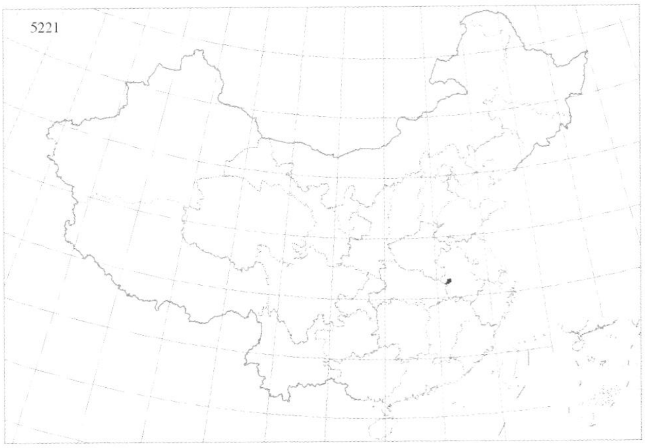

Aquifoliaceae 冬青科 *Ilex* 冬青属
Ilex dasyclada C. Y. Wu ex Y. R. Li 毛枝冬青
Shrubs, evergreen

17.3；11.0；21.6；17.3；148.0；0.0；832 ❙ 1049；598；44 ❙ 26.1；783；365.3

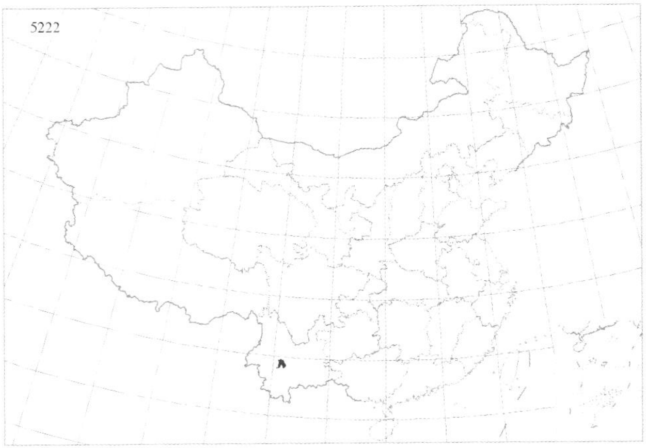

Aquifoliaceae 冬青科 *Ilex* 冬青属
Ilex dasyphylla Merr. 黄毛冬青
Shrubs or trees, evergreen

19.3 (12.6~23.0)；9.7 (3.8~15.1)；27.5 (17.9~30.1)；19.3 (12.6~23.0)；172.1 (91.6~216.2)；0.0 (0.0~1.5)；1015 (678~1247) ❙ 1661 (1057~2435)；656 (453~1428)；176 (42~259) ❙ 65.8 (12.9~130.7)；1011 (678~1247)；248.2 (148.9~349.8)

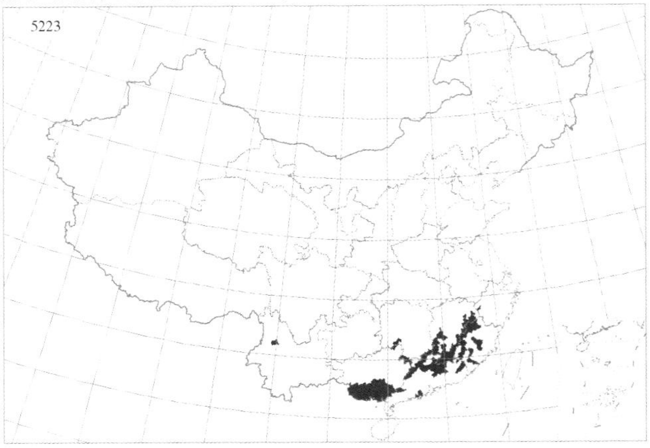

Aquifoliaceae 冬青科 *Ilex* 冬青属
Ilex dehongensis S. K. Chen et Y. X. Feng 德宏冬青
Trees, evergreen

17.7；10.9；21.8；17.7；152.7；0.0；849 ❙ 1616；946；51 ❙ 90.4；849；589.2

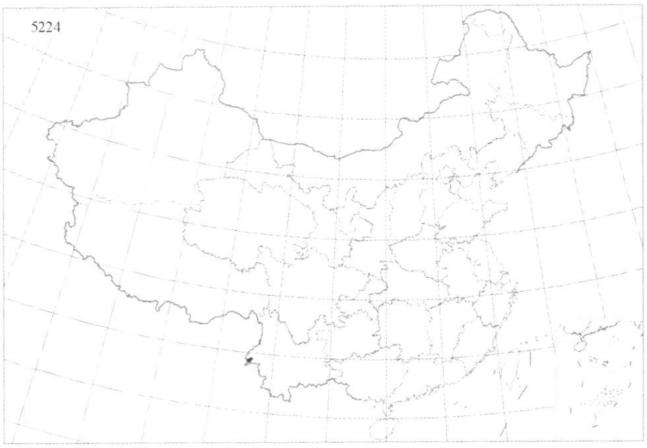

Aquifoliaceae 冬青科 *Ilex* 冬青属
Ilex delavayi Franch. 陷脉冬青
Shrubs or trees, evergreen

14.4 (5.4~19.2)；7.1 (-2.1~13.3)；20.1 (12.3~26.6)；14.4 (5.8~19.4)；114.5 (31.5~172.6)；1.4 (0.0~25.6)；744 (481~932) ❙ 1039 (741~1736)；595 (411~1007)；39 (12~107) ❙ 41.3 (-14.2~145.8)；720 (481~923)；309.8 (142.9~506.2)

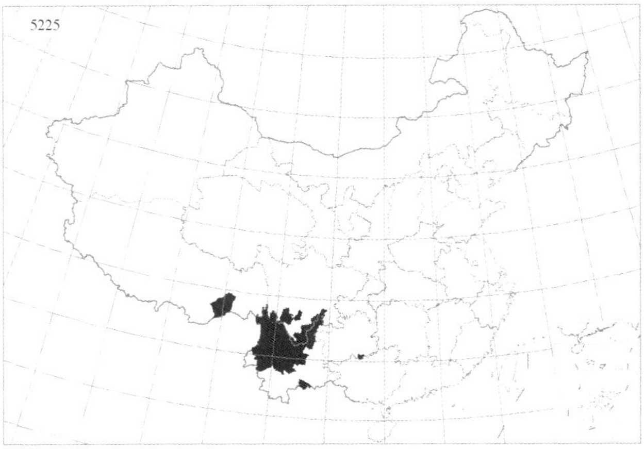

Aquifoliaceae 冬青科 *Ilex* 冬青属
Ilex delavayi var. comberiana S. Y. Hu 丽江陷脉冬青
Shrubs or trees, evergreen

12.0 (6.1~15.0)；5.2 (-1.5~8.2)；17.5 (12.7~20.0)；12.0 (6.3~15.0)；86.9 (35.2~119.9)；2.7 (0.0~21.8)；663 (496~748) ❙ 1028 (801~1164)；590 (496~651)；42 (21~59) ❙ 55.9 (36.1~78.3)；661 (496~748)；345.2 (282.9~384.1)

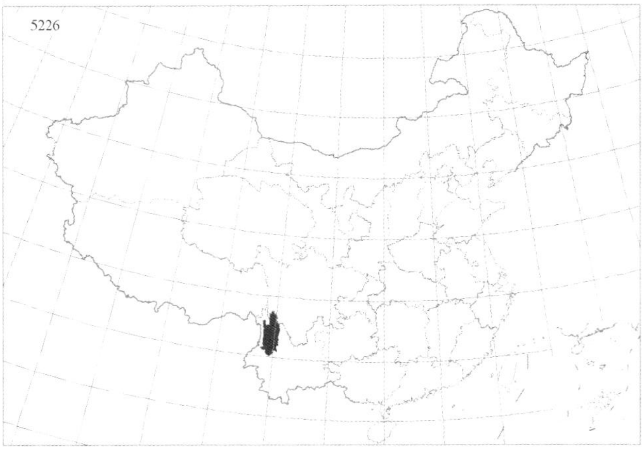

Aquifoliaceae 冬青科 *Ilex* 冬青属
Ilex delavayi var. exalata H. F. Comber 高山陷脉冬青
Shrubs or trees, evergreen

11.0；3.0；17.7；11.0；75.8；3.2；641 ❙ 1027；599；23 ❙ 60.4；641；283.7

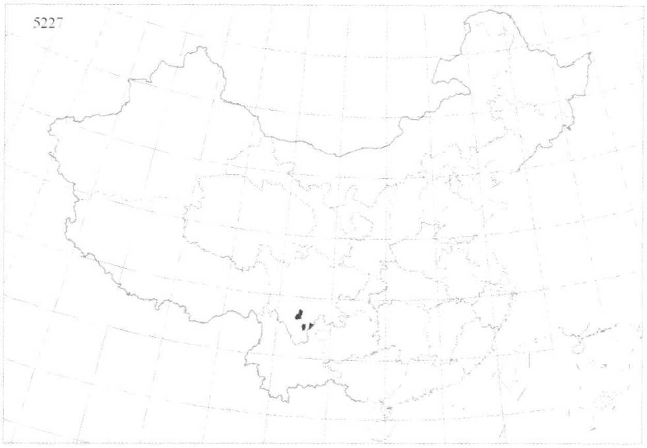

Aquifoliaceae 冬青科 *Ilex* 冬青属
Ilex delavayi var. linearifolia S. Y. Hu 线叶陷脉冬青
Shrubs or trees, evergreen

10.2；3.2；16.2；10.2；66.3；3.4；617 ❙ 988；586；32 ❙ 60.4；617；335.8

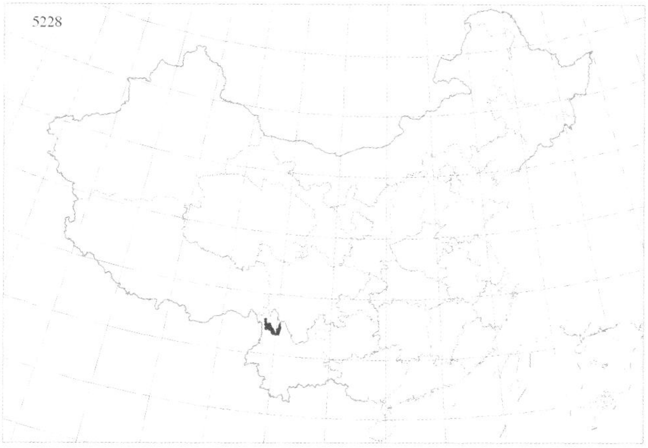

Aquifoliaceae 冬青科 *Ilex* 冬青属
Ilex delavayi var. muliensis D. Fang et Z. M. Tan 木里陷脉冬青
Shrubs or trees, evergreen

6.6；−1.3；13.3；6.8；39.6；19.7；514 ❙ 862；548；12 ❙ 67.7；514；311.5

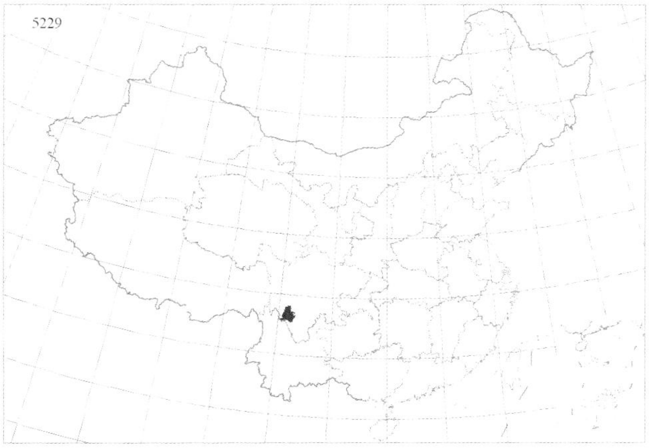

Aquifoliaceae 冬青科 *Ilex* 冬青属
Ilex denticulata Wall. ex Wight 细齿冬青
Trees, evergreen

18.3；10.6；23.9；18.3；159.7；0.0；897 ❙ 1440；662；83 ❙ 60.4；883；379.6

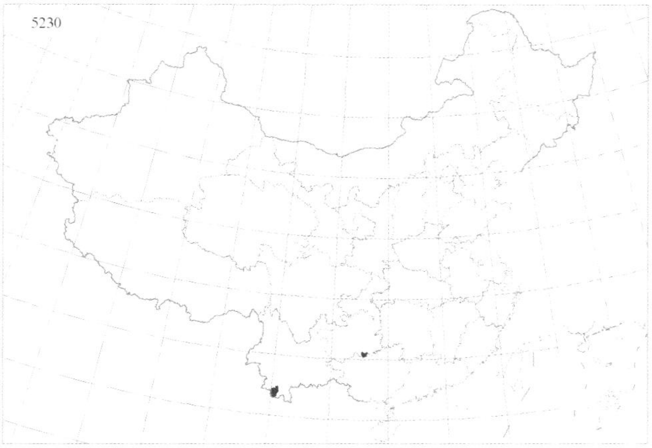

Aquifoliaceae 冬青科 *Ilex* 冬青属
Ilex dianguiensis C. J. Tseng 滇贵冬青
Small trees, evergreen

13.4；5.1；20.6；13.5；102.7；0.9；713 ❙ 978；536；45 ❙ 37.2；713；208.7

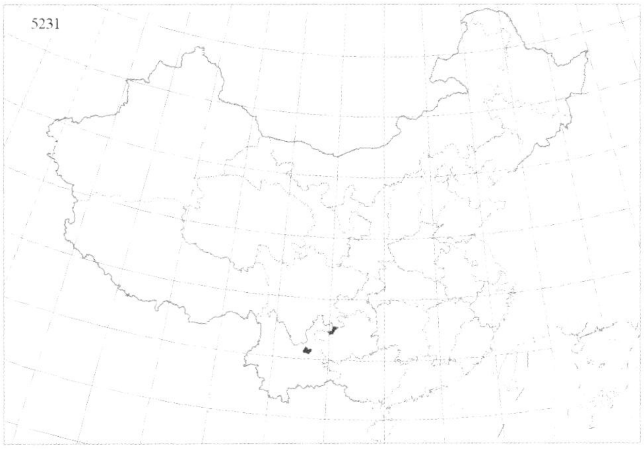

Aquifoliaceae 冬青科 *Ilex* 冬青属
Ilex dicarpa Y. R. Li 双果冬青
Small trees, evergreen

13.3；5.4；19.6；13.4；100.4；0.0；706 ❙ 1736；1007；51 ❙ 145.8；706；447.4

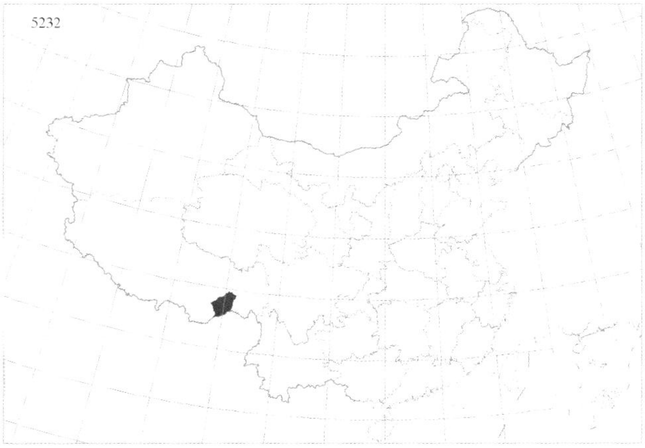

Aquifoliaceae 冬青科 *Ilex* 冬青属
Ilex dipyrena Wall. 双核枸骨
Trees, evergreen

15.0 (-1.8~21.0); 7.6 (-10.2~15.1); 20.8 (6.1~27.5); 15.1 (1.8~21.0); 123.5 (1.9~191.8); 2.5 (0.0~82.8); 771 (329~1053) ❙ 1122 (320~1751); 620 (217~1007); 47 (9~176) ❙ 45.7 (-14.2~145.8); 756 (320~1053); 319.8 (72.2~659.0)

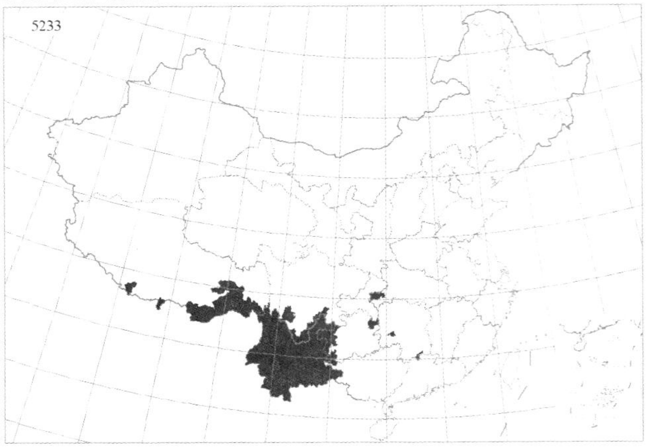

Aquifoliaceae 冬青科 *Ilex* 冬青属
Ilex dolichopoda Merr. et Chun 长柄冬青
Trees, evergreen

23.5; 18.4; 27.0; 23.5; 222.6; 0.0; 1259 ❙ 1638; 575; 58 ❙ 30.4; 1155; 596.5

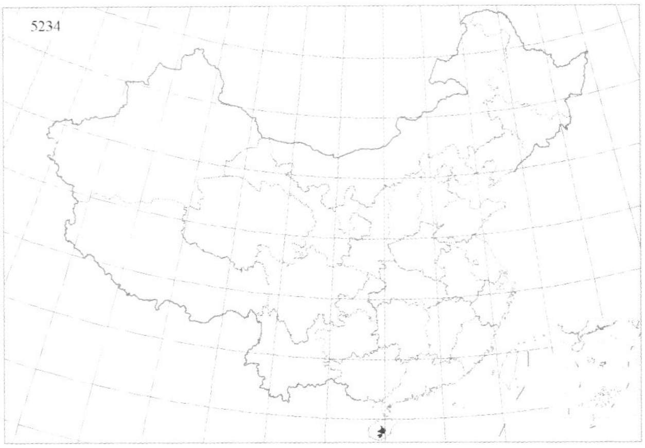

Aquifoliaceae 冬青科 *Ilex* 冬青属
Ilex dunniana H. Lév. 龙里冬青
Trees, evergreen

15.2 (6.2~22.2); 5.5 (-2.4~13.9); 24.0 (14.1~29.3); 15.3 (6.7~22.3); 124.8 (39.4~207.5); 1.4 (0.0~24.1); 810 (509~1190) ❙ 1229 (797~1679); 571 (411~850); 84 (19~225) ❙ 52.4 (-14.2~107.2); 798 (509~1190); 234.7 (134.6~384.7)

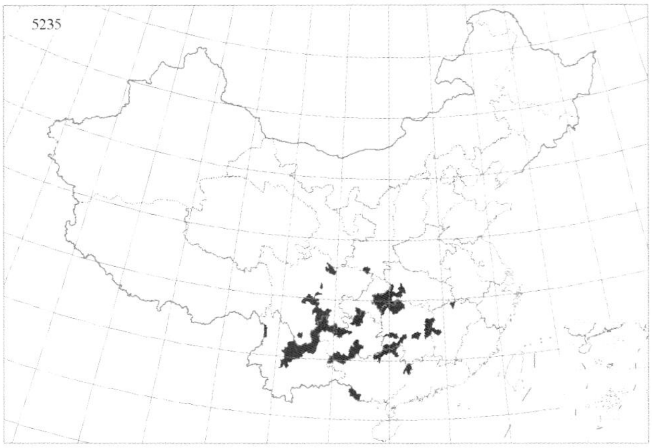

Aquifoliaceae 冬青科 *Ilex* 冬青属
Ilex editicostata Hu et T. Tang 显脉冬青
Shrubs or small trees, evergreen

17.8 (12.7~23.6); 7.6 (2.1~17.8); 27.0 (20.6~29.4); 17.8 (12.8~23.6); 154.3 (99.0~223.3); 0.2 (0.0~5.6); 939 (713~1271) ❙ 1566 (1133~2435); 601 (411~1428); 162 (43~259) ❙ 68.1 (-2.2~147.0); 935 (713~1230); 232.8 (140.5~627.0)

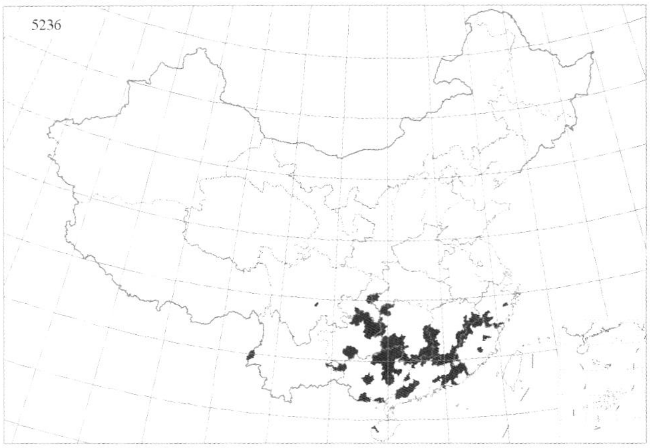

Aquifoliaceae 冬青科 *Ilex* 冬青属
Ilex elmerrilliana S. Y. Hu 厚叶冬青
Shrubs or small trees, evergreen

17.2 (11.4~23.6); 6.5 (-0.9~17.8); 27.2 (22.6~29.3); 17.3 (11.5~23.6); 147.9 (90.8~223.3); 0.6 (0.0~13.8); 914 (686~1271) ❙ 1584 (815~1997); 562 (352~891); 185 (26~259) ❙ 74.0 (-2.2~147.0); 907 (686~1153); 235.5 (151.0~439.7)

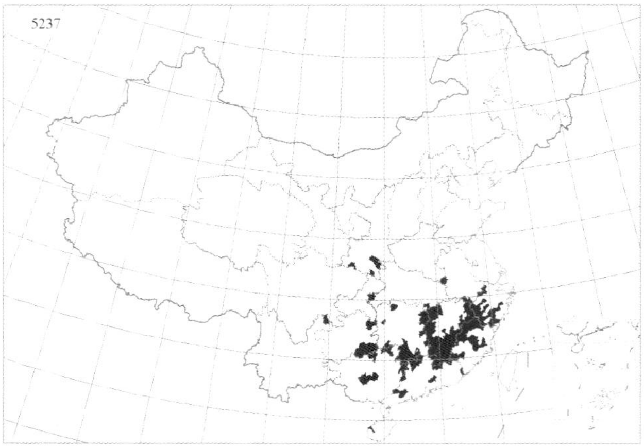

Aquifoliaceae 冬青科 *Ilex* 冬青属
Ilex estriata C. J. Tseng 平核冬青
Shrubs, evergreen

15.5; 5.5; 24.8; 15.6; 127.3; 0.6; 833 ❙ 1084; 531; 49 ❙ 33.4; 833; 191.9

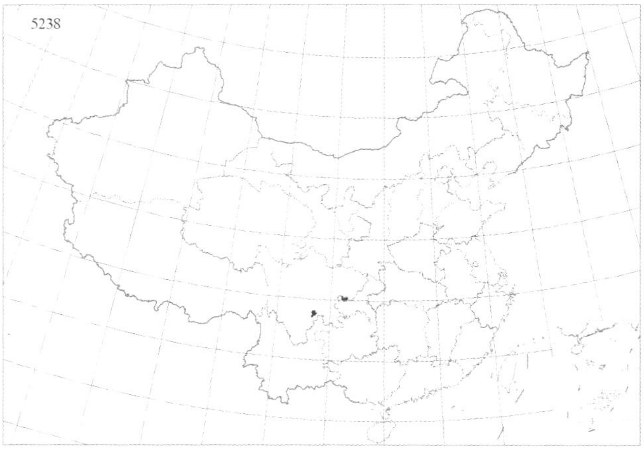

Aquifoliaceae 冬青科 *Ilex* 冬青属

Ilex euryoides C. J. Tseng 柃叶冬青

Small trees, evergreen

12.7；2.1；23.0；12 8；99.0；5.6；713 ▎1478；592；85 ▎107.2；713；234.7

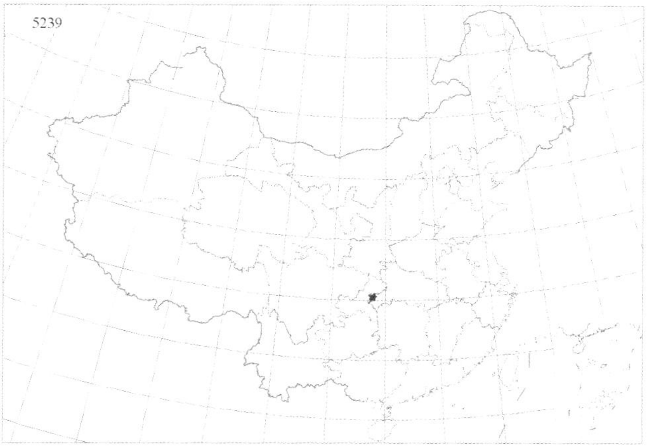

Aquifoliaceae 冬青科 *Ilex* 冬青属

Ilex excelsa (Wall.) Hook. f. 高冬青

Trees, evergreen

16.9 (11.3~21.0)；10.3 (4.3~15.1)；21.5 (17.0~27.7)；17.0 (11.3~21.0)；143.5 (76.6~191.8)；0.0 (0.0~0.7)；826 (644~1053) ▎1191 (797~1751)；664 (411~996)；49 (23~166) ▎44.0 (-14.2~112.2)；805 (644~1053)；372.2 (218.0~659.0)

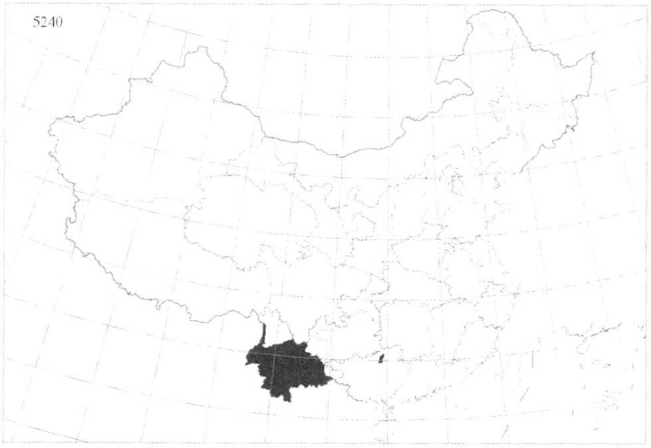

Aquifoliaceae 冬青科 *Ilex* 冬青属

Ilex excelsa var. hypotricha (Loes.) S. Y. Hu 毛背高冬青

Trees, evergreen

17.7；10.9；22.5；17.8；153.4；0.0；854 ▎1368；756；59 ▎60.1；854；491.5

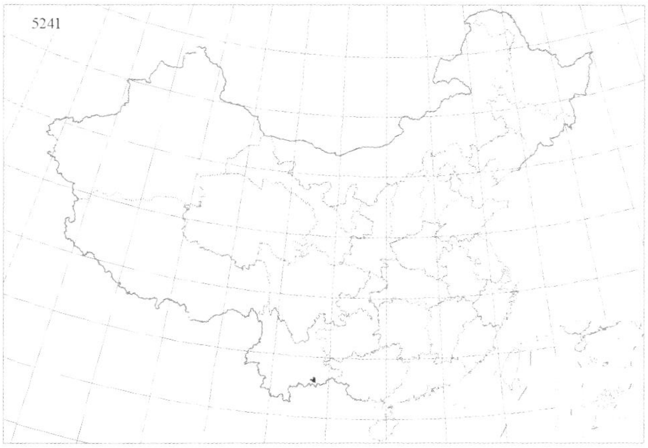

Aquifoliaceae 冬青科 *Ilex* 冬青属

Ilex fargesii Franch. 狭叶冬青

Trees, evergreen

12.7 (2.2~22.7)；2.3 (-7.0~15.0)；22.5 (9.9~28.6)；12.9 (3.8~22.8)；102.4 (15.5~213.1)；8.9 (0.0~49.1)；737 (413~1221) ▎1131 (632~2435)；541 (306~1428)；51 (14~142) ▎55.3 (0.0~125.9)；737 (413~1221)；252.5 (147.7~334.4)

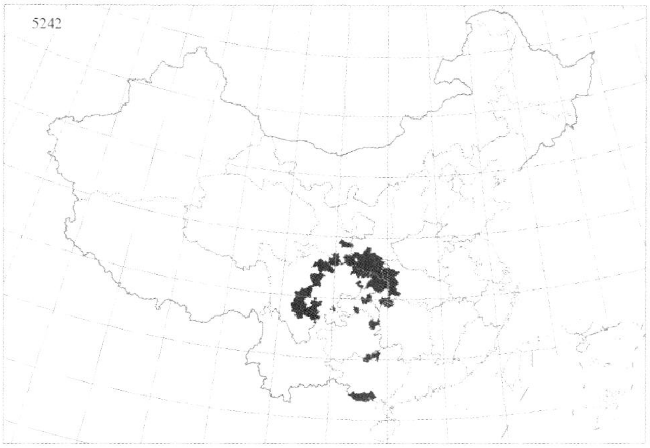

Aquifoliaceae 冬青科 *Ilex* 冬青属

Ilex fargesii var. angustifolia C. Y. Chang 线叶冬青

Trees, evergreen

10.4；-1.5；21.7；10.5；81.0；16.3；650 ▎800；369；22 ▎23.8；647；319.0

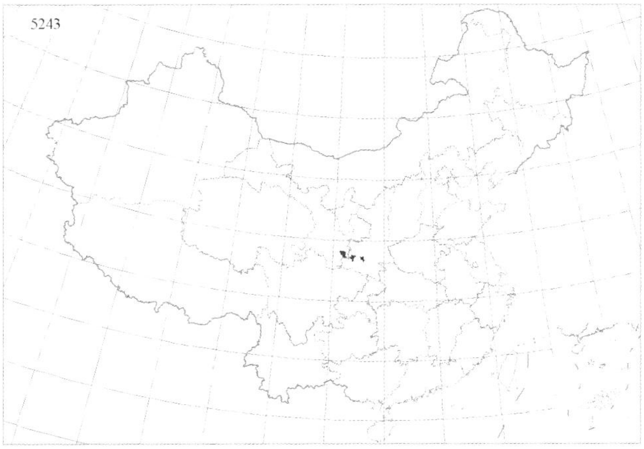

Aquifoliaceae 冬青科 *Ilex* 冬青属

Ilex fengqingensis C. Y. Wu ex Y. R. Li 凤庆冬青

Trees, evergreen

14.2；7.4；19.3；14.3；112.9；1.7；731 ▎1117；636；38 ▎53.8；726；358.5

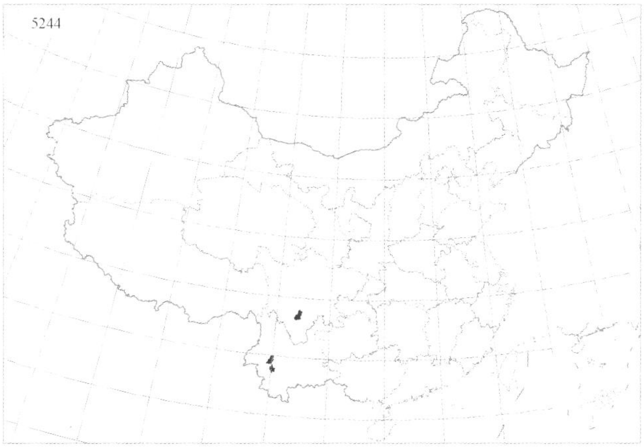

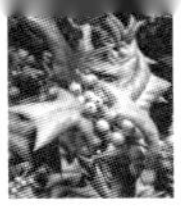

Aquifoliaceae 冬青科 *Ilex* 冬青属
Ilex ferruginea Hand.-Mazz. 锈毛冬青
Shrubs or trees, evergreen

16.0 (12.4~21.0); 7.2 (3.8~14.2); 23.4 (18.6~27.9); 16.1 (12.4~21.0); 133.0 (89.1~191.8); 0.2 (0.0~1.5); 820 (675~1053) ❙ 1190 (896~1710); 596 (473~818); 66 (29~186) ❙ 44.8 (12.7~72.9); 817 (675~1053); 246.9 (141.2~551.6)

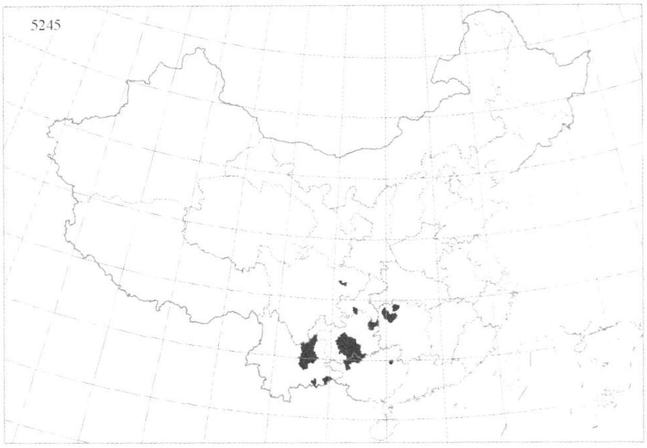

Aquifoliaceae 冬青科 *Ilex* 冬青属
Ilex ficifolia C. J. Tseng ex S. K. Chen et Y. X. Feng 硬叶冬青
Trees, evergreen

16.8 (11.4~19.4); 5.9 (2.1~11.0); 27.0 (18.8~29.3); 16.9 (11.4~19.4); 143.1 (80.7~173.4); 0.4 (0.0~5.6); 894 (652~1029) ❙ 1583 (874~1997); 563 (427~848); 184 (25~259) ❙ 77.5 (28.8~147.0); 891 (652~1029); 234.6 (166.9~313.7)

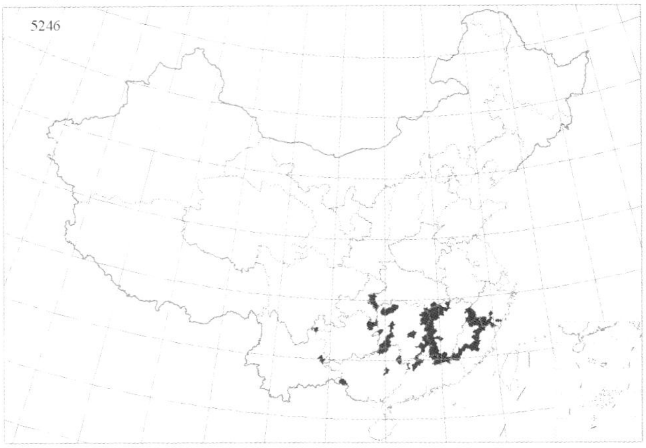

Aquifoliaceae 冬青科 *Ilex* 冬青属
Ilex ficoidea Hemsl. 榕叶冬青
Trees, evergreen

17.7 (2.0~24.8); 7.6 (-7.2~20.1); 26.8 (10.1~29.8); 17.7 (3.8~24.8); 153.4 (15.4~237.6); 0.5 (0.0~51.4); 937 (413~1385) ❙ 1529 (863~3495); 605 (392~1847); 151 (23~897) ❙ 64.4 (-11.9~246.5); 926 (413~1273); 228.9 (108.2~620.0)

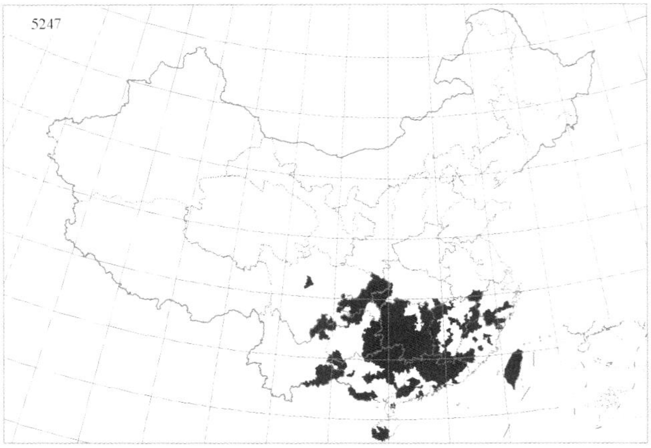

Aquifoliaceae 冬青科 *Ilex* 冬青属
Ilex formosana Maxim. 台湾冬青
Shrubs or trees, evergreen

17.5 (11.9~22.2); 7.8 (2.7~17.7); 26.0 (19.3~29.9); 17.5 (12.0~22.2); 150.4 (87.2~206.9); 0.2 (0.0~3.4); 907 (671~1177) ❙ 1480 (911~2993); 600 (392~1847); 144 (29~547) ❙ 63.3 (5.8~246.5); 896 (671~1177); 258.9 (108.2~659.0)

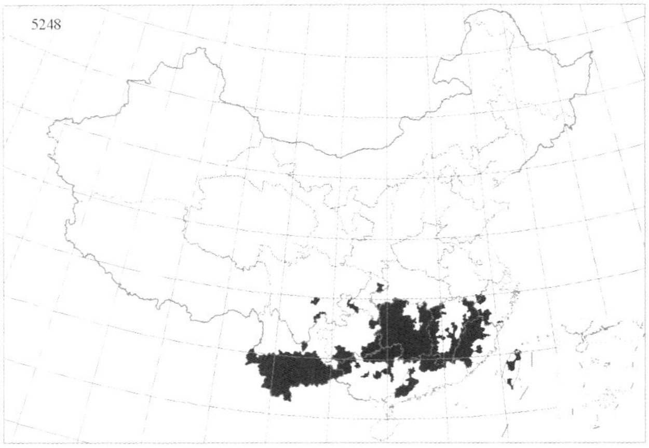

Aquifoliaceae 冬青科 *Ilex* 冬青属
Ilex formosana var. macropyrena S. Y. Hu 大核台湾冬青
Shrubs or trees, evergreen

19.7; 10.3; 27.6; 19.7; 176.3; 0.0; 1022 ❙ 1803; 680; 192 ❙ 76.3; 1022; 267.2

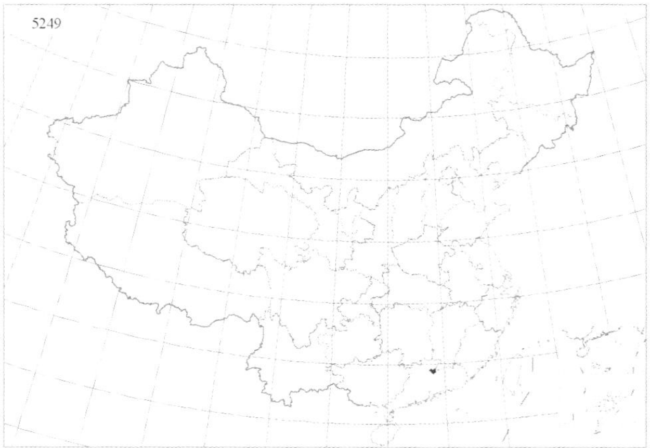

Aquifoliaceae 冬青科 *Ilex* 冬青属
Ilex forrestii H. F. Comber 滇西冬青
Shrubs or small trees, evergreen

12.9 (5.4~18.0); 5.9 (-2.1~10.0); 18.5 (12.3~28.1); 13.0 (5.8~18.0); 98.2 (31.5~156.3); 2.8 (0.0~25.6); 696 (481~958) ❙ 1066 (741~1454); 596 (426~792); 46 (18~70) ❙ 55.0 (11.7~99.9); 690 (481~958); 340.4 (128.7~487.2)

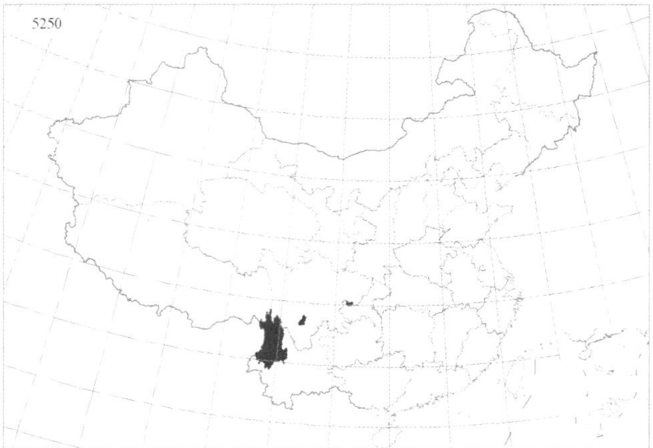

Aquifoliaceae 冬青科 *Ilex* 冬青属
Ilex forrestii var. glabra S. Y. Hu 无毛滇西冬青
Shrubs or small trees, evergreen

10.9；3.6；17.0；10.9；76.0；4.7；638 ❙ 1021；597；36 ❙ 62.1；622；323.0

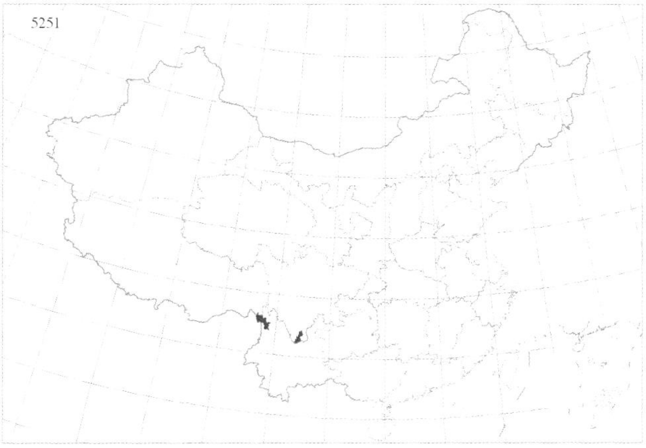

Aquifoliaceae 冬青科 *Ilex* 冬青属
Ilex fragilis Hook. f. 薄叶冬青
Shrubs or small trees, deciduous

14.6 (−1.8~21.0)；6.9 (−10.2~15.1)；20.6 (6.1~26.6)；14.7 (1.8~21.0)；118.5 (1.9~191.8)；3.1 (0.0~82.8)；758 (329~1053) ❙ 1122 (320~1751)；620 (217~996)；44 (10~129) ❙ 49.1 (−14.2~125.9)；745 (320~1053)；308.6 (61.6~659.0)

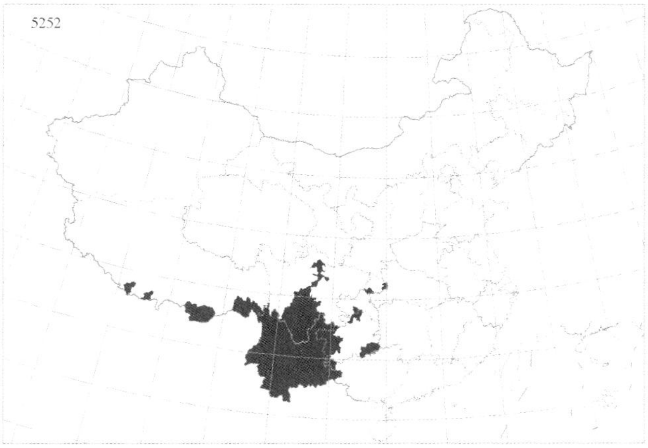

Aquifoliaceae 冬青科 *Ilex* 冬青属
Ilex franchetiana Loes. 康定冬青
Shrubs or small trees, evergreen

13.9 (2.2~19.6)；5.6 (−6.6~12.8)；21.0 (9.9~27.5)；14.0 (3.8~19.6)；110.7 (15.5~175.7)；3.0 (0.0~49.1)；739 (413~953) ❙ 1097 (797~1688)；581 (411~959)；49 (15~184) ❙ 50.2 (−14.2~125.9)；725 (413~935)；283.5 (133.6~627.0)

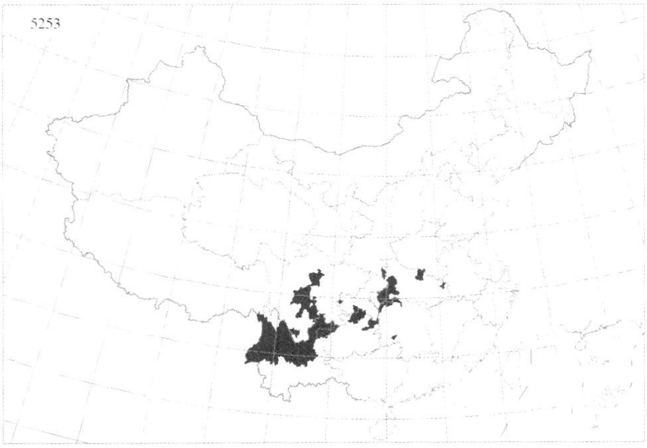

Aquifoliaceae 冬青科 *Ilex* 冬青属
Ilex franchetiana var. parvifolia S. Y. Hu 小叶康定冬青
Shrubs or small trees, evergreen

12.3 (10.2~14.9)；2.9 (0.9~5.3)；20.9 (16.9~23.6)；12.3 (10.3~15.0)；92.4 (68.2~119.6)；4.4 (0.0~8.7)；693 (620~782) ❙ 1194 (1006~1414)；657 (429~817)；37 (18~49) ❙ 71.9 (41.8~89.0)；693 (620~782)；257.8 (214.5~298.0)

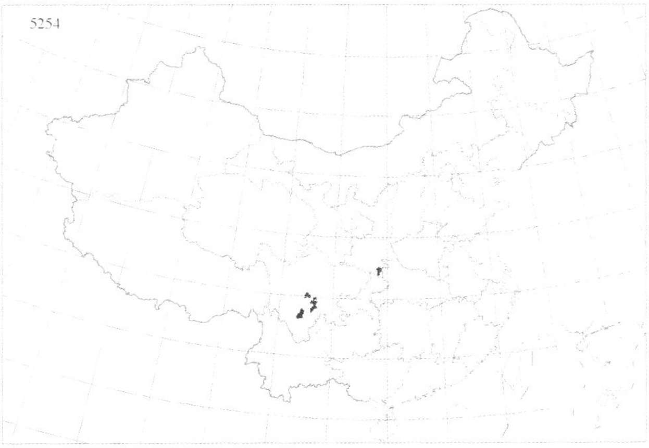

Aquifoliaceae 冬青科 *Ilex* 冬青属
Ilex fukienensis S. Y. Hu 福建冬青
Shrubs, evergreen

17.7 (16.6~19.1)；8.0 (5.9~10.0)；26.7 (25.8~27.6)；17.7 (16.6~19.1)；152.9 (138.9~169.1)；0.0 (0.0~0.0)；915 (862~973) ❙ 1797 (1702~1997)；620 (583~669)；218 (193~259) ❙ 97.1 (76.3~130.7)；915 (862~973)；289.6 (267.5~303.4)

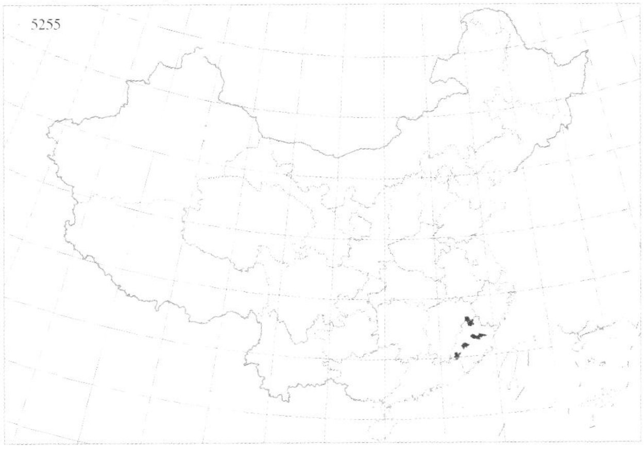

Aquifoliaceae 冬青科 *Ilex* 冬青属
Ilex georgei H. F. Comber 长叶枸骨
Shrubs or small trees, evergreen

15.6 (2.2~18.9)；8.9 (−6.6~12.5)；20.4 (9.9~23.9)；15.6 (3.8~19.0)；128.3 (15.5~167.8)；1.1 (0.0~49.1)；775 (413~929) ❙ 1066 (797~1454)；608 (411~816)；42 (19~70) ❙ 39.3 (−14.2~125.9)；746 (413~851)；333.4 (218.0~516.7)

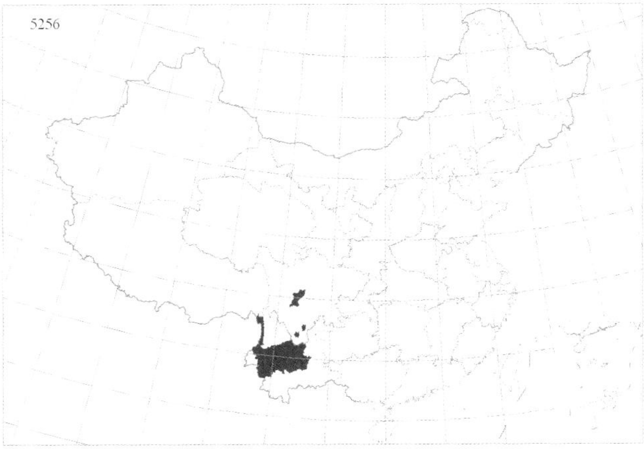

Aquifoliaceae 冬青科 *Ilex* 冬青属
Ilex gintungensis H. W. Li ex Y. R. Li 景东冬青
Shrubs or trees, evergreen

16.7; 10.3; 21.1; 16.7; 141.0; 0.0; 808 | 1066; 599; 47 | 32.1; 781; 385.1

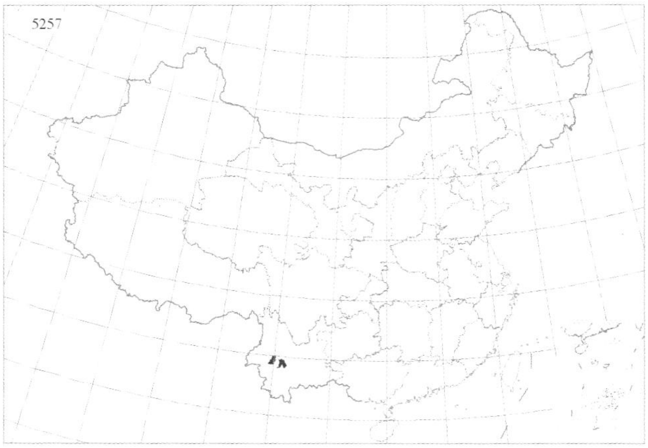

Aquifoliaceae 冬青科 *Ilex* 冬青属
Ilex glomerata King 团花冬青
Trees, evergreen

19.8 (15.5~22.7); 10.3 (4.8~15.0); 27.6 (25.1~28.6); 19.8 (15.6~22.8); 177.8 (127.2~213.1); 0.0 (0.0~0.2); 1041 (815~1221) | 1625 (1302~2435); 729 (505~1428); 153 (86~189) | 57.1 (12.9~99.5); 1040 (815~1221); 242.9 (185.5~327.7)

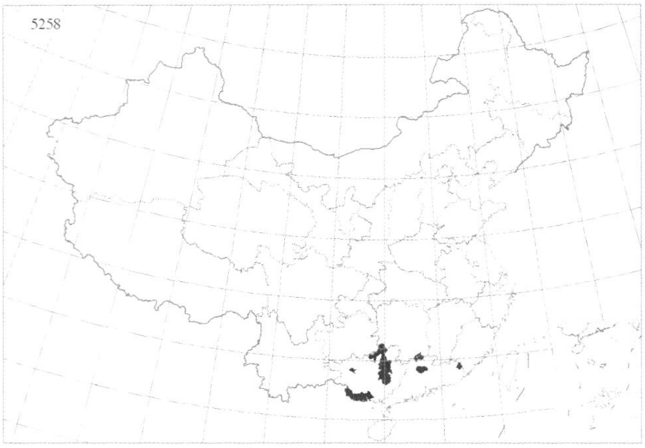

Aquifoliaceae 冬青科 *Ilex* 冬青属
Ilex godajam (Colebr. ex Wall.) Wall. ex Hook. f. 伞花冬青
Shrubs or trees, evergreen

19.8 (14.3~24.8); 11.8 (3.3~20.1); 26.0 (20.3~28.9); 19.8 (14.3~24.8); 177.6 (114.1~237.6); 0.1 (0.0~2.3); 1024 (765~1385) | 1500 (974~2355); 658 (392~996); 111 (37~308) | 50.2 (-12.0~172.5); 995 (717~1360); 349.0 (185.5~659.0)

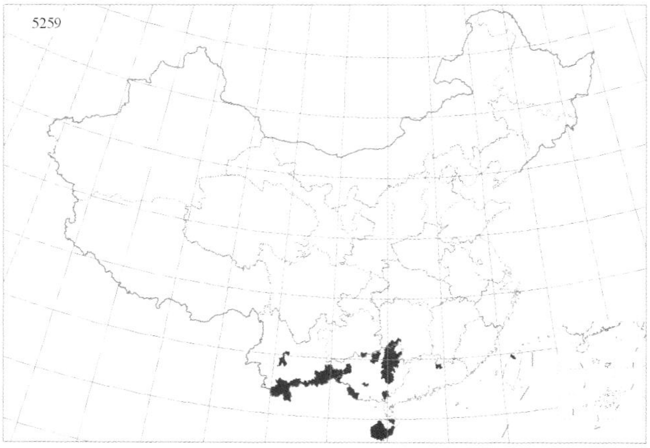

Aquifoliaceae 冬青科 *Ilex* 冬青属
Ilex goshiensis Hayata 海岛冬青
Shrubs or trees, evergreen

20.5 (16.2~24.1); 14.9 (11.0~19.1); 25.2 (20.1~28.2); 20.6 (16.2~24.1); 187.0 (134.2~229.7); 0.0 (0.0~0.0); 1045 (773~1322) | 2317 (1380~3495); 1050 (455~1847); 237 (47~897) | 134.1 (6.7~246.5); 996 (773~1147); 271.6 (119.4~620.0)

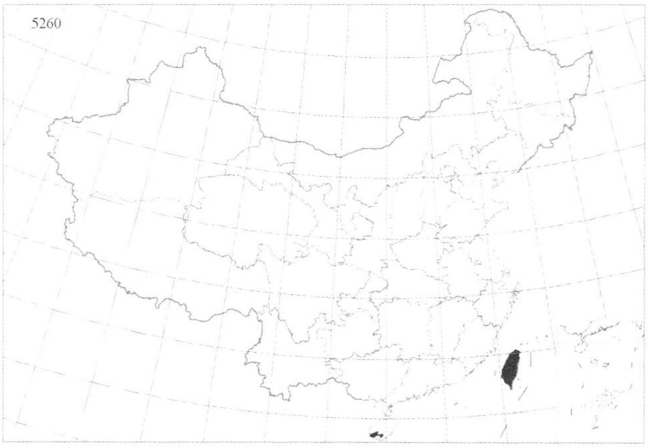

Aquifoliaceae 冬青科 *Ilex* 冬青属
Ilex graciliflora Champ. ex Benth. 纤花冬青
Trees, evergreen

22.2 (19.7~22.6); 14.3 (10.3~15.2); 28.6 (27.6~28.9); 22.2 (19.7~22.7); 206.9 (176.3~212.7); 0.0 (0.0~0.0); 1186 (1022~1216) | 1766 (1447~2037); 859 (680~1033); 116 (82~192) | 49.3 (20.8~76.3); 1186 (1022~1216); 167.6 (95.3~267.2)

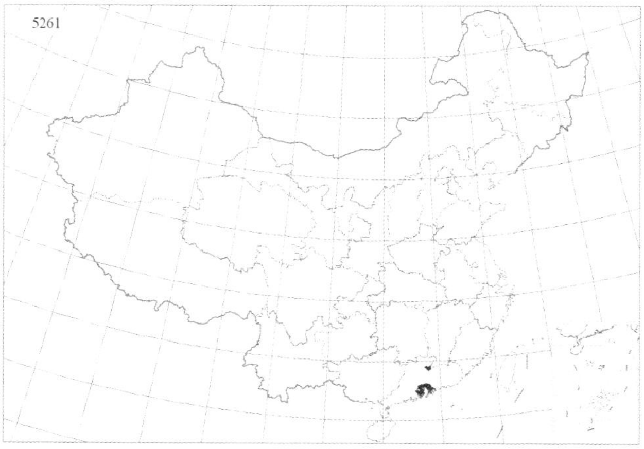

Aquifoliaceae 冬青科 *Ilex* 冬青属
Ilex gracilis C. J. Tseng 纤枝冬青
Small trees, evergreen

8.6; 1.3; 15.0; 8.7; 52.6; 8.8; 574 | 1041; 585; 45 | 81.5; 574; 352.3

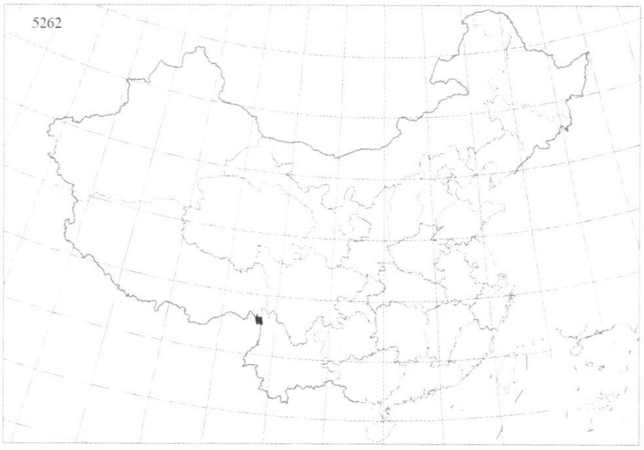

Aquifoliaceae 冬青科 *Ilex* 冬青属
Ilex guangnanensis C. J. Tseng et Y. R. Li 广南冬青
Trees, evergreen

17.2；9.1；23.1；17.3；147.5；0.0；847 ❙ 1161；633；50 ❙ 37.1；847；314.4

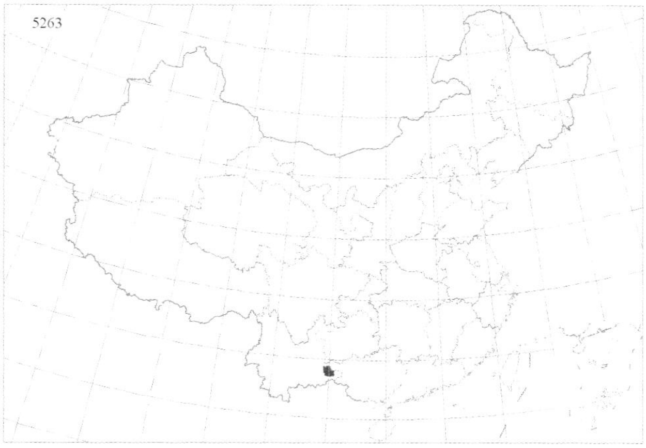

Aquifoliaceae 冬青科 *Ilex* 冬青属
Ilex guizhouensis C. J. Tseng 贵州冬青
Shrubs or trees, evergreen

15.6；5.5；24.6；15.6；127.5；0.0；808 ❙ 1206；546；69 ❙ 49.2；808；189.4

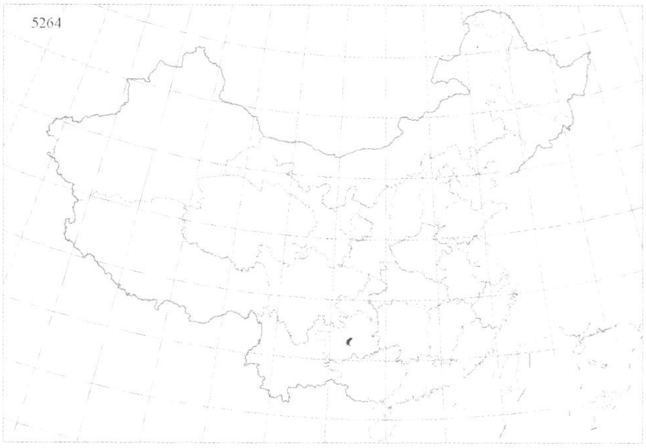

Aquifoliaceae 冬青科 *Ilex* 冬青属
Ilex hainanensis Merr. 海南冬青
Trees, evergreen

20.0 (15.0~24.8)；11.9 (4.4~20.1)；26.5 (21.4~29.0)；20.1 (15.0~24.8)；181.1 (121.0~237.6)；0.0 (0.0~0.6)；1045 (797~1385) ❙ 1488 (958~2435)；704 (392~1428)；105 (41~194) ❙ 43.7 (-11.9~99.5)；1029 (797~1299)；296.0 (122.7~620.0)

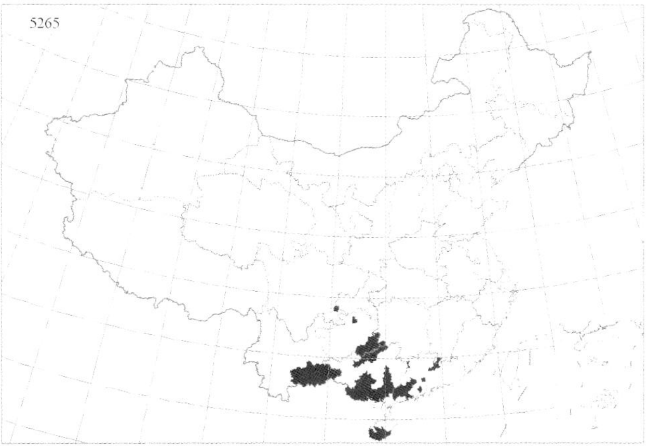

Aquifoliaceae 冬青科 *Ilex* 冬青属
Ilex hanceana Maxim. 青茶香
Shrubs or small trees, evergreen

19.4 (14.3~24.3)；9.9 (3.3~18.7)；27.6 (24.3~29.8)；19.5 (14.3~24.3)；173.7 (114.1~231.9)；0.1 (0.0~2.3)；1026 (765~1338) ❙ 1600 (1179~2435)；643 (392~1428)；162 (42~230) ❙ 58.2 (-11.9~106.4)；1009 (765~1221)；248.9 (139.6~590.7)

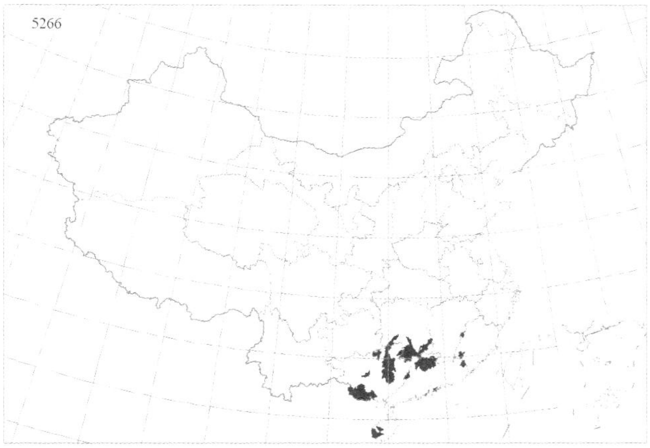

Aquifoliaceae 冬青科 *Ilex* 冬青属
Ilex hayatana Loes. 早田氏冬青
Small trees, evergreen

18.2；12.4；22.9；18.2；158.4；0.0；890 ❙ 2439；1056；253 ❙ 184.5；879；259.8

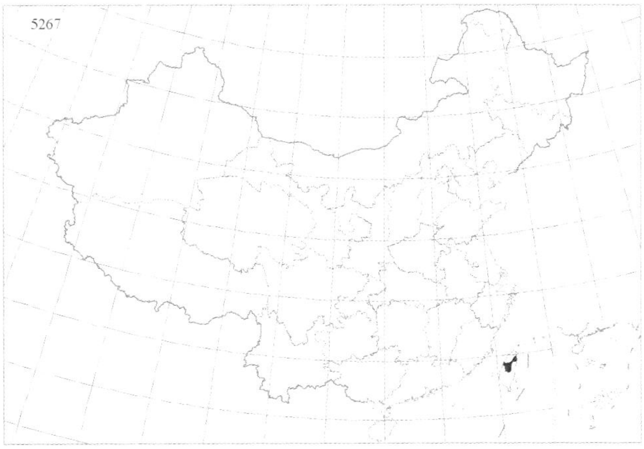

Aquifoliaceae 冬青科 *Ilex* 冬青属
Ilex hirsuta C. J. Tseng ex S. K. Chen et Y. X. Feng 硬毛冬青
Small trees, evergreen

16.5 (12.1~21.6)；4.9 (1.1~12.6)；27.4 (22.9~29.9)；16.5 (12.2~21.6)；139.0 (95.0~199.3)；0.6 (0.0~8.5)；888 (696~1150) ❙ 1419 (1087~1742)；496 (392~650)；160 (64~225) ❙ 60.6 (14.2~128.9)；871 (696~1148)；194.3 (134.6~285.1)

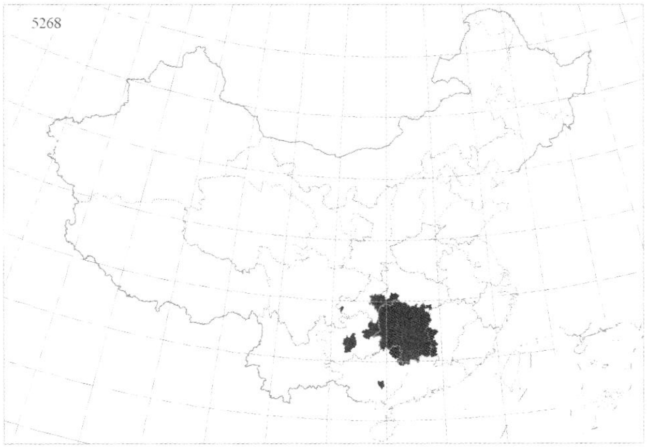

Aquifoliaceae 冬青科 *Ilex* 冬青属
Ilex hookeri King 贡山冬青
Trees, evergreen

13.4 (−0.7~19.6); 6.5 (−9.6~12.8); 18.6 (7.7~23.5); 13.6 (2.5~19.6); 106.5 (6.8~175.7); 5.3 (0.0~74.2); 715 (338~953) ❙ 1223 (370~1688); 695 (262~959); 50 (10~70) ❙ 69.1 (4.7~112.2); 713 (338~932); 397.3 (61.6~627.0)

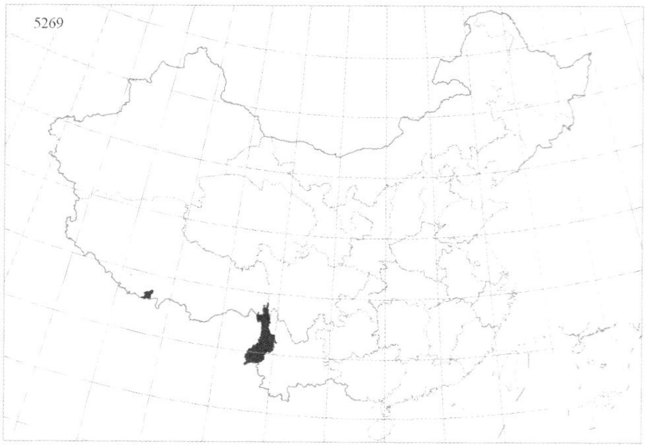

Aquifoliaceae 冬青科 *Ilex* 冬青属
Ilex huana C. J. Tseng ex S. K. Chen et Y. X. Feng 秀英冬青
Shrubs, evergreen

22.4 (15.0~24.7); 15.6 (4.4~19.8); 27.7 (25.2~28.7); 22.4 (15.0~24.7); 208.8 (121.0~236.5); 0.1 (0.0~0.6); 1218 (798~1376) ❙ 1579 (1141~1854); 579 (455~862); 89 (48~189) ❙ 31.1 (6.7~53.5); 1146 (798~1360); 387.8 (139.6~620.0)

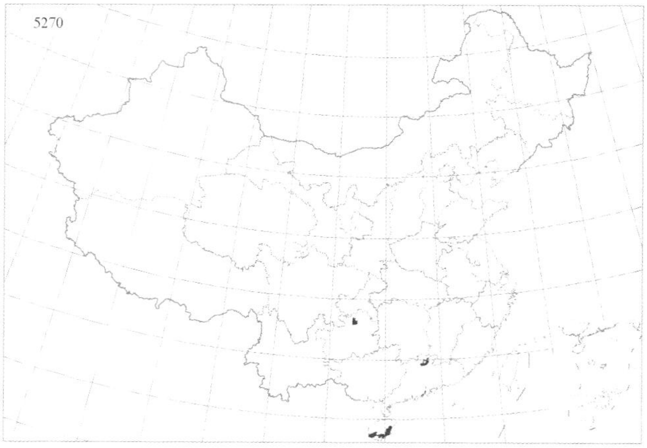

Aquifoliaceae 冬青科 *Ilex* 冬青属
Ilex hylonoma Hu et T. Tang 细刺枸骨
Trees, evergreen

17.0 (12.0~22.2); 6.3 (0.7~13.9); 26.8 (22.5~29.7); 17.0 (12.0~22.3); 145.1 (92.3~207.5); 0.6 (0.0~9.4); 899 (690~1190) ❙ 1469 (951~1776); 546 (400~816); 159 (26~225) ❙ 64.3 (9.3~128.9); 892 (690~1190); 223.1 (123.1~327.7)

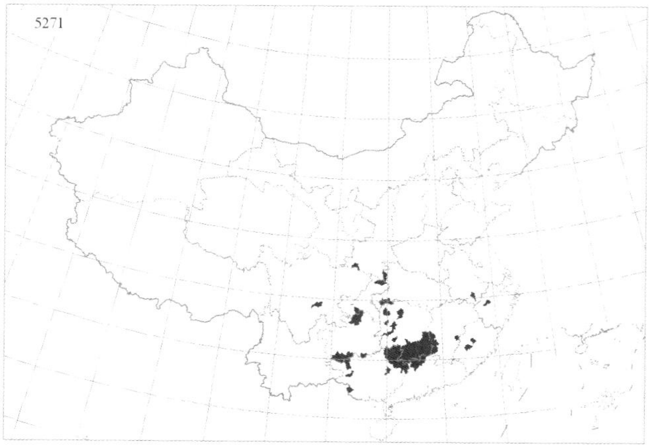

Aquifoliaceae 冬青科 *Ilex* 冬青属
Ilex hylonoma var. glabra S. Y. Hu 光叶细刺枸骨
Trees, evergreen

17.2 (14.6~19.1); 6.5 (3.3~10.2); 26.9 (24.6~29.9); 17.3 (14.7~19.2); 147.6 (117.7~170.1); 0.3 (0.0~1.9); 906 (779~981) ❙ 1485 (1250~1726); 557 (439~678); 167 (58~220) ❙ 64.9 (30.5~101.8); 896 (779~981); 242.9 (141.7~322.1)

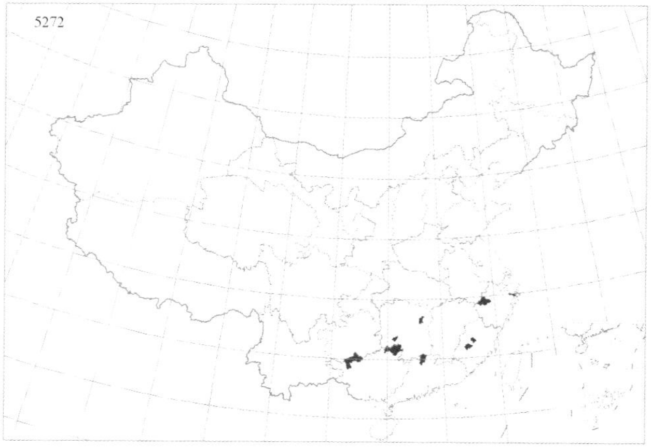

Aquifoliaceae 冬青科 *Ilex* 冬青属
Ilex integra Thunb. 全缘冬青
Small trees, evergreen

19.6 (16.2~24.1); 13.0 (4.4~18.9); 25.4 (20.1~28.9); 19.7 (16.2~24.1); 175.9 (134.2~229.7); 0.0 (0.0~0.6); 998 (773~1322) ❙ 2279 (1363~3495); 1016 (481~1847); 250 (47~897) ❙ 136.3 (25.2~246.5); 968 (773~1144); 245.6 (119.4~363.7)

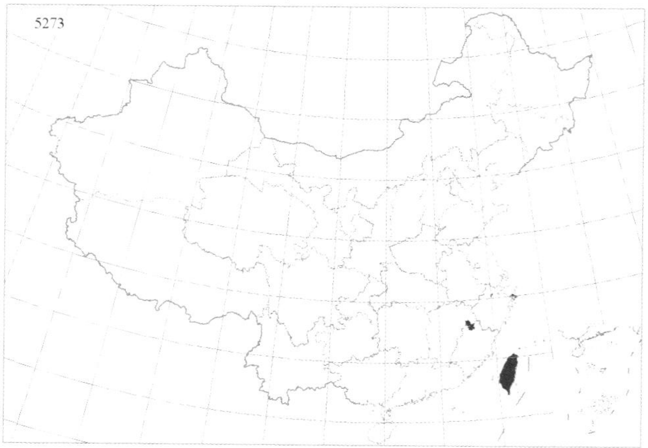

Aquifoliaceae 冬青科 *Ilex* 冬青属
Ilex intermedia Loes. et Diels 中型冬青
Trees, evergreen

16.1 (10.6~21.6); 5.2 (−0.2~13.1); 26.4 (19.6~29.3); 16.1 (10.6~21.6); 134.8 (79.4~199.6); 1.3 (0.0~12.1); 864 (643~1150) ❙ 1294 (958~1624); 522 (419~766); 103 (44~199) ❙ 51.8 (8.0~108.0); 862 (643~1148); 196.9 (108.2~382.6)

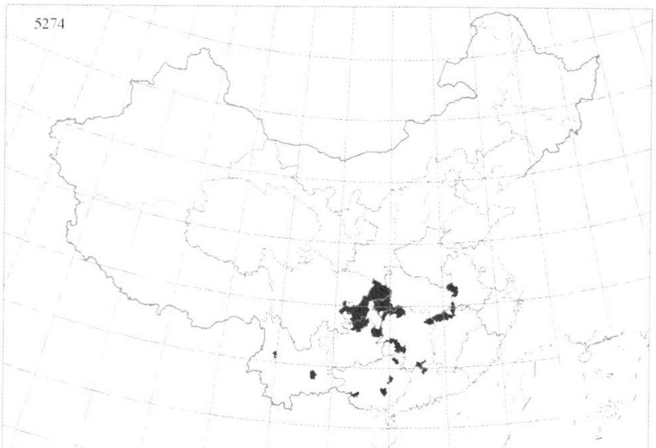

Aquifoliaceae 冬青科 *Ilex* 冬青属
Ilex intricata Hook. f. 错枝冬青
Small shrubs, evergreen

10.4 (1.8~16.7); 3.1 (-7.6~10.0); 16.4 (10.1~21.3); 10.7 (3.9~16.7); 77.1 (16.6~140.8); 11.7 (0.0~54.8); 625 (411~809) ❙ 1047 (579~1736); 598 (362~1007); 42 (9~69) ❙ 67.0 (34.7~145.8); 625 (411~809); 336.5 (201.9~447.4)

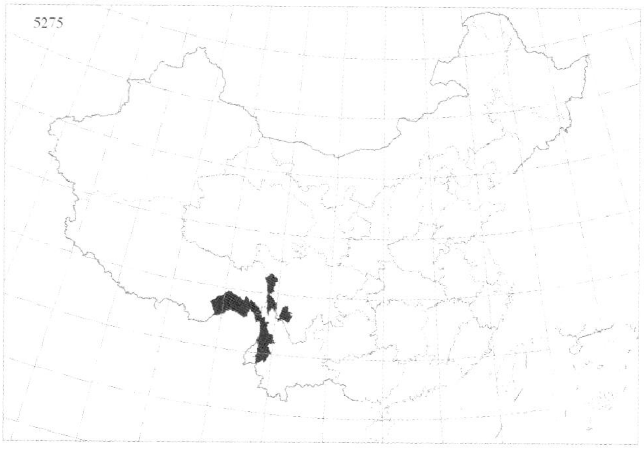

Aquifoliaceae 冬青科 *Ilex* 冬青属
Ilex jiaolingensis C. J. Tseng et H. H. Liu 蕉岭冬青
Trees, evergreen

20.1; 11.0; 27.8; 20.1; 181.7; 0.0; 1045 ❙ 1631; 592; 183 ❙ 56.1; 1045; 270.0

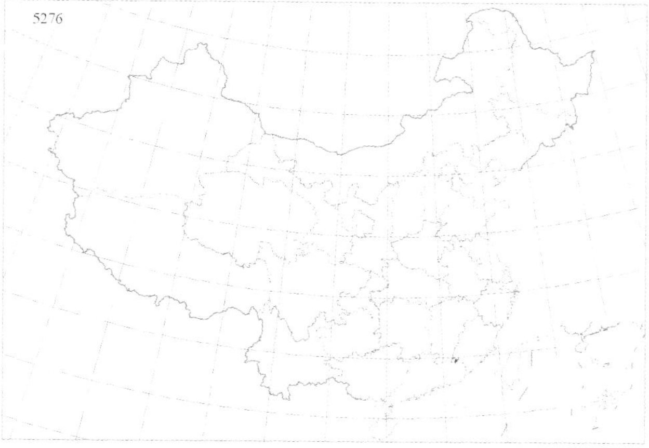

Aquifoliaceae 冬青科 *Ilex* 冬青属
Ilex jinyunensis Z. M. Tan 缙云冬青
Trees, evergreen

16.6 (10.6~18.7); 5.8 (-0.2~8.1); 26.8 (21.1~28.9); 16.6 (10.6~18.8); 140.7 (79.4~156.0); 1.0 (0.0~12.1); 890 (643~1002) ❙ 1156 (1028~1320); 436 (442~546); 63 (46~76) ❙ 32.2 (8.0~93.8); 889 (643~1002); 160.5 (108.2~284.3)

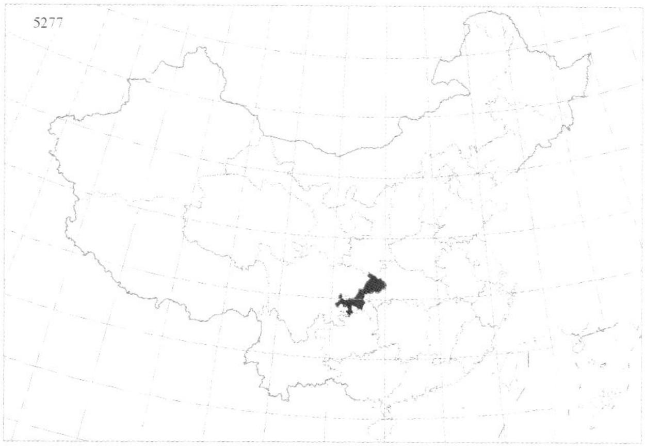

Aquifoliaceae 冬青科 *Ilex* 冬青属
Ilex jiuwanshanensis C. J. Tseng 九万山冬青
Shrubs, evergreen

16.6; 5.7; 26.6; 16.7; 140.6; 0.3; 876 ❙ 1573; 562; 186 ❙ 80.3; 876; 220.2

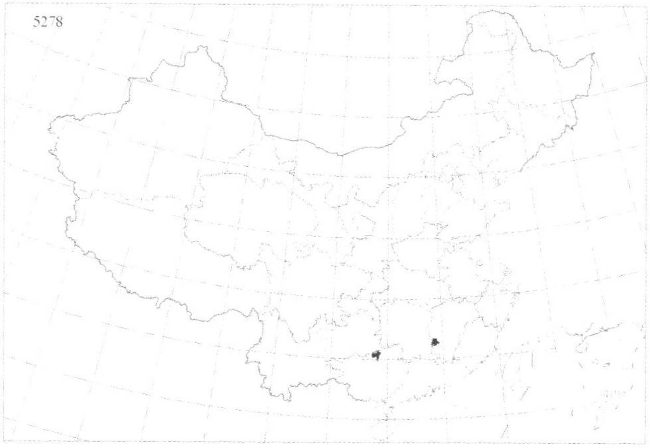

Aquifoliaceae 冬青科 *Ilex* 冬青属
Ilex kaushue S. Y. Hu 扣树
Trees, evergreen

19.0 (12.0~24.2); 9.8 (0.6~18.7); 26.8 (22.7~28.9); 19.1 (12.1~24.3); 169.4 (94.7~231.2); 0.7 (0.0~9.7); 1006 (697~1335) ❙ 1446 (1097~1781); 633 (447~813); 109 (48~194) ❙ 47.0 (-12.0~100.5); 990 (697~1202); 269.1 (158.7~590.7)

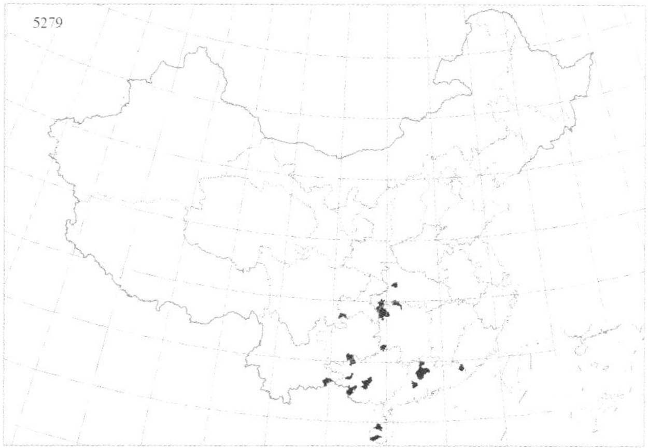

Aquifoliaceae 冬青科 *Ilex* 冬青属
Ilex kengii S. Y. Hu 皱柄冬青
Trees, evergreen

17.3 (14.3~24.3); 6.6 (3.3~18.7); 27.0 (24.3~29.4); 17.3 (14.3~24.3); 148.2 (114.1~231.9); 0.3 (0.0~2.3); 915 (765~1338) ❙ 1488 (1168~1876); 557 (392~891); 162 (42~230) ❙ 63.6 (-11.9~106.4); 909 (765~1150); 222.9 (151.0~455.2)

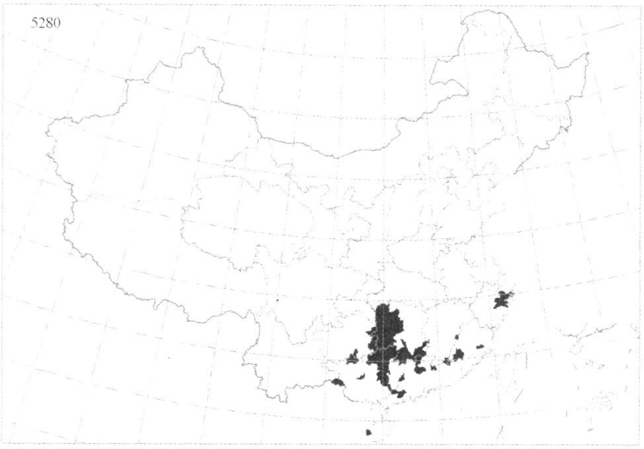

Aquifoliaceae 冬青科 *Ilex* 冬青属

Ilex kiangsiensis (S. Y. Hu) C. J. Tseng et B. W. Liu 江西满树星

Shrubs or small trees, deciduous

17.9 (13.6~19.4); 7.3 (4.2~9.4); 27.4 (19.0~29.3); 17.9 (13.6~19.4); 154.9 (103.6~173.4); 0.1 (0.0~0.8); 943 (707~1037) ❙ 1572 (957~1828); 553 (452~644); 183 (36~253) ❙ 66.7 (14.2~103.1); 938 (707~1029); 220.8 (143.8~261.3)

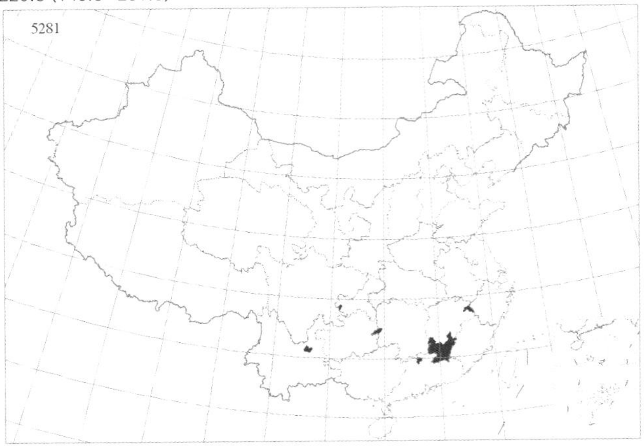

Aquifoliaceae 冬青科 *Ilex* 冬青属

Ilex kobuskiana S. Y. Hu 凸脉冬青

Shrubs or trees, evergreen

20.5 (16.4~24.3); 12.5 (4.8~18.7); 27.3 (25.7~28.4); 20.6 (16.5~24.3); 186.9 (138.1~231.9); 0.0 (0.0~0.2); 1086 (866~1338) ❙ 1513 (1179~1997); 527 (392~669); 126 (42~254) ❙ 44.7 (-11.9~130.7); 1012 (866~1142); 354.8 (169.7~590.7)

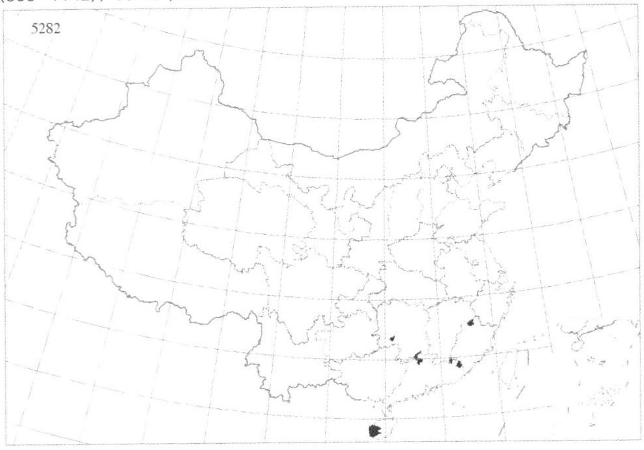

Aquifoliaceae 冬青科 *Ilex* 冬青属

Ilex kunmingensis H. W. Li 昆明冬青

Shrubs, evergreen

15.3; 8.7; 19.9; 15.3; 123.8; 0.0; 758 ❙ 1117; 651; 42 ❙ 46.7; 752; 333.3

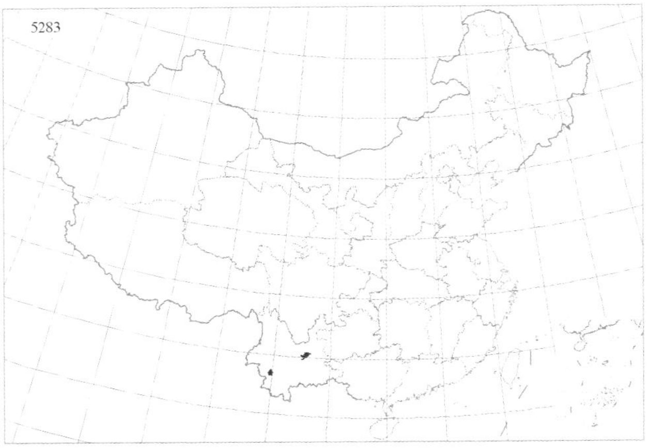

Aquifoliaceae 冬青科 *Ilex* 冬青属

Ilex kusanoi Hayata 兰屿冬青

Trees, deciduous

18.5; 14.1; 21.7; 18.5; 162.1; 0.0; 863 ❙ 2713; 1388; 188 ❙ 214.4; 863; 363.7

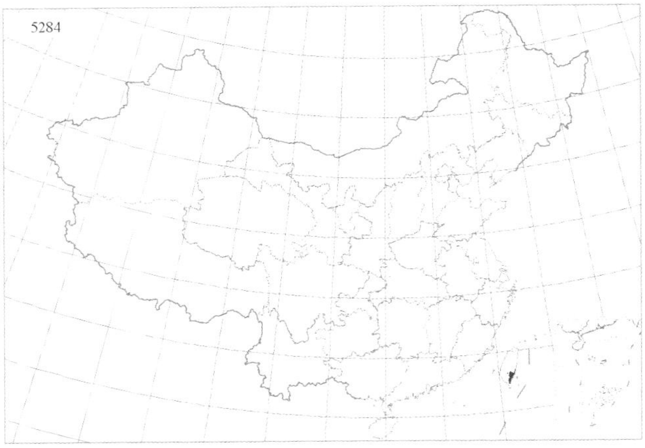

Aquifoliaceae 冬青科 *Ilex* 冬青属

Ilex kwangtungensis Merr. 广东冬青

Shrubs or small trees, evergreen

18.6 (2.9~24.8); 8.8 (-6.3~20.1); 27.1 (11.2~29.8); 18.6 (4.4~24.8); 163.7 (20.5~237.6); 0.3 (0.0~44.6); 977 (428~1385) ❙ 1592 (950~2435); 622 (392~1428); 163 (25~259) ❙ 64.7 (1.5~147.0); 967 (428~1239); 248.7 (111.7~627.0)

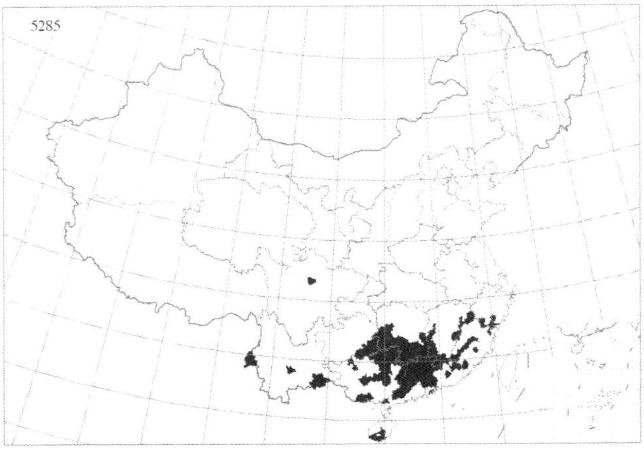

Aquifoliaceae 冬青科 *Ilex* 冬青属

Ilex lancilimba Merr. 剑叶冬青

Shrubs or small trees, evergreen

21.0 (17.5~24.8); 12.9 (7.1~20.1); 27.6 (25.8~29.0); 21.1 (17.6~24.8); 193.0 (150.9~237.6); 0.0 (0.0~0.0); 1115 (909~1385) ❙ 1587 (1284~1961); 633 (455~1044); 127 (41~229) ❙ 44.8 (1.5~99.7); 1077 (909~1212); 313.3 (197.5~620.0)

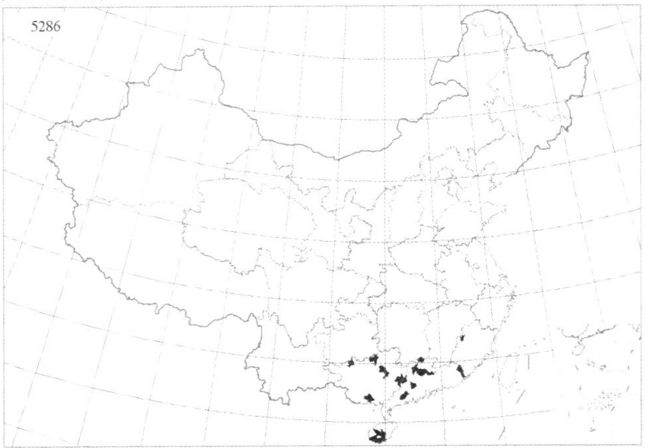

Aquifoliaceae 冬青科 *Ilex* 冬青属
Ilex latifolia Thurb. 大叶冬青
Trees, evergreen

16.7 (12.0~23.6); 5.6 (0.6~17.8); 27.3 (22.4~29.3); 16.8 (12.1~23.6); 142.1 (94.7~223.3); 0.9 (0.0~9.7); 898 (697~1271) ❙ 1490 (931~1997); 557 (323~860); 169 (44~259) ❙ 66.6 (-2.2~147.0); 893 (697~1190); 230.0 (144.4~551.6)

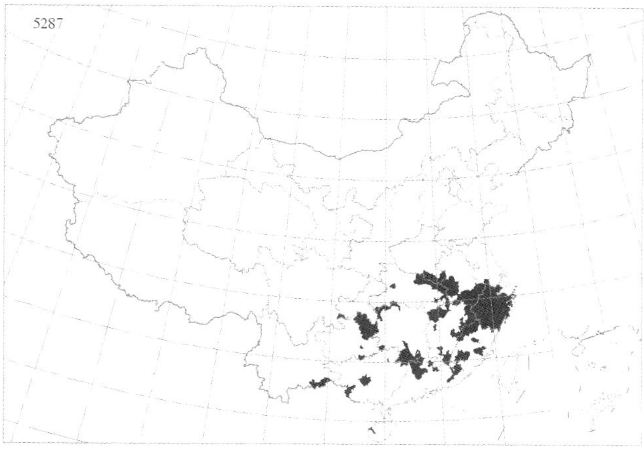

Aquifoliaceae 冬青科 *Ilex* 冬青属
Ilex latifrons Chun 阔叶冬青
Trees, evergreen

20.7 (17.2~22.7); 12.6 (9.5~15.0); 26.8 (22.4~28.9); 20.7 (17.3~22.8); 189.0 (147.0~213.1); 0.0 (0.0~0.0); 1075 (833~1221) ❙ 1584 (1254~2435); 840 (599~1428); 108 (48~194) ❙ 48.2 (12.9~99.5); 1075 (833~1221); 311.1 (201.4~551.6)

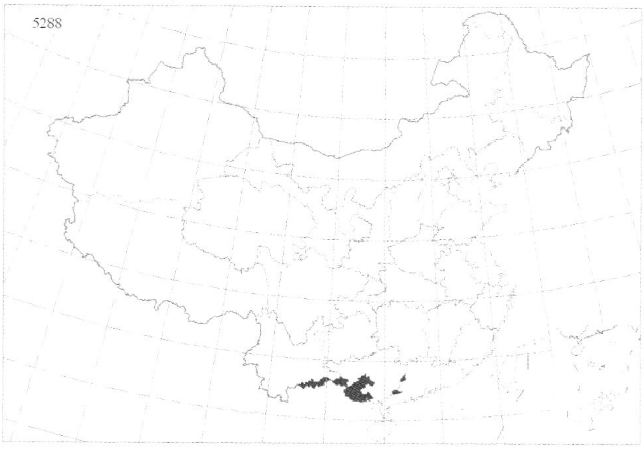

Aquifoliaceae 冬青科 *Ilex* 冬青属
Ilex liana S. Y. Hu 毛核冬青
Trees, evergreen

17.6; 10.6; 22.4; 17.6; 151.3; 0.0; 848 ❙ 1280; 722; 52 ❙ 50.6; 832; 405.5

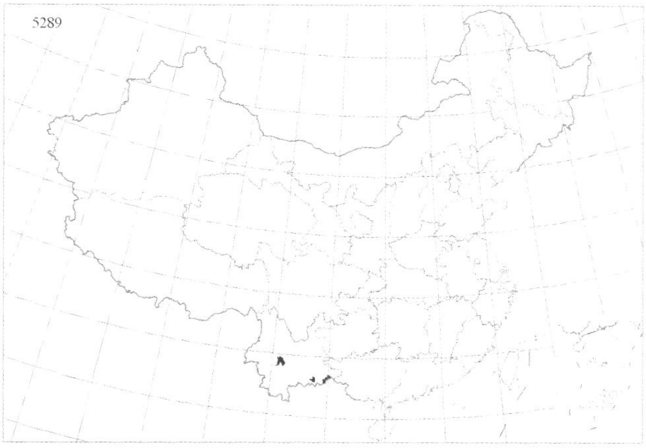

Aquifoliaceae 冬青科 *Ilex* 冬青属
Ilex liangii S. Y. Hu 保亭冬青
Shrubs or trees, evergreen

24.2 (23.9~24.7); 18.9 (18.2~19.8); 27.9 (27.2~28.4); 24.2 (23.9~24.7); 230.7 (226.8~236.5); 0.0 (0.0~0.0); 1327 (1294~1376) ❙ 1484 (1179~1723); 508 (392~619); 61 (42~110) ❙ 11.8 (-11.9~29.9); 1171 (1056~1326); 475.6 (352.9~620.0)

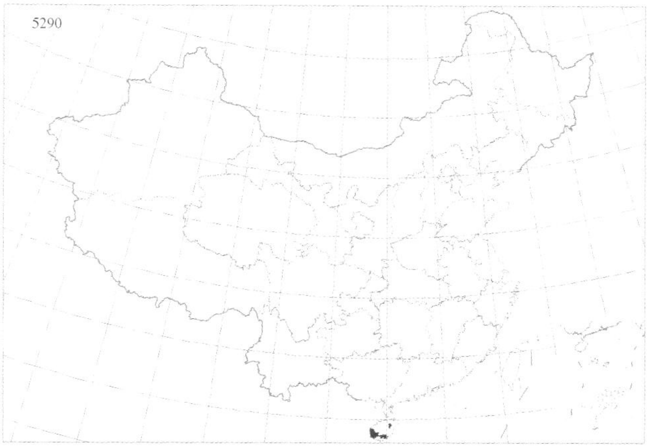

Aquifoliaceae 冬青科 *Ilex* 冬青属
Ilex lihuaensis T. R. Dudley. 溪畔冬青
Shrubs, evergreen

17.4; 7.4; 25.5; 17.4; 148.9; 0.0; 886 ❙ 1381; 637; 107 ❙ 55.9; 886; 245.3

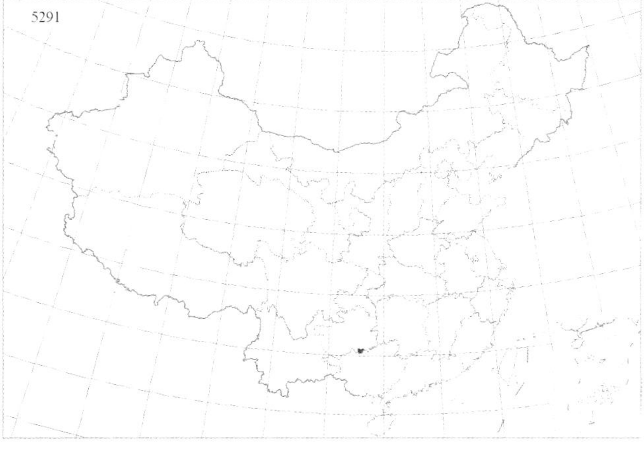

Aquifoliaceae 冬青科 *Ilex* 冬青属
Ilex limi C. J. Tseng 汝昌冬青
Trees, evergreen

18.2 (15.1~21.3); 8.3 (4.0~12.0); 27.1 (25.2~29.1); 18.2 (15.2~21.3); 158.8 (122.9~196.1); 0.0 (0.0~1.0); 947 (811~1137) ❙ 1718 (1427~1922); 603 (495~677); 205 (156~259) ❙ 82.8 (27.7~129.9); 945 (811~1136); 255.7 (183.9~305.7)

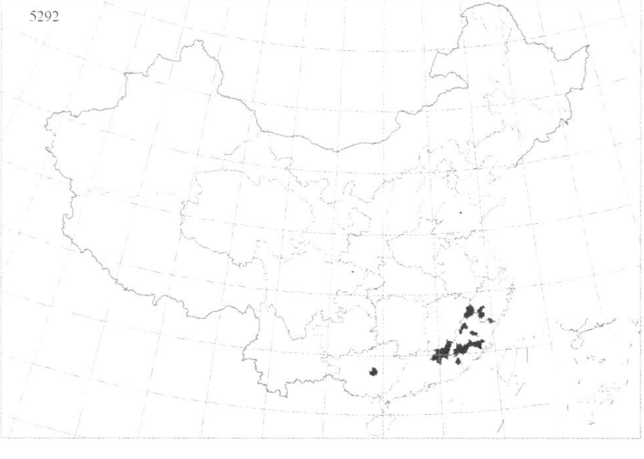

Aquifoliaceae 冬青科 *Ilex* 冬青属
Ilex litseifolia Hu et T. Tang 木姜冬青
Shrubs or small trees, evergreen

16.8 (14.3~23.6); 6.1 (2.5~17.8); 26.7 (23.5~29.1); 16.9 (14.3~23.6); 142.5 (112.6~223.3); 0.3 (0.0~3.9); 888 (756~1271) ❙ 1571 (1171~2005); 567 (411~824); 179 (44~259) ❙ 77.7 (-2.2~147.0); 885 (756~1198); 235.1 (164.9~439.7)

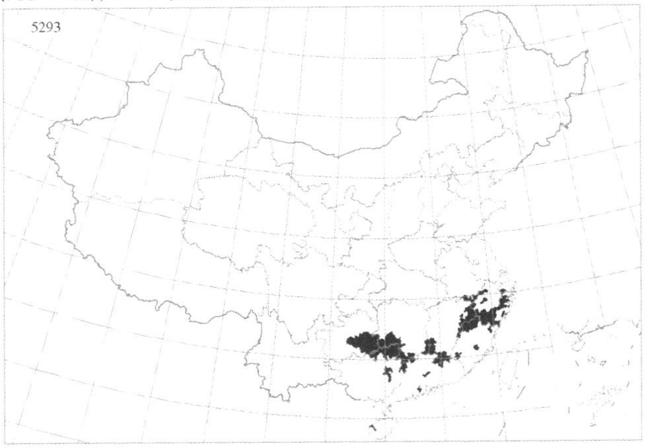

Aquifoliaceae 冬青科 *Ilex* 冬青属
Ilex lohfauensis Merr. 矮冬青
Shrubs or small trees, evergreen

17.8 (14.3~22.3); 7.4 (3.1~15.2); 27.3 (23.5~29.8); 17.8 (14.3~22.7); 154.1 (112.6~212.7); 0.2 (0.0~2.3); 939 (756~1216) ❙ 1599 (1171~2024); 583 (403~994); 181 (81~259) ❙ 71.4 (22.7~147.0); 934 (756~1216); 230.4 (111.7~305.7)

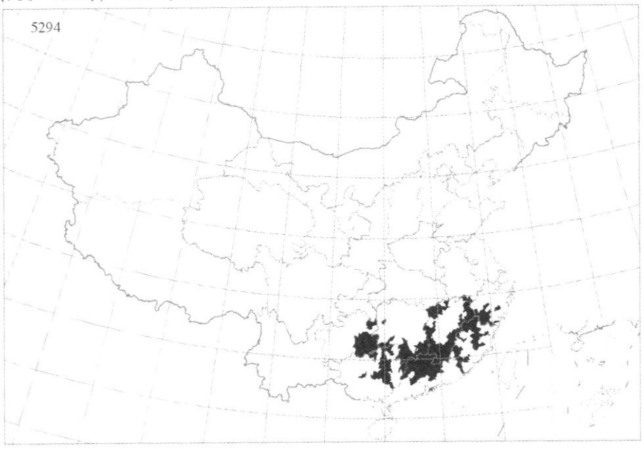

Aquifoliaceae 冬青科 *Ilex* 冬青属
Ilex longecaudata H. F. Comber 长尾冬青
Shrubs or trees, evergreen

16.1 (8.6~21.0); 9.2 (1.3~14.2); 20.9 (15.0~25.6); 16.1 (8.7~21.0); 133.6 (52.6~191.8); 0.3 (0.0~8.8); 796 (574~1053) ❙ 1276 (1018~1736); 711 (554~1007); 53 (42~70) ❙ 61.7 (25.7~145.8); 792 (574~1053); 400.0 (277.0~551.6)

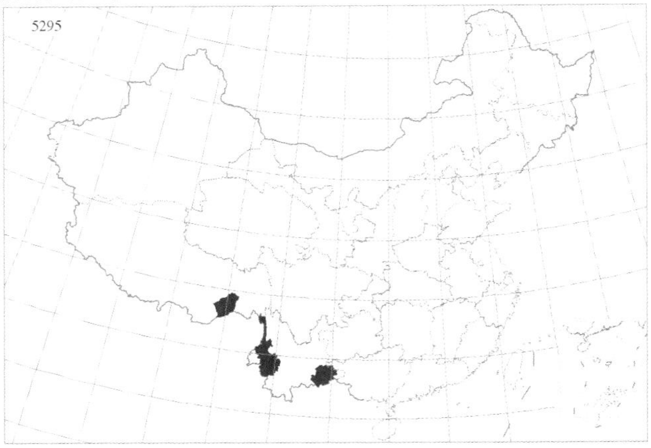

Aquifoliaceae 冬青科 *Ilex* 冬青属
Ilex longzhouensis C. J. Tseng 龙州冬青
Small trees, evergreen

20.1; 11.9; 26.1; 20.1; 181.5; 0.0; 1025 ❙ 1368; 754; 65 ❙ 35.1; 1025; 347.3

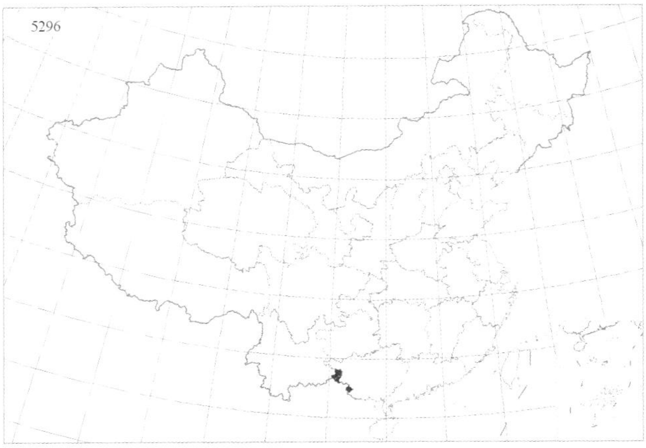

Aquifoliaceae 冬青科 *Ilex* 冬青属
Ilex lonicerifolia Hayata 忍冬叶冬青
Trees, evergreen

19.7 (15.3~24.1); 13.3 (3.6~18.9); 25.1 (20.1~28.2); 19.7 (15.4~24.1); 177.0 (125.7~229.7); 0.1 (0.0~1.4); 999 (773~1322) ❙ 2265 (1206~3495); 1031 (507~1847); 238 (47~897) ❙ 134.5 (25.2~246.5); 967 (773~1144); 241.2 (119.4~363.7)

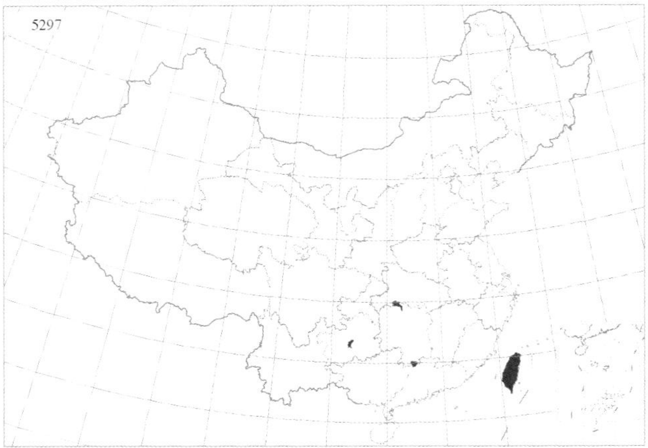

Aquifoliaceae 冬青科 *Ilex* 冬青属
Ilex ludianensis S. C. Huang ex Y. R. Li 鲁甸冬青
Shrubs, evergreen

12.4; 3.9; 19.3; 12.4; 89.4; 1.1; 678 ❙ 863; 496; 25 ❙ 27.3; 678; 190.3

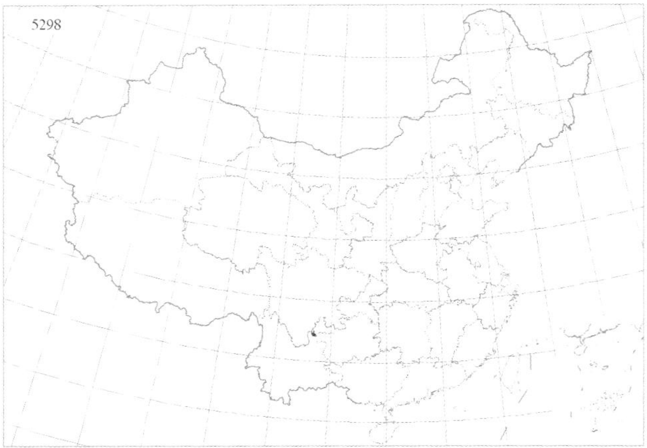

Aquifoliaceae 冬青科 *Ilex* 冬青属
Ilex machilifolia H. W. Li ex Y. R. Li 楠叶冬青
Trees, evergreen

17.6; 10.3; 22.8; 17.7; 152.4; 0.0; 856 ❙ 1306; 740; 53 ❙ 52.5; 856; 369.0

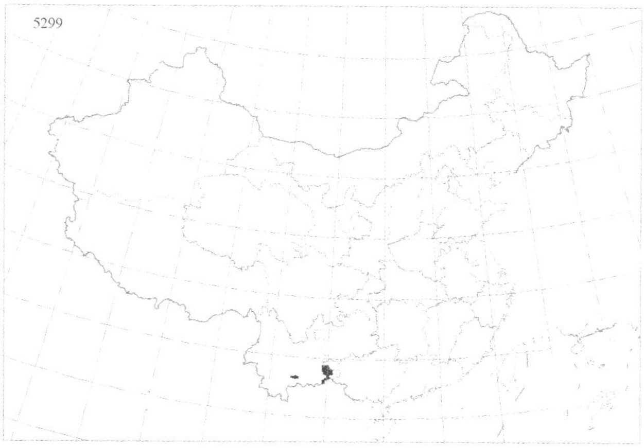

Aquifoliaceae 冬青科 *Ilex* 冬青属
Ilex maclurei Merr. 长圆叶冬青
Shrubs or small trees, evergreen

22.9; 15.2; 29.0; 23.0; 215.7; 0.0; 1239 ❙ 1477; 714; 97 ❙ 19.2; 1239; 203.1

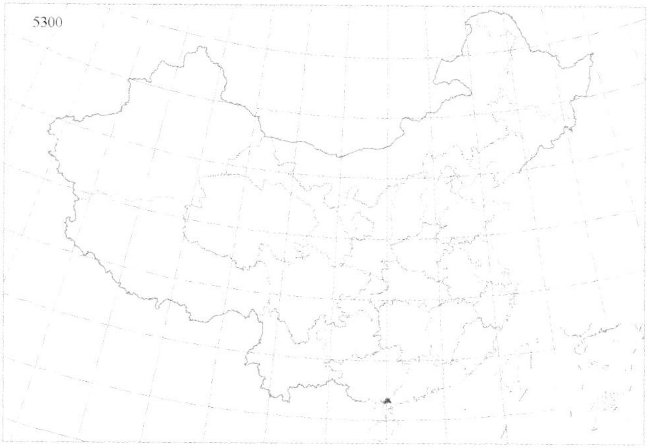

Aquifoliaceae 冬青科 *Ilex* 冬青属
Ilex macrocarpa Oliv. 大果冬青
Trees, deciduous

16.2 (3.8~22.4); 6.3 (-4.3~15.2); 25.0 (10.8~29.8); 16.2 (4.7~22.7); 135.8 (21.6~212.7); 1.2 (0.0~35.4); 854 (444~1216) ❙ 1289 (741~2037); 571 (357~1033); 103 (15~259) ❙ 51.3 (-14.2~130.7); 845 (444~1216); 243.3 (108.2~659.0)

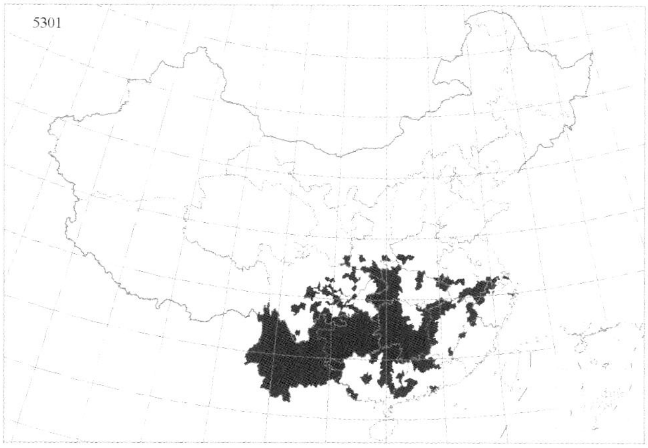

Aquifoliaceae 冬青科 *Ilex* 冬青属
Ilex macrocarpa var. reevesiae (S. Y. Hu) S. Y. Hu 柔毛冬青
Trees, deciduous

14.6; 3.9; 24.2; 14.6; 117.7; 2.8; 789 ❙ 929; 529; 26 ❙ 18.5; 782; 186.5

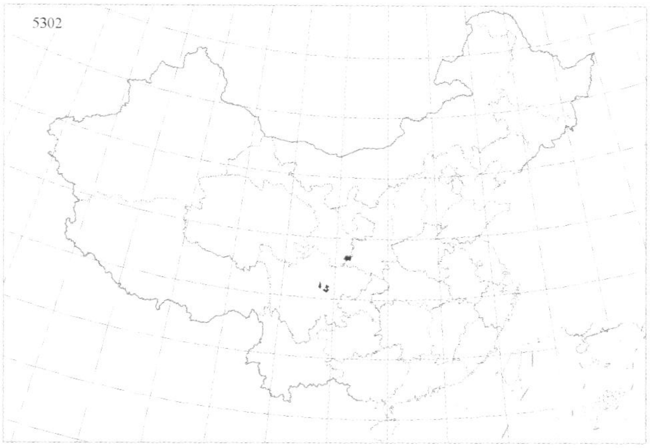

Aquifoliaceae 冬青科 *Ilex* 冬青属
Ilex macropoda Miq. 大柄冬青
Trees, deciduous

15.7 (9.3~23.6); 3.9 (-1.7~17.8); 27.0 (20.1~29.3); 15.8 (9.5~23.6); 131.6 (69.8~223.3); 2.6 (0.0~18.0); 862 (603~1271) ❙ 1427 (753~1997); 538 (376~685); 157 (44~254) ❙ 66.4 (-12.2~130.7); 855 (603~1148); 230.8 (147.1~439.7)

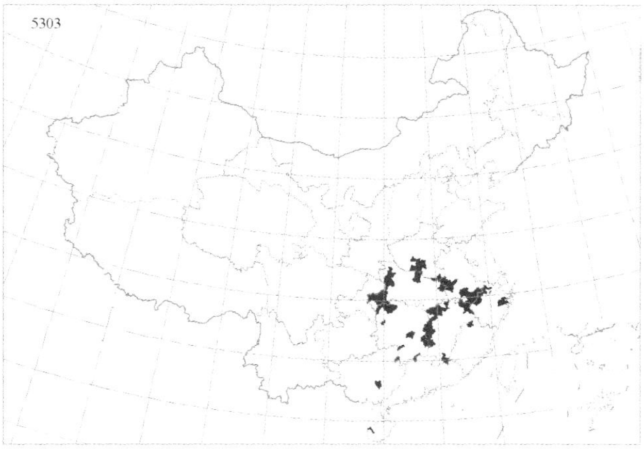

Aquifoliaceae 冬青科 *Ilex* 冬青属
Ilex macrostigma C. Y. Wu et Y. R. Li 大柱头冬青
Shrubs, evergreen

11.4; 4.5; 17.3; 11.5; 82.2; 4.4; 651 ❙ 1016; 582; 40 ❙ 58.8; 640; 324.6

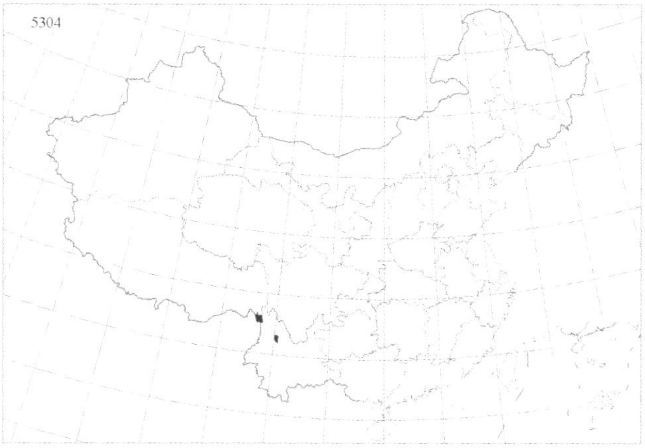

Aquifoliaceae 冬青科 *Ilex* 冬青属
Ilex manneiensis S. Y. Hu 红河冬青
Shrubs or small trees, evergreen

16.7 (10.2~21.0); 9.9 (2.4~14.2); 21.5 (16.9~25.6); 16.7 (10.3~21.0); 140.6 (68.2~191.8); 0.1 (0.0~5.1); 817 (620~1053) ❙ 1092 (907~1547); 624 (521~911); 44 (18~68) ❙ 33.6 (6.5~80.6); 797 (620~1053); 325.6 (234.5~551.6)

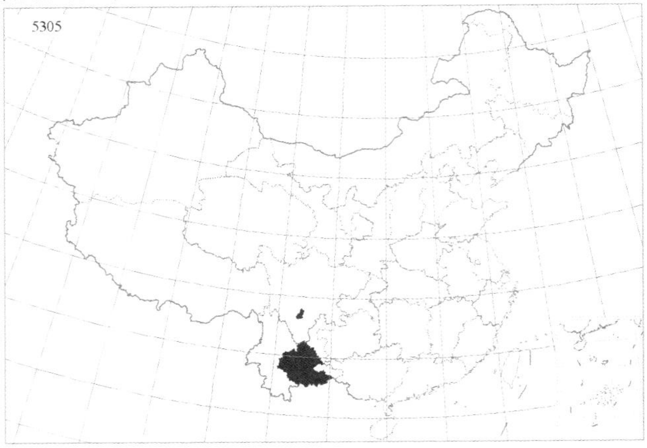

Aquifoliaceae 冬青科 *Ilex* 冬青属
Ilex marlipoensis H. W. Li ex Y. R. Li 麻栗坡冬青
Trees, evergreen

17.6; 9.9; 23.2; 17.7; 152.4; 0.0; 859 ❙ 1422; 813; 52 ❙ 65.6; 859; 359.7

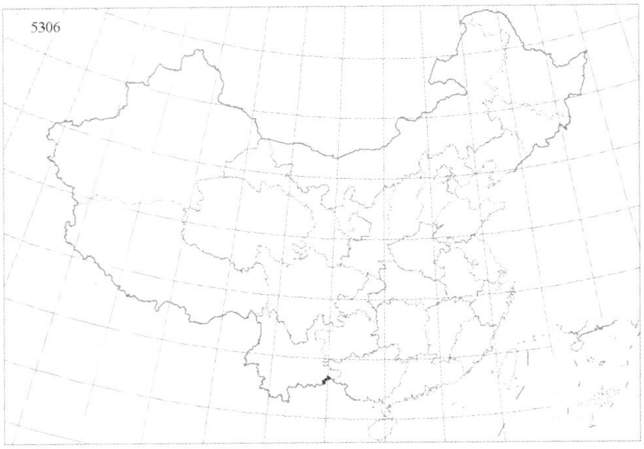

Aquifoliaceae 冬青科 *Ilex* 冬青属
Ilex maximowicziana Loes. 倒卵叶冬青
Small trees, evergreen

18.5; 14.1; 21.7; 18.5; 162.1; 0.0; 863 ❙ 2713; 1388; 188 ❙ 214.4; 863; 363.7

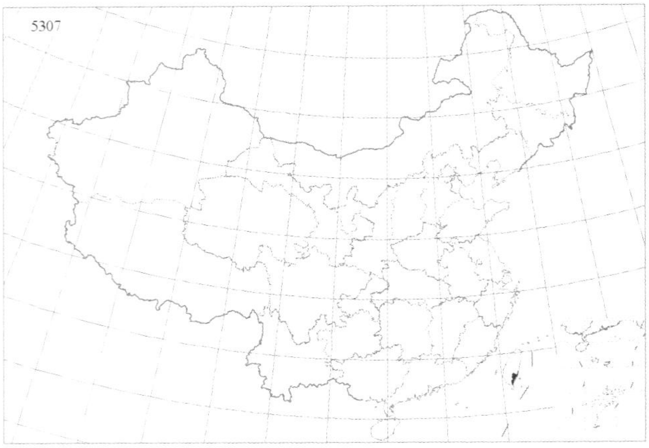

Aquifoliaceae 冬青科 *Ilex* 冬青属
Ilex medogensis Y. R. Li 墨脱冬青
Trees, evergreen

13.3; 5.4; 19.6; 13.4; 100.4; 0.0; 706 ❙ 1736; 1007; 51 ❙ 145.8; 706; 447.4

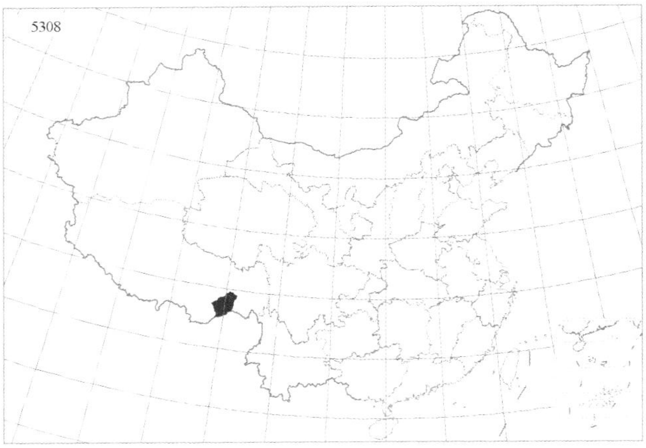

Aquifoliaceae 冬青科 *Ilex* 冬青属
Ilex melanophylla H. T. Chang 黑叶冬青
Shrubs, evergreen

17.5 (14.9~22.1); 6.6 (3.6~13.7); 27.5 (25.7~29.1); 17.5 (15.0~22.1); 150.7 (121.2~205.6); 0.1 (0.0~1.6); 927 (807~1182) ❙ 1529 (1211~1829); 535 (403~821); 184 (105~230) ❙ 65.8 (28.8~104.6); 919 (807~1182); 209.4 (151.0~270.5)

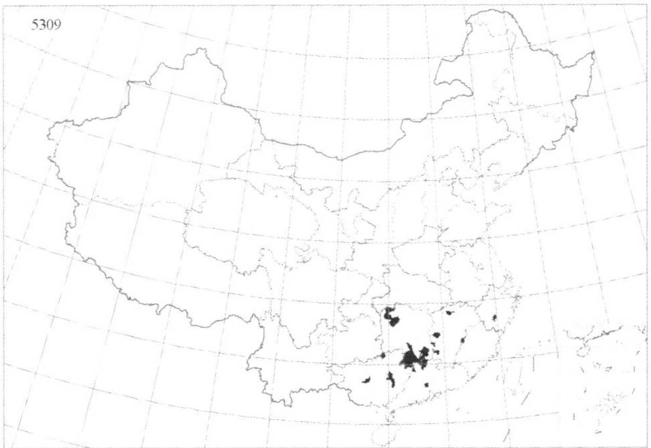

Aquifoliaceae 冬青科 *Ilex* 冬青属
Ilex melanotricha Merr. 黑毛冬青
Trees, evergreen

9.8 (5.3~14.8); 1.9 (-2.4~4.3); 16.9 (12.2~24.9); 9.9 (5.7~14.9); 67.0 (30.5~119.2); 8.5 (0.7~26.2); 615 (479~791) ❙ 1035 (961~1197); 561 (489~594); 42 (24~71) ❙ 70.8 (51.2~100.7); 615 (479~791); 288.2 (182.9~352.3)

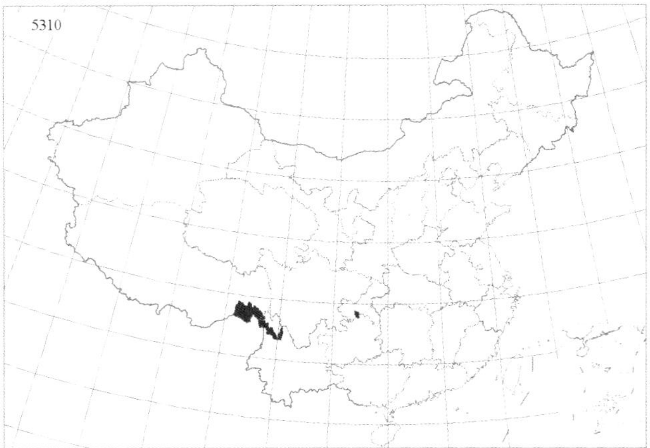

Aquifoliaceae 冬青科 *Ilex* 冬青属

Ilex memecylifolia Champ. ex Benth. 谷木叶冬青

Trees, evergreen

19.9 (16.6~22.7)；10.9 (5.6~15.2)；27.6 (25.2~29.8)；20.0 (16.6~22.8)；179.9 (138.9~2[illegible]3.1)；0.0 (0.0~0.0)；1046 (863~1221) ❙ 1707 (1268~2435)；704 (497~1428)；165 (75~259) ❙ 64.8 (12.9~130.7)；1044 (863~1221)；248.1 (122.7~327.7)

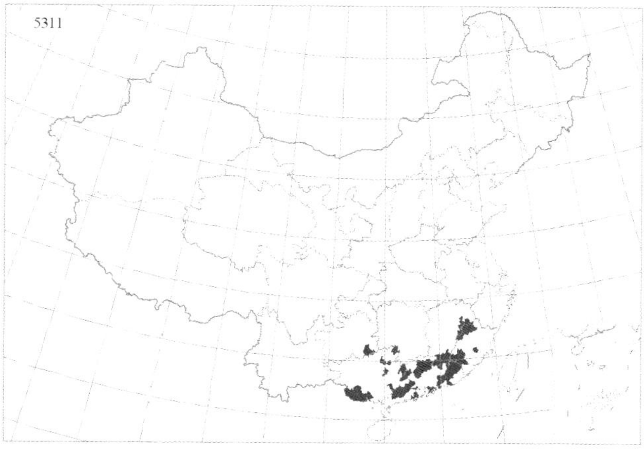

Aquifoliaceae 冬青科 *Ilex* 冬青属

Ilex metabaptista Loes. 河滩冬青

Shrubs or small trees, evergreen

15.7 (12.6~21.3)；5.3 (1.6~13.2)；25.2 (19.3~28.1)；15.8 (12.7~21.3)；129.8 (96.0~195.9)；0.7 (0.0~6.9)；828 (699~1111) ❙ 1247 (879~1900)；548 (443~1056)；83 (29~191) ❙ 51.0 (10.4~107.2)；828 (690~1111)；207.5 (118.7~322.1)

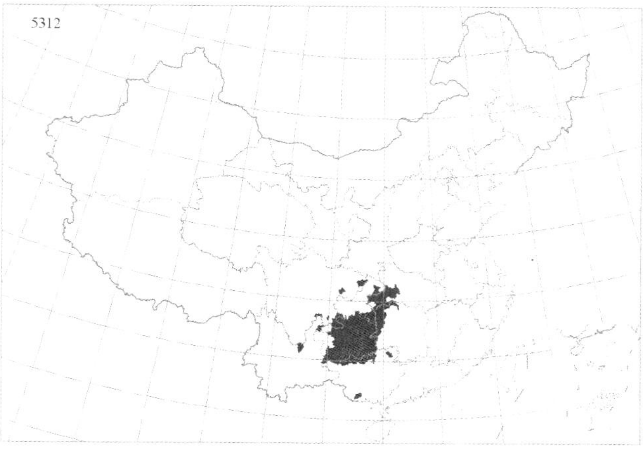

Aquifoliaceae 冬青科 *Ilex* 冬青属

Ilex metabaptista var. bodinieri (Loesener) G. Barriera 紫金牛叶冬青

Shrubs or small trees, evergreen

17.1 (15.0~19.1)；7.1 (4.5~10.1)；25.8 (24.9~26.6)；17.1 (15.0~19.1)；145.7 (120.8~169.2)；0.1 (0.0~0.5)；884 (794~969) ❙ 1248 (1089~1368)；581 (499~701)；81 (51~129) ❙ 41.6 (27.5~59.2)；884 (794~969)；224.3 (181.2~282.2)

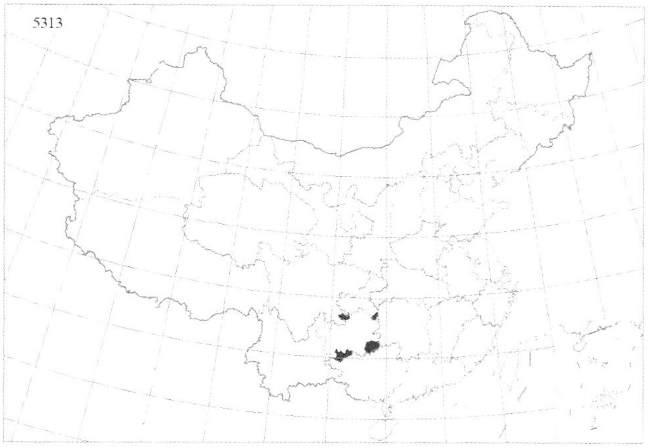

Aquifoliaceae 冬青科 *Ilex* 冬青属

Ilex micrococca Maxim. 小果冬青

Trees, deciduous

16.9 (2.6~24.1)；7.2 (-7.0~18.9)；25.5 (10.8~29.9)；16.9 (4.3~24.1)；144.1 (20.0~229.7)；0.8 (0.0~48.2)；890 (427~1322) ❙ 1395 (741~3495)；600 (376~1847)；123 (11~897) ❙ 57.4 (-14.2~246.5)；881 (427~1247)；244.2 (108.2~659.0)

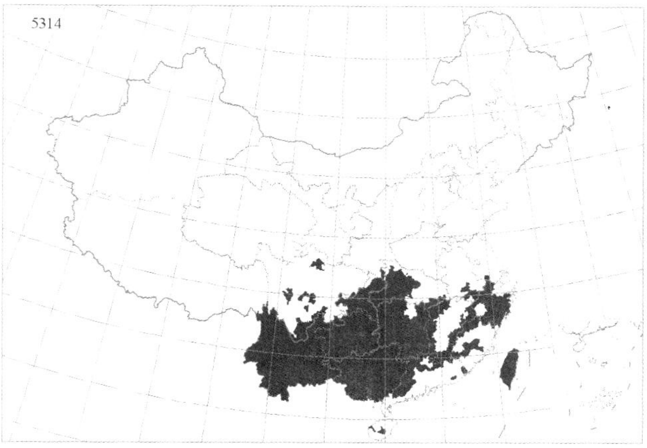

Aquifoliaceae 冬青科 *Ilex* 冬青属

Ilex micropyrena C. Y. Wu ex Y. R. Li 小核冬青

Shrubs, evergreen

8.6；1.3；15.0；8.7；52.6；8.8；574 ❙ 1041；585；45 ❙ 81.5；574；352.3

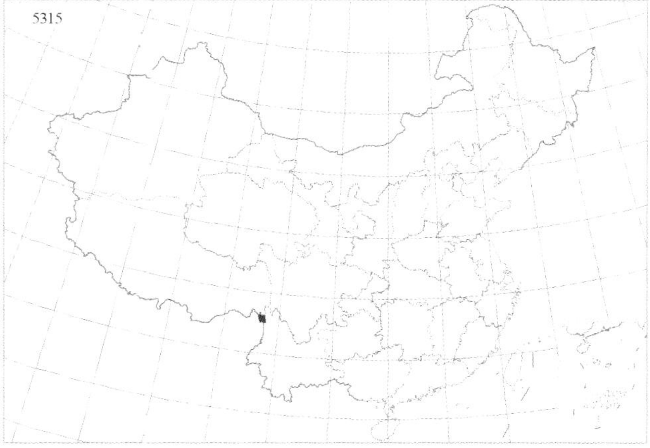

Aquifoliaceae 冬青科 *Ilex* 冬青属

Ilex miguensis S. Y. Hu 米谷冬青

Shrubs, evergreen

5.3；-2.4；12.2；5.7；30.5；26.2；479 ❙ 961；558；31 ❙ 100.7；479；234.2

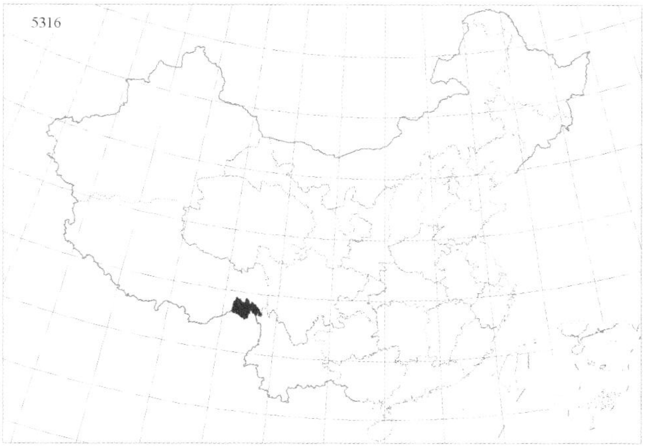

Aquifoliaceae 冬青科 *Ilex* 冬青属
Ilex nanchuanensis Z. M. Tan 南川冬青
Shrubs, evergreen

14.8；4.3；24.9；14.9；119.2；0.7；791 ❙ 1197；489；71 ❙ 51.2；791；182.9

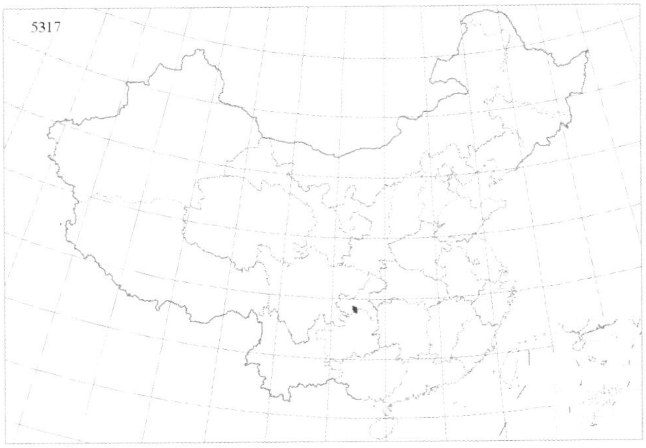

Aquifoliaceae 冬青科 *Ilex* 冬青属
Ilex nanningensis Hand.-Mazz. 南宁冬青
Trees, evergreen

21.9 (19.5~24.0)；13.9 (10.6~19.1)；27.9 (26.9~28.6)；21.9 (19.6~24.0)；203.2 (175.1~228.1)；0.0 (0.0~0.0)；1163 (1003~1302) ❙ 1724 (1302~2435)；877 (539~1428)；117 (48~191) ❙ 48.9 (12.9~99.5)；1149 (1003~1221)；310.2 (232.5~620.0)

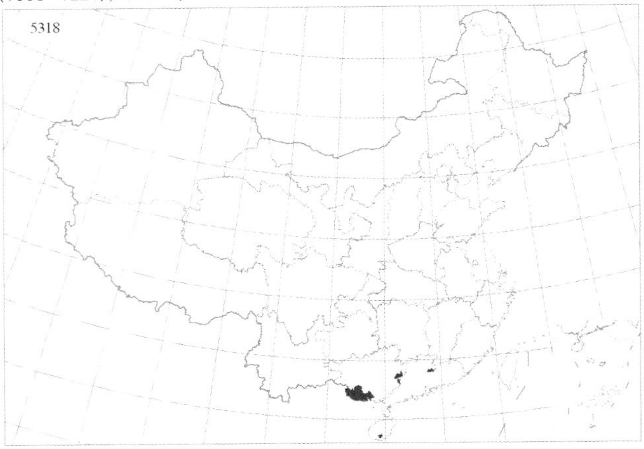

Aquifoliaceae 冬青科 *Ilex* 冬青属
Ilex ningdeensis C. J. Tseng 宁德冬青
Trees, evergreen

17.0；7.7；26.1；17.0；144.5；0.0；870 ❙ 1718；591；198 ❙ 97.5；870；267.7

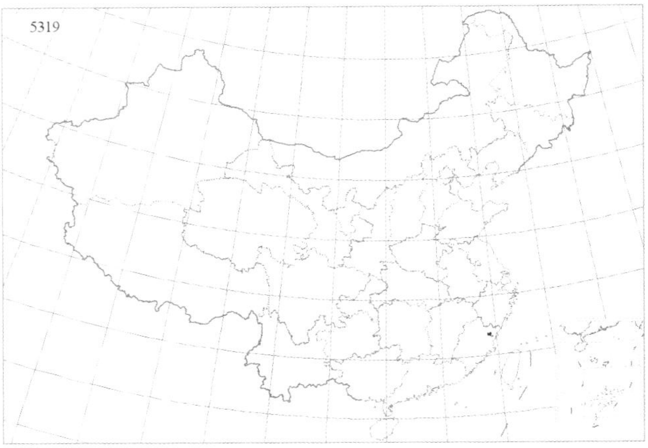

Aquifoliaceae 冬青科 *Ilex* 冬青属
Ilex nitidissima C. J. Tseng 亮叶冬青
Small trees, evergreen

18.3；8.6；26.6；18.3；160.1；0.0；950 ❙ 1631；697；166 ❙ 72.0；950；245.7

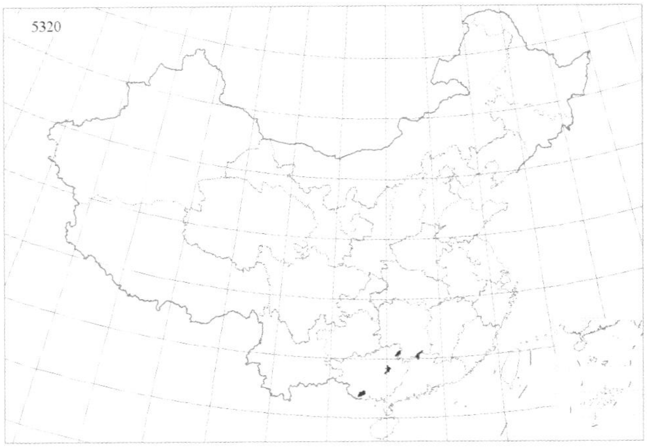

Aquifoliaceae 冬青科 *Ilex* 冬青属
Ilex nothofagifolia Kingdon-Ward 小圆叶冬青
Small trees, evergreen

15.1；8.3；20.3；15.1；123.8；2.2；782 ❙ 1370；792；50 ❙ 80.0；780；439.5

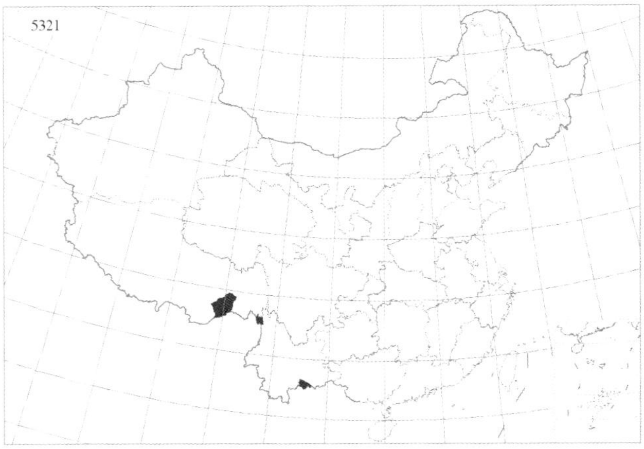

Aquifoliaceae 冬青科 *Ilex* 冬青属
Ilex nubicola C. Y. Wu ex Y. R. Li 云中冬青
Shrubs or trees, evergreen

19.1；13.0；23.3；19.1；169.7；0.0；917 ❙ 1428；835；56 ❙ 55.7；917；506.2

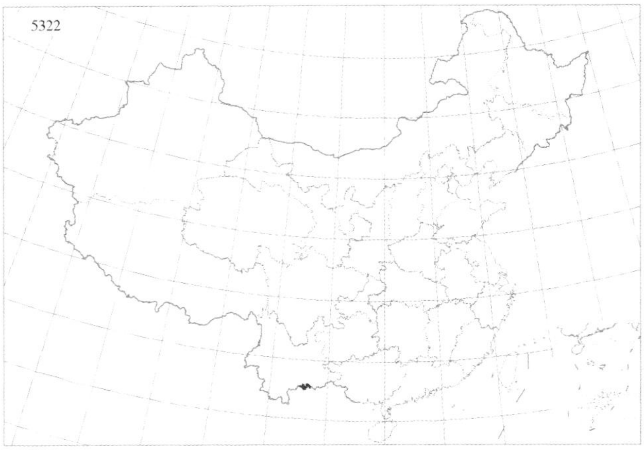

Aquifoliaceae 冬青科 *Ilex* 冬青属
Ilex nuculicava S. Y. Hu 洼皮冬青
Trees, evergreen

23.3 (18.3~24.8); 17.1 (8.1~20.1); 28.0 (25.7~28.8); 23.4 (18.4~24.8); 220.7 (160.7~237.6); 0.0 (0.0~0.0); 1268 (965~1385) ❙ 1525 (868~1990); 571 (312~827); 85 (41~181) ❙ 21.3 (-34.3~83.4); 1174 (868~1360); 392.1 (152.6~620.0)

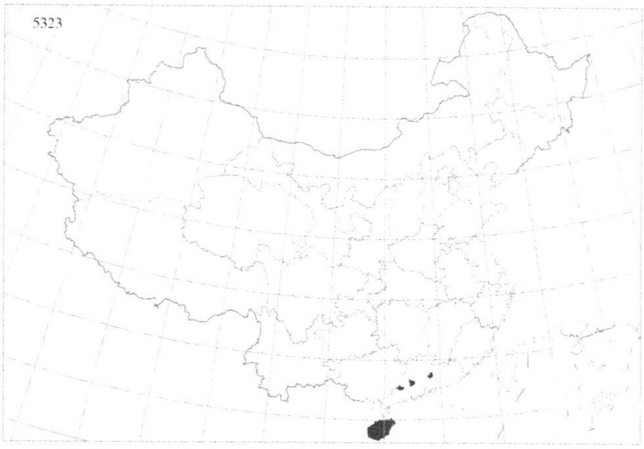

Aquifoliaceae 冬青科 *Ilex* 冬青属
Ilex nuculicava var. auctumnalis S. Y. Hu 秋花洼皮冬青
Trees, evergreen

24.2 (23.1~24.8); 19.0 (17.5~20.1); 27.7 (27.2~28.4); 24.2 (23.2~24.8); 230.5 (218.4~237.6); 0.0 (0.0~0.0); 1327 (1232~1385) ❙ 1442 (1179~1588); 496 (392~544); 48 (41~58) ❙ 8.9 (-11.9~20.7); 1139 (1056~1212); 527.1 (455.2~620.0)

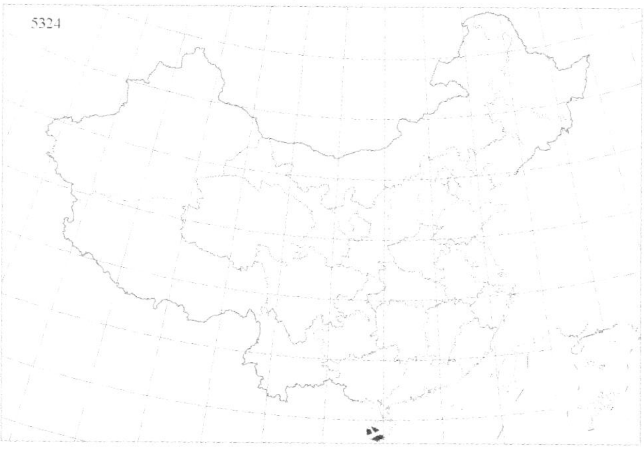

Aquifoliaceae 冬青科 *Ilex* 冬青属
Ilex nuculicava var. glabra S. Y. Hu 光枝洼皮冬青
Trees, evergreen

24.8; 20.1; 28.0; 24.8; 237.6; 0.0; 1385 ❙ 1405; 473; 41 ❙ 1.5; 1137; 527.5

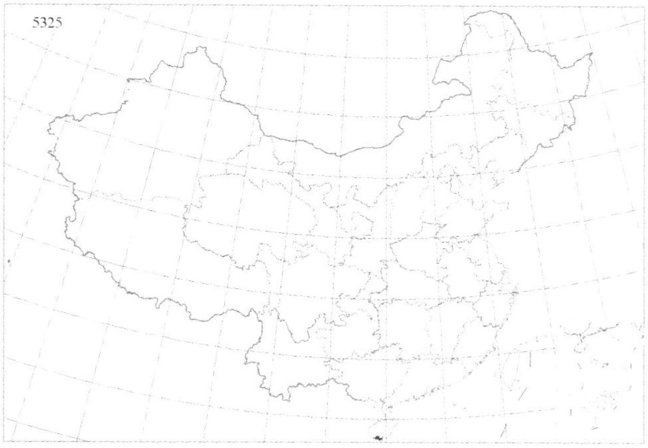

Aquifoliaceae 冬青科 *Ilex* 冬青属
Ilex oblonga C. J. Tseng 长圆果冬青
Trees, evergreen

17.9; 7.9; 26.4; 17.9; 154.9; 0.0; 927 ❙ 1599; 645; 168 ❙ 73.3; 927; 238.8

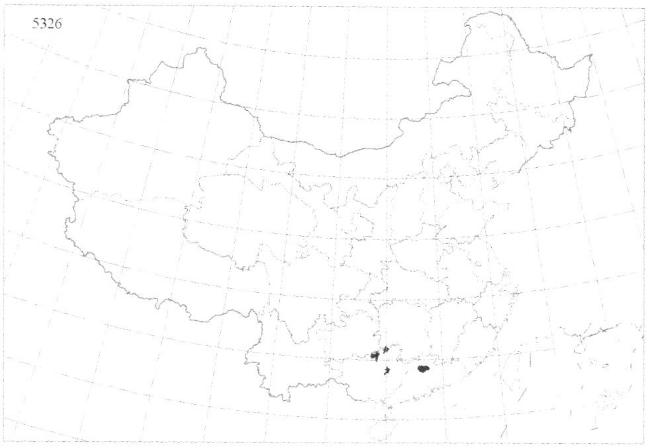

Aquifoliaceae 冬青科 *Ilex* 冬青属
Ilex occulta C. J. Tseng 隐脉冬青
Shrubs, evergreen

22.1; 14.4; 28.0; 22.2; 206.0; 0.0; 1174 ❙ 2153; 1185; 123 ❙ 83.1; 1174; 297.5

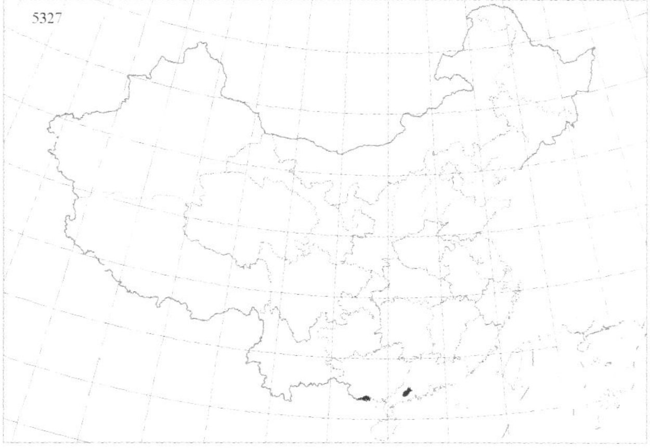

Aquifoliaceae 冬青科 *Ilex* 冬青属
Ilex oligodonta Merr. et Chun 疏齿冬青
Shrubs, evergreen

17.8 (16.5~21.0); 7.4 (5.2~12.0); 27.3 (25.7~29.0); 17.9 (16.5~21.1); 154.7 (138.3~193.0); 0.0 (0.0~0.0); 936 (866~1114) ❙ 1667 (1410~2024); 571 (403~824); 205 (173~259) ❙ 78.6 (47.4~130.7); 929 (866~1114); 231.0 (151.0~303.4)

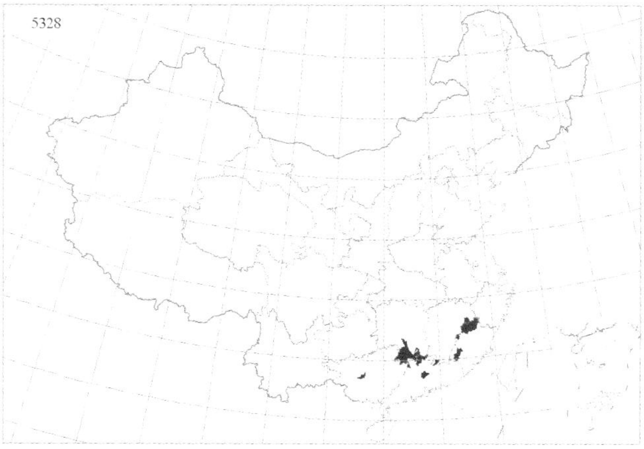

Aquifoliaceae 冬青科 *Ilex* 冬青属
Ilex omeiensis Hu et T. Tang 峨眉冬青
Shrubs or trees, evergreen

15.6 (13.8~18.0); 6.6 (4.5~11.6); 23.5 (19.3~25.7); 15.7 (13.8~18.1); 127.9 (105.9~156.9); 0.1 (0.0~0.5); 807 (714~869) ❙ 1274 (963~1590); 738 (556~946); 48 (29~59) ❙ 57.9 (12.2~89.0); 797 (690~866); 219.0 (166.3~293.0)

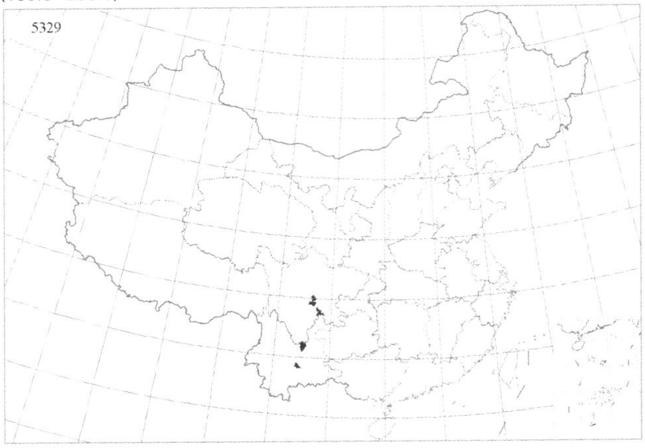

Aquifoliaceae 冬青科 *Ilex* 冬青属
Ilex pedunculosa Miq. 具柄冬青
Shrubs or trees, evergreen

16.2 (9.3~23.6); 5.2 (-1.7~17.8); 26.4 (20.1~29.3); 16.2 (9.5~23.6); 136.4 (69.8~223.3); 1.7 (0.0~18.0); 870 (603~1271) ❙ 1496 (842~2680); 565 (375~1186); 158 (19~259) ❙ 72.5 (-2.2~246.5); 866 (603~1190); 243.2 (164.1~439.7)

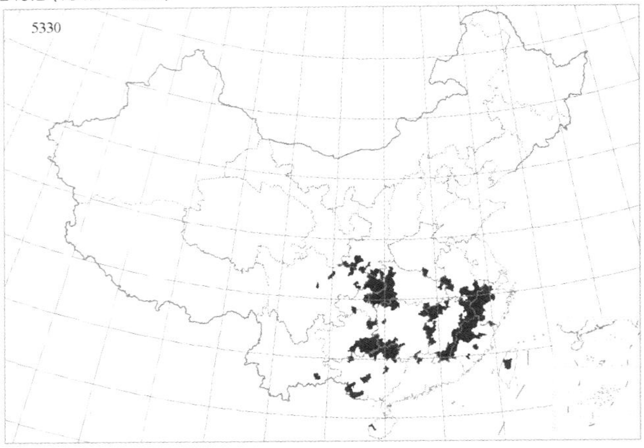

Aquifoliaceae 冬青科 *Ilex* 冬青属
Ilex peiradena S. Y. Hu 上思冬青
Shrubs, evergreen

21.3 (17.6~22.7); 12.8 (7.3~15.0); 27.9 (26.3~28.6); 21.4 (17.6~22.8); 196.5 (151.6~213.1); 0.0 (0.0~0.0); 1130 (909~1221) ❙ 1684 (1302~2435); 869 (561~1428); 123 (86~168) ❙ 49.3 (12.9~99.5); 1130 (909~1221); 274.0 (225.9~327.7)

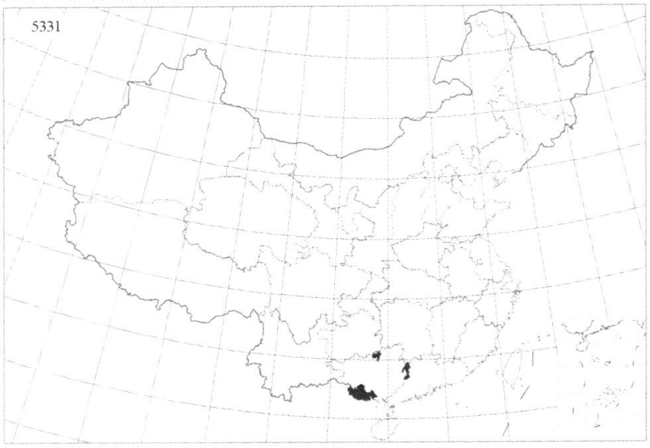

Aquifoliaceae 冬青科 *Ilex* 冬青属
Ilex pentagona S. K. Chen, Y. X. Feng et C. F. Liang 五棱苦丁茶
Trees, evergreen

16.5 (14.9~21.6); 6.1 (3.6~13.1); 25.9 (22.7~27.9); 16.6 (15.0~21.6); 139.7 (121.0~199.6); 0.6 (0.0~1.6); 872 (798~1143) ❙ 1373 (1141~1451); 613 (486~813); 95 (48~123) ❙ 58.7 (27.0~79.3); 872 (798~1143); 242.3 (182.7~359.7)

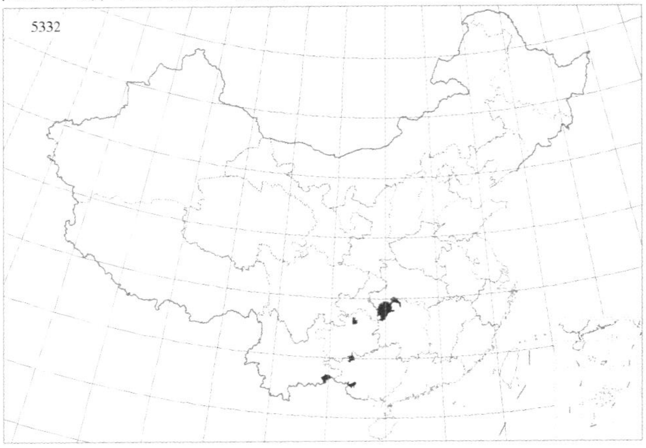

Aquifoliaceae 冬青科 *Ilex* 冬青属
Ilex perlata C. Chen et S. C. Huang ex Y. R. Li 巨叶冬青
Shrubs or trees, evergreen

20.0; 13.6; 24.4; 20.1; 180.7; 0.0; 985 ❙ 1475; 827; 62 ❙ 50.1; 985; 528.9

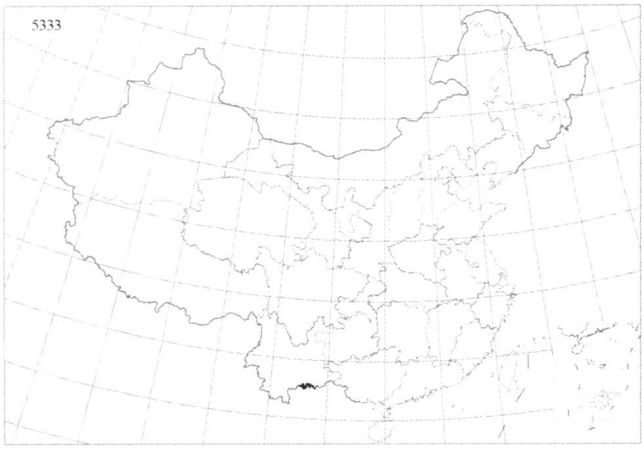

Aquifoliaceae 冬青科 *Ilex* 冬青属
Ilex pernyi Franch. 猫儿刺
Shrubs or trees, evergreen

13.9 (2.2~21.0); 2.8 (-7.0~11.9); 24.3 (9.9~29.3); 14.0 (3.8~21.1); 113.3 (15.5~193.0); 5.7 (0.0~49.1); 780 (413~1114) ❙ 1221 (644~1997); 504 (316~821); 96 (12~254) ❙ 56.7 (-8.1~147.0); 776 (413~1114); 236.1 (128.7~491.5)

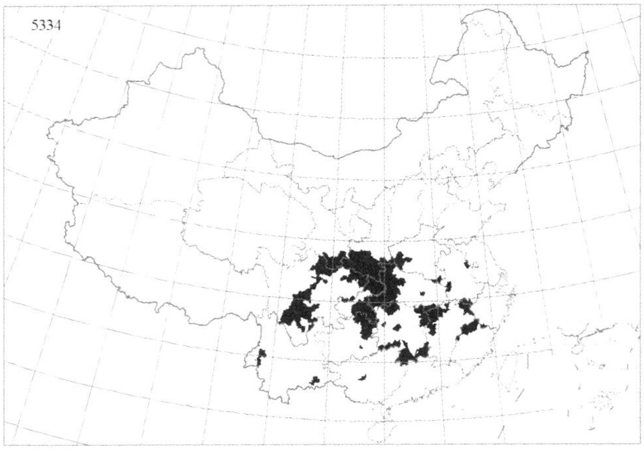

Aquifoliaceae 冬青科 *Ilex* 冬青属
Ilex perryana S. Y. Hu 皱叶冬青
Shrubs, evergreen

11.6 (5.3~16.2); 4.5 (-2.4~9.4); 17.4 (12.2~20.8); 11.7 (5.7~16.3); 85.0 (30.5~135.1); 5.1 (0.0~26.2); 655 (479~790) ❙ 1122 (741~1454); 629 (426~792); 49 (24~70) ❙ 71.8 (39.8~100.7); 655 (479~790); 345.9 (234.2~487.2)

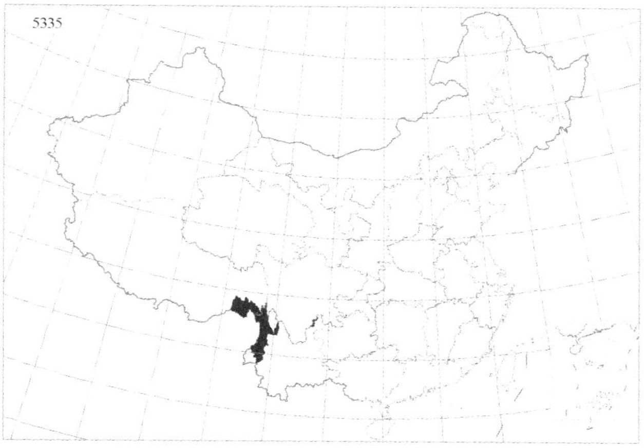

Aquifoliaceae 冬青科 *Ilex* 冬青属
Ilex pingheensis C. J. Tseng 平和冬青
Shrubs, evergreen

19.4; 11.5; 26.5; 19.5; 173.5; 0.0; 980 ❙ 1587; 562; 151 ❙ 62.0; 980; 273.2

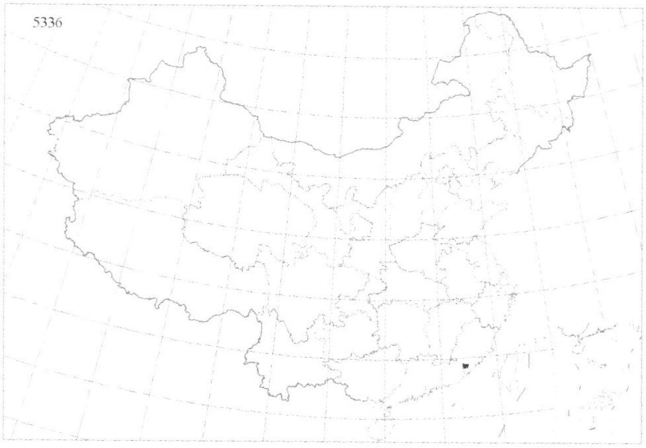

Aquifoliaceae 冬青科 *Ilex* 冬青属
Ilex pingnanensis S. Y. Hu 平南冬青
Shrubs or trees, evergreen

21.1 (17.6~22.2); 12.2 (7.3~13.9); 28.2 (26.3~28.8); 21.1 (17.6~22.3); 193.4 (151.6~207.5); 0.0 (0.0~0.0); 1117 (909~1190) ❙ 1535 (1374~1647); 675 (599~726); 134 (87~168) ❙ 38.3 (15.4~64.7); 1117 (909~1190); 232.1 (203.4~327.7)

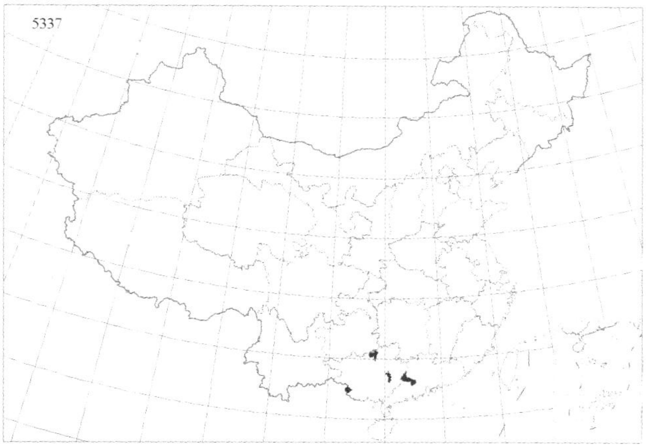

Aquifoliaceae 冬青科 *Ilex* 冬青属
Ilex polyneura (Hand.-Mazz.) S. Y. Hu 多脉冬青
Trees, deciduous

15.6 (5.4~21.8); 8.4 (-2.1~15.1); 21.0 (12.3~27.7); 15.7 (5.8~21.9); 128.6 (31.5~202.2); 0.7 (0.0~25.6); 788 (481~1150) ❙ 1140 (741~2148); 641 (411~1234); 45 (15~124) ❙ 45.1 (-14.2~113.0); 772 (481~1150); 332.8 (142.9~659.0)

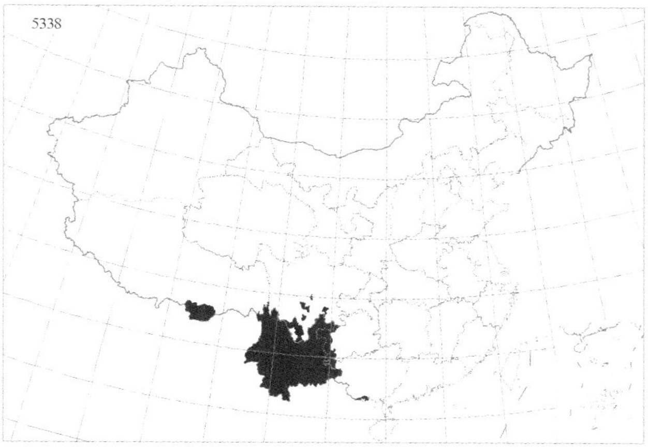

Aquifoliaceae 冬青科 *Ilex* 冬青属
Ilex polypyrena C. J. Tseng et B. W. Liu 多核冬青
Trees, evergreen

21.9 (21.3~22.7); 13.9 (13.2~15.0); 27.9 (27.3~28.6); 22.0 (21.3~22.8); 203.6 (195.9~213.1); 0.0 (0.0~0.0); 1162 (1111~1221) ❙ 1876 (1302~2435); 1047 (670~1428); 116 (86~130) ❙ 61.1 (12.9~99.5); 1162 (1111~1221); 289.6 (235.1~318.8)

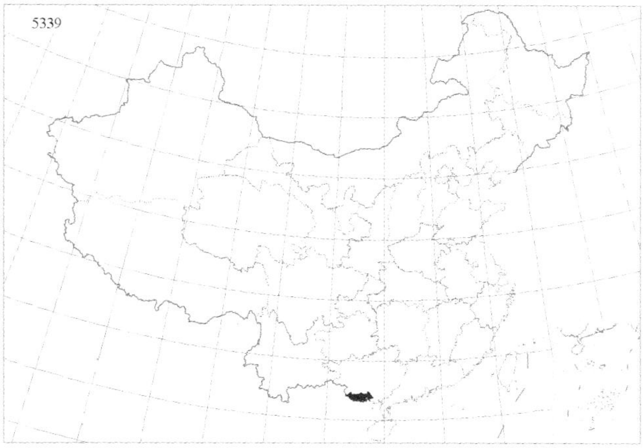

Aquifoliaceae 冬青科 *Ilex* 冬青属
Ilex pseudomachilifolia C. Y. Wu ex Y. R. Li 假楠叶冬青
Trees, evergreen

17.7; 10.9; 22.5; 17.8; 153.4; 0.0; 854 ❙ 1368; 756; 59 ❙ 60.1; 854; 491.5

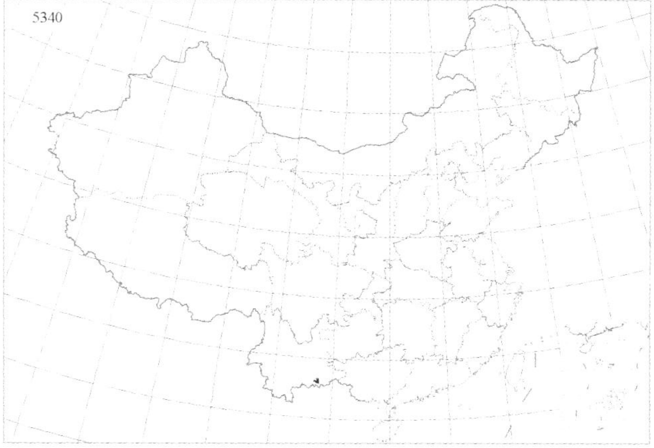

Aquifoliaceae 冬青科 *Ilex* 冬青属

Ilex pubescens Hook. et Arn. 毛冬青

Shrubs or small trees, evergreen

18.5 (11.2~24.8); 9.5 (2.5~20.1); 26.3 (16.7~30.1); 18.5 (11.3~24.8); 162.5 (75.8~237.6); 0.1 (0.0~3.9); 963 (642~1385) ❙ 1548 (797~3495); 631 (312~1847); 154 (23~897) ❙ 62.5 (-34.3~246.5); 947 (642~1360); 263.2 (119.4~620.0)

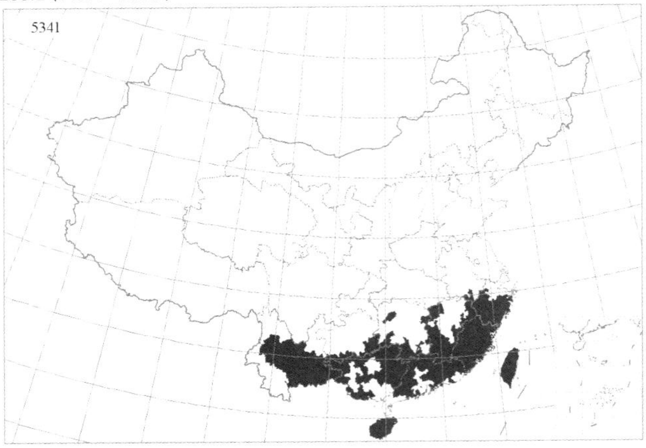

Aquifoliaceae 冬青科 *Ilex* 冬青属

Ilex pubescens var. kwangsiensis Handel-Mazzetti 广西毛冬青

Shrubs or small trees, evergreen

18.5; 10.0; 25.1; 18.6; 163.1; 0.0; 940 ❙ 1234; 645; 65 ❙ 32.9; 939; 295.1

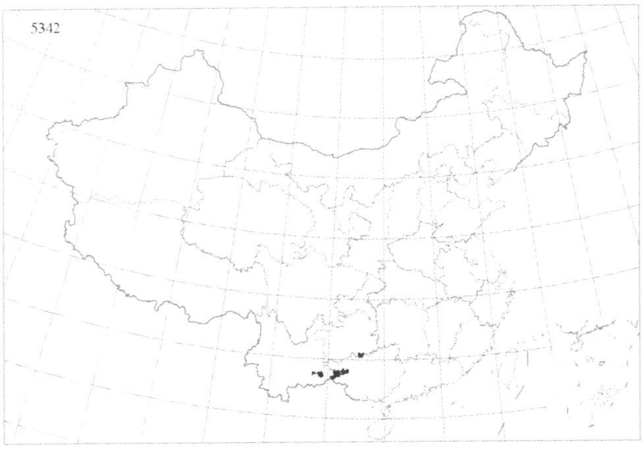

Aquifoliaceae 冬青科 *Ilex* 冬青属

Ilex pubigera (C. Y. Wu ex Y. R. Li) S. K. Chen et Y. X. Feng 有毛冬青

Trees, evergreen

17.8; 8.9; 25.0; 17.8; 153.7; 0.0; 900 ❙ 1479; 649; 138 ❙ 63.6; 900; 276.0

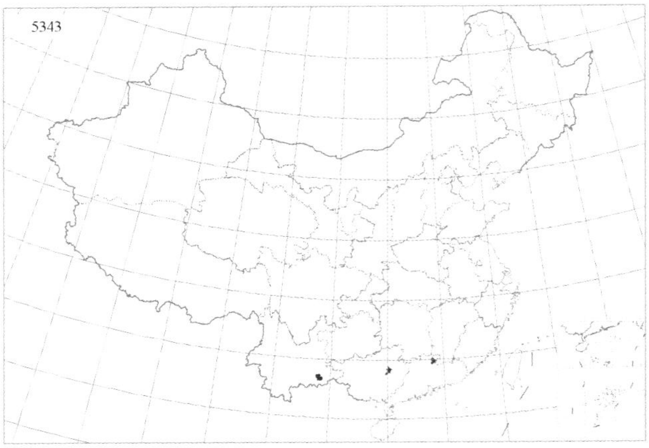

Aquifoliaceae 冬青科 *Ilex* 冬青属

Ilex pubilimba Merr. et Chun 毛叶冬青

Trees, evergreen

21.5 (18.0~24.8); 13.6 (7.5~20.1); 27.8 (25.8~28.6); 21.5 (18.0~24.8); 198.5 (156.4~237.6); 0.0 (0.0~0.0); 1145 (913~1385) ❙ 1586 (1084~2435); 716 (392~1428); 106 (41~194) ❙ 39.5 (-11.9~99.5); 1107 (913~1326); 317.2 (130.1~620.0)

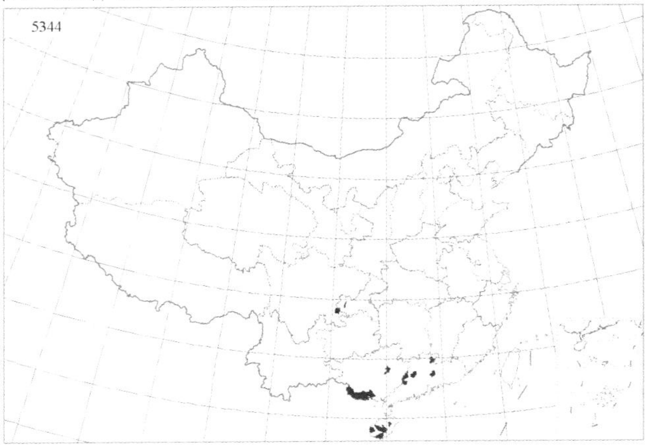

Aquifoliaceae 冬青科 *Ilex* 冬青属

Ilex punctatilimba C. Y. Wu ex Y. R. Li 点叶冬青

Shrubs or small trees, evergreen

11.8; 4.9; 17.4; 11.9; 87.1; 4.7; 668 ❙ 1033; 587; 43 ❙ 58.5; 652; 356.4

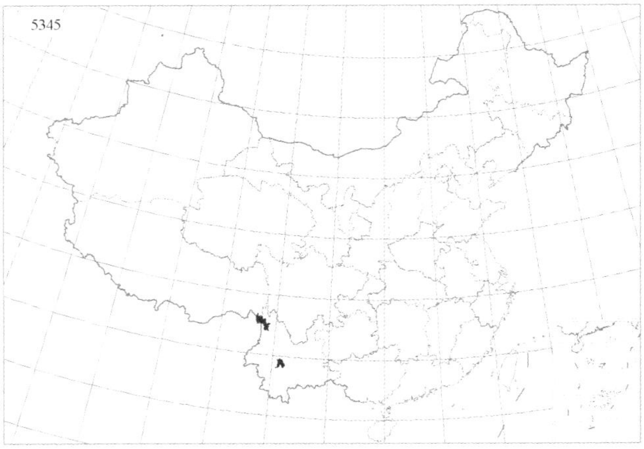

Aquifoliaceae 冬青科 *Ilex* 冬青属

Ilex pyrifolia C. J. Tseng 梨叶冬青

Small trees, evergreen

12.7; 3.6; 21.1; 12.8; 95.3; 1.5; 702 ❙ 1102; 605; 30 ❙ 57.1; 702; 298.1

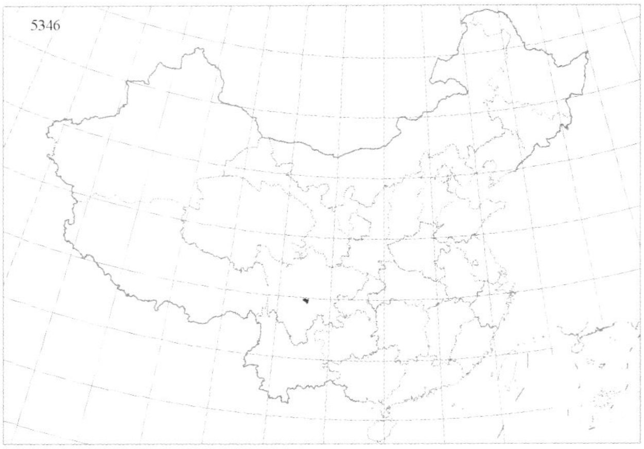

Aquifoliaceae 冬青科 *Ilex* 冬青属
Ilex qianlingshanensis C. J. Tseng 黔灵山冬青
Trees, evergreen

15.6; 5.5; 24.6; 15.6; 127.5; 0.0; 808 ❙ 1206; 546; 69 ❙ 49.2; 808; 189.4

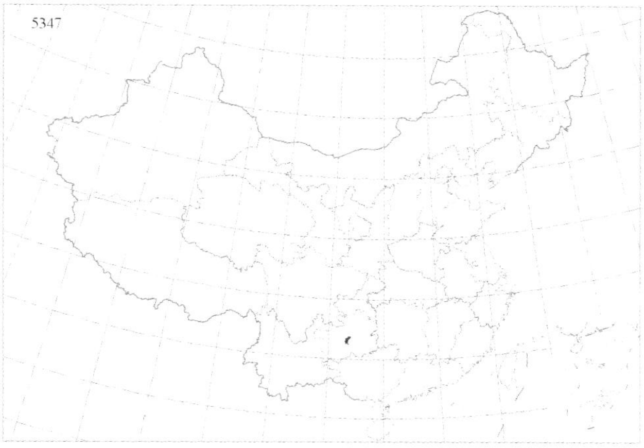

Aquifoliaceae 冬青科 *Ilex* 冬青属
Ilex qingyuanensis C. Z. Zheng 庆元冬青
Trees, evergreen

17.7; 8.4; 26.6; 17.8; 153.3; 0.0; 917 ❙ 1770; 626; 200 ❙ 95.7; 917; 293.8

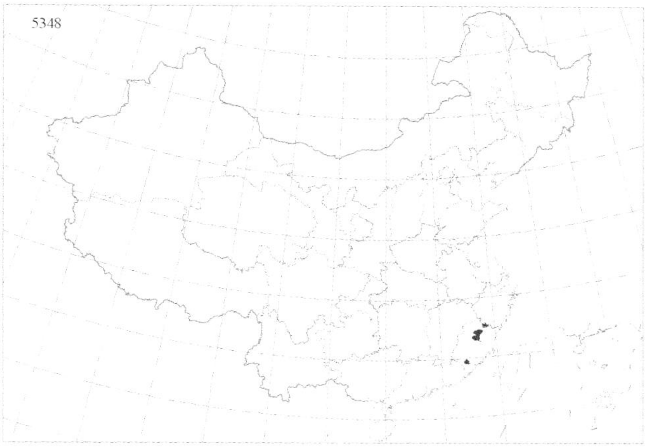

Aquifoliaceae 冬青科 *Ilex* 冬青属
Ilex rarasanensis Sasaki 拉拉山冬青
Trees, evergreen

18.8; 12.4; 24.6; 18.9; 166.6; 0.0; 921 ❙ 2281; 825; 325 ❙ 149.1; 921; 272.4

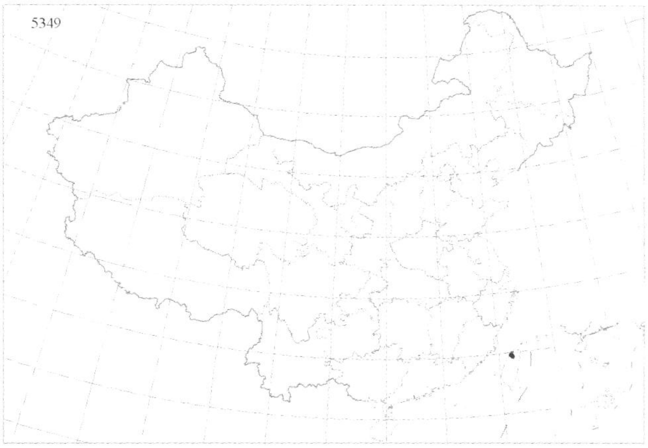

Aquifoliaceae 冬青科 *Ilex* 冬青属
Ilex reticulata C. J. Tseng 网脉冬青
Shrubs, evergreen

19.0; 8.7; 27.9; 19.1; 168.7; 0.0; 1002 ❙ 1710; 705; 186 ❙ 70.6; 1002; 236.8

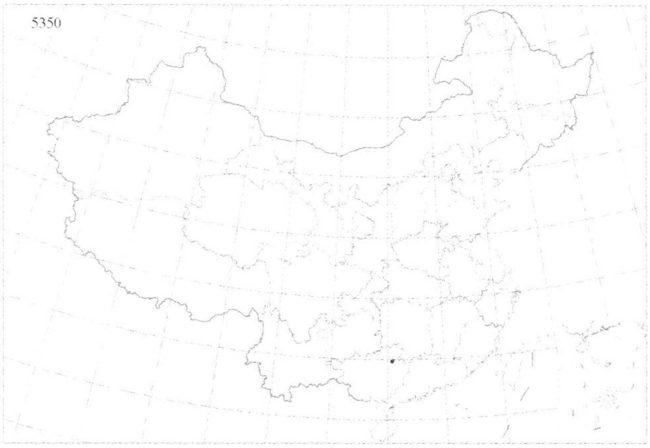

Aquifoliaceae 冬青科 *Ilex* 冬青属
Ilex retusifolia S. Y. Hu 微凹冬青
Shrubs, evergreen

17.0; 6.2; 26.7; 17.0; 144.4; 0.0; 891 ❙ 1624; 594; 191 ❙ 82.3; 891; 240.7

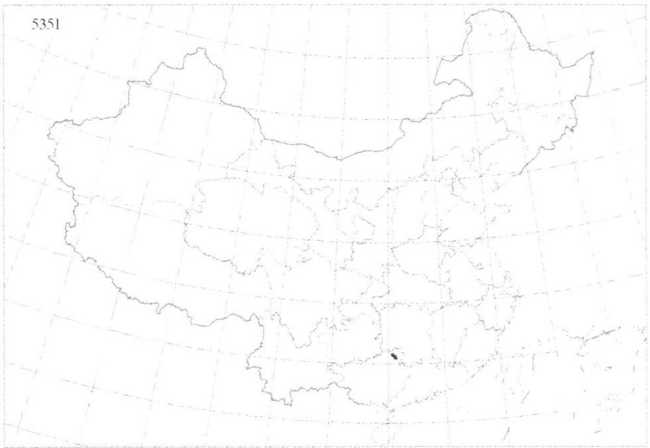

Aquifoliaceae 冬青科 *Ilex* 冬青属
Ilex robusta C. J. Tseng 粗枝冬青
Shrubs, evergreen

21.1; 12.0; 28.4; 21.2; 194.2; 0.0; 1121 ❙ 1578; 662; 168 ❙ 40.8; 1121; 218.0

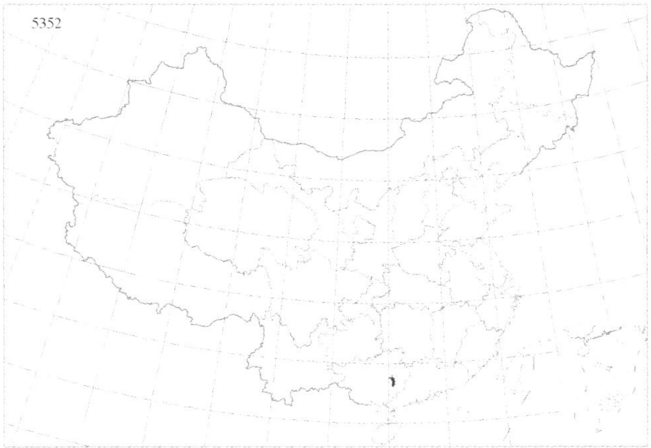

Aquifoliaceae 冬青科 *Ilex* 冬青属

Ilex robustinervosa C. J. Tseng ex S. K. Chen et Y. X. Feng 粗脉冬青

Shrubs or small trees, evergreen

22.1；14.6；27.9；22.1；205.4；0.0；1167 ❙ 1955；930；115 ❙ 67.5；1167；255.6

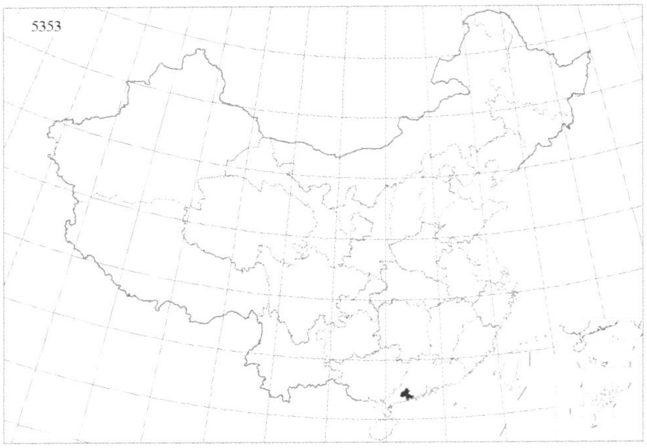

Aquifoliaceae 冬青科 *Ilex* 冬青属

Ilex rockii S. Y. Hu 高山冬青

Small shrubs, evergreen

12.6 (5.4~16.2)；5.0 (−2.1~9.4)；18.8 (12.3~24.6)；12.6 (5.8~16.3)；94.6 (31.5~135.1)；3.2 (0.0~25.6)；689 (481~822) ❙ 1204 (741~1736)；676 (426~1007)；51 (28~69) ❙ 74.9 (49.1~145.8)；689 (481~822)；334.4 (189.1~447.4)

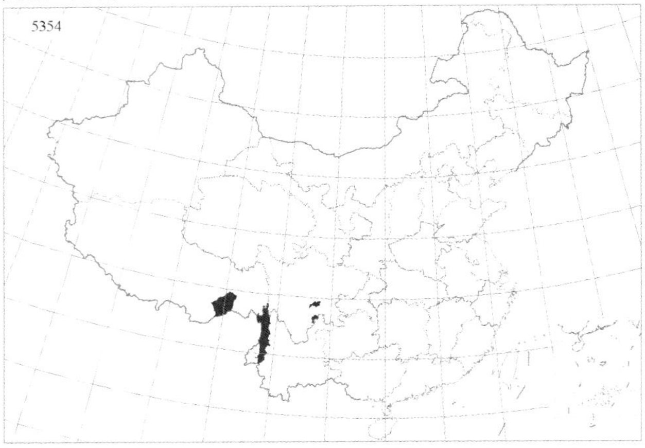

Aquifoliaceae 冬青科 *Ilex* 冬青属

Ilex rotunda Thunb. 铁冬青

Shrubs or trees, evergreen

18.0 (3.8~24.8)；8.3 (−4.3~20.1)；26.6 (10.8~29.9)；18.1 (4.7~24.8)；157.1 (21.6~237.6)；0.4 (0.0~35.4)；949 (444~1385) ❙ 1495 (736~3495)；610 (299~1847)；147 (16~897) ❙ 58.9 (−34.3~246.5)；934 (444~1360)；251.1 (111.7~659.0)

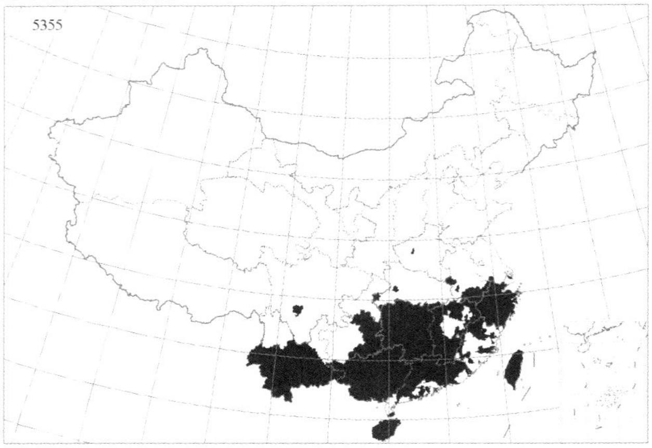

Aquifoliaceae 冬青科 *Ilex* 冬青属

Ilex salicina Hand.-Mazz. 柳叶冬青

Shrubs, evergreen

21.3 (17.4~22.7)；12.9 (7.4~15.0)；27.7 (25.5~28.6)；21.4 (17.4~22.8)；196.5 (148.9~213.1)；0.0 (0.0~0.0)；1127 (886~1221) ❙ 1675 (1284~2435)；902 (637~1428)；110 (73~130) ❙ 48.4 (12.9~99.5)；1127 (886~1221)；287.5 (235.1~327.7)

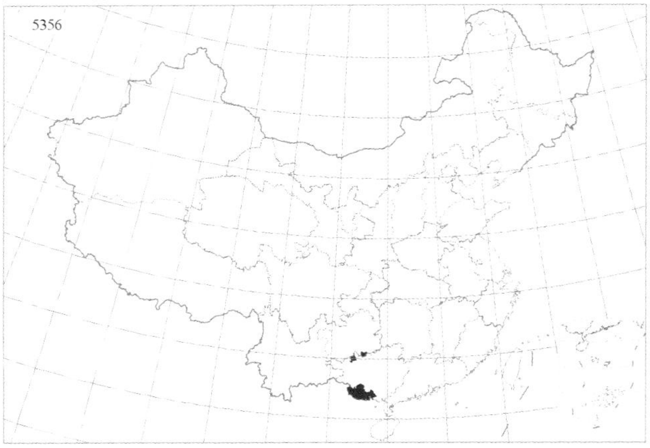

Aquifoliaceae 冬青科 *Ilex* 冬青属

Ilex serrata Thunb. 落霜红

Shrubs, deciduous

16.3 (2.2~18.6)；4.9 (−6.6~9.3)；27.1 (9.9~29.9)；16.4 (3.8~18.7)；137.4 (15.5~164.0)；1.4 (0.0~49.1)；882 (413~976) ❙ 1567 (932~1950)；581 (427~938)；180 (19~259) ❙ 78.6 (28.8~147.0)；874 (413~965)；222.1 (141.7~300.0)

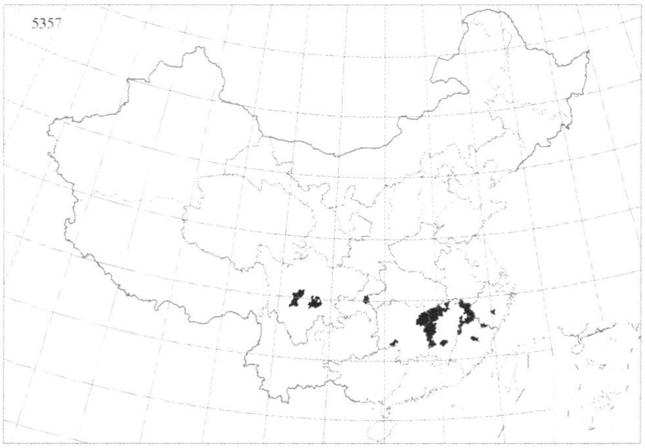

Aquifoliaceae 冬青科 *Ilex* 冬青属

Ilex shennongjiaensis T. R. Dudley 神农架冬青

Trees, evergreen

9.3；−1.7；20.1；9.5；69.8；18.0；603 ❙ 1214；516；71 ❙ 101.4；603；279.5

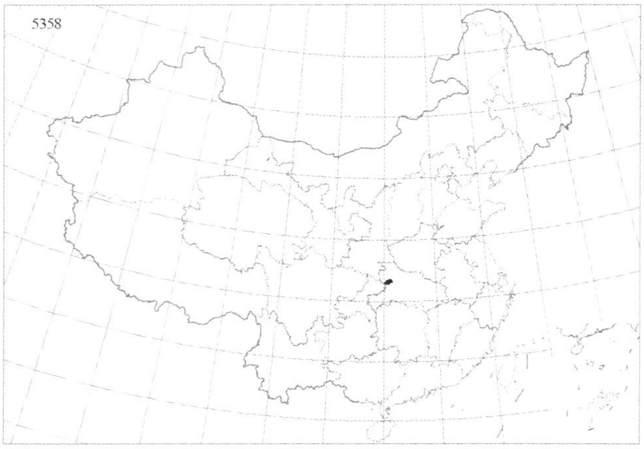

Aquifoliaceae 冬青科 *Ilex* 冬青属
Ilex shimeica Tam. 石枚冬青
Trees, evergreen

24.6；19.7；28.1；24.7；236.2；0.0；1374 ❙ 1677；550；73 ❙ 22.1；1255；521.4

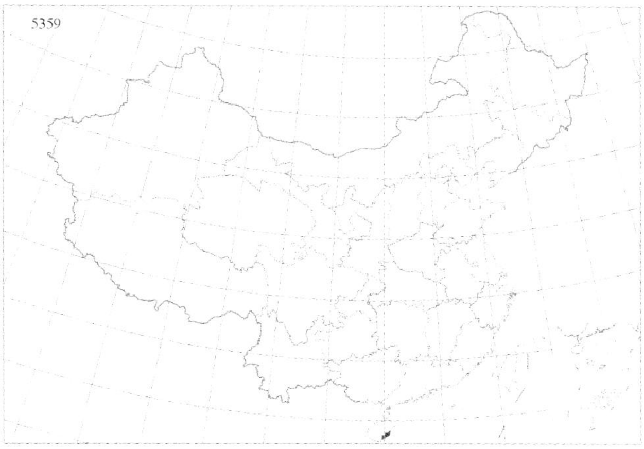

Aquifoliaceae 冬青科 *Ilex* 冬青属
Ilex sikkimensis Kurz 锡金冬青
Trees, evergreen

15.3 (8.6~19.6)；8.4 (1.3~12.8)；20.1 (15.0~23.5)；15.4 (8.7~19.6)；125.3 (52.6~175.7)；0.9 (0.0~8.8)；778 (574~953) ❙ 1482 (1041~1736)；849 (585~1007)；55 (42~70) ❙ 92.0 (61.2~145.8)；774 (574~932)；477.4 (352.3~627.0)

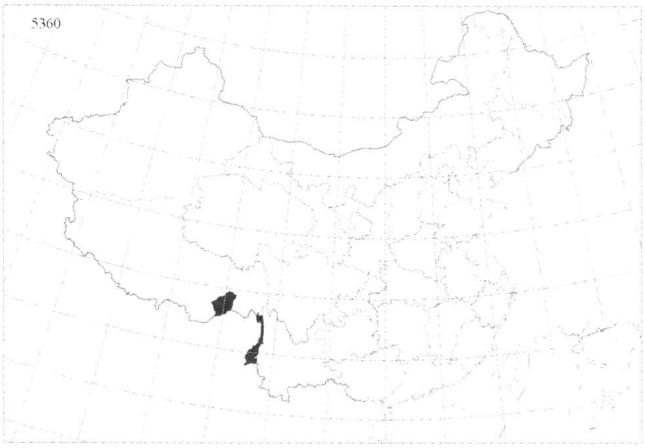

Aquifoliaceae 冬青科 *Ilex* 冬青属
Ilex sinica (Loes.) S. Y. Hu 中华冬青
Trees, evergreen

17.7 (14.3~21.0)；9.8 (3.3~15.1)；23.9 (21.4~28.7)；17.8 (14.3~21.0)；153.6 (114.1~191.8)；0.2 (0.0~2.3)；882 (765~1053) ❙ 1444 (946~1776)；695 (497~996)；95 (38~196) ❙ 63.9 (15.5~101.8)；877 (763~1053)；379.0 (203.9~659.0)

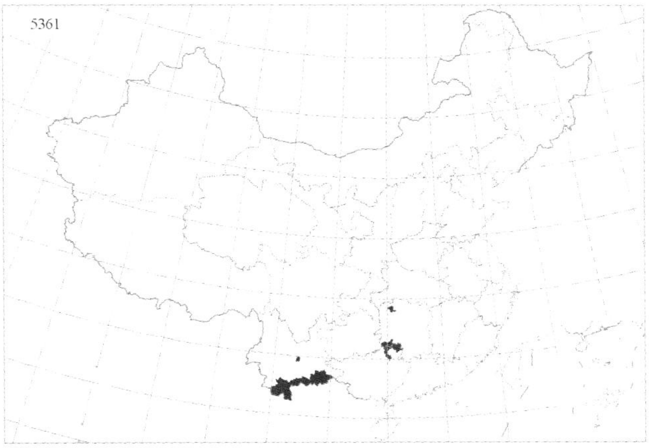

Aquifoliaceae 冬青科 *Ilex* 冬青属
Ilex sterrophylla Merr. et Chun 华南冬青
Trees, evergreen

19.0 (14.3~24.0)；9.3 (3.3~19.1)；27.4 (24.3~29.8)；19.0 (14.3~24.0)；168.4 (114.1~228.1)；0.1 (0.0~2.3)；1004 (765~1302) ❙ 1570 (1117~2435)；627 (403~1428)；163 (48~221) ❙ 58.5 (6.7~106.4)；988 (765~1221)；247.1 (151.0~620.0)

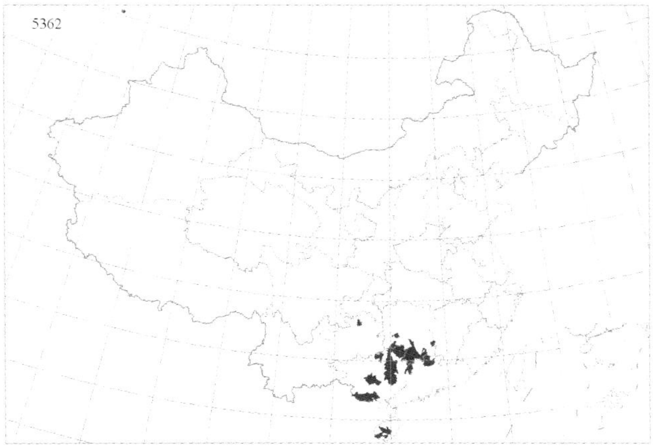

Aquifoliaceae 冬青科 *Ilex* 冬青属
Ilex stewardii S. Y. Hu 黔桂冬青
Shrubs or small trees, evergreen

19.1 (15.2~22.8)；10.1 (4.4~15.1)；26.4 (21.4~28.9)；19.2 (15.2~22.8)；170.3 (123.1~213.6)；0.0 (0.0~0.6)；994 (797~1230) ❙ 1473 (1074~2435)；722 (452~1428)；108 (47~189) ❙ 48.8 (12.9~99.5)；994 (797~1230)；268.4 (189.5~659.0)

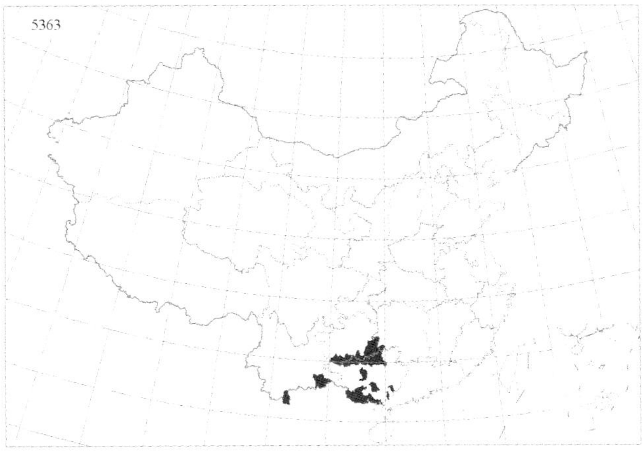

Aquifoliaceae 冬青科 *Ilex* 冬青属
Ilex strigillosa T. R. Dudley 粗毛冬青
Small trees, evergreen

22.1；15.2；27.4；22.1；205.8；0.0；1190 ❙ 1775；656；124 ❙ 55.4；1108；385.0

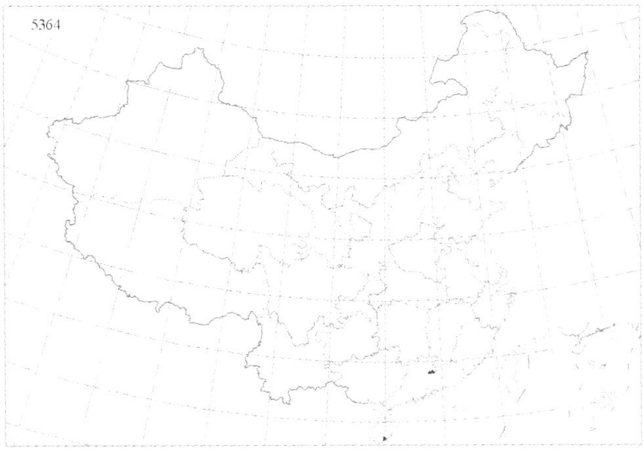

Celastraceae 卫矛科 *Celastrus* 南蛇藤属

Celastrus cuneatus (Rehder et Wils.) C. Y. Cheng et T. C. Kao 小南蛇藤

Scandent shrubs

15.3 (13.0~17.1); 2.9 (1.9~4.0); 27.0 (23.8~29.4); 15.4 (13.1~17.1); 127.3 (102.9~146.1); 3.0 (1.0~6.1); 855 (730~952) ❙ 1053 (849~1235); 459 (388~509); 86 (65~123) ❙ 24.3 (-5.3~64.6); 844 (730~921); 182.0 (111.9~262.4)

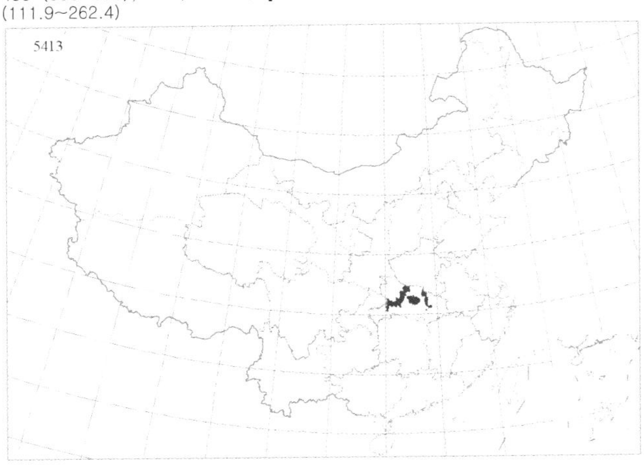

Celastraceae 卫矛科 *Celastrus* 南蛇藤属

Celastrus flagellaris Rupr. 刺苞南蛇藤

Scandent shrubs, deciduous

10.3 (-0.8~16.8); -5.4 (-25.0~5.1); 24.3 (18.4~28.0); 11.8 (6.4~16.8); 97.0 (45.8~142.2); 33.0 (0.0~118.4); 721 (501~904) ❙ 783 (409~1489); 423 (258~691); 43 (8~196) ❙ 8.1 (-36.5~82.7); 659 (409~904); 209.2 (120.1~335.3)

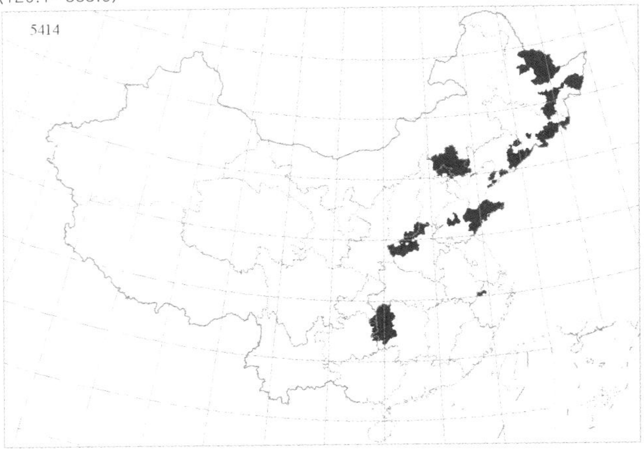

Celastraceae 卫矛科 *Celastrus* 南蛇藤属

Celastrus gemmatus Loes. 大芽南蛇藤

Scandent shrubs

14.3 (2.0~21.7); 2.9 (-12.7~13.5); 24.8 (9.9~30.1); 14.6 (3.8~21.7); 120.1 (15.4~200.8); 7.8 (0.0~65.0); 804 (413~1151) ❙ 1157 (367~2024); 494 (235~902); 98 (10~259) ❙ 41.8 (-37.7~130.7); 776 (367~1151); 219.5 (70.3~506.2)

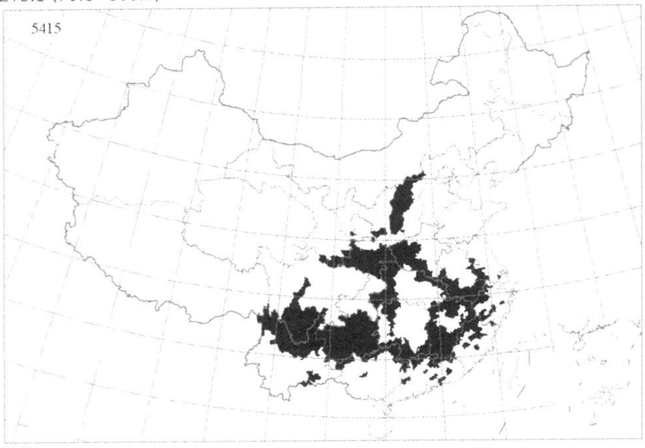

Celastraceae 卫矛科 *Celastrus* 南蛇藤属

Celastrus glaucophyllus Rehder et Wils. 灰叶南蛇藤

Scandent shrubs, deciduous

13.8 (1.8~21.0); 4.0 (-7.1~13.0); 22.6 (9.9~29.4); 13.9 (3.8~21.1); 110.7 (15.5~193.0); 4.8 (0.0~53.3); 759 (404~1114) ❙ 1138 (583~1736); 531 (317~1007); 70 (12~214) ❙ 51.2 (-14.2~145.8); 747 (404~1114); 249.8 (135.3~506.2)

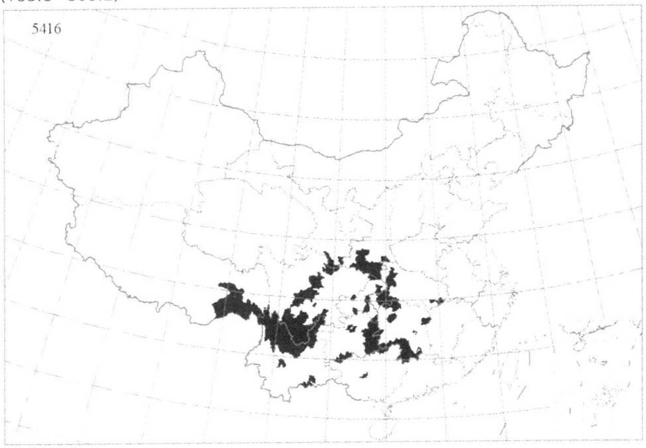

Celastraceae 卫矛科 *Celastrus* 南蛇藤属

Celastrus hirsutus H. F. Comber 硬毛南蛇藤

Scandent shrubs

16.1 (8.6~22.0); 9.1 (1.3~14.2); 21.4 (15.0~28.8); 16.2 (8.7~22.0); 135.0 (52.6~203.8); 1.0 (0.0~8.8); 831 (574~1170) ❙ 1351 (1007~1647); 739 (578~909); 63 (35~132) ❙ 65.5 (26.1~99.9); 826 (574~1170); 402.0 (203.4~551.6)

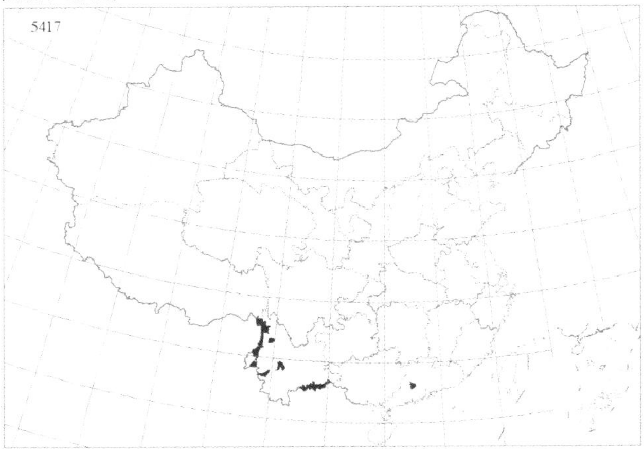

Celastraceae 卫矛科 *Celastrus* 南蛇藤属

Celastrus homaliifolius Hsu. 小果南蛇藤

Scandent shrubs, evergreen

14.9; 5.3; 23.6; 15.0; 119.6; 0.0; 782 ❙ 1410; 816; 49 ❙ 80.4; 782; 214.5

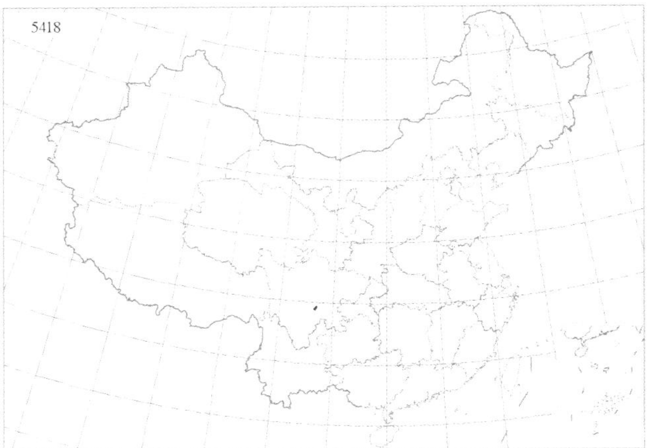

Celastraceae 卫矛科 *Celastrus* 南蛇藤属

Celastrus hookeri Prain 滇边南蛇藤

Scandent shrubs

16.8 (-1.1~22.4); 7.3 (-9.6~14.8); 24.9 (6.2~29.8); 16.9 (2.0~22.4); 144.2 (2.5~208.9); 2.2 (0.0~75.6); 887 (334~1191) ❙ 1411 (321~2037); 617 (193~1033); 112 (10~197) ❙ 58.7 (-4.0~113.0); 886 (321~1191); 236.1 (61.6~491.5)

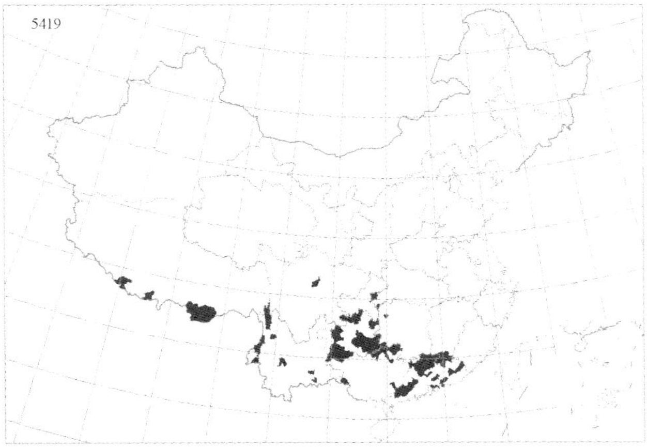

Celastraceae 卫矛科 *Celastrus* 南蛇藤属

Celastrus hypoleucoides P. L. Chiu 薄叶南蛇藤

Scandent shrubs

15.6 (14.3~17.7); 4.4 (2.5~6.9); 26.2 (23.5~28.6); 15.7 (14.3~17.7); 128.7 (112.6~152.4); 0.8 (0.0~3.9); 835 (756~932) ❙ 1355 (1101~1801); 507 (419~598); 133 (68~242) ❙ 62.4 (35.3~109.2); 829 (756~927); 207.1 (134.6~284.0)

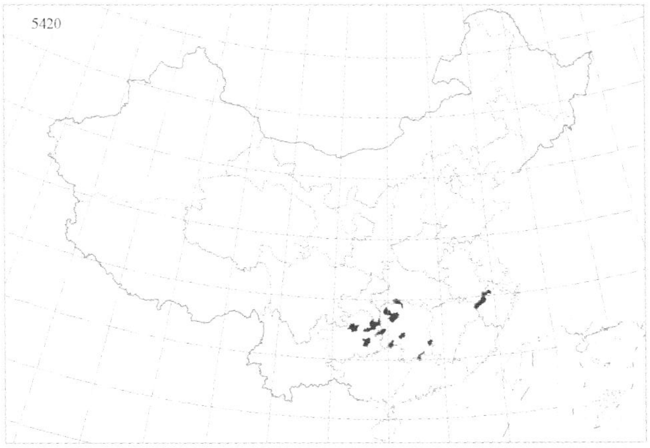

Celastraceae 卫矛科 *Celastrus* 南蛇藤属

Celastrus hypoleucus (Oliv.) Warb. ex Loesen. 粉背南蛇藤

Scandent shrubs

14.0 (0.0~19.[illegible]); 1.8 (-11.4~9.3); 25.5 (9.7~29.7); 14.1 (3.3~19.1); 116.3 (12.2~169.2); 7.8 (0.0~72.2); 795 (395~988) ❙ 1085 (512~1742); 441 (258~666); 91 (11~228) ❙ 35.2 (-35.9~128.9); 768 (395~988); 215.0 (121.6~364.1)

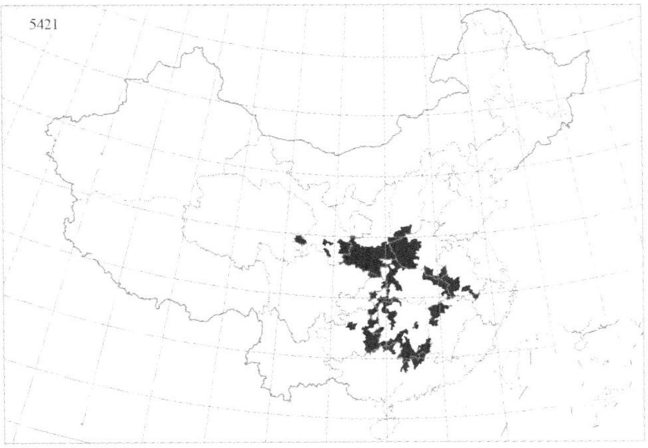

Celastraceae 卫矛科 *Celastrus* 南蛇藤属

Celastrus kusanoi Hayata 圆叶南蛇藤

Scandent shrubs, deciduous

20.3 (15.5~24.2); 12.2 (4.8~18.9); 27.0 (20.1~29.8); 20.4 (15.6~24.3); 184.4 (127.2~231.2); 0.0 (0.0~0.2); 1057 (773~1335) ❙ 1778 (1176~3495); 782 (447~1847); 170 (47~897) ❙ 73.0 (-12.0~246.5); 1041 (773~1269); 250.4 (119.4~538.2)

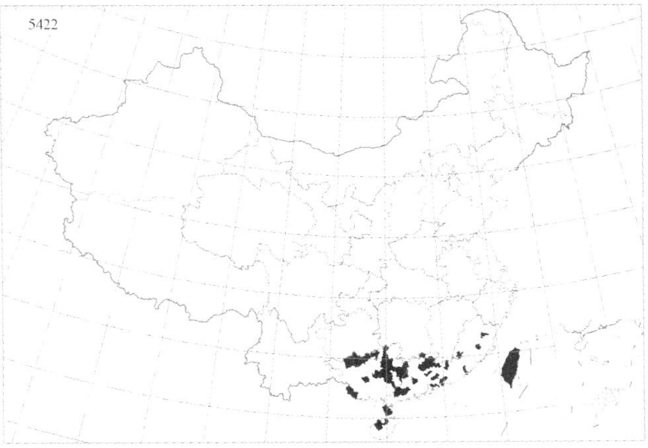

Celastraceae 卫矛科 *Celastrus* 南蛇藤属

Celastrus monospermus Roxb. 独子藤

Scandent shrubs, evergreen

19.4 (3.6~24.0); 11.2 (1.3~19.1); 26.0 (15.0~29.8); 19.5 (8.7~24.0); 173.8 (52.6~228.1); 0.1 (0.0~8.8); 1005 (574~1302) ❙ 1556 (1007~2435); 718 (471~1428); 118 (44~194) ❙ 56.0 (12.9~145.8); 998 (574~1257); 305.4 (163.7~659.0)

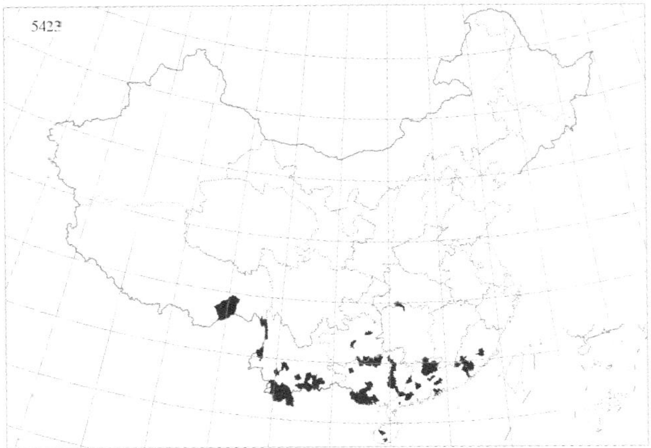

Celastraceae 卫矛科 *Celastrus* 南蛇藤属

Celastrus oblanceifolius C. H. Wang et P. C. Tsoong 窄叶南蛇藤

Scandent shrubs

15.6 (2.2~22.4); 3.9 (-17.9~14.8); 26.3 (15.8~29.5); 16.0 (6.7~22.4); 135.8 (45.2~209.2); 7.5 (0.0~83.9); 868 (513~1198) ❙ 1375 (457~2037); 540 (273~1033); 141 (8~259) ❙ 56.7 (-28.3~147.0); 858 (457~1198); 230.7 (123.3~491.5)

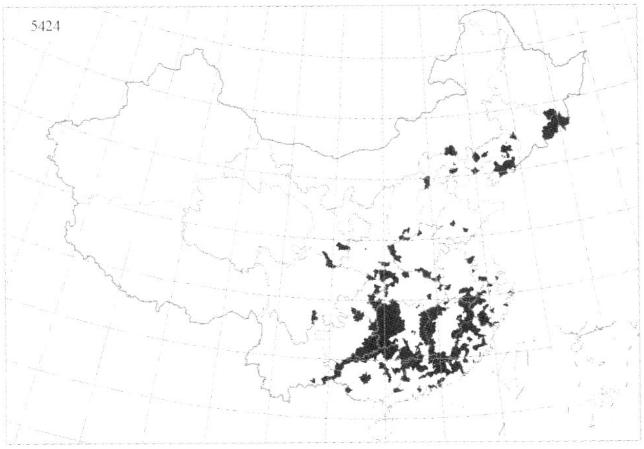

Celastraceae 卫矛科 *Celastrus* 南蛇藤属
Celastrus orbiculatus Thunberg 南蛇藤
Scandent shrubs, deciduous

13.3 (2.1~22.4); -0.5 (-19.2~15.2); 25.8 (14.0~30.1); 13.9 (5.6~22.7); 116.3 (32.4~212.7); 16.6 (0.0~85.4); 802 (467~1216) ❙ 972 (341~2014); 444 (190~994); 84 (8~259) ❙ 18.0 (-50.3~130.7); 731 (341~1216); 191.2 (61.2~551.3)

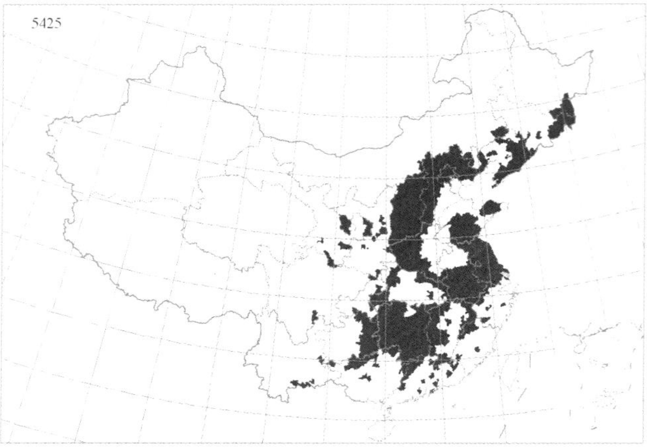

Celastraceae 卫矛科 *Celastrus* 南蛇藤属
Celastrus paniculatus Willd. 灯油藤
Scandent shrubs, evergreen

19.2 (5.3~24.8); 12.2 (-2.4~20.1); 24.4 (12.2~28.6); 19.2 (5.7~24.8); 171.3 (30.5~237.6); 0.8 (0.0~26.2); 975 (479~1385) ❙ 1449 (961~2992); 719 (455~1847); 83 (31~209) ❙ 51.6 (-1.4~166.5); 951 (479~1254); 377.9 (223.0~659.0)

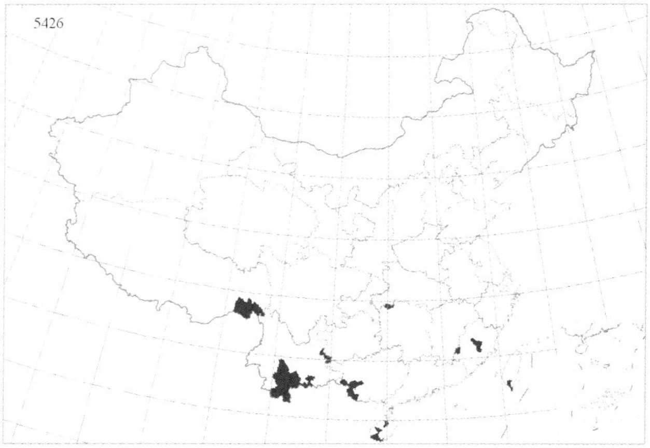

Celastraceae 卫矛科 *Celastrus* 南蛇藤属
Celastrus punctatus Thunb. 东南南蛇藤
Scandent shrubs, deciduous

17.6 (14.9~21.1); 7.4 (2.6~15.3); 27.1 (20.1~30.1); 17.7 (14.9~21.1); 152.4 (120.8~193.7); 0.3 (0.0~3.4); 928 (773~1104) ❙ 1796 (1258~2859); 653 (425~1412); 222 (110~402) ❙ 96.6 (14.0~246.5); 918 (773~1047); 242.7 (148.9~363.7)

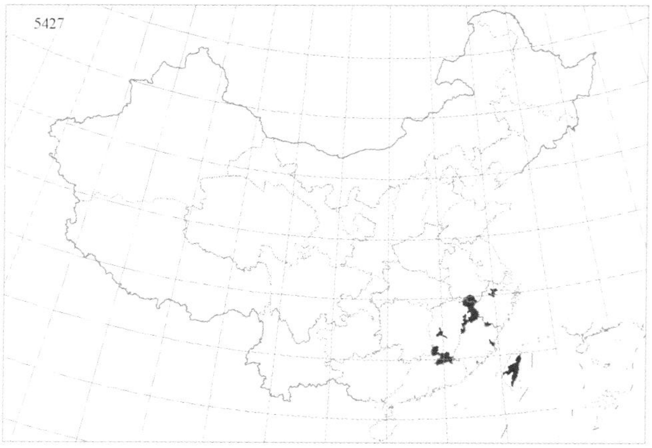

Celastraceae 卫矛科 *Celastrus* 南蛇藤属
Celastrus rosthornianus Loes. 短梗南蛇藤
Scandent shrubs

14.9 (2.2~21.9); 4.2 (-6.6~15.1); 24.6 (9.5~29.3); 15.0 (3.8~22.0); 123.3 (15.5~203.8); 4.0 (0.0~49.1); 814 (411~1176) ❙ 1211 (512~1922); 525 (258~835); 101 (11~259) ❙ 48.6 (-35.3~129.9); 797 (411~1176); 232.8 (121.6~659.0)

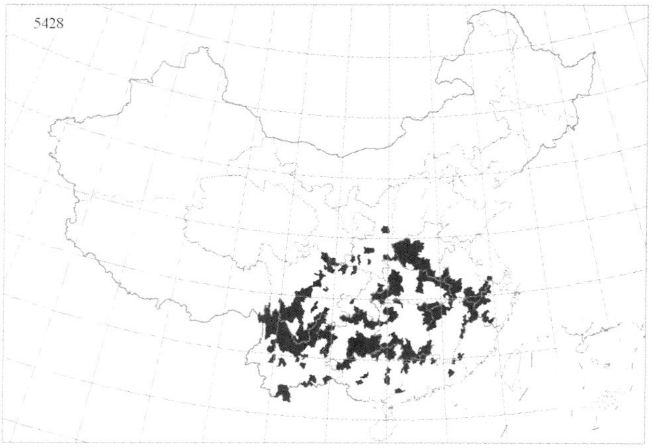

Celastraceae 卫矛科 *Celastrus* 南蛇藤属
Celastrus rosthornianus var. loeseneri (Rehder et Wils.) C. Y. Wu 宽叶短梗南蛇藤
Scandent shrubs

13.4 (6.2~21.0); 2.4 (-4.2~14.2); 23.5 (15.0~26.8); 13.5 (7.0~21.0); 110.2 (45.2~191.8); 8.8 (0.0~29.9); 753 (522~1053) ❙ 993 (634~1522); 463 (290~835); 48 (15~107) ❙ 30.9 (-16.6~81.5); 750 (522~1053); 265.9 (146.0~551.6)

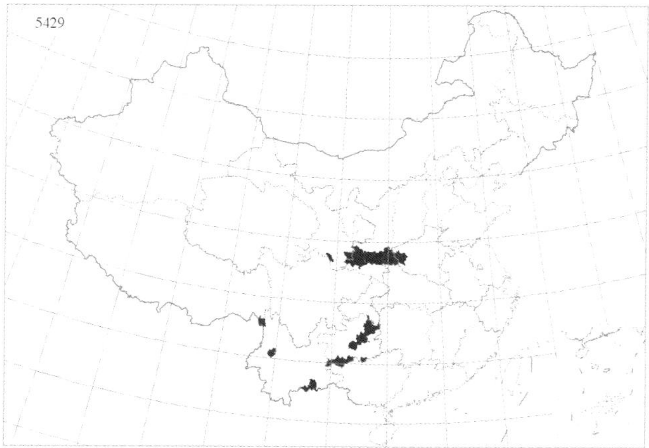

Celastraceae 卫矛科 *Celastrus* 南蛇藤属
Celastrus rugosus Rehder et Wils. 皱叶南蛇藤
Scandent shrubs, deciduous

11.2 (1.8~19.1); 2.2 (-7.1~10.5); 19.1 (9.9~26.6); 11.4 (3.8~19.1); 85.9 (15.5~169.2); 11.6 (0.0~53.3); 669 (404~969) ❙ 1073 (583~1736); 565 (348~1007); 40 (12~94) ❙ 65.0 (5.5~145.8); 663 (404~969); 267.7 (169.5~447.4)

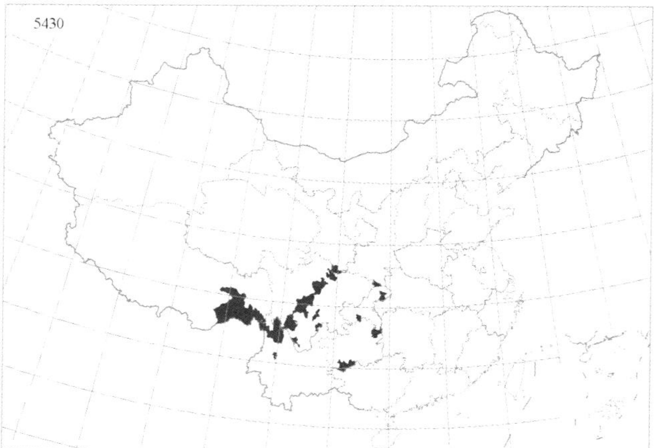

Celastraceae 卫矛科 *Celastrus* 南蛇藤属

Celastrus stylosus Wall. 显柱南蛇藤

Scandent shrubs

15.8 (-0.7~22.7); 5.5 (-9.6~15.0); 25.2 (7.7~29.1); 15.9 (2.5~22.8); 132.3 (6.8~213.1); 1.9 (0.0~74.2); 845 (338~1221) ❙ 1324 (370~2435); 569 (262~1428); 104 (9~259) ❙ 57.3 (4.7~145.8); 843 (338~1221); 229.3 (61.6~570.1)

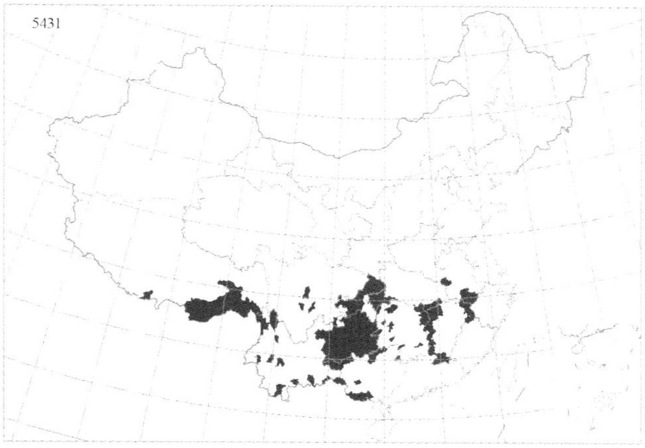

Celastraceae 卫矛科 *Celastrus* 南蛇藤属

Celastrus stylosus var. puberulus (Hsu.) C. Y. Cheng et T. C. Kao 毛脉显柱南蛇藤

Scandent shrubs

15.2 (14.4~16.1); 2.9 (2.1~3.9); 27.0 (26.3~27.8); 15.3 (14.5~16.1); 126.1 (118.7~134.4); 2.9 (1.1~4.9); 838 (801~878) ❙ 1600 (1444~1689); 582 (527~617); 196 (169~212) ❙ 91.1 (64.3~105.7); 838 (801~878); 274.4 (240.8~319.9)

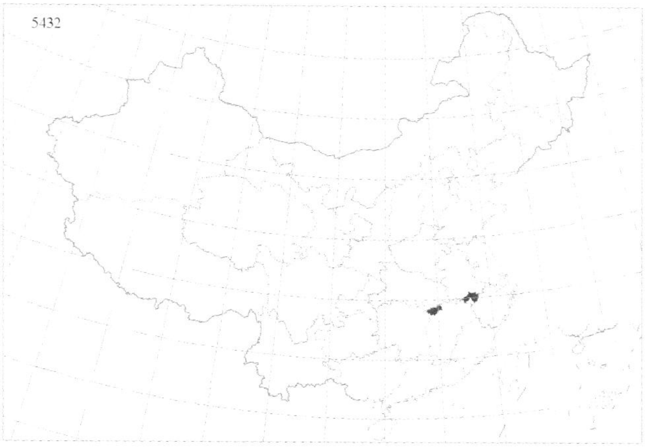

Celastraceae 卫矛科 *Celastrus* 南蛇藤属

Celastrus tonkinensis Pitard 皱果南蛇藤

Scandent shrubs, evergreen

18.5 (1.8~24.3); 9.4 (-7.1~20.1); 26.5 (9.9~29.8); 18.6 (3.8~24.8); 163.9 (15.7~237.6); 1.1 (0.0~53.3); 977 (404~1385) ❙ 1512 (583~2992); 652 (348~1847); 126 (13~303) ❙ 56.9 (-6.9~219.3); 970 (404~1269); 241.7 (95.3~659.0)

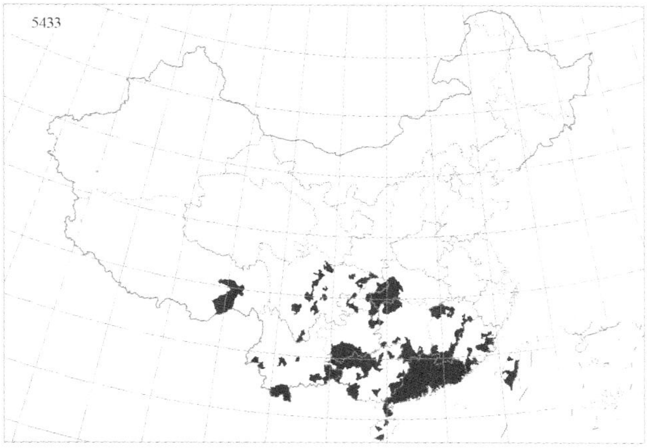

Celastraceae 卫矛科 *Celastrus* 南蛇藤属

Celastrus vaniotii (Lév.) Rehder 长序南蛇藤

Scandent shrubs

15.1 (3.5~21.0); 5.1 (-6.5~12.0); 24.1 (12.6~28.2); 15.2 (5.0~21.1); 123.5 (26.3~193.0); 1.6 (0.0~43.4); 800 (446~1114) ❙ 1217 (731~1662); 559 (346~821); 73 (19~192) ❙ 53.0 (16.9~107.2); 799 (446~1114); 218.0 (146.0~432.9)

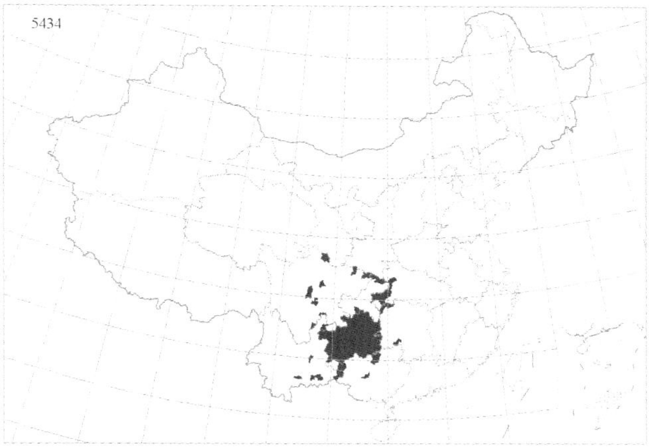

Celastraceae 卫矛科 *Celastrus* 南蛇藤属

Celastrus virens (Wang et Tang) C. Y. Cheng et T. C. Kao 绿独子藤

Scandent shrubs, evergreen

19.6; 13.8; 23.1; 19.6; 174.9; 0.0; 940 ❙ 1506; 783; 64 ❙ 60.9; 923; 560.7

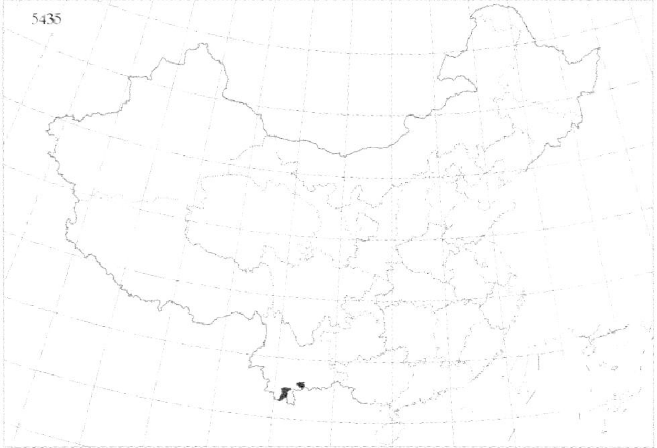

Celastraceae 卫矛科 *Euonymus* 卫矛属

Euonymus acanthocarpus Franch. 刺果卫矛

Shrubs, deciduous

15.5 (-1.8~22.7); 5.0 (-10.2~15.0); 25.0 (6.1~30.1); 15.5 (1.8~22.8); 128.3 (1.9~213.1); 2.3 (0.0~82.8); 832 (329~1221) ❙ 1309 (320~2435); 540 (217~1428); 116 (10~259) ❙ 56.7 (-33.2~130.7); 822 (320~1221); 230.9 (61.6~551.6)

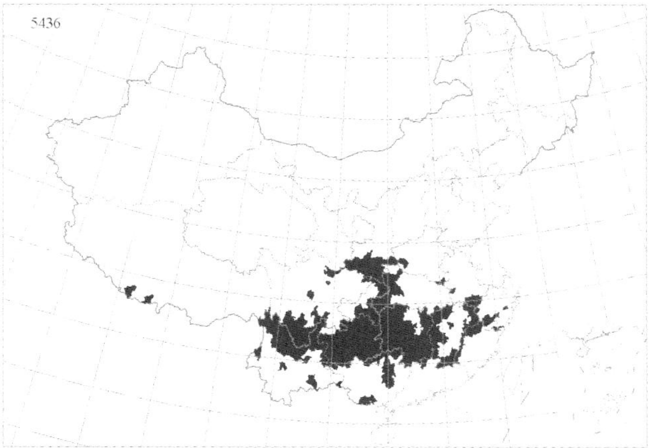

Celastraceae 卫矛科 *Euonymus* 卫矛属

Euonymus acanthoxanthus Pitard 三脉卫矛

Shrubs, evergreen

18.9; 9.5; 26.6; 18.9; 167.3; 0.0; 969 ❙ 1236; 617; 60 ❙ 27.5; 969; 254.3

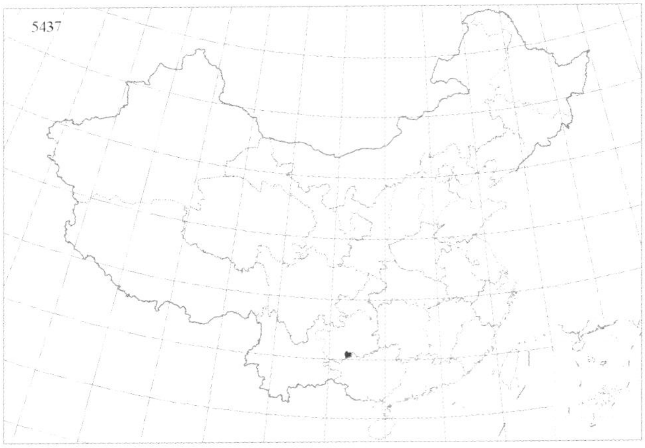

Celastraceae 卫矛科 *Euonymus* 卫矛属

Euonymus actinocarpus Loes. 星刺卫矛

Shrubs, evergreen

16.7 (1.9~22.4); 6.5 (-9.2~15.2); 25.8 (9.9~29.1); 16.9 (3.8~22.7); 144.8 (15.5~212.7); 3.3 (0.0~57.8); 897 (413~1216) ❙ 1429 (525~2037); 581 (270~1033); 141 (10~221) ❙ 59.3 (0.0~125.9); 893 (413~1216); 229.6 (151.0~491.5)

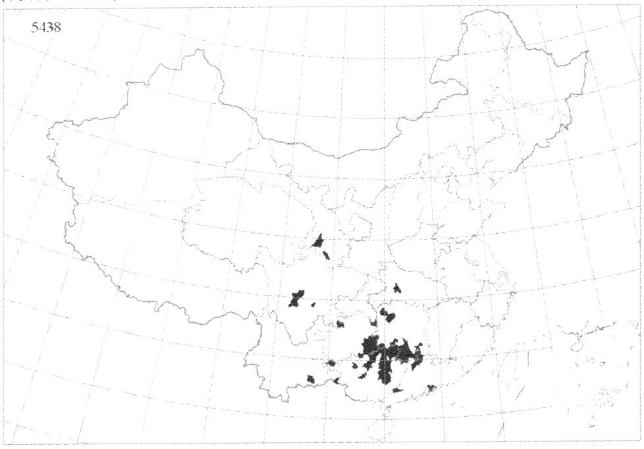

Celastraceae 卫矛科 *Euonymus* 卫矛属

Euonymus aculeatus Hemsl. 小千金

Shrubs, evergreen

15.9 (9.3~22.2); 5.5 (-1.7~13.9); 25.2 (20.1~29.1); 15.9 (9.5~22.3); 131.8 (69.8~207.5); 0.8 (0.0~18.0); 835 (603~1190) ❙ 1274 (985~1662); 547 (403~821); 95 (43~221) ❙ 53.2 (15.4~107.2); 834 (603~1190); 216.7 (146.0~491.5)

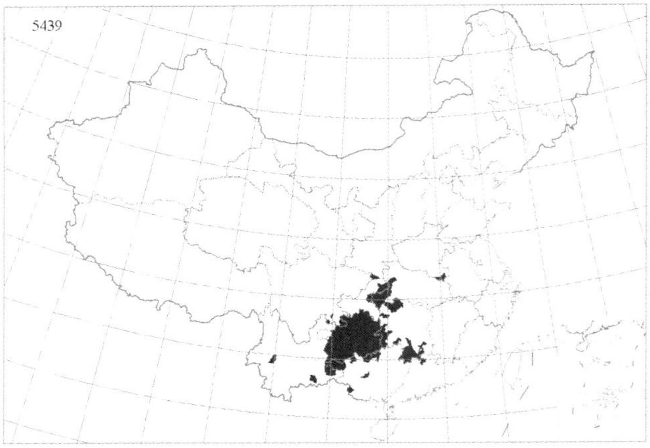

Celastraceae 卫矛科 *Euonymus* 卫矛属

Euonymus aculeolus C. Y. Cheng 微刺卫矛

Shrubs

15.3; 8.7; 20.5; 15.4; 124.3; 0.0; 761 ❙ 976; 564; 38 ❙ 28.3; 717; 309.8

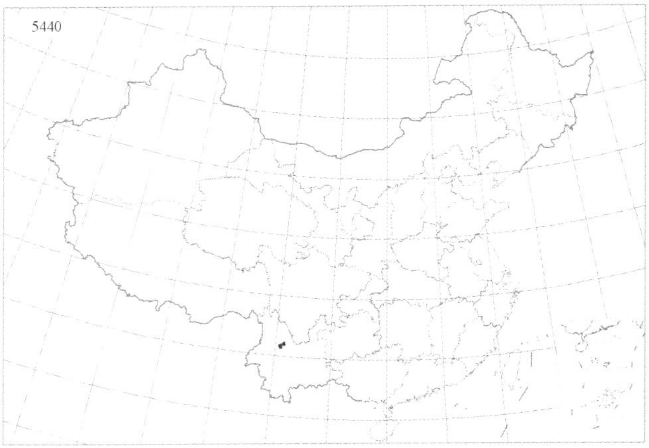

Celastraceae 卫矛科 *Euonymus* 卫矛属

Euonymus alatus (Thunb.) Sieber 卫矛

Shrubs, deciduous

12.9 (-4.0~21.3); -0.9 (-28.2~14.2); 25.4 (10.7~29.7); 13.5 (3.8~21.3); 112.7 (16.2~196.1); 17.9 (0.0~146.2); 787 (411~1130) ❙ 910 (63~2024); 431 (45~911); 69 (2~259) ❙ 13.5 (-84.9~147.0); 698 (63~1130); 189.1 (8.4~551.6)

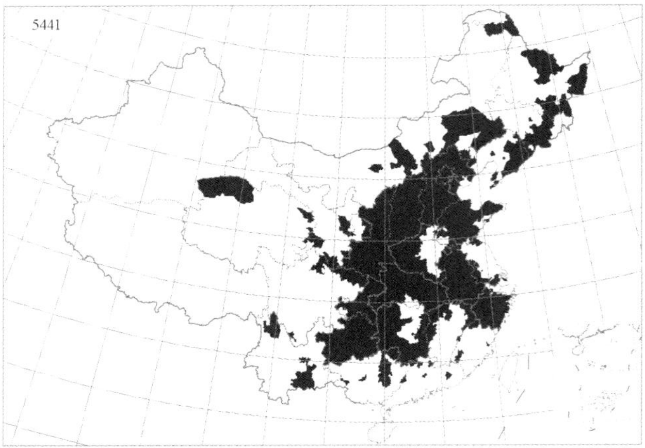

Celastraceae 卫矛科 *Euonymus* 卫矛属

Euonymus balansae Sprague 刺猬卫矛

Scandent shrubs, evergreen

18.0; 10.6; 23.2; 18.0; 156.2; 0.0; 879 ❙ 1308; 733; 54 ❙ 48.5; 879; 377.6

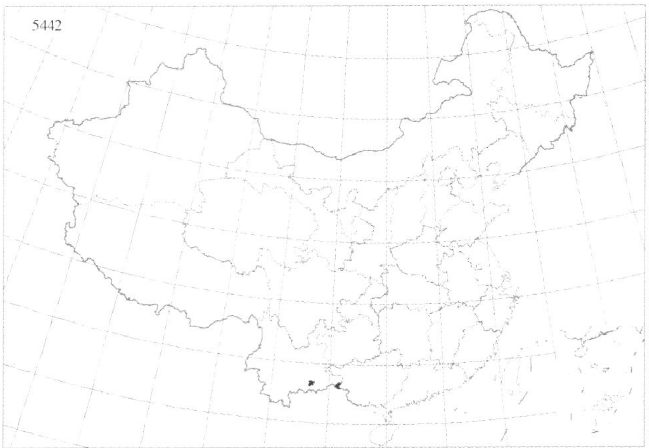

Celastraceae 卫矛科 *Euonymus* 卫矛属
Euonymus bockii Loes. ex Diels 南川卫矛
Shrubs, evergreen

17.3 (11.1~22.2); 3.5 (1.8~13.9); 24.4 (19.4~28.3); 17.3 (11.1~22.3); 148.3 (79.8~207.5); 0.5 (0.0~6.9); 885 (648~1190) ❙ 1290 (999~1581); 680 (489~850); 63 (30~106) ❙ 47.5 (15.4~80.4); 885 (648~1190); 274.8 (182.9~491.5)

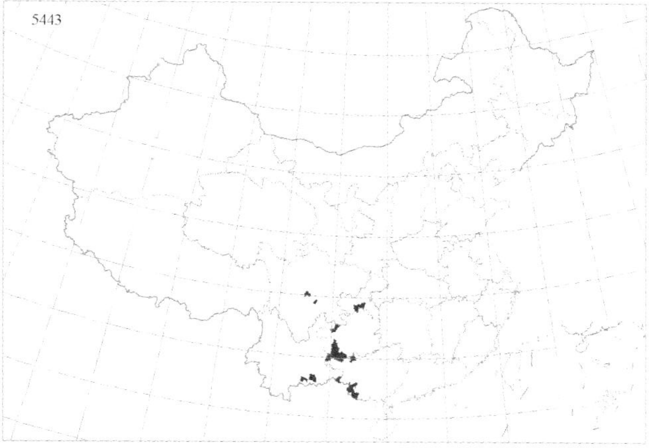

Celastraceae 卫矛科 *Euonymus* 卫矛属
Euonymus bullatus Wallich ex Loddiges 凸脉卫矛
Small trees, evergreen

19.0; 12.9; 22.8; 19.0; 168.2; 0.0; 914 ❙ 1257; 654; 50 ❙ 37.2; 868; 467.8

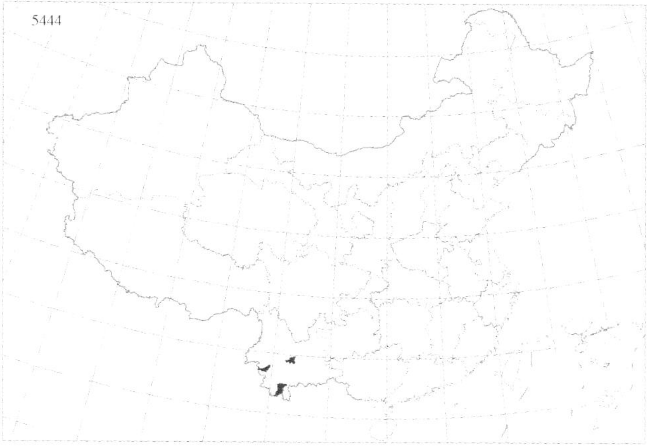

Celastraceae 卫矛科 *Euonymus* 卫矛属
Euonymus carnosus Hemsl. 肉花卫矛
Shrubs, evergreen

16.0 (5.7~22.6); 4.5 (-3.5~16.3); 27.0 (14.2~29.4); 16.1 (6.4~22.6); 135.0 (38.1~211.5); 2.2 (0.0~28.1); 876 (504~1197) ❙ 1509 (753~3495); 576 (384~1412); 178 (24~897) ❙ 71.9 (-12.2~224.4); 870 (504~1090); 224.8 (111.9~363.7)

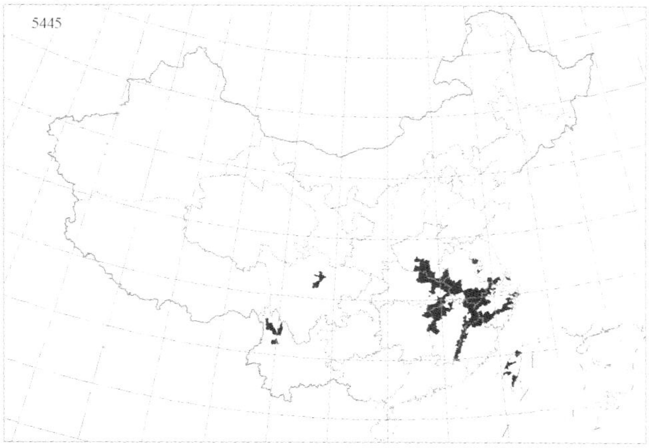

Celastraceae 卫矛科 *Euonymus* 卫矛属
Euonymus centidens H. Lév. 百齿卫矛
Shrubs

16.9 (11.1~22.4); 6.1 (1.6~14.0); 26.9 (19.4~29.8); 17.0 (11.1~22.4); 143.9 (79.8~209.1); 0.5 (0.0~6.9); 898 (648~1204) ❙ 1475 (871~1997); 560 (299~1044); 154 (29~259) ❙ 64.4 (-6.7~130.7); 892 (648~1204); 223.2 (130.1~551.3)

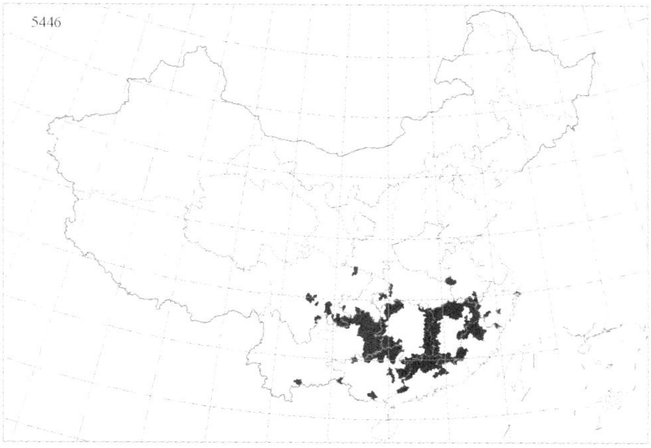

Celastraceae 卫矛科 *Euonymus* 卫矛属
Euonymus chengii J. S. Ma 静容卫矛
Shrubs, evergreen

21.9; 14.6; 27.9; 22.0; 203.7; 0.0; 1158 ❙ 1787; 877; 105 ❙ 54.4; 1158; 248.3

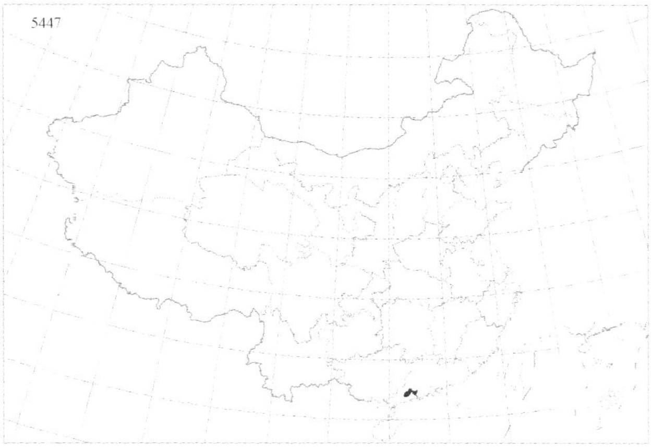

Celastraceae 卫矛科 *Euonymus* 卫矛属
Euonymus chenmoui W. C. Cheng 陈谋卫矛
Small shrubs

15.9 (14.4~17.6); 3.7 (2.1~7.4); 27.8 (26.3~29.3); 16.0 (14.5~17.6); 133.6 (118.7~151.2); 1.9 (0.0~4.9); 875 (801~941) ❙ 1518 (1211~1689); 565 (474~685); 182 (134~220) ❙ 74.1 (28.8~105.7); 874 (801~941); 244.8 (198.6~319.9)

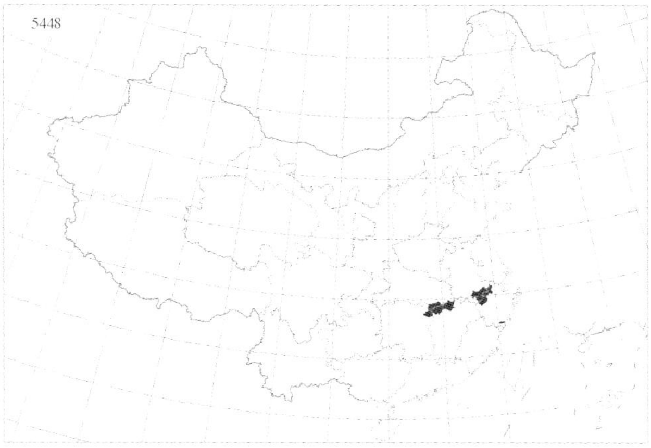

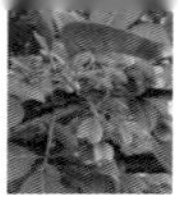

Celastraceae 卫矛科 *Euonymus* 卫矛属

Euonymus chloranthoides Yang 缙云卫矛

Small shrubs, evergreen

16.5 (10.2~18.7); 5.7 (-0.7~8.1); 26.7 (21.0~28.9); 16.5 (10.4~18.8); 139.2 (77.2~166.0); 1.3 (0.0~13.6); 883 (634~1002) ❙ 1158 (1028~1320); 489 (442~549); 62 (46~76) ❙ 33.6 (8.0~93.8); 883 (634~1002); 163.9 (108.2~305.6)

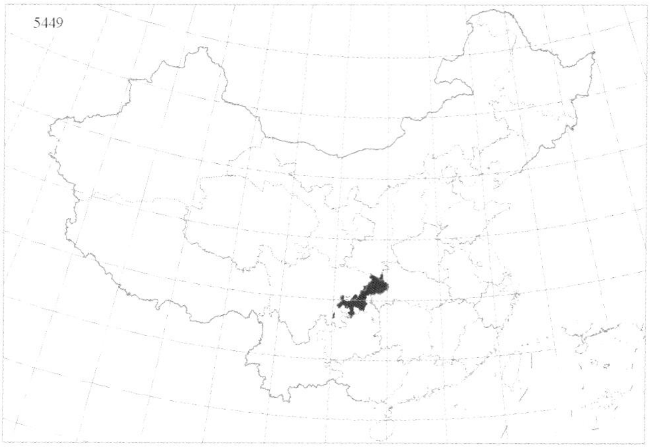

Celastraceae 卫矛科 *Euonymus* 卫矛属

Euonymus chui Hand.-Mazz. 隐刺卫矛

Scandent shrubs, deciduous

15.0 (6.0~18.7); 4.8 (-3.2~8.7); 24.5 (14.4~28.9); 15.0 (6.6~18.8); 122.8 (39.7~166.0); 2.4 (0.0~26.7); 816 (508~1002) ❙ 1144 (879~1414); 529 (442~817); 55 (23~92) ❙ 43.8 (8.0~101.2); 815 (508~1002); 195.6 (108.2~352.3)

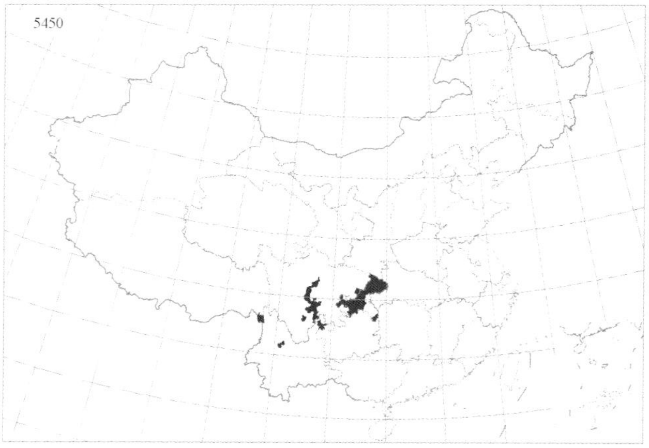

Celastraceae 卫矛科 *Euonymus* 卫矛属

Euonymus clivicola W. W. Sm. 岩坡卫矛

Shrubs, deciduous

9.0 (-0.7~14.9); 1.1 (-9.6~7.8); 15.8 (7.7~23.6); 9.5 (2.5~15.0); 64.7 (6.8~119.6); 16.5 (0.0~74.2); 587 (338~782) ❙ 945 (370~1410); 545 (262~816); 35 (9~69) ❙ 58.7 (4.7~100.7); 586 (338~782); 284.1 (61.6~407.1)

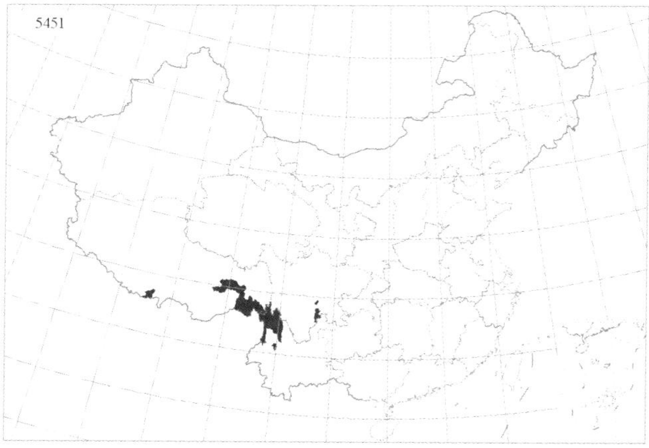

Celastraceae 卫矛科 *Euonymus* 卫矛属

Euonymus cornutus Hemsl. 角翅卫矛

Shrubs, deciduous

12.4 (-0.7~19.4); 2.5 (-9.6~12.0); 21.5 (7.7~27.8); 12.6 (2.5~19.4); 98.5 (6.8~172.7); 8.7 (0.0~74.2); 711 (338~994) ❙ 985 (370~1852); 495 (262~817); 43 (10~209) ❙ 39.8 (-14.2~121.7); 699 (338~994); 248.8 (61.6~407.1)

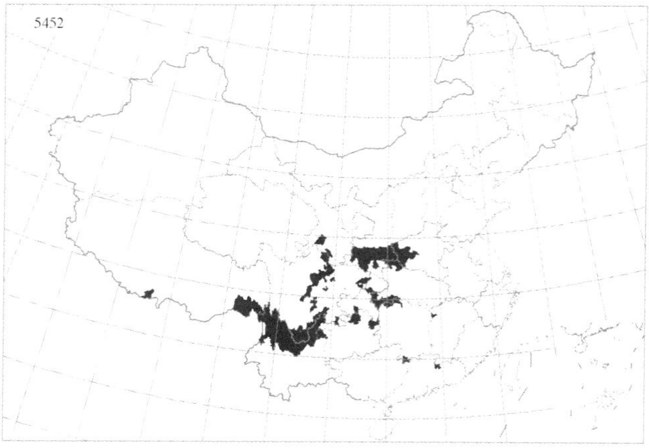

Celastraceae 卫矛科 *Euonymus* 卫矛属

Euonymus dielsianus Loes. et Diels 裂果卫矛

Shrubs or trees, evergreen

16.1 (6.0~22.3); 5.7 (-3.2~14.7); 25.5 (14.4~29.0); 16.2 (6.6~22.4); 135.0 (39.7~208.3); 1.2 (0.0~26.7); 852 (508~1188) ❙ 1309 (793~1961); 563 (377~946); 100 (28~230) ❙ 54.3 (-2.0~107.2); 850 (508~1188); 217.4 (108.2~506.2)

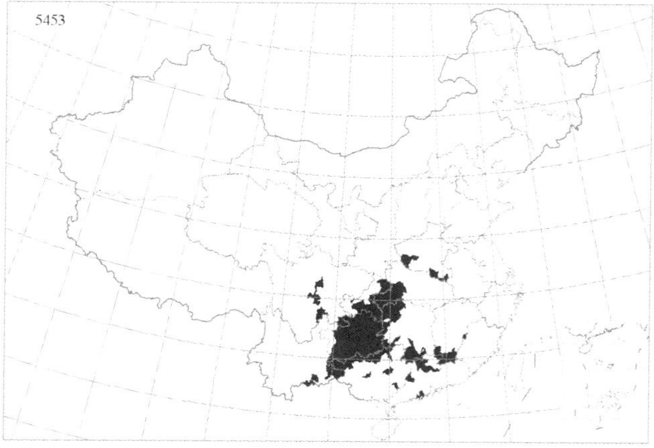

Celastraceae 卫矛科 *Euonymus* 卫矛属

Euonymus distichus H. Lév. 双歧卫矛

Small shrubs

16.0 (14.0~19.2); 5.8 (3.6~13.3); 25.4 (22.3~27.4); 16.0 (14.1~19.4); 132.6 (109.4~172.6); 0.3 (0.0~1.6); 839 (748~932) ❙ 1253 (958~1590); 596 (465~946); 73 (36~192) ❙ 49.9 (12.0~89.0); 837 (748~923); 203.8 (130.1~452.0)

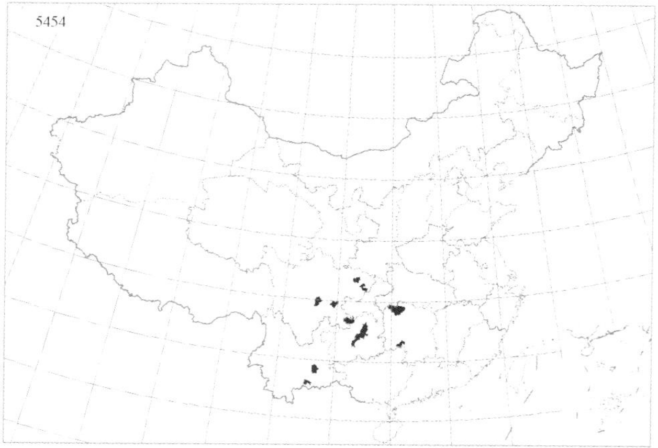

Celastraceae 卫矛科 *Euonymus* 卫矛属
Euonymus dolichopus Merr. ex J. S. Ma 长梗卫矛
Shrubs, evergreen

21.3; 13.2; 27.3; 21.3; 195.9; 0.0; 1111 ❙ 1900; 1056; 125 ❙ 71.0; 1111; 297.2

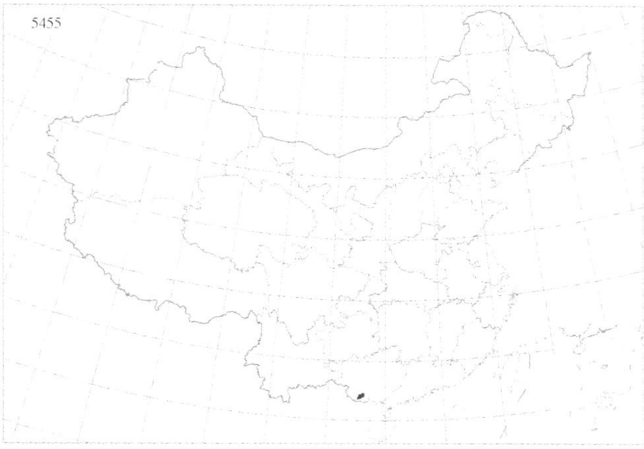

Celastraceae 卫矛科 *Euonymus* 卫矛属
Euonymus echinatus Wall. 棘刺卫矛
Scandent shrubs, evergreent or semievergreen

16.0 (-1.8~22.2); 5.5 (-10.2~17.7); 25.8 (6.1~29.8); 16.1 (1.8~22.2); 134.6 (1.9~206.9); 1.6 (0.0~82.8); 857 (329~1177) ❙ 1404 (320~2992); 556 (217~1847); 132 (10~402) ❙ 63.7 (-3.0~246.5); 852 (320~1177); 215.6 (61.6~383.4)

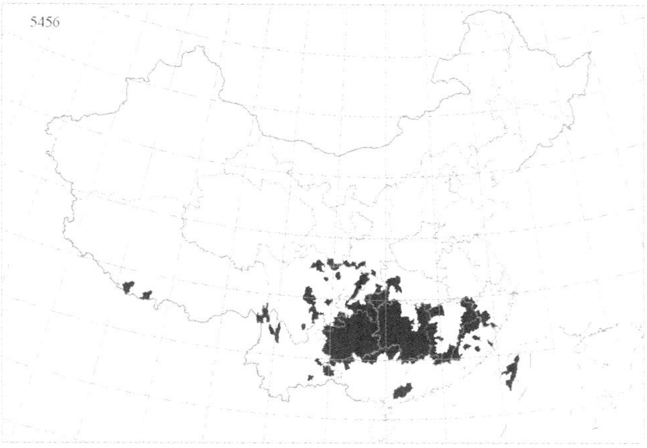

Celastraceae 卫矛科 *Euonymus* 卫矛属
Euonymus euscaphis Hand.-Mazz. 鸭椿卫矛
Shrubs, deciduous

16.4 (12.0~19.1); 5.0 (0.6~10.0); 27.1 (23.1~29.2); 16.4 (12.1~19.1); 137.8 (94.7~169.1); 0.8 (0.0~9.7); 878 (697~984) ❙ 1605 (1118~1997); 553 (427~677) 198 (75~259) ❙ 83.2 (28.8~130.7); 873 (697~973); 239.3 (169.7~303.4)

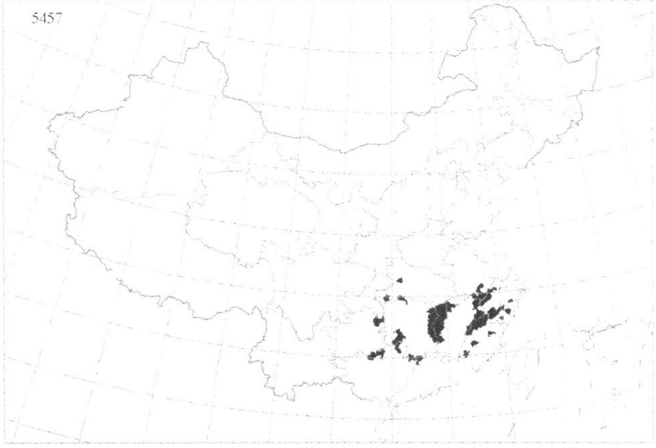

Celastraceae 卫矛科 *Euonymus* 卫矛属
Euonymus ficoides C. Y. Cheng ex J. S. Ma 榕叶卫矛
Shrubs, evergreen

17.4; 9.5; 23.0; 17.4; 149.0; 0.0; 848 ❙ 1279; 715; 50 ❙ 50.8; 848; 336.8

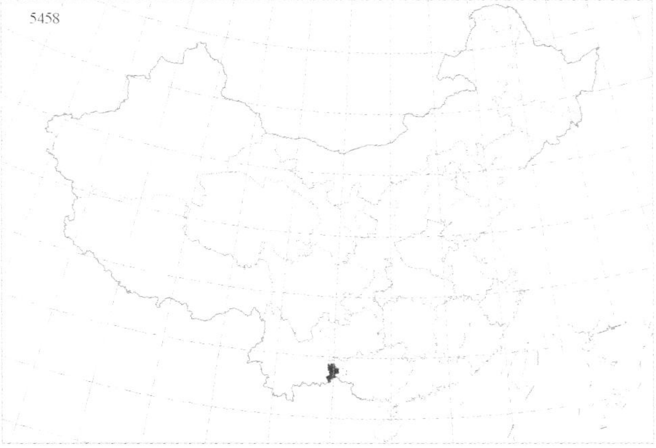

Celastraceae 卫矛科 *Euonymus* 卫矛属
Euonymus fimbriatus Wall. et Roxb. 遂叶卫矛
Shrubs

10.6 (-1.1~18.2); -1.1 (-9.1~11.8); 21.7 (6.2~25.1); 11.1 (2.0~18.2); 87.5 (2.5~158.4); 19.9 (0.0~75.6); 669 (334~873) ❙ 772 (321~1059); 358 (193~604); 29 (15~44) ❙ 15.3 (-4.0~52.6); 664 (321~847); 244.1 (79.3~318.2)

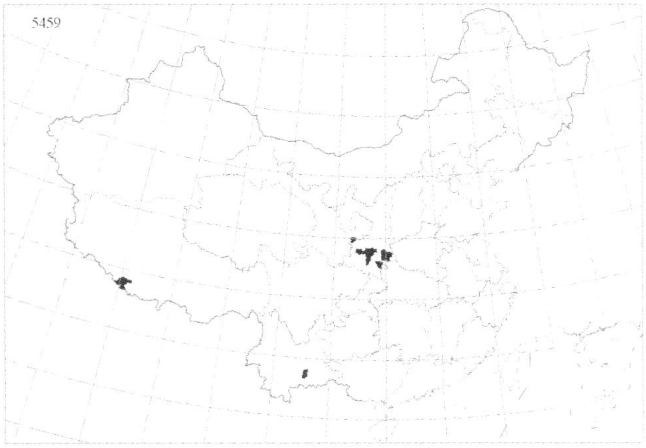

Celastraceae 卫矛科 *Euonymus* 卫矛属
Euonymus fortunei (Turcz.) Hand.-Mazz. 扶芳藤
Shrubs, semievergreen

15.0 (3.9~24.1); 2.9 (-13.7~18.9); 26.2 (18.6~29.9); 15.2 (7.3~24.1); 128.0 (52.8~229.7); 7.4 (0.0~71.7); 843 (535~1322) ❙ 1145 (367~2759); 493 (235~1594); 102 (8~259) ❙ 33.6 (-40.9~202.5); 806 (367~1221); 201.0 (101.9~572.9)

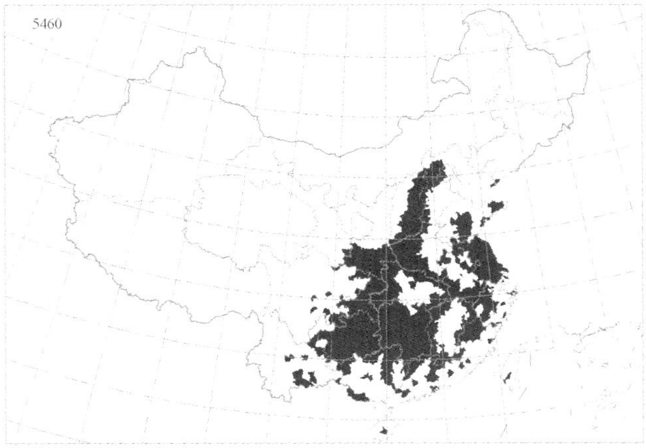

Celastraceae 卫矛科 *Euonymus* 卫矛属
Euonymus frigidus Wall. ex Roxb 冷地卫矛
Shrubs or small trees, deciduous

11.3 (-1.3~22.0); 1.5 (-13.5~13.5); 20.0 (7.7~28.8); 11.7 (2.5~22.0); 90.3 (6.8~203.8); 14.3 (0.0~86.3); 678 (338~1170) ❙ 963 (320~1736); 501 (190~1007); 41 (3~132) ❙ 43.3 (-40.9~145.8); 658 (320~1170); 258.6 (61.2~506.2)

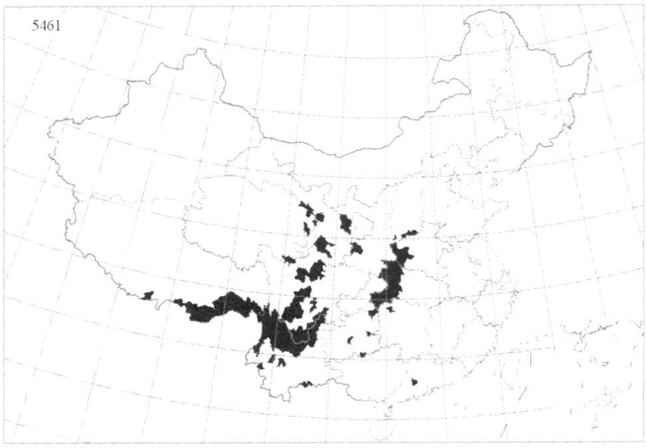

Celastraceae 卫矛科 *Euonymus* 卫矛属
Euonymus gibber Hance 流苏卫矛
Shrubs or small trees, evergreen

22.5 (17.7~24.8); 16.5 (7.7~20.1); 27.0 (21.7~28.4); 22.5 (17.8~24.8); 210.5 (153.6~237.6); 0.0 (0.0~0.0); 1208 (863~1385) ❙ 1594 (1179~2713); 631 (392~1388); 87 (41~188) ❙ 41.4 (-11.9~214.4); 1079 (863~1212); 434.6 (232.5~620.0)

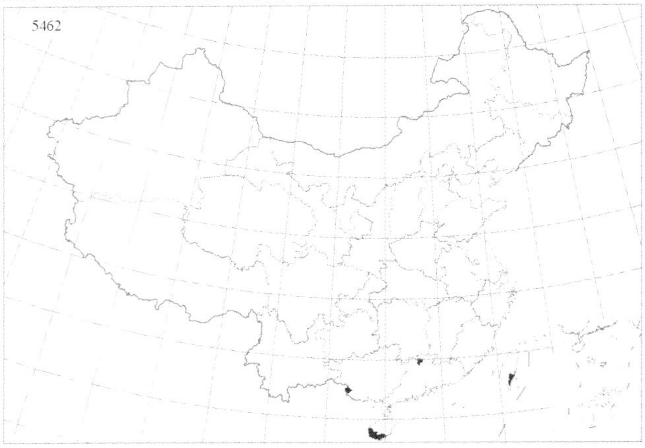

Celastraceae 卫矛科 *Euonymus* 卫矛属
Euonymus giraldii Loes. ex Diels 纤齿卫矛
Shrubs or small trees, deciduous

10.3 (0.6~15.5); -3.8 (-13.7~4.5); 23.0 (9.5~28.0); 11.1 (3.4~15.6); 89.2 (12.7~128.0); 25.4 (0.5~71.7); 690 (394~880) ❙ 663 (341~1451); 352 (190~817); 23 (8~120) ❙ -2.4 (-40.9~121.7); 590 (341~839); 193.5 (61.2~344.4)

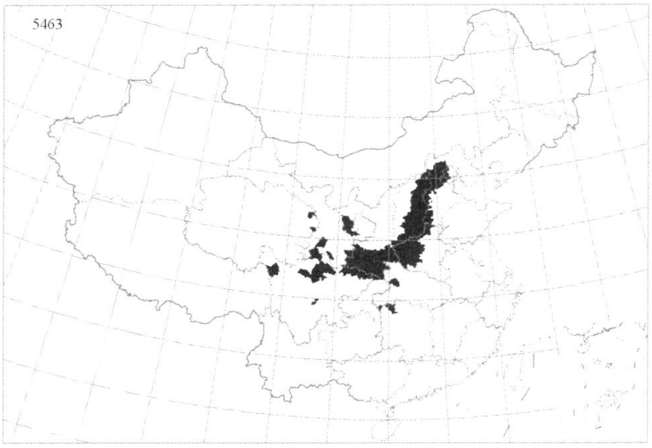

Celastraceae 卫矛科 *Euonymus* 卫矛属
Euonymus glaber Roxb. 帽果卫矛
Shrubs, evergreen

20.0; 13.0; 25.1; 20.0; 180.1; 0.0; 1002 ❙ 1525; 797; 82 ❙ 52.8; 1002; 426.7

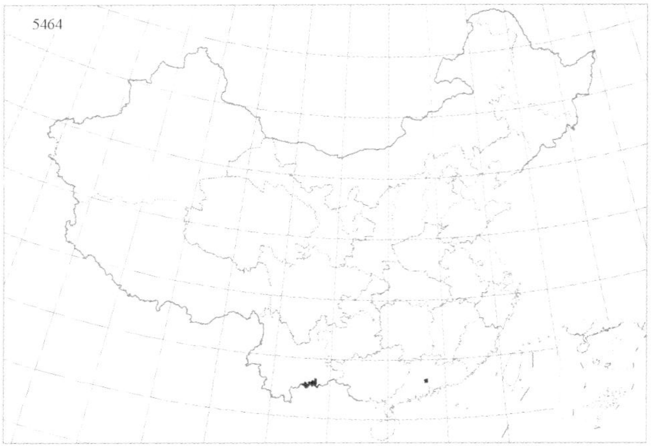

Celastraceae 卫矛科 *Euonymus* 卫矛属
Euonymus gracillimus Hemsl. 纤细卫矛
Shrubs, deciduous

20.9; 12.4; 27.8; 20.9; 191.1; 0.0; 1094 ❙ 1809; 773; 153 ❙ 65.2; 1094; 256.5

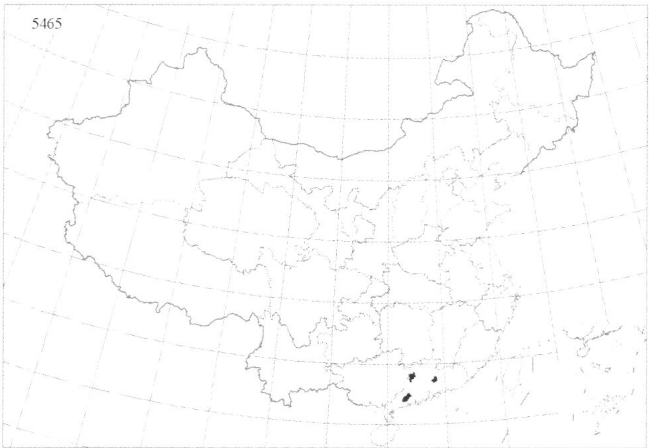

Celastraceae 卫矛科 *Euonymus* 卫矛属
Euonymus grandiflorus Wall. 大花卫矛
Shrubs or trees, deciduous

13.6 (-1.8~22.3); 2.9 (-10.5~14.0); 23.3 (6.1~29.4); 13.8 (1.8~22.4); 111.7 (1.9~208.2); 8.4 (0.0~82.8); 772 (329~1198) ❙ 1048 (255~1852); 505 (191~1007); 68 (1~224) ❙ 35.4 (-35.3~145.8); 751 (255~1198); 220.2 (60.4~506.2)

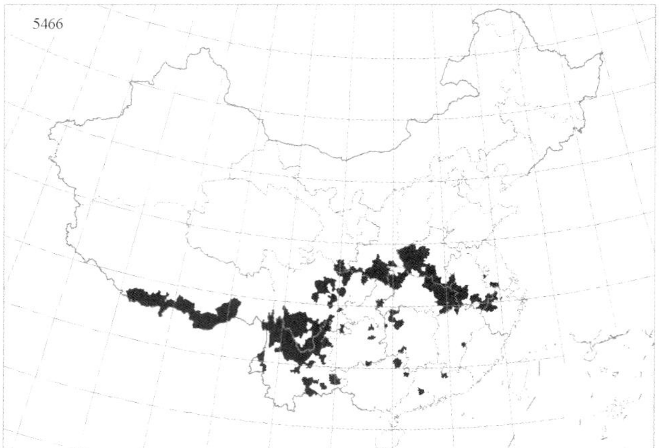

Celastraceae 卫矛科 *Euonymus* 卫矛属
Euonymus hainanensis Chun et How 海南卫矛
Shrubs, evergreen

24.5; 19.7; 27.7; 24.5; 234.1; 0.0; 1354 ❙ 1514; 515; 49 ❙ 12.0; 1166; 547.4

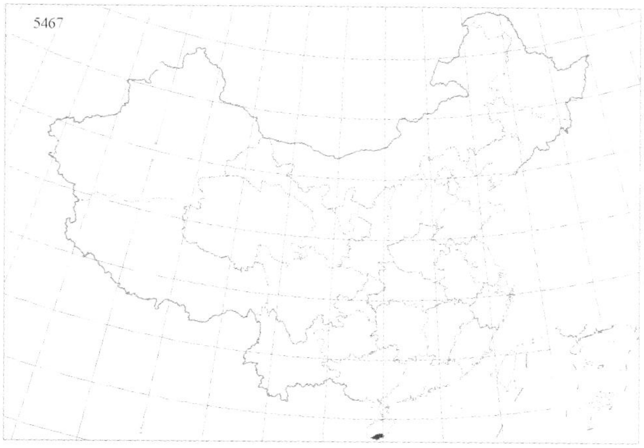

Celastraceae 卫矛科 *Euonymus* 卫矛属
Euonymus hamiltonianus Wall. et Roxb. 西南卫矛
Small trees, deciduous

14.9 (2.2~20.1); 3.8 (-6.8~13.3); 25.1 (9.9~29.6); 15.0 (3.8~20.1); 123.4 (15.5~181.1); 4.2 (0.0~49.1); 817 (413~1048) ❙ 1216 (459~1997); 509 (253~1007); 101 (10~259) ❙ 48.5 (-25.8~147.0); 806 (413~1048); 222.2 (108.2~506.2)

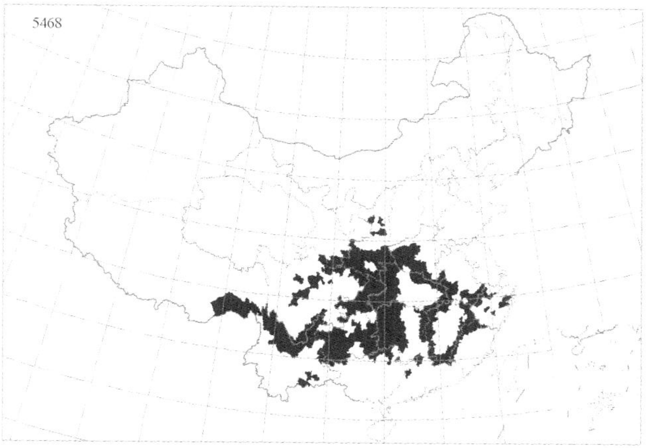

Celastraceae 卫矛科 *Euonymus* 卫矛属
Euonymus hui J. S. Ma 秀英卫矛
Trees, deciduous

11.1; 1.8; 19.4; 11.1; 79.8; 6.9; 648 ❙ 1076; 584; 30 ❙ 66.0; 648; 246.3

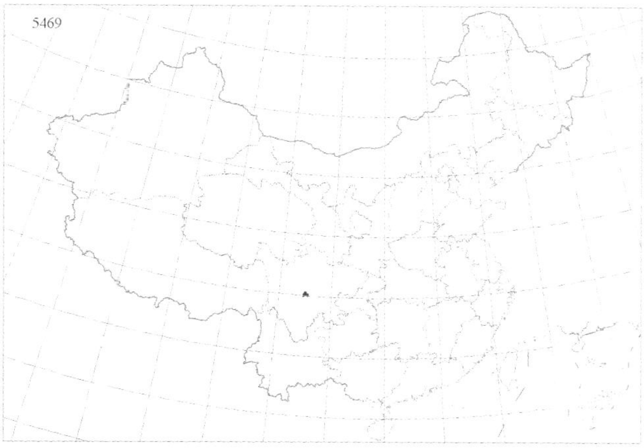

Celastraceae 卫矛科 *Euonymus* 卫矛属
Euonymus hukuangensis C. Y. Cheng ex J. S. Ma 湖广卫矛
Shrubs or small trees, evergreen

17.1; 6.7; 26.4; 17.2; 146.1; 0.3; 892 ❙ 1512; 581; 157 ❙ 69.8; 892; 216.5

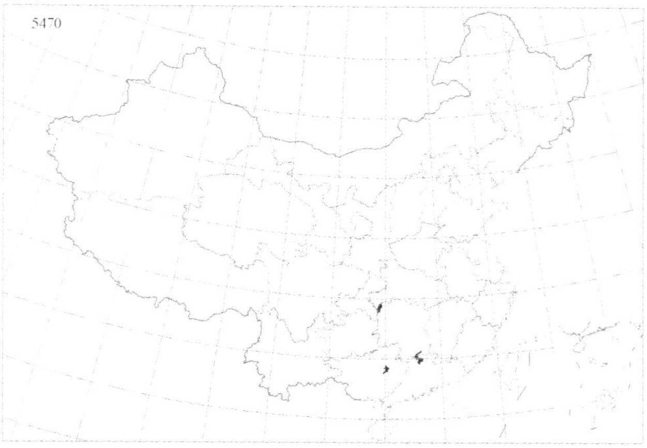

Celastraceae 卫矛科 *Euonymus* 卫矛属
Euonymus hupehensis (Loes.) Loes. 湖北卫矛
Shrubs, evergreen

16.1; 7.2; 23.7; 16.1; 133.9; 0.4; 818 ❙ 1309; 619; 72 ❙ 59.9; 818; 284.4

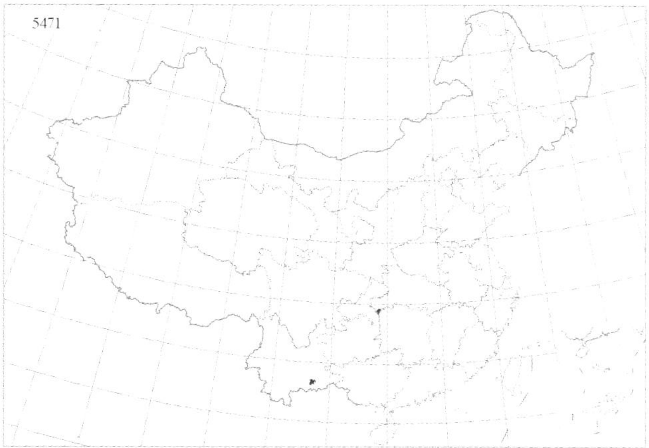

Celastraceae 卫矛科 *Euonymus* 卫矛属
Euonymus kachinensis Prain 克钦卫矛
Shrubs, deciduous

8.6; 1.3; 15.0; 8.7; 52.6; 8.8; 574 ❙ 1041; 585; 45 ❙ 81.5; 574; 352.3

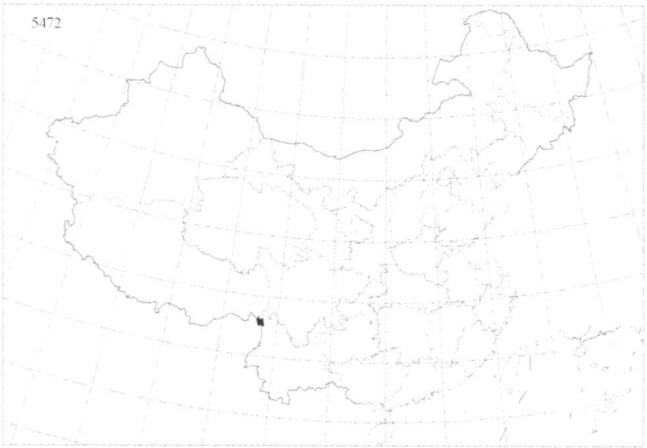

Celastraceae 卫矛科 *Euonymus* 卫矛属
Euonymus kengmaensis C. Y. Cheng ex J. S. Ma 耿马卫矛
Shrubs or small trees, evergreen

17.4 (15.5~19.5); 10.4 (5.5~13.1); 22.1 (20.3~25.5); 17.4 (15.6~19.5); 149.0 (127.3~173.8); 0.0 (0.0~0.0); 844 (768~942) ❙ 1357 (1021~1688); 759 (496~959); 53 (42~89) ❙ 61.0 (25.6~112.2); 835 (768~939); 421.4 (188.8~627.0)

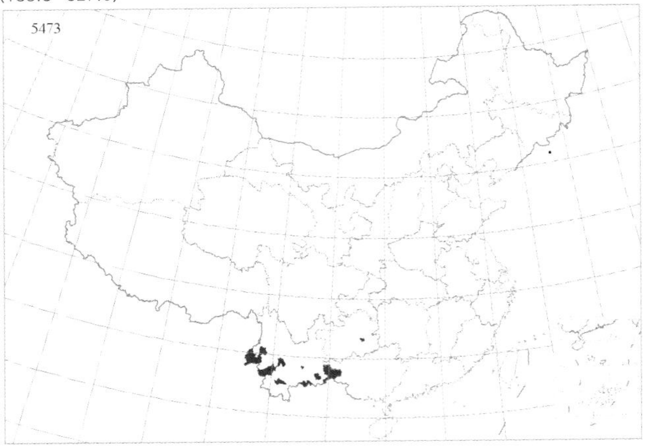

Celastraceae 卫矛科 *Euonymus* 卫矛属
Euonymus kweichowensis Chen H. Wang 贵州卫矛
Shrubs, deciduous

15.2; 4.2; 25.7; 15.2; 123.2; 0.9; 810 ❙ 1231; 483; 88 ❙ 52.1; 810; 196.6

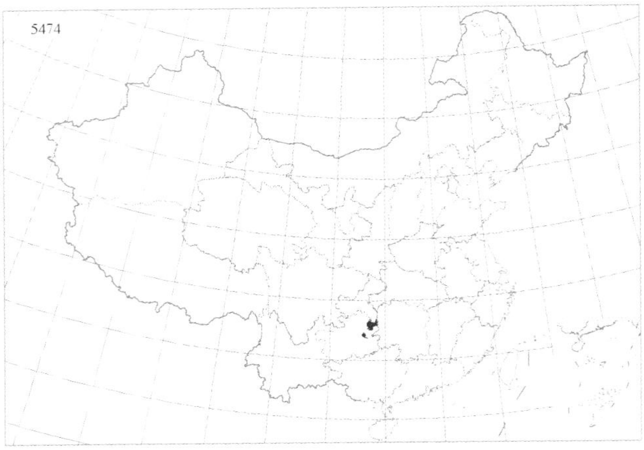

Celastraceae 卫矛科 *Euonymus* 卫矛属
Euonymus laxicymosus C. Y. Cheng ex J. S. Ma 稀序卫矛
Shrubs, evergreen

19.7; 11.9; 25.4; 19.7; 176.7; 0.0; 996 ❙ 1632; 867; 74 ❙ 63.2; 996; 323.1

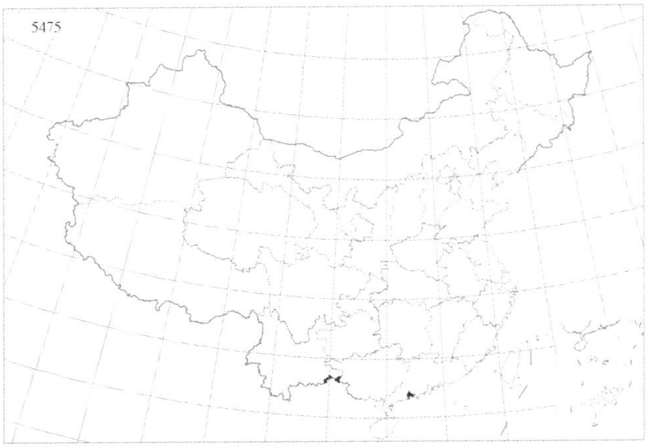

Celastraceae 卫矛科 *Euonymus* 卫矛属
Euonymus laxiflorus Champ. et Benth. 疏花卫矛
Shrubs or small trees, deciduous

18.3 (8.6~24.8); 8.8 (1.3~20.1); 26.6 (15.0~29.8); 18.4 (8.7~24.8); 161.0 (52.6~237.6); 0.2 (0.0~8.8); 963 (574~1385) ❙ 1503 (868~3495); 629 (312~1428); 140 (41~897) ❙ 58.0 (-34.3~224.4); 952 (574~1257); 249.6 (122.7~627.0)

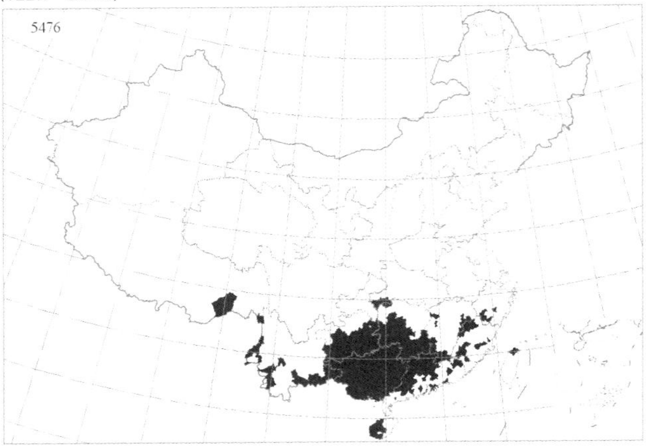

Celastraceae 卫矛科 *Euonymus* 卫矛属
Euonymus lichiangensis W. W. Sm. 丽江卫矛
Shrubs, evergreen

9.8; 2.8; 15.8; 9.9; 66.2; 7.7; 602 ❙ 942; 567; 29 ❙ 56.8; 602; 317.5

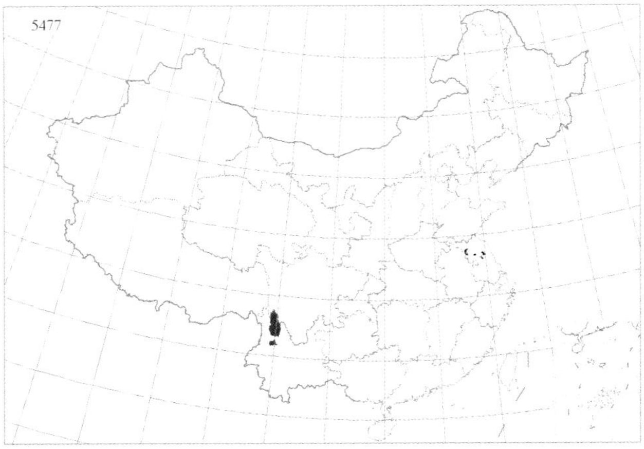

Celastraceae 卫矛科 *Euonymus* 卫矛属
Euonymus lucidus D. Don 垂序卫矛
Shrubs or trees, evergreen

-1.2; -9.9; 6.9; 2.1; 4.4; 78.5; 334 ❙ 345; 240; 15 ❙ 0.9; 329; 66.9

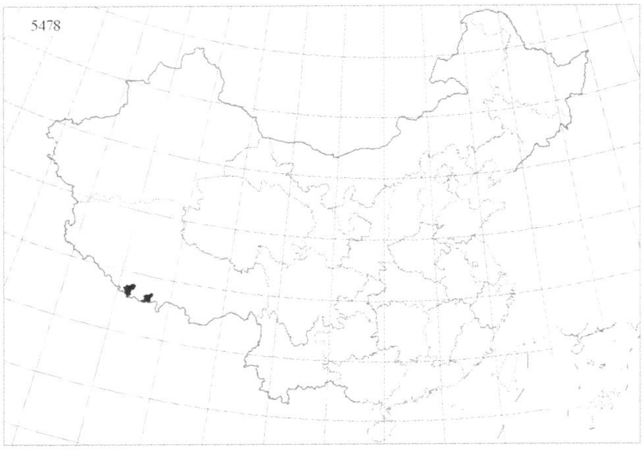

Celastraceae 卫矛科 *Euonymus* 卫矛属
Euonymus lushanensis F. H. Chen et M. C. Wang 庐山卫矛
Shrubs, deciduous

16.9；4.5；29.1；17.0；144.0；0.5；931 ❙ 1560；630；171 ❙ 67.5；931；205.0

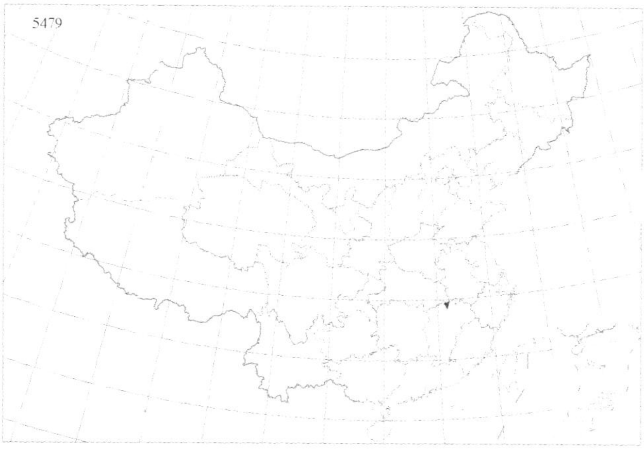

Celastraceae 卫矛科 *Euonymus* 卫矛属
Euonymus maackii Rupr. 白杜
Small trees

10.7 (−5.8~21.6)；−4.6 (−29.8~13.2)；24.5 (8.5~29.9)；12.1 (2.7~21.6)；99.8 (8.3~199.2)；30.7 (0.0~157.4)；735 (363~1139) ❙ 800 (195~1953)；401 (133~858)；58 (4~259) ❙ 5.8 (−67.0~112.2)；647 (195~1139)；188.7 (43.7~346.5)

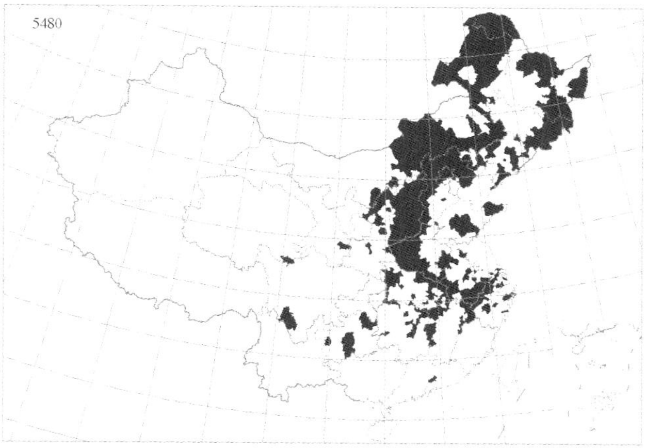

Celastraceae 卫矛科 *Euonymus* 卫矛属
Euonymus macropterus Rupr. 黄心卫矛
Shrubs

10.3 (2.6~15.1)；−5.0 (−15.9~2.0)；24.0 (18.7~28.0)；11.4 (7.2~15.2)；93.5 (52.0~128.0)；29.6 (5.4~80.4)；707 (535~880) ❙ 663 (377~1437)；384 (241~691)；27 (8~140) ❙ −5.7 (−37.2~101.4)；614 (377~821)；195.5 (109.4~335.3)

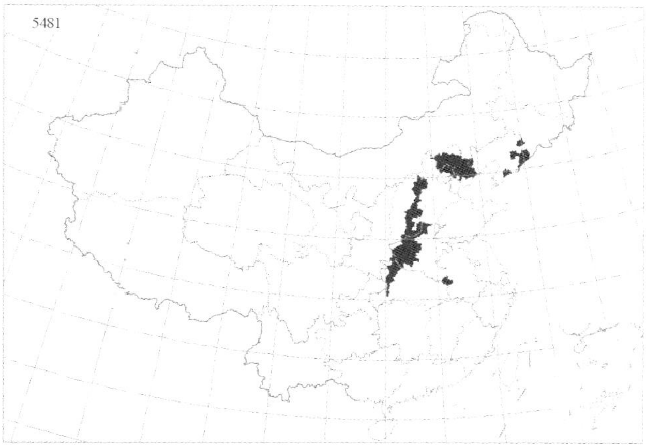

Celastraceae 卫矛科 *Euonymus* 卫矛属
Euonymus microcarpus (Oliv.) Sprague 小果卫矛
Shrubs

11.8 (1.2~16.9)；0.1 (−7.1~6.2)；22.8 (9.0~28.1)；11.9 (3.3~16.9)；93.8 (11.9~143.4)；12.1 (0.0~56.5)；703 (390~924) ❙ 909 (578~1492)；420 (290~753)；38 (12~111) ❙ 30.7 (−16.6~125.9)；699 (390~924)；253.6 (130.1~334.4)

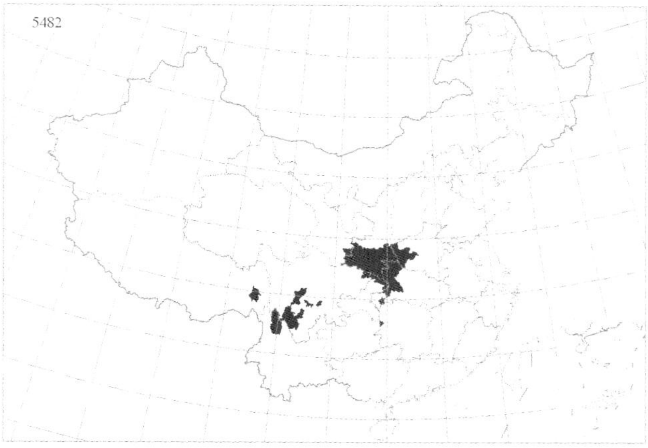

Celastraceae 卫矛科 *Euonymus* 卫矛属
Euonymus myrianthus Hemsl. 大果卫矛
Shrubs, evergreen

16.3 (9.3~22.7)；5.5 (−2.5~15.0)；26.3 (18.8~29.8)；16.4 (9.5~22.8)；137.4 (69.8~213.1)；1.2 (0.0~20.6)；871 (603~1221) ❙ 1405 (815~2435)；544 (323~1428)；138 (19~259) ❙ 61.2 (−5.3~130.7)；863 (603~1221)；223.5 (133.5~452.0)

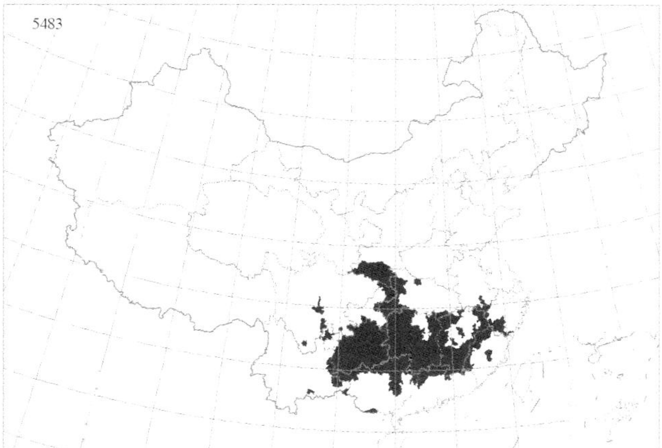

Celastraceae 卫矛科 *Euonymus* 卫矛属
Euonymus nanoides Loes. et Rehder 小卫矛
Small shrubs

7.0 (0.8~14.0)；−7.9 (−13.9~2.4)；20.1 (10.1~26.9)；8.7 (3.7~14.0)；66.7 (15.1~120.0)；42.6 (5.1~71.7)；596 (409~830) ❙ 530 (341~1036)；308 (190~606)；15 (8~39) ❙ −7.8 (−41.6~121.7)；480 (341~679)；173.8 (51.3~346.5)

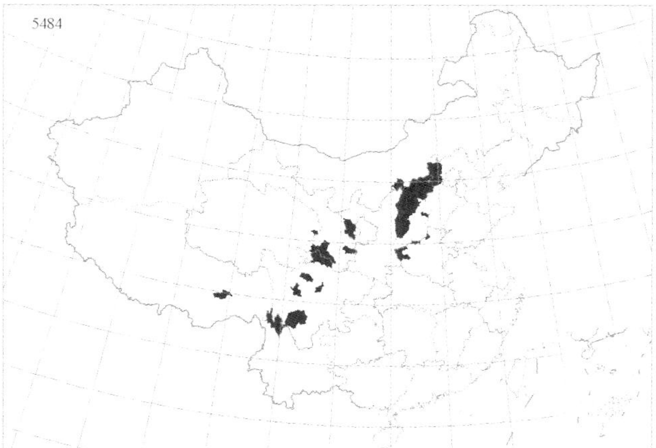

Celastraceae 卫矛科 *Euonymus* 卫矛属
Euonymus nanus M. Bieb. 矮卫矛
Small shrubs

6.7 (-1.0~14.5); -8.5 (-16.1~-0.8); 20.3 (8.5~28.2); 8.7 (2.7~14.6); 66.4 (8.3~126.9); 45.5 (11.9~81.1); 593 (363~874) ❙ 456 (84~856); 255 (53~391); 12 (1~25) ❙ -20.5 (-87.1~82.4); 430 (84~638); 152.5 (7.5~323.3)

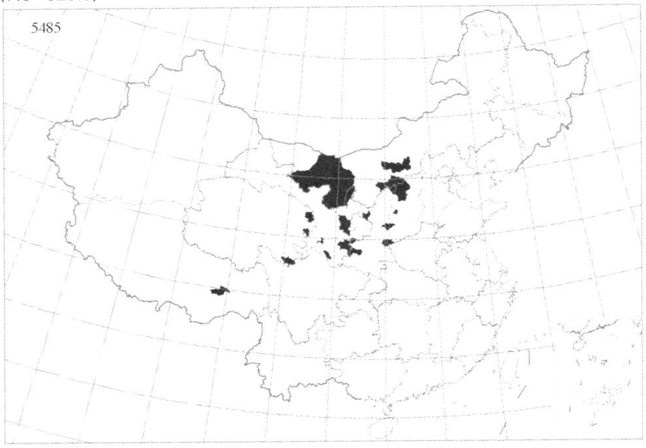

Celastraceae 卫矛科 *Euonymus* 卫矛属
Euonymus nitidus Benth. 中华卫矛
Shrubs or trees, evergreen

17.7 (9.1~24.0); 7.6 (-0.7~19.1); 26.7 (18.1~29.8); 17.7 (9.2~24.0); 153.5 (64.1~228.1); 0.6 (0.0~14.1); 935 (595~1302) ❙ 1532 (822~2435); 618 (323~1428); 150 (19~259) ❙ 64.9 (-1.6~133.8); 930 (595~1257); 243.0 (120.3~620.0)

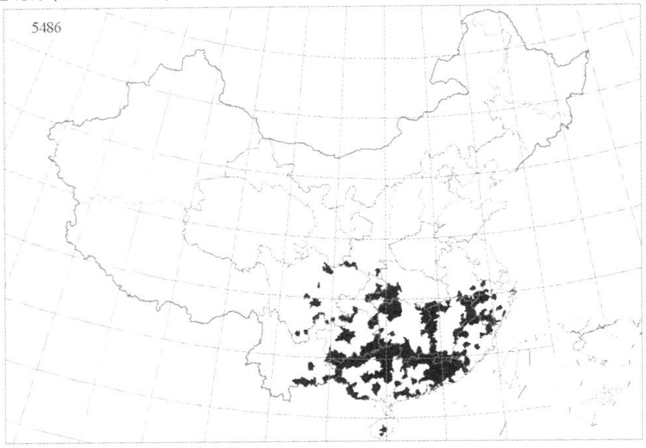

Celastraceae 卫矛科 *Euonymus* 卫矛属
Euonymus oxyphyllus Miq. 垂丝卫矛
Shrubs

14.1 (8.7~19.8); 0.9 (-8.3~14.1); 26.3 (20.1~29.3); 14.3 (9.5~19.8); 119.6 (69.8~177.9); 9.9 (0.0~40.2); 815 (603~1010) ❙ 1068 (604~2713); 488 (298~1388); 94 (23~341) ❙ 29.4 (-26.3~214.4); 772 (603~1010); 200.0 (136.4~363.7)

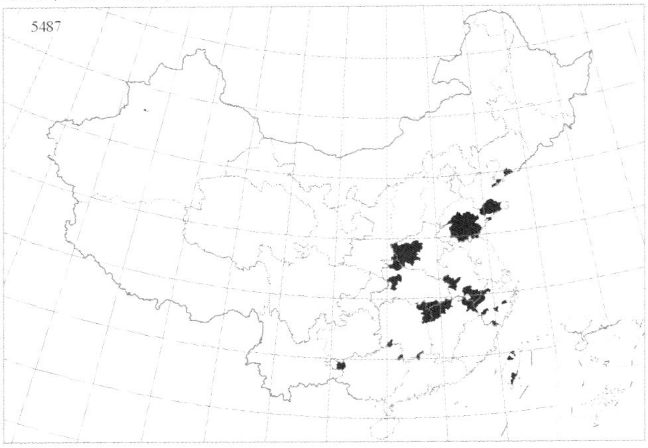

Celastraceae 卫矛科 *Euonymus* 卫矛属
Euonymus pallidifolius Hayata 淡绿叶卫矛
Shrubs, evergreen

22.2; 17.7; 25.5; 22.2; 206.7; 0.0; 1123 ❙ 2992; 1847; 107 ❙ 166.5; 1123; 296.9

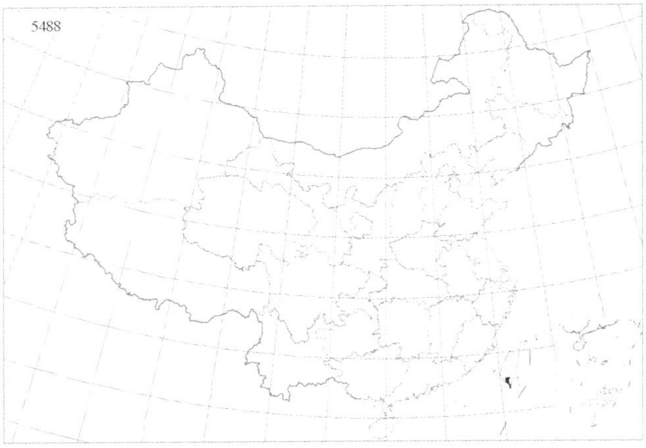

Celastraceae 卫矛科 *Euonymus* 卫矛属
Euonymus parasimilis C. Y. Cheng ex J. S. Ma 碧江卫矛
Trees, evergreen

11.3; 4.3; 17.0; 11.3; 76.6; 0.7; 644 ❙ 1243; 696; 58 ❙ 93.0; 644; 384.7

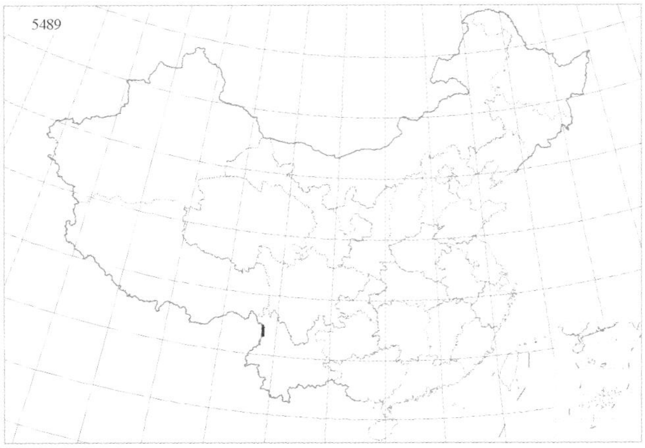

Celastraceae 卫矛科 *Euonymus* 卫矛属
Euonymus percoriaceus C. Y. Wu ex J. S. Ma 西畴卫矛
Shrubs or small trees, evergreen

17.2; 9.5; 22.7; 17.3; 147.0; 0.0; 838 ❙ 1254; 699; 48 ❙ 49.7; 838; 336.3

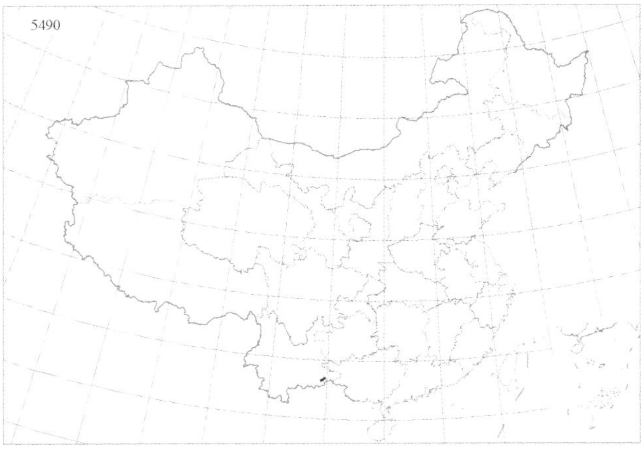

Celastraceae 卫矛科 *Euonymus* 卫矛属
Euonymus phellomanus Loes. ex Diels 栓翅卫矛
Shrubs

10.0 (−1.0~16.9); −3.5 (−12.7~4.3); 22.4 (8.5~29.1); 10.9 (2.7~16.9); 86.1 (8.3~143.8); 25.1 (0.7~79.5); 674 (363~933) ❙ 728 (341~1495); 354 (190~596); 32 (8~183) ❙ 8.4 (−40.9~128.6); 606 (341~932); 203.9 (61.2~357.7)

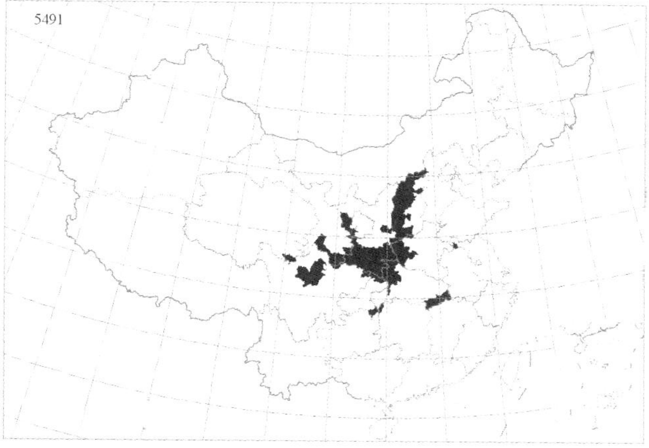

Celastraceae 卫矛科 *Euonymus* 卫矛属
Euonymus pittosporoides C. Y. Cheng ex J. S. 海桐卫矛
Small trees

18.7 (16.3~21.3); 10.8 (9.1~14.2); 24.4 (21.4~27.3); 18.7 (16.4~21.3); 164.5 (136.4~195.9); 0.0 (0.0~0.0); 934 (797~1111) ❙ 1377 (1074~1900); 753 (582~1056); 65 (47~125) ❙ 46.9 (27.5~71.0); 934 (797~1111); 347.1 (254.3~551.6)

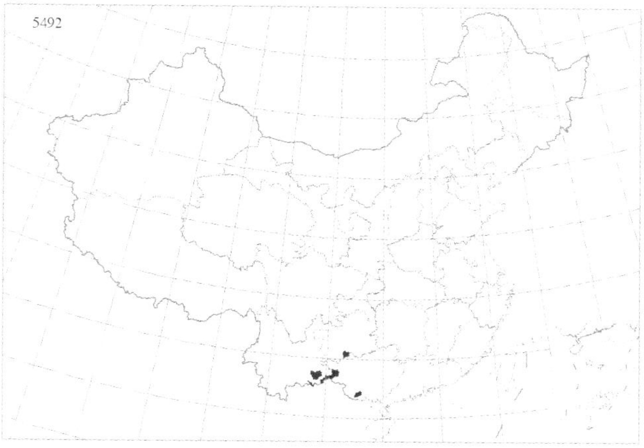

Celastraceae 卫矛科 *Euonymus* 卫矛属
Euonymus potingensis Chun et F. C. How ex J. S. Ma 保亭卫矛
Shrubs, evergreen

24.0; 19.1; 27.2; 24.0; 228.1; 0.0; 1302 ❙ 1551; 539; 48 ❙ 19.1; 1147; 620.0

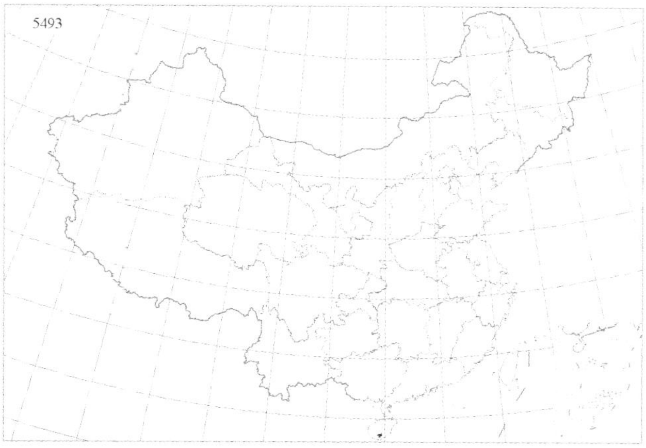

Celastraceae 卫矛科 *Euonymus* 卫矛属
Euonymus prismatomerioides C. Y. Wu ex J. S. Ma 显脉卫矛
Shrubs, evergreen

17.2; 9.5; 22.7; 17.3; 147.0; 0.0; 838 ❙ 1254; 699; 48 ❙ 49.7; 838; 336.3

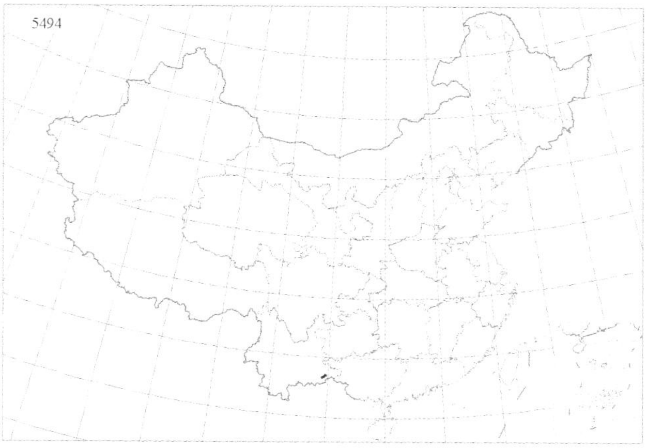

Celastraceae 卫矛科 *Euonymus* 卫矛属
Euonymus pseudovagans Pitard 假游藤卫矛
Shrubs, evergreen

21.3 (18.6~22.7); 13.7 (12.7~15.0); 26.8 (22.5~28.6); 21.3 (18.7~22.8); 195.8 (163.9~213.1); 0.0 (0.0~0.0); 1107 (888~1221) ❙ 1734 (1302~2435); 971 (670~1428); 100 (56~130) ❙ 56.9 (12.9~99.5); 1107 (888~1221); 346.9 (235.1~551.3)

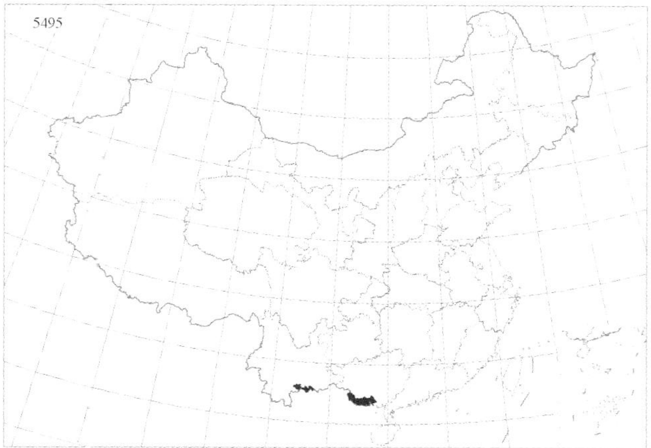

Celastraceae 卫矛科 *Euonymus* 卫矛属
Euonymus rehderianus Loes. 短翅卫矛
Shrubs or small trees, deciduous

14.4 (7.3~19.1); 4.9 (−2.0~10.1); 22.8 (15.6~27.7); 14.5 (7.5~19.1); 116.2 (48.2~169.2); 2.4 (0.0~20.4); 774 (537~992) ❙ 1236 (1001~1620); 619 (489~817); 66 (22~196) ❙ 60.8 (30.0~89.0); 774 (537~992); 242.2 (160.8~322.4)

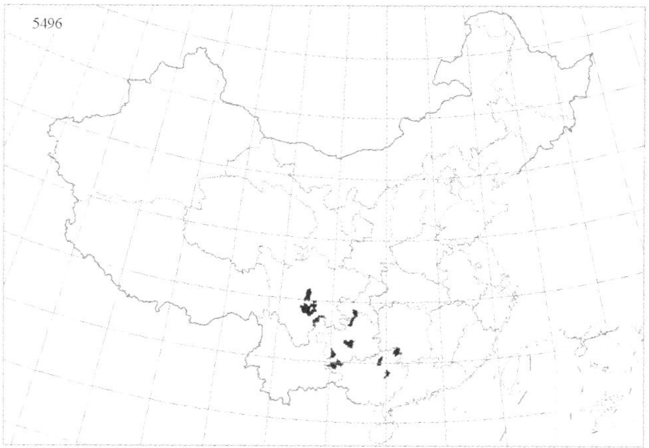

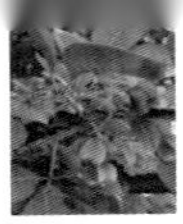

Celastraceae 卫矛科 *Euonymus* 卫矛属
Euonymus sachalinensis (F. Schmidt) Maxim. 东北卫矛
Shrubs, deciduous

6.3 (1.4~9.0); -12.6 (-18.0~-9.1); 22.5 (18.4~24.8); 9.1 (6.4~10.8); 74.4 (45.8~91.5); 58.5 (42.6~88.8); 622 (501~701) ❙ 847 (674~1073); 532 (413~691); 31 (23~40) ❙ 37.4 (-3.8~74.1); 620 (501~678); 285.7 (164.3~335.3)

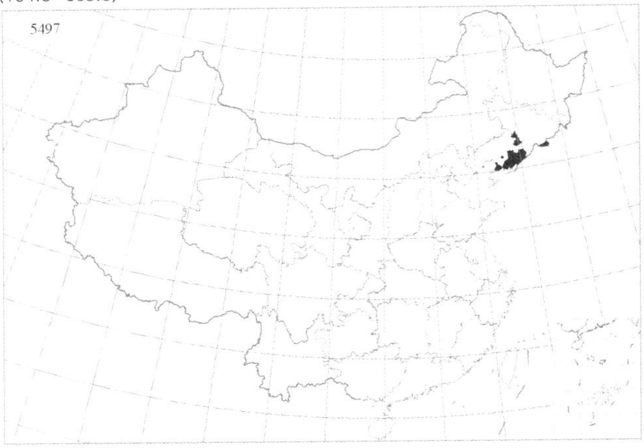

Celastraceae 卫矛科 *Euonymus* 卫矛属
Euonymus salicifolius Loes. 柳叶卫矛
Shrubs, evergreen

18.2 (14.4~22.4); 11.7 (7.5~14.0); 22.7 (19.1~28.6); 18.3 (14.4~22.4); 159.3 (113.1~209.1); 0.0 (0.0~0.0); 898 (727~1204) ❙ 1434 (1049~1888); 800 (598~1044); 67 (44~124) ❙ 60.3 (26.1~99.9); 889 (727~1204); 422.4 (235.1~551.3)

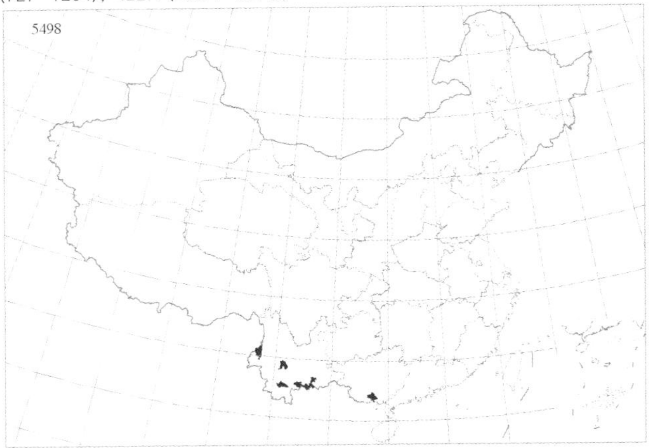

Celastraceae 卫矛科 *Euonymus* 卫矛属
Euonymus sanguineus Loes. ex Diels 石枣子
Shrubs or small trees, deciduous

10.3 (-1.5~17.6); -1.3 (-10.5~6.7); 21.1 (6.2~28.1); 10.8 (2.0~17.6); 83.8 (2.5~151.4); 19.3 (0.0~82.2); 666 (332~940) ❙ 855 (311~1736); 414 (190~1007); 36 (7~120) ❙ 30.8 (-40.9~145.8); 647 (311~940); 228.0 (61.2~447.4)

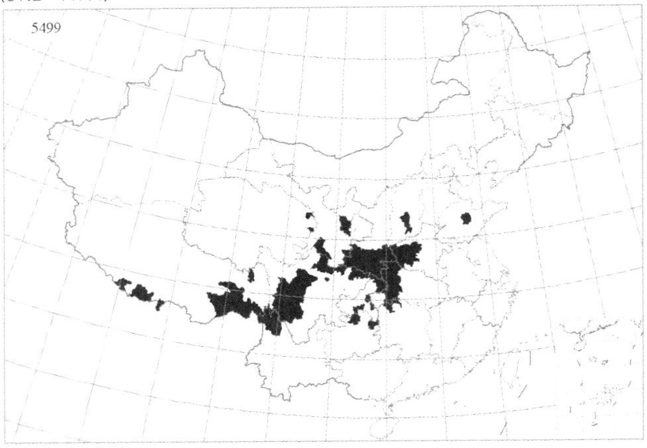

Celastraceae 卫矛科 *Euonymus* 卫矛属
Euonymus schensianus Maxim. 陕西卫矛
Shrubs, deciduous

12.0 (3.5~16.9); 0.0 (-7.4~6.0); 23.5 (12.6~28.0); 12.2 (5.0~17.0); 97.2 (26.3~143.4); 12.6 (0.0~43.4); 715 (446~907) ❙ 880 (341~1280); 390 (190~532); 38 (8~94) ❙ 23.1 (-40.9~101.4); 702 (341~907); 240.3 (61.2~334.4)

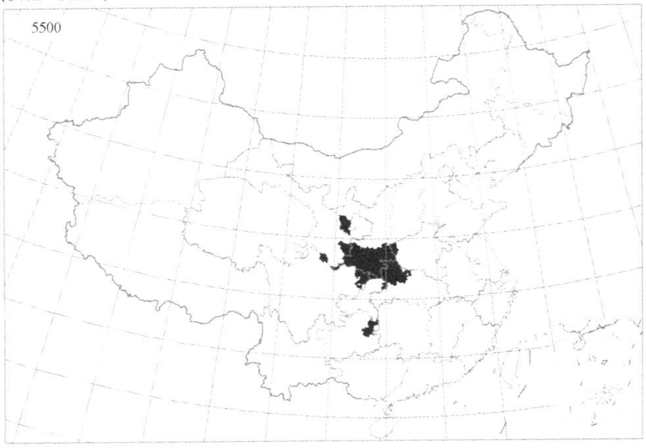

Celastraceae 卫矛科 *Euonymus* 卫矛属
Euonymus semenovii Regel et Herd. 中亚卫矛
Small shrubs, deciduous

5.9 (-2.7~14.8); -7.6 (-20.9~3.8); 17.8 (6.2~27.1); 7.8 (2.0~14.8); 56.4 (2.5~119.2); 44.4 (1.4~103.9); 558 (334~820) ❙ 574 (225~1358); 315 (65~651); 18 (4~88) ❙ 7.9 (-65.1~128.6); 480 (225~795); 209.6 (61.2~357.7)

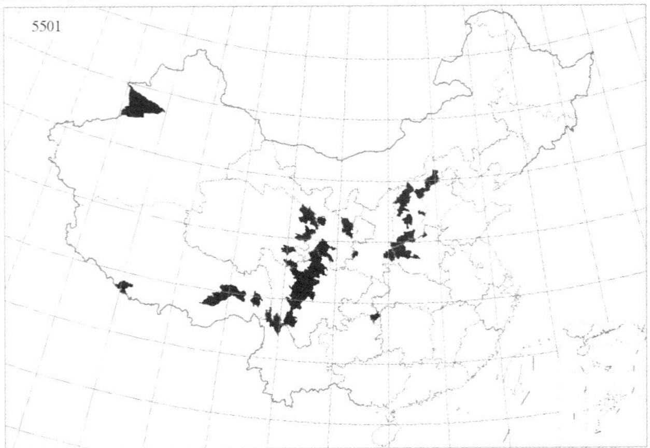

Celastraceae 卫矛科 *Euonymus* 卫矛属
Euonymus serratifolius Beddome 印度卫矛
Shrubs, evergreen

16.0; 9.3; 20.4; 16.0; 132.4; 0.0; 780 ❙ 1348; 762; 56 ❙ 72.8; 780; 395.4

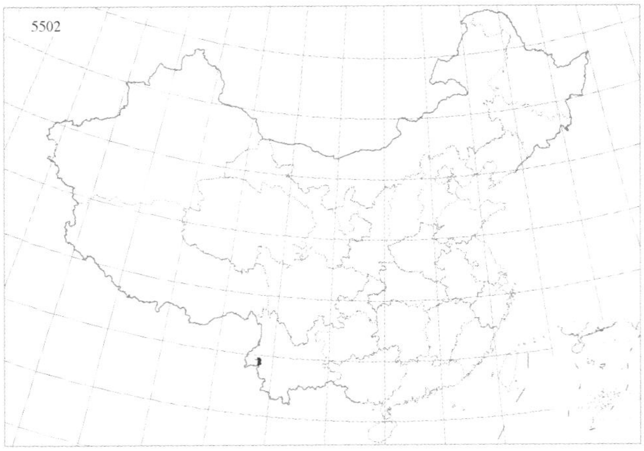

Celastraceae 卫矛科 *Euonymus* 卫矛属
Euonymus spraguei Hayata 疏刺卫矛
Scandent shrubs, deciduous

18.9 (16.2~21.0); 10.4 (6.9~15.3); 26.1 (20.1~28.5); 18.9 (16.2~21.0); 167.2 (134.2~192.3); 0.0 (0.0~0.0); 967 (773~1081) ❙ 1851 (1381~2680); 794 (556~1412); 169 (107~245) ❙ 95.6 (50.6~246.5); 963 (773~1081); 248.0 (207.7~289.1)

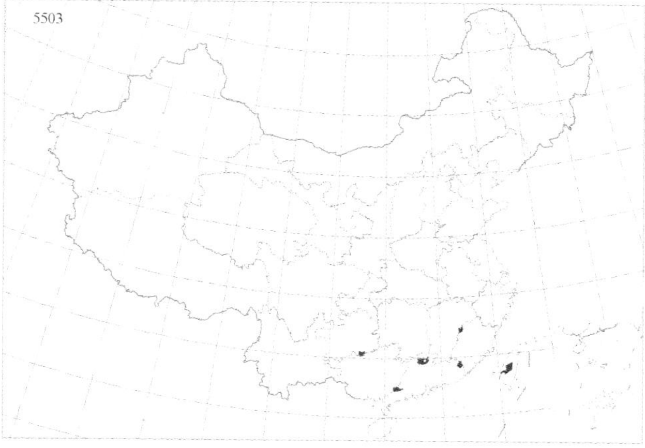

Celastraceae 卫矛科 *Euonymus* 卫矛属
Euonymus subcordatus J. S. Ma 近心叶卫矛
Shrubs

22.2; 13.9; 28.3; 22.3; 207.5; 0.0; 1190 ❙ 1374; 713; 87 ❙ 15.4; 1190; 327.7

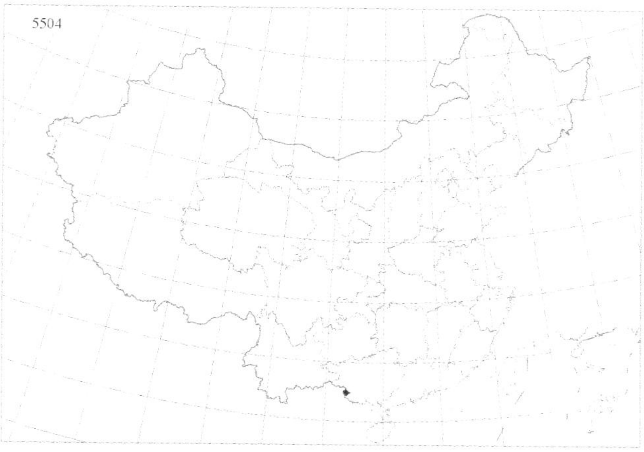

Celastraceae 卫矛科 *Euonymus* 卫矛属
Euonymus szechuanensis C. H. Wang 四川卫矛
Shrubs

12.1 (7.3~14.9) 2.7 (-2.0~5.3); 20.7 (15.6~23.6); 12.2 (7.5~15.0); 91.0 (48.2~119.6); 5.1 (0.0~20.4); 686 (537~782) ❙ 1129 (806~1414); 631 (448~817); 35 (16~49) ❙ 65.3 (8.2~89.0); 686 (537~782); 254.8 (214.5~292.6)

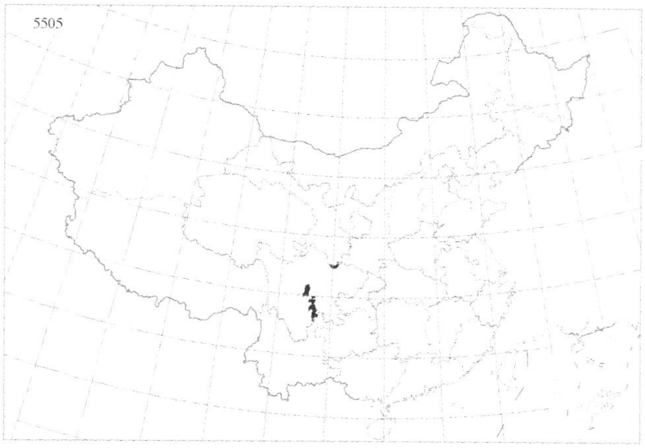

Celastraceae 卫矛科 *Euonymus* 卫矛属
Euonymus tashiroi Maxim. 菱叶卫矛
Small shrubs, evergreen

20.3; 15.9; 23.6; 20.4; 184.4; 0.0; 993 ❙ 2853; 1618; 147 ❙ 190.4; 993; 330.3

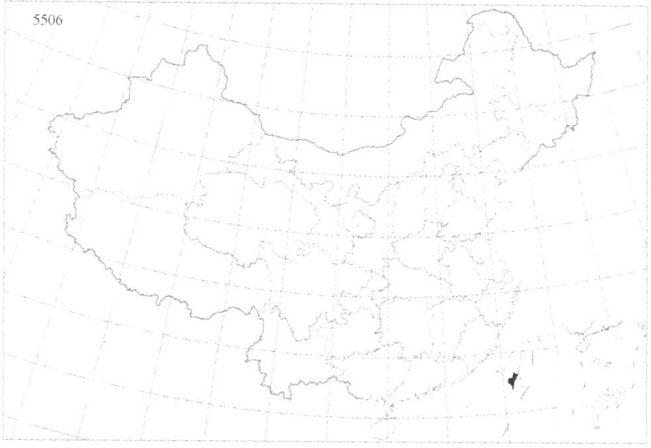

Celastraceae 卫矛科 *Euonymus* 卫矛属
Euonymus tenuiserratus C. Y. Cheng ex J. S. Ma 柔齿卫矛
Shrubs

18.2; 11.8; 22.5; 18.2; 158.4; 0.0; 873 ❙ 1059; 604; 44 ❙ 21.3; 847; 294.5

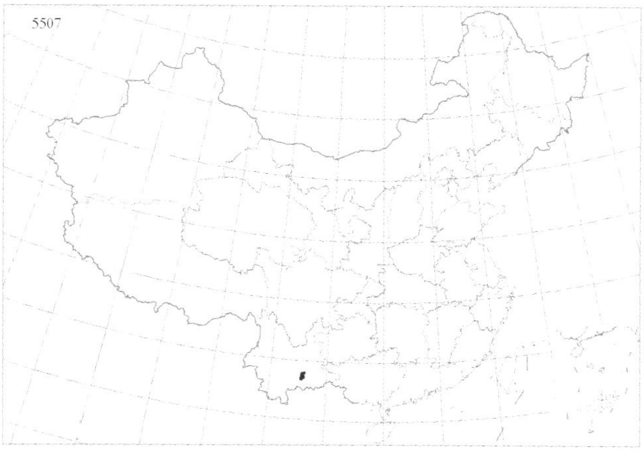

Celastraceae 卫矛科 *Euonymus* 卫矛属
Euonymus ternifolius Hand.-Mazz. 韩氏卫矛
Shrubs, deciduous

11.0; 3.7; 17.0; 11.1; 74.6; 1.8; 640 ❙ 990; 619; 15 ❙ 54.7; 638; 296.0

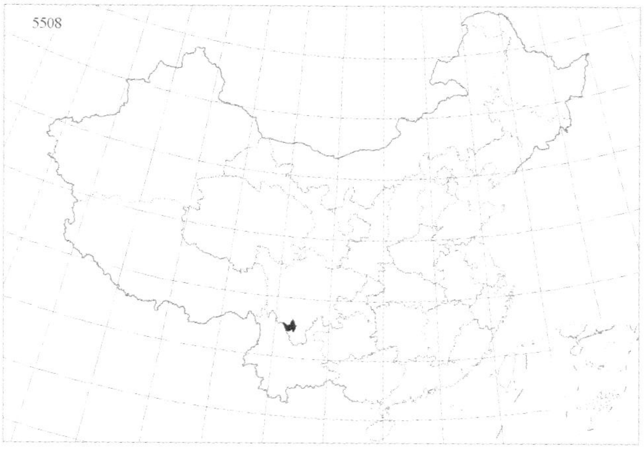

Celastraceae 卫矛科 *Euonymus* 卫矛属

Euonymus theacola C. Y. Cheng 茶色卫矛

Shrubs or small trees, evergreen

17.6 (14.9~19.1); 10.0 (8.2~13.0); 23.2 (20.1~26.6); 17.7 (15.0~19.1); 151.9 (119.8~169.7); 0.0 (0.0~0.0); 868 (748~969) ▎1227 (956~1428); 681 (573~835); 51 (34~60) ▎41.5 (26.1~65.6); 861 (716~969); 349.0 (254.3~506.2)

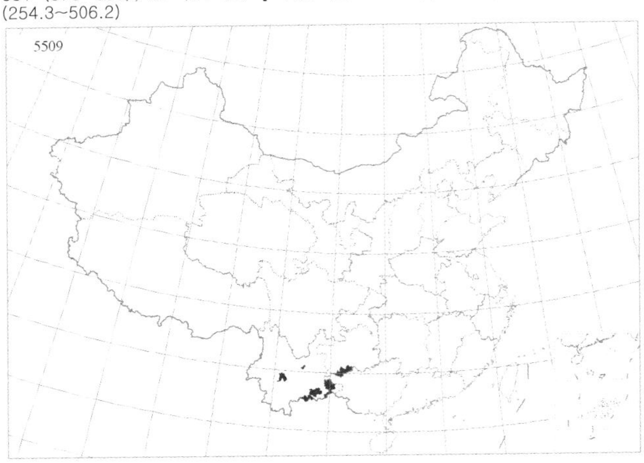

Celastraceae 卫矛科 *Euonymus* 卫矛属

Euonymus theifolius Wall. 茶叶卫矛

Shrubs, evergreen

11.2 (5.3~15.4); 3.4 (-2.4~8.2); 17.7 (12.2~23.5); 11.3 (5.7~15.4); 81.0 (30.5~125.4); 6.6 (0.0~26.2); 647 (479~770) ▎1198 (801~1736); 669 (496~1007); 41 (19~84) ▎85.0 (27.6~145.8); 638 (479~735); 335.3 (220.9~487.2)

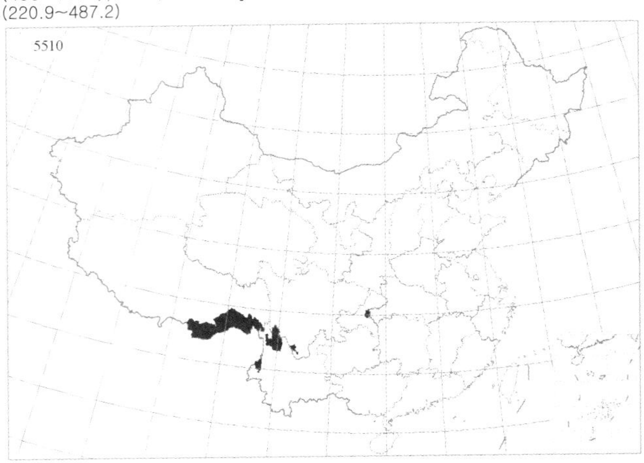

Celastraceae 卫矛科 *Euonymus* 卫矛属

Euonymus tingens Wall. 染用卫矛

Trees

11.0 (-1.1~18.5); 3.0 (-9.1~11.0); 17.7 (6.2~25.9); 11.4 (2.0~18.3); 84.7 (2.5~160.0); 11.8 (0.0~75.6); 655 (334~928) ▎1002 (321~1431); 570 (193~817); 37 (11~107) ▎53.5 (-4.0~113.0); 642 (321~928); 275.5 (79.3~372.3)

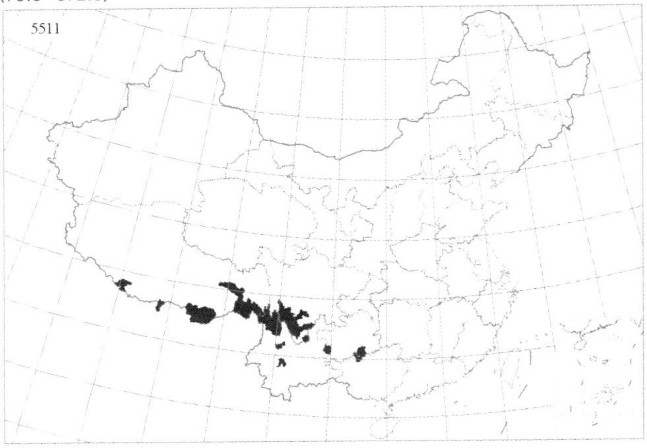

Celastraceae 卫矛科 *Euonymus* 卫矛属

Euonymus tonkinensis (Loes.) Loes. 北部湾卫矛

Shrubs, evergreen

23.0 (21.8~24.1); 16.5 (14.0~19.1); 27.8 (25.7~29.0); 23.1 (21.9~24.1); 216.9 (202.2~229.4); 0.0 (0.0~0.0); 1234 (1127~1321) ▎1638 (1187~2148); 732 (539~1234); 87 (48~124) ▎33.8 (-6.9~86.7); 1189 (1055~1294); 389.1 (225.6~620.0)

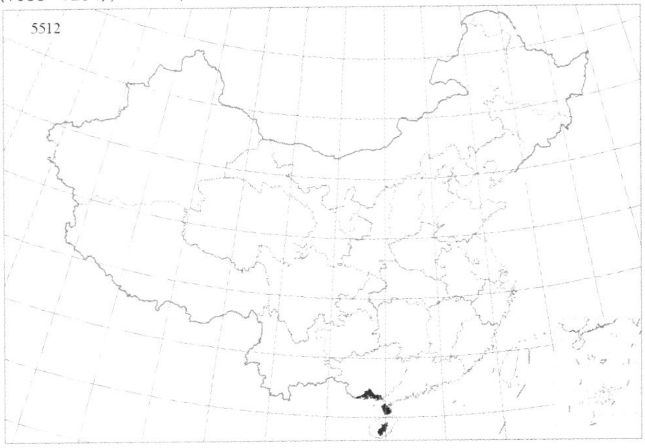

Celastraceae 卫矛科 *Euonymus* 卫矛属

Euonymus tsoi Merr. 狭叶卫矛

Small shrubs, evergreen

19.9 (15.8~22.0); 11.0 (4.1~15.2); 28.0 (26.9~28.8); 20.3 (15.9~22.7); 183.2 (131.5~212.7); 0.1 (0.0~0.9); 1071 (859~1216) ▎1606 (1387~1965); 683 (513~994); 138 (94~182) ▎50.7 (29.9~66.9); 1071 (859~1216); 214.2 (197.5~235.5)

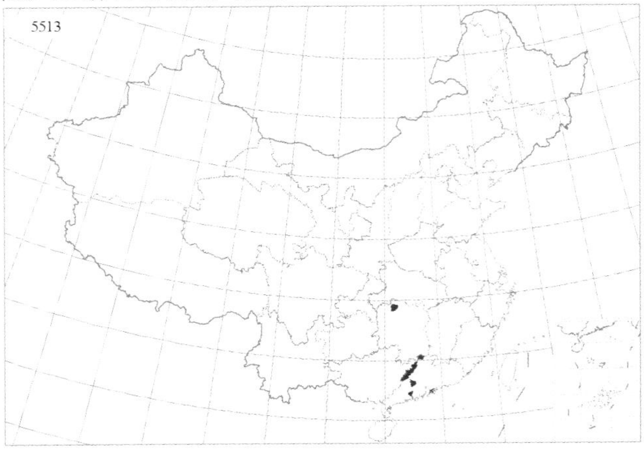

Celastraceae 卫矛科 *Euonymus* 卫矛属

Euonymus vaganoides C. Y. Cheng ex J. S. Ma 拟游藤卫矛

Scandent shrubs, evergreen

17.4; 10.7; 21.9; 17.4; 148.8; 0.0; 836 ▎1151; 635; 50 ▎37.5; 822; 395.2

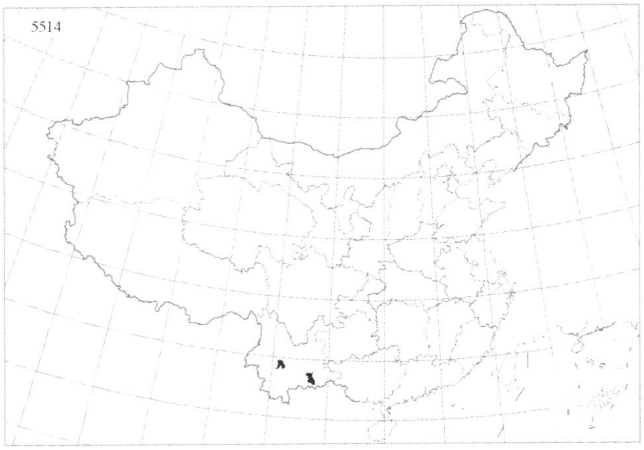

Celastraceae 卫矛科 *Euonymus* 卫矛属
Euonymus vagans Wall. ex Roxb. 游藤卫矛
Small shrubs

13.5 (−1.1~21.5); 5.7 (−9.1~13.6); 19.9 (6.2~28.2); 13.6 (2.0~21.6); 106.4 (2.5~199.1); 3.8 (0.0~75.6); 727 (334~1131) ❙ 1040 (321~1736); 580 (193~1007); 44 (13~216) ❙ 44.0 (−14.2~145.8); 706 (321~1131); 283.4 (79.3~491.5)

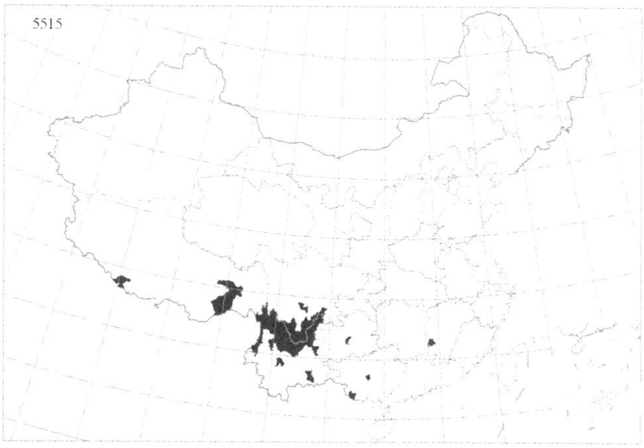

Celastraceae 卫矛科 *Euonymus* 卫矛属
Euonymus venosus Hemsl. 曲脉卫矛
Shrubs or small trees

12.1 (5.4~17.7); 0.5 (−4.2~10.9); 23.2 (12.3~28.1); 12.2 (5.8~17.8); 96.8 (31.5~153.4); 10.9 (0.0~29.4); 713 (481~924) ❙ 950 (634~1466); 438 (290~812); 41 (16~106) ❙ 34.2 (−16.6~106.2); 710 (481~924); 251.0 (147.3~491.5)

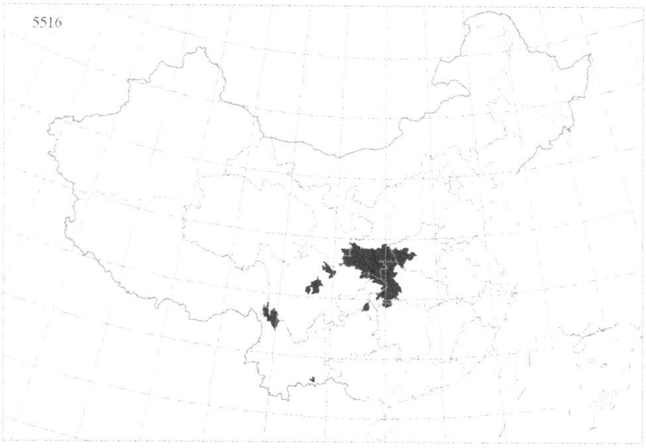

Celastraceae 卫矛科 *Euonymus* 卫矛属
Euonymus verrucocarpus C. Y. Cheng ex J. S. Ma 瘤果卫矛
Shrubs or small trees, deciduous

15.2; 8.4; 19.8; 15.2; 122.8; 0.0; 754 ❙ 1401; 777; 63 ❙ 86.3; 754; 441.3

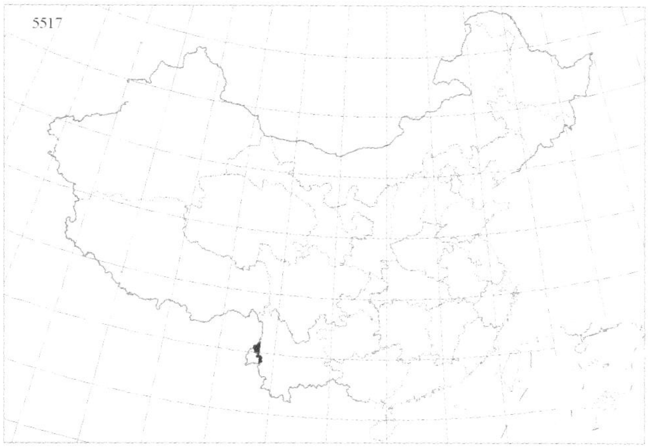

Celastraceae 卫矛科 *Euonymus* 卫矛属
Euonymus verrucosoides Loes. 疣点卫矛
Shrubs, deciduous

11.8 (−1.0~17.7); −0.1 (−11.8~7.7); 22.9 (8.5~29.2); 12.1 (2.7~17.8); 96.1 (8.3~153.6); 13.9 (0.0~79.5); 718 (363~936) ❙ 947 (512~1736); 441 (257~1007); 51 (9~195) ❙ 34.8 (−35.3~145.8); 698 (363~930); 228.6 (121.6~447.4)

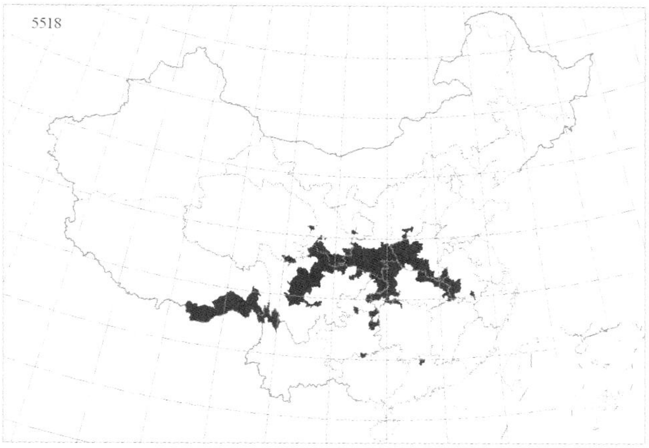

Celastraceae 卫矛科 *Euonymus* 卫矛属
Euonymus verrucosus Scop. 瘤枝卫矛
Shrubs, deciduous

4.3 (−0.8~14.5); −14.8 (−25.0~−0.2); 20.9 (17.9~28.2); 8.2 (6.4~14.6); 63.0 (45.8~125.3); 71.3 (10.4~118.4); 581 (501~864) ❙ 670 (341~1073); 392 (190~691); 22 (8~40) ❙ 15.5 (−40.9~74.1); 562 (341~697); 272.7 (61.2~346.5)

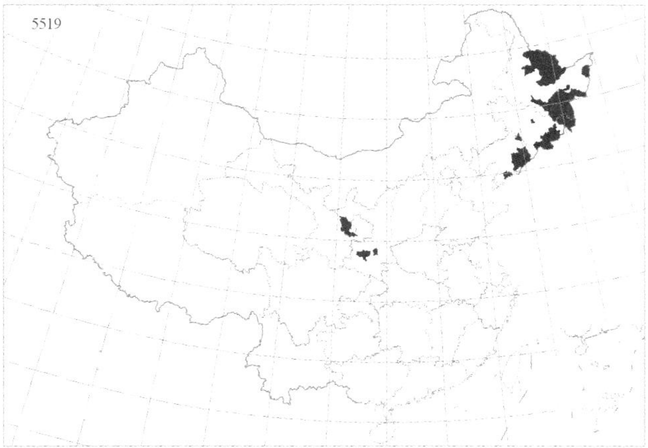

Celastraceae 卫矛科 *Euonymus* 卫矛属
Euonymus viburnoides Prain 荚蒾卫矛
Shrubs or small trees, deciduous

15.4 (7.3~21.6); 7.5 (−2.0~13.1); 21.7 (15.6~28.5); 15.5 (7.5~21.7); 127.7 (48.2~199.8); 2.1 (0.0~20.4); 803 (537~1153) ❙ 1197 (946~1703); 655 (510~911); 54 (19~194) ❙ 51.3 (15.5~101.6); 792 (537~1153); 307.8 (183.1~551.3)

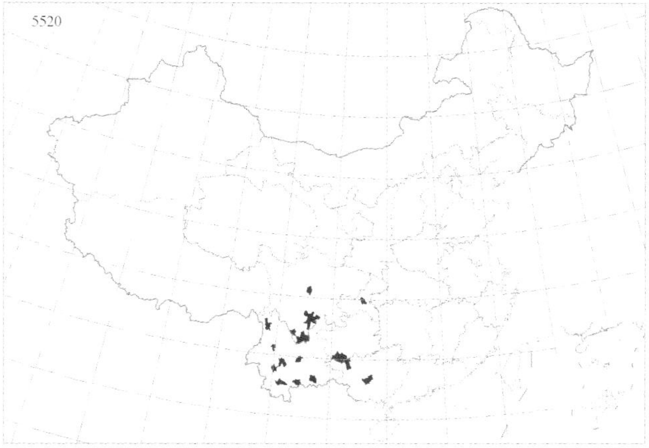

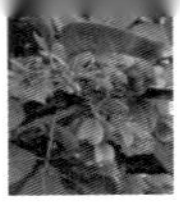

Celastraceae 卫矛科 *Euonymus* 卫矛属
Euonymus wilsonii Sprague 长刺卫矛
Scandent shrubs

15.0 (5.7~19.1); 5.8 (-3.5~11.8); 22.8 (14.2~27.2); 15.1 (6.4~19.2); 122.5 (38.1~170.1); 1.9 (0.0~28.1); 789 (504~988) ❙ 1197 (874~1740); 619 (460~817); 60 (25~187) ❙ 52.7 (18.3~93.2); 788 (504~988); 245.8 (149.6~491.5)

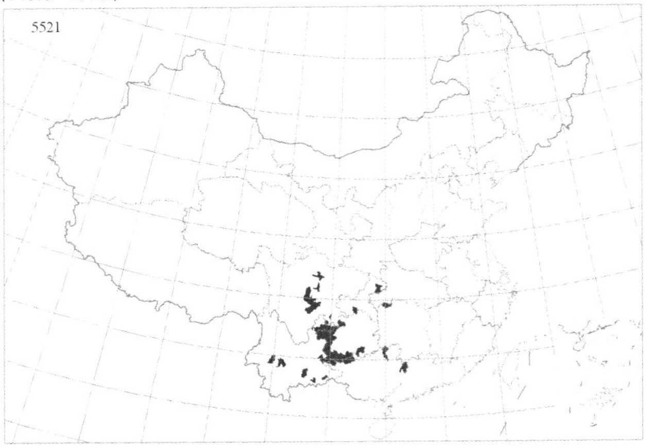

Celastraceae 卫矛科 *Euonymus* 卫矛属
Euonymus wui J. S. Ma 征镒卫矛
Shrubs, deciduous

18.4 (16.3~20.5); 12.1 (9.1~15.1); 22.4 (20.3~23.7); 18.4 (16.4~20.5); 160.9 (136.4~185.9); 0.0 (0.0~0.0); 886 (789~992) ❙ 1397 (1190~1607); 714 (655~813); 56 (45~77) ❙ 57.7 (47.6~65.6); 873 (780~986); 477.0 (347.3~659.0)

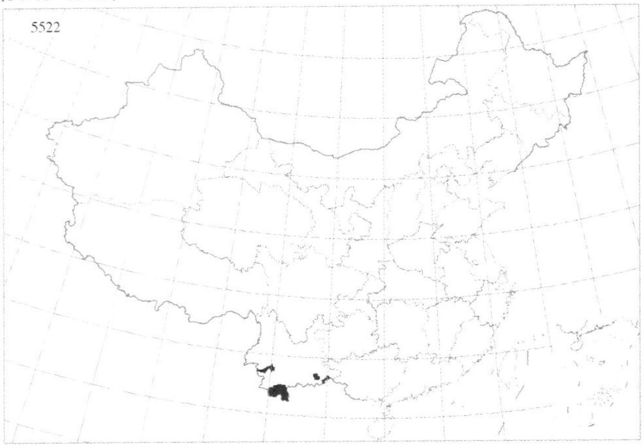

Celastraceae 卫矛科 *Euonymus* 卫矛属
Euonymus yunnanensis Franch. 云南卫矛
Shrubs, evergreen

13.9 (1.8~19.3); 6.4 (-7.1~13.1); 20.0 (9.9~26.6); 14.0 (3.8~19.4); 110.3 (15.7~173.2); 2.8 (0.0~53.3); 732 (404~939) ❙ 964 (583~1143); 561 (348~637); 32 (13~57) ❙ 33.7 (-14.2~100.7); 703 (404~912); 271.6 (142.9~372.3)

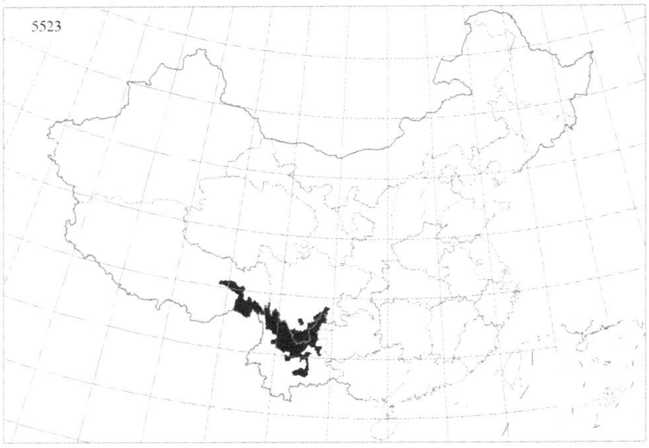

Celastraceae 卫矛科 *Glyptopetalum* 沟瓣属
Glyptopetalum aquifolium (Loes. et Rehder) C. Y. Cheng et Q. S. Ma 冬青沟瓣
Shrubs, evergreen

12.6; 3.4; 21.1; 12.7; 95.1; 2.9; 700 ❙ 1317; 752; 42 ❙ 88.1; 700; 255.1

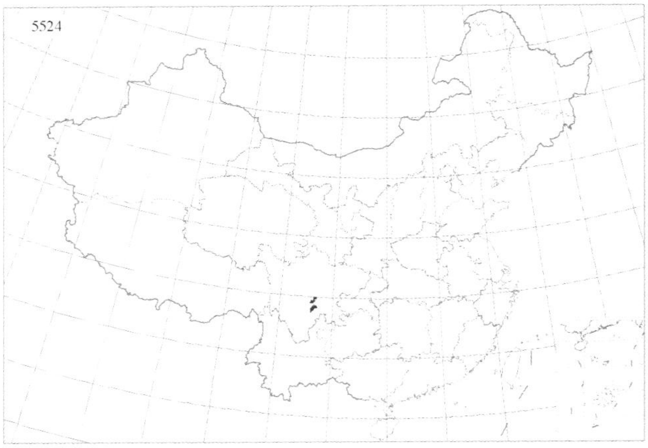

Celastraceae 卫矛科 *Glyptopetalum* 沟瓣属
Glyptopetalum feddei (Lév.) D. Hou 罗甸沟瓣
Shrubs, evergreen

18.5 (15.1~19.8); 9.4 (5.0~11.1); 25.8 (24.1~26.6); 18.6 (15.2~19.8); 162.9 (121.9~177.9); 0.0 (0.0~0.0); 942 (788~1010) ❙ 1283 (1182~1438); 671 (545~820); 65 (43~97) ❙ 36.7 (16.9~57.8); 942 (788~1010); 295.7 (221.1~371.7)

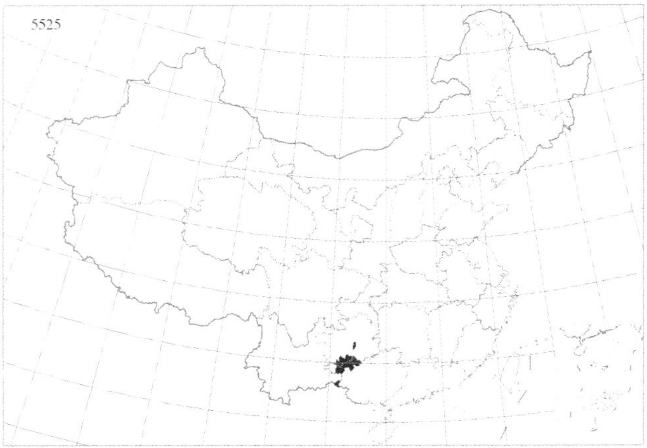

Celastraceae 卫矛科 *Glyptopetalum* 沟瓣属
Glyptopetalum fengii (Chun et How) D. Hou 海南沟瓣
Shrubs, evergreen

23.9; 18.7; 27.5; 23.9; 226.8; 0.0; 1294 ❙ 1380; 455; 48 ❙ 6.7; 1112; 455.2

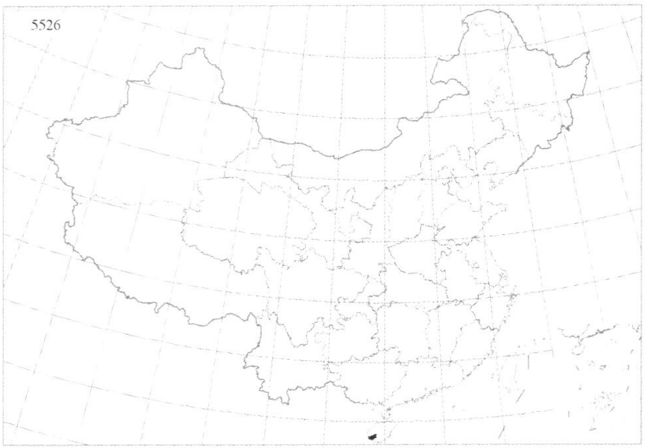

Celastraceae 卫矛科 *Glyptopetalum* 沟瓣属

Glyptopetalum geloniifolium (Chun et How) C. Y. Cheng 白树沟瓣

Shrubs, evergreen

22.8；15.9；28.0；22.8；214.0；0.0；1238 ▎1403；628；65 ▎15.1；1105；376.1

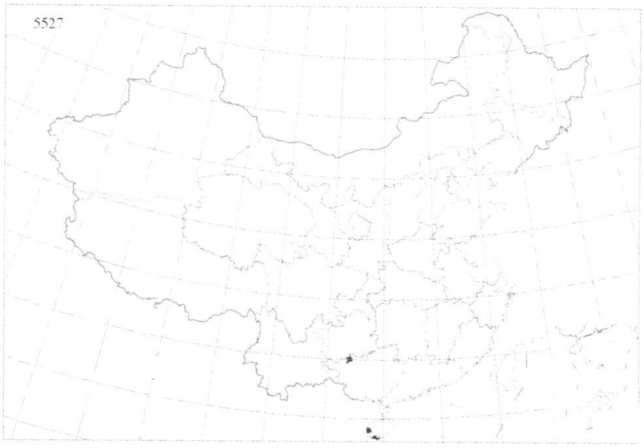

Celastraceae 卫矛科 *Glyptopetalum* 沟瓣属

Glyptopetalum ilicifolium (Franch.) C. Y. Cheng et Q. S. Ma 刺叶沟瓣

Shrubs, evergreen

13.6 (5.4~18.9)；5.9 (-2.1~12.0)；20.1 (12.3~26.6)；13.7 (5.8~19.0)；106.2 (31.5~167.8)；1.9 (0.0~25.6)；725 (481~929) ▎962 (741~1141)；560 (411~641)；30 (15~60) ▎34.6 (-14.2~70.8)；695 (481~912)；267.9 (142.9~351.7)

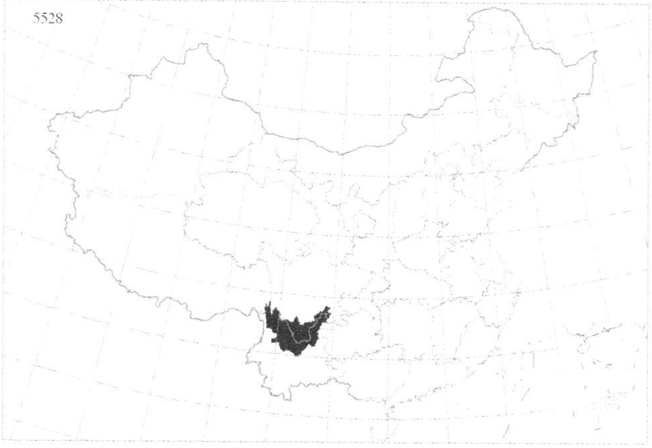

Celastraceae 卫矛科 *Glyptopetalum* 沟瓣属

Glyptopetalum longepedunculatum Tardieu 细梗沟瓣

Small trees, evergreen

19.1；11.0；25.0；19.1；169.6；0.0；945 ▎1438；820；57 ▎52.2；945；371.7

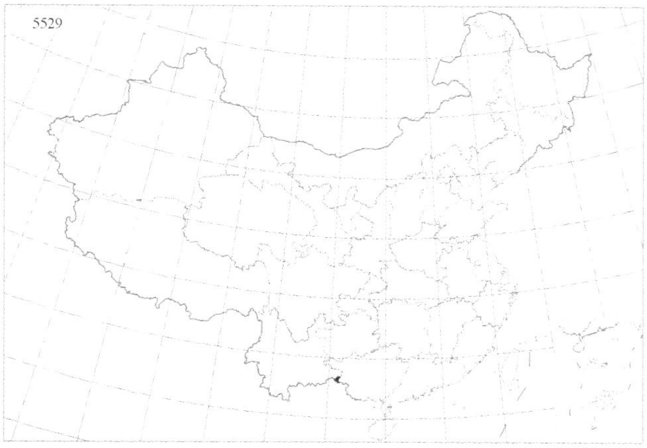

Celastraceae 卫矛科 *Glyptopetalum* 沟瓣属

Glyptopetalum longipedicellatum (Merr. et Chun) C. Y. Cheng 长梗沟瓣

Shrubs or small trees, evergreen

19.6 (17.6~24.0)；10.7 (7.0~19.1)；26.9 (25.0~27.7)；19.6 (17.6~24.0)；175.3 (151.7~228.1)；0.0 (0.0~0.0)；1020 (920~1302) ▎1518 (1438~1590)；619 (507~820)；127 (48~191) ▎50.9 (19.1~69.4)；994 (920~1147)；326.9 (228.7~620.0)

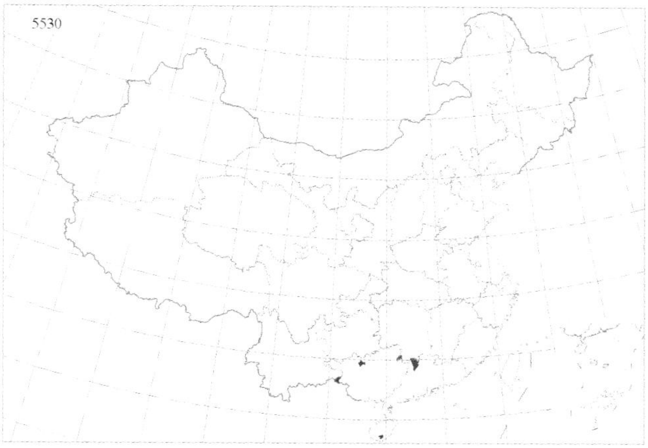

Celastraceae 卫矛科 *Glyptopetalum* 沟瓣属

Glyptopetalum rhytidophyllum (Chun et How) C. Y. Cheng 皱叶沟瓣

Shrubs, evergreen

20.5 (18.1~22.2)；12.6 (9.5~15.1)；26.4 (23.6~28.3)；20.6 (18.2~22.3)；187.0 (158.1~207.5)；0.0 (0.0~0.0)；1055 (901~1190) ▎1371 (1182~1607)；723 (634~850)；71 (43~106) ▎31.1 (5.8~62.5)；1055 (901~1190)；366.3 (281.7~659.0)

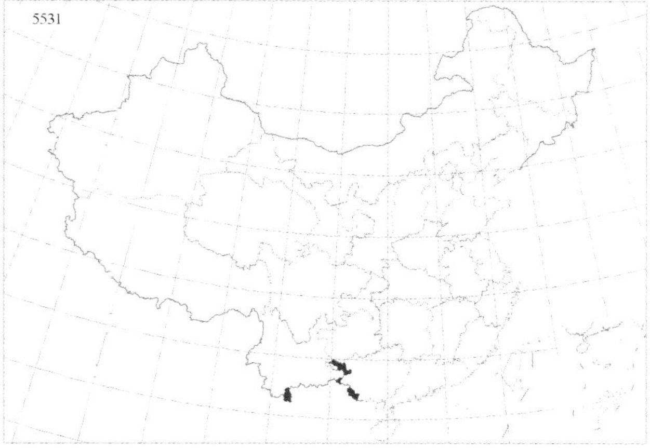

Celastraceae 卫矛科 *Glyptopetalum* 沟瓣属

Glyptopetalum sclerocarpum (Kurz.) Laws 硬果沟瓣

Shrubs or trees, evergreen

19.5；13.9；23.1；19.6；174.8；0.0；938 ▎1577；806；70 ▎68.4；937；605.2

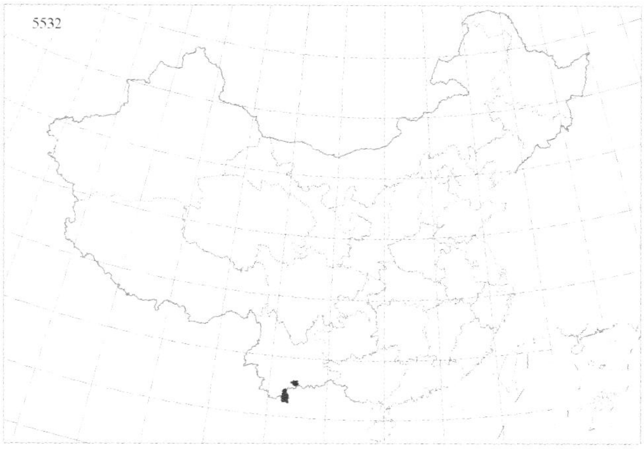

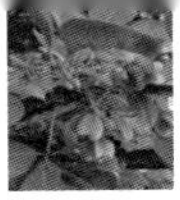

Celastraceae 卫矛科 *Gymnosporia* 裸实属

Gymnosporia berberoides W. W. Sm. 小檗裸实

Shrubs, evergreen

13.9 (5.4~19.1); 7.0 (-2.1~13.0); 19.3 (12.3~23.9); 14.0 (5.8~19.1); 110.6 (31.5~169.7); 3.3 (0.0~25.6); 729 (481~929) ❙ 975 (741~1454); 571 (411~835); 32 (12~70) ❙ 36.4 (-14.2~99.9); 685 (481~917); 312.4 (218.0~506.2)

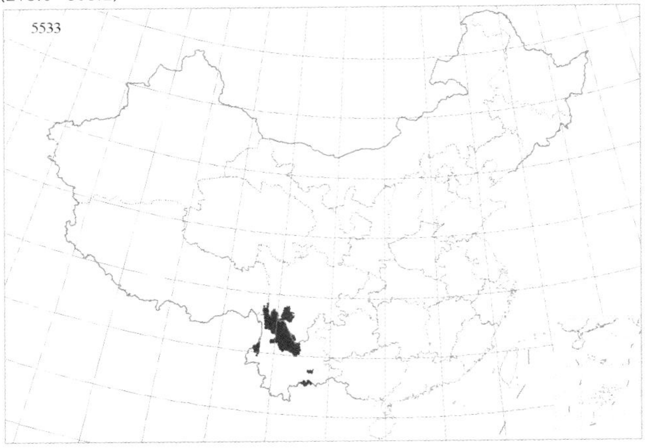

Celastraceae 卫矛科 *Gymnosporia* 裸实属

Gymnosporia diversifolia Maxim. 变叶裸实

Shrubs or small trees, evergreen

22.3 (17.1~24.8); 16.2 (11.2~20.1); 26.9 (21.7~28.9); 22.3 (17.1~24.8); 207.9 (145.0~237.6); 0.0 (0.0~0.0); 1182 (816~1385) ❙ 1759 (868~2759); 794 (312~1594); 114 (41~308) ❙ 59.0 (-34.3~214.4); 1099 (816~1329); 309.7 (154.2~620.0)

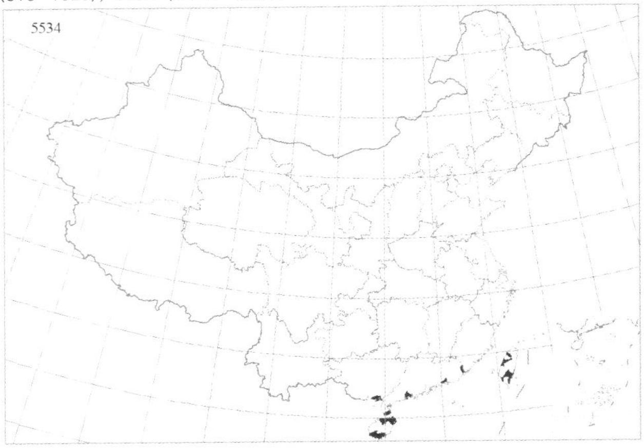

Celastraceae 卫矛科 *Gymnosporia* 裸实属

Gymnosporia emarginata (Willdenow) Thwaites 台湾裸实

Small shrubs, evergreen

18.5; 14.1; 21.7; 18.5; 162.1; 0.0; 863 ❙ 2713; 1388; 188 ❙ 214.4; 863; 363.7

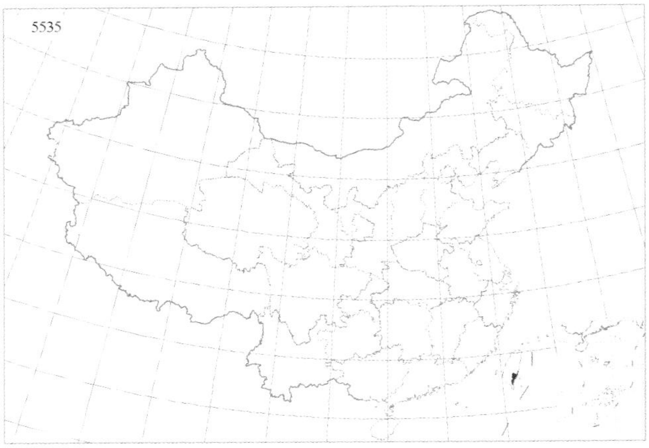

Celastraceae 卫矛科 *Gymnosporia* 裸实属

Gymnosporia esquirolii H. Lév. 贵州裸实

Small shrubs, evergreen

18.4; 11.0; 23.8; 18.4; 161.1; 0.0; 907 ❙ 1186; 633; 54 ❙ 31.4; 895; 328.2

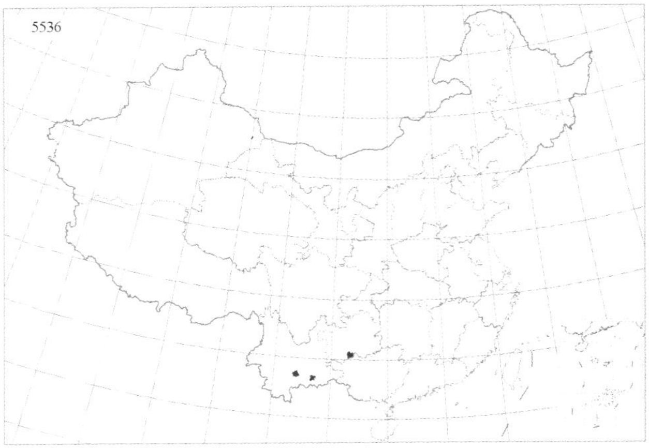

Celastraceae 卫矛科 *Gymnosporia* 裸实属

Gymnosporia graciliramula (S. J. Pei et Y. H. Li) Q. R. Liu et Funston 细梗裸实

Shrubs, evergreen

18.4 (16.3~20.5); 11.7 (9.1~15.1); 23.0 (20.3~26.6); 18.5 (16.4~20.5); 161.8 (136.6~185.9); 0.0 (0.0~0.0); 898 (789~992) ❙ 1291 (1049~1607); 690 (598~774); 56 (44~77) ❙ 43.9 (21.7~65.9); 884 (780~986); 417.3 (314.4~659.0)

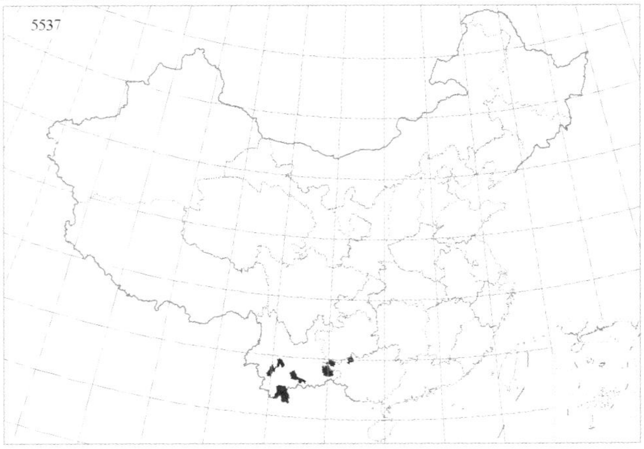

Celastraceae 卫矛科 *Gymnosporia* 裸实属

Gymnosporia jinyangensis (C. Y. Cheng) Q. R. Liu et Funston 金阳美登木

Shrubs, evergreen

12.0; 3.9; 18.6; 12.0; 85.0; 1.5; 665 ❙ 942; 544; 29 ❙ 41.8; 665; 230.3

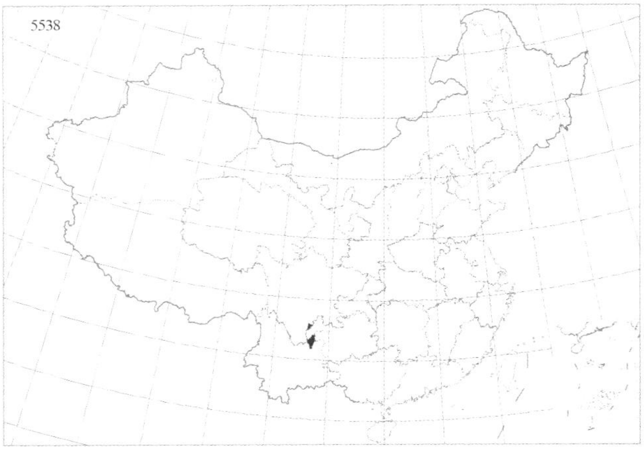

Celastraceae 卫矛科 *Gymnosporia* 裸实属
Gymnosporia orbiculata (C. Y. Wu ex S. J. Pei et Y. H. Li) Q. R. Liu et Funston 圆叶裸实
Shrubs, evergreen

14.3 (5.4~19.1); 7.4 (-2.1~13.0); 19.6 (12.3~23.9); 14.3 (5.8~19.1); 114.2 (31.5~169.7) 2.5 (0.0~25.6); 742 (481~929) ❙ 961 (741~1428); 567 (411~835); 30 (15~56) ❙ 32.0 (-14.2~67.9); 692 (481~917); 300.8 (218.0~506.2)

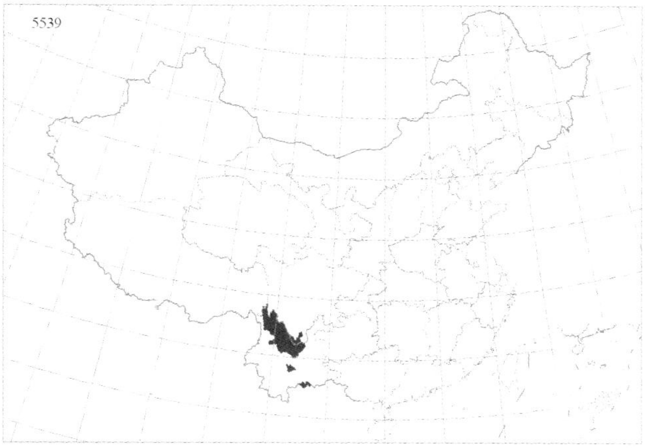

Celastraceae 卫矛科 *Gymnosporia* 裸实属
Gymnosporia royleana Laws. 被子裸实
Shrubs, evergreen

2.1; -8.3; 11.2; 5.3; 31.6; 66.1; 438 ❙ 598; 360; 20 ❙ 8.1; 340; 172.7

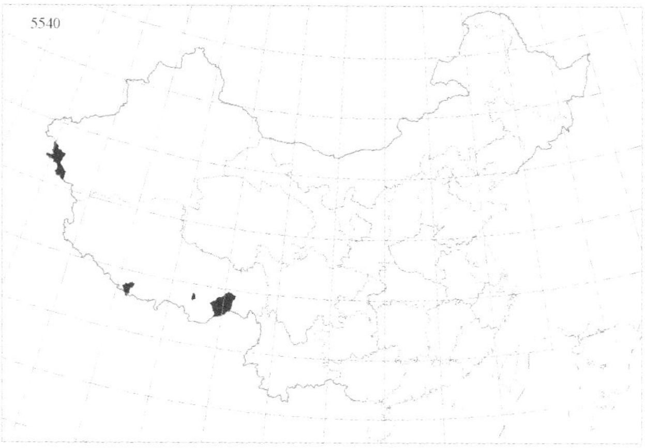

Celastraceae 卫矛科 *Gymnosporia* 裸实属
Gymnosporia rufa (Wall.) Laws. 淡红美登木
Trees, evergreen

-1.8; -10.2; 6.1; 1.8; 1.9; 82.8; 329 ❙ 320; 217; 20 ❙ -3.0; 320; 72.2

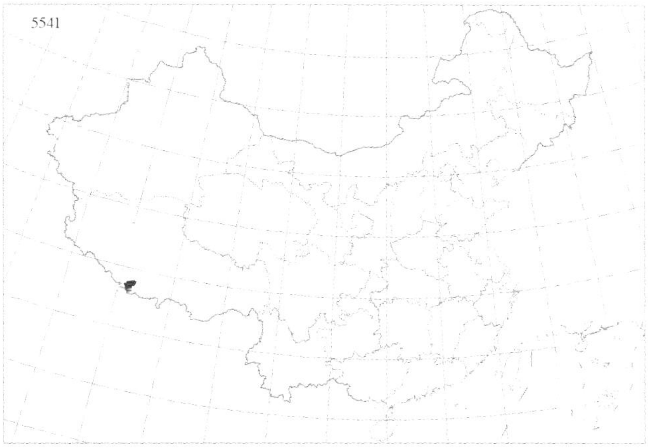

Celastraceae 卫矛科 *Gymnosporia* 裸实属
Gymnosporia tiaoloshanensis Chun et F. C. How 吊罗裸实
Small shrubs, evergreen

24.1; 18.9; 27.8; 24.1; 229.8; 0.0; 1319 ❙ 1595; 539; 70 ❙ 20.7; 1206; 501.3

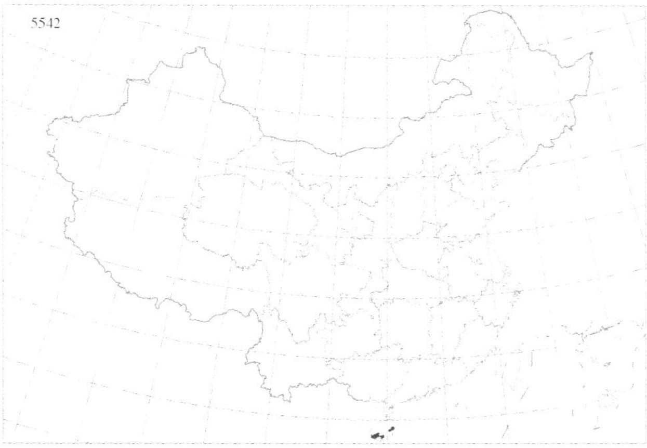

Celastraceae 卫矛科 *Gymnosporia* 裸实属
Gymnosporia variabilis (Hemsl.) Loes. 刺茶裸实
Shrubs, evergreen

16.2 (10.6~20.5); 5.6 (-0.2~14.9); 26.2 (21.1~28.9); 16.3 (10.6~20.5); 136.7 (79.4~185.9); 1.4 (0.0~12.1); 868 (643~1018) ❙ 1178 (1028~1478); 509 (442~705); 67 (46~96) ❙ 38.1 (8.0~107.2); 867 (643~1018); 190.5 (108.2~570.1)

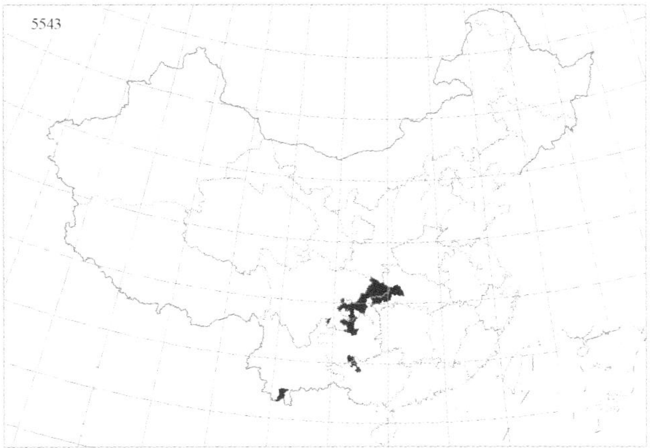

Celastraceae 卫矛科 *Loeseneriella* 翅子藤属
Loeseneriella concinna A. C. Sm. 程香仔树
Vines

21.8 (20.4~22.6); 13.7 (11.9~15.3); 28.3 (27.7~29.0); 21.9 (20.5~22.7); 202.4 (185.5~212.7); 0.0 (0.0~0.0); 1158 (1057~1216) ❙ 1732 (1371~2148); 817 (565~1234); 130 (104~173) ❙ 49.4 (17.4~86.7); 1156 (1057~1216); 234.7 (163.7~318.8)

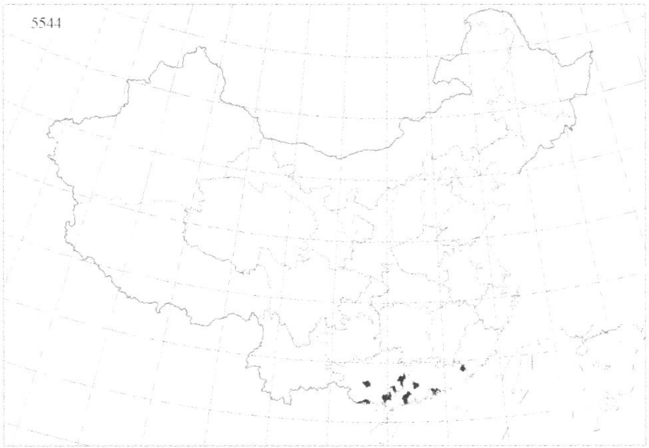

Celastraceae 卫矛科 *Loeseneriella* 翅子藤属
Loeseneriella griseoramula S. Y. Pao 灰枝翅子藤
Vines

21.3; 12.6; 27.9; 21.3; 196.0; 0.0; 1123 ❙ 1188; 634; 56 ❙ 5.8; 1121; 315.6

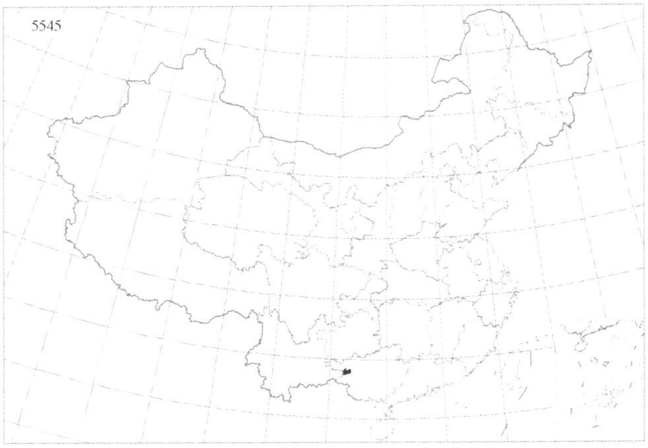

Celastraceae 卫矛科 *Loeseneriella* 翅子藤属
Loeseneriella lenticellata C. Y. Wu 皮孔翅子藤
Vines

18.7 (17.1~20.5); 12.2 (8.5~15.1); 23.1 (21.6~26.0); 18.8 (17.1~20.5); 165.2 (145.1~185.9); 0.0 (0.0~0.0); 906 (832~992) ❙ 1368 (1028~1751); 729 (566~996); 59 (44~77) ❙ 50.8 (12.5~88.8); 888 (783~986); 422.6 (262.1~659.0)

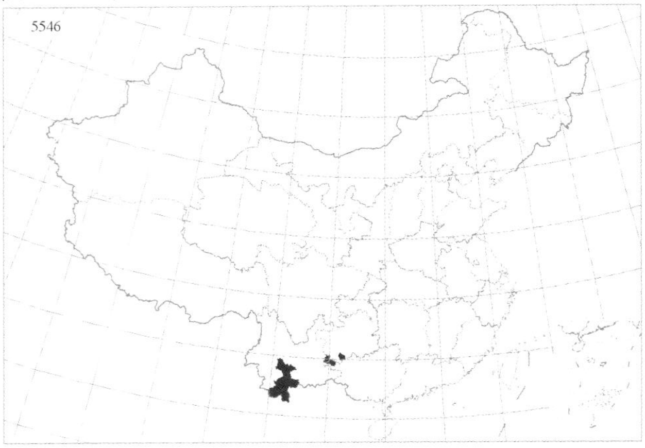

Celastraceae 卫矛科 *Loeseneriella* 翅子藤属
Loeseneriella merrilliana A. C. Sm. 翅子藤
Vines

21.0 (16.3~24.8); 14.1 (9.1~20.1); 25.8 (20.3~28.7); 21.0 (16.4~24.8); 192.0 (136.6~237.6); 0.0 (0.0~0.0); 1090 (789~1385) ❙ 1381 (868~1767); 636 (312~860); 62 (41~130) ❙ 30.8 (-34.3~80.6); 1013 (780~1299); 415.0 (248.2~659.0)

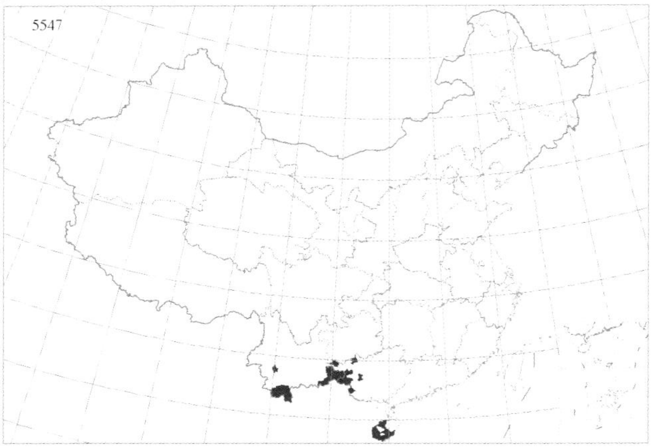

Celastraceae 卫矛科 *Loeseneriella* 翅子藤属
Loeseneriella yunnanensis (Hu) A. C. Sm. 云南翅子藤
Vines, evergreen

18.8 (17.5~20.5); 12.5 (10.5~15.1); 22.9 (22.1~25.0); 18.9 (17.5~20.5); 166.2 (149.9~185.9); 0.0 (0.0~0.0); 907 (846~992) ❙ 1293 (958~1607); 674 (531~836); 55 (44~77) ❙ 42.0 (13.3~66.9); 888 (823~986); 411.6 (294.5~659.0)

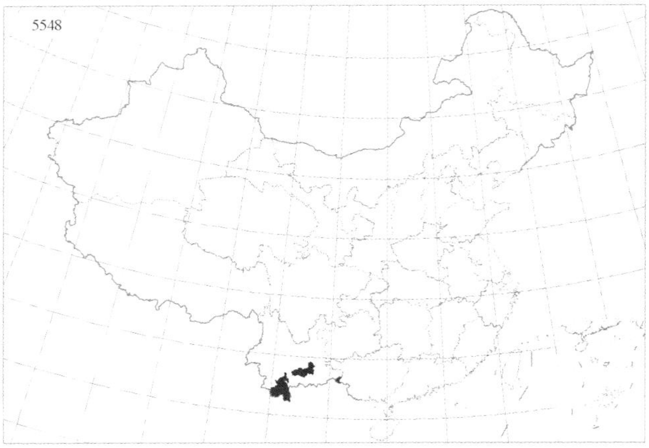

Celastraceae 卫矛科 *Maytenus* 美登木属
Maytenus austroyunnanensis S. J. Pei et Y. H. Li 滇南美登木
Shrubs

19.1 (17.1~20.5); 13.3 (11.0~15.1); 22.5 (20.8~23.7); 19.1 (17.1~20.5); 169.2 (145.0~185.9); 0.0 (0.0~0.0); 914 (815~992) ❙ 1458 (1352~1607); 719 (655~783); 59 (45~77) ❙ 59.7 (47.6~65.9); 895 (802~986); 539.7 (438.9~659.0)

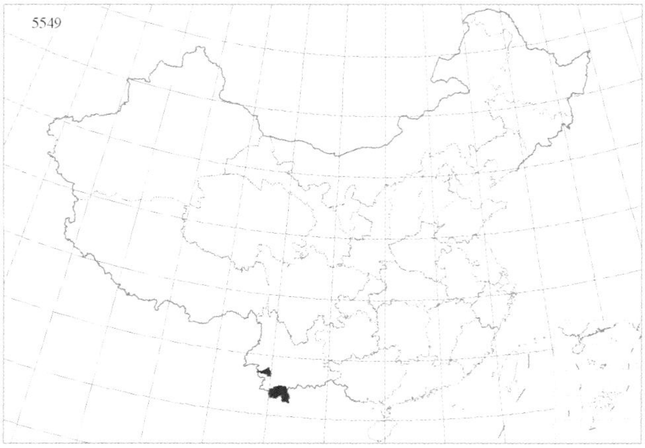

Celastraceae 卫矛科 *Maytenus* 美登木属
Maytenus confertiflora J. Y. Luo et X. X. Chen 密花美登木
Shrubs

21.9 (21.5~22.2); 13.6 (13.1~13.9); 27.9 (27.4~28.4); 21.9 (21.6~22.3); 203.0 (199.1~207.5); 0.0 (0.0~0.0); 1161 (1131~1190) ❙ 1448 (1302~1581); 760 (670~850); 99 (86~116) ❙ 24.8 (12.9~39.9); 1161 (1131~1190); 304.3 (273.1~335.1)

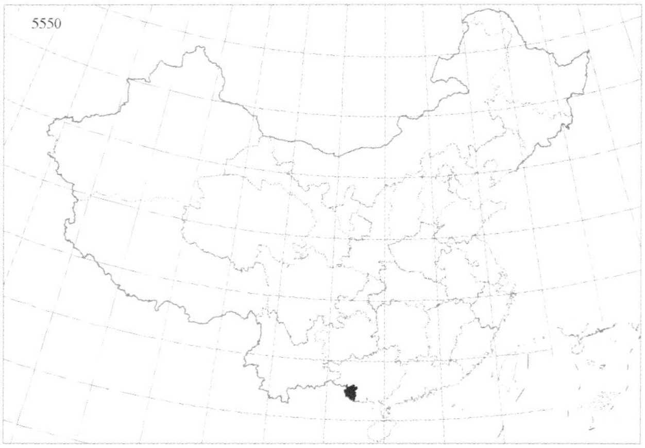

Celastraceae 卫矛科 *Maytenus* 美登木属
Maytenus guangxiensis C. Y. Cheng et W. L. Sha 广西美登木
Shrubs

21.2; 12.5; 27.9; 21.3; 195.5; 0.0; 1124 ▎1445; 748; 109 ▎28.7; 1124; 274.1

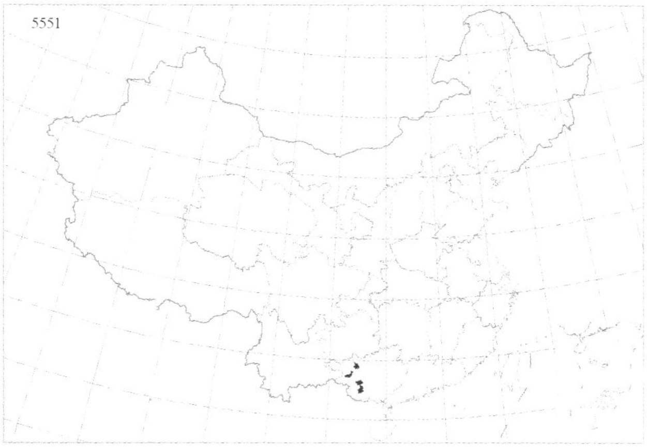

Celastraceae 卫矛科 *Maytenus* 美登木属
Maytenus hainanensis (Merr. et Chun) C. Y. Cheng 海南美登木
Shrubs

20.8; 15.3; 24.2; 20.8; 189.5; 0.0; 1064 ▎1407; 633; 52 ▎38.3; 966; 512.2

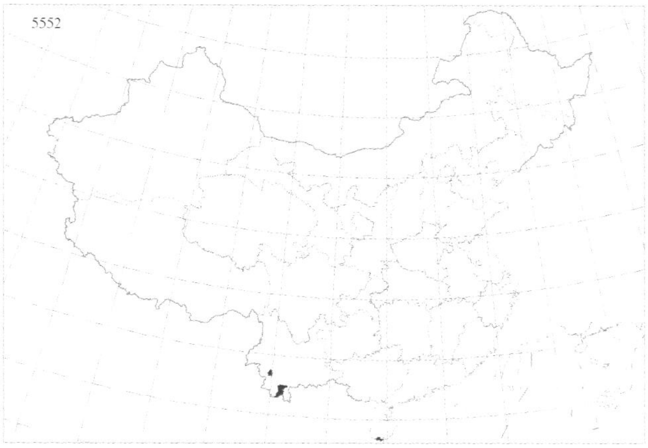

Celastraceae 卫矛科 *Maytenus* 美登木属
Maytenus hookeri Loes. 美登木
Shrubs

18.8 (17.1~20.5); 13.0 (11.0~15.1); 22.3 (20.8~23.7); 18.9 (17.1~20.5); 166.2 (145.0~185.9); 0.0 (0.0~0.0); 903 (815~992) ▎1445 (1352~1607); 743 (655~803); 56 (42~77) ▎60.3 (47.6~68.0); 885 (802~986); 538.0 (438.9~659.0)

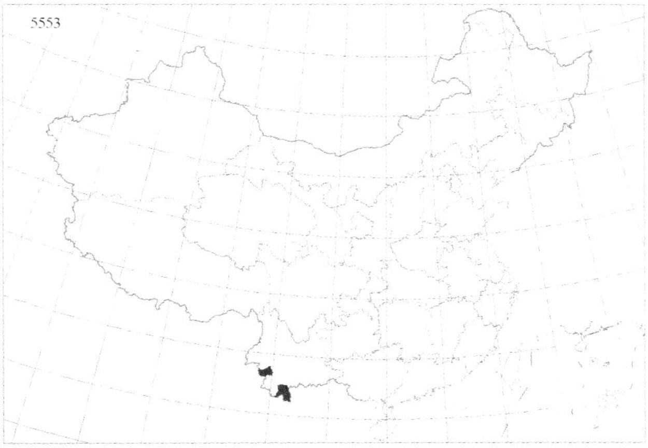

Celastraceae 卫矛科 *Maytenus* 美登木属
Maytenus inflata S. J. Pei et Y. H. Li 胀果美登木
Shrubs

20.0; 14.6; 23.2; 20.1; 180.8; 0.0; 963 ▎1523; 681; 67 ▎58.4; 942; 581.0

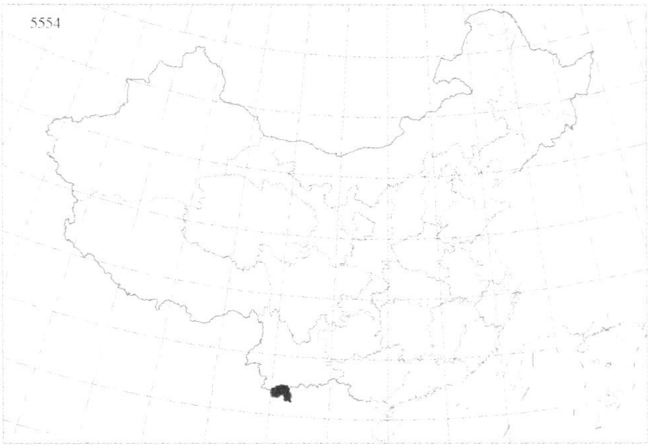

Celastraceae 卫矛科 *Microtropis* 假卫矛属
Microtropis biflora Merr. et Freem. 双花假卫矛
Shrubs

20.2 (15.5~22.5); 11.6 (4.8~15.0); 27.2 (24.6~28.7); 20.2 (15.6~22.6); 182.5 (127.2~210.9); 0.0 (0.0~0.2); 1057 (815~1206) ▎1770 (1293~2037); 829 (591~1033); 130 (58~187) ▎68.2 (43.5~93.2); 1057 (815~1206); 221.0 (139.6~316.1)

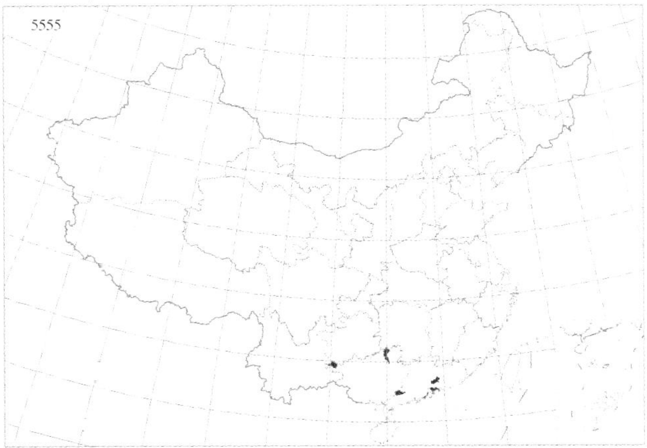

Celastraceae 卫矛科 *Microtropis* 假卫矛属
Microtropis discolor (Wall.) Wall. ex Arn. 异色假卫矛
Shrubs or small trees, evergreen

18.0 (8.6~21.0); 11.7 (1.3~15.1); 22.3 (15.0~25.6); 18.1 (8.7~21.0); 157.5 (52.6~191.8); 0.6 (0.0~8.8); 882 (574~1053) ▎1359 (1028~1607); 724 (566~911); 57 (44~77) ▎54.8 (12.5~81.5); 867 (574~1053); 468.0 (339.5~659.0)

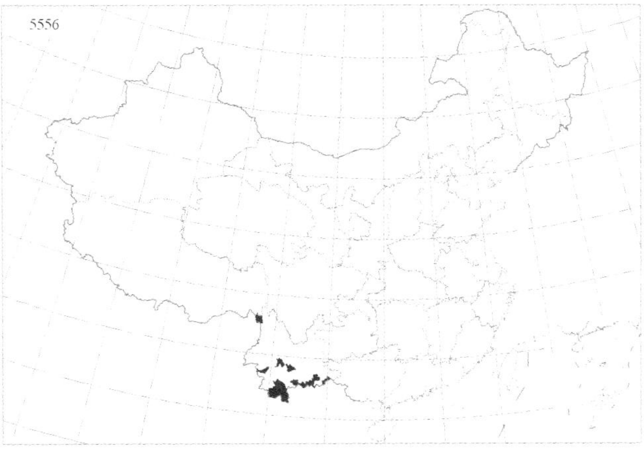

Celastraceae 卫矛科 *Microtropis* 假卫矛属
Microtropis fokienensis Dunn 福建假卫矛
Shrubs

17.9 (14.1~24.1); 7.7 (1.6~18.9); 27.1 (20.1~29.3); 17.9 (14.2~24.1); 155.4 (114.1~229.7); 0.5 (0.0~6.5); 944 (765~1322) ❙ 1759 (1190~3495); 662 (403~1847); 209 (47~897) ❙ 88.2 (25.2~246.5); 933 (765~1191); 237.4 (119.4~363.7)

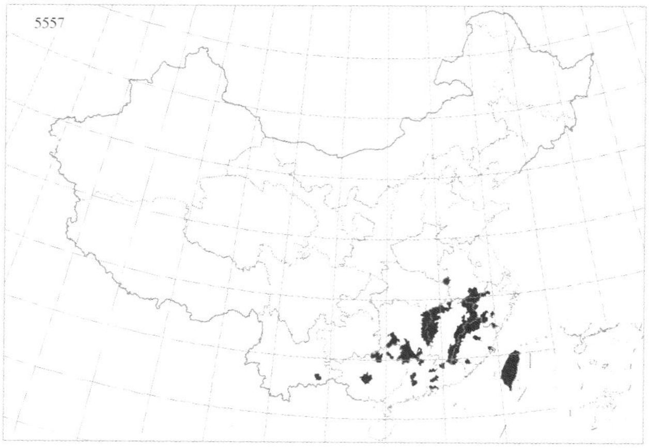

Celastraceae 卫矛科 *Microtropis* 假卫矛属
Microtropis gracilipes Merr. et F. P. Metcalf 密花假卫矛
Shrubs

18.4 (14.3~22.0); 8.7 (3.3~14.1); 26.9 (23.0~29.8); 18.5 (14.3~22.0); 161.5 (114.1~203.8); 0.1 (0.0~2.3); 958 (765~1170) ❙ 1567 (1182~2005); 606 (403~902); 163 (43~254) ❙ 64.3 (16.9~130.7); 954 (765~1170); 240.3 (151.0~354.5)

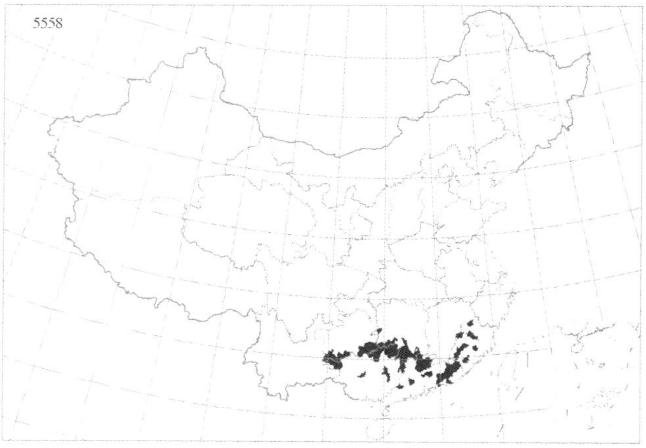

Celastraceae 卫矛科 *Microtropis* 假卫矛属
Microtropis henryi Merr. et Freem. 滇东假卫矛
Shrubs

18.6; 11.8; 23.2; 18.6; 162.6; 0.0; 907 ❙ 1356; 740; 59 ❙ 49.8; 907; 475.5

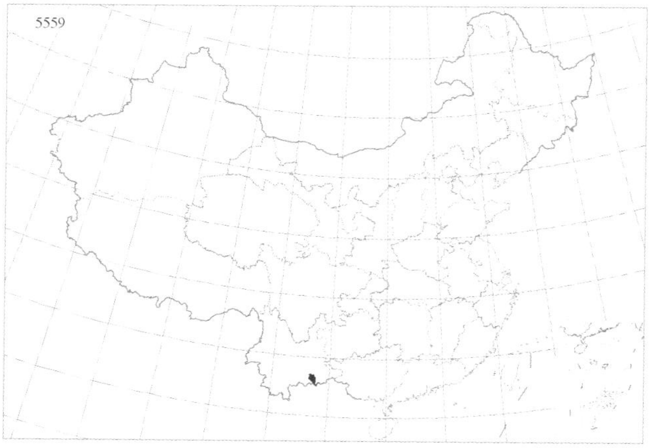

Celastraceae 卫矛科 *Microtropis* 假卫矛属
Microtropis hexandra Merr. et Freem. 六蕊假卫矛
Shrubs

19.3; 12.8; 23.9; 19.4; 172.6; 0.0; 947 ❙ 1388; 772; 59 ❙ 47.1; 944; 498.4

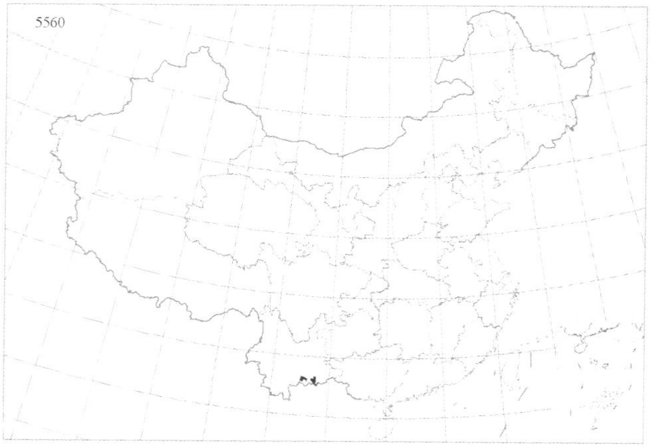

Celastraceae 卫矛科 *Microtropis* 假卫矛属
Microtropis japonica (Franch. et Sav.) Hall f. Meded. 日本假卫矛
Shrubs or small trees, evergreen

18.5; 14.1; 21.7; 18.5; 162.1; 0.0; 863 ❙ 2713; 1388; 188 ❙ 214.4; 863; 363.7

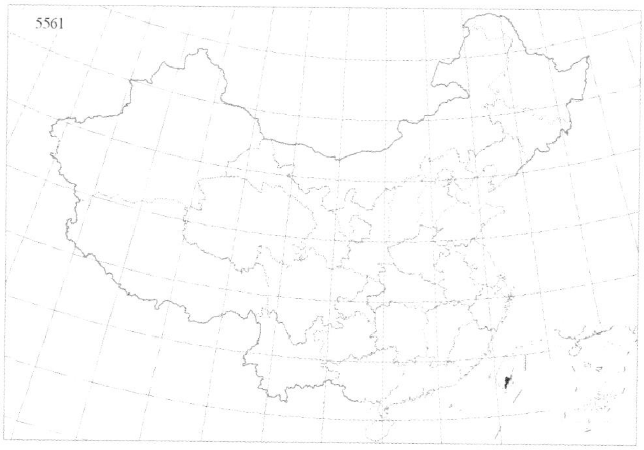

Celastraceae 卫矛科 *Microtropis* 假卫矛属
Microtropis macrophylla Merr. et Freem. 大叶假卫矛
Shrubs

15.5; 8.2; 21.1; 15.6; 126.9; 0.0; 780 ❙ 1552; 882; 55 ❙ 103.0; 780; 469.5

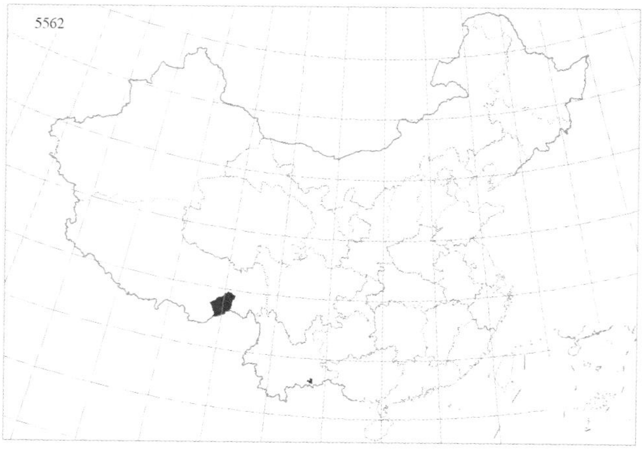

Celastraceae 卫矛科 *Microtropis* 假卫矛属
Microtropis obliquinervia Merr. et Freem. 斜脉假卫矛
Shrubs or small trees

16.9 (14.3~22.5); 6.3 (3.3~15.0); 26.4 (21.4~28.8); 16.9 (14.3~22.6); 143.7 (112.6~210.9); 0.4 (0.0~2.3); 891 (756~1206) ❙ 1413 (1190~1907); 565 (419~862); 130 (48~198) ❙ 59.4 (27.5~101.8); 889 (756~1206); 217.3 (139.6~347.3)

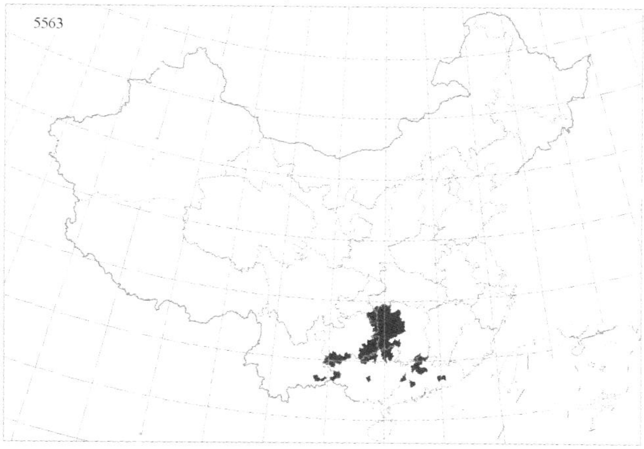

Celastraceae 卫矛科 *Microtropis* 假卫矛属
Microtropis obscurinervia Merr. et Freem. 隐脉假卫矛
Shrubs, evergreen

24.0 (22.3~24.8); 17.9 (14.7~20.1); 28.1 (27.2~28.7); 23.8 (22.4~24.8); 225.8 (208.3~237.6); 0.0 (0.0~0.0); 1301 (1188~1385) ❙ 1544 (1179~1965); 629 (392~994); 61 (41~104) ❙ 19.7 (−11.9~61.6); 1160 (1056~1216); 419.5 (204.4~620.0)

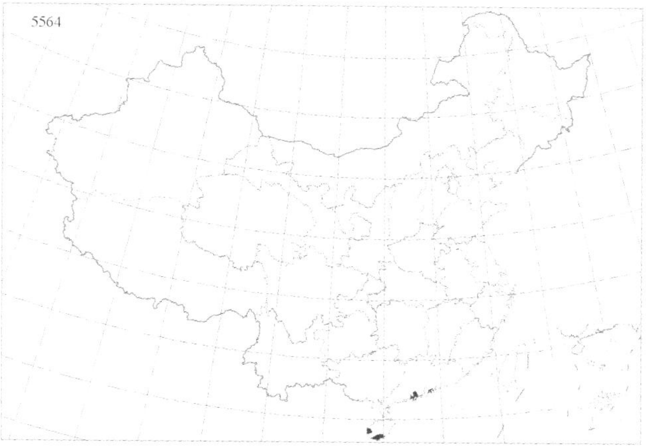

Celastraceae 卫矛科 *Microtropis* 假卫矛属
Microtropis oligantha Merr. et F. L. Freeman 逢春假卫矛
Shrubs

18.3; 12.0; 22.7; 18.4; 161.1; 0.0; 883 ❙ 1326; 757; 55 ❙ 50.6; 880; 458.8

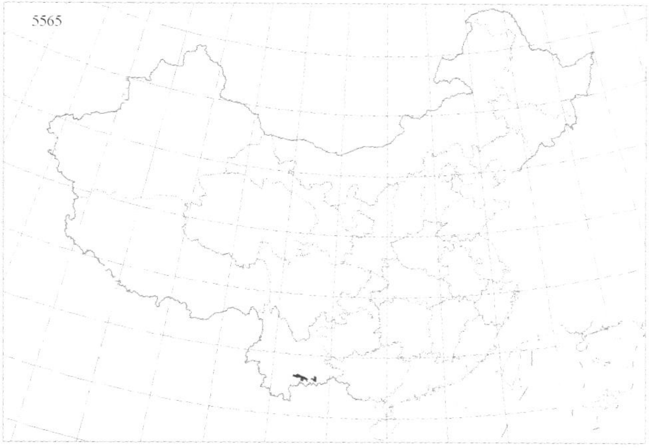

Celastraceae 卫矛科 *Microtropis* 假卫矛属
Microtropis osmanthoides (Hand.-Mazz.) Hand.-Mazz. 木樨假卫矛
Shrubs

22.0 (21.3~22.7); 14.1 (13.2~15.0); 27.9 (27.3~28.6); 22.0 (21.3~22.8); 204.0 (195.9~213.1); 0.0 (0.0~0.0); 1163 (1111~1221) ❙ 1838 (1302~2435); 1002 (670~1428); 111 (75~130) ❙ 57.8 (12.9~99.5); 1163 (1111~1221); 278.2 (215.3~318.8)

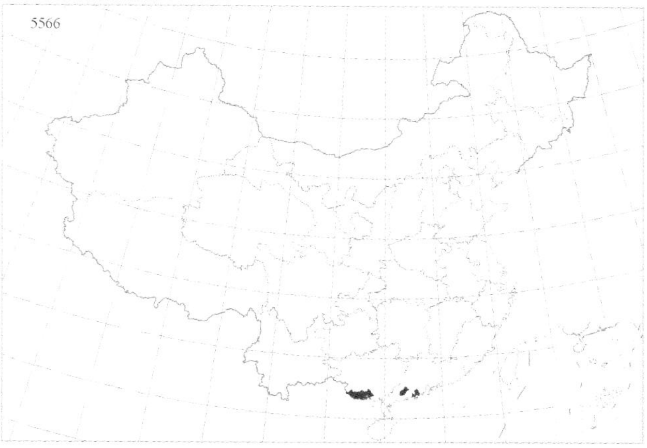

Celastraceae 卫矛科 *Microtropis* 假卫矛属
Microtropis paucinervia Merr. et Chun ex Merr. et Freem. 少脉假卫矛
Shrubs or small trees

22.0 (19.1~24.1); 14.4 (11.0~19.1); 27.7 (25.0~28.8); 22.1 (19.1~24.2); 204.7 (169.6~230.2); 0.0 (0.0~0.0); 1168 (945~1329) ❙ 1506 (1188~2435); 814 (480~1428); 93 (48~130) ❙ 38.0 (−1.4~99.5); 1156 (945~1329); 301.9 (139.6~620.0)

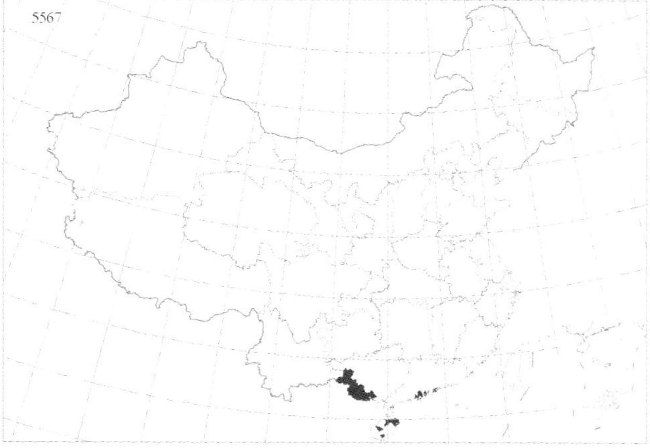

Celastraceae 卫矛科 *Microtropis* 假卫矛属
Microtropis petelotii Merr. et Freem. 广序假卫矛
Shrubs or trees

17.9 (16.1~21.0); 9.4 (5.3~14.2); 24.4 (21.4~25.8); 17.9 (16.1~21.0); 155.0 (133.1~191.8); 0.0 (0.0~0.0); 892 (813~1053) ❙ 1326 (1161~1522); 684 (523~835); 73 (48~129) ❙ 48.9 (30.5~65.6); 892 (813~1053); 340.1 (222.1~551.6)

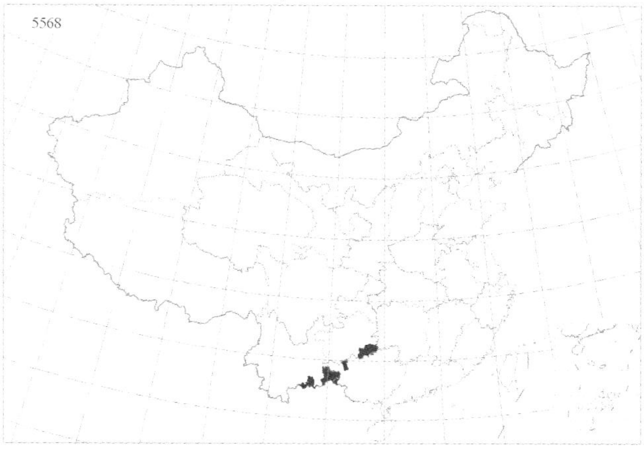

Celastraceae 卫矛科 *Microtropis* 假卫矛属
Microtropis pyramidalis C. Y. Cheng et T. C. Kao 塔蕾假卫矛
Small shrubs

19.6 (16.3~22.2); 12.1 (9.1~14.2); 25.0 (21.4~28.3); 19.6 (16.4~22.3); 175.8 (136.4~207.5); 0.0 (0.0~0.0); 992 (797~1190) ❙ 1465 (1190~1900); 803 (656~1056); 74 (50~125) ❙ 48.8 (15.4~71.0); 992 (797~1190); 397.8 (297.2~551.6)

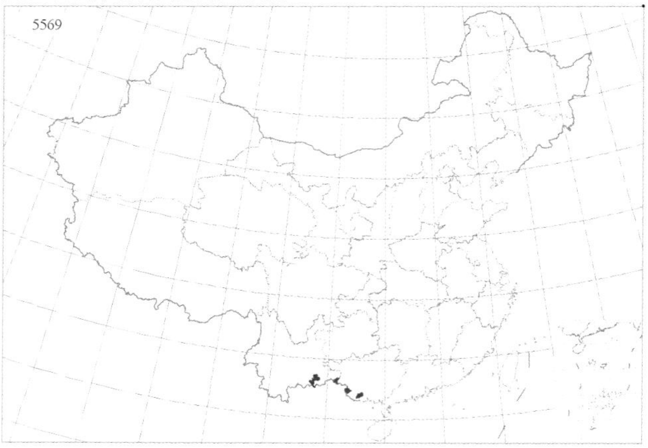

Celastraceae 卫矛科 *Microtropis* 假卫矛属
Microtropis reticulata Dunn 网脉假卫矛
Small shrubs

22.1 (21.6~22.5); 14.3 (12.6~15.2); 28.5 (28.2~28.7); 22.2 (21.6~22.7); 206.6 (199.2~212.7); 0.0 (0.0~0.0); 1182 (1139~1216) ❙ 1895 (1625~2037); 928 (795~1033); 128 (97~173) ❙ 60.3 (40.9~71.8); 1182 (1139~1216); 195.7 (139.6~240.0)

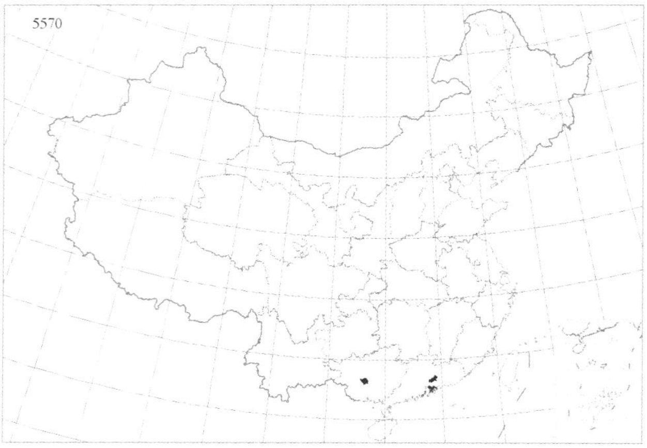

Celastraceae 卫矛科 *Microtropis* 假卫矛属
Microtropis semipaniculata C. Y. Cheng et T. C. Kao 复序假卫矛
Shrubs or small trees

19.1; 11.0; 25.0; 19.1; 169.6; 0.0; 945 ❙ 1438; 820; 57 ❙ 52.2; 945; 371.7

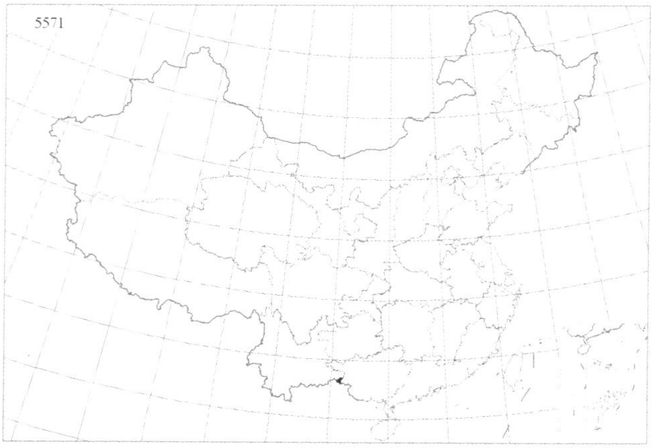

Celastraceae 卫矛科 *Microtropis* 假卫矛属
Microtropis submembranacea Merr. et Freem. 灵香假卫矛
Shrubs

19.3 (16.3~24.7); 12.3 (7.7~19.8); 24.4 (21.4~28.2); 19.3 (16.4~24.7); 171.6 (136.4~236.5); 0.0 (0.0~0.0); 979 (797~1376) ❙ 1436 (1007~1776); 673 (534~860); 74 (44~182) ❙ 48.9 (15.4~80.6); 958 (797~1294); 391.8 (232.5~572.9)

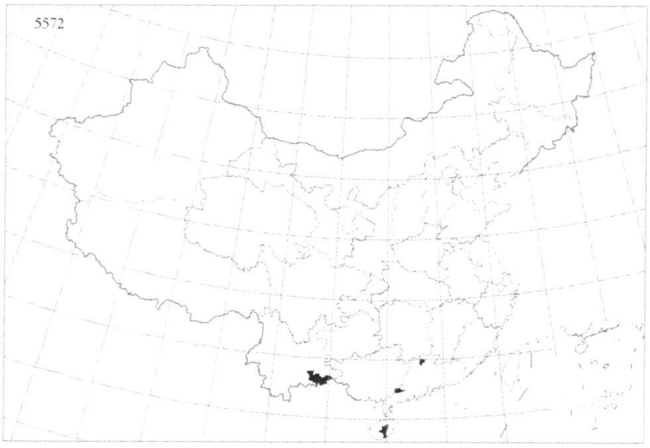

Celastraceae 卫矛科 *Microtropis* 假卫矛属
Microtropis tetragona Merr. et Freem. 方枝假卫矛
Shrubs or small trees

18.3 (8.6~21.3); 9.1 (1.3~13.8); 25.8 (15.0~28.6); 18.3 (8.7~21.3); 159.9 (52.6~196.1); 0.4 (0.0~8.8); 944 (574~1130) ❙ 1482 (1041~1758); 660 (523~911); 134 (45~189) ❙ 58.1 (22.7~85.5); 942 (574~1130); 281.6 (185.5~551.3)

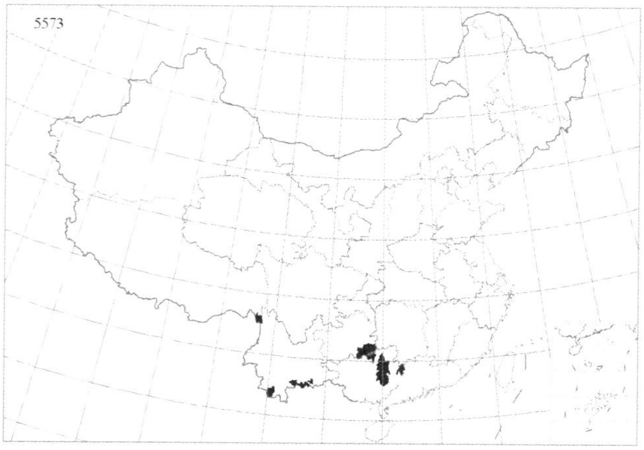

Celastraceae 卫矛科 *Microtropis* 假卫矛属
Microtropis thyrsiflora C. Y. Cheng et T. C. Kao 大序假卫矛
Shrubs or small trees

21.3; 12.2; 28.3; 21.3; 196.0; 0.0; 1131 ❙ 1636; 788; 178 ❙ 44.7; 1131; 219.4

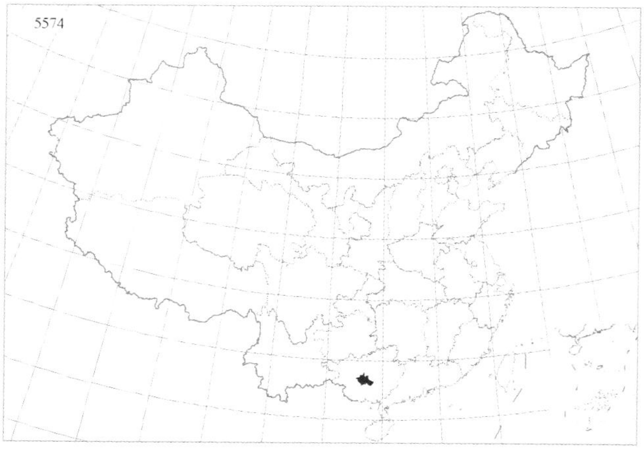

Celastraceae 卫矛科 *Microtropis* 假卫矛属

Microtropis triflora Merr. et Freem. 三花假卫矛

Shrubs

15.8 (11.7~22.8); 5.9 (2.1~14.7); 24.8 (19.6~28.9); 15.9 (11.7~22.8); 131.1 (84.6~213.6); 0.7 (0.0~5.6); 833 (661~1230) ❙ 1265 (942~1647); 583 (444~932); 76 (39~142) ❙ 52.7 (30.9~107.2); 833 (661~1230); 211.7 (146.0~491.5)

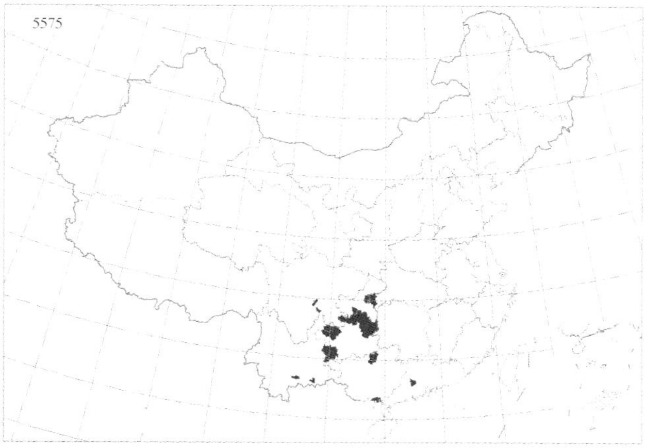

Celastraceae 卫矛科 *Microtropis* 假卫矛属

Microtropis yunnanensis (Hu) C. Y. Cheng et T. C. Kao 云南假卫矛

Shrubs or small trees

17.9 (15.6~21.0); 9.7 (5.5~14.2); 24.1 (21.4~26.4); 17.9 (15.6~21.0); 154.9 (127.5~191.8); 0.0 (0.0~0.0); 890 (797~1053) ❙ 1310 (1161~1547); 714 (546~911); 58 (43~69) ❙ 47.9 (16.9~74.2); 890 (797~1053); 335.2 (189.4~551.6)

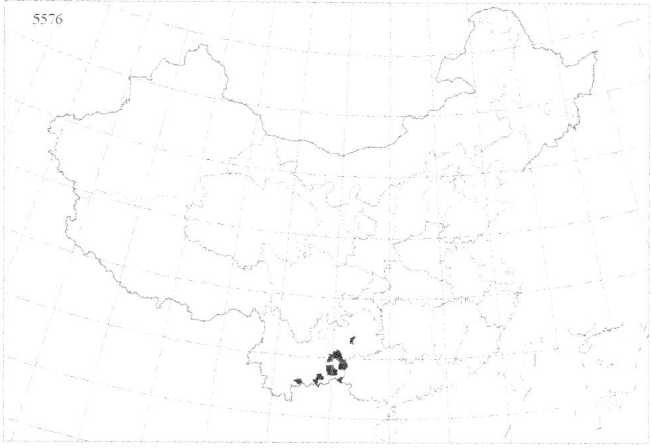

Celastraceae 卫矛科 *Monimopetalum* 永瓣藤属

Monimopetalu m chinense Rehder 永瓣藤

Scandent shrubs, deciduous

16.6 (15.5~17.8); 4.6 (3.3~6.2); 28.1 (27.0~29.1); 16.7 (15.5~17.9); 140.6 (127.5~154.6); 0.6 (0.0~1.7); 905 (841~961) ❙ 1529 (931~1828); 514 (323~601) 198 (113~253) ❙ 69.1 (-1.6~102.5); 884 (790~961); 231.5 (198.4~273.6)

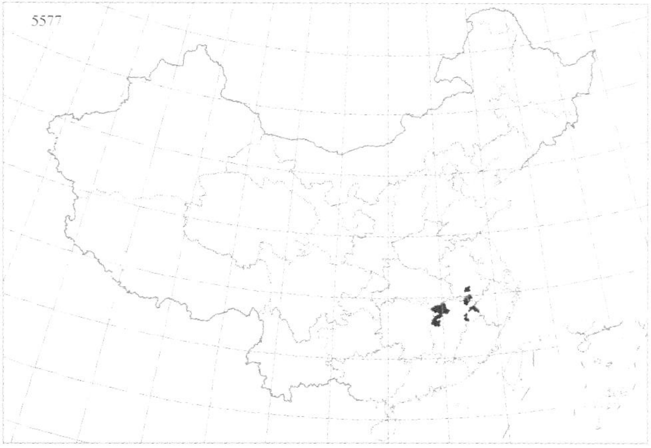

Celastraceae 卫矛科 *Pleurostylia* 盾柱属

Pleurostylia opposita (Wall.) Alston 盾柱

Shrubs or small trees, evergreen

24.2 (23.6~24.8); 18.7 (17.8~20.1); 28.1 (27.5~28.7); 24.2 (23.6~24.8); 231.0 (223.3~237.6); 0.0 (0.0~0.0); 1331 (1271~1385) ❙ 1263 (868~1588); 437 (312~534); 52 (41~71) ❙ -5.3 (-34.3~15.4); 1088 (868~1212); 438.9 (288.3~527.5)

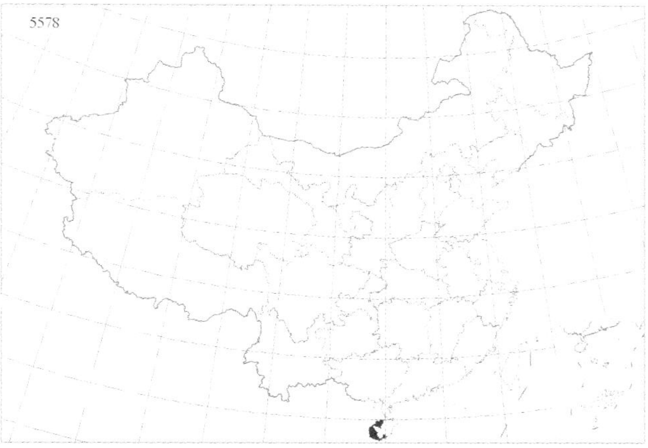

Celastraceae 卫矛科 *Pristimera* 扁蒴藤属

Pristimera arborea (Roxb.) A. C. Sm. 二籽扁蒴藤

Vines

20.8 (19.2~22.2); 13.6 (11.0~15.1); 26.0 (22.3~28.6); 20.9 (19.2~22.3); 190.6 (17C.5~207.5); 0.0 (0.0~0.0); 1063 (909~1190) ❙ 1506 (1374~1607, 745 (655~850); 94 (58~145) ❙ 43.3 (15.4~64.9); 1055 (880~1190); 393.4 (208.5~659.0)

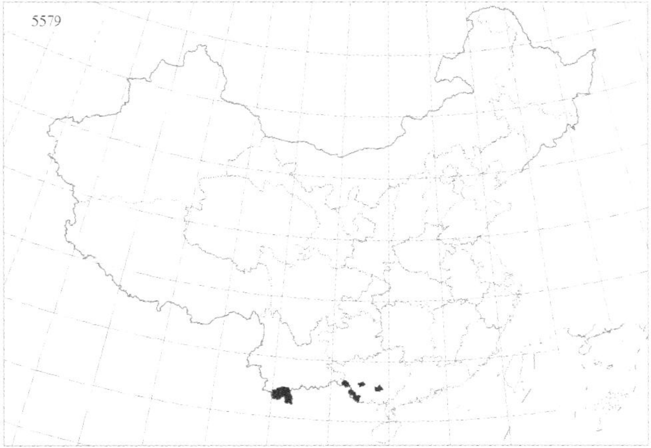

Celastraceae 卫矛科 *Pristimera* 扁蒴藤属

Pristimera cambodiana (Pierre) A. C. Sm. 风车果

Vines

20.1 (16.3~22.7); 13.3 (10.2~15.1); 24.8 (20.3~28.6); 20.2 (16.4~22.8); 182.1 (136.6~213.1); 0.0 (0.0~0.0); 1015 (789~1221) ❙ 1586 (1143~2435); 839 (637~1428); 83 (47~130) ❙ 56.6 (15.4~99.5); 1003 (780~1221); 400.3 (256.6~659.0)

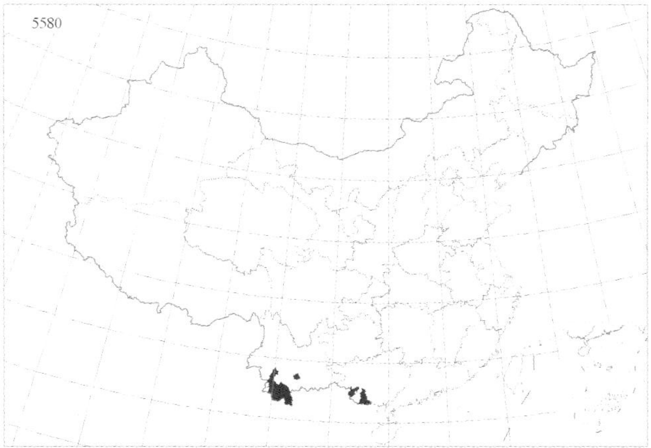

Celastraceae 卫矛科 *Pristimera* 扁蒴藤属
Pristimera indica (Willd.) A. C. Sm. 扁蒴藤
Vines

24.0 (23.5~24.8); 18.0 (16.4~20.1); 28.3 (27.5~29.0); 24.0 (23.5~24.8); 228.5 (222.3~237.6); 0.0 (0.0~0.0); 1314 (1271~1385) ❙ 1224 (868~1405); 455 (312~551); 62 (41~88) ❙ -6.8 (-34.3~6.7); 1124 (868~1269); 376.6 (225.6~527.5)

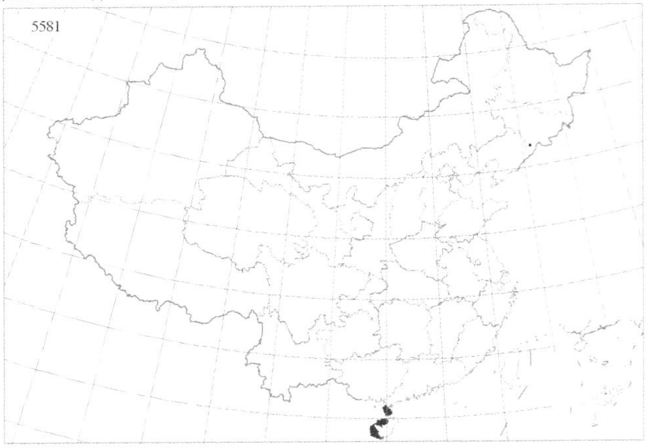

Celastraceae 卫矛科 *Pristimera* 扁蒴藤属
Pristimera setulosa A. C. Sm. 毛扁蒴藤
Vines

19.9 (17.3~22.2); 13.1 (9.9~15.1); 24.6 (22.3~28.5); 20.0 (17.3~22.3); 179.8 (147.1~207.5); 0.0 (0.0~0.0); 993 (833~1190) ❙ 1507 (1368~1751); 796 (655~996); 76 (55~128) ❙ 53.8 (15.4~88.8); 988 (833~1190); 455.6 (256.6~659.0)

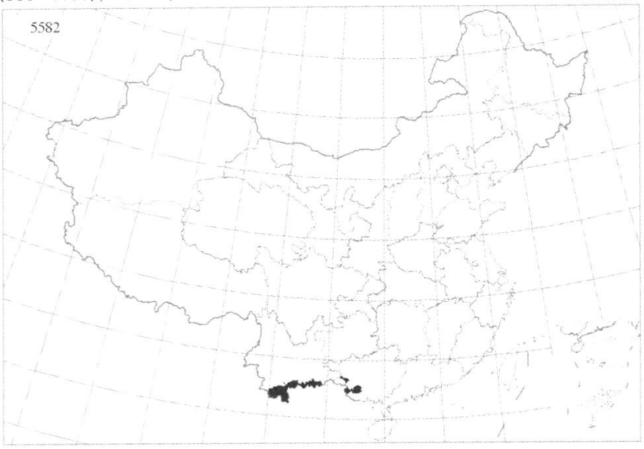

Celastraceae 卫矛科 *Salacia* 五层龙属
Salacia amplifolia Merr. 阔叶五层龙
Shrubs

21.3; 15.5; 25.3; 21.3; 195.5; 0.0; 1120 ❙ 1386; 615; 50 ❙ 30.8; 996; 509.5

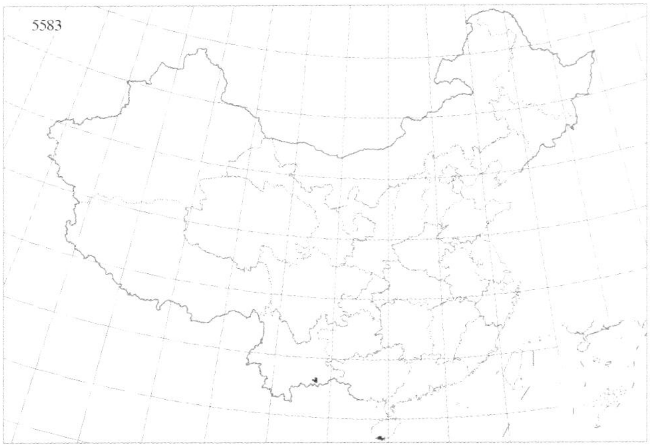

Celastraceae 卫矛科 *Salacia* 五层龙属
Salacia aurantiaca C. Y. Wu 橙果五层龙
Scandent shrubs

19.6 (17.3~21.0); 13.5 (9.9~15.1); 23.5 (22.3~25.6); 19.6 (17.3~21.0); 175.2 (147.1~191.8); 0.0 (0.0~0.0); 949 (833~1053) ❙ 1504 (1428~1607); 759 (655~860); 63 (55~77) ❙ 59.3 (44.5~80.6); 938 (833~1053); 535.7 (413.5~659.0)

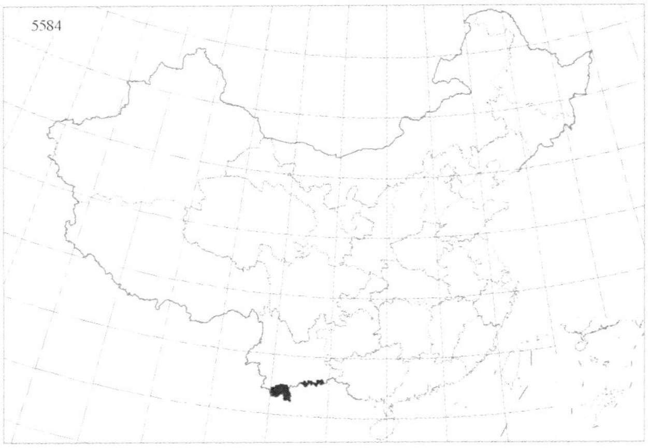

Celastraceae 卫矛科 *Salacia* 五层龙属
Salacia chinensis L. 五层龙
Scandent shrubs

23.3 (19.2~24.8); 16.7 (9.2~20.1); 28.3 (27.2~29.2); 23.4 (19.2~24.8); 220.2 (170.8~237.6); 0.0 (0.0~0.0); 1268 (1002~1385) ❙ 1364 (868~1723); 533 (312~714); 79 (41~171) ❙ 8.7 (-34.3~49.1); 1127 (868~1326); 364.8 (178.7~620.0)

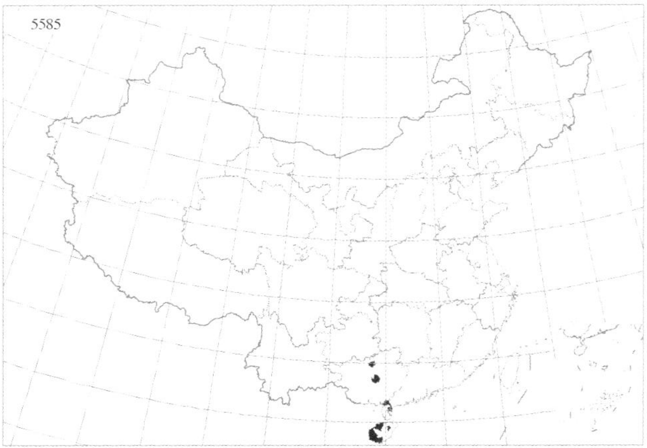

Celastraceae 卫矛科 *Salacia* 五层龙属
Salacia cochinchinensis Lour. 柳叶五层龙
Shrubs

20.0 (19.1~21.0); 14.2 (13.0~15.1); 23.7 (22.3~25.6); 20.1 (19.1~21.0); 180.8 (169.7~191.8); 0.0 (0.0~0.0); 972 (909~1053) ❙ 1504 (1428~1607); 739 (655~835); 65 (56~77) ❙ 55.1 (44.5~64.9); 959 (880~1053); 560.1 (506.2~659.0)

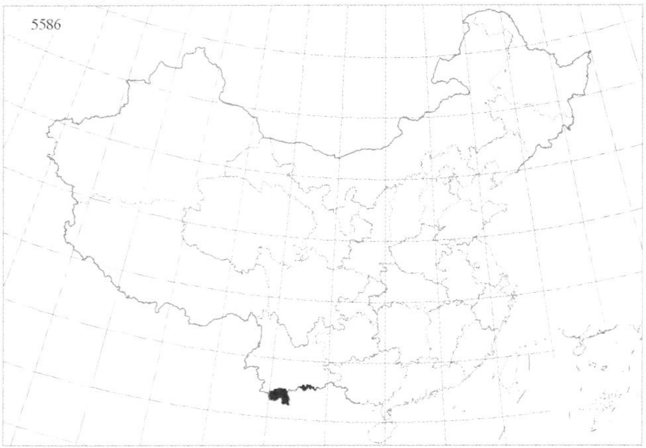

Celastraceae 卫矛科 *Salacia* 五层龙属
Salacia confertiflora Merr. 密花五层龙
Scandent shrubs

24.4; 19.6; 27.6; 24.4; 232.9; 0.0; 1343 ❙ 1478; 506; 45 ❙ 10.3; 1142; 573.8

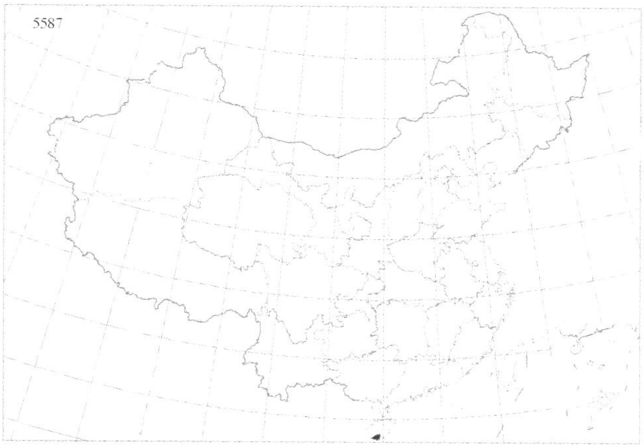

Celastraceae 卫矛科 *Salacia* 五层龙属
Salacia glaucifolia C. Y. Wu 粉叶五层龙
Scandent shrubs

18.2; 11.8; 22.5; 18.2; 158.6; 0.0; 871 ❙ 1457; 834; 61 ❙ 67.2; 871; 521.4

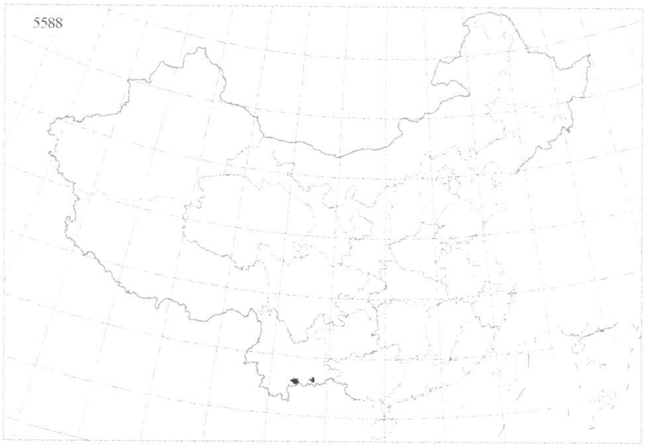

Celastraceae 卫矛科 *Salacia* 五层龙属
Salacia hainanensis Chun et F. C. How 海南五层龙
Scandent shrubs

24.0; 19.1; 27.2; 24.0; 228.1; 0.0; 1302 ❙ 1551; 539; 48 ❙ 19.1; 1147; 620.0

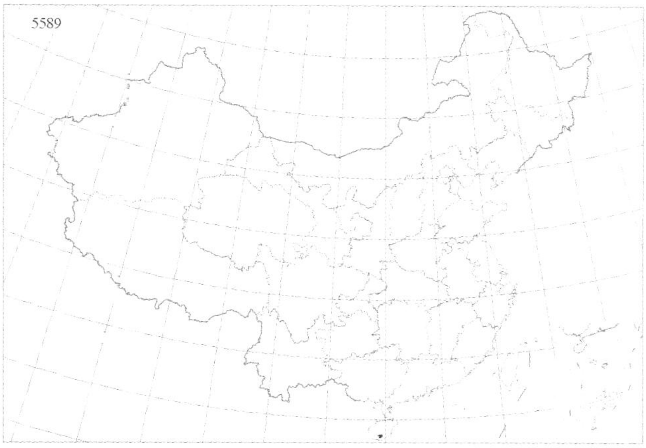

Celastraceae 卫矛科 *Salacia* 五层龙属
Salacia obovatilimba S. Y. Pao 河口五层龙
Shrubs

19.6; 13.1; 23.9; 19.6; 174.9; 0.0; 958 ❙ 1544; 793; 67 ❙ 62.5; 957; 541.4

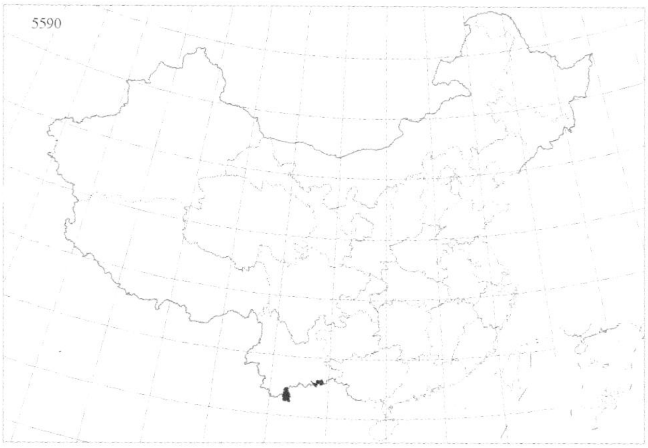

Celastraceae 卫矛科 *Salacia* 五层龙属
Salacia polysperma Hu 多籽五层龙
Scandent shrubs

19.2 (14.4~21.6); 12.6 (7.5~15.1); 23.6 (19.1~28.5); 19.2 (14.4~21.7); 170.6 (113.1~199.8); 0.0 (0.0~0.0); 951 (727~1153) ❙ 1525 (1454~1625); 758 (655~860); 81 (55~173) ❙ 63.0 (40.9~99.9); 942 (727~1153); 490.4 (237.8~659.0)

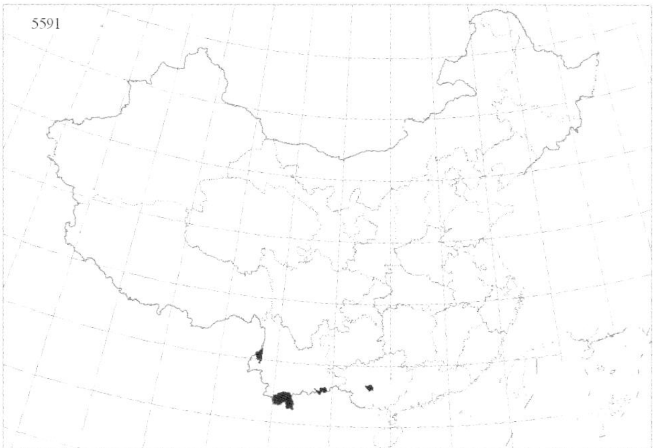

Celastraceae 卫矛科 *Salacia* 五层龙属
Salacia sessiliflora Hand.-Mazz. 无柄五层龙
Shrubs

18.6 (14.3~22.7); 9.5 (3.3~15.0); 25.9 (21.4~28.7); 18.7 (14.3~22.8); 163.9 (114.1~213.1); 0.1 (0.0~2.3); 958 (765~1221) ❙ 1492 (1161~2435); 717 (497~1428); 113 (48~196) ❙ 56.7 (12.9~101.8); 958 (765~1221); 277.7 (189.5~551.3)

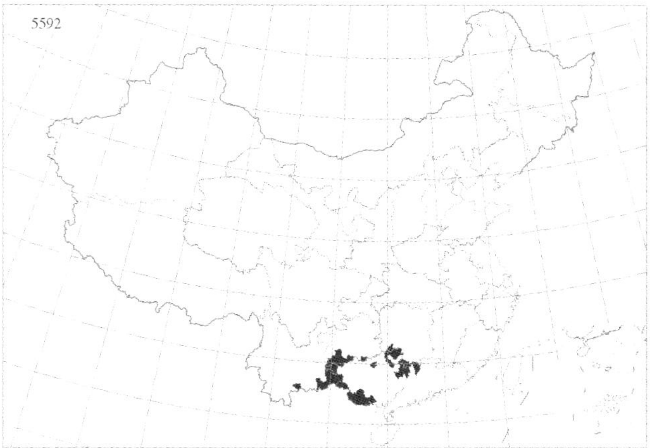

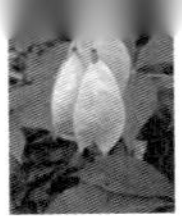

S

Celastraceae 卫矛科 *Tripterygium* 雷公藤属
Tripterygium wilfordii Hook. f. 雷公藤
Scandent shrubs, deciduous

15.6 (-0.2~21.1); 4.5 (-23.8~14.6); 25.6 (12.3~29.9); 15.9 (5.8~21.1); 132.5 (31.5~193.1); 5.0 (0.0~112.6); 851 (481~1114) ❙ 1368 (561~3495); 543 (323~924); 138 (15~897) ❙ 60.1 (-14.2~224.4); 839 (481~1114); 237.5 (134.1~551.3)

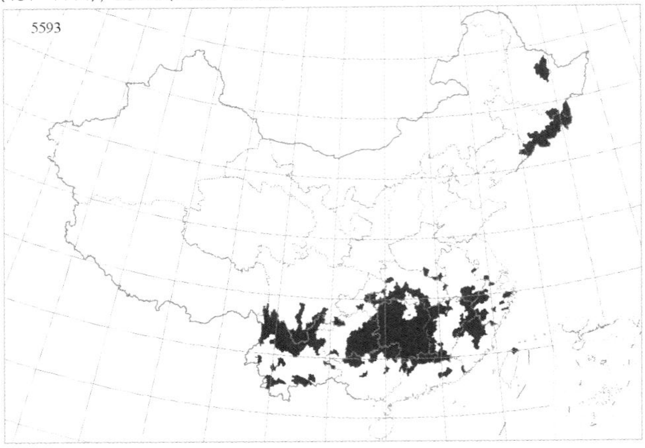

Salvadoraceae 刺茉莉科 *Azima* 刺茉莉属
Azima sarmentosa (Blume) Benth. et Hook. 刺茉莉
Shrubs

24.0 (22.2~24.8); 18.4 (17.2~20.1); 28.0 (25.7~28.8); 24.0 (22.3~24.8); 228.4 (207.2~237.6); 0.0 (0.0~0.0); 1310 (1127~1385) ❙ 1488 (1176~1813); 544 (444~712); 71 (41~113) ❙ 14.1 (-12.0~47.8); 1202 (1055~1331); 425.2 (248.2~620.0)

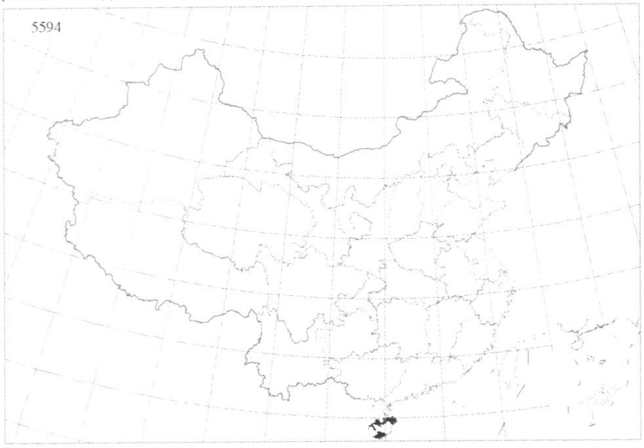

S

Staphyleaceae 省沽油科 *Euscaphis* 野鸦椿属
Euscaphis japonica (Thunb.) Dippel 野鸦椿
Shrubs or small trees, deciduous

16.4 (5.4~24.7); 5.6 (-7.1~19.8); 26.3 (14.2~30.1); 16.5 (6.4~24.7); 139.7 (38.1~236.5); 2.3 (0.0~40.6); 885 (504~1376) ❙ 1321 (525~3495); 550 (270~1428); 126 (10~897) ❙ 49.1 (-22.2~224.4); 875 (504~1360); 222.3 (108.2~620.0)

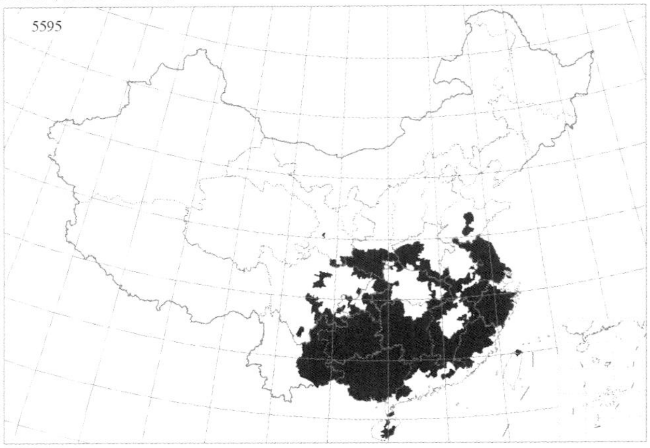

Staphyleaceae 省沽油科 *Staphylea* 省沽油属
Staphylea bumalda DC. 省沽油
Shrubs, deciduous

11.5 (3.3~18.2); -2.7 (-17.7~8.4); 24.4 (17.5~29.3); 12.3 (6.6~18.3); 100.3 (45.4~159.1); 22.0 (0.0~77.3); 735 (507~964) ❙ 811 (367~1950); 396 (235~932); 53 (8~227) ❙ 8.1 (-40.9~147.0); 651 (367~961); 191.3 (70.3~335.3)

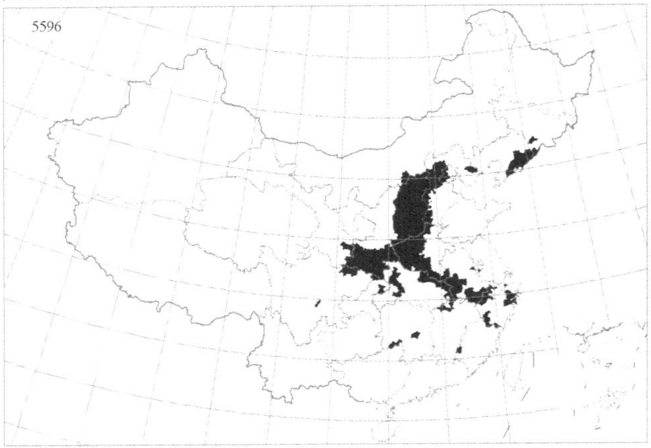

Staphyleaceae 省沽油科 *Staphylea* 省沽油属
Staphylea forrestii Balf. f. 嵩明省沽油
Trees

16.1 (13.6~18.9); 7.8 (4.4~12.5); 22.9 (19.0~27.5); 16.1 (13.6~19.0); 133.8 (103.6~167.5); 0.0 (0.0~0.6); 814 (707~916) ❙ 1172 (939~1564); 598 (524~816); 63 (26~182) ❙ 44.2 (6.5~80.4); 796 (696~916); 265.7 (172.1~383.4)

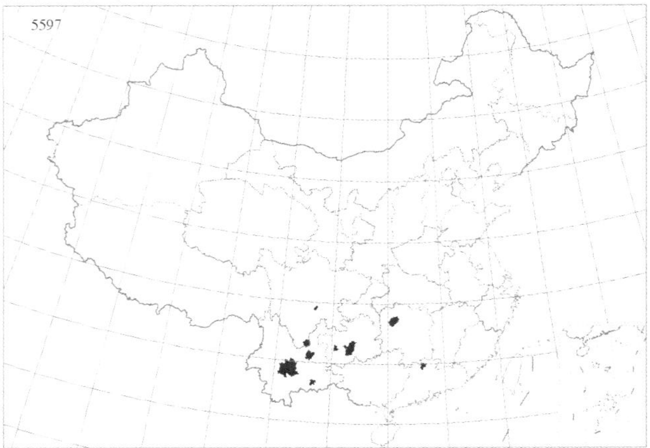

Staphyleaceae 省沽油科 *Staphylea* 省沽油属
Staphylea holocarpa Hemsl. 膀胱果
Shrubs or small trees, deciduous

13.1 (3.0~19.4); 1.1 (-9.8~10.1); 24.3 (12.2~29.3); 13.4 (4.8~19.4); 108.9 (25.1~173.4); 10.7 (0.0~49.3); 761 (441~1029) ❙ 1010 (189~1745); 456 (114~932); 66 (4~220) ❙ 31.1 (-70.2~108.0); 725 (189~1029); 218.1 (33.5~487.2)

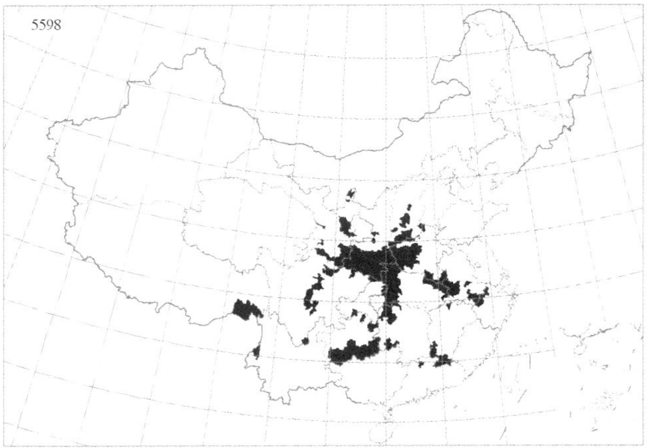

Staphyleaceae 省沽油科 *Staphylea* 省沽油属
Staphylea holocarpa var. rosea Rehder et E. H. Wilson 玫红省沽油
Shrubs or small trees, deciduous

11.4 (7.3~14.6); −0.6 (−2.9~2.9); 22.7 (15.6~26.8); 11.6 (7.5~14.7); 91.3 (48.2~122.7); 13.5 (3.4~20.9); 694 (537~834) ❙ 841 (622~1220); 411 (298~688); 35 (19~53) ❙ 22.9 (−13.4~87.3); 685 (537~817); 241.5 (151.3~323.3)

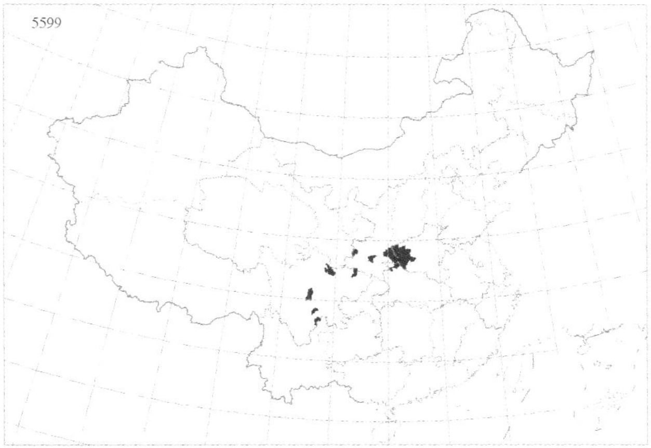

Staphyleaceae 省沽油科 *Staphylea* 省沽油属
Staphylea shweliensis W. W. Sm. 腺齿省沽油
Shrubs or trees

11.8; 5.0; 17.3; 11.9; 83.0; 0.5; 658 ❙ 1238; 691; 58 ❙ 87.9; 658; 377.1

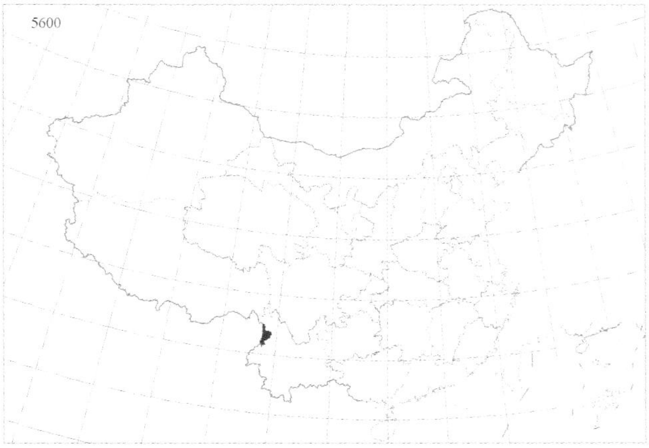

Staphyleaceae 省沽油科 *Staphylea* 省沽油属
Staphylea yuanjiangensis K. M. Feng et T. Z. Hsu 元江省沽油
Small trees, deciduous

19.3; 13.1; 23.5; 19.4; 173.2; 0.0; 939 ❙ 1143; 637; 50 ❙ 21.7; 903; 346.9

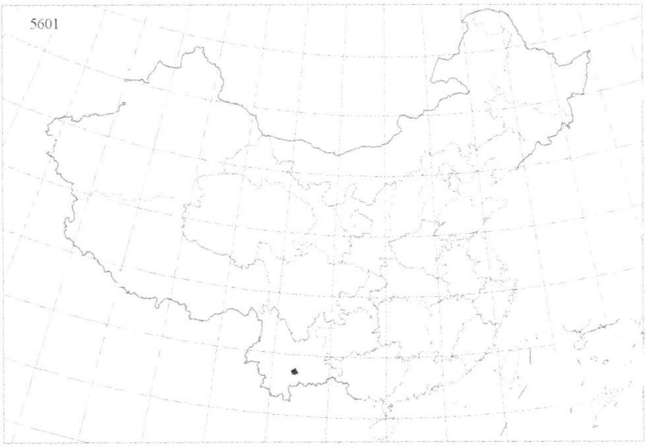

Staphyleaceae 省沽油科 *Turpinia* 山香圆属
Turpinia affinis Merr. et L. M. Perry 硬毛山香圆
Trees, evergreen

17.0 (8.6~22.7); 7.4 (−0.2~15.0); 25.5 (15.0~29.4); 17.1 (8.7~22.8); 145.6 (52.6~213.1); 0.7 (0.0~12.1); 890 (574~1221) ❙ 1338 (1001~2435); 603 (392~1428); 98 (29~209) ❙ 51.7 (5.8~107.2); 884 (574~1221); 253.9 (108.2~551.6)

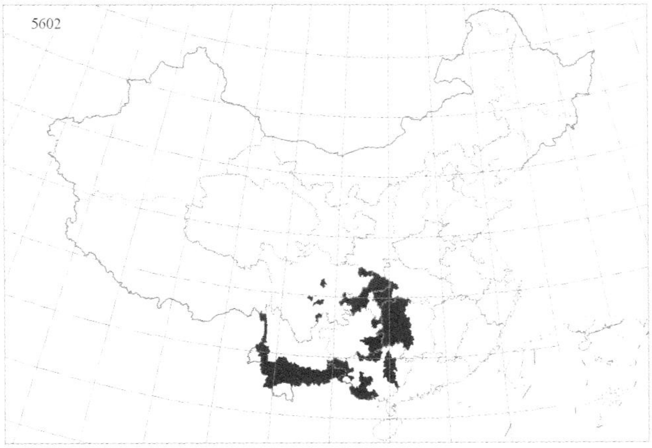

Staphyleaceae 省沽油科 *Turpinia* 山香圆属
Turpinia arguta Seem. 锐尖山香圆
Shrubs, deciduous

18.7 (10.6~23.6); 8.8 (−0.2~16.7); 27.4 (21.1~30.1); 18.7 (10.6~23.6); 164.7 (79.4~223.5); 0.3 (0.0~12.1); 986 (643~1281) ❙ 1545 (1070~2435); 609 (356~1428); 159 (43~259) ❙ 58.7 (−6.9~147.0); 979 (643~1273); 227.0 (95.3~411.1)

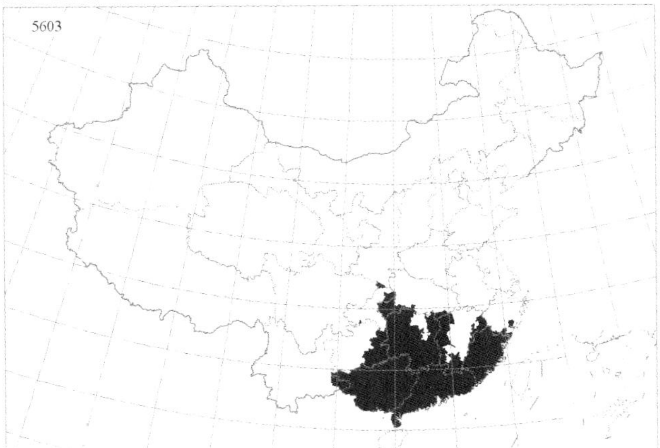

Staphyleaceae 省沽油科 *Turpinia* 山香圆属
Turpinia arguta var. pubescens T. Z. Hsu 绒毛锐尖山香圆
Shrubs, deciduous

18.4 (14.2~22.2); 8.6 (3.6~13.7); 27.1 (23.5~29.8); 18.5 (14.2~22.2); 162.0 (112.5~206.4); 0.1 (0.0~1.7); 969 (756~1190) ❙ 1562 (1175~2024); 601 (448~858); 157 (70~259) ❙ 61.5 (29.9~111.3); 967 (756~1190); 226.6 (152.9~279.5)

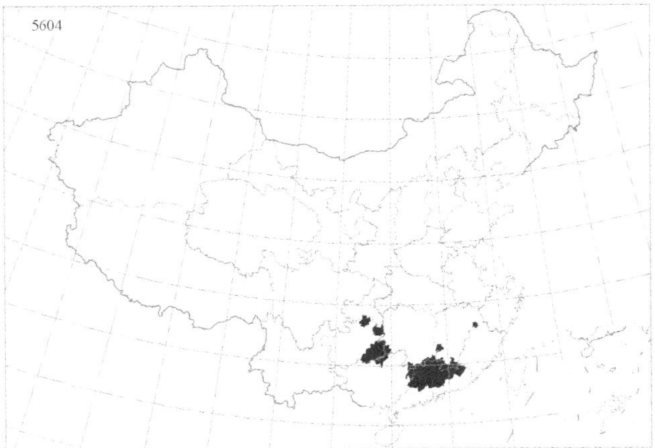

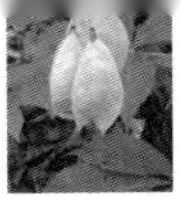

Staphyleaceae 省沽油科 *Turpinia* 山香圆属
Turpinia cochinchinensis (Lour.) Merr. 越南山香圆
Trees, deciduous

17.6 (7.3~23.2); 10.3 (-2.0~15.8); 23.2 (15.0~28.9); 17.7 (7.5~23.2); 153.1 (48.2~218.9); 0.5 (0.0~20.4); 886 (537~1257) ❙ 1373 (939~1965); 721 (454~996); 70 (27~174) ❙ 56.0 (6.5~112.2); 875 (537~1257); 363.3 (142.9~659.0)

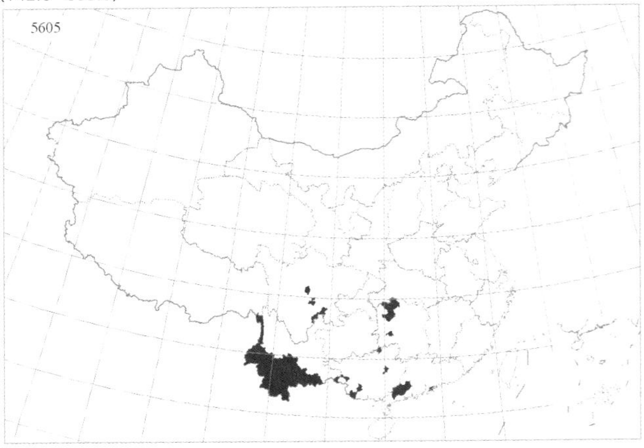

Staphyleaceae 省沽油科 *Turpinia* 山香圆属
Turpinia indochinensis Merr. 疏脉山香圆
Shrubs or small trees, evergreen

20.2 (16.2~24.0); 13.2 (9.1~19.1); 25.5 (20.1~28.6); 20.2 (16.2~24.0); 182.7 (134.2~228.1); 0.0 (0.0~0.0); 1034 (773~1302) ❙ 1899 (1074~3495); 920 (539~1428); 170 (47~897) ❙ 87.9 (12.9~246.5); 1017 (773~1221); 309.9 (119.4~620.0)

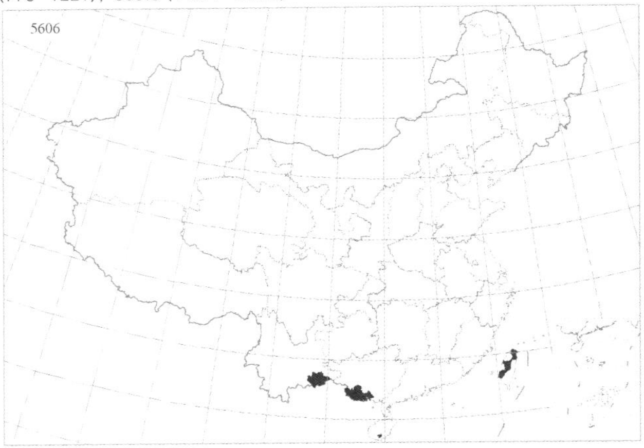

Staphyleaceae 省沽油科 *Turpinia* 山香圆属
Turpinia macrosperma C. C. Huang 大籽山香圆
Trees

11.0; 3.4; 17.3; 11.0; 76.5; 4.4; 640 ❙ 1389; 796; 48 ❙ 113.7; 640; 399.9

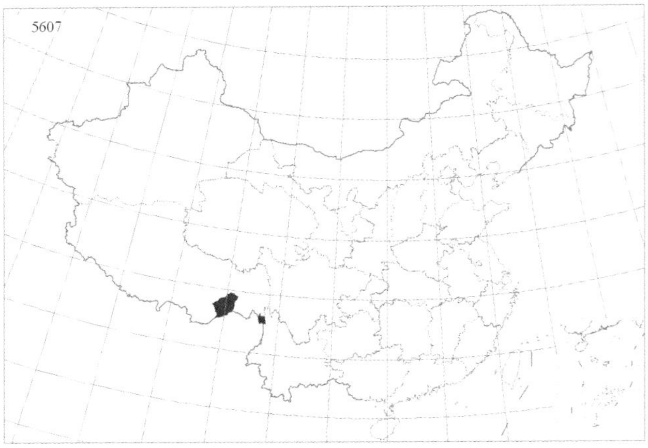

Staphyleaceae 省沽油科 *Turpinia* 山香圆属
Turpinia montana (Blume) Kurz 山香圆
Small trees

18.8 (14.3~24.8); 10.1 (3.5~20.1); 26.0 (20.3~29.2); 18.9 (14.3~24.8); 166.6 (112.6~237.6); 0.1 (0.0~1.5); 976 (756~1385) ❙ 1508 (868~2435); 667 (312~1428); 119 (41~259) ❙ 56.0 (-34.3~128.9); 959 (756~1257); 292.9 (139.6~659.0)

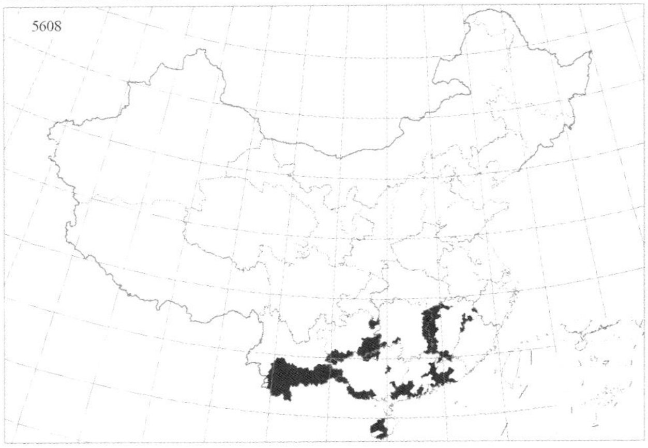

Staphyleaceae 省沽油科 *Turpinia* 山香圆属
Turpinia ovalifolia Elmer 卵叶山香圆
Trees

18.5; 14.1; 21.7; 18.5; 162.1; 0.0; 863 ❙ 2713; 1388; 188 ❙ 214.4; 863; 363.7

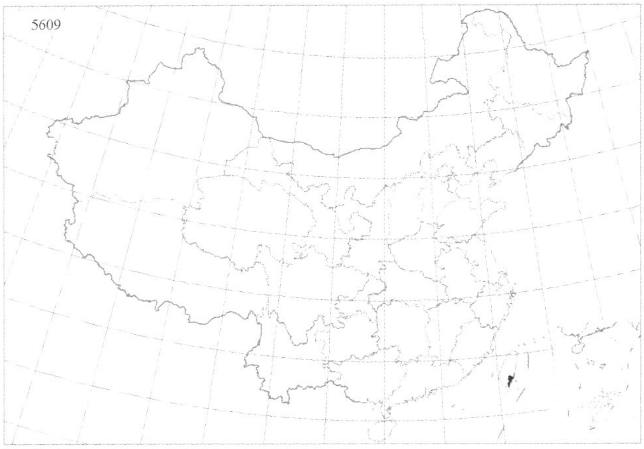

Staphyleaceae 省沽油科 *Turpinia* 山香圆属
Turpinia pomifera (Roxburgh) Candolle 大果山香圆
Trees

17.8 (8.6~22.2); 10.3 (1.3~15.1); 23.4 (15.0~28.6); 17.9 (8.7~22.2); 154.7 (52.6~206.5); 0.2 (0.0~8.8); 890 (574~1188) ❙ 1352 (939~1758); 701 (473~1007); 76 (37~189) ❙ 53.1 (5.8~145.8); 881 (574~1188); 362.6 (170.4~659.0)

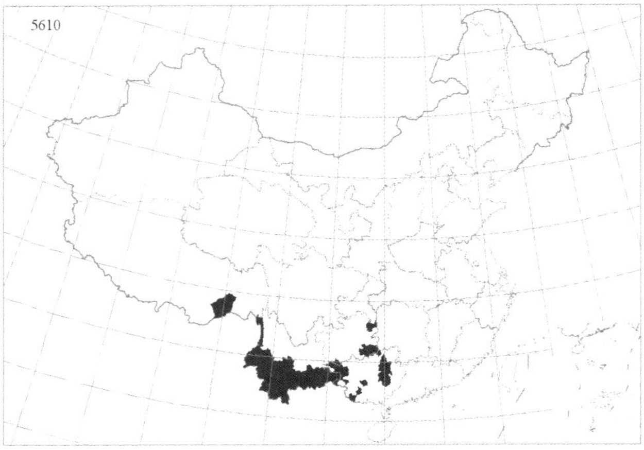

Staphyleaceae 省沽油科 *Turpinia* 山香圆属
Turpinia pomifera var. minor C. C. Huang 山麻风树
Trees

17.1 (8.6~21.0); 8.3 (1.3~15.1); 23.7 (15.0~26.6); 17.1 (8.7~21.0); 145.7 (52.6~191.8); 0.5 (0.0~8.8); 862 (574~1053) ❙ 1328 (1041~1751); 655 (452~996); 74 (45~129) ❙ 55.0 (27.5~93.0); 860 (574~1053); 348.6 (170.4~659.0)

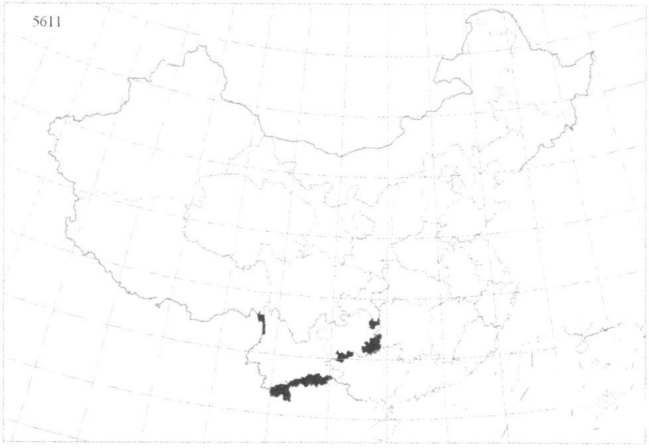

Staphyleaceae 省沽油科 *Turpinia* 山香圆属
Turpinia robusta Craib 粗壮山香圆
Trees

20.0; 14.6; 23.2; 20.1; 180.8; 0.0; 963 ❙ 1523; 681; 67 ❙ 58.4; 942; 581.0

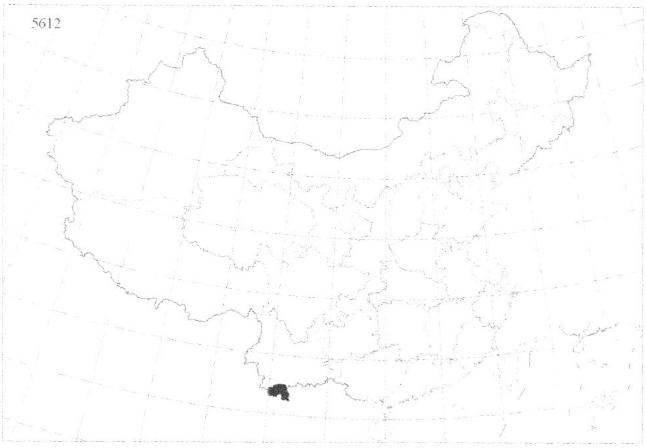

Staphyleaceae 省沽油科 *Turpinia* 山香圆属
Turpinia simplicifolia Merr. 亮叶山香圆
Shrubs or small trees

21.7 (17.6~24.8); 14.5 (7.3~20.1); 27.1 (25.2~28.1); 21.7 (17.6~24.8); 200.8 (151.6~237.6); 0.0 (0.0~0.0); 1148 (909~1385) ❙ 1625 (1405~2035); 750 (473~1056); 91 (41~142) ❙ 43.6 (1.5~71.8); 1098 (909~1185); 379.1 (225.9~620.0)

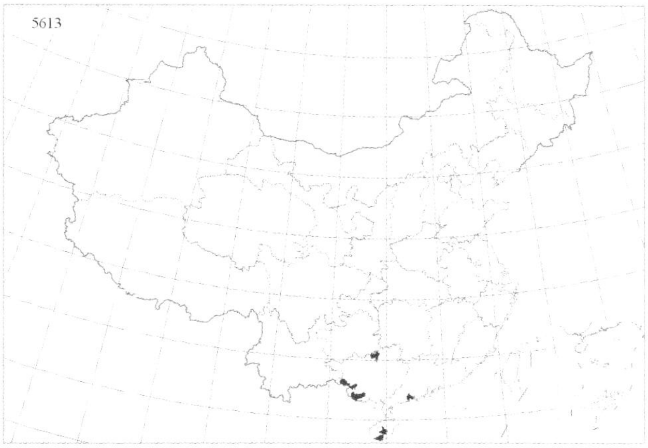

Staphyleaceae 省沽油科 *Turpinia* 山香圆属
Turpinia subsessilifolia C. Y. Wu 心叶山香圆
Shrubs or small trees

17.2; 9.5; 22.7; 17.3; 147.0; 0.0; 838 ❙ 1254; 699; 48 ❙ 49.7; 838; 336.3

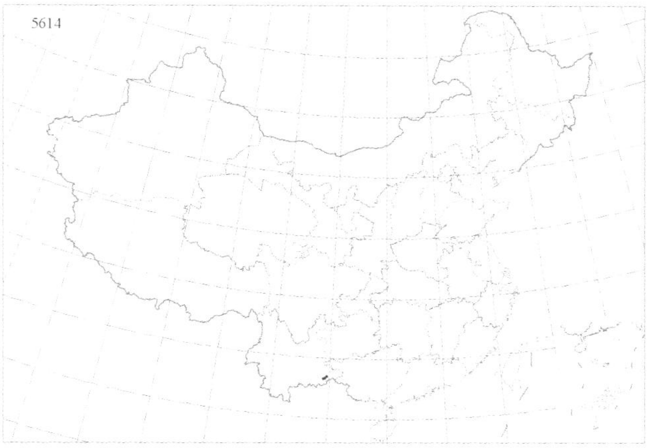

Icacinaceae 茶茱萸科 *Apodytes* 柴龙树属
Apodytes dimidiata E. Mey. ex Arn. 柴龙树
Shrubs or trees

21.9 (16.3~24.8); 15.3 (10.2~20.1); 26.5 (20.3~28.7); 21.9 (16.4~24.8); 203.2 (136.6~237.6); 0.0 (0.0~0.0); 1151 (789~1385) ❙ 1424 (868~1900); 644 (312~1056); 75 (41~173) ❙ 27.3 (-34.3~71.0); 1053 (780~1190); 413.7 (237.8~659.0)

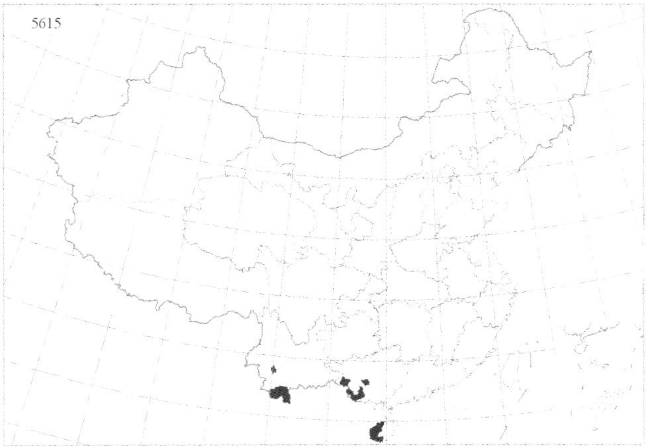

Icacinaceae 茶茱萸科 *Gomphandra* 须蕊木属
Gomphandra mollis Merr. 毛粗丝木
Shrubs or small trees

19.2 (17.2~21.6); 10.9 (7.4~14.2); 25.4 (22.4~28.2); 19.2 (17.3~21.6); 170.3 (147.0~199.6); 0.0 (0.0~0.0); 968 (833~1143) ❙ 1427 (1254~1567); 752 (637~860); 87 (48~141) ❙ 48.5 (27.0~80.6); 967 (833~1143); 344.9 (232.6~551.6)

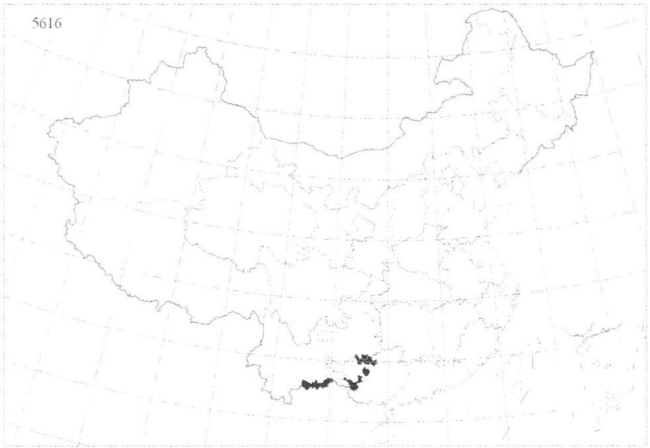

Icacinaceae 茶茱萸科 *Gomphandra* 须蕊木属
Gomphandra tetrandra (Wall.) Sleumer 粗丝木
Shrubs or small trees

19.0 (14.3~24.8); 10.9 (3.6~20.1); 25.4 (20.3~28.9); 19.1 (14.3~24.8); 169.0 (112.6~237.6); 0.0 (0.0~1.6); 977 (756~1385) ❙ 1441 (1161~2435); 694 (455~1428); 85 (41~189) ❙ 49.3 (1.5~99.5); 961 (756~1326); 340.0 (188.6~659.0)

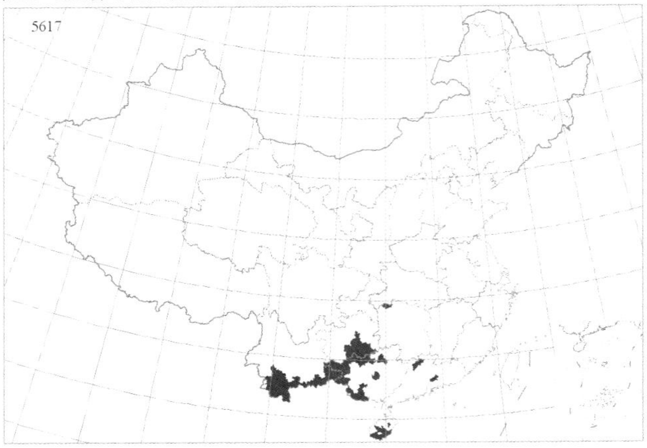

Icacinaceae 茶茱萸科 *Gonocaryum* 琼榄属
Gonocaryum calleryanum (Baill.) Becc. 台湾琼榄
Shrubs

20.2; 15.1; 24.3; 20.2; 182.6; 0.0; 991 ❙ 2899; 1386; 281 ❙ 194.8; 991; 307.3

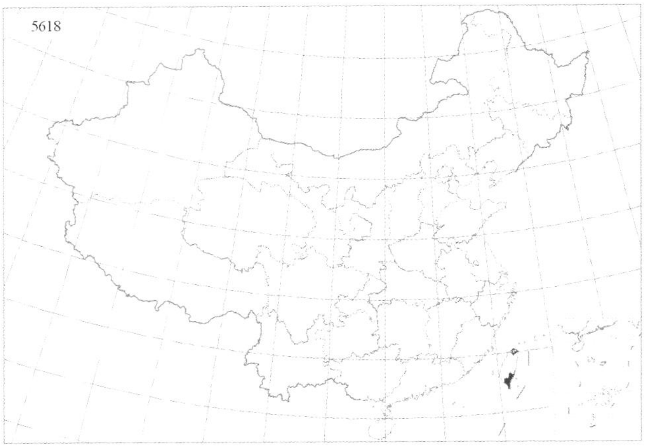

Icacinaceae 茶茱萸科 *Gonocaryum* 琼榄属
Gonocaryum lobbianum (Miers) Kurz 琼榄
Shrubs or small trees

21.8 (17.2~24.8); 15.9 (9.5~20.1); 26.0 (22.4~28.6); 21.9 (17.3~24.8); 202.8 (147.0~237.6); 0.0 (0.0~0.0); 1151 (833~1385) ❙ 1489 (1254~1767); 639 (455~860); 62 (41~110) ❙ 33.0 (-1.4~80.6); 1081 (833~1326); 472.0 (336.3~620.0)

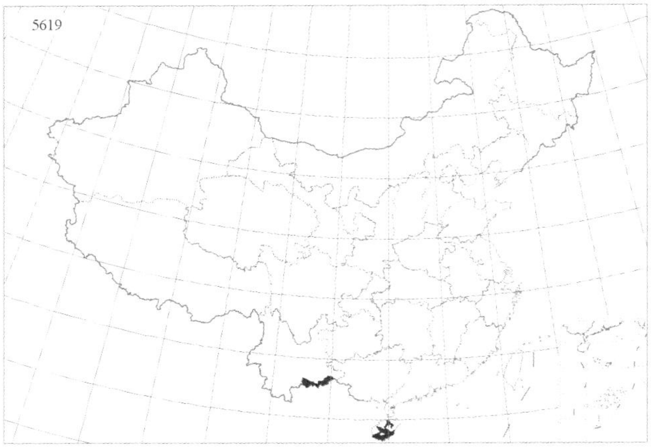

Icacinaceae 茶茱萸科 *Hosiea* 无须藤属
Hosiea sinensis (Oliv.) Hemsl. et E. H. Wilson 无须藤
Climbing shrubs

14.2 (9.3~16.9); 3.3 (-1.7~5.3); 24.7 (19.4~28.3); 14.3 (9.5~16.9); 114.7 (69.8~143.4); 3.2 (0.0~18.0); 782 (603~924) ❙ 1332 (1024~1950); 568 (445~817); 98 (29~223) ❙ 71.3 (19.6~147.0); 782 (603~924); 224.7 (164.0~292.6)

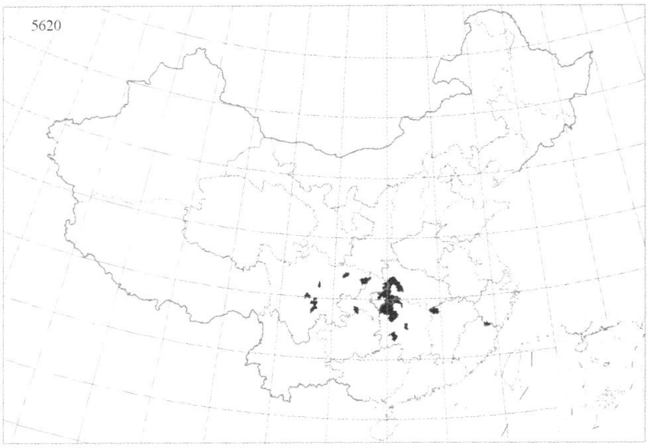

Icacinaceae 茶茱萸科 *Iodes* 微花藤属
Iodes balansae Gagnep. 大果微花藤
Vines

19.9 (14.4~22.7); 11.6 (5.2~15.0); 26.0 (19.1~29.1); 19.9 (14.4~22.8); 178.7 (113.1~213.1); 0.0 (0.0~0.0); 1021 (727~1221) ❙ 1482 (931~2435); 804 (323~1428); 87 (43~173) ❙ 45.9 (-1.6~99.9); 1016 (727~1221); 335.7 (198.4~551.6)

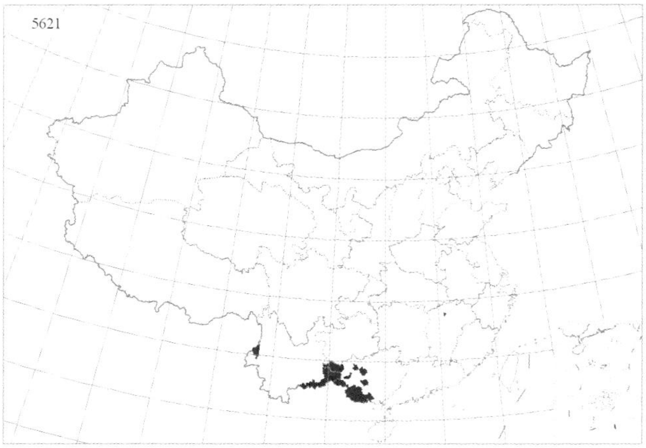

Icacinaceae 茶茱萸科 *Iodes* 微花藤属
Iodes cirrhosa Turcz. 微花藤
Vines

19.9 (18.1~22.8); 12.1 (9.4~15.1); 25.6 (22.1~28.9); 20.0 (18.1~22.8); 179.6 (157.3~213.6); 0.0 (0.0~0.0); 1008 (863~1230) ❙ 1410 (1172~1751); 747 (634~996); 76 (49~141) ❙ 40.9 (5.8~88.8); 1005 (863~1230); 373.4 (211.1~659.0)

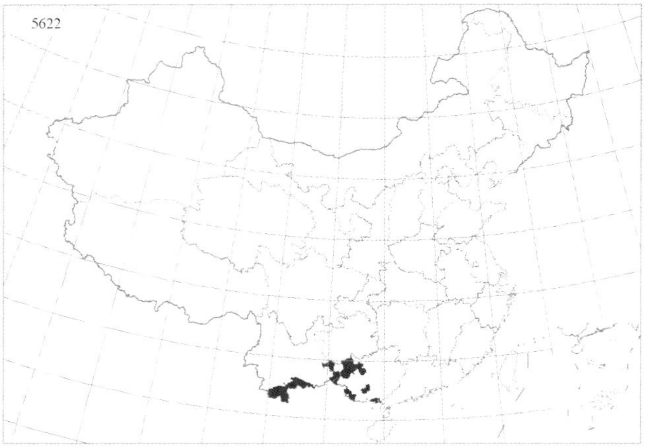

Icacinaceae 茶茱萸科 *Iodes* 微花藤属

Iodes seguinii (Levl.) Rehd. 瘤枝微花藤

Vines

18.8 (15.4~22.2); 0.3 (5.8~14.2); 25.3 (21.4~28.6); 18.8 (15.4~22.3); 165.9 (124.8~207.5); 0.0 (0.0~0.0); 955 (790~1190) ▌ 1346 (1007~1581); 715 (539~860); 73 (43~144) ▌ 42.3 (12.9~80.6); 954 (790~1190); 295.2 (171.2~551.6)

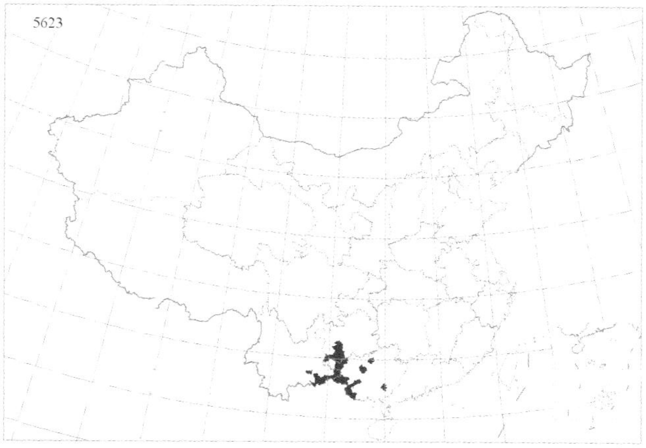

Icacinaceae 茶茱萸科 *Iodes* 微花藤属

Iodes vitiginea (Hance) Hemsl. 小果微花藤

Vines

20.4 (16.8~24.8); 12.6 (7.4~20.1); 26.4 (21.4~28.8); 20.5 (16.8~24.8); 185.9 (141.5~237.6); 0.0 (0.0~0.0); 1059 (813~1385) ▌ 1443 (1161~2435); 720 (444~1428); 85 (41~194) ▌ 37.9 (-12.0~99.5); 1039 (813~1269); 336.3 (188.6~659.0)

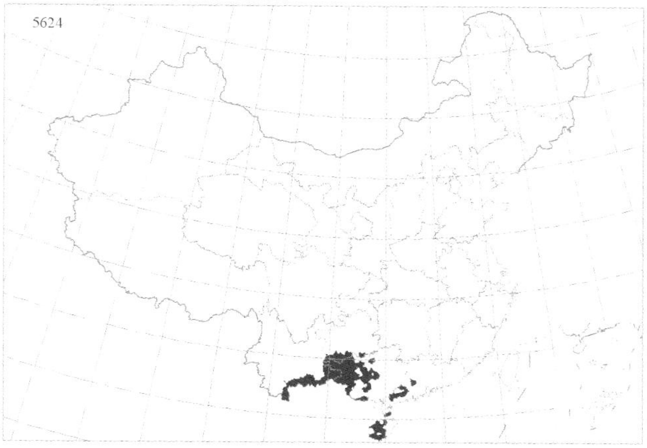

Icacinaceae 茶茱萸科 *Mappianthus* 定心藤属

Mappianthus icdoides Hand.-Mazz. 定心藤

Vines

19.3 (14.2~24.8); 10.9 (3.8~20.1); 26.2 (20.1~29.4); 19.4 (14.2~24.8); 172.6 (110.5~237.6); 0.0 (0.0~1.5); 995 (729~1385) ▌ 1535 (958~2435); 682 (435~1428) 126 (41~235) ▌ 55.1 (-2.2~115.7); 987 (729~1326); 293.2 (163.7~659.0)

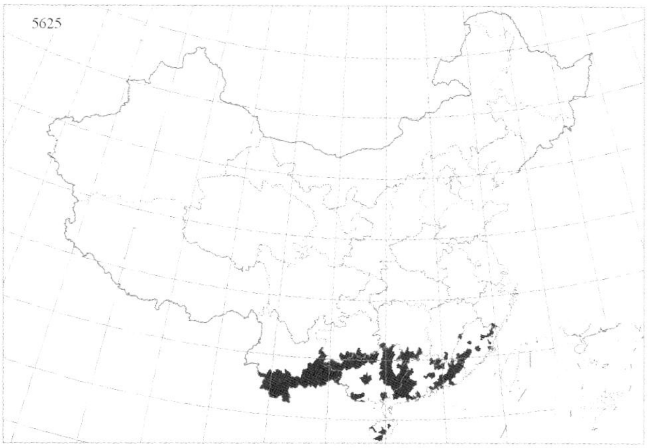

Icacinaceae 茶茱萸科 *Natsiatopsis* 麻核藤属

Natsiatopsis thunbergiifolia Kurz 麻核藤

Climbing shrubs

20.4; 15.1; 23.6; 20.5; 185.8; 0.0; 989 ▌ 1607; 702; 77 ▌ 62.5; 986; 659.0

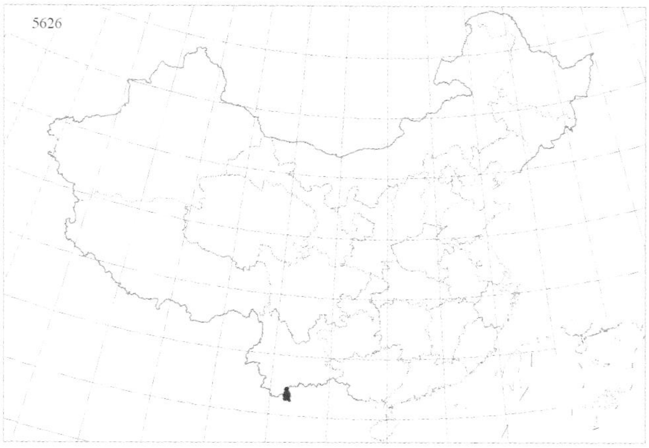

Icacinaceae 茶茱萸科 *Natsiatum* 薄核藤属

Natsiatum herpeticum Buch.-Ham. ex Arn. 薄核藤

Climbing shrubs

13.2; 6.4; 18.4; 13.3; 99.4; 0.0; 695 ▌ 1337; 727; 69 ▌ 92.4; 695; 407.1

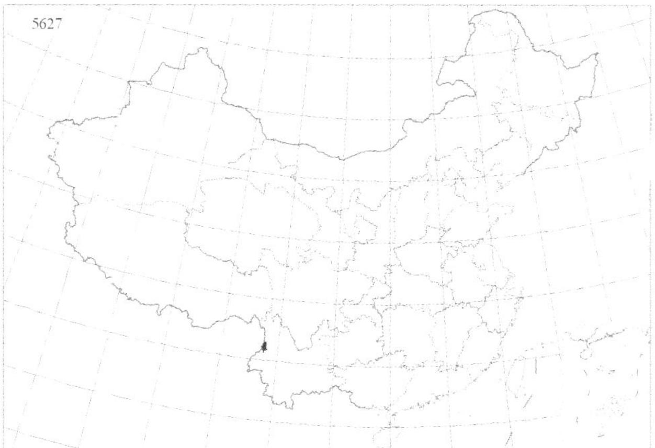

Icacinaceae 茶茱萸科 *Nothapodytes* 南柴龙树属

Nothapodytes collina C. Y. Wu 厚叶假柴龙树

Trees

20.0; 14.6; 23.2; 20.1; 180.8; 0.0; 963 ▌ 1523; 681; 67 ▌ 58.4; 942; 581.0

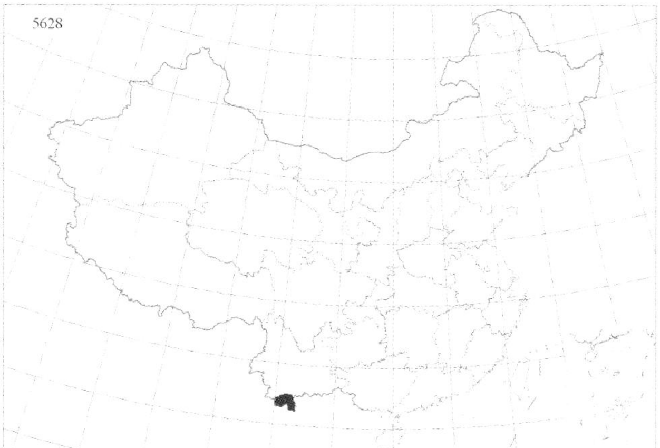

Icacinaceae 茶茱萸科 *Nothapodytes* 南柴龙树属
Nothapodytes nimmoniana (J. Graham) Mabb. 臭味假柴龙树
Trees

18.1; 11.0; 23.1; 18.1; 157.3; 0.0; 875 ❙ 1670; 907; 85 ❙ 91.8; 875; 350.5

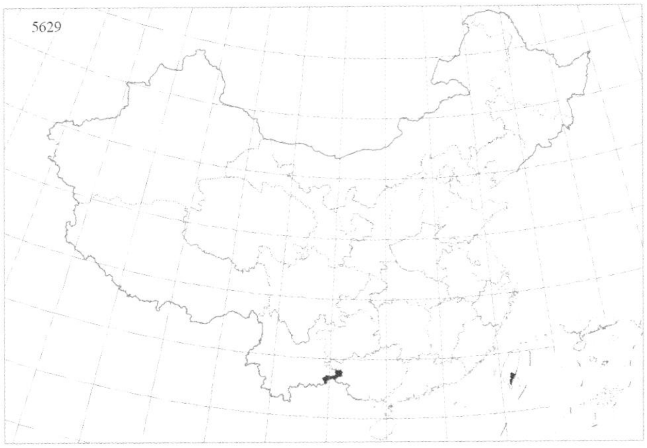

Icacinaceae 茶茱萸科 *Nothapodytes* 南柴龙树属
Nothapodytes obscura C. Y. Wu 薄叶假柴龙树
Shrubs or small trees

18.5 (17.2~21.0); 11.3 (9.1~15.1); 23.5 (22.3~25.6); 18.5 (17.3~21.0); 162.0 (147.0~191.8); 0.0 (0.0~0.0); 902 (833~1053) ❙ 1359 (958~1607); 745 (531~860); 56 (44~77) ❙ 50.8 (13.3~80.6); 899 (823~1053); 423.5 (311.9~659.0)

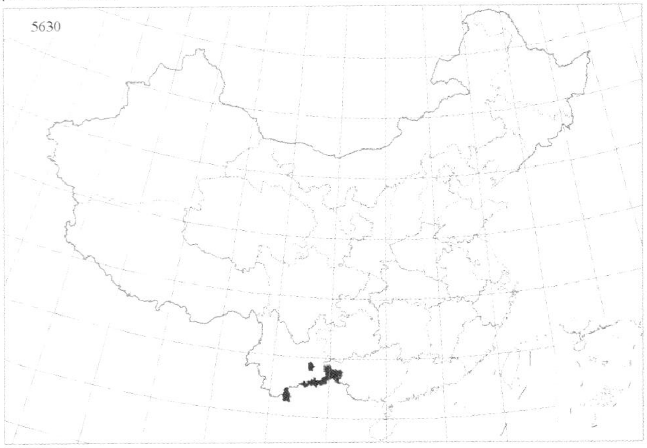

Icacinaceae 茶茱萸科 *Nothapodytes* 南柴龙树属
Nothapodytes obtusifolia (Merr.) R. A. Howard 假柴龙树
Shrubs or trees

24.8; 20.1; 28.0; 24.8; 237.6; 0.0; 1385 ❙ 1405; 473; 41 ❙ 1.5; 1137; 527.5

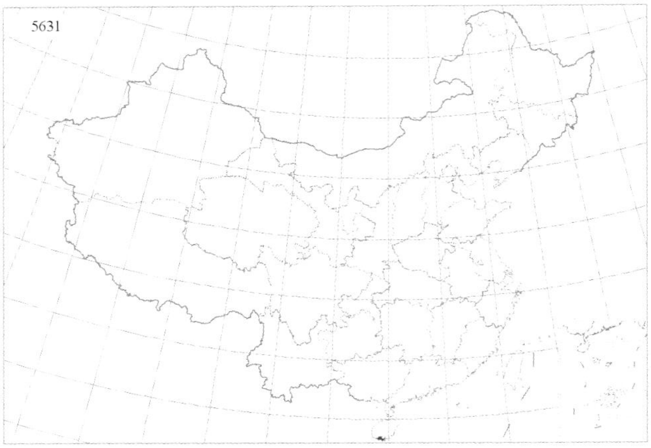

Icacinaceae 茶茱萸科 *Nothapodytes* 南柴龙树属
Nothapodytes pittosporoides (Oliv.) Sleumer 马比木
Small shrubs

15.5 (9.3~20.3); 4.8 (-1.7~10.9); 25.5 (19.4~29.3); 15.6 (9.5~20.4); 128.7 (69.8~184.5); 1.7 (0.0~18.0); 833 (603~1072) ❙ 1310 (711~1777); 541 (357~817); 103 (15~209) ❙ 58.5 (8.0~107.2); 832 (603~1072); 211.5 (108.2~292.6)

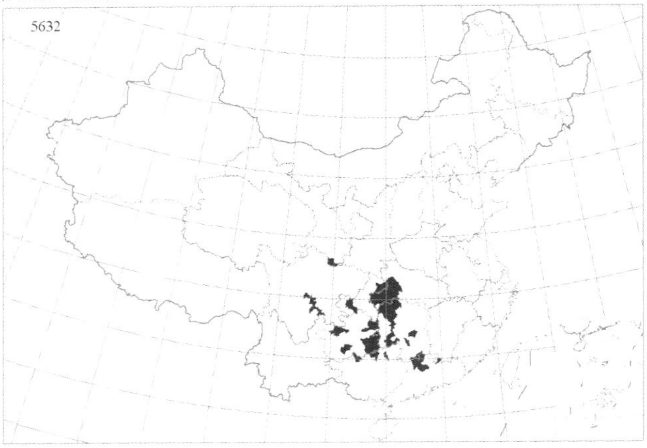

Icacinaceae 茶茱萸科 *Nothapodytes* 南柴龙树属
Nothapodytes tomentosa C. Y. Wu 毛假柴龙树
Shrubs

16.8 (14.5~21.6); 9.8 (7.8~12.6); 22.0 (19.6~28.5); 16.9 (14.5~21.7); 142.3 (113.5~199.8); 0.0 (0.0~0.0); 840 (730~1153) ❙ 1078 (946~1625); 596 (531~795); 61 (27~173) ❙ 27.9 (13.3~43.0); 817 (683~1153); 292.8 (234.5~382.6)

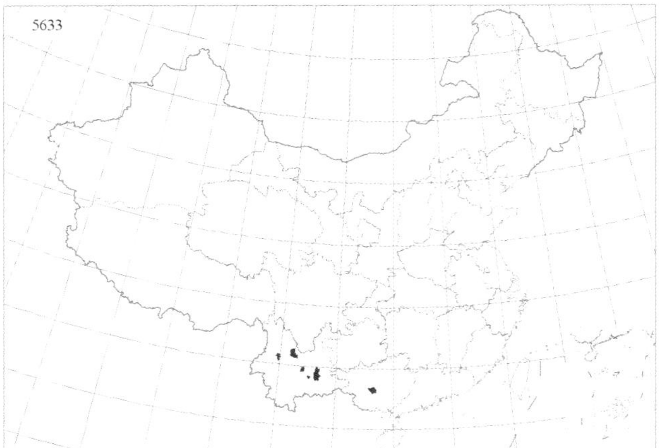

Icacinaceae 茶茱萸科 *Pittosporopsis* 假海桐属
Pittosporopsis kerrii Craib 假海桐
Shrubs or small trees

19.0 (17.9~20.5); 13.2 (11.9~15.1); 22.7 (21.6~23.7); 19.1 (18.0~20.5); 169.2 (155.8~185.9); 0.0 (0.0~0.0); 913 (854~992) ❙ 1485 (1276~1751); 787 (655~996); 60 (42~77) ❙ 62.9 (36.8~88.8); 901 (833~986); 506.6 (427.9~659.0)

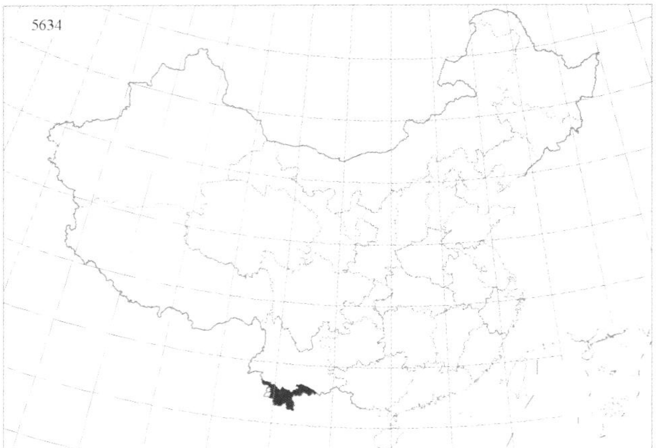

Icacinaceae 茶茱萸科 *Platea* 肖榄属

Platea latifolia Blume 阔叶肖榄

Trees

21.2 (17.3~24.8); 13.4 (7.4~20.1); 27.3 (22.4~28.6); 21.2 (17.3~24.8); 194.9 (147.1~237.6); 0.0 (0.0~0.0); 1120 (833~1385) ❙ 1605 (1179~2435); 730 (392~1428); 117 (41~189) ❙ 45.7 (-11.9~99.5); 1084 (833~1299); 339.2 (185.5~620.0)

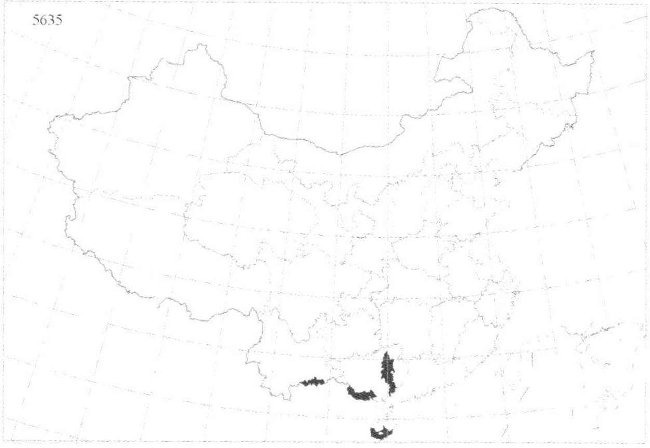

Icacinaceae 茶茱萸科 *Pyrenacantha* 刺核藤属

Pyrenacantha volubilis Wight 刺核藤

Vines

23.9 (22.2~24.8); 19.0 (17.2~20.1); 27.3 (25.7~28.0); 23.9 (22.3~24.8); 227.3 (207.2~237.6); 0.0 (0.0~0.0); 1297 (1127~1385) ❙ 1518 (1380~1666); 520 (455~597); 50 (41~58) ❙ 18.1 (1.5~47.8); 1133 (1055~1212); 537.6 (455.2~620.0)

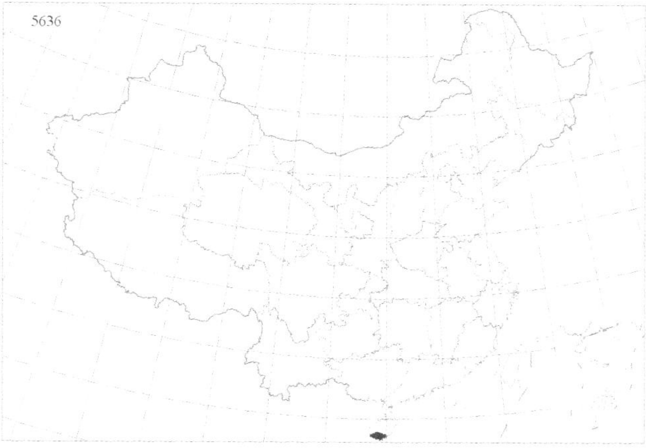

Aceraceae 槭树科 *Acer* 槭属

Acer acutum W. P. Fang 锐角枫

Small trees, deciduous

16.0 (13.4~18.2); 4.1 (1.5~6.5); 27.4 (25.2~29.1); 16.0 (13.4~18.2); 133.9 (108.3~158.8); 1.7 (0.0~7.3); 872 (755~984) ❙ 1657 (1363~1922); 584 (481~666); 211 (129~259) ❙ 90.0 (53.5~112.2); 868 (755~927); 240.0 (183.4~284.0)

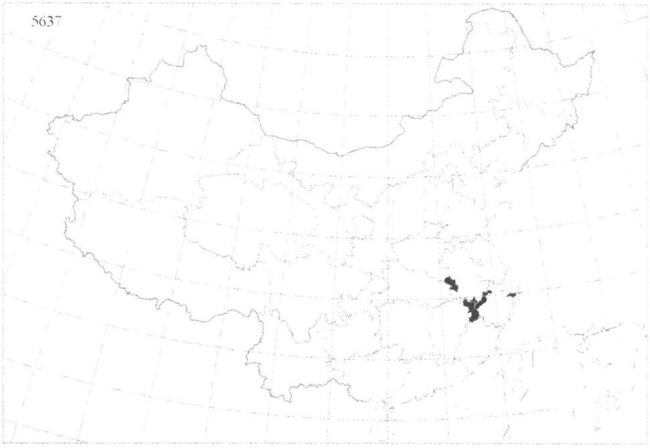

Aceraceae 槭树科 *Acer* 槭属

Acer albopurpurascens Hayata 紫白枫

Trees, evergreen

20.6 (16.2~24.1); 14.3 (11.0~18.9); 25.7 (20.1~28.3); 20.6 (16.2~24.1); 187.1 (134.2~229.7); 0.0 (0.0~0.0); 1048 (773~1322) ❙ 2246 (1498~3495); 1038 (742~1847); 224 (47~897) ❙ 124.3 (25.2~246.5); 1019 (773~1177); 243.3 (119.4~363.7)

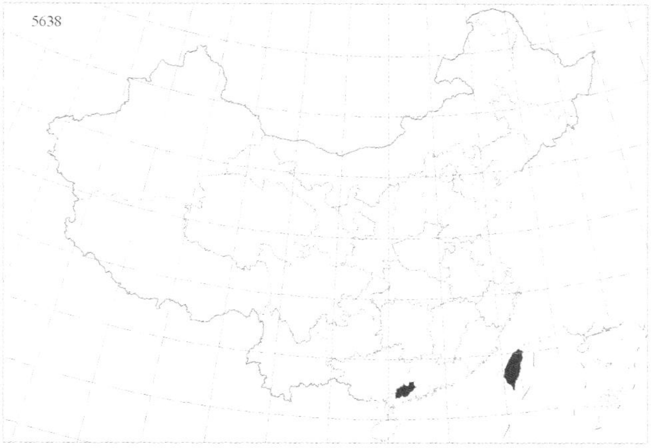

Aceraceae 槭树科 *Acer* 槭属

Acer amplum Rehder 阔叶枫

Trees, deciduous

15.7 (1.2~21.2); 5.5 (-7.1~14.2); 24.9 (9.0~29.8); 15.8 (3.3~21.3); 131.6 (11.9~195.2); 3.1 (0.0~56.5); 842 (390~1126) ❙ 1450 (578~1997); 586 (345~996); 142 (11~259) ❙ 72.7 (5.5~130.7); 837 (390~1126); 270.9 (130.1~627.0)

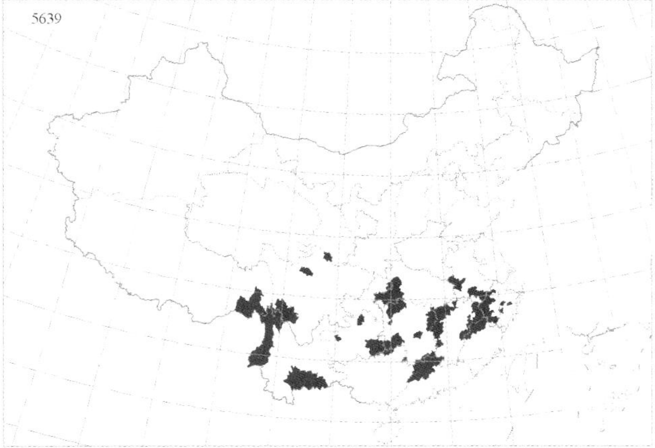

Aceraceae 槭树科 *Acer* 槭属

Acer amplum subsp. bodinieri (H. Lév.) Y. S. Chen 建水阔叶枫

Trees, deciduous

15.7 (13.3~18.2); 5.4 (3.3~11.8); 25.2 (21.4~27.2); 15.8 (13.4~18.2); 130.2 (101.4~158.4); 0.7 (0.0~2.3); 826 (715~878) ❙ 1362 (1059~1742); 544 (450~756); 121 (44~221) ❙ 65.2 (21.3~128.9); 825 (715~878); 237.6 (170.4~491.5)

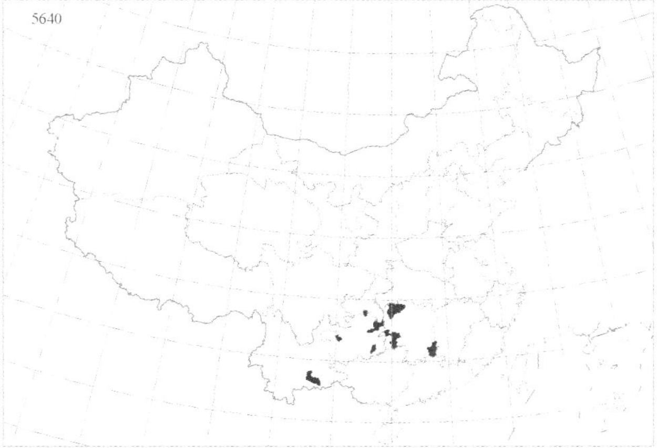

A

Aceraceae 槭树科 *Acer* 槭属
Acer amplum subsp. catalpifolium (Rehder) Y. S. Chen 梓叶枫
Trees, deciduous

13.5 (2.2~17.0); 3.8 (-6.6~6.4); 22.2 (9.9~26.7); 13.6 (3.8~17.0); 107.0 (15.5~144.4); 4.6 (0.0~49.1); 746 (413~891) ❙ 1220 (932~1624); 680 (546~938); 47 (19~191) ❙ 65.9 (12.0~125.9); 746 (413~891); 213.8 (133.8~318.7)

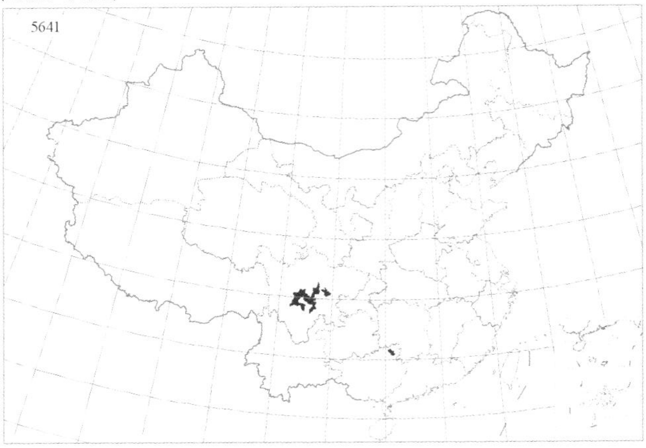

Aceraceae 槭树科 *Acer* 槭属
Acer amplum subsp. tientaiense (C. K. Schneid.) Y. S. Chen 天台阔叶枫
Trees, deciduous

16.7 (15.6~17.6); 5.4 (4.3~7.2); 27.7 (26.3~29.3); 16.8 (15.7~17.7); 141.8 (128.8~151.9); 0.3 (0.0~0.7); 899 (843~941) ❙ 1756 (1502~1997); 636 (587~685); 212 (160~254) ❙ 96.1 (59.6~130.7); 899 (843~941); 249.4 (202.6~303.4)

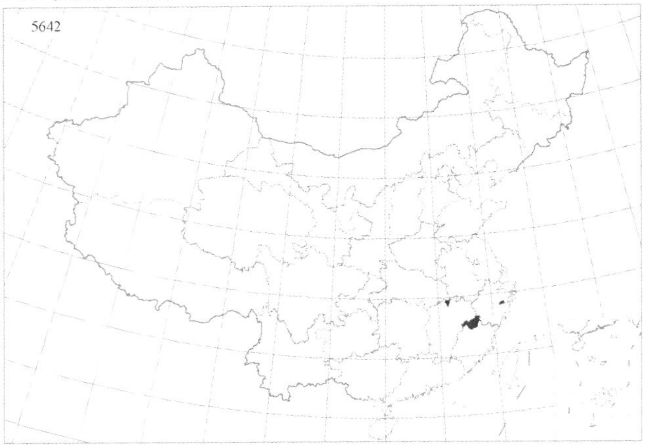

Aceraceae 槭树科 *Acer* 槭属
Acer barbinerve Maxim. 簇毛枫
Small trees, deciduous

3.6 (0.8~7.6); -16.4 (-22.8~-10.8); 20.7 (18.4~23.4); 7.9 (6.4~9.8); 60.2 (45.8~83.2); 76.6 (51.4~105.6); 567 (501~660) ❙ 762 (537~1073); 475 (340~691); 27 (9~40) ❙ 34.2 (-18.7~74.1); 561 (501~624); 307.0 (158.5~346.5)

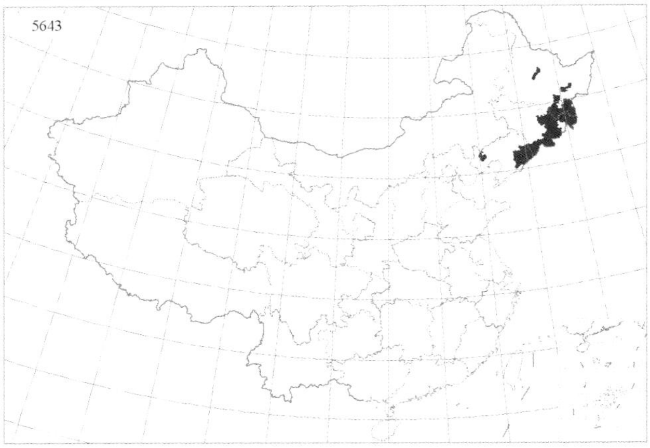

Aceraceae 槭树科 *Acer* 槭属
Acer buergerianum Miq. 三角枫
Trees, deciduous

16.1 (7.4~22.1); 4.3 (-5.6~15.3); 27.1 (18.7~30.1); 16.1 (8.2~22.1); 136.3 (58.8~205.6); 3.0 (0.0~29.4); 879 (561~1182) ❙ 1285 (70~2993); 507 (23~1412); 137 (11~547) ❙ 45.5 (-91.1~203.4); 858 (70~1182); 202.0 (110.1~332.5)

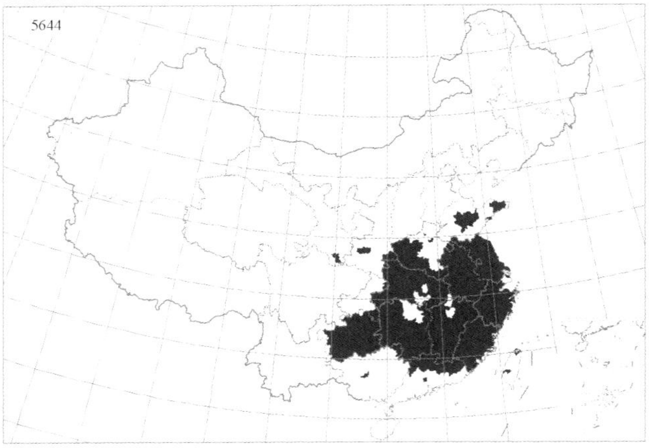

Aceraceae 槭树科 *Acer* 槭属
Acer buergerianum var. jiujiangense Z. X. Yu 九江三角枫
Trees, deciduous

16.9; 4.4; 29.1; 16.9; 143.8; 0.6; 932 ❙ 1546; 602; 177 ❙ 66.0; 932; 206.2

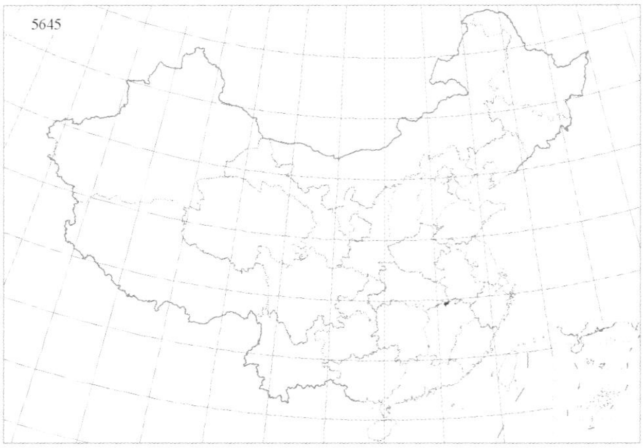

Aceraceae 槭树科 *Acer* 槭属
Acer buergerianum var. kaiscianense (Pamp.) W. P. Fang 界山三角枫
Trees, deciduous

10.9; -0.1; 21.0; 11.0; 83.0; 11.6; 660 ❙ 677; 338; 15 ❙ 2.1; 641; 256.3

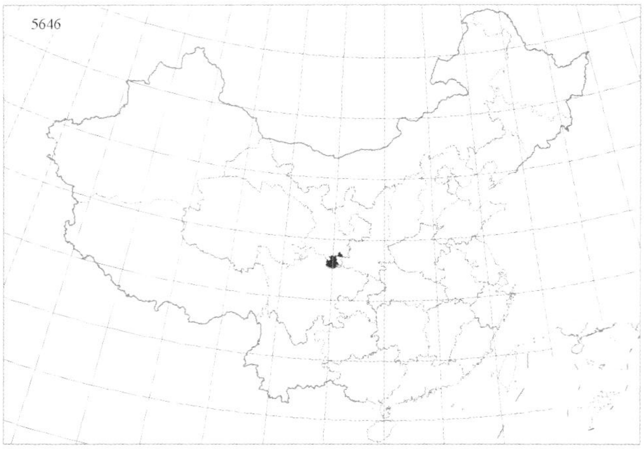

Aceraceae 槭树科 *Acer* 槭属

Acer buergerianum var. yentangense W. P. Fang et M. Y. Fang 雁荡三角枫

Trees, deciduous

16.3; 5.5; 27.3; 16.4; 136.6; 0.3; 869 ❙ 1626; 596; 187 ❙ 87.1; 869; 228.5

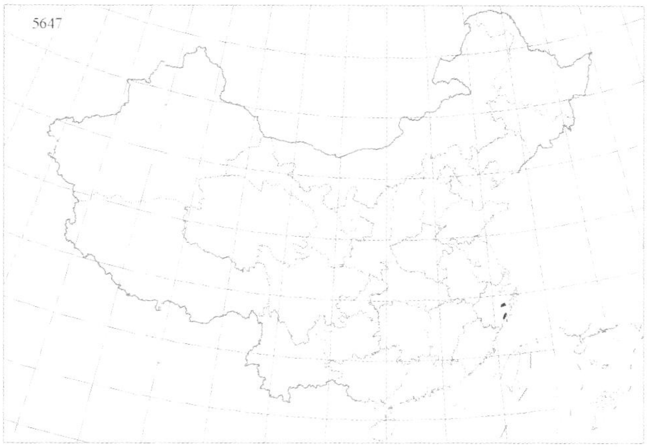

Aceraceae 槭树科 *Acer* 槭属

Acer caesium Wall. ex Brandis 深灰枫

Trees, deciduous

9.7 (2.0~16.7); 0.0 (-7.7~9.7); 18.5 (10.1~25.3); 10.1 (3.8~16.7); 73.8 (15.4~140.9); 17.0 (0.0~51.4); 621 (413~819) ❙ 898 (341~1358); 468 (190~727); 30 (7~88) ❙ 46.8 (-40.9~128.6); 609 (341~795); 279.1 (61.2~407.1)

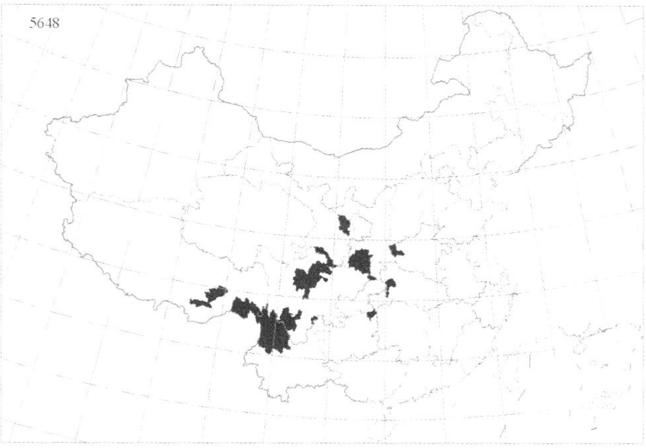

Aceraceae 槭树科 *Acer* 槭属

Acer campbellii Hook. f. et Thomson ex Hiern 藏南枫

Trees, deciduous

2.9 (-1.8~9.6); -5.0 (-10.2~2.4); 10.1 (6.1~15.7); 4.6 (1.8~9.6); 22.8 (1.9~60.8); 46.8 (5.4~82.8); 432 (329~600) ❙ 623 (320~1007); 370 (193~578); 27 (11~40) ❙ 35.9 (-4.0~100.7); 428 (320~600); 192.2 (72.2~351.7)

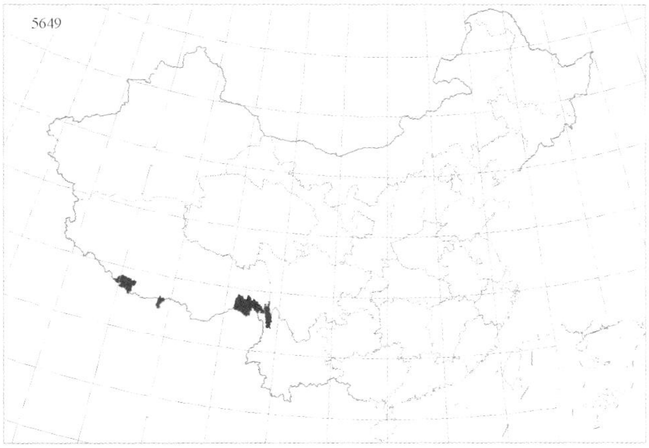

Aceraceae 槭树科 *Acer* 槭属

Acer campbellii var. serratifolium Banerji 重齿藏南枫

Trees, deciduous

14.0 (-1.8~19.3); 7.2 (-10.2~13.1); 19.3 (6.1~24.3); 14.2 (1.8~19.4); 114.1 (1.9~173.2); 5.1 (0.0~82.8); 737 (329~939) ❙ 1066 (320~1456); 606 (217~836); 44 (11~69) ❙ 45.6 (-3.0~100.7); 717 (320~903); 335.7 (72.2~438.9)

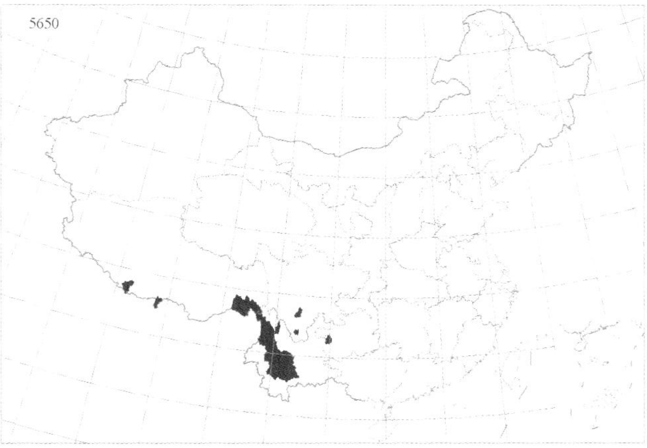

Aceraceae 槭树科 *Acer* 槭属

Acer cappadocicum Gled. 青皮枫

Trees, deciduous

14.4 (-1.1~18.9); 5.5 (-9.1~12.0); 22.2 (6.2~26.9); 14.5 (2.0~19.0); 116.1 (2.5~167.8); 2.2 (0.0~75.6); 763 (334~929) ❙ 1095 (321~1446); 545 (193~817); 57 (13~129) ❙ 44.0 (-14.2~113.0); 753 (321~918); 237.8 (79.3~384.1)

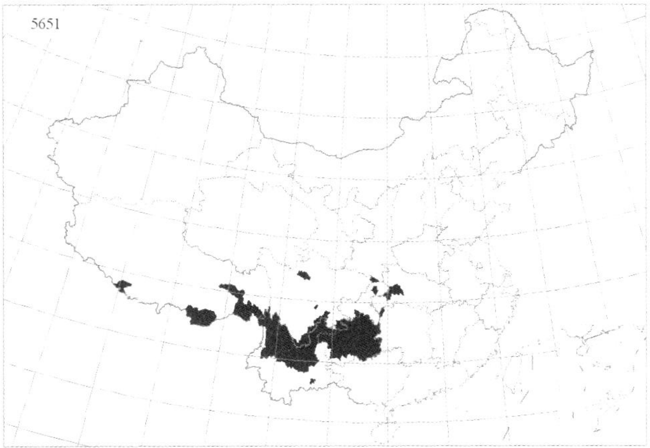

Aceraceae 槭树科 *Acer* 槭属

Acer cappadocicum subsp. sinicum Rehder 小叶青皮枫

Trees, deciduous

10.6 (2.2~18.1); 0.7 (-7.4~12.0); 19.7 (9.5~25.9); 11.0 (3.9~18.1); 81.4 (15.8~157.3); 13.7 (0.0~48.6); 651 (411~863) ❙ 977 (341~1431); 492 (190~817); 41 (8~100) ❙ 50.4 (-40.9~121.7); 640 (341~863); 242.1 (61.2~432.9)

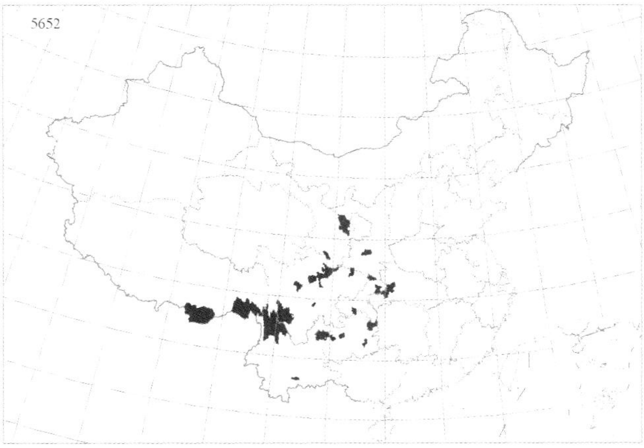

Aceraceae 槭树科 *Acer* 槭属
Acer caudatifolium Hayata 尖尾枫
Trees, deciduous

19.5 (17.1~21.7); 13.3 (10.2~15.3); 24.7 (21.7~27.0); 19.6 (17.1~21.7); 174.7 (145.0~200.2); 0.0 (0.0~0.0); 966 (816~1118) ❙ 2370 (1836~2993); 1086 (711~1412); 234 (110~547) ❙ 150.0 (64.2~214.4); 959 (816~1075); 264.7 (202.8~363.7)

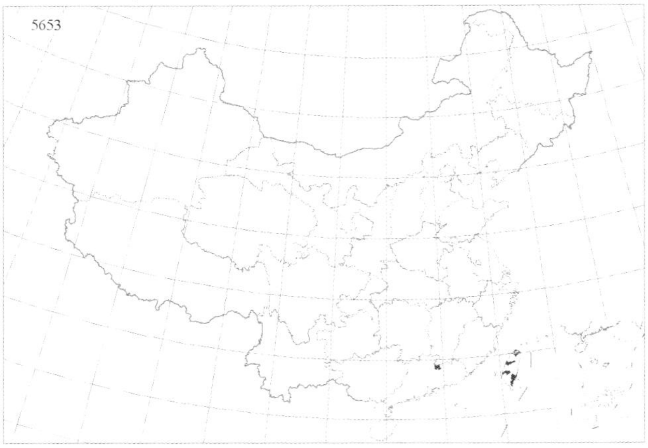

Aceraceae 槭树科 *Acer* 槭属
Acer caudatum Wall. 长尾枫
Trees, deciduous

10.7 (−1.8~16.7); 0.1 (−10.2~9.7); 20.3 (6.1~28.8); 11.0 (1.8~16.7); 83.7 (1.9~140.9); 15.4 (0.0~82.8); 664 (329~920) ❙ 941 (320~1736); 476 (190~1007); 43 (7~153) ❙ 43.3 (−40.9~145.8); 642 (320~899); 247.5 (61.2~447.4)

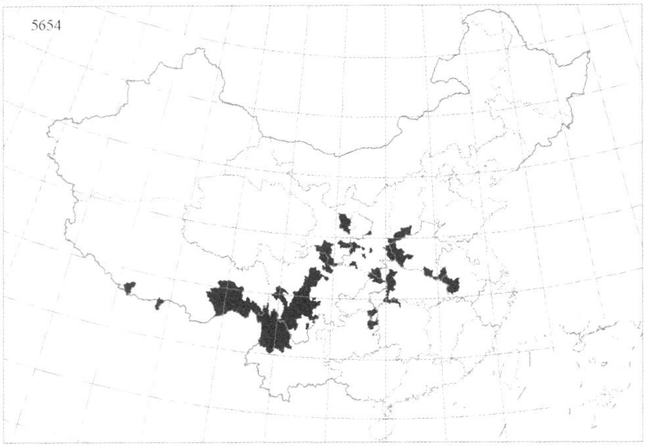

Aceraceae 槭树科 *Acer* 槭属
Acer ceriferum Rehder 杈叶枫
Trees, deciduous

13.1 (6.2~23.6); 1.2 (−6.7~17.8); 24.3 (15.6~29.3); 13.2 (7.0~23.6); 106.5 (45.2~223.3); 8.9 (0.0~35.7); 756 (522~1271) ❙ 1049 (373~1820); 461 (201~835); 69 (7~250) ❙ 38.5 (−37.6~108.0); 739 (373~1091); 241.2 (88.1~506.2)

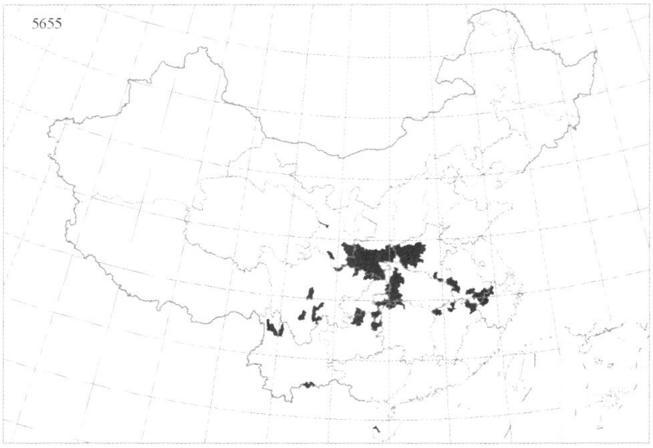

Aceraceae 槭树科 *Acer* 槭属
Acer chienii Hu et W. C. Cheng 怒江枫
Trees, deciduous

11.8 (5.4~16.2); 4.9 (−2.1~9.4); 17.4 (12.3~20.8); 11.9 (5.8~16.3); 86.2 (31.5~135.1); 3.6 (0.0~25.6); 661 (481~790) ❙ 1117 (741~1348); 630 (426~762); 49 (24~69) ❙ 69.1 (50.8~93.0); 661 (481~790); 352.0 (273.5~407.1)

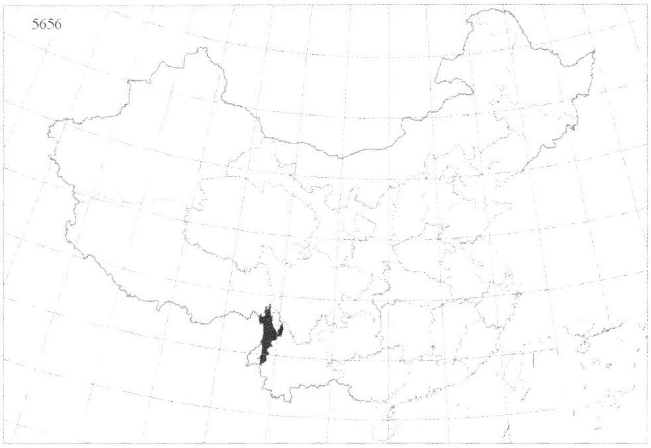

Aceraceae 槭树科 *Acer* 槭属
Acer chingii Hu 黔桂枫
Trees, deciduous

15.6 (−1.8~21.3); 5.3 (−10.2~12.3); 24.9 (6.1~28.7); 15.8 (1.8~21.3); 132.0 (1.9~196.1); 3.9 (0.0~82.8); 843 (329~1130) ❙ 1303 (320~1776); 542 (217~1007); 112 (7~196) ❙ 53.9 (−3.0~145.8); 842 (320~1130); 215.6 (72.2~447.4)

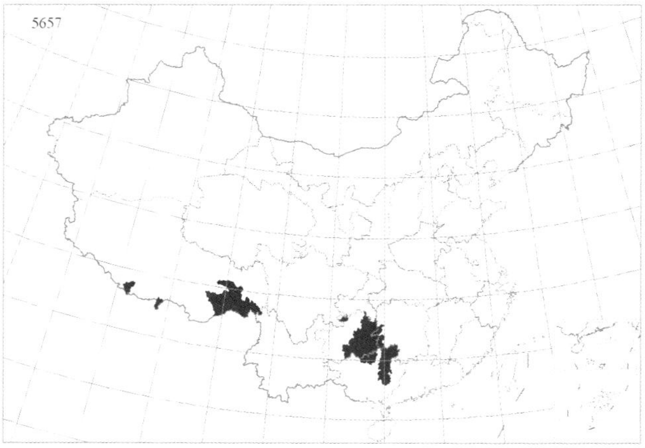

Aceraceae 槭树科 *Acer* 槭属
Acer chunii W. P. Fang 乳源枫
Small trees, deciduous

15.7; 5.8; 24.8; 15.8; 132.0; 2.2; 854 ❙ 1433; 579; 123 ❙ 72.3; 832; 230.2

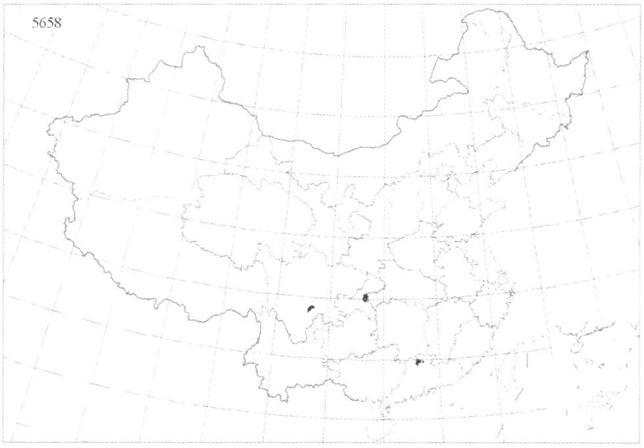

Aceraceae 槭树科 *Acer* 槭属
Acer confertifolium Merr. et F. P. Metcalf 密叶枫
Trees, evergreen

18.6 (16.2~21.1); 10.9 (7.6~15.3); 25.6 (20.1~28.6); 18.6 (16.2~21.1); 163.7 (134.2~193.7); 0.0 (0.0~0.0); 938 (773~1107) ❙ 1931 (1258~2993); 753 (425~1412); 221 (110~547) ❙ 110.9 (14.0~246.5); 935 (773~1107); 267.9 (206.0~363.7)

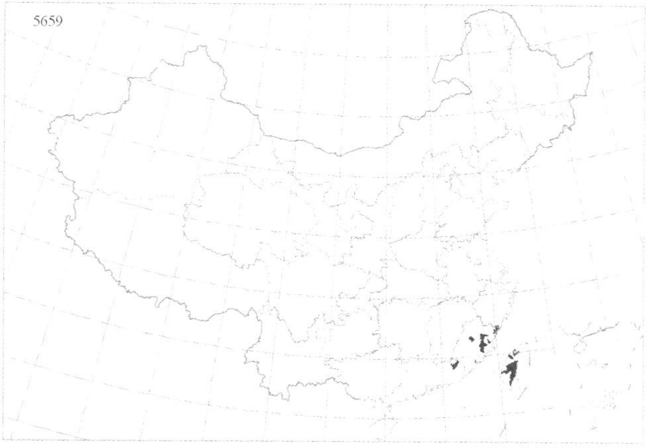

Aceraceae 槭树科 *Acer* 槭属
Acer cordatum Pax 紫果枫
Trees, deciduous

16.9 (10.6~21.7); 6.1 (-0.2~14.3); 27.0 (21.1~29.9); 17.0 (10.6~21.7); 144.5 (79.4~200.5); 0.6 (0.0~12.1); 902 (643~1140) ❙ 1501 (1072~1997); 544 (392~858); 165 (46~259) ❙ 67.1 (28.8~147.0); 894 (643~1140); 219.0 (134.6~551.6)

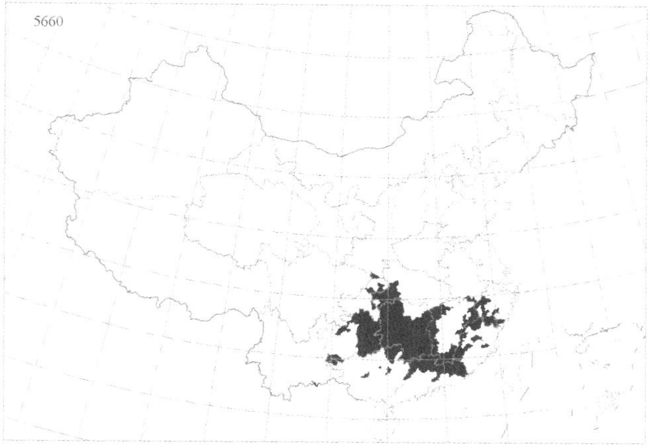

Aceraceae 槭树科 *Acer* 槭属
Acer coriaceifolium H. Lév. 樟叶枫
Trees, evergreen

16.9 (11.7~24.1); 6.5 (2.6~18.9); 26.3 (18.8~29.9); 16.9 (11.7~24.1); 143.6 (82.6~229.7); 0.3 (0.0~3.9); 890 (658~1322) ❙ 1426 (916~2992); 565 (411~1847); 132 (31~308) ❙ 60.6 (8.0~246.5); 885 (658~1190); 211.0 (108.2~371.7)

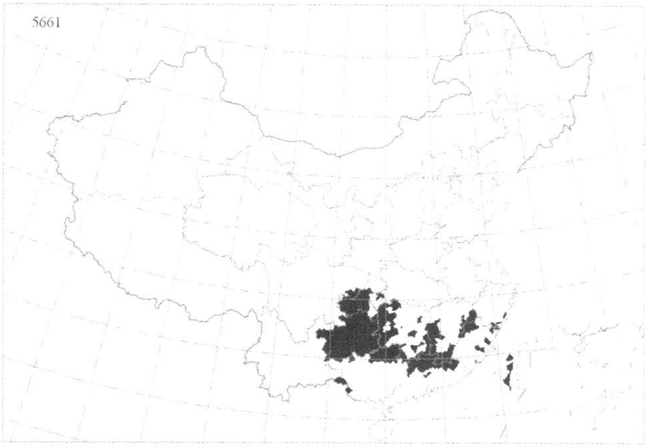

Aceraceae 槭树科 *Acer* 槭属
Acer crassum Hu et W. C. Cheng 厚叶枫
Trees, evergreen

18.1 (16.3~21.0); 11.0 (9.1~14.2); 23.1 (21.4~25.6); 18.1 (16.4~21.0); 157.2 (136.4~191.8); 0.0 (0.0~0.0); 879 (797~1053) ❙ 1321 (1074~1547); 743 (582~911); 54 (47~68) ❙ 50.4 (32.4~80.6); 879 (797~1053); 407.0 (277.0~551.6)

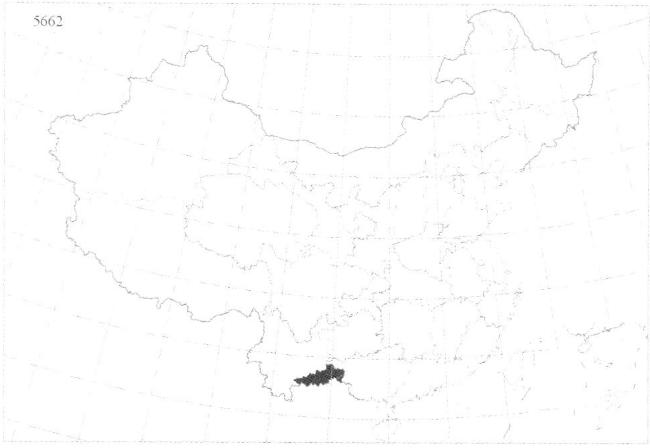

Aceraceae 槭树科 *Acer* 槭属
Acer davidii Franch. 青榨枫
Trees, deciducus

14.3 (-0.8~22.7); 2.9 (-15.9~15.0); 24.7 (9.2~29.9); 14.6 (2.9~22.8); 121.7 (10.1~213.1); 9.3 (0.0~80.4); 811 (374~1221) ❙ 1128 (317~2435); 495 (190~1428); 95 (5~259) ❙ 37.0 (-40.9~147.0); 778 (317~1221); 218.3 (61.2~551.6)

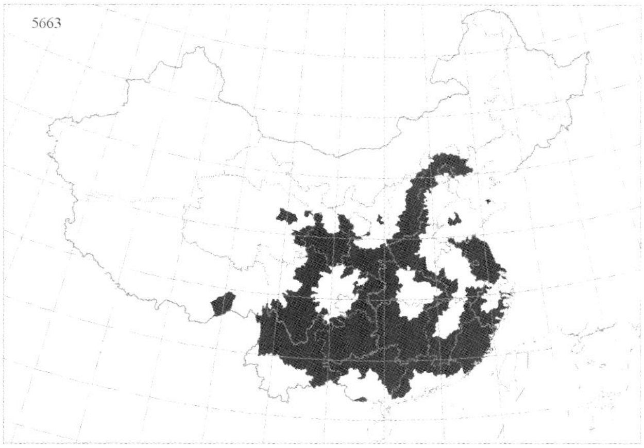

Aceraceae 槭树科 *Acer* 槭属
Acer davidii subsp. grosseri (Pax) P. C. de Jong 葛萝枫
Trees, deciduous

11.5 (2.5~18.7); -2.7 (-15.9~8.6); 24.5 (11.2~29.2); 12.3 (4.4~18.8); 100.7 (20.5~165.2); 21.7 (0.0~80.4); 738 (428~984) ❙ 788 (367~1922); 386 (201~666); 50 (7~259) ❙ 4.5 (-40.9~121.7); 649 (367~976); 191.2 (70.3~334.4)

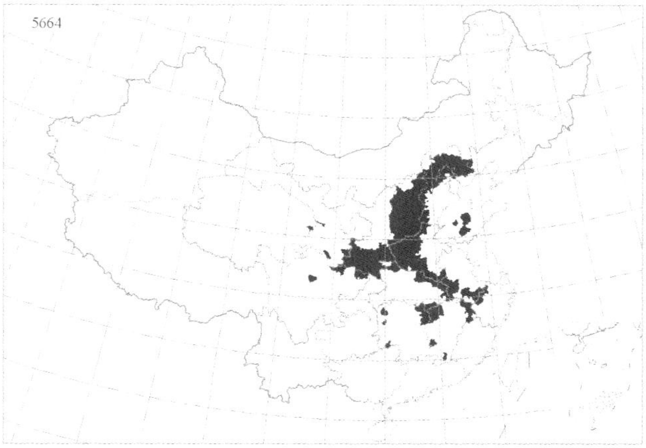

Aceraceae 槭树科 *Acer* 槭属

Acer duplicatoserratum var. chinense C. S. Chang 中华重齿枫

Small trees, deciduous

17.1 (15.9~18.2); 5.3 (3.8~6.9); 28.3 (26.7~29.3); 17.1 (16.0~18.2); 145.7 (132.7~158.8); 0.3 (0.0~1.2); 924 (868~984) ❙ 1660 (1211~1950); 583 (474~685); 208 (134~259) ❙ 79.8 (28.8~121.9); 918 (868~961); 232.9 (183.9~279.5)

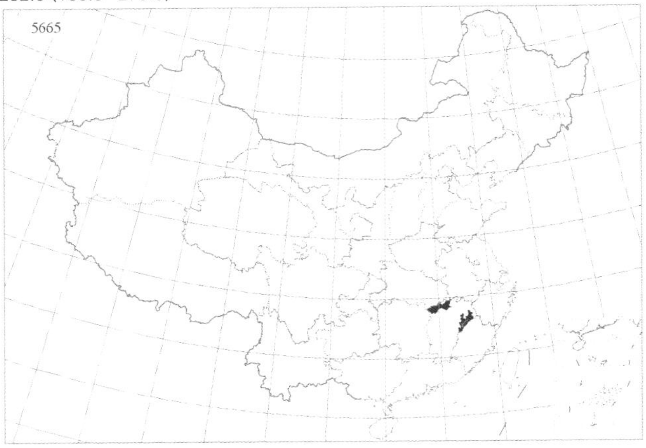

Aceraceae 槭树科 *Acer* 槭属

Acer elegantulum W. P. Fang et P. L. Chiu 秀丽枫

Trees, deciduous

16.4 (13.4~23.6); 4.7 (1.5~17.8); 27.5 (25.2~29.3); 16.4 (13.4~23.6); 138.3 (108.3~223.3); 1.4 (0.0~7.3); 889 (755~1271) ❙ 1585 (1211~1922); 574 (427~685); 193 (44~259) ❙ 79.8 (-2.2~129.9); 882 (755~1099); 248.1 (183.4~439.7)

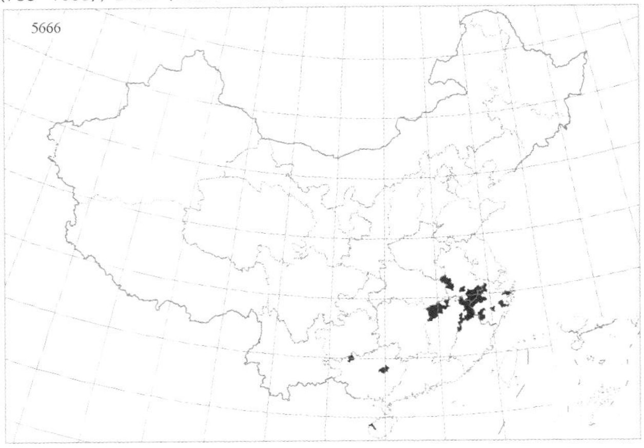

Aceraceae 槭树科 *Acer* 槭属

Acer erianthum Schwerin 毛花枫

Trees, deciduous

13.4 (2.2~22.2); 2.8 (-7.1~14.7); 23.0 (9.9~28.3); 13.5 (3.8~22.2); 108.1 (15.5~206.9); 7.2 (0.0~49.1); 751 (413~1177) ❙ 1035 (626~1876); 497 (306~891); 53 (14~199) ❙ 38.9 (-22.2~125.9); 740 (413~1177); 238.4 (139.1~352.3)

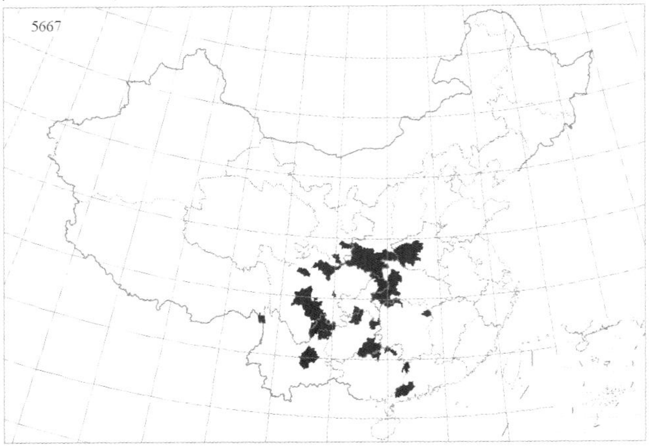

Aceraceae 槭树科 *Acer* 槭属

Acer fabri Hance 罗浮枫

Trees, evergreen

17.6 (6.0~24.0); 7.6 (-3.2~19.1); 26.3 (14.4~29.9); 17.6 (6.6~24.0); 151.7 (39.7~228.1); 0.5 (0.0~26.7); 925 (508~1302) ❙ 1406 (916~2435); 591 (392~1428); 124 (27~258) ❙ 53.0 (5.8~128.9); 918 (508~1247); 231.2 (108.2~620.0)

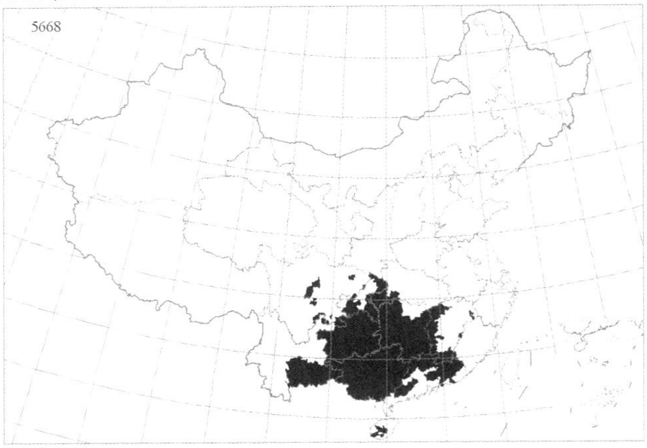

Aceraceae 槭树科 *Acer* 槭属

Acer fenzelianum Hand.-Mazz. 河口枫

Trees, deciduous

19.3; 12.7; 23.8; 19.3; 171.6; 0.0; 942 ❙ 1439; 803; 61 ❙ 53.4; 942; 516.4

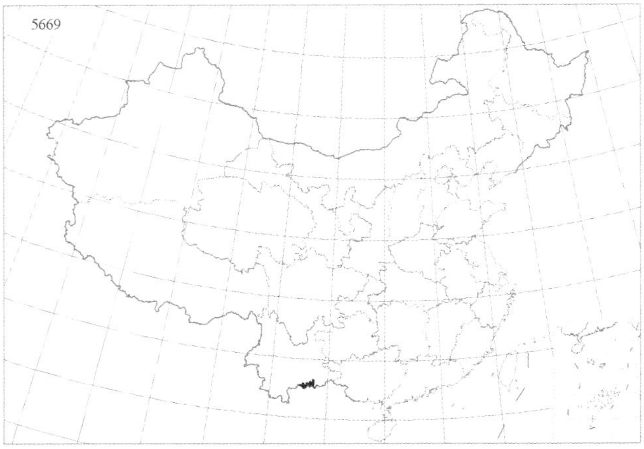

Aceraceae 槭树科 *Acer* 槭属

Acer flabellatum Rehder 扇叶枫

Trees, deciduous

15.1 (2.6~22.7); 5.9 (-7.0~15.0); 23.2 (11.2~29.2); 15.2 (4.3~22.8); 124.6 (20.0~213.1); 2.5 (0.0~48.2); 807 (427~1221) ❙ 1271 (741~2435); 598 (376~1428); 89 (15~253) ❙ 58.5 (6.5~128.9); 801 (427~1221); 278.6 (143.4~551.6)

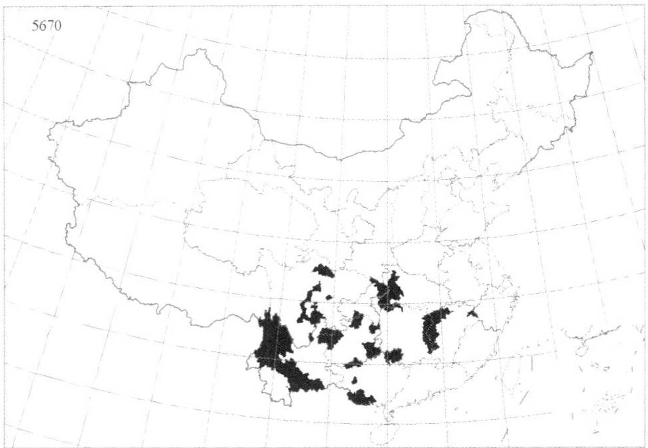

Aceraceae 槭树科 *Acer* 槭属
Acer forrestii Diels 丽江枫
Trees, deciduous

13.1 (2.0~18.9); 5.5 (−7.2~12.0); 19.3 (10.1~26.6); 13.2 (3.8~19.0); 101.7 (15.4~167.8); 3.9 (0.0~51.4); 707 (413~929) ▎ 964 (731~1243); 561 (346~696); 3[illegible] (12~58) ▎ 39.6 (−14.2~128.6); 679 (413~912); 278.8 (142.9~384.7)

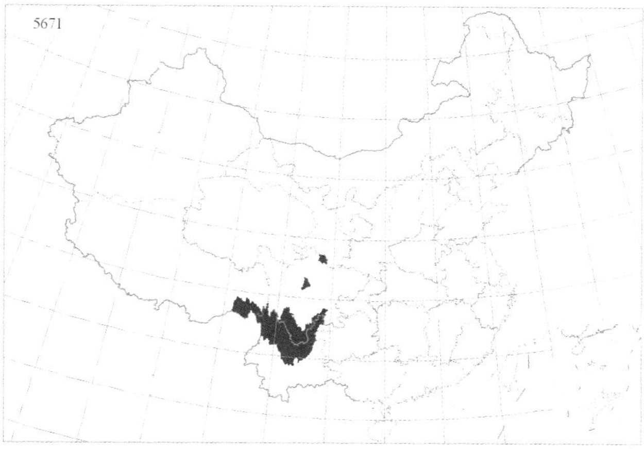

Aceraceae 槭树科 *Acer* 槭属
Acer fulvescens Rehder 黄毛枫
Trees, deciduous

10.2 (2.0~16.7); 0.2 (−7.2~9.7); 19.5 (10.1~25.4); 10.6 (3.8~16.7); 78.6 (15.4~140.9); 15.3 (0.0~51.6); 642 (413~819) ▎ 924 (510~1358); 468 (311~651); 27 (11~88) ▎ 48.2 (5.0~128.6); 636 (413~795); 279.4 (171.0~351.7)

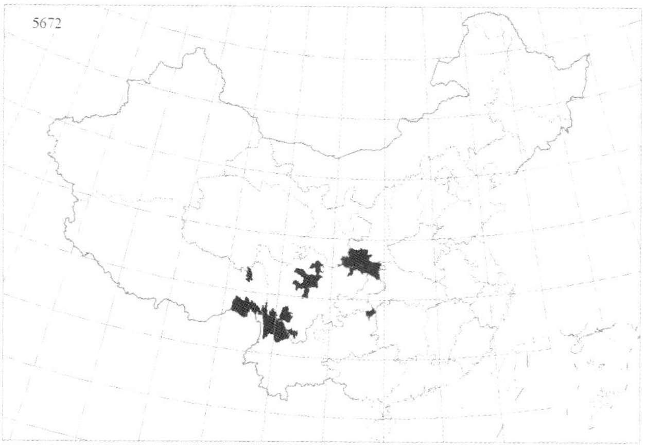

Aceraceae 槭树科 *Acer* 槭属
Acer gracilifolium W. P. Fang et C. C. Fu 长叶枫
Trees, evergreen

10.3; −0.1; 19.8; 10.3; 75.8; 11.7; 634 ▎ 711; 357; 15 ▎ 12.1; 634; 279.5

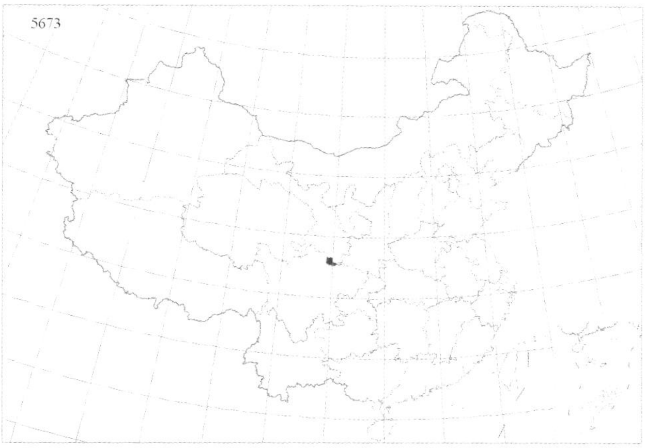

Aceraceae 槭树科 *Acer* 槭属
Acer griseum (Franch.) Pax 血皮枫
Trees, deciduous

11.9 (3.3~16.2); −1.8 (−12.7~4.4); 24.4 (17.5~28.1); 12.3 (6.6~16.2); 100.9 (45.4~135.6); 17.9 (0.6~65.0); 737 (507~891) ▎ 752 (427~1492); 368 (258~613); 36 (10~140) ▎ 2.3 (−36.5~106.2); 648 (427~891); 193.9 (70.3~334.4)

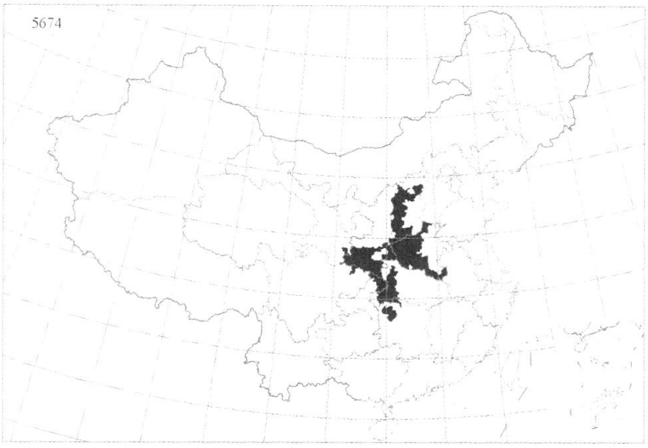

Aceraceae 槭树科 *Acer* 槭属
Acer henryi Pax 三叶枫
Trees, deciduous

14.5 (6.0~21.0); 2.3 (−4.9~11.9); 26.1 (14.4~29.4); 14.6 (6.6~21.1); 120.4 (39.7~193.0); 5.9 (0.0~29.4); 816 (508~1114) ▎ 1116 (512~1997); 474 (258~821); 100 (14~259) ▎ 36.2 (−35.9~130.7); 796 (508~1114); 206.0 (121.6~354.5)

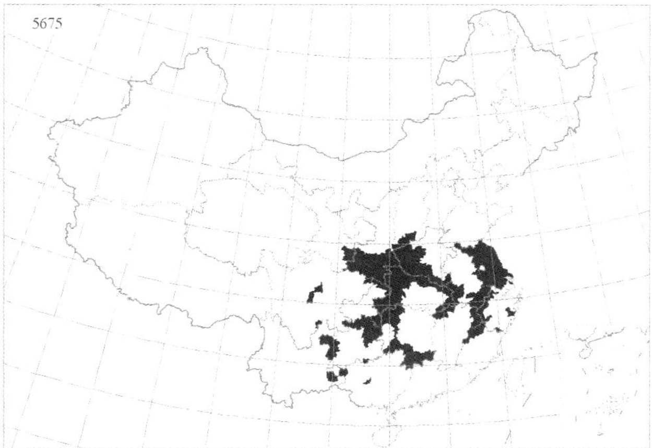

Aceraceae 槭树科 *Acer* 槭属
Acer hilaense Hu et Cheng 海拉枫
Trees, evergreen

12.7; 5.6; 18.2; 12.7; 96.7; 4.4; 692 ▎ 1066; 589; 50 ▎ 58.1; 692; 379.1

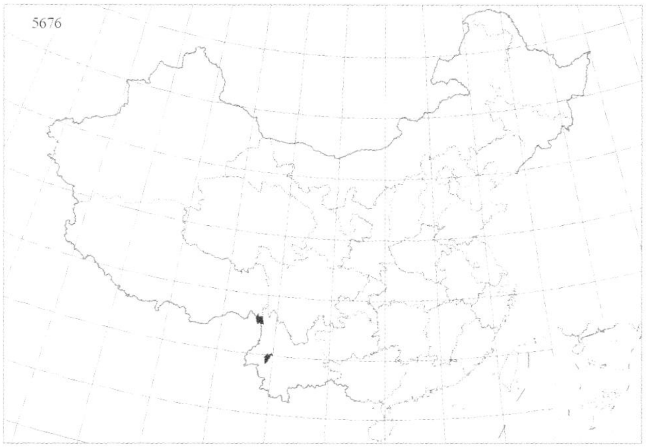

Aceraceae 槭树科 *Acer* 槭属

Acer komarovii Pojark. 小楷槭

Small trees, deciduous

5.3 (0.8~12.8); -13.9 (-22.8~-3.0); 21.7 (18.4~26.5); 8.8 (6.4~13.2); 69.1 (45.8~112.9); 65.3 (18.7~105.6); 607 (501~810) ❙ 702 (466~1073); 445 (312~691); 23 (11~40) ❙ 18.3 (-37.6~74.1); 563 (466~659); 278.7 (117.3~346.5)

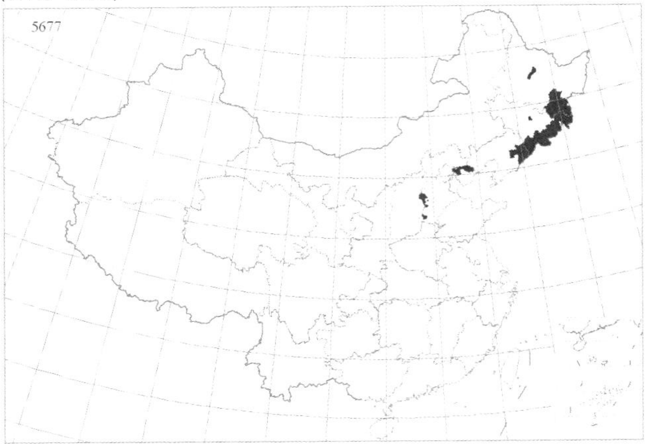

Aceraceae 槭树科 *Acer* 槭属

Acer kungshanense W. P. Fang et C. Y. Chang 贡山枫

Trees, deciduous

11.5 (5.4~17.2); 4.5 (-2.1~9.1); 17.3 (12.3~23.1); 11.6 (5.8~17.3); 83.2 (31.5~147.5); 4.6 (0.0~25.6); 655 (481~847) ❙ 1013 (741~1161); 579 (426~633); 40 (24~54) ❙ 56.0 (37.1~81.5); 655 (481~847); 340.9 (273.5~382.6)

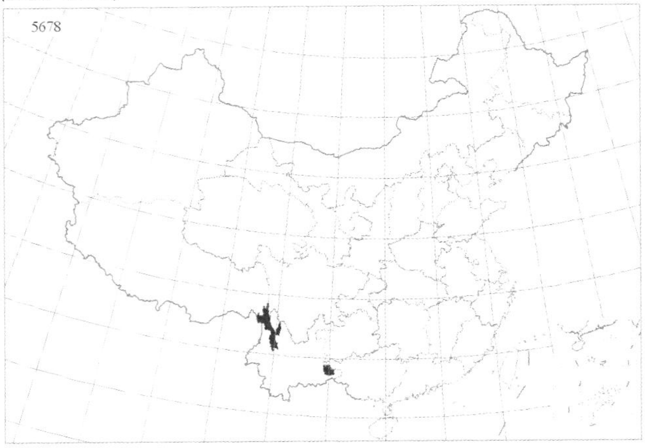

Aceraceae 槭树科 *Acer* 槭属

Acer kwangnanense Hu et Cheng 广南枫

Trees, evergreen

18.0 (16.3~21.0); 10.9 (9.0~14.2); 23.0 (21.4~25.6); 18.1 (16.4~21.0); 157.1 (136.4~191.8); 0.0 (0.0~0.0); 879 (797~1053) ❙ 1339 (1018~1751); 753 (554~996); 55 (47~77) ❙ 52.2 (25.7~88.8); 878 (797~1053); 407.7 (277.0~551.6)

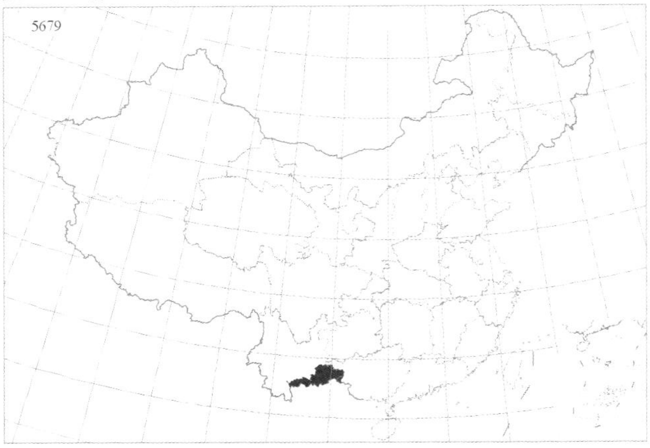

Aceraceae 槭树科 *Acer* 槭属

Acer kweilinense W. P. Fang et M. Y. Fang 桂林枫

Trees, deciduous

17.5 (14.3~21.3); 7.5 (3.3~12.3); 26.0 (21.4~28.7); 17.5 (14.3~21.3); 150.7 (114.1~196.1); 0.2 (0.0~2.3); 909 (765~1130) ❙ 1442 (1074~1776); 613 (419~860); 133 (43~198) ❙ 59.6 (16.9~101.8); 908 (765~1130); 247.2 (170.4~413.5)

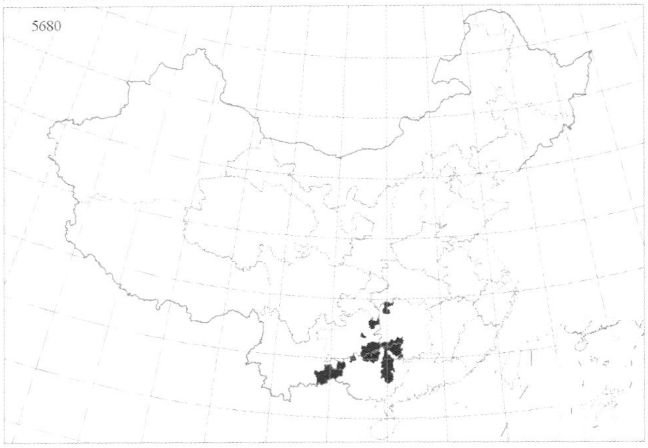

Aceraceae 槭树科 *Acer* 槭属

Acer laevigatum Wall. 光叶枫

Trees, evergreen

16.4 (2.2~23.1); 6.8 (-7.0~17.5); 24.9 (9.9~28.9); 16.5 (3.8~23.2); 138.9 (15.5~218.4); 1.5 (0.0~49.1); 866 (413~1232) ❙ 1309 (829~2435); 612 (400~1428); 81 (19~188) ❙ 52.3 (8.0~125.9); 865 (413~1221); 233.4 (108.2~551.6)

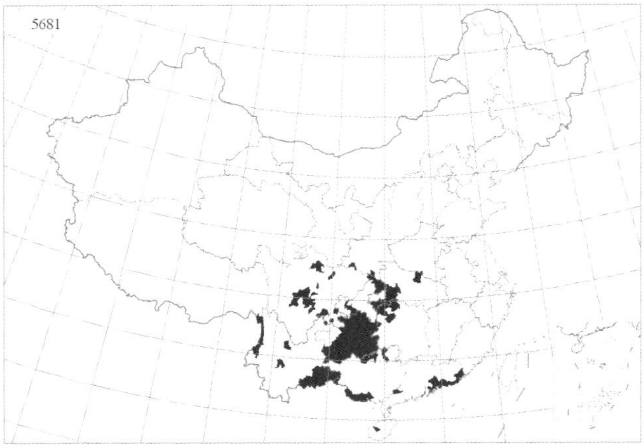

Aceraceae 槭树科 *Acer* 槭属

Acer laevigatum var. salweenense (W. W. Sm.) J. M. Cowan ex Fang 怒江光叶枫

Trees, evergreen

12.4; 4.9; 18.4; 12.4; 92.2; 3.2; 688 ❙ 1149; 638; 51 ❙ 70.6; 688; 350.5

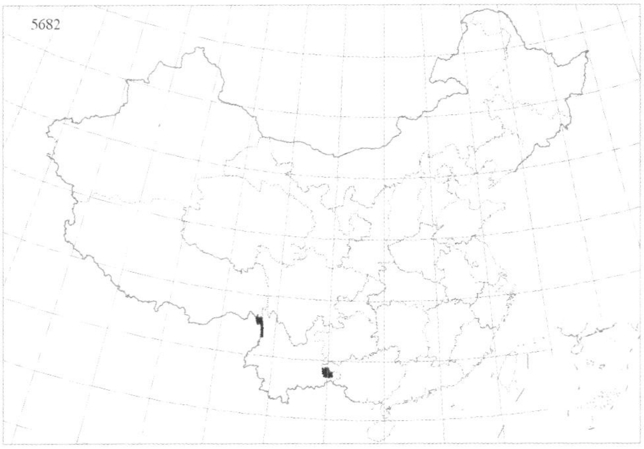

Aceraceae 槭树科 *Acer* 槭属
Acer laurinum Hasskarl 十蕊枫
Trees, evergreen

20.1 (8.6~24.8); 12.8 (1.3~20.1); 25.3 (15.0~29.0); 20.1 (8.7~24.8); 181.4 (52.6~237.6); 0.1 (0.0~8.8); 1033 (574~1385) ❙ 1443 (868~2435); 719 (312~1428); 84 (41~191) ❙ 42.5 (-34.3~145.8); 1002 (574~1299); 351.5 (199.4~659.0)

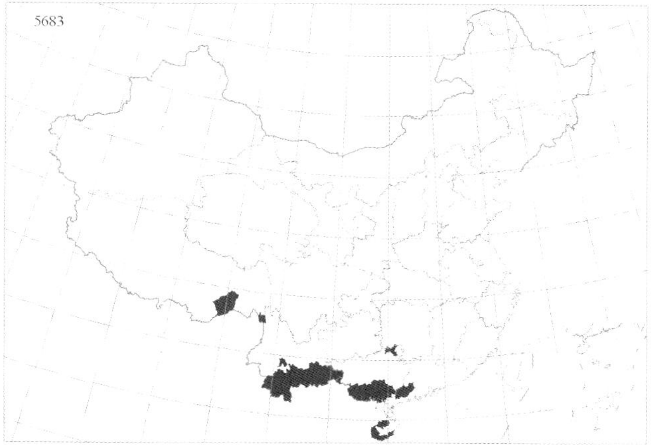

Aceraceae 槭树科 *Acer* 槭属
Acer laxiflorum Pax 疏花枫
Trees, deciduous

11.6 (2.0~21.6); 2.4 (-7.2~13.3); 19.7 (9.5~28.5); 11.9 (3.8~21.7); 89.9 (15.4~199.8); 10.6 (0.0~51.4); 686 (411~1153) ❙ 1095 (729~1703); 562 (346~835); 55 (11~196) ❙ 63.4 (18.6~128.6); 685 (411~1153); 254.9 (150.8~506.2)

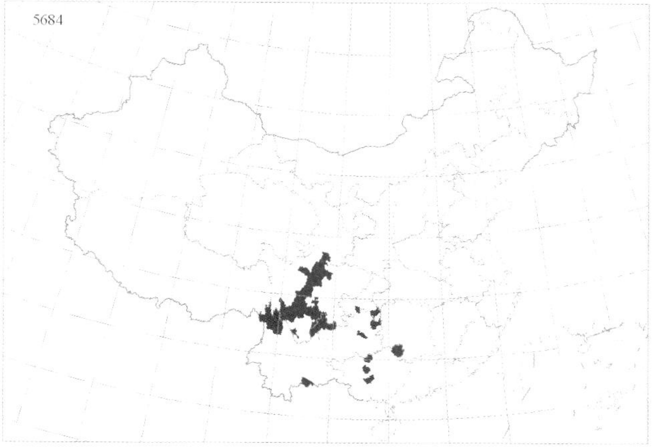

Aceraceae 槭树科 *Acer* 槭属
Acer leipoense W. P. Fang et T. P. Soong 雷波枫
Trees, deciduous

10.8; 1.6; 19.1; 10.9; 78.4; 8.0; 641 ❙ 1045; 569; 33 ❙ 64.4; 641; 263.2

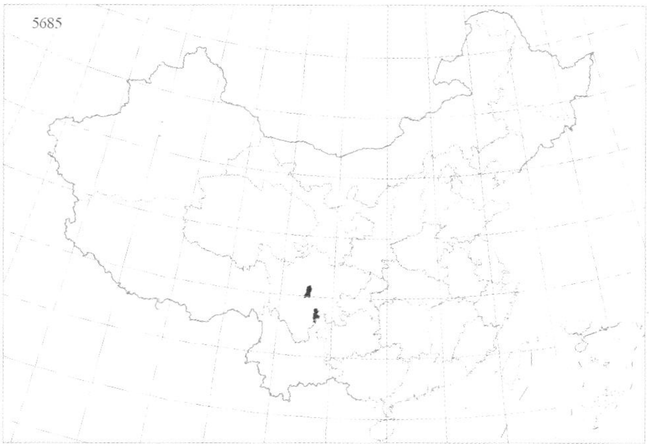

Aceraceae 槭树科 *Acer* 槭属
Acer linganense W. P. Fang et P. L. Chiu 临安枫
Small trees, deciduous

15.6 (13.4~18.2); 3.5 (1.5~6.5); 27.2 (25.2~29.1); 15.6 (13.4~18.2); 129.7 (108.3~158.8); 2.3 (0.0~7.3); 854 (755~984) ❙ 1620 (1317~1922); 582 (529~666); 202 (122~259) ❙ 89.9 (53.3~123.6); 851 (755~927); 253.1 (183.9~319.9)

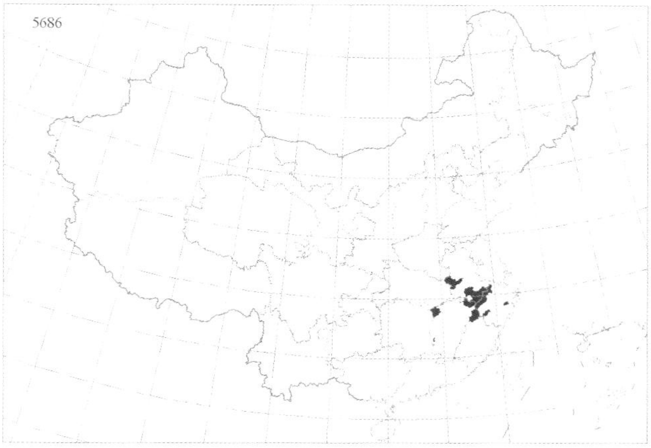

Aceraceae 槭树科 *Acer* 槭属
Acer longipes Franch. ex Rehder 长柄枫
Trees, deciduous

14.8 (9.3~18.2); 3.1 (-2.5~6.9); 25.9 (15.7~29.7); 14.9 (9.5~18.2); 122.0 (60.8~158.8); 3.6 (0.0~20.6); 820 (600~984) ❙ 1365 (741~1997); 524 (368~685); 139 (24~259) ❙ 66.3 (-2.0~130.7); 816 (600~961); 232.5 (151.3~351.7)

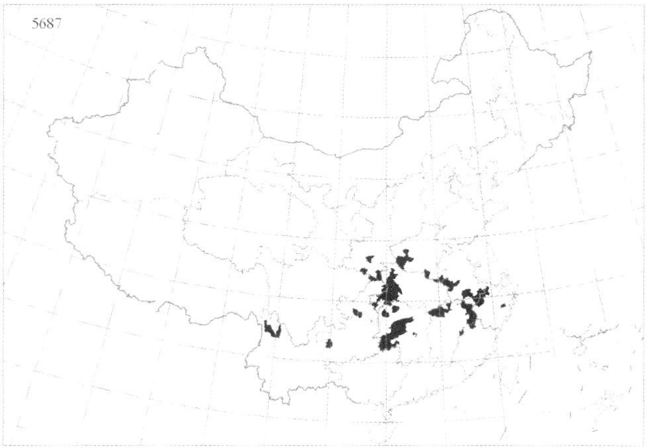

Aceraceae 槭树科 *Acer* 槭属
Acer lucidum F. P. Metcalf 亮叶枫
Trees, evergreen

17.8 (10.6~22.2); 7.6 (-0.2~13.9); 27.0 (21.1~29.8); 17.8 (10.6~22.3); 154.5 (79.4~207.5); 0.4 (0.0~12.1); 937 (643~1190) ❙ 1458 (1028~2005); 569 (442~850); 135 (46~259) ❙ 56.0 (8.0~130.7); 936 (643~1190); 217.6 (108.2~359.7)

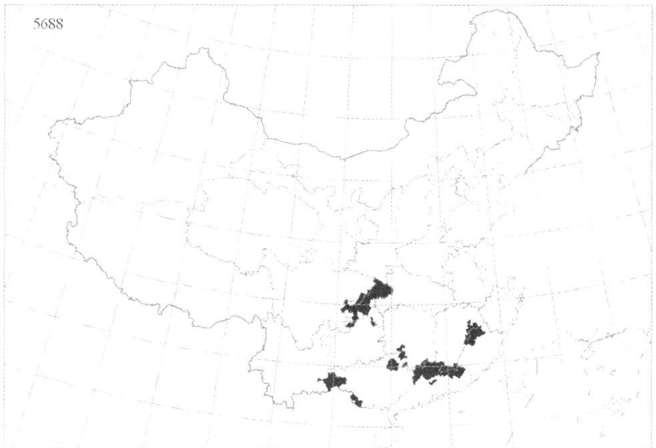

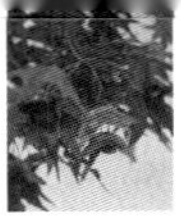

Aceraceae 槭树科 *Acer* 槭属

Acer paihengii W. P. Fang 富宁枫

Small trees, deciduous

18.9; 10.6; 25.0; 19.0; 167.5; 0.0; 939 ❙ 1292; 729; 49 ❙ 37.6; 939; 342.5

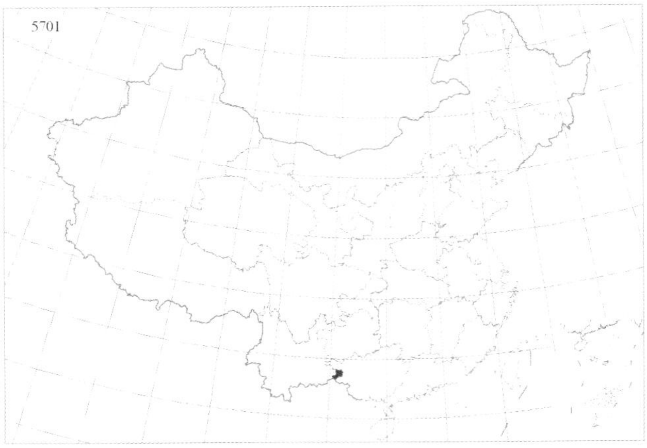

Aceraceae 槭树科 *Acer* 槭属

Acer palmatum Thunb. 鸡爪槭

Small trees, deciduous

15.4 (10.2~21.7); 3.2 (−2.9~14.0); 26.9 (18.8~29.4); 15.4 (10.3~21.7); 128.8 (75.8~200.5); 4.0 (0.0~20.5); 853 (634~1140) ❙ 1272 (512~2993); 500 (258~924); 133 (15~547) ❙ 48.1 (−35.3~203.4); 831 (512~1140); 205.9 (111.9~319.9)

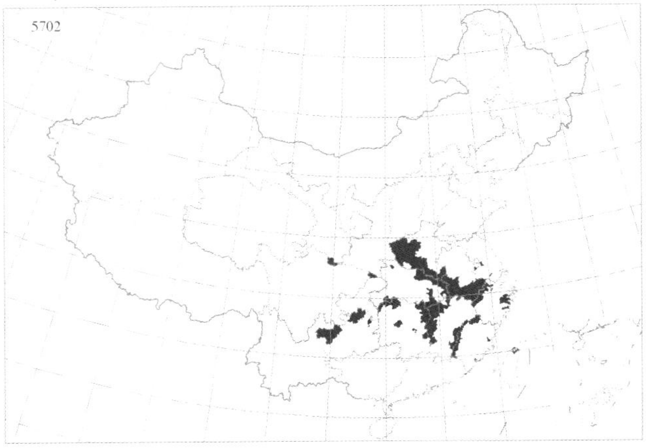

Aceraceae 槭树科 *Acer* 槭属

Acer pauciflorum Fang 稀花槭

Trees, deciduous

16.0 (13.4~23.6); 4.2 (1.5~17.8); 27.6 (25.2~29.4); 16.1 (13.4~23.6); 134.7 (108.3~223.3); 1.7 (0.0~7.3); 879 (755~1271) ❙ 1528 (1211~1828); 567 (444~685); 182 (44~253) ❙ 75.4 (−2.2~108.0); 873 (755~1091); 241.4 (111.9~439.7)

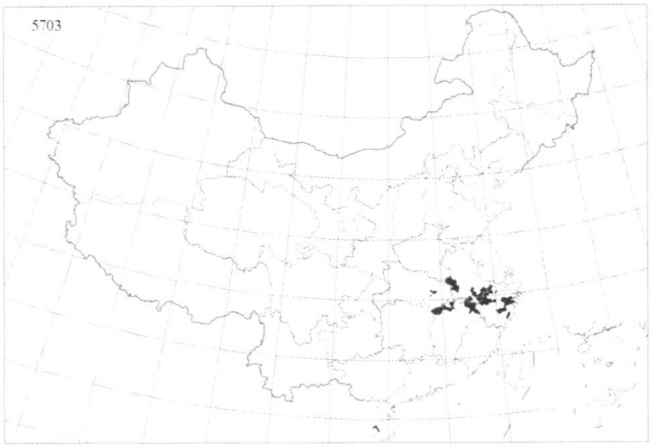

Aceraceae 槭树科 *Acer* 槭属

Acer paxii Franch. 金沙槭

Trees, evergreen

13.2 (2.0~18.9); 5.6 (−7.2~12.0); 19.4 (10.1~25.5); 13.4 (3.8~19.0); 104.6 (15.4~167.8); 5.8 (0.0~51.4); 712 (413~929) ❙ 963 (797~1381); 556 (400~641); 31 (12~107) ❙ 40.1 (−14.2~128.6); 677 (413~886); 282.5 (214.6~331.6)

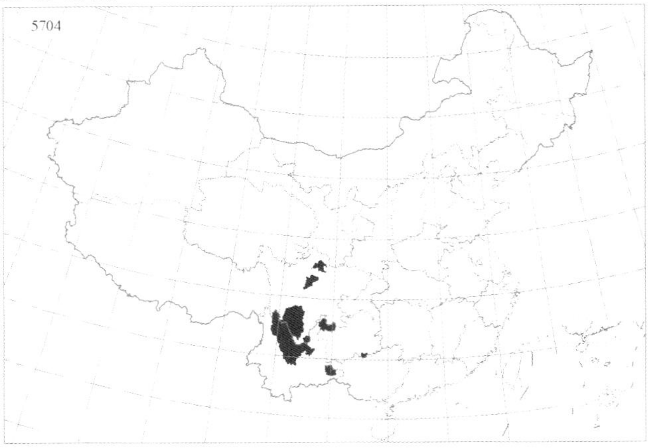

Aceraceae 槭树科 *Acer* 槭属

Acer pectinatum Wall. ex G. Nicholson 篦齿槭

Trees, deciduous

10.2 (−1.8~19.1); 2.4 (−10.2~13.0); 16.8 (6.1~25.3); 10.7 (1.8~19.1); 78.1 (1.9~169.7); 15.5 (0.0~82.8); 628 (329~917) ❙ 964 (320~1736); 553 (217~1007); 29 (7~88) ❙ 52.0 (−3.0~145.8); 610 (320~917); 287.8 (72.2~506.2)

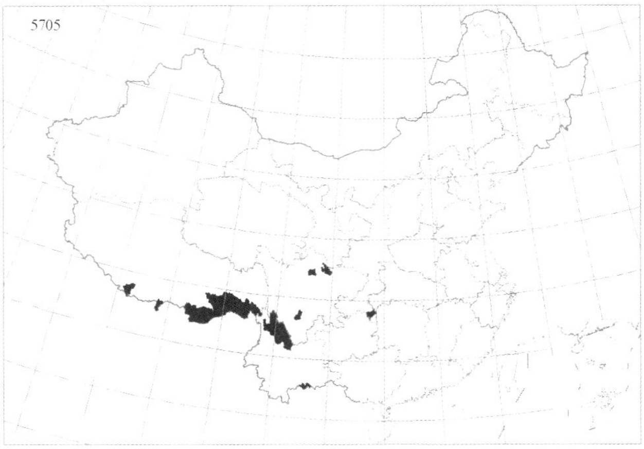

Aceraceae 槭树科 *Acer* 槭属

Acer pectinatum subsp. taronense (Hand.-Mazz.) E. Murray 独龙槭

Trees, deciduous

8.0 (2.7~13.2); 0.2 (−7.1~6.4); 14.7 (11.1~18.4); 8.3 (4.4~13.3); 52.5 (20.5~99.4); 16.1 (0.0~48.0); 557 (429~695) ❙ 998 (741~1337); 561 (426~727); 36 (18~69) ❙ 79.4 (52.8~101.2); 557 (429~695); 312.6 (200.8~407.1)

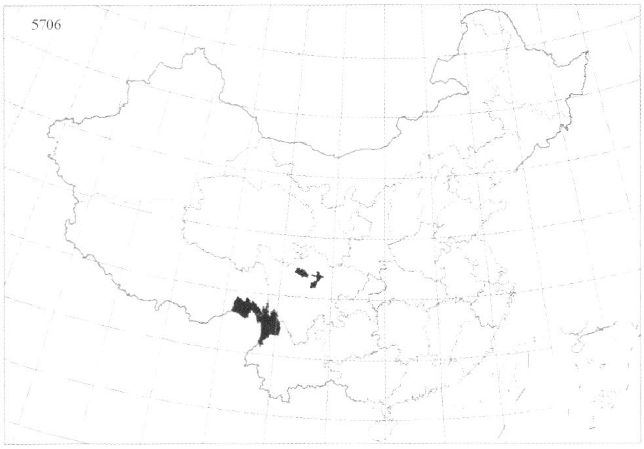

Aceraceae 槭树科 *Acer* 槭属

Acer pentaphyllum Diels 五小叶枫

Trees, evergreen

4.2; -4.1; 11.3; 5.1; 25.6; 34.8; 457 ❙ 926; 565; 16 ❙ 105.1; 457; 275.0

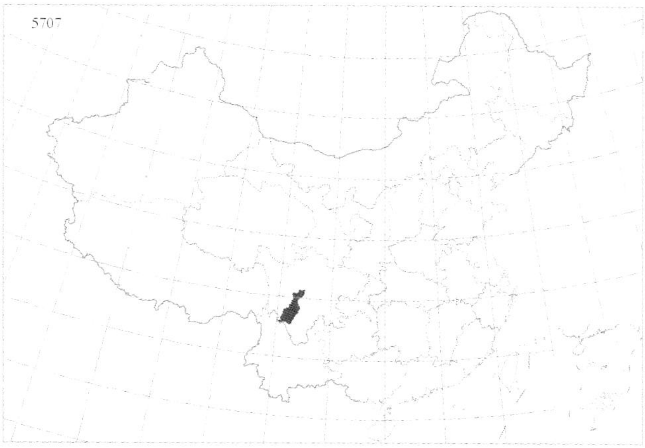

Aceraceae 槭树科 *Acer* 槭属

Acer pictum subsp. macropterum (W. P. Fang) Ohashi 大翅色木枫

Trees, deciduous

7.5 (1.8~18.6); -1.3 (-7.7~12.7); 15.2 (9.5~25.0); 8.2 (3.8~18.7); 54.6 (15.4~163.9); 24.6 (0.0~53.3); 555 (404~888) ❙ 935 (583~1547); 517 (323~911); 26 (11~71) ❙ 71.4 (27.3~128.6); 552 (404~888); 276.3 (169.5~551.3)

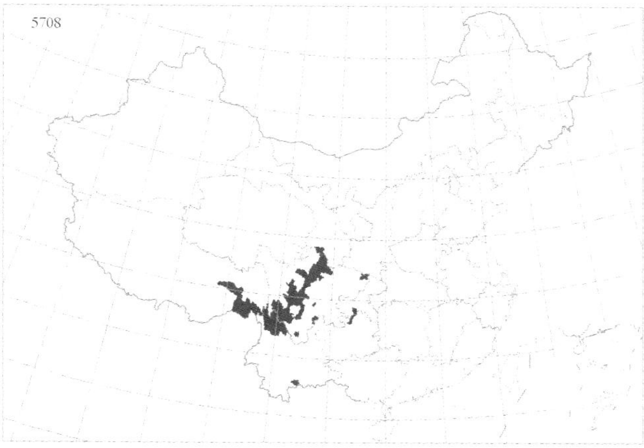

Aceraceae 槭树科 *Acer* 槭属

Acer pictum subsp. mono (Maxim.) Ohashi 五角枫

Trees, deciducus

10.6 (-0.2~17.7); -4.0 (-23.8~5.5); 23.9 (9.5~29.3); 11.8 (3.8~17.7); 95.8 (15.4~152.4); 27.7 (0.0~112.6); 719 (404~964) ❙ 811 (70~1828); 414 (23~816); 52 (5~253) ❙ 12.0 (-91.1~128.6); 657 (70~961); 199.4 (33.5~357.7)

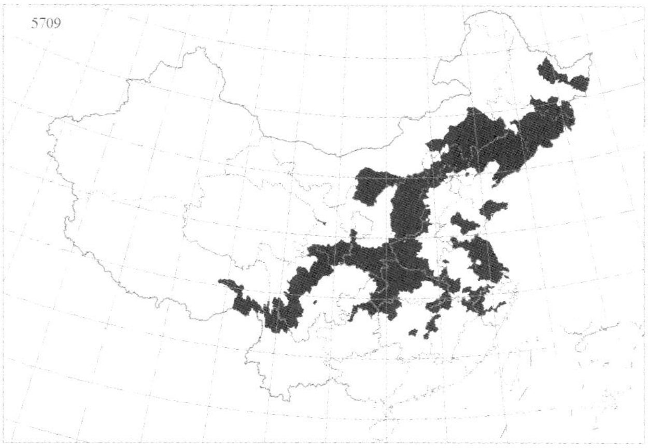

Aceraceae 槭树科 *Acer* 槭属

Acer pictum subsp. pubigerum (W. P. Fang) Y. S. Chen 江南色木枫

Trees, evergreen

15.5 (14.7~16.2); 3.4 (2.5~4.4); 27.5 (26.8~28.3); 15.6 (14.8~16.2); 129.1 (121.1~136.1); 2.0 (0.6~3.9); 858 (815~898) ❙ 1441 (1317~1581); 556 (529~587); 164 (122~196) ❙ 68.3 (53.3~87.6); 858 (815~898); 239.6 (185.5~284.0)

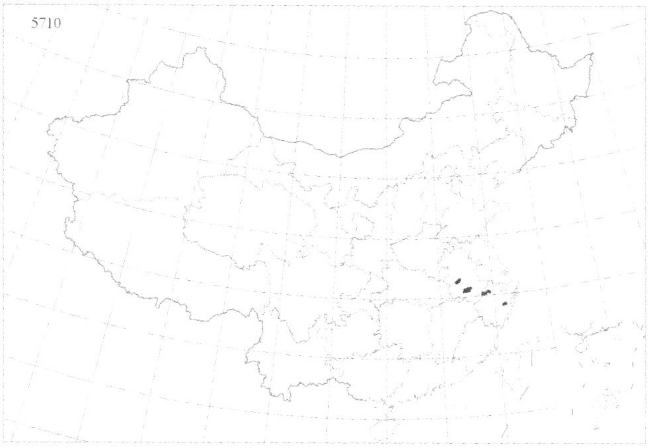

Aceraceae 槭树科 *Acer* 槭属

Acer pictum subsp. tricuspis (Rehder) Ohashi 三尖色木枫

Trees, deciduous

11.5 (7.4~15.7); -0.6 (-4.2~5.3); 23.1 (18.7~25.6); 11.7 (8.2~15.7); 92.8 (58.8~128.4); 13.9 (0.0~29.4); 695 (561~824) ❙ 839 (634~1410); 382 (290~816); 31 (16~84) ❙ 20.6 (-16.6~80.4); 690 (561~824); 259.9 (182.9~334.4)

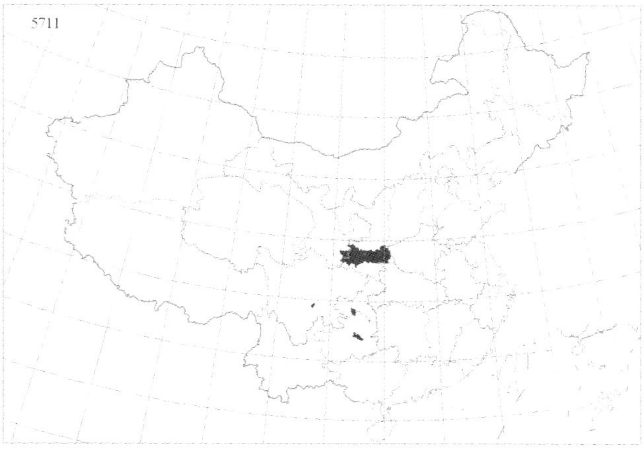

Aceraceae 槭树科 *Acer* 槭属

Acer pilosum Maxim. 疏毛枫

Shrubs or small trees, deciduous

9.6; -2.8; 21.5; 10.1; 78.3; 21.7; 641 ❙ 655; 303; 18 ❙ 1.2; 640; 220.6

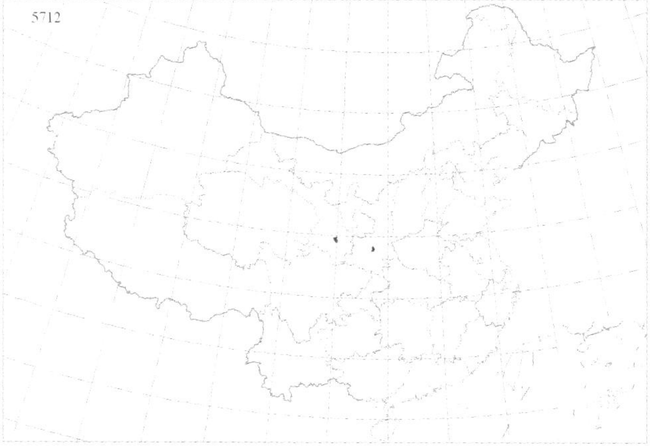

Aceraceae 槭树科 *Acer* 槭属

Acer pilosum var. stenolobum (Rehder) W. P. Fang 细裂枫

Small trees, deciduous

9.1 (3.3~13.5); -6.2 (-12.7~-1.6); 22.6 (17.5~27.1); 10.3 (6.6~13.7); 82.4 (45.4~117.5); 33.2 (14.6~65.0); 657 (507~820) ❙ 486 (132~640); 268 (81~350); 14 (2~26) ❙ -25.7 (-80.1~-2.4); 486 (132~640); 145.5 (13.6~268.0)

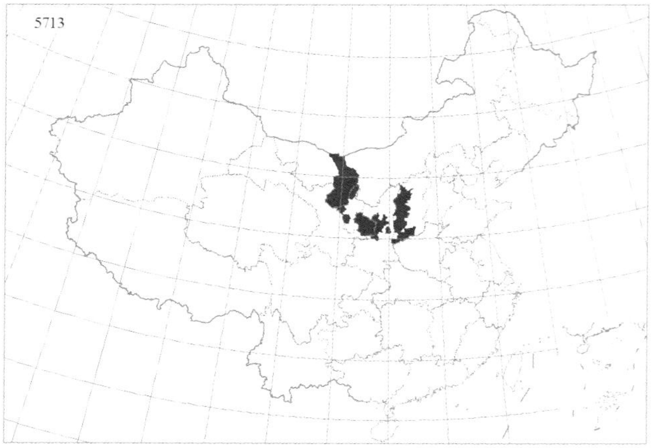

Aceraceae 槭树科 *Acer* 槭属

Acer pinnatinervium Merr. 楠叶枫

Trees, evergreen

13.0; 6.1; 18.3; 13.0; 100.3; 4.4; 703 ❙ 1045; 591; 44 ❙ 53.8; 678; 358.8

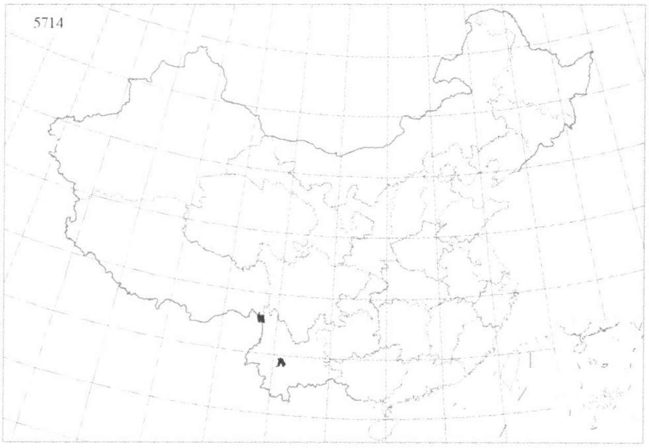

Aceraceae 槭树科 *Acer* 槭属

Acer poliophyllum W. P. Fang et Y. T. Wu 灰叶枫

Small trees, evergreen

16.9; 8.5; 23.2; 16.9; 142.9; 0.0; 834 ❙ 1314; 710; 62 ❙ 57.5; 834; 301.4

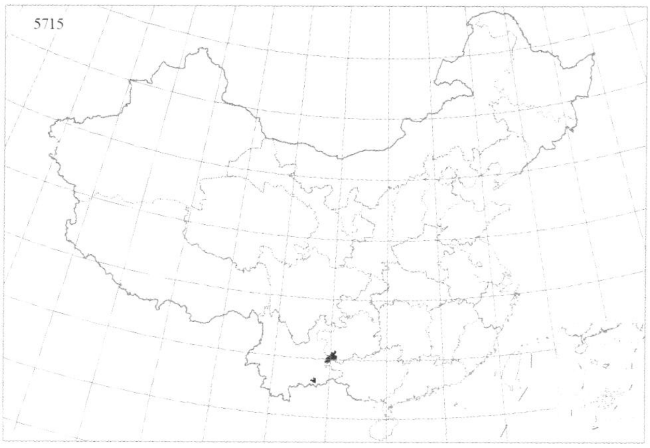

Aceraceae 槭树科 *Acer* 槭属

Acer pseudosieboldianum (Pax) Kom. 紫花枫

Shrubs or small trees, deciduous

5.0 (1.8~9.5); -14.3 (-18.8~-7.2); 21.6 (18.5~25.0); 8.6 (6.7~10.8); 67.7 (47.6~91.3); 67.0 (35.9~88.8); 598 (513~702) ❙ 744 (484~1073); 462 (333~691); 26 (7~40) ❙ 25.3 (-27.9~74.1); 586 (484~698); 286.2 (140.8~346.5)

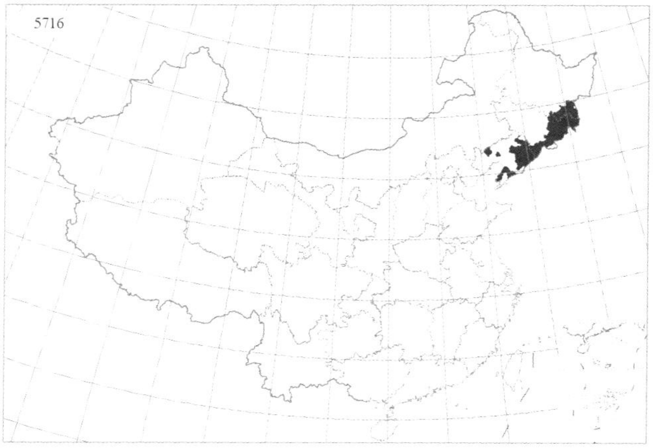

Aceraceae 槭树科 *Acer* 槭属

Acer pubinerve Rehder 毛脉枫

Trees, deciduous

17.3 (14.1~22.0); 6.5 (1.6~13.5); 27.1 (23.5~29.8); 17.3 (14.2~22.0); 148.2 (112.6~203.8); 0.5 (0.0~6.5); 915 (756~1170) ❙ 1566 (1177~1922); 576 (427~820); 177 (57~259) ❙ 71.9 (27.5~129.9); 912 (756~1170); 239.3 (174.9~371.7)

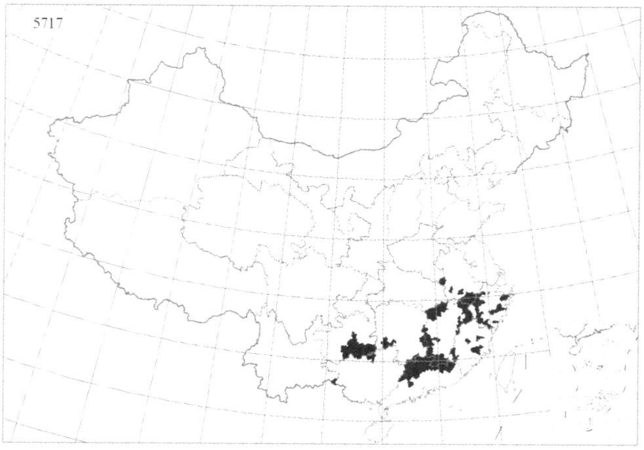

Aceraceae 槭树科 *Acer* 槭属

Acer pubipetiolatum Hu et Cheng 毛柄枫

Trees, deciduous

15.1 (8.6~18.2); 8.4 (1.3~11.9); 20.0 (15.0~22.7); 15.2 (8.7~18.2); 122.4 (52.6~158.7); 0.4 (0.0~8.8); 760 (574~868) ❙ 1195 (1019~1736); 670 (553~1007); 51 (35~69) ❙ 58.4 (26.1~145.8); 754 (574~854); 390.4 (335.1~516.7)

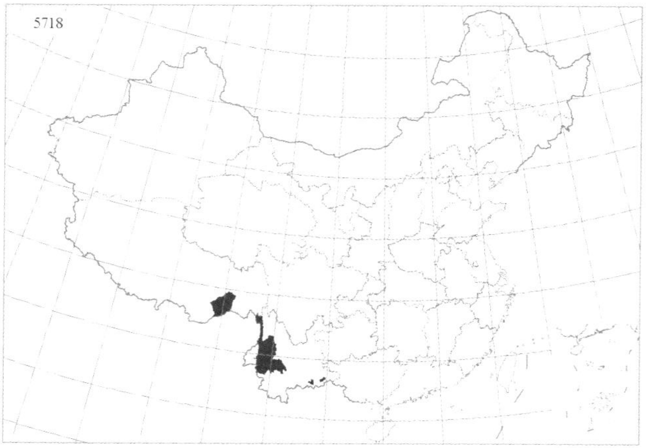

Aceraceae 槭树科 *Acer* 槭属
Acer pubipetiolatum var. pingpienense W. P. Fang et W. K. Hu 屏边毛柄枫
Trees, deciduous

16.4 (14.3~21.0); 6.8 (3.8~14.2); 24.8 (22.1~25.9); 16.5 (14.3~21.0); 138.1 (112.6~191.8); 0.4 (0.0~1.5); 847 (756~1053) ❙ 1300 (1218~1522); 532 (473~818); 89 (55~107) ❙ 53.9 (44.5~71.9); 847 (756~1053); 283.1 (170.4~551.6)

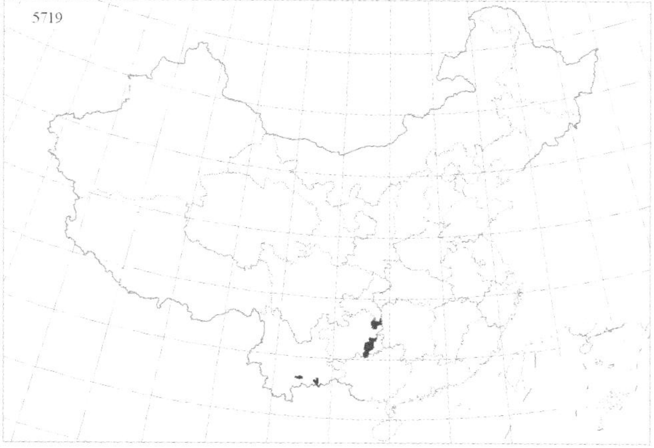

Aceraceae 槭树科 *Acer* 槭属
Acer shenkanense W. P. Fang et C. C. Fu 陕甘枫
Trees

10.7 (1.6~16.9); 0.2 (-8.0~10.3); 20.3 (9.8~28.1); 11.0 (3.7~16.9); 82.8 (15.2~143.4); 14.1 (0.0~56.2); 659 (408~924) ❙ 931 (562~1280); 460 (288~646); 37 (9~94) ❙ 43.6 (-23.0~128.6); 652 (408~924); 250.4 (162.7~383.4)

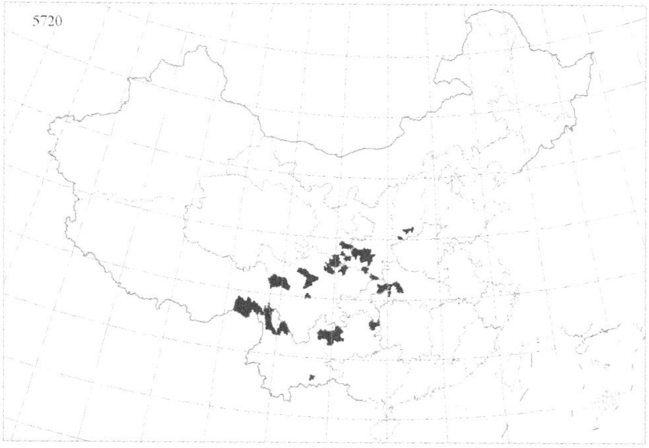

Aceraceae 槭树科 *Acer* 槭属
Acer shihweii F. Chun et W. P. Fang 平坝枫
Trees, evergreen

15.5; 5.7; 24.2; 15.6; 126.7; 0.0; 801 ❙ 1248; 600; 64 ❙ 55.8; 801; 172.1

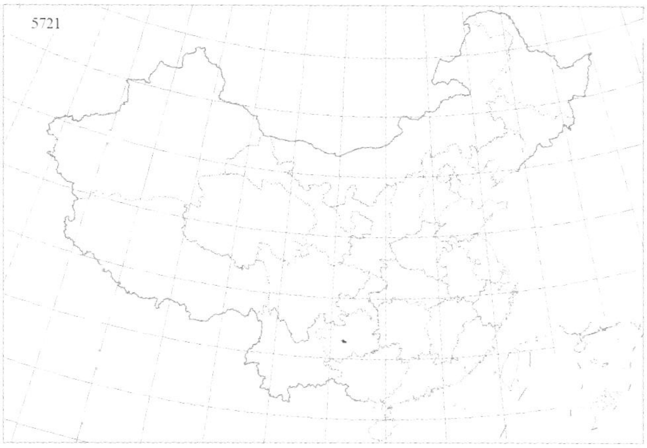

Aceraceae 槭树科 *Acer* 槭属
Acer sikkimense Miq. 锡金枫
Trees, deciduous

14.6 (0.3~21.0); 7.5 (-8.6~14.2); 20.1 (8.3~26.6); 14.7 (2.9~21.0); 117.7 (9.6~191.8); 2.1 (0.0~65.4); 753 (368~1053) ❙ 1092 (389~1751); 625 (246~1007); 41 (11~77) ❙ 45.9 (-14.2~145.8); 732 (368~1053); 333.3 (142.5~551.6)

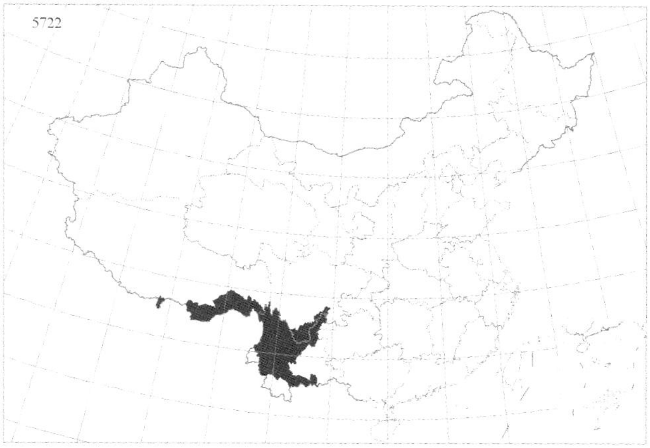

Aceraceae 槭树科 *Acer* 槭属
Acer sinense Pax 中华枫
Trees, deciduous

17.0 (2.6~23.6); 6.8 (-7.0~16.7); 26.0 (11.2~29.8); 17.0 (4.3~23.6); 145.8 (20.0~223.5); 1.7 (0.0~48.2); 903 (427~1281) ❙ 1378 (626~2435); 593 (316~1428); 118 (20~259) ❙ 53.3 (-22.2~145.8); 895 (427~1273); 236.6 (139.1~659.0)

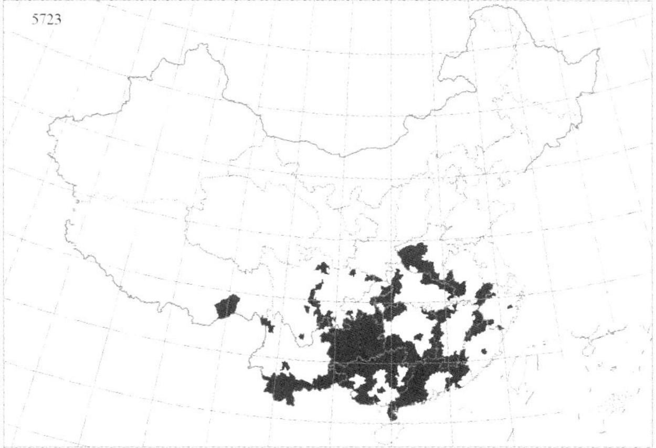

Aceraceae 槭树科 *Acer* 槭属
Acer sino-oblongum F. P. Metcalf 滨海枫
Trees, evergreen

21.3 (17.6~22.5); 13.3 (7.3~15.2); 27.9 (25.0~29.0); 21.4 (17.6~22.7); 196.6 (151.6~212.7); 0.0 (0.0~0.0); 1121 (909~1216) ❙ 1648 (1188~2037); 752 (539~1033); 123 (56~167) ❙ 47.1 (5.8~79.9); 1119 (909~1216); 228.6 (139.6~371.7)

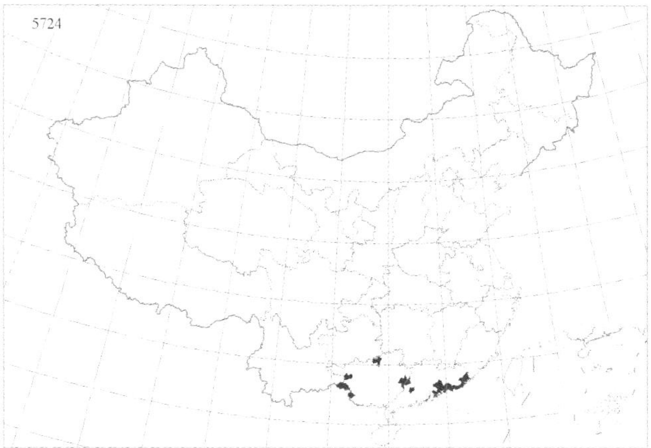

Aceraceae 槭树科　*Acer* 槭属

Acer sinopurpurascens Cheng 天目枫

Trees, deciduous

15.5 (13.4~17.2); 3.5 (1.5~5.9); 27.2 (25.2~29.3); 15.6 (13.4~17.2); 129.1 (108.3~146.8); 2.2 (0.0~7.3); 852 (755~941) ❙ 1508 (1211~1791); 572 (427~685); 174 (105~220) ❙ 77.8 (28.8~115.7); 850 (755~941); 244.4 (185.5~319.9)

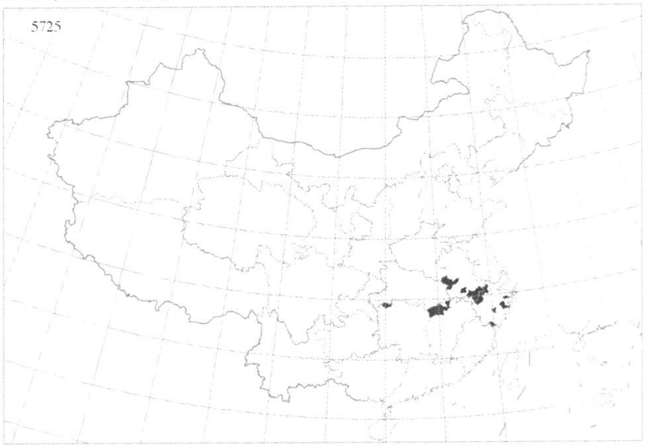

Aceraceae 槭树科　*Acer* 槭属

Acer stachyophyllum Hiern 毛叶枫

Trees, deciduous

7.9 (−1.1~14.9); −1.0 (−9.6~5.3); 16.0 (6.2~24.4); 8.5 (2.0~15.0); 57.0 (2.5~119.6); 21.3 (0.0~75.6); 566 (334~782) ❙ 988 (321~1435); 531 (193~817); 34 (10~88) ❙ 74.9 (−4.0~128.6); 565 (321~782); 260.7 (61.6~363.7)

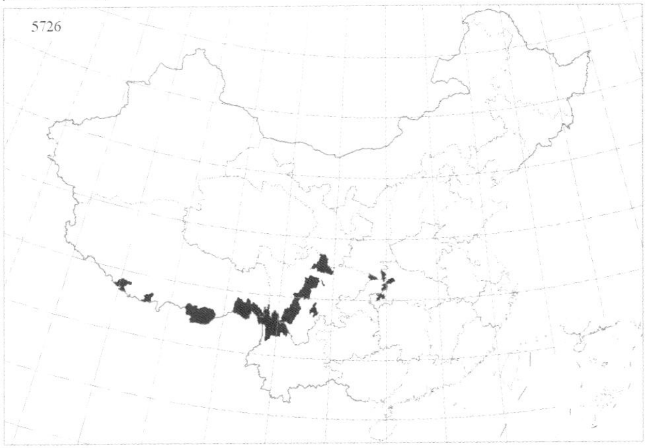

Aceraceae 槭树科　*Acer* 槭属

Acer stachyophyllum subsp. betulifolium (Maxim.) P. C. de Jong 四蕊枫

Trees, deciduous

10.0 (−1.1~16.2); −1.0 (−10.2~8.6); 20.3 (6.2~26.8); 10.5 (2.0~16.2); 79.1 (2.5~134.6); 18.0 (0.0~75.6); 646 (334~854) ❙ 888 (321~1492); 430 (190~817); 35 (7~113) ❙ 38.5 (−40.9~128.6); 639 (321~854); 250.7 (61.2~372.3)

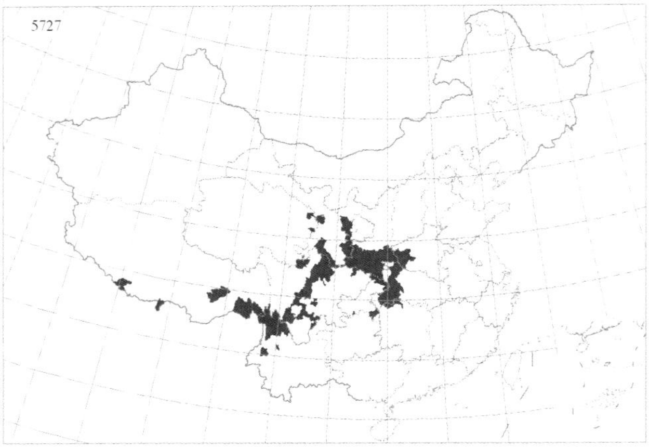

Aceraceae 槭树科　*Acer* 槭属

Acer sterculiaceum Wallich 苹婆枫

Trees, deciduous

10.3 (−1.8~14.5); 3.3 (−10.2~7.8); 16.3 (6.1~19.7); 10.7 (1.8~14.5); 74.7 (1.9~113.5); 10.4 (0.0~82.8); 621 (329~730) ❙ 971 (320~1164); 561 (217~651); 38 (20~58) ❙ 54.5 (−3.0~89.0); 619 (320~730); 312.1 (72.2~382.6)

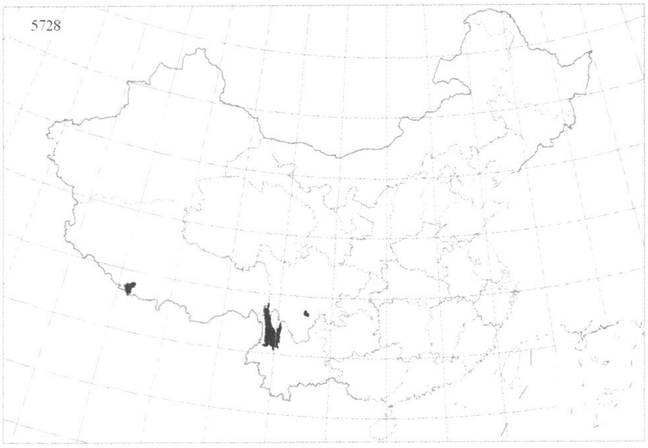

Aceraceae 槭树科　*Acer* 槭属

Acer sterculiaceum subsp. franchetii (Pax) A. E. Murray 房县枫

Trees, deciduous

14.0 (5.4~21.0); 4.8 (−5.2~14.2); 22.1 (12.3~29.3); 14.1 (5.8~21.0); 113.2 (31.5~191.8); 4.2 (0.0~29.4); 753 (481~1053) ❙ 1057 (520~1852); 538 (245~860); 50 (12~204) ❙ 40.8 (−36.2~106.2); 740 (481~1053); 270.7 (131.4~551.6)

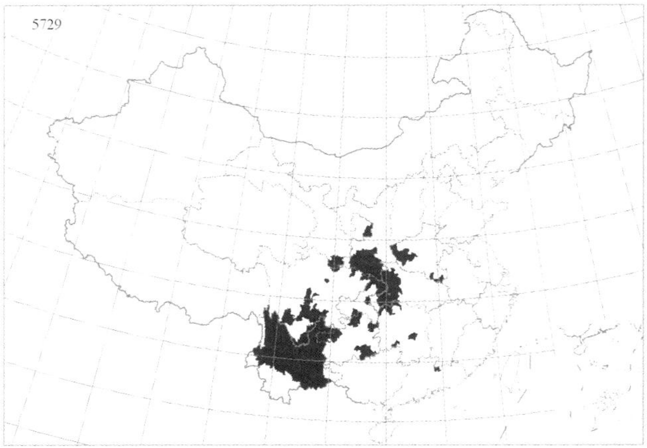

Aceraceae 槭树科　*Acer* 槭属

Acer sutchuenense Franch. 四川枫

Small trees, deciduous

11.5 (6.2~15.3); 1.4 (−2.4~5.3); 21.1 (14.1~26.5); 11.6 (6.7~15.4); 87.6 (39.4~125.7); 8.4 (0.0~24.1); 676 (509~832) ❙ 1181 (964~1441); 568 (421~816); 54 (23~113) ❙ 75.7 (31.7~101.4); 676 (509~832); 254.3 (213.4~292.6)

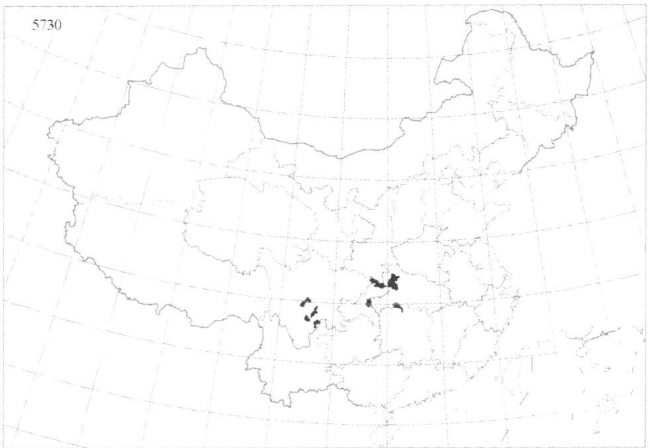

Aceraceae 槭树科 *Acer* 槭属

Acer sycopseoides Chun 角叶枫

Small trees, evergreen

17.6 (15.6~19.2); 7.8 (5.5~11.0); 25.8 (24.4~27.9); 17.7 (15.7~19.2); 152.2 (128.3~170.8); 0.0 (0.0~0.0); 907 (810~1002) ❙ 1432 (1272~1710); 649 (552~820); 115 (57~195) ❙ 58.1 (47.9~86.9); 907 (810~1002); 258.2 (215.4~371.7)

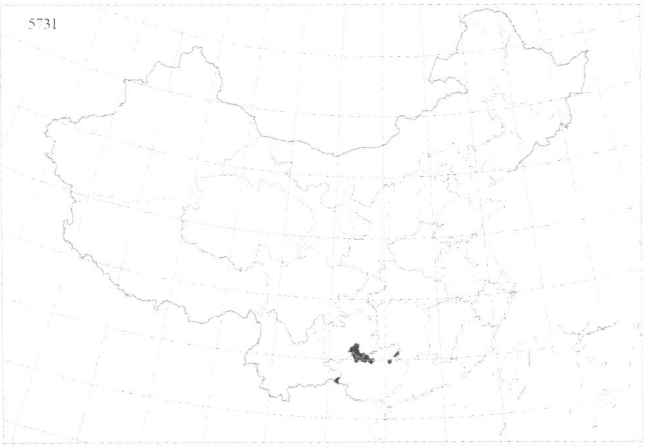

Aceraceae 槭树科 *Acer* 槭属

Acer tataricum subsp. ginnala (Maxim.) Wesmael 茶条枫

Shrubs or small trees, deciduous

10.4 (-4.0~17.4); -4.8 (-28.2~5.4); 24.2 (13.1~29.2); 11.8 (4.9~17.5); 96.1 (27.4~149.9); 30.5 (0.0~146.2); 721 (445~950) ❙ 757 (111~1893); 390 (74~691); 50 (2~246) ❙ 2.9 (-82.3~129.9); 643 (111~936); 190.5 (13.9~346.5)

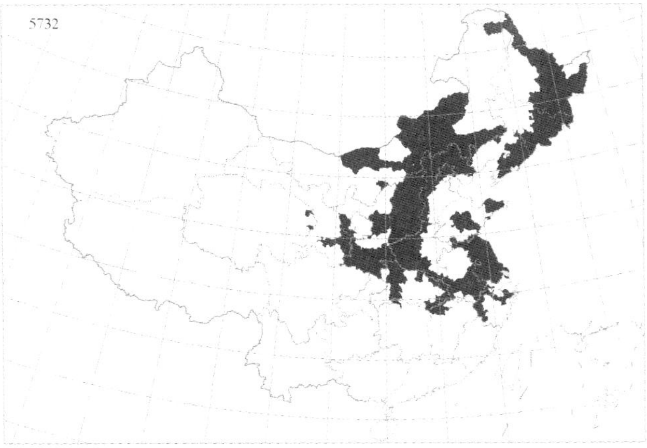

Aceraceae 槭树科 *Acer* 槭属

Acer tataricum subsp. semenovii (Regel et Herder) A. E. Murray 天山枫

Shrubs or small trees, deciduous

2.9 (-1.3~9.0); -15.7 (-20.9~-9.3); 17.3 (13.2~23.2); 7.0 (4.6~10.7); 50.4 (26.7~88.9); 74.7 (40.7~103.9); 537 (440~703) ❙ 310 (225~406); 143 (65~211); 27 (12~48) ❙ -40.1 (-65.1~-7.6); 310 (225~406); 216.4 (165.7~297.3)

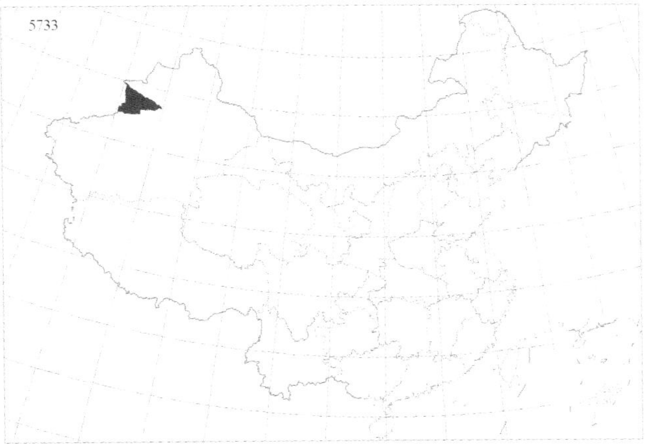

Aceraceae 槭树科 *Acer* 槭属

Acer tataricum subsp. theiferum (W. P. Fang) Y. S. Chen et P. C. de Jong 苦条枫

Shrubs or small trees, deciduous

16.0 (2.5~21.0); 4.0 (-9.7~12.2); 27.2 (13.1~29.8); 16.1 (5.0~21.0); 136.4 (27.9~192.6); 4.1 (0.0~56.4); 882 (449~1113) ❙ 1316 (516~1893); 509 (281~711); 143 (9~242) ❙ 48.1 (-22.2~129.9); 864 (449~1107); 212.7 (139.1~319.9)

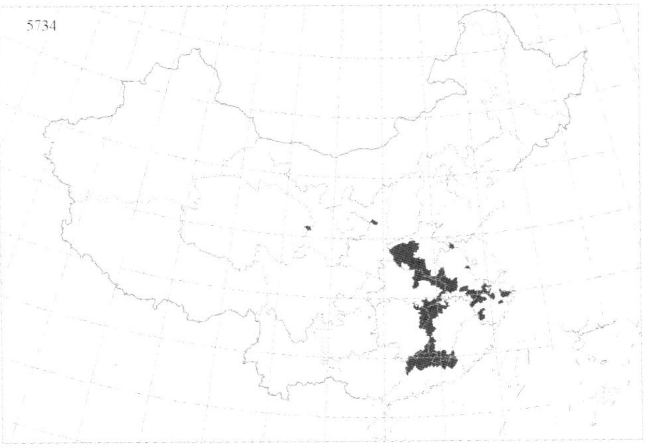

Aceraceae 槭树科 *Acer* 槭属

Acer tegmentosum Maxim. 青楷枫

Trees, deciduous

6.6 (-0.2~15.2); -11.9 (-23.8~2.3); 22.5 (18.5~27.4); 9.5 (6.7~15.3); 76.7 (47.6~127.7); 57.3 (4.3~112.6); 638 (513~860) ❙ 708 (409~1198); 455 (258~691); 24 (8~109) ❙ 12.7 (-24.3~74.1); 596 (409~860); 262.6 (136.6~346.5)

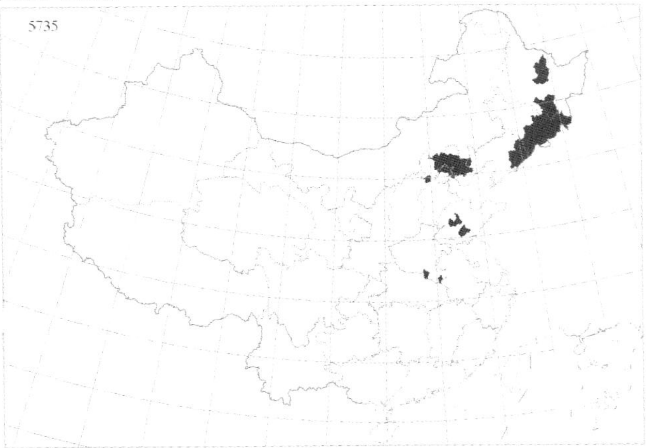

Aceraceae 槭树科 *Acer* 槭属

Acer tenellum Pax 薄叶枫

Trees, deciduous

14.4 (13.4~15.3); 3.3 (2.6~4.3); 25.0 (23.5~26.5); 14.4 (13.4~15.4); 115.5 (104.5~125.7); 2.2 (0.7~4.0); 783 (735~832) ❙ 1306 (1148~1448); 532 (475~600); 84 (65~113) ❙ 66.9 (51.2~95.9); 783 (735~832); 218.8 (182.9~258.0)

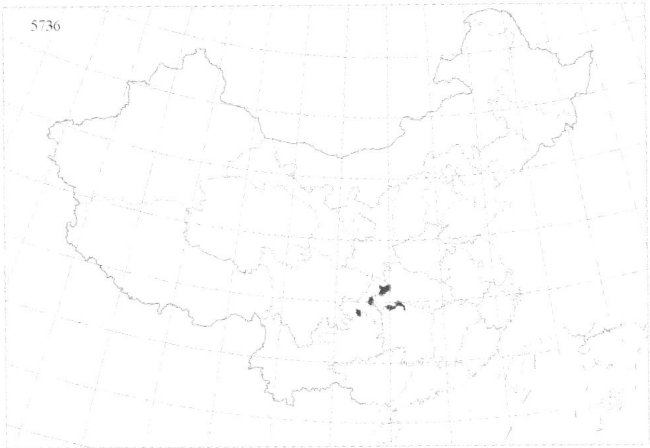

Aceraceae 槭树科 *Acer* 槭属
Acer tenellum var. septemlobum (W. P. Fang et T. P. Soong) W. P. Fang et T. P. Soong 七裂薄叶枫
Trees, deciduous

14.8；4.3；24.9；14.9；119.2；0.7；791 ❙ 1197；489；71 ❙ 51.2；791；182.9

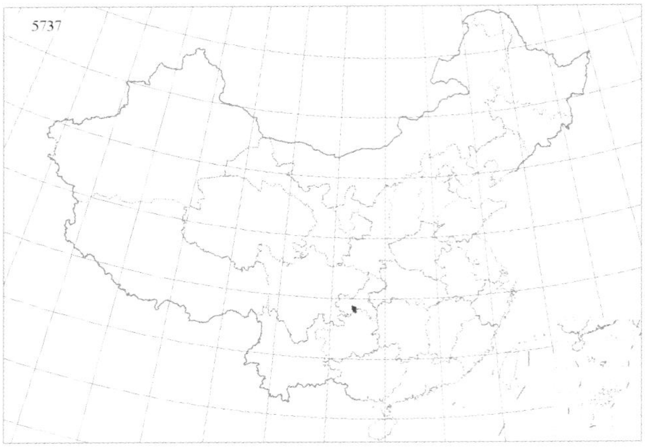

Aceraceae 槭树科 *Acer* 槭属
Acer thomsonii Miq. 巨果枫
Trees, deciduous

11.8 (−1.5~20.5)；4.8 (−10.5~15.1)；17.2 (7.1~23.7)；12.9 (2.2~20.5)；107.3 (4.7~185.9)；24.7 (0.0~82.2)；710 (332~992) ❙ 1117 (311~1607)；556 (227~817)；42 (11~77) ❙ 47.0 (−6.3~113.0)；696 (311~986)；385.3 (62.8~659.0)

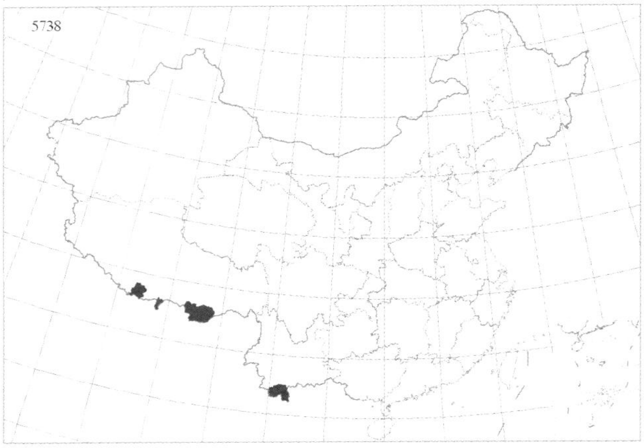

Aceraceae 槭树科 *Acer* 槭属
Acer tibetense W. P. Fang 察隅枫
Trees, deciduous

5.3；−2.4；12.2；5.7；30.5；26.2；479 ❙ 961；558；31 ❙ 100.7；479；234.2

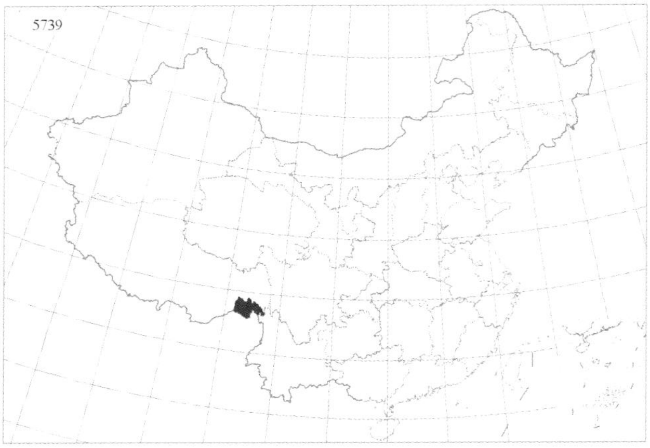

Aceraceae 槭树科 *Acer* 槭属
Acer tonkinense Lecomte 粗柄枫
Trees, deciduous

18.8 (13.3~22.2)；10.0 (5.4~13.9)；25.8 (19.6~28.6)；18.9 (13.4~22.3)；166.7 (100.4~207.5)；0.0 (0.0~0.0)；968 (706~1190) ❙ 1440 (1074~1740)；707 (497~1007)；106 (43~196) ❙ 51.0 (5.8~145.8)；968 (706~1190)；302.0 (205.4~491.5)

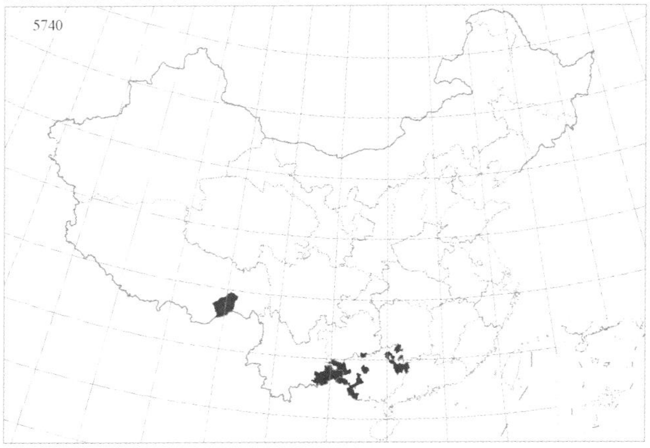

Aceraceae 槭树科 *Acer* 槭属
Acer triflorum Kom. 三花枫
Trees, deciduous

3.8 (1.7~7.0)；−16.2 (−20.1~−11.0)；20.8 (18.5~22.8)；8.0 (6.7~9.3)；61.1 (47.6~76.0)；75.5 (51.4~93.6)；571 (513~624) ❙ 747 (561~1073)；463 (333~691)；26 (16~40) ❙ 30.4 (−2.6~74.1)；569 (513~624)；310.8 (264.2~346.5)

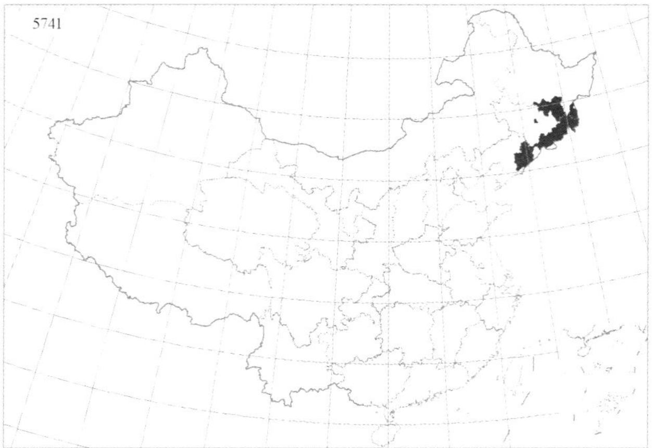

Aceraceae 槭树科 *Acer* 槭属
Acer truncatum Bunge 元宝枫
Trees, deciduous

10.0 (−0.2~17.1)；−5.4 (−23.8~4.7)；23.8 (17.5~29.4)；11.2 (6.4~17.1)；91.9 (45.4~146.1)；31.5 (0.3~112.6)；702 (506~952) ❙ 648 (304~1689)；371 (190~691)；27 (5~212) ❙ −7.6 (−48.6~105.7)；601 (304~941)；186.7 (61.2~338.4)

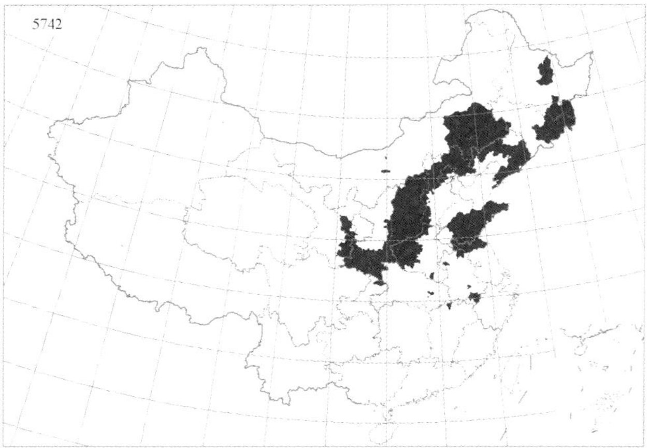

Aceraceae 槭树科 *Acer* 槭属

Acer tsinglingense W. P. Fang et C. C. Hsieh, 秦岭枫

Trees, deciduous

12.3 (6.2~16.2); -0.7 (-4.9~5.5); 24.4 (15.8~28.3); 12.5 (7.0~16.2); 102.0 (45.2~134.7); 13.5 (0.0~29.9); 743 (522~887) ❙ 779 (520~1489); 378 (245~644); 41 (15~196) ❙ 6.0 (-36.2~89.0); 677 (520~887); 203.0 (128.6~318.2)

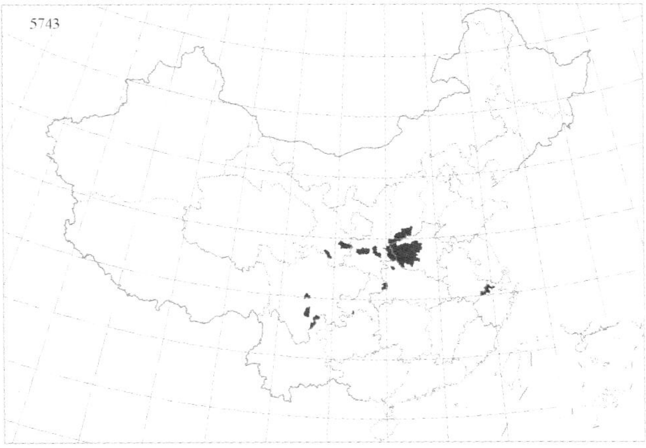

Aceraceae 槭树科 *Acer* 槭属

Acer tutcheri Duthie 岭南枫

Trees, deciduous

19.1 (14.3~24.1); 9.9 (0.5~18.9); 27.2 (20.1~29.8); 19.2 (14.3~24.1); 170.6 (113.2~229.7); 0.2 (0.0~9.2); 1002 (761~1322) ❙ 1728 (687~3495); 700 (393~1847); 185 (47~897) ❙ 75.0 (-23.1~246.5); 992 (687~1216); 236.5 (119.4~363.7)

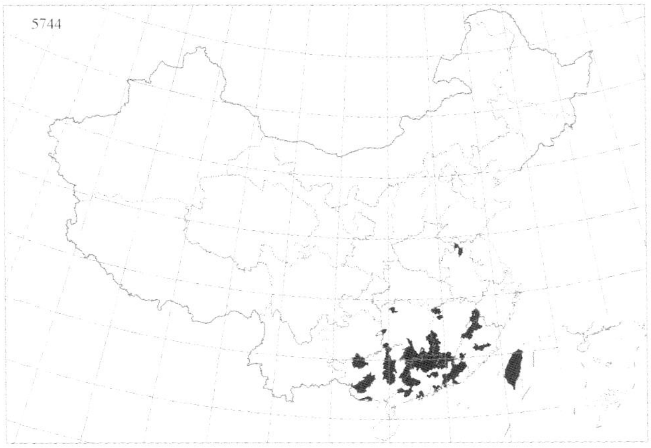

Aceraceae 槭树科 *Acer* 槭属

Acer ukurunduense Trautv. et C. A. Mey. 花楷枫

Trees, deciduous

3.3 (-0.2~6.5); -16.9 (-23.8~-6.5); 20.6 (18.3~22.8); 7.8 (6.7~9.2); 58.9 (47.6~75.2); 78.8 (38.9~112.6); 564 (513~624) ❙ 713 (558~1073); 438 (298~691). 25 (13~40) ❙ 26.0 (-4.3~74.1); 561 (513~624); 308.1 (229.4~346.5)

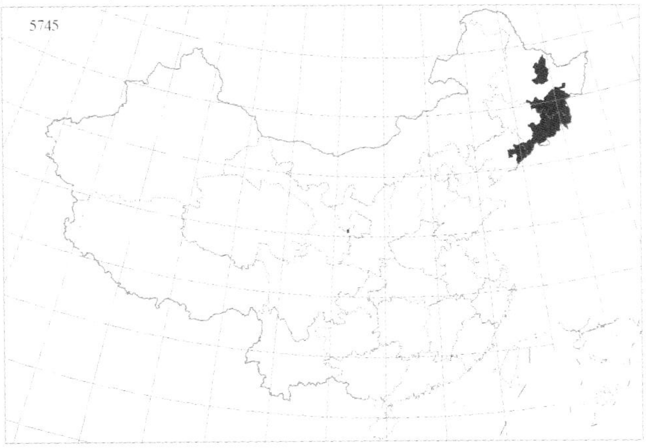

Aceraceae 槭树科 *Acer* 槭属

Acer wangchii W. P. Fang 天峨枫

Trees, evergreen

16.9 (10.6~19.1); 6.3 (-0.2~10.1); 26.7 (21.1~28.9); 17.0 (10.6~19.2); 144.2 (79.4~170.1); 0.8 (0.0~12.1); 898 (643~1002) ❙ 1181 (1028~1381); 513 (442~701); 64 (46~107) ❙ 33.2 (8.0~93.8); 898 (643~1002); 177.4 (108.2~322.1)

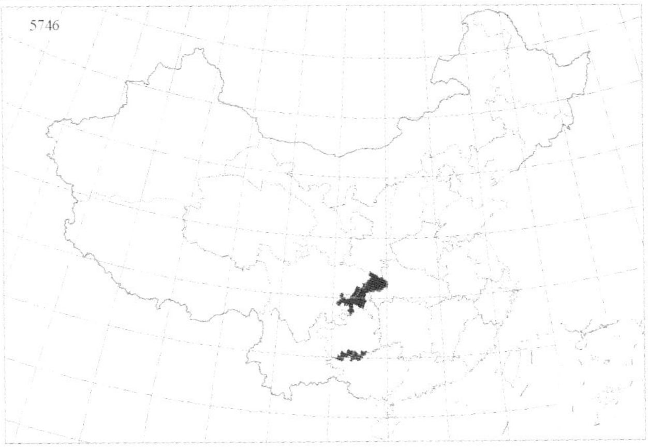

Aceraceae 槭树科 *Acer* 槭属

Acer wardii W. W. Sm. 滇藏枫

Trees, deciduous

8.2 (0.7~13.2); 0.8 (-8.8~6.4); 14.6 (9.5~18.4); 8.6 (3.3~13.3); 54.4 (12.7~99.4); 15.4 (0.0~63.9); 563 (389~695) ❙ 975 (514~1337); 563 (314~727); 37 (10~69) ❙ 71.4 (32.3~100.7); 563 (389~695); 319.9 (234.2~407.1)

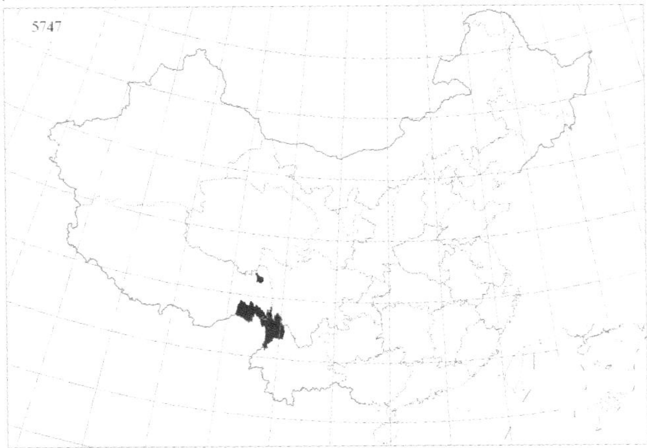

Aceraceae 槭树科 *Acer* 槭属

Acer wilsonii Rehder 三峡枫

Trees, evergreen

17.3 (8.6~22.3); 7.4 (1.3~15.1); 26.1 (15.0~29.8); 17.4 (8.7~22.3); 149.1 (52.6~208.1); 0.5 (0.0~8.8); 909 (574~1202) ❙ 1459 (1007~1997); 585 (403~996); 137 (41~254) ❙ 61.4 (15.4~147.0); 905 (574~1202); 251.6 (146.0~659.0)

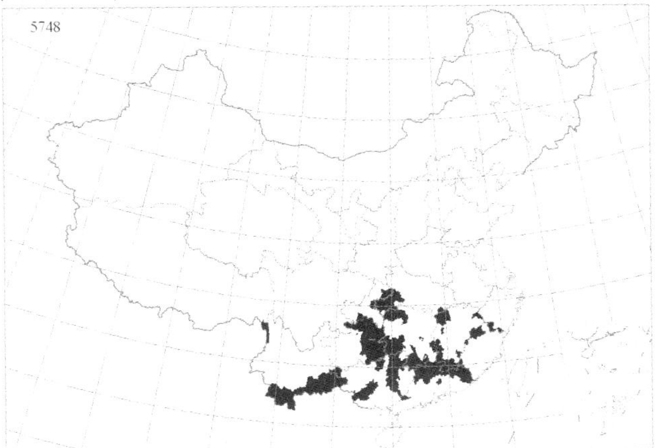

Aceraceae 槭树科 *Acer* 槭属
Acer yinkunii W. P. Fang 都安枫
Small trees, evergreen

20.0; 10.5; 27.6; 20.1; 180.9; 0.0; 1049 ❙ 1555; 755; 144 ❙ 48.3; 1049; 230.3

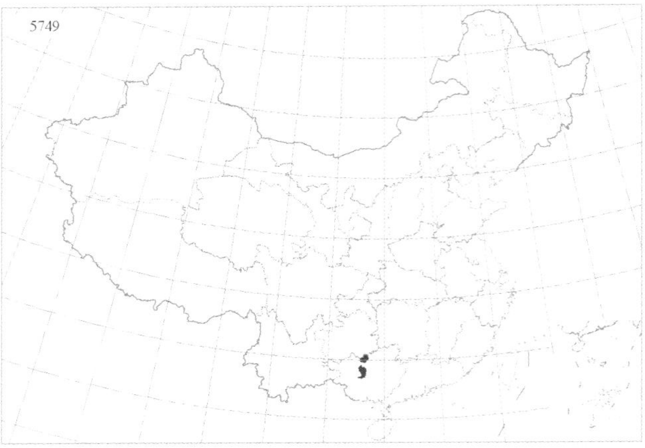

Aceraceae 槭树科 *Acer* 槭属
Acer yui W. P. Fang 川甘枫
Trees, deciduous

4.3; -6.1; 13.9; 5.7; 33.2; 40.5; 473 ❙ 671; 324; 16 ❙ 42.8; 473; 276.6

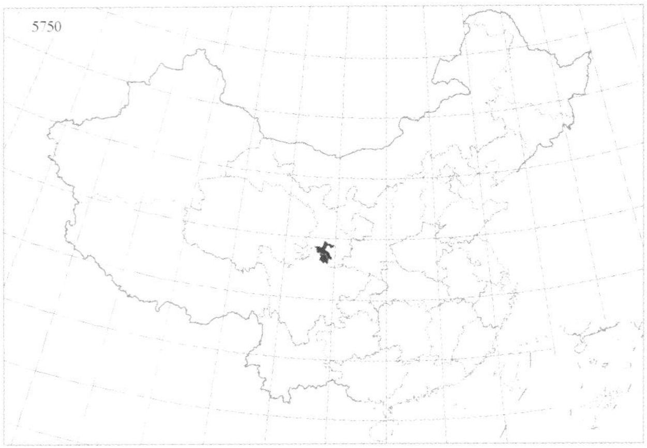

Aceraceae 槭树科 *Dipteronia* 金钱槭属
Dipteronia dyeriana Henry 云南金钱枫
Trees, deciduous

17.7 (16.3~21.0); 10.1 (8.5~14.2); 23.1 (21.4~26.0); 17.7 (16.4~21.0); 152.5 (136.4~191.8); 0.0 (0.0~0.0); 867 (797~1053) ❙ 1300 (1074~1522); 720 (582~860); 54 (47~68) ❙ 50.3 (30.0~80.6); 867 (797~1053); 372.6 (277.0~551.6)

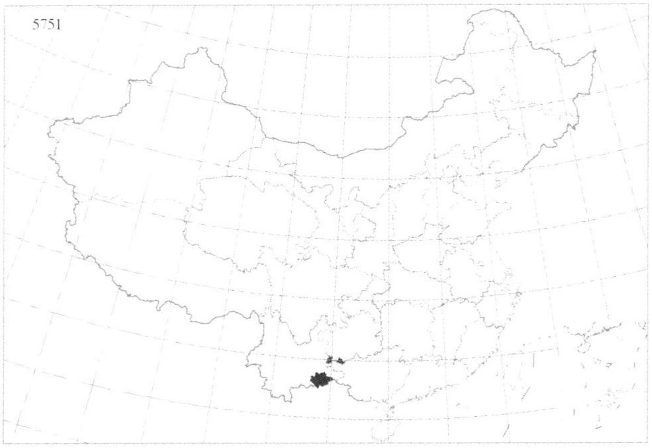

Aceraceae 槭树科 *Dipteronia* 金钱槭属
Dipteronia sinensis Oliv. 金钱枫
Small trees, deciduous

12.1 (1.6~17.5); 0.2 (-8.0~7.6); 23.3 (9.8~28.4); 12.3 (3.7~17.6); 98.2 (15.2~150.7); 12.1 (0.0~56.2); 723 (408~924) ❙ 942 (373~1562); 429 (201~688); 47 (7~193) ❙ 32.5 (-37.6~125.9); 706 (373~924); 232.9 (88.1~334.4)

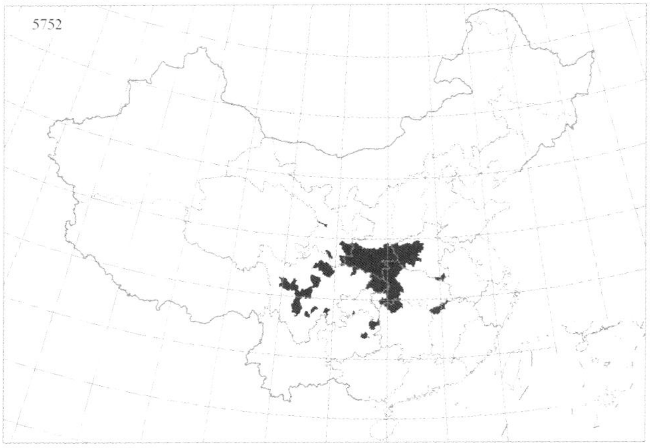

Dipentodontaceae 十齿花科 *Dipentodon* 十萼花属
Dipentodon sinicus Dunn 十齿花
Shrubs or small trees, semievergreen

16.0 (8.6~21.0); 6.9 (1.3~14.2); 23.6 (15.0~26.6); 16.0 (8.7~21.0); 132.9 (52.6~191.8); 0.5 (0.0~8.8); 826 (574~1053) ❙ 1268 (964~1736); 626 (452~1007); 74 (39~142) ❙ 55.2 (16.9~145.8); 825 (574~1053); 277.4 (179.5~551.6)

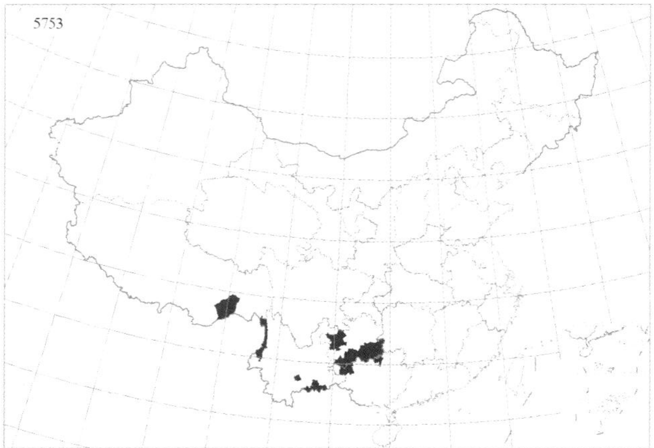

Dipentodontaceae 十齿花科 *Perrottetia* 核子木属
Perrottetia arisanensis Hayata 台湾核子木
Shrubs or small trees, deciduous

19.4 (16.4~24.1); 14.0 (11.5~18.9); 23.7 (20.3~28.0); 19.4 (16.5~24.1); 173.1 (137.8~229.7); 0.0 (0.0~0.0); 959 (783~1322) ❙ 2445 (1751~2859); 1163 (900~1594); 216 (47~402) ❙ 167.1 (32.5~235.6); 921 (783~1095); 244.7 (154.2~303.6)

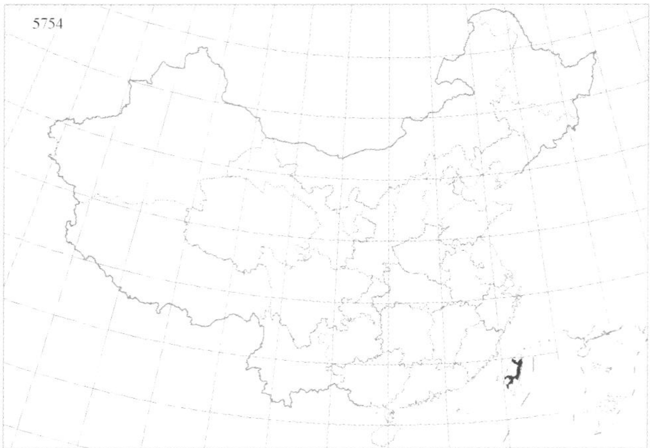

D

Dipentodontaceae 十齿花科 *Perrottetia* 核子木属
Perrottetia racemosa (Oliv.) Loes. 核子木
Shrubs, deciduous

15.4 (7.3~19.1); 4.6 (-2.0~11.0); 25.6 (15.6~29.4); 15.5 (7.5~19.1); 127.1 (48.2~169.6); 1.5 (0.0~20.4); 827 (537~972) ❙ 1292 (936~1583); 531 (392~938); 104 (23~209) ❙ 57.1 (18.6~107.2); 822 (537~969); 205.0 (134.6~371.7)

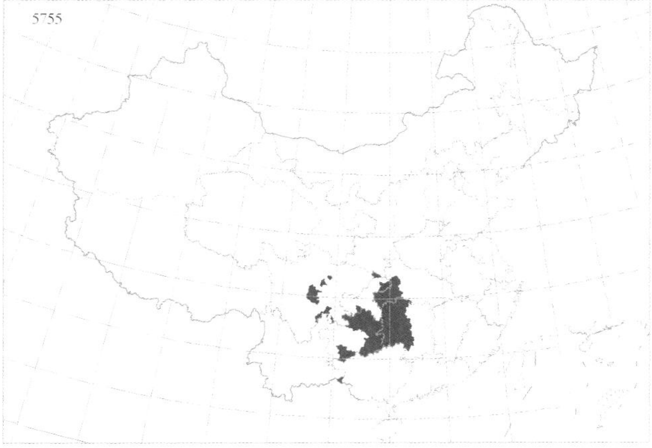

Nitrariaceae 白刺科 *Nitraria* 白刺属
Nitraria pamirica L. I. Vassiljeva 帕米尔白刺
Shrubs

7.0 (3.4~11.6); -10.8 (-13.7~-6.7); 20.9 (17.0~25.6); 9.4 (6.8~12.7); 74.7 (47.4~109.3); 50.6 (30.1~66.5); 638 (521~791) ❙ 153 (35~357); 76 (22~136); 12 (0~45) ❙ -72.9 (-95.6~-34.9); 153 (35~357); 54.9 (1.7~190.1)

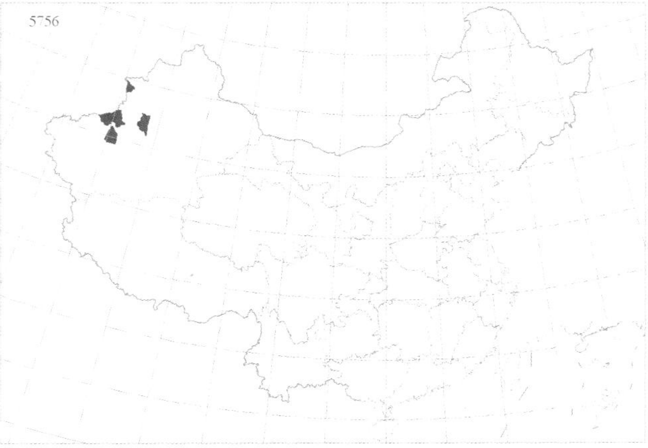

Nitrariaceae 白刺科 *Nitraria* 白刺属
Nitraria roborowskii Kom. 大白刺
Shrubs, deciduous

7.0 (-4.3~12.4); -10.5 (-20.1~-5.2); 21.7 (9.6~27.5); 9.4 (2.5~13.2); 76.5 (9.7~114.1); 52.2 (24.3~121.1); 640 (329~822) ❙ 146 (21~445); 84 (15~266); 6 (0~30) ❙ -74.8 (-97.3~0.0); 146 (21~437); 40.8 (1.6~220.8)

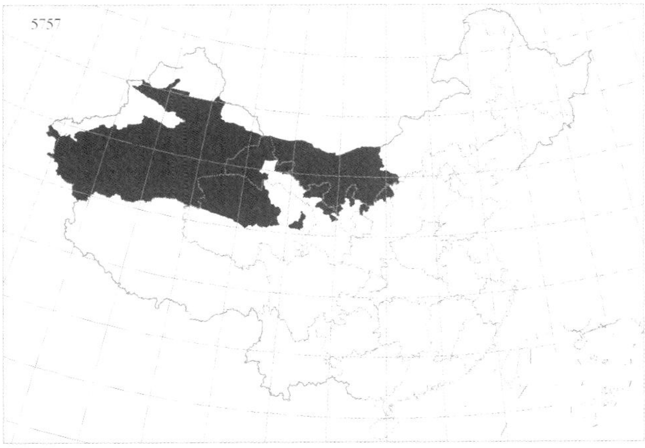

Nitrariaceae 白刺科 *Nitraria* 白刺属
Nitraria sibirica Pall. 小果白刺
Shrubs, deciduous

6.5 (-4.3~15.2); -10.7 (-25.9~2.3); 21.3 (8.7~28.1); 9.1 (2.5~15.3); 72.2 (7.7~127.5); 54.2 (4.1~128.6); 621 (329~864) ❙ 287 (21~1034); 176 (15~486); 10 (0~120) ❙ -51.8 (-97.3~48.8); 285 (21~864); 90.9 (1.6~297.3)

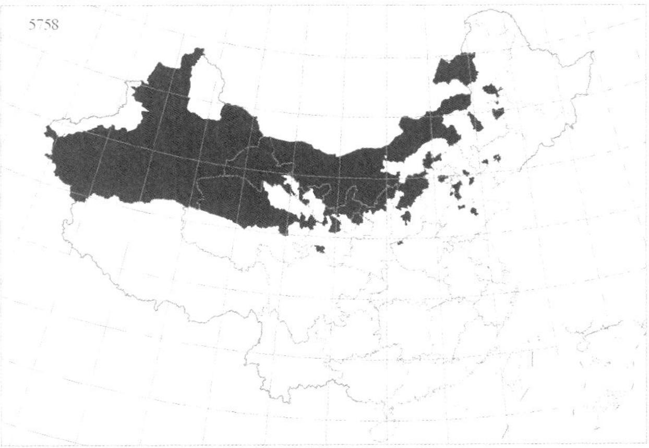

Nitrariaceae 白刺科 *Nitraria* 白刺属
Nitraria sphaerocarpa Maxim. 泡泡刺
Shrubs, deciduous

6.9 (-4.1~12.5); -10.9 (-20.3~-5.1); 21.7 (9.8~27.5); 9.5 (2.5~13.2); 77.4 (9.9~114.1); 53.6 (23.1~117.9); 647 (330~822) ❙ 116 (21~450); 64 (15~211); 7 (0~48) ❙ -79.3 (-97.3~3.9); 116 (21~406); 42.6 (1.6~261.4)

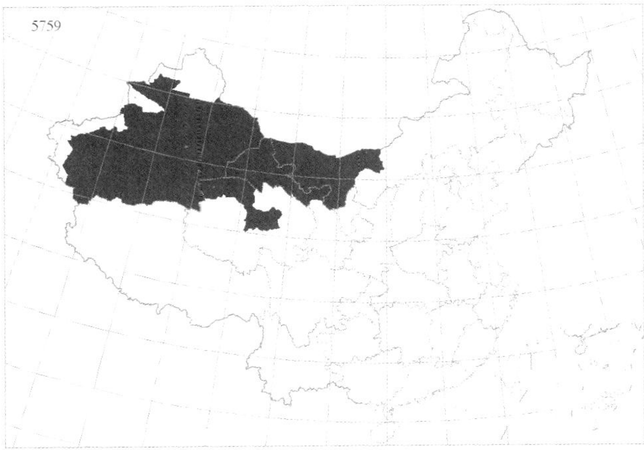

Nitrariaceae 白刺科 *Nitraria* 白刺属
Nitraria tangutorum Bobrov 白刺
Shrubs, deciduous

5.5 (-5.9~12.5); -12.0 (-21.5~-5.1); 20.3 (5.4~27.5); 8.5 (1.3~13.2); 67.2 (0.4~114.1); 60.9 (23.1~130.5); 602 (300~822) ❙ 187 (21~678); 103 (15~343); 9 (0~51) ❙ -64.7 (-97.3~61.8); 182 (21~449); 69.9 (1.6~344.4)

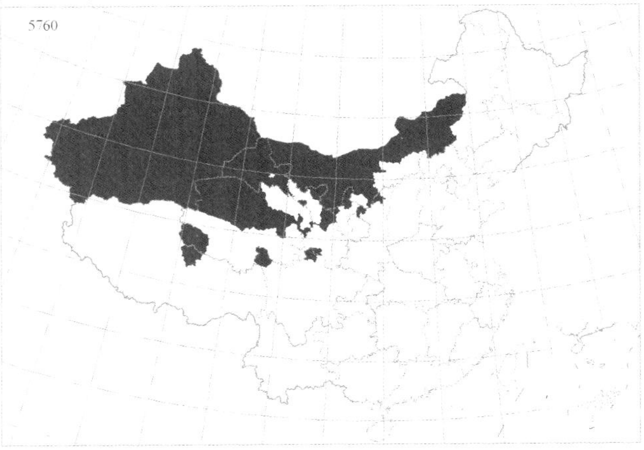

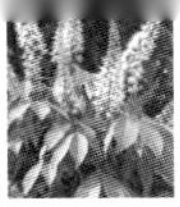

Tapisciaceae 瘿椒树科 *Tapiscia* 瘿椒树属

Tapiscia sinensis Oliv. 瘿椒树

Trees, deciduous

16.0 (7.3~22.2); 6.0 (-2.5~15.1); 24.8 (15.6~29.9); 16.0 (7.5~22.3); 133.9 (48.2~207.5); 1.6 (0.0~20.6); 840 (537~1190) ❙ 1307 (741~1997); 572 (352~996); 102 (17~259) ❙ 55.5 (0.0~130.7); 834 (537~1190); 254.8 (128.7~659.0)

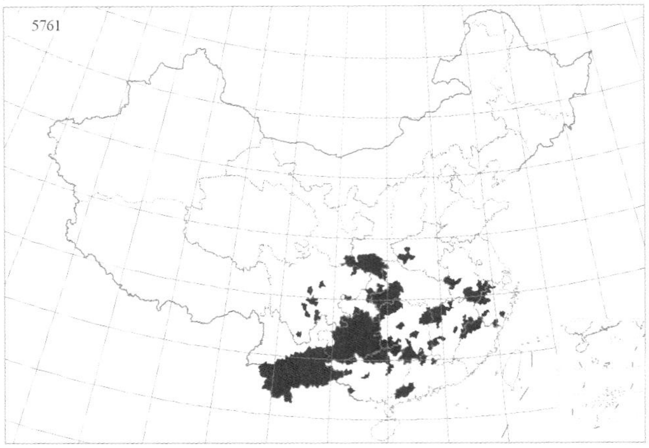

Tapisciaceae 瘿椒树科 *Tapiscia* 瘿椒树属

Tapiscia yunnanensis W. C. Cheng et C. D. Chu 云南瘿椒树

Trees, deciduous

15.7 (11.1~20.4); 6.3 (1.1~15.1); 23.9 (19.4~27.5); 15.7 (11.3~20.5); 130.1 (80.8~185.8); 1.2 (0.0~8.5); 816 (652~989) ❙ 1308 (956~1607); 621 (475~816); 73 (20~140) ❙ 61.1 (27.3~107.2); 810 (652~986); 281.6 (173.7~659.0)

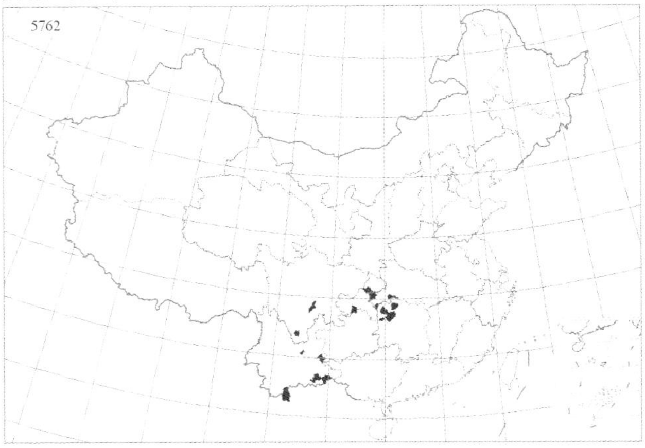

Hippocastanaceae 七叶树科 *Aesculus* 七叶树属

Aesculus assamica Griff. 长柄七叶树

Trees, deciduous

18.1 (15.1~22.2); 10.4 (5.0~15.1); 23.9 (20.3~28.6); 18.1 (15.2~22.3); 157.7 (121.9~207.5); 0.0 (0.0~0.0); 897 (767~1190) ❙ 1323 (939~1751); 698 (517~996); 63 (37~138) ❙ 47.9 (6.5~88.8); 887 (717~1190); 345.9 (189.5~659.0)

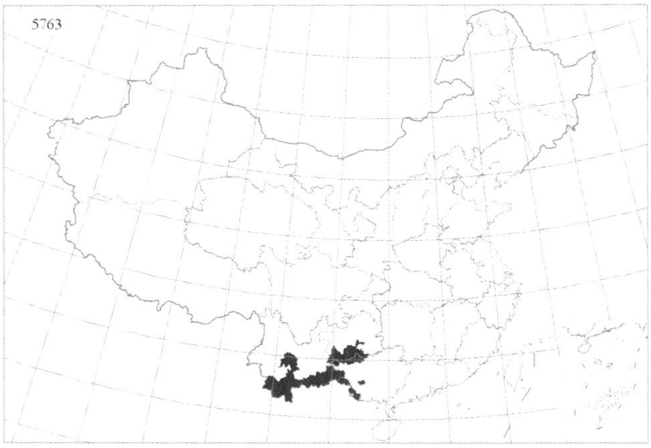

Hippocastanaceae 七叶树科 *Aesculus* 七叶树属

Aesculus chinensis Bunge 七叶树

Trees, deciduous

12.5 (3.5~16.9); 0.0 (-6.5~5.5); 24.2 (12.6~28.9); 12.7 (5.0~16.9); 102.3 (26.3~143.4); 12.0 (0.0~43.4); 741 (446~924) ❙ 919 (520~1801); 413 (258~602); 53 (12~242) ❙ 24.4 (-35.9~109.2); 710 (446~924); 230.3 (128.6~334.4)

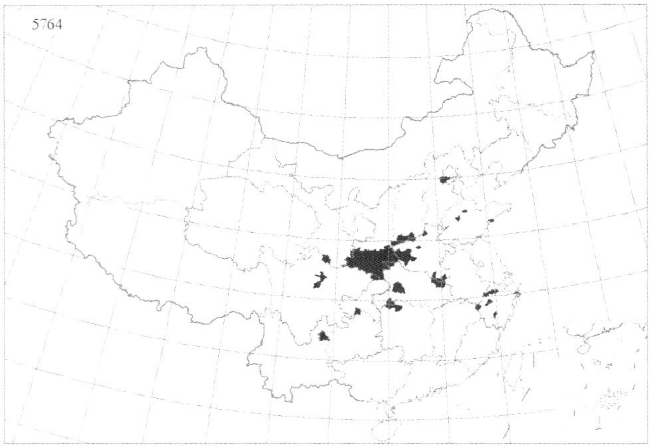

Hippocastanaceae 七叶树科 *Aesculus* 七叶树属

Aesculus chinensis var. wilsonii (Rehder) Turland et N. H. Xia 天师栗

Trees, deciduous

14.4 (5.7~17.8); 3.5 (-3.5~7.7); 24.6 (14.2~28.8); 14.5 (6.4~17.8); 116.5 (38.1~153.6); 3.3 (0.0~28.1); 786 (504~936) ❙ 1206 (590~1671); 508 (349~816); 92 (16~228) ❙ 53.6 (-33.3~107.2); 783 (504~936); 215.0 (146.0~292.6)

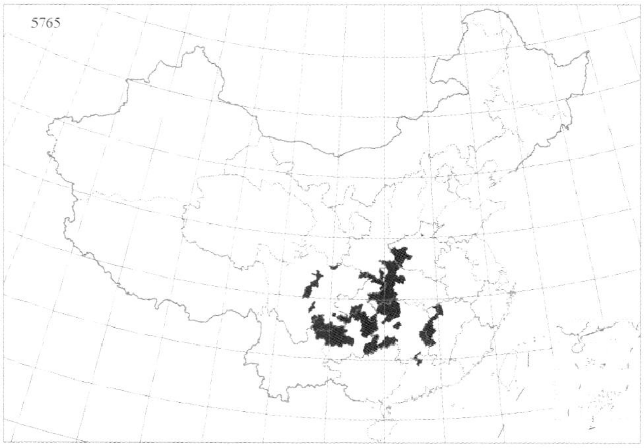

Hippocastanaceae 七叶树科 *Handeliodendron* 掌叶木属

Handeliodendron bodinieri (H. Lév.) Rehder 掌叶木

Shrubs or trees, deciduous

18.7 (16.2~22.2); 9.3 (6.3~13.9); 26.2 (23.4~28.4); 18.7 (16.2~22.3); 164.8 (134.5~207.5); 0.0 (0.0~0.0); 961 (826~1190) ❙ 1371 (1182~1546); 655 (561~723); 102 (43~184) ❙ 43.6 (15.4~64.7); 961 (826~1190); 270.8 (215.4~354.5)

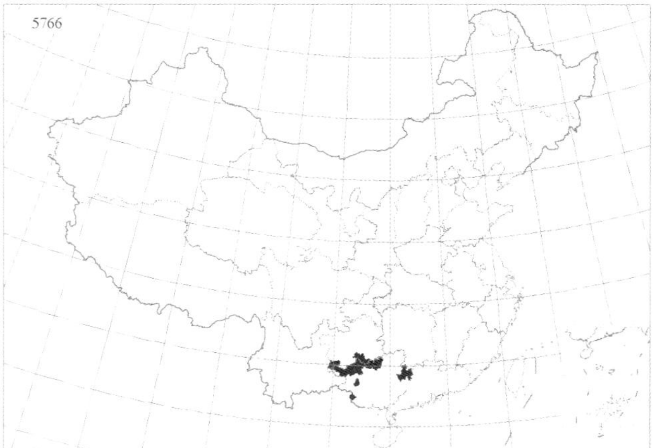

Sapindaceae 无患子科 *Allophylus* 异木患属
Allophylus caudatus Radlk. 波叶异木患
Shrubs or small trees

20.1 (16.9~22.7); 12.5 (9.9~15.0); 25.6 (21.4~28.6); 20.1 (16.9~22.8); 181.7 (142.7~213.1); 0.0 (0.0~0.0); 1028 (813~1221) ❙ 1605 (1178~2435); 893 (646~1428); 85 (51~130) ❙ 56.2 (15.4~99.5); 1028 (813~1221); 365.8 (275.8~551.6)

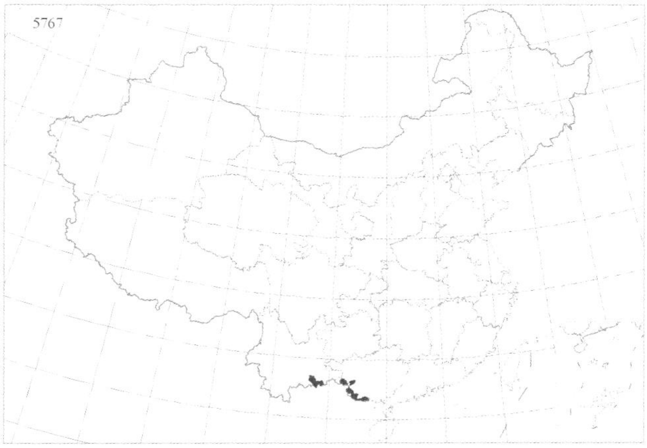

Sapindaceae 无患子科 *Allophylus* 异木患属
Allophylus chartaceus (Kurz) Radlk. 大叶异木患
Shrubs

13.3; 5.4; 19.6; 13.4; 100.4; 0.0; 706 ❙ 1736; 1007; 51 ❙ 145.8; 706; 447.4

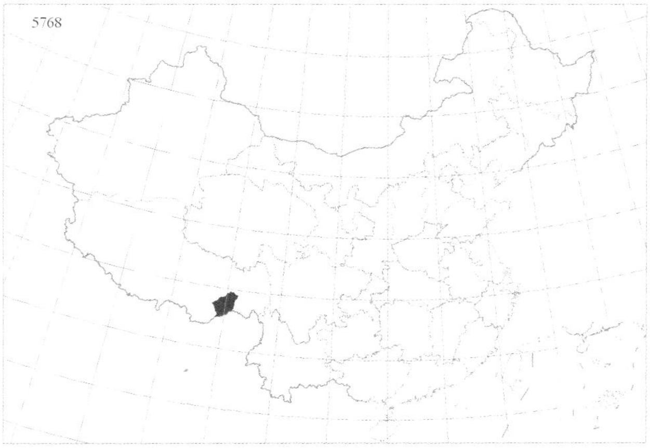

Sapindaceae 无患子科 *Allophylus* 异木患属
Allophylus cobbe var. velutinus Corner 滇南异木患
Shrubs

20.2; 13.3; 25.0; 20.3; 183.1; 0.0; 1016 ❙ 1502; 773; 72 ❙ 50.7; 1016; 487.9

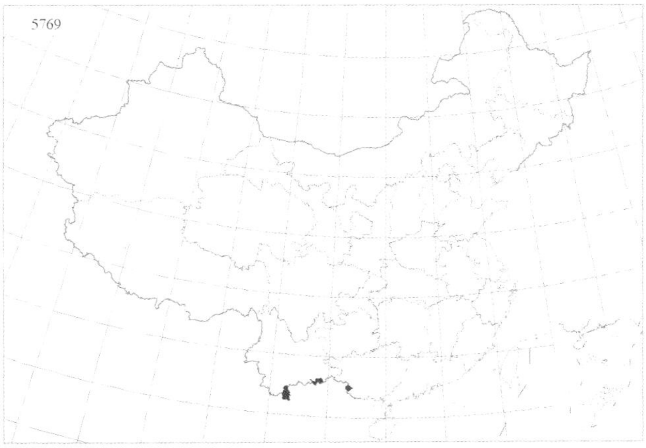

Sapindaceae 无患子科 *Allophylus* 异木患属
Allophylus dimorphus Radlk. 五叶异木患
Shrubs

24.6; 19.7; 28.1; 24.7; 236.2; 0.0; 1374 ❙ 1677; 550; 73 ❙ 22.1; 1255; 521.4

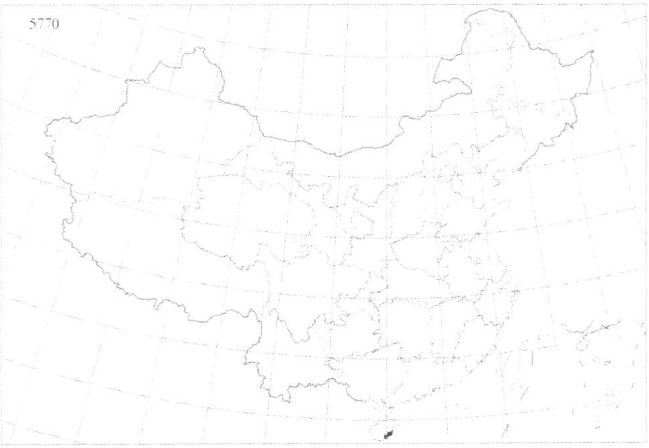

Sapindaceae 无患子科 *Allophylus* 异木患属
Allophylus hirsutus Radlk. 云南异木患
Shrubs

20.0; 14.6; 23.2; 20.1; 180.8; 0.0; 963 ❙ 1523; 681; 67 ❙ 58.4; 942; 581.0

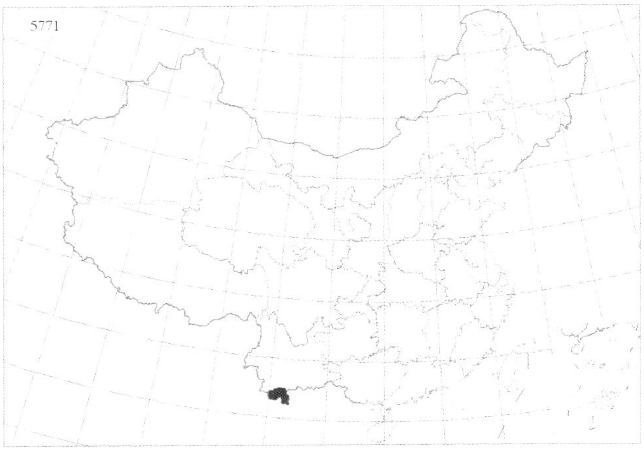

Sapindaceae 无患子科 *Allophylus* 异木患属
Allophylus longipes Radlk. 长柄异木患
Shrubs or small trees

18.1 (16.0~21.0); 11.4 (8.5~15.1); 22.5 (20.3~26.0); 18.1 (16.0~21.0); 157.7 (132.4~191.8); 0.0 (0.0~0.0); 876 (780~1053) ❙ 1396 (1074~1751); 770 (582~996); 54 (41~77) ❙ 59.8 (30.0~112.2); 867 (780~1053); 442.5 (277.0~659.0)

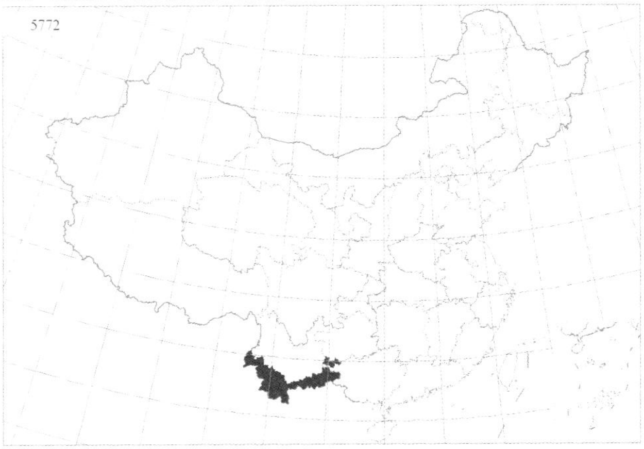

Sapindaceae 无患子科 *Allophylus* 异木患属

Allophylus timoriensis (DC.) Blume 海滨异木患

Shrubs

23.7 (22.2~24.8); 18.6 (17.2~20.1); 27.4 (25.5~28.5); 23.8 (22.2~24.8); 225.3 (206.7~237.6); 0.0 (0.0~0.0); 1283 (1123~1385) ❙ 1653 (1179~2992); 635 (392~1847); 64 (41~115) ❙ 30.9 (-11.9~166.5); 1158 (1055~1360); 497.3 (296.9~620.0)

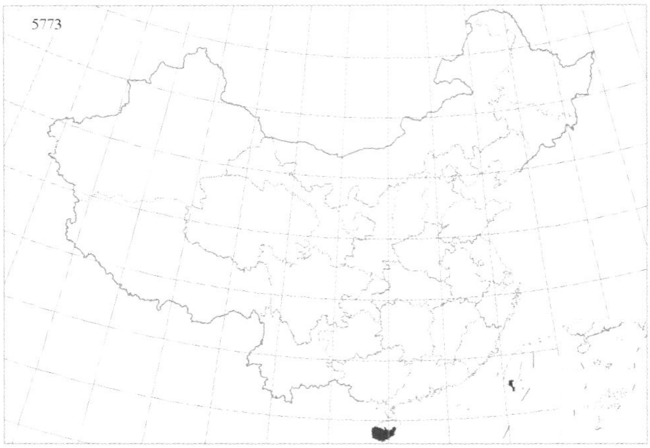

Sapindaceae 无患子科 *Allophylus* 异木患属

Allophylus viridis Radlk. 异木患

Shrubs

23.7 (19.1~24.8); 17.6 (11.0~20.1); 28.1 (25.0~29.2); 23.7 (19.1~24.8); 224.7 (169.6~237.6); 0.0 (0.0~0.0); 1290 (945~1385) ❙ 1409 (868~1813); 547 (312~820); 70 (41~113) ❙ 9.9 (-34.3~52.2); 1170 (868~1329); 384.8 (178.7~620.0)

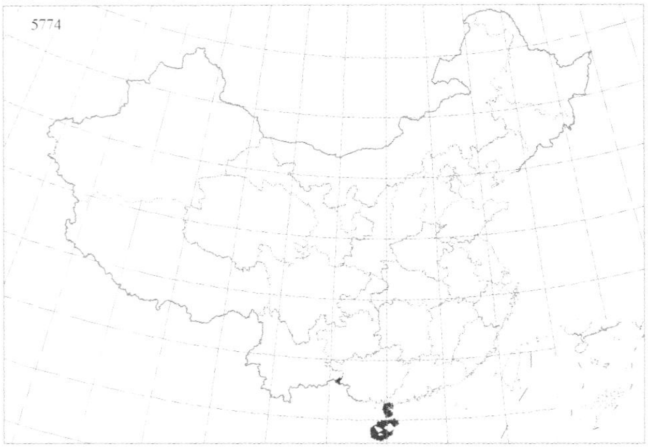

Sapindaceae 无患子科 *Amesiodendron* 细子龙属

Amesiodendron chinense (Merr.) Hu 细子龙

Trees, evergreen

21.6 (17.7~24.8); 14.6 (8.5~20.1); 26.9 (22.5~28.6); 21.7 (17.8~24.8); 200.2 (153.4~237.6); 0.0 (0.0~0.0); 1144 (854~1385) ❙ 1456 (1179~1854); 642 (392~850); 74 (41~169) ❙ 29.1 (-11.9~69.8); 1087 (854~1360); 402.7 (189.5~620.0)

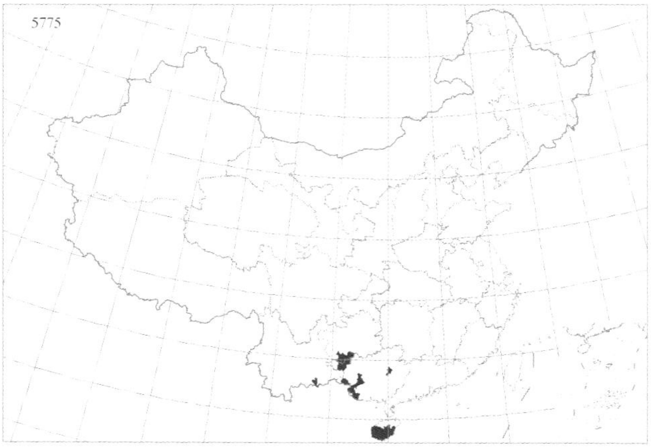

Sapindaceae 无患子科 *Arytera* 滨木患属

Arytera littoralis Blume 滨木患

Shrubs or small trees, evergreen

21.8 (17.2~24.8); 14.6 (9.1~20.1); 27.2 (21.6~29.1); 21.8 (17.3~24.8); 201.8 (147.1~237.6); 0.0 (0.0~0.0); 1145 (833~1385) ❙ 1639 (868~2435); 760 (312~1428); 100 (41~188) ❙ 44.8 (-34.3~99.5); 1119 (833~1360); 308.4 (95.3~659.0)

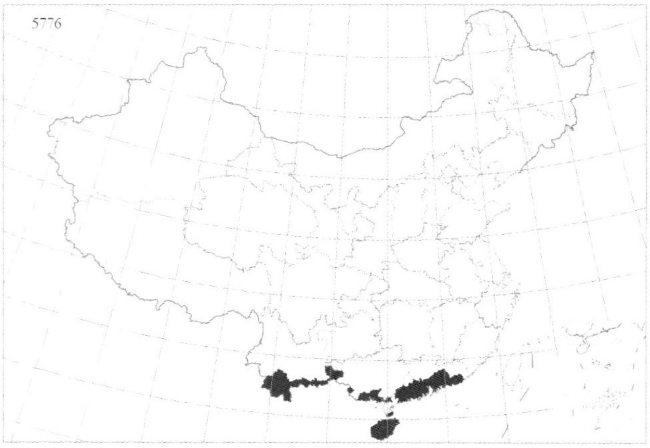

Sapindaceae 无患子科 *Boniodendron* 黄梨木属

Boniodendron minus (Hemsl.) T. C. Chen 黄梨木

Small trees, evergreen

19.3 (14.3~22.9); 9.7 (3.7~15.2); 27.5 (23.1~29.8); 19.3 (14.3~22.9); 171.6 (113.2~214.8); 0.0 (0.0~1.4); 1008 (761~1232) ❙ 1578 (1161~2024); 621 (400~902); 162 (49~221) ❙ 57.6 (15.4~128.9); 1002 (761~1232); 236.5 (151.0~411.1)

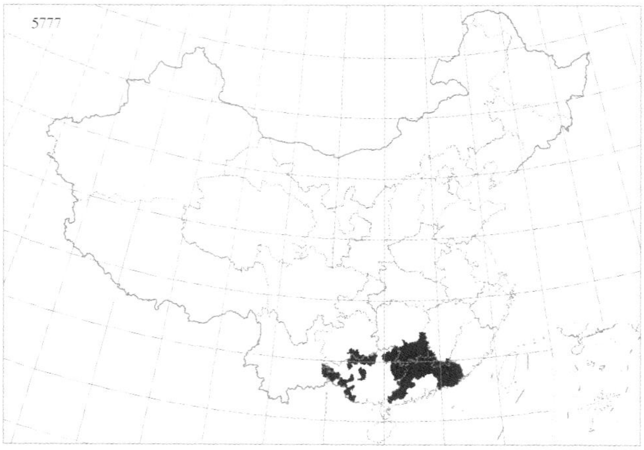

Sapindaceae 无患子科 *Delavaya* 茶条木属

Delavaya toxocarpa Franch. 茶条木

Shrubs or small trees

15.7 (5.4~22.2); 8.2 (-2.1~14.2); 21.4 (12.3~28.4); 15.7 (5.8~22.3); 129.3 (31.5~207.5); 0.8 (0.0~25.6); 798 (481~1190) ❙ 1071 (741~1547); 605 (411~911); 45 (15~130) ❙ 35.3 (-14.2~80.6); 782 (481~1190); 300.6 (142.9~551.6)

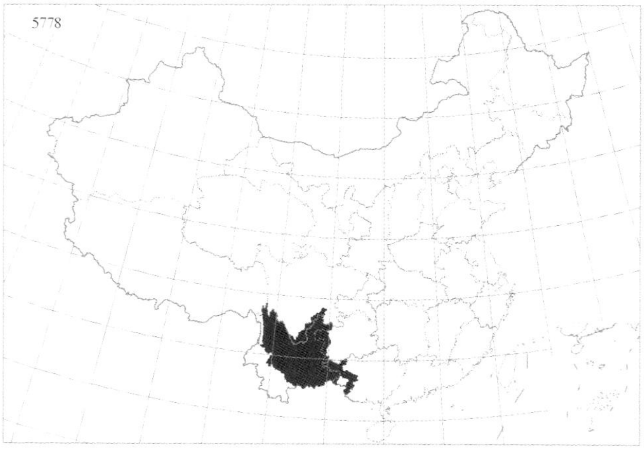

Sapindaceae 无患子科 *Dimocarpus* 龙眼属

Dimocarpus confinis (F. C. How et C. N. Ho) H. S. Lo 龙荔

Trees, evergreen

19.0 (14.9~22.3); 10.3 (4.6~14.7); 25.8 (21.4~28.9); 19.0 (15.0~22.3); 168.1 (119.9~208.1); 0.0 (0.0~0.4); 974 (781~1202) ❙ 1403 (1074~1876); 697 (494~891); 90 (43~194) ❙ 45.1 (5.8~85.5); 974 (781~1202); 283.5 (189.5~551.6)

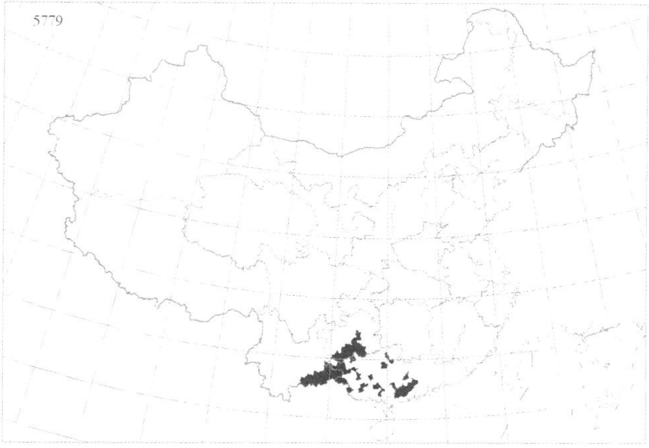

Sapindaceae 无患子科 *Dimocarpus* 龙眼属

Dimocarpus fumatus subsp. calcicola C. Y. Wu 灰岩肖韶子

Trees

18.4 (16.8~21.0); 9.6 (7.8~11.8); 25.2 (22.7~28.2); 18.5 (16.8~21.1); 161.6 (141.5~192.6); 0.0 (0.0~0.0); 931 (836~1112) ❙ 1342 (1236~1703); 713 (617~813); 73 (48~194) ❙ 44.6 (27.5~65.6); 931 (836~1112); 265.5 (189.5~359.7)

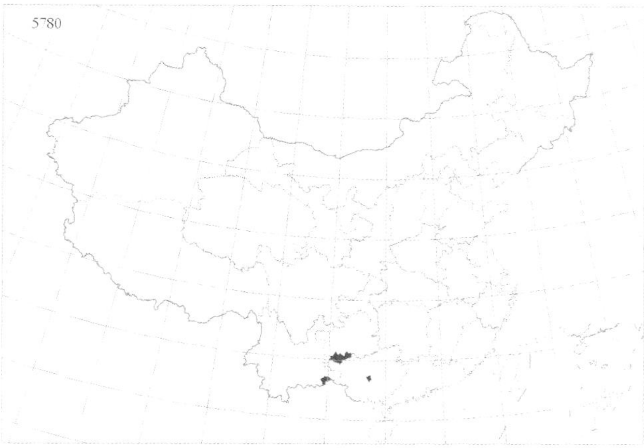

Sapindaceae 无患子科 *Dimocarpus* 龙眼属

Dimocarpus longan Lour. 龙眼

Trees, evergreen

20.4 (14.4~24.8); 12.5 (5.1~20.1); 26.7 (19.1~29.1); 20.4 (14.4~24.8); 185.0 (113.1~237.6); 0.0 (0.0~0.0); 1060 (727~1385) ❙ 1515 (797~2435); 696 (312~1428); 107 (23~255) ❙ 44.2 (-34.3~112.2); 1045 (717~1360); 285.8 (95.3~659.0)

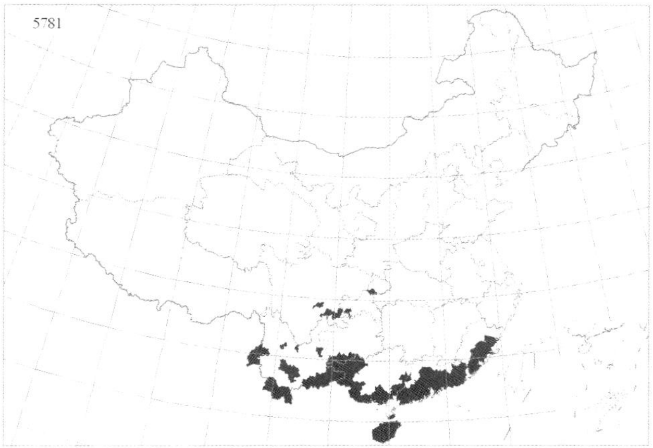

Sapindaceae 无患子科 *Dimocarpus* 龙眼属

Dimocarpus yunnanensis (W. T. Wang) C. Y. Wu et T. L. Ming 滇龙眼

Trees, evergreen

19.3; 12.7; 23.8; 19.3; 171.6; 0.0; 942 ❙ 1439; 803; 61 ❙ 53.4; 942; 516.4

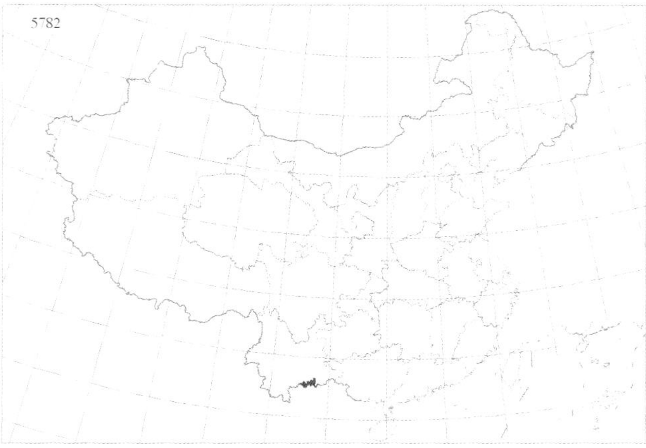

Sapindaceae 无患子科 *Dodonaea* 车桑仔属

Dodonaea viscosa Jacquin 车桑子

Shrubs or small trees, evergreen

17.6 (5.4~24.8); 10.5 (-2.4~20.1); 23.2 (12.3~29.2); 17.6 (5.8~24.8); 152.0 (31.5~237.6); 0.6 (0.0~25.6); 902 (481~1385) ❙ 1363 (741~3495); 671 (312~1847); 85 (12~897) ❙ 52.3 (-34.3~246.5); 877 (481~1360); 314.8 (111.7~659.0)

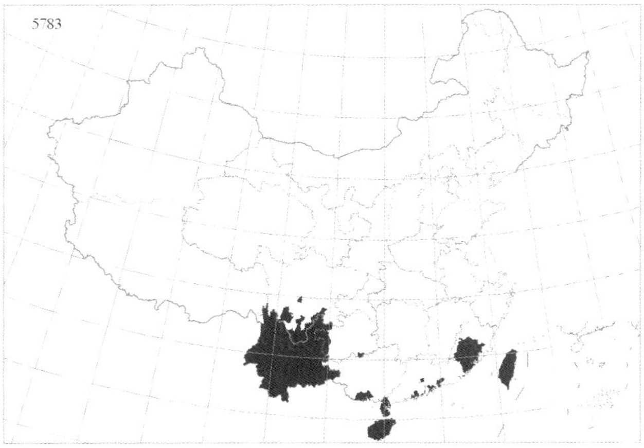

Sapindaceae 无患子科 *Eurycorymbus* 伞花树属

Eurycorymbus cavaleriei (H. Lév.) Rehder et Hand.-Mazz. 伞花木

Trees, deciduous

17.5 (8.6~24.1); 7.5 (-1.7~18.9); 26.7 (15.0~29.8); 17.6 (8.7~24.1); 151.5 (52.6~229.7); 0.4 (0.0~18.0); 920 (574~1322) ❙ 1553 (1041~3495); 564 (356~1594); 179 (45~897) ❙ 70.1 (-5.0~224.4); 911 (574~1131); 234.3 (123.9~383.4)

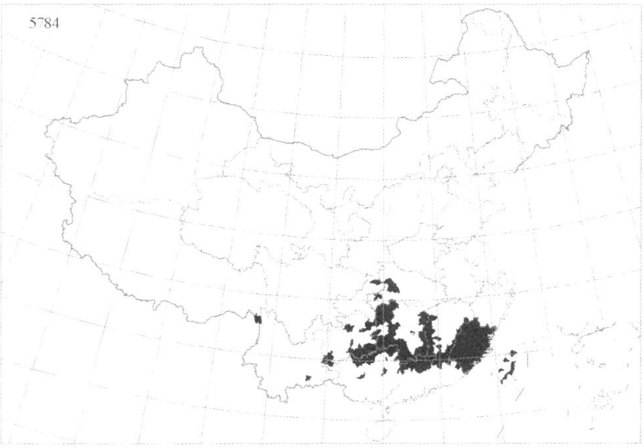

Sapindaceae 无患子科 *Harpullia* 假山萝属
Harpullia cupanioides Roxb. 假山萝
Trees

21.8 (17.2~24.8); 15.8 (9.5~20.1); 26.0 (21.6~29.2); 21.8 (17.3~24.8); 202.0 (147.0~237.6); 0.0 (0.0~0.0); 1138 (833~1385) ❙ 1469 (868~1854); 643 (312~996); 67 (41~115) ❙ 33.5 (-34.3~88.8); 1071 (833~1360); 440.3 (178.7~659.0)

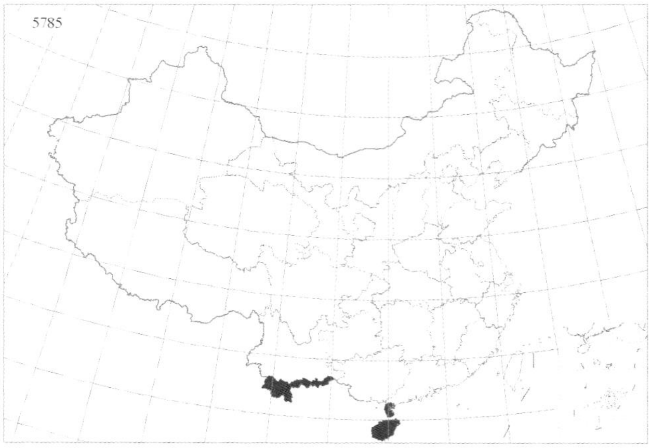

Sapindaceae 无患子科 *Koelreuteria* 栾树属
Koelreuteria bipinnata Franch. 复羽叶栾树
Trees, deciduous

16.2 (2.2~22.4); 6.8 (-6.6~15.1); 24.3 (9.9~29.9); 16.2 (3.8~22.4); 135.4 (15.5~208.9); 1.0 (0.0~49.1); 841 (413~1191) ❙ 1272 (797~2037); 583 (376~1033); 91 (17~242) ❙ 51.6 (-14.2~129.9); 833 (413~1191); 257.4 (108.2~659.0)

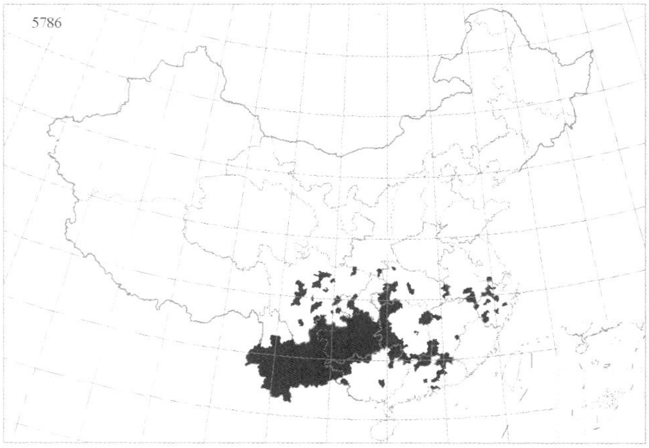

Sapindaceae 无患子科 *Koelreuteria* 栾树属
Koelreuteria elegans subsp. formosana (Hayata) F. G. Meyer 台湾栾树
Trees

19.9 (17.1~22.2); 13.8 (11.2~17.7); 25.2 (21.8~27.4); 20.0 (17.1~22.2); 179.6 (145.0~206.7); 0.0 (0.0~0.0); 995 (816~1123) ❙ 2509 (1836~2993); 1049 (750~1847); 317 (107~551) ❙ 154.8 (64.2~203.4); 989 (816~1123); 260.8 (202.8~296.9)

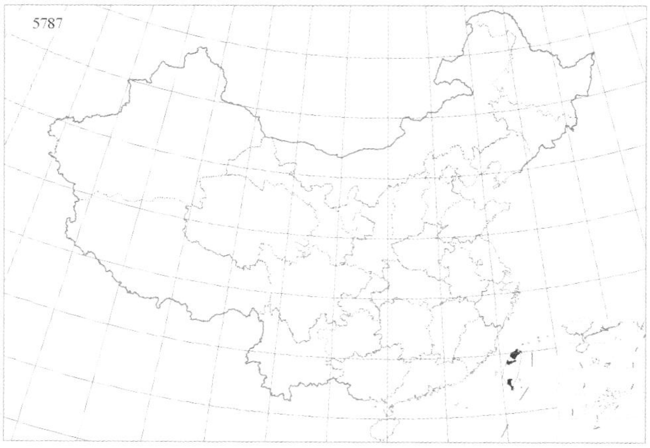

Sapindaceae 无患子科 *Koelreuteria* 栾树属
Koelreuteria paniculata Laxm. 栾树
Shrubs or trees, deciduous

13.5 (2.0~21.4); 0.0 (-15.9~13.6); 25.8 (9.5~29.8); 13.9 (3.8~21.5); 116.7 (15.4~197.7); 14.3 (0.0~80.4); 806 (411~1128) ❙ 940 (349~1997); 433 (201~753); 81 (7~259) ❙ 14.8 (-51.1~147.0); 718 (349~1094); 189.9 (70.3~357.7)

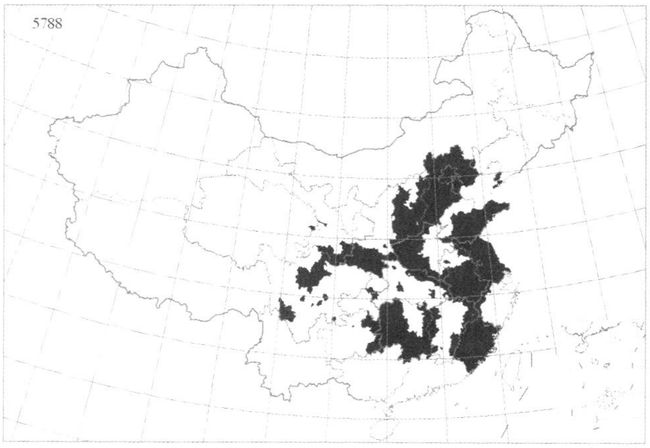

Sapindaceae 无患子科 *Lepisanthes* 鳞花木属
Lepisanthes hainanensis H. S. Lo 鳞花木
Trees

23.9 (22.2~24.8); 19.0 (17.2~20.1); 27.3 (25.7~28.0); 23.9 (22.3~24.8); 227.3 (207.2~237.6); 0.0 (0.0~0.0); 1297 (1127~1385) ❙ 1518 (1380~1666); 520 (455~597); 50 (41~58) ❙ 18.1 (1.5~47.8); 1133 (1055~1212); 537.6 (455.2~620.0)

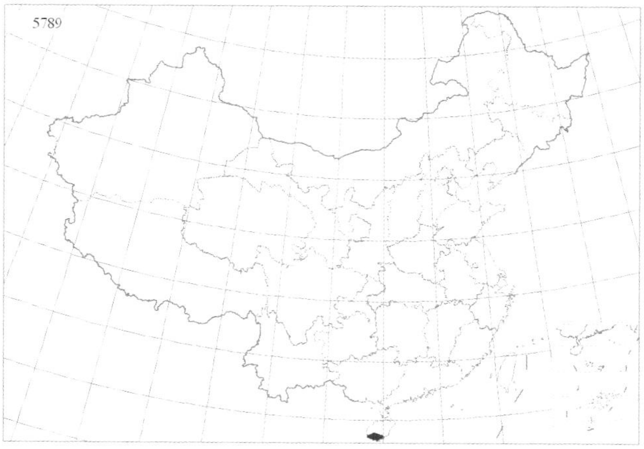

Sapindaceae 无患子科 *Lepisanthes* 鳞花木属
Lepisanthes rubiginosa (Roxb.) Leenh. 赤才
Shrubs or small trees, evergreen

23.2 (21.0~24.8); 16.4 (12.6~20.1); 28.2 (25.7~29.2); 23.2 (21.0~24.8); 218.4 (192.3~237.6); 0.0 (0.0~0.0); 1252 (1081~1385) ❙ 1530 (868~1876); 648 (312~997); 91 (41~173) ❙ 23.1 (-34.3~64.3); 1188 (868~1360); 343.8 (178.7~620.0)

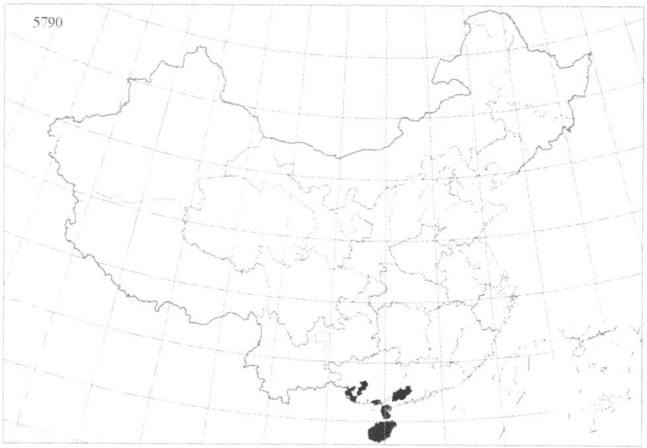

Sapindaceae 无患子科 *Lepisanthes* 鳞花木属
Lepisanthes senegalensis (Juss. ex Poir.) Leenh. 滇赤才
Shrubs or trees, evergreen

18.8 (16.8~21.6); 12.7 (10.4~15.1); 22.6 (20.8~27.9); 18.9 (16.9~21.6); 166.4 (142.6~199.6); 0.0 (0.0~0.0); 909 (811~1143) ❙ 1442 (1275~1607); 744 (655~816); 56 (41~102) ❙ 59.6 (27.0~74.2); 890 (802~1143); 481.8 (335.1~659.0)

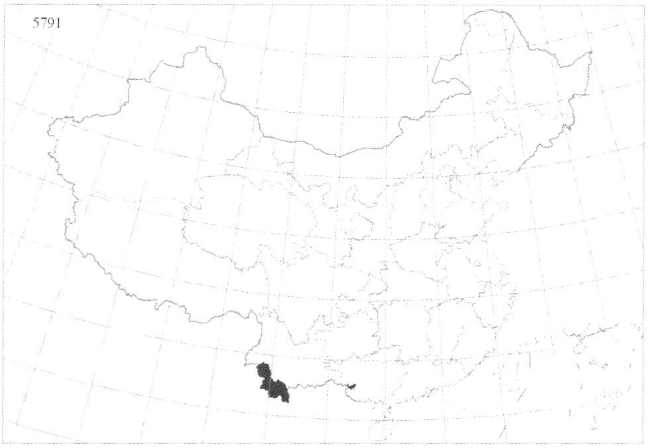

Sapindaceae 无患子科 *Lepisanthes* 鳞花木属
Lepisanthes unilocularis Leenh. 爪耳木
Shrubs

24.3; 19.4; 27.7; 24.4; 232.2; 0.0; 1339 ❙ 1393; 464; 45 ❙ 4.1; 1124; 491.4

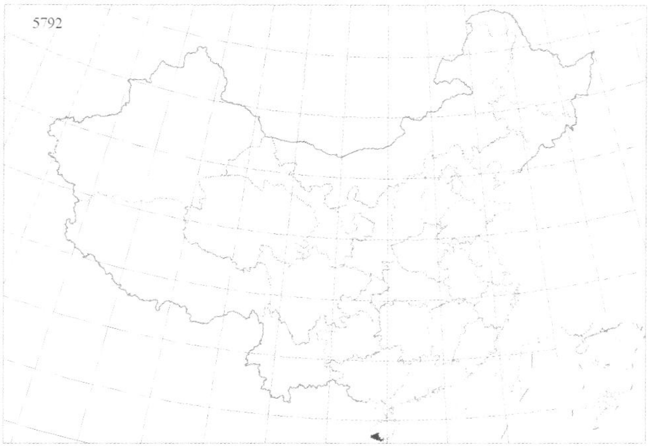

Sapindaceae 无患子科 *Mischocarpus* 柄果木属
Mischocarpus hainanensis H. S. Lo 海南柄果木
Shrubs or small trees, evergreen

24.4 (23.9~24.8); 19.4 (18.7~20.1); 27.8 (27.2~28.2); 24.4 (23.9~24.8); 233.0 (226.8~237.6); 0.0 (0.0~0.0); 1346 (1294~1385) ❙ 1538 (1380~1767); 513 (455~567); 57 (41~88) ❙ 14.3 (1.5~28.8); 1181 (1112~1299); 529.1 (455.2~620.0)

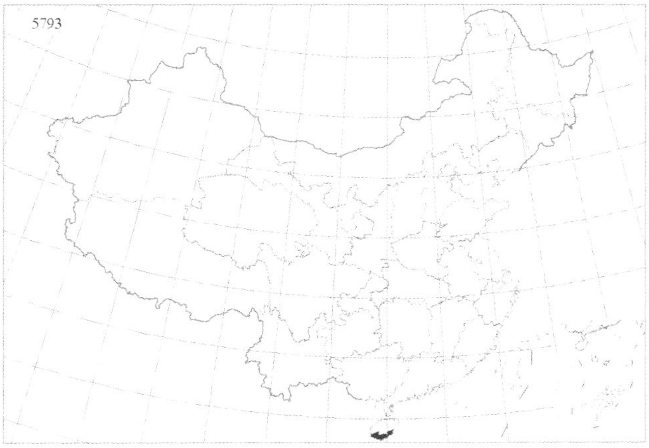

Sapindaceae 无患子科 *Mischocarpus* 柄果木属
Mischocarpus pentapetalus (Roxb.) Radlk. 褐叶柄果木
Trees, evergreen

20.8 (16.3~24.8); 13.3 (8.5~20.1); 26.7 (20.8~29.1); 20.9 (16.4~24.8); 190.5 (136.4~237.6); 0.0 (0.0~0.0); 1080 (797~1385) ❙ 1584 (868~2435); 735 (312~1428); 110 (41~197) ❙ 48.0 (-34.3~99.5); 1062 (797~1247); 295.8 (95.3~659.0)

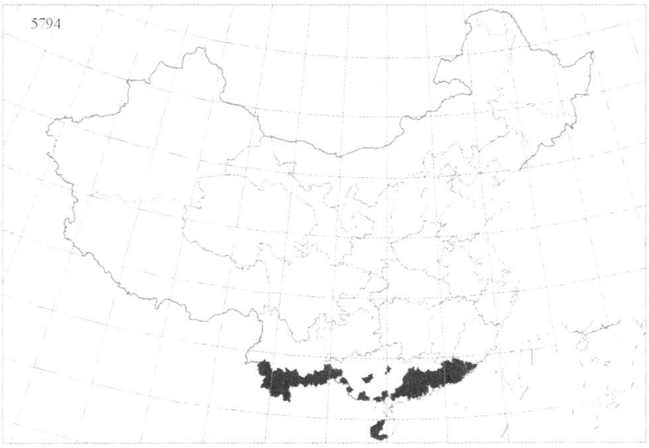

Sapindaceae 无患子科 *Mischocarpus* 柄果木属
Mischocarpus sundaicus Blume 柄果木
Trees, evergreen

22.6 (19.1~24.8); 16.0 (9.8~20.1); 27.5 (25.0~29.0); 22.7 (19.1~24.8); 212.1 (169.6~237.6); 0.0 (0.0~0.0); 1213 (945~1385) ❙ 1649 (1182~2435); 741 (455~1428); 83 (41~130) ❙ 37.2 (-1.4~99.5); 1162 (945~1360); 391.4 (199.6~620.0)

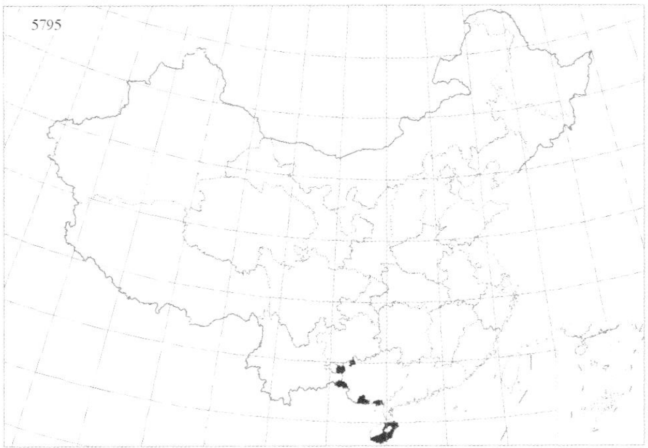

Sapindaceae 无患子科 *Nephelium* 韶子属
Nephelium chryseum Blume 韶子
Trees, evergreen

19.9 (16.3~22.7); 13.1 (9.1~15.1); 24.7 (21.4~28.8); 19.9 (16.4~22.8); 179.3 (136.4~213.1); 0.0 (0.0~0.0); 996 (797~1221) ❙ 1558 (1163~2435); 814 (641~1428); 79 (41~132) ❙ 56.8 (12.9~99.5); 989 (797~1221); 407.4 (203.4~659.0)

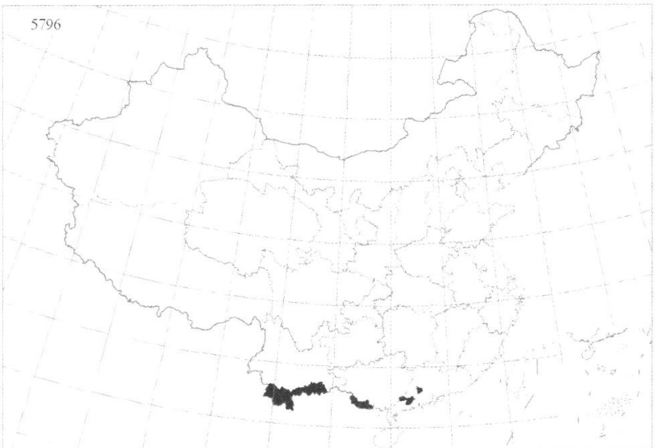

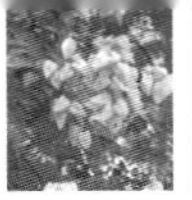

Sapindaceae 无患子科 *Nephelium* 韶子属

Nephelium topengii (Merr.) H. S. Lo 海南韶子

Trees, evergreen

23.8 (21.8~24.8); 17.9 (13.7~20.1); 28.0 (27.2~28.7); 23.9 (21.8~24.8); 226.4 (202.2~237.6); 0.0 (0.0~0.0); 1302 (1154~1385) ▎1382 (868~1767); 513 (312~713); 66 (41~110) ▎6.3 (-34.3~29.9); 1152 (868~1326); 436.8 (288.3~620.0)

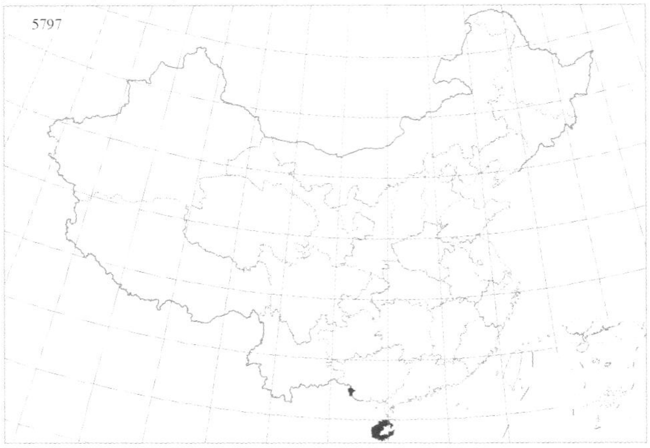

Sapindaceae 无患子科 *Paranephelium* 假韶子属

Paranephelium hainanense H. S. Lo 海南假韶子

Trees, evergreen

23.9 (22.2~24.8); 18.9 (17.2~20.1); 27.3 (25.7~28.2); 23.9 (22.3~24.8); 227.1 (207.2~237.6); 0.0 (0.0~0.0); 1296 (1127~1385) ▎1617 (1405~1767); 553 (473~610); 59 (41~88) ▎25.7 (1.5~47.8); 1169 (1055~1299); 559.0 (494.7~620.0)

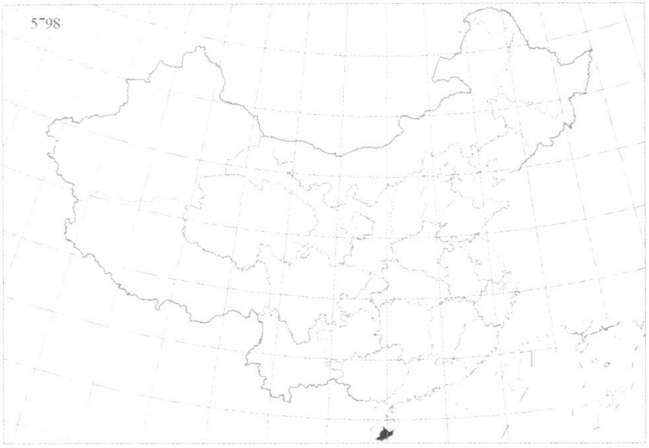

Sapindaceae 无患子科 *Pavieasia* 檀栗属

Pavieasia kwangsiensis H. S. Lo 广西檀栗

Trees, evergreen

22.2; 13.9; 28.3; 22.3; 207.5; 0.0; 1190 ▎1374; 713; 87 ▎15.4; 1190; 327.7

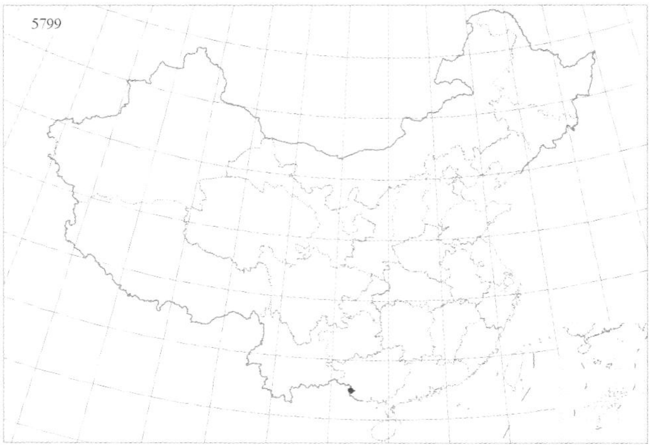

Sapindaceae 无患子科 *Pavieasia* 檀栗属

Pavieasia yunnanensis H. S. Lo 云南檀栗

Trees, evergreen

18.6 (17.2~21.0); 11.9 (9.5~14.2); 23.2 (22.4~25.6); 18.6 (17.3~21.0); 163.4 (147.0~191.8); 0.0 (0.0~0.0); 900 (833~1053) ▎1452 (1254~1751); 826 (699~996); 59 (48~77) ▎61.8 (36.8~88.8); 899 (833~1053); 464.0 (336.3~551.6)

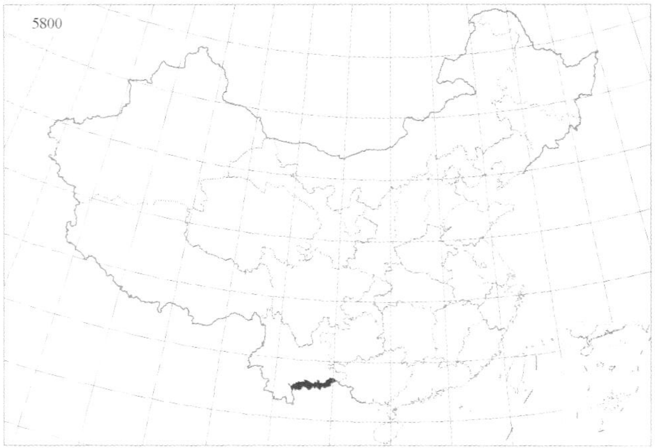

Sapindaceae 无患子科 *Pometia* 番龙眼属

Pometia pinnata J. R. Forst. et G. Forst. 番龙眼

Trees, evergreen

18.5 (16.3~22.2); 12.2 (9.1~17.7); 22.9 (20.3~27.4); 18.6 (16.4~22.2); 163.0 (136.4~206.7); 0.0 (0.0~0.0); 900 (783~1123) ▎1653 (1074~3495); 840 (582~1847); 111 (41~897) ▎82.8 (32.4~224.4); 891 (783~1123); 424.5 (202.8~659.0)

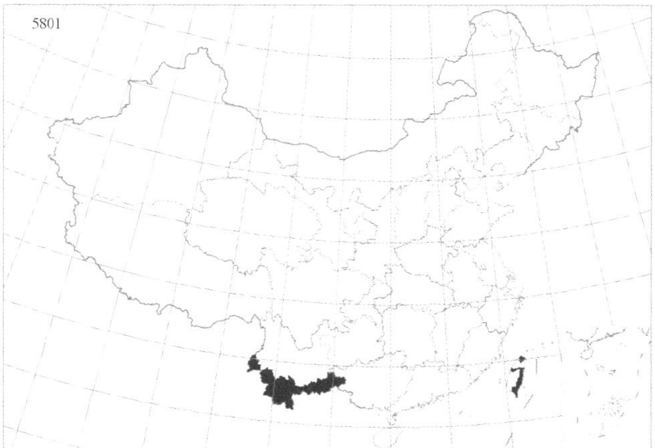

Sapindaceae 无患子科 *Sapindus* 无患子属

Sapindus delavayi (Franch.) Radlk. 川滇无患子

Trees, deciduous

15.4 (5.4~21.0); 7.8 (-2.1~15.1); 21.4 (12.3~27.8); 15.5 (5.8~21.0); 126.8 (31.5~191.8); 1.1 (0.0~25.6); 788 (481~1053) ▎1126 (741~1751); 624 (393~996); 45 (12~107) ▎43.9 (-14.2~112.2); 773 (481~1053); 314.3 (133.6~659.0)

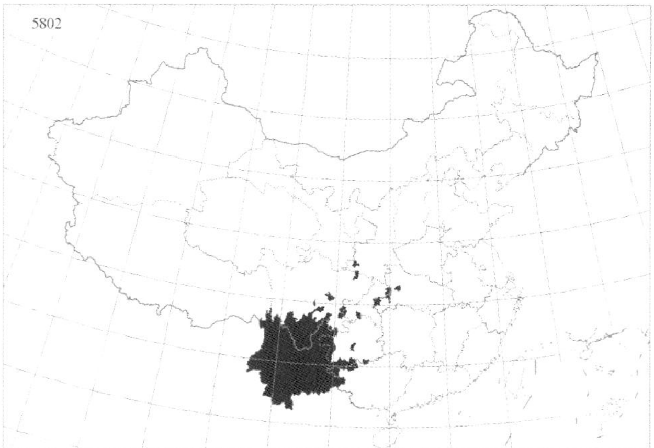

Sapindaceae 无患子科 *Sapindus* 无患子属
Sapindus rarak DC. 毛瓣无患子
Trees, deciduous

17.9 (15.2~21.0); 11.5 (8.7~15.1); 22.1 (20.1~25.6); 18.0 (15.2~21.0); 155.7 (122.5~191.8); 0.0 (0.0~0.0); 865 (753~1053) ❙ 1331 (1007~1751); 736 (539~996); 52 (41~77) ❙ 53.9 (12.5~112.2); 851 (746~1053); 432.9 (277.0~659.0)

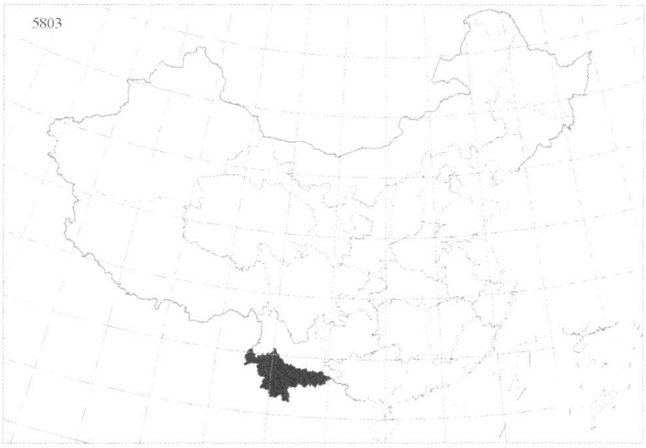

Sapindaceae 无患子科 *Sapindus* 无患子属
Sapindus rarak var. velutinus C. Y. Wu 石屏无患子
Trees, deciduous

18.0 (14.4~21.0); 11.7 (7.5~15.1); 22.1 (19.1~25.6); 18.1 (14.4~21.0); 156.7 (113.1~191.8); 0.0 (0.0~0.0); 869 (727~1053) ❙ 1367 (1028~1751); 755 (566~996); 54 (41~77) ❙ 57.7 (12.5~112.2); 855 (727~1053); 448.6 (294.5~659.0)

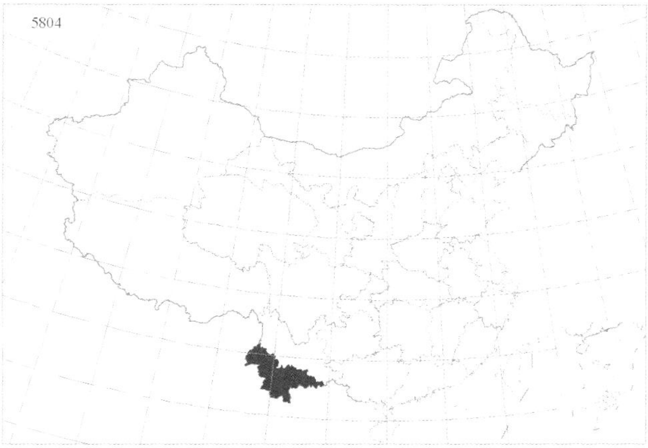

Sapindaceae 无患子科 *Sapindus* 无患子属
Sapindus saponaria L. 无患子
Trees, deciduous

17.6 (9.3~24.8); 7.1 (-3.3~20.1); 27.1 (17.0~29.9); 17.7 (9.5~24.8); 153.8 (69.8~237.6); 1.9 (0.0~22.0); 946 (603~1385) ❙ 1396 (626~3495); 581 (312~1847); 135 (15~897) ❙ 48.0 (-34.3~246.5); 931 (603~1360); 228.4 (95.3~659.0)

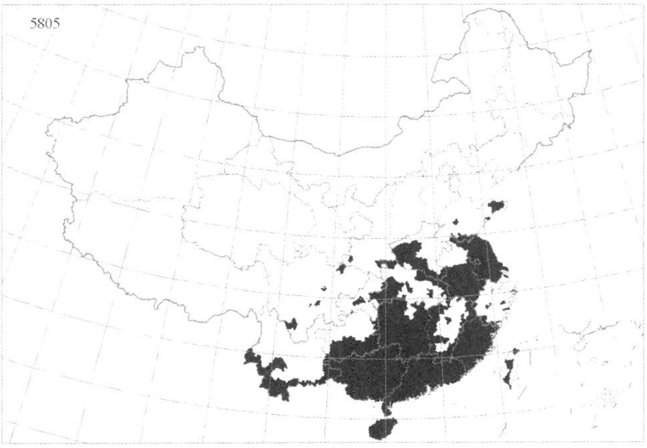

Sapindaceae 无患子科 *Sapindus* 无患子属
Sapindus tomentosus Kurz 绒毛无患子
Trees

15.6; 8.9; 20.3; 15.7; 127.9; 0.0; 770 ❙ 1316; 719; 60 ❙ 72.4; 770; 435.3

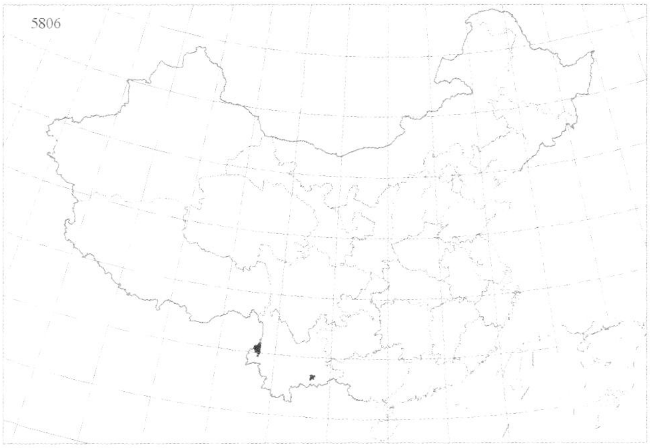

Sapindaceae 无患子科 *Xanthoceras* 文冠果属
Xanthoceras sorbifolium Bunge 文冠果
Shrubs or small trees, deciduous

9.1 (0.0~14.8); -6.5 (-16.1~1.5); 23.0 (12.1~28.6); 10.5 (4.0~14.9); 85.0 (20.9~128.9); 35.4 (6.5~81.2); 671 (403~886) ❙ 525 (63~856); 306 (45~453); 16 (2~53) ❙ -21.7 (-84.9~52.6); 519 (63~817); 150.8 (8.4~323.3)

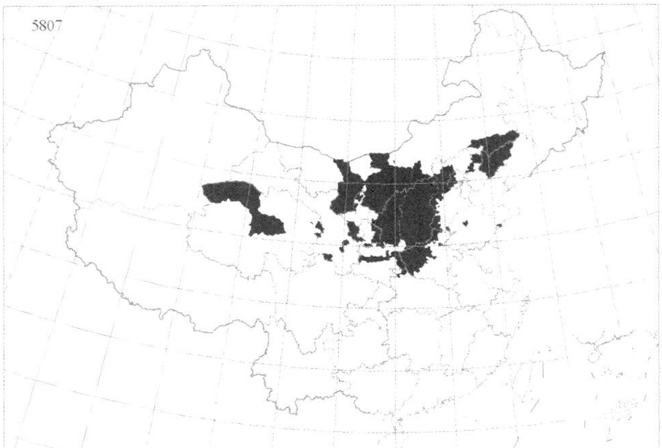

Sapindaceae 无患子科 *Xerospermum* 干果木属
Xerospermum bonii (Lecomte) Radlk. 干果木
Trees

19.7 (16.9~22.7); 13.2 (10.3~15.1); 24.1 (21.4~28.6); 19.7 (16.9~22.8); 176.6 (142.7~213.1); 0.0 (0.0~0.0); 973 (813~1221) ❙ 1527 (1143~2435); 829 (637~1428); 73 (48~130) ❙ 56.7 (15.4~99.5); 965 (813~1221); 434.7 (281.7~659.0)

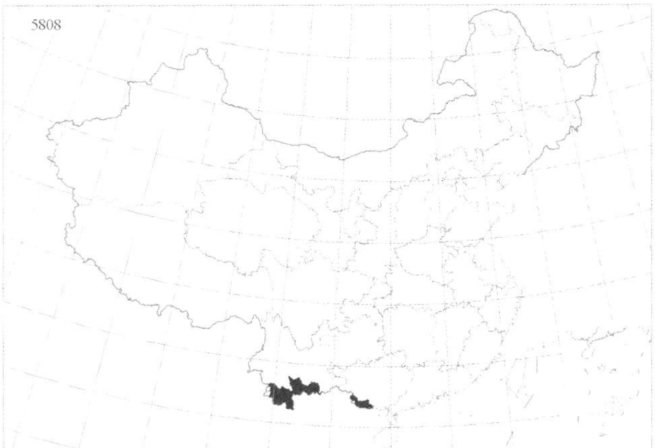

Sabiaceae 清风藤科 *Meliosma* 泡花树属
Meliosma alba (Schlech.) Walp. 珂南树
Trees

14.5 (6.2~21.0); 3.0 (-2.9~14.2); 25.2 (14.1~29.3); 14.5 (6.7~21.0); 119.9 (39.4~191.8); 5.8 (0.0~24.1); 806 (509~1053) ❙ 1166 (626~1997); 502 (316~932); 89 (19~259) ❙ 44.3 (-22.2~130.7); 791 (509~1053); 231.6 (139.1~551.6)

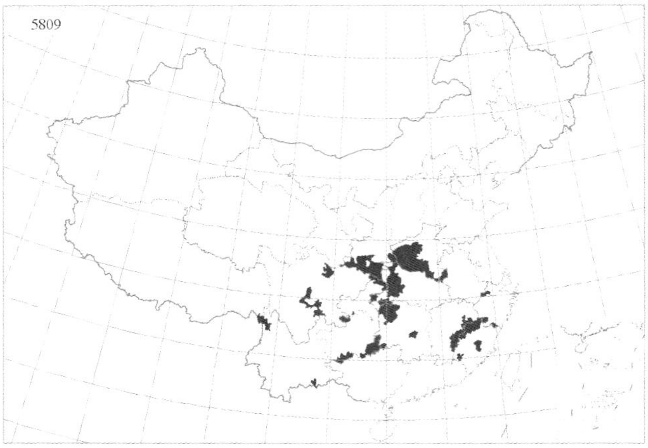

Sabiaceae 清风藤科 *Meliosma* 泡花树属
Meliosma angustifolia Merr. 狭叶泡花树
Trees, evergreen

21.6 (17.2~24.8); 13.9 (6.4~20.1); 27.5 (22.4~29.8); 21.6 (17.3~24.8); 199.2 (147.0~237.6); 0.0 (0.0~0.0); 1140 (833~1385) ❙ 1557 (868~2435); 710 (312~1428); 104 (41~197) ❙ 38.6 (-34.3~99.5); 1112 (833~1360); 308.3 (152.9~620.0)

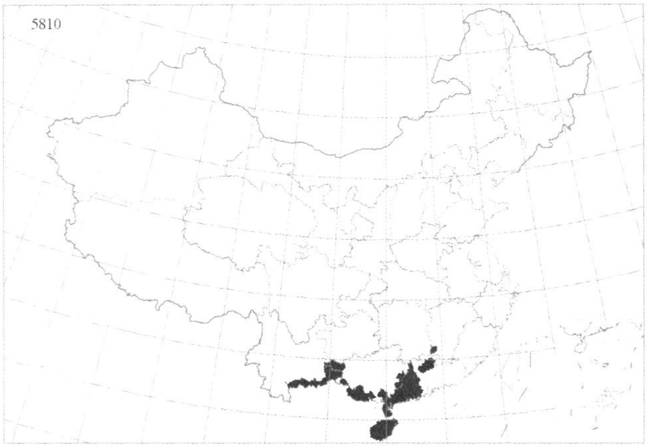

Sabiaceae 清风藤科 *Meliosma* 泡花树属
Meliosma arnottiana (Wight) Walp. 南亚泡花树
Trees, evergreen

15.7 (5.3~21.8); 8.5 (-2.4~15.1); 21.1 (12.2~27.7); 15.7 (5.7~21.8); 129.6 (30.5~202.2); 0.9 (0.0~26.2); 791 (479~1154) ❙ 1132 (741~1751); 633 (411~996); 46 (15~142) ❙ 43.6 (-4.2~112.2); 776 (479~1154); 331.7 (142.9~659.0)

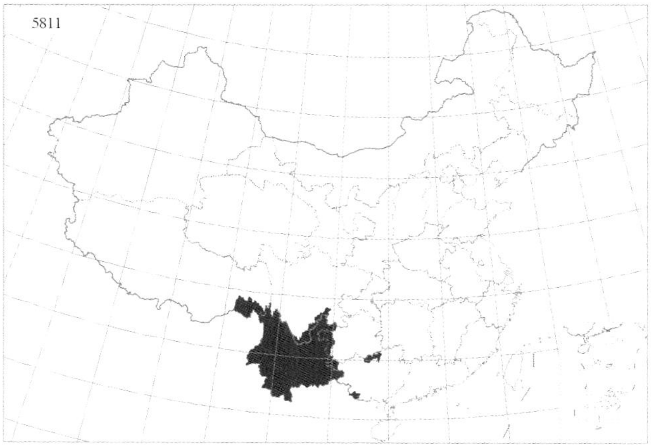

Sabiaceae 清风藤科 *Meliosma* 泡花树属
Meliosma bifida Law 双裂泡花树
Trees

19.4; 12.6; 24.1; 19.4; 172.6; 0.0; 954 ❙ 1445; 787; 64 ❙ 52.3; 954; 521.5

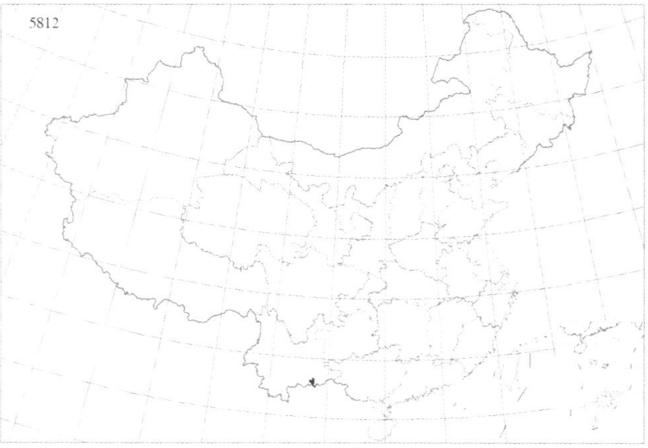

Sabiaceae 清风藤科 *Meliosma* 泡花树属
Meliosma callicarpifolia Hayata 紫珠叶泡花树
Trees, evergreen

21.5; 16.5; 25.2; 21.5; 198.0; 0.0; 1071 ❙ 2718; 1630; 109 ❙ 153.3; 1071; 252.3

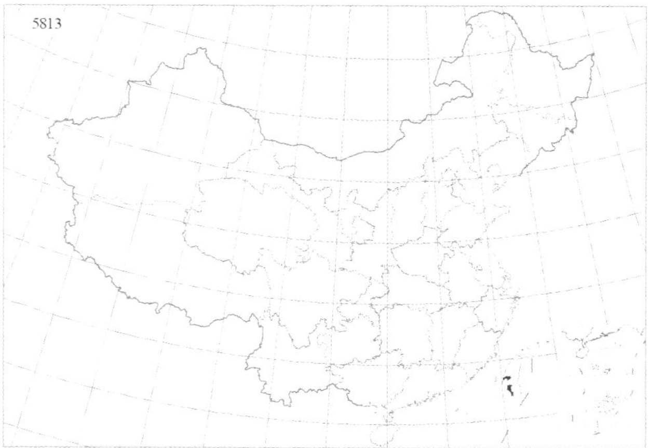

Sabiaceae 清风藤科 *Meliosma* 泡花树属
Meliosma cuneifolia Franch. 泡花树
Shrubs or trees, deciduous

14.0 (1.8~19.3); 3.6 (-7.7~13.1); 23.4 (9.9~29.3); 14.1 (3.8~19.4); 114.4 (15.4~173.2); 5.8 (0.0~53.3); 774 (404~1002) ❙ 1065 (341~1736); 513 (190~1007); 58 (8~214) ❙ 38.4 (-40.9~145.8); 757 (341~1002); 225.8 (61.2~491.5)

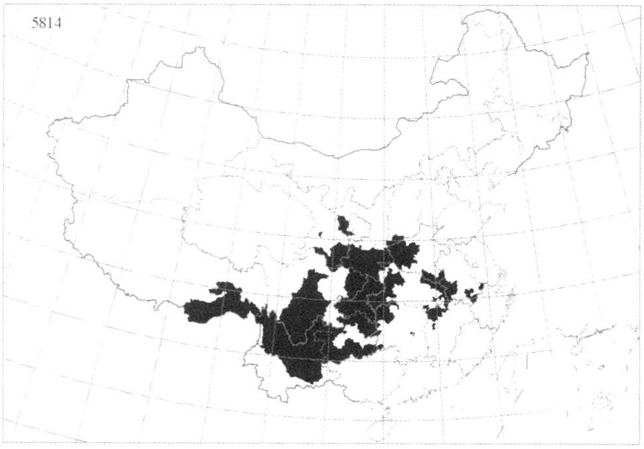

Sabiaceae 清风藤科 *Meliosma* 泡花树属
Meliosma cuneifolia var. glabriuscula Cufod. 光叶泡花树
Shrubs or trees, deciduous

12.8 (5.3~16.8); 3.0 (-3.8~8.6); 21.6 (12.2~29.0); 12.9 (5.7~16.9); 98.9 (30.5~143.2); 5.2 (0.0~26.7); 718 (479~931) ❙ 1077 (548~1502); 544 (264~817); 59 (12~175) ❙ 50.9 (-28.9~106.2); 705 (479~931); 250.6 (153.3~366.3)

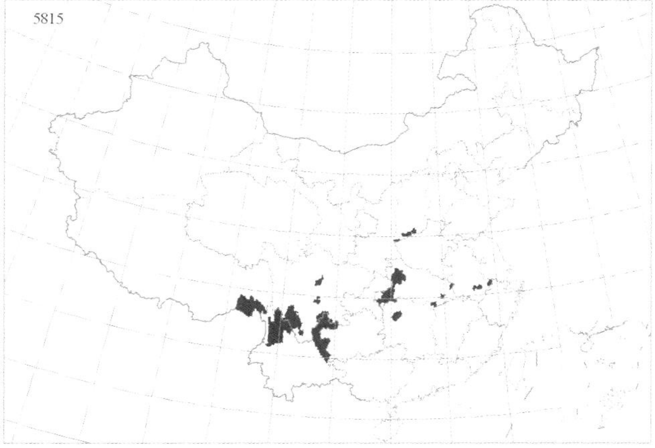

Sabiaceae 清风藤科 *Meliosma* 泡花树属
Meliosma dilleniifolia (Wall. ex Wight et Arn.) Walp. 重齿泡花树
Trees

6.6 (-1.8~15.4); -1.4 (-10.2~8.2); 13.4 (6.1~21.1); 8.0 (1.8~15.4); 53.9 (1.9~125.4); 34.6 (0.0~82.8); 532 (329~770) ❙ 886 (320~1736); 527 (193~1007); 30 (10~51) ❙ 52.2 (-4.0~145.8); 518 (320~706); 240.9 (61.6~447.4)

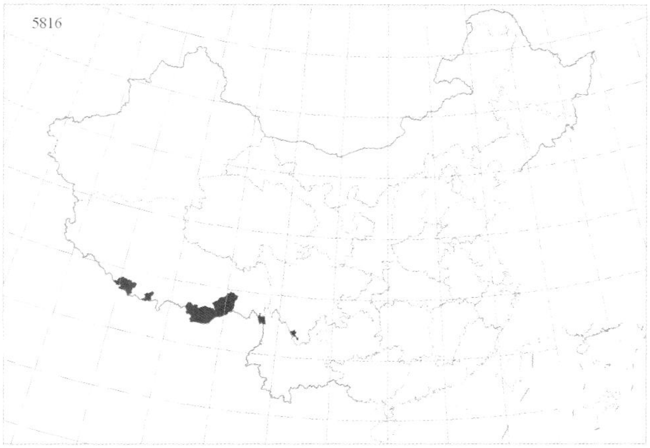

Sabiaceae 清风藤科 *Meliosma* 泡花树属
Meliosma dumicola W. W. Sm. 灌丛泡花树
Trees, evergreen

18.9 (13.3~24.3); 12.2 (5.4~18.7); 23.7 (19.1~28.6); 19.0 (13.4~24.3); 167.8 (100.4~231.9); 0.0 (0.0~0.0); 964 (706~1338) ❙ 1342 (1007~1736); 658 (392~1007); 59 (42~110) ❙ 43.8 (-11.9~145.8); 937 (706~1326); 408.5 (277.0~572.9)

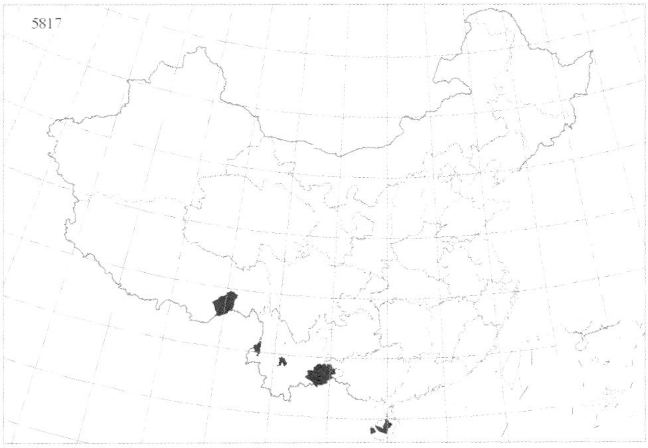

Sabiaceae 清风藤科 *Meliosma* 泡花树属
Meliosma flexuosa Pamp. 垂枝泡花树
Trees

16.2 (2.9~21.0); 4.9 (-6.3~12.2); 26.8 (11.2~29.9); 16.3 (4.4~21.1); 136.7 (20.5~193.0); 1.6 (0.0~44.6); 875 (428~1114) ❙ 1436 (793~1997); 526 (377~821); 156 (16~259) ❙ 64.5 (-2.0~130.7); 867 (428~1114); 218.1 (134.6~319.9)

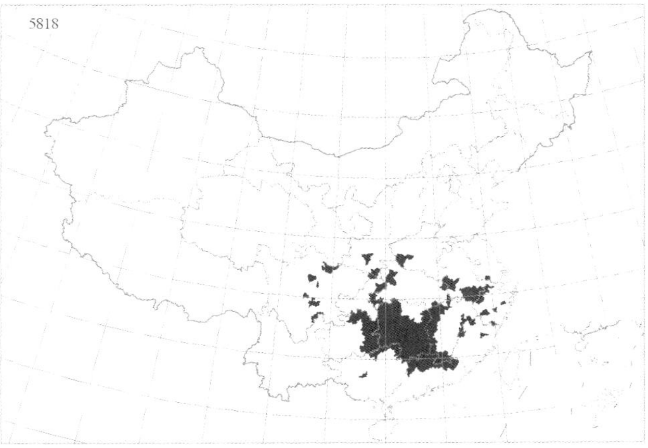

Sabiaceae 清风藤科 *Meliosma* 泡花树属
Meliosma fordii Hemsl. 香皮树
Trees

19.7 (14.3~24.8); 10.7 (3.3~20.1); 27.4 (22.4~29.8); 19.8 (14.3~24.8); 177.3 (112.6~237.6); 0.1 (0.0~2.3); 1037 (756~1385) ❙ 1576 (1168~2435); 662 (403~1428); 140 (41~230) ❙ 53.3 (-6.9~104.6); 1031 (756~1299); 242.6 (95.3~620.0)

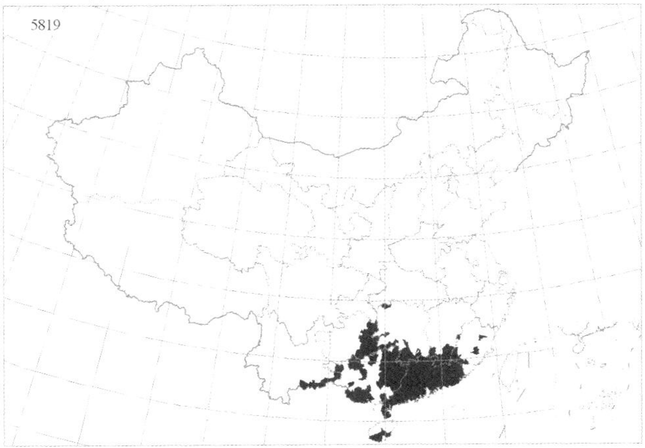

Sabiaceae 清风藤科 *Meliosma* 泡花树属
Meliosma fordii var. sinii (Diels) Law 辛氏泡花树
Trees

16.4 (15.9~16.9); 5.7 (4.7~6.8); 25.8 (25.5~26.3); 16.4 (16.0~17.0); 137.4 (131.7~143.8); 0.1 (0.0~0.3); 855 (838~871) ❙ 1293 (1234~1368); 521 (452~579); 116 (99~129) ❙ 51.3 (45.6~59.2); 855 (838~871); 225.5 (206.2~246.1)

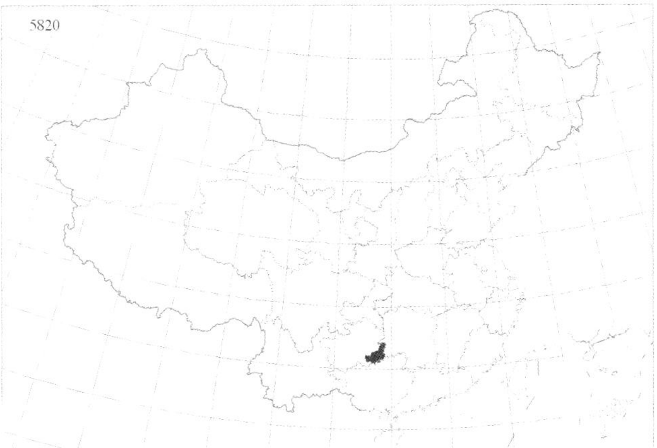

Sabiaceae 清风藤科 *Meliosma* 泡花树属

Meliosma glandulosa Cufod. 腺毛泡花树

Trees, evergreen

17.5 (14.3~21.6); 6.9 (3.3~13.2); 26.9 (23.5~29.0); 17.5 (14.3~21.6); 150.3 (112.6~199.2); 0.2 (0.0~2.3); 920 (756~1139) ❙ 1501 (1185~1961); 566 (403~858); 163 (57~230) ❙ 63.9 (22.7~104.6); 915 (756~1139); 220.8 (151.0~371.7)

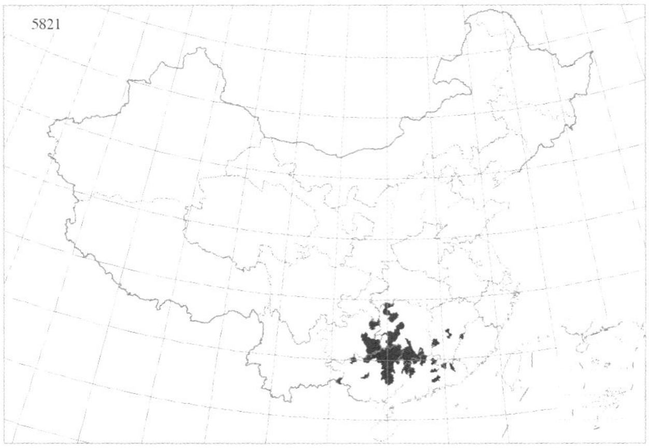

Sabiaceae 清风藤科 *Meliosma* 泡花树属

Meliosma henryi Diels 贵州泡花树

Trees

16.2 (9.3~22.2); 6.0 (-1.7~13.9); 25.1 (20.1~28.6); 16.2 (9.5~22.3); 136.8 (69.8~207.5); 2.3 (0.0~18.0); 856 (603~1190) ❙ 1315 (1118~1519); 601 (473~848); 86 (46~161) ❙ 56.4 (15.4~107.2); 856 (603~1190); 244.9 (163.0~371.7)

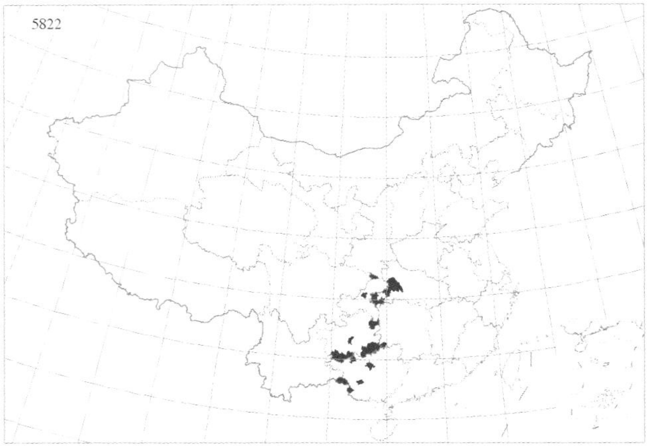

Sabiaceae 清风藤科 *Meliosma* 泡花树属

Meliosma laui Merr. 华南泡花树

Trees

21.3 (17.2~24.8); 13.8 (7.3~20.1); 26.9 (22.4~28.9); 21.3 (17.3~24.8); 195.9 (147.0~237.6); 0.0 (0.0~0.0); 1121 (833~1385) ❙ 1609 (1254~2435); 751 (455~1428); 96 (41~181) ❙ 45.5 (1.5~99.5); 1090 (833~1360); 366.8 (194.5~620.0)

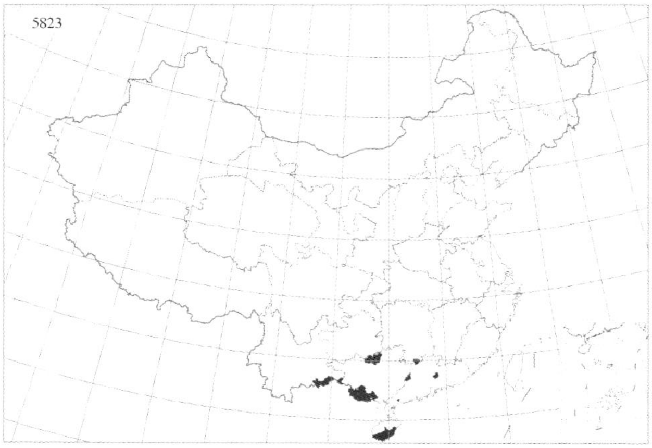

Sabiaceae 清风藤科 *Meliosma* 泡花树属

Meliosma longipes Merr. 疏枝泡花树

Shrubs or trees, evergreen

20.2 (17.2~22.7); 11.8 (7.3~15.0); 26.5 (22.4~28.6); 20.2 (17.3~22.8); 182.7 (147.0~213.1); 0.0 (0.0~0.0); 1047 (833~1221) ❙ 1604 (1254~2435); 847 (639~1428); 104 (48~166) ❙ 54.0 (12.9~99.5); 1047 (833~1221); 316.7 (225.9~551.6)

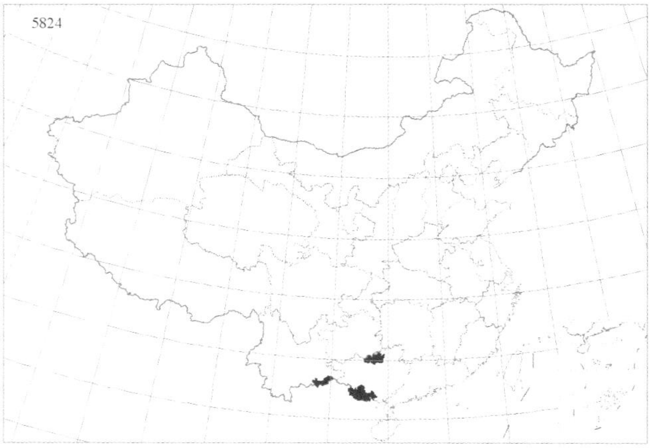

Sabiaceae 清风藤科 *Meliosma* 泡花树属

Meliosma myriantha (Leconte) Law 多花泡花树

Trees, deciduous

15.6 (9.3~21.0); 3.6 (-3.7~11.9); 26.8 (20.1~29.7); 15.7 (9.5~21.1); 131.9 (69.8~193.0); 4.1 (0.0~23.0); 858 (603~1114) ❙ 1338 (604~1997); 518 (371~821); 147 (26~259) ❙ 54.6 (-19.2~130.7); 842 (603~1114); 213.6 (134.1~319.9)

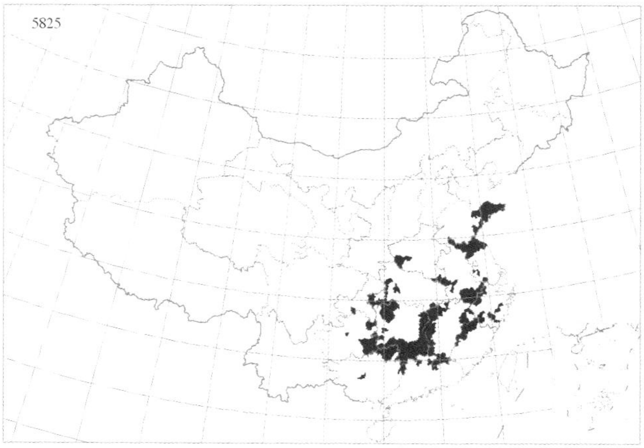

Sabiaceae 清风藤科 *Meliosma* 泡花树属

Meliosma myriantha var. discolor Dunn 异色泡花树

Trees, deciduous

16.9 (12.7~21.0); 5.8 (-0.2~12.2); 27.2 (23.0~29.8); 16.9 (12.8~21.0); 144.3 (99.0~192.6); 1.0 (0.0~11.7); 904 (713~1113) ❙ 1579 (793~1950); 571 (398~713); 182 (44~259) ❙ 76.0 (0.0~147.0); 901 (713~1107); 238.3 (169.7~319.9)

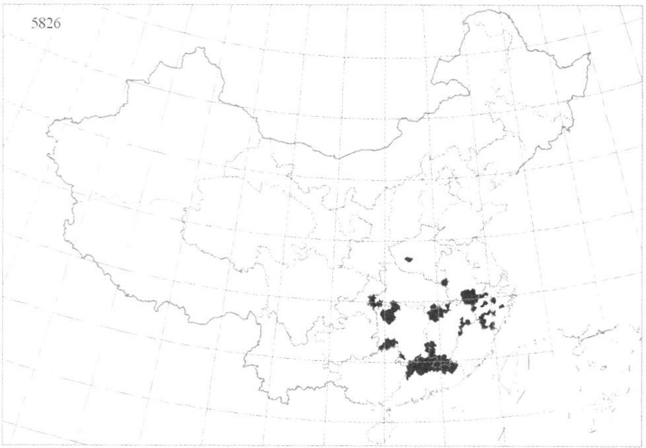

Sabiaceae 清风藤科 *Meliosma* 泡花树属

Meliosma myriantha var. pilosa (Leconte) Law 柔毛泡花树

Trees, deciduous

15.4 (10.2~17.8)；3.8 (-0.7~6.0)；26.5 (21.0~29.3)；15.5 (10.4~17.8)；127.4 (77.2~154.1)；2.0 (0.0~13.6)；838 (634~960) ❙ 1463 (1006~1922)；539 (403~685)；159 (29~259) ❙ 74.9 (28.8~123.6)；834 (634~941)；232.9 (151.0~319.9)

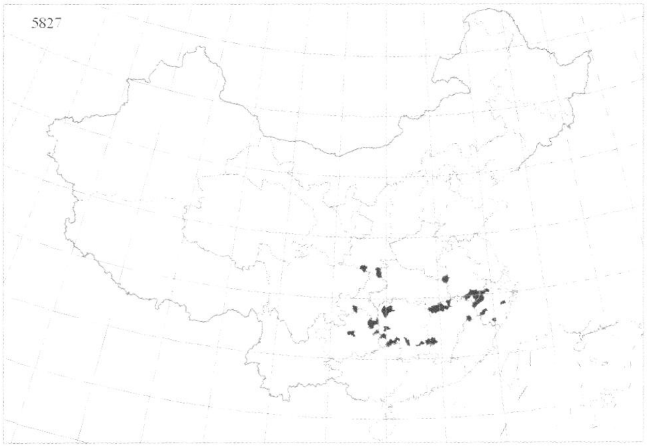

Sabiaceae 清风藤科 *Meliosma* 泡花树属

Meliosma oldhamii Miq. ex Maxim. 红柴枝

Trees, deciduous

15.2 (7.4~21.0)；3.4 (-4.2~11.9)；26.3 (18.7~29.8)；15.2 (8.2~21.1)；126.7 (58.8~193.0)；4.1 (0.0~29.4)；837 (561~1114) ❙ 1238 (512~1922)；493 (258~821)；120 (17~259) ❙ 47.1 (-35.3~129.9)；823 (512~1114)；213.5 (121.6~334.4)

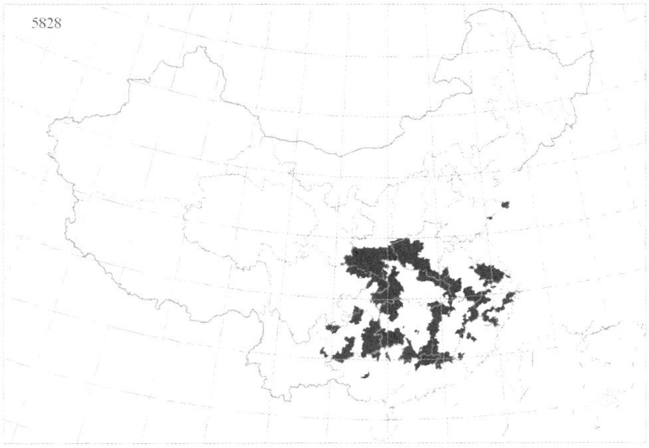

Sabiaceae 清风藤科 *Meliosma* 泡花树属

Meliosma oldhamii var. glandulifera Cufod. 有腺泡花树

Trees, deciduous

16.5 (14.3~18.2)；4.8 (2.6~7.7)；27.6 (24.2~29.8)；16.6 (14.3~18.3)；139.6 (113.2~159.4)；0.5 (0.0~3.5)；892 (761~986) ❙ 1442 (1171~1742)；490 (392~600)；172 (88~225) ❙ 62.1 (34.4~128.9)；878 (761~972)；196.2 (134.6~299.5)

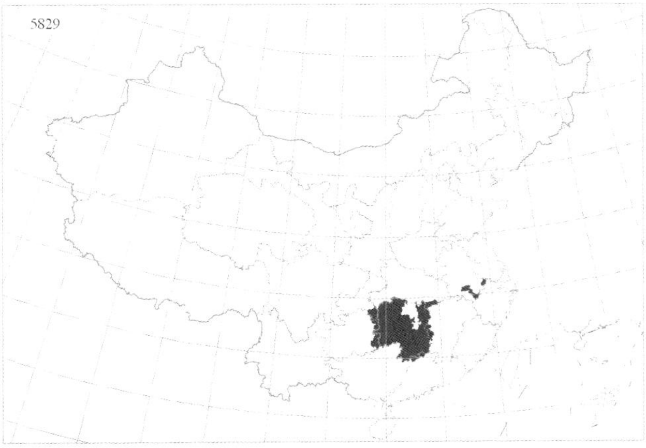

Sabiaceae 清风藤科 *Meliosma* 泡花树属

Meliosma parviflora Lecomte 细花泡花树

Shrubs or trees, deciduous

14.5 (2.2~17.1)；2.2 (-7.0~7.1)；26.2 (9.9~28.9)；14.6 (3.8~17.1)；120.6 (15.5~145.2)；5.9 (0.0~49.1)；820 (413~933) ❙ 1081 (512~1663)；490 (258~946)；92 (19~220) ❙ 32.2 (-35.3~125.9)；800 (413~914)；185.6 (121.6~318.0)

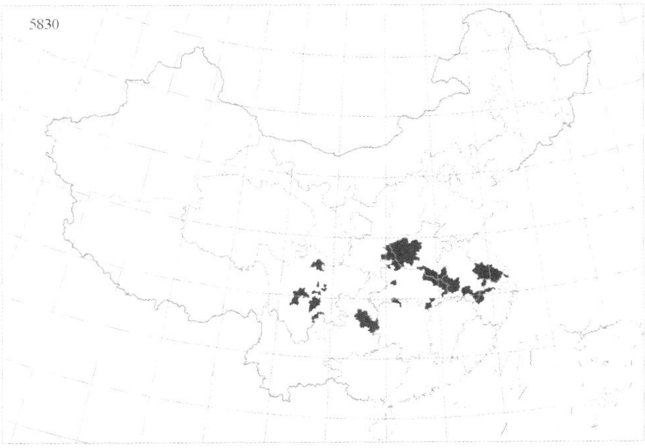

Sabiaceae 清风藤科 *Meliosma* 泡花树属

Meliosma paupera Hand.-Mazz. 狭序泡花树

Trees

18.4 (14.3~22.7)；8.6 (4.0~15.0)；26.9 (22.7~29.8)；18.5 (14.3~22.8)；161.9 (112.6~213.1)；0.0 (0.0~1.1)；962 (756~1221) ❙ 1564 (1218~2435)；627 (452~1428)；157 (48~200) ❙ 62.7 (17.4~99.5)；960 (756~1221)；240.2 (174.9~359.7)

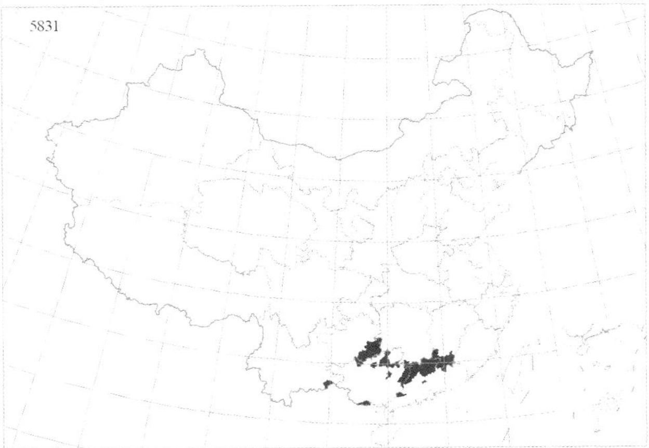

Sabiaceae 清风藤科 *Meliosma* 泡花树属

Meliosma pinnata (Roxb.) Maxim. 羽叶泡花树

Trees

13.3；5.4；19.6；13.4；100.4；0.0；706 ❙ 1736；1007；51 ❙ 145.8；706；447.4

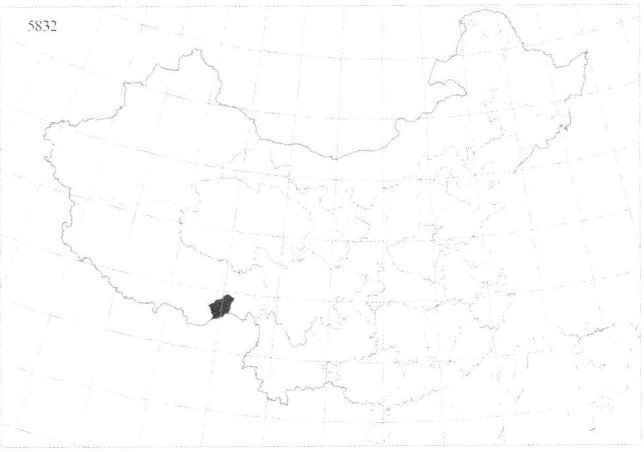

Sabiaceae 清风藤科 *Meliosma* 泡花树属
Meliosma rhoifolia Maxim. 漆叶泡花树
Trees, evergreen

17.4 (14.3~21.7); 7.2 (3.8~15.3); 26.6 (21.8~29.0); 17.4 (14.3~21.7); 148.9 (112.6~200.2); 0.1 (0.0~1.5); 907 (756~1118) | 1658 (1136~3495); 588 (403~1045); 201 (68~897) | 82.8 (30.7~235.6); 903 (756~1114); 235.8 (151.0~305.7)

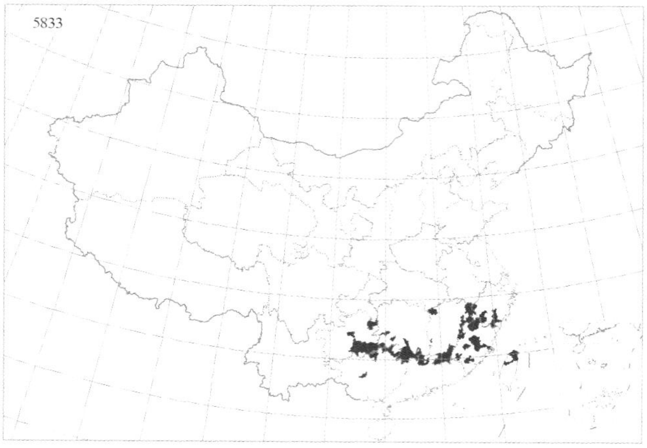

Sabiaceae 清风藤科 *Meliosma* 泡花树属
Meliosma rhoifolia var. barbulata (Cufod.) Law 腋毛泡花树
Trees, evergreen

14.6 (6.1~21.3); 3.0 (-4.9~12.3); 25.3 (12.7~29.8); 14.7 (6.3~21.3); 121.9 (35.2~196.1); 6.5 (0.0~29.4); 814 (496~1130) | 1207 (520~1997); 491 (258~932); 107 (15~259) | 46.3 (-35.9~130.7); 794 (496~1130); 231.2 (128.6~491.5)

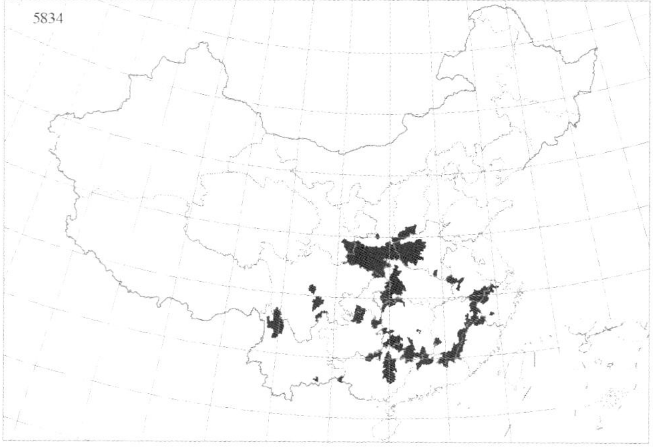

Sabiaceae 清风藤科 *Meliosma* 泡花树属
Meliosma rigida Siebold et Zucc. 笔罗子
Trees

17.8 (10.2~24.7); 8.0 (-2.9~19.8); 26.8 (21.4~29.8); 17.9 (10.6~24.7); 155.9 (83.1~236.5); 1.1 (0.0~20.5); 941 (656~1376) | 1505 (626~3495); 593 (299~1428); 154 (29~897) | 60.5 (-22.2~224.4); 928 (626~1230); 246.1 (123.9~659.0)

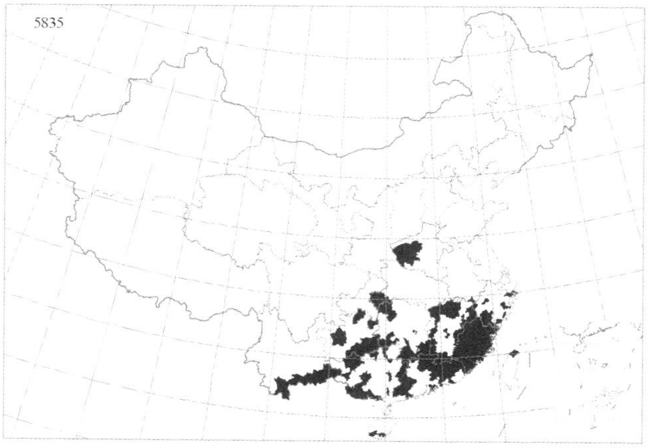

Sabiaceae 清风藤科 *Meliosma* 泡花树属
Meliosma rigida var. pannosa (Hand.-Mazz.) Law 毡毛泡花树
Trees

16.7 (14.3~19.0); 5.9 (3.3~9.4); 26.8 (23.5~29.3); 16.7 (14.3~19.1); 141.3 (112.6~168.8); 0.3 (0.0~1.9); 883 (756~992) | 1577 (1211~1997); 561 (403~713); 181 (84~259) | 78.9 (45.6~147.0); 880 (756~992); 235.0 (151.0~303.4)

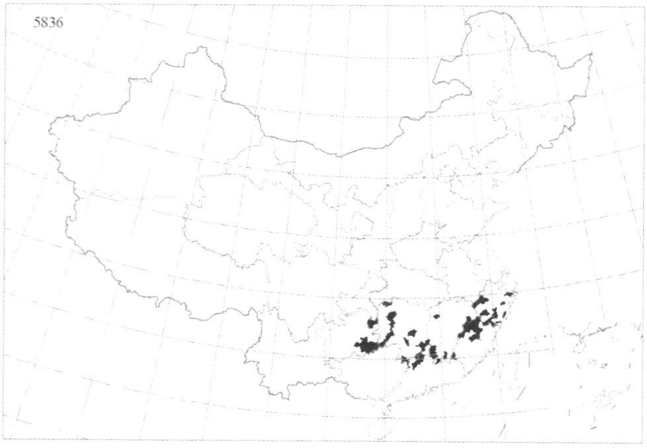

Sabiaceae 清风藤科 *Meliosma* 泡花树属
Meliosma squamulata Hance 樟叶泡花树
Trees

19.4 (14.3~24.8); 10.7 (3.8~20.1); 26.8 (20.1~29.8); 19.4 (14.3~24.8); 173.1 (112.6~237.6); 0.0 (0.0~1.5); 1011 (756~1385) | 1673 (868~3495); 687 (312~1847); 159 (41~897) | 68.6 (-34.3~246.5); 995 (756~1360); 272.1 (119.4~620.0)

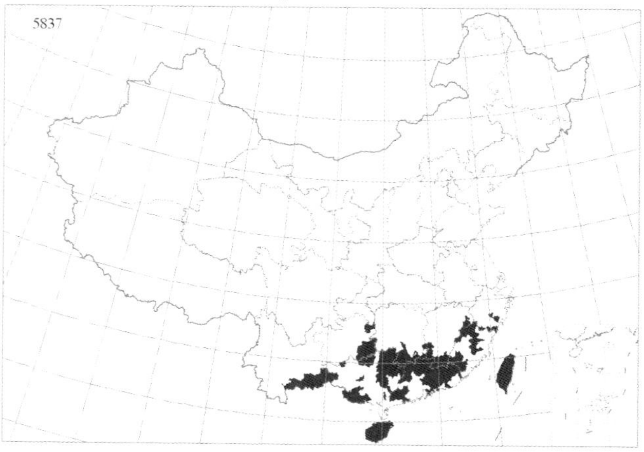

Sabiaceae 清风藤科 *Meliosma* 泡花树属
Meliosma thomsonii King ex Brandis 西南泡花树
Trees

16.0 (5.3~21.0); 8.9 (-2.4~14.2); 21.4 (12.2~26.9); 16.1 (5.7~21.0); 135.3 (30.5~191.8); 2.8 (0.0~26.2); 819 (479~1053) | 1358 (961~1751); 756 (501~1007); 57 (31~117) | 69.4 (21.7~145.8); 815 (479~1053); 404.3 (234.2~551.6)

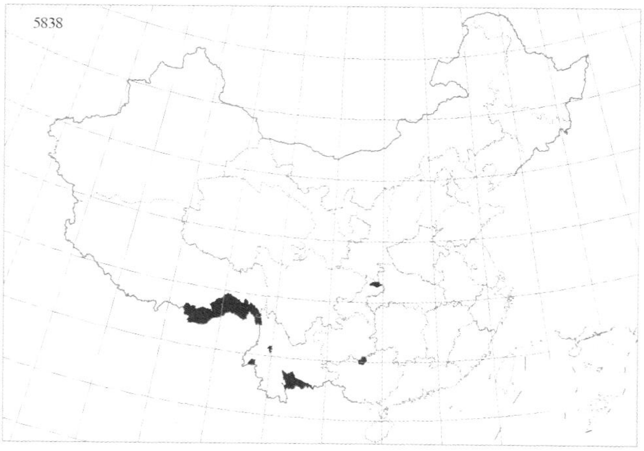

Sabiaceae 清风藤科 *Meliosma* 泡花树属
Meliosma thorelii Lecomte 山様叶泡花树
Trees

18.8 (13.2~24.8); 10.9 (3.8~20.1); 25.0 (18.4~29.8); 18.9 (13.3~24.8); 166.6 (99.4~237.6); 0.0 (0.0~1.5); 959 (695~1385) ❙ 1454 (868~2435); 691 (312~1428); 100 (41~235) ❙ 52.9 (-34.3~112.2); 942 (695~1326); 332.9 (158.7~659.0)

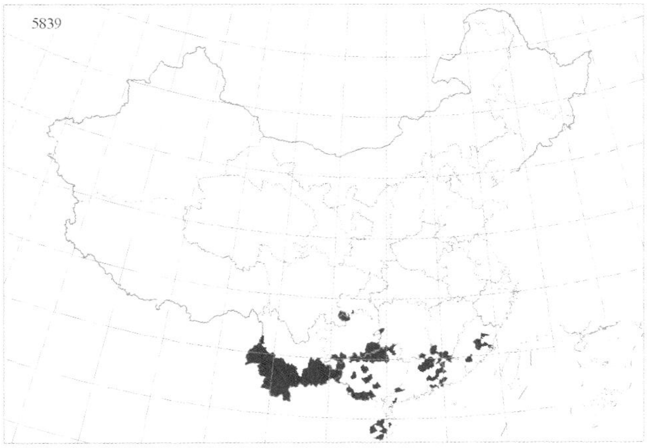

Sabiaceae 清风藤科 *Meliosma* 泡花树属
Meliosma velutina Rehder et E. H. Wilson 毛泡花树
Trees

19.2 (17.2~22.2); 11.2 (7.1~15.1); 25.3 (22.3~28.5); 19.3 (17.3~22.2); 171.0 (147.0~206.5); 0.0 (0.0~0.0); 969 (833~1188) ❙ 1463 (1254~1607); 699 (507~860); 99 (48~182) ❙ 52.2 (24.6~80.6); 964 (833~1188); 360.9 (197.5~659.0)

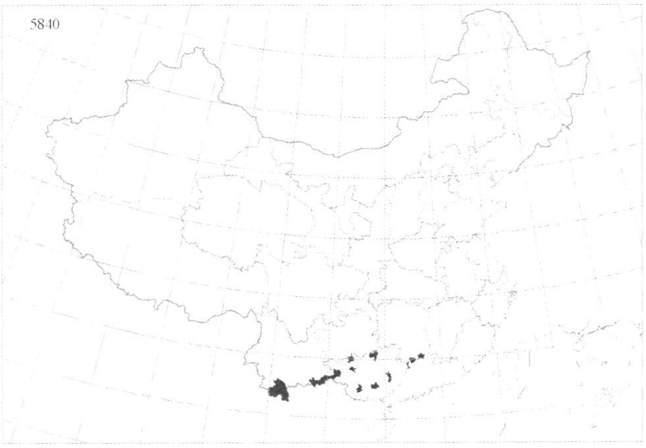

Sabiaceae 清风藤科 *Meliosma* 泡花树属
Meliosma yunnanensis Franch. 云南泡花树
Trees

14.2 (-1.8~20.5); 6.6 (-10.2~15.1); 20.4 (6.1~26.6); 14.4 (1.8~20.5); 114.3 (1.9~185.9); 2.8 (0.0~82.8); 744 (329~992) ❙ 1037 (320~1736); 582 (217~1007); 40 (12~77) ❙ 40.6 (-14.2~145.8); 725 (320~986); 280.3 (72.2~659.0)

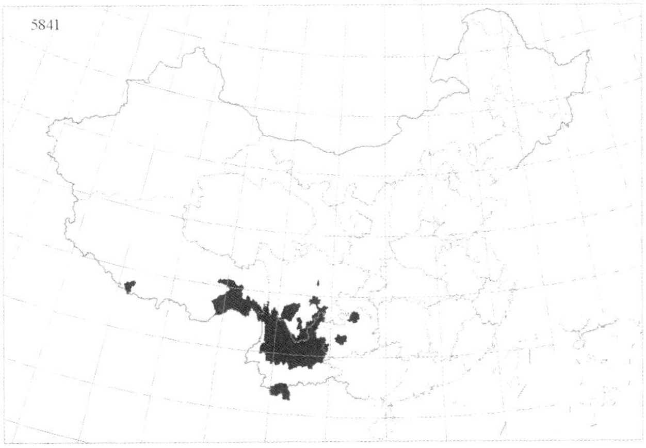

Sabiaceae 清风藤科 *Sabia* 清风藤属
Sabia campanulata (L. Chen) Y. F. Wu 钟花清风藤
Vines, deciduous

15.3 (-1.8~21.0); 4.2 (-10.2~14.2); 25.7 (6.1~29.3); 15.4 (1.8~21.0); 128.4 (1.9~191.8); 4.4 (0.0~82.8); 839 (329~1053) ❙ 1434 (320~1950); 544 (193~1007); 153 (10~259) ❙ 69.8 (-4.0~145.8); 836 (320~1053); 243.1 (61.6~551.6)

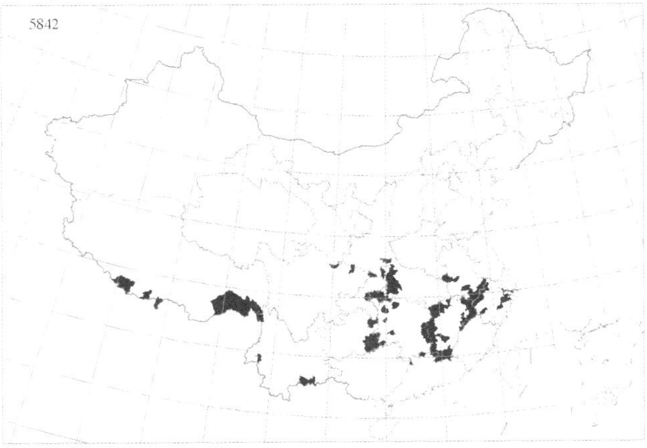

Sabiaceae 清风藤科 *Sabia* 清风藤属
Sabia campanulata subsp. ritchieae (Rehder et E. H. Wilson) Y. F. Wu 鄂西清风藤
Vines, deciduous

15.6 (3.0~19.4); 3.9 (-7.7~9.4); 26.6 (12.4~29.3); 15.6 (4.8~19.4); 130.6 (25.1~173.4); 3.2 (0.0~49.3); 854 (441~1029) ❙ 1427 (656~1997); 534 (303~713); 155 (15~259) ❙ 66.6 (-12.2~147.0); 850 (441~1029); 231.0 (147.1~332.5)

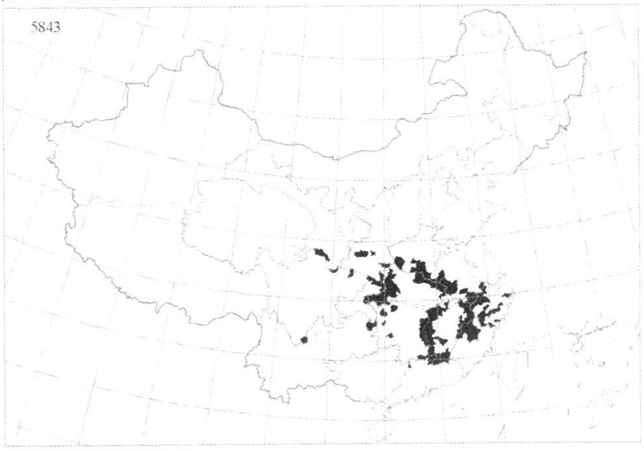

Sabiaceae 清风藤科 *Sabia* 清风藤属
Sabia coriacea Rehder et E. H. Wilson 革叶清风藤
Vines, evergreen

18.4 (14.3~22.2); 8.3 (3.7~14.2); 27.4 (24.2~29.1); 18.4 (14.3~22.3); 160.8 (113.2~207.2); 0.0 (0.0~1.4); 962 (761~1186) ❙ 1665 (1182~2014); 588 (403~954); 195 (43~259) ❙ 74.2 (16.9~130.7); 956 (761~1186); 239.9 (151.0~354.5)

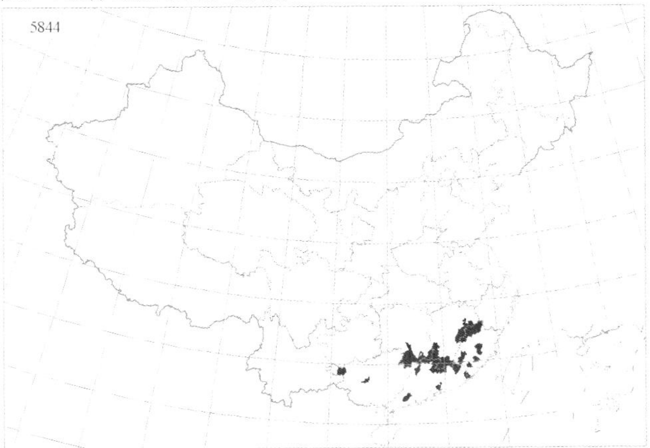

Sabiaceae 清风藤科 *Sabia* 清风藤属

Sabia dielsii H. Lév. 平伐清风藤

Vines, deciduous

17.4 (6.6~21.6); 9.7 (−1.3~15.1); 23.3 (13.3~28.5); 17.5 (6.8~21.7); 150.2 (39.6~199.8); 0.4 (0.0~19.7); 872 (514~1153) ❙ 1278 (862~1751); 668 (473~996); 64 (12~194) ❙ 47.5 (5.8~93.0); 862 (514~1153); 342.1 (176.4~659.0)

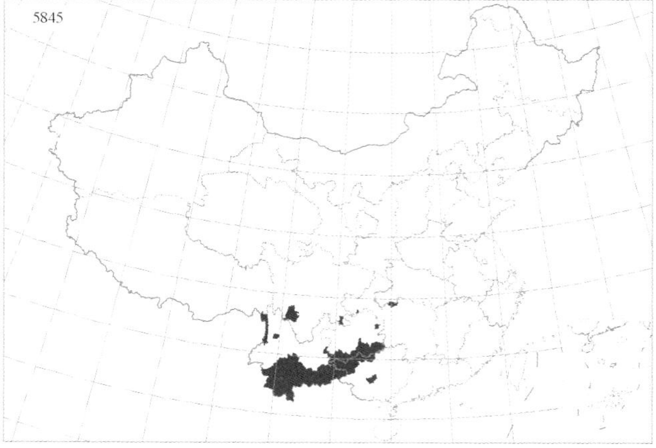

Sabiaceae 清风藤科 *Sabia* 清风藤属

Sabia emarginata Lecomte 凹萼清风藤

Vines, deciduous

15.7 (10.6~18.7); 4.5 (−0.2~8.8); 26.1 (21.1~29.8); 15.7 (10.6~18.8); 130.1 (79.4~165.2); 1.6 (0.0~12.1); 844 (643~986) ❙ 1392 (1077~1740); 529 (392~709); 136 (46~220) ❙ 66.0 (30.4~107.2); 834 (643~968); 215.7 (145.3~285.1)

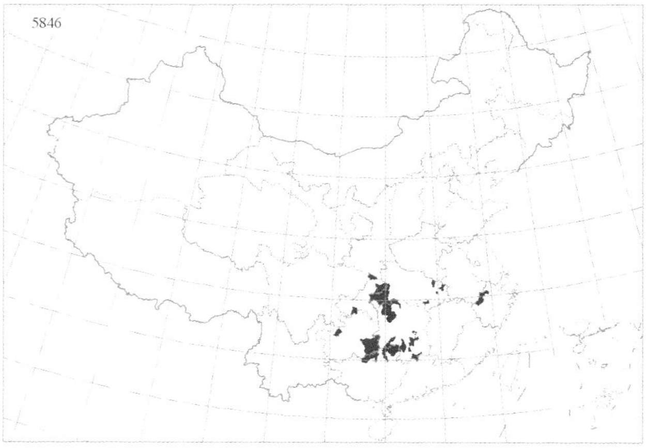

Sabiaceae 清风藤科 *Sabia* 清风藤属

Sabia fasciculata Lecomte ex L. Chen 簇花清风藤

Vines, evergreen

18.5 (8.6~22.4); 10.7 (1.3~15.1); 24.5 (15.0~28.6); 18.6 (8.7~22.4); 163.1 (52.6~208.9); 0.2 (0.0~8.8); 935 (574~1191) ❙ 1435 (1007~2037); 709 (513~1033); 93 (41~193) ❙ 54.3 (5.8~93.0); 930 (574~1191); 345.1 (163.7~659.0)

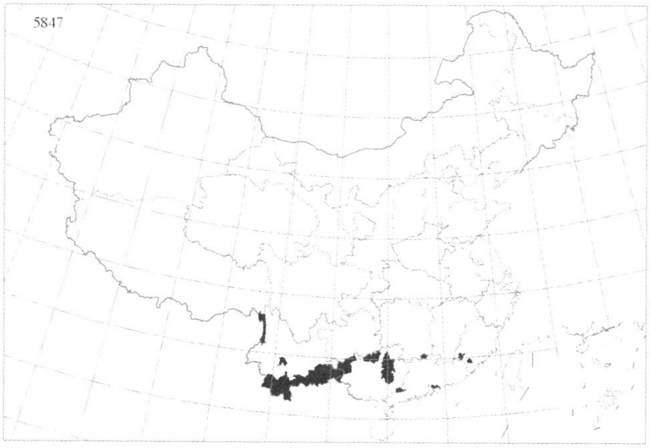

Sabiaceae 清风藤科 *Sabia* 清风藤属

Sabia japonica (Stapf ex Anon) L. Chen 清风藤

Vines, deciduous

16.1 (7.4~23.6); 4.7 (−4.2~16.7); 26.9 (18.7~30.1); 16.2 (8.2~23.6); 137.3 (58.8~223.5); 3.2 (0.0~29.4); 879 (561~1281) ❙ 1350 (512~2037); 517 (258~1033); 145 (17~259) ❙ 52.8 (−35.3~147.0); 865 (512~1269); 221.8 (111.9~371.7)

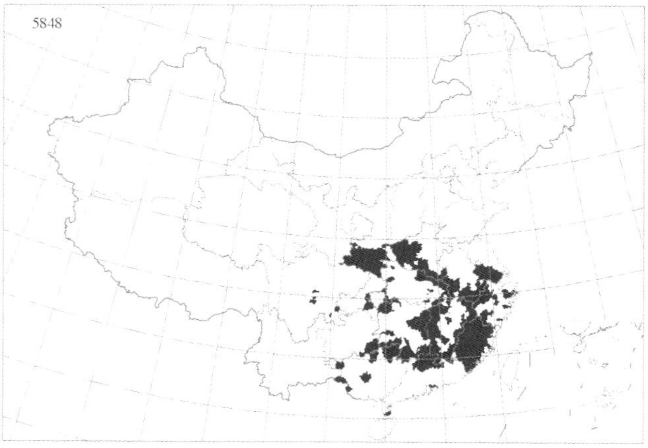

Sabiaceae 清风藤科 *Sabia* 清风藤属

Sabia japonica var. sinensis (Stapf ex Anon) L. Chen 中华清风藤

Vines, deciduous

19.1 (16.6~21.0); 9.3 (6.0~12.2); 27.6 (26.2~29.8); 19.1 (16.6~21.0); 169.7 (138.9~192.6); 0.0 (0.0~0.0); 997 (866~1113) ❙ 1647 (1486~1997); 595 (505~711); 188 (161~254) ❙ 66.0 (35.6~130.7); 994 (866~1107); 241.5 (174.9~303.4)

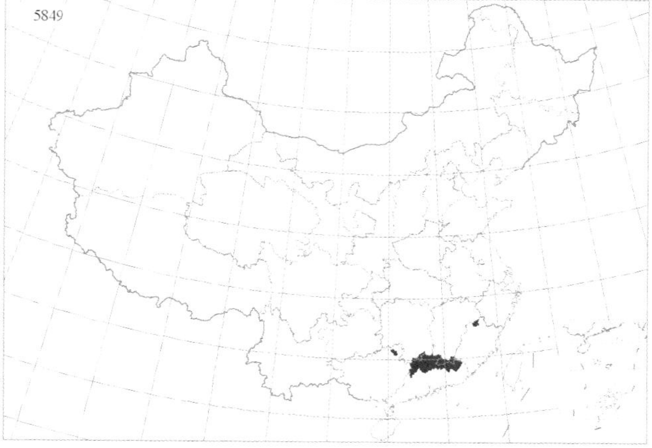

Sabiaceae 清风藤科 *Sabia* 清风藤属

Sabia limoniacea Wall. ex Hook. f. et Thomson 柠檬清风藤

Vines, evergreen

20.3 (10.6~24.8); 11.6 (−0.2~20.1); 27.7 (21.1~29.8); 20.4 (10.6~24.8); 184.7 (79.4~237.6); 0.2 (0.0~12.1); 1077 (643~1385) ❙ 1545 (868~2435); 661 (312~1428); 120 (33~198) ❙ 44.7 (−34.3~100.8); 1065 (643~1360); 241.6 (95.3~659.0)

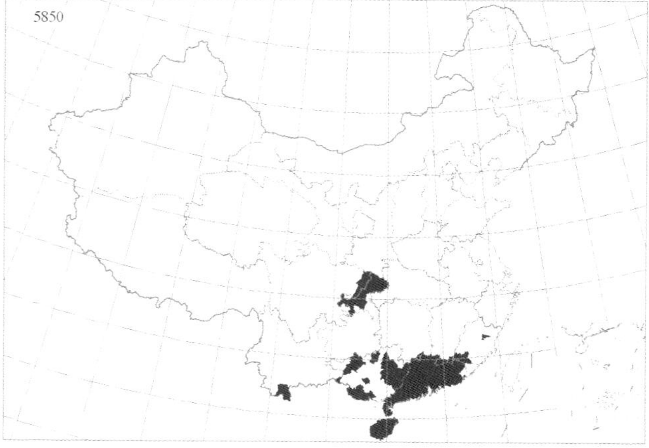

Sabiaceae 清风藤科 *Sabia* 清风藤属
Sabia nervosa Chun ex Y. F. Wu 长脉清风藤
Vines, evergreen

19.2 (17.6~21.3); 9.5 (6.9~13.2); 27.4 (25.8~28.6); 19.3 (17.6~21.3); 171.2 (151.6~196.1); 0.0 (0.0~0.0); 1004 (906~1130) | 1568 (1284~1776); 648 (505~828); 163 (73~194) | 56.9 (22.7~85.5); 1004 (906~1130); 231.9 (185.5~322.1)

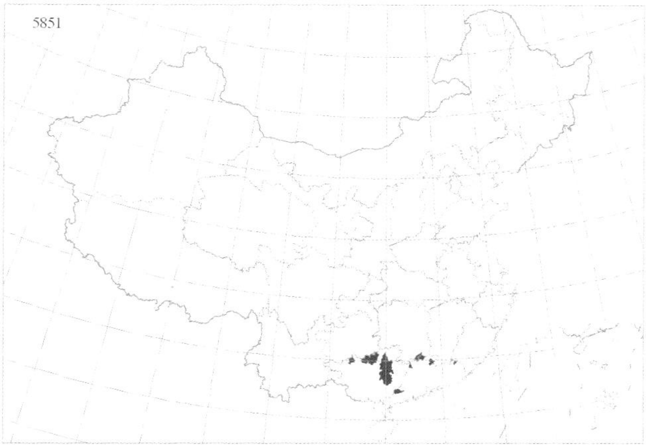

Sabiaceae 清风藤科 *Sabia* 清风藤属
Sabia paniculata Edgew. ex Hook. f. et Thomson 锥序清风藤
Vines, evergreen

17.4; 7.4; 25.5; 17.4; 148.9; 0.0; 886 | 1381; 637; 107 | 55.9; 886; 245.3

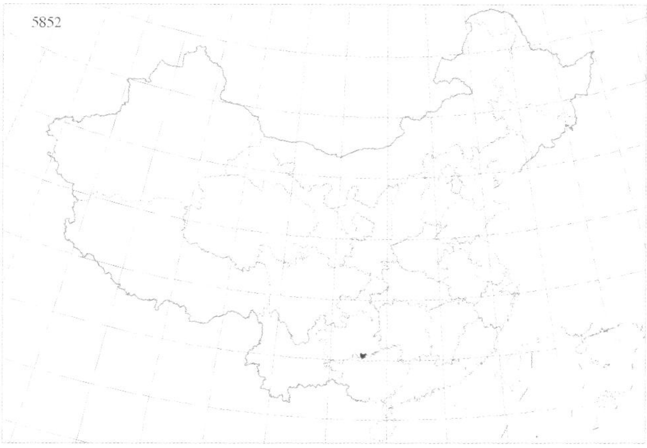

Sabiaceae 清风藤科 *Sabia* 清风藤属
Sabia parviflora Wall. ex Roxb. 小花清风藤
Vines, evergreen

17.4 (8.6~22.2); 7.8 (1.3~15.1); 25.8 (15.0~29.8); 17.5 (8.7~22.3); 149.8 (52.6~207.5); 0.3 (0.0~8.8); 904 (574~1190) | 1465 (939~2005); 611 (403~1007); 136 (37~259) | 62.6 (5.8~147.0); 898 (574~1190); 269.3 (146.0~659.0)

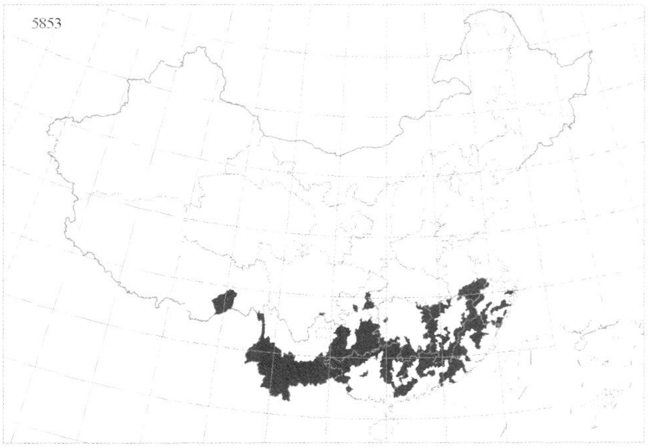

Sabiaceae 清风藤科 *Sabia* 清风藤属
Sabia purpurea subsp. dumicola (W. W. Smith) Water 紫花清风藤
Vines, deciduous

16.6 (14.4~18.2); 10.1 (7.5~11.9); 20.9 (19.1~22.0); 16.7 (14.4~18.2); 139.9 (113.1~158.7); 0.0 (0.0~0.0); 805 (727~868) | 1235 (1048~1454); 691 (553~816); 52 (42~70) | 53.6 (26.1~99.9); 794 (727~847); 410.5 (335.1~516.7)

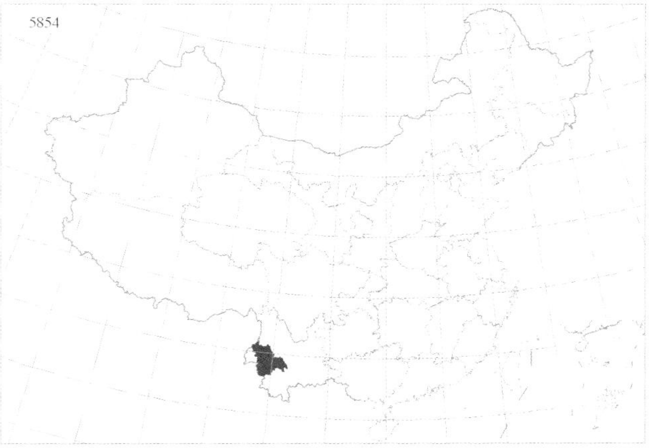

Sabiaceae 清风藤科 *Sabia* 清风藤属
Sabia schumanniana Diels 四川清风藤
Vines, deciduous

14.2 (6.1~21.6); 3.0 (-2.9~12.6); 24.6 (12.7~29.0); 14.3 (6.3~21.7); 115.8 (35.2~199.8); 4.8 (0.0~21.8); 786 (496~1153) | 1108 (626~1828); 497 (316~816); 70 (15~253) | 41.6 (-22.2~107.2); 773 (496~1153); 219.3 (118.7~354.5)

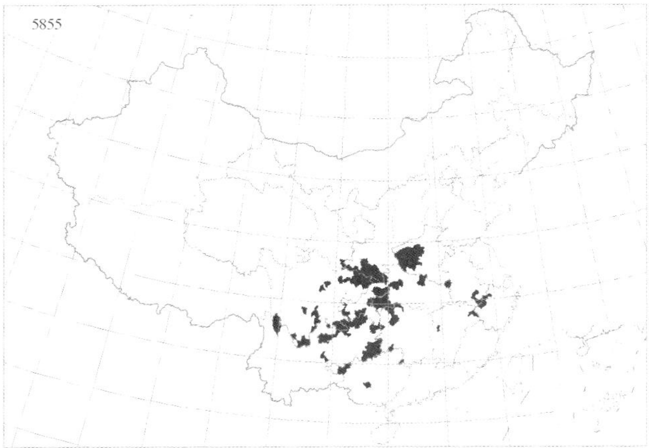

Sabiaceae 清风藤科 *Sabia* 清风藤属
Sabia schumanniana subsp. pluriflora (Rehder et E. H. Wilson) Y. F. Wu 多花清风藤
Vines, deciduous

13.1 (10.2~15.1); 0.4 (-2.9~4.7); 24.8 (18.8~27.5); 13.2 (10.4~15.2); 107.8 (77.2~128.0); 10.3 (0.3~20.5); 763 (634~865) | 807 (626~1164); 397 (316~565); 42 (29~65) | 6.9 (-22.2~83.5); 719 (626~821); 187.3 (139.1~305.6)

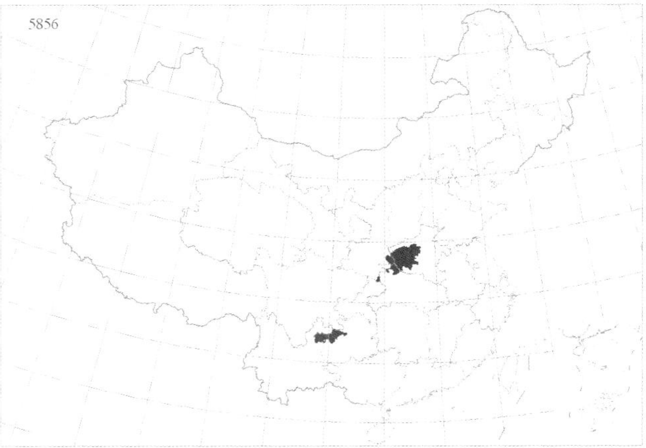

Sabiaceae 清风藤科 *Sabia* 清风藤属
Sabia swinhoei Hemsl. 尖叶清风藤
Vines, evergreen

17.4 (11.1~24.1); 7.1 (1.1~18.9); 26.7 (19.4~30.1); 17.4 (11.1~24.1); 149.2 (79.8~229.7); 0.4 (0.0~8.5); 915 (648~1322) ❙ 1493 (999~3495); 586 (356~1847); 152 (29~897) ❙ 63.9 (-5.0~246.5); 905 (648~1221); 222.4 (108.2~551.6)

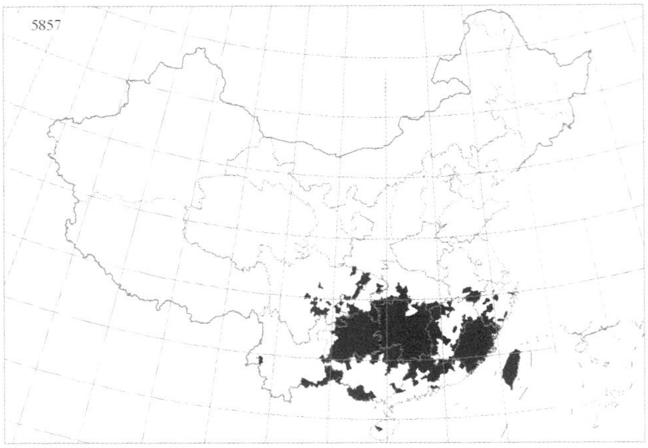

Sabiaceae 清风藤科 *Sabia* 清风藤属
Sabia transarisanensis Hayata 阿里山清风藤
Vines, deciduous

19.8; 13.8; 24.9; 19.9; 178.4; 0.0; 985 ❙ 2414; 1104; 255 ❙ 149.8; 974; 238.3

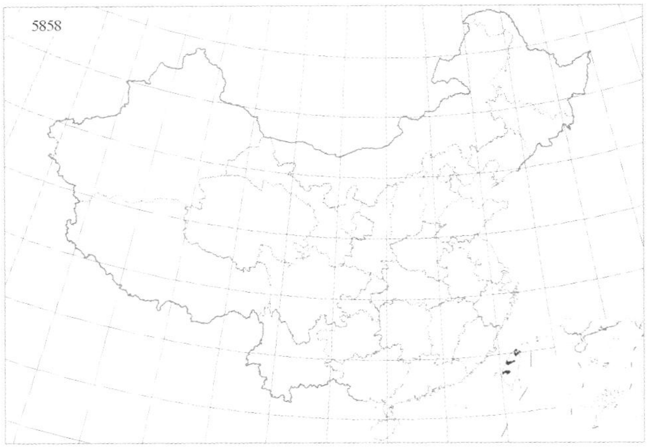

Sabiaceae 清风藤科 *Sabia* 清风藤属
Sabia yunnanensis Franch. 云南清风藤
Vines, deciduous

14.4 (1.8~21.0); 4.9 (-7.2~14.2); 22.7 (9.9~29.3); 14.5 (3.8~21.0); 117.1 (15.4~191.8); 4.0 (0.0~53.3); 776 (404~1053) ❙ 1091 (583~1751); 550 (316~1007); 63 (12~228) ❙ 42.1 (-22.2~145.8); 759 (404~1053); 257.9 (133.6~551.6)

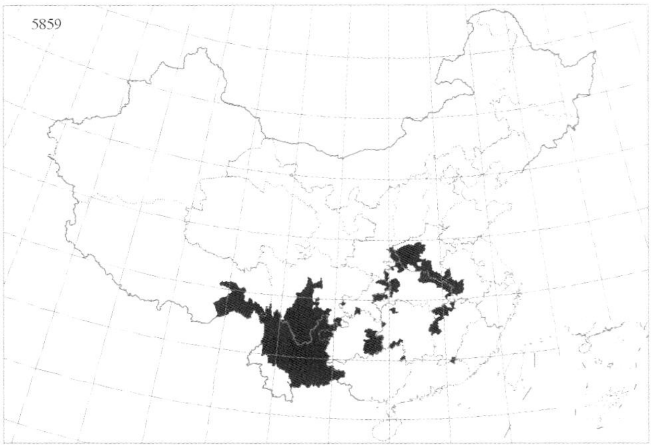

Sabiaceae 清风藤科 *Sabia* 清风藤属
Sabia yunnanensis subsp. latifolia (Rehder et E. H. Wilson) Y. F. Wu 阔叶清风藤
Vines, deciduous

13.5 (2.2~16.6); 1.2 (-6.6~8.6); 24.9 (9.9~28.8); 13.6 (3.8~16.7); 111.2 (15.5~140.6); 8.7 (0.0~49.1); 782 (413~920) ❙ 1007 (626~1571); 480 (316~816); 67 (19~157) ❙ 31.0 (-22.2~125.9); 755 (413~910); 198.6 (139.1~330.0)

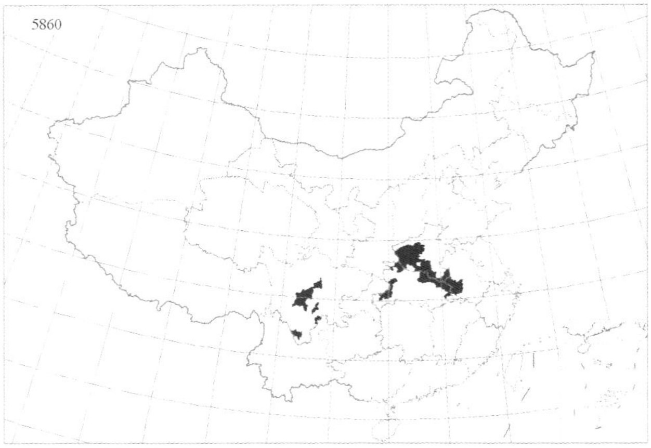

Rhamnaceae 鼠李科 *Alphitonia* 麦珠子属
Alphitonia incana (Roxburgh) Teijsmann et Binnendijk ex Kurz 麦珠子
Trees, evergreen

23.9 (22.2~24.8); 18.6 (17.2~20.1); 27.6 (25.7~28.5); 23.9 (22.3~24.8); 227.3 (207.2~237.6); 0.0 (0.0~0.0); 1300 (1127~1385) ❙ 1547 (1179~1854); 533 (392~622); 64 (41~115) ❙ 19.5 (-11.9~47.8); 1175 (1055~1360); 502.0 (352.9~620.0)

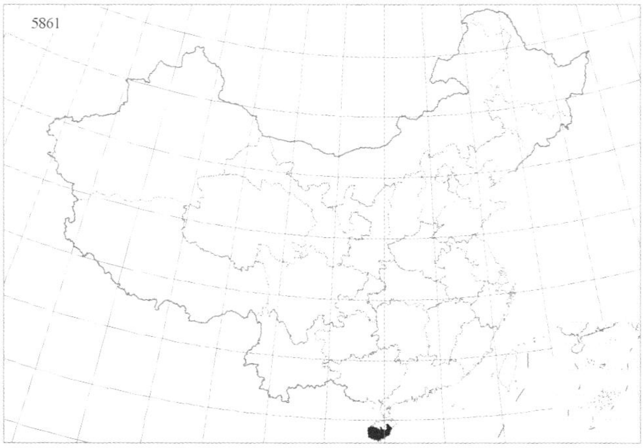

Rhamnaceae 鼠李科 *Berchemia* 勾儿茶属
Berchemia annamensis Pitard 越南勾儿茶
Climbing shrubs, deciduous

19.0 (15.5~22.0); 11.0 (4.8~14.2); 25.0 (21.4~28.8); 19.0 (15.6~22.0); 168.3 (127.2~203.8); 0.0 (0.0~0.2); 958 (797~1170) ❙ 1485 (1074~2005); 736 (582~911); 95 (47~181) ❙ 55.4 (29.9~93.2); 957 (797~1170); 332.3 (203.4~551.6)

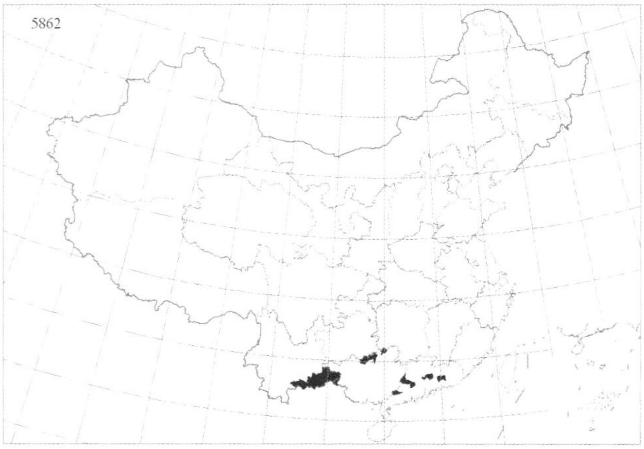

Rhamnaceae 鼠李科 *Berchemia* 勾儿茶属
Berchemia barbigera C. Y. Wu ex Y. L. Chen 腋毛勾儿茶
Scandent shrubs, deciduous

14.8 (14.1~15.5); 2.5 (1.6~3.3); 26.8 (26.2~27.6); 14.9 (14.2~15.6); 122.7 (116.7~128.6); 3.9 (1.9~6.5); 824 (796~851) ❙ 1526 (1331~1689); 585 (546~617); 86 (129~220) ❙ 85.4 (56.6~105.7); 824 (796~851); 271.8 (235.9~319.9)

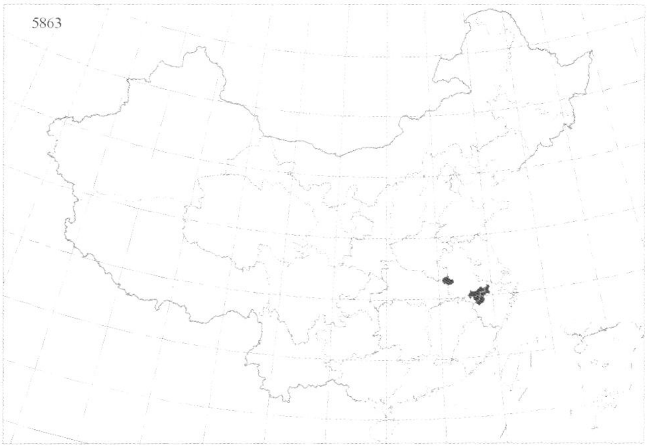

Rhamnaceae 鼠李科 *Berchemia* 勾儿茶属
Berchemia brachycarpa C. Y. Wu ex Y. L. Chen et P. K. Chou 短果勾儿茶
Climbing shrubs, deciduous

18.5; 12.1; 22.9; 18.6; 162.9; 0.0; 903 ❙ 1388; 778; 58 ❙ 54.1; 901; 480.4

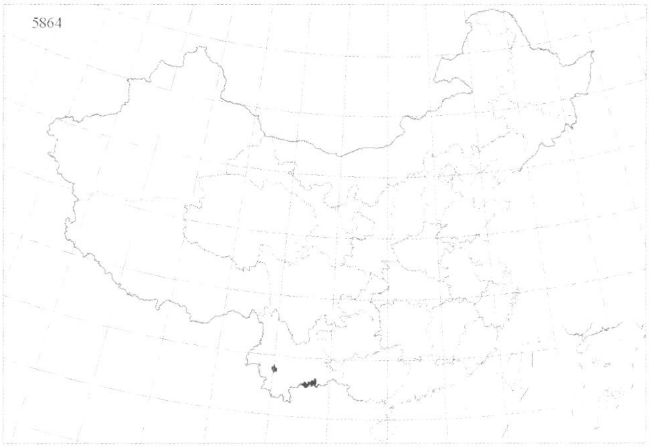

Rhamnaceae 鼠李科 *Berchemia* 勾儿茶属
Berchemia edgeworthii M. A. Lawson 腋花勾儿茶
Shrubs, deciduous

2.8 (−1.8~10.8); −6.0 (−10.2~3.9); 10.6 (6.1~16.6); 4.5 (1.8~10.9); 21.3 (1.9~71.8); 47.3 (1.3~82.8); 428 (329~633) ❙ 674 (320~968); 405 (193~594); 16 (9~33) ❙ 55.9 (−4.0~128.6); 427 (320~633); 241.5 (61.6~357.7)

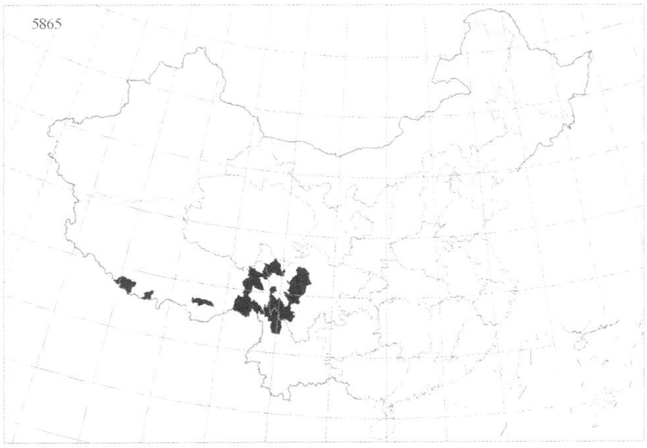

Rhamnaceae 鼠李科 *Berchemia* 勾儿茶属
Berchemia flavescens Wall. 黄背勾儿茶
Shrubs, deciduous

12.0 (−1.8~17.2); 0.4 (−10.2~8.2); 22.8 (6.1~28.8); 12.3 (1.8~17.2); 97.5 (1.9~146.2); 12.8 (0.0~82.8); 725 (329~924) ❙ 972 (320~1630); 458 (193~817); 58 (7~228) ❙ 35.4 (−37.6~121.7); 708 (320~924); 221.4 (61.6~384.7)

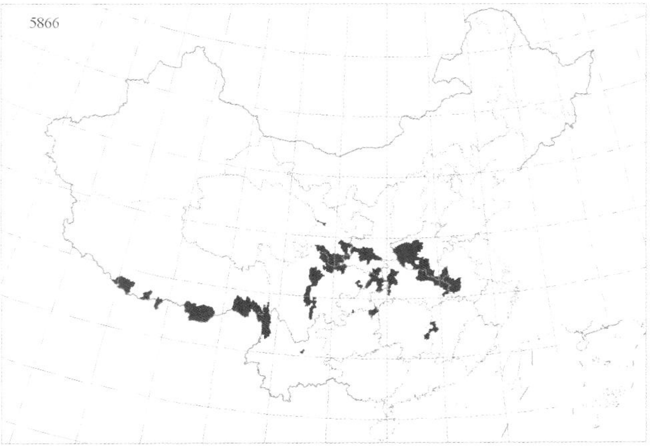

Rhamnaceae 鼠李科 *Berchemia* 勾儿茶属
Berchemia floribunda Wall. 多花勾儿茶
Shrubs, deciduous

16.1 (1.8~24.8); 6.0 (−12.5~20.1); 25.1 (9.9~29.8); 16.2 (3.8~24.8); 137.9 (15.5~237.6); 4.3 (0.0~62.6); 869 (404~1385) ❙ 1275 (377~2993); 560 (241~1428); 104 (8~547) ❙ 46.1 (−38.2~203.4); 851 (377~1273); 243.6 (95.3~659.0)

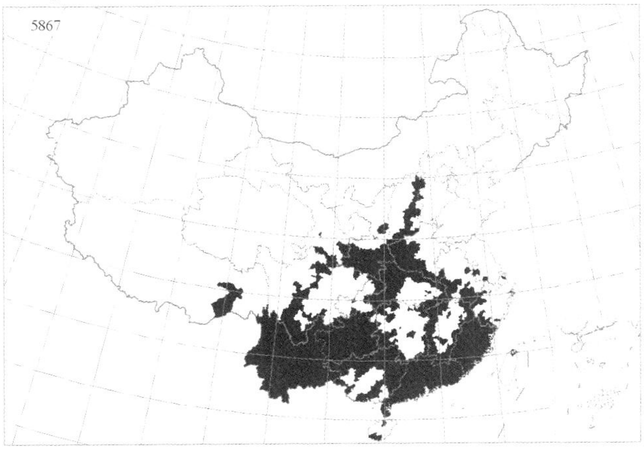

Rhamnaceae 鼠李科 *Berchemia* 勾儿茶属
Berchemia floribunda var. oblongifolia Y. L. Chen et P. K. Chou 矩叶勾儿茶
Shrubs, deciduous

15.7 (14.4~17.5); 4.3 (2.1~7.1); 26.6 (25.6~27.8); 15.7 (14.5~17.6); 130.1 (118.7~150.6); 1.8 (0.0~4.9); 842 (801~920) ❙ 1749 (1542~1997); 629 (582~677); 211 (181~259) ❙ 107.8 (84.6~130.7); 842 (801~920); 273.8 (204.3~319.9)

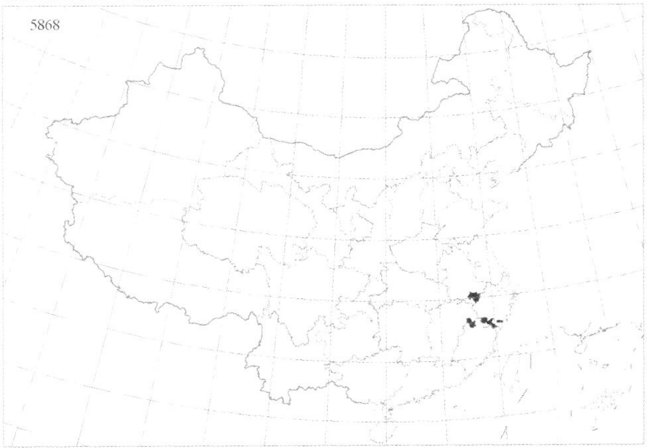

Rhamnaceae 鼠李科 *Berchemia* 勾儿茶属

Berchemia hirtella Tsai et K. M. Feng 大果勾儿茶

Scandent shrubs, deciduous

15.1 (5.4~20.5); 7.6 (−2.1~15.1); 20.8 (12.3~26.6); 15.1 (5.8~20.5); 122.5 (31.5~185.9); 1.4 (0.0~25.6); 773 (481~992) ❙ 1103 (741~1688); 624 (411~959); 40 (15~77) ❙ 43.2 (−14.2~112.2); 750 (481~986); 325.7 (142.9~659.0)

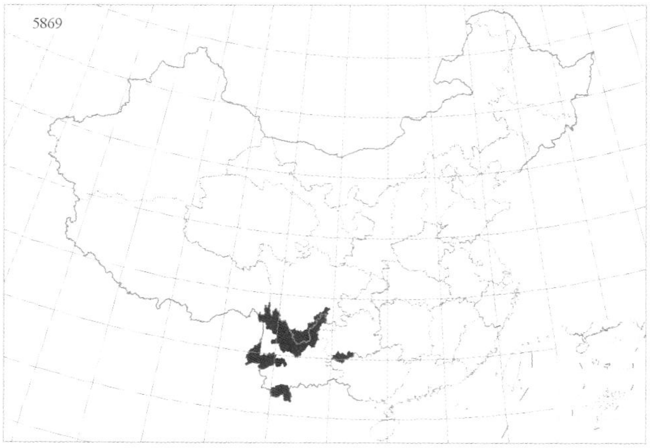

Rhamnaceae 鼠李科 *Berchemia* 勾儿茶属

Berchemia hirtella var. glabrescens C. Y. Wu ex Y. L. Chen 大老鼠耳

Scandent shrubs, deciduous

18.3 (16.8~19.1); 9.2 (7.8~10.1); 25.5 (23.6~26.6); 18.4 (16.8~19.1); 160.2 (141.5~169.2); 0.0 (0.0~0.0); 927 (836~969) ❙ 1271 (1236~1334); 677 (617~720); 57 (51~64) ❙ 37.7 (27.5~59.6); 927 (836~969); 242.9 (189.5~282.2)

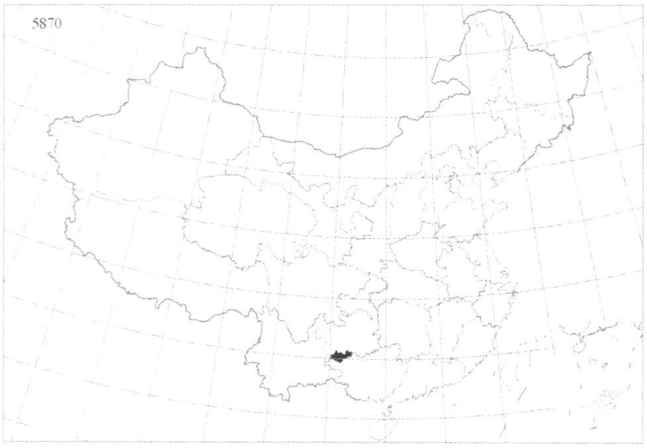

Rhamnaceae 鼠李科 *Berchemia* 勾儿茶属

Berchemia hispida (Tsai et K. M. Feng) Y. L. Chen et P. K. Chou 毛背勾儿茶

Climbing shrubs, deciduous

13.8 (6.0~17.8); 4.3 (−3.2~9.9); 22.2 (14.4~27.0); 13.8 (6.6~17.9); 108.6 (39.7~154.2); 2.8 (0.0~26.7); 742 (508~929) ❙ 1149 (879~1422); 604 (472~817); 50 (28~71) ❙ 56.1 (18.3~101.2); 742 (508~929); 223.1 (149.6~359.7)

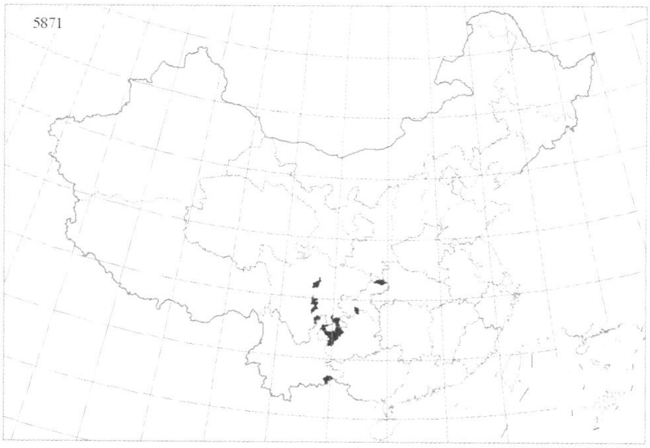

Rhamnaceae 鼠李科 *Berchemia* 勾儿茶属

Berchemia hispida var. glabrata Y. L. Chen et P. K. Chou 光轴勾儿茶

Climbing shrubs, deciduous

14.4 (11.7~17.8); 5.2 (2.7~8.0); 22.3 (19.6~27.0); 14.4 (11.7~17.9); 113.7 (84.6~154.2); 0.8 (0.0~3.8); 757 (661~929) ❙ 1092 (964~1410); 577 (501~816); 53 (39~67) ❙ 45.1 (18.3~80.4); 757 (661~929); 204.6 (149.6~298.7)

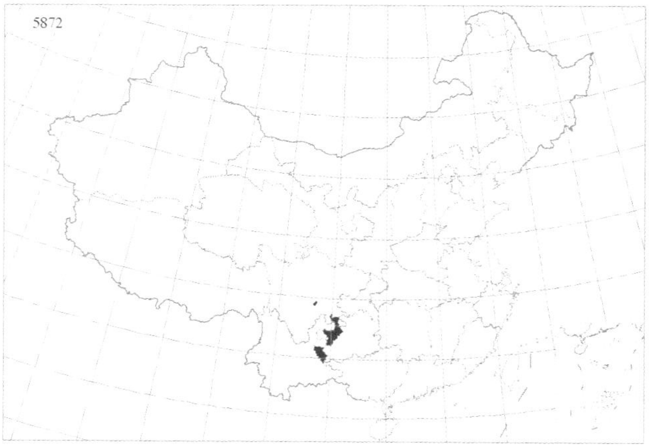

Rhamnaceae 鼠李科 *Berchemia* 勾儿茶属

Berchemia huana Rehder 大叶勾儿茶

Shrubs, deciduous

16.4 (14.3~19.4); 4.8 (2.1~9.4); 27.4 (24.2~29.3); 16.4 (14.3~19.4); 138.4 (113.2~173.4); 1.2 (0.0~4.9); 885 (761~1029) ❙ 1561 (1002~1922); 567 (448~685); 189 (105~259) ❙ 76.9 (14.0~129.9); 884 (761~1029); 237.8 (134.1~319.9)

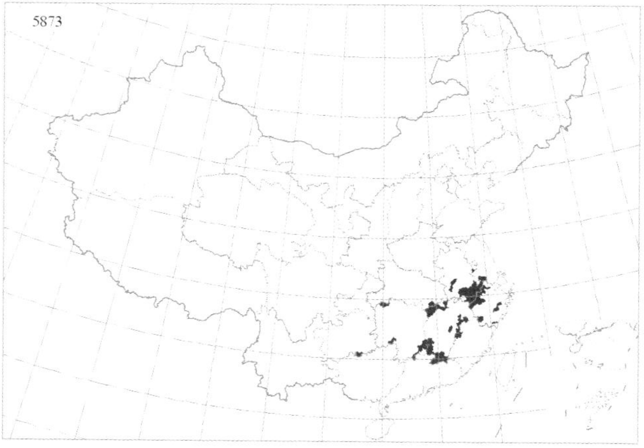

Rhamnaceae 鼠李科 *Berchemia* 勾儿茶属

Berchemia huana var. glabrescens Cheng ex Y. L. Chen 脱毛大叶勾儿茶

Shrubs, deciduous

15.2 (14.4~16.2); 2.9 (2.1~4.0); 27.2 (26.3~28.3); 15.3 (14.5~16.2); 126.4 (118.7~136.1); 2.9 (1.0~4.9); 843 (801~898) ❙ 1542 (1331~1689); 570 (513~617); 192 (153~220) ❙ 83.4 (55.4~105.7); 843 (801~898); 273.3 (225.2~319.9)

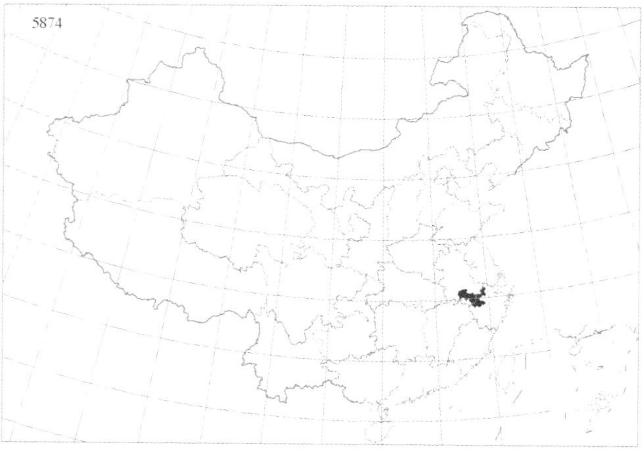

Rhamnaceae 鼠李科 *Berchemia* 勾儿茶属
Berchemia kulingensis C. K. Schneid. 牯岭勾儿茶
Shrubs, deciduous

16.2 (9.3~22.3); 4.8 (-1.7~13.5); 27.0 (20.1~29.9); 16.2 (9.5~22.3); 136.0 (69.8~208.1); 1.4 (0.0~18.0); 873 (603~1202) ❙ 1467 (856~1997); 547 (403~821); 163 (46~254) ❙ 68.8 (-0.9~130.7); 867 (603~1202); 224.3 (131.2~319.9)

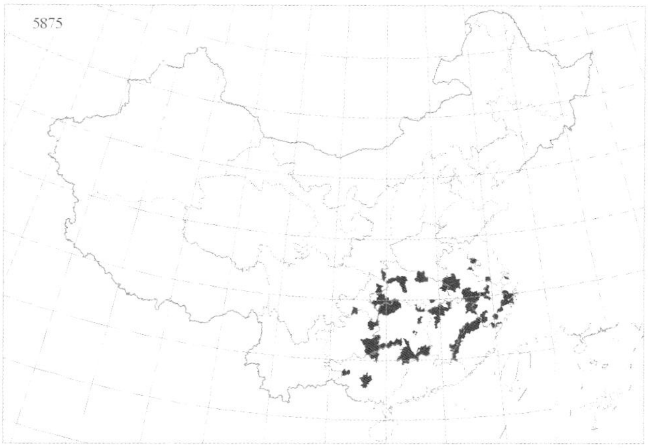

Rhamnaceae 鼠李科 *Berchemia* 勾儿茶属
Berchemia lineata (L.) DC. 铁包金
Shrubs, deciduous

20.1 (13.2~24.1); 11.6 (3.4~18.9); 27.2 (19.6~29.8); 20.1 (13.2~24.1); 181.5 (98.5~229.7); 0.0 (0.0~2.1); 1050 (700~1322) ❙ 1638 (978~3495); 713 (386~1847); 151 (37~897) ❙ 58.2 (-6.9~246.5); 1042 (700~1273); 229.5 (95.3~411.1)

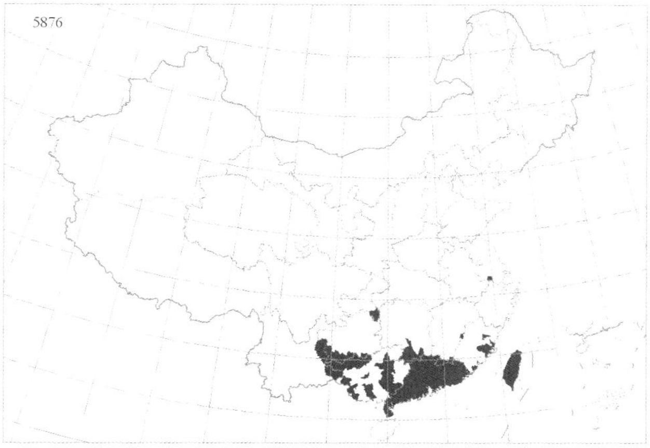

Rhamnaceae 鼠李科 *Berchemia* 勾儿茶属
Berchemia longipedicellata Y. L. Chen et P. K. Chou 细梗勾儿茶
Shrubs, deciduous

2.8; -6.0; 10.7; 4.3; 20.3; 46.5; 423 ❙ 544; 332; 11 ❙ 29.0; 423; 195.7

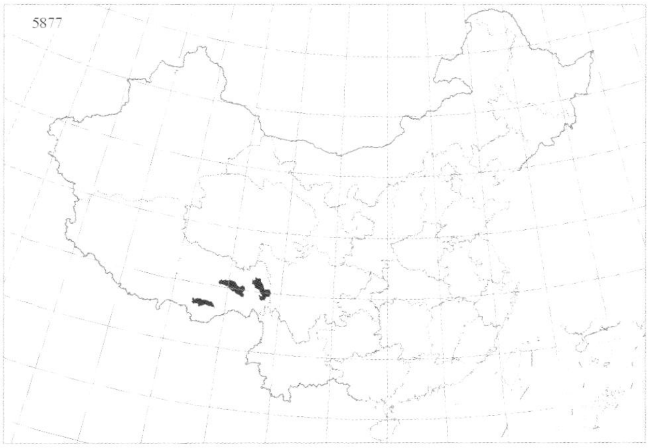

Rhamnaceae 鼠李科 *Berchemia* 勾儿茶属
Berchemia longipes Y. L. Chen et P. K. Chou 长梗勾儿茶
Scandent shrubs, deciduous

17.2; 9.5; 22.7; 17.3; 147.0; 0.0; 838 ❙ 1254; 699; 48 ❙ 49.7; 838; 336.3

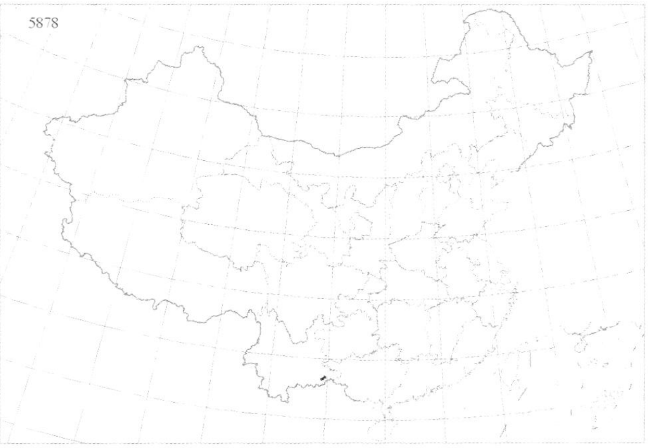

Rhamnaceae 鼠李科 *Berchemia* 勾儿茶属
Berchemia omeiensis Feng ex Y. L. Chen et P. K. Chou 峨眉勾儿茶
Shrubs, deciduous

14.5 (10.6~17.1); 3.9 (-0.2~5.5); 24.7 (19.4~29.4); 14.5 (10.6~17.1); 116.5 (79.4~146.5); 2.0 (0.0~12.1); 785 (643~951) ❙ 1229 (1072~1492); 537 (475~816); 73 (30~143) ❙ 57.7 (30.1~100.5); 784 (643~929); 204.7 (140.2~292.6)

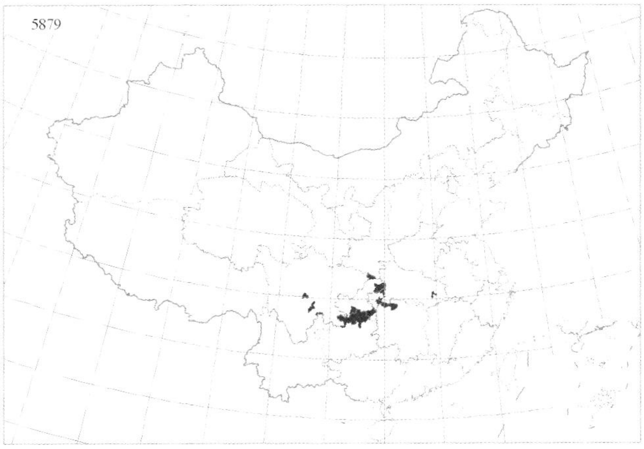

Rhamnaceae 鼠李科 *Berchemia* 勾儿茶属
Berchemia polyphylla Wall. ex M. A. Lawson 多叶勾儿茶
Shrubs, deciduous

16.0 (2.6~24.7); 6.2 (-7.1~19.8); 24.8 (11.1~29.0); 16.1 (4.3~24.7); 134.3 (20.0~236.5); 1.6 (0.0~48.2); 846 (427~1376) ❙ 1213 (711~1829); 555 (350~860); 77 (15~230) ❙ 44.9 (0.0~128.9); 842 (427~1299); 229.8 (108.2~572.9)

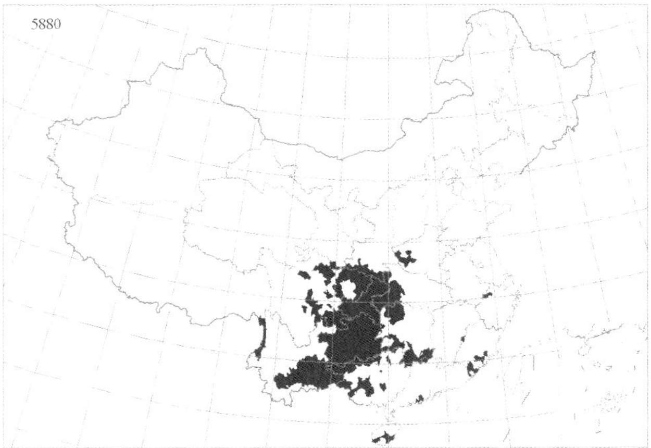

Rhamnaceae 鼠李科 *Berchemia* 勾儿茶属
Berchemia polyphylla var. leioclada Hand.-Mazz 光枝勾儿茶
Shrubs, deciduous

19.7 (9.3~24.7); 10.7 (-1.7~19.8); 27.3 (20.1~29.8); 19.7 (9.5~24.7); 176.9 (69.8~236.5); 0.5 (0.0~18.0); 1039 (603~1376) ❙ 1545 (999~2037); 643 (356~1033); 135 (43~202) ❙ 50.5 (-6.9~107.2); 1032 (603~1299); 228.1 (95.3~572.9)

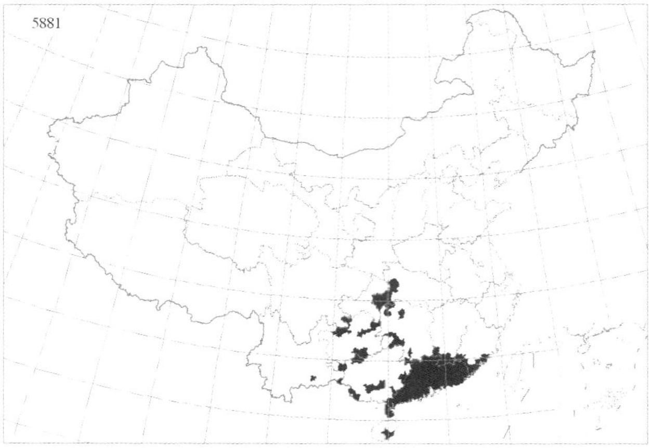

Rhamnaceae 鼠李科 *Berchemia* 勾儿茶属
Berchemia polyphylla var. trichophylla Hand.-Mazz. 毛叶勾儿茶
Shrubs, deciduous

17.7 (16.0~19.1); 8.9 (7.8~10.1); 24.6 (21.8~26.6); 17.8 (16.0~19.1); 153.3 (132.2~169.2); 0.0 (0.0~0.0); 893 (793~969) ❙ 1236 (1107~1334); 663 (607~720); 57 (51~64) ❙ 39.1 (27.5~59.6); 893 (793~969); 268.0 (226.6~298.7)

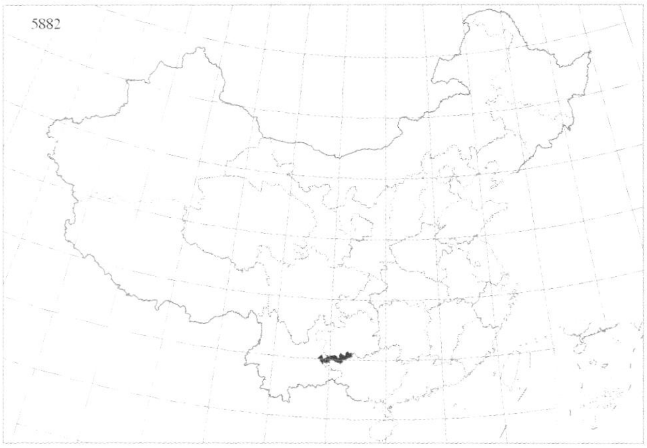

Rhamnaceae 鼠李科 *Berchemia* 勾儿茶属
Berchemia sinica C. K. Schneid. 勾儿茶
Shrubs, deciduous

13.5 (2.7~19.1); 1.5 (-12.5~10.1); 24.7 (11.1~29.2); 13.8 (4.4~19.1); 112.1 (20.5~169.2); 9.0 (0.0~62.6); 776 (429~969) ❙ 1016 (373~1770); 459 (201~816); 69 (7~223) ❙ 30.2 (-37.6~112.2); 750 (373~969); 206.1 (88.1~341.6)

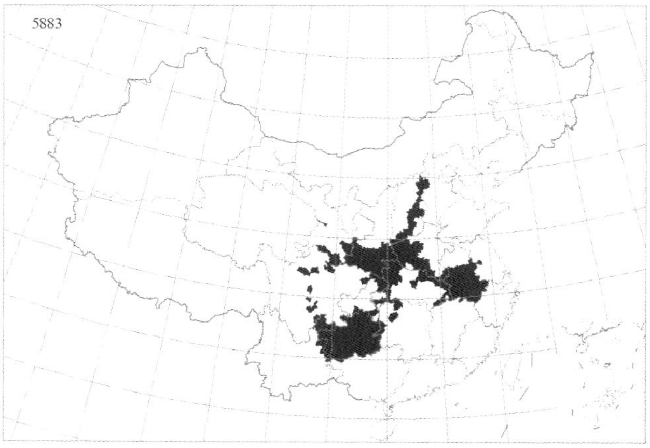

Rhamnaceae 鼠李科 *Berchemia* 勾儿茶属
Berchemia yunnanensis Franchet 云南勾儿茶
Shrubs, deciduous

11.3 (0.3~21.6); 2.8 (-9.9~12.6); 18.5 (9.2~28.5); 11.7 (3.3~21.7); 88.2 (11.9~199.8); 12.2 (0.0~67.5); 664 (391~1153) ❙ 973 (373~1736); 534 (201~1007); 37 (7~173) ❙ 49.4 (-37.6~145.8); 652 (373~1153); 265.7 (88.1~447.4)

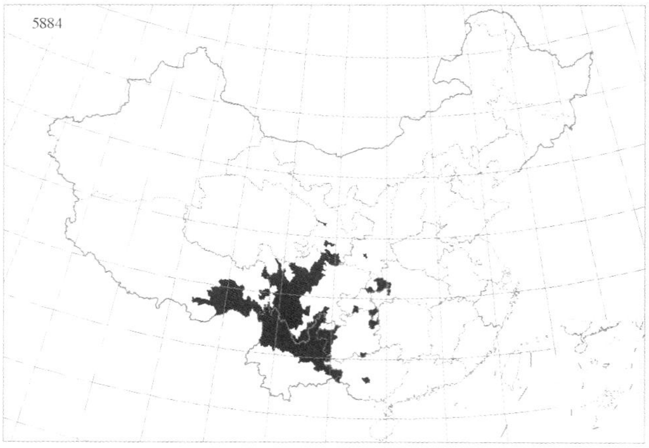

Rhamnaceae 鼠李科 *Berchemiella* 小勾儿茶属
Berchemiella wilsonii (C. K. Schneid.) Nakai 小勾儿茶
Shrubs, deciduous

14.0 (12.0~15.4); 2.1 (0.6~3.4); 25.6 (22.9~27.4); 14.1 (12.1~15.4); 114.1 (94.7~126.7); 5.3 (1.7~9.7); 785 (696~851) ❙ 1428 (993~1770); 566 (420~622); 147 (46~223) ❙ 81.4 (36.2~112.2); 785 (696~851); 252.6 (183.4~284.0)

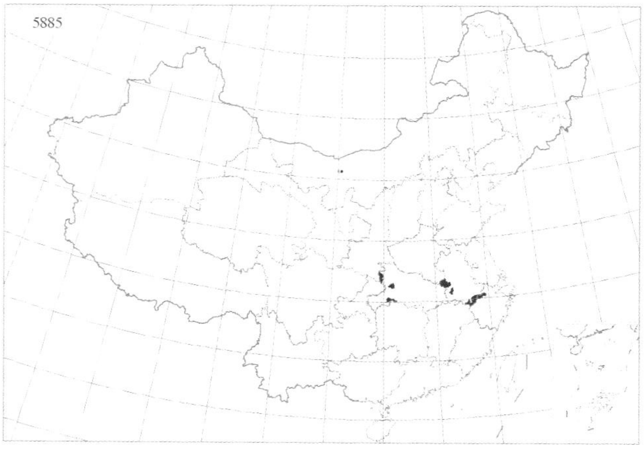

Rhamnaceae 鼠李科 *Berchemiella* 小勾儿茶属
Berchemiella wilsonii var. pubipetiolata H. Qian 毛柄小勾儿茶
Shrubs, deciduous

14.6; 2.3; 26.5; 14.6; 119.9; 4.6; 812 ❙ 1460; 599; 152 ❙ 79.9; 812; 234.9

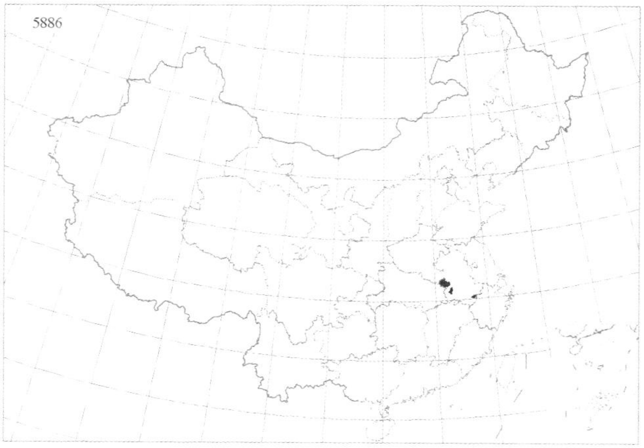

Rhamnaceae 鼠李科 *Berchemiella* 小勾儿茶属
Berchemiella yunnanensis Y. L. Chen et P. K. Chou 滇小勾儿茶
Trees, deciduous

18.9; 10.6; 25.0; 19.0; 167.5; 0.0; 939 ❙ 1292; 729; 49 ❙ 37.6; 939; 342.5

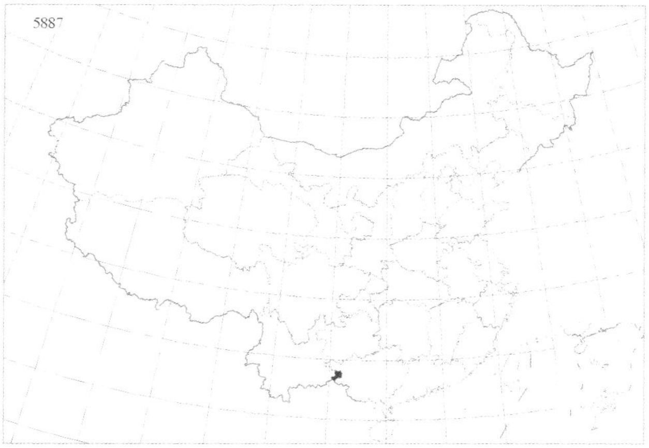

Rhamnaceae 鼠李科 *Colubrina* 蛇藤属
Colubrina asiatica (L.) Brongn. 蛇藤
Scandent shrubs, evergreen

21.5 (16.2~24.8); 14.2 (10.6~20.1); 27.5 (20.1~29.1); 21.6 (16.2~24.8); 198.8 (134.2~237.6); 0.0 (0.0~0.0); 1131 (773~1385) ❙ 1907 (1269~3495); 873 (455~1847); 157 (41~897) ❙ 73.8 (-1.0~246.5); 1115 (773~1269); 229.2 (95.3~527.5)

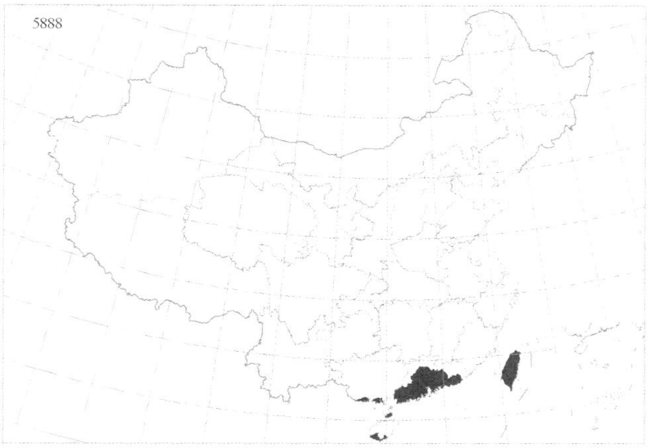

Rhamnaceae 鼠李科 *Colubrina* 蛇藤属
Colubrina javanica Miquel 毛蛇藤
Shrubs, evergreen

17.9 (16.9~19.3); 11.4 (10.3~13.1); 22.3 (21.4~23.6); 17.9 (16.9~19.4); 155.3 (142.7~173.2); 0.0 (0.0~0.0); 863 (813~939) ❙ 1087 (958~1178); 605 (531~646); 47 (44~51) ❙ 26.1 (13.3~44.9); 846 (813~903); 340.0 (294.5~406.7)

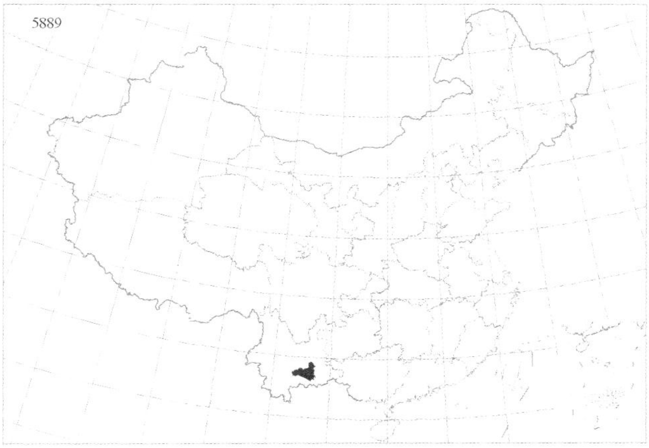

Rhamnaceae 鼠李科 *Gouania* 咀签属
Gouania javanica Miq. 毛咀签
Climbing shrubs

19.7 (16.3~24.8); 11.8 (7.3~20.1); 25.7 (21.4~28.9); 19.7 (16.4~24.8); 177.0 (136.4~237.6); 0.0 (0.0~0.0); 1009 (797~1385) ❙ 1389 (1018~1777); 700 (392~996); 81 (41~205) ❙ 39.7 (-11.9~98.6); 990 (797~1212); 335.2 (189.5~659.0)

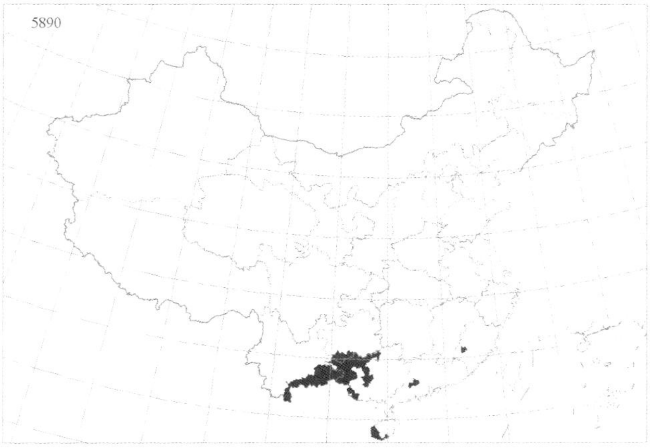

Rhamnaceae 鼠李科 *Gouania* 咀签属
Gouania leptostachya DC. 咀签
Climbing shrubs

18.3 (14.4~22.2); 11.6 (7.5~15.1); 22.7 (19.1~28.3); 18.3 (14.4~22.3); 159.7 (113.1~207.5); 0.0 (0.0~0.0); 888 (727~1190) ❙ 1340 (939~1751); 740 (524~996); 55 (37~130) ❙ 51.5 (6.5~112.2); 874 (727~1190); 418.6 (248.2~659.0)

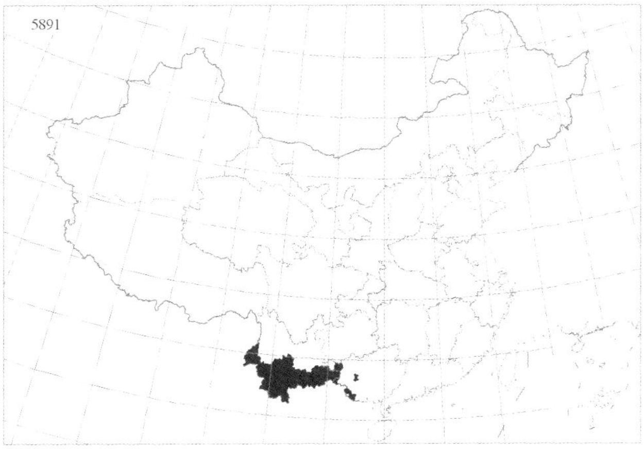

Rhamnaceae 鼠李科 *Gouania* 咀签属
Gouania leptostachya var. macrocarpa Pitard 大果咀签
Climbing shrubs

21.0; 14.2; 25.6; 21.0; 191.8; 0.0; 1053 ❙ 1522; 818; 68 ❙ 44.5; 1053; 551.6

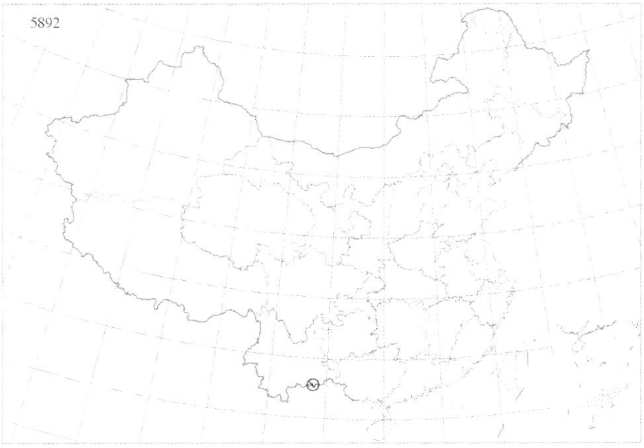

Rhamnaceae 鼠李科 *Gouania* 咀签属
Gouania leptostachya var. tonkinensis Pitard 越南咀签
Climbing shrubs

18.1 (16.3~21.0); 11.3 (9.0~14.2); 22.9 (21.4~25.6); 18.2 (16.4~21.0); 158.0 (136.4~191.8); 0.0 (0.0~0.0); 879 (797~1053) ❙ 1286 (1007~1751); 724 (539~996); 53 (44~77) ❙ 46.2 (19.0~88.8); 874 (797~1053); 391.1 (277.0~551.6)

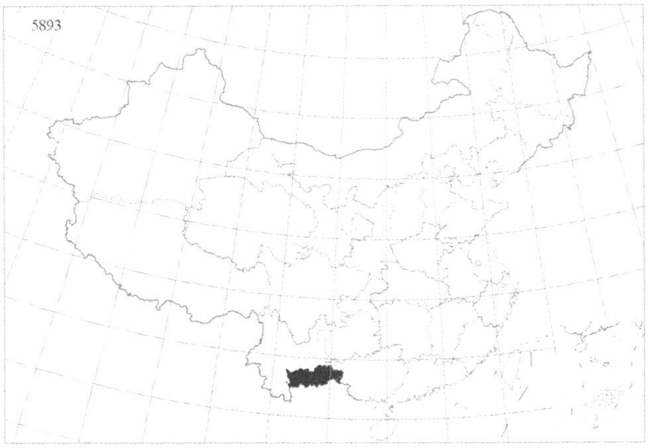

Rhamnaceae 鼠李科 *Hovenia* 枳椇属
Hovenia acerba Lindl. 枳椇
Trees, deciduous

17.1 (3.2~24.8); 7.1 (-6.4~20.1); 26.0 (11.4~29.9); 17.1 (4.7~24.8); 146.4 (22.7~237.6); 1.2 (0.0~43.8); 906 (436~1385) ❙ 1335 (526~2037); 560 (259~1033); 116 (14~259) ❙ 48.1 (-39.7~145.8); 893 (436~1360); 233.0 (95.3~659.0)

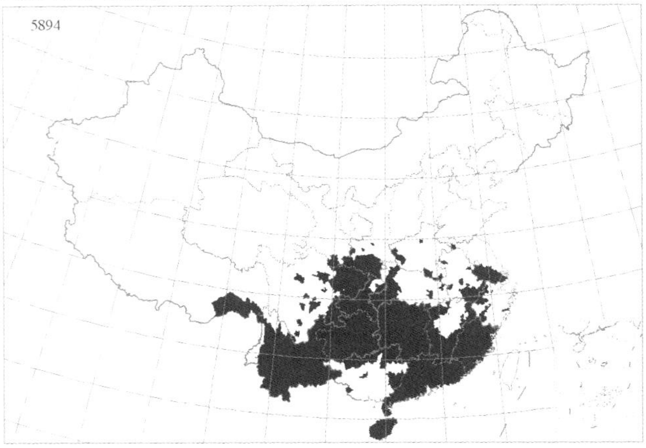

Rhamnaceae 鼠李科 *Hovenia* 枳椇属
Hovenia acerba var. kiukiangensis (Hu et Cheng) C. Y. Wu ex Y. L. Chen 俅江枳椇
Trees, deciduous

15.7 (5.3~21.0); 9.0 (-2.4~14.2); 20.6 (12.2~25.6); 15.7 (5.7~21.0); 132.0 (30.5~191.8); 3.7 (0.0~26.2); 803 (479~1053) ❙ 1257 (961~1547); 713 (558~911); 52 (31~68) ❙ 59.6 (34.8~100.7); 803 (479~1053); 428.5 (234.2~551.6)

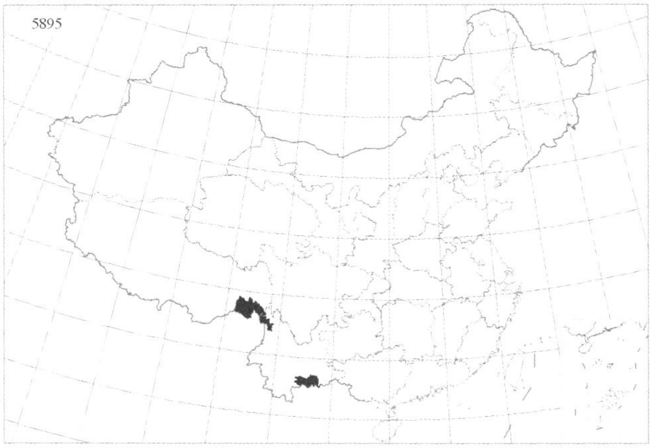

Rhamnaceae 鼠李科 *Hovenia* 枳椇属
Hovenia trichocarpa Chun et Tsiang 毛果枳椇
Trees, deciduous

16.6 (12.1~21.0); 5.3 (1.1~11.9); 27.0 (22.9~29.8); 16.6 (12.2~21.1); 140.1 (95.0~193.0); 0.8 (0.0~8.5); 886 (696~1114) ❙ 1511 (1117~1997); 540 (392~821); 176 (83~254) ❙ 71.2 (28.5~130.7); 880 (696~1114); 225.6 (111.9~318.0)

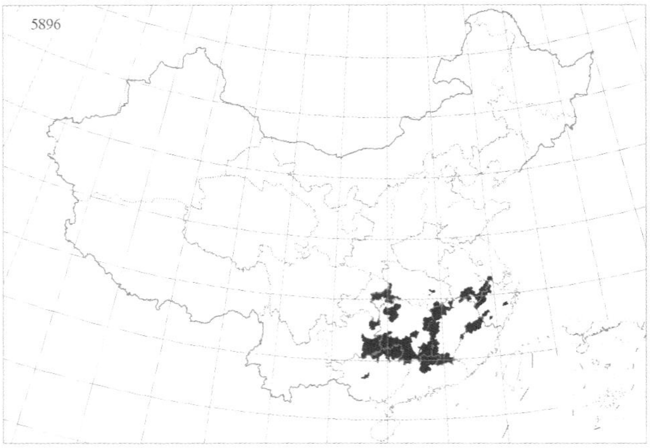

Rhamnaceae 鼠李科 *Hovenia* 枳椇属
Hovenia trichocarpa var. robusta (Nakai et Y. Kimura) Y. L. Chou et P. K. Chou 光叶毛果枳椇
Trees, deciduous

15.0 (3.9~23.0); 2.6 (-13.7~15.1); 26.1 (18.7~29.8); 15.2 (7.3~23.0); 130.0 (52.8~216.2); 10.0 (0.0~71.7); 853 (535~1247) ❙ 1078 (367~2435); 491 (235~1428); 88 (8~254) ❙ 23.4 (-40.9~130.7); 794 (367~1247); 202.8 (101.9~411.1)

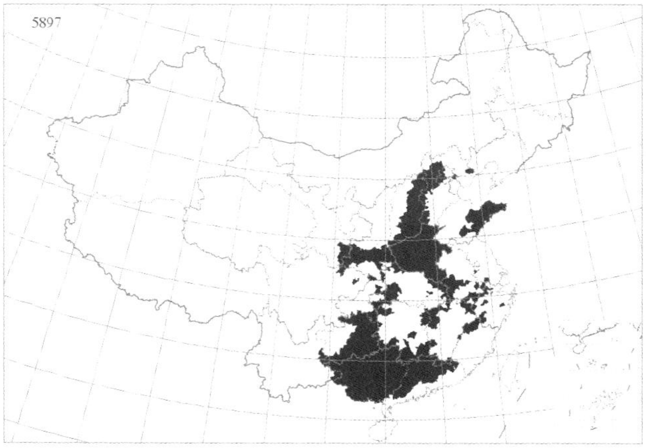

Rhamnaceae 鼠李科 *Paliurus* 马甲子属
Paliurus hemsleyanus Rehder 铜钱树
Trees, evergreen

15.5 (7.4~21.8); 3.8 (-4.2~14.2); 26.3 (17.0~29.8); 15.6 (8.2~21.9); 131.2 (58.8~202.6); 4.9 (0.0~29.4); 855 (561~1153) ❙ 1226 (512~1876); 498 (258~891); 114 (14~230) ❙ 41.9 (-35.3~108.0); 840 (512~1153); 221.3 (121.6~334.4)

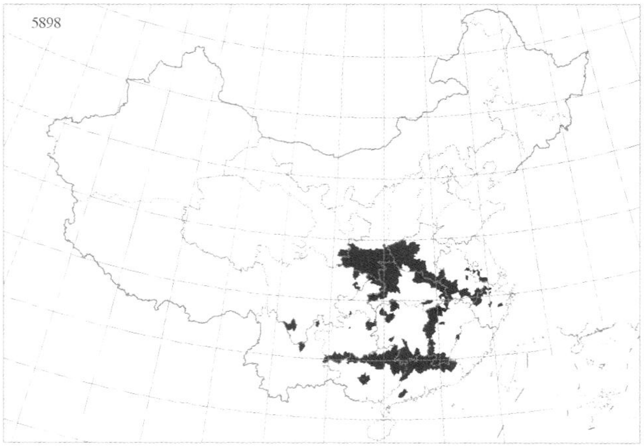

Rhamnaceae 鼠李科 *Paliurus* 马甲子属
Paliurus hirsutus Hemsl. 硬毛马甲子
Shrubs or small trees, evergreen

17.2 (14.4~22.0); 6.0 (2.1~13.5); 27.6 (25.8~29.5); 17.3 (14.5~22.0); 148.5 (118.7~203.8); 1.2 (0.0~4.9); 928 (801~1170) ❙ 1603 (1002~1997); 598 (448~858); 187 (114~254) ❙ 74.5 (14.0~130.7); 927 (801~1170); 242.9 (134.1~319.9)

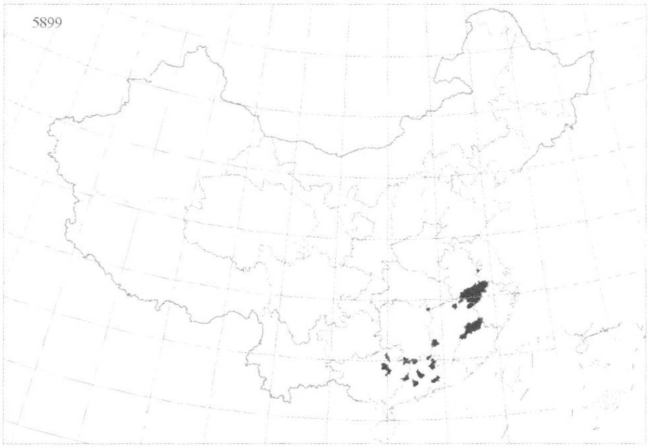

Rhamnaceae 鼠李科 *Paliurus* 马甲子属
Paliurus orientalis (Franchet) Hemsley 短柄铜钱树
Shrubs or small trees, evergreen

14.3 (3.0~24.8); 7.2 (-5.6~20.1); 19.9 (10.5~28.5); 14.4 (4.3~24.8); 114.9 (20.1~237.6); 2.8 (0.0~43.9); 755 (423~1385) ❙ 1007 (628~1625); 581 (410~818); 34 (11~173) ❙ 36.0 (-14.2~92.4); 717 (423~1153); 309.2 (190.3~551.6)

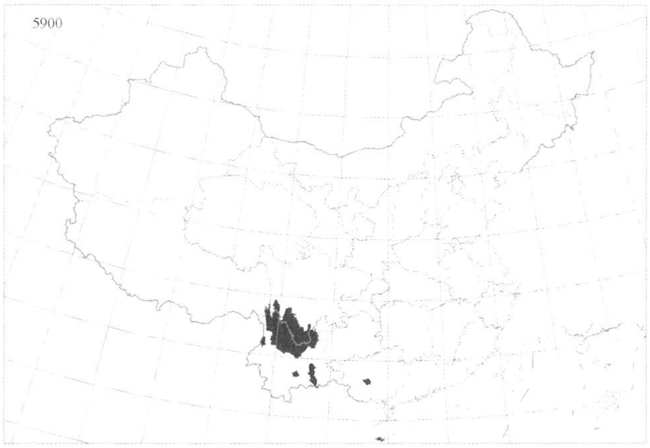

Rhamnaceae 鼠李科 *Paliurus* 马甲子属
Paliurus ramosissimus (Lour.) Poir. 马甲子
Shrubs, deciduous

17.4 (9.1~23.8); 7.5 (-1.5~18.1); 26.1 (18.1~29.8); 17.4 (9.2~23.8); 149.5 (64.1~225.8); 0.5 (0.0~15.5); 914 (595~1297) ❙ 1351 (724~3495); 597 (299~1428); 111 (16~897) ❙ 48.2 (-6.7~246.5); 909 (595~1269); 216.0 (108.2~491.5)

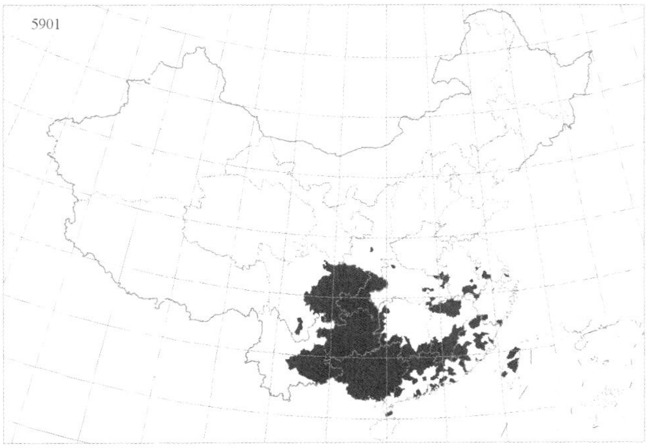

Rhamnaceae 鼠李科 *Rhamnella* 猫乳属
Rhamnella caudata Merr. 尾叶猫乳
Shrubs or small trees, deciduous

17.7; 7.7; 26.6; 17.8; 153.6; 0.0; 916 ❙ 1564; 556; 182 ❙ 70.7; 916; 232.5

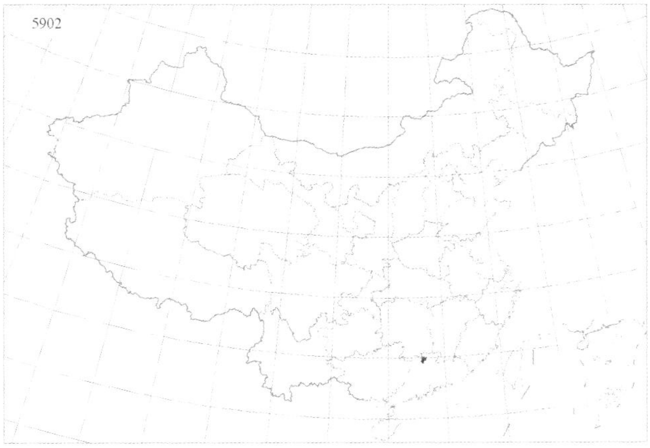

Rhamnaceae 鼠李科 *Rhamnella* 猫乳属
Rhamnella forrestii W. W. Sm. 川滇猫乳
Shrubs, deciduous

5.5 (0.6~14.2); -2.7 (-9.4~6.2); 12.6 (9.5~20.1); 6.4 (3.4~14.2); 37.2 (12.7~110.5); 30.9 (0.0~65.4); 493 (394~729) ❙ 803 (535~1147); 488 (326~636); 19 (0~50) ❙ 63.6 (27.1~125.9); 493 (394~729); 261.5 (169.5~319.8)

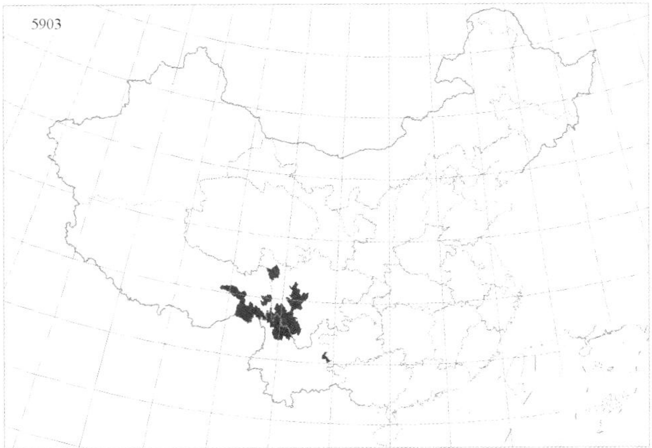

Rhamnaceae 鼠李科 *Rhamnella* 猫乳属
Rhamnella franguloides (Maxim.) Weberbauer 猫乳
Shrubs or small trees, deciduous

13.8 (3.9~18.4); 0.1 (-13.7~6.7); 26.5 (18.7~29.7); 14.1 (7.3~18.4); 117.8 (52.8~161.1); 12.2 (0.0~71.7); 816 (535~993) ❙ 928 (367~1828); 431 (235~685); 84 (8~253) ❙ 11.8 (-40.9~112.2); 749 (367~961); 178.2 (101.9~319.9)

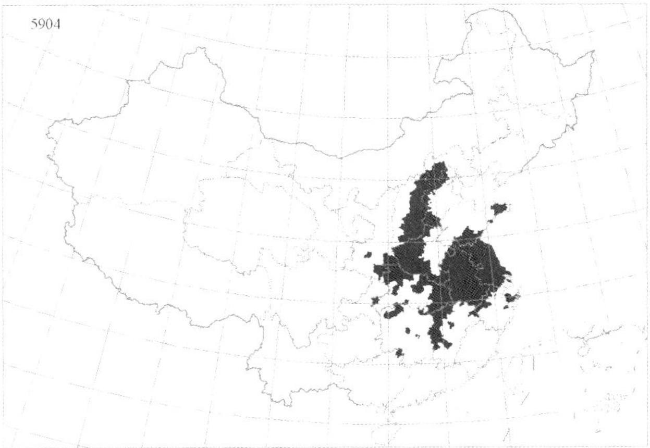

Rhamnaceae 鼠李科 *Rhamnella* 猫乳属
Rhamnella gilgitica Mansf. et Melch 西藏猫乳
Shrubs, deciduous

4.1 (2.2~6.1); -3.9 (-6.3~-1.5); 11.3 (9.5~12.7); 5.1 (3.9~6.3); 25.5 (15.8~35.2); 35.2 (21.8~48.6); 452 (411~496) ❙ 733 (535~961); 450 (326~558); 19 (10~31) ❙ 61.5 (27.1~100.7); 452 (411~496); 245.1 (208.8~282.9)

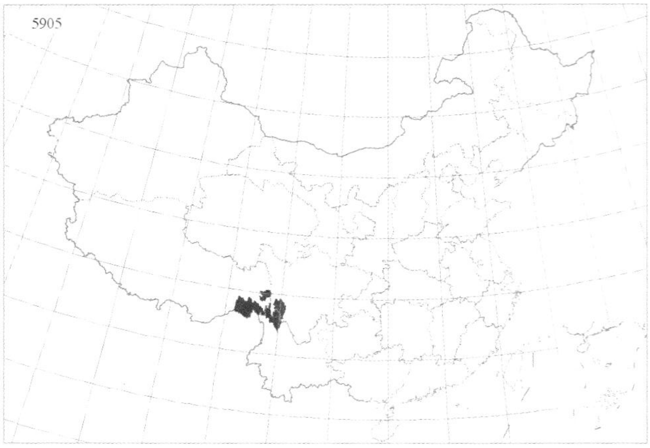

Rhamnaceae 鼠李科 *Rhamnella* 猫乳属
Rhamnella julianae C. K. Schneid. 毛背猫乳
Shrubs, deciduous

13.5 (6.0~16.4); 3.1 (-3.2~8.0); 23.2 (14.1~27.5); 13.6 (6.6~16.4); 107.7 (39.4~136.6); 5.2 (0.0~26.7); 753 (508~877) ❙ 1145 (862~1444); 502 (376~727); 70 (23~140) ❙ 54.7 (7.8~101.2); 753 (508~877); 241.2 (170.4~407.1)

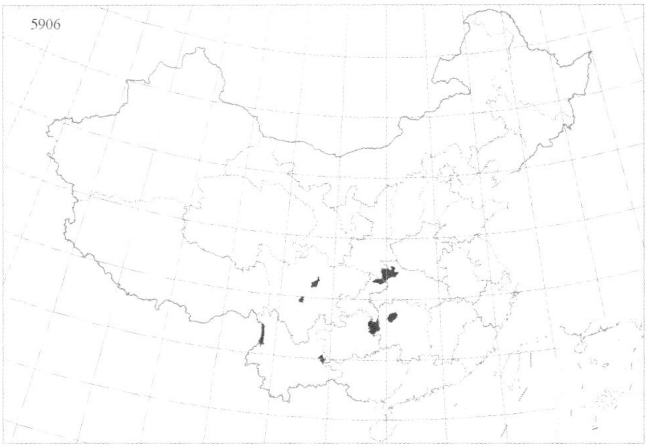

Rhamnaceae 鼠李科 *Rhamnella* 猫乳属
Rhamnella martini (H. Lév.) C. K. Schneid. 多脉猫乳
Shrubs or small trees, deciduous

15.2 (5.3~19.1); 5.7 (-2.4~11.8); 23.5 (12.2~27.5); 15.2 (5.7~19.1); 123.8 (30.5~169.2); 1.4 (0.0~26.2); 793 (479~969) ❙ 1170 (916~1564); 554 (444~756); 68 (15~182) ❙ 48.3 (13.3~106.2); 791 (479~969); 229.3 (146.0~491.5)

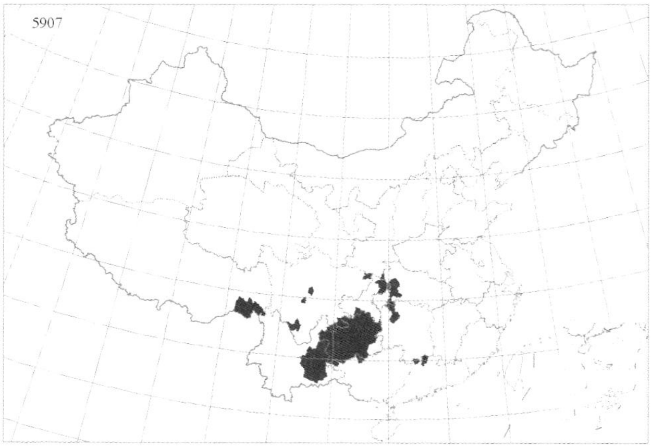

Rhamnaceae 鼠李科 *Rhamnella* 猫乳属
Rhamnella rubrinervis (H. Lév.) Rehder 苞叶木
Shrubs or small trees, evergreen

19.0 (15.0~24.8); 10.6 (4.5~20.1); 25.6 (21.4~29.0); 19.1 (15.0~24.8); 168.8 (120.8~237.6); 0.0 (0.0~0.5); 974 (794~1385) ❙ 1383 (1018~1776); 680 (455~996); 83 (41~194) ❙ 43.3 (1.5~88.8); 963 (794~1299); 322.7 (160.7~659.0)

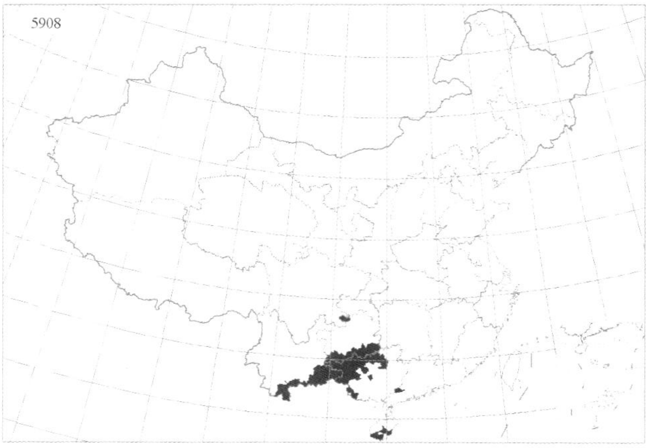

Rhamnaceae 鼠李科 *Rhamnella* 猫乳属
Rhamnella wilsonii C. K. Schneid. 卵叶猫乳
Shrubs or small trees, deciduous

5.1 (2.0~17.9); -3.4 (-7.2~11.6); 12.4 (9.9~22.1); 6.2 (3.8~17.9); 37.8 (15.4~155.1); 36.3 (0.0~52.8); 489 (410~858) ❙ 846 (520~1098); 491 (317~634); 20 (10~46) ❙ 78.7 (26.6~128.6); 486 (410~836); 278.1 (201.9~357.7)

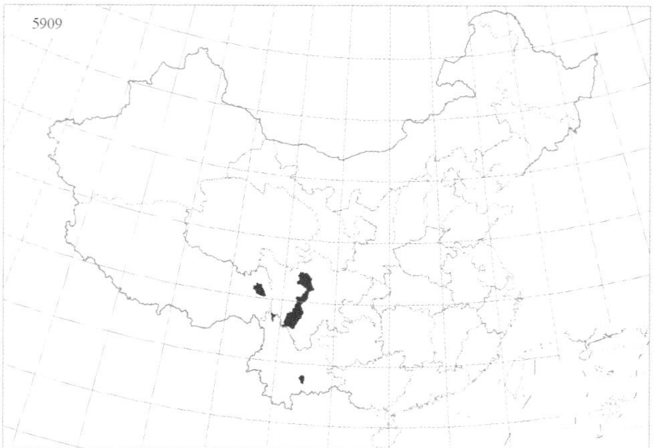

Rhamnaceae 鼠李科 *Rhamnus* 鼠李属
Rhamnus arguta Maxim. 锐齿鼠李
Shrubs, deciduous

9.5 (-4.6~15.5); -6.4 (-28.8~2.1); 23.7 (16.5~28.0); 11.0 (5.0~15.6); 90.0 (34.4~131.3); 35.3 (4.3~149.1); 694 (448~881) ❙ 581 (249~1077); 355 (179~601); 21 (5~97) ❙ -16.4 (-54.1~56.9); 571 (249~881); 171.3 (70.3~323.4)

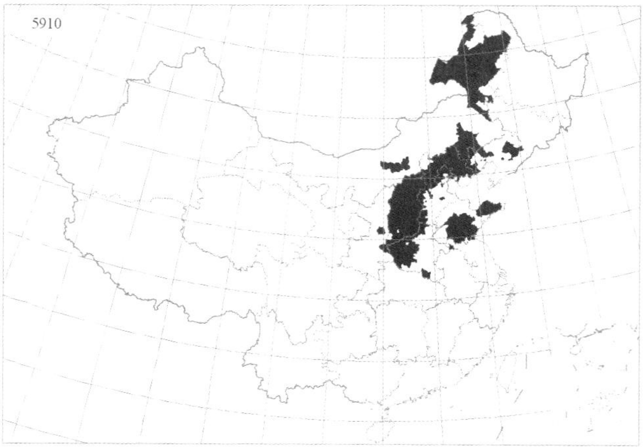

Rhamnaceae 鼠李科 *Rhamnus* 鼠李属
Rhamnus arguta var. velutina Hand.-Mazz. 毛背锐齿鼠李
Shrubs or small trees, deciduous

8.9 (3.3~13.5); -6.6 (-12.7~-1.6); 22.6 (17.5~27.1); 10.2 (6.6~13.7); 81.9 (45.4~117.5); 34.8 (14.6~65.0); 656 (507~820) ❙ 522 (377~702); 302 (241~409); 16 (8~27) ❙ -19.9 (-37.2~12.7); 520 (377~649); 152.5 (50.7~228.3)

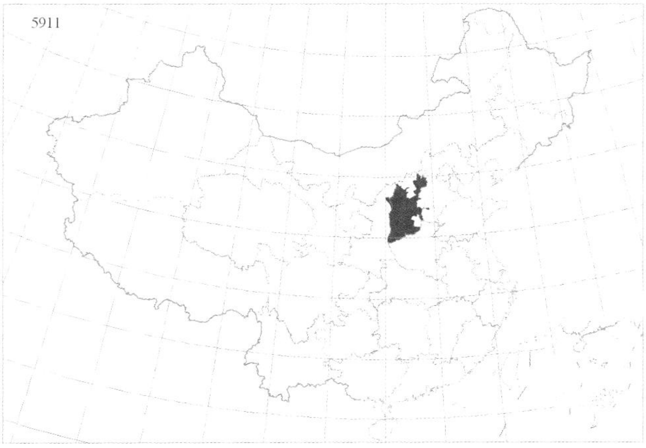

Rhamnaceae 鼠李科 *Rhamnus* 鼠李属
Rhamnus aurea Heppeler 铁马鞭
Shrubs, deciduous

14.3 (5.4~18.9); 7.0 (-2.1~12.0); 20.1 (12.3~26.6); 14.3 (5.8~19.0); 113.1 (31.5~167.8); 1.2 (0.0~25.6); 739 (481~929) ❙ 973 (741~1164); 563 (411~651); 36 (15~58) ❙ 33.3 (-14.2~78.3); 714 (481~912); 276.1 (142.9~382.6)

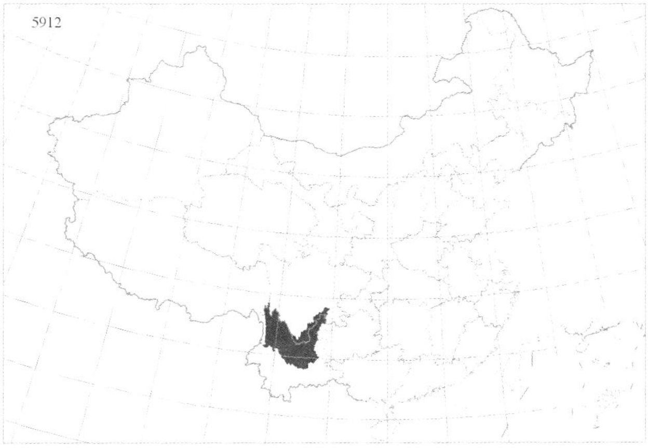

Rhamnaceae 鼠李科 *Rhamnus* 鼠李属
Rhamnus bodinieri H. Lév. 陷脉鼠李
Shrubs, evergreen

17.2 (14.2~21.0); 8.8 (5.7~14.2); 23.7 (20.1~26.9); 17.2 (14.2~21.0); 146.9 (110.5~191.8); 0.0 (0.0~0.0); 858 (729~1053) ❙ 1254 (958~1522); 664 (531~860); 61 (44~117) ❙ 46.3 (13.3~80.6); 858 (729~1053); 294.3 (171.2~551.6)

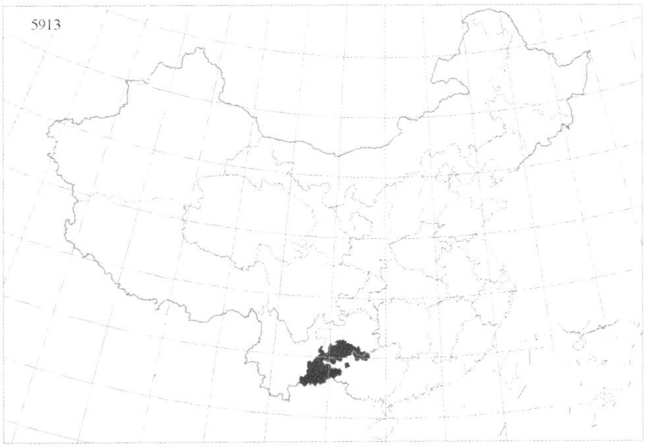

Rhamnaceae 鼠李科 *Rhamnus* 鼠李属
Rhamnus brachypoda C. Y. Wu ex Y. L. Chen 山绿柴
Shrubs, deciduous

19.3 (14.2~23.6); 10.0 (3.3~17.8); 27.4 (24.2~29.8); 19.4 (14.2~23.6); 172.6 (112.5~223.5); 0.1 (0.0~2.3); 1017 (761~1281) ❙ 1600 (1179~2037); 648 (392~1033); 154 (43~254) ❙ 59.5 (-6.9~130.7); 1012 (761~1273); 228.8 (95.3~439.7)

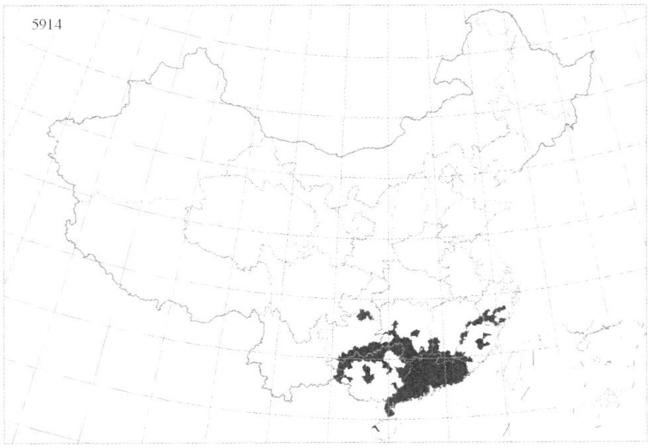

Rhamnaceae 鼠李科 *Rhamnus* 鼠李属
Rhamnus bungeana J. J. Vassiljev 卵叶鼠李
Shrubs, deciduous

10.6 (4.2~15.3); -4.2 (-12.5~3.6); 23.9 (18.9~27.3); 11.5 (7.4~15.4); 94.0 (53.4~125.7); 25.8 (1.4~62.6); 711 (538~842) ❙ 637 (367~1401); 363 (235~579); 26 (11~113) ❙ -10.7 (-37.7~101.3); 590 (367~832); 173.8 (109.4~274.[illegible])

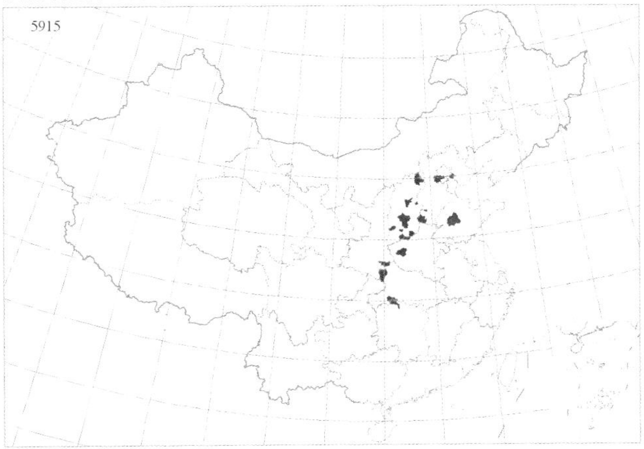

Rhamnaceae 鼠李科 *Rhamnus* 鼠李属
Rhamnus coriophylla Hand.-Mazz. 革叶鼠李
Shrubs, deciduous

18.7 (16.1~22.2); 9.6 (6.3~14.2); 25.9 (23.4~28.3); 18.7 (16.1~22.3); 164.4 (133.6~207.5); 0.0 (0.0~0.0); 954 (822~1190) ❙ 1364 (1232~1876); 696 (594~891); 79 (49~136) ❙ 43.6 (15.4~63.2); 954 (822~1190); 266.3 (189.5~371.7)

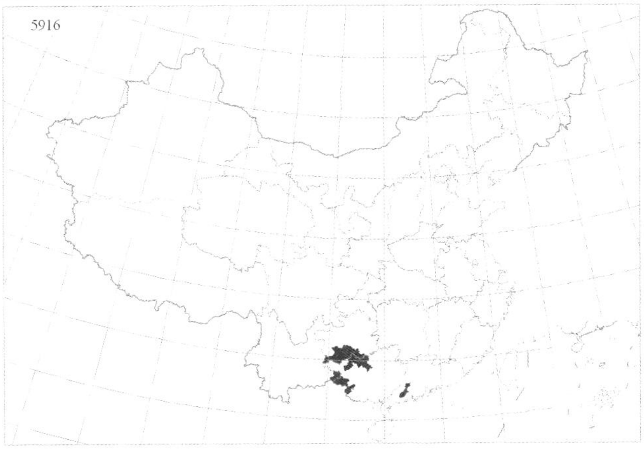

Rhamnaceae 鼠李科 *Rhamnus* 鼠李属
Rhamnus coriophylla var. acutidens Y. L. Chen et P. K. Chou 锐齿革叶鼠李
Shrubs, deciduous

16.2; 6.3; 24.4; 16.2; 134.5; 0.0; 826 ❙ 1342; 611; 95 ❙ 62.5; 826; 215.4

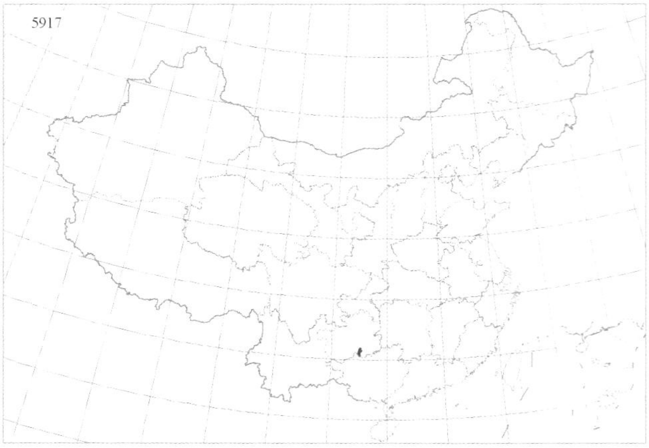

Rhamnaceae 鼠李科 *Rhamnus* 鼠李属
Rhamnus crenata Siebold et Zucc. 长叶冻绿
Shrubs or small trees, deciduous

17.5 (9.3~23.6); 7.1 (-2.9~17.7); 26.8 (18.8~30.1); 17.5 (9.5~23.6); 151.5 (69.8~223.5); 1.3 (0.0~20.5); 931 (603~1281) ❙ 1432 (626~3495); 582 (316~1847); 140 (24~897) ❙ 54.4 (-22.2~246.5); 923 (603~1273); 222.9 (95.3~551.6)

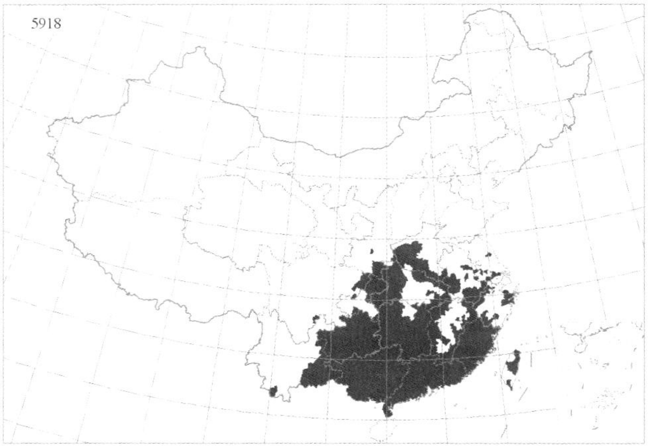

Rhamnaceae 鼠李科 *Rhamnus* 鼠李属
Rhamnus diamantiaca Nakai 金刚鼠李
Shrubs, deciduous

2.8 (-4.0~8.2); -18.3 (-28.2~-10.3); 20.8 (17.6~24.6); 7.8 (5.3~10.3); 59.9 (38.6~87.1); 86.1 (47.7~146.2); 567 (463~681) ❙ 682 (470~1073); 425 (309~691); 23 (8~40) ❙ 19.9 (-19.0~74.1); 558 (463~681); 291.5 (189.1~346.5)

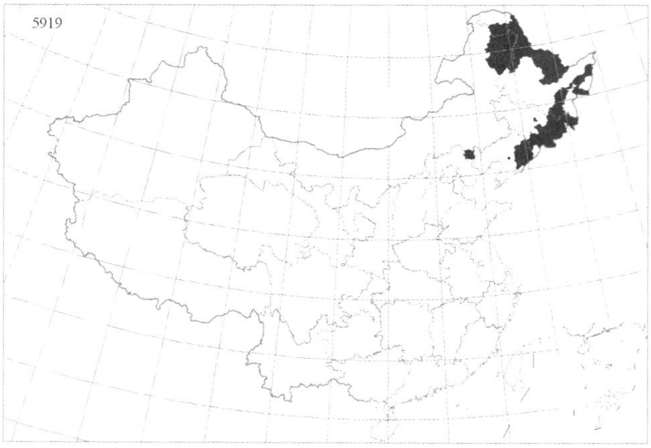

Rhamnaceae 鼠李科 *Rhamnus* 鼠李属
Rhamnus dumetorum C. K. Schneid. 刺鼠李
Shrubs, deciduous

11.0 (1.8~19.2); -0.4 (-12.7~10.9); 21.3 (9.9~28.4); 11.7 (3.8~19.2); 90.4 (15.4~170.8); 17.3 (0.0~65.0); 683 (404~961) ❙ 924 (367~1770); 465 (235~817); 48 (9~225) ❙ 34.7 (-37.7~128.6); 643 (367~961); 227.8 (70.3~411.1)

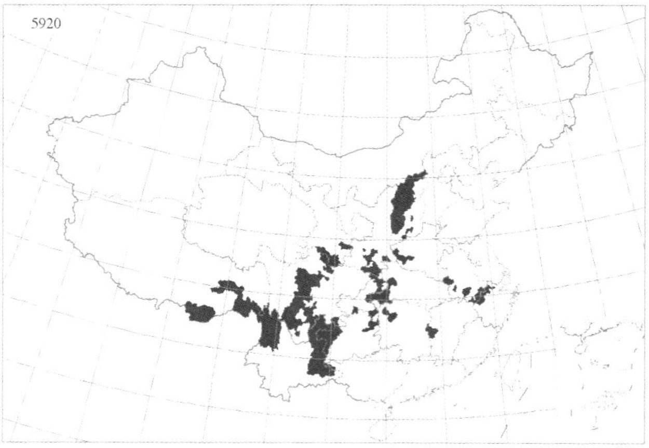

Rhamnaceae 鼠李科 *Rhamnus* 鼠李属
Rhamnus dumetorum var. crenoserrata Rehder et E. H. Wilson 圆齿刺鼠李
Shrubs, deciduous

7.0 (1.8~12.0); -2.2 (-7.1~2.4); 15.3 (9.9~23.7); 7.5 (3.8~12.0); 48.6 (15.5~94.7); 24.6 (5.4~53.3); 538 (404~703) ❙ 874 (527~1087); 474 (317~601); 26 (9~45) ❙ 64.9 (15.7~125.9); 538 (404~703); 280.4 (169.5~352.3)

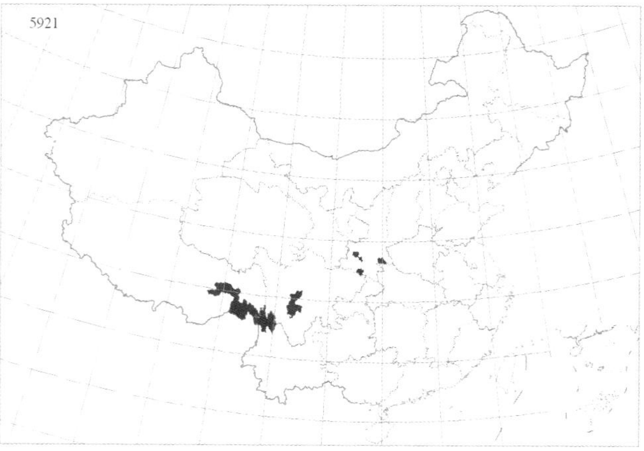

Rhamnaceae 鼠李科 *Rhamnus* 鼠李属
Rhamnus erythroxylum Pall. 柳叶鼠李
Shrubs or small trees, deciduous

9.3 (-1.1~15.1); -6.2 (-24.4~1.8); 23.1 (12.4~28.5); 10.6 (4.8~15.2); 86.4 (25.1~128.7); 34.5 (5.6~122.1); 677 (441~886) ❙ 519 (101~880); 290 (61~409); 18 (2~58) ❙ -23.4 (-83.8~48.8); 512 (101~821); 145.3 (13.6~282.7)

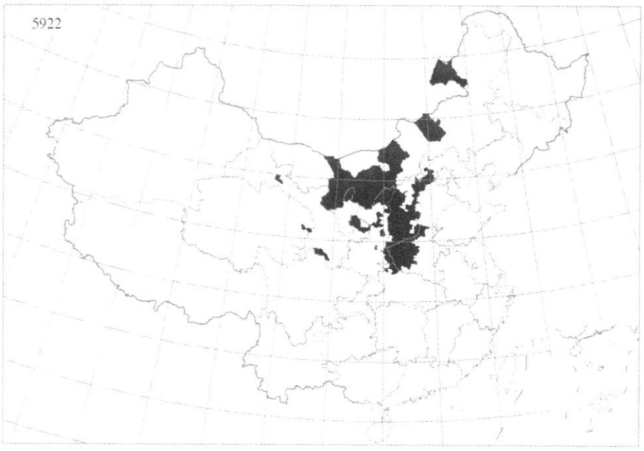

Rhamnaceae 鼠李科 *Rhamnus* 鼠李属
Rhamnus esquirolii H. Léveillé 贵州鼠李
Shrubs or small trees, deciduous

16.1 (9.3~21.3); 6.0 (−1.7~12.6); 25.3 (18.8~28.9); 16.2 (9.5~21.3); 134.9 (69.8~196.0); 0.6 (0.0~18.0); 849 (603~1123) ❙ 1215 (862~1740); 554 (376~816); 74 (29~187) ❙ 44.2 (5.8~107.2); 849 (603~1121); 205.4 (108.2~487.2)

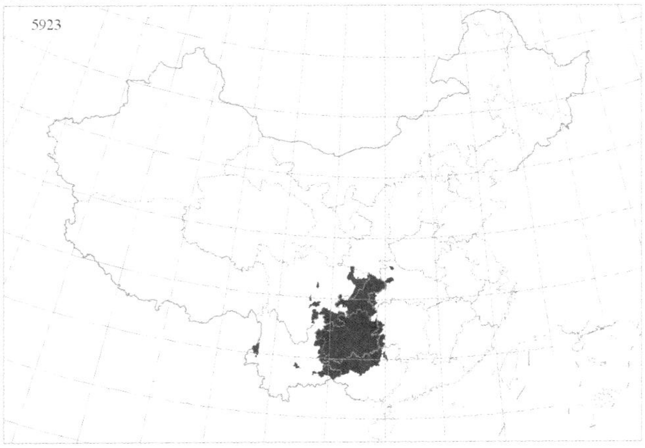

Rhamnaceae 鼠李科 *Rhamnus* 鼠李属
Rhamnus esquirolii var. glabrata Y. L. Chen et P. K. Chou 木子花
Shrubs or small trees, deciduous

15.8 (13.3~19.1); 6.3 (3.6~10.1); 24.0 (22.2~26.0); 15.9 (13.3~19.1); 130.8 (101.9~169.2); 0.3 (0.0~1.8); 816 (718~966) ❙ 1202 (999~1334); 604 (501~720); 61 (51~69) ❙ 47.7 (30.0~64.7); 816 (718~966); 199.7 (153.9~283.9)

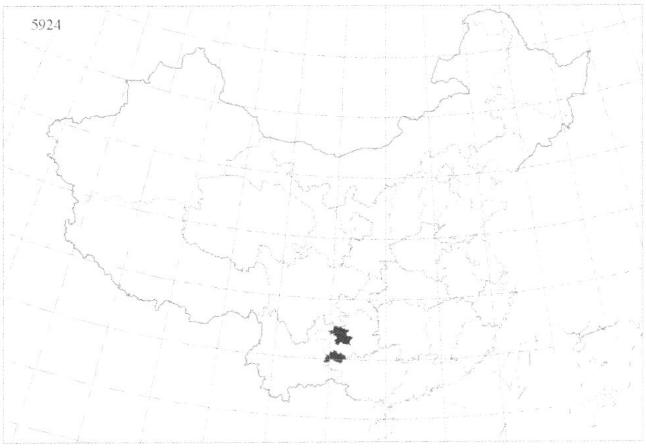

Rhamnaceae 鼠李科 *Rhamnus* 鼠李属
Rhamnus flavescens Y. L. Chen et P. K. Chou 淡黄鼠李
Shrubs, deciduous

3.8 (−1.2~14.2); −4.8 (−11.7~6.2); 11.4 (8.1~20.1); 5.2 (2.6~14.2); 28.0 (8.0~110.5); 41.8 (0.0~82.3); 454 (347~729) ❙ 742 (491~1147); 443 (301~636); 19 (9~50) ❙ 63.7 (27.1~128.6); 454 (347~729); 251.9 (111.0~319.8)

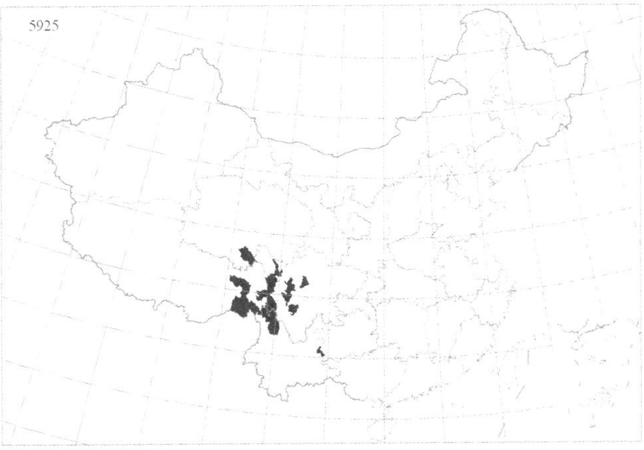

Rhamnaceae 鼠李科 *Rhamnus* 鼠李属
Rhamnus formosana Matsum. 台湾鼠李
Shrubs, deciduous

20.0 (16.2~24.1); 14.6 (11.0~18.9); 24.2 (20.1~28.0); 20.0 (16.2~24.1); 179.8 (134.2~229.7); 0.0 (0.0~0.0); 993 (773~1322) ❙ 2506 (1751~2993); 1289 (924~1847); 193 (47~547) ❙ 162.4 (32.5~246.5); 963 (773~1123); 256.6 (154.2~363.7)

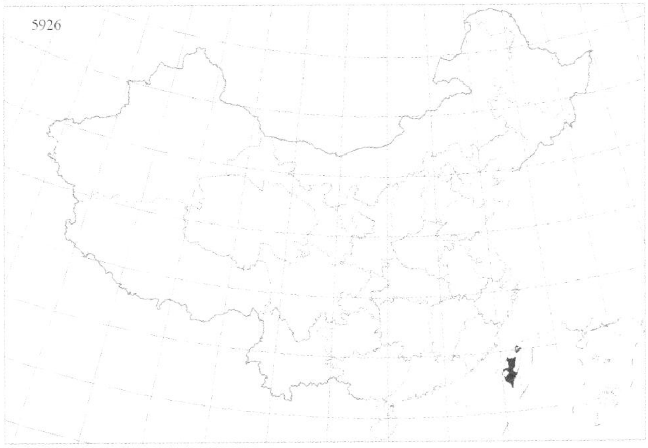

Rhamnaceae 鼠李科 *Rhamnus* 鼠李属
Rhamnus fulvotincta Metcalf 黄鼠李
Shrubs, deciduous

18.1 (14.2~20.7); 8.0 (3.6~12.0); 26.9 (24.4~29.0); 18.1 (14.2~20.8); 157.4 (112.5~189.3); 0.2 (0.0~1.7); 947 (765~1108) ❙ 1486 (1133~2005); 621 (486~824); 137 (58~196) ❙ 57.0 (28.9~85.5); 947 (765~1108); 225.6 (160.7~322.1)

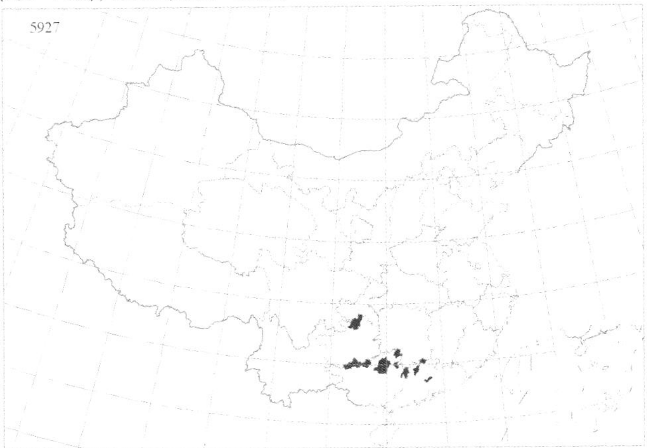

Rhamnaceae 鼠李科 *Rhamnus* 鼠李属
Rhamnus gilgiana Heppeler 川滇鼠李
Shrubs, deciduous

13.4 (5.3~19.8); 5.4 (−2.4~12.0); 20.0 (12.2~26.6); 13.4 (5.7~19.8); 103.9 (30.5~177.9); 3.1 (0.0~26.2); 717 (479~1010) ❙ 985 (741~1265); 553 (382~673); 36 (12~75) ❙ 39.7 (−14.2~101.4); 696 (479~1010); 272.3 (142.9~366.3)

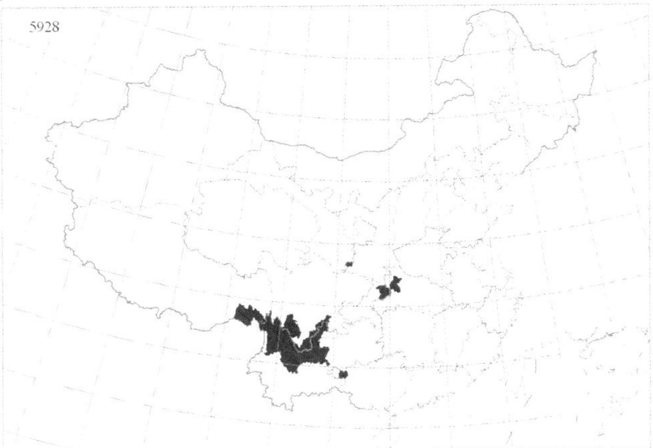

Rhamnaceae 鼠李科 *Rhamnus* 鼠李属

Rhamnus globosa Bunge 圆叶鼠李

Shrubs or small trees, deciduous

12.4 (1.9~21.0); -1.4 (-16.1~11.9); 25.1 (11.7~29.4); 13.0 (4.3~21.1); 107.1 (21.1~193.0); 18.1 (0.0~81.1); 764 (427~1114) ❙ 875 (202~1893); 409 (134~821); 70 (6~228) ❙ 11.6 (-65.5~129.9); 692 (202~1114); 184.0 (52.7~334.4)

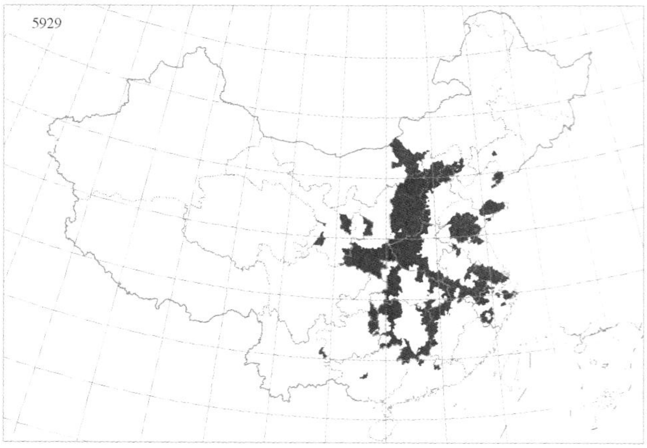

Rhamnaceae 鼠李科 *Rhamnus* 鼠李属

Rhamnus grandiflora C. Y. Wu ex Y. L. Chen 大花鼠李

Shrubs, deciduous

15.7 (12.6~19.1); 5.5 (1.6~9.8); 24.8 (22.2~27.0); 15.7 (12.7~19.2); 129.3 (98.8~170.1); 0.5 (0.0~6.9); 821 (711~981) ❙ 1230 (999~1466); 545 (452~678); 84 (53~129) ❙ 50.5 (18.3~106.2); 821 (711~981); 209.8 (149.6~322.1)

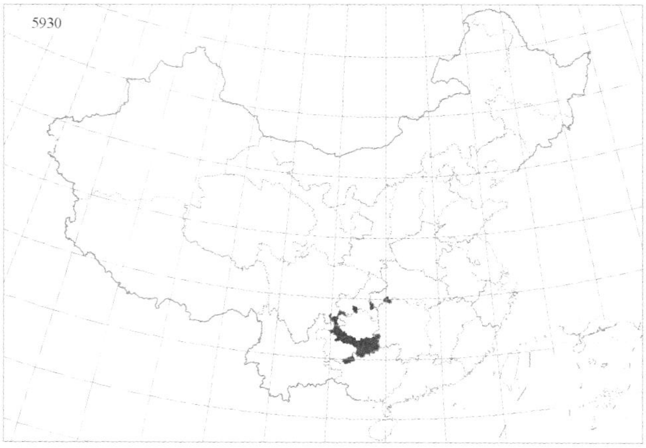

Rhamnaceae 鼠李科 *Rhamnus* 鼠李属

Rhamnus hainanensis Merr. et Chun 海南鼠李

Shrubs, deciduous

22.3 (19.1~24.8); 15.5 (9.7~20.1); 27.3 (25.0~28.6); 22.3 (19.1~24.8); 207.8 (169.6~237.6); 0.0 (0.0~0.0); 1186 (945~1385) ❙ 1523 (1380~1725); 622 (455~820); 85 (41~69) ❙ 30.4 (1.5~56.8); 1105 (945~1176); 426.3 (236.1~620.0)

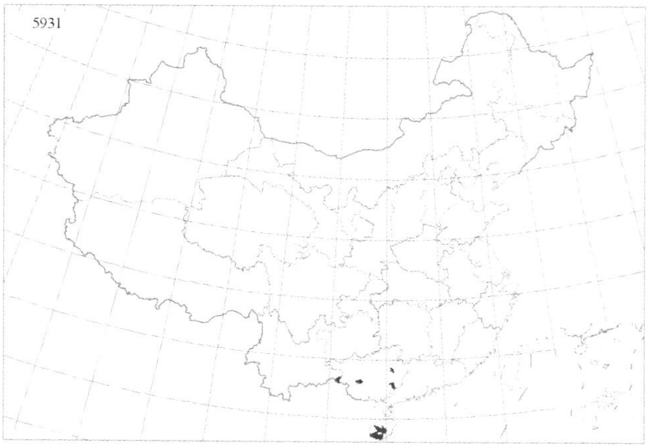

Rhamnaceae 鼠李科 *Rhamnus* 鼠李属

Rhamnus hemsleyana C. K. Schneid. 亮叶鼠李

Trees, evergreen

13.6 (5.4~19.1); 3.2 (-2.6~9.8); 23.3 (12.3~29.0); 13.7 (5.8~19.2); 109.0 (31.5~170.1); 4.8 (0.0~25.6); 754 (481~981) ❙ 1080 (738~1496); 505 (345~817); 58 (12~171) ❙ 43.9 (-2.0~101.4); 752 (481~981); 244.9 (146.0~384.1)

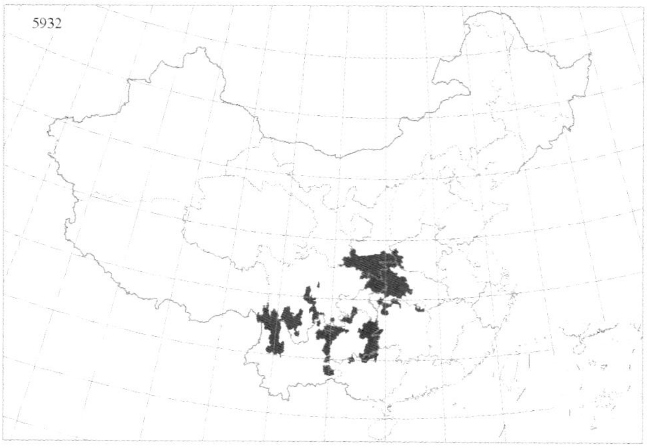

Rhamnaceae 鼠李科 *Rhamnus* 鼠李属

Rhamnus hemsleyana var. yunnanensis C. Y. Wu ex Y. L. Chen et P. K. Chou 高山亮叶鼠李

Trees, evergreen

11.3 (5.4~16.1); 4.2 (-2.1~8.7); 17.2 (12.3~21.9); 11.4 (5.8~16.1); 81.5 (31.5~133.0); 5.0 (0.0~25.6); 647 (481~797) ❙ 999 (741~1164); 585 (426~651); 34 (12~58) ❙ 55.7 (27.3~81.5); 640 (481~746); 331.2 (273.5~382.6)

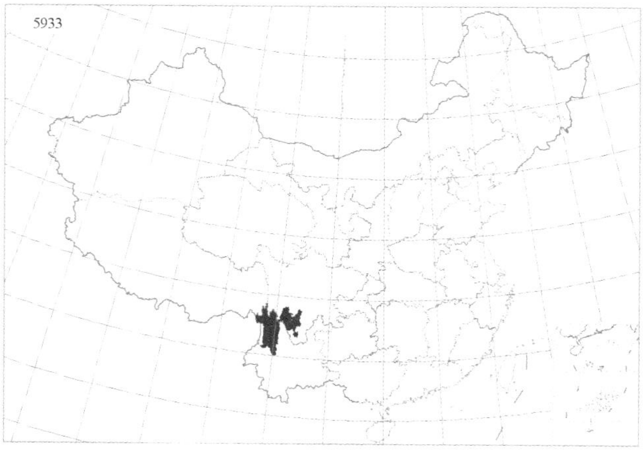

Rhamnaceae 鼠李科 *Rhamnus* 鼠李属

Rhamnus henryi C. K. Schneid. 毛叶鼠李

Trees, deciduous

15.8 (1.2~21.6); 7.9 (-7.1~14.2); 22.2 (9.0~28.5); 15.9 (3.3~21.7); 132.4 (11.9~199.8); 2.1 (0.0~56.5); 815 (390~1153) ❙ 1234 (578~1751); 651 (345~1007); 61 (15~194) ❙ 53.2 (5.8~145.8); 812 (390~1153); 332.0 (130.1~551.6)

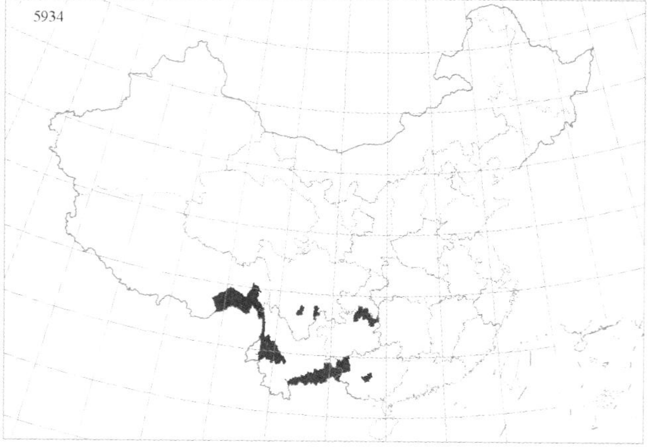

Rhamnaceae 鼠李科 *Rhamnus* 鼠李属
Rhamnus heterophylla Oliv. 异叶鼠李
Shrubs, deciduous

14.9 (5.7~18.7); 4.4 (-3.5~8.6); 24.5 (14.2~28.9); 14.9 (6.4~18.8); 121.4 (38.1~166.0); 2.6 (0.0~28.1); 801 (504~1002) ❙ 1114 (644~1410); 515 (321~816); 59 (14~129) ❙ 40.2 (-8.1~101.4); 801 (504~1002); 206.0 (108.2~366.3)

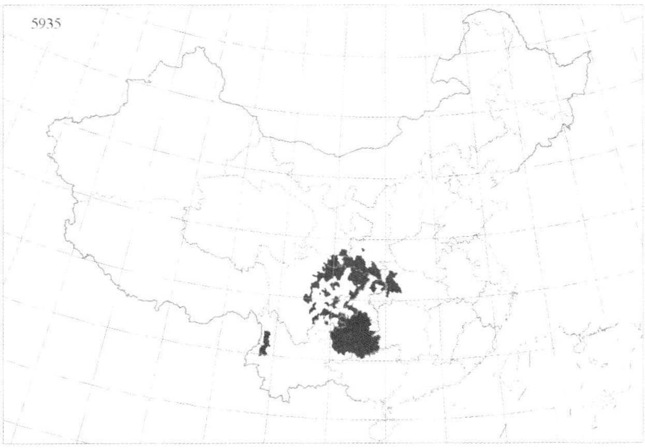

Rhamnaceae 鼠李科 *Rhamnus* 鼠李属
Rhamnus hupehensis C. K. Schneid. 湖北鼠李
Shrubs, deciduous

12.6; 1.3; 23.5; 12.7; 100.1; 8.1; 721 ❙ 1279; 538; 81 ❙ 78.2; 721; 256.9

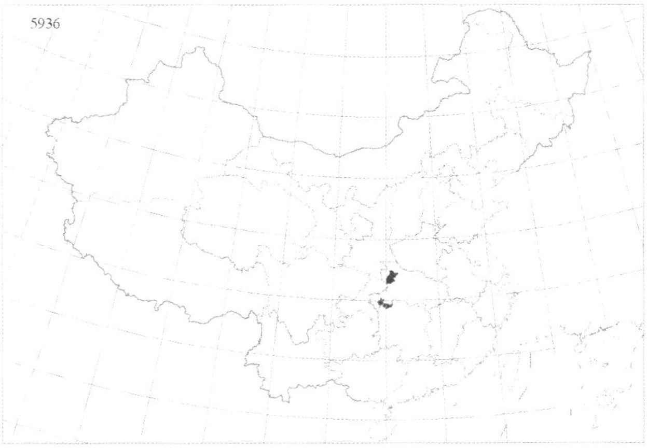

Rhamnaceae 鼠李科 *Rhamnus* 鼠李属
Rhamnus iteinophylla C. K. Schneid. 桃叶鼠李
Shrubs, deciduous

13.7 (9.3~17.5); 3.7 (-2.5~10.5); 22.8 (20.1~26.5); 13.8 (9.5~17.5); 110.8 (69.8~149.9); 5.9 (0.0~20.6); 745 (603~847) ❙ 1169 (862~1492); 532 (376~633); 61 (29~113) ❙ 59.1 (7.8~107.2); 743 (603~847); 267.5 (213.4~314.4)

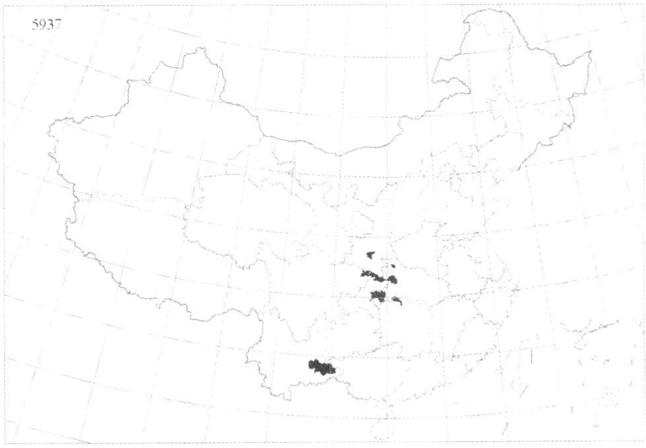

Rhamnaceae 鼠李科 *Rhamnus* 鼠李属
Rhamnus koraiensis C. K. Schneid. 朝鲜鼠李
Shrubs, deciduous

9.1 (2.2~15.5); -7.3 (-18.0~2.1); 23.5 (19.2~27.8); 10.7 (6.9~15.6); 88.0 (50.3~131.3); 37.8 (4.3~83.8); 681 (525~881) ❙ 764 (455~1077); 467 (306~691); 34 (8~97) ❙ 13.6 (-32.9~74.1); 650 (455~881); 219.4 (136.4~335.3)

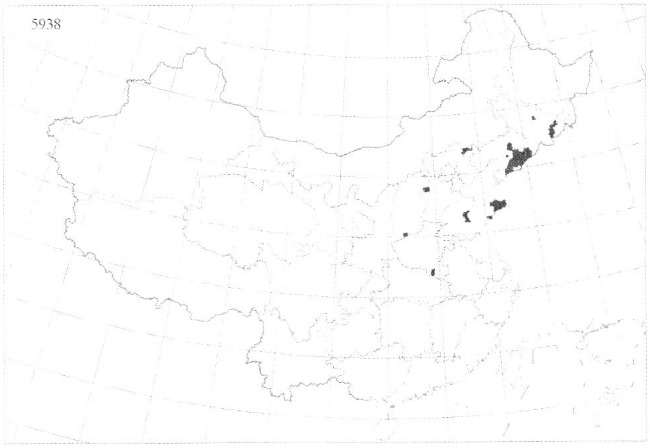

Rhamnaceae 鼠李科 *Rhamnus* 鼠李属
Rhamnus kwangsiensis Y. L. Chen et P. K. Chou 广西鼠李
Shrubs, deciduous

19.4; 8.9; 28.7; 19.5; 173.7; 0.0; 1038 ❙ 1776; 718; 188 ❙ 71.1; 1038; 203.9

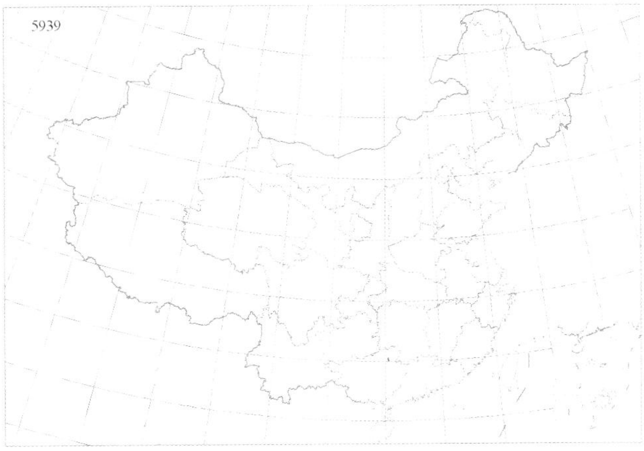

Rhamnaceae 鼠李科 *Rhamnus* 鼠李属
Rhamnus lamprophylla C. K. Schneid. 钩齿鼠李
Shrubs or small trees, deciduous

16.6 (9.3~21.3); 6.4 (-1.7~12.6); 25.7 (20.1~29.8); 16.7 (9.5~21.3); 140.5 (69.8~196.0); 0.7 (0.0~18.0); 871 (603~1123) ❙ 1410 (1018~1997); 573 (419~848); 125 (42~259) ❙ 62.3 (5.8~130.7); 870 (603~1121); 231.2 (146.0~411.1)

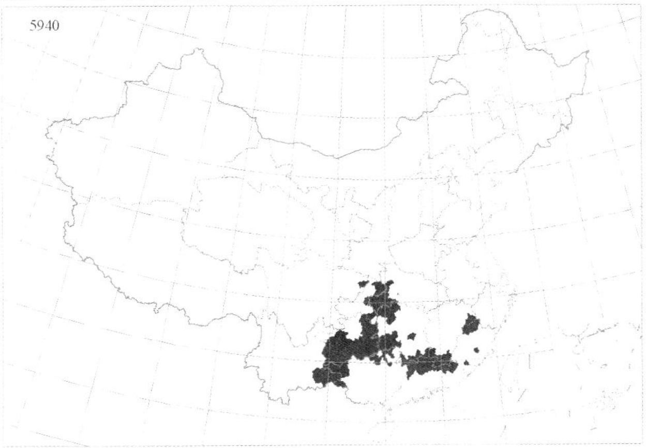

Rhamnaceae 鼠李科 *Rhamnus* 鼠李属
Rhamnus leptophylla C. K. Schneid. 薄叶鼠李
Shrubs or small trees, deciduous

15.6 (3.5~22.4); 4.8 (-9.3~14.9); 25.4 (12.6~29.9); 15.7 (5.0~22.4); 131.5 (26.3~208.9); 3.7 (0.0~46.8); 844 (446~1191) ❙ 1242 (422~2037); 527 (258~1033); 108 (8~259) ❙ 45.9 (-35.9~130.7); 825 (422~1191); 221.3 (111.9~570.1)

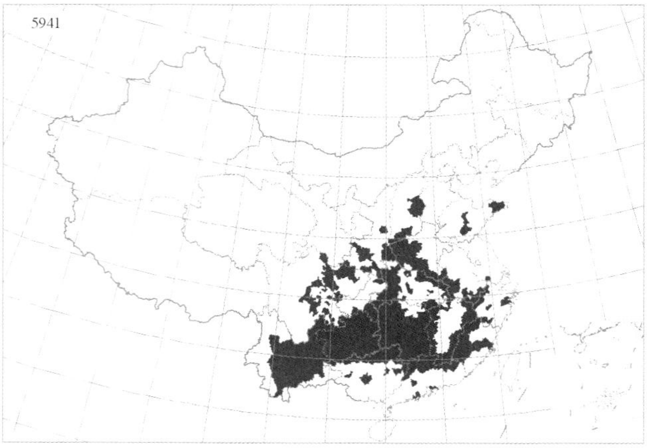

Rhamnaceae 鼠李科 *Rhamnus* 鼠李属
Rhamnus liukiuensis (E. H. Wilson) Koidzumi 琉球鼠李
Shrubs, deciduous

17.8; 12.1; 22.8; 17.9; 154.4; 0.0; 852 ❙ 2859; 959; 402 ❙ 235.6; 852; 265.8

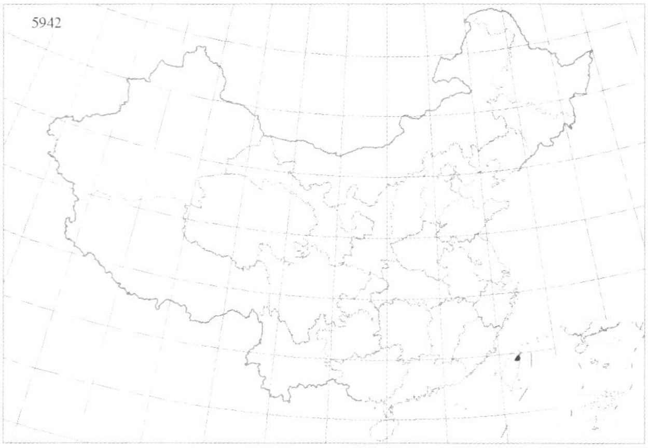

Rhamnaceae 鼠李科 *Rhamnus* 鼠李属
Rhamnus longipes Merr. et Chun 长柄鼠李
Shrubs or small trees, deciduous

20.7 (16.5~24.3); 12.2 (5.8~19.1); 27.6 (22.4~29.8); 20.7 (16.5~24.3); 188.4 (138.2~231.9); 0.0 (0.0~0.0); 1084 (833~1338) ❙ 1628 (1161~2037); 700 (392~1033); 139 (42~203) ❙ 51.5 (-11.9~95.5); 1078 (833~1273); 243.1 (95.3~620.0)

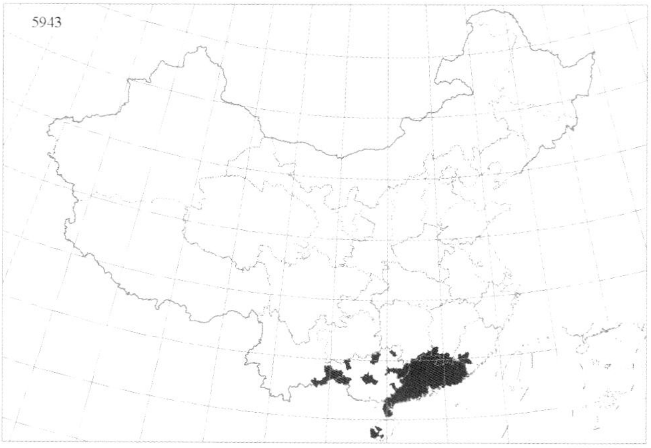

Rhamnaceae 鼠李科 *Rhamnus* 鼠李属
Rhamnus maximovicziana J. J. Vassiljev 黑桦树
Shrubs, deciduous

8.1 (2.0~14.5); -7.2 (-16.1~-0.8); 21.8 (9.9~28.2); 9.7 (3.8~14.6); 77.0 (15.4~126.9); 38.8 (11.9~81.1); 638 (413~874) ❙ 487 (111~950); 278 (74~546); 14 (2~35) ❙ -20.8 (-82.3~128.6); 454 (111~699); 158.0 (13.6~357.7)

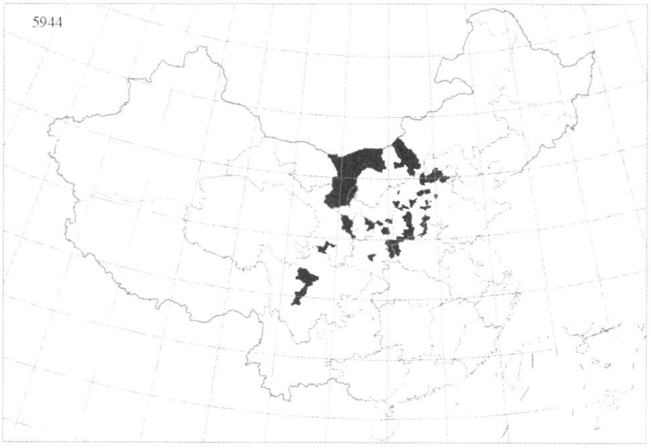

Rhamnaceae 鼠李科 *Rhamnus* 鼠李属
Rhamnus minuta Grubov 矮小鼠李
Shrubs, deciduous

-2.8; -18.9; 11.1; 3.2; 15.3; 108.6; 360 ❙ 104; 35; 16 ❙ -71.8; 104; 19.3

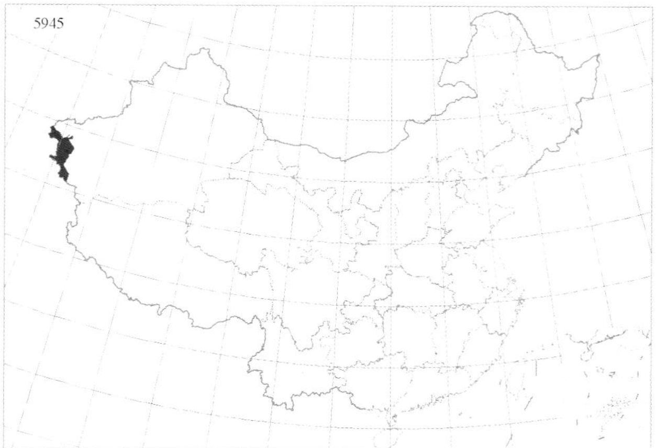

Rhamnaceae 鼠李科 *Rhamnus* 鼠李属
Rhamnus nakaharae (Hayata) Hayata 台中鼠李
Shrubs, deciduous

20.3 (17.1~22.2); 14.6 (11.2~17.7); 25.0 (21.8~27.0); 20.3 (17.1~22.2); 184.1 (145.0~206.7); 0.0 (0.0~0.0); 1012 (816~1123) ❙ 2529 (1836~2993); 1252 (924~1847); 226 (107~547) ❙ 153.2 (64.2~203.4); 1004 (816~1123); 250.1 (202.8~296.9)

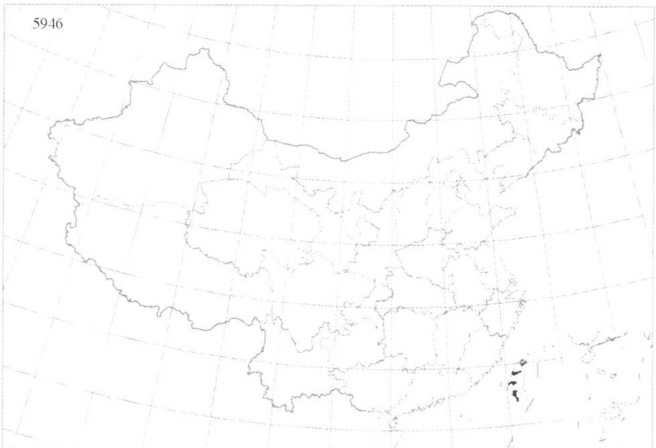

Rhamnaceae 鼠李科 *Rhamnus* 鼠李属

Rhamnus napalensis (Wall.) Lawson 尼泊尔鼠李

Shrubs or trees, deciduous

18.0（12.7~24.7）；8.4（1.9~19.8）；26.4（19.6~29.8）；18.1（12.8~24.7）；157.0 (99.0~236.5)；0.2 (0.0~6.1)；939 (706~1376) ▌ 1528 (1018~2030)；625（403~1033）；143（41~259）▌ 63.7（-1.0~145.8）；934（706~1326）；258.1 (95.3~659.0)

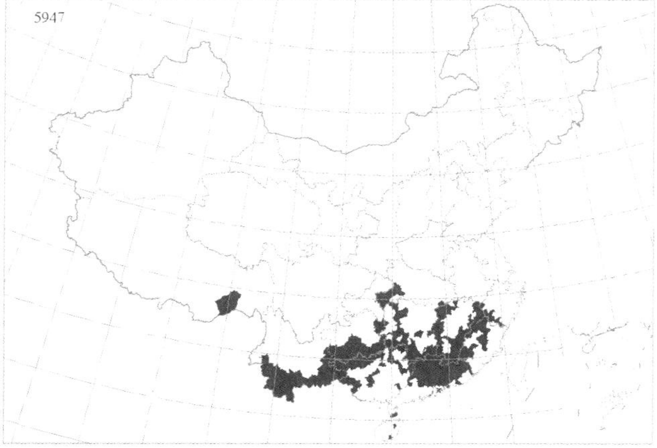

Rhamnaceae 鼠李科 *Rhamnus* 鼠李属

Rhamnus nigricans Hand.-Mazz. 黑背鼠李

Shrubs or small trees, deciduous

15.9（13.6~18.9）；9.3（7.2~12.5）；20.7（18.8~23.2）；15.9（13.7~19.0）；131.3（104.1~167.5）；0.0（0.0~0.0）；782（706~914）▌ 1057（907~1454）；593（521~792）；46（27~70）▌ 35.8（6.5~99.9）；751（683~851）；348.2（264.0~487.2）

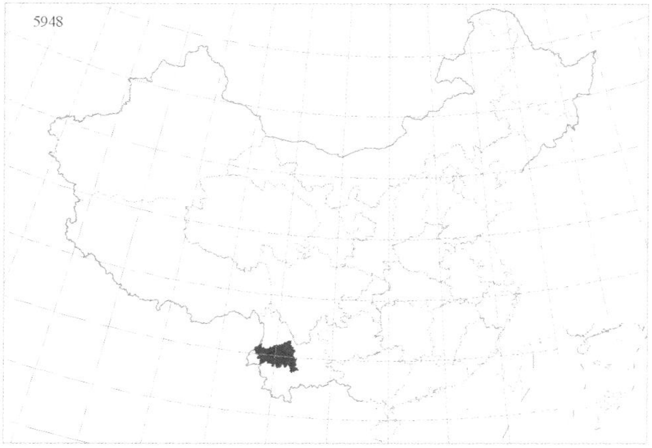

Rhamnaceae 鼠李科 *Rhamnus* 鼠李属

Rhamnus ninglangensis Y. L. Chen 宁蒗鼠李

Shrubs, deciduous

7.1；-0.4；13.6；7.3；43.6；17.4；529 ▌ 838；518；19 ▌ 58.8；527；292.1

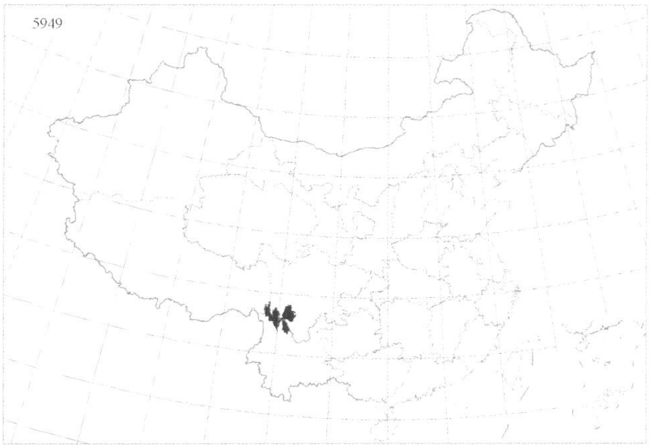

Rhamnaceae 鼠李科 *Rhamnus* 鼠李属

Rhamnus parvifolia Bunge 小叶鼠李

Shrubs, deciduous

9.6（-1.2~24.1）；-6.2（-21.5~18.9）；23.6（8.5~29.5）；11.0（2.6~24.1）；90.3（8.1~229.7）；35.0（0.0~108.6）；695（347~1322）▌ 597（137~2759）；352（92~1594）；23（3~303）▌ -14.5（-77.5~246.5）；561（137~1095）；166.2（25.4~346.5）

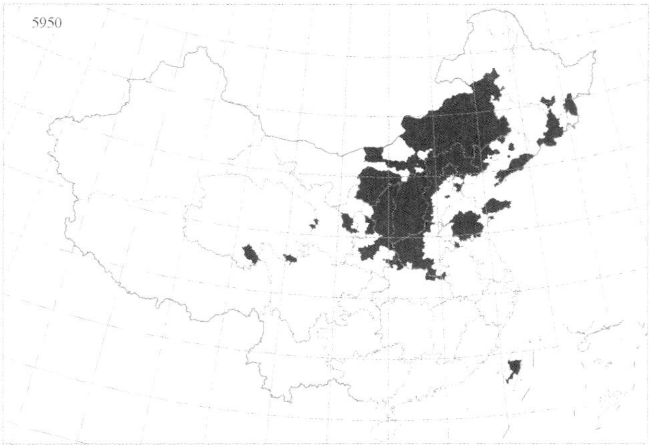

Rhamnaceae 鼠李科 *Rhamnus* 鼠李属

Rhamnus procumbens Edgew. 蔓生鼠李

Small shrubs, deciduous

-1.7；-10.0；6.2；1.8；2.3；82.6；329 ▌ 470；250；68 ▌ 43.7；322；63.1

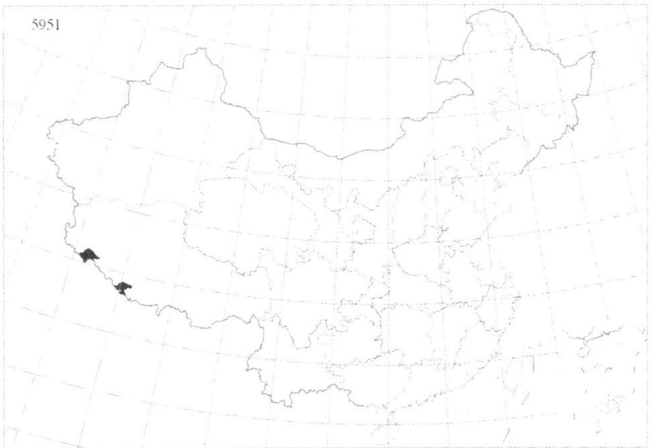

Rhamnaceae 鼠李科 *Rhamnus* 鼠李属

Rhamnus prostrata Jacquin ex R. N. Parker 平卧鼠李

Shrubs, deciduous

-1.8；-10.7；7.0；2.0；4.2；84.7；330 ▌ 483；244；78 ▌ 46.8；326；52.8

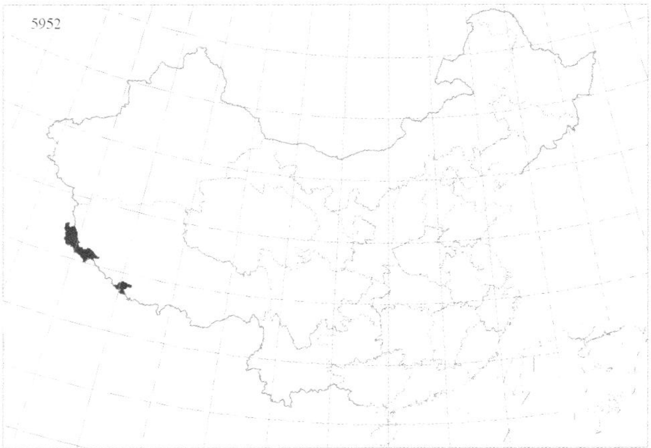

Rhamnaceae 鼠李科 *Rhamnus* 鼠李属
Rhamnus rhododendriphylla Y. L. Chen et P. K. Chou 杜鹃叶鼠李
Shrubs, deciduous

20.0; 10.9; 27.3; 20.0; 180.1; 0.0; 1045 | 1568; 714; 128 | 50.2; 1045; 234.7

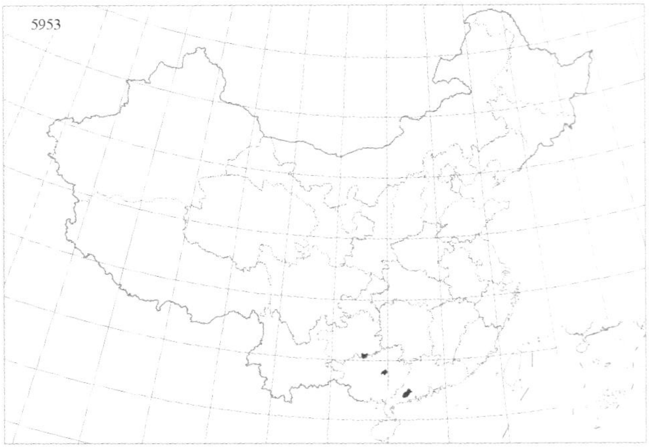

Rhamnaceae 鼠李科 *Rhamnus* 鼠李属
Rhamnus rosthornii E. Pritz. ex Diels 小冻绿树
Shrubs or small trees, deciduous

15.3 (2.9~21.3); 5.4 (-8.2~15.1); 24.1 (11.2~28.8); 15.3 (4.4~21.3); 125.9 (20.5~196.1); 2.3 (0.0~45.0); 808 (428~1137) | 1179 (374~1607); 549 (205~860); 67 (8~171) | 46.8 (-29.3~121.7); 807 (374~1136); 235.6 (81.2~659.0)

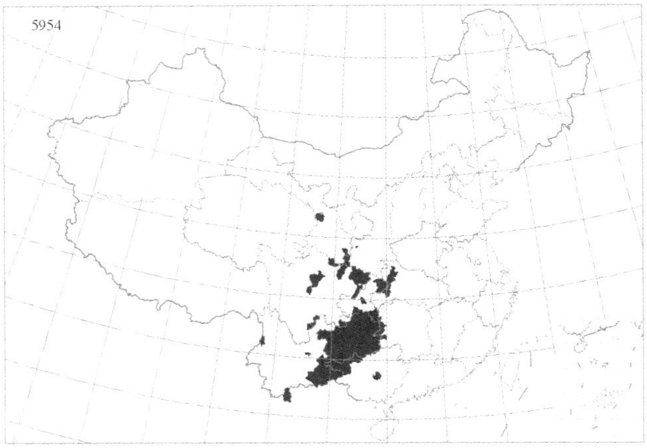

Rhamnaceae 鼠李科 *Rhamnus* 鼠李属
Rhamnus rugulosa Hemsl. ex Forbes et Hemsl. 皱叶鼠李
Shrubs, deciduous

14.3 (4.2~22.1); 1.9 (-12.5~13.7); 25.8 (15.8~29.4); 14.5 (7.0~22.1); 120.8 (45.2~205.6); 8.8 (0.0~62.6); 816 (522~1182) | 1131 (377~1922); 461 (241~779); 103 (8~259) | 35.7 (-37.2~128.9); 784 (377~1182); 200.7 (109.4~383.4)

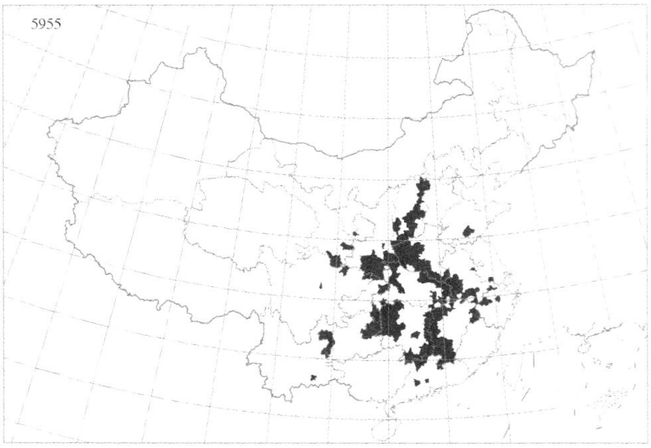

Rhamnaceae 鼠李科 *Rhamnus* 鼠李属
Rhamnus rugulosa var. chekiangensis (Cheng) Y. L. Chen et P. K. Chou 浙江鼠李
Shrubs, deciduous

16.2 (15.4~16.9); 4.2 (3.4~4.8); 28.3 (27.4~29.1); 16.3 (15.4~16.9); 136.0 (126.7~143.3); 0.9 (0.2~1.7); 889 (851~925) | 1423 (1318~1536); 507 (465~582); 188 (140~213) | 60.2 (42.6~73.7); 889 (851~925); 208.4 (183.4~234.0)

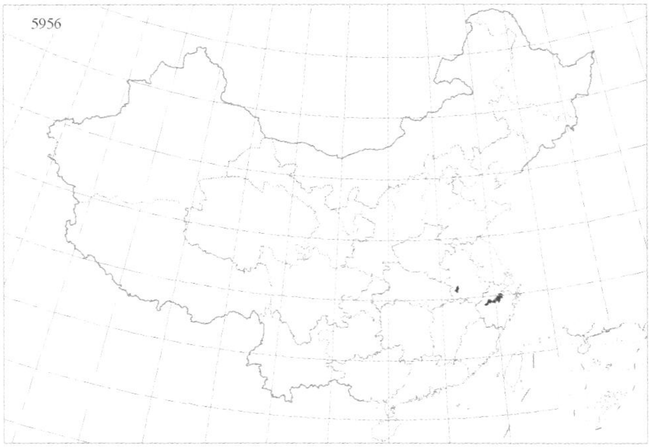

Rhamnaceae 鼠李科 *Rhamnus* 鼠李属
Rhamnus rugulosa var. glabrata Y. L. Chen et P. K. Chou 脱毛皱叶鼠李
Shrubs, deciduous

12.6 (9.3~14.2); 1.4 (-1.7~3.3); 23.4 (20.1~26.2); 12.6 (9.5~14.3); 99.2 (69.8~116.7); 7.8 (2.1~18.0); 720 (603~796) | 1288 (1148~1492); 541 (475~613); 79 (46~129) | 80.2 (51.4~101.4); 720 (603~796); 254.2 (214.7~284.3)

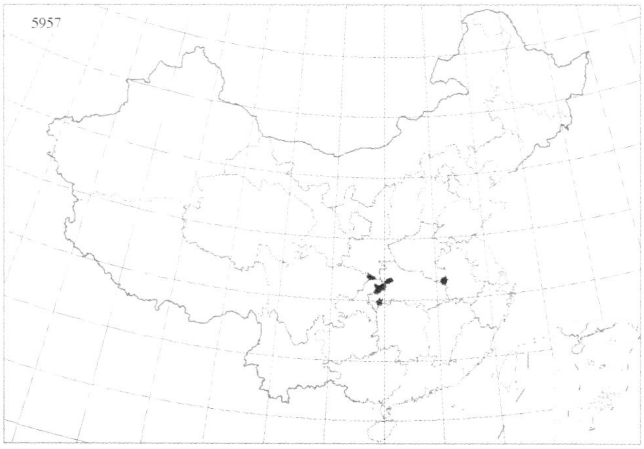

Rhamnaceae 鼠李科 *Rhamnus* 鼠李属
Rhamnus sargentiana C. K. Schneid. 多脉鼠李
Shrubs or trees, deciduous

12.3 (2.2~17.9); 3.1 (-6.6~10.7); 20.5 (9.9~26.6); 12.4 (3.8~18.0); 93.6 (15.5~155.7); 5.9 (0.0~49.1); 693 (413~912) | 1032 (711~1736); 543 (357~1007); 37 (15~83) | 51.0 (4.4~145.8); 684 (413~912); 268.6 (142.9~447.4)

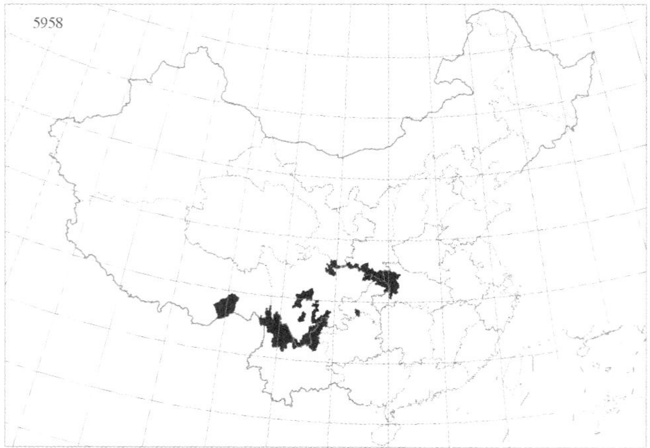

Rhamnaceae 鼠李科 *Rhamnus* 鼠李属

Rhamnus schneideri H. Léveillé et Vaniot 长梗鼠李

Shrubs, deciduous

7.3 (2.2~12.8); -9.6 (-20.0~1.6); 22.2 (18.5~27.2); 9.6 (6.7~13.3); 76.9 (47.6~114.7); 48.5 (6.9~89.7); 635 (513~827) | 599 (367~1466); 382 (235~691); 21 (9~105) | -5.3 (-47.7~106.2); 545 (367~729); 215.9 (91.4~335.3)

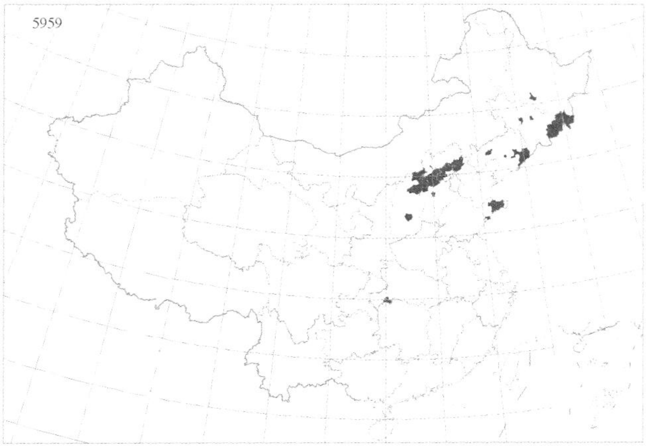

Rhamnaceae 鼠李科 *Rhamnus* 鼠李属

Rhamnus schneideri var. manshurica (Nakai) Nakai 东北鼠李

Shrubs, deciduous

5.9 (2.2~11.0); -12.1 (-20.0~-5.7); 21.6 (18.5~25.5); 8.8 (6.7~12.0); 69.9 (47.6~101.7); 58.6 (29.5~89.7); 607 (513~757) | 555 (367~1073); 367 (235~691); 15 (8~40) | -8.4 (-41.6~74.1); 517 (367~661); 215.7 (91.4~335.3)

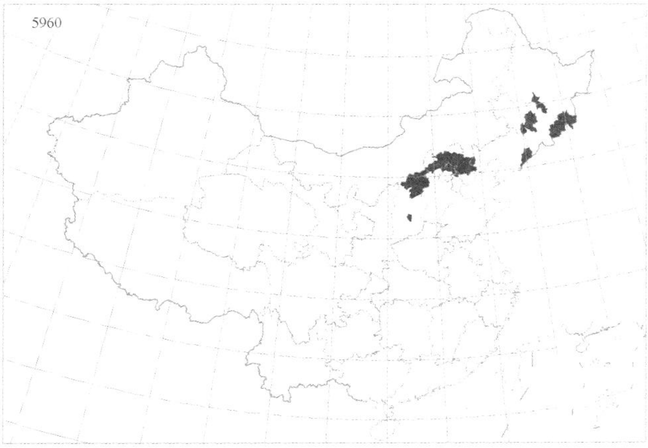

Rhamnaceae 鼠李科 *Rhamnus* 鼠李属

Rhamnus songorica Gontsch. 新疆鼠李

Shrubs, deciduous

2.9 (-1.3~9.0); -15.7 (-20.9~-9.3); 17.3 (13.2~23.2); 7.0 (4.6~10.7); 50.4 (26.7~88.9); 74.7 (40.7~103.9); 537 (440~703) | 310 (225~406); 143 (65~211); 27 (12~48) | -40.1 (-65.1~-7.6); 310 (225~406); 216.4 (165.7~297.3)

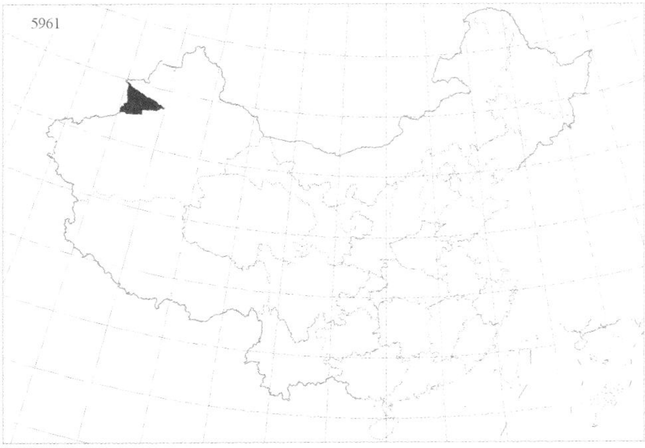

Rhamnaceae 鼠李科 *Rhamnus* 鼠李属

Rhamnus subapetala Merr. 紫背鼠李

Shrubs or trees, deciduous

18.0 (16.3~19.2); 9.9 (8.7~11.0); 24.0 (21.4~26.6); 18.0 (16.4~19.2); 156.3 (136.4~170.5); 0.0 (0.0~0.0); 890 (797~981) | 1303 (1074~1504); 713 (582~860); 60 (47~97) | 46.7 (30.5~80.6); 890 (797~981); 340.8 (277.0~491.5)

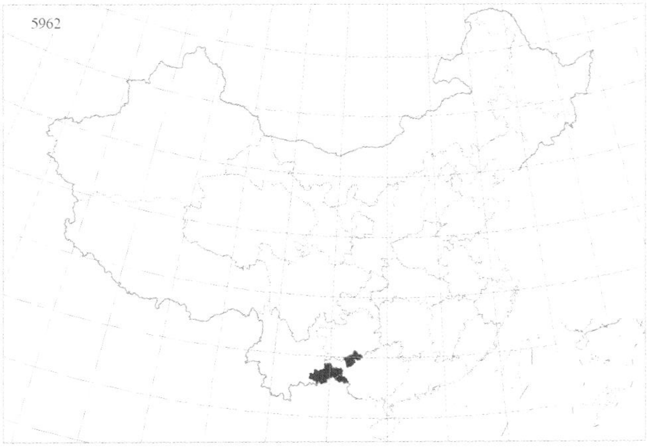

Rhamnaceae 鼠李科 *Rhamnus* 鼠李属

Rhamnus tangutica J. J. Vassiljev 甘青鼠李

Shrubs or trees, deciduous

6.5 (-1.3~16.8); -5.0 (-13.5~6.1); 17.1 (8.5~26.7); 7.8 (2.6~16.9); 54.8 (8.1~142.4); 35.6 (0.0~86.3); 552 (347~897) | 683 (72~1246); 348 (46~601); 19 (3~81) | 27.8 (-89.1~128.6); 529 (72~897); 238.6 (12.0~364.1)

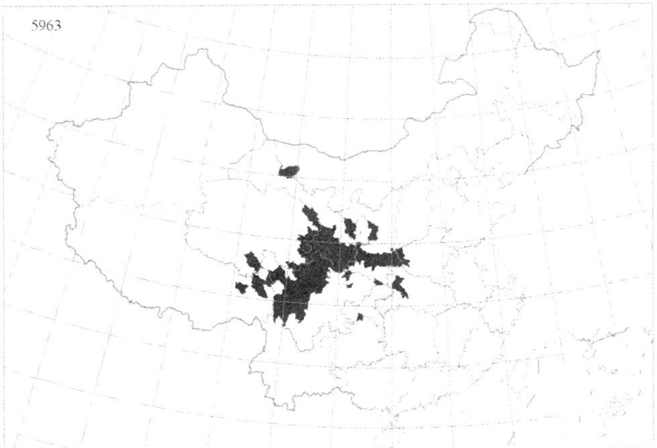

Rhamnaceae 鼠李科 *Rhamnus* 鼠李属

Rhamnus tzekweiensis Y. L. Chen et P. K. Chou 鄂西鼠李

Shrubs, deciduous

13.4; 2.1; 24.3; 13.5; 107.0; 5.3; 748 | 1161; 489; 83 | 55.4; 748; 249.2

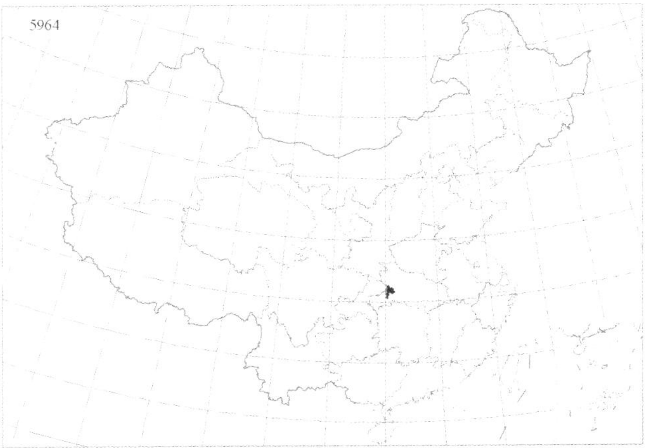

Rhamnaceae 鼠李科 *Rhamnus* 鼠李属

Rhamnus ussuriensis J. J. Vassiljev 乌苏里鼠李

Shrubs, deciduous

5.0 (-4.6~12.1); -14.0 (-28.8~-1.2); 21.5 (16.6~26.3); 8.6 (5.0~12.8); 68.0 (34.4~108.4); 67.7 (14.2~149.1); 601 (448~795) ❙ 537 (132~1073); 345 (81~691); 17 (2~65) ❙ -10.4 (-80.1~74.1); 487 (132~724); 204.1 (13.6~346.5)

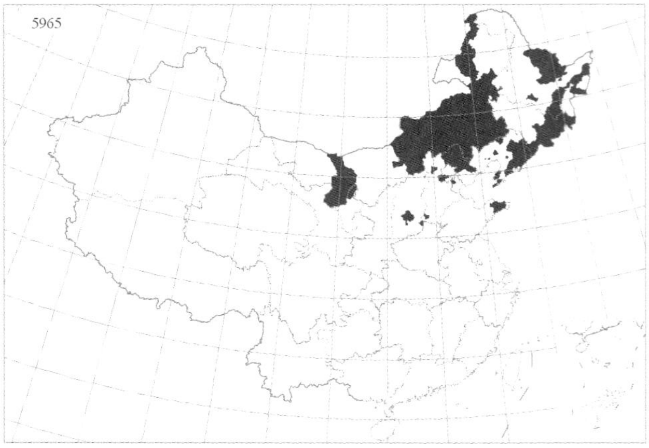

Rhamnaceae 鼠李科 *Rhamnus* 鼠李属

Rhamnus utilis J. Decaisne 冻绿

Shrubs or small trees, deciduous

13.2 (-5.0~22.7); 0.6 (-29.4~15.2); 24.6 (11.2~30.1); 13.9 (4.3~22.8); 115.2 (20.0~213.1); 15.9 (0.0~154.8); 784 (427~1221) ❙ 1023 (290~2435); 473 (190~1428); 76 (5~259) ❙ 27.4 (-49.3~147.0); 738 (290~1221); 212.4 (61.2~551.6)

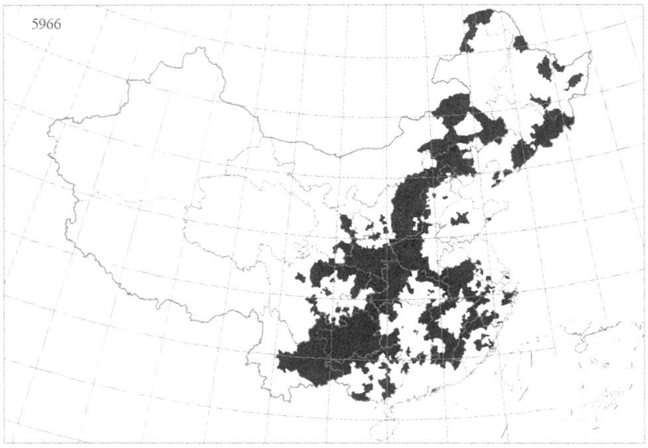

Rhamnaceae 鼠李科 *Rhamnus* 鼠李属

Rhamnus utilis var. hypochrysa (C. K. Schneid.) Rehder 毛冻绿

Shrubs or small trees, deciduous

15.0 (7.2~19.1); 4.0 (-7.6~10.1); 24.7 (20.1~28.3); 15.1 (8.7~19.1); 125.1 (66.3~169.2); 5.0 (0.0~39.8); 809 (588~969) ❙ 1132 (565~1580); 544 (306~816); 67 (17~181) ❙ 39.8 (-20.0~101.4); 798 (565~969); 209.7 (133.8~318.0)

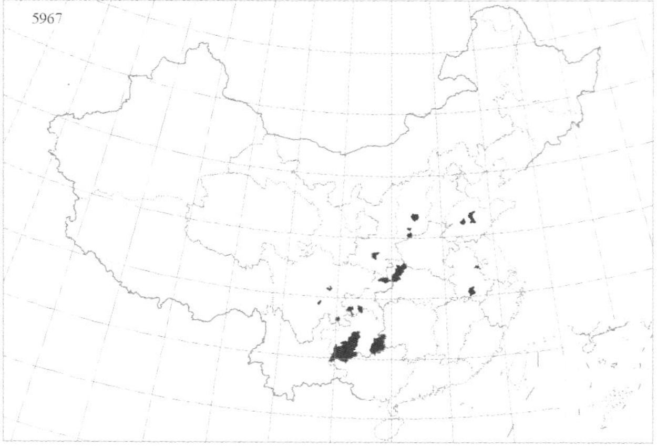

Rhamnaceae 鼠李科 *Rhamnus* 鼠李属

Rhamnus utilis var. szechuanensis Y. L. Chen et P. K. Chou 高山冻绿

Shrubs or small trees, deciduous

5.3 (2.6~9.1); -5.7 (-7.7~-0.7); 15.4 (11.2~19.9); 6.6 (4.3~9.2); 42.1 (20.0~64.1); 37.5 (14.1~49.3); 505 (427~595) ❙ 705 (341~1023); 361 (190~535); 17 (8~28) ❙ 43.9 (-40.9~121.7); 472 (341~595); 222.8 (61.2~357.7)

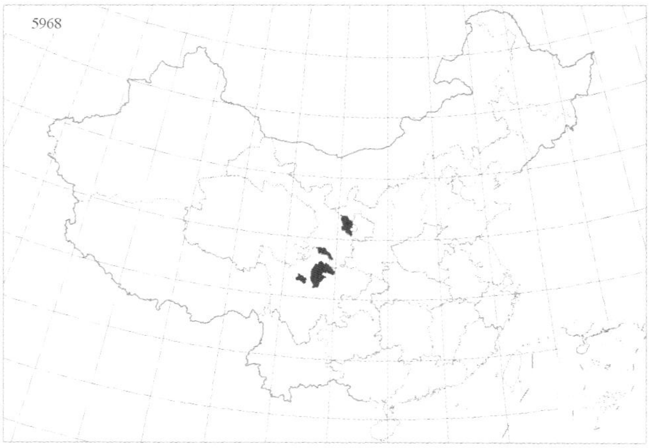

Rhamnaceae 鼠李科 *Rhamnus* 鼠李属

Rhamnus virgata Roxb. 帚枝鼠李

Shrubs or trees, deciduous

14.2 (-1.1~21.2); 6.5 (-9.9~15.2); 20.4 (6.2~28.7); 14.4 (2.0~22.7); 116.0 (2.5~212.7); 4.6 (0.0~75.6); 751 (334~1216) ❙ 1045 (321~1965); 584 (193~994); 43 (7~177) ❙ 40.0 (-14.2~121.7); 734 (321~1216); 272.0 (79.3~491.5)

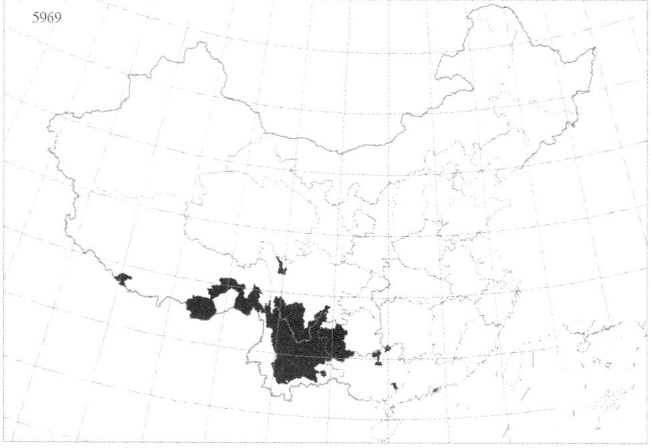

Rhamnaceae 鼠李科 *Rhamnus* 鼠李属

Rhamnus virgata var. hirsuta (Wight et Arn.) Y. L. Chen et P. K. Chou 糙毛帚枝鼠李

Shrubs or trees, deciduous

8.6; 0.4; 15.5; 9.3; 66.3; 23.5; 593 ❙ 895; 537; 19 ❙ 60.6; 563; 259.1

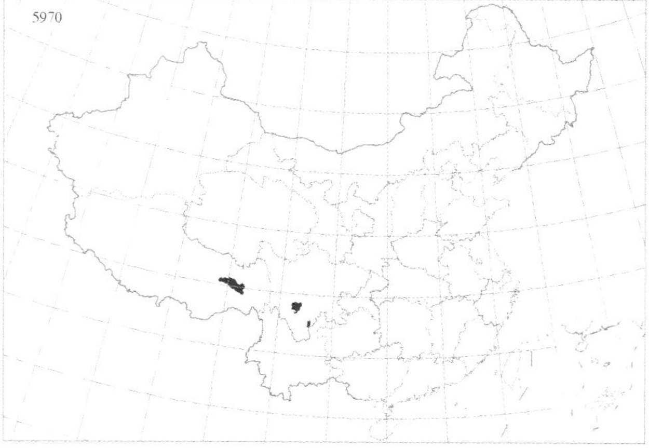

Rhamnaceae 鼠李科 *Rhamnus* 鼠李属

Rhamnus wilsonii C. K. Schneid. 山鼠李

Shrubs, deciduous

16.8 (12.7~21.3); 5.7 (1.5~12.6); 27.0 (22.7~29.6); 16.8 (12.8~21.3) 142.4 (99.0~193.0); 0.7 (0.0~7.3); 892 (713~1123) | 1544 (1117~1997) 563 (403~848); 177 (48~259) | 73.6 (5.8~130.7); 888 (713~1121); 234.6 (147.8~359.7)

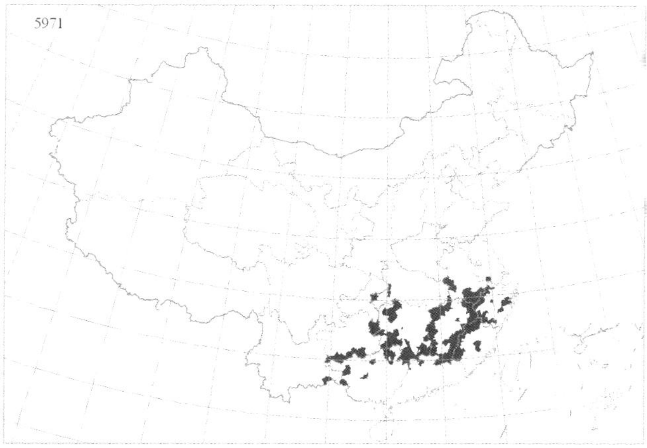

Rhamnaceae 鼠李科 *Rhamnus* 鼠李属

Rhamnus wilsonii var. pilosa Rehder 毛山鼠李

Shrubs, deciduous

15.8 (13.4~18.2); 3.7 (1.5~6.5); 27.4 (25.2~29.3); 15.8 (13.4~18.2); 132.0 (108.3~158.8); 2.2 (0.0~7.3); 866 (755~984) | 1624 (1364~1922); 594 (529~685); 198 (129~259) | 88.0 (55.4~112.2); 863 (755~941); 245.4 (183.4~319.9)

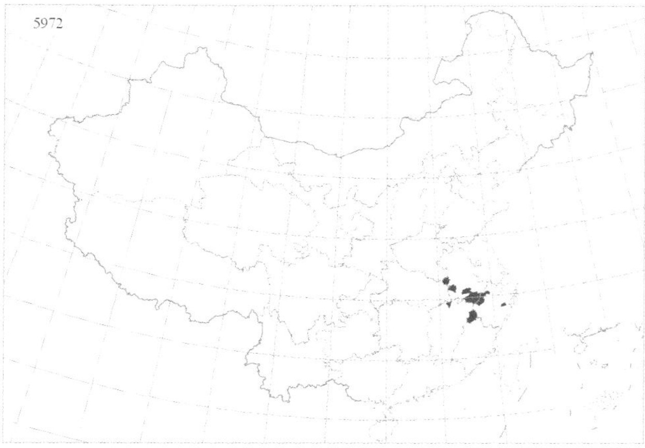

Rhamnaceae 鼠李科 *Rhamnus* 鼠李属

Rhamnus wumingensis Y. L. Chen et P. K. Chou 武鸣鼠李

Shrubs, deciduous

21.1; 11.8; 28.5; 21.1; 193.7; 0.0; 1127 | 1591; 742; 149 | 42.5; 1127; 256.4

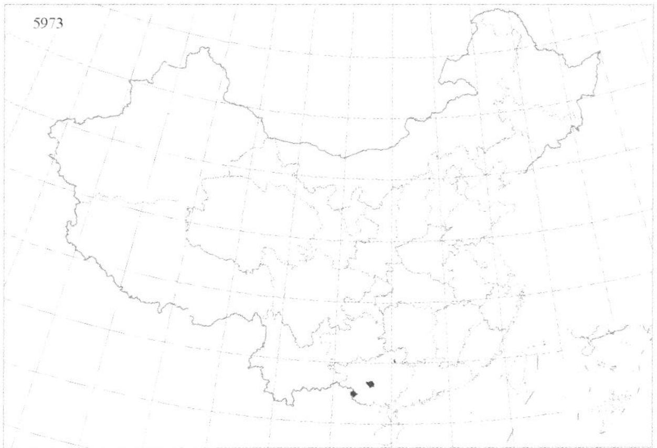

Rhamnaceae 鼠李科 *Rhamnus* 鼠李属

Rhamnus xizangensis Y. L. Chen et P. K. Chou 西藏鼠李

Shrubs or small trees, evergreen

11.4 (1.8~17.3); 4.1 (-7.1~11.0); 17.2 (9.9~21.6); 11.7 (3.8~17.3); 89.6 (15.7~148.0); 12.6 (0.0~53.3); 657 (404~832) | 1079 (583~1736); 617 (348~1007); 40 (13~51) | 67.2 (26.1~145.8); 645 (404~792); 338.3 (169.5~447.4)

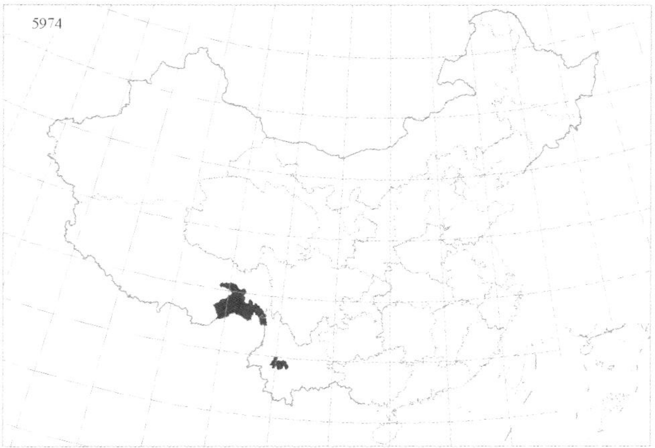

Rhamnaceae 鼠李科 *Sageretia* 雀梅藤属

Sageretia camelliifolia Y. L. Chen et P. K. Chou 茶叶雀梅藤

Shrubs

20.3 (17.5~22.2); 12.0 (9.5~13.9); 26.3 (22.3~28.6); 20.3 (17.5~22.3); 183.8 (149.9~207.5); 0.0 (0.0~0.0); 1051 (846~1190) | 1319 (958~1518); 701 (531~773); 86 (44~138) | 25.6 (13.3~43.5); 1047 (823~1190); 312.6 (272.3~335.1)

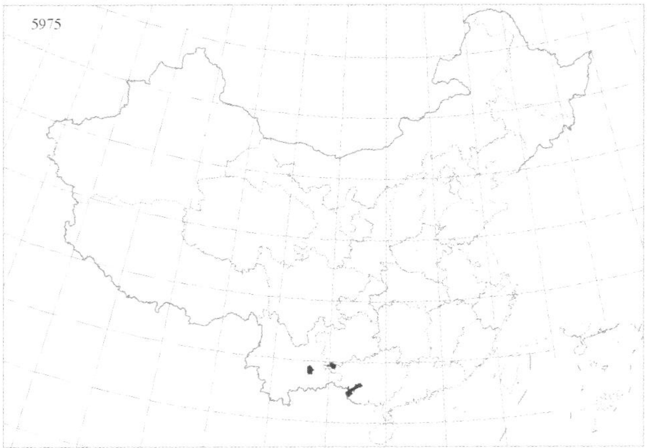

Rhamnaceae 鼠李科 *Sageretia* 雀梅藤属

Sageretia gracilis J. R. Drummond et Sprague 纤细雀梅藤

Shrubs, evergreen

14.6 (1.8~20.0); 6.6 (-7.1~12.0); 21.1 (9.9~28.4); 14.7 (3.8~20.0); 117.6 (15.7~180.0); 2.0 (0.0~53.3); 759 (404~1058) | 1034 (583~1590); 564 (348~820); 45 (12~166) | 37.1 (-14.2~100.7); 742 (404~1058); 259.4 (142.9~406.7)

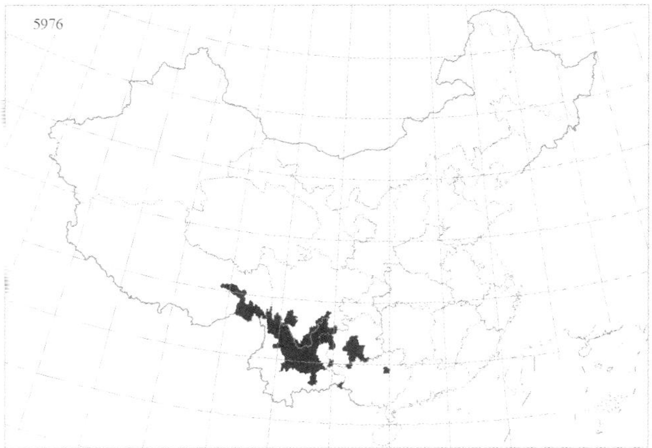

Rhamnaceae 鼠李科 *Sageretia* 雀梅藤属
Sageretia hamosa (Wallich) Brongniart 钩枝雀梅藤
Shrubs, evergreen

17.6 (5.3~24.7); 7.6 (-2.4~19.8); 26.7 (12.2~29.8); 17.7 (5.7~24.7); 152.7 (30.5~236.5); 0.5 (0.0~26.2); 929 (479~1376) ❙ 1553 (961~3495); 596 (427~924); 160 (30~897) ❙ 68.0 (15.4~224.4); 925 (479~1212); 248.2 (149.6~659.0)

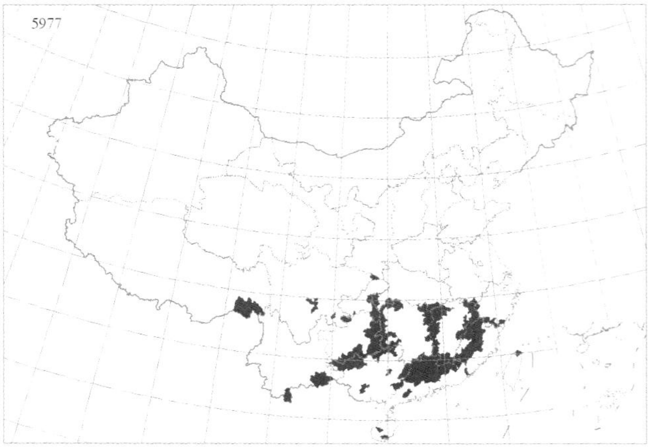

Rhamnaceae 鼠李科 *Sageretia* 雀梅藤属
Sageretia horrida Pax et K. Hoffm. 凹叶雀梅藤
Shrubs, evergreen

3.4 (0.0~6.1); -5.1 (-8.7~-1.5); 11.1 (8.1~14.4); 4.7 (2.7~6.6); 23.5 (8.0~39.7); 41.8 (21.8~67.9); 439 (363~508) ❙ 676 (510~1023); 396 (311~558); 18 (10~31) ❙ 52.6 (20.7~101.2); 439 (363~508); 215.8 (111.0~282.9)

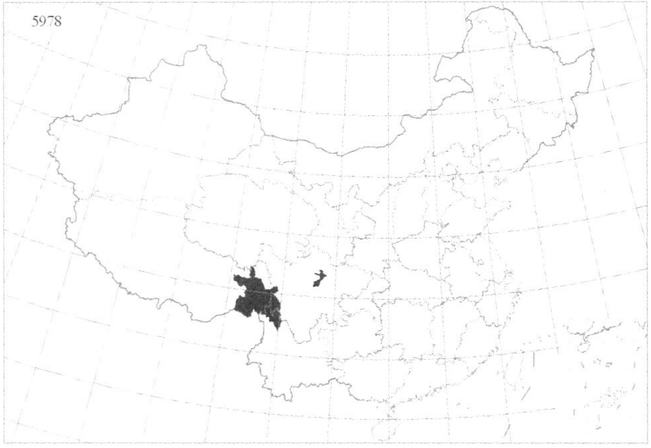

Rhamnaceae 鼠李科 *Sageretia* 雀梅藤属
Sageretia laxiflora Hand.-Mazz. 疏花雀梅藤
Shrubs, evergreen

17.6 (15.4~21.3); 7.6 (4.1~12.6); 26.5 (23.6~29.7); 17.7 (15.5~21.3); 152.4 (126.5~196.0); 0.1 (0.0~0.9); 918 (820~1123) ❙ 1291 (1112~1619); 583 (403~820); 95 (43~228) ❙ 41.4 (5.8~77.0); 911 (820~1121); 249.1 (175.3~371.7)

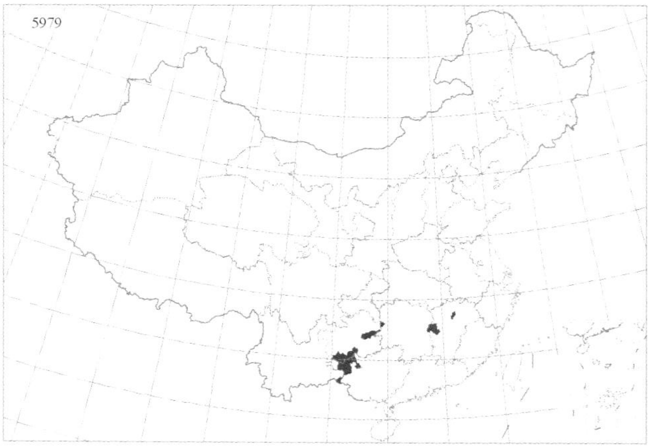

Rhamnaceae 鼠李科 *Sageretia* 雀梅藤属
Sageretia lijiangensis G. S. Fan et S. K. Chen 丽江雀梅藤
Shrubs, evergreen

12.2 (6.1~15.3); 5.3 (-1.5~8.7); 17.9 (12.7~20.5); 12.3 (6.3~15.4); 91.7 (35.2~124.3); 4.6 (0.0~21.8); 671 (496~761) ❙ 940 (801~991); 563 (496~594); 29 (21~38) ❙ 41.8 (28.3~61.4); 649 (496~717); 303.8 (282.9~319.8)

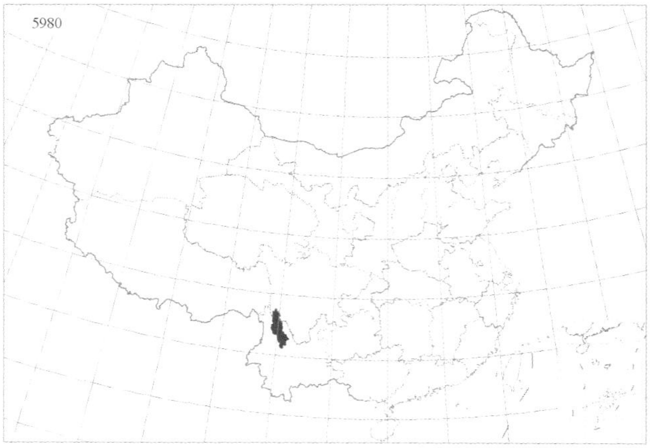

Rhamnaceae 鼠李科 *Sageretia* 雀梅藤属
Sageretia lucida Merr. 亮叶雀梅藤
Shrubs, evergreen

17.4 (5.3~24.7); 7.7 (-3.5~19.8); 26.0 (12.2~29.8); 17.5 (5.7~24.7); 151.2 (30.5~236.5); 1.4 (0.0~28.1); 928 (479~1376) ❙ 1414 (711~3495); 604 (312~1056); 115 (15~897) ❙ 53.7 (-34.3~224.4); 922 (479~1326); 231.2 (95.3~659.0)

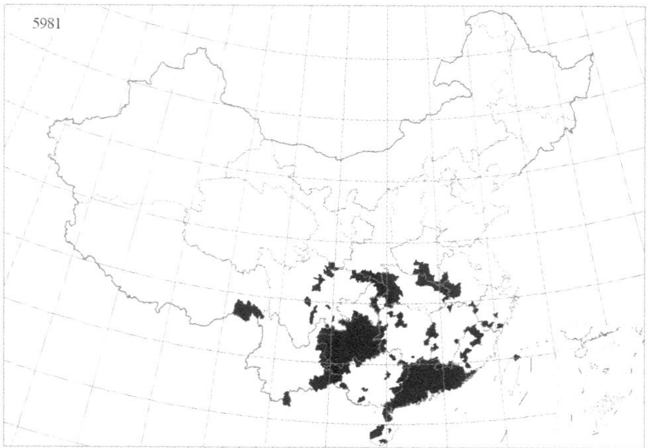

Rhamnaceae 鼠李科 *Sageretia* 雀梅藤属
Sageretia melliana Hand.-Mazz. 刺藤子
Shrubs, evergreen

16.9 (14.3~20.0); 5.7 (2.1~11.5); 27.3 (22.7~30.1); 16.9 (14.3~20.0); 143.5 (112.6~180.0); 0.6 (0.0~4.9); 900 (756~1040) ❙ 1557 (1133~1997); 559 (427~813); 183 (48~259) ❙ 73.4 (28.8~130.7); 894 (756~1040); 235.3 (148.9~359.7)

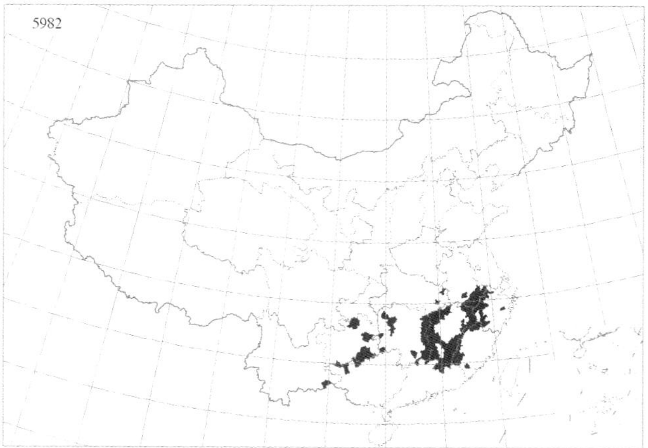

Rhamnaceae 鼠李科 *Sagaretia* 雀梅藤属
Sageretia omeiensis C. K. Schneid. 峨眉雀梅藤
Shrubs, evergreen

14.9 (14.2~15.6); 4.3 (3.6~5.3); 24.9 (23.6~26.5); 14.9 (14.2~15.7); 119.6 (112.5~129.2); 0.8 (0.0~1.7); 793 (765~840) ▎1233 (1133~1410); 534 (486~816); 73 (49~111) ▎55.6 (42.1~80.4); 793 (765~840); 192.2 (182.9~214.5)

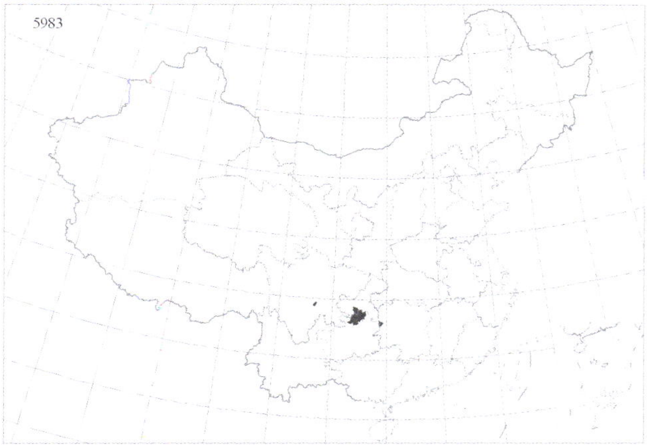

Rhamnaceae 鼠李科 *Sagaretia* 雀梅藤属
Sageretia paucicostata Maxim. 少脉雀梅藤
Shrubs or small trees, evergreen

10.0 (0.9~18.5); −3.8 (−12.7~10.5); 22.4 (9.5~29.5); 10.8 (3.6~18.3); 86.6 (14.0~160.0); 26.3 (0.0~65.0); 679 (403~932) ▎655 (367~1358); 348 (201~699); 23 (7~109) ▎−0.3 (−45.2~128.6); 580 (367~858); 187.9 (50.7~357.7)

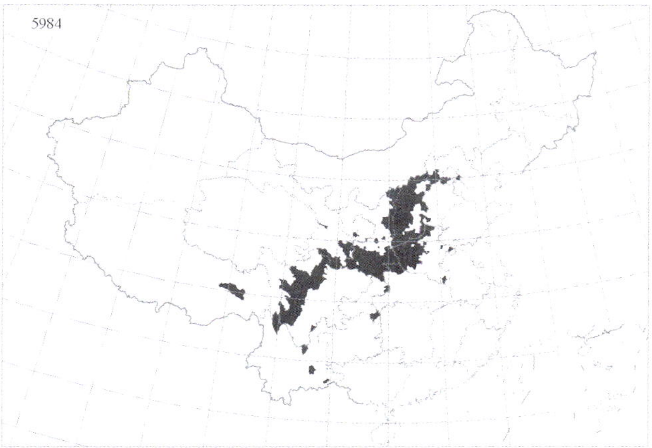

Rhamnaceae 鼠李科 *Sagaretia* 雀梅藤属
Sageretia pedicellata C. Z. Gao 南丹雀梅藤
Shrubs, evergreen

18.2; 8.7; 25.9; 18.3; 159.3; 0.0; 928 ▎1372; 670; 97 ▎47.9; 928; 280.9

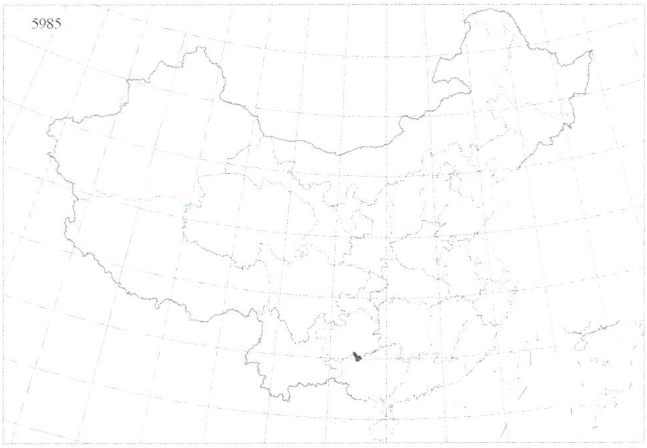

Rhamnaceae 鼠李科 *Sagaretia* 雀梅藤属
Sageretia randaiensis Hayata 峦大雀梅藤
Small shrubs, evergreen

18.7; 12.8; 23.7; 18.7; 164.6; 0.0; 924 ▎2473; 1047; 289 ▎176.4; 913; 258.7

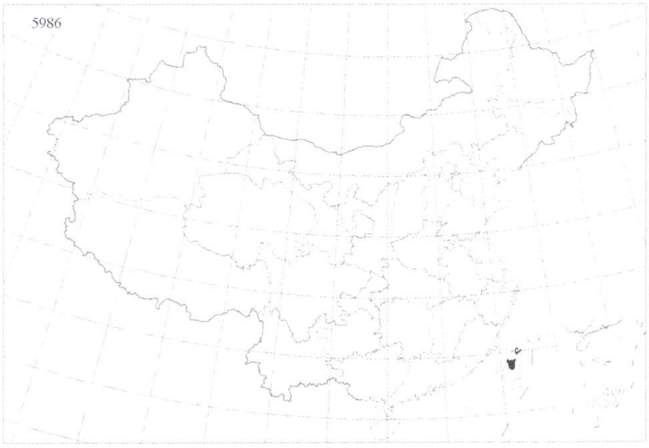

Rhamnaceae 鼠李科 *Sagaretia* 雀梅藤属
Sageretia rugosa Hance 皱叶雀梅藤
Shrubs, evergreen

16.6 (11.7~22.2); 6.5 (1.1~13.9); 25.5 (18.8~29.8); 16.6 (11.7~22.3); 140.1 (82.6~207.5); 0.5 (0.0~8.5); 870 (658~1190) ▎1334 (916~1852); 568 (403~860); 107 (31~230) ▎53.7 (5.8~107.2); 869 (658~1190); 222.9 (146.0~413.5)

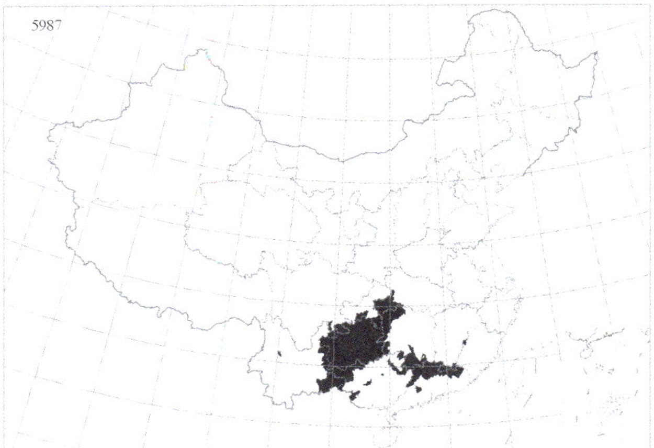

Rhamnaceae 鼠李科 *Sagaretia* 雀梅藤属
Sageretia subcaudata C. K. Schneid. 尾叶雀梅藤
Shrubs, evergreen

14.8 (6.0~20.2); 3.0 (−3.2~11.1); 25.9 (14.4~29.2); 14.9 (6.6~20.3); 124.4 (39.7~183.3); 6.0 (0.0~26.7); 827 (508~1056) ▎1169 (512~1852); 492 (258~816); 95 (17~221) ▎40.3 (−35.3~108.0); 805 (508~1056); 216.1 (121.6~491.5)

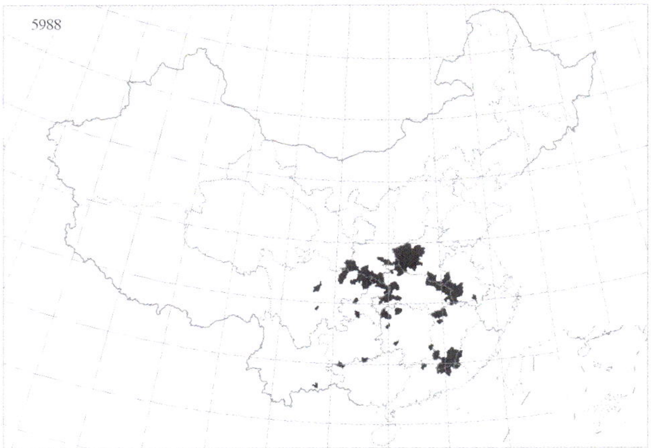

Rhamnaceae 鼠李科 *Sageretia* 雀梅藤属

Sageretia thea (Osbeck) M. C. Johnston 雀梅藤

Shrubs, evergreen

17.0 (2.2~24.8); 6.9 (-6.3~20.1); 26.2 (9.5~30.1); 17.1 (3.9~24.8); 146.5 (15.8~237.6); 1.7 (0.0~48.6); 909 (411~1385) ❙ 1366 (512~3495); 571 (258~1847); 135 (11~897) ❙ 50.5 (-35.3~246.5); 891 (411~1360); 241.0 (110.1~659.0)

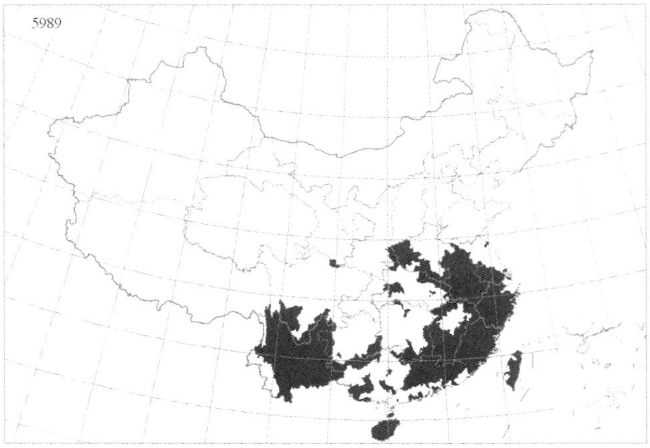

Rhamnaceae 鼠李科 *Sageretia* 雀梅藤属

Sageretia thea var. cordiformis Y. L. Chen et P. K. Chou 心叶雀梅藤

Shrubs, evergreen

19.2; 13.5; 22.8; 19.3; 171.3; 0.0; 926 ❙ 1437; 672; 63 ❙ 55.0; 910; 531.6

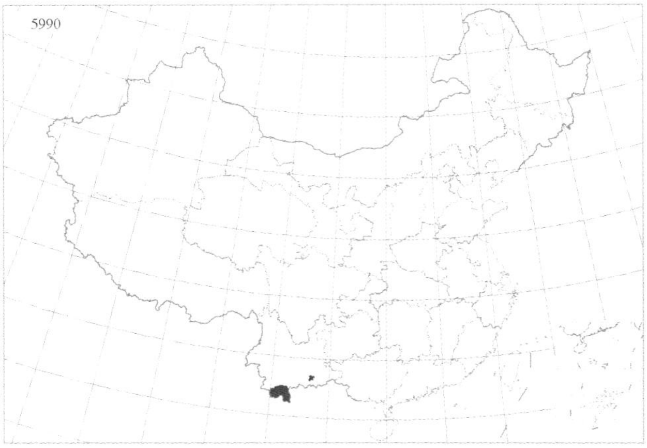

Rhamnaceae 鼠李科 *Sageretia* 雀梅藤属

Sageretia thea var. tomentosa (C. K. Schneid.) Y. L. Chen et P. K. Chou 毛叶雀梅藤

Shrubs, evergreen

18.0 (6.1~21.4); 8.6 (-1.5~13.6); 26.7 (12.7~29.0); 18.1 (6.3~21.5); 157.3 (35.2~197.7); 0.4 (0.0~21.8); 941 (496~1128) ❙ 1592 (801~1997); 567 (356~756); 189 (15~259) ❙ 71.5 (-5.0~147.0); 932 (496~1107); 258.8 (123.9~491.5)

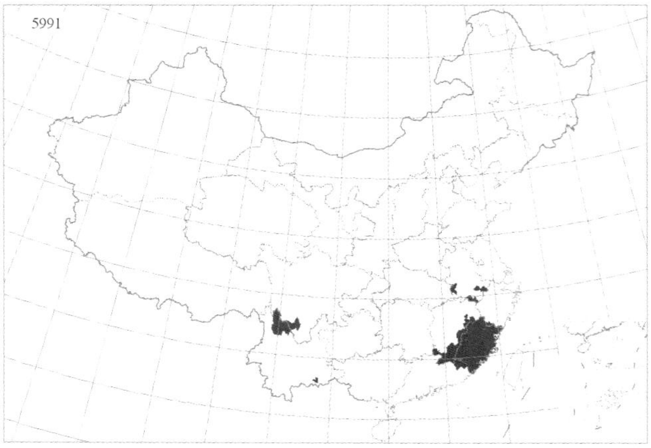

Rhamnaceae 鼠李科 *Sageretia* 雀梅藤属

Sageretia yilinii G. S. Fan et S. K. Chen 脱毛雀梅藤

Shrubs, evergreen

13.8; 7.2; 19.1; 13.8; 105.7; 0.0; 712 ❙ 1013; 591; 37 ❙ 42.6; 695; 332.2

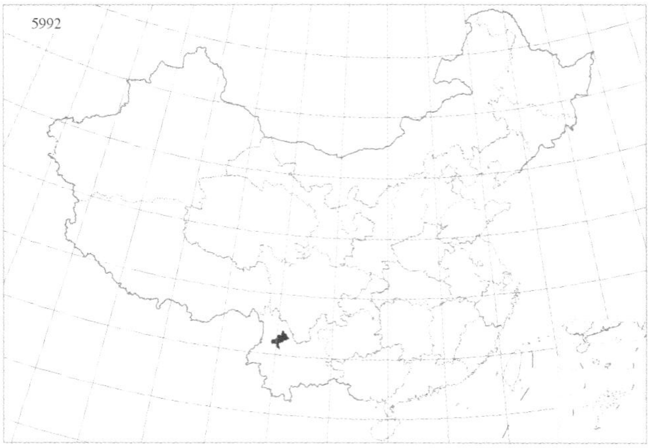

Rhamnaceae 鼠李科 *Scutia* 对刺藤属

Scutia myrtina (N. L. Burman) Kurz 对刺藤

Shrubs, evergreen

19.0 (16.4~22.2); 12.5 (9.6~15.1); 23.0 (20.6~28.3); 19.0 (16.4~22.3); 168.6 (137.3~207.5); 0.0 (0.0~0.0); 929 (795~1190) ❙ 1543 (1374~1688); 829 (655~959); 60 (42~87) ❙ 69.0 (15.4~112.2); 918 (795~1190); 524.3 (327.7~659.0)

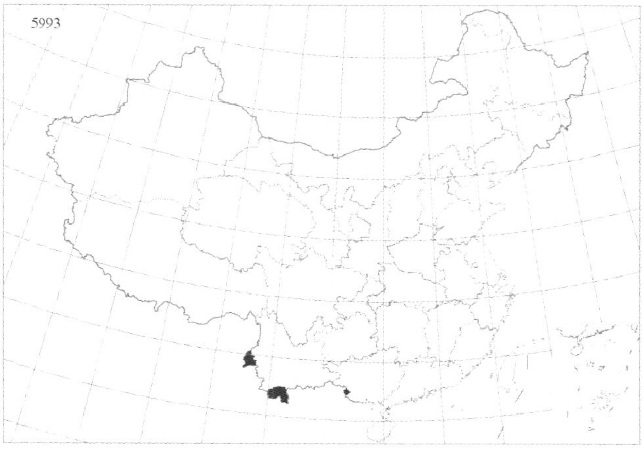

Rhamnaceae 鼠李科 *Ventilago* 翼核果属

Ventilago calyculata Tulasne 毛果翼核果

Scandent shrubs, evergreen

17.6 (12.5~22.2); 10.3 (3.6~15.1); 22.8 (18.4~28.3); 17.6 (12.5~22.3); 151.4 (92.0~207.5); 0.0 (0.0~1.4); 865 (685~1190) ❙ 1291 (939~1751); 702 (524~996); 55 (37~87) ❙ 50.1 (6.5~99.9); 854 (685~1190); 366.6 (186.7~659.0)

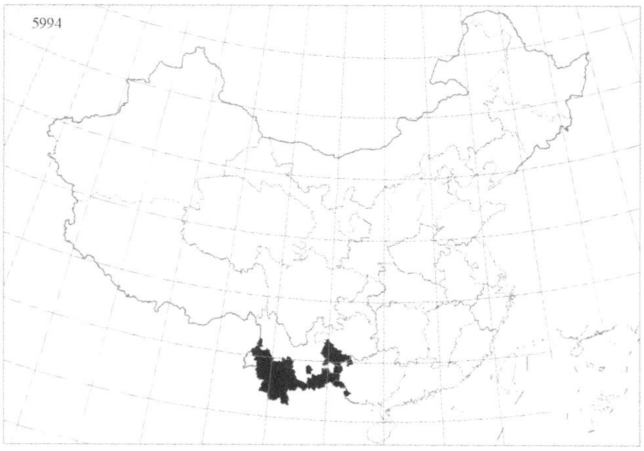

Rhamnaceae 鼠李科 *Ventilago* 翼核果属
Ventilago calyculata var. trichoclada Y. L. Chen et P. K. Chou 毛枝翼核果
Scandent shrubs, evergreen

19.5; 10.5; 26.5; 19.5; 174.0; 0.0; 996 | 1233; 662; 58 | 23.9; 996; 338.3

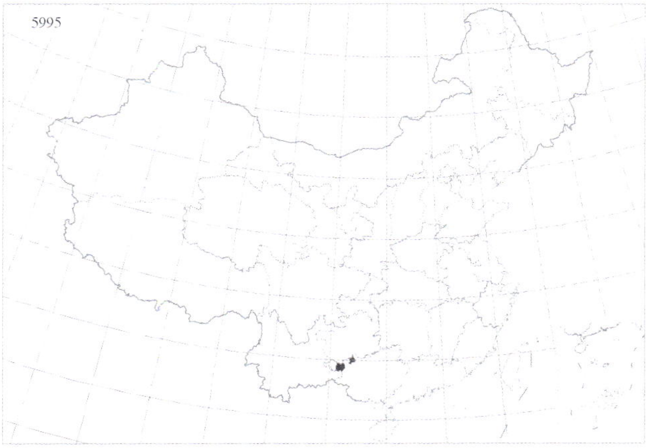

Rhamnaceae 鼠李科 *Ventilago* 翼核果属
Ventilago elegans Hemsl. 台湾翼核果
Scandent shrubs

20.1 (16.4~24.1); 15.0 (11.5~18.9); 24.0 (20.3~28.0); 20.1 (16.5~24.1); 181.2 (137.8~229.7); 0.0 (0.0~0.0); 998 (783~1322) | 2618 (1751~2993); 1312 (924~1847); 220 (47~547) | 173.1 (32.5~219.3); 960 (783~1123); 272.0 (154.2~363.7)

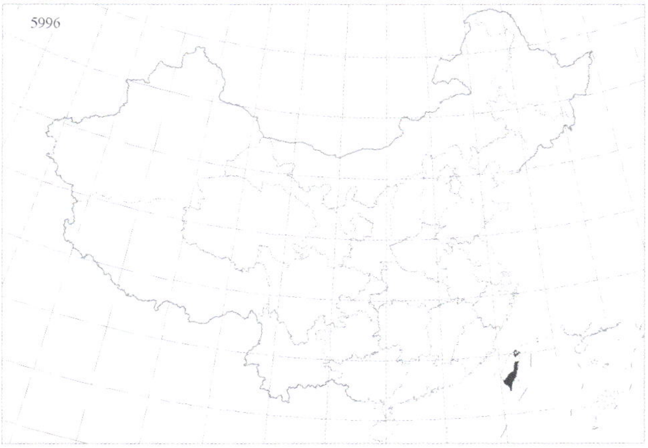

Rhamnaceae 鼠李科 *Ventilago* 翼核果属
Ventilago inaequilateralis Merr. et Chun 海南翼核果
Scandent shrubs

19.0 (14.2~24.8); 11.6 (5.6~20.1); 24.3 (20.8~28.4); 19.0 (14.2~24.8); 167.9 (110.9~237.6); 0.0 (0.0~0.0); 955 (734~1385) | 1339 (1007~1751); 677 (392~996); 60 (41~165) | 42.6 (-11.9~88.8); 929 (734~1190); 376.9 (189.5~659.0)

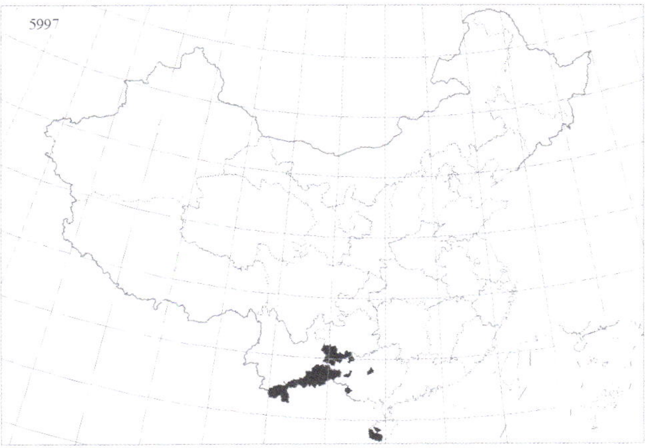

Rhamnaceae 鼠李科 *Ventilago* 翼核果属
Ventilago leiocarpa Benth. 翼核果
Scandent shrubs

20.8 (16.2~24.8); 12.9 (5.8~20.1); 27.2 (20.1~29.3); 20.8 (16.2~24.8); 189.9 (134.2~237.6); 0.0 (0.0~0.0); 1087 (773~1385) | 1673 (868~3495); 736 (312~1847); 137 (41~897) | 56.9 (-34.3~246.5); 1071 (773~1360); 266.2 (95.3~659.0)

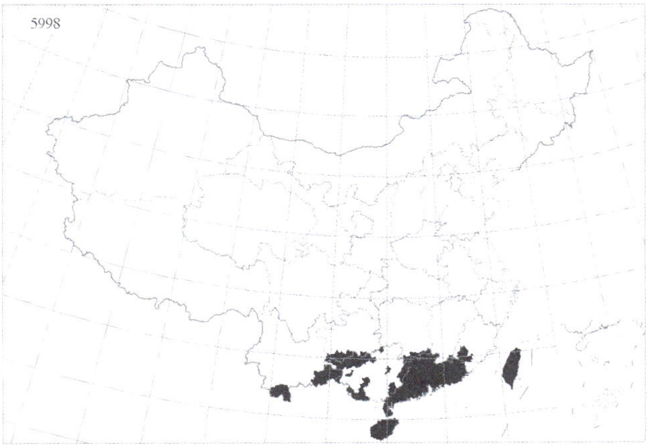

Rhamnaceae 鼠李科 *Ventilago* 翼核果属
Ventilago leiocarpa var. pubescens Y. L. Chen et P. K. Chou 毛叶翼核果
Scandent shrubs

18.7 (17.9~19.8); 9.5 (7.1~11.1); 26.1 (25.0~27.2); 18.8 (17.9~19.8); 165.5 (154.8~177.9); 0.0 (0.0~0.0); 954 (913~1010) | 1325 (1182~1551); 675 (588~820); 80 (43~169) | 39.1 (16.9~69.8); 954 (913~1010); 279.4 (189.5~371.7)

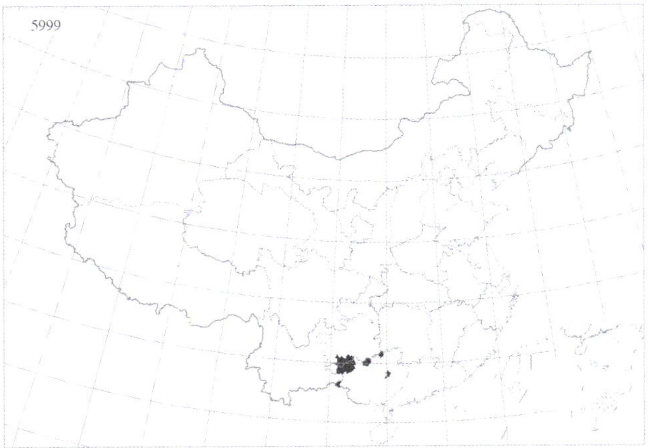

Rhamnaceae 鼠李科 *Ventilago* 翼核果属
Ventilago oblongifolia Blume 矩叶翼核果
Scandent shrubs

19.8; 13.2; 24.2; 19.8; 177.7; 0.0; 985 | 1365; 701; 72 | 37.9; 958; 405.7

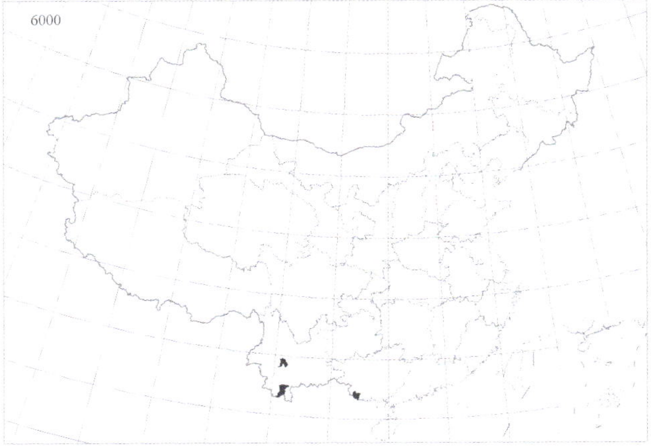

Rhamnaceae 鼠李科 *Ziziphus* 枣属
Ziziphus attopensis Pierre 毛果枣
Scandent shrubs

18.8 (16.3~22.2); 12.0 (9.1~15.1); 23.3 (20.8~28.5); 18.8 (16.4~22.3); 165.7 (136.4~207.5); 0.0 (0.0~0.0); 917 (797~1190) ❙ 1406 (1074~1751); 758 (582~996); 59 (41~128) ❙ 54.4 (14.8~88.8); 909 (797~1190); 435.2 (256.6~659.0)

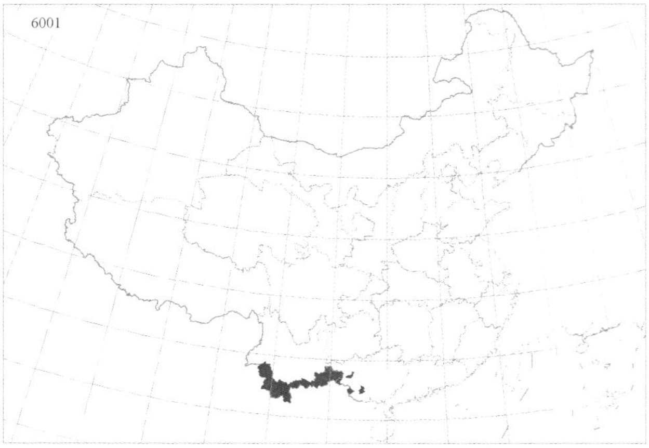

Rhamnaceae 鼠李科 *Ziziphus* 枣属
Ziziphus fungii Merr. 褐果枣
Shrubs

19.5 (16.3~24.8); 13.4 (9.1~20.1); 23.6 (21.4~28.6); 19.6 (16.4~24.8); 174.6 (136.4~237.6); 0.0 (0.0~0.0); 966 (797~1385) ❙ 1360 (1028~1751); 700 (444~996); 57 (41~88) ❙ 43.6 (-2.2~88.8); 928 (783~1254); 446.3 (339.5~659.0)

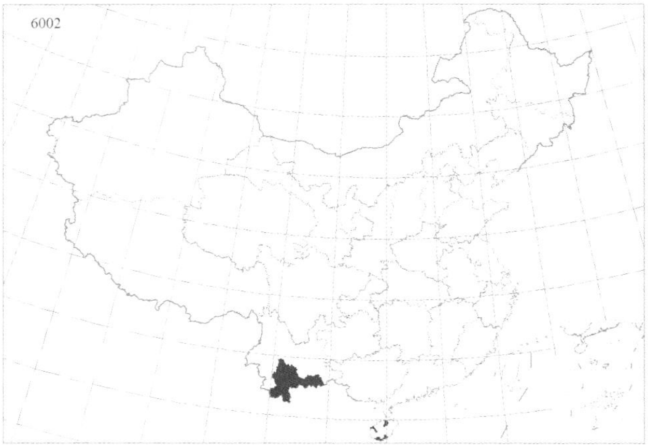

Rhamnaceae 鼠李科 *Ziziphus* 枣属
Ziziphus incurva Roxb. 印度枣
Shrubs or trees

16.1 (-1.1~22.2); 8.7 (-9.1~15.1); 21.6 (6.2~28.5); 16.1 (2.0~22.3); 134.7 (2.5~207.5); 1.5 (0.0~75.6); 812 (334~1190) ❙ 1159 (321~1758); 640 (193~996); 52 (11~189) ❙ 42.9 (-14.2~112.2); 797 (321~1190); 328.3 (79.3~659.0)

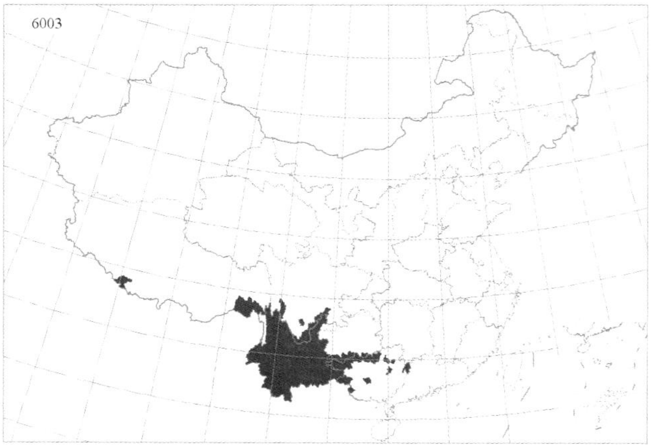

Rhamnaceae 鼠李科 *Ziziphus* 枣属
Ziziphus jujuba Mill. 枣
Trees, deciduous

13.9 (-4.3~24.8); 1.4 (-20.9~20.1); 24.9 (9.6~29.9); 14.5 (2.5~24.8); 121.7 (9.7~237.6); 14.9 (0.0~121.1); 812 (329~1385) ❙ 992 (21~1893); 441 (15~996); 81 (0~242) ❙ 18.4 (-97.3~129.9); 713 (21~1360); 196.2 (1.6~620.0)

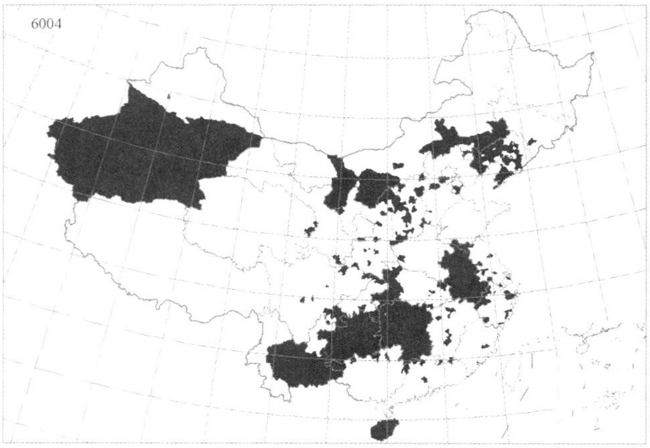

Rhamnaceae 鼠李科 *Ziziphus* 枣属
Ziziphus jujuba var. inermis (Bunge) Rehd. 无刺枣
Trees, deciduous

13.9 (1.8~19.1); 1.1 (-15.9~10.7); 25.4 (9.9~29.9); 14.3 (3.8~19.1); 119.0 (15.4~169.2); 12.1 (0.0~80.4); 811 (413~1002) ❙ 980 (349~1742); 470 (227~946); 66 (7~228) ❙ 20.5 (-51.1~128.9); 720 (349~1002); 183.9 (76.1~331.6)

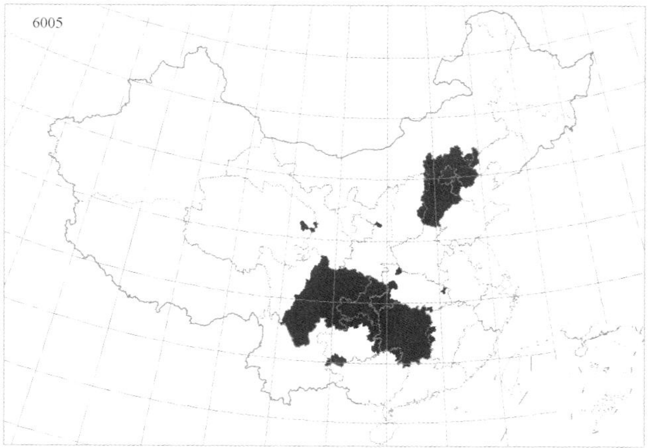

Rhamnaceae 鼠李科 *Ziziphus* 枣属
Ziziphus jujuba var. spinosa (Bunge) Hu ex H. F. Chow 酸枣
Shrubs, deciduous

11.6 (0.4~15.9); -3.6 (-15.9~2.9); 25.2 (10.5~28.6); 12.3 (3.6~16.0); 103.2 (15.5~134.9); 23.4 (2.4~80.4); 758 (406~901) ❙ 618 (349~1200); 368 (227~560); 27 (7~120) ❙ -17.9 (-51.1~52.6); 603 (349~901); 169.9 (58.7~334.4)

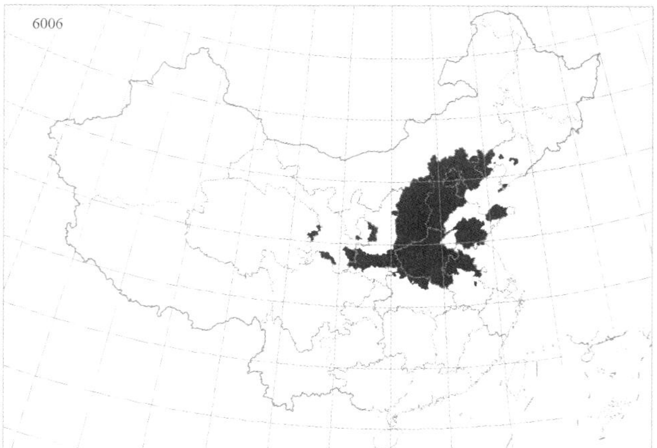

Rhamnaceae 鼠李科 *Ziziphus* 枣属

Ziziphus laui Merr. 球枣

Shrubs or trees

24.5; 19.4; 28.2; 24.6; 234.8; 0.0; 1361 ❙ 1292; 432; 41 ❙ -5.2; 1097; 491.4

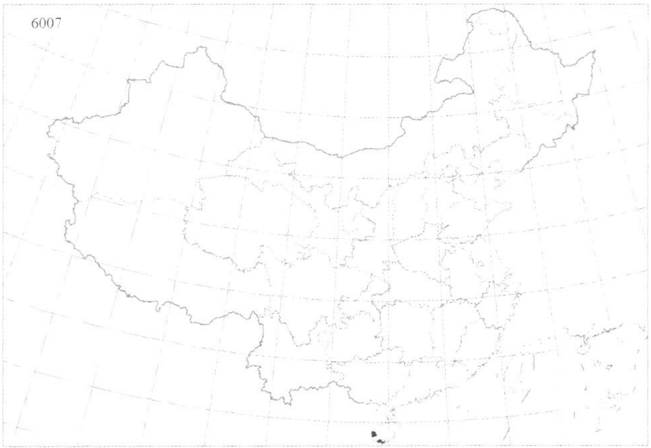

Rhamnaceae 鼠李科 *Ziziphus* 枣属

Ziziphus mairei Dode 大果枣

Trees

12.5; 5.6; 18.0; 12.7; 99.5; 8.5; 688 ❙ 919; 521; 37 ❙ 36.7; 683; 297.5

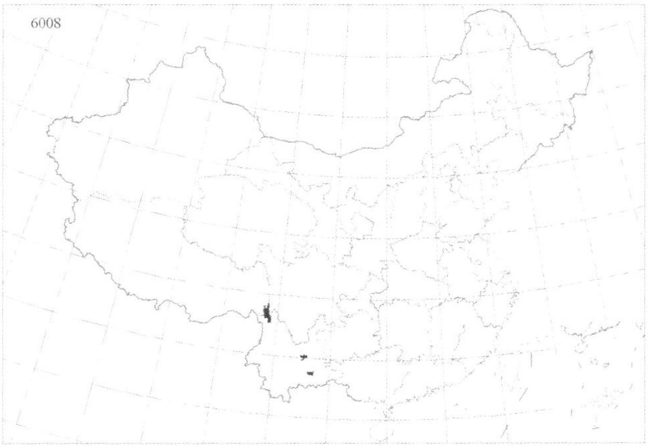

Rhamnaceae 鼠李科 *Ziziphus* 枣属

Ziziphus mauritiana Lam. 滇刺枣

Shrubs or trees, evergreen

17.2 (8.6~23.6); 10.2 (1.3~17.8); 22.2 (15.0~29.0); 17.2 (8.7~23.6); 146.9 (52.6~223.5); 0.1 (0.0~8.8); 858 (574~1281) ❙ 1232 (797~2435); 678 (411~1428); 55 (20~173) ❙ 43.8 (-14.2~112.2); 839 (574~1269); 339.1 (157.7~659.0)

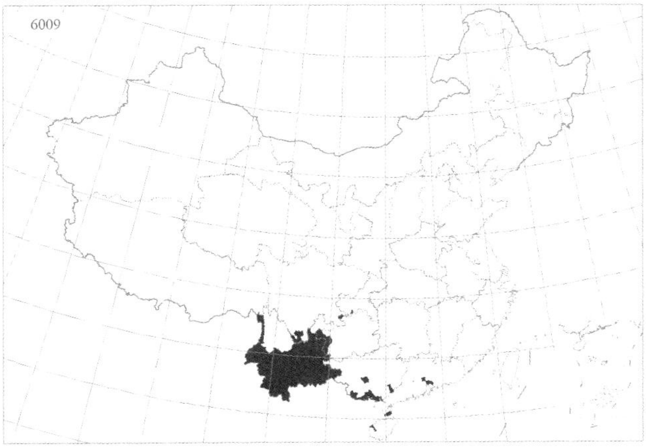

Rhamnaceae 鼠李科 *Ziziphus* 枣属

Ziziphus montana W. W. Sm. 山枣

Shrubs or trees

13.1 (2.2~18.9); 6.0 (-6.6~12.0); 18.6 (9.5~23.9); 13.2 (3.8~19.0); 102.4 (15.5~167.8); 5.3 (0.0~49.1); 702 (411~929) ❙ 940 (628~1057); 557 (410~627); 30 (11~46) ❙ 37.7 (-14.2~125.9); 670 (411~799); 280.0 (205.2~372.3)

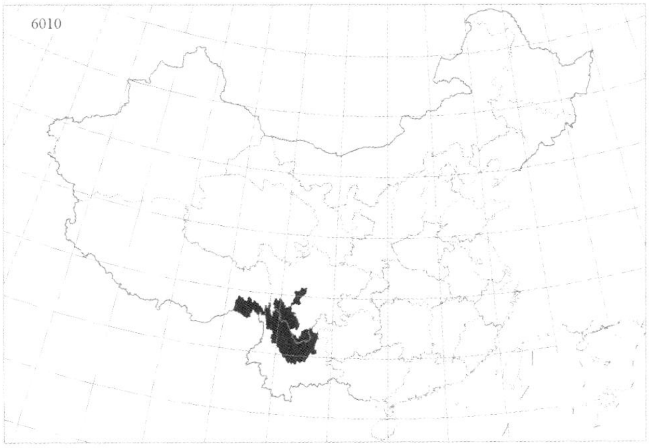

Rhamnaceae 鼠李科 *Ziziphus* 枣属

Ziziphus oenopolia (L.) Mill. 小果枣

Shrubs

20.4 (18.3~22.3); 12.9 (9.4~15.1); 25.8 (22.3~28.9); 20.4 (18.4~22.3); 185.4 (160.9~208.1); 0.0 (0.0~0.0); 1036 (909~1202) ❙ 1434 (1188~1607); 730 (634~850); 77 (56~136) ❙ 40.1 (5.8~64.9); 1029 (880~1202); 387.3 (201.4~659.0)

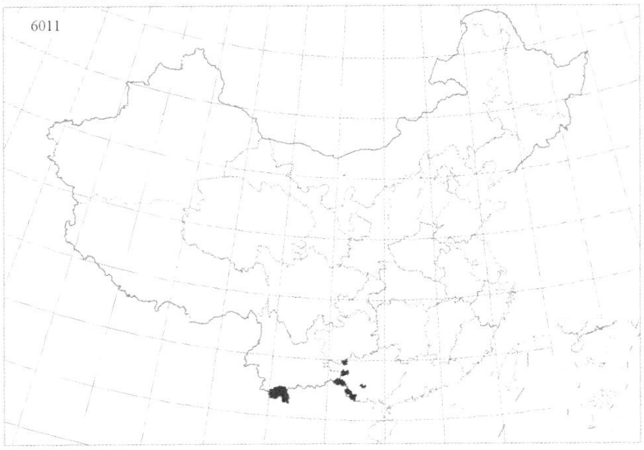

Rhamnaceae 鼠李科 *Ziziphus* 枣属

Ziziphus rugosa Lam. 皱枣

Shrubs or small trees, evergreen

19.6 (16.3~24.8); 13.5 (10.2~20.1); 23.8 (20.3~28.7); 19.7 (16.4~24.8); 176.1 (136.6~237.6); 0.0 (0.0~0.0); 982 (789~1385) ❙ 1297 (868~1767); 663 (312~996); 54 (37~88) ❙ 35.5 (-34.3~88.8); 923 (744~1299); 428.6 (278.6~659.0)

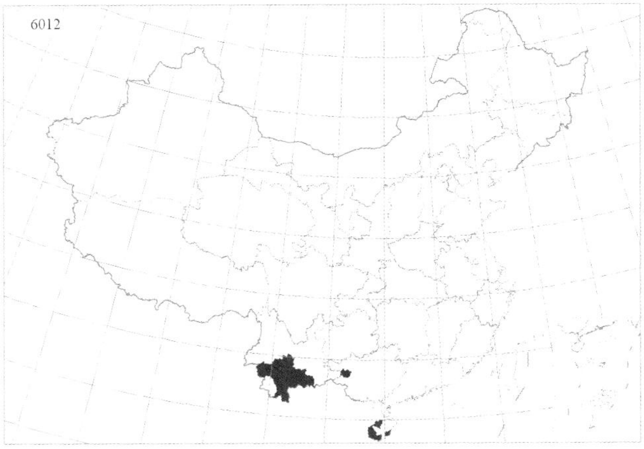

Rhamnaceae 鼠李科 *Ziziphus* 枣属
Ziziphus xiangchengensis Y. L. Chen et P. K. Chou 蜀枣
Shrubs or trees

7.0; -1.0; 13.8; 7.7; 47.4; 22.9; 531 ❙ 809; 514; 13 ❙ 51.7; 530; 266.1

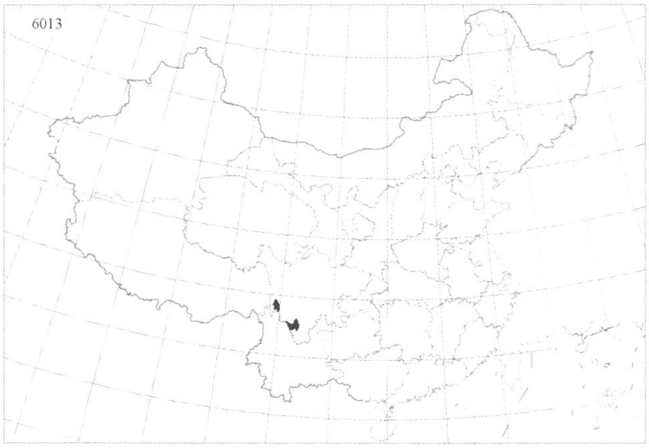

Leeaceae 火筒树科 *Leea* 火筒树属
Leea aequata L. 圆腺火筒树
Shrubs or small trees

19.4 (18.2~21.0); 13.6 (12.4~15.1); 23.1 (21.6~25.6); 19.5 (18.2~21.0); 173.6 (158.6~191.8); 0.0 (0.0~0.0); 934 (861~1053) ❙ 1499 (1276~1751); 775 (655~996); 60 (41~77) ❙ 60.8 (36.8~88.8); 917 (839~1053); 515.8 (427.9~659.0)

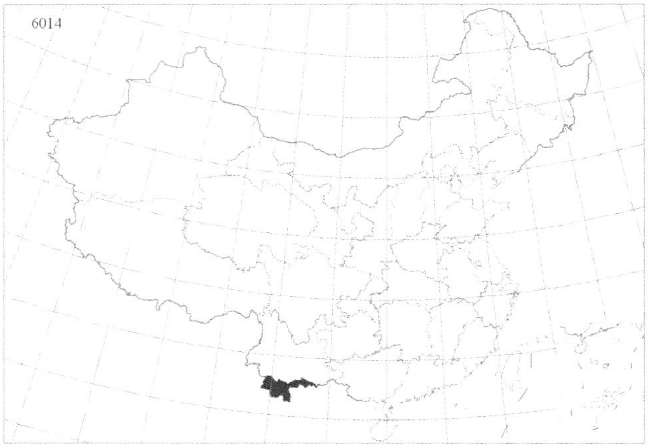

Leeaceae 火筒树科 *Leea* 火筒树属
Leea asiatica (L.) Ridsdale 单羽火筒树
Shrubs or small trees

18.9 (17.3~20.5); 13.0 (11.0~15.1); 22.7 (21.6~23.7); 19.0 (17.3~20.5); 167.7 (148.0~185.9); 0.0 (0.0~0.0); 907 (832~992) ❙ 1383 (1028~1751); 741 (566~996); 56 (41~77) ❙ 52.4 (12.5~88.8); 884 (783~986); 454.7 (339.5~659.0)

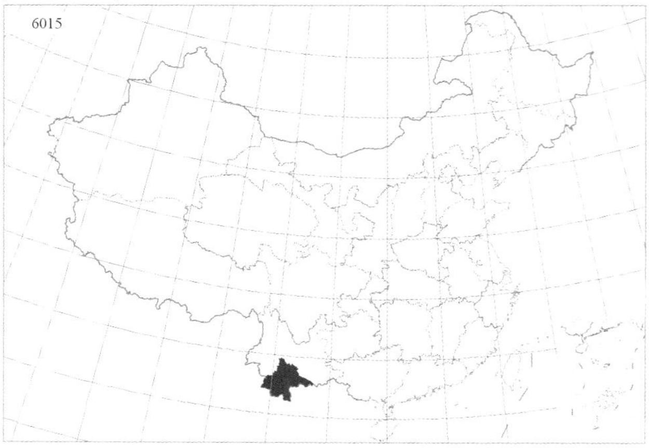

Leeaceae 火筒树科 *Leea* 火筒树属
Leea compactiflora Kurz 密花火筒树
Shrubs

18.2 (13.3~21.0); 11.9 (5.4~15.1); 22.1 (19.6~25.6); 18.2 (13.4~21.0); 158.4 (100.4~191.8); 0.0 (0.0~0.0); 875 (706~1053) ❙ 1413 (1028~1751); 781 (566~1007); 54 (41~77) ❙ 62.3 (12.5~145.8); 859 (706~1053); 462.3 (339.5~659.0)

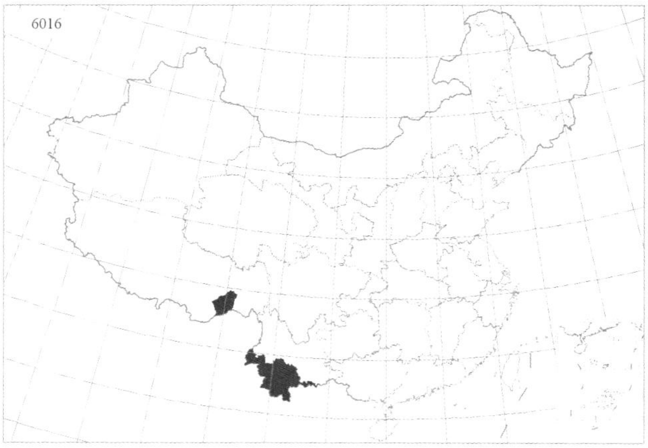

Leeaceae 火筒树科 *Leea* 火筒树属
Leea glabra C. L. Li 光叶火筒树
Shrubs

19.7 (18.6~22.2); 13.2 (12.7~13.7); 24.2 (22.5~28.5); 19.8 (18.7~22.2); 177.1 (163.9~206.5); 0.0 (0.0~0.0); 970 (888~1188) ❙ 1520 (1276~1751); 864 (741~996); 75 (51~128) ❙ 58.1 (34.7~88.8); 969 (888~1188); 456.0 (256.6~551.3)

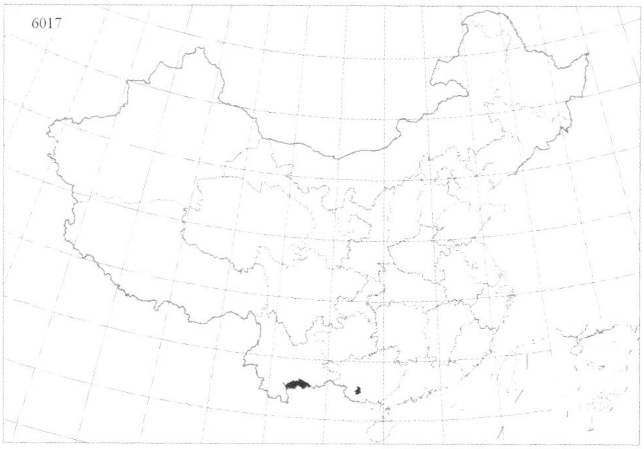

Leeaceae 火筒树科 *Leea* 火筒树属
Leea guineensis G. Don. 台湾火筒树
Shrubs or small trees

19.6 (16.2~24.1); 14.1 (11.0~18.9); 23.4 (20.1~28.0); 19.6 (16.2~24.1); 175.3 (134.2~229.7); 0.0 (0.0~0.0); 955 (773~1322) ❙ 2026 (1345~2992); 1082 (655~1847); 114 (47~308) ❙ 116.2 (32.5~246.5); 933 (773~1123); 380.0 (154.2~659.0)

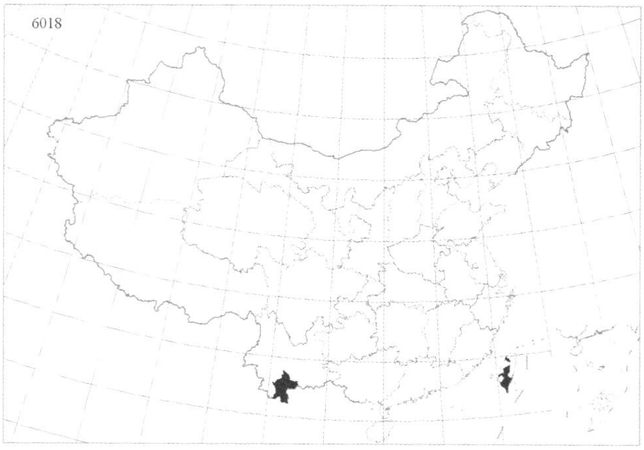

Leeaceae 火筒树科 *Leea* 火筒树属
Leea indica (N. L. Burman) Merr. 火筒树
Shrubs or small trees

19.6 (16.0~24.8); 13.0 (7.8~20.1); 24.4 (20.1~28.8); 19.7 (16.0~24.8); 175.9 (132.4~237.6); 0.0 (0.0~0.0); 986 (773~1385) ❙ 1592 (939~3495); 792 (392~1847); 100 (37~897) ❙ 64.6 (−12.0~246.5); 955 (744~1326); 369.3 (119.4~659.0)

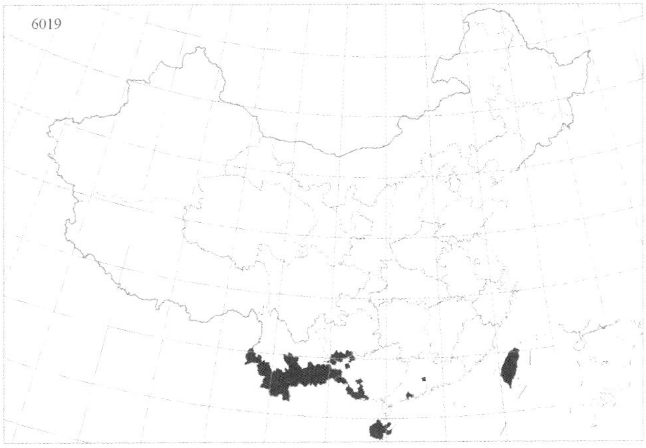

Leeaceae 火筒树科 *Leea* 火筒树属
Leea longifolia Merr. 窄叶火筒树
Shrubs

24.1; 18.8; 27.9; 24.2; 229.9; 0.0; 1322 ❙ 1302; 441; 44 ❙ −1.5; 1099; 469.4

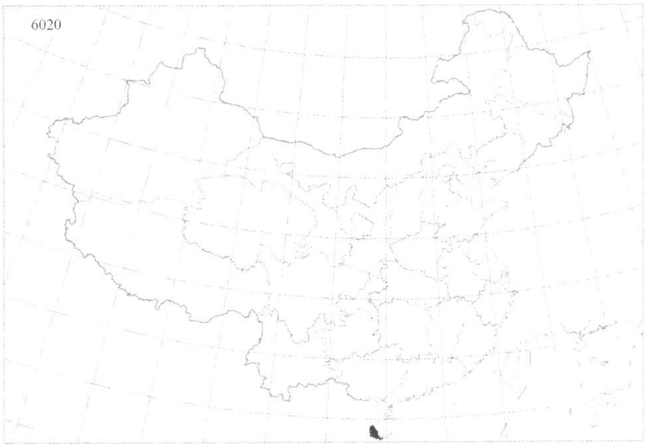

Leeaceae 火筒树科 *Leea* 火筒树属
Leea macrophylla Roxb. ex Hornem. 大叶火筒树
Shrubs or small trees

18.3 (13.3~20.5); 12.0 (5.4~15.1); 22.2 (19.6~23.7); 18.3 (13.4~20.5); 160.0 (100.4~185.9); 0.0 (0.0~0.0); 885 (706~992) ❙ 1553 (1434~1736); 828 (655~1007); 54 (41~77) ❙ 77.8 (47.6~145.8); 868 (706~986); 523.6 (439.5~659.0)

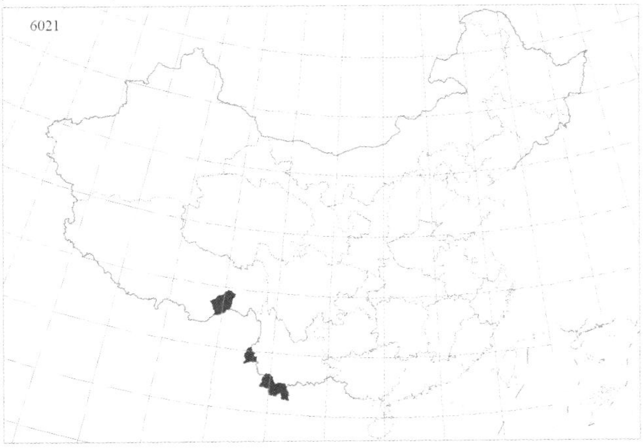

Leeaceae 火筒树科 *Leea* 火筒树属
Leea philippinensis Merr. 菲律宾火筒树
Trees

18.5; 14.1; 21.7; 18.5; 162.1; 0.0; 863 ❙ 2713; 1388; 188 ❙ 214.4; 863; 363.7

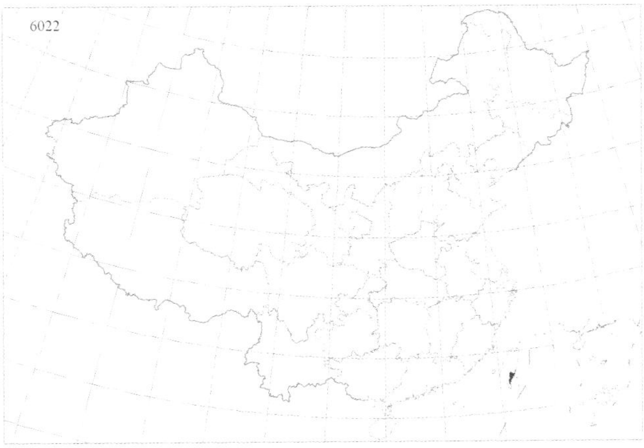

Vitaceae 葡萄科 *Ampelocissus* 酸蔹藤属
Ampelocissus artemisiifolia Planch. 酸蔹藤
Vines

14.0 (6.1~18.9); 7.0 (−1.5~12.0); 19.6 (12.7~24.3); 14.1 (6.3~19.0); 111.7 (35.2~167.8); 3.0 (0.0~21.8); 737 (496~929) ❙ 926 (797~991); 555 (411~603); 24 (12~38) ❙ 29.3 (−14.2~67.7); 677 (496~786); 286.8 (218.0~322.5)

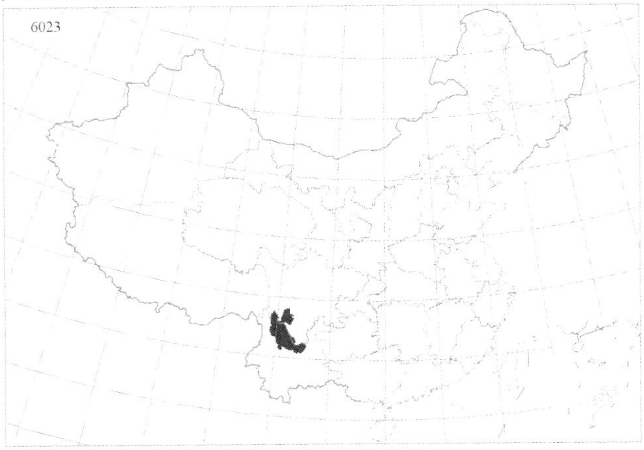

Vitaceae 葡萄科 *Ampelocissus* 酸蔹藤属
Ampelocissus hoabinhensis C. L. Li 红河酸蔹藤
Vines

17.6; 9.9; 23.2; 17.7; 152.4; 0.0; 859 ❙ 1422; 813; 52 ❙ 65.6; 859; 359.7

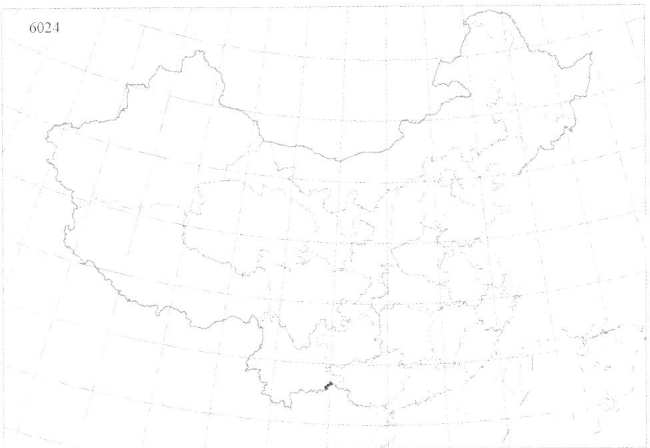

Vitaceae 葡萄科 *Ampelocissus* 酸蔹藤属
Ampelocissus sikkimensis (Lawson) Planch. 锡金酸蔹藤
Vines

16.9；10.3；21.4；16.9；142.7；0.0；813 ❙ 1178；646；51 ❙ 44.9；813；383.4

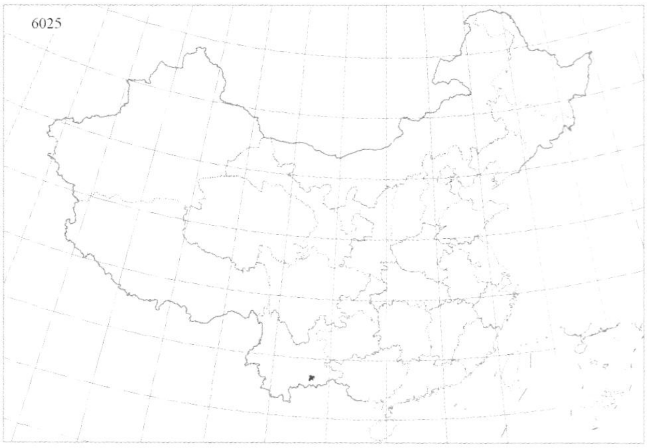

Vitaceae 葡萄科 *Ampelocissus* 酸蔹藤属
Ampelocissus xizangensis C. L. Li 西藏酸蔹藤
Vines

-1.1；-9.1；6.2；2.0；2.5；75.6；334 ❙ 321；193；33 ❙ -4.0；321；79.3

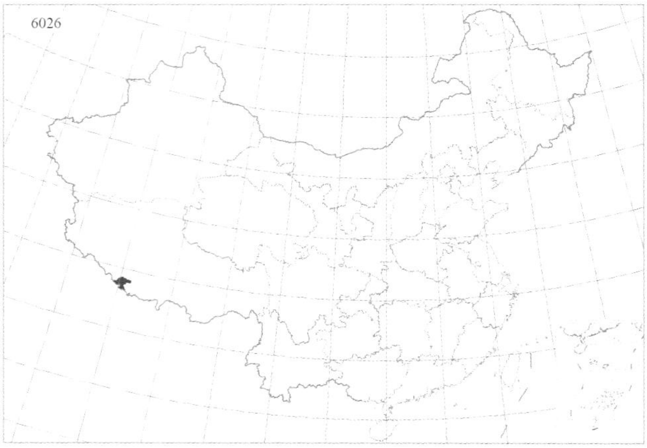

Vitaceae 葡萄科 *Ampelopsis* 蛇葡萄属
Ampelopsis acerifolia W. T. Wang 槭叶蛇葡萄
Vines, deciduous

15.1；4.4；25.4；15.2；122.8；0.6；817 ❙ 899；512；19 ❙ 10.0；817；197.9

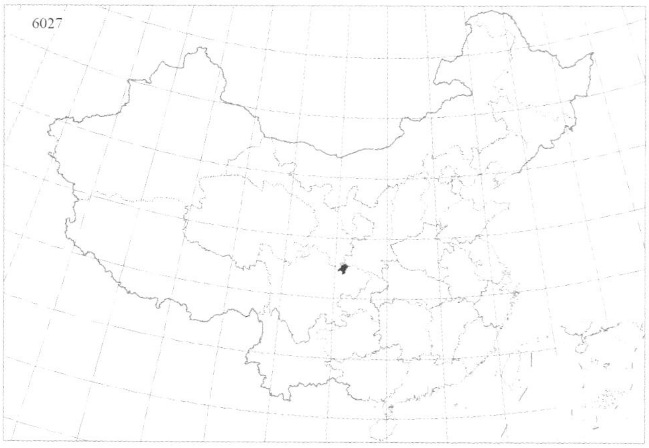

Vitaceae 葡萄科 *Ampelopsis* 蛇葡萄属
Ampelopsis aconitifolia Bunge 乌头叶蛇葡萄
Vines, deciduous

11.3 (0.2~22.2)；-3.7 (-20.8~13.9)；24.7 (9.9~29.3)；12.2 (3.8~22.3)；100.9 (15.5~207.5)；25.0 (0.0~104.9)；745 (413~1190) ❙ 653 (132~1776)；372 (81~816)；34 (2~197) ❙ -12.7 (-80.1~125.9)；594 (132~1190)；169.5 (13.6~357.7)

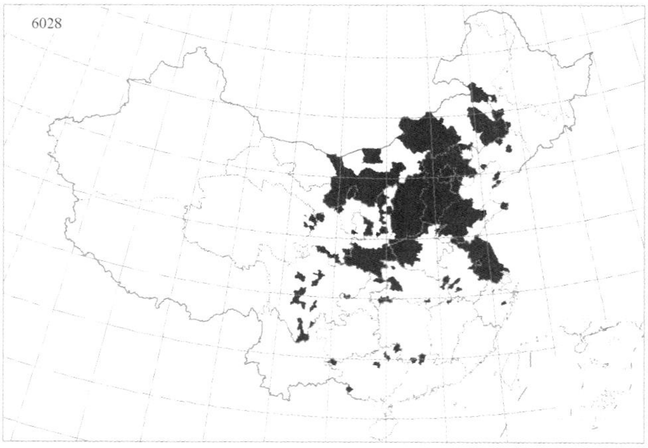

Vitaceae 葡萄科 *Ampelopsis* 蛇葡萄属
Ampelopsis aconitifolia var. palmiloba (Carrière) Rehder 掌裂草葡萄
Vines, deciduous

20.4 (17.9~22.2)；10.9 (7.4~13.9)；28.2 (27.1~28.7)；20.5 (18.0~22.3)；185.9 (155.6~207.5)；0.0 (0.0~0.0)；1086 (938~1190) ❙ 1553 (1374~1776)；674 (579~718)；154 (87~188) ❙ 45.2 (15.4~85.5)；1084 (938~1190)；234.4 (199.4~327.7)

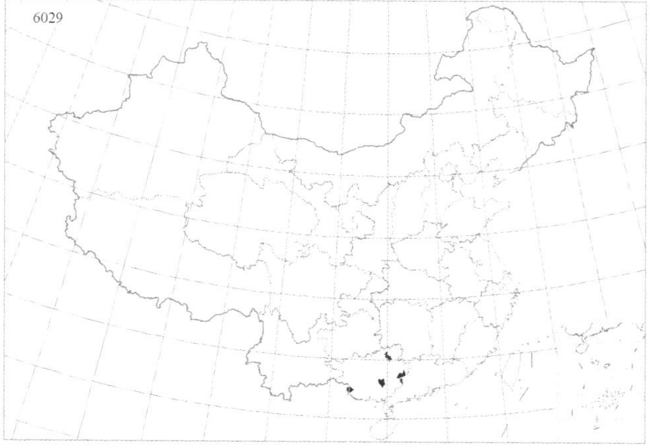

Vitaceae 葡萄科 *Ampelopsis* 蛇葡萄属
Ampelopsis acutidentata W. T. Wang 尖齿蛇葡萄
Vines, deciduous

6.4 (2.7~16.0)；-1.6 (-6.0~8.0)；13.4 (10.4~21.8)；7.1 (4.2~16.0)；45.1 (18.9~132.2)；27.6 (0.0~46.9)；516 (419~793) ❙ 741 (535~1107)；438 (326~607)；24 (10~52) ❙ 43.7 (27.1~61.4)；516 (419~793)；254.4 (201.9~298.7)

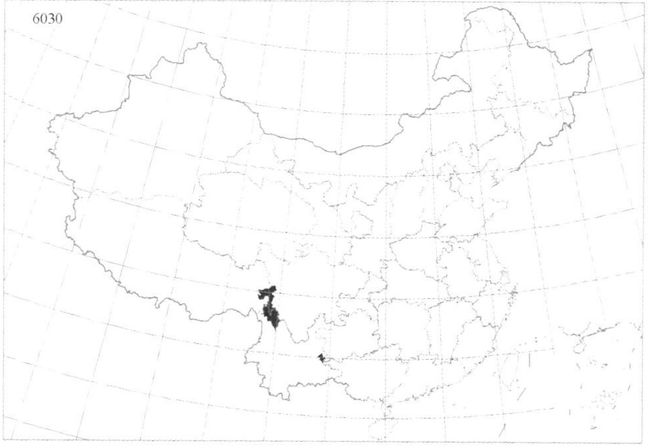

Vitaceae 葡萄科 *Ampelopsis* 蛇葡萄属
Ampelopsis bodinieri (H. Lév. et Vaniot) Rehder 蓝果蛇葡萄
Vines, deciduous

14.7 (2.2~20.4); 4.2 (-12.3~15.1); 24.3 (9.5~29.9); 14.8 (3.8~20.5); 121.6 (15.5~185.8); 4.2 (0.0~61.5); 802 (411~989) ❙ 1160 (466~1922); 503 (285~996); 91 (11~259) ❙ 44.1 (-41.7~128.9); 782 (411~986); 228.6 (116.6~659.0)

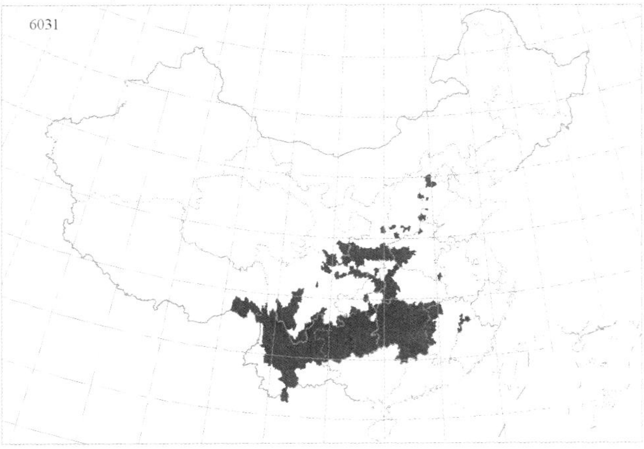

Vitaceae 葡萄科 *Ampelopsis* 蛇葡萄属
Ampelopsis bodinieri var. cinerea (Gagnep.) Rehder 灰毛蛇葡萄
Vines, deciduous

11.5 (6.2~16.2); -0.9 (-4.4~4.4); 23.2 (15.8~27.5); 11.7 (7.0~16.2); 92.7 (45.2~135.6); 14.5 (0.6~29.9); 698 (522~877) ❙ 848 (583~1451); 379 (290~591); 36 (15~140) ❙ 21.6 (-20.0~101.4); 688 (522~877); 256.9 (151.3~334.4)

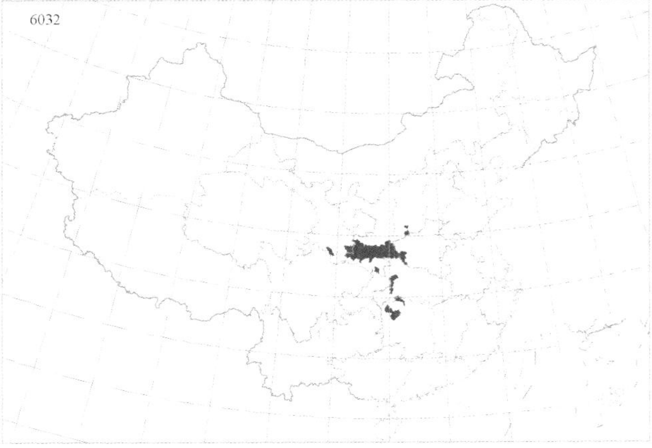

Vitaceae 葡萄科 *Ampelopsis* 蛇葡萄属
Ampelopsis cantoniensis (Hook. et Arn.) Planch. 广东蛇葡萄
Vines, deciduous

18.2 (12.0~24.8); 8.6 (0.6~20.1); 26.8 (20.1~29.8); 18.3 (12.1~24.8); 159.6 (94.7~237.6); 0.3 (0.0~9.7); 955 (697~1385) ❙ 1556 (868~2993); 609 (312~1428); 159 (41~547) ❙ 65.2 (-34.3~246.5); 945 (697~1294); 259.8 (123.9~659.0)

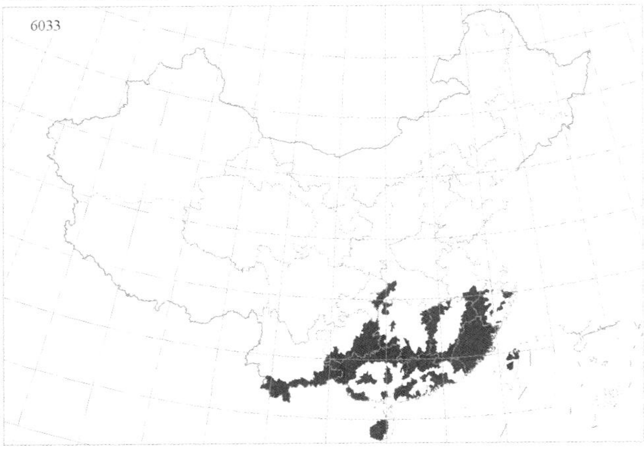

Vitaceae 葡萄科 *Ampelopsis* 蛇葡萄属
Ampelopsis chaffanjonii (H. Lév.) Rehder 羽叶蛇葡萄
Vines, deciduous

16.5 (11.1~22.2); 6.0 (0.9~14.7); 26.1 (19.4~29.3); 16.5 (11.3~22.2); 139.1 (80.8~206.9); 0.7 (0.0~8.7); 873 (652~1177) ❙ 1399 (976~1991); 567 (403~902); 125 (32~259) ❙ 60.6 (12.2~128.9); 869 (652~1177); 221.4 (135.1~551.6)

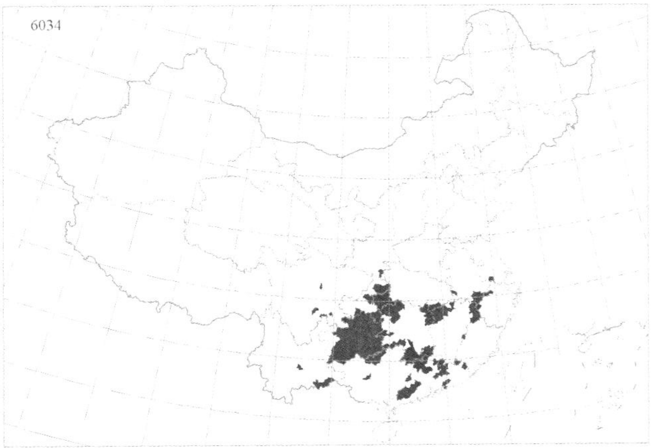

Vitaceae 葡萄科 *Ampelopsis* 蛇葡萄属
Ampelopsis delavayana Planch. 三裂蛇葡萄
Vines, deciduous

15.3 (2.2~21.7); 5.0 (-11.4~15.1); 24.6 (9.9~29.9); 15.4 (3.8~21.7); 128.1 (15.5~200.5); 3.6 (0.0~57.0); 825 (413~1140) ❙ 1190 (460~1950); 535 (258~996); 88 (10~259) ❙ 43.8 (-35.3~125.9); 811 (413~1140); 241.1 (108.2~659.0)

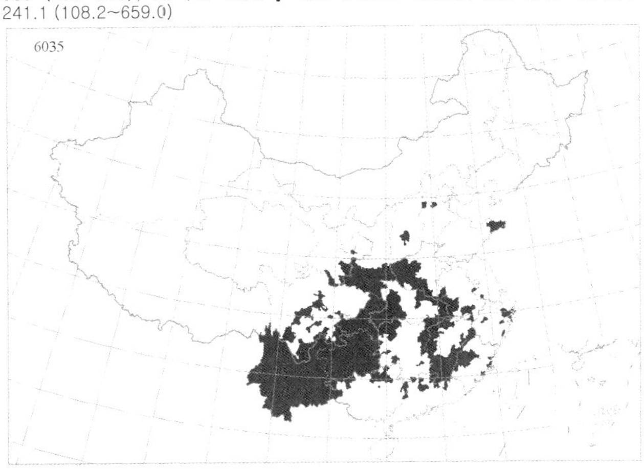

Vitaceae 葡萄科 *Ampelopsis* 蛇葡萄属
Ampelopsis delavayana var. glabra (Diels et Gilg) C. L. Li 掌裂蛇葡萄
Vines, deciduous

12.0 (0.2~16.8); -3.0 (-20.8~4.2); 25.5 (11.4~28.9); 12.7 (4.7~16.9); 106.0 (22.7~143.4); 21.4 (0.8~104.9); 770 (436~930) ❙ 657 (132~1571); 379 (81~666); 36 (2~157) ❙ -14.5 (-80.1~108.0); 612 (132~913); 168.7 (13.6~357.7)

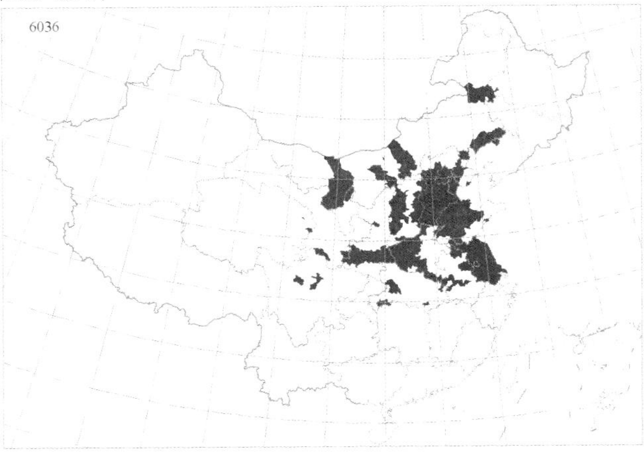

Vitaceae 葡萄科 *Ampelopsis* 蛇葡萄属

Ampelopsis delavayana var. setulosa (Diels et Gilg) C. L. Li 毛三裂蛇葡萄

Vines, deciduous

13.6 (3.5~20.7); 2.3 (−6.5~11.1); 24.1 (12.6~29.0); 13.7 (5.0~20.8); 111.9 (26.3~189.3); 8.1 (0.0~43.4); 766 (446~1108) ❙ 1032 (634~1740); 467 (290~720); 54 (16~187) ❙ 34.5 (−16.6~101.4); 763 (446~1108); 236.9 (126.1~354.5)

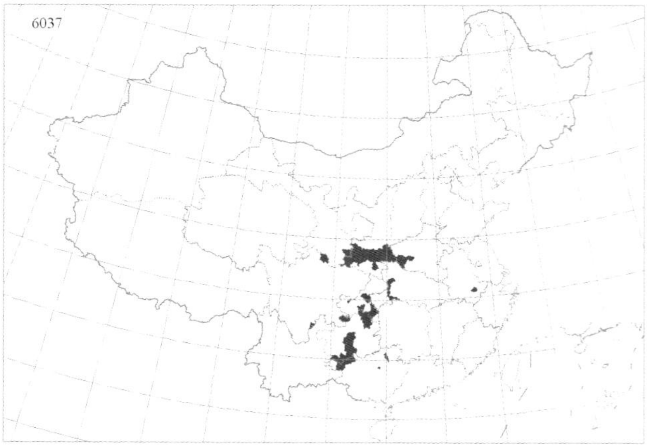

Vitaceae 葡萄科 *Ampelopsis* 蛇葡萄属

Ampelopsis glandulosa (Wall.) Momiy. 蛇葡萄

Vines, deciduous

16.8 (9.8~24.8); 6.0 (−2.9~20.1); 26.6 (18.8~29.6); 16.8 (10.2~24.8); 143.4 (79.2~237.6); 1.8 (0.0~20.5); 903 (641~1385) ❙ 1362 (626~1997); 543 (312~1044); 131 (25~259) ❙ 51.8 (−34.3~133.8); 888 (626~1360); 228.1 (111.9~620.0)

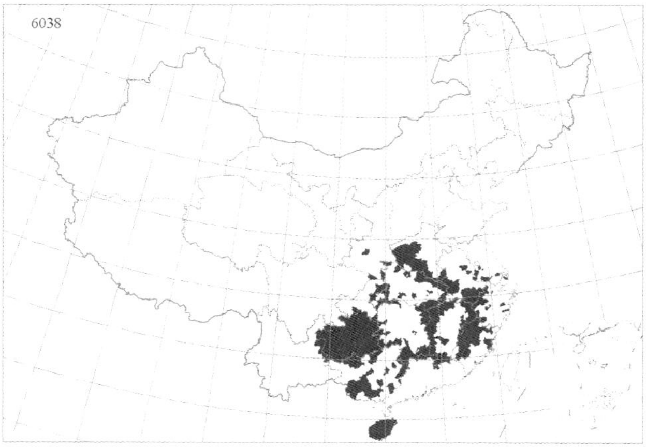

Vitaceae 葡萄科 *Ampelopsis* 蛇葡萄属

Ampelopsis glandulosa var. brevipedunculata (Maxim.) Momiy. 东北蛇葡萄

Vines, deciduous

15.2 (1.7~24.8); 3.2 (−20.5~20.1); 26.1 (18.5~29.4); 15.8 (6.7~24.8); 135.2 (47.6~237.6); 11.6 (0.0~93.6); 872 (513~1385) ❙ 1254 (460~3495); 547 (289~1847); 117 (8~897) ❙ 42.1 (−34.3~246.5); 850 (460~1360); 230.9 (111.9~620.0)

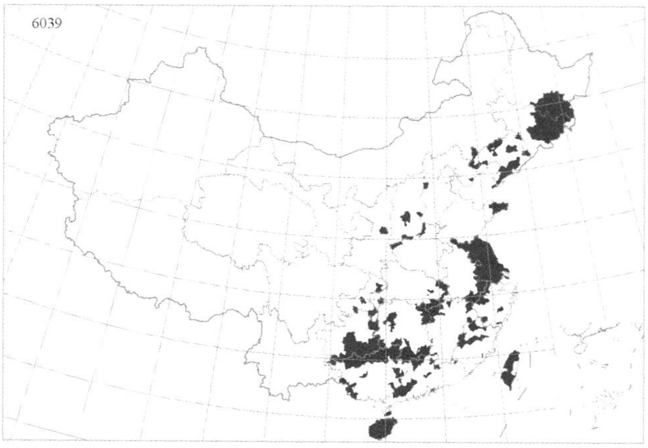

Vitaceae 葡萄科 *Ampelopsis* 蛇葡萄属

Ampelopsis glandulosa var. heterophylla (Thunberg) Momiyama 异叶蛇葡萄

Vines, deciduous

15.7 (5.5~22.4); 3.6 (−15.3~14.8); 26.9 (14.4~29.8); 15.8 (6.6~22.4); 134.0 (39.7~208.9); 4.9 (0.0~68.0); 869 (508~1191) ❙ 1273 (541~2037); 497 (302~1033); 131 (9~254) ❙ 45.1 (−36.5~130.7); 838 (508~1191); 203.7 (120.1~371.7)

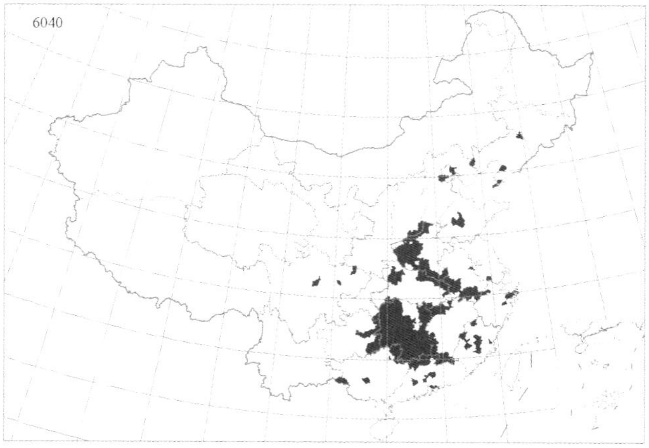

Vitaceae 葡萄科 *Ampelopsis* 蛇葡萄属

Ampelopsis glandulosa var. kulingensis (Rehder) Momiy. 牯岭蛇葡萄

Vines, deciduous

16.7 (13.5~20.3); 6.2 (2.1~12.2); 26.3 (23.0~28.8); 16.7 (13.6~20.4); 141.1 (106.7~184.2); 0.6 (0.0~4.9); 875 (744~1052) ❙ 1523 (1136~1997); 586 (413~720); 160 (51~257) ❙ 74.2 (19.5~132.1); 873 (744~1007); 243.1 (160.1~332.5)

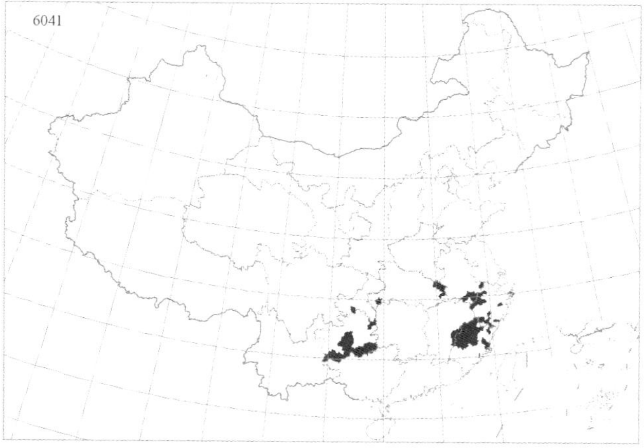

Vitaceae 葡萄科 *Ampelopsis* 蛇葡萄属

Ampelopsis gongshanensis C. L. Li 贡山蛇葡萄

Vines, deciduous

8.6; 1.3; 15.0; 8.7; 52.6; 8.8; 574 ❙ 1041; 585; 45 ❙ 81.5; 574; 352.3

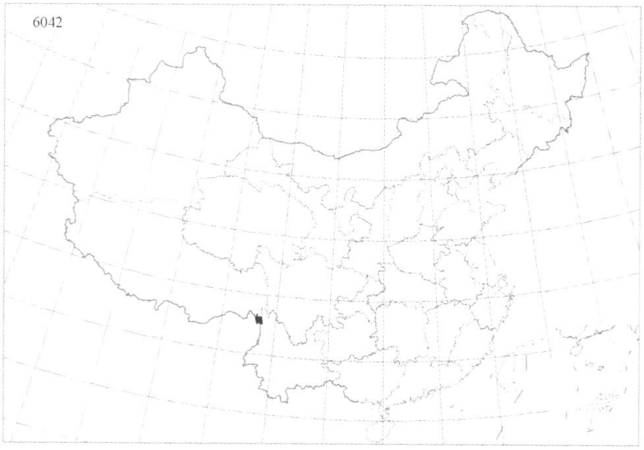

Vitaceae 葡萄科 *Ampelopsis* 蛇葡萄属
Ampelopsis grossedentata (Hand.-Mazz.) W. T. Wang 显齿蛇葡萄
Vines, deciduous

17.7 (13.9~24.3); 7.6 (3.1~18.7); 26.6 (20.8~29.8); 17.7 (14.0~24.3); 153.1 (110.5~231.9); 0.2 (0.0~2.6); 927 (734~1338) ❙ 1477 (868~2148); 580 (312~1234); 146 (41~259) ❙ 60.2 (-34.3~130.7); 917 (734~1190); 244.3 (134.6~659.0)

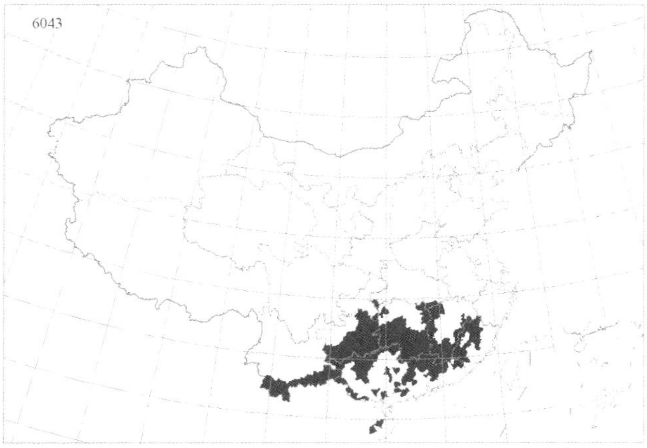

Vitaceae 葡萄科 *Ampelopsis* 蛇葡萄属
Ampelopsis humulifolia Bunge 葎叶蛇葡萄
Vines, deciduous

11.4 (-5.8~21.6); -3.2 (-29.8~12.6); 24.6 (15.3~29.6); 12.4 (4.4~21.7); 102.6 (28.0~199.8); 25.3 (0.0~157.4); 745 (426~1153) ❙ 316 (202~1997); 404 (134~795); 57 (6~259) ❙ 6.3 (-65.5~130.7); 663 (202~1153); 192.0 (52.7~334.4)

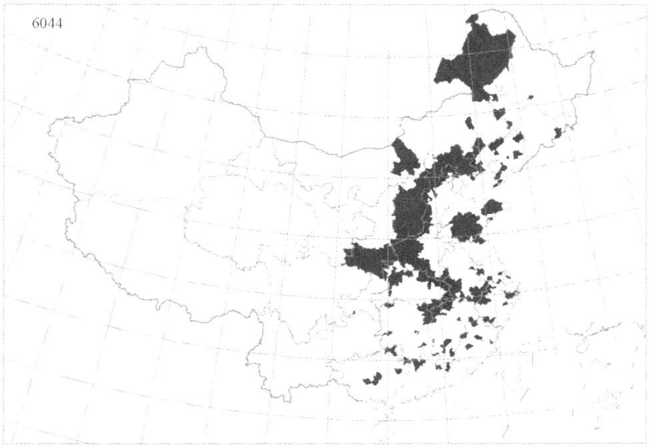

Vitaceae 葡萄科 *Ampelopsis* 蛇葡萄属
Ampelopsis japonica (Thunb.) Makino 白蔹
Vines, deciduous

14.2 (1.4~22.3); 0.8 (-21.5~13.5); 26.5 (18.4~30.1); 14.6 (6.4~22.3); 123.1 (45.8~208.1); 12.2 (0.0~97.2); 832 (501~1202) ❙ 1052 (450~1997); 468 (258~821); 97 (9~259) ❙ 25.1 (-43.4~147.0); 783 (450~1202); 201.2 (110.1~346.5)

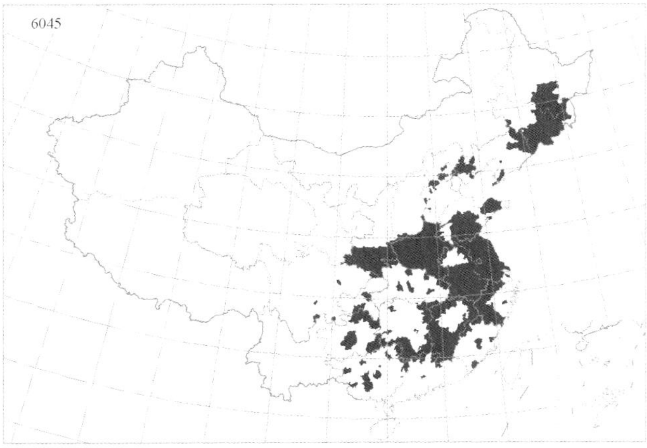

Vitaceae 葡萄科 *Ampelopsis* 蛇葡萄属
Ampelopsis megalophylla Diels et Gilg 大叶蛇葡萄
Vines, deciduous

15.5 (7.3~22.2); 5.2 (-4.2~14.2); 24.7 (15.6~28.8); 15.5 (7.5~22.3); 129.4 (48.2~207.5); 3.5 (0.0~29.4); 827 (537~1190) ❙ 1244 (632~1852); 536 (290~996); 96 (15~230) ❙ 50.1 (-16.6~107.2); 823 (537~1190); 247.4 (130.1~551.6)

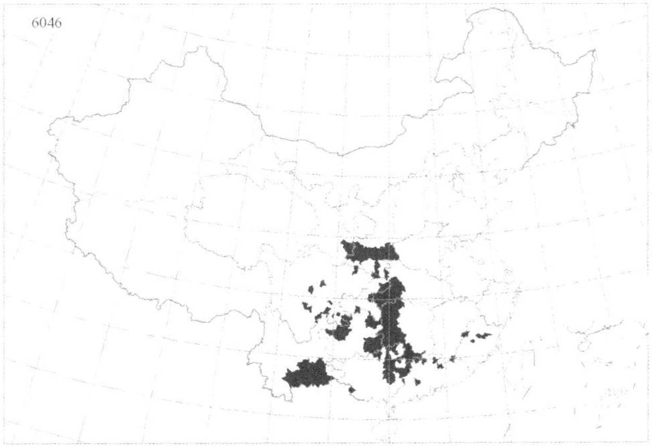

Vitaceae 葡萄科 *Ampelopsis* 蛇葡萄属
Ampelopsis megalophylla var. jiangxiensis (W. T. Wang) C. L. Li 柔毛大叶蛇葡萄
Vines, deciduous

18.6 (17.3~19.4); 8.4 (6.4~9.4); 27.6 (26.9~28.5); 18.6 (17.3~19.4); 163.8 (147.9~173.4); 0.0 (0.0~0.0); 973 (907~1029) ❙ 1665 (1579~1745); 594 (541~644); 194 (187~197) ❙ 71.3 (57.0~78.8); 973 (907~1029); 239.6 (208.1~261.3)

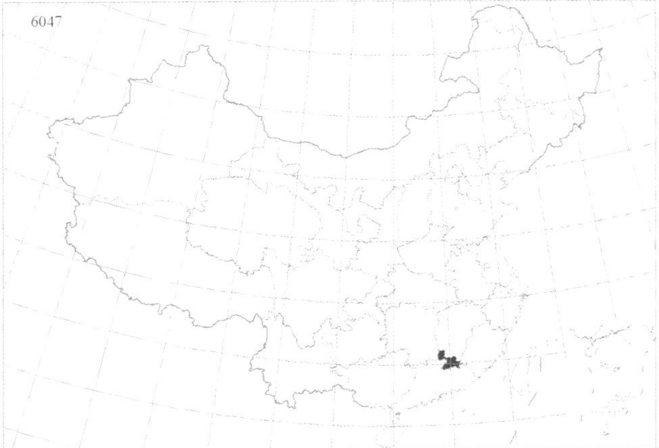

Vitaceae 葡萄科 *Ampelopsis* 蛇葡萄属
Ampelopsis mollifolia W. T. Wang 毛叶蛇葡萄
Vines, deciduous

11.0; 3.7; 17.0; 11.1; 74.6; 1.8; 640 ❙ 990; 619; 15 ❙ 54.7; 638; 296.0

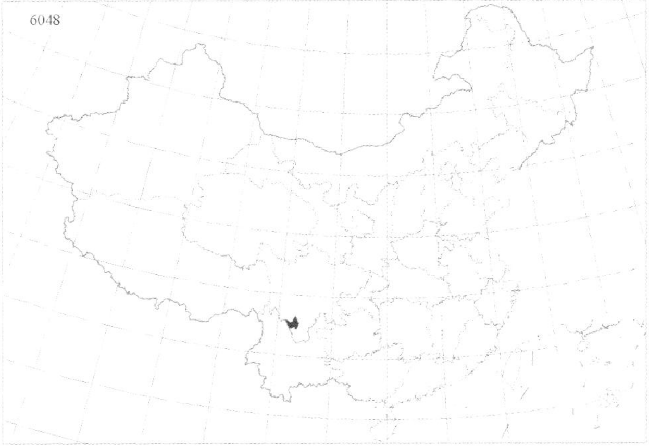

Vitaceae 葡萄科 *Ampelopsis* 蛇葡萄属
Ampelopsis rubifolia (Wall.) Planch. 毛枝蛇葡萄
Vines, deciduous

16.4 (14.3~20.7); 5.7 (3.3~12.7); 26.1 (22.5~29.1); 16.4 (14.3~20.8); 137.2 (112.6~189.3); 0.3 (0.0~2.3); 861 (756~1108) ❙ 1365 (1168~1997); 530 (419~911); 129 (63~254) ❙ 58.7 (28.9~130.7); 859 (756~1108); 221.6 (160.7~551.3)

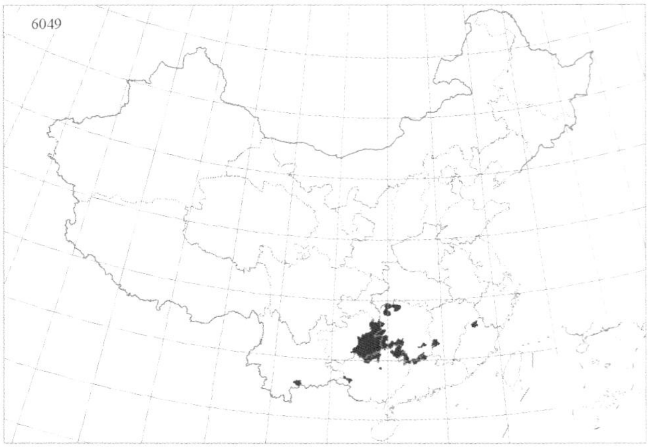

Vitaceae 葡萄科 *Cayratia* 乌蔹莓属
Cayratia ciliifera (Merr.) Chun 节毛乌蔹莓
Vines

24.1; 18.8; 27.9; 24.2; 229.9; 0.0; 1322 ❙ 1302; 441; 44 ❙ -1.5; 1099; 469.4

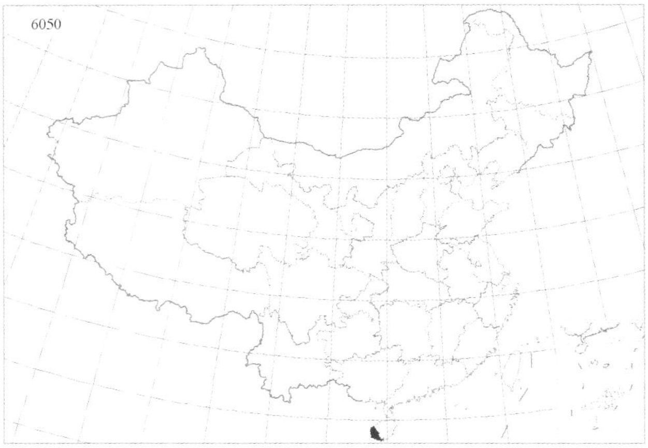

Vitaceae 葡萄科 *Cayratia* 乌蔹莓属
Cayratia cordifolia C. Y. Wu 心叶乌蔹莓
Vines

18.3 (16.0~21.0); 11.6 (8.0~14.2); 22.9 (21.4~25.6); 18.3 (16.0~21.0); 159.8 (132.2~191.8); 0.0 (0.0~0.0); 889 (793~1053) ❙ 1292 (1107~1522); 721 (607~835); 55 (48~68) ❙ 45.2 (34.8~60.1); 888 (793~1053); 441.4 (298.7~551.6)

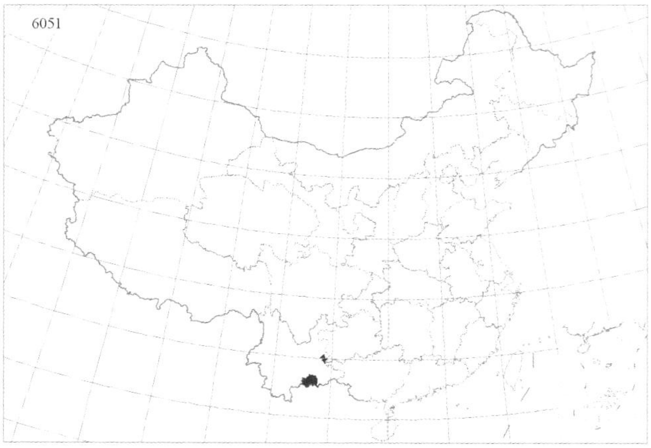

Vitaceae 葡萄科 *Cayratia* 乌蔹莓属
Cayratia geniculata (Blume) Gagnep. 膝曲乌蔹莓
Vines

16.8 (9.3~24.8); 8.3 (-1.7~20.1); 23.8 (15.7~28.8); 16.8 (9.5~24.8); 142.1 (60.8~237.6); 0.6 (0.0~18.0); 863 (600~1385) ❙ 1275 (707~2435); 624 (312~1428); 68 (15~173) ❙ 49.6 (-34.3~145.8); 848 (600~1221); 303.8 (146.0~659.0)

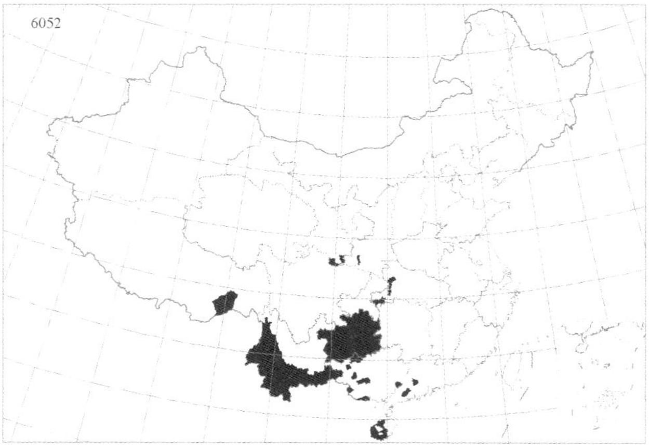

Vitaceae 葡萄科 *Cayratia* 乌蔹莓属
Cayratia japonica var. mollis (Wall.) Mcmiy. 毛乌蔹莓
Vines

18.7 (11.3~21.0); 12.4 (4.3~15.1); 22.9 (17.0~27.6); 18.7 (11.3~21.0); 164.9 (76.6~191.8); 0.0 (0.0~0.7); 912 (644~1053) ❙ 1501 (1243~1751); 760 (618~996); 70 (41~194) ❙ 65.6 (44.5~93.0); 898 (644~1053); 489.3 (236.9~659.0)

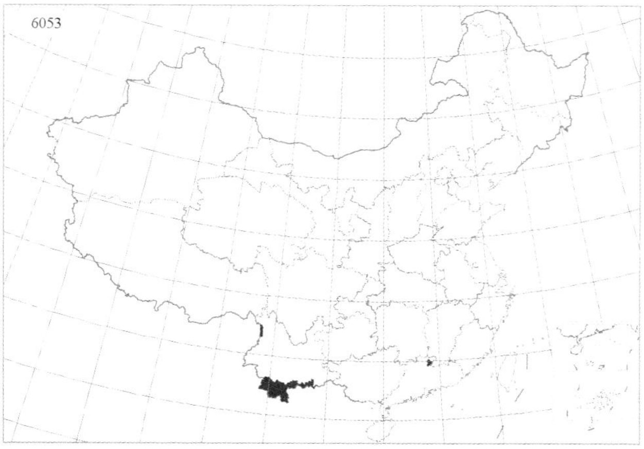

Vitaceae 葡萄科 *Cayratia* 乌蔹莓属
Cayratia japonica var. pseudotrifolia (W. T. Wang) C. L. Li 尖叶乌蔹莓
Vines

12.9 (7.4~17.6); 1.2 (-4.2~5.3); 24.1 (18.7~29.3); 13.1 (8.2~17.6); 104.4 (58.8~151.3); 8.5 (0.0~29.4); 742 (561~961) ❙ 1016 (620~1679); 432 (301~613); 56 (13~195) ❙ 36.3 (-8.1~107.2); 741 (561~961); 238.1 (146.0~334.4)

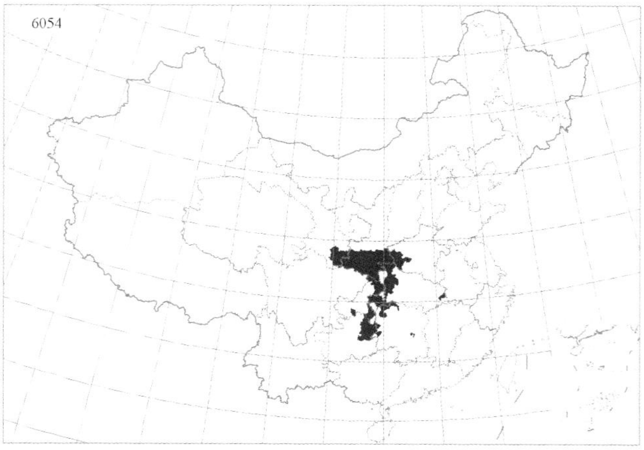

Vitaceae 葡萄科 *Cayratia* 乌蔹莓属

Cayratia oligocarpa (H. Léveillé et Vaniot) Gagnepain 华中乌蔹莓

Vines

15.1 (12.8~18.3); 5.1 (2.5~12.7); 24.1 (20.1~29.9); 15.1 (12.9~18.7); 122.7 (95.9~163.9); 1.2 (0.0~3.9); 805 (699~976) ❙ 1249 (885~1922); 563 (439~911); 106 (32~259) ❙ 53.8 (26.5~111.3); 786 (699~910); 249.3 (141.7~551.3)

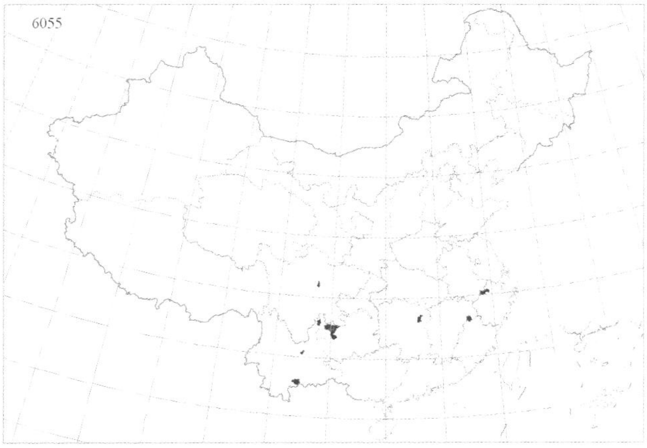

Vitaceae 葡萄科 *Cayratia* 乌蔹莓属

Cayratia pedata (Lam.) Juss. ex Gagnep. 乌足乌蔹莓

Vines

19.1 (17.2~22.2); 11.0 (9.1~13.9); 25.0 (22.7~28.3); 19.1 (17.3~22.3); 169.6 (147.0~207.5); 0.0 (0.0~0.0); 963 (838~1190) ❙ 1304 (1161~1438); 720 (633~820); 57 (48~87) ❙ 37.6 (5.8~65.6); 963 (838~1190); 338.2 (314.4~371.7)

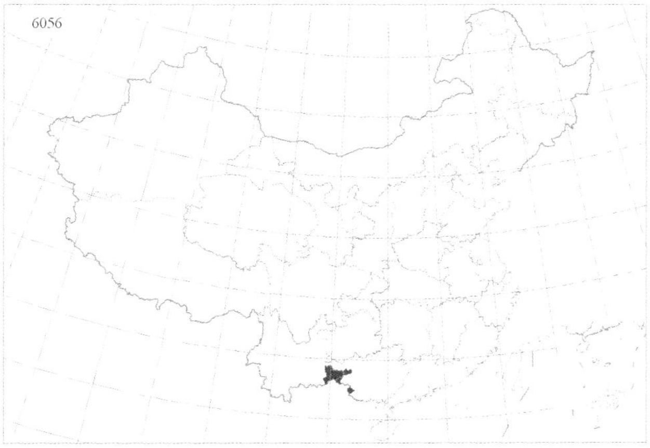

Vitaceae 葡萄科 *Cayratia* 乌蔹莓属

Cayratia timorensis var. mekongensis (C. Y. Wu) C. L. Li 澜沧乌蔹莓

Vines

18.9 (17.1~20.5); 13.1 (11.0~15.1); 22.3 (20.8~23.7); 18.9 (17.1~20.5); 167.2 (145.0~185 9); 0.0 (0.0~0.0); 903 (815~992) ❙ 1461 (1352~1607); 736 (655~803); 53 (41~77) ❙ 62.1 (47.6~74.2); 878 (802~986); 507.5 (427.9~659.0)

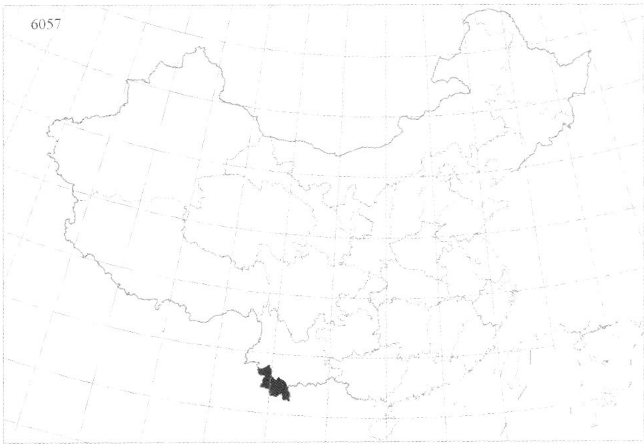

Vitaceae 葡萄科 *Cayratia* 乌蔹莓属

Cayratia trifolia (L.) Domin 三叶乌蔹莓

Vines

20.1 (18.2~24.8); 14.6 (12.4~20.1); 23.3 (21.6~28.0); 20.1 (18.2~24.8); 181.2 (158.6~237.6); 0.0 (0.0~0.0); 985 (861~1385) ❙ 1483 (1405~1607); 684 (473~785); 55 (41~77) ❙ 53.9 (1.5~74.2); 930 (839~1137); 517.7 (427.9~659.0)

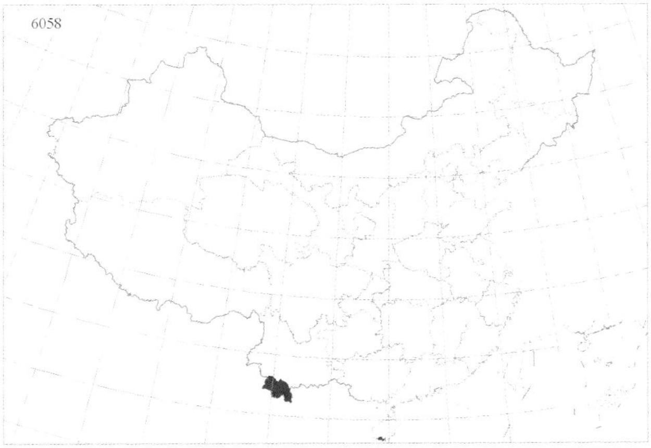

Vitaceae 葡萄科 *Cissus* 白粉藤属

Cissus adnata Roxb. 贴生白粉藤

Vines

20.0; 14.6; 23.2; 20.1; 180.8; 0.0; 963 ❙ 1523; 681; 67 ❙ 58.4; 942; 581.0

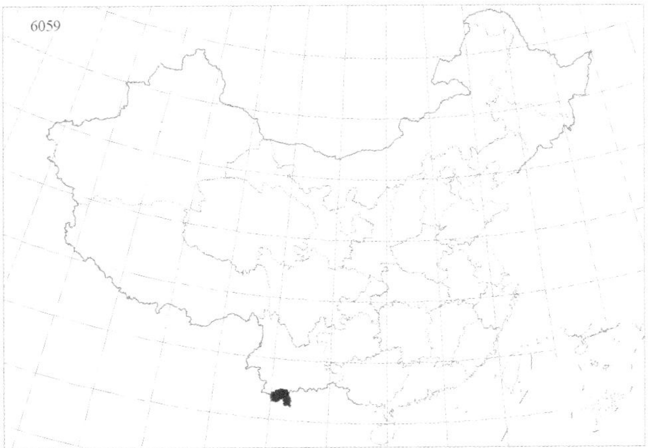

Vitaceae 葡萄科 *Cissus* 白粉藤属

Cissus aristata Blume 毛叶苦郎藤

Vines

18.0 (12.7~24.8); 8.6 (0.9~20.1); 26.2 (19.6~29.1); 18.1 (12.8~24.8); 157.2 (99.0~237.6); 0.3 (0.0~8.5); 938 (706~1385) ❙ 1491 (807~1997); 617 (349~1007); 134 (32~254) ❙ 60.3 (-11.9~145.8); 929 (706~1212); 264.4 (126.1~659.0)

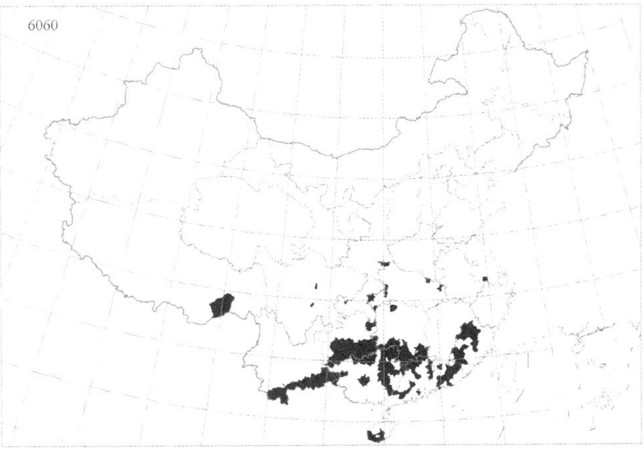

Vitaceae 葡萄科 *Cissus* 白粉藤属
Cissus austroyunnanensis Y. H. Li et Y. Zhang 滇南青紫葛
Vines

20.5; 14.9; 23.7; 20.5; 185.9; 0.0; 992 ❙ 1464; 655; 66 ❙ 47.6; 959; 570.1

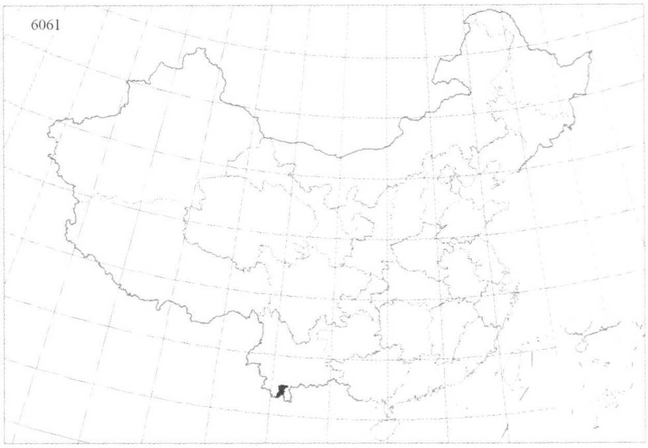

Vitaceae 葡萄科 *Cissus* 白粉藤属
Cissus elongata Roxb. 五叶白粉藤
Vines

24.0 (22.2~24.8); 18.4 (17.2~20.1); 28.0 (25.7~28.8); 24.0 (22.3~24.8); 228.3 (207.2~237.6); 0.0 (0.0~0.0); 1310 (1127~1385) ❙ 1501 (868~1854); 534 (312~712); 74 (41~115) ❙ 15.0 (-34.3~47.8); 1191 (868~1360); 440.3 (248.2~620.0)

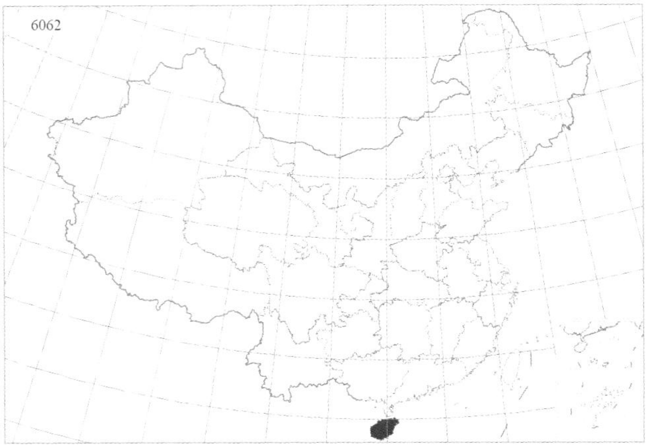

Vitaceae 葡萄科 *Cissus* 白粉藤属
Cissus hexangularis Thorel ex Planch. 翅茎白粉藤
Vines

22.1 (17.9~24.8); 14.7 (7.4~20.1); 27.8 (25.7~29.2); 22.1 (18.0~24.8); 205.3 (155.6~237.6); 0.0 (0.0~0.0); 1178 (928~1385) ❙ 1510 (868~2148); 651 (312~1234); 94 (41~188) ❙ 30.2 (-34.3~86.7); 1115 (868~1360); 340.0 (163.7~620.0)

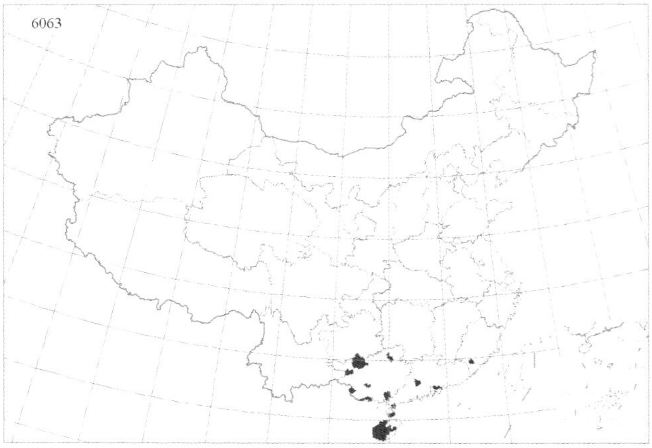

Vitaceae 葡萄科 *Cissus* 白粉藤属
Cissus pteroclada Hayata 翼茎白粉藤
Vines

21.1 (17.9~24.8); 13.1 (7.4~20.1); 27.5 (22.3~28.8); 21.1 (18.0~24.8); 193.3 (155.6~237.6); 0.0 (0.0~0.0); 1105 (909~1385) ❙ 1599 (1258~2435); 692 (425~1428); 138 (41~194) ❙ 46.0 (1.5~99.5); 1082 (880~1221); 299.7 (203.4~620.0)

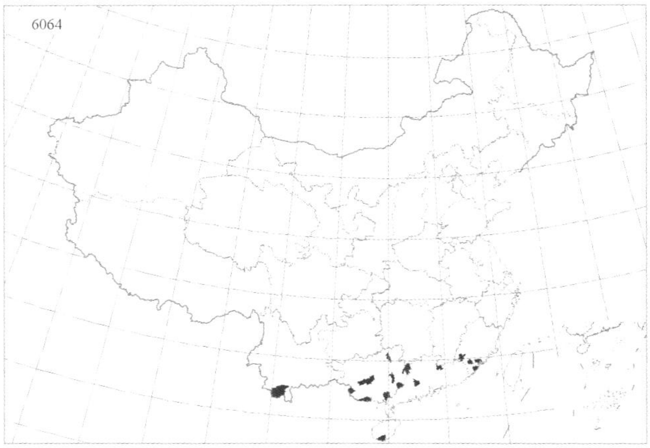

Vitaceae 葡萄科 *Cissus* 白粉藤属
Cissus repanda Vahl 大叶白粉藤
Large climbers

18.5 (16.3~21.0); 12.1 (9.1~15.1); 22.8 (21.4~25.6); 18.6 (16.4~21.0); 162.9 (136.4~191.8); 0.0 (0.0~0.0); 894 (797~1053) ❙ 1378 (1074~1751); 744 (582~996); 56 (41~77) ❙ 54.1 (21.7~88.8); 884 (797~1053); 436.8 (277.0~659.0)

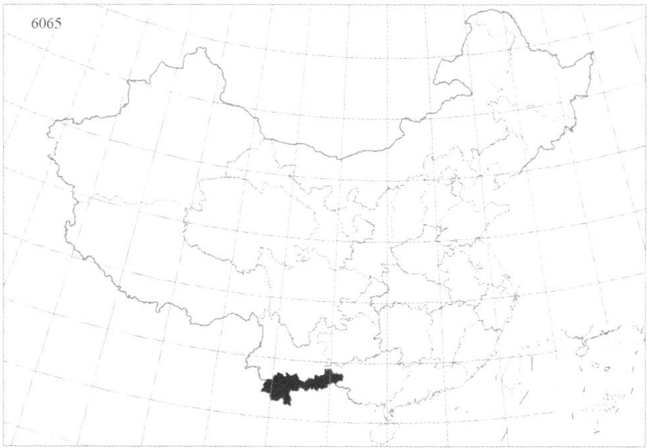

Vitaceae 葡萄科 *Cissus* 白粉藤属
Cissus repanda var. subferruginea (Merr. et Chun) C. L. Li 海南大叶白粉藤
Vines, deciduous

24.7; 20.0; 27.9; 24.8; 237.1; 0.0; 1380 ❙ 1496; 503; 50 ❙ 8.4; 1175; 511.1

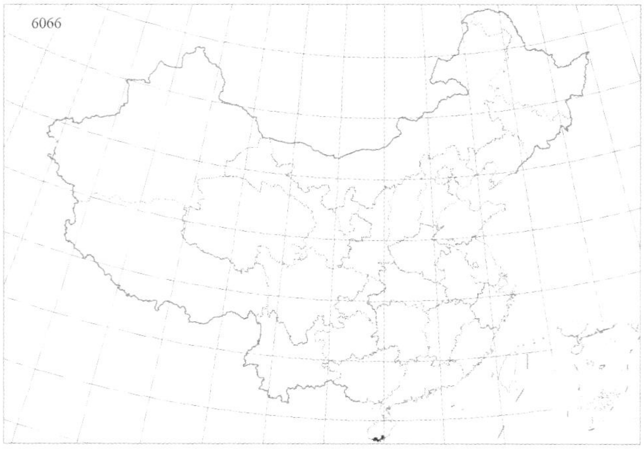

Vitaceae 葡萄科 *Cissus* 白粉藤属
Cissus subtetragona Planch. 四棱白粉藤
Vines

18.9 (16.3~22.7); 12.3 (9.1~15.1); 23.4 (21.4~28.6); 19.0 (16.4~22.8); 167.4 (136.4~213.1); 0.0 (0.0~0.0); 924 (797~1221) ❙ 1438 (1074~2435); 783 (582~1428); 61 (41~130) ❙ 55.3 (15.4~99.5); 915 (797~1221); 423.0 (277.0~659.0)

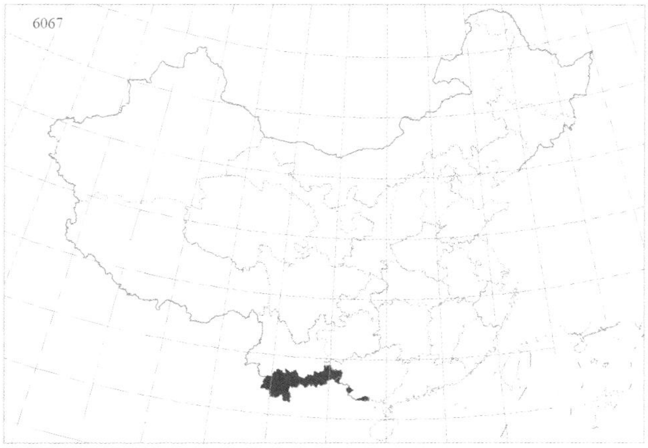

Vitaceae 葡萄科 *Parthenocissus* 地锦属
Parthenocissus dalzielii Gagnep. 异叶地锦
Vines, deciduous

17.1 (2.9~23.2); 6.6 (-6.3~17.5); 26.5 (11.2~29.8); 17.1 (4.4~23.2); 146.1 (20.5~218.9); 0.8 (0.0~44.6); 906 (428~1257) ❙ 1433 (808~2037); 553 (378~1044); 147 (24~259) ❙ 58.7 (-6.5~130.7); 895 (428~1257); 225.6 (120.3~551.6)

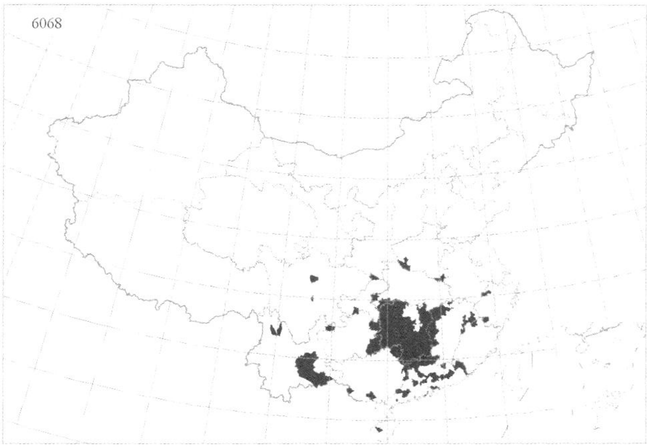

Vitaceae 葡萄科 *Parthenocissus* 地锦属
Parthenocissus feddei (H. Lév.) C. L. Li 长柄地锦
Vines, deciduous

16.6 (10.2~24.2); 5.9 (-2.9~18.1); 26.3 (20.1~29.3); 16.6 (10.6~24.3); 141.3 (83.1~231.2); 1.8 (0.0~20.5); 887 (656~1335) ❙ 1376 (626~2435); 569 (316~1428); 126 (29~254) ❙ 55.5 (-22.2~137.5); 879 (626~1221); 222.5 (128.7~538.2)

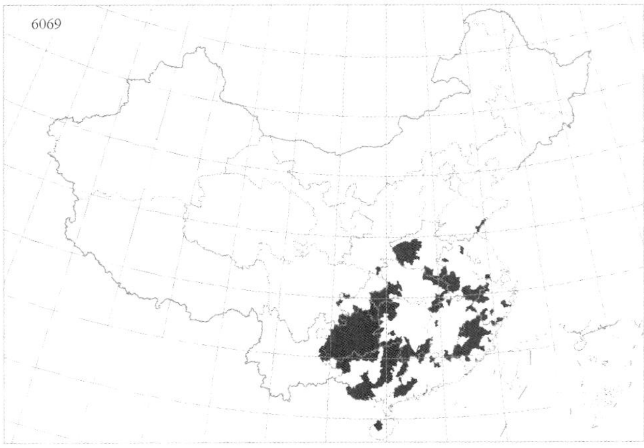

Vitaceae 葡萄科 *Parthenocissus* 地锦属
Parthenocissus henryana (Hemsl.) Graebn. ex Diels et Gilg 花叶地锦
Vines, deciduous

14.8 (9.3~18.9); 3.6 (-2.5~9.5); 25.3 (19.8~28.9); 14.8 (9.5~18.9); 121.8 (69.8~167.3); 3.9 (0.0~20.6); 809 (603~1002) ❙ 1122 (711~1559); 480 (352~816); 67 (15~196) ❙ 39.5 (-5.3~107.2); 809 (603~1002); 211.2 (108.2~334.4)

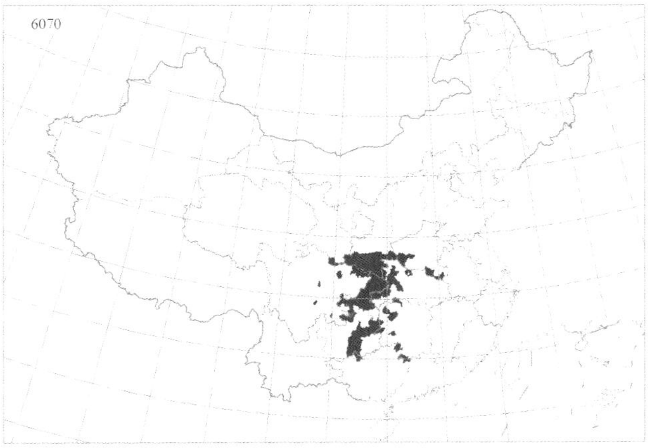

Vitaceae 葡萄科 *Parthenocissus* 地锦属
Parthenocissus laetevirens Rehder 绿叶地锦
Vines, deciduous

16.3 (10.2~22.2); 4.8 (-2.9~14.7); 27.2 (22.4~30.1); 16.4 (10.6~22.2); 138.6 (83.1~206.9); 1.9 (0.0~20.5); 887 (656~1177) ❙ 1400 (738~1997); 531 (349~891); 157 (27~259) ❙ 57.7 (-14.0~130.7); 880 (656~1177); 216.0 (134.1~332.5)

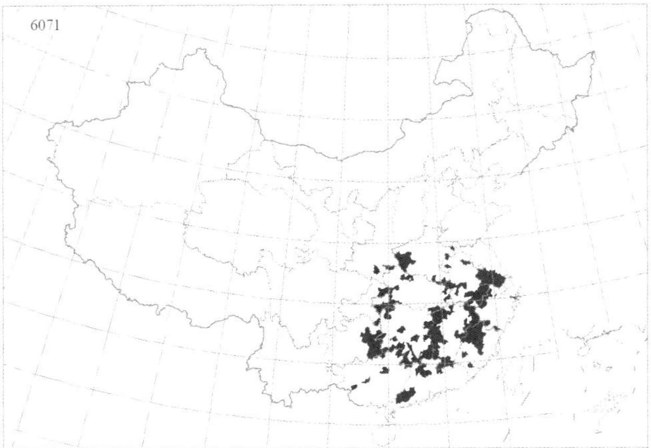

Vitaceae 葡萄科 *Parthenocissus* 地锦属
Parthenocissus semicordata (Wall.) Planch. 三叶地锦
Vines, deciduous

15.1 (-1.8~22.2); 5.0 (-10.2~15.1); 24.0 (6.1~29.9); 15.2 (1.8~22.2); 125.3 (1.9~206.9); 4.0 (0.0~82.8); 811 (329~1177) ❙ 1179 (320~1953); 532 (193~1007); 86 (9~253) ❙ 45.0 (-22.2~145.8); 798 (320~1177); 247.4 (72.2~659.0)

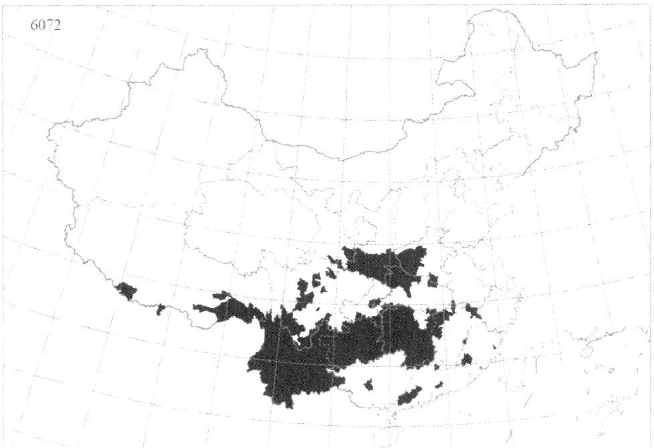

Vitaceae 葡萄科 *Parthenocissus* 地锦属
Parthenocissus suberosa Hand.-Mazz. 栓翅地锦
Vines, deciduous

16.6 (14.6~20.3); 5.3 (2.5~10.9); 27.0 (23.9~29.3); 16.6 (14.7~20.4); 139.8 (117.7~184.4); 0.6 (0.0~3.9); 885 (779~1065) ❙ 1476 (1211~1847); 567 (427~775); 157 (68~253) ❙ 67.1 (28.8~109.2); 883 (779~1065); 233.2 (184.7~322.1)

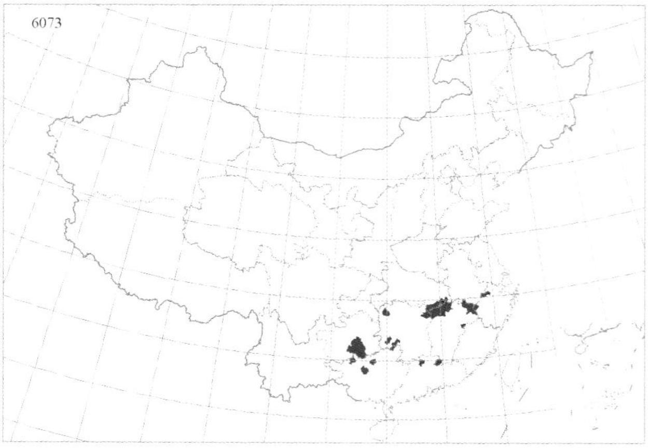

Vitaceae 葡萄科 *Parthenocissus* 地锦属
Parthenocissus tricuspidata (Siebold et Zucc.) Planch. 地锦
Vines, deciduous

14.6 (3.9~23.8); 2.5 (-14.5~18.1); 25.7 (16.7~30.1); 14.8 (7.3~23.8); 123.8 (52.8~225.8); 8.4 (0.0~71.7); 823 (535~1297) ❙ 1101 (367~3495); 489 (235~1847); 96 (8~897) ❙ 31.4 (-43.4~246.5); 783 (367~1216); 201.4 (101.9~382.6)

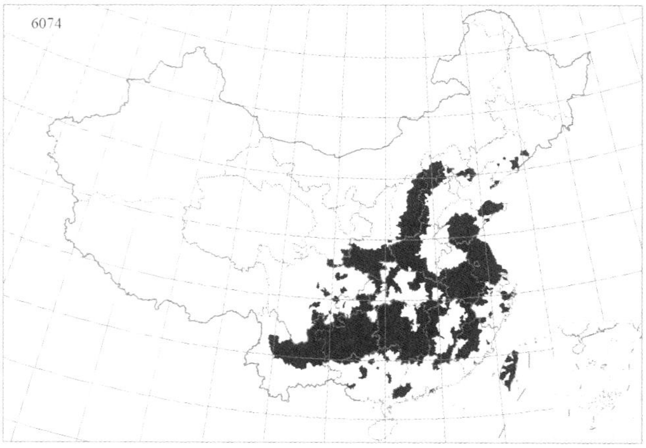

Vitaceae 葡萄科 *Tetrastigma* 崖爬藤属
Tetrastigma apiculatum Gagnep. 草崖藤
Vines

19.8 (18.2~21.6); 13.7 (12.4~15.1); 23.6 (21.6~27.9); 19.9 (18.2~21.6); 178.2 (158.6~199.6); 0.0 (0.0~0.0); 967 (861~1143) ❙ 1503 (1451~1607); 728 (655~785); 70 (52~102) ❙ 56.8 (27.0~74.2); 950 (839~1143); 490.9 (335.1~659.0)

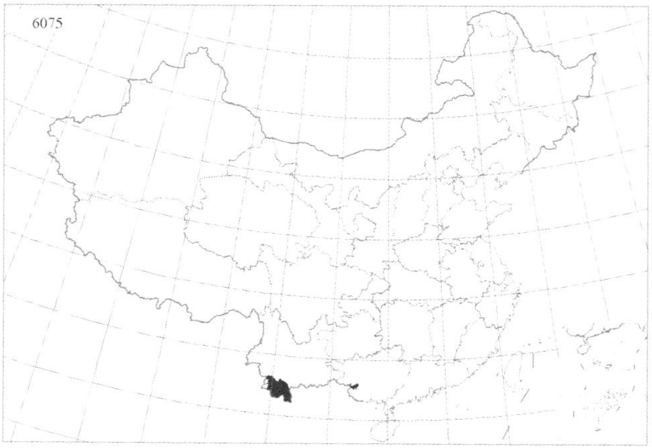

Vitaceae 葡萄科 *Tetrastigma* 崖爬藤属
Tetrastigma campylocarpum (Kurz) Planch. 多花崖爬藤
Vines

18.5 (16.3~20.5); 12.5 (10.2~15.1); 22.3 (20.3~23.7); 18.6 (16.4~20.5); 163.0 (136.6~185.9); 0.0 (0.0~0.0); 888 (789~992) ❙ 1367 (1028~1751); 735 (566~996); 54 (41~77) ❙ 53.8 (12.5~88.8); 864 (780~986); 450.4 (339.5~659.0)

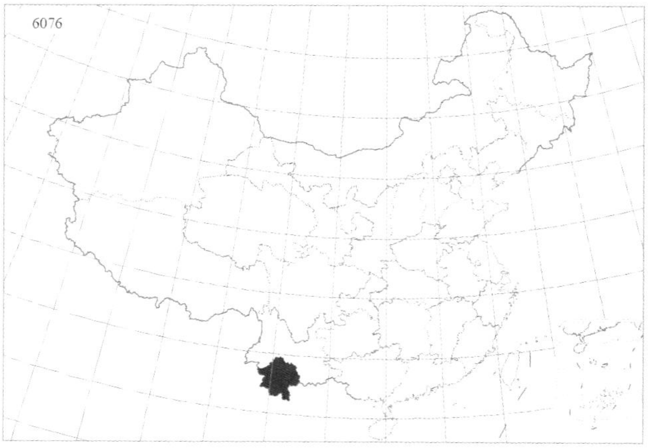

Vitaceae 葡萄科 *Tetrastigma* 崖爬藤属
Tetrastigma caudatum Merr. et Chun 尾叶崖爬藤
Vines

20.4 (16.3~24.8); 12.4 (7.1~20.1); 26.6 (21.4~29.0); 20.4 (16.4~24.8); 185.3 (136.4~237.6); 0.0 (0.0~0.0); 1065 (797~1385) ❙ 1501 (1007~1900); 672 (455~1056); 110 (41~189) ❙ 42.5 (1.5~85.5); 1045 (797~1326); 316.3 (157.7~620.0)

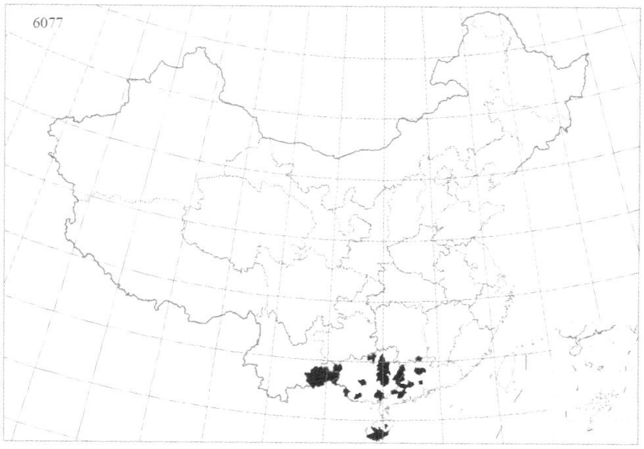

Vitaceae 葡萄科 *Tetrastigma* 崖爬藤属
Tetrastigma cauliflorum Merr. 茎花崖爬藤
Large climbers

21.1 (16.3~24.8); 15.2 (9.1~20.1); 25.2 (21.4~28.4); 21.2 (16.4~24.8); 194.3 (136.4~237.6); 0.0 (0.0~0.0); 1090 (797~1385) ❙ 1455 (1163~1767); 654 (392~996); 61 (41~92) ❙ 37.2 (-11.9~88.8); 1029 (797~1299); 488.8 (304.0~659.0)

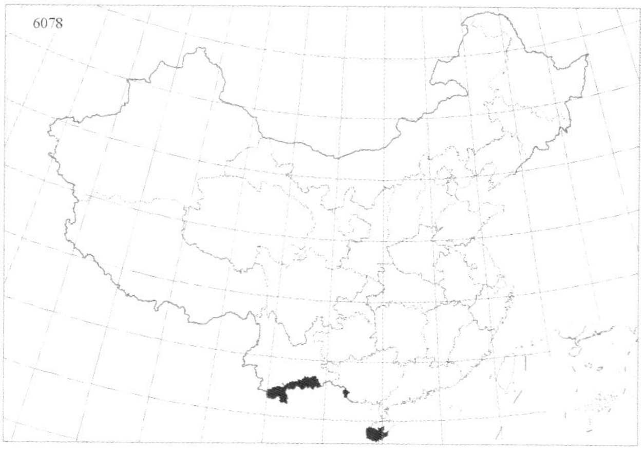

Vitaceae 葡萄科 *Tetrastigma* 崖爬藤属
Tetrastigma ceratopetalum C. Y. Wu 角花崖爬藤
Vines

17.7 (14.7~2[illegible].9); 9.4 (3.8~14.2); 24.2 (21.4~28.6); 17.7 (14.8~22.0); 152.6 (118.7~203.8); 0.2 (0.0~1.5); 888 (791~1176) ❙ 1287 (1074~1522); 660 (463~860); 67 (47~138) ❙ 46.0 (5.8~80.6); 887 (791~1176); 335.6 (170.4~551.6)

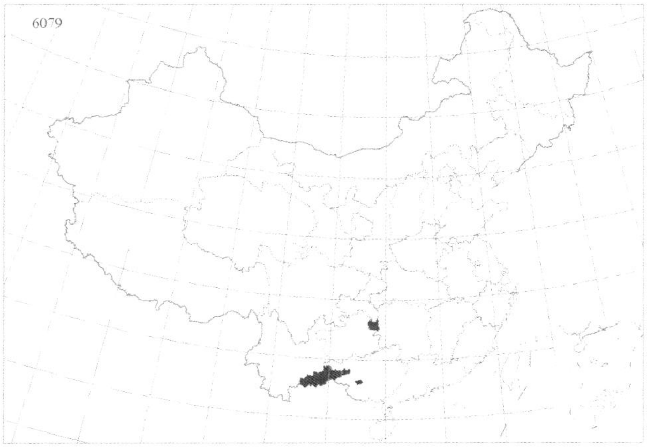

Vitaceae 葡萄科 *Tetrastigma* 崖爬藤属
Tetrastigma cruciatum Craib et Gagnep. 十字崖爬藤
Vines

19.3 (18.2~20.5); 13.5 (12.4~15.1); 22.8 (21.6~23.7); 19.3 (18.2~20.5); 171.9 (158.6~185.9); 0.0 (0.0~0.0); 923 (861~992) ❙ 1497 (1276~1751); 771 (655~996); 59 (41~77) ❙ 62.3 (36.8~88.8); 905 (839~986); 512.5 (427.9~659.0)

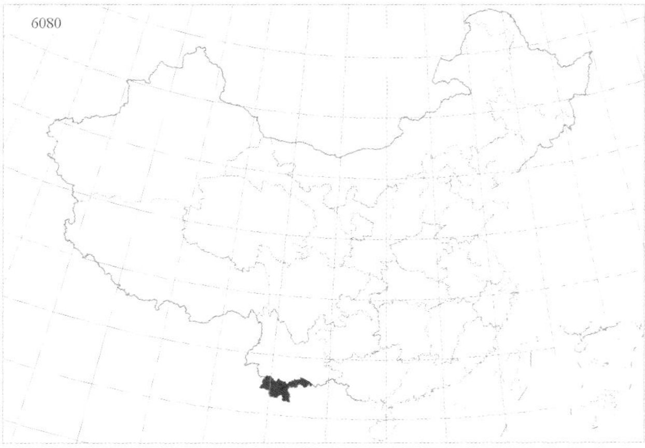

Vitaceae 葡萄科 *Tetrastigma* 崖爬藤属
Tetrastigma delavayi Gagnep. 七小叶崖爬藤
Vines

17.5 (8.6~21.3); 10.7 (1.3~15.1); 22.2 (15.0~27.9); 17.6 (8.7~21.3); 150.8 (52.6~196.0); 0.1 (0.0~8.8); 858 (574~1123) ❙ 1294 (958~1751); 713 (531~996); 53 (41~77) ❙ 51.8 (5.8~112.2); 848 (574~1121); 408.5 (254.3~659.0)

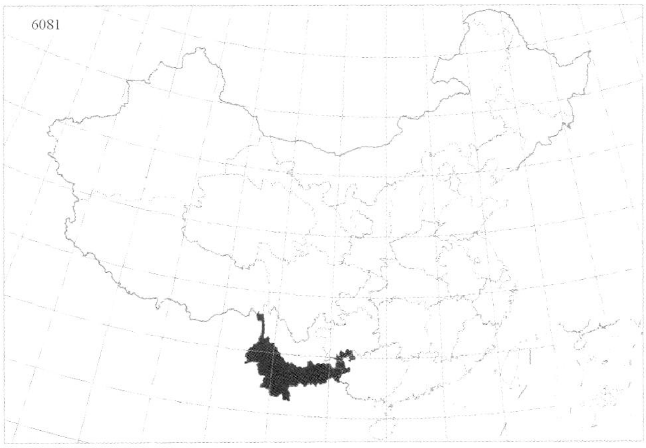

Vitaceae 葡萄科 *Tetrastigma* 崖爬藤属
Tetrastigma erubescens Planch. 红枝崖爬藤
Vines

20.3 (16.2~24.8); 13.7 (7.3~20.1); 25.4 (20.1~29.0); 20.3 (16.2~24.8); 184.0 (134.2~237.6); 0.0 (0.0~0.0); 1034 (773~1385) ❙ 1852 (1178~3495); 853 (473~1847); 154 (41~897) ❙ 85.4 (1.5~246.5); 1002 (773~1360); 348.6 (119.4~659.0)

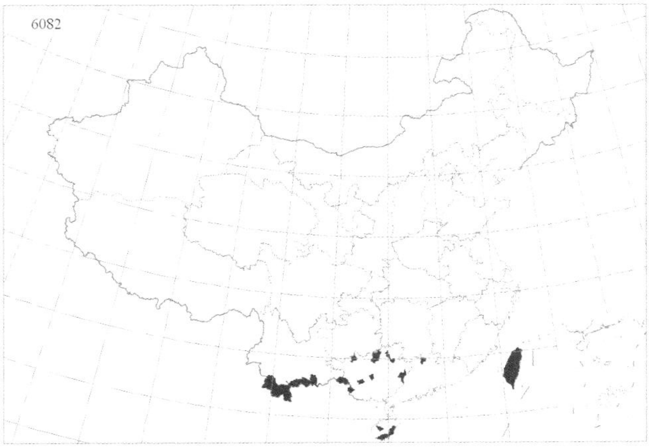

Vitaceae 葡萄科 *Tetrastigma* 崖爬藤属
Tetrastigma erubescens var. monophyllum Gagnep. 单叶红枝崖爬藤
Vines

18.8 (16.3~21.8); 12.0 (9.1~15.1); 23.7 (21.4~28.1); 18.9 (16.4~21.9); 166.6 (136.4~202.6); 0.0 (0.0~0.0); 925 (797~1151) ❙ 1450 (1163~1876); 785 (641~996); 66 (48~124) ❙ 56.8 (34.8~88.8); 925 (797~1151); 427.4 (221.6~659.0)

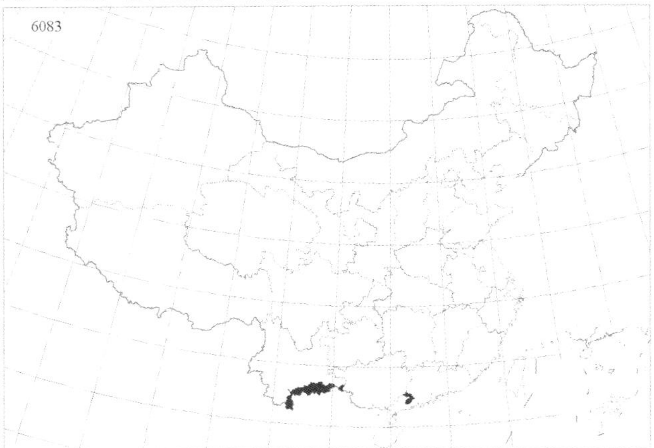

Vitaceae 葡萄科 *Tetrastigma* 崖爬藤属
Tetrastigma formosanum (Hemsl.) Gagnep. 台湾崖爬藤
Vines, evergreen

20.4 (16.2~24.1); 15.1 (11.0~18.9); 24.5 (20.1~28.2); 20.4 (16.2~24.1); 184.9 (134.2~229.7); 0.0 (0.0~0.0); 1027 (773~1322) ❙ 2443 (1751~2992); 1298 (900~1847); 167 (47~402) ❙ 151.5 (32.5~246.5); 976 (773~1144); 244.2 (119.4~363.7)

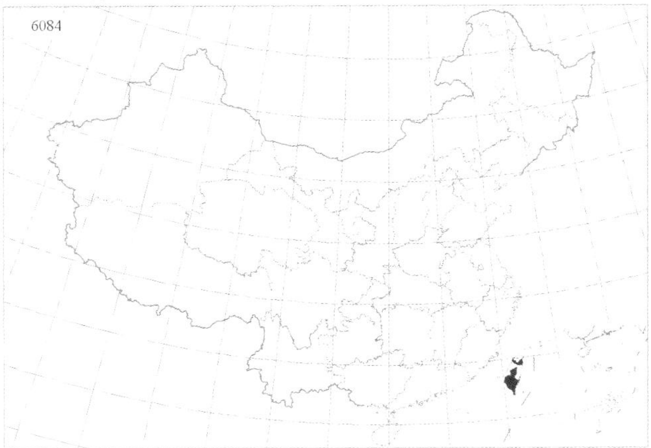

Vitaceae 葡萄科 *Tetrastigma* 崖爬藤属
Tetrastigma funingense C. L. Li 富宁崖爬藤
Vines

18.6 (17.2~21.0); 11.8 (9.5~14.2); 23.4 (22.4~25.6); 18.7 (17.3~21.0); 163.8 (147.0~191.8); 0.0 (0.0~0.0); 904 (833~1053) ❙ 1436 (1254~1751); 816 (699~996); 58 (48~77) ❙ 59.4 (36.8~88.8); 903 (833~1053); 451.9 (336.3~551.6)

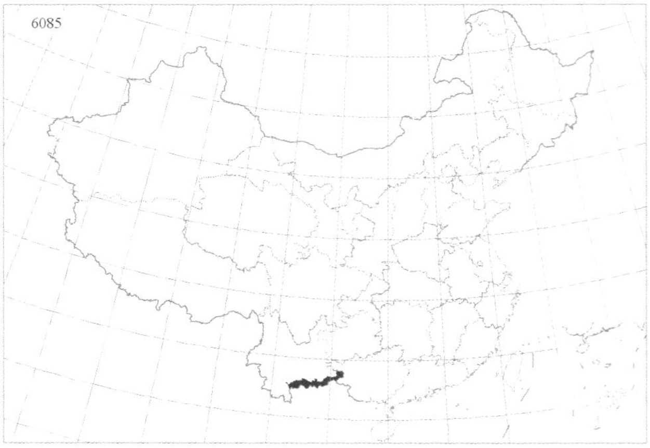

Vitaceae 葡萄科 *Tetrastigma* 崖爬藤属
Tetrastigma godefroyanum Planch. 柄果崖爬藤
Vines

22.8; 16.6; 27.1; 22.8; 213.7; 0.0; 1217 ❙ 1225; 537; 70 ❙ 6.7; 1037; 400.7

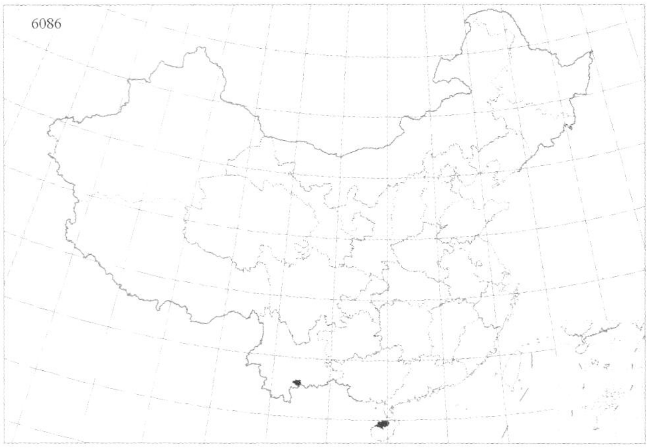

Vitaceae 葡萄科 *Tetrastigma* 崖爬藤属
Tetrastigma hemsleyanum Diels et Gilg 三叶崖爬藤
Vines

19.5 (16.4~24.1); 14.1 (11.5~18.9); 23.9 (20.3~28.0); 19.5 (16.5~24.1); 174.5 (137.8~229.7); 0.0 (0.0~0.0); 960 (783~1322) ❙ 2565 (1751~2993); 1210 (900~1847); 239 (47~547) ❙ 174.8 (32.5~235.6); 937 (783~1123); 268.9 (154.2~363.7)

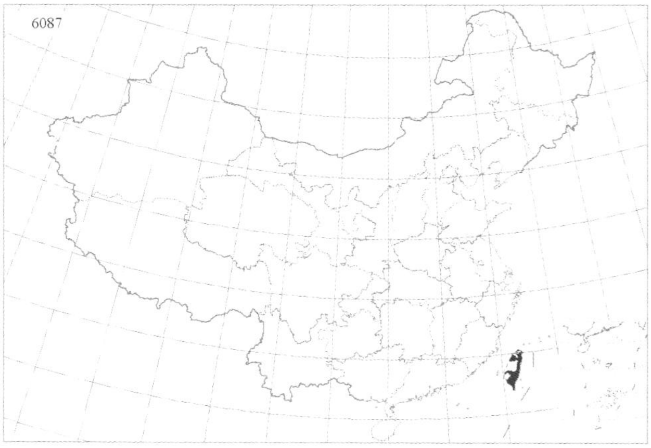

Vitaceae 葡萄科 *Tetrastigma* 崖爬藤属
Tetrastigma henryi Gagnep. 蒙自崖爬藤
Vines

18.0 (13.2~21.6); 11.3 (6.4~15.1); 22.5 (18.4~28.5); 18.1 (13.3~21.7); 156.7 (99.4~199.8); 0.0 (0.0~0.0); 876 (695~1153) ❙ 1339 (958~1751); 739 (531~996); 56 (41~173) ❙ 53.6 (5.8~112.2); 865 (695~1153); 418.2 (237.8~659.0)

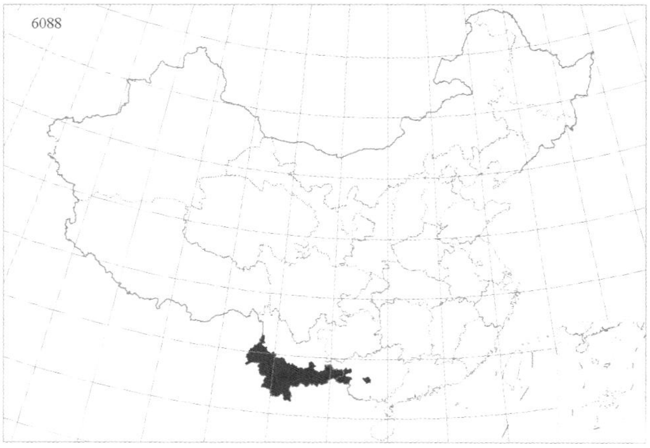

Vitaceae 葡萄科 *Tetrastigma* 崖爬藤属
Tetrastigma hypoglaucum Planch. 叉须崖爬藤
Vines

15.2 (−1.8~21.0); 6.8 (−10.2~15.1); 22.2 (6.1~27.5); 15.3 (1.8~21.0); 125.3 (1.9~192.3); 2.1 (0.0~82.8); 790 (329~1081) ❙ 1164 (320~1776); 597 (193~1007); 59 (12~161) ❙ 48.0 (−14.2~145.8); 780 (320~1081); 280.3 (72.2~659.0)

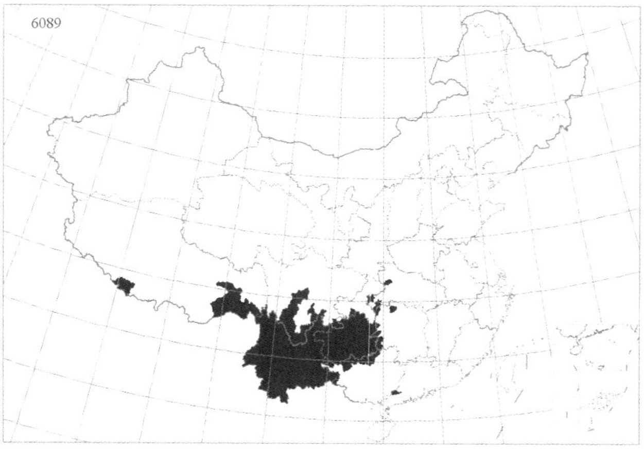

Vitaceae 葡萄科 *Tetrastigma* 崖爬藤属
Tetrastigma jinghongense C. L. Li 景洪崖爬藤
Vines

20.5; 14.9; 23.7; 20.5; 185.9; 0.0; 992 ❙ 1464; 655; 66 ❙ 47.6; 959; 570.1

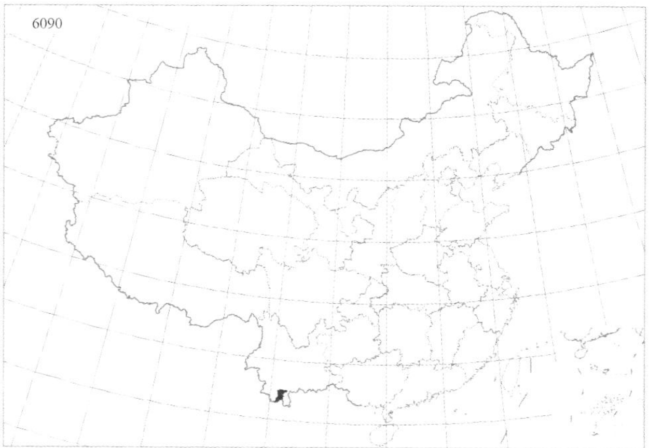

Vitaceae 葡萄科 *Tetrastigma* 崖爬藤属
Tetrastigma jinxiuense C. L. Li 金秀崖爬藤
Vines

18.5；9.7；25.4；18.6；163.0；0.0；929 | 1494；746；113 | 61.0；929；307.8

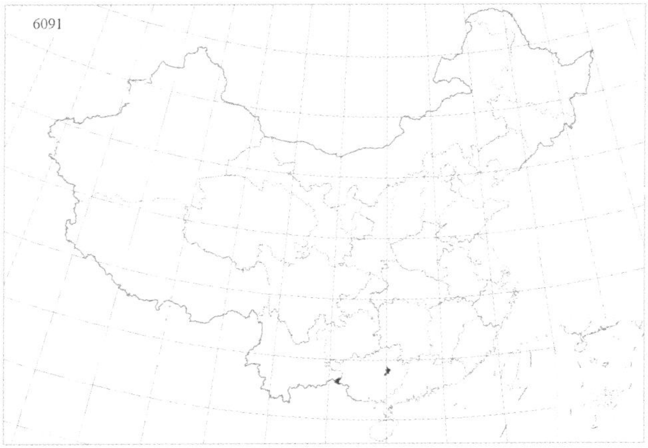

Vitaceae 葡萄科 *Tetrastigma* 崖爬藤属
Tetrastigma kwangsiense C. L. Li 广西崖爬藤
Vines

20.7；12.5；26.7；20.7；188.5；0.0；1068 | 1406；766；72 | 33.8；1068；349.7

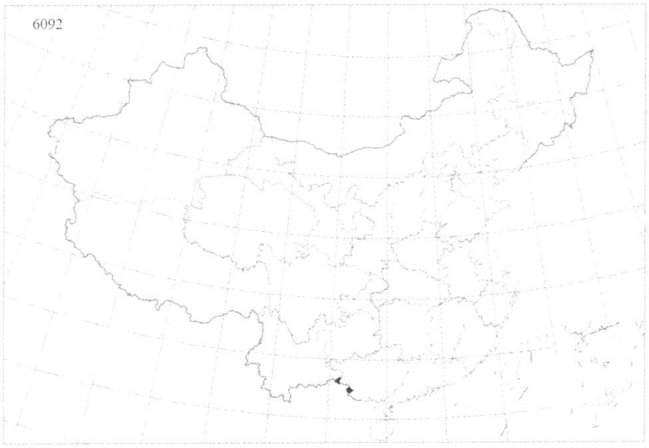

Vitaceae 葡萄科 *Tetrastigma* 崖爬藤属
Tetrastigma lanyuense Chang 兰屿崖爬藤
Vines

18.5；14.1；21.7；18.5；162.1；0.0；863 | 2713；1388；188 | 214.4；863；363.7

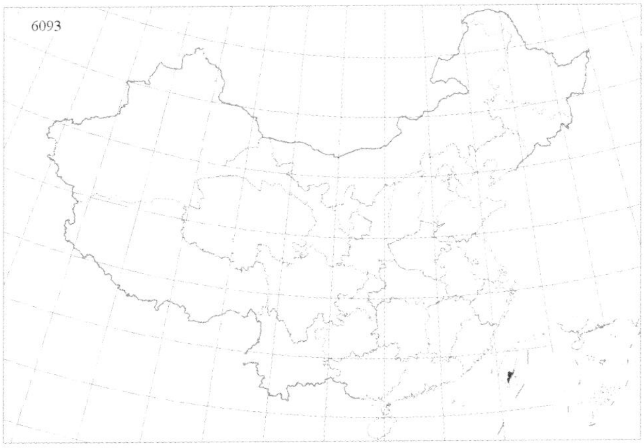

Vitaceae 葡萄科 *Tetrastigma* 崖爬藤属
Tetrastigma lenticellatum C. Y. Wu ex W. T. Wang 显孔崖爬藤
Vines

19.5 (18.2~20.5)；13.9 (12.4~15.1)；22.7 (21.6~23.7)；19.5 (18.2~20.5)；173.9 (158.6~185.9)；0.0 (0.0~0.0)；931 (861~992) | 1494 (1434~1607)；705 (655~785)；56 (41~77) | 60.9 (47.6~74.2)；901 (839~986)；531.1 (439.5~659.0)

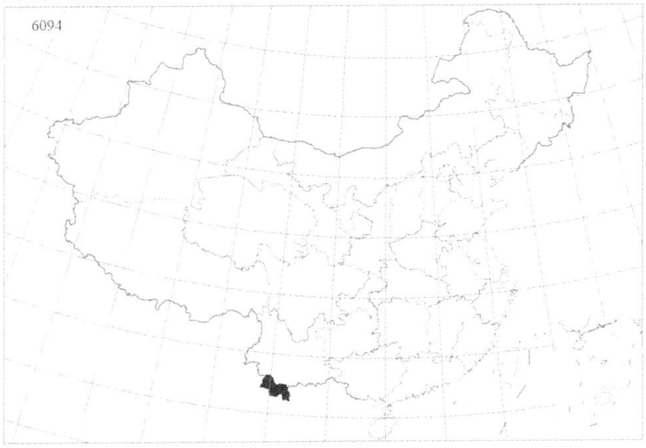

Vitaceae 葡萄科 *Tetrastigma* 崖爬藤属
Tetrastigma lincangense C. L. Li 临沧崖爬藤
Vines

16.3；10.2；20.3；16.4；136.6 0.0；789 | 1232；703；47 | 56.2；780；372.4

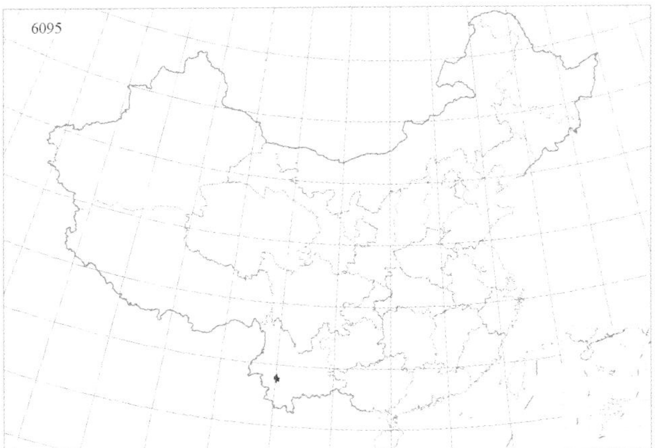

Vitaceae 葡萄科 *Tetrastigma* 崖爬藤属
Tetrastigma longipedunculatum C. L. Li 长梗崖爬藤
Vines

21.6；13.1；27.9；21.6；199.6；0.0；1143 | 1451；766；102 | 27.0；1143；335.1

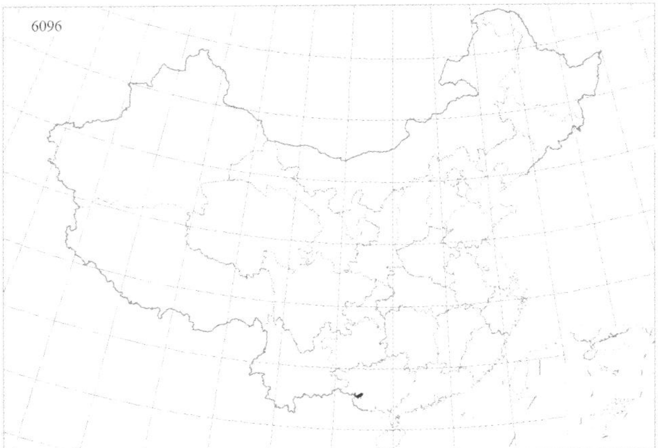

Vitaceae 葡萄科 *Tetrastigma* 崖爬藤属
Tetrastigma obtectum var. glabrum (H. Lév.) Gagnep. 无毛崖爬藤
Vines

19.5 (16.2~24.1); 14.1 (11.0~18.9); 23.9 (20.1~28.0); 19.6 (16.2~24.1); 174.8 (134.2~229.7); 0.0 (0.0~0.0); 970 (773~1322) ❙ 2520 (1751~2993); 1212 (924~1847); 225 (47~547) ❙ 170.2 (32.5~246.5); 943 (773~1123); 251.5 (154.2~303.6)

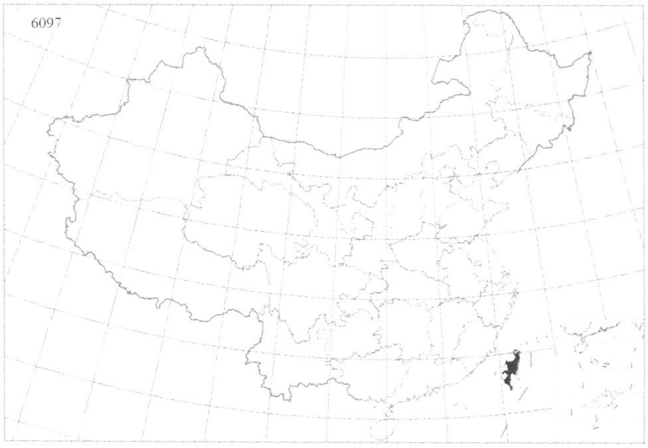

Vitaceae 葡萄科 *Tetrastigma* 崖爬藤属
Tetrastigma papillatum (Hance) C. Y. Wu 海南崖爬藤
Vines

20.9 (18.3~24.8); 12.9 (9.2~20.1); 27.0 (25.0~28.8); 21.0 (18.4~24.8); 191.4 (160.9~237.6); 0.0 (0.0~0.0); 1097 (924~1385) ❙ 1392 (1188~1567); 670 (473~848); 80 (41~141) ❙ 29.0 (-1.4~55.9); 1069 (924~1331); 332.9 (232.6~538.2)

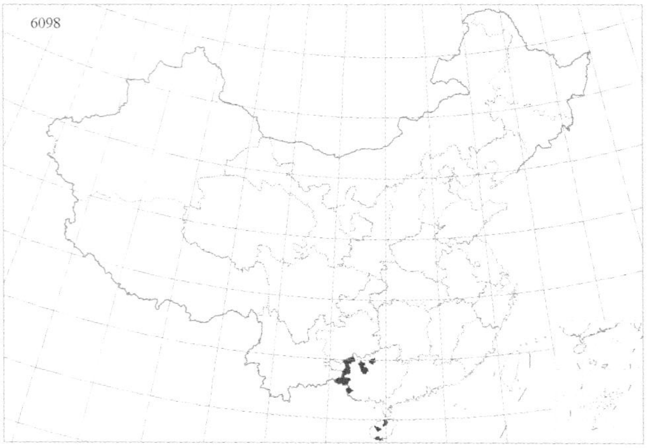

Vitaceae 葡萄科 *Tetrastigma* 崖爬藤属
Tetrastigma pseudocruciatum C. L. Li 过山崖爬藤
Vines

22.6 (19.1~24.8); 15.7 (10.7~20.1); 27.8 (23.6~29.0); 22.7 (19.1~24.8); 212.3 (169.6~237.6); 0.0 (0.0~0.0); 1216 (945~1385) ❙ 1467 (868~2035); 642 (312~997); 85 (41~194) ❙ 22.4 (-34.3~71.8); 1132 (868~1299); 365.2 (160.7~659.0)

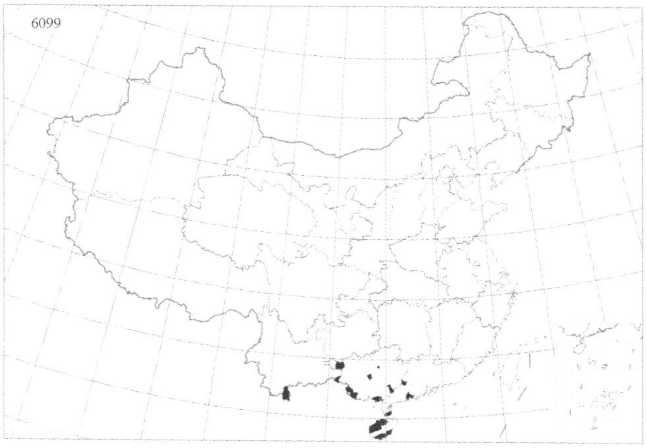

Vitaceae 葡萄科 *Tetrastigma* 崖爬藤属
Tetrastigma pubinerve Merr. et Chun 毛脉崖爬藤
Vines

21.8 (19.1~24.8); 14.3 (10.5~20.1); 27.4 (23.6~28.9); 21.9 (19.1~24.8); 202.6 (169.6~237.6); 0.0 (0.0~0.0); 1154 (945~1385) ❙ 1573 (1374~1900); 747 (473~1056); 102 (41~194) ❙ 37.6 (1.5~71.0); 1122 (945~1257); 368.5 (194.5~659.0)

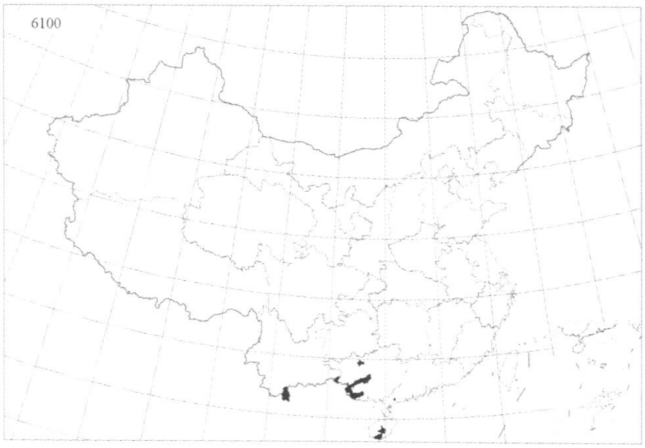

Vitaceae 葡萄科 *Tetrastigma* 崖爬藤属
Tetrastigma retinervium var. pubescens C. L. Li 柔毛网脉崖爬藤
Vines

19.8 (17.3~22.2); 11.3 (8.7~13.9); 26.3 (21.6~28.3); 19.9 (17.3~22.3); 178.5 (148.0~207.5); 0.0 (0.0~0.0); 1016 (832~1190) ❙ 1358 (1049~1581); 726 (598~850); 79 (44~106) ❙ 34.3 (5.8~55.9); 1013 (783~1190); 311.0 (275.8~371.7)

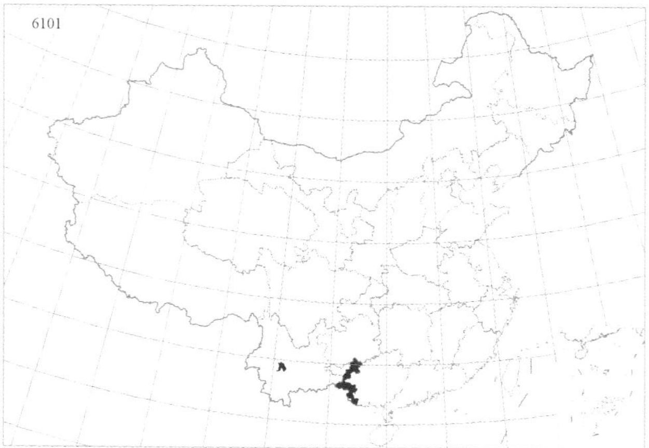

Vitaceae 葡萄科 *Tetrastigma* 崖爬藤属
Tetrastigma rumicispermum (Lawson) Planch. 喜马拉雅崖爬藤
Vines

16.9 (-1.8~21.0); 10.3 (-10.2~15.1); 21.3 (6.1~25.6); 17.0 (1.8~21.0); 145.1 (1.9~191.8); 2.1 (0.0~82.8); 833 (329~1053) ❙ 1308 (320~1751); 725 (217~1007); 52 (20~77) ❙ 57.7 (-3.0~145.8); 821 (320~1053); 416.7 (72.2~659.0)

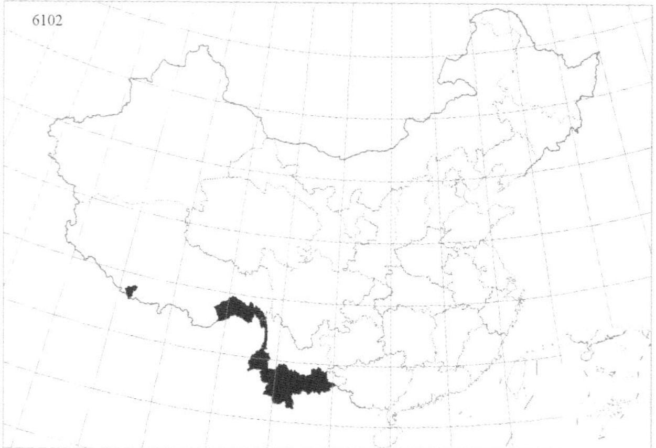

Vitaceae 葡萄科 *Tetrastigma* 崖爬藤属

Tetrastigma rumicispermum var. lasiogynum (W. T. Wang) C. L. Li 锈毛喜马拉雅崖爬藤 Vines

17.6 (10.8~21.0); 11.2 (3.9~15.1); 21.9 (16.6~25.6); 17.6 (10.9~21.0); 151.8 (71.8~191.8); 0.0 (0.0~1.3); 855 (633~1053) ❙ 1343 (968~1751); 744 (539~996); 53 (24~77) ❙ 57.8 (12.5~112.2); 842 (633~1053); 432.7 (277.0~659.0)

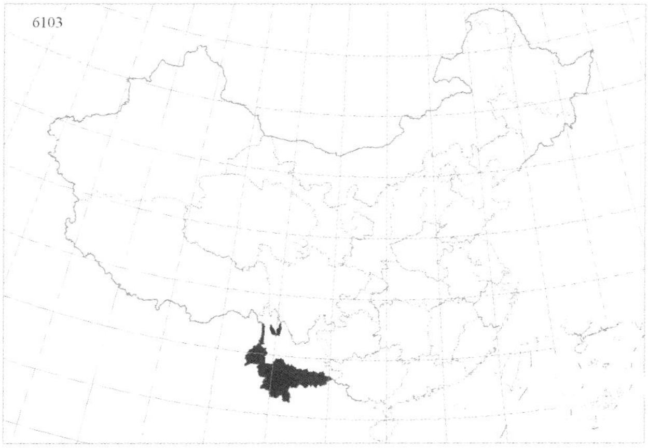

Vitaceae 葡萄科 *Tetrastigma* 崖爬藤属

Tetrastigma serrulatum var. puberulum W. T. Wang 毛狭叶崖爬藤 Vines

5.9 (-1.1~11.3); -1.8 (-9.1~4.3); 12.6 (6.2~17.0); 6.8 (2.0~11.3); 39.8 (2.5~76.6); 28.3 (0.7~75.6); 506 (334~644) ❙ 860 (321~1243); 493 (193~696); 36 (13~58) ❙ 63.9 (-4.0~100.7); 504 (321~644); 261.9 (79.3~384.7)

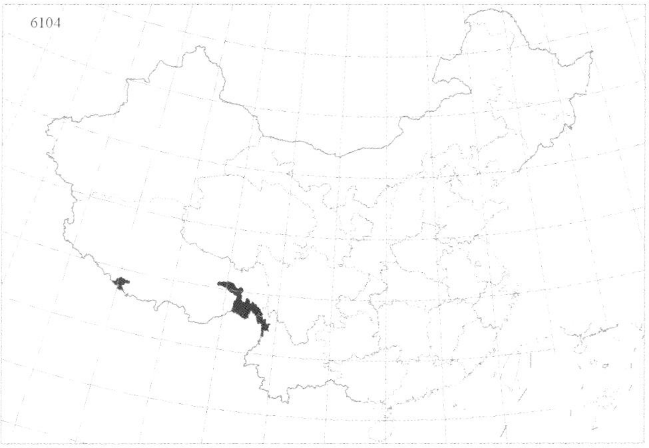

Vitaceae 葡萄科 *Tetrastigma* 崖爬藤属

Tetrastigma sichouense C. L. Li 西畴崖爬藤 Vines

18.5 (16.3~21.7); 12.0 (9.1~15.1); 22.8 (20.8~28.4); 18.5 (16.4~21.8); 162.3 (136.4~201.3); 0.0 (0.0~0.0); 894 (797~1162) ❙ 1332 (939~1751); 726 (524~996); 54 (39~97) ❙ 49.3 (6.5~88.8); 877 (783~1162); 422.5 (273.6~659.0)

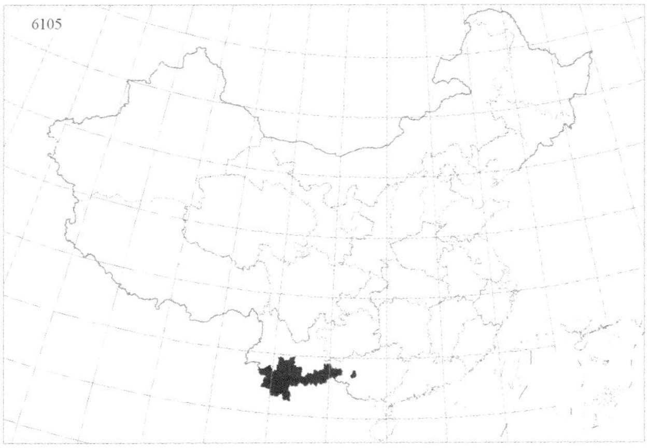

Vitaceae 葡萄科 *Tetrastigma* 崖爬藤属

Tetrastigma sichouense var. megalocarpum C. L. Li 大果西畴崖爬藤 Vines

20.0 (13.3~24.8); 11.6 (3.8~20.1); 26.6 (19.6~29.0); 20.1 (13.4~24.8); 180.8 (100.4~237.6); 0.0 (0.0~1.5); 1037 (706~1385) ❙ 1517 (939~2435); 713 (455~1428); 116 (39~196) ❙ 47.4 (1.5~145.8); 1030 (706~1326); 289.6 (139.6~659.0)

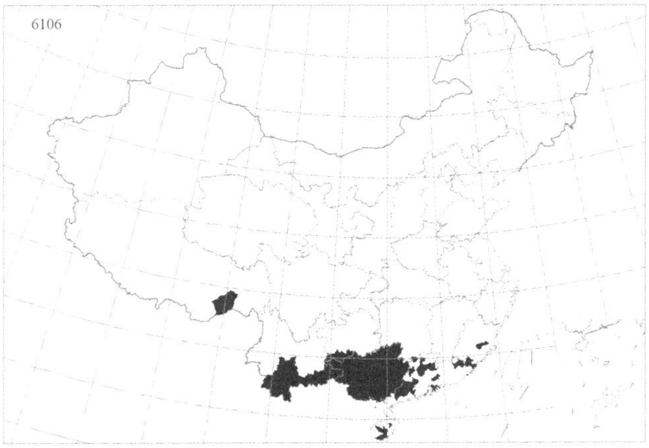

Vitaceae 葡萄科 *Tetrastigma* 崖爬藤属

Tetrastigma subtetragonum C. L. Li 红花崖爬藤 Vines

19.4; 12.6; 24.1; 19.4; 172.6; 0.0; 954 ❙ 1445; 787; 64 ❙ 52.3; 954; 521.5

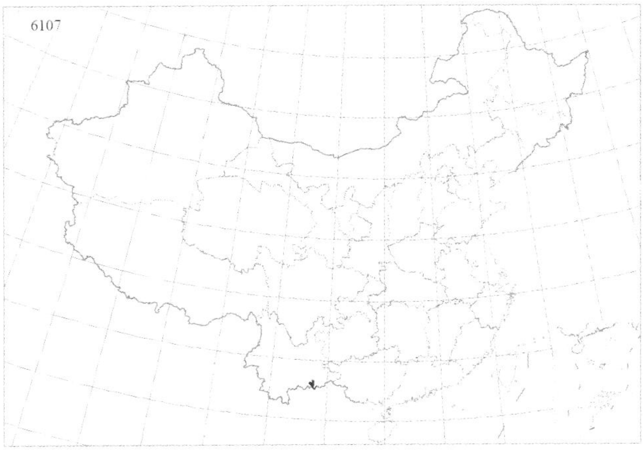

Vitaceae 葡萄科 *Tetrastigma* 崖爬藤属

Tetrastigma tonkinense Gagnep. 越南崖爬藤 Vines

20.4; 11.7; 27.0; 20.4; 185.1; 0.0; 1056 ❙ 1440; 749; 107 ❙ 36.8; 1056; 285.3

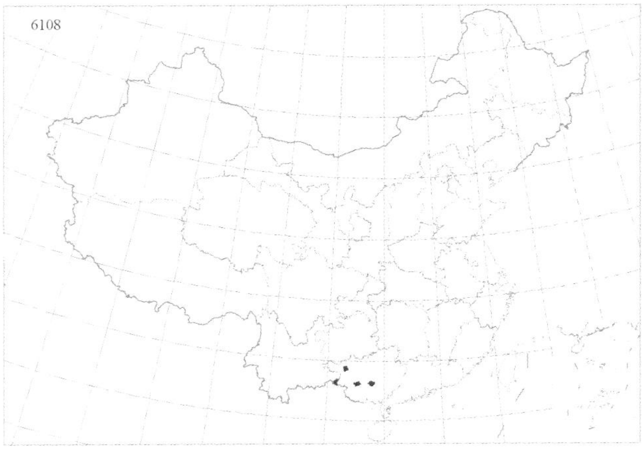

Vitaceae 葡萄科 *Tetrastigma* 崖爬藤属
Tetrastigma tsaianum C. Y. Wu 蔡氏崖爬藤
Vines

17.8；11.1；22.0；17.8；154.1；0.0；855 ❙ 1530；909；51 ❙ 78.9；855；482.8

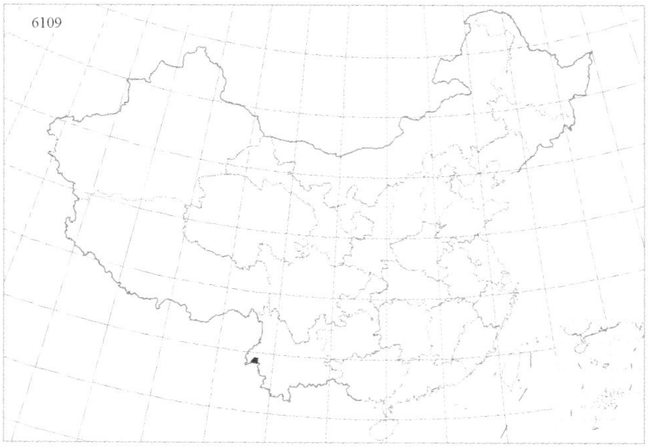

Vitaceae 葡萄科 *Tetrastigma* 崖爬藤属
Tetrastigma venulosum C. Y. Wu 马关崖爬藤
Vines

18.8；12.0；23.5；18.8；165.5；0.0；914 ❙ 1455；818；60 ❙ 60.2；914；490.7

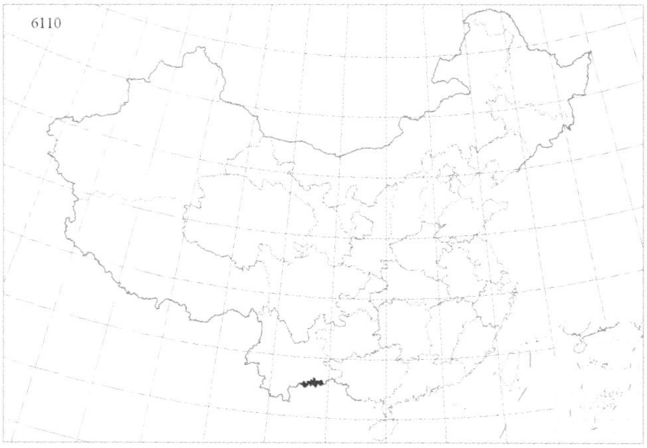

Vitaceae 葡萄科 *Tetrastigma* 崖爬藤属
Tetrastigma xishuangbannaense C. L. Li 西双版纳崖爬藤
Vines

20.0；14.6；23.2；20.1；180.8；0.0；963 ❙ 1523；681；67 ❙ 58.4；942；581.0

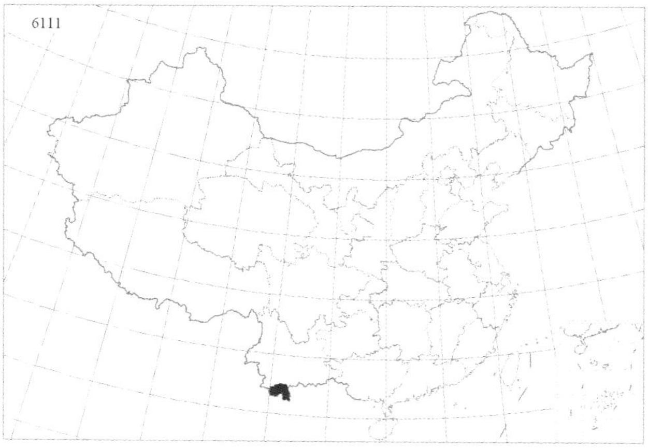

Vitaceae 葡萄科 *Tetrastigma* 崖爬藤属
Tetrastigma yiwuense C. L. Li 易武崖爬藤
Vines

18.9；12.5；23.2；19.0；167.5；0.0；914 ❙ 1028；566；45 ❙ 12.5；851；339.5

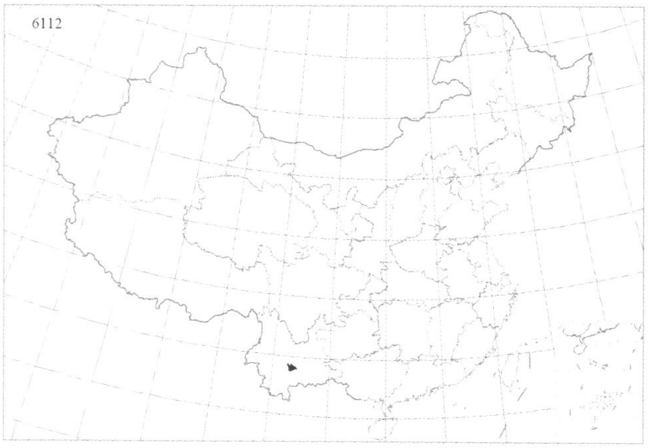

Vitaceae 葡萄科 *Vitis* 葡萄属
Vitis amurensis Rupr. 山葡萄
Vines, deciduous

9.8 (−0.2~18.6)；−6.3 (−23.8~9.3)；24.2 (17.5~28.9)；11.4 (6.4~18.7)；93.5 (45.4~164.0)；35.4 (0.0~112.6)；710 (507~965) ❙ 684 (132~1702)；390 (81~691)；39 (2~220) ❙ −4.3 (−80.1~101.3)；625 (132~965)；190.0 (13.6~346.5)

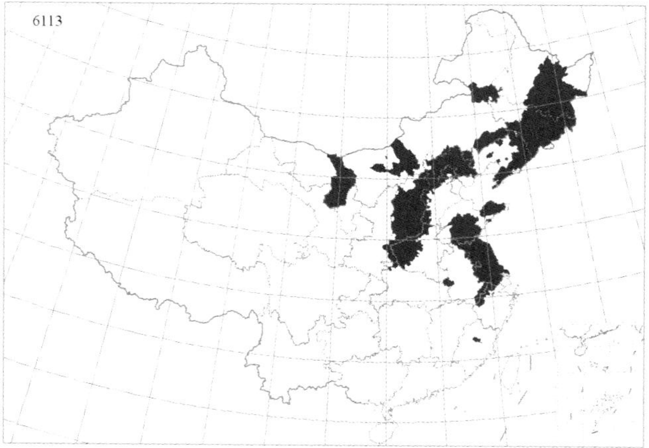

Vitaceae 葡萄科 *Vitis* 葡萄属
Vitis balansana Planch. 小果葡萄
Vines, deciduous

20.3 (5.3~24.8)；12.5 (−2.4~20.1)；26.3 (12.2~29.1)；20.3 (5.7~24.8)；184.3 (30.5~237.6)；0.4 (0.0~26.2)；1063 (479~1385) ❙ 1537 (868~2435)；701 (312~1428)；105 (31~259) ❙ 47.9 (−34.3~130.7)；1036 (479~1360)；333.9 (139.6~659.0)

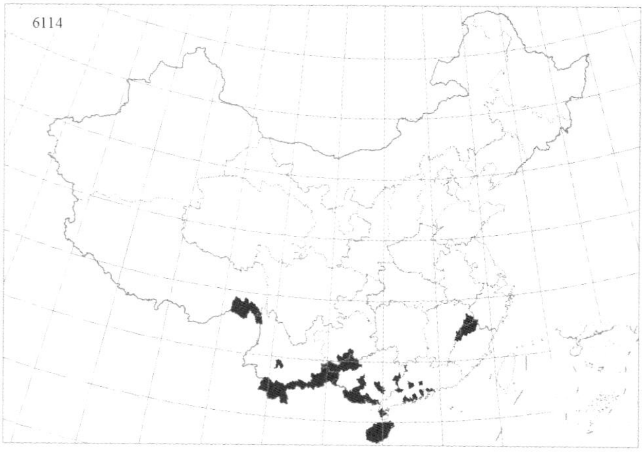

Vitaceae 葡萄科 *Vitis* 葡萄属
Vitis balansana var. ficifolioides (W. T. Wang) C. L. Li 龙州葡萄
Vines, deciduous

20.6；11.5；28.0；20.7；187.8；0.0；1090 ❙ 1535；665；140 ❙ 43.5；1090；282.3

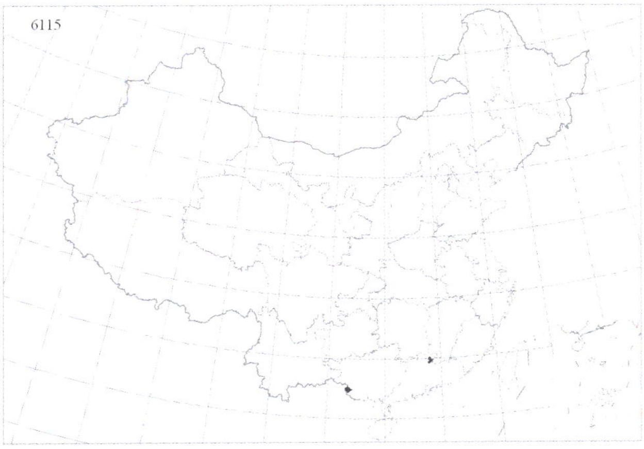

Vitaceae 葡萄科 *Vitis* 葡萄属
Vitis bellula (Rehder) W. T. Wang 美丽葡萄
Vines, deciduous

16.0 (6.6~21.0)；6.3 (-1.3~14.2)；24.5 (13.3~29.3)；16.0 (6.8~21.0)；133.6 (39.6~191.8)；1.3 (0.0~19.7)；837 (514~1072) ❙ 1319 (817~1777)；599 (378~996)；95 (12~192) ❙ 57.8 (-2.0~101.8)；835 (514~1072)；269.9 (148.9~551.6)

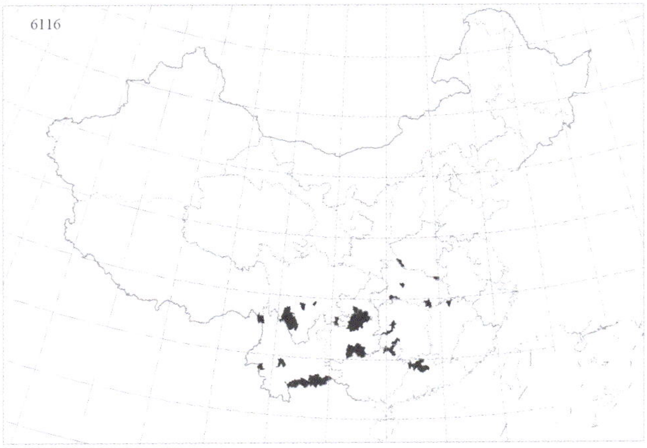

Vitaceae 葡萄科 *Vitis* 葡萄属
Vitis bryoniifolia Bunge 蘡薁
Vines, deciduous

14.6 (3.8~24.1)；2.3 (-13.7~18.0)；25.8 (10.8~30.1)；14.9 (4.7~24.2)；125.5 (21.6~230.2)；10.0 (0.0~71.7)；832 (444~1329) ❙ 1100 (367~2037)；464 (235~1056)；104 (8~259) ❙ 28.9 (-41.7~132.1)；773 (367~1329)；199.8 (101.9~572.9)

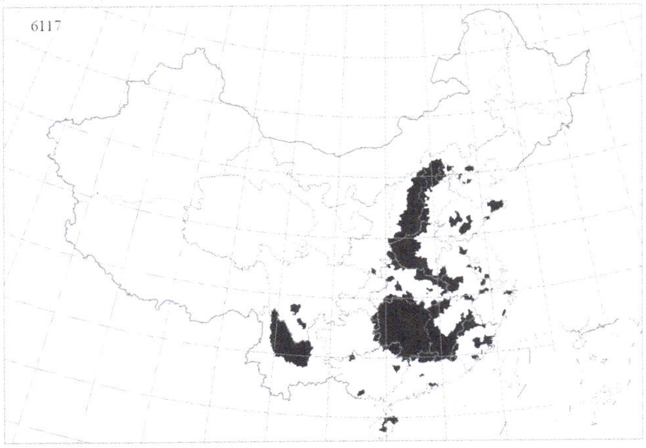

Vitaceae 葡萄科 *Vitis* 葡萄属
Vitis bryoniifolia var. ternata (W. T. Wang) C. L. Li 三出蘡薁
Vines, deciduous

14.9 (13.2~16.5)；1.7 (-0.9~4.4)；27.6 (26.2~28.9)；15.0 (13.3~16.6)；125.6 (112.8~139.8)；6.0 (0.6~13.7)；852 (787~910) ❙ 944 (687~1489)；469 (393~565)；99 (38~197) ❙ 10.5 (-23.1~82.7)；826 (687~910)；174.3 (134.1~284.0)

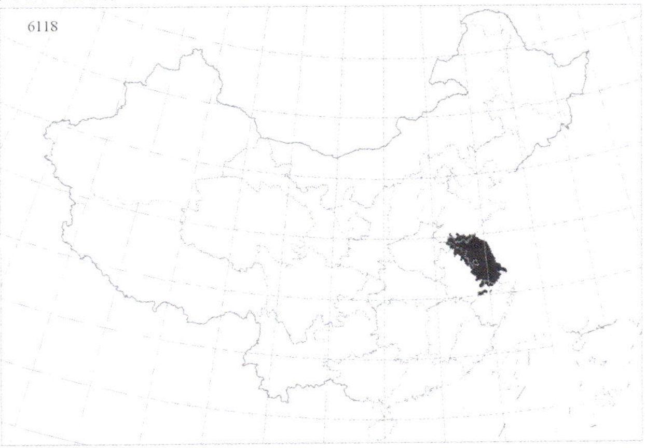

Vitaceae 葡萄科 *Vitis* 葡萄属
Vitis chunganensis Hu 东南葡萄
Vines, deciduous

17.3 (14.3~21.0)；6.6 (3.1~11.9)；27.1 (23.5~30.1)；17.4 (14.3~21.1)；148.7 (112.6~193.0)；0.2 (0.0~2.3)；914 (756~1114) ❙ 1576 (1187~1997)；562 (403~821)；183 (60~259) ❙ 72.9 (27.5~130.7)；907 (756~1114)；237.4 (148.9~332.5)

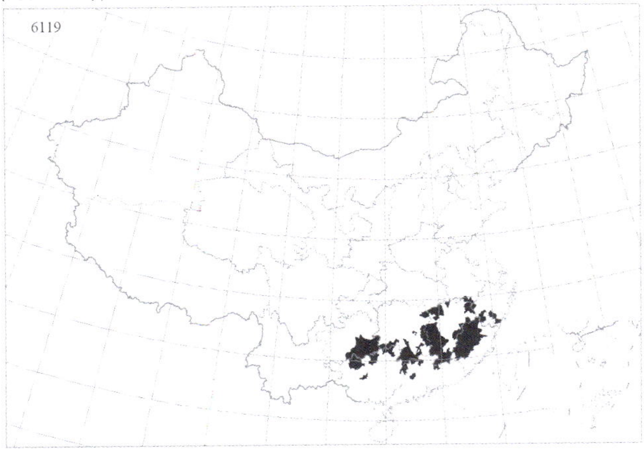

Vitaceae 葡萄科 *Vitis* 葡萄属
Vitis chungii F. P. Metcalf 闽赣葡萄
Vines, deciduous

18.3 (15.0~22.0)；8.5 (3.9~13.5)；27.1 (24.7~29.0)；18.3 (15.1~22.0)；159.6 (120.8~203.8)；0.0 (0.0~1.1)；950 (789~1170) ❙ 1713 (1371~1997)；604 (507~726)；200 (132~259) ❙ 82.3 (17.4~147.0)；948 (789~1170)；267.4 (203.4~332.5)

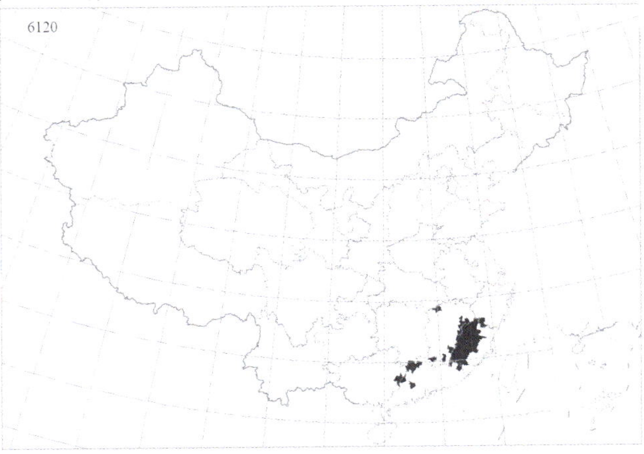

Vitaceae 葡萄科 *Vitis* 葡萄属
Vitis davidii (Rom. Caill.) Foex. 刺葡萄
Vines, deciduous

15.7 (2.2~21.0); 4.6 (-6.6~14.2); 26.0 (9.9~29.5); 15.7 (3.8~21.0); 131.1 (15.5~191.8); 2.4 (0.0~49.1); 846 (413~1053) ❙ 1303 (512~1997); 526 (258~938); 120 (14~259) ❙ 53.7 (-35.3~130.7); 836 (413~1053); 217.0 (118.7~551.6)

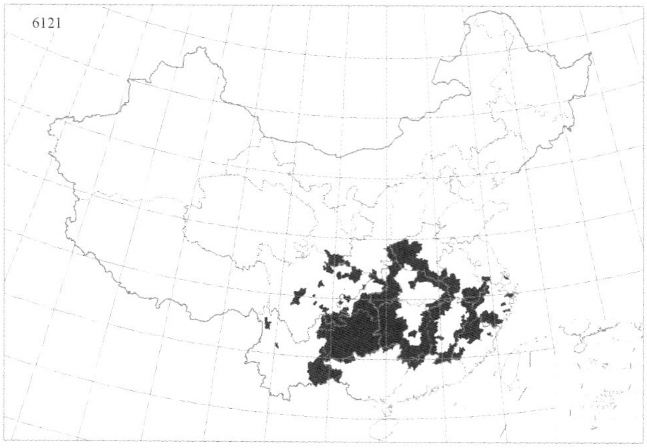

Vitaceae 葡萄科 *Vitis* 葡萄属
Vitis davidii var. cyanocarpa (Gagnep.) Gagnep. 蓝果刺葡萄
Vines, deciduous

14.7 (9.3~17.6); 4.4 (-1.7~9.9); 24.0 (18.8~28.0); 14.7 (9.5~17.7); 120.1 (69.8~152.4); 3.4 (0.0~18.0); 786 (603~866) ❙ 1341 (942~1770); 598 (421~860); 112 (38~223) ❙ 70.3 (31.7~112.2); 786 (603~866); 288.0 (200.5~413.5)

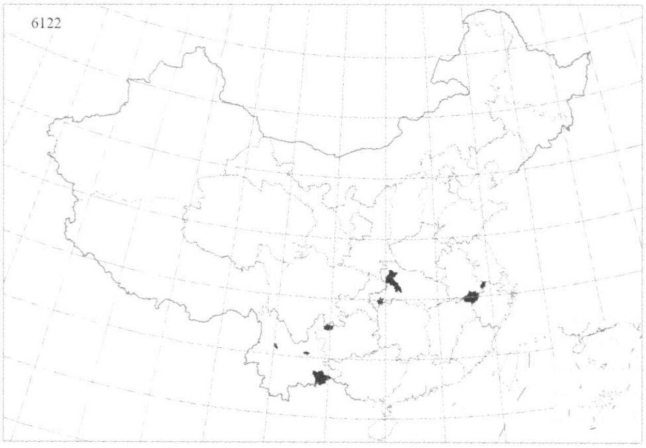

Vitaceae 葡萄科 *Vitis* 葡萄属
Vitis davidii var. ferruginea Merr. et Chun 锈毛刺葡萄
Vines, deciduous

17.1 (14.6~19.1); 6.7 (3.8~10.0); 26.5 (24.6~27.8); 17.1 (14.7~19.1); 145.5 (117.7~169.1); 0.2 (0.0~1.5); 892 (779~973) ❙ 1761 (1564~1997); 604 (544~669); 216 (176~259) ❙ 98.0 (66.9~130.7); 892 (779~973); 261.9 (215.5~303.4)

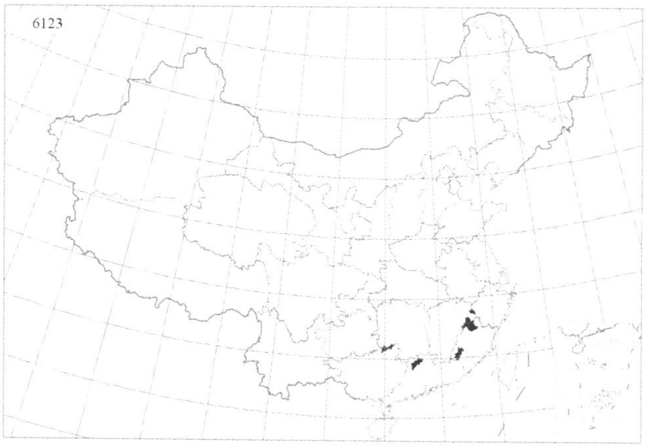

Vitaceae 葡萄科 *Vitis* 葡萄属
Vitis erythrophylla W. T. Wang 红叶葡萄
Vines, deciduous

16.1 (15.3~17.0); 4.5 (3.9~5.7); 27.2 (25.6~28.3); 16.1 (15.4~17.0); 134.3 (124.9~144.3); 0.6 (0.0~1.1); 868 (810~912) ❙ 1697 (1536~1821); 576 (512~671); 219 (200~242) ❙ 96.1 (73.0~124.8); 868 (810~912); 244.5 (212.3~273.1)

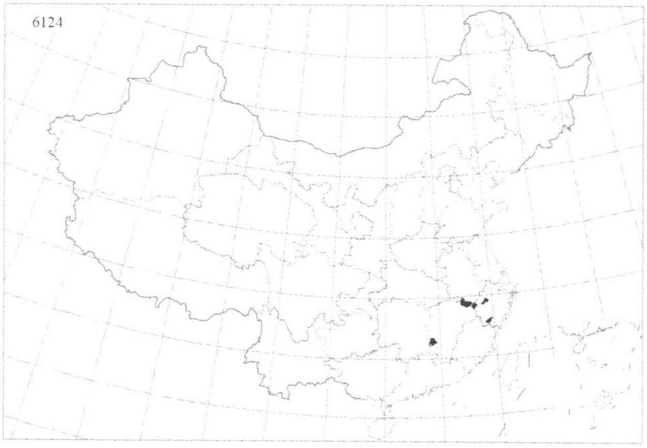

Vitaceae 葡萄科 *Vitis* 葡萄属
Vitis fengqinensis C. L. Li 凤庆葡萄
Vines, deciduous

16.1; 9.6; 20.6; 16.2; 133.9; 0.0; 785 ❙ 1083; 600; 50 ❙ 38.0; 779; 405.0

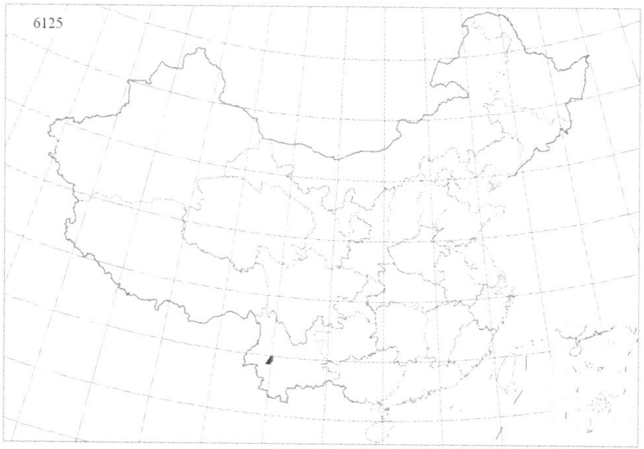

Vitaceae 葡萄科 *Vitis* 葡萄属
Vitis flexuosa Thunberg 葛藟葡萄
Vines, deciduous

15.5 (2.0~24.1); 4.7 (-7.2~19.1); 25.4 (9.5~29.6); 15.6 (3.8~24.1); 130.9 (15.4~229.7); 3.7 (0.0~51.4); 844 (411~1322) ❙ 1192 (457~2992); 537 (258~1847); 96 (11~259) ❙ 40.9 (-44.9~202.5); 821 (411~1221); 229.7 (108.2~659.0)

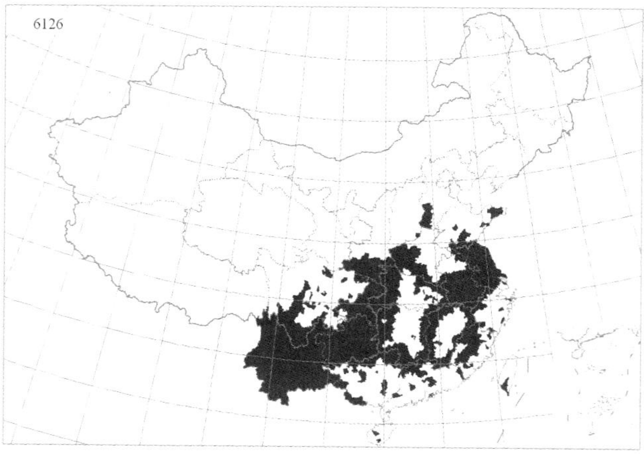

Vitaceae 葡萄科 *Vitis* 葡萄属
Vitis hancockii Hance 菱叶葡萄
Vines, deciduous

16.8 (14.4~18.8); 5.4 (2.1~9.3); 27.7 (25.8~30.1); 16.8 (14.5~18.9); 142.5 (118.7~166.5); 0.6 (0.0~4.9); 902 (801~1024) ❙ 1680 (1039~1997) 577 (381~685); 213 (124~259) ❙ 87.2 (7.5~130.7); 887 (801~965); 246.3 (147.5~319.9)

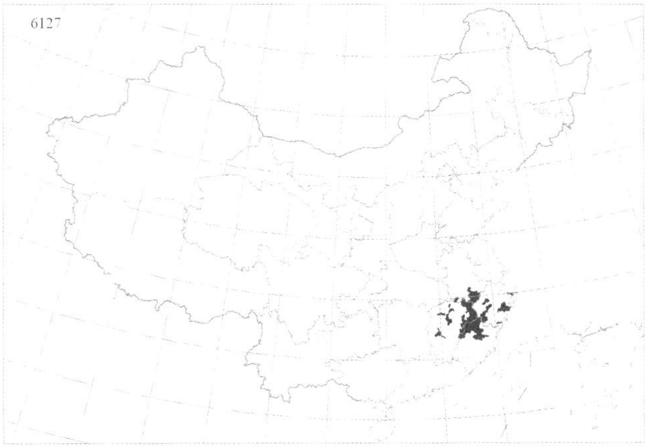

Vitaceae 葡萄科 *Vitis* 葡萄属
Vitis heyneana Roem. et Schult. 毛葡萄
Vines, deciduous

14.9 (−1.1~24.1); 4.2 (−12.5~18.9); 24.5 (6.2~29.4); 15.1 (2.0~24.1); 125.3 (2.5~229.7); 6.0 (0.0~75.6); 817 (334~1322) ❙ 1165 (321~2993); 534 (193~1847); 90 (8~547) ❙ 41.5 (−37.2~246.5); 798 (321~1221); 236.7 (61.6~659.0)

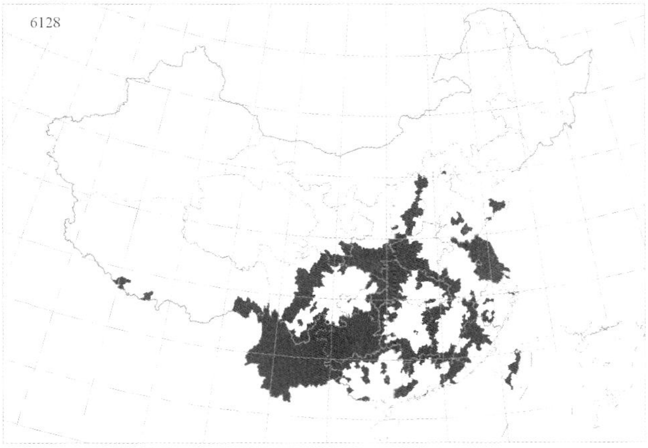

Vitaceae 葡萄科 *Vitis* 葡萄属
Vitis heyneana subsp. ficifolia (Bunge) C. L. Li 桑叶葡萄
Vines, deciduous

12.0 (2.5~17.1); −2.3 (−13.7~4.9); 25.1 (13.1~29.4); 12.6 (5.0~17.1); 104.2 (27.9~146.1); 19.5 (0.1~71.7); 756 (449~952) ❙ 741 (341~1621); 393 (190~635); 48 (8~205) ❙ −3.1 (−43.4~108.0); 673 (341~930); 173.9 (61.2~334.4)

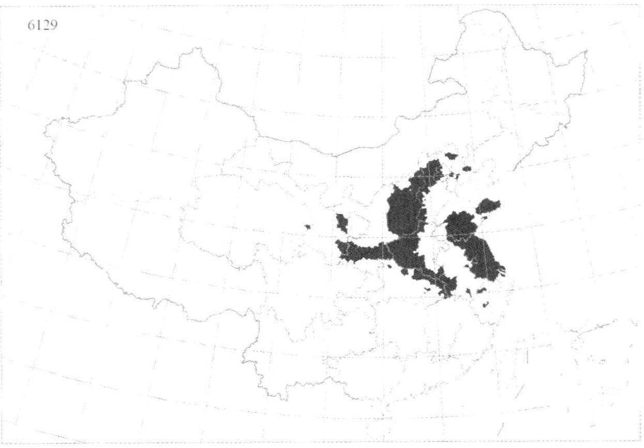

Vitaceae 葡萄科 *Vitis* 葡萄属
Vitis hui Cheng 庐山葡萄
Vines, deciduous

16.5 (15.7~17.4); 4.3 (3.5~5.1); 28.5 (27.5~29.5); 16.6 (15.7~17.5); 139.9 (130.5~149.7); 0.8 (0.0~1.8); 909 (862~959) ❙ 1541 (1444~1610); 592 (527~685); 180 (160~204) ❙ 69.7 (59.6~80.0); 909 (862~959); 225.7 (202.6~268.6)

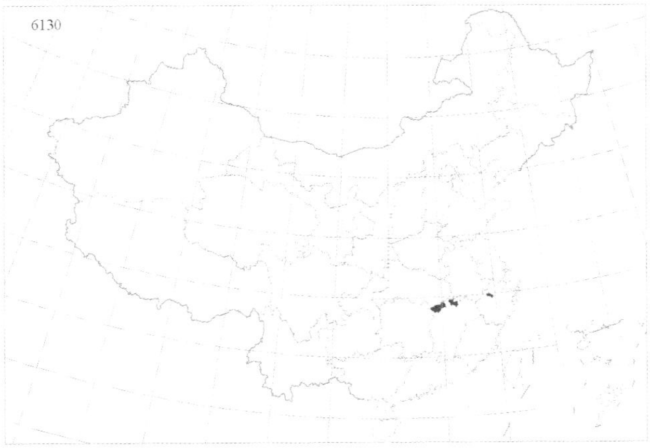

Vitaceae 葡萄科 *Vitis* 葡萄属
Vitis jinggangensis W. T. Wang 井冈葡萄
Vines, deciduous

16.0; 4.7; 26.7; 16.0; 132.8; 0.6; 853 ❙ 1557; 546; 176 ❙ 83.0; 853; 209.4

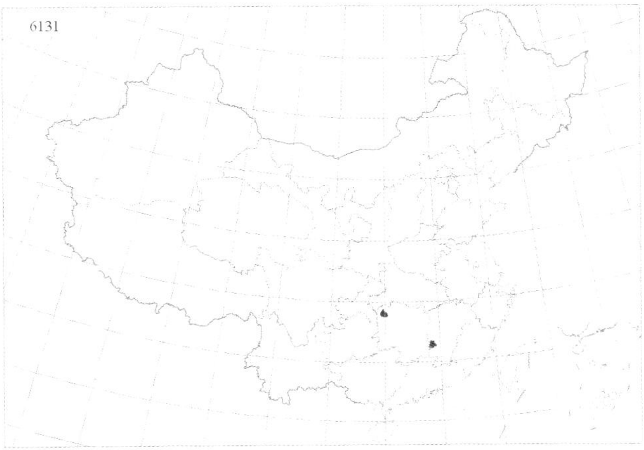

Vitaceae 葡萄科 *Vitis* 葡萄属
Vitis longquanensis P. L. Chiu 龙泉葡萄
Vines, deciduous

17.2 (15.6~18.2); 5.9 (4.9~6.9); 27.9 (25.9~29.1); 17.3 (15.6~18.2); 147.3 (127.5~158.8); 0.0 (0.0~0.1); 923 (824~984) ❙ 1839 (1759~1922); 595 (535~677); 248 (227~259) ❙ 100.0 (78.7~129.9); 913 (824~961); 240.5 (183.9~279.5)

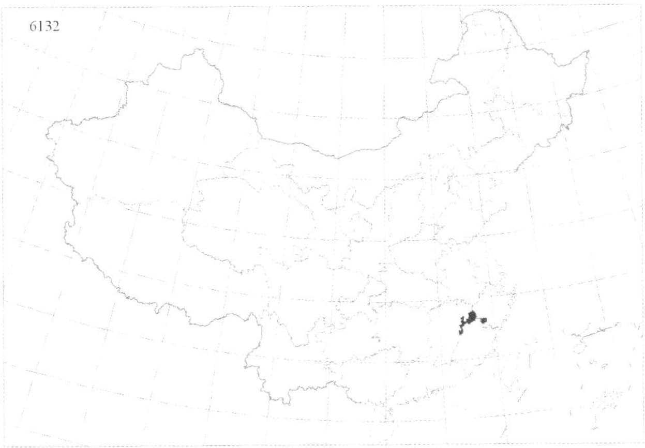

Vitaceae 葡萄科 *Vitis* 葡萄属

Vitis luochengensis W. T. Wang 罗城葡萄

Vines, deciduous

18.1; 8.0; 26.8; 18.2; 158.1; 0.0; 939 ❙ 1518; 581; 165 ❙ 62.2; 939; 244.7

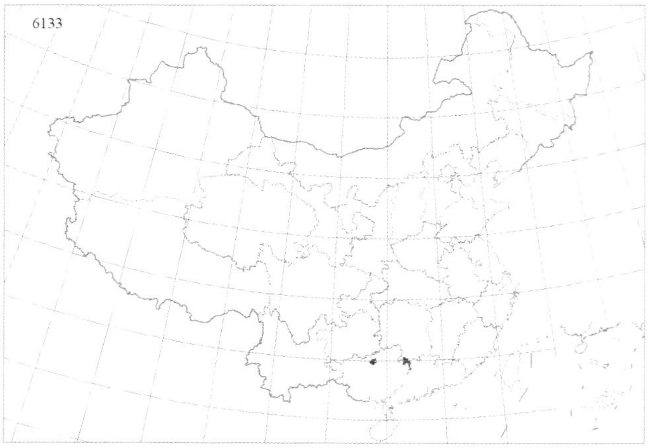

Vitaceae 葡萄科 *Vitis* 葡萄属

Vitis menghaiensis C. L. Li 勐海葡萄

Vines, deciduous

19.2; 13.8; 22.3; 19.2; 170.5; 0.0; 909 ❙ 1499; 686; 58 ❙ 64.9; 880; 513.9

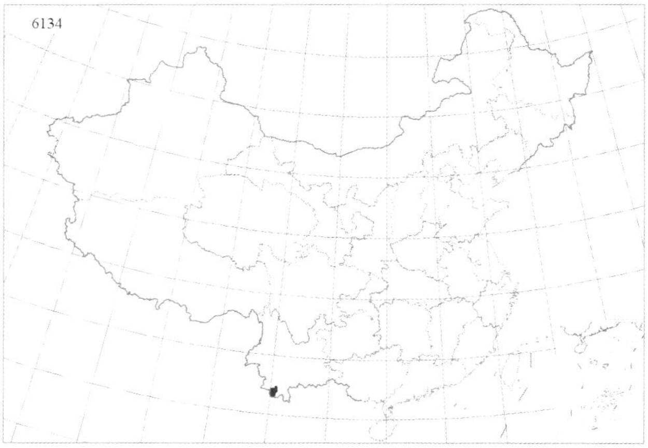

Vitaceae 葡萄科 *Vitis* 葡萄属

Vitis mengziensis C. L. Li 蒙自葡萄

Vines, deciduous

16.9; 10.3; 21.4; 16.9; 142.7; 0.0; 813 ❙ 1178; 646; 51 ❙ 44.9; 813; 383.4

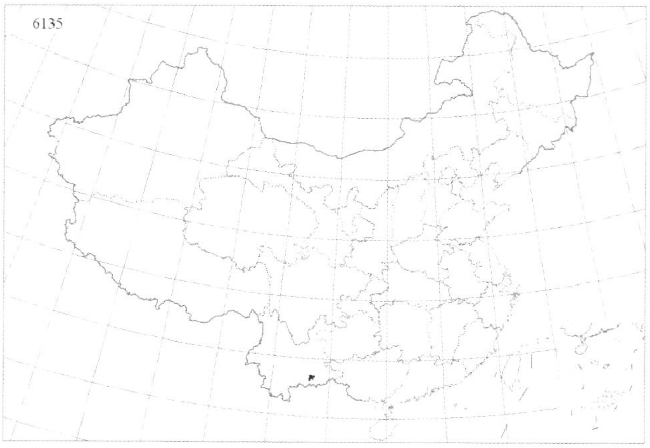

Vitaceae 葡萄科 *Vitis* 葡萄属

Vitis piasezkii Maxim. 变叶葡萄

Vines, deciduous

11.3 (1.9~19.4); -2.7 (-13.7~9.4); 24.0 (11.7~28.7); 12.0 (4.3~19.5); 97.5 (21.1~173.7); 21.4 (0.0~71.7); 723 (427~1038) ❙ 708 (341~1776); 359 (190~718); 33 (8~196) ❙ -2.5 (-40.9~101.4); 627 (341~1038); 188.2 (61.2~334.4)

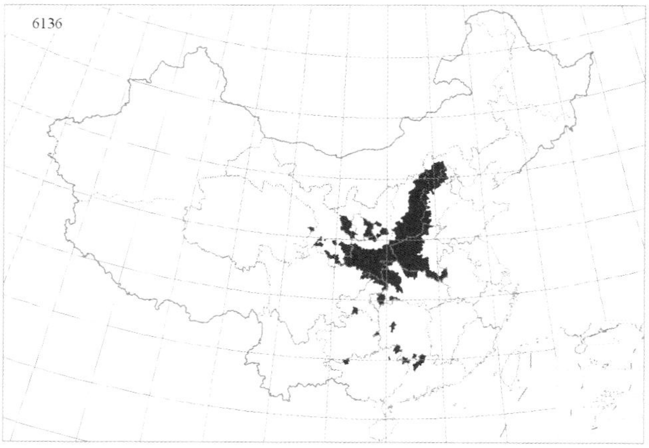

Vitaceae 葡萄科 *Vitis* 葡萄属

Vitis pilosonerva F. P. Metcalf 毛脉葡萄

Vines, deciduous

18.8 (17.7~20.6); 8.9 (7.1~11.6); 27.4 (26.6~28.0); 18.9 (17.7~20.7); 166.3 (152.6~187.9); 0.0 (0.0~0.0); 979 (916~1081) ❙ 1680 (1503~1779); 585 (513~642); 199 (182~220) ❙ 71.9 (50.5~93.8); 979 (916~1081); 245.6 (197.5~300.0)

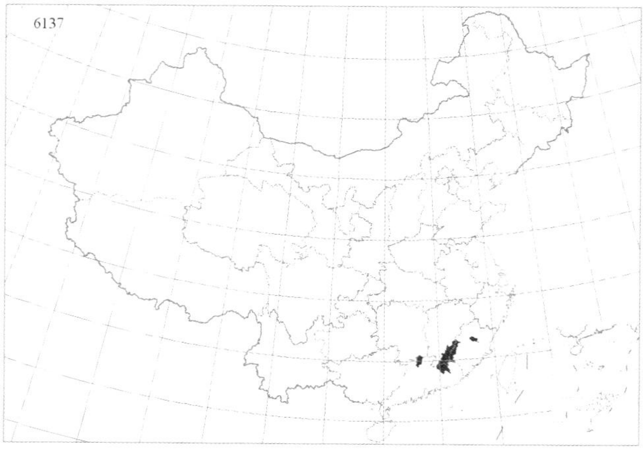

Vitaceae 葡萄科 *Vitis* 葡萄属

Vitis pseudoreticulata W. T. Wang 华东葡萄

Vines, deciduous

16.5 (12.3~21.0); 4.8 (-2.3~11.9); 27.6 (24.4~29.9); 16.6 (12.3~21.1); 140.3 (100.7~193.0); 1.6 (0.0~17.4); 896 (728~1114) ❙ 1469 (774~1922); 527 (377~821); 179 (32~259) ❙ 63.8 (-2.0~129.9); 888 (728~1114); 214.5 (111.9~300.0)

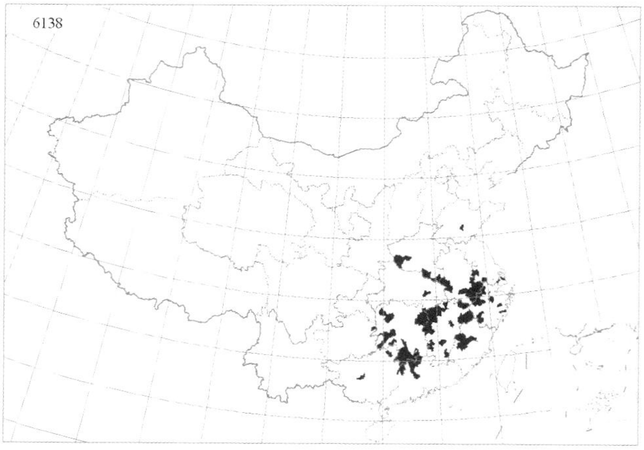

Vitaceae 葡萄科 *Vitis* 葡萄属
Vitis retordii Romanet du Caillaud ex Planchon 绵毛葡萄
Vines, deciduous

19.4 (14.6~23.0); 11.1 (3.8~17.7); 26.0 (21.4~29.0); 19.5 (14.7~23.`); 174.2 (117.7~217.0); 0.0 (0.0~1.5); 997 (779~1217) ❙ 1501 (1074~2037); 715 (544~1033); 106 (41~191) ❙ 51.3 (15.4~101.8); 993 (779~1216); 303.1 (139.6~659.0)

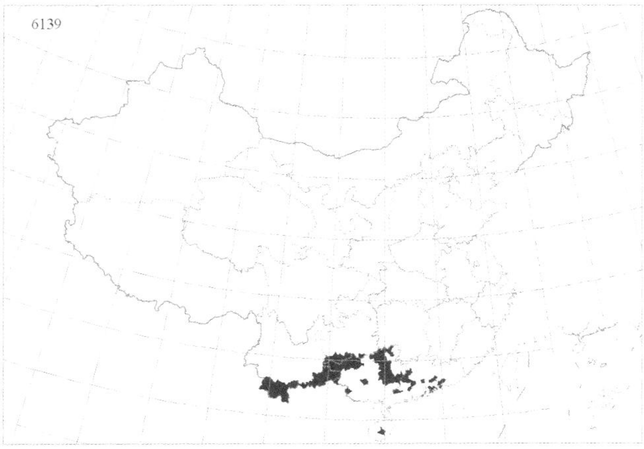

Vitaceae 葡萄科 *Vitis* 葡萄属
Vitis romanetii Rom. Caill. 秋葡萄
Vines, deciduous

13.6 (7.4~20.7); 1.1 (-4.2~10.5); 25.5 (15.7~29.8); 13.7 (8.2~20.8); 113.2 (58.8~189.3); 9.1 (0.0~29.4); 788 (561~1113) ❙ 971 (512~1705); 434 (258~666); 67 (15~220) ❙ 22.5 (-35.3~108.0); 760 (512~1032); 210.0 (111.9~351.7)

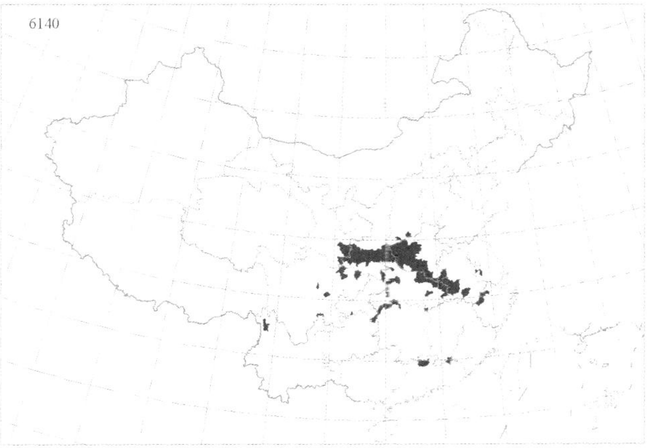

Vitaceae 葡萄科 *Vitis* 葡萄属
Vitis shenxiensis C. L. Li 陕西葡萄
Vines, deciduous

11.6; -1.2; 23.9; 11.8; 94.6; 14.5; 705 ❙ 769; 337; 30 ❙ 9.1; 703; 240.2

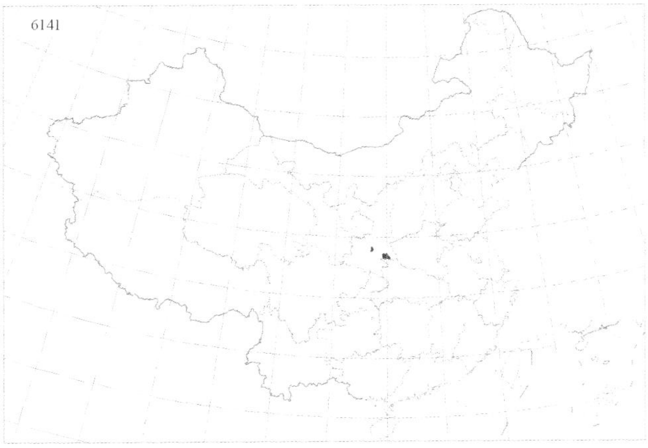

Vitaceae 葡萄科 *Vitis* 葡萄属
Vitis silvestrii Pamp. 湖北葡萄
Vines, deciduous

14.5 (9.3~19.1); 3.0 (-1.7~10.1); 25.1 (19.4~28.1); 14.5 (9.5~19.1); 120.1 (69.8~169.2); 5.7 (0.0~18.0); 804 (603~969) ❙ 1001 (807~1334); 471 (349~720); 55 (25~75) ❙ 26.1 (-6.5~10`.4); 797 (603~969); 223.3 (133.5~292.6)

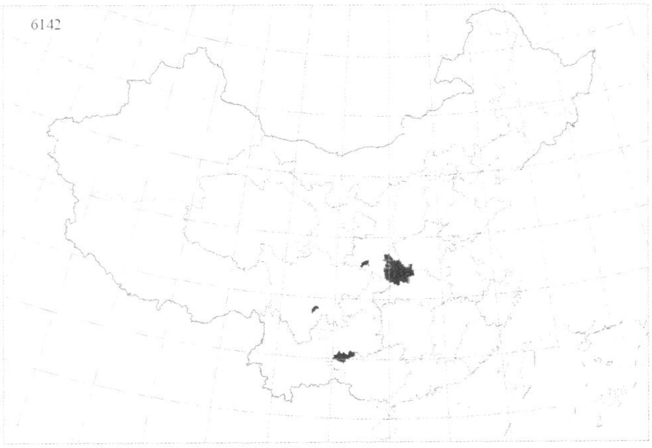

Vitaceae 葡萄科 *Vitis* 葡萄属
Vitis sinocinerea W. T. Wang 小叶葡萄
Vines, deciduous

15.7 (5.4~21.1); 6.4 (-2.1~13.2); 23.8 (12.3~30.1); 15.7 (5.8~21.1); 129.5 (31.5~193.7); 0.9 (0.0~25.6); 824 (481~1104) ❙ 1278 (741~1997); 560 (411~762); 113 (12~259) ❙ 54.2 (-14.2~130.7); 811 (481~1094); 254.0 (135.1~406.0)

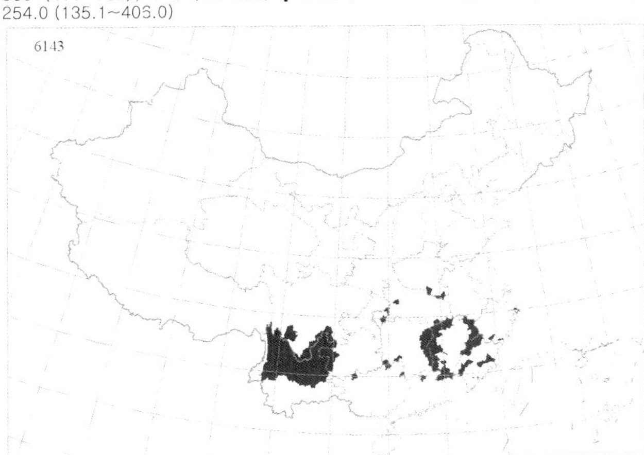

Vitaceae 葡萄科 *Vitis* 葡萄属
Vitis tsoi Merr. 狭叶葡萄
Vines, deciduous

18.2 (14.3~21.2); 8.3 (3.7~15.2); 27.4 (24.2~29.1); 18.4 (14.3~22.7); 160.6 (113.2~212.7); 0.1 (0.0~1.4); 960 (761~1216) ❙ 1716 (1215~1997); 601 (425~994); 203 (104~259) ❙ 80.0 (14.0~130.7); 955 (761~1216); 252.0 (183.9~332.5)

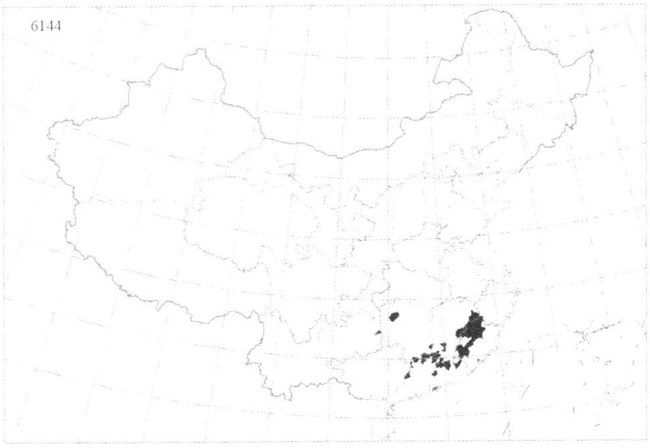

Vitaceae 葡萄科 *Vitis* 葡萄属
Vitis wenchowensis C. Ling 温州葡萄
Vines, deciduous

16.8；6.4；27.0；16.8；141.6；0.0；879 ❙ 1736；645；186 ❙ 98.0；879；219.0

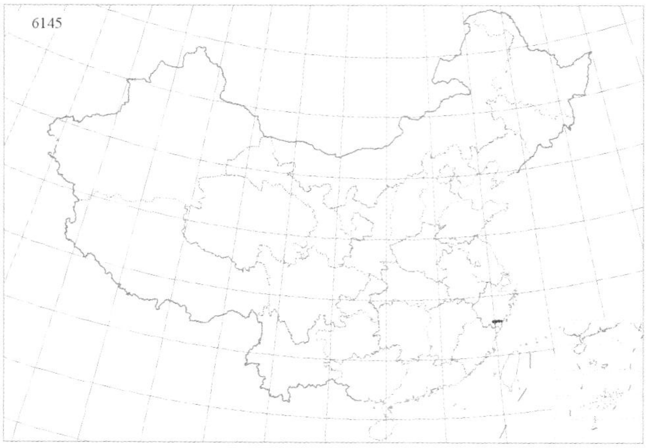

Vitaceae 葡萄科 *Vitis* 葡萄属
Vitis wilsoniae H. J. Veitch 网脉葡萄
Vines, deciduous

15.0 (5.2~22.3)；3.6 (−10.7~14.2)；25.5 (18.1~29.2)；15.1 (7.7~22.3)；124.0 (57.5~208.1)；3.7 (0.0~54.8)；818 (550~1202) ❙ 1238 (478~1997)；524 (298~818)；102 (14~254) ❙ 51.0 (−29.0~147.0)；809 (478~1202)；213.8 (135.3~551.6)

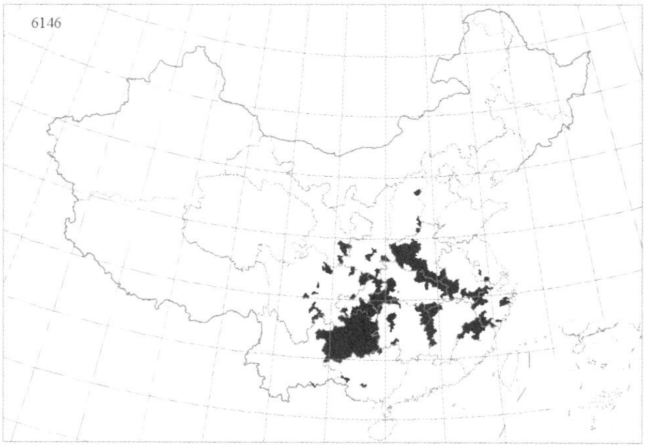

Vitaceae 葡萄科 *Vitis* 葡萄属
Vitis wuhanensis C. L. Li 武汉葡萄
Vines, deciduous

17.1 (15.1~18.8)；4.6 (1.7~6.9)；28.9 (27.4~30.1)；17.1 (15.2~18.9)；146.4 (127.7~166.5)；1.1 (0.0~5.4)；939 (860~1024) ❙ 1428 (1042~1745)；529 (453~685)；168 (89~237) ❙ 51.8 (21.1~80.8)；912 (860~965)；187.3 (111.9~249.0)

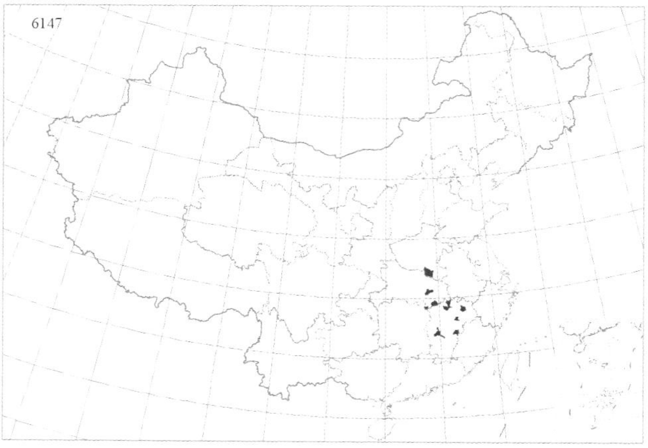

Vitaceae 葡萄科 *Vitis* 葡萄属
Vitis zhejiang-adstricta P. L. Chiu 浙江蘡薁
Vines, deciduous

15.9；4.1；27.3；15.9；132.3；1.3；862 ❙ 1663；560；220 ❙ 92.8；862；241.8

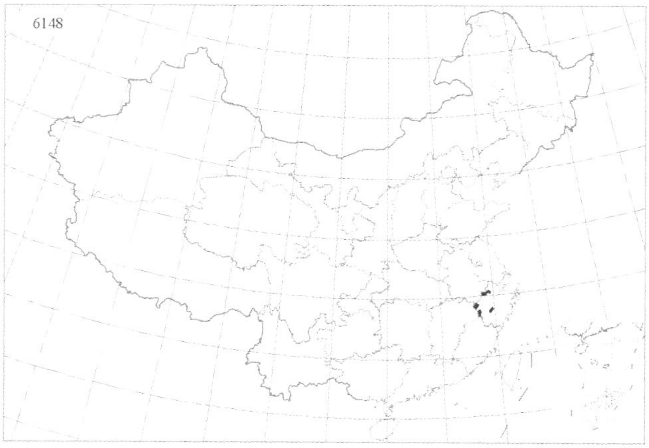

Vitaceae 葡萄科 *Yua* 俞藤属
Yua austro-orientalis (F. P. Metcalf) C. L. Li 大果俞藤
Vines

18.0 (14.3~22.2)；7.6 (3.5~15.1)；27.3 (22.5~29.8)；18.0 (14.3~22.3)；156.5 (113.2~207.5)；0.2 (0.0~1.5)；948 (761~1190) ❙ 1552 (1061~1990)；583 (427~891)；169 (59~231) ❙ 64.6 (10.5~137.5)；946 (761~1190)；237.6 (130.1~659.0)

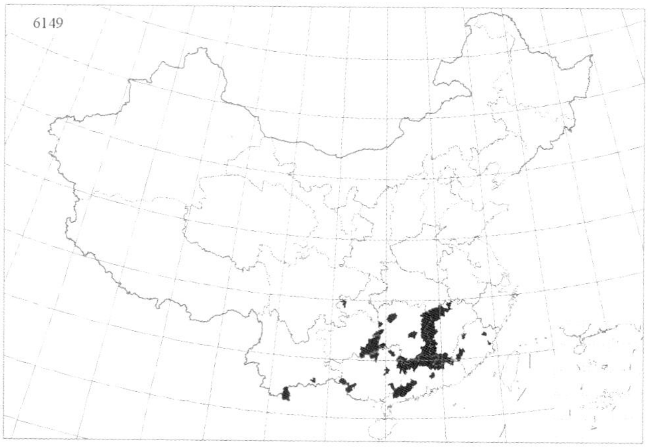

Vitaceae 葡萄科 *Yua* 俞藤属
Yua thomsonii (Lawson) C. L. Li 俞藤
Vines

15.6 (2.6~22.2)；4.7 (−7.0~13.9)；25.6 (11.2~29.8)；15.7 (4.3~22.3)；130.2 (20.0~207.5)；2.2 (0.0~48.2)；842 (427~1190) ❙ 1286 (626~1950)；529 (316~821)；113 (19~259) ❙ 52.9 (−22.2~137.5)；834 (427~1190)；212.2 (108.2~383.4)

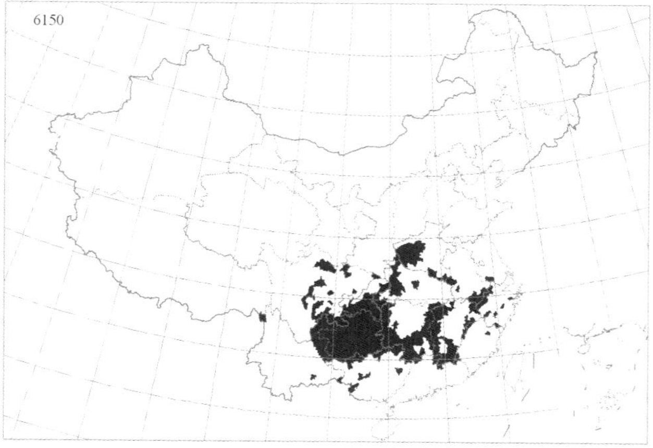

Vitaceae 葡萄科 *Yua* 俞藤属
Yua thomsonii var. glancescens (Diels et Gilg) C. L. Li 华西俞藤
Vines, evergreen

13.0 (2.6~19.2); 4.0 (-7.0~11.0); 20.7 (11.2~27.5); 13.1 (4.3~19.2); 103.4 (20.0~170.8); 7.6 (0.0~48.2); 733 (427~1002) ‖ 1229 (829~1496); 648 (400~848); 58 (23~142) ‖ 71.4 (32.5~101.6); 733 (427~1002); 271.0 (171.0~407.1)

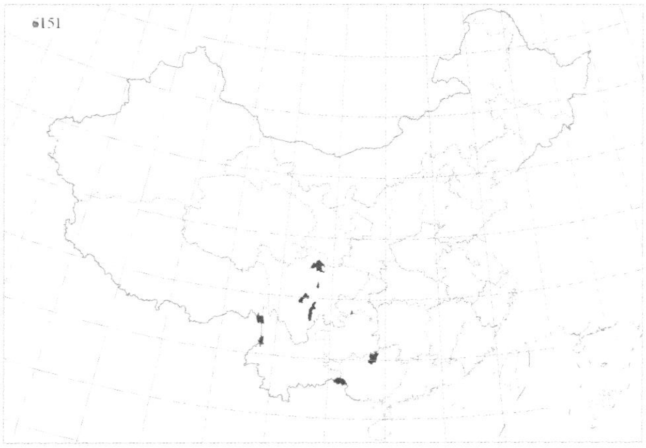

Elaeocarpaceae 杜英科 *Elaeocarpus* 杜英属
Elaeocarpus angustifolius Blume 圆果杜英
Trees, evergreen

23.2 (18.5~24.8); 17.8 (13.9~20.1); 27.0 (21.7~28.5); 23.2 (18.5~24.8); 218.8 (162.1~237.6); 0.0 (0.0~0.0); 1243 (863~1385) ‖ 1604 (1179~2713); 607 (392~1388); 73 (41~188) ‖ 33.8 (-11.9~214.4); 1139 (863~1360); 493.8 (327.7~620.0)

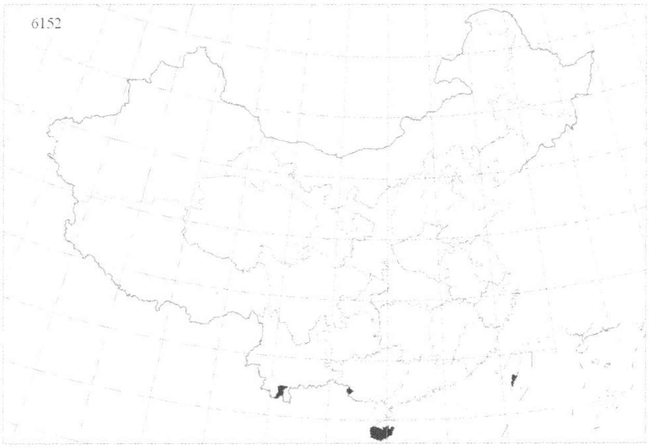

Elaeocarpaceae 杜英科 *Elaeocarpus* 杜英属
Elaeocarpus argenteus Merr. 腺叶杜英
Trees, evergreen

18.5; 14.1; 21.7; 18.5; 162.1; 0.0; 863 ‖ 2713; 1388; 188 ‖ 214.4; 863; 353.7

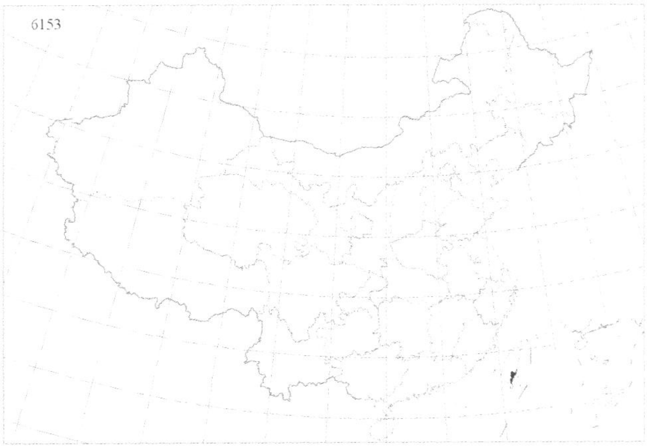

Elaeocarpaceae 杜英科 *Elaeocarpus* 杜英属
Elaeocarpus atropunctatus H. T. Chang 黑腺杜英
Trees, evergreen

21.8; 13.7; 28.1; 21.8; 201.8; 0.0; 1151 ‖ 1675; 768; 124 ‖ 45.5; 1151; 221.6

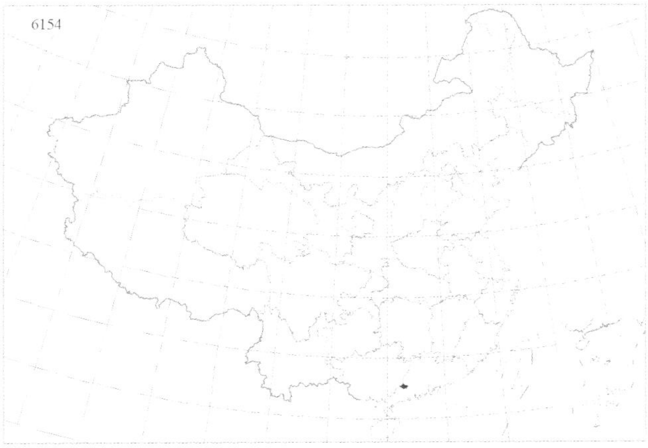

Elaeocarpaceae 杜英科 *Elaeocarpus* 杜英属
Elaeocarpus auricomus C. Y. Wu ex H. T. Chang 金毛杜英
Small trees, evergreen

18.8; 12.0; 23.5; 18.8; 165.5; 0.0; 914 ‖ 1455; 818; 60 ‖ 60.2; 914; 490.7

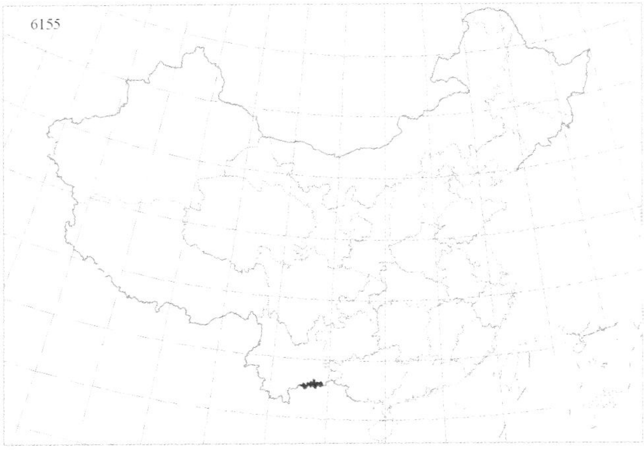

Elaeocarpaceae 杜英科 *Elaeocarpus* 杜英属
Elaeocarpus austroyunnanensis Hu 滇南杜英
Trees, evergreen

18.9 (16.4~21.0); 12.5 (9.6~15.1); 23.1 (20.6~27.5); 19.0 (16.4~21.0); 167.5 (137.3~191.8); 0.0 (0.0~0.0); 916 (795~1072) ‖ 1465 (1276~1688); 767 (655~959); 58 (41~89) ‖ 61.1 (24.6~112.2); 902 (795~1072); 486.0 (255.0~659.0)

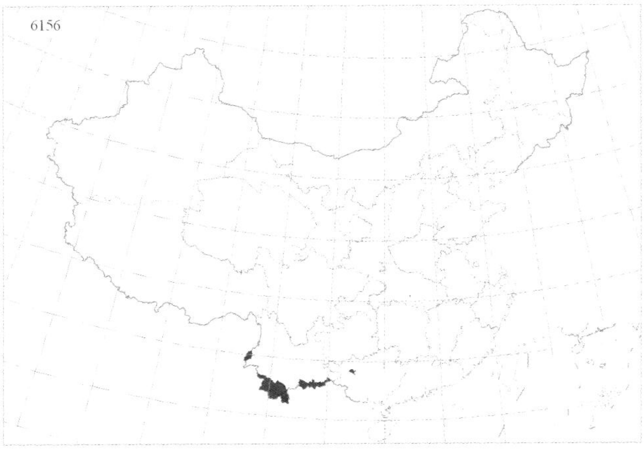

Elaeocarpaceae 杜英科 *Elaeocarpus* 杜英属

Elaeocarpus bachmaensis Gagnep. 少花杜英

Trees, evergreen

21.5 (18.9~22.7); 13.3 (10.6~15.0); 27.6 (25.0~28.6); 21.5 (19.0~22.8); 198.2 (167.5~213.1); 0.0 (0.0~0.0); 1130 (939~1221) ❙ 1695 (1292~2435); 935 (670~1428); 102 (49~130) ❙ 49.4 (12.9~99.5); 1130 (939~1221); 295.8 (235.1~342.5)

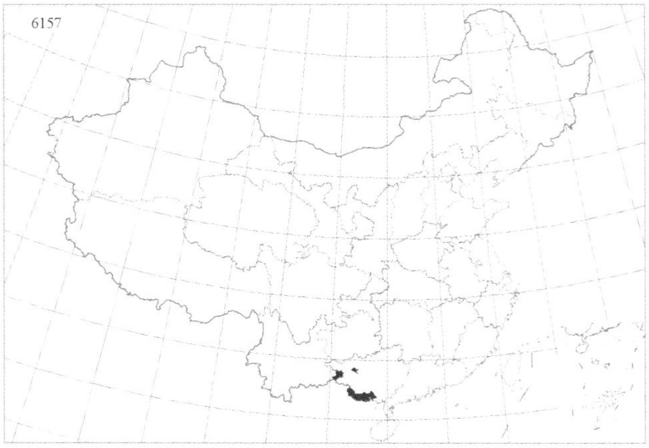

Elaeocarpaceae 杜英科 *Elaeocarpus* 杜英属

Elaeocarpus balansae A. Candolle 大叶杜英

Trees, evergreen

18.7 (17.3~21.0); 12.2 (9.9~14.2); 23.3 (22.4~25.6); 18.8 (17.3~21.0); 165.2 (147.1~191.8); 0.0 (0.0~0.0); 909 (833~1053) ❙ 1474 (1368~1547); 836 (756~911); 60 (55~68) ❙ 63.0 (44.5~80.6); 909 (833~1053); 502.8 (413.5~551.6)

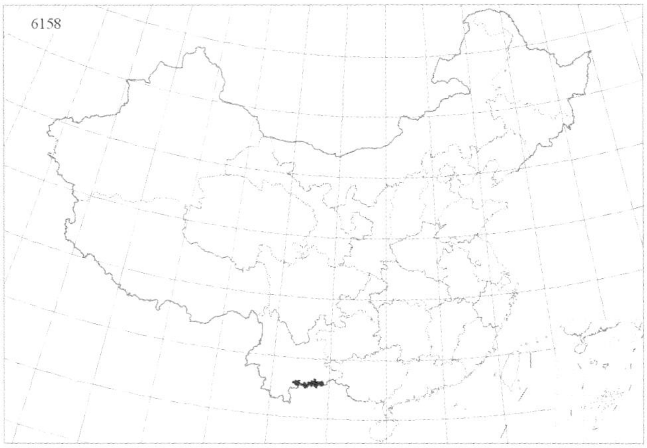

Elaeocarpaceae 杜英科 *Elaeocarpus* 杜英属

Elaeocarpus braceanus Watt ex C. B. Clarke 滇藏杜英

Trees, evergreen

17.7 (13.3~20.5); 11.4 (5.4~15.1); 21.8 (19.1~23.7); 17.7 (13.4~20.5); 152.9 (100.4~185.9); 0.0 (0.0~0.0); 856 (706~992) ❙ 1399 (1049~1736); 772 (594~1007); 55 (42~77) ❙ 64.6 (21.7~145.8); 845 (706~986); 469.6 (346.9~659.0)

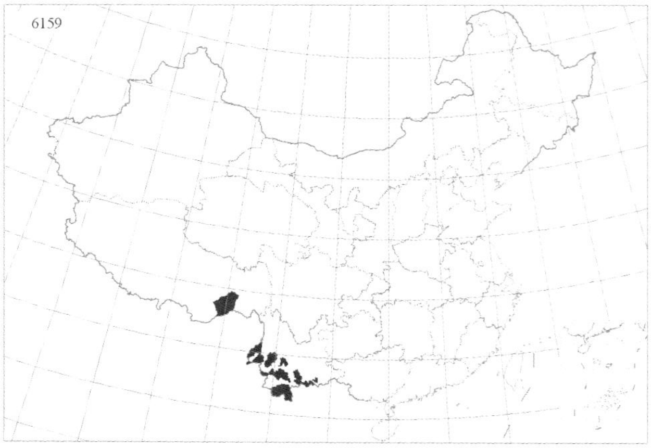

Elaeocarpaceae 杜英科 *Elaeocarpus* 杜英属

Elaeocarpus brachystachyus H. T. Chang 短穗杜英

Trees, evergreen

8.6; 1.3; 15.0; 8.7; 52.6; 8.8; 574 ❙ 1041; 585; 45 ❙ 81.5; 574; 352.3

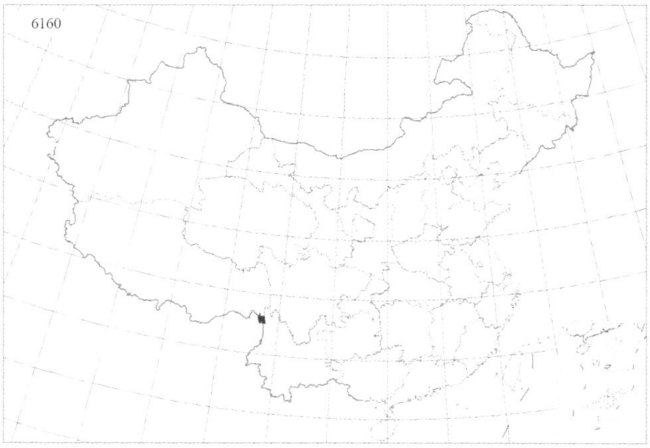

Elaeocarpaceae 杜英科 *Elaeocarpus* 杜英属

Elaeocarpus brachystachyus var. fengii C. Chen et Y. Tang 贡山杜英

Trees, evergreen

8.6; 1.3; 15.0; 8.7; 52.6; 8.8; 574 ❙ 1041; 585; 45 ❙ 81.5; 574; 352.3

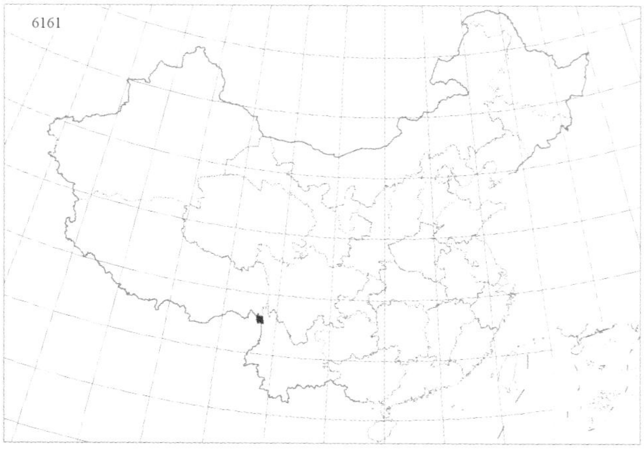

Elaeocarpaceae 杜英科 *Elaeocarpus* 杜英属

Elaeocarpus chinensis (Gardner et Champion) Hook. f. ex Bentham 华杜英

Small trees, evergreen

17.8 (12.7~24.7); 7.6 (2.1~19.8); 27.1 (21.4~30.1); 17.8 (12.8~24.7); 154.1 (99.0~236.5); 0.2 (0.0~5.6); 939 (713~1376) ❙ 1520 (871~2435); 577 (299~1428); 161 (42~259) ❙ 63.0 (−11.9~147.0); 925 (713~1221); 228.6 (120.3~620.0)

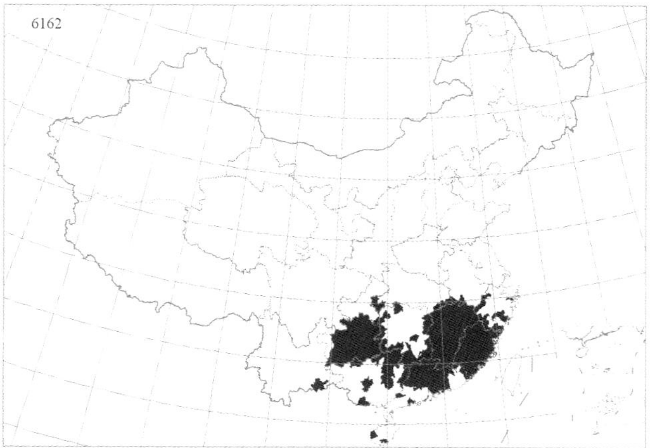

Elaeocarpaceae 杜英科 *Elaeocarpus* 杜英属

Elaeocarpus decandrus Merr. 缘瓣杜英

Trees, evergreen

18.1; 10.9; 23.1; 18.1; 157.2; 0.0; 886 ❙ 1334; 732; 56 ❙ 50.9; 886; 431.7

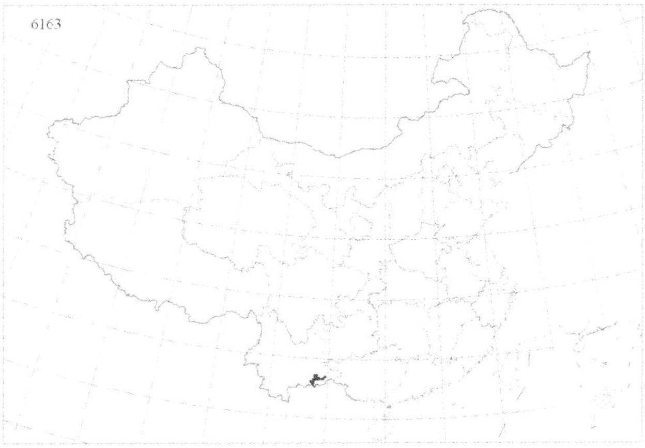

Elaeocarpaceae 杜英科 *Elaeocarpus* 杜英属

Elaeocarpus decipiens Hemsley 杜英

Trees, evergreen

18.3 (14.3~23.6); 8.4 (2.6~17.8); 27.3 (19.1~30.1); 18.3 (14.3~23.6); 160.1 (112.6~223.3); 0.1 (0.0~3.5); 961 (727~1271) ❙ 1582 (871~2499); 595 (299~1428); 176 (44~303) ❙ 66.4 (-6.7~219.3); 945 (727~1221); 247.4 (123.9~570.1)

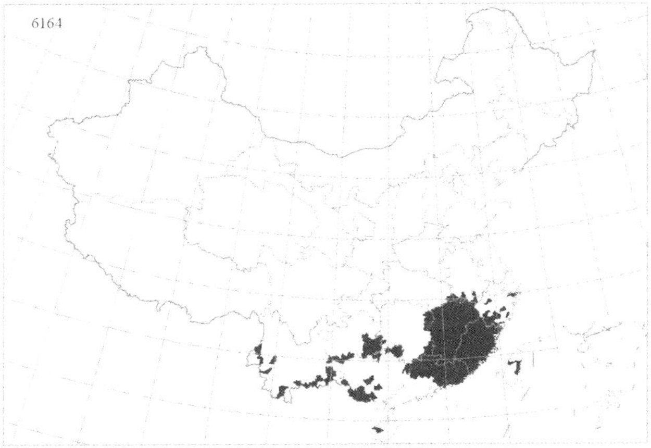

Elaeocarpaceae 杜英科 *Elaeocarpus* 杜英属

Elaeocarpus dubius A. DC. 显脉杜英

Trees, evergreen

20.8 (13.3~24.8); 12.3 (3.8~20.1); 27.7 (19.6~29.8); 20.8 (13.4~24.8); 189.7 (100.4~237.6); 0.0 (0.0~1.5); 1098 (706~1385) ❙ 1620 (868~2435); 684 (312~1428); 139 (41~230) ❙ 49.8 (-34.3~145.8); 1083 (706~1360); 258.1 (95.3~659.0)

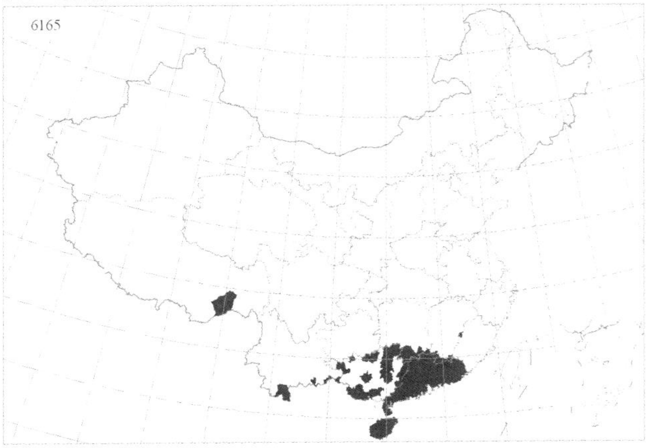

Elaeocarpaceae 杜英科 *Elaeocarpus* 杜英属

Elaeocarpus duclouxii Gagnep. 冬桃

Trees, evergreen

17.2 (2.6~23.2); 6.5 (-7.0~15.8); 27.0 (11.2~30.1); 17.3 (4.3~23.2); 147.6 (20.0~218.9); 0.5 (0.0~48.2); 916 (427~1257) ❙ 1432 (731~2037); 554 (299~1026); 142 (19~259) ❙ 56.5 (-6.7~128.9); 904 (427~1257); 206.7 (108.2~570.1)

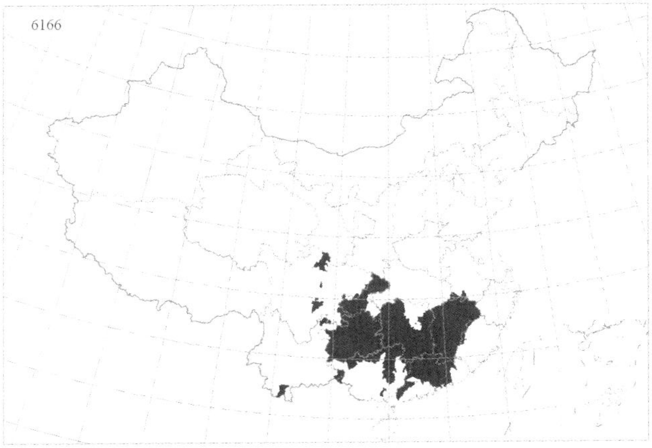

Elaeocarpaceae 杜英科 *Elaeocarpus* 杜英属

Elaeocarpus duclouxii var. funingensis Y. C. Hsu et Y. Tang 富宁杜英

Trees, evergreen

18.9; 10.6; 25.0; 19.0; 167.5; 0.0; 939 ❙ 1292; 729; 49 ❙ 37.6; 939; 342.5

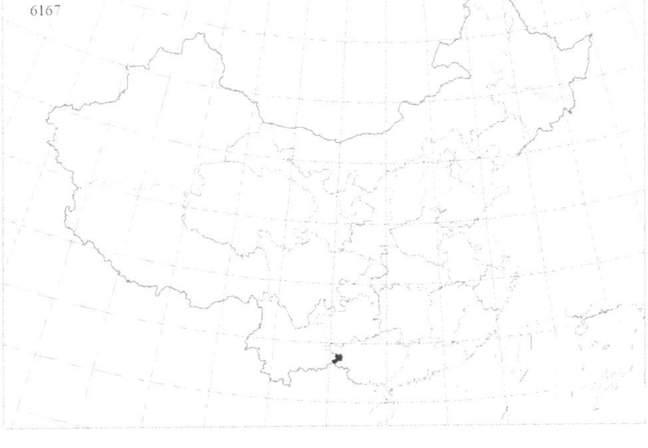

Elaeocarpaceae 杜英科 *Elaeocarpus* 杜英属

Elaeocarpus glabripetalus Merr. 秃瓣杜英

Trees, evergreen

17.8 (14.3~22.2); 7.1 (3.1~14.9); 27.5 (21.4~30.1); 17.8 (14.3~22.3); 153.8 (113.2~207.5); 0.2 (0.0~2.3); 941 (761~1190) ❙ 1547 (931~2024); 566 (323~954); 174 (42~259) ❙ 65.2 (-1.6~147.0); 926 (761~1190); 227.0 (134.6~570.1)

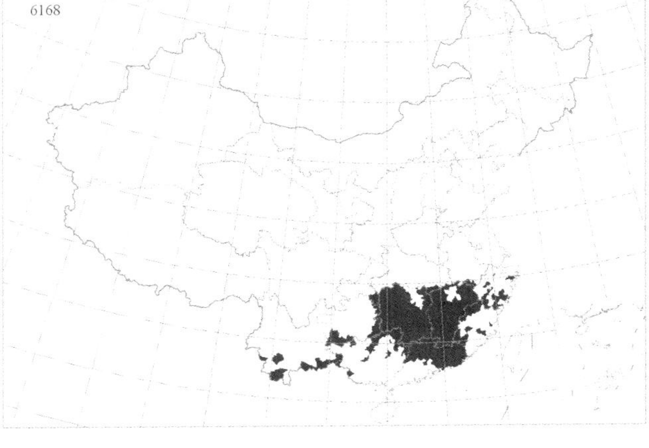

Elaeocarpaceae 杜英科 *Elaeocarpus* 杜英属

Elaeocarpus glabripetalus var. alatus (Kunth) H. T. Chang 棱枝杜英

Trees, evergreen

19.0 (17.7~21.0); 12.2 (9.2~14.2); 23.6 (21.6~27.5); 19.0 (17.8~21.0); 168.3 (153.4~191.8); 0.0 (0.0~0.0); 927 (854~1053) ❙ 1439 (1323~1522); 762 (678~835); 67 (42~135) ❙ 55.6 (44.5~68.0); 915 (833~1053); 461.1 (256.4~551.6)

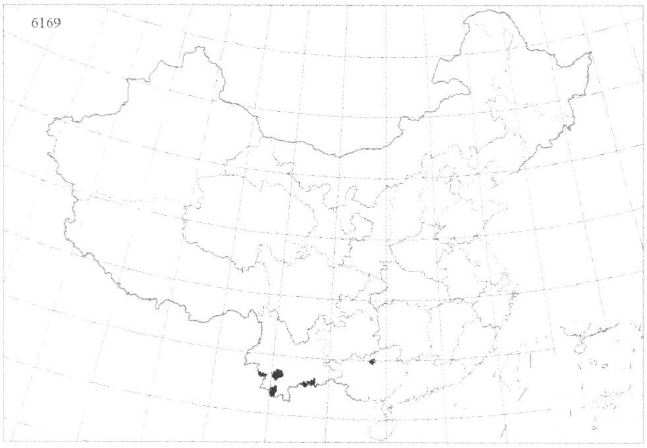

Elaeocarpaceae 杜英科 *Elaeocarpus* 杜英属

Elaeocarpus gymnogynus H. T. Chang 秃蕊杜英

Trees, evergreen

19.6 (17.6~21.7); 10.4 (7.3~13.4); 27.2 (26.3~28.3); 19.7 (17.6~21.7); 176.1 (151.6~200.7); 0.0 (0.0~0.0); 1021 (909~1147) ❙ 1564 (1437~1776); 711 (639~828); 128 (117~142) ❙ 53.5 (41.1~64.7); 1021 (909~1147); 236.8 (209.2~256.4)

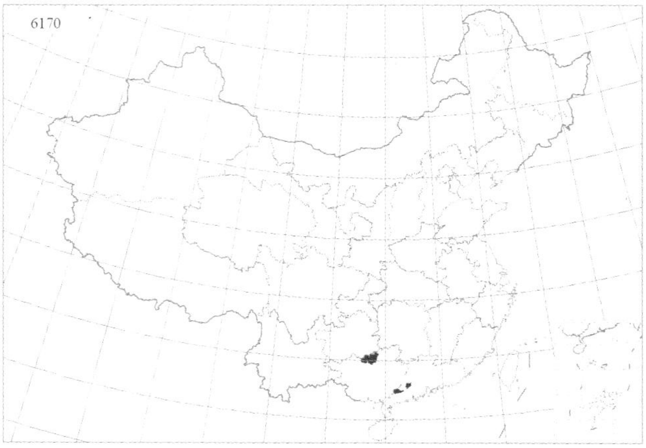

Elaeocarpaceae 杜英科 *Elaeocarpus* 杜英属

Elaeocarpus hainanensis Oliv. 水石榕

Shrubs or small trees, evergreen

22.6 (17.2~24.8); 16.3 (9.5~20.1); 27.1 (22.5~28.8); 22.6 (17.3~24.8); 211.8 (147.0~237.6); 0.0 (0.0~0.0); 1208 (838~1385) ❙ 1563 (868~2435); 677 (312~1428); 77 (41~130) ❙ 31.9 (-34.3~99.5); 1131 (838~1360); 418.4 (235.1~620.0)

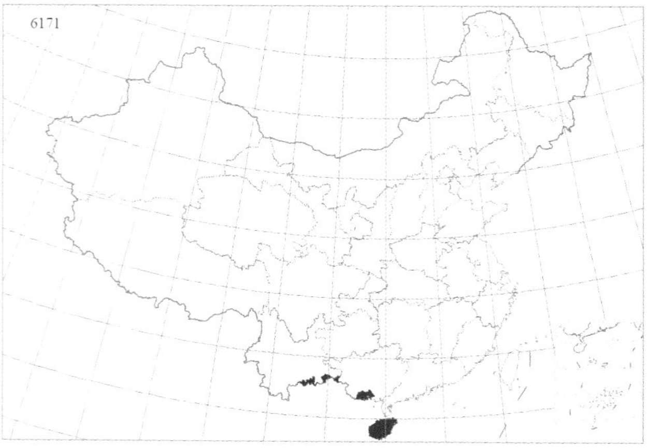

Elaeocarpaceae 杜英科 *Elaeocarpus* 杜英属

Elaeocarpus hainanensis var. brachyphyllus Merr. 短叶水石榕

Shrubs or small trees, evergreen

23.9 (23.0~24.2); 17.9 (17.5~18.3); 28.2 (26.7~28.8); 23.9 (23.1~24.3); 227.2 (217.0~231.2); 0.0 (0.0~0.0); 1303 (1217~1335) ❙ 1467 (868~1813); 542 (312~712); 82 (44~113) ❙ 12.8 (-34.3~41.8); 1205 (868~1331); 386.0 (248.2~572.9)

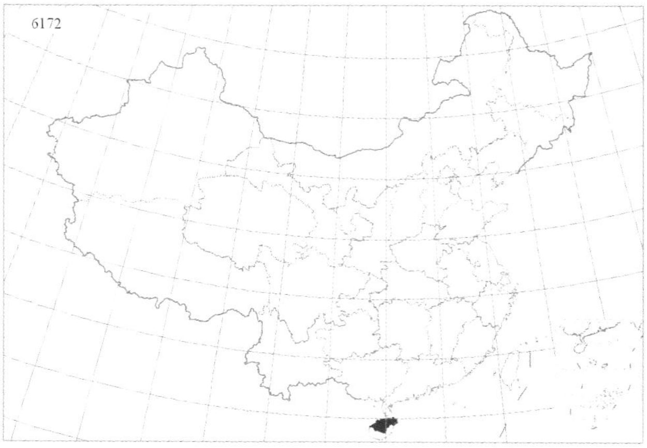

Elaeocarpaceae 杜英科 *Elaeocarpus* 杜英属

Elaeocarpus harmandii Pierre 肿柄杜英

Trees, evergreen

8.6; 1.3; 15.0; 8.7; 52.6; 8.8; 574 ❙ 1041; 585; 45 ❙ 81.5; 574; 352.3

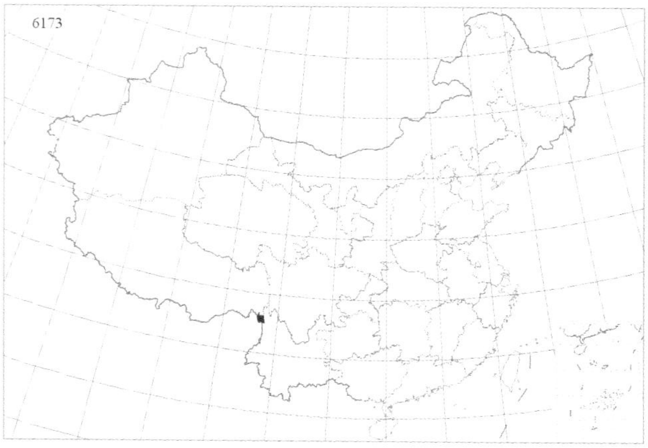

Elaeocarpaceae 杜英科 *Elaeocarpus* 杜英属

Elaeocarpus howii Merr. et Chun 锈毛杜英

Trees, evergreen

22.2 (16.9~24.8); 16.4 (9.5~20.1); 26.3 (21.4~28.5); 22.2 (16.9~24.8); 206.7 (142.7~237.6); 0.0 (0.0~0.0); 1177 (813~1385) ❙ 1485 (1178~1854); 593 (392~818); 60 (41~115) ❙ 29.1 (-11.9~65.6); 1087 (813~1360); 486.4 (336.3~620.0)

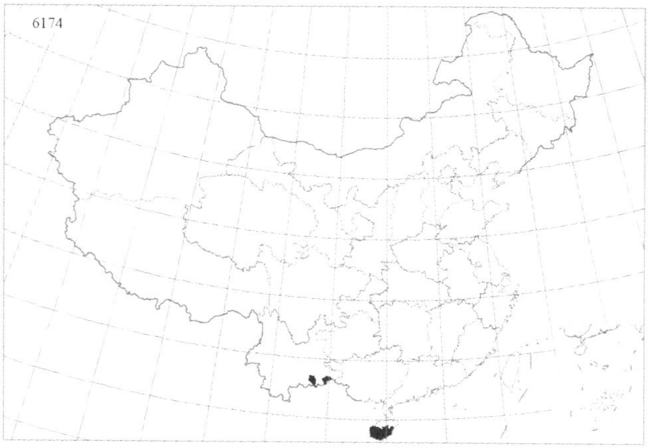

Elaeocarpaceae 杜英科 *Elaeocarpus* 杜英属

Elaeocarpus japonicus Siebold et Zuccarini 薯豆

Trees, evergreen

17.8 (3.5~23.6)；7.9 (-6.5~17.8)；26.7 (12.6~30.1)；17.8 (5.0~23.6)；154.3 (26.3~223.5)；0.4 (0.0~43.4)；939 (446~1281) | 1504 (731~3495)；599 (346~1847)；149 (19~897) | 61.5 (-6.9~246.5)；928 (446~1273)；230.0 (95.3~659.0)

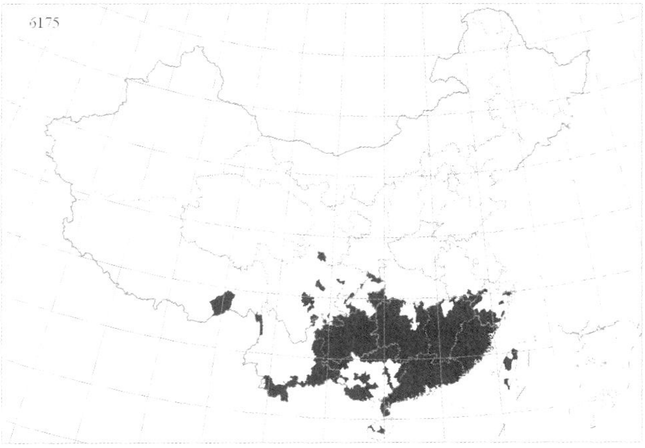

Elaeocarpaceae 杜英科 *Elaeocarpus* 杜英属

Elaeocarpus japonicus var. lantsangensis (Hu) H. T. Chang 澜沧杜英

Trees, evergreen

15.5 (8.6~19.1)；5.0 (1.3~13.0)；25.2 (15.0~27.7)；15.5 (8.7~19.1)；126.6 (52.6~169.7)；0.6 (0.0~8.8)；820 (574~917) | 1223 (1041~1500)；504 (444~835)；89 (45~146) | 49.7 (30.1~93.0)；819 (574~917)；213.4 (146.0~506.2)

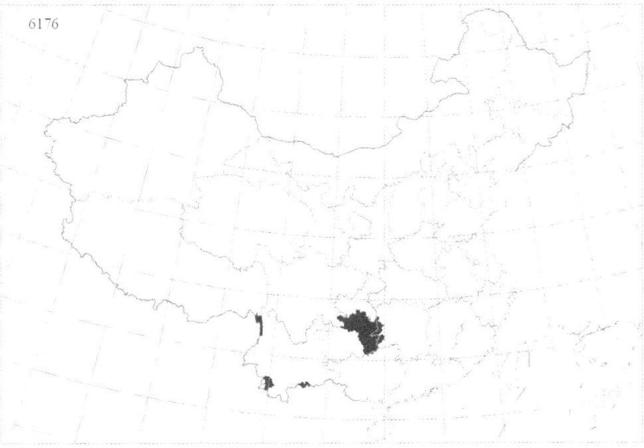

Elaeocarpaceae 杜英科 *Elaeocarpus* 杜英属

Elaeocarpus lacunosus Wall. ex Kurz 多沟杜英

Trees, evergreen

13.4 (8.6~16.2)；6.4 (1.3~9.4)；18.8 (15.0~20.8)；13.5 (8.7~16.3)；102.7 (52.6~135.1)；1.2 (0.0~8.8)；706 (574~790) | 1280 (1041~1735)；712 (585~1007)；57 (45~69) | 82.3 (50.8~145.8)；706 (574~790)；379.4 (335.1~447.4)

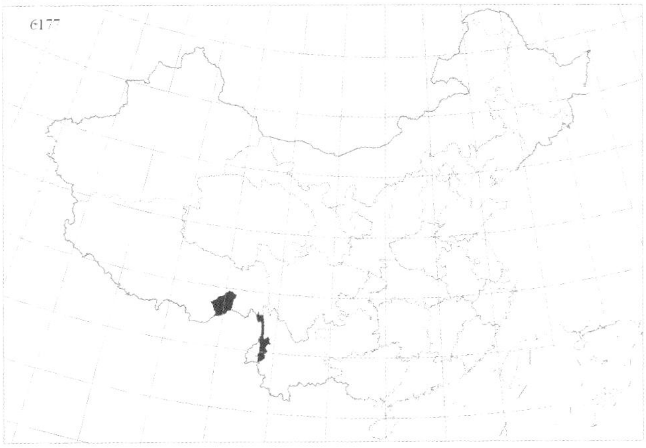

Elaeocarpaceae 杜英科 *Elaeocarpus* 杜英属

Elaeocarpus lanceifolius Roxb. 披针叶杜英

Trees, evergreen

18.4 (8.6~22.3)；9.2 (1.3~14.2)；25.9 (15.0~28.9)；18.4 (8.7~22.3)；161.0 (52.6~208.1)；0.1 (0.0~8.8)；949 (574~1202) | 1488 (1041~2005)；656 (435~1007)；130 (42~230) | 58.2 (16.9~145.8)；946 (574~1202)；278.0 (170.8~551.6)

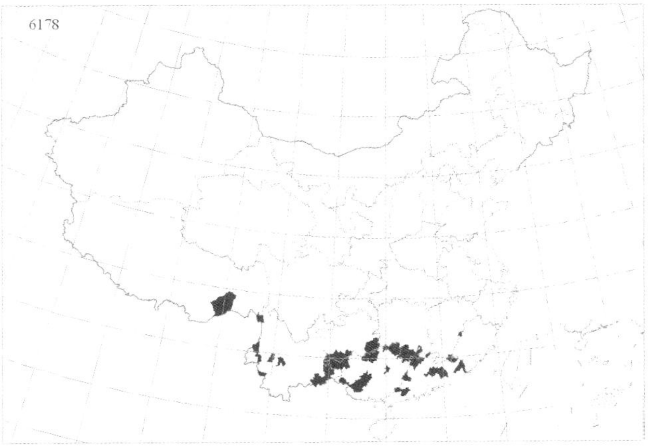

Elaeocarpaceae 杜英科 *Elaeocarpus* 杜英属

Elaeocarpus laoticus Gagnep. 老挝杜英

Trees, evergreen

18.7 (17.6~21.0)；12.0 (9.9~14.2)；23.4 (22.1~25.6)；18.7 (17.7~21.0)；164.9 (152.4~191.8)；0.0 (0.0~0.0)；909 (854~1053) | 1415 (1336~1522)；799 (756~835)；58 (52~68) | 56.1 (44.5~65.6)；909 (854~1053)；468.4 (359.7~551.6)

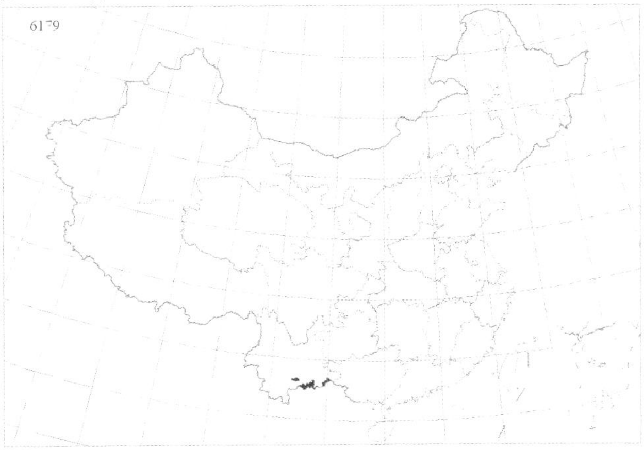

Elaeocarpaceae 杜英科 *Elaeocarpus* 杜英属

Elaeocarpus limitaneus Hand.-Mazz. 灰毛杜英

Small trees, evergreen

19.6 (14.3~24.8)；11.1 (3.8~20.1)；25.7 (22.4~28.9)；19.7 (14.3~24.8)；176.1 (112.6~237.6)；0.1 (0.0~1.5)；1041 (756~1385) | 1476 (868~2435)；628 (312~1428)；107 (41~191) | 45.2 (-34.3~101.8)；1004 (756~1360)；317.6 (178.6~620.0)

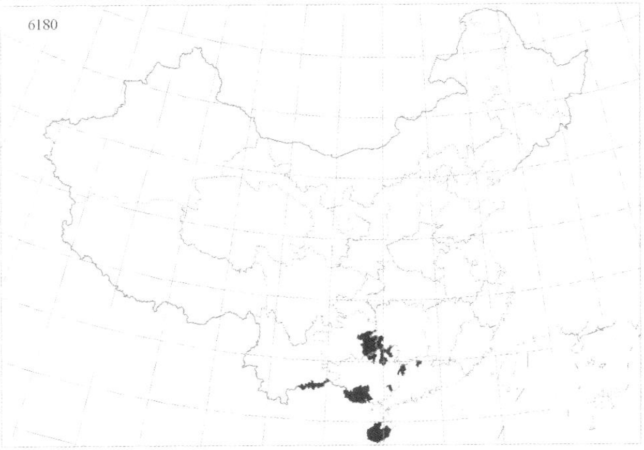

Elaeocarpaceae 杜英科 *Elaeocarpus* 杜英属
Elaeocarpus longlingensis Y. C. Hsu et Y. Tang 龙陵杜英
Trees, evergreen

16.0；9.3；20.4；16.0；132.4；0.0；780 ❙ 1348；762；56 ❙ 72.8；780；395.4

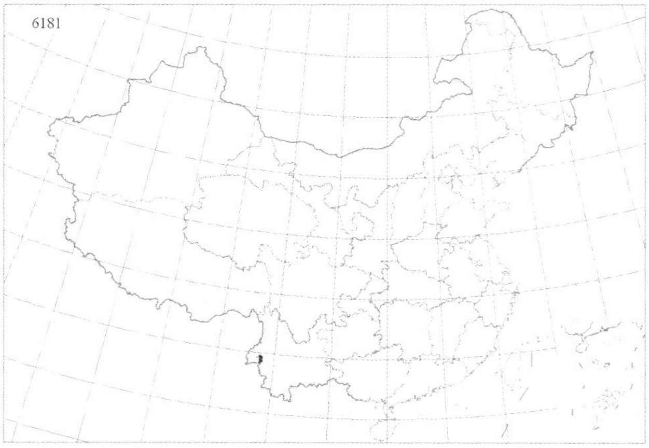

Elaeocarpaceae 杜英科 *Elaeocarpus* 杜英属
Elaeocarpus multiflorus (Turcz.) Fern.-Vill. 繁花杜英
Trees, evergreen

20.2 (16.2~24.1)；14.5 (11.0~18.9)；25.0 (20.1~28.2)；20.2 (16.2~24.1)；182.9 (134.2~229.7)；0.0 (0.0~0.0)；1019 (773~1322) ❙ 2402 (1498~3495)；1106 (750~1847)；255 (47~897) ❙ 146.2 (25.2~246.5)；983 (773~1144)；245.0 (119.4~363.7)

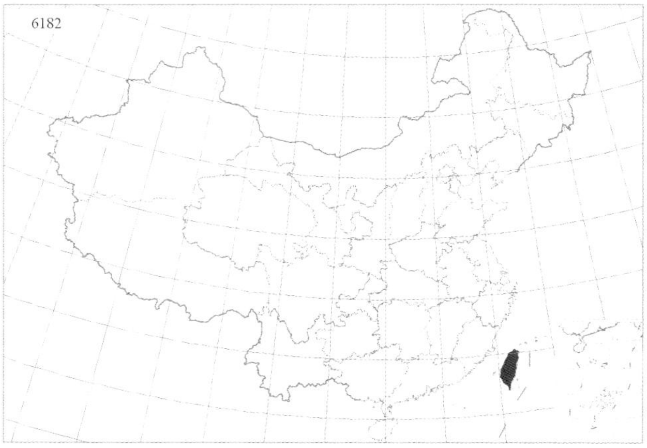

Elaeocarpaceae 杜英科 *Elaeocarpus* 杜英属
Elaeocarpus nitentifolius Merr. et Chun 绢毛杜英
Trees, evergreen

21.2 (17.6~24.8)；12.9 (7.1~20.1)；28.0 (22.5~29.8)；21.2 (17.6~24.8)；194.8 (151.6~237.6)；0.0 (0.0~0.0)；1121 (854~1385) ❙ 1646 (868~2435)；707 (312~1428)；138 (41~197) ❙ 48.5 (-34.3~99.5)；1106 (854~1360)；247.8 (95.3~620.0)

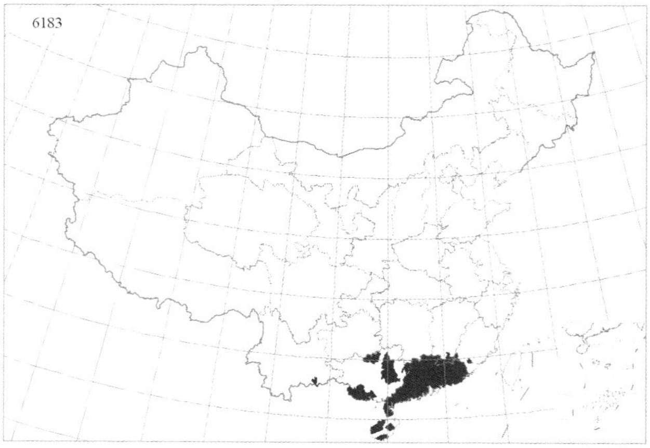

Elaeocarpaceae 杜英科 *Elaeocarpus* 杜英属
Elaeocarpus oblongilimbus H. T. Chang 长圆叶杜英
Trees, evergreen

19.3；12.7；23.8；19.3；171.6；0.0；942 ❙ 1439；803；61 ❙ 53.4；942；516.4

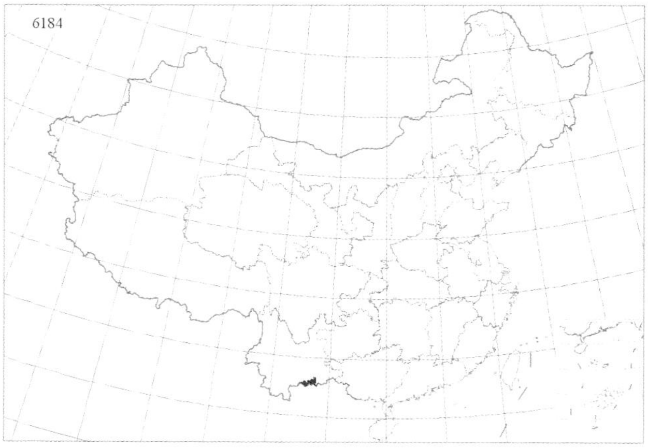

Elaeocarpaceae 杜英科 *Elaeocarpus* 杜英属
Elaeocarpus petiolatus (Jack) Wall. 长柄杜英
Trees, evergreen

20.2 (13.3~24.8)；11.9 (3.8~20.1)；26.8 (19.6~29.2)；20.2 (13.4~24.8)；182.5 (100.4~237.6)；0.0 (0.0~1.5)；1063 (706~1385) ❙ 1505 (868~2435)；662 (312~1428)；105 (41~194) ❙ 44.6 (-34.3~145.8)；1037 (706~1360)；286.8 (178.7~659.0)

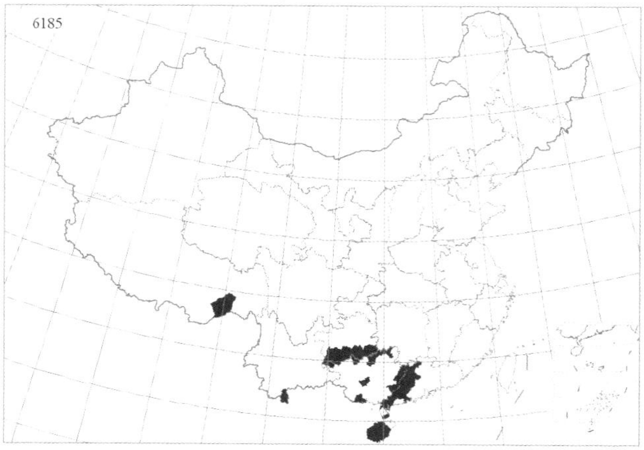

Elaeocarpaceae 杜英科 *Elaeocarpus* 杜英属
Elaeocarpus poilanei Gagnep. 滇越杜英
Trees, evergreen

22.3 (17.6~24.8)；15.8 (7.3~20.1)；27.0 (22.5~28.7)；22.4 (17.6~24.8)；208.5 (151.6~237.6)；0.0 (0.0~0.0)；1186 (854~1385) ❙ 1535 (868~2435)；687 (312~1428)；79 (41~142) ❙ 32.5 (-34.3~99.5)；1093 (854~1294)；422.8 (225.9~620.0)

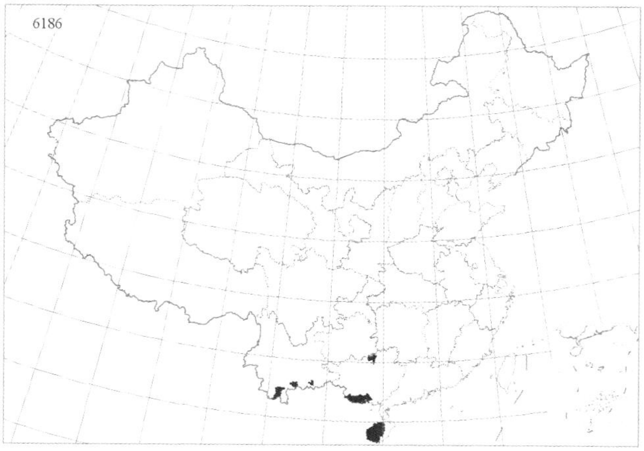

Elaeocarpaceae 杜英科 *Elaeocarpus* 杜英属

Elaeocarpus prunifolioides Hu 假樱叶杜英

Trees, evergreen

17.7 (8.6~20.5); 11.7 (1.3~15.1); 21.7 (15.0~23.7); 17.8 (8.7~20.5); 154.2 (52.6~185.9); 0.8 (0.0~8.8); 866 (574~992) ❙ 1413 (1041~1607); 743 (585~911); 55 (42~77) ❙ 63.8 (26.1~81.5); 849 (574~986); 478.4 (352.3~659.0)

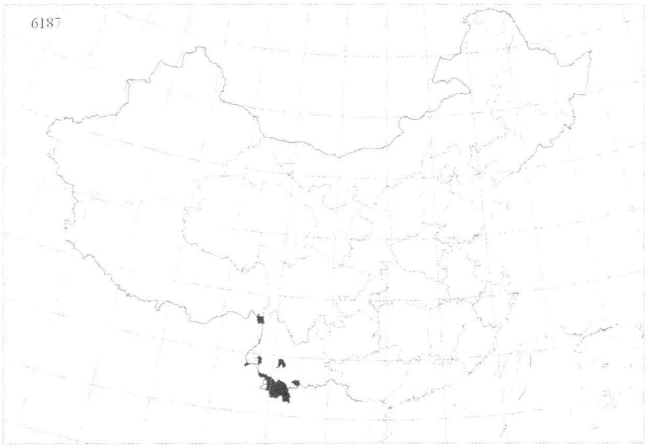

Elaeocarpaceae 杜英科 *Elaeocarpus* 杜英属

Elaeocarpus rugosus Roxb. 毛果杜英

Trees, evergreen

22.2 (13.3~24.8); 16.6 (5.4~20.1); 26.0 (19.6~28.5); 22.2 (13.4~24.8); 206.4 (100.4~237.6); 0.0 (0.0~0.0); 1173 (706~1385) ❙ 1539 (1179~1854); 618 (392~1007); 61 (41~115) ❙ 36.9 (-11.9~145.8); 1077 (706~1360); 525.3 (428.6~659.0)

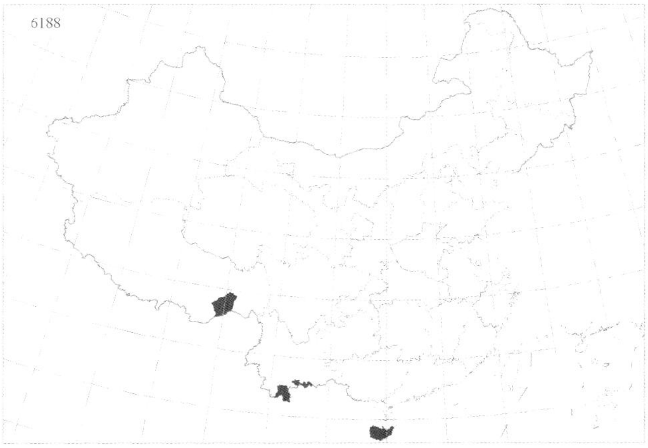

Elaeocarpaceae 杜英科 *Elaeocarpus* 杜英属

Elaeocarpus sikkimensis Masters 大果杜英

Trees, evergreen

17.1 (13.3~19.5); 10.6 (5.4~13.8); 21.2 (19.6~23.3); 17.1 (13.4~19.5); 145.3 (100.4~173.8); 0.0 (0.0~0.0); 829 (706~942) ❙ 1489 (1232~1736); 827 (686~1007); 50 (42~62) ❙ 81.4 (56.2~145.8); 818 (706~921); 481.0 (372.4~627.0)

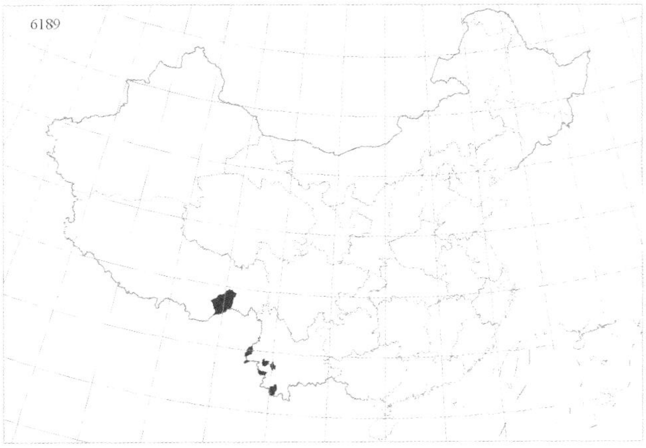

Elaeocarpaceae 杜英科 *Elaeocarpus* 杜英属

Elaeocarpus sphaerocarpus H. T. Chang 阔叶圆果杜英

Trees, evergreen

19.0; 13.4; 22.6; 19.1; 169.1; 0.0; 919 ❙ 1413; 661; 63 ❙ 53.3; 901; 537.0

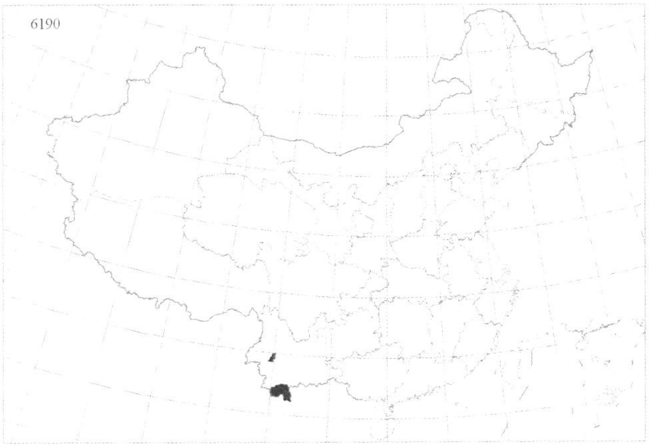

Elaeocarpaceae 杜英科 *Elaeocarpus* 杜英属

Elaeocarpus subpetiolatus H. T. Chang 屏边杜英

Trees, evergreen

18.2; 11.8; 22.5; 18.2; 158.6; 0.0; 871 ❙ 1457; 834; 61 ❙ 67.2; 871; 521.4

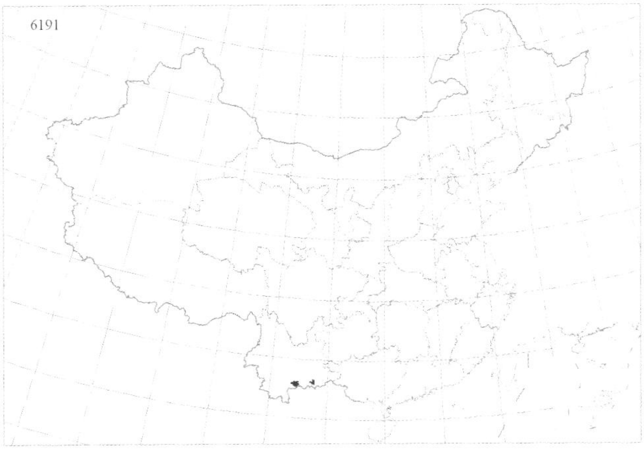

Elaeocarpaceae 杜英科 *Elaeocarpus* 杜英属

Elaeocarpus sylvestris (Lour.) Poiret 山杜英

Small trees, evergreen

18.8 (14.3~24.8); 9.2 (2.5~20.1); 27.4 (20.1~30.1); 18.9 (14.3~24.8); 166.8 (112.6~237.6); 0.1 (0.0~3.9); 993 (756~1385) ❙ 1566 (868~3495); 614 (299~1847); 161 (41~897) ❙ 59.7 (-34.3~246.5); 977 (755~1331); 239.6 (95.3~659.0)

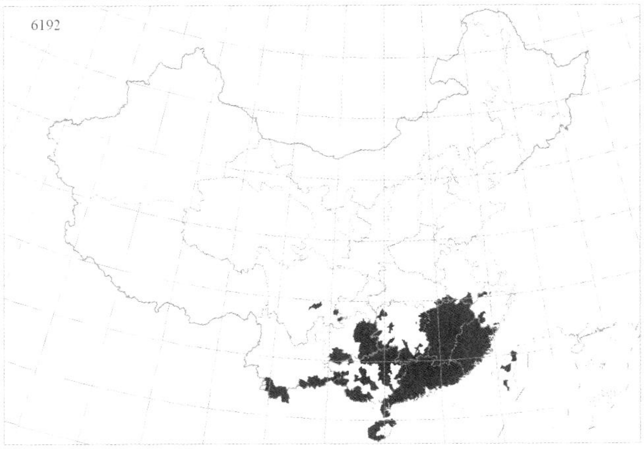

Elaeocarpaceae 杜英科 *Elaeocarpus* 杜英属

Elaeocarpus varunua Buch.-Ham. 滇印杜英

Trees, evergreen

20.2 (8.6~23.4); 11.8 (1.3~16.1); 26.7 (15.0~29.2); 20.2 (8.7~23.5); 182.5 (52.6~221.7); 0.2 (0.0~8.8); 1052 (574~1273) ❙ 1589 (1041~2435); 737 (547~1428); 123 (42~189) ❙ 53.0 (9.0~145.8); 1050 (574~1273); 281.8 (157.7~659.0)

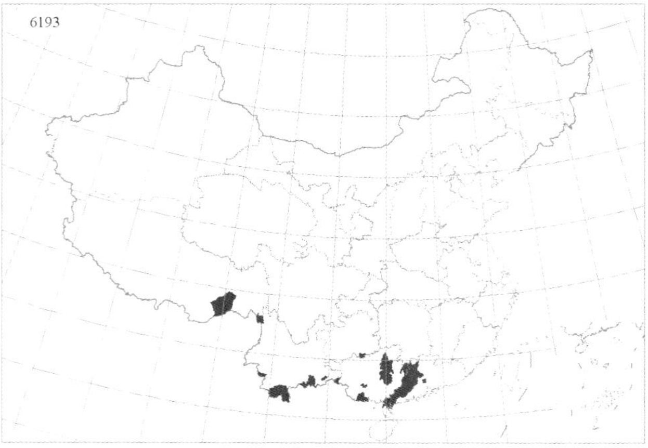

Elaeocarpaceae 杜英科 *Sloanea* 猴欢喜属

Sloanea changii Coode 樟叶猴欢喜

Trees, evergreen

19.6; 11.5; 25.5; 19.6; 175.7; 0.0; 1002 ❙ 1309; 715; 61 ❙ 34.1; 1002; 334.8

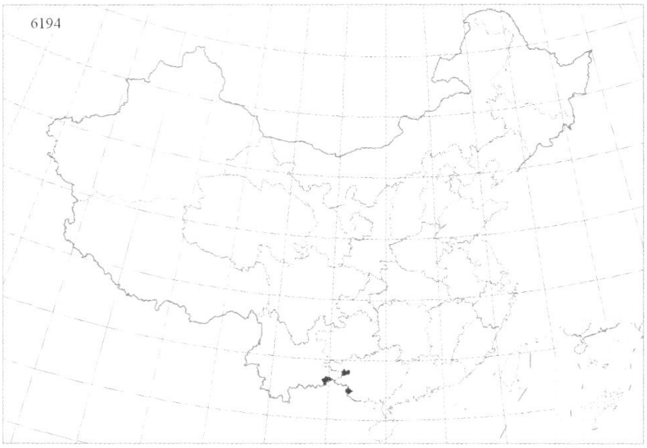

Elaeocarpaceae 杜英科 *Sloanea* 猴欢喜属

Sloanea chingiana Hu 白色猴欢喜

Trees, evergreen

19.5 (14.6~23.6); 10.0 (3.8~17.8); 27.4 (24.6~28.6); 19.6 (14.7~23.6); 175.1 (117.7~223.3); 0.1 (0.0~1.5); 1028 (779~1271) ❙ 1547 (1188~1758); 667 (444~832); 155 (44~194) ❙ 53.0 (-2.2~101.8); 1020 (779~1176); 244.1 (185.5~439.7)

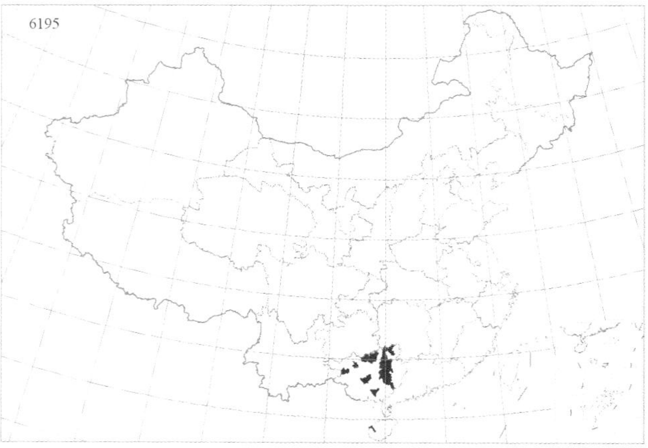

Elaeocarpaceae 杜英科 *Sloanea* 猴欢喜属

Sloanea cordifolia K. M. Feng ex H. T. Chang 心叶猴欢喜

Trees, evergreen

17.6; 9.9; 23.2; 17.7; 152.4; 0.0; 859 ❙ 1422; 813; 52 ❙ 65.6; 859; 359.7

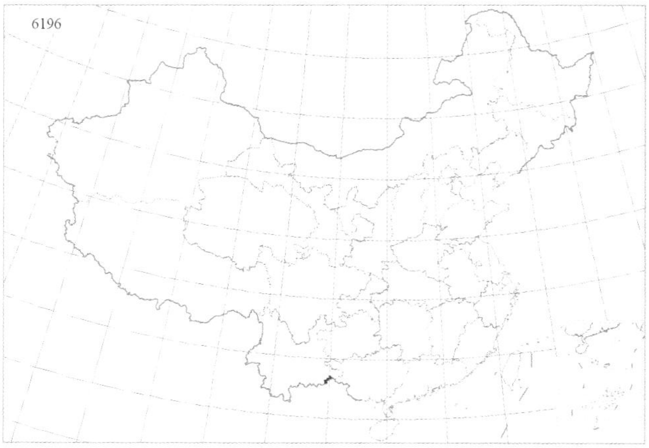

Elaeocarpaceae 杜英科 *Sloanea* 猴欢喜属

Sloanea dasycarpa (Benth.) Hemsl. 毛果猴欢喜

Trees, evergreen

18.2 (13.3~22.2); 11.6 (5.4~17.7); 23.5 (19.1~27.6); 18.2 (13.4~22.2); 158.9 (100.4~206.7); 0.0 (0.0~0.0); 902 (706~1123) ❙ 2174 (1422~3495); 934 (512~1847); 246 (51~897) ❙ 139.8 (55.7~246.5); 902 (706~1123); 380.9 (212.3~659.0)

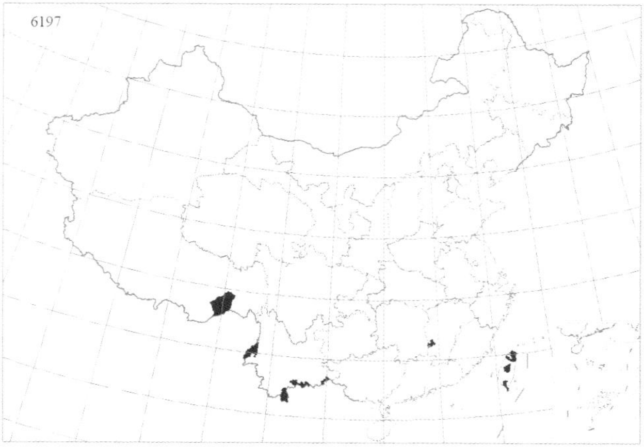

Elaeocarpaceae 杜英科 *Sloanea* 猴欢喜属

Sloanea hainanensis Merr. et Chun 海南猴欢喜

Trees, evergreen

23.9 (22.2~24.8); 18.6 (17.2~20.1); 27.6 (25.7~28.7); 23.9 (22.3~24.8); 227.0 (207.2~237.6); 0.0 (0.0~0.0); 1298 (1127~1385) ❙ 1486 (1176~1767); 514 (392~610); 58 (41~92) ❙ 15.1 (-12.0~47.8); 1154 (1055~1299); 510.7 (411.8~620.0)

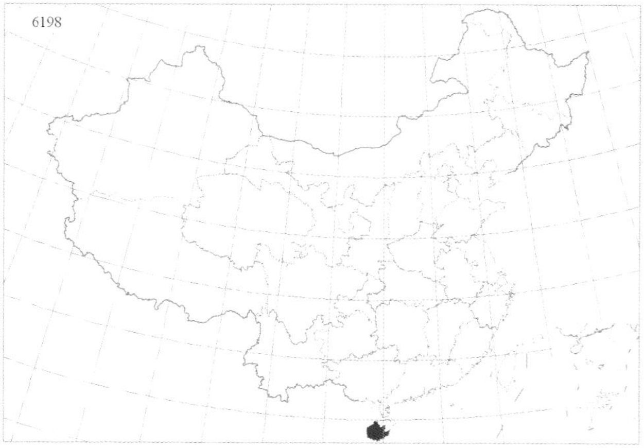

Elaeocarpaceae 杜英科 *Sloanea* 猴欢喜属

Sloanea hemsleyana (Ito) Rehder et E. H. Wilson 仿栗

Trees, evergreen

15.4 (7.3~21.0); 5.4 (-2.6~14.2); 24.4 (15.6~29.2); 15.5 (7.5~21.0); 127.1 (48.2~191.8); 1.5 (0.0~20.9); 814 (537~1053) ❙ 1234 (711~1745); 562 (357~911); 80 (15~197) ❙ 52.1 (5.0~107.2); 812 (537~1053); 231.5 (146.0~551.6)

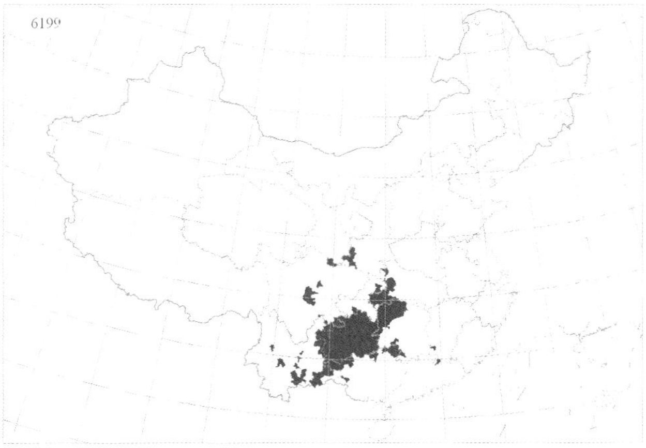

Elaeocarpaceae 杜英科 *Sloanea* 猴欢喜属

Sloanea integrifolia Chun et F. C. How 全缘叶猴欢喜

Trees, evergreen

20.0 (18.2~23.6); 11.1 (8.7~17.8); 27.3 (25.9~27.9); 20.1 (18.3~23.6); 180.7 (159.3~223.3); 0.0 (0.0~0.0); 1044 (928~1271) ❙ 1426 (1188~1697); 628 (444~705); 112 (44~194) ❙ 38.4 (-2.2~71.6); 1022 (928~1121); 290.5 (236.9~439.7)

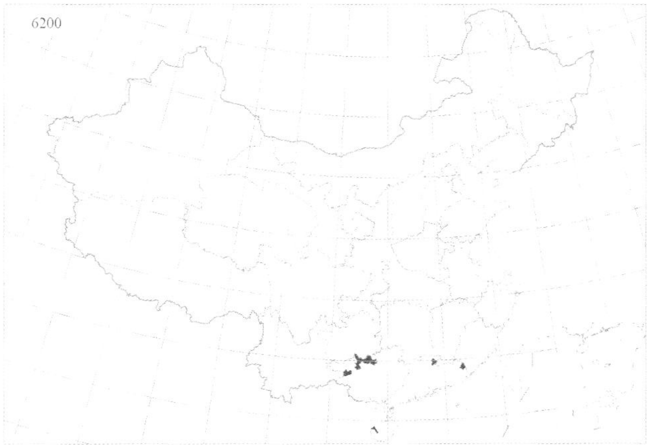

Elaeocarpaceae 杜英科 *Sloanea* 猴欢喜属

Sloanea leptocarpa Diels 薄果猴欢喜

Trees, evergreen

18.3 (14.3~22.0); 8.7 (3.3~14.9); 26.6 (22.0~29.1); 18.4 (14.3~22.0); 160.6 (112.6~203.8); 0.1 (0.0~2.3); 952 (756~1170) ❙ 1428 (1048~1829); 611 (356~911); 128 (43~230) ❙ 51.0 (-5.0~104.6); 948 (756~1170); 237.6 (108.2~570.1)

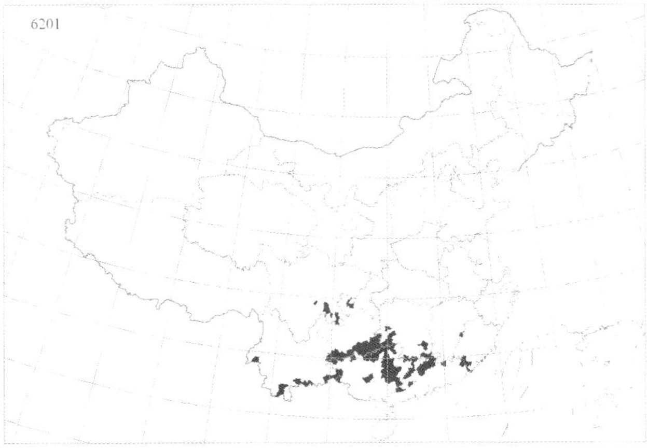

Elaeocarpaceae 杜英科 *Sloanea* 猴欢喜属

Sloanea mollis Gagnep. 滇越猴欢喜

Trees, evergreen

18.9 (8.6~22.7); 11.2 (1.3~15.0); 24.7 (15.0~28.6); 19.0 (8.7~22.8); 168.2 (52.6~213.1); 0.4 (0.0~8.8); 976 (574~1221) ❙ 1483 (1041~2435); 810 (585~1428); 82 (44~146) ❙ 52.5 (12.9~99.5); 973 (574~1221); 342.0 (235.1~551.6)

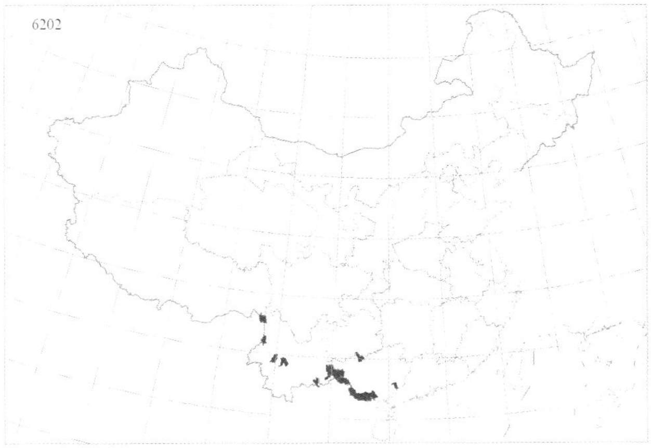

Elaeocarpaceae 杜英科 *Sloanea* 猴欢喜属

Sloanea sinensis (Hance) Hemsl. 猴欢喜

Trees, evergreen

18.1 (3.5~24.8); 8.1 (-6.5~20.1); 27.1 (12.6~30.1); 18.1 (5.0~24.8); 157.6 (26.3~237.6); 0.4 (0.0~43.4); 956 (446~1385) ❙ 1540 (731~3495); 602 (299~1847); 158 (19~897) ❙ 63.0 (-34.3~246.5); 940 (446~1273); 225.0 (95.3~620.0)

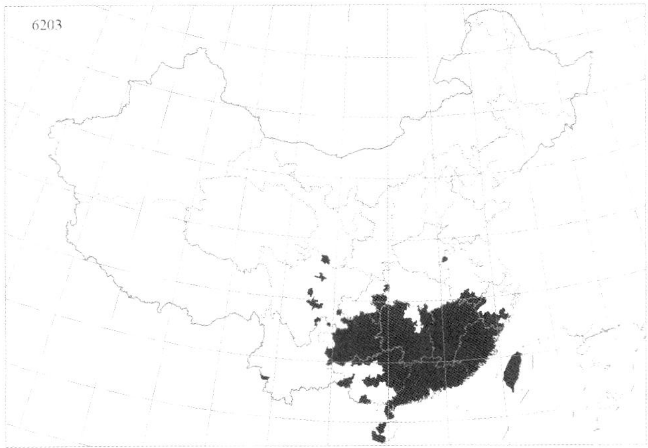

Elaeocarpaceae 杜英科 *Sloanea* 猴欢喜属

Sloanea sterculiacea (Benth.) Rehder et E. H. Wilson 贡山猴欢喜

Trees, evergreen

15.6 (8.6~21.0); 8.8 (1.3~14.2); 20.3 (15.0~25.6); 15.7 (8.7~21.0); 128.7 (52.6~191.8); 0.7 (0.0~8.8); 788 (574~1053) ❙ 1255 (1041~1547); 699 (585~911); 56 (44~70) ❙ 61.4 (26.1~99.9); 782 (574~1053); 398.6 (295.5~551.6)

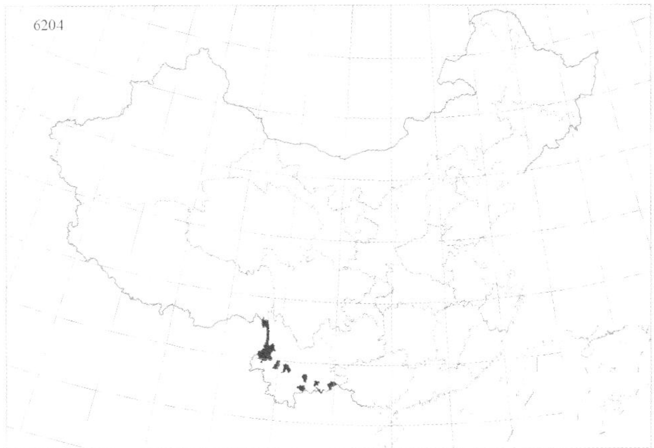

Elaeocarpaceae 杜英科 *Sloanea* 猴欢喜属
Sloanea sterculiacea var. assamica (Benth.) Coode 长叶猴欢喜
Trees, evergreen

17.1; 9.5; 22.6; 17.1; 145.4; 0.0; 840 ❙ 1276; 687; 65 ❙ 52.1; 828; 348.7

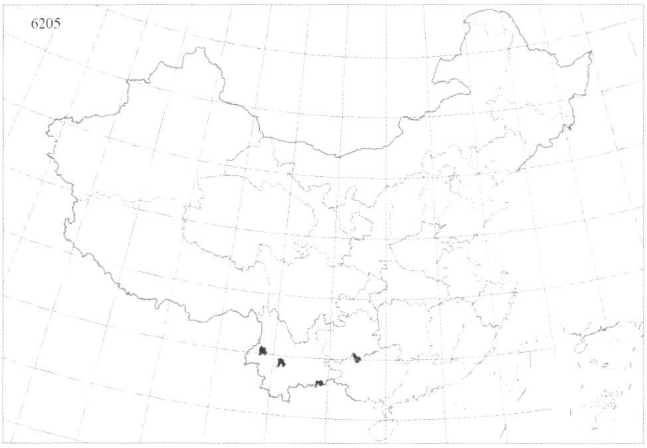

Elaeocarpaceae 杜英科 *Sloanea* 猴欢喜属
Sloanea tomentosa (Benth.) Rehder et E. H. Wilson 绒毛猴欢喜
Trees, evergreen

21.3 (17.4~22.7); 13.5 (11.0~15.1); 26.9 (21.4~28.6); 21.3 (17.4~22.8); 195.6 (149.3~213.1); 0.0 (0.0~0.0); 1109 (834~1221) ❙ 1685 (1302~2435); 895 (670~1428); 113 (48~194) ❙ 52.5 (12.9~99.5); 1108 (834~1221); 326.2 (213.5~659.0)

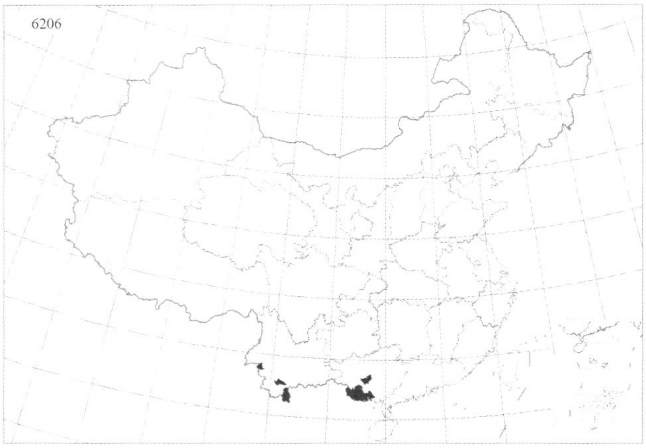

Elaeocarpaceae 杜英科 *Sloanea* 猴欢喜属
Sloanea xichouensis K. M. Feng 西畴猴欢喜
Trees, evergreen

17.2; 9.5; 22.7; 17.3; 147.0; 0.0; 838 ❙ 1254; 699; 48 ❙ 49.7; 838; 336.3

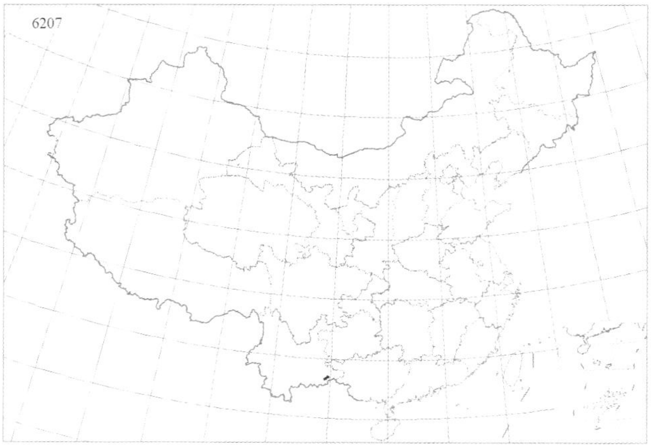

Tiliaceae 椴树科 *Berrya* 六翅木属
Berrya cordifolia (Willd.) Burret 六翅木
Trees

22.2; 17.7; 25.5; 22.2; 206.7; 0.0; 1123 ❙ 2992; 1847; 107 ❙ 166.5; 1123; 296.9

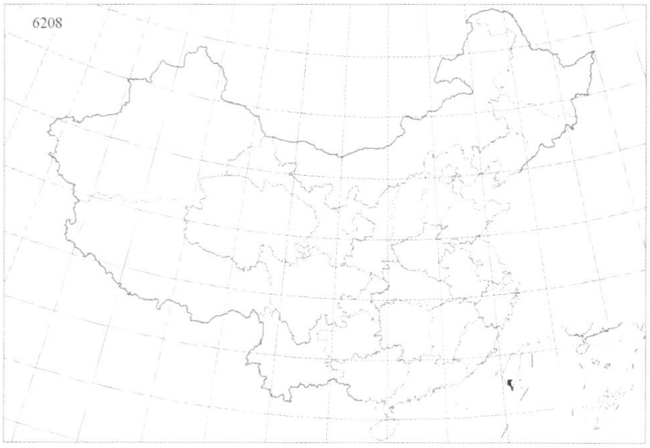

Tiliaceae 椴树科 *Burretiodendron* 柄翅果属
Burretiodendron esquirolii (H. Lév.) Rehder 柄翅果
Small trees, deciduous

18.6 (16.3~21.3); 10.8 (7.8~14.2); 24.4 (21.4~28.8); 18.7 (16.4~21.3); 164.2 (136.4~196.1); 0.0 (0.0~0.0); 929 (797~1137) ❙ 1311 (958~1751); 707 (531~996); 72 (43~174) ❙ 41.4 (13.3~88.8); 925 (797~1136); 338.3 (188.6~551.6)

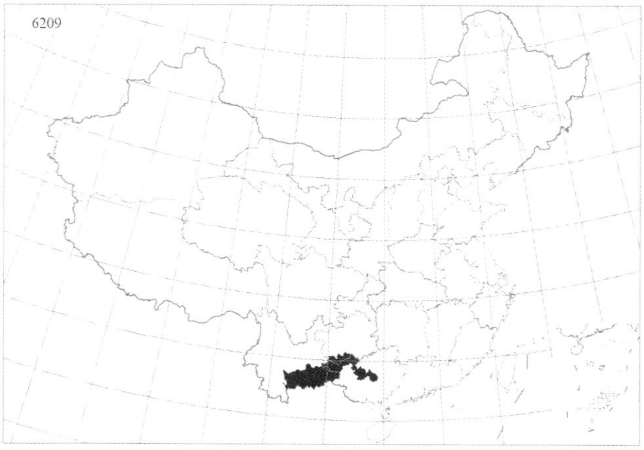

Tiliaceae 椴树科 *Burretiodendron* 柄翅果属
Burretiodendron kydiifolium Hsu et R. Zhuge 元江柄翅果
Trees, deciduous

18.7; 12.5; 22.9; 18.8; 165.2; 0.0; 901 ❙ 1240; 705; 53 ❙ 38.3; 883; 389.9

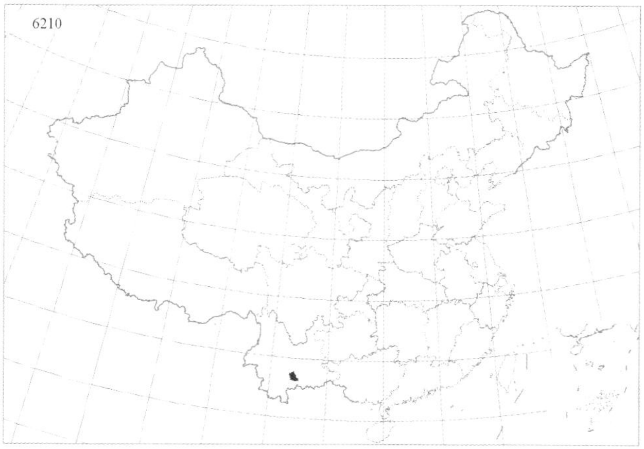

Tiliaceae 椴树科 *Colona* 一担柴属

Colona floribunda (Wall. ex Kurz) Craib 一担柴

Small trees

17.9 (14.4~21.0); 11.6 (7.5~15.1); 22.0 (19.1~25.6); 17.9 (14.4~21.0); 154.9 (113.1~191.8); 0.0 (0.0~0.0); 862 (727~1053) ❙ 1310 (939~1751); 725 (524~996); 52 (37~77) ❙ 52.1 (6.5~112.2); 843 (717~1053); 430.3 (289 9~659.0)

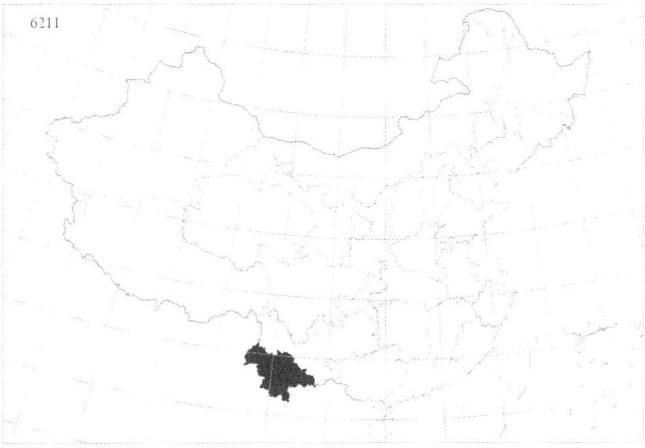

Tiliaceae 椴树科 *Colona* 一担柴属

Colona thorelii (Gagnep.) Burret 狭叶一担柴

Small trees, deciduous

19.1; 13.0; 23.1; 19.1; 169.6; 0.0; 922 ❙ 1487; 729; 58 ❙ 61.3; 920; 575.3

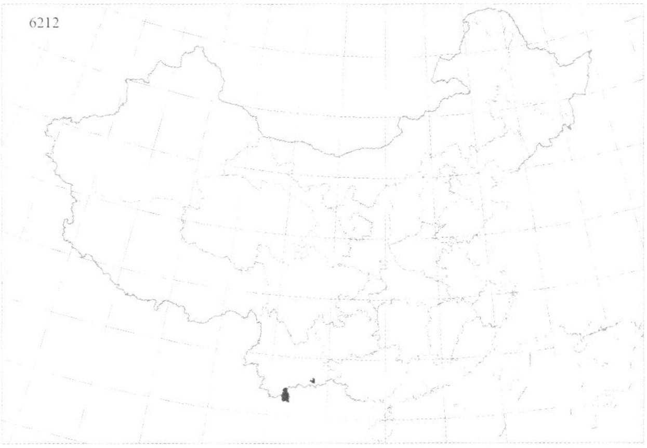

Tiliaceae 椴树科 *Craigia* 滇桐属

Craigia kwangsiensis Hsue 桂滇桐

Trees, deciduous

19.8; 11.1; 26.4; 19.8; 177.9; 0.0; 1010 ❙ 1182; 673; 43 ❙ 16.9; 1010 354.5

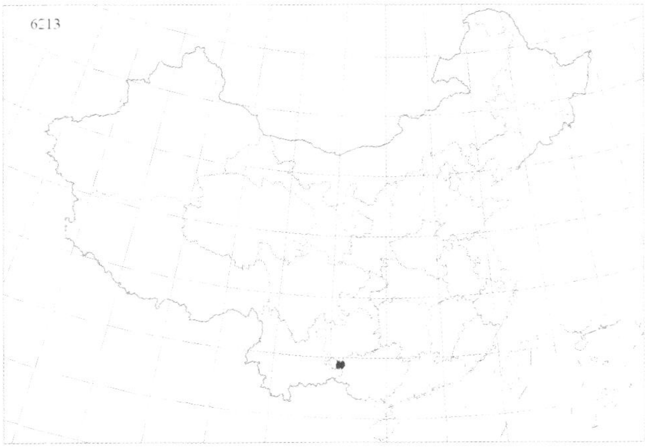

Tiliaceae 椴树科 *Craigia* 滇桐属

Craigia yunnanensis W. W. Sm. et W. E. Evans 滇桐

Trees, deciduous

17.6 (13.2~21.0); 9.9 (6.3~14.2); 23.2 (18.4~26.6); 17.7 (13.3~21.0); 152.0 (99.4~191.8); 0.0 (0.0~0.0); 873 (695~1053) ❙ 1366 (1074~1688); 756 (582~959); 57 (42~95) ❙ 58.0 (16.9~112.2); 872 (695~1053); 380.9 (189.5~627.0)

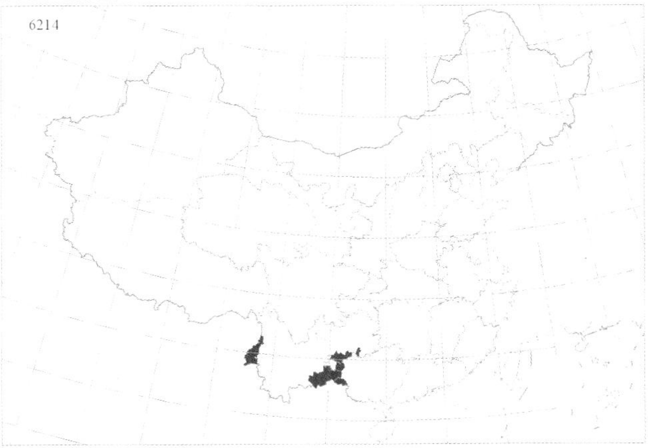

Tiliaceae 椴树科 *Diplodiscus* 海南椴属

Diplodiscus trichospermus (Merr.) Y. Tang, M. G. Gilbert et Dorr 海南椴

Shrubs or small trees

22.9 (21.3~24.8); 16.0 (12.5~20.1); 28.0 (27.2~28.7); 23.0 (21.3~24.8); 215.8 (195.9~237.6); 0.0 (0.0~0.0); 1238 (1111~1385) ❙ 1410 (868~1900); 632 (312~1056); 82 (41~130) ❙ 15.4 (-34.3~71.0); 11[illegible]7 (868~1[illegible]90); 372.7 (248.2~620.0)

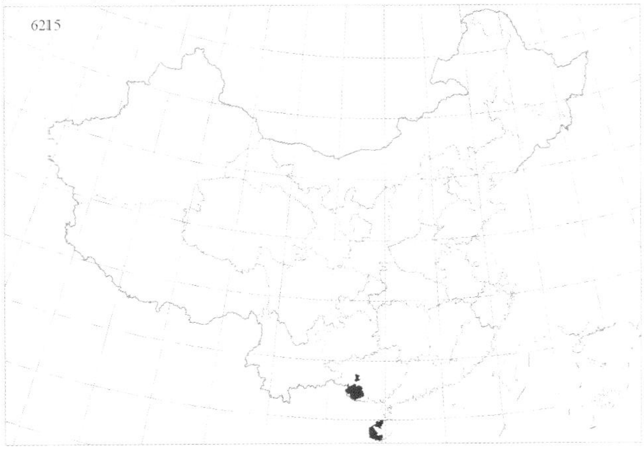

Tiliaceae 椴树科 *Excentrodendron* 蚬木属

Excentrodendron obconicum (Chun et F. C. How) H. T. Chang et R. H. Miao 长蒴蚬木

Trees, deciduous

22.2; 13.9; 28.3; 22.3; 207.5; 0.0; 1190 ❙ 1374; 7[illegible]3; 87 ❙ 15.4; 1190; 327.7

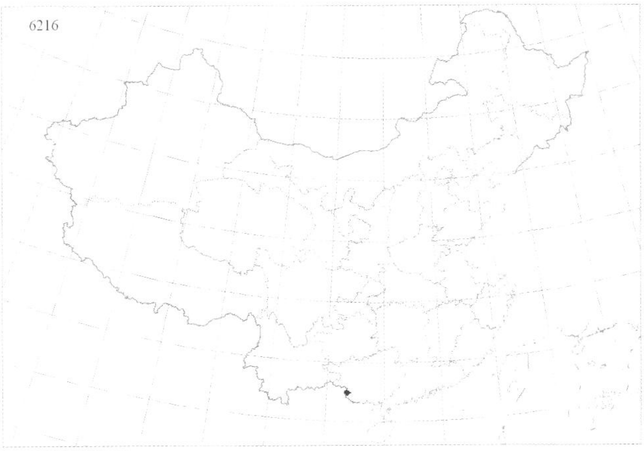

Tiliaceae 椴树科 *Excentrodendron* 蚬木属

Excentrodendron tonkinense (A. Chev.) H. T. Chang et R. H. Miao 蚬木

Trees, deciduous

19.9 (16.3~22.2); 11.8 (9.1~14.2); 26.0 (21.4~28.6); 20.0 (16.4~22.3); 179.6 (136.4~207.5); 0.0 (0.0~0.0); 1021 (797~1190) ▎1399 (1161~1662); 751 (633~860); 84 (48~173) ▎38.6 (5.8~80.6); 1021 (797~1190); 330.4 (220.1~551.6)

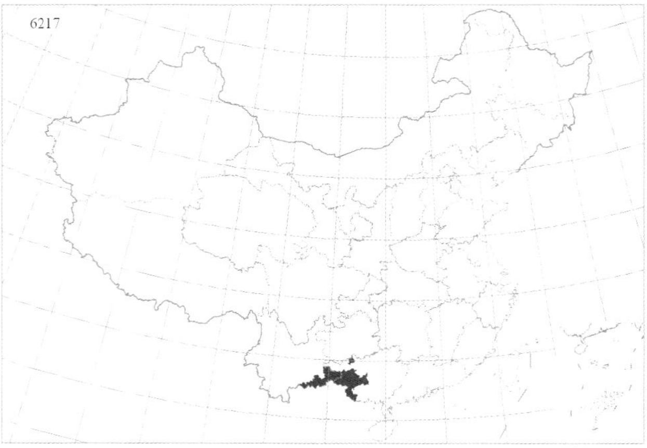

Tiliaceae 椴树科 *Grewia* 扁担杆属

Grewia abutilifolia W. Vent ex Juss. 苘麻叶扁担杆

Shrubs or small trees

19.9 (14.4~24.8); 12.5 (2.1~20.1); 25.5 (19.1~29.1); 19.9 (14.4~24.8); 179.3 (113.1~237.6); 0.1 (0.0~4.6); 1020 (727~1385) ▎1579 (939~3495); 772 (392~1847); 106 (37~897) ▎57.0 (-11.9~246.5); 1003 (717~1329); 306.4 (95.3~659.0)

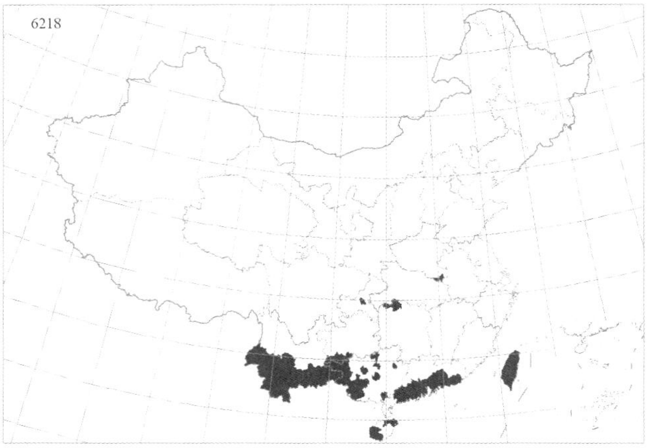

Tiliaceae 椴树科 *Grewia* 扁担杆属

Grewia acuminata Juss. 密齿扁担杆

Shrubs

20.0; 14.6; 23.2; 20.1; 180.8; 0.0; 963 ▎1523; 681; 67 ▎58.4; 942; 581.0

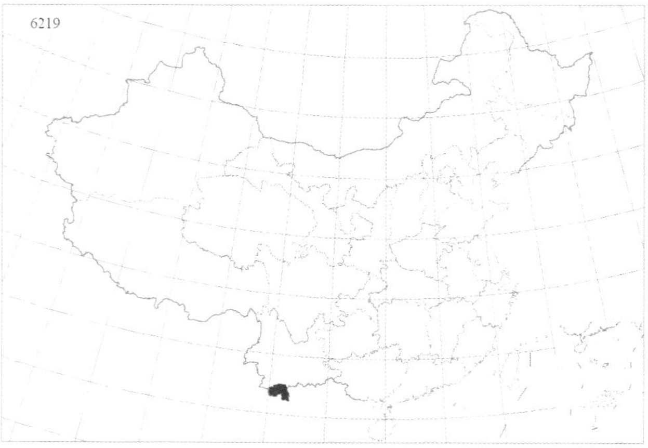

Tiliaceae 椴树科 *Grewia* 扁担杆属

Grewia angustisepala H. T. Chang 狭萼扁担杆

Shrubs

17.4; 11.0; 21.4; 17.4; 149.3; 0.0; 834 ▎1406; 816; 48 ▎68.7; 834; 434.4

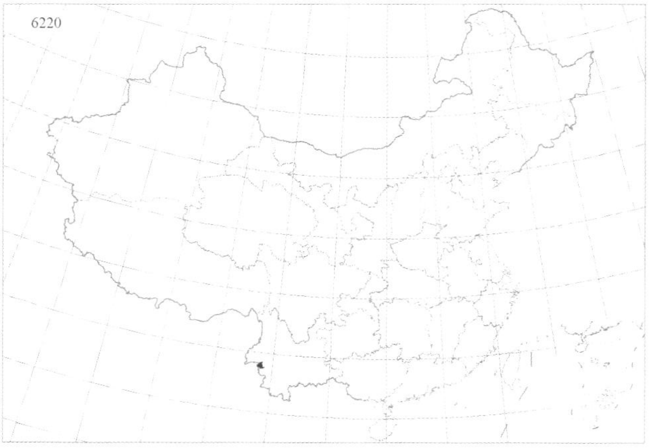

Tiliaceae 椴树科 *Grewia* 扁担杆属

Grewia biloba G. Don 扁担杆

Shrubs or small trees

14.8 (2.0~24.1); 3.1 (-13.7~18.9); 25.4 (9.9~30.1); 15.1 (3.8~24.1); 126.5 (15.4~229.7); 8.3 (0.0~71.7); 830 (413~1322) ▎1129 (367~2993); 494 (235~1847); 100 (8~551) ▎33.8 (-40.9~219.3); 788 (367~1190); 213.1 (70.3~551.6)

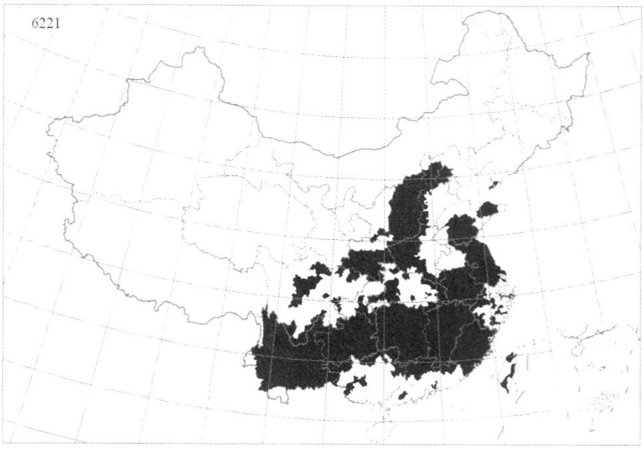

Tiliaceae 椴树科 *Grewia* 扁担杆属

Grewia biloba var. microphylla (Maxim.) Hand.-Mazz. 小叶扁担杆

Shrubs or small trees

16.0 (2.0~24.1); 7.8 (-7.2~18.9); 23.4 (9.9~28.2); 16.2 (3.8~24.1); 138.3 (15.4~229.7); 5.6 (0.0~51.4); 859 (413~1322) ▎1671 (731~3495); 758 (346~1847); 147 (19~897) ▎95.5 (10.0~246.5); 842 (413~1144); 229.5 (119.4~363.7)

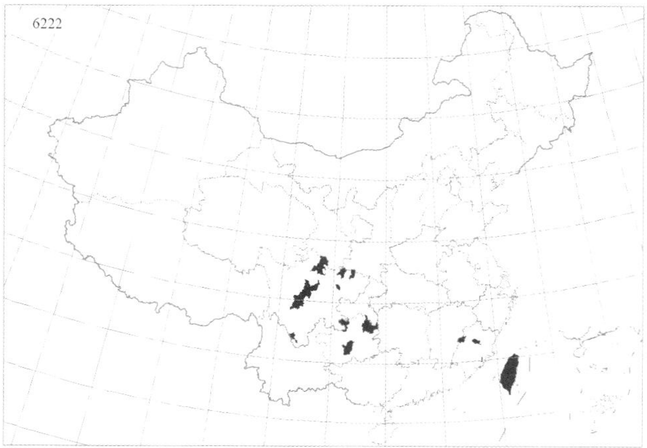

Tiliaceae 椴树科 *Grewia* 扁担杆属

Grewia biloba var. parviflora (Bunge) Hand.-Mazz. 小花扁担杆

Shrubs or small trees

14.6 (2.0~21.8); 2.4 (-15.9~13.6); 25.9 (9.9~30.1); 14.9 (3.8~21.8); 125.8 (15.4~201.6); 9.6 (0.0~80.4); 837 (413~1160) ❙ 1107 (377~1997); 469 (241~770); 104 (8~259) ❙ 30.3 (-47.7~137.5); 784 (377~1160); 209.3 (109.4~357.7)

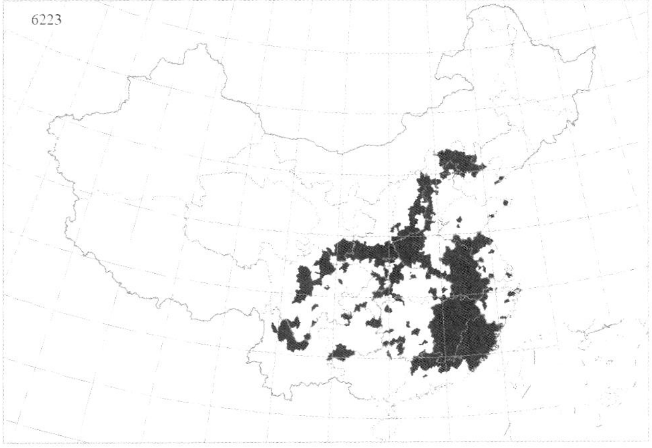

Tiliaceae 椴树科 *Grewia* 扁担杆属

Grewia brachypoda C. Y. Wu 短柄扁担杆

Shrubs

16.4 (11.6~19.6); 9.3 (3.1~12.8); 21.4 (18.4~24.3); 16.4 (11.5~19.6); 136.5 (80.9~175.7); 0.2 (0.0~2.8); 811 (654~953) ❙ 1223 (797~1688); 712 (411~959); 39 (19~70) ❙ 51.7 (-14.2~112.2); 766 (654~932); 387.3 (218.0~627.0)

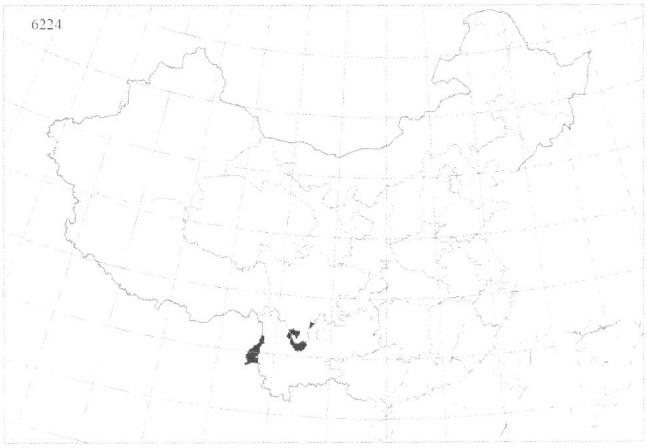

Tiliaceae 椴树科 *Grewia* 扁担杆属

Grewia celtidifolia Juss. 朴叶扁担杆

Shrubs

18.1 (12.4~24.8); 1[illegible].3 (4.7~20.1); 22.9 (18.6~28.8); 18.1 (12.4~24.8); 157.4 (89.1~237.6); 0.0 (0.0~0.3); 906 (675~1385) ❙ 1260 (868~2992); 668 (312~1847); 56 (26~244) ❙ 40.9 (-34.3~191.8); 877 (675~1360); 357.9 (189.5~659.0)

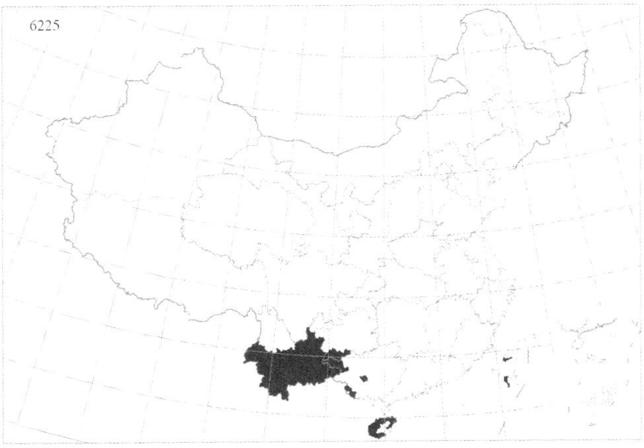

Tiliaceae 椴树科 *Grewia* 扁担杆属

Grewia chuniana Burret 崖县扁担杆

Shrubs

24.8; 20.1; 28.0; 24.8; 237.6; 0.0; 1385 ❙ 1405; 473; 41 ❙ 1.5; 1137; 527.5

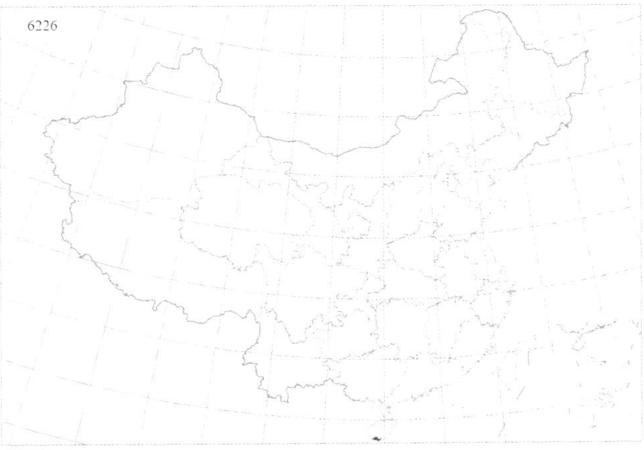

Tiliaceae 椴树科 *Grewia* 扁担杆属

Grewia concolor Merr. 同色扁担杆

Shrubs

23.2 (17.1~24.8); 16.8 (6.5~20.1); 28.0 (26.8~28.8); 23.2 (17.1~24.8); 218.7 (145.5~237.6); 0.0 (0.0~0.0); 1267 (894~1385) ❙ 1450 (868~1829); 513 (312~712); 91 (41~230) ❙ 18.4 (-34.3~104.6); 1129 (868~1331); 385.4 (243.2~620.0)

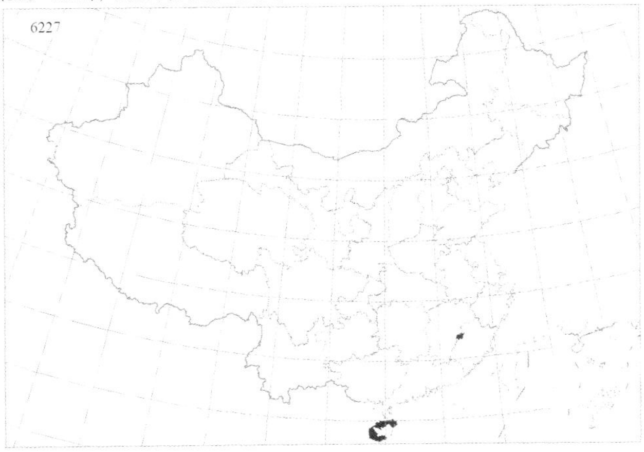

Tiliaceae 椴树科 *Grewia* 扁担杆属

Grewia cuspidatoserrata Burret 复齿扁担杆

Shrubs

17.8; 11.3; 22.1; 17.8; 153.2; 0.0; 852 ❙ 1124; 631; 47 ❙ 32.2; 840; 345.1

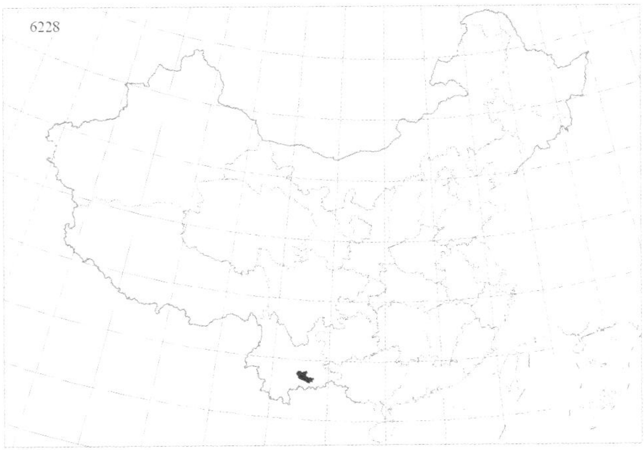

Tiliaceae 椴树科 *Grewia* 扁担杆属
Grewia falcata C. Y. Wu 镰叶扁担杆
Shrubs or small trees

18.1 (14.4~22.2); 10.1 (6.4~13.9); 24.1 (19.1~28.5); 18.2 (14.4~22.3); 157.8 (113.1~207.5); 0.0 (0.0~0.0); 909 (727~1190) ❙ 1515 (1178~1745); 739 (541~959); 105 (42~197) ❙ 68.0 (15.4~112.2); 905 (727~1190); 377.5 (208.1~627.0)

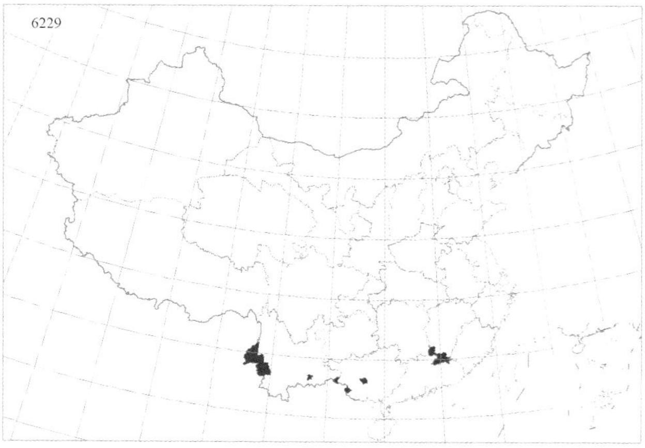

Tiliaceae 椴树科 *Grewia* 扁担杆属
Grewia henryi Burret 黄麻叶扁担杆
Shrubs or small trees

18.5 (14.6~22.2); 10.4 (3.8~15.1); 24.7 (21.3~28.5); 18.5 (14.7~22.3); 162.4 (117.7~207.5); 0.0 (0.0~1.5); 926 (779~1190) ❙ 1379 (939~1873); 667 (477~996); 94 (37~200) ❙ 49.0 (5.8~101.8); 918 (744~1190); 333.1 (197.5~659.0)

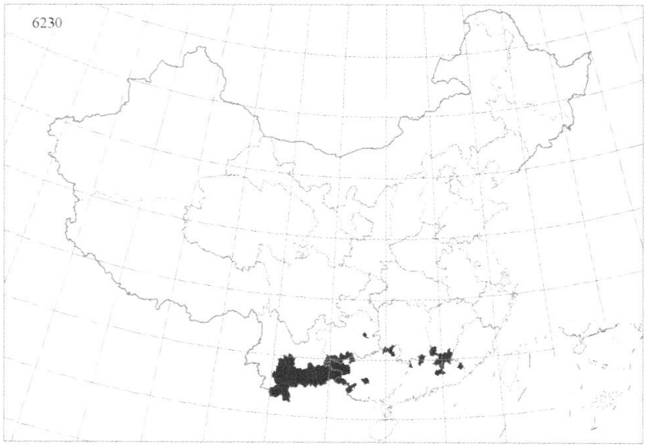

Tiliaceae 椴树科 *Grewia* 扁担杆属
Grewia kwangtungensis H. T. Chang 广东扁担杆
Shrubs

17.8; 7.1; 27.5; 17.8; 154.0; 0.0; 933 ❙ 1503; 513; 182 ❙ 61.1; 933; 197.5

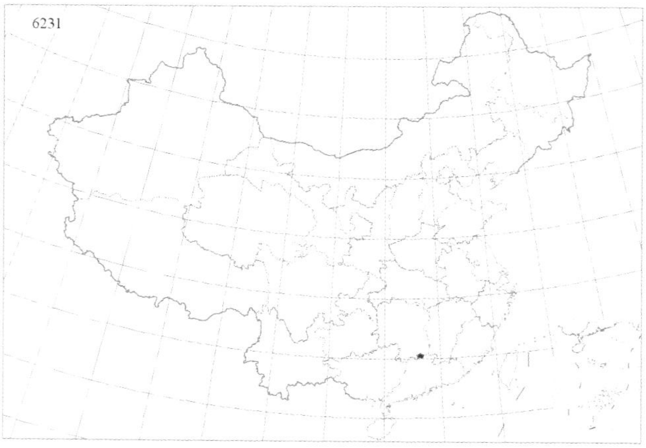

Tiliaceae 椴树科 *Grewia* 扁担杆属
Grewia macropetala Burret 长瓣扁担杆
Shrubs

18.5; 8.7; 26.4; 18.5; 162.3; 0.0; 948 ❙ 1405; 670; 107 ❙ 48.2; 948; 259.6

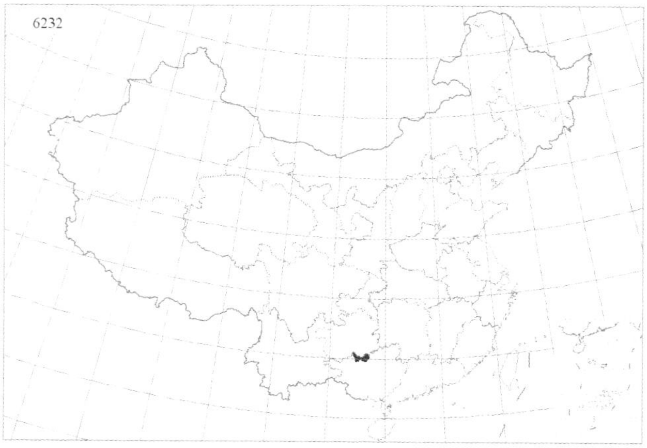

Tiliaceae 椴树科 *Grewia* 扁担杆属
Grewia multiflora Jussieu 光叶扁担杆
Shrubs or small trees

19.0 (13.6~21.9); 12.4 (7.2~15.1); 23.4 (18.8~28.6); 19.0 (13.7~22.0); 167.9 (104.1~203.8); 0.0 (0.0~0.0); 938 (706~1176) ❙ 1394 (1042~1607); 718 (592~916); 66 (44~138) ❙ 48.9 (26.1~64.9); 921 (706~1176); 447.0 (272.3~659.0)

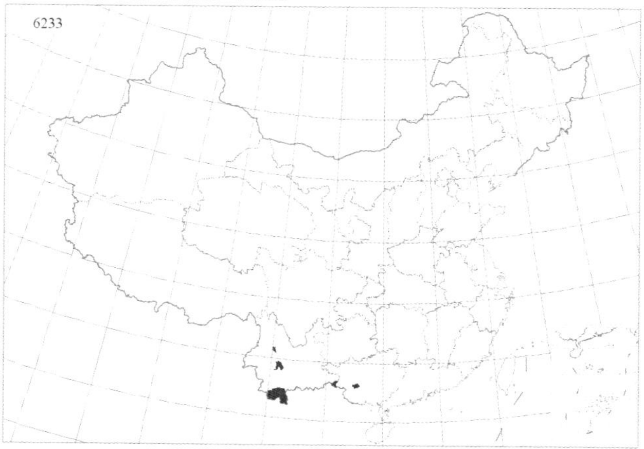

Tiliaceae 椴树科 *Grewia* 扁担杆属
Grewia oligandra Pierre 寡蕊扁担杆
Shrubs

21.6 (19.0~24.3); 13.3 (9.9~18.7); 27.9 (25.2~28.9); 21.6 (19.0~24.3); 199.6 (168.4~231.9); 0.0 (0.0~0.0); 1144 (952~1338) ❙ 1506 (1179~1953); 716 (392~1056); 115 (42~194) ❙ 32.5 (-11.9~71.5); 1119 (952~1202); 283.1 (157.7~538.2)

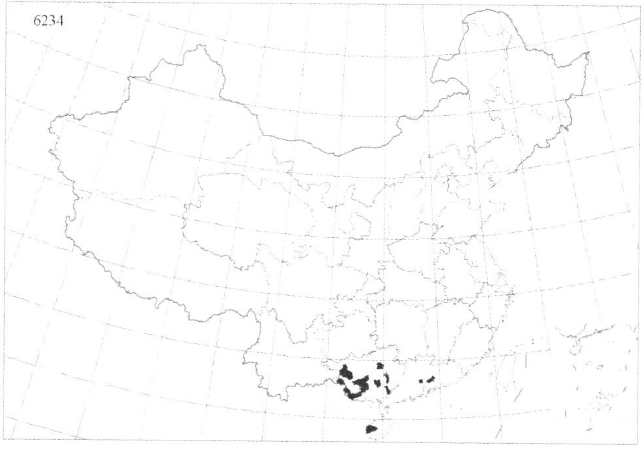

Tiliaceae 椴树科 *Grewia* 扁担杆属

Grewia permagna C. Y. Wu ex H. T. Chang 大叶扁担杆

Shrubs or small trees

19.0 (17.4~20.5); 13.1 (10.5~15.1); 22.7 (21.4~23.7); 19.0 (17.4~20.5); 168.3 (149.3~185.9); 0.0 (0.0~0.0); 914 (834~992) ❙ 1387 (958~1607); 678 (531~816); 58 (44~77) ❙ 51.4 (13.3~68.7); 896 (823~986); 497.9 (311.9~659.0)

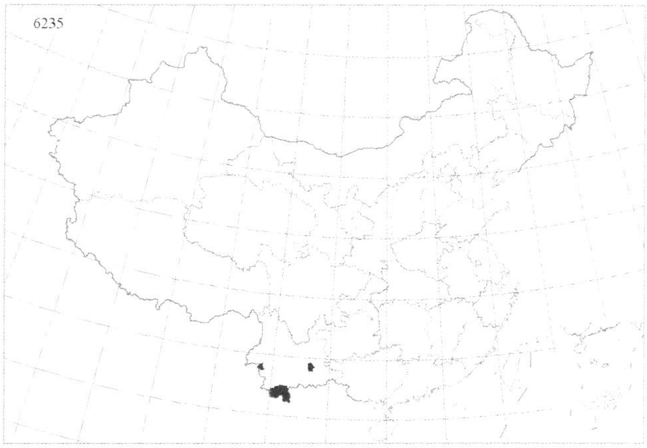

Tiliaceae 椴树科 *Grewia* 扁担杆属

Grewia retusifolia Pierre 钝叶扁担杆

Shrubs or small trees

22.3; 13.5; 28.9; 22.3; 208.1; 0.0; 1202 ❙ 1446; 710; 136 ❙ 20.3; 1202; 201.4

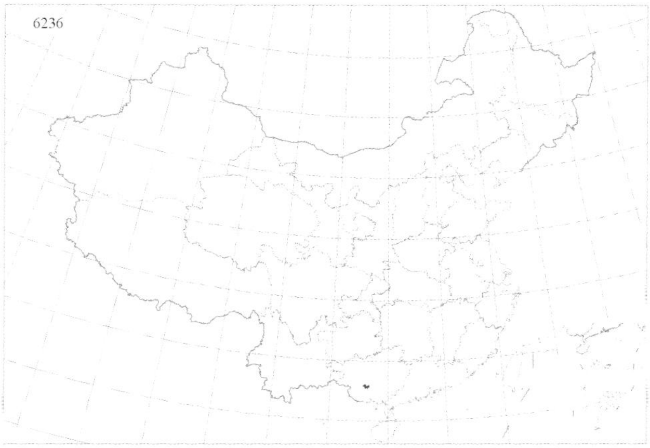

Tiliaceae 椴树科 *Grewia* 扁担杆属

Grewia rhombifolia Kaneh. et Sasaki 菱叶扁担杆

Shrubs

20.6 (17.1~24.1); 14.9 (11.2~18.9); 25.4 (21.8~28.2); 20.7 (17.1~24.1); 188.2 (145.0~229.7); 0.0 (0.0~0.0); 1051 (816~1322) ❙ 2299 (1498~2993); 1151 (859~1847); 198 (47~547) ❙ 130.1 (25.2~235.6); 996 (816~1144); 235.9 (119.4~298.9)

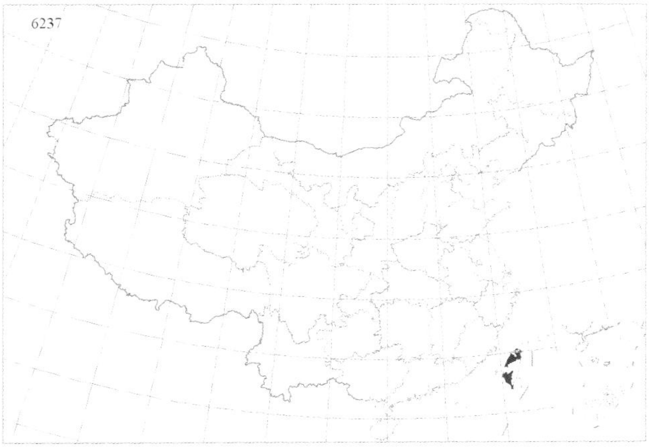

Tiliaceae 椴树科 *Grewia* 扁担杆属

Grewia sessiliflora Gagnep. 无柄扁担杆

Shrubs

22.2 (20.4~24.8); 14.7 (11.9~20.1); 28.1 (25.7~29.1); 22.3 (20.4~24.8); 207.0 (184.9~237.6); 0.0 (0.0~0.0); 1182 (1039~1385) ❙ 1684 (1179~2037); 757 (392~1033); 114 (41~188) ❙ 43.4 (-11.9~83.4); 1161 (1039~1329); 249.0 (95.3~620.0)

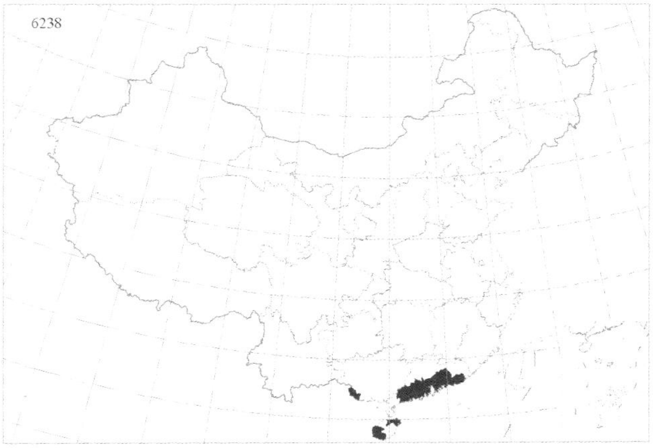

Tiliaceae 椴树科 *Grewia* 扁担杆属

Grewia tiliifolia Vahl 椴叶扁担杆

Trees

15.4 (8.6~21.0); 8.0 (1.3~14.2); 21.0 (15.0~26.5); 15.4 (8.7~21.0); 125.5 (52.6~191.8); 0.7 (0.0~8.8); 780 (574~1053) ❙ 1247 (1041~1522); 662 (511~860); 61 (45~92) ❙ 61.2 (34.7~93.0); 780 (574~1053); 372.5 (180.4~551.6)

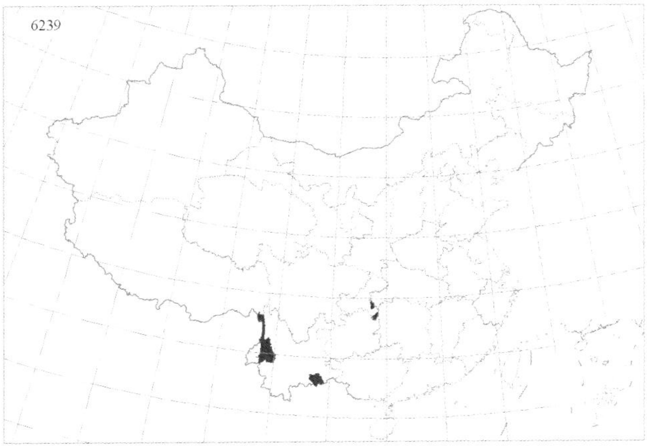

Tiliaceae 椴树科 *Grewia* 扁担杆属

Grewia urenifolia (Pierre) Gagnep. 稔叶扁担杆

Shrubs or small trees

18.9; 10.0; 26.0; 19.0; 167.6; 0.0; 976 ❙ 1363; 649; 103 ❙ 41.4; 975; 285.7

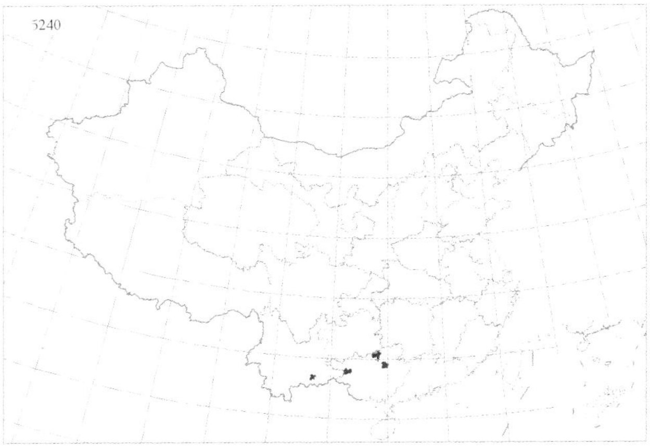

Tiliaceae 椴树科 *Microcos* 破布叶属
Microcos chungii (Merr.) Chun 海南破布叶
Trees

23.6 (20.4~24.8); 18.0 (14.9~20.1); 27.6 (23.6~28.8); 23.7 (20.5~24.8); 224.2 (185.8~237.6); 0.0 (0.0~0.0); 1279 (989~1385) ❙ 1505 (868~1854); 547 (312~712); 73 (41~115) ❙ 18.8 (-34.3~62.5); 1171 (868~1360); 456.9 (248.2~659.0)

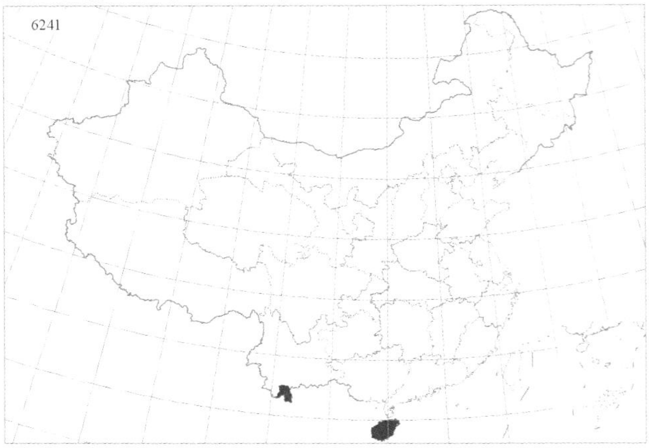

Tiliaceae 椴树科 *Microcos* 破布叶属
Microcos paniculata L. 破布叶
Shrubs or small trees

20.0 (14.4~24.8); 12.8 (6.4~20.1); 25.2 (19.1~29.1); 20.0 (14.4~24.8); 180.3 (113.1~237.6); 0.0 (0.0~0.0); 1025 (727~1385) ❙ 1451 (868~2435); 705 (312~1428); 85 (37~197) ❙ 43.8 (-34.3~112.2); 998 (717~1360); 358.8 (139.6~659.0)

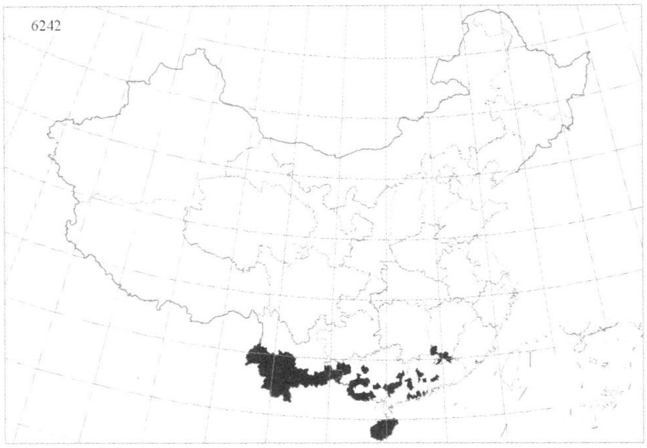

Tiliaceae 椴树科 *Tilia* 椴属
Tilia amurensis Rupr. 紫椴
Trees, deciduous

6.2 (-5.8~14.2); -12.7 (-29.8~-0.7); 22.6 (15.3~27.8); 9.4 (4.4~14.4); 75.6 (28.0~124.3); 61.2 (12.6~157.4); 635 (426~859) ❙ 574 (367~1073); 368 (235~691); 17 (4~65) ❙ -8.7 (-40.9~74.1); 543 (367~729); 208.4 (97.2~346.5)

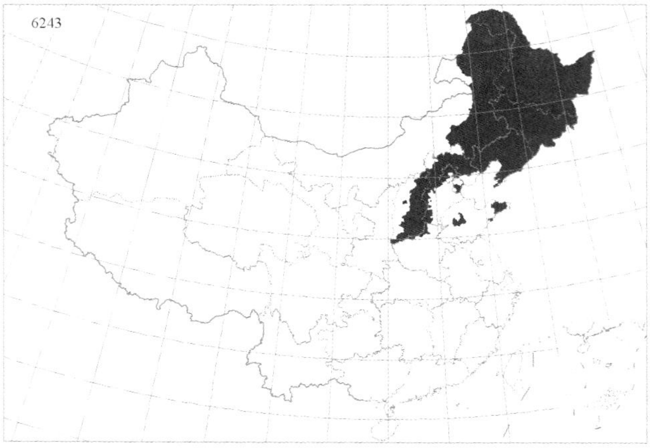

Tiliaceae 椴树科 *Tilia* 椴属
Tilia callidonta H. T. Chang 美齿椴
Trees, deciduous

8.8; 1.6; 15.0; 8.9; 55.9; 9.5; 576 ❙ 925; 556; 28 ❙ 60.7; 576; 318.1

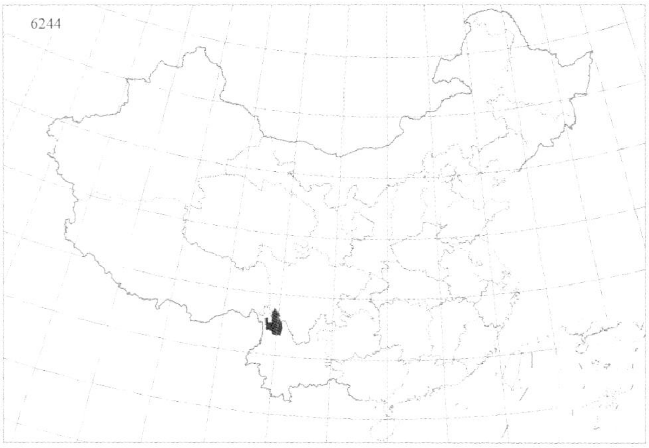

Tiliaceae 椴树科 *Tilia* 椴属
Tilia chinensis Maxim. 华椴
Trees, deciduous

11.6 (1.9~18.9); 0.8 (-9.2~12.0); 21.6 (9.5~27.8); 11.9 (3.8~19.0); 92.9 (15.4~167.8); 12.9 (0.0~57.8); 695 (411~929) ❙ 913 (341~1492); 450 (190~816); 38 (7~120) ❙ 34.0 (-40.9~128.6); 668 (341~881); 246.8 (61.2~384.7)

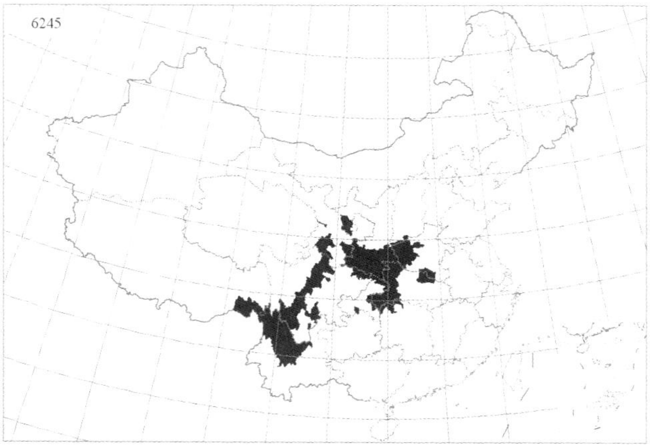

Tiliaceae 椴树科 *Tilia* 椴属
Tilia chinensis var. intonsa (Rehder et Wils) Hsu et Zhuge 多毛椴
Trees, deciduous

10.7 (2.2~18.9); 2.2 (-6.6~12.0); 18.0 (9.5~25.6); 11.0 (3.8~19.0); 81.8 (15.5~167.8); 13.0 (0.0~49.1); 646 (411~929) ❙ 941 (634~1220); 534 (290~688); 24 (11~59) ❙ 52.9 (-16.6~125.9); 617 (411~844); 267.8 (150.8~322.5)

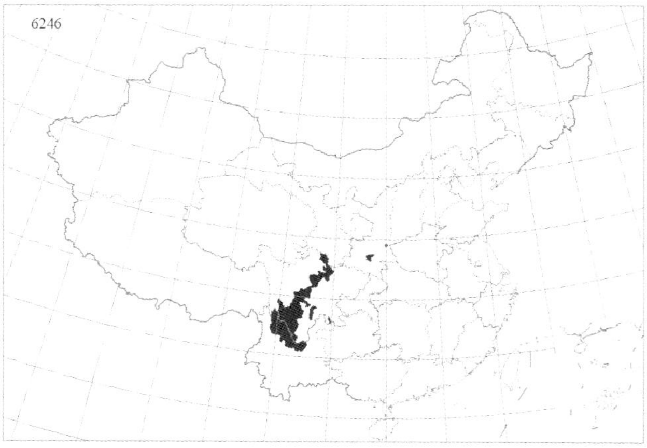

Tiliaceae 椴树科 *Tilia* 椴属
Tilia chinensis var. investita (V. Engl.) Rehder 秃华椴
Trees, deciduous

11.7 (-0.7~16.7); 4.7 (-9.6~9.7); 17.4 (7.7~22.1); 11.9 (2.5~16.7); 87.0 (6.8~140.9); 6.4 (0.0~74.2); 658 (338~819) ❙ 990 (370~1337); 575 (262~727); 38 (10~69) ❙ 51.0 (4.7~93.0); 642 (338~748); 321.9 (61.6~407.1)

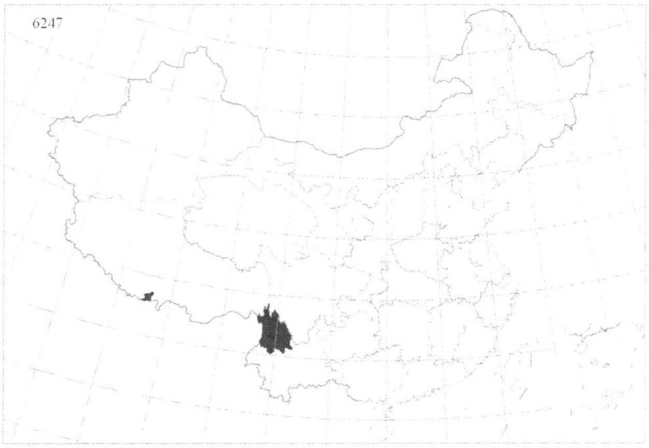

Tiliaceae 椴树科 *Tilia* 椴属
Tilia chingiana Hu et W. C. Cheng 短毛椴
Trees, deciduous

16.2 (14.4~17.4); 4.1 (2.1~6.2); 27.7 (24.6~29.3); 16.2 (14.5~17.4); 135.8 (117.7~149.4); 1.2 (0.0~4.9); 881 (779~950) ❙ 1542 (1211~1689); 538 (427~685); 193 (134~228) ❙ 75.5 (28.8~105.7); 878 (779~950); 239.2 (186.9~319.9)

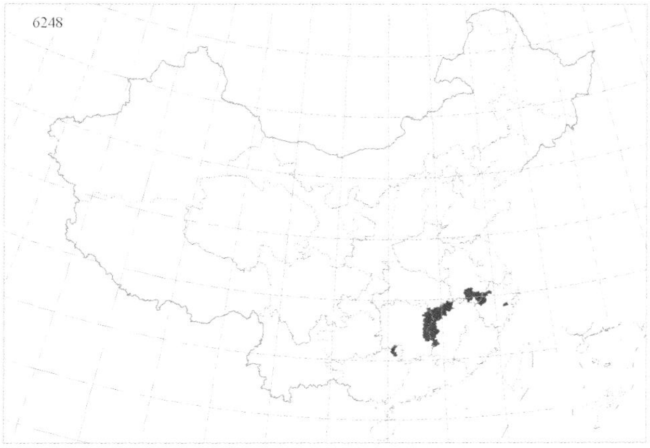

Tiliaceae 椴树科 *Tilia* 椴属
Tilia endochrysea Hand.-Mazz. 白毛椴
Trees, deciduous

16.8 (14.1~19.4); 5.4 (1.6~9.4); 27.6 (24.2~29.9); 16.9 (14.2~19.4); 143.3 (113.2~173.4); 0.6 (0.0~6.5); 903 (761~1037) ❙ 1572 (1215~1997); 536 (392~713); 194 (105~259) ❙ 74.7 (40.8~147.0); 892 (761~1029); 222.1 (134.6~319.9)

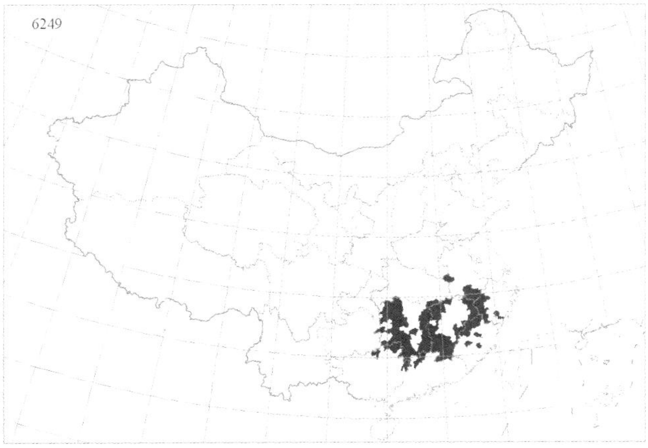

Tiliaceae 椴树科 *Tilia* 椴属
Tilia henryana Szyszyl. 毛糯米椴
Trees, deciduous

14.5 (8.2~18.2); 1.4 (-6.9~6.9); 26.8 (21.6~29.3); 14.6 (9.6~18.2); 122.5 (73.3~158.6); 7.5 (0.0~34.4); 833 (622~969) ❙ 1065 (512~1828); 454 (258~685); 99 (12~253) ❙ 26.7 (-36.5~112.2); 775 (512~941); 193.0 (120.1~319.9)

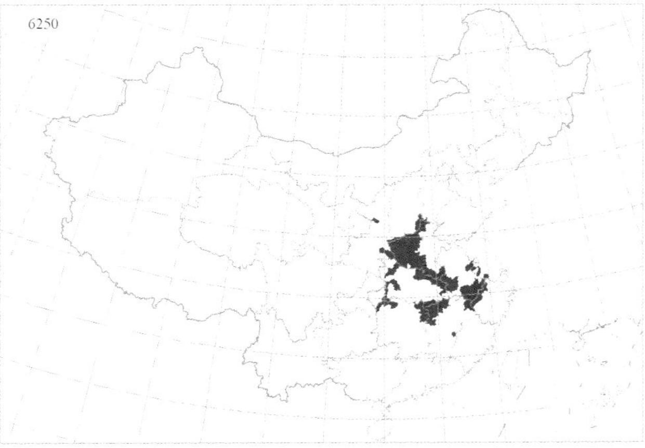

Tiliaceae 椴树科 *Tilia* 椴属
Tilia henryana var. subglabra V. Engl. 糯米椴
Trees, deciduous

14.5 (10.2~17.2); 1.2 (-2.9~5.1); 26.9 (22.4~29.4); 14.6 (10.6~17.2); 122.2 (83.1~146.8); 7.8 (0.0~20.5); 833 (656~952) ❙ 1021 (512~1828); 448 (258~685); 92 (19~253) ❙ 21.7 (-36.5~112.2); 762 (512~941); 192.6 (111.9~319.9)

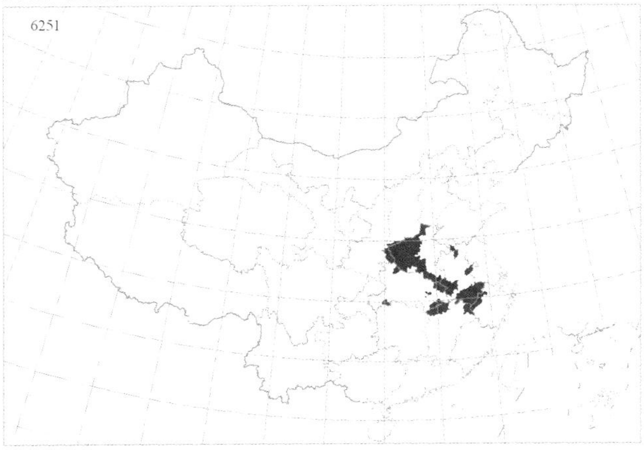

Tiliaceae 椴树科 *Tilia* 椴属
Tilia japonica (Miq.) Simonk. 华东椴
Trees, deciduous

15.3 (11.1~18.2); 2.8 (-3.3~6.5); 27.1 (23.0~29.3); 15.4 (11.6~18.2); 128.7 (95.6~158.8); 4.6 (0.0~22.0); 853 (709~984) ❙ 1375 (670~1922); 529 (392~685); 160 (27~259) ❙ 59.6 (-13.1~123.6); 845 (670~961); 216.9 (136.4~319.9)

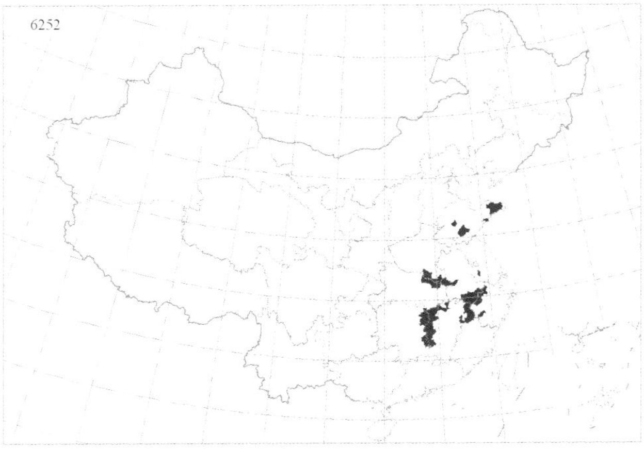

Tiliaceae 椴树科 *Tilia* 椴属

Tilia jiaodongensis S. B. Liang 胶东椴

Trees, deciduous

11.7 (11.1~12.1); -1.9 (-3.3~-1.2); 23.8 (23.0~24.6); 12.0 (11.6~12.4); 98.4 (95.6~103.4); 17.0 (14.2~22.0); 720 (709~746) ❙ 696 (670~759); 414 (392~466); 42 (31~65) ❙ -3.6 (-9.5~3.8); 691 (670~724); 167.8 (155.2~189.3)

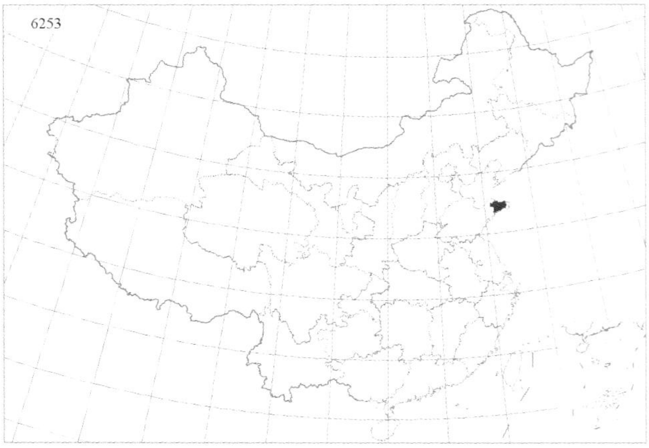

Tiliaceae 椴树科 *Tilia* 椴属

Tilia kueichouensis Hu 黔椴

Trees, deciduous

15.0 (12.8~15.9); 4.6 (3.5~5.5); 25.0 (21.0~26.0); 15.1 (12.9~15.9); 121.6 (95.9~131.1); 0.5 (0.0~1.7); 799 (699~833) ❙ 1143 (885~1255); 491 (475~506); 69 (34~79) ❙ 43.1 (26.5~62.4); 799 (699~833); 183.3 (148.9~276.9)

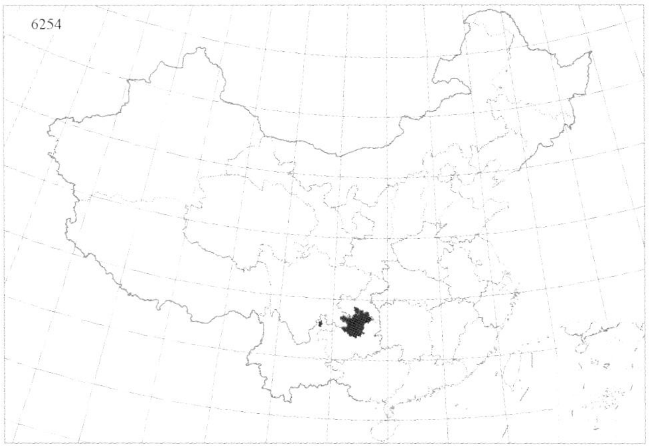

Tiliaceae 椴树科 *Tilia* 椴属

Tilia likiangensis H. T. Chang 丽江椴

Trees, deciduous

10.8; 3.9; 16.6; 10.9; 71.8; 1.3; 633 ❙ 968; 594; 24 ❙ 52.8; 633; 319.8

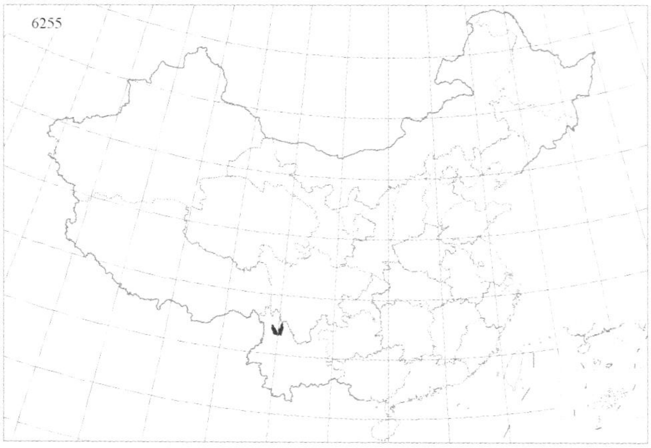

Tiliaceae 椴树科 *Tilia* 椴属

Tilia mandshurica Rupr. et Maxim. 糠椴

Trees, deciduous

7.5 (-4.0~15.5); -10.5 (-28.2~2.3); 23.2 (17.6~28.5); 10.0 (5.3~15.5); 82.2 (38.6~130.2); 51.4 (4.1~146.2); 662 (463~883) ❙ 635 (384~1073); 403 (249~691); 21 (7~113) ❙ -2.6 (-40.9~74.1); 587 (384~883); 221.7 (106.8~346.5)

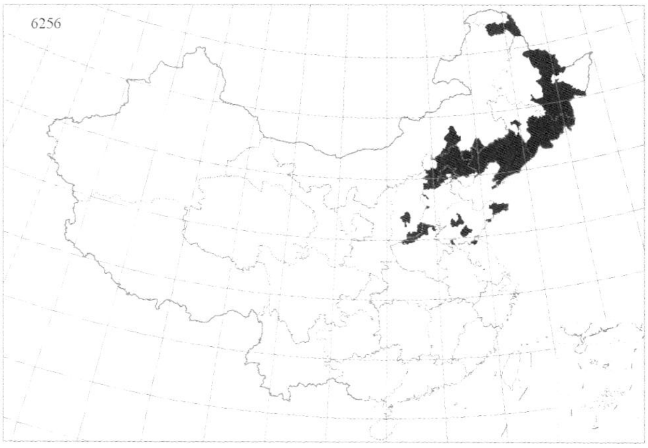

Tiliaceae 椴树科 *Tilia* 椴属

Tilia mandshurica var. megaphylla (Nakai) Liou et Li 棱果辽椴

Trees, deciduous

3.2 (1.8~8.4); -16.3 (-18.7~-8.9); 20.1 (18.5~23.2); 7.6 (6.7~10.2); 57.1 (47.6~84.0); 78.1 (42.8~88.8); 556 (513~661) ❙ 638 (558~1018); 386 (330~648); 22 (16~36) ❙ 14.4 (-3.6~54.0); 552 (513~661); 300.3 (265.9~333.1)

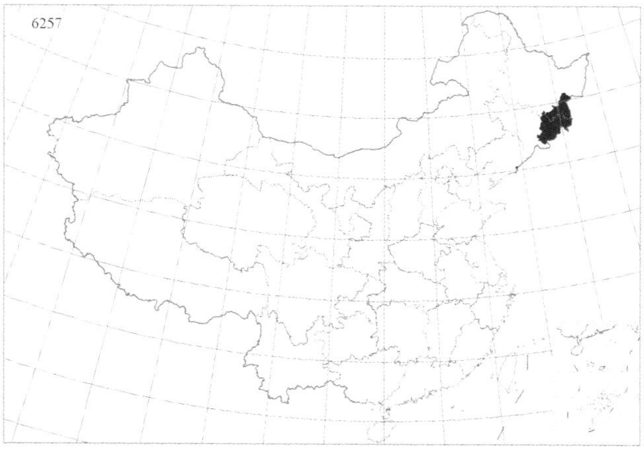

Tiliaceae 椴树科 *Tilia* 椴属

Tilia mandshurica var. ovalis (Nakai) Liou et Li 卵果糠椴

Trees, deciduous

3.7; -15.8; 20.5; 7.8; 59.1; 74.4; 560 ❙ 837; 525; 31 ❙ 48.1; 560; 322.3

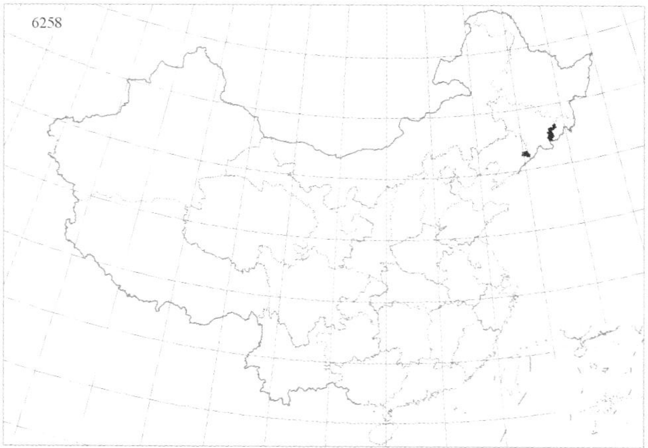

Tiliaceae 椴树科 *Tilia* 椴属

Tilia mandshurica var. tuberculata Liou et Li 瘤果糠椴

Trees, deciduous

3.0 (1.4~8.2); -17.0 (-19.2~-10.3); 20.2 (18.4~24.3); 7.6 (6.4~10.3); 56.7 (45.8~87.0); 80.4 (47.7~88.8); 555 (501~678) ❙ 659 (558~853); 402 (340~527); 24 (16~35) ❙ 19.1 (-3.6~57.4); 552 (501~678); 299.8 (219.8~346.5)

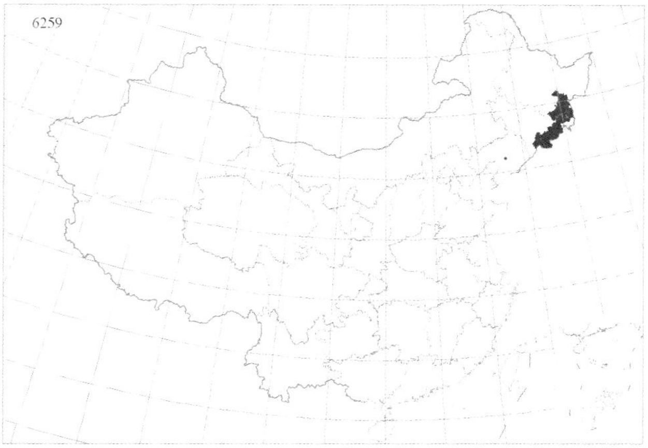

Tiliaceae 椴树科 *Tilia* 椴属

Tilia membranacea H. T. Chang 膜叶椴

Trees, deciduous

16.5 (15.5~17.1); 4.3 (3.5~5.2); 28.2 (26.6~29.3); 16.5 (15.5~17.1); 138.9 (128.0~145.7); 0.7 (0.0~1.5); 899 (839~941) ❙ 1530 (1403~1630); 534 (467~685); 189 (120~228) ❙ 70.3 (51.1~91.5); 894 (839~941); 225.3 (202.6~268.8)

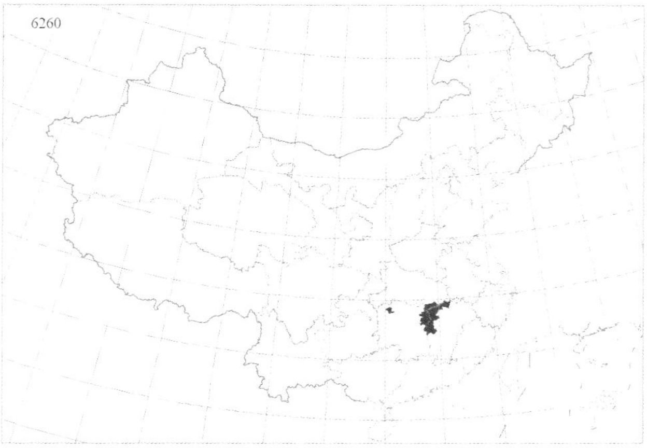

Tiliaceae 椴树科 *Tilia* 椴属

Tilia miqueliana Maxim. 南京椴

Trees, deciduous

15.0 (10.2~19.9); 2.1 (-2.9~10.6); 27.1 (22.4~29.3); 15.1 (10.6~19.9); 127.2 (83.1~179.3); 6.5 (0.0~20.5); 851 (656~1036) ❙ 1083 (512~1828); 463 (258~685); 137 (19~253) ❙ 25.8 (-36.5~128.9); 797 (512~1036); 192.4 (120.1~284.3)

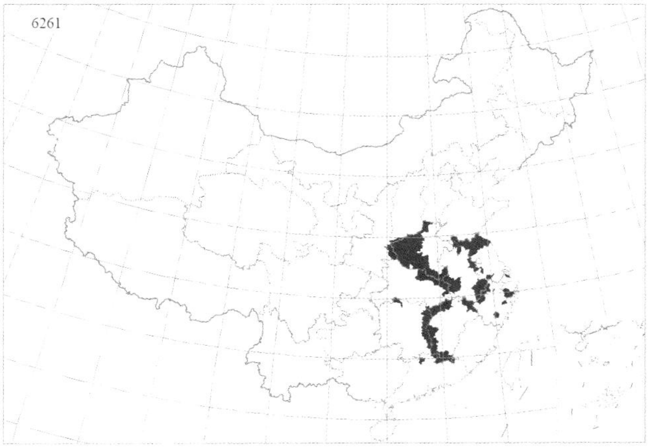

Tiliaceae 椴树科 *Tilia* 椴属

Tilia mongolica Maxim. 蒙椴

Trees, deciduous

9.8 (0.2~15.7); -5.9 (-20.8~2.5); 23.8 (17.5~28.5); 11.1 (6.4~15.7); 91.3 (45.4~132.3); 32.8 (3.4~104.9); 700 (506~889) ❙ 577 (256~1200); 345 (169~542); 21 (5~109) ❙ -17.8 (-57.7~41.4); 567 (256~881); 165.1 (61.2~287.6)

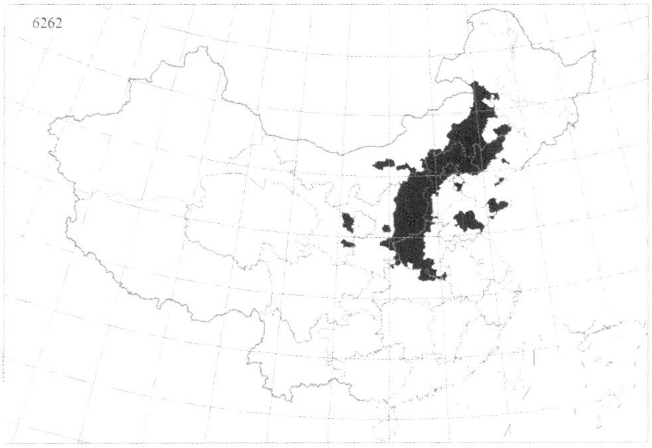

Tiliaceae 椴树科 *Tilia* 椴属

Tilia nobilis Rehder et E. H. Wilson 大叶椴

Trees, deciduous

12.0 (5.7~16.6); 2.3 (-3.5~6.6); 20.7 (12.7~26.8); 12.1 (6.3~16.6); 91.5 (35.2~139.7); 7.0 (0.0~28.1); 691 (496~866) ❙ 1009 (793~1590); 559 (377~946); 35 (21~59) ❙ 47.9 (-2.0~92.9); 690 (496~866); 227.6 (147.3~318.7)

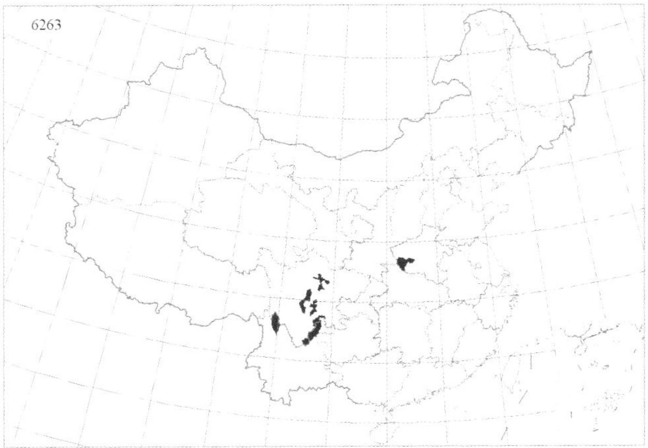

Tiliaceae 椴树科 *Tilia* 椴属

Tilia oliveri Szyszyl. 鄂椴

Trees, deciduous

13.7 (7.1~18.2); 0.9 (-10.6~7.5); 25.7 (17.7~29.5); 13.9 (8.2~18.2); 114.8 (58.8~158.8); 9.8 (0.0~50.1); 794 (561~984) ❙ 1035 (358~1997); 431 (200~816); 89 (8~259) ❙ 28.2 (-45.2~130.7); 744 (358~961); 197.2 (144.7~334.4)

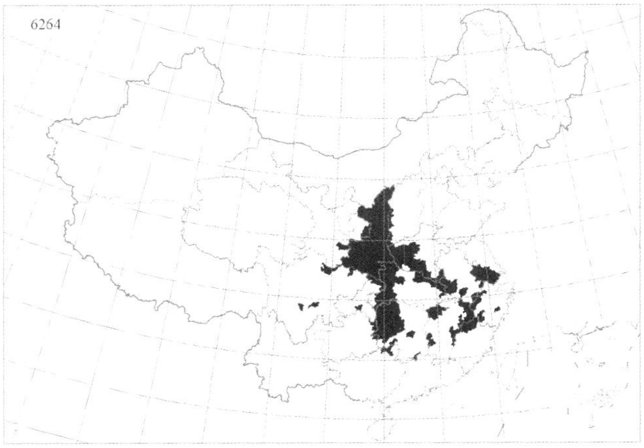

Tiliaceae 椴树科 *Tilia* 椴属

Tilia oliveri var. cinerascens Rehder et E. H. Wilson 灰背椴

Trees, deciduous

15.1 (10.6~16.4); 4.0 (-0.2~5.3); 25.8 (21.1~27.5); 15.1 (10.6~16.4); 123.2 (79.4~136.6); 1.4 (0.0~12.1); 814 (643~877) ❙ 1269 (1165~1451); 504 (444~600); 90 (46~140) ❙ 56.5 (34.4~93.8); 814 (643~877); 193.8 (146.0~284.3)

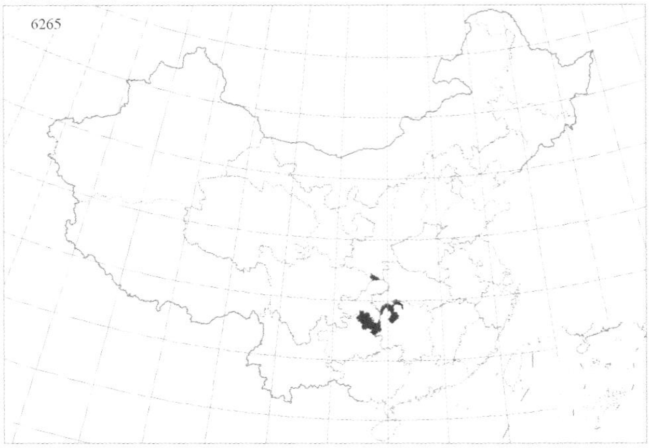

Tiliaceae 椴树科 *Tilia* 椴属

Tilia paucicostata Maxim. 少脉椴

Trees, deciduous

10.9 (1.9~18.9); -1.4 (-12.7~12.0); 22.0 (9.5~29.2); 11.6 (3.8~19.0); 92.2 (15.5~167.8); 20.4 (0.0~65.0); 689 (411~936) ❙ 768 (341~1770); 399 (190~651); 33 (7~223) ❙ 11.4 (-40.9~125.9); 624 (341~926); 214.1 (61.2~384.1)

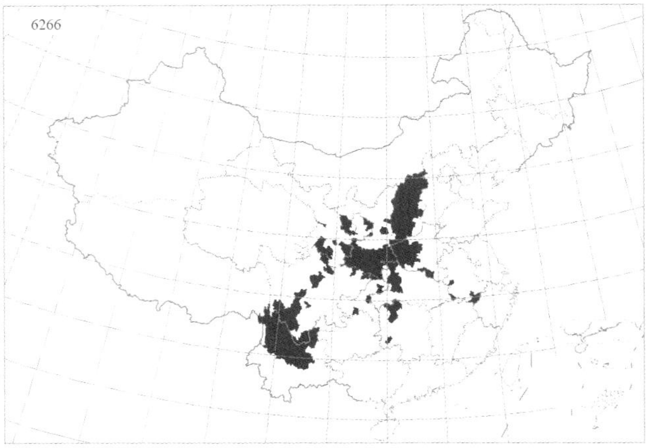

Tiliaceae 椴树科 *Tilia* 椴属

Tilia paucicostata var. dictyoneura (V. Engl. ex C. K. Schneid.) H. T. Chang et E. W. Miao 红皮椴

Trees, deciduous

11.7 (3.0~15.7); -2.3 (-10.6~2.5); 24.4 (12.4~29.5); 12.1 (4.8~15.8); 99.3 (25.1~136.0); 18.6 (3.5~50.1); 733 (441~932) ❙ 662 (358~1200); 328 (200~536); 25 (8~109) ❙ -8.7 (-45.2~83.5); 618 (358~890); 179.8 (44.7~334.4)

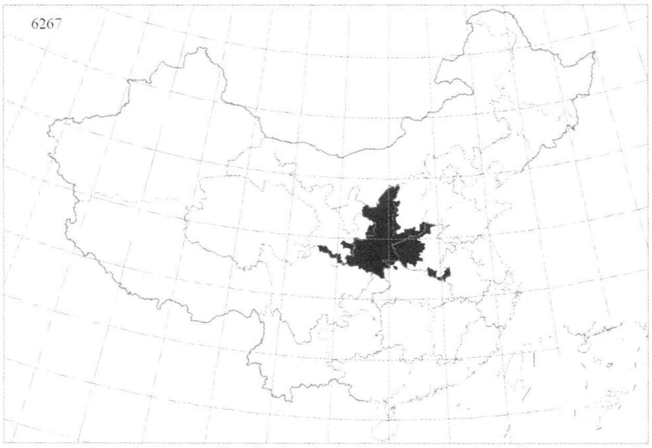

Tiliaceae 椴树科 *Tilia* 椴属

Tilia paucicostata var. yunnanensis Diels 毛少脉椴

Trees, deciduous

7.7 (2.2~21.0); 0.1 (-6.3~14.2); 14.2 (9.5~25.6); 8.3 (3.9~21.0); 56.1 (15.8~191.8); 22.8 (0.0~48.6); 568 (411~1053) ❙ 923 (628~1522); 551 (410~818); 28 (11~68) ❙ 66.0 (44.5~121.7); 568 (411~1053); 311.2 (208.8~551.6)

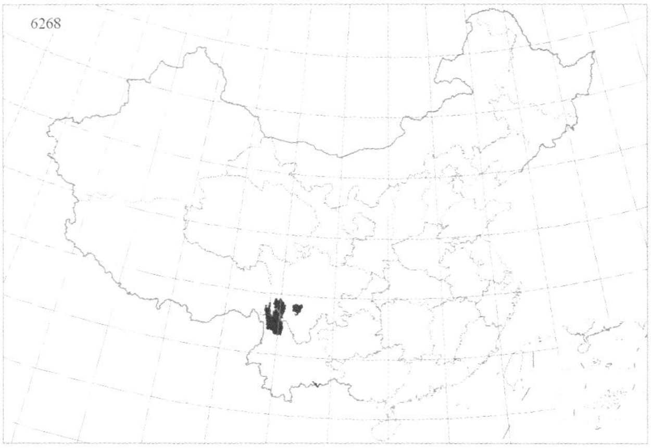

Tiliaceae 椴树科 *Tilia* 椴属

Tilia taishanensis S. B. Liang 泰山椴

Trees, deciduous

13.3 (11.7~15.5); -1.7 (-3.4~2.1); 26.6 (25.0~27.8); 13.6 (12.2~15.6); 115.9 (101.5~131.3); 15.5 (4.3~20.4); 814 (743~881) ❙ 734 (636~1077); 455 (427~488); 35 (23~97) ❙ -9.9 (-23.9~22.3); 707 (636~881); 154.7 (140.5~180.5)

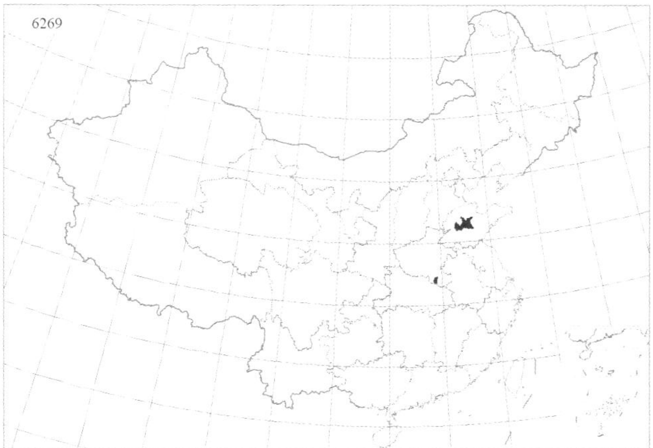

Tiliaceae 椴树科 *Tilia* 椴属

Tilia tuan Szyszyl. 椴树

Trees, deciduous

15.8 (3.5~22.2); 6.3 (-6.5~15.1); 24.1 (10.8~29.4); 15.8 (4.7~22.3); 131.1 (21.6~207.5); 1.2 (0.0~43.4); 826 (444~1190) ❙ 1290 (731~1997); 577 (346~996); 99 (15~259) ❙ 56.0 (-14.2~130.7); 817 (444~1190); 261.2 (134.1~659.0)

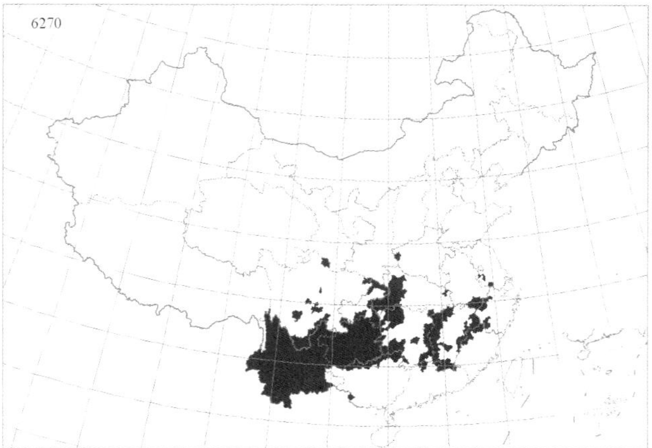

Tiliaceae 椴树科 *Tilia* 椴属
Tilia tuan var. chenmoui (Cheng) Y. Tang 长苞椴
Trees, deciduous

14.3; 7.7; 19.6; 14.4; 112.7; 0.0; 730 ❙ 1000; 581; 39 ❙ 37.3; 714; 315.4

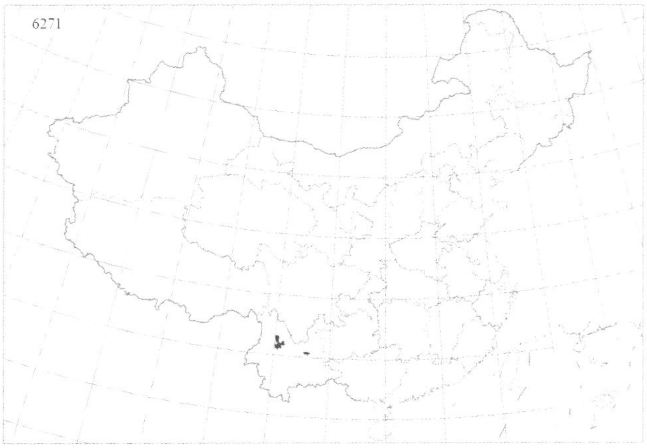

Tiliaceae 椴树科 *Tilia* 椴属
Tilia tuan var. chinensis (Szyszyl.) Rehder et E. H. Wilson 毛芽椴
Trees, deciduous

15.4 (11.4~19.1); 6.2 (2.6~12.0); 23.3 (18.6~27.2); 15.4 (11.4~19.2); 125.5 (80.7~170.1); 0.4 (0.0~4.0); 797 (652~981) ❙ 1127 (797~1496); 549 (411~720); 64 (20~161) ❙ 41.5 (-14.2~71.9); 788 (652~981); 227.5 (146.0~331.6)

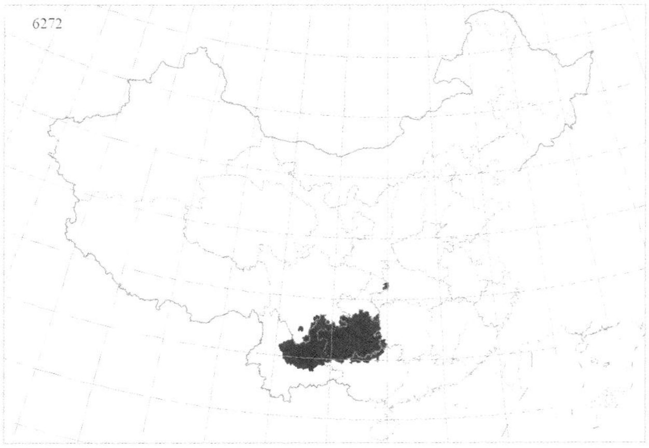

Malvaceae 锦葵科 *Abutilon* 苘麻属
Abutilon gebauerianum Hand.-Mazz. 滇西苘麻
Shrubs, deciduous

14.5 (8.6~20.5); 7.7 (1.3~14.9); 19.4 (15.0~23.7); 14.5 (8.7~20.5); 114.8 (52.6~185.9); 0.9 (0.0~8.8); 742 (574~992) ❙ 1214 (952~1464); 661 (564~792); 55 (28~70) ❙ 65.1 (28.1~99.9); 731 (574~959); 384.6 (275.9~570.1)

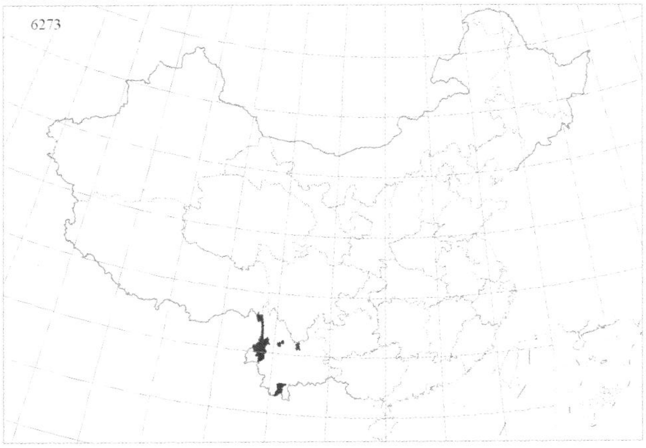

Malvaceae 锦葵科 *Abutilon* 苘麻属
Abutilon paniculatum Hand.-Mazz. 圆锥苘麻
Shrubs, deciduous

11.3 (5.4~21.6); 3.9 (-2.1~13.3); 17.4 (12.3~28.5); 11.4 (5.8~21.7); 84.1 (31.5~199.8); 8.2 (0.0~25.6); 672 (481~1153) ❙ 1026 (741~1625); 598 (426~795); 40 (12~173) ❙ 55.4 (27.6~81.5); 664 (481~1153); 317.3 (237.8~452.0)

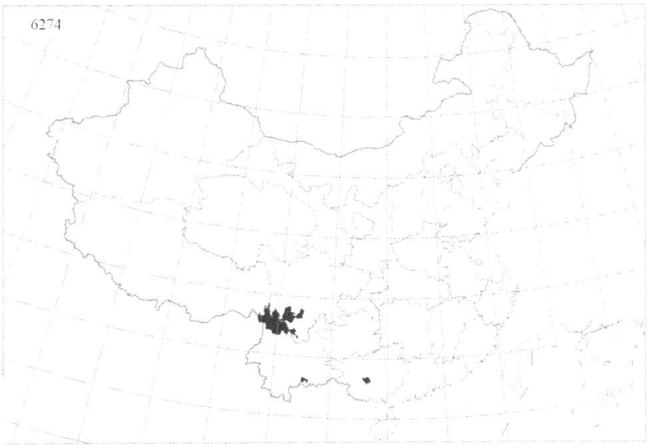

Malvaceae 锦葵科 *Abutilon* 苘麻属
Abutilon sinense Oliv. 华苘麻
Shrubs, deciduous

15.1 (5.4~21.6); 7.3 (-2.1~12.6); 21.3 (12.3~28.9); 15.1 (5.8~21.7); 122.3 (31.5~199.8); 0.9 (0.0~25.6); 779 (481~1153) ❙ 1044 (741~1625); 577 (411~795); 45 (15~173) ❙ 35.7 (-14.2~93.0); 762 (481~1153); 275.7 (108.2~407.1)

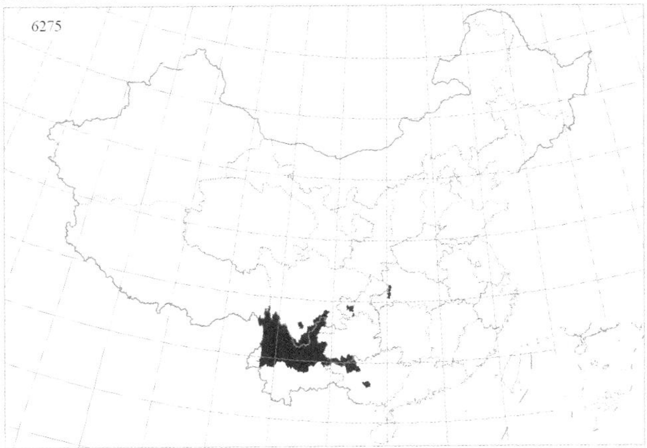

Malvaceae 锦葵科 *Abutilon* 苘麻属
Abutilon sinense var. edentatum K. M. Feng 无齿华苘麻
Shrubs, deciduous

13.2; 6.4; 18.4; 13.3; 99.4; 0.0; 695 ❙ 1337; 727; 69 ❙ 92.4; 695; 407.1

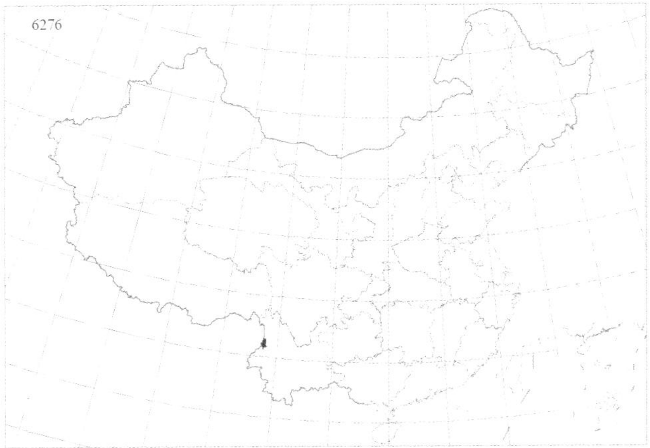

Malvaceae 锦葵科 *Cenocentrum* 大萼葵属
Cenocentrum tonkinense Gagnep. 大萼葵
Shrubs, deciduous

18.5 (16.3~21.0); 11.6 (9.0~15.1); 23.3 (21.4~26.6); 18.5 (16.4~21.0); 162.2 (136.4~191.8); 0.0 (0.0~0.0); 902 (797~1053) ❙ 1328 (958~1751); 714 (531~996); 57 (44~77) ❙ 47.3 (13.3~88.8); 897 (797~1053); 414.8 (189.5~659.0)

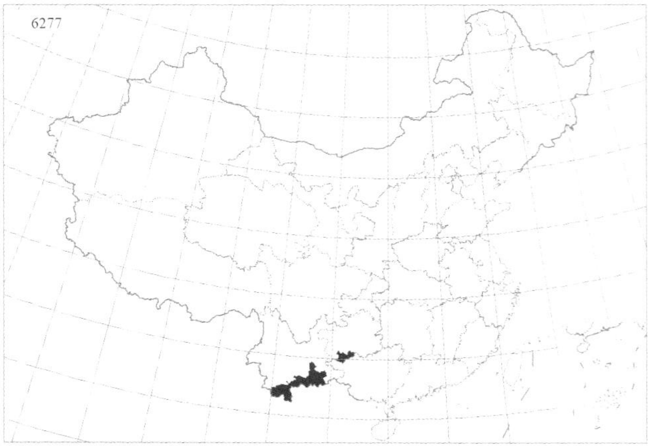

Malvaceae 锦葵科 *Hibiscus* 木槿属
Hibiscus aridicola J. Anthony 旱地木槿
Shrubs, deciduous

10.7 (5.4~16.7); 3.4 (-2.1~9.7); 16.9 (12.3~22.1); 10.8 (5.8~16.7); 76.9 (31.5~140.9); 7.9 (0.0~25.6); 633 (481~819) ❙ 920 (741~1007); 557 (426~610); 23 (12~40) ❙ 47.9 (14.4~67.9); 610 (481~721); 299.4 (255.4~351.7)

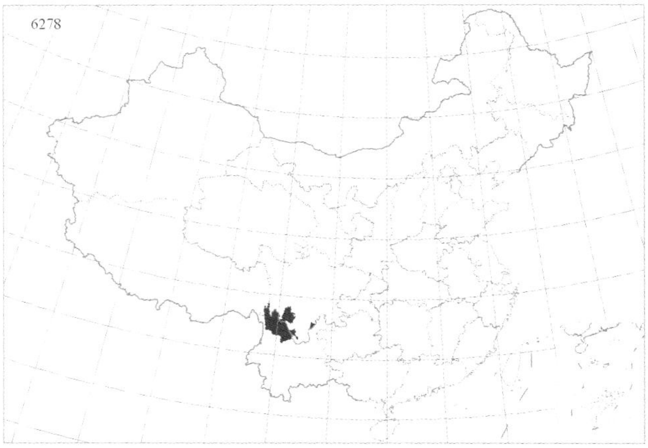

Malvaceae 锦葵科 *Hibiscus* 木槿属
Hibiscus austroyunnanensis Wu et Feng 滇南芙蓉
Scandent shrubs

16.3; 9.6; 20.9; 16.3; 135.5; 0.0; 807 ❙ 1617; 846; 54 ❙ 105.4; 793; 480.6

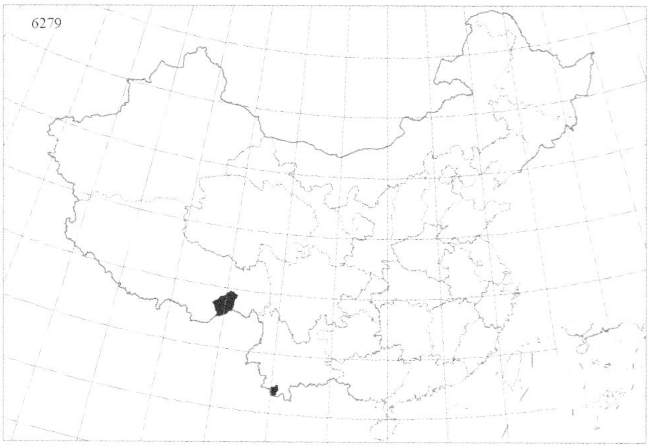

Malvaceae 锦葵科 *Hibiscus* 木槿属
Hibiscus grewiifolius Hassk. 樟叶槿
Small trees, evergreen

23.9 (22.2~24.8); 18.6 (17.2~20.1); 27.6 (25.7~28.5); 23.9 (22.3~24.8); 227.0 (207.2~237.6); 0.0 (0.0~0.0); 1298 (1127~1385) ❙ 1531 (1179~1854); 525 (392~622); 60 (41~115) ❙ 18.5 (-11.9~47.8); 1161 (1055~1360); 515.5 (428.6~620.0)

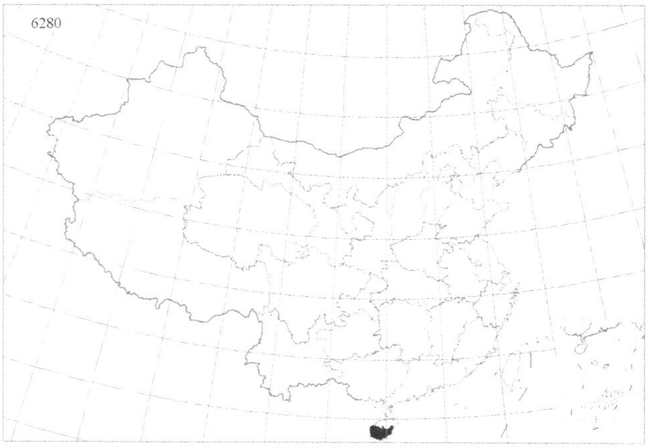

Malvaceae 锦葵科 *Hibiscus* 木槿属
Hibiscus hamabo Sieb. et Zucc. 海滨木槿
Shrubs or trees, deciduous

16.5; 5.3; 27.7; 16.5; 138.0; 0.0; 888 ❙ 1363; 481; 183 ❙ 53.5; 888; 222.6

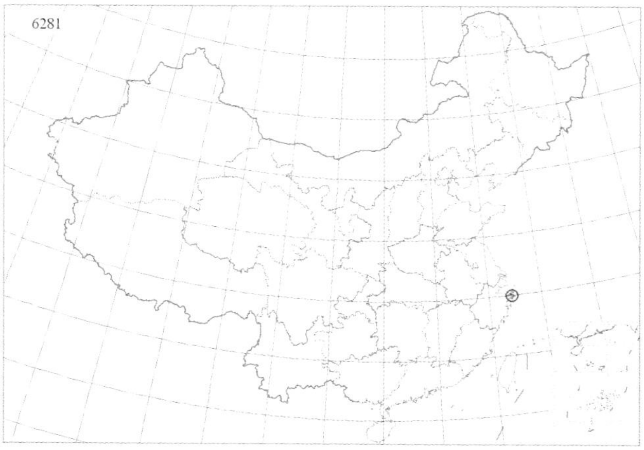

Malvaceae 锦葵科 *Hibiscus* 木槿属
Hibiscus indicus (Burm. f.) Hochr. 美丽芙蓉
Shrubs, deciduous

17.1 (5.4~24.7); 9.3 (-2.1~19.8); 23.3 (12.3~29.8); 17.2 (5.8~24.7); 146.5 (31.5~236.5); 0.4 (0.0~25.6); 878 (481~1376) ❙ 1244 (741~2992); 621 (312~1847); 77 (15~197) ❙ 42.1 (-34.3~166.5); 858 (481~1299); 299.3 (139.6~659.0)

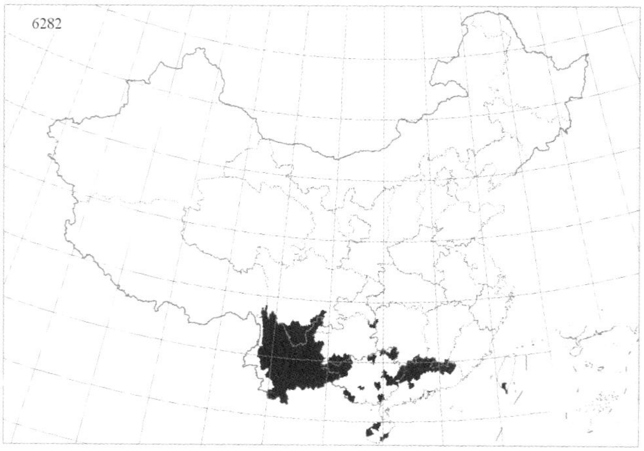

Malvaceae 锦葵科 *Hibiscus* 木槿属
Hibiscus labordei H. Lév. 贵州芙蓉
Shrubs, deciduous

16.0 (14.3~19.2); 5.3 (4.0~9.2); 25.7 (23.5~27.5); 16.0 (14.3~19.2); 132.4 (112.6~170.8); 0.3 (0.0~1.1); 840 (756~1002) ❙ 1261 (1168~1495); 505 (444~678); 104 (81~192) ❙ 50.2 (34.4~76.9); 840 (756~1002); 215.2 (178.6~256.4)

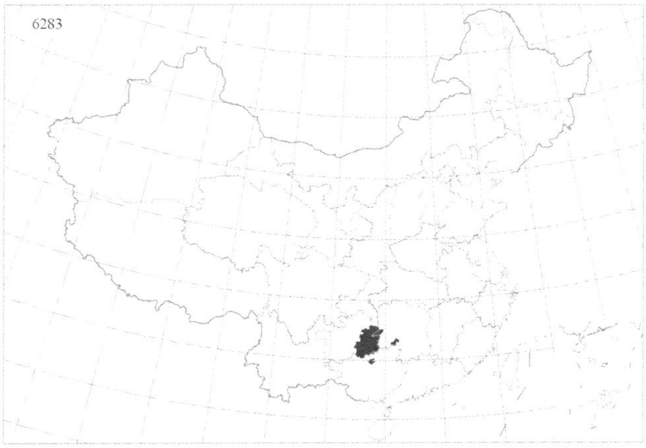

Malvaceae 锦葵科 *Hibiscus* 木槿属
Hibiscus leviseminus M. G. Gilbert, Y. Tang et Dorr 光籽木槿
Shrubs, deciduous

15.6 (9.2~19.2); 4.5 (-2.6~9.2); 26.1 (20.7~29.3); 15.7 (9.6~19.2); 129.9 (72.1~170.8); 1.7 (0.0~20.9); 842 (615~1002) ❙ 1324 (689~1679); 512 (322~685); 121 (17~221) ❙ 56.7 (0.0~106.4); 840 (615~1002); 209.8 (146.0~323.3)

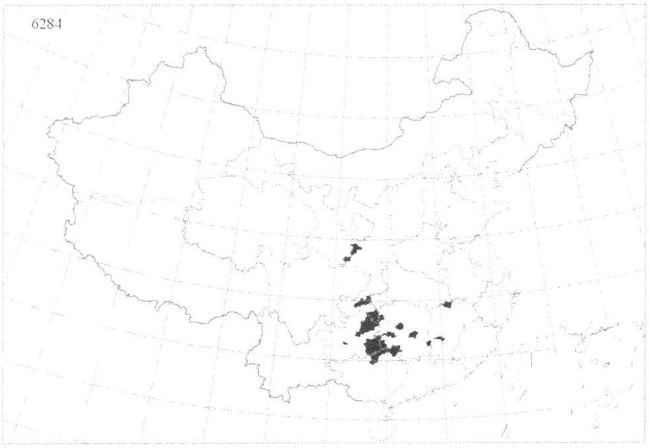

Malvaceae 锦葵科 *Hibiscus* 木槿属
Hibiscus macrophyllus Roxb. ex Harnemann 大叶木槿
Trees

18.7 (16.3~20.5); 12.8 (10.2~15.1); 22.2 (20.3~23.7); 18.7 (16.4~20.5); 164.5 (136.6~185.9); 0.0 (0.0~0.0); 893 (789~992) ❙ 1431 (1232~1607); 735 (655~803); 53 (41~77) ❙ 60.4 (47.1~74.2); 869 (780~986); 487.5 (372.4~659.0)

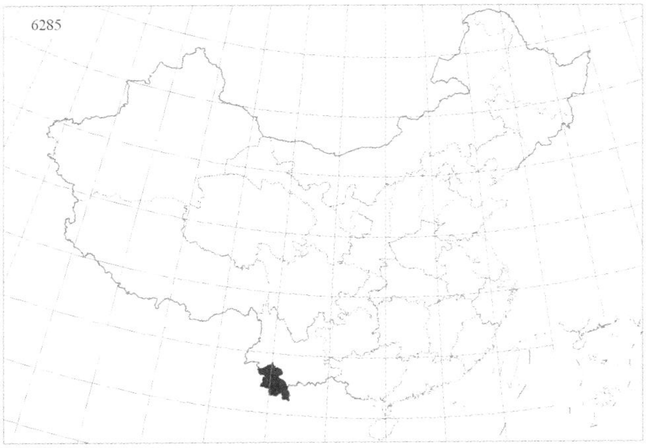

Malvaceae 锦葵科 *Hibiscus* 木槿属
Hibiscus mutabilis L. 木芙蓉
Shrubs or small trees, deciduous

17.9 (9.3~24.8); 7.6 (-3.3~20.1); 27.1 (18.8~30.1); 17.9 (9.5~24.8); 155.8 (69.8~237.6); 0.9 (0.0~20.2); 954 (603~1385) ❙ 1430 (510~2713); 580 (312~1428); 136 (15~257) ❙ 51.3 (-39.1~214.4); 941 (510~1360); 221.3 (87.8~620.0)

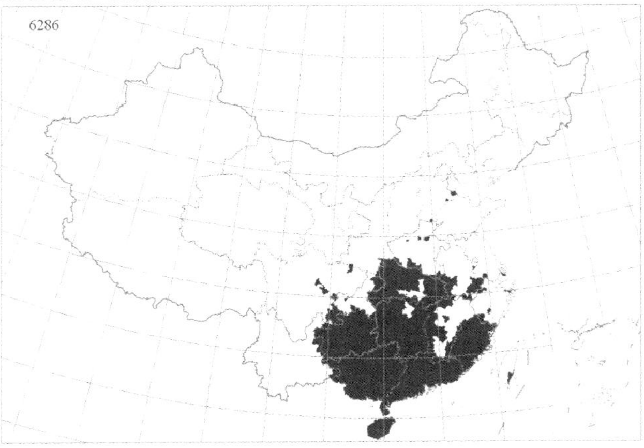

Malvaceae 锦葵科 *Hibiscus* 木槿属
Hibiscus paramutabilis L. H. Bailey 庐山芙蓉
Shrubs or small trees, deciduous

16.5 (14.3~18.0); 5.0 (3.3~8.5); 27.2 (24.3~29.3); 16.5 (14.3~18.0); 138.8 (114.1~156.4); 0.6 (0.0~2.3); 883 (765~964) ❙ 1509 (1211~1679); 539 (442~685); 180 (134~221) ❙ 71.6 (28.8~106.4); 881 (765~961); 225.6 (166.9~268.8)

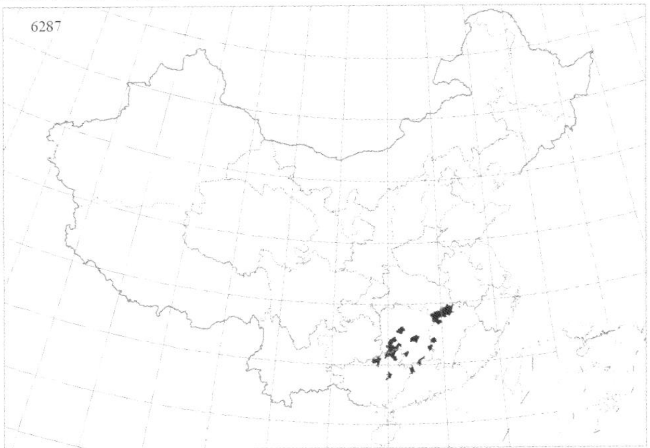

Malvaceae 锦葵科 *Hibiscus* 木槿属
Hibiscus rosa-sinensis L. 朱槿
Shrubs, evergreen

18.9 (11.7~24.8); 10.3 (2.7~20.1); 26.1 (18.8~29.8); 18.9 (11.7~24.8); 167.4 (82.6~237.6); 0.1 (0.0~3.8); 989 (658~1385) ❙ 1462 (868~2037); 647 (312~1056); 106 (27~254) ❙ 49.6 (-34.3~130.7); 978 (658~1360); 259.5 (95.3~659.0)

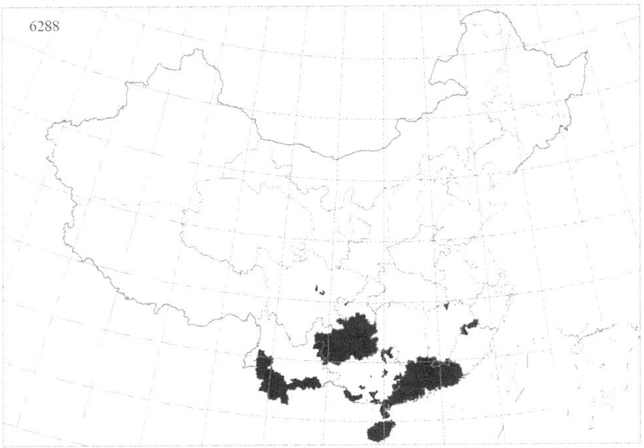

Malvaceae 锦葵科 *Hibiscus* 木槿属
Hibiscus syriacus L. 木槿
Shrubs, deciduous

17.1 (7.3~24.8); 7.0 (-6.9~20.1); 26.1 (15.6~30.1); 17.2 (7.5~24.8); 147.2 (48.2~237.6); 1.5 (0.0~34.4); 913 (537~1385) ❙ 1349 (70~3495); 579 (23~1847); 117 (11~897) ❙ 48.4 (-91.1~246.5); 896 (70~1360); 238.1 (95.3~659.0)

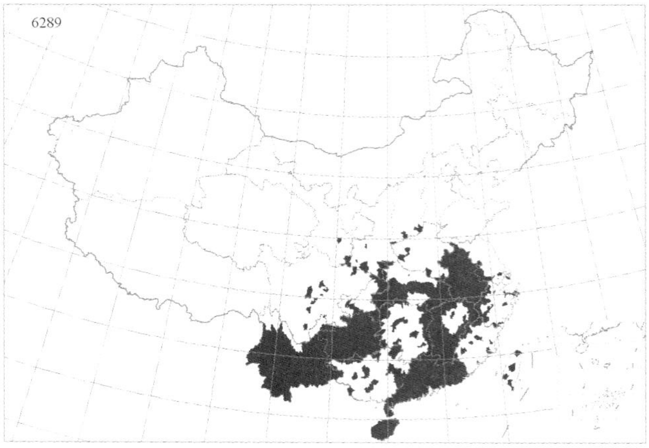

Malvaceae 锦葵科 *Hibiscus* 木槿属
Hibiscus taiwanensis S. Y. Hu 台湾芙蓉
Shrubs, deciduous

20.8; 15.3; 25.0; 20.8; 189.2; 0.0; 1019 ❙ 2445; 1412; 110 ❙ 140.0; 1019; 207.7

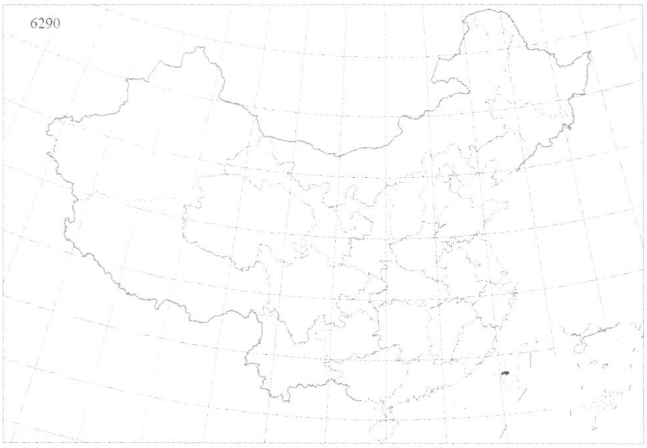

Malvaceae 锦葵科 *Hibiscus* 木槿属
Hibiscus tiliaceus L. 黄槿
Shrubs or trees, evergreen

22.0 (10.2~24.8); 15.1 (2.4~20.1); 27.4 (16.9~29.2); 22.0 (10.3~24.8); 204.1 (68.2~237.6); 0.1 (0.0~5.1); 1164 (620~1385) ❙ 1798 (868~3495); 812 (312~1847); 137 (18~897) ❙ 60.6 (-34.3~246.5); 1129 (620~1360); 268.7 (111.7~620.0)

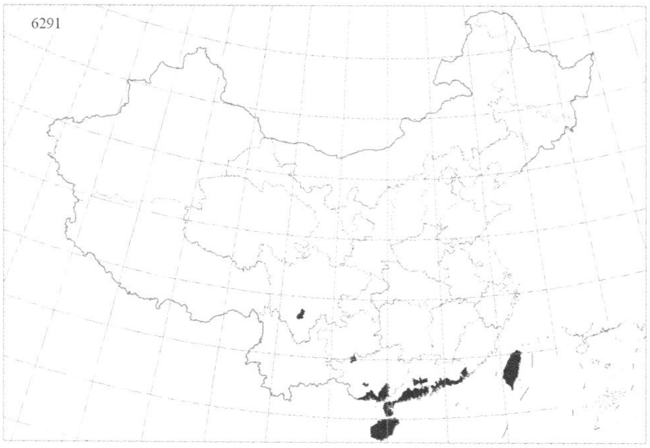

Malvaceae 锦葵科 *Hibiscus* 木槿属
Hibiscus yunnanensis S. Y. Hu 云南芙蓉
Herbs or subshrubs

19.3; 13.1; 23.6; 19.4; 173.2; 0.0; 939 ❙ 1143; 637; 50 ❙ 21.7; 903; 346.9

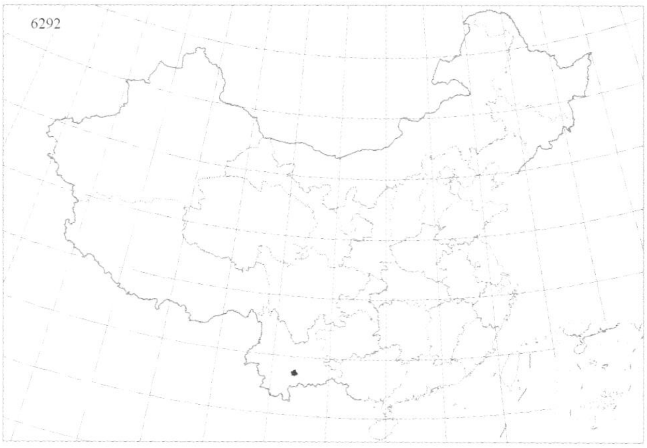

Malvaceae 锦葵科 *Kydia* 翅果麻属
Kydia calycina Roxb. 翅果麻
Trees

18.0 (15.4~21.0); 11.5 (8.1~15.1); 22.3 (20.3~25.6); 18.1 (15.5~21.0); 157.0 (125.7~191.8); 0.0 (0.0~0.0); 871 (767~1053) ❙ 1341 (958~1751); 741 (531~996); 52 (41~77) ❙ 54.1 (12.5~112.2); 860 (767~1053); 429.6 (274.9~659.0)

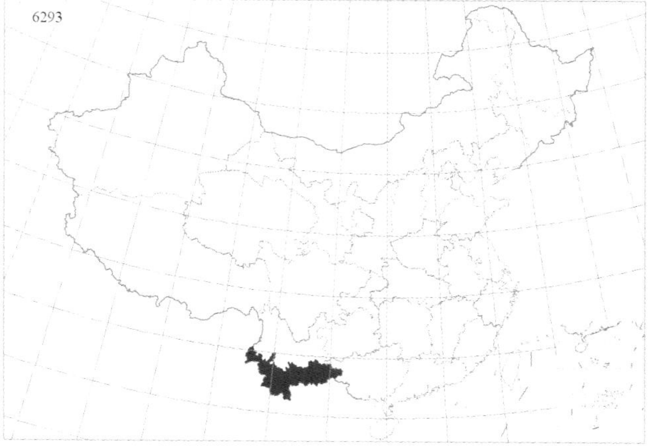

Malvaceae 锦葵科 *Kydia* 翅果麻属
Kydia glabrescens Mast. 光叶翅果麻
Trees

18.5 (16.3~21.0); 11.8 (9.1~15.1); 23.1 (21.4~26.4); 18.6 (16.4~21.0); 162.8 (136.4~191.8); 0.0 (0.0~0.0); 899 (797~1053) ❙ 1371 (1074~1751); 748 (582~996); 57 (43~77) ❙ 52.9 (16.9~88.8); 892 (797~1053); 426.1 (277.0~659.0)

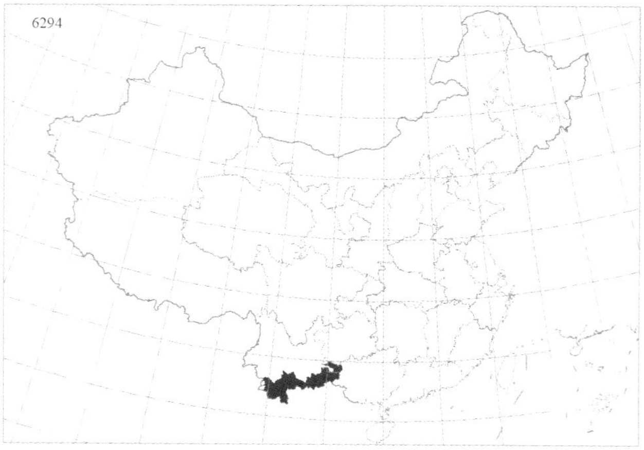

Malvaceae 锦葵科 *Nayariophyton* 枣叶槿属

Nayariophyton zizyphifolium (Griffith) D. G. Long et A.G. Miller 枣叶槿

Trees

17.8；11.5；21.8；17.8；153.7；0.0；851 ❙ 1334；775；52 ❙ 56.4；835；450.4

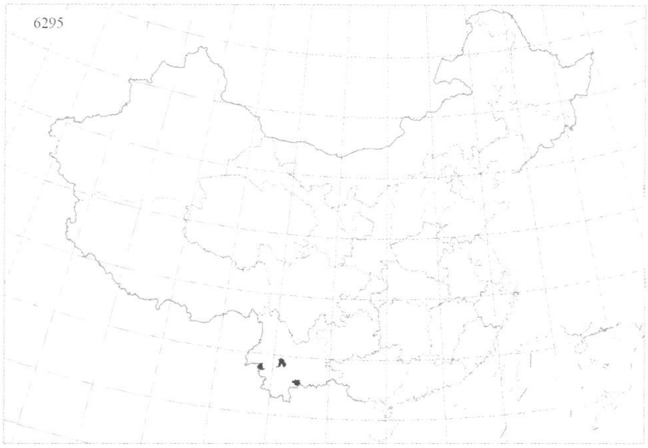

Malvaceae 锦葵科 *Sida* 黄花稔属

Sida alnifolia L. 桤叶黄花稔

Shrubs or subshrubs

20.0 (14.4~24.8)；12.1 (5.3~20.1)；26.5 (19.1~29.1)；20.0 (14.4~24.8)；180.2 (113.1~237.6)；0.0 (0.0~0.0)；1031 (727~1385) ❙ 1635 (868~3495)；713 (312~1847)；146 (41~897) ❙ 61.3 (-34.3~246.5)；1017 (727~1331)；272.9 (95.3~659.0)

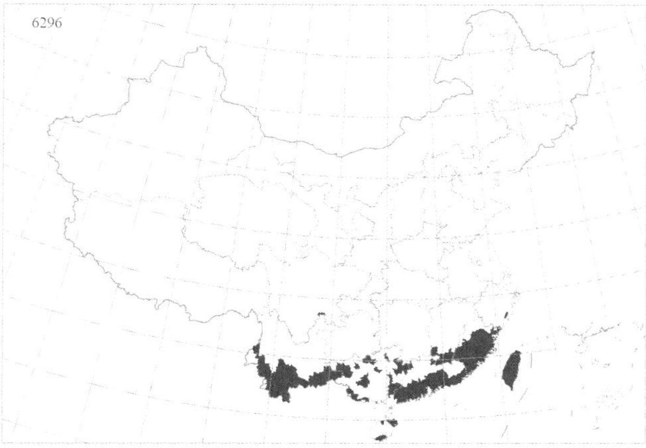

Malvaceae 锦葵科 *Sida* 黄花稔属

Sida alnifolia var. microphylla (Cav.) S. Y. Hu 小叶黄花稔

Shrubs or subshrubs

19.0 (16.0~22.4)；10.3 (5.6~14.8)；27.2 (23.5~28.9)；19.1 (16.0~22.4)；168.6 (132.5~208.9)；0.0 (0.0~0.0)；981 (838~1191) ❙ 1519 (1244~2030)；586 (411~1033)；182 (51~257) ❙ 56.4 (14.0~111.5)；974 (838~1191)；251.2 (163.7~452.0)

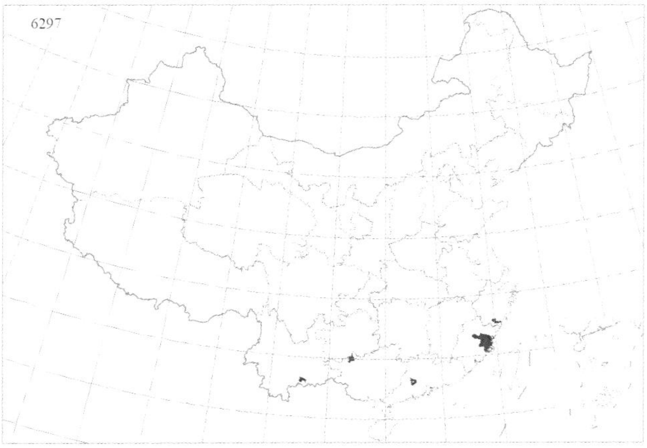

Malvaceae 锦葵科 *Sida* 黄花稔属

Sida alnifolia var. obovata (Wall. ex Mast.) S. Y. Hu 倒卵叶黄花稔

Shrubs or subshrubs

22.5；15.0；28.7；22.5；210.3；0.0；1202 ❙ 1535；708；108 ❙ 27.8；1202；198.3

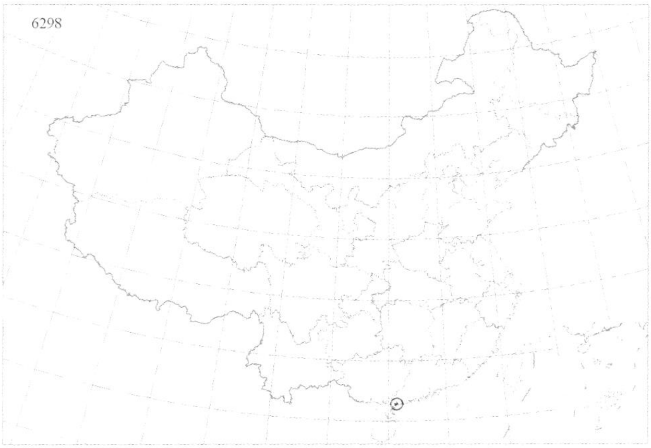

Malvaceae 锦葵科 *Sida* 黄花稔属

Sida chinensis Retz. 中华黄花稔

Small shrubs

17.7 (5.4~24.8)；11.3 (-2.1~20.1)；22.2 (12.3~28.5)；17.7 (5.8~24.8)；153.0 (31.5~237.6)；0.7 (0.0~25.6)；896 (481~1385) ❙ 1386 (741~3495)；688 (392~1033)；80 (21~897) ❙ 56.5 (-11.9~224.4)；867 (481~1360)；412.5 (163.7~659.0)

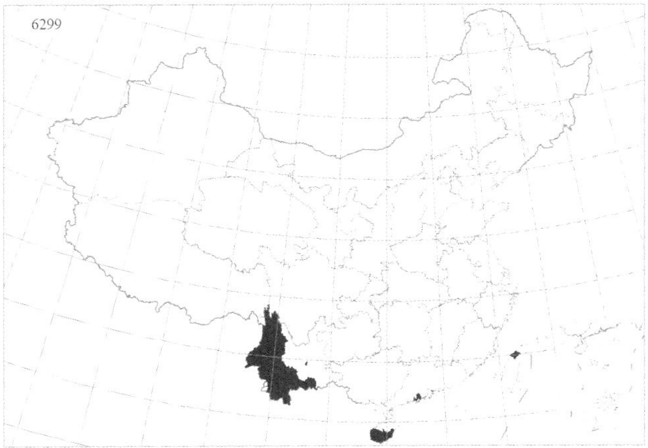

Malvaceae 锦葵科 *Sida* 黄花稔属

Sida cordifolia L. 心叶黄花稔

Subshrubs

19.2 (5.4~24.8)；12.1 (-2.1~20.1)；24.8 (12.3~29.2)；19.3 (5.8~24.8)；171.5 (31.5~237.6)；0.3 (0.0~25.6)；998 (481~1385) ❙ 1459 (741~3495)；701 (312~1847)；100 (15~897) ❙ 47.5 (-34.3~246.5)；973 (481~1360)；282.3 (95.3~659.0)

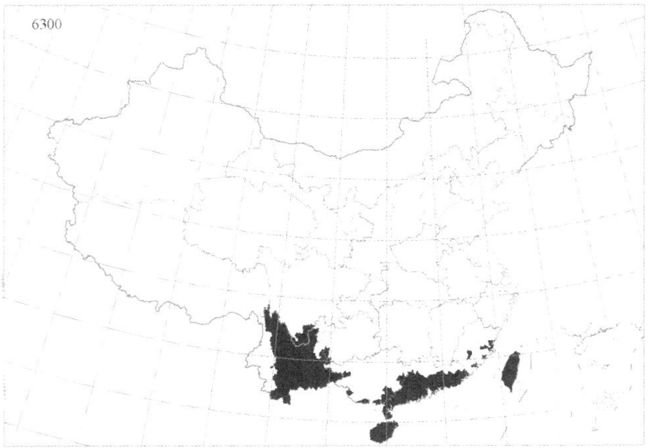

Malvaceae 锦葵科 *Sida* 黄花稔属
Sida orientalis Cav. 东方黄花稔
Subshrubs

17.4 (11.6~22.8); 10.8 (3.1~17.7); 22.9 (18.6~27.8); 17.5 (11.5~22.8); 149.9 (80.9~213.7); 0.2 (0.0~2.8); 892 (654~1209) ❙ 1847 (797~3495); 821 (411~1847); 217 (23~897) ❙ 100.0 (-14.2~246.5); 864 (654~1131); 248.8 (179.2~296.9)

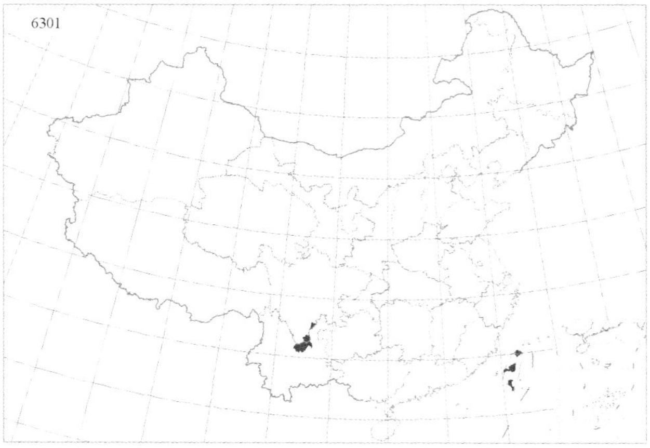

Malvaceae 锦葵科 *Sida* 黄花稔属
Sida rhombifolia L. 白背黄花稔
Subshrubs

18.7 (6.1~24.8); 10.4 (-1.5~20.1); 25.6 (12.7~29.8); 18.8 (6.3~24.8); 165.8 (35.2~237.6); 0.2 (0.0~21.8); 975 (496~1385) ❙ 1472 (797~3495); 662 (312~1847); 121 (12~897) ❙ 52.3 (-34.3~246.5); 960 (496~1360); 269.5 (95.3~659.0)

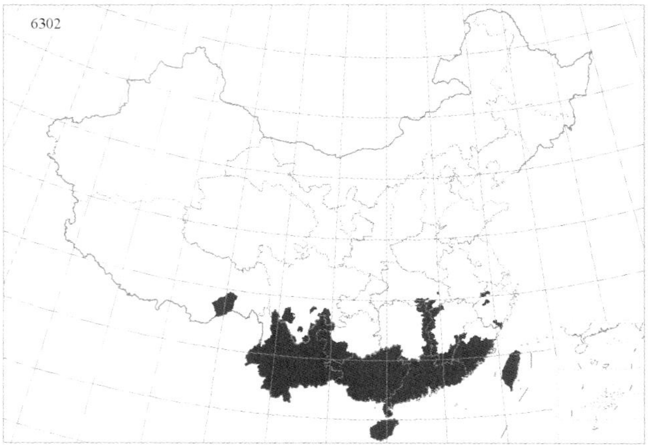

Malvaceae 锦葵科 *Sida* 黄花稔属
Sida subcordata Span. 榛叶黄花稔
Subshrubs

21.5 (16.3~24.8); 14.1 (9.1~20.1); 27.2 (21.4~29.1); 21.5 (16.4~24.8); 198.2 (136.4~237.6); 0.0 (0.0~0.0); 1128 (797~1385) ❙ 1611 (868~2037); 731 (312~1033); 104 (41~191) ❙ 44.5 (-34.3~95.5); 1105 (797~1360); 298.7 (95.3~659.0)

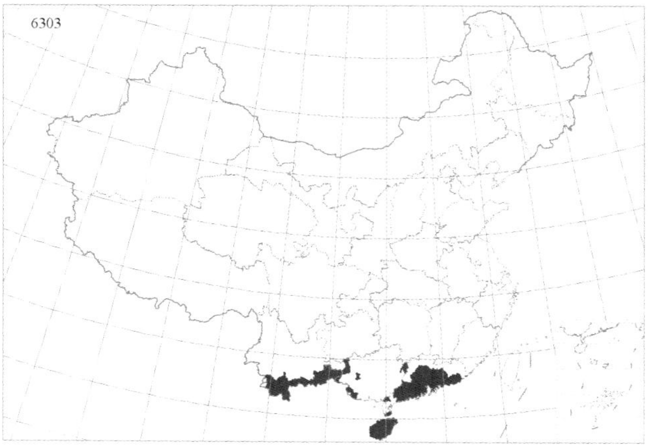

Malvaceae 锦葵科 *Sida* 黄花稔属
Sida szechuensis Matsuda 拔毒散
Subshrubs

15.5 (2.2~21.3); 8.0 (-6.6~15.1); 21.2 (9.9~27.9); 15.5 (3.8~21.3); 127.5 (15.5~196.0); 1.2 (0.0~49.1); 787 (413~1123) ❙ 1145 (741~1751); 635 (411~996); 47 (12~187) ❙ 46.2 (-14.2~125.9); 773 (413~1121); 316.6 (144.6~659.0)

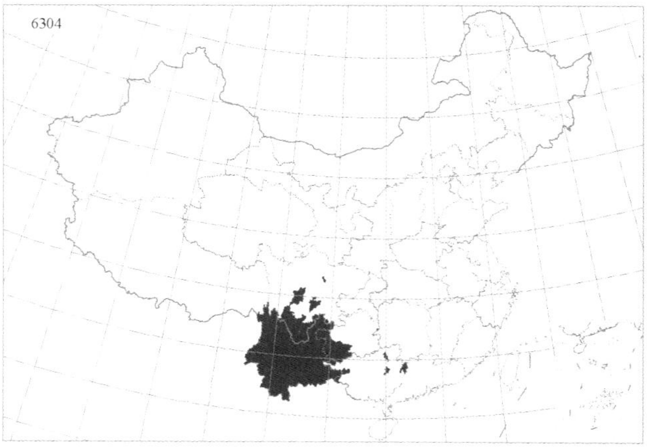

Malvaceae 锦葵科 *Sida* 黄花稔属
Sida yunnanensis S. Y. Hu 云南黄花稔
Subshrubs

16.7 (5.4~24.2); 9.5 (-2.4~18.3); 22.0 (12.3~29.0); 16.7 (5.8~24.2); 141.4 (31.5~230.4); 1.0 (0.0~25.6); 845 (481~1331) ❙ 1160 (741~2005); 626 (411~996); 53 (12~181) ❙ 38.4 (-14.2~92.9); 826 (481~1331); 318.9 (189.5~659.0)

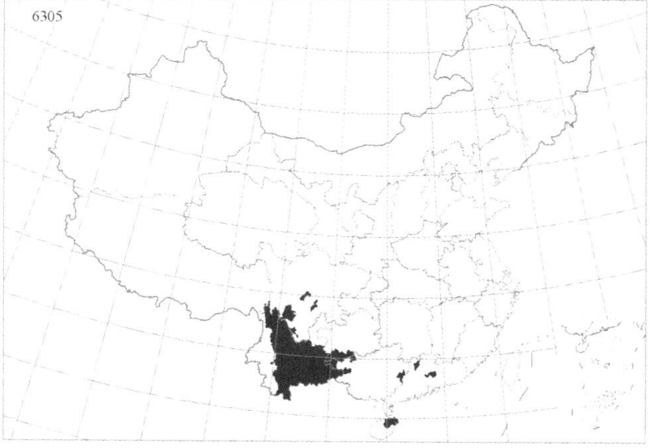

Malvaceae 锦葵科 *Thespesia* 桐棉属
Thespesia lampas (Cav.) Dalzell et A. Gibson 白脚桐棉
Shrubs, evergreen

21.6 (16.8~24.8); 15.5 (10.4~20.1); 25.7 (20.8~28.8); 21.6 (16.9~24.8); 199.5 (142.6~237.6); 0.0 (0.0~0.0); 1126 (811~1385) ❙ 1492 (1179~1854); 654 (392~850); 70 (41~144) ❙ 36.2 (-11.9~74.2); 1070 (802~1360); 435.4 (157.7~659.0)

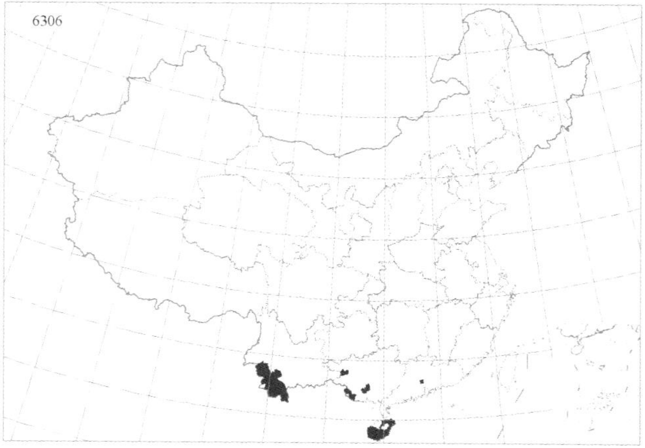

Malvaceae 锦葵科 *Thespesia* 桐棉属
Thespesia populnea (L.) Solander ex Corrêa 桐棉
Shrubs or trees, evergreen

22.8 (19.3~24.8)；16.2 (13.2~20.1)；27.8 (22.8~29.0)；22.9 (19.3~24.8)；214.2 (171.9~237.6)；0.0 (0.0~0.0)；1221 (912~1385) ❙ 1827 (1269~2992)；863 (473~1847)；100 (41~142) ❙ 53.9 (-1.4~202.5)；1135 (912~1329)；292.6 (139.6~620.0)

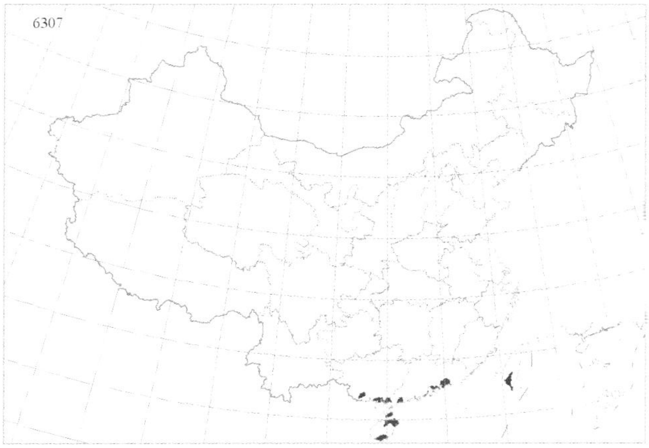

Malvaceae 锦葵科 *Urena* 梵天花属
Urena procumbens L. 梵天花
Small shrubs

19.0 (12.7~24.8)；9.7 (1.9~20.1)；27.2 (20.8~29.8)；19.1 (12.8~24.8)；168.8 (99.0~237.6)；0.1 (0.0~6.1)；1000 (713~1385) ❙ 1589 (868~3495)；620 (312~1847)；161 (41~897) ❙ 61.7 (-34.3~224.4)；989 (713~1360)；242.5 (95.3~620.0)

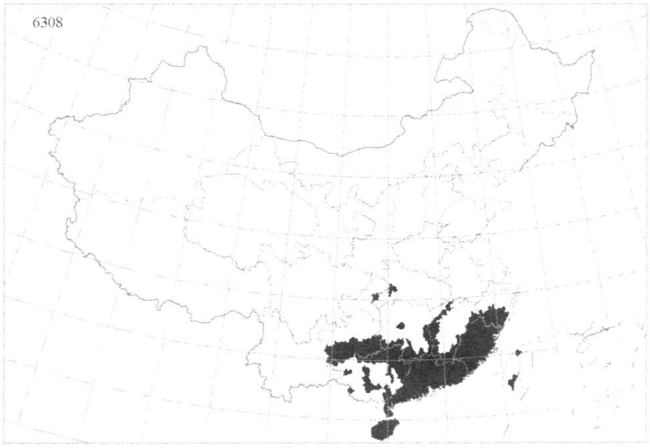

Malvaceae 锦葵科 *Urena* 梵天花属
Urena procumbens var. microphylla K. M. Feng 小叶梵天花
Small shrubs

17.0；6.8；27.1；17.0；144.6；0.0；891 ❙ 1692；615；186 ❙ 90.5；891；220.0

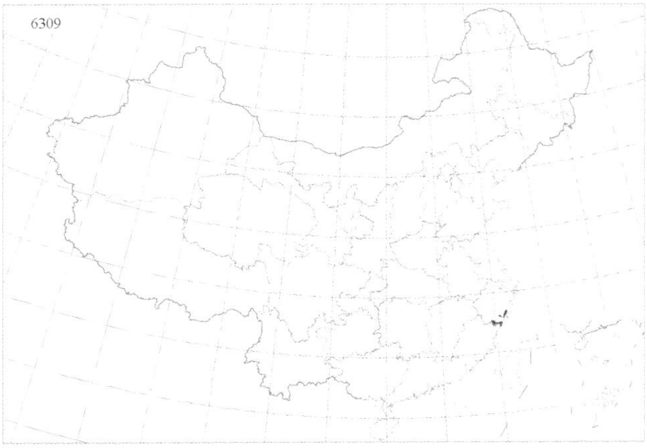

Malvaceae 锦葵科 *Wissadula* 隔蒴苘属
Wissadula periplocifolia (L.) C. Presl ex Thwaites 隔蒴苘
Small shrubs

23.8 (22.2~24.8)；18.4 (17.2~20.1)；27.6 (25.7~28.7)；23.8 (22.3~24.8)；226.1 (207.2~237.6)；0.0 (0.0~0.0)；1291 (1127~1385) ❙ 1388 (868~1725)；486 (312~610)；53 (41~71) ❙ 8.4 (-34.3~47.8)；1102 (868~1212)；490.4 (288.3~620.0)

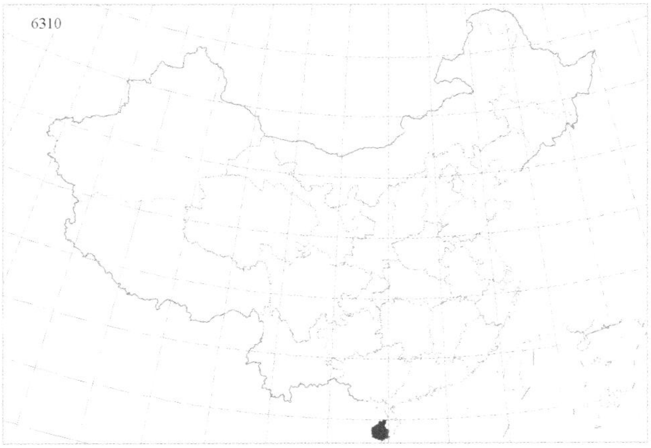

Bombacaceae 木棉科 *Bombax* 木棉属
Bombax cambodiense Pierre 澜沧木棉
Trees, deciduous

18.2；12.4；21.6；18.2；158.6；0.0；861 ❙ 1500；785；52 ❙ 74.2；839；439.5

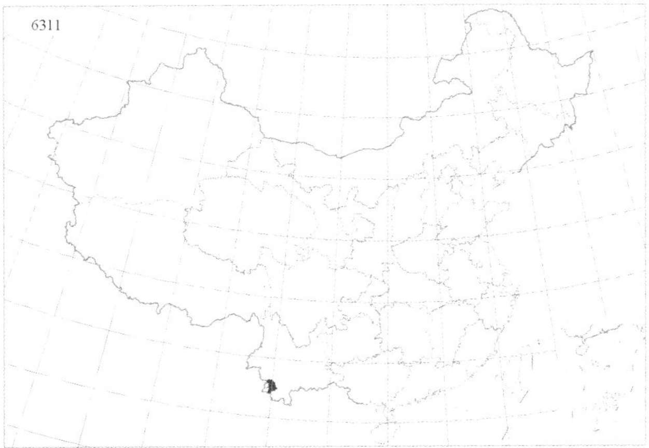

Bombacaceae 木棉科 *Bombax* 木棉属
Bombax ceiba L. 木棉
Trees, deciduous

18.7 (5.4~24.8)；11.0 (-2.1~20.1)；24.8 (12.3~29.3)；18.7 (5.8~24.8)；165.1 (31.5~237.6)；0.4 (0.0~25.6)；972 (481~1385) ❙ 1445 (741~3495)；694 (312~1847)；102 (15~897) ❙ 50.6 (-34.3~246.5)；945 (481~1360)；296.2 (119.4~659.0)

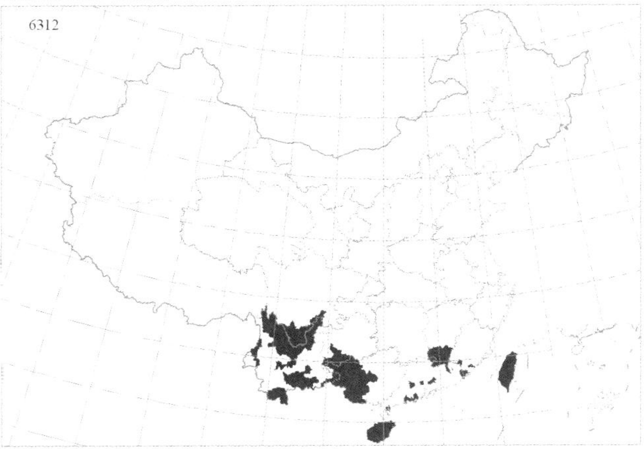

Bombacaceae 木棉科 *Bombax* 木棉属

Bombax insigne Wallich var. tenebrosum (Dunn) A.Robyns 长果木棉

Trees, deciduous

18.1 (16.4~20.4); 11.7 (9.6~15.1); 22.1 (20.6~23.6); 18.1 (16.4~20.5); 157.3 (137.3~185.8); 0.0 (0.0~0.0); 870 (795~989) ❙ 1462 (958~1688); 788 (531~959); 57 (44~77) ❙ 68.6 (13.3~112.2); 863 (795~986); 508.2 (311.9~659.0)

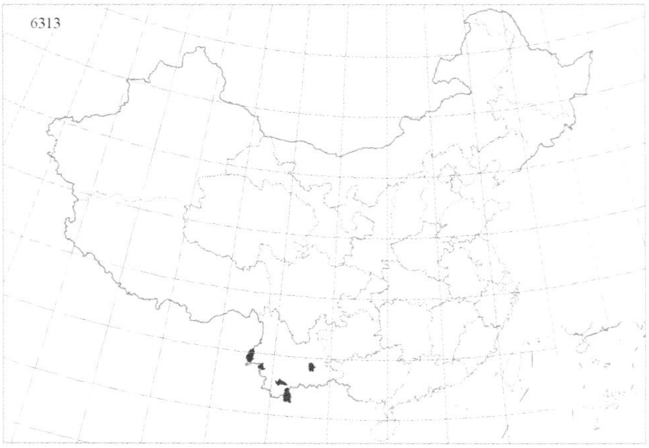

Sterculiaceae 梧桐科 *Ambroma* 昂天莲属

Ambroma augustum (L.) L. f. 昂天莲

Shrubs

20.5 (13.3~24.8); 13.2 (5.4~20.1); 25.7 (19.1~28.9); 20.5 (13.4~24.8); 186.2 (100.4~237.6); 0.0 (0.0~0.0); 1057 (706~1385) ❙ 1491 (868~2435); 731 (312~1428); 80 (41~194) ❙ 44.0 (-34.3~145.8); 1029 (706~1360); 370.4 (139.6~659.0)

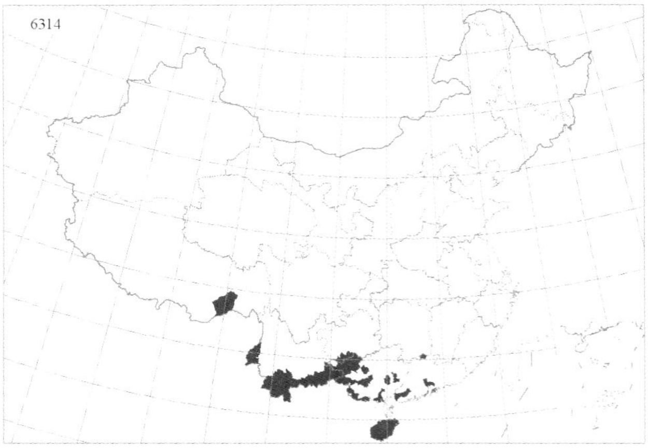

Sterculiaceae 梧桐科 *Byttneria* 刺果藤属

Byttneria grandifolia DC. 刺果藤

Large vines

20.5 (14.4~24.8); 13.1 (7.3~20.1); 26.1 (19.1~29.0); 20.6 (14.4~24.8); 186.7 (113.1~237.6); 0.0 (0.0~0.0); 1061 (727~1385) ❙ 1536 (868~2435); 740 (312~1428); 95 (41~198) ❙ 47.0 (-34.3~112.2); 1040 (727~1299); 339.2 (111.7~659.0)

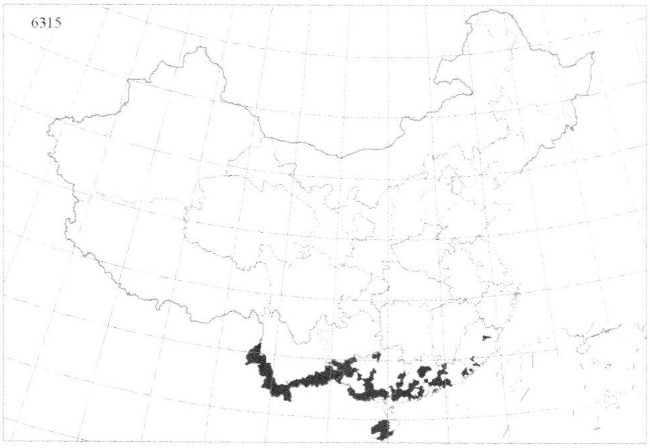

Sterculiaceae 梧桐科 *Byttneria* 刺果藤属

Byttneria integrifolia Lace 全缘刺果藤

Large vines

20.5 (18.6~24.2); 14.4 (12.7~17.6); 24.6 (22.3~28.8); 20.6 (18.7~24.2); 187.0 (163.9~230.4); 0.0 (0.0~0.0); 1025 (888~1331) ❙ 1542 (1451~1751); 757 (581~996); 78 (58~102) ❙ 53.7 (10.8~88.8); 1016 (880~1331); 489.7 (284.3~659.0)

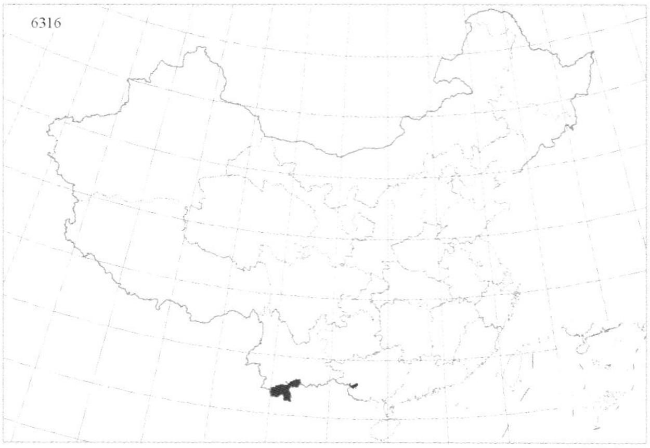

Sterculiaceae 梧桐科 *Byttneria* 刺果藤属

Byttneria pilosa Roxb. 粗毛刺果藤

Vines

17.8 (14.4~20.5); 11.6 (7.5~15.1); 21.7 (19.1~23.7); 17.9 (14.4~20.5); 154.6 (113.1~185.9); 0.0 (0.0~0.0); 860 (727~992) ❙ 1438 (1180~1688); 780 (622~959); 54 (41~77) ❙ 67.8 (47.1~112.2); 846 (727~986); 476.7 (335.1~659.0)

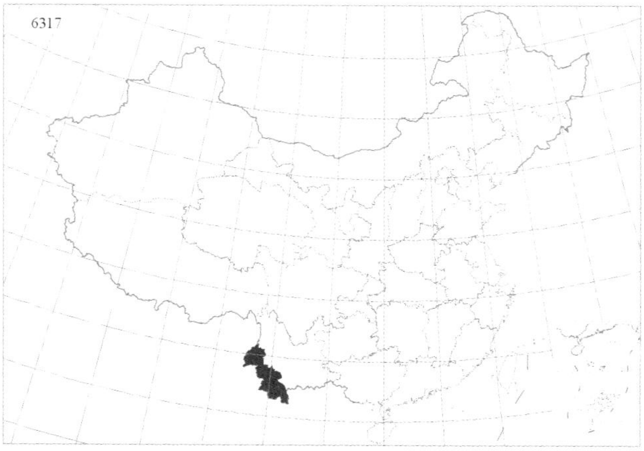

Sterculiaceae 梧桐科 *Commersonia* 山麻树属

Commersonia bartramia (L.) Merr. 山麻树

Trees

22.7 (17.3~24.8); 16.5 (9.9~20.1); 27.2 (22.4~28.7); 22.7 (17.3~24.8); 213.0 (147.1~237.6); 0.0 (0.0~0.0); 1214 (833~1385) ❙ 1563 (868~2435); 679 (312~1428); 76 (41~130) ❙ 31.5 (-34.3~99.5); 1121 (833~1326); 427.1 (235.1~620.0)

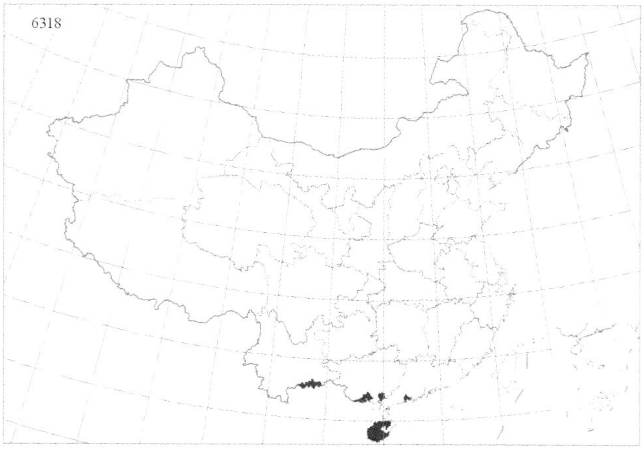

Sterculiaceae 梧桐科 *Eriolaena* 火绳树属
Eriolaena candollei Wall. 南火绳
Trees

19.2 (17.3~21.7); 12.9 (10.7~15.1); 23.3 (21.6~28.4); 19.2 (17.3~21.8); 170.8 (148.0~201.3); 0.0 (0.0~0.0); 932 (832~1162) ❙ 1382 (1028~1751); 740 (566~996); 60 (41~106) ❙ 48.9 (12.5~88.8); 911 (783~1162); 436.6 (273.6~659.0)

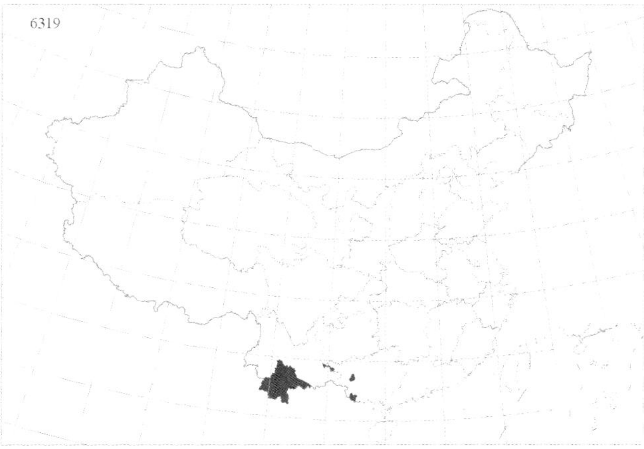

Sterculiaceae 梧桐科 *Eriolaena* 火绳树属
Eriolaena glabrescens Aug. Candolle 光叶火绳
Trees

18.3 (16.3~20.5); 12.3 (10.2~15.1); 22.2 (20.3~23.7); 18.4 (16.4~20.5); 160.7 (136.6~185.9); 0.0 (0.0~0.0); 880 (789~992) ❙ 1332 (939~1751); 723 (524~996); 52 (37~77) ❙ 51.3 (6.5~88.8); 854 (744~986); 437.6 (301.1~659.0)

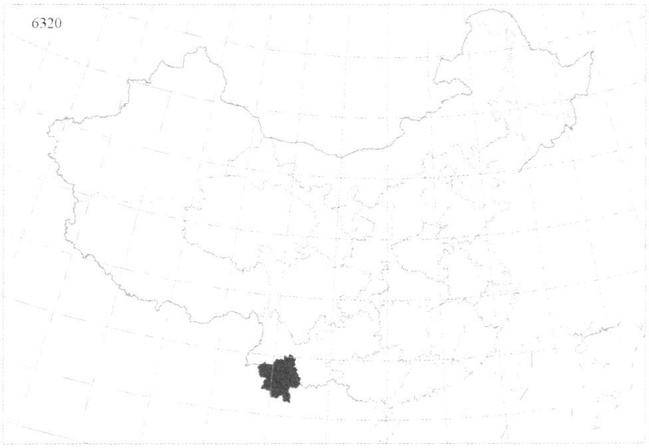

Sterculiaceae 梧桐科 *Eriolaena* 火绳树属
Eriolaena kwangsiensis Hand.-Mazz. 桂火绳
Shrubs or trees

19.2 (16.3~22.3); 11.8 (9.1~15.1); 24.6 (20.3~28.9); 19.3 (16.4~22.3); 171.4 (136.6~208.[illegible]); 0.0 (0.0~0.0); 961 (789~1202) ❙ 1336 (939~1751); 713 (524~996); 68 (37~141) ❙ 40.0 (5.8~88.8); 948 (744~1202); 359.5 (201.4~659.0)

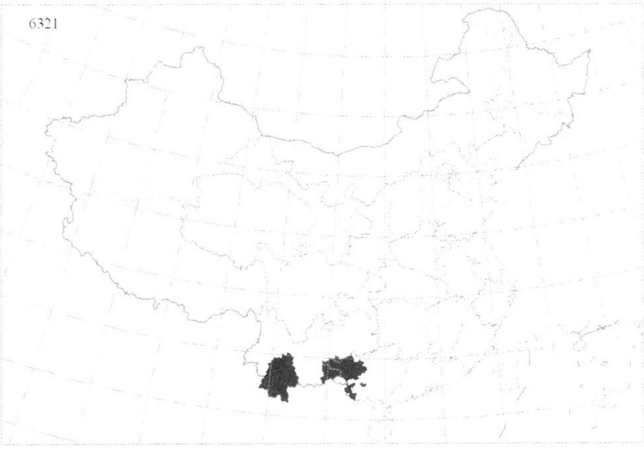

Sterculiaceae 梧桐科 *Eriolaena* 火绳树属
Eriolaena quinquelocularis (Wight et Arn.) Wight 五室火绳
Trees

17.2 (8.6~20.5); 10.9 (1.3~15.1); 21.4 (15.0~23.7); 17.3 (8.7~20.5); 147.4 (52.6~185.9); 0.2 (0.0~8.8); 839 (574~992) ❙ 1352 (1028~1751); 743 (553~996); 54 (41~77) ❙ 61.8 (12.5~112.2); 825 (574~986); 444.0 (335.1~659.0)

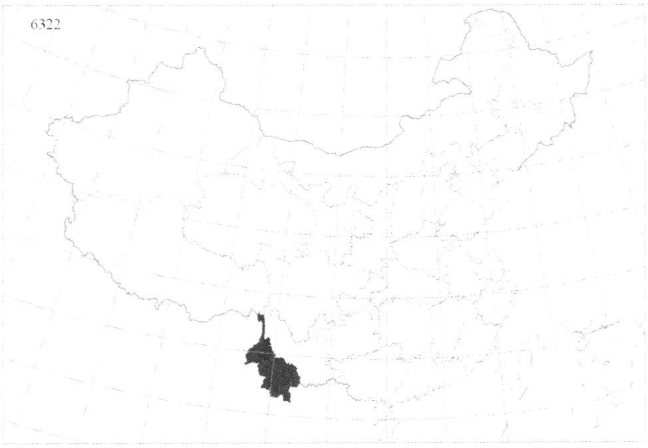

Sterculiaceae 梧桐科 *Eriolaena* 火绳树属
Eriolaena spectabilis (DC.) Planch. ex Mast. 火绳树
Shrubs or small trees, deciduous

16.8 (11.2~22.2); 9.6 (2.7~15.1); 21.9 (16.7~28.5); 16.8 (11.3~22.3); 141.6 (75.8~207.5); 0.1 (0.0~3.7); 831 (642~1190) ❙ 1174 (797~1751); 652 (411~996); 50 (20~173) ❙ 41.3 (-14.2~112.2); 814 (642~1190); 340.7 (185.3~659.0)

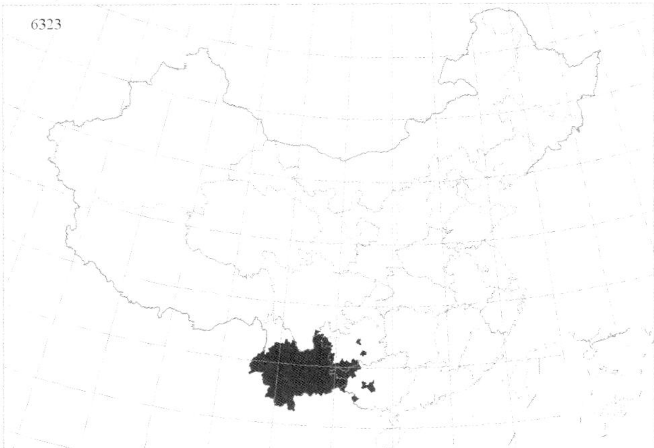

Sterculiaceae 梧桐科 *Firmiana* 梧桐属
Firmiana colorata (Roxburgh) R. Brown 火桐
Trees, deciduous

18.5 (14.4~24.7); 12.2 (7.5~19.8); 22.6 (19.1~28.2); 18.5 (14.4~24.7); 162.4 (113.1~236.5); 0.0 (0.0~0.0); 904 (727~1376) ❙ 1389 (1007~1767); 739 (534~996); 55 (41~88) ❙ 55.1 (12.5~112.2); 884 (727~1299); 448.6 (277.0~659.0)

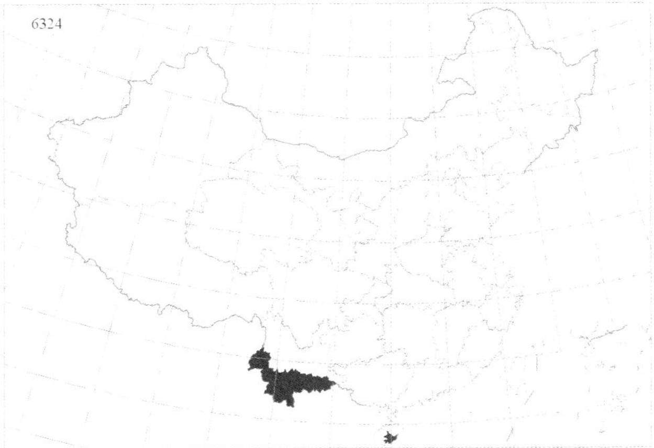

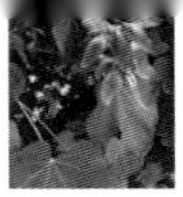

Sterculiaceae 梧桐科 *Firmiana* 梧桐属
Firmiana danxiaensis Hsue et H. S. Kiu 丹霞梧桐
Trees, deciduous

19.2; 8.9; 28.5; 19.3; 171.4; 0.0; 1019 ❙ 1542; 531; 188 ❙ 51.9; 995; 204.6

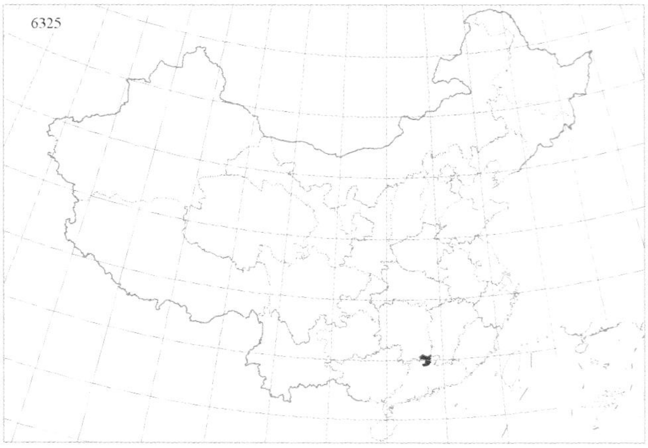

Sterculiaceae 梧桐科 *Firmiana* 梧桐属
Firmiana hainanensis Kosterm. 海南梧桐
Trees, deciduous

24.1 (23.0~24.8); 18.6 (17.7~20.1); 28.1 (26.7~28.7); 24.1 (23.1~24.8); 229.7 (217.0~237.6); 0.0 (0.0~0.0); 1321 (1217~1385) ❙ 1400 (868~1854); 480 (312~622); 64 (41~115) ❙ 6.1 (-34.3~41.8); 1136 (868~1360); 458.6 (288.3~572.9)

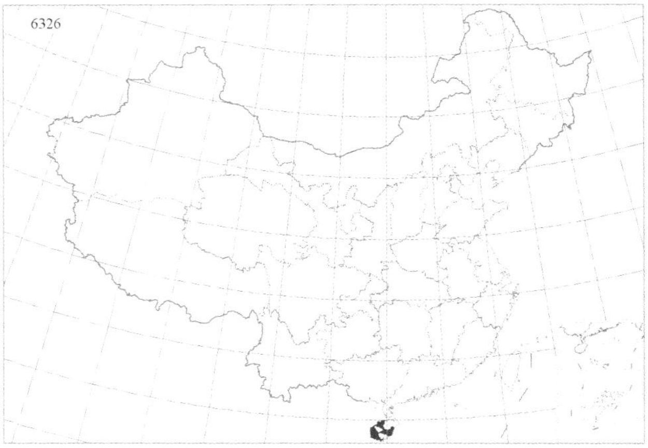

Sterculiaceae 梧桐科 *Firmiana* 梧桐属
Firmiana kwangsiensis Hsu 广西火桐
Trees, deciduous

19.1; 10.9; 25.2; 19.2; 170.3; 0.0; 953 ❙ 1365; 773; 60 ❙ 43.3; 953; 365.5

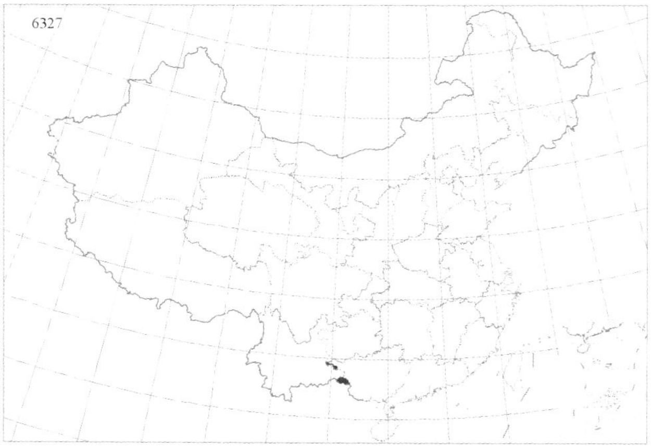

Sterculiaceae 梧桐科 *Firmiana* 梧桐属
Firmiana major (W. W. Sm.) Hand.-Mazz. 云南梧桐
Trees, deciduous

15.0 (5.4~19.6); 7.9 (-2.1~12.8); 20.3 (12.3~26.6); 15.0 (5.8~19.6); 120.8 (31.5~175.7); 0.9 (0.0~25.6); 760 (481~953) ❙ 1063 (741~1688); 610 (411~959); 40 (15~70) ❙ 40.7 (-14.2~112.2); 737 (481~932); 320.9 (142.9~627.0)

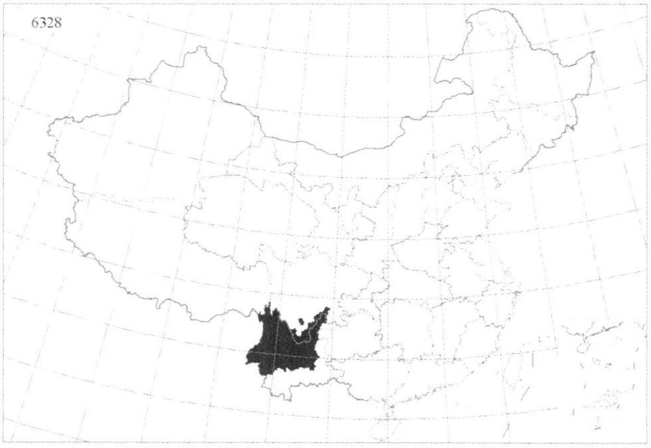

Sterculiaceae 梧桐科 *Firmiana* 梧桐属
Firmiana pulcherrima H. H. Hsue 美丽火桐
Trees, deciduous

23.1 (19.2~24.8); 17.1 (11.0~20.1); 27.3 (25.2~28.5); 23.1 (19.2~24.8); 217.3 (170.5~237.6); 0.0 (0.0~0.0); 1241 (952~1385) ❙ 1623 (1380~1900); 636 (455~1056); 76 (41~125) ❙ 32.5 (1.5~71.0); 1166 (952~1360); 459.3 (294.6~620.0)

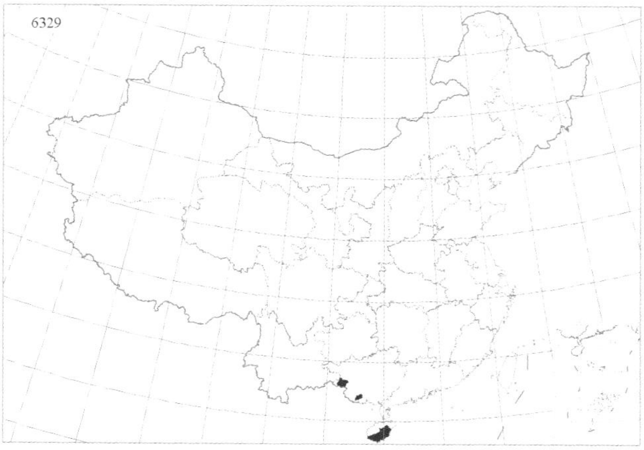

Sterculiaceae 梧桐科 *Firmiana* 梧桐属
Firmiana simplex (L.) W. Wight 梧桐
Trees, deciduous

17.6 (9.3~24.8); 7.3 (-6.2~20.1); 26.9 (18.6~29.9); 17.6 (9.5~24.8); 152.9 (69.8~237.6); 1.4 (0.0~32.8); 942 (603~1385) ❙ 1412 (481~2993); 571 (264~1847); 134 (11~547) ❙ 51.2 (-42.0~203.4); 925 (481~1360); 222.4 (95.3~620.0)

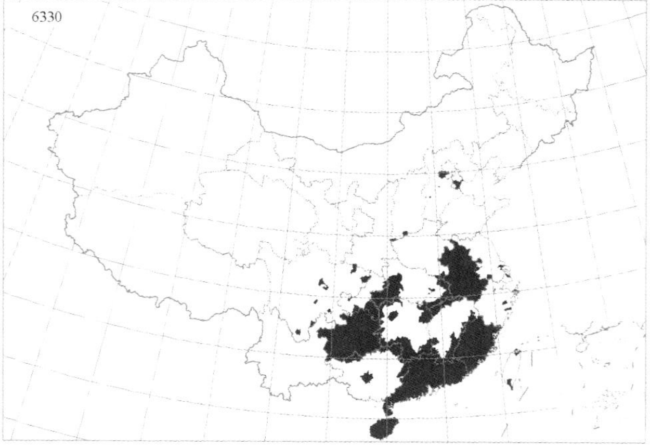

Sterculiaceae 梧桐科 *Helicteres* 山芝麻属
Helicteres angustifolia L. 山芝麻
Small shrubs

19.8 (14.3~24.8); 11.4 (3.3~20.1); 26.7 (19.1~30.1); 19.9 (14.3~24.8); 178.2 (113.1~237.6); 0.0 (0.0~2.3); 1030 (727~1385) ❙ 1589 (868~3495); 638 (312~1847); 140 (37~897) ❙ 56.6 (-34.3~246.5); 1015 (727~1360); 273.8 (95.3~659.0)

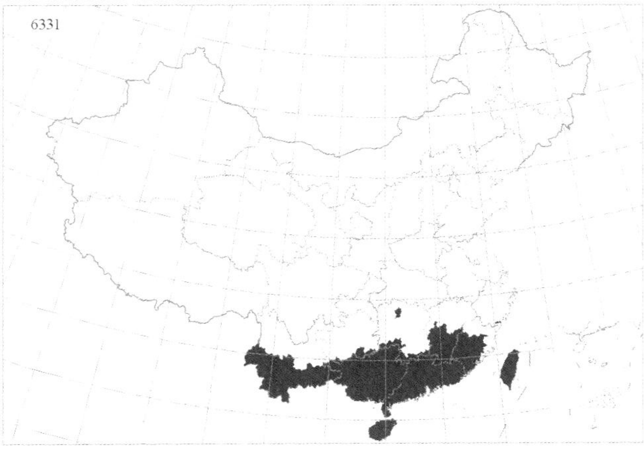

Sterculiaceae 梧桐科 *Helicteres* 山芝麻属
Helicteres elongata Wall. ex Mast. 长序山芝麻
Shrubs

18.2 (14.4~21.8); 11.3 (7.5~15.1); 23.0 (19.1~28.5); 18.3 (14.4~21.8); 159.4 (113.1~201.6); 0.0 (0.0~0.0); 893 (727~1162) ❙ 1326 (939~1900); 727 (524~1056); 58 (37~173) ❙ 49.1 (5.8~112.2); 881 (727~1162); 397.9 (208.5~659.0)

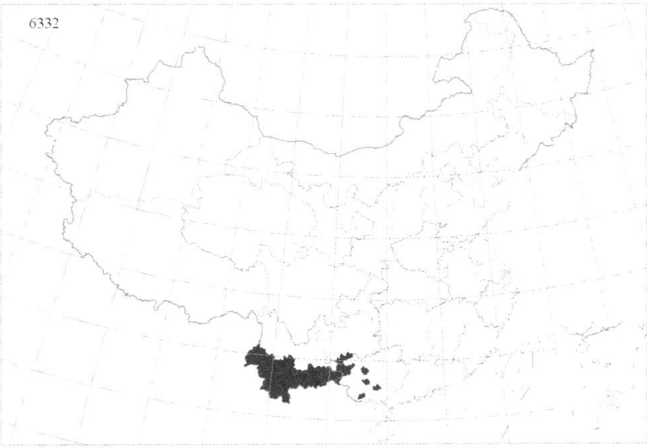

Sterculiaceae 梧桐科 *Helicteres* 山芝麻属
Helicteres glabriuscula Wall. ex Mast. 细齿山芝麻
Shrubs

18.9 (16.3~22.3); 11.8 (9.0~15.1); 23.9 (20.3~28.9); 19.0 (16.4~22.3); 167.4 (136.4~208.1); 0.0 (0.0~0.0); 936 (789~1202) ❙ 1341 (958~1751); 725 (531~996); 52 (41~136) ❙ 44.3 (5.8~88.8); 928 (780~1202); 382.0 (201.4~659.0)

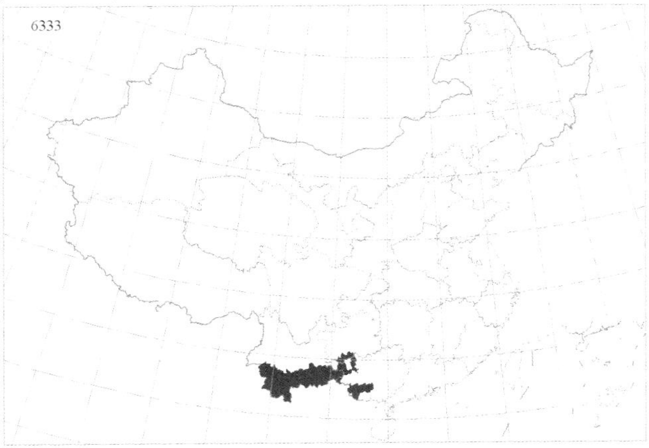

Sterculiaceae 梧桐科 *Helicteres* 山芝麻属
Helicteres hirsuta Lour. 雁婆麻
Shrubs

22.3 (15.9~24.8); 15.2 (6.9~20.1); 27.7 (23.0~29.0); 22.4 (16.0~24.8); 208.4 (131.6~237.6); 0.0 (0.0~0.0); 1196 (802~1385) ❙ 1524 (868~2435); 685 (312~1428); 85 (41~144) ❙ 28.9 (-34.3~99.5); 1140 (802~1360); 328.8 (157.7~620.0)

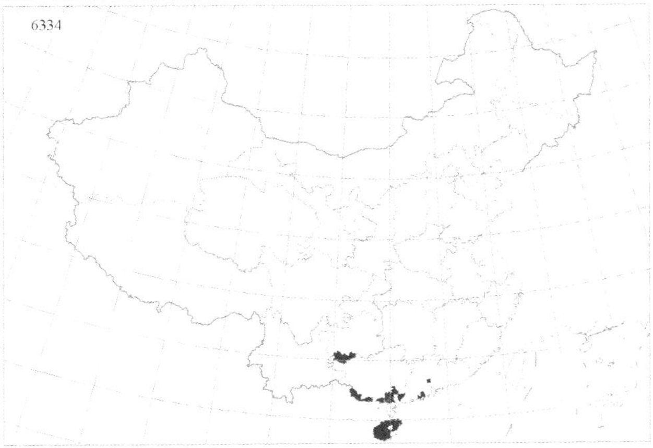

Sterculiaceae 梧桐科 *Helicteres* 山芝麻属
Helicteres isora L. 火索麻
Shrubs

19.3 (16.3~24.8); 13.1 (9.0~20.1); 23.5 (20.3~28.7); 19.4 (16.4~24.8); 172.2 (136.4~237.6); 0.0 (0.0~0.0); 962 (789~1385) ❙ 1309 (868~1767); 662 (312~996); 52 (37~88) ❙ 39.0 (-34.3~88.8); 910 (744~1299); 426.7 (277.0~659.0)

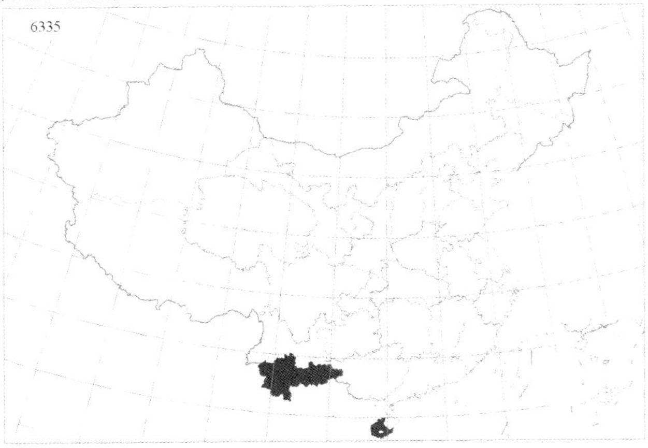

Sterculiaceae 梧桐科 *Helicteres* 山芝麻属
Helicteres lanceolata DC. 剑叶山芝麻
Shrubs

19.7 (15.3~24.8); 11.3 (4.2~20.1); 26.4 (21.4~30.1); 19.7 (15.4~24.8); 177.0 (125.8~237.6); 0.0 (0.0~0.8); 1021 (797~1385) ❙ 1499 (868~2435); 658 (312~1428); 118 (41~222) ❙ 48.6 (-34.3~103.1); 990 (783~1360); 319.4 (148.9~659.0)

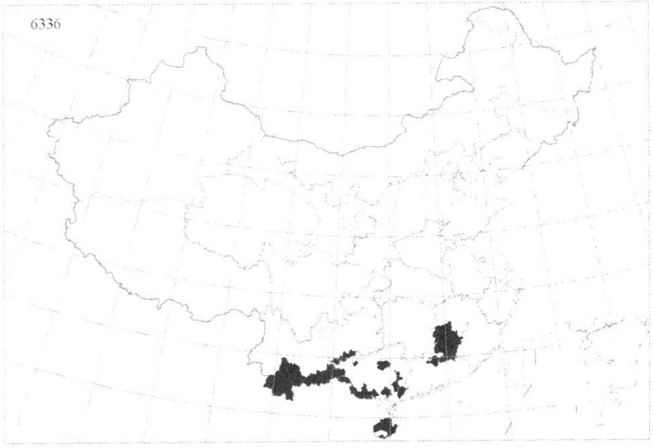

Sterculiaceae 梧桐科 *Helicteres* 山芝麻属
Helicteres obtusa Wall. ex Masters 钝叶山芝麻
Shrubs or subshrubs

19.3; 13.1; 23.6; 19.4; 173.2; 0.0; 939 ❙ 1143; 637; 50 ❙ 21.7; 903; 346.9

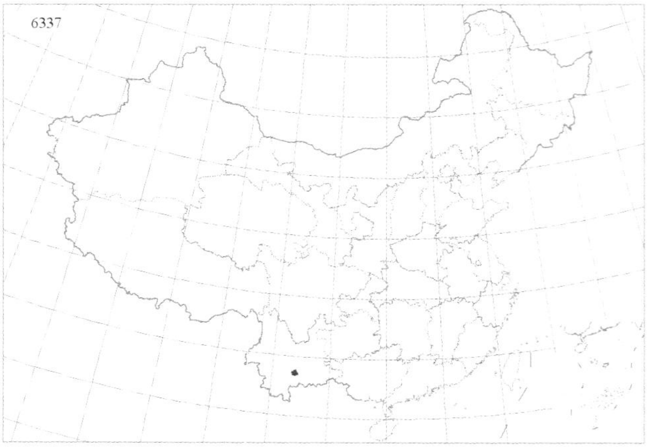

Sterculiaceae 梧桐科 *Helicteres* 山芝麻属
Helicteres plebeja Kurz 矮山芝麻
Shrubs

19.6; 11.8; 25.1; 19.6; 174.9; 0.0; 977 ❙ 1408; 771; 63 ❙ 44.0; 977; 433.9

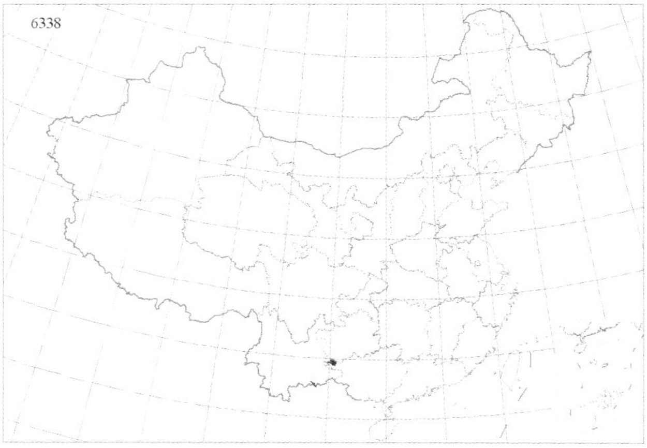

Sterculiaceae 梧桐科 *Helicteres* 山芝麻属
Helicteres viscida Blume 黏毛山芝麻
Shrubs

20.4 (16.3~24.7); 14.4 (9.1~19.8); 24.5 (21.4~29.0); 20.5 (16.4~24.7); 185.8 (136.4~236.5); 0.0 (0.0~0.0); 1034 (797~1376) ❙ 1394 (868~1751); 671 (312~997); 60 (41~110) ❙ 38.5 (-34.3~88.8); 976 (797~1326); 437.1 (199.6~659.0)

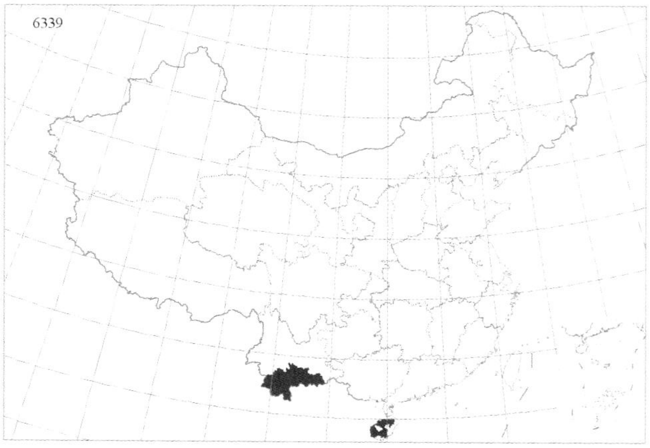

Sterculiaceae 梧桐科 *Heritiera* 银叶树属
Heritiera angustata Pierre 长柄银叶树
Trees, evergreen

22.4 (19.2~24.8); 17.1 (13.8~20.1); 25.8 (22.3~28.0); 22.4 (19.2~24.8); 208.7 (170.5~237.6); 0.0 (0.0~0.0); 1164 (909~1385) ❙ 1541 (1380~1725); 607 (455~818); 58 (41~77) ❙ 35.2 (1.5~64.9); 1071 (880~1212); 555.6 (455.2~659.0)

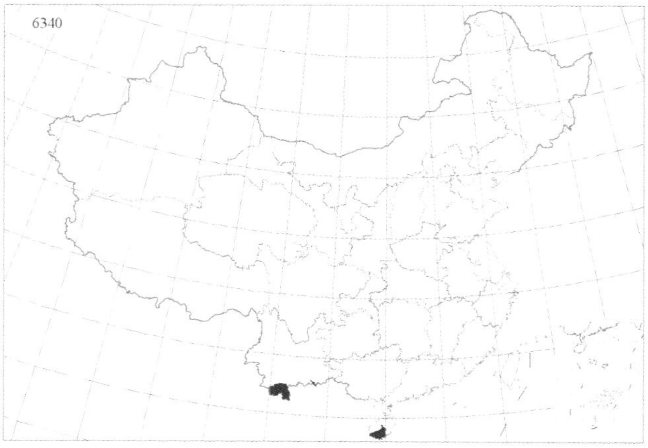

Sterculiaceae 梧桐科 *Heritiera* 银叶树属
Heritiera littoralis Aiton 银叶树
Trees, evergreen

22.6 (19.3~24.8); 15.9 (13.5~20.1); 27.7 (22.8~28.9); 22.6 (19.3~24.8); 211.3 (171.9~237.6); 0.0 (0.0~0.0); 1201 (912~1385) ❙ 1823 (1309~2992); 854 (455~1847); 101 (41~149) ❙ 55.4 (-1.4~202.5); 1162 (912~1294); 296.4 (95.3~620.0)

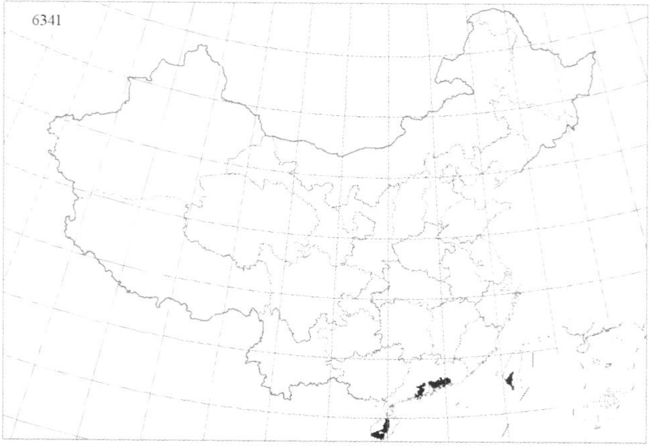

Sterculiaceae 梧桐科 *Heritiera* 银叶树属
Heritiera parvifolia Merr. 蝴蝶树
Trees, evergreen

23.9 (22.2~24.8); 18.5 (17.2~20.1); 27.7 (25.7~28.7); 23.9 (22.3~24.8); 226.9 (207.2~237.6); 0.0 (0.0~0.0); 1296 (1127~1385) ❙ 1396 (868~1767); 484 (312~610); 56 (41~88) ❙ 8.5 (-34.3~47.8); 1107 (868~1299); 490.9 (288.3~620.0)

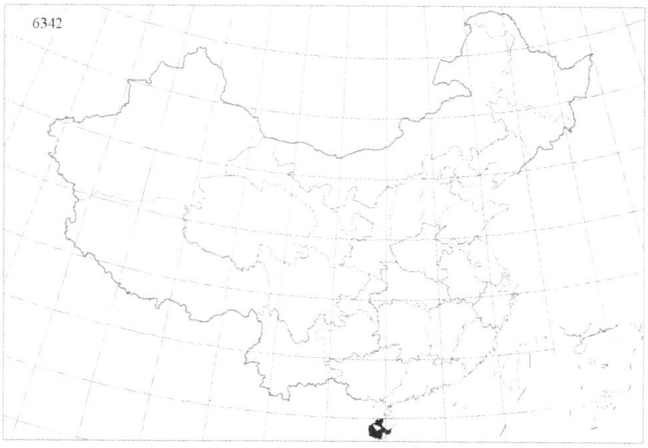

Sterculiaceae 梧桐科 *Kleinhovia* 鹧鸪麻属
Kleinhovia hospita L. 鹧鸪麻
Trees

23.1 (18.5~24.8)；17.3 (12.6~20.1)；27.4 (21.7~28.8)；23.2 (18.5~24.8)；218.2 (162.1~237.6)；0.0 (0.0~0.0)；1242 (863~1385) ❙ 1702 (868~2992)；761 (312~1847)；87 (41~188) ❙ 42.3 (-34.3~214.4)；1150 (863~1360)；363.8 (119.4~620.0)

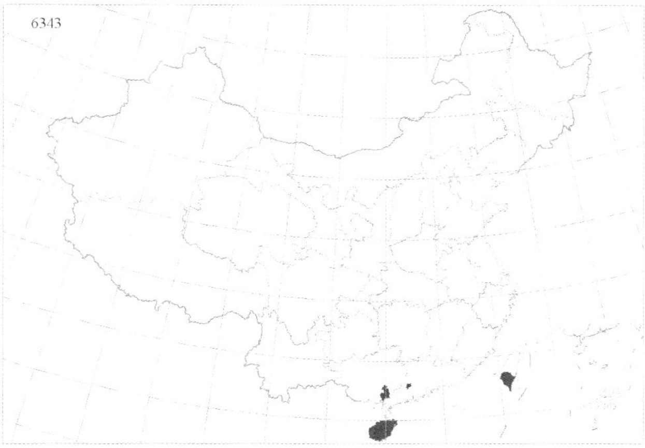

Sterculiaceae 梧桐科 *Melhania* 梅蓝属
Melhania hamiltoniana Wall. 梅蓝
Small shrubs

18.7；12.4；23.0；18.8；165.2；0.0；904 ❙ 1090；612；47 ❙ 20.7；864；327.3

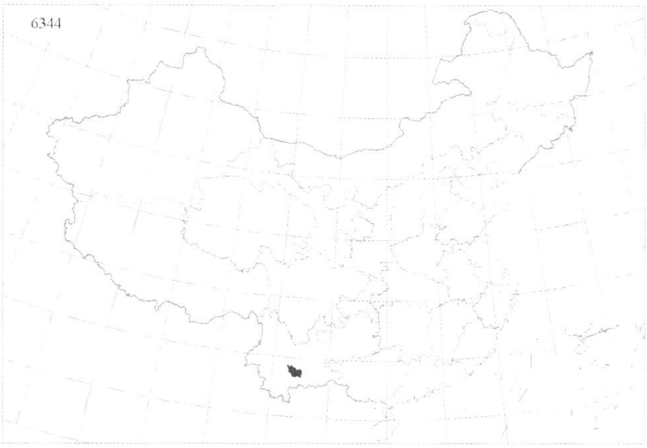

Sterculiaceae 梧桐科 *Paradombeya* 平当树属
Paradombeya sinensis Dunn 平当树
Shrubs or small trees

15.3 (11.0~19.3)；8.1 (2.7~13.1)；20.8 (16.6~26.6)；15.3 (11.0~19.4)；124.0 (73.1~173.2)；0.2 (0.0~3.7)；770 (636~939) ❙ 1020 (797~1454)；579 (411~792)；42 (23~70) ❙ 33.5 (-14.2~99.9)；748 (636~912)；293.7 (142.9~487.2)

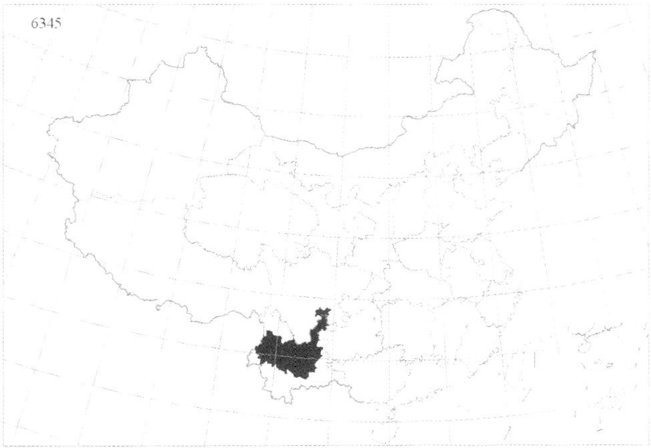

Sterculiaceae 梧桐科 *Pterospermum* 翅子树属
Pterospermum heterophyllum Hance 翻白叶树
Trees

20.0 (14.6~24.8)；11.8 (3.8~20.1)；26.6 (20.3~29.0)；20.1 (14.7~24.8)；180.8 (117.7~237.6)；0.0 (0.0~1.5)；1040 (779~1385) ❙ 1519 (868~2037)；667 (312~1044)；122 (41~255) ❙ 48.6 (-34.3~104.6)；1022 (779~1331)；298.6 (157.7~659.0)

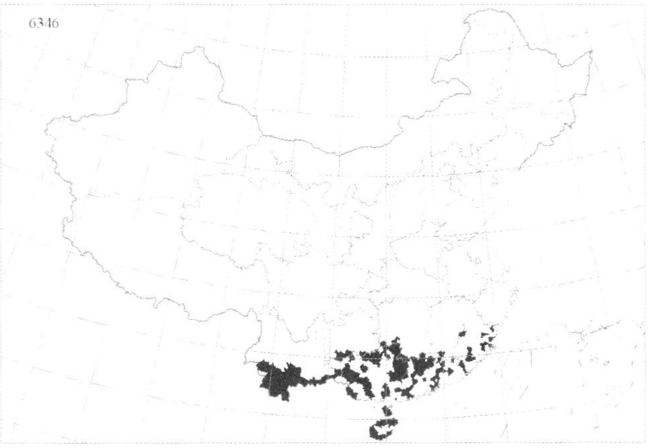

Sterculiaceae 梧桐科 *Pterospermum* 翅子树属
Pterospermum kingtungense C. Y. Wu ex Hsue 景东翅子树
Trees

18.2；11.9；22.3；18.3；159.0；0.0；875 ❙ 1135；642；47 ❙ 29.8；829；367.7

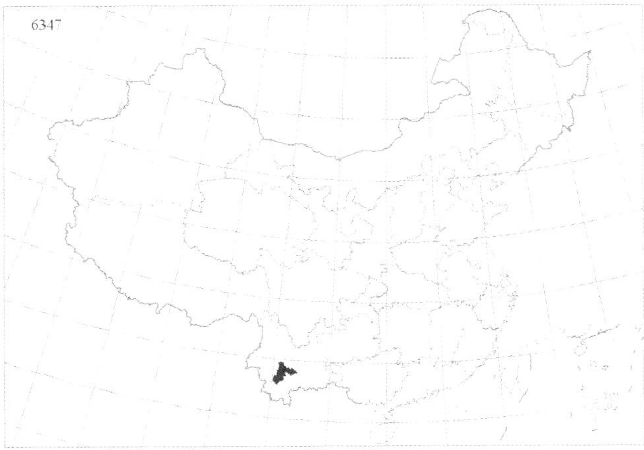

Sterculiaceae 梧桐科 *Pterospermum* 翅子树属
Pterospermum lanceifolium Roxb. 窄叶半枫荷
Trees

20.1 (13.3~24.8)；13.5 (5.4~20.1)；24.7 (19.6~28.8)；20.1 (13.4~24.8)；181.7 (100.4~237.6)；0.0 (0.0~0.0)；1024 (706~1385) ❙ 1431 (868~2148)；704 (312~1234)；68 (41~136) ❙ 43.1 (-34.3~145.8)；985 (706~1360)；400.0 (203.4~659.0)

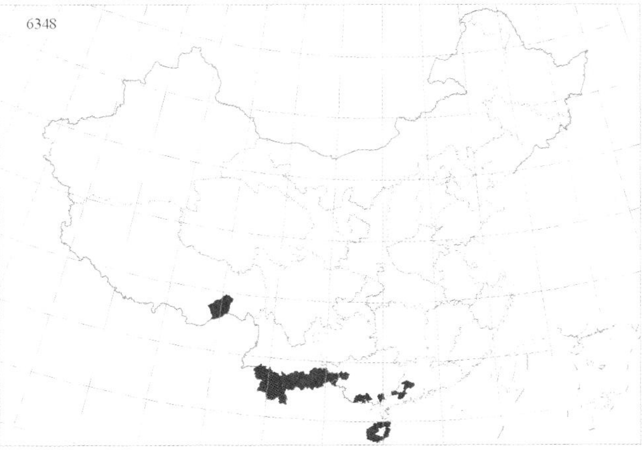

Sterculiaceae 梧桐科 *Pterospermum* 翅子树属
Pterospermum menglunense Hsue 勐仑翅子树
Trees

19.5 (18.6~20.5); 13.9 (12.7~15.1); 23.0 (22.3~23.7); 19.6 (18.7~20.5); 174.6 (163.9~185.9); 0.0 (0.0~0.0); 935 (888~992) ▎1560 (1464~1751); 787 (655~996); 67 (58~77) ▎67.1 (47.6~88.8); 922 (880~986); 539.4 (427.9~659.0)

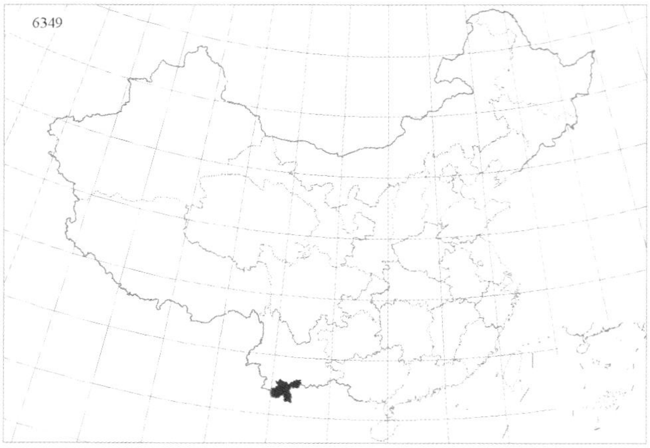

Sterculiaceae 梧桐科 *Pterospermum* 翅子树属
Pterospermum niveum Vidal 台湾翅子树
Trees

19.2; 13.8; 23.8; 19.2; 170.6; 0.0; 925 ▎2853; 1156; 367 ▎208.9; 925; 312.6

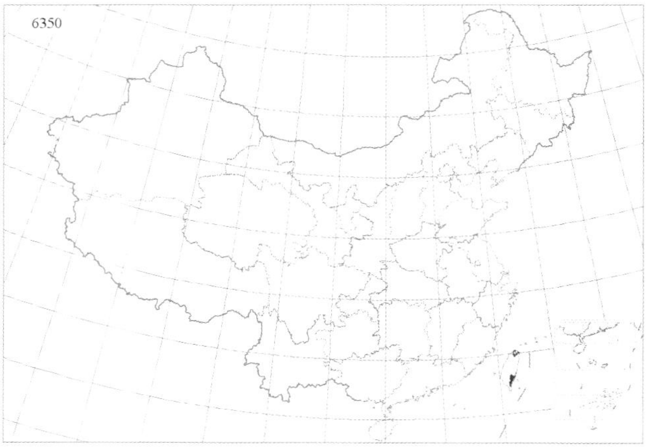

Sterculiaceae 梧桐科 *Pterospermum* 翅子树属
Pterospermum proteus Burkill 变叶翅子树
Small trees

18.5 (16.3~21.0); 12.1 (9.1~15.1); 22.7 (21.4~25.6); 18.5 (16.4~21.0); 162.0 (136.4~191.8); 0.0 (0.0~0.0); 891 (797~1053) ▎1310 (958~1751); 713 (531~996); 56 (44~77) ▎46.8 (13.3~88.8); 881 (797~1053); 429.3 (277.0~659.0)

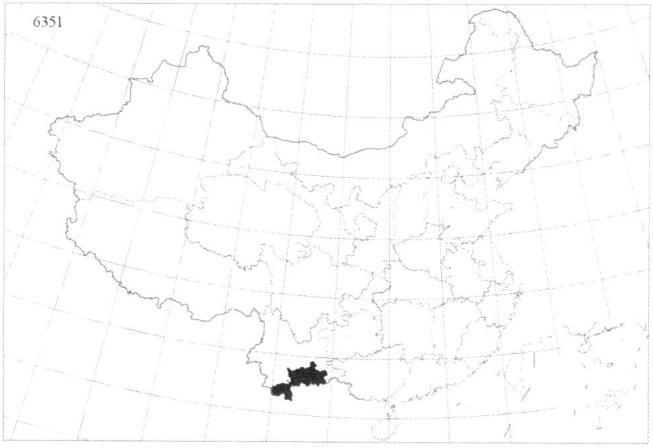

Sterculiaceae 梧桐科 *Pterospermum* 翅子树属
Pterospermum truncatolobatum Gagnep. 截裂翅子树
Trees

20.0 (17.3~22.2); 12.8 (9.9~14.2); 25.2 (22.3~28.4); 20.0 (17.3~22.3); 180.4 (147.1~207.5); 0.0 (0.0~0.0); 1010 (833~1190) ▎1420 (1163~1581); 778 (641~860); 75 (48~116) ▎42.4 (15.4~80.6); 1009 (833~1190); 403.9 (273.1~551.6)

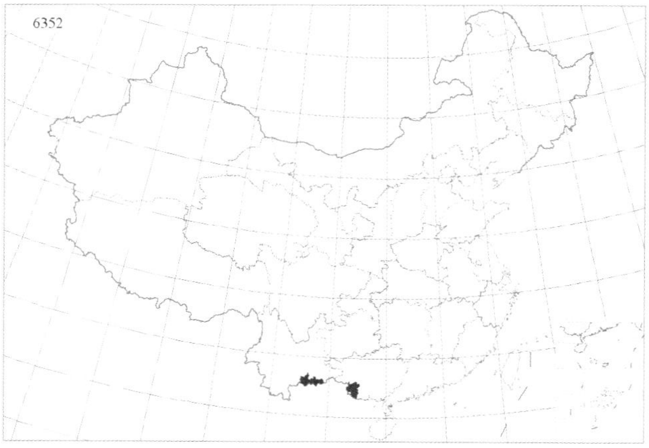

Sterculiaceae 梧桐科 *Pterospermum* 翅子树属
Pterospermum yunnanense Hsue 云南翅子树
Small trees

20.0; 14.6; 23.2; 20.1; 180.8; 0.0; 963 ▎1523; 681; 67 ▎58.4; 942; 581.0

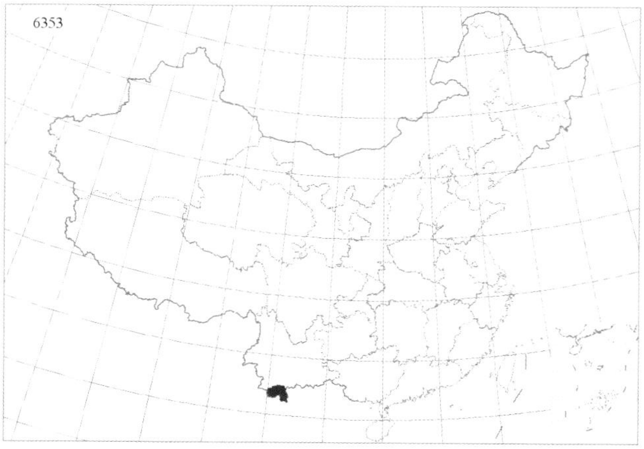

Sterculiaceae 梧桐科 *Pterygota* 翅苹婆属
Pterygota alata (Roxb.) R. Br. 翅苹婆
Trees

23.8 (20.4~24.7); 18.1 (15.1~19.8); 27.8 (23.6~28.8); 23.8 (20.5~24.7); 226.1 (185.8~236.5); 0.0 (0.0~0.0); 1294 (989~1376) ▎1452 (868~1813); 529 (312~712); 72 (42~113) ▎13.2 (-34.3~62.5); 1165 (868~1331); 436.6 (248.2~659.0)

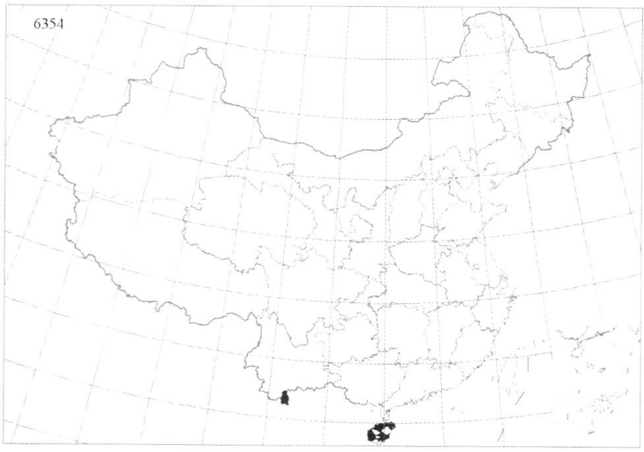

Sterculiaceae 梧桐科 *Reevesia* 梭罗树属
Reevesia botingensis H. H. Hsue 保亭梭罗
Trees

24.3; 19.5; 27.6; 24.4; 232.3; 0.0; 1339 ❙ 1569; 536; 53 ❙ 17.2; 1180; 557.4

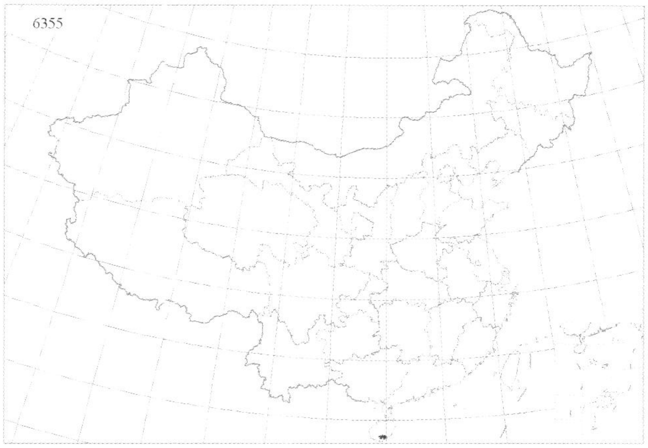

Sterculiaceae 梧桐科 *Reevesia* 梭罗树属
Reevesia formosana Sprague 台湾梭罗
Trees, deciduous

19.8 (17.1~22.2); 14.6 (11.2~17.7); 23.8 (21.7~27.0); 19.8 (17.1~22.2); 177.2 (145.0~206.7); 0.0 (0.0~0.0); 967 (816~1123) ❙ 2536 (1836~2992); 1381 (1033~1847); 157 (107~244) ❙ 167.9 (64.2~214.4); 958 (816~1123); 279.4 (202.8~363.7)

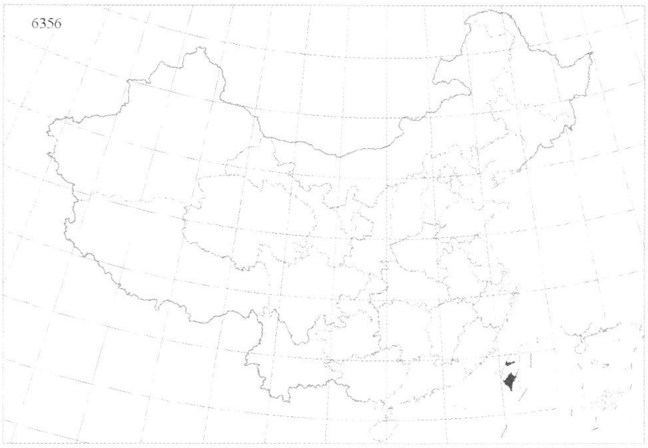

Sterculiaceae 梧桐科 *Reevesia* 梭罗树属
Reevesia glaucophylla H. H. Hsue 瑶山梭罗
Trees

18.1 (14.6~22.2); 7.8 (3.6~13.9); 26.9 (24.4~28.7); 18.1 (14.7~22.3); 157.4 (117.7~207.5); 0.1 (0.0~1.6); 947 (779~1190) ❙ 1506 (1268~1776); 602 (452~728); 157 (87~197) ❙ 60.2 (15.4~101.8); 947 (779~1190); 228.1 (180.3~327.7)

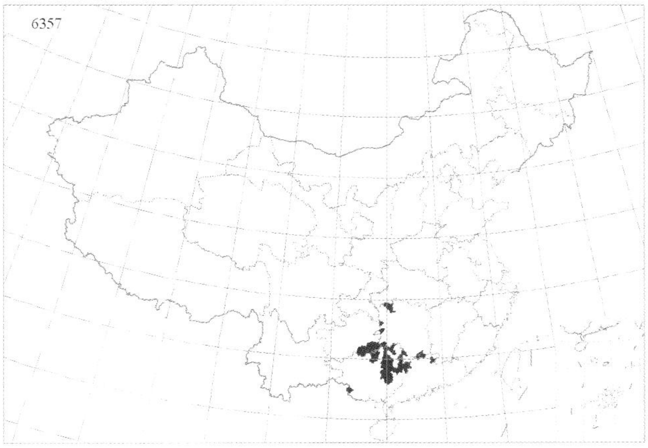

Sterculiaceae 梧桐科 *Reevesia* 梭罗树属
Reevesia lancifolia H. L. Li 剑叶梭罗
Trees

23.0; 17.7; 26.7; 23.1; 217.0; 0.0; 1217 ❙ 1725; 610; 68 ❙ 41.8; 1163; 572.9

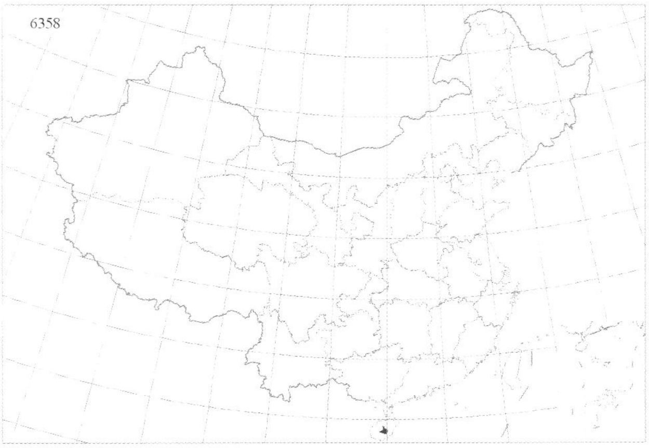

Sterculiaceae 梧桐科 *Reevesia* 梭罗树属
Reevesia lofouensis Chun et H. H. Hsue 罗浮梭罗
Trees

21.6; 13.2; 28.2; 21.6; 199.2; 0.0; 1139 ❙ 1953; 858; 167 ❙ 71.5; 1139; 240.0

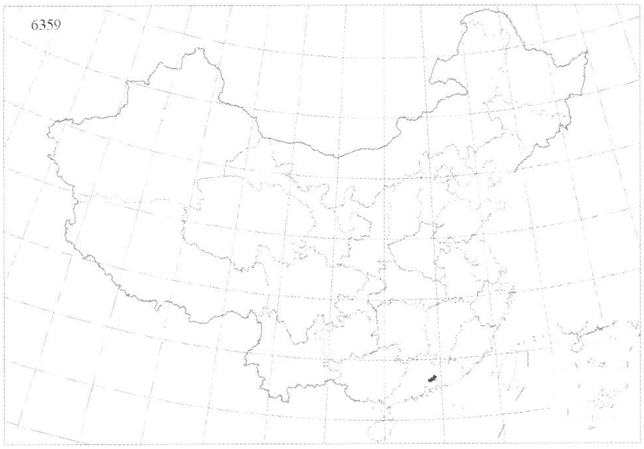

Sterculiaceae 梧桐科 *Reevesia* 梭罗树属
Reevesia longipetiolata Merr. et Chun 长柄梭罗
Trees

22.6 (21.3~24.7); 15.2 (12.9~19.8); 28.2 (27.3~28.9); 22.7 (21.3~24.7); 212.1 (195.9~236.5); 0.0 (0.0~0.0); 1216 (1111~1376) ❙ 1652 (868~2435); 806 (312~1428); 101 (44~137) ❙ 36.9 (-34.3~99.5); 1170 (868~1326); 305.2 (139.6~548.0)

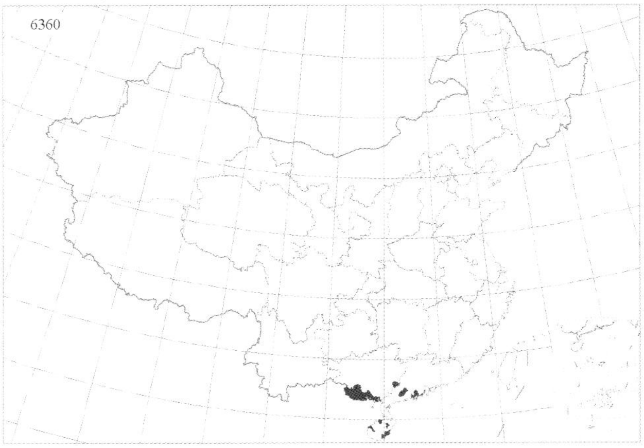

Sterculiaceae 梧桐科 *Reevesia* 梭罗树属
Reevesia orbicularifolia Hsue 圆叶梭罗
Trees

17.7；9.8；23.5；17.8；153.6；0.0；871 ❙ 1282；719；50 ❙ 47.5；871；338.2

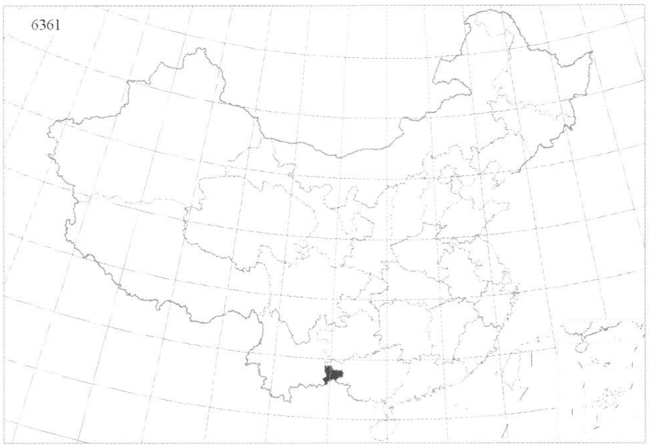

Sterculiaceae 梧桐科 *Reevesia* 梭罗树属
Reevesia pubescens Mast. 梭罗树
Trees, evergreen

17.2 (2.0~24.8)；8.7 (-7.2~20.1)；24.3 (10.1~29.6)；17.3 (3.8~24.8)；148.1 (15.4~237.6)；0.6 (0.0~51.4)；896 (413~1385) ❙ 1321 (741~2435)；622 (312~1428)；89 (15~253) ❙ 48.2 (-34.3~145.8)；881 (413~1326)；276.9 (108.2~659.0)

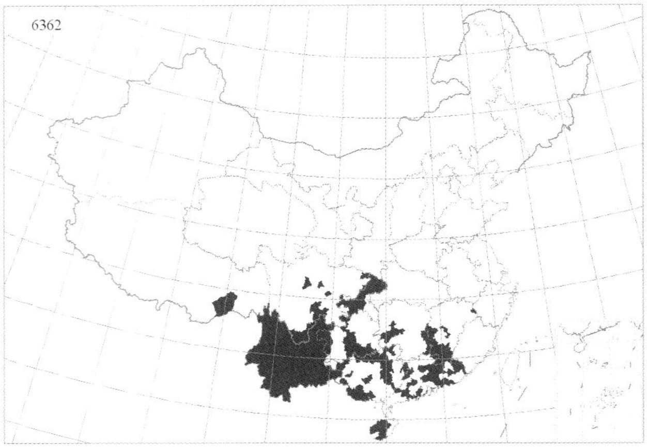

Sterculiaceae 梧桐科 *Reevesia* 梭罗树属
Reevesia pubescens var. siamensis (Craib) J. Anthony 泰梭罗
Trees, evergreen

18.6 (16.1~20.5)；12.5 (9.6~15.1)；22.5 (20.6~23.7)；18.6 (16.2~20.5)；163.3 (133.9~185.9)；0.0 (0.0~0.0)；896 (785~992) ❙ 1330 (958~1607)；655 (531~756)；59 (44~77) ❙ 47.7 (13.3~64.9)；880 (779~986)；491.9 (311.9~659.0)

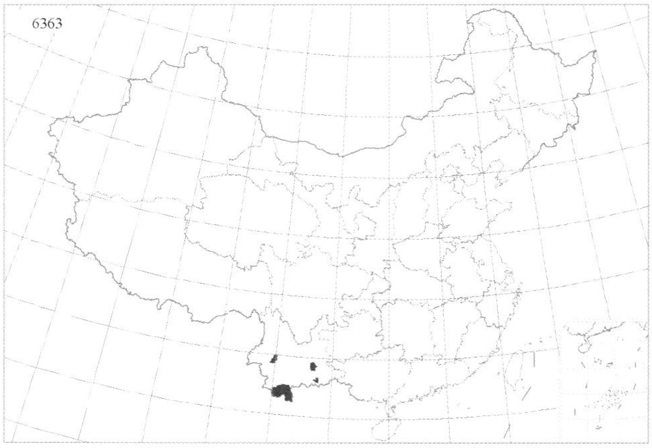

Sterculiaceae 梧桐科 *Reevesia* 梭罗树属
Reevesia pubescens var. xuefengensis C. J. Qi 雪峰山梭罗
Trees, evergreen

16.3；4.6；27.4；16.3；136.5；0.5；876 ❙ 1391；437；181 ❙ 59.1；847；167.0

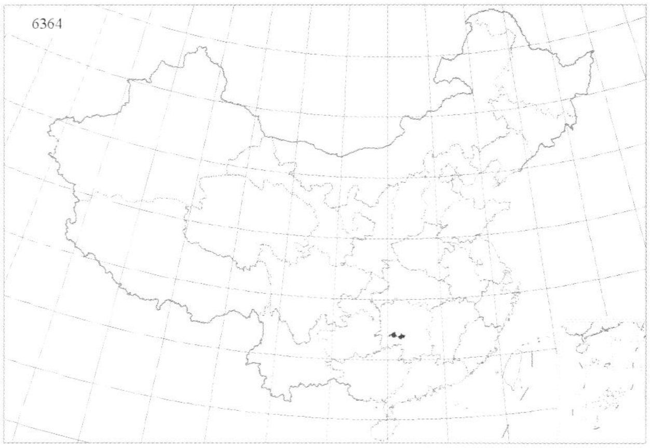

Sterculiaceae 梧桐科 *Reevesia* 梭罗树属
Reevesia pycnantha Ling 密花梭罗
Trees

17.6 (14.3~19.9)；7.1 (2.6~10.6)；27.3 (24.2~30.1)；17.6 (14.3~19.9)；151.9 (113.2~179.3)；0.3 (0.0~3.5)；926 (761~1036) ❙ 1732 (1452~1997)；590 (453~713)；214 (172~259) ❙ 88.2 (41.7~147.0)；919 (761~1036)；255.4 (148.9~332.5)

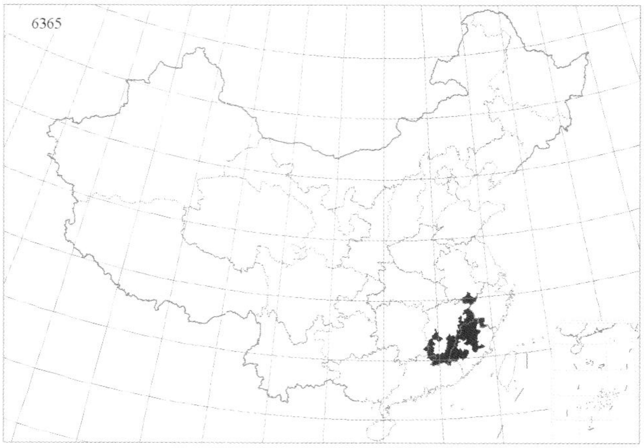

Sterculiaceae 梧桐科 *Reevesia* 梭罗树属
Reevesia rotundifolia Chun 粗齿梭罗
Trees

21.9 (21.3~22.7)；13.9 (13.2~15.0)；27.9 (27.3~28.6)；22.0 (21.3~22.8)；203.6 (195.9~213.1)；0.0 (0.0~0.0)；1162 (1111~1221) ❙ 1876 (1302~2435)；1047 (670~1428)；116 (86~130) ❙ 61.1 (12.9~99.5)；1162 (1111~1221)；289.6 (235.1~318.8)

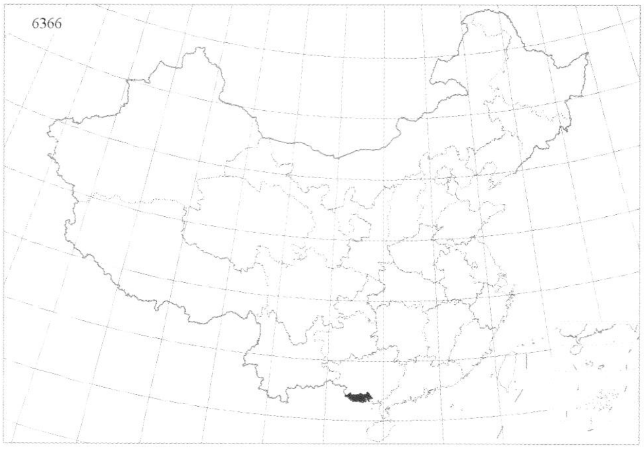

Sterculiaceae 梧桐科 *Reevesia* 梭罗树属
Reevesia rubronervia H. H. Hsue 红脉梭罗
Trees

19.4 (18.1~20.5); 13.3 (10.6~15.1); 23.4 (22.1~25.0); 19.5 (18.1~20.5); 173.4 (157.3~185.9); 0.0 (0.0~0.0); 938 (863~992) ▎1440 (1292~1607); 709 (655~774); 61 (49~77) ▎53.5 (37.6~64.9); 926 (863~986); 503.7 (342.5~659.0)

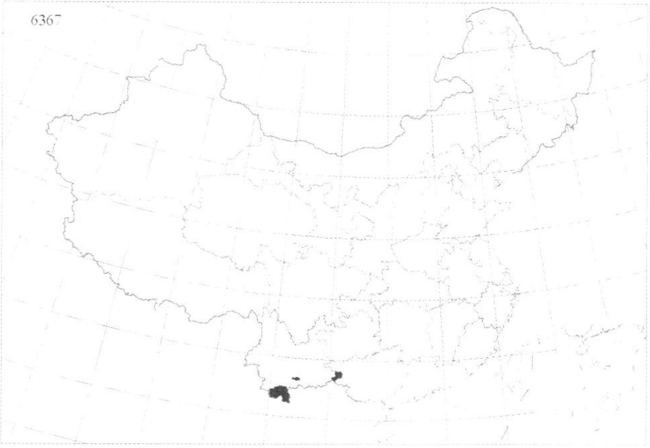

Sterculiaceae 梧桐科 *Reevesia* 梭罗树属
Reevesia shangszeensis H. H. Hsue 上思梭罗
Trees

21.9 (21.3~22.7); 13.9 (13.2~15.0); 27.9 (27.3~28.6); 22.0 (21.3~22.8); 203.6 (195.9~213.1); 0.0 (0.0~0.0); 1162 (1111~1221) ▎1876 (1302~2435); 1047 (670~1428); 116 (86~130) ▎61.1 (12.9~99.5); 1162 (1111~1221); 289.6 (235.1~318.8)

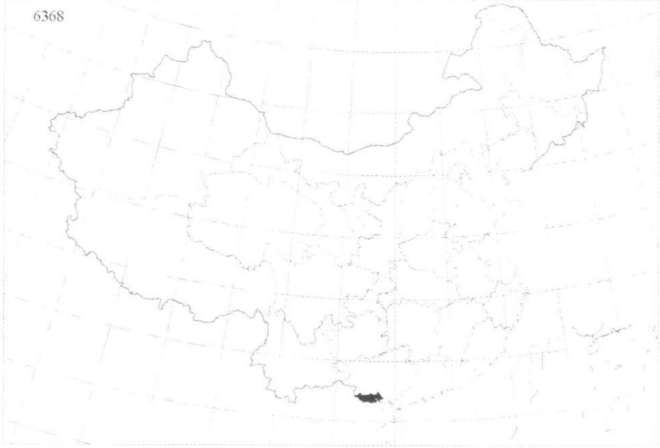

Sterculiaceae 梧桐科 *Reevesia* 梭罗树属
Reevesia tomentosa H. L. Li 绒果梭罗
Trees

19.1 (17.6~21.2); 10.3 (7.3~13.0); 26.8 (25.9~27.8); 19.2 (17.6~21.3); 170.1 (151.6~195.4); 0.0 (0.0~0.0); 980 (906~1115) ▎1605 (1494~1745); 600 (539~720); 166 (136~193) ▎64.4 (39.6~89.4); 980 (906~1115); 261.2 (225.9~300.5)

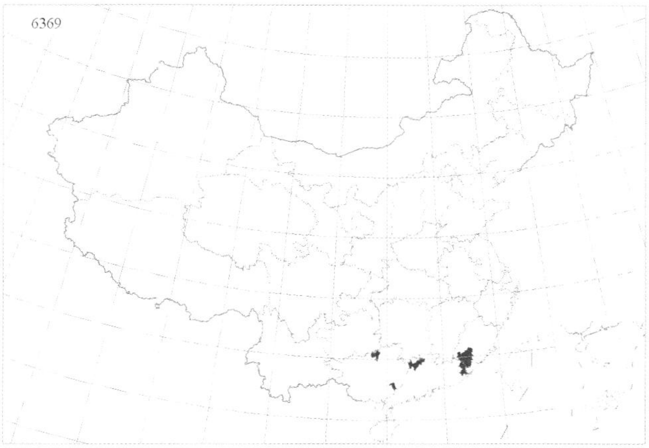

Sterculiaceae 梧桐科 *Sterculia* 苹婆属
Sterculia brevissima H. H. Hsue ex Y. Tang, M. G. Gilbert et Dorr 短柄苹婆
Shrubs or small trees, evergreen

19.0 (16.4~21.0); 13.1 (9.6~15.1); 22.7 (20.6~25.6); 19.1 (16.4~21.0); 169.1 (137.3~191.8); 0.0 (0.0~0.0); 915 (795~1053) ▎1504 (1276~1751); 793 (655~996); 59 (41~77) ▎65.1 (36.8~112.2); 900 (795~1053); 514.4 (395.0~659.0)

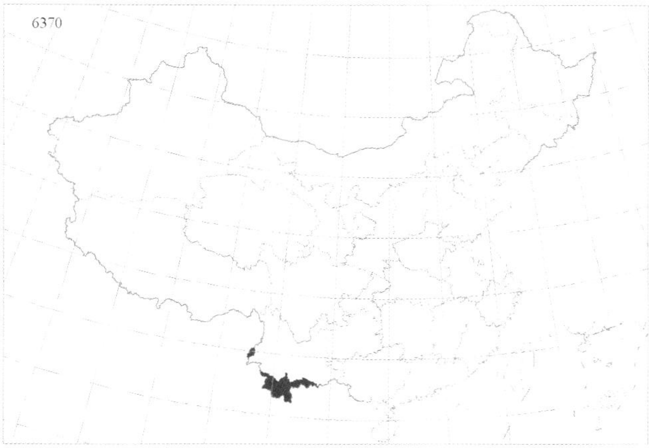

Sterculiaceae 梧桐科 *Sterculia* 苹婆属
Sterculia ceramica R. Br. 台湾苹婆
Small trees, evergreen

20.2; 15.4; 23.8; 20.2; 182.5; 0.0; 979 ▎2727; 1560; 133 ▎180.8; 979; 280.1

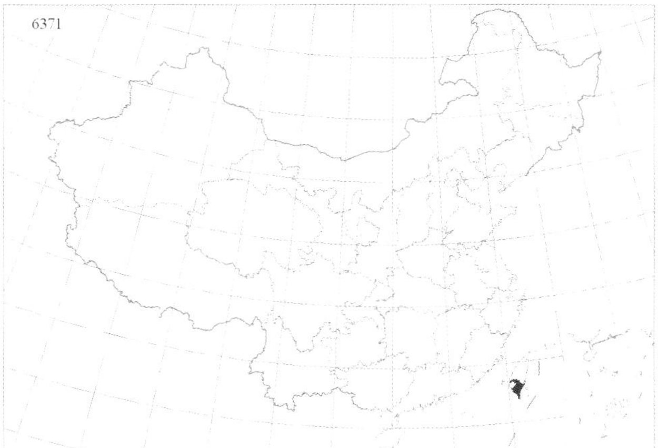

Sterculiaceae 梧桐科 *Sterculia* 苹婆属
Sterculia cinnamomifolia Tsai et Mao 樟叶苹婆
Shrubs, evergreen

17.7; 10.9; 22.5; 17.8; 153.4; 0.0; 854 ▎1368; 756; 59 ▎60.1; 854; 491.5

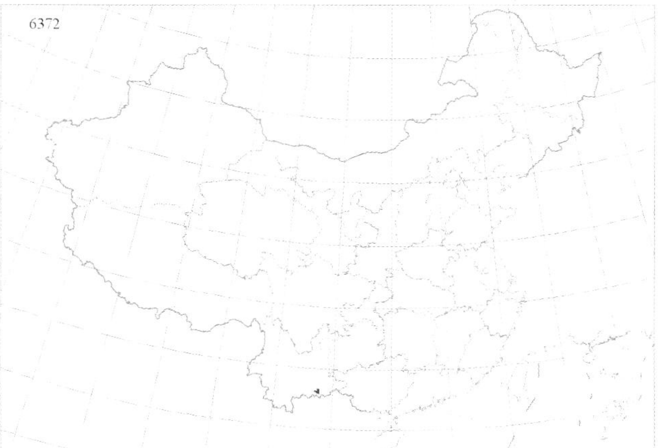

Sterculiaceae 梧桐科 *Sterculia* 苹婆属

Sterculia euosma W. W. Sm. 粉苹婆

Trees, evergreen

19.2 (13.3~22.2); 10.3 (5.4~13.9); 26.2 (19.1~29.0); 19.3 (13.4~22.3); 171.1 (100.4~207.5); 0.0 (0.0~0.0); 993 (706~1190) ❙ 1419 (1236~1736); 721 (617~1007); 97 (51~177) ❙ 45.3 (14.8~145.8); 993 (706~1190); 286.0 (160.7~491.5)

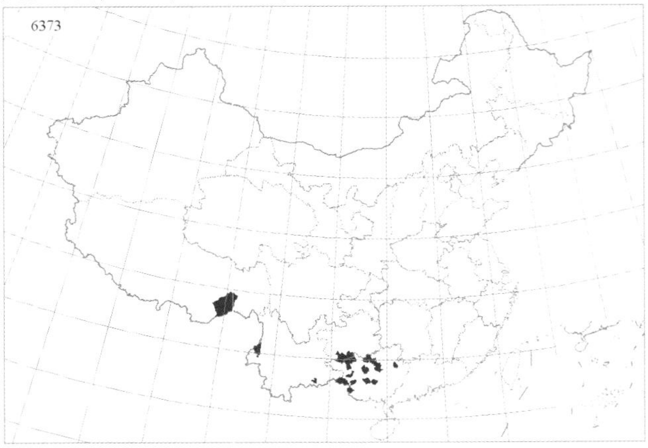

Sterculiaceae 梧桐科 *Sterculia* 苹婆属

Sterculia gengmaensis H. H. Hsue ex Y. Tang, M. G. Gilbert et Dorr 绿花苹婆

Shrubs, evergreen

19.6; 13.1; 23.8; 19.6; 175.2; 0.0; 961 ❙ 1444; 801; 57 ❙ 50.9; 950; 534.2

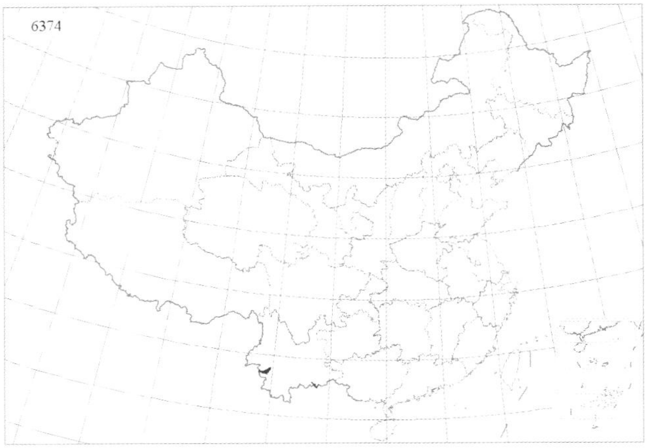

Sterculiaceae 梧桐科 *Sterculia* 苹婆属

Sterculia hainanensis Merr. et Chun 海南苹婆

Shrubs or small trees, evergreen

22.0 (19.1~24.8); 14.6 (11.0~20.1); 27.5 (25.0~28.6); 22.0 (19.1~24.8); 204.4 (169.6~237.6); 0.0 (0.0~0.0); 1168 (945~1385) ❙ 1650 (1182~2435); 820 (473~1428); 93 (41~130) ❙ 42.2 (1.5~99.5); 1143 (945~1360); 347.6 (235.1~620.0)

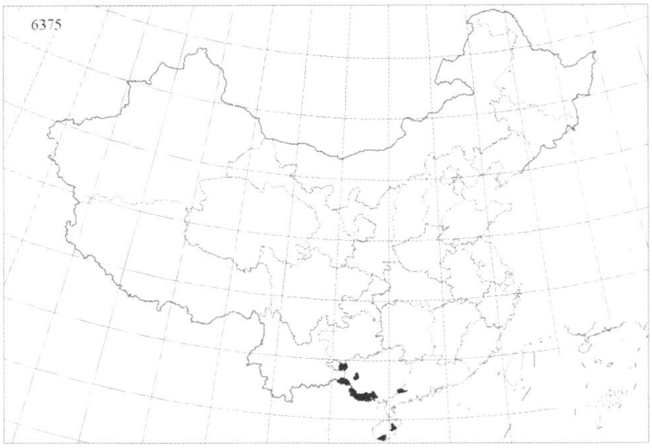

Sterculiaceae 梧桐科 *Sterculia* 苹婆属

Sterculia henryi Hemsl. 蒙自苹婆

Shrubs or small trees, evergreen

18.6; 11.8; 23.2; 18.6; 162.6; 0.0; 907 ❙ 1356; 740; 59 ❙ 49.8; 907; 475.5

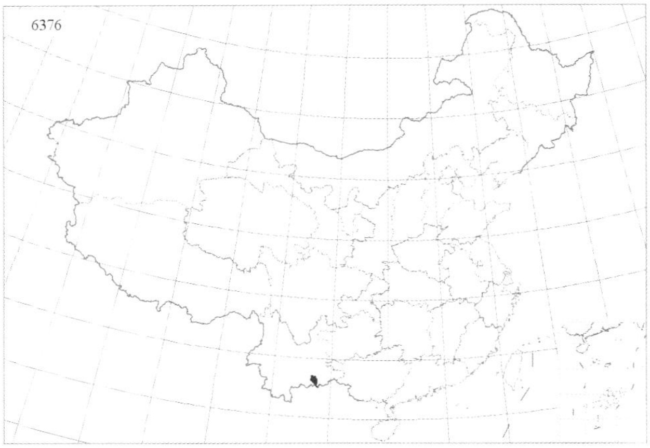

Sterculiaceae 梧桐科 *Sterculia* 苹婆属

Sterculia henryi var. cuneata Chun et H. H. Hsue 大围山苹婆

Shrubs or small trees, evergreen

19.3; 12.7; 23.8; 19.3; 171.6; 0.0; 942 ❙ 1439; 803; 61 ❙ 53.4; 942; 516.4

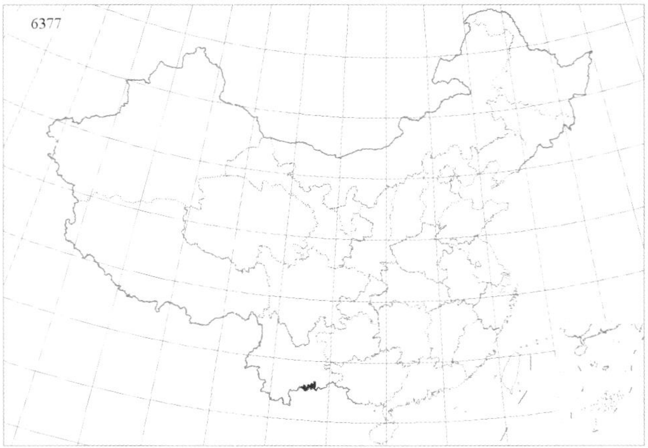

Sterculiaceae 梧桐科 *Sterculia* 苹婆属

Sterculia hymenocalyx K. Schum. 膜萼苹婆

Shrubs, evergreen

19.8; 13.5; 24.1; 19.8; 178.0; 0.0; 968 ❙ 1408; 798; 58 ❙ 45.7; 964; 503.3

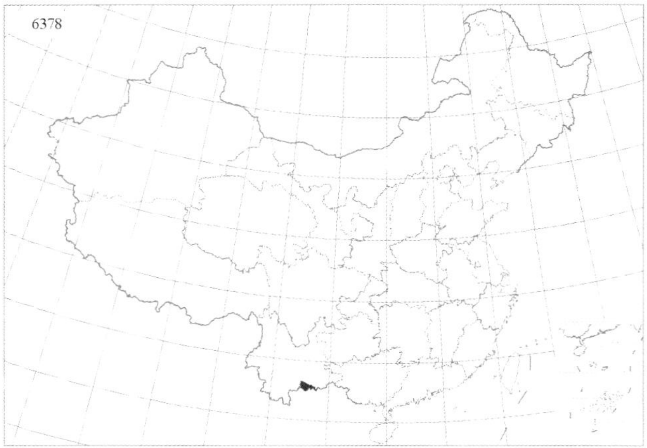

Sterculiaceae 梧桐科 *Sterculia* 苹婆属

Sterculia impressinervis Hsue 凹脉苹婆

Shrubs or trees, evergreen

17.4；9.7；22.9；17.5；149.7；0.0；848 ▎1338；756；50 ▎57.6；848；348.0

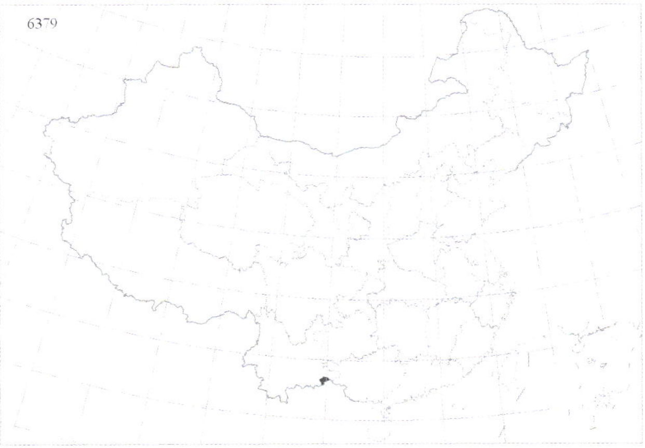

Sterculiaceae 梧桐科 *Sterculia* 苹婆属

Sterculia kingtungensis H. H. Hsue ex Y. Tang, M. G. Gilbert et Dorr 大叶苹婆

Trees, evergreen

17.3；11.0；21.6；17.3；148.0；0.0；832 ▎1049；598；44 ▎26.1；783；365.3

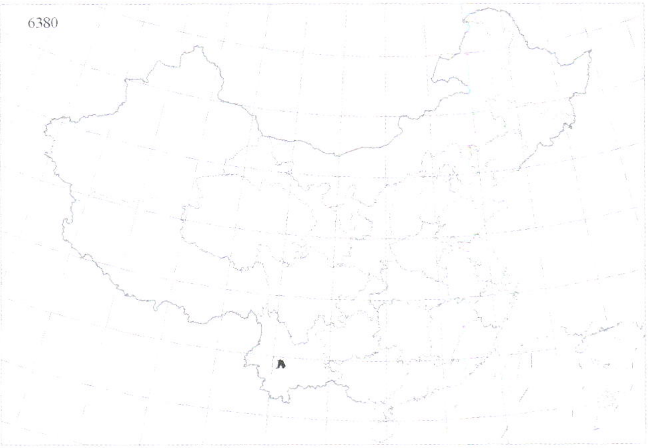

Sterculiaceae 梧桐科 *Sterculia* 苹婆属

Sterculia lanceifolia Roxb. 西蜀苹婆

Shrubs or trees, evergreen

18.4 (13.3~21.0)；11.8 (5.3~15.1)；22.9 (19.6~26.6)；18.4 (13.4~21.0)；160.9 (100.4~191.8)；0.0 (0.0~0.0)；892 (706~1053) ▎1336 (1007~1751)；729 (539~1007)；55 (44~77) ▎50.6 (12.5~145.8)；877 (706~1053)；413.5 (214.5~659.0)

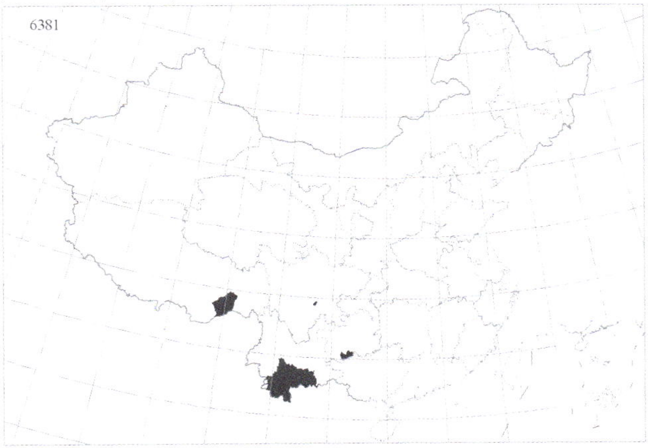

Sterculiaceae 梧桐科 *Sterculia* 苹婆属

Sterculia lanceolata Cav. 假苹婆

Trees, evergreen

19.7 (11.9~24.8)；11.7 (2.9~20.1)；26.1 (20.0~29.1)；19.8 (12.0~24.8)；177.4 (87.2~237.6)；0.0 (0.0~3.4)；1023 (671~1385) ▎1438 (868~2435)；682 (312~1428)；95 (35~197) ▎41.3 (-34.3~99.5)；1007 (671~1360)；290.3 (120.3~659.0)

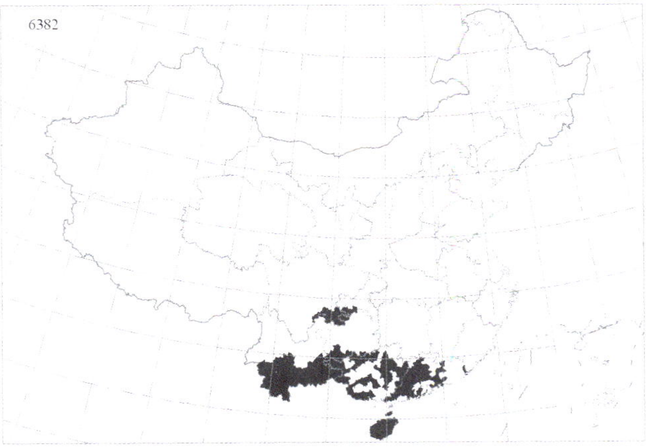

Sterculiaceae 梧桐科 *Sterculia* 苹婆属

Sterculia micrantha Chun et H. H. Hsue 小花苹婆

Trees, evergreen

19.2；12.6；23.6；19.2；169.9；0.0；942 ▎1285；708；56 ▎35.3；918；458.4

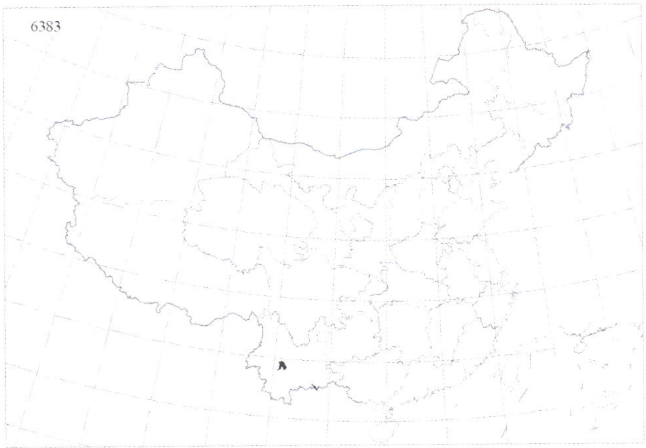

Sterculiaceae 梧桐科 *Sterculia* 苹婆属

Sterculia monosperma Ventenat 苹婆

Trees, evergreen

20.2 (14.6~24.6)；12.6 (6.5~19.5)；26.3 (20.1~29.1)；20.3 (14.6~24.7)；183.5 (114.8~235.9)；0.0 (0.0~0.0)；1043 (741~1372) ▎1607 (917~3495)；762 (356~1847)；125 (19~897) ▎55.6 (-5.0~246.5)；1030 (692~1299)；267.9 (95.3~659.0)

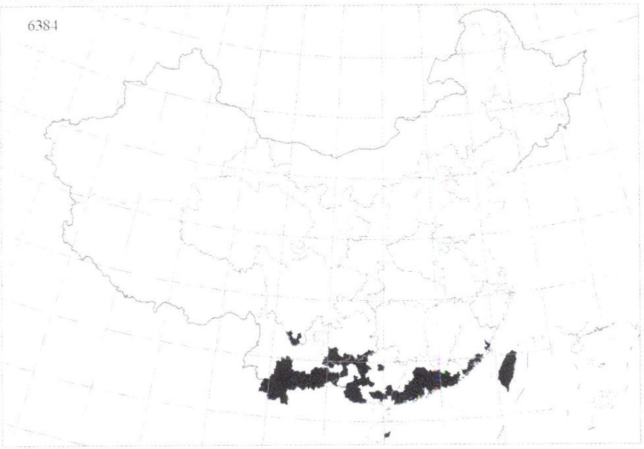

Sterculiaceae 梧桐科 *Sterculia* 苹婆属
Sterculia pexa Pierre 家麻树
Trees, evergreen

18.5 (16.0~22.2); 11.6 (9.0~15.1); 23.1 (20.3~28.5); 18.5 (16.0~22.3); 162.0 (132.4~207.5); 0.0 (0.0~0.0); 903 (780~1190) ❙ 1339 (939~1751); 735 (524~996); 59 (37~194) ❙ 49.0 (5.8~112.2); 890 (744~1190); 402.6 (213.5~659.0)

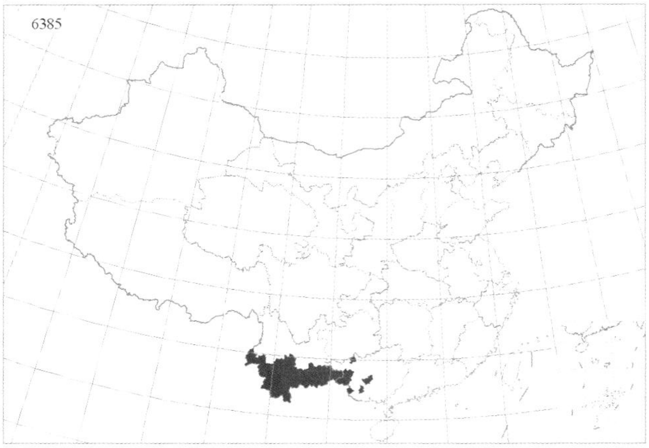

Sterculiaceae 梧桐科 *Sterculia* 苹婆属
Sterculia pinbienensis Tsai et Mao 屏边苹婆
Shrubs, evergreen

20.1; 12.9; 25.2; 20.1; 181.4; 0.0; 1013 ❙ 1490; 808; 78 ❙ 48.2; 1013; 441.6

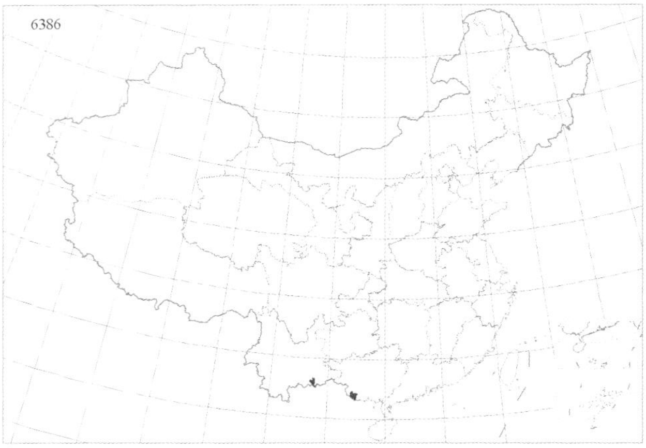

Sterculiaceae 梧桐科 *Sterculia* 苹婆属
Sterculia principis Gagnep. 基苹婆
Shrubs, evergreen

19.1; 13.2; 23.4; 19.3; 171.1; 0.0; 925 ❙ 1352; 788; 53 ❙ 46.3; 920; 479.1

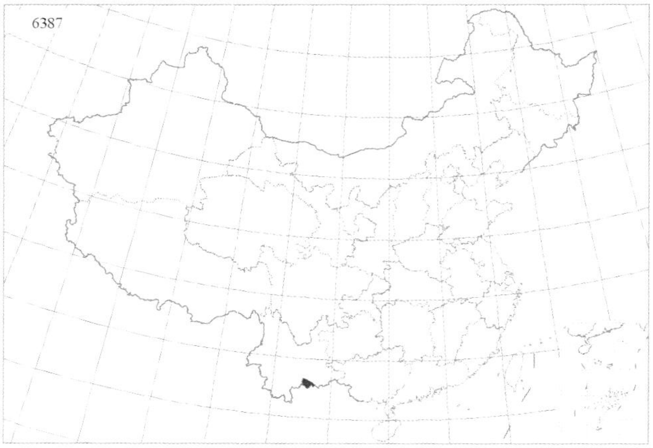

Sterculiaceae 梧桐科 *Sterculia* 苹婆属
Sterculia scandens Hemsl. 河口苹婆
Shrubs, evergreen

19.3; 12.7; 23.8; 19.3; 171.6; 0.0; 942 ❙ 1439; 803; 61 ❙ 53.4; 942; 516.4

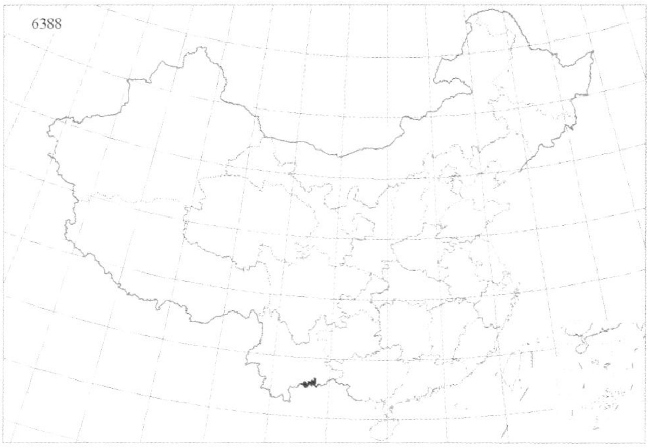

Sterculiaceae 梧桐科 *Sterculia* 苹婆属
Sterculia subnobilis Hsue 罗浮苹婆
Trees, evergreen

22.0 (21.6~22.4); 14.0 (13.2~14.8); 28.3 (27.8~28.5); 22.0 (21.6~22.4); 204.5 (199.2~208.9); 0.0 (0.0~0.0); 1169 (1139~1191) ❙ 1896 (1600~2037); 910 (811~1033); 134 (110~167) ❙ 62.3 (34.7~71.8); 1169 (1139~1191); 222.6 (163.7~273.3)

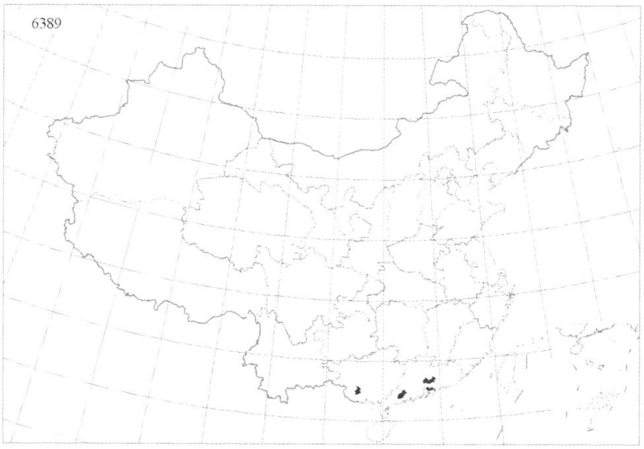

Sterculiaceae 梧桐科 *Sterculia* 苹婆属
Sterculia subracemosa Chun et H. H. Hsue 信宜苹婆
Shrubs, evergreen

21.6; 13.6; 27.6; 21.7; 199.9; 0.0; 1136 ❙ 1575; 770; 106 ❙ 39.8; 1136; 290.9

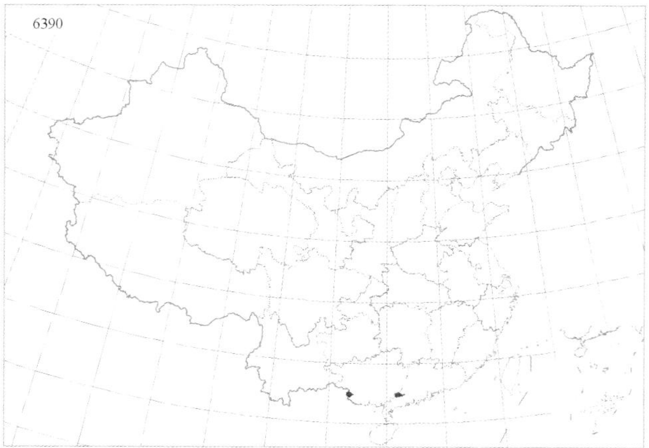

Sterculiaceae 梧桐科 *Sterculia* 苹婆属
Sterculia tonkinensis Aug. DC. 北越苹婆
Shrubs or small trees, evergreen

19.4; 12.6; 24.1; 19.4; 172.6; 0.0; 954 ❙ 1445; 787; 64 ❙ 52.3; 954; 521.5

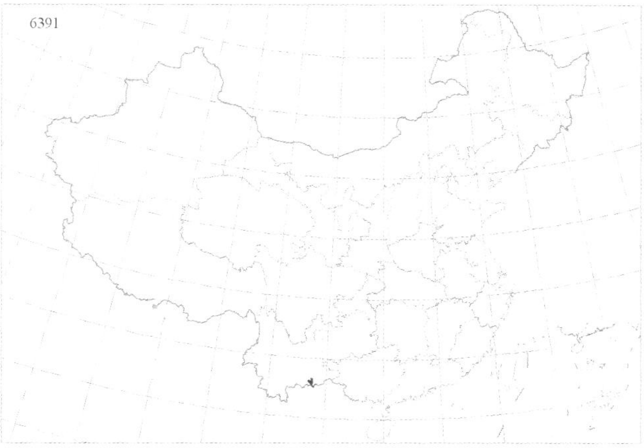

Sterculiaceae 梧桐科 *Sterculia* 苹婆属
Sterculia villosa Roxb. 绒毛苹婆
Trees, evergreen

18.1 (16.0~20.5); 11.9 (9.3~15.1); 22.1 (20.3~23.7); 18.1 (16.0~20.5); 157.8 (132.4~185.9); 0.0 (0.0~0.0); 871 (780~992) ❙ 1340 (939~1751); 743 (524~996); 52 (37~77) ❙ 54.1 (6.5~112.2); 850 (744~986); 441.3 (293.0~659.0)

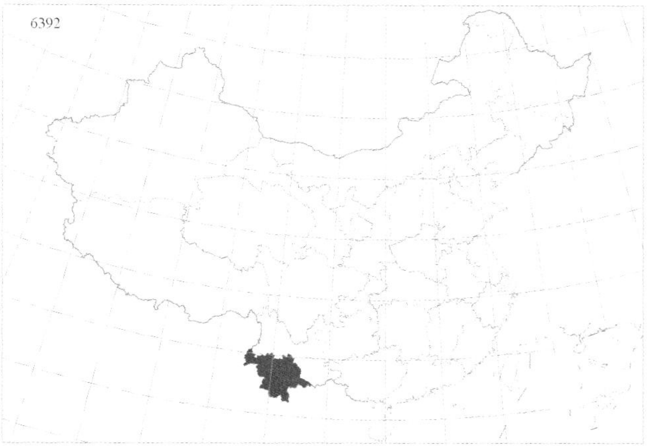

Sterculiaceae 梧桐科 *Waltheria* 蛇婆子属
Waltheria indica L. 蛇婆子
Subshrubs

20.2 (14.4~24.8); 13.4 (7.5~20.1); 25.3 (19.1~29.0); 20.2 (14.4~24.8); 182.7 (113.1~237.6); 0.0 (0.0~0.0); 1032 (727~1385) ❙ 1562 (868~3495); 736 (312~1847); 110 (37~897) ❙ 55.3 (-34.3~246.5); 997 (727~1360); 337.5 (95.3~659.0)

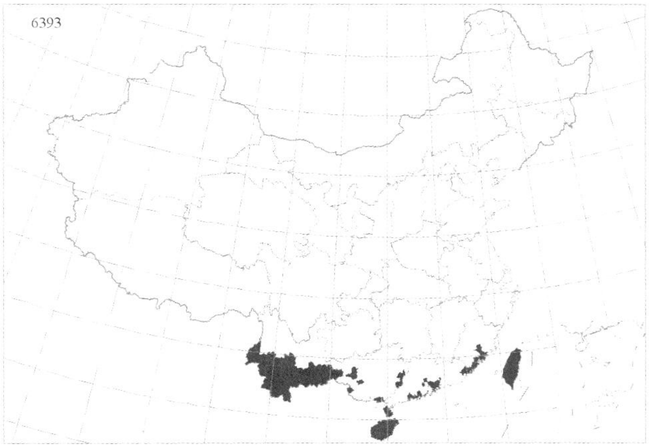

Dilleniaceae 五桠果科 *Dillenia* 五桠果属
Dillenia indica L. 五桠果
Trees, evergreen

19.3 (16.9~22.2); 12.5 (9.5~15.1); 24.1 (21.4~28.3); 19.4 (16.9~22.2); 172.3 (142.7~206.9); 0.0 (0.0~0.0); 955 (813~1177) ❙ 1425 (1007~1876); 754 (539~996); 68 (44~125) ❙ 49.6 (5.8~88.8); 948 (813~1177); 388.8 (209.2~659.0)

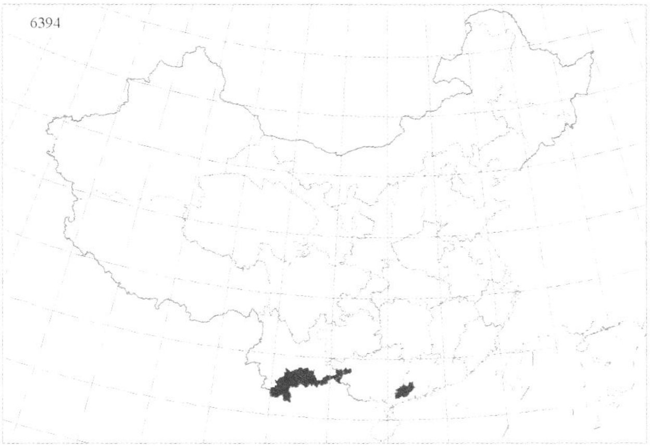

Dilleniaceae 五桠果科 *Dillenia* 五桠果属
Dillenia pentagyna Roxb. 小花五桠果
Trees, deciduous

23.8 (20.5~24.8); 18.2 (14.9~20.1); 27.8 (23.7~28.8); 23.8 (20.5~24.8); 226.2 (185.9~237.6); 0.0 (0.0~0.0); 1294 (992~1385) ❙ 1499 (868~1854); 540 (312~712); 73 (41~115) ❙ 16.6 (-34.3~47.8); 1180 (868~1360); 445.8 (248.2~620.0)

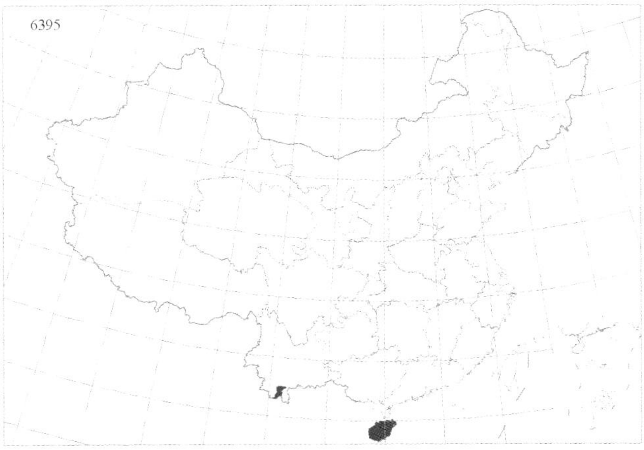

Dilleniaceae 五桠果科 *Dillenia* 五桠果属
Dillenia turbinata Finet et Gagnep. 大花五桠果
Trees, evergreen

21.8 (17.2~24.8); 14.6 (9.5~20.1); 27.1 (22.3~28.9); 21.8 (17.3~24.8); 202.1 (147.0~237.6); 0.0 (0.0~0.0); 1149 (833~1385) ❙ 1543 (868~2435); 721 (312~1428); 96 (41~194) ❙ 36.4 (-34.3~99.5); 1112 (833~1360); 359.9 (157.7~659.0)

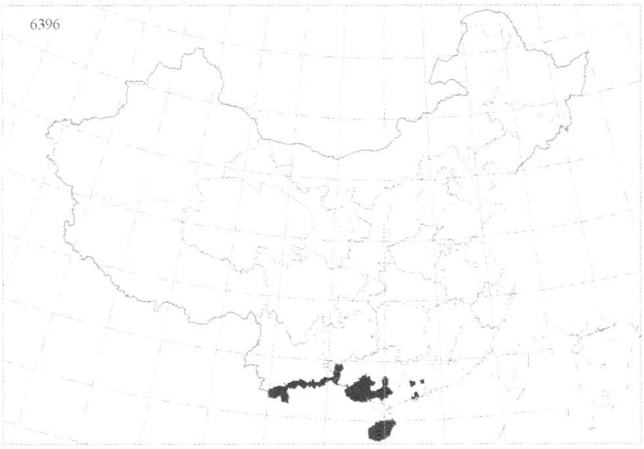

Dilleniaceae 五桠果科 *Tetracera* 锡叶藤属

Tetracera sarmentosa (L.) Vahl 锡叶藤

Large climbers, evergreen

21.9 (16.9~24.8); 14.2 (7.1~20.1); 28.0 (22.5~29.8); 22.0 (17.0~24.8); 203.7 (143.6~237.6); 0.0 (0.0~0.0); 1165 (862~1385) ❙ 1608 (868~2435); 728 (312~1428); 116 (41~198) ❙ 39.1 (-34.3~100.8); 1145 (862~1360); 266.6 (95.3~659.0)

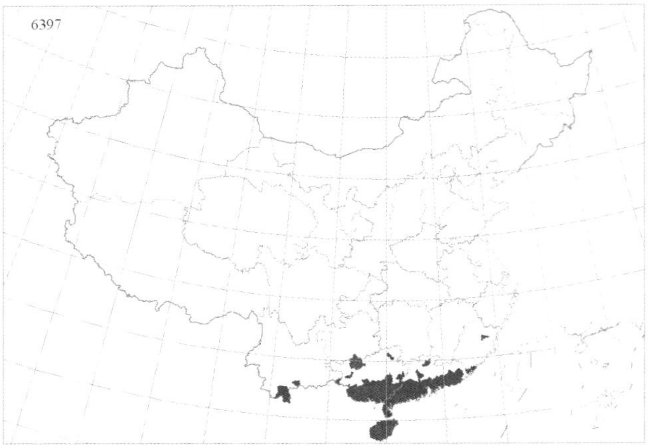

Dilleniaceae 五桠果科 *Tetracera* 锡叶藤属

Tetracera scandens sensu Merr. 毛果锡叶藤

Vines, evergreen

19.9 (15.5~21.8); 11.9 (4.8~15.1); 26.0 (23.6~29.0); 19.9 (15.6~21.8); 178.8 (127.2~202.0); 0.0 (0.0~0.2); 1016 (815~1168) ❙ 1487 (1371~1607); 618 (565~702); 120 (66~177) ❙ 49.6 (17.4~93.2); 1000 (815~1134); 387.3 (218.1~659.0)

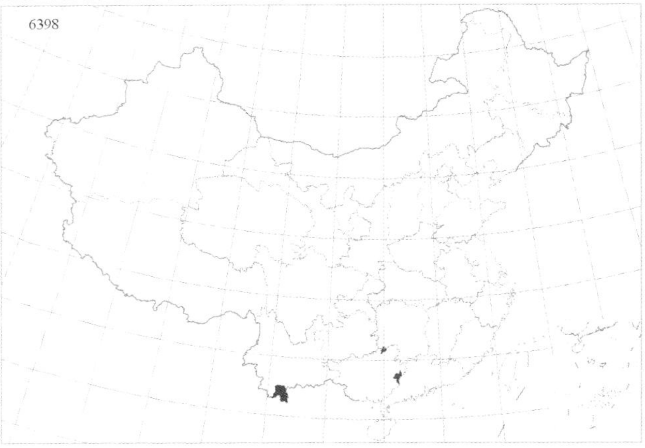

Actinidiaceae 猕猴桃科 *Actinidia* 猕猴桃属

Actinidia arguta (Siebold et Zucc.) Planch. ex Miq. 软枣猕猴桃

Large climbers, deciduous

12.8 (1.4~19.4); 0.2 (-20.5~12.5); 24.3 (9.9~30.1); 13.5 (3.8~19.4); 110.5 (15.5~173.4); 15.9 (0.0~93.6); 769 (413~1037) ❙ 1043 (341~1997); 479 (190~817); 80 (8~259) ❙ 33.7 (-40.9~130.7); 723 (341~1029); 230.6 (61.2~491.5)

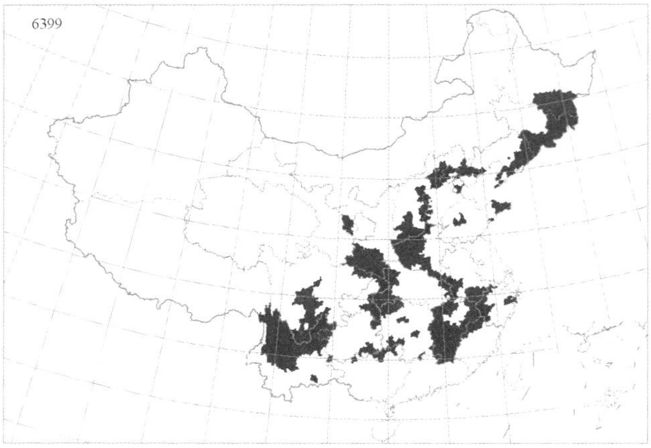

Actinidiaceae 猕猴桃科 *Actinidia* 猕猴桃属

Actinidia arguta var. giraldii (Diels) Vorosch. 陕西猕猴桃

Vines, deciduous

13.7 (9.3~19.2); 2.4 (-1.7~9.2); 24.5 (20.1~27.5); 13.8 (9.5~19.2); 110.6 (69.8~170.8); 5.7 (0.0~18.0); 764 (603~1002) ❙ 1320 (724~1638); 542 (316~678); 102 (24~220) ❙ 73.5 (2.5~108.0); 764 (603~1002); 251.5 (213.4~285.1)

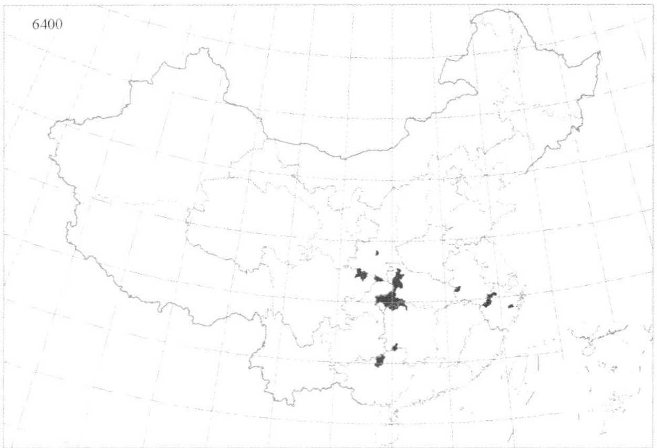

Actinidiaceae 猕猴桃科 *Actinidia* 猕猴桃属

Actinidia callosa Lindl. 硬齿猕猴桃

Large climbers, deciduous

16.4 (2.6~24.0); 6.7 (-7.0~19.1); 24.9 (11.2~29.3); 16.4 (4.3~24.0); 138.2 (20.0~228.1); 1.0 (0.0~48.2); 858 (427~1302) ❙ 1376 (798~3495); 594 (377~1847); 124 (15~897) ❙ 60.2 (-2.0~235.6); 851 (427~1190); 253.8 (149.5~620.0)

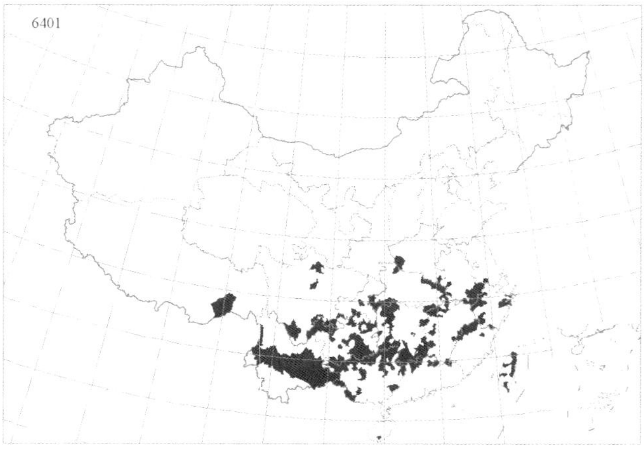

Actinidiaceae 猕猴桃科 *Actinidia* 猕猴桃属

Actinidia callosa var. acuminata C. F. Liang 尖叶猕猴桃

Large climbers, deciduous

15.7; 5.2; 25.6; 15.8; 129.5; 0.2; 827 ❙ 1359; 522; 134 ❙ 63.3; 827; 205.5

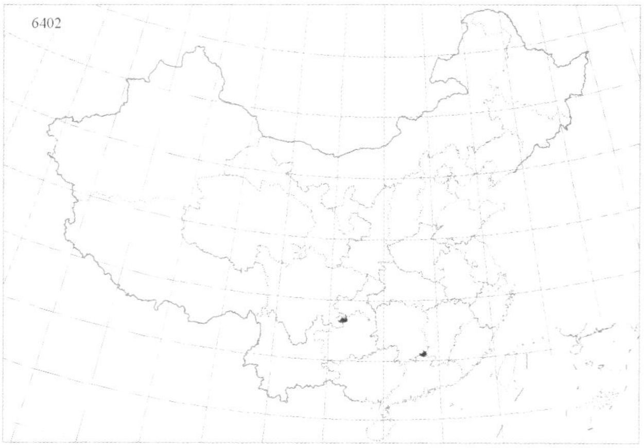

Actinidiaceae 猕猴桃科 *Actinidia* 猕猴桃属
Actinidia callosa var. discolor C. F. Liang 异色猕猴桃
Large climbers, deciduous

17.6 (14.3~24.0); 6.9 (2.1~19.1); 27.6 (24.2~30.1); 17.7 (14.3~24.0); 152.3 (113.2~228.1); 0.3 (0.0~4.9); 935 (761~1302) ❙ 1565 (1070~2024); 554 (356~902); 188 (43~259) ❙ 68.6 (-5.0~147.0); 921 (761~1179); 231.0 (123.9~620.0)

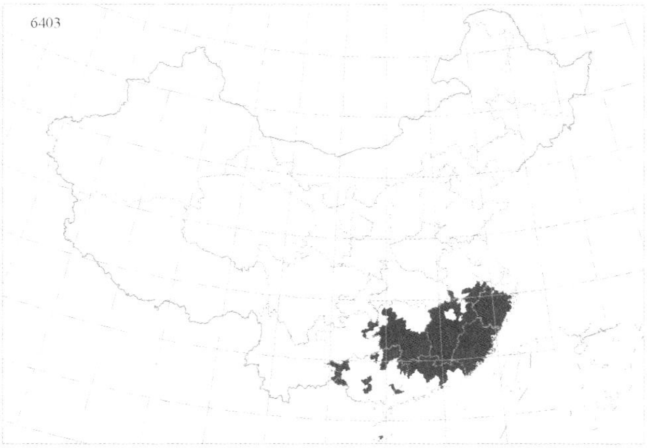

Actinidiaceae 猕猴桃科 *Actinidia* 猕猴桃属
Actinidia callosa var. henryi Maxim. 京梨猕猴桃
Large climbers, deciduous

16.1 (2.6~22.2); 5.0 (-7.0~13.9); 26.4 (11.2~30.1); 16.2 (4.3~22.3); 136.3 (20.0~207.5); 1.9 (0.0~48.2); 870 (427~1190) ❙ 1364 (626~1997); 532 (299~850); 135 (23~259) ❙ 56.8 (-22.2~137.5); 858 (427~1190); 216.8 (134.6~384.7)

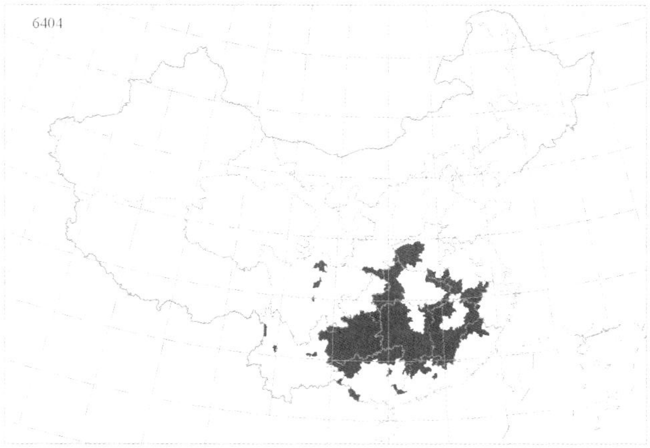

Actinidiaceae 猕猴桃科 *Actinidia* 猕猴桃属
Actinidia callosa var. strigillosa C. F. Liang 毛叶硬齿猕猴桃
Large climbers, deciduous

17.7 (14.3~21.6); 7.2 (2.0~12.6); 27.0 (24.6~28.8); 17.7 (14.3~21.7); 153.1 (117.5~199.8); 0.3 (0.0~5.4); 931 (779~1153) ❙ 1623 (1211~1997); 613 (496~809); 182 (73~254) ❙ 75.8 (30.9~130.7); 927 (779~1153); 247.0 (188.8~322.1)

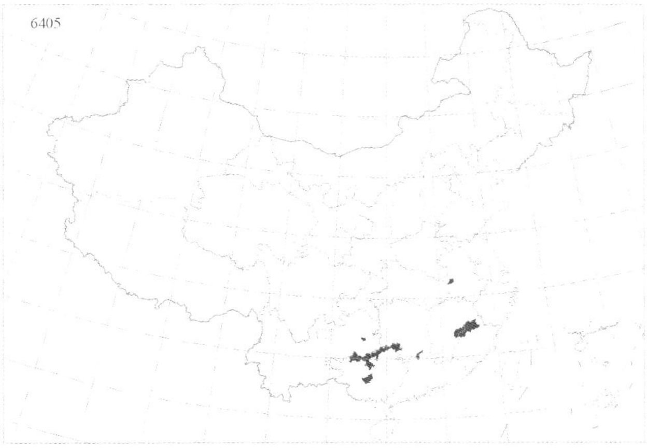

Actinidiaceae 猕猴桃科 *Actinidia* 猕猴桃属
Actinidia chengkouensis C. Y. Chang 城口猕猴桃
Vines, deciduous

13.9 (10.6~16.2); 3.0 (-0.2~5.2); 24.3 (21.1~26.7); 13.9 (10.6~16.3); 110.7 (79.4~135.2); 4.0 (0.0~12.1); 763 (643~869) ❙ 1208 (1148~1265); 509 (475~541); 60 (33~84) ❙ 59.6 (44.1~93.8); 763 (643~869); 232.2 (165.6~284.3)

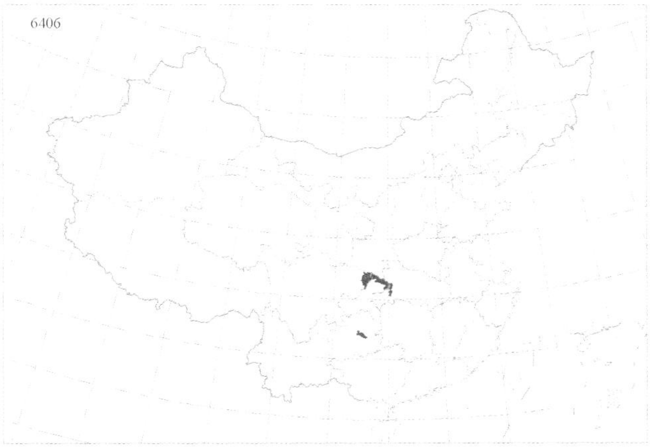

Actinidiaceae 猕猴桃科 *Actinidia* 猕猴桃属
Actinidia chinensis Planch. 中华猕猴桃
Vines, deciduous

15.6 (2.6~21.7); 4.3 (-7.0~15.3); 26.2 (11.2~30.1); 15.7 (4.3~21.7); 130.8 (20.0~200.2); 3.0 (0.0~48.2); 850 (427~1118) ❙ 1329 (512~2859); 517 (258~1412); 133 (15~402) ❙ 55.7 (-35.3~235.6); 835 (427~1114); 217.1 (121.6~334.4)

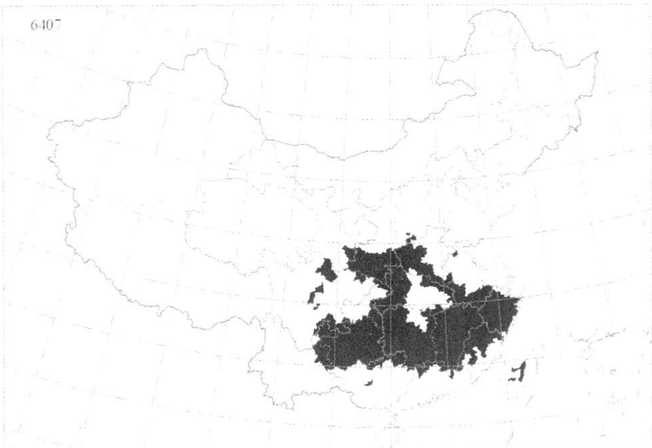

Actinidiaceae 猕猴桃科 *Actinidia* 猕猴桃属
Actinidia chinensis var. deliciosa (A. Chev.) A Chev. 美味猕猴桃
Vines, deciduous

14.2 (2.6~19.1); 3.6 (-7.0~10.1); 23.8 (11.2~28.1); 14.2 (4.3~19.1); 114.5 (20.0~169.2); 4.0 (0.0~48.2); 771 (427~969) ❙ 1123 (632~1572); 516 (306~817); 65 (15~196) ❙ 46.1 (-13.6~107.2); 770 (427~969); 222.8 (130.1~321.8)

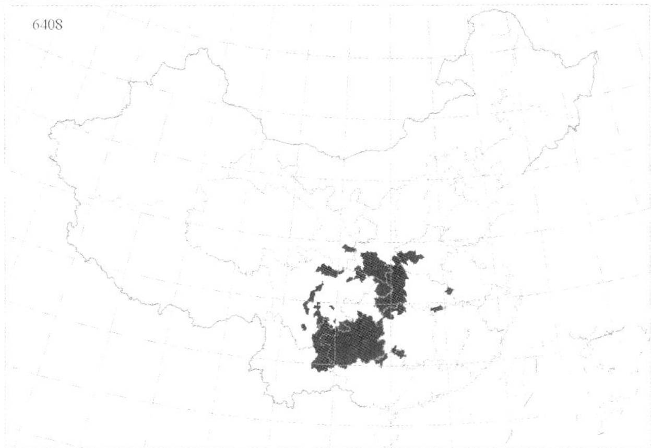

Actinidiaceae 猕猴桃科 *Actinidia* 猕猴桃属
Actinidia chrysantha C. F. Liang 金花猕猴桃
Large climbers, deciduous

17.1 (14.3~21.0); 6.0 (3.3~11.9); 27.4 (24.2~29.8); 17.2 (14.3~21.1); 146.5 (113.2~193.0); 0.2 (0.0~2.3); 911 (761~1114) ❙ 1503 (1117~1922); 515 (392~821); 183 (57~259) ❙ 65.5 (28.5~128.9); 898 (761~1114); 210.2 (134.6~371.7)

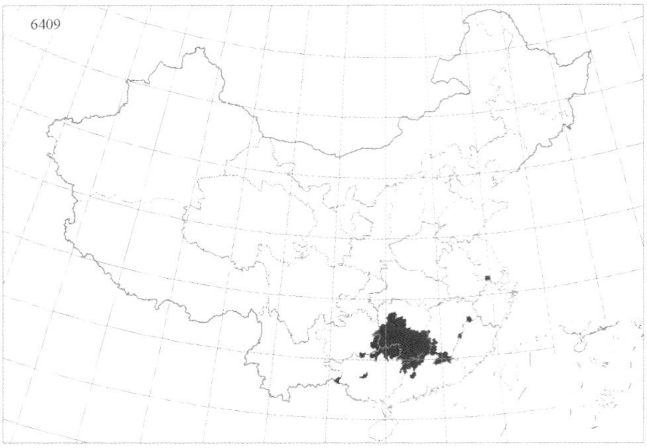

Actinidiaceae 猕猴桃科 *Actinidia* 猕猴桃属
Actinidia cylindrica C. F. Liang 柱果猕猴桃
Vines, evergreen

17.9 (15.2~19.2); 8.1 (4.1~9.8); 26.3 (25.5~27.5); 18.0 (15.2~19.2); 156.0 (123.4~170.8); 0.1 (0.0~0.9); 926 (812~1002) ❙ 1355 (1221~1496); 631 (473~678); 99 (58~142) ❙ 46.5 (30.9~64.7); 926 (812~1002); 269.3 (211.8~322.1)

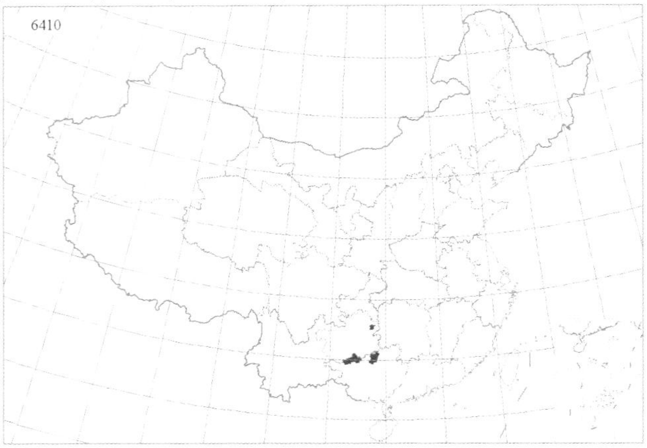

Actinidiaceae 猕猴桃科 *Actinidia* 猕猴桃属
Actinidia cylindrica var. reticulata C. F. Liang 网脉猕猴桃
Vines, evergreen

14.3; 4.0; 23.5; 14.3; 112.6; 1.1; 756 ❙ 1299; 539; 100 ❙ 71.9; 756; 225.8

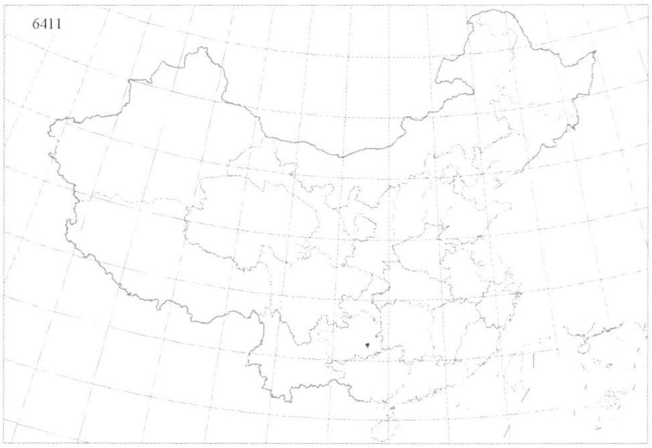

Actinidiaceae 猕猴桃科 *Actinidia* 猕猴桃属
Actinidia eriantha Benth. 毛花猕猴桃
Large climbers, deciduous

17.6 (14.3~22.0); 7.0 (2.6~13.6); 27.5 (23.5~30.1); 17.7 (14.3~22.0); 152.2 (112.6~203.8); 0.2 (0.0~3.6); 933 (756~1170) ❙ 1541 (871~1997); 537 (299~821); 186 (62~259) ❙ 66.3 (-6.7~147.0); 916 (755~1170); 224.1 (123.9~332.5)

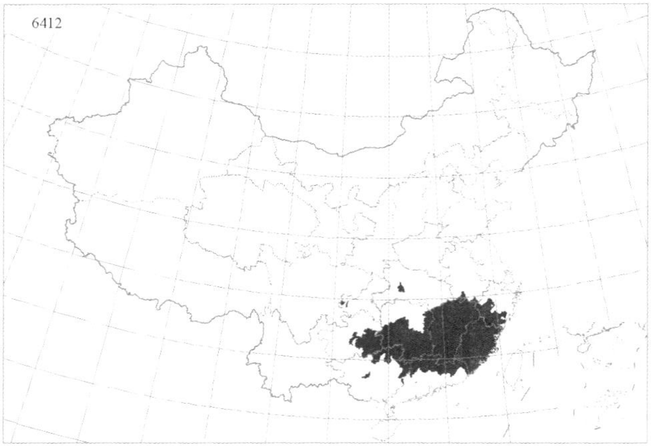

Actinidiaceae 猕猴桃科 *Actinidia* 猕猴桃属
Actinidia farinosa C. F. Liang 粉毛猕猴桃
Vines, evergreen

19.8; 11.1; 26.4; 19.8; 177.9; 0.0; 1010 ❙ 1182; 673; 43 ❙ 16.9; 1010; 354.5

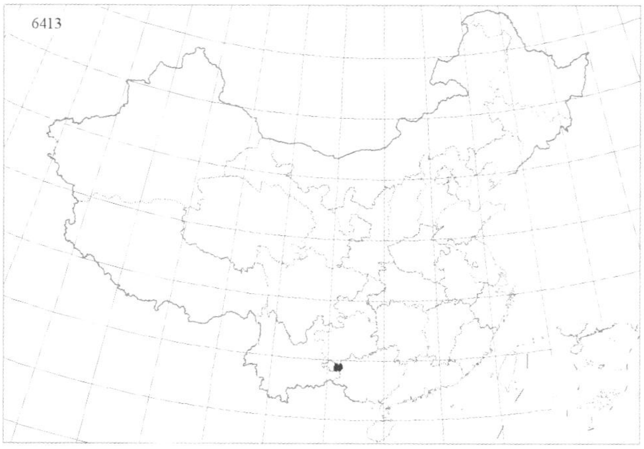

Actinidiaceae 猕猴桃科 *Actinidia* 猕猴桃属
Actinidia fasciculoides C. F. Liang 簇花猕猴桃
Large climbers, deciduous

18.3 (15.3~22.2); 9.4 (4.2~13.9); 25.4 (22.7~28.5); 18.3 (15.4~22.3); 160.0 (125.8~207.5); 0.1 (0.0~0.8); 930 (818~1190) ❙ 1377 (1161~1660); 679 (492~820); 92 (43~228) ❙ 50.2 (15.4~103.1); 929 (818~1190); 316.0 (215.0~371.7)

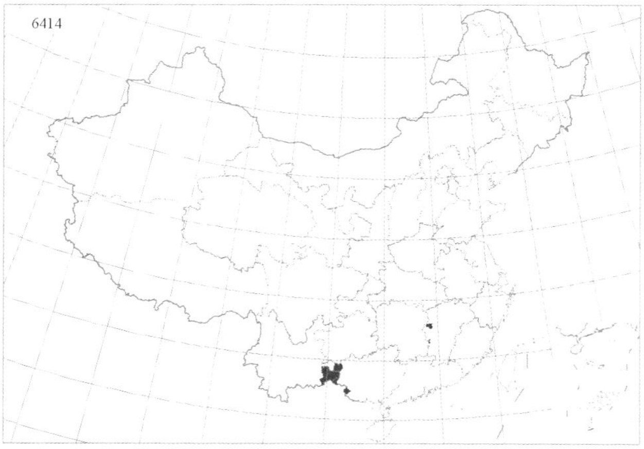

Actinidiaceae 猕猴桃科 *Actinidia* 猕猴桃属

Actinidia fasciculoides var. cuneata C. F. Liang 楔叶猕猴桃

Vines, deciduous

19.7; 10.8; 26.8; 19.8; 177.2; 0.0; 1014 | 1512; 734; 122 | 49.2; 1014; 297.8

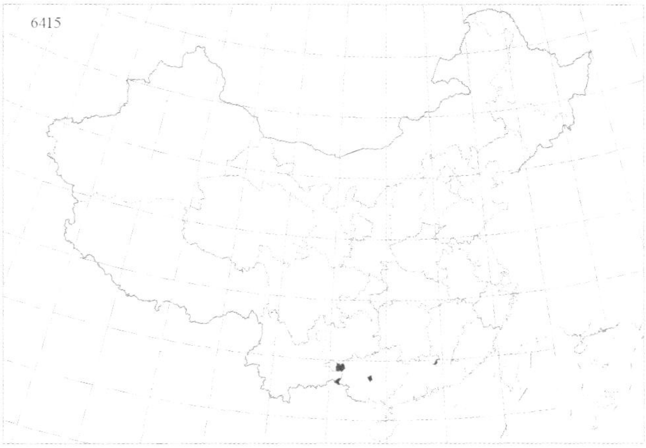

Actinidiaceae 猕猴桃科 *Actinidia* 猕猴桃属

Actinidia fasciculoides var. orbiculata C.F.Liang 圆叶猕猴桃

Vines, deciduous

22.2; 13.9; 28.3; 22.3; 207.5; 0.0; 1190 | 1374; 713; 87 | 15.4; 1190; 327.7

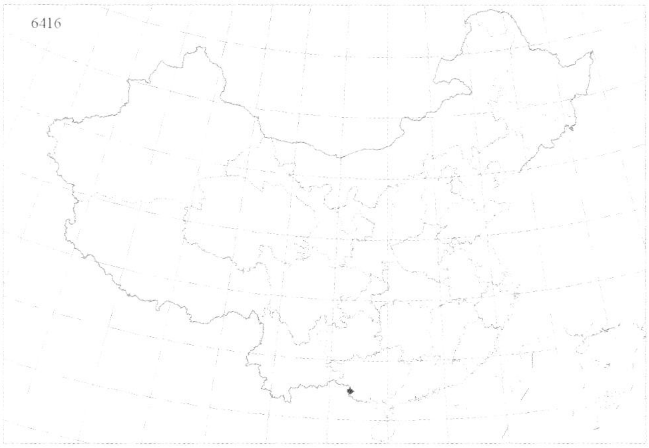

Actinidiaceae 猕猴桃科 *Actinidia* 猕猴桃属

Actinidia fortunatii Finet et Gagnep. 条叶猕猴桃

Vines, deciduous or semideciduous

18.0 (14.3~22.7); 7.5 (3.3~15.0); 27.3 (23.5~29.8); 18.0 (14.3~22.8); 156.7 (112.6~213.1); 0.2 (0.0~2.3); 952 (756~1221) | 1513 (1188~2435); 587 (392~1428); 163 (56~221) | 60.0 (5.8~128.9); 943 (756~1221); 218.6 (134.6~335.1)

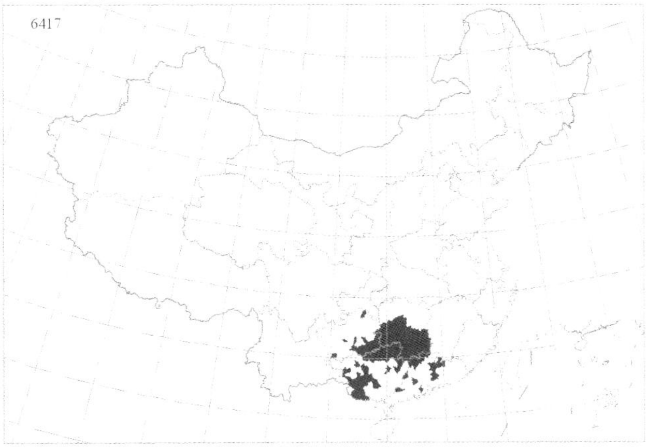

Actinidiaceae 猕猴桃科 *Actinidia* 猕猴桃属

Actinidia fulvicoma Hance 黄毛猕猴桃

Vines, semievergreen

18.6 (14.3~22.2); 8.7 (3.3~14.1); 27.2 (21.9~29.8); 18.7 (14.3~22.3); 164.3 (113.2~207.5); 0.1 (0.0~2.3); 975 (761~1190) | 1543 (871~2024); 616 (299~954); 160 (43~259) | 59.3 (-6.7~130.7); 969 (755~1190); 239.6 (145.3~411.1)

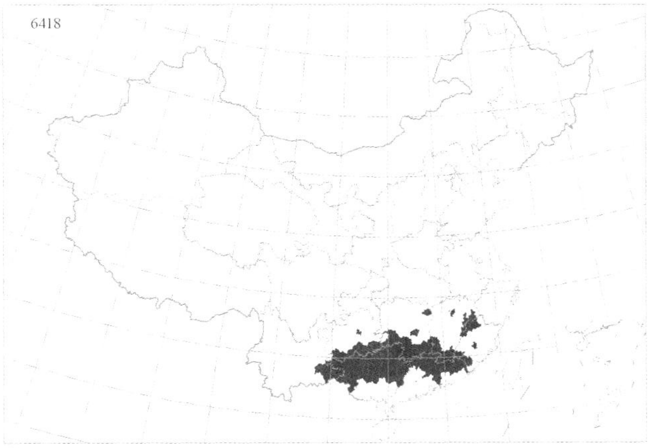

Actinidiaceae 猕猴桃科 *Actinidia* 猕猴桃属

Actinidia fulvicoma var. cinerascens (C. F. Liang) J. Q. Li et Soejarto 灰毛猕猴桃

Vines, semievergreen

18.8 (14.6~22.0); 9.1 (3.8~14.2); 27.3 (24.6~29.8); 18.9 (14.7~22.1); 166.6 (117.7~205.1); 0.0 (0.0~1.5); 984 (779~1170) | 1684 (1486~2014); 624 (474~954); 186 (116~254) | 72.1 (35.6~130.7); 982 (779~1170); 243.1 (174.9~303.4)

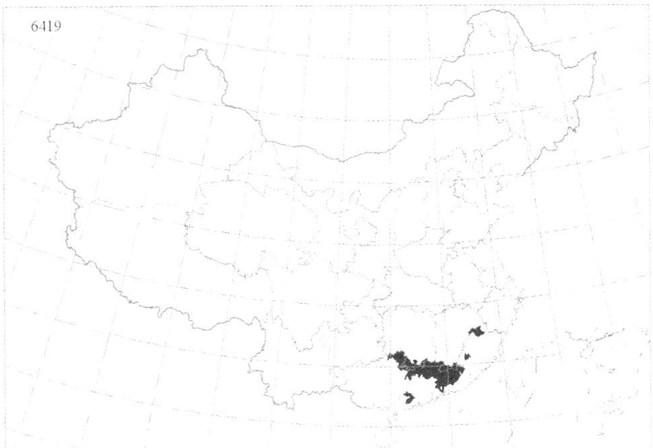

Actinidiaceae 猕猴桃科 *Actinidia* 猕猴桃属

Actinidia fulvicoma var. hirsuta Finet et Gagnepain 糙毛猕猴桃

Vines, semievergreen

17.9 (14.3~22.3); 7.4 (3.3~13.8); 27.1 (22.5~29.8); 17.9 (14.3~22.3); 154.9 (112.6~208.1); 0.1 (0.0~2.3); 940 (756~1202) | 1494 (1177~1873); 572 (392~996); 160 (43~242) | 60.0 (16.9~128.9); 931 (756~1202); 226.4 (134.6~551.3)

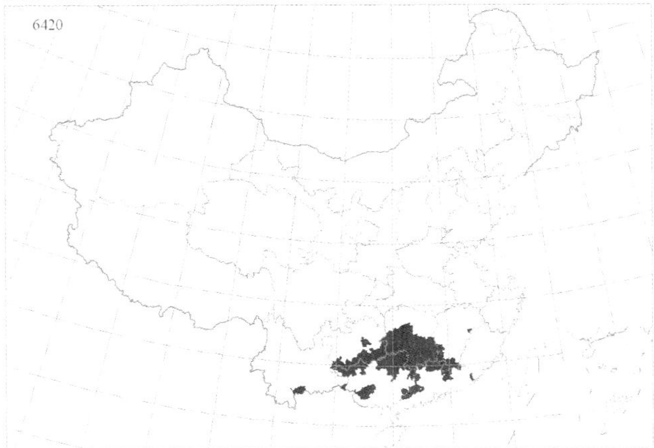

Actinidiaceae 猕猴桃科 *Actinidia* 猕猴桃属
Actinidia fulvicoma var. pachyphylla (Dunn) H. L. Li 厚叶猕猴桃
Vines, semievergreen

18.9 (17.1~20.8); 9.1 (6.4~12.2); 27.5 (26.8~28.5); 19.0 (17.1~20.8); 167.8 (145.5~189.7); 0.0 (0.0~0.0); 989 (894~1085) ❙ 1727 (1579~2005); 634 (541~827); 191 (156~230) ❙ 74.8 (57.0~104.6); 989 (894~1085); 246.1 (208.1~270.5)

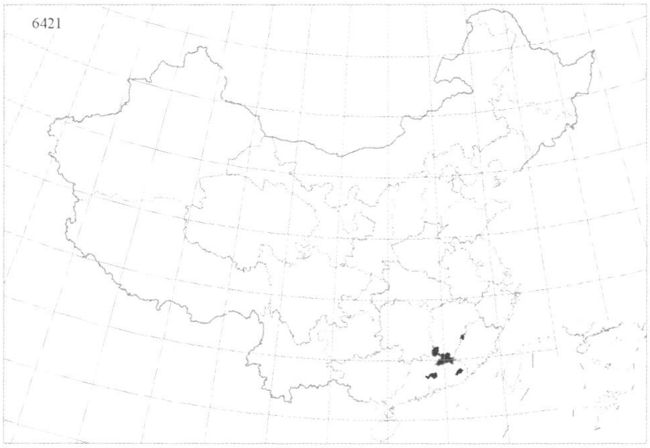

Actinidiaceae 猕猴桃科 *Actinidia* 猕猴桃属
Actinidia glaucocallosa C. Y. Wu 粉叶猕猴桃
Large climbers, deciduous

16.5 (11.4~18.9); 10.0 (2.7~12.5); 21.0 (18.8~23.2); 16.6 (11.4~19.0); 139.2 (80.7~167.5); 0.2 (0.0~3.7); 806 (652~914) ❙ 1212 (874~1454); 680 (494~816); 49 (25~70) ❙ 50.8 (12.5~99.9); 791 (652~851); 396.0 (208.2~516.7)

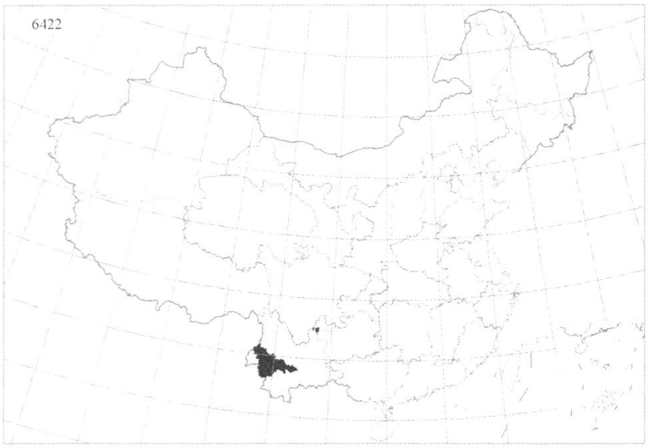

Actinidiaceae 猕猴桃科 *Actinidia* 猕猴桃属
Actinidia grandiflora C. F. Liang 大花猕猴桃
Large climbers, deciduous

13.0; 3.6; 21.5; 13.0; 98.9; 2.8; 713 ❙ 1085; 592; 36 ❙ 53.2; 713; 252.1

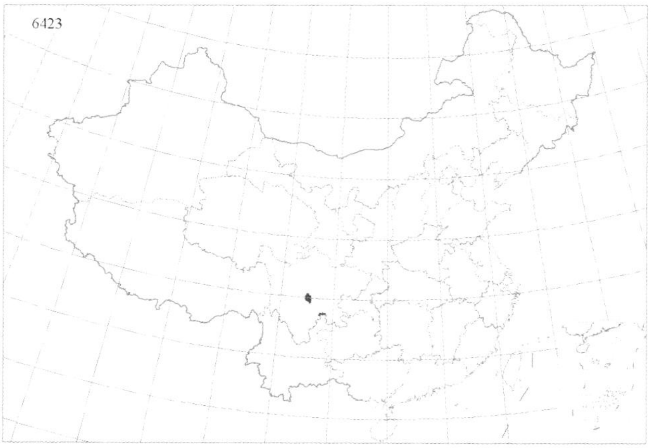

Actinidiaceae 猕猴桃科 *Actinidia* 猕猴桃属
Actinidia hemsleyana Dunn 长叶猕猴桃
Vines, deciduous

17.1 (15.0~19.1); 7.0 (4.0~10.5); 26.7 (23.0~29.1); 17.2 (15.1~19.1); 146.0 (120.8~169.1); 0.0 (0.0~1.0); 893 (789~984) ❙ 1756 (1321~1997); 612 (495~713); 210 (64~259) ❙ 97.3 (50.3~147.0); 891 (789~973); 265.0 (183.9~332.5)

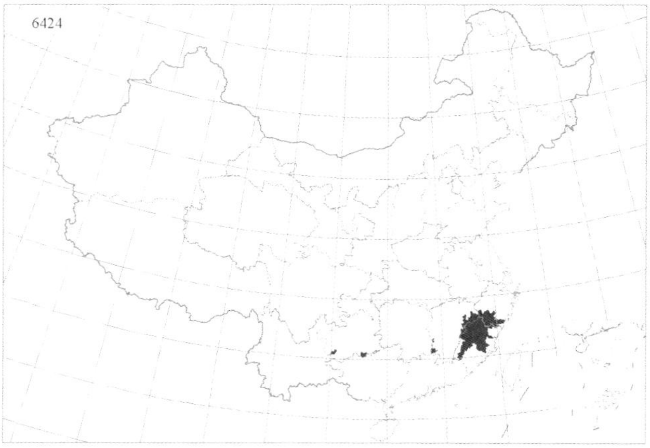

Actinidiaceae 猕猴桃科 *Actinidia* 猕猴桃属
Actinidia henryi Dunn 蒙自猕猴桃
Vines, deciduous

18.1 (14.2~23.2); 9.1 (3.3~15.8); 25.6 (21.4~28.9); 18.2 (14.2~23.2); 158.1 (112.5~218.9); 0.1 (0.0~2.3); 928 (756~1257) ❙ 1408 (1007~1776); 643 (419~996); 112 (43~198) ❙ 52.4 (5.8~101.8); 926 (756~1257); 276.6 (152.4~551.6)

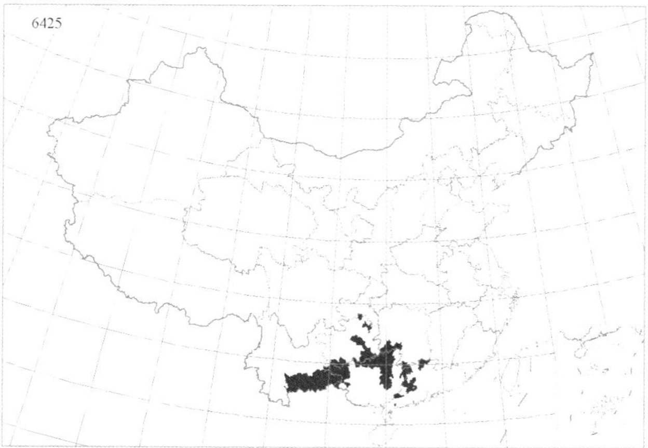

Actinidiaceae 猕猴桃科 *Actinidia* 猕猴桃属
Actinidia holotricha Finet et Gagnep. 全毛猕猴桃
Large climbers, deciduous

14.6; 6.2; 21.6; 14.6; 115.6; 0.4; 757 ❙ 1110; 620; 44 ❙ 46.1; 757; 291.0

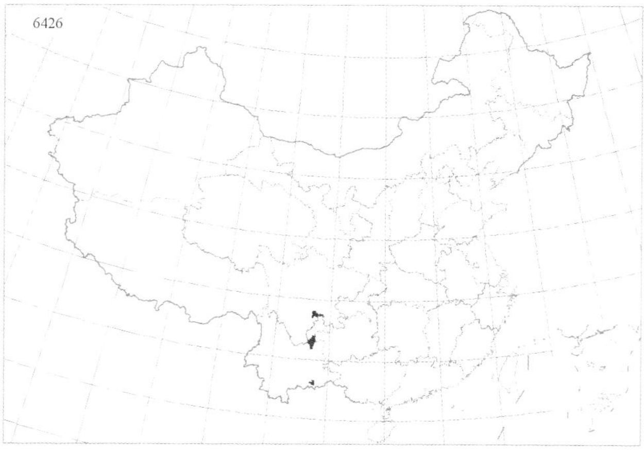

Actinidiaceae 猕猴桃科 *Actinidia* 猕猴桃属

Actinidia indochinensis Merr. 中越猕猴桃

Large climbers, deciduous

18.6 (14.6~22.7); 10.1 (3.8~15.0); 25.3 (21.4~28.6); 18.6 (14.7~22.8); 163.6 (117.7~213.1); 0.1 (0.0~1.5); 950 (779~1221) | 1508 (1074~2435); 758 (486~1428); 101 (47~191) | 58.9 (15.4~101.8); 950 (779~1221); 310.9 (190.5~551.6)

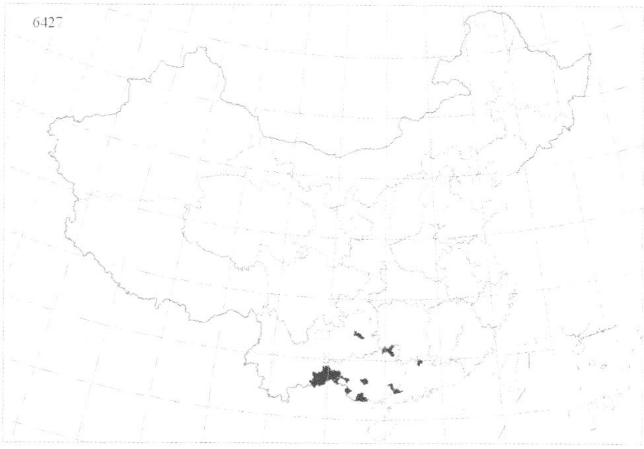

Actinidiaceae 猕猴桃科 *Actinidia* 猕猴桃属

Actinidia kolomikta (Maxim. et Rupr.) Maxim. 狗枣猕猴桃

Vines, deciduous

9.9 (-0.2~18.3); -4.7 (-23.8~8.1); 22.9 (9.9~29.1); 11.3 (3.8~18.4); 89.6 (15.5~160.7); 30.5 (0.0~112.6); 689 (413~965) | 796 (377~1770); 430 (241~817); 33 (8~223) | 16.3 (-37.6~125.9); 636 (377~965); 224.5 (109.4~346.5)

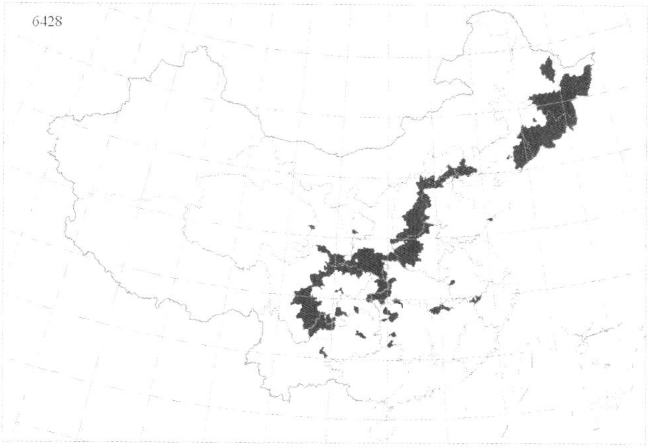

Actinidiaceae 猕猴桃科 *Actinidia* 猕猴桃属

Actinidia laevissima C. F. Liang 滑叶猕猴桃

Vines, deciduous

15.0; 4.0; 25.7; 15.1; 122.0; 1.1; 806 | 1247; 484; 89 | 54.7; 806; 189.9

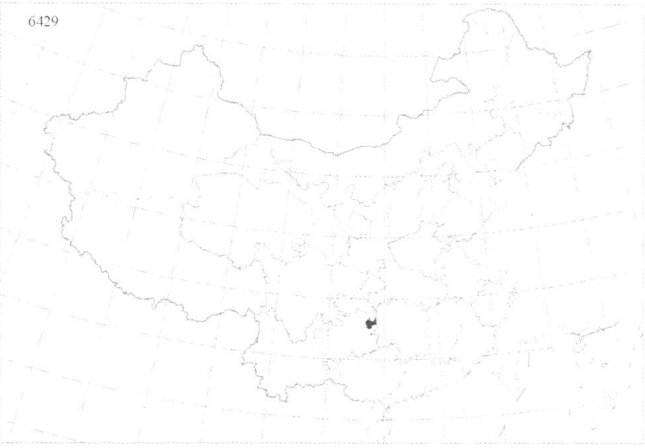

Actinidiaceae 猕猴桃科 *Actinidia* 猕猴桃属

Actinidia lanceolata Dunn 小叶猕猴桃

Vines, deciduous

17.1 (14.3~21.0); 5.9 (2.1~12.2); 27.7 (24.2~30.1); 17.1 (14.3~21.0); 145.7 (113.2~192.6); 0.3 (0.0~4.9); 912 (761~1107) | 1590 (871~1997); 550 (299~713); 197 (81~259) | 75.3 (-6.7~147.0); 897 (755~1107); 232.6 (147.5~332.5)

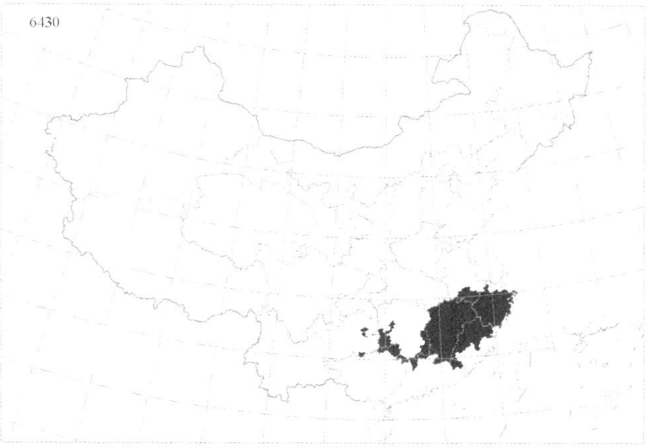

Actinidiaceae 猕猴桃科 *Actinidia* 猕猴桃属

Actinidia latifolia (Gardner et Champ.) Merr. 阔叶猕猴桃

Large climbers, deciduous

18.8 (12.1~24.8); 9.0 (1.1~20.1); 27.5 (17.9~30.1); 18.8 (12.2~24.8); 166.1 (91.6~237.6); 0.2 (0.0~8.5); 993 (678~1385) | 1537 (862~2992); 601 (312~1847); 158 (41~259) | 56.7 (-34.3~246.5); 979 (678~1273); 233.7 (95.3~620.0)

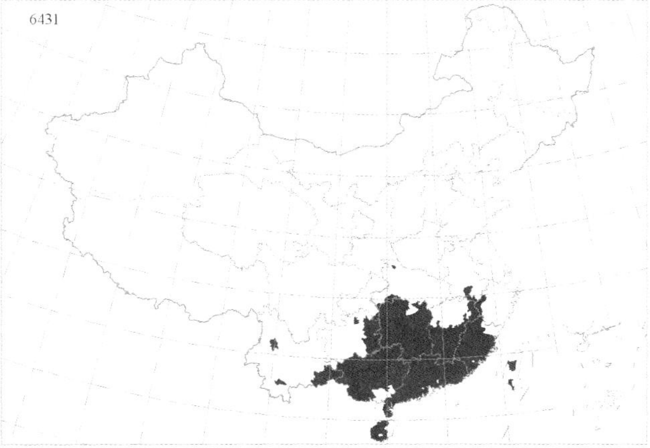

Actinidiaceae 猕猴桃科 *Actinidia* 猕猴桃属

Actinidia latifolia var. mollis (Dunn) Hand.-Mazz. 长绒猕猴桃

Vines, deciduous

18.3 (16.3~21.0); 11.5 (9.1~14.2); 23.0 (21.4~25.6); 18.3 (16.4~21.0); 159.9 (136.4~191.8); 0.0 (0.0~0.0); 886 (797~1053) | 1350 (1074~1751); 759 (582~996); 56 (47~77) | 52.4 (21.7~88.8); 882 (797~1053); 408.4 (277.0~551.6)

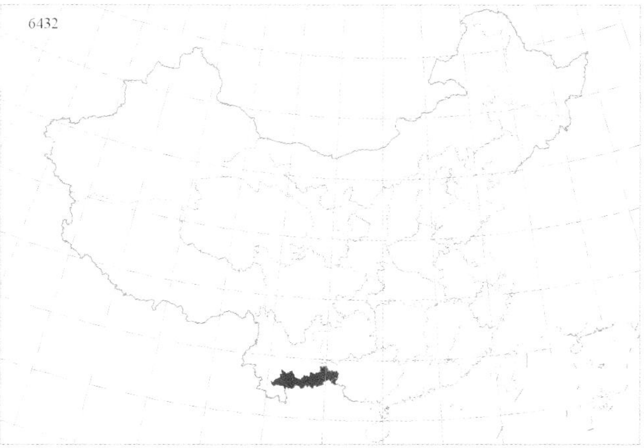

Actinidiaceae 猕猴桃科 *Actinidia* 猕猴桃属
Actinidia liangguangensis C. F. Liang 两广猕猴桃
Large climbers, evergreen

18.6 (14.6~21.4); 8.8 (3.8~13.1); 26.9 (24.6~28.2); 18.6 (14.7~21.4); 163.6 (117.7~196.9); 0.1 (0.0~1.5); 968 (779~1121) ❙ 1565 (1486~1781); 603 (505~734); 169 (137~191) ❙ 63.2 (44.0~101.8); 968 (779~1121); 239.2 (188.8~263.5)

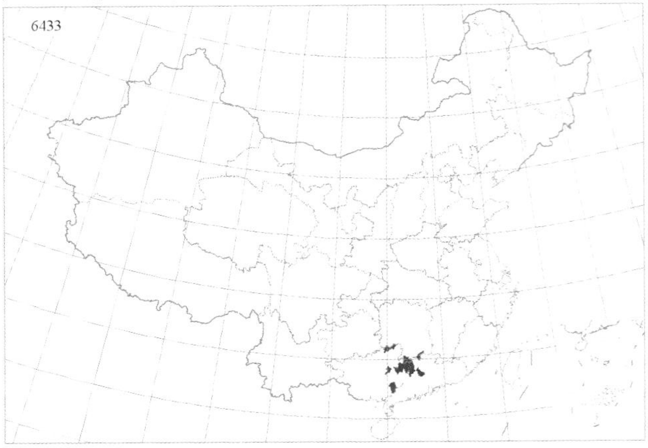

Actinidiaceae 猕猴桃科 *Actinidia* 猕猴桃属
Actinidia macrosperma C. F. Liang 大籽猕猴桃
Vines, deciduous

16.0 (13.4~18.2); 3.9 (1.5~7.7); 27.8 (25.2~29.3); 16.1 (13.4~18.2); 134.6 (108.3~158.8); 1.7 (0.0~7.3); 880 (755~984) ❙ 1486 (994~1922); 554 (448~685); 180 (79~259) ❙ 69.1 (13.5~111.3); 878 (755~961); 223.9 (134.1~319.9)

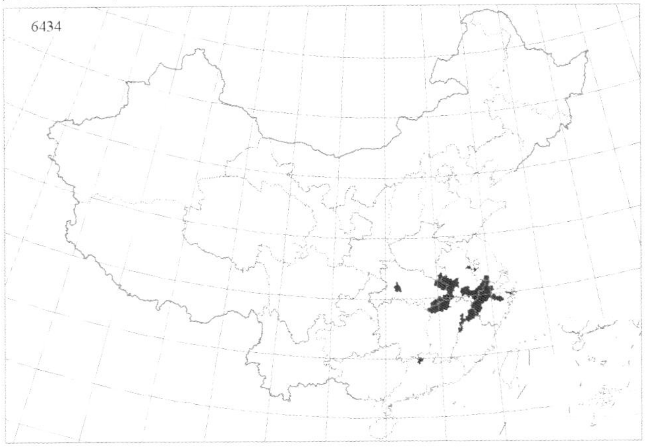

Actinidiaceae 猕猴桃科 *Actinidia* 猕猴桃属
Actinidia macrosperma var. mumoides C. F. Liang 梅叶猕猴桃
Vines, deciduous

16.6 (15.0~18.5); 4.1 (1.8~6.6); 28.6 (26.8~29.7); 16.6 (15.0~18.5); 140.8 (123.9~162.3); 1.3 (0.0~5.0); 914 (829~1004) ❙ 1461 (994~1732); 558 (464~685); 170 (102~239) ❙ 59.9 (13.5~95.5); 908 (829~950); 215.1 (147.5~299.5)

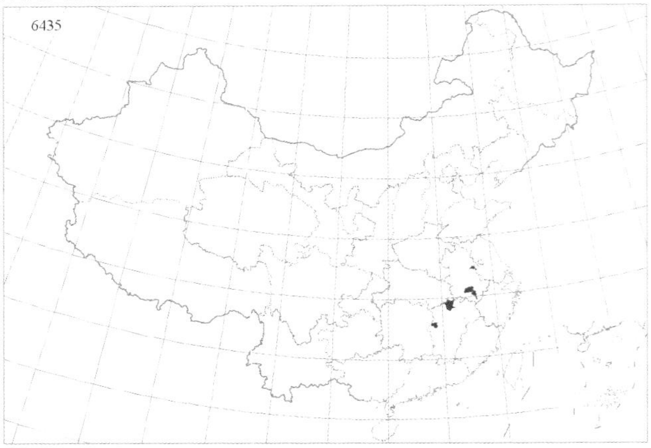

Actinidiaceae 猕猴桃科 *Actinidia* 猕猴桃属
Actinidia melanandra Franch. 黑蕊猕猴桃
Vines, deciduous

14.3 (5.4~19.2); 3.1 (-4.9~12.0); 24.7 (12.3~29.6); 14.4 (5.8~19.2); 117.6 (31.5~170.8); 5.1 (0.0~29.4); 793 (481~1002) ❙ 1177 (512~1997); 501 (258~817); 97 (15~259) ❙ 47.1 (-35.9~147.0); 773 (481~1002); 234.8 (121.6~491.5)

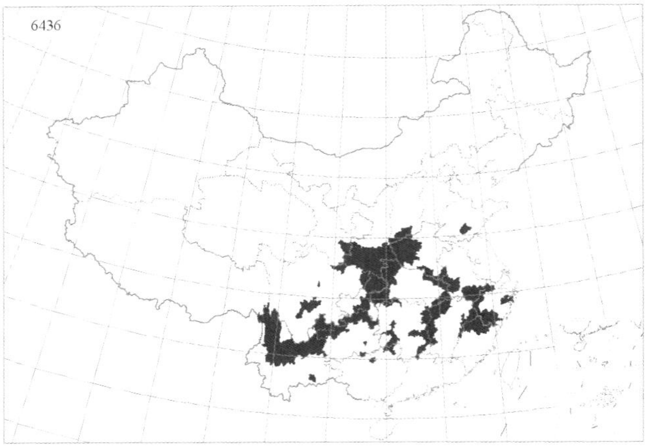

Actinidiaceae 猕猴桃科 *Actinidia* 猕猴桃属
Actinidia melanandra var. glabrescens C. F. Liang 无髯猕猴桃
Vines, deciduous

15.5 (14.4~17.7); 3.3 (2.1~5.3); 27.2 (25.7~29.3); 15.6 (14.5~17.7); 129.6 (118.7~152.4); 2.6 (0.0~4.9); 854 (801~964) ❙ 1596 (1448~1689); 584 (490~617); 188 (105~212) ❙ 87.6 (57.7~105.7); 853 (801~961); 253.7 (166.9~319.9)

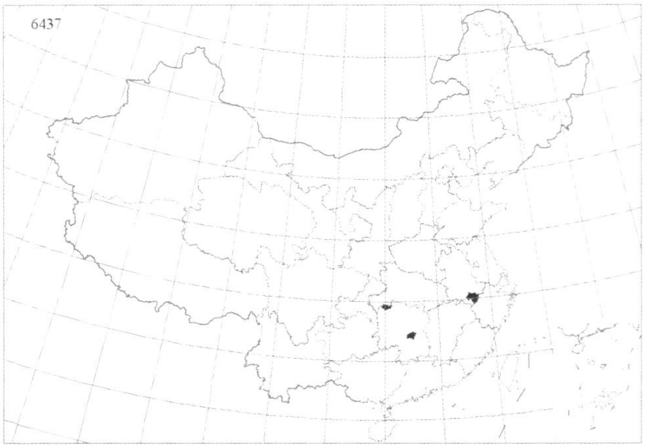

Actinidiaceae 猕猴桃科 *Actinidia* 猕猴桃属
Actinidia melliana Hand.-Mazz. 美丽猕猴桃
Vines, evergreen

19.3 (14.6~23.9); 9.6 (3.8~18.7); 27.5 (24.6~29.8); 19.3 (14.7~23.9); 172.0 (117.7~226.8); 0.0 (0.0~1.5); 1013 (779~1294) ❙ 1606 (1371~1961); 631 (455~891); 168 (48~242) ❙ 60.4 (6.7~106.4); 1007 (779~1172); 243.5 (166.9~572.9)

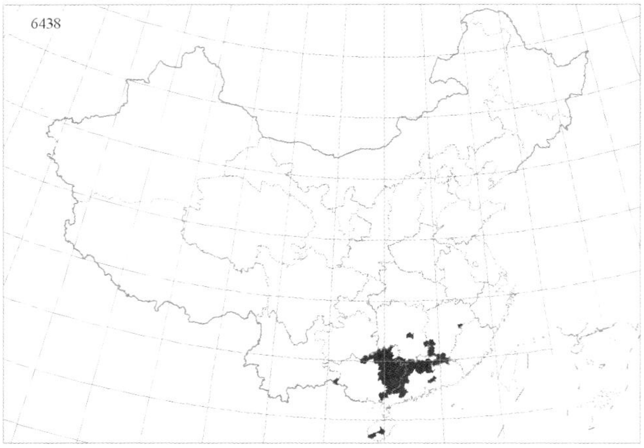

Actinidiaceae 猕猴桃科 *Actinidia* 猕猴桃属
Actinidia obovata Chun ex C. F. Liang 倒卵叶猕猴桃
Large climbers, deciduous

12.5 (1.8~16.8); 3.1 (-7.1~7.8); 20.7 (9.9~25.8); 12.9 (3.8~16.8); 101.3 (15.7~141.5); 10.8 (0.0~53.3); 712 (404~836) ❙ 1213 (583~1736); 622 (348~1007); 56 (13~88) ❙ 69.9 (44.2~145.8); 712 (404~836); 243.2 (160.8~447.4)

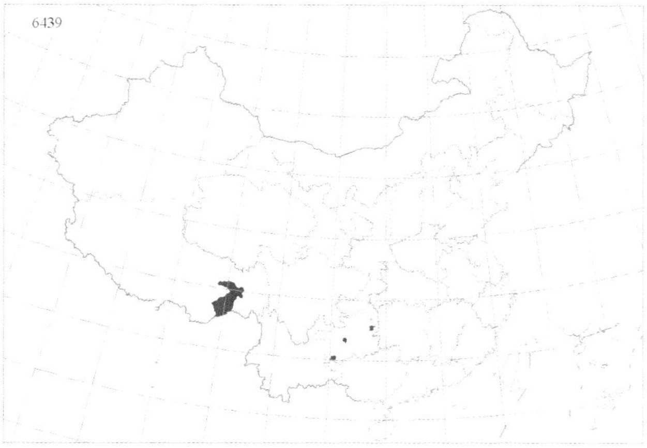

Actinidiaceae 猕猴桃科 *Actinidia* 猕猴桃属
Actinidia pilosula (Finet et Gagnep.) Stapf ex Hand.-Mazz. 贡山猕猴桃
Vines, deciduous

12.5 (5.3~16.7); 5.3 (-2.4~10.0); 18.2 (12.2~21.3); 12.6 (5.7~16.7); 94.7 (30.5~140.8); 4.2 (0.0~26.2); 681 (479~809) ❙ 1109 (741~1454); 614 (426~792); 50 (28~70) ❙ 64.2 (26.5~100.7); 681 (479~809); 352.7 (234.2~487.2)

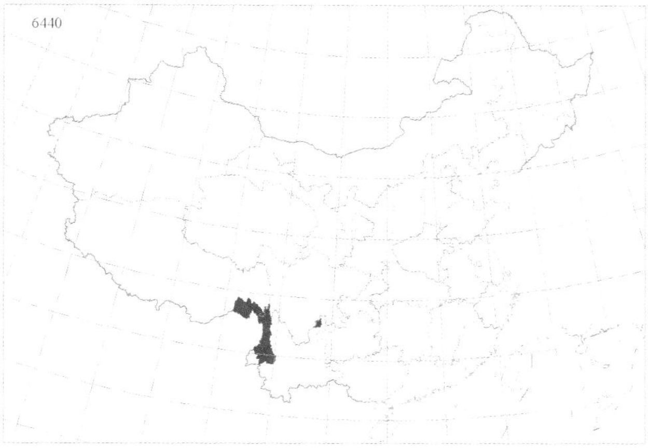

Actinidiaceae 猕猴桃科 *Actinidia* 猕猴桃属
Actinidia polygama (Siebold et Zucc.) Maxim. 葛枣猕猴桃
Large climbers, deciduous

11.5 (1.4~18.2); -2.4 (-20.5~8.5); 24.1 (9.9~29.3); 12.5 (3.8~18.2); 101.1 (15.5~158.1); 22.3 (0.0~93.6); 735 (413~965) ❙ 960 (377~1959); 450 (241~695); 68 (8~250) ❙ 29.0 (-37.2~137.5); 698 (377~961); 224.5 (109.4~346.5)

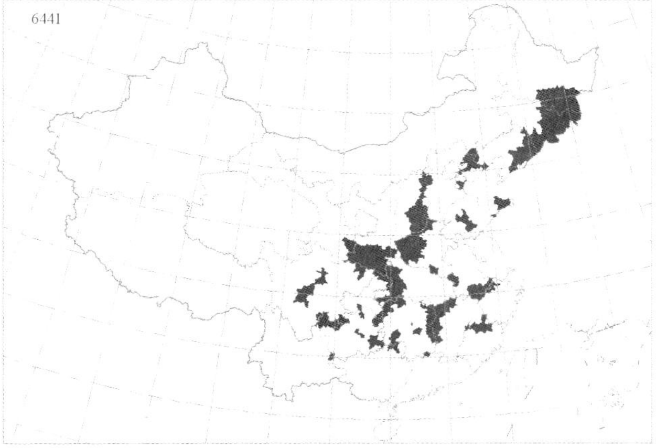

Actinidiaceae 猕猴桃科 *Actinidia* 猕猴桃属
Actinidia rubricaulis Dunn 红茎猕猴桃
Vines, semievergreen

16.3 (5.7~21.0); 6.2 (-3.5~14.2); 25.4 (14.2~29.3); 16.3 (6.4~21.0); 137.0 (38.1~191.8); 1.0 (0.0~28.1); 857 (504~1053) ❙ 1264 (797~1865); 570 (411~996); 88 (23~258) ❙ 48.5 (-14.2~112.2); 855 (504~1053); 232.4 (108.2~551.6)

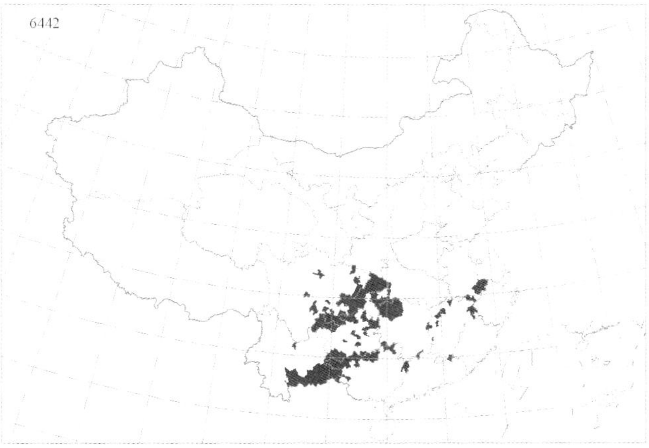

Actinidiaceae 猕猴桃科 *Actinidia* 猕猴桃属
Actinidia rubricaulis var. coriacea (Finet et Gagnep.) C. F. Liang 革叶猕猴桃
Vines, semievergreen

15.7 (5.7~21.3); 4.2 (-3.5~12.6); 26.3 (14.2~29.3); 15.7 (6.4~21.3); 131.8 (38.1~196.0); 3.2 (0.0~28.1); 854 (504~1123) ❙ 1302 (626~1950); 522 (316~765); 120 (25~259) ❙ 52.3 (-22.2~121.9); 842 (504~1121); 216.7 (139.1~354.5)

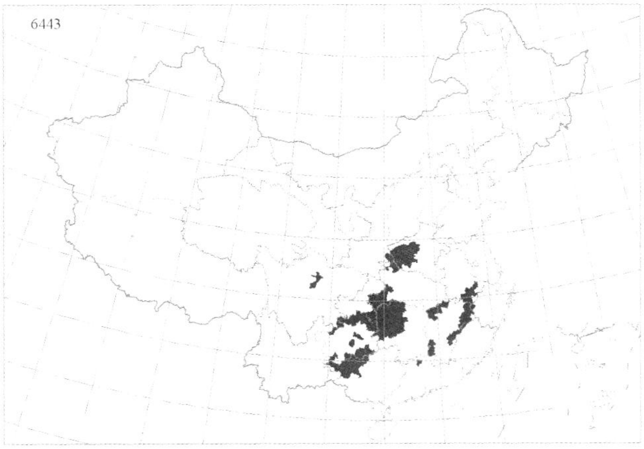

Actinidiaceae 猕猴桃科 *Actinidia* 猕猴桃属
Actinidia rubus H. Lév. 昭通猕猴桃
Vines, deciduous

13.1 (9.0~16.3); 3.7 (0.6~6.3); 21.4 (16.5~25.4); 13.1 (9.0~16.3); 99.7 (58.7~136.0); 2.2 (0.0~10.3); 715 (584~844) ❙ 992 (863~1220); 534 (472~688); 42 (25~64) ❙ 40.2 (22.6~101.6); 715 (584~844); 233.5 (179.0~293.1)

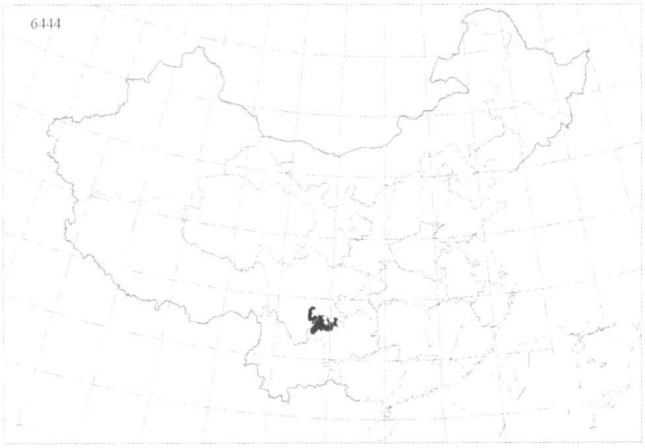

Actinidiaceae 猕猴桃科 *Actinidia* 猕猴桃属
Actinidia rudis Dunn 糙叶猕猴桃
Climbing shrubs, deciduous

18.6 (16.3~21.0); 12.2 (9.1~15.1); 22.8 (21.4~25.6); 18.6 (16.4~21.0); 163.3 (136.4~191.8); 0.0 (0.0~0.0); 896 (797~1053) ❙ 1420 (1074~1751); 753 (582~996); 57 (41~77) ❙ 58.6 (32.4~88.8); 885 (797~1053); 463.1 (277.0~659.0)

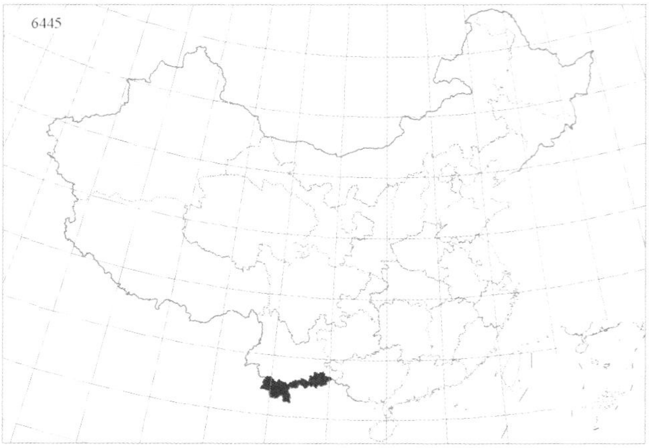

Actinidiaceae 猕猴桃科 *Actinidia* 猕猴桃属
Actinidia rudis var. glabricaulis C. Y. Wu 光茎猕猴桃
Climbing shrubs, deciduous

19.0 (18.1~19.3); 12.9 (12.0~13.5); 23.0 (22.1~23.6); 19.0 (18.1~19.4); 168.2 (157.3~173.2); 0.0 (0.0~0.0); 911 (863~939) ❙ 1404 (1143~1751); 809 (637~996); 59 (50~77) ❙ 54.2 (21.7~88.8); 902 (863~927); 451.2 (346.9~551.3)

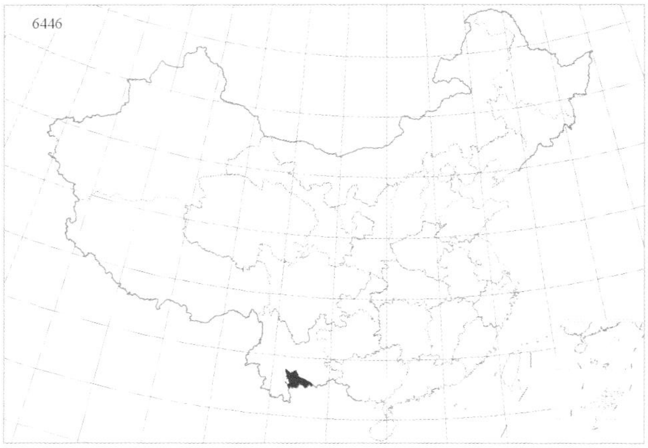

Actinidiaceae 猕猴桃科 *Actinidia* 猕猴桃属
Actinidia rufotricha C. Y. Wu 红毛猕猴桃
Vines, semievergreen

18.0 (16.3~20.3); 9.5 (7.8~10.9); 24.4 (21.4~27.7); 18.0 (16.4~20.4); 156.6 (136.4~184.4); 0.0 (0.0~0.0); 900 (797~1065) ❙ 1319 (1074~1567); 693 (561~860); 72 (47~168) ❙ 46.9 (30.5~80.6); 900 (797~1065); 307.3 (226.6~413.5)

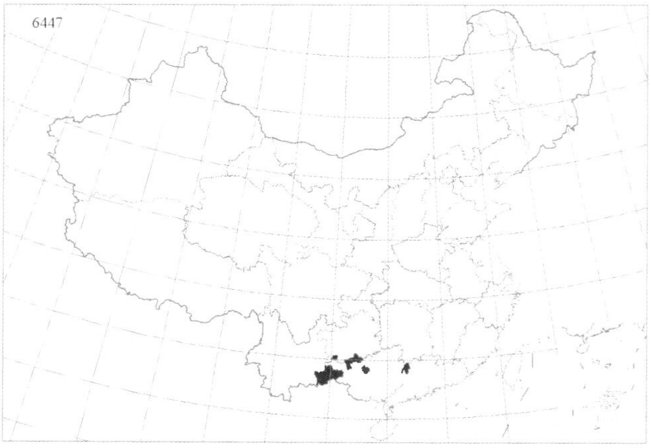

Actinidiaceae 猕猴桃科 *Actinidia* 猕猴桃属
Actinidia rufotricha var. glomerata C. F. Liang 密花猕猴桃
Vines, semievergreen

18.3; 9.3; 25.4; 18.4; 160.2; 0.0; 925 ❙ 1282; 681; 64 ❙ 39.3; 925; 289.6

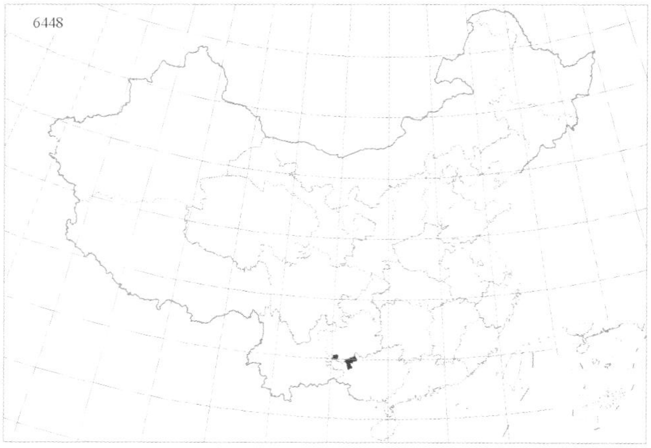

Actinidiaceae 猕猴桃科 *Actinidia* 猕猴桃属
Actinidia sabiifolia Dunn 清风藤猕猴桃
Vines, deciduous

18.1 (14.6~22.7); 7.6 (2.8~15.0); 27.5 (24.6~29.3); 18.1 (14.7~22.8); 157.8 (117.7~213.1); 0.2 (0.0~2.7); 961 (779~1221) ❙ 1690 (1302~2435); 658 (448~1428); 187 (57~259) ❙ 77.7 (12.9~130.7); 959 (779~1221); 254.4 (170.8~371.7)

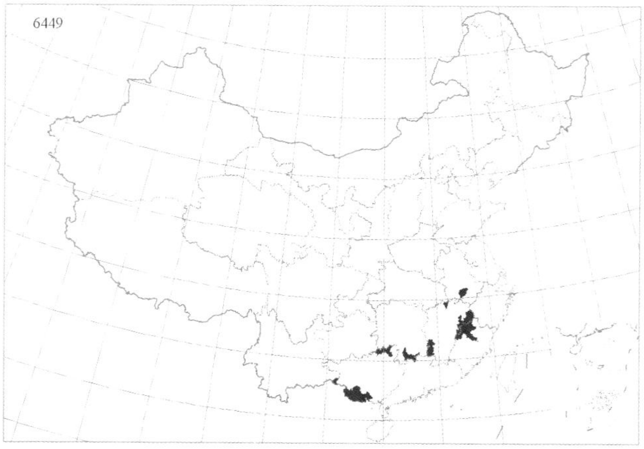

Actinidiaceae 猕猴桃科 *Actinidia* 猕猴桃属
Actinidia sorbifolia C. F. Liang 花楸猕猴桃
Large climbers, deciduous

15.5 (14.3~17.4); 4.9 (3.3~7.8); 25.3 (23.5~26.7); 15.6 (14.3~17.4); 127.4 (112.6~149.1); 0.8 (0.0~2.3); 819 (756~903) ❙ 1374 (1185~1624); 549 (454~720); 124 (53~191) ❙ 68.3 (35.9~101.8); 819 (756~903); 216.6 (154.2~258.0)

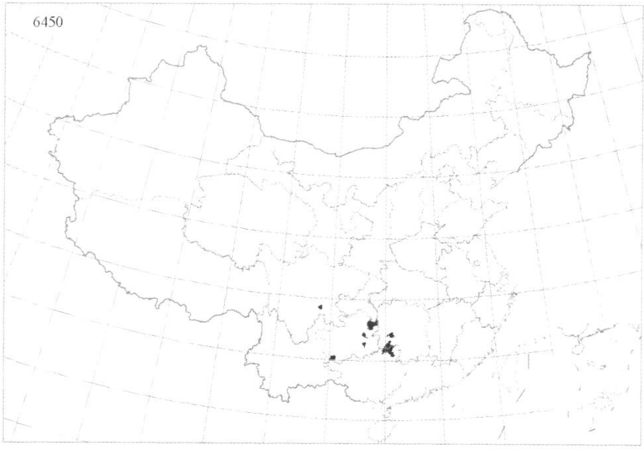

Actinidiaceae 猕猴桃科 *Actinidia* 猕猴桃属

Actinidia stellatopilosa C. Y. Chang 星毛猕猴桃

Large climbers, deciduous

10.6; −0.2; 21.1; 10.6; 79.4; 12.1; 643 ❙ 1246; 525; 46 ❙ 93.8; 643; 284.3

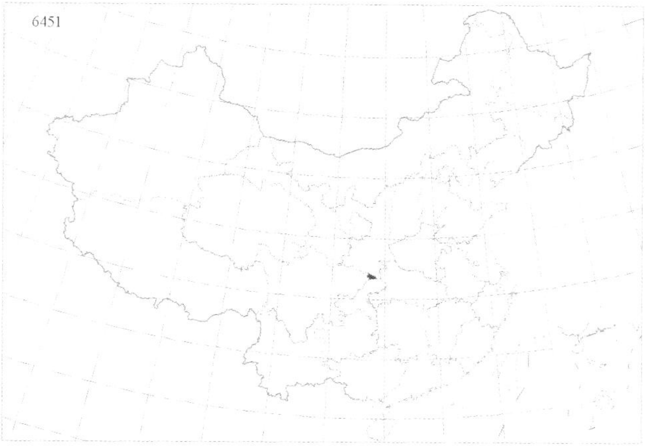

Actinidiaceae 猕猴桃科 *Actinidia* 猕猴桃属

Actinidia styracifolia C. F. Liang 安息香猕猴桃

Vines, deciduous

17.1 (15.0~18.6); 6.5 (4.2~9.9); 27.0 (24.3~28.7); 17.1 (15.1~18.7); 145.7 (120.8~164.0); 0.1 (0.0~0.8); 900 (789~975) ❙ 1769 (1273~1997); 600 (452~713); 221 (124~259) ❙ 97.3 (46.2~147.0); 899 (789~975); 256.9 (186.2~305.7)

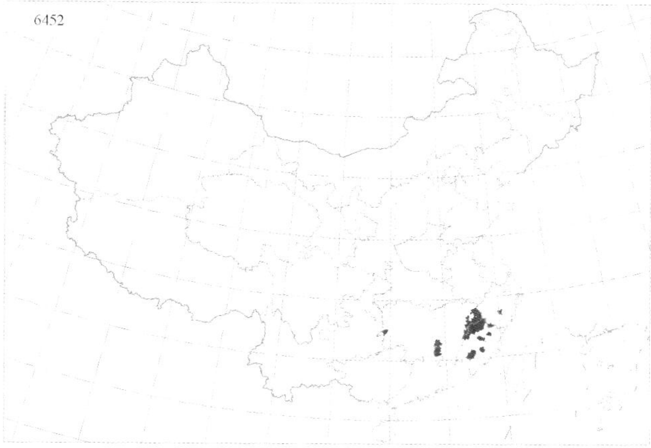

Actinidiaceae 猕猴桃科 *Actinidia* 猕猴桃属

Actinidia suberifolia C. Y. Wu 栓叶猕猴桃

Large climbers, deciduous

18.6; 11.8; 23.2; 18.6; 162.6; 0.0; 907 ❙ 1356; 740; 59 ❙ 49.8; 907; 475.5

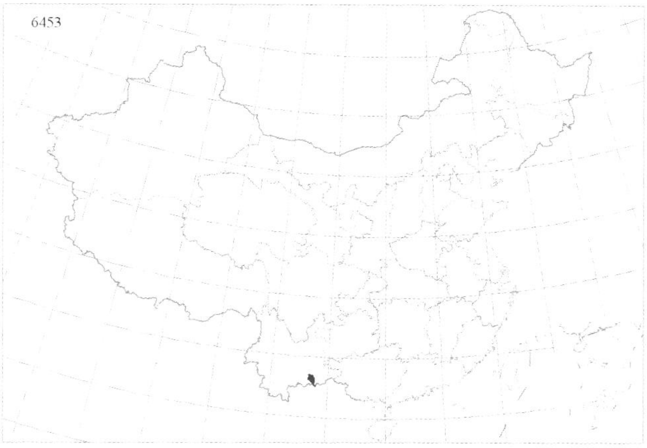

Actinidiaceae 猕猴桃科 *Actinidia* 猕猴桃属

Actinidia tetramera Maxim. 四萼猕猴桃

Vines, deciduous

10.5 (2.0~15.2); −1.7 (−9.7~5.3); 21.8 (9.9~27.5); 10.9 (3.8~15.3); 85.3 (15.4~128.0); 19.0 (0.0~56.4); 672 (413~865) ❙ 811 (341~1492); 390 (190~816); 32 (7~109) ❙ 24.1 (−40.9~128.6); 635 (341~858); 223.9 (61.2~357.7)

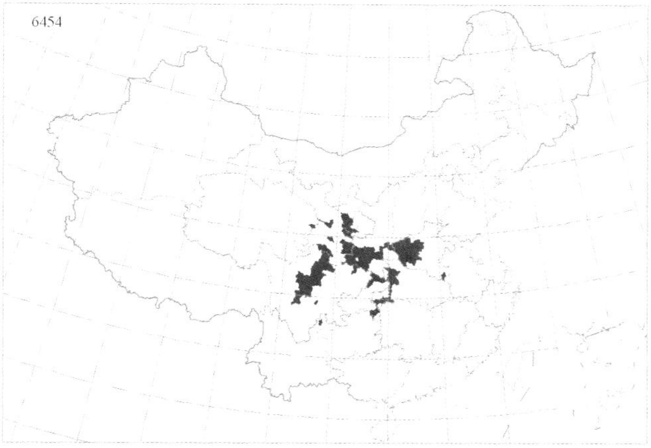

Actinidiaceae 猕猴桃科 *Actinidia* 猕猴桃属

Actinidia trichogyna Franch. 毛蕊猕猴桃

Vines, deciduous

15.3 (10.6~21.2); 4.4 (−0.2~12.8); 25.5 (21.1~28.7); 15.3 (10.6~21.3); 126.0 (79.4~195.4); 2.1 (0.0~12.1); 825 (643~1115) ❙ 1468 (1014~1922); 573 (475~720); 128 (20~259) ❙ 78.2 (27.3~111.3); 822 (643~1115); 235.7 (161.9~330.0)

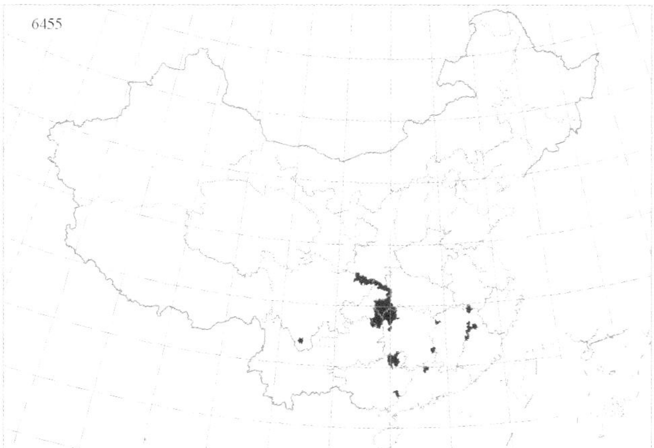

Actinidiaceae 猕猴桃科 *Actinidia* 猕猴桃属

Actinidia ulmifolia C. F. Liang 榆叶猕猴桃

Vines, deciduous

16.3 (13.8~19.2); 7.9 (5.3~13.8); 23.5 (19.3~26.5); 16.4 (13.8~19.2); 136.5 (105.9~170.5); 0.0 (0.0~0.0); 831 (714~909) ❙ 1132 (963~1499); 591 (528~686); 51 (29~60) ❙ 36.0 (20.8~64.9); 821 (690~900); 264.9 (153.3~513.9)

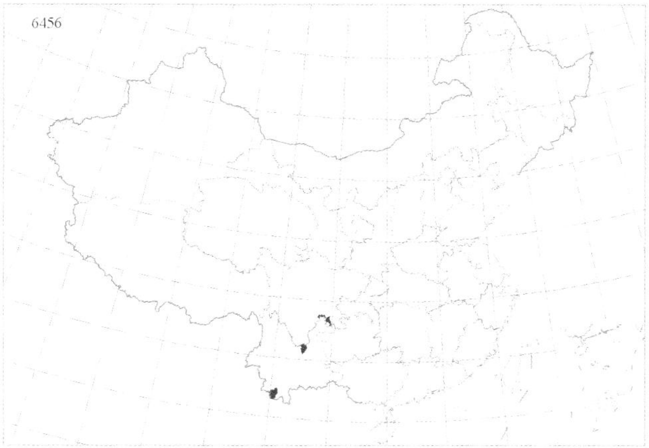

Actinidiaceae 猕猴桃科 *Actinidia* 猕猴桃属

Actinidia umbelloides C. F. Liang 伞花猕猴桃

Large climbers, deciduous

17.4 (14.4~19.6); 11.1 (7.5~13.9); 21.4 (19.1~22.7); 17.4 (14.4~19.6); 148.8 (113.1~175.1); 0.0 (0.0~0.0); 836 (727~933) ❙ 1306 (1049~1500); 710 (594~816); 51 (41~70) ❙ 56.4 (26.1~99.9); 818 (727~891); 427.3 (335.1~530.1)

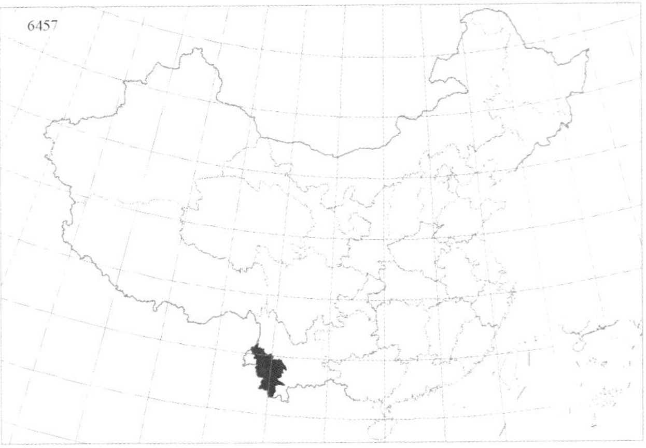

Actinidiaceae 猕猴桃科 *Actinidia* 猕猴桃属

Actinidia valvata Dunn 对萼猕猴桃

Climbing shrubs, deciduous

15.8 (10.2~18.5); 3.8 (-2.9~8.6); 27.3 (22.4~29.7); 15.8 (10.6~18.5); 132.0 (83.1~162.1); 2.2 (0.0~20.5); 864 (656~984) ❙ 1383 (622~1997); 527 (298~713); 158 (26~259) ❙ 59.9 (-13.4~147.0); 861 (622~961); 212.1 (131.2~319.9)

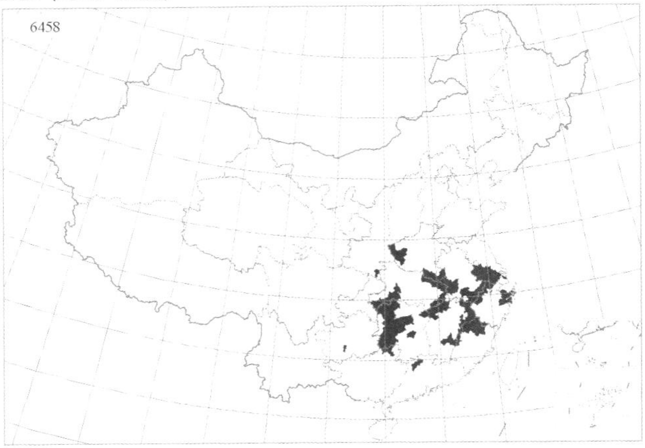

Actinidiaceae 猕猴桃科 *Actinidia* 猕猴桃属

Actinidia venosa Rehder 显脉猕猴桃

Vines, deciduous

12.9 (-0.7~18.5); 4.7 (-9.6~11.9); 19.7 (7.7~26.9); 13.0 (2.5~18.3); 100.4 (6.8~160.0); 5.7 (0.0~74.2); 705 (338~906) ❙ 1077 (370~1736); 599 (262~1007); 42 (9~107) ❙ 54.3 (4.7~145.8); 699 (338~905); 291.2 (61.6~516.7)

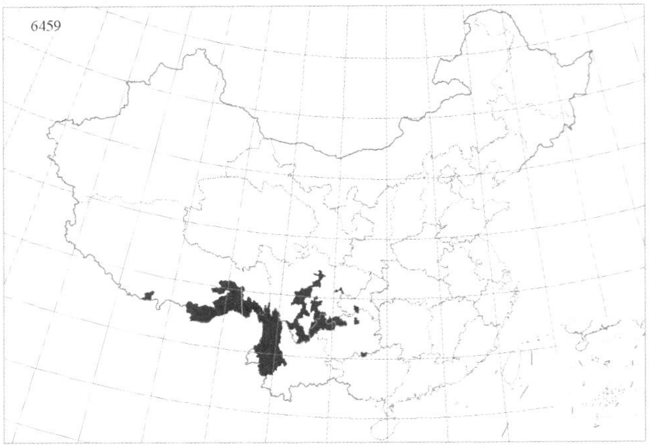

Actinidiaceae 猕猴桃科 *Actinidia* 猕猴桃属

Actinidia vitifolia C. Y. Wu 葡萄叶猕猴桃

Vines, deciduous

13.8 (11.1~16.3); 4.4 (2.2~6.4); 22.3 (18.4~25.6); 13.9 (11.3~16.3); 107.5 (80.7~136.0); 1.2 (0.0~5.3); 741 (652~845) ❙ 1019 (863~1337); 536 (472~727); 47 (25~69) ❙ 38.3 (19.4~92.4); 741 (652~845); 227.6 (154.4~407.1)

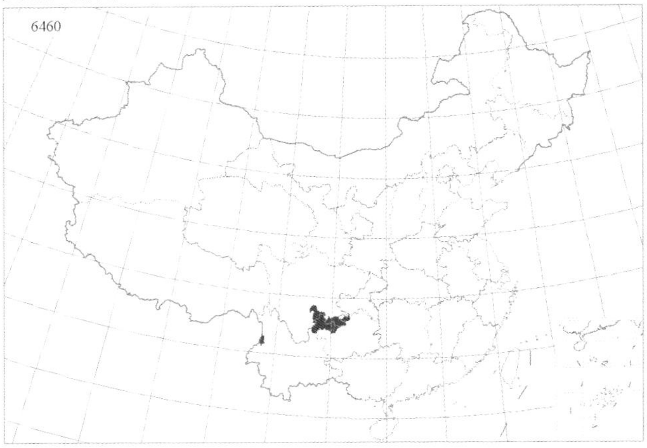

Actinidiaceae 猕猴桃科 *Actinidia* 猕猴桃属

Actinidia zhejiangensis C. F. Liang 浙江猕猴桃

Vines

16.9 (15.0~18.2); 5.9 (5.0~7.1); 27.3 (24.7~29.1); 16.9 (15.1~18.2); 142.9 (120.8~158.8); 0.0 (0.0~0.0); 898 (789~984) ❙ 1832 (1698~1950); 613 (535~713); 235 (181~259) ❙ 105.3 (78.7~147.0); 889 (789~927); 233.8 (183.9~283.9)

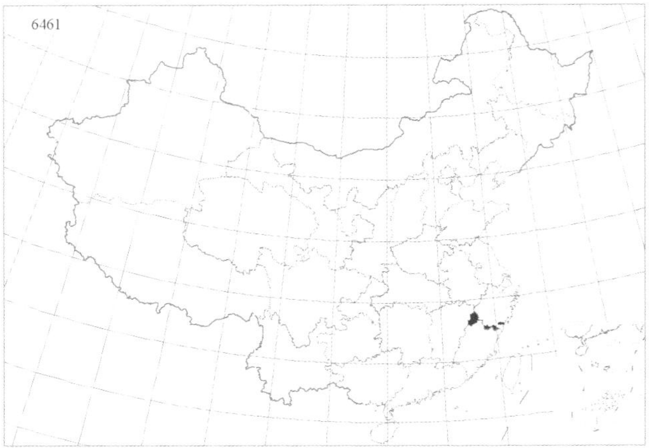

Actinidiaceae 猕猴桃科 *Clematoclethra* 藤山柳属

Clematoclethra scandens (Franch.) Maxim. 藤山柳

Vines, deciduous

13.0 (2.2~17.0); 2.9 (-6.6~6.2); 22.4 (9.9~27.6); 13.1 (3.8~17.0); 101.6 (15.5~144.4); 4.8 (0.0~49.1); 727 (413~902) ❙ 1132 (632~1660); 531 (306~817); 67 (14~216) ❙ 55.8 (-8.1~125.9); 726 (413~902); 236.0 (147.3~322.4)

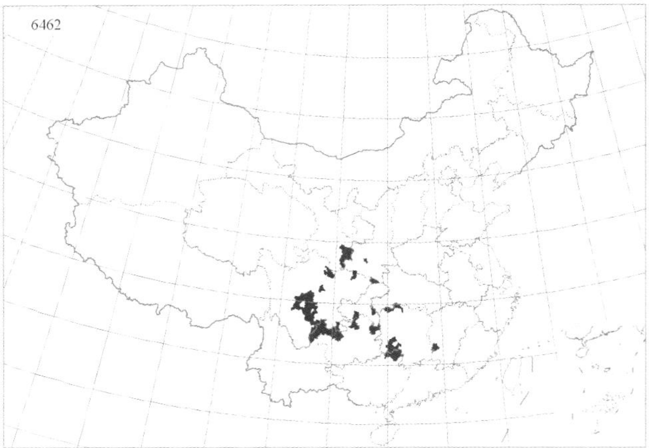

Actinidiaceae 猕猴桃科 *Clematoclethra* 藤山柳属

Clematoclethra scandens subsp. actinidioides (Maxim.) Y. C. Tang et Q. Y. Xiang 猕猴桃藤山柳 Vines, deciduous

11.0 (0.8~17.0); -0.3 (-10.6~6.2); 21.5 (9.8~27.5); 11.4 (3.6~17.0); 87.7 (14.0~144.4); 15.2 (0.0~66.0); 679 (403~891) ❙ 889 (341~1624); 432 (190~817); 37 (8~191) ❙ 32.9 (-40.9~128.6); 659 (341~891); 234.6 (61.2~357.7)

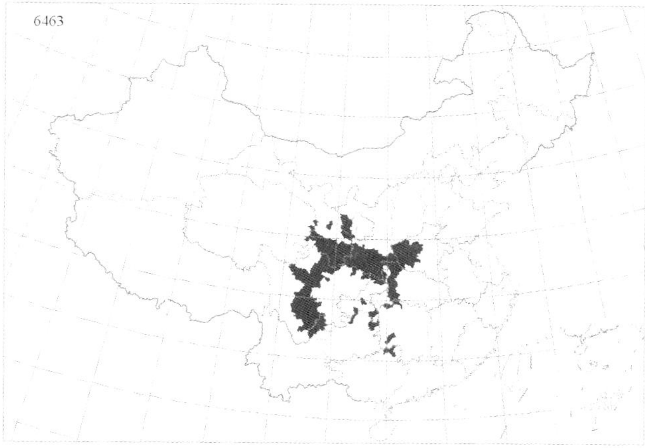

Actinidiaceae 猕猴桃科 *Clematoclethra* 藤山柳属

Clematoclethra scandens subsp. hemsleyi (Baill.) Y. C. Tang et Q. Y. Xiang 繁花藤山柳 Vines, deciduous

11.1 (7.4~14.0); -0.6 (-4.2~2.9); 22.4 (18.7~24.6); 11.3 (8.2~14.1); 87.8 (58.8~111.4); 14.3 (2.8~29.4); 677 (561~772) ❙ 900 (738~1105); 410 (332~504); 26 (22~31) ❙ 33.4 (5.8~52.6); 677 (561~772); 282.2 (236.9~334.4)

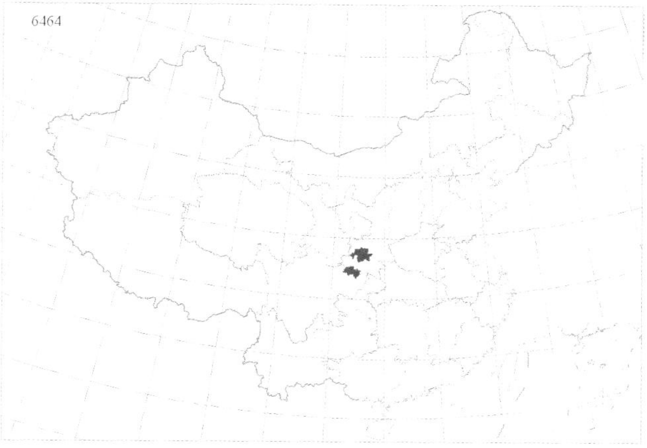

Actinidiaceae 猕猴桃科 *Clematoclethra* 藤山柳属

Clematoclethra scandens subsp. tomentella (Franch.) Y. C. Tang et Q. Y. Xiang 绒毛藤山柳 Vines, deciduous

9.8 (2.2~14.9); 0.6 (-6.6~5.3); 18.3 (9.9~24.9); 10.2 (3.8~15.0); 73.1 (15.5~119.6); 14.3 (0.0~49.1); 625 (413~791) ❙ 1134 (922~1410); 607 (460~816); 38 (19~71) ❙ 85.0 (51.2~125.9); 625 (413~791); 248.6 (182.9~293.1)

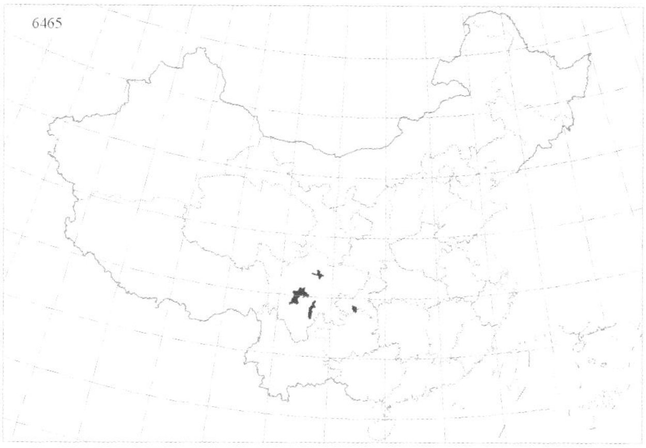

Actinidiaceae 猕猴桃科 *Saurauia* 水东哥属

Saurauia cerea Griff. ex Dyer 蜡质水东哥

Trees

18.6 (13.3~21.0); 12.4 (5.4~15.1); 22.7 (19.6~25.6); 18.6 (13.4~21.0); 163.3 (100.4~191.8); 0.0 (0.0~0.0); 897 (706~1053) ❙ 1374 (1028~1751); 750 (566~1007); 57 (44~77) ❙ 54.1 (12.5~145.8); 881 (706~1053); 451.2 (339.5~659.0)

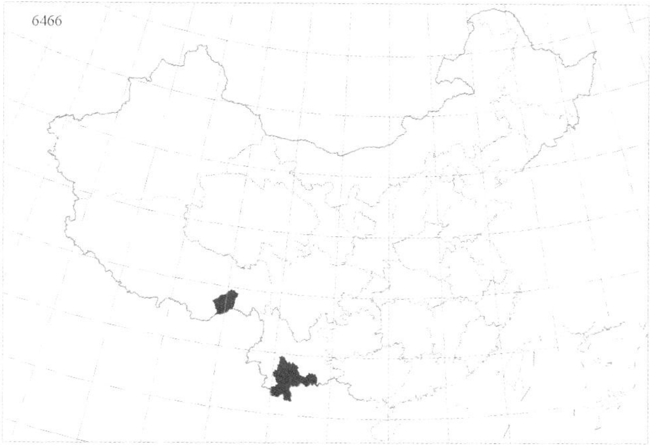

Actinidiaceae 猕猴桃科 *Saurauia* 水东哥属

Saurauia erythrocarpa C. F. Liang et Y. S. Wang 红果水东哥

Trees

14.6; 7.9; 19.6; 14.6; 118.2; 2.9; 757 ❙ 1272; 727; 52 ❙ 70.0; 757; 417.9

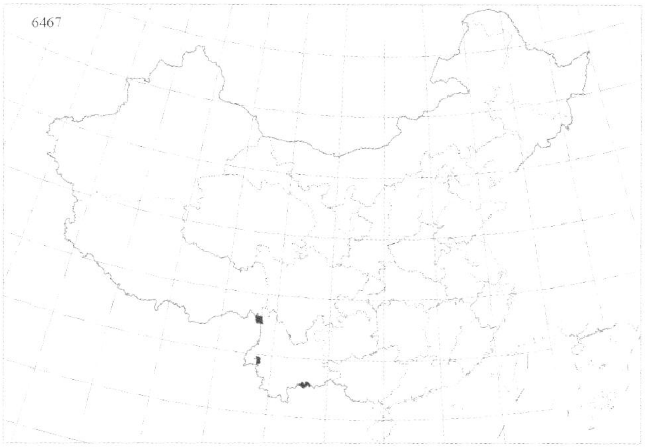

Actinidiaceae 猕猴桃科 *Saurauia* 水东哥属

Saurauia erythrocarpa var. grosseserrata C. F. Liang et Y. S. Wang 粗齿水东哥 Shrubs

19.1; 13.0; 23.3; 19.1; 169.7; 0.0; 917 ❙ 1428; 835; 56 ❙ 55.7; 917; 506.2

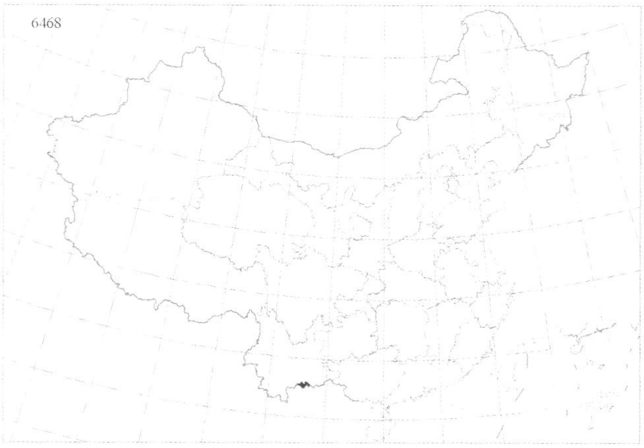

Actinidiaceae 猕猴桃科 *Saurauia* 水东哥属
Saurauia griffithii Dyer 绵毛水东哥
Trees

13.3；5.4；19.6；13.4；100.4；0.0；706 ❙ 1736；1007；51 ❙ 145.8；706；447.4

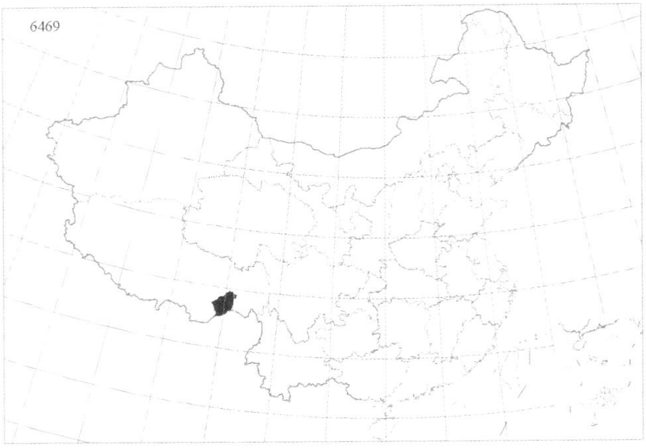

Actinidiaceae 猕猴桃科 *Saurauia* 水东哥属
Saurauia griffithii var. annamica Gagnep. 越南水东哥
Trees

13.3；5.4；19.6；13.4；100.4；0.0；706 ❙ 1736；1007；51 ❙ 145.8；706；447.4

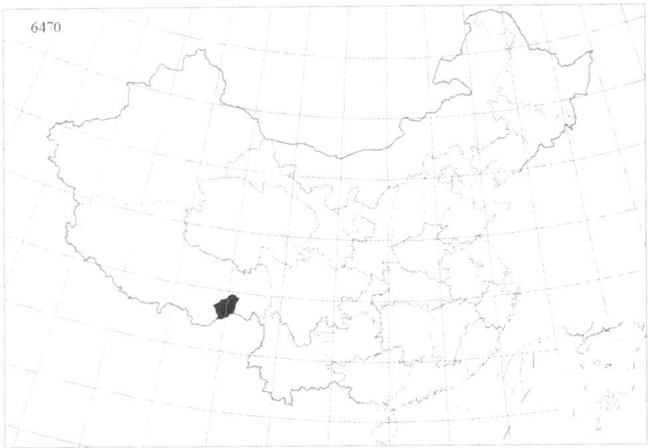

Actinidiaceae 猕猴桃科 *Saurauia* 水东哥属
Saurauia macrotricha Kurz ex Dyer 长毛水东哥
Shrubs or small trees

17.9 (8.6~20.5)；11.7 (1.3~15.1)；21.8 (15.0~23.7)；18.0 (8.7~20.5)；156.3 (52.6~185.9)；0.7 (0.0~8.8)；874 (574~992) ❙ 1500 (1041~1688)；796 (585~959)；53 (41~77) ❙ 72.8 (47.6~112.2)；858 (574~986)；516.3 (352.3~659.0)

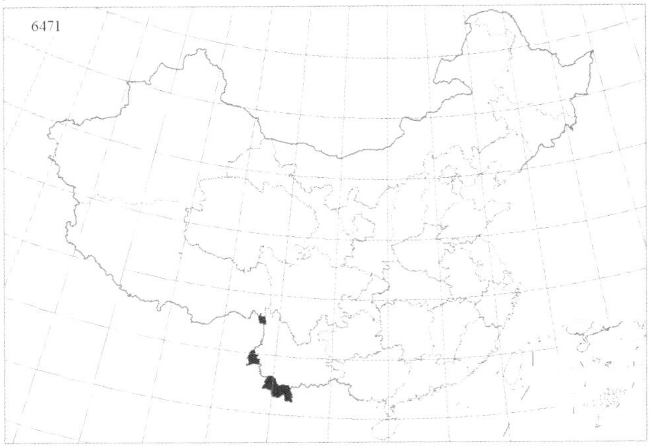

Actinidiaceae 猕猴桃科 *Saurauia* 水东哥属
Saurauia miniata C. F. Liang et Y. S. Wang 朱毛水东哥
Shrubs or small trees

18.4 (16.3~21.0)；11.3 (9.1~14.2)；23.3 (21.4~26.4)；18.4 (16.4~21.0)；160.9 (136.4~191.8)；0.0 (0.0~0.0)；895 (797~1053) ❙ 1336 (1074~1751)；754 (582~996)；55 (43~77) ❙ 49.6 (16.9~88.8)；892 (797~1053)；401.0 (277.0~551.6)

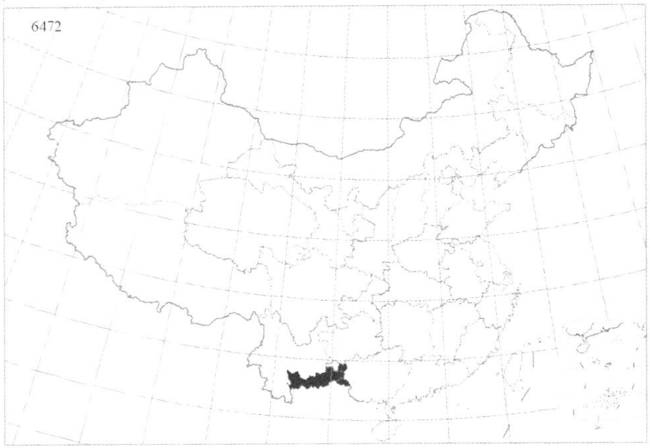

Actinidiaceae 猕猴桃科 *Saurauia* 水东哥属
Saurauia napaulensis DC. 尼泊尔水东哥
Trees

16.0 (5.4~21.5)；8.6 (−2.1~15.1)；21.5 (12.3~27.7)；16.0 (5.8~21.6)；132.7 (31.5~199.1)；0.6 (0.0~25.6)；803 (481~1131) ❙ 1141 (741~1751)；637 (411~996)；47 (15~111) ❙ 42.4 (−14.2~112.2)；787 (481~1131)；327.8 (149.6~659.0)

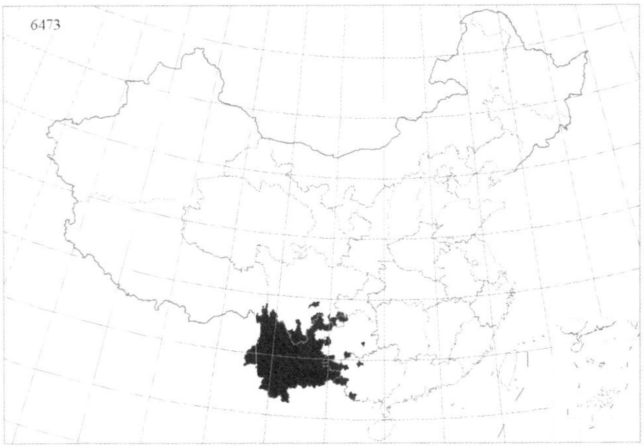

Actinidiaceae 猕猴桃科 *Saurauia* 水东哥属
Saurauia polyneura C. F. Liang et Y. S. Wang 多脉水东哥
Small trees

15.2 (5.3~19.1)；8.6 (−2.4~13.0)；20.0 (12.2~23.3)；15.3 (5.7~19.1)；124.8 (30.5~169.7)；1.4 (0.0~26.2)；765 (479~917) ❙ 1211 (961~1736)；680 (553~1007)；51 (31~69) ❙ 60.1 (26.1~145.8)；756 (479~917)；394.5 (234.2~516.7)

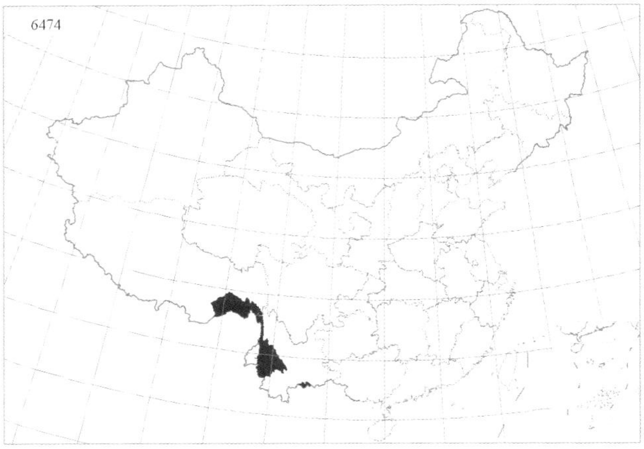

Actinidiaceae 猕猴桃科 *Saurauia* 水东哥属

Saurauia polyneura var. paucinervis J. Q. Li et D. D. Soejarto 少脉水东哥

Trees

5.3; −2.4; 12.2; 5.7; 30.5; 26.2; 479 ▎961; 558; 31 ▎100.7; 479; 234.2

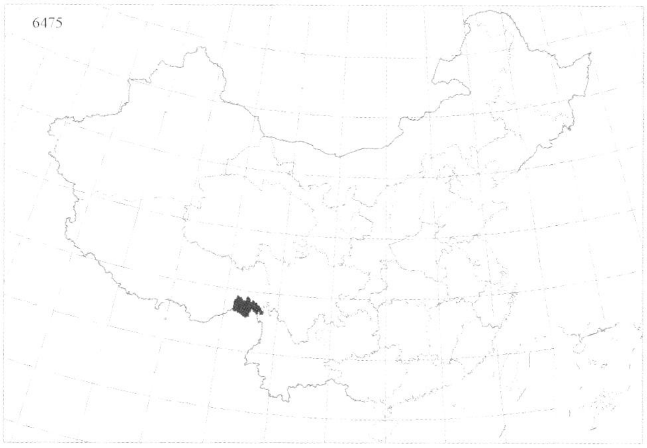

Actinidiaceae 猕猴桃科 *Saurauia* 水东哥属

Saurauia punduana Wall. 大花水东哥

Trees

17.1 (8.6~21.3); 10.2 (1.3~14.2); 22.1 (15.0~27.9); 17.2 (8.7~21.3); 146.8 (52.6~196.0); 0.9 (0.0~8.8); 862 (574~1123) ▎1325 (1041~1736); 743 (585~1007); 54 (45~68) ▎58.3 (5.8~145.8); 861 (574~1121); 430.2 (315.6~551.6)

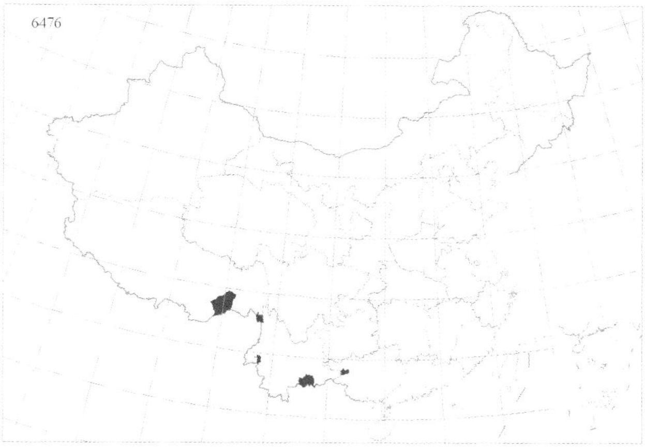

Actinidiaceae 猕猴桃科 *Saurauia* 水东哥属

Saurauia rubricalyx C. F. Liang et Y. S. Wang 红萼水东哥

Shrubs

13.3; 5.4; 19.6; 13.4; 100.4; 0.0; 706 ▎1736; 1007; 51 ▎145.8; 706; 447.4

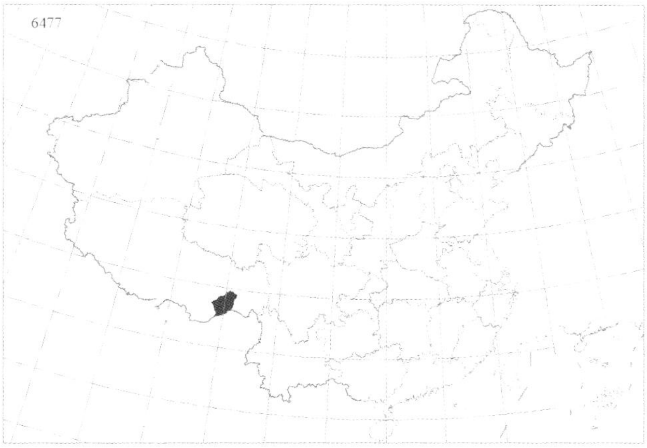

Actinidiaceae 猕猴桃科 *Saurauia* 水东哥属

Saurauia sinohirsuta J. Q. Li et Soejarto 糙毛水东哥

Trees

13.3; 5.4; 19.6; 13.4; 100.4; 0.0; 706 ▎1736; 1007; 51 ▎145.8; 706; 447.4

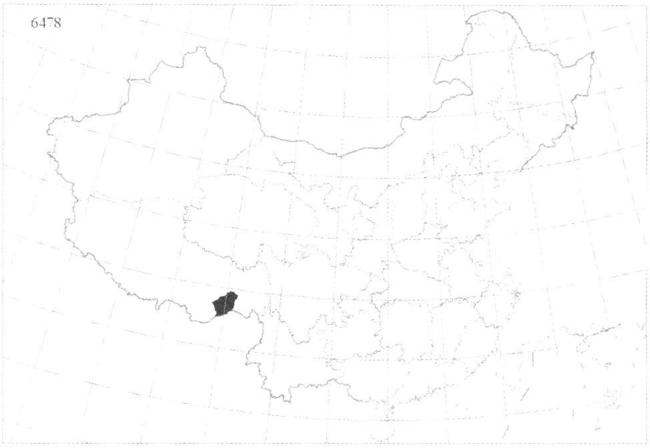

Actinidiaceae 猕猴桃科 *Saurauia* 水东哥属

Saurauia thyrsiflora C. F. Liang et Y. S. Wang 聚锥水东哥

Shrubs or small trees

19.3 (15.9~22.3); 10.7 (6.8~14.2); 25.8 (21.4~28.9); 19.3 (16.0~22.3); 172.1 (131.6~208.1); 0.0 (0.0~0.0); 988 (797~1202) ▎1377 (1074~1703); 7[illegible]9 (566~860); 89 (43~194) ▎40.3 (5.8~80.6); 988 (797~1202); 300.0 (201.4~551.6)

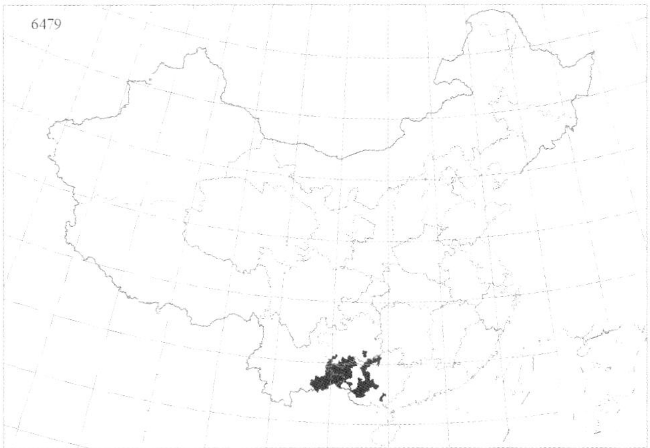

Actinidiaceae 猕猴桃科 *Saurauia* 水东哥属

Saurauia tristyla DC. 水东哥

Shrubs or small trees

20.5 (14.6~24.8); 12.3 (3.8~20.1); 27.1 (20.1~29.8); 20.5 (14.7~24.8); 186.4 (117.7~237.6); 0.0 (0.0~1.5); 1070 (773~1385) ▎1573 (868~2993); 708 (312~1847); 125 (41~547) ▎48.9 (−34.3~246.5); 1059 (773~1360); 268.1 (95.3~659.0)

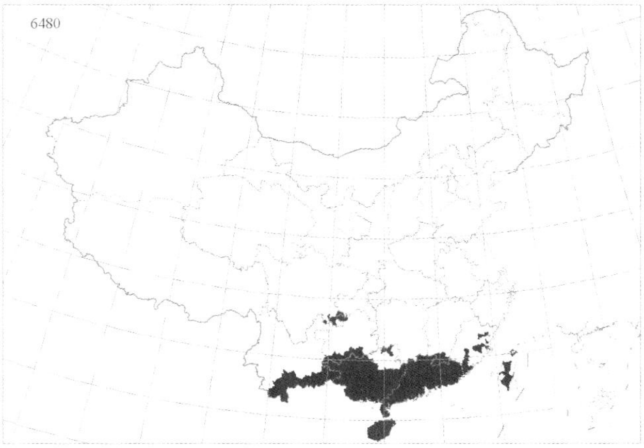

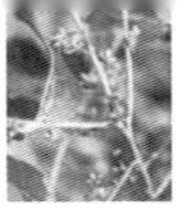

Actinidiaceae 猕猴桃科 *Saurauia* 水东哥属

Saurauia yunnanensis C. F. Liang et Y. S. Wang 云南水东哥

Shrubs or small trees

17.3 (6.1~21.0); 10.9 (−1.5~15.1); 21.8 (12.7~25.6); 17.4 (6.3~21.0); 150.0 (35.2~191.8); 1.6 (0.0~21.8); 860 (496~1053) ❙ 1364 (801~1751); 739 (496~996); 55 (21~77) ❙ 58.9 (36.8~88.8); 854 (496~1053); 468.3 (282.9~659.0)

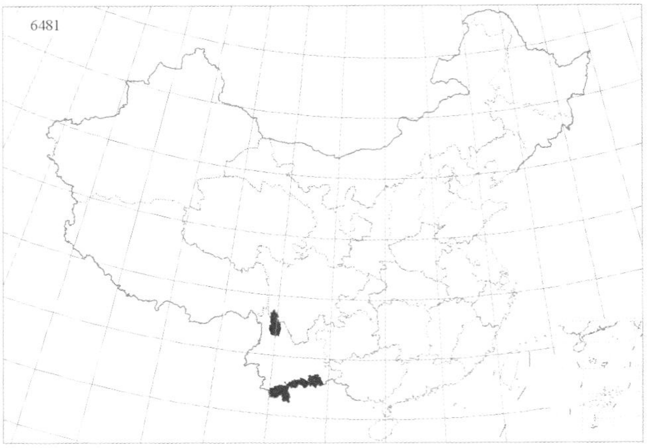

Ochnaceae 金莲木科 *Campylospermum* 赛金莲木属

Campylospermum serratum (Gaertner) Bittrich et M. C. E. Amaral 齿叶赛金莲木

Shrubs or small trees

24.1 (23.6~24.8); 18.6 (17.8~20.1); 28.0 (27.2~28.7); 24.1 (23.6~24.8); 229.8 (223.3~237.6); 0.0 (0.0~0.0); 1321 (1271~1385) ❙ 1257 (868~1551); 437 (312~539); 50 (41~71) ❙ −4.7 (−34.3~19.1); 1079 (868~1147); 456.8 (288.3~620.0)

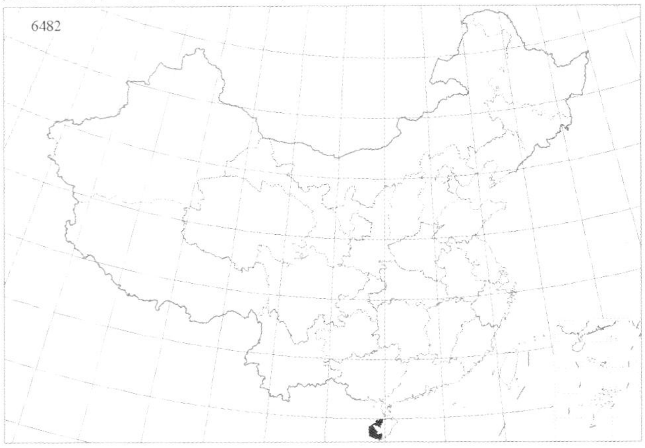

Ochnaceae 金莲木科 *Campylospermum* 赛金莲木属

Campylospermum striatum (Tieghem) M. C. E. Amaral 赛金莲木

Shrubs

23.9 (22.2~24.8); 18.6 (17.2~20.1); 27.5 (25.7~28.5); 23.9 (22.3~24.8); 226.9 (207.2~237.6); 0.0 (0.0~0.0); 1296 (1127~1385) ❙ 1581 (1242~1854); 546 (444~622); 66 (41~115) ❙ 22.3 (−2.2~47.8); 1186 (1055~1360); 506.2 (352.9~620.0)

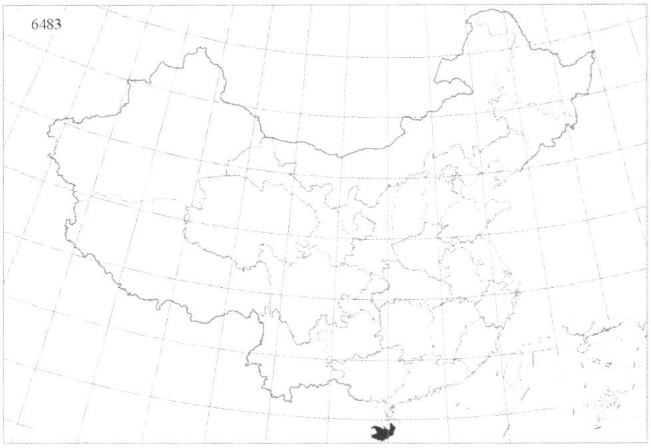

Ochnaceae 金莲木科 *Ochna* 金莲木属

Ochna integerrima (Lour.) Merr. 金莲木

Shrubs or small trees, deciduous

23.0 (18.7~24.8); 16.3 (8.8~20.1); 28.0 (25.7~29.0); 23.1 (18.8~24.8); 216.7 (165.2~237.6); 0.0 (0.0~0.0); 1241 (968~1385) ❙ 1609 (868~2435); 702 (312~1428); 92 (41~130) ❙ 30.9 (−34.3~99.5); 1177 (868~1360); 361.2 (203.1~620.0)

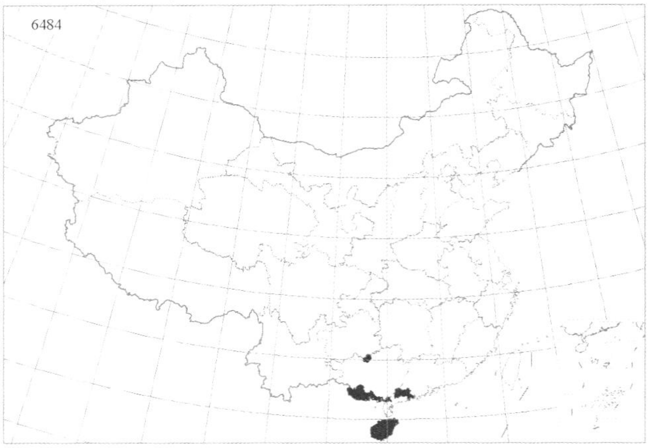

Ochnaceae 金莲木科 *Sauvagesia* 合柱金莲木属

Sauvagesia rhodoleuca (Diels) M. C. E. Amaral 合柱金莲木

Small shrubs

19.2 (15.5~21.3); 9.4 (4.8~12.3); 27.4 (25.1~28.6); 19.3 (15.6~21.3); 171.1 (127.2~196.1); 0.0 (0.0~0.2); 1005 (815~1130) ❙ 1555 (1371~1758); 633 (517~728); 165 (117~189) ❙ 56.1 (22.7~93.2); 1004 (815~1130); 230.8 (185.5~263.5)

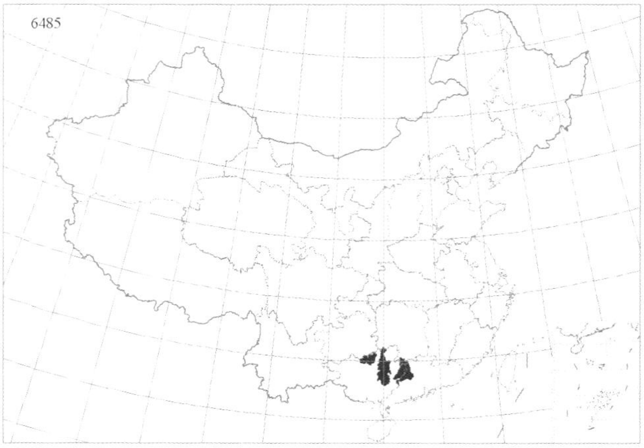

Sladeniaceae 肋果茶科 *Sladenia* 肋果茶属

Sladenia celastrifolia Kurz 肋果茶

Trees

16.8 (13.6~20.5); 10.2 (6.2~15.1); 21.4 (18.8~26.6); 16.9 (13.6~20.5); 142.7 (103.6~185.9); 0.0 (0.0~0.0); 823 (706~992) ❙ 1149 (911~1751); 641 (524~996); 48 (28~77) ❙ 39.2 (6.5~88.8); 804 (690~986); 354.3 (230.7~659.0)

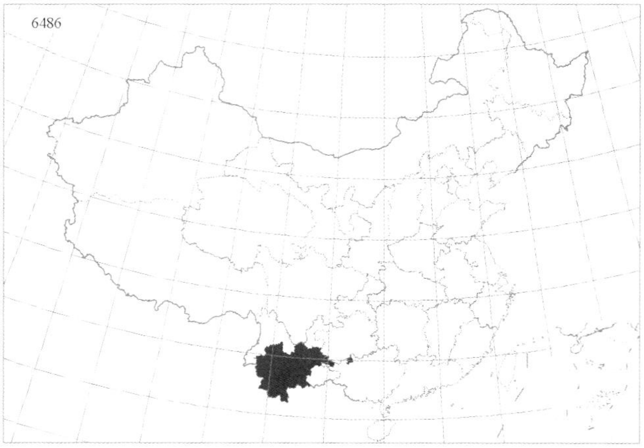

Pentaphylacaceae 五列木科 *Pentaphylax* 五列木属
Pentaphylax euryoides Gardner et Champ. 五列木
Shrubs or trees, evergreen

19.5 (14.3~24.8); 10.3 (3.3~20.1); 27.2 (22.5~29.8); 19.5 (14.3~24.8); 174.5 (112.6~237.6); 0.1 (0.0~2.3); 1027 (756~1385) ❙ 1557 (868~2435); 652 (312~1428); 139 (41~197) ❙ 53.5 (-34.3~101.8); 1011 (756~1326); 259.3 (122.7~620.0)

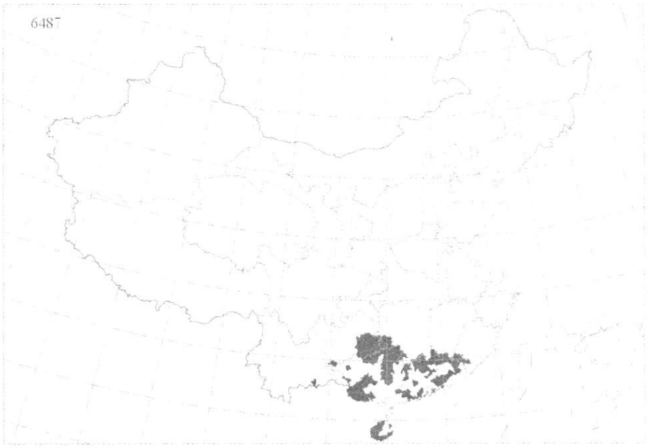

Theaceae 山茶科 *Adinandra* 杨桐属
Adinandra bockiana E. Pritz. 川杨桐
Shrubs or small trees, evergreen

17.4 (10.6~22.7); 7.1 (-0.2~15.0); 26.6 (21.1~29.9); 17.4 (10.6~22.8); 149.3 (79.4~213.1); 0.3 (0.0~12.1); 916 (643~1221) ❙ 1407 (1028~2435); 579 (439~1428); 123 (46~254) ❙ 53.9 (8.0~130.7); 914 (643~1221); 212.5 (108.2~327.7)

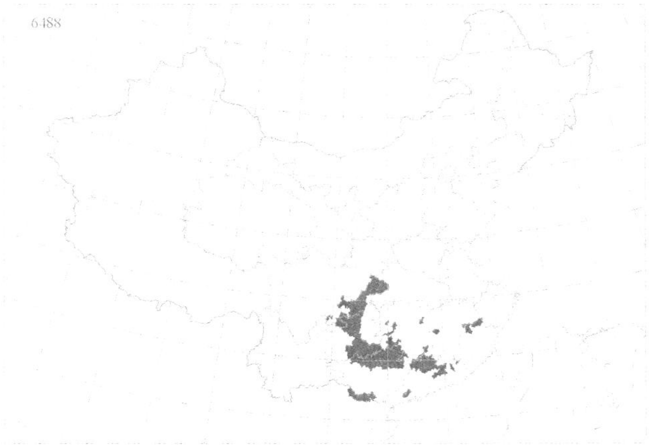

Theaceae 山茶科 *Adinandra* 杨桐属
Adinandra bockiana var. acutifolia (Hand.-Mazz.) Kobuski 尖叶川杨桐
Shrubs or small trees, evergreen

17.6 (11.7~22.7); 7.5 (2.7~15.0); 26.8 (18.8~29.9); 17.6 (11.7~22.8); 151.9 (82.6~213.1); 0.2 (0.0~3.8); 925 (658~1221) ❙ 1499 (916~2435); 560 (356~1428); 167 (31~259) ❙ 63.2 (-5.0~137.5); 915 (658~1221); 226.0 (123.9~335.1)

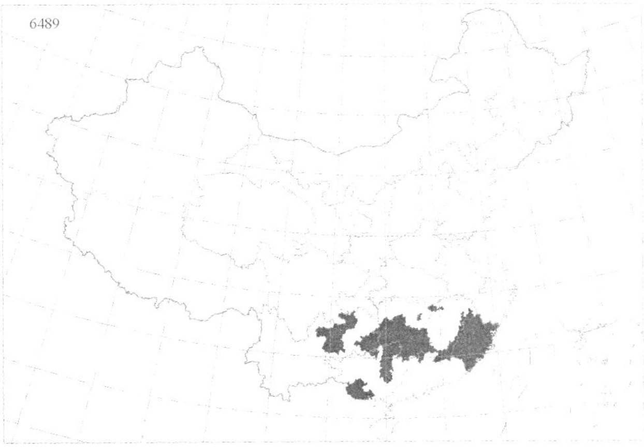

Theaceae 山茶科 *Adinandra* 杨桐属
Adinandra elegans How et Ko ex H. T. Chang 长梗杨桐
Shrubs, evergreen

22.4; 15.0; 28.1; 22.4; 209.4; 0.0; 1193 ❙ 1913; 912; 113 ❙ 60.4; 1193; 255.7

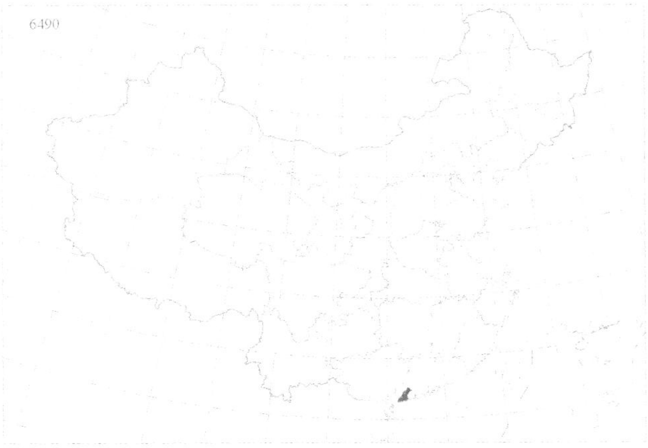

Theaceae 山茶科 *Adinandra* 杨桐属
Adinandra epunctata Merr. et Chun 无腺杨桐
Trees, evergreen

24.1; 18.2; 28.4; 24.2; 230.1; 0.0; 1326 ❙ 1723; 619; 110 ❙ 29.9; 1326; 352.9

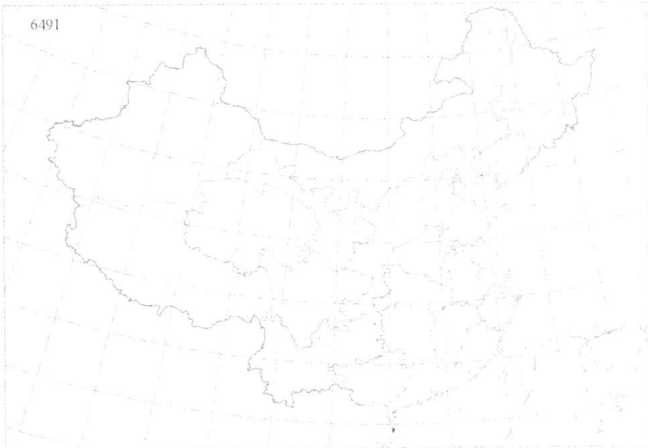

Theaceae 山茶科 *Adinandra* 杨桐属
Adinandra filipes Merr. ex Kobuski 细梗杨桐
Shrubs, evergreen

21.5 (19.0~22.7); 13.4 (10.2~15.0); 27.6 (25.8~28.6); 21.5 (19.0~22.8); 198.6 (168.4~213.1); 0.0 (0.0~0.0); 1133 (958~1221) ❙ 1786 (1250~2435); 994 (670~1428); 108 (60~130) ❙ 56.7 (12.9~99.5); 1133 (958~1221); 289.8 (235.1~318.8)

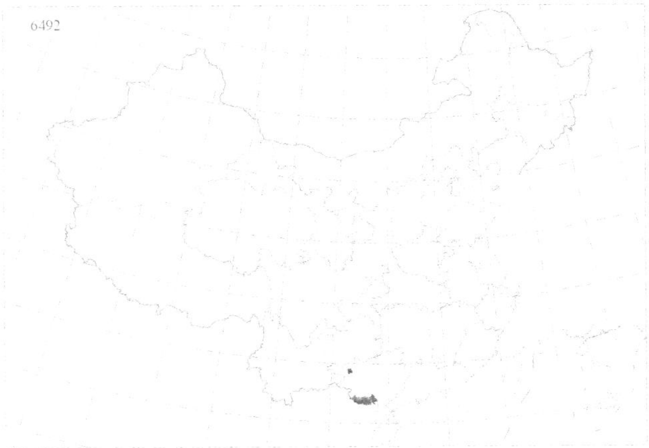

Theaceae 山茶科 *Adinandra* 杨桐属
Adinandra formosana Hayata 台湾杨桐
Shrubs or small trees, evergreen

20.5 (16.2~24.1); 14.8 (11.0~18.9); 25.1 (20.1~28.2); 20.5 (16.2~24.1); 185.9 (134.2~229.7); 0.0 (0.0~0.0); 1037 (773~1322) ❙ 2510 (1751~3495); 1175 (821~1847); 270 (47~897) ❙ 152.5 (32.5~246.5); 994 (773~1144); 244.5 (119.4~363.7)

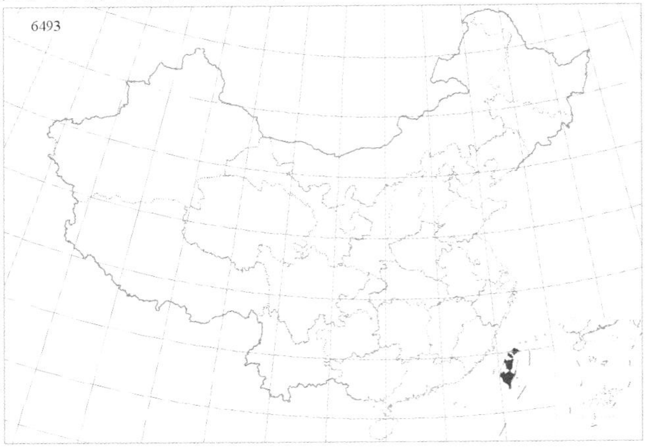

Theaceae 山茶科 *Adinandra* 杨桐属
Adinandra glischroloma Hand.-Mazz. 两广杨桐
Shrubs or small trees, evergreen

19.6 (14.8~24.1); 10.2 (4.2~18.0); 27.5 (22.5~29.8); 19.6 (14.9~24.2); 175.2 (119.2~230.2); 0.0 (0.0~0.8); 1025 (791~1329) ❙ 1628 (1197~2014); 655 (448~954); 166 (59~259) ❙ 60.6 (-1.4~130.7); 1023 (791~1329); 246.8 (170.8~491.5)

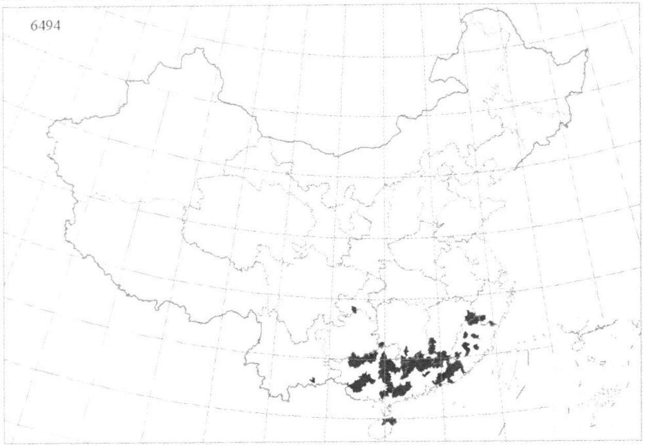

Theaceae 山茶科 *Adinandra* 杨桐属
Adinandra glischroloma var. jubata (H. L. Li) Kobuski 长毛杨桐
Shrubs or small trees, evergreen

21.3 (18.4~22.2); 13.3 (9.9~14.7); 27.7 (26.2~28.4); 21.3 (18.4~22.2); 195.5 (161.4~206.9); 0.0 (0.0~0.0); 1115 (928~1177) ❙ 1802 (1675~2014); 814 (614~954); 136 (114~185) ❙ 62.3 (45.5~80.5); 1115 (928~1177); 251.4 (217.4~300.5)

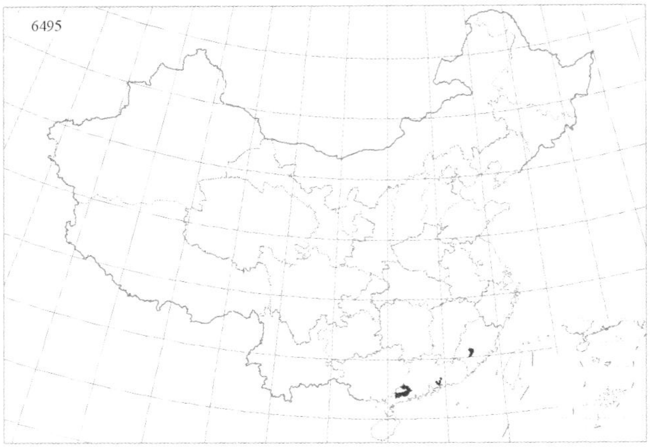

Theaceae 山茶科 *Adinandra* 杨桐属
Adinandra glischroloma var. macrosepala (F. P. Metcalf) Kobuski 大萼杨桐
Shrubs or small trees, evergreen

17.9 (14.3~22.0); 7.9 (4.0~13.8); 27.0 (23.5~29.1); 18.0 (14.3~22.0); 155.6 (112.6~204.5); 0.0 (0.0~1.1); 936 (756~1172) ❙ 1708 (1211~1997); 612 (490~778); 197 (89~259) ❙ 84.0 (24.9~147.0); 934 (756~1172); 254.4 (183.9~332.5)

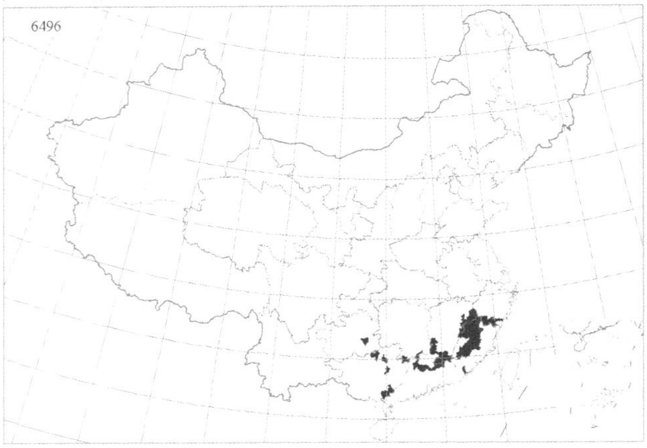

Theaceae 山茶科 *Adinandra* 杨桐属
Adinandra grandis L. K. Ling 大杨桐
Trees, evergreen

18.3; 10.2; 25.2; 18.4; 160.7; 0.2; 920 ❙ 1360; 690; 101 ❙ 47.9; 917; 382.4

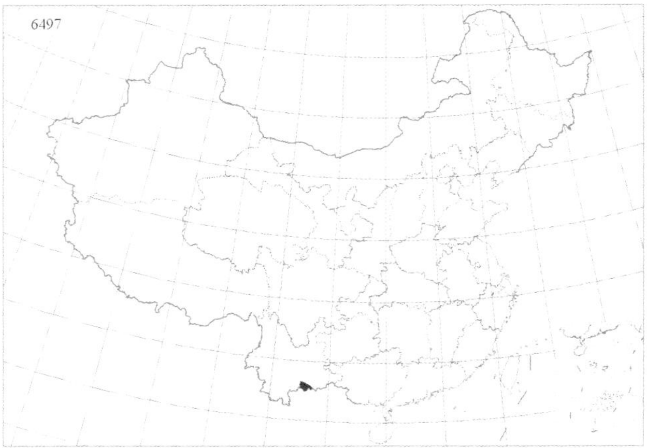

Theaceae 山茶科 *Adinandra* 杨桐属
Adinandra hainanensis Merr. 海南杨桐
Shrubs or trees, evergreen

22.4 (19.1~24.8); 14.9 (9.2~20.1); 28.0 (25.0~29.2); 22.4 (19.1~24.8); 208.9 (169.6~237.6); 0.0 (0.0~0.0); 1196 (945~1385) ❙ 1580 (868~2435); 724 (312~1428); 103 (41~146) ❙ 33.1 (-34.3~99.5); 1163 (868~1360); 300.9 (157.7~620.0)

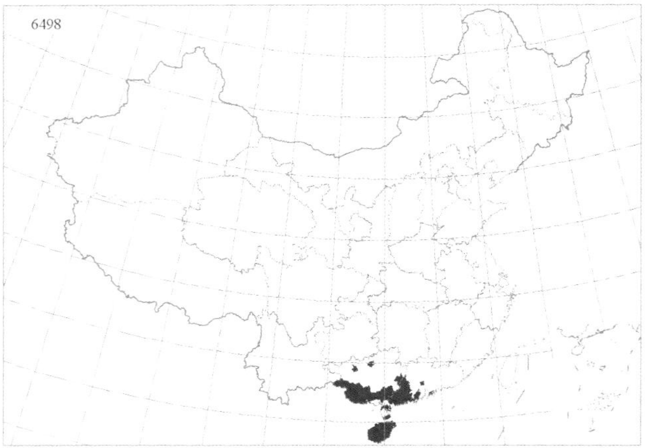

Theaceae 山茶科 *Adinandra* 杨桐属

Adinandra hirta Gagnep. 粗毛杨桐

Shrubs or trees, evergreen

17.3 (11.9~22.2); 8.6 (2.9~15.1); 24.4 (19.6~28.5); 17.3 (12.0~22.2); 148.1 (87.2~206.5); 0.1 (0.0~3.4); 877 (671~1188) ❙ 1348 (1001~1776); 646 (448~1007); 81 (35~142) ❙ 54.4 (27.0~145.8); 873 (671~1188); 318.5 (188.8~659.0)

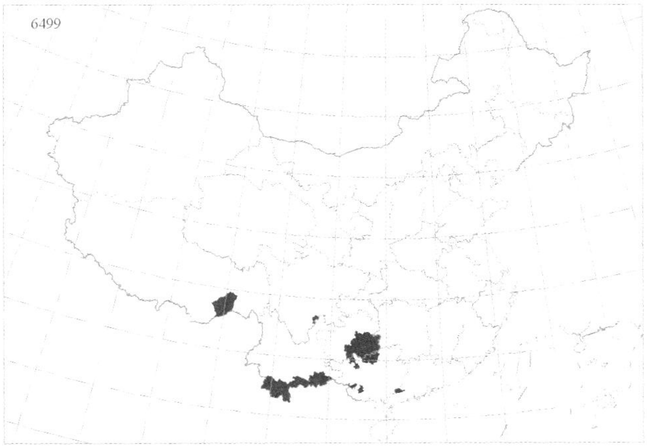

Theaceae 山茶科 *Adinandra* 杨桐属

Adinandra howii Merr. et Chun 保亭杨桐

Small trees, evergreen

23.8 (22.2~24.7); 18.6 (17.2~19.8); 27.4 (25.7~28.4); 23.8 (22.3~24.7); 225.8 (207.2~236.5); 0.0 (0.0~0.0); 1287 (1127~1376) ❙ 1670 (1551~1767); 578 (534~619); 71 (48~110) ❙ 30.5 (15.4~47.8); 1200 (1055~1326); 529.9 (352.9~620.0)

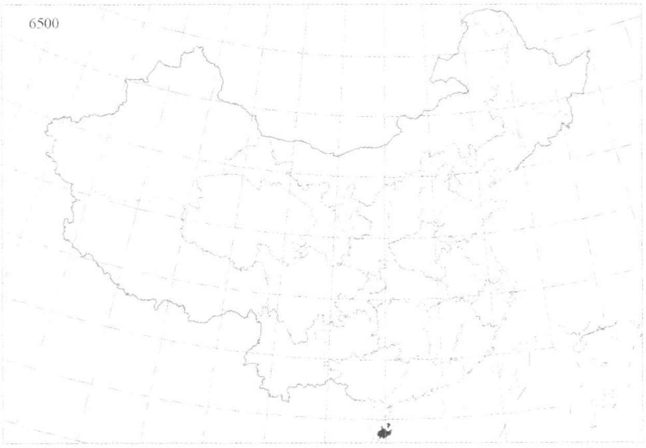

Theaceae 山茶科 *Adinandra* 杨桐属

Adinandra lancipetala L. K. Ling 狭瓣杨桐

Shrubs or trees, evergreen

21.9 (21.3~23.0); 13.6 (12.4~15.1); 28.3 (27.3~29.0); 22.0 (21.3~23.0); 203.8 (195.7~2[illegible]6.2); 0.0 (0.0~0.0); 1169 (1111~1247) ❙ 1619 (1302~2435); 836 (617~1428); 124 (86~171) ❙ 38.5 (12.9~99.5); 1169 (1111~1247); 254.1 (199.4~335.1)

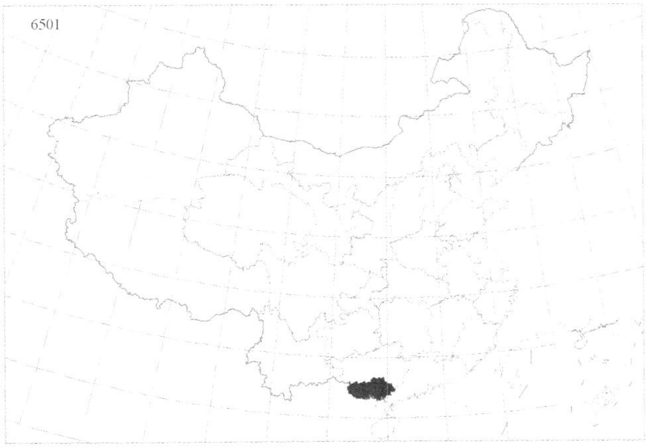

Theaceae 山茶科 *Adinandra* 杨桐属

Adinandra lasiostyla Hayata 毛柱杨桐

Trees, evergreen

19.5; 14.0; 23.9; 19.5; 174.1; 0.0; 954 ❙ 2344; 1220; 166 ❙ 152.6; 943; 263.9

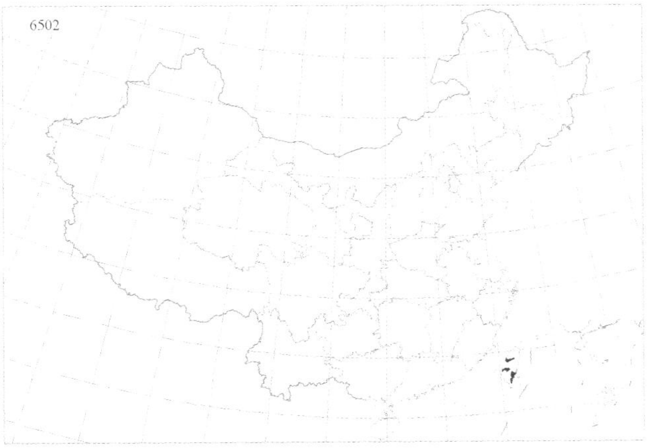

Theaceae 山茶科 *Adinandra* 杨桐属

Adinandra latifolia L. K. Ling 阔叶杨桐

Trees, evergreen

8.6; 1.3; 15.0; 8.7; 52.6; 8.8; 574 ❙ 1041; 585; 45 ❙ 81.5; 574; 352.3

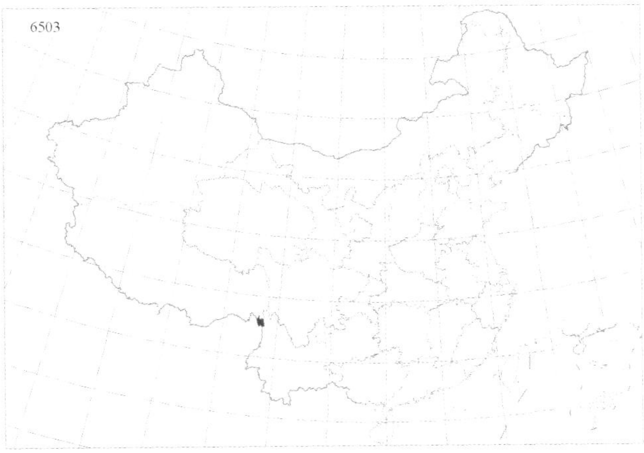

Theaceae 山茶科 *Adinandra* 杨桐属

Adinandra megaphylla Hu 大叶杨桐

Trees, evergreen

18.5 (16.3~21.8); 11.1 (9.1~14.2); 23.8 (21.4~27.7); 18.5 (16.4~21.9); 162.4 (136.4~202.2); 0.0 (0.0~0.0); 910 (797~1150) ❙ 1363 (1074~2148); 768 (582~1234); 59 (43~124) ❙ 49.6 (16.9~86.7); 909 (797~1150); 386.2 (277.0~551.6)

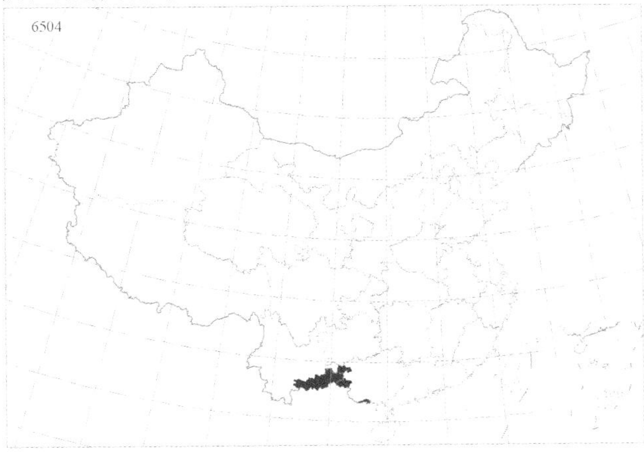

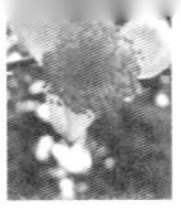

Theaceae 山茶科 *Adinandra* 杨桐属

Adinandra millettii (Hook. et Arn.) Benth. et Hook. f. ex Hance 杨桐

Shrubs or small trees, evergreen

18.5 (14.3~24.1); 8.4 (2.7~18.2); 27.6 (22.2~29.9); 18.5 (14.3~24.2); 162.4 (112.6~230.1); 0.2 (0.0~3.2); 978 (756~1326) ❙ 1569 (1039~2435); 597 (356~1428); 169 (51~259) ❙ 62.2 (-6.9~147.0); 968 (756~1326); 222.6 (95.3~352.9)

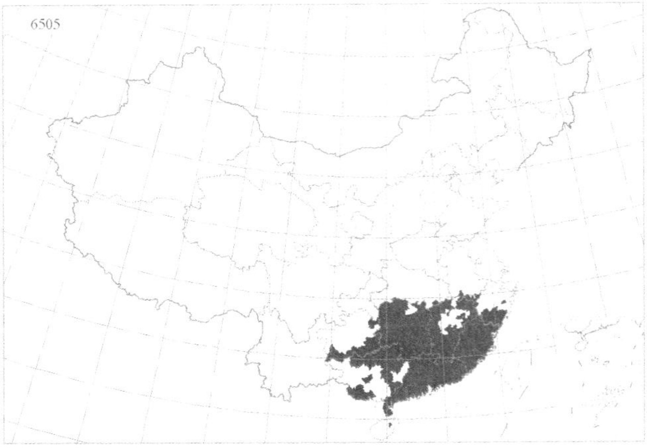

Theaceae 山茶科 *Adinandra* 杨桐属

Adinandra nigroglandulosa L. K. Ling 腺叶杨桐

Trees, evergreen

18.3 (17.2~21.0); 11.2 (9.5~14.2); 23.3 (22.4~25.6); 18.4 (17.3~21.0); 160.2 (147.0~191.8); 0.0 (0.0~0.0); 892 (833~1053) ❙ 1416 (1254~1522); 797 (699~860); 57 (48~68) ❙ 59.4 (44.5~80.6); 892 (833~1053); 443.1 (336.3~551.6)

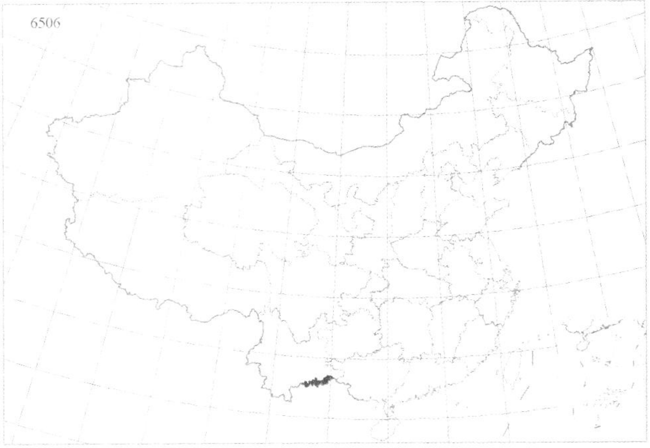

Theaceae 山茶科 *Adinandra* 杨桐属

Adinandra nitida Merr. ex H. L. Li 亮叶杨桐

Shrubs or trees, evergreen

19.7 (14.3~23.4); 10.5 (3.8~16.1); 27.3 (22.5~29.8); 19.7 (14.3~23.5); 176.8 (112.6~221.7); 0.0 (0.0~1.5); 1036 (756~1273) ❙ 1578 (1159~2435); 700 (461~1428); 139 (59~194) ❙ 53.4 (11.9~101.8); 1034 (756~1273); 239.9 (174.9~491.5)

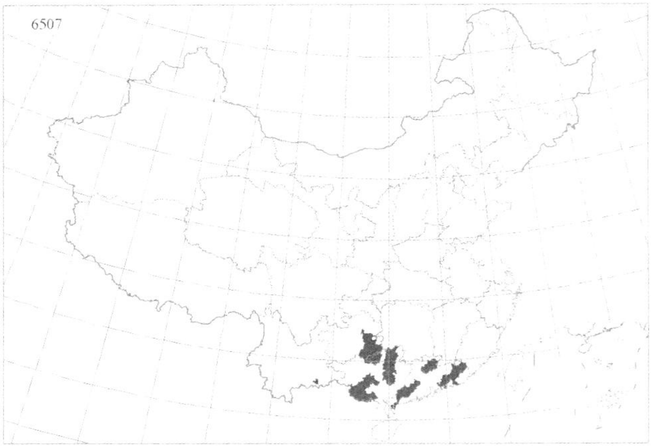

Theaceae 山茶科 *Adinandra* 杨桐属

Adinandra pingbianensis L. K. Ling 屏边杨桐

Trees, evergreen

19.3; 12.7; 23.8; 19.3; 171.6; 0.0; 942 ❙ 1439; 803; 61 ❙ 53.4; 942; 516.4

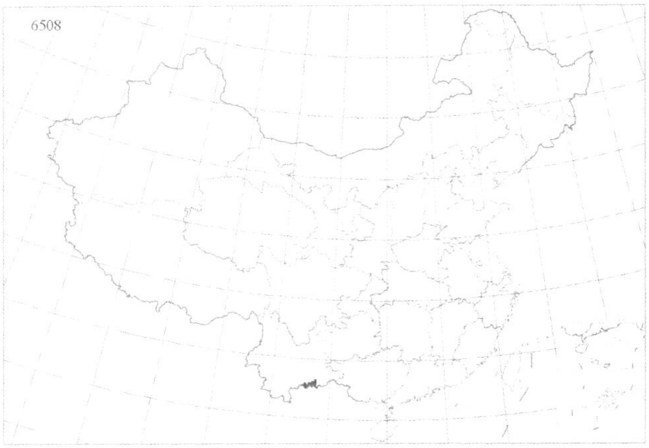

Theaceae 山茶科 *Anneslea* 茶梨属

Anneslea fragrans Wall. 茶梨

Trees, evergreen

19.5 (14.3~24.7); 11.0 (3.3~19.8); 26.3 (19.1~29.8); 19.5 (14.3~24.7); 174.3 (112.6~236.5); 0.0 (0.0~2.3); 1008 (727~1376) ❙ 1553 (907~2992); 690 (448~1847); 128 (29~259) ❙ 55.2 (-6.9~214.4); 1001 (709~1273); 274.3 (95.3~659.0)

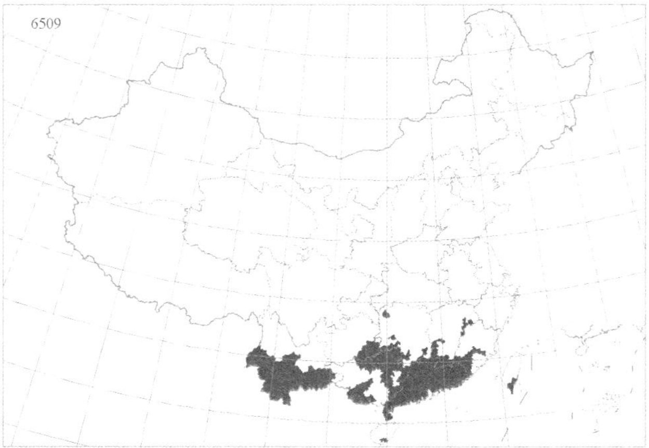

Theaceae 山茶科 *Apterosperma* 圆籽荷属

Apterosperma oblata H. T. Chang 圆籽荷

Shrubs or small trees, evergreen

21.6 (19.1~22.6); 13.5 (9.3~14.9); 28.0 (26.9~28.8); 21.6 (19.1~22.6); 199.6 (169.2~211.7); 0.0 (0.0~0.0); 1140 (988~1213) ❙ 1663 (1498~1876); 784 (561~891); 119 (75~178) ❙ 46.2 (31.8~64.3); 1140 (988~1213); 227.0 (166.0~273.3)

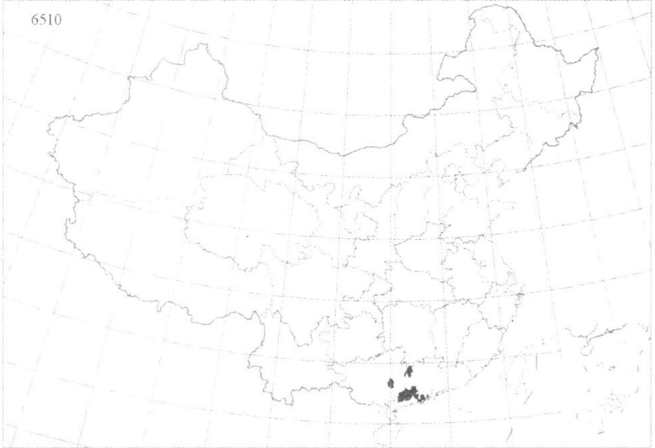

Theaceae 山茶科 *Camellia* 山茶属
Camellia amplexifolia Merr. et Chun 抱茎短蕊茶
Shrubs or small trees, evergreen

24.3；19.5；27.6；24.4；232.3；0.0；1339 ❙ 1569；536；53 ❙ 17.2；1180；557.4

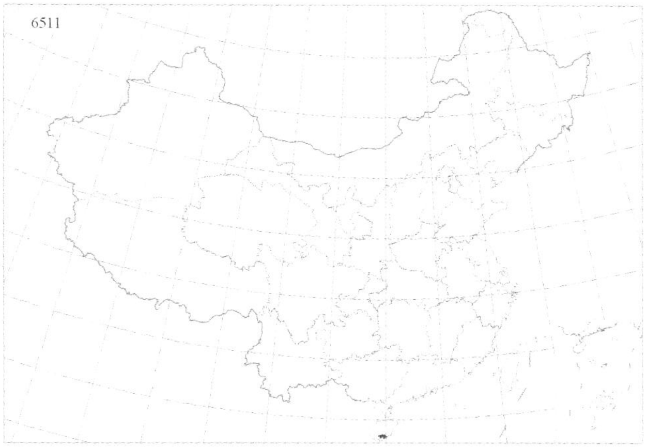

Theaceae 山茶科 *Camellia* 山茶属
Camellia anlungensis H. T. Chang 安龙瘤果茶
Shrubs or small trees, evergreen

18.0 (16.3~22.3)；9.4 (5.9~13.5)；24.7 (21.4~28.9)；18.1 (16.4~22.3)；156.9 (136.4~208.1)；0.0 (0.0~0.0)；907 (797~1202) ❙ 1312 (1018~1847)；675 (554~860)；75 (47~242) ❙ 45.3 (20.3~101.9)；906 (797~1202)；298.5 (189.5~413.5)

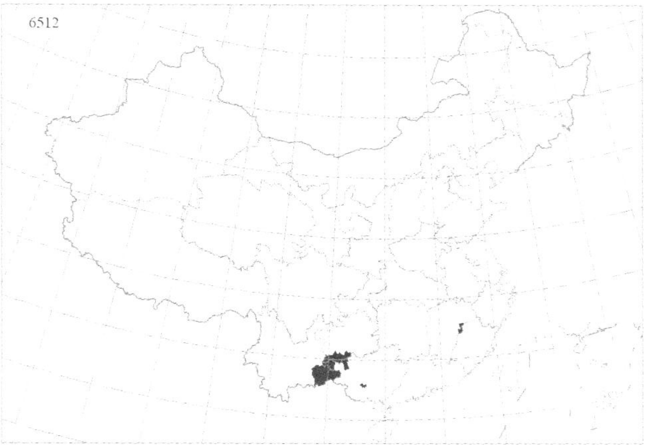

Theaceae 山茶科 *Camellia* 山茶属
Camellia anlungensis var. acutiperulata (Hung T. Chang et C. X. Ye) T. L. Ming 尖苞瘤果茶
Shrubs or trees, evergreen

18.1；9.5；24.6；18.2；158.1；0.0；901 ❙ 1293；723；58 ❙ 43.5；901；316.1

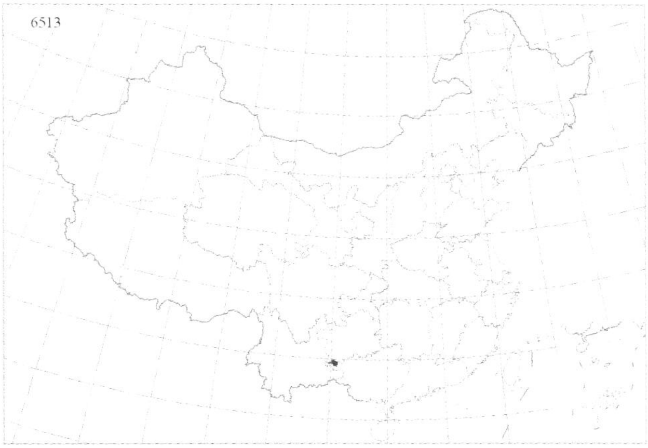

Theaceae 山茶科 *Camellia* 山茶属
Camellia assimiloides Sealy 大萼毛蕊茶
Shrubs, evergreen

18.1 (14.6~21.3)；8.1 (3.8~13.5)；27.0 (24.6~27.9)；18.2 (14.7~21.3)；158.4 (117.7~195.6)；0.1 (0.0~1.5)；947 (779~1107) ❙ 1641 (1483~1991)；594 (481~902)；184 (142~218) ❙ 73.7 (49.1~101.8)；947 (779~1107)；236.2 (188.8~272.2)

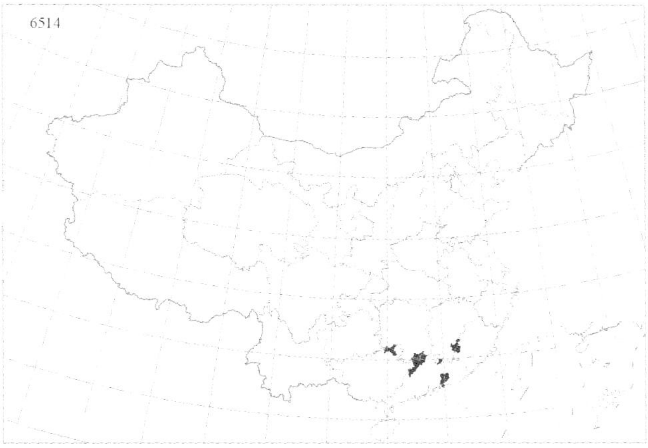

Theaceae 山茶科 *Camellia* 山茶属
Camellia azalea C. F. Wei 杜鹃叶山茶
Shrubs, evergreen

21.8；14.2；27.8；21.9；202.6；0.0；1150 ❙ 1876；891；116 ❙ 63.2；1150；273.3

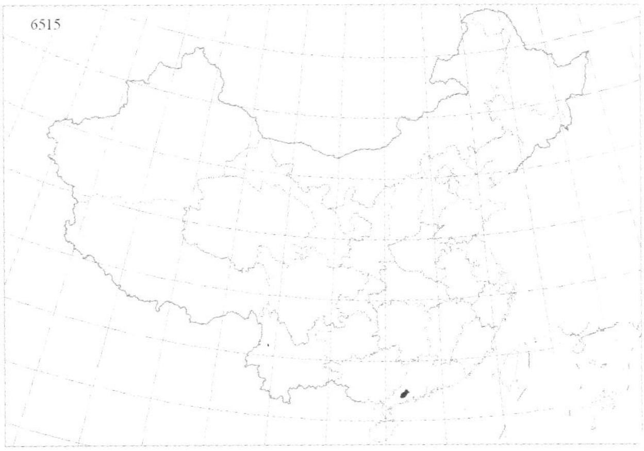

Theaceae 山茶科 *Camellia* 山茶属
Camellia brevistyla (Hayata) Cohen-Stuart 短柱油茶
Shrubs or small trees, evergreen

18.1 (12.1~23.0)；8.2 (1.1~15.3)；27.2 (22.8~30.1)；18.1 (12.2~23.0)；157.9 (95.0~216.2)；0.2 (0.0~8.5)；949 (696~1247) ❙ 1597 (1039~2859)；587 (356~1412)；183 (51~402) ❙ 70.8 (-5.0~235.6)；941 (696~1247)；242.1 (123.9~332.5)

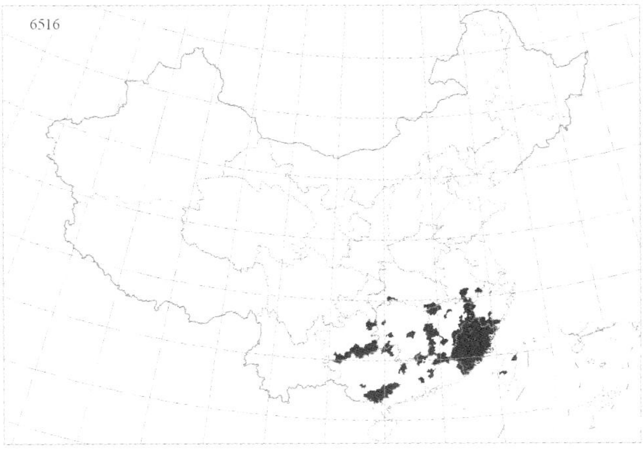

Theaceae 山茶科 *Camellia* 山茶属

Camellia candida H. T. Chang 白毛蕊茶

Small trees, evergreen

17.6; 9.9; 23.2; 17.7; 152.4; 0.0; 859 ❙ 1422; 813; 52 ❙ 65.6; 859; 359.7

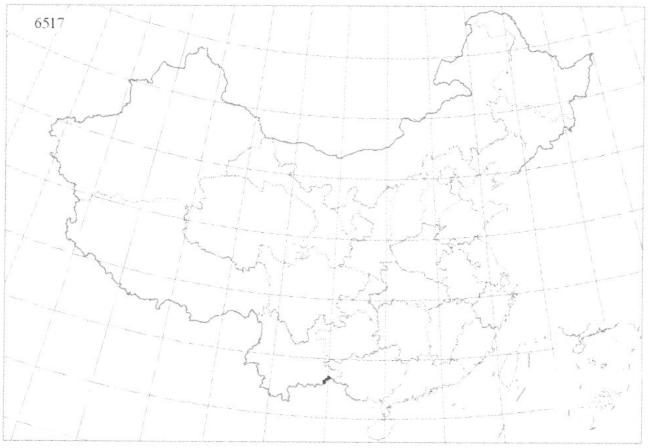

Theaceae 山茶科 *Camellia* 山茶属

Camellia caudata Wall. 长尾毛蕊茶

Shrubs or small trees, evergreen

20.2 (12.6~24.8); 11.9 (1.6~20.1); 27.0 (19.6~29.8); 20.2 (12.7~24.8); 182.9 (98.8~237.6); 0.1 (0.0~6.9); 1055 (706~1385) ❙ 1655 (868~3495); 721 (312~1847); 144 (41~897) ❙ 59.9 (-34.3~246.5); 1042 (706~1273); 256.0 (95.3~620.0)

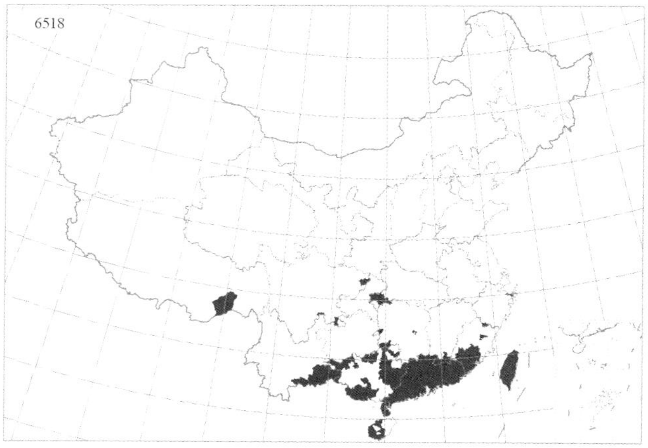

Theaceae 山茶科 *Camellia* 山茶属

Camellia caudata var. gracilis Wall. 小长尾毛蕊茶

Shrubs or small trees, evergreen

21.2; 12.6; 27.8; 21.2; 194.7; 0.0; 1114 ❙ 1395; 666; 101 ❙ 24.9; 1114; 273.0

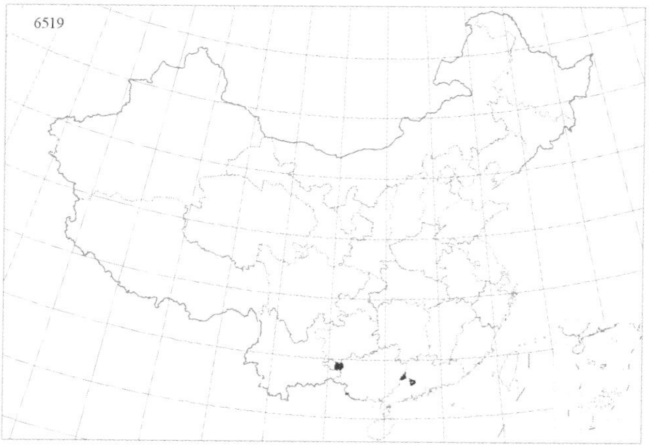

Theaceae 山茶科 *Camellia* 山茶属

Camellia chekiangoleosa Hu 浙江山茶

Small trees, evergreen

16.6 (13.4~19.0); 5.5 (1.5~9.3); 27.2 (24.0~29.4); 16.7 (13.4~19.1); 140.8 (106.7~168.6); 0.6 (0.0~7.3); 889 (744~989) ❙ 1692 (1198~1997); 590 (445~713); 205 (91~259) ❙ 91.2 (29.7~147.0); 885 (744~989); 241.9 (111.9~305.7)

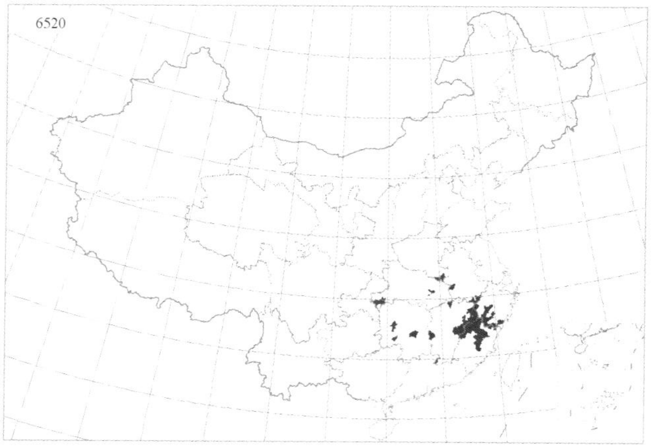

Theaceae 山茶科 *Camellia* 山茶属

Camellia chrysanthoides H. T. Chang 薄叶金花茶

Shrubs, evergreen

22.2; 13.9; 28.3; 22.3; 207.5; 0.0; 1190 ❙ 1374; 713; 87 ❙ 15.4; 1190; 327.7

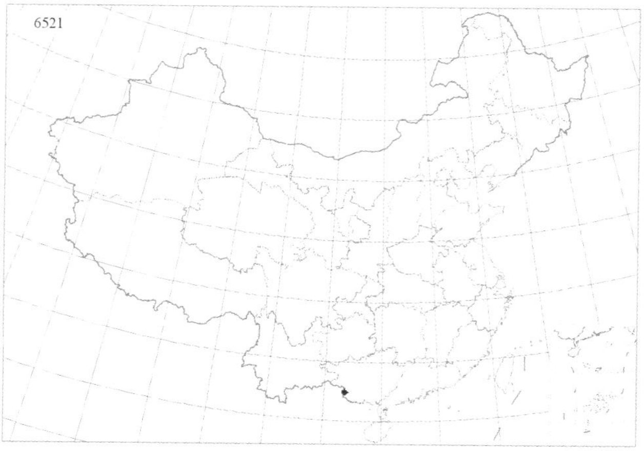

Theaceae 山茶科 *Camellia* 山茶属

Camellia cordifolia (F. P. Metcalf) Nakai 心叶毛蕊茶

Shrubs or small trees, evergreen

19.2 (14.3~22.7); 9.9 (3.3~15.0); 26.9 (21.4~29.8); 19.2 (14.3~22.8); 170.6 (114.1~213.1); 0.1 (0.0~2.3); 999 (765~1221) ❙ 1549 (1007~2435); 664 (442~1428); 140 (44~221) ❙ 55.9 (5.8~106.4); 997 (765~1221); 248.6 (95.3~551.6)

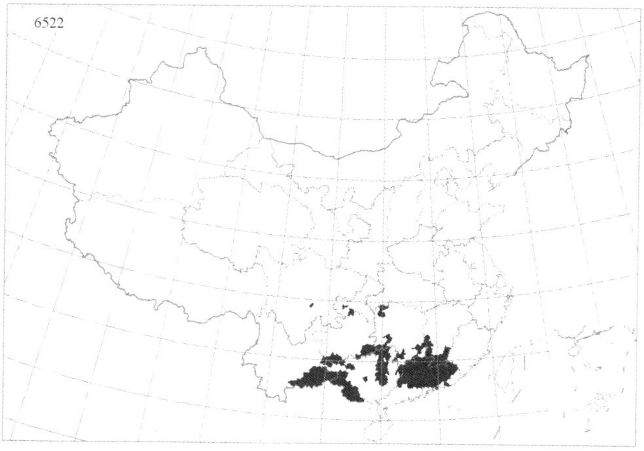

Theaceae 山茶科 *Camellia* 山茶属
Camellia cordifolia var. glabrisepala T. L. Ming 光萼心叶毛蕊茶
Shrubs or small trees, evergreen

17.2；9.1；23.1；17.3；147.5；0.0；847 ❙ 1161；633；50 ❙ 37.1；847；314.4

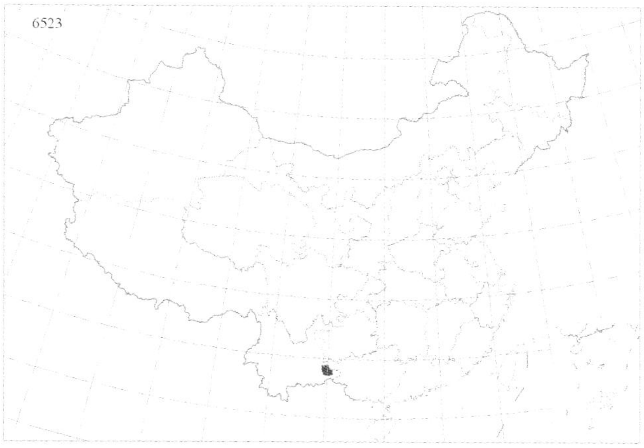

Theaceae 山茶科 *Camellia* 山茶属
Camellia costata H. T. Chang 突肋茶
Shrubs or trees, evergreen

17.3 (14.2~21.0)；7.9 (4.0~14.2)；25.1 (20.8~27.7)；17.4 (14.2~21.0)；148.6 (110.9~191.8)；0.1 (0.0~1.1)；888 (734~1053) ❙ 1355 (1236~1590)；631 (523~835)；96 (51~176) ❙ 53.4 (27.5~71.9)；888 (734~1053)；269.3 (215.4~551.6)

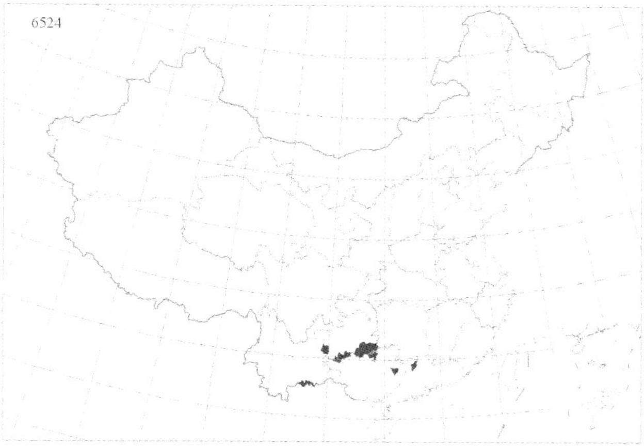

Theaceae 山茶科 *Camellia* 山茶属
Camellia costei H. Lév. 贵州连蕊茶
Shrubs or small trees, evergreen

16.2 (11.7~22.2)；5.6 (0.6~13.9)；25.8 (19.6~29.3)；16.2 (11.7~22.3)；135.3 (84.6~207.5)；0.6 (0.0~9.7)；855 (661~1190) ❙ 1338 (942~1990)；557 (444~827)；111 (39~233) ❙ 56.9 (15.4~107.2)；854 (661~1190)；216.1 (142.5~411.1)

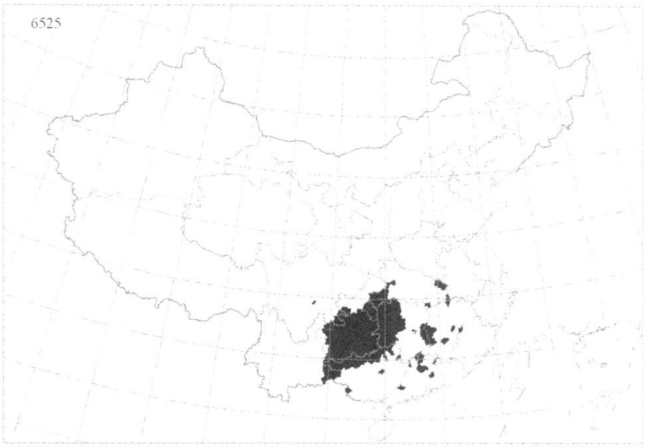

Theaceae 山茶科 *Camellia* 山茶属
Camellia crapnelliana Tutcher 红皮糙果茶
Small trees, evergreen

19.7 (15.0~23.0)；10.7 (4.0~15.2)；27.4 (24.3~29.0)；19.8 (15.1~23.0)；177.4 (120.8~216.2)；0.0 (0.0~1.0)；1037 (789~1247) ❙ 1727 (1302~2435)；725 (522~1428)；170 (86~254) ❙ 70.0 (12.9~147.0)；1037 (789~1247)；258.7 (157.7~335.1)

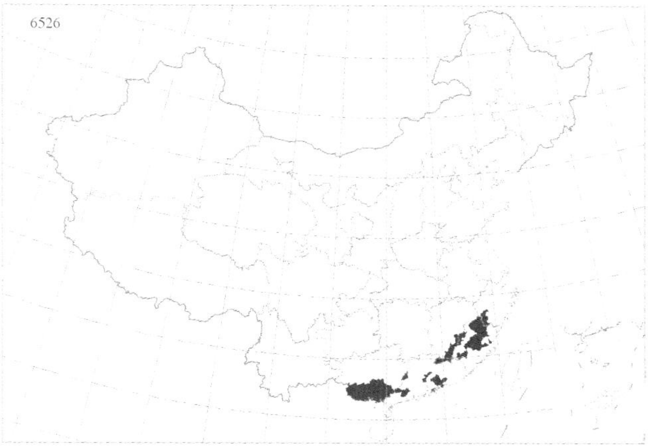

Theaceae 山茶科 *Camellia* 山茶属
Camellia crassicolumna H. T. Chang 厚轴茶
Small trees, evergreen

18.1 (16.3~21.0)；11.3 (9.1~14.2)；22.9 (21.4~25.6)；18.2 (16.4~21.0)；158.4 (136.4~191.8)；0.0 (0.0~0.0)；881 (797~1053) ❙ 1293 (1028~1522)；726 (566~860)；53 (45~68) ❙ 47.2 (12.5~80.6)；873 (797~1053)；405.8 (314.4~551.6)

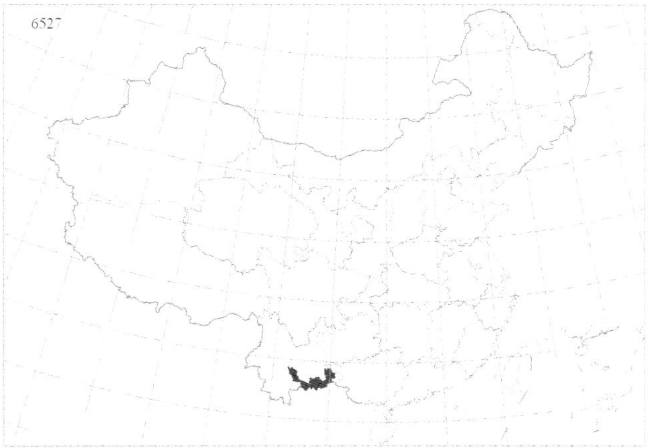

Theaceae 山茶科 *Camellia* 山茶属
Camellia crassicolumna var. multiplex T. L. Ming 光萼厚轴茶
Small trees, evergreen

16.3；9.1；21.4；16.4；136.4；0.0；797 ❙ 1190；656；50 ❙ 49.4；797；347.3

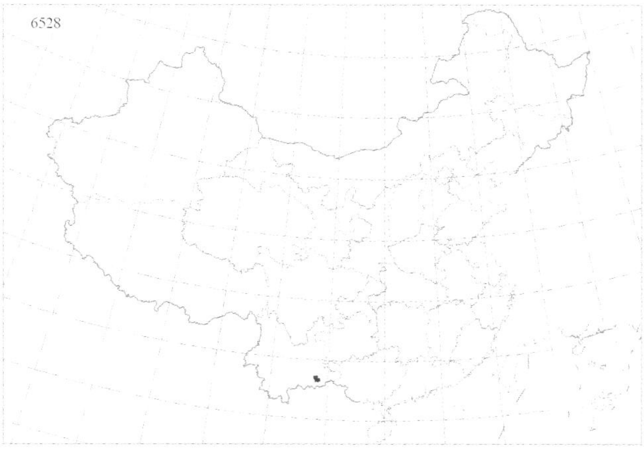

Theaceae 山茶科 *Camellia* 山茶属
Camellia cupiformis T. L. Ming 滇南连蕊茶
Shrubs or small trees, evergreen

16.3; 10.2; 20.3; 16.4; 136.6; 0.0; 789 ❙ 1232; 703; 47 ❙ 56.2; 780; 372.4

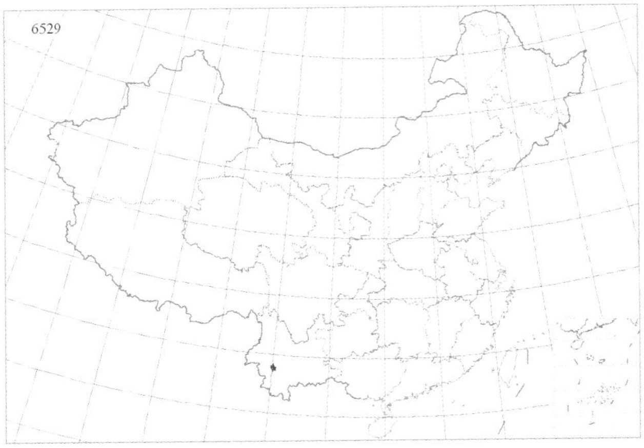

Theaceae 山茶科 *Camellia* 山茶属
Camellia cuspidata (Kochs) Wright ex Gard. 连蕊茶
Shrubs or trees, evergreen

16.6 (9.3~22.3); 5.5 (-1.7~14.2); 27.1 (18.8~30.1); 16.7 (9.5~22.4); 140.9 (69.8~208.2); 0.9 (0.0~18.0); 891 (603~1198) ❙ 1468 (807~2024); 541 (299~911); 161 (24~259) ❙ 65.2 (-6.7~147.0); 879 (603~1198); 220.0 (111.7~551.6)

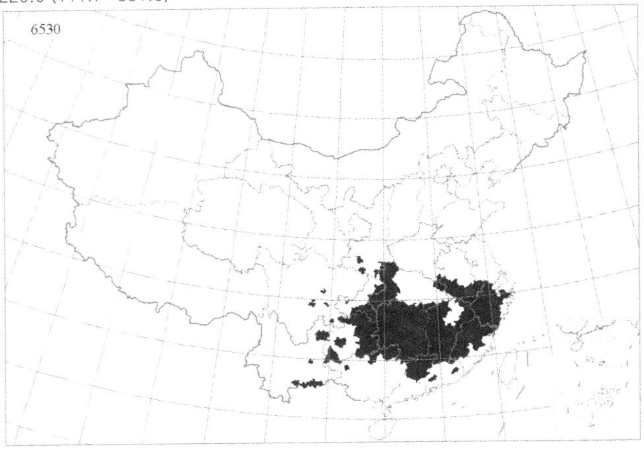

Theaceae 山茶科 *Camellia* 山茶属
Camellia cuspidata var. chekiangensis Sealy 浙江连蕊茶
Shrubs or trees, evergreen

17.2 (14.3~20.7); 6.8 (3.3~11.7); 27.1 (24.3~28.7); 17.3 (14.3~20.8); 147.6 (114.1~189.5); 0.1 (0.0~2.3); 909 (765~1094) ❙ 1645 (1363~1997); 612 (481~728); 188 (97~259) ❙ 82.3 (38.7~147.0); 909 (765~1094); 240.5 (176.3~303.4)

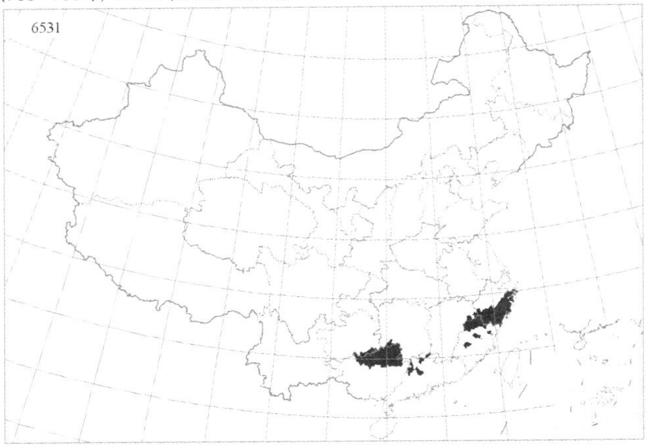

Theaceae 山茶科 *Camellia* 山茶属
Camellia cuspidata var. grandiflora Sealy 大花连蕊茶
Shrubs or trees, evergreen

18.2 (16.4~20.2); 8.0 (4.8~12.0); 27.3 (26.3~28.3); 18.2 (16.5~20.2); 158.7 (138.1~183.0); 0.0 (0.0~0.2); 950 (881~1045) ❙ 1553 (1409~1638); 544 (445~662); 183 (156~222) ❙ 63.7 (56.1~72.3); 948 (878~1045); 231.8 (169.7~270.0)

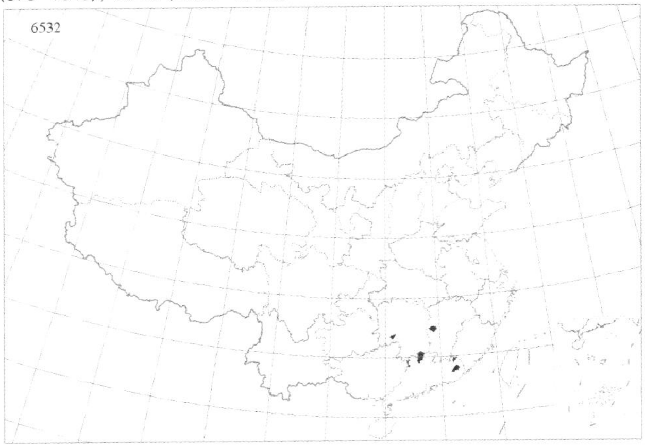

Theaceae 山茶科 *Camellia* 山茶属
Camellia cuspidata var. trichandra (H. T. Chang) Ming 毛丝连蕊茶
Shrubs or trees, evergreen

19.6 (17.6~21.8); 10.6 (7.3~13.7); 26.6 (25.5~27.9); 19.7 (17.6~21.8); 175.8 (151.6~202.2); 0.0 (0.0~0.0); 1014 (909~1154) ❙ 1299 (1188~1496); 659 (634~678); 80 (56~142) ❙ 30.0 (5.8~64.7); 1013 (909~1154); 291.1 (225.9~318.5)

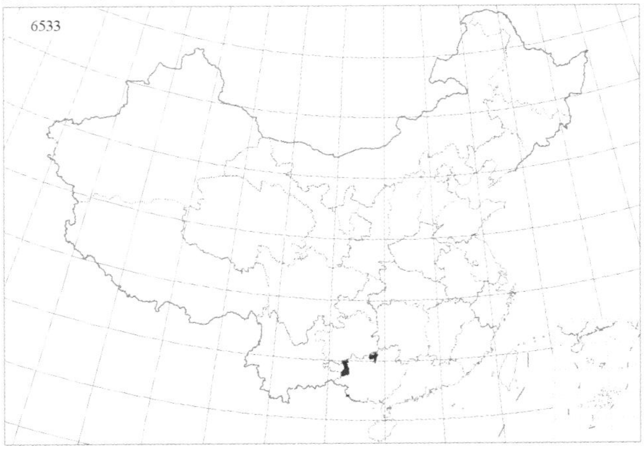

Theaceae 山茶科 *Camellia* 山茶属
Camellia drupifera Lour. 越南油茶
Shrubs or small trees, evergreen

22.2; 14.7; 28.0; 22.2; 206.9; 0.0; 1177 ❙ 1771; 842; 114 ❙ 50.4; 1177; 244.0

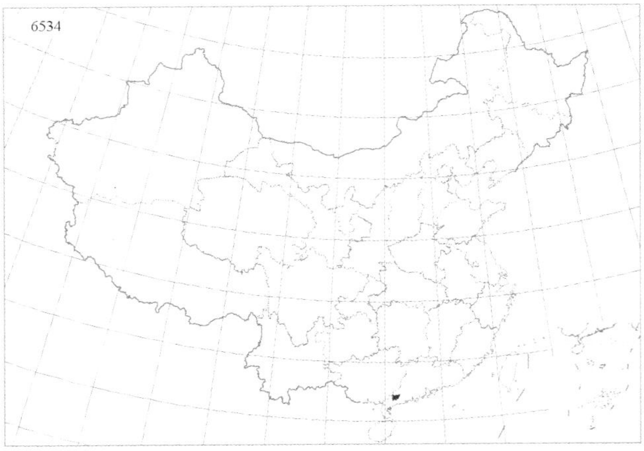

Theaceae 山茶科 *Camellia* 山茶属
Camellia edithae Hance 东南山茶
Shrubs or small trees, evergreen

18.4 (16.6~21.0); 8.8 (6.0~12.2); 27.1 (25.2~28.5); 18.5 (16.6~21.0); 161.4 (138.9~192.6); 0.0 (0.0~0.0); 954 (862~1107) | 1712 (1559~1997); 606 (541~669); 199 (156~254) | 80.7 (41.1~130.7); 954 (862~1107); 267.6 (208.1~305.7)

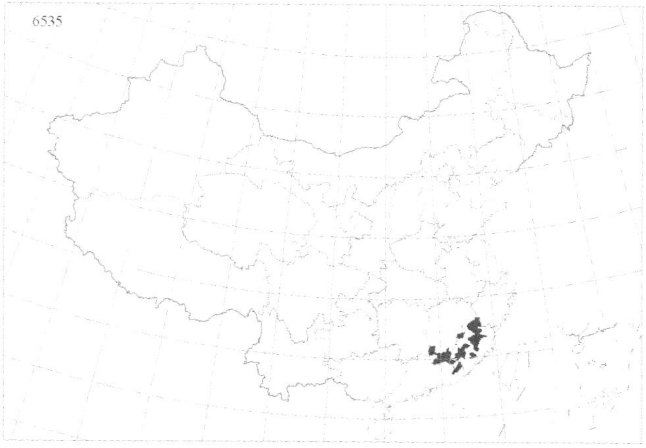

Theaceae 山茶科 *Camellia* 山茶属
Camellia elongata (Rehder et E. H. Wilson) Rehder 长管连蕊茶
Shrubs or small trees, evergreen

15.7 (14.3~17.5); 5.3 (3.8~7.6); 25.4 (23.5~26.6); 15.8 (14.3~17.6); 129.8 (112.6~150.7); 0.4 (0.0~1.5); 829 (756~912) | 1265 (1090~1577); 579 (473~932); 81 (49~106) | 53.2 (22.7~80.4); 829 (756~912); 194.9 (130.1~246.1)

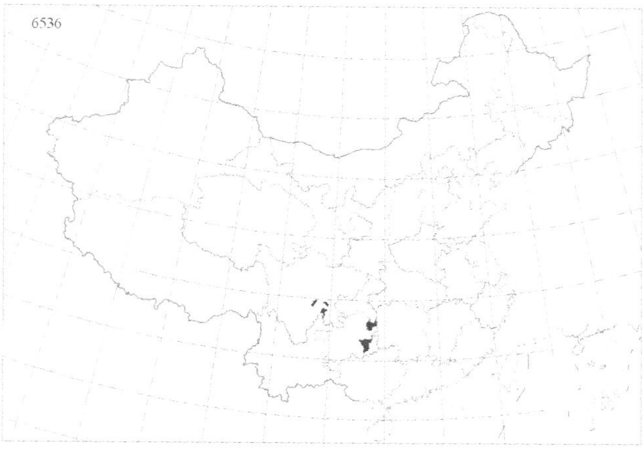

Theaceae 山茶科 *Camellia* 山茶属
Camellia euphlebia Merr. ex Sealy 显脉金花茶
Shrubs or small trees, evergreen

22.0 (21.3~22.7); 13.9 (13.2~15.0); 27.9 (27.3~28.6); 22.0 (21.3~22.8); 204.1 (195.9~2[illegible]3.1); 0.0 (0.0~0.0); 1166 (1111~1221) | 1804 (1302~2435); 999 (670~1428); 112 (86~130) | 54.6 (12.9~99.5); 1166 (1111~1221); 295 0 (235.1~327.7)

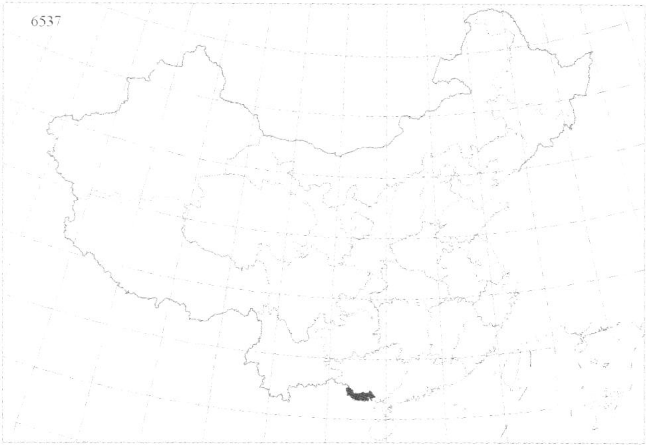

Theaceae 山茶科 *Camellia* 山茶属
Camellia euryoides Lindl. 柃叶连蕊茶
Shrubs or small trees, evergreen

18.0 (15.2~22.5); 7.4 (3.5~15.0); 27.7 (25.2~29.5); 18.0 (15.3~22.6); 156.3 (124.1~210.9); 0.2 (0.0~1.5); 951 (810~1206) | 1628 (1211~2030); 572 (443~1033); 195 (97~259) | 72.1 (28.8~130.7); 945 (810~1206); 235.5 (139.6~305.7)

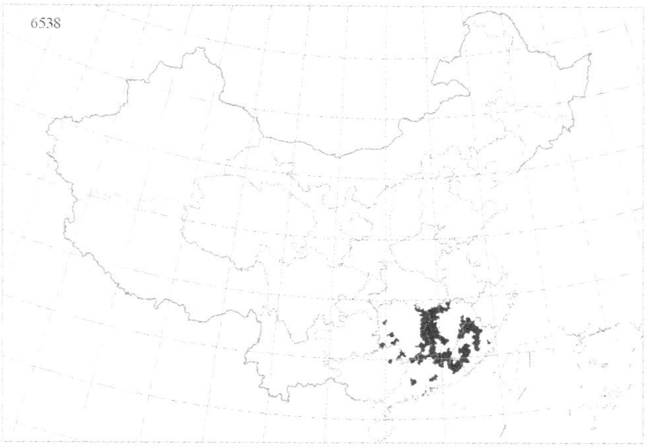

Theaceae 山茶科 *Camellia* 山茶属
Camellia euryoides var. nokoensis (Hayata) Ming 毛蕊柃叶连蕊茶
Shrubs or small trees, evergreen

16.4 (10.6~22.2); 5.5 (-0.2~17.7); 26.6 (20.3~28.9); 16.4 (10.6~22.2); 137.7 (79.4~206.7); 0.7 (0.0~12.1); 874 (643~1123) | 1361 (1028~2992); 534 (419~1847); 117 (46~303) | 56.9 (8.0~219.3); 872 (643~1123); 139.4 (108.2~303.6)

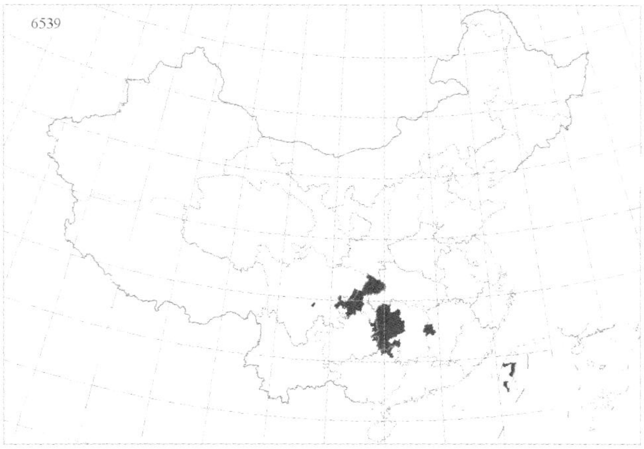

Theaceae 山茶科 *Camellia* 山茶属
Camellia fangchengensis S. Y. Liang et Y. C. Zhong 防城茶
Small trees, evergreen

21.8; 14.0; 27.7; 21.9; 202.2; 0.0; 1150 | 2148; 1234; 124 | 86.7; 1150; 318.8

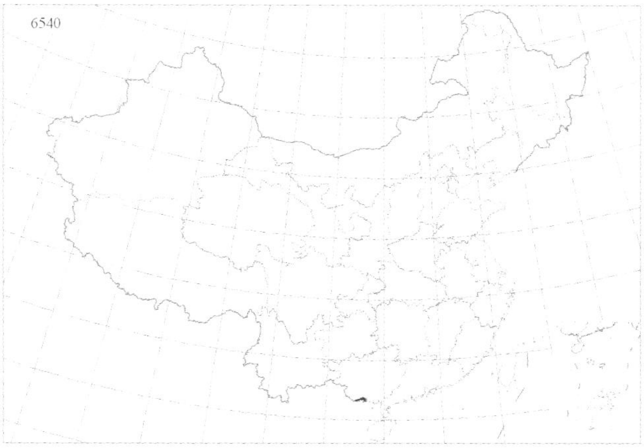

Theaceae 山茶科 *Camellia* 山茶属
Camellia fascicularis H. T. Chang 云南金花茶
Shrubs or small trees, evergreen

18.2 (16.9~21.0); 11.4 (9.9~14.2); 22.9 (21.4~25.6); 18.2 (16.9~21.0); 158.3 (142.7~191.8); 0.0 (0.0~0.0); 883 (813~1053) ❙ 1347 (1163~1522); 745 (641~860); 56 (48~68) ❙ 53.0 (34.8~80.6); 883 (813~1053); 449.4 (383.4~551.6)

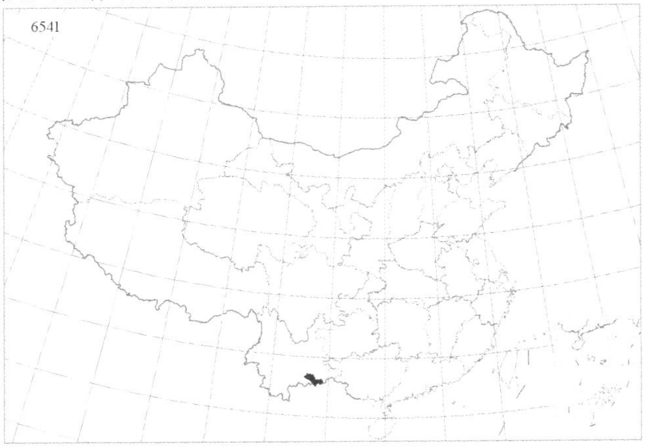

Theaceae 山茶科 *Camellia* 山茶属
Camellia flavida H. T. Chang 淡黄金花茶
Shrubs, evergreen

21.6 (19.2~22.3); 13.5 (12.6~13.9); 27.5 (22.3~28.9); 21.6 (19.2~22.3); 199.7 (170.5~208.1); 0.0 (0.0~0.0); 1138 (909~1202) ❙ 1513 (1374~1625); 770 (686~850); 113 (58~173) ❙ 34.0 (15.4~64.9); 1134 (880~1202); 303.4 (201.4~513.9)

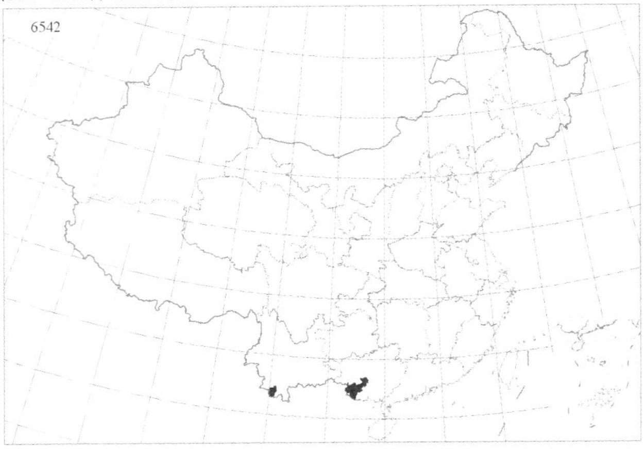

Theaceae 山茶科 *Camellia* 山茶属
Camellia fluviatilis Hand.-Mazz. 窄叶油茶
Shrubs, evergreen

21.7 (17.7~24.5); 13.8 (7.7~18.9); 27.9 (25.8~29.0); 21.7 (17.8~24.5); 200.5 (153.6~234.4); 0.0 (0.0~0.0); 1160 (913~1360) ❙ 1497 (868~1900); 581 (312~1056); 125 (42~182) ❙ 32.4 (-34.3~71.0); 1093 (868~1360); 299.1 (218.1~455.2)

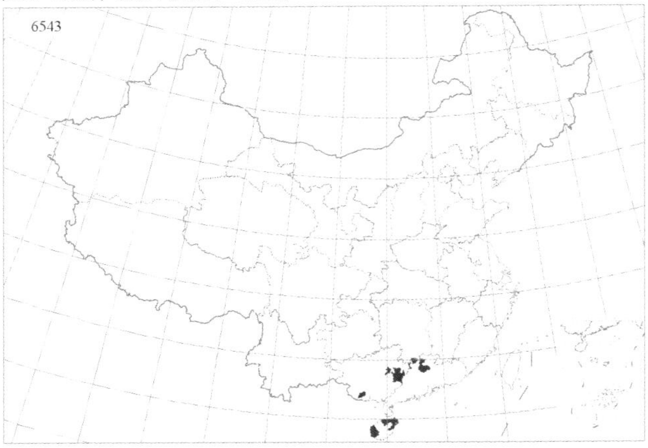

Theaceae 山茶科 *Camellia* 山茶属
Camellia fluviatilis var. megalantha (H. T. Chang) Ming 大花窄叶油茶
Shrubs, evergreen

17.6; 5.5; 29.3; 17.6; 151.3; 0.0; 958 ❙ 991; 351; 121 ❙ 3.2; 796; 176.5

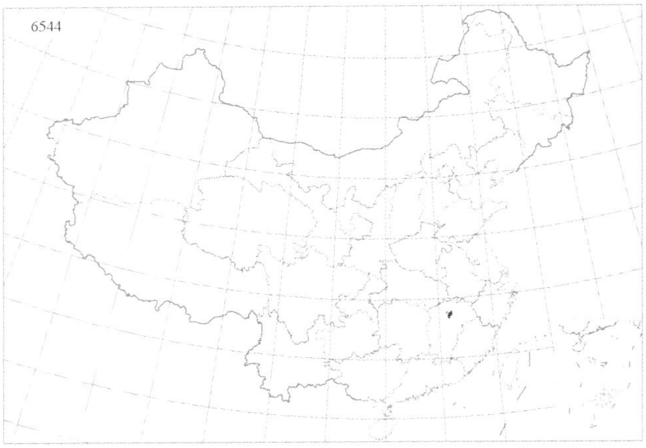

Theaceae 山茶科 *Camellia* 山茶属
Camellia forrestii (Diels) Cohen-Stuart 云南连蕊茶
Shrubs or small trees, evergreen

16.3 (11.2~19.3); 9.7 (4.6~13.3); 21.0 (16.7~23.9); 16.3 (11.3~19.4); 136.0 (75.8~173.2); 0.0 (0.0~0.4); 800 (642~939) ❙ 1089 (797~1547); 620 (411~911); 45 (23~70) ❙ 36.5 (-14.2~99.9); 777 (642~923); 336.9 (218.0~551.3)

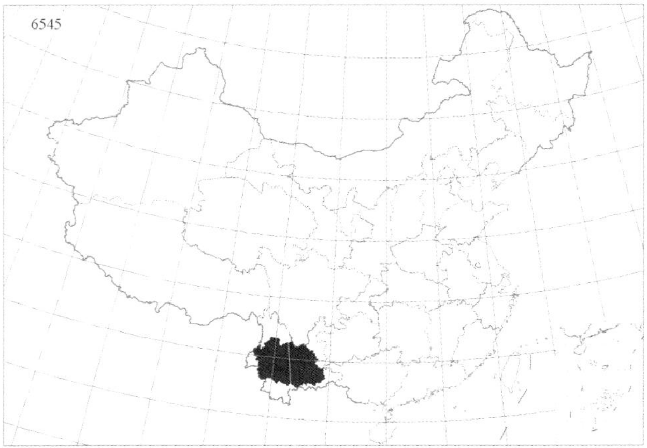

Theaceae 山茶科 *Camellia* 山茶属
Camellia forrestii var. acutisepala (Tsai et K. M. Feng) H. T. Chang 尖萼云南连蕊茶
Shrubs or small trees, evergreen

15.9 (14.6~17.3); 9.1 (7.9~11.0); 20.7 (19.6~21.6); 15.9 (14.6~17.3); 130.8 (115.8~148.0); 0.0 (0.0~0.0); 779 (736~832) ❙ 1058 (956~1190); 605 (573~656); 43 (34~50) ❙ 35.7 (26.1~49.4); 761 (716~797); 333.7 (272.5~405.0)

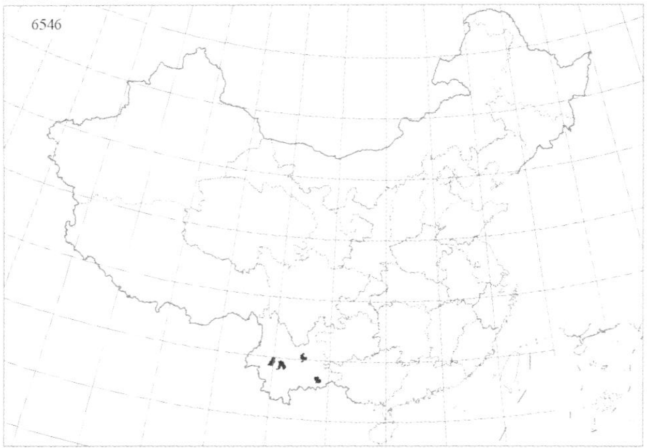

Theaceae 山茶科 *Camellia* 山茶属

Camellia fraterna Hance 毛花连蕊茶

Shrubs or small trees, evergreen

17.4 (12.0~24.1); 6.7 (0.6~18.3); 27.7 (23.1~30.1); 17.5 (12.1~24.2); 150.2 (94.7~230.2); 0.6 (0.0~9.7); 930 (697~1327) | 1559 (871~2435); 560 (299~1428); 188 (64~259) | 69.3 (-6.7~147.0); 914 (697~1294); 235.9 (123.9~474.7)

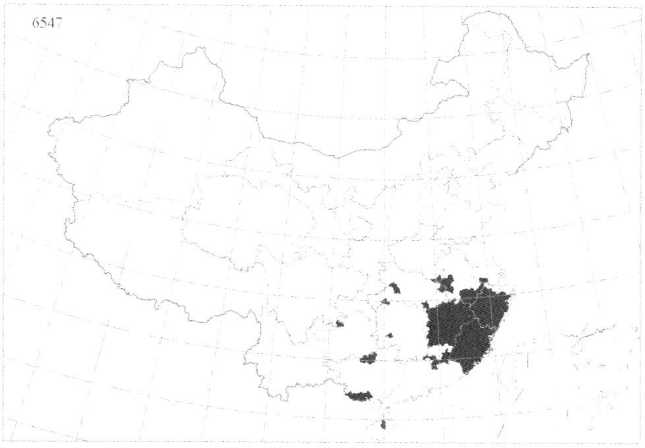

Theaceae 山茶科 *Camellia* 山茶属

Camellia furfuracea (Merr.) Cohen-Stuart 糙果茶

Shrubs or small trees, evergreen

20.4 (15.0~24.8); 11.6 (5.0~20.1); 27.7 (24.7~29.0); 20.5 (15.1~24.8); 185.8 (120.8~237.6); 0.0 (0.0~0.0); 1078 (789~1385) | 1673 (1302~2435); 710 (473~1428); 153 (41~242) | 57.1 (1.5~147.0); 1069 (789~1247); 260.0 (188.8~620.0)

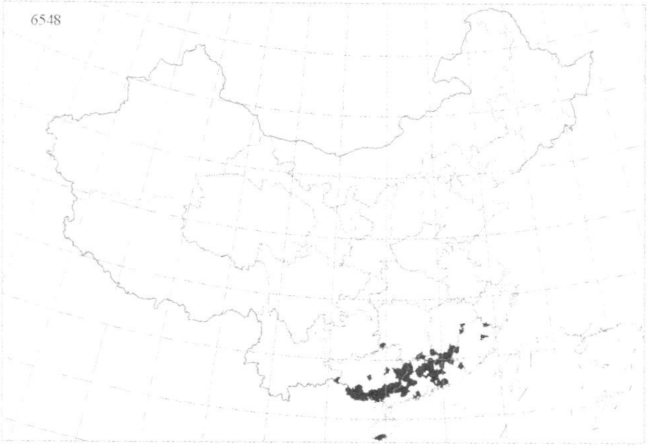

Theaceae 山茶科 *Camellia* 山茶属

Camellia furfuracea var. latipetiolate (Chi) Ming 阔柄糙果茶

Shrubs or small trees, evergreen

21.0; 13.2; 26.9; 21.0; 192.3; 0.0; 1081 | 1776; 828; 125 | 64.3; 1081; 254.1

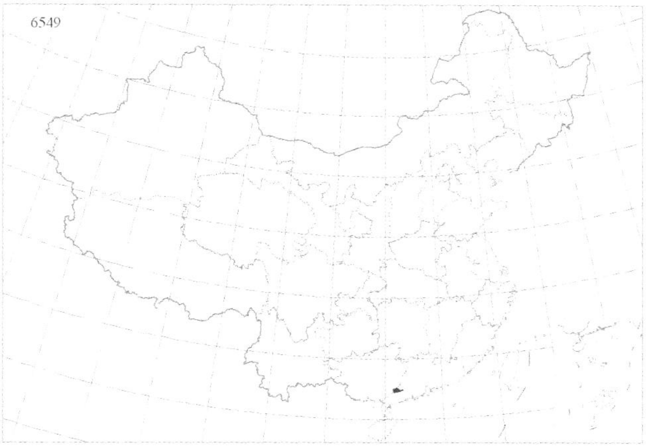

Theaceae 山茶科 *Camellia* 山茶属

Camellia gaudichaudii (Gagnep.) Sealy 硬叶糙果茶

Small trees, evergreen

21.6; 13.7; 27.7; 21.6; 199.6; 0.0; 1150 | 1889; 851; 145 | 66.3; 1150; 276.1

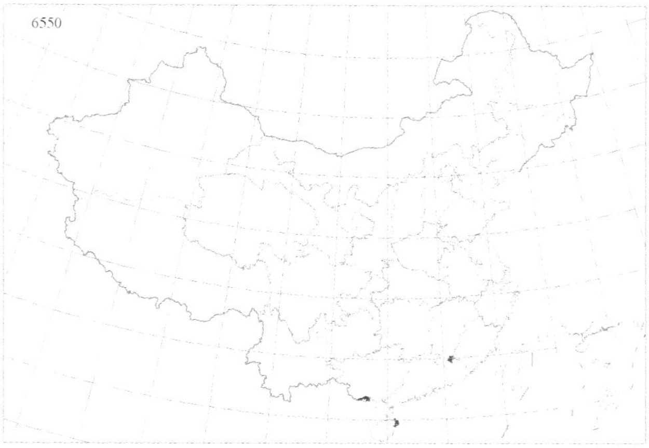

Theaceae 山茶科 *Camellia* 山茶属

Camellia gilbertii (A. Chev.) Sealy 中越短蕊茶

Shrubs, evergreen

21.0; 14.2; 25.6; 21.0; 191.8; 0.0; 1053 | 1522; 818; 68 | 44.5; 1053; 551.6

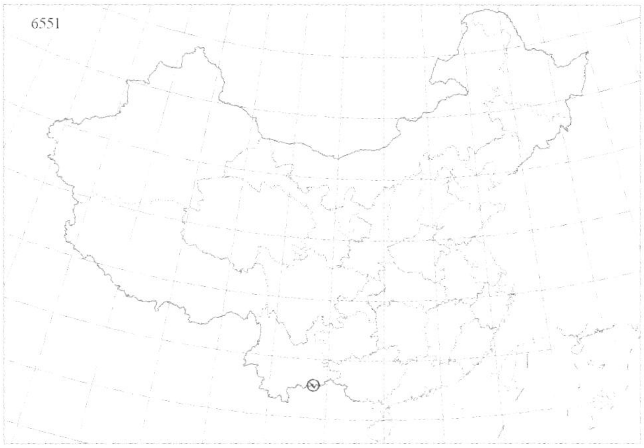

Theaceae 山茶科 *Camellia* 山茶属

Camellia glabricostata Ming 秃肋连蕊茶

Shrubs or trees, evergreen

21.8; 13.7; 27.7; 21.8; 202.2; 0.0; 1154 | 1302; 670; 86 | 12.9; 1154; 304.0

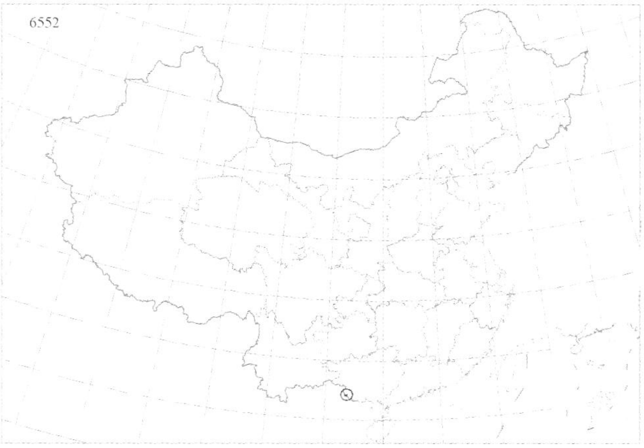

Theaceae 山茶科 *Camellia* 山茶属
Camellia gracilipes Merr. ex Sealy 狭叶长梗茶
Shrubs, evergreen

21.8; 14.0; 27.7; 21.9; 202.2; 0.0; 1150 ❙ 2148; 1234; 124 ❙ 86.7; 1150; 318.8

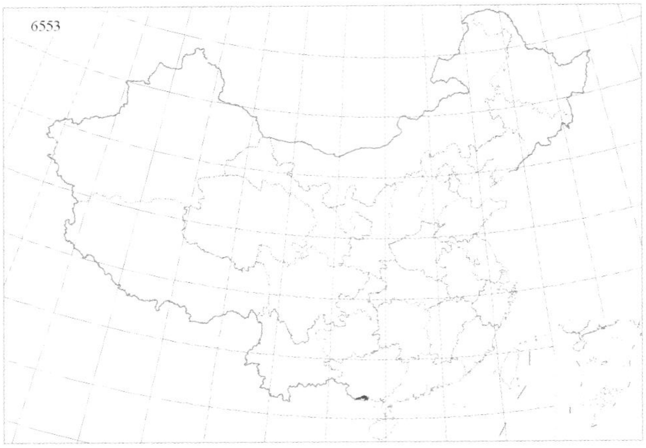

Theaceae 山茶科 *Camellia* 山茶属
Camellia grandibracteata H. T. Chang et F. L. Yu 大苞茶
Trees, evergreen

17.1; 10.8; 21.4; 17.2; 145.8; 0.0; 823 ❙ 1102; 622; 46 ❙ 33.9; 792; 394.7

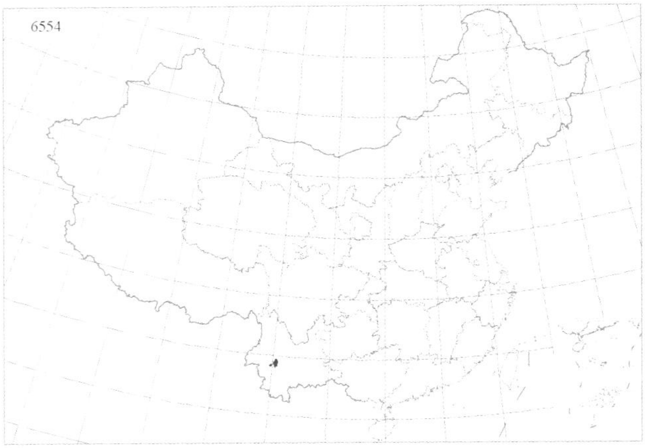

Theaceae 山茶科 *Camellia* 山茶属
Camellia granthamiana Sealy 大苞白山茶
Trees, evergreen

21.8 (20.4~22.4); 14.0 (12.0~15.2); 28.2 (27.7~28.7); 21.9 (20.5~22.7); 202.7 (185.5~212.7); 0.0 (0.0~0.0); 1155 (1057~1216) ❙ 1831 (1462~2037); 837 (599~1033); 131 (104~162) ❙ 58.2 (29.9~72.1); 1155 (1057~1216); 206.0 (163.7~266.9)

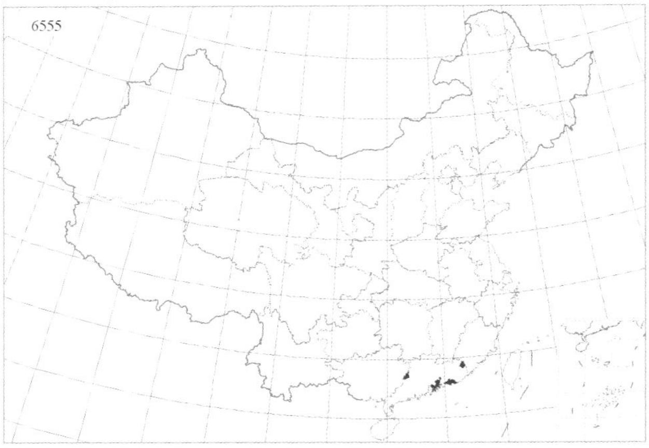

Theaceae 山茶科 *Camellia* 山茶属
Camellia grijsii Hance 长瓣短柱茶
Shrubs, evergreen

16.1 (10.6~18.7); 5.2 (-0.2~9.3); 26.4 (21.1~29.1); 16.1 (10.6~18.8); 134.8 (79.4~166.0); 1.7 (0.0~12.1); 862 (643~1002) ❙ 1296 (845~1847); 515 (375~816); 100 (24~242) ❙ 51.9 (7.8~107.2); 862 (643~1002); 198.2 (108.2~302.2)

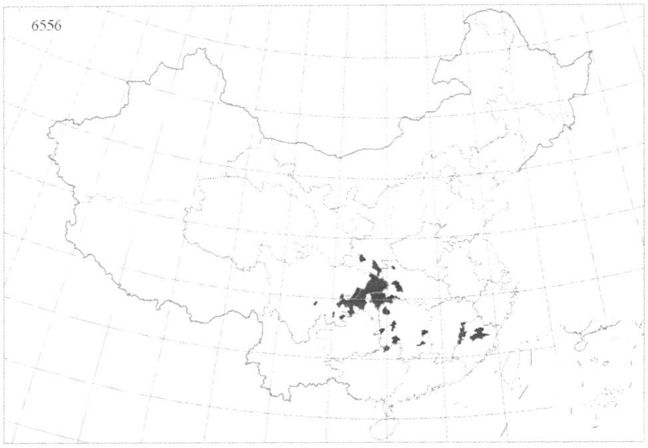

Theaceae 山茶科 *Camellia* 山茶属
Camellia grijsii var. shensiensis (H. T. Chang) Ming 小叶短柱茶
Shrubs, evergreen

15.8 (12.0~18.7); 4.7 (0.2~8.1); 26.3 (22.5~28.9); 15.8 (12.0~18.8); 132.9 (92.3~166.0); 3.3 (0.0~10.8); 859 (690~1002) ❙ 1063 (845~1410); 467 (375~816); 52 (24~66) ❙ 25.7 (7.8~80.4); 859 (690~1002); 184.0 (108.2~286.7)

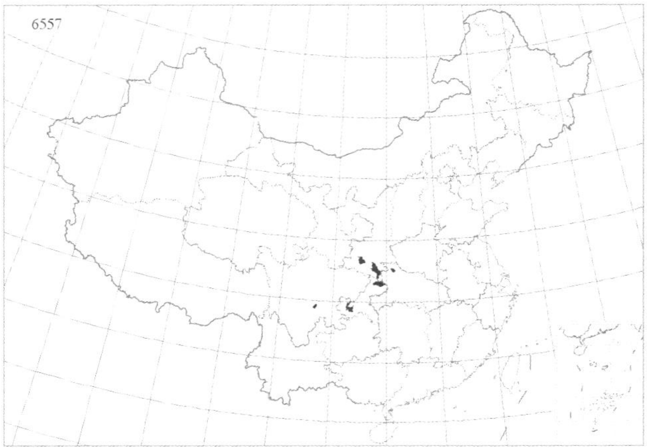

Theaceae 山茶科 *Camellia* 山茶属
Camellia gymnogyna H. T. Chang 秃房茶
Shrubs or trees, evergreen

17.1 (10.6~22.2); 7.7 (-0.2~14.7); 25.4 (18.6~28.9); 17.1 (10.6~22.2); 146.1 (79.4~206.9); 0.5 (0.0~12.1); 885 (643~1177) ❙ 1242 (896~2148); 587 (442~1234); 69 (31~191) ❙ 41.2 (6.5~93.8); 880 (643~1177); 243.2 (108.2~551.6)

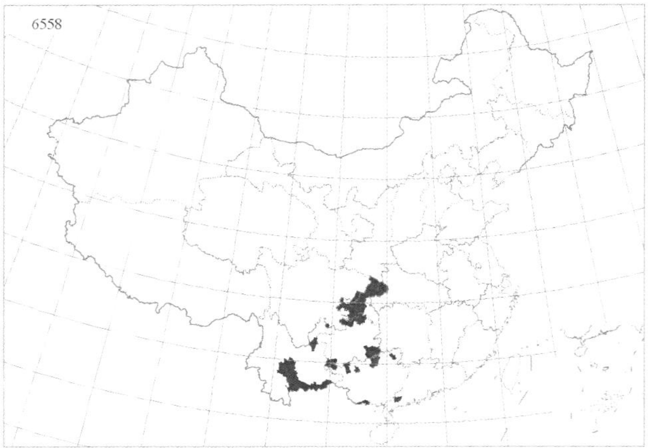

Theaceae 山茶科 *Camellia* 山茶属

Camellia hekouensis C. J. Wang et G. S. Fan 河口长梗茶

Shrubs or trees, evergreen

21.0; 14.2; 25.6; 21.0; 191.8; 0.0; 1053 ❙ 1522; 818; 68 ❙ 44.5; 1053; 551.6

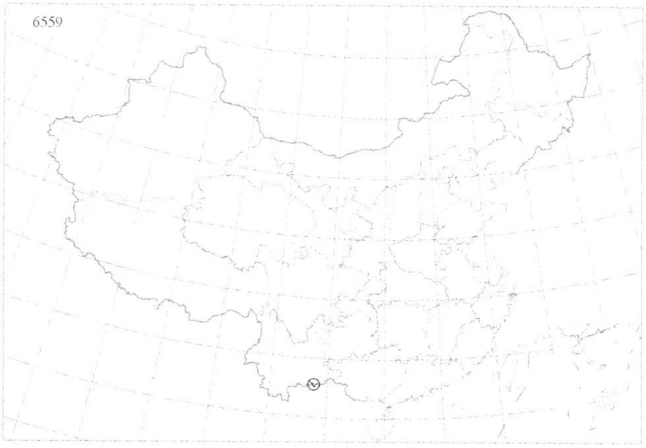

Theaceae 山茶科 *Camellia* 山茶属

Camellia hongkongensis Seem. 香港山茶

Trees, evergreen

22.0 (19.5~23.6); 14.3 (10.6~16.7); 28.3 (26.9~29.2); 22.1 (19.6~23.6); 205.0 (175.1~223.5); 0.0 (0.0~0.0); 1170 (1003~1281) ❙ 1670 (1187~2037); 771 (539~1033); 116 (75~191) ❙ 43.3 (-6.9~95.5); 1168 (1003~1273); 208.9 (116.6~275.2)

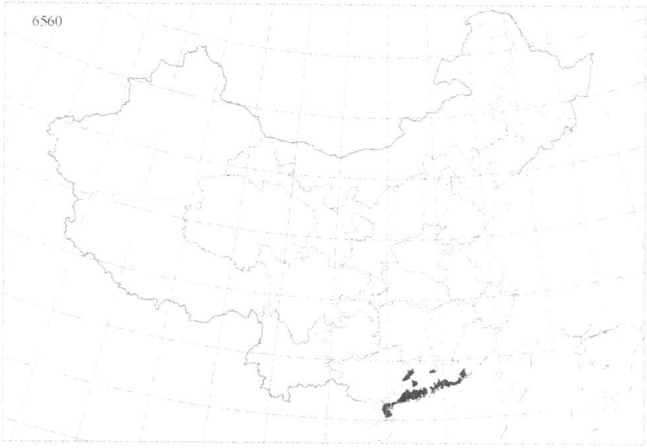

Theaceae 山茶科 *Camellia* 山茶属

Camellia huana T. L. Ming et W. J. Zhang 贵州金花茶

Shrubs, evergreen

18.1 (15.9~19.1); 9.1 (6.9~10.1); 25.3 (23.0~26.6); 18.2 (16.0~19.2); 158.0 (131.6~170.1); 0.0 (0.0~0.0); 918 (802~981) ❙ 1265 (1232~1321); 668 (617~701); 60 (51~73) ❙ 38.6 (27.5~64.7); 918 (802~981); 256.8 (189.5~322.1)

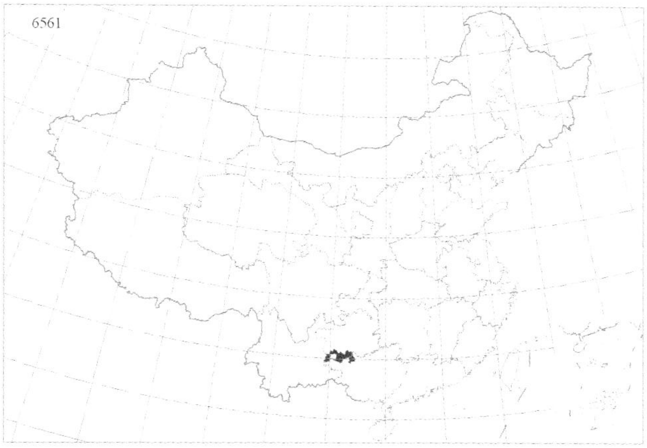

Theaceae 山茶科 *Camellia* 山茶属

Camellia ilicifolia Y. K. Li ex H. T. Chang 冬青叶瘤果茶

Shrubs or small trees, evergreen

16.6; 6.5; 25.6; 16.7; 140.0; 0.0; 859 ❙ 1235; 570; 88 ❙ 43.4; 859; 213.2

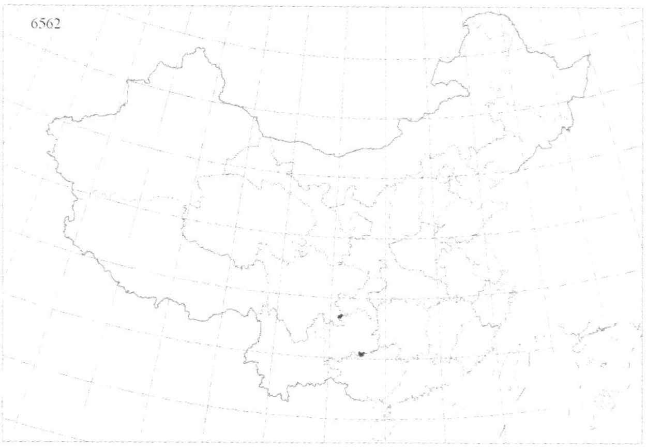

Theaceae 山茶科 *Camellia* 山茶属

Camellia ilicifolia var. neriifolia (H. T. Chang) T. L. Ming 狭叶瘤果茶

Shrubs or small trees, evergreen

15.9; 5.5; 25.6; 15.9; 131.1; 0.0; 833 ❙ 1089; 503; 69 ❙ 30.9; 833; 181.2

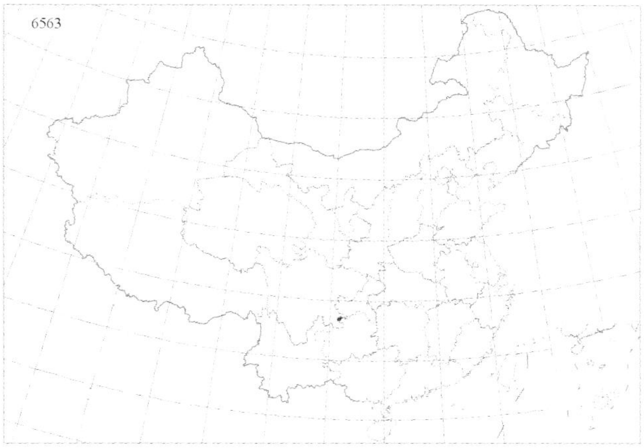

Theaceae 山茶科 *Camellia* 山茶属

Camellia impressinervis H. T. Chang et S. Y. Liang 凹脉金花茶

Shrubs, evergreen

22.2; 13.9; 28.3; 22.3; 207.5; 0.0; 1190 ❙ 1374; 713; 87 ❙ 15.4; 1190; 327.7

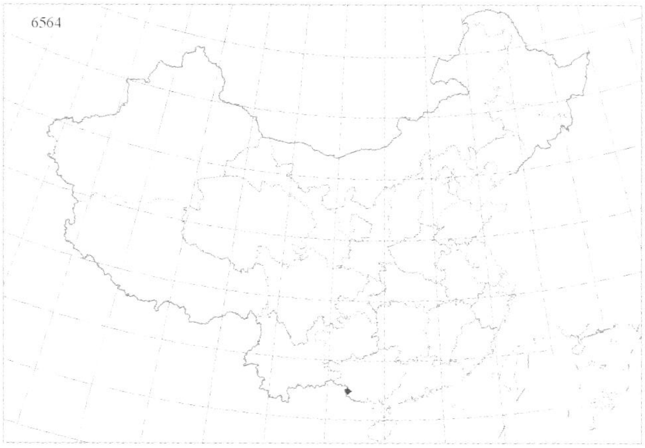

Theaceae 山茶科 *Camellia* 山茶属
Camellia indochinensis Merr. 柠檬金花茶
Shrubs, evergreen

19.7 (16.8~22.7); 11.5 (7.8~15.0); 26.0 (22.6~28.6); 19.8 (16.8~22.8); 177.4 (141.5~213.1); 0.0 (0.0~0.0); 1012 (836~1221) ❙ 1495 (1232~2435); 813 (617~1428); 79 (51~130) ❙ 47.2 (15.4~99.5); 1011 (836~1221); 302.9 (189.5~506.2)

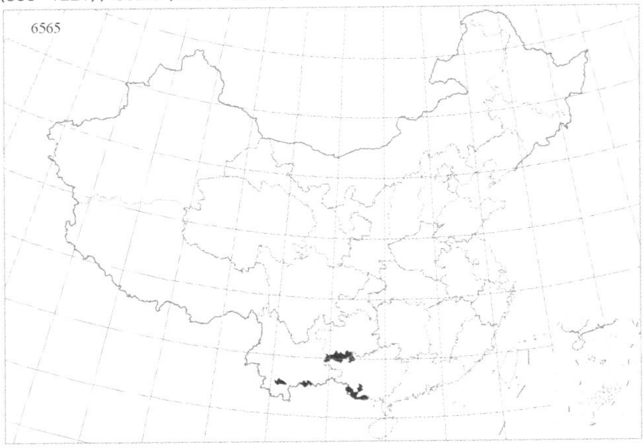

Theaceae 山茶科 *Camellia* 山茶属
Camellia indochinensis var. tunghinensis (H. T. Chang) T. L. Ming et W. J. Zhang 东兴金花茶
Shrubs, evergreen

22.3; 14.5; 28.1; 22.3; 207.7; 0.0; 1186 ❙ 2291; 1331; 127 ❙ 93.1; 1186; 309.7

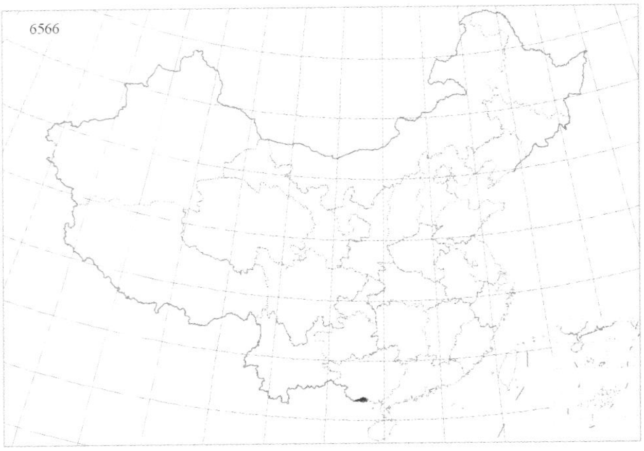

Theaceae 山茶科 *Camellia* 山茶属
Camellia japonica L. 山茶
Shrubs or small trees, evergreen

18.3 (11.9~24.8); 9.4 (-2.0~20.1); 26.1 (18.8~29.5); 18.3 (12.2~24.8); 160.6 (100.7~237.6); 0.5 (0.0~17.1); 961 (706~1385) ❙ 1568 (747~3495); 651 (312~1847); 149 (29~897) ❙ 65.8 (-34.3~246.5); 935 (690~1360); 281.8 (119.4~620.0)

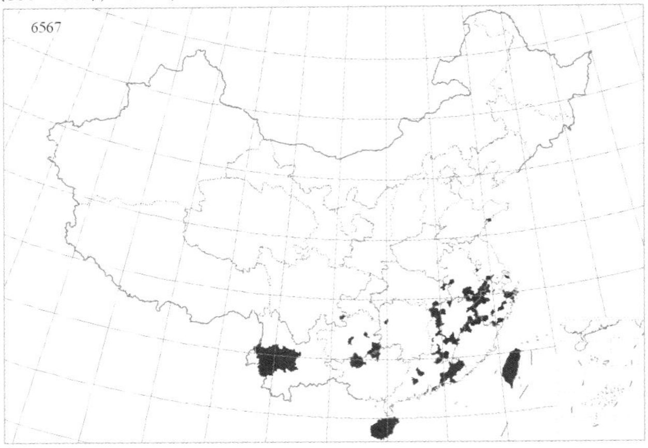

Theaceae 山茶科 *Camellia* 山茶属
Camellia kissii Wall. 落瓣油茶
Shrubs or small trees, evergreen

19.2 (11.3~24.3); 11.8 (3.8~19.1); 24.6 (17.0~28.9); 19.2 (11.3~24.3); 170.9 (76.6~231.9); 0.0 (0.0~1.5); 977 (644~1338) ❙ 1503 (939~2435); 743 (392~1428); 95 (39~191) ❙ 55.4 (-11.9~112.2); 964 (644~1331); 353.9 (163.7~659.0)

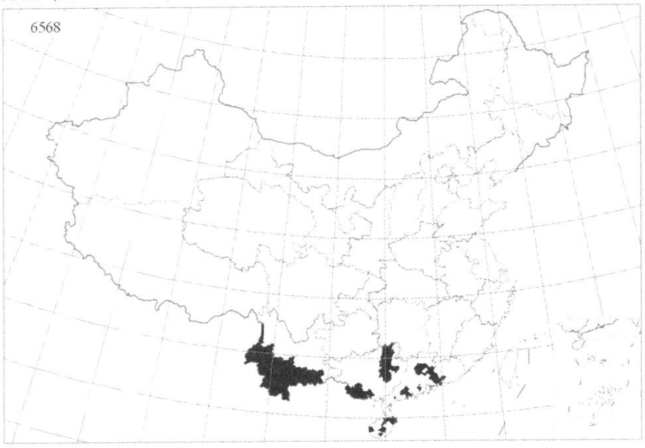

Theaceae 山茶科 *Camellia* 山茶属
Camellia kissii var. confusa (Craib) Ming 大叶落瓣油茶
Shrubs or small trees, evergreen

20.8 (17.9~22.7); 13.8 (11.9~15.1); 25.6 (21.6~28.6); 20.8 (18.0~22.8); 190.1 (155.8~213.1); 0.0 (0.0~0.0); 1061 (854~1221) ❙ 1635 (1302~2435); 854 (655~1428); 88 (41~130) ❙ 54.9 (12.9~99.5); 1049 (833~1221); 392.9 (235.1~659.0)

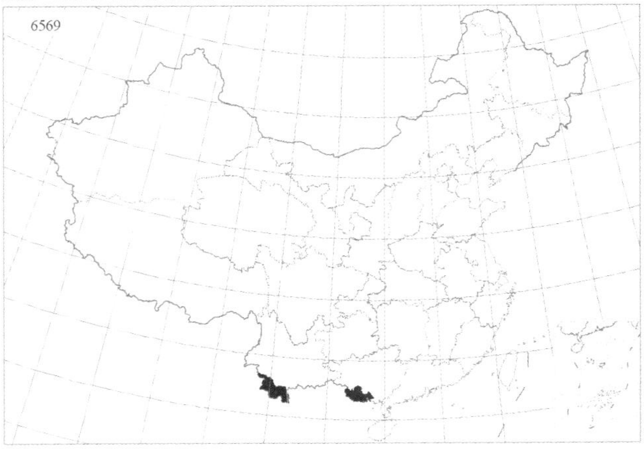

Theaceae 山茶科 *Camellia* 山茶属
Camellia kwangsiensis (H. T. Chang et B. H. Chen) T. L. Ming 广西茶
Shrubs or small trees, evergreen

17.9 (16.3~21.8); 9.9 (7.8~13.7); 24.0 (21.4~27.7); 18.0 (16.4~21.8); 155.8 (136.4~202.2); 0.0 (0.0~0.0); 893 (797~1154) ❙ 1301 (1018~1590); 673 (547~860); 73 (43~177) ❙ 46.7 (12.9~80.6); 893 (797~1154); 333.9 (228.7~491.5)

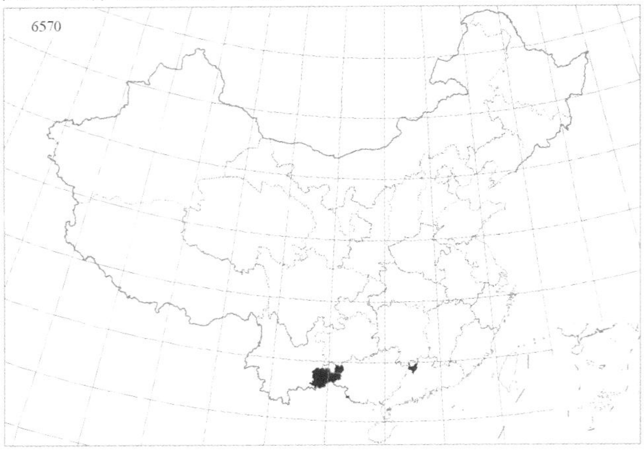

Theaceae 山茶科 *Camellia* 山茶属
Camellia kwangsiensis var. kwangnanica (H. T. Chang et B. H. Chen) T. L. Ming 毛萼广西茶 Small trees, evergreen

17.2; 9.1; 23.1; 17.3; 147.5; 0.0; 847 | 1161; 633; 50 | 37.1; 847; 314.4

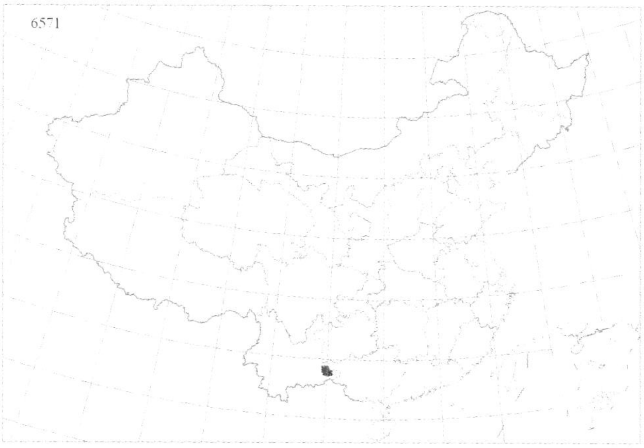

Theaceae 山茶科 *Camellia* 山茶属
Camellia lawii Sealy 四川毛蕊茶
Shrubs, evergreen

16.6 (10.6~18.7); 5.8 (-0.2~8.1); 26.8 (21.1~28.9); 16.6 (10.6~18.8); 140.2 (79.4~166.0); 0.9 (0.0~12.1); 887 (643~1002) | 1165 (1028~1410); 495 (442~816); 62 (46~76) | 33.5 (8.0~93.8); 887 (643~1002); 161.2 (108.2~284.3)

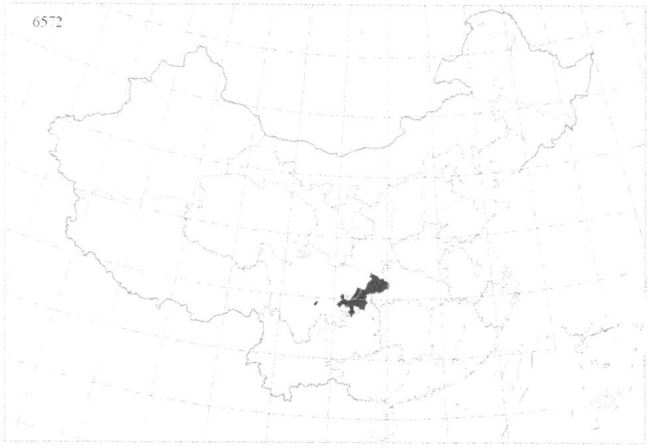

Theaceae 山茶科 *Camellia* 山茶属
Camellia leptophylla S. Y. Liang ex H. T. Chang 膜叶茶
Shrubs or small trees, evergreen

22.2; 13.9; 28.3; 22.3; 207.5; 0.0; 1190 | 1374; 713; 87 | 15.4; 1190; 327.7

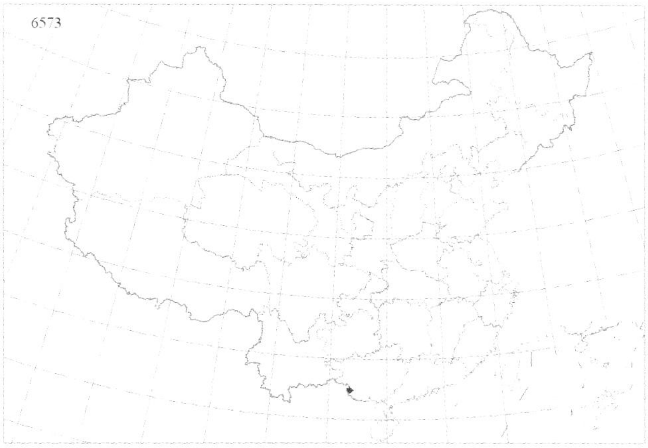

Theaceae 山茶科 *Camellia* 山茶属
Camellia longicalyx H. T. Chang 长萼连蕊茶
Small trees, evergreen

18.8; 8.9; 27.5; 18.9; 166.5; 0.0; 997 | 1574; 601; 171 | 62.3; 997; 266.0

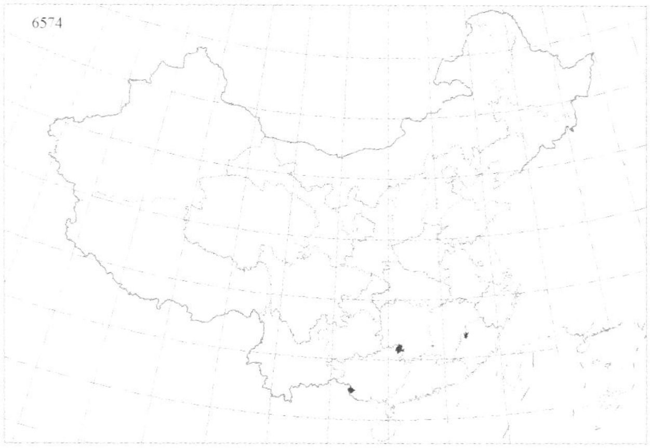

Theaceae 山茶科 *Camellia* 山茶属
Camellia longipedicellata (Hu) H. T. Chang et D. Fang 长梗茶
Shrubs, evergreen

20.3; 10.9; 27.8; 20.4; 184.6; 0.0; 1068 | 1576; 762; 153 | 47.5; 1068; 218.5

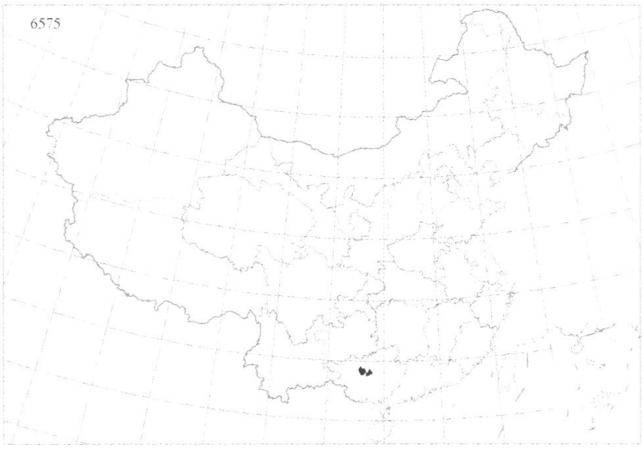

Theaceae 山茶科 *Camellia* 山茶属
Camellia longissima H. T. Chang et S. Y. Liang ex H. T. Chang 超长梗茶 Shrubs or small trees, evergreen

22.2; 13.9; 28.3; 22.3; 207.5; 0.0; 1190 | 1374; 713; 87 | 15.4; 1190; 327.7

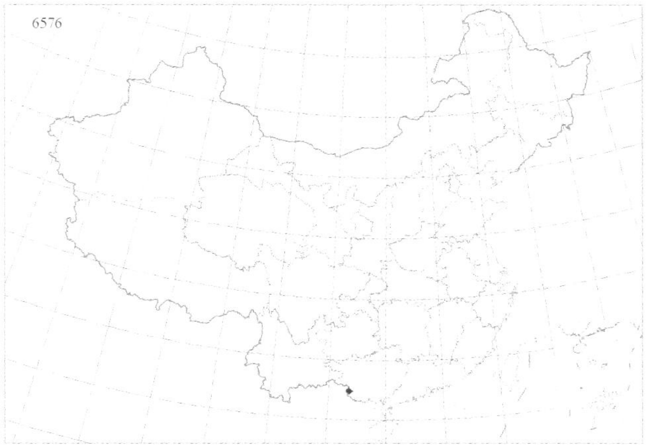

Theaceae 山茶科 *Camellia* 山茶属
Camellia lutchuensis T. Ito 台湾连蕊茶
Shrubs or trees, evergreen

20.2 (16.2~24.1); 14.5 (11.0~18.9); 25.0 (20.1~28.2); 20.2 (16.2~24.1); 182.9 (134.2~229.7); 0.0 (0.0~0.0); 1019 (773~1322) ❙ 2402 (1498~3495); 1106 (750~1847); 255 (47~897) ❙ 146.2 (25.2~246.5); 983 (773~1144); 245.0 (119.4~363.7)

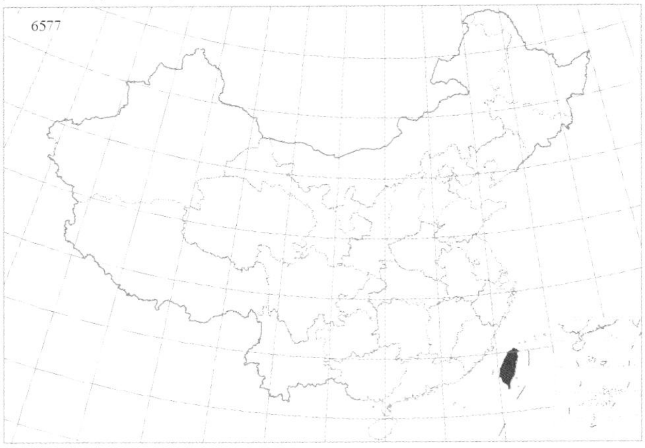

Theaceae 山茶科 *Camellia* 山茶属
Camellia luteoflora Y. K. Li ex H. T. Chang et F. A. Zeng 小黄花茶
Shrubs or small trees, evergreen

15.9; 5.5; 25.6; 15.9; 131.1; 0.0; 833 ❙ 1089; 503; 69 ❙ 30.9; 833; 181.2

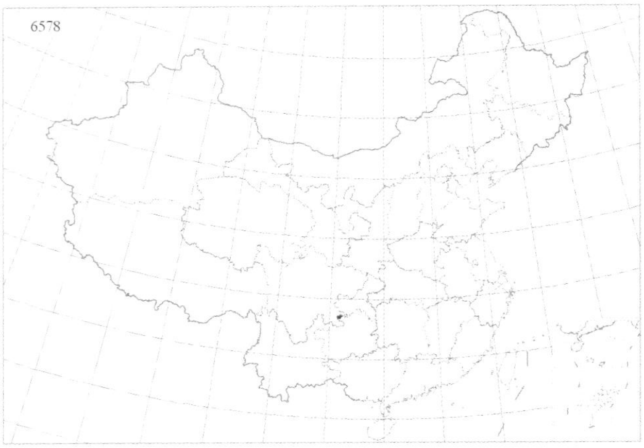

Theaceae 山茶科 *Camellia* 山茶属
Camellia mairei (H. Lév.) Melch. 毛蕊山茶
Shrubs or small trees, evergreen

17.3 (11.7~22.7); 7.9 (2.7~15.0); 25.3 (18.8~28.9); 17.4 (11.7~22.8); 148.8 (82.6~213.1); 0.2 (0.0~3.8); 899 (658~1221) ❙ 1379 (879~2435); 645 (445~1428); 103 (31~197) ❙ 53.5 (12.9~101.8); 898 (658~1221); 245.9 (153.9~551.6)

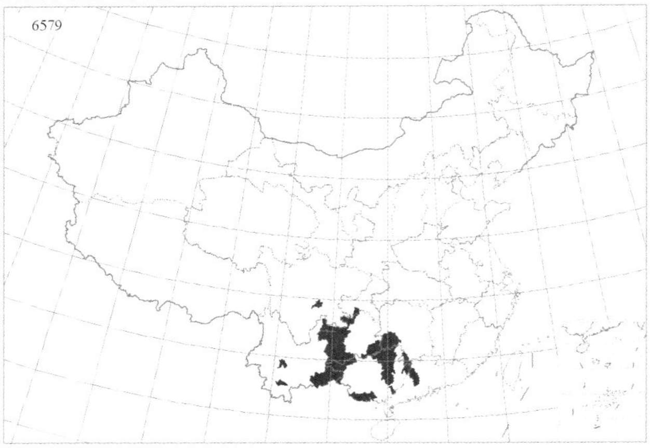

Theaceae 山茶科 *Camellia* 山茶属
Camellia mairei var. lapidea (Y. C. Wu) Sealy 石果毛蕊山茶
Shrubs or small trees, evergreen

19.7 (14.3~22.7); 10.4 (3.3~15.0); 27.3 (23.6~28.9); 19.8 (14.3~22.8); 177.5 (114.1~213.1); 0.1 (0.0~2.3); 1041 (765~1221) ❙ 1568 (1089~2435); 728 (478~1428); 140 (64~189) ❙ 51.9 (12.9~99.5); 1041 (765~1221); 245.9 (181.2~335.1)

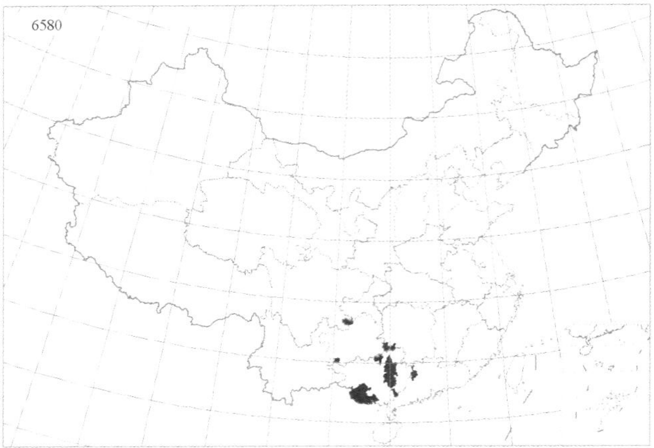

Theaceae 山茶科 *Camellia* 山茶属
Camellia mairei var. velutina Sealy 滇南毛蕊山茶
Shrubs or small trees, evergreen

16.0 (11.1~21.0); 6.5 (1.8~14.2); 24.3 (18.6~28.9); 16.0 (11.1~21.0); 133.0 (79.8~191.8); 0.7 (0.0~6.9); 829 (648~1053) ❙ 1251 (879~1624); 588 (419~911); 79 (18~198) ❙ 51.5 (8.0~101.8); 827 (648~1053); 249.6 (108.2~551.6)

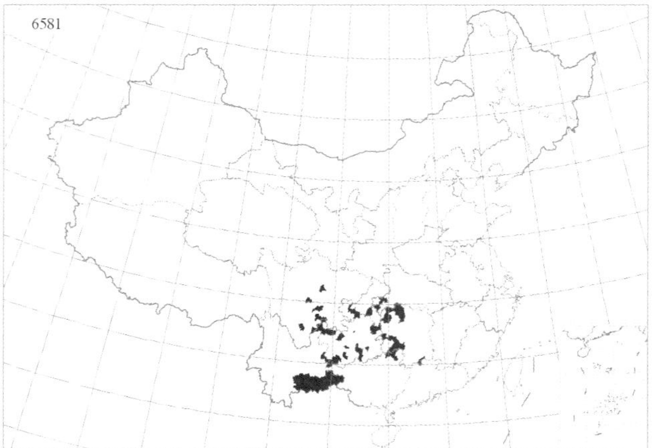

Theaceae 山茶科 *Camellia* 山茶属
Camellia melliana Hand.-Mazz. 广东毛蕊茶
Shrubs, evergreen

20.6; 12.6; 26.8; 20.6; 187.6; 0.0; 1061 ❙ 1826; 824; 144 ❙ 71.7; 1061; 307.1

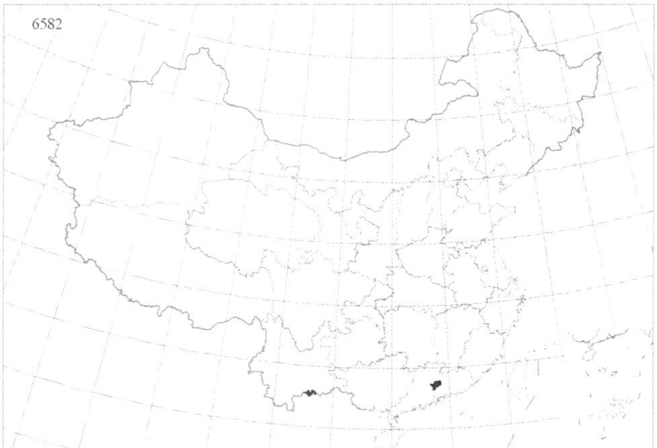

Theaceae 山茶科 *Camellia* 山茶属
Camellia micrantha S. Y. Liang et Y. C. Zhong 小花金花茶
Shrubs, evergreen

22.2；13.9；28.3；22.3；207.5；0.0；1190 ❙ 1374；713；87 ❙ 15.4；1190；327.7

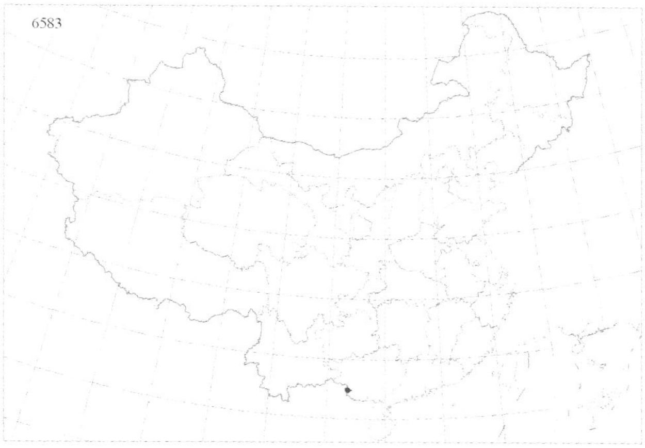

Theaceae 山茶科 *Camellia* 山茶属
Camellia mileensis T. L. Ming 弥勒糙果茶
Shrubs, evergreen

17.5；10.5；22.3；17.5；149.9；0.0；846 ❙ 958；531；44 ❙ 13.3；823；311.9

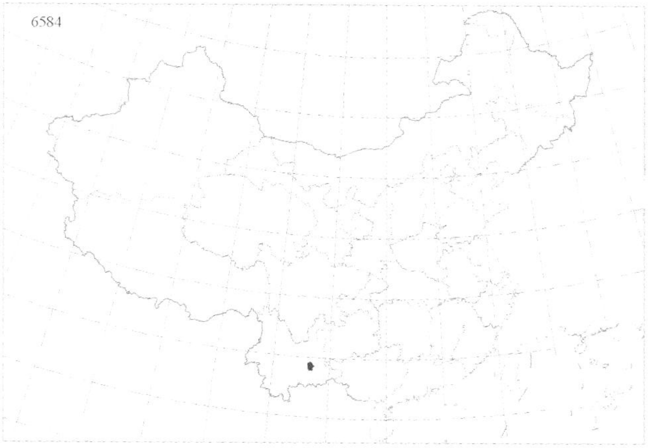

Theaceae 山茶科 *Camellia* 山茶属
Camellia mileensis var. microphylla T. L. Ming 小叶弥勒糙果茶
Shrubs, evergreen

17.5；10.5；22.3；17.5；149.9；0.0；846 ❙ 958；531；44 ❙ 13.3；823；311.9

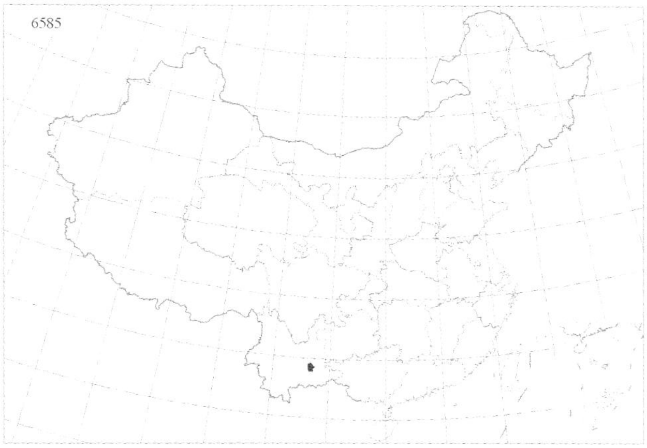

Theaceae 山茶科 *Camellia* 山茶属
Camellia oleifera Abel 油茶
Shrubs or trees, evergreen

17.1 (9.6~24.8)；7.1 (-1.9~20.1)；26.2 (15.7~30.1)；17.1 (9.6~24.8)；146.3 (60.8~237.6)；0.8 (0.0~16.9)；905 (600~1385) ❙ 1388 (584~2435)；572 (299~1428)；126 (15~259) ❙ 54.1 (-34.3~147.0)；892 (584~1360)；238.2 (108.2~659.0)

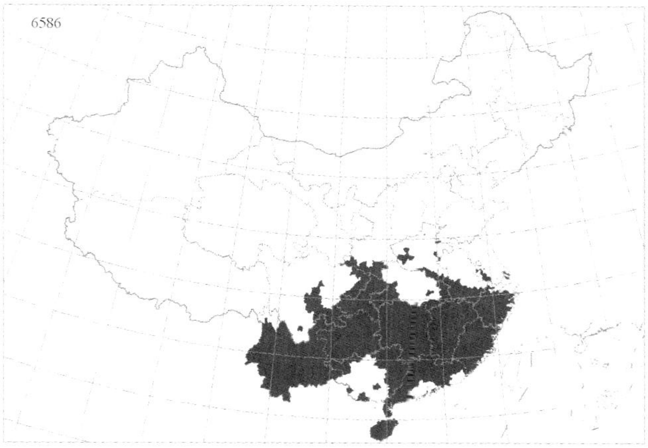

Theaceae 山茶科 *Camellia* 山茶属
Camellia pachyandra Hu 滇南离蕊茶
Trees, evergreen

18.5；11.4；23.3；18.6；162.7；0.0；899 ❙ 1436；811；50 ❙ 60.1；889；438.4

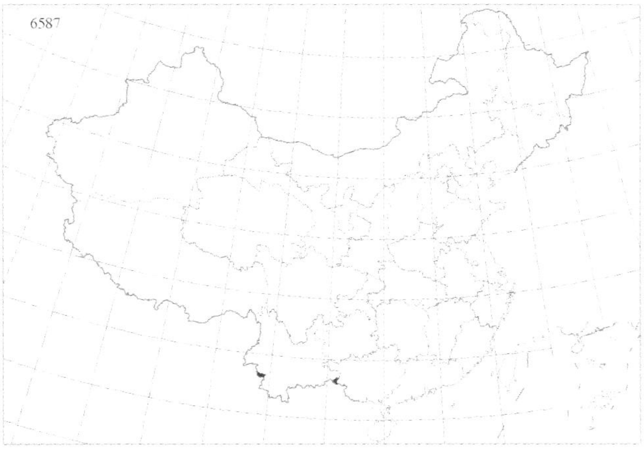

Theaceae 山茶科 *Camellia* 山茶属
Camellia parviflora Merr. et Chun ex Sealy 细花短蕊茶
Shrubs, evergreen

23.9 (22.2~24.8)；18.6 (17.2~20.1)；27.6 (25.7~28.7)；23.9 (22.3~24.8)；227.4 (207.2~237.6)；0.0 (0.0~0.0)；1301 (1127~1385) ❙ 1502 (1176~1854)；519 (392~622)；60 (41~115) ❙ 16.0 (-12.0~47.8)；1160 (1055~1360)；506.9 (411.8~620.0)

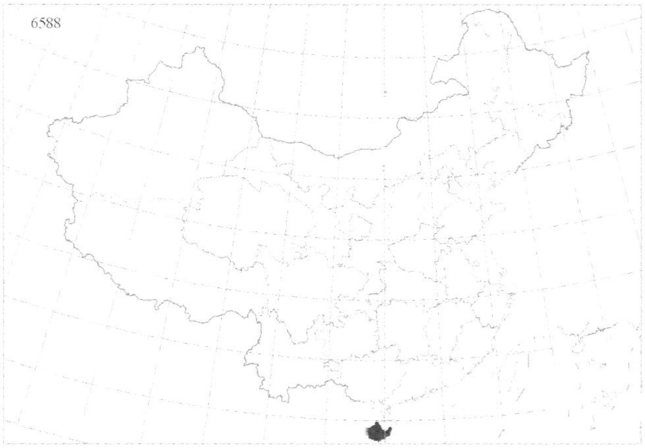

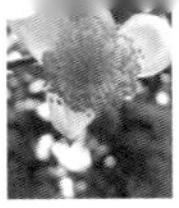

Theaceae 山茶科 *Camellia* 山茶属
Camellia parvimuricata H. T. Chang 小瘤果茶
Shrubs, evergreen

15.8 (12.7~18.7); 5.1 (2.1~8.8); 25.8 (23.0~26.9); 15.8 (12.8~18.8); 130.4 (99.0~165.2); 0.6 (0.0~5.6); 836 (713~968) ❙ 1255 (1090~1478); 513 (450~670); 98 (49~174) ❙ 50.8 (22.7~107.2); 836 (713~968); 196.1 (130.1~239.6)

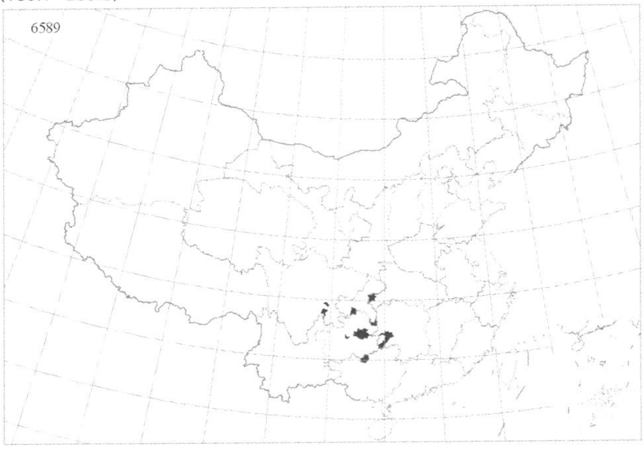

Theaceae 山茶科 *Camellia* 山茶属
Camellia petelotii (Merr.) Sealy 金花茶
Shrubs, evergreen

22.0 (21.3~22.7); 13.6 (12.5~15.0); 28.3 (27.3~28.9); 22.0 (21.3~22.8); 204.2 (195.9~213.1); 0.0 (0.0~0.0); 1171 (1111~1221) ❙ 1669 (1302~2435); 874 (670~1428); 122 (86~138) ❙ 42.5 (12.9~99.5); 1171 (1111~1221); 262.5 (201.4~327.7)

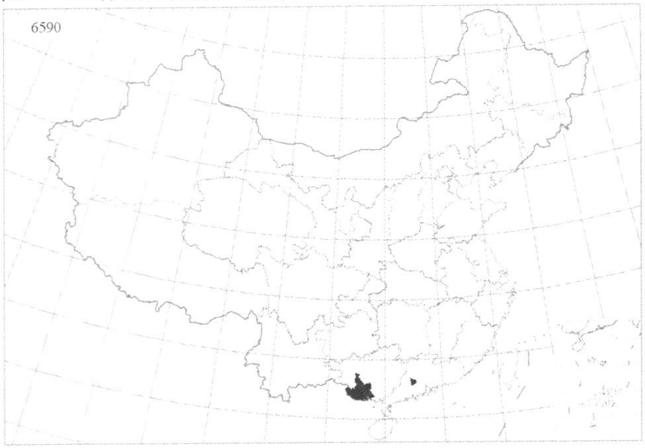

Theaceae 山茶科 *Camellia* 山茶属
Camellia pilosperma S. Y. Liang 毛籽短蕊茶
Shrubs, evergreen

20.6; 11.3; 27.9; 20.6; 187.6; 0.0; 1086 ❙ 1504; 690; 147 ❙ 38.9; 1086; 250.6

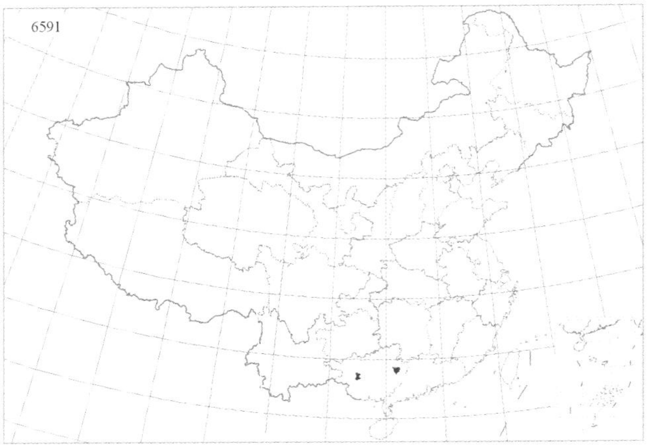

Theaceae 山茶科 *Camellia* 山茶属
Camellia pingguoensis D. Fang 平果金花茶
Shrubs, evergreen

21.3; 12.6; 27.9; 21.4; 196.4; 0.0; 1127 ❙ 1439; 750; 109 ❙ 27.8; 1127; 265.9

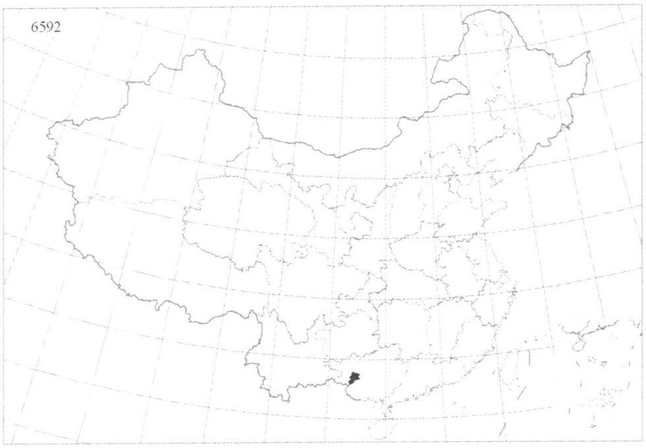

Theaceae 山茶科 *Camellia* 山茶属
Camellia pitardii Cohen-Stuart 西南山茶
Shrubs or small trees, evergreen

15.6 (5.4~21.6); 6.3 (-2.1~13.3); 23.8 (12.3~28.9); 15.7 (5.8~21.6); 129.0 (31.5~199.3); 0.7 (0.0~25.6); 817 (481~1150) ❙ 1186 (741~1740); 562 (411~911); 75 (15~207) ❙ 45.7 (-14.2~106.2); 809 (481~1148); 234.7 (108.2~551.3)

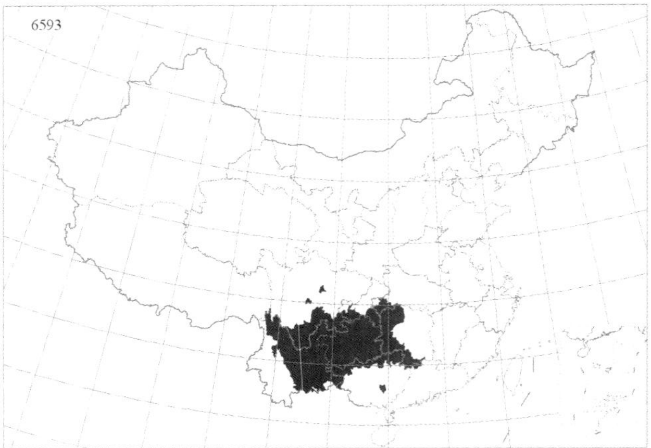

Theaceae 山茶科 *Camellia* 山茶属
Camellia pitardii var. compressa (H. T. Chang et X. K. Wen) T. L. Ming 多变西南山茶
Shrubs or small trees, evergreen

15.2; 3.7; 26.0; 15.2; 123.6; 1.3; 817 ❙ 1445; 562; 122 ❙ 76.8; 817; 206.3

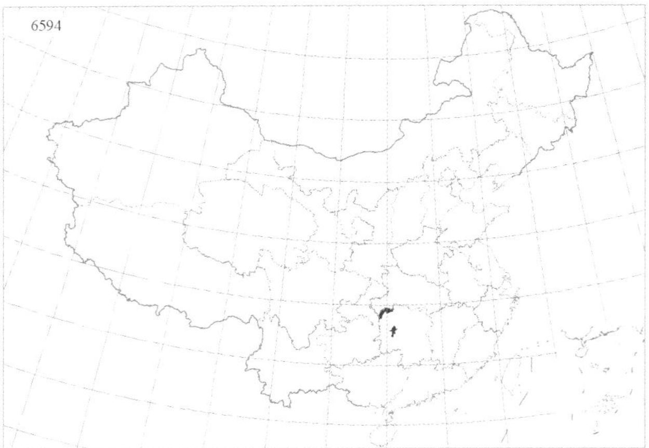

Theaceae 山茶科 *Camellia* 山茶属
Camellia polyodonta How ex Hu 多齿山茶
Small trees, evergreen

17.9 (14.3~23.4); 7.6 (3.3~16.1); 27.1 (24.3~29.2); 18.0 (14.3~23.5); 155.7 (114.1~221.7); 0.1 (0.0~2.3); 945 (765~1273) ❙ 1501 (1035~1740); 562 (403~821); 170 (59~211) ❙ 60.7 (11.9~101.8); 938 (765~1273); 211.8 (151.0~258.0)

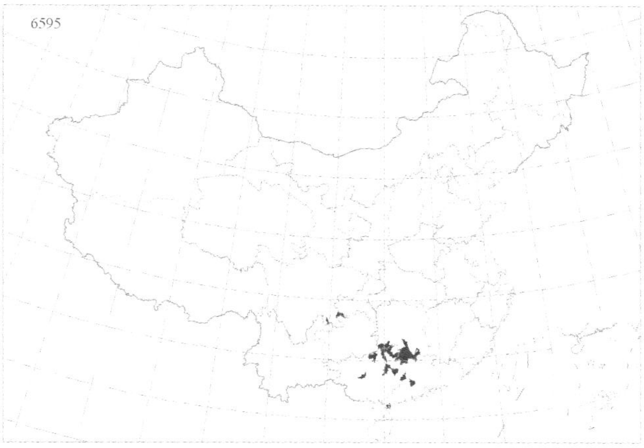

Theaceae 山茶科 *Camellia* 山茶属
Camellia polyodonta var. longicaudata (H. T. Chang et S. Ye Liang) T. L. Ming 长尾多齿山茶
Shrubs or small trees, evergreen

19.8 (16.4~23.4); 10.2 (5.4~16.1); 27.8 (25.8~29.2); 19.9 (16.5~23.5); 178.6 (137.7~221.7); 0.0 (0.0~0.0); 1047 (864~1273) ❙ 1575 (1371~1758); 663 (518~728); 164 (82~189) ❙ 52.4 (11.9~85.5); 1046 (864~1273); 217.3 (178.7~243.9)

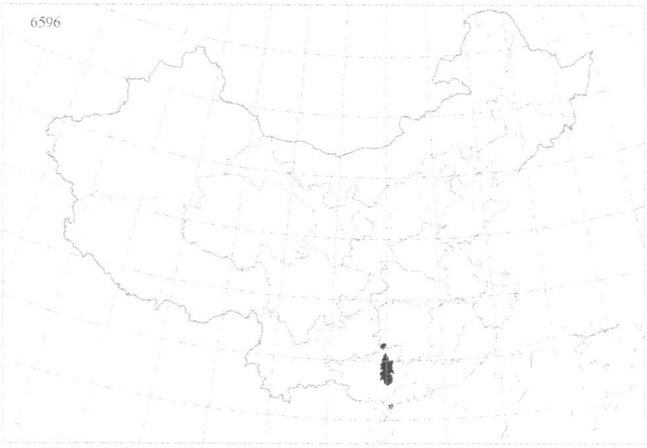

Theaceae 山茶科 *Camellia* 山茶属
Camellia ptilophylla H. T. Chang 毛叶茶
Small trees, evergreen

20.6 (14.6~22.7); 12.1 (3.8~15.2); 27.9 (24.6~28.9); 20.8 (14.7~22.7); 189.2 (117.7~212.7); 0.1 (0.0~1.5); 1100 (779~1216) ❙ 1776 (1462~2037); 781 (544~1033); 146 (97~203) ❙ 63.5 (29.9~101.8); 1100 (779~1216); 201.8 (116.6~260.7)

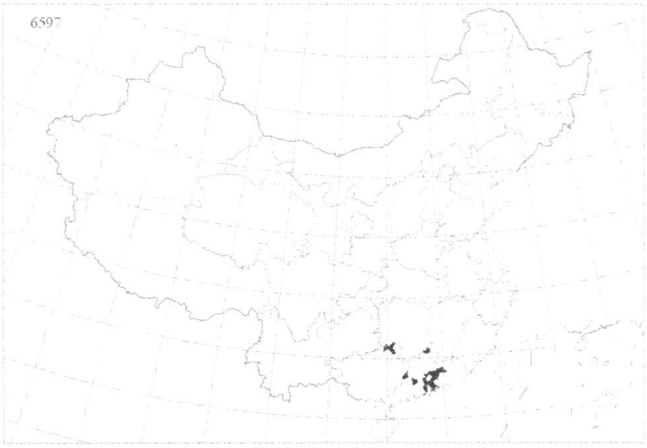

Theaceae 山茶科 *Camellia* 山茶属
Camellia pubifurfuracea Y. C. Zhong 毛糙果茶
Shrubs or small trees, evergreen

18.0; 8.5; 25.8; 18.0; 156.4; 0.0; 913 ❙ 1551; 672; 169 ❙ 69.8; 913; 243.9

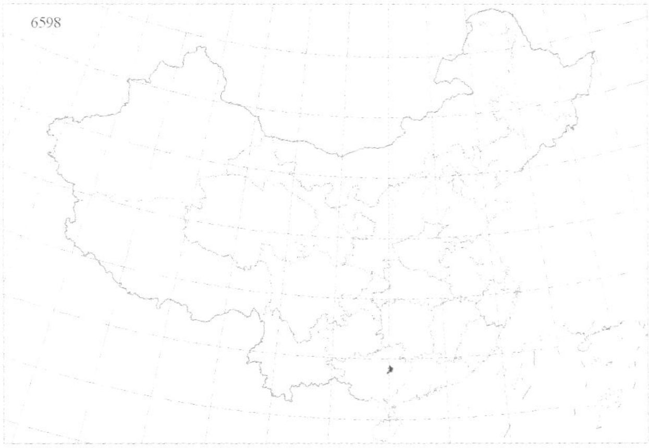

Theaceae 山茶科 *Camellia* 山茶属
Camellia pubipetala Y. Wan et S. Z. Huang 毛瓣金花茶
Small trees, evergreen

21.8; 13.1; 28.2; 21.8; 201.7; 0.0; 1159 ❙ 1485; 770; 120 ❙ 28.0; 1159; 303.7

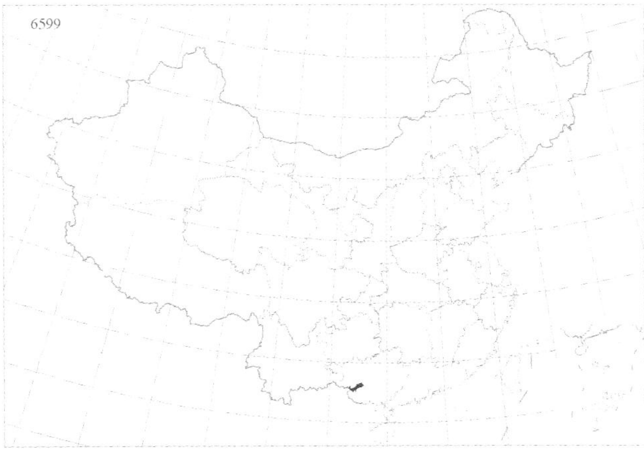

Theaceae 山茶科 *Camellia* 山茶属
Camellia punctata (Kochs) Cohen-Stuart 斑枝毛蕊茶
Shrubs or small trees, evergreen

15.9; 6.2; 24.6; 15.9; 130.8; 0.0; 822 ❙ 1318; 753; 50 ❙ 60.3; 822; 189.1

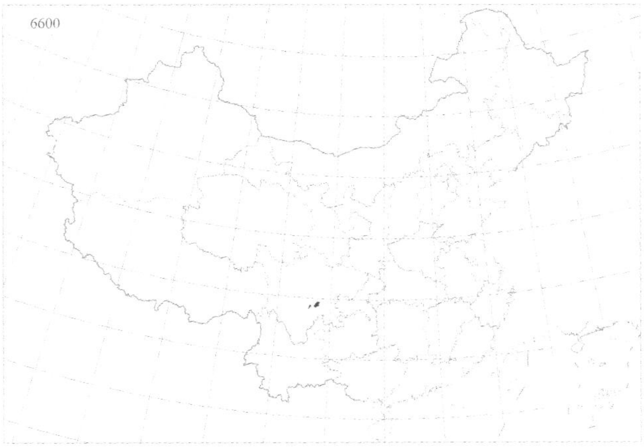

Theaceae 山茶科　*Camellia* 山茶属

Camellia pyxidiacea Z. R. Xu, F. P. Chen et C. Y. Deng 三江瘤果茶

Shrubs or trees, evergreen

16.6 (15.1~18.2); 7.7 (6.2~9.0); 23.5 (21.8~25.5); 16.6 (15.2~18.2); 139.3 (122.6~158.8); 0.0 (0.0~0.0); 832 (772~918) ❙ 1266 (1107~1334); 675 (607~720); 59 (52~64) ❙ 52.5 (39.1~69.4); 832 (772~918); 229.7 (189.5~298.7)

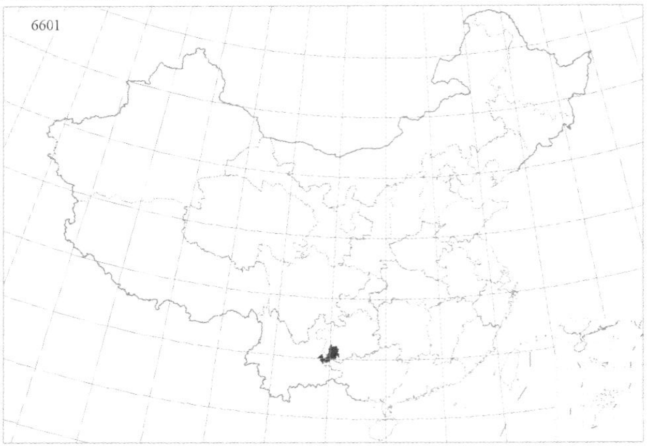

Theaceae 山茶科　*Camellia* 山茶属

Camellia pyxidiacea var. rubituberculata (H. T. Chang ex M. J. Lin et Q. M. Lu) T. L. Ming 红花三江瘤果茶

Shrubs or trees, evergreen

16.5; 7.3; 24.0; 16.6; 139.0; 0.0; 833 ❙ 1274; 671; 59 ❙ 52.9; 833; 208.5

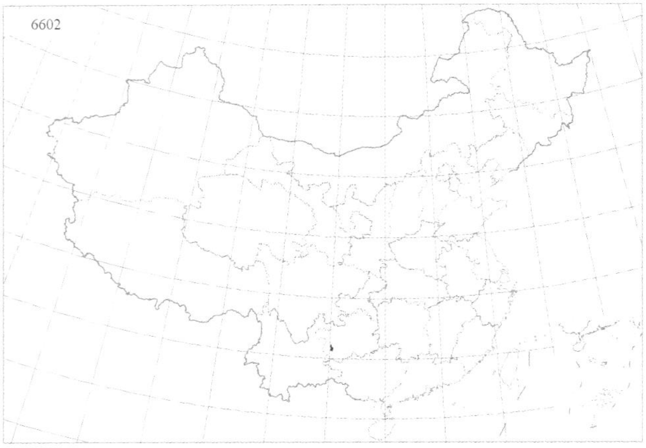

Theaceae 山茶科　*Camellia* 山茶属

Camellia reticulata Lindl. 滇山茶

Small trees, evergreen

15.7 (5.4~22.0); 7.5 (-2.1~15.1); 22.3 (12.3~28.9); 15.7 (5.8~22.1); 129.1 (31.5~205.1); 0.7 (0.0~25.6); 800 (481~1170) ❙ 1170 (741~2014); 610 (411~996); 58 (15~198) ❙ 46.5 (-14.2~112.2); 788 (481~1170); 290.1 (108.2~659.0)

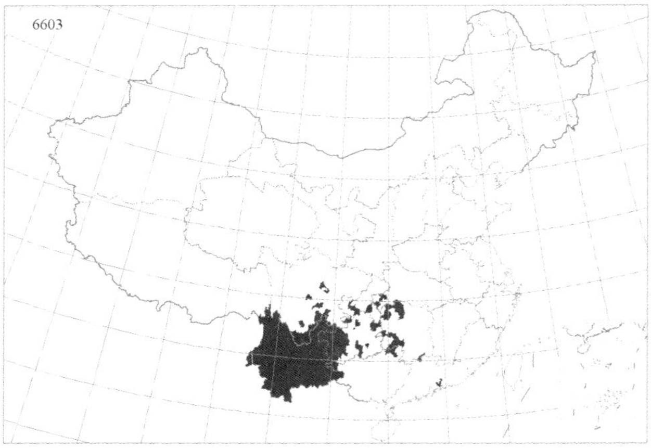

Theaceae 山茶科　*Camellia* 山茶属

Camellia rhytidocarpa H. T. Chang et S. Y. Liang 皱果茶

Shrubs or trees, evergreen

17.2 (14.3~21.3); 6.5 (3.3~12.6); 26.7 (24.3~29.4); 17.2 (14.3~21.3); 146.8 (114.1~196.0); 0.2 (0.0~2.3); 902 (765~1123) ❙ 1414 (1089~1662); 534 (392~821); 153 (56~211) ❙ 57.7 (5.8~101.8); 892 (765~1121); 223.9 (145.3~322.1)

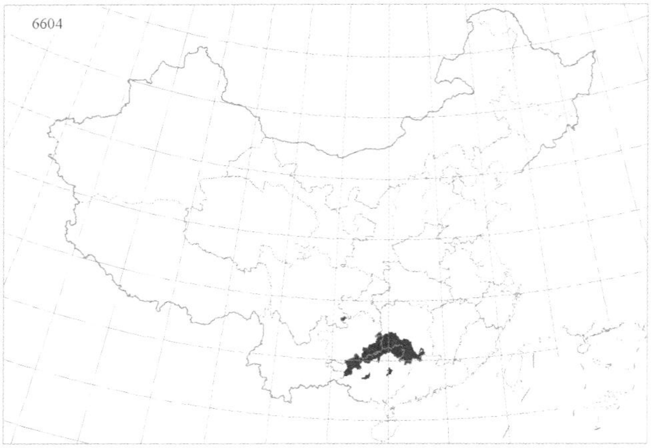

Theaceae 山茶科　*Camellia* 山茶属

Camellia rhytidocarpa var. microphylla Y. C. Zhong ex Ming et Y. C. Zhang 小叶皱果茶

Shrubs or trees, evergreen

16.3 (14.3~18.8); 4.9 (2.6~6.9); 27.1 (24.3~30.1); 16.4 (14.3~18.9); 137.3 (114.1~166.5); 0.7 (0.0~3.4); 878 (765~1024) ❙ 1533 (1175~1865); 538 (448~652); 180 (84~258) ❙ 74.6 (41.7~115.7); 867 (765~961); 226.1 (148.9~318.0)

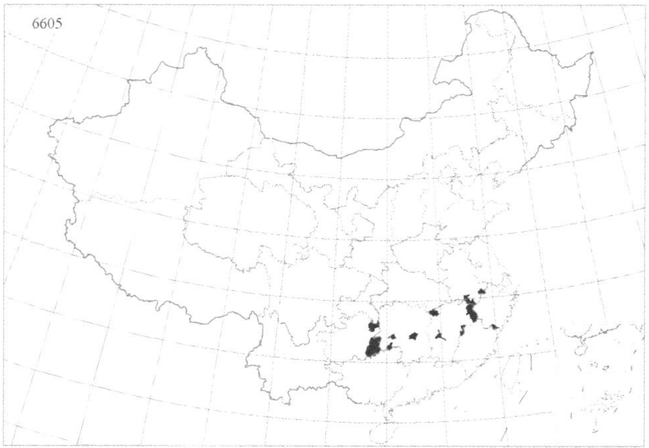

Theaceae 山茶科　*Camellia* 山茶属

Camellia rosthorniana Hand.-Mazz. 川鄂连蕊茶

Shrubs or small trees, evergreen

16.2 (10.6~22.0); 5.4 (-0.2~13.5); 26.2 (21.1~29.9); 16.2 (10.6~22.0); 135.6 (79.4~203.8); 0.9 (0.0~12.1); 862 (643~1170) ❙ 1264 (1028~1671); 520 (439~816); 96 (32~221) ❙ 48.1 (8.0~106.4); 860 (643~1170); 193.8 (108.2~284.3)

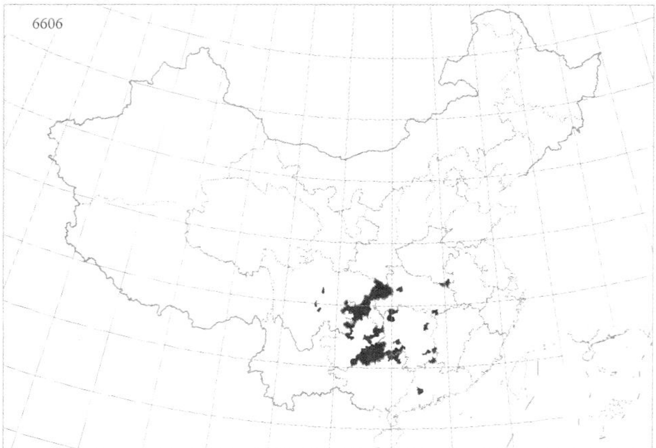

Theaceae 山茶科 *Camellia* 山茶属

Camellia salicifolia Champ. ex Benth. 柳叶毛蕊茶

Shrubs or small trees, evergreen

19.7 (14.6~24.1); 10.7 (3.8~18.9); 27.4 (20.1~29.8); 19.8 (14.7~24.1); 177.1 (117.7~229.7); 0.0 (0.0~1.5); 1030 (773~1322) ❙ 1727 (1371~2992); 694 (476~1847); 172 (47~259) ❙ 70.1 (20.8~246.5); 1027 (773~1216); 237.9 (95.3~332.5)

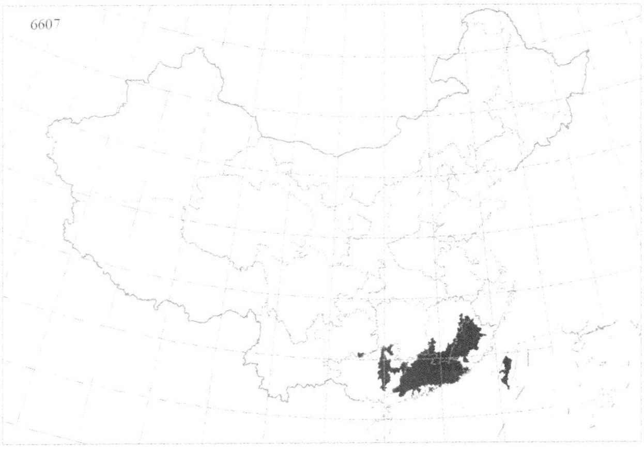

Theaceae 山茶科 *Camellia* 山茶属

Camellia saluenensis Stapf ex Bean 怒江山茶

Shrubs or small trees, evergreen

14.6 (5.4~19.3); 7.3 (-2.1~13.3); 20.3 (12.3~26.6); 14.6 (5.8~19.4); 116.6 (31.5~173.2); 1.1 (0.0~25.6); 750 (481~939) ❙ 1040 (741~1454); 592 (411~816); 39 (15~71) ❙ 39.9 (-14.2~99.9); 727 (481~923); 304.7 (142.9~516.7)

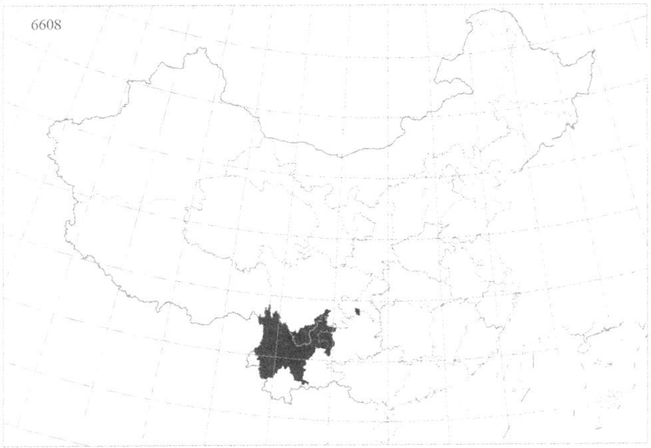

Theaceae 山茶科 *Camellia* 山茶属

Camellia semiserrata C. W. Chi 南山茶

Small trees, evergreen

20.4 (14.6~23.2); 11.2 (3.8~15.8); 28.0 (24.6~29.1); 20.4 (14.7~23.2); 185.1 (117.7~218.9); 0.0 (0.0~1.5); 1079 (779~1257) ❙ 1627 (1371~2005); 688 (466~1044); 153 (94~217) ❙ 52.5 (17.4~101.8); 1078 (779~1257); 227.3 (157.7~273.3)

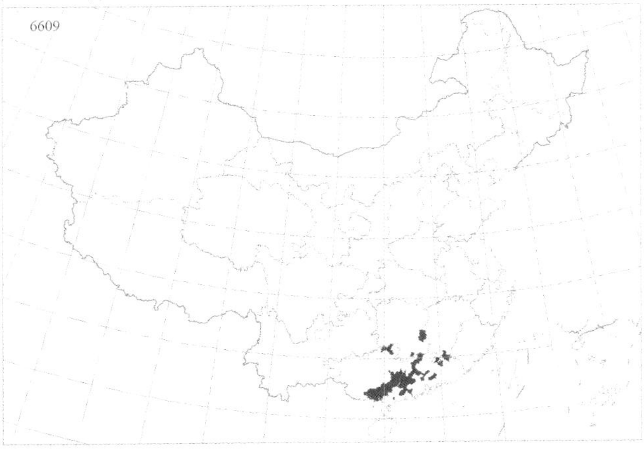

Theaceae 山茶科 *Camellia* 山茶属

Camellia semiserrata var. magnocarpa Hu et T. C. Huang ex Hu 大果南山茶

Small trees, evergreen

20.3; 11.3; 27.7; 20.4; 184.7; 0.0; 1068 ❙ 1597; 706; 151 ❙ 50.5; 1067; 245.3

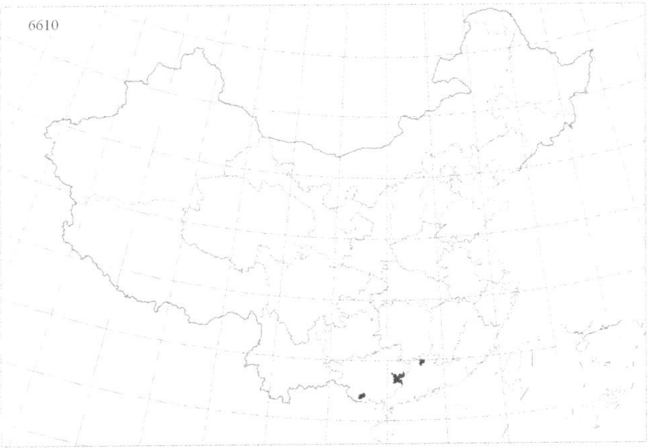

Theaceae 山茶科 *Camellia* 山茶属

Camellia sinensis (L.) Kuntze 茶

Shrubs or small trees, evergreen

17.0 (1.8~24.8); 6.9 (-7.1~20.1); 26.1 (9.9~30.1); 17.1 (3.8~24.8); 145.9 (15.7~237.6); 1.4 (0.0~53.3); 907 (404~1385) ❙ 1347 (583~2148); 572 (299~1234); 119 (13~259) ❙ 49.4 (-34.3~147.0); 894 (404~1360); 230.6 (95.3~659.0)

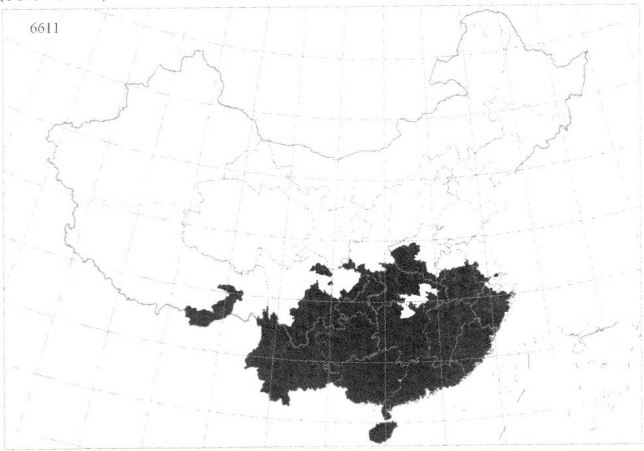

Theaceae 山茶科 *Camellia* 山茶属

Camellia sinensis var. assamica (Mast.) Kitamura 普洱茶

Shrubs or small trees, evergreen

17.9 (3.8~24.7); 10.2 (-4.3~19.8); 23.8 (10.8~28.9); 17.9 (4.7~24.7); 155.9 (21.6~236.5); 0.8 (0.0~35.4); 906 (444~1376) ❙ 1367 (862~1758); 664 (442~996); 77 (12~216) ❙ 53.0 (-2.2~121.7); 891 (444~1257); 356.3 (108.2~659.0)

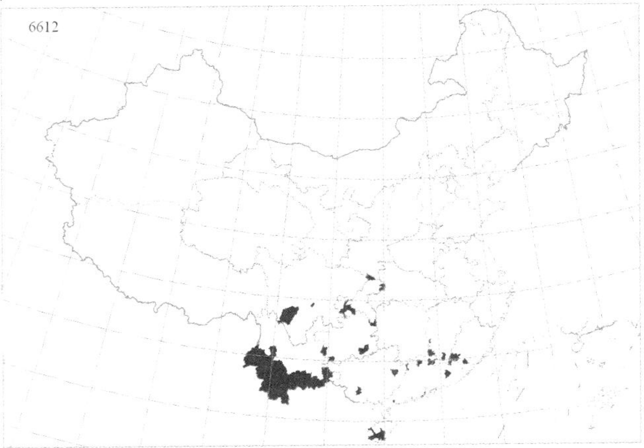

Theaceae 山茶科 *Camellia* 山茶属

Camellia sinensis var. dehungensis (H. T. Chang et B. H. Chen) T. L. Ming 德宏茶
Shrubs or small trees, evergreen

18.8 (16.7~20.5); 12.6 (10.0~15.1); 22.6 (21.3~23.7); 18.9 (16.7~20.5); 166.2 (140.8~185.9); 0.0 (0.0~0.0); 906 (809~992) ❙ 1481 (1090~1616); 771 (594~946); 57 (42~77) ❙ 63.5 (34.7~90.4); 894 (809~986); 529.1 (406.0~659.0)

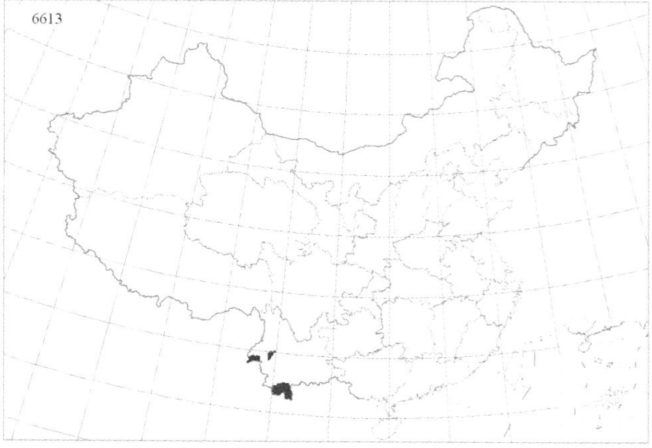

Theaceae 山茶科 *Camellia* 山茶属

Camellia sinensis var. pubilimba H. T. Chang 白毛茶
Shrubs or small trees, evergreen

18.6 (14.6~22.2); 11.1 (3.8~14.2); 24.1 (21.4~28.6); 18.6 (14.7~22.2); 163.5 (117.7~206.5); 0.0 (0.0~1.5); 922 (779~1188) ❙ 1375 (1007~2148); 722 (539~1234); 78 (41~191) ❙ 49.6 (12.5~101.8); 914 (779~1188); 355.0 (199.4~551.6)

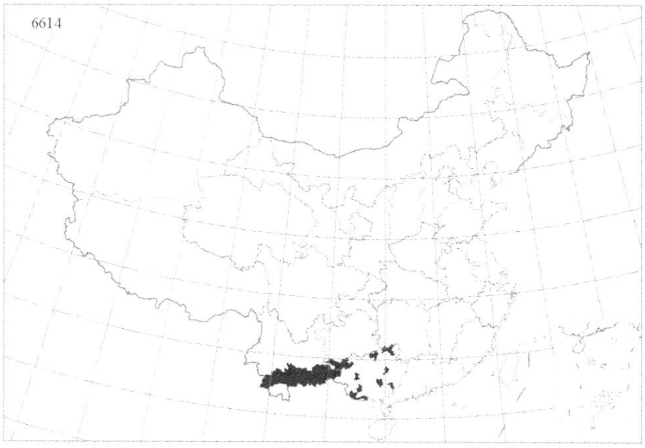

Theaceae 山茶科 *Camellia* 山茶属

Camellia stuartiana Sealy 五室连蕊茶
Small trees, evergreen

19.2 (18.1~21.0); 13.0 (12.0~14.2); 23.4 (22.1~25.6); 19.3 (18.1~21.0); 171.4 (157.3~191.8); 0.0 (0.0~0.0); 932 (863~1053) ❙ 1375 (1143~1547); 786 (637~911); 57 (50~68) ❙ 48.0 (21.7~74.2); 925 (863~1053); 473.5 (346.9~551.6)

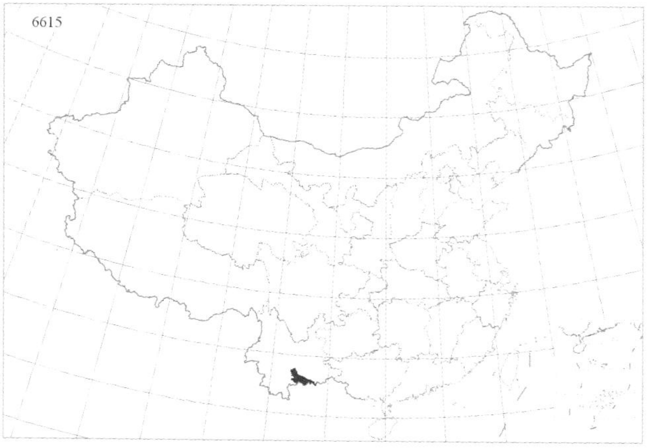

Theaceae 山茶科 *Camellia* 山茶属

Camellia subintegra P. C. Huang ex H. T. Chang 全缘叶山茶
Small trees, evergreen

16.9 (14.6~17.9); 5.8 (3.8~8.1); 27.1 (24.6~28.7); 16.9 (14.7~17.9); 143.1 (117.7~155.4); 0.2 (0.0~1.5); 894 (779~948) ❙ 1525 (1197~1624); 514 (411~594); 187 (64~228) ❙ 71.3 (31.7~101.8); 884 (779~918); 221.7 (150.2~263.5)

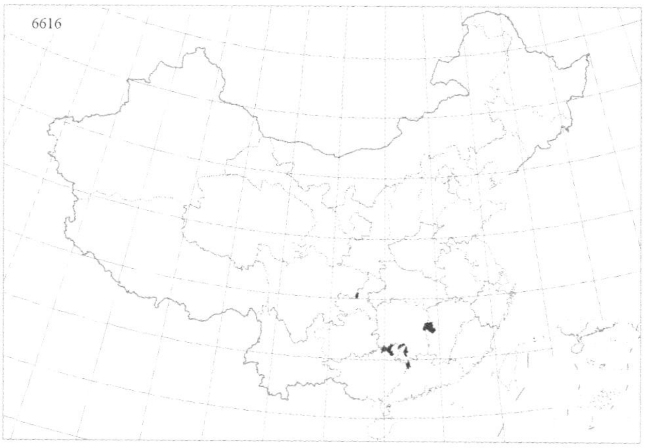

Theaceae 山茶科 *Camellia* 山茶属

Camellia synaptica Sealy 川滇连蕊茶
Shrubs or small trees, evergreen

14.5 (5.4~19.1); 6.5 (-2.1~12.0); 21.0 (12.3~27.8); 14.5 (5.8~19.1); 115.1 (31.5~169.2); 1.1 (0.0~25.6); 753 (481~966) ❙ 1044 (741~1428); 569 (411~816); 48 (15~180) ❙ 39.4 (-14.2~80.4); 734 (481~966); 262.9 (142.9~384.1)

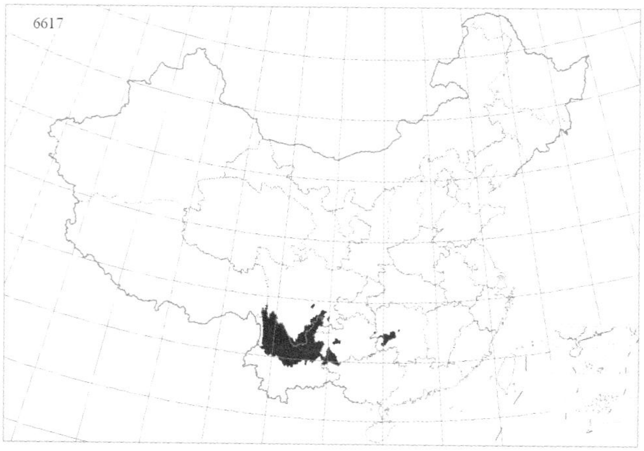

Theaceae 山茶科 *Camellia* 山茶属

Camellia synaptica var. parviovata (H. T. Chang) Ming 毛蕊川滇连蕊茶
Shrubs or small trees, evergreen

14.9; 5.3; 23.6; 15.0; 119.6; 0.0; 782 ❙ 1410; 816; 49 ❙ 80.4; 782; 214.5

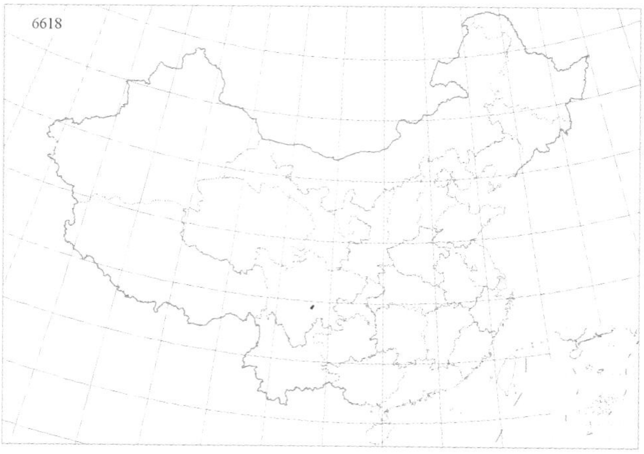

Theaceae 山茶科 *Camellia* 山茶属
Camellia szechuanensis C. W. Chi 四川离蕊茶
Shrubs or small trees, evergreen

14.4; 4.9; 23.2; 14.5; 114.5; 0.2; 765 ▎1412; 817; 48 ▎84.7; 765; 216.0

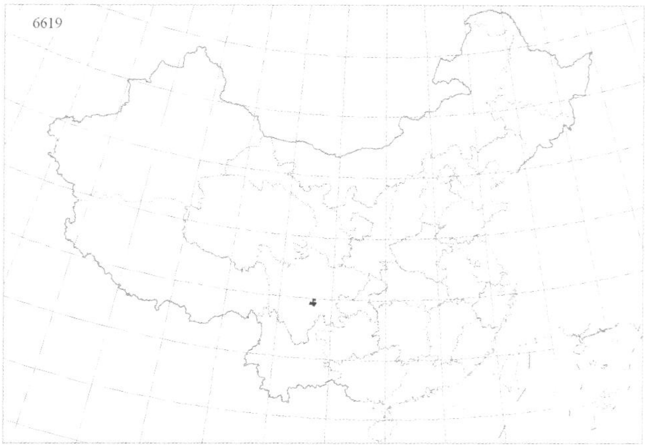

Theaceae 山茶科 *Camellia* 山茶属
Camellia szemaoensis H. T. Chang 斑叶离蕊茶
Shrubs or small trees, evergreen

19.7; 14.0; 23.2; 19.8; 177.4; 0.0; 950 ▎1479; 715; 64 ▎56.1; 925; 499.0

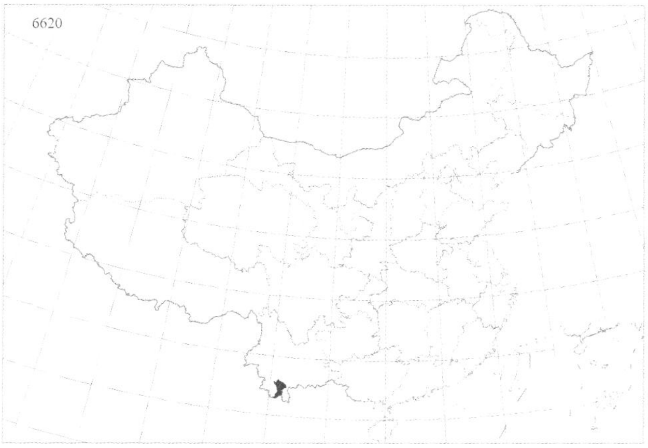

Theaceae 山茶科 *Camellia* 山茶属
Camellia tachangensis F. C. Zhang 大厂茶
Trees, evergreen

16.7 (14.2~19.2); 8.5 (5.6~10.7); 22.9 (20.1~26.0); 16.8 (14.2~19.2); 141.3 (110.5~170.8); 0.0 (0.0~0.0); 834 (729~966) ▎1209 (1007~1422); 663 (554~813); 54 (47~64) ▎45.5 (21.9~70.6); 834 (729~966); 287.2 (203.5~411.1)

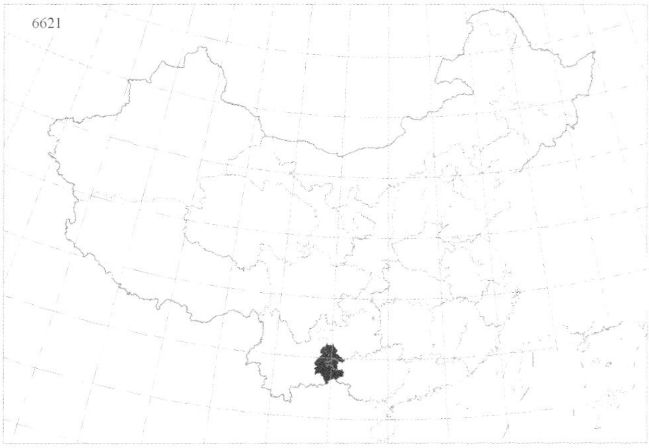

Theaceae 山茶科 *Camellia* 山茶属
Camellia taliensis (W. W. Sm.) Melch. 大理茶
Trees, evergreen

16.5 (11.2~21.0); 9.8 (4.6~14.2); 21.3 (16.7~25.6); 16.6 (11.3~21.0); 138.7 (75.8~191.8); 0.0 (0.0~0.4); 810 (642~1053) ▎1136 (797~1751); 646 (411~996); 46 (23~77) ▎40.2 (-14.2~112.2); 790 (642~1053); 352.3 (218.0~627.0)

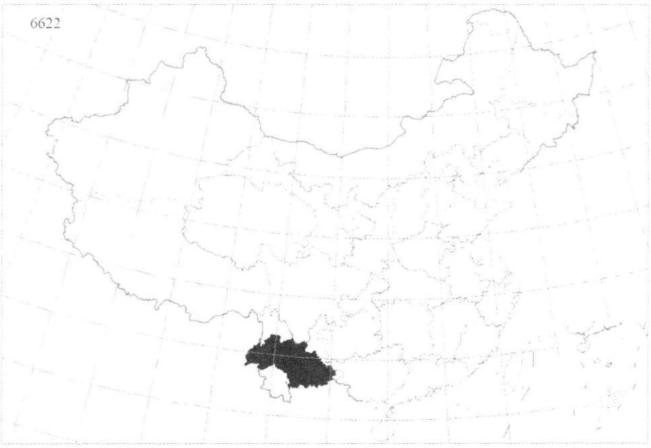

Theaceae 山茶科 *Camellia* 山茶属
Camellia tenii Sealy 小糙果茶
Shrubs, evergreen

15.0; 8.4; 20.2; 15.1; 121.2; 0.0; 753 ▎950; 570; 27 ▎26.3; 683; 307.7

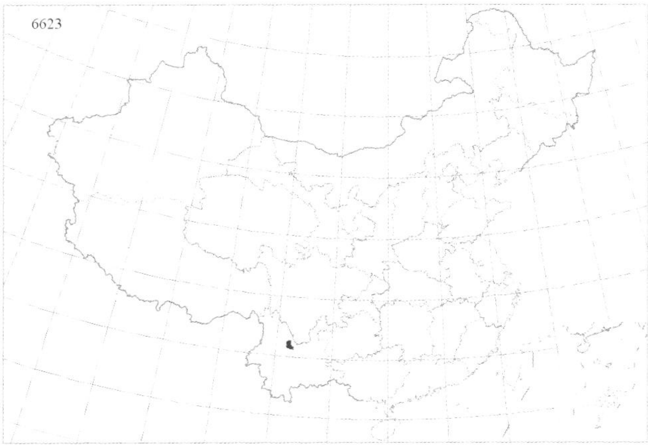

Theaceae 山茶科 *Camellia* 山茶属
Camellia transarisanensis (Hayata) Cohen-Stuart 毛萼连蕊茶
Shrubs, evergreen

18.1 (15.3~23.3); 8.7 (3.6~17.7); 26.5 (20.1~30.1); 18.1 (15.4~23.3); 157.7 (124.7~219.7); 0.2 (0.0~1.4); 945 (773~1248) ▎1677 (1101~2759); 736 (439~1594); 156 (68~308) ▎80.5 (25.2~246.5); 914 (773~1144); 210.1 (141.7~322.1)

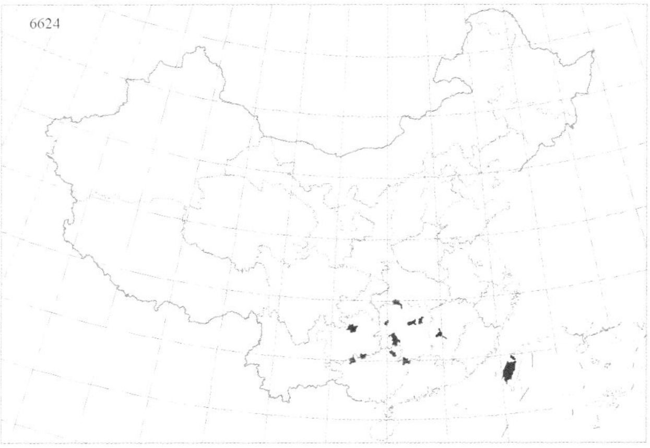

Theaceae 山茶科　*Camellia* 山茶属
Camellia trichoclada (Rehder) Chien 毛枝连蕊茶
Shrubs, evergreen

17.4 (15.9~18.8); 7.4 (5.9~9.6); 26.8 (25.8~27.6); 17.5 (16.0~18.9); 149.7 (131.9~166.3); 0.0 (0.0~0.0); 905 (831~970) ❙ 1776 (1520~1997); 609 (551~669); 214 (172~254) ❙ 96.7 (65.3~130.7); 905 (831~970); 266.2 (216.1~303.4)

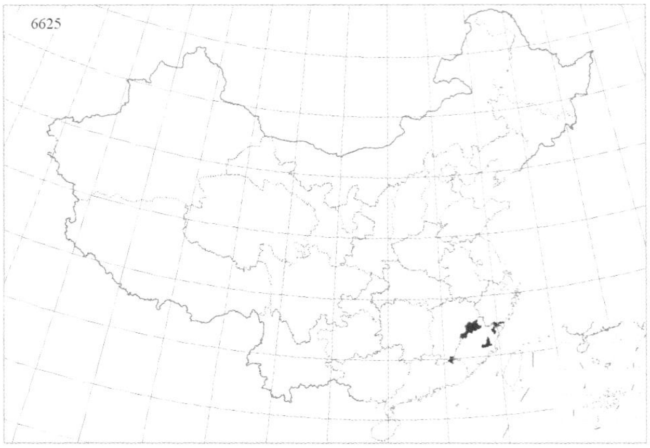

Theaceae 山茶科　*Camellia* 山茶属
Camellia tsaii Hu 窄叶连蕊茶
Shrubs or small trees, evergreen

17.5 (14.4~21.0); 10.9 (7.5~14.2); 21.7 (19.1~25.6); 17.5 (14.4~21.0); 149.9 (113.1~191.8); 0.0 (0.0~0.0); 845 (727~1053) ❙ 1427 (1007~1688); 814 (561~959); 54 (42~70) ❙ 69.5 (31.4~112.2); 840 (727~1053); 481.4 (274.9~627.0)

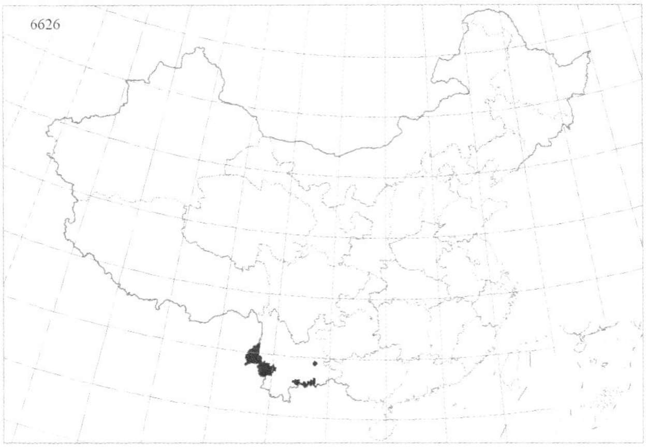

Theaceae 山茶科　*Camellia* 山茶属
Camellia tsingpienensis Hu 屏边连蕊茶
Shrubs or small trees, evergreen

18.8 (16.1~21.6); 10.1 (5.3~14.2); 25.6 (21.4~28.5); 18.8 (16.1~21.7); 165.6 (133.1~199.8); 0.0 (0.0~0.0); 957 (813~1153) ❙ 1462 (1178~1740); 723 (523~860); 111 (51~194) ❙ 53.5 (35.6~85.5); 957 (813~1153); 314.3 (206.0~551.6)

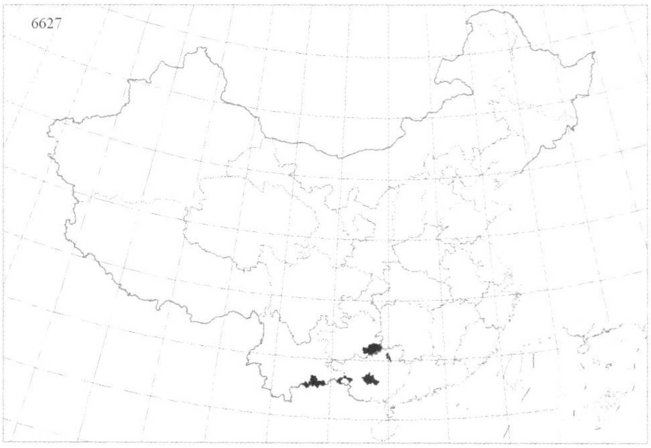

Theaceae 山茶科　*Camellia* 山茶属
Camellia tsingpienensis var. pubisepala H. T. Chang 毛萼屏边连蕊茶
Shrubs or small trees, evergreen

17.3 (14.3~21.6); 8.3 (4.0~14.2); 24.6 (22.2~28.5); 17.3 (14.3~21.7); 148.2 (112.6~199.8); 0.1 (0.0~1.1); 876 (756~1153) ❙ 1348 (1161~1740); 667 (452~860); 85 (48~187) ❙ 54.3 (37.1~85.5); 876 (756~1153); 288.8 (189.5~551.6)

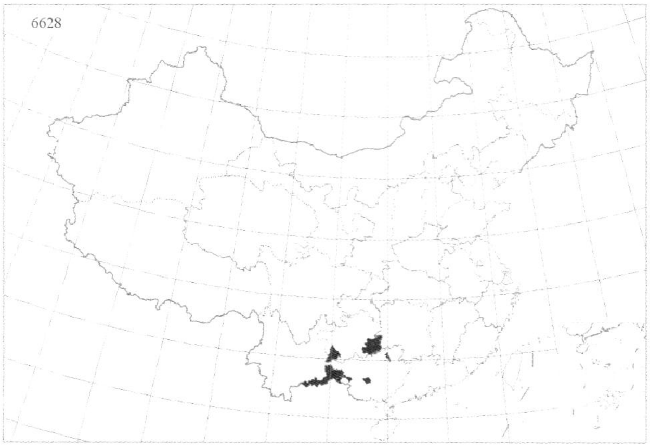

Theaceae 山茶科　*Camellia* 山茶属
Camellia tuberculata Chien 瘤果茶
Shrubs, evergreen

16.5 (10.6~18.7); 5.9 (-0.2~8.1); 26.5 (19.6~28.9); 16.5 (10.6~18.8); 139.3 (79.4~166.0); 0.9 (0.0~12.1); 881 (643~1002) ❙ 1156 (1028~1320); 505 (442~753); 62 (20~76) ❙ 33.4 (8.0~93.8); 880 (643~1002); 166.0 (108.2~331.6)

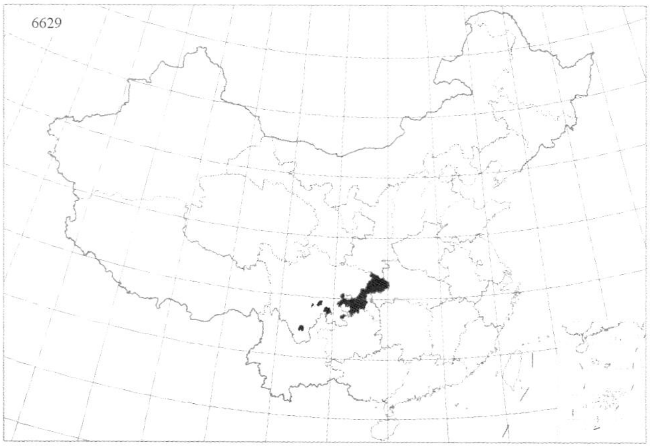

Theaceae 山茶科　*Camellia* 山茶属
Camellia villicarpa Chien 小果毛蕊茶
Shrubs, evergreen

14.8; 4.6; 24.0; 14.8; 118.3; 0.8; 786 ❙ 1159; 621; 45 ❙ 48.3; 785; 175.5

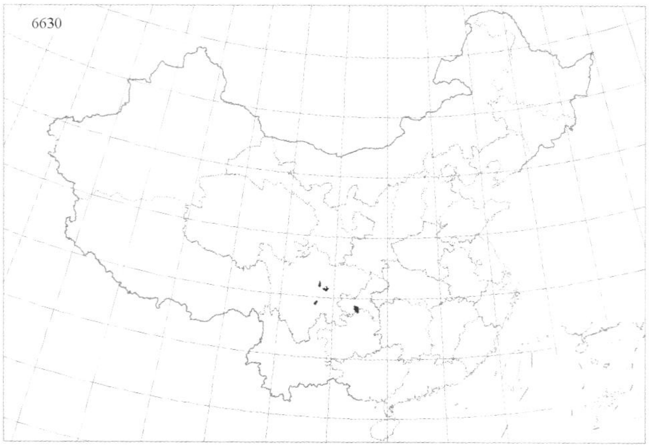

Theaceae 山茶科　*Camellia* 山茶属
Camellia viridicalyx H. T. Chang et S. Y. Liang 绿萼连蕊茶
Shrubs, evergreen

16.9 (14.6~19.1); 6.3 (3.8~9.3); 26.3 (24.6~27.7); 16.9 (14.7~19.1); 143.6 (117.7~169.2); 0.3 (0.0~1.5); 885 (779~988) ❙ 1543 (1494~1624); 567 (476~639); 178 (142~197) ❙ 75.8 (51.7~101.8); 884 (779~988); 233.6 (180.3~258.0)

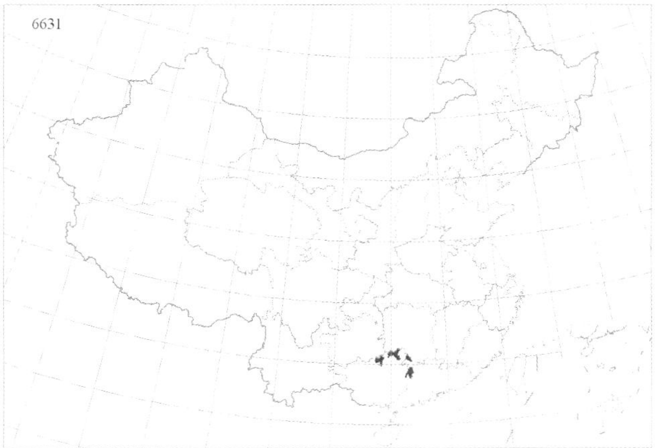

Theaceae 山茶科　*Camellia* 山茶属
Camellia viridicalyx var. linearifolia Ming 线叶连蕊茶
Shrubs, evergreen

18.9; 9.5; 26.6; 18.9; 167.3; 0.0; 969 ❙ 1236; 617; 60 ❙ 27.5; 969; 254.3

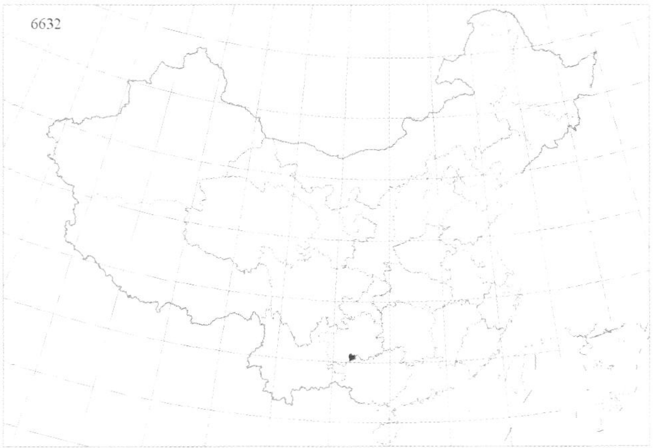

Theaceae 山茶科　*Camellia* 山茶属
Camellia wardii Kobuski 滇缅离蕊茶
Shrubs or small trees, evergreen

16.7 (11.1~19.6); 9.6 (1.8~12.8); 21.4 (19.1~23.5); 16.7 (11.1~19.6); 141.1 (79.8~175.7); 0.9 (0.0~6.9); 823 (648~953) ❙ 1503 (1076~1688); 864 (584~959); 52 (30~70) ❙ 83.2 (61.2~112.2); 818 (648~932); 488.3 (246.3~627.0)

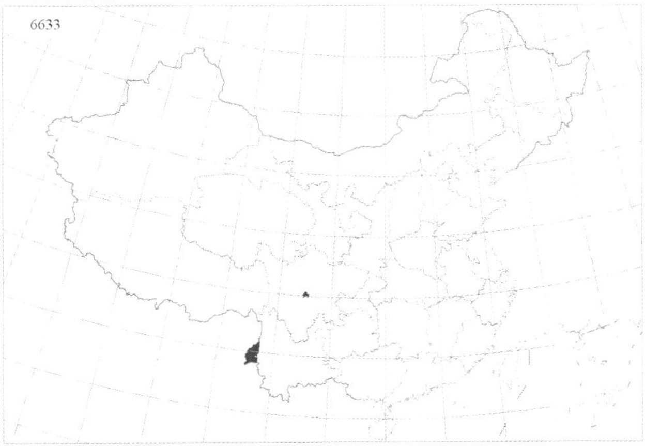

Theaceae 山茶科　*Camellia* 山茶属
Camellia wardii var. muricatula (H. T. Chang) Ming 毛滇缅离蕊茶
Shrubs or small trees, evergreen

18.0 (16.4~19.6); 11.2 (9.6~12.8); 22.1 (20.6~23.5); 18.0 (16.4~19.6); 156.0 (137.3~175.7); 0.0 (0.0~0.0); 868 (795~953) ❙ 1582 (1530~1688); 922 (897~959); 52 (42~62) ❙ 83.3 (61.2~112.2); 861 (795~932); 528.8 (451.2~627.0)

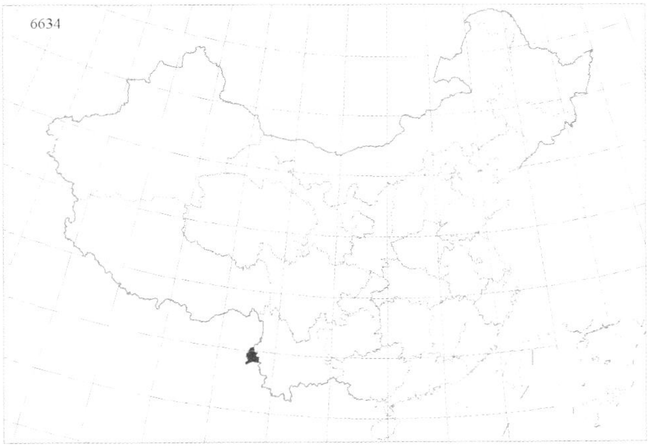

Theaceae 山茶科　*Camellia* 山茶属
Camellia xanthochroma K. M. Feng et L. S. Xie 黄花短蕊茶
Shrubs, evergreen

23.6; 17.5; 28.3; 23.7; 224.3; 0.0; 1295 ❙ 1582; 667; 69 ❙ 23.6; 1172; 333.6

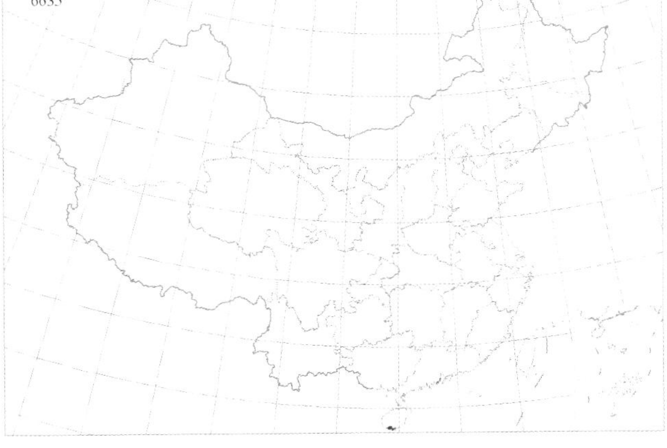

Theaceae 山茶科　*Camellia* 山茶属
Camellia yunnanensis (Pit. ex Diels) Cohen-Stuart 猴子木
Shrubs or small trees, evergreen

15.7 (5.4~21.0); 8.5 (-2.1~15.1); 21.0 (12.3~27.5); 15.7 (5.8~21.0); 129.3 (31.5~191.8); 0.7 (0.0~25.6); 789 (481~1053) ❙ 1123 (741~1751); 631 (411~996); 44 (15~135) ❙ 42.7 (-14.2~112.2); 771 (481~1053); 333.1 (142.9~659.0)

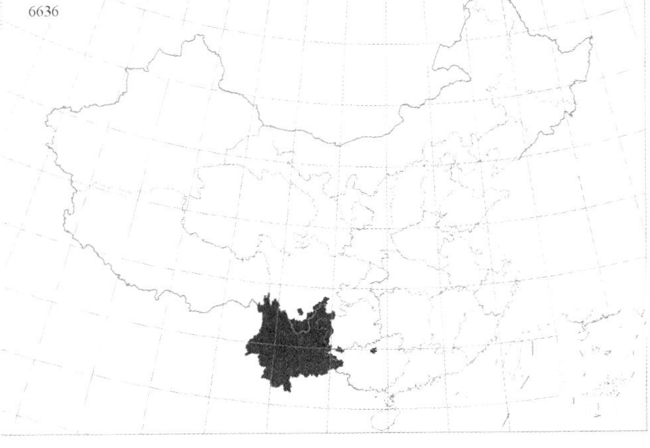

Theaceae 山茶科 *Camellia* 山茶属

Camellia yunnanensis var. camellioides (Hu) Ming 毛果猴子木

Shrubs or small trees, evergreen

15.9 (11.2~21.0); 9.3 (4.6~14.2); 20.8 (16.7~25.6); 16.0 (11.3~21.0); 131.7 (75.8~191.8); 0.0 (0.0~0.4); 787 (642~1053) ❙ 1041 (797~1547); 594 (411~911); 43 (23~68) ❙ 32.6 (-14.2~74.2); 760 (642~1053); 320.9 (218.0~551.6)

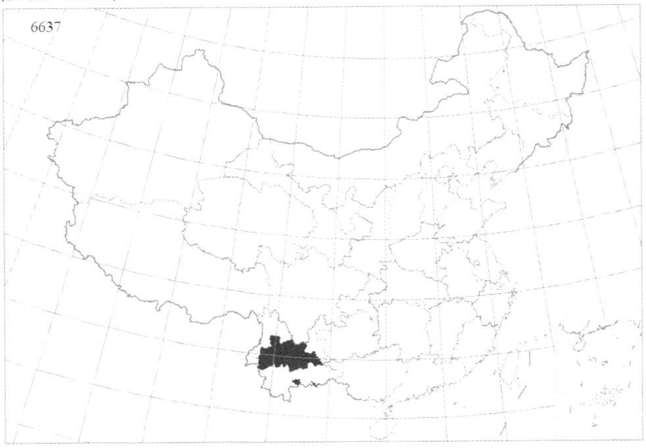

Theaceae 山茶科 *Cleyera* 红淡比属

Cleyera incornuta Y. C. Wu 凹脉红淡比

Shrubs or small trees, evergreen

18.0 (14.6~21.8); 7.6 (3.8~14.2); 27.3 (24.6~29.3); 18.1 (14.7~21.9); 157.0 (117.7~202.6); 0.1 (0.0~1.5); 952 (779~1150) ❙ 1549 (1141~1876); 602 (473~891); 169 (67~228) ❙ 63.7 (14.8~101.8); 951 (779~1150); 224.3 (166.9~278.6)

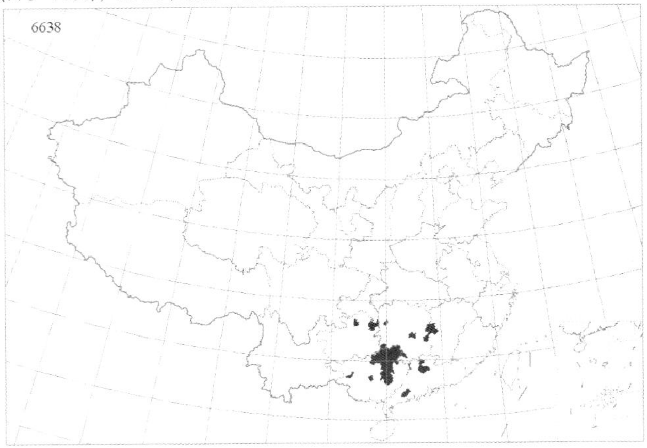

Theaceae 山茶科 *Cleyera* 红淡比属

Cleyera japonica Thunb. 红淡比

Shrubs or trees, evergreen

17.7 (8.6~24.1); 7.4 (1.3~19.1); 27.3 (15.0~30.1); 17.8 (8.7~24.1); 153.9 (52.6~229.7); 0.4 (0.0~8.8); 941 (574~1322) ❙ 1556 (1001~3495); 591 (356~1594); 172 (35~897) ❙ 66.9 (-5.0~246.5); 931 (574~1221); 230.9 (95.3~627.0)

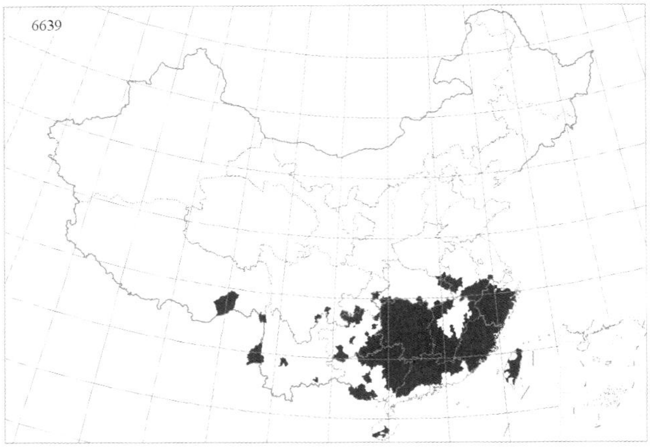

Theaceae 山茶科 *Cleyera* 红淡比属

Cleyera japonica var. wallichiana (DC.) Sealy 大花红淡比

Shrubs or small trees, evergreen

11.0; 3.4; 17.3; 11.0; 76.5; 4.4; 640 ❙ 1389; 796; 48 ❙ 113.7; 640; 399.9

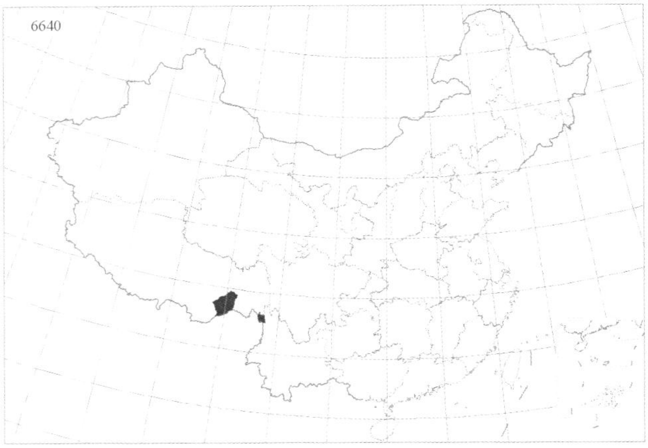

Theaceae 山茶科 *Cleyera* 红淡比属

Cleyera lipingensis (Hand.-Mazz.) Ming 齿叶红淡比

Shrubs or small trees, evergreen

15.9 (12.7~21.7); 5.5 (2.1~15.3); 25.5 (21.8~27.5); 16.0 (12.8~21.7); 132.1 (99.0~200.2); 0.6 (0.0~5.6); 838 (713~1118) ❙ 1351 (1048~2382); 561 (444~1045); 108 (58~308) ❙ 61.5 (30.1~191.8); 838 (713~1075); 208.5 (146.0~284.4)

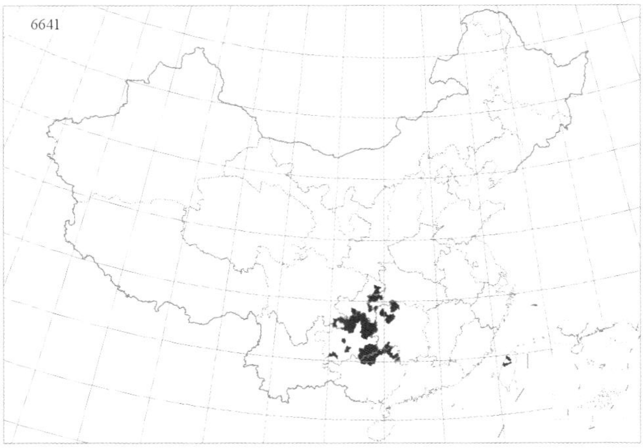

Theaceae 山茶科 *Cleyera* 红淡比属

Cleyera longicarpa (Yamam.) L. K. Ling 长果红淡比

Trees, evergreen

20.2 (16.2~24.1); 14.5 (11.0~18.9); 25.0 (20.1~28.2); 20.2 (16.2~24.1); 182.9 (134.2~229.7); 0.0 (0.0~0.0); 1019 (773~1322) ❙ 2402 (1498~3495); 1106 (750~1847); 255 (47~897) ❙ 146.2 (25.2~246.5); 983 (773~1144); 245.0 (119.4~363.7)

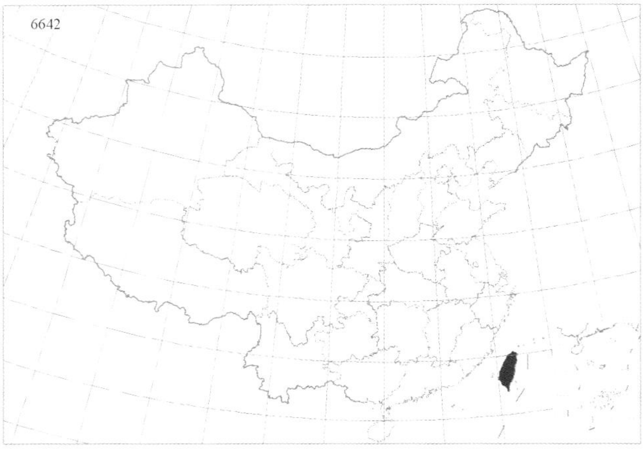

Theaceae 山茶科 *Cleyera* 红淡比属
Cleyera obovata H. T. Chang 倒卵叶红淡比
Shrubs or small trees, evergreen

21.8 (21.0~22.7); 13.9 (13.2~15.0); 27.9 (26.9~28.6); 21.9 (21.0~22.8); 202.2 (192.3~213.1); 0.0 (0.0~0.0); 1152 (1081~1221) ❙ 1800 (1302~2435); 929 (670~1428); 117 (86~130) ❙ 56.1 (12.9~99.5); 1152 (1081~1221); 263.6 (209.2~318.8)

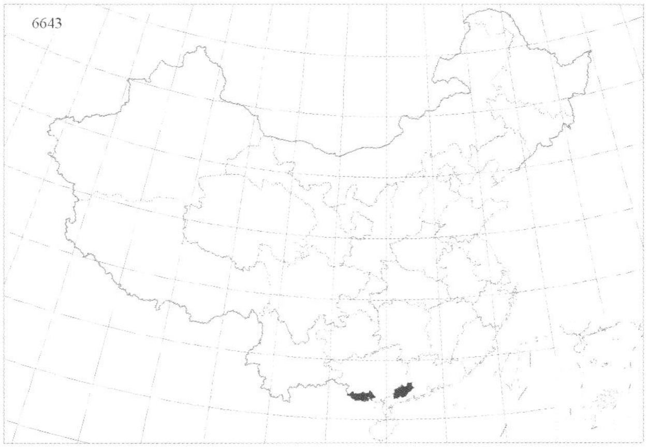

Theaceae 山茶科 *Cleyera* 红淡比属
Cleyera obscurinervia (Merr. et Chun) H. T. Chang 隐脉红淡比
Trees, evergreen

22.3 (21.3~24.8); 14.6 (12.4~20.1); 28.0 (25.7~29.0); 22.3 (21.3~24.8); 208.2 (195.7~237.6); 0.0 (0.0~0.0); 1191 (1111~1385) ❙ 1589 (1179~2435); 762 (392~1428); 109 (41~173) ❙ 34.0 (-11.9~99.5); 1157 (1055~1247); 317.5 (199.4~620.0)

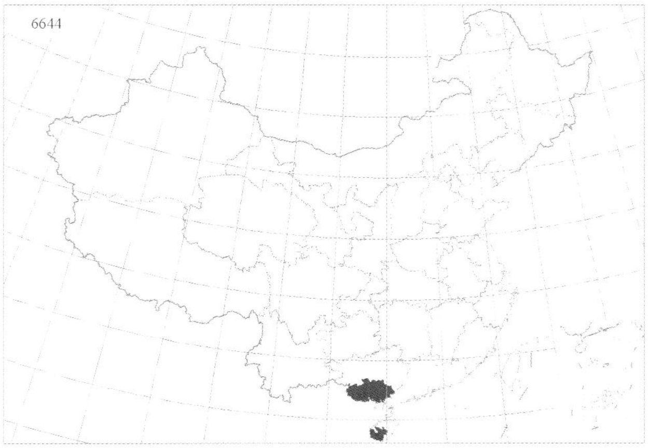

Theaceae 山茶科 *Cleyera* 红淡比属
Cleyera pachyphylla Chun ex H. T. Chang 厚叶红淡比
Shrubs or small trees, evergreen

18.3 (14.3~23.2); 7.9 (3.3~15.8); 27.5 (24.2~30.1); 18.3 (14.3~23.2); 159.9 (113.2~218.9); 0.1 (0.0~2.3); 963 (761~1257) ❙ 1637 (1363~1997); 593 (392~891); 193 (57~259) ❙ 71.5 (15.4~130.7); 954 (761~1257); 236.2 (145.3~371.7)

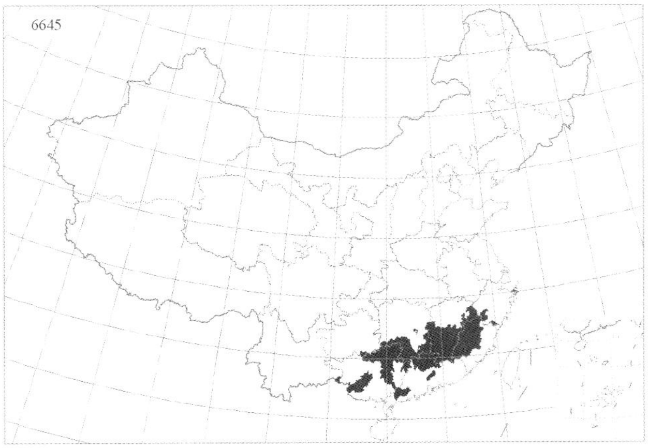

Theaceae 山茶科 *Cleyera* 红淡比属
Cleyera parvifolia (Kobuski) Hu ex L. K. Ling 小叶红淡比
Shrubs or small trees, evergreen

20.6 (17.7~22.0); 12.1 (7.7~14.1); 27.7 (25.9~23.5); 20.6 (17.8~22.1); 187.6 (153.6~205.1); 0.0 (0.0~0.0); 1074 (916~1170) ❙ 1886 (1462~2993); 744 (539~954); 199 (136~547) ❙ 76.8 (29.9~203.4); 1074 (916~1170); 243.4 (213.4~267.2)

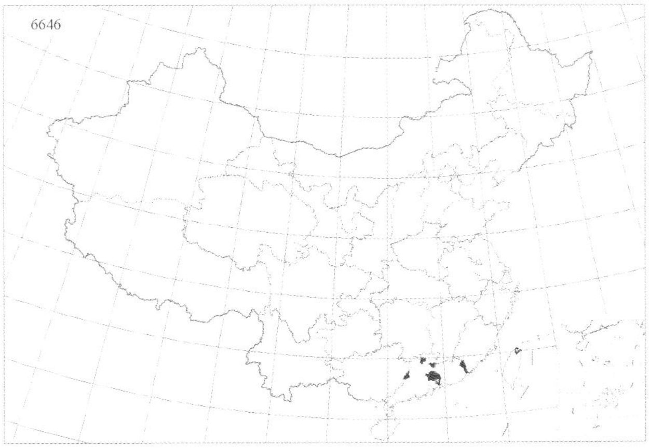

Theaceae 山茶科 *Cleyera* 红淡比属
Cleyera yangchunensis L. K. Ling 阳春红淡比
Shrubs or small trees, evergreen

21.8; 14.2; 27.8; 21.9; 202.6; 0.0; 1150 ❙ 1876; 891; 116 ❙ 63.2; 1150; 273.3

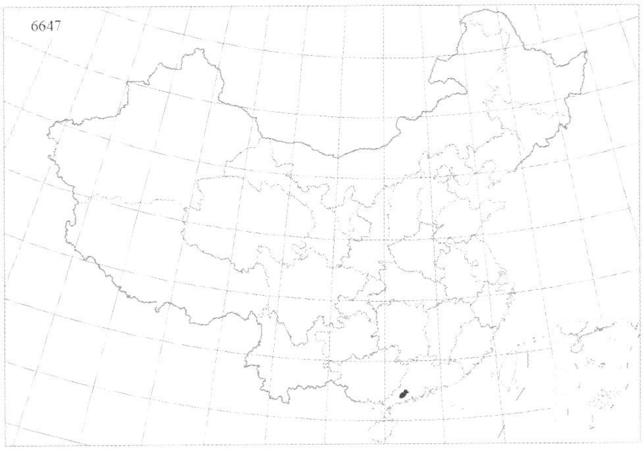

Theaceae 山茶科 *Eurya* 柃属
Eurya acuminata DC. 尖叶柃
Shrubs or small trees, evergreen

16.3 (8.6~21.0); 9.0 (1.3~14.2); 21.6 (15.0~25.6); 16.3 (8.7~21.0); 136.4 (52.6~191.8); 0.9 (0.0~8.8); 814 (574~1053) ❙ 1329 (1041~1736); 742 (582~1007); 53 (45~68) ❙ 65.4 (32.4~145.8); 814 (574~1053); 396.0 (277.0~551.6)

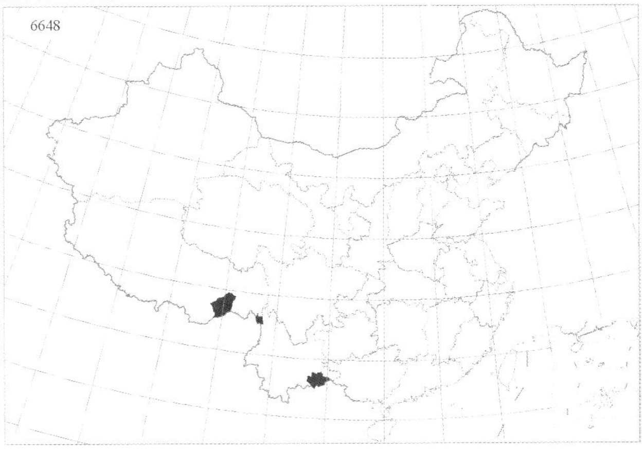

Theaceae 山茶科 *Eurya* 柃属

Eurya acuminatissima Merr. et Chun 尖叶毛柃

Shrubs or small trees, evergreen

18.1 (14.3~23.2); 7.8 (3.3~15.8); 27.3 (23.5~29.8); 18.2 (14.3~23.2); 158.6 (112.6~218.9); 0.2 (0.0~2.3); 960 (756~1257) ❙ 1519 (1182~2024); 591 (392~1056); 158 (43~230) ❙ 59.3 (15.4~128.9); 954 (756~1257); 220.9 (134.6~371.7)

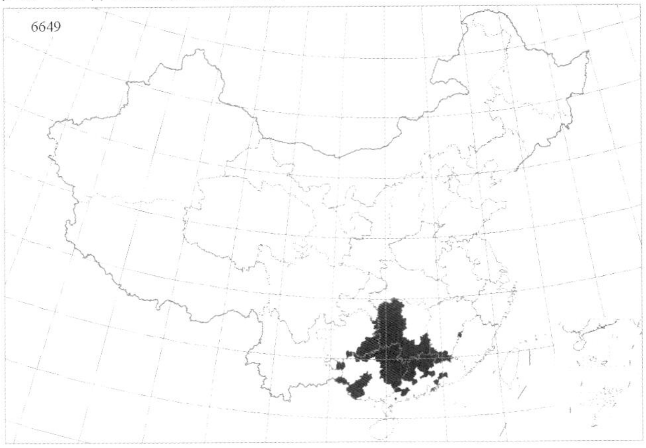

Theaceae 山茶科 *Eurya* 柃属

Eurya acuminoides Hu et L. K. Ling 川黔尖叶柃

Shrubs, evergreen

15.6 (11.1~18.7); 5.6 (1.8~8.1); 25.0 (19.3~28.9); 15.7 (11.1~18.8); 129.1 (79.8~166.0); 0.8 (0.0~6.9); 835 (648~1002) ❙ 1120 (963~1414); 540 (442~817); 55 (29~69) ❙ 36.6 (8.0~89.0); 834 (648~1002); 181.0 (108.2~278.8)

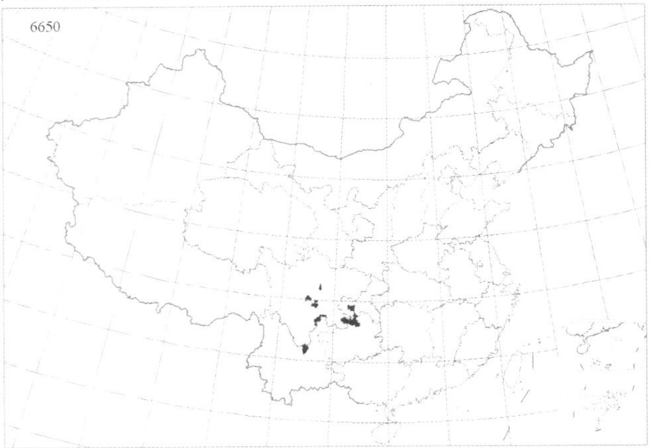

Theaceae 山茶科 *Eurya* 柃属

Eurya acutisepala Hu et L. K. Ling 尖萼毛柃

Shrubs or small trees, evergreen

17.5 (14.2~22.2); 7.1 (3.3~14.2); 26.8 (20.8~29.9); 17.5 (14.2~22.3); 150.4 (110.9~207.5); 0.2 (0.0~2.3); 917 (734~1190) ❙ 1508 (1049~1961); 569 (392~911); 164 (44~230) ❙ 65.1 (15.4~147.0); 908 (734~1190); 232.3 (134.6~551.6)

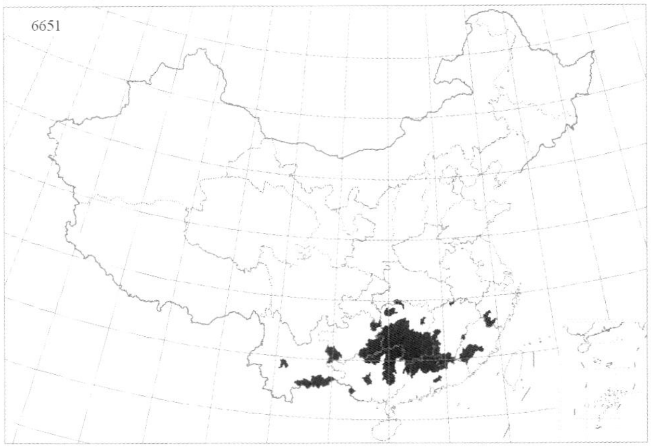

Theaceae 山茶科 *Eurya* 柃属

Eurya alata Kobuski 翅柃

Shrubs, evergreen

16.5 (9.3~22.2); 5.2 (−2.9~14.7); 26.9 (20.1~29.9); 16.5 (9.5~22.2); 140.0 (69.8~206.9); 1.9 (0.0~20.5); 886 (603~1177) ❙ 1442 (512~1997); 538 (258~891); 156 (21~259) ❙ 62.8 (−35.3~147.0); 875 (512~1177); 222.5 (121.6~332.5)

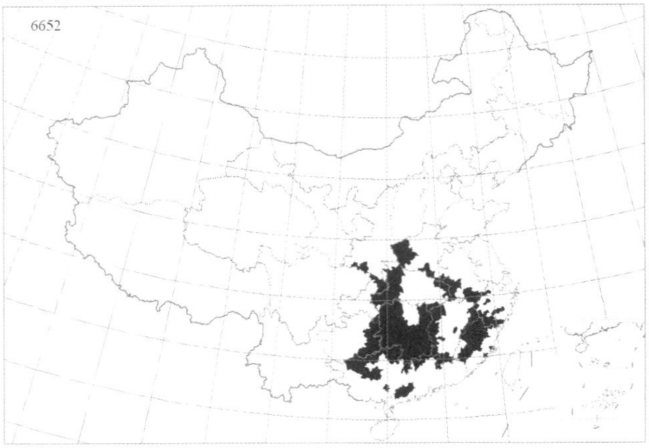

Theaceae 山茶科 *Eurya* 柃属

Eurya amplexifolia Dunn 穿心柃

Shrubs or small trees, evergreen

19.2 (17.1~21.7); 10.5 (6.5~14.0); 27.0 (25.2~28.6); 19.3 (17.1~21.7); 171.4 (145.5~200.5); 0.0 (0.0~0.0); 989 (864~1140) ❙ 1625 (1258~1829); 591 (425~682); 177 (125~230) ❙ 65.9 (14.0~104.6); 987 (864~1140); 264.1 (197.5~305.7)

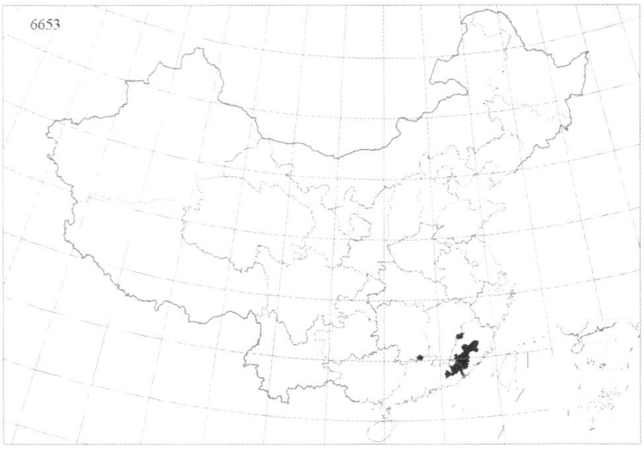

Theaceae 山茶科 *Eurya* 柃属

Eurya auriformis H. T. Chang 耳叶柃

Shrubs, evergreen

20.4 (19.1~21.5); 11.4 (9.3~12.8); 27.9 (27.2~28.5); 20.5 (19.1~21.5); 185.5 (169.2~198.2); 0.0 (0.0~0.0); 1068 (988~1138) ❙ 1614 (1462~1990); 635 (561~827); 163 (136~188) ❙ 51.4 (29.9~83.4); 1068 (988~1138); 244.1 (216.5~266.9)

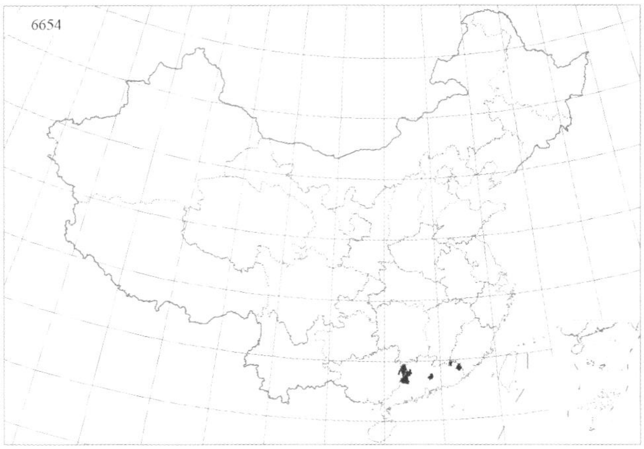

Theaceae 山茶科 *Eurya* 柃属

Eurya bifidostyla K. M. Feng et P. I Mao 双柱柃

Small trees, evergreen

18.6; 12.0; 23.0; 18.6; 163.4; 0.0; 907 ❙ 1414; 782; 58 ❙ 56.8; 903; 494.0

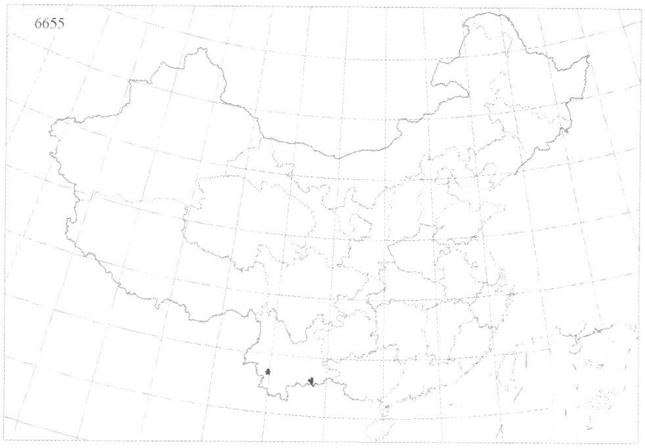

Theaceae 山茶科 *Eurya* 柃属

Eurya brevistyla Kobuski 短柱柃

Shrubs or small trees, evergreen

15.0 (2.6~19.9); 4.6 (-7.0~13.1); 24.4 (10.8~29.3); 15.1 (4.3~19.9); 123.7 (20.0~179.3); 3.3 (0.0~48.2); 806 (427~1036) ❙ 1210 (512~1997); 527 (258~836); 94 (16~259) ❙ 49.4 (-35.3~132.1); 795 (427~1036); 240.4 (121.6~432.9)

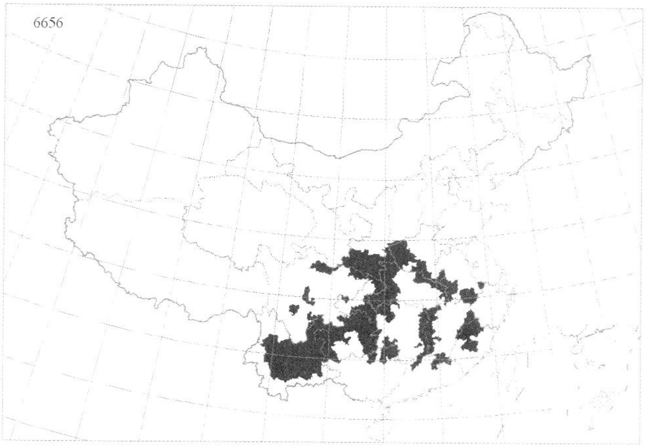

Theaceae 山茶科 *Eurya* 柃属

Eurya cavinervis Vesque 云南凹脉柃

Shrubs or small trees, evergreen

16.8 (5.3~22.7); 9.6 (-2.4~15.0); 22.2 (12.2~28.6); 16.8 (5.7~22.8); 143.0 (30.5~213.1); 0.9 (0.0~26.2); 859 (479~1221) ❙ 1291 (801~2435); 695 (496~1428); 74 (21~189) ❙ 51.3 (6.5~145.8); 845 (479~1221); 336.6 (185.5~551.3)

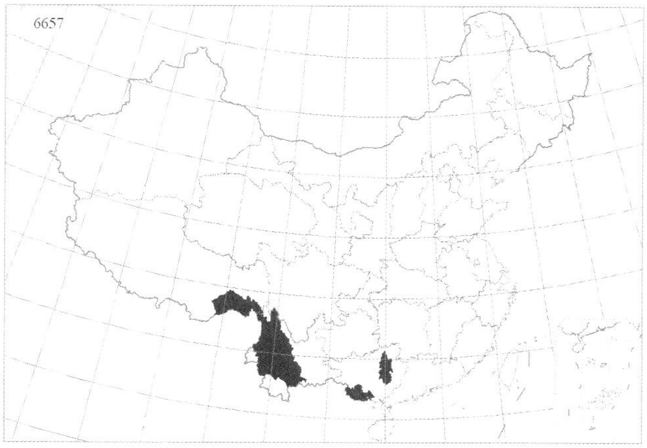

Theaceae 山茶科 *Eurya* 柃属

Eurya chinensis R. Brown 米碎花

Shrubs, evergreen

19.9 (14.3~24.8); 10.9 (3.3~20.1); 27.6 (20.1~29.8); 19.9 (14.3~24.8); 179.1 (113.2~237.6); 0.0 (0.0~2.3); 1047 (761~1385) ❙ 1691 (1174~3495); 689 (386~1847); 171 (41~897) ❙ 64.5 (-6.9~246.5); 1036 (761~1326); 234.4 (95.3~620.0)

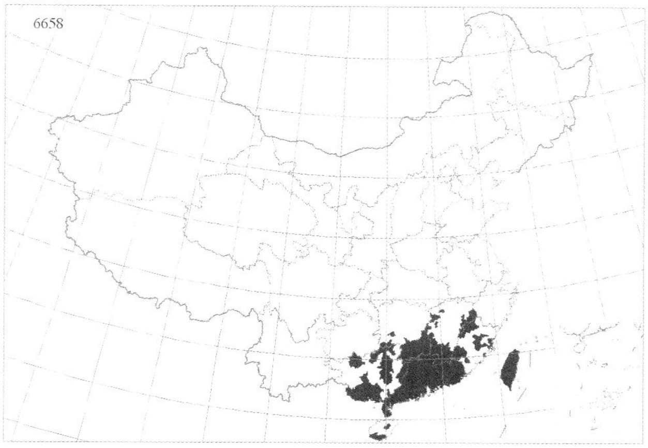

Theaceae 山茶科 *Eurya* 柃属

Eurya chinensis var. glabra Hu et L. K. Ling 光枝米碎花

Shrubs, evergreen

19.6 (14.3~23.4); 10.0 (3.3~16.1); 28.0 (24.2~29.8); 19.7 (14.3~23.5); 176.1 (113.2~221.7); 0.0 (0.0~2.3); 1038 (761~1273) ❙ 1625 (1371~2024); 626 (400~954); 171 (82~231) ❙ 58.1 (9.0~128.9); 1031 (761~1273); 216.6 (95.3~275.2)

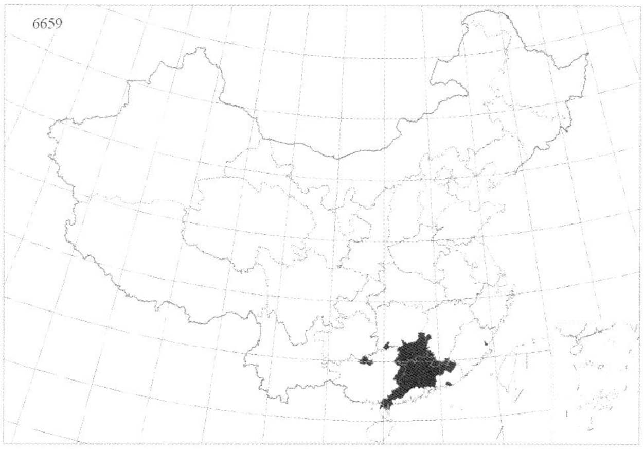

Theaceae 山茶科 *Eurya* 柃属

Eurya chuekiangensis Hu 大果柃

Shrubs, evergreen

8.4; 1.1; 14.7; 8.6; 53.2; 11.9; 566 ❙ 1082; 613; 44 ❙ 91.7; 566; 323.7

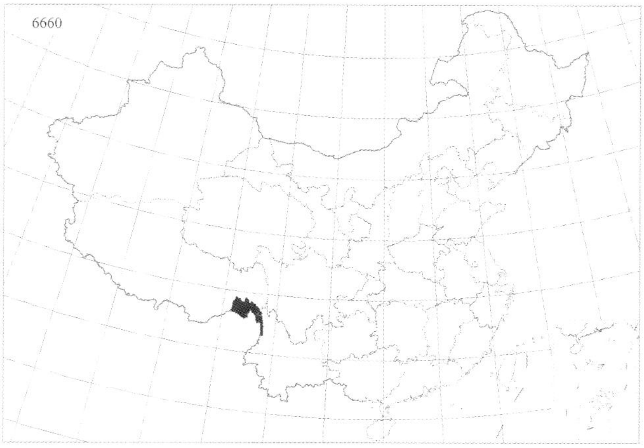

Theaceae 山茶科 *Eurya* 柃属

Eurya ciliata Merr. 华南毛柃

Shrubs or small trees, evergreen

20.1 (14.6~24.8); 11.4 (3.8~20.1); 27.3 (21.4~29.0); 20.1 (14.7~24.8); 181.8 (117.7~237.6); 0.0 (0.0~1.5); 1061 (779~1385) ❙ 1571 (868~2435); 645 (312~1428); 136 (41~225) ❙ 51.0 (-34.3~101.8); 1035 (779~1360); 295.4 (183.3~620.0)

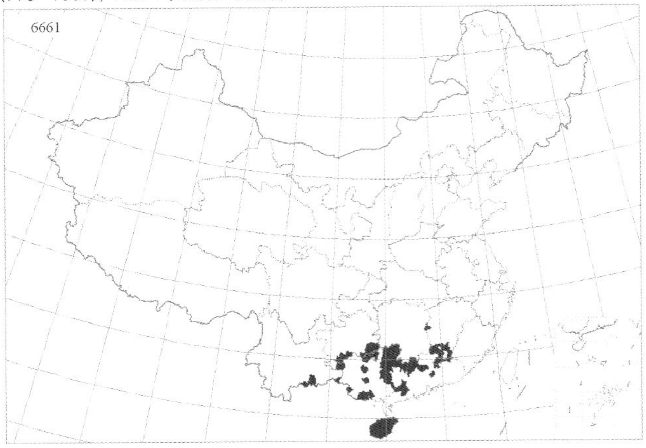

Theaceae 山茶科 *Eurya* 柃属

Eurya crassilimba H. T. Chang 厚叶柃

Shrubs, evergreen

13.3; 5.4; 19.6; 13.4; 100.4; 0.0; 706 ❙ 1736; 1007; 51 ❙ 145.8; 706; 447.4

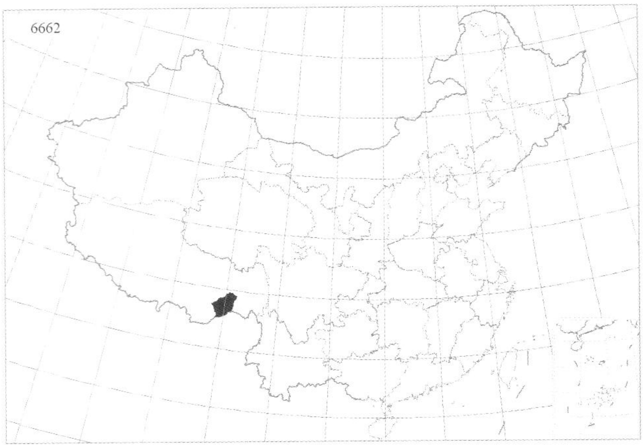

Theaceae 山茶科 *Eurya* 柃属

Eurya crenatifolia (Yamam.) Kobuski 钝齿柃

Shrubs, evergreen

18.0; 12.3; 23.0; 18.1; 157.1; 0.0; 874 ❙ 2784; 943; 417 ❙ 219.4; 874; 276.9

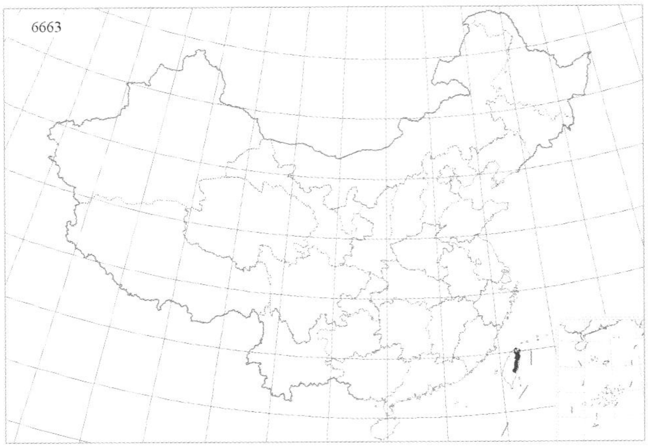

Theaceae 山茶科 *Eurya* 柃属

Eurya cuneata Kobuski 楔叶柃

Shrubs or small trees, evergreen

21.7 (19.0~24.7); 14.0 (9.1~19.8); 27.9 (26.7~29.8); 21.7 (19.0~24.7); 200.4 (168.2~236.5); 0.0 (0.0~0.0); 1158 (989~1376) ❙ 1625 (1380~1803); 576 (455~680); 130 (48~194) ❙ 42.9 (6.7~76.3); 1086 (989~1212); 364.1 (174.9~620.0)

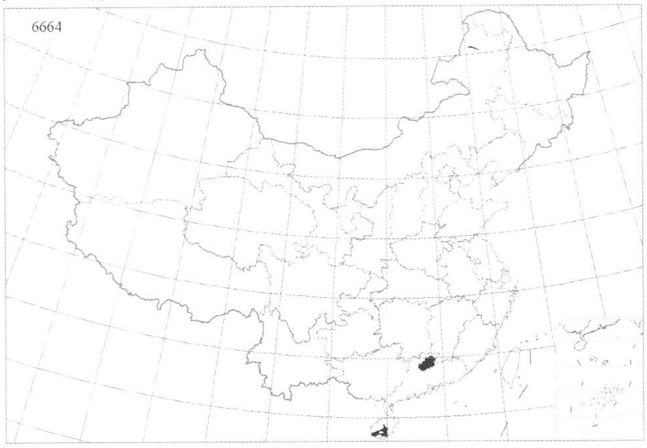

Theaceae 山茶科 *Eurya* 柃属

Eurya cuneata var. glabra Kobuski 光枝楔叶柃

Shrubs or small trees, evergreen

23.9; 18.9; 27.4; 23.9; 227.1; 0.0; 1297 ❙ 1561; 534; 56 ❙ 20.7; 1159; 535.7

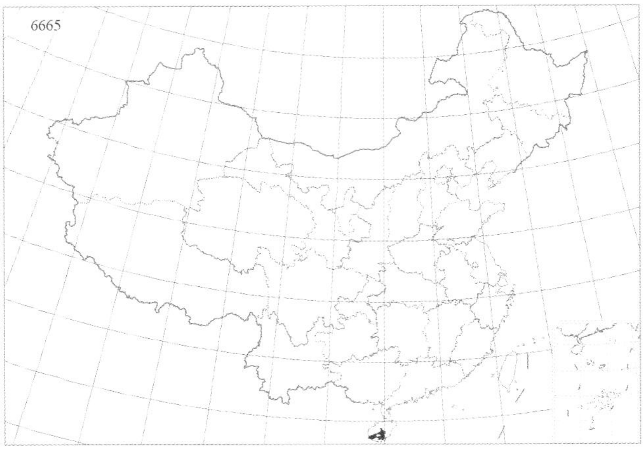

Theaceae 山茶科 *Eurya* 柃属

Eurya disticha Chun 秃小耳柃

Small shrubs, evergreen

20.0 (18.2~21.8); 11.1 (8.7~14.2); 27.1 (25.9~28.5); 20.1 (18.3~21.9); 180.8 (159.3~202.6); 0.0 (0.0~0.0); 1037 (928~1150) ❙ 1490 (1284~1876); 719 (599~891); 105 (73~142) ❙ 43.3 (29.9~64.3); 1037 (928~1150); 273.0 (226.9~322.1)

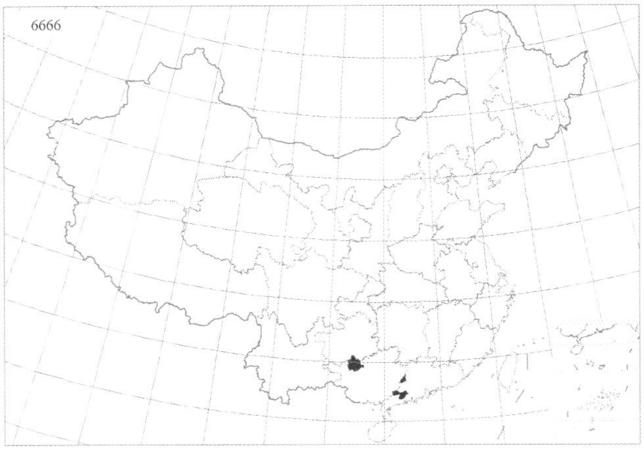

Theaceae 山茶科 *Eurya* 柃属

Eurya distichophylla Hemsl. 二列叶柃

Shrubs or small trees, evergreen

19.2 (14.3~23.2); 9.8 (3.7~15.8); 27.5 (24.2~29.8); 19.3 (14.3~23.2); 171.7 (113.2~218.9); 0.1 (0.0~1.5); 1012 (761~1257) ❙ 1587 (1039~2037); 636 (381~1033); 156 (57~230) ❙ 57.8 (7.5~128.9); 1007 (761~1257); 224.7 (95.3~371.7)

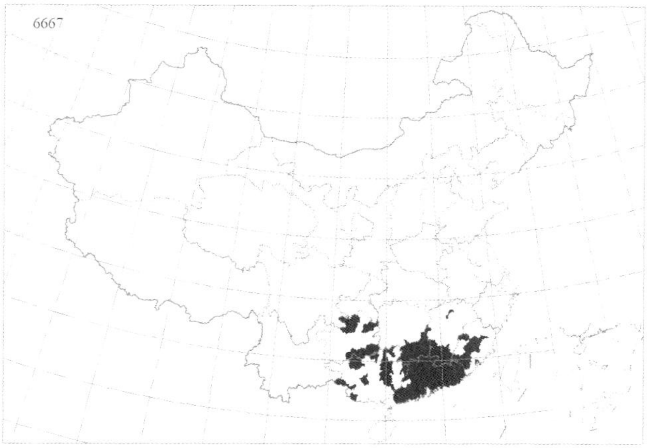

Theaceae 山茶科 *Eurya* 柃属

Eurya emarginata (Thunb.) Makino 滨柃

Shrubs, evergreen

19.0 (16.4~22.2); 10.8 (5.3~17.7); 26.9 (20.3~28.9); 19.0 (16.5~22.2); 168.6 (137.8~206.7); 0.0 (0.0~0.0); 976 (783~1123) ❙ 1873 (1174~3495); 685 (386~1847); 271 (107~897) ❙ 93.3 (7.3~224.4); 966 (783~1123); 236.8 (188.7~363.7)

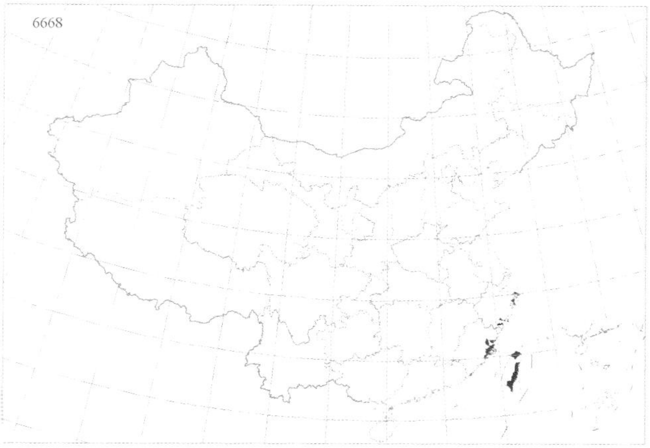

Theaceae 山茶科 *Eurya* 柃属

Eurya fangii Rehder 川柃

Shrubs or small trees, evergreen

13.7 (2.2~18.7); 4.1 (-6.6~8.1); 22.5 (9.9~28.9); 13.9 (3.8~18.8); 110.6 (15.5~166.0); 5.6 (0.0~49.1); 768 (413~1002) ❙ 1141 (932~1583); 591 (447~938); 46 (18~71) ❙ 54.8 (8.0~125.9); 767 (413~1002); 205.2 (108.2~351.7)

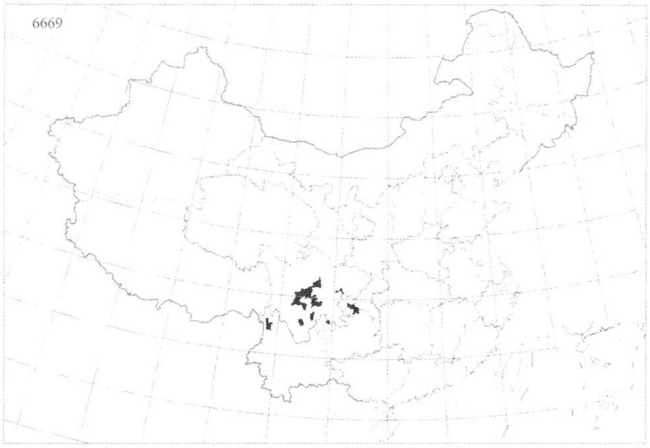

Theaceae 山茶科 *Eurya* 柃属

Eurya fangii var. megaphylla P. S. Hsu 大叶柃

Shrubs or small trees, evergreen

0.6; 2.0; 18.6; 10.8; 75.4; 6.1; 636 ❙ 1121; 628; 30 ❙ 75.9; 636; 305.7

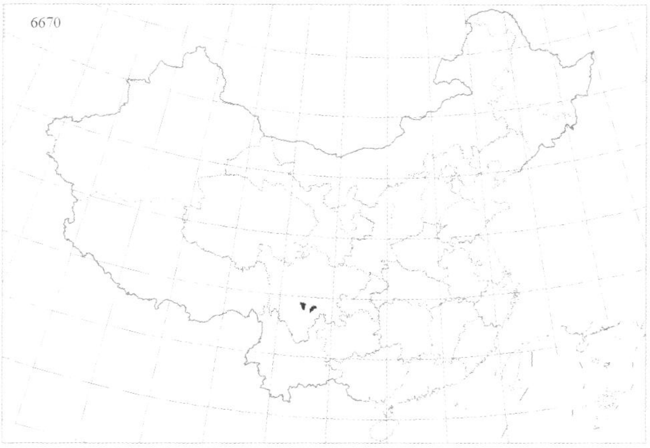

Theaceae 山茶科 *Eurya* 柃属

Eurya glaberrima Hayata 光柃

Shrubs or small trees, evergreen

18.8; 13.0; 23.7; 18.9; 166.4; 0.0; 912 ❙ 2553; 1090; 274 ❙ 182.7; 912; 252.6

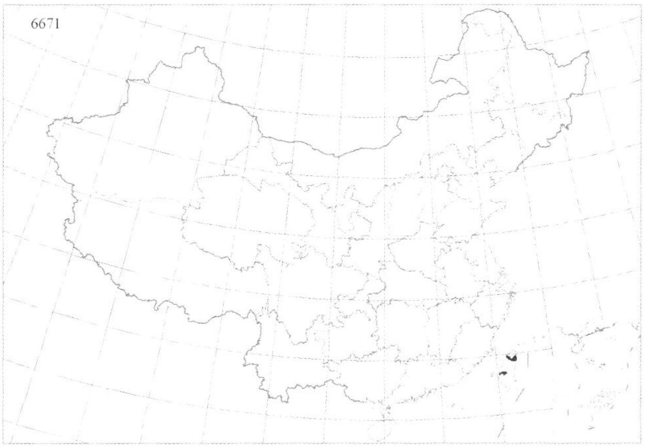

Theaceae 山茶科 *Eurya* 柃属

Eurya glandulosa Merr. 腺柃

Shrubs, evergreen

19.2 (14.6~22.3); 10.2 (3.8~14.3); 26.9 (24.6~28.8); 19.2 (14.7~22.4); 171.0 (117.7~208.7); 0.1 (0.0~1.5); 995 (779~1197) ❙ 1662 (1284~2014); 686 (539~954); 156 (73~205) ❙ 68.2 (30.9~101.8); 995 (779~1197); 261.1 (161.4~322.1)

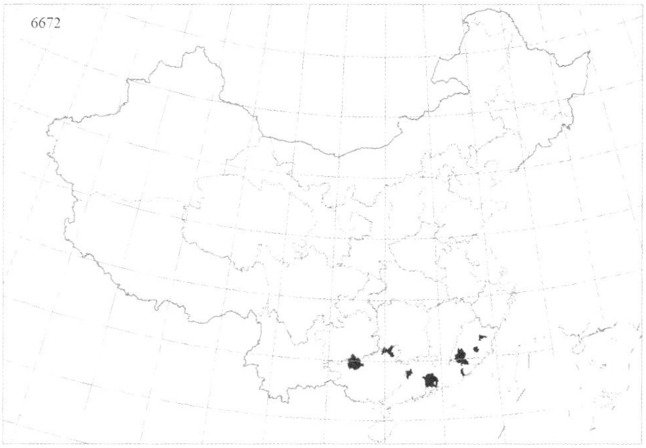

Theaceae 山茶科 *Eurya* 柃属

Eurya glandulosa var. cuneiformis H. T. Chang 楔基腺柃

Shrubs, evergreen

20.8 (17.9~22.0); 12.8 (8.1~14.3); 27.6 (26.3~28.5); 20.8 (17.9~22.1); 189.9 (155.4~205.1); 0.0 (0.0~0.0); 1079 (918~1170) ❙ 1635 (1375~2014); 658 (476~954); 147 (123~182) ❙ 51.8 (30.2~79.9); 1079 (918~1170); 238.8 (188.9~273.2)

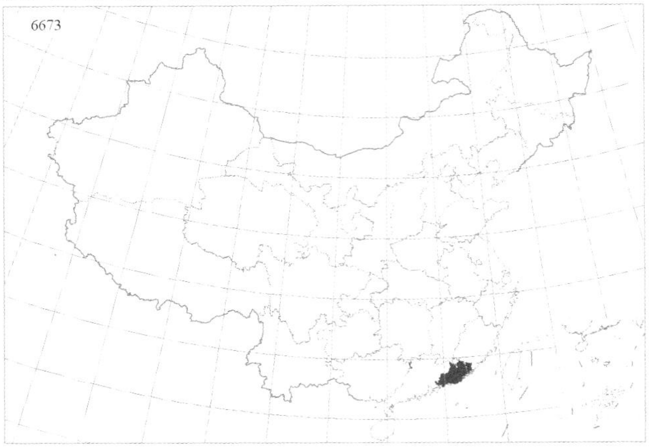

Theaceae 山茶科 *Eurya* 柃属

Eurya glandulosa var. dasyclados (Kobuski) H. T. Chang 粗枝腺柃

Shrubs, evergreen

19.9 (17.3~22.4); 11.3 (8.5~14.5); 27.2 (25.2~28.7); 19.9 (17.3~22.4); 178.8 (147.6~209.2); 0.0 (0.0~0.0); 1027 (864~1198) ❙ 1676 (1447~2005); 665 (577~891); 163 (82~205) ❙ 64.8 (20.8~98.6); 1027 (864~1198); 259.0 (164.9~305.7)

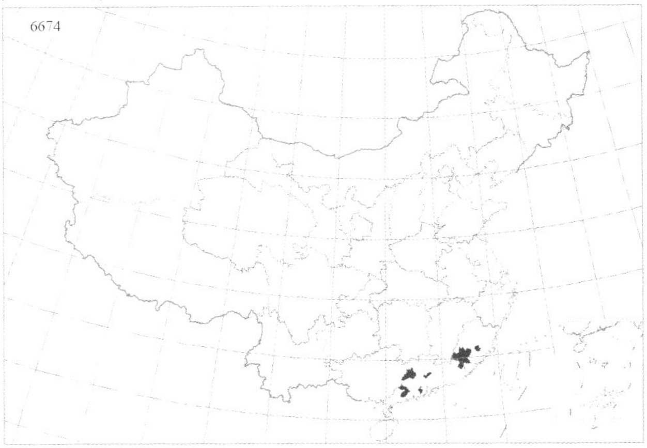

Theaceae 山茶科 *Eurya* 柃属

Eurya gnaphalocarpa Hayata 灰毛柃

Shrubs or small trees, evergreen

18.4 (16.2~21.7); 13.1 (11.0~15.3); 22.7 (20.1~27.0); 18.5 (16.2~21.7); 161.4 (134.2~200.2); 0.0 (0.0~0.0); 895 (773~1118) ❙ 2426 (1836~2713); 1169 (947~1412); 202 (110~303) ❙ 179.3 (64.2~246.5); 888 (773~1075); 274.7 (202.8~363.7)

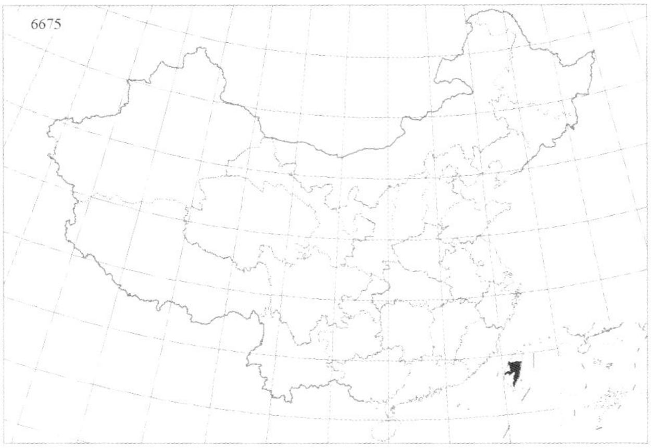

Theaceae 山茶科 *Eurya* 柃属

Eurya groffii Merr. 岗柃

Shrubs or small trees, evergreen

18.1 (5.4~24.3); 9.6 (-2.1~18.7); 25.2 (12.3~29.8); 18.2 (5.8~24.3); 158.7 (31.5~231.9); 0.4 (0.0~25.6); 944 (481~1338) ❙ 1390 (741~2435); 650 (392~1428); 100 (15~230) ❙ 48.0 (-14.2~145.8); 937 (481~1273); 260.5 (95.3~659.0)

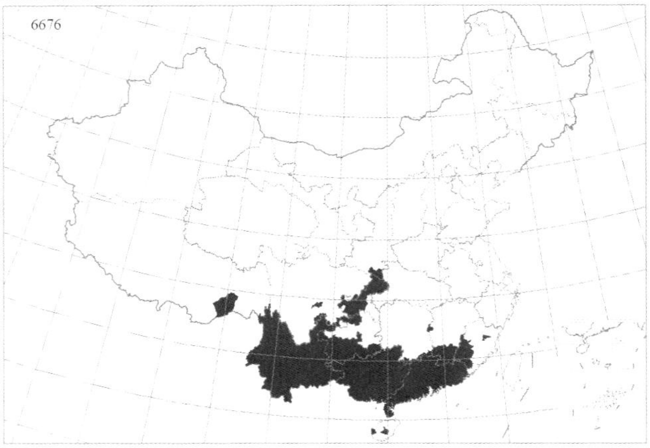

Theaceae 山茶科 *Eurya* 柃属

Eurya gungshanensis Hu et L. K. Ling 贡山柃

Small trees, evergreen

11.1; 3.7; 17.2; 11.1; 76.5; 3.2; 641 ❙ 1340; 762; 51 ❙ 106.8; 641; 394.8

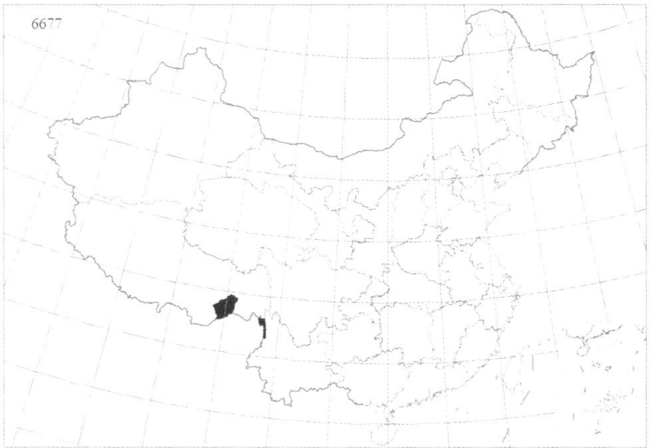

Theaceae 山茶科 *Eurya* 柃属

Eurya hainanensis (Kobuski) H. T. Chang 海南柃

Shrubs or small trees, evergreen

23.2 (21.0~24.8); 17.0 (13.2~20.1); 27.7 (25.7~28.5); 23.2 (21.0~24.8); 218.5 (192.3~237.6); 0.0 (0.0~0.0); 1247 (1081~1385) ❙ 1606 (1179~1876); 626 (392~891); 82 (41~125) ❙ 30.0 (-11.9~64.3); 1164 (1055~1360); 413.8 (209.2~620.0)

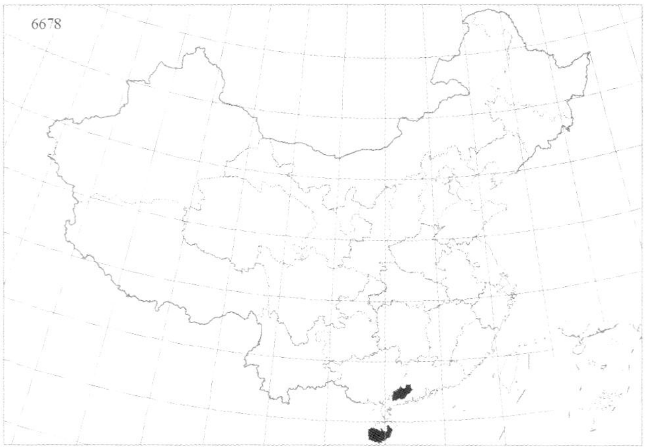

Theaceae 山茶科 *Eurya* 柃属
Eurya handel-mazzettii H. T. Chang 丽江柃
Shrubs or small trees, evergreen

14.8 (−1.8~19.6); 7.6 (−10.2~13.1); 20.3 (6.1~27.1); 14.9 (1.8~19.6); 119.8 (1.9~175.7); 2.1 (0.0~82.8); 757 (329~953) ❙ 1061 (320~1736); 607 (217~1007); 40 (10~70) ❙ 40.8 (−14.2~145.8); 738 (320~932); 310.8 (61.6~627.0)

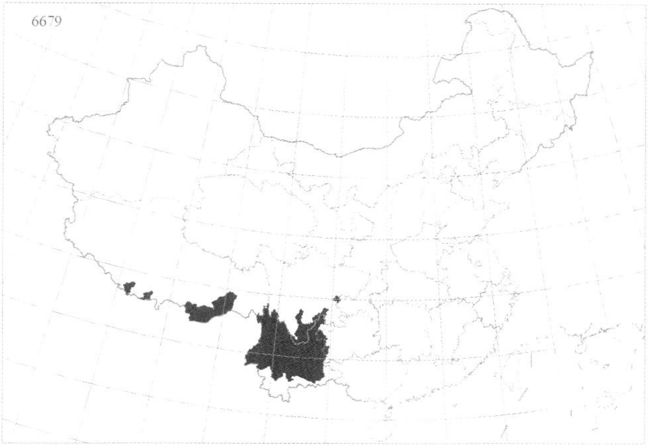

Theaceae 山茶科 *Eurya* 柃属
Eurya hebeclados Ling 微毛柃
Shrubs or small trees, evergreen

17.7 (10.6~22.7); 7.2 (−0.2~15.0); 27.4 (19.6~30.1); 17.8 (10.6~22.8); 153.8 (79.4~213.1); 0.5 (0.0~12.1); 943 (643~1221) ❙ 1522 (871~2435); 567 (299~1428); 169 (29~259) ❙ 62.9 (−6.7~147.0); 931 (643~1221); 222.9 (108.2~371.7)

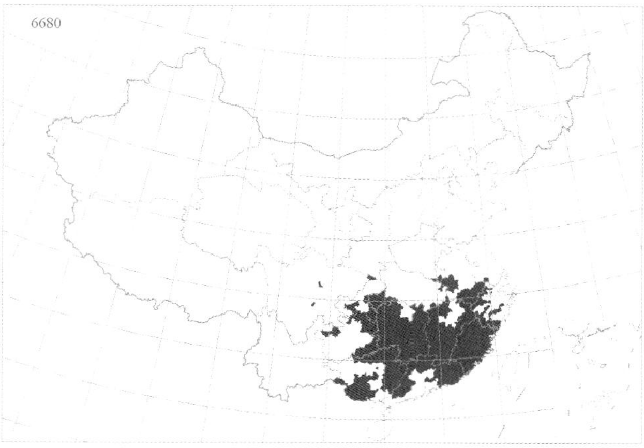

Theaceae 山茶科 *Eurya* 柃属
Eurya henryi Hemsl. 披针叶毛柃
Shrubs, evergreen

17.4 (14.2~21.0); 9.8 (5.0~14.2); 23.0 (20.8~25.6); 17.4 (14.2~21.0); 148.8 (110.9~191.8); 0.0 (0.0~0.0); 856 (734~1053) ❙ 1290 (955~1547); 708 (506~911); 58 (44~107) ❙ 51.1 (19.0~80.6); 854 (734~1053); 364.3 (215.5~551.6)

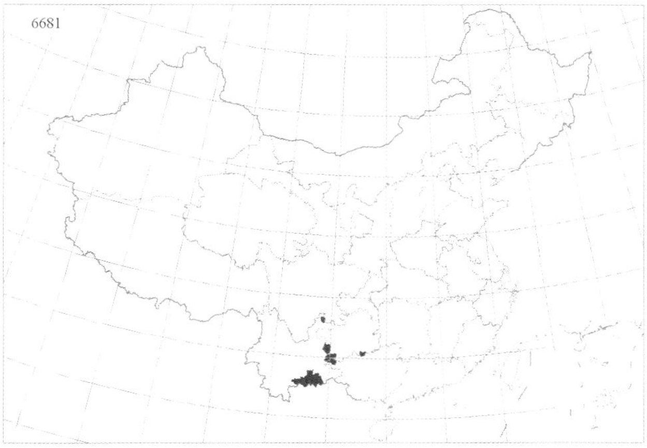

Theaceae 山茶科 *Eurya* 柃属
Eurya hupehensis P. S. Hsu 鄂柃
Small trees, evergreen

12.5; 1.2; 23.4; 12.6; 98.8; 7.9; 714 ❙ 1160; 491; 79 ❙ 62.5; 714; 254.9

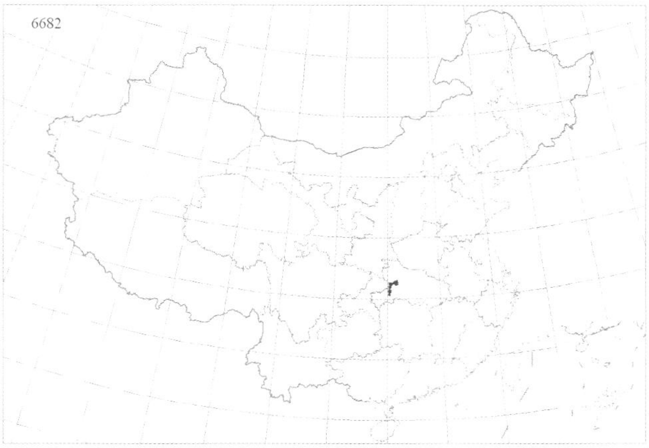

Theaceae 山茶科 *Eurya* 柃属
Eurya impressinervis Kobuski 凹脉柃
Shrubs or small trees, evergreen

18.9 (8.6~22.3); 9.5 (1.3~14.1); 26.8 (15.0~29.8); 19.0 (8.7~22.3); 167.8 (52.6~208.1); 0.1 (0.0~8.8); 988 (574~1202) ❙ 1559 (1041~1991); 651 (465~902); 152 (43~197) ❙ 58.9 (16.9~101.8); 986 (574~1202); 247.2 (174.9~413.5)

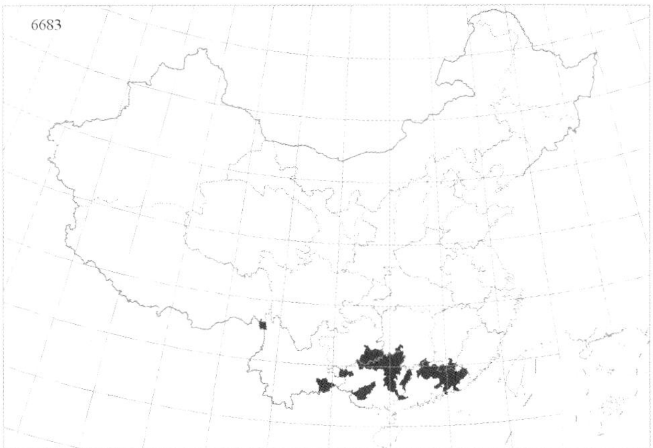

Theaceae 山茶科 *Eurya* 柃属
Eurya inaequalis P. S. Hsu 偏心叶柃
Shrubs or small trees, evergreen

18.6 (16.9~21.0); 12.3 (10.3~14.2); 22.9 (21.4~25.6); 18.6 (16.9~21.0); 163.5 (142.7~191.8); 0.0 (0.0~0.0); 898 (813~1053) ❙ 1352 (1163~1547); 765 (641~911); 56 (48~68) ❙ 50.7 (34.8~74.2); 897 (813~1053); 472.0 (383.4~551.6)

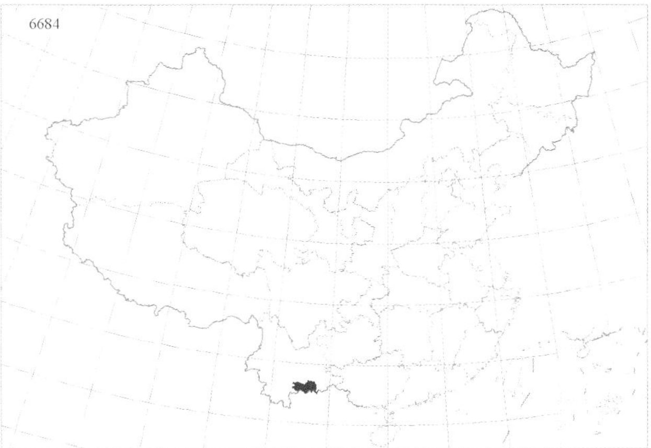

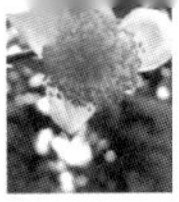

Theaceae 山茶科 *Eurya* 柃属

Eurya japonica Thunb. 柃木

Shrubs, evergreen

19.3 (14.6~22.2); 10.9 (2.8~17.7); 26.8 (20.1~28.7); 19.4 (14.7~22.7); 173.2 (117.7~212.7); 0.2 (0.0~3.1); 1006 (773~1216) ❙ 1864 (1060~3495); 772 (447~1847); 196 (104~897) ❙ 87.4 (26.4~246.5); 1005 (773~1216); 247.1 (194.3~363.7)

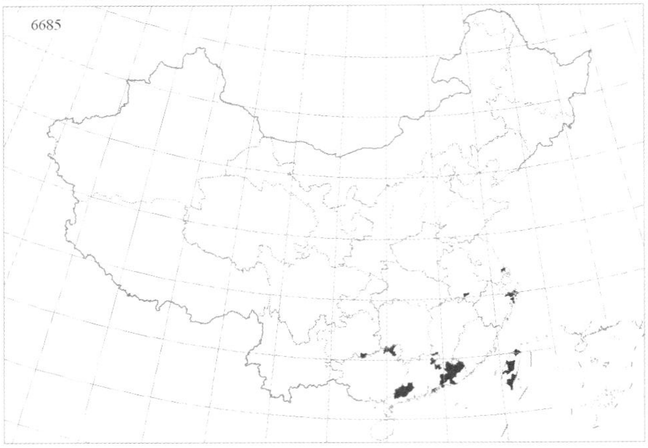

Theaceae 山茶科 *Eurya* 柃属

Eurya jintungensis Hu et L. K. Ling 景东柃

Shrubs or small trees, evergreen

17.2 (14.4~19.6); 10.8 (7.5~13.1); 21.4 (19.1~23.5); 17.3 (14.4~19.6); 147.3 (113.1~175.7); 0.0 (0.0~0.0); 832 (727~953) ❙ 1339 (1014~1688); 759 (568~959); 52 (42~70) ❙ 61.0 (26.1~112.2); 820 (727~932); 436.1 (318.7~627.0)

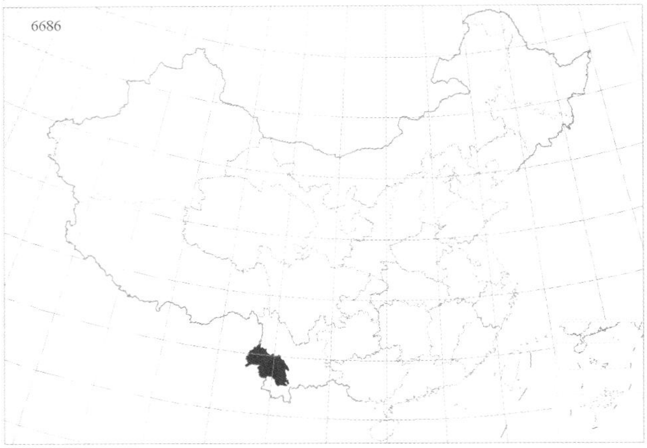

Theaceae 山茶科 *Eurya* 柃属

Eurya kueichowensis Hu et L. K. Ling ex P. T. Li, J. S. China 贵州毛柃

Shrubs or small trees, evergreen

16.1 (10.2~21.6); 6.0 (-0.7~12.6); 25.3 (20.0~28.9); 16.2 (10.4~21.7); 134.9 (77.2~199.8); 0.7 (0.0~13.6); 847 (634~1153) ❙ 1206 (879~1703); 542 (442~809); 74 (35~194) ❙ 43.4 (6.5~101.8); 846 (634~1153); 208.1 (108.2~365.3)

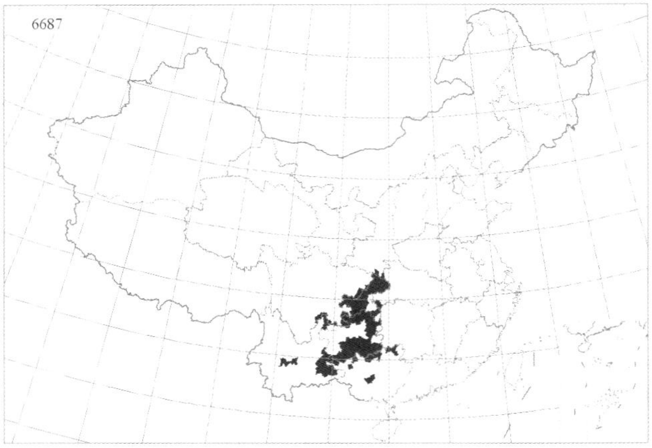

Theaceae 山茶科 *Eurya* 柃属

Eurya lanciformis Kobuski 披针叶柃

Trees, evergreen

20.4 (14.6~22.7); 11.7 (3.8~15.0); 27.4 (24.6~28.6); 20.5 (14.7~22.8); 186.0 (117.7~213.1); 0.1 (0.0~1.5); 1085 (779~1221) ❙ 1711 (1302~2435); 864 (544~1428); 132 (86~191) ❙ 60.3 (12.9~101.8); 1085 (779~1221); 279.2 (235.1~327.7)

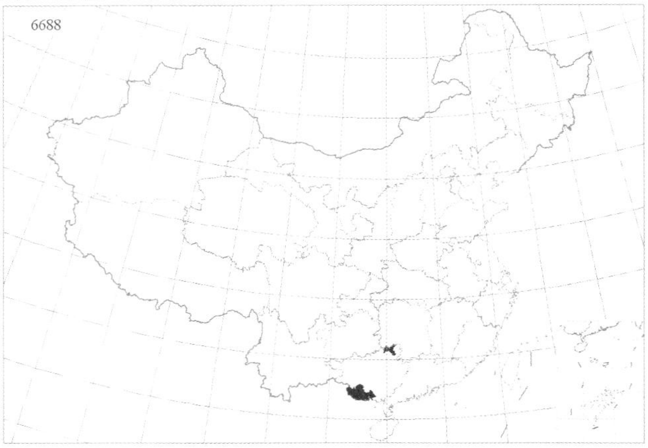

Theaceae 山茶科 *Eurya* 柃属

Eurya leptophylla Hayata 菠叶柃

Shrubs, evergreen

19.1 (16.2~21.7); 13.3 (11.0~15.3); 24.0 (20.1~27.0); 19.1 (16.2~21.7); 169.6 (134.2~200.2); 0.0 (0.0~0.0); 943 (773~1118) ❙ 2467 (1836~2993); 1120 (924~1412); 253 (110~547) ❙ 169.2 (64.2~246.5); 934 (773~1075); 248.5 (202.8~289.1)

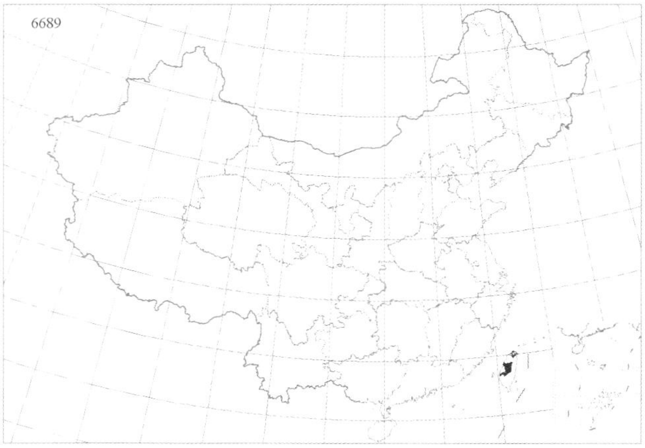

Theaceae 山茶科 *Eurya* 柃属

Eurya loquaiana Dunn 细枝柃

Shrubs or small trees, evergreen

17.9 (10.6~24.7); 8.2 (-0.2~19.8); 26.5 (18.6~30.1); 17.9 (10.6~24.7); 155.6 (79.4~236.5); 0.3 (0.0~12.1); 942 (643~1376) ❙ 1435 (797~2680); 592 (299~1428); 133 (23~259) ❙ 53.4 (-34.3~246.5); 930 (643~1360); 230.8 (95.3~620.0)

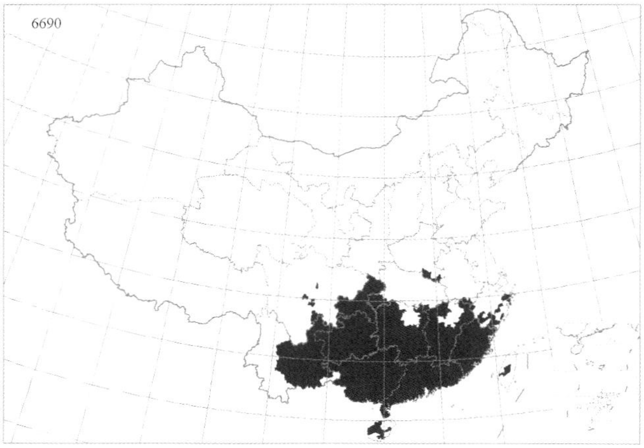

Theaceae 山茶科 *Eurya* 柃属

Eurya loquaiana var. aureopunctata H. T. Chang 金叶细枝柃

Shrubs or small trees, evergreen

18.5 (14.2~22.4); 9.2 (3.6~14.3); 26.3 (20.1~29.8); 18.5 (14.2~22.5); 162.4 (110.5~209.7); 0.1 (0.0~1.4); 959 (729~1205) ❙ 1524 (976~2024); 650 (445~954); 137 (42~259) ❙ 59.8 (12.2~137.5); 955 (729~1205); 255.9 (95.3~551.6)

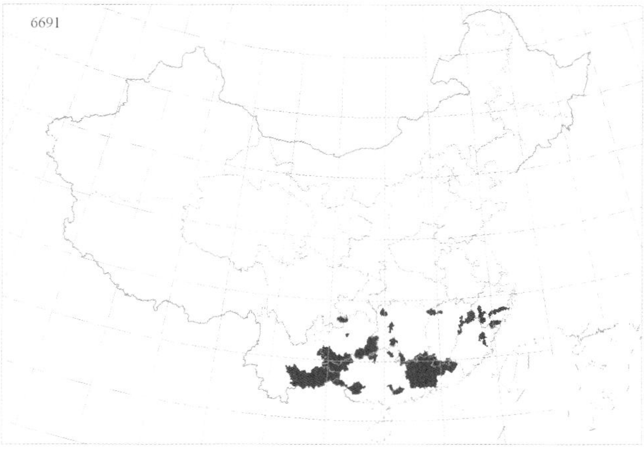

Theaceae 山茶科 *Eurya* 柃属

Eurya lunglingensis Hu et L. K. Ling 隆林耳叶柃

Shrubs, evergreen

18.1; 9.5; 24.6; 18.2; 158.1; 0.0; 901 ❙ 1293; 723; 58 ❙ 43.5; 901; 316.1

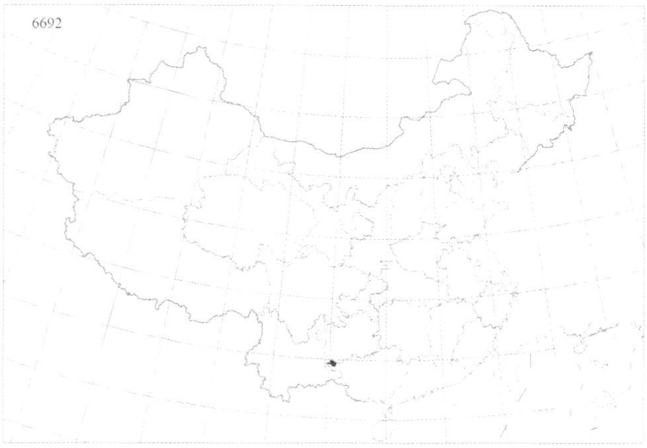

Theaceae 山茶科 *Eurya* 柃属

Eurya macartneyi Champ. 黑柃

Shrubs or small trees, evergreen

20.1 (14.3~24.3); 11.0 (3.3~18.7); 27.8 (24.3~29.8); 20.1 (14.3~24.3); 181.6 (114.1~231.9); 0.1 (0.0~2.3); 1061 (765~1338) ❙ 1651 (868~2435); 671 (312~1428); 158 (42~259) ❙ 58.0 (-34.3~130.7); 1051 (765~1273); 237.2 (95.3~538.2)

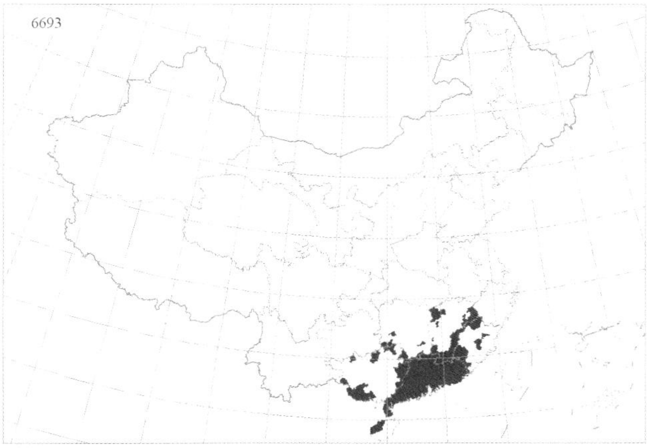

Theaceae 山茶科 *Eurya* 柃属

Eurya magniflora P. I Mao et P. X. He 大花柃

Small trees, evergreen

17.3; 9.9; 22.4; 17.3; 147.1; 0.0; 833 ❙ 1504; 860; 55 ❙ 80.6; 833; 413.5

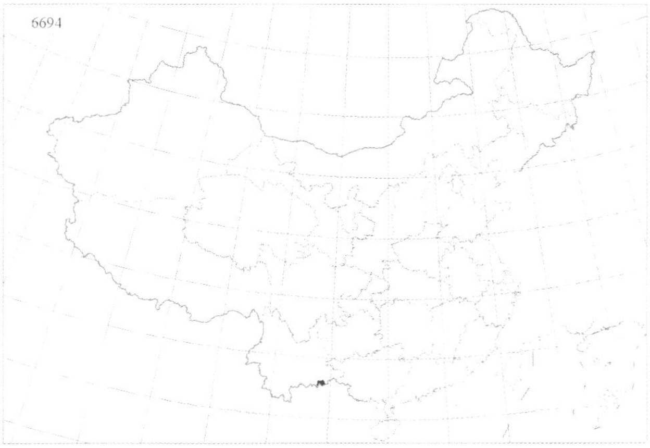

Theaceae 山茶科 *Eurya* 柃属

Eurya marlipoensis Hu 麻栗坡柃

Shrubs, evergreen

17.6; 9.9; 23.2; 17.7; 152.4; 0.0; 859 ❙ 1422; 813; 52 ❙ 65.6; 859; 359.7

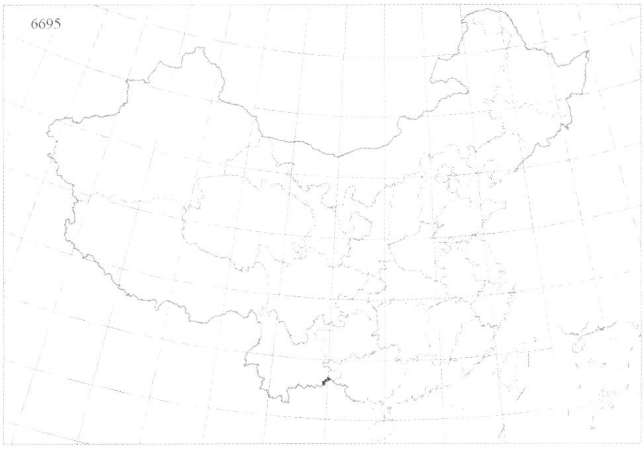

Theaceae 山茶科 *Eurya* 柃属

Eurya megatrichocarpa H. T. Chang 大果毛柃

Shrubs or small trees, evergreen

21.4 (16.3~22.7); 13.0 (4.8~15.0); 27.9 (27.0~28.6); 21.5 (16.3~22.8); 197.8 (136.2~213.1); 0.0 (0.0~0.2); 1141 (871~1221) ❙ 1703 (1273~2435); 909 (452~1428); 115 (86~130) ❙ 49.2 (12.9~99.5); 1141 (871~1221); 278.1 (186.2~327.7)

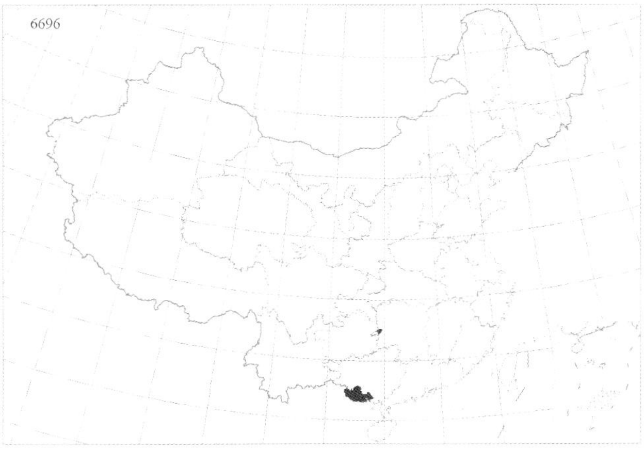

Theaceae 山茶科 *Eurya* 柃属
Eurya metcalfiana Kobuski 从化柃
Shrubs, evergreen

17.8 (14.3~22.5); 7.7 (3.3~15.0); 27.1 (24.3~29.0); 17.9 (14.3~22.6); 154.7 (114.1~210.9); 0.1 (0.0~2.3); 934 (765~1206) ❙ 1678 (1317~2005); 593 (403~862); 202 (97~258) ❙ 80.9 (30.7~130.7); 930 (765~1206); 241.9 (139.6~305.7)

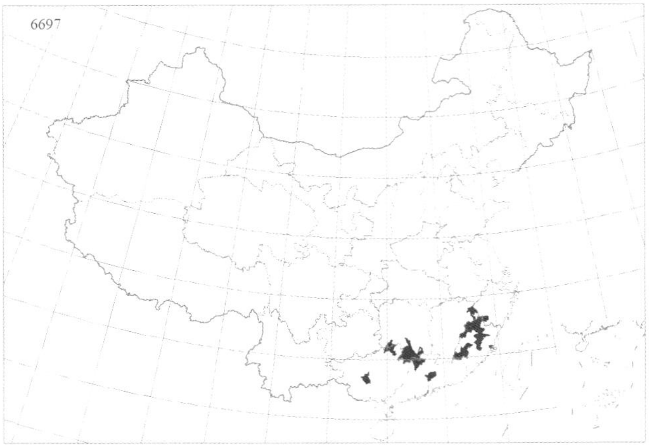

Theaceae 山茶科 *Eurya* 柃属
Eurya muricata Dunn 格药柃
Shrubs or small trees, evergreen

17.5 (13.4~23.2); 6.9 (1.5~15.8); 27.4 (20.3~30.1); 17.5 (13.4~23.2); 151.0 (105.4~218.9); 0.4 (0.0~7.3); 929 (745~1257) ❙ 1532 (871~2030); 561 (299~1033); 174 (33~259) ❙ 65.9 (-6.7~147.0); 916 (745~1257); 227.4 (95.3~530.1)

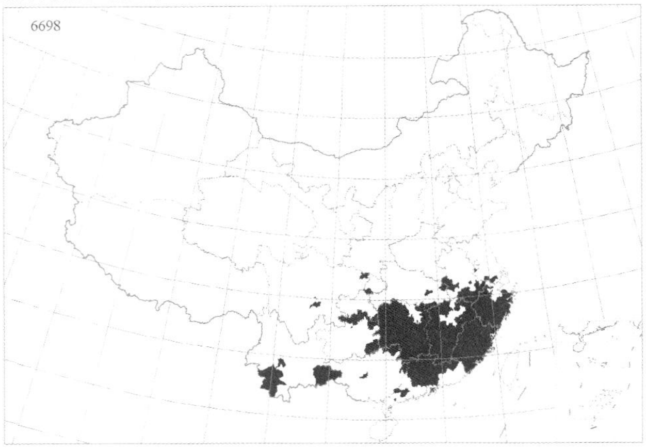

Theaceae 山茶科 *Eurya* 柃属
Eurya muricata var. huana (Kobuski) L. K. Ling 毛枝格药柃
Shrubs or small trees, evergreen

16.7 (12.1~19.2); 5.2 (1.1~13.8); 27.3 (20.8~29.9); 16.7 (12.2~19.4); 141.0 (95.0~172.6); 0.5 (0.0~8.5); 893 (696~992) ❙ 1424 (1018~1846); 498 (381~771); 164 (44~256) ❙ 59.9 (7.5~128.9); 875 (696~977); 204.9 (134.6~513.9)

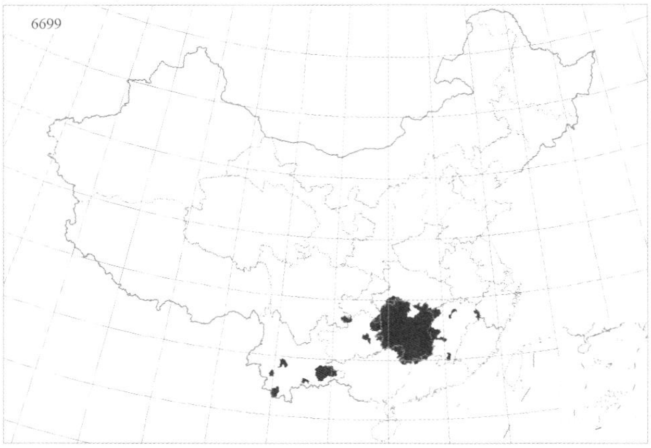

Theaceae 山茶科 *Eurya* 柃属
Eurya nitida Korth. 细齿叶柃
Shrubs or small trees, evergreen

17.9 (9.3~24.8); 8.1 (-1.7~20.1); 26.7 (18.8~30.1); 17.9 (9.5~24.8); 155.8 (69.8~237.6); 0.6 (0.0~18.0); 947 (603~1385) ❙ 1460 (798~2992); 591 (312~1847); 135 (21~259) ❙ 55.7 (-34.3~166.5); 937 (603~1326); 227.4 (95.3~620.0)

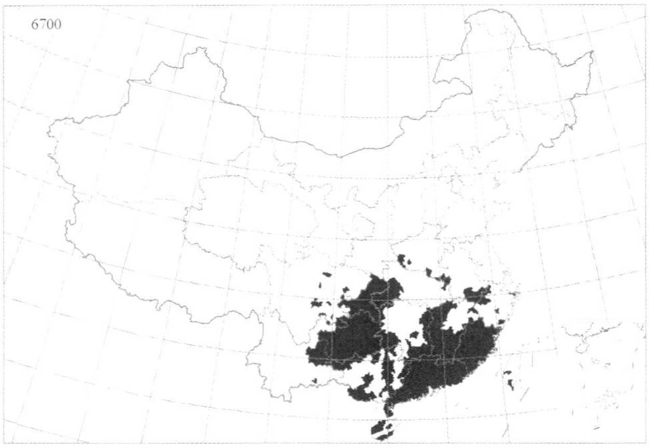

Theaceae 山茶科 *Eurya* 柃属
Eurya obliquifolia Hemsl. 斜基叶柃
Shrubs or small trees, evergreen

17.4 (8.6~21.0); 10.9 (1.3~14.2); 21.7 (15.0~25.6); 17.4 (8.7~21.0); 149.4 (52.6~191.8); 0.2 (0.0~8.8); 844 (574~1053) ❙ 1279 (939~1751); 723 (524~996); 52 (37~77) ❙ 51.9 (6.5~112.2); 829 (574~1053); 411.8 (277.0~627.0)

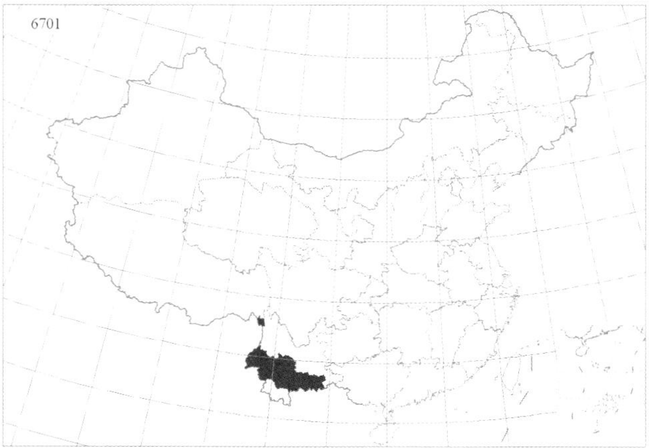

Theaceae 山茶科 *Eurya* 柃属
Eurya oblonga Yang 矩圆叶柃
Shrubs or small trees, evergreen

15.6 (6.0~21.0); 6.4 (-3.2~14.2); 23.5 (14.4~26.9); 15.7 (6.6~21.0); 129.5 (39.7~191.8); 1.3 (0.0~26.7); 813 (508~1053) ❙ 1214 (885~1590); 631 (476~946); 59 (28~129) ❙ 50.5 (10.0~101.2); 813 (508~1053); 273.4 (133.5~551.6)

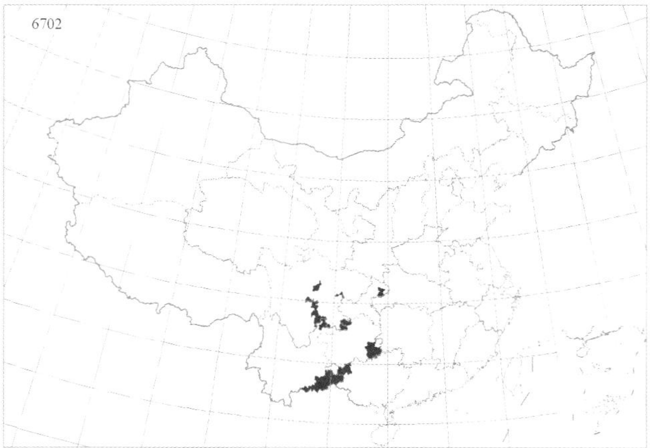

Theaceae 山茶科 *Eurya* 柃属

Eurya obtusifolia H. T. Chang 钝叶柃

Shrubs or small trees, evergreen

15.7 (10.2~21.6); 4.5 (-0.7~12.6); 26.2 (19.4~28.9); 15.7 (10.4~21.7); 130.3 (77.2~199.8); 1.7 (0.0~13.6); 846 (634~1153) ❙ 1337 (1006~1852); 532 (429~816); 115 (30~242) ❙ 59.4 (8.0~112.2); 845 (634~1153); 210.9 (108.2~438.9)

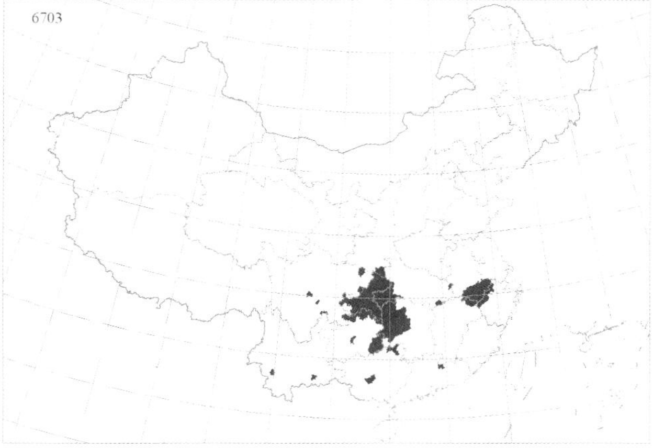

Theaceae 山茶科 *Eurya* 柃属

Eurya obtusifolia var. aurea (H. Lév.) Ming 金叶柃

Shrubs, evergreen

16.2 (2.2~21.0); 7.3 (-6.6~14.2); 23.8 (9.9~28.9); 16.2 (3.8~21.0); 135.2 (15.5~191.8); 0.6 (0.0~49.1); 830 (413~1053) ❙ 1194 (797~1776); 573 (411~996); 71 (19~196) ❙ 44.5 (-14.2~125.9); 821 (413~1053); 260.9 (108.2~551.6)

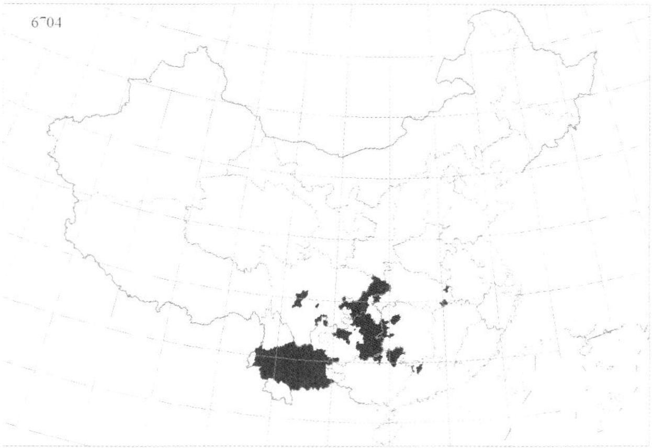

Theaceae 山茶科 *Eurya* 柃属

Eurya ovatifolia H. T. Chang 卵叶柃

Shrubs or small trees, evergreen

23.7; 18.3; 27.5; 23.8; 225.5; 0.0; 1288 ❙ 1631; 577; 72 ❙ 27.0; 1211; 489.7

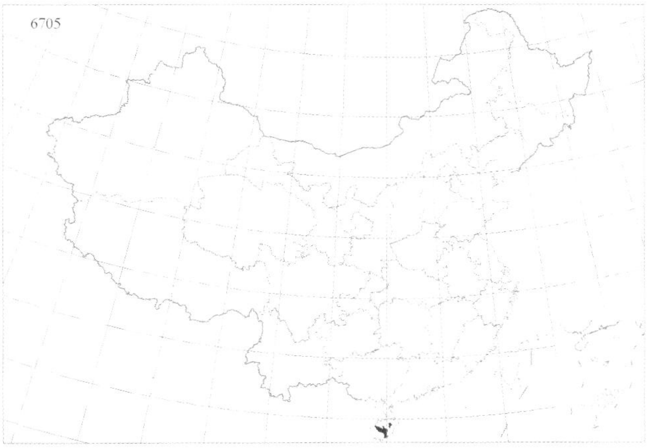

Theaceae 山茶科 *Eurya* 柃属

Eurya paratetragonoclada Hu 滇四角柃

Shrubs or small trees, evergreen

15.6 (5.4~21.0); 9.0 (-2.1~14.2); 20.5 (12.3~25.6); 15.7 (5.8~21.0); 128.9 (31.5~191.8); 0.7 (0.0~25.6); 783 (481~1053) ❙ 1093 (741~1547); 620 (411~911); 46 (23~70) ❙ 40.9 (-14.2~99.9); 758 (481~1053); 350.4 (218.0~551.6)

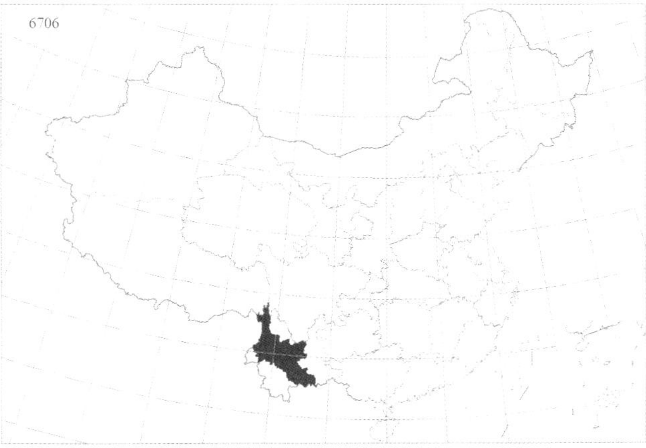

Theaceae 山茶科 *Eurya* 柃属

Eurya patentipila Chun 长毛柃

Shrubs, evergreen

19.2 (14.6~22.3); 9.6 (3.8~14.2); 27.4 (24.6~29.0); 19.3 (14.7~22.3); 171.2 (117.7~208.1); 0.0 (0.0~1.5); 1006 (779~1202) ❙ 1578 (1250~1991); 633 (403~902); 165 (60~230) ❙ 58.0 (20.3~104.6); 1003 (779~1202); 237.2 (151.0~335.1)

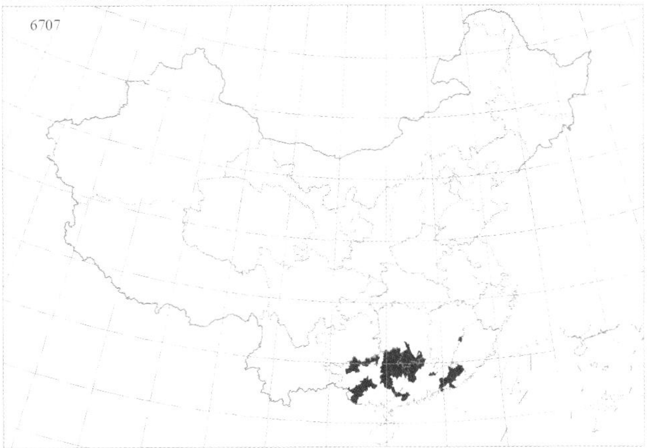

Theaceae 山茶科 *Eurya* 柃属

Eurya pentagyna H. T. Chang 五柱柃

Shrubs or small trees, evergreen

17.8; 8.5; 25.9; 17.8; 154.3; 0.4; 947 ❙ 1580; 567; 152 ❙ 74.1; 908; 343.9

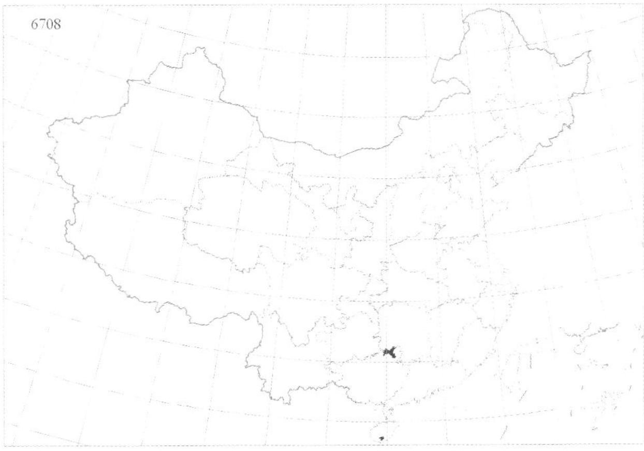

Theaceae 山茶科 *Eurya* 柃属
Eurya perserrata Kobuski 尖齿叶柃
Shrubs, evergreen

8.6; 1.3; 15.0; 8.7; 52.6; 8.8; 574 ❙ 1041; 585; 45 ❙ 81.5; 574; 352.3

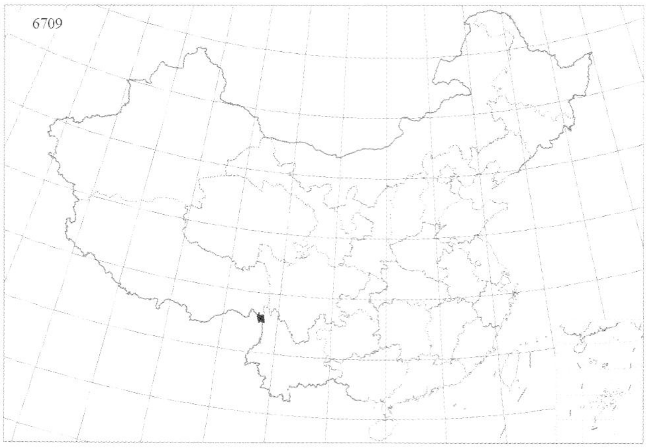

Theaceae 山茶科 *Eurya* 柃属
Eurya persicifolia Gagnep. 坚桃叶柃
Shrubs or small trees, evergreen

18.3 (17.2~21.0); 11.2 (9.5~14.2); 23.3 (22.4~25.6); 18.4 (17.3~21.0); 160.2 (147.0~191.8); 0.0 (0.0~0.0); 892 (833~1053) ❙ 1416 (1254~1522); 797 (699~860); 57 (48~68) ❙ 59.4 (44.5~80.6); 892 (833~1053); 443.1 (336.3~551.6)

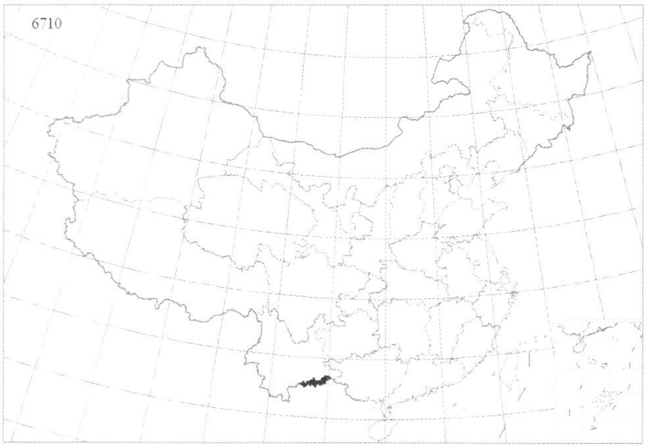

Theaceae 山茶科 *Eurya* 柃属
Eurya pittosporifolia Hu 海桐叶柃
Small trees, evergreen

20.5; 14.9; 23.7; 20.5; 185.9; 0.0; 992 ❙ 1464; 655; 66 ❙ 47.6; 959; 570.1

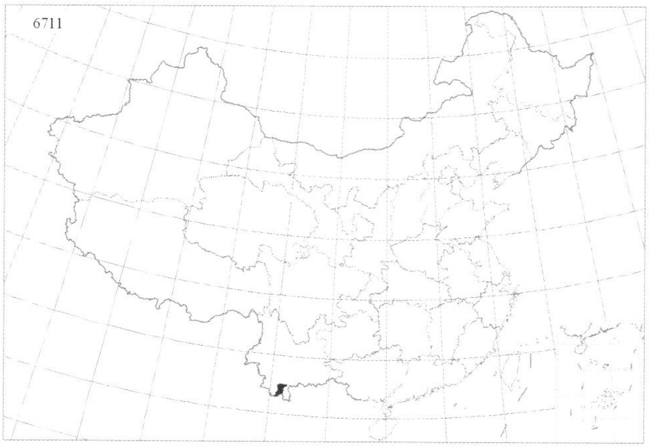

Theaceae 山茶科 *Eurya* 柃属
Eurya polyneura Chun 多脉柃
Shrubs, evergreen

20.1 (14.6~22.2); 11.4 (3.8~14.7); 27.2 (24.6~28.3); 20.2 (14.7~22.2); 182.2 (117.7~206.9); 0.2 (0.0~1.5); 1054 (779~1177) ❙ 1608 (1284~1876); 740 (544~891); 121 (73~191) ❙ 54.0 (30.9~101.8); 1054 (779~1177); 252.0 (209.2~322.1)

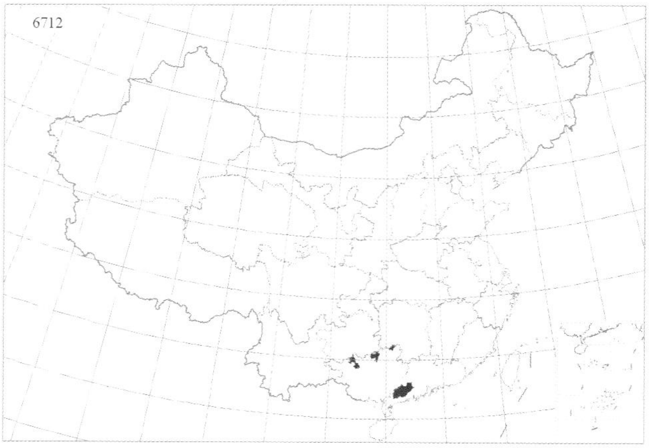

Theaceae 山茶科 *Eurya* 柃属
Eurya prunifolia P. S. Hsu 桃叶柃
Small trees, evergreen

19.3; 12.7; 23.8; 19.3; 171.6; 0.0; 942 ❙ 1439; 803; 61 ❙ 53.4; 942; 516.4

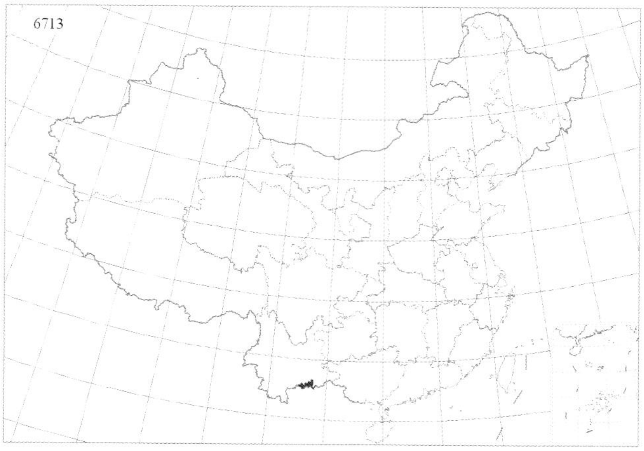

Theaceae 山茶科 *Eurya* 柃属
Eurya pseudocerasifera Kobuski 肖樱叶柃
Shrubs or trees, evergreen

16.6 (5.3~21.0); 10.1 (-2.4~15.1); 21.1 (12.2~25.6); 16.6 (5.7~21.0); 140.1 (30.5~191.8); 0.4 (0.0~26.2); 816 (479~1053) ❙ 1187 (797~1751); 666 (411~1007); 48 (23~77) ❙ 46.0 (-14.2~145.8); 794 (479~1053); 374.2 (218.0~659.0)

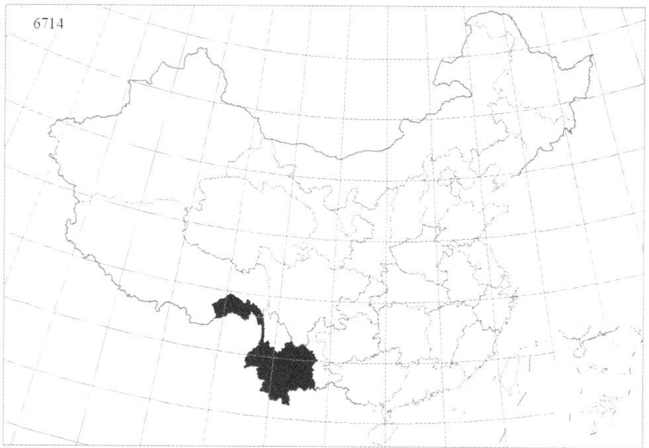

Theaceae 山茶科 *Eurya* 柃属
Eurya pyracanthifolia P. S. Hsu 火棘叶柃
Shrubs, evergreen

16.6 (13.2~19.6); 9.8 (6.4~12.8); 21.0 (18.4~23.5); 16.7 (13.3~19.6); 139.9 (99.4~175.7); 0.0 (0.0~0.0); 816 (695~953) ❙ 1459 (965~1688); 837 (555~959); 55 (40~70) ❙ 79.4 (29.3~112.2); 812 (695~932); 470.5 (242.1~627.0)

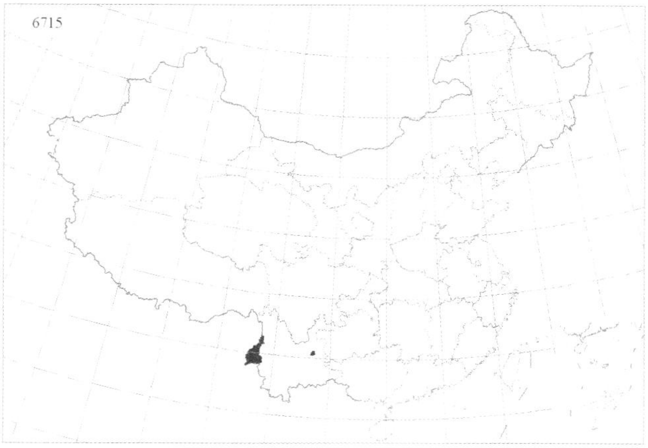

Theaceae 山茶科 *Eurya* 柃属
Eurya quinquelocularis Kobuski 大叶五室柃
Shrubs or small trees, evergreen

19.4 (13.3~22.7); 10.6 (4.5~15.0); 26.2 (19.6~28.9); 19.4 (13.4~22.8); 173.0 (100.4~213.1); 0.0 (0.0~0.5); 1003 (706~1221) ❙ 1509 (1074~2435); 753 (499~1428); 112 (47~194) ❙ 51.6 (12.9~145.8); 1003 (706~1221); 287.9 (181.2~551.6)

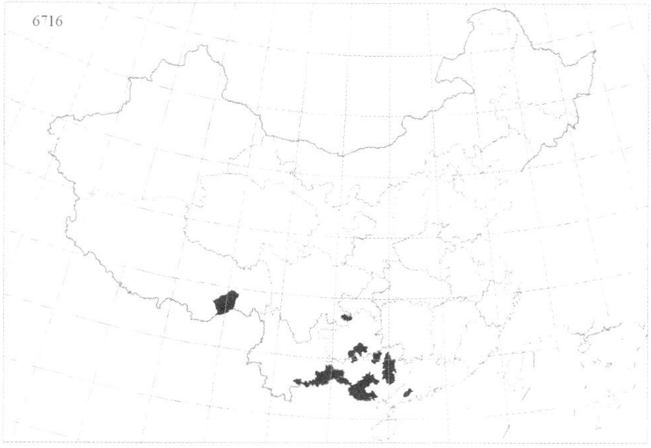

Theaceae 山茶科 *Eurya* 柃属
Eurya rengechiensis Yamam. 莲华柃
Shrubs or small trees, evergreen

19.4; 13.3; 24.4; 19.4; 172.6; 0.0; 967 ❙ 2109; 1039; 182 ❙ 128.0; 946; 242.1

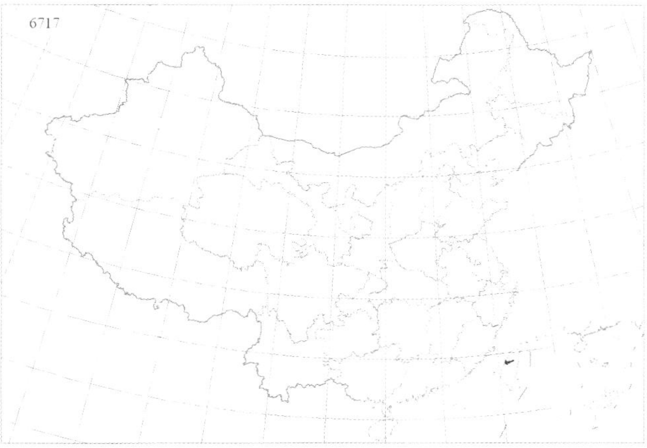

Theaceae 山茶科 *Eurya* 柃属
Eurya rubiginosa H. T. Chang 红褐柃
Shrubs, evergreen

17.4 (14.4~21.8); 6.7 (2.1~13.4); 27.3 (24.6~29.3); 17.5 (14.5~21.8); 150.4 (117.7~201.9); 0.7 (0.0~4.9); 926 (779~1158) ❙ 1650 (1117~2005); 611 (473~848); 196 (72~254) ❙ 79.3 (28.5~130.7); 925 (779~1158); 246.4 (179.2~313.7)

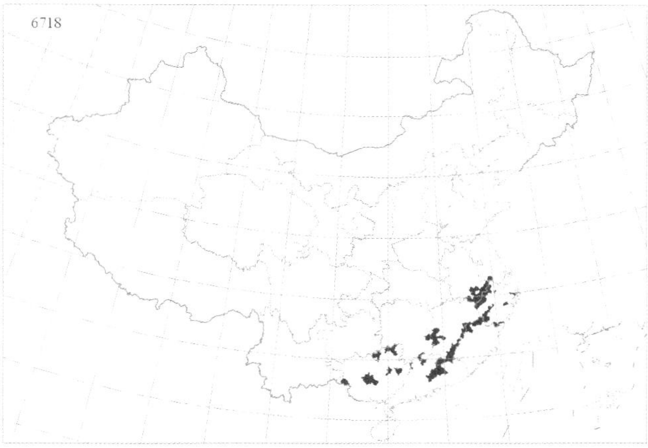

Theaceae 山茶科 *Eurya* 柃属
Eurya rubiginosa var. attenuata H. T. Chang 窄基红褐柃
Shrubs, evergreen

18.3 (14.2~22.4); 8.5 (2.1~14.3); 27.2 (20.1~29.8); 18.4 (14.2~22.5); 160.8 (110.5~209.7); 0.2 (0.0~4.9); 963 (729~1205) ❙ 1603 (1070~2024); 610 (356~954); 177 (48~259) ❙ 68.4 (-5.0~147.0); 958 (729~1205); 247.1 (95.3~551.6)

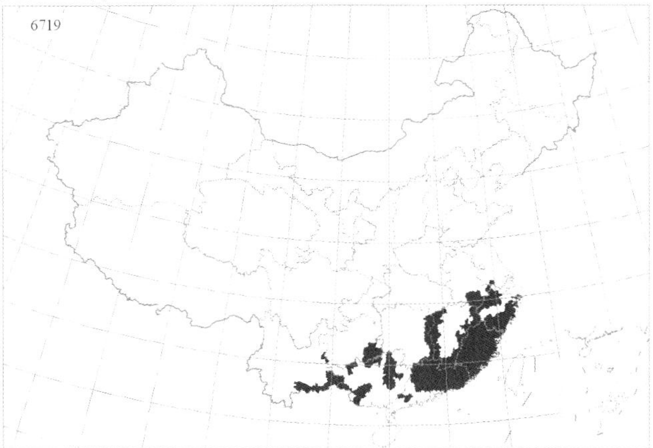

Theaceae 山茶科 *Eurya* 柃属
Eurya rugosa Hu 皱叶柃
Shrubs, evergreen

15.1; 5.3; 24.0; 15.1; 121.7; 0.0; 789 ❙ 1076; 586; 48 ❙ 36.4; 789; 212.0

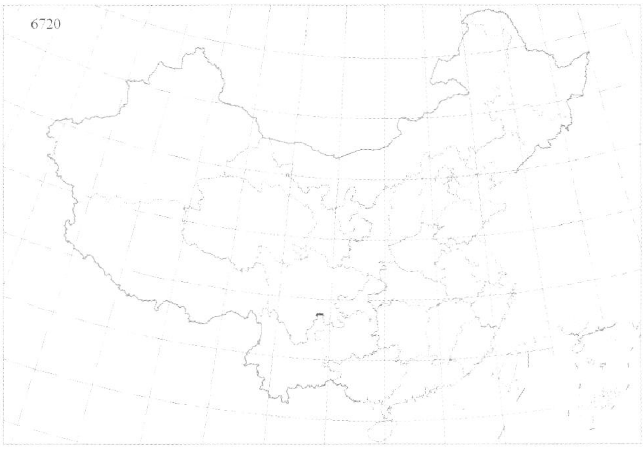

Theaceae 山茶科 *Eurya* 柃属
Eurya saxicola H. T. Chang 岩柃
Shrubs, evergreen

18.6 (14.3~24.8); 8.8 (2.1~20.1); 27.3 (24.3~29.8); 18.7 (14.3~24.8); 164.5 (114.1~237.6); 0.3 (0.0~4.9); 985 (765~1385) ❙ 1653 (1179~1997); 599 (392~809); 184 (41~259) ❙ 71.3 (−11.9~147.0); 968 (765~1212); 270.7 (174.9~620.0)

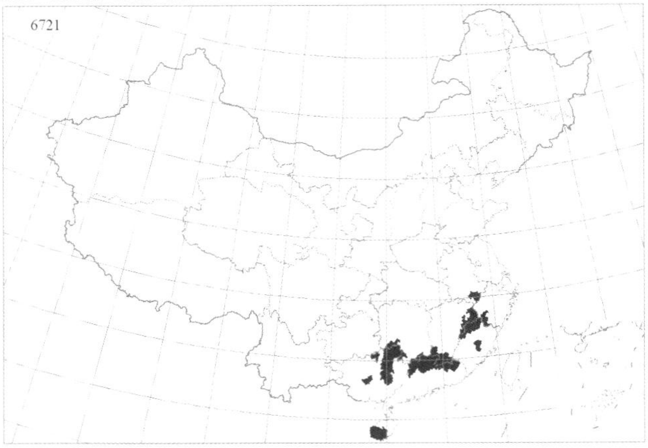

Theaceae 山茶科 *Eurya* 柃属
Eurya semiserrulata H. T. Chang 半持柃
Shrubs or small trees, evergreen

14.8 (2.2~19.1); 4.8 (−6.6~10.5); 23.8 (9.9~29.0); 14.9 (3.8~19.2); 120.0 (15.5~170.1); 1.8 (0.0~49.1); 793 (413~981) ❙ 1218 (863~1630); 554 (450~817); 88 (19~228) ❙ 54.0 (5.5~125.9); 789 (413~981); 228.9 (147.3~347.3)

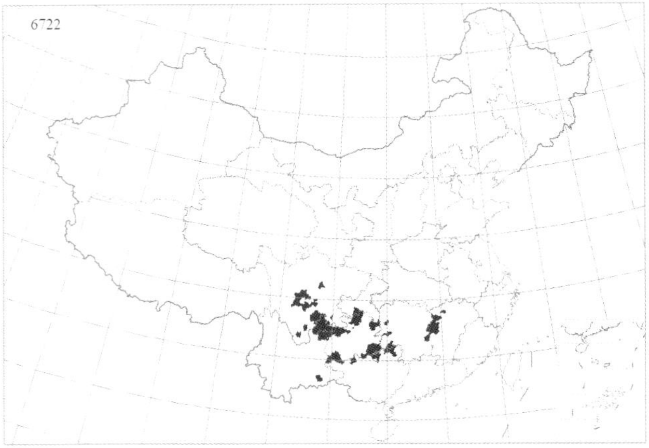

Theaceae 山茶科 *Eurya* 柃属
Eurya stenophylla Merr. 窄叶柃
Shrubs, evergreen

17.5 (12.9~22.7); 7.5 (1.9~15.0); 26.4 (22.2~28.9); 17.6 (12.9~22.8); 151.1 (101.2~213.1); 0.3 (0.0~5.9); 925 (718~1221) ❙ 1393 (999~2435); 604 (444~1428); 118 (48~202) ❙ 51.1 (12.9~101.8); 925 (718~1221); 219.1 (140.5~335.1)

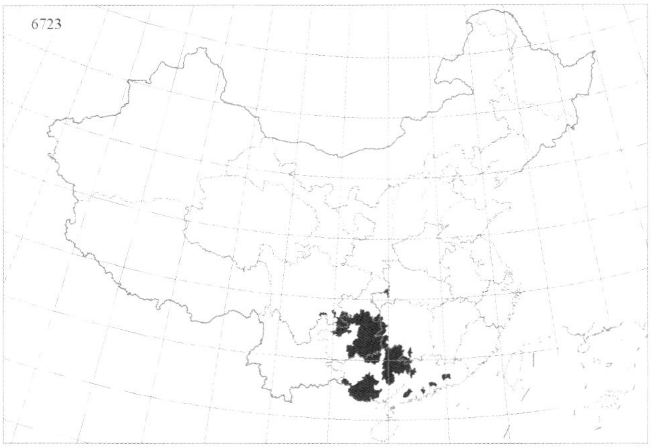

Theaceae 山茶科 *Eurya* 柃属
Eurya stenophylla var. caudata H. T. Chang 长尾窄叶柃
Shrubs, evergreen

21.4 (15.5~23.4); 12.9 (4.8~16.1); 28.2 (25.1~29.2); 21.4 (15.6~23.5); 196.8 (127.2~221.7); 0.0 (0.0~0.2); 1133 (815~1273) ❙ 1699 (1302~2435); 768 (547~1428); 138 (82~195) ❙ 51.2 (9.0~99.5); 1132 (815~1273); 225.6 (95.3~327.7)

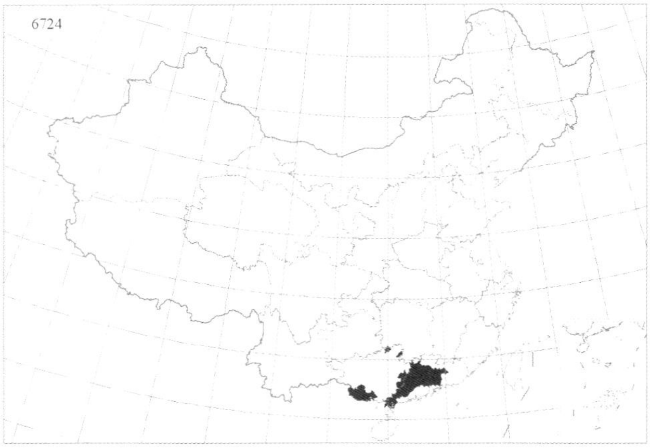

Theaceae 山茶科 *Eurya* 柃属
Eurya stenophylla var. pubescens T. L. Ming 毛窄叶柃
Shrubs, evergreen

21.0 (15.5~22.7); 12.8 (4.8~15.0); 27.2 (23.3~28.6); 21.0 (15.6~22.8); 192.4 (127.2~213.1); 0.0 (0.0~0.2); 1099 (815~1221) ❙ 1696 (1302~2435); 892 (591~1428); 114 (56~177) ❙ 55.1 (12.9~99.5); 1099 (815~1221); 296.5 (211.1~506.2)

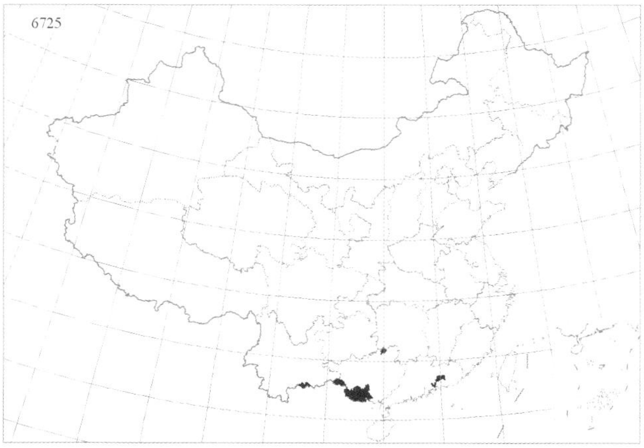

Theaceae 山茶科 *Eurya* 柃属
Eurya strigillosa Hayata 台湾毛柃
Small trees, evergreen

19.6 (16.2~22.2); 14.1 (11.0~17.7); 23.9 (20.1~27.0); 19.6 (16.2~22.2); 175.1 (134.2~206.7); 0.0 (0.0~0.0); 970 (773~1123) ❙ 2467 (1836~2992); 1305 (1033~1847); 165 (107~245) ❙ 161.8 (64.2~246.5); 961 (773~1123); 255.6 (202.8~296.9)

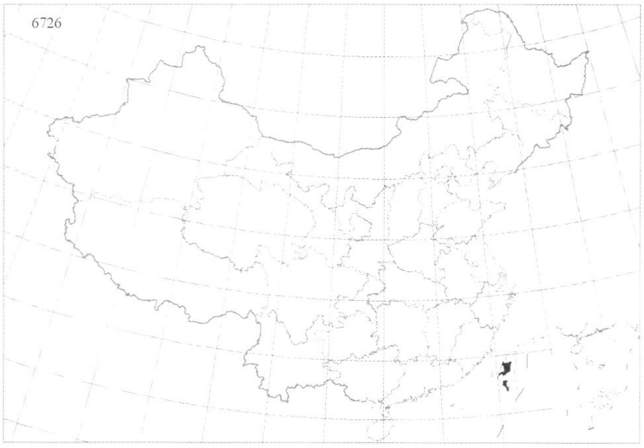

Theaceae 山茶科 *Eurya* 柃属

Eurya tetragonoclada Merr. et Chun 四角柃

Shrubs or trees, evergreen

18.0 (10.6~22.4); 8.9 (-0.2~14.3); 25.6 (19.3~29.9); 18.0 (10.6~22.5); 156.8 (79.4~209.7); 0.5 (0.0~12.1); 931 (643~1205) ❙ 1425 (797~2024); 627 (399~1056); 121 (23~221) ❙ 52.9 (-14.2~128.9); 921 (643~1205); 256.8 (95.3~551.6)

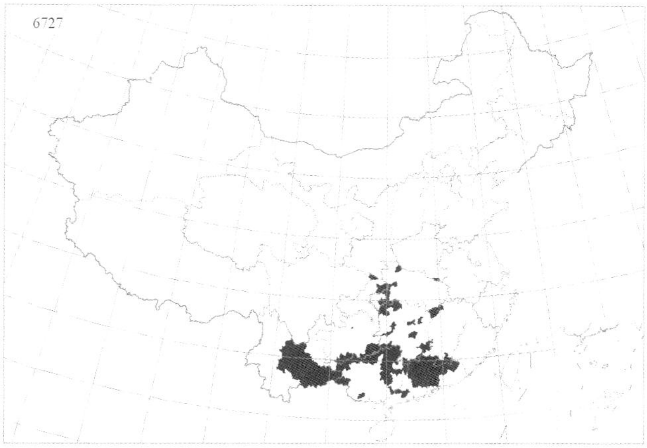

Theaceae 山茶科 *Eurya* 柃属

Eurya trichocarpa Korthals 毛果柃

Shrubs or small trees, evergreen

20.0 (8.6~24.8); 11.8 (1.3~20.1); 26.5 (15.0~29.8); 20.1 (8.7~24.8); 180.7 (52.6~237.6); 0.1 (0.0~8.8); 1047 (574~1385) ❙ 1566 (868~2435); 690 (312~1428); 120 (41~197) ❙ 52.6 (-34.3~145.8); 1026 (574~1360); 315.9 (163.7~627.0)

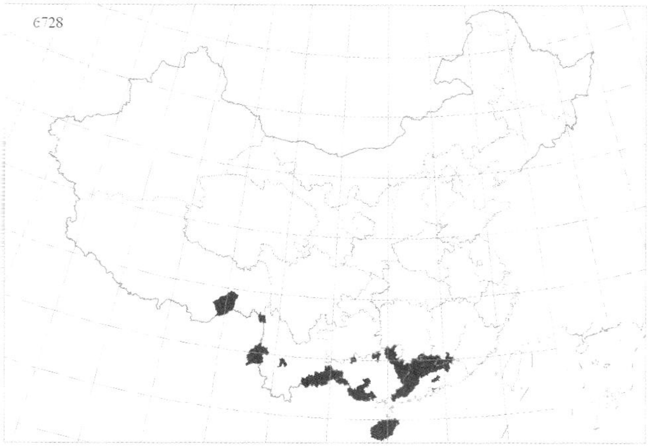

Theaceae 山茶科 *Eurya* 柃属

Eurya tsaii H. T. Chang 怒江柃

Shrubs or small trees, evergreen

14.6 (1.8~19.6); 7.5 (-7.1~13.3); 19.8 (9.9~24.3); 14.7 (3.8~19.6); 119.6 (15.7~175.7); 4.5 (0.0~53.3); 759 (404~953) ❙ 1258 (583~1736); 726 (348~1007); 45 (13~70) ❙ 68.1 (4.4~145.8); 738 (404~932); 399.5 (169.5~627.0)

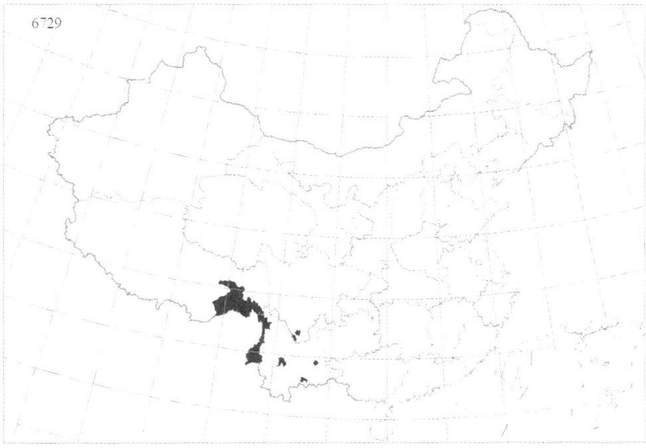

Theaceae 山茶科 *Eurya* 柃属

Eurya velutina Chun 信宜毛柃

Trees, evergreen

21.0; 13.2; 26.9; 21.0; 192.3; 0.0; 1081 ❙ 1776; 828; 125 ❙ 64.3; 1081; 254.1

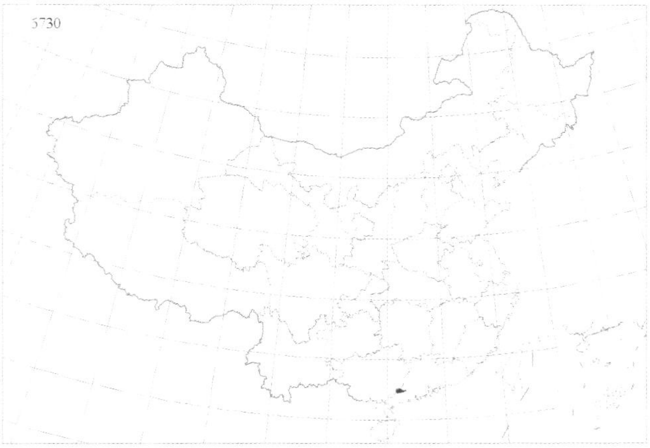

Theaceae 山茶科 *Eurya* 柃属

Eurya weissiae Chun 单耳柃

Shrubs, evergreen

18.7 (14.3~22.4); 8.9 (3.3~14.3); 27.4 (23.5~29.8); 18.8 (14.3~22.5); 165.2 (112.6~209.7); 0.1 (0.0~2.3); 983 (756~1205) ❙ 1664 (1218~2024); 623 (445~954); 180 (73~259) ❙ 70.6 (30.2~129.9); 981 (756~1205); 234.3 (95.3~322.1)

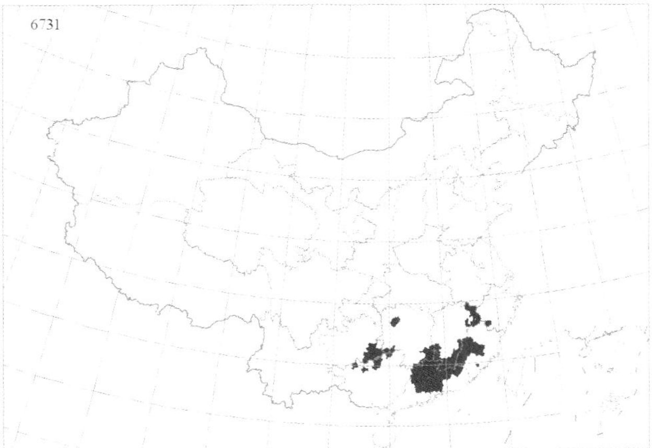

Theaceae 山茶科 *Euryodendron* 猪血木属

Euryodendron excelsum H. T. Chang 猪血木

Trees, evergreen

20.4; 11.6; 27.4; 20.4; 185.1; 0.0; 1064 ❙ 1585; 729; 136 ❙ 49.6; 1064; 247.5

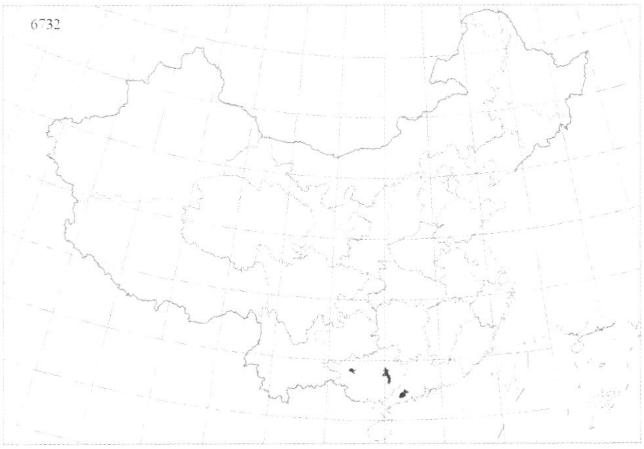

Theaceae 山茶科 *Polyspora* 大头茶属
Polyspora axillaris (Roxb. ex Ker Gawl.) Sweet 大头茶
Trees, evergreen

21.1 (16.2~24.3); 13.3 (8.5~19.1); 27.3 (20.1~29.1); 21.1 (16.2~24.3); 193.5 (134.2~231.9); 0.0 (0.0~0.0); 1102 (773~1338) ❙ 1809 (1179~3495); 817 (392~1847); 155 (42~897) ❙ 67.9 (-11.9~246.5); 1090 (773~1326); 249.7 (95.3~620.0)

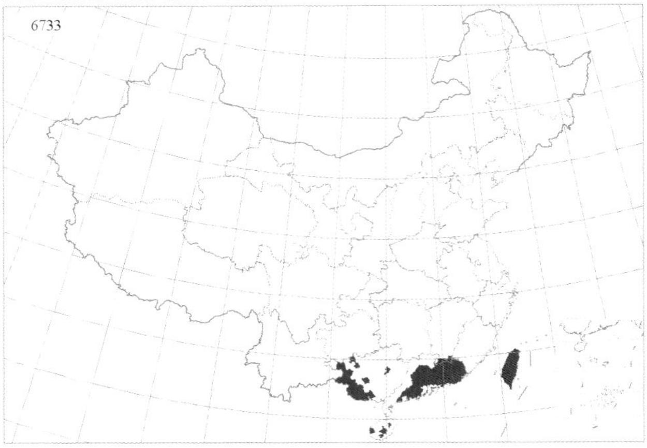

Theaceae 山茶科 *Polyspora* 大头茶属
Polyspora chrysandra (Cowan) Hu ex Barthol. et Ming 黄药大头茶
Small trees, evergreen

15.6 (5.4~21.0); 8.1 (-2.1~15.1); 21.3 (12.3~26.6); 15.6 (5.8~21.0); 128.2 (31.5~191.8); 0.7 (0.0~25.6); 787 (481~1053) ❙ 1131 (741~1751); 624 (411~996); 47 (15~77) ❙ 43.9 (-14.2~112.2); 773 (481~1053); 318.4 (142.9~659.0)

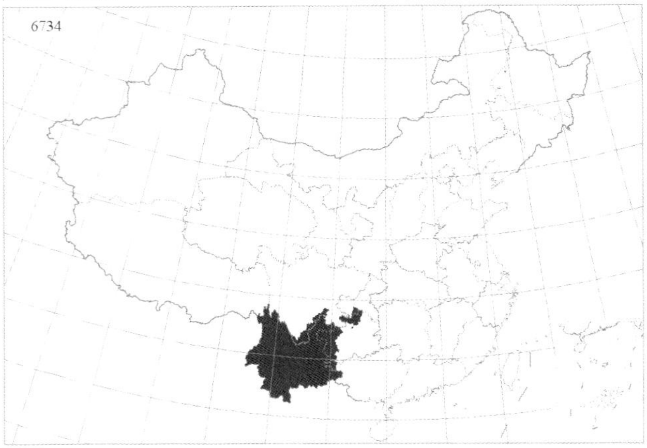

Theaceae 山茶科 *Polyspora* 大头茶属
Polyspora hainanensis (H. T. Chang) C. X. Ye ex Barthol. et Ming 海南大头茶
Trees, evergreen

24.2 (23.6~24.8); 18.9 (17.8~20.1); 28.0 (27.2~28.7); 24.3 (23.6~24.8); 231.2 (223.3~237.6); 0.0 (0.0~0.0); 1333 (1271~1385) ❙ 1351 (868~1767); 462 (312~567); 55 (41~88) ❙ 1.2 (-34.3~28.8); 1118 (868~1299); 471.2 (288.3~620.0)

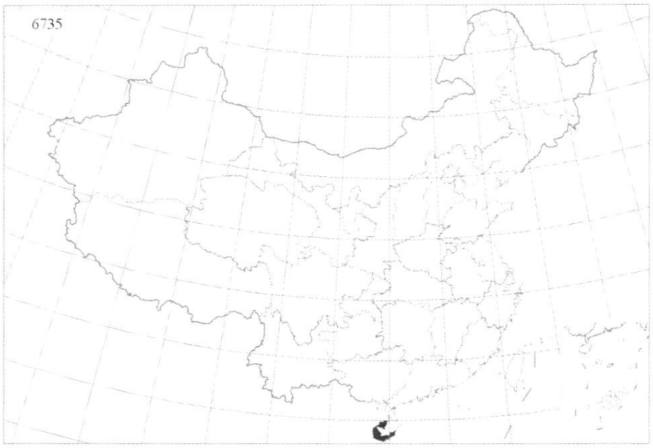

Theaceae 山茶科 *Polyspora* 大头茶属
Polyspora longicarpa (H. T. Chang) , C. X. Ye ex Barthol. et Ming 长果大头茶
Trees, evergreen

17.5 (13.1~21.0); 11.1 (6.4~14.2); 21.7 (18.4~25.6); 17.5 (13.2~21.0); 150.1 (98.1~191.8); 0.0 (0.0~0.0); 846 (692~1053) ❙ 1279 (939~1751); 717 (524~996); 52 (37~77) ❙ 51.3 (6.5~112.2); 828 (692~1053); 417.3 (289.9~627.0)

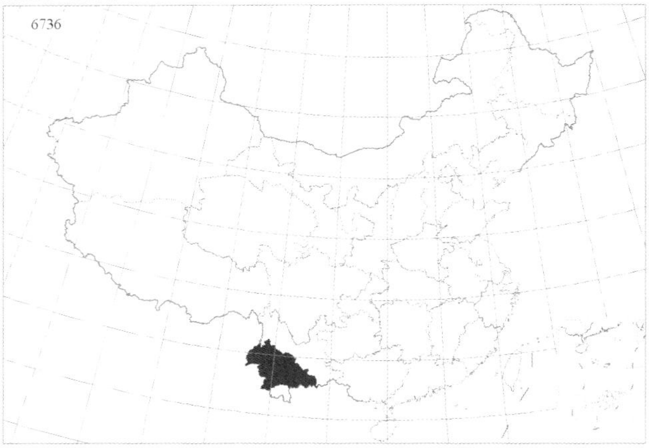

Theaceae 山茶科 *Polyspora* 大头茶属
Polyspora speciosa (Kochs) Barthol. et T. L. Ming 四川大头茶
Trees, evergreen

17.0 (10.6~21.3); 7.1 (-0.2~14.2); 25.9 (19.4~28.9); 17.1 (10.6~21.3); 145.6 (79.4~196.1); 0.5 (0.0~12.1); 892 (643~1130) ❙ 1307 (885~1776); 590 (442~932); 92 (30~197) ❙ 47.1 (8.0~101.8); 892 (643~1130); 222.6 (108.2~551.6)

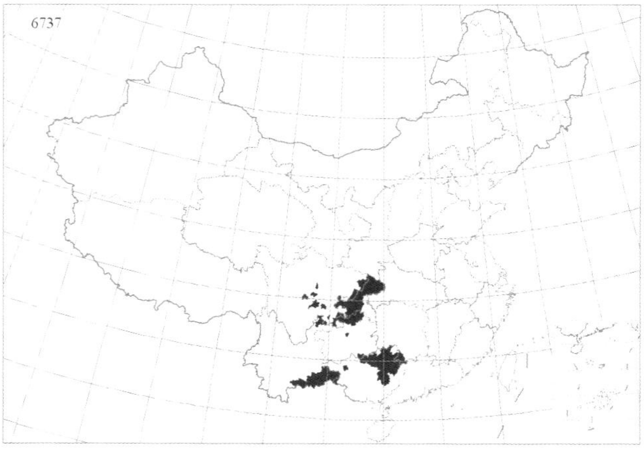

Theaceae 山茶科 *Pyrenaria* 核果茶属
Pyrenaria diospyricarpa Kurz 叶萼核果茶
Trees, evergreen

18.2 (16.3~21.0); 11.6 (9.0~15.1); 22.5 (21.4~25.6); 18.2 (16.4~21.0); 158.4 (136.4~191.8); 0.0 (0.0~0.0); 877 (797~1053) ❙ 1330 (958~1607); 698 (531~860); 52 (41~77) ❙ 51.4 (13.3~80.6); 864 (797~1053); 438.5 (277.0~659.0)

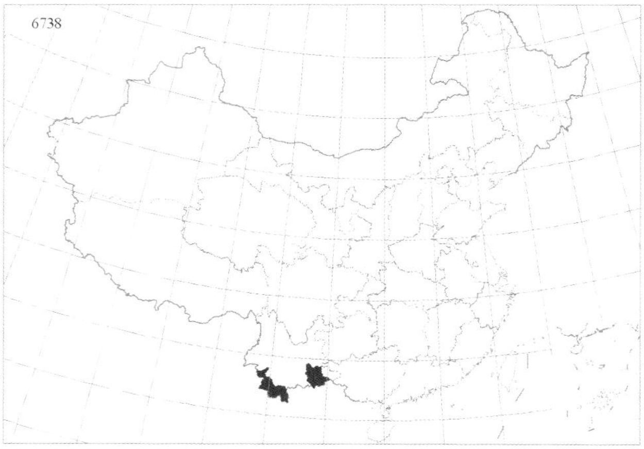

Theaceae 山茶科 *Pyrenaria* 核果茶属

Pyrenaria hirta Keng 粗毛核果茶

Trees, evergreen

18.0 (13.5~22.3); 7.3 (2.6~13.5); 27.6 (21.4~29.9); 18.0 (13.6~22.3); 156.7 (106.7~208.1); 0.3 (0.0~4.1); 956 (744~1202) ❙ 1520 (1190~2024); 566 (392~821); 171 (50~230) ❙ 60.2 (20.3~128.9); 947 (744~1202); 212.2 (134.6~347.3)

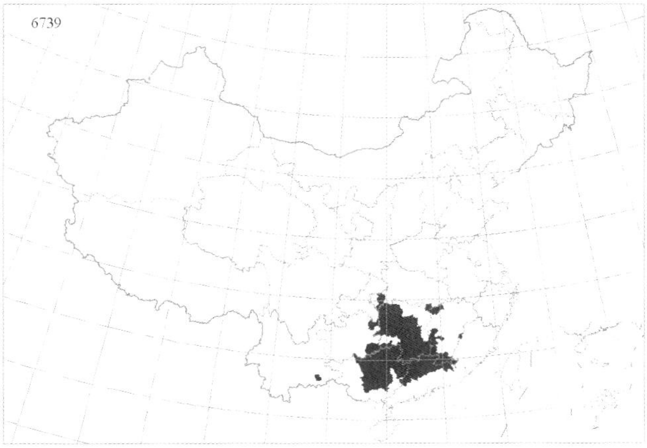

Theaceae 山茶科 *Pyrenaria* 核果茶属

Pyrenaria jonquieriana subsp. multisepala (Merr. et Chun) S. X. Yang 多瓣核果茶

Trees, evergreen

24.1 (23.0~24.8); 18.9 (17.7~20.1); 27.8 (26.7~28.5); 24.1 (23.1~24.8); 229.5 (217.0~237.6); 0.0 (0.0~0.0); 1318 (1217~1385) ❙ 1491 (1179~1854); 509 (392~622); 58 (41~115) ❙ 13.3 (−11.9~41.8); 1160 (1056~1360); 499.2 (428.6~620.0)

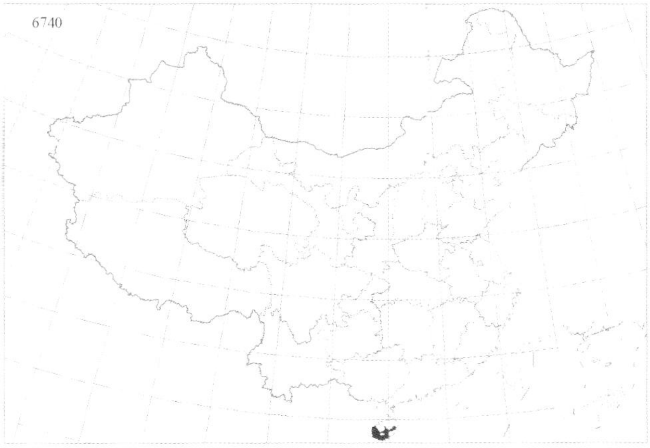

Theaceae 山茶科 *Pyrenaria* 核果茶属

Pyrenaria khasiana R. N. Pao 印藏核果茶

Trees, evergreen

13.3; 5.4; 19.6; 13.4; 100.4; 0.0; 706 ❙ 1736; 1007; 51 ❙ 145.8; 706; 447.4

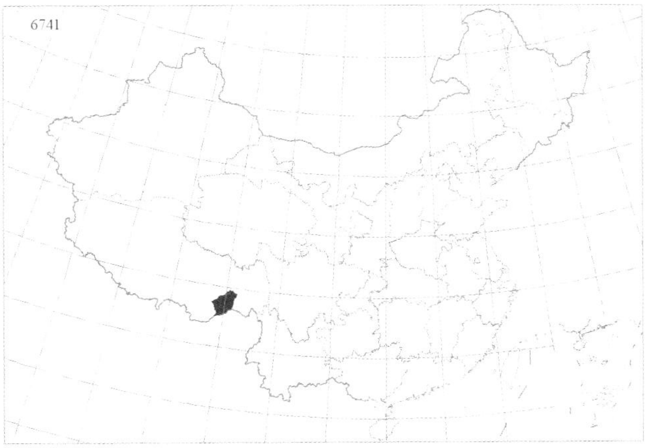

Theaceae 山茶科 *Pyrenaria* 核果茶属

Pyrenaria kwangsiensis H. T. Chang 广西核果茶

Trees, evergreen

19.9; 10.1; 27.9; 19.9; 179.0; 0.0; 1044 ❙ 1473; 587; 165 ❙ 41.6; 1044; 212.7

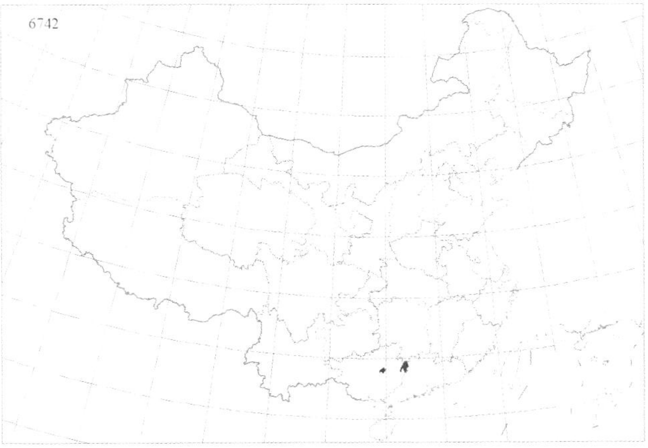

Theaceae 山茶科 *Pyrenaria* 核果茶属

Pyrenaria menglaensis G. D. Tao 勐腊核果茶

Trees, evergreen

20.4; 15.1; 23.6; 20.5; 185.8; 0.0; 989 ❙ 1607; 702; 77 ❙ 62.5; 986; 659.0

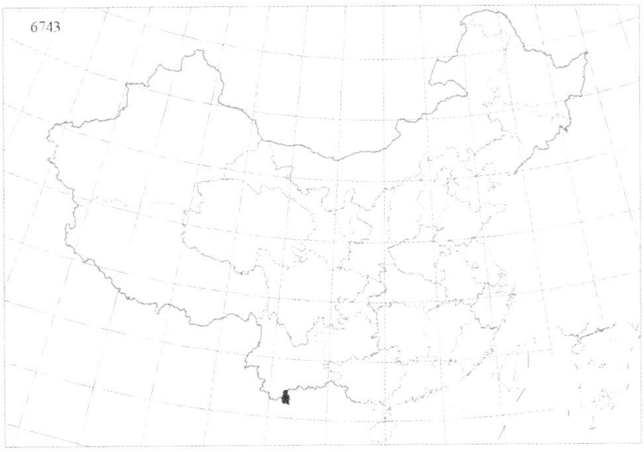

Theaceae 山茶科 *Pyrenaria* 核果茶属

Pyrenaria microcarpa Keng 小果核果茶

Small trees, evergreen

19.0 (14.3~24.8); 9.7 (3.1~20.1); 27.3 (21.8~30.1); 19.1 (14.3~24.8); 168.8 (113.2~237.6); 0.1 (0.0~2.3); 998 (761~1385) ❙ 1628 (868~2993); 593 (312~1045); 180 (41~547) ❙ 66.6 (−34.3~235.6); 980 (761~1360); 259.4 (123.9~620.0)

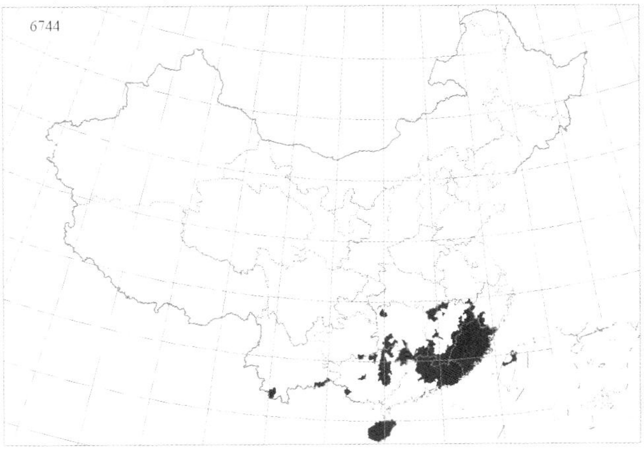

Theaceae 山茶科 *Pyrenaria* 核果茶属

Pyrenaria microcarpa var. ovalifolia (Li) Ming et S. X. Yang ex S. X. Yang 卵叶核果茶 Trees, evergreen

24.0 (22.2~24.8); 18.4 (17.2~20.1); 28.0 (25.7~28.8); 24.0 (22.3~24.8); 228.3 (207.2~237.6); 0.0 (0.0~0.0); 1310 (1127~1385) ❙ 1501 (868~1854); 534 (312~712); 74 (41~115) ❙ 15.0 (-34.3~47.8); 1191 (868~1360); 440.3 (248.2~620.0)

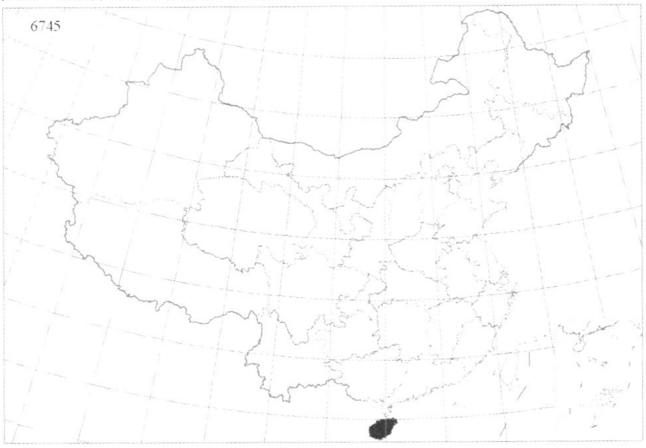

Theaceae 山茶科 *Pyrenaria* 核果茶属

Pyrenaria oblongicarpa H. T. Chang 长核果茶

Trees, evergreen

19.4; 12.6; 24.1; 19.4; 172.6; 0.0; 954 ❙ 1445; 787; 64 ❙ 52.3; 954; 521.5

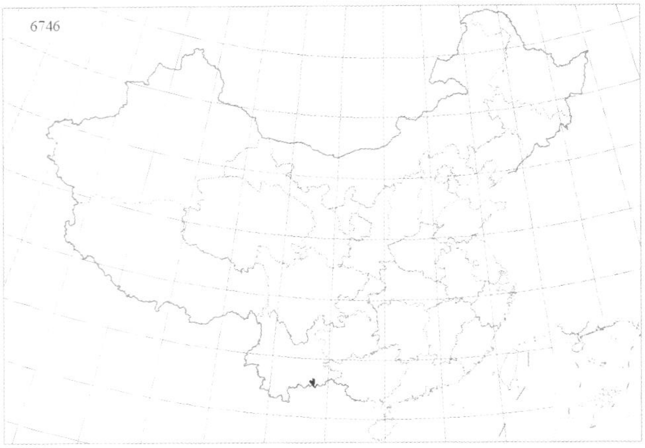

Theaceae 山茶科 *Pyrenaria* 核果茶属

Pyrenaria pingpienensis (H. T. Chang) S. X. Yang et Ming ex S. X. Yang 屏边核果茶 Trees, evergreen

18.1 (15.9~21.0); 10.7 (5.5~14.2); 23.5 (22.4~25.6); 18.1 (15.9~21.0); 157.1 (131.1~191.8); 0.0 (0.0~0.0); 884 (833~1053) ❙ 1392 (1089~1547); 775 (503~911); 59 (48~69) ❙ 57.7 (30.9~80.6); 884 (833~1053); 423.9 (181.2~551.6)

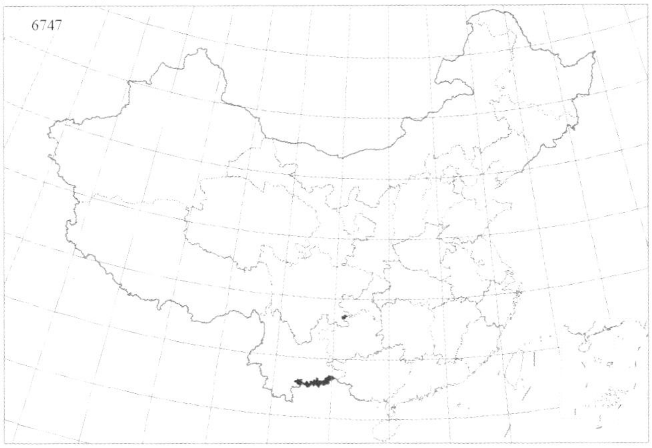

Theaceae 山茶科 *Pyrenaria* 核果茶属

Pyrenaria sophiae (Hu) S. X. Yang et Ming ex S. X. Yang 云南核果茶 Shrubs or small trees, evergreen

18.8 (17.9~19.3); 12.7 (11.6~13.5); 22.9 (22.1~23.6); 18.9 (17.9~19.4); 166.3 (155.1~173.2); 0.0 (0.0~0.0); 904 (858~939) ❙ 1368 (1098~1751); 790 (634~996); 57 (46~77) ❙ 51.4 (21.7~88.8); 894 (836~927); 442.7 (295.5~551.3)

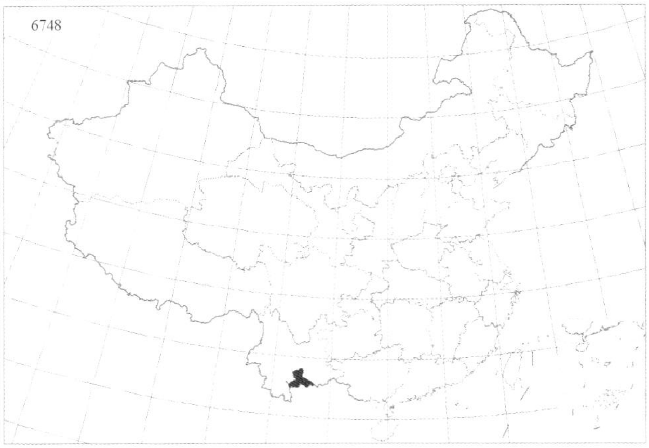

Theaceae 山茶科 *Pyrenaria* 核果茶属

Pyrenaria spectabilis (Champ.) C. Y. Wu et S. X. Yang ex S. X. Yang 大果核果茶 Trees, evergreen

19.1 (14.6~22.7); 9.7 (3.8~15.0); 27.3 (22.1~29.4); 19.2 (14.7~22.8); 170.2 (117.7~213.1); 0.0 (0.0~1.5); 1003 (779~1221) ❙ 1617 (1098~2435); 666 (421~1428); 163 (46~230) ❙ 62.4 (12.9~104.6); 997 (779~1221); 247.8 (163.7~327.7)

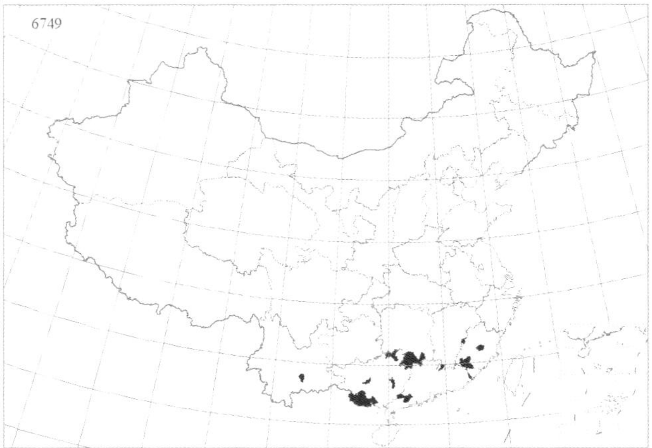

Theaceae 山茶科 *Pyrenaria* 核果茶属

Pyrenaria spectabilis var. greeniae (Chun) S. X. Yang 长柱核果茶

Small trees, evergreen

18.9 (14.3~24.3); 9.6 (3.7~19.1); 27.0 (19.6~29.8); 19.0 (14.3~24.3); 167.5 (113.2~231.9); 0.0 (0.0~1.5); 991 (730~1338) ❙ 1647 (868~3495); 646 (312~1847); 173 (42~897) ❙ 68.6 (-34.3~246.5); 974 (730~1326); 245.1 (108.2~620.0)

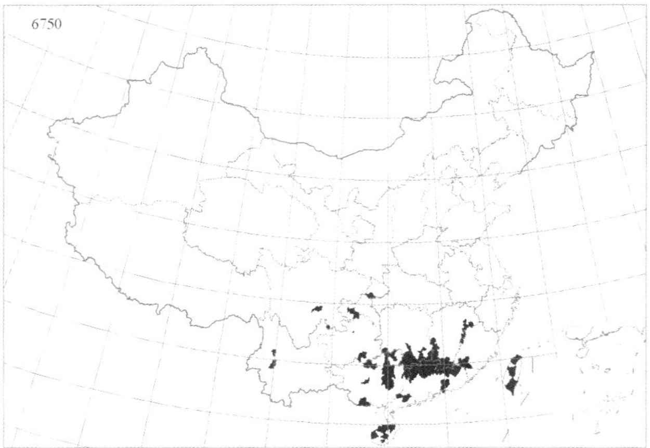

Theaceae 山茶科 *Pyrenaria* 核果茶属
Pyrenaria wuana (H. T. Chang) S. X. Yang 长萼核果茶
Shrubs or small trees, evergreen

20.3; 10.9; 28.2; 20.4; 184.3; 0.0; 1073 ❙ 1653; 700; 159 ❙ 54.3; 1073; 226.1

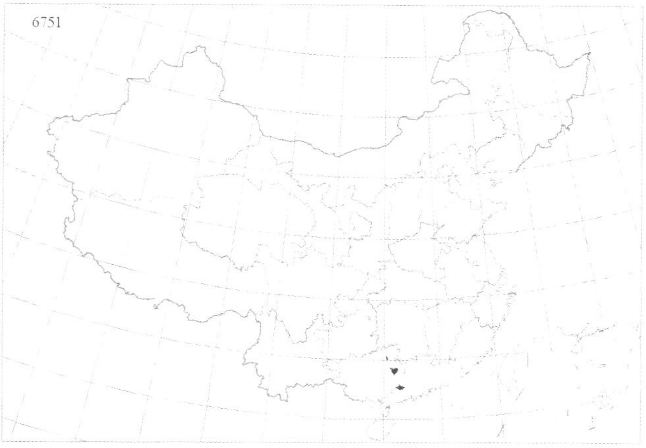

Theaceae 山茶科 *Schima* 木荷属
Schima argentea E. Pritz. 银木荷
Trees, evergreen

16.7 (5.4~22.7); 7.2 (-2.1~15.1); 24.9 (12.3~30.1); 16.7 (5.8~22.8); 141.3 (31.5~213.1); 0.5 (0.0~25.6); 872 (481~1221) ❙ 1310 (741~2435); 585 (392~1428); 102 (15~259) ❙ 50.5 (-14.2~128.9); 862 (481~1221); 248.6 (108.2~659.0)

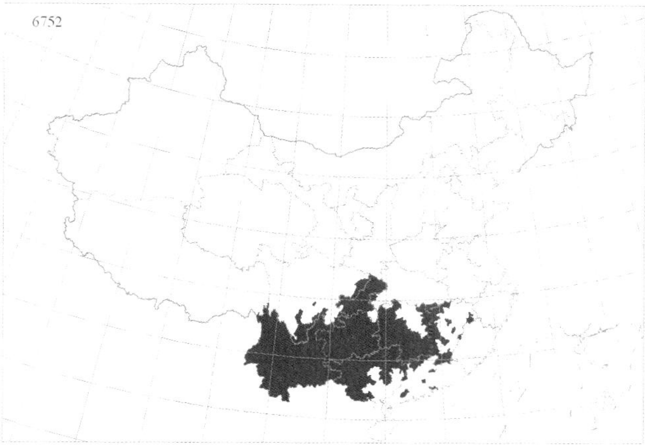

Theaceae 山茶科 *Schima* 木荷属
Schima brevipedicellata H. T. Chang 短梗木荷
Trees, evergreen

17.7 (15.3~21.0); 7.9 (4.0~14.2); 26.4 (21.4~29.3); 17.8 (15.4~21.0); 153.5 (125.8~191.8); 0.1 (0.0~1.0); 920 (813~1053) ❙ 1509 (1163~1776); 641 (473~996); 145 (48~222) ❙ 64.2 (28.8~103.1); 916 (813~1053); 303.9 (189.7~551.6)

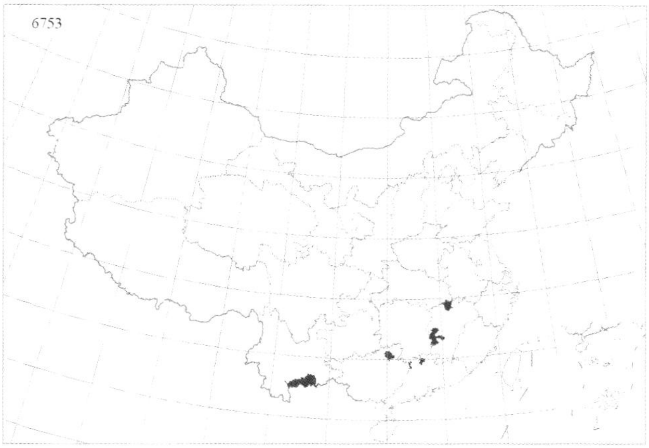

Theaceae 山茶科 *Schima* 木荷属
Schima crenata Korth. 钝齿木荷
Trees, evergreen

22.6 (17.7~24.7); 16.5 (7.7~19.8); 27.2 (25.7~28.2); 22.6 (17.8~24.7); 211.6 (153.6~236.5); 0.0 (0.0~0.0); 1209 (916~1376) ❙ 1677 (1551~1876); 613 (534~891); 87 (48~182) ❙ 41.0 (15.4~70.7); 1135 (916~1299); 476.0 (232.5~620.0)

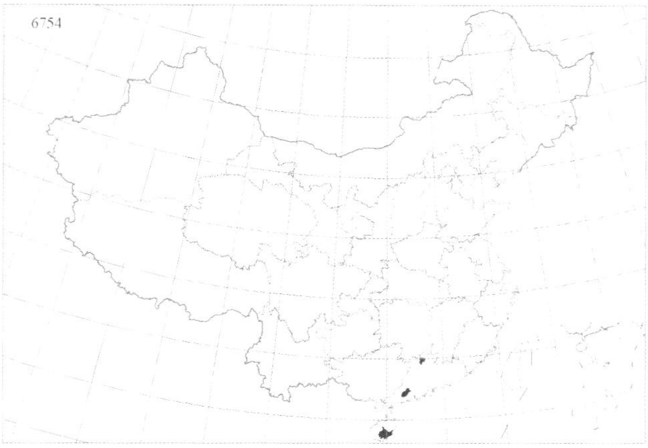

Theaceae 山茶科 *Schima* 木荷属
Schima khasiana Dyer 印度木荷
Trees, evergreen

16.9 (8.6~21.0); 10.2 (1.3~14.2); 21.6 (15.0~25.6); 17.0 (8.7~21.0); 143.9 (52.6~191.8); 0.3 (0.0~8.8); 830 (574~1053) ❙ 1306 (1041~1751); 736 (582~1007); 54 (44~77) ❙ 58.7 (21.7~145.8); 823 (574~1053); 407.2 (277.0~551.6)

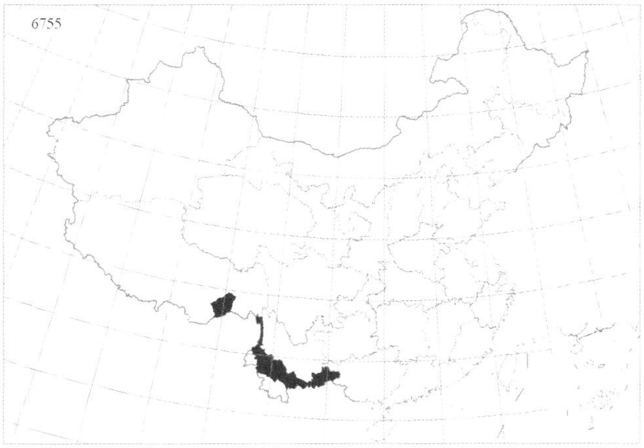

Theaceae 山茶科 *Schima* 木荷属
Schima multibracteata H. T. Chang 多苞木荷
Trees, evergreen

19.9 (17.9~22.2); 10.2 (7.4~13.9); 27.8 (25.8~28.6); 19.9 (18.0~22.3); 178.7 (155.6~207.5); 0.0 (0.0~0.0); 1045 (913~1190) ❙ 1578 (1371~1758); 673 (562~728); 168 (87~189) ❙ 52.6 (15.4~85.5); 1044 (913~1190); 229.0 (185.5~327.7)

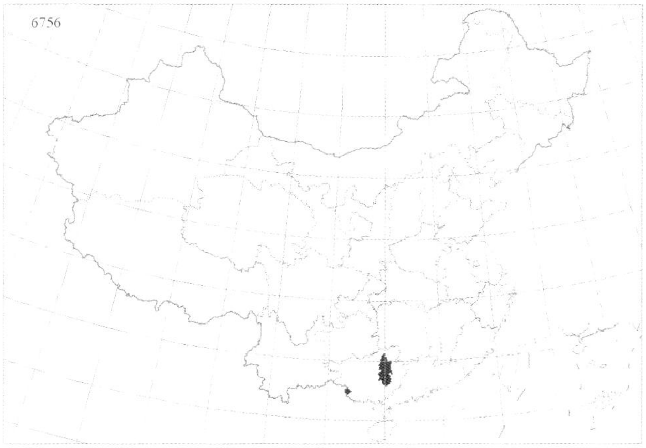

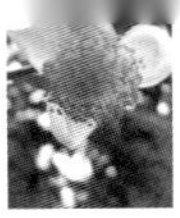

Theaceae 山茶科　*Schima* 木荷属

Schima noronhae Reinw. ex Blume 南洋木荷

Trees, evergreen

17.0 (14.4~18.9); 10.1 (5.0~12.5); 22.0 (19.1~23.6); 17.0 (14.4~19.0); 144.5 (113.1~167.5); 0.0 (0.0~0.0); 831 (727~914) ❙ 1110 (939~1454); 618 (506~792); 48 (37~70) ❙ 34.9 (6.5~99.9); 794 (727~854); 362.6 (236.5~491.5)

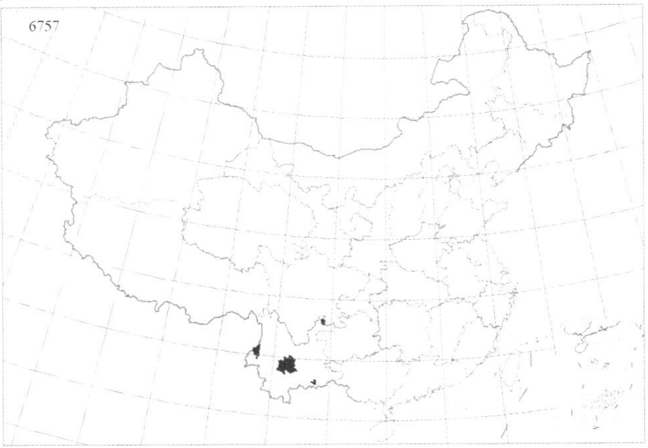

Theaceae 山茶科　*Schima* 木荷属

Schima parviflora Cheng et H. T. Chang ex H. T. Chang 小花木荷

Trees, evergreen

16.0 (11.7~21.0); 7.8 (1.1~14.2); 22.8 (18.4~26.6); 16.1 (11.7~21.0); 133.8 (82.6~191.8); 0.7 (0.0~8.5); 813 (658~1053) ❙ 1264 (916~1751); 659 (488~1007); 60 (31~113) ❙ 56.3 (12.5~145.8); 808 (658~1053); 311.7 (153.9~551.6)

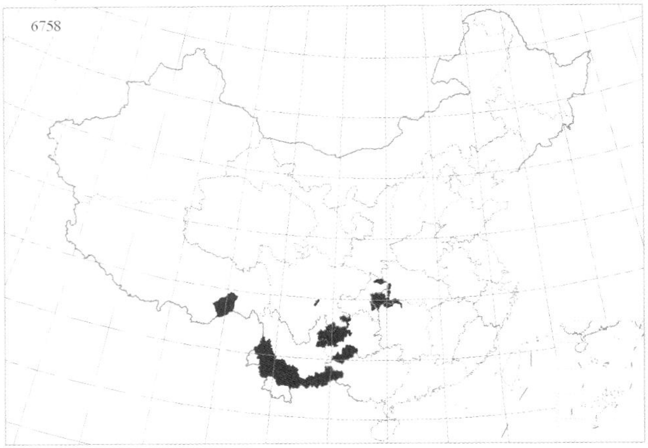

Theaceae 山茶科　*Schima* 木荷属

Schima remotiserrata H. T. Chang 疏齿木荷

Trees, evergreen

16.9 (14.3~20.7); 5.7 (3.3~10.9); 27.4 (24.2~29.8); 17.0 (14.3~20.8); 143.8 (113.2~189.3); 0.3 (0.0~2.3); 902 (761~1113) ❙ 1499 (1168~1997); 503 (392~679); 183 (89~259) ❙ 66.7 (34.4~130.7); 888 (761~1072); 207.0 (134.6~303.4)

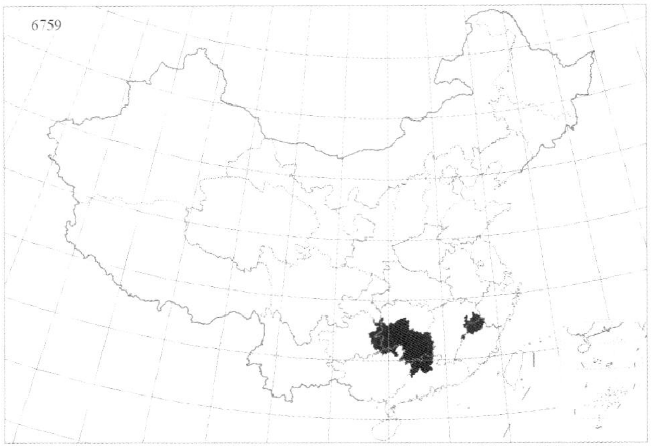

Theaceae 山茶科　*Schima* 木荷属

Schima sericans (Hand.-Mazz.) Ming 贡山木荷

Trees, evergreen

16.7 (5.3~21.0); 10.1 (-2.4~14.2); 21.3 (12.2~25.6); 16.8 (5.7~21.0); 141.7 (30.5~191.8); 0.7 (0.0~26.2); 821 (479~1053) ❙ 1238 (939~1751); 696 (524~1007); 52 (31~77) ❙ 52.3 (6.5~145.8); 806 (479~1053); 390.5 (234.2~551.6)

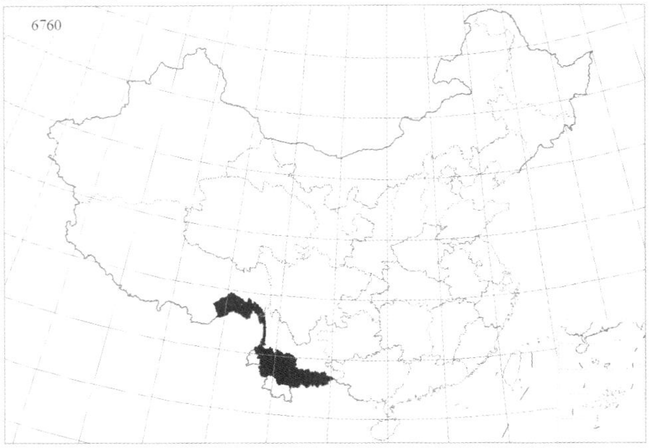

Theaceae 山茶科　*Schima* 木荷属

Schima sericans var. paracrenata (H. T. Chang) Ming 独龙木荷

Trees, evergreen

11.1; 4.0; 16.8; 11.1; 76.2; 3.2; 638 ❙ 1207; 669; 57 ❙ 89.0; 638; 381.4

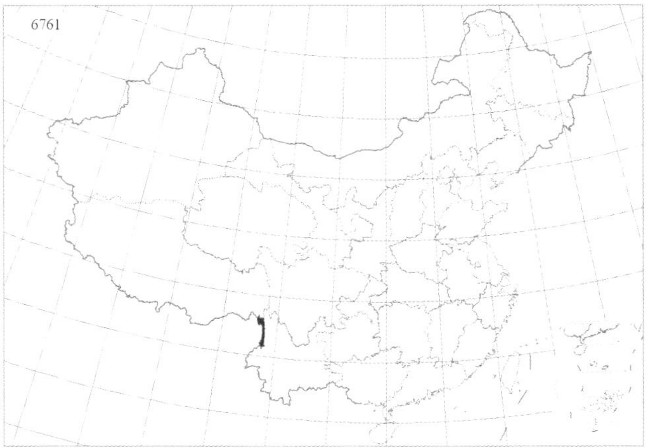

Theaceae 山茶科　*Schima* 木荷属

Schima sinensis (Hemsl. et E. H. Wilson) Airy Shaw 华木荷

Trees, evergreen

15.1 (9.0~21.0); 4.9 (0.6~14.2); 24.5 (16.5~28.1); 15.1 (9.0~21.0); 122.9 (58.7~191.8); 1.1 (0.0~10.3); 803 (584~1053) ❙ 1242 (874~1624); 538 (419~911); 90 (25~198) ❙ 55.0 (10.0~107.2); 802 (584~1053); 232.1 (133.5~551.6)

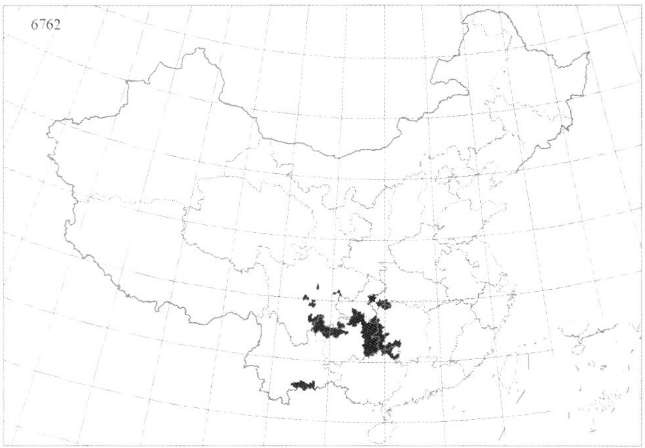

Theaceae 山茶科 *Schima* 木荷属

Schima superba Gardner et Champ. 木荷

Trees, evergreen

18.0 (11.7~24.8); 8.0 (0.7~20.1); 27.1 (18.6~30.1); 18.0 (11.7~24.8); 157.0 (82.6~237.6); 0.5 (0.0~9.4); 955 (658~1385) ❙ 1507 (868~3495); 603 (312~1847); 154 (31~897) ❙ 59.6 (-34.3~246.5); 944 (658~1360); 226.2 (95.3~620.0)

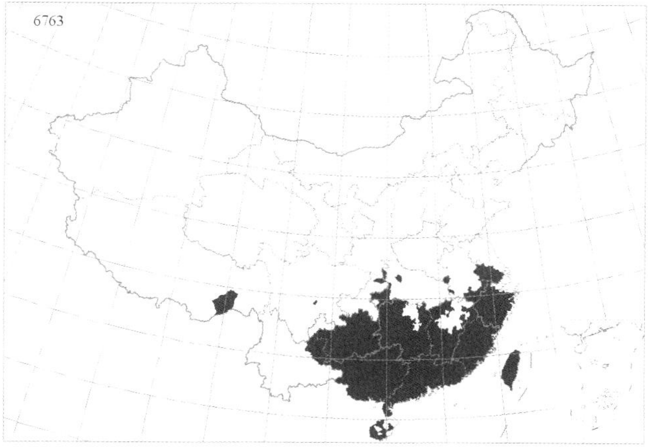

Theaceae 山茶科 *Schima* 木荷属

Schima villosa Hu 毛木荷

Trees, evergreen

19.3; 12.7; 23.8; 19.3; 171.6; 0.0; 942 ❙ 1439; 803; 61 ❙ 53.4; 942; 516.4

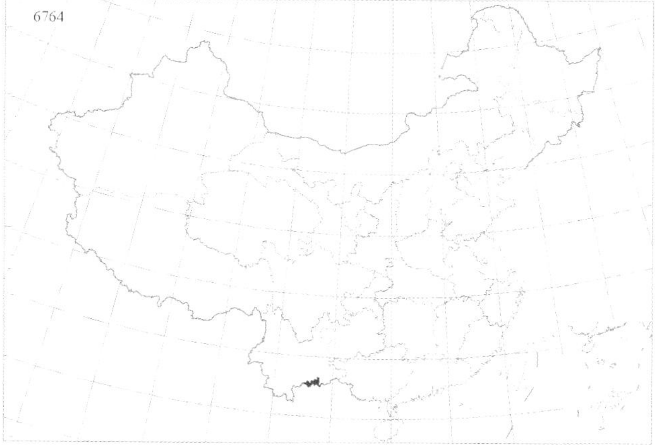

Theaceae 山茶科 *Schima* 木荷属

Schima wallichii (DC.) Korth. 红木荷

Trees, evergreen

17.1 (11.0~22.3); 9.7 (3.8~15.1); 22.5 (16.6~28.9); 17.1 (11.0~22.3); 145.3 (73.1~208.1); 0.0 (0.0~1.5); 851 (636~1202) ❙ 1219 (797~1751); 662 (411~996); 58 (23~191) ❙ 43.7 (-14.2~112.2); 837 (636~1202); 330.6 (171.2~659.0)

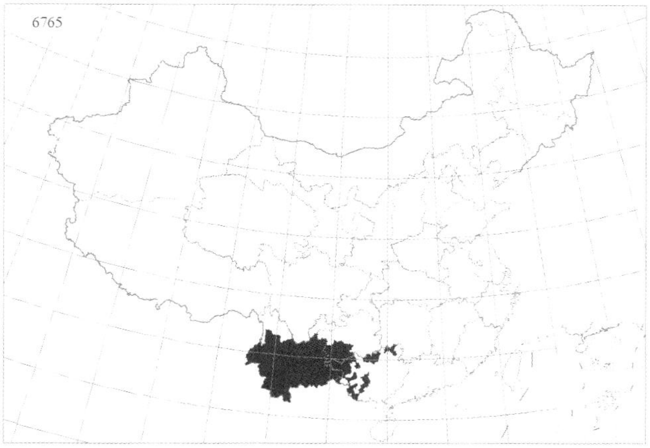

Theaceae 山茶科 *Stewartia* 紫茎属

Stewartia calcicola Ming et J. Li ex J. Li 云南紫茎

Trees, evergreen

18.5 (16.3~21.0); 11.7 (9.1~15.1); 23.2 (21.4~25.6); 18.5 (16.4~21.0); 162.3 (136.4~191.8); 0.0 (0.0~0.0); 898 (797~1053) ❙ 1375 (1074~1607); 745 (582~911); 57 (47~77) ❙ 53.3 (32.4~80.6); 894 (797~1053); 440.6 (277.0~659.0)

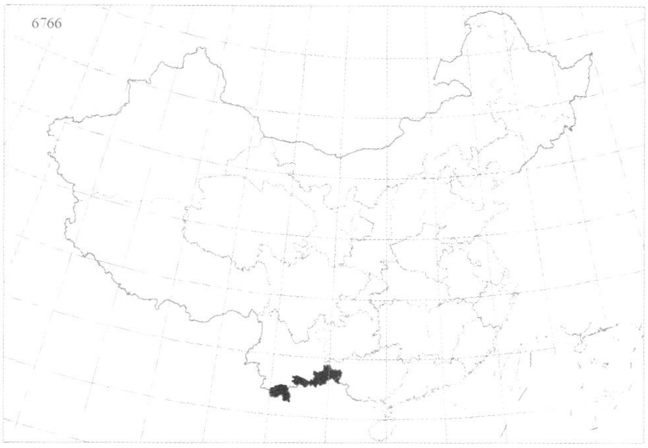

Theaceae 山茶科 *Stewartia* 紫茎属

Stewartia cordifolia (H. L. Li) J. Li et Ming ex J. Li 心叶紫茎

Trees, evergreen

18.6 (14.3~21.3); 8.7 (4.0~12.3); 26.9 (23.5~28.6); 18.6 (14.3~21.3); 163.7 (112.6~196.1); 0.1 (0.0~1.1); 970 (756~1130) ❙ 1503 (1268~1758); 644 (523~848); 149 (72~189) ❙ 55.8 (22.7~85.5); 970 (756~1130); 230.3 (185.5~313.7)

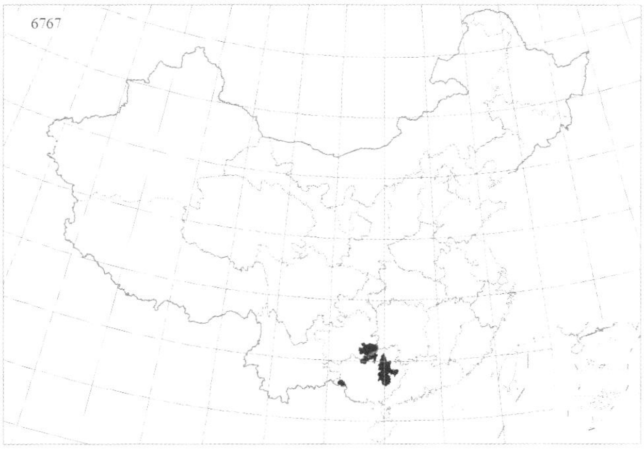

Theaceae 山茶科 *Stewartia* 紫茎属

Stewartia crassifolia (S. Z. Yan) J. Li et Ming 厚叶紫茎

Trees, evergreen

17.9 (14.3~19.9); 7.5 (3.7~10.6); 27.1 (24.2~28.5); 17.9 (14.3~19.9); 155.0 (113.2~179.3); 0.2 (0.0~1.4); 936 (761~1036) ❙ 1653 (1486~1745); 576 (505~644); 197 (176~221) ❙ 78.0 (57.0~128.9); 936 (761~1036); 236.0 (188.8~270.8)

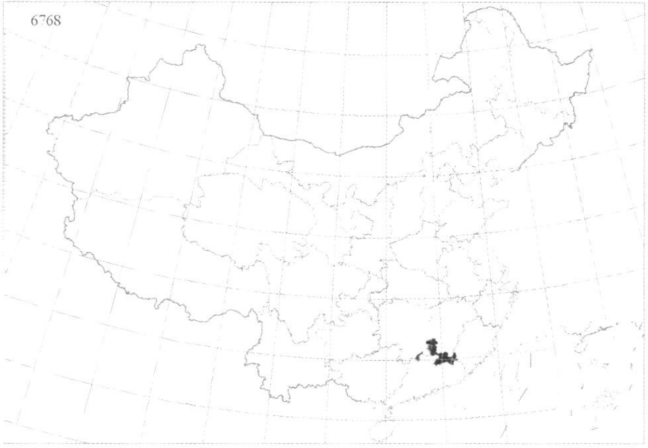

Theaceae 山茶科 *Stewartia* 紫茎属

Stewartia densivillosa (Hu et H. T. Chang et C. X. Ye) J. Li et Ming ex J. Li 狭萼紫茎

Shrubs, evergreen

18.9；10.6；25.0；19.0；167.5；0.0；939 ❙ 1292；729；49 ❙ 37.6；939；342.5

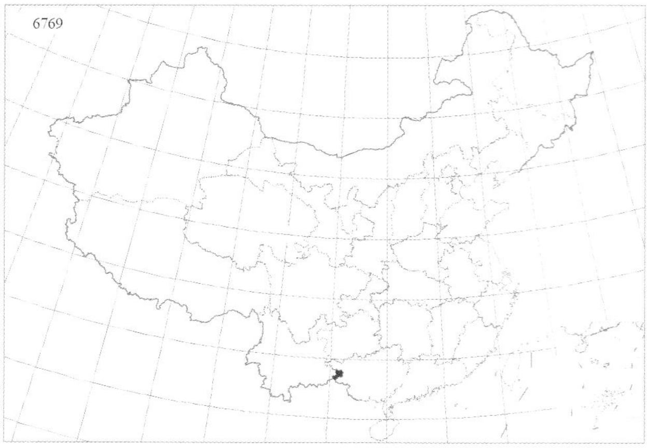

Theaceae 山茶科 *Stewartia* 紫茎属

Stewartia laotica (Gagnep.) J. Li et Ming ex J. Li 老挝紫茎

Small trees, evergreen

17.7；10.4；22.9；17.7；152.9；0.0；857 ❙ 1395；785；56 ❙ 62.8；857；425.6

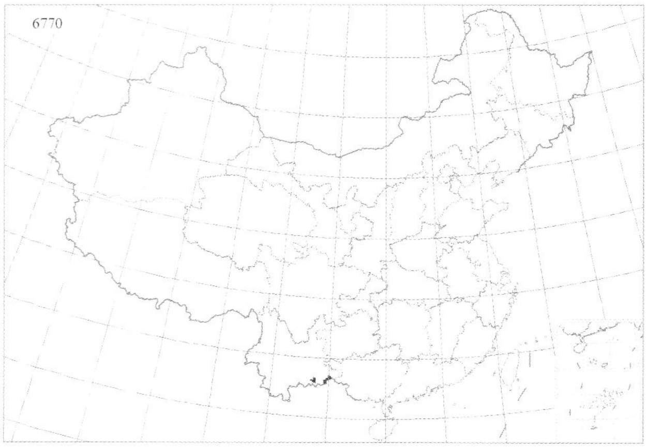

Theaceae 山茶科 *Stewartia* 紫茎属

Stewartia medogensis J. Li et Ming 墨脱紫茎

Small trees, evergreen

13.3；5.4；19.6；13.4；100.4；0.0；706 ❙ 1736；1007；51 ❙ 145.8；706；447.4

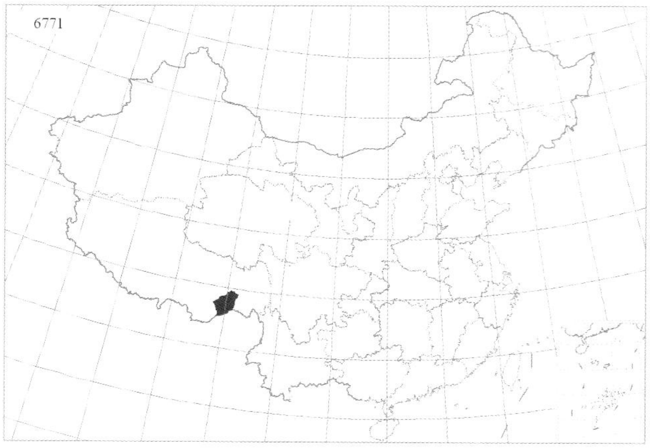

Theaceae 山茶科 *Stewartia* 紫茎属

Stewartia micrantha (Chun) Sealy 小花紫茎

Shrubs, evergreen

20.2；11.9；27.4；20.2；182.8；0.0；1039 ❙ 1592；596；159 ❙ 53.6；1039；262.3

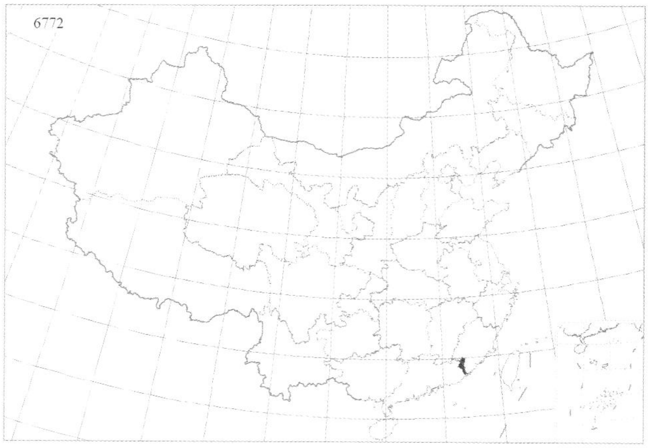

Theaceae 山茶科 *Stewartia* 紫茎属

Stewartia obovata (Chun ex H. T. Chang) J. Li et Ming ex J. Li 钝叶紫茎

Trees, evergreen

20.6 (14.6~22.7)；11.9 (3.8~15.0)；27.5 (24.6~28.9)；20.6 (14.7~22.8)；187.5 (117.7~213.1)；0.1 (0.0~1.5)；1089 (779~1221) ❙ 1648 (1260~2435)；831 (544~1428)；119 (58~191) ❙ 53.5 (12.9~101.8)；1089 (779~1221)；259.5 (122.7~327.7)

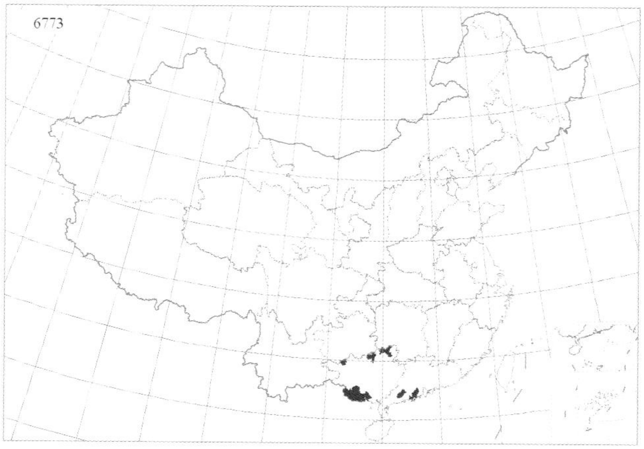

Theaceae 山茶科 *Stewartia* 紫茎属

Stewartia pteropetiolata W. C. Cheng 翅柄紫茎

Trees, evergreen

17.7 (14.4~21.0)；11.3 (7.5~15.1)；22.0 (19.1~25.6)；17.8 (14.4~21.0)；153.3 (113.1~191.8)；0.0 (0.0~0.0)；857 (727~1053) ❙ 1310 (939~1751)；725 (524~996)；52 (37~77) ❙ 52.9 (6.5~112.2)；840 (717~1053)；424.3 (277.0~659.0)

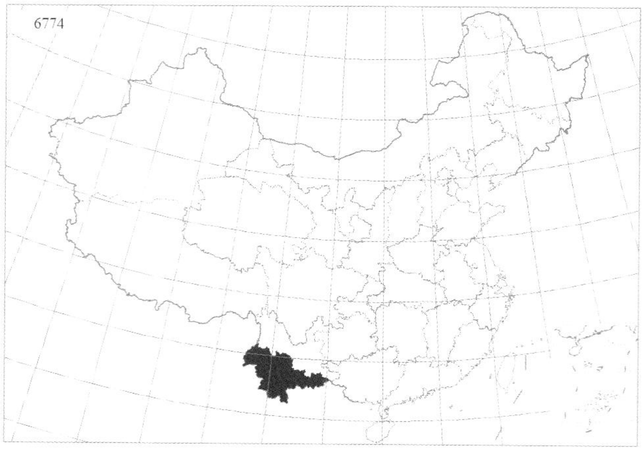

Theaceae 山茶科 *Stewartia* 紫茎属
Stewartia rostrata Spongberg 长喙紫茎
Small trees, deciduous

15.3 (9.3~17.2); 3.2 (-1.7~5.0); 27.0 (20.1~29.3); 15.3 (9.5~17.2); 126.6 (69.8~146.8); 2.9 (0.0~18.0); 843 (603~941) ▎1445 (1198~1663); 556 (427~685); 163 (71~220) ▎72.3 (28.8~108.0); 841 (603~941); 237.4 (181.4~299.5)

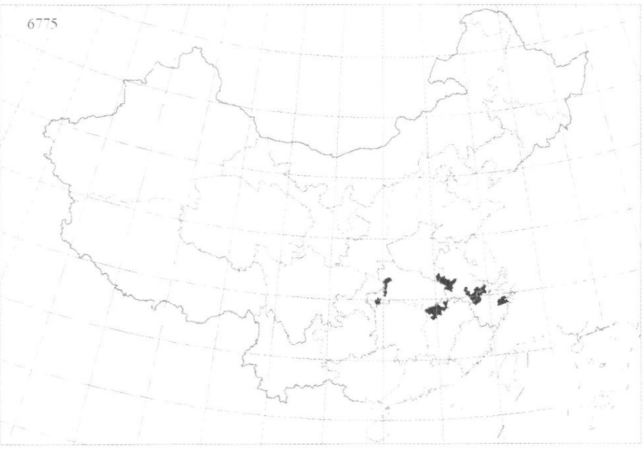

Theaceae 山茶科 *Stewartia* 紫茎属
Stewartia rubiginosa H. T. Chang 红皮紫茎
Trees, deciduous

18.4 (14.6~21.6); 8.3 (3.8~12.6); 27.1 (24.6~28.5); 18.5 (14.7~21.7); 162.0 (117.7~199.8); 0.2 (0.0~1.5); 973 (779~1153) ▎1597 (1469~1777); 636 (465~821); 183 (171~194) ▎66.6 (35.6~101.8); 973 (779~1153); 228.4 (188.8~258.0)

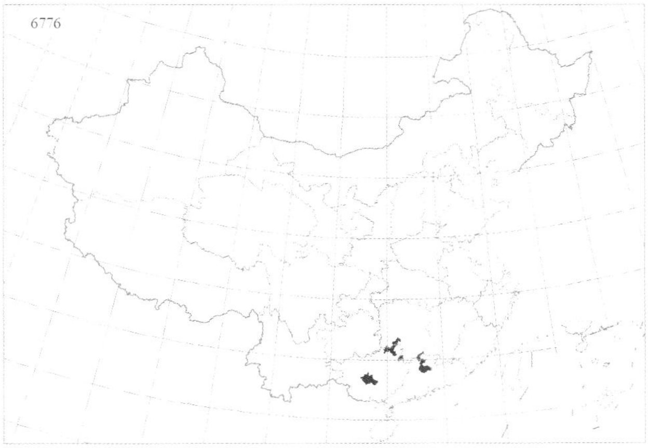

Theaceae 山茶科 *Stewartia* 紫茎属
Stewartia rubiginosa var. damingshanica (J. Li) Ming 大明紫茎
Trees, deciduous

21.6; 12.6; 28.5; 21.7; 199.8; 0.0; 1153 ▎1625; 795; 173 ▎40.9; 1153; 237.8

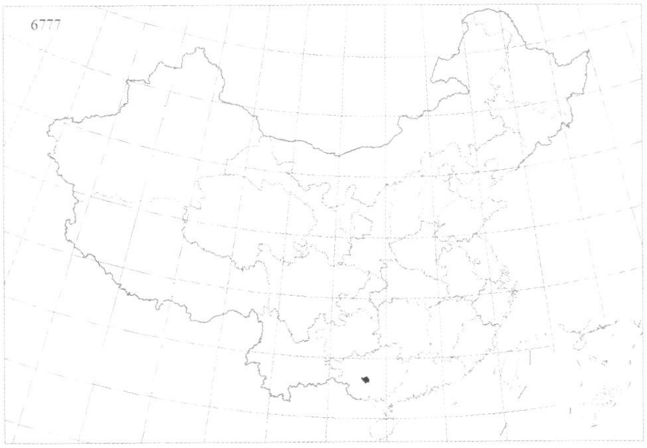

Theaceae 山茶科 *Stewartia* 紫茎属
Stewartia sichuanensis (S. Z. Yan) J. Li et Ming ex J. Li 四川紫茎
Shrubs or small trees, evergreen

15.4; 6.1; 23.9; 15.5; 125.7; 0.1; 807 ▎1070; 555; 53 ▎33.2; 802; 204.3

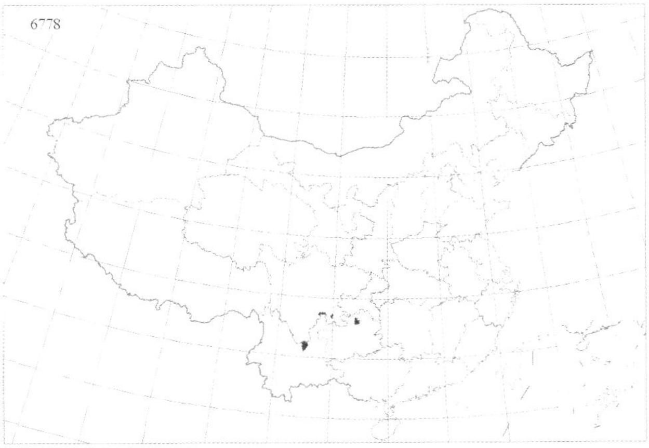

Theaceae 山茶科 *Stewartia* 紫茎属
Stewartia sinensis (S. Z. Yan) Ming et J. Li ex J. Li 紫茎
Small trees, deciduous

16.2 (9.0~21.0); 5.8 (-2.5~14.2); 25.6 (16.5~30.1); 16.2 (9.0~21.1); 136.5 (58.7~193.0); 1.8 (0.0~20.6); 857 (584~1114) ▎1420 (793~1997); 585 (366~1007); 133 (19~259) ▎66.1 (0.0~147.0); 850 (584~1114); 271.2 (148.9~627.0)

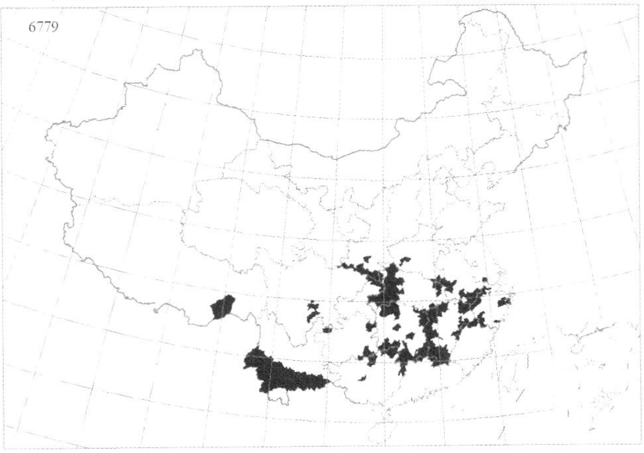

Theaceae 山茶科 *Stewartia* 紫茎属
Stewartia sinensis var. acutisepala (P. L. Chiu et G. R. Zhong) Ming et J. Li ex J. Li 尖萼紫茎
Small trees, deciduous

15.8 (15.0~17.5); 5.1 (4.0~7.1); 26.3 (24.7~27.8); 15.8 (15.1~17.6); 130.4 (120.8~150.6); 0.3 (0.0~1.0); 837 (789~920) ▎1765 (1581~1950); 641 (587~713); 204 (181~227) ▎111.6 (84.6~147.0); 837 (789~920); 244.2 (204.3~283.9)

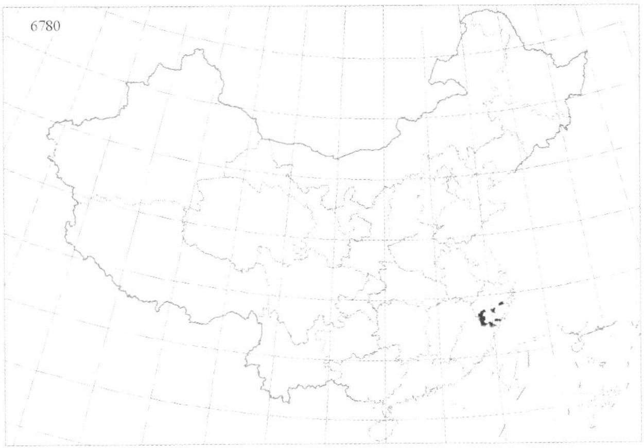

Theaceae 山茶科 *Stewartia* 紫茎属

Stewartia sinensis var. brevicalyx (S. Z. Yan) Ming et J. Li ex J. Li 短萼紫茎

Small trees, deciduous

14.5; 3.8; 24.6; 14.6; 116.7; 1.9; 791 ❙ 1621; 711; 155 ❙ 107.0; 791; 331.4

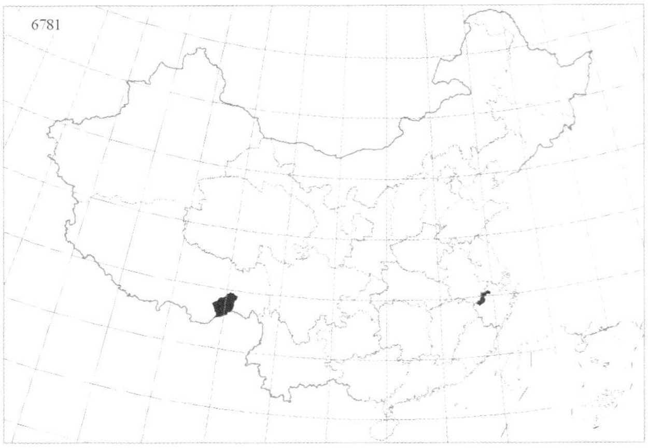

Theaceae 山茶科 *Stewartia* 紫茎属

Stewartia sinensis var. shensiensis (H. T. Chang) Ming et J. Li ex J. Li 陕西紫茎

Small trees, evergreen

12.4 (9.4~14.0); 0.8 (-2.5~2.9); 23.6 (21.0~25.8); 12.5 (9.8~14.1); 98.8 (73.9~113.4); 9.2 (2.8~20.6); 721 (619~786) ❙ 945 (760~1246); 425 (352~528); 31 (17~51) ❙ 31.9 (0.0~93.8); 721 (619~786); 267.0 (189.3~334.4)

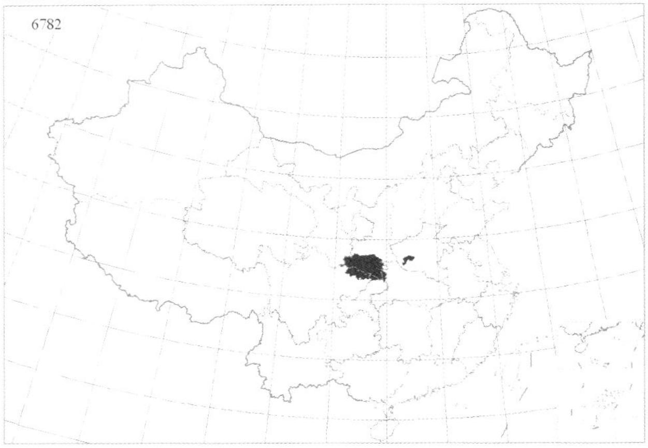

Theaceae 山茶科 *Stewartia* 紫茎属

Stewartia sinii (Y. C. Wu) Sealy 黄毛紫茎

Trees, evergreen

20.0 (17.9~21.8); 10.5 (7.4~14.0); 27.7 (25.8~28.6); 20.0 (18.0~21.9); 180.6 (155.6~202.2); 0.0 (0.0~0.0); 1052 (913~1150) ❙ 1665 (1371~2148); 747 (562~1234); 169 (124~194) ❙ 59.1 (22.7~86.7); 1051 (913~1150); 232.1 (185.5~318.8)

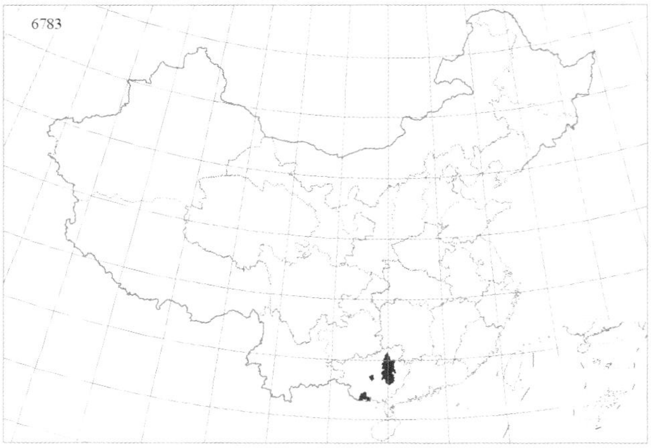

Theaceae 山茶科 *Stewartia* 紫茎属

Stewartia villosa Merr. 柔毛紫茎

Trees, evergreen

20.4 (14.6~22.7); 11.3 (3.8~15.0); 27.8 (24.6~29.8); 20.4 (14.7~22.8); 184.9 (117.7~213.1); 0.0 (0.0~1.5); 1071 (779~1221) ❙ 1726 (1302~2435); 732 (507~1428); 162 (86~197) ❙ 62.0 (12.9~101.8); 1070 (779~1221); 238.4 (120.3~327.7)

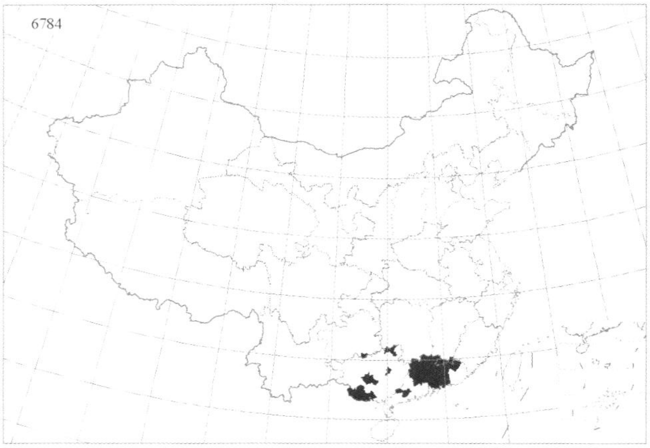

Theaceae 山茶科 *Stewartia* 紫茎属

Stewartia villosa var. kwangtungensis (Chun) J. Li et Ming 广东柔毛紫茎

Trees, evergreen

21.3 (17.4~22.7); 12.9 (7.4~15.0); 27.7 (25.5~28.6); 21.3 (17.4~22.8); 196.0 (148.9~213.1); 0.0 (0.0~0.0); 1124 (886~1221) ❙ 1722 (1302~2435); 895 (637~1428); 121 (86~194) ❙ 53.6 (12.9~99.5); 1124 (886~1221); 278.4 (235.1~327.7)

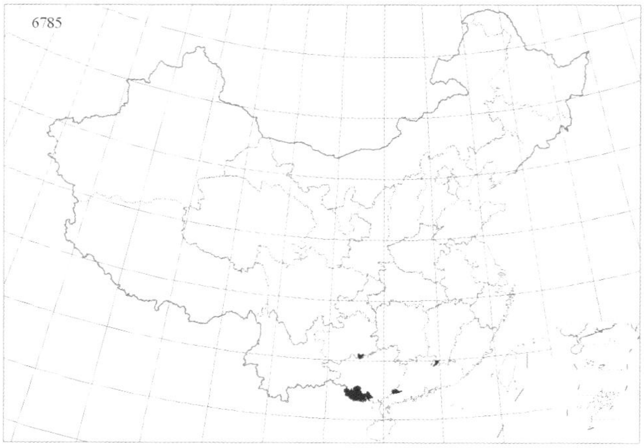

Theaceae 山茶科 *Ternstroemia* 厚皮香属

Ternstroemia biangulipes H. T. Chang 角柄厚皮香

Small trees, evergreen

13.6 (8.6~20.4); 6.7 (1.3~15.1); 18.8 (15.0~23.6); 13.6 (8.7~20.5); 104.7 (52.6~185.8); 1.6 (0.0~8.8); 722 (574~989) ❙ 1403 (1041~1736); 751 (585~1007); 61 (45~77) ❙ 95.9 (62.5~145.8); 722 (574~986); 456.3 (352.3~659.0)

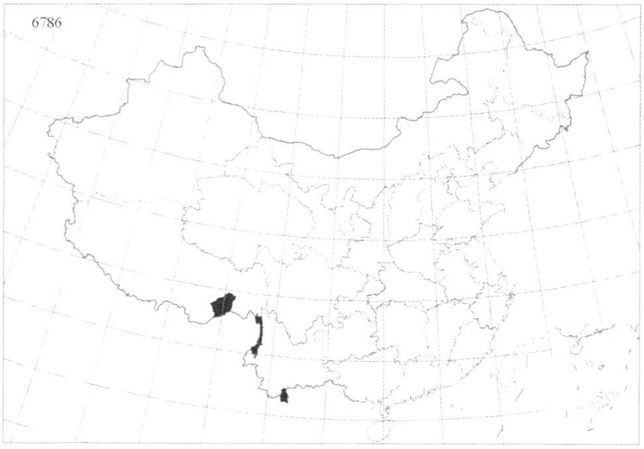

Theaceae 山茶科 *Ternstroemia* 厚皮香属
Ternstroemia conicocarpa L. K. Ling 锥果厚皮香
Shrubs or small trees, evergreen

16.7 (14.3~19.2); 5.9 (3.3~9.2); 26.6 (24.3~28.1); 16.8 (14.3~19.2); 141.6 (114.1~170.8); 0.3 (0.0~2.3); 882 (765~1002) ❙ 1507 (1415~1588); 527 (419~678); 181 (135~198) ❙ 71.9 (49.1~99.4); 877 (765~1002); 224.2 (177.4~257.1)

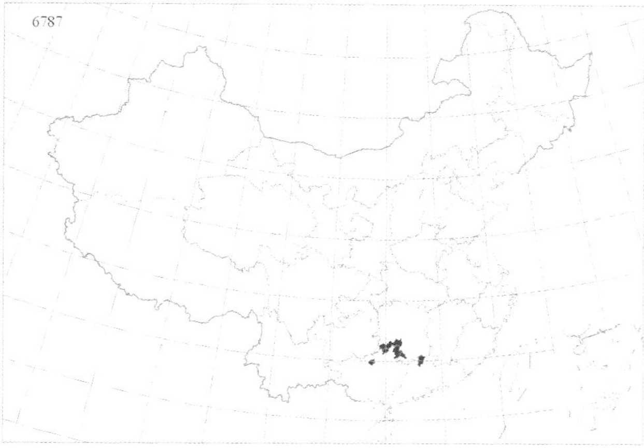

Theaceae 山茶科 *Ternstroemia* 厚皮香属
Ternstroemia gymnanthera (Wight et Arn.) Beddome 厚皮香
Small trees, evergreen

17.7 (5.4~24.7); 8.3 (-2.1~19.8); 25.9 (12.3~30.1); 17.7 (5.8~24.7); 153.0 (31.5~236.5); 0.4 (0.0~25.6); 927 (481~1376) ❙ 1436 (741~2993); 605 (312~1428); 130 (15~547) ❙ 56.1 (-34.3~246.5); 915 (481~1273); 249.5 (95.3~659.0)

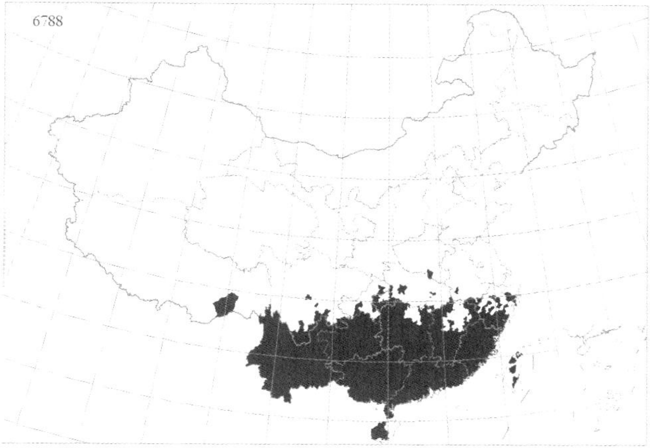

Theaceae 山茶科 *Ternstroemia* 厚皮香属
Ternstroemia gymnanthera var. wightii (Choisy) Hand.-Mazz. 阔叶厚皮香
Small trees, evergreen

16.4 (5.4~22.7); 7.7 (-2.1~15.1); 23.8 (12.3~29.8); 16.5 (5.8~22.8); 138.5 (31.5~213.1); 0.5 (0.0~25.6); 849 (481~1221) ❙ 1288 (741~2435); 595 (392~1428); 100 (15~221) ❙ 51.6 (-14.2~128.9); 835 (481~1221); 283.1 (134.6~659.0)

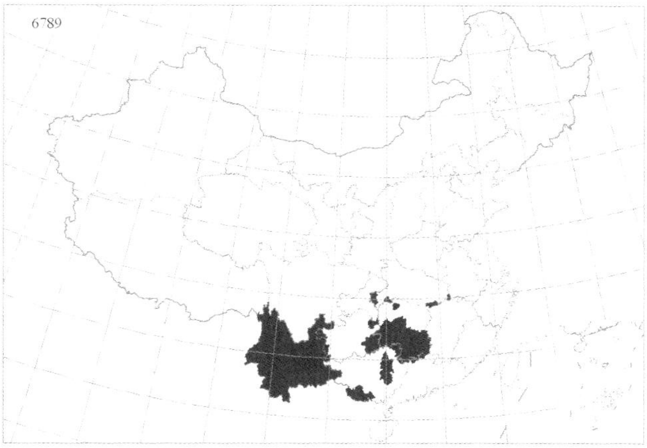

Theaceae 山茶科 *Ternstroemia* 厚皮香属
Ternstroemia hainanensis H. T. Chang 海南厚皮香
Trees, evergreen

22.8; 15.9; 27.9; 22.8; 214.1; 0.0; 1222 ❙ 1988; 1039; 101 ❙ 64.3; 1161; 358.2

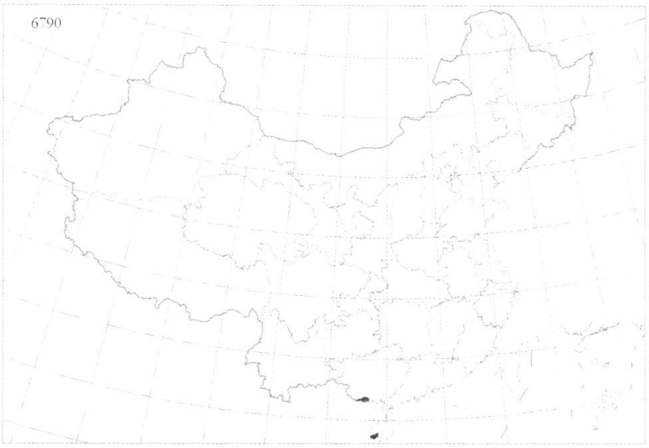

Theaceae 山茶科 *Ternstroemia* 厚皮香属
Ternstroemia insignis Y. C. Wu 大果厚皮香
Trees, evergreen

18.3 (15.5~21.3); 9.9 (4.8~14.2); 24.9 (21.4~28.6); 18.4 (15.6~21.3); 160.6 (127.2~196.1); 0.0 (0.0~0.2); 924 (797~1130) ❙ 1406 (1007~1758); 693 (539~911); 101 (44~189) ❙ 52.5 (19.0~93.2); 923 (797~1130); 309.2 (185.5~551.6)

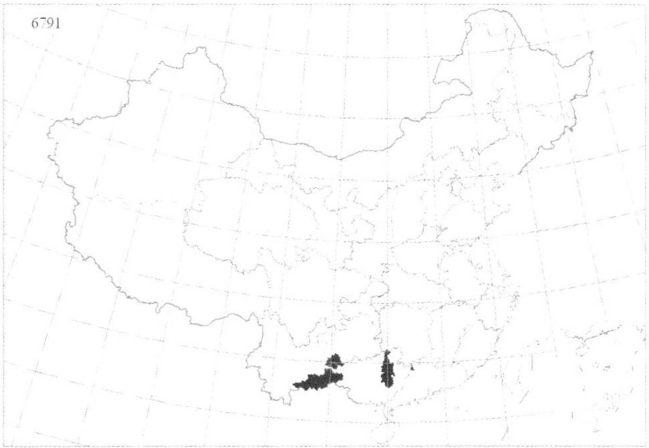

Theaceae 山茶科 *Ternstroemia* 厚皮香属
Ternstroemia japonica (Thunberg) Thunberg 日本厚皮香
Shrubs or trees, evergreen

20.2 (16.2~24.1); 14.5 (11.0~18.9); 25.0 (20.1~28.2); 20.2 (16.2~24.1); 182.9 (134.2~229.7); 0.0 (0.0~0.0); 1019 (773~1322) ❙ 2402 (1498~3495); 1106 (750~1847); 255 (47~897) ❙ 146.2 (25.2~246.5); 983 (773~1144); 245.0 (119.4~363.7)

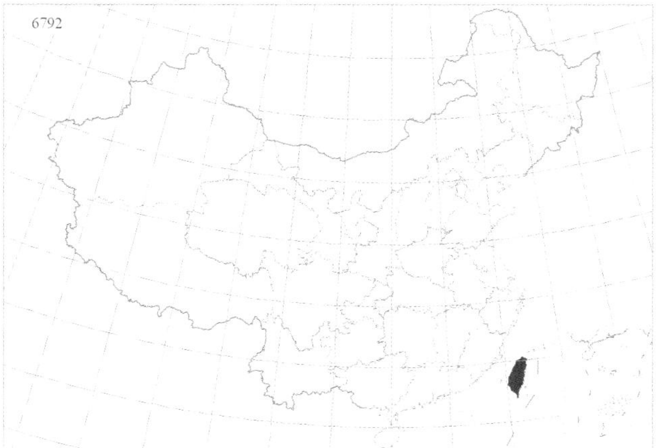

Theaceae 山茶科 *Ternstroemia* 厚皮香属

Ternstroemia kwangtungensis Merr. 厚叶厚皮香

Shrubs or small trees, evergreen

19.5 (14.6~24.8); 10.3 (3.8~20.1); 27.5 (24.3~29.2); 19.6 (14.7~24.8); 174.8 (117.7~237.6); 0.0 (0.0~1.5); 1032 (779~1385) ❙ 1656 (868~2435); 631 (312~1428); 167 (41~259) ❙ 64.1 (-34.3~137.5); 1010 (779~1360); 277.5 (183.9~620.0)

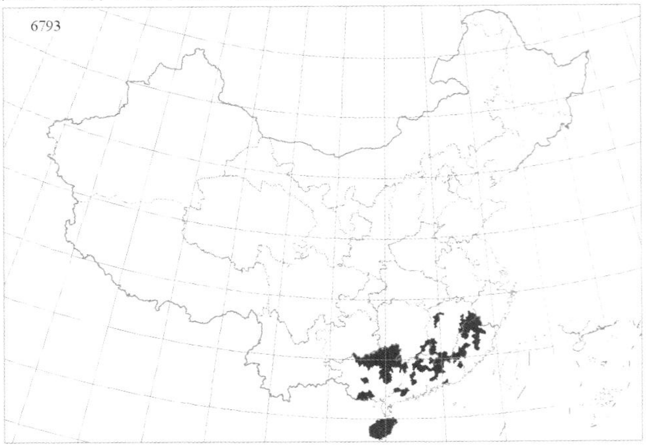

Theaceae 山茶科 *Ternstroemia* 厚皮香属

Ternstroemia luteoflora L. K. Ling 尖萼厚皮香

Small trees, evergreen

18.9 (12.6~23.6); 9.2 (1.6~17.5); 27.4 (21.4~29.8); 19.0 (12.7~23.6); 168.2 (98.8~223.5); 0.2 (0.0~6.9); 1000 (711~1281) ❙ 1544 (1018~2435); 635 (392~1428); 148 (43~259) ❙ 55.8 (-6.9~128.9); 995 (711~1273); 227.4 (95.3~538.2)

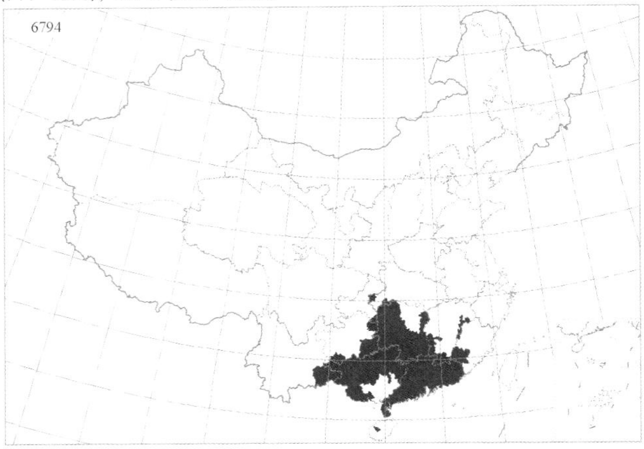

Theaceae 山茶科 *Ternstroemia* 厚皮香属

Ternstroemia microphylla Merr. 小叶厚皮香

Shrubs or small trees, evergreen

22.5 (18.0~24.8); 15.3 (8.2~20.1); 28.0 (25.7~29.0); 22.5 (18.0~24.8); 210.4 (156.4~237.6); 0.0 (0.0~0.0); 1207 (913~1385) ❙ 1604 (1174~2435); 712 (386~1428); 106 (41~195) ❙ 34.3 (-11.9~99.5); 1161 (913~1360); 328.4 (188.7~620.0)

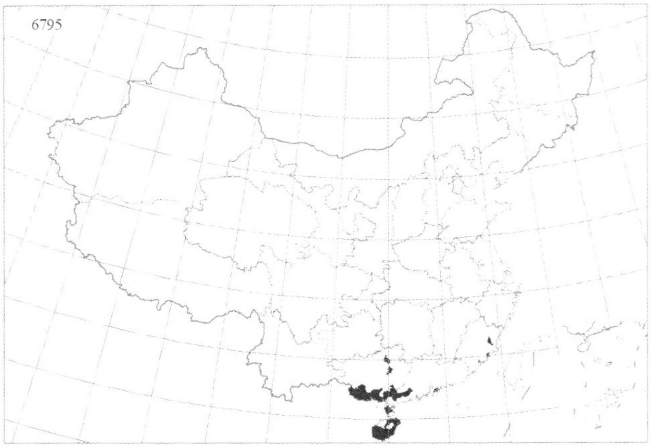

Theaceae 山茶科 *Ternstroemia* 厚皮香属

Ternstroemia nitida Merr. 亮叶厚皮香

Shrubs or small trees, evergreen

17.8 (12.7~24.3); 7.4 (2.1~18.7); 27.2 (23.0~30.1); 17.9 (12.8~24.3); 154.5 (99.0~231.9); 0.2 (0.0~5.6); 942 (713~1338) ❙ 1578 (868~2024); 565 (312~845); 179 (42~259) ❙ 69.4 (-34.3~130.7); 929 (713~1188); 240.9 (148.9~572.9)

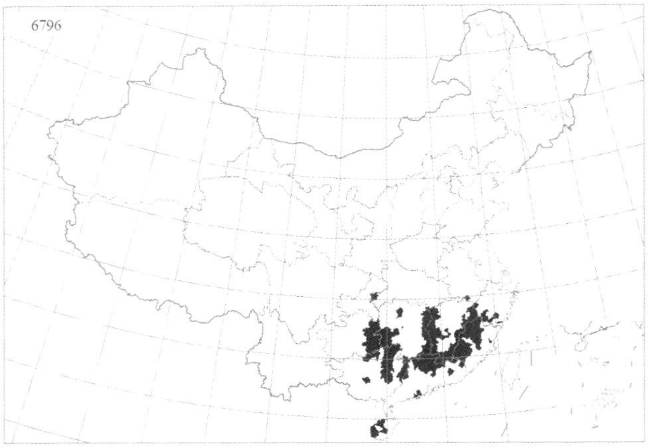

Theaceae 山茶科 *Ternstroemia* 厚皮香属

Ternstroemia sichuanensis L. K. Ling 四川厚皮香

Shrubs or trees, evergreen

16.7 (11.9~18.7); 6.3 (2.9~8.1); 26.5 (20.0~28.9); 16.8 (12.0~18.8); 141.4 (87.2~166.0); 0.3 (0.0~3.4); 889 (671~1002) ❙ 1146 (1001~1410); 509 (442~816); 62 (35~71) ❙ 30.8 (8.0~80.4); 889 (671~1002); 156.7 (108.2~278.8)

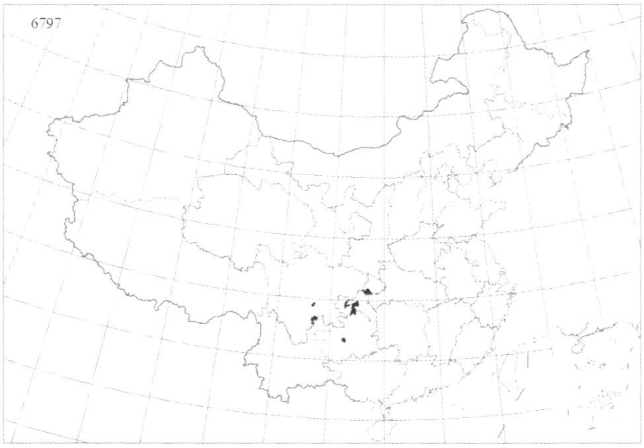

Theaceae 山茶科 *Ternstroemia* 厚皮香属

Ternstroemia simaoensis L. K. Ling 思茅厚皮香

Shrubs or trees, evergreen

18.7 (17.1~20.5); 12.9 (11.0~14.9); 22.2 (20.8~23.7); 18.7 (17.1~20.5); 164.8 (145.0~185.9); 0.0 (0.0~0.0); 893 (815~992) ❙ 1461 (1352~1500); 752 (655~836); 58 (47~66) ❙ 64.0 (47.6~74.2); 873 (802~959); 464.2 (395.0~570.1)

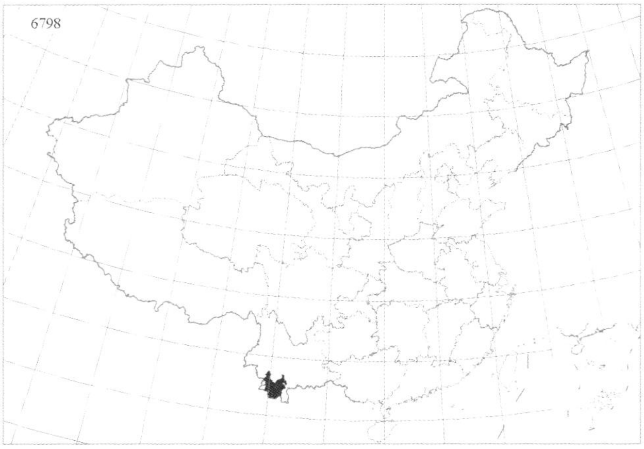

Theaceae 山茶科 *Ternstroemia* 厚皮香属
Ternstroemia yunnanensis L. K. Ling 云南厚皮香
Trees, evergreen

19.3；12.7；23.8；19.3；171.6；0.0；942 ❙ 1439；803；61 ❙ 53.4；942；516.4

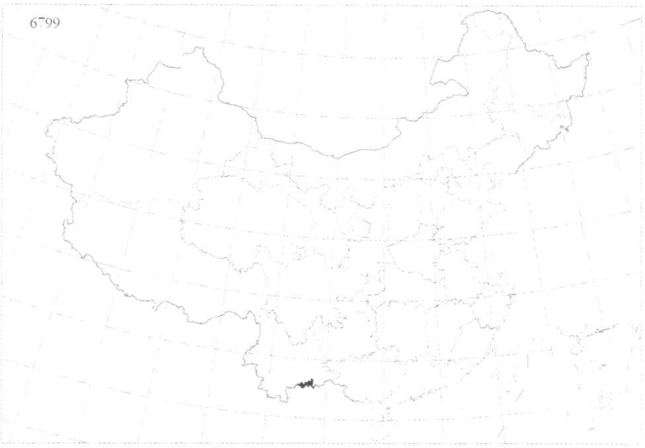

Clusiaceae 藤黄科 *Calophyllum* 红厚壳属
Calophyllum blancoi Planch. et Triana 兰屿红厚壳
Trees, evergreen

18.5；14.1；21.7；18.5；162.1；0.0；863 ❙ 2713；1388；188 ❙ 214.4；863；363.7

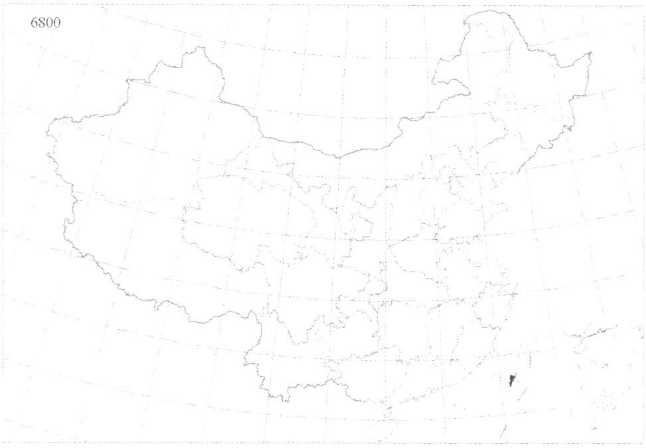

Clusiaceae 藤黄科 *Calophyllum* 红厚壳属
Calophyllum inophyllum L. 红厚壳
Trees, evergreen

21.9 (16.2~24.8)；16.2 (11.0~20.1)；26.3 (20.1~29.2)；22.0 (16.2~24.8)；203.8 (134.2~237.6)；0.0 (0.0~0.0)；1150 (773~1385) ❙ 1923 (868~3495)；817 (312~1847)；157 (41~897) ❙ 78.8 (-34.3~246.5)；1081 (773~1360)；350.6 (119.4~659.0)

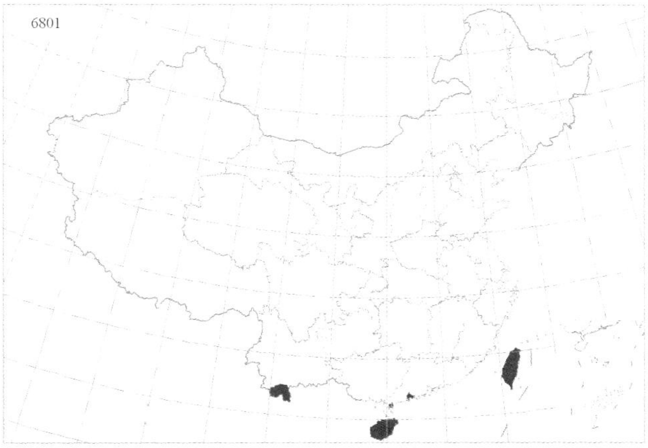

Clusiaceae 藤黄科 *Calophyllum* 红厚壳属
Calophyllum membranaceum Gardner et Champ. 薄叶红厚壳
Shrubs or small trees, evergreen

21.5 (17.6~24.8)；13.3 (7.1~20.1)；28.0 (25.0~29.8)；21.5 (17.6~24.8)；198.2 (151.6~237.6)；0.0 (0.0~0.0)；1139 (906~1385) ❙ 1632 (868~2435)；706 (312~1428)；133 (41~197) ❙ 44.8 (-34.3~99.5)；1123 (868~1360)；252.1 (95.3~620.0)

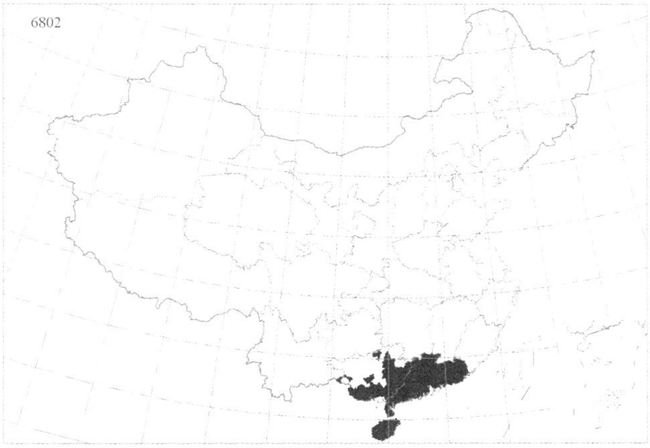

Clusiaceae 藤黄科 *Calophyllum* 红厚壳属
Calophyllum polyanthum Wall. ex Choisy 滇南红厚壳
Trees, evergreen

18.8；13.2；22.3；18.9；166.7；0.0；908 ❙ 1451；711；61 ❙ 60.1；891；510.3

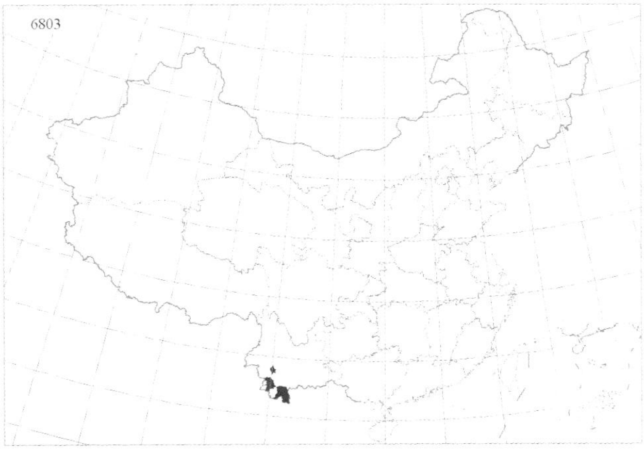

Clusiaceae 藤黄科 *Cratoxylum* 黄牛木属
Cratoxylum cochinchinense (Lour.) Blume 黄牛木
Trees, deciduous

21.7 (18.0~24.8)；13.7 (8.5~20.1)；27.9 (22.3~29.1)；21.7 (18.0~24.8)；200.6 (156.4~237.6)；0.0 (0.0~0.0)；1148 (888~1385) ❙ 1595 (868~2435)；735 (312~1428)；118 (41~194) ❙ 39.9 (-34.3~99.5)；1131 (868~1360)；273.8 (95.3~659.0)

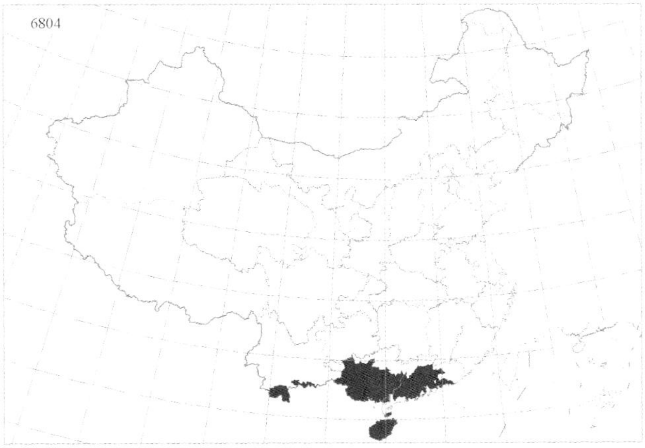

Clusiaceae 藤黄科 *Cratoxylum* 黄牛木属
Cratoxylum formosum (Jack) Dyer 越南黄牛木
Shrubs or trees, deciduous

20.6 (16.3~22.7); 13.5 (10.2~15.1); 25.5 (20.3~28.6); 20.6 (16.4~22.8); 187.6 (136.6~213.1); 0.0 (0.0~0.0); 1047 (789~1221) ❙ 1596 (1232~2435); 850 (655~1428); 84 (41~130) ❙ 52.5 (12.9~99.5); 1039 (780~1221); 394.6 (235.1~659.0)

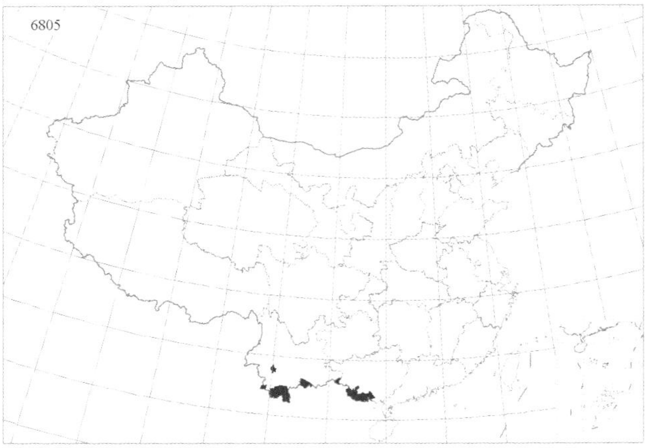

Clusiaceae 藤黄科 *Garcinia* 藤黄属
Garcinia bracteata C. Y. Wu ex Y. H. Li 大苞藤黄
Trees, evergreen

19.4 (16.1~22.2); 12.0 (9.5~15.1); 24.6 (20.3~28.3); 19.4 (16.2~22.3); 173.2 (133.9~207.5); 0.0 (0.0~0.0); 973 (785~1190) ❙ 1381 (1083~1607); 748 (594~909); 68 (42~130) ❙ 43.7 (5.8~78.9); 970 (779~1190); 391.4 (248.2~659.0)

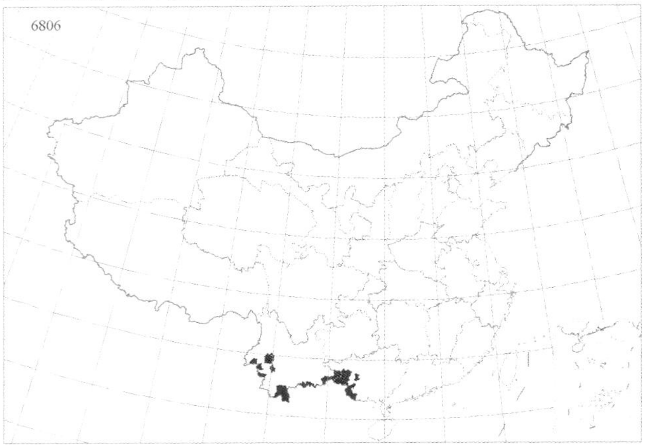

Clusiaceae 藤黄科 *Garcinia* 藤黄属
Garcinia cowa Roxb. 云树
Trees, evergreen

18.6 (16.3~21.0); 12.3 (9.6~15.1); 22.5 (20.3~25.6); 18.6 (16.4~21.0); 163.0 (136.6~191.8); 0.0 (0.0~0.0); 894 (789~1053) ❙ 1496 (1232~1688); 819 (655~959); 57 (42~77) ❙ 68.2 (44.5~112.2); 884 (780~1053); 510.8 (372.4~659.0)

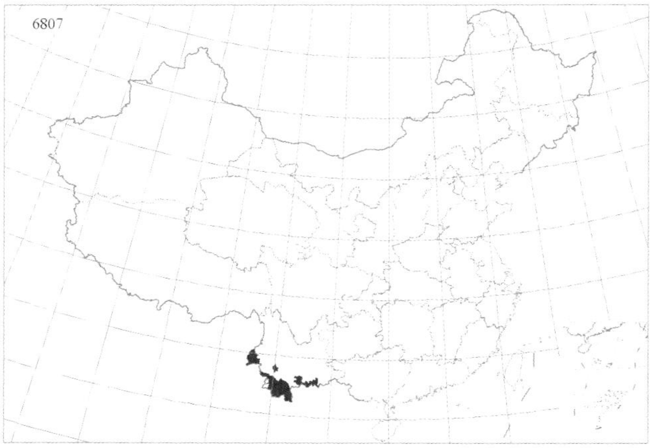

Clusiaceae 藤黄科 *Garcinia* 藤黄属
Garcinia erythrosepala Y. H. Li 红萼藤黄
Trees, evergreen

16.4; 9.6; 20.6; 16.4; 137.3; 0.0; 795 ❙ 1688; 959; 62 ❙ 112.2; 795; 627.0

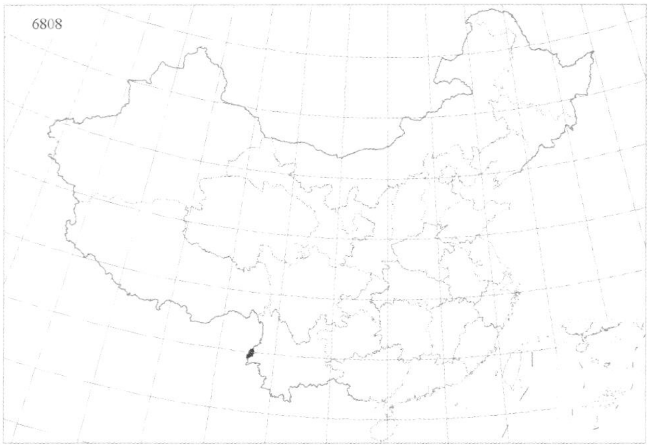

Clusiaceae 藤黄科 *Garcinia* 藤黄属
Garcinia esculenta Y. H. Li 山木瓜
Trees, evergreen

15.6; 8.6; 20.2; 15.6; 129.1; 2.2; 790 ❙ 1476; 849; 50 ❙ 87.4; 785; 512.8

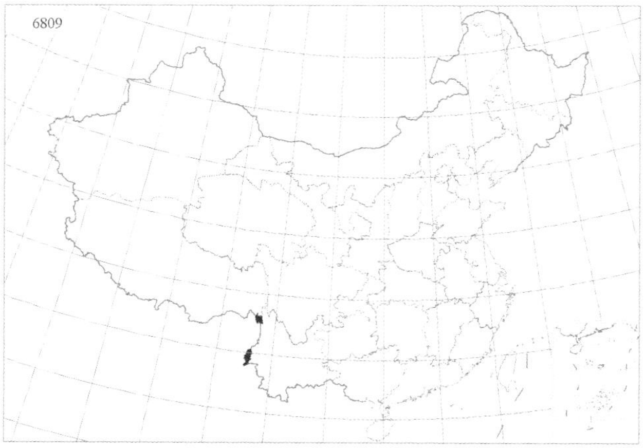

Clusiaceae 藤黄科 *Garcinia* 藤黄属
Garcinia kwangsiensis Merr. ex F. N. Wei 广西藤黄
Small trees, evergreen

21.9 (21.3~22.7); 13.9 (13.2~15.0); 27.9 (27.3~28.6); 22.0 (21.3~22.8); 203.6 (195.9~213.1); 0.0 (0.0~0.0); 1162 (1111~1221) ❙ 1876 (1302~2435); 1047 (670~1428); 116 (86~130) ❙ 61.1 (12.9~99.5); 1162 (1111~1221); 289.6 (235.1~318.8)

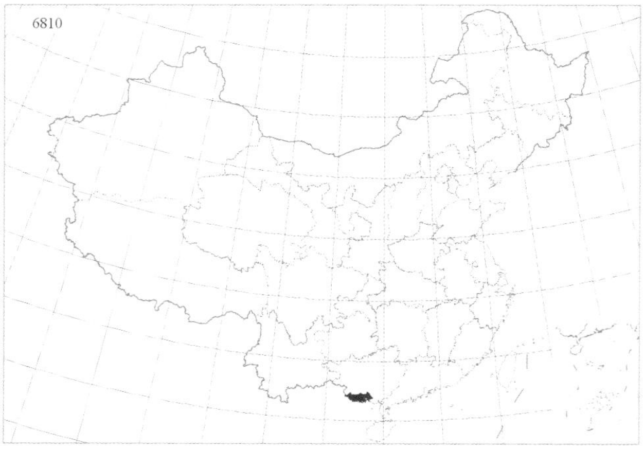

Clusiaceae 藤黄科 *Garcinia* 藤黄属
Garcinia lancilimba C. Y. Wu ex Y. H. Li 长裂藤黄
Small trees, evergreen

19.8; 14.3; 23.0; 19.9; 178.2; 0.0; 950 ❙ 1482; 671; 62 ❙ 56.3; 920; 542.0

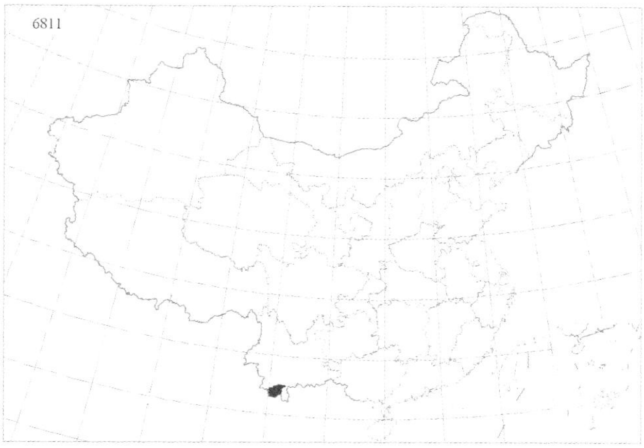

Clusiaceae 藤黄科 *Garcinia* 藤黄属
Garcinia linii C. E. Chang 兰屿福木
Small trees, evergreen

13.5; 14.1; 21.7; 18.5; 162.1; 0.0; 863 ❙ 2713; 1388; 188 ❙ 214.4; 833; 363.7

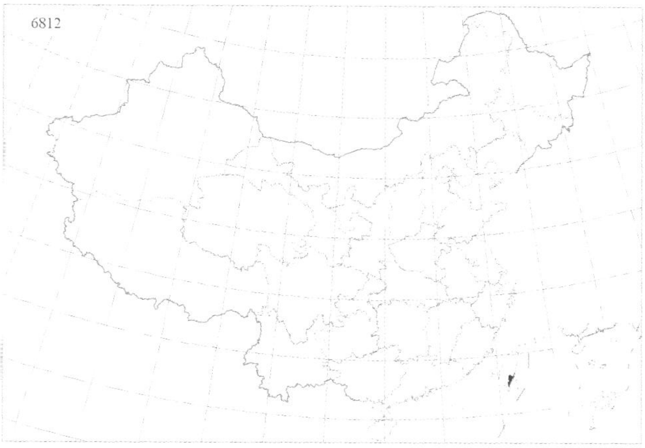

Clusiaceae 藤黄科 *Garcinia* 藤黄属
Garcinia mangostana L. 莽吉柿
Small trees, evergreen

20.2 (16.2~24.1); 14.5 (11.0~18.9); 25.0 (20.1~28.2); 20.2 (16.2~24.1); 182.9 (134.2~229.7); 0.0 (0.0~0.0); 1019 (773~1322) ❙ 2402 (1498~3495); 1106 (750~1847); 255 (47~897) ❙ 146.2 (25.2~246.5); 983 (773~1144); 245.0 (119.4~363.7)

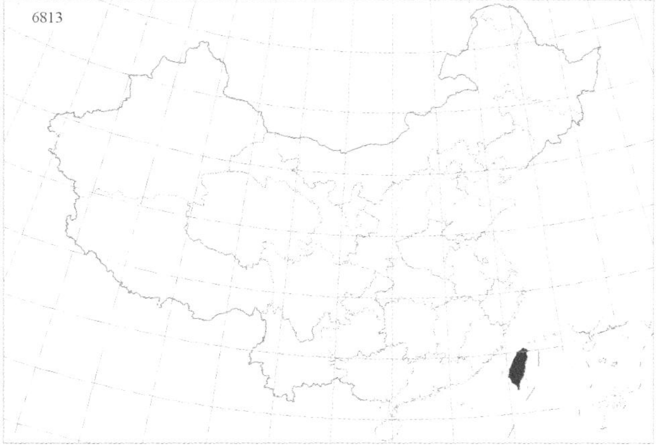

Clusiaceae 藤黄科 *Garcinia* 藤黄属
Garcinia multiflora Champ. ex Benth. 木竹子
Trees, evergreen

19.8 (14.3~24.8); 11.0 (3.3~20.1); 27.2 (20.1~30.1); 19.9 (14.3~24.8); 178.6 (112.6~237.6); 0.0 (0.0~2.3); 1040 (756~1385) ❙ 1624 (868~3495); 683 (312~1847); 151 (41~897) ❙ 58.9 (-34.3~246.5); 1026 (756~1360); 256.5 (95.3~659.0)

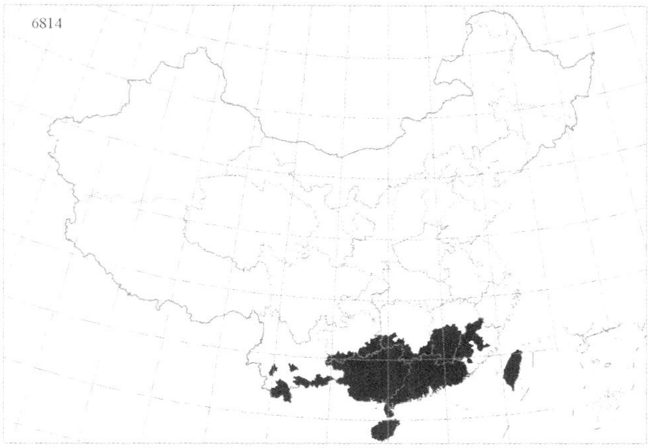

Clusiaceae 藤黄科 *Garcinia* 藤黄属
Garcinia nujiangensis C. Y. Wu et Y. H. Li 怒江藤黄
Trees, evergreen

15.1 (8.6~19.6); 8.0 (1.3~12.8); 20.1 (15.0~23.5); 15.2 (8.7~19.6); 123.7 (52.6~175.7); 1.8 (0.0~8.8); 775 (574~953) ❙ 1523 (1041~1736); 882 (585~1007); 51 (44~62) ❙ 98.2 (61.2~145.8); 771 (574~932); 493.4 (352.3~627.0)

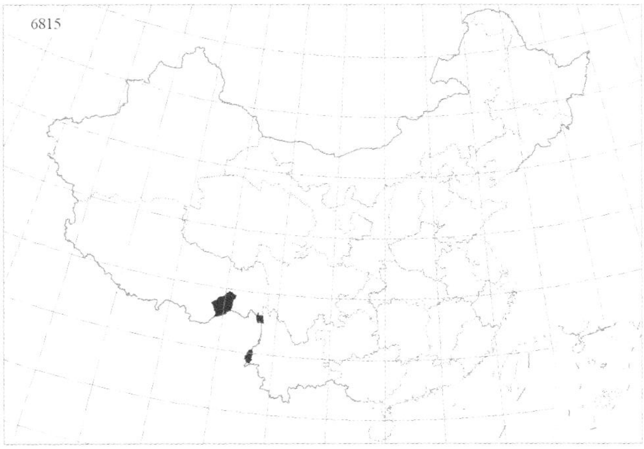

Clusiaceae 藤黄科 *Garcinia* 藤黄属
Garcinia oblongifolia Champ. ex Benth. 岭南山竹子
Shrubs or trees, evergreen

21.2 (14.6~24.8); 13.0 (3.8~20.1); 27.7 (23.4~29.0); 21.2 (14.7~24.8); 194.9 (117.7~237.6); 0.0 (0.0~1.5); 1126 (779~1385) ❙ 1609 (868~2435); 711 (312~1428); 123 (41~197) ❙ 44.9 (-34.3~101.8); 1103 (779~1360); 274.1 (95.3~620.0)

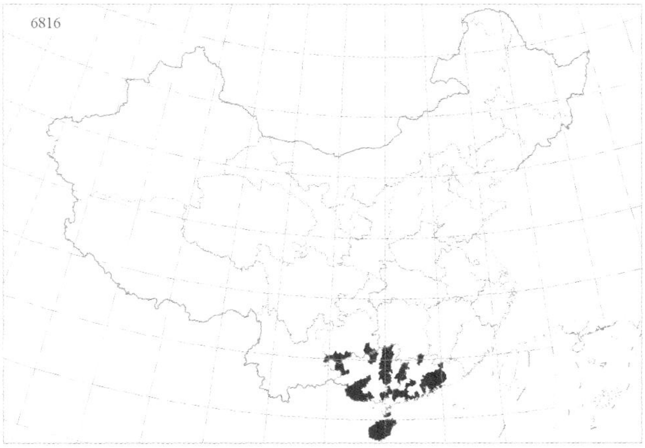

Clusiaceae 藤黄科 *Garcinia* 藤黄属
Garcinia oligantha Merr. 单花山竹子
Shrubs, evergreen

24.0 (22.2~24.8); 18.4 (17.2~20.1); 28.0 (25.7~28.8); 24.0 (22.3~24.8); 228.3 (207.2~237.6); 0.0 (0.0~0.0); 1310 (1127~1385) ❙ 1501 (868~1854); 534 (312~712); 74 (41~115) ❙ 15.0 (-34.3~47.8); 1191 (868~1360); 440.3 (248.2~620.0)

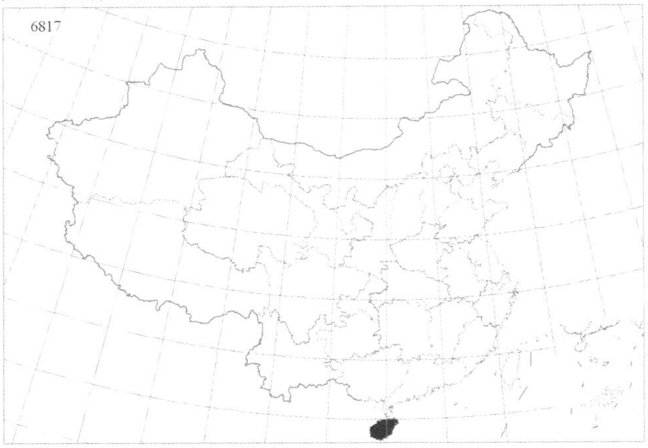

Clusiaceae 藤黄科 *Garcinia* 藤黄属
Garcinia paucinervis Chun ex F. C. How 金丝李
Trees, evergreen

20.3 (17.2~22.2); 11.6 (8.8~13.9); 27.0 (22.7~28.6); 20.4 (17.3~22.3); 184.7 (147.0~207.5); 0.0 (0.0~0.0); 1058 (838~1190) ❙ 1433 (1172~1703); 745 (651~850); 102 (43~194) ❙ 36.2 (12.9~65.6); 1058 (838~1190); 286.0 (204.5~411.1)

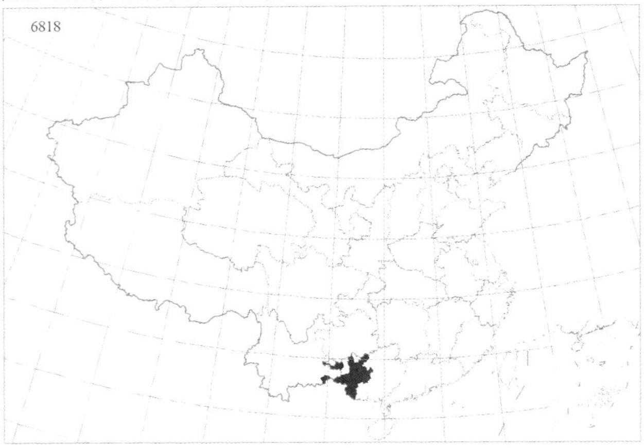

Clusiaceae 藤黄科 *Garcinia* 藤黄属
Garcinia pedunculata Roxb. ex Buch.-Ham. 大果藤黄
Trees, evergreen

16.9 (13.3~19.6); 10.1 (5.4~13.0); 21.5 (18.8~23.5); 17.0 (13.4~19.6); 143.5 (100.4~175.7); 0.0 (0.0~0.0); 837 (706~953) ❙ 1498 (1042~1736); 869 (592~1007); 50 (42~62) ❙ 81.3 (47.6~145.8); 830 (706~932); 481.2 (372.3~627.0)

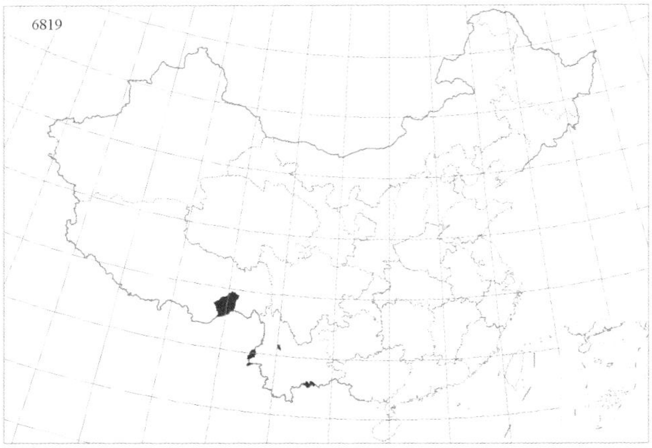

Clusiaceae 藤黄科 *Garcinia* 藤黄属
Garcinia subelliptica Merr. 菲岛福木
Trees, evergreen

20.8 (18.5~24.1); 15.7 (13.5~18.9); 24.8 (21.7~28.0); 20.8 (18.5~24.1); 189.9 (162.1~229.7); 0.0 (0.0~0.0); 1041 (863~1322) ❙ 2642 (1751~2993); 1384 (924~1847); 203 (47~547) ❙ 163.8 (32.5~214.4); 996 (863~1123); 265.7 (154.2~363.7)

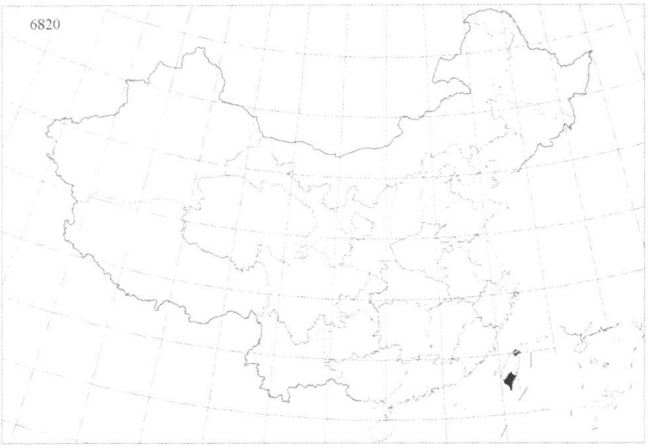

Clusiaceae 藤黄科 *Garcinia* 藤黄属
Garcinia subfalcata Y. H. Li et F. N. Wei 尖叶藤黄
Trees, evergreen

21.7 (21.0~22.7); 13.2 (11.8~15.0); 28.1 (27.3~28.6); 21.7 (21.1~22.8); 200.5 (192.6~213.1); 0.0 (0.0~0.0); 1149 (1111~1221) ❙ 1780 (1302~2435); 943 (670~1428); 141 (86~194) ❙ 54.6 (12.9~99.5); 1149 (1111~1221); 261.5 (206.0~318.8)

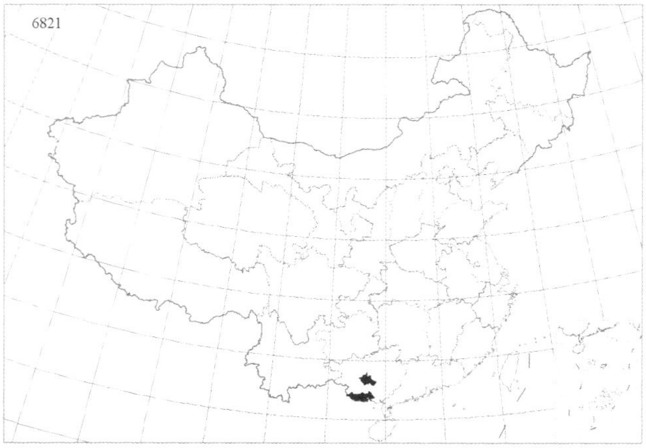

Clusiaceae 藤黄科 *Garcinia* 藤黄属
Garcinia tetralata C. Y. Wu ex Y. H. Li 双籽藤黄
Trees, evergreen

19.1 (17.9~20.5); 13.3 (11.9~15.1); 22.7 (21.6~23.7); 19.2 (18.0~20.5); 170.0 (155.8~185.9); 0.0 (0.0~0.0); 918 (854~992) ❙ 1484 (1366~1607); 771 (655~911); 59 (42~77) ❙ 62.0 (47.6~74.2); 903 (833~986); 560.5 (505.1~659.0)

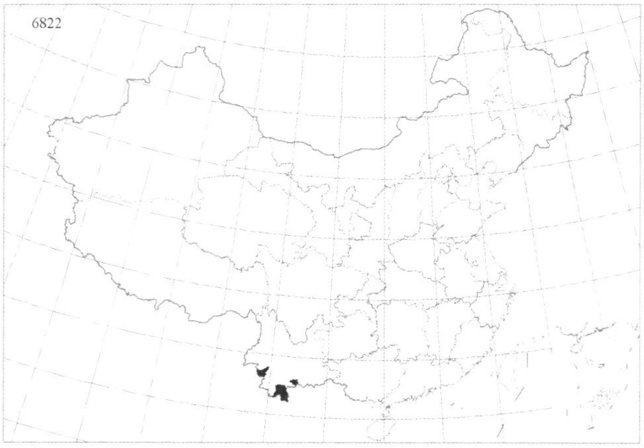

Clusiaceae 藤黄科 *Garcinia* 藤黄属
Garcinia xanthochymus Hook. f. ex T. Anderson 大叶藤黄
Trees, evergreen

18.5 (16.3~21.0); 12.0 (9.1~15.1); 22.9 (20.3~25.6); 18.5 (16.4~21.0); 162.1 (136.4~191.8); 0.0 (0.0~0.0); 896 (789~1053) ❙ 1354 (1049~1607); 723 (598~835); 56 (44~77) ❙ 51.1 (26.1~65.6); 886 (780~1053); 445.6 (336.3~659.0)

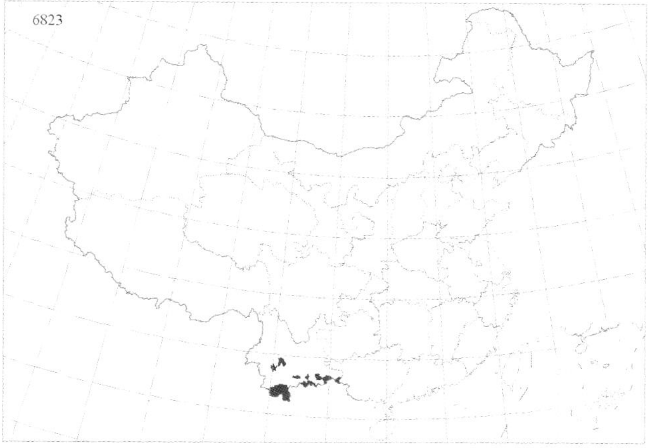

Clusiaceae 藤黄科 *Garcinia* 藤黄属
Garcinia xishuanbannaensis Y. H. Li 版纳藤黄
Trees, evergreen

19.6 (18.6~20.5); 14.0 (12.7~15.1); 23.1 (22.3~23.7); 19.6 (18.7~20.5); 175.7 (163.9~185.9); 0.0 (0.0~0.0); 942 (888~992) ❙ 1478 (1276~1607); 739 (655~911); 63 (51~77) ❙ 57.2 (36.8~74.2); 927 (880~986); 549.3 (452.0~659.0)

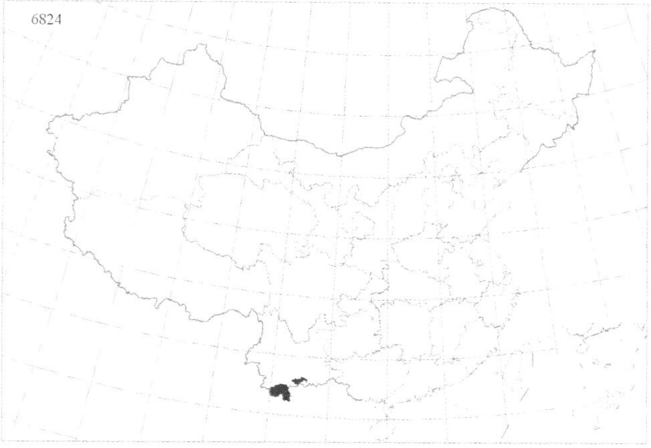

Clusiaceae 藤黄科 *Garcinia* 藤黄属
Garcinia yunnanensis H. H. Hu 云南藤黄
Trees

17.9; 11.9; 21.6; 18.0; 155.8; 0.0; 854 ❙ 1435; 803; 42 ❙ 68.0; 833; 505.1

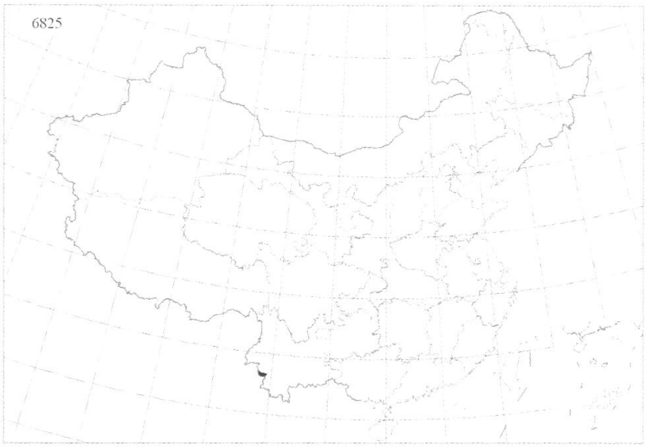

Clusiaceae 藤黄科 *Hypericum* 金丝桃属
Hypericum acmosepalum N. Robson 尖萼金丝桃
Shrubs

14.8 (2.9~21.3); 6.2 (-6.4~14.2); 22.0 (11.2~27.9); 14.8 (4.4~21.3); 119.9 (20.5~196.0); 2.2 (0.0~44.6); 777 (428~1123) ❙ 1099 (741~1547); 578 (411~911); 50 (12~106) ❙ 42.8 (-14.2~21.7); 764 (428~1121); 264.6 (142.9~551.6)

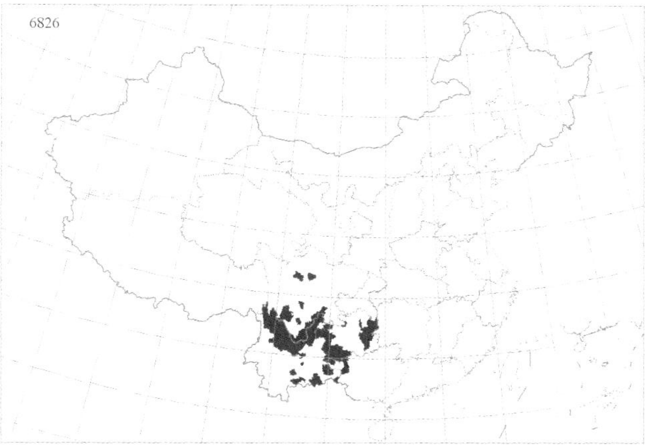

Clusiaceae 藤黄科 *Hypericum* 金丝桃属
Hypericum addingtonii N. Robson 碟花金丝桃
Shrubs

12.3; 5.5; 17.8; 12.3; 91.1; 2.9; 673 ❙ 1075; 595; 51 ❙ 61.6; 670; 366.7

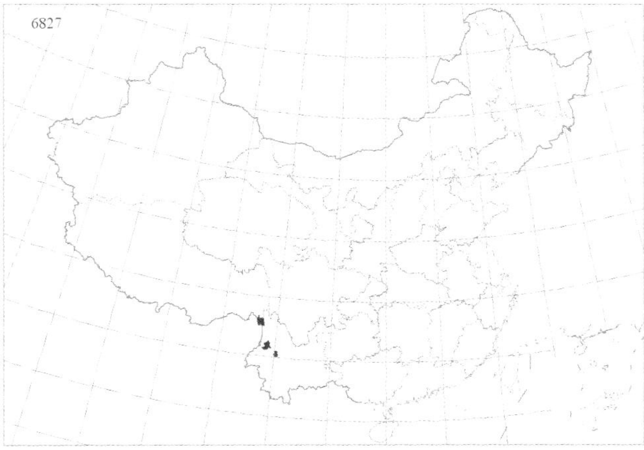

Clusiaceae 藤黄科 *Hypericum* 金丝桃属
Hypericum ascyron subsp. gebleri (Ledeb.) N. Robson 短柱黄海棠
Herbs or subshrubs

2.9 (-0.2~8.8); -18.9 (-23.8~-7.7); 21.6 (13.4~24.3); 8.1 (6.4~10.4); 62.3 (45.8~87.0); 86.9 (40.0~112.6); 586 (501~678) ❙ 586 (416~966); 365 (291~616); 18 (7~35) ❙ 0.4 (-33.8~55.1); 548 (416~678); 249.3 (121.2~333.3)

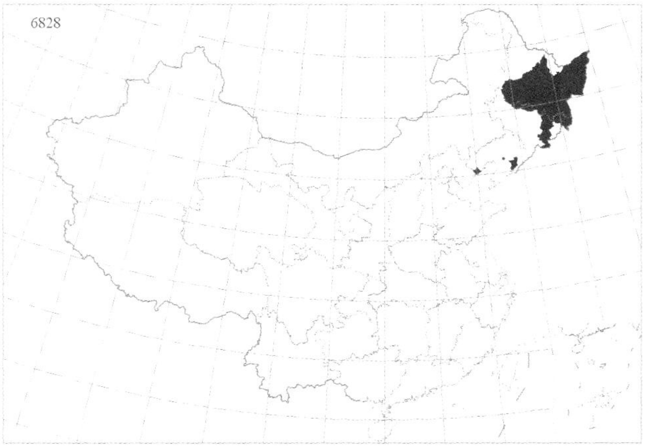

Clusiaceae 藤黄科 *Hypericum* 金丝桃属
Hypericum augustinii N. Robson 无柄金丝桃
Shrubs

13.9 (-1.1~20.5); 5.6 (-9.1~14.9); 20.6 (6.2~26.0); 14.2 (2.0~20.5); 114.1 (2.5~185.9); 6.9 (0.0~75.6); 748 (334~992) ❙ 1124 (321~1464); 598 (193~720); 51 (31~73) ❙ 50.1 (-4.0~100.7); 743 (321~966); 248.3 (79.3~570.1)

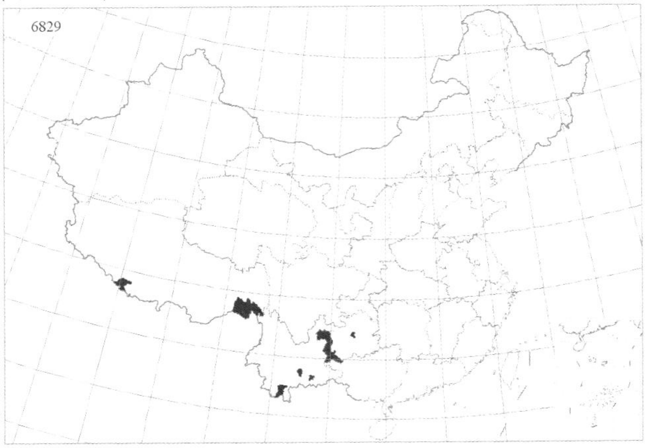

Clusiaceae 藤黄科 *Hypericum* 金丝桃属
Hypericum beanii N. Robson 栽秧花
Shrubs

15.0 (3.0~19.1); 5.8 (-5.6~11.0); 22.9 (10.5~27.4); 15.1 (4.3~19.1); 123.2 (20.1~169.2); 2.9 (0.0~43.9); 790 (423~966) ❙ 1194 (628~1454); 583 (410~792); 68 (11~184) ❙ 52.1 (26.1~99.9); 788 (423~966); 248.4 (148.9~487.2)

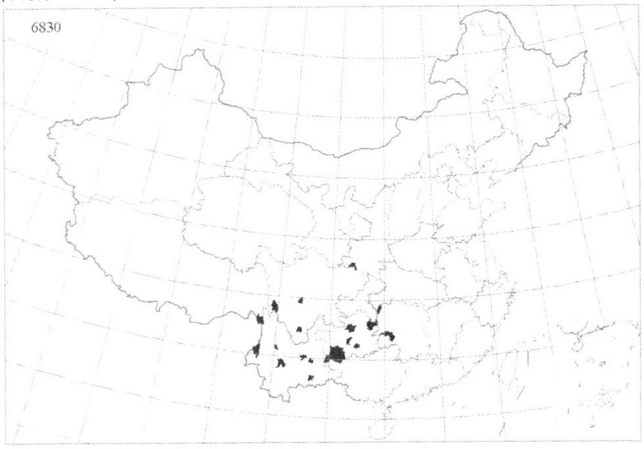

Clusiaceae 藤黄科 *Hypericum* 金丝桃属
Hypericum bellum H. L. Li 美丽金丝桃
Shrubs

6.1 (1.8~11.9); -2.0 (-7.1~4.3); 13.2 (9.5~20.0); 6.9 (3.8~12.0); 40.4 (15.5~87.2); 26.3 (0.7~53.3); 508 (404~671) ❙ 850 (527~1243); 494 (317~696); 26 (9~58) ❙ 67.4 (18.2~125.9); 508 (404~671); 271.8 (168.2~384.7)

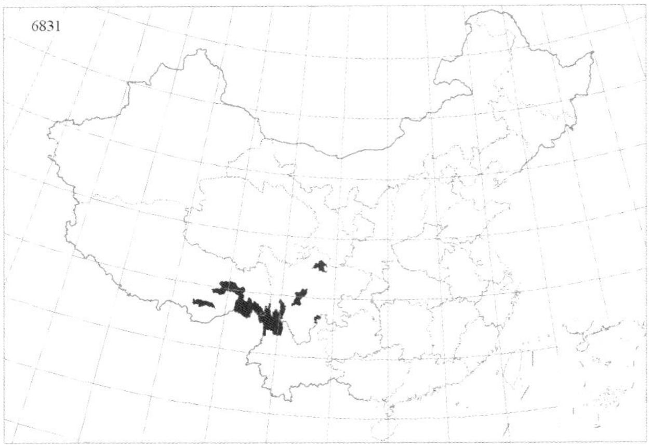

Clusiaceae 藤黄科 *Hypericum* 金丝桃属
Hypericum choisyanum Wall. ex N. Robson 多蕊金丝桃
Shrubs

7.4 (-1.8~17.3); -0.2 (-10.5~11.0); 13.9 (6.1~22.4); 8.7 (1.8~17.3); 61.0 (1.9~148.0); 32.0 (0.0~82.8); 554 (329~833) ❙ 792 (311~1504); 468 (193~860); 32 (10~55) ❙ 34.4 (-6.3~81.5); 547 (311~833); 247.2 (61.6~413.5)

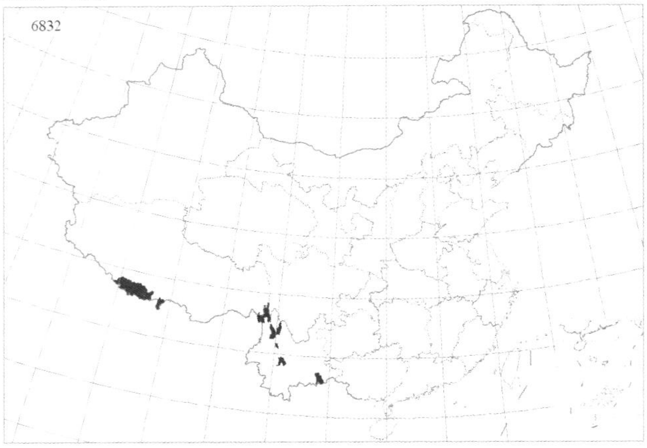

Clusiaceae 藤黄科 *Hypericum* 金丝桃属
Hypericum cohaerens N. Robson 连柱金丝桃
Shrubs

14.5; 3.9; 24.5; 14.5; 115.5; 1.2; 780 ❙ 1156; 483; 75 ❙ 47.6; 780; 211.7

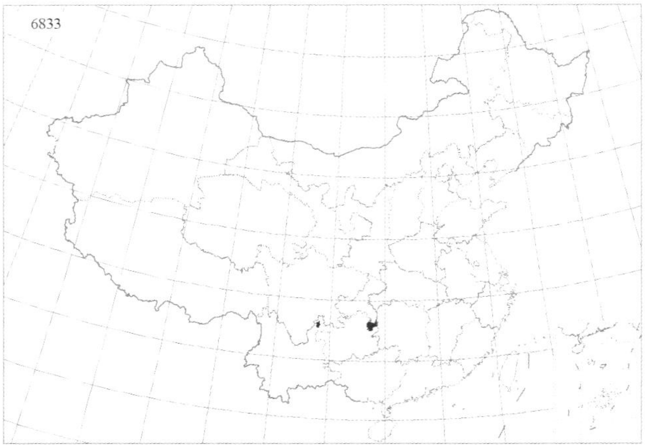

Clusiaceae 藤黄科 *Hypericum* 金丝桃属
Hypericum curvisepalum N. Robson 弯萼金丝桃
Shrubs

14.8 (5.4~19.1); 5.9 (-2.1~10.1); 22.4 (12.3~27.7); 14.8 (5.8~19.1); 119.1 (31.5~169.2); 1.5 (0.0~25.6); 782 (481~966) ❙ 1097 (741~1334); 591 (426~720); 49 (20~76) ❙ 42.4 (16.1~101.6); 779 (481~966); 236.5 (142.5~382.6)

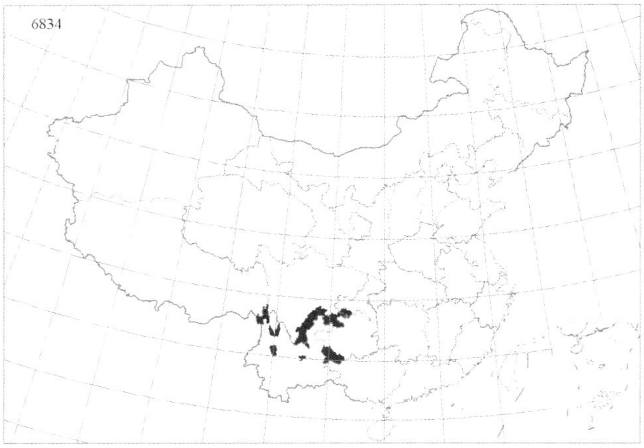

Clusiaceae 藤黄科 *Hypericum* 金丝桃属

Hypericum elatoides R. Keller 岐山金丝桃

Subshrubs

12.5; -0.8; 25.0; 12.5; 102.9; 13.2; 744 ❙ 675; 308; 19 ❙ -9.3; 675; 211.8

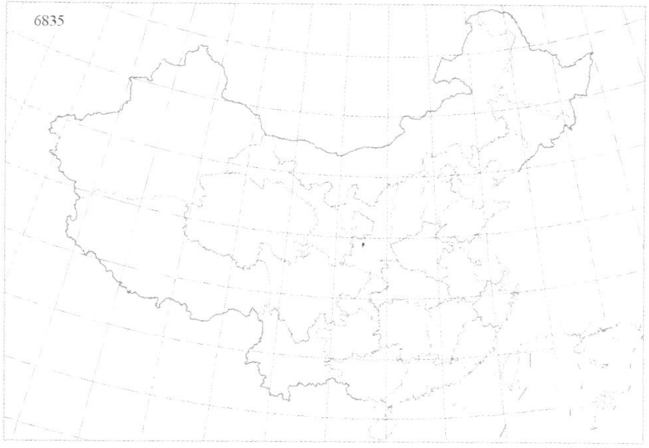

Clusiaceae 藤黄科 *Hypericum* 金丝桃属

Hypericum elongatum Ledeb. 延伸金丝桃

Herbs or subshrubs

4.6 (-2.8~10.2); -14.6 (-20.9~-8.1); 20.5 (10.9~27.5); 8.4 (3.5~12.1); 63.0 (16.3~107.6); 70.3 (36.6~109.2); 603 (390~813) ❙ 196 (45~450); 90 (27~211); 18 (3~51) ❙ -64.9 (-94.0~3.9); 195 (45~406); 87.7 (7.6~297.3)

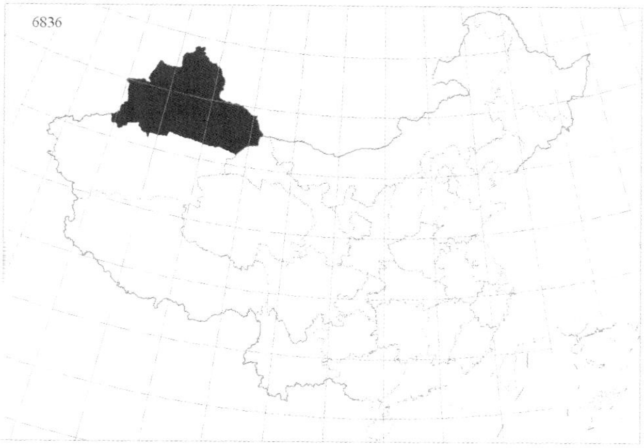

Clusiaceae 藤黄科 *Hypericum* 金丝桃属

Hypericum formosanum Maxim. 台湾金丝桃

Shrubs

19.9; 13.5; 25.9; 19.9; 179.1; 0.0; 987 ❙ 2993; 924; 547 ❙ 203.4; 987; 261.4

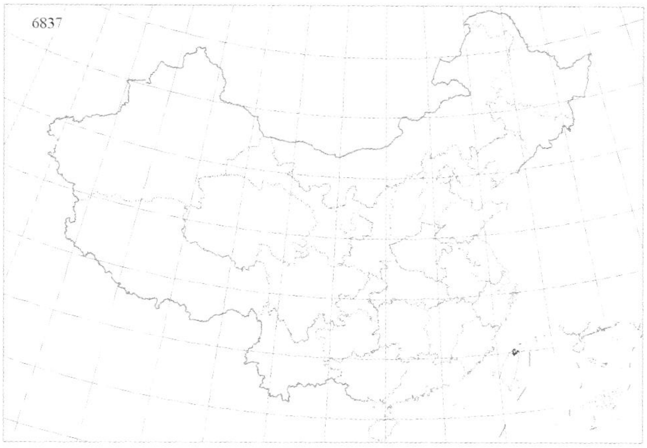

Clusiaceae 藤黄科 *Hypericum* 金丝桃属

Hypericum forrestii (Chitt.) N. Robson 川滇金丝桃

Shrubs

9.0 (2.2~17.7); 0.8 (-6.6~7.5); 16.1 (9.5~27.8); 9.4 (3.8~17.7); 64.9 (15.5~152.2); 16.5 (0.0~49.1); 596 (411~937) ❙ 977 (579~1454); 557 (365~792); 32 (10~70) ❙ 68.8 (12.1~125.9); 594 (411~937); 278.2 (126.1~487.2)

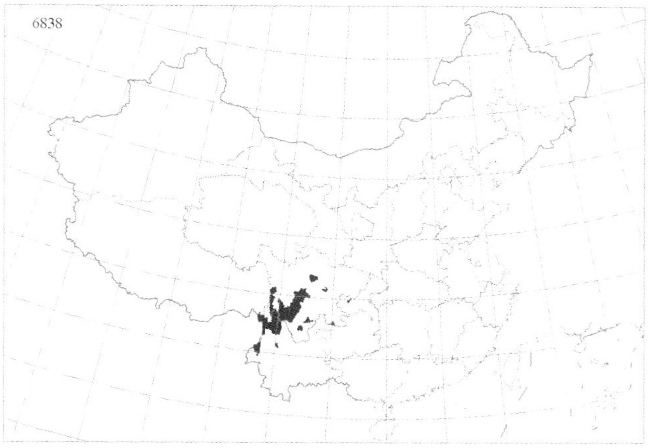

Clusiaceae 藤黄科 *Hypericum* 金丝桃属

Hypericum geminiflorum Hemsl. 双花金丝桃

Shrubs

19.4 (16.2~24.1); 14.1 (11.0~18.9); 23.7 (20.1~28.0); 19.5 (16.2~24.1); 173.7 (134.2~229.7); 0.0 (0.0~0.0); 961 (773~1322) ❙ 2537 (1751~2993); 1228 (924~1847); 222 (47~547) ❙ 174.2 (32.5~246.5); 936 (773~1123); 261.7 (154.2~363.7)

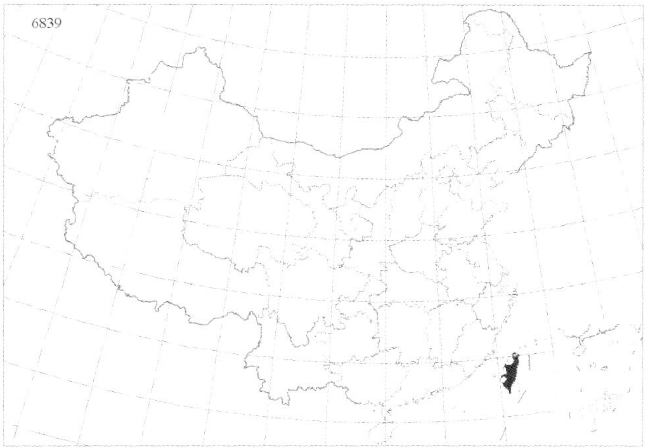

Clusiaceae 藤黄科 *Hypericum* 金丝桃属

Hypericum henryi H. Lév. et Vaniot 西南金丝桃

Shrubs

16.3 (11.1~21.0); 8.8 (1.8~14.2); 22.0 (18.8~25.6); 16.3 (11.1~21.0); 136.0 (79.8~191.8); 0.3 (0.0~6.9); 814 (648~1053) ❙ 1169 (911~1522); 655 (534~835); 46 (20~68) ❙ 43.7 (5.5~66.0); 794 (648~1053); 328.1 (160.8~551.6)

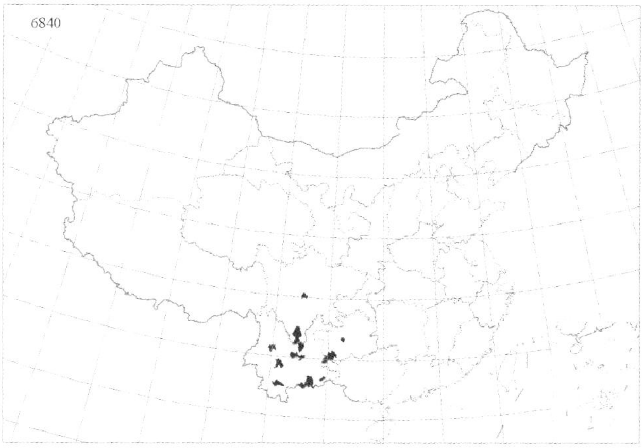

Clusiaceae 藤黄科 *Hypericum* 金丝桃属
Hypericum henryi subsp. hancockii N. Robson 蒙自金丝桃
Shrubs

18.4 (16.9~19.8); 11.6 (10.3~13.0); 23.2 (21.4~26.4); 18.5 (16.9~19.8); 161.5 (142.7~177.9); 0.0 (0.0~0.0); 897 (813~1010) ❙ 1341 (1178~1547); 764 (646~911); 54 (43~63) ❙ 50.4 (16.9~74.2); 897 (813~1010); 457.4 (354.5~551.3)

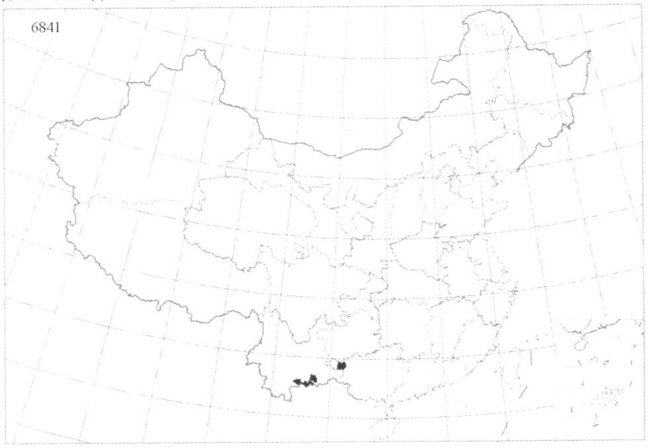

Clusiaceae 藤黄科 *Hypericum* 金丝桃属
Hypericum henryi subsp. uraloides (Rehder) N. Robson 岷江金丝桃
Shrubs

18.1; 11.7; 22.4; 18.1; 157.4; 0.0; 867 ❙ 1338; 767; 56 ❙ 54.1; 859; 446.

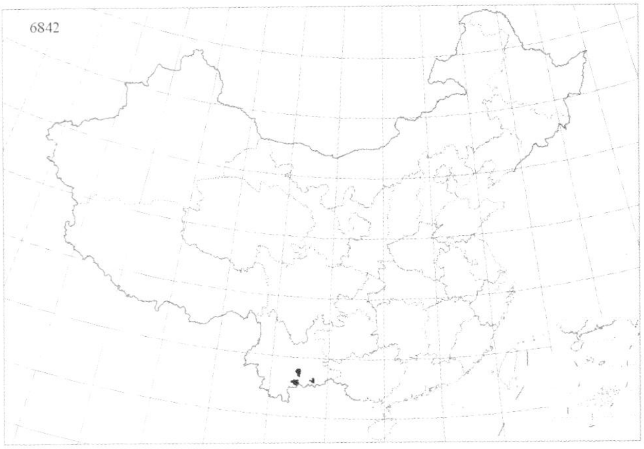

Clusiaceae 藤黄科 *Hypericum* 金丝桃属
Hypericum hookerianum Wight et Arn. 短柱金丝桃
Shrubs

13.9 (1.8~20.7); 4.8 (-7.2~13.0); 21.7 (9.9~29.0); 14.1 (3.8~20.8); 114.2 (15.4~189.3); 7.2 (0.0~53.3); 766 (404~1108) ❙ 1119 (473~1486); 592 (287~835); 50 (7~176) ❙ 47.7 (5.0~128.6); 766 (404~1108); 256.6 (133.8~506.2)

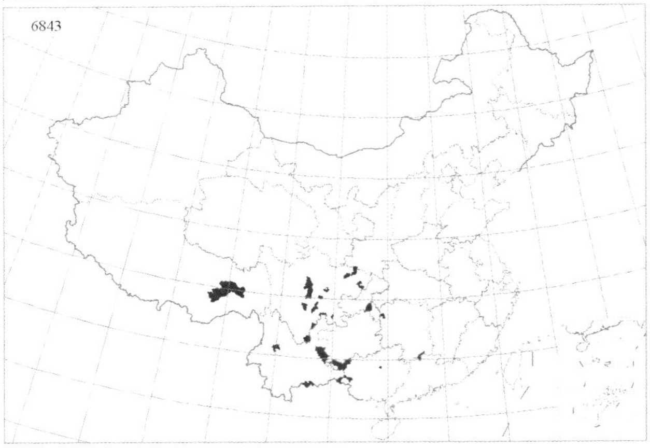

Clusiaceae 藤黄科 *Hypericum* 金丝桃属
Hypericum kingdonii N. Robson 察隅遍地金
Herbs or subshrubs

5.3; -2.4; 12.2; 5.7; 30.5; 26.2; 479 ❙ 961; 558; 31 ❙ 100.7; 479; 234.2

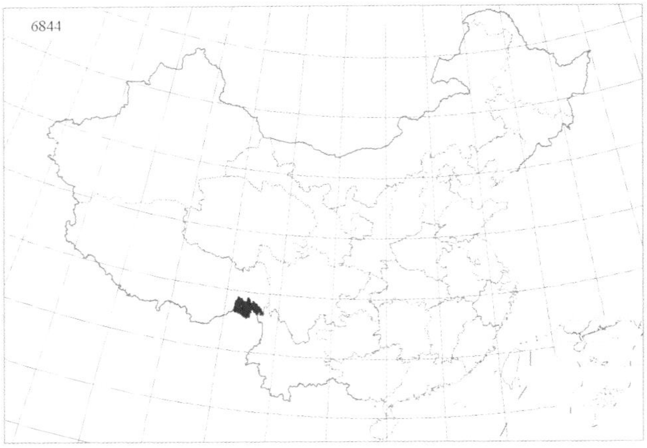

Clusiaceae 藤黄科 *Hypericum* 金丝桃属
Hypericum kouytchense H. Lév. 贵州金丝桃
Shrubs

15.5 (11.7~19.1); 5.5 (2.7~10.1); 24.5 (18.8~27.7); 15.6 (11.7~19.1); 127.2 (82.6~169.2); 0.4 (0.0~3.8); 813 (658~969) ❙ 1195 (916~1381); 550 (444~723); 71 (31~107) ❙ 47.4 (16.1~71.9); 813 (658~969); 201.6 (142.5~316.1)

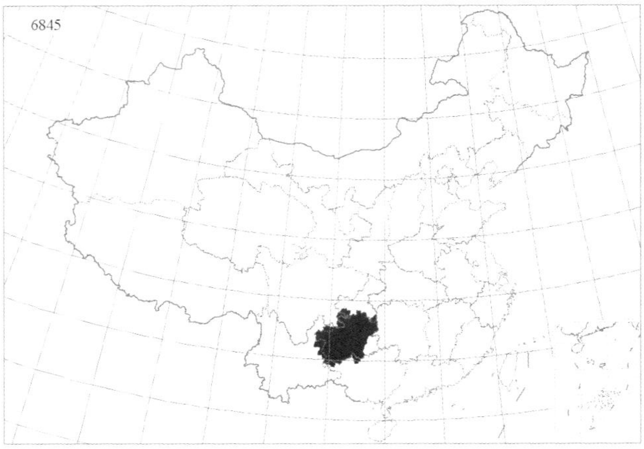

Clusiaceae 藤黄科 *Hypericum* 金丝桃属
Hypericum lagarocladum N. Robson 纤枝金丝桃
Shrubs

15.2 (6.2~19.1); 5.6 (-2.4~10.5); 23.5 (13.3~27.2); 15.2 (6.7~19.1); 124.4 (39.4~169.2); 2.1 (0.0~24.1); 800 (509~969) ❙ 1201 (862~1440); 573 (450~816); 71 (12~161) ❙ 51.7 (5.5~92.9); 792 (509~969); 238.1 (160.8~330.0)

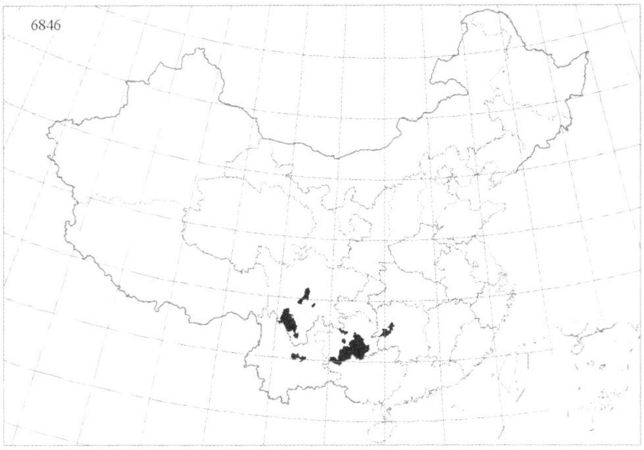

Clusiaceae 藤黄科 *Hypericum* 金丝桃属
Hypericum lancasteri N. Robson 展萼金丝桃
Shrubs

14.7 (13.6~17.3); 8.0 (6.2~11.0); 19.9 (18.8~21.6); 14.8 (13.7~17.3); 117.2 (104.1~148.0); 0.0 (0.0~0.0); 744 (706~832) ❙ 1021 (963~1049); 590 (570~598); 42 (26~54) ❙ 37.7 (26.1~47.6); 729 (696~783); 334.0 (271.3~382.6)

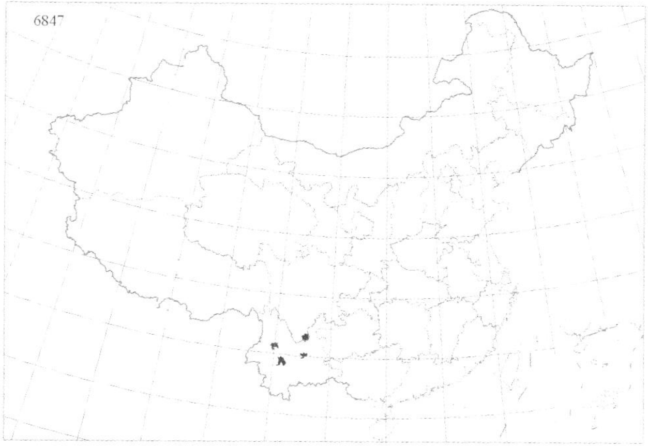

Clusiaceae 藤黄科 *Hypericum* 金丝桃属
Hypericum latisepalum (N. Robson) N. Robson 宽萼金丝桃
Shrubs

11.6; 4.6; 17.3; 11.6; 83.0; 4.4; 652 ❙ 1043; 577; 49 ❙ 62.3; 652; 367.4

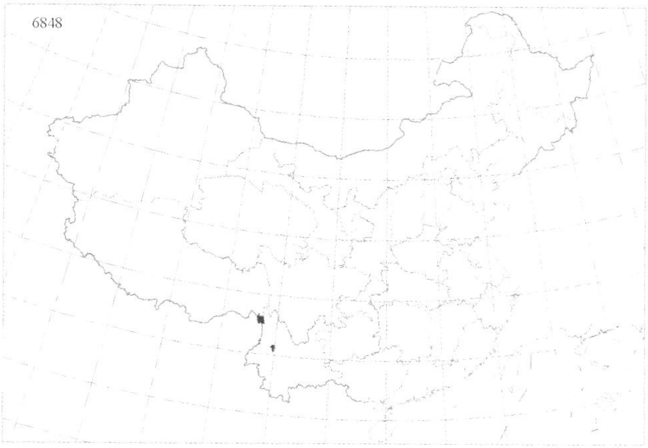

Clusiaceae 藤黄科 *Hypericum* 金丝桃属
Hypericum longistylum Oliv. 长柱金丝桃
Shrubs

14.7 (9.3~17.0); 2.5 (-2.5~4.8); 26.4 (19.8~29.2); 14.8 (9.5~17.0); 121.6 (69.8~144.9); 4.3 (0.2~20.6); 825 (603~941) ❙ 1218 (711~1770); 503 (345~666); 107 (15~223) ❙ 48.1 (-5.3~112.2); 822 (603~932); 214.6 (133.5~319.9)

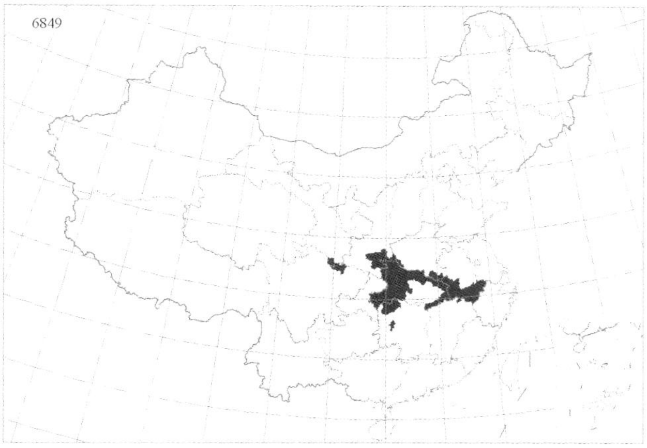

Clusiaceae 藤黄科 *Hypericum* 金丝桃属
Hypericum maclarenii N. Robson 康定金丝桃
Shrubs

4.1; -4.9; 12.2; 5.2; 27.6; 37.9; 461 ❙ 977; 541; 24 ❙ 113.5; 461; 224.6

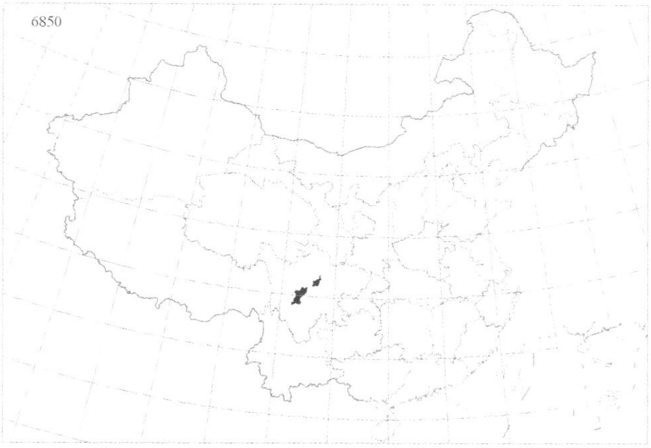

Clusiaceae 藤黄科 *Hypericum* 金丝桃属
Hypericum monogynum L. 金丝桃
Shrubs, semievergreen

16.0 (2.2~22.1); 4.5 (-6.6~14.6); 26.8 (9.9~30.1); 16.1 (3.8~22.1); 135.1 (15.5~205.6); 2.3 (0.0~49.1); 871 (413~1182) ❙ 1318 (626~3495); 513 (316~959); 132 (19~897) ❙ 51.2 (-22.2~235.6); 855 (413~1182); 204.5 (110.1~354.5)

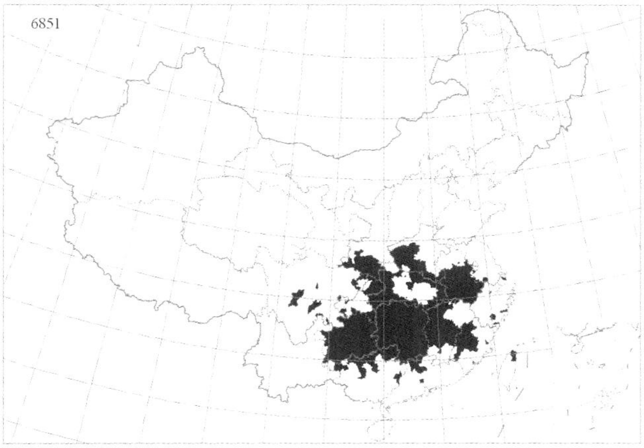

Clusiaceae 藤黄科 *Hypericum* 金丝桃属
Hypericum nakamurae (Masam.) N. Robson 清水金丝桃
Shrubs

16.4; 11.5; 20.3; 16.5; 137.8; 0.0; 783 ❙ 2499; 947; 303 ❙ 219.3; 783; 303.6

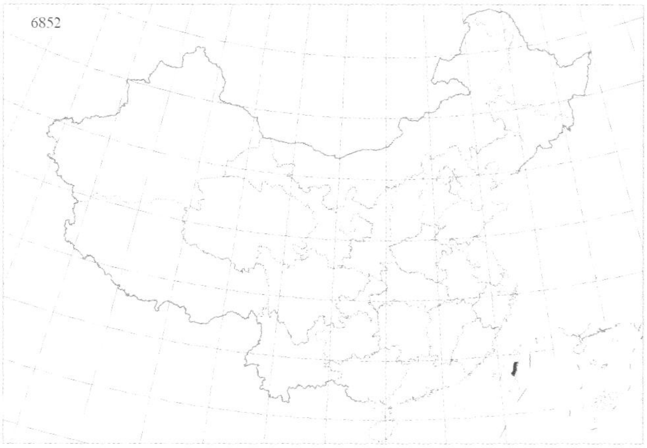

Clusiaceae 藤黄科 *Hypericum* 金丝桃属
Hypericum patulum Thunb. 金丝梅
Shrubs, semievergreen or green

15.1 (2.0~22.8); 4.4 (-7.2~16.7); 25.0 (9.5~29.4); 15.2 (3.8~22.8); 125.9 (15.4~213.7); 3.6 (0.0~51.4); 824 (411~1209) ❙ 1258 (626~2445); 542 (316~1412); 100 (11~259) ❙ 53.3 (-22.2~147.0); 816 (411~1207); 221.0 (108.2~506.2)

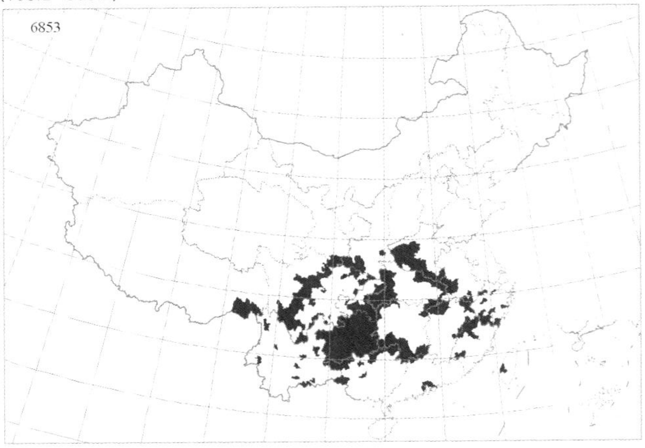

Clusiaceae 藤黄科 *Hypericum* 金丝桃属
Hypericum prattii Hemsl. 大叶金丝桃
Shrubs

13.4 (2.2~17.7); 3.0 (-6.6~7.4); 23.0 (9.9~27.8); 13.7 (3.8~17.7); 110.6 (15.5~152.2); 9.5 (0.0~49.1); 766 (413~937) ❙ 1091 (932~1154); 536 (458~640); 55 (19~81) ❙ 50.3 (18.0~125.9); 766 (413~937); 199.7 (126.1~258.0)

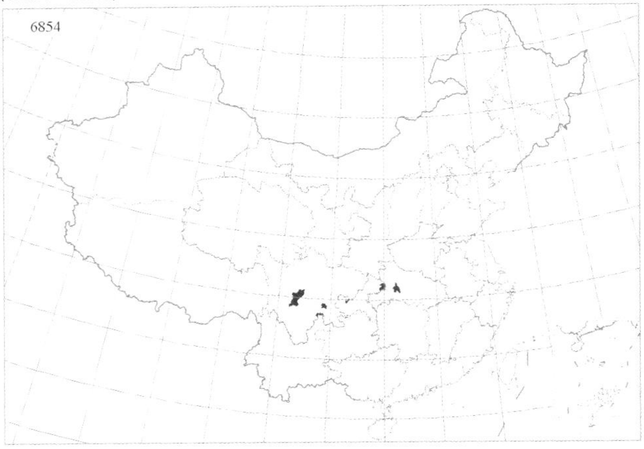

Clusiaceae 藤黄科 *Hypericum* 金丝桃属
Hypericum przewalskii Maxim. 突脉金丝桃
Subshrubs

3.4 (-1.0~13.0); -8.0 (-12.5~1.9); 13.4 (8.5~23.8); 5.6 (2.7~13.1); 33.1 (8.3~102.9); 52.1 (6.1~79.5); 468 (363~730) ❙ 669 (417~1214); 354 (250~531); 19 (6~83) ❙ 41.8 (-3.9~128.6); 467 (363~730); 255.2 (168.7~315.4)

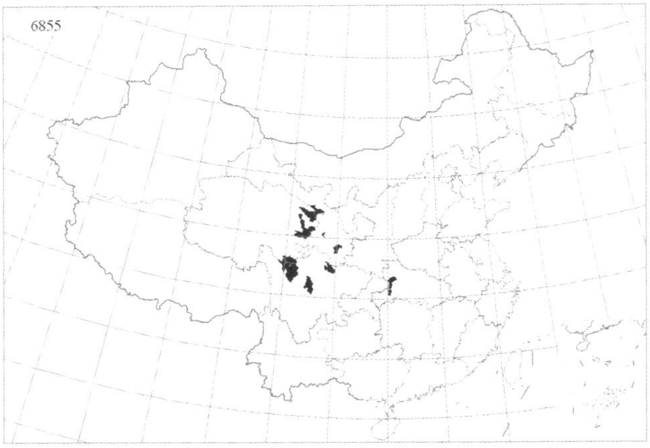

Clusiaceae 藤黄科 *Hypericum* 金丝桃属
Hypericum pseudohenryi N. Robson 北栽秧花
Shrubs, evergreen

11.5 (2.0~19.3); 3.6 (-7.4~13.1); 18.2 (9.5~26.6); 11.8 (3.8~19.4); 88.9 (15.4~173.2); 10.2 (0.0~51.9); 665 (411~939) ❙ 964 (729~1260); 553 (411~678); 30 (11~58) ❙ 50.8 (-14.2~128.6); 646 (411~924); 275.6 (142.9~383.4)

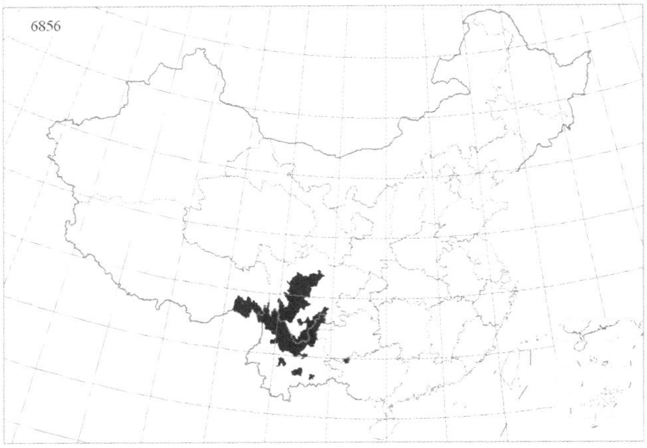

Clusiaceae 藤黄科 *Hypericum* 金丝桃属
Hypericum reptans Hook. f. et Thomson ex Dyer 匍枝金丝桃
Small shrubs

8.4 (5.3~13.3); 0.8 (-2.4~5.4); 14.8 (12.2~19.6); 8.5 (5.7~13.4); 54.5 (30.5~100.4); 13.8 (0.0~26.2); 563 (479~706) ❙ 1087 (741~1736); 628 (426~1007); 39 (21~58) ❙ 89.4 (54.0~145.8); 563 (479~706); 329.2 (234.2~447.4)

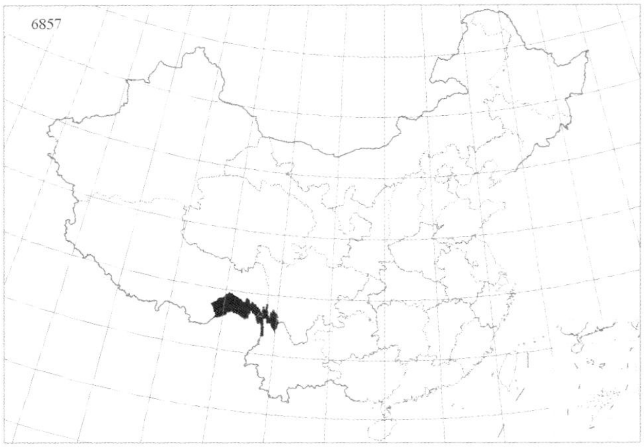

Clusiaceae 藤黄科 *Hypericum* 金丝桃属
Hypericum stellatum N. Robson 星萼金丝桃
Shrubs

12.4 (2.2~15.9); 2.1 (-6.6~5.5); 22.0 (9.9~26.5); 12.7 (3.8~16.0); 98.3 (15.5~131.6); 9.5 (0.0~49.1); 716 (413~855) ❙ 1123 (932~1265); 509 (486~550); 54 (19~71) ❙ 62.8 (30.9~125.9); 716 (413~855); 215.8 (167.5~284.3)

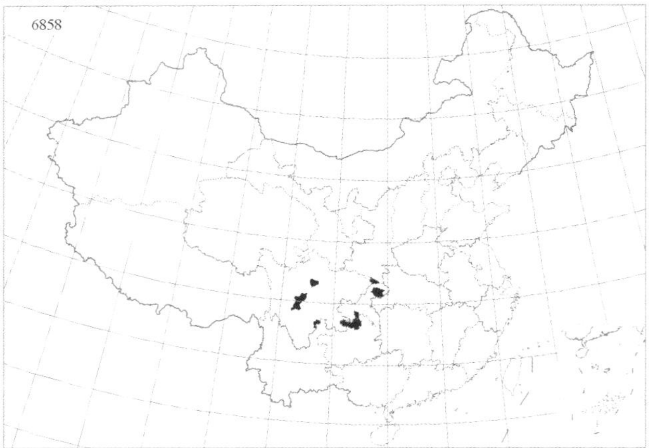

Clusiaceae 藤黄科 *Hypericum* 金丝桃属
Hypericum subalatum Hayata 方茎金丝桃
Shrubs

18.0；12.3；23.0；18.1；157.1；0.0；874 ❙ 2784；943；417 ❙ 219.4；874；276.9

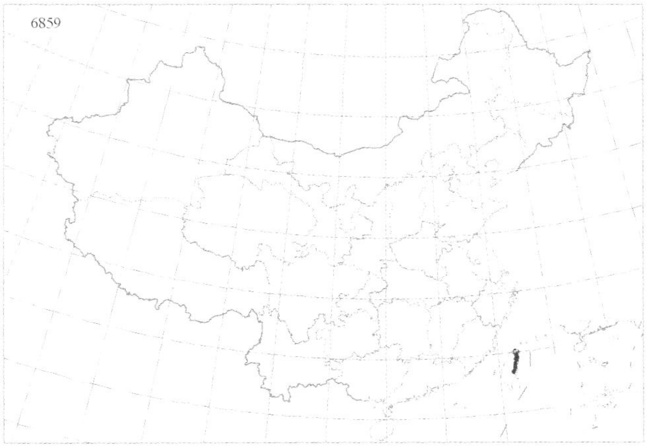

Clusiaceae 藤黄科 *Hypericum* 金丝桃属
Hypericum subsessile N. Robson 近无柄金丝桃
Shrubs

13.1；5.5；19.5；13.2；98.9；0.8；699 ❙ 1044；602；31 ❙ 49.4；692；342.1

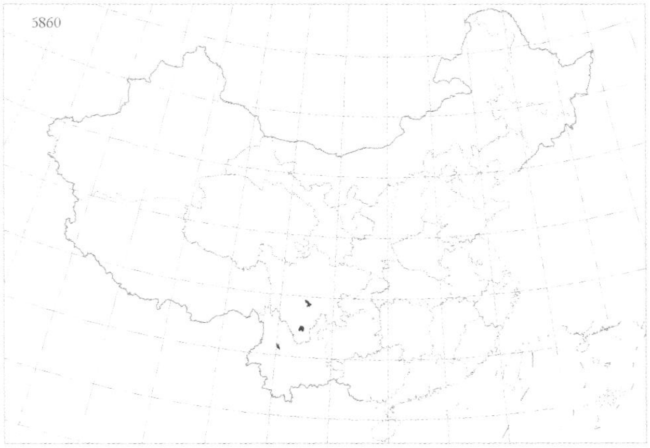

Clusiaceae 藤黄科 *Hypericum* 金丝桃属
Hypericum uralum Buch.-Ham. ex D. Don 匙萼金丝桃
Shrubs

14.1 (−1.8~19.8)；6.3 (−10.2~12.7)；20.3 (6.1~26.6)；14.2 (1.8~19.8)；112.7 (1.9~177.9)；3.3 (0.0~82.8)；744 (329~1010) ❙ 1033 (320~1736)；591 (217~1007)；37 (15~64) ❙ 40.5 (−14.2~145.8)；723 (320~1010)；280.0 (72.2~551.3)

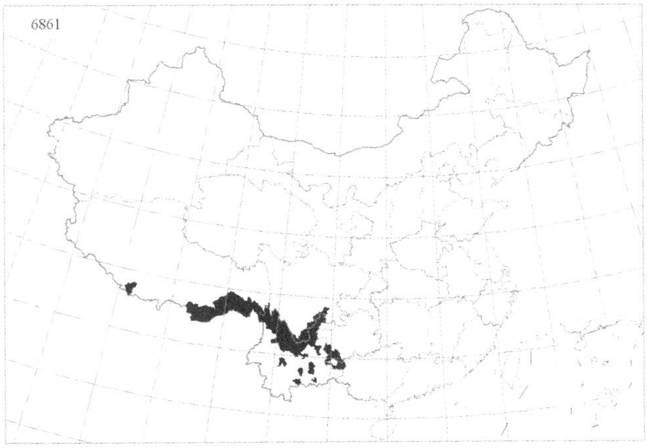

Clusiaceae 藤黄科 *Hypericum* 金丝桃属
Hypericum wardianum N. Robson 漾濞金丝桃
Shrubs

14.5；7.8；19.6；14.5；113.5；0.0；730 ❙ 1044；570；54 ❙ 43.0；730；382.6

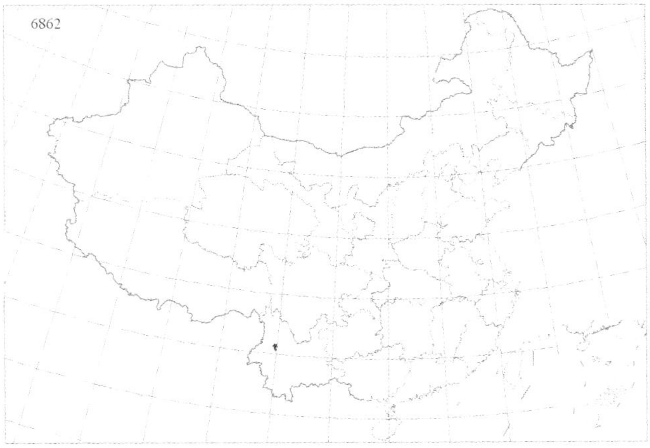

Clusiaceae 藤黄科 *Hypericum* 金丝桃属
Hypericum wightianum Wall. ex Wight et Arn. 遍地金
Herbs or subshrubs

16.9 (14.9~18.1)；6.7 (5.2~7.8)；26.3 (23.6~28.2)；17.0 (15.0~18.2)；143.4 (119.6~158.0)；0.0 (0.0~0.0)；887 (782~962) ❙ 1108 (985~1410)；553 (470~816)；61 (49~72) ❙ 25.9 (12.7~80.4)；887 (782~962)；161.7 (130.1~214.5)

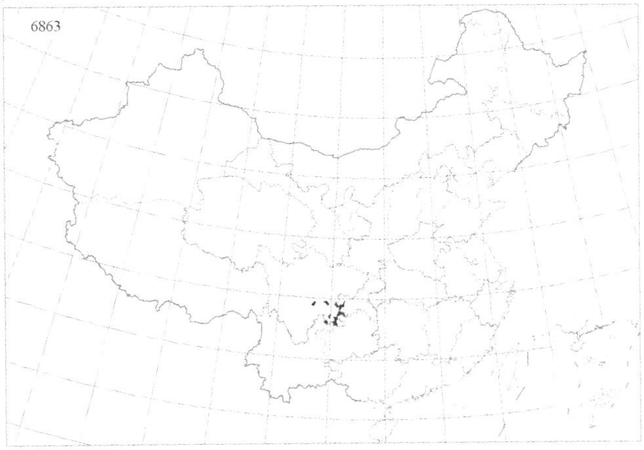

Clusiaceae 藤黄科 *Hypericum* 金丝桃属
Hypericum wilsonii N. Robson 川鄂金丝桃
Shrubs

13.5 (12.7~14.8)；2.7 (1.9~4.3)；24.0 (23.0~24.9)；13.6 (12.8~14.9)；107.1 (99.0~119.2)；4.1 (0.7~6.1)；747 (713~791) ❙ 1319 (1197~1476)；540 (489~592)；79 (65~88) ❙ 77.2 (51.2~107.2)；747 (713~791)；234.6 (182.9~274.9)

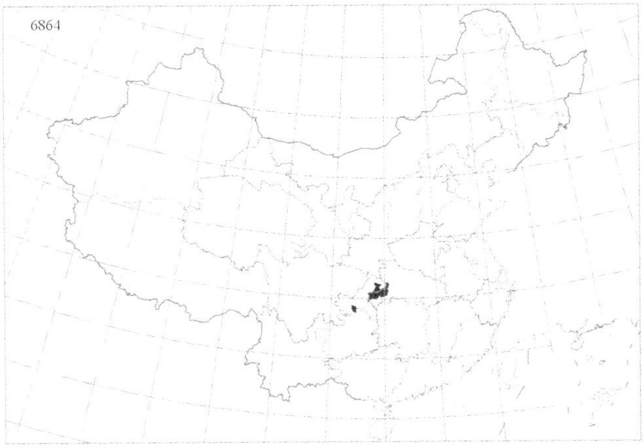

Clusiaceae 藤黄科 *Lianthus* 惠林花属
Lianthus ellipticifolius (H. L. Li) N. Robson 惠林花
Shrubs

8.6；1.3；15.0；8.7；52.6；8.8；574 ❙ 1041；585；45 ❙ 81.5；574；352.3

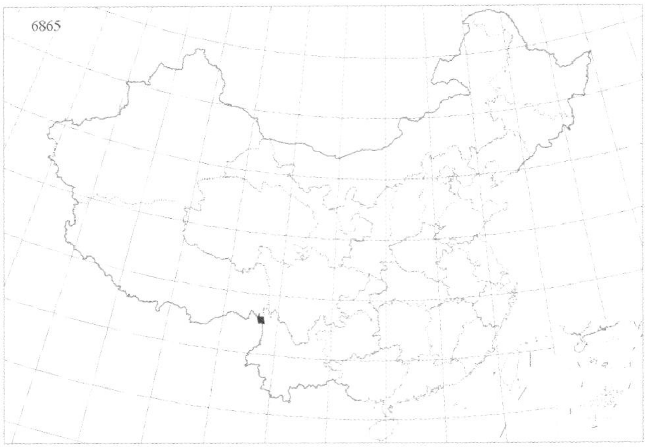

Clusiaceae 藤黄科 *Mammea* 格脉树属
Mammea yunnanensis (H. L. Li) Kosterm. 格脉树
Trees, evergreen

19.1；13.4；22.6；19.1；169.5；0.0；923 ❙ 1434；687；63 ❙ 55.5；909；533.8

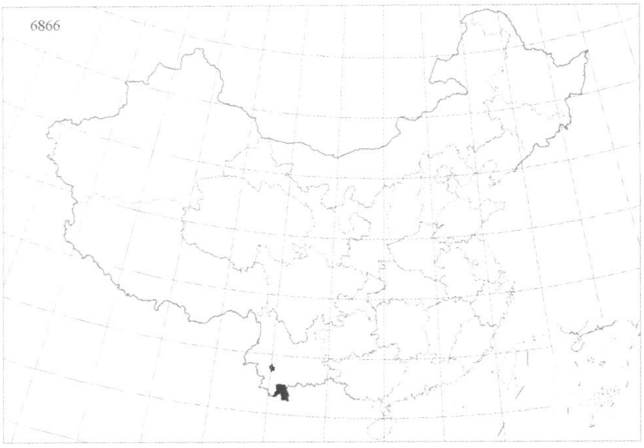

Clusiaceae 藤黄科 *Mesua* 铁力木属
Mesua ferrea L. 铁力木
Trees, evergreen

18.8 (13.1~21.3)；12.0 (6.4~14.9)；23.4 (18.4~28.5)；18.8 (13.2~21.3)；166.0 (98.1~196.1)；0.0 (0.0~0.0)；932 (692~1132) ❙ 1479 (1164~1776)；758 (634~909)；76 (42~154) ❙ 59.9 (33.2~78.9)；920 (692~1132)；417.5 (219.1~570.1)

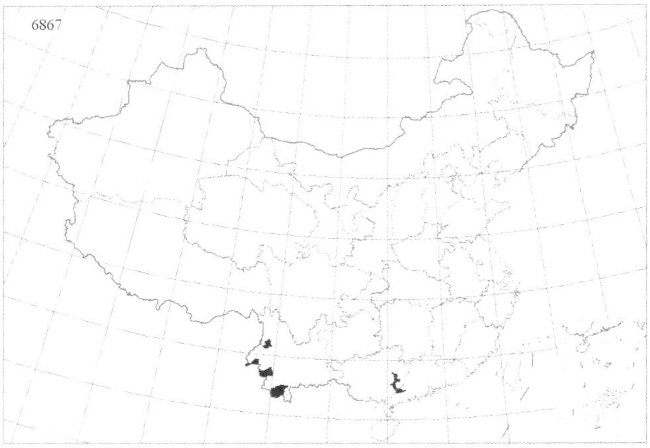

Dipterocarpaceae 龙脑香科 *Dipterocarpus* 龙脑香属
Dipterocarpus retusus Blume 东京龙脑香
Trees, evergreen

17.7 (13.3~21.0)；11.1 (5.4~14.2)；22.3 (19.6~25.6)；17.8 (13.4~21.0)；153.2 (100.4~191.8)；0.0 (0.0~0.0)；865 (706~1053) ❙ 1525 (1178~1751)；865 (646~1007)；60 (51~77) ❙ 78.5 (44.5~145.8)；865 (706~1053)；498.4 (383.4~627.0)

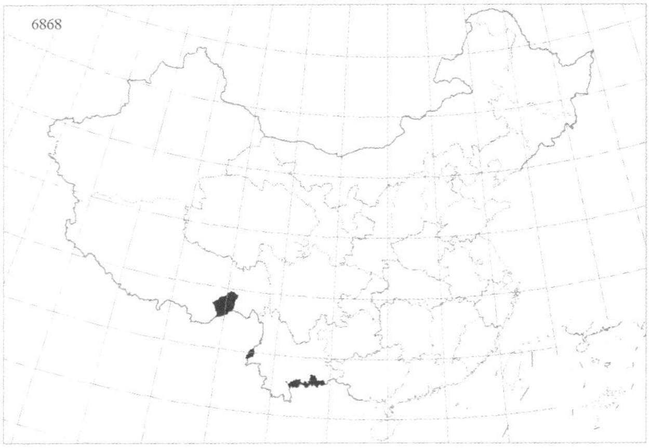

Dipterocarpaceae 龙脑香科 *Dipterocarpus* 龙脑香属
Dipterocarpus turbinatus Gaertn. 羯布罗香
Trees, evergreen

17.5 (13.3~20.5)；11.0 (5.4~15.1)；21.8 (19.6~23.7)；17.5 (13.4~20.5)；150.4 (100.4~185.9)；0.0 (0.0~0.0)；859 (706~992) ❙ 1611 (1464~1736)；844 (655~1007)；64 (51~77) ❙ 92.0 (47.6~145.8)；852 (706~986)；568.6 (447.4~659.0)

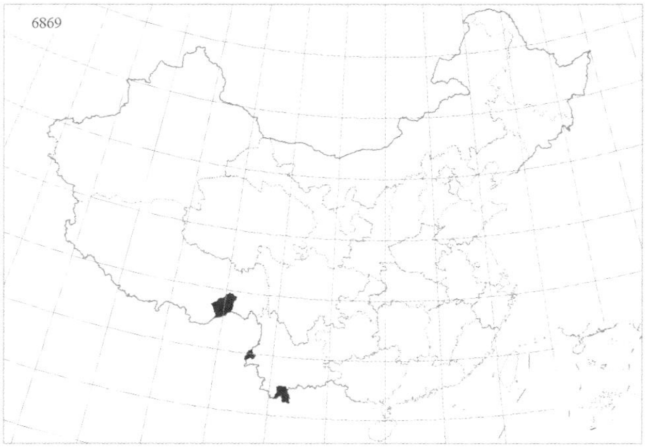

Dipterocarpaceae 龙脑香科 *Hopea* 坡垒属
Hopea chinensis Hand.-Mazz. 狭叶坡垒
Trees, evergreen

20.4 (17.6~22.7)；12.3 (7.3~15.0)；26.4 (22.5~28.6)；20.4 (17.6~22.8)；184.7 (151.6~213.1)；0.0 (0.0~0.0)；1050 (854~1221) ❙ 1643 (1293~2435)；873 (639~1428)；110 (58~174) ❙ 57.1 (12.9~99.5)；1050 (854~1221)；335.2 (200.0~551.6)

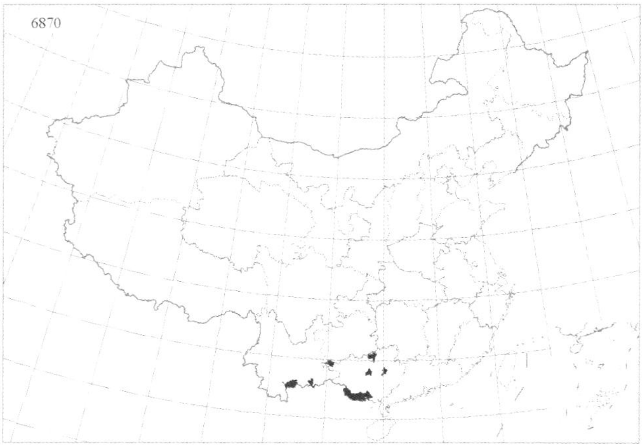

Dipterocarpaceae 龙脑香科 *Hopea* 坡垒属
Hopea hainanensis Merr. et Chun 坡垒
Trees, evergreen

24.1 (23.0~24.8); 18.7 (17.7~20.1); 28.0 (26.7~28.7); 24.2 (23.1~24.8); 230.0 (217.0~237.6); 0.0 (0.0~0.0); 1323 (1217~1385) ❙ 1370 (868~1767); 470 (312~610); 58 (41~88) ❙ 3.8 (-34.3~41.8); 1120 (868~1299); 465.9 (288.3~572.9)

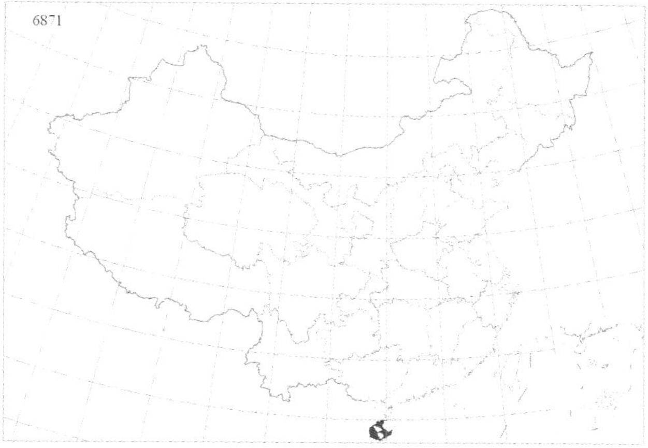

Dipterocarpaceae 龙脑香科 *Hopea* 坡垒属
Hopea reticulata Tardieu 铁凌
Trees, evergreen

23.9; 19.1; 27.2; 23.9; 227.4; 0.0; 1298 ❙ 1552; 536; 50 ❙ 20.9; 1138; 558.2

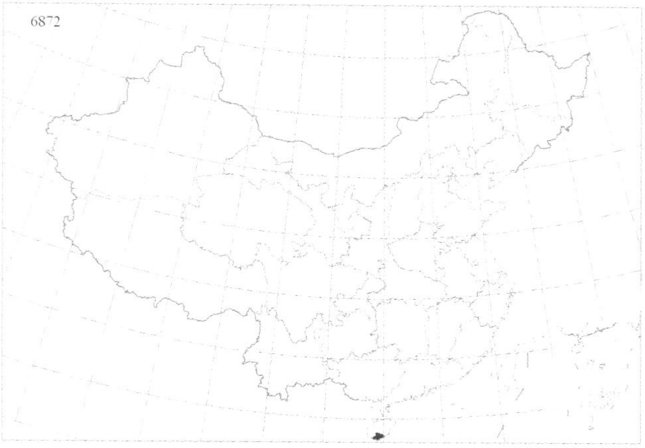

Dipterocarpaceae 龙脑香科 *Hopea* 坡垒属
Hopea shingkeng (Dunn) Bor 西藏坡垒
Trees, evergreen

13.3; 5.4; 19.6; 13.4; 100.4; 0.0; 706 ❙ 1736; 1007; 51 ❙ 145.8; 706; 447.4

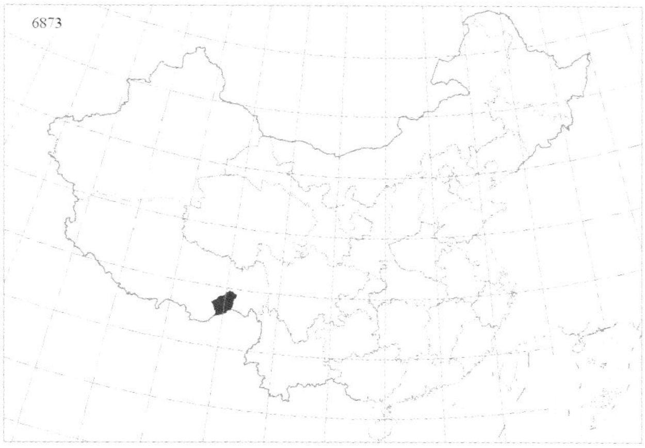

Dipterocarpaceae 龙脑香科 *Parashorea* 柳安属
Parashorea chinensis H. Wang 望天树
Trees, evergreen

19.5 (16.9~22.2); 12.2 (9.5~15.1); 24.6 (21.4~28.4); 19.5 (16.9~22.3); 174.2 (142.7~207.5); 0.0 (0.0~0.0); 972 (813~1190) ❙ 1417 (1163~1751); 764 (641~996); 71 (48~141) ❙ 47.3 (14.8~88.8); 970 (813~1190); 410.1 (232.6~659.0)

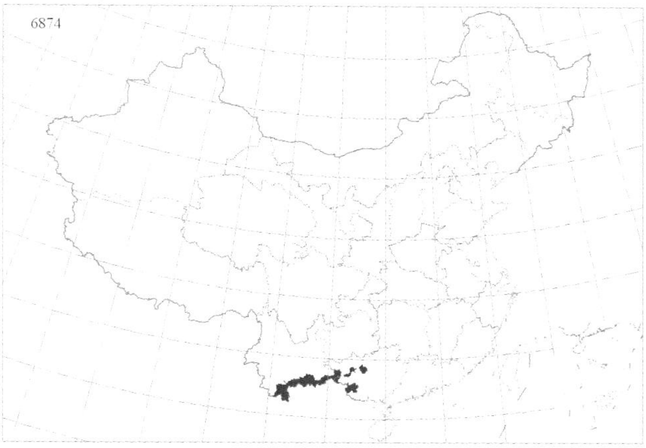

Dipterocarpaceae 龙脑香科 *Shorea* 娑罗双属
Shorea assamica Dyer 云南娑罗双
Trees, evergreen

14.9; 7.5; 20.1; 14.9; 118.9; 0.0; 751 ❙ 1712; 983; 57 ❙ 129.0; 751; 537.2

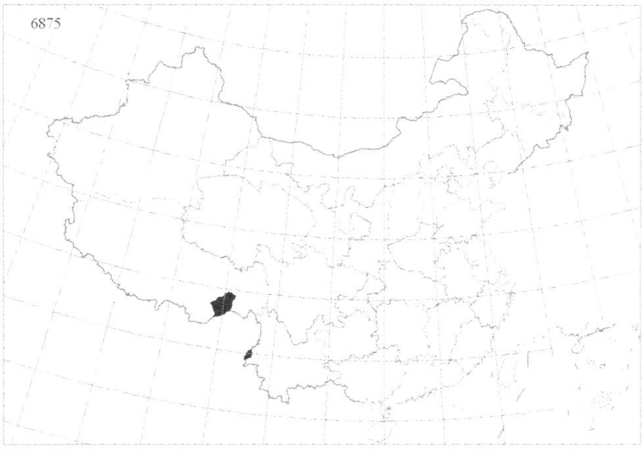

Dipterocarpaceae 龙脑香科 *Shorea* 娑罗双属
Shorea robusta C. F. Gaertner 娑罗双
Trees, evergreen

20.4; 15.1; 23.6; 20.5; 185.8; 0.0; 989 ❙ 1607; 702; 77 ❙ 62.5; 986; 659.0

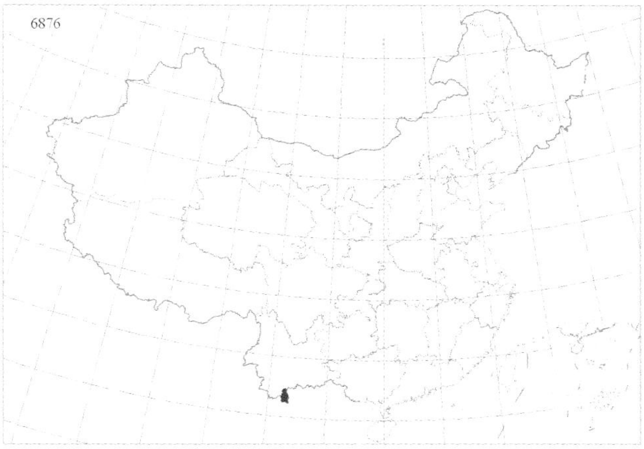

Dipterocarpaceae 龙脑香科 *Vatica* 青梅属
Vatica guangxiensis S. L. Mo 广西青梅
Trees, evergreen

18.6; 11.9; 23.1; 18.7; 164.2; 0.0; 909 ❙ 1577; 827; 66 ❙ 75.7; 909; 552.6

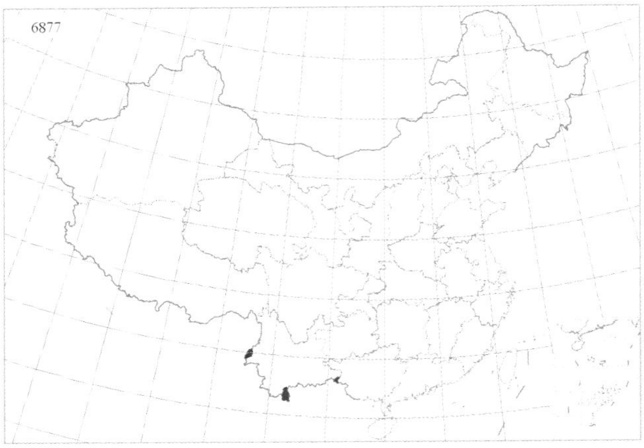

Dipterocarpaceae 龙脑香科 *Vatica* 青梅属
Vatica lanceifolia (Roxburgh) Blume 西藏青梅
Trees, evergreen

5.3; -2.4; 12.2; 5.7; 30.5; 26.2; 479 ❙ 961; 558; 31 ❙ 100.7; 479; 234.2

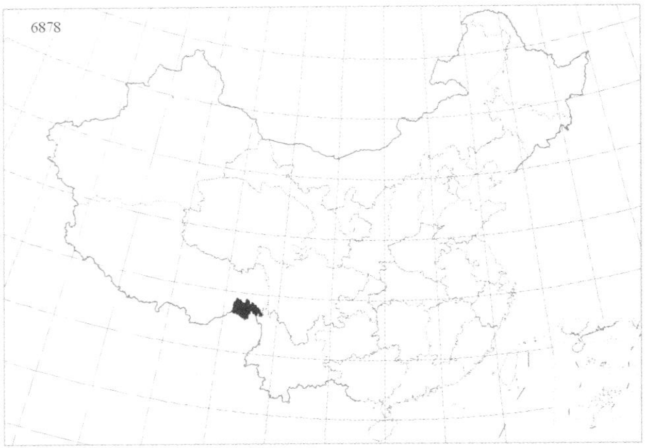

Dipterocarpaceae 龙脑香科 *Vatica* 青梅属
Vatica mangachapoi Blanco 青梅
Trees, evergreen

23.9 (22.2~24.8); 18.5 (17.2~20.1); 27.7 (25.7~28.7); 23.9 (22.3~24.8); 227.0 (207.2~237.6); 0.0 (0.0~0.0); 1297 (1127~1385) ❙ 1420 (868~1767); 493 (312~610); 56 (41~88) ❙ 10.1 (-34.3~47.8); 1119 (868~1299); 495.2 (288.3~620.0)

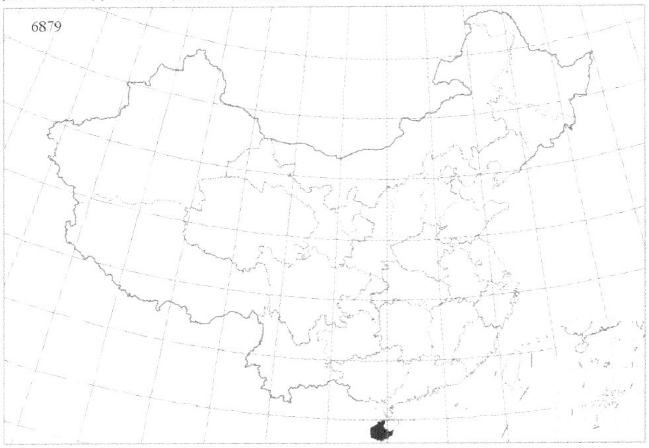

Tamaricaceae 柽柳科 *Myricaria* 水柏枝属
Myricaria bracteata Royle 宽苞水柏枝
Shrubs, deciduous

7.7 (-4.8~16.9); -7.7 (-20.9~7.6); 21.0 (7.0~28.1); 9.6 (1.9~16.9); 75.6 (3.6~143.4); 43.1 (0.0~121.6); 636 (328~924) ❙ 426 (21~1337); 231 (15~753); 17 (0~109) ❙ -31.3 (-97.3~125.9); 382 (21~924); 136.2 (1.6~407.1)

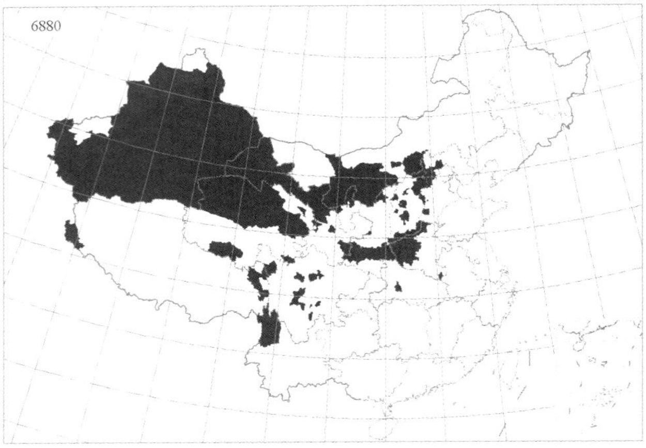

Tamaricaceae 柽柳科 *Myricaria* 水柏枝属
Myricaria elegans Royle 秀丽水柏枝
Shrubs or small trees, deciduous

3.1 (-5.9~12.4); -11.5 (-20.1~4.3); 15.7 (5.4~26.8); 6.6 (1.3~13.2); 48.0 (0.4~114.1); 70.0 (0.7~133.0); 512 (294~811) ❙ 164 (21~1431); 97 (15~817); 11 (0~103) ❙ -59.4 (-97.3~113.0); 135 (21~672); 43.2 (2.6~363.7)

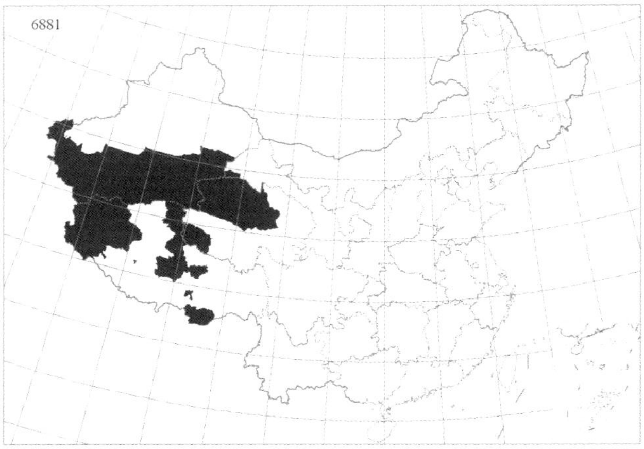

Tamaricaceae 柽柳科 *Myricaria* 水柏枝属
Myricaria laxiflora (Franch.) P. Y. Zhang et Y. J. Zhang 疏花水柏枝
Shrubs, deciduous

14.3 (13.0~16.9); 2.9 (1.9~4.8); 25.2 (23.8~28.1); 14.4 (13.1~16.9); 116.1 (102.9~143.4); 3.4 (0.2~6.1); 791 (730~924) ❙ 1170 (1121~1265); 503 (475~535); 76 (65~83) ❙ 48.7 (24.4~64.6); 791 (730~924); 225.5 (166.0~262.4)

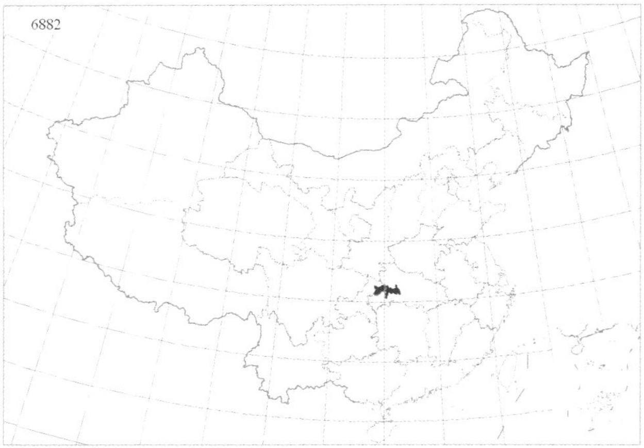

Tamaricaceae 柽柳科 *Myricaria* 水柏枝属

Myricaria paniculata P. Y. Zhang et Y. J. Zhang 三春水柏枝

Shrubs, deciduous

8.4 (-4.8~18.9); -1.8 (-18.0~12.5); 17.3 (7.4~26.4); 9.6 (1.9~19.0); 70.7 (4.3~167.8); 29.0 (0.0~121.6); 599 (332~929) ❙ 733 (63~1358); 414 (45~663); 22 (1~88) ❙ 23.3 (-84.9~125.9); 557 (63~851); 240.1 (8.4~382.6)

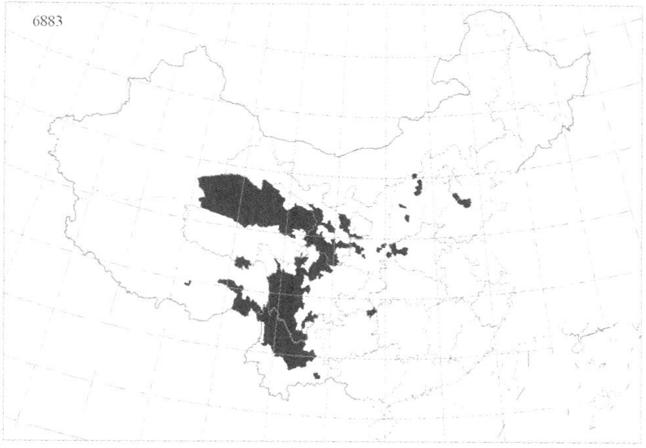

Tamaricaceae 柽柳科 *Myricaria* 水柏枝属

Myricaria platyphylla Maxim. 宽叶水柏枝

Shrubs, deciduous

7.6 (4.7~9.5); -9.6 (-13.8~-6.5); 22.7 (21.1~24.2); 9.6 (8.2~10.9); 78.1 (63.8~89.2); 46.8 (32.3~67.3); 637 (583~689) ❙ 315 (111~496); 189 (74~276); 7 (2~12) ❙ -50.5 (-82.3~-26.5); 315 (111~496); 50.7 (13.6~90.0)

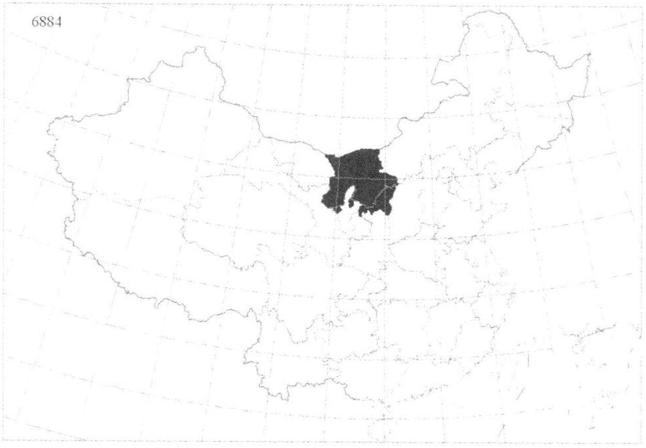

Tamaricaceae 柽柳科 *Myricaria* 水柏枝属

Myricaria prostrata Hook. f. et Thomson 匍匐水柏枝

Dwarf shrubs, deciduous

-0.6 (-5.9~12.0); -13.4 (-18.9~-5.2); 11.1 (5.4~25.8); 3.9 (1.3~12.8); 21.4 (0.4~109.4); 87.9 (24.3~133.0); 398 (294~786) ❙ 237 (34~662); 149 (16~379); 10 (0~103) ❙ -33.4 (-95.6~91.4); 207 (34~463); 88.9 (7.7~290.1)

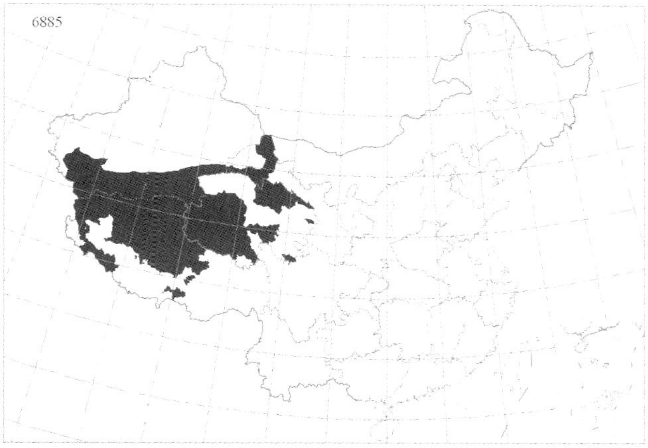

Tamaricaceae 柽柳科 *Myricaria* 水柏枝属

Myricaria pulcherrima Batalin 心叶水柏枝

Shrubs or subshrubs, deciduous

9.1 (-4.3~12.4); -8.5 (-20.1~-5.2); 23.1 (9.6~26.5); 11.0 (2.5~13.2); 92.2 (9.7~114.1); 42.2 (24.3~121.1); 712 (329~822) ❙ 41 (26~71); 22 (16~33); 3 (0~10) ❙ -93.4 (-96.7~-78.5); 41 (26~71); 16.1 (1.6~65.1)

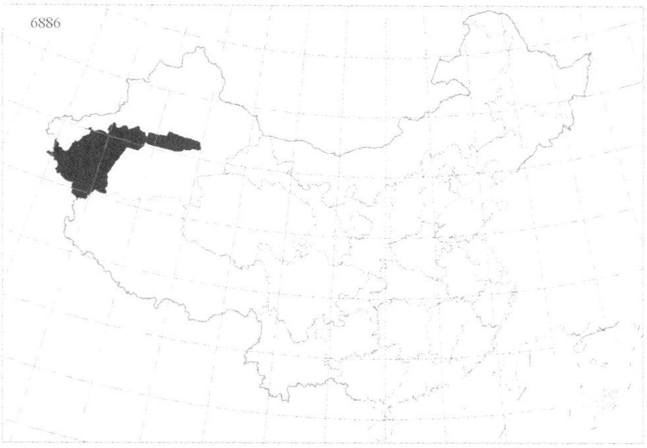

Tamaricaceae 柽柳科 *Myricaria* 水柏枝属

Myricaria rosea W. W. Sm. 卧生水柏枝

Shrubs, deciduous

1.6 (-4.8~10.8); -7.2 (-16.1~3.9); 9.7 (6.1~16.6); 4.0 (1.6~10.9); 18.7 (1.9~71.8); 58.4 (1.3~118.7); 406 (300~633) ❙ 518 (220~968); 327 (160~594); 15 (0~33) ❙ 21.7 (-26.8~100.7); 385 (220~633); 164.0 (55.5~319.8)

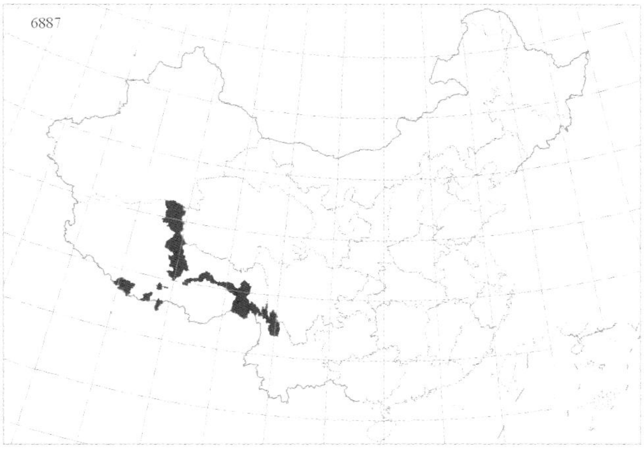

Tamaricaceae 柽柳科 *Myricaria* 水柏枝属

Myricaria squamosa Desv. 具鳞水柏枝

Shrubs, deciduous

2.7 (-5.9~15.9); -9.9 (-20.3~6.2); 13.7 (5.4~26.8); 5.5 (1.3~15.9); 34.5 (0.4~130.8); 62.0 (0.0~130.5); 471 (299~822) ❙ 488 (23~1318); 278 (16~753); 15 (1~98) ❙ 8.6 (-97.2~128.6); 368 (23~822); 191.7 (3.0~364.1)

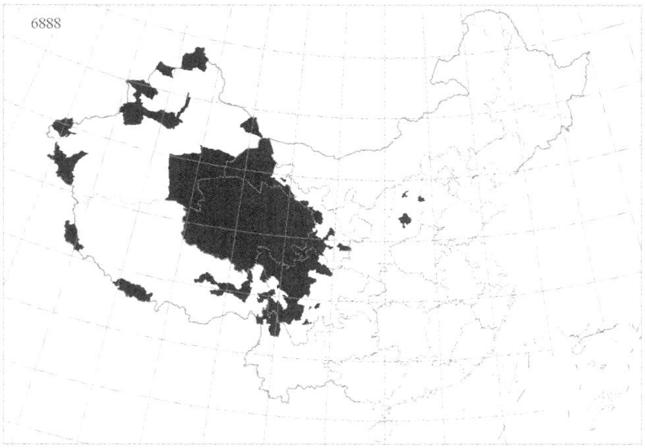

Tamaricaceae 柽柳科 *Reaumuria* 琵琶柴属
Reaumuria kaschgarica Rupr. 五柱红砂
Dwarf subshrubs

2.1 (-5.9~12.4); -13.2 (-20.1~-5.2); 15.4 (5.4~26.8); 6.1 (1.3~13.2); 43.2 (0.4~114.1); 77.4 (24.3~133.0); 493 (294~811) ❙ 129 (21~386); 78 (15~235); 6 (0~26) ❙ -67.6 (-97.3~0.0); 128 (21~386); 39.3 (2.8~208.7)

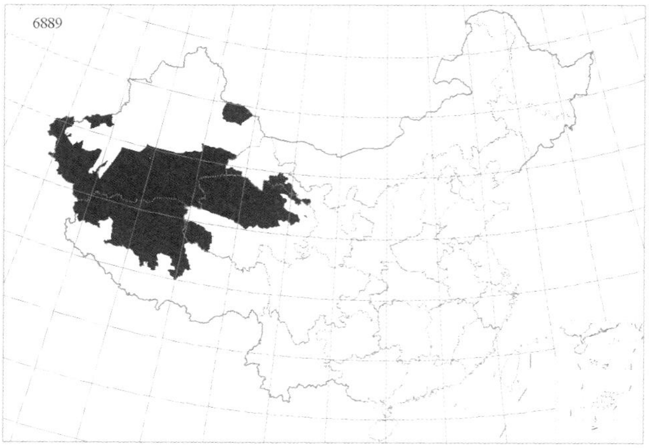

Tamaricaceae 柽柳科 *Reaumuria* 琵琶柴属
Reaumuria soongarica (Pall.) Maxim. 红砂
Shrubs

5.7 (-4.8~12.5); -11.3 (-25.5~-4.8); 20.1 (7.4~27.5); 8.5 (1.9~13.2); 66.7 (4.3~114.1); 57.8 (23.1~128.7); 600 (329~822) ❙ 196 (21~590); 111 (15~453); 8 (0~48) ❙ -62.8 (-97.3~35.3); 193 (21~590); 72.2 (1.6~297.3)

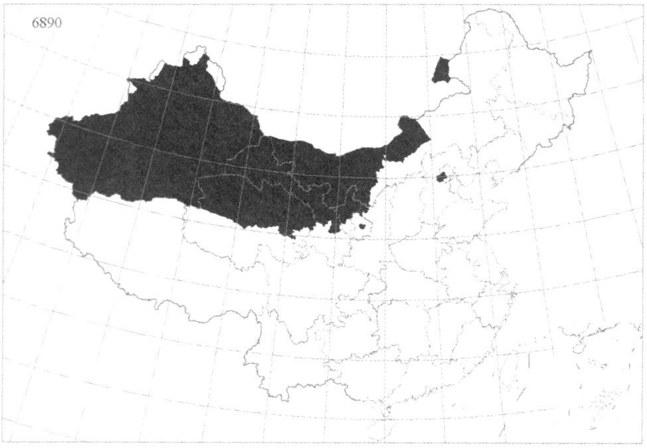

Tamaricaceae 柽柳科 *Reaumuria* 琵琶柴属
Reaumuria trigyna Maxim. 黄花红砂
Subshrubs

7.6 (4.7~8.8); -9.6 (-13.8~-7.0); 22.8 (20.7~24.2); 9.6 (8.2~10.3); 78.4 (63.8~84.9); 47.0 (35.3~67.3); 639 (583~671) ❙ 201 (84~416); 122 (53~265); 4 (1~12) ❙ -68.3 (-87.1~-31.0); 201 (84~416); 56.5 (7.5~116.9)

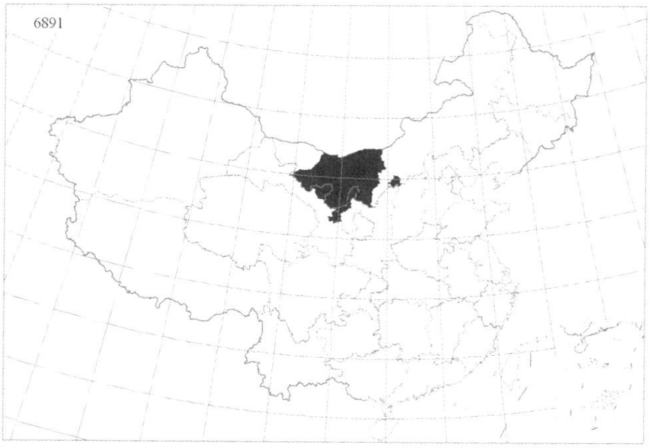

Tamaricaceae 柽柳科 *Tamarix* 柽柳属
Tamarix androssowii Litv. 白花柽柳
Shrubs or small trees, deciduous

8.3 (-2.3~14.6); -8.8 (-17.2~0.9); 22.4 (10.7~27.5); 10.2 (3.2~14.7); 83.8 (14.3~124.7); 43.9 (8.6~102.0); 673 (364~852) ❙ 102 (21~725); 60 (15~382); 4 (0~42) ❙ -83.1 (-97.3~-14.9); 102 (21~725); 34.3 (1.6~161.1)

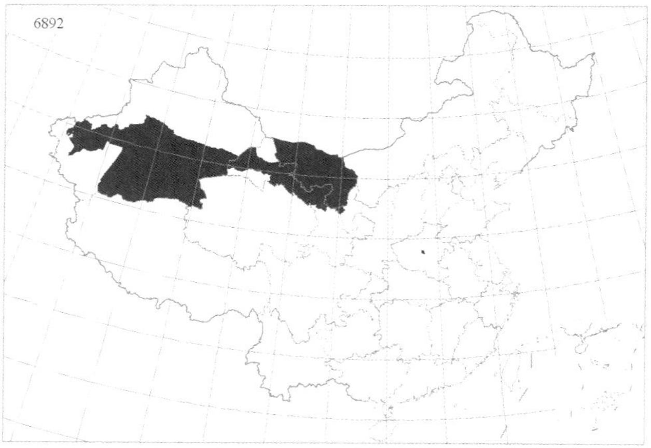

Tamaricaceae 柽柳科 *Tamarix* 柽柳属
Tamarix aphylla (L.) H. Karst. 无叶柽柳
Shrubs or trees, deciduous

21.8 (18.5~24.1); 16.6 (14.1~18.9); 25.8 (21.7~28.2); 21.9 (18.5~24.1); 202.4 (162.1~229.7); 0.0 (0.0~0.0); 1124 (863~1322) ❙ 2270 (1625~2992); 1361 (951~1847); 100 (47~188) ❙ 111.9 (32.5~214.4); 1051 (863~1144); 222.7 (119.4~363.7)

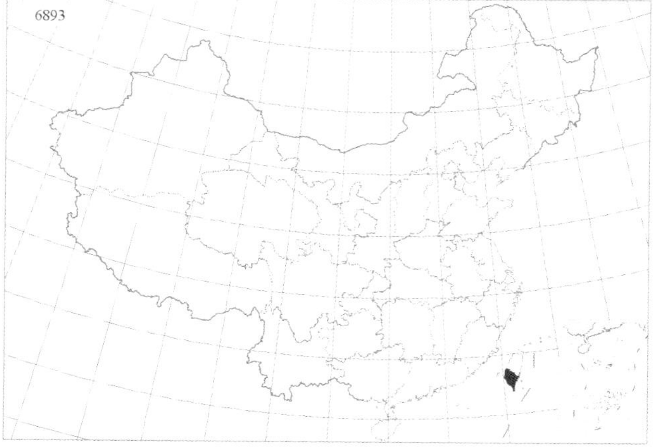

Tamaricaceae 柽柳科 *Tamarix* 柽柳属
Tamarix arceuthoides Bunge 密花柽柳
Shrubs or small trees, deciduous

6.2 (-4.1~12.4); -10.9 (-20.9~-5.2); 20.3 (8.5~26.8); 8.9 (2.5~13.2); 70.3 (8.1~114.1); 55.3 (24.3~117.9); 616 (330~822) ❙ 130 (21~491); 72 (15~301); 7 (0~48) ❙ -74.6 (-97.3~41.8); 128 (21~406); 55.3 (1.6~297.3)

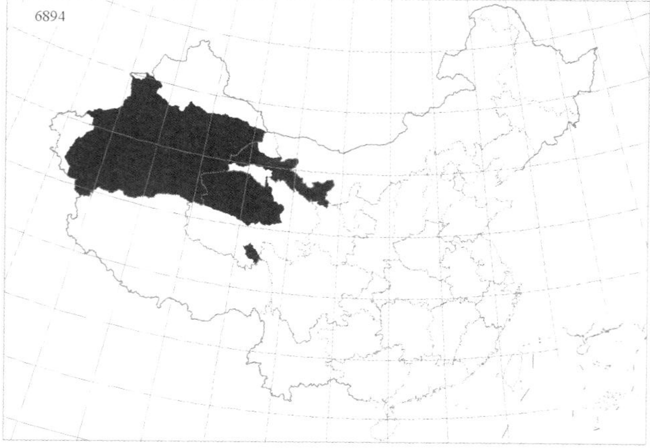

Tamaricaceae 柽柳科 *Tamarix* 柽柳属
Tamarix austromongolica Nakai 甘蒙柽柳
Shrubs or trees, deciduous

8.0 (-2.3~14.9); -7.8 (-17.2~-0.2); 21.9 (10.7~28.6); 9.8 (3.2~14.9); 78.7 (14.3~129.4); 41.7 (10.2~102.0); 648 (364~889) ❙ 366 (21~652); 221 (15~441); 11 (0~36) ❙ -43.3 (-97.3~14.7); 365 (21~652); 96.3 (2.1~256.0)

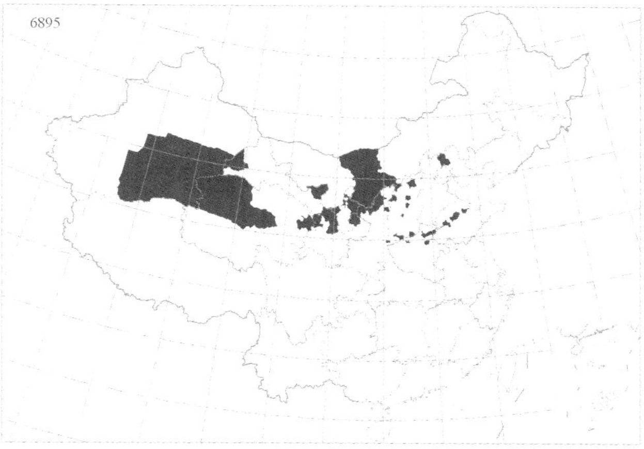

Tamaricaceae 柽柳科 *Tamarix* 柽柳属
Tamarix chinensis Lour. 柽柳
Shrubs or trees, deciduous

12.3 (-2.6~22.5); -2.0 (-16.0~15.0); 25.2 (8.7~29.9); 13.1 (2.5~22.6); 109.2 (7.7~210.9); 20.5 (0.0~98.0); 775 (349~1207) ❙ 737 (35~1784); 399 (23~862); 40 (2~220) ❙ -6.4 (-94.9~101.4); 638 (35~1207); 165.5 (7.4~322.1)

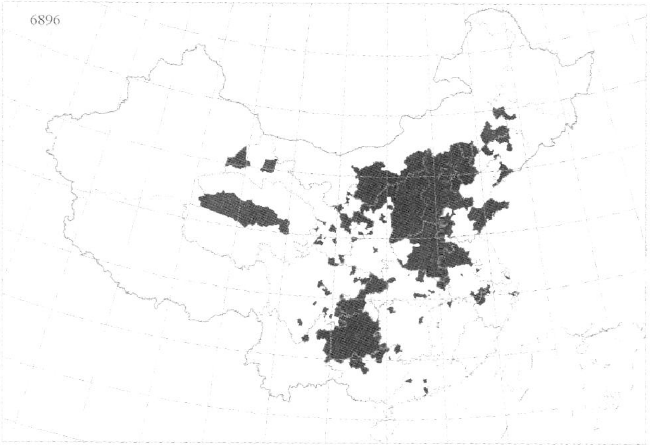

Tamaricaceae 柽柳科 *Tamarix* 柽柳属
Tamarix elongata Ledeb. 长穗柽柳
Tall shrubs, deciduous

6.7 (-5.9~12.5); -11.0 (-19.5~-5.1); 21.4 (5.4~27.5); 9.3 (1.3~13.2); 75.6 (0.4~114.1); 55.2 (23.1~130.5); 638 (300~822) ❙ 125 (21~310); 68 (15~190); 8 (0~48) ❙ -77.6 (-97.3~-10.5); 125 (21~310); 43.5 (1.6~189.0)

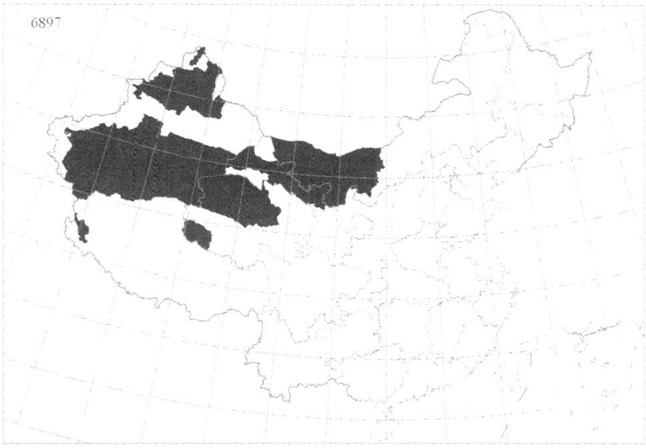

Tamaricaceae 柽柳科 *Tamarix* 柽柳属
Tamarix gansuensis H. Z. Zhang 甘肃柽柳
Shrubs, deciduous

6.6 (-4.8~12.2); -10.3 (-18.9~-5.7); 20.7 (7.4~26.8); 9.2 (1.9~13.1); 74.1 (4.3~113.8); 54.5 (26.7~121.6); 631 (330~822) ❙ 78 (21~287); 50 (15~188); 3 (0~12) ❙ -83.8 (-97.3~-13.5); 78 (21~287); 19.1 (1.6~112.4)

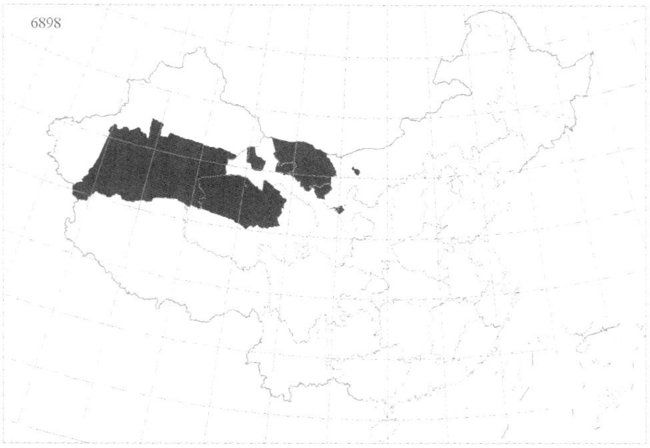

Tamaricaceae 柽柳科 *Tamarix* 柽柳属
Tamarix gracilis Willd. 翠枝柽柳
Shrubs, deciduous

3.9 (-4.8~8.6); -12.7 (-18.4~-8.6); 18.4 (7.4~24.0); 7.3 (1.9~10.3); 55.2 (4.3~84.8); 67.8 (41.2~121.6); 544 (332~667) ❙ 184 (63~308); 107 (45~188); 7 (1~30) ❙ -63.2 (-87.1~-13.5); 184 (63~308); 47.5 (7.5~112.4)

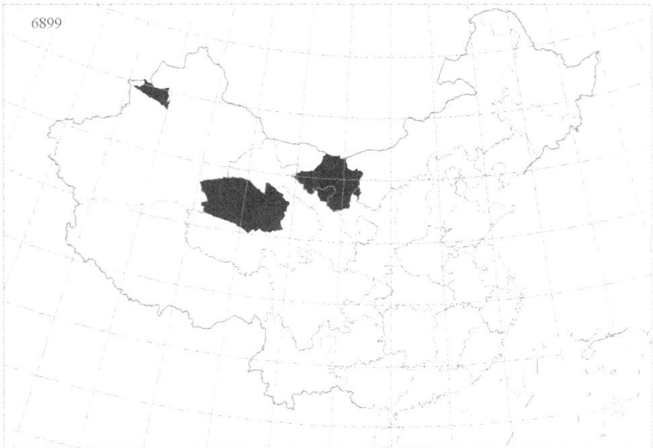

Tamaricaceae 柽柳科 *Tamarix* 柽柳属
Tamarix hispida Willd. 刚毛柽柳
Shrubs or small trees, deciduous

6.1 (-4.1~12.5); -11.6 (-20.9~-5.1); 20.9 (9.8~27.5); 9.0 (2.5~13.2); 72.2 (9.9~114.1); 57.8 (23.1~117.9); 626 (330~822) ❙ 143 (21~450); 73 (15~211); 10 (0~51) ❙ -74.1 (-97.3~3.9); 142 (21~406); 57.5 (1.6~297.3)

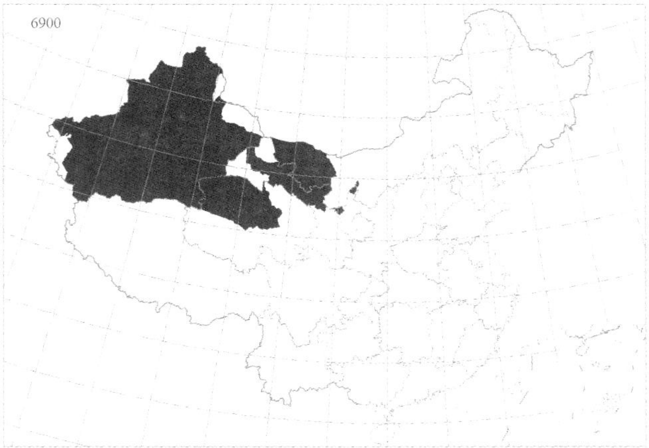

Tamaricaceae 柽柳科 *Tamarix* 柽柳属

Tamarix hohenackeri Bunge 多花柽柳

Shrubs or small trees, deciduous

6.1 (-2.5~12.1); -10.7 (-18.4~-6.0); 20.3 (10.1~26.5); 8.7 (2.9~13.1); 68.3 (12.0~113.4); 54.9 (27.2~102.0); 602 (358~811) ❙ 140 (21~308); 82 (15~182); 5 (0~30) ❙ -74.1 (-97.3~-24.8); 140 (21~308); 44.1 (1.6~147.7)

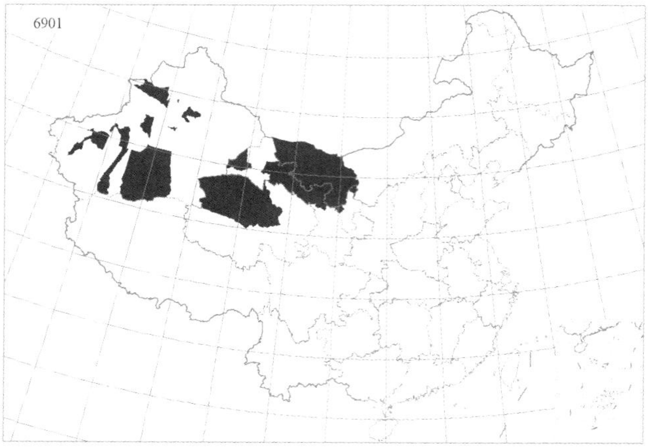

Tamaricaceae 柽柳科 *Tamarix* 柽柳属

Tamarix jintaensis P. Y. Zhang et M. T. Liu 金塔柽柳

Shrubs, deciduous

8.2; -8.9; 23.1; 10.0; 82.9; 44.3; 662 ❙ 72; 46; 3 ❙ -89.1; 72; 12.0

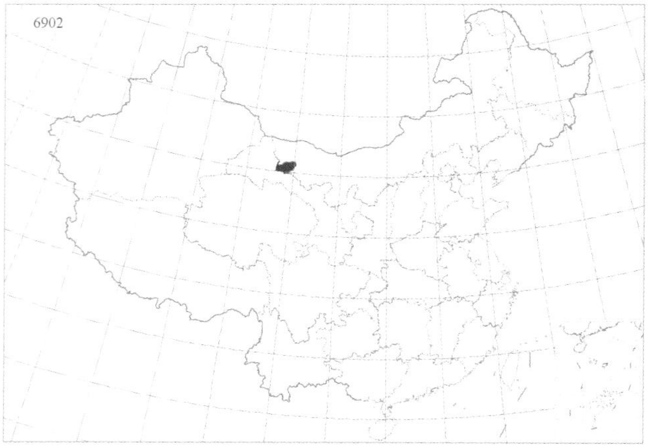

Tamaricaceae 柽柳科 *Tamarix* 柽柳属

Tamarix karelinii Bunge 盐地柽柳

Shrubs or small trees, deciduous

5.3 (-2.5~11.8); -12.6 (-19.5~-7.3); 20.6 (10.1~27.5); 8.5 (2.9~13.0); 67.4 (12.0~113.8); 63.2 (31.4~104.2); 603 (358~822) ❙ 158 (28~308); 83 (20~182); 10 (0~30) ❙ -71.5 (-96.6~-24.8); 158 (28~308); 53.5 (2.0~147.7)

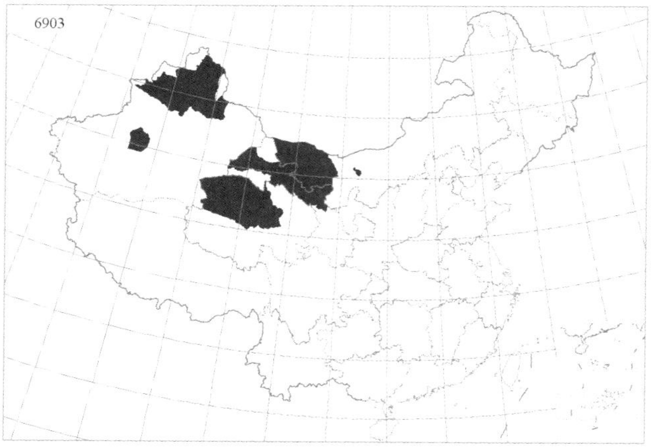

Tamaricaceae 柽柳科 *Tamarix* 柽柳属

Tamarix laxa Willd. 短穗柽柳

Shrubs, deciduous

6.2 (-4.8~12.4); -11.6 (-19.5~-5.2); 21.3 (7.4~27.5); 9.1 (1.9~13.2); 73.0 (4.3~114.1); 57.4 (24.3~121.6); 626 (330~822) ❙ 162 (21~416); 95 (15~270); 7 (0~48) ❙ -71.6 (-97.3~-13.5); 162 (21~416); 48.3 (1.6~189.0)

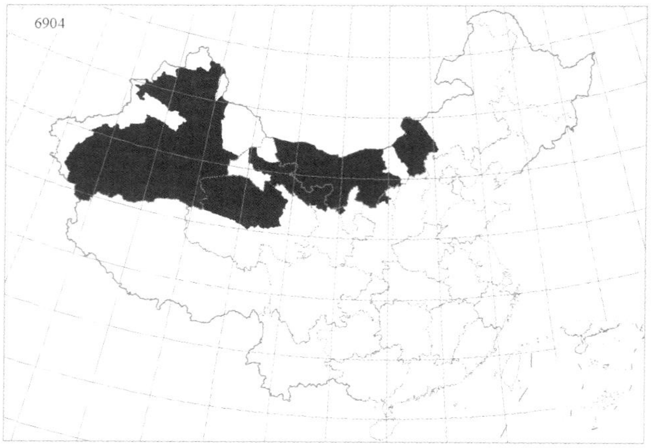

Tamaricaceae 柽柳科 *Tamarix* 柽柳属

Tamarix leptostachya Bunge 细穗柽柳

Shrubs, deciduous

8.3 (-2.5~12.5); -9.8 (-18.4~-5.1); 23.1 (10.1~27.5); 10.5 (2.9~13.2); 87.2 (12.0~114.1); 47.1 (23.1~101.3); 691 (358~822) ❙ 102 (21~300); 55 (15~168); 6 (0~21) ❙ -83.2 (-97.3~-37.4); 102 (21~300); 33.7 (1.6~120.2)

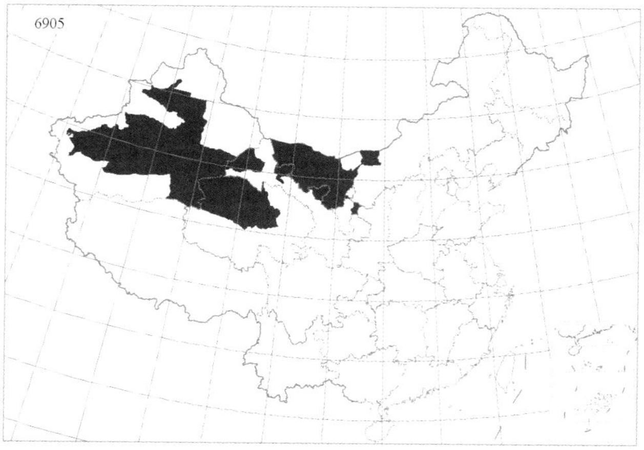

Tamaricaceae 柽柳科 *Tamarix* 柽柳属

Tamarix ramosissima Ledeb. 多枝柽柳

Shrubs or small trees, deciduous

8.7 (-4.1~15.4); -8.0 (-18.9~1.2); 23.1 (8.7~28.6); 10.5 (2.5~15.4); 86.6 (7.7~131.9); 41.9 (7.0~117.9); 684 (330~894) ❙ 307 (21~756); 182 (15~453); 12 (0~61) ❙ -55.8 (-97.3~0.0); 307 (21~756); 83.0 (1.6~269.8)

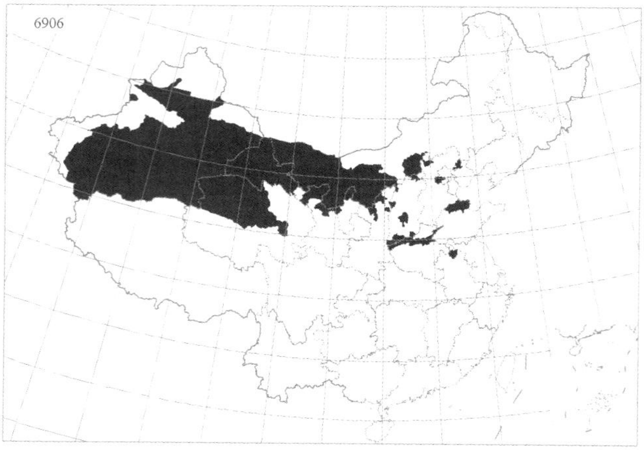

Tamaricaceae 柽柳科 *Tamarix* 柽柳属
Tamarix sachensis P. Y. Zhang et M. T. Liu 莎车柽柳
Shrubs, deciduous

11.7; -5.7; 25.6; 12 6; 107.3; 26.1; 778 | 46; 20; 5 | -94.1; 46; 39.3

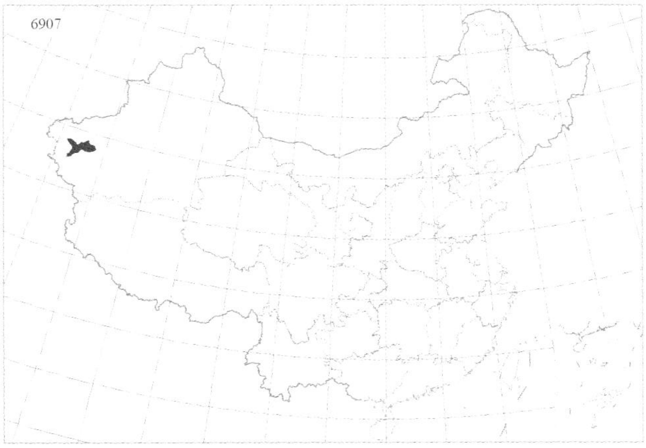

Tamaricaceae 柽柳科 *Tamarix* 柽柳属
Tamarix taklamakanensis M. T. Liu 沙生柽柳
Shrubs or small trees, deciduous

10.6 (-4.1~12.4); -7.5 (-18.9~-5.3); 25.0 (9.8~26.8); 12.1 (2.5~13.2); 103.8 (9.9~114.1); 35.5 (24.7~117.9); 766 (330~820) | 33 (21~51); 20 (15~31); 2 (0~4) | -95.3 (-97.3~-84.7); 33 (21~51); 6.2 (1.6~24.8)

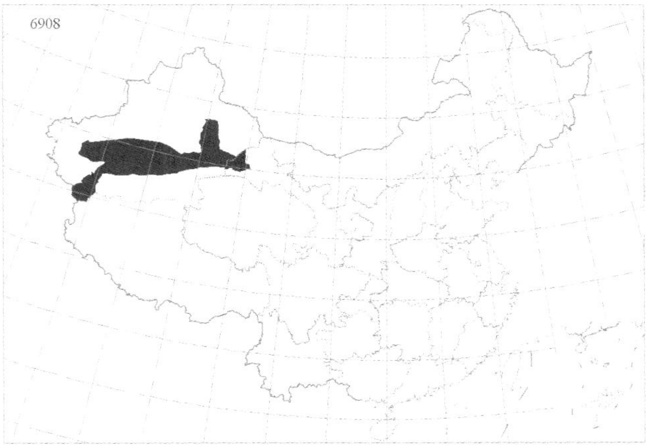

Tamaricaceae 柽柳科 *Tamarix* 柽柳属
Tamarix tarimensis P. Y. Zhang et M. T. Liu 塔里木柽柳
Shrubs

3.1 (-4.1~11.3); -13.5 (-18.9~-7.3); 17.1 (9.8~25.9); 7.0 (2.5~12.6); 53.0 (9.9~109.0); 75.7 (32.6~117.9); 533 (330~788) | 46 (21~76); 32 (15~56); 2 (0~4) | -88.7 (-97.3~-79.3); 46 (21~76); 7.5 (3.8~13.7)

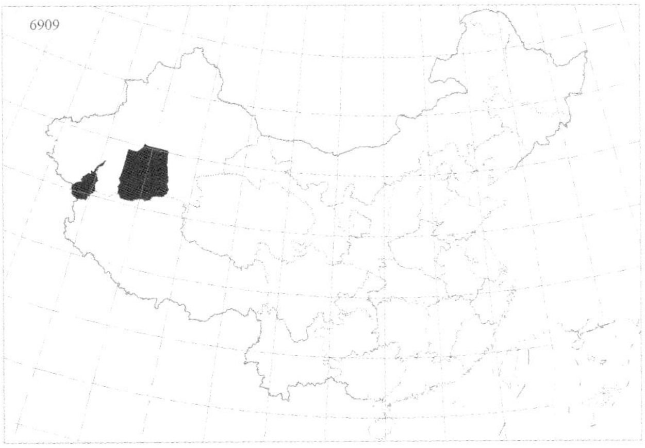

Cistaceae 半日花科 *Helianthemum* 半日花属
Helianthemum songaricum Schrenk 半日花
Shrubs, evergreen

5.7 (-1.4~9.3); -11.6 (-20.9~-8.7); 20.4 (11.5~26.5); 8.5 (3.5~11.8); 66.2 (17.2~104.4); 56.9 (40.7~103.9); 592 (375~786) | 267 (77~445); 151 (48~276); 13 (2~48) | -53.7 (-87.2~-24.8); 267 (77~445); 92.1 (13.6~297.3)

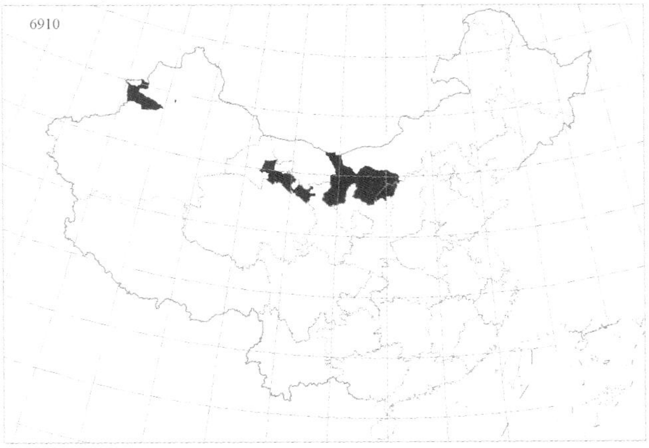

Violaceae 堇菜科 *Rinorea* 三角车属
Rinorea bengalensis (Wall.) Kuntze 三角车
Shrubs or small trees

23.0 (20.8~24.8); 16.8 (12.3~20.1); 27.5 (25.7~28.5); 23.1 (20.8~24.8); 216.9 (190.0~237.6); 0.0 (0.0~0.0); 1236 (1080~1385) | 1543 (1374~1854); 623 (455~850); 73 (41~115) | 25.5 (1.5~47.8); 1156 (1055~1360); 454.0 (275.8~620.0)

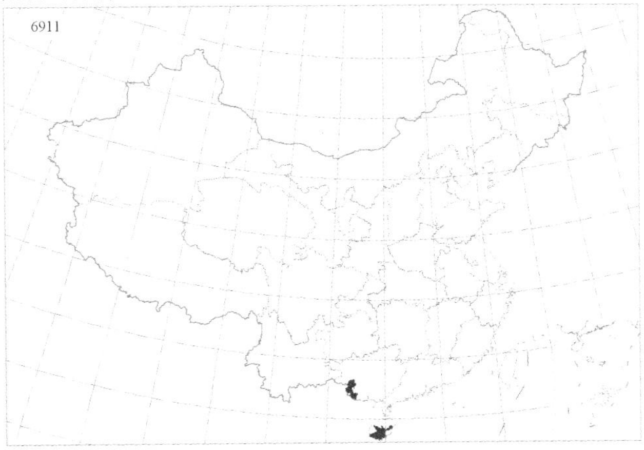

Violaceae 堇菜科 *Rinorea* 三角车属
Rinorea erianthera C. Y. Wu et C. Ho 毛蕊三角车
Shrubs, deciduous

10.8; 2.7; 17.8; 10.9; 74.5; 4.0; 636 | 1022; 588; 27 | 60.9; 636; 267.3

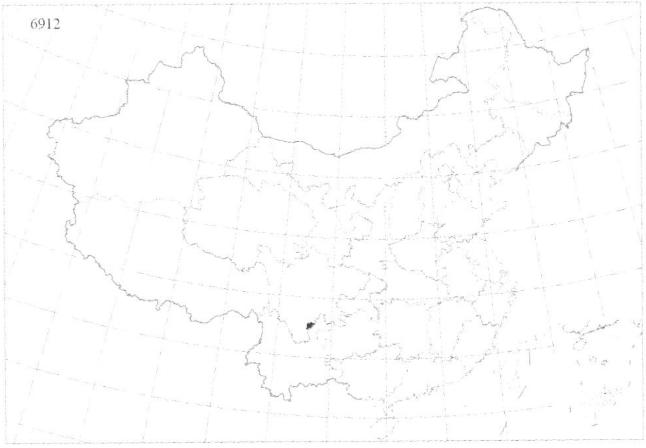

Violaceae 堇菜科 *Rinorea* 三角车属
Rinorea virgata (Thwaites) Kuntze 鳞隔堇
Shrubs

23.9 (23.0~24.8); 18.3 (17.5~20.1); 27.9 (26.7~28.7); 23.9 (23.1~24.8); 227.0 (217.0~237.6); 0.0 (0.0~0.0); 1299 (1217~1385) ❙ 1308 (868~1725); 460 (312~610); 53 (41~71) ❙ 1.3 (-34.3~41.8); 1089 (868~1163); 461.1 (288.3~572.9)

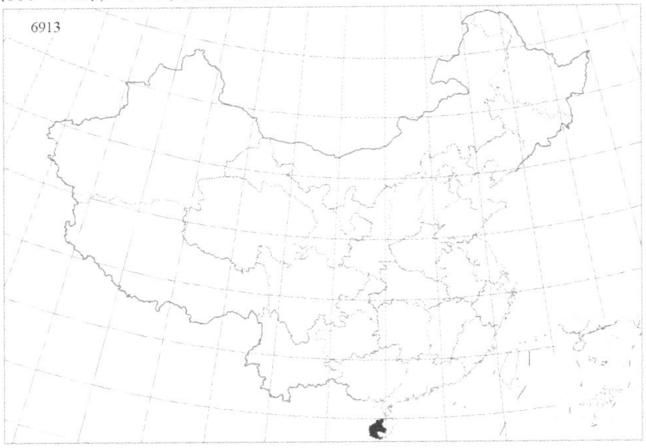

Flacourtiaceae 大风子科 *Bennettiodendron* 山桂花属
Bennettiodendron leprosipes (Clos) Merr. 山桂花
Shrubs or trees, evergreen

19.8 (15.5~24.8); 11.0 (4.8~20.1); 26.7 (20.6~29.8); 19.8 (15.5~24.8); 177.6 (126.3~237.6); 0.0 (0.0~0.2); 1028 (768~1385) ❙ 1524 (989~2435); 687 (473~1428); 123 (41~215) ❙ 49.6 (1.5~99.5); 1019 (768~1326); 291.9 (174.9~659.0)

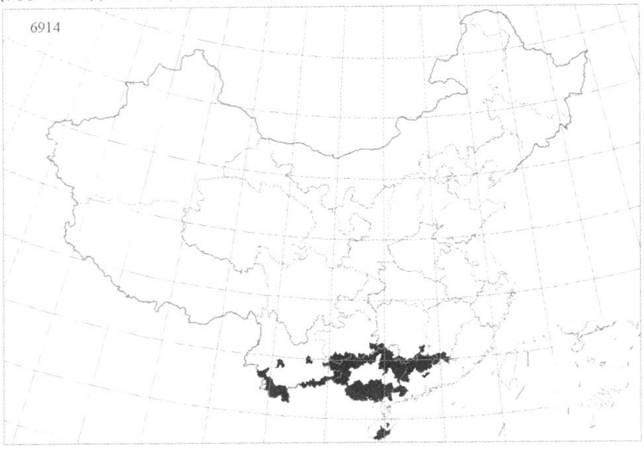

Flacourtiaceae 大风子科 *Carrierea* 山羊角树属
Carrierea calycina Franch. 山羊角树
Trees, deciduous

14.9 (6.0~22.0); 4.8 (-3.2~14.2); 24.0 (14.4~28.4); 14.9 (6.6~22.0); 122.2 (39.7~204.5); 3.2 (0.0~26.7); 799 (508~1172) ❙ 1271 (958~1528); 592 (421~860); 75 (27~192) ❙ 61.0 (13.3~107.2); 798 (508~1172); 250.0 (147.3~551.6)

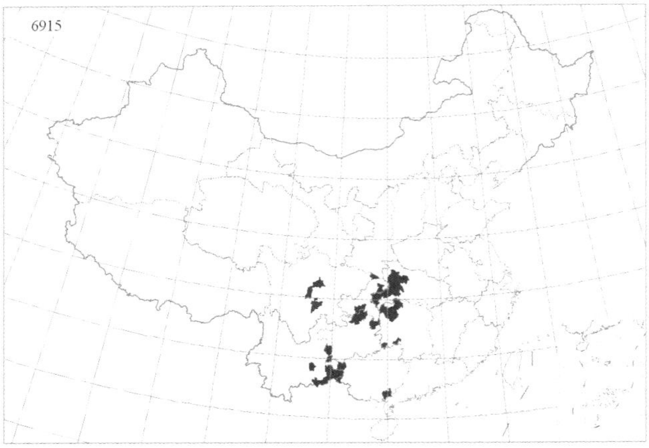

Flacourtiaceae 大风子科 *Carrierea* 山羊角树属
Carrierea dunniana H. Lév. 贵州嘉丽树
Trees, deciduous

16.7 (14.9~19.2); 7.5 (4.7~13.8); 24.3 (21.4~26.6); 16.7 (15.0~19.2); 140.4 (119.9~170.5); 0.0 (0.0~0.3); 846 (781~969) ❙ 1290 (1098~1564); 591 (511~686); 82 (46~182) ❙ 52.7 (27.5~70.7); 842 (781~969); 262.6 (194.1~513.9)

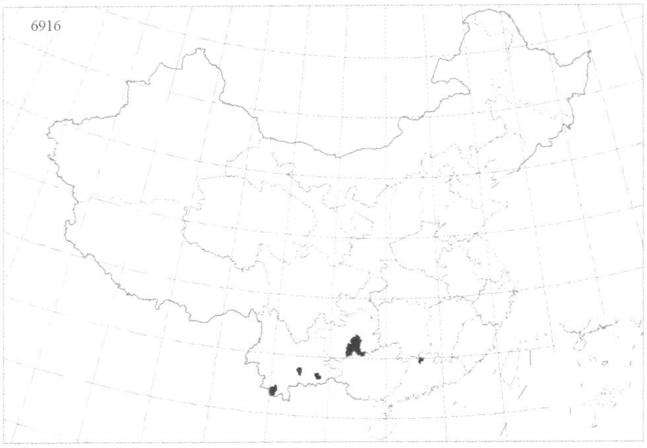

Flacourtiaceae 大风子科 *Casearia* 脚骨脆属
Casearia flexuosa Craib 云南脚骨脆
Shrubs

18.2 (17.2~21.0); 10.6 (9.5~14.2); 23.5 (22.4~25.6); 18.2 (17.3~21.0); 158.3 (147.0~191.8); 0.0 (0.0~0.0); 890 (833~1053) ❙ 1394 (1254~1522); 778 (699~860); 57 (48~68) ❙ 57.3 (43.5~80.6); 890 (833~1053); 411.4 (316.1~551.6)

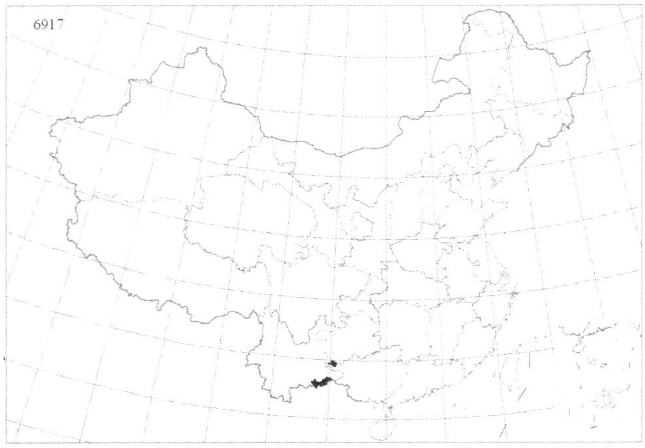

Flacourtiaceae 大风子科 *Casearia* 脚骨脆属
Casearia glomerata Roxb. 球花脚骨脆
Shrubs or trees, evergreen

20.6 (13.3~24.1); 12.5 (5.4~19.1); 27.4 (19.6~29.0); 20.7 (13.4~24.2); 188.5 (100.4~230.2); 0.0 (0.0~0.0); 1081 (706~1327) ❙ 1626 (1182~2435); 713 (419~1428); 133 (43~217) ❙ 52.2 (-1.4~145.8); 1076 (706~1326); 269.4 (139.6~659.0)

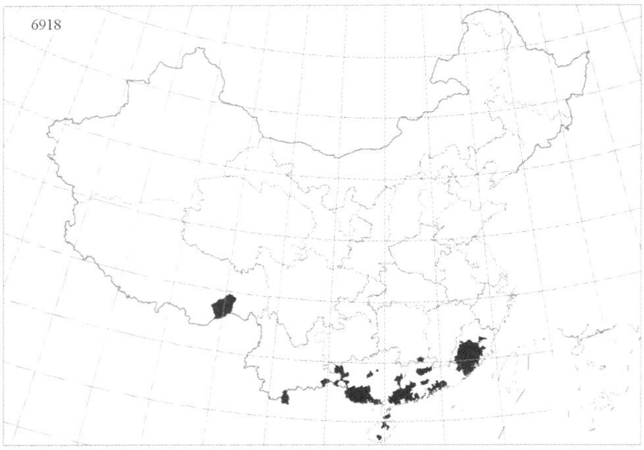

Flacourtiaceae 大风子科 *Casearia* 脚骨脆属
Casearia graveolens Dalzell 香味脚骨脆
Trees, deciduous

18.2 (16.1~21.0); 12.0 (9.6~15.1); 22.4 (20.3~25.6); 18.3 (16.2~21.0); 159.4 (133.9~191.3); 0.0 (0.0~0.0); 880 (785~1053) ❙ 1349 (1049~1607); 725 (594~860); 56 (44~77) ❙ 53.1 (26.1~80.6); 865 (779~1053); 446.5 (359.7~659.0)

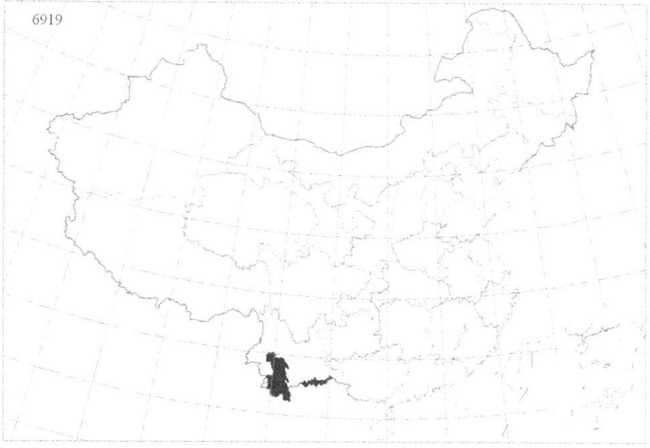

Flacourtiaceae 大风子科 *Casearia* 脚骨脆属
Casearia kurzii C. B. Clarke 印度脚骨脆
Small trees

18.5 (16.3~21.0); 12.3 (9.9~15.1); 22.4 (20.3~25.6); 18.5 (16.4~21.0); 162.2 (136.6~191.8); 0.0 (0.0~0.0); 890 (789~1053) ❙ 1387 (1049~1607); 736 (598~860); 56 (44~77) ❙ 56.1 (26.1~80.6); 871 (780~1053); 466.2 (363.2~659.0)

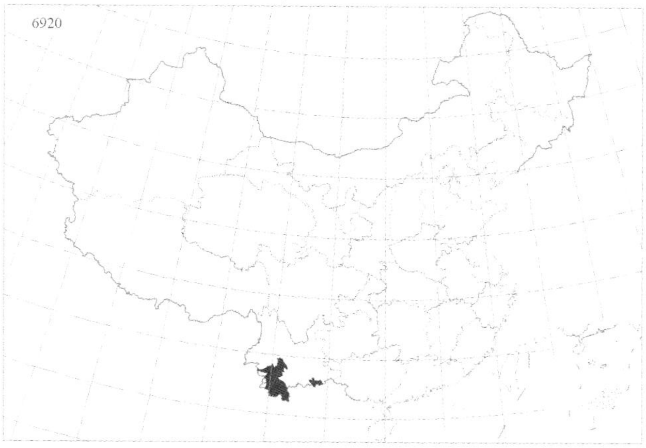

Flacourtiaceae 大风子科 *Casearia* 脚骨脆属
Casearia membranacea Hance 膜叶脚骨脆
Trees, evergreen

21.1 (15.5~24.8); 13.2 (4.8~20.1); 27.2 (20.1~29.8); 21.1 (15.6~24.8); 193.3 (127.2~237.6); 0.0 (0.0~0.2); 1110 (773~1385) ❙ 1632 (868~2993); 726 (312~1847); 128 (41~547) ❙ 51.4 (-34.3~246.5); 1079 (773~1360); 302.2 (174.9~620.0)

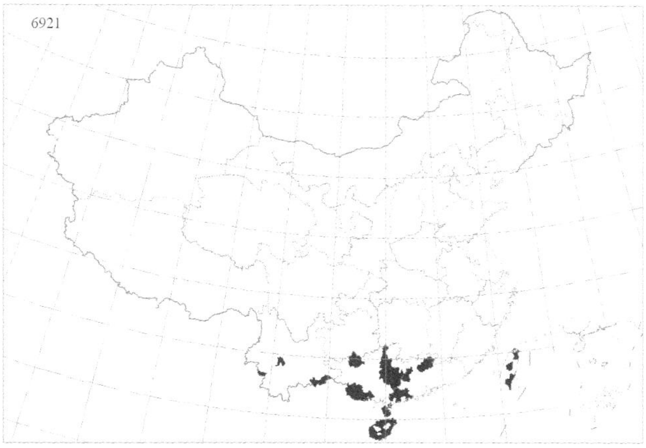

Flacourtiaceae 大风子科 *Casearia* 脚骨脆属
Casearia tardieuae Lescot et Sleumer 石生脚骨脆
Trees

18.1 (16.1~22.2); 10.8 (9.5~13.9); 23.3 (20.6~28.3); 18.1 (16.2~22.3); 157.8 (133.9~207.5); 0.0 (0.0~0.0); 896 (785~1190) ❙ 1246 (1049~1438); 695 (598~820); 55 (44~87) ❙ 40.1 (15.4~65.6); 884 (779~1190); 365.7 (327.7~405.0)

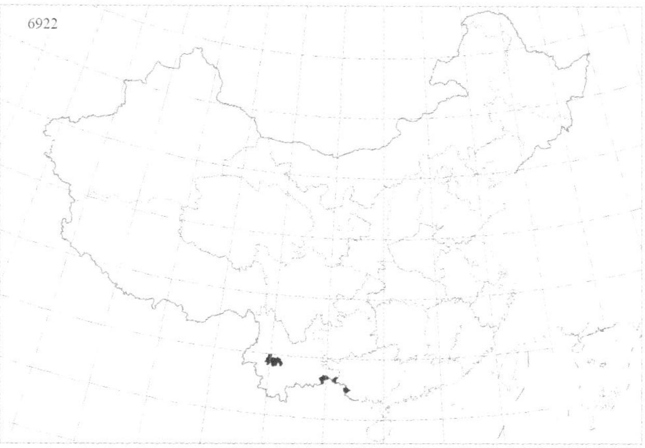

Flacourtiaceae 大风子科 *Casearia* 脚骨脆属
Casearia velutina Blume 毛叶脚骨脆
Shrubs or small trees

20.0 (16.3~23.2); 12.2 (7.1~17.5); 25.9 (20.3~29.0); 20.0 (16.4~23.2); 180.2 (136.6~218.9); 0.0 (0.0~0.0); 1023 (789~1257) ❙ 1577 (1049~2435); 760 (513~1428); 110 (42~193) ❙ 55.2 (5.8~99.5); 1017 (780~1257); 32[illegible].4 (160.7~659.0)

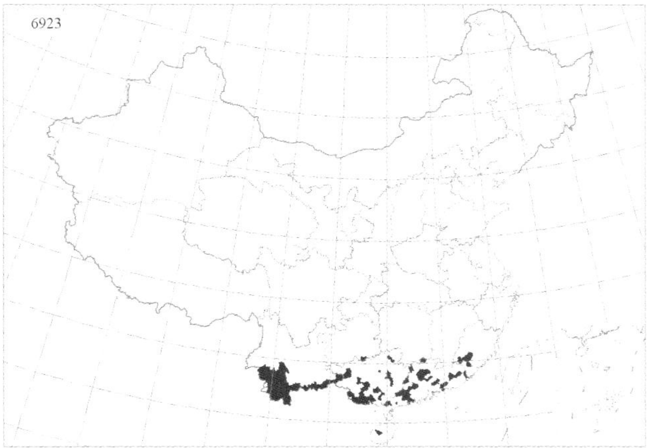

Flacourtiaceae 大风子科 *Flacourtia* 刺篱木属
Flacourtia indica (Burm. f.) Merr. 刺篱木
Shrubs or small trees, deciduous

23.5 (19.7~24.8); 17.1 (11.5~20.1); 28.2 (25.7~29.0); 23.5 (19.7~24.8); 221.9 (176.5~237.6); 0.0 (0.0~0.0); 1273 (1001~1385) ❙ 1530 (868~1888); 605 (312~1044); 88 (41~163) ❙ 21.1 (-34.3~62.9); 1189 (868~1360); 367.2 (145.3~620.0)

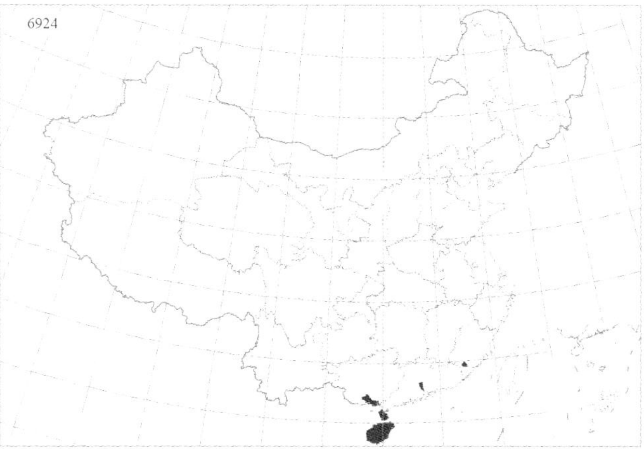

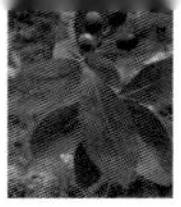

Flacourtiaceae 大风子科 *Flacourtia* 刺篱木属

Flacourtia jangomas (Lour.) Raeusch. 云南刺篱木

Shrubs or small trees, deciduous

22.9 (20.5~24.8); 17.1 (12.6~20.1); 27.0 (23.7~28.3); 23.0 (20.5~24.8); 215.7 (185.9~237.6); 0.0 (0.0~0.0); 1224 (992~1385) ❙ 1452 (1188~1666); 575 (455~713); 57 (41~87) ❙ 19.9 (1.5~47.8); 1117 (959~1212); 487.7 (315.6~620.0)

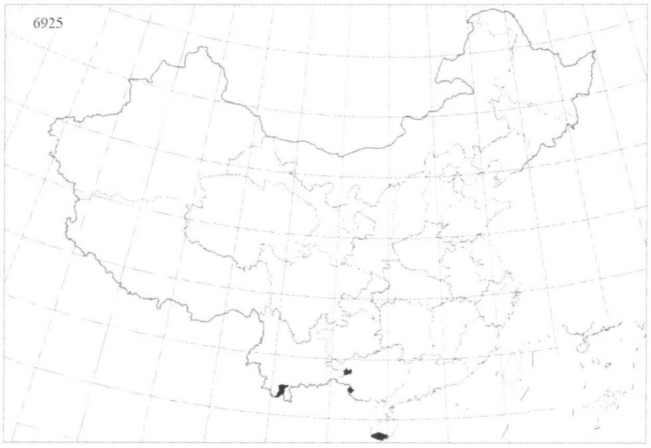

Flacourtiaceae 大风子科 *Flacourtia* 刺篱木属

Flacourtia mollis Hook. f. et Thomson 毛叶刺篱木

Small trees, deciduous

20.3; 14.5; 23.8; 20.3; 183.5; 0.0; 986 ❙ 1523; 715; 67 ❙ 54.9; 970; 573.6

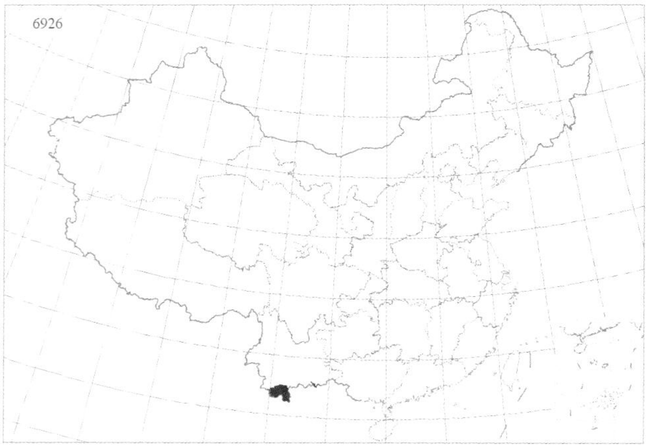

Flacourtiaceae 大风子科 *Flacourtia* 刺篱木属

Flacourtia ramontchi L'Hér. 大果刺篱木

Trees, deciduous

19.4 (14.4~23.0); 11.5 (5.3~15.1); 25.3 (19.1~29.0); 19.5 (14.4~23.0); 173.6 (113.1~216.2); 0.0 (0.0~0.0); 987 (727~1247) ❙ 1426 (1028~1745); 735 (523~997); 83 (41~169) ❙ 46.6 (5.8~112.2); 981 (727~1247); 345.8 (189.5~659.0)

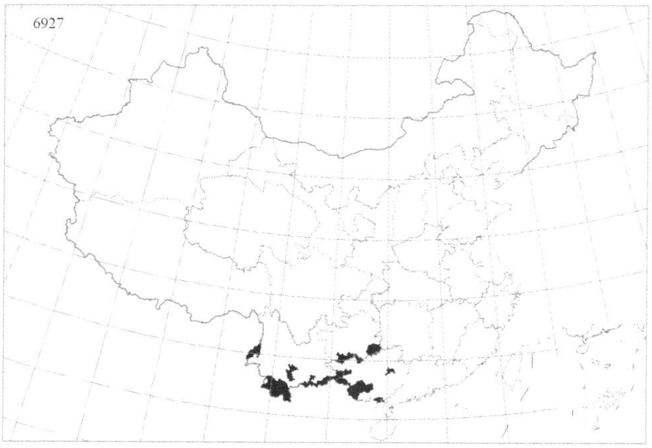

Flacourtiaceae 大风子科 *Flacourtia* 刺篱木属

Flacourtia rukam Zoll. et Moritzi 大叶刺篱木

Trees, deciduous

21.3 (17.2~24.8); 13.4 (7.4~20.1); 27.4 (21.7~29.2); 21.4 (17.3~24.8); 196.3 (147.0~237.6); 0.0 (0.0~0.0); 1123 (838~1385) ❙ 1617 (868~2992); 768 (312~1847); 112 (41~189) ❙ 46.5 (-34.3~214.4); 1091 (838~1273); 299.3 (152.9~659.0)

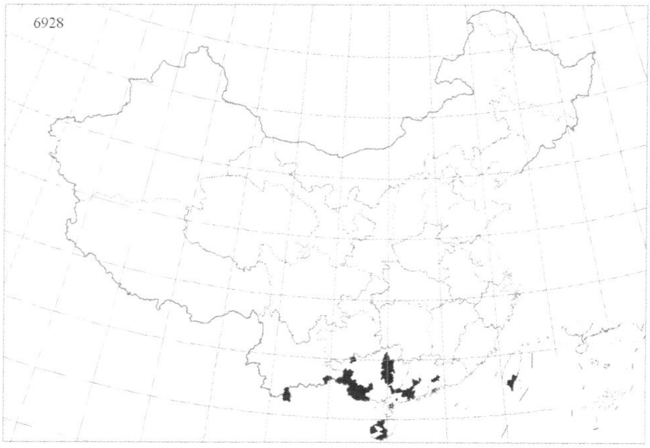

Flacourtiaceae 大风子科 *Gynocardia* 马蛋果属

Gynocardia odorata R. Br. 马蛋果

Shrubs or trees, evergreen

18.7 (13.3~21.0); 12.8 (5.4~15.2); 23.7 (19.6~28.7); 19.3 (13.4~22.7); 171.3 (100.4~212.7); 0.0 (0.0~0.0); 955 (706~1216) ❙ 1574 (1368~1965); 807 (655~1007); 67 (51~104) ❙ 67.8 (44.5~145.8); 947 (706~1216); 493.0 (204.4~659.0)

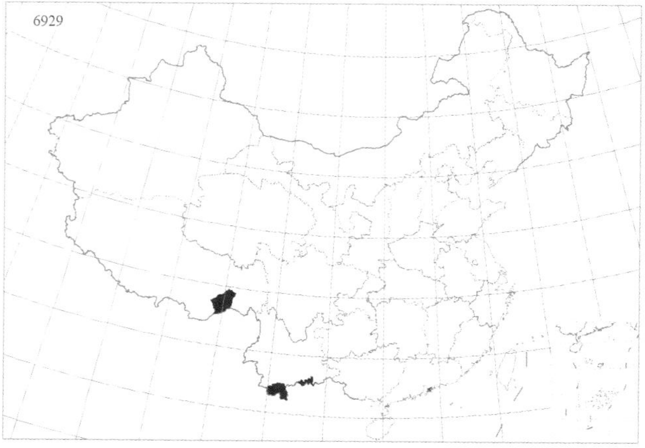

Flacourtiaceae 大风子科 *Homalium* 天料木属

Homalium breviracemosum F. C. How et W. C. Ko 短穗天料木

Shrubs

22.3 (21.8~22.7); 14.6 (14.0~15.0); 28.3 (27.7~28.7); 22.3 (21.9~22.8); 207.7 (202.2~213.1); 0.0 (0.0~0.0); 1186 (1150~1221) ❙ 1873 (1447~2435); 1006 (777~1428); 104 (75~130) ❙ 58.1 (20.8~99.5); 1186 (1150~1221); 235.4 (139.6~318.8)

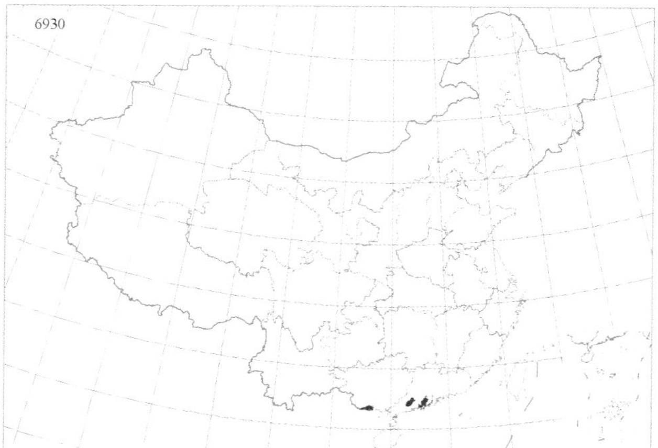

Flacourtiaceae 大风子科 *Homalium* 天料木属
Homalium ceylanicum (Gardner) Benth. 斯里兰卡天料木
Trees

22.2 (13.3~24.8); 15.4 (5.4~20.1); 27.2 (19.6~29.0); 22.2 (13.4~24.8); 206.8 (100.4~237.6); 0.0 (0.0~0.0); 1182 (706~1385) ❙ 1525 (868~2037); 655 (312~1033); 86 (41~144) ❙ 31.7 (-34.3~145.8); 1124 (706~1360); 362.6 (145.3~659.0)

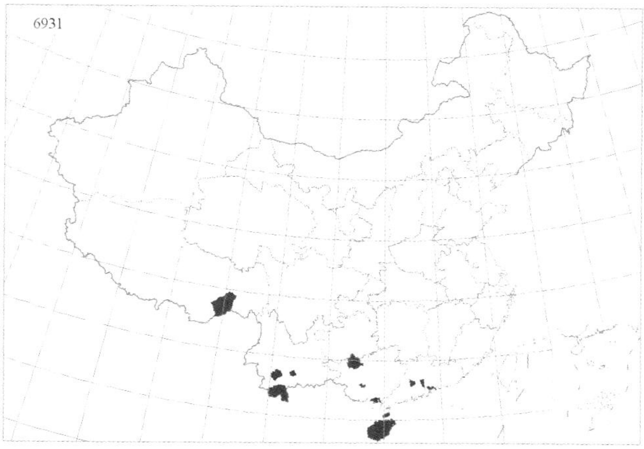

Flacourtiaceae 大风子科 *Homalium* 天料木属
Homalium cochinchinense (Lour.) Druce 天料木
Shrubs or small trees

20.2 (14.3~24.8); 11.8 (3.7~20.1); 27.4 (20.1~29.8); 20.3 (14.3~24.8); 183.6 (113.2~237.6); 0.0 (0.0~1.5); 1061 (761~1385) ❙ 1689 (868~3495); 703 (312~1847); 160 (41~897) ❙ 62.4 (-34.3~246.5); 1047 (761~1273); 242.4 (95.3~620.0)

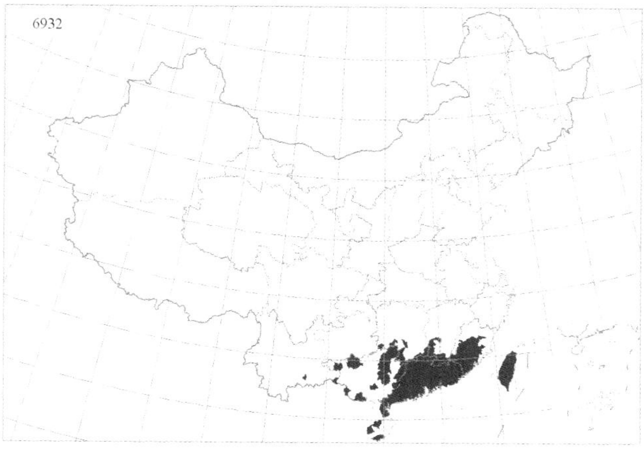

Flacourtiaceae 大风子科 *Homalium* 天料木属
Homalium kainantense Masam. 阔瓣天料木
Trees

22.9 (21.3~24.8); 16.4 (13.2~20.1); 27.7 (26.7~28.6); 23.0 (21.3~24.8); 215.7 (195.9~237.6); 0.0 (0.0~0.0); 1233 (1111~1385) ❙ 1800 (1405~2435); 846 (473~1428); 87 (41~130) ❙ 47.8 (1.5~99.5); 1164 (1111~1221); 402.2 (215.3~620.0)

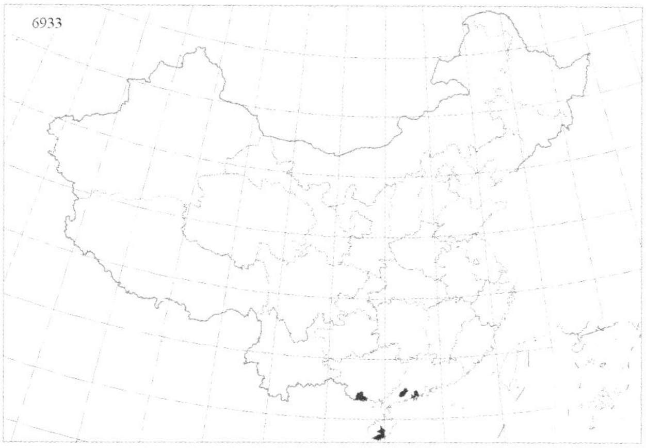

Flacourtiaceae 大风子科 *Homalium* 天料木属
Homalium kwangsiense F. C. How et W. C. Ko 广西天料木
Trees

22.2; 13.9; 28.3; 22.3; 207.5; 0.0; 1190 ❙ 1374; 713; 87 ❙ 15.4; 1190; 327.7

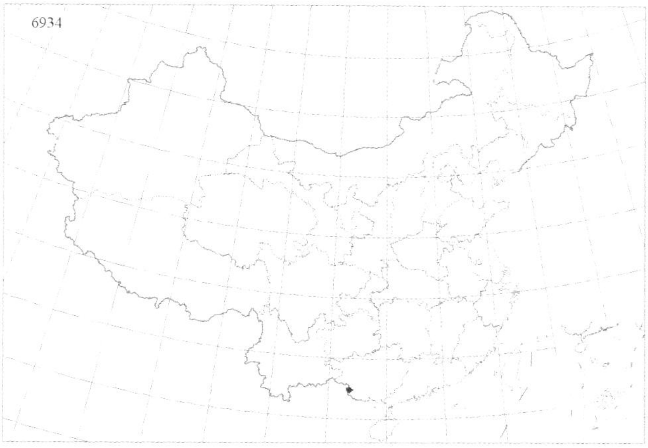

Flacourtiaceae 大风子科 *Homalium* 天料木属
Homalium mollissimum Merr. 毛天料木
Shrubs or small trees

23.0 (17.0~24.8); 16.6 (6.2~20.1); 27.9 (26.7~28.8); 23.0 (17.0~24.8); 216.5 (144.4~237.6); 0.0 (0.0~0.0); 1249 (891~1385) ❙ 1592 (1179~1876); 594 (392~891); 90 (41~191) ❙ 29.8 (-11.9~82.3); 1153 (891~1326); 412.5 (203.4~620.0)

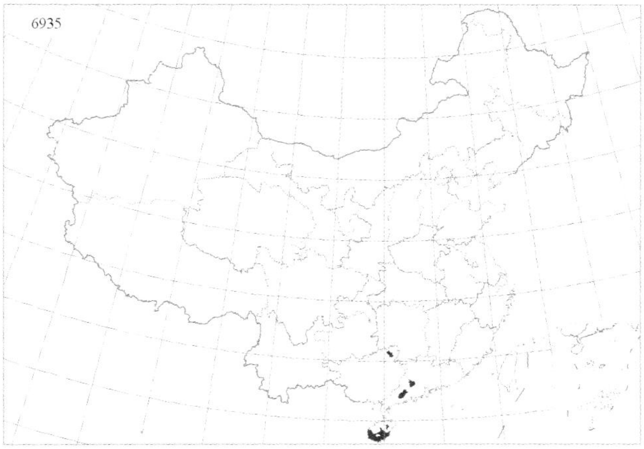

Flacourtiaceae 大风子科 *Homalium* 天料木属
Homalium paniculiflorum F. C. How et W. C. Ko 广南天料木
Shrubs or small trees

22.3 (20.4~24.8); 14.8 (11.9~20.1); 28.2 (26.7~29.1); 22.3 (20.4~24.8); 207.8 (184.9~237.6); 0.0 (0.0~0.0); 1188 (1039~1385) ❙ 1675 (868~2037); 749 (312~1033); 114 (41~188) ❙ 42.2 (-34.3~83.4); 1155 (868~1299); 244.1 (95.3~548.0)

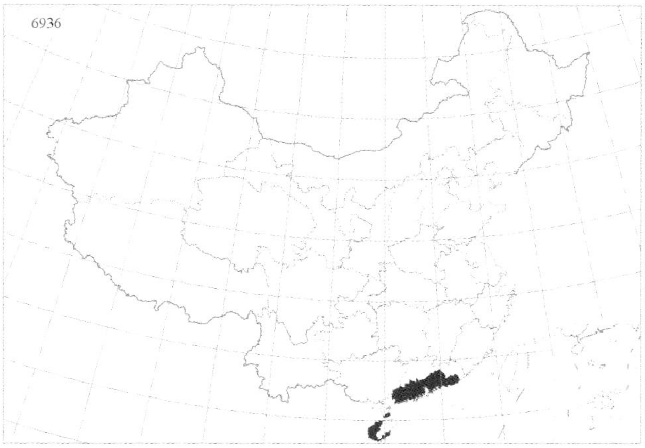

Flacourtiaceae 大风子科 *Homalium* 天料木属
Homalium phanerophlebium F. C. How et W. C. Ko 显脉天料木
Trees, evergreen

22.7 (17.1~24.8); 15.8 (8.5~20.1); 27.9 (23.4~29.8); 22.7 (17.1~24.8); 212.4 (145.1~237.6); 0.0 (0.0~0.0); 1222 (845~1385) ❙ 1582 (868~2035); 601 (312~969); 103 (41~194) ❙ 31.7 (-34.3~85.1); 1144 (845~1360); 367.3 (174.9~620.0)

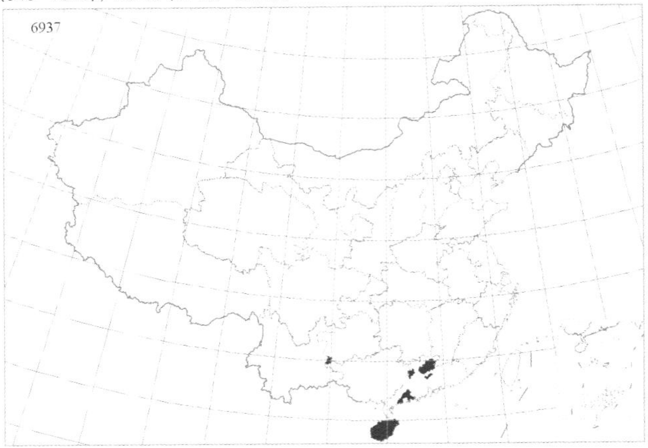

Flacourtiaceae 大风子科 *Homalium* 天料木属
Homalium sabiifolium F. C. How et W. C. Ko 窄叶天料木
Shrubs

21.5; 13.0; 28.0; 21.5; 198.4; 0.0; 1141 ❙ 1863; 1013; 119 ❙ 62.7; 1141; 300.8

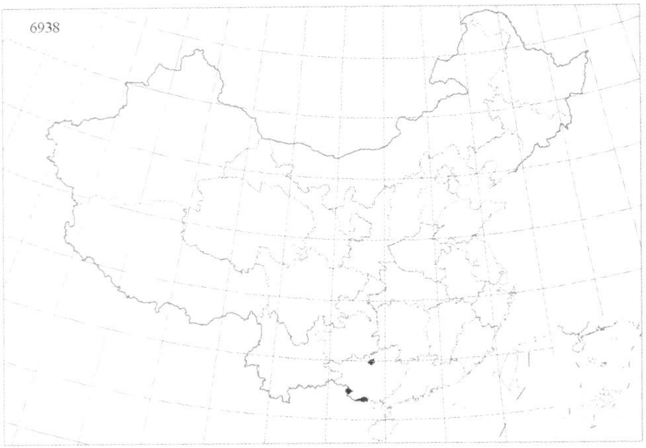

Flacourtiaceae 大风子科 *Homalium* 天料木属
Homalium stenophyllum Merr. et Chun 海南天料木
Trees

23.6 (21.8~24.8); 17.7 (14.0~20.1); 27.8 (26.7~28.6); 23.6 (21.9~24.8); 223.5 (202.2~237.6); 0.0 (0.0~0.0); 1282 (1150~1385) ❙ 1796 (1405~2435); 777 (473~1428); 83 (41~130) ❙ 42.0 (1.5~99.5); 1194 (1137~1326); 455.3 (300.5~620.0)

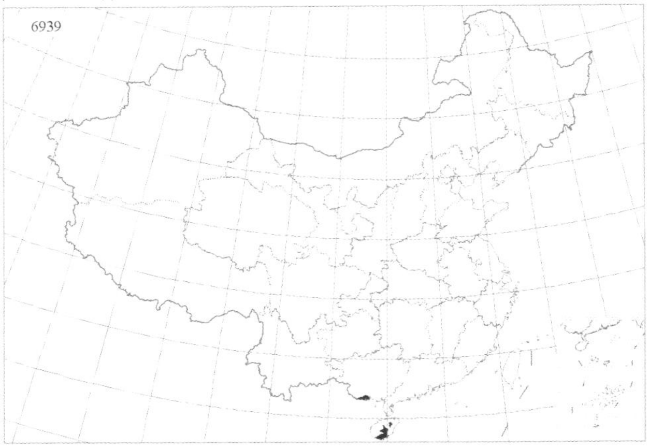

Flacourtiaceae 大风子科 *Hydnocarpus* 大风子属
Hydnocarpus annamensis (Gagnep.) Lescot et Sleumer 大叶龙角
Trees, evergreen

20.6 (17.7~22.2); 13.4 (10.9~15.1); 25.7 (22.5~28.5); 20.6 (17.8~22.3); 187.6 (153.4~207.5); 0.0 (0.0~0.0); 1047 (854~1190) ❙ 1523 (1302~1900); 815 (670~1056); 86 (56~128) ❙ 47.1 (12.9~74.2); 1047 (854~1190); 422.7 (256.6~659.0)

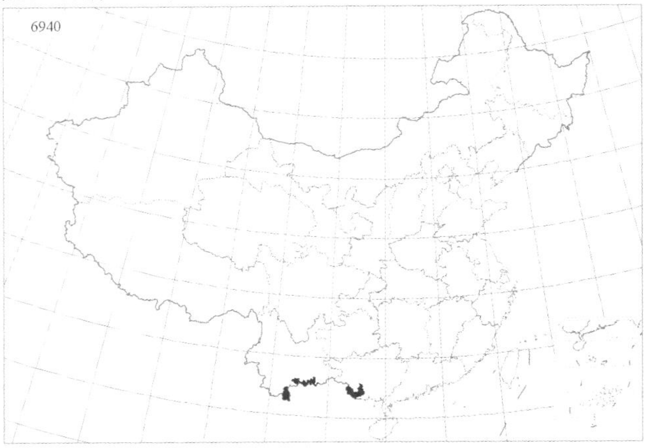

Flacourtiaceae 大风子科 *Hydnocarpus* 大风子属
Hydnocarpus anthelminthicus Pierre 泰国大风子
Trees, evergreen

21.9; 13.6; 28.1; 22.0; 203.5; 0.0; 1166 ❙ 1491; 779; 110 ❙ 28.1; 1166; 293.9

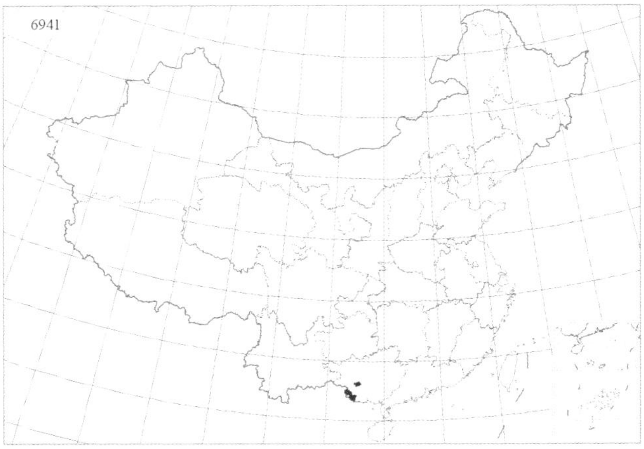

Flacourtiaceae 大风子科 *Hydnocarpus* 大风子属
Hydnocarpus hainanensis (Merr.) Sleumer 海南大风子
Trees, evergreen

23.0 (17.7~24.8); 16.7 (10.9~20.1); 27.4 (22.5~28.8); 23.0 (17.8~24.8); 216.0 (153.4~237.6); 0.0 (0.0~0.0); 1234 (854~1385) ❙ 1483 (868~1854); 607 (312~850); 75 (41~115) ❙ 21.9 (-34.3~60.1); 1150 (854~1360); 420.1 (248.2~620.0)

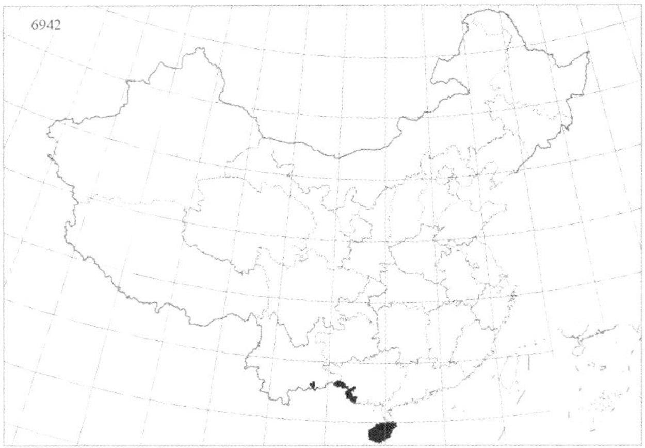

Flacourtiaceae 大风子科 *Idesia* 山桐子属
Idesia polycarpa Maxim. 山桐子
Trees, deciduous

16.6 (2.6~22.4); 6.0 (-7.0~17.7); 26.3 (11.2~30.1); 16.7 (4.3~22.5); 142.2 (20.0~209.7); 2.0 (0.0~48.2); 891 (427~1205) ❙ 1384 (520~3495); 558 (258~1847); 134 (12~897) ❙ 55.3 (-35.9~224.4); 877 (427~1205); 220.9 (95.3~491.5)

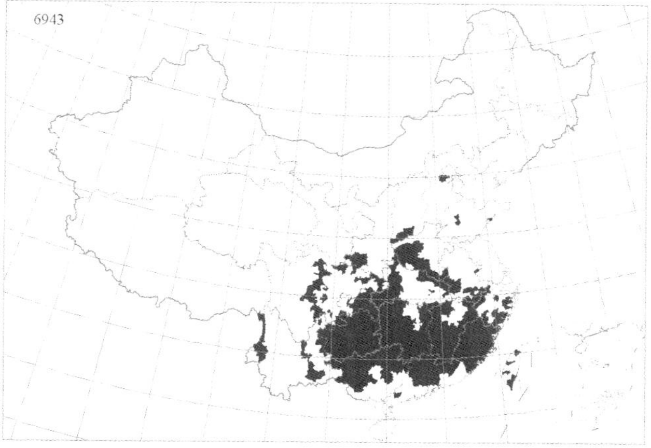

Flacourtiaceae 大风子科 *Idesia* 山桐子属
Idesia polycarpa var. fujianensis (G. S. Fan) S. S. Lai 福建山桐子
Trees, deciduous

17.7; 7.7; 27.0; 17.8; 153.4; 0.0; 924 ❙ 1667; 609; 187 ❙ 81.8; 924; 251.0

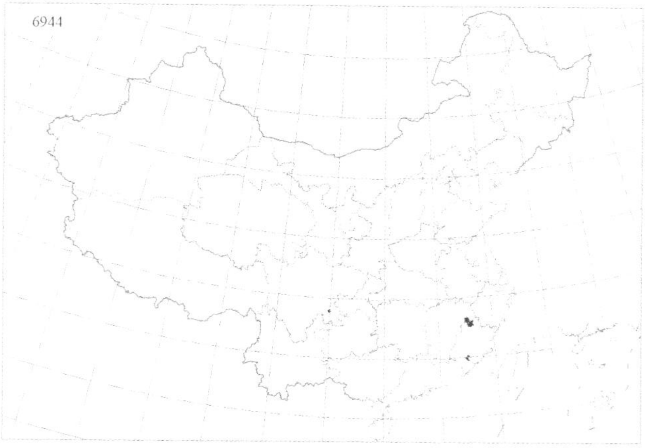

Flacourtiaceae 大风子科 *Idesia* 山桐子属
Idesia polycarpa var. vestita Diels 毛叶山桐子
Trees, deciduous

15.4 (3.8~19.4); 4.1 (-4.3~9.4); 26.0 (10.8~30.1); 15.4 (4.7~19.4); 128.4 (21.6~173.4); 3.4 (0.0~35.4); 841 (444~1037) ❙ 1348 (626~1997); 516 (299~816); 132 (15~259) ❙ 59.7 (-22.2~137.5); 825 (444~1029); 225.1 (111.9~332.5)

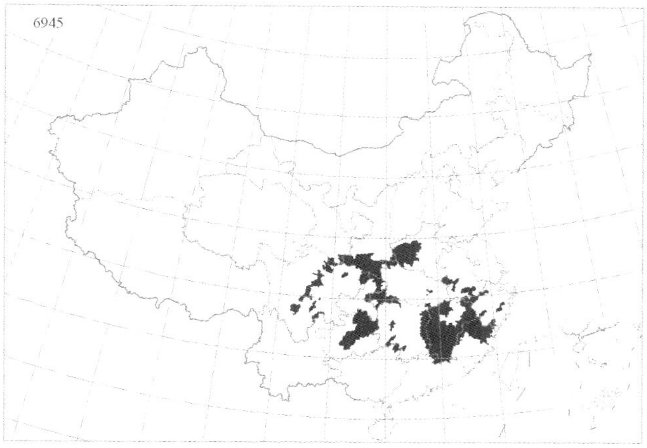

Flacourtiaceae 大风子科 *Itoa* 栀子皮属
Itoa orientalis Hemsl. 栀子皮
Trees, evergreen

18.0 (13.3~24.6); 9.3 (3.2~19.5); 25.0 (19.1~28.9); 18.0 (13.2~24.7); 156.5 (100.9~235.9); 0.1 (0.0~2.5); 919 (715~1372) ❙ 1316 (928~1776); 656 (452~996); 80 (23~189) ❙ 44.1 (5.8~99.9); 913 (715~1299); 285.5 (120.2~659.0)

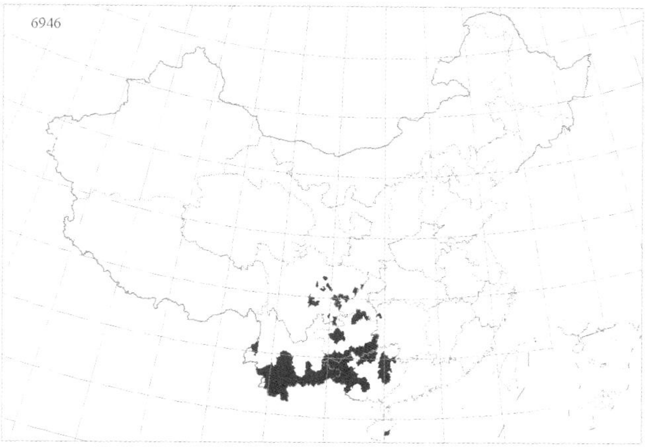

Flacourtiaceae 大风子科 *Itoa* 栀子皮属
Itoa orientalis var. glabrescens C. Y. Wu ex G. S. Fan 光叶栀子皮
Trees, evergreen

19.6; 11.2; 25.8; 19.6; 175.5; 0.0; 986 ❙ 1400; 776; 74 ❙ 42.0; 986; 294.6

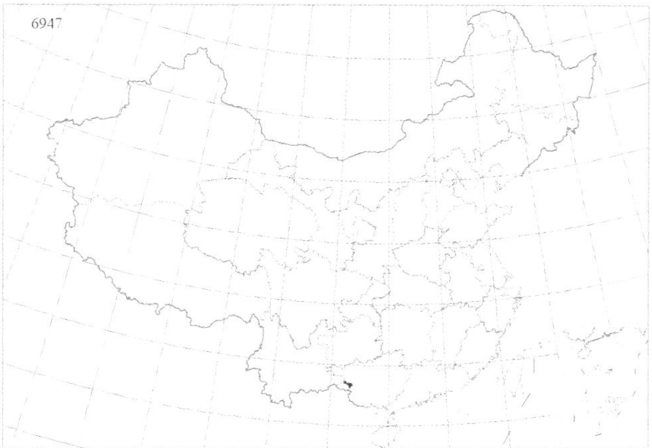

Flacourtiaceae 大风子科 *Poliothyrsis* 山拐枣属
Poliothyrsis sinensis Oliv. 山拐枣
Trees, deciduous

14.7 (5.7~20.0); 3.0 (-4.9~10.6); 25.6 (14.2~29.4); 14.7 (6.4~20.0); 121.0 (38.1~180.0); 4.3 (0.0~28.1); 812 (504~1040) ❙ 1198 (622~1922); 492 (298~720); 101 (14~259) ❙ 47.0 (-22.2~123.6); 801 (504~1040); 211.1 (134.6~320.6)

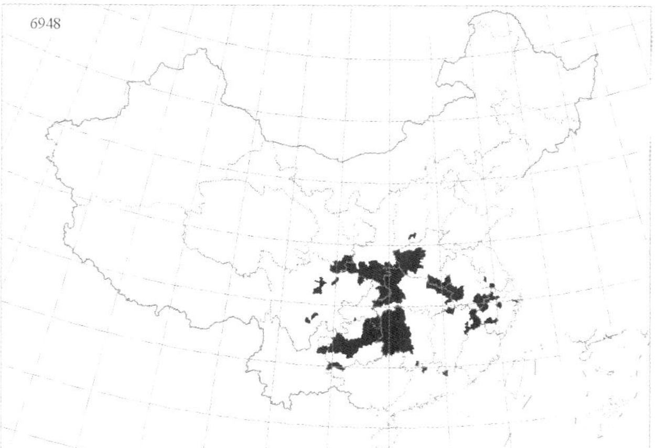

Flacourtiaceae 大风子科 *Scolopia* 箣柊属
Scolopia buxifolia Gagnep. 黄杨叶箣柊
Shrubs or small trees, evergreen

22.4 (17.9~24.8); 15.2 (7.4~20.1); 28.0 (25.7~29.0); 22.5 (18.0~24.8); 209.5 (155.6~237.6); 0.0 (0.0~0.0); 1210 (913~1385) ❙ 1546 (868~1854); 606 (312~997); 110 (41~189) ❙ 30.4 (-34.3~85.5); 1139 (868~1360); 350.5 (185.5~620.0)

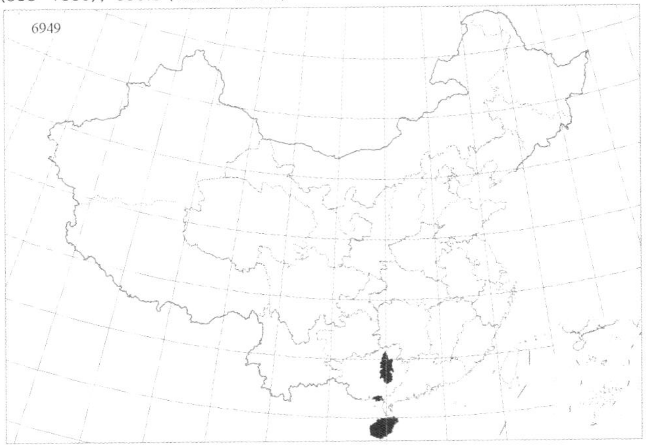

Flacourtiaceae 大风子科 *Scolopia* 箣柊属
Scolopia chinensis (Lour.) Clos 箣柊
Shrubs or small trees, evergreen

19.8 (15.0~24.5); 11.2 (4.9~18.9); 27.4 (24.3~29.0); 19.9 (15.1~24.5); 178.8 (120.8~234.4); 0.0 (0.0~0.1); 1039 (789~1360) ❙ 1609 (1070~2435); 653 (356~1428); 161 (48~257) ❙ 58.3 (-5.0~147.0); 1032 (789~1360); 254.9 (123.9~455.2)

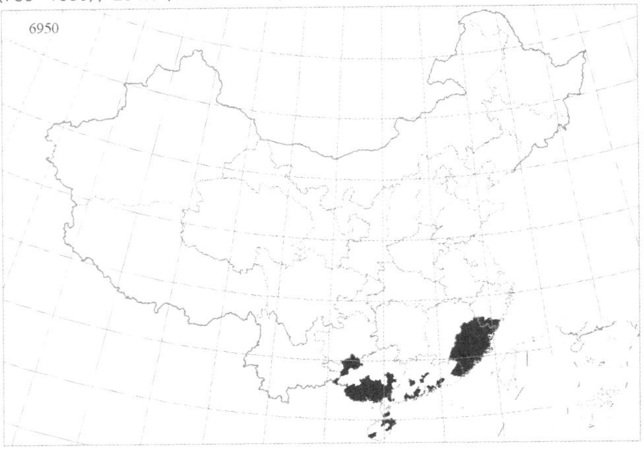

Flacourtiaceae 大风子科 *Scolopia* 箣柊属
Scolopia oldhamii Hance 台湾箣柊
Shrubs or small trees, evergreen

20.4 (17.9~24.1); 14.7 (11.6~18.9); 25.3 (21.7~28.0); 20.5 (18.0~24.1); 185.4 (155.6~229.7); 0.0 (0.0~0.0); 1023 (863~1322) ❙ 2674 (1751~3495); 1137 (750~1847); 346 (47~897) ❙ 166.8 (32.5~224.4); 997 (863~1123); 259.9 (154.2~363.7)

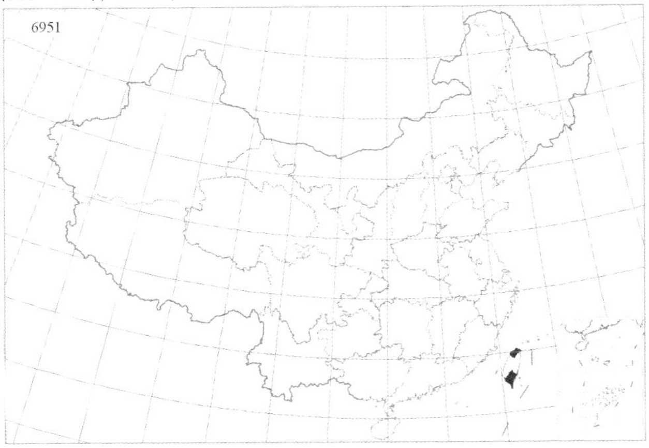

Flacourtiaceae 大风子科 *Scolopia* 箣柊属
Scolopia saeva (Hance) Hance 广东箣柊
Shrubs or small trees, evergreen

21.4 (16.9~24.8); 13.5 (6.5~20.1); 27.8 (22.6~29.2); 21.4 (17.0~24.8); 197.1 (143.6~237.6); 0.0 (0.0~0.0); 1129 (862~1385) ❙ 1626 (868~2435); 695 (312~1428); 129 (41~230) ❙ 46.1 (-34.3~104.6); 1111 (862~1360); 262.9 (95.3~620.0)

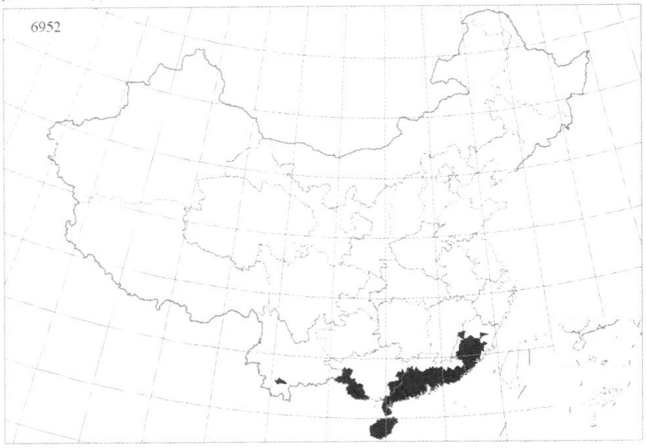

Flacourtiaceae 大风子科 *Xylosma* 柞木属
Xylosma congesta (Lour.) Merr. 柞木
Small trees, evergreen

17.6 (-1.1~23.6); 7.3 (-9.1~16.7); 26.9 (6.2~30.1); 17.6 (2.0~23.6); 152.3 (2.5~223.5); 0.9 (0.0~75.6); 936 (334~1281) ❙ 1439 (321~2499); 579 (193~1428); 140 (15~303) ❙ 54.7 (-6.9~219.3); 926 (321~1273); 220.8 (79.3~659.0)

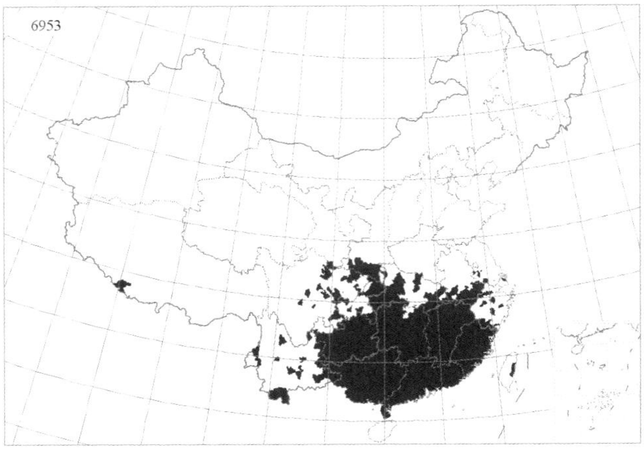

Flacourtiaceae 大风子科 *Xylosma* 柞木属
Xylosma controversa Clos 南岭柞木
Shrubs or small trees, evergreen

18.4 (13.3~24.7); 9.5 (3.2~19.8); 25.8 (19.6~29.8); 18.5 (13.2~24.7); 161.7 (100.9~236.5); 0.1 (0.0~2.5); 952 (723~1376) ❙ 1402 (939~2435); 652 (448~1428); 107 (32~216) ❙ 48.1 (-1.4~103.1); 946 (717~1326); 267.8 (120.0~659.0)

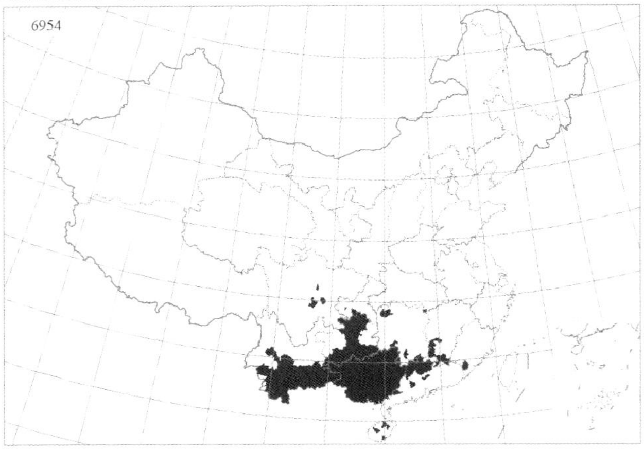

S

Flacourtiaceae 大风子科 *Xylosma* 柞木属
Xylosma longifolia Clos 长叶柞木
Shrubs or small trees, evergreen

19.3 (10.8~24.7); 11.2 (3.9~19.8); 25.9 (16.6~29.8); 19.3 (10.9~24.7); 171.9 (71.8~236.5); 0.0 (0.0~1.3); 992 (633~1376) ❙ 1513 (911~2435); 672 (359~1428); 124 (24~257) ❙ 53.6 (-6.9~137.5); 983 (633~1299); 283.0 (95.3~659.0)

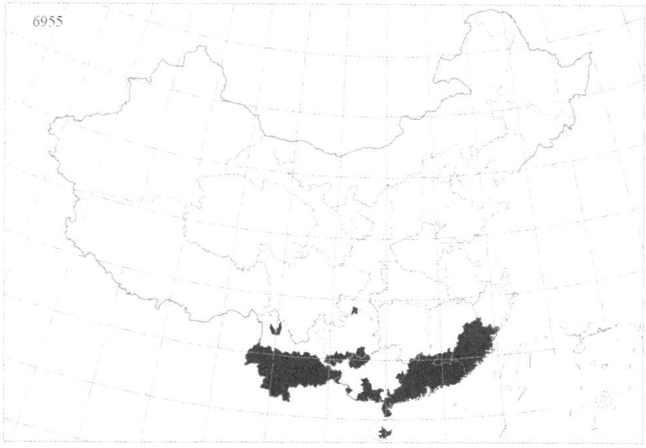

Stachyuraceae 旌节花科 *Stachyurus* 旌节花属
Stachyurus chinensis Franch. 中国旌节花
Shrubs, deciduous

15.5 (1.8~22.2); 4.8 (-7.7~13.9); 25.4 (9.9~29.9); 15.6 (3.8~22.3); 129.5 (15.5~207.5); 3.2 (0.0~53.3); 839 (404~1190) ❙ 1296 (527~1997); 526 (290~848); 119 (9~259) ❙ 54.4 (-22.2~147.0); 830 (404~1190); 228.7 (123.9~491.5)

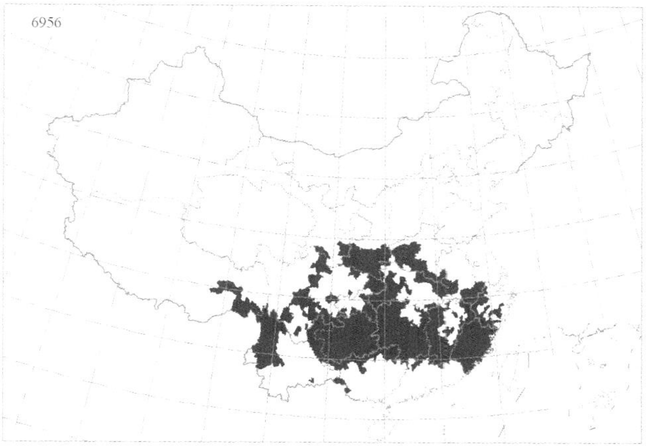

Stachyuraceae 旌节花科 *Stachyurus* 旌节花属
Stachyurus himalaicus Hook. f. et Thomson ex Benth. 西域旌节花
Shrubs or small trees, deciduous

15.8 (1.8~22.2); 6.0 (-7.1~17.7); 24.5 (9.9~29.8); 15.8 (3.8~22.2); 131.7 (15.7~206.7); 1.7 (0.0~53.3); 834 (404~1123) ❙ 1267 (527~2992); 558 (316~1847); 101 (9~402) ❙ 51.7 (-22.2~246.5); 820 (404~1123); 242.4 (108.2~551.6)

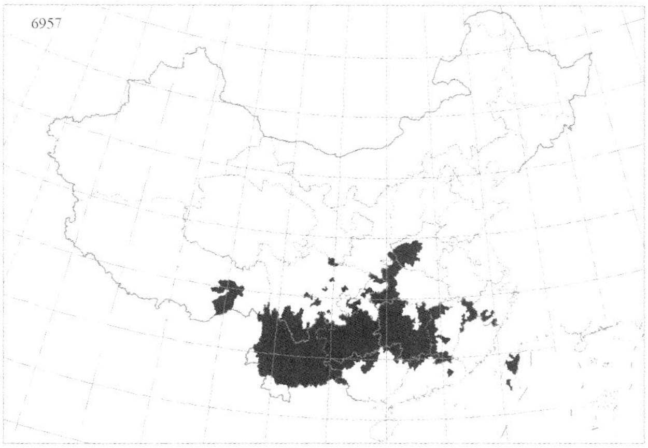

Stachyuraceae 旌节花科 *Stachyurus* 旌节花属
Stachyurus obovatus (Rehder) Cheng 倒卵叶旌节花
Shrubs or small trees, evergreen

14.9 (11.4~18.1); 4.9 (2.6~8.1); 24.1 (18.8~27.3); 15.0 (11.4~18.2); 120.2 (80.7~158.1); 0.8 (0.0~4.1); 792 (652~949) ❙ 1090 (863~1492); 540 (472~816); 58 (25~113) ❙ 38.1 (18.3~100.5); 792 (652~949); 194.6 (135.1~276.9)

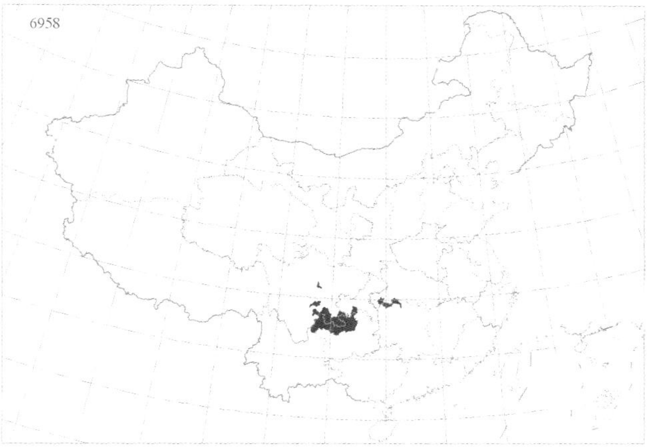

Stachyuraceae 旌节花科 *Stachyurus* 旌节花属
Stachyurus retusus Y. C. Yang 凹叶旌节花
Shrubs

13.4 (11.4~15.9); 4.2 (2.7~6.2); 21.6 (18.8~24.6); 13.5 (11.4~15.9); 102.8 (80.7~130.8); 1.3 (0.0~3.7); 724 (652~822) ❙ 1004 (863~1410); 552 (472~816); 40 (25~55) ❙ 38.2 (23.3~80.4); 724 (652~822); 227.9 (189.1~278.8)

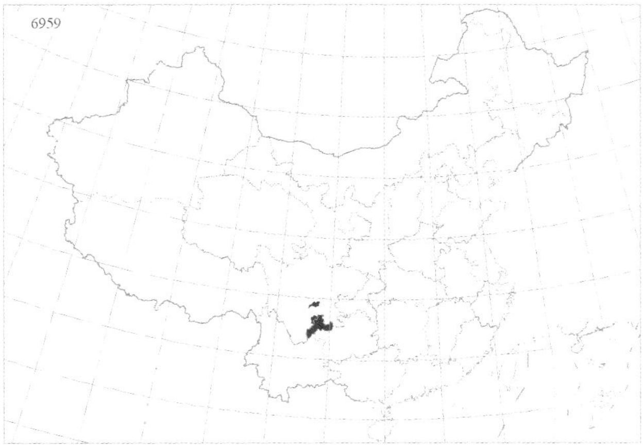

Stachyuraceae 旌节花科 *Stachyurus* 旌节花属
Stachyurus salicifolius Franch. 柳叶旌节花
Shrubs, evergreen

13.9 (10.1~18.5); 4.5 (2.2~10.5); 22.2 (16.9~25.7); 13.9 (10.2~18.3); 108.0 (68.1~160.0); 1.3 (0.0~5.3); 743 (618~906) ❙ 1090 (863~1590); 581 (472~946); 43 (18~76) ❙ 46.8 (5.5~89.0); 736 (618~866); 231.1 (166.3~330.0)

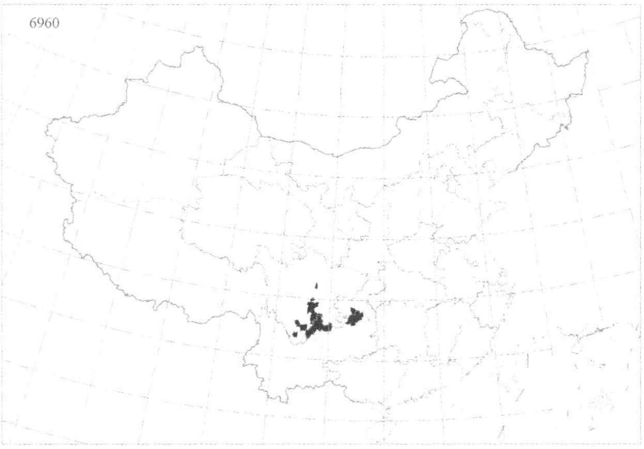

Stachyuraceae 旌节花科 *Stachyurus* 旌节花属
Stachyurus yunnanensis Franch. 云南旌节花
Shrubs, evergreen

15.1 (8.6~21.0); 5.2 (−1.7~14.2); 24.0 (15.0~28.1); 15.1 (8.7~21.0); 122.9 (52.6~191.8); 1.1 (0.0~18.0); 797 (574~1053) ❙ 1188 (863~1522); 557 (444~835); 68 (18~131) ❙ 49.5 (23.3~101.4); 797 (574~1053); 224.3 (146.0~551.6)

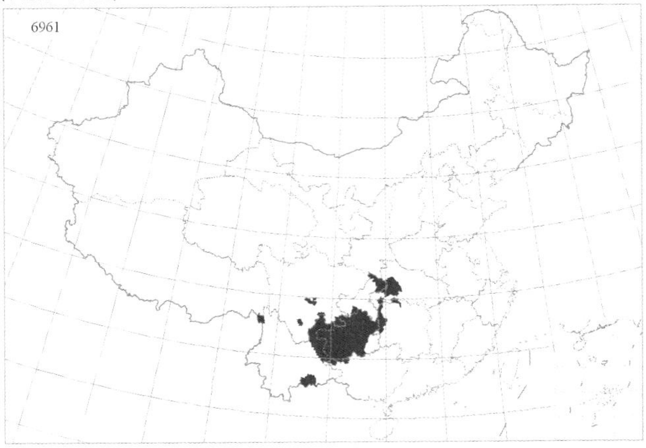

Passifloraceae 西番莲科 *Adenia* 蒴莲属
Adenia cardiophylla (Masters) Engl. 三开瓢
Vines

19.0 (16.4~21.0); 13.0 (9.6~15.1); 22.8 (20.6~25.6); 19.0 (16.4~21.0); 168.6 (137.3~191.8); 0.0 (0.0~0.0); 918 (795~1053) ❙ 1438 (1049~1688); 752 (598~959); 58 (41~77) ❙ 57.5 (21.7~112.2); 897 (783~1053); 513.4 (346.9~659.0)

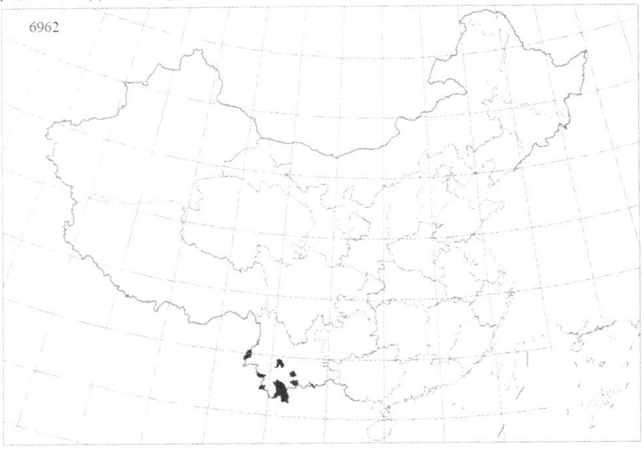

Passifloraceae 西番莲科 *Adenia* 蒴莲属
Adenia heterophylla (Blume) Koord. 异叶蒴莲
Vines

17.8 (16.0~20.5); 11.5 (9.3~15.1); 21.8 (20.3~23.7); 17.9 (16.0~20.5); 154.2 (132.4~185.9); 0.0 (0.0~0.0); 857 (780~992) ❙ 1381 (1049~1688); 762 (594~959); 53 (42~77) ❙ 61.2 (26.1~112.2); 843 (779~986); 459.4 (346.6~659.0)

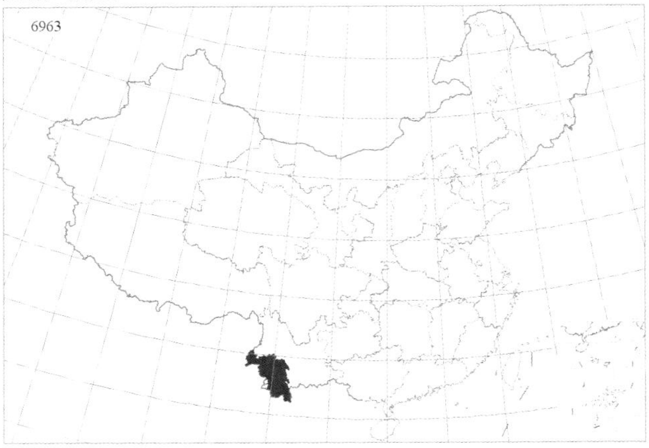

Passifloraceae 西番莲科 *Adenia* 蒴莲属
Adenia penangiana (Wall. ex G. Don) W. J. de Wilde 滇南蒴莲
Vines

20.4; 15.1; 23.6; 20.5; 185.8; 0.0; 989 ❙ 1607; 702; 77 ❙ 62.5; 986; 659.0

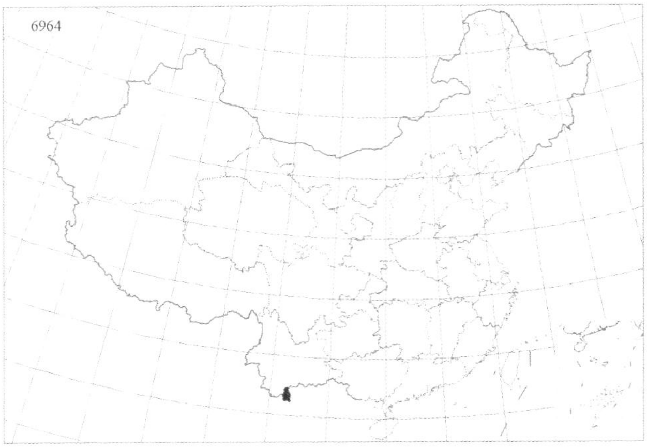

Passifloraceae 西番莲科 *Passiflora* 西番莲属
Passiflora eberhardtii Gagnep. 心叶西番莲
Vines

18.5; 10.1; 24.7; 18.5; 162.3; 0.0; 921 ❙ 1325; 724; 60 ❙ 44.3; 921; 343.3

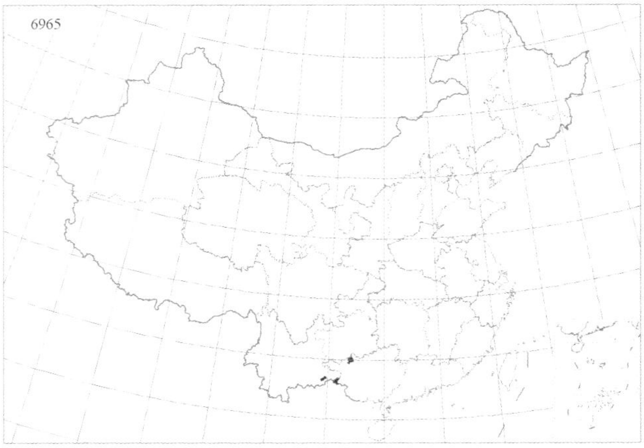

Passifloraceae 西番莲科 *Passiflora* 西番莲属
Passiflora jianfengensis S. M. Hwang et Q. Huang 尖峰西番莲
Vines

22.5 (20.8~23.9); 15.1 (12.2~18.7); 28.2 (27.5~28.9); 22.6 (20.8~23.9); 211.0 (189.7~226.8); 0.0 (0.0~0.0); 1209 (1085~1294) ❙ 1582 (1269~1990); 666 (455~827); 119 (48~181) ❙ 32.7 (−1.0~83.4); 1170 (1085~1269); 274.4 (194.5~455.2)

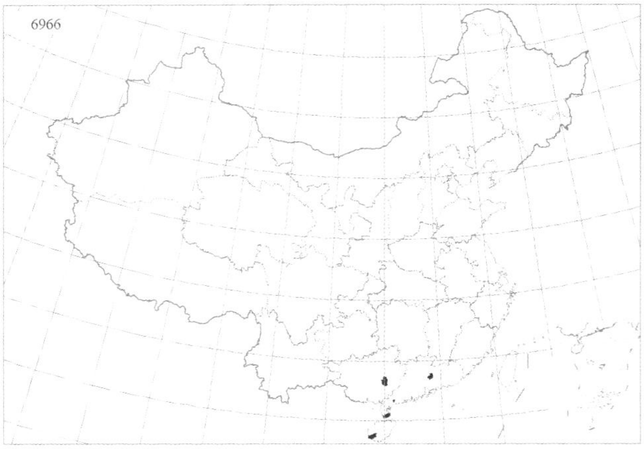

Passifloraceae 西番莲科 *Passiflora* 西番莲属
Passiflora jugorum W. W. Sm. 山峰西番莲
Vines

17.4 (14.4~21.0); 10.4 (7.5~14.2); 22.1 (19.1~25.6); 17.5 (14.4~21.0); 149.6 (113.1~191.8); 0.0 (0.0~0.0); 849 (727~1053) ❙ 1367 (1049~1688); 773 (594~959); 54 (42~70) ❙ 61.6 (26.1~112.2); 842 (727~1053); 439.1 (314.4~627.0)

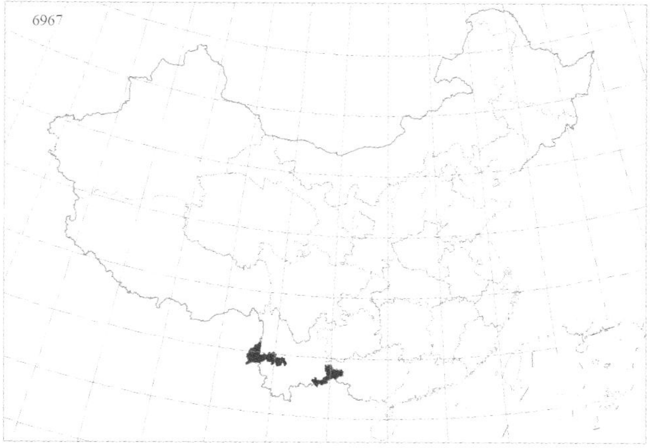

Passifloraceae 西番莲科 *Passiflora* 西番莲属
Passiflora siamica Craib 长叶西番莲
Vines

20.3; 14.0; 24.5; 20.4; 184.6; 0.0; 1005 ❙ 1538; 679; 95 ❙ 53.8; 989; 492.4

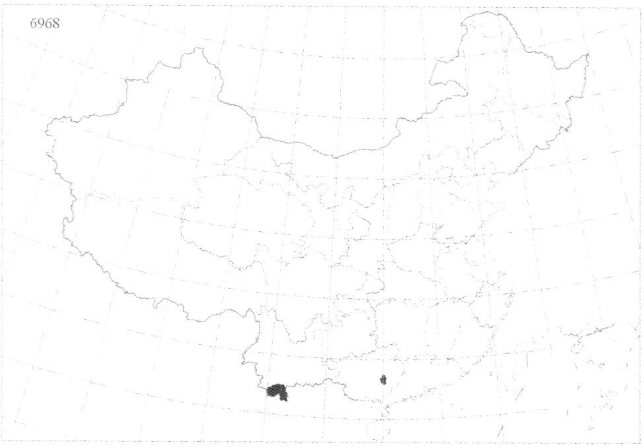

Passifloraceae 西番莲科 *Passiflora* 西番莲属
Passiflora wilsonii Hemsl. 镰叶西番莲
Herbs or subshrubs

17.7 (12.8~21.0); 10.7 (3.3~14.2); 22.5 (19.6~25.6); 17.8 (12.9~21.0); 153.4 (97.1~191.8); 0.1 (0.0~2.2); 866 (702~1053) ❙ 1454 (942~1751); 826 (489~1007); 55 (42~77) ❙ 69.0 (34.2~145.8); 863 (702~1053); 452.2 (219.7~627.0)

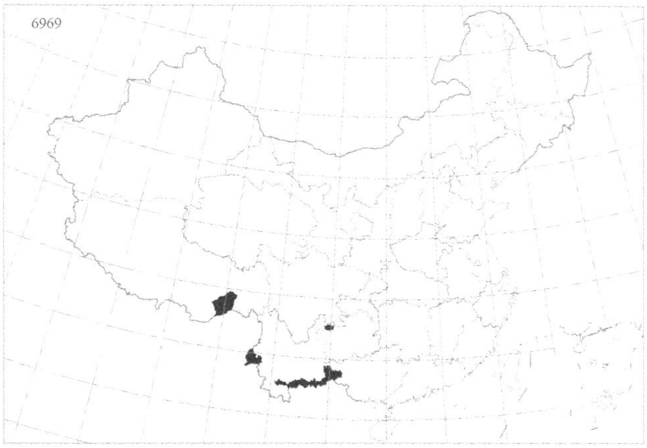

Tetramelaceae 四数木科 *Tetrameles* 四数木属
Tetrameles nudiflora R. Br. 四数木
Trees, deciduous

18.7 (16.4~20.5); 12.4 (9.6~15.1); 22.5 (20.6~23.7); 18.7 (16.4~20.5); 164.3 (137.3~185.9); 0.0 (0.0~0.0); 898 (795~992) ❙ 1532 (1366~1688); 836 (655~959); 56 (42~77) ❙ 71.7 (47.6~112.2); 887 (795~986); 539.9 (451.2~659.0)

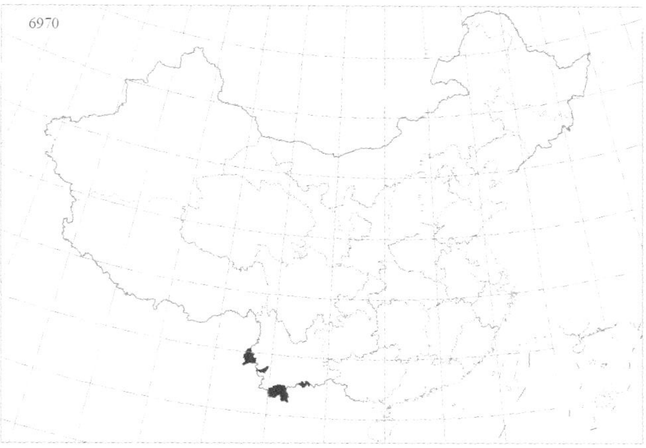

Ancistrocladaceae 钩枝藤科 *Ancistrocladus* 钩枝藤属
Ancistrocladus tectorius (Lour.) Merr. 钩枝藤
Scandent shrubs

23.5 (21.3~24.8); 17.4 (13.2~20.1); 27.9 (25.7~28.8); 23.6 (21.3~24.8); 222.7 (195.9~237.6); 0.0 (0.0~0.0); 1276 (1111~1385) ❙ 1535 (868~2148); 611 (312~1234); 80 (41~125) ❙ 21.3 (-34.3~86.7); 1182 (868~1360); 412.3 (248.2~620.0)

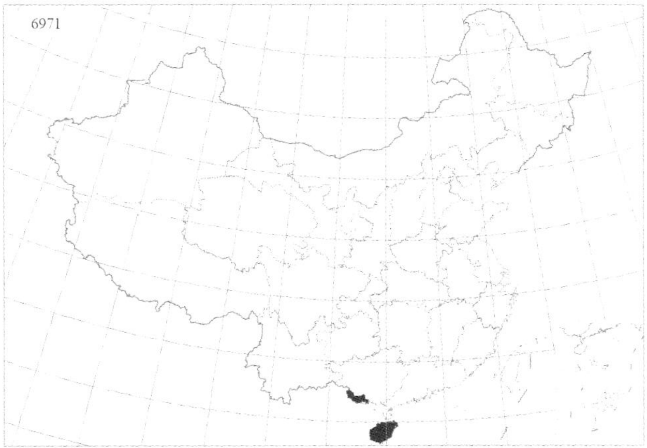

Thymelaeaceae 瑞香科 *Aquilaria* 沉香属
Aquilaria sinensis (Lour.) Spreng. 土沉香
Trees

21.5 (16.2~24.8); 13.9 (7.8~20.1); 27.6 (20.1~29.2); 21.6 (16.2~24.8); 198.7 (134.2~237.6); 0.0 (0.0~0.0); 1133 (773~1385) ❙ 1749 (868~3495); 776 (312~1847); 144 (41~897) ❙ 58.3 (-34.3~246.5); 1113 (773~1360); 256.3 (95.3~659.0)

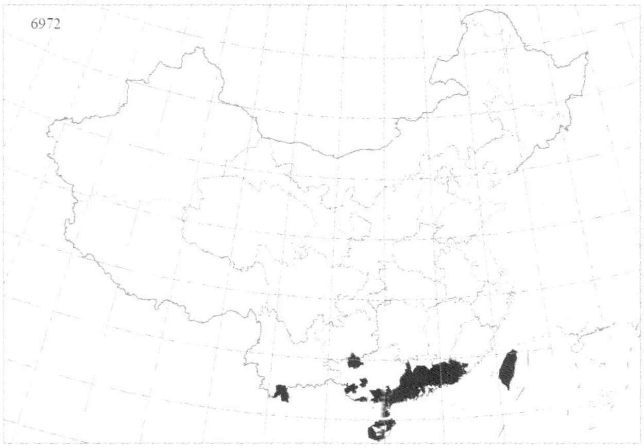

Thymelaeaceae 瑞香科 *Aquilaria* 沉香属
Aquilaria yunnanensis S. C. Huang 云南沉香
Small trees

19.3；13.7；22.7；19.4；172.3；0.0；932 ❙ 1474；709；63 ❙ 58.7；916；556.0

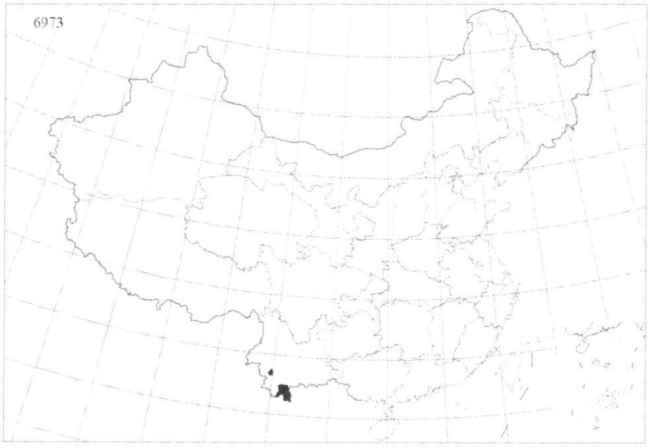

Thymelaeaceae 瑞香科 *Daphne* 瑞香属
Daphne acutiloba Rehder 尖瓣瑞香
Shrubs, evergreen

15.6 (5.4~21.0)；6.4 (-3.2~14.2)；23.6 (12.3~29.2)；15.7 (5.8~21.0)；128.9 (31.5~191.8)；1.1 (0.0~26.7)；811 (481~1053) ❙ 1191 (741~1751)；572 (411~996)；70 (22~197) ❙ 47.6 (-14.2~101.4)；802 (481~1053)；245.3 (143.8~551.6)

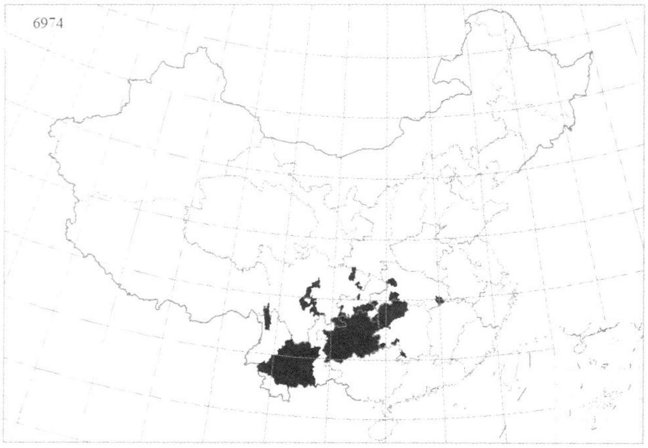

Thymelaeaceae 瑞香科 *Daphne* 瑞香属
Daphne altaica Pall 阿尔泰瑞香
Shrubs, deciduous

3.7；-15.1；19.9；7.8；59.1；74.3；577 ❙ 309；106；45 ❙ -46.5；307；157.3

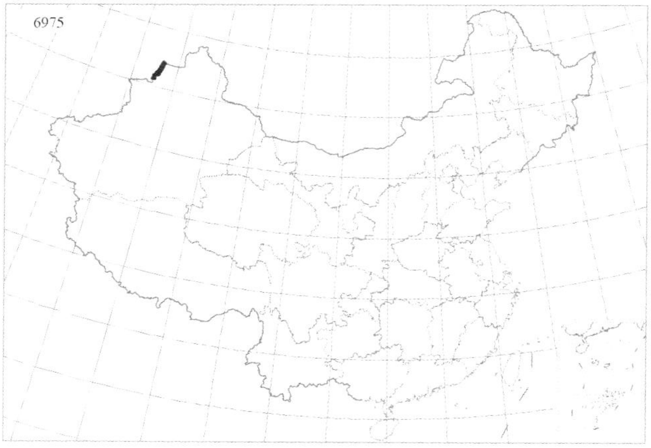

Thymelaeaceae 瑞香科 *Daphne* 瑞香属
Daphne arisanensis Hayata 台湾瑞香
Subshrubs

18.9 (16.4~21.7)；13.5 (11.2~15.3)；23.2 (20.3~27.0)；18.9 (16.5~21.7)；166.9 (137.8~200.2)；0.0 (0.0~0.0)；920 (783~1118) ❙ 2375 (1836~2713)；1165 (947~1412)；193 (110~303) ❙ 165.9 (64.2~219.3)；911 (783~1075)；271.8 (202.8~363.7)

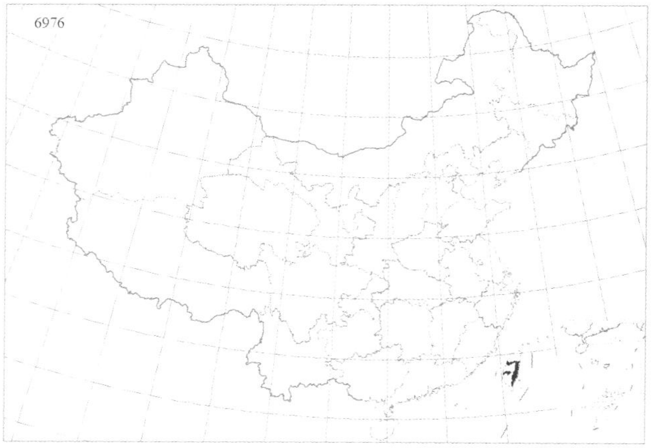

Thymelaeaceae 瑞香科 *Daphne* 瑞香属
Daphne aurantiaca Diels 橙黄瑞香
Dwarf shrubs

13.3 (2.2~18.9)；6.2 (-6.3~12.5)；18.8 (9.5~24.3)；13.4 (3.9~19.0)；104.8 (15.8~167.8)；5.2 (0.0~48.6)；710 (411~929) ❙ 948 (579~1164)；557 (365~651)；32 (10~59) ❙ 36.4 (-14.2~78.3)；674 (411~851)；300.2 (201.9~384.1)

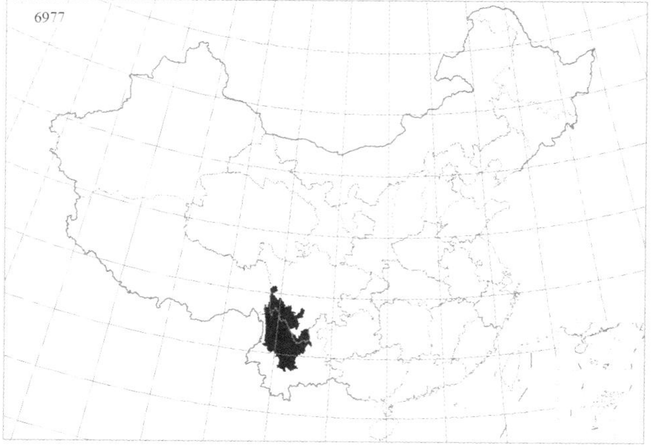

Thymelaeaceae 瑞香科 *Daphne* 瑞香属
Daphne axillaris (Merr. et Chun) Chun et C. F. Wei 腋花瑞香
Shrubs

24.2；19.3；27.6；24.2；230.9；0.0；1327 ❙ 1445；489；46 ❙ 9.1；1132；534.2

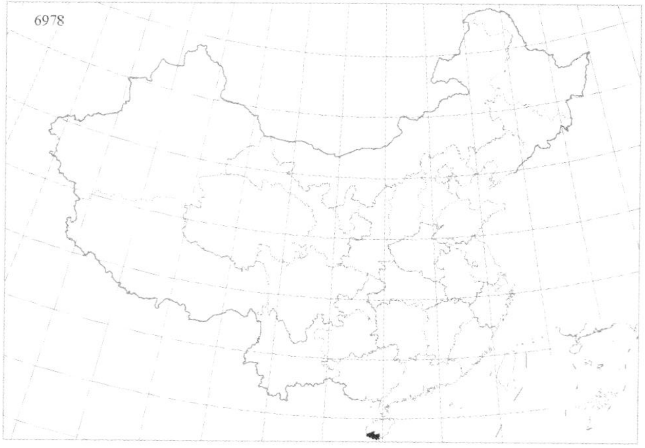

Thymelaeaceae 瑞香科 *Daphne* 瑞香属
Daphne bholua Buch.-Ham. ex D. Don 藏东瑞香
Shrubs, evergreen

5.7 (-1.8~13.3)；-2.2 (-10.2~5.4)；12.7 (6.1~19.6)；6.9 (1.8~13.4)；41.9 (1.9~100.4)；32.9 (0.0~82.8)；504 (329~706) | 822 (320~1736)；489 (217~1007)；29 (10~51) | 50.5 (-3.0~145.8)；502 (320~706)；252.6 (61.6~447.4)

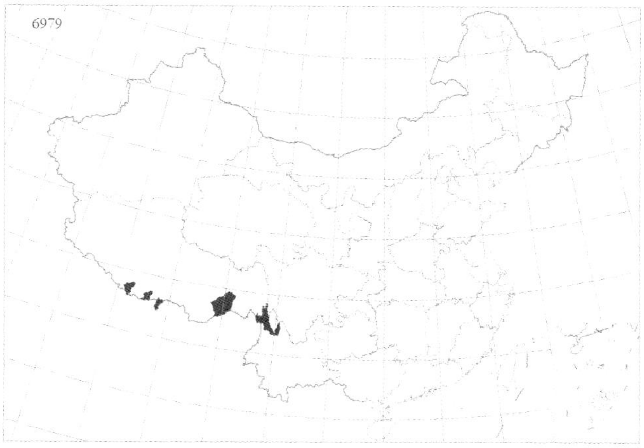

Thymelaeaceae 瑞香科 *Daphne* 瑞香属
Daphne bholua var. glacialis (W. W. Sm. et Cave) B. L. Burtt 落叶瑞香
Shrubs, evergreen

-0.2；-9.1；8.0；2.7；8.2；69.8；353 | 380；254；0 | 2.4；353；102.0

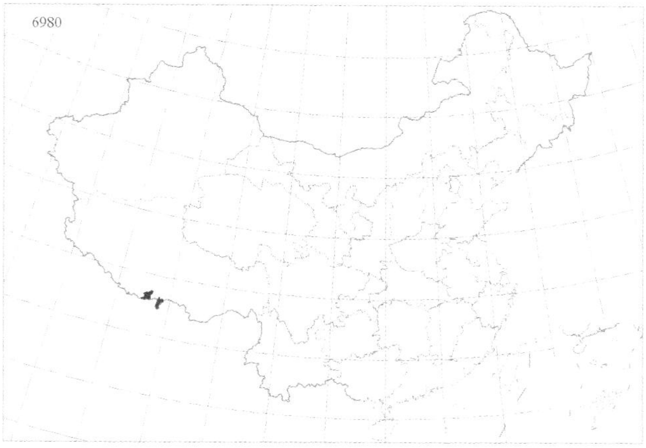

Thymelaeaceae 瑞香科 *Daphne* 瑞香属
Daphne brevituba H. F. Zhou 短管瑞香
Shrubs, evergreen

14.3；7.6；19.7；14.3；111.9；0.0；728 | 991；580；35 | 36.1；707；296.9

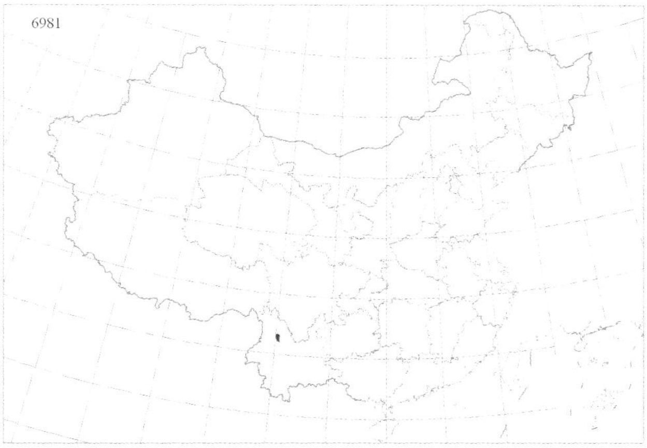

Thymelaeaceae 瑞香科 *Daphne* 瑞香属
Daphne championii Benth. 长柱瑞香
Shrubs, evergreen

19.1 (14.3~22.4)；9.4 (3.7~15.2)；27.5 (24.2~29.8)；19.2 (14.3~22.7)；169.9 (113.2~212.7)；0.0 (0.0~1.4)；1001 (761~1216) | 1641 (931~2024)；634 (323~1056)；174 (60~259) | 64.6 (-1.6~128.9)；993 (761~1216)；241.4 (159.4~322.1)

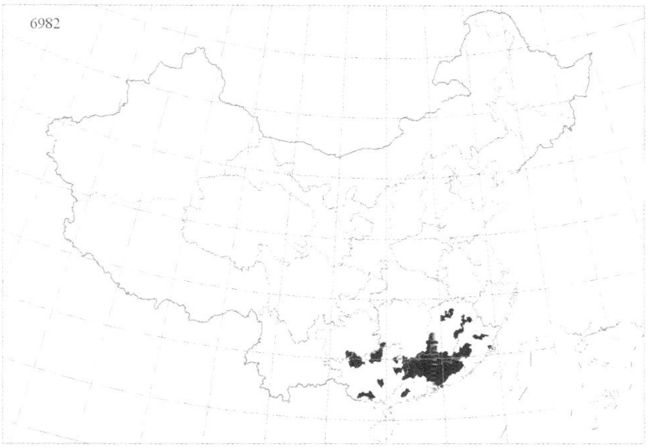

Thymelaeaceae 瑞香科 *Daphne* 瑞香属
Daphne depauperata H. F. Zhou 少花瑞香
Shrubs, evergreen

14.5；7.8；19.6；14.5；113.5；0.0；730 | 1044；570；54 | 43.0；730；382.6

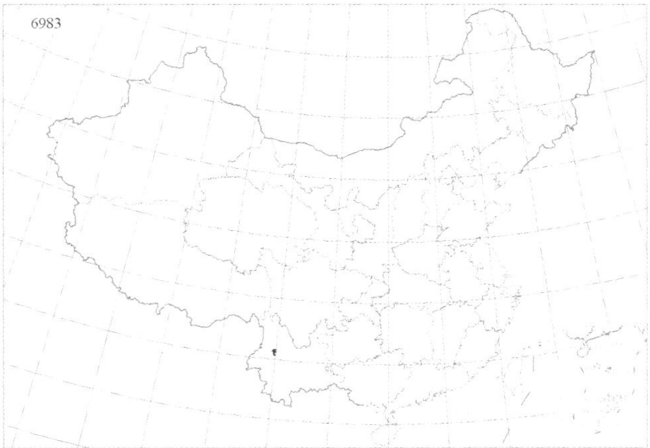

Thymelaeaceae 瑞香科 *Daphne* 瑞香属
Daphne emeiensis C. Y. Chang 峨眉瑞香
Shrubs, evergreen

15.0；5.3；23.8；15.1；120.6；0.0；785 | 1243；701；49 | 58.4；785；213.3

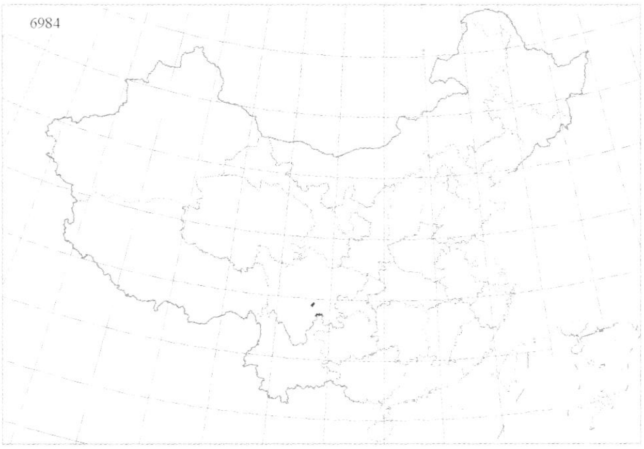

Thymelaeaceae 瑞香科 *Daphne* 瑞香属
Daphne erosiloba C. Y. Chang 啮蚀瓣瑞香
Shrubs, deciduous

0.6; -9.4; 9.5; 3.4; 12.7; 65.4; 394 ❙ 581; 347; 12 ❙ 47.6; 394; 307.8

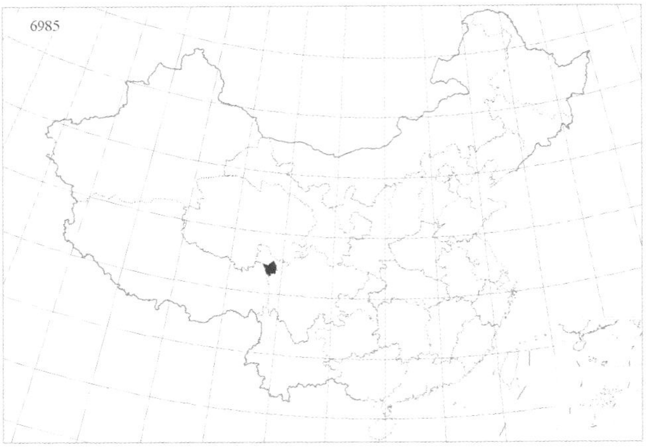

Thymelaeaceae 瑞香科 *Daphne* 瑞香属
Daphne esquirolii H. Lév. 穗花瑞香
Shrubs, deciduous

14.4 (6.1~18.0); 6.9 (-1.5~11.6); 20.2 (12.7~23.8); 14.4 (6.3~18.1); 114.0 (35.2~156.9); 1.0 (0.0~21.8); 739 (496~869) ❙ 977 (801~1147); 563 (479~636); 37 (18~50) ❙ 33.0 (12.2~61.4); 726 (496~815); 258.3 (190.3~309.3)

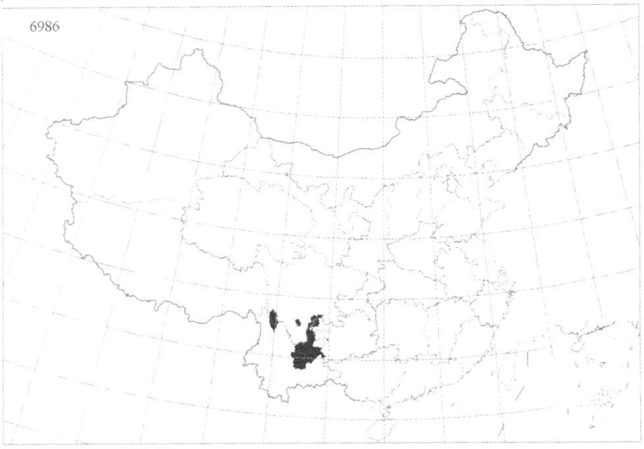

Thymelaeaceae 瑞香科 *Daphne* 瑞香属
Daphne feddei H. Lév. 滇瑞香
Shrubs, evergreen

15.3 (5.4~21.4); 7.0 (-2.1~13.5); 22.3 (12.3~28.9); 15.3 (5.8~21.4); 125.2 (31.5~196.9); 1.1 (0.0~25.6); 794 (481~1121) ❙ 1076 (741~1751); 564 (411~996); 47 (15~137) ❙ 36.9 (-14.2~95.9); 781 (481~1121); 255.4 (108.2~551.3)

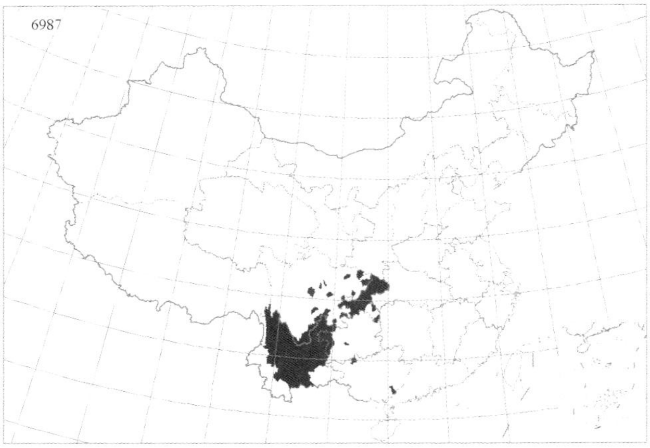

Thymelaeaceae 瑞香科 *Daphne* 瑞香属
Daphne gemmata E. Pritz. et Diels 川西瑞香
Shrubs, deciduous

7.9 (2.9~13.3); -1.5 (-6.3~3.2); 16.3 (10.8~22.1); 8.4 (4.4~13.2); 56.1 (20.5~100.9); 21.2 (2.5~44.6); 564 (428~723) ❙ 978 (890~1081); 530 (460~606); 25 (16~32) ❙ 79.4 (35.4~121.7); 564 (428~723); 230.4 (144.6~272.3)

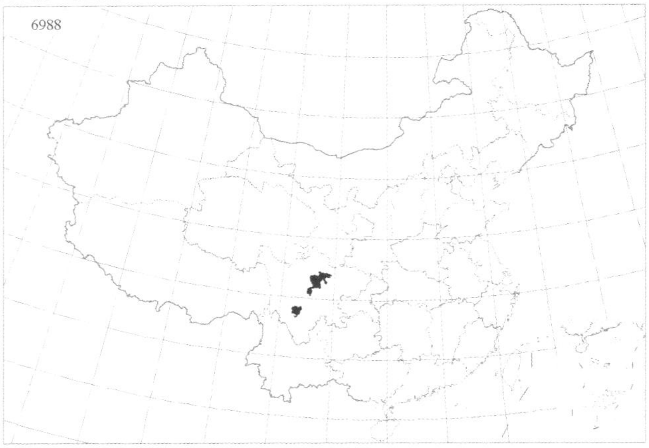

Thymelaeaceae 瑞香科 *Daphne* 瑞香属
Daphne genkwa Siebold et Zucc. 芫花
Shrubs, deciduous

14.9 (3.3~21.4); 2.5 (-12.7~14.1); 26.4 (10.8~30.1); 15.1 (4.7~21.5); 127.2 (21.6~197.7); 7.8 (0.0~65.0); 842 (444~1128) ❙ 1135 (367~2713); 471 (235~1388); 111 (8~308) ❙ 32.5 (-37.7~214.4); 803 (367~1099); 194.9 (70.3~363.7)

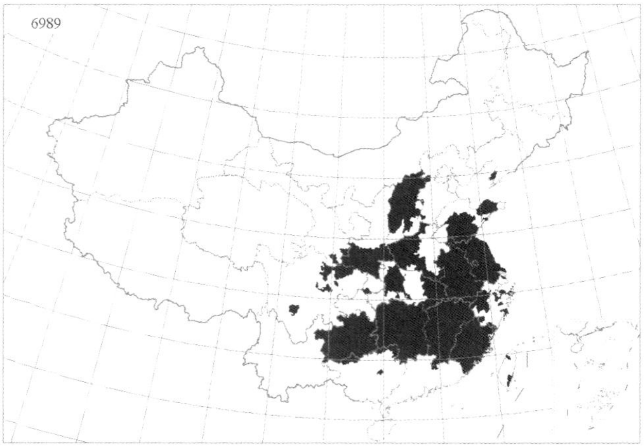

Thymelaeaceae 瑞香科 *Daphne* 瑞香属
Daphne giraldii Nitsche 黄瑞香
Shrubs, deciduous

9.9 (-2.1~15.8); -4.0 (-16.2~3.2); 22.4 (9.2~28.0); 10.8 (2.9~15.8); 85.9 (10.1~133.1); 27.0 (2.5~99.3); 677 (359~891) ❙ 638 (53~1335); 329 (33~606); 24 (1~109) ❙ -4.4 (-92.4~101.4); 580 (53~891); 183.3 (10.3~344.4)

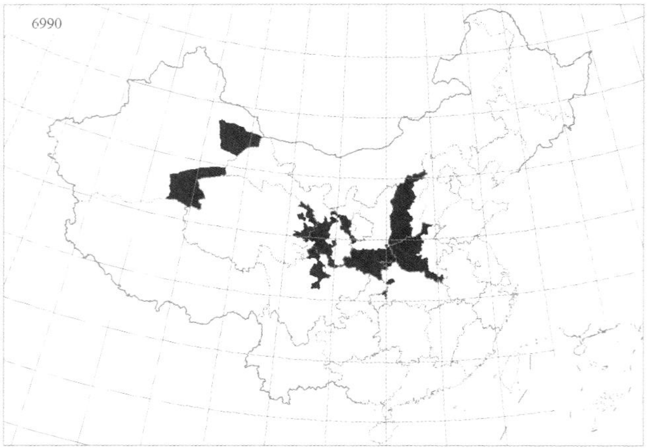

Thymelaeaceae 瑞香科 *Daphne* 瑞香属
Daphne tripartita H. F. Zhou ex C. Y. Chang 九龙瑞香
Shrubs, evergreen

3.8；-4.3；10.8；4.7；21.6；35.4；444 ▎984；601；16 ▎121.7；444；265.2

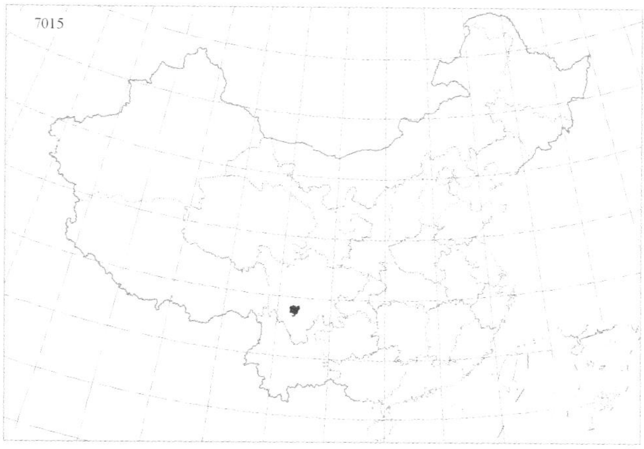

Thymelaeaceae 瑞香科 *Daphne* 瑞香属
Daphne wangiana (Hamaya) Halda 少丝毛瑞香
Shrubs, evergreen

5.3；-2.4；12.2；5.7；30.5；26.2；479 ▎961；558；31 ▎100.7；479；234.2

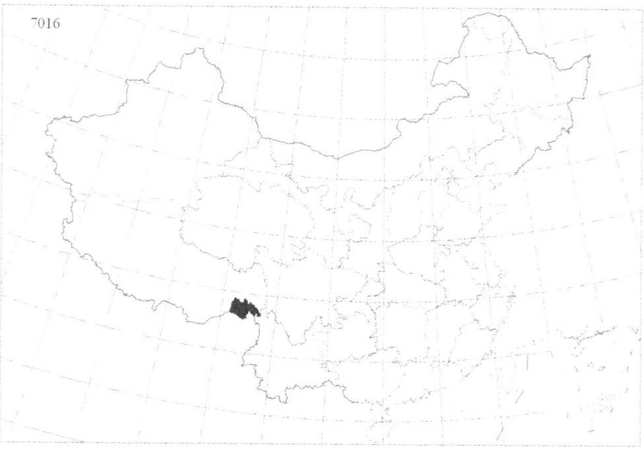

Thymelaeaceae 瑞香科 *Daphne* 瑞香属
Daphne xichouensis H. F. Zhou 西畴瑞香
Shrubs, evergreen

17.4；9.7；22.9；17.5；149.7；0.0；848 ▎1338；756；50 ▎57.6；848；348.0

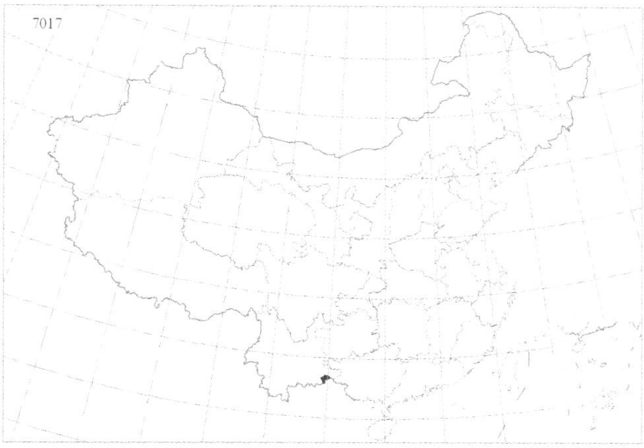

Thymelaeaceae 瑞香科 *Daphne* 瑞香属
Daphne yunnanensis H. F. Zhou 云南瑞香
Shrubs, evergreen

14.4；7.5；19.1；14.4；113.1；0.0；727 ▎1454；792；70 ▎99.9；727；487.2

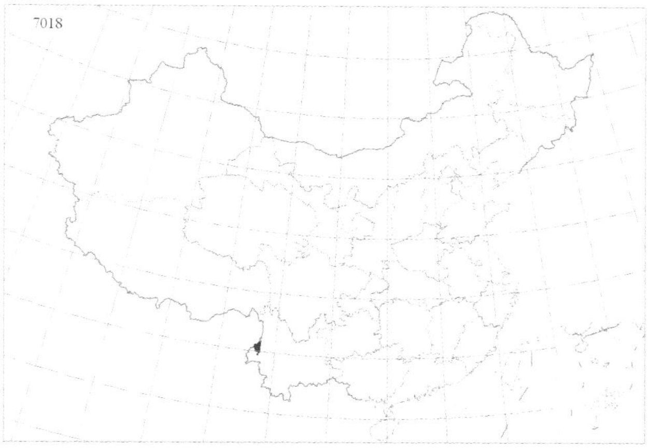

Thymelaeaceae 瑞香科 *Edgeworthia* 结香属
Edgeworthia albiflora Nakai 白结香
Shrubs, deciduous

14.9；7.4；20.8；14.9；119.2；0.0；756 ▎1017；624；20 ▎34.8；706；300.3

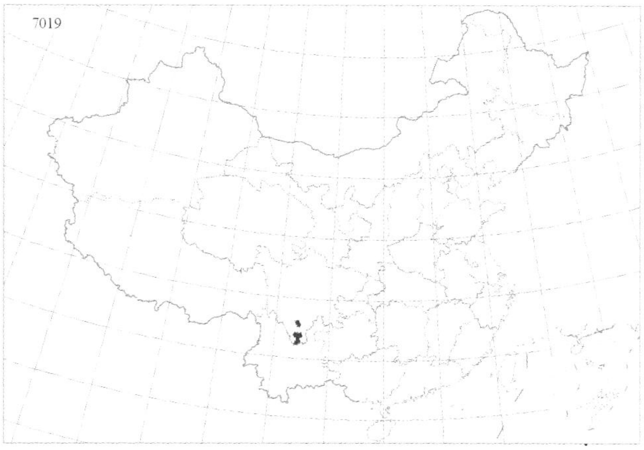

Thymelaeaceae 瑞香科 *Edgeworthia* 结香属
Edgeworthia chrysantha Lindl. 结香
Shrubs, deciduous

16.0 (7.4~21.0)；5.5 (-4.2~13.1)；25.6 (15.0~30.1)；16.1 (8.2~21.0)；134.5 (52.6~192.6)；1.6 (0.0~29.4)；853 (561~1113) ▎1309 (626~1997)；542 (316~1007)；120 (20~259) ▎53.2 (-22.2~145.8)；841 (561~1107)；232.0 (111.9~516.7)

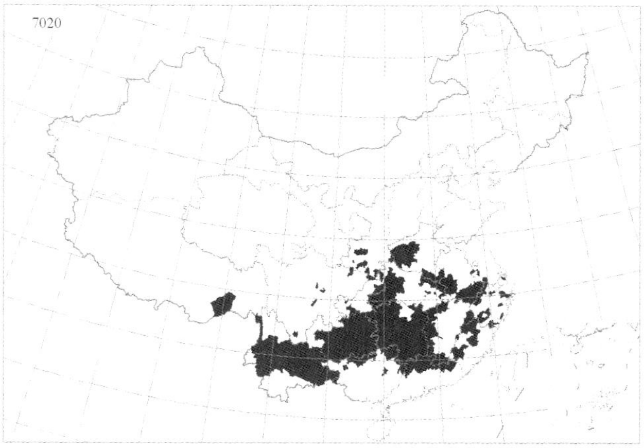

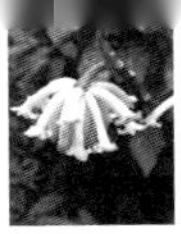

Thymelaeaceae 瑞香科 *Edgeworthia* 结香属
Edgeworthia eriosolenoides Feng et S. C. Huang 西畴结香
Shrubs, deciduous

17.2; 9.5; 22.7; 17.3; 147.0; 0.0; 838 ❙ 1254; 699; 48 ❙ 49.7; 838; 336.3

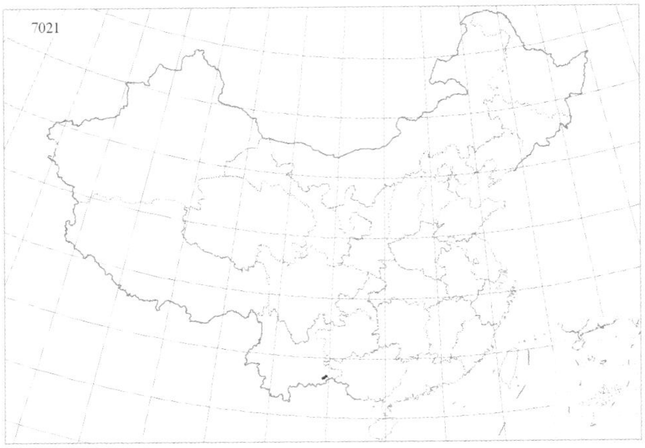

Thymelaeaceae 瑞香科 *Eriosolena* 毛花瑞香属
Eriosolena composita (L. f.) Tiegh. 毛花瑞香
Shrubs

16.5 (8.6~21.0); 9.8 (1.3~15.1); 21.2 (15.0~25.6); 16.5 (8.7~21.0); 138.4 (52.6~191.8); 0.1 (0.0~8.8); 812 (574~1053) ❙ 1177 (797~1751); 660 (411~1007); 48 (23~77) ❙ 45.2 (−14.2~145.8); 794 (574~1053); 363.9 (207.5~659.0)

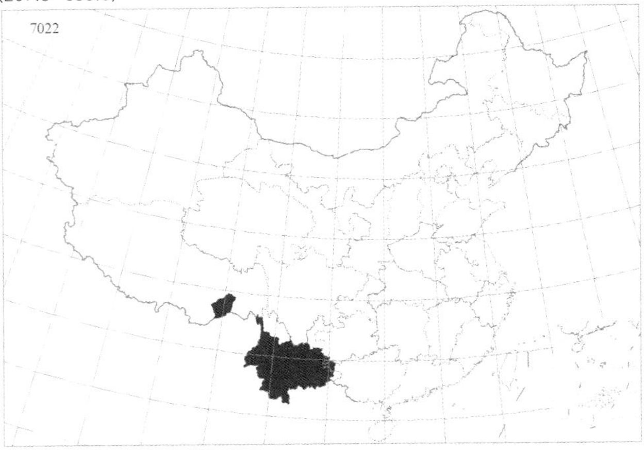

Thymelaeaceae 瑞香科 *Rhamnoneuron* 鼠皮树属
Rhamnoneuron balansae (Drake) Gilg 鼠皮树
Shrubs

18.7; 11.7; 23.5; 18.7; 164.1; 0.0; 914 ❙ 1465; 812; 61 ❙ 61.7; 914; 485.5

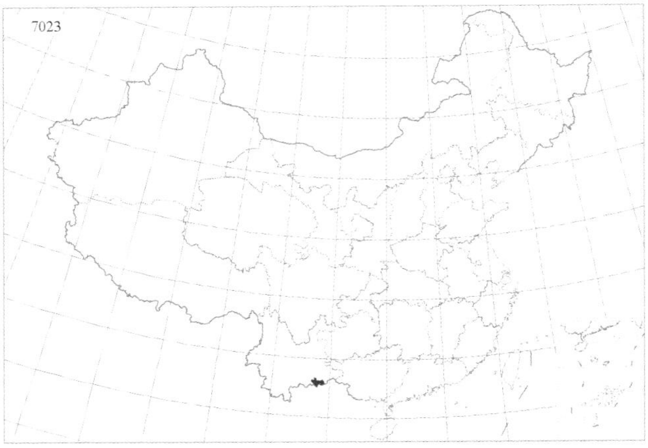

Thymelaeaceae 瑞香科 *Wikstroemia* 荛花属
Wikstroemia alternifolia Batalin 互生叶荛花
Shrubs

3.5; −6.5; 12.6; 5.0; 26.3; 43.4; 446 ❙ 731; 346; 19 ❙ 63.8; 446; 284.1

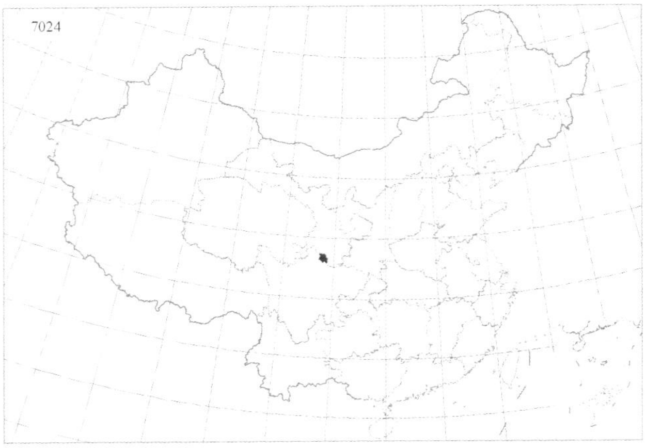

Thymelaeaceae 瑞香科 *Wikstroemia* 荛花属
Wikstroemia angustifolia Hemsl. 岩杉树
Shrubs

14.0 (9.1~16.8); 2.6 (−2.3~6.5); 24.7 (18.1~28.1); 14.1 (9.2~16.9); 114.3 (64.1~142.7); 5.4 (0.0~17.6); 780 (595~902) ❙ 1044 (738~1368); 458 (350~579); 63 (17~129) ❙ 33.9 (−5.3~74.9); 777 (595~869); 221.4 (133.5~294.3)

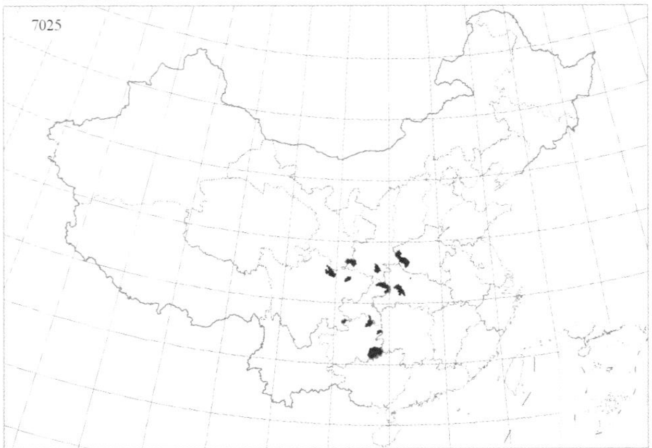

Thymelaeaceae 瑞香科 *Wikstroemia* 荛花属
Wikstroemia anhuiensis D. C. Zhang et X. P. Zhang 安徽荛花
Shrubs

15.0; 2.7; 26.8; 15.0; 123.7; 3.2; 826 ❙ 1663; 595; 212 ❙ 101.3; 826; 256.1

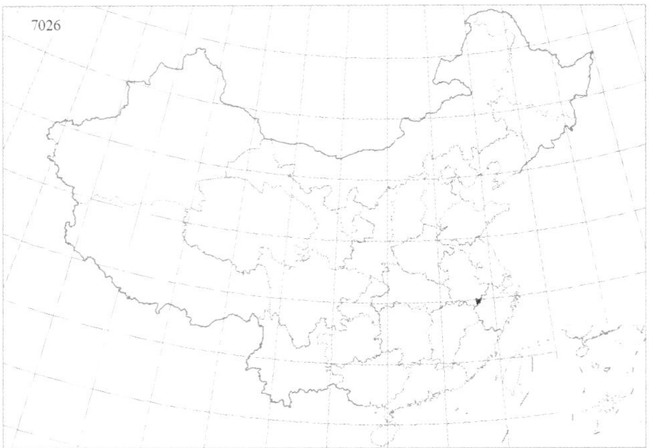

Thymelaeaceae 瑞香科 *Wikstroemia* 荛花属
Wikstroemia baimashanensis S. C. Huang 白马山荛花
Shrubs

5.4; -2.1; 12.3; 5.8; 31.5; 25.6; 481 | 741; 426; 28 | 54.0; 481; 273.5

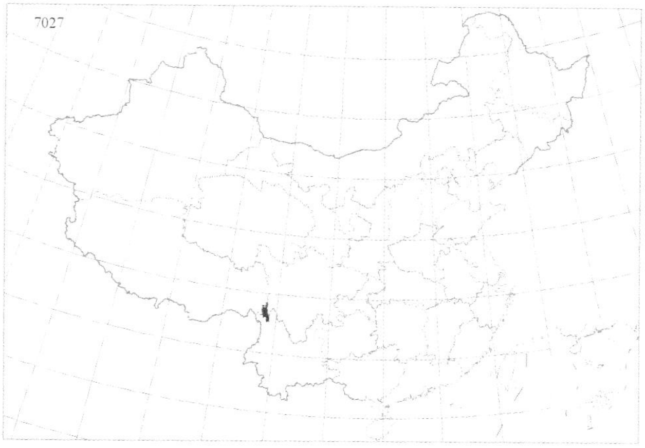

Thymelaeaceae 瑞香科 *Wikstroemia* 荛花属
Wikstroemia canescens Wallich ex Meisner 荛花
Shrubs

10.8 (-1.1~16.8); -0.7 (-9.1~4.3); 21.5 (6.2~29.1); 11.4 (2.0~16.9); 89.8 (2.5~143.0); 20.3 (0.7~75.6); 696 (334~930) | 1086 (321~1770); 466 (193~628); 102 (10~223) | 51.3 (-4.0~112.2); 695 (321~930); 218.4 (79.3~273.6)

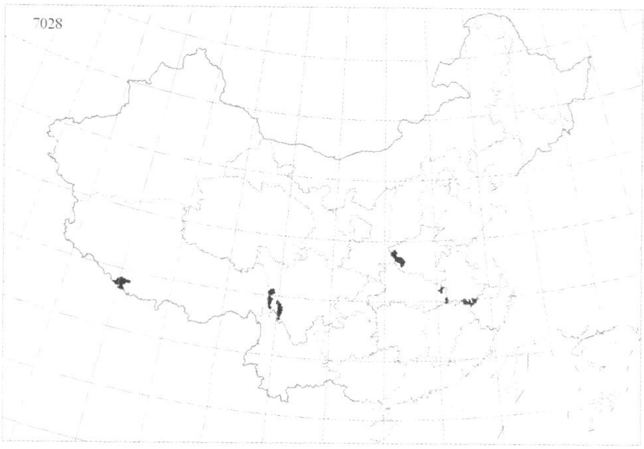

Thymelaeaceae 瑞香科 *Wikstroemia* 荛花属
Wikstroemia capitata Rehder et E. H. Wilson 头序荛花
Small shrubs

13.8 (3.8~18.6); 2.8 (-4.3~7.9); 24.3 (10.8~28.8); 13.9 (4.7~18.7); 111.8 (21.6~164.0); 5.6 (0.0~35.4); 774 (444~992) | 1112 (806~1492); 482 (349~613); 60 (16~96) | 46.7 (5.6~121.7); 774 (444~992); 216.8 (120.7~279.5)

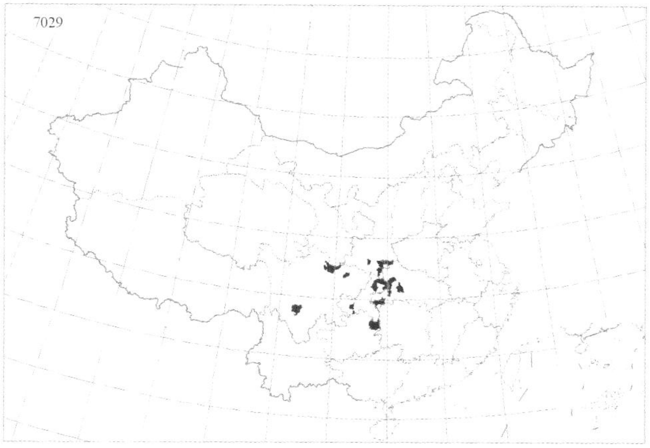

Thymelaeaceae 瑞香科 *Wikstroemia* 荛花属
Wikstroemia capitatoracemosa S. C. Huang 短总序荛花
Shrubs

4.8 (0.7~13.7); -3.7 (-8.8~6.2); 12.1 (9.5~19.7); 5.9 (3.3~13.7); 33.8 (12.7~104.7); 36.2 (0.0~63.9); 475 (389~714) | 790 (514~963); 47 (314~593); 20 (10~31) | 68.7 (32.3~128.6); 473 (389~696); 262.8 (234.2~299.8)

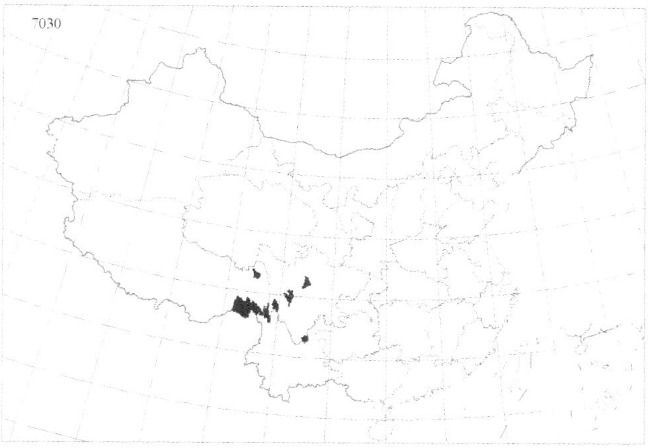

Thymelaeaceae 瑞香科 *Wikstroemia* 荛花属
Wikstroemia chamaedaphne Meisn. 河朔荛花
Shrubs

9.4 (0.8~14.7); -5.7 (-15.9~1.5); 22.8 (10.7~28.3); 10.5 (3.8~14.7); 85.1 (16.2~126.3); 31.9 (7.1~80.4); 672 (411~873) | 576 (367~892); 329 (201~499); 17 (7~42) | -13.1 (-47.7~94.4); 552 (367~764); 187.1 (44.7~334.4)

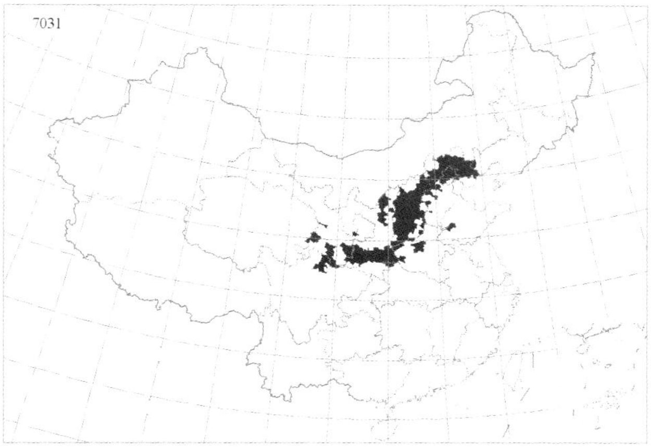

Thymelaeaceae 瑞香科 *Wikstroemia* 荛花属
Wikstroemia chui Merr. 窄叶荛花
Shrubs

24.3; 19.5; 27.6; 24.4; 232.3; 0.0; 1339 | 1569; 536; 53 | 17.2; 1180; 557.4

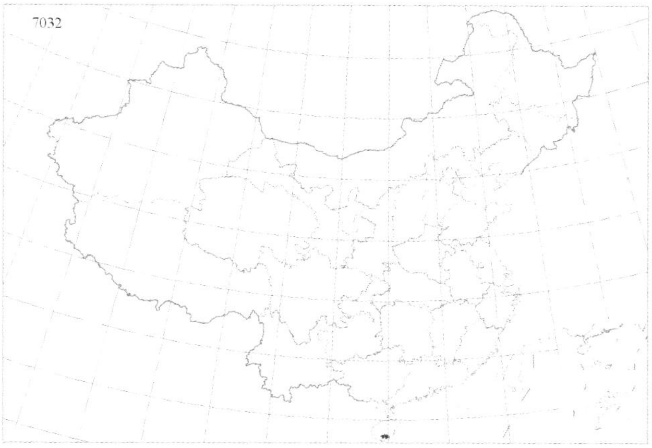

Thymelaeaceae 瑞香科 *Wikstroemia* 荛花属
Wikstroemia cochlearifolia S. C. Huang 匙叶荛花
Shrubs

15.3; 4.7; 24.9; 15.3; 124.0; 0.3; 816 ❙ 905; 527; 20 ❙ 11.0; 816; 181.8

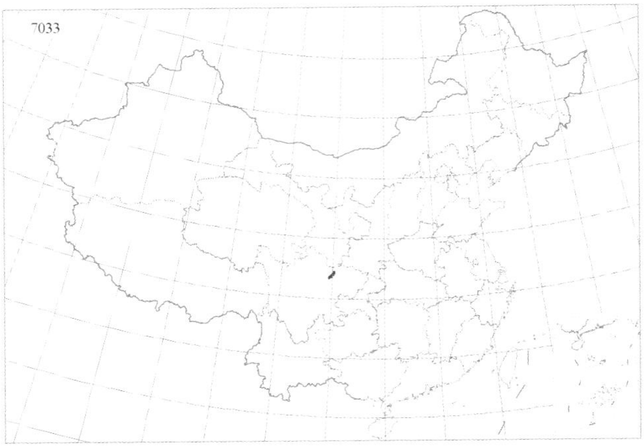

Thymelaeaceae 瑞香科 *Wikstroemia* 荛花属
Wikstroemia delavayi Lecomte 澜沧荛花
Shrubs

15.0 (2.2~19.6); 8.4 (-6.6~13.9); 19.9 (9.5~23.9); 15.1 (3.8~19.6); 123.8 (15.5~175.1); 3.1 (0.0~49.1); 765 (411~939) ❙ 1090 (729~1500); 618 (411~836); 41 (11~65) ❙ 44.5 (-14.2~125.9); 735 (411~903); 349.1 (208.8~530.1)

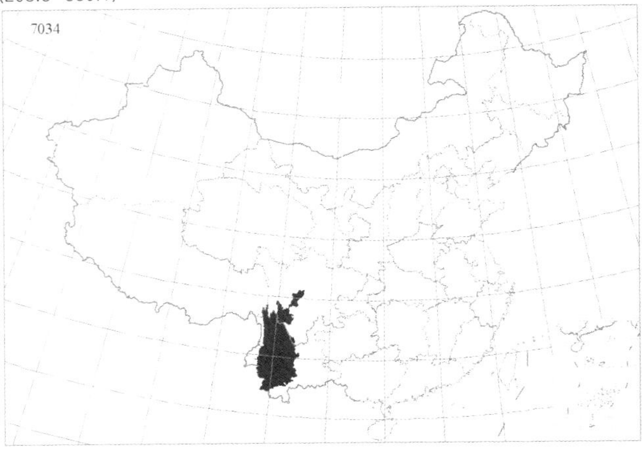

Thymelaeaceae 瑞香科 *Wikstroemia* 荛花属
Wikstroemia dolichantha Diels 一把香
Shrubs

15.1 (2.2~19.6); 7.1 (-6.3~13.9); 21.5 (9.5~29.3); 15.2 (3.9~19.6); 123.3 (15.8~175.1); 1.8 (0.0~48.6); 779 (411~994) ❙ 1112 (579~1852); 593 (323~911); 58 (10~253) ❙ 43.1 (-14.2~121.7); 762 (411~994); 290.5 (142.9~551.3)

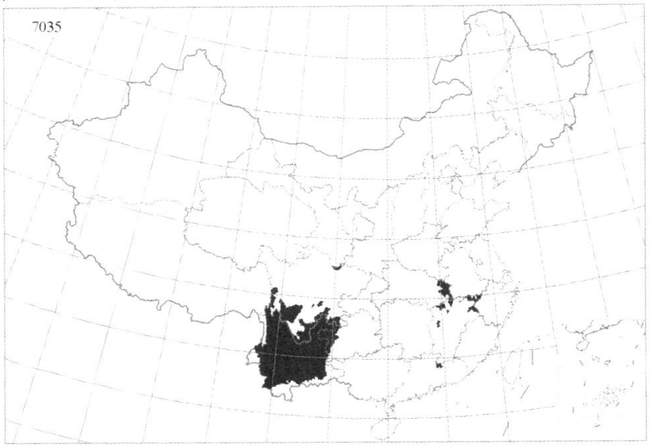

Thymelaeaceae 瑞香科 *Wikstroemia* 荛花属
Wikstroemia fargesii (Lecomte) Domke 城口荛花
Shrubs

10.6; -0.2; 21.1; 10.6; 79.4; 12.1; 643 ❙ 1246; 525; 46 ❙ 93.8; 643; 284.3

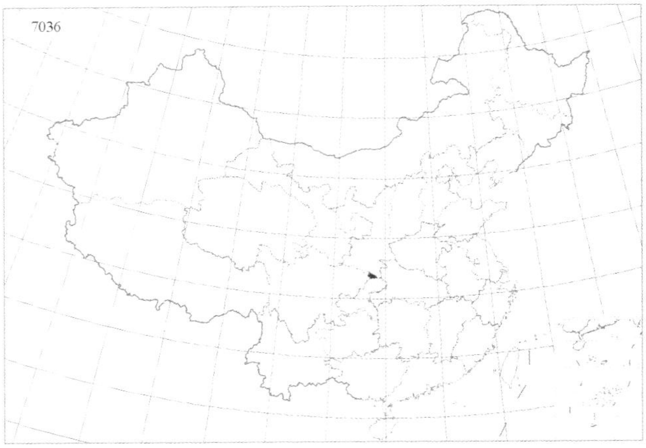

Thymelaeaceae 瑞香科 *Wikstroemia* 荛花属
Wikstroemia glabra Cheng 光叶荛花
Shrubs

17.1 (10.6~21.8); 6.2 (-0.2~13.4); 27.1 (21.1~30.1); 17.1 (10.6~21.8); 146.5 (79.4~201.9); 0.9 (0.0~12.1); 912 (643~1158) ❙ 1625 (1197~2005); 589 (442~858); 189 (46~259) ❙ 79.3 (31.6~130.7); 907 (643~1158); 241.1 (148.9~332.5)

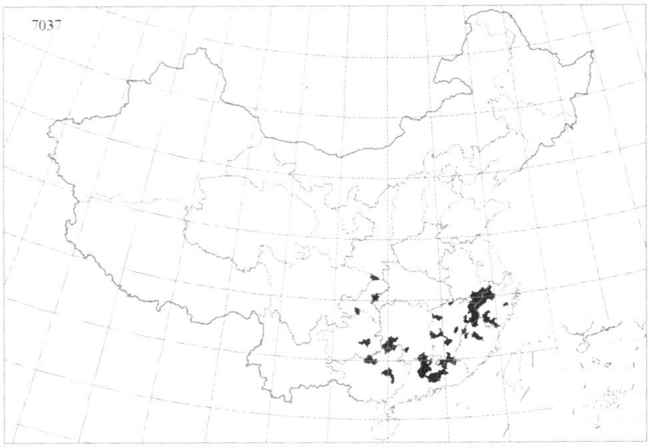

Thymelaeaceae 瑞香科 *Wikstroemia* 荛花属
Wikstroemia gracilis Hemsl. 纤细荛花
Shrubs

12.3 (7.3~15.9); 0.5 (-3.5~4.2); 23.4 (15.6~27.5); 12.5 (7.5~16.0); 98.9 (48.2~132.7); 10.5 (0.8~21.2); 723 (537~868) ❙ 1121 (598~1660); 467 (297~606); 87 (20~216) ❙ 54.6 (-14.9~103.1); 704 (537~868); 233.0 (163.5~279.5)

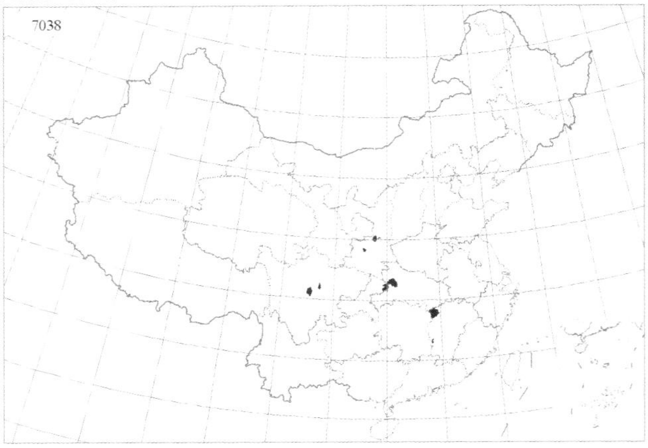

Thymelaeaceae 瑞香科 *Wikstroemia* 荛花属
Wikstroemia hainanensis Merr. 海南荛花
Shrubs

23.8 (22.2~24.8); 18.4 (17.2~20.1); 27.8 (25.7~28.7); 23.9 (22.3~24.8); 226.7 (207.2~237.6); 0.0 (0.0~0.0); 1295 (1127~1385) ❙ 1417 (868~1767); 495 (312~610); 61 (41~92) ❙ 10.0 (-34.3~47.8); 1135 (868~1299); 482.6 (288.3~620.0)

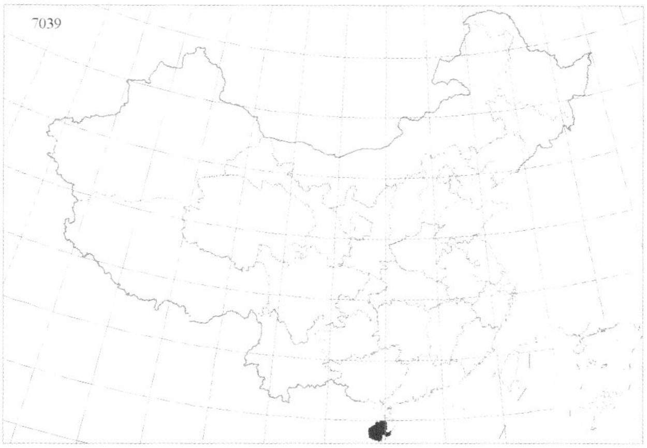

Thymelaeaceae 瑞香科 *Wikstroemia* 荛花属
Wikstroemia haoi Domke 武都荛花
Shrubs

8.4; -2.0; 18.0; 9.0; 63.9; 22.1; 589 ❙ 721; 362; 16 ❙ 25.6; 589; 259.5

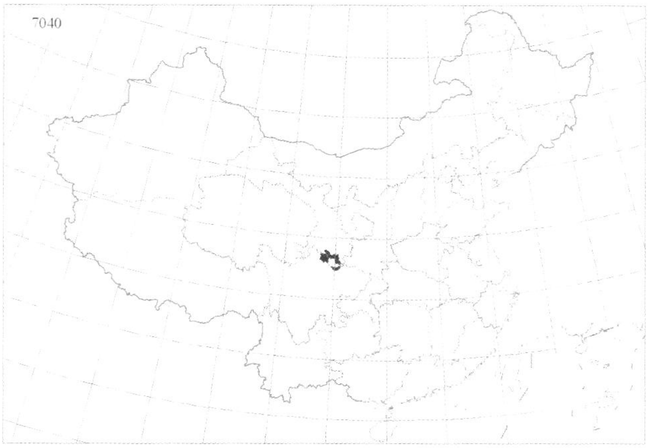

Thymelaeaceae 瑞香科 *Wikstroemia* 荛花属
Wikstroemia huidongensis C. Y. Chang 会东荛花
Shrubs, evergreen

13.7; 6.2; 19.7; 13.7; 104.7; 0.0; 714 ❙ 963; 593; 26 ❙ 34.8; 696; 271.3

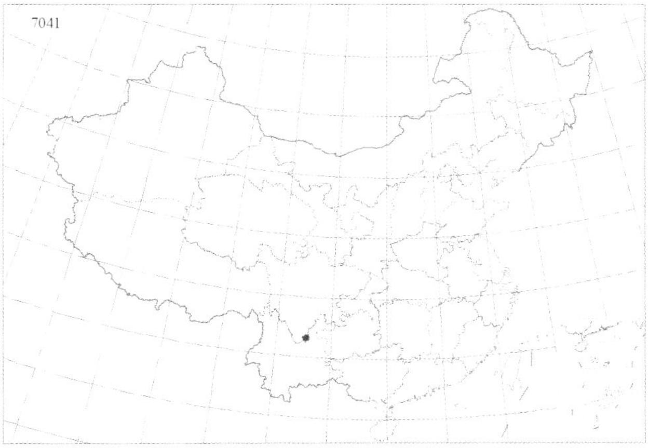

Thymelaeaceae 瑞香科 *Wikstroemia* 荛花属
Wikstroemia indica (L.) C. A. Mey. 了哥王
Shrubs

18.6 (2.6~24.8); 9.3 (-7.0~20.1); 26.6 (11.2~30.1); 18.6 (4.3~24.8); 163.9 (20.0~237.6); 0.4 (0.0~48.2); 979 (427~1385) ❙ 1510 (685~3495); 624 (299~1594); 141 (12~897) ❙ 56.2 (-34.3~235.6); 968 (427~1360); 237.8 (95.3~620.0)

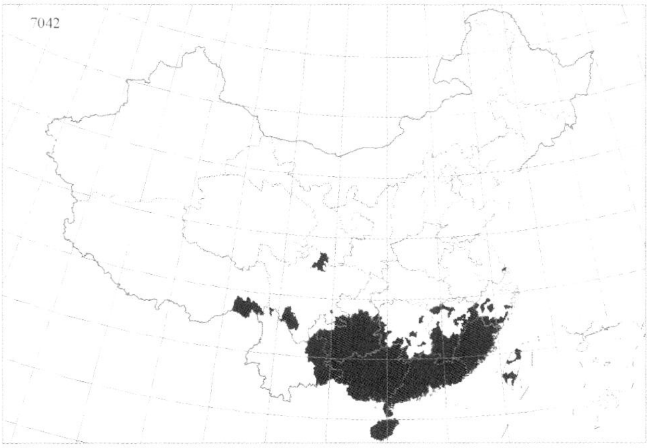

Thymelaeaceae 瑞香科 *Wikstroemia* 荛花属
Wikstroemia leptophylla W. W. Sm. 细叶荛花
Shrubs

7.8; 0.4; 14.2; 8.0; 48.9; 14.3; 548 ❙ 877; 546; 19 ❙ 60.7; 548; 304.7

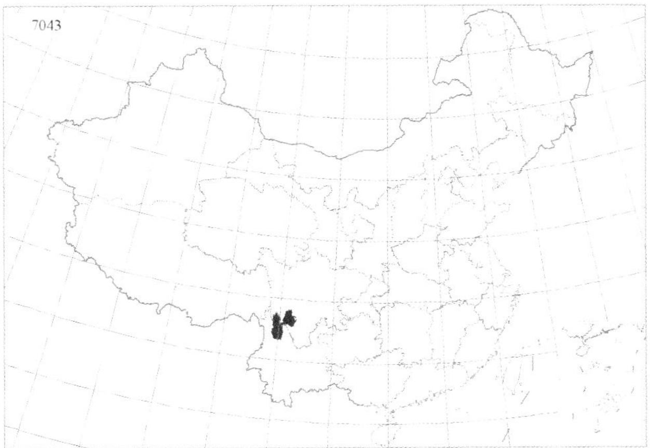

Thymelaeaceae 瑞香科 *Wikstroemia* 荛花属
Wikstroemia liangii Merr. et Chun 大叶荛花
Shrubs

24.0; 18.7; 27.8; 24.1; 229.1; 0.0; 1314 ❙ 1637; 579; 79 ❙ 24.5; 1237; 486.5

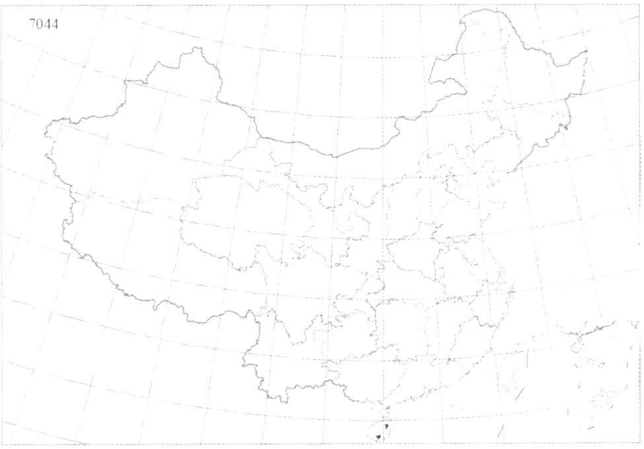

Thymelaeaceae 瑞香科 *Wikstroemia* 荛花属
Wikstroemia lichiangensis W. W. Sm. 丽江荛花
Shrubs

7.0 (-1.2~14.5); -0.9 (-11.7~7.8); 13.8 (8.5~19.7); 8.0 (2.6~14.5); 52.0 (8.1~113.5); 27.1 (0.0~82.3); 537 (347~730) ❙ 856 (491~1057); 505 (301~627); 26 (10~54) ❙ 62.3 (36.1~128.6); 535 (347~730); 301.1 (208.8~382.6)

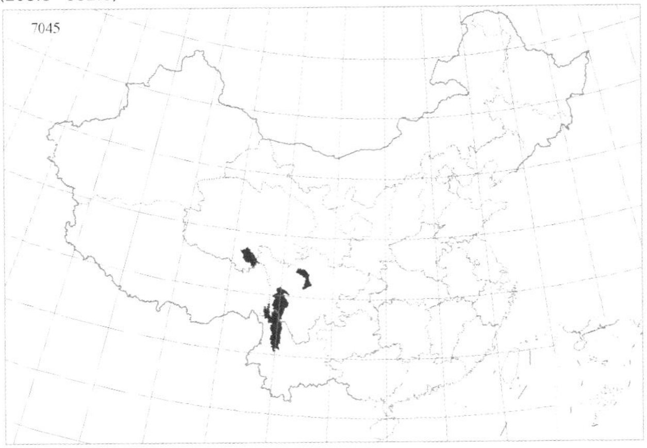

Thymelaeaceae 瑞香科 *Wikstroemia* 荛花属
Wikstroemia ligustrina Rehder 白腊叶荛花
Shrubs

6.3 (-0.2~13.6); -7.0 (-12.7~0.9); 18.2 (8.8~25.8); 7.9 (3.0~13.6); 57.4 (11.1~113.4); 41.6 (8.7~72.9); 559 (363~786) ❙ 628 (298~1006); 347 (215~507); 19 (2~49) ❙ 15.7 (-37.7~128.6); 508 (298~786); 225.1 (109.4~357.7)

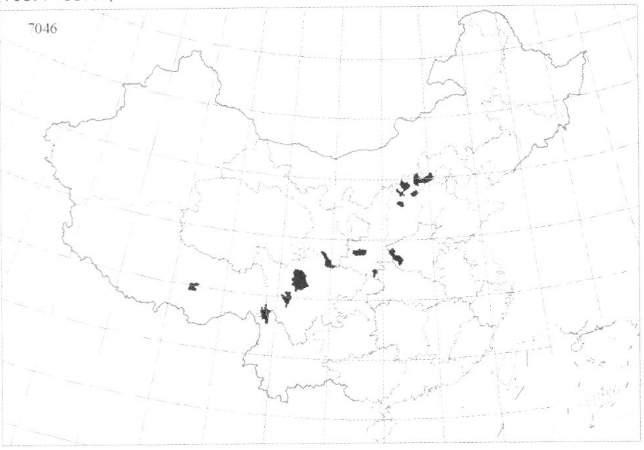

Thymelaeaceae 瑞香科 *Wikstroemia* 荛花属
Wikstroemia linearifolia H. F. Zhou ex C. Y. Chang 线叶荛花
Shrubs

1.6; -8.0; 9.8; 3.7; 15.2; 56.2; 408 ❙ 687; 419; 12 ❙ 68.4; 408; 251.8

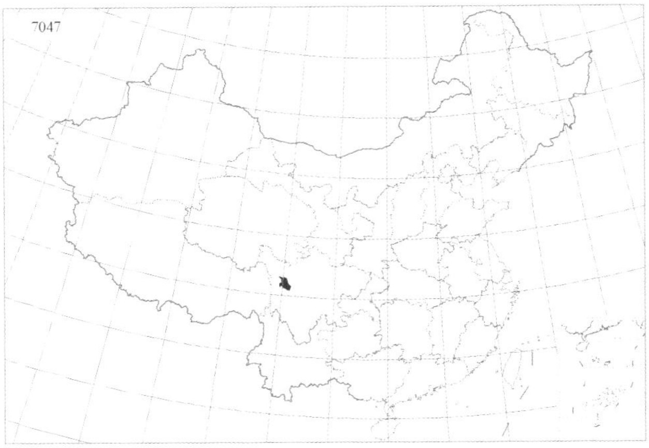

Thymelaeaceae 瑞香科 *Wikstroemia* 荛花属
Wikstroemia linoides Hemsl. 亚麻荛花
Subshrubs

14.6; 3.5; 24.8; 14.6; 117.4; 2.2; 791 ❙ 1034; 526; 52 ❙ 30.7; 791; 183.5

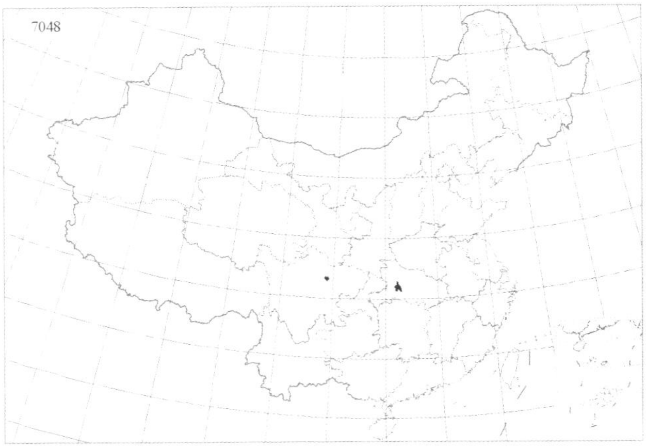

Thymelaeaceae 瑞香科 *Wikstroemia* 荛花属
Wikstroemia longipaniculata S. C. Huang 长锥序荛花
Shrubs

22.2; 13.9; 28.3; 22.3; 207.5; 0.0; 1190 ❙ 1374; 713; 87 ❙ 15.4; 1190; 327.7

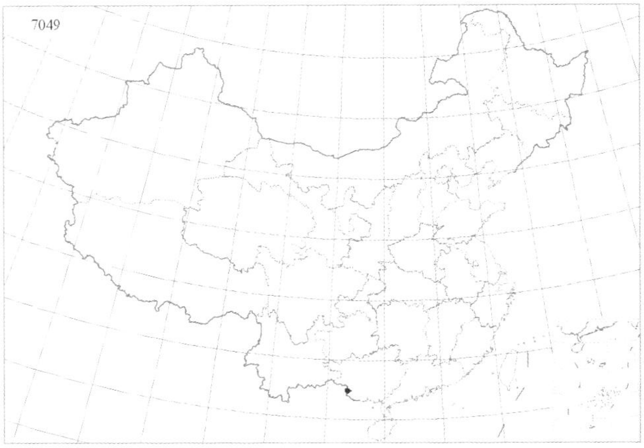

Thymelaeaceae 瑞香科 *Wikstroemia* 荛花属
Wikstroemia lungtzeensis S. C. Huang 隆子荛花
Shrubs

4.4; -4.3; 12.1; 5.3; 28.5; 34.6; 455 ❙ 538; 337; 9 ❙ 18.2; 455; 168.2

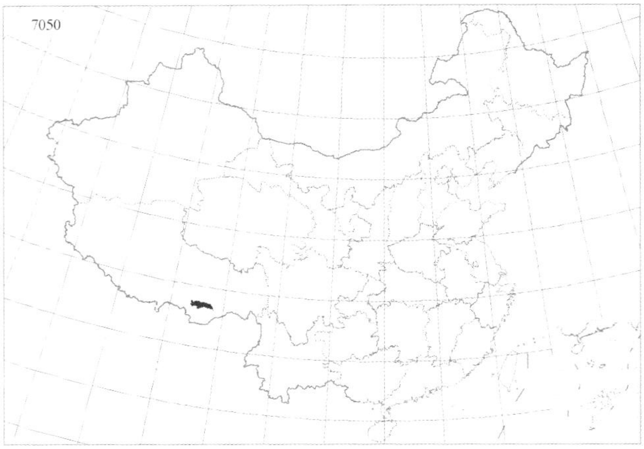

Thymelaeaceae 瑞香科 *Wikstroemia* 荛花属
Wikstroemia micrantha Hemsl. 小黄构
Shrubs

15.6 (2.0~22.2); 5.1 (−7.2~14.7); 25.3 (9.9~29.2); 15.7 (3.8~22.3); 131.1 (15.4~207.5); 3.3 (0.0~51.4); 844 (413~1190) ❙ 1239 (622~1876); 527 (298~891); 89 (15~228) ❙ 47.9 (−13.4~128.9); 842 (413~1190); 222.5 (108.2~491.5)

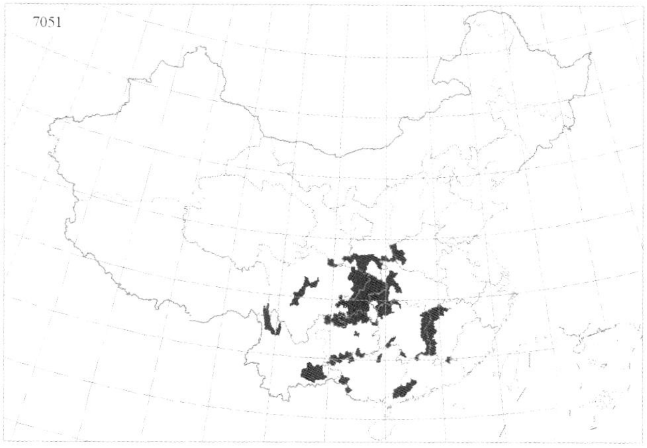

Thymelaeaceae 瑞香科 *Wikstroemia* 荛花属
Wikstroemia monnula Hance 北江荛花
Shrubs

18.7 (12.7~22.4); 9.1 (2.1~14.3); 27.3 (23.0~29.8); 18.7 (12.8~22.5); 164.9 (99.0~209.7); 0.1 (0.0~5.6); 979 (713~1205) ❙ 1632 (1070~2024); 604 (356~954); 185 (65~259) ❙ 69.1 (−5.0~147.0); 973 (713~1205); 237.1 (95.3~332.5)

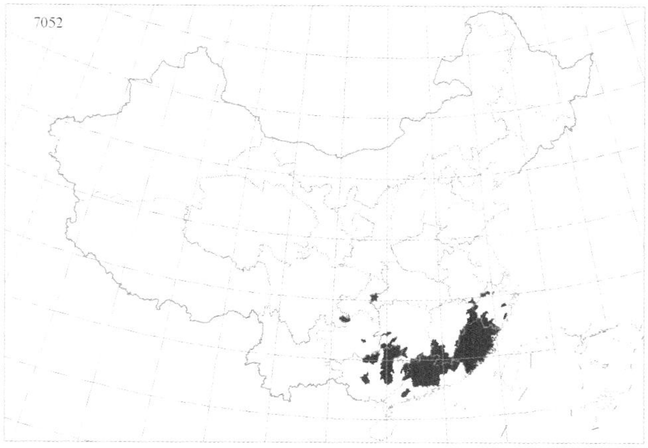

Thymelaeaceae 瑞香科 *Wikstroemia* 荛花属
Wikstroemia mononectaria Hayata 独鳞荛花
Shrubs

19.6; 13.6; 24.8; 19.6; 175.7; 0.0; 960 ❙ 2575; 1099; 301 ❙ 168.4; 960; 256.0

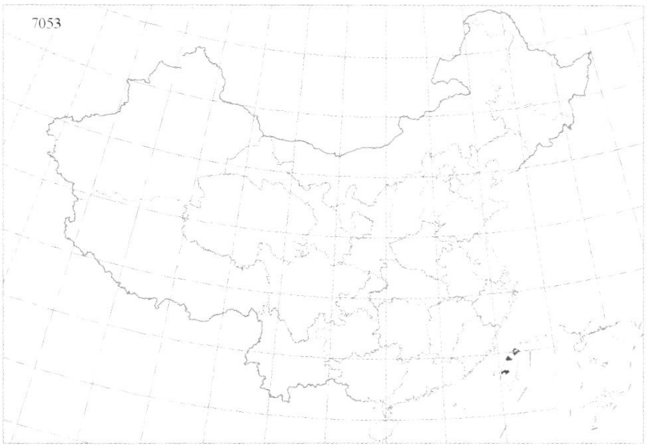

Thymelaeaceae 瑞香科 *Wikstroemia* 荛花属
Wikstroemia nutans Champion ex Bentham 细轴荛花
Shrubs

20.1 (12.5~24.8); 11.4 (3.6~20.1); 27.3 (20.1~29.8); 20.2 (12.5~24.8); 182.1 (92.0~237.6); 0.0 (0.0~1.4); 1059 (685~1385) ❙ 1588 (868~2992); 687 (312~1847); 132 (41~259) ❙ 51.8 (−34.3~202.5); 1048 (685~1360); 244.5 (95.3~620.0)

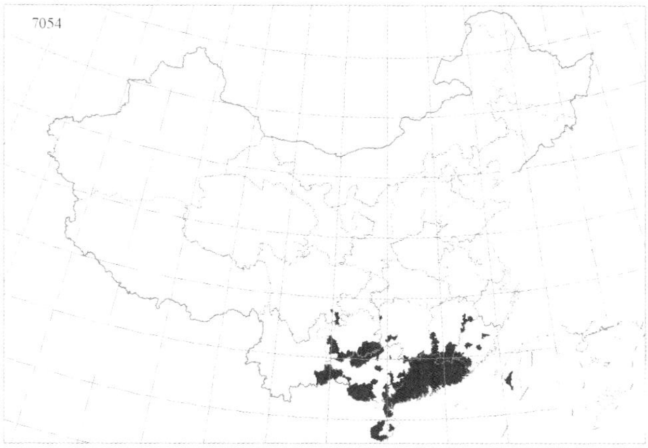

Thymelaeaceae 瑞香科 *Wikstroemia* 荛花属
Wikstroemia pachyrachis S. L. Tsai 粗轴荛花
Shrubs

23.2 (17.4~24.7); 16.9 (7.4~19.8); 27.9 (25.5~28.7); 23.3 (17.4~24.7); 219.4 (148.9~236.5); 0.0 (0.0~0.0); 1260 (886~1376) ❙ 1498 (868~2035); 594 (312~969); 80 (42~113) ❙ 20.4 (−34.3~71.8); 1152 (868~1329); 367.4 (139.6~620.0)

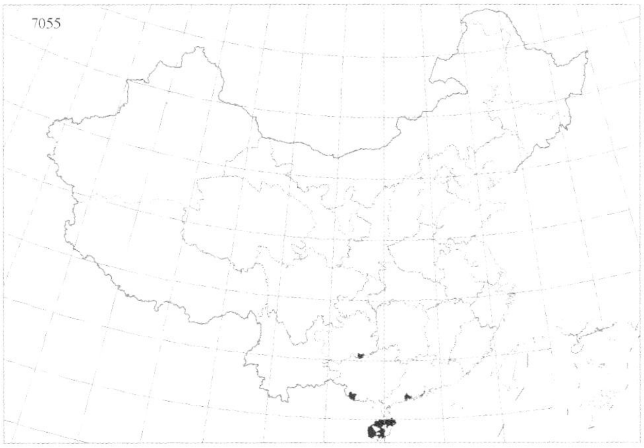

Thymelaeaceae 瑞香科 *Wikstroemia* 荛花属
Wikstroemia pampaninii Rehder et E. H. Wilson 鄂北荛花
Shrubs

8.4 (3.3~14.5); −6.2 (−12.7~1.8); 21.7 (17.5~26.4); 9.8 (6.6~14.6); 76.3 (45.4~120.7); 34.9 (5.7~65.0); 631 (507~821) ❙ 585 (341~880); 309 (190~399); 21 (8~58) ❙ −7.8 (−40.9~52.6); 559 (341~821); 182.2 (61.2~318.2)

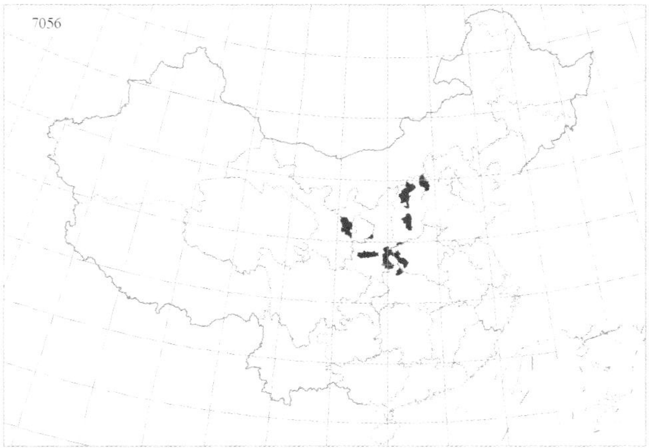

Thymelaeaceae 瑞香科 *Wikstroemia* 荛花属
Wikstroemia pilosa Cheng 多毛荛花
Shrubs

16.3 (14.1~18.2); 4.5 (1.6~7.8); 27.8 (25.3~29.5); 16.4 (14.2~18.3); 137.4 (116.7~159.6); 0.9 (0.0~6.5); 888 (791~966) ❙ 1481 (1039~1997); 542 (381~685); 171 (84~254) ❙ 67.2 (7.5~130.7); 881 (791~965); 217.2 (111.9~318.0)

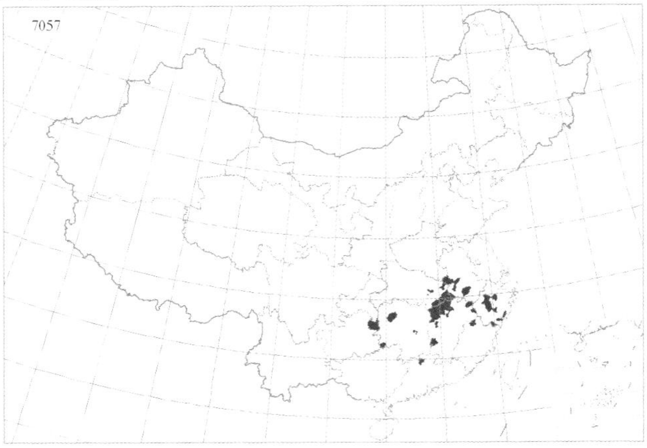

Thymelaeaceae 瑞香科 *Wikstroemia* 荛花属
Wikstroemia salicina (Lév.) Lév. et Blin 柳状荛花
Shrubs

12.7; 4.7; 19.2; 12.7; 93.0; 0.3; 687 ❙ 922; 546; 26 ❙ 34.2; 687; 232.7

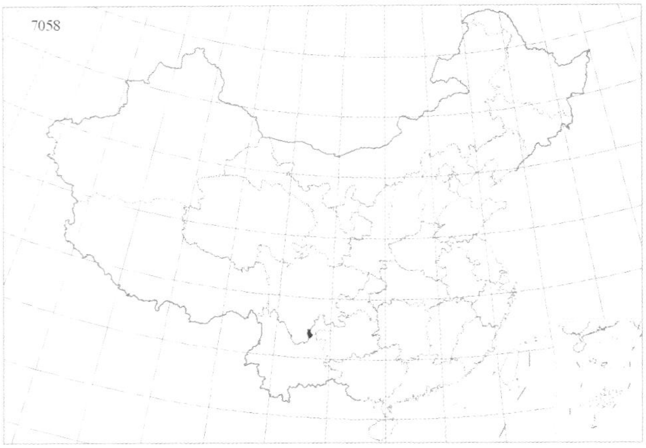

Thymelaeaceae 瑞香科 *Wikstroemia* 荛花属
Wikstroemia scytophylla Diels 革叶荛花
Shrubs

9.9 (0.7~18.5); 1.8 (-8.8~10.7); 16.8 (9.5~24.3); 10.3 (3.3~18.3); 74.0 (12.7~160.0); 14.9 (0.0~63.9); 617 (389~906) ❙ 851 (510~1096); 503 (311~627); 24 (10~53) ❙ 40.0 (4.4~77.4); 600 (389~786); 246.7 (169.5~351.7)

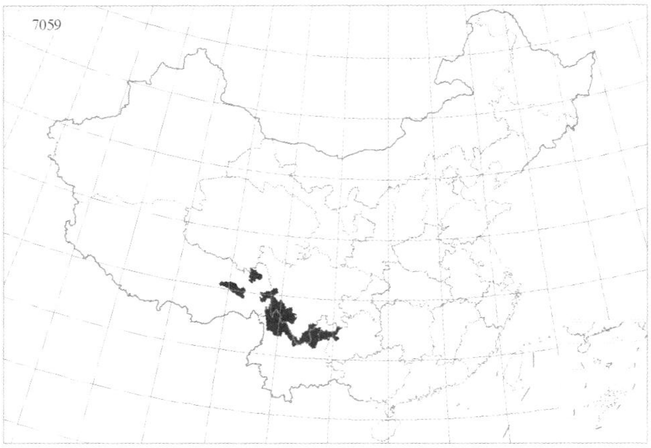

Thymelaeaceae 瑞香科 *Wikstroemia* 荛花属
Wikstroemia sinoparviflora Yin Z. Wang et M. G. Gilbert 小花荛花
Shrubs

10.3; -0.1; 19.8; 10.3; 75.8; 11.7; 634 ❙ 711; 357; 15 ❙ 12.1; 634; 279.5

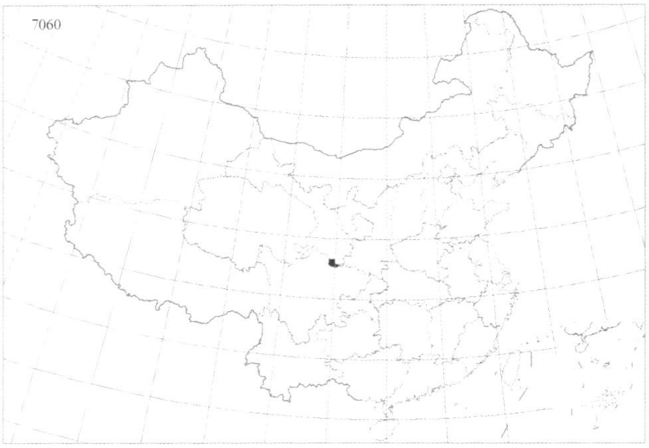

Thymelaeaceae 瑞香科 *Wikstroemia* 荛花属
Wikstroemia stenophylla E. Pritz. 轮叶荛花
Shrubs, evergreen

10.0 (2.0~16.0); -0.4 (-7.2~5.5); 19.5 (9.9~26.3); 10.5 (3.8~16.0); 78.9 (15.4~132.4); 18.7 (0.0~51.4); 645 (413~847) ❙ 1019 (731~1410); 482 (346~816); 47 (17~123) ❙ 65.7 (5.0~128.6); 645 (413~847); 247.5 (181.2~357.7)

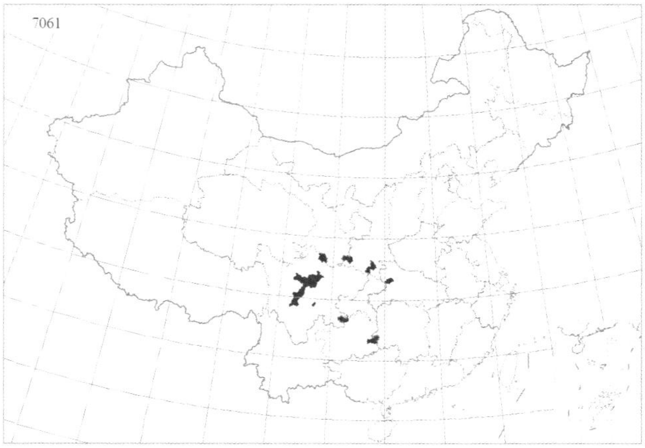

Thymelaeaceae 瑞香科 *Wikstroemia* 荛花属
Wikstroemia subcyclolepidota L. P. Liu et Y. S. Lian 亚环鳞荛花
Shrubs, deciduous

9.5; -4.3; 22.2; 10.2; 80.6; 26.2; 648 ❙ 588; 294; 18 ❙ -9.2; 588; 174.0

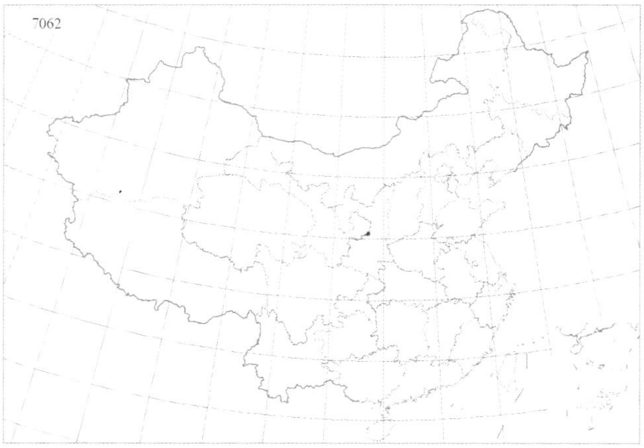

Thymelaeaceae 瑞香科 *Wikstroemia* 荛花属
Wikstroemia techinensis S. C. Huang 德钦荛花
Shrubs

5.4; -2.1; 12.3; 5.8; 31.5; 25.6; 481 | 741; 426; 28 | 54.0; 481; 273.5

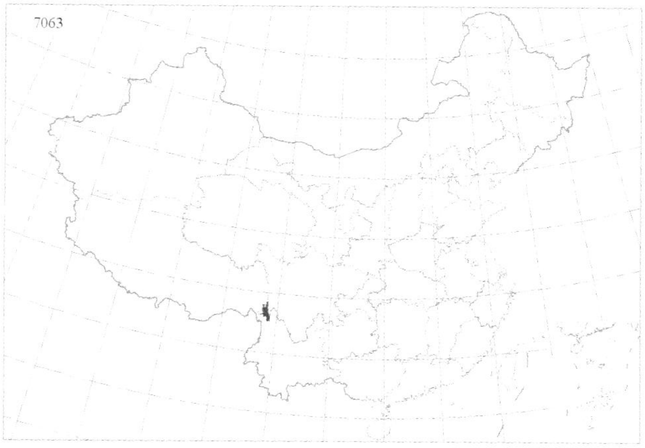

Thymelaeaceae 瑞香科 *Wikstroemia* 荛花属
Wikstroemia trichotoma (Thunb.) Makino 白花荛花
Shrubs, evergreen

18.0 (14.3~24.1); 7.7 (2.7~18.9); 27.5 (21.8~29.8); 18.1 (14.3~24.1); 157.2 (114.1~229.7); 0.3 (0.0~3.2); 952 (765~1322) | 1692 (1211~3495); 605 (420~1847); 210 (47~897) | 78.6 (14.0~235.6); 941 (765~1123); 232.3 (140.5~332.5)

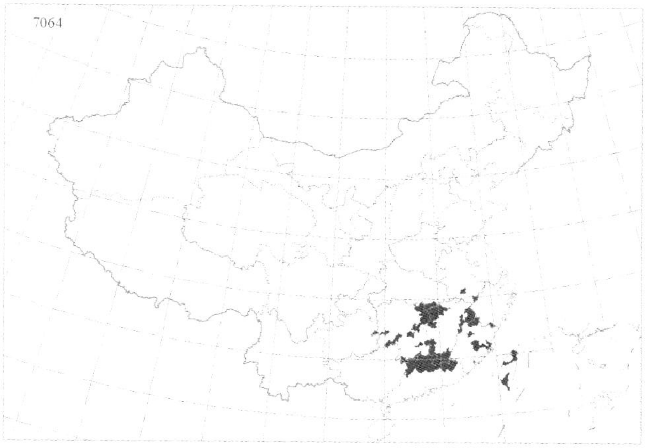

Thymelaeaceae 瑞香科 *Wikstroemia* 荛花属
Wikstroemia vaccinium (H. Lév.) Rehder 平伐荛花
Shrubs, deciduous

15.9; 6.0; 24.5; 16.0; 131.8; 0.0; 820 | 1267; 573; 77 | 54.5; 820; 235.5

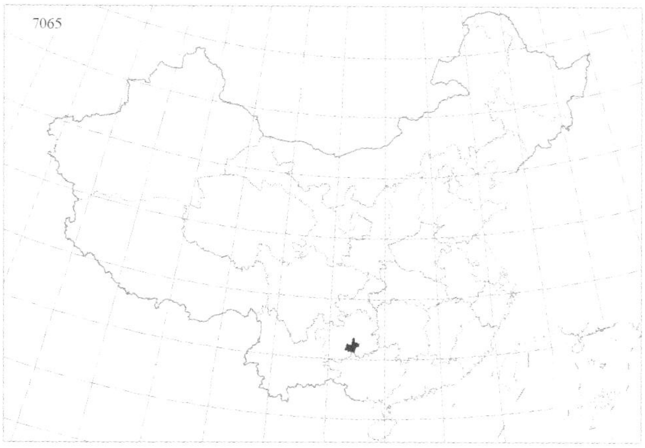

Elaeagnaceae 胡颓子科 *Elaeagnus* 胡颓子属
Elaeagnus angustata (Rehder) C. Y. Chang 狭叶木半夏
Shrubs, deciduous

11.2 (2.2~14.5); 2.5 (-6.6~7.3); 18.7 (9.9~23.3); 11.4 (3.8~14.6); 83.7 (15.5~115.0); 8.6 (0.0~49.1); 653 (413~739) | 1030 (932~1202); 558 (468~636); 35 (18~83) | 60.9 (32.4~125.9); 652 (413~739); 271.5 (230.7~318.7)

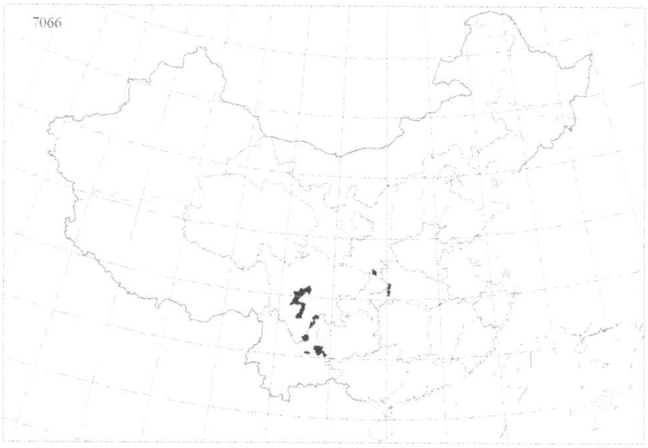

Elaeagnaceae 胡颓子科 *Elaeagnus* 胡颓子属
Elaeagnus angustifolia L. 沙枣
Trees, deciduous

10.2 (-4.3~17.1); -5.5 (-20.9~4.4); 23.9 (9.6~29.4); 11.5 (2.5~17.1); 95.2 (9.7~146.1); 32.4 (0.6~121.1); 723 (329~952) | 502 (21~1663); 269 (15~666); 32 (0~212) | -32.9 (-97.3~108.0); 465 (21~930); 121.9 (1.6~299.5)

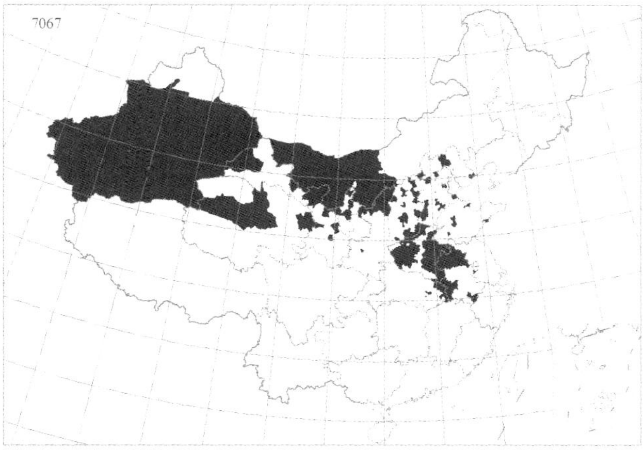

Elaeagnaceae 胡颓子科 *Elaeagnus* 胡颓子属
Elaeagnus angustifolia var. orientalis (L.) Kuntze 东方沙枣
Shrubs or small trees, deciduous

7.0 (4.7~8.0); -10.8 (-13.8~-8.7); 22.8 (21.1~24.3); 9.4 (8.2~10.2); 76.6 (63.8~85.9); 52.7 (43.5~67.3); 632 (583~679) | 222 (49~416); 142 (34~265); 5 (1~12) | -64.2 (-92.8~-31.0); 222 (49~416); 46.0 (5.1~90.0)

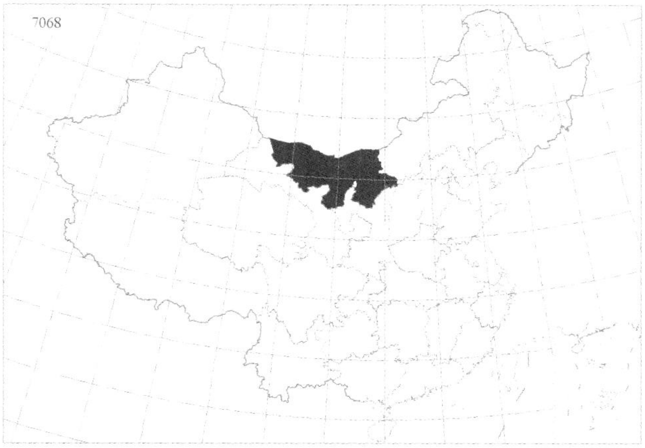

Elaeagnaceae 胡颓子科 *Elaeagnus* 胡颓子属
Elaeagnus argyi H. Lév. 佘山羊奶子
Shrubs, deciduous or evergreen

12.9 (−1.1~21.7); 0.0 (−16.0~15.3); 24.7 (6.2~30.1); 13.4 (2.0~21.7); 110.0 (2.5~200.2); 14.6 (0.0~81.2); 776 (334~1153) ❙ 943 (63~2859); 443 (45~1186); 74 (2~402) ❙ 20.2 (−84.9~246.5); 711 (63~1153); 197.0 (8.4~447.4)

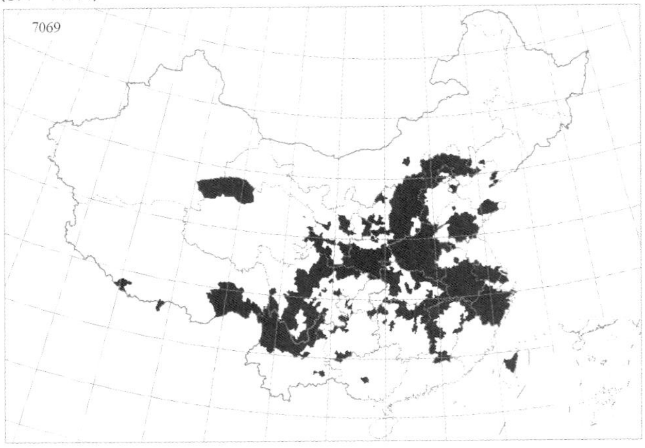

Elaeagnaceae 胡颓子科 *Elaeagnus* 胡颓子属
Elaeagnus bambusetorum Hand.-Mazz. 竹生羊奶子
Shrubs

17.4; 11.0; 21.8; 17.4; 149.1; 0.0; 836 ❙ 1130; 629; 48 ❙ 35.3; 820; 385.2

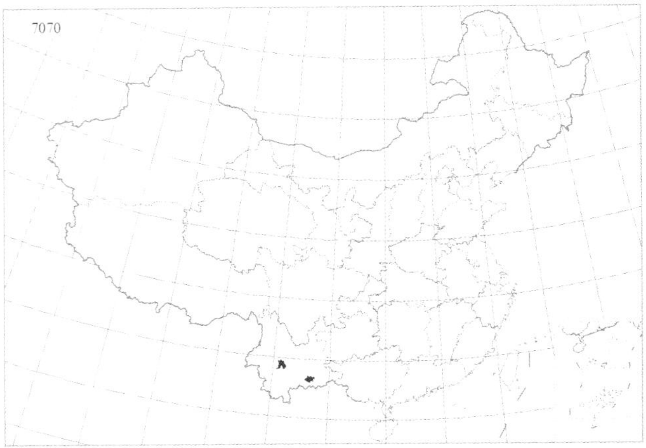

Elaeagnaceae 胡颓子科 *Elaeagnus* 胡颓子属
Elaeagnus bockii Diels 长叶胡颓子
Shrubs, evergreen

14.7 (2.2~22.4); 4.1 (−6.6~14.3); 24.5 (9.9~29.1); 14.8 (3.8~22.5); 123.0 (15.5~209.7); 6.1 (0.0~49.1); 817 (413~1205) ❙ 1173 (551~2024); 525 (245~954); 71 (12~191) ❙ 43.6 (−36.2~125.9); 812 (413~1205); 231.1 (95.3~411.1)

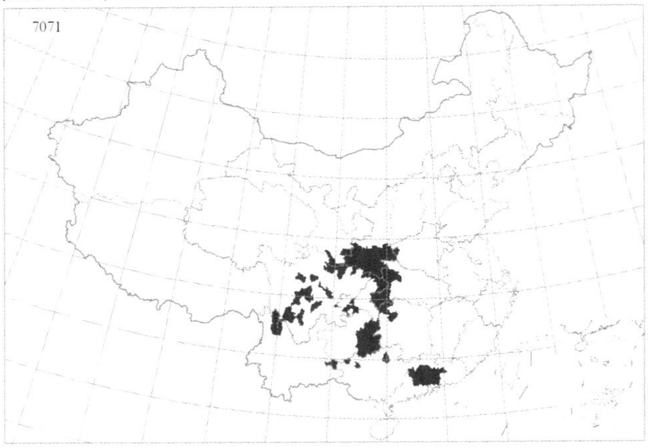

Elaeagnaceae 胡颓子科 *Elaeagnus* 胡颓子属
Elaeagnus bockii var. muliensis C. Y. Chang 木里胡颓子
Shrubs, evergreen

6.4; −1.4; 13.1; 7.0; 40.5; 23.0; 514 ❙ 891; 546; 19 ❙ 77.0; 514; 290.6

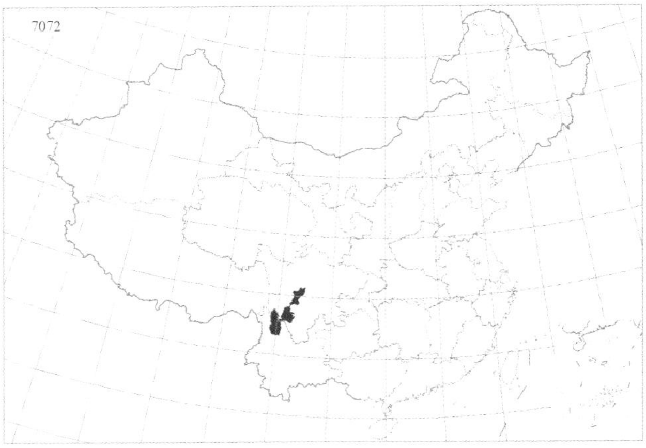

Elaeagnaceae 胡颓子科 *Elaeagnus* 胡颓子属
Elaeagnus calcarea Z. R. Xu 石山胡颓子
Shrubs, semievergreen

17.4; 7.4; 25.5; 17.4; 148.9; 0.0; 886 ❙ 1381; 637; 107 ❙ 55.9; 886; 245.3

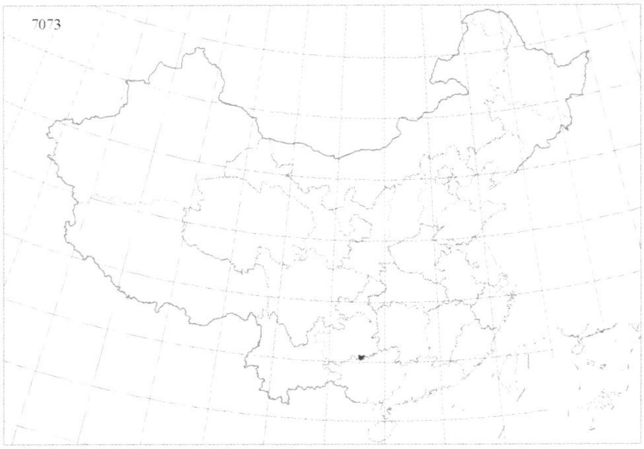

Elaeagnaceae 胡颓子科 *Elaeagnus* 胡颓子属
Elaeagnus cinnamomifolia W. K. Hu et H. F. Chow 樟叶胡颓子
Climbing shrubs, evergreen

18.5; 8.7; 26.6; 18.5; 162.6; 0.0; 953 ❙ 1599; 684; 173 ❙ 67.8; 953; 235.1

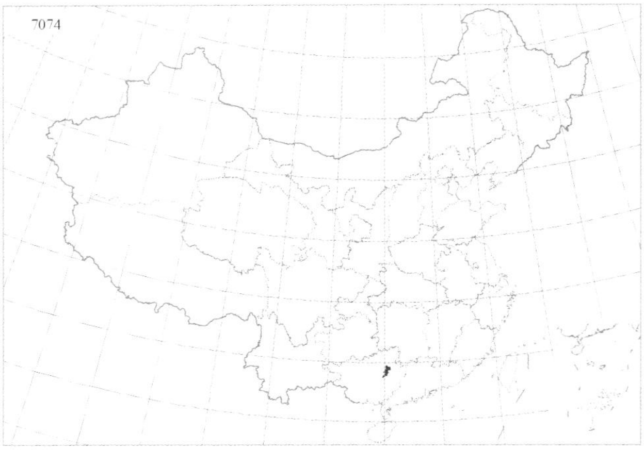

Elaeagnaceae 胡颓子科 *Elaeagnus* 胡颓子属

Elaeagnus conferta Roxb. 密花胡颓子

Scandent shrubs, evergreen

20.5 (15.5~22.4); 12.6 (8.6~15.1); 26.4 (20.4~29.1); 20.5 (15.6~22.5); 185.9 (127.3~209.7); 0.0 (0.0~0.0); 1057 (768~1205) ❙ 1636 (1049~2024); 775 (598~959); 117 (42~194) ❙ 55.6 (15.4~112.2); 1053 (768~1205); 298.7 (95.3~659.0)

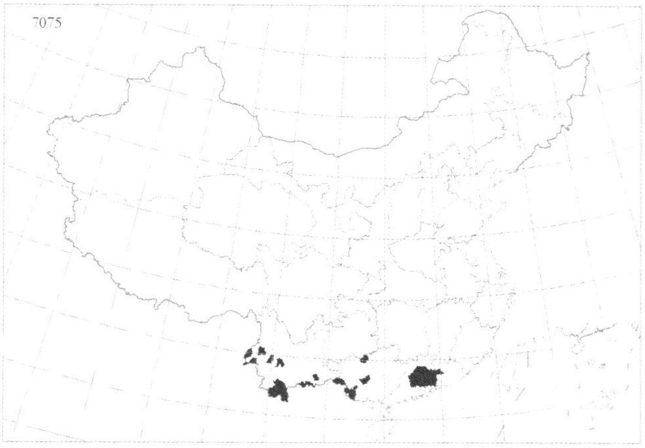

Elaeagnaceae 胡颓子科 *Elaeagnus* 胡颓子属

Elaeagnus courtoisii Belval 毛木半夏

Shrubs, deciduous

16.1 (13.4~18.4); 3.9 (1.5~6.7); 27.8 (25.2~29.5); 16.1 (13.4~18.4); 135.0 (108.3~161.1); 1.6 (0.0~7.3); 882 (755~993) ❙ 1510 (1037~1959); 564 (440~685); 177 (81~253) ❙ 72.2 (17.0~124.8); 876 (755~942); 225.6 (111.9~319.9)

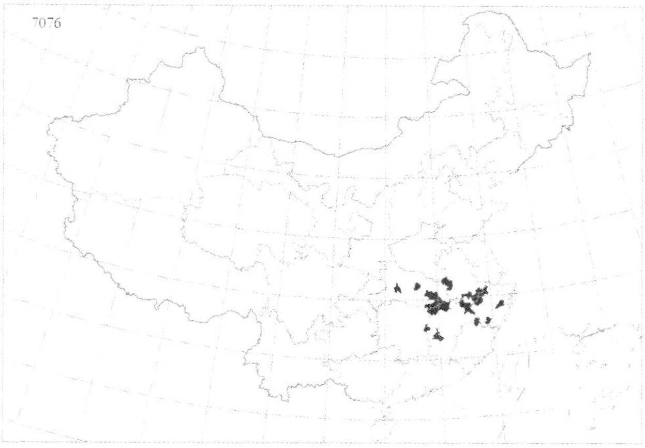

Elaeagnaceae 胡颓子科 *Elaeagnus* 胡颓子属

Elaeagnus delavayi Lecomte 长柄胡颓子

Shrubs, evergreen

13.7 (5.4~18.9); 6.0 (-2.1~12.0); 20.0 (12.3~26.6); 13.7 (5.8~19.0); 106.6 (31.5~167.8); [illegible].9 (0.0~25.6); 724 (481~929) ❙ 956 (741~1141); 557 (411~641); 30 ([illegible]5~57) ❙ 33.7 (-14.2~70.8); 693 (481~912); 269.3 (142.9~351.7)

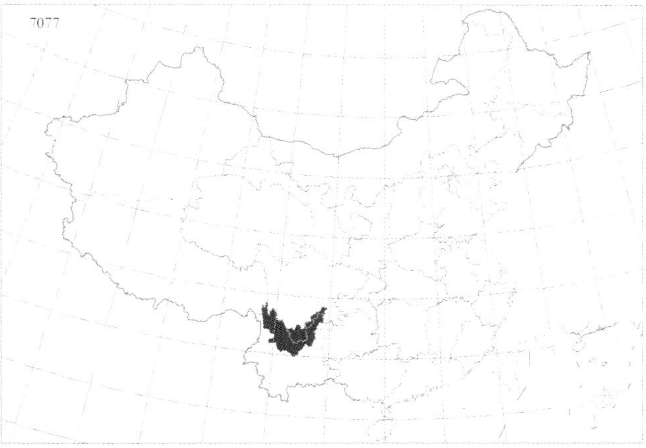

Elaeagnaceae 胡颓子科 *Elaeagnus* 胡颓子属

Elaeagnus difficilis Servettaz 巴东胡颓子

Shrubs, evergreen

15.7 (9.3~21.6); 4.9 (-2.0~13.2); 25.7 (18.8~29.5); 15.7 (9.5~21.6); 130.2 (69.8~199.2); 1.5 (0.0~18.5); 839 (603~1139) ❙ 1347 (815~2005); 541 (352~858); 117 (19~259) ❙ 60.5 (7.2~130.7); 836 (603~1139); 219.5 (108.2~354.5)

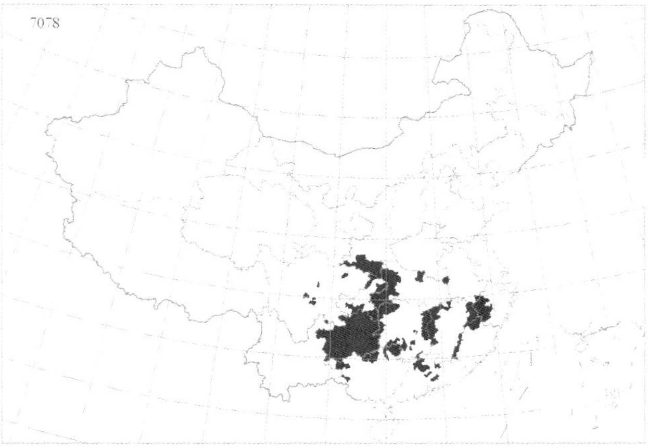

Elaeagnaceae 胡颓子科 *Elaeagnus* 胡颓子属

Elaeagnus difficilis var. brevistyla W. K. Hu et H. F. Chow 短柱胡颓子

Shrubs, evergreen

14.8; 4.3; 24.9; 14.9; 119.2; 0.7; 791 ❙ 1197; 489; 71 ❙ 51.2; 791; 182.9

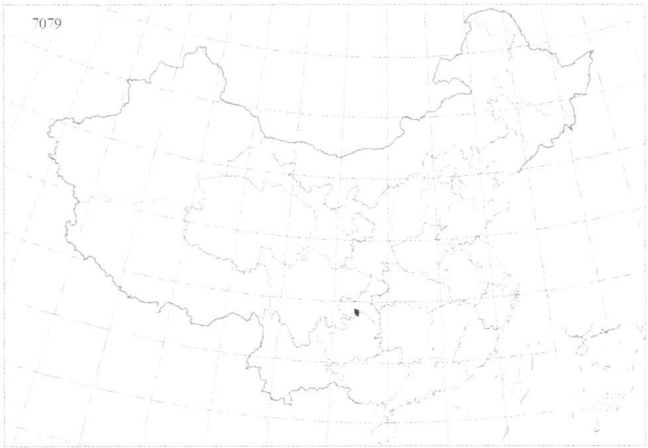

Elaeagnaceae 胡颓子科 *Elaeagnus* 胡颓子属

Elaeagnus formosana Nakai 台湾胡颓子

Shrubs, evergreen

17.2 (11.9~21.1); 7.7 (-2.0~14.6); 26.3 (20.3~27.8); 17.3 (12.1~21.1); 149.0 (98.7~193.1); 2.2 (0.0~17.1); 902 (720~1077) ❙ 1840 (689~3495); 668 (393~1388); 253 (34~897) ❙ 100.2 (-13.7~224.4); 894 (689~1077); 228.1 (136.4~363.7)

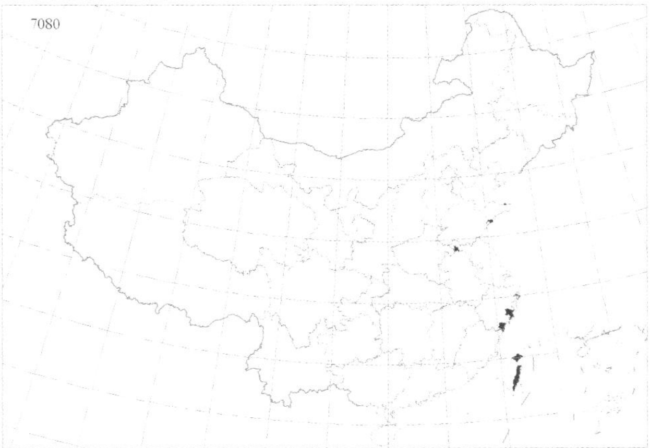

Elaeagnaceae 胡颓子科 *Elaeagnus* 胡颓子属
Elaeagnus formosensis Hatusima 蓬莱胡颓子
Scandent shrubs

21.0; 16.3; 24.5; 21.1; 192.6; 0.0; 1055 ❙ 2554; 1499; 117 ❙ 154.0; 998; 266.8

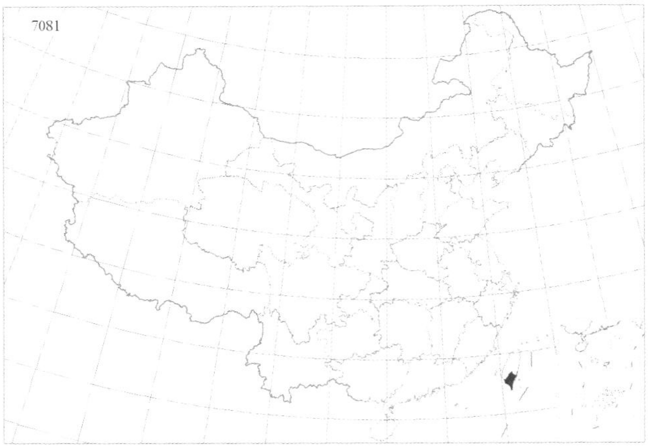

Elaeagnaceae 胡颓子科 *Elaeagnus* 胡颓子属
Elaeagnus geniculata D. Fang 膝柱胡颓子
Climbing shrubs, evergreen

22.2; 13.9; 28.3; 22.3; 207.5; 0.0; 1190 ❙ 1374; 713; 87 ❙ 15.4; 1190; 327.7

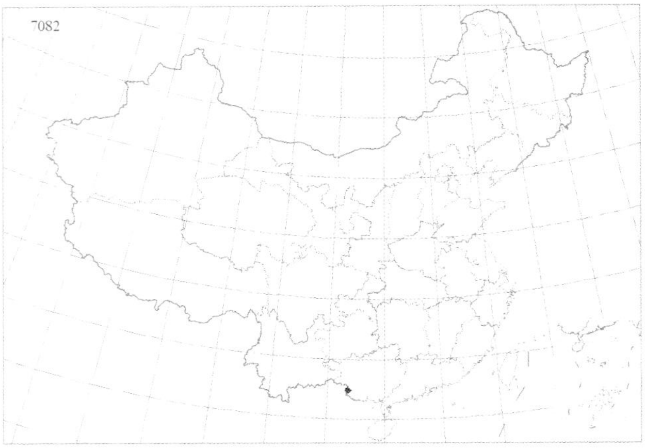

Elaeagnaceae 胡颓子科 *Elaeagnus* 胡颓子属
Elaeagnus glabra Thunb. 蔓胡颓子
Shrubs, evergreen

17.4 (9.3~24.1); 6.8 (−2.9~18.9); 27.1 (19.6~29.9); 17.4 (9.5~24.1); 150.3 (69.8~229.7); 1.1 (0.0~20.5); 929 (603~1322) ❙ 1462 (626~2993); 571 (299~1594); 151 (29~547) ❙ 58.1 (−22.2~246.5); 918 (603~1273); 216.4 (95.3~371.7)

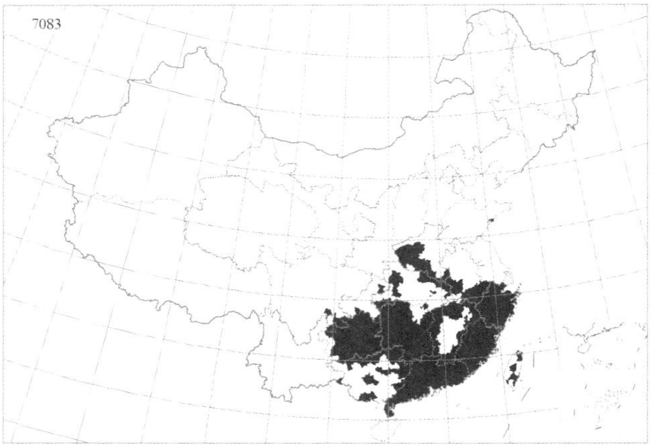

Elaeagnaceae 胡颓子科 *Elaeagnus* 胡颓子属
Elaeagnus gonyanthes Benth. 角花胡颓子
Climbing shrubs, evergreen

19.9 (14.3~24.8); 10.8 (3.7~20.1); 27.5 (22.4~29.7); 19.9 (14.3~24.8); 178.9 (113.2~237.6); 0.0 (0.0~1.4); 1049 (761~1385) ❙ 1593 (1179~2435); 652 (392~1428); 150 (41~221) ❙ 54.2 (−11.9~128.9); 1031 (761~1360); 266.7 (139.6~659.0)

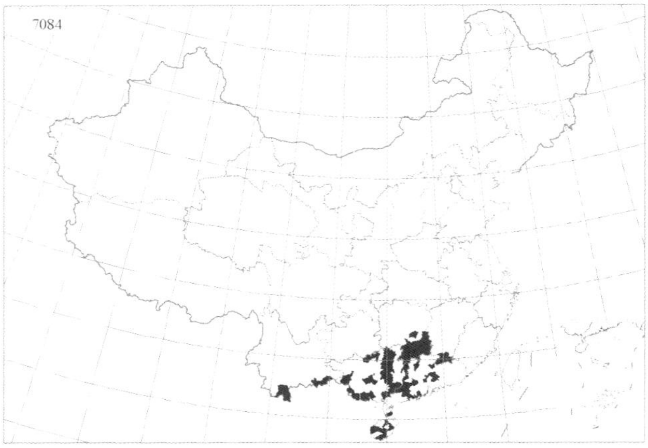

Elaeagnaceae 胡颓子科 *Elaeagnus* 胡颓子属
Elaeagnus grandifolia Hayata 慈恩胡颓子
Scandent shrubs

18.3; 13.0; 22.4; 18.3; 160.1; 0.0; 880 ❙ 2566; 1309; 181 ❙ 195.2; 880; 257.6

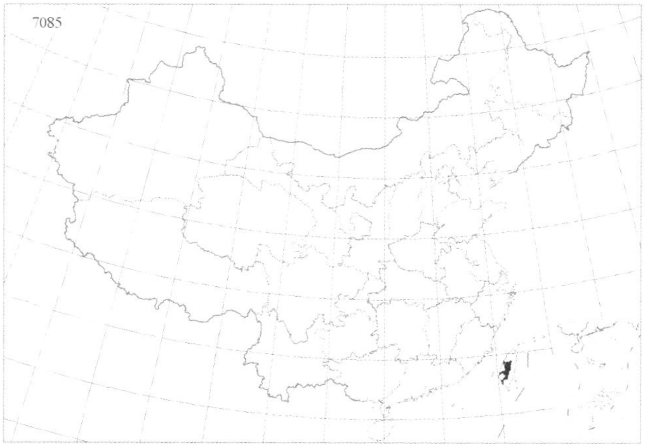

Elaeagnaceae 胡颓子科 *Elaeagnus* 胡颓子属
Elaeagnus griffithii Servettaz 钟花胡颓子
Shrubs

18.0; 10.7; 23.0; 18.0; 156.2; 0.0; 873 ❙ 1408; 801; 54 ❙ 61.6; 873; 414.3

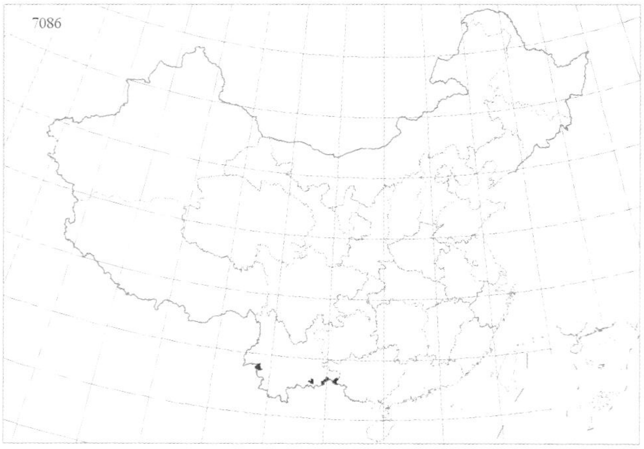

Elaeagnaceae 胡颓子科 *Elaeagnus* 胡颓子属
Elaeagnus grijsii Hance 多毛羊奶子
Small shrubs, deciduous or semievergreen

16.8 (13.9~18.5); 7.3 (3.1~10.4); 25.8 (24.3~26.3); 16.9 (14.0~18.5); 143.1 (110.5~162.3); 0.7 (0.0~2.6); 871 (760~928) ❙ 1573 (1439~1711); 587 (549~614); 147 (91~202) ❙ 81.0 (68.2~94.5); 871 (760~928); 247.5 (185.4~300.5)

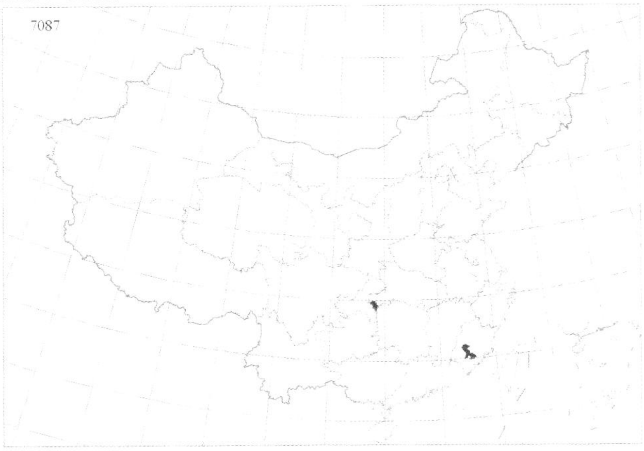

Elaeagnaceae 胡颓子科 *Elaeagnus* 胡颓子属
Elaeagnus guizhouensis C. Y. Chang 贵州羊奶子
Shrubs, deciduous or semievergreen

15.6; 4.4; 26.3; 15.7; 128.8; 0.6; 837 ❙ 1221; 468; 92 ❙ 46.1; 837; 199.5

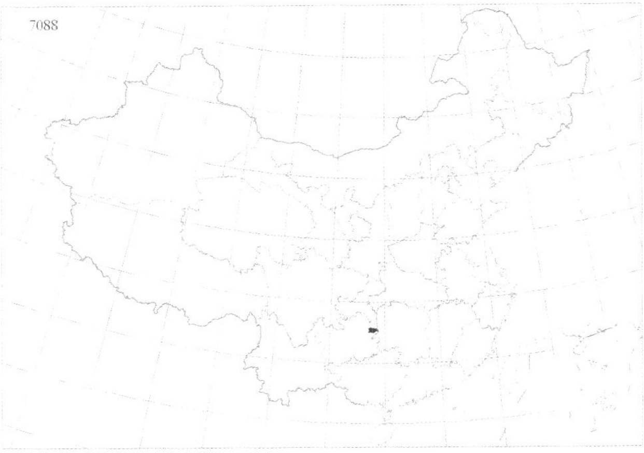

Elaeagnaceae 胡颓子科 *Elaeagnus* 胡颓子属
Elaeagnus henryi Warb. 宜昌胡颓子
Climbing shrubs, evergreen

15.8 (5.3~21.8); 5.0 (-4.2~13.4); 25.8 (12.2~29.5); 15.8 (5.7~21.8); 132.2 (30.5~201.9); 2.4 (0.0~29.4); 849 (479~1158) ❙ 1350 (701~2005); 539 (299~853); 123 (17~259) ❙ 58.5 (-6.7~130.7); 846 (479~1158); 228.5 (108.2~383.4)

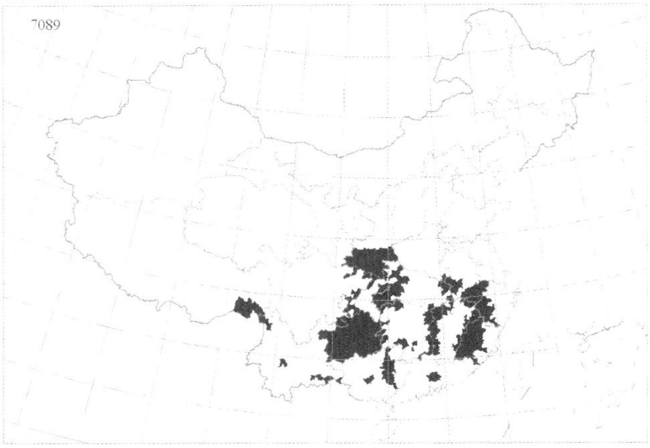

Elaeagnaceae 胡颓子科 *Elaeagnus* 胡颓子属
Elaeagnus heterophylla D. Fang et D. R. Liang 异叶胡颓子
Shrubs, evergreen

18.0; 8.5; 25.8; 18.0; 156.4; 0.0; 913 ❙ 1551; 672; 169 ❙ 69.8; 913; 243.9

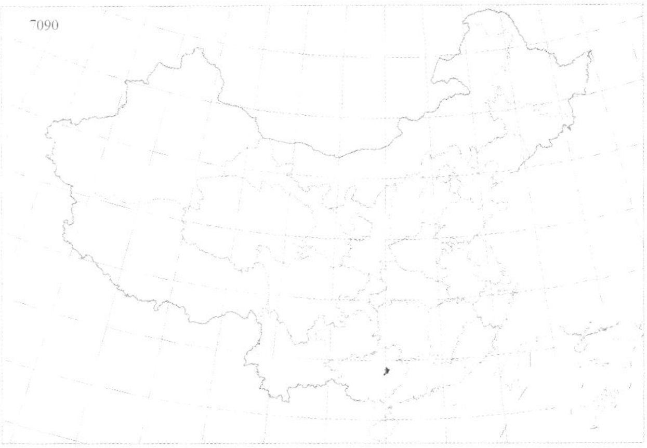

Elaeagnaceae 胡颓子科 *Elaeagnus* 胡颓子属
Elaeagnus jiangxiensis C. Y. Chang 江西羊奶子
Shrubs, deciduous or semievergreen

16.2; 5.0; 26.8; 16.2; 135.1; 0.4; 860 ❙ 1611; 524; 208 ❙ 88.0; 860; 217.4

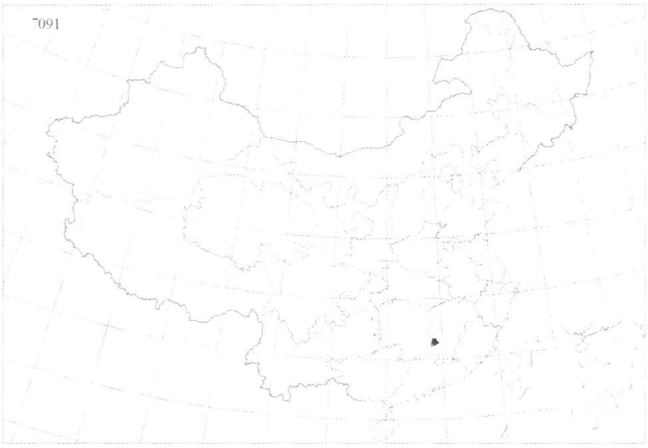

Elaeagnaceae 胡颓子科 *Elaeagnus* 胡颓子属
Elaeagnus jingdongensis C. Y. Chang 景东羊奶子
Shrubs, deciduous or semievergreen

17.3; 11.0; 21.6; 17.3; 148.0; 0.0; 832 ❙ 1049; 598; 44 ❙ 26.1; 783; 365.3

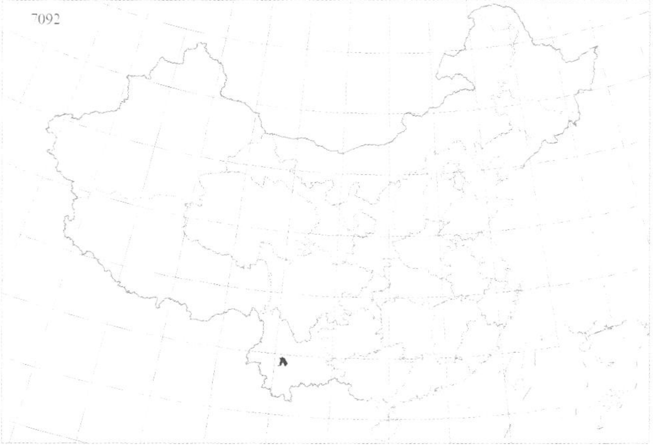

Elaeagnaceae 胡颓子科 *Elaeagnus* 胡颓子属
Elaeagnus lanceolata Warb. 披针叶胡颓子
Shrubs, evergreen

14.3 (1.4~20.1); 4.0 (-7.6~13.3); 23.6 (9.2~29.7); 14.4 (3.5~20.1); 116.2 (13.4~181.6); 4.1 (0.0~55.6); 778 (402~1052) ❙ 1143 (518~1740); 526 (262~860); 71 (10~221) ❙ 47.4 (-41.4~121.7); 770 (402~1052); 234.5 (142.9~452.0)

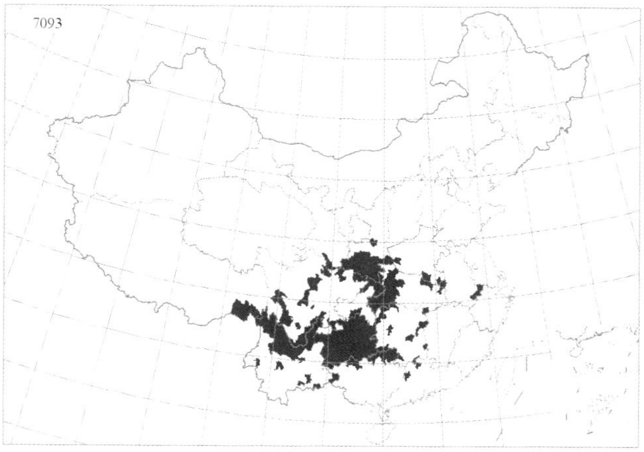

Elaeagnaceae 胡颓子科 *Elaeagnus* 胡颓子属
Elaeagnus lanpingensis C. Y. Chang 兰坪胡颓子
Shrubs or small trees, evergreen

9.8; 2.8; 15.8; 9.8; 62.8; 4.8; 605 ❙ 1088; 618; 46 ❙ 79.9; 605; 345.9

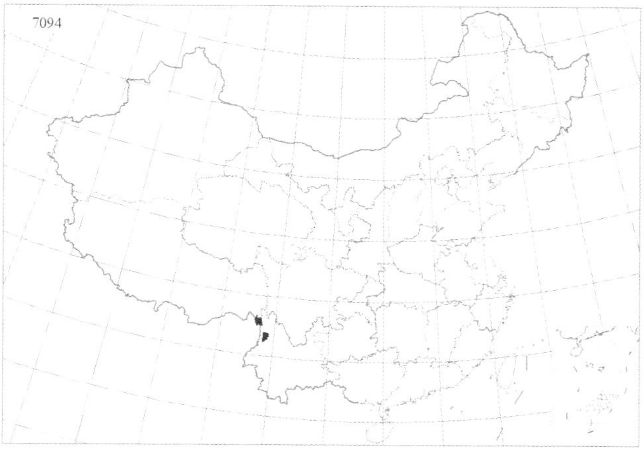

Elaeagnaceae 胡颓子科 *Elaeagnus* 胡颓子属
Elaeagnus liuzhouensis C. Y. Chang 柳州胡颓子
Shrubs, evergreen

19.4; 9.5; 27.6; 19.5; 173.7; 0.0; 1019 ❙ 1427; 649; 123 ❙ 40.9; 1019; 224.9

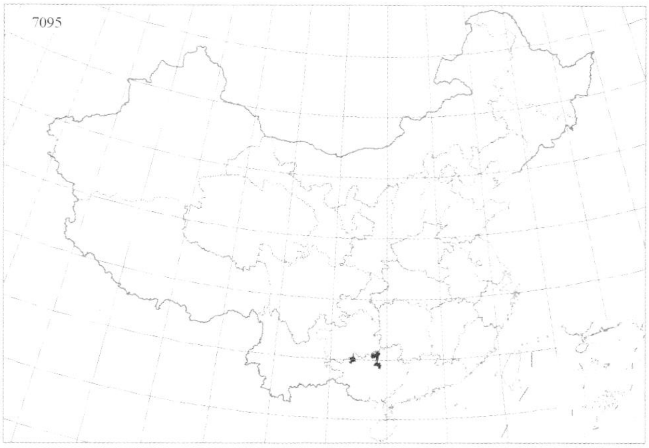

Elaeagnaceae 胡颓子科 *Elaeagnus* 胡颓子属
Elaeagnus longiloba C. Y. Chang 长裂胡颓子
Shrubs, evergreen

13.3; 3.6; 22.2; 13.3; 101.9; 1.7; 718 ❙ 999; 501; 53 ❙ 39.0; 718; 183.0

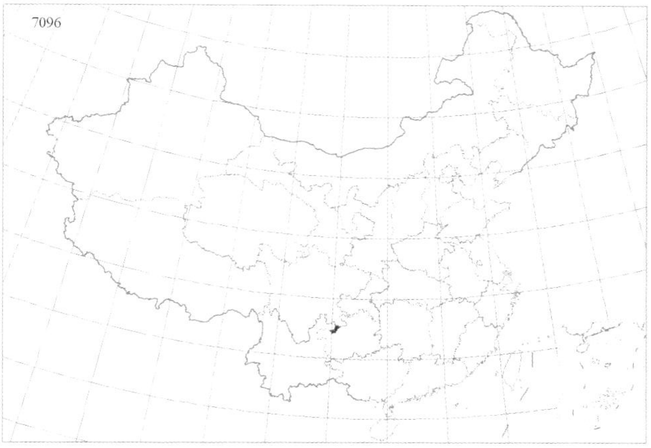

Elaeagnaceae 胡颓子科 *Elaeagnus* 胡颓子属
Elaeagnus loureiroi Champion ex Bentham 鸡柏紫藤
Shrubs, evergreen

19.6 (13.6~23.4); 11.4 (4.8~16.1); 26.3 (18.8~29.2); 19.7 (13.7~23.5); 176.8 (104.1~221.7); 0.0 (0.0~0.2); 1018 (706~1273) ❙ 1561 (939~2435); 737 (512~1428); 117 (39~200) ❙ 53.4 (6.5~99.5); 1012 (706~1273); 272.4 (139.6~506.2)

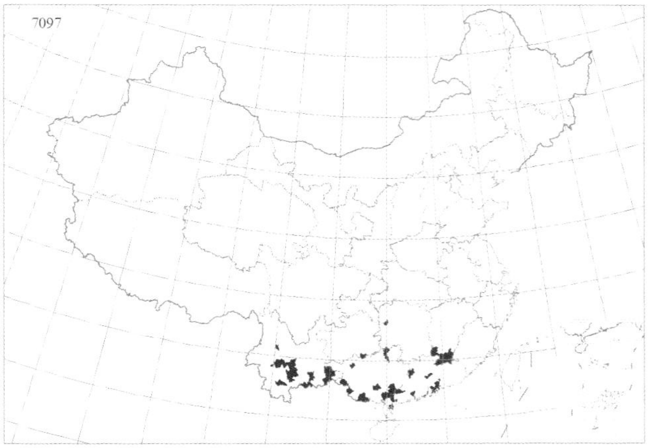

Elaeagnaceae 胡颓子科 *Elaeagnus* 胡颓子属
Elaeagnus luoxiangensis C. Y. Chang 罗香胡颓子
Shrubs, evergreen

18.0; 8.5; 25.8; 18.0; 156.4; 0.0; 913 ❙ 1551; 672; 169 ❙ 69.8; 913; 243.9

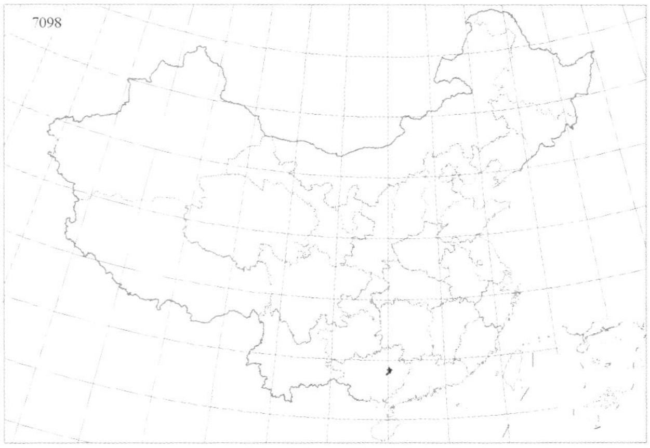

Elaeagnaceae 胡颓子科 *Elaeagnus* 胡颓子属
Elaeagnus luxiensis C. Y. Chang 潞西胡颓子
Shrubs, evergreen

17.0; 10.4; 21.3; 17.1; 144.8; 0.0; 822 ❙ 1309; 756; 51 ❙ 59.3; 806; 414.5

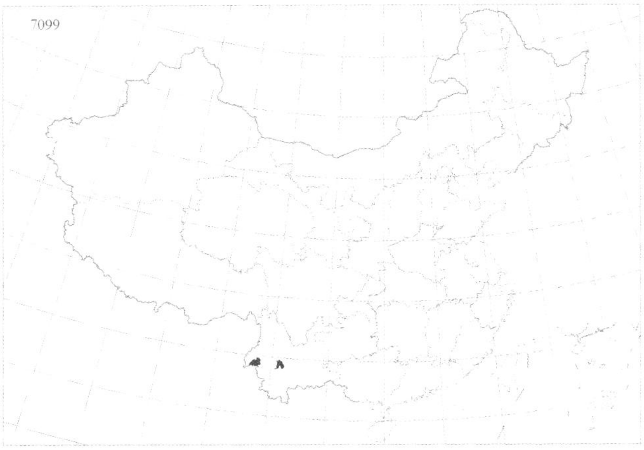

Elaeagnaceae 胡颓子科 *Elaeagnus* 胡颓子属
Elaeagnus macrantha Rehder 大花胡颓子
Shrubs, evergreen

19.6 (19.0~20.5); 14.0 (13.0~15.1); 23.1 (22.3~23.7); 19.7 (19.1~20.5); 176.2 (168.8~185.9); 0.0 (0.0~0.0); 943 (909~992) ❙ 1498 (1428~1607); 731 (655~835); 64 (56~77) ❙ 59.1 (47.6~64.9); 927 (880~986); 535.4 (427.9~659.0)

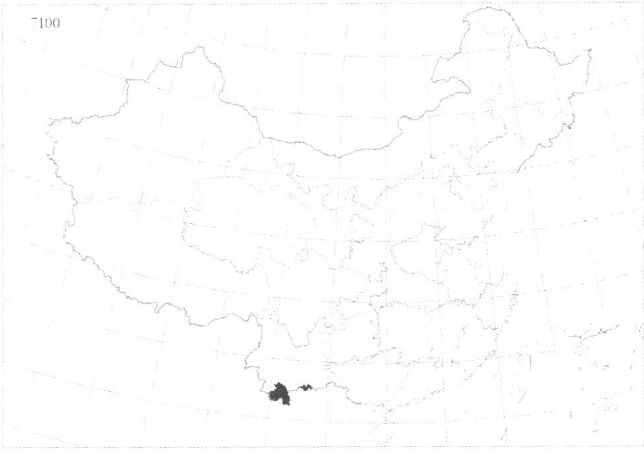

Elaeagnaceae 胡颓子科 *Elaeagnus* 胡颓子属
Elaeagnus magna Rehder 银果牛奶子
Shrubs, deciduous

15.9 (9.3~21.8); 4.7 (-1.7~14.0); 26.3 (20.1~29.5); 15.9 (9.5~21.9); 132.5 (69.8~202.2); 1.6 (0.0~18.0); 853 (603~1150) ❙ 1341 (807~2148); 515 (349~1234); 129 (37~230) ❙ 57.7 (-5.3~107.2); 846 (603~1150); 216.6 (111.9~371.7)

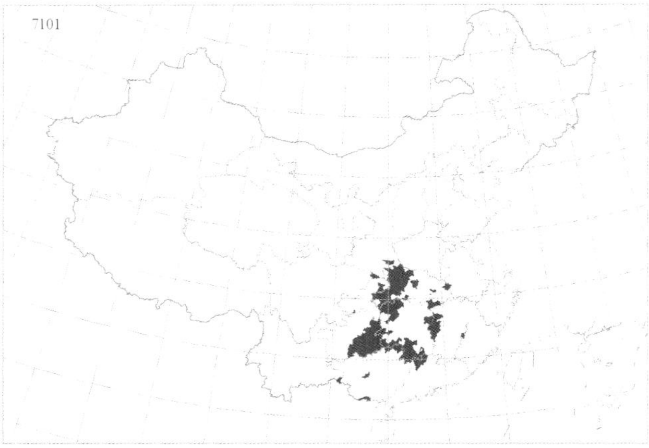

Elaeagnaceae 胡颓子科 *Elaeagnus* 胡颓子属
Elaeagnus micrantha C. Y. Chang 小花羊奶子
Shrubs, deciduous or semievergreen

14.2; 7.3; 19.4; 14.2; 110.6; 0.0; 724 ❙ 991; 586; 38 ❙ 37.0; 724; 282.6

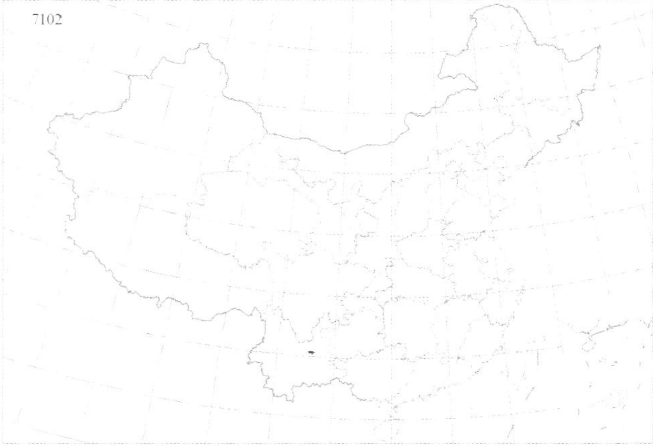

Elaeagnaceae 胡颓子科 *Elaeagnus* 胡颓子属
Elaeagnus mollis Diels 翅果油树
Shrubs or trees, deciduous

10.7 (7.2~14.6); -4.0 (-7.6~0.4); 23.9 (20.4~28.2); 11.4 (8.7~14.7); 93.0 (66.3~125.5); 23.8 (9.3~39.8); 706 (588~866) ❙ 563 (428~892); 296 (258~389); 19 (12~29) ❙ -19.2 (-35.9~44.0); 551 (428~702); 170.3 (125.5~311.0)

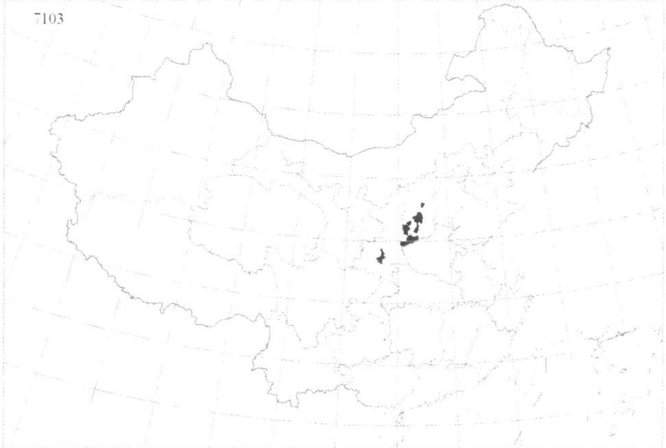

Elaeagnaceae 胡颓子科 *Elaeagnus* 胡颓子属
Elaeagnus multiflora Thunb. 木半夏
Shrubs or small trees, deciduous

14.1 (2.5~18.9); 1.2 (-13.7~12.0); 26.0 (12.3~29.9); 14.3 (5.0~19.0); 119.2 (27.9~167.8); 10.0 (0.0~71.7); 814 (449~1002) ❙ 1015 (367~1950); 451 (235~816); 92 (8~258) ❙ 23.0 (-43.4~147.0); 751 (367~1002); 191.9 (101.9~351.7)

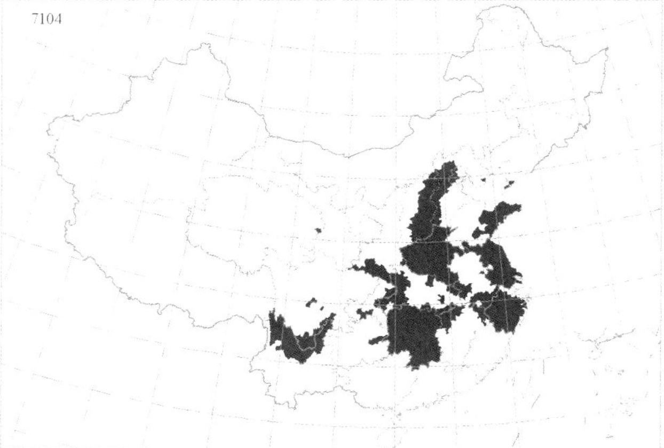

Elaeagnaceae 胡颓子科 *Elaeagnus* 胡颓子属
Elaeagnus multiflora var. obovoidea C. Y. Chang 倒果木半夏
Shrubs or small trees, deciduous

15.8 (13.4~18.2); 3.0 (1.5~6.5); 28.0 (25.2~29.2); 15.8 (13.4~18.2); 132.7 (108.3~158.8); 2.8 (0.0~7.3); 881 (755~984) ▎1308 (879~1922); 516 (363~666); 151 (87~259) ▎49.2 (-1.4~112.2); 872 (755~927); 189.8 (110.1~319.9)

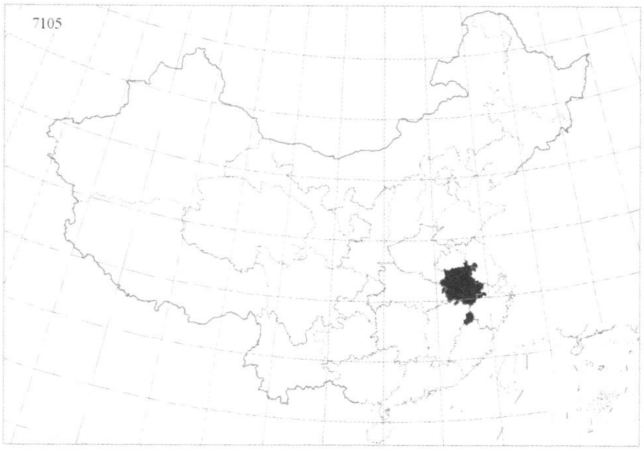

Elaeagnaceae 胡颓子科 *Elaeagnus* 胡颓子属
Elaeagnus multiflora var. siphonantha (Nakai) C. Y. Chang 长萼木半夏
Shrubs or small trees, deciduous

18.4; 8.5; 26.9; 18.4; 161.2; 0.0; 951 ▎1655; 600; 188 ▎73.9; 951; 239.6

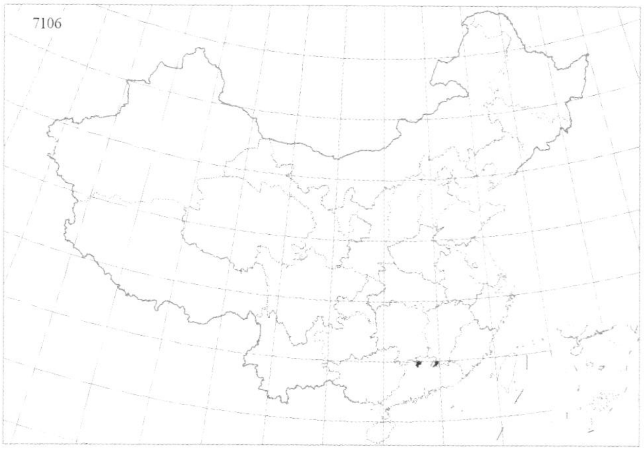

Elaeagnaceae 胡颓子科 *Elaeagnus* 胡颓子属
Elaeagnus multiflora var. tenuipes C. Y. Chang 细枝木半夏
Shrubs or small trees, deciduous

14.9; 5.3; 23.6; 15.0; 119.6; 0.0; 782 ▎1410; 816; 49 ▎80.4; 782; 214.5

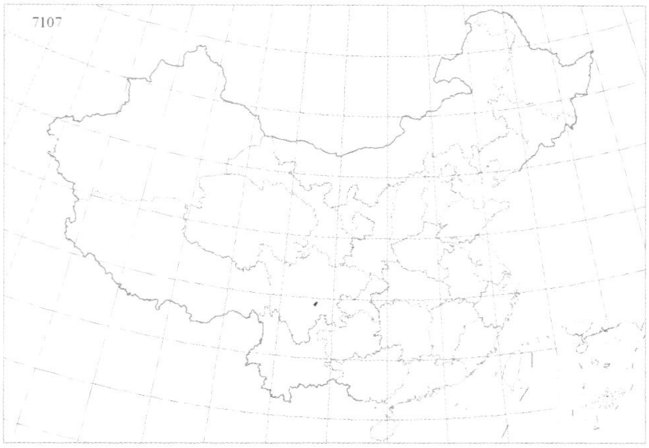

Elaeagnaceae 胡颓子科 *Elaeagnus* 胡颓子属
Elaeagnus nanchuanensis C. Y. Chang 南川牛奶子
Shrubs, deciduous

15.6 (10.8~18.7); 5.2 (1.8~9.4); 25.3 (18.5~28.9); 15.6 (10.9~18.8); 128.0 (76.1~166.0); 0.8 (0.0~6.9); 830 (639~1002) ▎1131 (924~1358); 516 (447~688); 61 (26~88) ▎38.0 (7.2~87.2); 829 (639~1002); 178.8 (108.2~318.5)

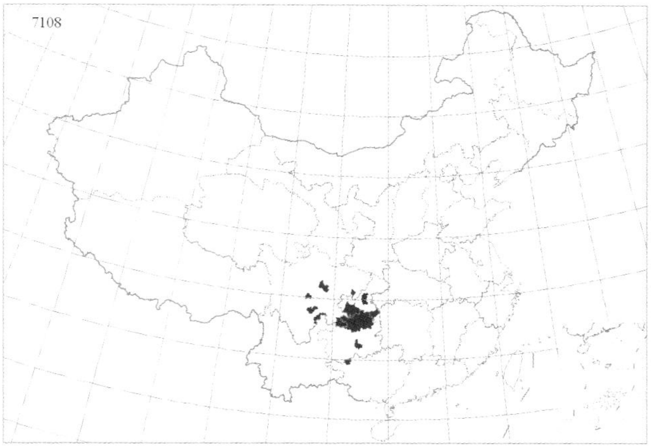

Elaeagnaceae 胡颓子科 *Elaeagnus* 胡颓子属
Elaeagnus obovatifolia D. Fang 弄化胡颓子
Shrubs, evergreen

19.1; 11.0; 25.0; 19.1; 169.6; 0.0; 945 ▎1438; 820; 57 ▎52.2; 945; 371.7

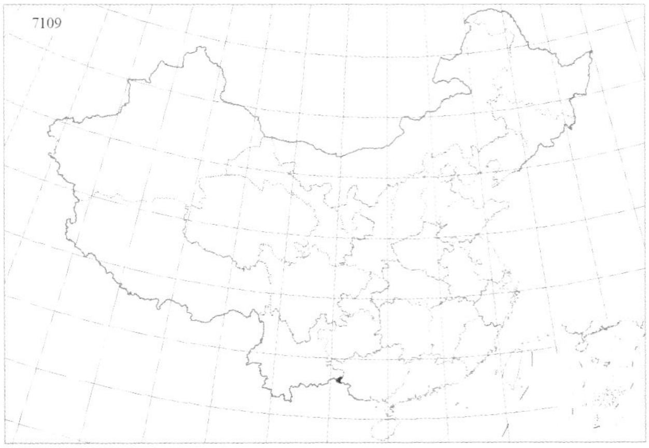

Elaeagnaceae 胡颓子科 *Elaeagnus* 胡颓子属
Elaeagnus obtusa C. Y. Chang 钝叶胡颓子
Climbing shrubs

17.7; 6.9; 27.5; 17.7; 152.4; 0.0; 927 ▎1486; 505; 176 ▎60.2; 927; 188.8

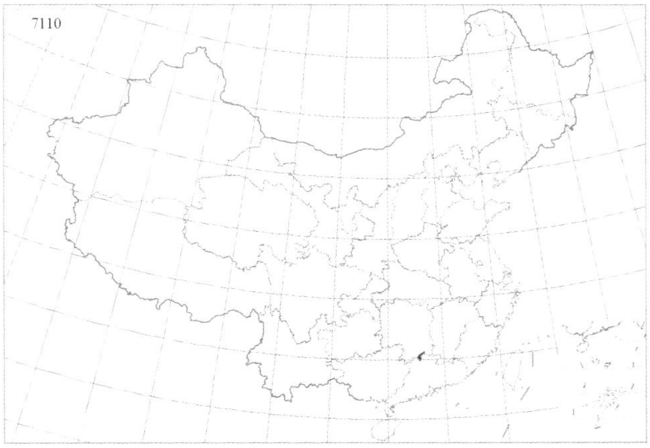

Elaeagnaceae 胡颓子科 *Elaeagnus* 胡颓子属

Elaeagnus oldhamii Maxim. 福建胡颓子

Shrubs or small trees, evergreen

20.3 (12.6~24.1); 13.3 (1.6~18.9); 26.4 (20.1~29.0); 20.3 (12.7~24.1); 183.7 (98.8~229.7); 0.2 (0.0~6.9); 1042 (711~1322) ❙ 1950 (1174~3495); 844 (386~1847); 213 (47~897) ❙ 94.5 (-1.0~246.5); 1015 (711~1269); 234.2 (119.4~363.7)

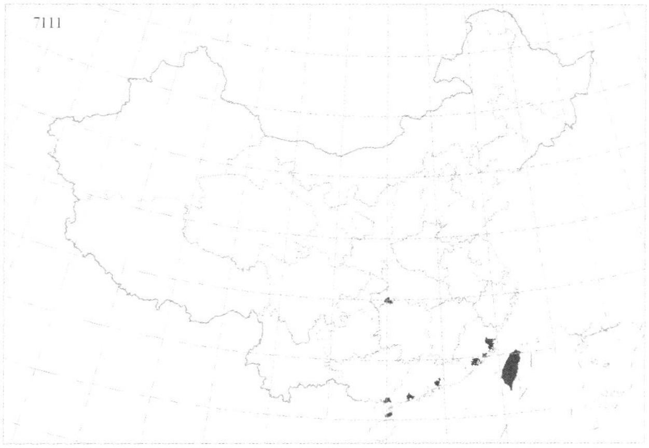

Elaeagnaceae 胡颓子科 *Elaeagnus* 胡颓子属

Elaeagnus ovata Servettaz 卵叶胡颓子

Shrubs, deciduous

16.0 (15.8~16.1); 4.2 (3.8~4.7); 27.8 (27.5~28.0); 16.1 (15.8~16.2); 133.5 (131.2~134.9); 0.8 (0.3~1.3); 877 (869~885) ❙ 1041 (996~1069); 402 (370~441); 143 (139~148) ❙ 18.7 (14.7~21.2); 873 (853~885); 174.2 (131.2~220.8)

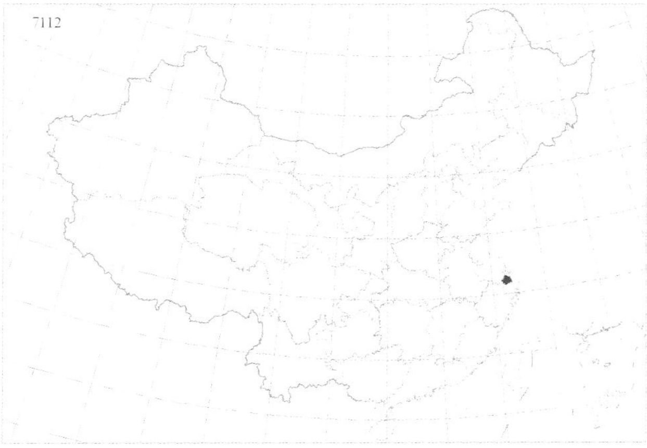

Elaeagnaceae 胡颓子科 *Elaeagnus* 胡颓子属

Elaeagnus oxycarpa Schltdl. 尖果沙枣

Trees, deciduous

6.2 (-4.3~12.5); -11.8 (-20.9~-5.1); 21.0 (9.6~27.5); 9.1 (2.5~13.2); 72.8 (9.7~114.1); 58.4 (23.1~121.1); 629 (329~822) ❙ 139 (21~450); 69 (15~211); 11 (0~51) ❙ -75.0 (-97.3~3.9); 138 (21~406); 57.7 (1.6~297.3)

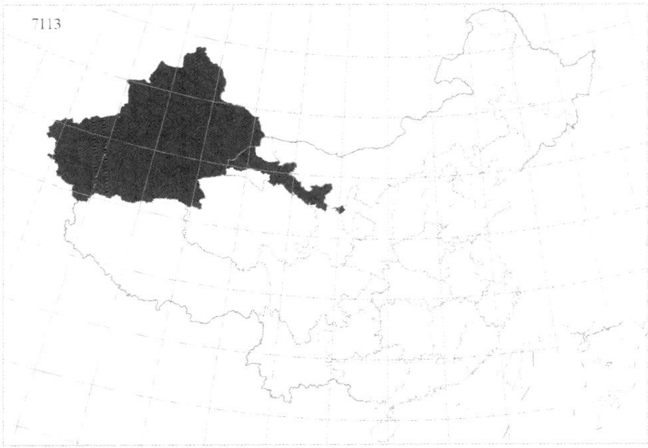

Elaeagnaceae 胡颓子科 *Elaeagnus* 胡颓子属

Elaeagnus pallidiflora C. Y. Chang 白花胡颓子

Shrubs, evergreen

11.6; 4.8; 17.3; 11.8; 89.0; 8.7; 668 ❙ 911; 526; 34 ❙ 41.5; 641; 313.9

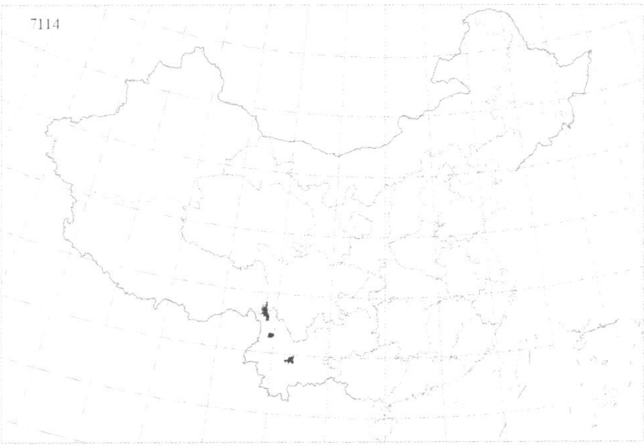

Elaeagnaceae 胡颓子科 *Elaeagnus* 胡颓子属

Elaeagnus pilostyla C. Y. Chang 毛柱胡颓子

Shrubs, evergreen

16.9 (14.8~18.3); 9.5 (4.3~11.8); 22.4 (20.6~24.9); 16.9 (14.9~18.3); 142.9 (119.2~159.8); 0.1 (0.0~0.7); 827 (785~882) ❙ 1102 (939~1197); 590 (489~656); 50 (39~71) ❙ 33.8 (6.5~51.2); 804 (779~847); 318.1 (182.9~405.0)

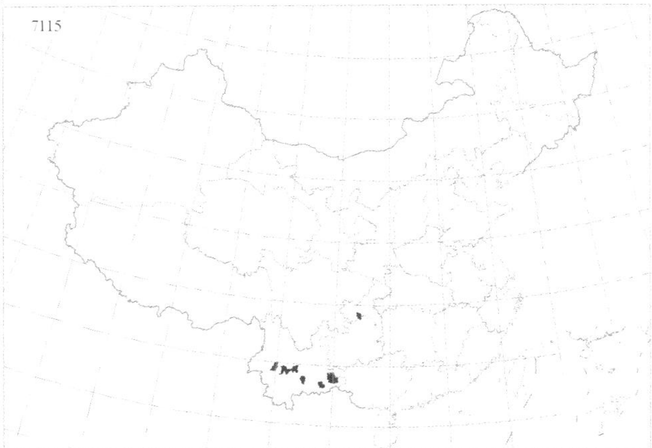

Elaeagnaceae 胡颓子科 *Elaeagnus* 胡颓子属

Elaeagnus pingnanensis C. Y. Chang 平南胡颓子

Climbing shrubs, evergreen

21.1; 12.0; 28.4; 21.2; 194.2; 0.0; 1121 ❙ 1578; 662; 168 ❙ 40.8; 1121; 218.0

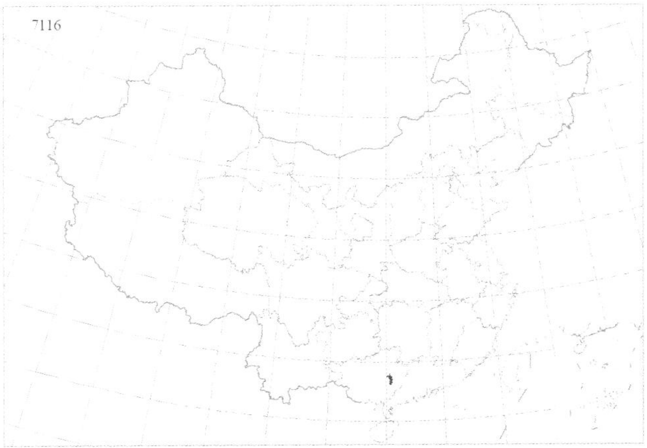

Elaeagnaceae 胡颓子科 *Elaeagnus* 胡颓子属

Elaeagnus pungens Thunb. 胡颓子

Shrubs, evergreen

16.2 (7.2~22.2); 4.7 (-7.6~14.6); 27.0 (15.6~29.9); 16.3 (7.5~22.3); 136.9 (48.2~207.5); 2.1 (0.0~39.8); 878 (537~1190) ❙ 1360 (503~3495); 521 (299~924); 150 (15~897) ❙ 54.5 (-22.3~224.4); 867 (503~1190); 205.9 (110.1~332.5)

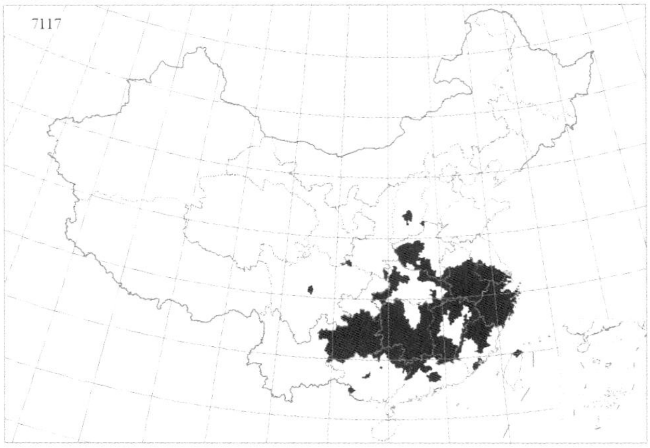

Elaeagnaceae 胡颓子科 *Elaeagnus* 胡颓子属

Elaeagnus retrostyla C. Y. Chang 卷柱胡颓子

Shrubs, evergreen

13.3; 3.6; 22.2; 13.3; 101.9; 1.7; 718 ❙ 999; 501; 53 ❙ 39.0; 718; 183.0

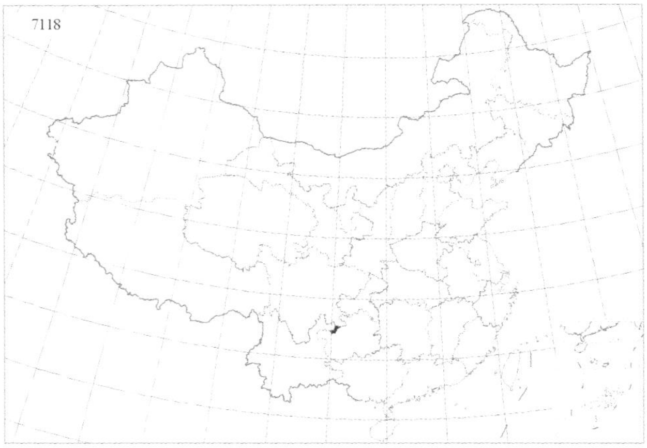

Elaeagnaceae 胡颓子科 *Elaeagnus* 胡颓子属

Elaeagnus sarmentosa Rehder 攀缘胡颓子

Shrubs, evergreen

17.6 (13.8~21.0); 10.3 (4.8~14.9); 22.8 (19.3~25.6); 17.6 (13.8~21.0); 151.2 (105.9~191.8); 0.0 (0.0~0.2); 864 (714~1053) ❙ 1337 (963~1574); 700 (554~860); 64 (29~177) ❙ 54.7 (32.4~93.2); 857 (690~1053); 385.8 (230.7~570.1)

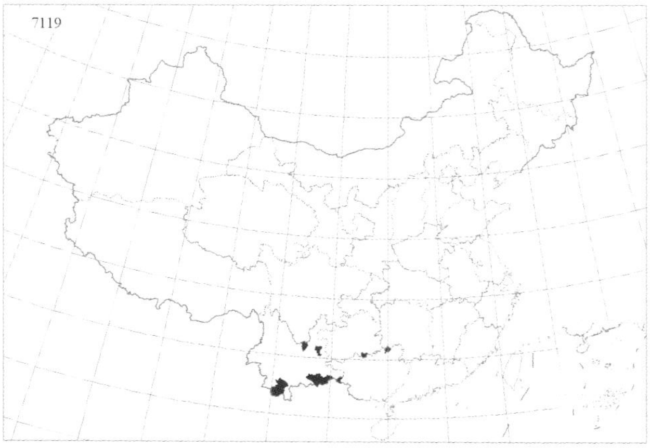

Elaeagnaceae 胡颓子科 *Elaeagnus* 胡颓子属

Elaeagnus schlechtendalii Servettaz 小胡颓子

Small shrubs, evergreen

19.7; 10.1; 27.7; 19.8; 177.4; 0.0; 1031 ❙ 1510; 614; 165 ❙ 46.4; 1031; 253.0

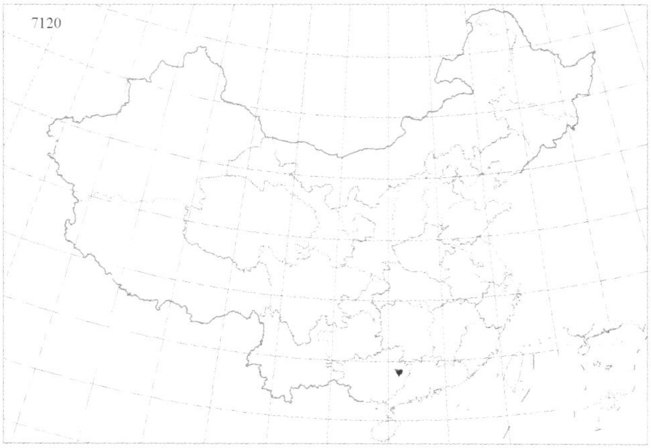

Elaeagnaceae 胡颓子科 *Elaeagnus* 胡颓子属

Elaeagnus stellipila Rehder 星毛羊奶子

Shrubs, deciduous

15.6 (10.6~18.7); 5.0 (-0.2~8.1); 25.6 (19.3~29.3); 15.7 (10.6~18.8); 128.9 (79.4~166.0); 0.9 (0.0~12.1); 834 (643~1002) ❙ 1223 (963~1660); 535 (442~946); 78 (29~216) ❙ 48.0 (8.0~107.2); 832 (643~1002); 186.6 (108.2~292.6)

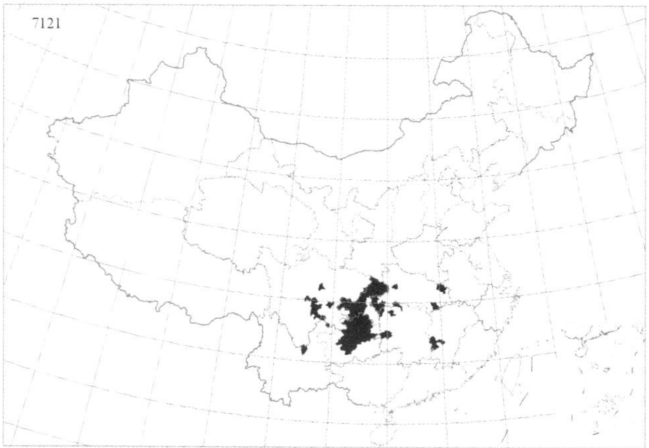

Elaeagnaceae 胡颓子科 *Elaeagnus* 胡颓子属

Elaeagnus taliensis C. Y. Chang 大理胡颓子

Shrubs, evergreen

13.6; 7.2; 18.8; 13.7; 104.1; 0.0; 706 ❙ 1042; 592; 46 ❙ 47.6; 706; 372.3

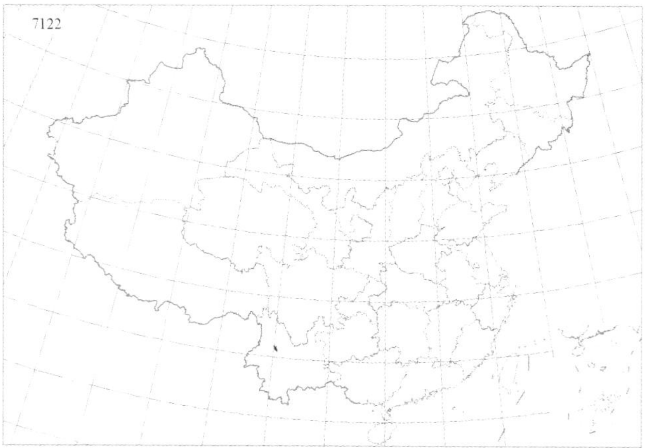

Elaeagnaceae 胡颓子科 *Elaeagnus* 胡颓子属
Elaeagnus tarokoensis S. Y. Lu et Y. P. Yang 太鲁阁胡颓子
Shrubs

16.4; 11.5; 20.3; 16.5; 137.8; 0.0; 783 ❙ 2499; 947; 303 ❙ 219.3; 783; 303.6

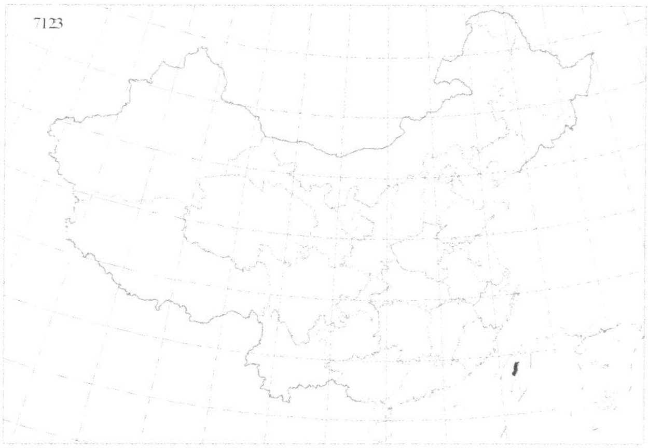

Elaeagnaceae 胡颓子科 *Elaeagnus* 胡颓子属
Elaeagnus thunbergii Servettaz 阿里胡颓子
Shrubs, evergreen

18.4 (16.2~20.8); 12.6 (11.0~15.3); 23.4 (20.1~25.9); 18.5 (16.2~20.8); 161.9 (134.2~189.2); 0.0 (0.0~0.0); 895 (773~1019) ❙ 2603 (2288~2993); 1057 (900~1412); 310 (110~547) ❙ 193.3 (140.0~246.5); 895 (773~1019); 267.9 (207.7~298.9)

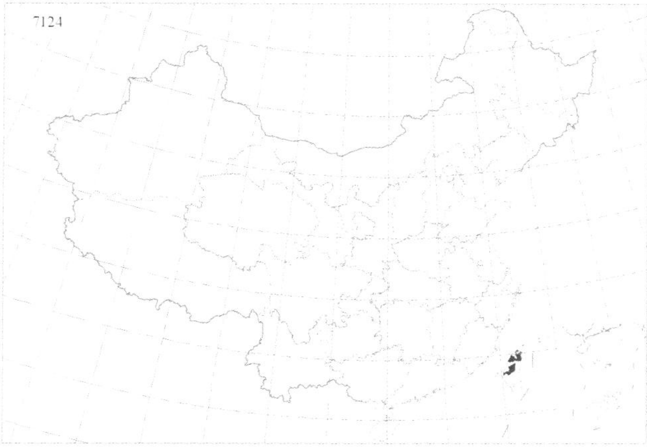

Elaeagnaceae 胡颓子科 *Elaeagnus* 胡颓子属
Elaeagnus tonkinensis Servettaz 越南胡颓子
Shrubs, evergreen

17.5; 10.5; 22.3; 17.5; 149.9; 0.0; 846 ❙ 958; 531; 44 ❙ 13.3; 823; 311.9

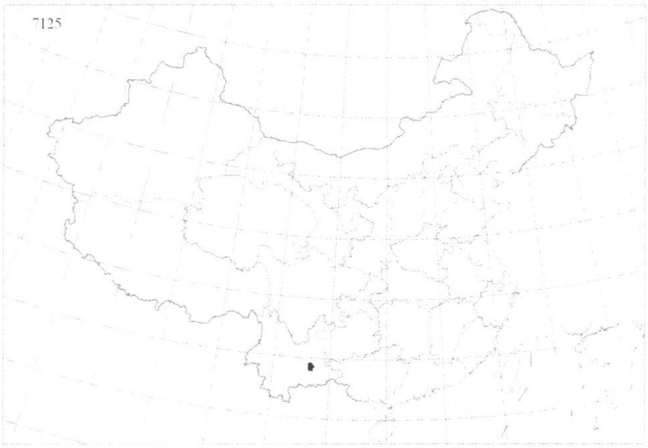

Elaeagnaceae 胡颓子科 *Elaeagnus* 胡颓子属
Elaeagnus triflora Roxburgh 菲律宾胡颓子
Scandent shrubs, deciduous

18.5; 14.1; 21.7; 18.5; 162.1; 0.0; 863 ❙ 2713; 1388; 188 ❙ 214.4; 863; 363.7

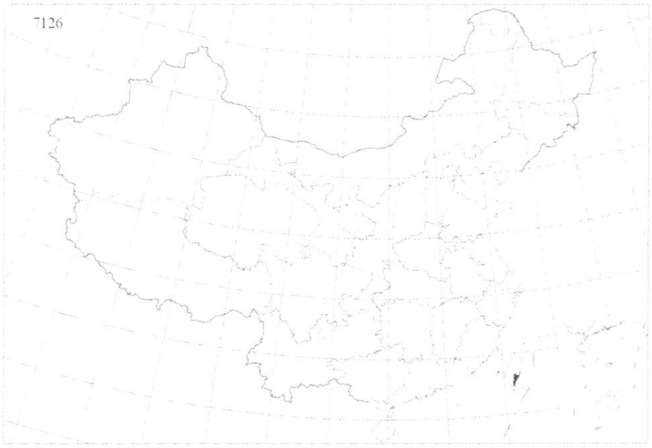

Elaeagnaceae 胡颓子科 *Elaeagnus* 胡颓子属
Elaeagnus tubiflora C. Y. Chang 管花胡颓子
Shrubs, evergreen

18.7; 10.5; 24.8; 18.8; 165.2; 0.0; 935 ❙ 1302; 743; 48 ❙ 41.3; 935; 357.1

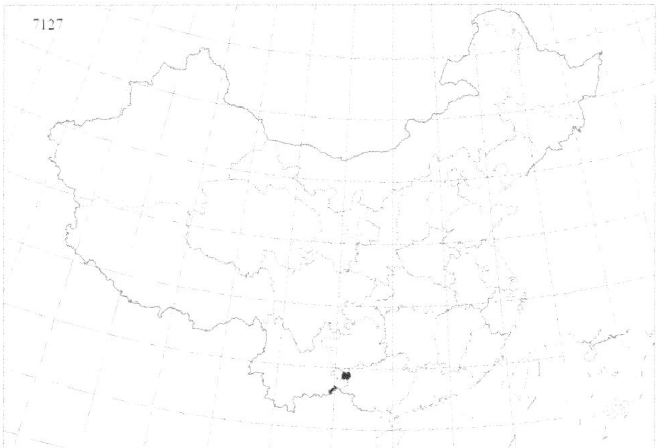

Elaeagnaceae 胡颓子科 *Elaeagnus* 胡颓子属
Elaeagnus tutcheri Dunn 香港胡颓子
Shrubs, evergreen

19.9 (16.5~22.4); 10.5 (5.2~14.8); 28.0 (26.2~29.8); 20.0 (16.5~22.5); 179.5 (138.3~209.7); 0.0 (0.0~0.0); 1050 (875~1205) ❙ 1685 (1410~2037); 655 (403~1033); 173 (110~211) ❙ 61.1 (30.2~95.5); 1045 (866~1205); 213.3 (95.3~275.2)

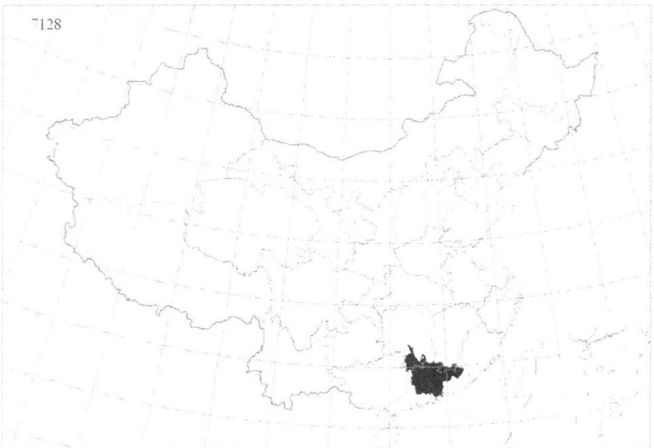

Elaeagnaceae 胡颓子科 *Elaeagnus* 胡颓子属
Elaeagnus viridis Servettaz 绿叶胡颓子
Shrubs, evergreen

13.6 (9.3~17.4); 2.9 (-1.7~7.4); 23.6 (20.1~25.9); 13.7 (9.5~17.4); 108.9 (69.8~148.9); 5.2 (0.0~18.0); 750 (603~886) ❙ 1252 (911~1478); 542 (390~637); 78 (40~107) ❙ 67.9 (20.2~107.2); 750 (603~886); 236.9 (185.4~279.5)

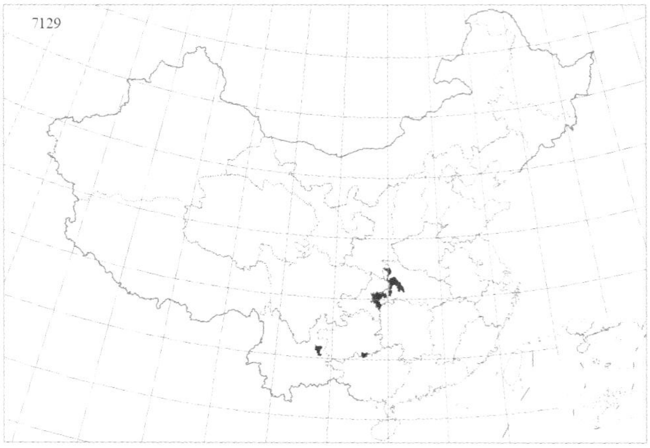

Elaeagnaceae 胡颓子科 *Elaeagnus* 胡颓子属
Elaeagnus wenshanensis C. Y. Chang 文山胡颓子
Shrubs, evergreen

15.5 (14.2~16.9); 6.4 (3.3~10.3); 23.4 (21.4~24.9); 15.5 (14.3~16.9); 127.2 (113.3~142.7); 0.6 (0.0~2.1); 796 (776~813) ❙ 1176 (1048~1265); 563 (489~656); 60 (50~71) ❙ 47.9 (30.8~63.0); 796 (776~813); 262.6 (182.9~383.4)

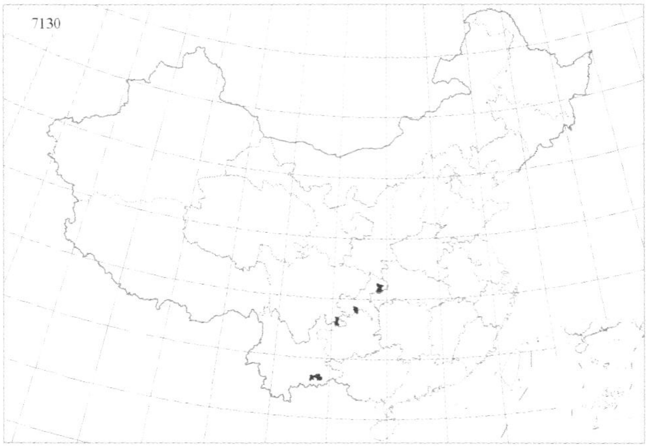

Elaeagnaceae 胡颓子科 *Elaeagnus* 胡颓子属
Elaeagnus wushanensis C. Y. Chang 巫山牛奶子
Shrubs, deciduous

13.2 (10.6~15.3); 2.2 (-0.2~4.2); 23.8 (21.1~25.9); 13.3 (10.6~15.4); 104.5 (79.4~125.2); 5.3 (0.8~12.1); 739 (643~822) ❙ 1276 (1047~1492); 536 (475~613); 69 (29~105) ❙ 73.2 (41.3~106.2); 739 (643~822); 244.7 (178.4~285.1)

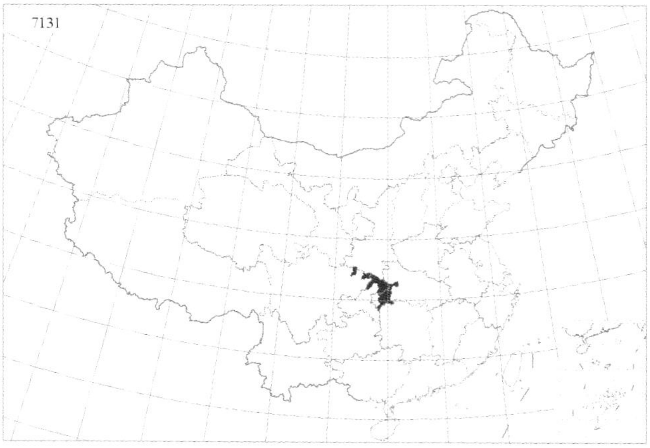

Elaeagnaceae 胡颓子科 *Elaeagnus* 胡颓子属
Elaeagnus xichouensis C. Y. Chang 西畴胡颓子
Shrubs, evergreen

17.2; 9.5; 22.7; 17.3; 147.0; 0.0; 838 ❙ 1254; 699; 48 ❙ 49.7; 838; 336.3

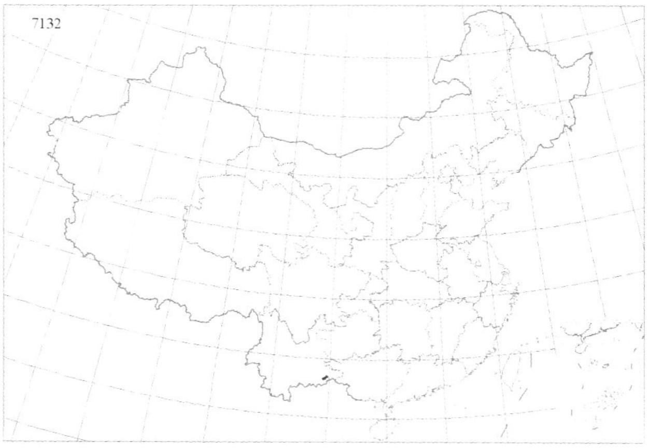

Elaeagnaceae 胡颓子科 *Elaeagnus* 胡颓子属
Elaeagnus xingwenensis C. Y. Chang 兴文胡颓子
Shrubs, evergreen

16.2; 6.2; 25.6; 16.3; 135.7; 0.0; 845 ❙ 1055; 522; 64 ❙ 24.9; 845; 154.4

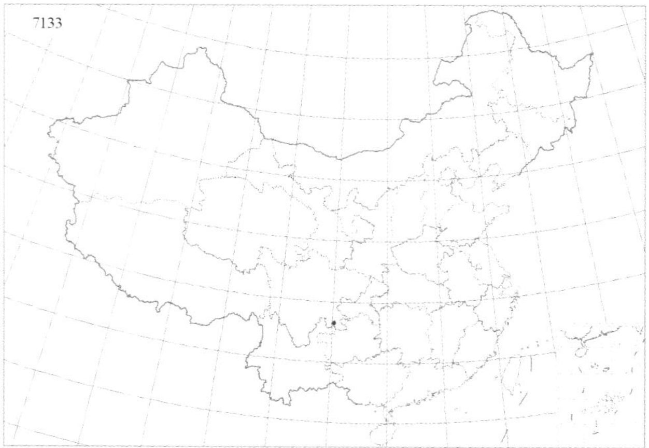

Elaeagnaceae 胡颓子科 *Elaeagnus* 胡颓子属
Elaeagnus xizangensis C. Y. Chang 西藏胡颓子
Shrubs, evergreen

13.3; 5.4; 19.6; 13.4; 100.4; 0.0; 706 ❙ 1736; 1007; 51 ❙ 145.8; 706; 447.4

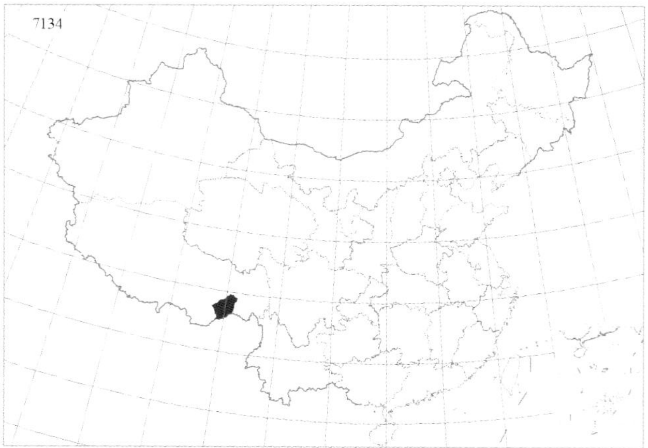

Elaeagnaceae 胡颓子科 *Elaeagnus* 胡颓子属

Elaeagnus yunnanensis Servettaz 云南胡颓子

Shrubs, evergreen

14.4; 7.5; 19.1; 14.4; 113.1; 0.0; 727 ❙ 1454; 792; 70 ❙ 99.9; 727; 487.2

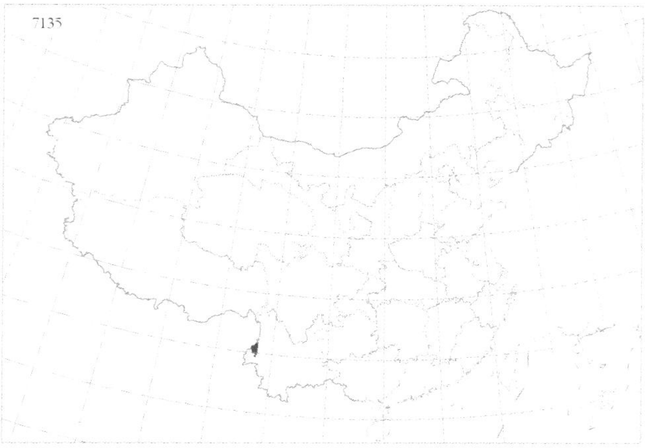

Elaeagnaceae 胡颓子科 *Hippophae* 沙棘属

Hippophae goniocarpa Y. S. Lian et al. ex Swenson et Bartish 棱果沙棘

Shrubs or small trees

0.5; -10.4; 10.2; 3.6; 15.2; 68.7; 401 ❙ 660; 343; 16 ❙ 60.1; 401; 281.6

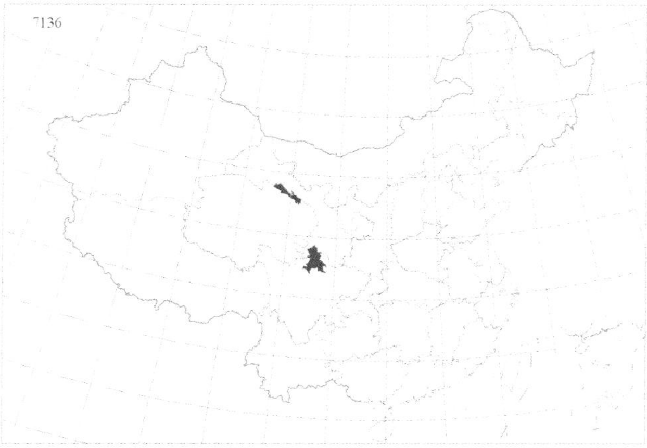

Elaeagnaceae 胡颓子科 *Hippophae* 沙棘属

Hippophae gyantsensis (Rousi) Y. S. Lian 江孜沙棘

Shrubs or trees, deciduous

4.2 (1.2~12.2); -5.1 (-8.6~4.3); 12.3 (9.7~18.5); 5.7 (3.6~12.2); 32.8 (15.5~87.4); 42.5 (0.7~61.3); 461 (394~672) ❙ 522 (252~1431); 341 (196~817); 8 (0~31) ❙ 2.4 (-35.9~113.0); 382 (252~672); 187.3 (93.6~363.7)

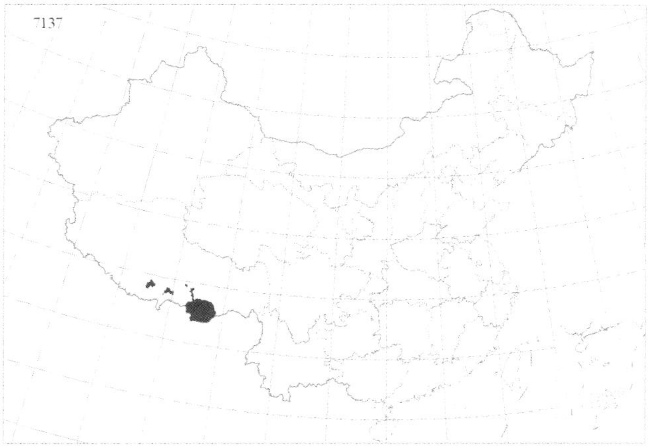

Elaeagnaceae 胡颓子科 *Hippophae* 沙棘属

Hippophae litangensis Y. S. Lian et Xue L. Chen ex Swenson et Bartish 理塘沙棘

Shrubs or small trees

1.4; -7.6; 9.2; 3.5; 13.4; 55.6; 402 ❙ 699; 454; 10 ❙ 73.9; 402; 238.3

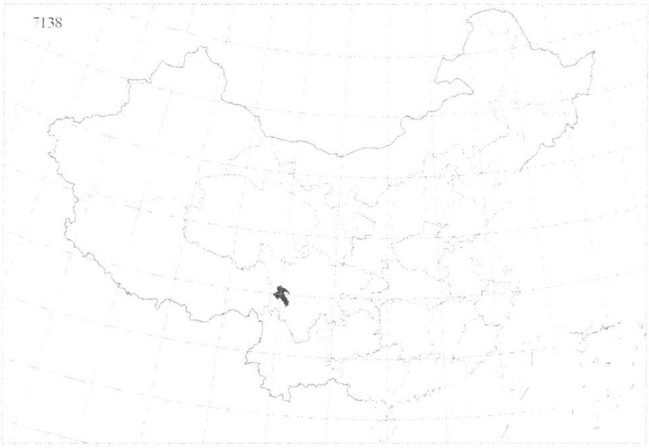

Elaeagnaceae 胡颓子科 *Hippophae* 沙棘属

Hippophae pleurocarpa S. W. Liu et T. N. He 肋果沙棘

Shrubs or small trees, deciduous

-0.2 (-5.5~10.3); -11.4 (-17.3~-0.1); 9.8 (5.8~19.8); 3.4 (1.3~10.3); 15.1 (1.2~75.8); 77.3 (11.7~127.1); 389 (299~634) ❙ 507 (252~730); 302 (153~482); 11 (4~17) ❙ 29.4 (-44.6~87.3); 368 (252~634); 216.0 (46.7~364.1)

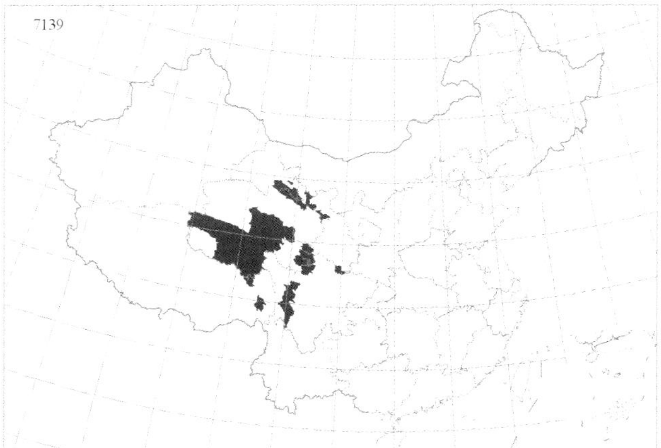

Elaeagnaceae 胡颓子科 *Hippophae* 沙棘属

Hippophae rhamnoides L. 沙棘

Shrubs or trees, deciduous

8.4 (-5.9~15.7); -6.6 (-20.9~4.3); 21.5 (5.4~28.3); 10.0 (1.3~15.8); 80.0 (0.4~133.5); 38.9 (0.7~133.0); 662 (294~898) ❙ 509 (26~1431); 313 (16~817); 19 (0~108) ❙ -19.6 (-96.7~128.6); 474 (26~887); 160.6 (2.8~364.1)

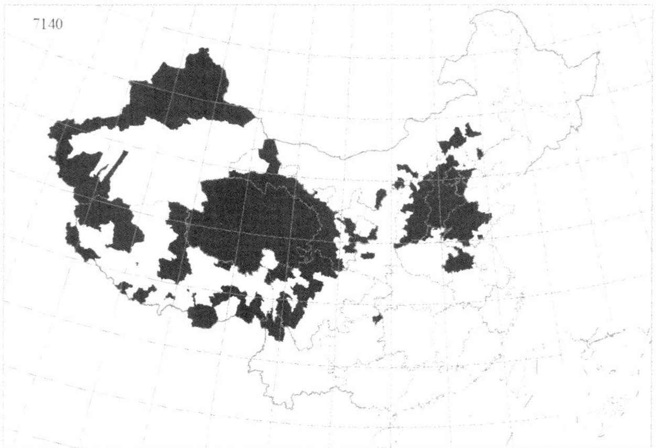

Elaeagnaceae 胡颓子科 *Hippophae* 沙棘属

Hippophae rhamnoides subsp. sinensis Rousi 中国沙棘

Shrubs or trees, deciduous

6.4 (-5.9~14.5); -7.7 (-18.4~1.1); 18.9 (5.4~28.0); 8.3 (1.3~14.6); 61.9 (0.4~127.2); 44.9 (8.1~130.5); 578 (299~880) ❙ 524 (63~1023); 299 (45~601); 13 (2~28) ❙ -5.0 (-84.9~128.6); 472 (63~707); 172.6 (8.4~364.1)

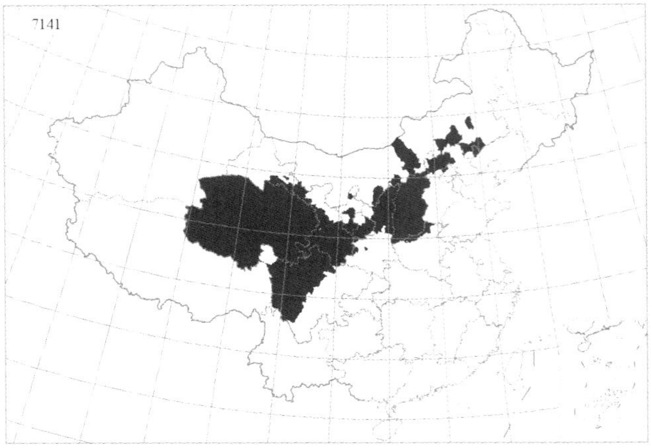

Elaeagnaceae 胡颓子科 *Hippophae* 沙棘属

Hippophae rhamnoides subsp. turkestanica Rousi 中亚沙棘

Shrubs or small trees, deciduous

-1.8; -12.0; 8.6; 2.4; 7.9; 88.8; 333 ❙ 509; 233; 98 ❙ 52.9; 333; 32.2

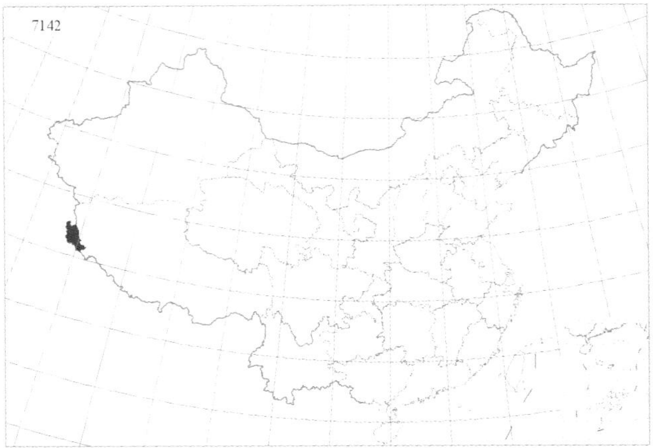

Elaeagnaceae 胡颓子科 *Hippophae* 沙棘属

Hippophae rhamnoides subsp. yunnanensis Rousi 云南沙棘

Shrubs or trees, deciduous

3.7 (0.6~7.3); -5.2 (-9.4~-1.3); 11.4 (9.2~15.6); 4.9 (3.4~7.5); 24.6 (12.7~48.2); 40.2 (19.7~65.4); 447 (394~537) ❙ 803 (473~1023); 461 (287~601); 18 (7~31) ❙ 79.2 (5.0~128.6); 447 (394~537); 258.0 (169.5~341.6)

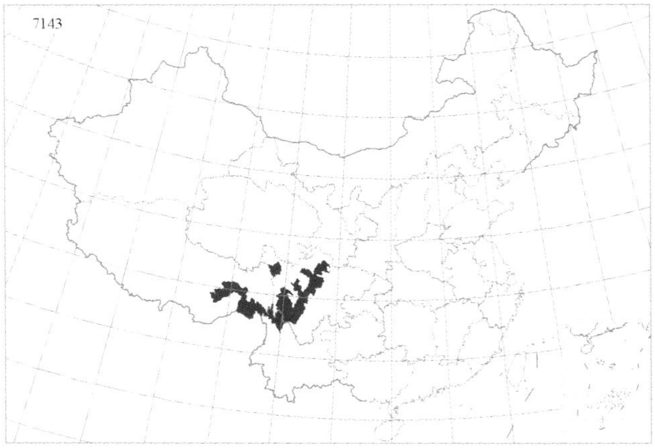

Elaeagnaceae 胡颓子科 *Hippophae* 沙棘属

Hippophae salicifolia D. Don 柳叶沙棘

Shrubs or small trees, deciduous

5.5; -2.4; 12.3; 7.1; 45.0; 38.2; 503 ❙ 876; 505; 32 ❙ 54.5; 496; 221.5

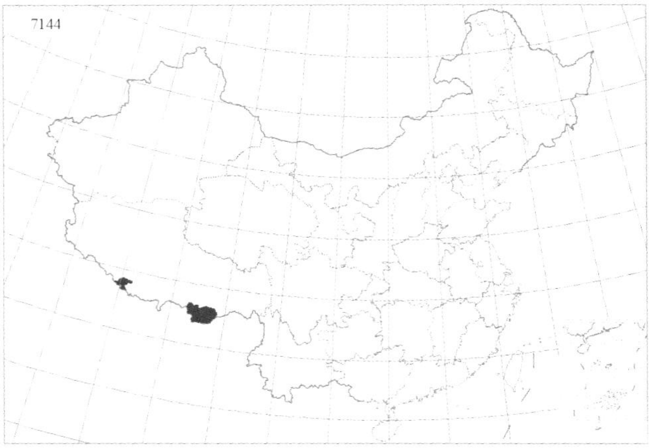

Elaeagnaceae 胡颓子科 *Hippophae* 沙棘属

Hippophae tibetana Schltdl. 西藏沙棘

Small shrubs, deciduous

0.0 (-5.7~6.1); -11.4 (-18.7~-2.4); 10.3 (5.8~17.3); 3.6 (1.3~7.6); 16.4 (1.2~53.6); 75.4 (26.2~133.0); 394 (294~544) ❙ 482 (74~961); 279 (46~558); 12 (4~103) ❙ 21.0 (-75.0~100.7); 368 (74~479); 214.1 (16.7~364.1)

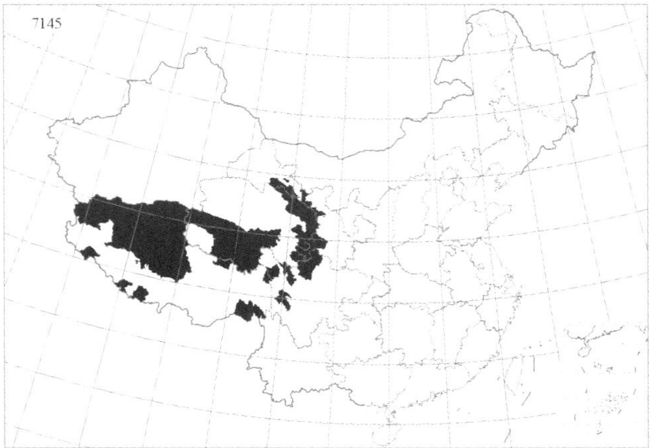

Lythraceae 千屈菜科 *Duabanga* 八宝树属

Duabanga grandiflora (Roxb. ex DC.) Walp. 八宝树

Trees

18.1 (14.4~22.2); 11.5 (7.5~15.1); 22.5 (19.1~28.3); 18.2 (14.4~22.3); 158.0 (113.1~207.5); 0.0 (0.0~0.0); 879 (727~1190) ❙ 1330 (939~1751); 736 (524~996); 54 (37~106) ❙ 51.8 (6.5~112.2); 865 (727~1190); 419.6 (277.0~659.0)

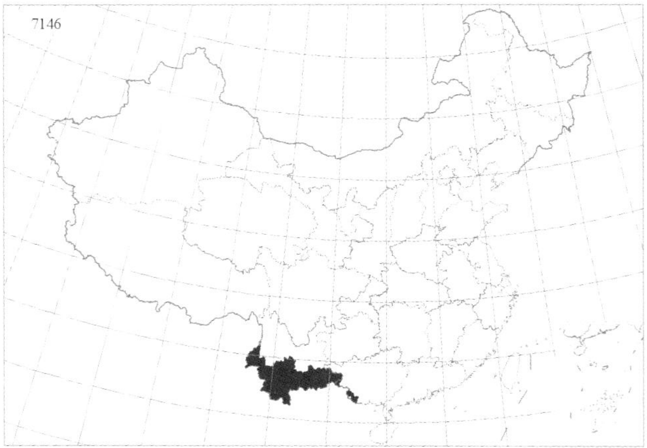

Lythraceae 千屈菜科 *Lagerstroemia* 紫薇属

Lagerstroemia anhuiensis X. H. Guo et S. B. Zhou 安徽紫薇

Shrubs or small trees

16.2; 3.5; 28.3; 16.2; 136.1; 1.5; 898 ❙ 1396; 529; 153 ❙ 55.4; 898; 225.2

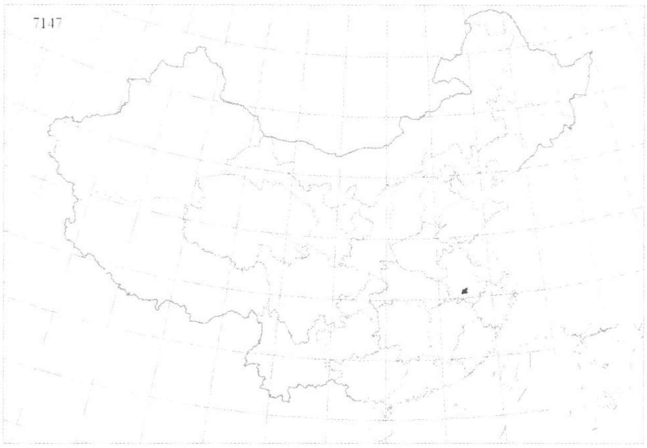

Lythraceae 千屈菜科 *Lagerstroemia* 紫薇属

Lagerstroemia balansae Koehne 毛萼紫薇

Shrubs or small trees, deciduous

23.9 (23.1~24.8); 18.6 (17.5~20.1); 27.8 (27.2~28.4); 24.0 (23.2~24.8); 227.6 (218.4~237.6); 0.0 (0.0~0.0); 1304 (1232~1385) ❙ 1339 (1179~1487); 462 (392~544); 46 (41~53) ❙ 3.0 (-11.9~20.7); 1108 (1056~1142); 483.2 (439.7~538.2)

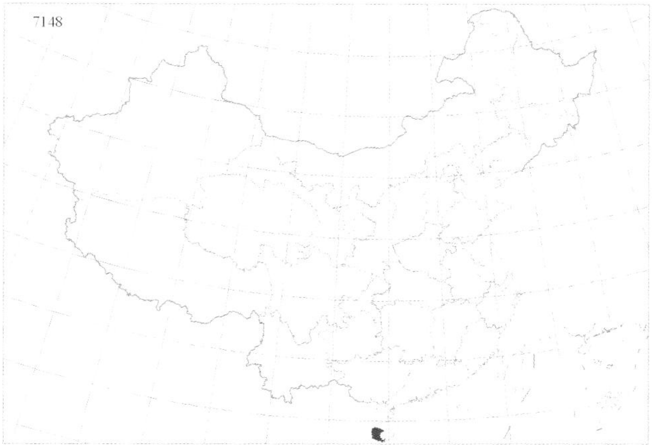

Lythraceae 千屈菜科 *Lagerstroemia* 紫薇属

Lagerstroemia caudata Chun et F. C. How ex S. K. Lee et L. F. Lau 尾叶紫薇

Trees, deciduous

17.4 (14.6~22.2); 6.6 (3.8~14.2); 27.2 (24.6~29.3); 17.4 (14.7~22.3); 149.6 (117.7~207.5); 0.3 (0.0~1.5); 924 (779~1190) ❙ 1536 (1171~1876); 588 (419~891); 168 (84~221) ❙ 67.3 (15.4~106.4); 922 (779~1190); 230.5 (170.4~327.7)

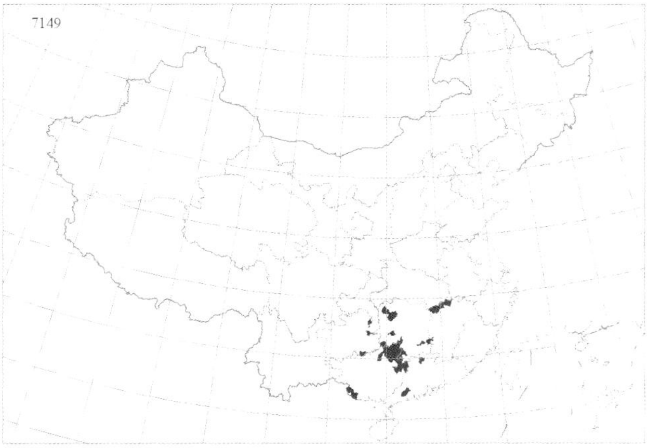

Lythraceae 千屈菜科 *Lagerstroemia* 紫薇属

Lagerstroemia excelsa (Dode) Chun ex S. K. Lee et L. F. Lau 川黔紫薇

Trees, deciduous

14.9 (11.7~16.3); 3.7 (-0.9~5.9); 25.5 (22.9~27.5); 15.0 (11.8~16.4); 121.7 (94.6~136.3); 2.3 (0.0~13.5); 806 (701~877) ❙ 1282 (814~1526); 509 (352~648); 100 (26~181) ❙ 59.0 (7.2~107.2); 806 (701~877); 205.1 (146.0~276.4)

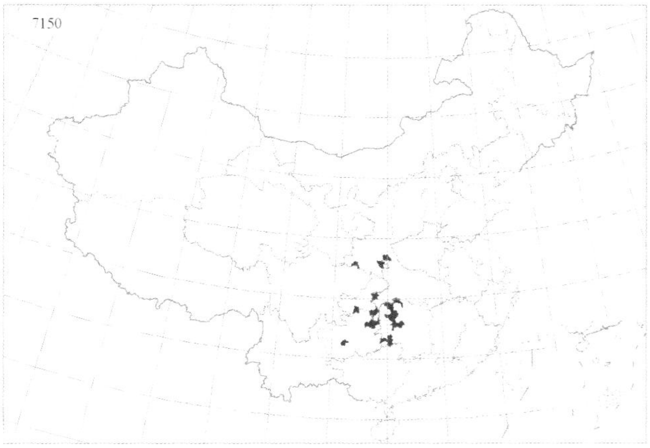

Lythraceae 千屈菜科 *Lagerstroemia* 紫薇属

Lagerstroemia fordii Oliv. et Koehne 广东紫薇

Trees, deciduous

20.3 (19.2~22.4); 12.7 (11.0~15.2); 27.5 (26.5~28.7); 20.7 (19.2~22.7); 188.5 (170.3~212.7); 0.0 (0.0~0.0); 1074 (971~1216) ❙ 1784 (1587~2030); 746 (562~1033); 147 (104~188) ❙ 66.1 (61.6~73.0); 1074 (971~1216); 245.0 (163.7~290.3)

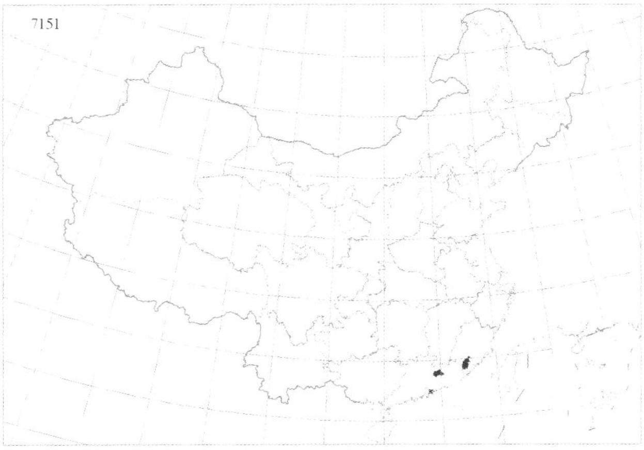

Lythraceae 千屈菜科 *Lagerstroemia* 紫薇属

Lagerstroemia glabra (Koehne) Koehne 光紫薇

Trees, deciduous

18.0 (13.0~21.8); 7.9 (1.9~14.2); 27.0 (23.8~27.8); 18.1 (13.1~21.9); 158.0 (102.9~202.6); 1.0 (0.0~6.1); 956 (730~1150) ❙ 1603 (1202~2005); 639 (500~891); 155 (83~189) ❙ 67.4 (51.7~85.1); 956 (730~1150); 244.3 (197.5~273.3)

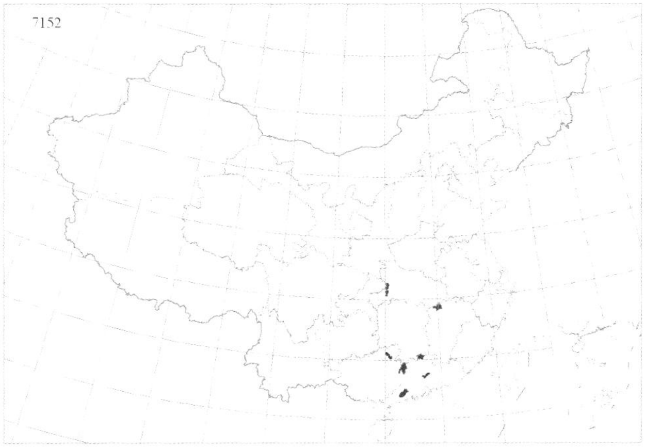

Lythraceae 千屈菜科 *Lagerstroemia* 紫薇属

Lagerstroemia guilinensis S. K. Lee et L. F. Lau 桂林紫薇

Shrubs, deciduous

19.2; 8.9; 28.1; 19.3; 171.2; 0.0; 1016 ❙ 1698; 691; 184 ❙ 67.0; 1016; 226.5

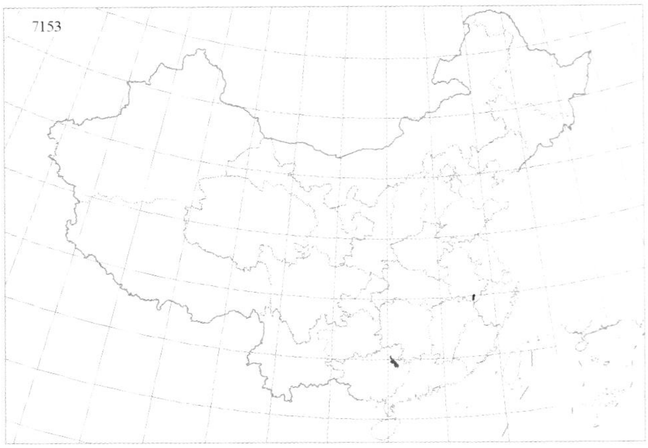

Lythraceae 千屈菜科 *Lagerstroemia* 紫薇属

Lagerstroemia indica L. 紫薇

Shrubs or small trees, deciduous

17.4 (2.2~24.8); 6.9 (-6.6~20.1); 26.8 (9.9~30.1); 17.4 (3.8~24.8); 150.2 (15.5~237.6); 1.4 (0.0~49.1); 931 (413~1385) ❙ 1432 (522~2435); 579 (277~1428); 138 (15~259) ❙ 54.9 (-34.6~147.0); 921 (413~1273); 220.5 (95.3~527.5)

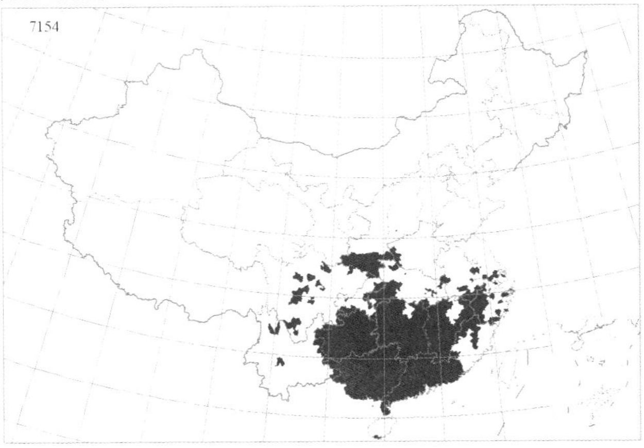

Lythraceae 千屈菜科 *Lagerstroemia* 紫薇属

Lagerstroemia intermedia Koehne 云南紫薇

Trees, deciduous

18.4 (16.3~20.5); 12.5 (10.2~15.1); 22.0 (20.3~23.7); 18.5 (16.4~20.5); 161.5 (136.6~185.9); 0.0 (0.0~0.0); 882 (789~992) ❙ 1425 (1232~1607); 739 (655~816); 52 (41~77) ❙ 61.8 (47.6~74.2); 862 (780~986); 484.6 (372.4~659.0)

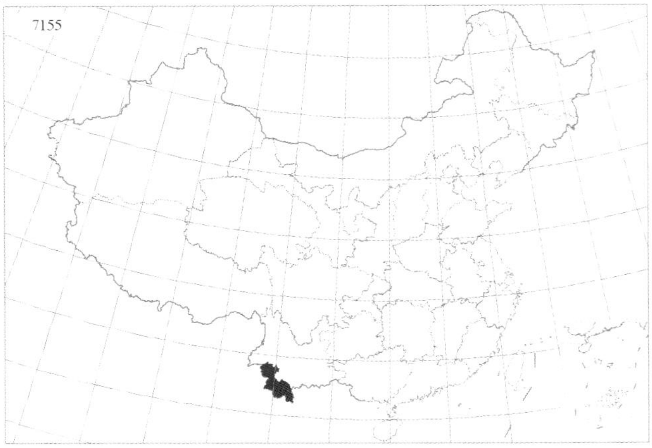

Lythraceae 千屈菜科 *Lagerstroemia* 紫薇属

Lagerstroemia limii Merr. 福建紫薇

Shrubs or small trees, deciduous

16.7 (12.7~21.7); 5.5 (2.1~14.0); 27.8 (23.0~29.1); 16.8 (12.8~21.7); 142.2 (99.0~200.5); 1.0 (0.0~5.6); 905 (713~1140) ❙ 1491 (1244~1773); 535 (411~659); 185 (85~223) ❙ 67.1 (14.0~112.2); 897 (713~1140); 227.6 (177.4~284.0)

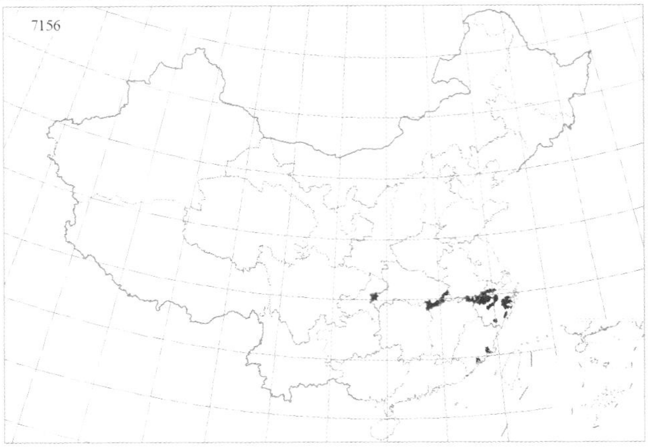

Lythraceae 千屈菜科 *Lagerstroemia* 紫薇属

Lagerstroemia subcostata Koehne 南紫薇

Shrubs or trees, deciduous

17.9 (9.3~24.1); 7.5 (-1.7~18.9); 27.4 (20.1~29.8); 18.0 (9.5~24.1); 156.8 (69.8~229.7); 1.1 (0.0~18.0); 959 (603~1322) ❙ 1591 (753~2993); 624 (323~1847); 169 (27~547) ❙ 67.9 (-12.2~246.5); 951 (603~1273); 225.7 (95.3~363.7)

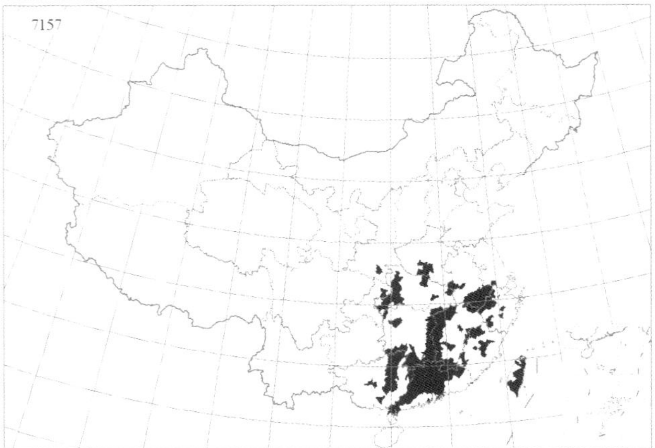

Lythraceae 千屈菜科 *Lagerstroemia* 紫薇属

Lagerstroemia suprareticulata S. K. Lee et L. F. Lau 网脉紫薇

Trees, deciduous

21.9; 13.3; 28.4; 22.0; 203.7; 0.0; 1172 ❙ 1499; 754; 130 ❙ 28.2; 1172; 282.7

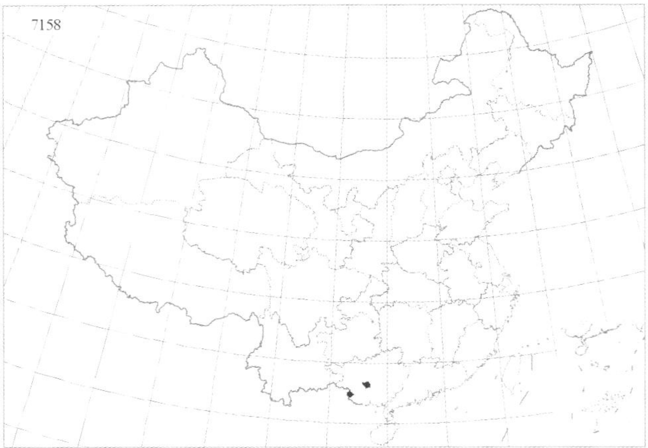

Lythraceae 千屈菜科 *Lagerstroemia* 紫薇属
Lagerstroemia tomentosa C. Presl 绒毛紫薇
Trees, deciduous

19.5; 13.8; 23.0; 19.6; 175.2; 0.0; 940 ❙ 1505; 742; 67 ❙ 60.4; 926; 513.0

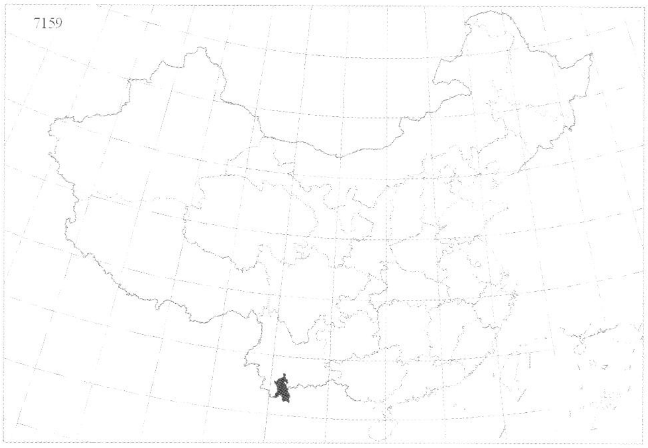

Lythraceae 千屈菜科 *Lagerstroemia* 紫薇属
Lagerstroemia venusta Wall. ex C. B. Clarke 西双紫薇
Trees, deciduous

20.4; 15.1; 23.6; 20.5; 185.8; 0.0; 989 ❙ 1607; 702; 77 ❙ 62.5; 986; 659.0

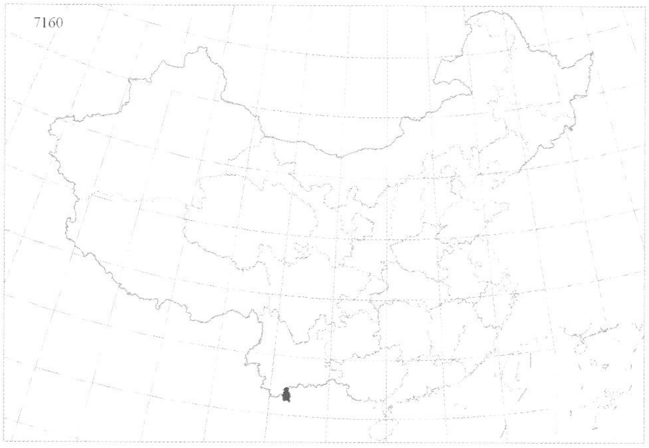

Lythraceae 千屈菜科 *Lagerstroemia* 紫薇属
Lagerstroemia villosa Wall. ex Kurz 毛紫薇
Trees, deciduous

19.2; 13.8; 22.3; 19.2; 170.5; 0.0; 909 ❙ 1499; 686; 58 ❙ 64.9; 880; 513.9

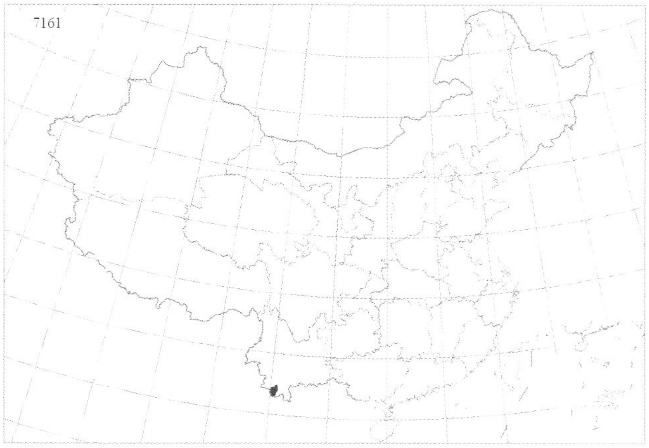

Lythraceae 千屈菜科 *Pemphis* 水芫花属
Pemphis acidula J. R. Forst. et G. Forst. 水芫花
Small shrubs, deciduous

20.4 (16.2~24.1); 14.7 (11.0~18.9); 25.2 (20.1~28.5); 20.4 (16.2~24.2); 185.2 (134.2~230.2); 0.0 (0.0~0.0); 1034 (773~1329) ❙ 2374 (1498~3495); 1087 (712~1847); 249 (47~897) ❙ 141.0 (25.2~246.5); 999 (773~1329); 245.1 (119.4~363.7)

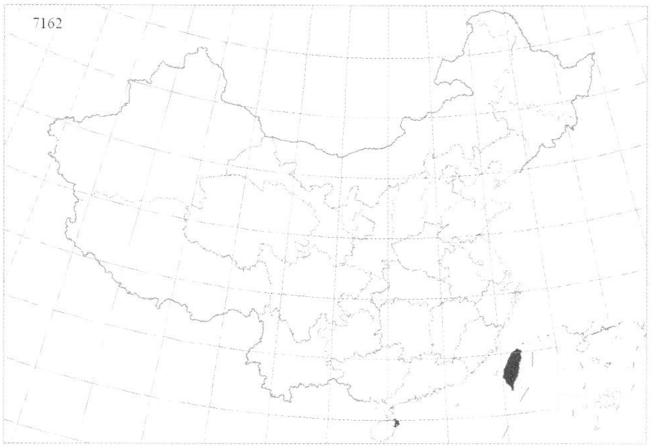

Lythraceae 千屈菜科 *Punica* 石榴属
Punica granatum L. 石榴
Shrubs or trees, deciduous

16.3 (2.2~24.8); 6.5 (-6.6~20.1); 25.3 (9.5~29.9); 16.5 (3.8~24.8); 139.4 (15.5~237.6); 2.5 (0.0~49.1); 882 (411~1385) ❙ 1313 (70~1997); 552 (23~994); 101 (11~254) ❙ 51.2 (-91.1~130.7); 862 (70~1360); 235.9 (111.9~620.0)

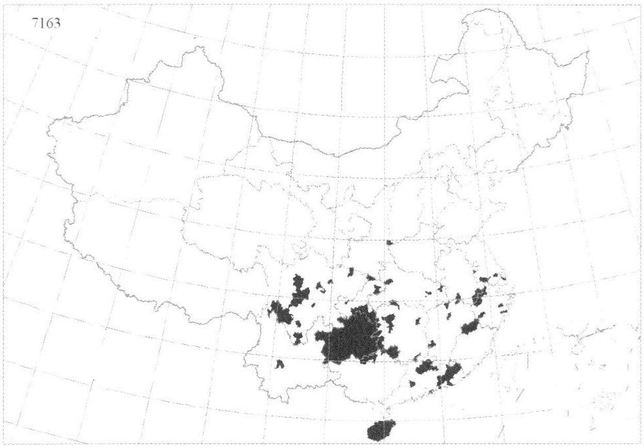

Lythraceae 千屈菜科 *Sonneratia* 海桑属
Sonneratia alba Sm. 杯萼海桑
Trees, evergreen

24.0 (22.2~24.8); 18.4 (17.2~20.1); 28.0 (25.7~28.8); 24.0 (22.3~24.8); 228.3 (207.2~237.6); 0.0 (0.0~0.0); 1310 (1127~1385) ❙ 1501 (868~1854); 534 (312~712); 74 (41~115) ❙ 15.0 (-34.3~47.8); 1191 (868~1360); 440.3 (248.2~620.0)

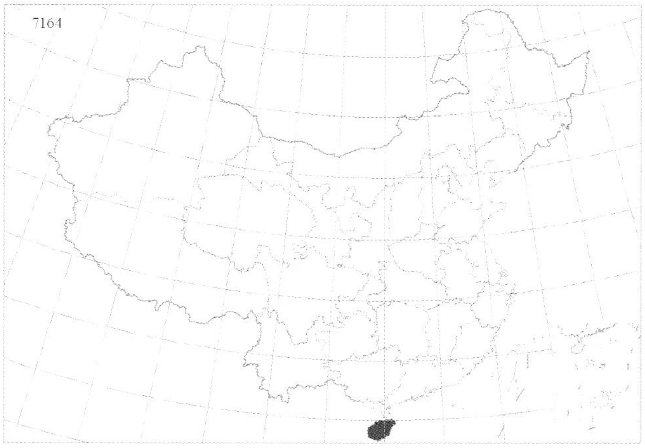

Lythraceae 千屈菜科 *Sonneratia* 海桑属

Sonneratia apetala Buchanan-Hamilton 无瓣海桑

Trees

22.4; 14.8; 28.4; 22.4; 208.9; 0.0; 1191 ❙ 2030; 1033; 110 ❙ 70.4; 1191; 163.7

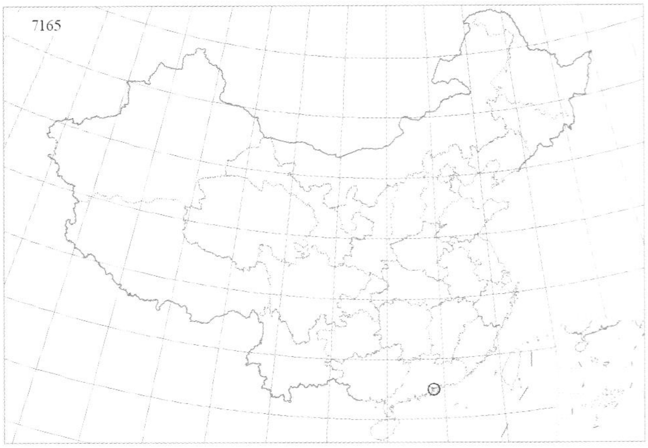

Lythraceae 千屈菜科 *Sonneratia* 海桑属

Sonneratia caseolaris (L.) Engl. 海桑

Trees

23.8 (22.2~24.8); 18.0 (14.8~20.1); 28.0 (25.7~28.8); 23.9 (22.3~24.8); 226.6 (207.2~237.6); 0.0 (0.0~0.0); 1299 (1127~1385) ❙ 1542 (868~2030); 573 (312~1033); 77 (41~115) ❙ 19.3 (-34.3~70.4); 1193 (868~1360); 411.7 (116.6~620.0)

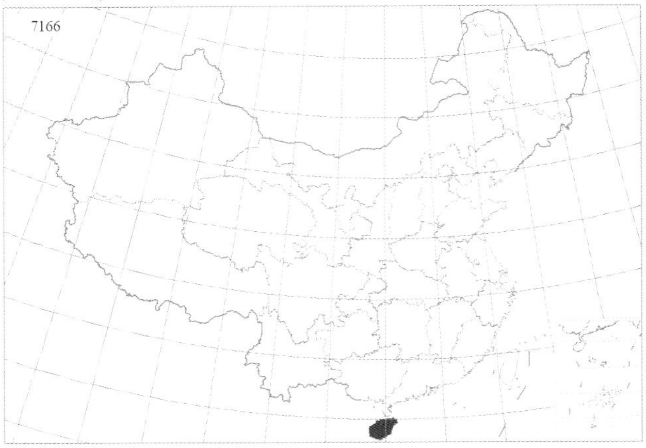

Lythraceae 千屈菜科 *Sonneratia* 海桑属

Sonneratia ovata Backer 桑海桑

Trees

24.1; 18.0; 28.5; 24.2; 230.2; 0.0; 1329 ❙ 1813; 712; 113 ❙ 36.4; 1329; 248.2

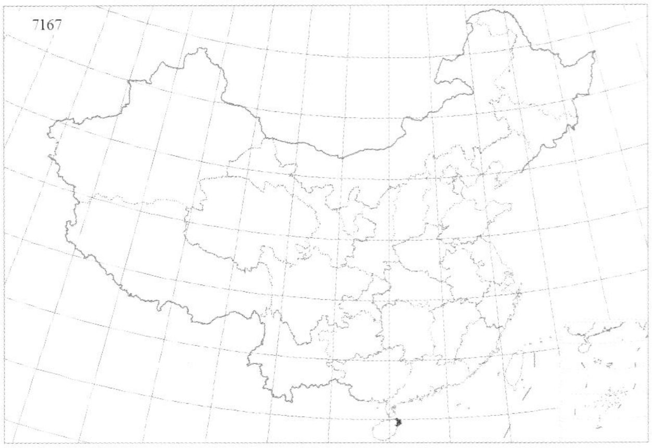

Lythraceae 千屈菜科 *Sonneratia* 海桑属

Sonneratia × gulngai N. C. Duke et B. R. Jackes 拟海桑

Trees, evergreen

24.3; 18.4; 28.5; 24.4; 232.3; 0.0; 1345 ❙ 1834; 667; 114 ❙ 36.4; 1345; 338.4

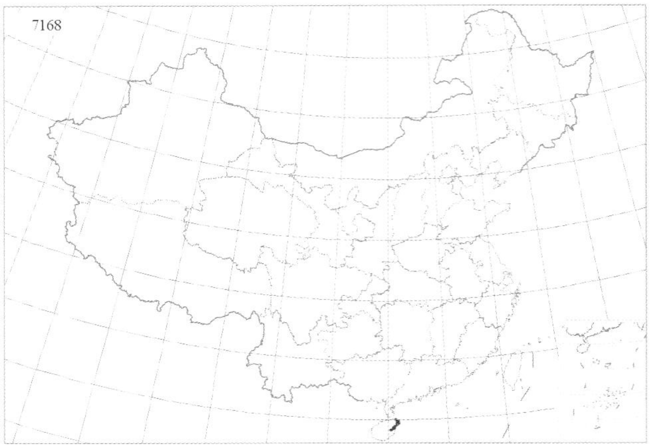

Lythraceae 千屈菜科 *Sonneratia* 海桑属

Sonneratia × hainanensis W. C. Ko, E. Y. Chen et W. Y. Chen 海南海桑

Trees, evergreen

24.1; 18.0; 28.5; 24.2; 230.2; 0.0; 1329 ❙ 1813; 712; 113 ❙ 36.4; 1329; 248.2

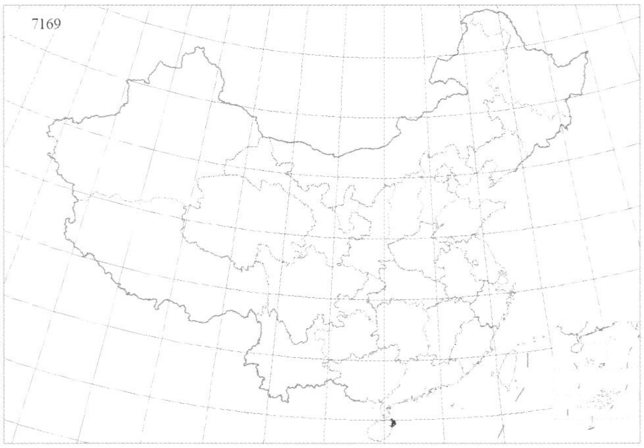

Lythraceae 千屈菜科 *Woodfordia* 虾子花属

Woodfordia fruticosa (L.) Kurz 虾子花

Shrubs, deciduous

16.8 (12.4~21.7); 9.4 (3.8~15.1); 22.3 (17.9~28.9); 16.8 (12.4~21.7); 142.2 (89.1~200.7); 0.0 (0.0~1.5); 836 (675~1147) ❙ 1202 (797~1751); 634 (411~996); 64 (20~209) ❙ 43.6 (-14.2~112.2); 816 (675~1147); 330.6 (163.7~659.0)

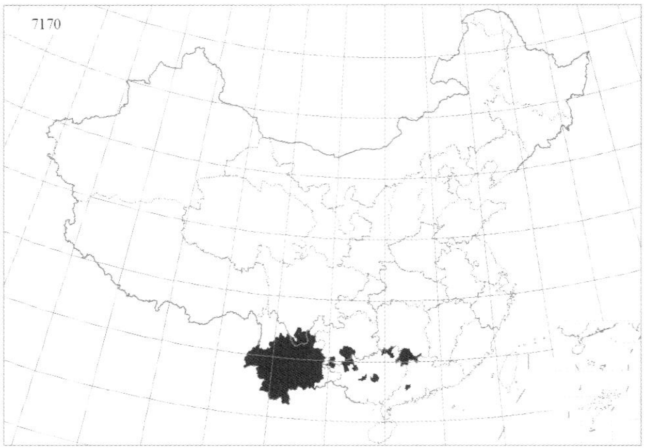

Crypteroniaceae 隐翼科 *Crypteronia* 隐翼属
Crypteronia paniculata Blume 隐翼木
Trees, evergreen

19.3 (17.7~21.0); 13.3 (10.9~15.1); 23.2 (21.6~25.6); 19.4 (17.8~21.0); 172.4 (153.4~191.8); 0.0 (0.0~0.0); 934 (854~1053) ❙ 1515 (1368~1751); 810 (655~996); 64 (42~77) ❙ 62.7 (44.5~88.8); 927 (833~1053); 543.6 (491.5~659.0)

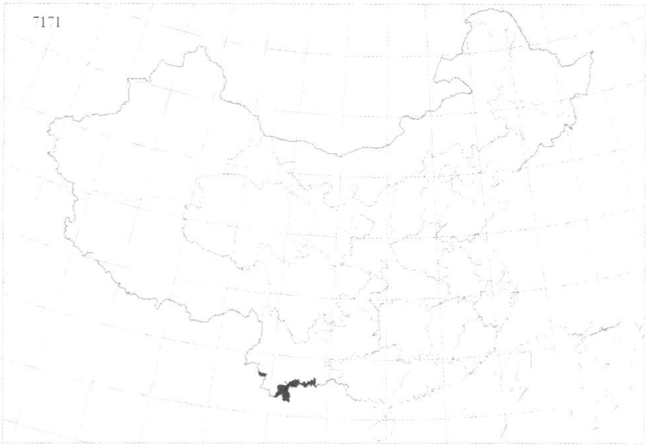

Lecythidaceae 玉蕊科 *Barringtonia* 玉蕊属
Barringtonia asiatica (L.) Kurz 滨玉蕊
Trees, evergreen

20.4 (16.2~24.6); 14.7 (11.0~19.5); 25.1 (20.1~28.2); 20.5 (16.2~24.7); 185.5 (134.2~235.9); 0.0 (0.0~0.0); 1036 (773~1372) ❙ 2372 (1498~3495); 1080 (567~1847); 248 (47~897) ❙ 140.7 (25.2~246.5); 998 (773~1299); 259.4 (119.4~548.0)

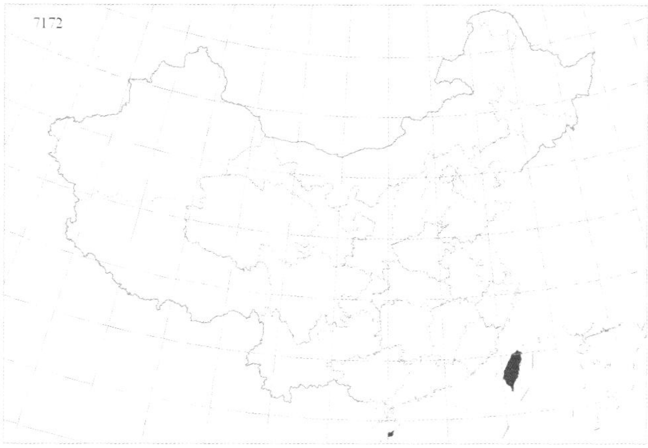

Lecythidaceae 玉蕊科 *Barringtonia* 玉蕊属
Barringtonia fusicarpa H. H. Hu 梭果玉蕊
Trees, evergreen

19.2 (17.3~21.0); 13.1 (9.9~15.1); 23.4 (22.4~25.6); 19.3 (17.3~21.0); 171.4 (147.1~191.8); 0.0 (0.0~0.0); 932 (833~1053) ❙ 1496 (1276~1751); 808 (655~996); 64 (51~77) ❙ 61.2 (36.8~88.8); 927 (833~1053); 523.3 (413.5~659.0)

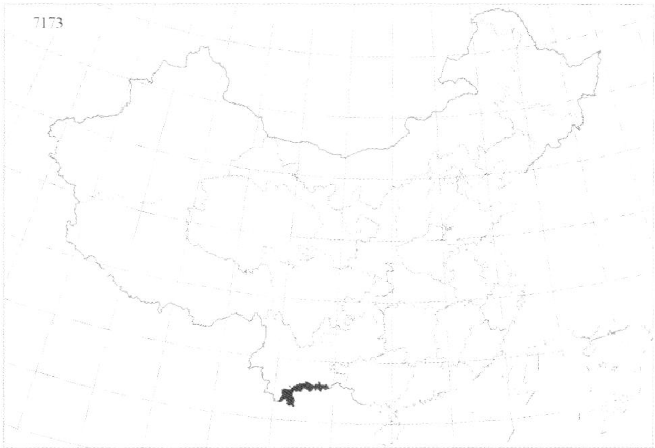

Lecythidaceae 玉蕊科 *Barringtonia* 玉蕊属
Barringtonia racemosa (L.) Blume ex DC. 玉蕊
Trees, evergreen

21.6 (16.2~24.8); 15.9 (11.0~20.1); 26.0 (20.1~28.6); 21.6 (16.2~24.8); 199.1 (134.2~237.6); 0.0 (0.0~0.0); 1123 (773~1385) ❙ 2121 (1309~3495); 914 (455~1847); 192 (41~897) ❙ 102.7 (−1.4~246.5); 1066 (773~1360); 323.2 (119.4~620.0)

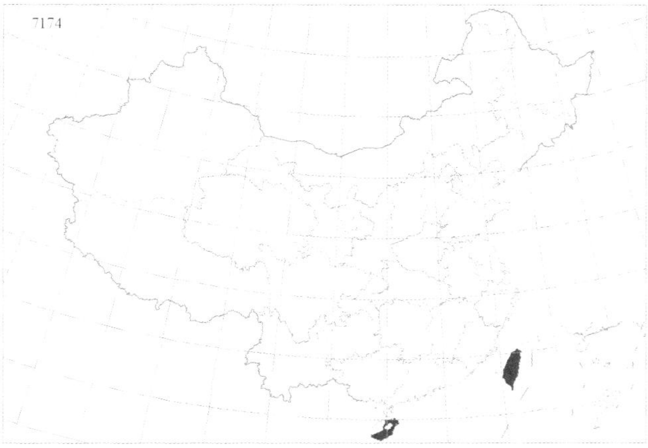

Rhizophoraceae 红树科 *Bruguiera* 木榄属
Bruguiera cylindrica (L.) Blume 柱果木榄
Trees, evergreen

24.2; 17.6; 28.8; 24.2; 230.4; 0.0; 1331 ❙ 1475; 581; 100 ❙ 10.8; 1331; 284.3

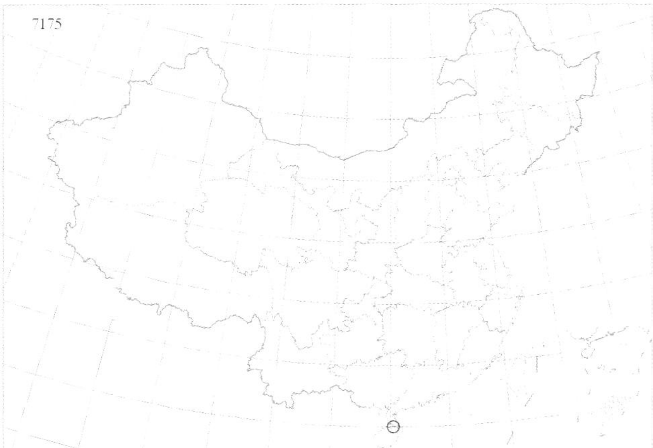

Rhizophoraceae 红树科 *Bruguiera* 木榄属
Bruguiera gymnorhiza (L.) Poir. 木榄
Trees, evergreen

21.5 (16.4~24.2); 14.9 (11.2~18.9); 26.8 (20.3~29.2); 21.6 (16.5~24.2); 198.7 (137.8~230.4); 0.0 (0.0~0.0); 1123 (783~1331) ❙ 2023 (1269~3495); 935 (455~1847); 182 (47~897) ❙ 88.7 (−1.4~235.6); 1103 (783~1331); 247.6 (119.4~363.7)

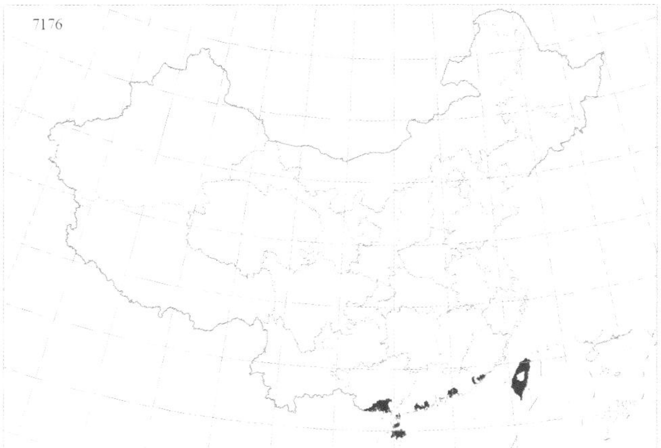

Rhizophoraceae 红树科 *Bruguiera* 木榄属
Bruguiera sexangula (Lour.) Poir. 海莲
Trees, evergreen

24.2 (23.1~24.8); 18.4 (17.5~20.1); 28.2 (27.2~28.8); 24.2 (23.2~24.8); 230.4 (218.4~237.6); 0.0 (0.0~0.0); 1328 (1232~1385) | 1443 (868~1854); 514 (312~712); 74 (41~115) | 8.6 (-34.3~36.4); 1189 (868~1360); 411.1 (248.2~548.0)

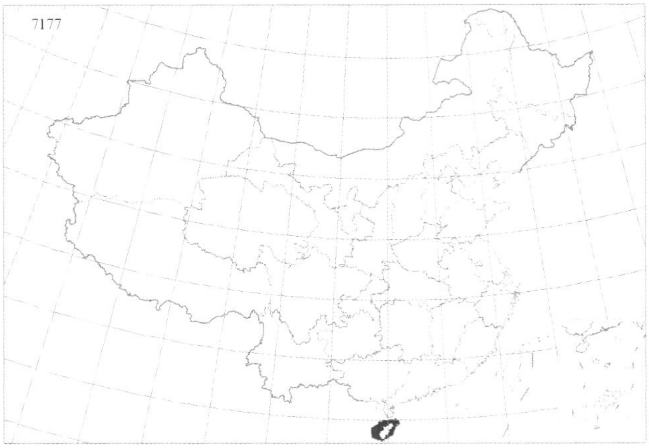

Rhizophoraceae 红树科 *Carallia* 竹节树属
Carallia brachiata (Lour.) Merr. 竹节树
Trees, evergreen

21.6 (16.1~24.8); 14.4 (9.6~20.1); 27.2 (20.6~29.2); 21.7 (16.2~24.8); 199.8 (133.9~237.6); 0.0 (0.0~0.0); 1136 (785~1385) | 1604 (868~2435); 742 (312~1428); 104 (41~191) | 42.6 (-34.3~99.5); 1112 (779~1360); 296.6 (95.3~659.0)

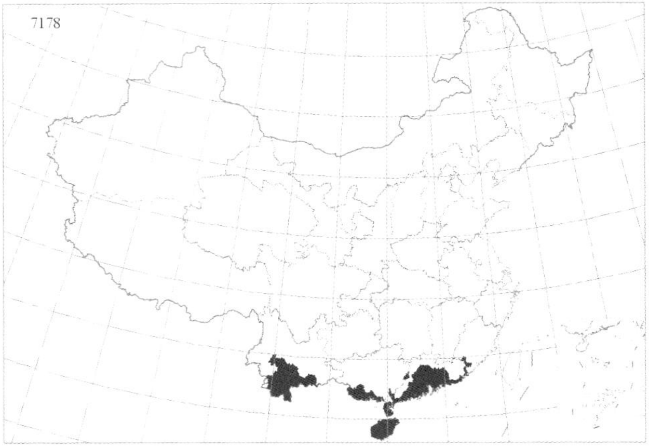

Rhizophoraceae 红树科 *Carallia* 竹节树属
Carallia diplopetala Hand.-Mazz. 锯叶竹节树
Shrubs or trees, evergreen

19.3 (16.3~23.0); 12.1 (7.3~15.1); 24.4 (21.4~29.0); 19.3 (16.4~23.0); 172.1 (136.4~216.2); 0.0 (0.0~0.0); 963 (797~1247) | 1459 (1018~2435); 779 (554~1428); 77 (41~169) | 51.9 (12.9~99.5); 956 (797~1247); 377.1 (199.6~659.0)

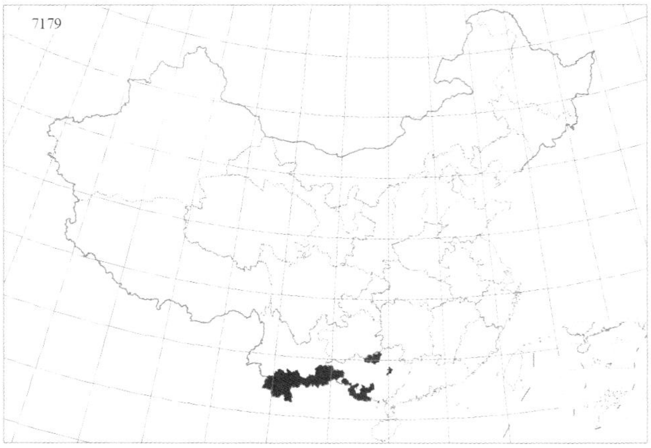

Rhizophoraceae 红树科 *Carallia* 竹节树属
Carallia garciniifolia F. C. How et C. N. Ho 大叶竹节树
Trees, evergreen

17.0; 9.4; 22.5; 17.1; 145.2; 0.0; 841 | 1321; 743; 57 | 58.1; 841; 355.7

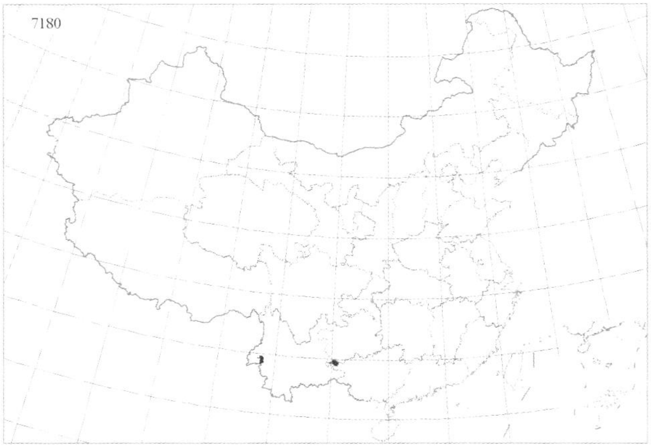

Rhizophoraceae 红树科 *Carallia* 竹节树属
Carallia pectinifolia W. C. Ko 旁杞木
Shrubs or small trees, evergreen

21.7 (21.0~22.9); 13.6 (12.2~15.2); 28.1 (26.9~28.8); 21.8 (21.0~22.9); 201.5 (192.3~214.8); 0.0 (0.0~0.0); 1150 (1081~1232) | 1651 (1462~1876); 753 (599~891); 124 (105~142) | 43.7 (28.6~64.3); 1150 (1081~1232); 231.4 (207.5~273.3)

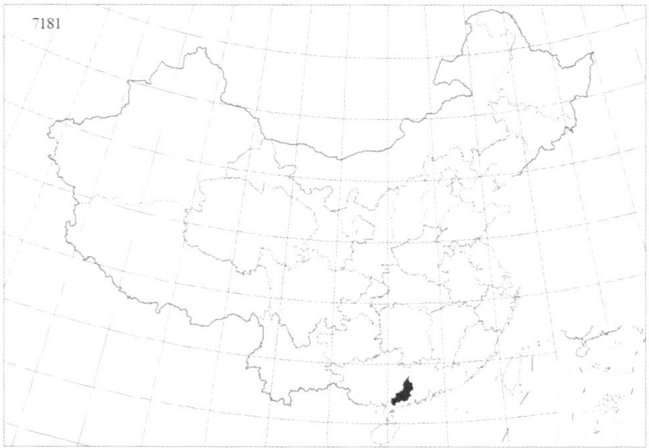

Rhizophoraceae 红树科 *Ceriops* 角果木属
Ceriops tagal (Perr.) C. B. Rob. 角果木
Shrubs or trees, evergreen

21.6 (16.2~24.8); 15.7 (11.0~20.1); 26.2 (20.1~28.9); 21.6 (16.2~24.8); 199.3 (134.2~237.6); 0.0 (0.0~0.0); 1127 (773~1385) | 2086 (1176~3495); 922 (447~1847); 194 (41~897) | 99.4 (-12.0~246.5); 1076 (773~1331); 288.3 (119.4~620.0)

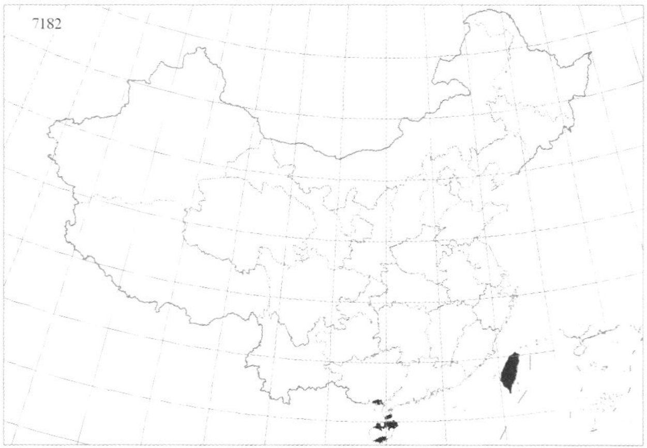

Rhizophoraceae 红树科 *Kandelia* 秋茄树属
Kandelia obovata Sheue et al. 秋茄树
Shrubs or trees, evergreen

21.0 (16.4~24.8)；13.8 (7.6~20.1)；27.3 (20.3~29.2)；21.1 (16.5~24.8)；192.8 (137.8~237.6)；0.0 (0.0~0.0)；1099 (783~1385) ❙ 1752 (1070~3495)；741 (356~1847)；179 (41~897) ❙ 65.2 (-6.9~235.6)；1067 (783~1331)；241.3 (119.4~620.0)

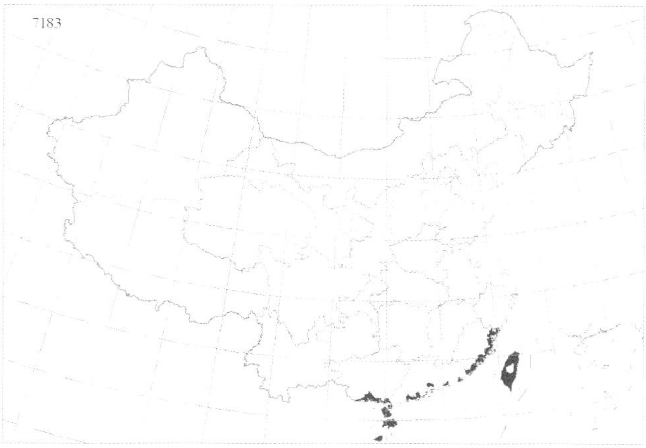

Rhizophoraceae 红树科 *Pellacalyx* 山红树属
Pellacalyx yunnanensis H. H. Hu 山红树
Trees, evergreen

19.2 (16.9~21.0)；13.2 (10.3~15.1)；23.2 (21.4~25.6)；19.2 (16.9~21.0)；170.9 (142.7~191.8)；0.0 (0.0~0.0)；927 (813~1053) ❙ 1464 (1178~1751)；775 (646~996)；63 (51~77) ❙ 58.0 (36.8~88.8)；920 (813~1053)；519.3 (383.4~659.0)

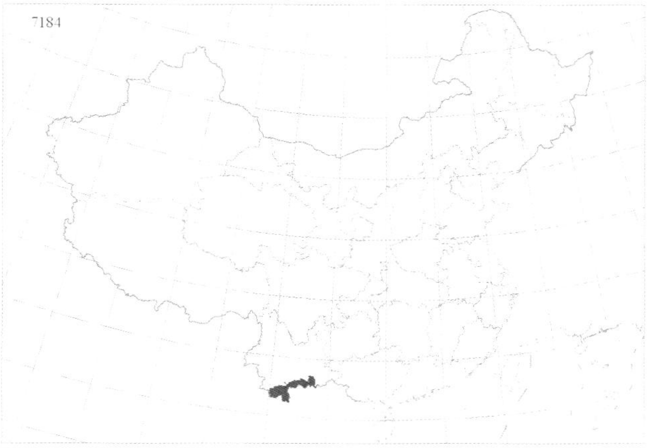

Rhizophoraceae 红树科 *Rhizophora* 红树属
Rhizophora apiculata Blume 红树
Shrubs or trees, evergreen

23.9 (22.8~24.8)；17.7 (14.7~20.1)；28.4 (27.2~29.0)；24.0 (22.8~24.8)；227.7 (213.6~237.6)；0.0 (0.0~0.0)；1313 (1230~1385) ❙ 1532 (868~1813)；620 (312~997)；89 (41~113) ❙ 17.0 (-34.3~39.9)；1226 (868~1331)；350.9 (199.6~620.0)

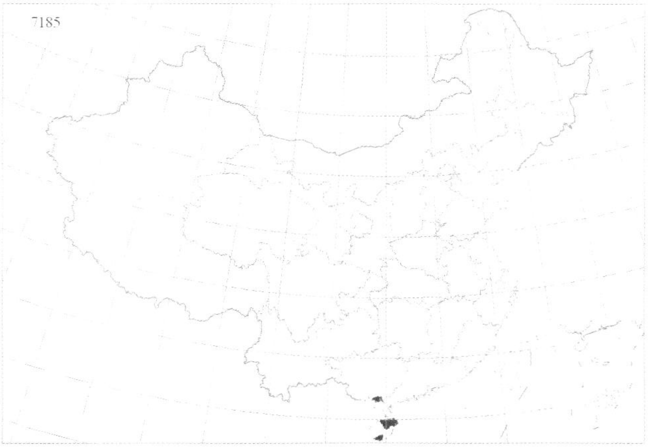

Rhizophoraceae 红树科 *Rhizophora* 红树属
Rhizophora mucronata Lam. ex Poiret 红茄苳
Trees, evergreen

22.4；16.2；27.1；22.5；209.7；0.0；1186 ❙ 1857；1052；91 ❙ 67.0；1126；217.7

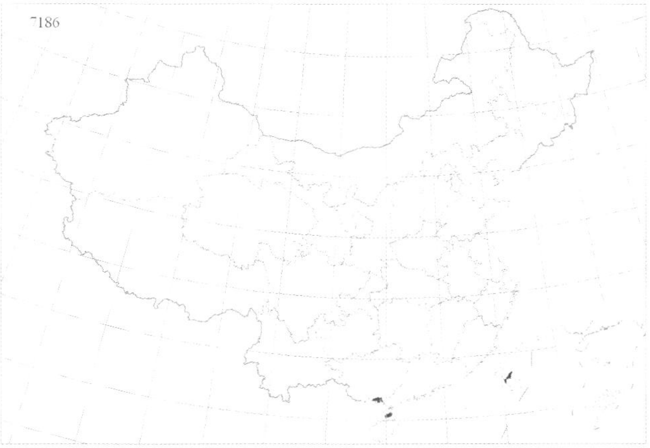

Rhizophoraceae 红树科 *Rhizophora* 红树属
Rhizophora stylosa Griff. 红海兰
Trees, evergreen

23.3 (21.3~24.2)；16.5 (13.5~18.3)；28.4 (27.5~29.0)；23.4 (21.3~24.2)；220.4 (195.6~230.4)；0.0 (0.0~0.0)；1265 (1107~1331) ❙ 1636 (1269~2035)；706 (480~969)；103 (75~142) ❙ 30.3 (-1.4~79.9)；1256 (1107~1331)；281.3 (203.1~474.7)

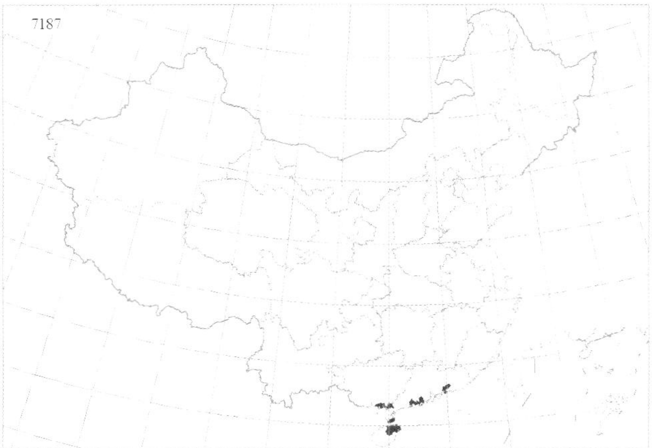

Nyssaceae 蓝果树科 *Camptotheca* 喜树属
Camptotheca acuminata Decne. 喜树
Trees, deciduous

16.8 (2.2~23.4)；6.9 (-6.6~16.1)；25.6 (9.9~30.1)；16.8 (3.8~23.5)；142.5 (15.5~221.7)；0.7 (0.0~49.1)；881 (413~1273) ❙ 1356 (584~2014)；569 (323~996)；125 (15~259) ❙ 54.4 (-34.1~147.0)；869 (413~1273)；243.9 (108.2~659.0)

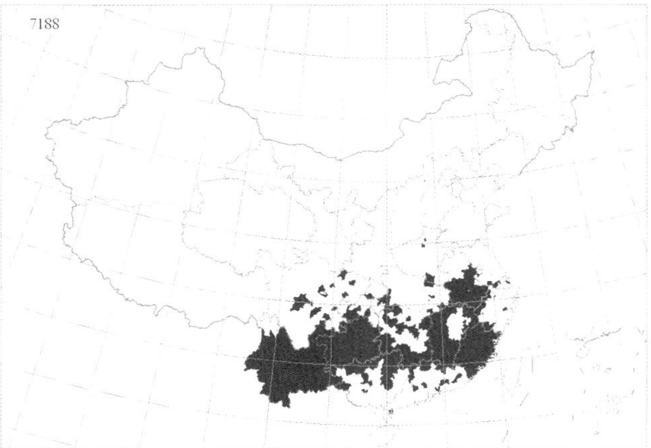

N

Nyssaceae 蓝果树科 *Davidia* 珙桐属

Davidia involucrata Baill. 珙桐

Trees, deciduous

13.4 (6.0~21.0); 3.2 (−3.2~14.2); 22.8 (14.4~27.5); 13.4 (6.6~21.0); 104.9 (39.7~191.8); 4.1 (0.0~26.7); 739 (508~1053) ❙ 1149 (677~1522); 537 (336~818); 63 (14~140) ❙ 56.2 (2.4~107.2); 739 (508~1053); 244.8 (146.0~551.6)

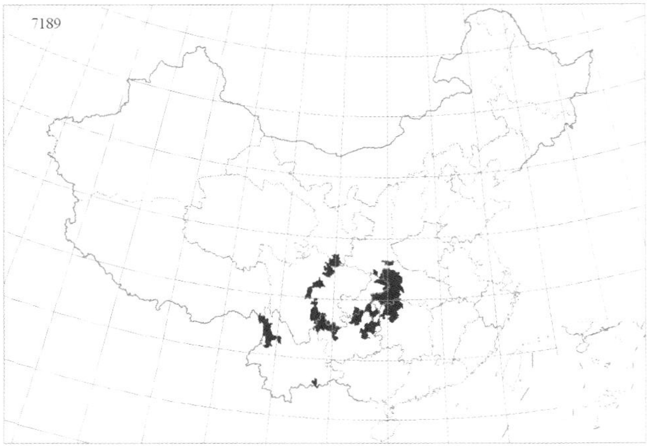

Nyssaceae 蓝果树科 *Davidia* 珙桐属

Davidia involucrata var. vilmoriniana (Dode) Wangerin 光叶珙桐

Trees, deciduous

13.1 (7.3~16.7); 2.9 (−2.0~10.3); 22.5 (15.0~26.5); 13.1 (7.5~16.8); 101.8 (48.2~141.5); 4.6 (0.0~20.4); 726 (537~839) ❙ 1205 (951~1492); 554 (420~816); 68 (27~113) ❙ 66.9 (17.2~107.2); 724 (537~839); 240.0 (147.3~352.3)

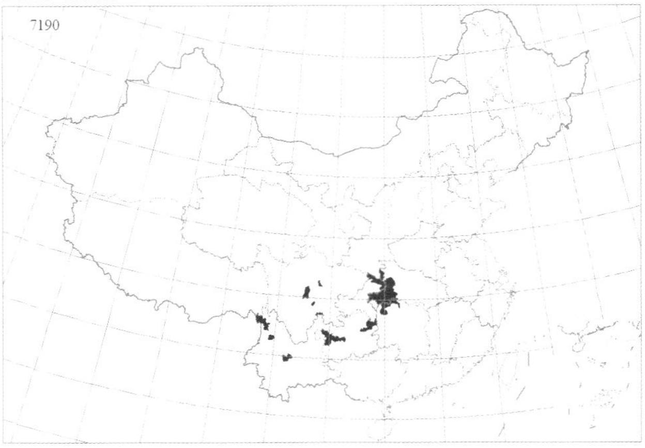

Nyssaceae 蓝果树科 *Nyssa* 蓝果树属

Nyssa javanica (Blume) Wangerin 华南蓝果树

Trees, deciduous

20.0 (8.6~24.0); 13.0 (1.3~19.1); 25.0 (15.0~28.6); 20.0 (8.7~24.0); 180.8 (52.6~228.1); 0.3 (0.0~8.8); 1027 (574~1302) ❙ 1503 (1041~2435); 775 (444~1428); 75 (44~142) ❙ 48.3 (−2.2~99.5); 1000 (574~1221); 400.6 (225.9~620.0)

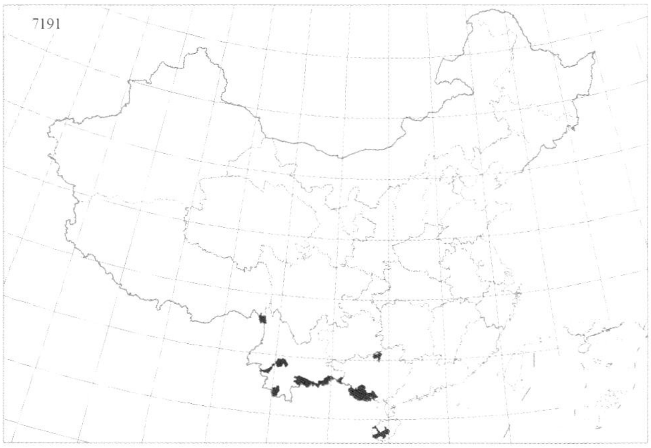

Nyssaceae 蓝果树科 *Nyssa* 蓝果树属

Nyssa leptophylla W. P. Fang et T. P. Chen 薄叶蓝果树

Small trees

15.8; 4.6; 26.3; 15.9; 131.1; 0.8; 843 ❙ 1459; 483; 183 ❙ 74.1; 831; 221.1

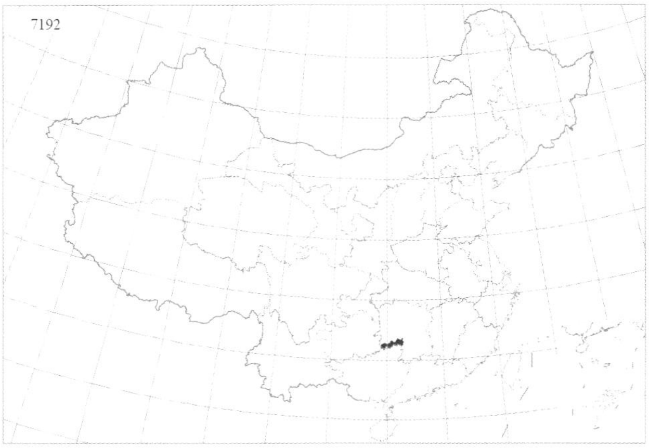

Nyssaceae 蓝果树科 *Nyssa* 蓝果树属

Nyssa shangszeensis W. P. Fang et Soong 上思蓝果树

Small trees, evergreen

21.9 (21.3~22.7); 13.9 (13.2~15.0); 27.9 (27.3~28.6); 22.0 (21.3~22.8); 203.6 (195.9~213.1); 0.0 (0.0~0.0); 1162 (1111~1221) ❙ 1876 (1302~2435); 1047 (670~1428); 116 (86~130) ❙ 61.1 (12.9~99.5); 1162 (1111~1221); 289.6 (235.1~318.8)

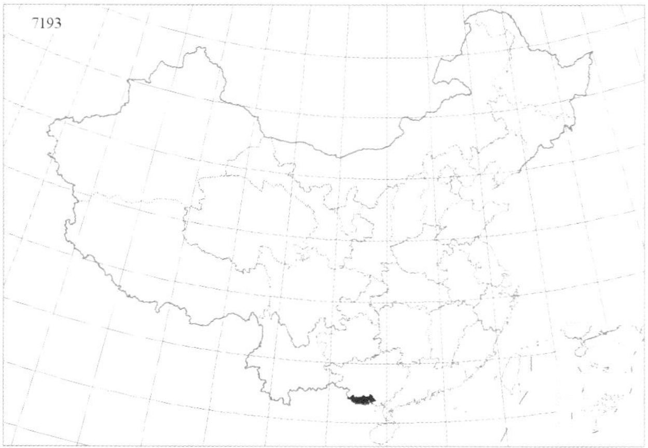

Nyssaceae 蓝果树科 *Nyssa* 蓝果树属

Nyssa shweliensis (W. W. Smith.) Airy-Shaw 瑞丽蓝果树

Trees, deciduous

16.6; 10.2; 21.0; 16.6; 139.5; 0.0; 807 ❙ 1280; 721; 56 ❙ 60.3; 791; 428.5

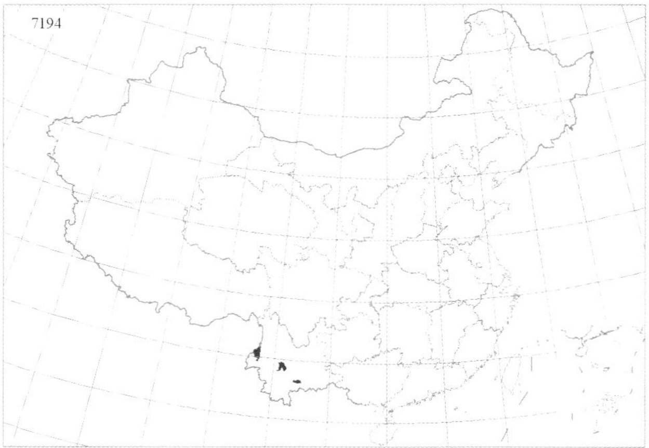

Nyssaceae 蓝果树科 *Nyssa* 蓝果树属
Nyssa sinensis Oliver 蓝果树
Trees, deciduous

17.4 (10.6~22.7); 7.0 (-0.2~15.1); 27.0 (21.1~30.1); 17.5 (10.6~22.8); 150.2 (79.4~213.1); 0.6 (0.0~12.1); 922 (643~1221) ❙ 1514 (969~2435); 568 (356~1428); 166 (46~259) ❙ 65.4 (-5.0~147.0); 912 (643~1221); 236.3 (123.9~659.0)

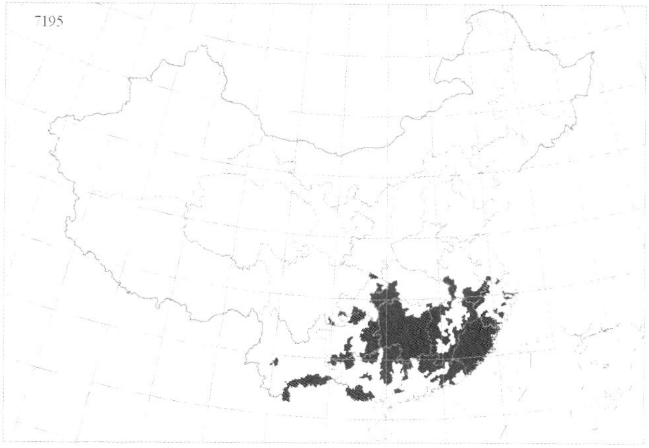

Nyssaceae 蓝果树科 *Nyssa* 蓝果树属
Nyssa wenshanensis W. P. Fang et Soong 文山蓝果树
Trees

18.6 (16.3~21.0); 12.0 (9.1~14.2); 23.1 (21.4~25.6); 18.6 (16.4~21.0); 163.0 (136.4~191.8); 0.0 (0.0~0.0); 902 (797~1053) ❙ 1411 (1190~1547); 795 (656~911); 59 (50~68) ❙ 56.8 (44.5~74.2); 902 (797~1053); 489.6 (347.3~551.6)

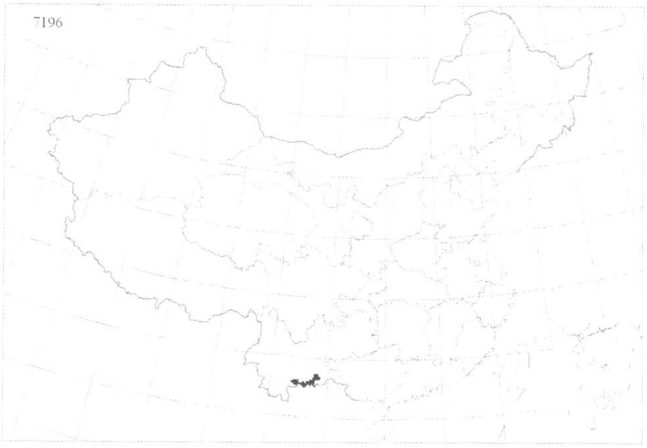

Nyssaceae 蓝果树科 *Nyssa* 蓝果树属
Nyssa yunnanensis W. Q. Yin ex H. N. Qin et Phengklai 云南蓝果树
Trees

19.3; 13.7; 22.5; 19.3; 171.7; 0.0; 921 ❙ 1488; 709; 59 ❙ 62.3; 893; 507.8

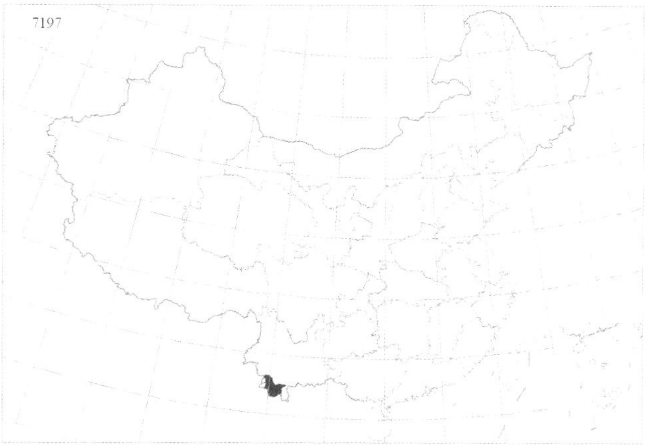

Alangiaceae 八角枫科 *Alangium* 八角枫属
Alangium alpinum (C. B. Clarke) W. W. Sm. et Cave 高山八角枫
Trees, deciduous

9.7 (-0.7~14.4); 2.2 (-9.6~7.5); 15.9 (7.7~19.6); 10.2 (2.5~14.4); 70.9 (6.8~113.1); 13.8 (0.0~74.2); 604 (338~727) ❙ 1197 (370~1736); 681 (262~1007); 46 (10~70) ❙ 91.4 (4.7~145.8); 604 (338~727); 342.3 (61.6~487.2)

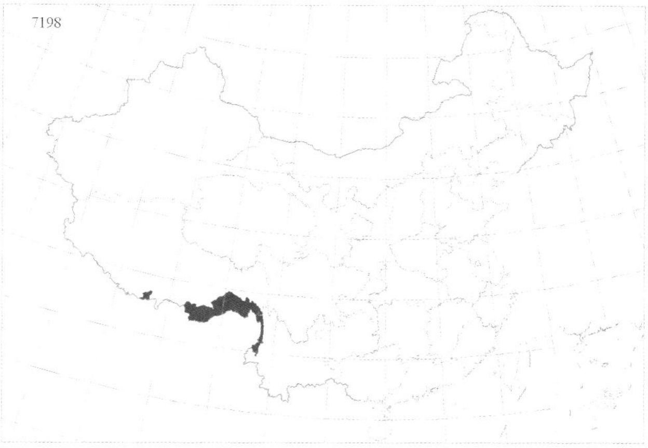

Alangiaceae 八角枫科 *Alangium* 八角枫属
Alangium barbatum Baillon ex Kuntze 髯毛八角枫
Shrubs or small trees, deciduous

19.3 (16.3~24.0); 12.6 (9.1~19.1); 24.0 (21.4~28.5); 19.3 (16.4~24.0); 172.1 (136.4~228.1); 0.0 (0.0~0.0); 959 (797~1302) ❙ 1473 (1074~1900); 763 (539~1056); 73 (47~188) ❙ 54.3 (19.1~88.8); 948 (797~1188); 429.0 (244.0~659.0)

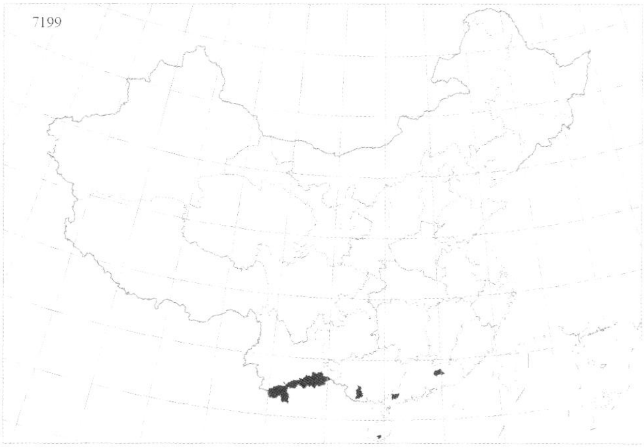

Alangiaceae 八角枫科 *Alangium* 八角枫属
Alangium chinense (Lour.) Harms 八角枫
Shrubs or trees, deciduous

16.2 (2.0~24.8); 5.6 (-12.5~20.1); 25.9 (10.1~30.1); 16.3 (3.8~24.8); 138.8 (15.4~237.6); 3.4 (0.0~62.6); 880 (413~1385) ❙ 1257 (377~2992); 551 (241~1847); 104 (8~402) ❙ 42.6 (-37.2~246.5); 861 (377~1273); 220.8 (95.3~659.0)

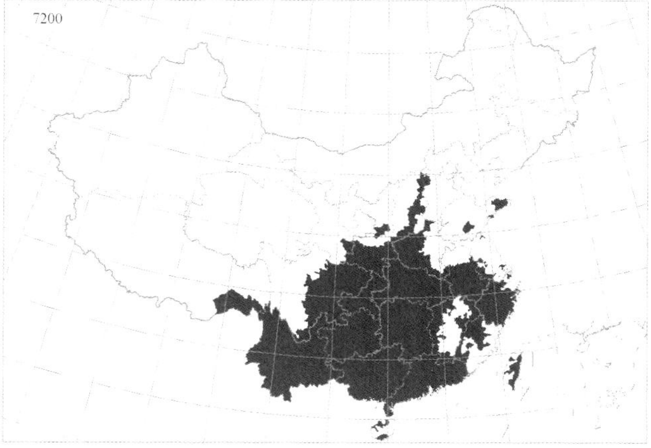

Alangiaceae 八角枫科 *Alangium* 八角枫属
Alangium chinense subsp. pauciflorum W. P. Fang 稀花八角枫
Shrubs or trees, deciduous

13.0 (7.4~19.2); 1.3 (-4.2~10.7); 24.1 (18.7~27.5); 13.1 (8.2~19.2); 105.2 (58.8~170.8); 9.1 (0.0~29.4); 743 (561~981) | 980 (634~1492); 425 (290~652); 47 (16~140) | 31.5 (-16.6~101.4); 741 (561~981); 248.6 (146.0~411.1)

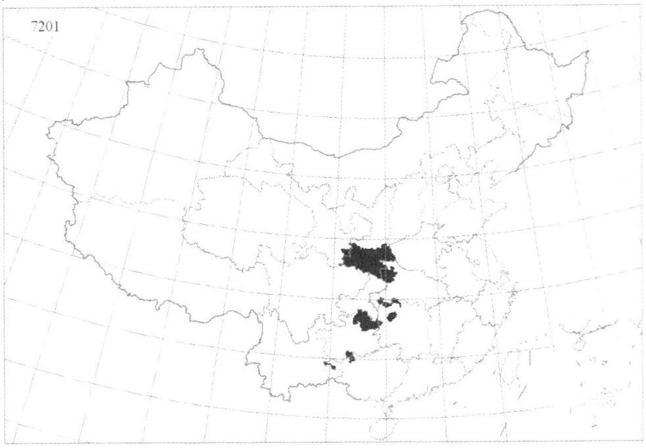

Alangiaceae 八角枫科 *Alangium* 八角枫属
Alangium chinense subsp. strigosum W. P. Fang 伏毛八角枫
Shrubs or small trees, deciduous

16.0 (8.6~21.0); 5.6 (-0.4~15.1); 25.3 (15.0~29.3); 16.0 (8.7~21.0); 133.4 (52.6~191.8); 1.3 (0.0~12.3); 843 (574~1053) | 1308 (848~1820); 566 (405~835); 104 (29~248) | 55.6 (6.2~112.2); 841 (574~1053); 255.2 (143.5~659.0)

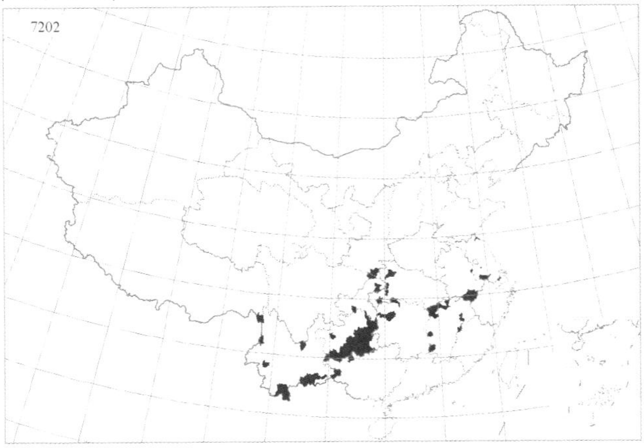

Alangiaceae 八角枫科 *Alangium* 八角枫属
Alangium chinense subsp. triangulare (Wangerin) W. P. Fang 深裂八角枫
Shrubs or trees, deciduous

14.9 (11.7~15.8); 3.7 (1.6~5.2); 25.6 (18.8~27.9); 15.0 (11.7~15.9); 121.3 (82.6~131.5); 1.7 (0.0~6.5); 813 (658~873) | 1317 (916~1770); 535 (451~622); 119 (31~223) | 61.3 (38.5~112.2); 813 (658~873); 217.1 (181.5~273.6)

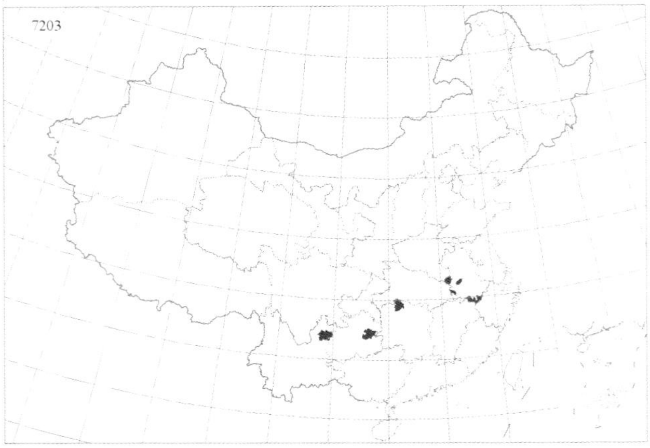

Alangiaceae 八角枫科 *Alangium* 八角枫属
Alangium faberi Oliv. 小花八角枫
Shrubs, deciduous

17.2 (10.6~24.8); 7.0 (-0.2~20.1); 26.4 (18.8~29.9); 17.3 (10.6~24.8); 147.7 (79.4~237.6); 0.5 (0.0~12.1); 911 (643~1385) | 1380 (916~2435); 570 (392~1428); 119 (29~225) | 52.5 (-12.0~145.8); 903 (643~1273); 223.1 (108.2~659.0)

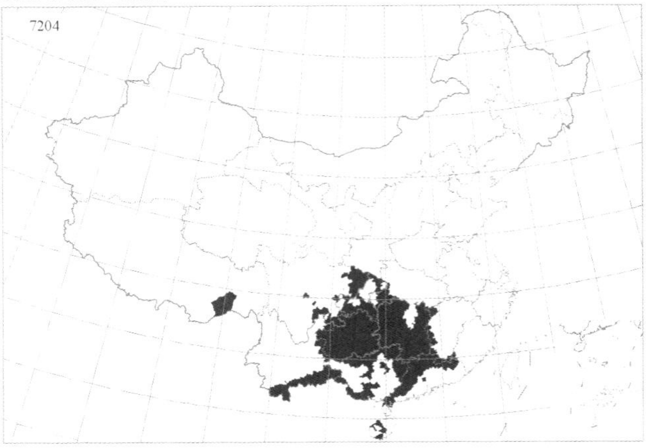

Alangiaceae 八角枫科 *Alangium* 八角枫属
Alangium faberi var. heterophyllum Yang 异叶八角枫
Shrubs, deciduous

15.9 (10.6~18.1); 5.4 (-0.2~8.1); 25.5 (21.1~27.8); 15.9 (10.6~18.2); 132.1 (79.4~158.1); 0.9 (0.0~12.1); 838 (643~949) | 1205 (985~1368); 530 (444~661); 77 (29~129) | 44.9 (18.0~93.8); 838 (643~949); 202.9 (126.1~284.3)

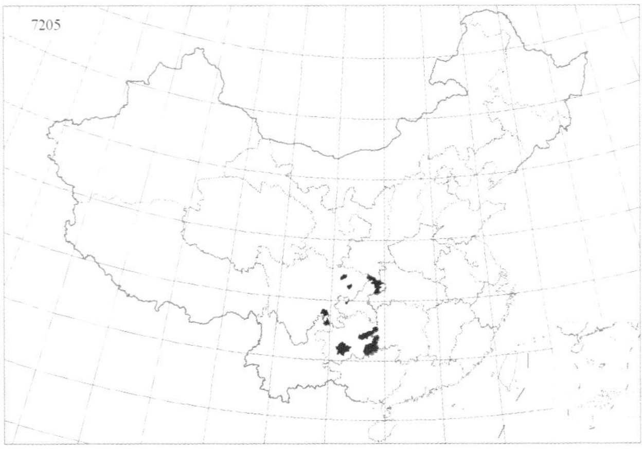

Alangiaceae 八角枫科 *Alangium* 八角枫属
Alangium faberi var. perforatum (H. Lév.) Rehder 小叶八角枫
Shrubs, deciduous

15.4 (14.2~16.8); 5.0 (3.6~6.6); 25.0 (23.5~26.4); 15.5 (14.2~16.9); 126.2 (112.5~142.7); 0.3 (0.0~1.7); 811 (756~869) | 1229 (1133~1368); 522 (448~648); 85 (63~129) | 51.6 (38.5~71.9); 811 (756~869); 207.0 (160.1~246.1)

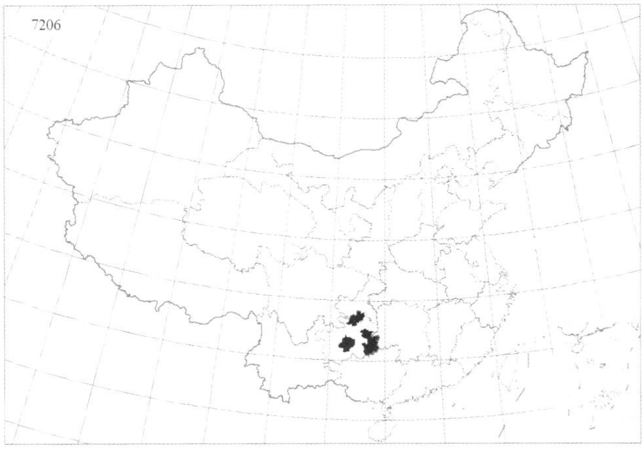

Alangiaceae 八角枫科 *Alangium* 八角枫属

Alangium faberi var. platyphyllum Chun et F. C. How 阔叶八角枫

Shrubs, deciduous

20.5 (16.1~24.0); 11.6 (5.3~19.1); 27.7 (25.5~29.0); 20.6 (16.1~24.0); 186.8 (133.1~228.1); 0.0 (0.0~0.0); 1084 (838~1302) ❙ 1596 (1187~2435); 736 (523~1428); 139 (48~194) ❙ 48.5 (-6.9~99.5); 1076 (838~1221); 261.8 (185.5~620.0)

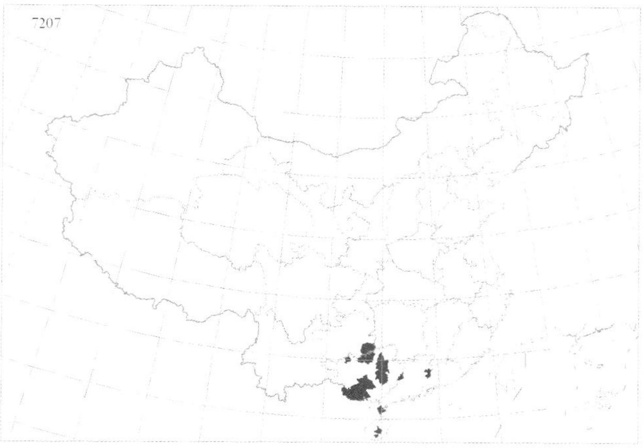

Alangiaceae 八角枫科 *Alangium* 八角枫属

Alangium kurzii Craib 毛八角枫

Small trees, deciduous

17.8 (11.7~24.8); 7.7 (1.1~20.1); 26.8 (18.8~29.9); 17.8 (11.7~24.8); 154.5 (82.6~237.6); 0.6 (0.0~8.5); 942 (658~1385) ❙ 1468 (753~2435); 601 (384~1428); 139 (31~259) ❙ 57.0 (-12.2~130.7); 934 (658~1326); 233.4 (95.3~659.0)

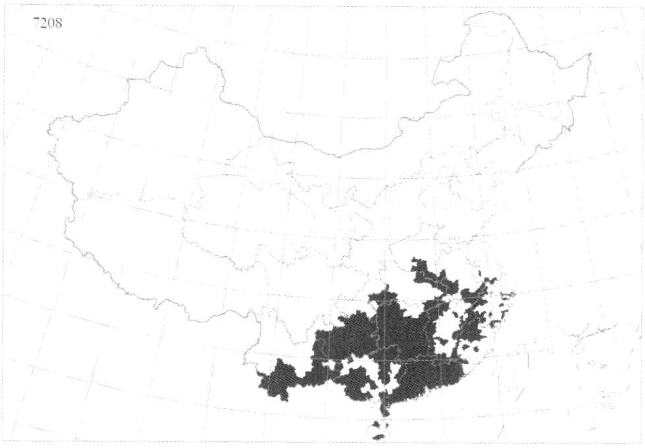

Alangiaceae 八角枫科 *Alangium* 八角枫属

Alangium kurzii var. handelii (Schnarf) W. P. Fang 云山八角枫

Small trees, deciduous

19.0 (13.4~23.6); 9.2 (1.1~16.7); 27.6 (22.5~29.8); 19.0 (13.4~23.6); 168.8 (108.3~223.5); 0.8 (0.0~7.9); 1009 (755~1281) ❙ 1564 (753~2037); 647 (384~1033); 147 (46~254) ❙ 56.7 (-12.2~147.0); 1007 (753~1273); 225.6 (95.3~551.6)

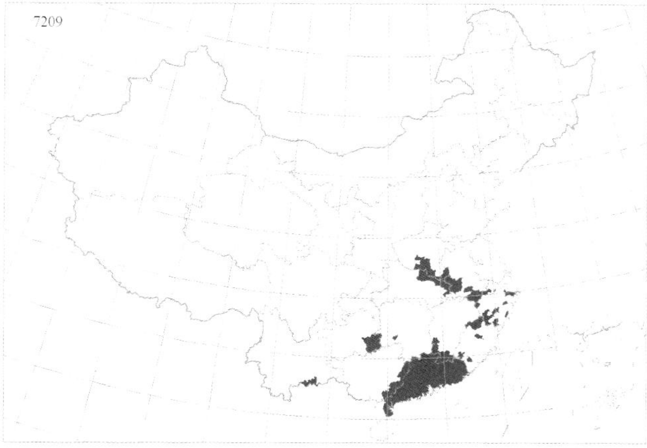

Alangiaceae 八角枫科 *Alangium* 八角枫属

Alangium kwangsiense Melch. 广西八角枫

Scandent shrubs, deciduous

20.1 (16.7~22.2); 11.2 (5.6~14.3); 27.6 (25.8~29.8); 20.2 (16.8~22.3); 182.0 (141.2~207.5); 0.0 (0.0~0.0); 1050 (885~1190) ❙ 1623 (1279~2005); 645 (497~902); 160 (73~197) ❙ 55.2 (14.8~86.3); 1049 (885~1190); 241.0 (174.9~327.7)

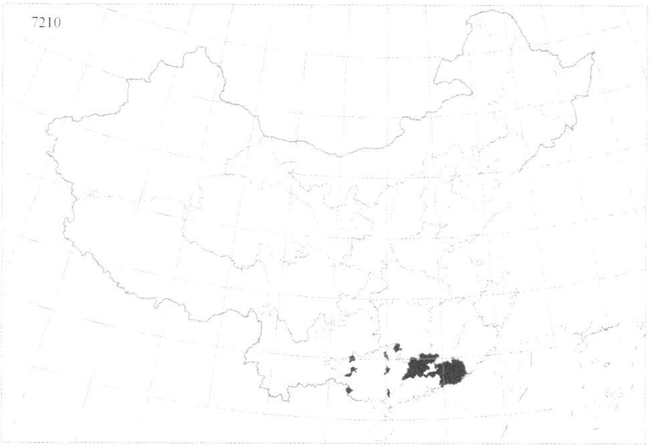

Alangiaceae 八角枫科 *Alangium* 八角枫属

Alangium platanifolium var. trilobum (Miq.) Ohwi 三裂瓜木

Shrubs, deciduous

13.9 (1.4~22.0); 1.2 (-19.2~15.3); 25.4 (9.9~30.1); 14.3 (3.8~22.0); 118.4 (15.7~203.8); 11.7 (0.0~88.8); 804 (404~1170) ❙ 1100 (367~2499); 475 (235~1045); 94 (8~303) ❙ 34.3 (-40.9~219.3); 762 (367~1170); 204.7 (101.9~346.5)

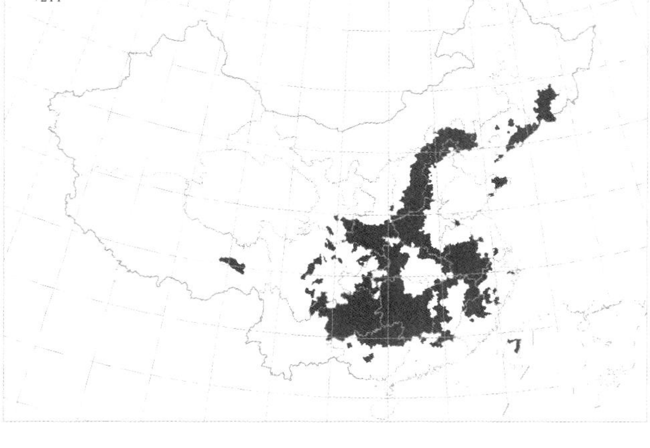

Alangiaceae 八角枫科 *Alangium* 八角枫属

Alangium qingchuanense M. Y. He 青川八角枫

Trees, deciduous

13.5; 2.9; 23.3; 13.5; 105.3; 3.0; 744 ❙ 806; 448; 16 ❙ 8.2; 744; 261.0

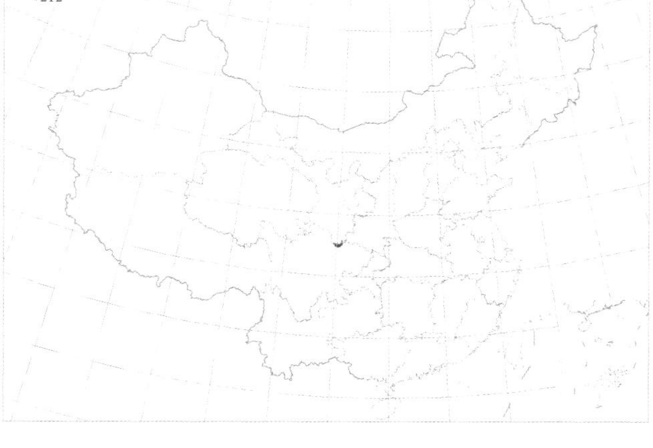

Alangiaceae 八角枫科 *Alangium* 八角枫属

Alangium salviifolium (L. f.) Wangerin 土坛树

Trees, deciduous

23.2 (21.3~24.8); 16.3 (13.2~20.1); 28.3 (27.2~29.2); 23.2 (21.3~24.8); 218.4 (195.9~237.6); 0.0 (0.0~0.0); 1251 (1111~1385) ❙ 1577 (1176~2035); 714 (444~1056); 84 (41~125) ❙ 27.0 (-12.0~71.8); 1196 (1091~1321); 312.3 (178.7~620.0)

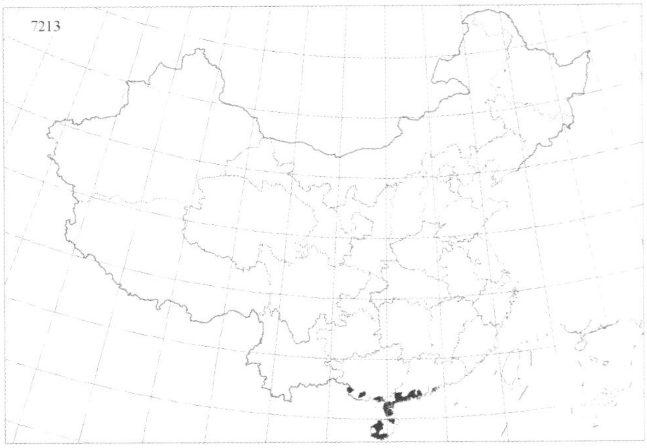

Alangiaceae 八角枫科 *Alangium* 八角枫属

Alangium yunnanense C. Y. Wu ex W. P. Fang et al. 云南八角枫

Shrubs or small trees, deciduous

17.1; 10.5; 21.6; 17.2; 146.1; 0.0; 831 ❙ 1144; 643; 48 ❙ 39.6; 789; 356.0

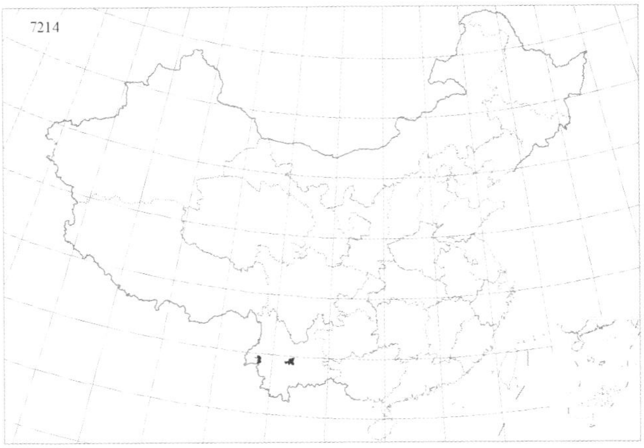

Combretaceae 使君子科 *Anogeissus* 榆绿木属

Anogeissus acuminata (Roxb. ex DC.) Guillaumin et al. 榆绿木

Trees, evergreen

19.1 (16.4~20.5); 13.3 (9.6~15.1); 22.6 (20.6~23.7); 19.1 (16.4~20.5); 169.7 (137.3~185.9); 0.0 (0.0~0.0); 919 (795~992) ❙ 1550 (1464~1688); 755 (655~959); 65 (58~77) ❙ 70.4 (47.6~112.2); 902 (795~986); 559.6 (427.9~659.0)

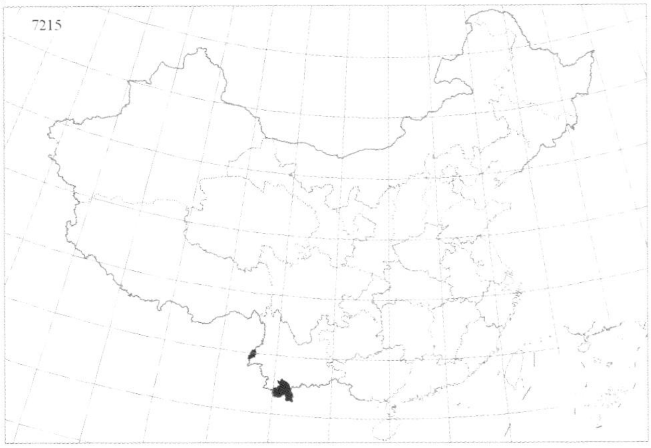

Combretaceae 使君子科 *Combretum* 风车藤属

Combretum alfredii Hance 风车子

Vines

20.3 (14.3~23.6); 11.3 (3.7~16.7); 27.9 (24.2~29.8); 20.4 (14.3~23.6); 184.7 (113.2~223.5); 0.0 (0.0~1.5); 1073 (761~1281) ❙ 1637 (1187~2037); 688 (411~1033); 153 (73~225) ❙ 54.0 (-6.9~128.9); 1071 (761~1273); 221.1 (95.3~327.7)

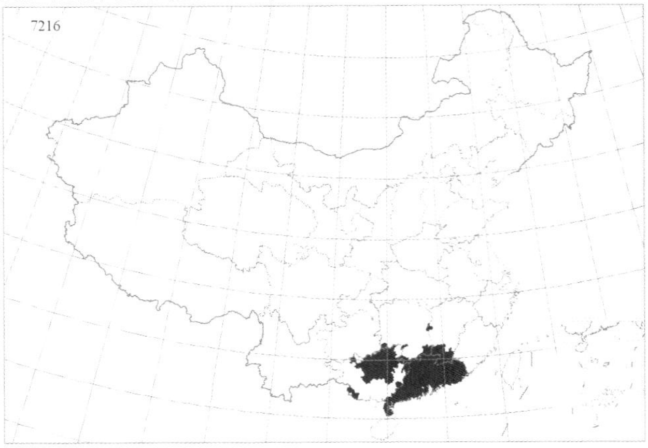

Combretaceae 使君子科 *Combretum* 风车藤属

Combretum griffithii Van Heurck et Müll. Arg. 西南风车子

Vines

18.6 (14.4~22.2); 12.1 (7.5~15.1); 22.8 (19.1~28.3); 18.6 (14.4~22.3); 163.7 (113.1~207.5); 0.0 (0.0~0.0); 912 (727~1190) ❙ 1405 (1049~1607); 761 (598~909); 59 (42~87) ❙ 56.2 (15.4~99.9); 900 (727~1190); 463.1 (327.7~659.0)

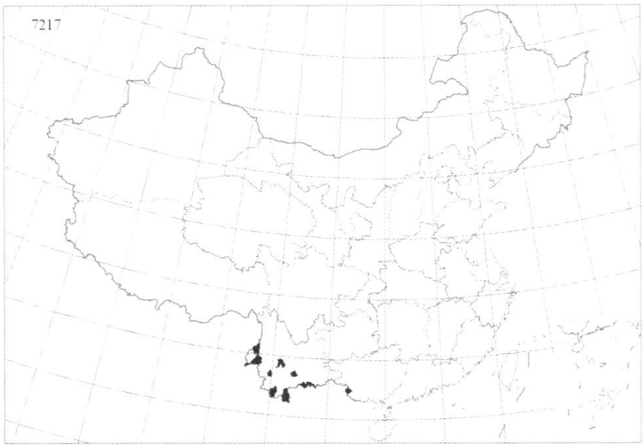

Combretaceae 使君子科 *Combretum* 风车藤属

Combretum griffithii var. yunnanense (Exell) Turland et C. Chen 云南风车子

Vines

19.0 (17.1~21.0); 13.0 (11.0~15.1); 22.7 (20.8~25.6); 19.0 (17.1~21.0); 167.9 (145.0~191.8); 0.0 (0.0~0.0); 913 (815~1053) ❙ 1430 (1049~1607); 751 (598~911); 57 (42~77) ❙ 56.7 (26.1~74.2); 893 (783~1053); 487.6 (363.2~659.0)

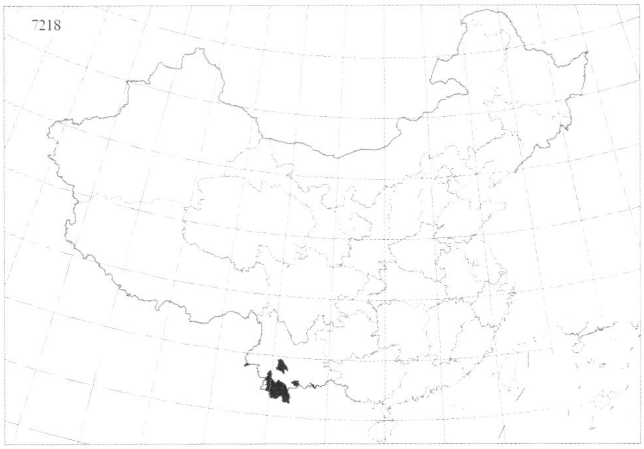

Combretaceae 使君子科 *Combretum* 风车藤属
Combretum latifolium Blume 阔叶风车子
Vines

19.6 (17.7~21.0); 13.7 (10.9~15.1); 23.4 (22.3~25.6); 19.7 (17.8~21.0); 176.0 (153.4~191.8); 0.0 (0.0~0.0); 951 (854~1053) ❙ 1492 (1368~1607); 732 (655~818); 65 (58~77) ❙ 57.4 (44.5~64.9); 937 (854~1053); 535.7 (427.9~659.0)

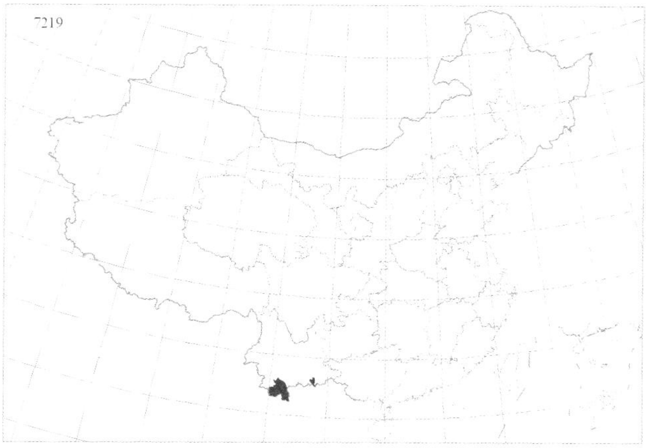

Combretaceae 使君子科 *Combretum* 风车藤属
Combretum pilosum Roxb. 长毛风车子
Vines

22.6 (17.4~24.8); 17.1 (11.0~20.1); 26.3 (21.4~28.7); 22.7 (17.4~24.8); 211.9 (149.3~237.6); 0.0 (0.0~0.0); 1198 (834~1385) ❙ 1373 (868~1607); 538 (312~816); 55 (41~77) ❙ 19.0 (-34.3~68.7); 1043 (834~1212); 492.9 (288.3~659.0)

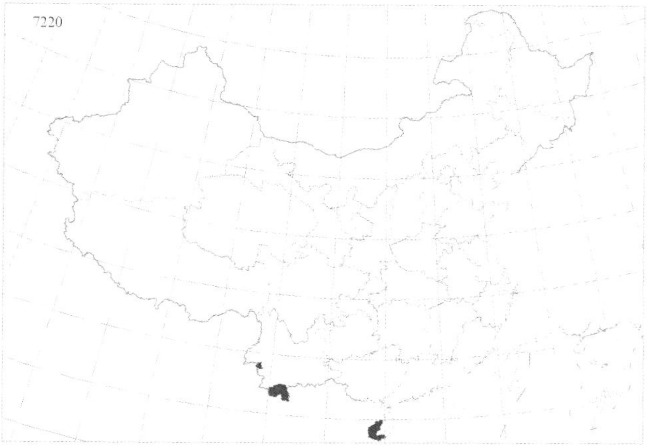

Combretaceae 使君子科 *Combretum* 风车藤属
Combretum punctatum Blume 盾鳞风车子
Vines

20.4 (14.4~24.8); 13.8 (7.5~20.1); 24.8 (19.1~28.6); 20.4 (14.4~24.8); 184.9 (113.1~237.6); 0.0 (0.0~0.0); 1043 (727~1385) ❙ 1590 (1269~2435); 818 (473~1428); 73 (41~130) ❙ 57.2 (-1.4~112.2); 1012 (727~1269); 450.3 (235.0~659.0)

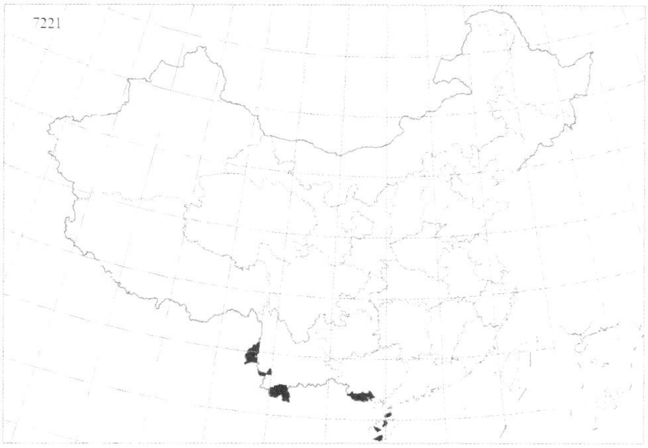

Combretaceae 使君子科 *Combretum* 风车藤属
Combretum punctatum var. squamosum (Roxb. ex G. Don) M. G. Gangopadhyay et Chatrabarty 水密花
Vines

21.6 (19.2~24.1); 15.9 (13.8~19.1); 25.2 (22.3~28.6); 21.7 (19.2~24.2); 199.9 (170.5~230.2); 0.0 (0.0~0.0); 1105 (909~1327) ❙ 1447 (1269~1607); 611 (480~702); 66 (41~88) ❙ 35.1 (-1.4~64.9); 1054 (880~1269); 497.1 (235.0~659.0)

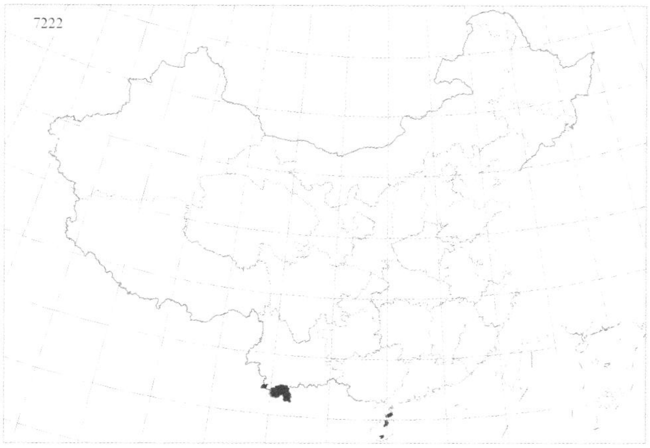

Combretaceae 使君子科 *Combretum* 风车藤属
Combretum roxburghii Sprengel 十蕊风车子
Vines

17.5 (14.4~20.3); 10.8 (7.5~13.9); 21.8 (19.1~27.7); 17.5 (14.4~20.4); 150.5 (113.1~184.4); 0.0 (0.0~0.0); 850 (727~1065) ❙ 1405 (1083~1688); 780 (594~959); 56 (41~141) ❙ 65.8 (34.7~112.2); 840 (727~1065); 447.0 (232.6~627.0)

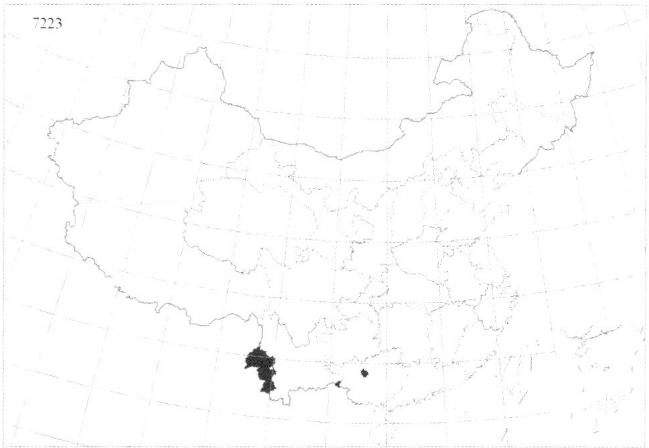

Combretaceae 使君子科 *Combretum* 风车藤属
Combretum sundaicum Miq. 榄形风车子
Vines

21.6 (18.6~24.8); 15.6 (12.7~20.1); 25.6 (22.3~28.3); 21.6 (18.7~24.8); 199.7 (163.9~237.6); 0.0 (0.0~0.0); 1114 (888~1385) ❙ 1487 (1374~1607); 665 (473~911); 66 (41~102) ❙ 37.0 (1.5~74.2); 1053 (880~1190); 515.9 (327.7~659.0)

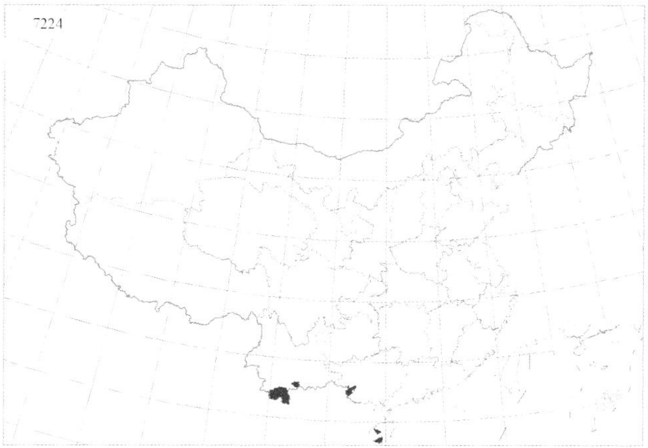

Combretaceae 使君子科 *Combretum* 风车藤属
Combretum wallichii DC. 石风车子
Vines, evergreen

17.3 (8.6~22.3); 8.7 (1.3~14.2); 24.1 (15.0~28.9); 17.3 (8.7~22.3); 147.8 (52.6~208.1); 0.2 (0.0~8.8); 886 (574~1202) ❙ 1306 (879~1625); 656 (463~850); 84 (28~191) ❙ 49.5 (5.8~101.8); 883 (574~1202); 303.7 (159.0~551.6)

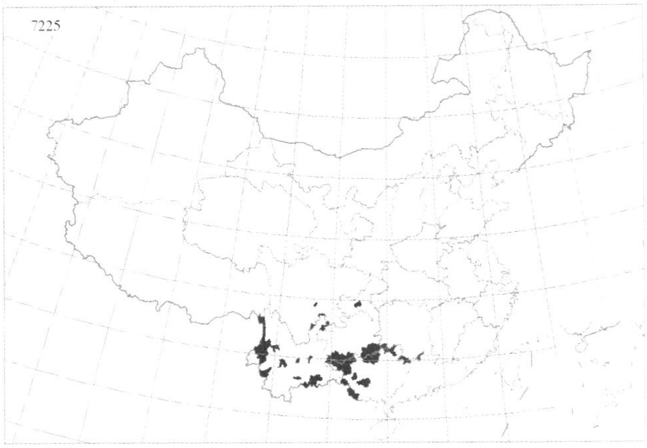

Combretaceae 使君子科 *Getonia* 萼翅藤属
Getonia floribunda Roxb. 萼翅藤
Vines, evergreen

16.4; 9.6; 20.6; 16.4; 137.3; 0.0; 795 ❙ 1688; 959; 62 ❙ 112.2; 795; 627.0

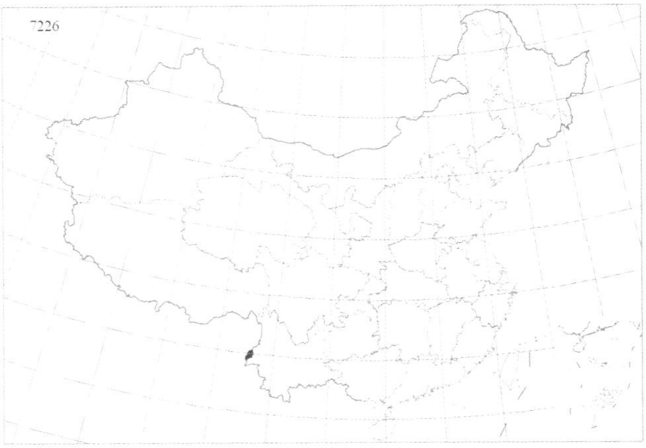

Combretaceae 使君子科 *Lumnitzera* 榄李属
Lumnitzera littorea (Jack) Voigt 红榄李
Trees, evergreen

24.7; 19.8; 27.9; 24.7; 236.5; 0.0; 1376 ❙ 1588; 534; 58 ❙ 15.4; 1212; 494.7

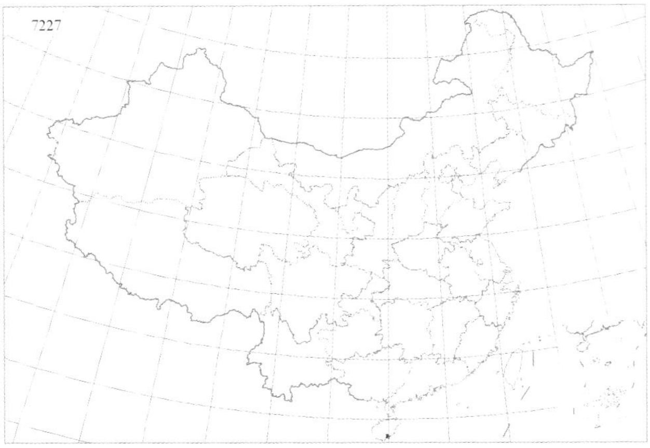

Combretaceae 使君子科 *Lumnitzera* 榄李属
Lumnitzera racemosa Willd. 榄李
Shrubs or small trees, evergreen

21.7 (16.2~24.8); 15.5 (11.0~20.1); 26.6 (20.1~29.0); 21.8 (16.2~24.8); 201.0 (134.2~237.6); 0.0 (0.0~0.0); 1138 (773~1385) ❙ 2002 (868~3495); 919 (312~1847); 174 (41~897) ❙ 87.8 (-34.3~246.5); 1079 (773~1331); 278.5 (119.4~527.5)

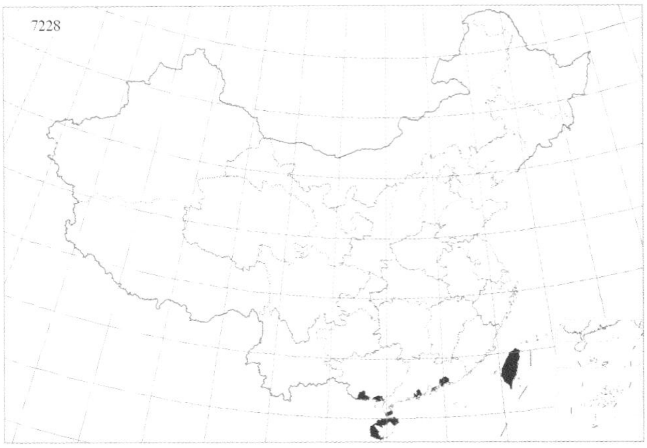

Combretaceae 使君子科 *Quisqualis* 使君子属
Quisqualis conferta (Jack) Exell 小花使君子
Vines

20.0; 14.6; 23.2; 20.1; 180.8; 0.0; 963 ❙ 1523; 681; 67 ❙ 58.4; 942; 581.0

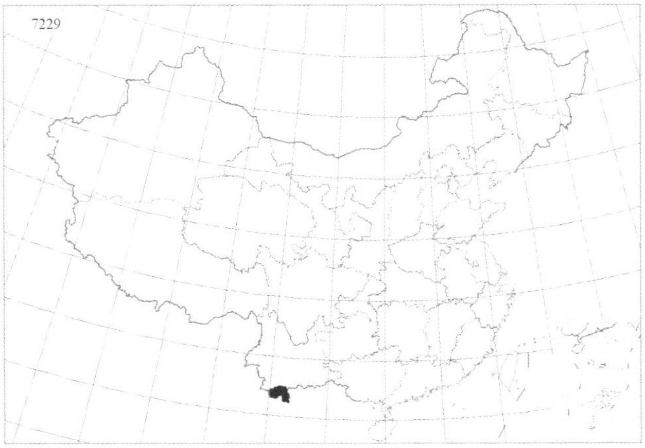

Combretaceae 使君子科 *Quisqualis* 使君子属
Quisqualis indica L. 使君子
Vines, evergreen

19.1 (3.0~24.8); 9.9 (-5.6~20.1); 26.9 (10.5~29.4); 19.1 (4.3~24.8); 169.8 (20.1~237.6); 0.2 (0.0~43.9); 998 (423~1385) ❙ 1483 (628~2435); 636 (312~1428); 131 (11~254) ❙ 50.2 (-34.3~130.7); 988 (423~1360); 253.4 (108.2~659.0)

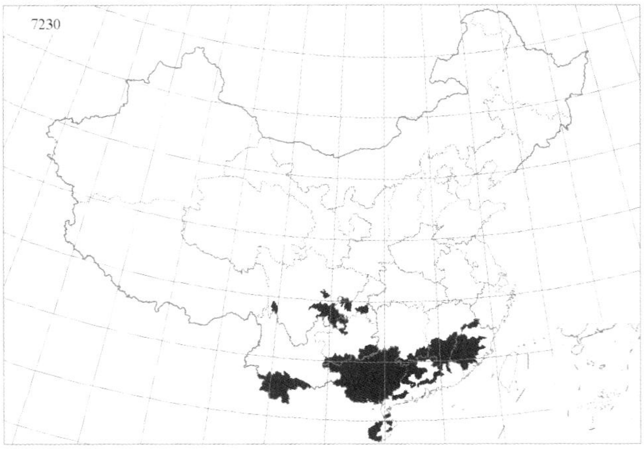

Combretaceae 使君子科 *Terminalia* 诃子属
Terminalia bellirica (Gaertn.) Roxb. 毗黎勒
Trees, deciduous

18.6 (16.3~21.0); 12.6 (10.2~15.1); 22.4 (20.3~25.6); 18.6 (16.4~21.0); 163.8 (136.6~191.8); 0.0 (0.0~0.0); 894 (789~1053) ❙ 1426 (1163~1751); 765 (641~996); 55 (41~77) ❙ 59.7 (34.8~88.8); 881 (780~1053); 491.9 (372.4~659.0)

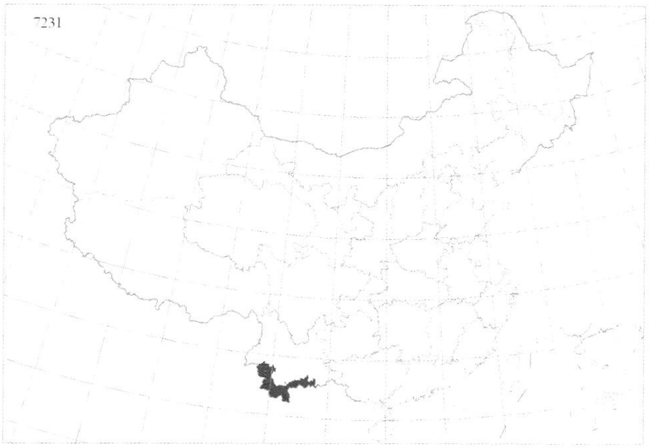

Combretaceae 使君子科 *Terminalia* 诃子属
Terminalia catappa L. 榄仁树
Trees, deciduous

22.0 (16.2~24.8); 16.0 (9.1~20.1); 26.7 (20.1~29.2); 22.1 (16.2~24.8); 204.6 (134.2~237.6); 0.0 (0.0~0.0); 1161 (773~1385) ❙ 1909 (868~3495); 830 (312~1847); 156 (41~897) ❙ 75.5 (-34.3~246.5); 1097 (773~1360); 323.8 (119.4~620.0)

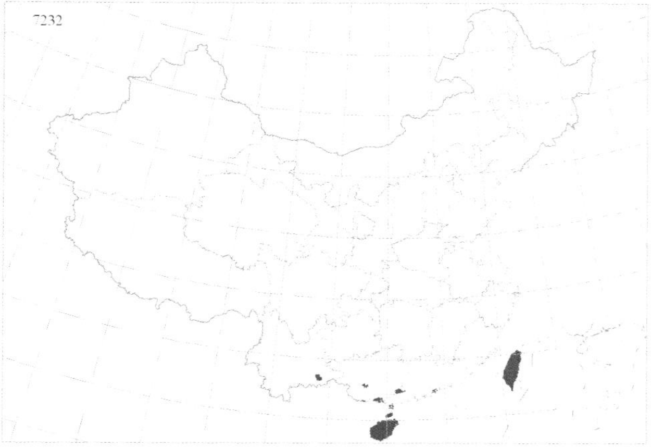

Combretaceae 使君子科 *Terminalia* 诃子属
Terminalia chebula Retz. 诃子
Trees, deciduous

17.6 (14.4~19.6); 11.1 (7.5~13.3); 21.7 (19.1~23.5); 17.6 (14.4~19.6); 151.4 (113.1~175.7); 0.0 (0.0~0.0); 850 (727~953) ❙ 1359 (1049~1559); 781 (598~916); 50 (42~70) ❙ 60.3 (26.1~99.9); 839 (727~932); 449.0 (365.3~516.7)

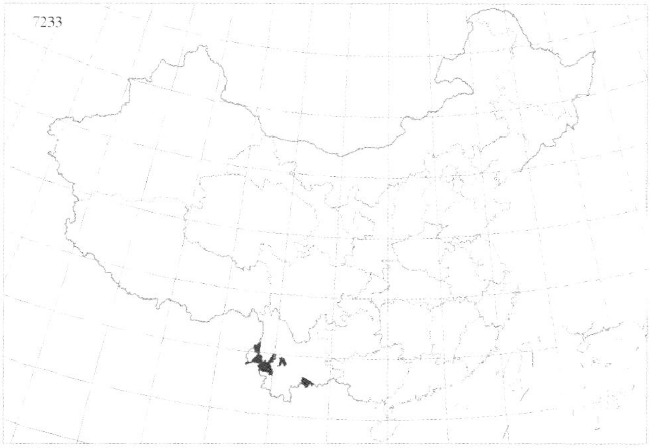

Combretaceae 使君子科 *Terminalia* 诃子属
Terminalia chebula var. tomentella (Kurz) C. B. Clarke 微毛诃子
Trees, deciduous

17.5 (16.0~19.5); 11.0 (9.3~12.7); 21.5 (20.4~23.3); 17.5 (16.0~19.5); 150.3 (132.4~173.8); 0.0 (0.0~0.0); 842 (780~942) ❙ 1384 (1275~1559); 795 (728~907); 48 (42~56) ❙ 64.6 (57.3~72.8); 832 (780~921); 447.8 (395.4~516.7)

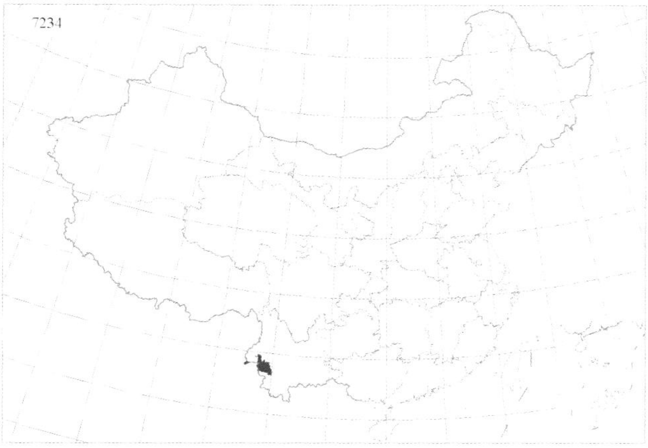

Combretaceae 使君子科 *Terminalia* 诃子属
Terminalia franchetii Gagnep. 滇榄仁
Shrubs or trees, deciduous

14.6 (5.4~21.6); 7.2 (-2.1~14.2); 20.4 (12.3~28.5); 14.6 (5.8~21.7); 117.0 (31.5~199.8); 1.6 (0.0~25.6); 757 (481~1153) ❙ 1050 (741~1625); 600 (411~911); 40 (12~173) ❙ 40.5 (-14.2~99.9); 734 (481~1153); 308.5 (142.9~551.6)

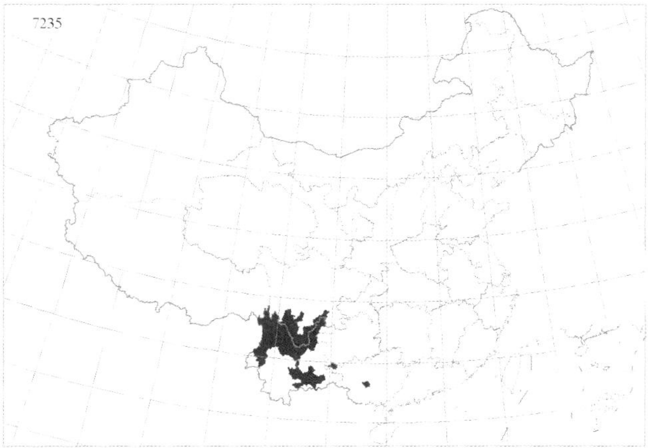

Combretaceae 使君子科 *Terminalia* 诃子属
Terminalia franchetii var. intricata (Hand.-Mazz.) Turland et C. Chen 错枝榄仁
Shrubs, deciduous

6.4 (2.7~11.6); -1.5 (-6.0~3.9); 13.3 (10.4~18.6); 7.0 (4.2~11.5); 41.3 (18.9~80.9); 24.3 (1.3~46.9); 512 (419~654) ❙ 770 (535~1007); 463 (326~594); 22 (10~40) ❙ 49.2 (27.1~67.9); 512 (419~654); 265.7 (201.9~351.7)

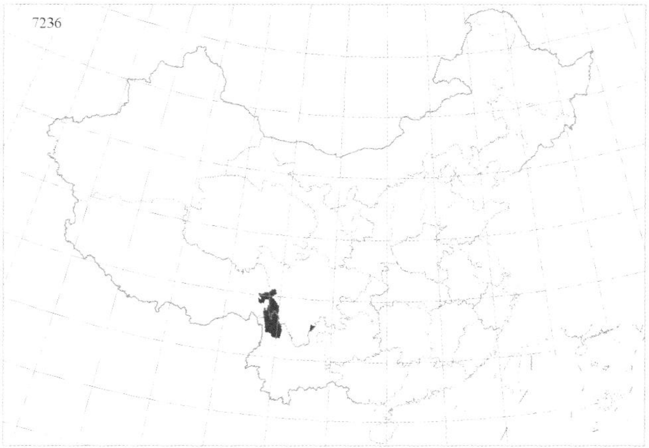

Combretaceae 使君子科 *Terminalia* 诃子属
Terminalia myriocarpa Van Heurck et Müll. Arg. 千果榄仁
Trees, evergreen

18.1 (13.2~22.5); 11.6 (5.4~15.1); 22.4 (18.4~28.7); 18.1 (13.3~22.6); 157.5 (99.4~210.9); 0.0 (0.0~0.0); 881 (695~1206) | 1352 (939~1758); 746 (524~1007); 55 (38~97) | 54.8 (6.5~145.8); 866 (695~1206); 424.1 (139.6~659.0)

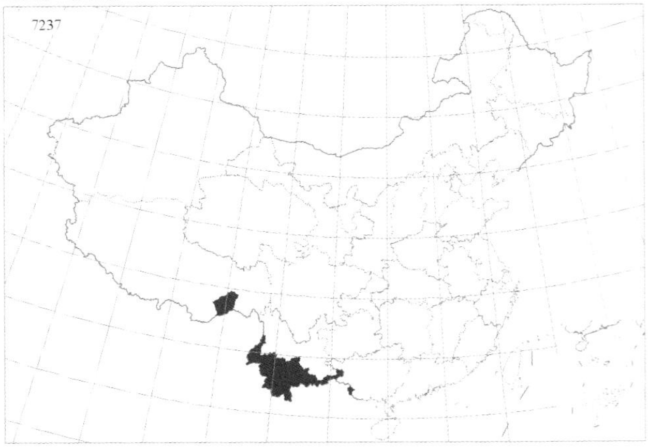

Combretaceae 使君子科 *Terminalia* 诃子属
Terminalia myriocarpa var. hirsuta Craib 硬毛千果榄仁
Trees, evergreen

13.2; 6.4; 18.4; 13.3; 99.4; 0.0; 695 | 1337; 727; 69 | 92.4; 695; 407.1

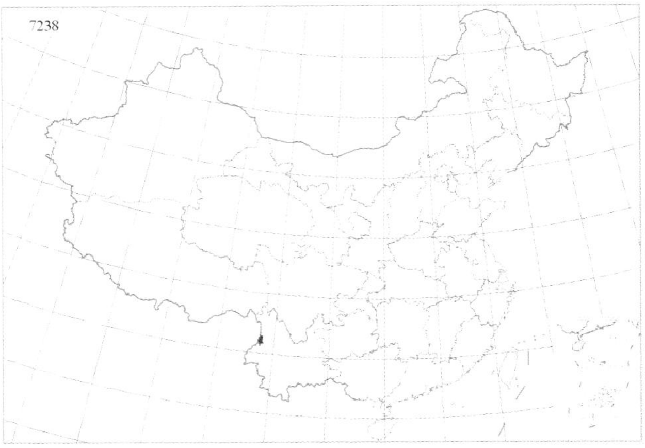

Combretaceae 使君子科 *Terminalia* 诃子属
Terminalia nigrovenulosa Pierre ex Lanessen 海南榄仁
Shrubs or trees

24.0 (22.2~24.8); 18.4 (17.2~20.1); 28.0 (25.7~28.8); 24.0 (22.3~24.8); 228.3 (207.2~237.6); 0.0 (0.0~0.0); 1310 (1127~1385) | 1501 (868~1854); 534 (312~712); 74 (41~115) | 15.0 (-34.3~47.8); 1191 (868~1360); 440.3 (248.2~620.0)

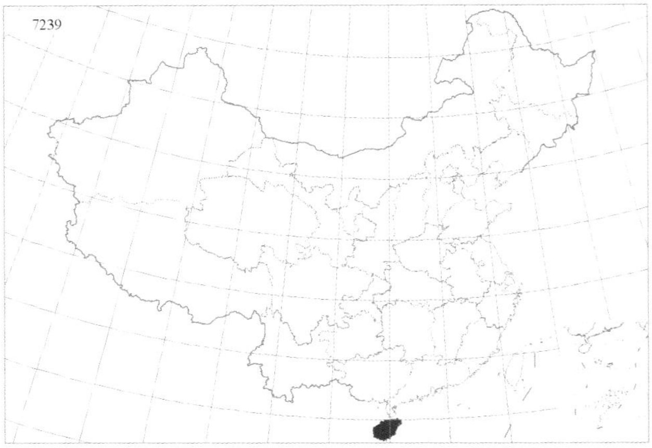

Myrtaceae 桃金娘科 *Baeckea* 岗松属
Baeckea frutescens L. 岗松
Shrubs or small trees, evergreen

21.0 (14.6~24.8); 12.5 (3.8~20.1); 28.0 (24.6~29.8); 21.1 (14.7~24.8); 192.8 (117.7~237.6); 0.0 (0.0~1.5); 1114 (779~1385) | 1612 (868~2435); 687 (312~1428); 141 (41~218) | 46.5 (-34.3~101.8); 1100 (779~1360); 247.5 (95.3~620.0)

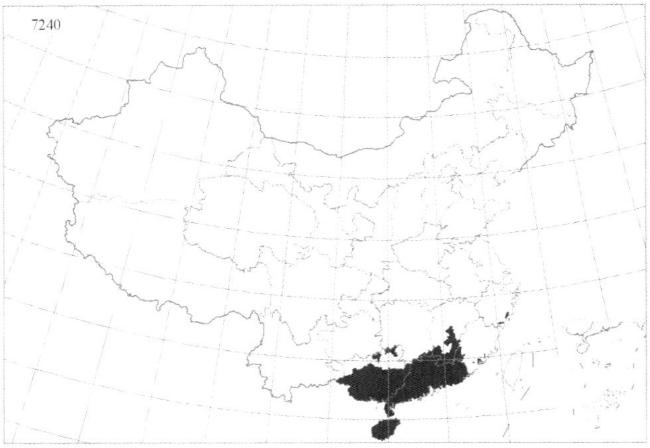

Myrtaceae 桃金娘科 *Decaspermum* 子楝树属
Decaspermum albociliatum Merr. et L. M. Perry 白毛子楝树
Shrubs, evergreen

24.4; 19.5; 27.8; 24.5; 233.5; 0.0; 1350 | 1635; 547; 65 | 21.1; 1219; 554.2

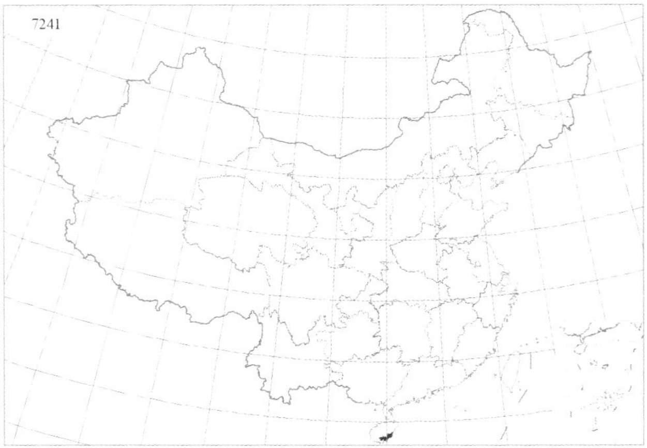

Myrtaceae 桃金娘科 *Decaspermum* 子楝树属
Decaspermum austrohainanicum H. T. Chang et R. H. Miao 琼南子楝树
Shrubs, evergreen

24.8; 20.1; 28.0; 24.8; 237.6; 0.0; 1385 | 1405; 473; 41 | 1.5; 1137; 527.5

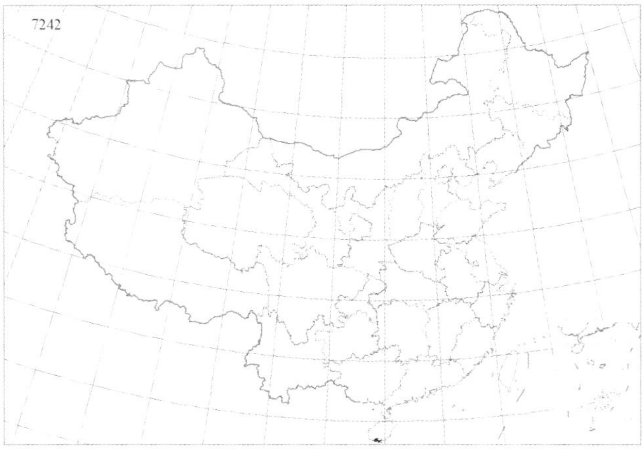

M

Myrtaceae 桃金娘科 *Decaspermum* 子楝树属
Decaspermum glabrum Chang et Miau 秃子楝树
Shrubs, evergreen

23.4; 16.1; 29.2; 23.5; 221.7; 0.0; 1273 ▌1425; 681; 82 ▌11.9; 1273; 178.7

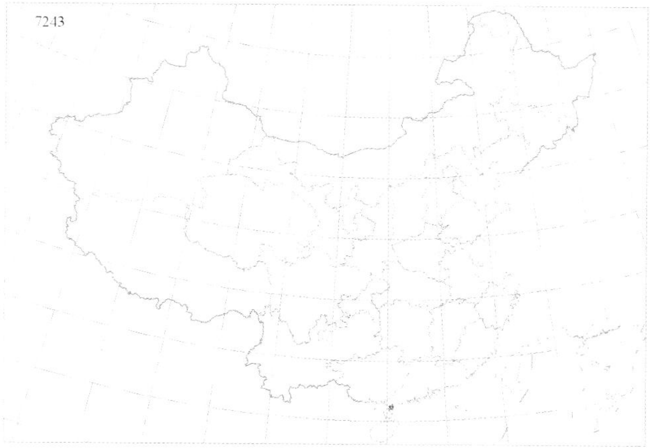

Myrtaceae 桃金娘科 *Decaspermum* 子楝树属
Decaspermum gracilentum (Hance) Merr. et L. M. Perry 子楝树
Shrubs or small trees, evergreen

21.1 (14.6~24.8); 12.9 (3.8~20.1); 27.7 (21.7~29.2); 21.1 (14.7~24.8); 193.7 (117.7~237.6); 0.0 (0.0~1.5); 1117 (779~1385) ▌1605 (868~2992); 722 (312~1847); 120 (41~195) ▌45.6 (-34.3~214.4); 1101 (779~1360); 257.6 (95.3~620.0)

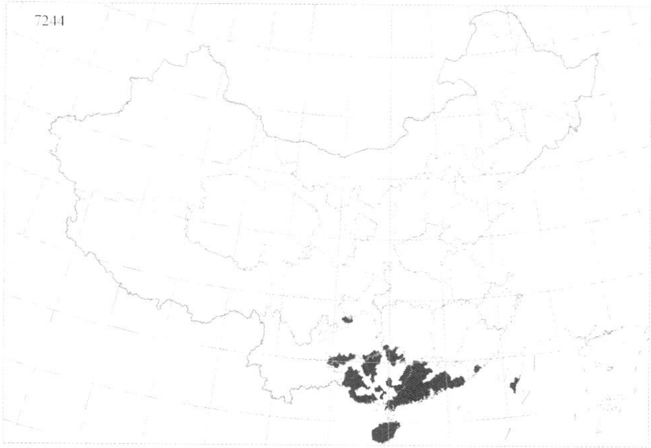

Myrtaceae 桃金娘科 *Decaspermum* 子楝树属
Decaspermum hainanense (Merr.) Merr. 海南子楝树
Trees, evergreen

23.9 (22.2~24.8); 18.9 (17.2~20.1); 27.3 (25.7~28.2); 23.9 (22.3~24.8); 227.0 (207.2~237.6); 0.0 (0.0~0.0); 1296 (1127~1385) ▌1583 (1380~1767); 539 (455~610); 58 (41~88) ▌23.0 (1.5~47.8); 1161 (1055~1299); 544.2 (455.2~620.0)

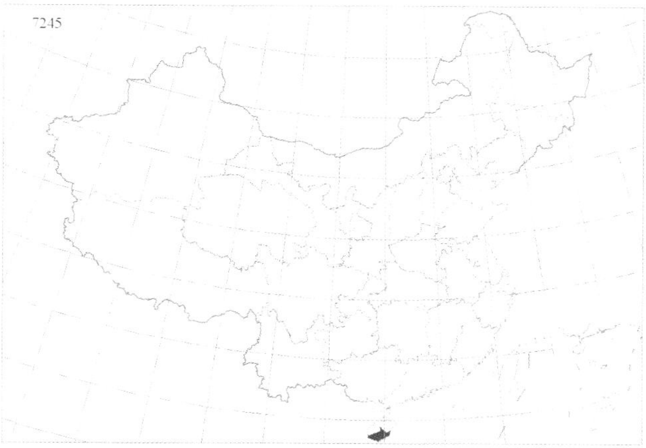

Myrtaceae 桃金娘科 *Decaspermum* 子楝树属
Decaspermum montanum Ridl. 柬埔寨子楝树
Trees, evergreen

24.0 (22.2~24.8); 18.4 (17.2~20.1); 28.0 (25.7~28.8); 24.0 (22.3~24.8); 228.3 (207.2~237.6); 0.0 (0.0~0.0); 1310 (1127~1385) ▌1501 (868~1854); 534 (312~712); 74 (41~115) ▌15.0 (-34.3~47.8); 1191 (868~1360); 440.3 (248.2~620.0)

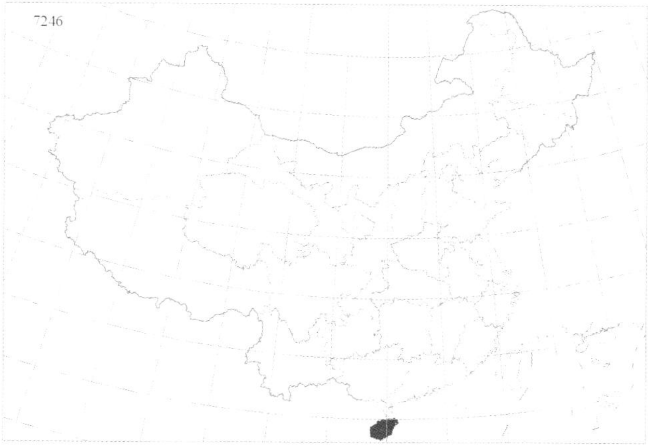

Myrtaceae 桃金娘科 *Decaspermum* 子楝树属
Decaspermum parviflorum (Lam.) A. J. Scott 五瓣子楝树
Trees, evergreen

19.3 (15.3~24.8); 12.2 (5.2~20.1); 24.4 (20.3~28.6); 19.3 (15.3~24.8); 171.6 (124.2~237.6); 0.0 (0.0~0.0); 969 (779~1385) ▌1423 (1014~2435); 719 (455~1428); 75 (41~189) ▌48.0 (1.5~99.5); 941 (751~1221); 383.3 (149.6~659.0)

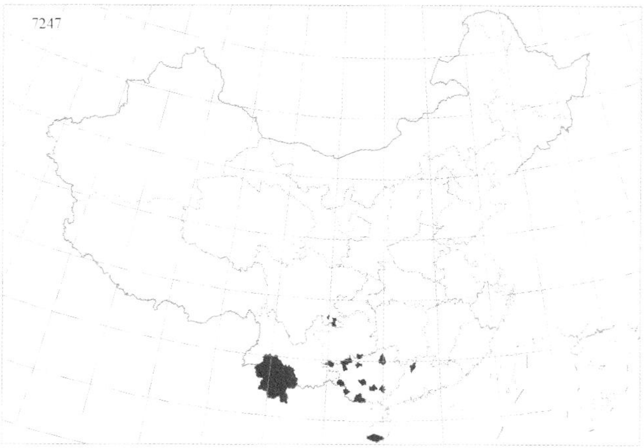

Myrtaceae 桃金娘科 *Decaspermum* 子楝树属
Decaspermum teretis (H. T. Chang et R. H. Miao) Craven 圆枝子楝树
Trees, evergreen

24.8; 20.1; 28.0; 24.8; 237.6; 0.0; 1385 ▌1405; 473; 41 ▌1.5; 1137; 527.5

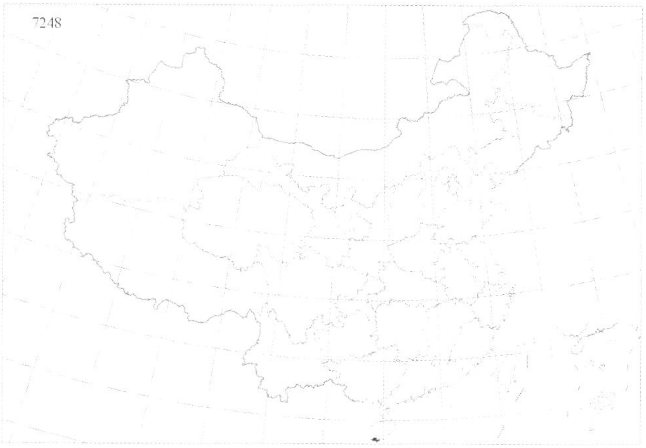

Myrtaceae 桃金娘科 *Psidium* 番石榴属
Psidium guajava L. 番石榴
Trees

19.1 (10.4~24.8); 11.9 (3.4~20.1); 24.4 (16.3~29.0); 19.2 (10.5~24.8); 169.9 (68.0~237.6); 0.0 (0.0~2.6); 978 (623~1385) ❙ 1339 (797~2435); 688 (312~1428); 70 (15~178) ❙ 38.6 (-34.3~112.2); 949 (615~1360); 347.8 (160.7~659.0)

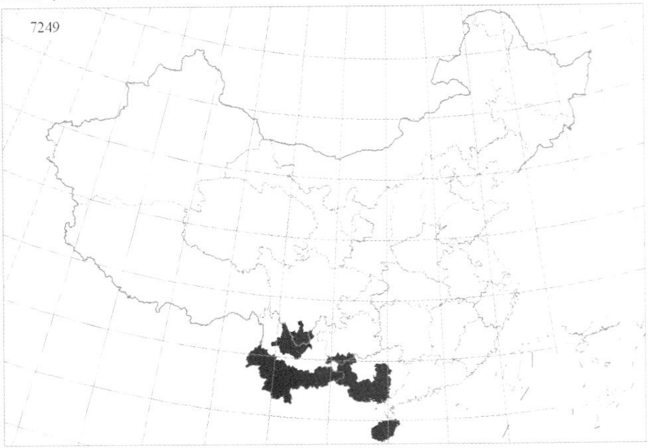

Myrtaceae 桃金娘科 *Rhodamnia* 玫瑰木属
Rhodamnia dumetorum (Poir.) Merr. et L. M. Perry 玫瑰木
Small trees

23.9 (21.9~24.8); 18.1 (13.2~20.1); 28.0 (25.7~28.8); 23.9 (22.0~24.8); 227.1 (203.8~237.6); 0.0 (0.0~0.0); 1303 (1127~1385) ❙ 1502 (868~1854); 546 (312~773); 77 (41~138) ❙ 15.7 (-34.3~47.8); 1191 (868~1360); 431.9 (248.2~620.0)

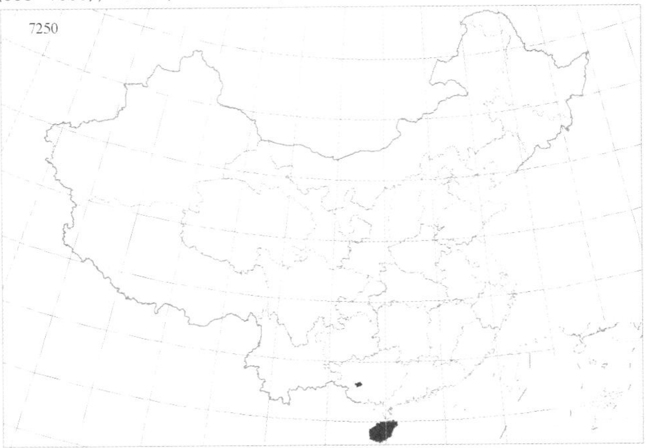

Myrtaceae 桃金娘科 *Rhodamnia* 玫瑰木属
Rhodamnia dumetorum var. hainanensis Merr. et L. M. Perry 海南玫瑰木
Trees

24.2; 19.3; 27.6; 24.2; 230.9; 0.0; 1327 ❙ 1445; 489; 46 ❙ 9.1; 1132; 534.2

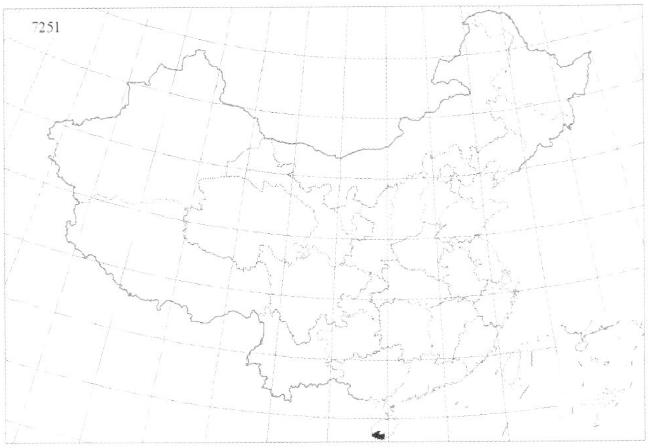

Myrtaceae 桃金娘科 *Rhodomyrtus* 桃金娘属
Rhodomyrtus tomentosa (Aiton) Hassk. 桃金娘
Shrubs

20.3 (14.3~24.8); 11.7 (3.7~20.1); 27.5 (20.1~30.1); 20.3 (14.3~24.8); 183.9 (113.2~237.6); 0.0 (0.0~1.5); 1065 (761~1385) ❙ 1621 (868~3495); 685 (312~1847); 150 (41~897) ❙ 54.9 (-34.3~246.5); 1050 (761~1360); 247.4 (95.3~620.0)

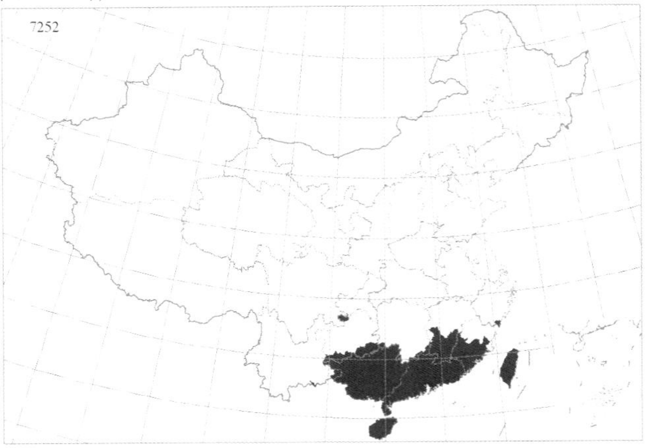

Myrtaceae 桃金娘科 *Syzygium* 蒲桃属
Syzygium acuminatissimum (Blume) DC. 肖蒲桃
Trees, evergreen

22.2 (18.5~24.8); 14.7 (10.6~20.1); 28.1 (21.7~29.1); 22.2 (18.5~24.8); 207.0 (162.1~237.6); 0.0 (0.0~0.0); 1183 (863~1385) ❙ 1704 (868~2713); 762 (312~1428); 117 (41~191) ❙ 45.8 (-34.3~214.4); 1158 (863~1360); 268.9 (95.3~620.0)

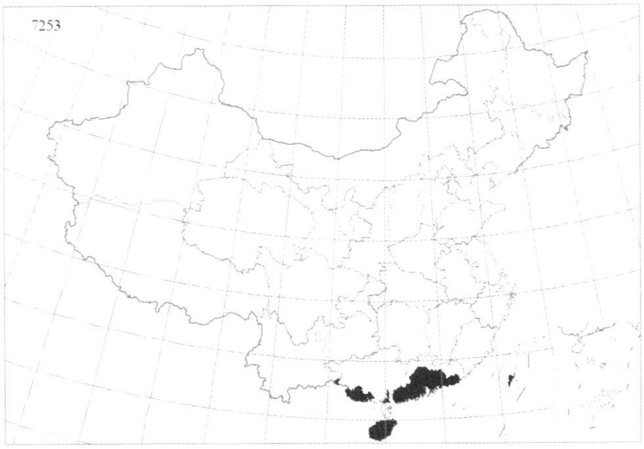

Myrtaceae 桃金娘科 *Syzygium* 蒲桃属
Syzygium album Q. F. Zheng 白果蒲桃
Trees, evergreen

20.4; 12.6; 27.4; 20.4; 184.9; 0.0; 1046 ❙ 1375; 476; 133 ❙ 31.5; 1046; 233.8

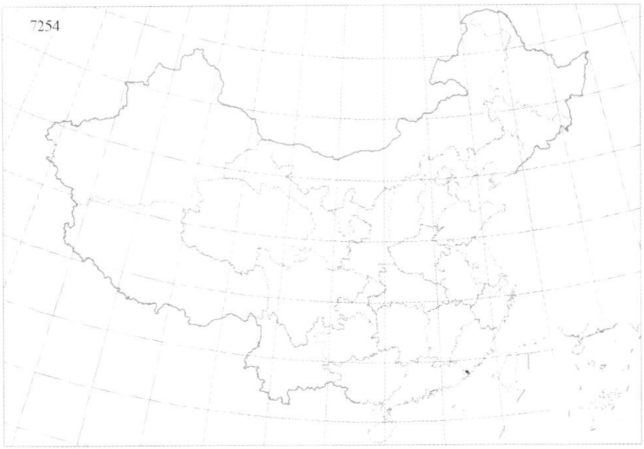

Myrtaceae 桃金娘科 *Syzygium* 蒲桃属

Syzygium araiocladum Merr. et L. M. Perry 线枝蒲桃

Small trees, evergreen

22.8 (21.3~24.7); 15.9 (12.9~19.8); 27.9 (25.7~28.7); 22.9 (21.3~24.7); 214.3 (195.9~236.5); 0.0 (0.0~0.0); 1226 (1111~1376) ❙ 1551 (868~2435); 727 (312~1428); 88 (42~145) ❙ 28.1 (-34.3~99.5); 1132 (868~1221); 364.4 (208.5~620.0)

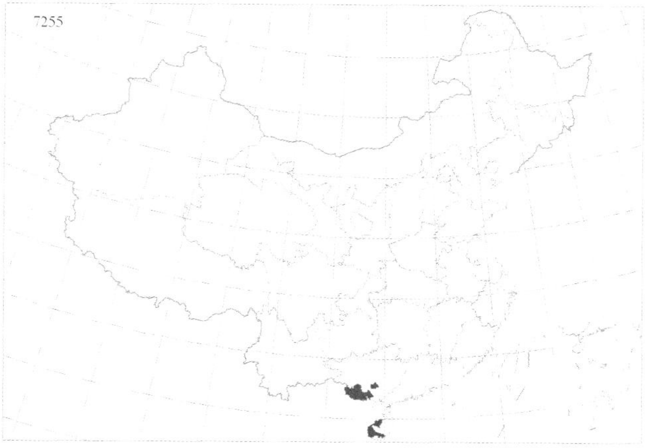

Myrtaceae 桃金娘科 *Syzygium* 蒲桃属

Syzygium austrosinense (Merr. et L. M. Perry) Chang et Miau 华南蒲桃

Shrubs or small trees, evergreen

19.8 (10.6~24.8); 10.8 (-0.2~20.1); 27.5 (21.1~29.8); 19.8 (10.6~24.8); 177.9 (79.4~237.6); 0.2 (0.0~12.1); 1047 (643~1385) ❙ 1558 (868~2435); 639 (312~1428); 131 (41~256) ❙ 51.1 (-34.3~130.7); 1036 (643~1360); 238.3 (95.3~620.0)

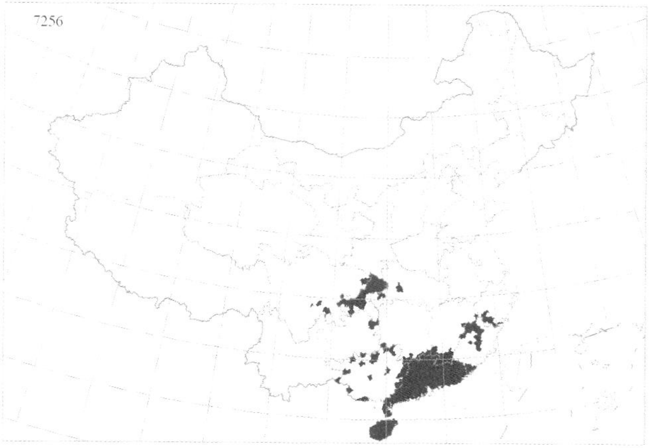

Myrtaceae 桃金娘科 *Syzygium* 蒲桃属

Syzygium austroyunnanense H. T. Chang et R. H. Miao 滇南蒲桃

Trees, evergreen

20.5; 14.2; 24.5; 20.5; 186.0; 0.0; 1013 ❙ 1522; 694; 87 ❙ 51 5; 997; 487.9

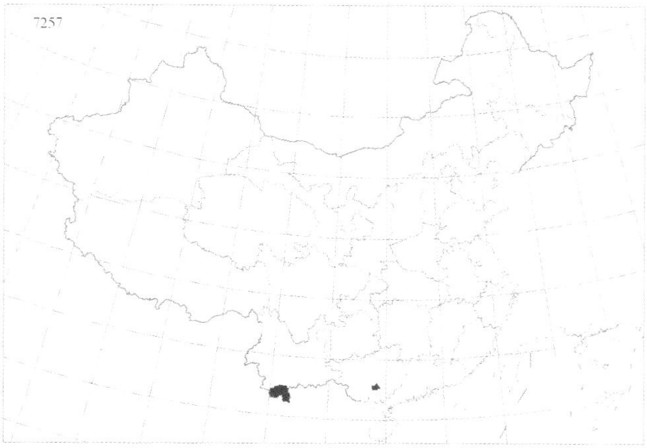

Myrtaceae 桃金娘科 *Syzygium* 蒲桃属

Syzygium balsameum (Wight) Wall. ex Walp. 香胶蒲桃

Shrubs or small trees, evergreen

18.9 (13.3~21.0); 12.7 (5.4~15.1); 23.0 (19.6~25.6); 18.9 (13.4~21.0); 166.9 (100.4~191.8); 0.0 (0.0~0.0); 930 (706~1053) ❙ 1565 (1464~1736); 774 (655~1007); 64 (51~77) ❙ 73.1 (44.5~145.8); 917 (706~1053); 548.4 (447.4~659.0)

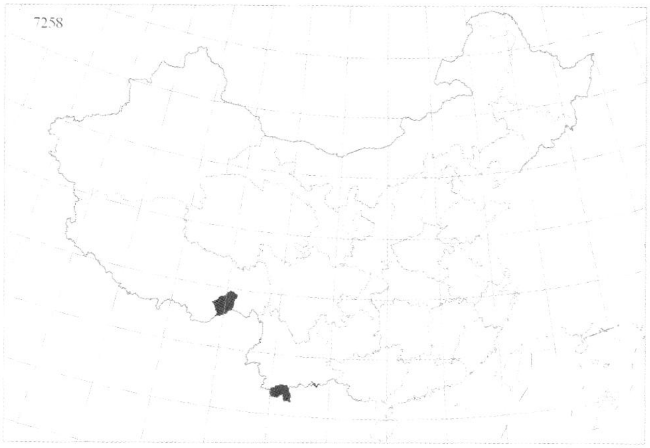

Myrtaceae 桃金娘科 *Syzygium* 蒲桃属

Syzygium baviense (Gagnep.) Merr. et L. M. Perry 短棒蒲桃

Shrubs or small trees, evergreen

21.0; 14.2; 25.6; 21.0; 191.8; 0.0; 1053 ❙ 1522; 818; 68 ❙ 44.5; 1053; 551.6

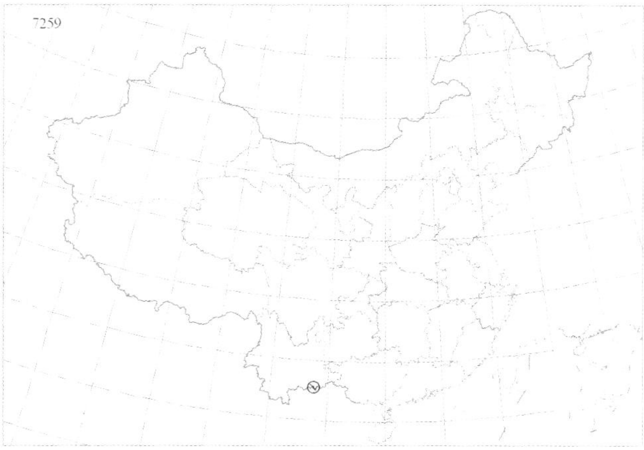

Myrtaceae 桃金娘科 *Syzygium* 蒲桃属

Syzygium boisianum (Gagnep.) Merr. et L. M. Perry 无柄蒲桃

Shrubs, evergreen

23.9 (22.2~24.8); 18.6 (17.2~20.1); 27.6 (25.7~28.5); 23.9 (22.3~24.8); 227.0 (207.2~237.6); 0.0 (0.0~0.0); 1298 (1127~1385) ❙ 1531 (1179~1854); 525 (392~622); 60 (41~115) ❙ 18.5 (-11.9~47.8); 1161 (1055~1360); 515.5 (428.6~620.0)

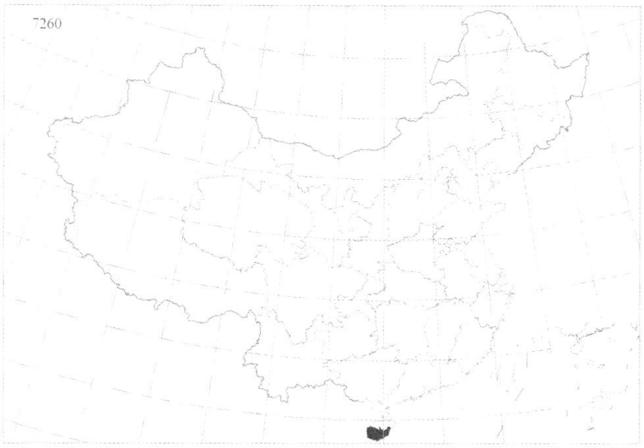

Myrtaceae 桃金娘科 *Syzygium* 蒲桃属

Syzygium brachythyrsum Merr. et L. M. Perry 短序蒲桃

Small trees, evergreen

20.5 (16.8~22.7); 12.8 (9.5~15.0); 26.0 (21.3~28.6); 20.5 (16.9~22.8); 186.1 (142.3~213.1); 0.0 (0.0~0.0); 1055 (815~1221) ❙ 1576 (984~2435); 855 (562~1428); 92 (43~145) ❙ 49.7 (12.9~99.5); 1052 (788~1221); 343.2 (208.5~570.1)

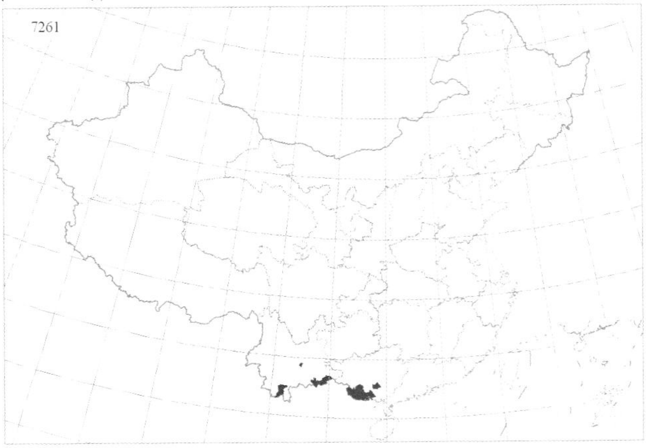

Myrtaceae 桃金娘科 *Syzygium* 蒲桃属

Syzygium bullockii (Hance) Merr. et L. M. Perry 黑嘴蒲桃

Shrubs or small trees, evergreen

22.2 (17.6~24.8); 14.6 (7.8~20.1); 28.2 (25.7~29.2); 22.2 (17.7~24.8); 207.0 (152.5~237.6); 0.0 (0.0~0.0); 1186 (906~1385) ❙ 1653 (868~2435); 732 (312~1428); 116 (41~188) ❙ 40.7 (-34.3~99.5); 1161 (868~1360); 259.4 (95.3~620.0)

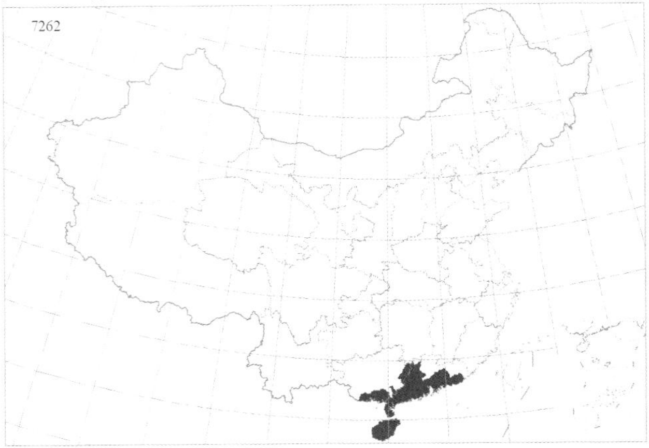

Myrtaceae 桃金娘科 *Syzygium* 蒲桃属

Syzygium buxifolioideum H. T. Chang et R. H. Miao 假赤楠

Shrubs or small trees, evergreen

24.0; 19.1; 27.2; 24.0; 228.1; 0.0; 1302 ❙ 1551; 539; 48 ❙ 19.1; 1147; 620.0

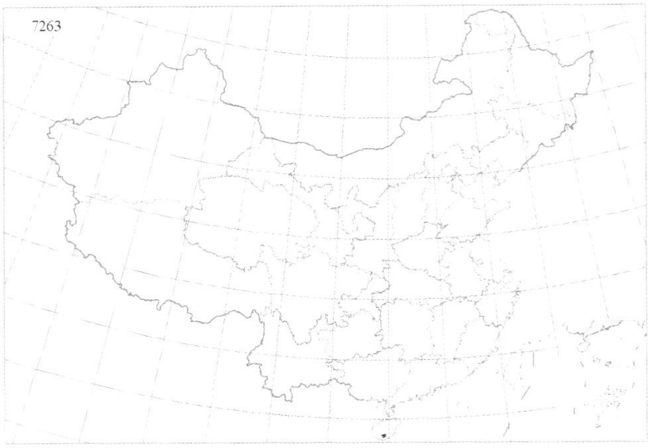

Myrtaceae 桃金娘科 *Syzygium* 蒲桃属

Syzygium buxifolium (Hook. et Arn.) Meyen et Schauer 赤楠

Shrubs or small trees, evergreen

18.2 (11.7~24.8); 8.2 (-0.8~20.1); 27.3 (19.6~30.1); 18.2 (11.7~24.8); 159.1 (84.6~237.6); 0.3 (0.0~13.4); 965 (661~1385) ❙ 1517 (798~3495); 591 (299~1847); 157 (27~897) ❙ 59.0 (-34.3~246.5); 953 (661~1360); 229.5 (95.3~620.0)

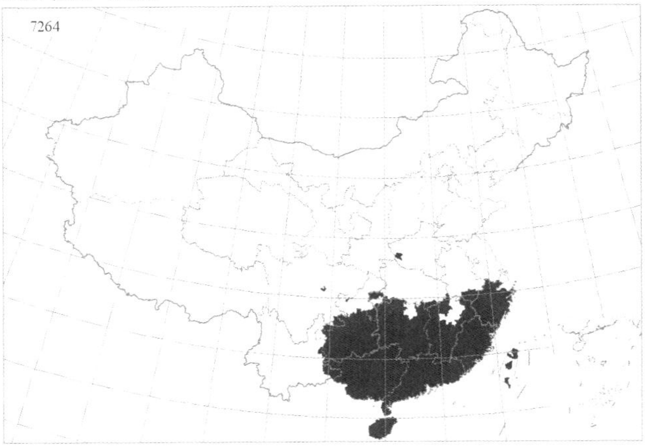

Myrtaceae 桃金娘科 *Syzygium* 蒲桃属

Syzygium cathayense Merr. et L. M. Perry 华夏蒲桃

Small trees, evergreen

17.0; 10.1; 22.1; 17.0; 146.9; 2.9; 904 ❙ 1599; 840; 82 ❙ 76.9; 903; 443.4

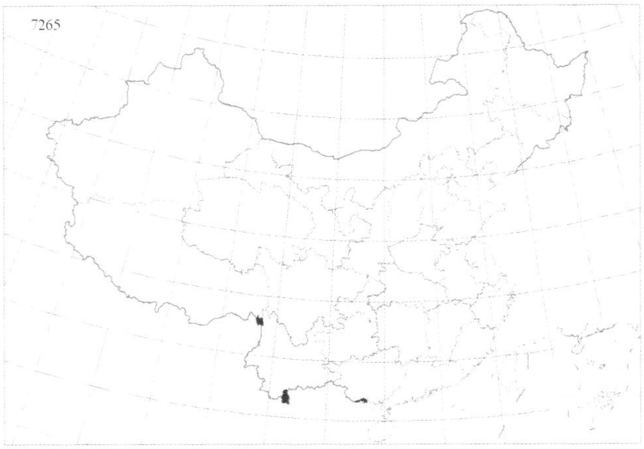

Myrtaceae 桃金娘科 *Syzygium* 蒲桃属

Syzygium championii (Benth.) Merr. et L. M. Perry 子凌蒲桃

Shrubs or trees, evergreen

21.8 (17.8~24.8); 14.0 (7.1~20.1); 28.0 (25.2~28.9); 21.9 (17.8~24.8); 202.8 (154.0~237.6); 0.0 (0.0~0.0); 1167 (913~1385) ❙ 1631 (868~2435); 714 (312~1428); 125 (41~194) ❙ 41.6 (-34.3~99.5); 1133 (868~1360); 295.9 (139.6~620.0)

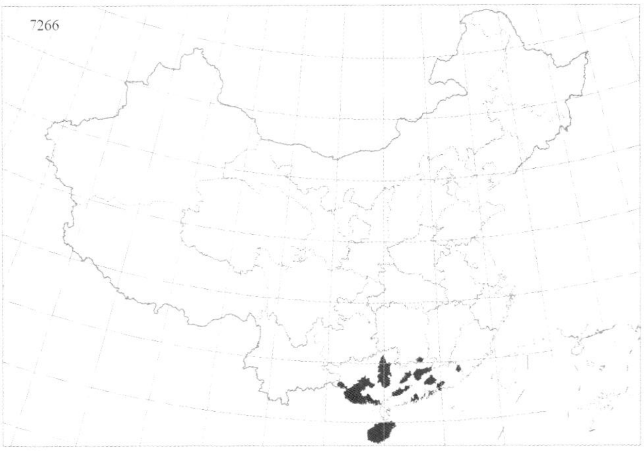

Myrtaceae 桃金娘科 *Syzygium* 蒲桃属
Syzygium chunianum Merr. et L. M. Perry 密脉蒲桃
Trees, evergreen

22.5 (21.3~24.6); 14.9 (12.4~19.5); 28.2 (26.7~29.0); 22.5 (21.3~24.7); 210.3 (195.7~235.9); 0.0 (0.0~0.0); 1206 (1111~1372) ❙ 1610 (1242~2435); 762 (444~1428); 112 (44~171) ❙ 33.9 (-2.2~99.5); 1186 (1091~1329); 306.2 (199.4~620.0)

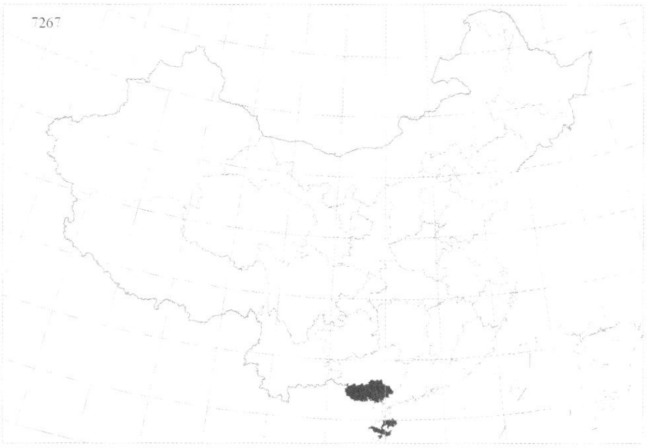

Myrtaceae 桃金娘科 *Syzygium* 蒲桃属
Syzygium cinereum (Kurz) Wall. ex Merr. et L. M. Perry 钝叶蒲桃
Small trees, evergreen

22.1; 14.0; 28.2; 22.1; 205.7; 0.0; 1177 ❙ 2018; 1139; 124 ❙ 71.8; 1177; 276.9

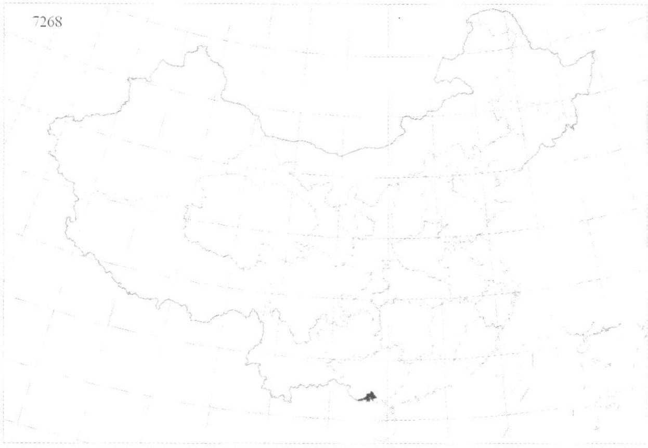

Myrtaceae 桃金娘科 *Syzygium* 蒲桃属
Syzygium claviflorum (Roxb.) Wall. ex Steudel 棒花蒲桃
Shrubs or small trees, evergreen

22.5 (18.2~24.8); 16.8 (11.9~20.1); 26.4 (21.7~28.8); 22.5 (18.2~24.8); 210.2 (158.7~237.6); 0.0 (0.0~0.0); 1187 (863~1385) ❙ 1599 (868~2993); 638 (312~1388); 91 (41~547) ❙ 40.7 (-34.3~214.4); 1102 (847~1360); 454.0 (248.2~659.0)

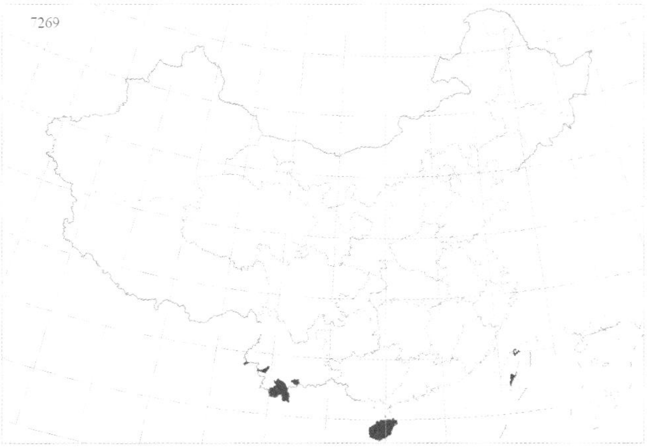

Myrtaceae 桃金娘科 *Syzygium* 蒲桃属
Syzygium congestiflorum H. T. Chang et R. H. Miao 团花蒲桃
Shrubs, evergreen

20.0; 14.6; 23.2; 20.1; 180.8; 0.0; 963 ❙ 1523; 681; 67 ❙ 58.4; 942; 581.0

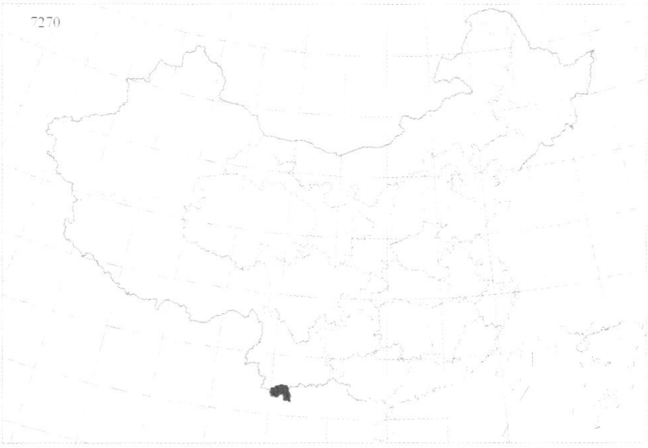

Myrtaceae 桃金娘科 *Syzygium* 蒲桃属
Syzygium conspersipunctatum (Merr. et L. M. Perry) Craven et Biffin 散点蒲桃
Trees, evergreen

23.9 (22.2~24.8); 18.6 (17.2~20.1); 27.6 (25.7~28.5); 23.9 (22.3~24.8); 227.3 (207.2~237.6); 0.0 (0.0~0.0); 1300 (1127~1385) ❙ 1547 (1179~1854); 533 (392~622); 64 (41~115) ❙ 19.5 (-11.9~47.8); 1175 (1055~1360); 502.0 (352.9~620.0)

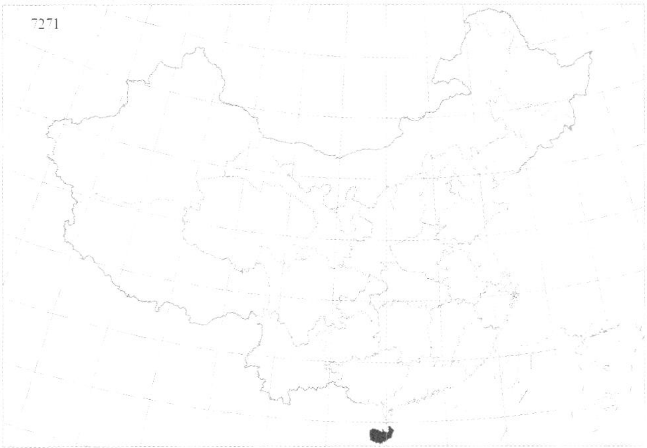

Myrtaceae 桃金娘科 *Syzygium* 蒲桃属
Syzygium cumini (L.) Skeels 乌墨
Trees, evergreen

20.6 (13.2~24.8); 13.1 (6.4~20.1); 26.4 (18.4~29.2); 20.7 (13.3~24.8); 188.3 (99.4~237.6); 0.0 (0.0~0.0); 1072 (695~1385) ❙ 1520 (868~2435); 726 (312~1428); 98 (39~231) ❙ 43.4 (-34.3~112.2); 1056 (695~1360); 303.9 (95.3~659.0)

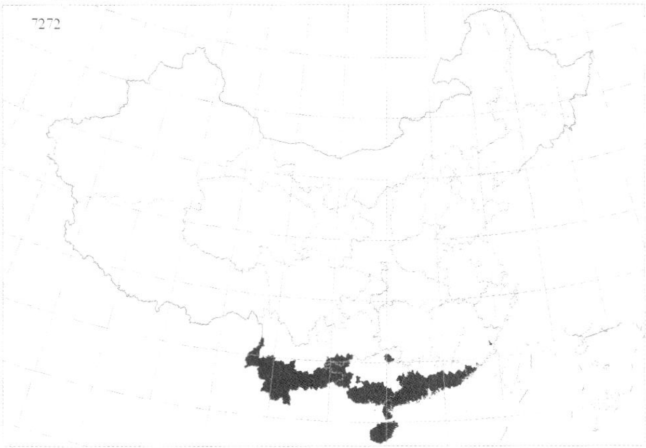

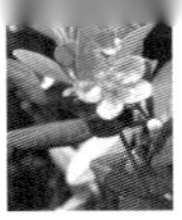

Myrtaceae 桃金娘科 *Syzygium* 蒲桃属
Syzygium cumini var. tsoi (Merr. et Chun) H. T. Chang et R. H. Miao 长萼乌墨
Trees, evergreen

23.9 (22.2~24.8); 18.5 (17.2~20.1); 27.8 (25.7~28.7); 24.0 (22.3~24.8); 227.8 (207.2~237.6); 0.0 (0.0~0.0); 1305 (1127~1385) ❙ 1457 (868~1854); 506 (312~622); 64 (41~115) ❙ 12.1 (-34.3~47.8); 1155 (868~1360); 479.8 (288.3~620.0)

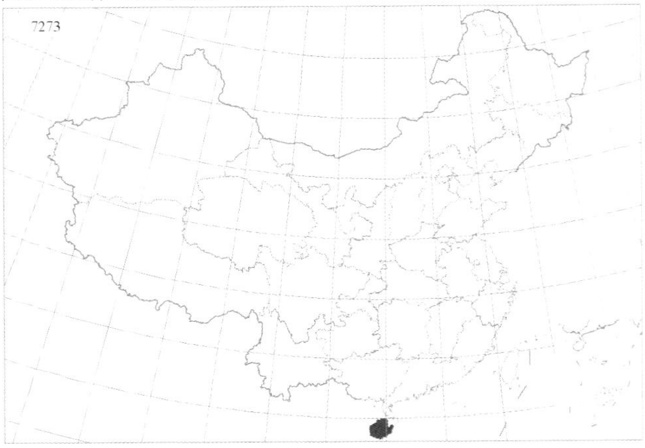

Myrtaceae 桃金娘科 *Syzygium* 蒲桃属
Syzygium densinervium var. insulare E. C. Chang 岛生蒲桃
Small trees, evergreen

18.5; 14.1; 21.7; 18.5; 162.1; 0.0; 863 ❙ 2713; 1388; 188 ❙ 214.4; 863; 363.7

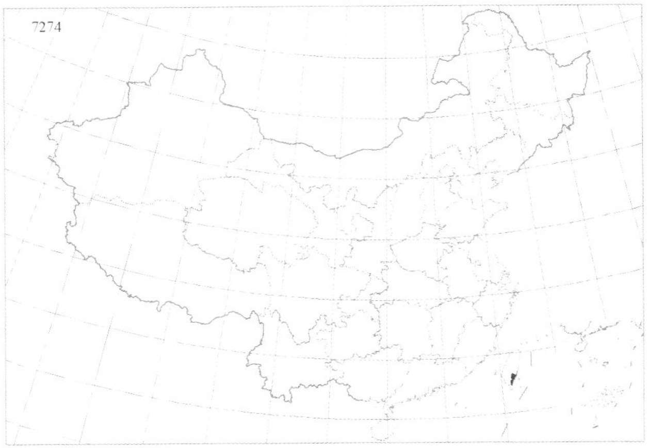

Myrtaceae 桃金娘科 *Syzygium* 蒲桃属
Syzygium euonymifolium (F. P. Metcalf) Merr. et L. M. Perry 卫矛叶蒲桃
Trees, evergreen

21.7 (18.7~24.8); 13.9 (9.7~20.1); 28.0 (25.7~29.0); 21.8 (18.8~24.8); 201.8 (165.4~237.6); 0.0 (0.0~0.0); 1156 (950~1385) ❙ 1675 (1302~2435); 750 (455~1428); 123 (41~192) ❙ 45.9 (1.5~99.5); 1133 (950~1239); 287.0 (157.7~620.0)

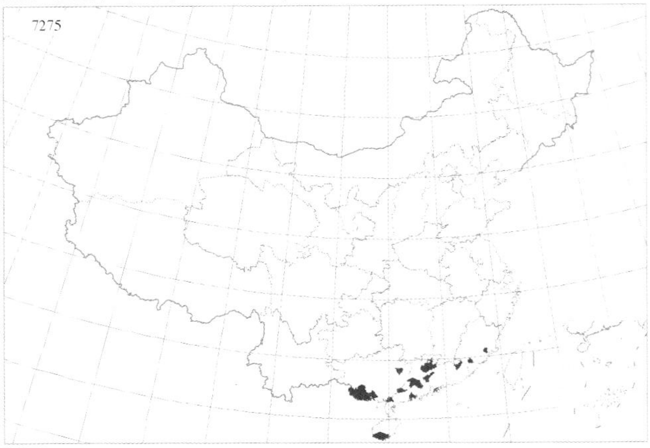

Myrtaceae 桃金娘科 *Syzygium* 蒲桃属
Syzygium euphlebium (Hayata) Mori 细叶蒲桃
Shrubs or trees, evergreen

20.3; 15.9; 23.6; 20.4; 184.4; 0.0; 993 ❙ 2853; 1618; 147 ❙ 190.4; 993; 330.3

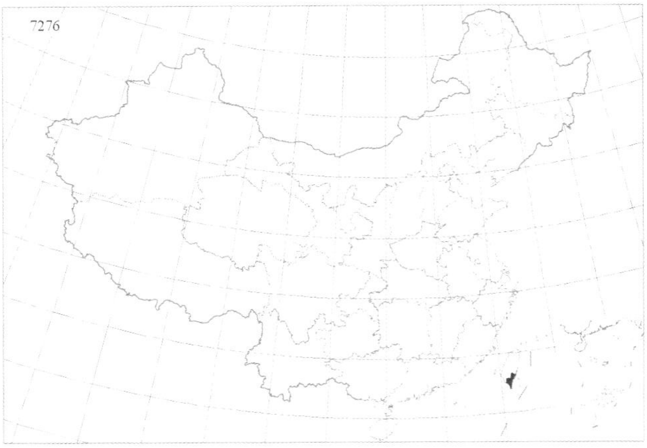

Myrtaceae 桃金娘科 *Syzygium* 蒲桃属
Syzygium fluviatile (Hemsl.) Merr. et L. M. Perry 水竹蒲桃
Shrubs, evergreen

22.3 (18.1~24.8); 15.4 (9.5~20.1); 27.4 (23.6~28.6); 22.4 (18.2~24.8); 208.3 (158.1~237.6); 0.0 (0.0~0.0); 1183 (901~1385) ❙ 1602 (1188~2435); 751 (444~1428); 87 (41~130) ❙ 36.6 (-2.2~99.5); 1144 (901~1329); 373.0 (215.3~659.0)

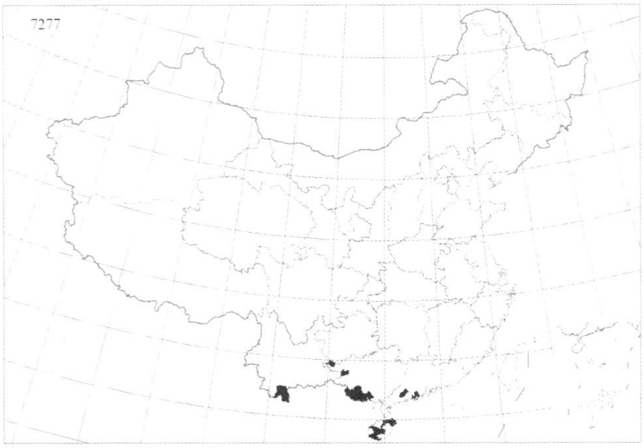

Myrtaceae 桃金娘科 *Syzygium* 蒲桃属
Syzygium formosanum (Hayata) Mori 台湾蒲桃
Shrubs or trees, evergreen

19.2 (16.2~22.2); 13.8 (11.0~17.7); 23.7 (20.1~27.0); 19.3 (16.2~22.2); 171.2 (134.2~206.7); 0.0 (0.0~0.0); 947 (773~1123) ❙ 2599 (1836~2993); 1237 (924~1847); 242 (107~547) ❙ 181.1 (64.2~246.5); 940 (773~1123); 282.6 (202.8~363.7)

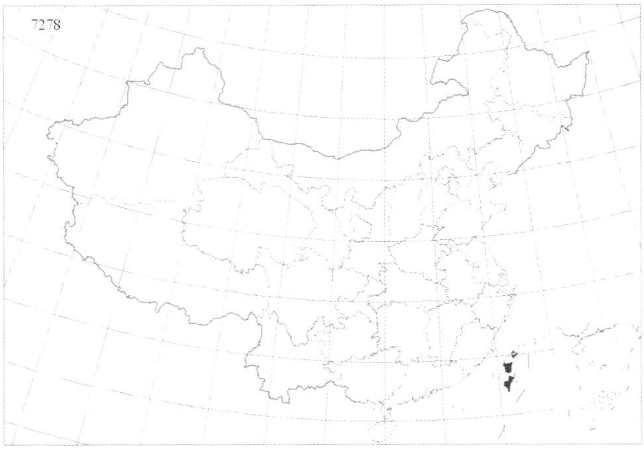

Myrtaceae 桃金娘科 *Syzygium* 蒲桃属
Syzygium forrestii Merr. et L. M. Perry 滇边蒲桃
Trees, evergreen

17.5 (8.6~21.0)；11.2 (1.3~15.1)；21.7 (15.0~25.6)；17.6 (8.7~21.0)；151.2 (52.6~191.8)；0.2 (0.0~8.8)；853 (574~1053) | 1330 (939~1751)；735 (524~996)；53 (39~77) | 56.9 (6.5~112.2)；838 (574~1053)；435.9 (294.5~659.0)

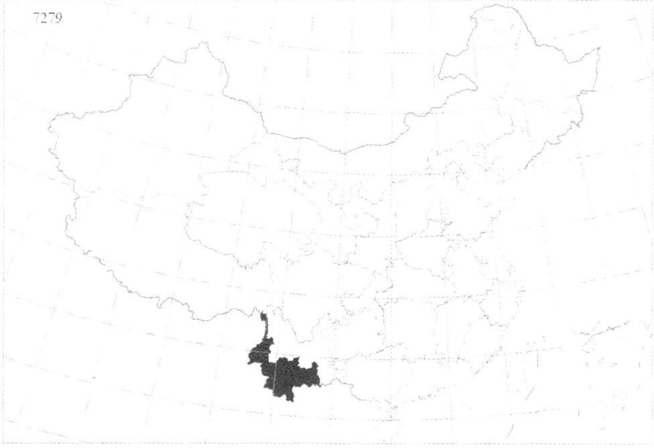

Myrtaceae 桃金娘科 *Syzygium* 蒲桃属
Syzygium fruticosum (Roxb.) DC. 簇花蒲桃
Trees, evergreen

18.2 (16.3~21.0)；11.4 (7.8~15.1)；23.0 (20.3~26.6)；18.3 (16.4~21.0)；159.4 (136.4~191.8)；0.0 (0.0~0.0)；887 (789~1053) | 1299 (958~1751)；714 (531~996)；54 (42~77) | 46.8 (12.5~88.8)；875 (780~1053)；395.8 (226.6~659.0)

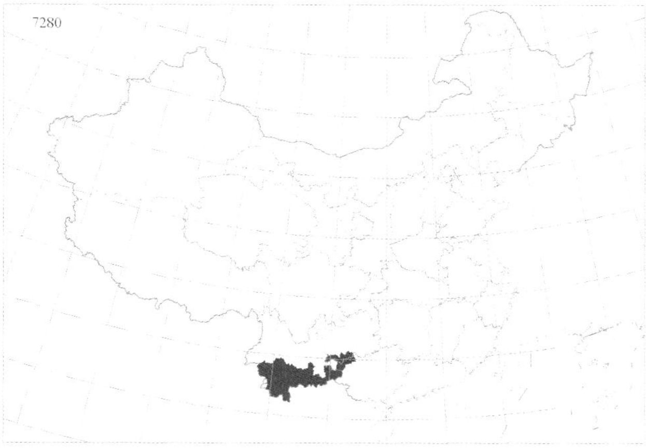

Myrtaceae 桃金娘科 *Syzygium* 蒲桃属
Syzygium globiflorum (Craib) Chantar. et J. Parnell 短药蒲桃
Shrubs or small trees, evergreen

21.6 (17.7~24.8)；15.0 (10.9~20.1)；26.1 (21.6~28.7)；21.6 (17.8~24.8)；199.3 (153.4~237.6)；0.0 (0.0~0.0)；1120 (854~1385) | 1505 (868~2435)；723 (312~1428)；77 (41~145) | 37.3 (-34.3~99.5)；1064 (839~1329)；413.5 (208.5~659.0)

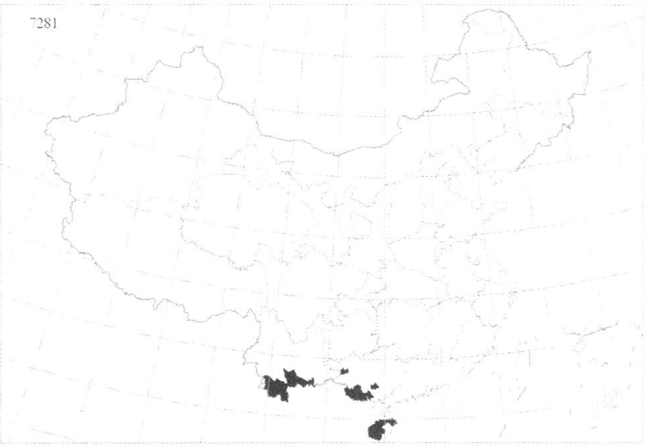

Myrtaceae 桃金娘科 *Syzygium* 蒲桃属
Syzygium gongshanense P. Y. Bai 贡山蒲桃
Trees, evergreen

8.6；1.3；15.0；8.7；52.6；8.8；574 | 1041；585；45 | 81.5；574；352.3

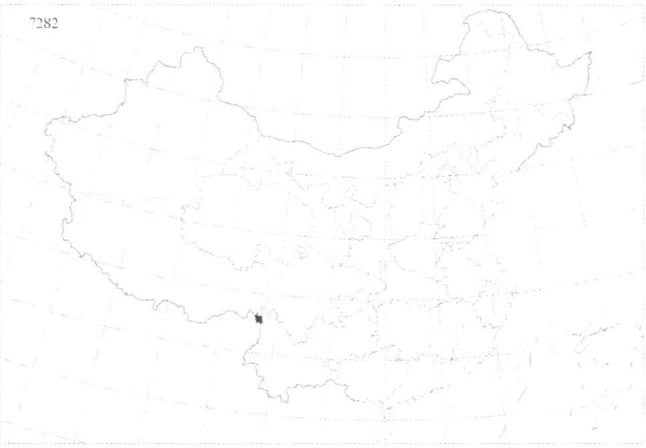

Myrtaceae 桃金娘科 *Syzygium* 蒲桃属
Syzygium grijsii (Hance) Merr. et L. M. Perry 轮叶蒲桃
Shrubs, evergreen

17.9 (13.4~22.5)；7.7 (1.5~15.0)；27.3 (24.2~29.8)；17.9 (13.4~22.6)；155.5 (108.3~210.9)；0.2 (0.0~7.3)；941 (755~1206) | 1589 (1070~2030)；562 (356~1033)；189 (84~259) | 70.6 (-5.0~147.0)；933 (755~1206)；236.7 (123.9~332.5)

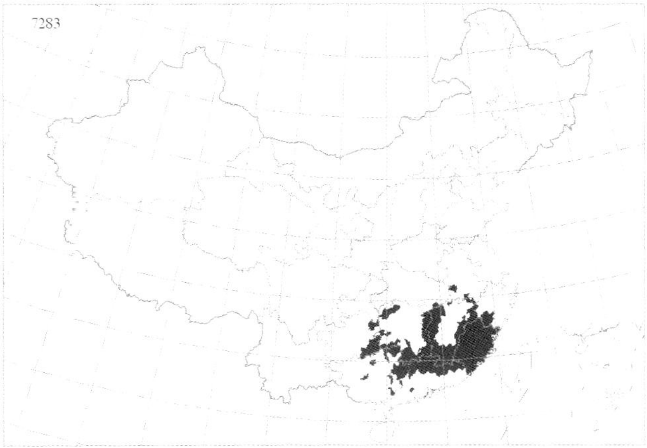

Myrtaceae 桃金娘科 *Syzygium* 蒲桃属
Syzygium guangxiense H. T. Chang et R. H. Miao 广西蒲桃
Shrubs, evergreen

21.6；13.1；27.9；21.6；199.6；0.0；1143 | 1451；766；102 | 27.0；1143；335.1

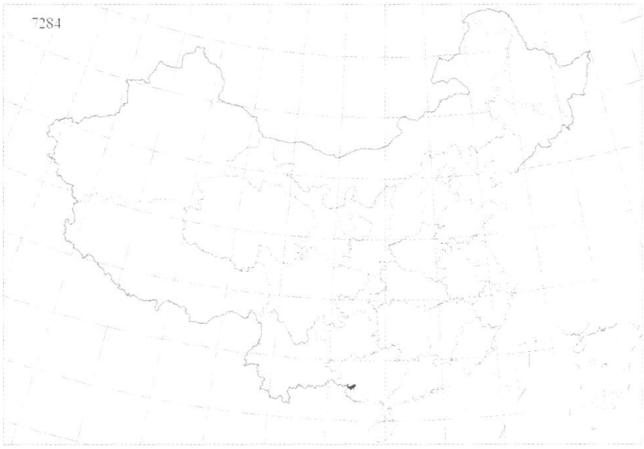

Myrtaceae 桃金娘科 *Syzygium* 蒲桃属
Syzygium hainanense H. T. Chang et R. H. Miao 海南蒲桃
Small trees, evergreen

23.6；17.8；27.8；23.6；223.3；0.0；1271 ❙ 1242；444；44 ❙ -2.2；1091；439.7

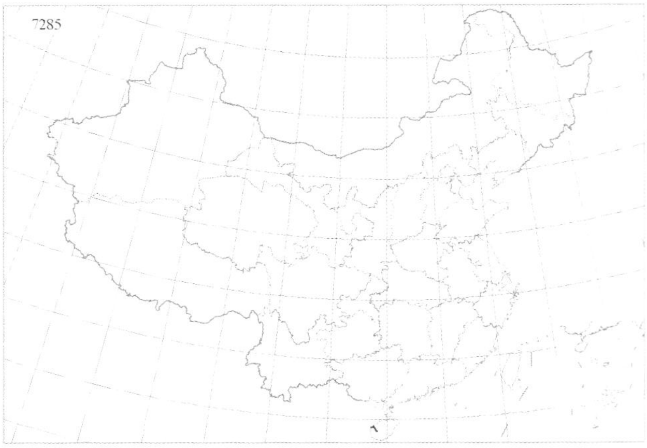

Myrtaceae 桃金娘科 *Syzygium* 蒲桃属
Syzygium hancei Merr. et L. M. Perry 红鳞蒲桃
Shrubs or trees, evergreen

21.1 (14.6~24.8)；12.7 (3.8~20.1)；27.9 (24.6~29.8)；21.1 (14.7~24.8)；193.7 (117.7~237.6)；0.0 (0.0~1.5)；1117 (779~1385) ❙ 1622 (868~2435)；698 (312~1428)；136 (41~202) ❙ 47.0 (-34.3~101.8)；1103 (779~1360)；252.7 (95.3~620.0)

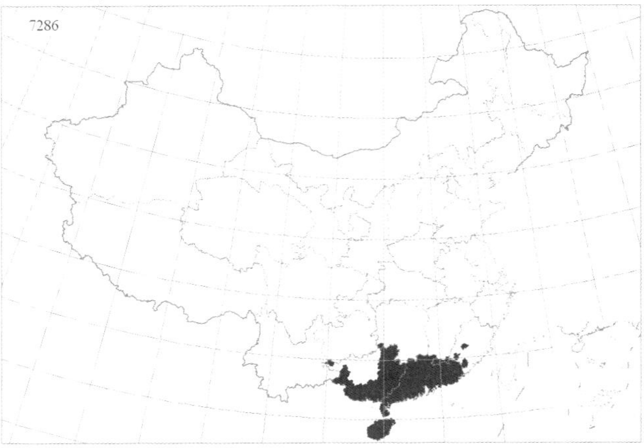

Myrtaceae 桃金娘科 *Syzygium* 蒲桃属
Syzygium handelii Merr. et L. M. Perry 贵州蒲桃
Shrubs, evergreen

18.7 (14.4~22.7)；8.9 (2.6~15.0)；27.0 (23.4~29.8)；18.7 (14.4~22.8)；164.9 (116.7~213.1)；0.2 (0.0~3.6)；978 (779~1221) ❙ 1529 (1133~2435)；649 (463~1428)；145 (51~227) ❙ 57.1 (20.3~129.9)；977 (779~1221)；242.8 (166.0~335.1)

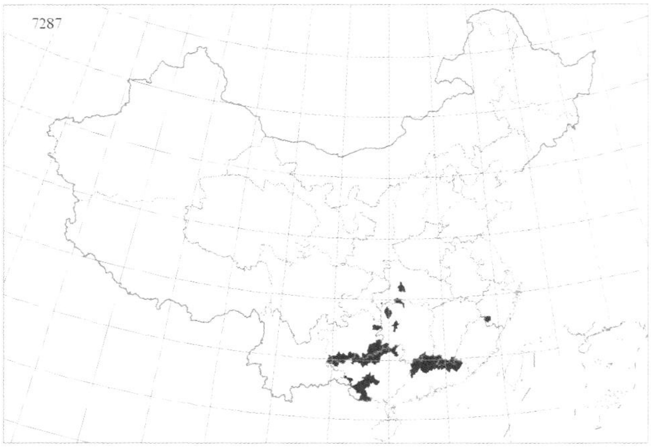

Myrtaceae 桃金娘科 *Syzygium* 蒲桃属
Syzygium howii Merr. et L. M. Perry 万宁蒲桃
Shrubs or small trees, evergreen

18.1 (14.6~24.6)；8.7 (3.8~19.5)；26.5 (24.6~28.2)；18.2 (14.7~24.7)；158.7 (117.7~235.9)；0.3 (0.0~1.5)；969 (779~1372) ❙ 1647 (1572~1767)；583 (544~618)；168 (88~194) ❙ 75.5 (28.8~101.8)；954 (779~1299)；308.1 (236.9~548.0)

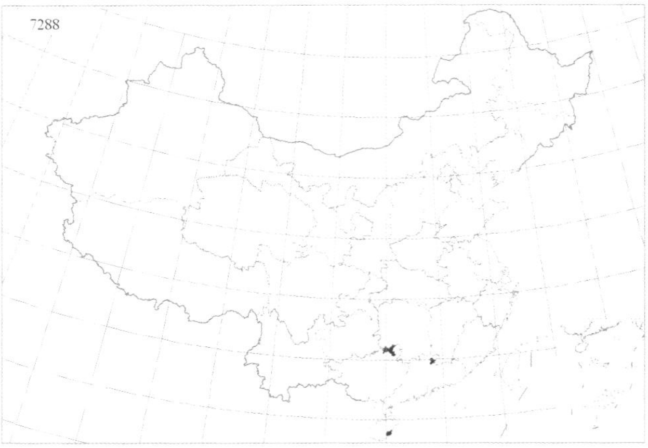

Myrtaceae 桃金娘科 *Syzygium* 蒲桃属
Syzygium imitans Merr. et L. M. Perry 桂南蒲桃
Trees, evergreen

22.0 (21.3~22.7)；13.9 (13.2~15.0)；28.1 (27.3~28.6)；22.1 (21.3~22.8)；204.7 (195.9~213.1)；0.0 (0.0~0.0)；1171 (1111~1221) ❙ 1751 (1302~2435)；959 (670~1428)；114 (86~130) ❙ 49.5 (12.9~99.5)；1171 (1111~1221)；288.3 (235.1~327.7)

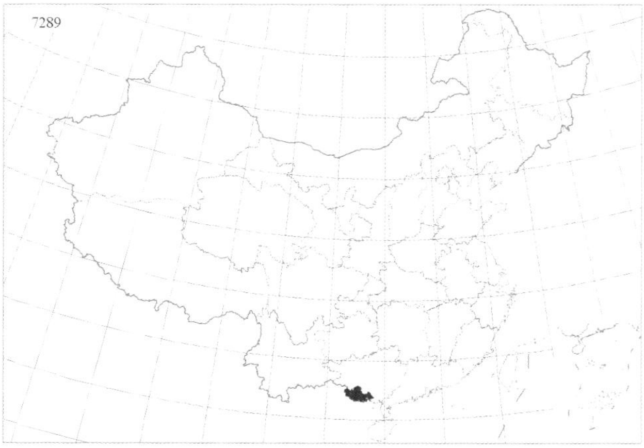

Myrtaceae 桃金娘科 *Syzygium* 蒲桃属
Syzygium infrarubiginosum H. T. Chang et R. H. Miao 褐背蒲桃
Trees, evergreen

24.3；18.7；28.4；24.3；231.9；0.0；1338 ❙ 1179；392；42 ❙ -11.9；1056；455.2

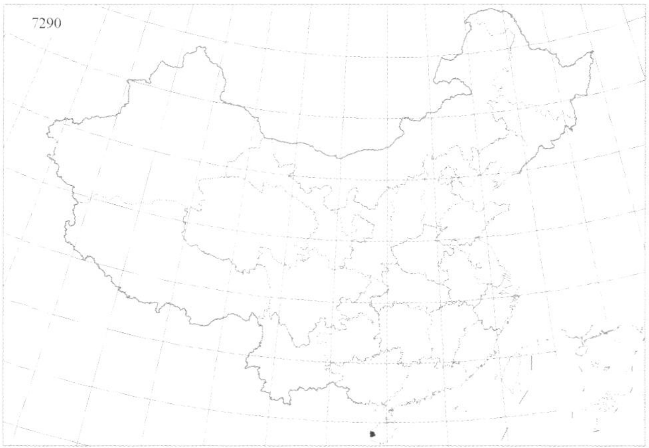

Myrtaceae 桃金娘科 *Syzygium* 蒲桃属
Syzygium jambos (L.) Alston 蒲桃
Trees, evergreen

21.0 (15.0~24.8); 13.4 (4.5~20.1); 26.9 (20.1~29.2); 21.0 (15.0~24.8); 192.3 (120.8~237.6); 0.0 (0.0~0.5); 1096 (773~1385) ❙ 1655 (868~3495); 767 (312~1847); 125 (20~897) ❙ 54.3 (-34.3~246.5); 1076 (773~1360); 281.4 (95.3~659.0)

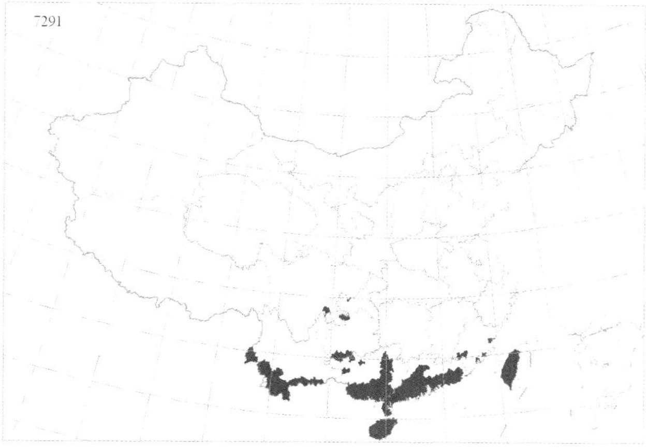

Myrtaceae 桃金娘科 *Syzygium* 蒲桃属
Syzygium jambos var. linearilimbum H. T. Chang et R. H. Miao 线叶蒲桃
Small trees, evergreen

21.0; 14.2; 25.6; 21.0; 191.8; 0.0; 1053 ❙ 1522; 818; 68 ❙ 44.5; 1053; 551.6

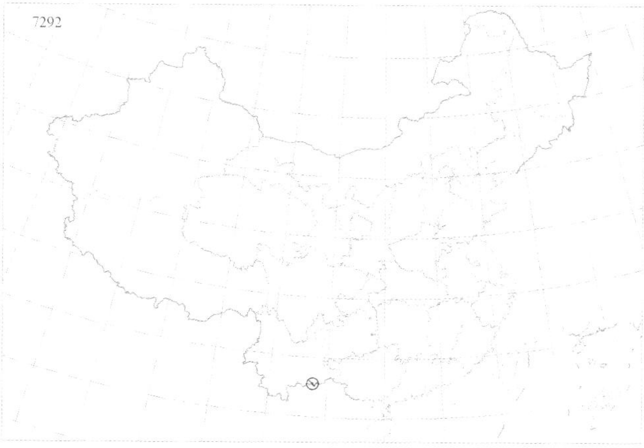

Myrtaceae 桃金娘科 *Syzygium* 蒲桃属
Syzygium jambos var. tripinnatum (Blanco) C. Chen et G. H. Zhu 大花赤楠
Trees, evergreen

18.5; 14. ; 21.7; 18.5; 162.1; 0.0; 863 ❙ 2713; 1388; 188 ❙ 214.4; 863; 363.7

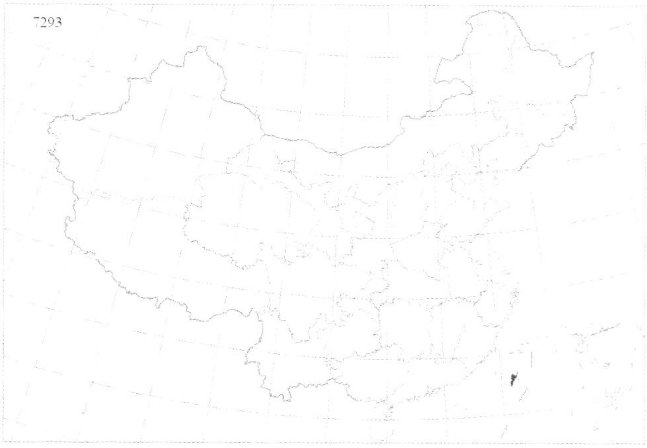

Myrtaceae 桃金娘科 *Syzygium* 蒲桃属
Syzygium jienfunicum H. T. Chang et R. H. Miao 尖峰蒲桃
Trees, evergreen

23.9; 18.7; 27.5; 23.9; 226.8; 0.0; 1294 ❙ 1380; 455; 48 ❙ 6.7; 1112; 455.2

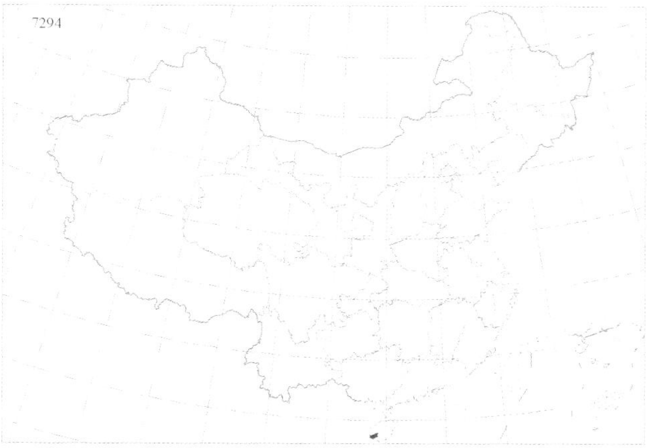

Myrtaceae 桃金娘科 *Syzygium* 蒲桃属
Syzygium kusukusense (Hayata) Mori 恒春蒲桃
Small trees, evergreen

20.3; 15.9; 23.6; 20.4; 184.4; 0.0; 993 ❙ 2853; 1618; 147 ❙ 190.4; 993; 330.3

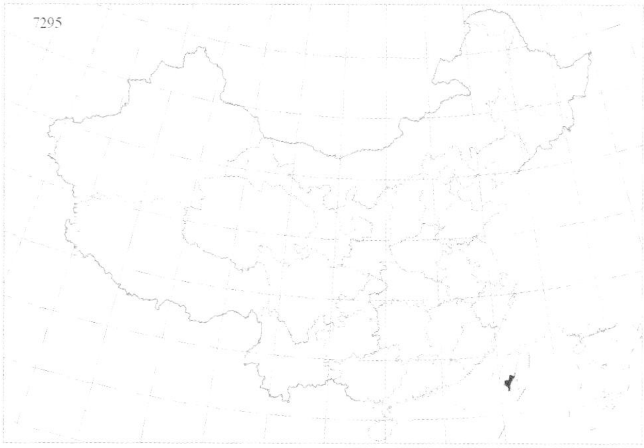

Myrtaceae 桃金娘科 *Syzygium* 蒲桃属
Syzygium kwangtungense (Merr.) Merr. et L. M. Perry 广东蒲桃
Small trees, evergreen

21.4 (17.3~23.4); 13.0 (6.9~16.1); 28.1 (26.2~29.2); 21.5 (17.3~23.5); 197.6 (148.1~221.7); 0.0 (0.0~0.0); 1135 (896~1273) ❙ 1646 (1302~2435); 760 (517~1428); 130 (75~191) ❙ 46.1 (9.0~99.5); 1134 (896~1273); 238.3 (139.6~327.7)

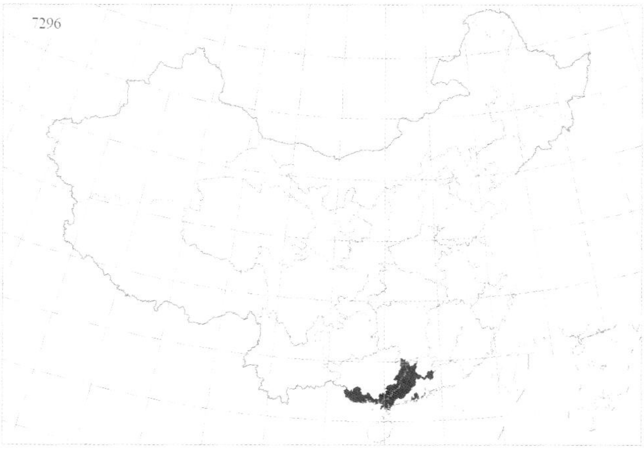

Myrtaceae 桃金娘科 *Syzygium* 蒲桃属
Syzygium laosense var. quocense (Gagnep.) Chang et Miau 少花老挝蒲桃
Small trees, evergreen

20.3; 14.5; 23.8; 20.3; 183.5; 0.0; 986 ❙ 1523; 715; 67 ❙ 54.9; 970; 573.6

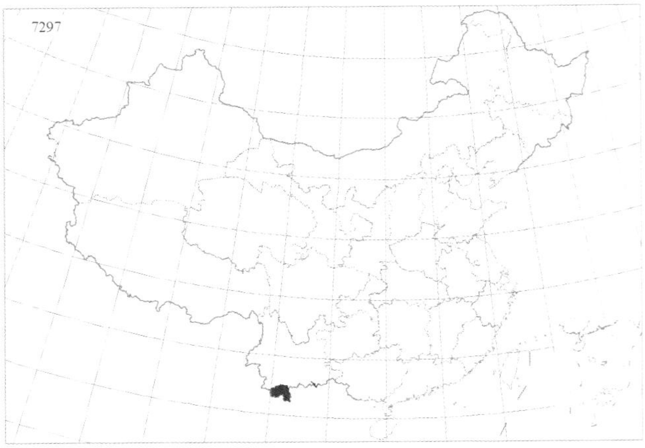

Myrtaceae 桃金娘科 *Syzygium* 蒲桃属
Syzygium lasianthifolium H. T. Chang et R. H. Miao 粗叶木蒲桃
Shrubs, evergreen

22.9; 15.2; 29.0; 23.0; 215.7; 0.0; 1239 ❙ 1477; 714; 97 ❙ 19.2; 1239; 203.1

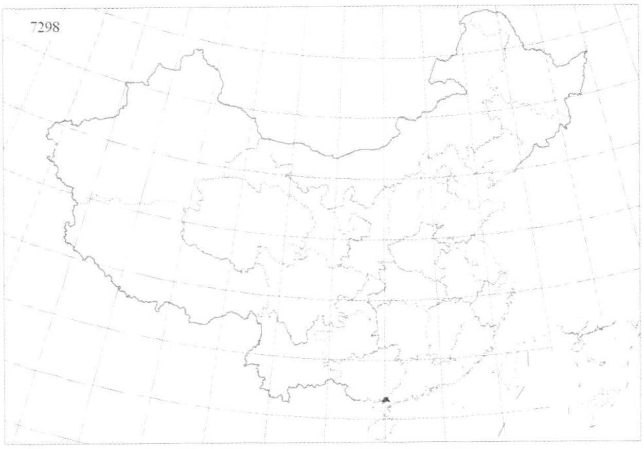

Myrtaceae 桃金娘科 *Syzygium* 蒲桃属
Syzygium levinei (Merr.) Merr. et L. M. Perry 山蒲桃
Trees, evergreen

22.4 (20.2~24.8); 14.9 (12.0~20.1); 28.2 (25.7~29.2); 22.4 (20.2~24.8); 209.2 (183.0~237.6); 0.0 (0.0~0.0); 1196 (1036~1385) ❙ 1650 (868~2435); 742 (312~1428); 112 (41~177) ❙ 39.0 (-34.3~99.5); 1171 (868~1360); 262.3 (95.3~620.0)

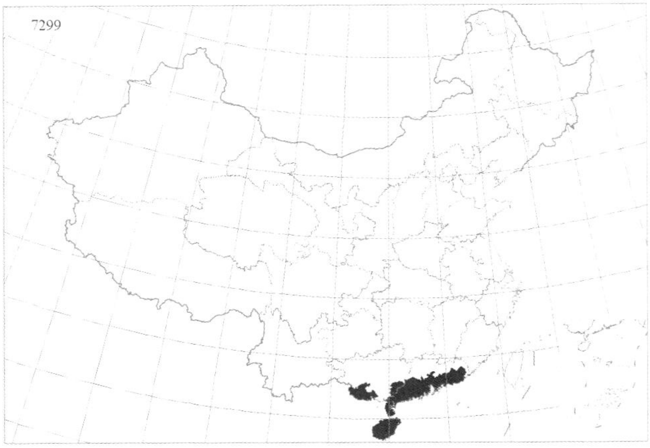

Myrtaceae 桃金娘科 *Syzygium* 蒲桃属
Syzygium lineatum (DC.) Merr. et L. M. Perry 长花蒲桃
Trees, evergreen

22.2; 13.9; 28.3; 22.3; 207.5; 0.0; 1190 ❙ 1374; 713; 87 ❙ 15.4; 1190; 327.7

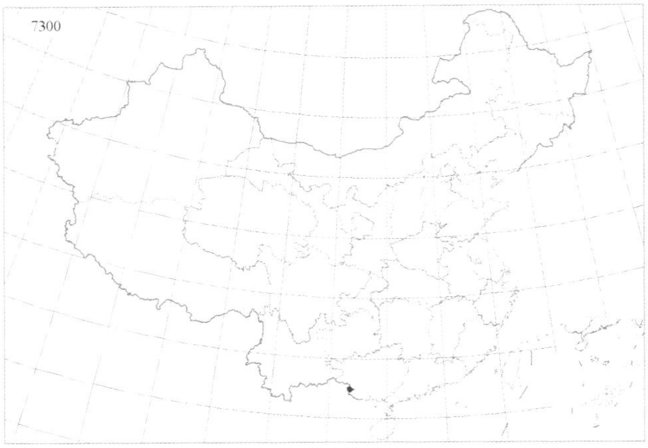

Myrtaceae 桃金娘科 *Syzygium* 蒲桃属
Syzygium malaccense (L.) Merr. et L. M. Perry 马六甲蒲桃
Trees, evergreen

20.3; 14.5; 23.8; 20.3; 183.5; 0.0; 986 ❙ 1523; 715; 67 ❙ 54.9; 970; 573.6

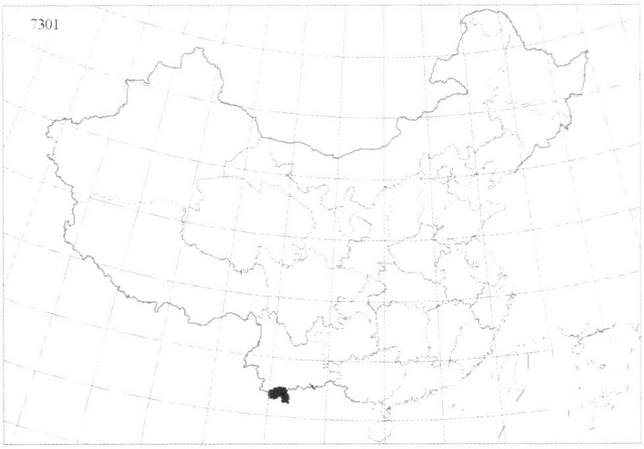

Myrtaceae 桃金娘科 *Syzygium* 蒲桃属
Syzygium megacarpum (Craib) Rathakrishnan et N. C. Nair 阔叶蒲桃
Trees, evergreen

19.8 (16.4~24.8); 13.4 (9.6~20.1); 24.1 (20.6~28.9); 19.8 (16.4~24.8); 178.2 (137.3~237.6); 0.0 (0.0~0.0); 993 (795~1385) ❙ 1520 (1178~2148); 781 (455~1234); 66 (41~142) ❙ 56.6 (1.5~112.2); 962 (795~1299); 469.3 (201.4~659.0)

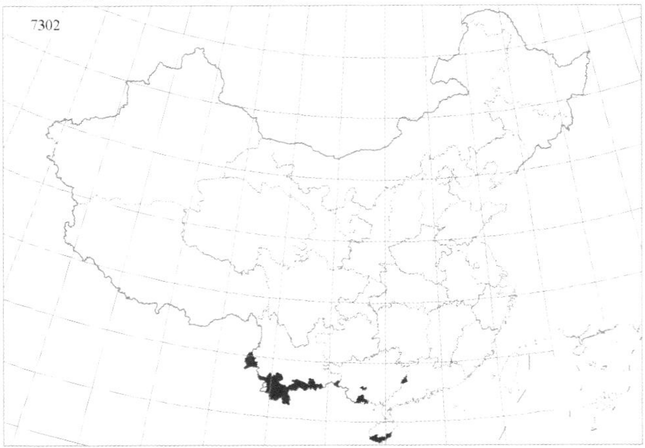

Myrtaceae 桃金娘科 *Syzygium* 蒲桃属

Syzygium melanophyllum H. T. Chang et R. H. Miao 黑长叶蒲桃

Trees, evergreen

20.0; 14.6 23.2; 20.1; 180.8; 0.0; 963 ❙ 1523; 681; 67 ❙ 58.4; 942; 581.0

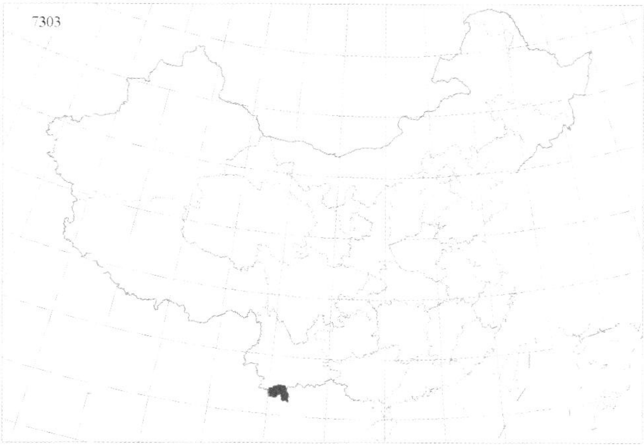

Myrtaceae 桃金娘科 *Syzygium* 蒲桃属

Syzygium myrsinifolium (Hance) Merr. et L. M. Perry 竹叶蒲桃

Shrubs or small trees, evergreen

23.7 (18.9~24.8); 18.0 (10.6~20.1); 27.8 (25.0~28.8); 23.8 (19.0~24.8); 225.2 (167.5~237.6); 0.0 (0.0~0.0); 1291 (939~1385) ❙ 1491 (868~1854); 543 (312~729); 72 (41~115) ❙ 16.1 (-34.3~47.8); 1179 (868~1360); 435.4 (248.2~620.0)

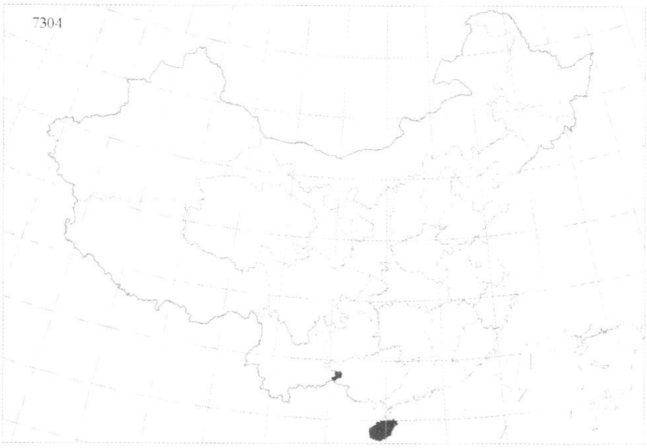

Myrtaceae 桃金娘科 *Syzygium* 蒲桃属

Syzygium myrsinifolium var. grandiflorum H. T. Chang et R. H. Miao 大花竹叶蒲桃

Shrubs or small trees, evergreen

23.9; 18.9; 27.3; 23.9; 227.3; 0.0; 1301 ❙ 1565; 542; 55 ❙ 21.6; 1150; 550.2

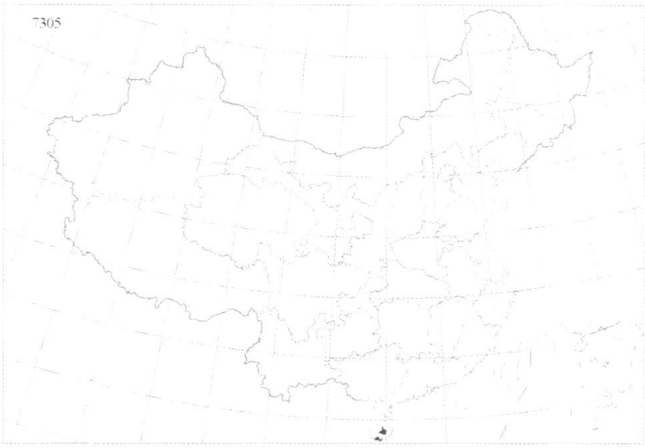

Myrtaceae 桃金娘科 *Syzygium* 蒲桃属

Syzygium nervosum DC. 水翁蒲桃

Trees, evergreen

21.3 (16.3~24.8); 13.8 (9.0~20.1); 27.1 (21.4~29.2); 21.4 (16.4~24.8); 196.4 (136.4~237.6); 0.0 (0.0~0.0); 1118 (797~1385) ❙ 1546 (868~2435); 738 (312~1428); 100 (41~178) ❙ 39.4 (-34.3~99.5); 1101 (797~1360); 292.4 (95.3~659.0)

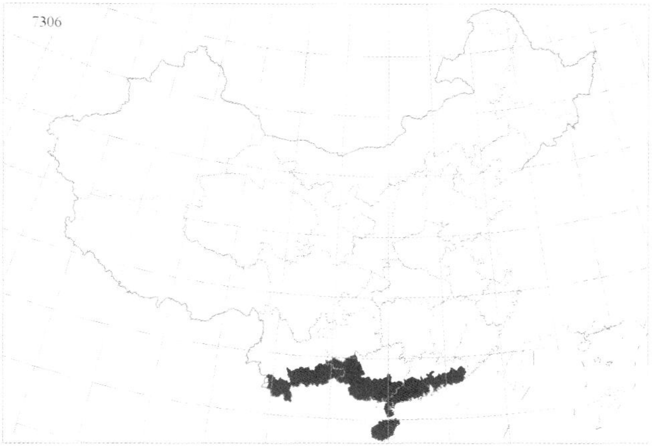

Myrtaceae 桃金娘科 *Syzygium* 蒲桃属

Syzygium oblancilimbum H. T. Chang et R. H. Miao 倒披针叶蒲桃

Dwarf shrubs, evergreen

17.1; 11.0; 20.8; 17.1; 145.0; 0.0; 815 ❙ 1352; 771; 47 ❙ 65.9; 802; 438.9

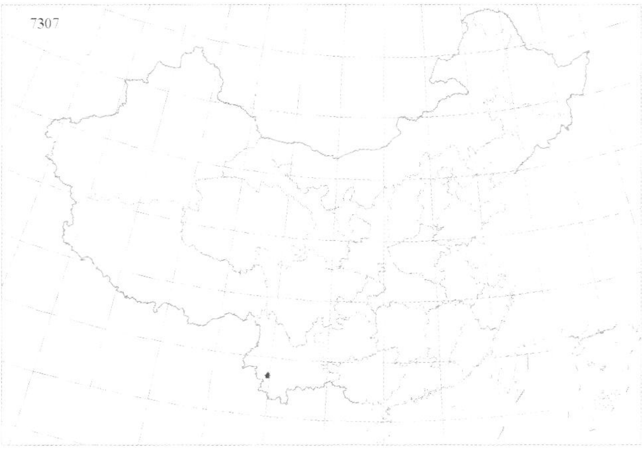

Myrtaceae 桃金娘科 *Syzygium* 蒲桃属

Syzygium oblatum (Roxb.) Wall. ex Steudel 高檐蒲桃

Trees, evergreen

18.9 (17.2~20.5); 12.7 (9.5~15.1); 22.9 (22.1~23.7); 18.9 (17.3~20.5); 167.2 (147.0~185.9); 0.0 (0.0~0.0); 910 (838~992) ❙ 1457 (1254~1607); 738 (655~836); 61 (48~77) ❙ 60.2 (47.6~66.9); 897 (838~986); 466.0 (336.3~659.0)

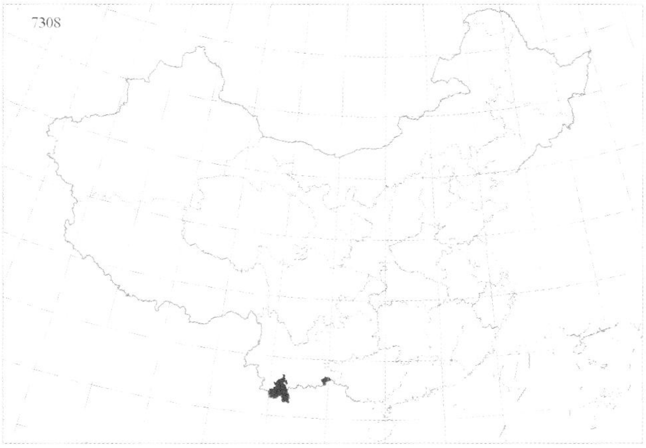

Myrtaceae 桃金娘科 *Syzygium* 蒲桃属
Syzygium odoratum (Lour.) DC. 香蒲桃
Trees, evergreen

22.4 (19.1~24.8); 15.3 (11.0~20.1); 27.8 (25.0~29.0); 22.5 (19.1~24.8); 210.0 (169.6~237.6); 0.0 (0.0~0.0); 1200 (945~1385) ❙ 1635 (868~2435); 774 (312~1428); 91 (41~130) ❙ 37.2 (-34.3~99.5); 1162 (868~1360); 328.0 (139.6~620.0)

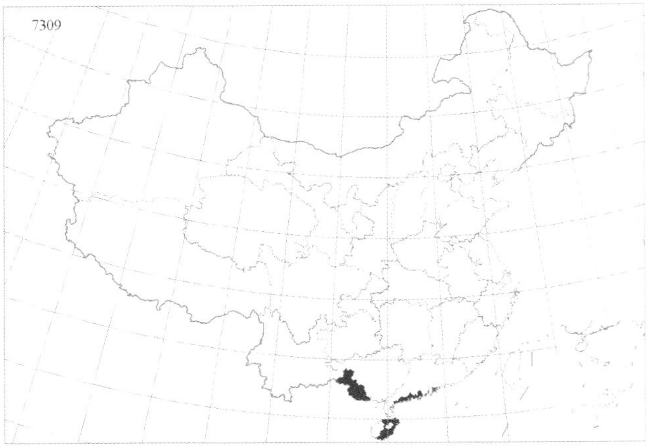

Myrtaceae 桃金娘科 *Syzygium* 蒲桃属
Syzygium paucivenium (C. B. Rob.) Merr. 圆顶蒲桃
Small trees, evergreen

18.5; 14.1; 21.7; 18.5; 162.1; 0.0; 863 ❙ 2713; 1388; 188 ❙ 214.4; 863; 363.7

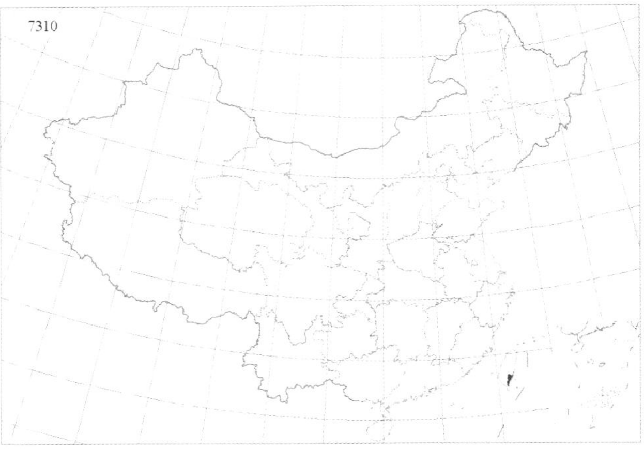

Myrtaceae 桃金娘科 *Syzygium* 蒲桃属
Syzygium polypetaloideum Merr. et L. M. Perry 假多瓣蒲桃
Shrubs, evergreen

19.7 (16.3~22.3); 11.9 (9.1~15.1); 25.3 (21.4~28.9); 19.7 (16.4~22.3); 176.5 (136.4~208.1); 0.0 (0.0~0.0); 996 (797~1202) ❙ 1440 (1074~1900); 770 (582~1056); 82 (47~173) ❙ 46.1 (5.8~88.8); 994 (797~1202); 368.2 (201.4~659.0)

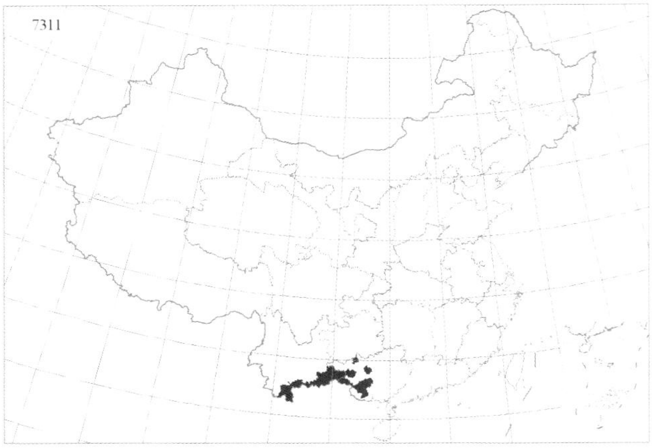

Myrtaceae 桃金娘科 *Syzygium* 蒲桃属
Syzygium rehderianum Merr. et L. M. Perry 红枝蒲桃
Shrubs or small trees, evergreen

20.3 (14.6~24.8); 11.4 (3.8~20.1); 27.6 (23.4~29.8); 20.3 (14.7~24.8); 184.1 (117.7~237.6); 0.0 (0.0~1.5); 1069 (779~1385) ❙ 1629 (1232~2435); 698 (473~1428); 144 (41~209) ❙ 53.8 (-1.4~101.8); 1062 (779~1329); 245.4 (95.3~620.0)

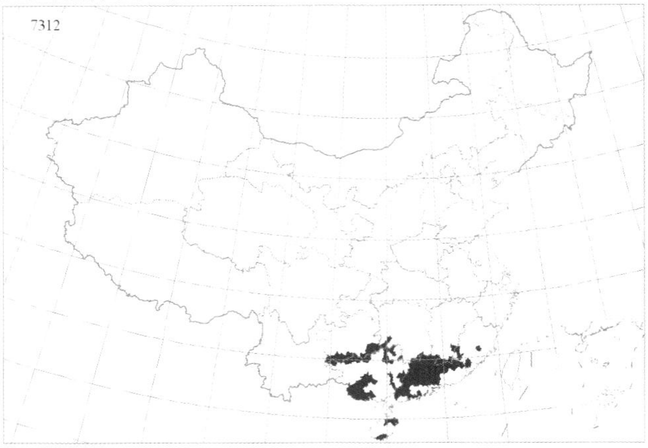

Myrtaceae 桃金娘科 *Syzygium* 蒲桃属
Syzygium rockii Merr. et L. M. Perry 滇西蒲桃
Trees, evergreen

19.3 (17.5~20.5); 13.5 (10.5~15.1); 22.9 (22.3~23.7); 19.4 (17.5~20.5); 172.2 (149.9~185.9); 0.0 (0.0~0.0); 929 (846~992) ❙ 1404 (958~1607); 670 (531~775); 61 (44~77) ❙ 50.6 (13.3~64.9); 908 (823~986); 496.6 (311.9~659.0)

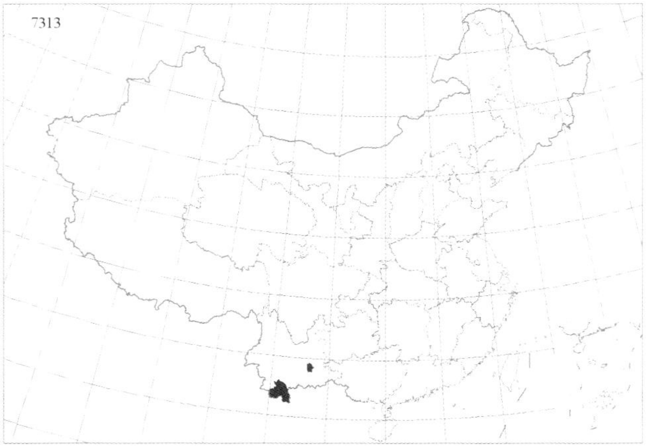

Myrtaceae 桃金娘科 *Syzygium* 蒲桃属
Syzygium rysopodum Merr. et L. M. Perry 皱萼蒲桃
Trees, evergreen

24.0 (22.2~24.8); 18.4 (17.2~20.1); 27.9 (25.7~28.8); 24.0 (22.3~24.8); 228.0 (207.2~237.6); 0.0 (0.0~0.0); 1308 (1127~1385) ❙ 1558 (1179~1854); 552 (392~712); 75 (41~115) ❙ 19.5 (-11.9~47.8); 1214 (1055~1360); 451.0 (248.2~620.0)

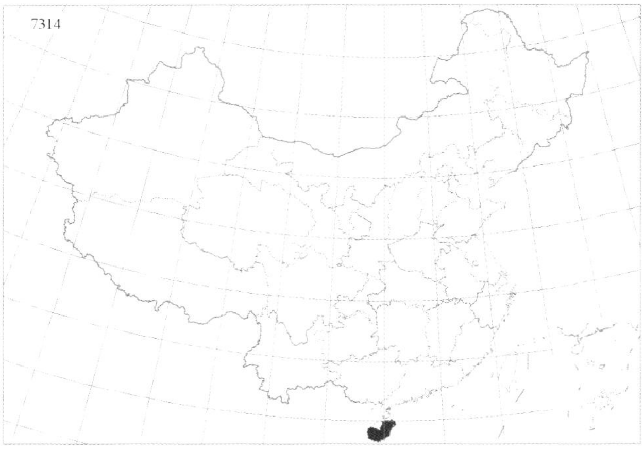

Myrtaceae 桃金娘科 *Syzygium* 蒲桃属
Syzygium salwinense Merr. et L. M. Perry 怒江蒲桃
Trees, evergreen

17.3 (13.2~19.6); 10.8 (6.4~13.8); 21.5 (18.4~23.5); 17.4 (13.3~19.6); 148.3 (99.4~175.7); 0.0 (0.0~0.0); 842 (695~953) ❙ 1489 (1049~1688); 841 (598~959); 56 (42~70) ❙ 78.0 (26.1~112.2); 831 (695~932); 499.8 (365.3~627.0)

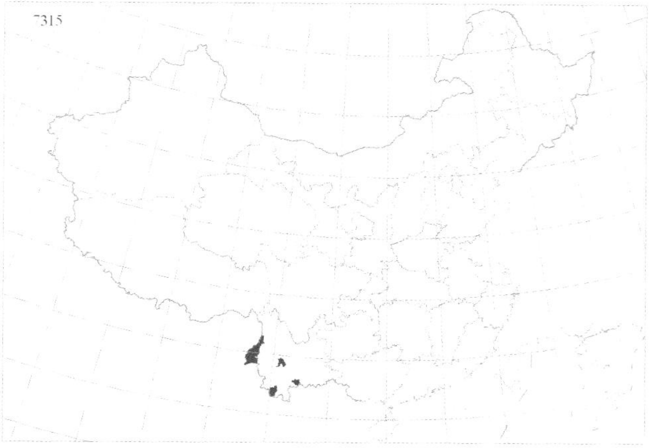

Myrtaceae 桃金娘科 *Syzygium* 蒲桃属
Syzygium samarangense (Blume) Merr. et L. M. Perry 洋蒲桃
Trees, evergreen

24.0 (22.2~24.8); 18.4 (17.2~20.1); 28.0 (25.7~28.8); 24.0 (22.3~24.8); 228.3 (207.2~237.6); 0.0 (0.0~0.0); 1310 (1127~1385) ❙ 1501 (868~1854); 534 (312~712); 74 (41~115) ❙ 15.0 (-34.3~47.8); 1191 (868~1360); 440.3 (248.2~620.0)

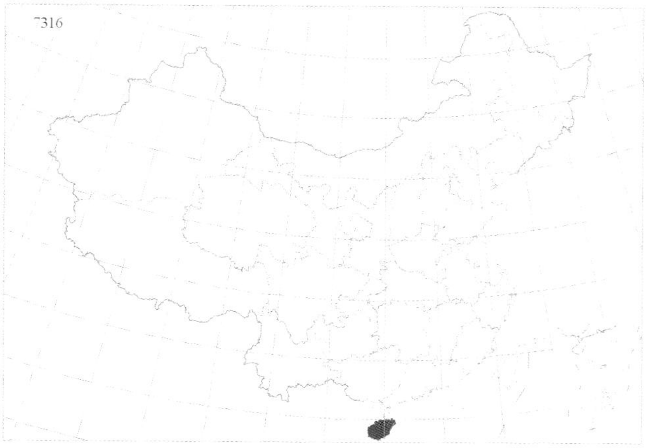

Myrtaceae 桃金娘科 *Syzygium* 蒲桃属
Syzygium saxatile H. T. Chang et R. H. Miao 石生蒲桃
Shrubs, evergreen

18.2; 10.3; 23.8; 18.2; 158.3; 0.0; 891 ❙ 1346; 760; 53 ❙ 51.0; 891; 354.0

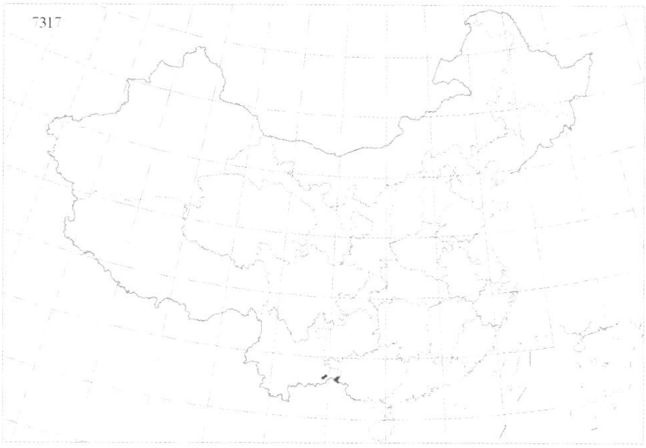

Myrtaceae 桃金娘科 *Syzygium* 蒲桃属
Syzygium sichuanense H. T. Chang et R. H. Miao 四川蒲桃
Small trees, evergreen

16.4 (10.6~18.7); 5.7 (-0.2~8.1); 26.6 (21.1~28.9); 16.5 (10.6~18.8); 138.8 (79.4~166.0); 1.2 (0.0~12.1); 881 (643~1002) ❙ 1173 (1028~1478); 491 (442~592); 64 (46~85) ❙ 35.9 (8.0~107.2); 881 (643~1002); 164.0 (108.2~284.3)

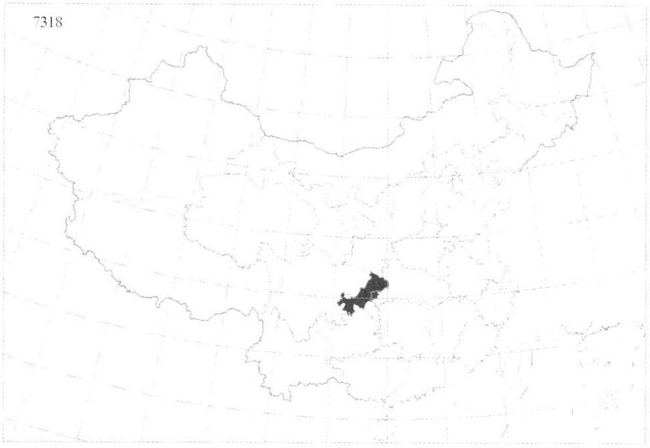

Myrtaceae 桃金娘科 *Syzygium* 蒲桃属
Syzygium simile (Merr.) Merr. 兰屿赤楠
Trees, evergreen

18.5; 14.1; 21.7; 18.5; 162.1; 0.0; 863 ❙ 2713; 1388; 188 ❙ 214.4; 863; 363.7

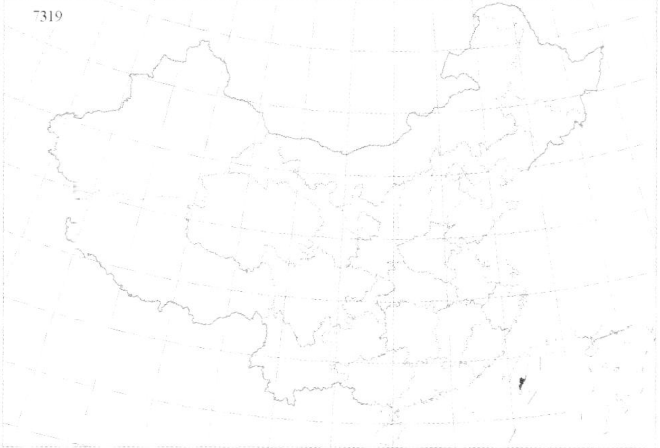

Myrtaceae 桃金娘科 *Syzygium* 蒲桃属
Syzygium stenocladum Merr. et L. M. Perry 纤枝蒲桃
Trees, evergreen

23.8 (22.2~24.8); 18.2 (17.2~20.1); 27.8 (25.7~28.7); 23.8 (22.3~24.8); 225.7 (207.2~237.6); 0.0 (0.0~0.0); 1288 (1127~1385) ❙ 1373 (868~1725); 484 (312~610); 60 (41~92) ❙ 7.5 (-34.3~47.8); 1119 (868~1294); 464.2 (288.3~590.7)

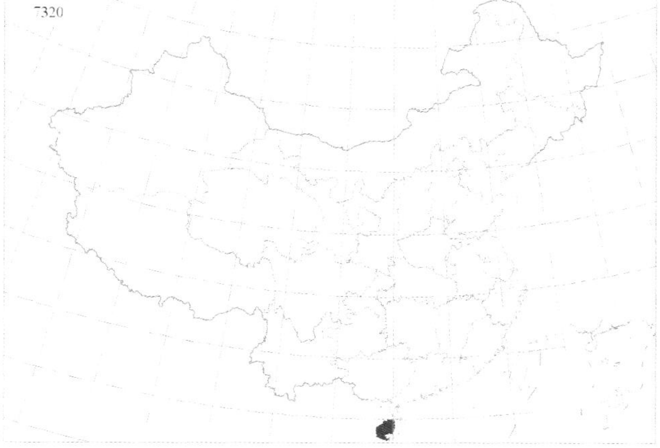

Myrtaceae 桃金娘科 *Syzygium* 蒲桃属
Syzygium sterrophyllum Merr. et L. M. Perry 硬叶蒲桃
Shrubs or small trees, evergreen

22.4 (17.2~24.8); 15.7 (9.5~20.1); 27.3 (22.4~28.9); 22.5 (17.3~24.8); 209.6 (147.0~237.6); 0.0 (0.0~0.0); 1199 (833~1385) ❙ 1533 (868~2435); 688 (312~1428); 82 (41~136) ❙ 29.9 (-34.3~99.5); 1137 (833~1360); 383.3 (201.4~620.0)

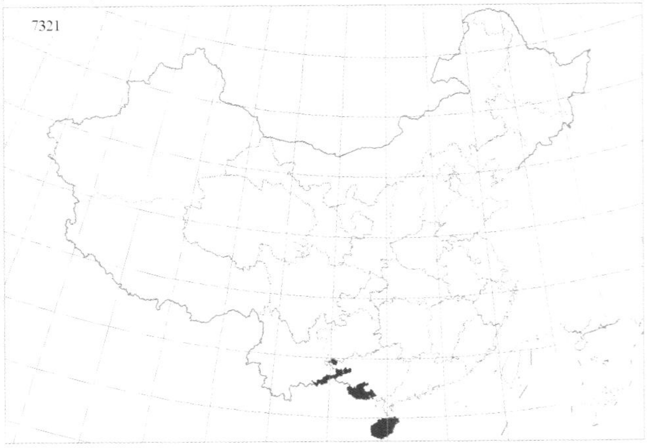

Myrtaceae 桃金娘科 *Syzygium* 蒲桃属
Syzygium szemaoense Merr. et L. M. Perry 思茅蒲桃
Shrubs or small trees, evergreen

18.4 (11.3~24.8); 10.6 (2.6~20.1); 24.5 (17.0~28.9); 18.4 (11.3~24.8); 161.2 (76.6~237.6); 0.2 (0.0~3.9); 936 (644~1385) ❙ 1467 (868~2024); 682 (312~996); 93 (41~256) ❙ 60.6 (-34.3~130.7); 905 (644~1230); 377.0 (166.0~659.0)

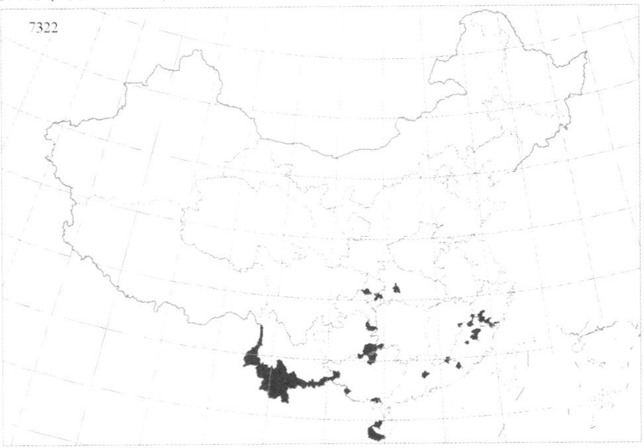

Myrtaceae 桃金娘科 *Syzygium* 蒲桃属
Syzygium taiwanicum H. T. Chang et R. H. Miao 台湾棒花蒲桃
Small trees, evergreen

18.5; 14.1; 21.7; 18.5; 162.1; 0.0; 863 ❙ 2713; 1388; 188 ❙ 214.4; 863; 363.7

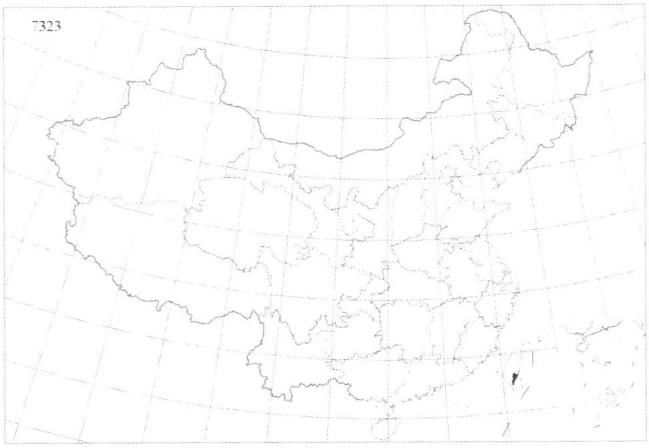

Myrtaceae 桃金娘科 *Syzygium* 蒲桃属
Syzygium tenuirhachis H. T. Chang et R. H. Miao 细轴蒲桃
Trees, evergreen

19.5; 10.9; 25.9; 19.5; 174.4; 0.0; 986 ❙ 1177; 662; 47 ❙ 19.4; 986; 382.8

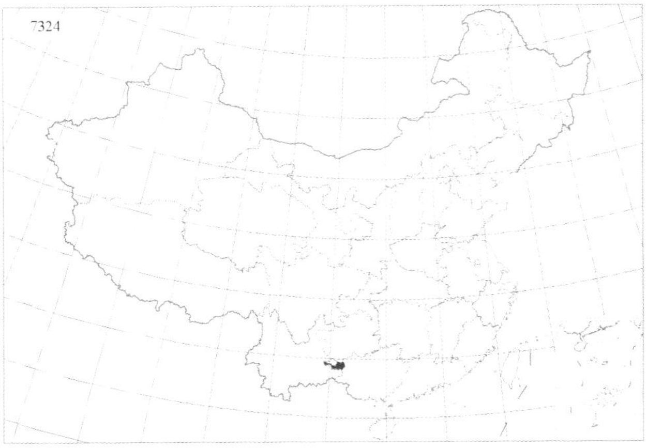

Myrtaceae 桃金娘科 *Syzygium* 蒲桃属
Syzygium tephrodes (Hance) Merr. et L. M. Perry 方枝蒲桃
Shrubs or small trees, evergreen

24.0 (22.2~24.8); 18.4 (17.2~20.1); 28.0 (25.7~28.8); 24.0 (22.3~24.8); 228.3 (207.2~237.6); 0.0 (0.0~0.0); 1310 (1127~1385) ❙ 1501 (868~1854); 534 (312~712); 74 (41~115) ❙ 15.0 (-34.3~47.8); 1191 (868~1360); 440.3 (248.2~620.0)

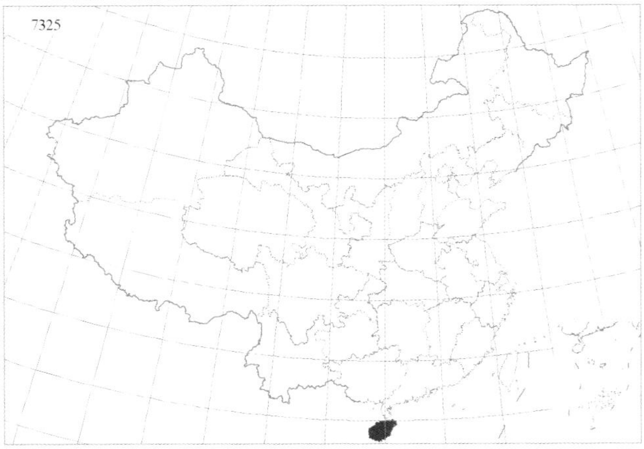

Myrtaceae 桃金娘科 *Syzygium* 蒲桃属
Syzygium tetragonum (Wight) Wall. ex Walp. 四角蒲桃
Trees, evergreen

18.8 (5.4~24.8); 11.5 (-2.1~20.1); 24.2 (12.3~28.9); 18.9 (5.8~24.8); 166.9 (31.5~237.6); 0.6 (0.0~25.6); 972 (481~1385) ❙ 1377 (741~1854); 688 (312~996); 77 (15~194) ❙ 45.1 (-34.3~112.2); 948 (481~1360); 369.2 (185.5~659.0)

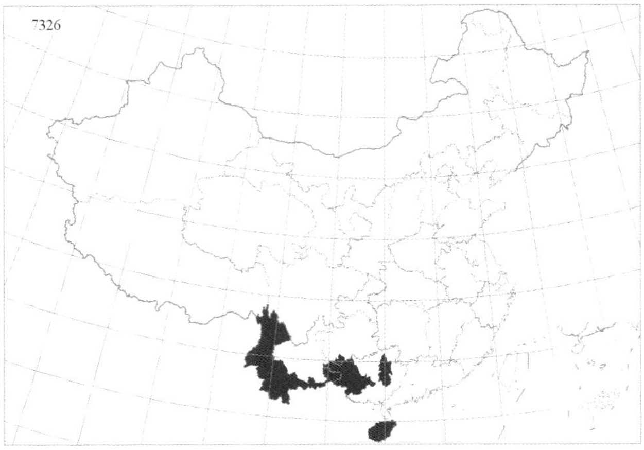

Myrtaceae 桃金娘科 *Syzygium* 蒲桃属
Syzygium thumra (Roxb.) Merr. et L. M. Perry 黑叶蒲桃
Trees, evergreen

19.5; 13.7; 23.0; 19.5; 173.9; 0.0; 936 ‖ 1485; 700; 65 ‖ 58.8; 920; 558.6

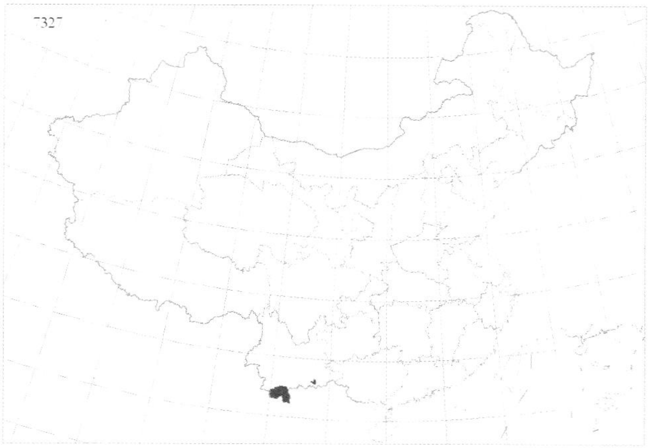

Myrtaceae 桃金娘科 *Syzygium* 蒲桃属
Syzygium toddalioides (Wight) Walpers 假乌墨
Small trees, evergreen

18.7; 12.5; 22.7; 18.7; 164.3; 0.0; 895 ‖ 1217; 649; 51 ‖ 35.5; 845; 372.3

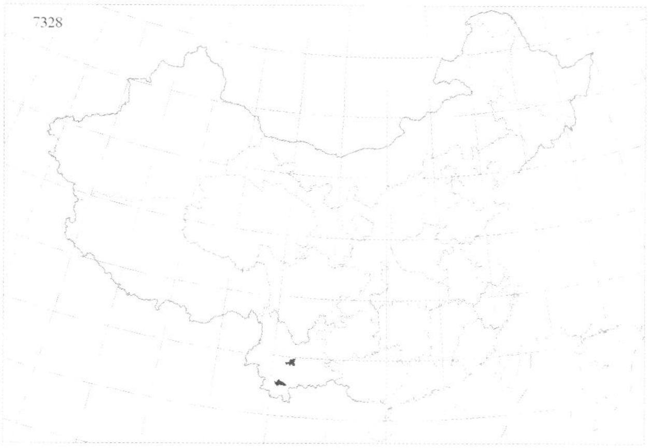

Myrtaceae 桃金娘科 *Syzygium* 蒲桃属
Syzygium tsoongii (Merr.) Merr. et L. M. Perry 狭叶蒲桃
Shrubs or small trees, evergreen

22.3 (14.6~24.8); 15.3 (3.8~20.1); 27.6 (24.6~28.8); 22.3 (14.7~24.8); 207.9 (117.7~237.6); 0.1 (0.0~1.5); 1205 (779~1385) ‖ 1563 (868~2148); 607 (312~1234); 101 (41~191) ‖ 33.8 (-34.3~101.8); 1125 (779~1360); 382.3 (235.1~620.0)

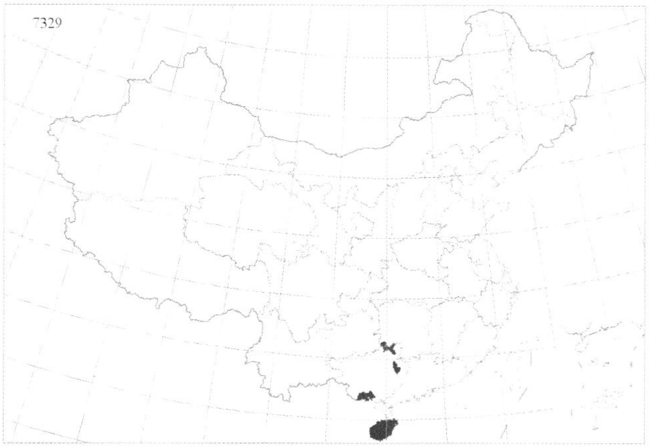

Myrtaceae 桃金娘科 *Syzygium* 蒲桃属
Syzygium vestitum Merr. et L. M. Perry 毛脉蒲桃
Small trees, evergreen

18.5 (17.2~21.0); 11.5 (9.5~14.2); 23.3 (22.4~25.6); 18.5 (17.3~21.0); 162.0 (147.0~191.8); 0.0 (0.0~0.0); 898 (833~1053) ‖ 1396 (1254~1522); 789 (699~860); 56 (48~68) ‖ 56.1 (36.8~80.6); 897 (833~1053); 444.4 (336.3~551.6)

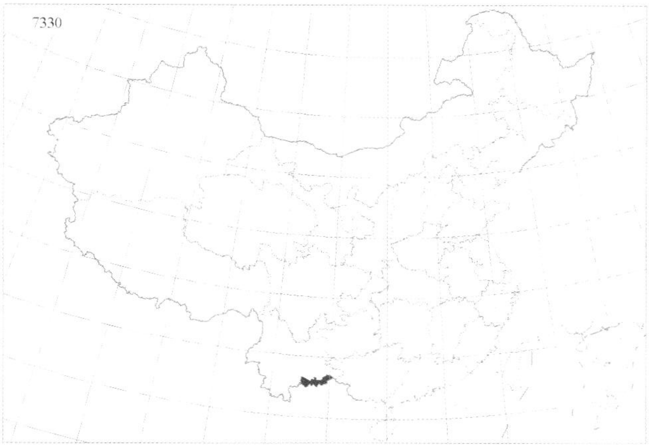

Myrtaceae 桃金娘科 *Syzygium* 蒲桃属
Syzygium wenshanense H. T. Chang et R. H. Miao 文山蒲桃
Small trees, evergreen

17.1; 9.6; 22.4; 17.1; 145.7; 0.0; 831 ‖ 1343; 757; 51 ‖ 61.3; 831; 364.2

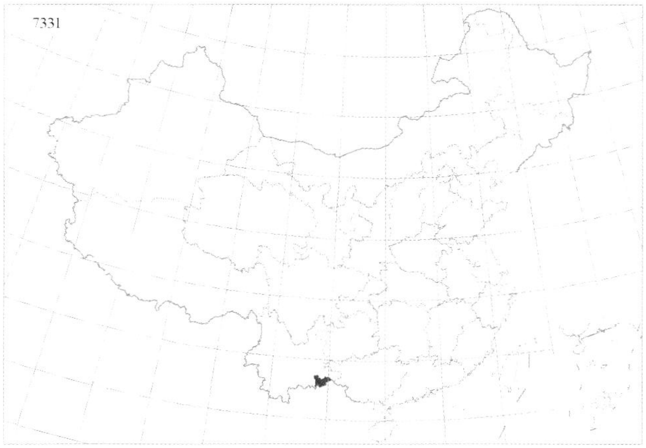

Myrtaceae 桃金娘科 *Syzygium* 蒲桃属
Syzygium xizangense Chang et Miau 西藏蒲桃
Trees, evergreen

13.3; 5.4; 19.6; 13.4; 100.4; 0.0; 706 ‖ 1736; 1007; 51 ‖ 145.8; 706; 447.4

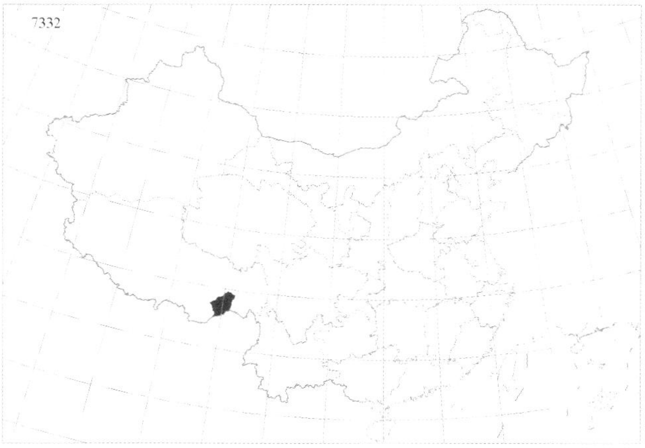

Myrtaceae 桃金娘科 *Syzygium* 蒲桃属
Syzygium yunnanense Merr. et L. M. Perry 云南蒲桃
Trees, evergreen

19.6 (18.6~21.0); 13.8 (12.7~15.1); 23.3 (22.3~25.6); 19.7 (18.7~21.0); 176.2 (163.9~191.8); 0.0 (0.0~0.0); 948 (888~1053) | 1539 (1428~1751); 797 (655~996); 66 (56~77) | 62.9 (44.5~88.8); 938 (880~1053); 536.8 (427.9~659.0)

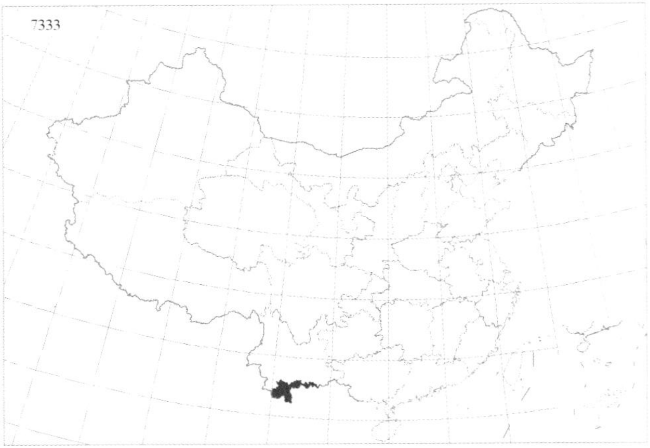

Myrtaceae 桃金娘科 *Syzygium* 蒲桃属
Syzygium zeylanicum (L.) DC. 锡兰蒲桃
Trees, evergreen

21.7 (21.3~22.4); 13.6 (12.6~14.2); 27.8 (27.3~28.6); 21.8 (21.3~22.4); 201.1 (195.9~209.1); 0.0 (0.0~0.0); 1147 (1111~1204) | 1800 (1188~2148); 972 (634~1234); 109 (56~125) | 56.7 (5.8~86.7); 1147 (1111~1204); 288.0 (235.1~318.8)

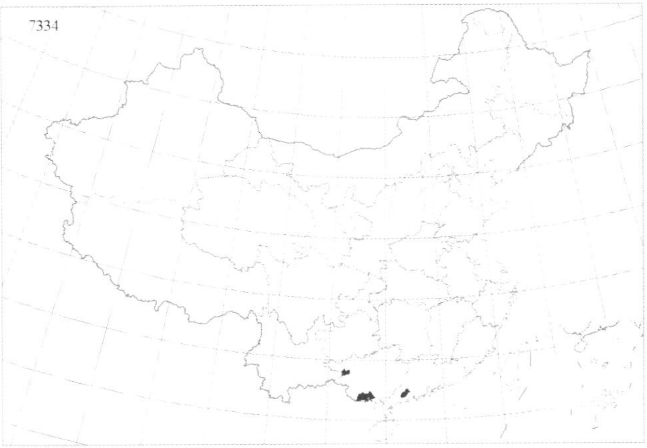

Melastomataceae 野牡丹科 *Allomorphia* 异形木属
Allomorphia balansae Cogn. 异形木
Shrubs

21.5 (17.6~24.3); 13.3 (7.3~19.1); 27.9 (25.8~28.9); 21.6 (17.6~24.3); 198.6 (151.6~231.9); 0.0 (0.0~0.0); 1145 (909~1338) | 1564 (868~2435); 703 (312~1428); 127 (42~194) | 38.5 (-34.3~99.5); 1107 (868~1326); 288.9 (185.5~620.0)

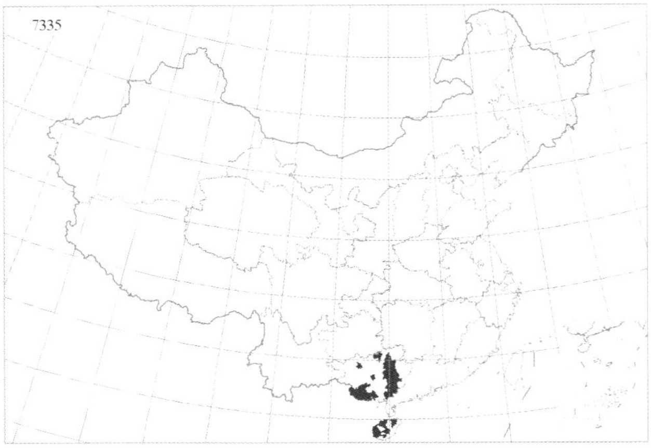

Melastomataceae 野牡丹科 *Allomorphia* 异形木属
Allomorphia baviensis Guillaumin 刺毛异形木
Shrubs

19.2 (17.2~22.2); 12.3 (9.1~15.1); 23.9 (22.3~28.3); 19.2 (17.3~22.3); 170.6 (147.0~207.5); 0.0 (0.0~0.0); 942 (833~1190) | 1453 (1161~1751); 780 (633~996); 64 (48~106) | 55.4 (15.4~88.8); 937 (833~1190); 443.1 (281.7~659.0)

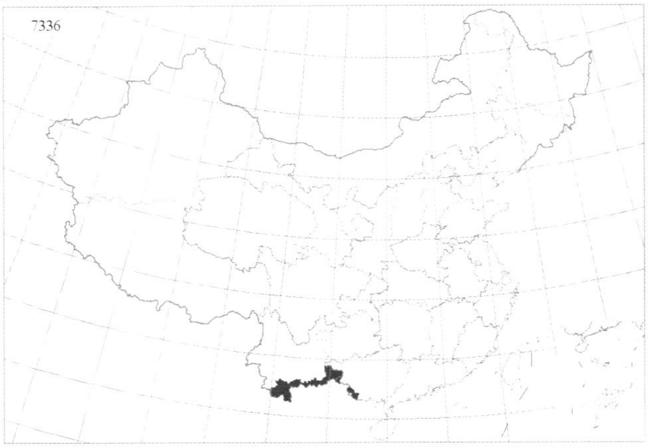

Melastomataceae 野牡丹科 *Allomorphia* 异形木属
Allomorphia urophylla Diels 尾叶异形木
Shrubs

19.0 (16.9~21.1); 11.1 (7.3~14.2); 25.0 (21.4~28.4); 19.1 (16.9~21.1); 169.0 (142.7~193.2); 0.0 (0.0~0.0); 953 (813~1115) | 1392 (1178~1522); 716 (561~848); 86 (51~168) | 46.7 (27.0~64.7); 951 (813~1107); 362.1 (225.9~551.6)

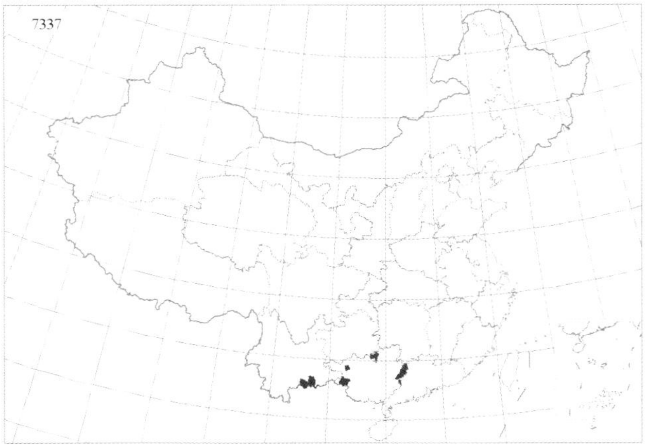

Melastomataceae 野牡丹科 *Astronia* 褐鳞木属
Astronia ferruginea Elmer 褐鳞木
Shrubs or small trees

19.1; 14.4; 22.6; 19.1; 169.6; 0.0; 920 | 2741; 1444; 181 | 200.7; 920; 304.1

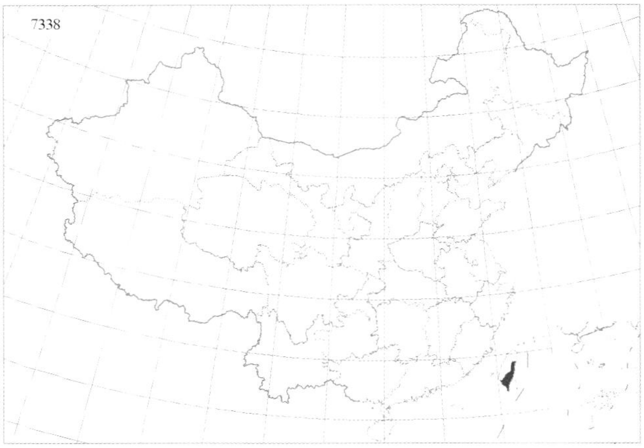

Melastomataceae 野牡丹科 *Barthea* 棱果花属
Barthea barthei (Hance ex Benth.) Krasser 棱果花
Shrubs

19.9 (14.6~22.8); 11.3 (3.8~15.5); 27.0 (20.1~28.9); 20.0 (14.7~22.8); 179.8 (117.7~214.1); 0.0 (0.0~1.5); 1041 (773~1225) ❙ 1735 (1302~2993); 794 (505~1428); 172 (82~547) ❙ 76.3 (12.9~246.5); 1040 (773~1225); 247.8 (163.7~363.7)

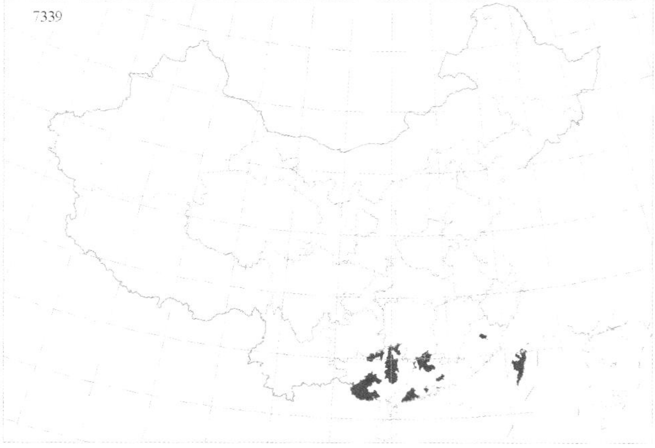

Melastomataceae 野牡丹科 *Barthea* 棱果花属
Barthea barthei var. valdealata C. Hansen 宽翅棱果花
Shrubs

21.9; 14.1; 27.9; 22.0; 203.7; 0.0; 1161 ❙ 2161; 1240; 126 ❙ 85.7; 1161; 305.5

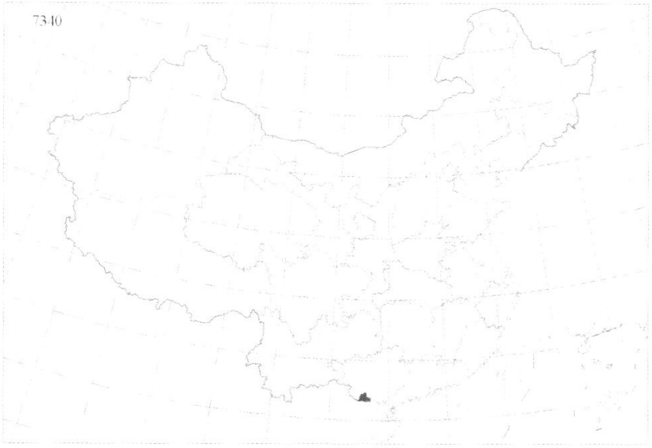

Melastomataceae 野牡丹科 *Blastus* 柏拉木属
Blastus auriculatus Y. C. Huang ex C. Chen 耳基柏拉木
Shrubs

21.0; 14.2; 25.6; 21.0; 191.8; 0.0; 1053 ❙ 1522; 818; 68 ❙ 44.5; 1053; 551.5

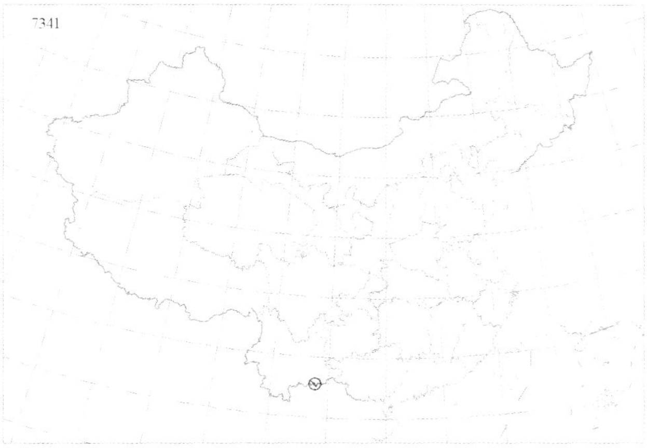

Melastomataceae 野牡丹科 *Blastus* 柏拉木属
Blastus borneensis Cogn. ex Boerlage 南亚柏拉木
Shrubs

19.2 (14.6~24.8); 10.4 (3.8~20.1); 26.7 (24.6~28.4); 19.2 (14.7~24.8); 171.2 (117.7~237.6); 0.3 (0.0~1.5); 1033 (779~1385) ❙ 1508 (1179~1697); 535 (392~618); 139 (41~194) ❙ 56.4 (-11.9~101.8); 945 (779~1137); 329.2 (236.9~527.5)

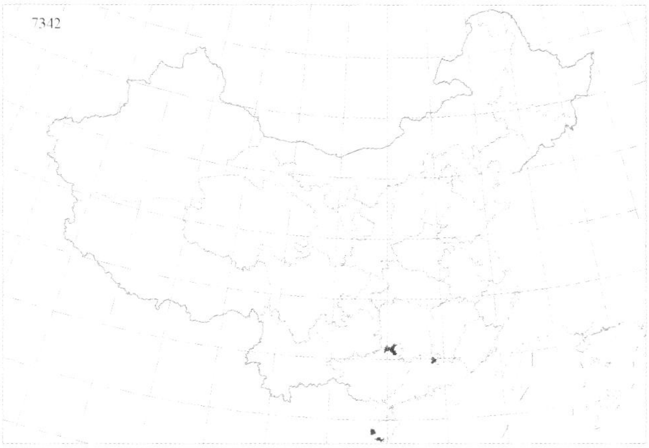

Melastomataceae 野牡丹科 *Blastus* 柏拉木属
Blastus brevissimus C. Chen 短柄柏拉木
Small shrubs

19.7; 10.1; 27.7; 19.8; 177.4; 0.0; 1031 ❙ 1510; 614; 165 ❙ 46.4; 1031; 253.0

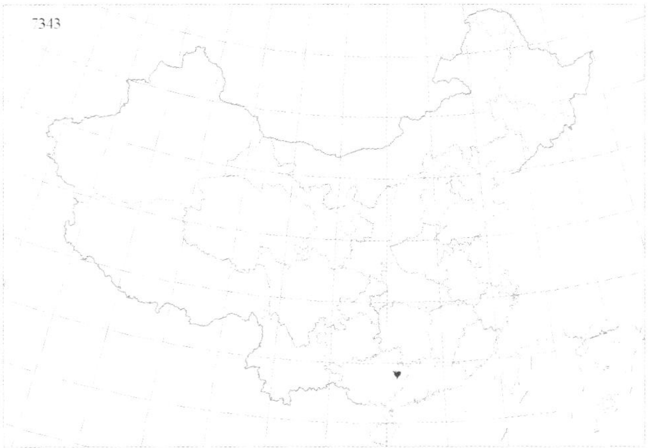

Melastomataceae 野牡丹科 *Blastus* 柏拉木属
Blastus cochinchinensis Lour. 柏拉木
Shrubs

19.6 (14.3~24.8); 10.5 (3.3~20.1); 27.4 (20.1~29.8); 19.6 (14.3~24.8); 175.9 (113.2~237.6); 0.1 (0.0~2.3); 1031 (761~1385) ❙ 1652 (1161~3495); 677 (392~1847); 163 (41~897) ❙ 62.8 (-11.9~246.5); 1021 (761~1326); 240.2 (95.3~620.0)

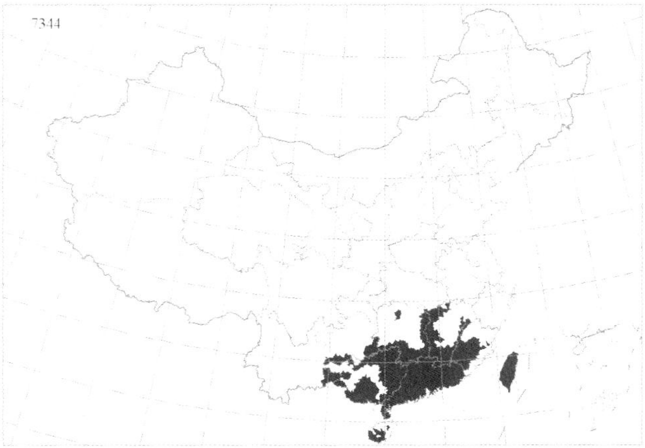

Melastomataceae 野牡丹科 *Blastus* 柏拉木属

Blastus mollissimus H. L. Li 密毛柏拉木

Shrubs

19.6 (17.9~21.3); 9.8 (7.4~12.3); 27.8 (25.8~28.6); 19.7 (18.0~21.3); 176.1 (155.6~196.1); 0.0 (0.0~0.0); 1032 (913~1130) ❙ 1596 (1371~1758); 669 (562~728); 175 (161~189) ❙ 56.0 (22.7~85.5); 1031 (913~1130); 220.0 (185.5~243.9)

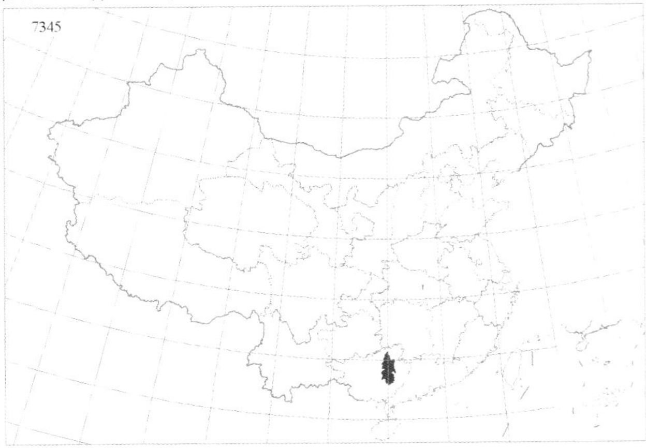

Melastomataceae 野牡丹科 *Blastus* 柏拉木属

Blastus pauciflorus (Bentham) Guillaumin 少花柏拉木

Shrubs

18.1 (13.7~23.4); 8.0 (3.3~16.1); 26.9 (20.5~29.8); 18.1 (13.7~23.5); 157.5 (104.8~221.7); 0.1 (0.0~2.3); 948 (722~1273) ❙ 1565 (976~2435); 603 (392~1428); 164 (38~259) ❙ 65.7 (11.9~132.1); 943 (717~1273); 237.5 (145.3~551.6)

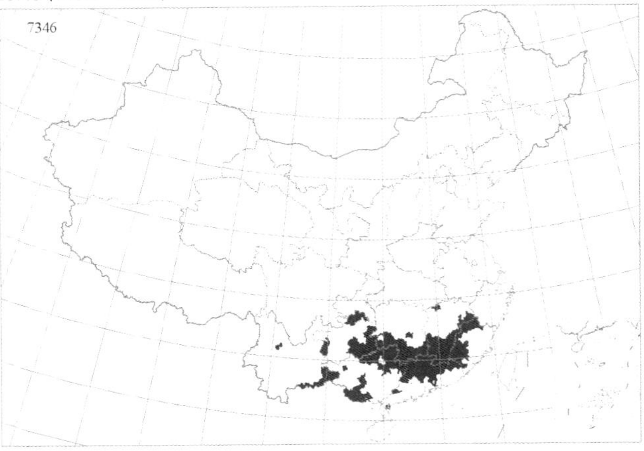

Melastomataceae 野牡丹科 *Blastus* 柏拉木属

Blastus setulosus Diels 刺毛柏拉木

Shrubs

19.8 (17.9~22.0); 10.2 (7.4~13.5); 27.8 (25.8~28.8); 19.9 (18.0~22.0); 178.4 (155.6~203.8); 0.0 (0.0~0.0); 1043 (913~1170) ❙ 1601 (1371~1758); 674 (562~728); 171 (132~189) ❙ 54.7 (22.7~85.5); 1042 (913~1170); 218.6 (185.5~243.9)

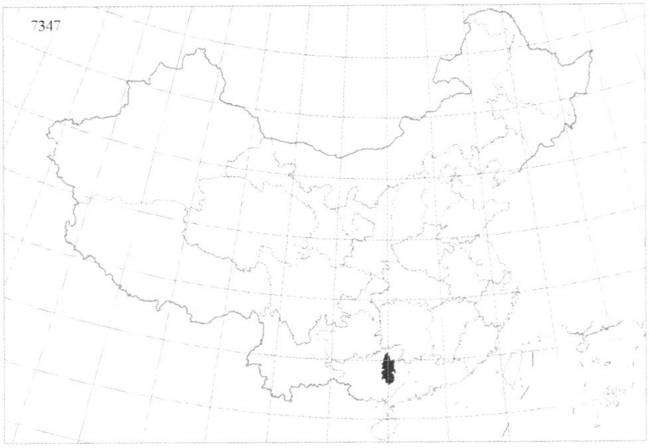

Melastomataceae 野牡丹科 *Blastus* 柏拉木属

Blastus tenuifolius Diels 薄叶柏拉木

Shrubs

19.8 (17.9~21.3); 10.1 (7.4~12.9); 27.8 (25.8~28.6); 19.8 (18.0~21.4); 177.8 (155.6~196.3); 0.0 (0.0~0.0); 1039 (913~1130) ❙ 1596 (1371~1758); 672 (562~728); 172 (137~189) ❙ 54.8 (22.7~85.5); 1038 (913~1130); 220.3 (185.5~243.9)

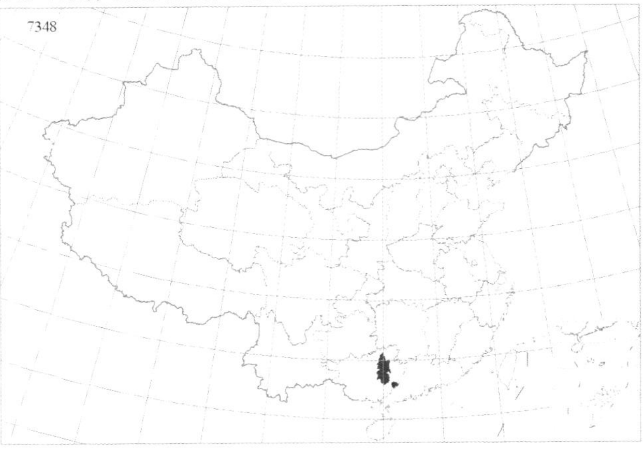

Melastomataceae 野牡丹科 *Blastus* 柏拉木属

Blastus tsaii H. L. Li 云南柏拉木

Shrubs

18.7; 11.7; 23.5; 18.7; 164.1; 0.0; 914 ❙ 1465; 812; 61 ❙ 61.7; 914; 485.5

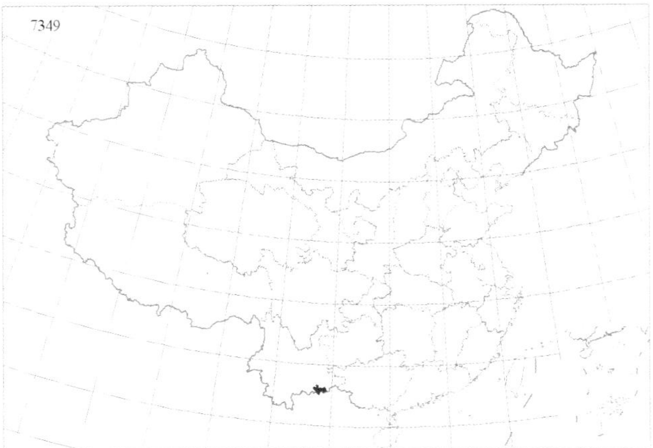

Melastomataceae 野牡丹科 *Bredia* 野海棠属

Bredia biglandularis C. Chen 双腺野海棠

Small shrubs

22.3; 14.5; 28.1; 22.3; 207.7; 0.0; 1186 ❙ 2291; 1331; 127 ❙ 93.1; 1186; 309.7

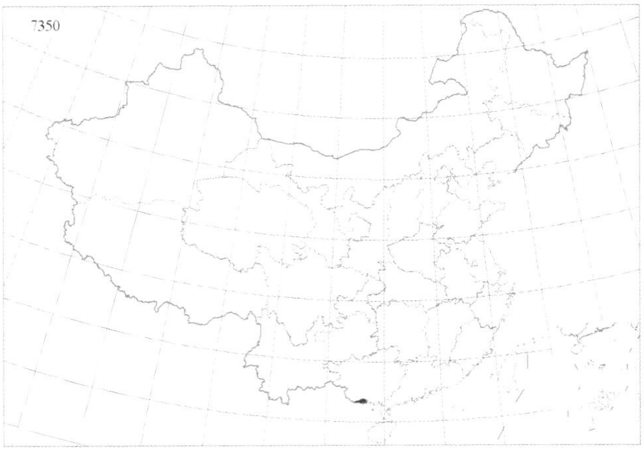

Melastomataceae 野牡丹科 *Bredia* 野海棠属

Bredia esquirolii (H. Leveille) Lauener 赤水野海棠

Subshrubs

16.6 (14.9~18.5); 5.8 (3.6~8.6); 26.6 (24.6~28.5); 16.7 (15.0~18.5); 140.3 (121.2~162.1); 0.3 (0.0~1.6); 879 (807~965) ❙ 1418 (1089~1773); 540 (450~753); 140 (50~233) ❙ 61.1 (30.9~83.7); 876 (807~953); 212.7 (181.2~261.7)

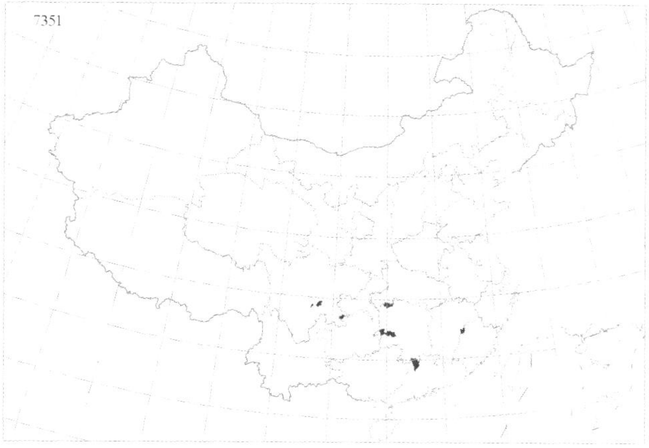

Melastomataceae 野牡丹科 *Bredia* 野海棠属

Bredia fordii (Hance) Diels 叶底红

Shrubs or subshrubs

18.8 (11.9~23.2); 9.4 (2.9~15.8); 27.1 (20.0~29.8); 18.8 (12.0~23.2); 166.2 (87.2~218.9); 0.1 (0.0~3.4); 981 (671~1257) ❙ 1626 (1001~2024); 621 (452~902); 167 (35~254) ❙ 66.7 (20.3~130.7); 980 (671~1257); 244.9 (146.0~332.5)

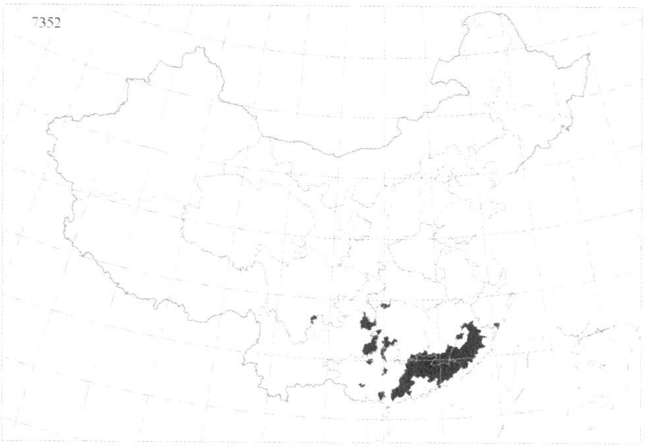

Melastomataceae 野牡丹科 *Bredia* 野海棠属

Bredia hirsuta var. scandens Ito et Matsum. 野海棠

Small shrubs

18.9; 13.2; 23.5; 18.9; 167.2; 0.0; 932 ❙ 2336; 1169; 180 ❙ 160.6; 921; 245.3

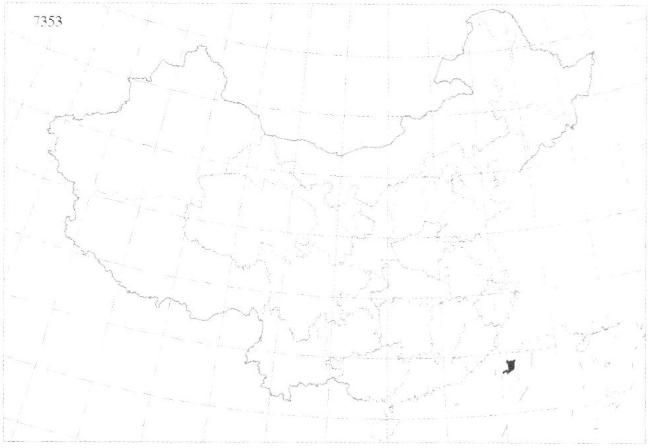

Melastomataceae 野牡丹科 *Bredia* 野海棠属

Bredia longiloba (Hand.-Mazz.) Diels 长萼野海棠

Subshrubs

16.3 (11.4~18.7); 4.9 (1.5~8.8); 26.6 (20.0~28.2); 16.3 (11.4~18.8); 137.5 (83.2~165.2); 1.9 (0.0~6.6); 880 (659~968) ❙ 1474 (828~1847); 563 (448~670); 163 (36~242) ❙ 67.9 (-6.0~101.9); 871 (659~968); 226.4 (172.8~268.9)

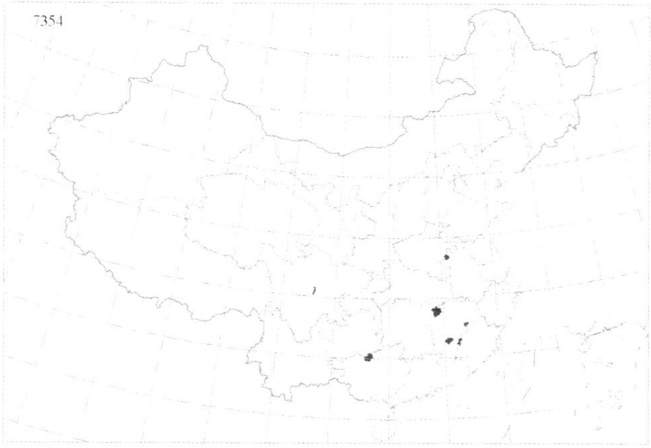

Melastomataceae 野牡丹科 *Bredia* 野海棠属

Bredia microphylla H. L. Li 小叶野海棠

Subshrubs

16.8; 6.2; 26.3; 16.9; 142.4; 0.1; 881 ❙ 1646; 631; 185 ❙ 87.0; 881; 235.4

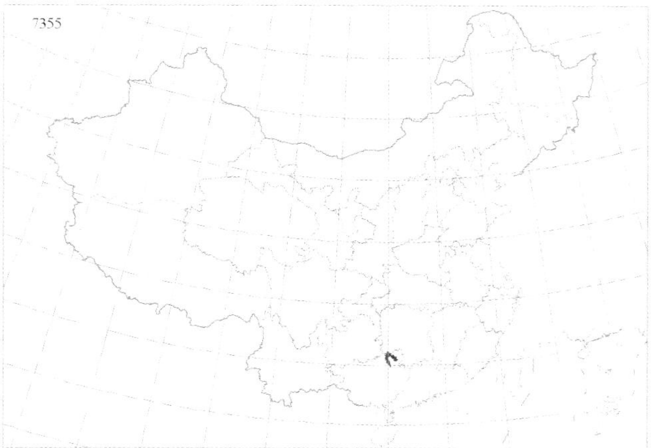

Melastomataceae 野牡丹科 *Bredia* 野海棠属

Bredia oldhamii Hook. f. 金石榴

Small shrubs

19.6 (16.2~24.1); 14.2 (11.0~18.9); 24.2 (20.1~28.0); 19.7 (16.2~24.1); 176.2 (134.2~229.7); 0.0 (0.0~0.0); 977 (773~1322) ❙ 2754 (1751~3495); 1168 (821~1847); 341 (47~897) ❙ 190.5 (32.5~246.5); 954 (773~1123); 265.2 (154.2~363.7)

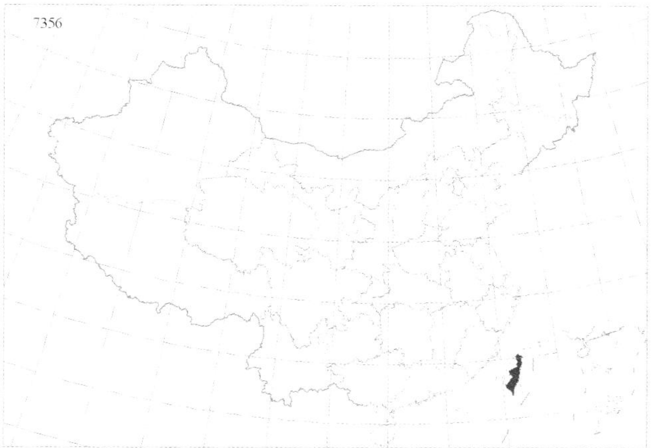

Melastomataceae 野牡丹科 *Bredia* 野海棠属

Bredia quadrangularis Cogn. 过路惊

Small shrubs

16.9 (12.6~21.3); 6.0 (0.7~12.3); 27.2 (23.0~29.2); 17.0 (12.6~21.3); 144.6 (98.8~196.1); 0.7 (0.0~9.4); 903 (711~1130) ❙ 1643 (976~1997); 582 (427~778); 196 (68~259) ❙ 83.2 (16.6~147.0); 899 (711~1130); 240.0 (158.9~332.5)

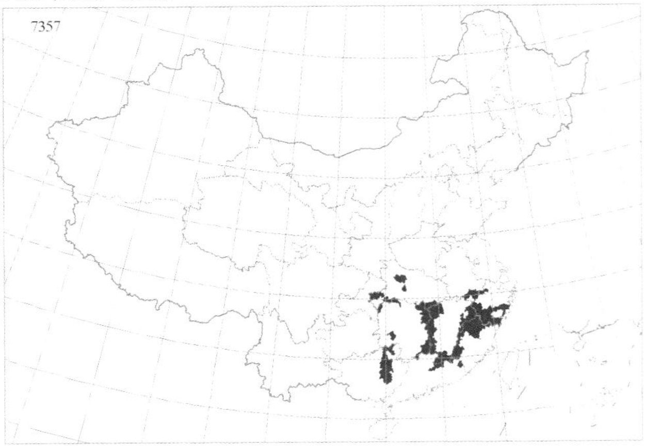

Melastomataceae 野牡丹科 *Bredia* 野海棠属

Bredia sessilifolia H. L. Li 短柄野海棠

Shrubs

20.0 (14.6~22.7); 11.2 (3.8~15.0); 27.1 (23.4~28.9); 20.1 (14.7~22.8); 180.9 (117.7~213.1); 0.1 (0.0~1.5); 1050 (779~1221) ❙ 1573 (1232~2435); 773 (544~1428); 120 (51~194) ❙ 50.9 (12.9~101.8); 1050 (779~1221); 259.1 (189.5~335.1)

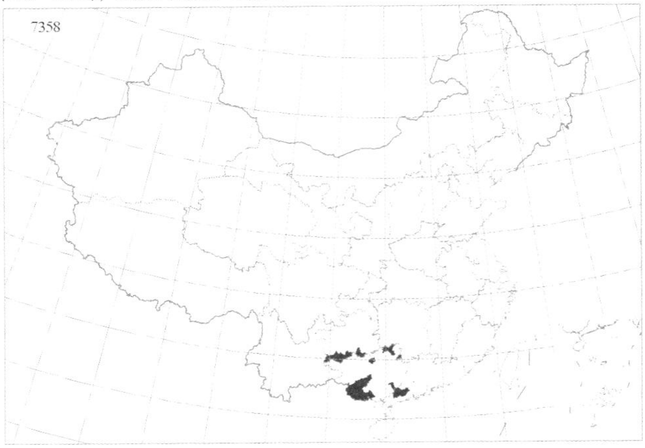

Melastomataceae 野牡丹科 *Bredia* 野海棠属

Bredia sinensis (Diels) H. L. Li 鸭脚茶

Shrubs

19.2 (14.3~23.2); 9.8 (3.7~15.8); 27.5 (24.2~29.8); 19.2 (14.3~23.2); 170.7 (113.2~218.9); 0.0 (0.0~1.5); 1006 (761~1257) ❙ 1656 (1090~2037); 631 (359~1033); 178 (75~259) ❙ 67.0 (-3.3~147.0); 1002 (761~1257); 233.1 (95.3~332.5)

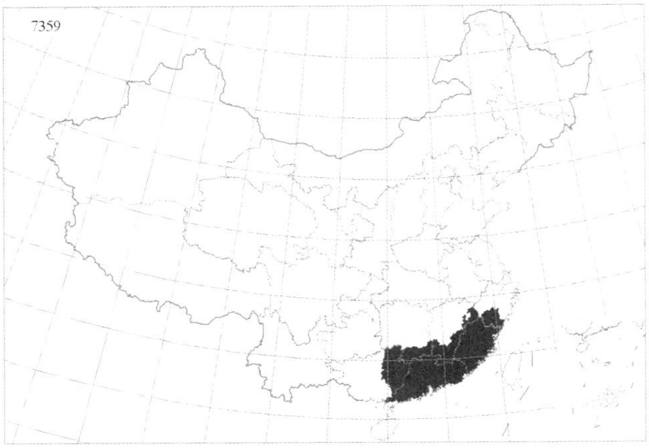

Melastomataceae 野牡丹科 *Cyphotheca* 药囊花属

Cyphotheca montana Diels 药囊花

Shrubs

18.1 (16.0~21.0); 11.9 (9.3~15.1); 22.2 (20.3~25.6); 18.2 (16.0~21.0); 158.2 (132.4~191.8); 0.0 (0.0~0.0); 873 (780~1053) ❙ 1313 (1028~1751); 724 (566~996); 54 (42~77) ❙ 50.5 (12.5~88.8); 858 (779~1053); 432.1 (294.5~659.0)

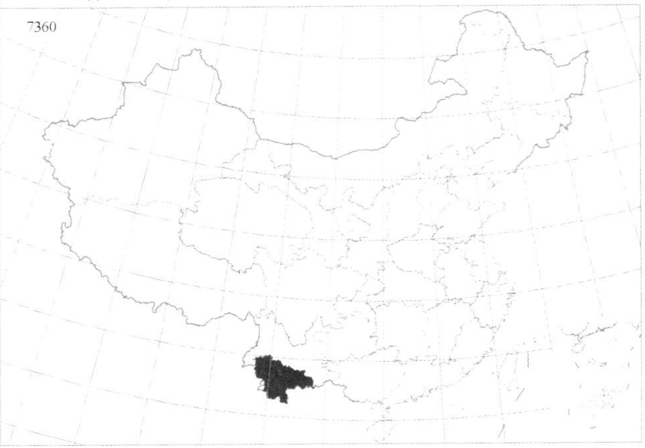

Melastomataceae 野牡丹科 *Diplectria* 藤牡丹属

Diplectria barbata (Wall. ex C. B. Clarke) Franken et M. C. Roos 藤牡丹

Scandent shrubss or vines

22.8 (18.0~24.8); 16.8 (8.5~20.1); 27.1 (25.2~28.4); 22.8 (18.0~24.8); 213.7 (156.4~237.6); 0.0 (0.0~0.0); 1219 (913~1385) ❙ 1453 (1179~1666); 550 (392~848); 63 (41~169) ❙ 22.3 (-11.9~69.8); 1082 (913~1212); 467.9 (243.9~620.0)

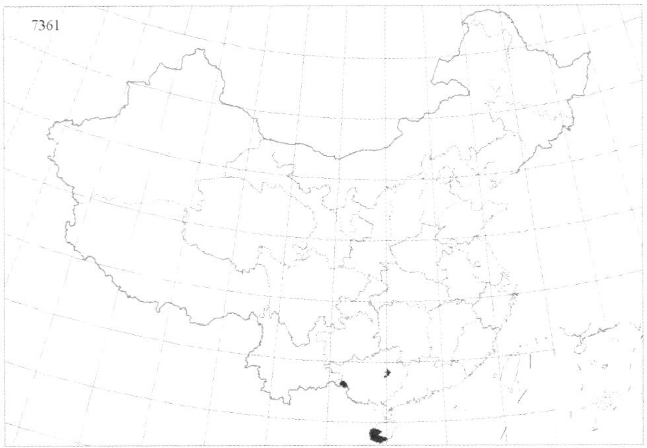

Melastomataceae 野牡丹科 *Fordiophyton* 异药花属

Fordiophyton breviscapum (C. Chen) Y. F. Deng et T. L. Wu 短葶无距花

Herbs or subshrubs

16.5 (14.6~17.7); 5.9 (3.8~7.8); 26.0 (24.6~27.5); 16.5 (14.7~17.7); 138.8 (117.7~152.5); 0.3 (0.0~1.5); 864 (779~927) ❙ 1562 (1486~1624); 556 (505~594); 182 (176~191) ❙ 81.9 (60.2~101.8); 864 (779~927); 241.0 (188.8~260.5)

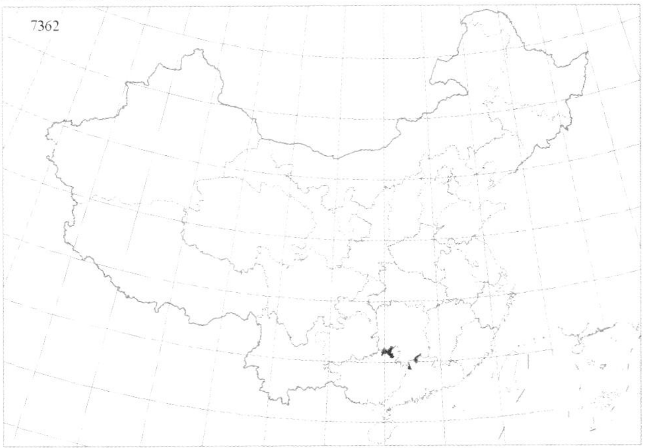

Melastomataceae 野牡丹科 *Fordiophyton* 异药花属
Fordiophyton degeneratum (C. Chen) Y. F. Deng et T. L. Wu 败蕊无距花
Herbs or subshrubs

21.7; 12.9 28.4; 21.7; 200.4; 0.0; 1155 ❙ 1512; 746; 133 ❙ 31.0; 1155; 251.2

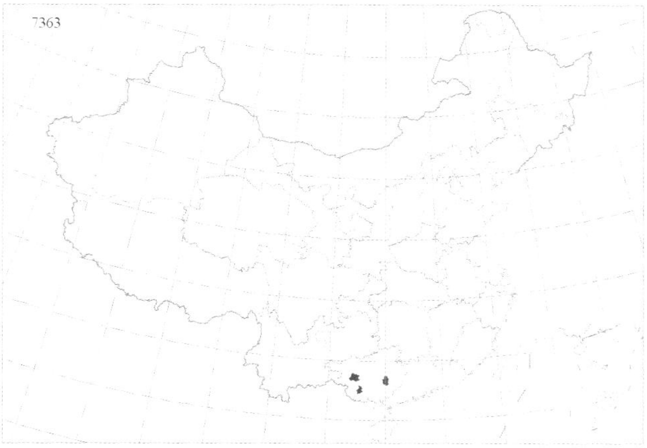

Melastomataceae 野牡丹科 *Medinilla* 酸脚杆属
Medinilla arboricola F. C. How 附生美丁花
Scandent shrubss or trees

24.0; 18.8; 27.6; 24.0; 228.4; 0.0; 1308 ❙ 1511; 519; 54 ❙ 16.1; 1145; 535.7

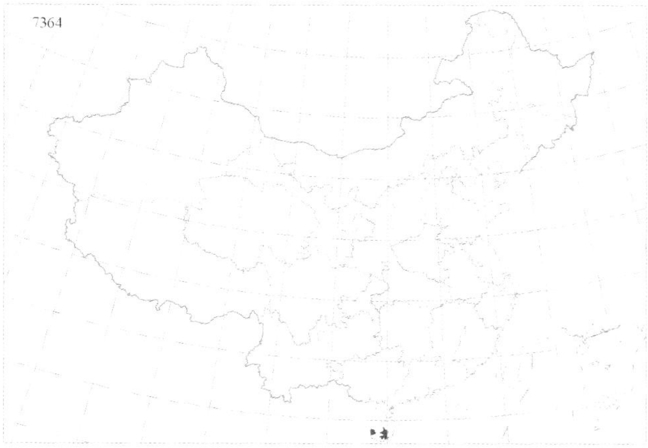

Melastomataceae 野牡丹科 *Medinilla* 酸脚杆属
Medinilla assamica (C. B. Clarke) C. Chen 顶花酸角杆
Shrubs

19.2 (17.2~24.3); 11.6 (7.3~18.7); 24.7 (22.4~28.4); 19.2 (17.3~24.3); 170.6 (147.0~231.9); 0.0 (0.0~0.0); 961 (833~1338) ❙ 1367 (1161~1547); 734 (392~911); 66 (42~142) ❙ 44.9 (-11.9~80.6); 943 (833~1140); 385.6 (225.9~551.6)

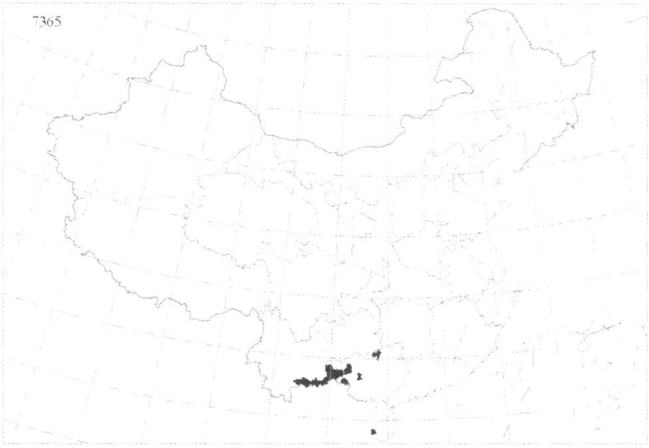

Melastomataceae 野牡丹科 *Medinilla* 酸脚杆属
Medinilla fengii (S. Y. Hu) C. Y. Wu et C. Chen 西畴酸角杆
Small shrubs

19.3 (16.2~24.1); 13.1 (9.5~18.9); 24.1 (20.1~28.0); 19.3 (16.2~24.1); 171.5 (134.2~229.7); 0.0 (0.0~0.0); 952 (773~1322) ❙ 2182 (1254~3495); 998 (699~1847); 226 (47~897) ❙ 132.3 (32.5~246.5); 939 (773~1123); 316.0 (154.2~551.6)

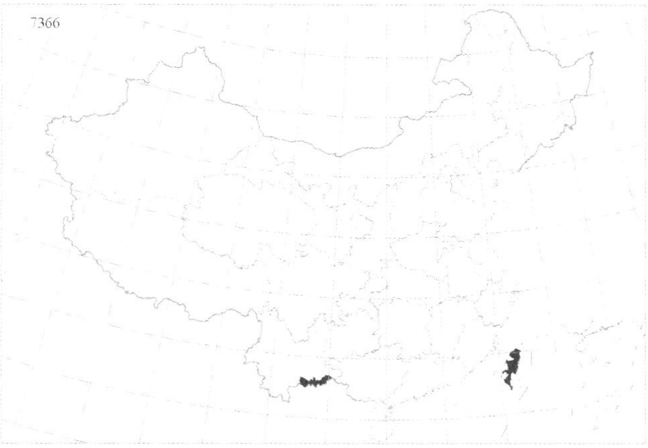

Melastomataceae 野牡丹科 *Medinilla* 酸脚杆属
Medinilla formosana Hayata 台湾酸角杆
Scandent shrubs

21.0; 16.3; 24.5; 21.1; 192.6; 0.0; 1055 ❙ 2554; 1499; 117 ❙ 154.0; 998; 266.8

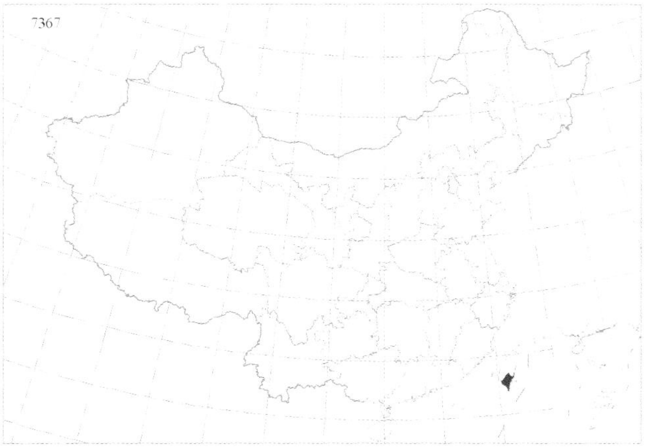

Melastomataceae 野牡丹科 *Medinilla* 酸脚杆属
Medinilla hayatana H. Keng 糠秕酸脚杆
Scandent shrubs

18.5; 14.1; 21.7; 18.5; 162.1; 0.0; 863 ❙ 2713; 1388; 188 ❙ 214.4; 863; 363.7

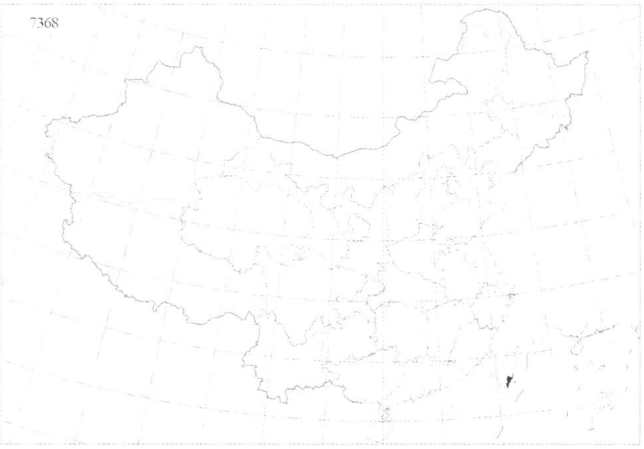

Melastomataceae 野牡丹科 *Medinilla* 酸脚杆属
Medinilla himalayana Hook. f. ex Triana 锥序酸角杆
Shrubs

19.6; 13.8; 23.1; 19.6; 174.9; 0.0; 940 ❙ 1506; 783; 64 ❙ 60.9; 923; 560.7

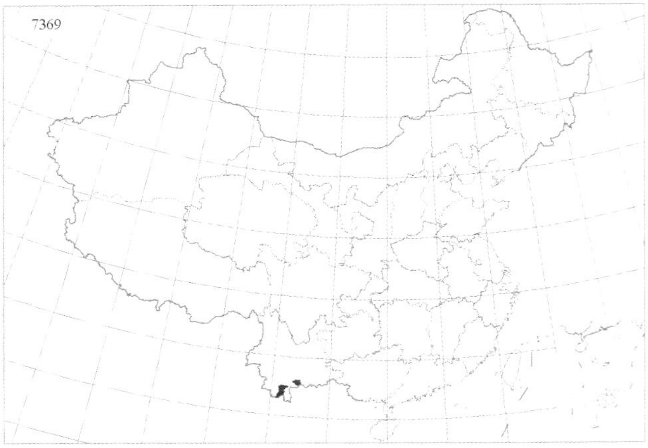

Melastomataceae 野牡丹科 *Medinilla* 酸脚杆属
Medinilla lanceata (Nayar) C. Chen 酸脚杆
Shrubs

22.5 (17.7~24.8); 16.7 (10.9~20.1); 26.5 (22.1~28.8); 22.5 (17.8~24.8); 210.4 (153.4~237.6); 0.0 (0.0~0.0); 1189 (854~1385) ❙ 1494 (868~1854); 619 (312~996); 70 (41~115) ❙ 28.9 (-34.3~88.8); 1107 (854~1360); 467.7 (248.2~659.0)

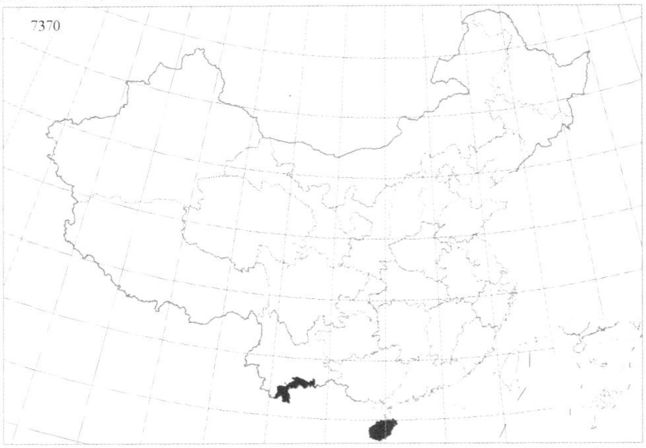

Melastomataceae 野牡丹科 *Medinilla* 酸脚杆属
Medinilla nana S. Y. Hu 矮酸脚杆
Small shrubs

17.6 (13.3~21.0); 10.4 (5.4~14.2); 22.8 (19.6~25.6); 17.6 (13.4~21.0); 151.7 (100.4~191.8); 0.0 (0.0~0.0); 866 (706~1053) ❙ 1462 (1254~1736); 827 (699~1007); 56 (48~68) ❙ 71.7 (44.5~145.8); 866 (706~1053); 443.7 (336.3~551.6)

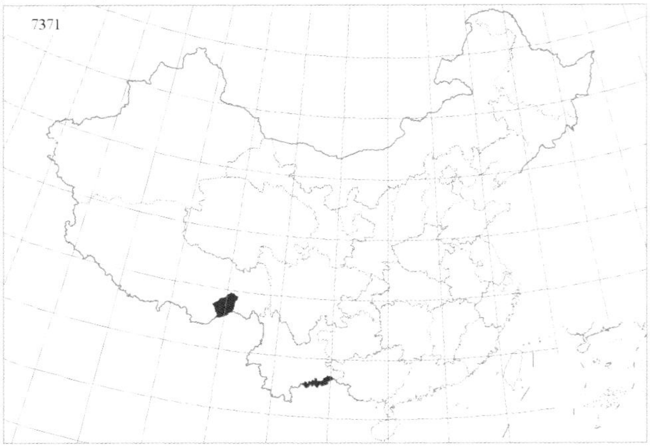

Melastomataceae 野牡丹科 *Medinilla* 酸脚杆属
Medinilla petelotii Merr. 沙巴酸脚杆
Shrubs

16.5 (8.6~21.0); 9.3 (1.3~14.2); 21.8 (15.0~25.6); 16.5 (8.7~21.0); 139.3 (52.6~191.8); 1.1 (0.0~8.8); 829 (574~1053) ❙ 1409 (1041~1736); 797 (585~1007); 54 (45~68) ❙ 72.9 (44.5~145.8); 829 (574~1053); 432.3 (336.3~551.6)

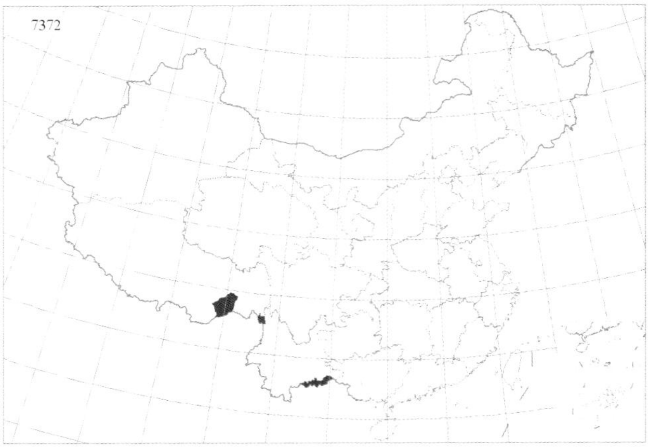

Melastomataceae 野牡丹科 *Medinilla* 酸脚杆属
Medinilla rubicunda (Jack) Blume 红花酸脚杆
Shrubs

18.1 (8.6~24.1); 11.7 (1.3~19.1); 22.6 (15.0~28.4); 18.1 (8.7~24.2); 158.0 (52.6~230.1); 0.8 (0.0~8.8); 926 (574~1326) ❙ 1432 (1041~1736); 688 (539~1007); 63 (44~110) ❙ 60.4 (19.1~145.8); 901 (574~1326); 455.2 (291.2~659.0)

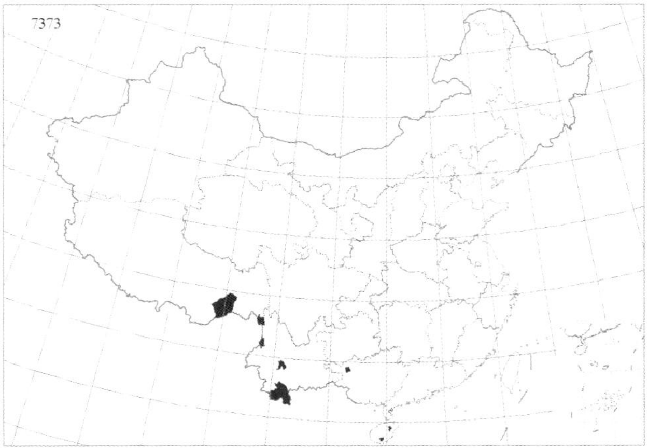

Melastomataceae 野牡丹科 *Medinilla* 酸脚杆属
Medinilla septentrionalis (W. W. Sm.) H. L. Li. 北酸角杆
Shrubs or small trees

19.3 (14.4~23.2); 11.8 (7.4~15.8); 24.7 (19.1~28.9); 19.3 (14.4~23.2); 171.9 (113.1~218.9); 0.0 (0.0~0.0); 972 (727~1257) ❙ 1490 (1028~2435); 775 (562~1428); 92 (42~194) ❙ 54.5 (5.8~112.2); 965 (727~1257); 361.8 (185.5~659.0)

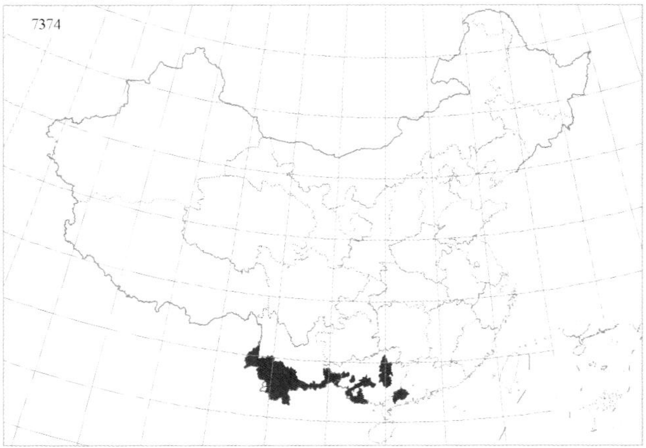

Melastomataceae 野牡丹科 *Melastoma* 野牡丹属
Melastoma dodecandrum Lour. 地菍
Small shrubs, evergreen

18.1 (12.7~23.6); 7.9 (2.1~16.7); 27.4 (21.6~30.1); 18.2 (12.8~23.6); 158.5 (99.0~223.5); 0.2 (0.0~5.6); 961 (713~1281) ❙ 1516 (871~2435); 591 (299~1428); 158 (43~259) ❙ 59.2 (-6.9~147.0); 951 (713~1273); 221.6 (95.3~411.1)

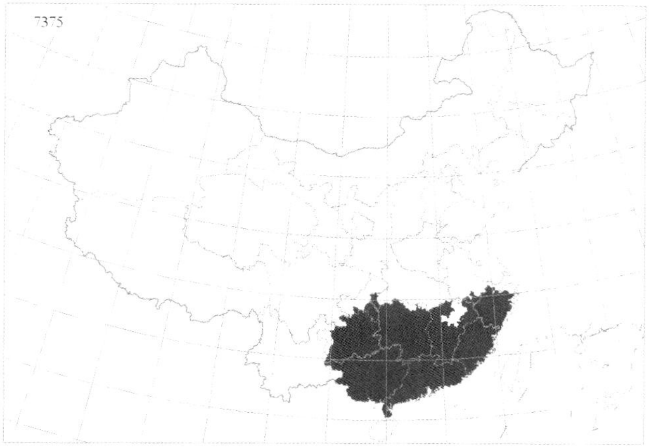

Melastomataceae 野牡丹科 *Melastoma* 野牡丹属
Melastoma imbricatum Wall. ex Triana 大野牡丹
Shrubs or small trees, evergreen

18.7 (16.1~22.2); 11.0 (5.3~15.1); 24.4 (22.4~28.3); 18.7 (16.1~22.3); 164.8 (133.1~207.5); 0.0 (0.0~0.0); 929 (833~1190) ❙ 1399 (1161~1751); 744 (523~996); 70 (48~129) ❙ 51.9 (5.8~88.8); 929 (833~1190); 395.5 (222.1~659.0)

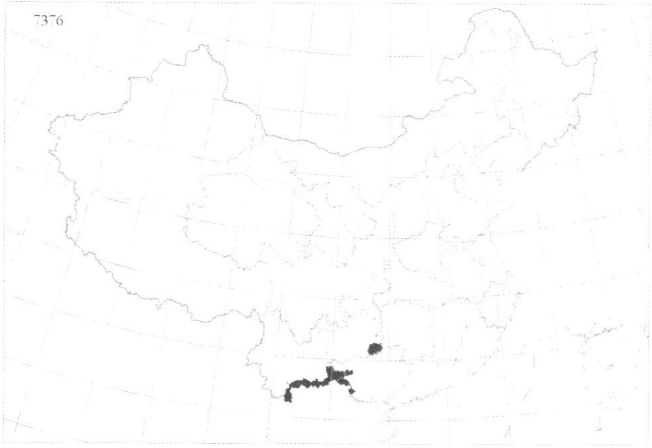

Melastomataceae 野牡丹科 *Melastoma* 野牡丹属
Melastoma intermedium Dunn 细叶野牡丹
Small shrubs, evergreen

21.0 (14.6~24.8); 12.9 (3.8~20.1); 27.6 (20.1~29.8); 21.0 (14.7~24.8); 192.2 (117.7~237.6); 0.0 (0.0~1.5); 1105 (773~1385) ❙ 1701 (868~3495); 733 (312~1847); 149 (41~897) ❙ 57.3 (-34.3~246.5); 1088 (773~1360); 250.3 (95.3~620.0)

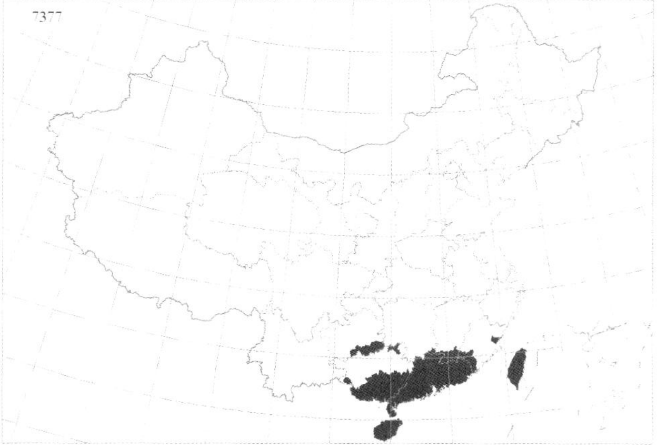

Melastomataceae 野牡丹科 *Melastoma* 野牡丹属
Melastoma malabathricum L. 野牡丹
Shrubs, evergreen

19.1 (-1.8~24.8); 10.6 (-10.2~20.1); 26.2 (6.1~29.8); 19.1 (1.8~24.8); 169.9 (1.9~237.6); 0.4 (0.0~82.8); 992 (329~1385) ❙ 1488 (320~3495); 670 (217~1847); 118 (10~897) ❙ 51.5 (-34.3~246.5); 982 (320~1360); 259.9 (72.2~659.0)

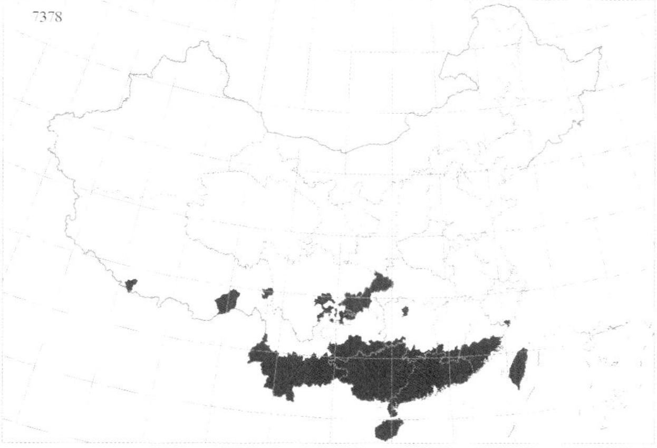

Melastomataceae 野牡丹科 *Melastoma* 野牡丹属
Melastoma sanguineum Sims 毛菍
Tall shrubs, evergreen

21.3 (17.6~24.8); 13.1 (7.1~20.1); 28.0 (25.7~29.8); 21.4 (17.7~24.8); 196.6 (152.5~237.6); 0.0 (0.0~0.0); 1129 (906~1385) ❙ 1637 (868~2435); 699 (312~1428); 136 (41~197) ❙ 46.4 (-34.3~99.5); 1112 (868~1299); 248.0 (95.3~620.0)

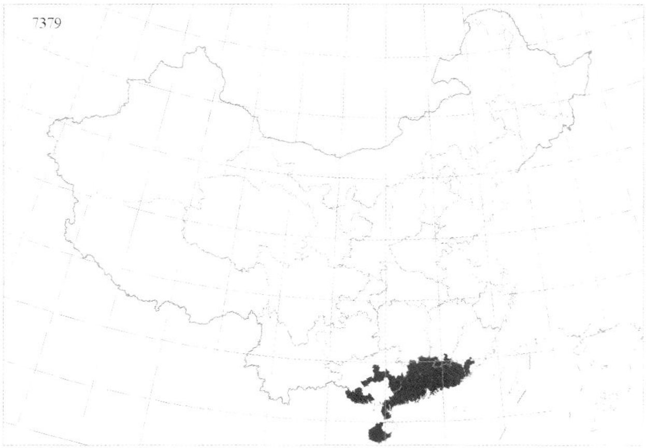

Melastomataceae 野牡丹科 *Melastoma* 野牡丹属
Melastoma sanguineum var. latisepalum C. Chen 宽萼毛菍
Subshrubs

23.9; 17.0; 28.1; 23.3; 219.8; 0.0; 1255 ❙ 1673; 724; 76 ❙ 34.1; 1164; 329.8

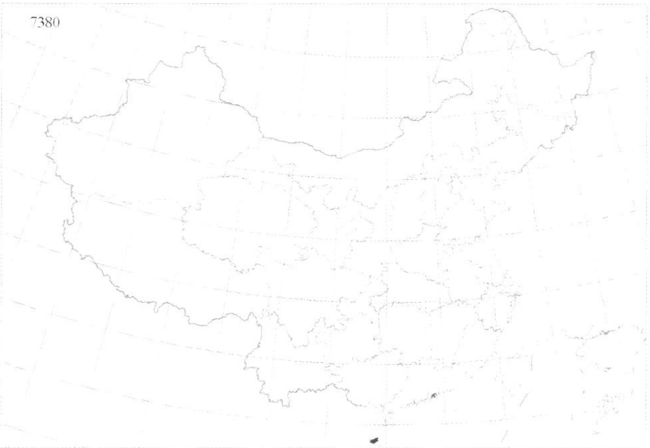

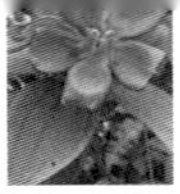

Melastomataceae 野牡丹科 *Memecylon* 谷木属

Memecylon caeruleum Jack 天蓝谷木

Shrubs or small trees

20.2 (13.3~24.3); 14.1 (5.4~18.7); 24.3 (19.6~28.4); 20.2 (13.4~24.3); 182.4 (100.4~231.9); 0.0 (0.0~0.0); 1043 (706~1338) ❙ 1424 (1179~1736); 637 (392~1007); 52 (42~66) ❙ 48.8 (-11.9~145.8); 939 (706~1091); 485.3 (439.7~570.1)

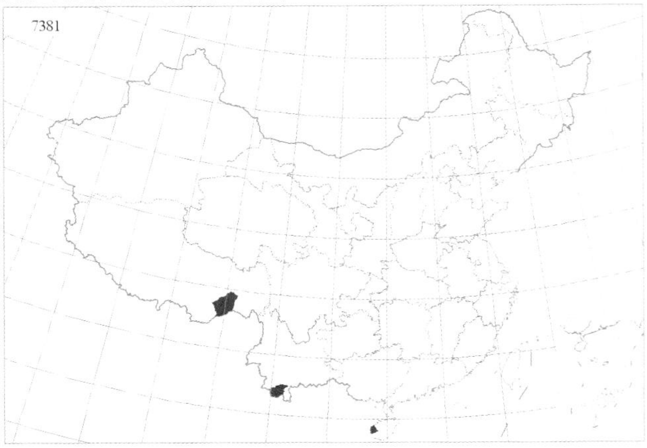

Melastomataceae 野牡丹科 *Memecylon* 谷木属

Memecylon hainanense Merr. et Chun 海南谷木

Shrubs or small trees

20.4 (16.3~24.8); 13.8 (8.8~20.1); 25.0 (21.4~28.6); 20.4 (16.4~24.8); 184.8 (136.4~237.6); 0.0 (0.0~0.0); 1052 (797~1385) ❙ 1401 (1074~1725); 631 (455~860); 59 (41~117) ❙ 36.9 (-1.4~80.6); 994 (797~1254); 436.9 (238.3~620.0)

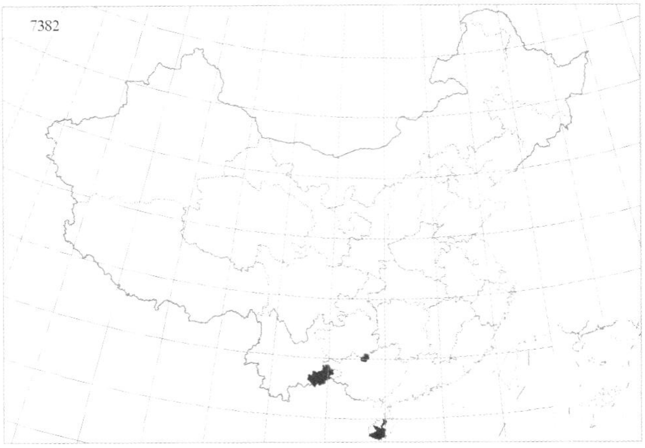

Melastomataceae 野牡丹科 *Memecylon* 谷木属

Memecylon lanceolatum Blanco 狭叶谷木

Shrubs or small trees

18.5; 14.1; 21.7; 18.5; 162.1; 0.0; 863 ❙ 2713; 1388; 188 ❙ 214.4; 863; 363.7

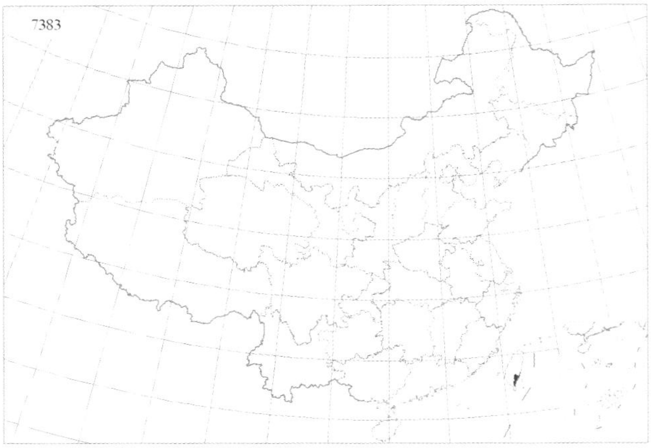

Melastomataceae 野牡丹科 *Memecylon* 谷木属

Memecylon ligustrifolium Champion ex Bentham 谷木

Shrubs or small trees

21.0 (17.3~24.8); 12.8 (7.1~20.1); 27.7 (22.3~29.0); 21.0 (17.3~24.8); 192.4 (147.6~237.6); 0.0 (0.0~0.0); 1106 (854~1385) ❙ 1614 (868~2435); 695 (312~1428); 134 (41~235) ❙ 47.6 (-34.3~99.5); 1083 (854~1360); 274.3 (111.7~620.0)

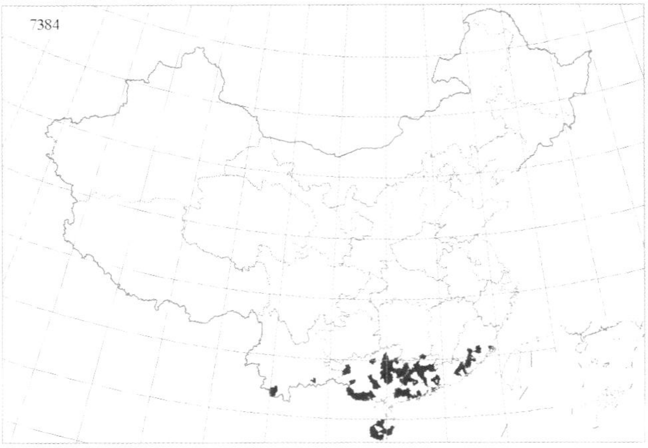

Melastomataceae 野牡丹科 *Memecylon* 谷木属

Memecylon ligustrifolium var. monocarpum C. Chen 单果谷木

Shrubs or small trees

17.7; 10.9; 22.5; 17.8; 153.4; 0.0; 854 ❙ 1368; 756; 59 ❙ 60.1; 854; 491.5

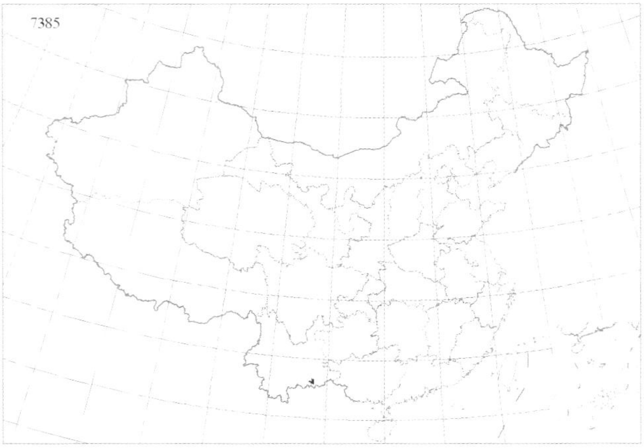

Melastomataceae 野牡丹科 *Memecylon* 谷木属

Memecylon luchuenense C. Chen 禄春谷木

Shrubs

18.6; 12.7; 22.5; 18.7; 163.9; 0.0; 888 ❙ 1547; 911; 63 ❙ 74.2; 888; 551.3

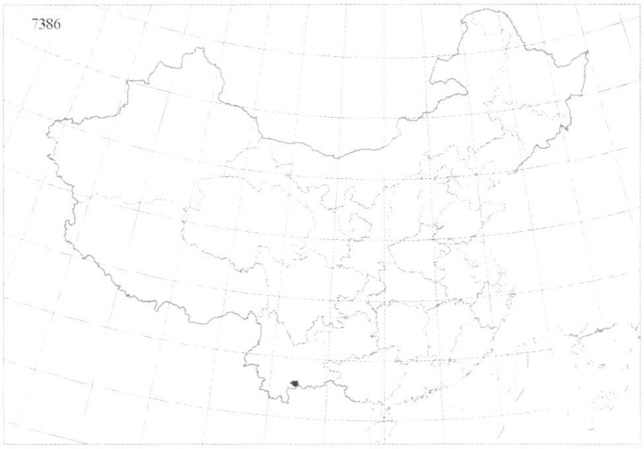

Melastomataceae 野牡丹科 *Memecylon* 谷木属

Memecylon nigrescens Hook. et Arn. 黑叶谷木

Shrubs or small trees

23.1 (21.3~24.8); 16.4 (12.7~20.1); 28.2 (27.2~29.0); 23.1 (21.3~24.8); 217.6 (195.6~237.6); 0.0 (0.0~0.0); 1250 (1107~1385) ❙ 1339 (1176~2030); 675 (447~1033); 96 (41~167) ❙ 32.8 (-12.0~79.9); 1169 (1107~1329); 331.8 (139.6~620.0)

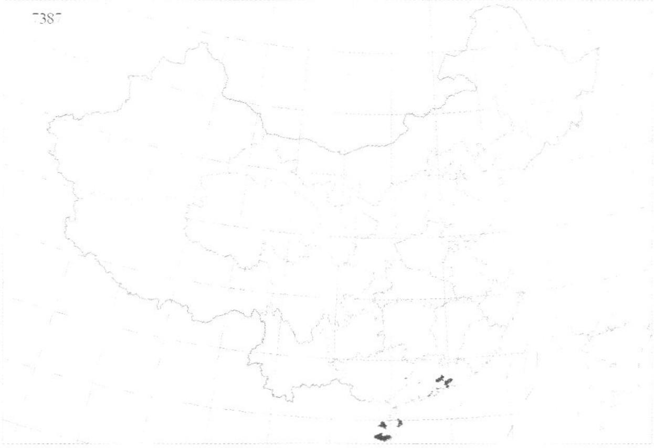

Melastomataceae 野牡丹科 *Memecylon* 谷木属

Memecylon octocostatum Merr. et Chun 棱果谷木

Shrubs

23.5 (22.3~24.8); 17.4 (14.7~20.1); 28.1 (27.5~28.6); 23.6 (22.3~24.8); 222.8 (208.2~237.6); 0.0 (0.0~0.0); 1277 (1185~1385) ❙ 1441 (1179~2035); 590 (392~969); 64 (41~113) ❙ 13.9 (-11.9~71.8); 1148 (1056~1269); 366.6 (215.3~527.5)

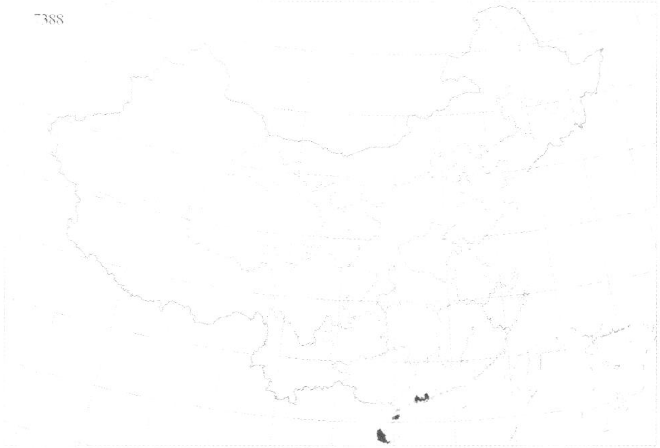

Melastomataceae 野牡丹科 *Memecylon* 谷木属

Memecylon pauciflorum Blume 少花谷木

Shrubs or small trees

23.4 (22.2~24.7); 16.7 (14.7~19.8); 28.4 (27.9~29.0); 23.5 (22.2~24.7); 221.4 (206.9~236.5); 0.0 (0.0~0.0); 1273 (1177~1376) ❙ 1530 (868~1840); 666 (312~889); 90 (58~114) ❙ 21.1 (-34.3~50.4); 1187 (868~1329); 285.6 (203.1~494.7)

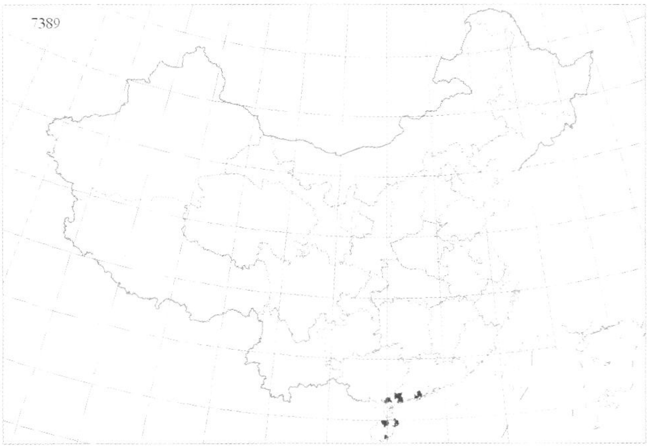

Melastomataceae 野牡丹科 *Memecylon* 谷木属

Memecylon polyanthum H. L. Li 滇谷木

Shrubs or small trees

19.5; 13.6; 23.3; 19.6; 175.1; 0.0; 945 ❙ 1480; 704; 67 ❙ 56.8; 933; 573.5

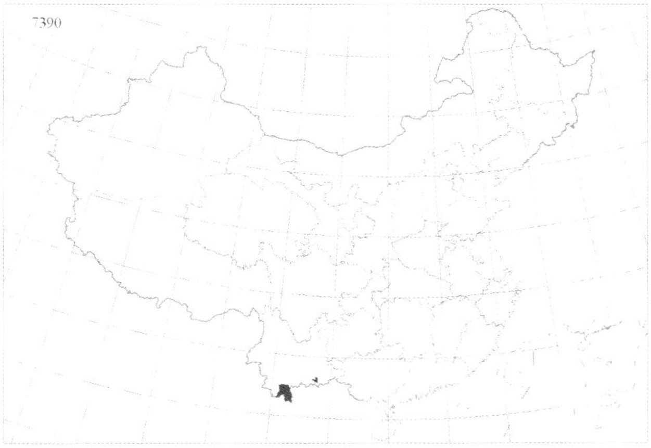

Melastomataceae 野牡丹科 *Memecylon* 谷木属

Memecylon scutellatum (Loureiro) Hook. et Arn. 细叶谷木

Shrubs

22.3 (18.3~24.7); 14.8 (9.2~19.8); 28.0 (25.5~29.2); 22.3 (18.4~24.7); 208.1 (160.9~236.5); 0.0 (0.0~0.0); 1194 (924~1376) ❙ 1551 (1176~2148); 705 (392~1234); 99 (42~167) ❙ 30.7 (-12.0~86.7); 1162 (924~1360); 313.1 (139.6~620.0)

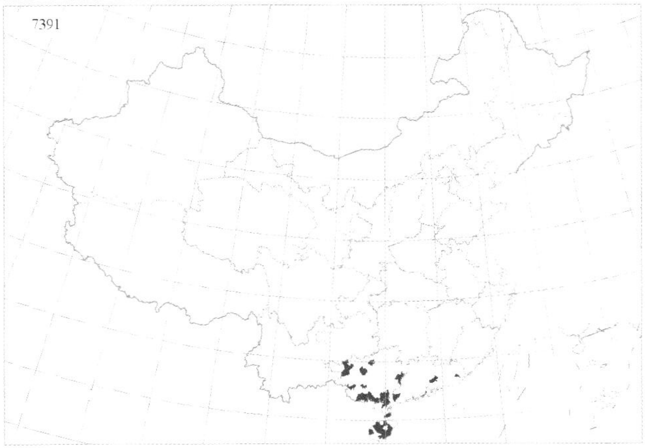

Melastomataceae 野牡丹科 *Osbeckia* 金锦香属

Osbeckia capitata Bentham ex Walpers 头序金锦香

Herbs or subshrubs

16.2; 10.0; 20.5; 16.2; 134.4; 0.0; 807 ❙ 1368; 695; 64 ❙ 70.0; 807; 506.2

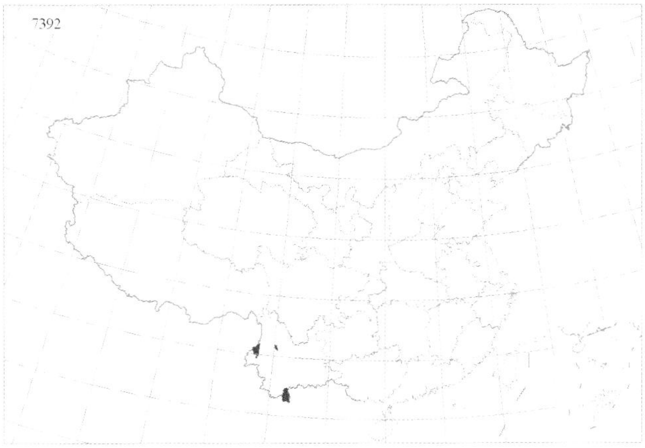

Melastomataceae 野牡丹科 *Osbeckia* 金锦香属

Osbeckia chinensis L. 金锦香

Shrubs

24.3; 19.4; 27.7; 24.4; 232.2; 0.0; 1339 ❙ 1393; 464; 45 ❙ 4.1; 1124; 491.4

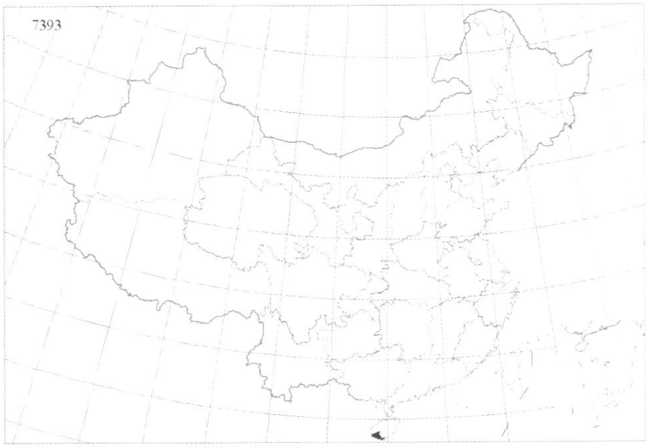

Melastomataceae 野牡丹科 *Osbeckia* 金锦香属

Osbeckia chinensis var. angustifolia (D. Don) C. Y. Wu et C. Chen 宽叶金锦香

Subshrubs

18.6 (14.6~20.5); 12.6 (7.1~15.1); 22.5 (20.3~23.7); 18.6 (14.6~20.5); 163.6 (114.8~185.9); 0.0 (0.0~0.0); 898 (741~992) ❙ 1387 (1014~1751); 729 (598~996); 55 (22~77) ❙ 53.7 (26.1~88.8); 870 (692~986); 462.7 (265.0~659.0)

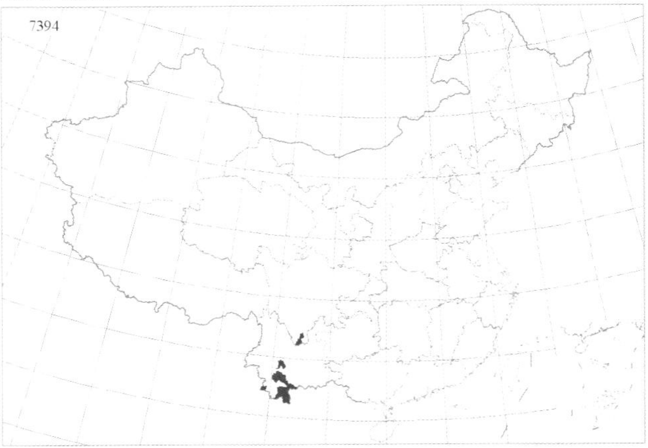

Melastomataceae 野牡丹科 *Osbeckia* 金锦香属

Osbeckia nepalensis Lindl. 蚂蚁花

Shrubs or subshrubs

17.2 (11.3~21.0); 10.7 (4.3~15.1); 21.7 (17.0~25.8); 17.3 (11.3~21.0); 147.3 (76.6~191.8); 0.0 (0.0~0.7); 839 (644~1053) ❙ 1238 (907~1751); 692 (521~1007); 50 (29~77) ❙ 47.8 (6.5~145.8); 821 (644~1053); 387.1 (234.5~659.0)

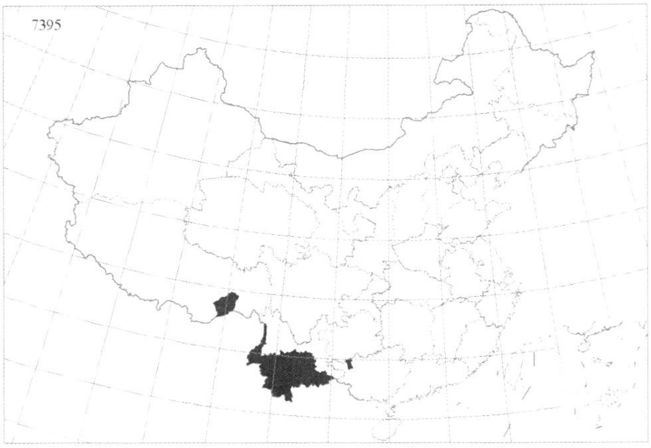

Melastomataceae 野牡丹科 *Osbeckia* 金锦香属

Osbeckia nepalensis var. albiflora Lindl. 白蚂蚁花

Shrubs or subshrubs

14.4; 7.5; 19.1; 14.4; 113.1; 0.0; 727 ❙ 1454; 792; 70 ❙ 99.9; 727; 487.2

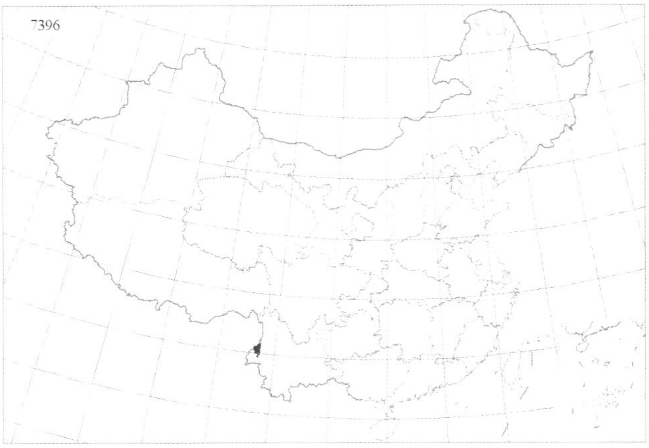

Melastomataceae 野牡丹科 *Osbeckia* 金锦香属

Osbeckia stellata Buchanan-Hamilton ex Kew Gawler 星毛金锦香

Herbs or subshrubs

17.6 (-1.8~24.8); 8.5 (-10.2~20.1); 25.5 (6.1~29.8); 17.7 (1.8~24.8); 152.7 (1.9~237.6); 0.6 (0.0~82.8); 924 (329~1385) ❙ 1405 (320~2680); 613 (193~1428); 116 (15~259) ❙ 53.2 (-34.3~246.5); 914 (320~1360); 253.7 (72.2~659.0)

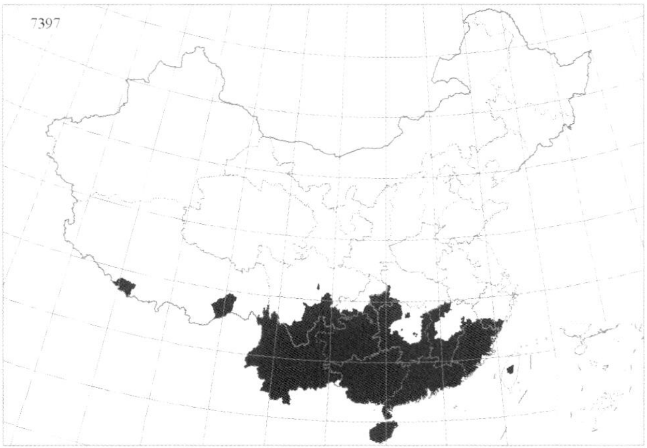

Melastomataceae 野牡丹科 *Oxyspora* 尖子木属

Oxyspora cernua (Roxb.) Hook. f. et Thomson ex Triana 墨脱尖子木

Shrubs

13.3; 5.4; 19.6; 13.4; 100.4; 0.0; 706 ❙ 1736; 1007; 51 ❙ 145.8; 706; 447.4

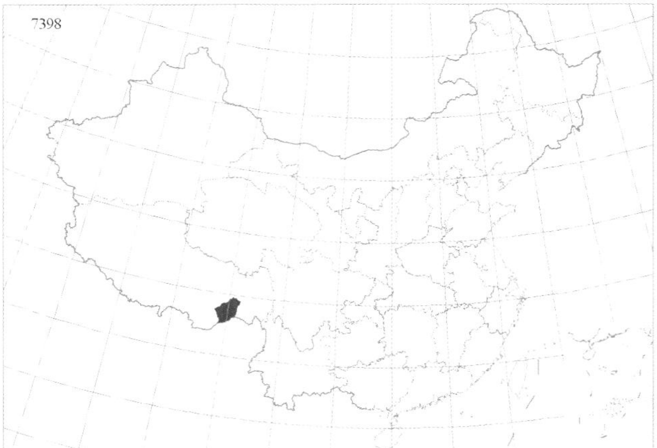

Melastomataceae 野牡丹科 *Oxyspora* 尖子木属
Oxyspora curtisii King 柑叶尖子木
Shrubs

18.6; 12.7; 22.5; 18.7; 163.9; 0.0; 888 ❙ 1547; 911; 63 ❙ 74.2; 888; 551.3

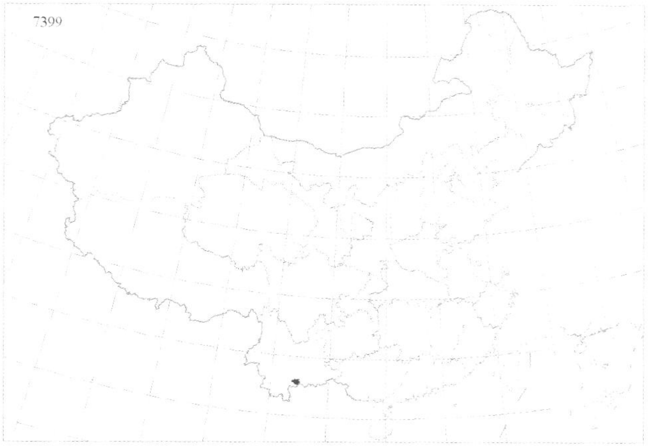

Melastomataceae 野牡丹科 *Oxyspora* 尖子木属
Oxyspora paniculata (D. Don) DC. 尖子木
Shrubs

17.7 (8.6~22.2); 10.2 (1.3~15.1); 23.2 (15.0~28.7); 17.7 (8.7~22.3); 152.9 (52.6~207.5); 0.1 (0.0~8.8); 880 (574~1190) ❙ 1312 (939~1776); 693 (497~1007); 71 (37~196) ❙ 50.0 (5.8~145.8); 871 (574~1190); 357.3 (189.5~659.0)

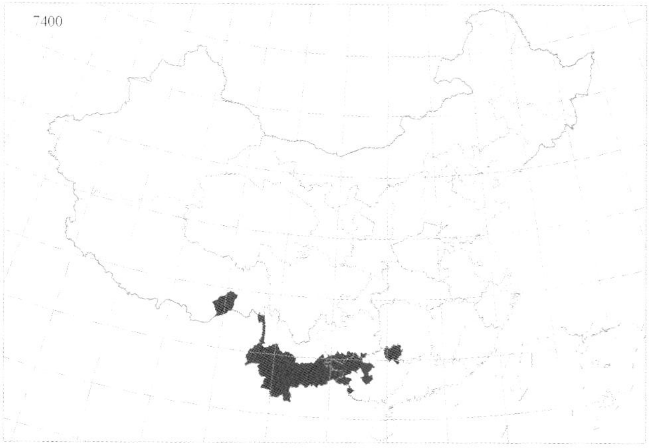

Melastomataceae 野牡丹科 *Oxyspora* 尖子木属
Oxyspora teretipetiolata (C. Y. Wu et C. Chen) W. H. Chen et Y. M. Shui 翅茎异形木 Shrubs

20.0; 13.6; 24.4; 20.1; 180.7; 0.0; 985 ❙ 1475; 827; 62 ❙ 50.1; 985; 528.9

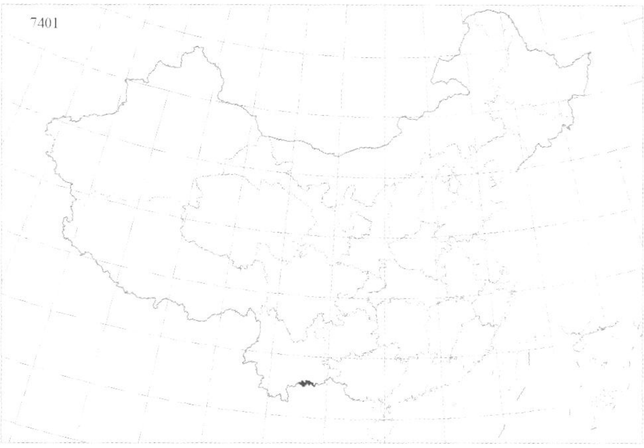

Melastomataceae 野牡丹科 *Oxyspora* 尖子木属
Oxyspora vagans (Roxburgh) Wallich 刚毛尖子木
Shrubs

18.7 (13.3~20.5); 12.8 (5.4~15.1); 22.4 (19.6~23.7); 18.7 (13.4~20.5); 164.5 (100.4~185.9); 0.0 (0.0~0.0); 901 (706~992) ❙ 1556 (1434~1751); 793 (655~1007); 59 (41~77) ❙ 74.9 (47.6~145.8); 881 (706~986); 522.1 (439.5~659.0)

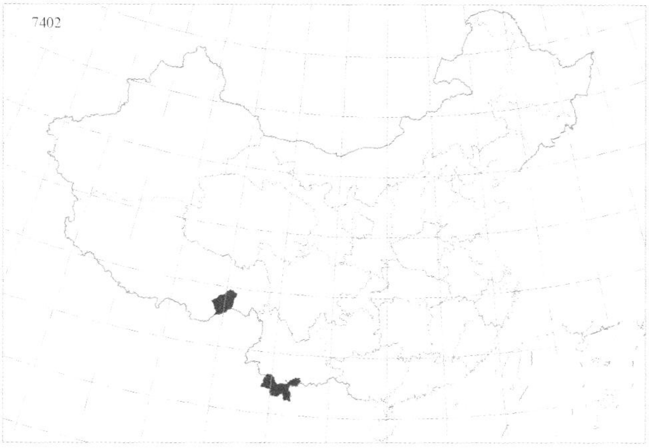

Melastomataceae 野牡丹科 *Oxyspora* 尖子木属
Oxyspora yunnanensis H. L. Li 滇尖子木
Shrubs

12.4; 5.5; 17.9; 12.4; 92.4; 3.2; 683 ❙ 1111; 626; 49 ❙ 66.9; 667; 367.4

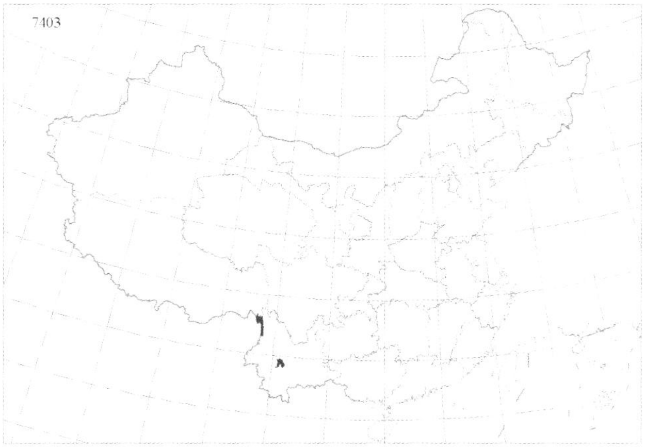

Melastomataceae 野牡丹科 *Phyllagathis* 锦香草属
Phyllagathis cavaleriei (H. Leveille et Vaniot) Guillaumin 锦香草 Herbs or subshrubs

14.3; 6.0; 21.2; 14.4; 113.3; 1.0; 743 ❙ 938; 517; 38 ❙ 26.1; 727; 290.5

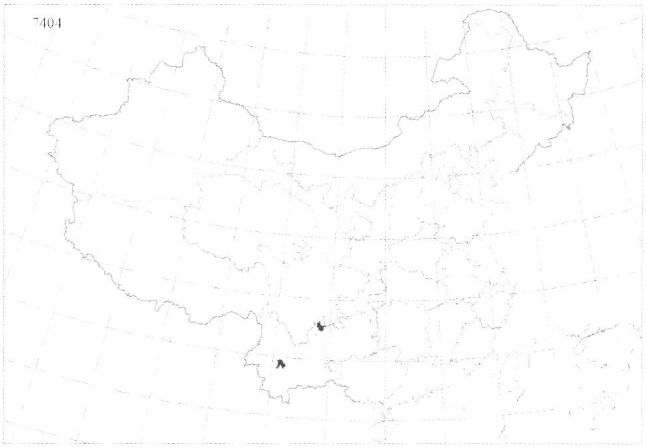

Melastomataceae 野牡丹科 *Phyllagathis* 锦香草属
Phyllagathis deltoidea C. Chen 三角齿锦香草
Small shrubs

21.5；13.6；27.4；21.6；199.1；0.0；1131 ❙ 1581；850；106 ❙ 39.9；1131；281.7

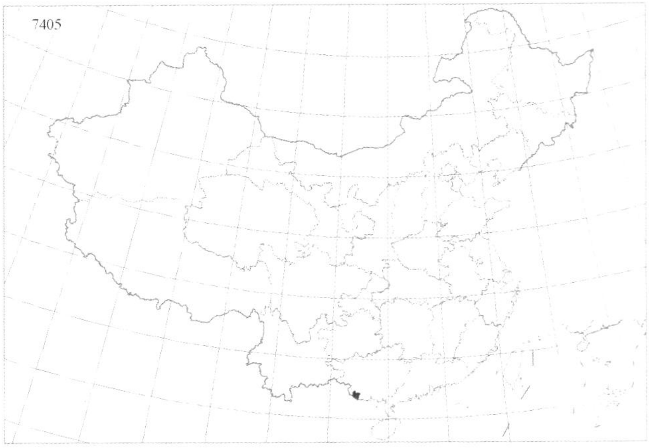

Melastomataceae 野牡丹科 *Phyllagathis* 锦香草属
Phyllagathis erecta (S. Y. Hu) C. Y. Wu ex C. Chen 直立锦香草
Subshrubs

20.1；13.7；24.6；20.2；182.4；0.0；1004 ❙ 1434；725；58 ❙ 42.9；982；449.4

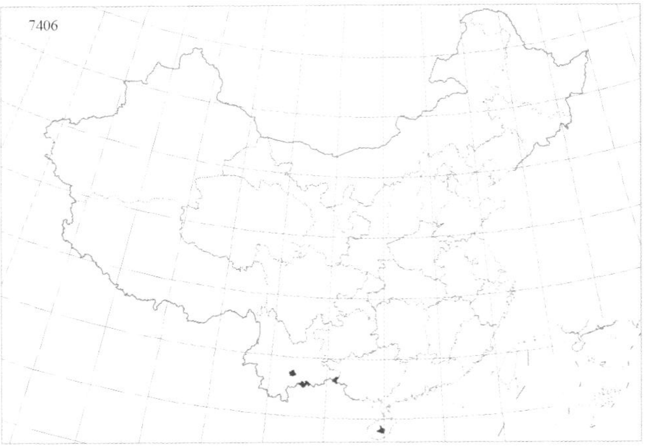

Melastomataceae 野牡丹科 *Phyllagathis* 锦香草属
Phyllagathis fengii C. Hansen 刚毛锦香草
Shrubs

18.2 (17.2~21.0)；10.9 (9.5~14.2)；23.3 (22.4~25.6)；18.2 (17.3~21.0)；158.4 (147.0~191.8)；0.0 (0.0~0.0)；887 (833~1053) ❙ 1414 (1254~1522)；789 (699~860)；57 (48~68) ❙ 60.1 (44.5~80.6)；887 (833~1053)；430.5 (336.3~551.6)

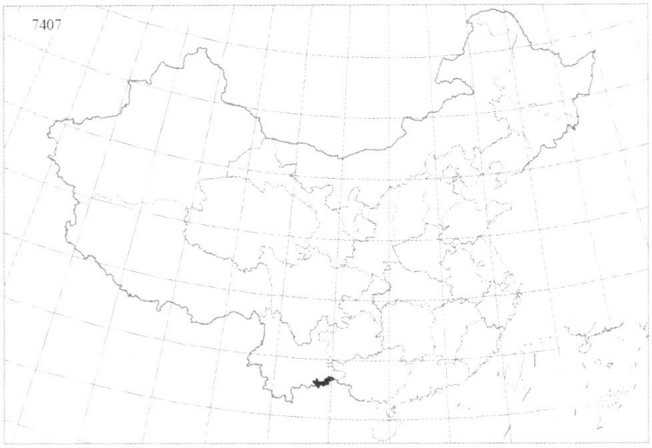

Melastomataceae 野牡丹科 *Phyllagathis* 锦香草属
Phyllagathis hainanensis (Merr. et Chun) C. Chen 海南锦香草
Small shrubs

23.7；18.6；27.3；23.7；225.0；0.0；1282 ❙ 1588；557；57 ❙ 24.2；1166；556.5

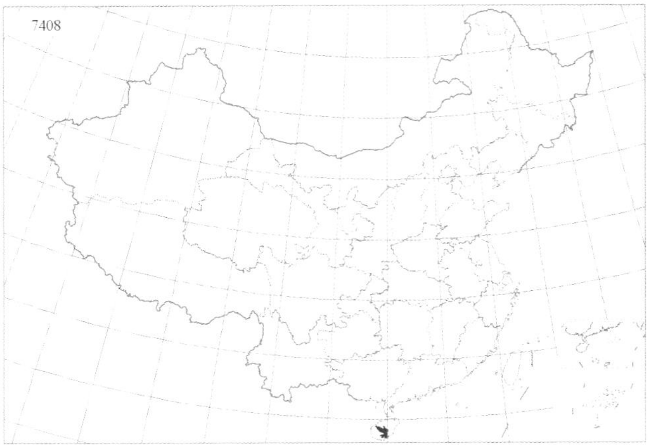

Melastomataceae 野牡丹科 *Phyllagathis* 锦香草属
Phyllagathis hispidissima (C. Chen) C. Chen 密毛锦香草
Small shrubs

18.4 (17.2~21.0)；10.7 (7.4~14.2)；24.0 (22.4~26.4)；18.4 (17.3~21.0)；161.0 (147.0~191.8)；0.0 (0.0~0.0)；906 (833~1053) ❙ 1383 (1182~1522)；762 (637~860)；61 (43~107) ❙ 53.6 (16.9~80.6)；906 (833~1053)；407.3 (245.3~551.6)

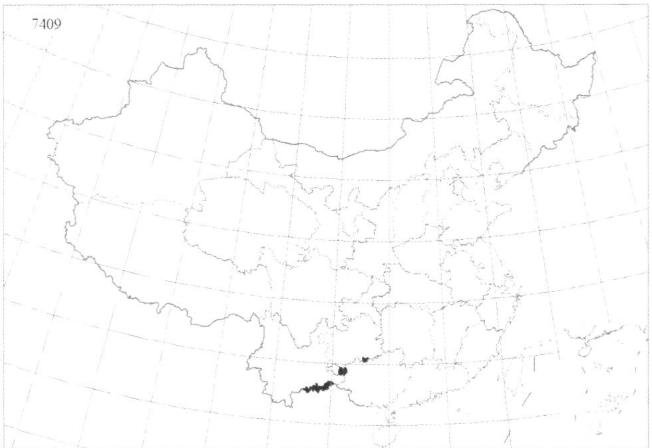

Melastomataceae 野牡丹科 *Phyllagathis* 锦香草属
Phyllagathis longearistata C. Chen 长芒锦香草
Shrubs

20.1；10.5；27.7；20.1；181.6；0.0；1052 ❙ 1459；723；111 ❙ 38.7；1052；248.4

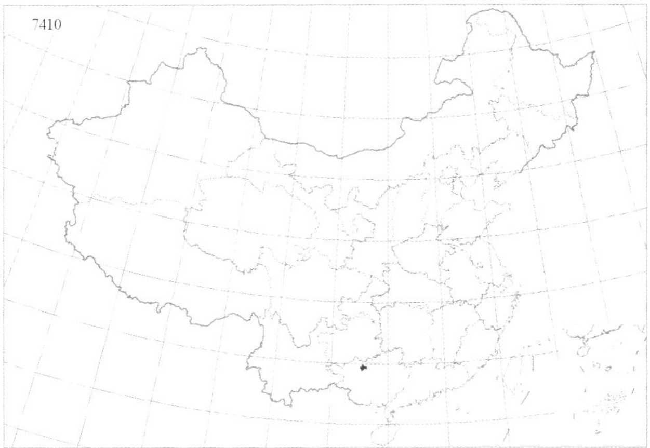

Melastomataceae 野牡丹科 *Phyllagathis* 锦香草属
Phyllagathis longiradiosa C. Chen 大叶熊巴掌
Shrubs or herbs

19.7; 11.9; 25.4; 19.8; 177.3; 0.0; 1012 ▎1439; 787; 71 ▎48.0; 1012; 370.6

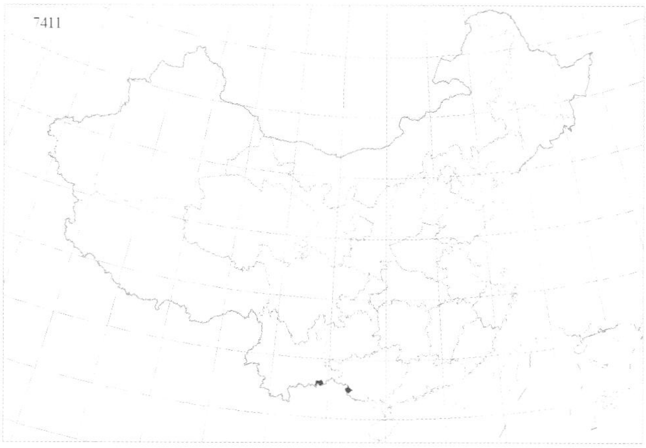

Melastomataceae 野牡丹科 *Phyllagathis* 锦香草属
Phyllagathis longiradiosa var. pulchella C. Chen 丽萼熊巴掌
Shrubs or herbs

21.6; 13.1; 27.9; 21.6; 199.6; 0.0; 1143 ▎1451; 766; 102 ▎27.0; 1143; 335.1

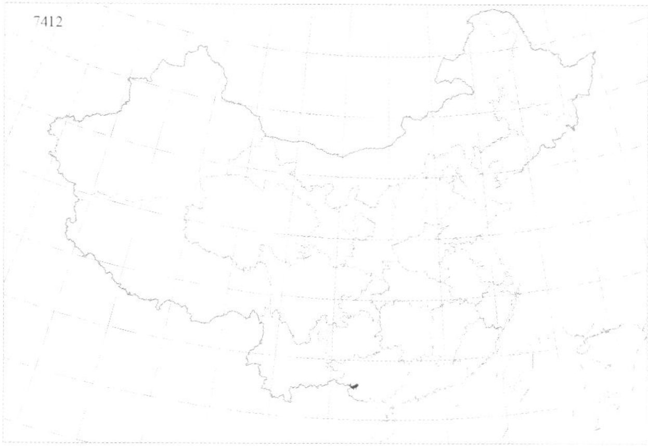

Melastomataceae 野牡丹科 *Phyllagathis* 锦香草属
Phyllagathis melastomatoides (Merr. et Chun) W. C. Ko 毛锦香草
Small shrubs

24.1 (23.0~24.8); 19.0 (17.7~20.1); 27.6 (26.7~28.4); 24.2 (23.1~24.8); 229.9 (217.0~237.6); 0.0 (0.0~0.0); 1321 (1217~1385) ▎1598 (1405~1725); 555 (473~619); 65 (41~110) ▎21.5 (1.5~41.8); 1197 (1137~1326); 513.6 (352.9~620.0)

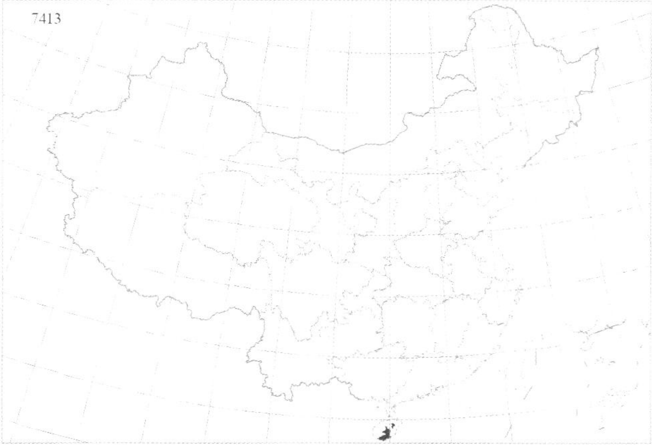

Melastomataceae 野牡丹科 *Phyllagathis* 锦香草属
Phyllagathis melastomatoides var. brevipes W. C. Ko 短柄锦香草
Shrubs

24.7; 19.8; 27.9; 24.7; 236.5; 0.0; 1376 ▎1588; 534; 58 ▎15.4; 1212; 494.7

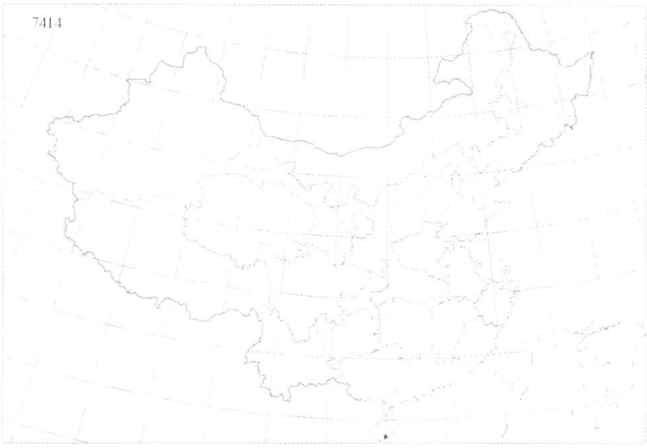

Melastomataceae 野牡丹科 *Phyllagathis* 锦香草属
Phyllagathis oligotricha Merr. 毛柄锦香草
Small shrubs

18.1 (14.6~21.6); 7.9 (3.8~12.6); 27.1 (24.6~29.4); 18.1 (14.7~21.7); 157.3 (117.7~199.8); 0.0 (0.0~1.5); 942 (779~1153) ▎1635 (1410~1997); 585 (403~821); 195 (161~254) ▎74.4 (40.9~130.7); 937 (779~1153); 240.2 (151.0~305.7)

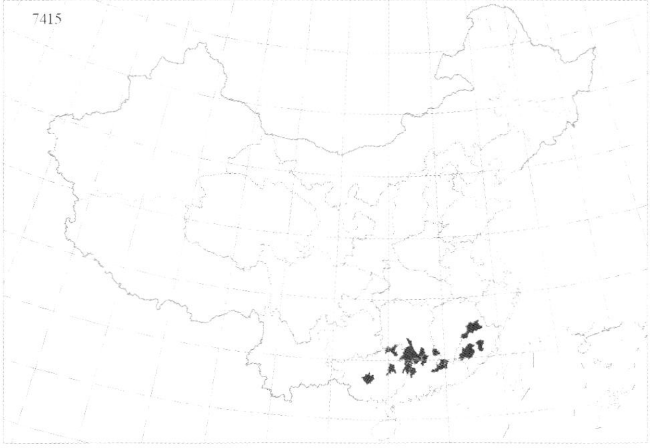

Melastomataceae 野牡丹科 *Phyllagathis* 锦香草属
Phyllagathis ovalifolia H. L. Li 卵叶锦香草
Shrubs

18.5; 10.9; 23.9; 18.5; 161.9; 0.0; 903 ▎1426; 802; 66 ▎58.0; 903; 402.6

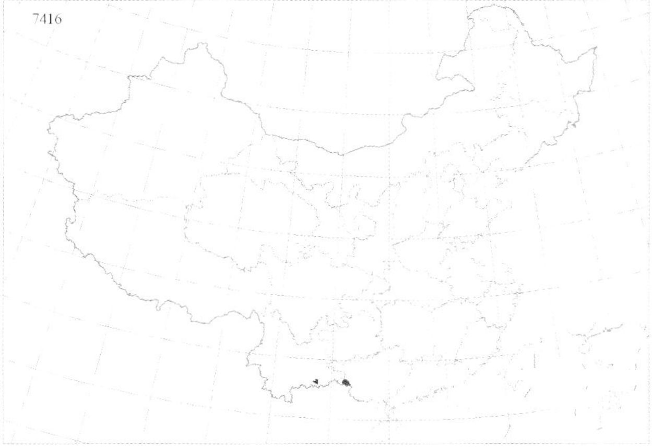

Melastomataceae 野牡丹科 *Phyllagathis* 锦香草属
Phyllagathis plagiopetala C. Chen 偏斜锦香草
Subshrubs

16.1; 4.9; 26.7; 16.2; 134.6; 0.4; 858 ❙ 1495; 481; 194 ❙ 74.4; 858; 207.8

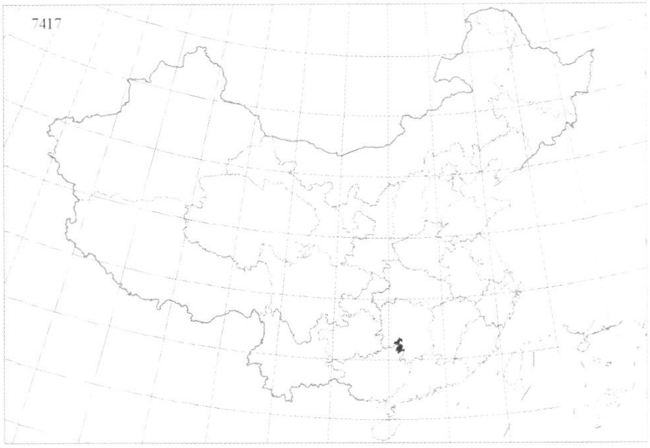

Melastomataceae 野牡丹科 *Phyllagathis* 锦香草属
Phyllagathis setotheca H. L. Li 刺蕊锦香草
Small shrubs

22.2 (21.3~23.4); 14.3 (13.2~16.1); 28.3 (27.3~29.2); 22.3 (21.3~23.5); 207.0 (195.9~221.7); 0.0 (0.0~0.0); 1184 (1111~1273) ❙ 1719 (1302~2435); 906 (670~1428); 107 (82~130) ❙ 45.5 (11.9~99.5); 1184 (1111~1273); 247.5 (122.7~327.7)

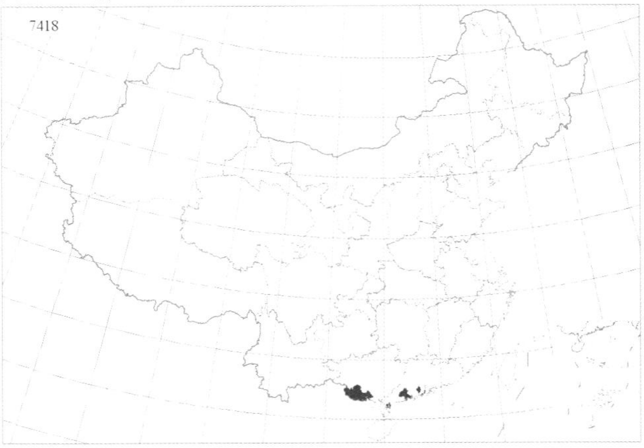

Melastomataceae 野牡丹科 *Phyllagathis* 锦香草属
Phyllagathis stenophylla (Merr. et Chun) H. L. Li 窄叶锦香草
Shrubs

24.3; 19.2; 28.0; 24.3; 232.1; 0.0; 1339 ❙ 1321; 440; 44 ❙ -1.2; 1102; 479.3

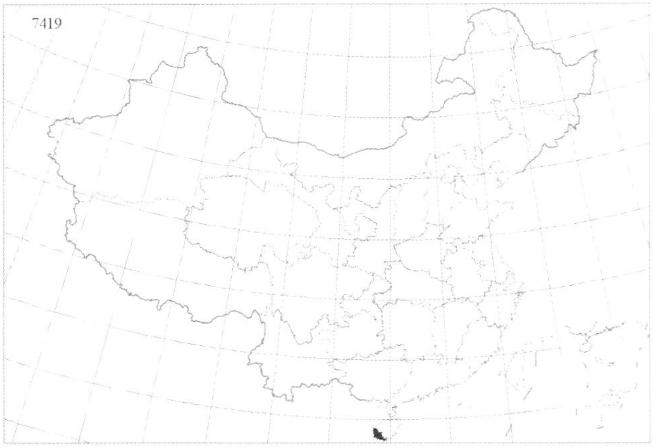

Melastomataceae 野牡丹科 *Phyllagathis* 锦香草属
Phyllagathis ternata C. Chen 三瓣锦香草
Subshrubs

21.0; 13.2; 26.9; 21.0; 192.3; 0.0; 1081 ❙ 1776; 828; 125 ❙ 64.3; 1081; 254.1

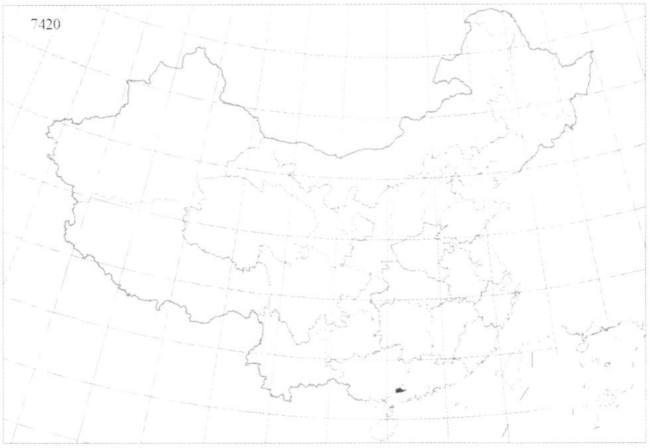

Melastomataceae 野牡丹科 *Phyllagathis* 锦香草属
Phyllagathis velutina (Diels) C. Chen 腺毛锦香草
Subshrubs

19.3; 12.1; 24.4; 19.3; 172.1; 0.0; 956 ❙ 1472; 816; 60 ❙ 55.0; 956; 455.6

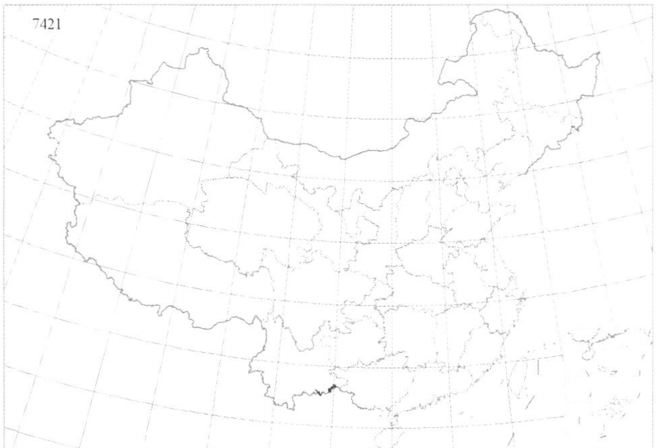

Melastomataceae 野牡丹科 *Plagiopetalum* 偏瓣花属
Plagiopetalum esquirolii (H. Léveillé) Rehder 偏瓣花
Shrubs

17.2 (8.6~21.0); 9.9 (1.3~14.2); 22.5 (15.0~27.1); 17.2 (8.7~21.0); 146.7 (52.6~191.8); 0.3 (0.0~8.8); 849 (574~1053) ❙ 1287 (1014~1751); 709 (497~996); 55 (37~196) ❙ 52.5 (12.5~93.0); 841 (574~1053); 368.0 (189.5~551.6)

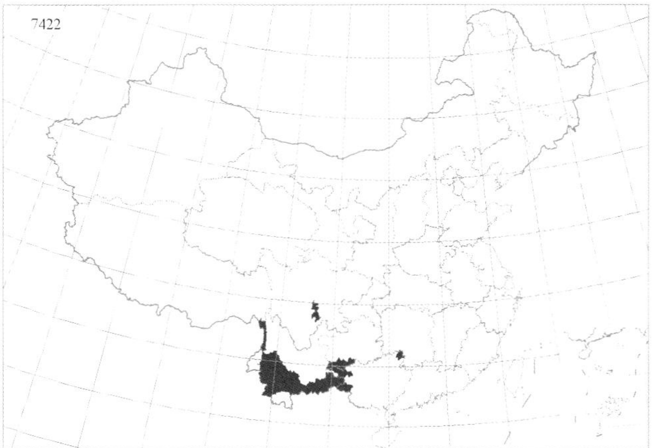

Melastomataceae 野牡丹科 *Sporoxeia* 八蕊花属
Sporoxeia clavicalcarata C. Chen 棒距八蕊花
Shrubs

18.6; 12.5; 22.9; 18.7; 164.3; 0.0; 896 | 1231; 705; 50 | 37.2; 887; 414.9

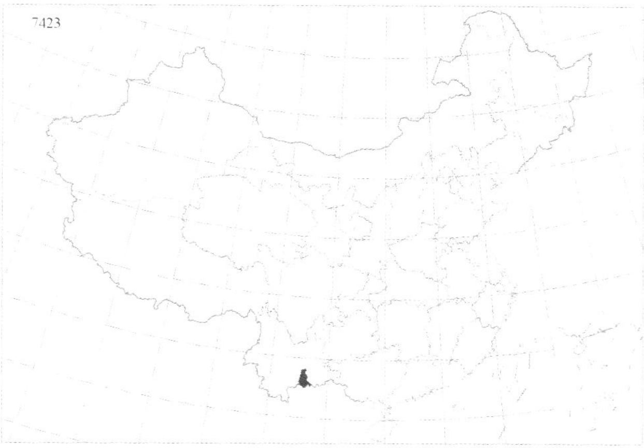

Melastomataceae 野牡丹科 *Sporoxeia* 八蕊花属
Sporoxeia sciadophila W. W. Sm. 八蕊花
Shrubs

17.0 (8.6~21.0); 10.1 (1.3~14.2); 21.9 (15.0~25.6); 17.0 (8.7~21.0); 144.9 (52.6~191.8); 0.6 (0.0~8.8); 845 (574~1053) | 1340 (1041~1547); 757 (585~911); 56 (45~69) | 60.6 (34.8~93.0); 844 (574~1053); 427.1 (342.5~551.6)

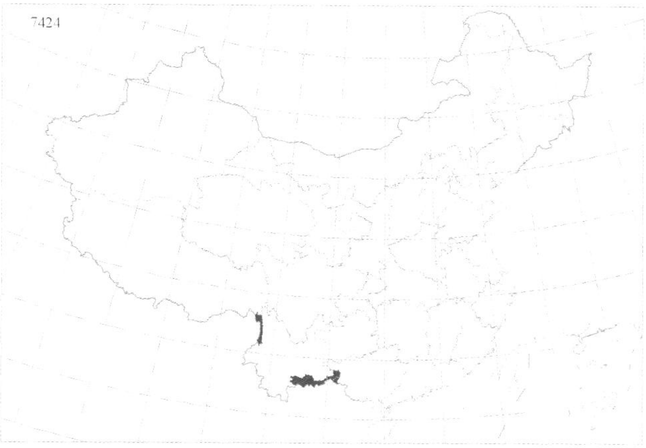

Melastomataceae 野牡丹科 *Styrophyton* 长穗花属
Styrophyton caudatum (Diels) S. Y. Hu 长穗花
Shrubs

18.5 (16.3~21.0); 11.6 (9.1~15.1); 23.1 (21.4~25.6); 18.5 (16.4~21.0); 162.0 (136.4~191.8); 0.0 (0.0~0.0); 896 (797~1053) | 1379 (1074~1751); 752 (582~996); 58 (47~77) | 53.8 (32.4~88.8); 893 (797~1053); 433.8 (277.0~659.0)

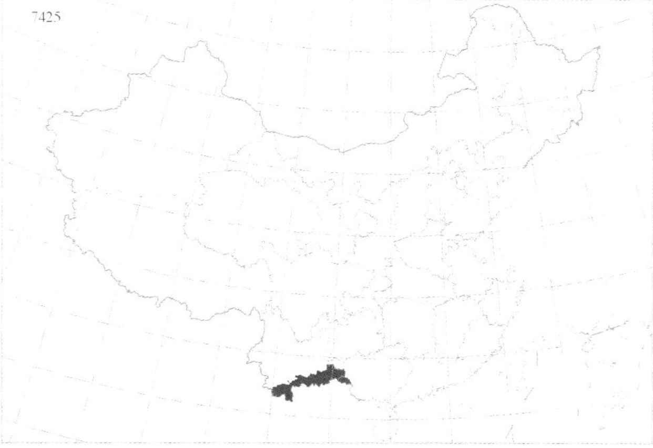

Araliaceae 五加科 *Aralia* 楤木属
Aralia armata (Wall. ex Don) Seem. 野楤头
Shrubs, deciduous

19.5 (13.3~24.7); 10.5 (3.3~19.8); 27.2 (19.1~29.8); 19.6 (13.4~24.7); 174.8 (100.4~236.5); 0.1 (0.0~2.3); 1023 (706~1376) | 1590 (1048~2435); 663 (403~1428); 142 (44~230) | 56.7 (-2.2~145.8); 1015 (706~1299); 256.4 (140.5~659.0)

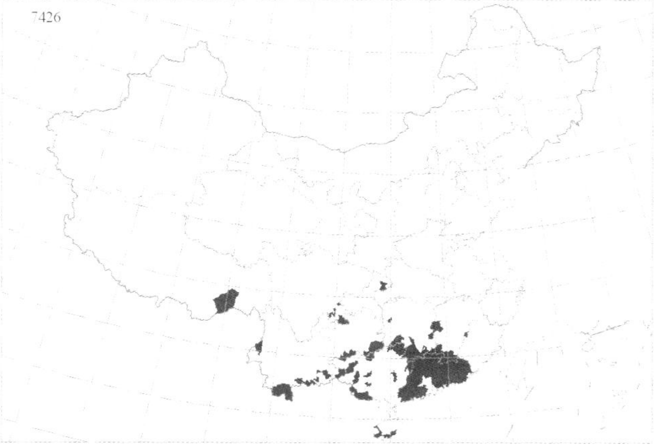

Araliaceae 五加科 *Aralia* 楤木属
Aralia bipinnata Blanco 台湾楤木
Shrubs or trees, deciduous

19.7 (16.4~24.1); 14.2 (11.2~18.9); 24.3 (20.3~28.0); 19.7 (16.5~24.1); 177.0 (137.8~229.7); 0.0 (0.0~0.0); 980 (783~1322) | 2487 (1751~2993); 1183 (900~1847); 232 (47~547) | 162.8 (32.5~235.6); 953 (783~1123); 251.1 (154.2~303.6)

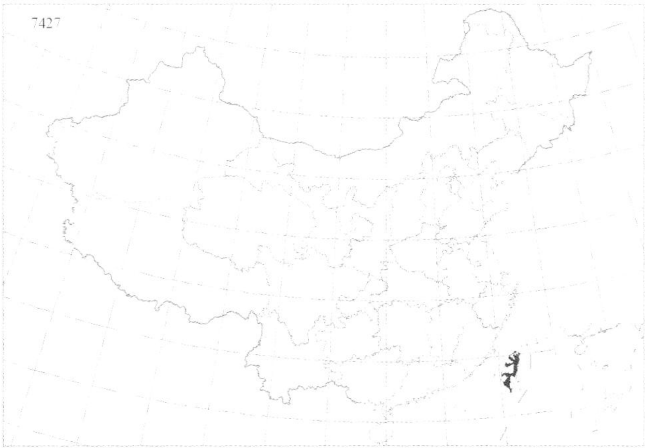

Araliaceae 五加科 *Aralia* 楤木属
Aralia chinensis L. 黄毛楤木
Shrubs or trees, deciduous

14.3 (0.8~23.6); 2.5 (-12.7~16.7); 25.3 (9.9~29.9); 14.6 (3.8~23.6); 121.4 (15.4~223.5); 8.6 (0.0~66.0); 813 (411~1281) | 1126 (341~2037); 482 (190~1033); 100 (5~259) | 36.4 (-47.7~147.0); 780 (341~1269); 206.7 (61.2~447.4)

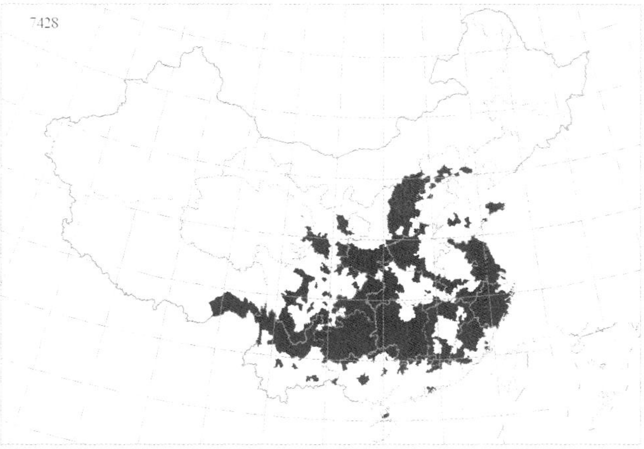

A

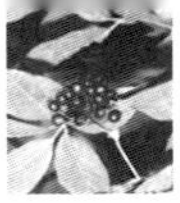

Araliaceae 五加科 *Aralia* 楤木属
Aralia cordata Thunb. 食用土当归
Herbs or subshrubs, deciduous

18.3; 13.4; 22.0; 18.3; 159.6; 0.0; 893 ❙ 2724; 1327; 218 ❙ 210.8; 893; 296.5

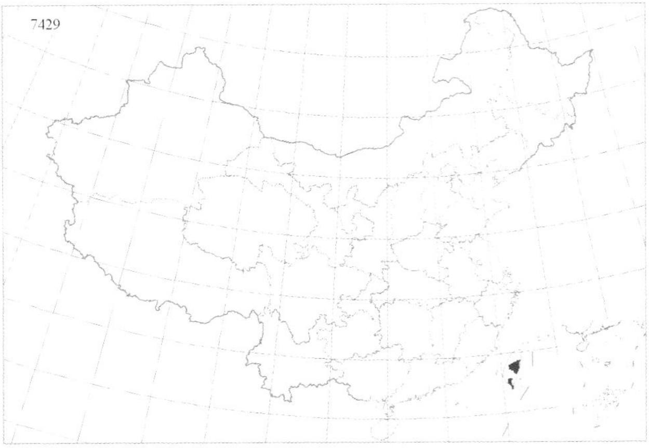

Araliaceae 五加科 *Aralia* 楤木属
Aralia dasyphylla Miq. 头序楤木
Shrubs or trees, deciduous

16.8 (10.2~22.5); 6.0 (-2.9~15.2); 26.8 (18.8~29.9); 16.9 (10.6~22.7); 144.2 (82.6~212.7); 1.4 (0.0~20.5); 903 (656~1216) ❙ 1416 (626~2037); 558 (316~1033); 140 (29~259) ❙ 56.7 (-22.2~130.7); 893 (626~1216); 212.2 (111.9~491.5)

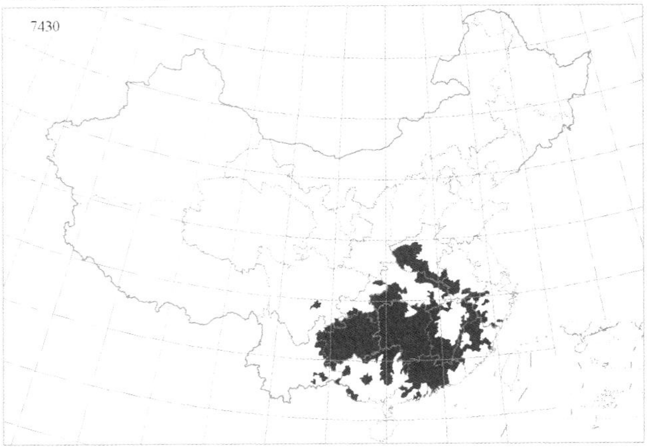

Araliaceae 五加科 *Aralia* 楤木属
Aralia debilis J. Wen 秀丽楤木
Shrubs, deciduous

19.0 (16.1~21.3); 9.0 (4.1~12.3); 27.7 (25.8~28.6); 19.1 (16.1~21.3); 168.8 (133.9~196.1); 0.1 (0.0~0.9); 1002 (883~1130) ❙ 1559 (1183~1781); 636 (490~728); 175 (161~189) ❙ 56.3 (22.7~85.5); 1001 (883~1130); 219.1 (185.5~260.5)

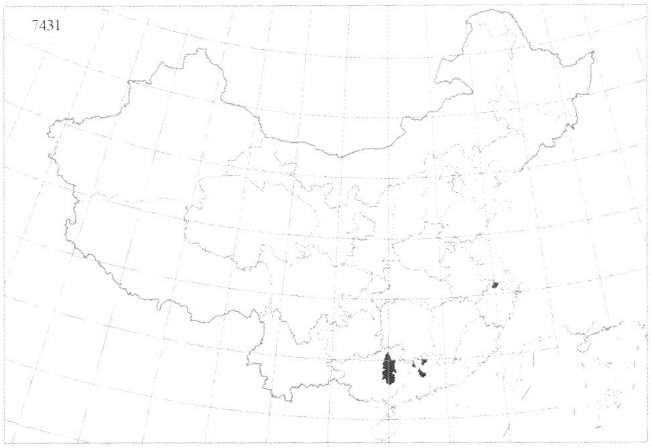

Araliaceae 五加科 *Aralia* 楤木属
Aralia decaisneana Hance 台湾毛楤木
Shrubs, deciduous

19.2 (14.1~24.8); 9.8 (1.6~20.1); 27.3 (20.1~29.8); 19.2 (14.2~24.8); 170.6 (112.6~237.6); 0.3 (0.0~6.5); 1010 (756~1385) ❙ 1612 (868~2993); 649 (312~1428); 157 (41~547) ❙ 62.1 (-34.3~246.5); 997 (756~1230); 256.7 (139.6~620.0)

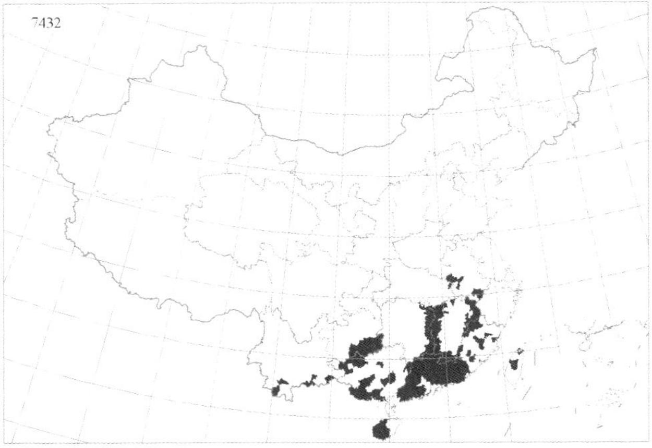

Araliaceae 五加科 *Aralia* 楤木属
Aralia echinocaulis Hand.-Mazz. 棘茎楤木
Small trees, deciduous

16.6 (3.4~21.7); 5.3 (-6.0~13.4); 27.1 (11.7~29.9); 16.7 (4.7~21.7); 140.7 (23.2~200.7); 1.0 (0.0~41.9); 892 (442~1147) ❙ 1456 (806~1997); 534 (299~821); 160 (16~259) ❙ 63.8 (-6.7~147.0); 882 (442~1147); 215.5 (134.1~365.3)

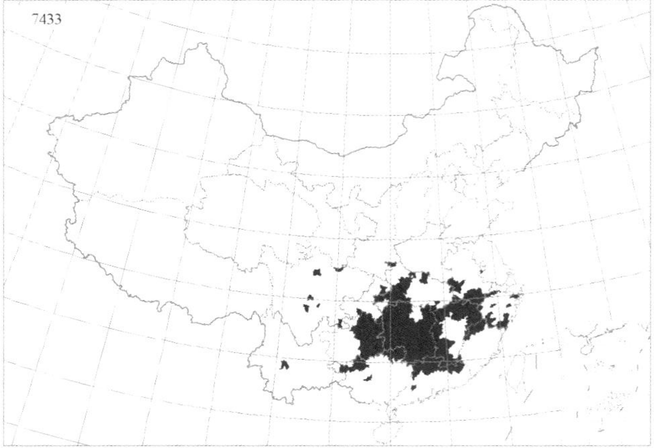

Araliaceae 五加科 *Aralia* 楤木属
Aralia elata (Miq.) Seem. 楤木
Shrubs or small trees, deciduous

13.0 (-0.8~22.1); -0.3 (-24.0~14.2); 24.9 (9.9~29.8); 13.8 (3.8~22.1); 114.6 (15.4~205.6); 18.6 (0.0~118.3); 789 (413~1182) ❙ 1085 (341~1997); 476 (190~818); 92 (8~259) ❙ 35.5 (-40.9~130.7); 750 (341~1182); 223.3 (61.2~551.6)

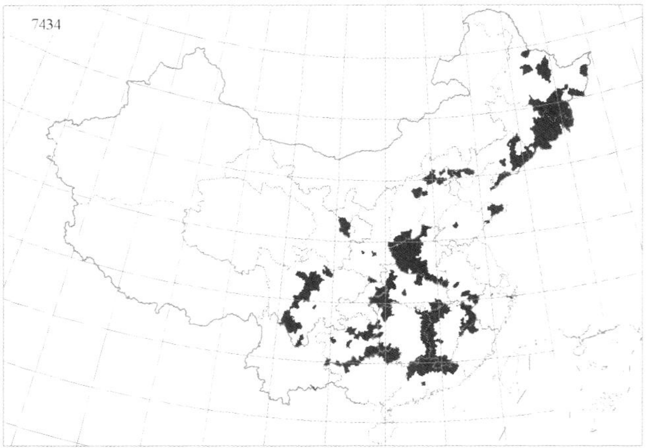

Araliaceae 五加科 *Aralia* 楤木属
Aralia foliolosa Seem. ex C. B. Clarke 小叶楤木
Small trees, deciduous

19.6 (14.3~22.7); 11.4 (3.3~15.1); 26.0 (19.˙~28.6); 19.7 (14.3~22.8); 176.5 (113.1~213.1); 0.1 (0.0~2.3); 1014 (727~1221) ❙ 1529 (1188~2435); 758 (419~1428); 105 (45~198) ❙ 52.9 (5.8~99.9); 1008 (727~1221); 338.2 (177.4~659.0)

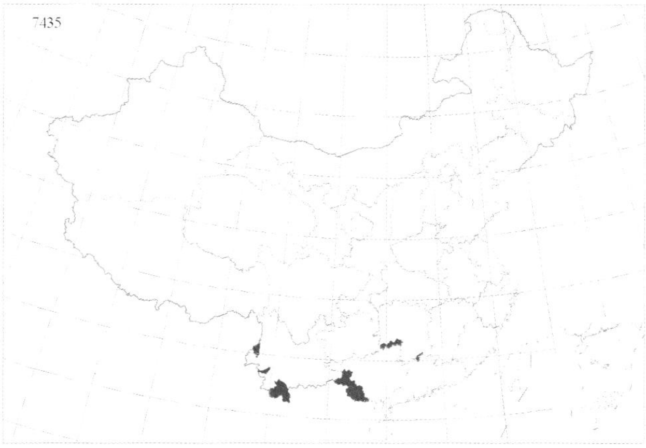

Araliaceae 五加科 *Aralia* 楤木属
Aralia gintungensis C. Y. Wu ex K. M. Feng 景东楤木
Shrubs or small trees, deciduous

16.6; 10.1; 21.0; 16.7; 140.2; 0.0; 806 ❙ 1198; 680; 50 ❙ 49.4; 782; 380.3

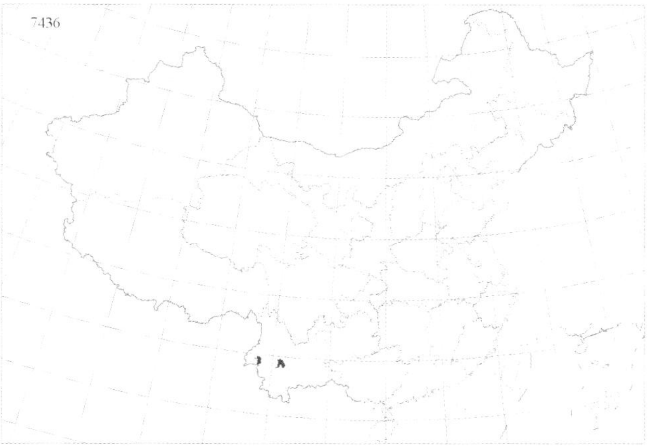

Araliaceae 五加科 *Aralia* 楤木属
Aralia scaberula G. Hoo 糙叶楤木
Shrubs, deciduous

16.6; 5.9; 26.5; 16.7; 140.0; 0.0; 872 ❙ 1973; 658; 252 ❙ 126.3; 872; 288.6

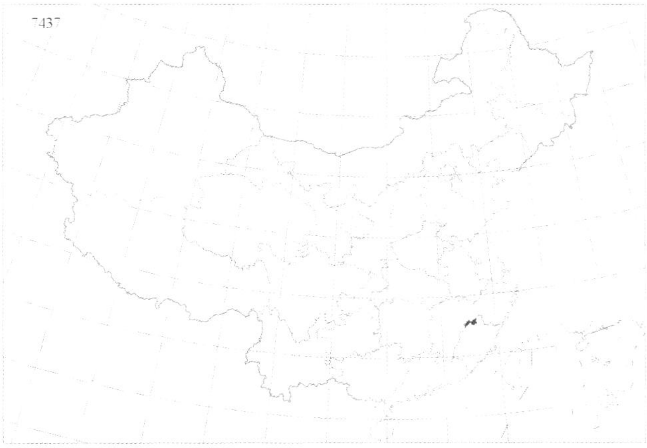

Araliaceae 五加科 *Aralia* 楤木属
Aralia searelliana Dunn 粗毛楤木
Small trees, deciduous

18.5 (17.3~21.0); 12.0 (10.5~14.2); 22.8 (21.6~25.6); 18.5 (17.3~21.0); 161.9 (148.0~191.8); 0.0 (0.0~0.0); 892 (832~1053) ❙ 1325 (958~1547); 738 (531~911); 57 (44~68) ❙ 48.2 (13.3~74.2); 879 (763~1053); 447.5 (311.9~551.6)

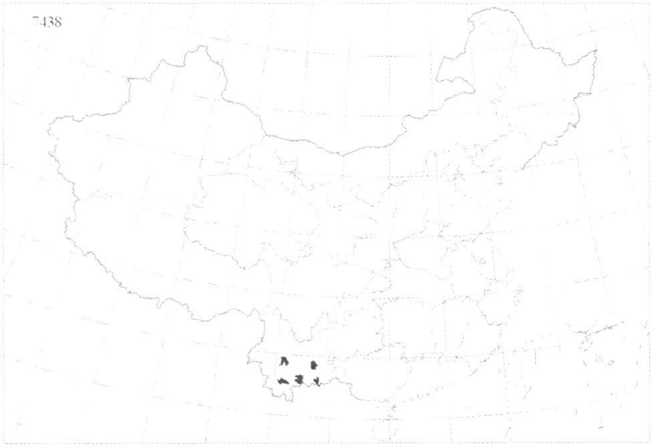

Araliaceae 五加科 *Aralia* 楤木属
Aralia spinifolia Merr. 长刺楤木
Shrubs, deciduous

18.6 (14.3~23.3); 8.5 (3.3~16.0); 27.6 (24.2~29.8); 18.7 (14.3~23.3); 164.1 (113.2~220.1); 0.1 (0.0~2.3); 982 (761~1264) ❙ 1627 (1211~2014); 604 (403~954); 184 (73~259) ❙ 67.5 (17.4~130.7); 977 (761~1264); 235.1 (151.0~322.1)

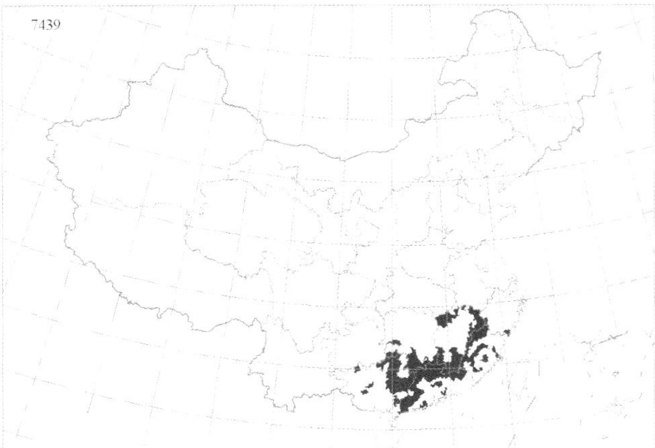

Araliaceae 五加科 *Aralia* 楤木属
Aralia stipulata Franch. 披针叶楤木
Shrubs or small trees

13.2; 6.5; 18.4; 13.2; 100.1; 1.8; 703 ❙ 1038; 595; 42 ❙ 50.0; 687; 355.6

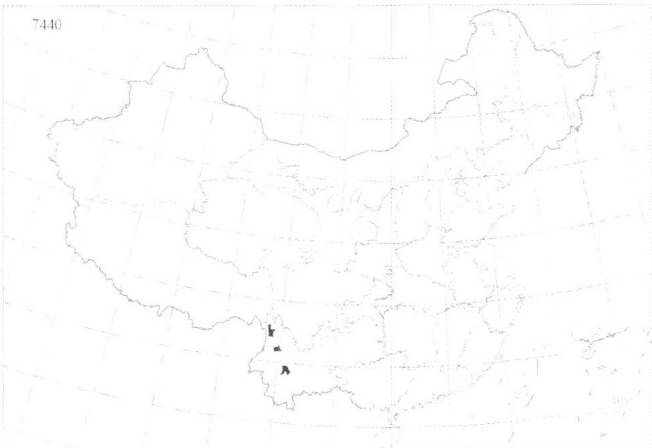

Araliaceae 五加科 *Aralia* 楤木属
Aralia thomsonii Seem. ex C. B. Clarke 云南楤木
Shrubs, deciduous

17.4 (8.6~20.5); 10.8 (1.3~15.1); 22.0 (15.0~26.4); 17.5 (8.7~20.5); 150.4 (52.6~185.9); 0.6 (0.0~8.8); 862 (574~1010) ❙ 1310 (1041~1607); 707 (585~909); 54 (43~77) ❙ 53.6 (16.9~93.0); 853 (574~1010); 431.6 (291.2~659.0)

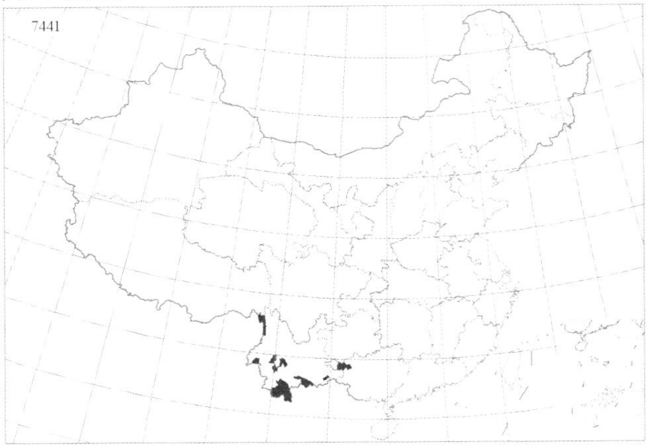

Araliaceae 五加科 *Aralia* 楤木属
Aralia undulata Hand.-Mazz. 波缘楤木
Shrubs or trees, deciduous

17.3 (11.9~22.2); 7.0 (-0.8~13.9); 26.4 (20.0~29.8); 17.3 (12.0~22.3); 148.7 (87.2~207.5); 0.8 (0.0~13.4); 912 (671~1190) ❙ 1434 (798~1852); 561 (377~850); 141 (22~197) ❙ 57.5 (-2.0~101.8); 909 (671~1190); 232.7 (151.3~327.7)

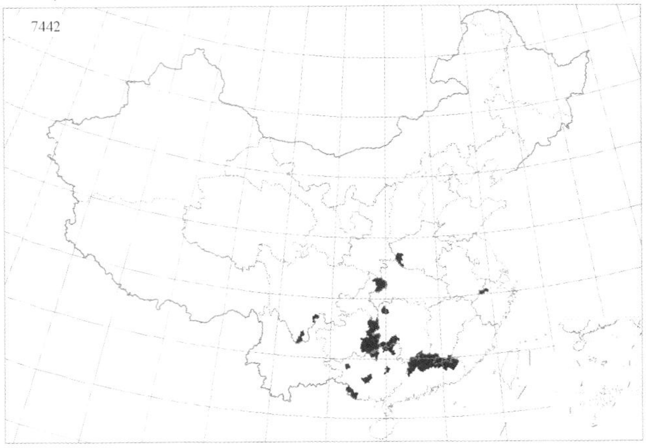

Araliaceae 五加科 *Aralia* 楤木属
Aralia vietnamensis Ha 偃毛楤木
Shrubs or small trees, deciduous

17.2; 9.5; 22.7; 17.3; 147.0; 0.0; 838 ❙ 1254; 699; 48 ❙ 49.7; 838; 336.3

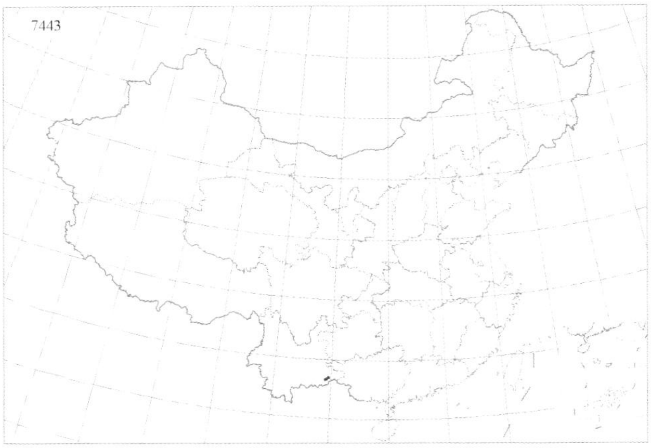

Araliaceae 五加科 *Brassaiopsis* 罗伞属
Brassaiopsis angustifolia K. M. Feng 狭叶罗伞
Shrubs

18.3 (17.0~19.3); 9.8 (7.6~13.1); 24.9 (22.1~26.6); 18.4 (17.0~19.4); 160.3 (143.7~173.2); 0.0 (0.0~0.0); 917 (854~969) ❙ 1255 (1143~1336); 674 (617~774); 56 (50~60) ❙ 37.4 (21.7~54.8); 912 (854~969); 274.2 (189.5~432.9)

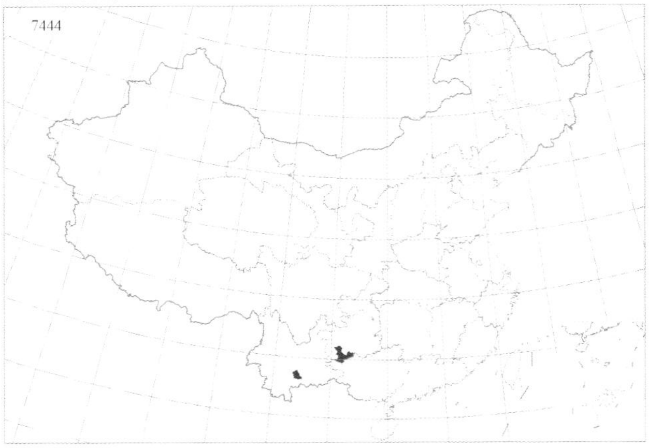

Araliaceae 五加科 *Brassaiopsis* 罗伞属
Brassaiopsis chengkangensis Hu 镇康罗伞
Trees

15.4 (8.6~17.8); 8.8 (1.3~11.1); 20.0 (15.0~22.0); 15.4 (8.7~17.8); 127.0 (52.6~154.1); 1.8 (0.0~8.8); 772 (574~855) ❙ 1282 (1041~1530); 736 (585~909); 48 (45~51) ❙ 66.6 (38.0~81.5); 768 (574~855); 422.7 (352.3~482.8)

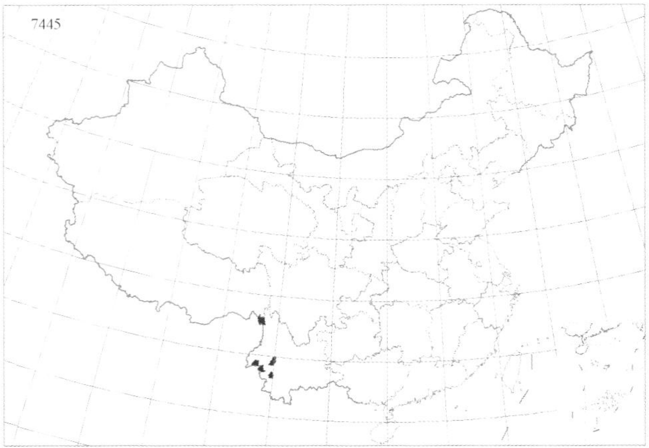

Araliaceae 五加科 *Brassaiopsis* 罗伞属
Brassaiopsis ciliata Dunn 纤齿罗伞
Shrubs

18.4 (12.2~22.2); 10.1 (4.3~14.2); 24.9 (18.5~28.5); 18.5 (12.2~22.3); 161.9 (87.4~207.5); 0.0 (0.0~0.7); 939 (672~1190) ❙ 1383 (1178~1703); 729 (591~911); 80 (31~194) ❙ 49.7 (12.9~113.0); 938 (672~1190); 317.2 (189.5~551.6)

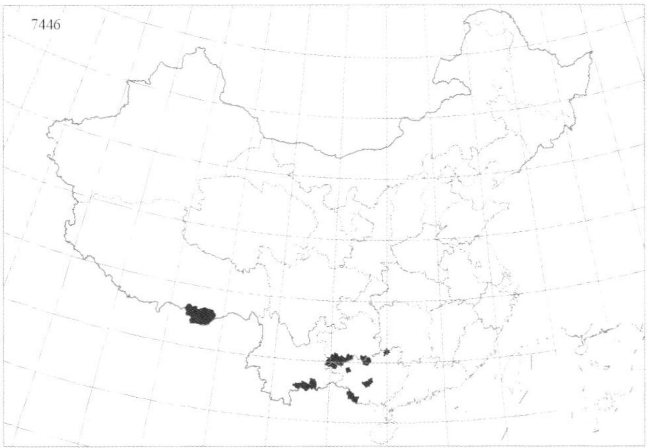

Araliaceae 五加科 *Brassaiopsis* 罗伞属
Brassaiopsis fatsioides Harms 盘叶罗伞
Shrubs or trees

15.7 (8.6~20.5); 7.4 (1.3~15.1); 22.4 (15.0~25.9); 15.7 (8.7~20.5); 129.2 (52.6~185.9); 0.7 (0.0~8.8); 802 (574~992) ‖ 1265 (985~1607); 684 (501~946); 53 (35~77) ‖ 58.0 (19.4~93.0); 798 (574~986); 302.8 (153.9~659.0)

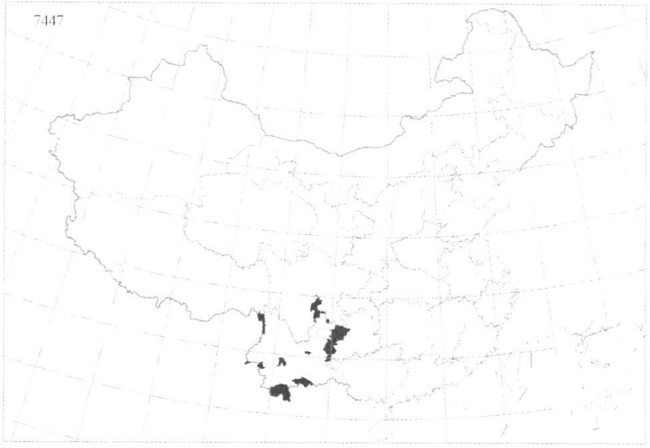

Araliaceae 五加科 *Brassaiopsis* 罗伞属
Brassaiopsis ferruginea (H. L. Li) C. Ho 锈毛罗伞
Shrubs

17.7 (13.3~21.9); 8.1 (3.6~14.2); 25.8 (21.4~28.6); 17.7 (13.3~22.0); 153.1 (101.9~203.8); 0.1 (0.0~1.8); 917 (718~1176) ‖ 1412 (999~1777); 650 (452~946); 114 (48~194) ‖ 54.6 (22.7~93.2); 916 (718~1176); 247.9 (152.1~551.6)

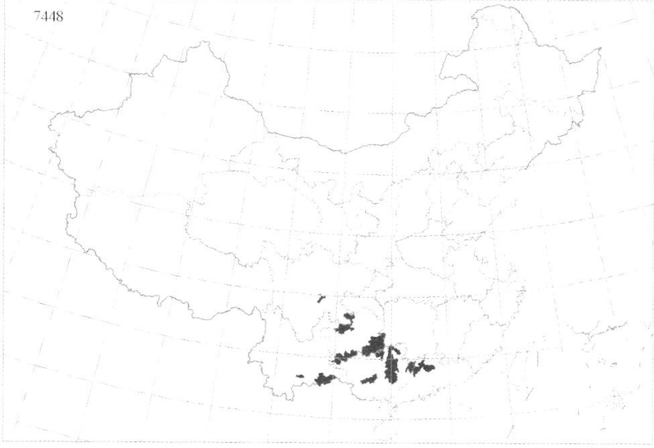

Araliaceae 五加科 *Brassaiopsis* 罗伞属
Brassaiopsis ficifolia Dunn 榕叶罗伞
Shrubs or trees

19.4; 12.6; 24.1; 19.4; 172.6; 0.0; 954 ‖ 1445; 787; 64 ‖ 52.3; 954; 521.5

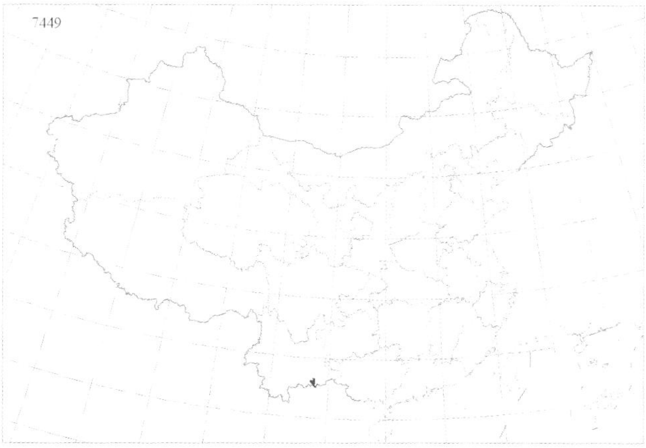

Araliaceae 五加科 *Brassaiopsis* 罗伞属
Brassaiopsis glomerulata (Blume) Regel 罗伞
Trees

19.0 (5.3~24.8); 10.7 (-2.4~20.1); 25.7 (12.2~28.8); 19.0 (5.7~24.8); 168.9 (30.5~237.6); 0.3 (0.0~26.2); 993 (479~1385) ‖ 1406 (868~1900); 650 (312~1056); 93 (31~204) ‖ 44.9 (-34.3~145.8); 970 (479~1360); 307.9 (140.5~659.0)

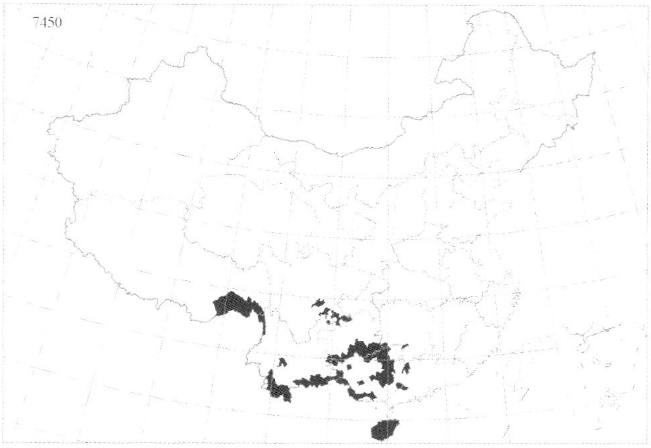

Araliaceae 五加科 *Brassaiopsis* 罗伞属
Brassaiopsis gracilis Hand.-Mazz. 细梗罗伞
Shrubs

18.2 (15.9~19.8); 9.6 (6.9~13.0); 24.8 (21.9~26.6); 18.3 (16.0~19.8); 159.1 (131.6~177.9); 0.0 (0.0~0.0); 913 (802~1010) ‖ 1289 (1074~155[illegible]); 692 (582~835); 63 (43~169) ‖ 41.8 (16.9~69.8); 913 (802~1010); 298.0 (189.5~506.2)

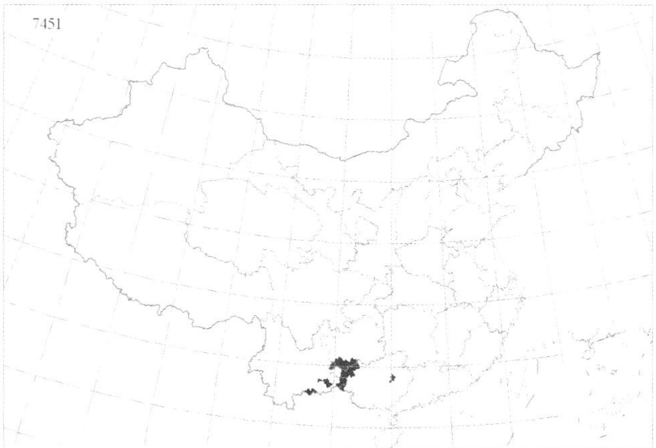

Araliaceae 五加科 *Brassaiopsis* 罗伞属
Brassaiopsis hainla (Buch.-Ham.) Seem. 浅裂罗伞
Trees

14.1 (5.3~18.2); 7.2 (-2.4~12.4); 19.3 (12.2~22.0); 14.2 (5.7~18.2); 112.0 (30.5~158.6); 2.0 (0.0~26.2); 730 (479~861) ‖ 1226 (956~1736); 697 (558~1007); 47 (31~69) ‖ 69.8 (25.4~145.8); 722 (479~855); 364.7 (234.2~482.8)

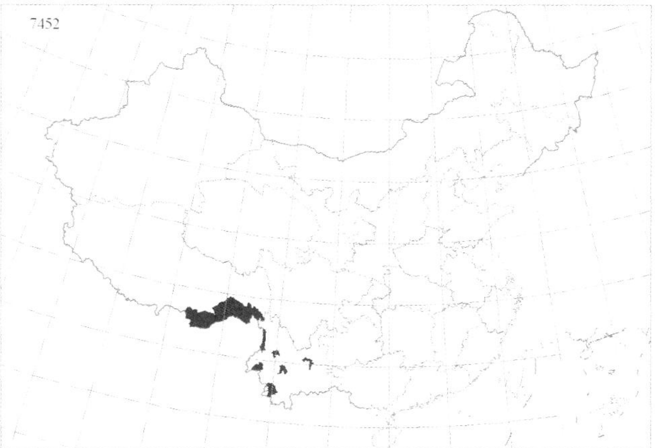

Araliaceae 五加科 *Brassaiopsis* 罗伞属
Brassaiopsis hispida Seem. 粗毛罗伞
Shrubs

12.1; 4.6; 18.1; 12.3; 94.8; 8.7; 680 ❙ 1355; 774; 47 ❙ 102.2; 680; 391.0

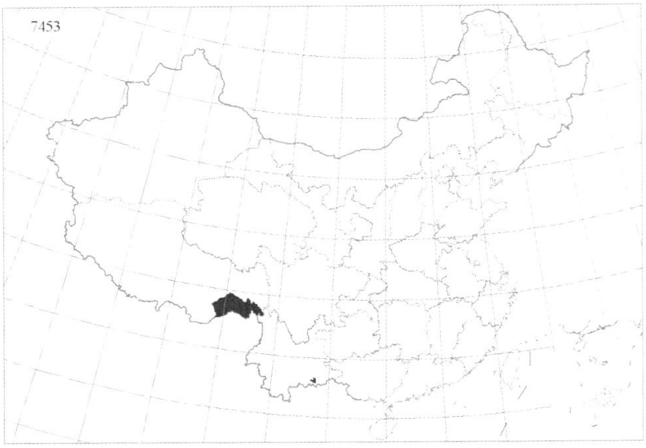

Araliaceae 五加科 *Brassaiopsis* 罗伞属
Brassaiopsis kwangsiensis C. Ho 广西罗伞
Shrubs

16.9 (8.6~19.2); 7.7 (1.3~11.0); 24.6 (15.0~26.6); 17.0 (8.7~19.2); 144.1 (52.6~170.5); 0.6 (0.0~8.8); 867 (574~969) ❙ 1303 (1041~1551); 643 (473~848); 76 (45~169) ❙ 51.4 (27.5~81.5); 867 (574~969); 268.3 (170.4~491.5)

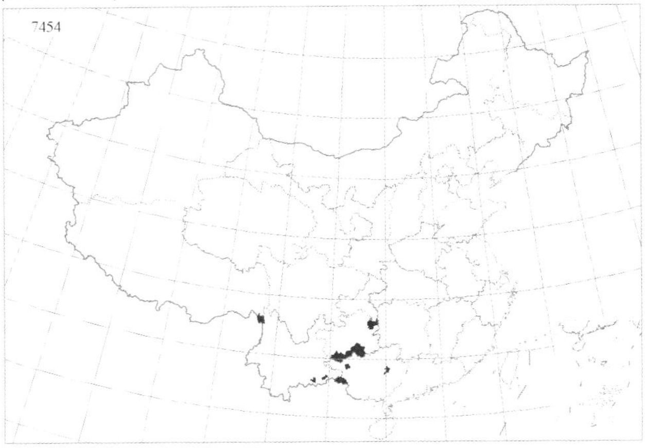

Araliaceae 五加科 *Brassaiopsis* 罗伞属
Brassaiopsis moumingensis (Y. R. Ling) C. B. Shang 茂名罗伞
Shrubs

23.2; 15.8; 28.9; 23.2; 218.9; 0.0; 1257 ❙ 1686; 824; 102 ❙ 34.1; 1257; 194.5

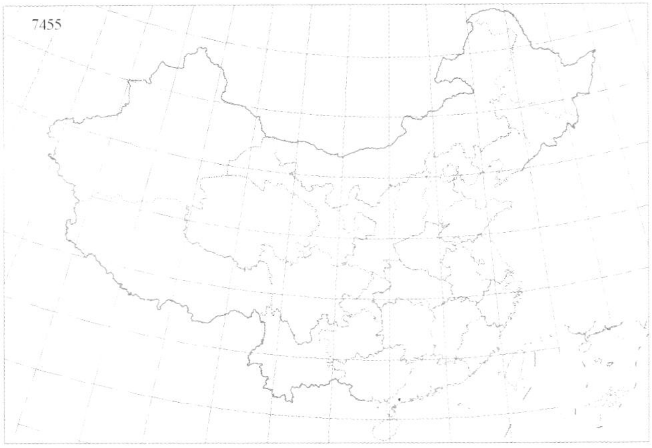

Araliaceae 五加科 *Brassaiopsis* 罗伞属
Brassaiopsis producta (Dunn) C. B. Shang 尖苞罗伞
Trees

17.4 (13.7~19.4); 8.5 (4.7~11.0); 24.4 (20.8~28.7); 17.4 (13.7~19.5); 148.9 (104.8~173.7); 0.0 (0.0~0.3); 876 (722~1038) ❙ 1315 (1143~1776); 685 (610~820); 70 (48~188) ❙ 50.5 (21.9~71.1); 876 (722~1038); 272.9 (186.7~491.5)

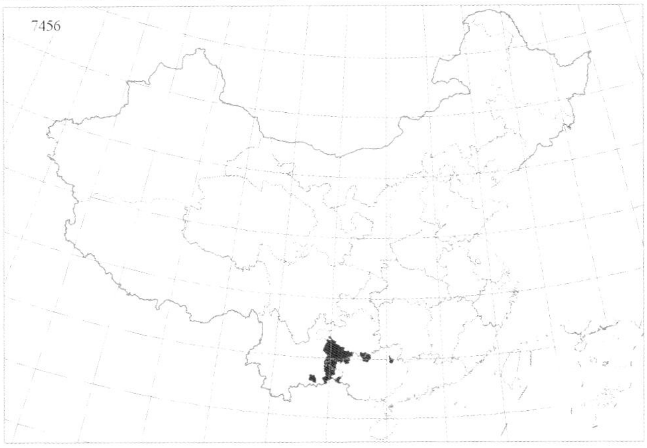

Araliaceae 五加科 *Brassaiopsis* 罗伞属
Brassaiopsis quercifolia G. Hoo 栎叶罗伞
Small trees

18.4 (16.2~21.9); 9.1 (6.3~13.2); 25.9 (24.4~28.6); 18.4 (16.2~22.0); 161.2 (134.5~203.8); 0.0 (0.0~0.0); 940 (826~1176) ❙ 1338 (1236~1710); 676 (611~773); 82 (51~186) ❙ 42.9 (27.5~70.6); 940 (826~1176); 249.2 (189.5~316.1)

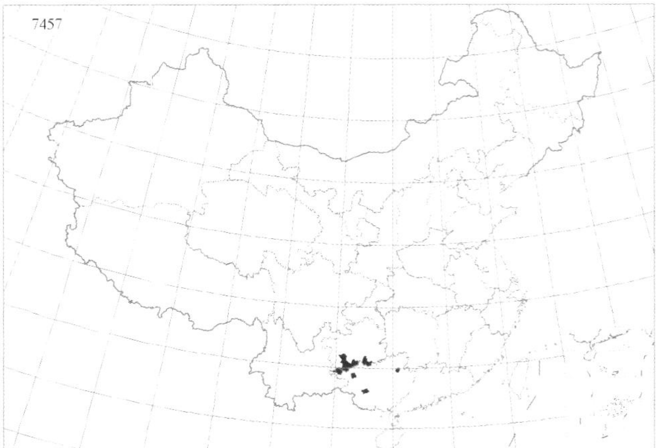

Araliaceae 五加科 *Brassaiopsis* 罗伞属
Brassaiopsis shweliensis W. W. Sm. 瑞丽罗伞
Trees

11.5; 4.0; 17.2; 12.2; 96.3; 17.8; 680 ❙ 1160; 661; 41 ❙ 67.3; 674; 353.1

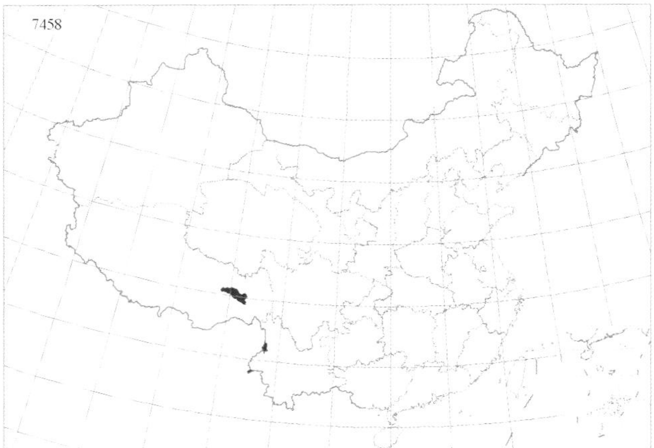

Araliaceae 五加科 *Brassaiopsis* 罗伞属

Brassaiopsis simplicifolia C. B. Clarke 单叶罗伞

Trees

13.3；5.4 19.6；13.4；100.4；0.0；706 ❙ 1736；1007；51 ❙ 145.8；706；447.4

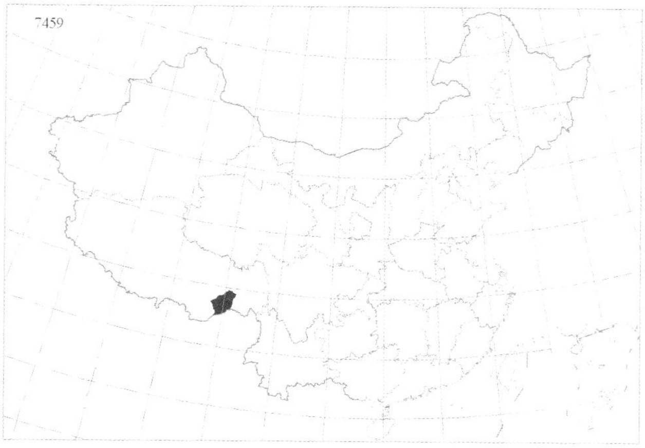

Araliaceae 五加科 *Brassaiopsis* 罗伞属

Brassaiopsis stellata K. M. Feng 星毛罗伞

Trees

19.8 (17.2~22.2)；11.8 (9.5~13.9)；25.6 (22.7~28.3)；19.9 (17.3~22.3)；178.3 (147.0~207.5)；0.0 (0.0~0.0)；1010 (838~1190) ❙ 1408 (1254~1581)；773 (670~850)；73 (48~106) ❙ 41.6 (12.9~65.6)；1010 (838~1190)；327.8 (281.7~371.7)

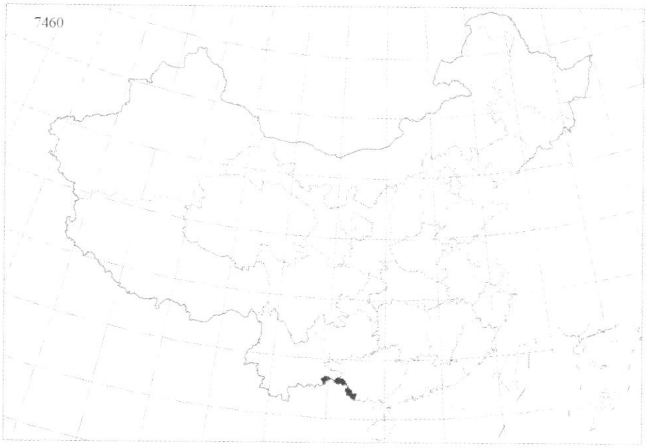

Araliaceae 五加科 *Brassaiopsis* 罗伞属

Brassaiopsis tibetana C. B. Shang 西藏罗伞

Trees

-1.8；-10.2；6.1；1.8；1.9；82.8；329 ❙ 320；217；20 ❙ -3.0；320；72.2

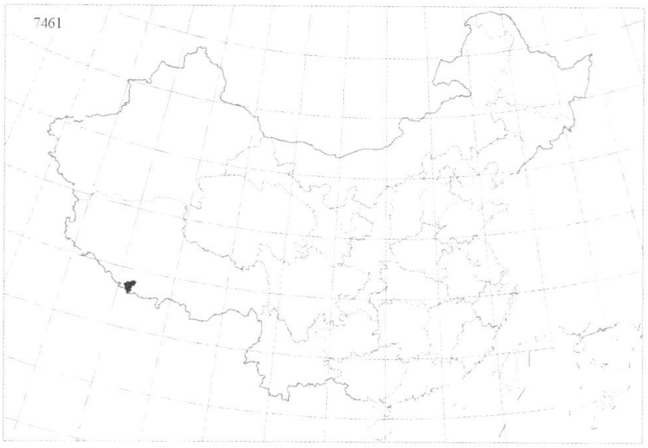

Araliaceae 五加科 *Brassaiopsis* 罗伞属

Brassaiopsis triloba K. M. Feng 三裂罗伞

Shrubs

18.9；10.6；25.0；19.0；167.5；0.0；939 ❙ 1292；729；49 ❙ 37.6；939；342.5

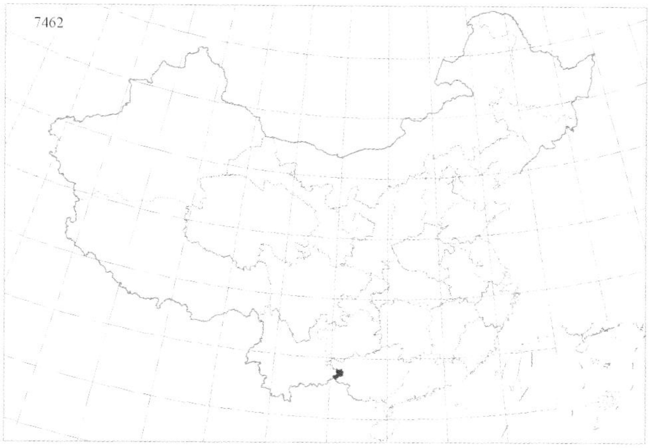

Araliaceae 五加科 *Brassaiopsis* 罗伞属

Brassaiopsis tripteris (H. Lév.) Rehder 显脉罗伞

Shrubs

20.3 (16.8~22.7)；11.3 (7.4~15.0)；27.4 (23.6~28.6)；20.3 (16.8~22.8)；183.7 (141.5~213.1)；0.0 (0.0~0.0)；1060 (836~1221) ❙ 1561 (1236~2435)；748 (562~1428)；129 (49~194) ❙ 47.6 (12.9~99.5)；1060 (836~1221)；252.8 (185.5~342.5)

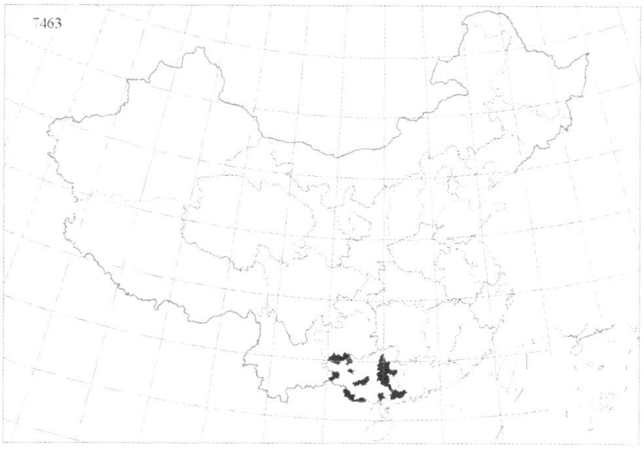

Araliaceae 五加科 *Chengiopanax* 人参木属

Chengiopanax fargesii (Franch.) C. B. Shang et J. Y. Huang 人参木

Trees, deciduous

15.4 (12.7~17.6)；4.3 (2.1~6.6)；25.7 (23.0~27.7)；15.4 (12.8~17.6)；126.5 (99.0~151.3)；1.5 (0.0~5.6)；825 (713~928) ❙ 1452 (1320~1526)；533 (465~600)；141 (64~197) ❙ 77.4 (52.8~107.2)；824 (713~922)；216.9 (173.7~257.1)

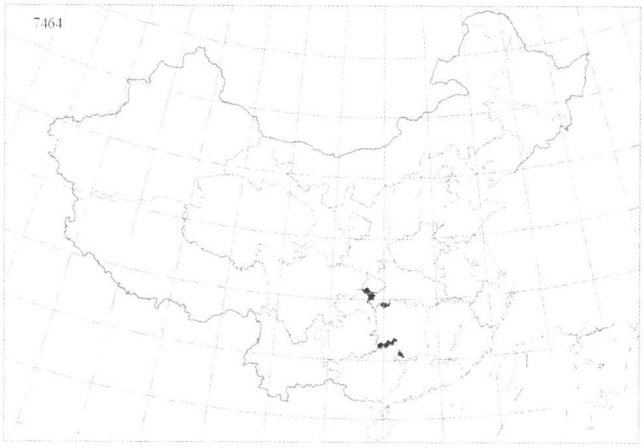

Araliaceae 五加科 *Dendropanax* 树参属
Dendropanax bilocularis C. N. Ho 双室树参
Shrubs

20.1 (17.7~22.2); 11.0 (7.4~14.7); 27.6 (25.0~28.6); 20.1 (17.8~22.2); 181.7 (153.6~206.9); 0.0 (0.0~0.0); 1052 (913~1177) ❙ 1615 (1292~1876); 705 (556~891); 153 (49~189) ❙ 54.4 (22.7~85.5); 1052 (913~1177); 235.0 (185.5~342.5)

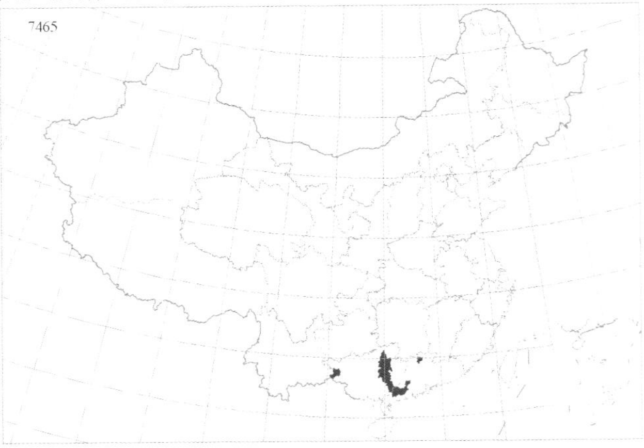

Araliaceae 五加科 *Dendropanax* 树参属
Dendropanax burmanicus Merr. 缅甸树参
Shrubs or small trees

12.3; 4.1; 19.3; 12.3; 90.2; 2.4; 689 ❙ 1277; 698; 55 ❙ 87.8; 689; 341.4

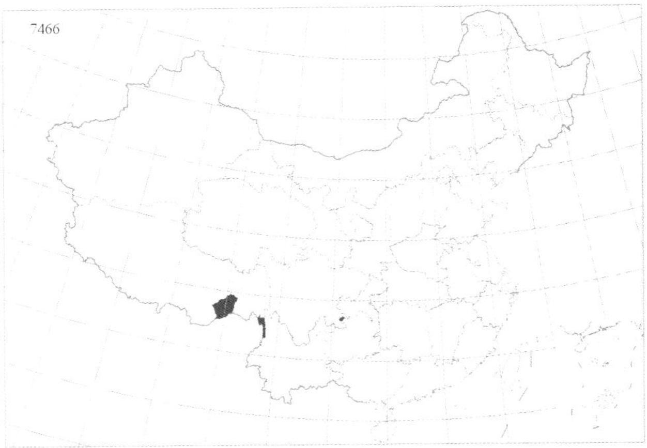

Araliaceae 五加科 *Dendropanax* 树参属
Dendropanax caloneurus (Harms) Merr. 榕叶树参
Small trees

17.3; 9.9; 22.4; 17.3; 147.1; 0.0; 833 ❙ 1504; 860; 55 ❙ 80.6; 833; 413.5

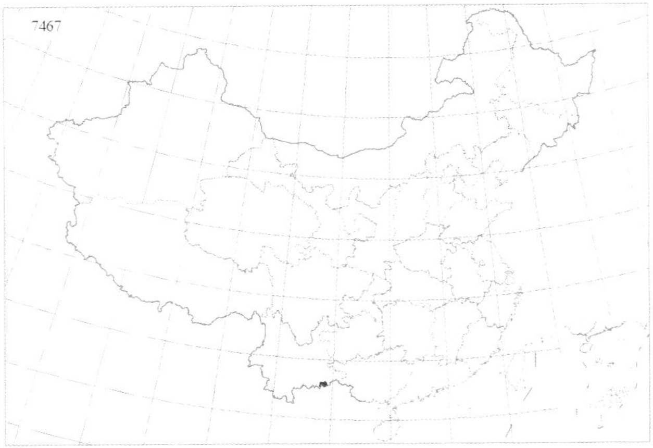

Araliaceae 五加科 *Dendropanax* 树参属
Dendropanax chevalieri (Vig.) Merr. 大果树参
Trees

18.4 (17.2~21.6); 10.6 (9.5~12.6); 24.1 (22.4~28.5); 18.5 (17.3~21.7); 161.6 (147.0~199.8); 0.0 (0.0~0.0); 914 (833~1153) ❙ 1435 (1254~1625); 791 (699~860); 74 (48~173) ❙ 58.2 (40.9~80.6); 914 (833~1153); 368.4 (237.8~491.5)

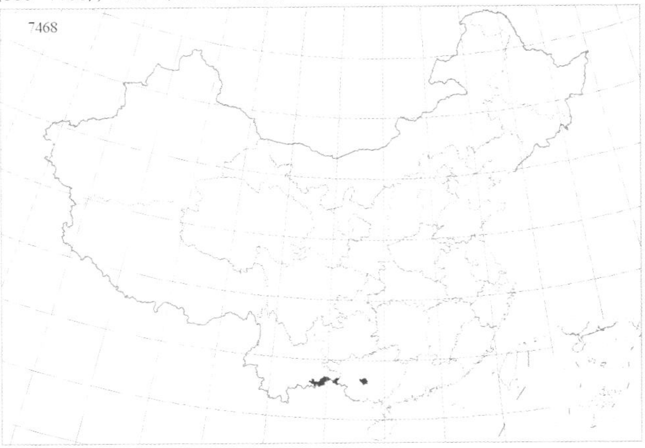

Araliaceae 五加科 *Dendropanax* 树参属
Dendropanax confertus H. L. Li 挤果树参
Shrubs or trees

19.3 (17.0~21.3); 9.5 (5.2~12.3); 27.7 (25.8~29.8); 19.3 (17.0~21.3); 172.1 (144.3~196.1); 0.0 (0.0~0.0); 1011 (902~1130) ❙ 1615 (1371~1852); 606 (494~728); 183 (142~222) ❙ 60.6 (22.7~86.3); 1007 (902~1130); 232.1 (174.9~271.3)

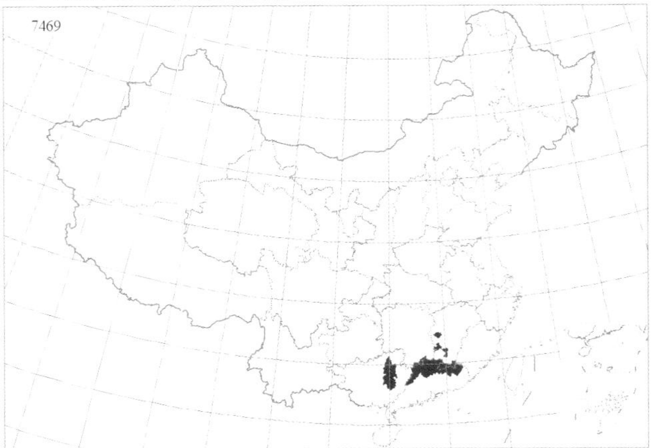

Araliaceae 五加科 *Dendropanax* 树参属
Dendropanax dentiger (Harms) Merr. 树参
Shrubs or trees

17.5 (2.2~24.0); 7.3 (-6.6~19.1); 26.8 (9.9~30.1); 17.5 (3.8~24.0); 150.8 (15.5~228.1); 0.6 (0.0~49.1); 924 (413~1302) ❙ 1515 (780~2993); 586 (356~1847); 158 (11~547) ❙ 65.0 (-5.0~235.6); 916 (413~1225); 229.2 (123.9~620.0)

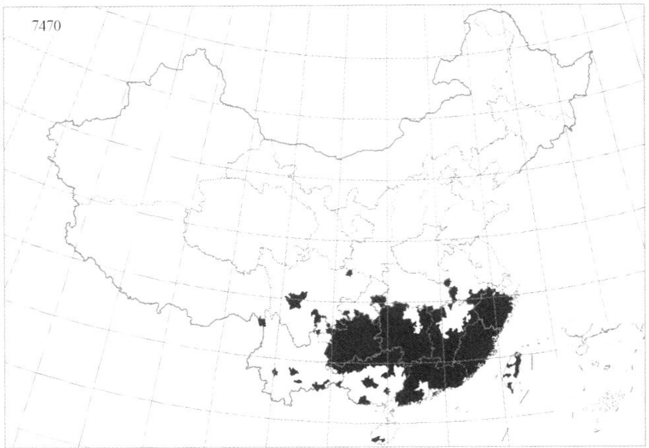

Araliaceae 五加科 *Dendropanax* 树参属
Dendropanax hainanensis (Merr. et Chun) Merr. et Chun 海南树参
Trees

18.9 (14.3~24.0); 9.3 (3.8~19.1); 27.1 (22.5~29.8); 18.9 (14.3~24.0); 167.5 (112.6~228.1); 0.1 (0.0~1.5); 992 (756~1302) ❙ 1543 (1188~2435); 615 (403~1428); 152 (44~230) ❙ 56.8 (-2.2~104.6); 981 (756~1221); 256.4 (151.0~620.0)

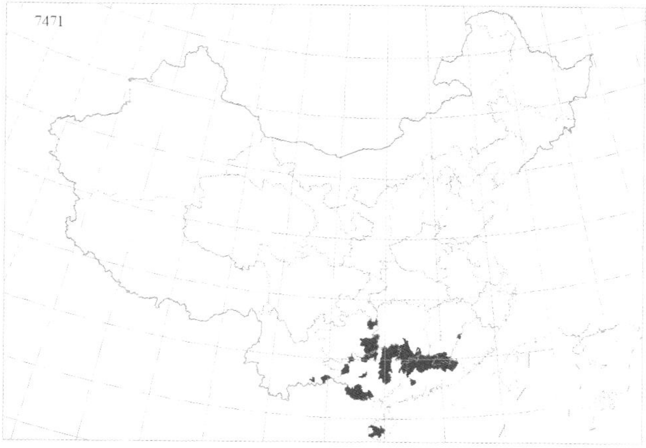

Araliaceae 五加科 *Dendropanax* 树参属
Dendropanax kwangsiensis H. L. Li 广西树参
Shrubs

19.1 (15.5~22.7); 9.4 (4.2~15.0); 27.5 (25.0~29.0); 19.2 (15.6~22.8); 169.9 (128.1~213.1); 0.0 (0.0~0.8); 1007 (830~1221) ❙ 1597 (1292~2435); 675 (403~1428); 158 (49~211) ❙ 59.0 (12.9~99.5); 998 (830~1221); 236.5 (151.0~342.5)

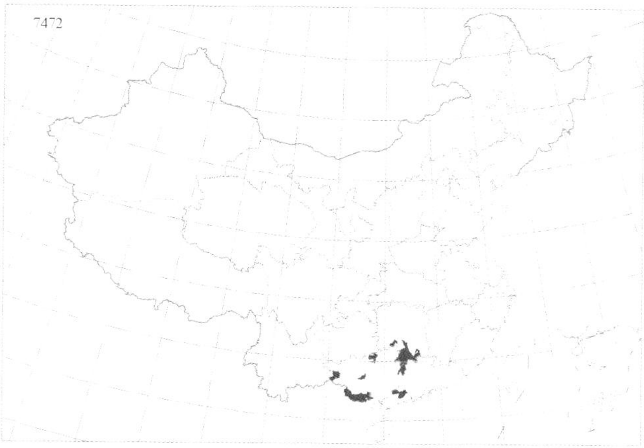

Araliaceae 五加科 *Dendropanax* 树参属
Dendropanax oligodontus Merr. et Chun 保亭树参
Shrubs

24.0; 19.1; 27.2; 24.0; 228.1; 0.0; 1302 ❙ 1551; 539; 48 ❙ 19.1; 1147; 620.0

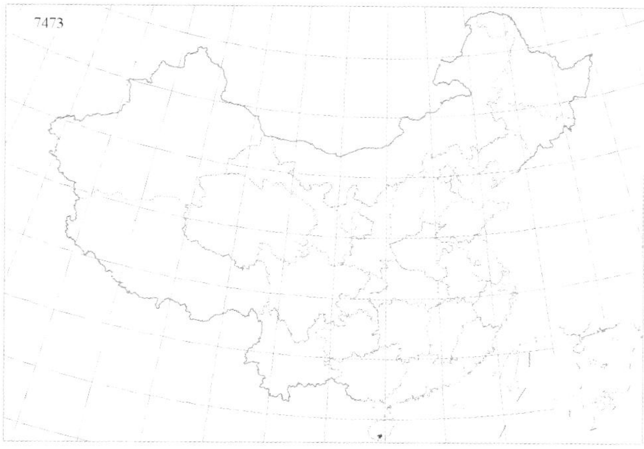

Araliaceae 五加科 *Dendropanax* 树参属
Dendropanax productus H. L. Li 长萼树参
Shrubs or small trees

19.1 (17.1~22.0); 9.3 (5.9~14.1); 27.7 (26.2~29.8); 19.2 (17.1~22.1); 170.2 (145.7~205.1); 0.0 (0.0~0.0); 1002 (906~1170) ❙ 1672 (1486~2014); 604 (505~954); 191 (142~259) ❙ 67.4 (35.6~111.3); 999 (906~1170); 241.1 (174.9~279.5)

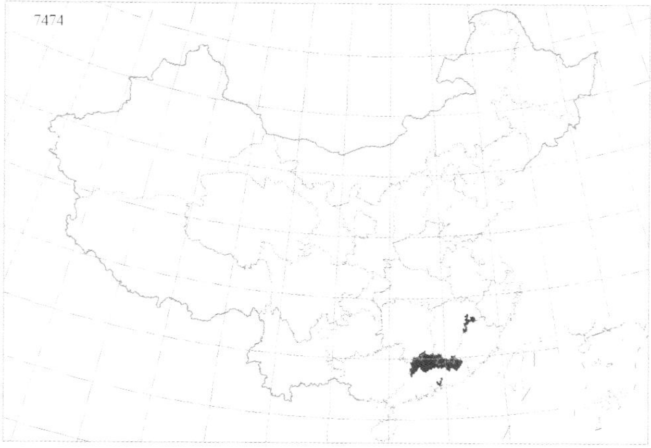

Araliaceae 五加科 *Dendropanax* 树参属
Dendropanax proteus (Champ. ex Benth.) Benth. 变叶树参
Shrubs

20.0 (14.2~23.6); 11.0 (3.8~16.7); 27.6 (20.1~29.8); 20.1 (14.2~23.6); 181.0 (110.5~223.5); 0.0 (0.0~1.5); 1054 (729~1281) ❙ 1630 (978~2435); 681 (403~1428); 152 (40~231) ❙ 56.2 (-6.9~104.6); 1051 (729~1273); 229.9 (95.3~335.1)

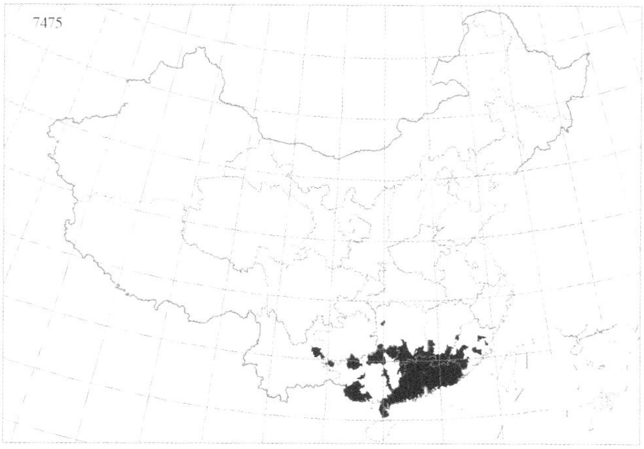

Araliaceae 五加科 *Diplopanax* 马蹄参属
Diplopanax stachyanthus Hand.-Mazz. 马蹄参
Trees

19.5 (14.6~23.0); 10.4 (3.8~17.7); 26.9 (22.5~28.9); 19.5 (14.7~23.1); 174.6 (117.7~217.0); 0.0 (0.0~1.5); 1020 (779~1225) ❙ 1572 (1221~2435); 720 (452~1428); 135 (48~194) ❙ 55.2 (12.9~101.8); 1019 (779~1225); 275.8 (185.5~572.9)

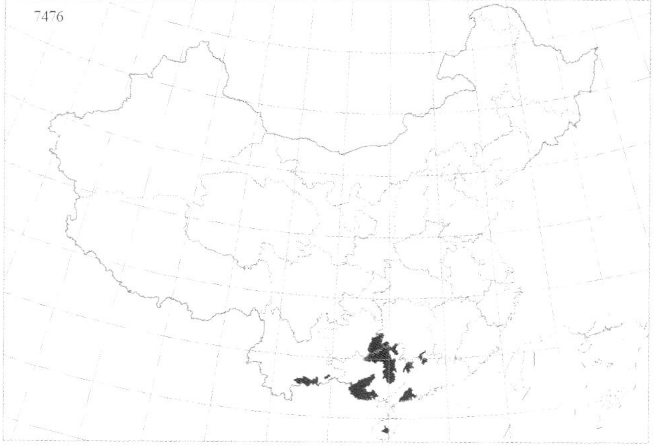

Araliaceae 五加科 *Eleutherococcus* 五加属

Eleutherococcus baoxinensis (X. P. Fang et C. K. Hsieh) P. S. Hsu et S. L. Pan 宝兴五加

Shrubs, deciduous

7.3; -2.0; 15.6; 7.5; 48.2; 20.4; 537 ❙ 1005; 531; 27 ❙ 87.3; 537; 272.3

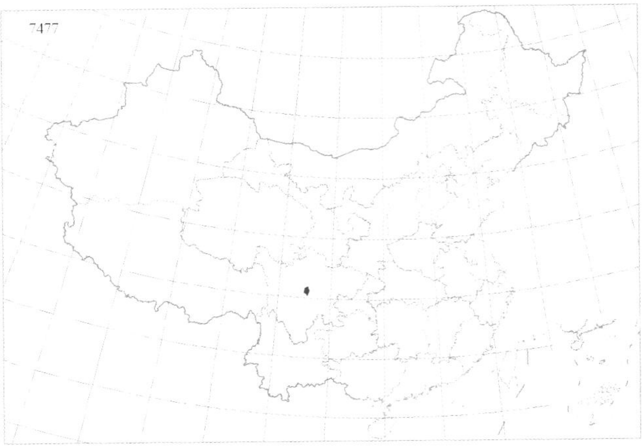

Araliaceae 五加科 *Eleutherococcus* 五加属

Eleutherococcus brachypus (Harms) Nakai 短柄五加

Shrubs, deciduous

8.9 (1.9~13.0); -4.7 (-9.2~1.1); 21.2 (11.7~26.6); 9.8 (4.3~13.3); 75.9 (21.1~113.3); 28.6 (8.1~57.8); 632 (427~800) ❙ 571 (341~892); 291 (190~389); 14 (8~29) ❙ -8.4 (-40.9~45.4); 543 (341~707); 198.1 (61.2~320.6)

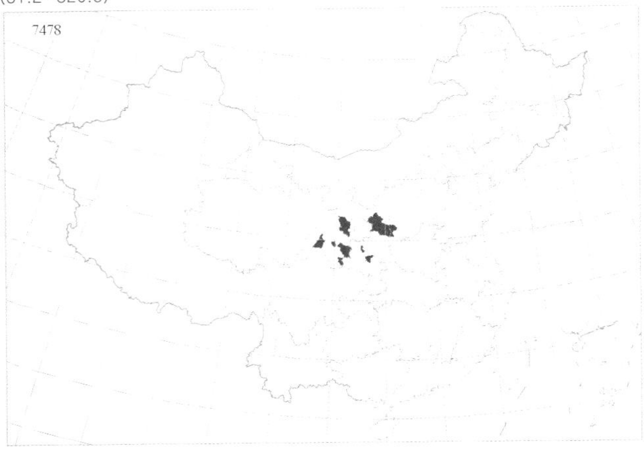

Araliaceae 五加科 *Eleutherococcus* 五加属

Eleutherococcus cissifolius (Griff. ex C. B. Clarke) Nakai 乌蔹莓五加

Shrubs, deciduous

4.5 (-1.8~14.3); -4.7 (-13.5~7.6); 12.5 (6.1~19.7); 6.1 (1.8~14.3); 35.9 (1.9~111.9); 41.7 (0.0~86.3); 478 (329~728) ❙ 704 (320~1736); 411 (193~1007); 22 (7~51) ❙ 40.9 (-14.6~145.8); 471 (320~707); 245.6 (72.2~447.4)

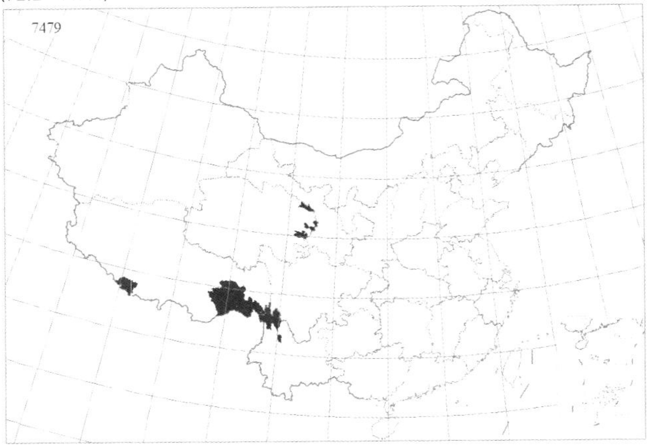

Araliaceae 五加科 *Eleutherococcus* 五加属

Eleutherococcus giraldii (Harms) Nakai 红毛五加

Shrubs, deciduous

9.0 (-1.3~15.5); -3.8 (-15.6~8.6); 20.7 (8.5~28.2); 9.9 (2.7~15.6); 76.2 (8.3~128.0); 27.4 (0.0~86.3); 636 (359~874) ❙ 709 (63~1492); 365 (45~622); 24 (2~105) ❙ 14.6 (-84.9~128.6); 585 (63~817); 219.3 (8.4~357.7)

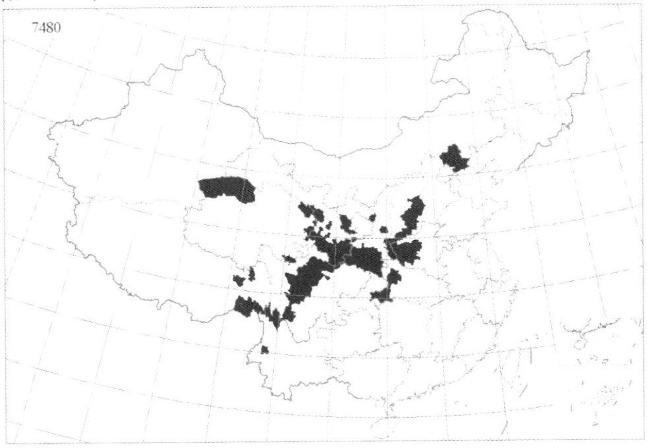

Araliaceae 五加科 *Eleutherococcus* 五加属

Eleutherococcus henryi Oliv. 糙叶五加

Shrubs, deciduous

12.9 (2.5~16.7); 0.2 (-12.7~5.3); 24.9 (11.2~28.9); 13.2 (4.4~16.8); 107.4 (20.5~142.4); 11.6 (0.0~65.5); 765 (428~924) ❙ 987 (450~1813); 443 (245~817); 70 (9~226) ❙ 28.9 (-36.2~123.6); 729 (428~924); 212.1 (127.2~334.4)

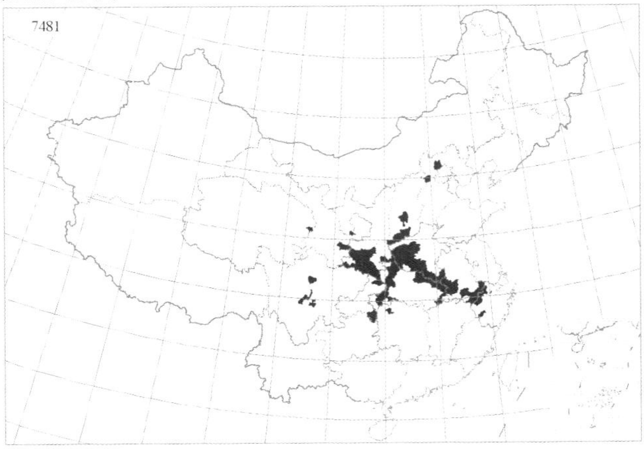

Araliaceae 五加科 *Eleutherococcus* 五加属

Eleutherococcus henryi var. faberi (Harms) S. Y. Hu 毛梗糙叶五加

Shrubs, deciduous

14.1 (10.2~17.7); 1.1 (-2.9~5.3); 26.4 (22.4~29.3); 14.2 (10.6~17.7); 118.2 (83.1~152.4); 8.1 (0.0~20.5); 810 (656~964) ❙ 1063 (626~1679); 458 (316~628); 95 (29~220) ❙ 30.1 (-22.2~101.3); 777 (626~961); 206.7 (134.1~318.0)

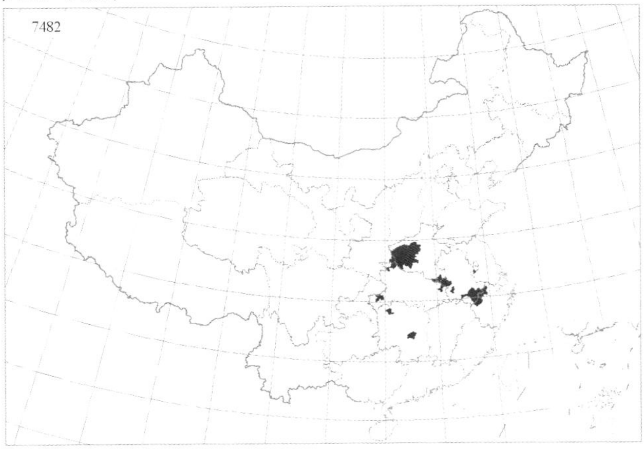

Araliaceae 五加科 *Eleutherococcus* 五加属
Eleutherococcus lasiogyne (Harms) S. Y. Hu 康定五加
Shrubs or trees, deciduous

4.7 (1.8~13.3); -4.2 (-7.4~5.4); 12.6 (9.9~19.6); 5.8 (3.8~13.4); 33.0 (15.4~100.4); 35.8 (0.0~53.3); 476 (404~706) ❙ 876 (473~1736); 491 (287~1007); 22 (7~51) ❙ 82.0 (5.0~145.8); 476 (404~706); 280.4 (169.5~447.4)

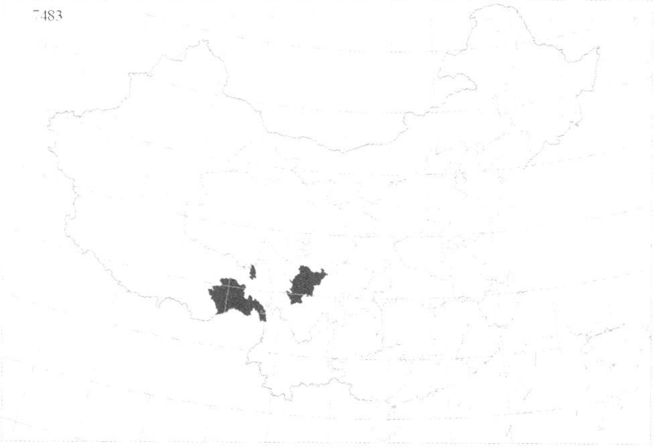

Araliaceae 五加科 *Eleutherococcus* 五加属
Eleutherococcus leucorrhizus Oliv. 藤五加
Shrubs, deciduous

13.8 (1.9~21.0); 2.2 (-11.4~11.9); 24.6 (11.2~29.3); 14.0 (4.3~21.1); 114.2 (20.5~193.0); 8.1 (0.0~57.8); 784 (427~1114) ❙ 1177 (341~1997); 475 (190~821); 104 (7~259) ❙ 48.1 (-40.9~130.7); 762 (341~1114); 221.6 (61.2~352.3)

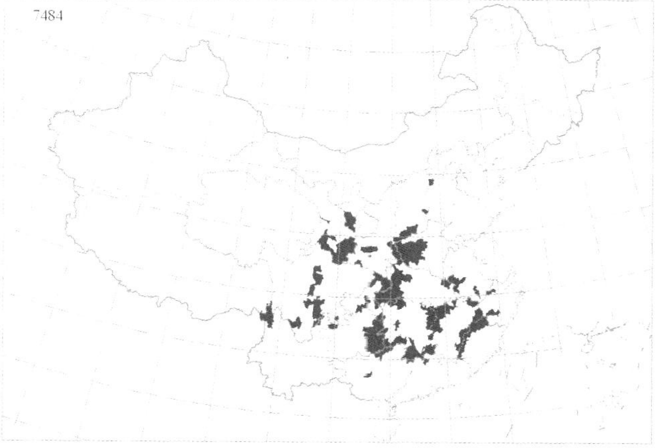

Araliaceae 五加科 *Eleutherococcus* 五加属
Eleutherococcus leucorrhizus var. fulvescens (Harms et Rehder) Nakai 糙叶藤五加
Shrubs, deciduous

11.5 (2.2~17.2); 0.8 (-7.4~5.7); 21.6 (9.9~28.4); 11.8 (3.8~17.2); 90.7 (15.5~146.8); 12.5 (0.0~49.1); 691 (413~925) ❙ 1136 (341~1821); 492 (190~817); 73 (8~225) ❙ 64.2 (-40.9~125.9); 684 (341~925); 234.9 (61.2~352.3)

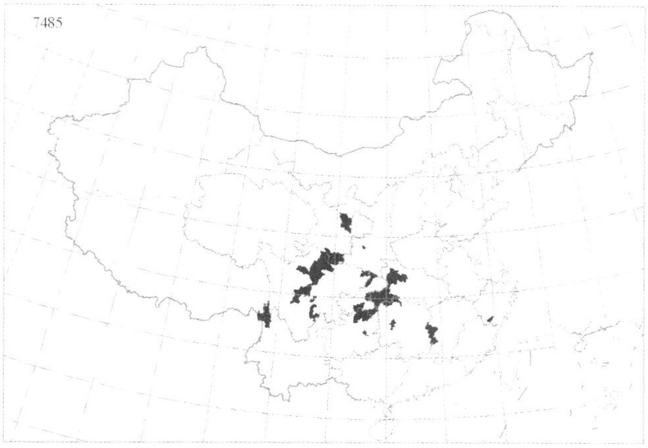

Araliaceae 五加科 *Eleutherococcus* 五加属
Eleutherococcus leucorrhizus var. scaberulus (Harms et Rehder) Nakai 狭叶藤五加
Shrubs, deciduous

15.0 (7.3~17.6); 4.2 (-2.0~7.4); 25.2 (15.6~29.3); 15.1 (7.5~17.7); 122.7 (48.2~151.9); 1.8 (0.0~20.4); 810 (537~950) ❙ 1304 (793~1997); 528 (377~685); 108 (27~254) ❙ 61.1 (0.0~130.7); 808 (537~950); 217.7 (146.0~303.4)

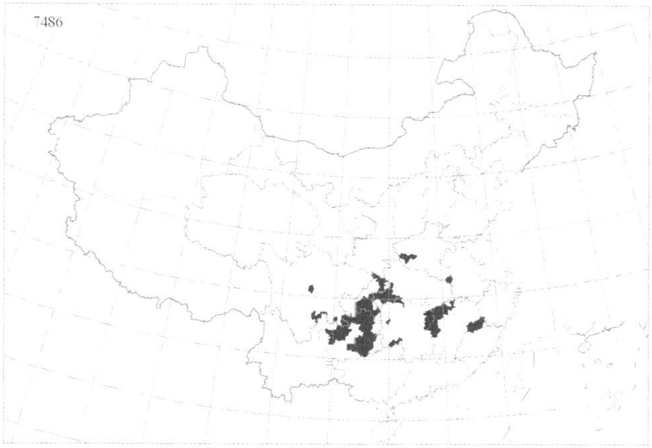

Araliaceae 五加科 *Eleutherococcus* 五加属
Eleutherococcus leucorrhizus var. setchuenensis (Harms) C. B. Shang et J. Y. Huang 蜀五加
Shrubs, deciduous

11.3 (2.2~15.5); -0.2 (-7.7~5.8); 22.1 (9.9~26.5); 11.5 (3.8~15.6); 89.4 (15.5~127.2); 13.7 (0.0~49.3); 683 (413~832) ❙ 947 (341~1526); 438 (190~817); 39 (8~181) ❙ 39.4 (-40.9~125.9); 676 (341~832); 246.1 (61.2~334.4)

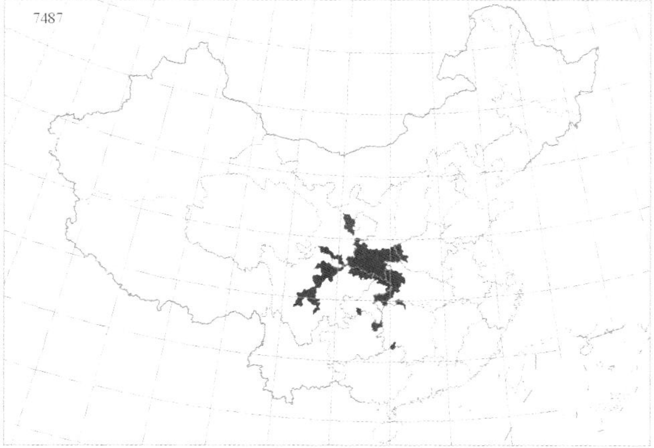

Araliaceae 五加科 *Eleutherococcus* 五加属
Eleutherococcus nodiflorus (Dunn) S. Y. Hu 细柱五加
Small trees, deciduous

15.5 (5.4~21.8); 4.0 (-7.7~13.4); 26.3 (12.3~30.1); 15.6 (5.8~21.8); 130.5 (31.5~201.9); 3.4 (0.0~40.0); 852 (481~1158) ❙ 1221 (503~1997); 507 (257~938); 115 (12~259) ❙ 42.7 (-36.5~147.0); 828 (481~1158); 204.2 (111.9~383.4)

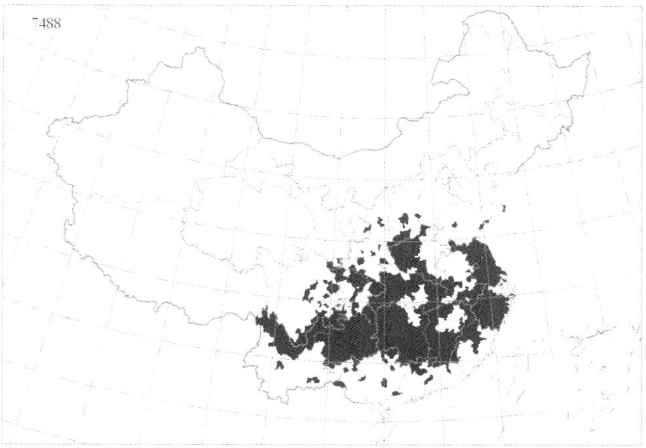

Araliaceae 五加科 *Eleutherococcus* 五加属
Eleutherococcus rehderianus (Harms) Nakai 匙叶五加
Shrubs, deciduous

12.8 (9.3~15.3); 1.3 (-3.3~5.2); 23.9 (20.1~27.5); 12.9 (9.5~15.4); 102.6 (69.8~128.7); 8.2 (0.0~22.7); 735 (603~864) ❙ 1031 (636~1214); 450 (308~528); 54 (15~83) ❙ 41.7 (-0.3~101.4); 734 (603~864); 243.9 (176.6~279.5)

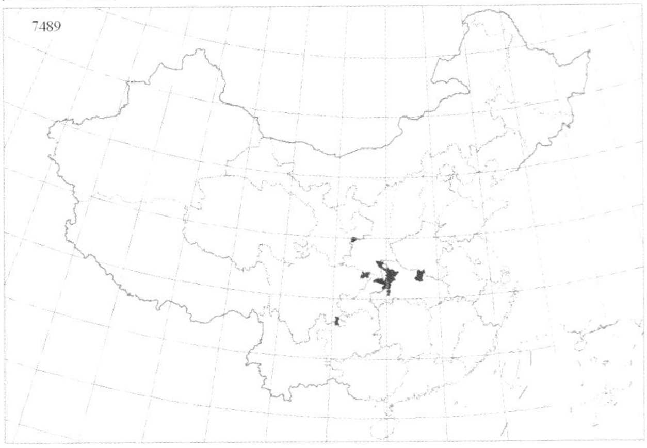

Araliaceae 五加科 *Eleutherococcus* 五加属
Eleutherococcus scandens H. Ohashi 匍匐五加
Scandent shrubs, deciduous

16.0 (13.4~18.8); 4.3 (1.5~9.4); 27.0 (25.2~29.2); 16.0 (13.4~18.9); 134.1 (108.3~166.4); 2.0 (0.0~7.3); 865 (755~967) ❙ 1626 (1296~1997); 586 (498~669); 199 (129~254) ❙ 88.4 (46.2~130.7); 864 (755~967); 259.6 (186.9~319.9)

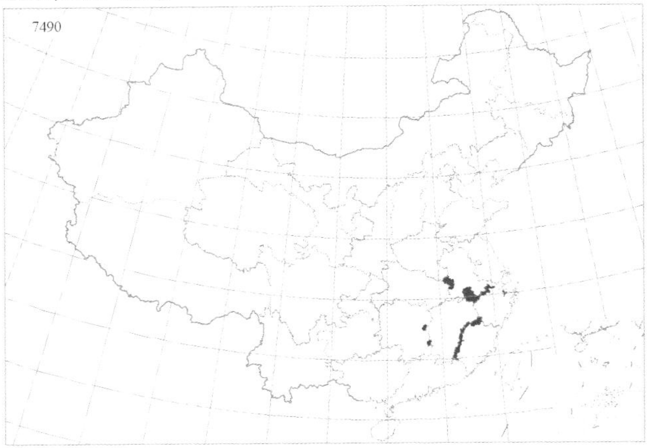

Araliaceae 五加科 *Eleutherococcus* 五加属
Eleutherococcus senticosus (Rupr. et Maxim.) Maxim. 刺五加
Shrubs, deciduous

8.3 (-1.5~15.8); -8.5 (-25.0~3.5); 23.1 (8.5~28.0); 10.3 (2.7~15.8); 84.0 (8.3~133.1); 43.6 (1.5~124.5); 669 (363~891) ❙ 593 (332~1200); 358 (200~691); 21 (6~109) ❙ -10.1 (-43.4~92.8); 557 (332~891); 189.1 (44.7~346.5)

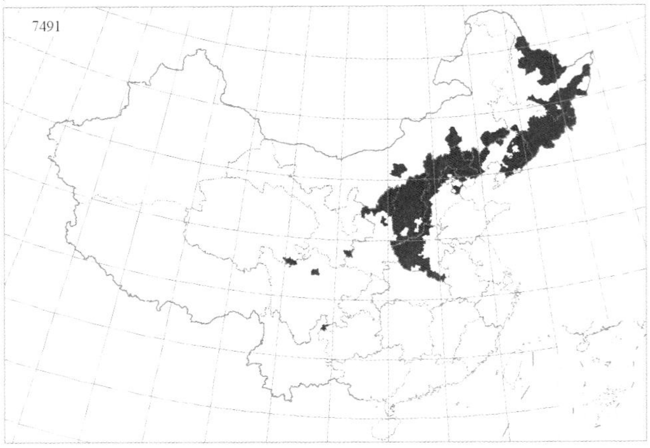

Araliaceae 五加科 *Eleutherococcus* 五加属
Eleutherococcus sessiliflorus (Rupr. et Maxim.) S. Y. Hu 无梗五加
Small trees, deciduous

6.3 (-0.8~13.0); -11.7 (-25.0~-1.3); 21.9 (17.5~26.6); 9.1 (6.4~13.3); 72.6 (45.4~113.3); 56.6 (14.7~118.4); 619 (501~800) ❙ 576 (202~1073); 362 (134~691); 17 (6~40) ❙ -6.2 (-65.5~74.1); 530 (202~698); 211.7 (50.7~346.5)

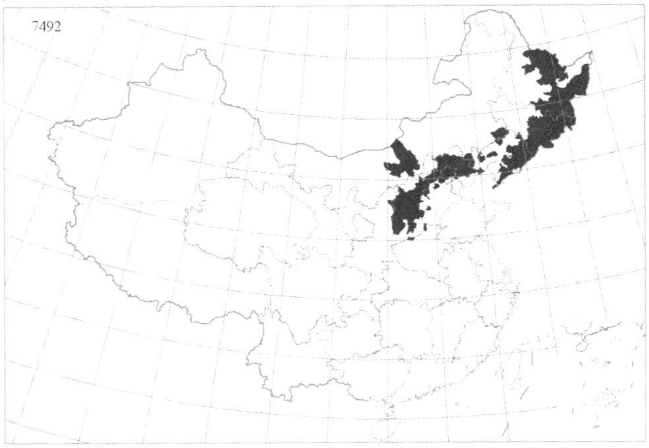

Araliaceae 五加科 *Eleutherococcus* 五加属
Eleutherococcus setulosus (Franch.) S. Y. Hu 细刺五加
Shrubs, deciduous

14.4 (7.3~16.5); 2.6 (-2.0~4.4); 25.7 (15.6~28.9); 14.4 (7.5~16.6); 117.3 (48.2~139.8); 4.3 (0.6~20.4); 803 (537~910) ❙ 1420 (1005~1801); 556 (493~606); 159 (27~242) ❙ 77.1 (45.3~109.2); 803 (537~910); 247.7 (171.0~299.5)

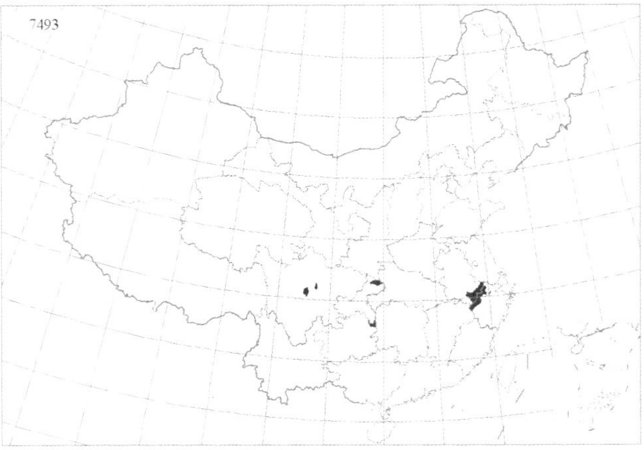

Araliaceae 五加科 *Eleutherococcus* 五加属
Eleutherococcus trifoliatus (L.) S. Y. Hu 白簕
Shrubs, deciduous

17.3 (2.2~24.5); 7.2 (-7.1~18.9); 26.4 (9.9~29.9); 17.4 (3.8~24.5); 149.7 (15.5~234.4); 1.5 (0.0~49.1); 923 (413~1360) ❙ 1414 (626~3495); 590 (316~1847); 133 (15~897) ❙ 53.9 (-22.2~246.5); 912 (413~1360); 225.0 (95.3~659.0)

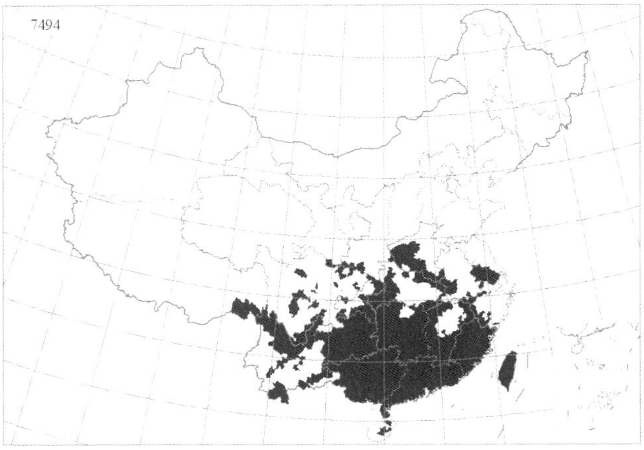

Araliaceae 五加科 *Eleutherococcus* 五加属
Eleutherococcus verticillatus (G. Hoo) H. Ohashi 轮伞五加
Shrubs, deciduous

0.3; -8.1; 8.1; 2.9; 9.1; 64.4; 369 ❙ 452; 271; 23 ❙ 20.1; 363; 124.4

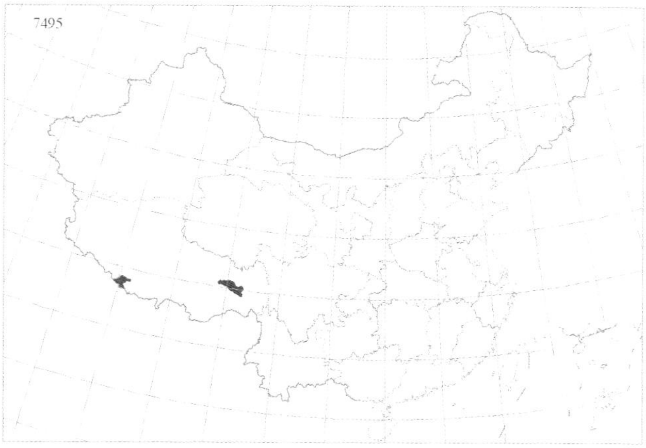

Araliaceae 五加科 *Eleutherococcus* 五加属
Eleutherococcus wilsonii (Harms) Nakai 狭叶五加
Shrubs, deciduous

7.7 (-1.7~15.1); -4.7 (-13.5~1.8); 19.1 (8.0~27.5); 8.9 (2.4~15.2); 67.2 (6.4~128.0); 33.8 (5.6~86.3); 604 (344~865) ❙ 650 (341~1023); 337 (190~546); 23 (5~53) ❙ 14.3 (-40.9~128.6); 554 (341~817); 213.6 (61.2~346.5)

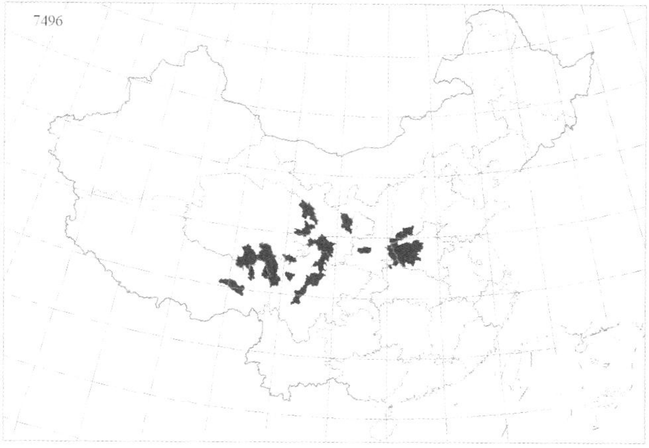

Araliaceae 五加科 *Eleutherococcus* 五加属
Eleutherococcus wilsonii var. pilosulus (Rehder) P. S. Hsu et S. L. Pan 毛狭叶五加
Shrubs, deciduous

3.2 (-1.3~7.0); -9.3 (-13.5~-6.5); 14.4 (9.2~19.9); 5.6 (2.9~8.5); 35.1 (10.1~64.1); 55.5 (37.9~86.3); 472 (359~577) ❙ 487 (341~621); 269 (190~321); 9 (5~13) ❙ 5.7 (-40.9~45.6); 430 (341~539); 205.5 (61.2~313.2)

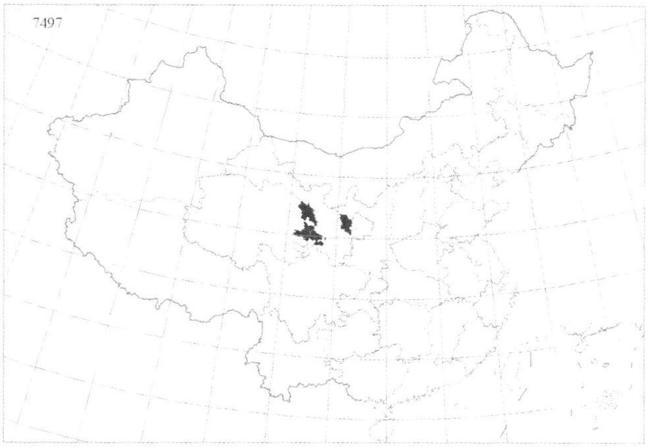

Araliaceae 五加科 *Fatsia* 八角金盘属
Fatsia japonica (Thunb.) Decaisne et Planchon 八角金盘
Shrubs, evergreen

15.6 (11.7~19.1); 5.6 (2.7~15.2); 24.7 (18.8~28.7); 15.7 (11.7~22.7); 129.0 (82.6~212.7); 0.4 (0.0~3.8); 821 (658~1216) ❙ 1224 (916~1965); 552 (444~994); 78 (31~158) ❙ 49.2 (23.8~71.9); 821 (658~1216); 203.8 (146.0~283.9)

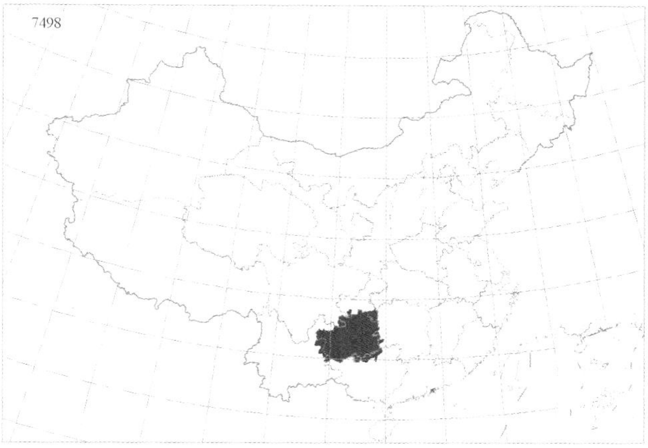

Araliaceae 五加科 *Fatsia* 八角金盘属
Fatsia polycarpa Hayata 多室八角金盘
Small trees, evergreen

19.2 (16.2~24.1); 13.7 (11.0~18.9); 23.8 (20.1~28.0); 19.3 (16.2~24.1); 171.3 (134.2~229.7); 0.0 (0.0~0.0); 954 (773~1322) ❙ 2467 (1751~2993); 1141 (924~1594); 238 (47~547) ❙ 170.6 (32.5~246.5); 924 (773~1095); 246.5 (154.2~303.6)

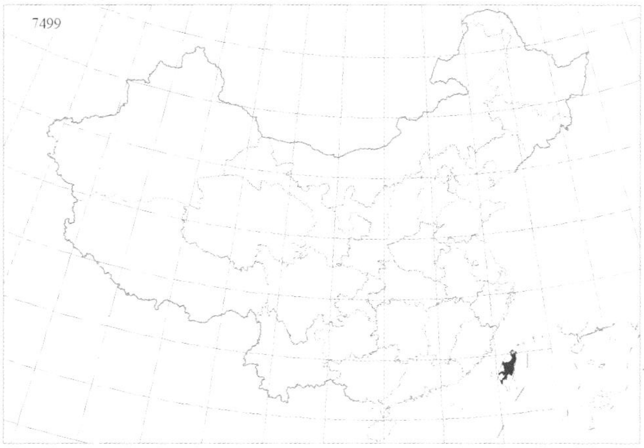

Araliaceae 五加科 *Gamblea* 吴茱叶五加属
Gamblea ciliata C. B. Clarke 萸叶五加
Shrubs or trees, evergreen

7.2 (-1.8~15.5); -1.8 (-10.5~5.4); 15.4 (6.1~26.3); 8.2 (1.8~15.6); 56.7 (1.9~128.1); 30.0 (0.0~82.8); 555 (329~830) ❙ 937 (311~1736); 499 (193~1007); 43 (9~192) ❙ 60.1 (-6.3~145.8); 551 (311~830); 213.2 (62.8~447.4)

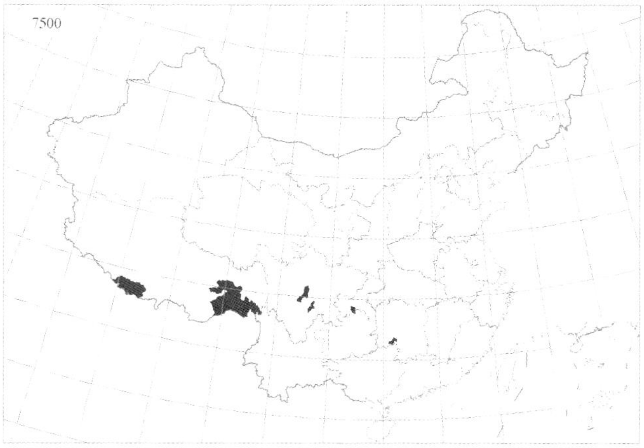

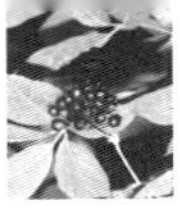

Araliaceae 五加科 *Gamblea* 吴茱叶五加属

Gamblea ciliata var. evodiifolia (Franchet) C. B. Shang et al. 吴茱萸五加

Shrubs or trees, evergreen

15.5 (−1.8~22.7); 5.1 (−10.5~15.0); 25.2 (6.1~29.7); 15.7 (1.8~22.8); 129.9 (1.9~213.1); 2.8 (0.0~82.8); 840 (329~1221) ❙ 1385 (311~2435); 558 (193~1428); 135 (9~259) ❙ 64.4 (−14.2~147.0); 831 (311~1221); 239.3 (61.6~551.6)

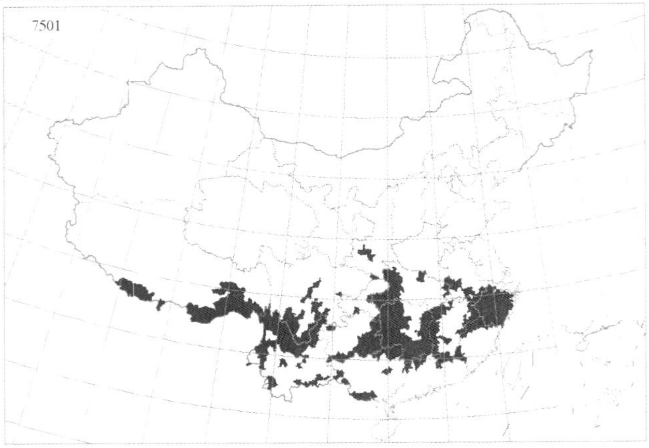

Araliaceae 五加科 *Hedera* 常春藤属

Hedera nepalensis var. sinensis (Tobler) Rehder 常春藤

Scandent shrubs, evergreen

16.1 (−1.8~23.6); 5.6 (−10.2~16.7); 25.9 (6.1~29.9); 16.2 (1.8~23.6); 137.3 (1.9~223.5); 3.2 (0.0~82.8); 875 (329~1281) ❙ 1305 (320~2037); 542 (193~1033); 123 (9~259) ❙ 49.2 (−22.2~147.0); 865 (320~1273); 223.7 (72.2~506.2)

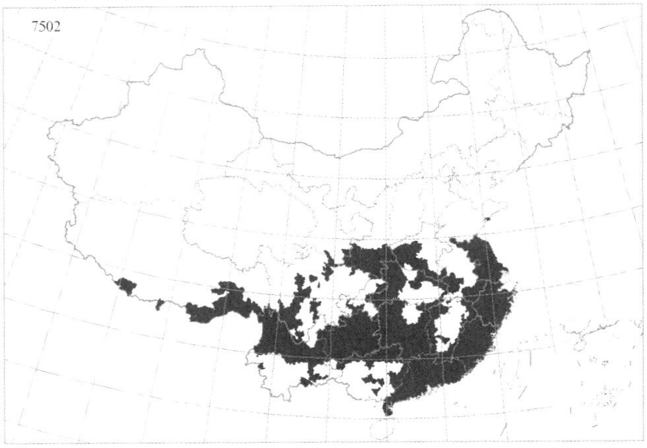

Araliaceae 五加科 *Heteropanax* 幌伞枫属

Heteropanax brevipedicellatus H. L. Li 短梗幌伞枫

Shrubs or small trees, evergreen

19.8 (15.9~22.9); 10.5 (5.4~15.2); 27.6 (25.2~29.8); 19.8 (16.0~22.9); 177.9 (131.9~214.8); 0.0 (0.0~0.0); 1036 (831~1232) ❙ 1669 (1250~2024); 668 (505~1056); 171 (60~259) ❙ 62.5 (17.4~115.7); 1034 (831~1232); 243.2 (152.9~335.1)

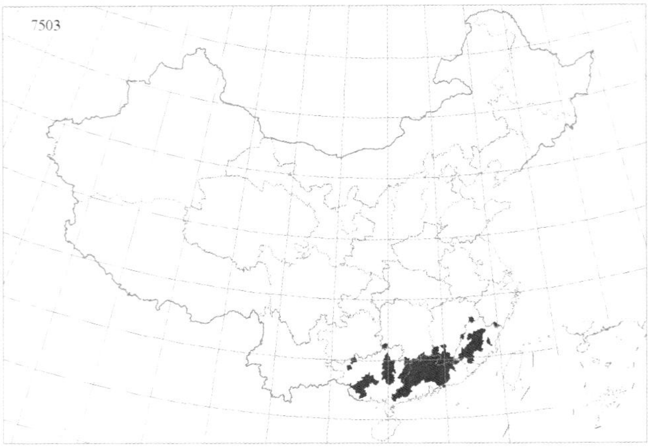

Araliaceae 五加科 *Heteropanax* 幌伞枫属

Heteropanax chinensis (Dunn) H. L. Li 华幌伞枫

Shrubs, evergreen

20.6 (17.3~22.7); 12.5 (6.5~15.1); 26.8 (22.3~28.9); 20.6 (17.3~22.8); 187.5 (147.9~213.1); 0.0 (0.0~0.0); 1068 (886~1221) ❙ 1644 (1302~2435); 800 (561~1428); 120 (58~197) ❙ 54.9 (12.9~99.5); 1063 (880~1221); 325.9 (201.4~659.0)

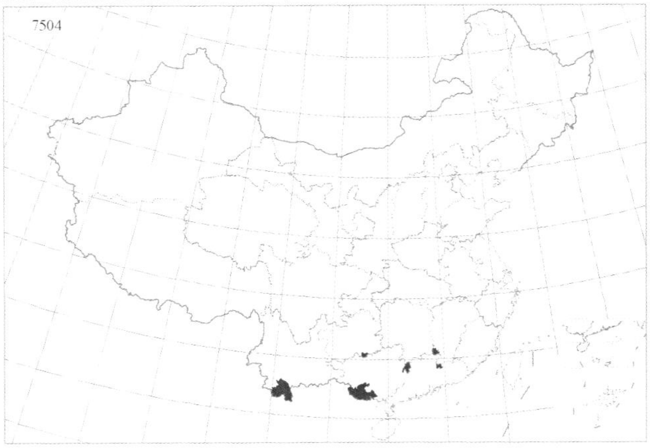

Araliaceae 五加科 *Heteropanax* 幌伞枫属

Heteropanax fragrans (D. Don) Seem. 幌伞枫

Trees, evergreen

21.3 (17.2~24.8); 13.4 (7.1~20.1); 27.7 (21.6~29.8); 21.4 (17.3~24.8); 196.5 (147.0~237.6); 0.0 (0.0~0.0); 1126 (832~1385) ❙ 1623 (868~2037); 691 (312~1033); 126 (41~197) ❙ 45.8 (−34.3~95.5); 1107 (783~1360); 270.3 (95.3~659.0)

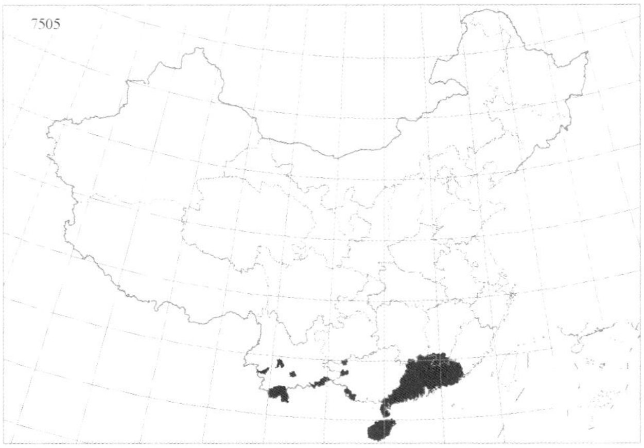

Araliaceae 五加科 *Heteropanax* 幌伞枫属

Heteropanax nitentifolius G. Hoo 亮叶幌伞枫

Trees, evergreen

21.0; 14.2; 25.6; 21.0; 191.8; 0.0; 1053 ❙ 1522; 818; 68 ❙ 44.5; 1053; 551.6

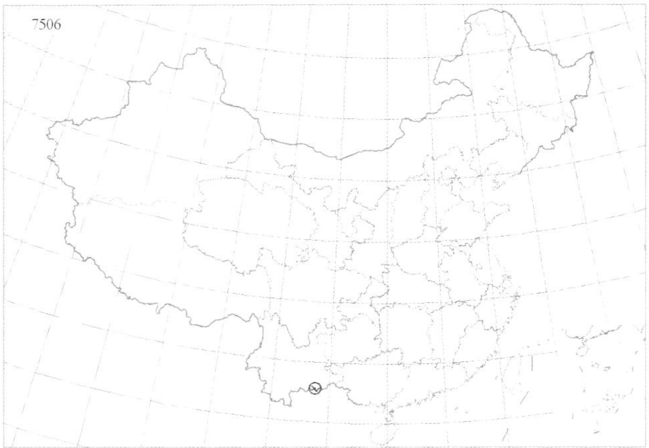

Araliaceae 五加科 *Heteropanax* 幌伞枫属

Heteropanax yunnanensis C. Ho 云南幌伞枫

Trees, evergreen

18.5; 12.6; 22.1; 18.5; 162.4; 0.0; 881 ❙ 1412; 770; 52 ❙ 60.6; 853; 421.3

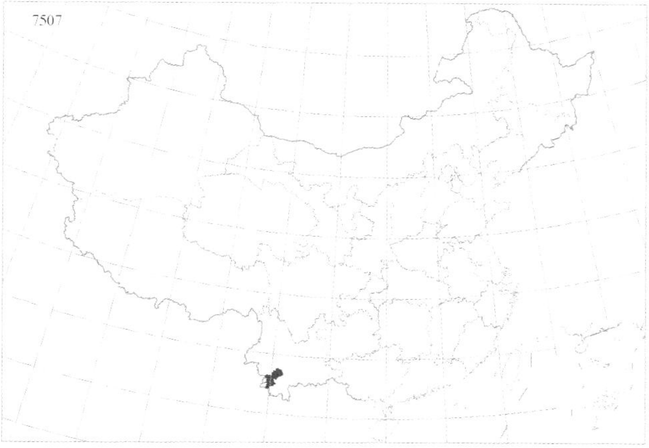

Araliaceae 五加科 *Kalopanax* 刺楸属

Kalopanax septemlobus (Thunb.) Koidz. 刺楸

Trees, deciduous

14.2 (1.4~21.0); 1.9 (−19.9~12.2); 25.6 (12.3~29.9); 14.6 (4.8~21.1); 121.0 (25.1~193.0); 9.5 (0.0~93.6); 814 (441~1114) ❙ 1088 (432~1997); 482 (257~1007); 93 (10~259) ❙ 32.5 (−47.7~145.8); 779 (432~1114); 208.7 (110.1~447.4)

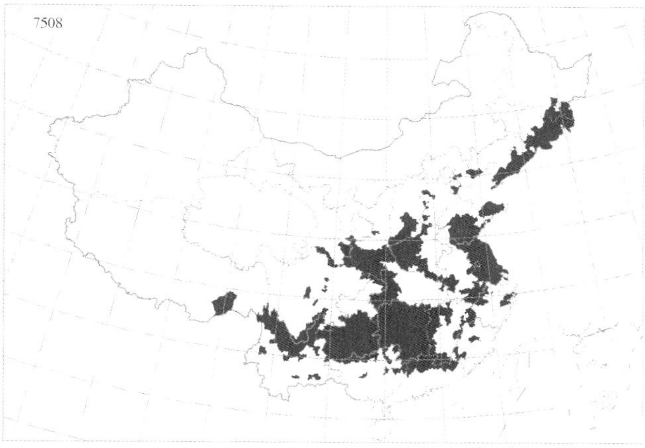

Araliaceae 五加科 *Macropanax* 大参属

Macropanax chienii G. Hoo 显脉大参

Small trees, evergreen

17.5; 11.1; 21.9; 17.5; 150.4; 0.0; 864 ❙ 1293; 628; 59 ❙ 47.5; 863; 444.9

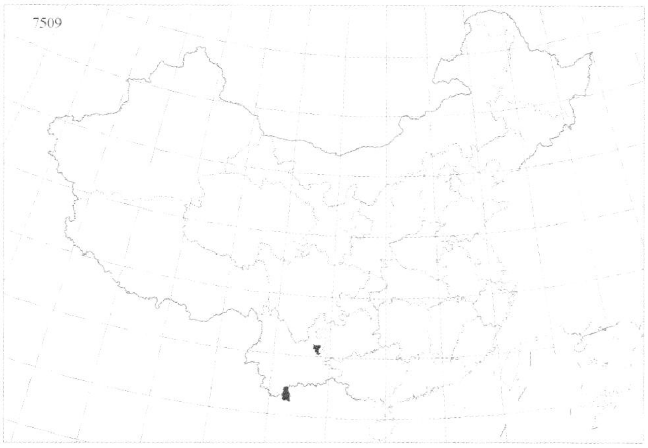

Araliaceae 五加科 *Macropanax* 大参属

Macropanax decandrus G. Hoo 十蕊大参

Trees, evergreen

23.5 (18.3~24.8); 17.5 (8.1~20.1); 27.9 (25.7~28.7); 23.5 (18.4~24.8); 222.0 (160.7~237.6); 0.0 (0.0~0.0); 1274 (965~1385) ❙ 1550 (1117~2037); 582 (392~1033); 73 (41~118) ❙ 22.5 (−12.0~71.8); 1174 (965~1360); 431.5 (152.6~620.0)

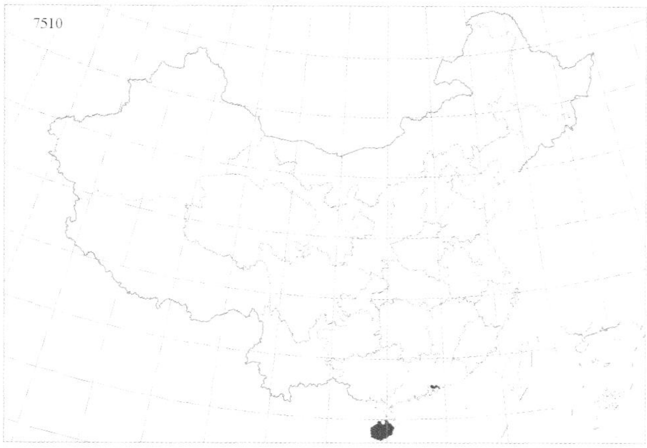

Araliaceae 五加科 *Macropanax* 大参属

Macropanax dispermus (Blume) Kuntze 大参

Trees, evergreen

20.7 (−1.8~24.8); 14.6 (−10.2~20.1); 25.0 (6.1~28.8); 20.8 (1.8~24.8); 191.2 (1.9~237.6); 2.4 (0.0~82.8); 1099 (329~1385) ❙ 1413 (320~1854); 609 (217~1007); 64 (20~115) ❙ 32.7 (−34.3~145.8); 1028 (320~1360); 430.8 (72.2~659.0)

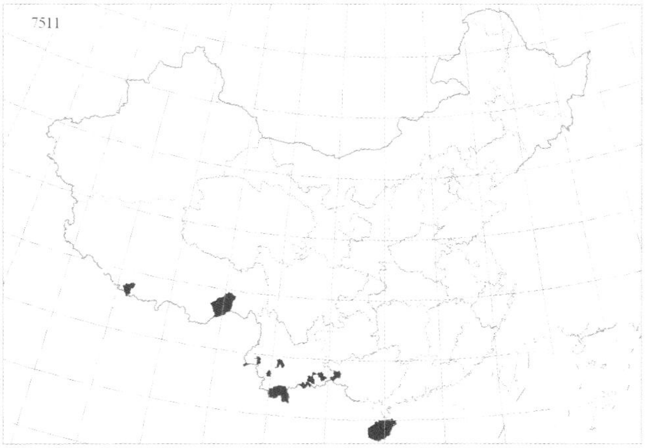

Araliaceae 五加科 *Macropanax* 大参属

Macropanax paucinervis C. B. Shang 疏脉大参

Trees, evergreen

19.1; 13.0; 23.3; 19.1; 169.7; 0.0; 917 ❙ 1428; 835; 56 ❙ 55.7; 917; 506.2

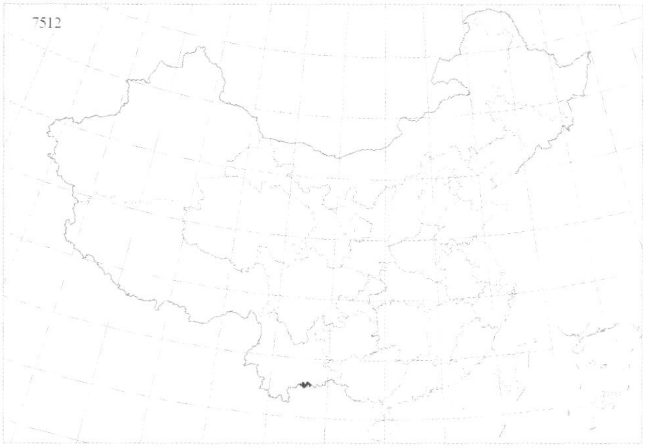

Araliaceae 五加科 *Macropanax* 大参属

Macropanax rosthornii (Harms) C. Y. Wu ex G. Hoo 短梗大参

Shrubs or small trees, evergreen

16.5 (3.5~21.0); 5.9 (−6.5~12.2); 26.1 (10.8~29.8); 16.5 (4.7~21.1); 139.6 (21.6~193.0); 1.4 (0.0~43.4); 877 (444~1114) ❙ 1470 (711~1997); 565 (346~946); 148 (12~259) ❙ 68.3 (12.1~130.7); 873 (444~1114); 228.3 (146.0~332.5)

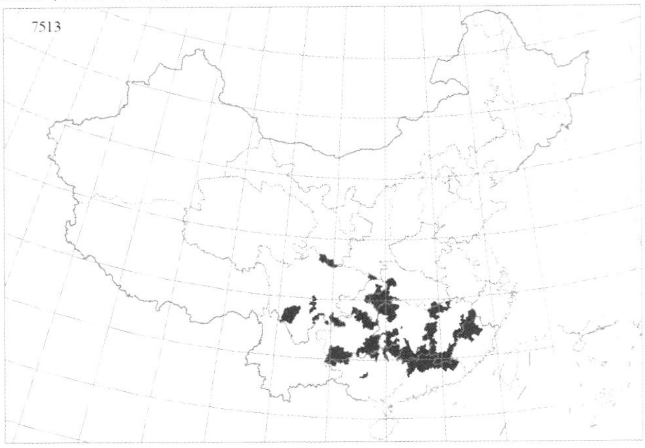

Araliaceae 五加科 *Macropanax* 大参属

Macropanax undulatus (Wall. ex G. Don) Seem. 波缘大参

Trees, evergreen

20.4 (5.3~24.8); 13.8 (−2.4~20.1); 25.2 (12.2~28.8); 20.5 (5.7~24.8); 186.5 (30.5~237.6); 0.9 (0.0~26.2); 1080 (479~1385) ❙ 1422 (868~1854); 626 (312~1007); 68 (31~123) ❙ 37.9 (−34.3~145.8); 1022 (479~1360); 410.5 (189.5~659.0)

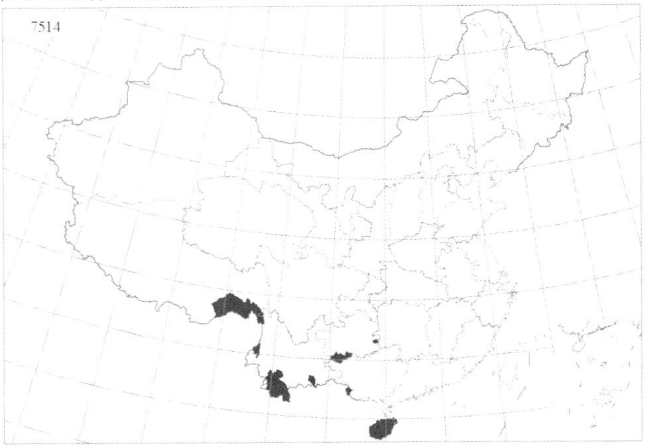

Araliaceae 五加科 *Merrilliopanax* 常春木属

Merrilliopanax alpinus (C. B. Clarke) C. B. Shang 西藏常春木

Small trees, evergreen

13.3; 5.4; 19.6; 13.4; 100.4; 0.0; 706 ❙ 1736; 1007; 51 ❙ 145.8; 706; 447.4

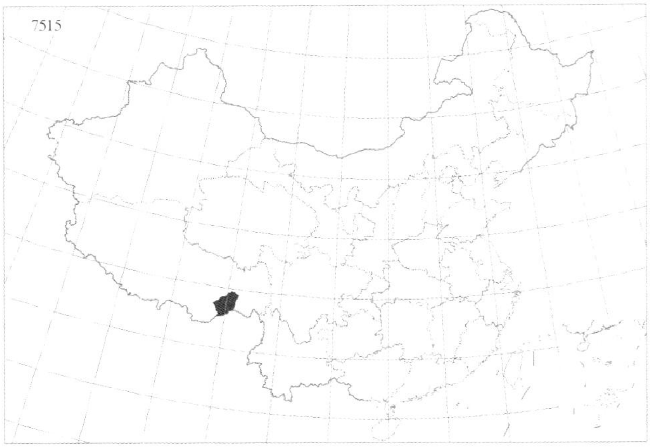

Araliaceae 五加科 *Merrilliopanax* 常春木属

Merrilliopanax listeri (King) H. L. Li 常春木

Small trees, evergreen

13.6 (5.4~19.6); 6.8 (−2.1~13.0); 18.9 (12.3~23.5); 13.7 (5.8~19.6); 106.7 (31.5~175.7); 2.6 (0.0~25.6); 722 (481~953) ❙ 1201 (741~1736); 695 (426~1007); 45 (15~70) ❙ 66.5 (14.4~145.8); 714 (481~932); 390.4 (273.5~589.2)

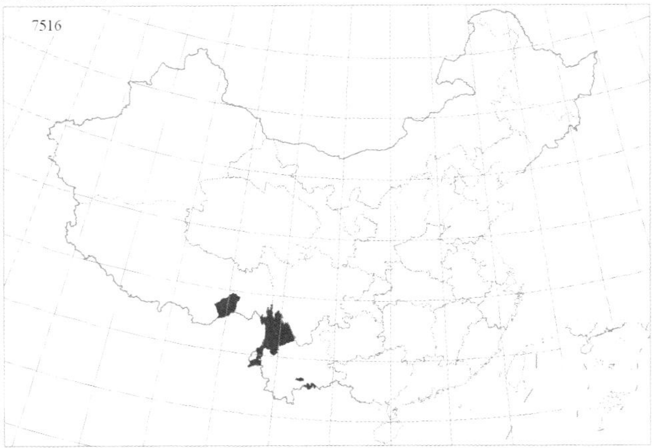

Araliaceae 五加科 *Merrilliopanax* 常春木属

Merrilliopanax membranifolius (W. W. Smith) C. B. Shang 长梗常春木

Shrubs or small trees, evergreen

7.7 (−1.7~19.6); −0.4 (−12.3~12.8); 14.6 (8.0~24.3); 8.6 (2.4~19.6); 59.3 (6.4~175.7); 26.3 (0.0~86.1); 562 (344~953) ❙ 920 (502~1616); 534 (301~946); 26 (8~70) ❙ 65.7 (5.5~128.6); 553 (344~932); 310.8 (130.1~589.2)

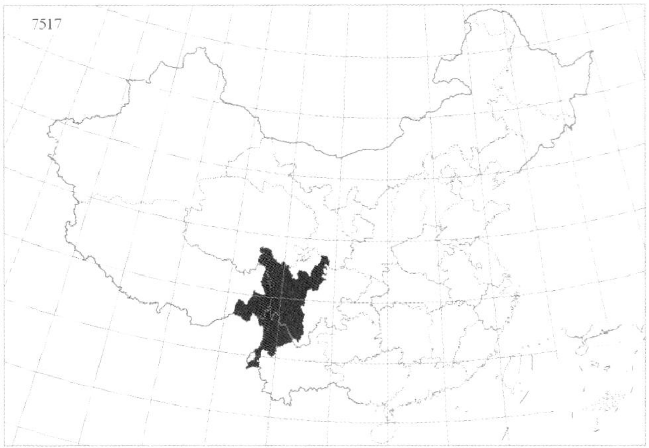

Araliaceae 五加科 *Metapanax* 梁王茶属

Metapanax davidii (Franch.) Frodin 异叶梁王茶

Shrubs or trees, evergreen

14.4 (6.0~21.0); 4.0 (−4.2~15.1); 23.9 (13.3~28.7); 14.4 (6.6~21.0); 116.3 (39.4~191.8); 3.3 (0.0~29.4); 778 (508~1053) ❙ 1178 (675~1733); 533 (331~818); 68 (12~218) ❙ 52.0 (−1.2~107.2); 777 (508~1053); 229.8 (118.7~659.0)

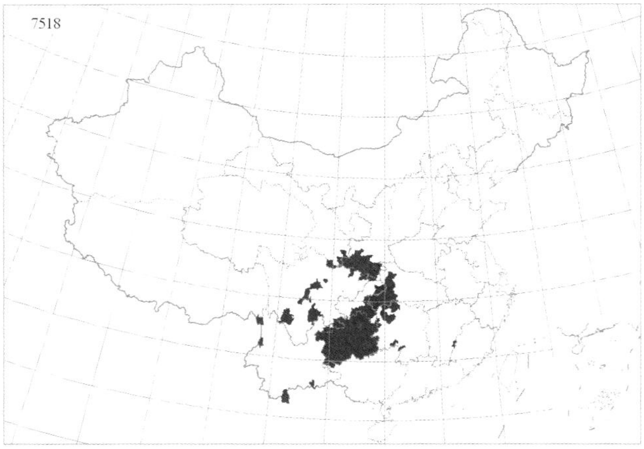

Araliaceae 五加科 *Metapanax* 梁王茶属
Metapanax delavayi (Franch.) J. Wen et Frodin 梁王茶
Shrubs, evergreen

14.5 (5.4~19.1); 6.0 (-2.1~11.7); 21.6 (12.3~26.6); 14.5 (5.8~19.1); 116.4 (31.5~169.2); 2.3 (0.0~25.6); 763 (481~969) ❙ 1116 (741~1368); 592 (426~756); 51 (19~94) ❙ 47.4 (27.3~81.5); 757 (481~969); 261.2 (153.9~491.5)

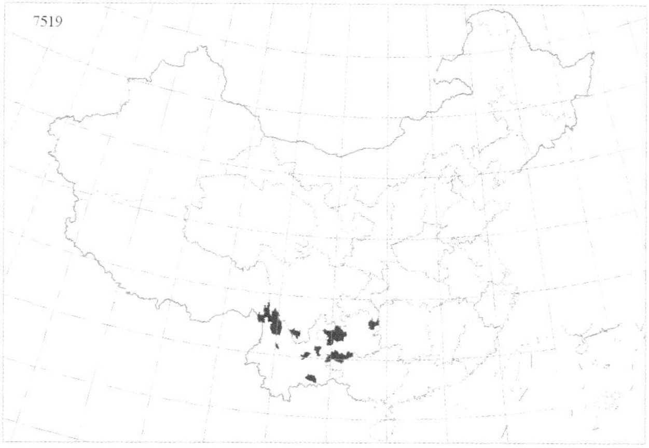

Araliaceae 五加科 *Oplopanax* 刺参属
Oplopanax elatus (Nakai) Nakai 刺参
Shrubs

4.0 (1.4~6.5); -15.3 (-18.7~-12.0); 20.6 (18.4~22.8); 7.9 (6.4~9.2); 60.5 (45.8~75.2); 72.2 (55.1~88.8); 566 (501~624) ❙ 786 (561~1073); 485 (333~691); 28 (16~40) ❙ 38.3 (-2.6~74.1); 564 (501~624); 317.1 (287.1~335.3)

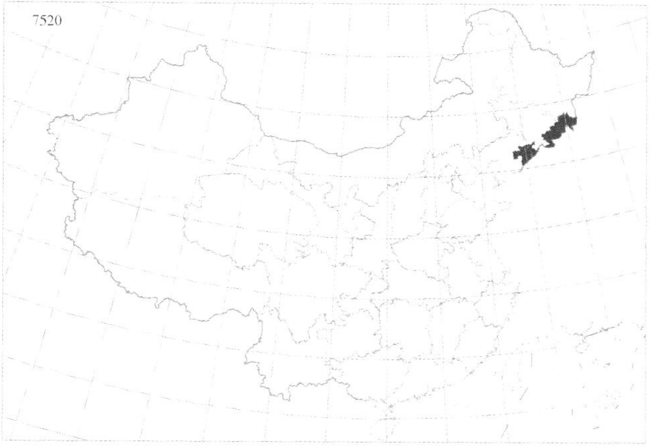

Araliaceae 五加科 *Osmoxylon* 兰屿加属
Osmoxylon pectinatum (Merr.) Philipson 兰屿加
Small trees

18.5; 14.1; 21.7; 18.5; 162.1; 0.0; 863 ❙ 2713; 1388; 188 ❙ 214.4; 863; 363.7

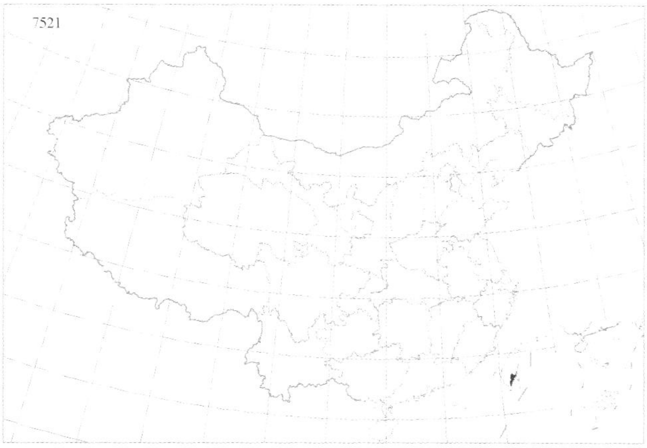

Araliaceae 五加科 *Pentapanax* 五叶参属
Pentapanax caesius (Hand.-Mazz.) C. B. Shang 圆叶羽叶参
Shrubs or small trees, deciduous

9.2 (2.2~16.9); 1.6 (-6.6~10.3); 15.5 (9.9~21.4); 9.5 (3.8~16.9); 65.6 (15.5~142.7); 14.7 (0.0~49.1); 591 (413~813) ❙ 959 (801~1178); 573 (496~646); 24 (12~51) ❙ 66.9 (35.0~125.9); 588 (413~813); 291.2 (221.3~383.4)

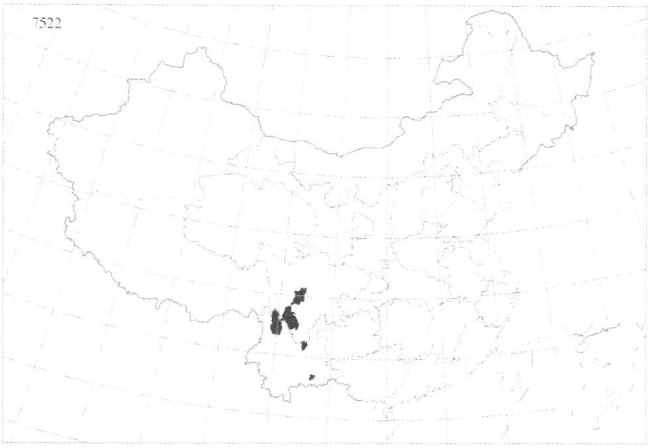

Araliaceae 五加科 *Pentapanax* 五叶参属
Pentapanax castanopsidicola Hayata 台湾羽叶参
Shrubs or small trees, evergreen

20.2; 14.2; 25.4; 20.3; 183.3; 0.0; 999 ❙ 2326; 1081; 226 ❙ 132.8; 998; 234.1

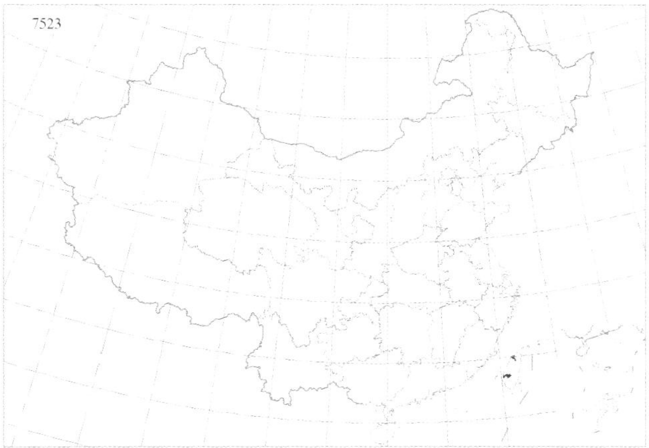

Araliaceae 五加科 *Pentapanax* 五叶参属
Pentapanax fragrans (D. Don) Ha 羽叶参
Shrubs or trees, evergreen

12.7 (-1.8~19.2); 5.1 (-10.2~13.3); 18.9 (6.1~26.6); 12.9 (1.8~19.4); 99.2 (1.9~172.6); 6.6 (0.0~82.8); 697 (329~932) ❙ 992 (320~1736); 574 (193~1007); 35 (11~70) ❙ 43.0 (-14.2~145.8); 676 (320~923); 287.1 (72.2~506.2)

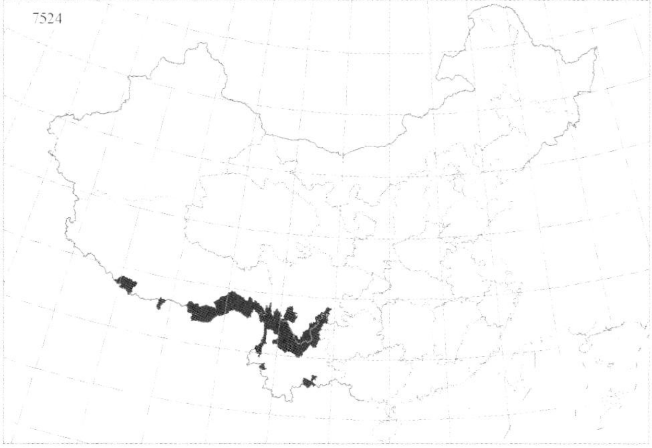

Araliaceae 五加科 *Pentapanax* 五叶参属
Pentapanax fragrans var. forrestii (W. W. Smith) C. B. Shang 全缘羽叶参
Shrubs or trees

5.8 (-1.8~13.3); -2.1 (-10.2~5.4); 12.7 (6.1~19.6); 7.0 (1.8~13.4); 43.3 (1.9~100.4); 33.0 (0.0~82.8); 508 (329~706) ❙ 873 (320~1736); 509 (193~1007); 32 (11~51) ❙ 56.9 (-4.0~145.8); 505 (320~706); 260.3 (72.2~447.4)

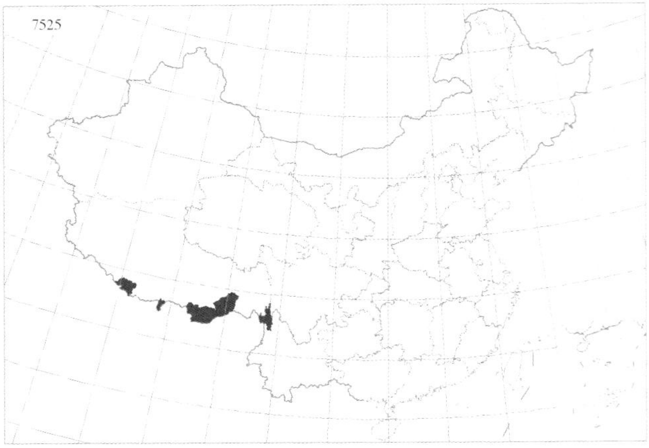

Araliaceae 五加科 *Pentapanax* 五叶参属
Pentapanax glabrifoliolatus C. B. Shang 光羽叶参
Shrubs or small trees

17.6; 9.9; 23.2; 17.7; 152.4; 0.0; 859 ❙ 1422; 813; 52 ❙ 65.6; 859; 359.7

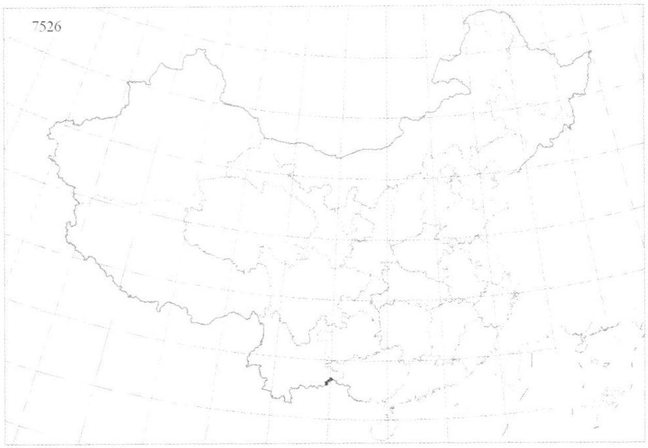

Araliaceae 五加科 *Pentapanax* 五叶参属
Pentapanax henryi Harms 锈毛羽叶参
Shrubs or small trees, evergreen

13.7 (2.2~18.9); 4.6 (-6.6~12.0); 21.8 (9.9~29.3); 13.8 (3.8~19.0); 109.4 (15.5~167.8); 3.9 (0.0~49.1); 748 (413~941) ❙ 1142 (741~1689); 565 (411~813); 72 (12~220) ❙ 53.8 (-14.2~125.9); 735 (413~941); 259.7 (142.9~383.4)

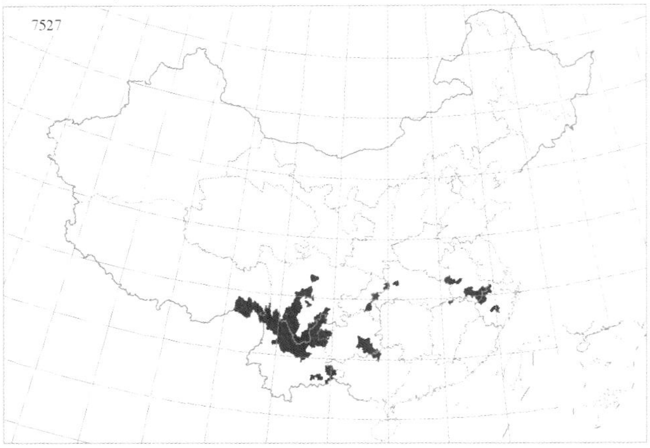

Araliaceae 五加科 *Pentapanax* 五叶参属
Pentapanax hypoglaucus (C. J. Qi et T. R. Cao) C. B. Shang et X. P. Li 粉背羽叶参
Shrubs, evergreen

16.2 (14.3~17.6); 5.0 (3.3~6.6); 26.7 (24.3~28.1); 16.3 (14.3~17.6); 135.8 (114.1~151.3); 0.6 (0.0~2.3); 864 (765~928) ❙ 1485 (1415~1526); 477 (419~530); 193 (181~198) ❙ 72.9 (55.1~99.4); 853 (765~922); 206.1 (177.4~257.1)

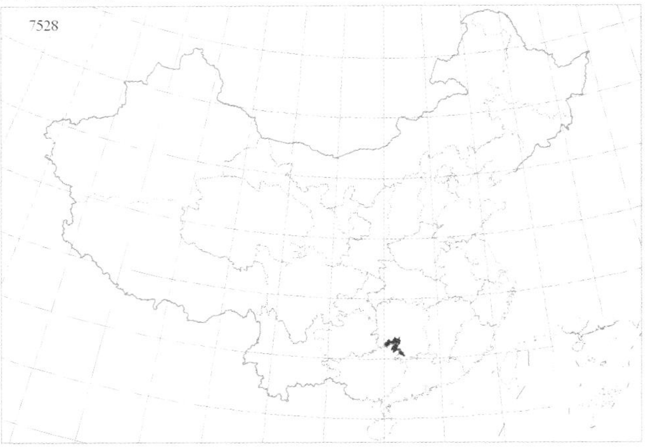

Araliaceae 五加科 *Pentapanax* 五叶参属
Pentapanax longipes (Merr.) C. B. Shang et C. F. Ji 独龙羽叶参
Shrubs, evergreen

8.6; 1.3; 15.0; 8.7; 52.6; 8.8; 574 ❙ 1041; 585; 45 ❙ 81.5; 574; 352.3

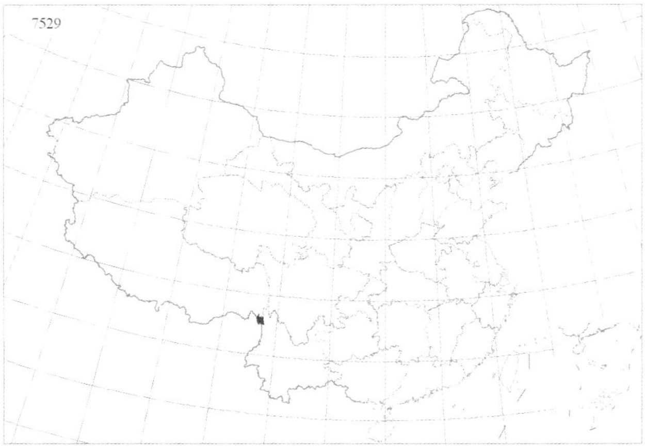

Araliaceae 五加科 *Pentapanax* 五叶参属
Pentapanax parasiticus (D. Don) Seem. 寄生羽叶参
Shrubs, evergreen

11.4 (0.3~16.9); 2.8 (-8.6~7.3); 19.0 (8.3~25.9); 11.7 (2.9~17.0); 87.5 (9.6~143.7); 10.1 (0.0~65.4); 671 (368~881) ❙ 1174 (389~1736); 679 (246~1007); 40 (11~60) ❙ 70.8 (0.0~145.8); 671 (368~881); 264.1 (142.5~447.4)

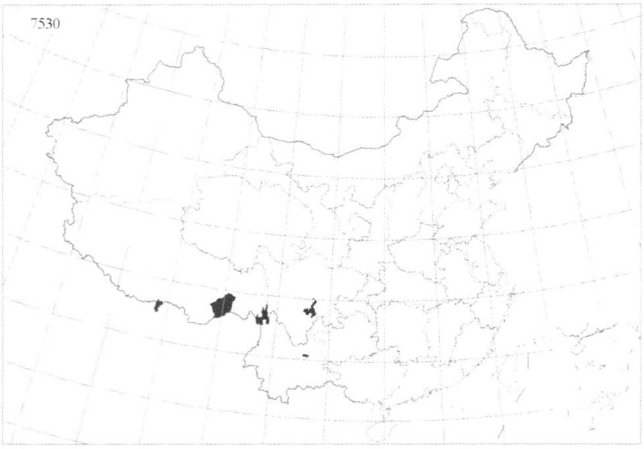

Araliaceae 五加科 *Pentapanax* 五叶参属

Pentapanax parasiticus var. khasianus C. B. Clarke 毛梗寄生羽叶参

Scandent shrubs, evergreen

13.8; 7.2; 19.1; 13.9; 111.1; 4.4; 745 ❙ 1235; 710; 50 ❙ 68.6; 745; 429.2

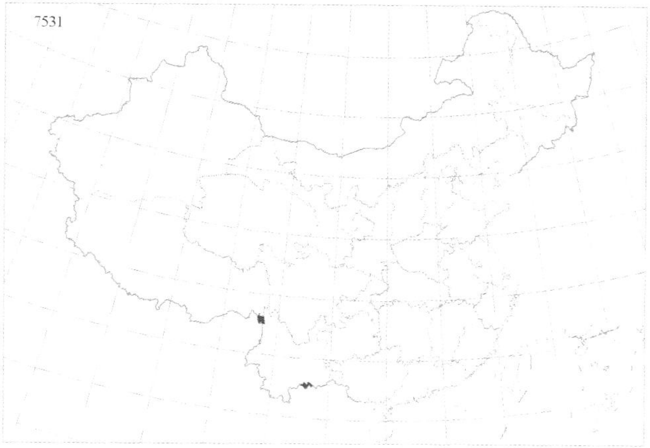

Araliaceae 五加科 *Pentapanax* 五叶参属

Pentapanax racemosus Seem. 总序羽叶参

Trees, evergreen

15.3 (12.2~19.1); 8.5 (4.3~13.0); 20.2 (17.9~23.3); 15.3 (12.2~19.1); 123.9 (87.4~169.7); 0.1 (0.0~0.7); 766 (672~917) ❙ 1270 (1014~1736); 721 (568~1007); 49 (31~70) ❙ 67.6 (26.1~145.8); 758 (672~917); 406.0 (318.7~506.2)

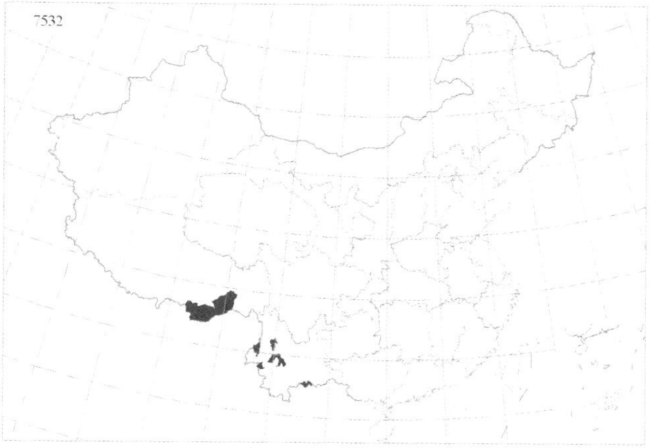

Araliaceae 五加科 *Pentapanax* 五叶参属

Pentapanax subcordatus Seem. 心叶羽叶参

Small trees, evergreen

14.4; 7.5; 19.1; 14.4; 113.1; 0.0; 727 ❙ 1454; 792; 70 ❙ 99.9; 727; 487.2

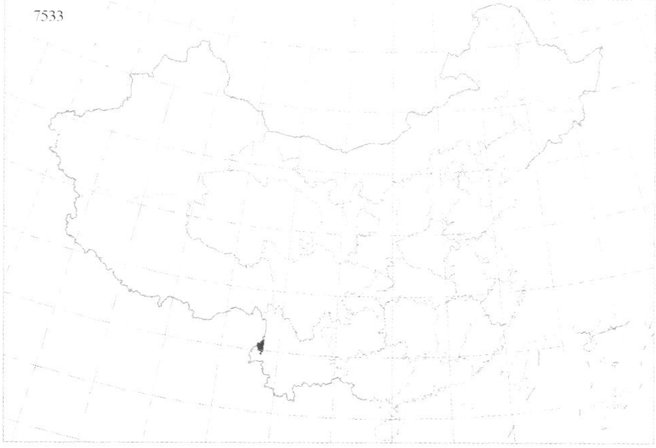

Araliaceae 五加科 *Pentapanax* 五叶参属

Pentapanax tomentellus (Franch.) C. B. Shang 马肠子树

Shrubs or small trees, evergreen

13.2 (2.2~18.0); 5.8 (-6.6~10.3); 19.2 (9.9~26.3); 13.4 (3.8~18.0); 105.1 (15.5~156.4); 5.7 (0.0~49.1); 715 (413~913) ❙ 1064 (741~1551); 600 (426~813); 46 (12~169) ❙ 51.0 (13.8~125.9); 703 (413~913); 305.2 (225.9~383.4)

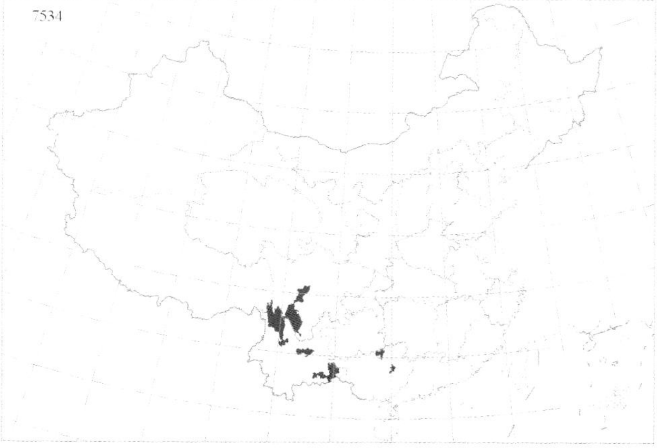

Araliaceae 五加科 *Pentapanax* 五叶参属

Pentapanax tomentellus var. distinctus C. B. Shang 离柱马肠子树

Shrubs or small trees, evergreen

8.6; 1.3; 15.0; 8.7; 52.6; 8.8; 574 ❙ 1041; 585; 45 ❙ 81.5; 574; 352.3

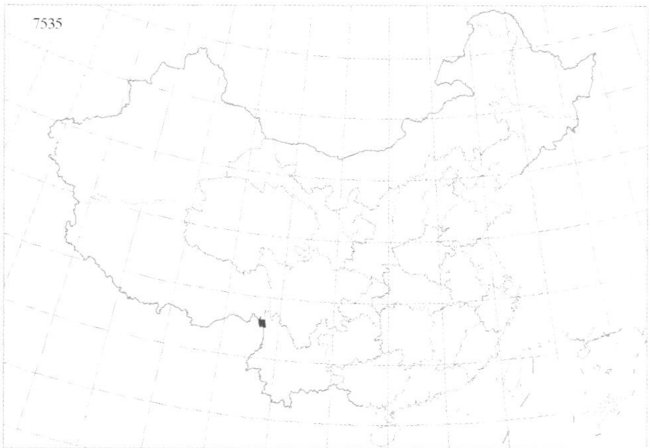

Araliaceae 五加科 *Pentapanax* 五叶参属

Pentapanax verticillatus Dunn 轮伞羽叶参

Shrubs, evergreen

16.3 (12.8~19.6); 8.4 (3.3~11.2); 22.4 (21.1~25.8); 16.4 (12.9~19.6); 137.2 (96.0~175.5); 0.5 (0.0~2.2); 818 (699~986) ❙ 1179 (879~1422); 653 (472~813); 51 (36~74) ❙ 43.4 (25.4~65.6); 816 (699~986); 323.9 (219.7~491.5)

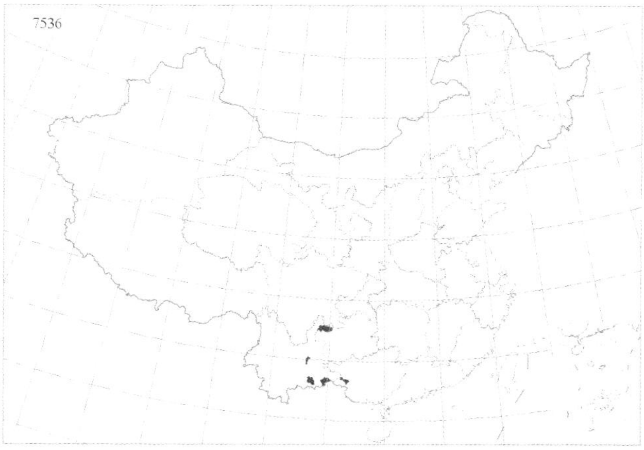

Araliaceae 五加科 *Pentapanax* 五叶参属
Pentapanax wilsonii (Harms) C. B. Shang 西南羽叶参
Shrubs, deciduous

13.1 (2.2~18.9); 5.4 (-6.3~12.0); 19.5 (9.5~26.6); 13.2 (3.9~19.0); 101.2 (15.8~167.8); 3.3 (0.0~48.6); 708 (411~929) ❙ 954 (729~1141); 560 (411~641); 29 (11~57) ❙ 37.1 (-14.2~77.4); 679 (411~912); 271.9 (142.9~351.7)

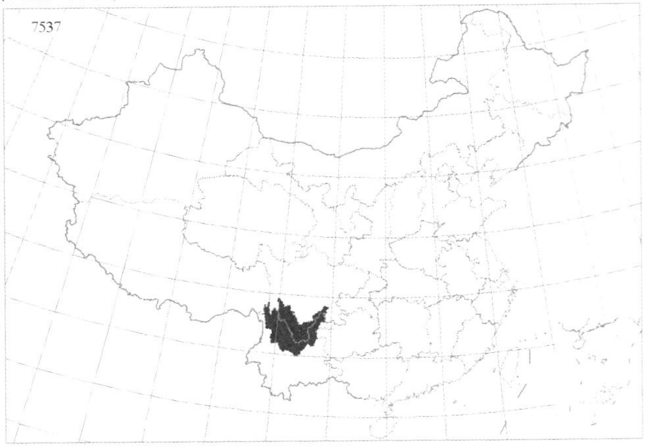

Araliaceae 五加科 *Pentapanax* 五叶参属
Pentapanax yunnanensis Franch. 云南羽叶参
Shrubs, evergreen

13.6 (5.4~18.9); 5.9 (-2.1~12.0); 19.9 (12.3~26.6); 13.6 (5.8~19.0); 105.4 (31.5~167.8); 2.0 (0.0~25.6); 721 (481~929) ❙ 954 (741~1141); 553 (411~627); 32 (15~57) ❙ 34.1 (-14.2~70.8); 692 (481~912); 267.8 (142.9~372.3)

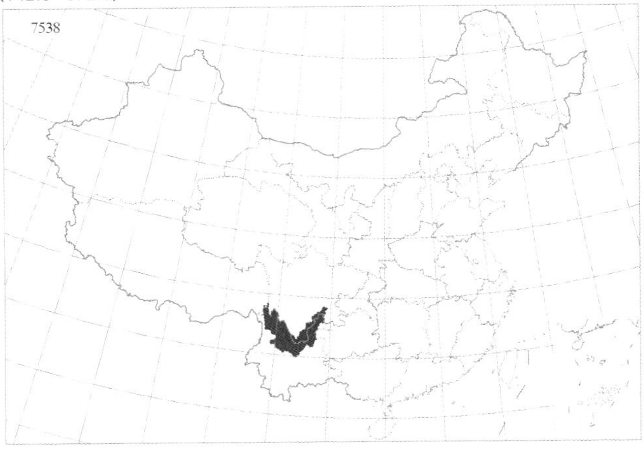

Araliaceae 五加科 *Schefflera* 鹅掌柴属
Schefflera arboricola (Hayata) Merr. 鹅掌藤
Scandent shrubs, evergreen

21.4 (16.2~24.8); 14.1 (7.4~20.1); 27.2 (20.1~29.0); 21.4 (16.2~24.8); 197.2 (134.2~237.6); 0.0 (0.0~0.0); 1125 (773~1385) ❙ 1823 (868~3495); 848 (312~1847); 157 (41~897) ❙ 69.0 (-34.3~246.5); 1087 (773~1299); 273.0 (119.4~620.0)

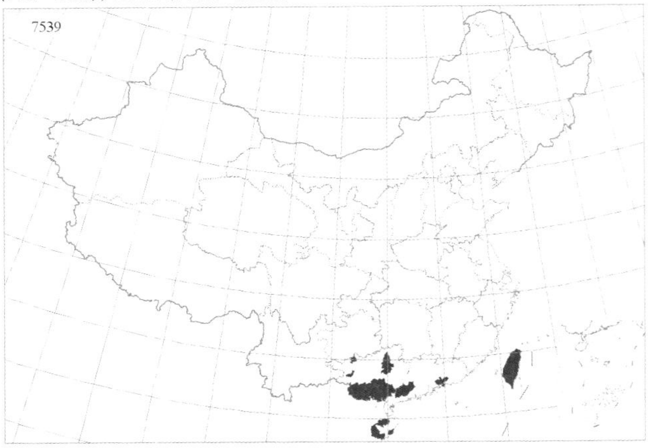

Araliaceae 五加科 *Schefflera* 鹅掌柴属
Schefflera bodinieri (H. Lév.) Rehder 短序鹅掌柴
Shrubs or trees, evergreen

19.1 (-1.8~24.8); 10.2 (-10.2~20.1); 26.7 (6.1~29.8); 19.1 (1.8~24.8); 169.9 (1.9~237.6); 0.5 (0.0~82.8); 1003 (329~1385) ❙ 1554 (320~3495); 669 (217~1847); 133 (20~897) ❙ 57.4 (-34.3~246.5); 993 (320~1360); 245.1 (72.2~659.0)

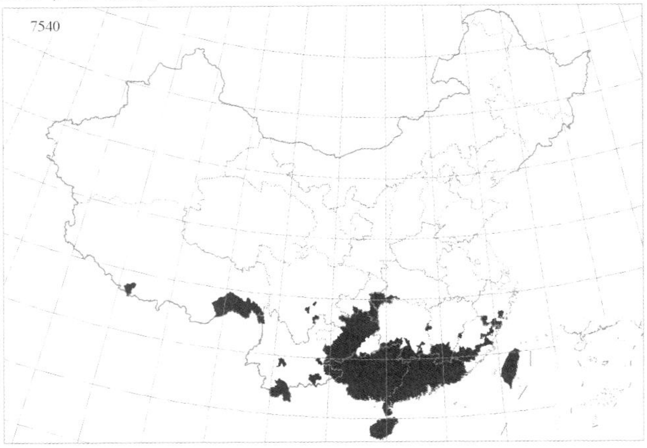

Araliaceae 五加科 *Schefflera* 鹅掌柴属
Schefflera brevipedicellata Harms 多核鹅掌柴
Shrubs or trees, evergreen

19.7; 13.4; 23.9; 19.8; 177.0; 0.0; 965 ❙ 1499; 759; 68 ❙ 55.7; 965; 567.4

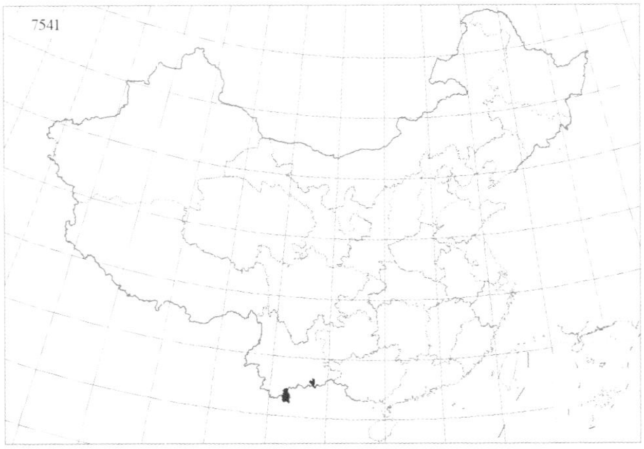

Araliaceae 五加科 *Schefflera* 鹅掌柴属
Schefflera chapana Harms 异叶鹅掌柴
Shrubs or small trees, evergreen

18.4; 12.0; 22.9; 18.5; 161.5; 0.0; 886 ❙ 1398; 796; 58 ❙ 57.9; 886; 498.8

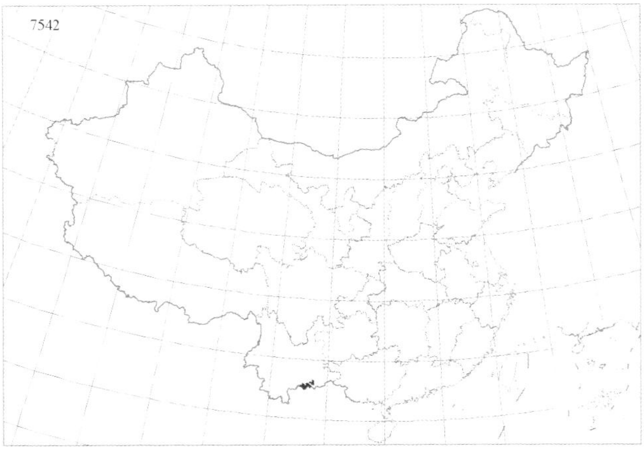

Araliaceae 五加科 *Schefflera* 鹅掌柴属

Schefflera chinensis (Dunn) H. L. Li 中华鹅掌柴

Trees, evergreen

16.1 (8.6~19.2); 9.4 (1.3~13.8); 20.9 (15.0~26.6); 16.1 (8.7~19.2); 134.5 (52.6~170.5); 0.9 (0.0~8.8); 808 (574~969) ❙ 1251 (946~1499); 680 (548~835); 53 (38~69) ❙ 57.1 (15.5~93.0); 794 (574~969); 394.0 (254.3~513.9)

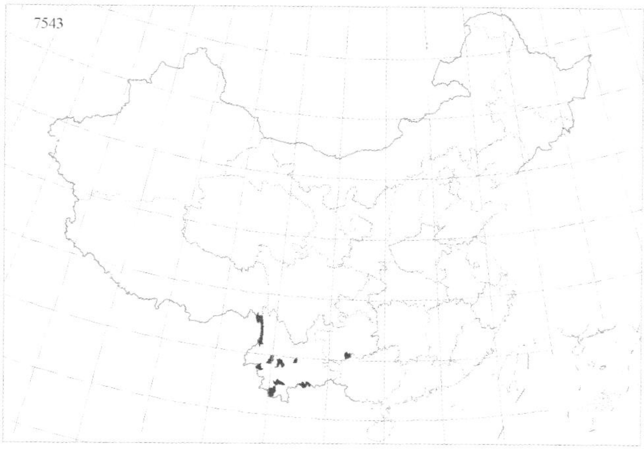

Araliaceae 五加科 *Schefflera* 鹅掌柴属

Schefflera delavayi (Franch.) Harms 穗序鹅掌柴

Shrubs or trees, evergreen

16.3 (3.8~22.2); 6.5 (-4.3~14.7); 25.0 (10.8~29.8); 16.3 (4.7~22.2); 136.6 (21.6~206.9); 0.9 (0.0~35.4); 858 (444~1177) ❙ 1321 (741~1961); 563 (400~938); 111 (15~259) ❙ 54.2 (-14.2~128.9); 851 (444~1177); 226.9 (108.2~432.9)

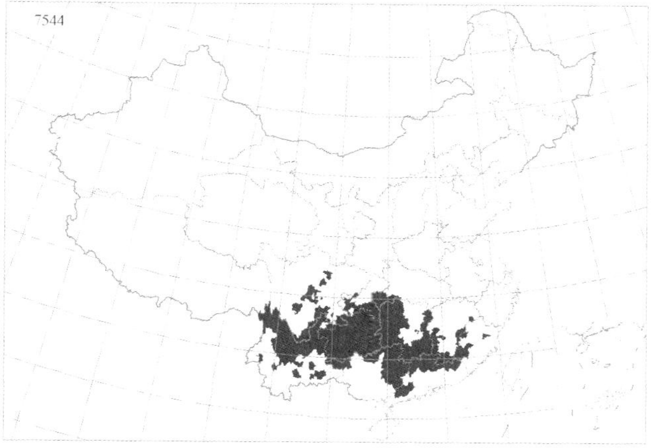

Araliaceae 五加科 *Schefflera* 鹅掌柴属

Schefflera elata (Buch.-Ham.) Harms 高鹅掌柴

Trees, evergreen

14.1; 7.3; 19.2; 14.1; 109.7; 0.3; 729 ❙ 1211; 671; 54 ❙ 68.9; 729; 384.1

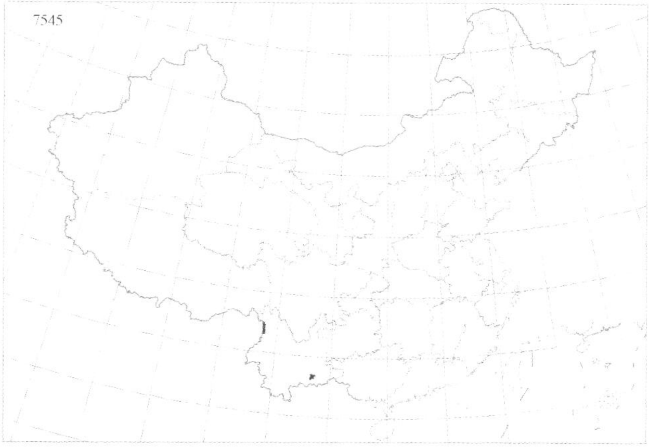

Araliaceae 五加科 *Schefflera* 鹅掌柴属

Schefflera elliptica (Blume) Harms 密脉鹅掌柴

Shrubs or trees, evergreen

17.1 (5.3~20.5); 10.0 (-2.4~15.1); 22.2 (12.2~26.6); 17.1 (5.7~20.5); 146.5 (30.5~185.9); 0.9 (0.0~26.2); 844 (479~992) ❙ 1341 (951~2713); 723 (531~1388); 59 (31~188) ❙ 60.1 (13.3~214.4); 836 (479~986); 379.8 (189.5~659.0)

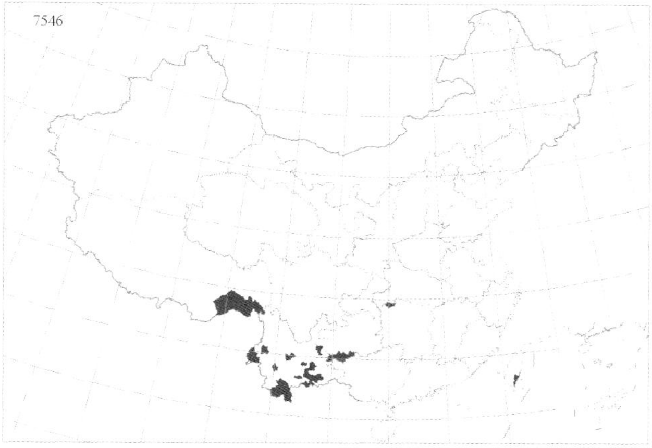

Araliaceae 五加科 *Schefflera* 鹅掌柴属

Schefflera fengii C. J. Tseng et G. Hoo 文山鹅掌柴

Shrubs or trees, evergreen

18.1 (16.3~19.3); 11.6 (9.1~13.3); 22.7 (21.4~23.6); 18.2 (16.4~19.4); 158.4 (136.4~173.2); 0.0 (0.0~0.0); 876 (797~939) ❙ 1269 (939~1547); 723 (524~911); 52 (39~63) ❙ 45.3 (6.5~80.5); 856 (783~923); 410.2 (316.6~551.3)

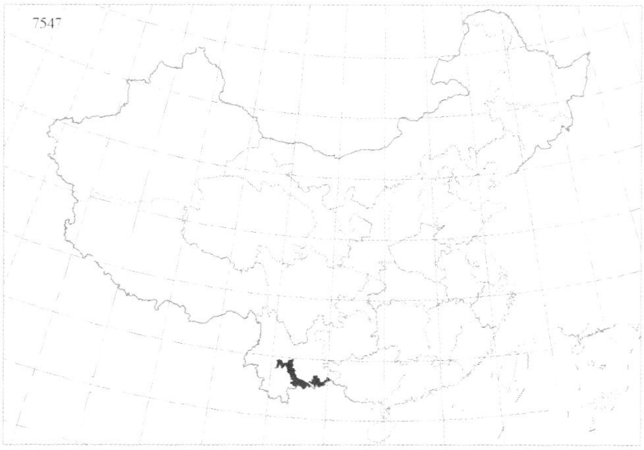

Araliaceae 五加科 *Schefflera* 鹅掌柴属

Schefflera hainanensis Merr. et Chun 海南鹅掌柴

Shrubs or trees, evergreen

21.8 (17.7~24.6); 15.0 (10.9~19.5); 26.9 (22.5~28.6); 21.9 (17.8~24.7); 202.7 (153.4~235.9); 0.0 (0.0~0.0); 1151 (854~1372) ❙ 1742 (1179~2435); 877 (392~1428); 92 (42~130) ❙ 52.8 (-11.9~99.5); 1111 (854~1299); 400.7 (235.1~572.9)

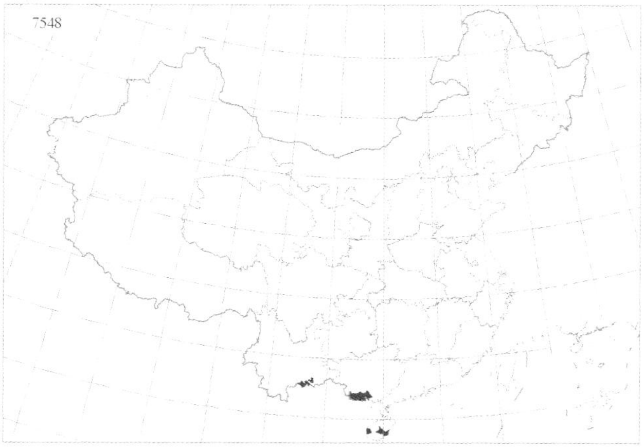

Araliaceae 五加科 *Schefflera* 鹅掌柴属
Schefflera heptaphylla (L.) Frodin 鹅掌柴
Trees, evergreen

20.4; 15.1; 23.6; 20.5; 185.8; 0.0; 989 ❙ 1607; 702; 77 ❙ 62.5; 986; 659.0

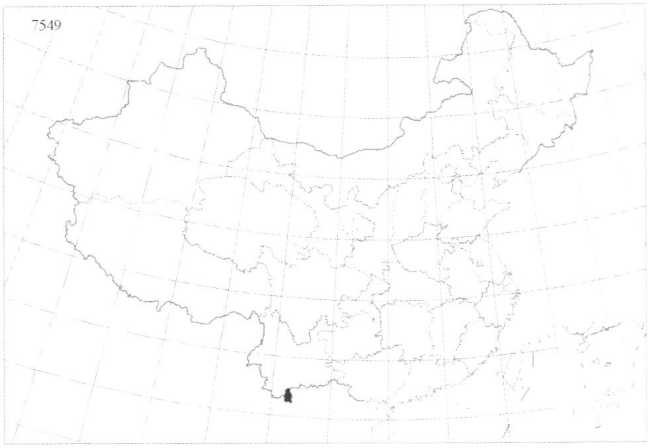

Araliaceae 五加科 *Schefflera* 鹅掌柴属
Schefflera hoi (Dunn) Vig. 红河鹅掌柴
Small trees, evergreen

14.3 (5.3~21.0); 6.8 (-2.4~14.2); 20.3 (12.2~26.6); 14.3 (5.7~21.0); 113.6 (30.5~191.8); 2.0 (0.0~26.2); 749 (479~1053) ❙ 1058 (741~1547); 603 (411~911); 39 (15~69) ❙ 43.0 (-14.2~100.7); 730 (479~1053); 304.7 (142.9~551.6)

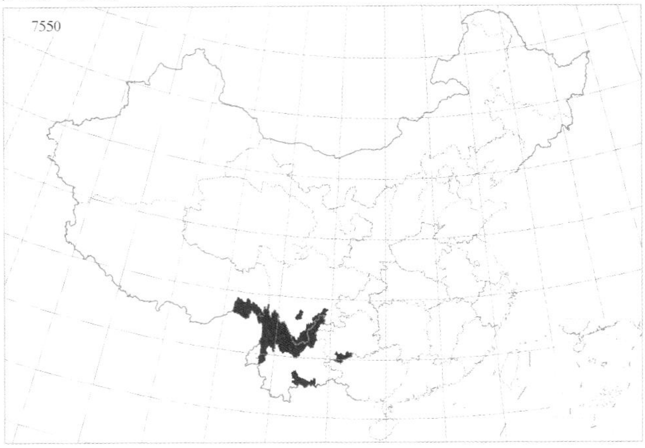

Araliaceae 五加科 *Schefflera* 鹅掌柴属
Schefflera hypoleuca (Kurz) Harms 白背鹅掌柴
Small trees, evergreen

18.4 (16.2~21.3); 8.7 (5.2~14.2); 26.5 (21.4~29.0); 18.4 (16.3~21.3); 160.9 (135.7~196.1); 0.0 (0.0~0.0); 950 (811~1130) ❙ 1479 (1049~1758); 608 (403~835); 152 (43~211) ❙ 56.1 (16.9~86.9); 943 (783~1130); 257.2 (151.0~551.6)

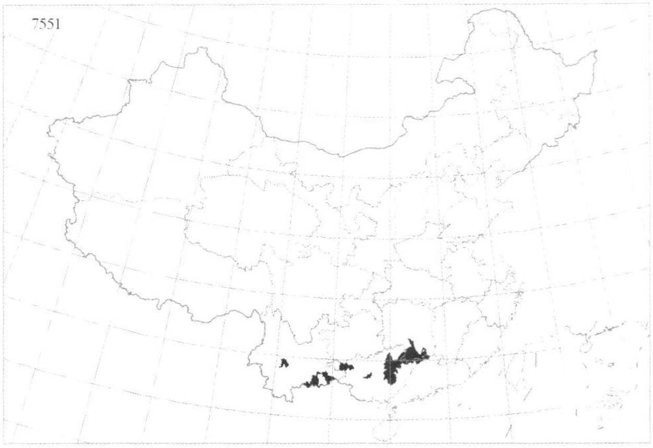

Araliaceae 五加科 *Schefflera* 鹅掌柴属
Schefflera hypoleucoides Harms 离柱鹅掌柴
Trees, evergreen

18.2 (16.1~21.0); 11.0 (9.1~14.2); 23.4 (20.6~26.4); 18.2 (16.2~21.0); 159.0 (133.9~191.8); 0.0 (0.0~0.0); 891 (785~1053) ❙ 1297 (1074~1547); 728 (582~911); 56 (43~76) ❙ 45.9 (16.9~80.6); 888 (779~1053); 385.8 (277.0~551.6)

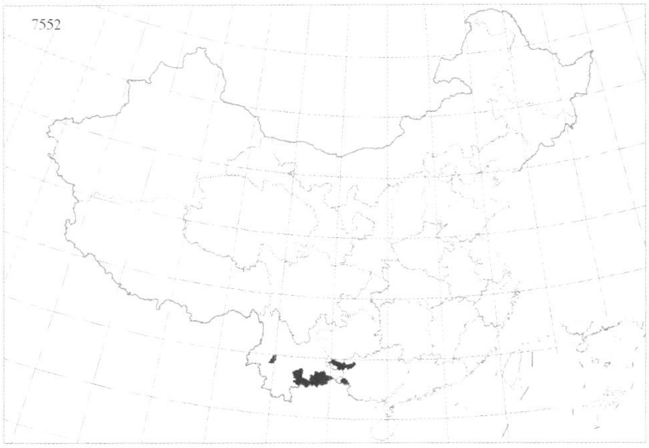

Araliaceae 五加科 *Schefflera* 鹅掌柴属
Schefflera insignis C. N. Ho 粉背鹅掌柴
Shrubs, evergreen

20.7; 12.4; 27.3; 20.7; 188.8; 0.0; 1076 ❙ 1918; 834; 154 ❙ 79.3; 1076; 274.3

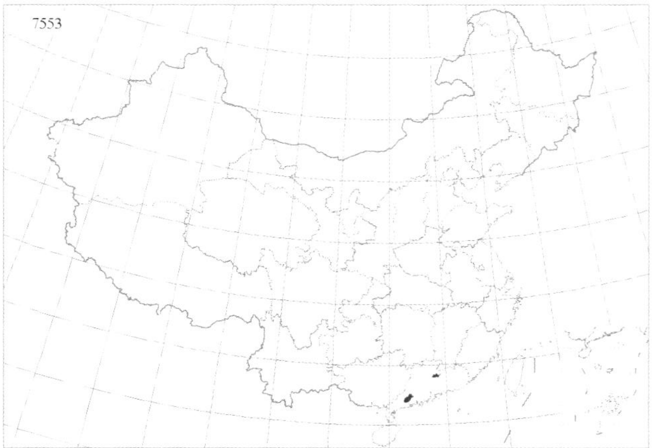

Araliaceae 五加科 *Schefflera* 鹅掌柴属
Schefflera khasiana (C. B. Clarke) R. Vig. 扁盘鹅掌柴
Trees, evergreen

12.0; 4.7; 17.7; 12.0; 87.8; 3.5; 669 ❙ 1368; 782; 49 ❙ 101.9; 669; 444.6

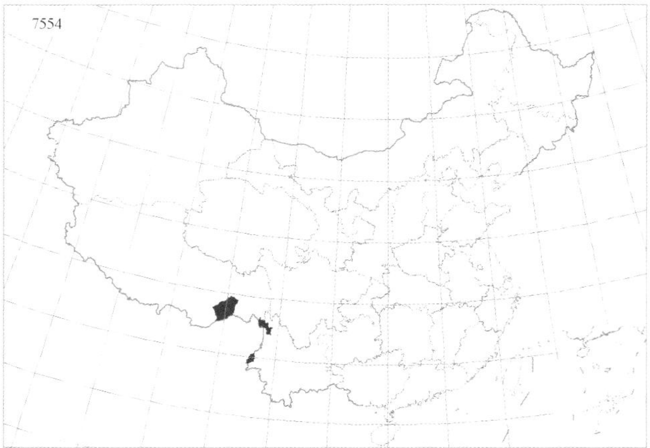

Araliaceae 五加科 *Schefflera* 鹅掌柴属

Schefflera leucantha R. Vig. 白花鹅掌柴

Shrubs or vines, evergreen

20.5 (8.6~22.7); 12.1 (1.3~15.0); 26.9 (15.0~28.9); 20.5 (8.7~22.8); 186.4 (52.6~213.1); 0.4 (0.0~8.8); 1089 (574~1221) ❙ 1590 (1041~2435); 833 (585~1428); 120 (45~194) ❙ 47.9 (12.9~99.5); 1089 (574~1221); 279.3 (201.4~384.7)

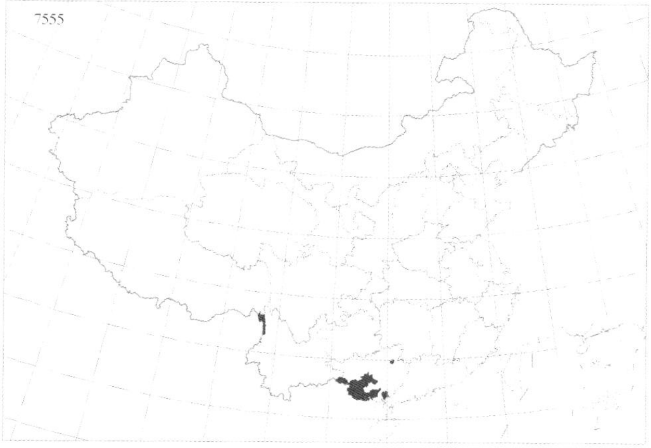

Araliaceae 五加科 *Schefflera* 鹅掌柴属

Schefflera lociana Grushv. et Skvortsova 谅山鹅掌柴

Trees, evergreen

21.1 (19.2~22.2); 12.9 (11.0~13.9); 27.1 (25.2~28.3); 21.2 (19.2~22.3); 193.8 (170.5~207.5); 0.0 (0.0~0.0); 1102 (952~1190) ❙ 1439 (1302~1581); 772 (670~850); 90 (72~106) ❙ 31.7 (12.9~55.9); 1102 (952~1190); 300.6 (275.8~327.7)

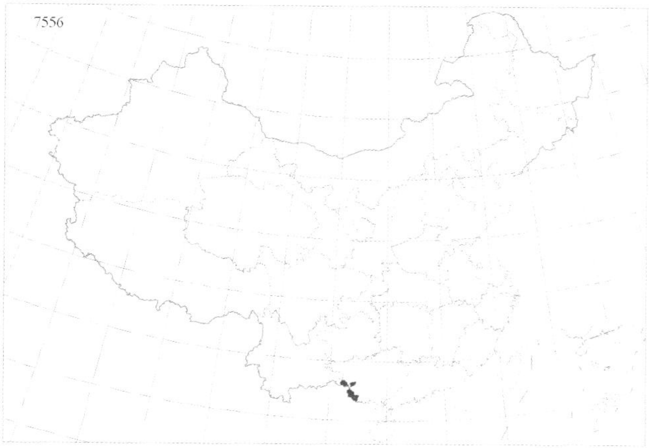

Araliaceae 五加科 *Schefflera* 鹅掌柴属

Schefflera macrophylla (Dunn) R. Vig. 大叶鹅掌柴

Trees, evergreen

17.3 (11.3~19.2); 11.1 (4.3~13.3); 21.7 (17.0~23.5); 17.4 (11.3~19.4); 148.7 (76.6~172.6); 0.1 (0.0~0.7); 846 (644~932) ❙ 1330 (1049~1547); 752 (598~911); 55 (44~63) ❙ 58.3 (26.1~93.0); 835 (644~923); 443.9 (365.3~551.3)

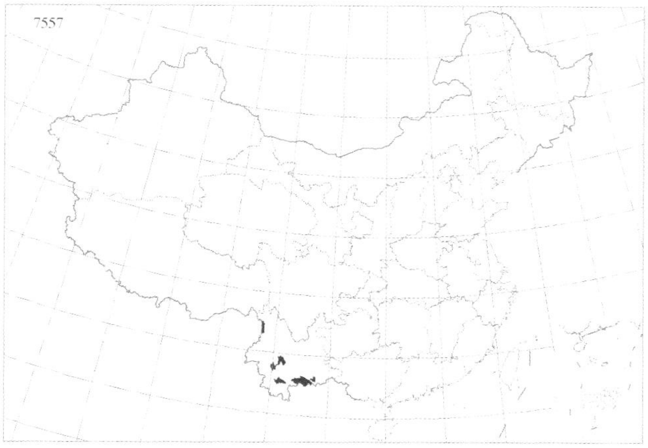

Araliaceae 五加科 *Schefflera* 鹅掌柴属

Schefflera marlipoensis C. J. Tseng et G. Hoo 麻栗坡鹅掌柴

Trees, evergreen

17.6; 9.9; 23.2; 17.7; 152.4; 0.0; 859 ❙ 1422; 813; 52 ❙ 65.6; 859; 359.7

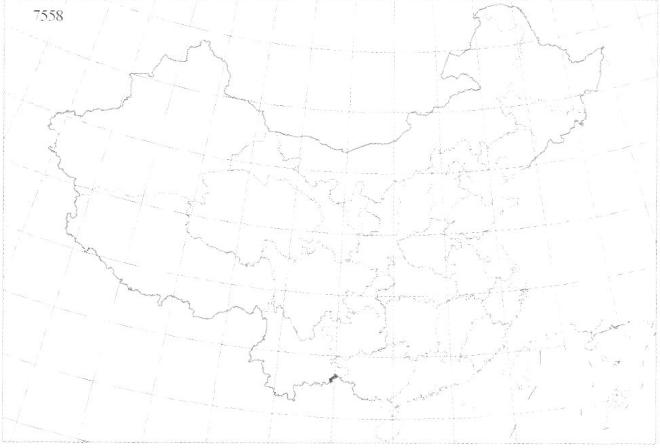

Araliaceae 五加科 *Schefflera* 鹅掌柴属

Schefflera metcalfiana Merr. ex H. L. Li 多叶鹅掌柴

Tall shrubs, evergreen

21.3; 13.2; 27.3; 21.3; 195.9; 0.0; 1111 ❙ 1900; 1056; 125 ❙ 71.0; 1111; 297.2

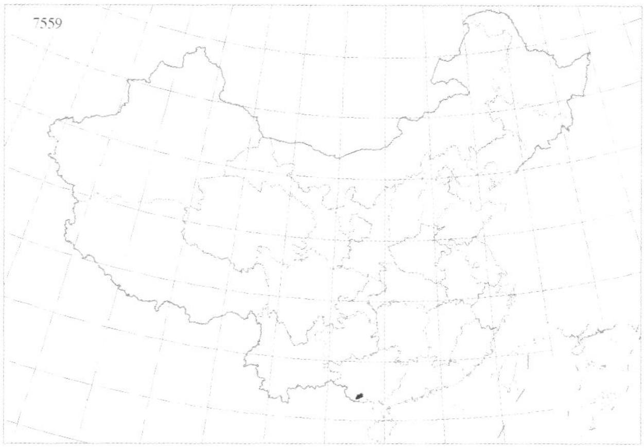

Araliaceae 五加科 *Schefflera* 鹅掌柴属

Schefflera minutistellata Merr. ex H. L. Li 星毛鹅掌柴

Shrubs or small trees, evergreen

18.7 (1.8~23.2); 9.1 (-7.1~15.8); 26.7 (9.9~29.8); 18.7 (3.8~23.2); 164.7 (15.7~218.9); 0.4 (0.0~53.3); 976 (404~1257) ❙ 1520 (583~2035); 630 (348~1007); 144 (13~221) ❙ 56.8 (12.9~145.8); 972 (404~1257); 243.0 (151.0~487.2)

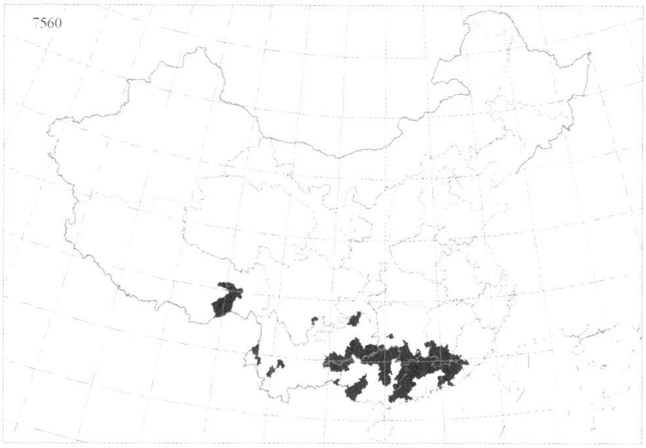

Araliaceae 五加科 *Schefflera* 鹅掌柴属

Schefflera multinervia H. L. Li 多脉鹅掌柴

Trees, evergreen

8.6; 1.3; 15.0; 8.7; 52.6; 8.8; 574 ❙ 1041; 585; 45 ❙ 81.5; 574; 352.3

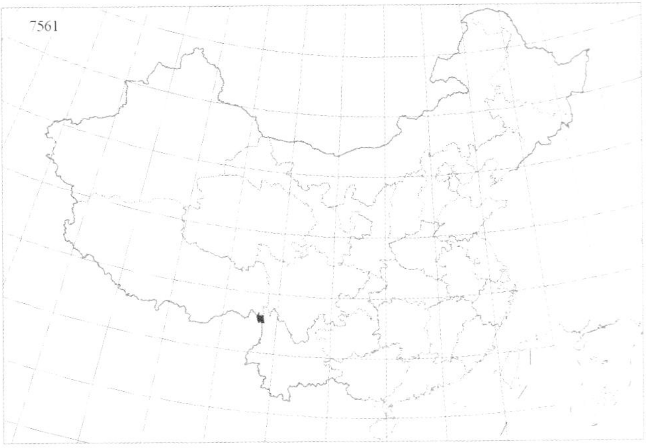

Araliaceae 五加科 *Schefflera* 鹅掌柴属

Schefflera napuoensis C. B. Shang 那坡鹅掌柴

Trees, evergreen

19.1; 11.0; 25.0; 19.1; 169.6; 0.0; 945 ❙ 1438; 820; 57 ❙ 52.2; 945; 371.7

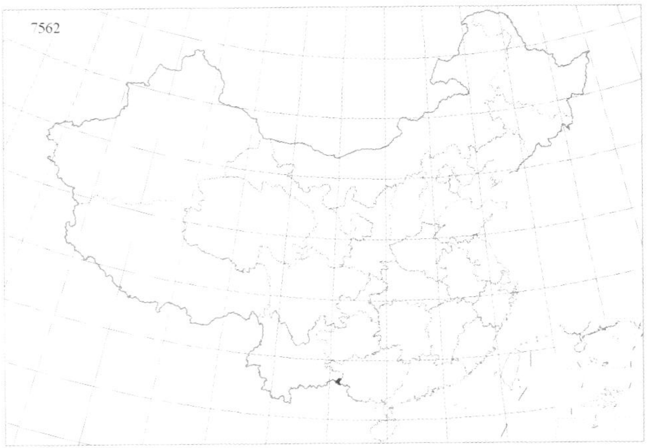

Araliaceae 五加科 *Schefflera* 鹅掌柴属

Schefflera parvifoliolata C. J. Tseng et C. Ho 小叶鹅掌柴

Shrubs, evergreen

17.6; 9.9; 23.2; 17.7; 152.4; 0.0; 859 ❙ 1422; 813; 52 ❙ 65.6; 859; 359.7

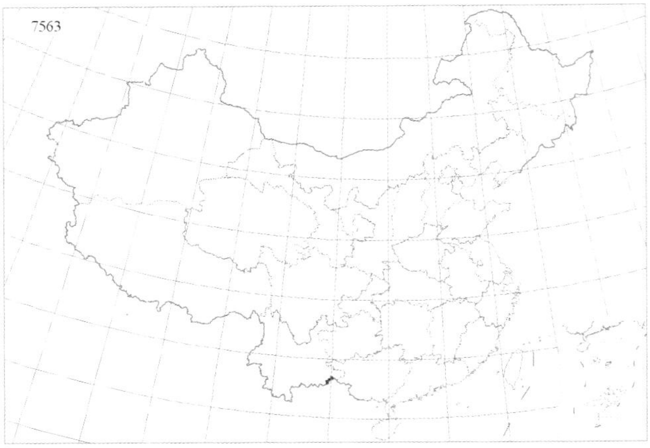

Araliaceae 五加科 *Schefflera* 鹅掌柴属

Schefflera pauciflora R. Vig. 球序鹅掌柴

Shrubs or trees, evergreen

19.1 (13.7~22.2); 10.6 (4.7~15.1); 25.8 (21.2~28.6); 19.2 (13.7~22.3); 170.2 (104.8~207.5); 0.0 (0.0~0.3); 978 (722~1190) ❙ 1434 (1143~1876); 711 (513~891); 94 (43~194) ❙ 47.4 (5.8~85.5); 976 (722~1190); 294.4 (186.7~659.0)

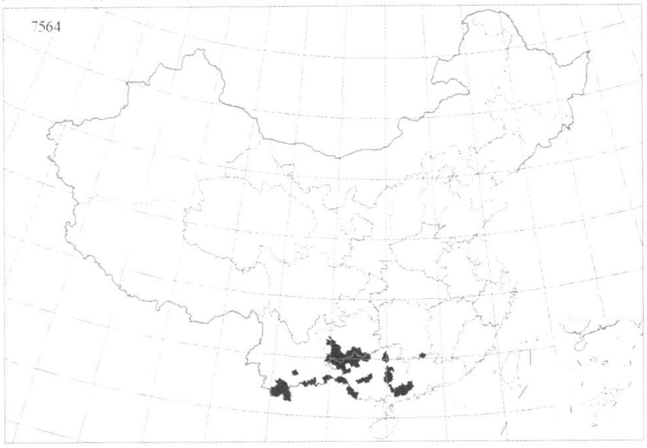

Araliaceae 五加科 *Schefflera* 鹅掌柴属

Schefflera pes-avis R. Vig. 樟叶鹅掌柴

Trees, evergreen

20.7; 12.5; 26.8; 20.7; 189.0; 0.0; 1071 ❙ 1429; 780; 79 ❙ 35.6; 1071; 320.7

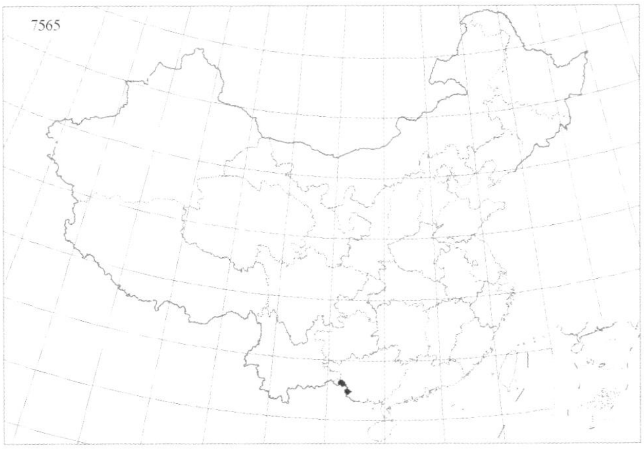

Araliaceae 五加科 *Schefflera* 鹅掌柴属

Schefflera petelotii Merr. 金平鹅掌柴

Shrubs, evergreen

18.5; 12.2; 22.8; 18.5; 162.3; 0.0; 886 ❙ 1448; 834; 59 ❙ 63.4; 886; 516.3

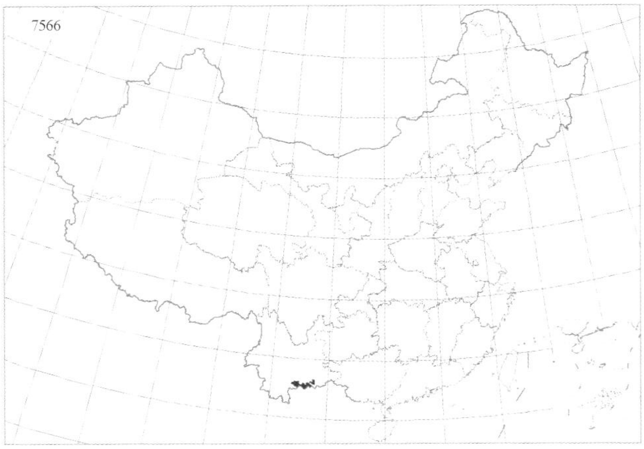

Araliaceae 五加科 *Schefflera* 鹅掌柴属

Schefflera rhododendrifolia (Griffich) Frodin 凹脉鹅掌柴

Trees, evergreen

15.9 (3.4~22.2); 8.1 (-5.5~14.7); 22.2 (11.4~28.5); 16.1 (4.7~22.2); 136.4 (23.3~206.9); 4.6 (0.0~41.8); 879 (431~1177) ❙ 1409 (527~1876); 723 (317~1007); 91 (9~194) ❙ 64.3 (22.4~145.8); 877 (431~1177); 292.2 (209.2~447.4)

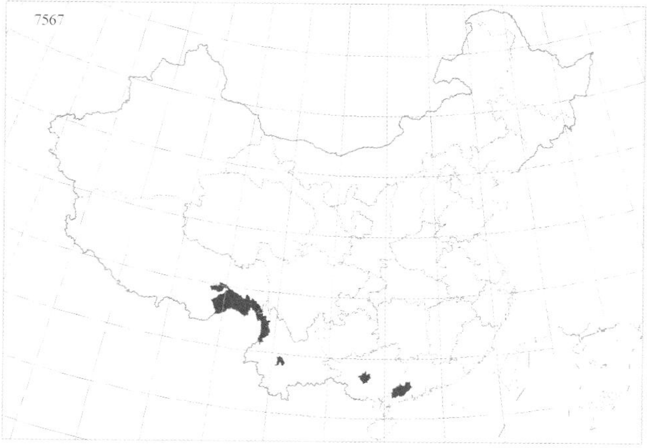

Araliaceae 五加科 *Schefflera* 鹅掌柴属

Schefflera shweliensis W. W. Sm. 瑞丽鹅掌柴

Shrubs or small trees, evergreen

14.7 (8.6~19.5); 8.0 (1.3~12.7); 19.5 (15.0~23.3); 14.7 (8.7~19.5); 118.4 (52.6~173.8); 1.6 (0.0~8.8); 753 (574~942) ❙ 1270 (1007~1559); 719 (578~907); 51 (40~70) ❙ 70.1 (26.1~99.9); 744 (574~921); 405.4 (351.7~487.2)

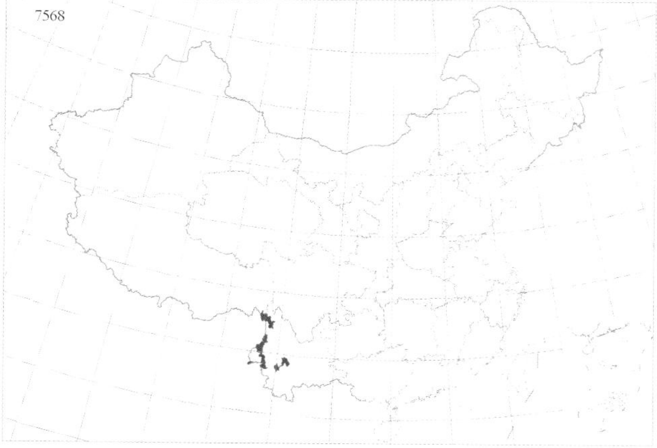

Araliaceae 五加科 *Schefflera* 鹅掌柴属

Schefflera taiwaniana (Nakai) Kaneh. 台湾鹅掌柴

Small trees, evergreen

18.6 (16.4~21.7); 12.8 (11.2~15.3); 23.4 (20.3~27.0); 18.6 (16.5~21.7); 163.7 (137.8~200.2); 0.0 (0.0~0.0); 909 (783~1118) ❙ 2396 (1836~2859); 1049 (900~1412); 248 (110~402) ❙ 170.5 (64.2~235.6); 902 (783~1075); 257.6 (202.8~303.6)

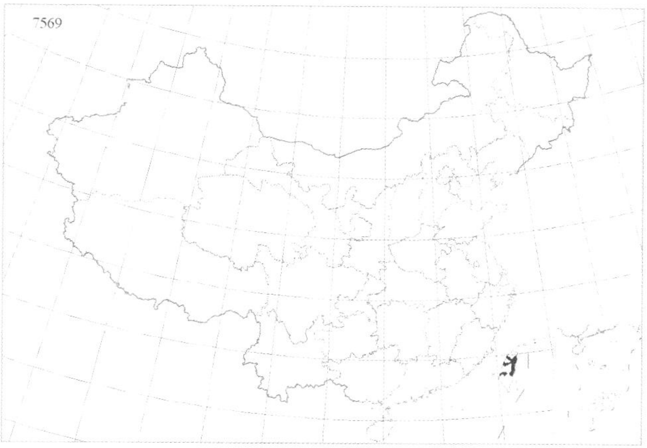

Araliaceae 五加科 *Schefflera* 鹅掌柴属

Schefflera wardii C. Marquand et Airy Shaw 西藏鹅掌柴

Shrubs, evergreen

7.0 (1.8~13.3); -1.2 (-7.1~5.4); 14.0 (9.9~19.6); 7.7 (3.8~13.4); 48.5 (15.7~100.4); 24.4 (0.0~53.3); 531 (404~706) ❙ 934 (473~1736); 540 (287~1007); 30 (7~58) ❙ 65.3 (5.0~145.8); 531 (404~706); 310.8 (169.5~447.4)

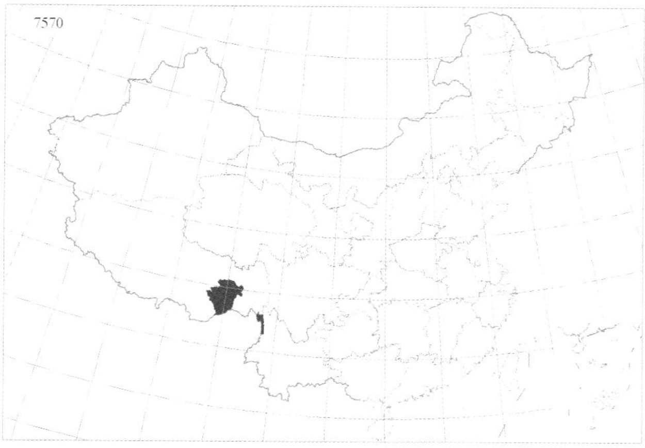

Araliaceae 五加科 *Sinopanax* 华参属

Sinopanax formosana (Hayata) H. L. Li 华参

Shrubs or small trees

20.2 (16.4~24.1); 14.9 (11.2~18.9); 24.4 (20.3~28.0); 20.2 (16.5~24.1); 182.9 (137.8~229.7); 0.0 (0.0~0.0); 1013 (783~1322) ❙ 2381 (1751~2992); 1292 (947~1847); 151 (47~303) ❙ 145.2 (32.5~219.3); 975 (783~1123); 242.7 (154.2~303.6)

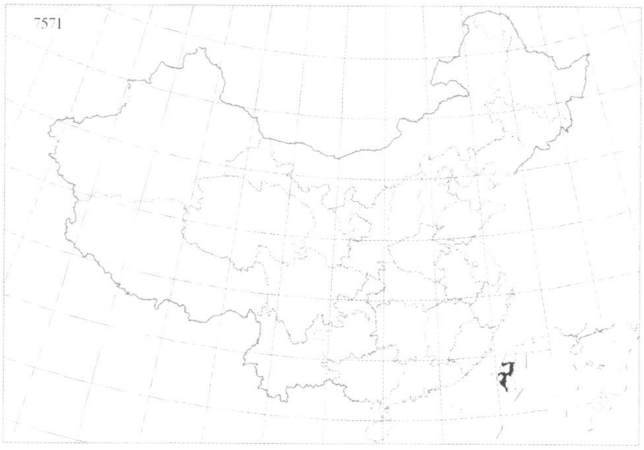

Araliaceae 五加科 *Tetrapanax* 通脱木属

Tetrapanax papyrifer (Hook.) K. Koch 通脱木

Shrubs or small trees, evergreen

15.7 (2.2~24.1); 5.0 (-6.6~18.9); 25.6 (9.9~30.1); 15.8 (3.8~24.1); 131.4 (15.5~229.7); 2.4 (0.0~49.1); 847 (413~1322) ❙ 1263 (626~3495); 526 (316~1594); 108 (15~897) ❙ 49.0 (-22.2~246.5); 835 (413~1182); 211.9 (108.2~491.5)

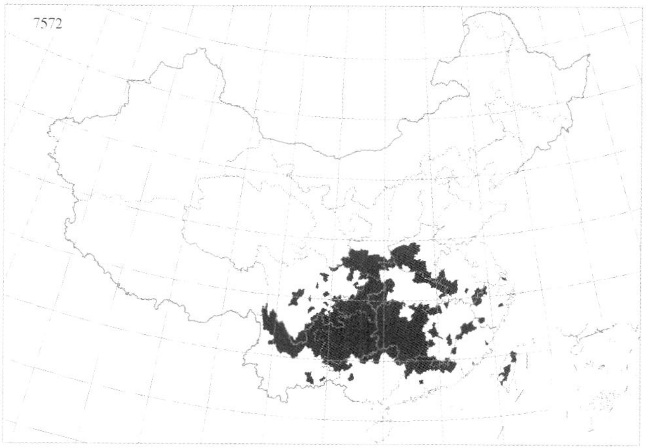

Araliaceae 五加科 *Trevesia* 刺通草属
Trevesia palmata (DC.) Vis. 刺通草
Small trees, evergreen

18.7 (5.3~22.3); 10.7 (-2.4~15.1); 24.7 (12.2~28.9); 18.8 (5.7~22.3); 165.6 (30.5~208.1); 0.5 (0.0~26.2); 953 (479~1202) | 1351 (879~1703); 710 (472~911); 76 (31~194) | 43.7 (12.9~100.7); 950 (479~1202); 332.6 (189.5~659.0)

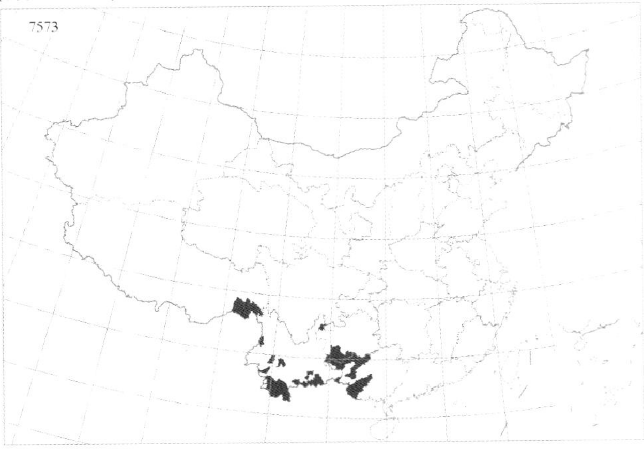

Araliaceae 五加科 *Tupidanthus* 多蕊木属
Tupidanthus calyptratus Hook. et Thomson 多蕊木
Small trees, evergreen

18.3 (13.3~20.5); 12.1 (5.4~15.1); 22.2 (19.1~23.7); 18.3 (13.4~20.5); 160.0 (100.4~185.9); 0.0 (0.0~0.0); 884 (706~992) | 1503 (1276~1736); 813 (655~1007); 55 (41~77) | 71.9 (36.8~145.8); 870 (706~986); 508.8 (427.9~659.0)

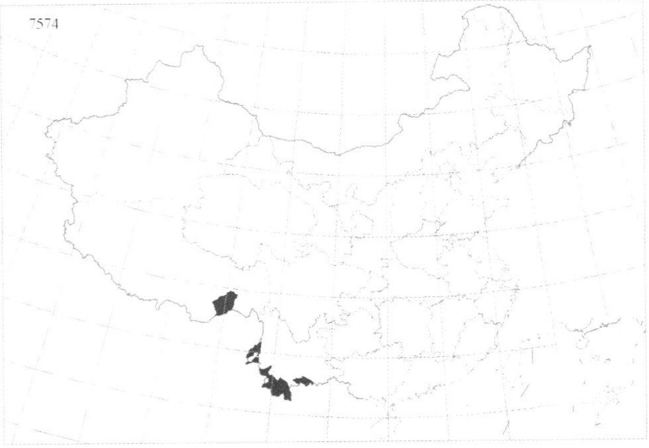

Cornaceae 山茱萸科 *Cornus* 山茱萸属
Cornus alba L. 红瑞木
Shrubs, deciduous

8.8 (-4.0~17.2); -7.3 (-28.2~5.0); 23.2 (13.1~29.1); 10.7 (5.0~17.2); 86.6 (27.9~146.8); 40.8 (0.0~146.2); 680 (448~941) | 666 (70~1571); 366 (23~691); 29 (4~193) | -1.9 (-91.1~108.0); 597 (70~936); 209.8 (61.2~337.1)

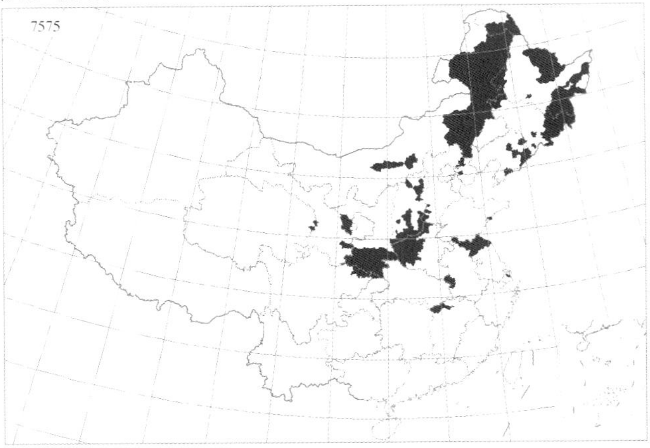

Cornaceae 山茱萸科 *Cornus* 山茱萸属
Cornus austrosinensis W. P. Fang et W. K. Hu 华南梾木
Shrubs or small trees, deciduous

17.9 (16.4~19.1); 8.0 (5.4~10.2); 26.5 (23.6~28.7); 18.0 (16.5~19.2); 155.8 (137.7~170.1); 0.0 (0.0~0.0); 925 (836~981) | 1382 (1236~1564); 585 (411~720); 124 (51~204) | 49.9 (27.5~70.7); 919 (836~981); 235.5 (170.8~322.1)

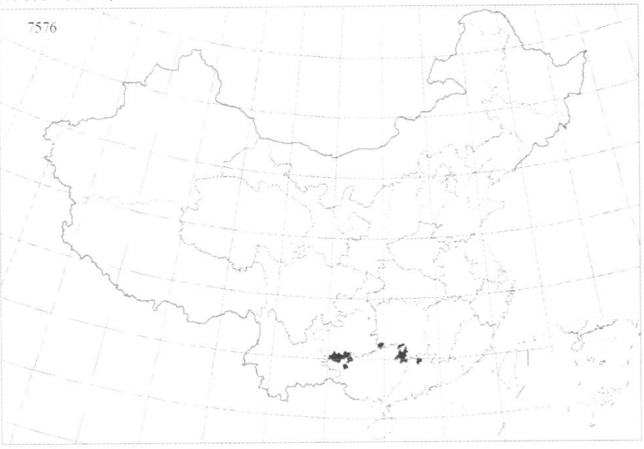

Cornaceae 山茱萸科 *Cornus* 山茱萸属
Cornus bretschneideri L. Henry 沙梾
Shrubs or small trees, deciduous

10.7 (0.8~16.9); -4.5 (-18.4~4.8); 24.3 (10.7~28.3); 11.7 (3.8~16.9); 96.6 (16.2~143.4); 27.3 (0.2~90.3); 729 (411~924) | 578 (183~1214); 348 (114~535); 20 (4~103) | -19.7 (-71.0~101.4); 552 (183~924); 170.2 (26.5~344.4)

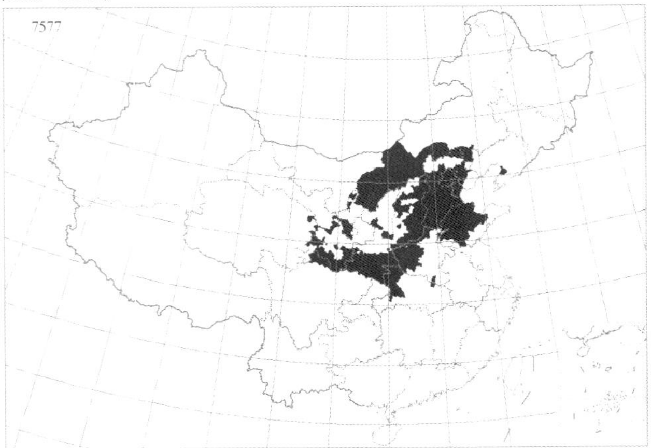

Cornaceae 山茱萸科 *Cornus* 山茱萸属
Cornus capitata Wallich 头状四照花
Trees, evergreen

15.2 (-1.1~21.5); 5.7 (-9.1~14.2); 23.5 (6.2~29.3); 15.3 (2.0~21.5); 125.4 (2.5~198.2); 2.4 (0.0~75.6); 811 (334~1138) | 1287 (321~1950); 565 (193~1007); 106 (11~259) | 58.6 (-14.2~147.0); 799 (321~1138); 249.2 (79.3~551.6)

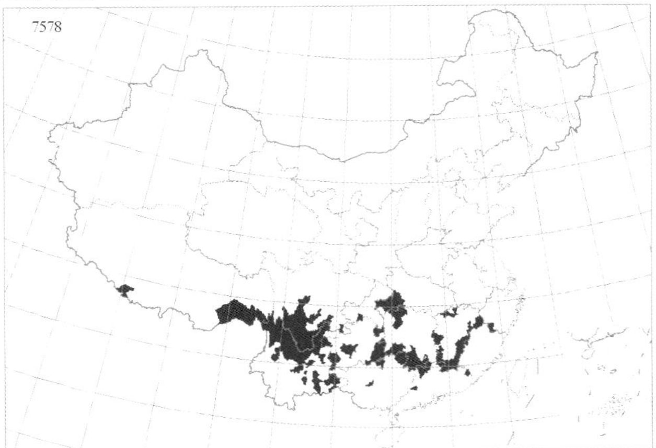

Atlas of Woody Plants in China

Distribution and Climate

Volume III

Edited by

Jingyun Fang

Zhiheng Wang

Zhiyao Tang

高等教育出版社 · 北京
HIGHER EDUCATION PRESS BEIJING

内容提要

准确翔实的物种分布资料是宏观生态学、生物地理学以及保护生物学等学科的研究基础，也是农业、林业生产实践的重要基础。北美、欧洲和日本在20世纪七八十年代就出版了详细的动、植物物种分布图，为这些区域的生物地理学、宏观生态学研究以及生物多样性保护实践提供了不可缺少的基础资料。中国是全球12个“巨大生物多样性国家”之一，拥有超过30 000种维管植物，其中包括约11 000种木本植物、3200种树木，物种数量远高于同处中高纬度的北美洲和欧洲。但中国目前还没有较为完善的植物分布图集。

本图集在忠实于标本记录的基础上，收集了我国目前已出版的全国、各省区和地方植物志、树木志，以及区域科学考察报告和学术论文中有关的木本植物分布资料，并在此基础上邀请全国各地20多位专家对现有资料进行修订和补充，最终编制了我国已知的全部11 405种木本植物的详尽分布图。图集中的物种名录采用了当前最新的分类系统，与国际标准一致。同时，图集还提供了物种的生活型以及物种分布区的13个气候和初级生产力指标，为生态学、生物地理学和保护生物学的基础研究和实践提供了重要的基础资料。

本图集是目前中国第一部具有完整木本植物名录、包含翔实分布信息和气候特征的专著，可作为生态学、植物学、地理学等相关学科的广大师生以及农林科技工作者的参考书。

图书在版编目（CIP）数据

中国木本植物分布图集 = Atlas of Woody Plants in China: Distribution and Climate: 中、英文 / 方精云，王志恒，唐志尧编著．— 北京：高等教育出版社，2009.9

ISBN 978-7-04-026859-1

I. 中… II. ① 方… ② 王… ③ 唐… III. 木本植物—中国—图集

Ⅳ．S717. 2-64

中国版本图书馆 CIP 数据核字（2009）第170374号

策划编辑 林金安 王 莉 **责任编辑** 王 莉 刘思涵 **封面设计** 张志奇
版式设计 张志奇 **责任排版** 高 瑾 张志奇 **责任印制** 陈伟光

出版发行 高等教育出版社
社 址 北京市西城区德外大街4号
邮政编码 100120
经 销 蓝色畅想图书发行有限公司
印 刷 涿州市星河印刷有限公司

购书热线 010-58581118
咨询电话 400-810-0598
网 址 http://www.hep.edu.cn
http://www.hep.com.cn
网上订购 http://www.landraco.com
http://www.landraco.com.cn

开 本 889×1194 1/16
印 张 41.75
字 数 1 300 000

版 次 2009年10月第1版
印 次 2009年10月第1次印刷
定 价 580.00元（全4册）

物 料 号 26859-002

Atlas of Woody Plants in China

Distribution and Climate

Edited by

Jingyun Fang

Zhiheng Wang

Zhiyao Tang

Other Contributors

Xiulian Chi, Zhaodi Guo, Yue Jiang, Lai Jiang, Jing Li, Liping Li, Xin Lin, Chang Liu, Jinglei Liu, Yining Liu, Xiujuan Qiao, Yanlin Ren, Zehao Shen, Kun Tan, Luying Tang, Jie Wang, Shaopeng Wang, Yetao Yan, Xinping Zhang, Yanming Zhang, Rui Zhou, Jiangling Zhu

Review Experts (Focal Areas)

Sizhao Fang (*Guizhou*), Ying Hai (*Xinjiang*), Shushen Lai (*Jiangxi*), Fazeng Li (*Shandong*), Guangzhao Li (*Guangxi*), Zhenyu Li (*Species checklist*), Cunzhu Liang (*Inner Mongolia and Ningxia*), Chudian Liu (*Fujian*), Anwar Mohammat (*Xinjiang*), Yi Ren (*Shaanxi*), Kun Sun (*Gansu*), Shizhou Sun (*North China and Tibet*), Zemin Wu (*Anhui*), Chengdong Xu (*Yunnan*), Qinzhou Yang (*Sichuan and Chongqing*), Huagu Ye (*Guangdong*), Yongzhong Ye (*Henan*), Xunlin Yu (*Hunan*), Zien Zhao (*Hubei*), Xingmin Zhou (*Qinghai*), Zhanjiang Zong (*Northeast China*)

中国木本植物分布图集

方精云　王志恒　唐志尧 编著

参与数据收集和数据库建立的人员（按姓氏拼音排序）

池秀莲、郭兆迪、江玥、姜来、李晶、李利平、林鑫、刘畅、刘晶磊、刘怿宁、乔秀娟、任艳林、沈泽昊、谭琨、汤璐瑛、汪洁、王少鹏、闫业涛、张新平、张延明、周睿、朱江玲

审校专家及负责区域（按姓氏拼音排序）

方嗣昭(贵州)、海鹰(新疆)、赖书绅(江西)、李法曾(山东)、李光照(广西)、李振宇(植物名录)、梁存柱(内蒙古、宁夏)、刘初钿(福建)、安尼瓦尔·买买提(新疆)、任毅(陕西)、孙坤(甘肃)、孙世洲(华北地区和西藏)、吴泽民(安徽)、徐成东(云南)、杨钦周(四川、重庆)、叶华谷(广东)、叶永忠(河南)、喻勋林(湖南)、赵子恩(湖北)、周兴民(青海)、宗占江(东北地区)

Biosketches of Editors 作者简介

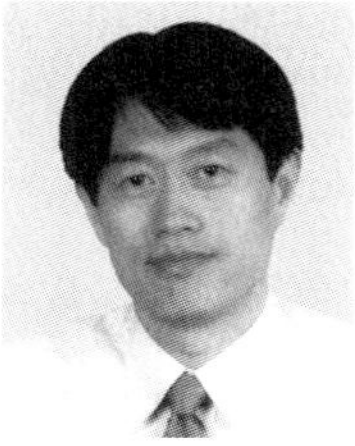

Jingyun Fang | 方精云

Dr. Jingyun Fang is a Cheung Kong Professor at Department of Ecology, Peking University. He obtained his PhD from Osaka City University (Japan) in 1989. His research interests cover vegetation ecology, biogeography of plants, carbon cycle and the application of remote sensing. He is an author of more than 230 publications including five books. He has received several awards including He-Liang-He-Li Science and Technology Progress Award, Cheung Kong Scholars Achievement Award, and National Natural Science Award of the State Council. He is a member of the Chinese Academy of Sciences and the Academy of Science for the Developing World (TWAS), and serves as the editorial board member for several journals including *Frontiers in Ecology and the Environment, Ecosystems, Global Environmental Change, Science China, Ecological Research, BMC Ecology* and others.

北京大学长江特聘教授、中科院院士，第三世界科学院院士。1989 年获日本大阪市立大学生物学博士学位。主要从事植被生态学、生物多样性、生态系统碳循环等方面的科研和教学工作。共发表论文 230 余篇，出版专著 5 部。曾获首届国家杰出青年科学基金 (1994)、教育部自然科学一等奖 (2003)，国家自然科学二等奖 (2004)，长江学者成就奖 (2006)，何梁何利科学技术进步奖 (2007) 等奖项以及全国野外科技工作先进个人称号 (2009)。是国内外多个专业杂志的编委或副主编，包括 *Frontiers in Ecology and the Environment, Ecosystems, Global Environmental Change, Science China, Ecological Research, BMC Ecology*。

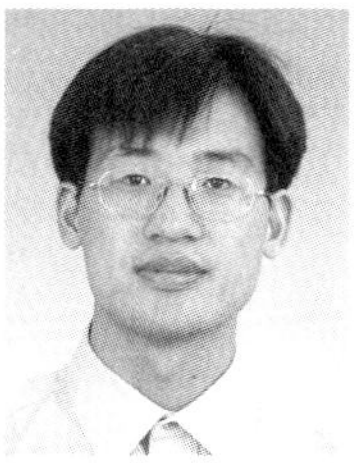

Zhiheng Wang | 王志恒

Dr. Zhiheng Wang is a postdoc researcher in Institute of Environmental Sciences, University of Zurich, Switzerland. He got his PhD at Peking University in 2008, and has research interests in all issues of macroecology. He has been working on the Database of China's Woody Plants for the last six years. His recent focus is on macro-scale patterns of species richness, species range size, species body size, patterns in community structure, and the mechanistic causes of such patterns. His current research involves testing the influence of contemporary climate, evolutionary history, human activities, and spatial scale on species diversity patterns using the synthesis of climatic, phylogenetic, and anthropogenic data. He has published 20 papers, including a recent paper in the Proceedings of the National Academy of Sciences, USA.

目前在瑞士苏黎世大学环境科学研究所从事博士后研究。2008 年毕业于北京大学，获理学博士学位。过去 6 年以来一直致力于“中国木本植物分布数据库”的建设。主要从事宏观生态学和全球变化生态学研究，内容包括物种多样性、物种分布区、个体大小和群落结构的大尺度格局及其进化和气候的形成机制，以及气候变化对物种多样性的影响。发表论文 20 余篇，包括一篇美国科学院院刊论文。

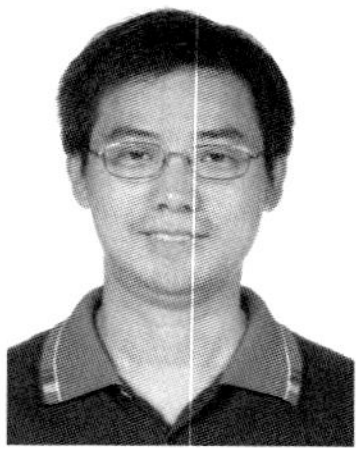

Zhiyao Tang | 唐志尧

Dr. Zhiyao Tang is an associate professor at the Department of Ecology, Peking University. He got his PhD at Peking University in 2003. His research interests include plant community ecology, biodiversity patterns, climate-vegetation interaction, and application of remote sensing in ecology. He has authored more than 40 peer-reviewed articles and two books.

北京大学生态学系副教授。2003 年毕业于北京大学，获理学博士学位。主要从事植被植物群落生态学、生物多样性、植被气候关系以及生态遥感等方面的科研与教学工作。发表论文 40 余篇，专著 2 部。

Preface | 前言

China is one of the earth's 12 mega-biodiversity countries, and the most diverse country in the northern hemisphere. With more than 30,000 species of vascular plants, more than 1/3 of which are woody plants, the number of plant species in China is higher than in Europe and North America combined. However, a comprehensive atlas of the distribution of woody plants in China has not been yet available though atlases of the species distribution in many western countries were compiled several decades ago. For example, an atlas of the United States trees was published in the 1970s. The atlas of vascular plants in Japan with a vertical resolution of 50 m was compiled in the 1970s. A project for mapping distribution of vascular plants in Europe was launched in 1965, and 13 volumes of the atlas have been published to date. These atlases of plants are of both scientific and practical importance, and provide an indispensable database for understanding the patterns and mechanisms of species distribution, utilizing the biological resources, protecting biological diversity, implementing regional development planning, and predicting ecological responses of plants to global environmental changes.

Why has China lacked of such atlas? This might be primarily due to there being too many species and limited information on the distributions of those species. These make some people think that it is not right time to compile the national atlas for China's plants. However, I don't agree with this opinion. Nothing is perfect, especially for the information on species distribution. For example, in 1817, Alexander von Humboldt drew the first isothermal map based on temperature data from only 58 climatic stations globally, which became the basis for climatic division and physiogeographical study for decades. Japan has completed the atlas for all its vascular plants with a vertical resolution of 50 m in the 1970s with limited information on distribution of plants. Compared with these, China has accumulated copious data on plant distribution through the efforts of the several generations of botanists and ecologists over the past half century. These hard-earned data provide an essential base to systematically compile the atlas of China's plants. Therefore, in my opinion, now it is the time to compile the atlas of China's plants, which is an urgent need for the conservation and sustainable use of biological resources.

The idea of compiling the atlas of China's plants initially came into my mind in the mid-1980s when I was a PhD student in Japan. My dissertation was to examine the relationships between climate and vegetation in China and I strongly felt that China was lacking of the species distribution data. I was very impressed when I saw the detailed spatial explicit distribution maps of vascular plants in Japan, and made a copy of the maps of major species and brought them back to China. They are still kept on my bookshelf now. In 1996, I was touched again when I saw the atlas of the United States trees and European species in McGill University, Canada, where I was a visiting scholar. These two shocks made me long hope that we should have an atlas for China's plants in the near future. In 2003 as the first step of compiling an atlas for all the plants of China, we initiated a project to compile the *Atlas for All the China's Woody Plants*, although without specific funds, because distribution information is relatively rich and the checklist is well-documented for the woody plants in China. As the basic data of the atlas, we first started to construct "*Database of China's Woody Plants*". We documented the distribution sites for all the species based on specimen records and all the validated records documented in publications. After six years' efforts, this atlas which documents more than 11,000 woody species was eventually published.

In this atlas, we provide distribution maps for more than 11,000 species at the level of counties. We also documented information on climate, primary production and life forms for each species. This atlas contains information on both species and environmental variables, compared with the previous atlases published in other countries or regions.

As above mentioned, we launched this project without specific financial support. However, we have been lucky to have creative human resources and excellent students, who have been dedicated to the collection and compilation of these basic data, which is difficult to yield articles but is critically significant in scientific research and practical use. Here, I would like especially to mention two major contributors of this atlas, Dr. Zhiheng Wang and Dr. Zhiyao Tang, who took major responsibility to complete the atlas. Zhiheng was my PhD student whose PhD project was to examine patterns of China's woody plants and the mechanisms that shape those patterns. He joined my group in his senior year, and I asked him to postpone his defense for his PhD twice to complete the construction of the database of China's woody plants. For this reason, he stayed in my group as a PhD student for eight years. During this period he led a group of students to overcome many problems that data compilations posed and spent six years and finally completed the database, which makes publishing this atlas possible. Zhiyao donated uncounted hours of his time to collect data, check distribution records, collate species name, design maps, and so on. In addition, many staff and students contributed their time in data collection, data input and species name collation. To all of them, I am grateful.

The atlas was based on many sources of information, including a Chinese version (*Flora Reipublicae Popularis Sinicae*) and English version of "*Flora of China*", and "*Seed Plants of China*" edited by Zhengyi Wu, "*Higher Plants of China*" edited by Liguo Fu, and nu-

merous local flora and species surveying reports. The compilation of this atlas would be impossible without these basic, most essential works. Here, I am greatly indebted to all the authors for their efforts. I would also like to thank the botanists and field biologists who reviewed and corrected the distribution records of species. I want to acknowledge many students in the Department of Ecology, Peking Universiy, who assisted this project by documenting distribution data. Mr. Jin'an Lin and Ms. Li Wang of Higher Education Press (China) spent lots of time editing, collating and publishing this atlas. I will like to express my many thanks for all these people. The compilation of this atlas was partly supported by the National Natural Science Foundation of China (Nos: 40638039, 90711002, 30721140306, 40871030, 40501025, 49971002, 39970044, 39425003).

In spite of our best efforts, errors still remain in this atlas due to our limited knowledge and limited information on species distribution. Please contact us (Email: zytang@urban.pku.edu.cn)and send us your comments so that we can update the atlas in the near future.

Jingyun Fang
Professor of Plant Ecology
September, 2009

中国是世界上物种多样性特别丰富的国家之一，拥有维管植物约30 000余种，其中1/3以上是木本植物，比北美和欧洲的总和还要多。但中国至今尚没有全面记述这些物种地理分布的图集。这应该说是件十分缺憾的事。早在几十年前，西方国家就已经完成了物种分布图的编制工作。比如，美国于20世纪70年代完成了全美木本植物分布图的编制；日本在20世纪70年代出版了所有维管植物的分布图，物种垂直分布范围的精度达50 m；欧洲也于1965年开始了维管植物分布图的编制工作。这些分布图为揭示物种的分布规律、开发和利用生物资源、开展生物多样性保护、进行区域发展规划，以及预测气候变化的生态影响等提供了不可或缺的基本资料，其科学价值和实践意义是不言而喻的。

为什么中国缺乏这类分布图？物种多、物种分布资料不足又散布于各类文献中，造成了编制工作难度大，这可能是最主要的原因。此外，国内对编制物种分布图的保守认识也是一个重要的限制因素。

据我所知，国内一些人认为，我国物种分布资料不足，目前尚不具备编制植物分布图集的条件。但我不同意这种说法。科学技术的进步是无止境的，大自然的很多现象和规律是逐渐被认知的，否则人类文明就不会进步到今天。1817年洪堡依据全球仅有的58个气象站的气温资料，编绘了世界首张全球等温线分布图，是气候分区和自然地理学研究的一个重要里程碑，成为相当长时期内开展相关研究的基石。又如，日本在30多年前就能出版垂直分布范围高达50 m精度的维管植物分布图，不能说他们当时已经掌握了所有物种准确的分布信息。因此，如果等物种分布资料齐全了才编制其分布图，那我们所待何时？事实上，经过广大植物学家、生态学家半个多世纪的努力，我国的物种分布资料已经有了相当深厚的积累，系统编制物种分布图的时机已经成熟，这也是时代发展的迫切要求。当然，随着知识的积累和认识水平的提高，这类分布图的信息需要进一步补充和完善。

编制中国植物分布图是我在学生时代就萌生的愿望。早在20世纪80年代的中后期，我在日本读书，博士期间的研究课题就是关于中国植物分布与气候关系的研究，实感中国物种分布数据之缺乏。当我看到日本详尽的物种分布图时，那种欣喜若狂至今还记忆深刻，当时我就不加思索地把主要物种的分布图全都复印了下来，带回国内，至今还保存着。1996年，我去加拿大McGill大学做访问学者，看到美国和欧洲也有详尽的物种分布图，又一次受到触动。于是，就产生了将来我们也要编制中国植物分布图的想法。2003年，在没有专项经费支持的情况下，我们决定先编制分布资料较多、种名相对稳定的木本植物分布图，于是启动了“中国木本植物分布图的编制”自设课题。次年，作为编制物种分布图的基础工作，课题组开始了“中国木本植物分布数据库”的建设。历经6年的艰苦工作，这本多达2 000余页的图集终于与读者见面了。

本图集的特点是为11 405个物种提供县级分布信息；每个物种的分布信息在忠实于标本记录的基础上，最大可能地吸收了所有物种分布研究的成果；提供物种分布区的气候资料和分布区初级生产力的平均状况。此外，我们也提供了绝大多数物种的主要生活型信息。本图集的这些特点，既克服了西方国家仅提供物种分布信息的局限性，也满足了人们对分布区气候和物种生活型等基础信息的需求，使其更具实用性。

前面讲过，我们是在没有专项经费的条件下开始此项工作的。我们虽然没有专项经费，但有丰厚的人力资源，有非常优秀的同学，他们愿意献身于这类发表成果难、但有重要科学和应用价值的基础性数据的收集和整理工作。这里，我要特别提到本图集的两位主要作者：王志恒和唐志尧，他们是本图集的主要贡献者。王志恒当时是我的直博生，我给他的博士论文题目就是中国木本植物多样性的分布格局及其机制的研究。他自大三就进入我的研究组，直到博士毕业，一共呆了8年。为了完成数据库的建设，我两度让他推延博士毕业。他带领一批同学，克服物种多、

物种名称杂乱、物种分布信息零散不全等困难，历时6年，终于建成了“中国木本植物分布数据库”，使本图集的编制成为可能。另一名主要作者唐志尧，在物种分布数据收集、分布区和种名校核、分布图制作与设计、编辑和出版等方面，做了主要的工作。可以说，他们二位是本图集编制出版的主要贡献者。此外，很多老师和同学在数据收集和录入、种名核校等方面也做了大量繁杂的工作。

本图集凝聚了几代人的心血和汗水，它是在中英文版《中国植物志》、吴征镒先生主编的《中国植物分布数据库》所记载的物种名录、傅立国主编的《中国高等植物》以及众多地方植物志和调查报告中所记载的物种分布信息的基础上编制出版的。因此，我要特别感谢这些作者们，没有他们的工作，就没有本图集的出版。北京大学生态学系的多位同学参与了数据的收集、整理和校对，以及分布图的制作等。数据库建成后邀请各地的有关专家对种名和分布区进行了校验和补充。高等教育出版社的林金安先生、王莉女士为本图集的制作、编辑和出版做了大量艰苦而有成效的工作。在此，对他们表示衷心的感谢。本图集在数据收集、整理和出版过程中，得到国家自然科学基金委的部分资助（编号：40638039, 90711002, 30721140306, 40871030, 40501025, 49971002, 39970044, 39425003）。

由于我们水平有限，以及现有资料的不足，本图集肯定存在一些纰漏，请读者们批评指正(请联系zytang@urban.pku.edu.cn)，以便在修订时改进。

方精云
2009年9月
于北京大学

How to Use This Atlas | 本图集使用说明

1 Arrangement of families, genera and species

The order of families, genera, and species were arranged as follows: the families are organized according to the phylogenetic system of *Flora of China*; within each family, the genera and species are arranged alphabetically. The atlas is published in three volumes, plus a separate volume of species checklist.

2 Arrangement of the climatic variables

The atlas lists mean and range of 12 climatic variables and vegetation net primary productivity. They are: mean annual temperature (MAT, °C), mean temperature of the coldest month (MTCM, °C), mean temperature of the warmest month (MTWM, °C), annual biotemperature (ABT, °C), warmth index (WI, °C), coldness index (CI, °C), potential evapotranspiration (PET, mm), annual precipitation (AP, mm), precipitation of warmest quarter (PWQ, mm), precipitation of coldest quarter (PCQ, mm), moisture index (Im), annual actual evaportranspiration (AET, mm), vegetation net primary productivity (NPP, $g \cdot a^{-1} \cdot m^{-2}$). For those species distributed in less than five counties, only the mean values of the climatic variables are listed. To save space, names of these variables and their units are not showed in the illustration of each species; these variables are listed as the following order, and three categories of variables (thermal index, humid/arid index, and integrative index) are separated by mark " ▌".

MAT (°C); MTCM (°C); MTWM (°C); ABT (°C); WI (°C); CI (°C); PET (mm) ▌AP(mm); PWQ (mm); PCQ (mm) ▌Im; AET (mm); NPP ($g \cdot a^{-1} \cdot m^{-2}$)

The following illustration explains the content of each map, using *Amentotaxus argotaenia* (Hance) Pilg. as an example. Following the illustration, MAT is 18.0 °C (Range is 9.3~22.9 °C); MTCM 8.0 °C (−1.7~15.2 °C); MTWM 26.8 °C (19.6 ~29.8 °C); ABT 18.0 °C (9.5~22.9 °C); WI 157.1 °C (69.8 ~214.8 °C); CI 0.7 °C (0.0 ~18.0 °C); PET 952 mm (603 ~1232 mm) ▌AP 1530 mm (711 ~2435 mm); PWQ 605 mm (357 ~1428 mm); PCQ 150 mm (15 ~250mm) ▌Im 61.9 (21.1 ~145.8); AET 947 mm (603 ~1232 mm); NPP 228.0 $g \cdot a^{-1} \cdot m^{-2}$ (134.6~447.4 $g \cdot a^{-1} \cdot m^{-2}$).

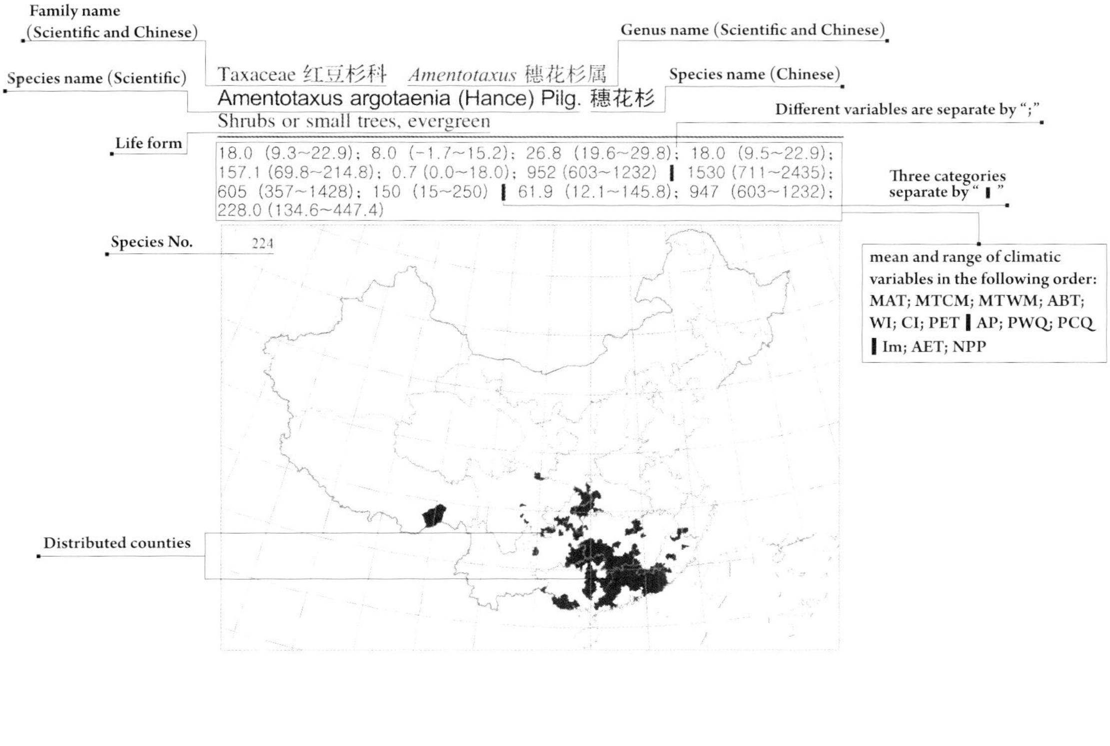

1 科、属、种排列说明

本图集科、属、种排列顺序如下：科序按 *Flora of China* 系统排列；科内（属、种）按字母顺序排列。本图集分三册出版，并附种名索引一册。

2 气候指标表示方法及使用说明

本图集给出每个种分布区的 12 个气候变量和植被净生产力的均值和范围。这些指标的意义和计算见第 xxvi~xxvii 页说明。对于分布区小于 5 个县域的物种，我们仅列出各气候指标的平均值。为节省空间，这些指标按以下顺序给出，并在各类气候指标（温度指标、水分指标和水热综合指标）之间隔以“▌”符号。另外，各指标都省去单位。顺序（含单位）如下：

年均温（℃）；最冷月均温（℃）；最热月均温（℃）；年生物温度（℃）；温暖指数（℃）；寒冷指数（℃）；潜在蒸散量（mm）▌年降水（mm）；最暖季降水（mm）；最冷季降水（mm）▌湿润指数；年实际蒸散量（mm）；植被净第一性生产力（$g \cdot a^{-1} \cdot m^{-2}$）

以穗花杉［*Amentotaxus argotaenia*（Hance）Pilg, 红豆杉科 Taxaceae, 穗花杉属 *Amentotaxus*］为例，用下面的示意图说明分布图及气候指标的含义。穗花杉在其分布区的气候指标（平均值和范围）如下：年均温为 18.0℃（范围为 9.3~22.9℃），最冷月均温 8.0℃（−1.7~15.2℃），最热月均温 26.8℃（19.6~29.8℃），年生物温度 18.0℃（9.5~22.9℃），温暖指数 157.1℃（69.8~214.8℃），寒冷指数 0.7℃（0.0~18.0℃）、潜在蒸散量 952 mm（603~1232 mm）▌年降水 1530 mm（711~2435 mm），最暖季降水 605 mm（357~1428 mm），最冷季降水 150 mm（15~250 mm）▌湿润指数 61.9（21.1~145.8），年实际蒸散量 947 mm（603~1232 mm），植被净第一性生产力 228.0 $g \cdot a^{-1} \cdot m^{-2}$（134.6~447.4 $g \cdot a^{-1} \cdot m^{-2}$）。

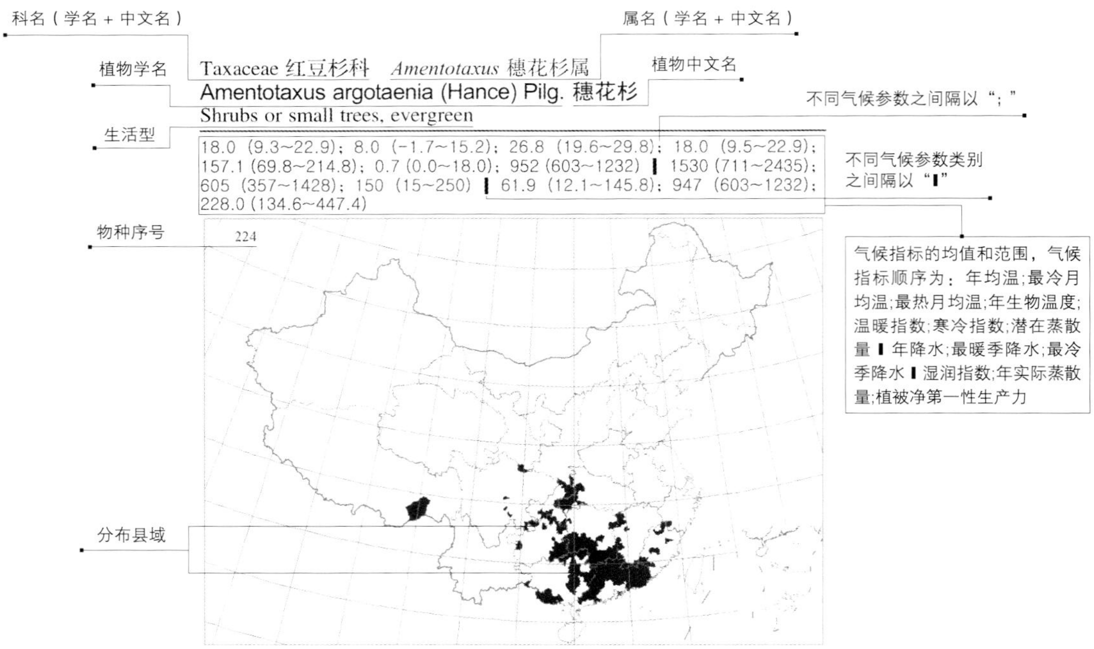

Contents | 目录

Cornaceae 山茱萸科 *Cornus* 山茱萸属
Cornus chinensis Wangerin 川鄂山茱萸
Trees, deciduous

13.3 (2.2~18.9); 3.6 (-6.6~12.0); 22.0 (9.9~29.1); 13.3 (3.8~19.0); 104.3 (15.5~167.8); 4.7 (0.0~49.1); 735 (413~933) ❙ 1099 (669~1813); 541 (317~1007); 59 (15~226) ❙ 51.0 (-14.2~145.8); 723 (413~932); 246.8 (142.9~447.4)

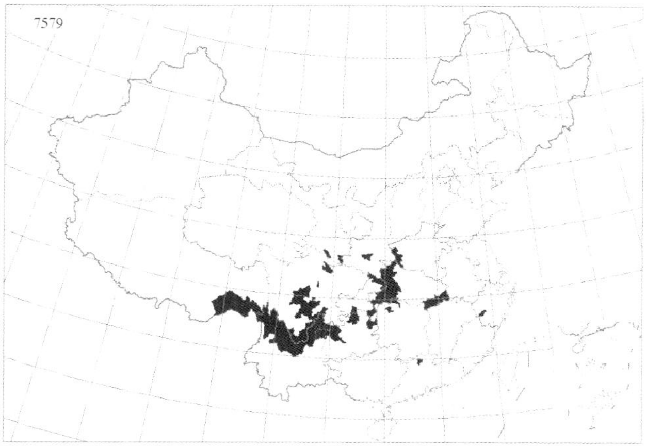

Cornaceae 山茱萸科 *Cornus* 山茱萸属
Cornus controversa Hemsl. 灯台树
Trees, deciduous

15.1 (1.8~21.4); 4.1 (-15.5~13.6); 25.3 (9.9~29.8); 15.2 (3.8~21.5); 126.7 (15.7~197.7); 4.9 (0.0~68.8); 828 (404~1128) ❙ 1244 (373~2680); 520 (201~1186); 111 (7~341) ❙ 49.8 (-37.6~246.5); 809 (373~1114); 227.8 (88.1~516.7)

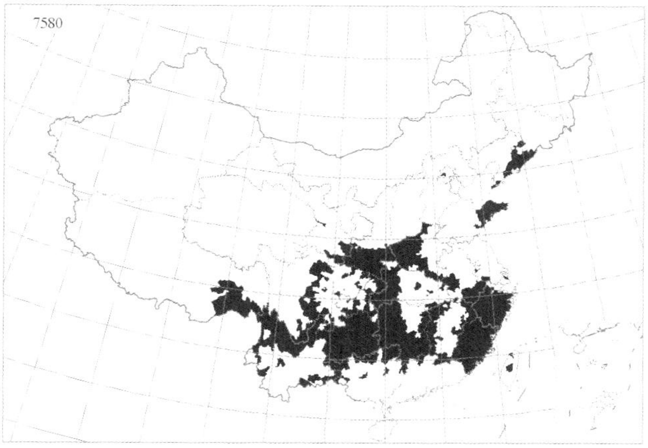

Cornaceae 山茱萸科 *Cornus* 山茱萸属
Cornus coreana Wangerin 朝鲜梾木
Trees, deciduous

6.7 (2.5~10.5); -11.4 (-17.7~-4.9); 22.3 (18.6~24.3); 9.2 (6.9~11.3); 74.9 (48.9~93.2); 54.6 (26.8~78.3); 626 (523~705) ❙ 663 (457~966); 442 (309~616); 13 (8~35) ❙ 6.0 (-16.4~56.9); 585 (457~678); 261.7 (180.8~323.4)

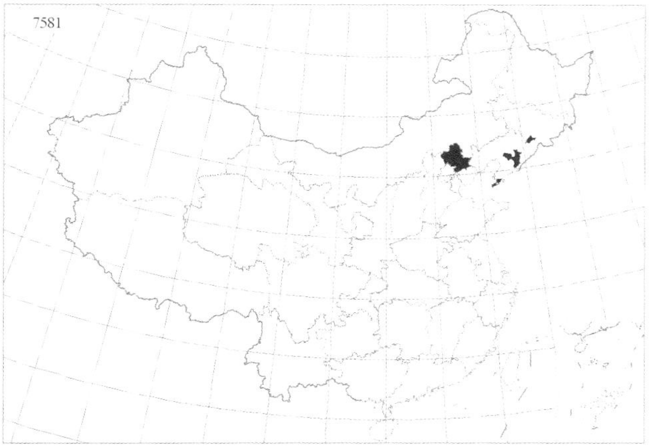

Cornaceae 山茱萸科 *Cornus* 山茱萸属
Cornus elliptica (Pojarkova) Q. Y. Xiang et Bofford 尖叶四照花
Trees, evergreen

16.1 (-1.1~21.4); 5.2 (-9.1~13.6); 26.2 (6.2~29.8); 16.1 (2.0~21.5); 136.0 (2.5~197.7); 2.5 (0.0~75.6); 864 (334~1128) ❙ 1354 (321~1997); 524 (193~1007); 135 (15~259) ❙ 56.3 (-22.2~147.0); 854 (321~1126); 219.6 (79.3~447.4)

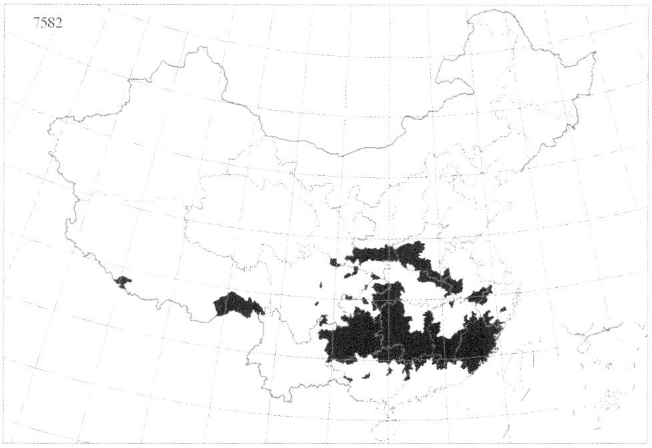

Cornaceae 山茱萸科 *Cornus* 山茱萸属
Cornus hemsleyi C. K. Schneid. et Wangerin 红椋子
Shrubs or small trees, deciduous

9.6 (-1.3~18.5); -2.2 (-12.3~10.5); 20.2 (8.3~28.0); 10.2 (2.5~18.3); 78.1 (7.0~160.0); 22.7 (0.0~82.3); 647 (347~906) ❙ 788 (341~1410); 415 (190~816); 29 (7~113) ❙ 27.1 (-40.9~128.6); 595 (341~832); 220.8 (51.9~357.7)

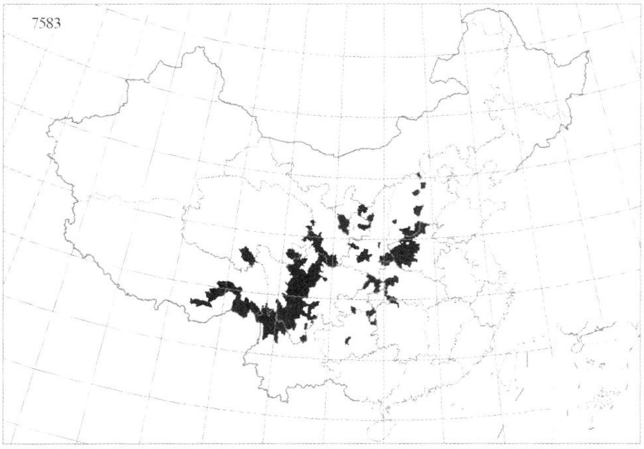

Cornaceae 山茱萸科 *Cornus* 山茱萸属
Cornus hongkongensis Hemsl. 香港四照花
Shrubs or trees, evergreen

17.5 (13.3~23.2); 7.5 (3.2~15.8); 26.3 (20.8~29.1); 17.5 (13.2~23.2); 150.5 (100.9~218.9); 0.2 (0.0~2.5); 916 (715~1257) ❙ 1458 (999~2435); 615 (403~1428); 129 (32~259) ❙ 59.6 (5.8~130.7); 914 (715~1257); 230.0 (146.0~551.3)

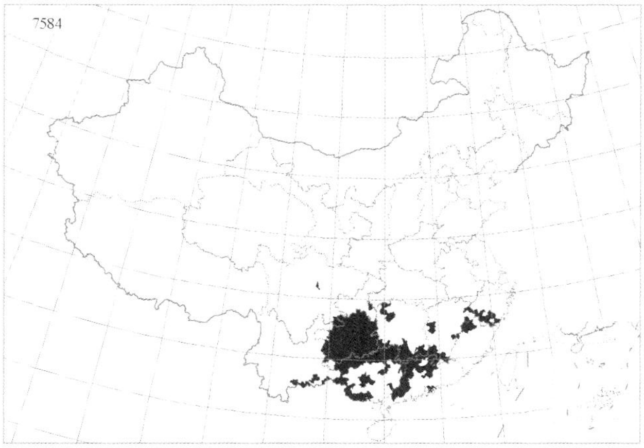

Cornaceae 山茱萸科 *Cornus* 山茱萸属

Cornus hongkongensis subsp. elegans (W. P. Fang et Y. T. Hsieh) Q. Y. Xiang 秀丽四照花 Shrubs or trees, evergreen

17.9 (15.0~21.4); 8.0 (3.3~13.6); 27.3 (24.3~29.4); 17.9 (15.1~21.5); 155.2 (120.8~197.7); 0.1 (0.0~1.9); 935 (789~1128) ❙ 1630 (1070~1997); 560 (356~713); 204 (130~259) ❙ 76.8 (-5.0~147.0); 925 (789~1094); 249.7 (123.9~332.5)

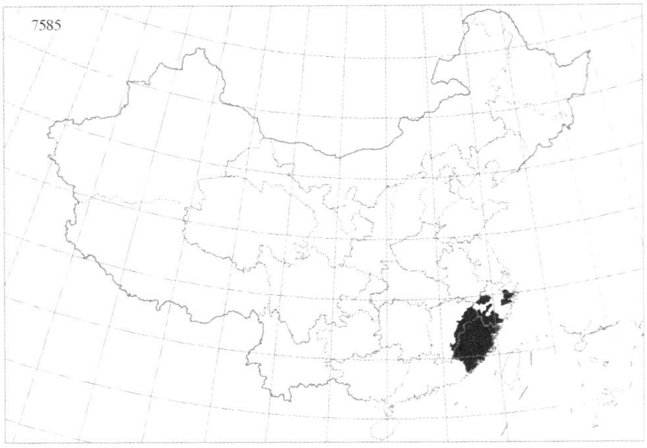

Cornaceae 山茱萸科 *Cornus* 山茱萸属

Cornus hongkongensis subsp. ferruginea (Y. C. Wu) Q. Y. Xiang 褐毛四照花 Trees, evergreen

17.7 (10.6~22.2); 7.4 (-0.2~14.1); 27.1 (21.1~28.9); 17.7 (10.6~22.3); 153.1 (79.4~207.2); 0.8 (0.0~12.1); 940 (643~1186) ❙ 1322 (1028~2014); 552 (442~954); 92 (46~197) ❙ 41.5 (8.0~99.4); 940 (643~1186); 182.2 (108.2~284.3)

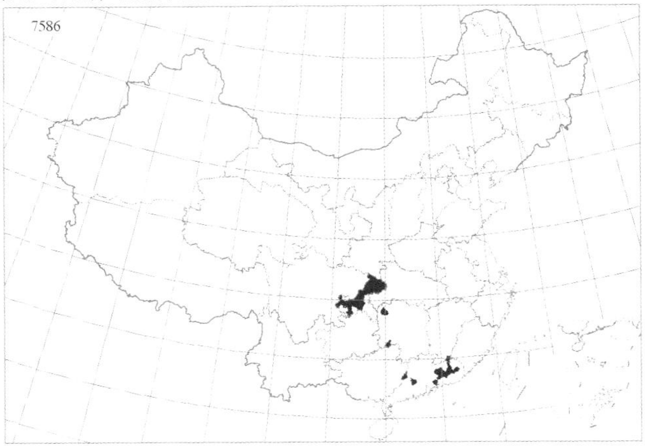

Cornaceae 山茱萸科 *Cornus* 山茱萸属

Cornus hongkongensis subsp. gigantea (Handel-Mazzetti) Q. Y. Xiang 大型四照花 Shrubs or small trees, evergreen

17.2 (14.7~21.0); 6.8 (4.4~11.9); 26.5 (23.6~29.0); 17.3 (14.8~21.1); 147.0 (117.8~193.0); 0.0 (0.0~0.6); 900 (779~1114) ❙ 1319 (985~1662); 536 (403~821); 127 (48~211) ❙ 46.4 (18.3~69.7); 889 (779~1114); 200.3 (149.6~282.2)

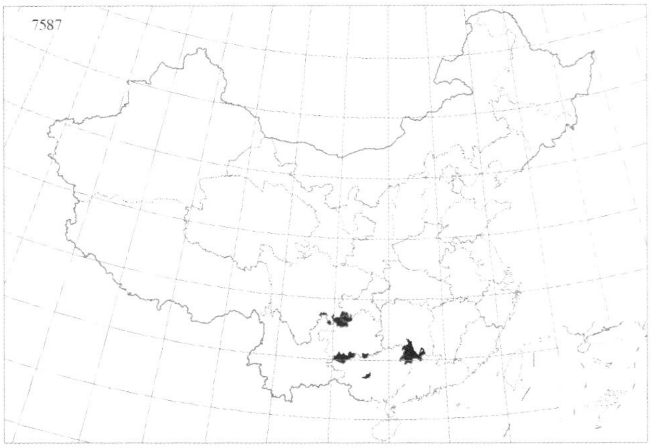

Cornaceae 山茱萸科 *Cornus* 山茱萸属

Cornus hongkongensis subsp. melanotricha (Pojarkova) Q. Y. Xiang 黑毛四照花 Shrubs or trees, evergreen

15.5 (7.3~19.2); 5.8 (-2.0~13.3); 24.4 (15.6~27.8); 15.6 (7.5~19.4); 128.1 (48.2~172.6); 1.0 (0.0~20.4); 816 (537~939) ❙ 1195 (955~1547); 575 (458~911); 63 (27~106) ❙ 47.4 (16.1~89.0); 816 (537~939); 228.2 (126.1~551.3)

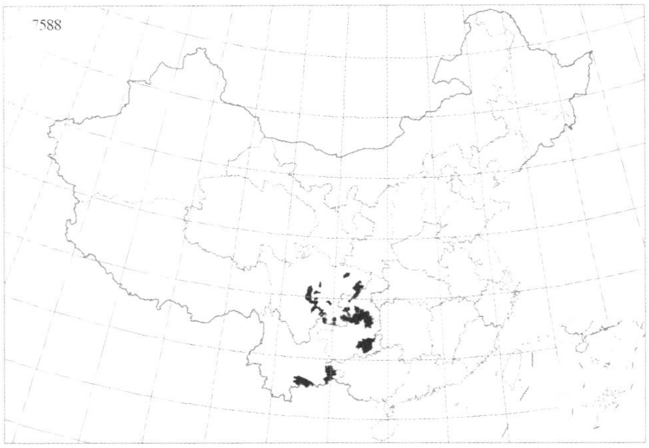

Cornaceae 山茱萸科 *Cornus* 山茱萸属

Cornus hongkongensis subsp. tonkinensis (W. P. Fang) Q. Y. Xiang 东京四照花 Shrubs or small trees, evergreen

16.3 (11.7~21.6); 7.7 (2.7~13.3); 23.4 (19.6~27.9); 16.3 (11.7~21.6); 136.5 (84.6~199.6); 0.4 (0.0~3.8); 830 (661~1143) ❙ 1182 (964~1547); 630 (488~911); 57 (39~102) ❙ 43.0 (19.4~74.2); 829 (661~1143); 277.0 (142.9~551.3)

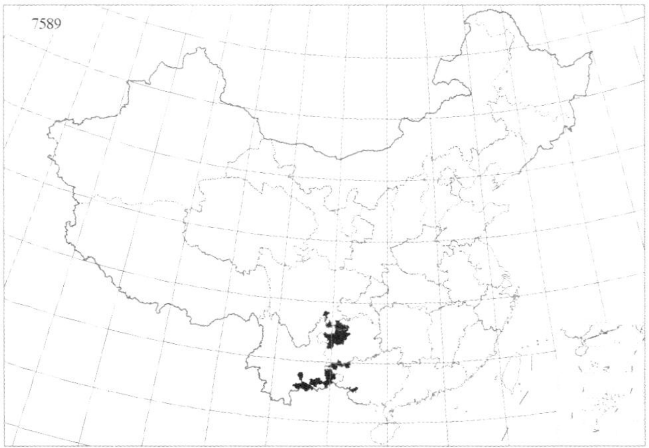

Cornaceae 山茱萸科 *Cornus* 山茱萸属

Cornus koehneana Wangerin 川陕梾木
Trees, deciduous

11.4 (2.0~14.9); -3.3 (-7.2~0.2); 24.5 (10.1~28.3); 12.0 (3.8~14.9); 98.5 (15.4~129.3); 21.5 (10.3~51.4); 735 (413~887) ❙ 616 (533~943); 348 (269~507); 22 (18~27) ❙ -13.1 (-34.1~128.6); 595 (413~649); 163.8 (118.4~270.7)

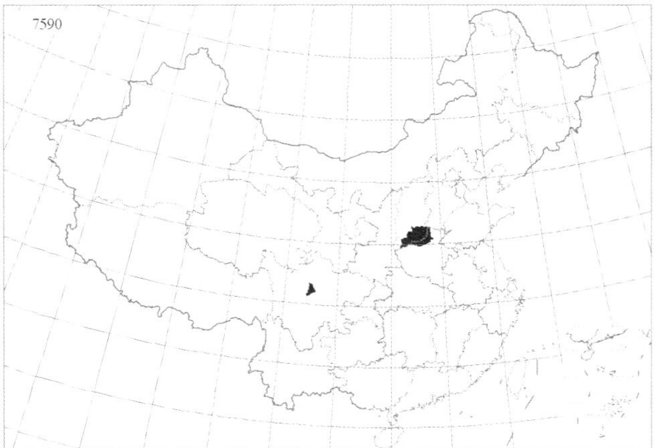

Cornaceae 山茱萸科 *Cornus* 山茱萸属
Cornus kousa subsp. chinensis (Osborn) Q. Y. Xiang 四照花
Trees or shrubs, deciduous

14.9 (7.4~22.2); 3.1 (-5.4~17.7); 25.9 (18.1~29.8); 15.0 (8.2~22.2); 124.0 (58.8~206.7); 4.8 (0.0~29.9); 824 (561~1123) ❙ 1192 (520~3495); 485 (258~1847); 107 (14~897) ❙ 43.5 (-36.5~235.6); 796 (520~1123); 208.0 (111.9~334.4)

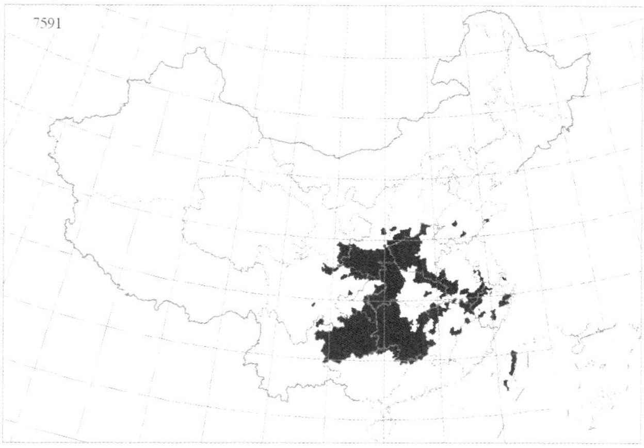

Cornaceae 山茱萸科 *Cornus* 山茱萸属
Cornus macrophylla Wall. 梾木
Trees, deciduous

14.1 (-0.7~21.7); 2.3 (-11.8~15.3); 25.0 (7.7~29.8); 14.3 (2.5~21.7); 117.5 (6.8~200.2); 7.7 (0.0~74.2); 796 (338~1118) ❙ 1120 (341~2680); 483 (190~1186); 91 (8~303) ❙ 39.8 (-40.9~246.5); 769 (338~1075); 208.7 (61.2~551.3)

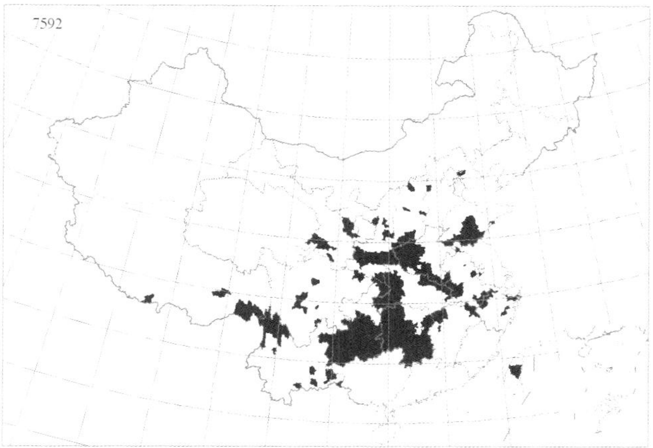

Cornaceae 山茱萸科 *Cornus* 山茱萸属
Cornus macrophylla var. stracheyi C. B. Clarke 密毛梾木
Shrubs or trees, deciduous

5.7 (-0.7~10.8); -2.2 (-9.6~3.9); 12.6 (7.7~16.6); 6.5 (2.5~10.9); 37.1 (6.8~71.8); 28.6 (1.3~74.2); 495 (338~633) ❙ 826 (370~1041); 492 (262~594); 25 (9~45) ❙ 64.7 (4.7~125.9); 495 (338~633); 266.3 (61.6~352.3)

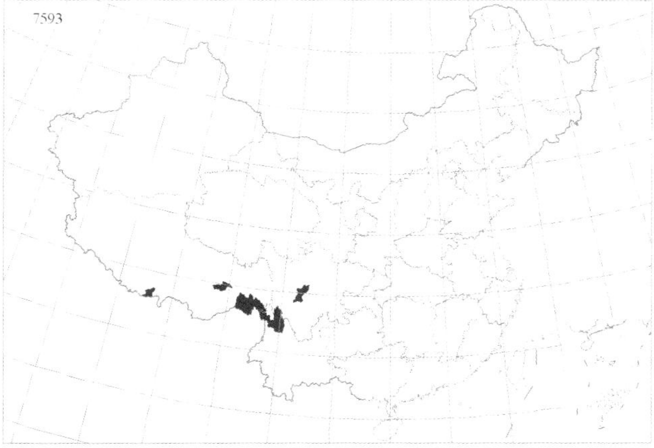

Cornaceae 山茱萸科 *Cornus* 山茱萸属
Cornus multinervosa (Pojarkova) Q. Y. Xiang 多脉四照花
Trees, deciduous

12.1 (2.2~18.0); 0.8 (-6.6~7.2); 22.8 (9.9~28.3); 12.3 (3.8~18.0); 96.5 (15.5~156.3); 10.6 (0.0~49.1); 710 (413~958) ❙ 939 (634~1621); 451 (290~817); 42 (16~195) ❙ 33.2 (-16.6~125.9); 707 (413~958); 250.6 (128.7~334.4)

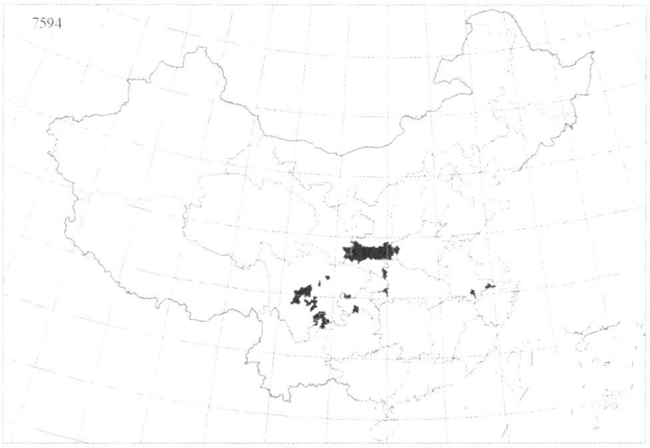

Cornaceae 山茱萸科 *Cornus* 山茱萸属
Cornus oblonga Wall. 长圆叶梾木
Trees, evergreen

14.3 (1.8~19.2); 5.2 (-7.1~13.8); 22.2 (9.9~28.9); 14.3 (3.8~19.2); 115.2 (15.5~170.5); 3.7 (0.0~53.3); 769 (404~1002) ❙ 1095 (583~1736); 553 (348~1007); 49 (12~113) ❙ 45.5 (-14.2~145.8); 756 (404~1002); 249.0 (108.2~513.9)

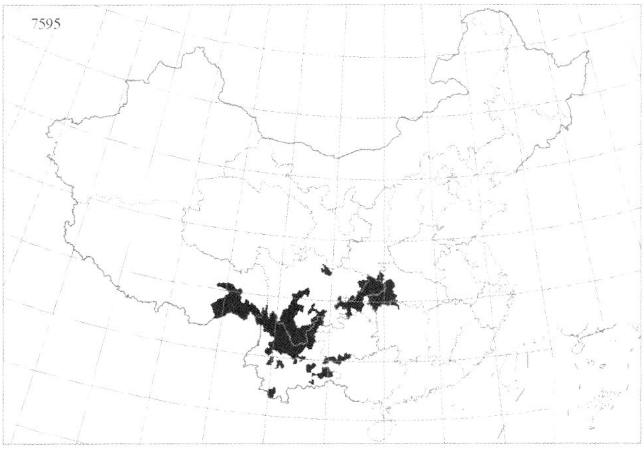

Cornaceae 山茱萸科 *Cornus* 山茱萸属
Cornus oblonga var. glabrescens W. P. Fang et W. K. Hu 无毛长圆叶梾木
Trees, evergreen

8.4 (5.3~16.1); 0.9 (-2.4~8.6); 14.8 (12.2~21.9); 8.6 (5.7~16.1); 56.3 (30.5~133.0); 14.8 (0.0~26.2); 568 (479~797) ❙ 913 (741~1041); 548 (426~630); 26 (12~45) ❙ 63.6 (27.3~100.7); 557 (479~719); 300.6 (234.2~352.3)

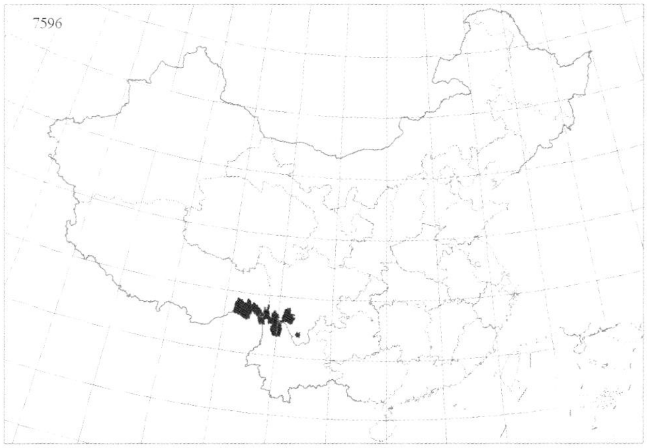

Cornaceae 山茱萸科 *Cornus* 山茱萸属

Cornus oblonga var. griffithii C. B. Clarke 毛叶梾木

Trees, evergreen

12.0 (5.3~18.5); 4.0 (−2.4~10.5); 18.8 (12.2~27.8); 12.1 (5.7~18.3); 91.0 (30.5~160.0); 6.6 (0.0~26.2); 681 (479~937) ❙ 1015 (801~1410); 583 (420~816); 31 (12~63) ❙ 52.6 (5.5~100.7); 665 (479~937); 292.2 (126.1~382.6)

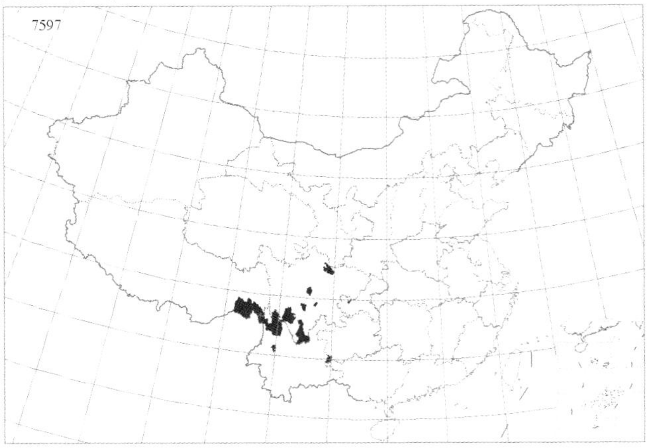

Cornaceae 山茱萸科 *Cornus* 山茱萸属

Cornus officinalis Sieb. et Zucc. 山茱萸

Shrubs or trees, deciduous

13.9 (5.4~18.7); 1.6 (−4.2~8.4); 25.5 (12.3~29.4); 14.0 (5.8~18.8); 115.4 (31.5~166.0); 8.0 (0.0~29.9); 797 (481~1002) ❙ 1032 (469~1922); 460 (316~661); 77 (13~259) ❙ 29.5 (−43.3~112.2); 770 (469~1002); 210.9 (108.2~352.3)

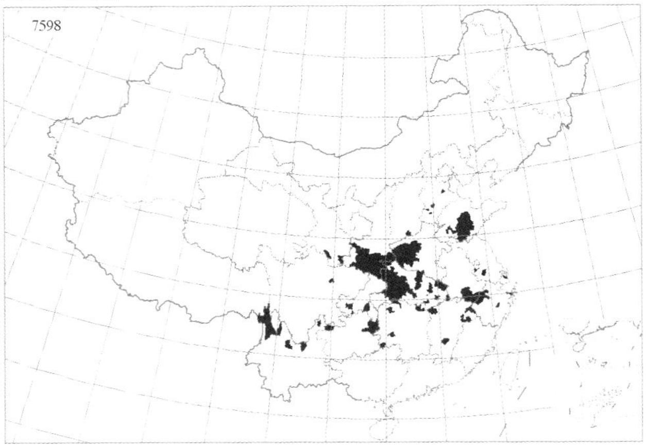

Cornaceae 山茱萸科 *Cornus* 山茱萸属

Cornus oligophlebia Merr. 樟叶梾木

Trees, deciduous

17.4; 9.7; 22.9; 17.5; 149.7; 0.0; 848 ❙ 1338; 756; 50 ❙ 57.6; 848; 348.0

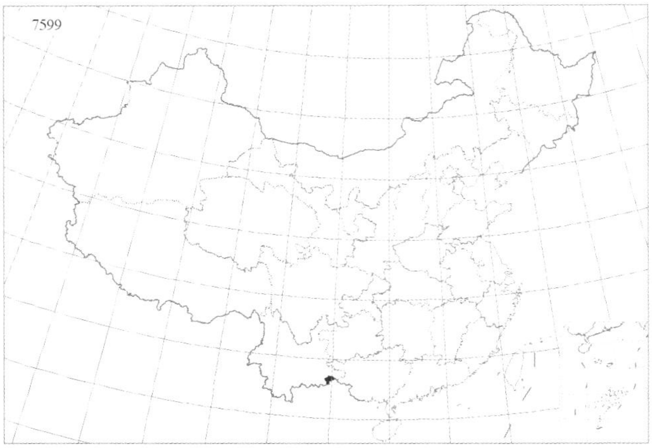

Cornaceae 山茱萸科 *Cornus* 山茱萸属

Cornus papillosa W. P. Fang et W. K. Hu 乳突梾木

Trees, deciduous

13.8; 5.3; 21.2; 13.9; 107.0; 0.2; 731 ❙ 1188; 705; 36 ❙ 61.9; 722; 244.4

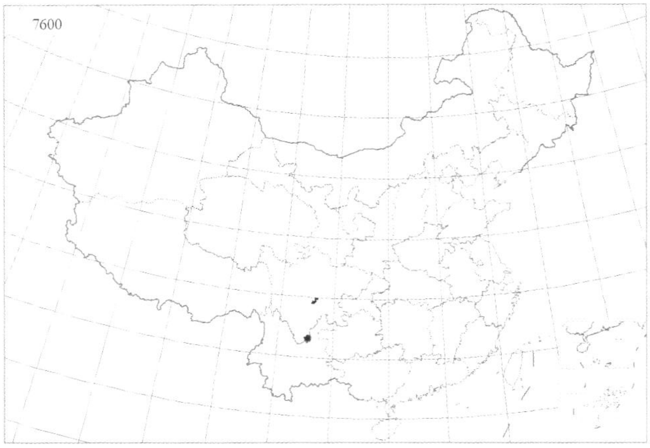

Cornaceae 山茱萸科 *Cornus* 山茱萸属

Cornus parviflora S. S. Chien 小花梾木

Shrubs or trees, deciduous

17.5 (15.1~19.6); 8.0 (5.0~11.2); 25.3 (24.1~26.6); 17.6 (15.2~19.6); 150.7 (121.9~175.5); 0.0 (0.0~0.0); 892 (788~986) ❙ 1307 (1227~1438); 639 (517~820); 80 (57~107) ❙ 47.0 (30.5~62.5); 892 (788~986); 269.9 (203.1~371.7)

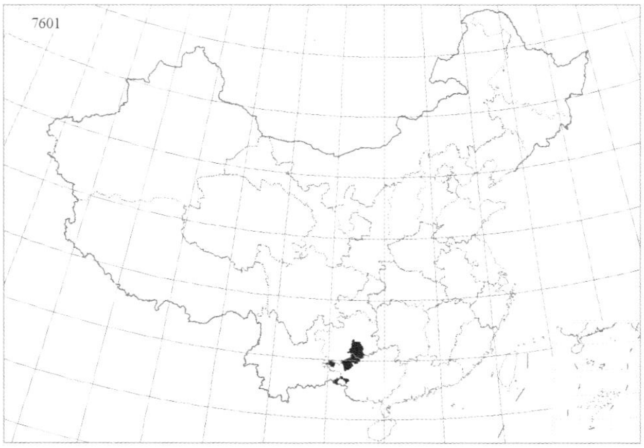

Cornaceae 山茱萸科 *Cornus* 山茱萸属

Cornus quinquenervis Franch. 小梾木

Shrubs, deciduous

15.1 (5.9~22.1); 4.3 (−6.9~13.7); 25.0 (17.9~29.2); 15.1 (7.3~22.1); 124.6 (51.7~205.6); 3.2 (0.0~41.0); 815 (529~1182) ❙ 1206 (447~1922); 520 (243~817); 91 (10~259) ❙ 47.5 (−22.2~128.9); 806 (447~1182); 220.4 (102.7~411.1)

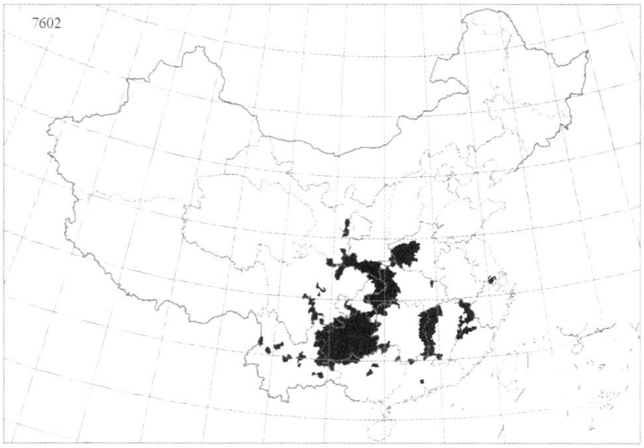

Cornaceae 山茱萸科 *Cornus* 山茱萸属

Cornus schindleri Wangerin 康定株木

Shrubs or small trees, deciduous

6.7 (-0.7~15.6); -1.8 (-9.6~6.1); 14.1 (7.7~24.6); 7.4 (2.5~15.6); 46.4 (6.8~127.5); 25.4 (0.0~74.2); 527 (338~808) ❙ 939 (370~1736); 532 (262~1007); 26 (9~69) ❙ 79.0 (4.7~145.8); 527 (338~808); 276.8 (61.6~447.4)

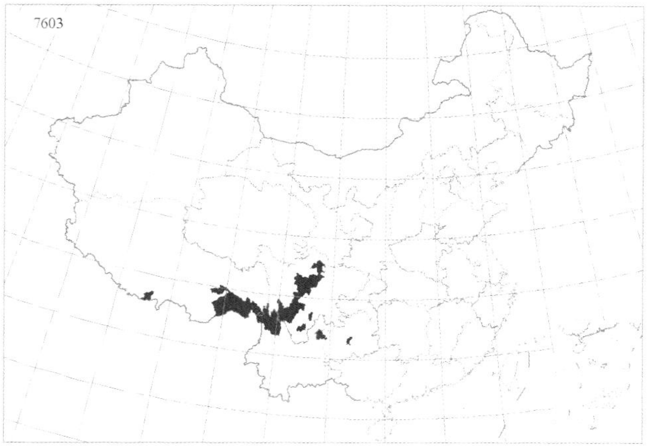

Cornaceae 山茱萸科 *Cornus* 山茱萸属

Cornus schindleri subsp. poliophylla (C. K. Schneid. et Wangerin) Q. Y. Xiang 灰叶株木

Shrubs or small trees, deciduous

10.5 (2.0~16.7); 0.4 (-7.2~10.3); 19.7 (9.9~28.0); 10.8 (3.8~16.8); 80.0 (15.4~141.5); 13.6 (0.0~51.4); 649 (413~888) ❙ 1020 (585~1483); 512 (297~817); 42 (15~183) ❙ 60.9 (-15.1~128.6); 645 (413~883); 256.1 (150.8~357.7)

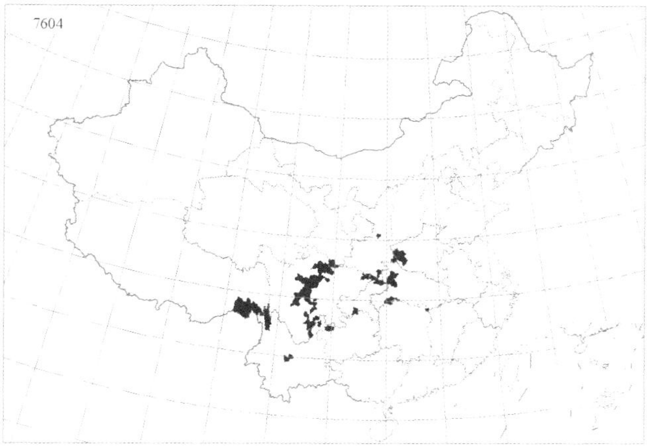

Cornaceae 山茱萸科 *Cornus* 山茱萸属

Cornus ulotricha C. K. Schneid. et Wangerin 卷毛株木

Trees, deciduous

10.7 (2.9~15.9); -0.6 (-6.7~5.5); 21.2 (11.2~25.8); 11.0 (4.4~15.9); 83.9 (20.5~131.1); 15.4 (0.0~44.6); 665 (428~833) ❙ 920 (373~1478); 438 (201~592); 42 (7~85) ❙ 41.5 (-37.6~121.7); 653 (373~833); 242.1 (88.1~351.7)

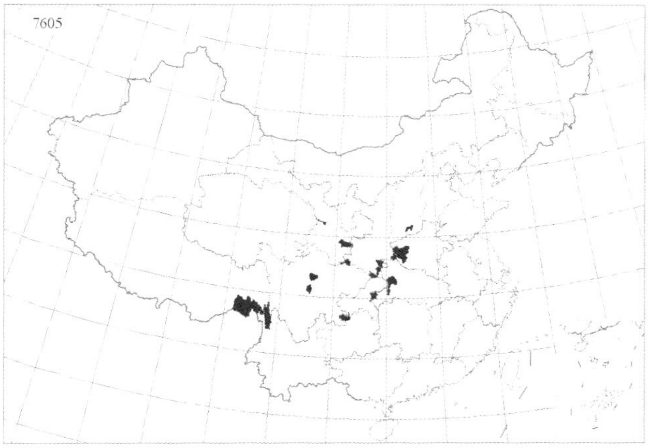

Cornaceae 山茱萸科 *Cornus* 山茱萸属

Cornus walteri Wangerin 毛株

Trees, deciduous

12.2 (1.9~19.2); -1.5 (-13.7~9.9); 24.7 (11.2~29.3); 12.7 (4.3~19.2); 105.0 (20.5~170.8); 18.0 (0.0~71.7); 755 (427~1002) ❙ 835 (367~1997); 413 (235~816); 55 (8~259) ❙ 9.1 (-47.7~130.7); 674 (367~1002); 192.2 (99.3~407.1)

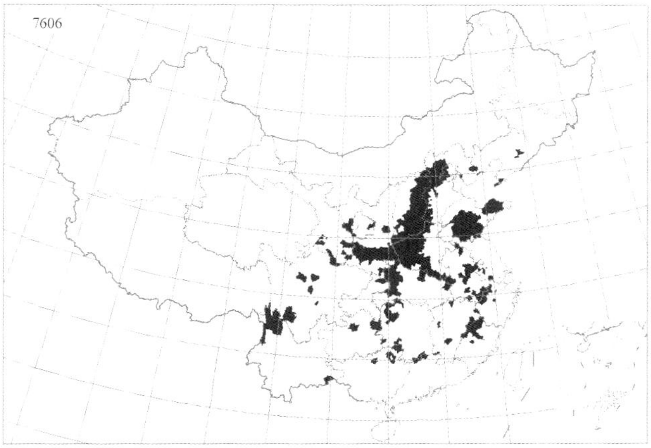

Cornaceae 山茱萸科 *Cornus* 山茱萸属

Cornus wilsoniana Wangerin 光皮株木

Trees, deciduous

16.1 (9.3~21.2); 4.8 (-2.9~12.5); 26.7 (20.1~29.7); 16.2 (9.5~21.2); 136.4 (69.8~194.6); 2.2 (0.0~21.0); 871 (603~1114) ❙ 1324 (626~1821); 515 (306~820); 129 (15~237) ❙ 51.7 (-22.2~128.9); 859 (603~1114); 205.6 (123.1~371.7)

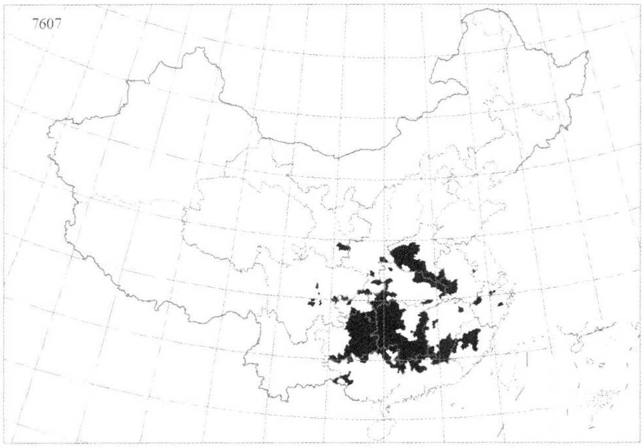

Aucubaceae 桃叶珊瑚科 *Aucuba* 桃叶珊瑚属

Aucuba albopunctifolia Wang 斑叶珊瑚

Shrubs or trees, evergreen

14.3 (9.3~17.8); 3.4 (-1.7~7.8); 24.7 (20.1~27.6); 14.3 (9.5~17.9); 116.0 (69.8~154.2); 4.1 (0.0~18.0); 781 (603~929) ❙ 1263 (1036~1674); 516 (462~611); 91 (46~181) ❙ 63.3 (18.3~101.4); 781 (603~929); 211.5 (149.6~284.3)

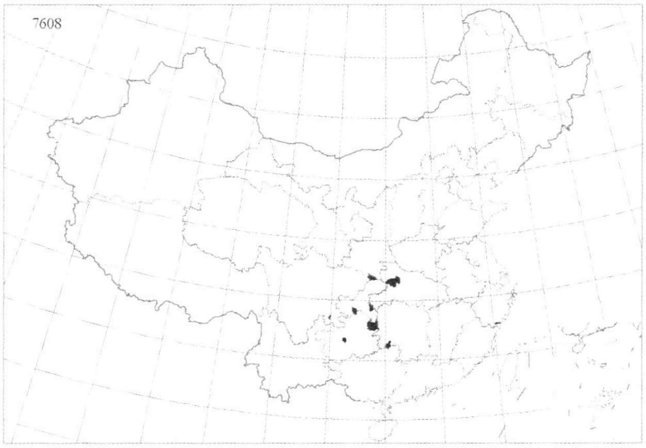

Aucubaceae 桃叶珊瑚科 *Aucuba* 桃叶珊瑚属
Aucuba albopunctifolia var. angustula W. P. Fang et Z. P. Song 窄斑叶珊瑚
Shrubs or trees, evergreen

15.3; 5.0; 25.2; 15.3; 124.0; 0.3; 804 ❙ 1646; 618; 163 ❙ 104.6; 804; 237.4

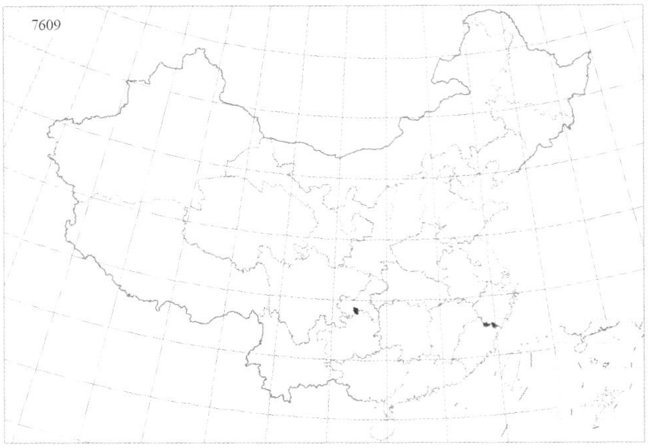

Aucubaceae 桃叶珊瑚科 *Aucuba* 桃叶珊瑚属
Aucuba chinensis Benth. 桃叶珊瑚
Shrubs, evergreen

17.9 (9.3~24.7); 8.5 (-1.7~19.8); 26.5 (20.1~29.2); 18.0 (9.5~24.7); 156.5 (69.8~236.5); 0.5 (0.0~18.0); 940 (603~1376) ❙ 1566 (807~3495); 619 (349~1847); 161 (24~897) ❙ 68.1 (-5.0~246.5); 931 (603~1257); 242.8 (119.4~620.0)

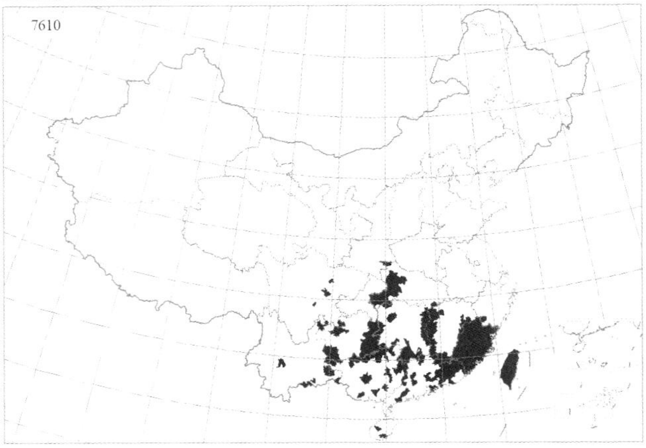

Aucubaceae 桃叶珊瑚科 *Aucuba* 桃叶珊瑚属
Aucuba chinensis var. angusta Wang 狭叶桃叶珊瑚
Shrubs or small trees, evergreen

19.1 (8.6~22.4); 11.2 (1.3~14.8); 25.4 (15.0~28.8); 19.1 (8.7~22.4); 170.7 (52.6~208.9); 1.0 (0.0~8.8); 1005 (574~1191) ❙ 1572 (1041~2037); 757 (538~1033); 106 (44~193) ❙ 57.0 (26.1~81.5); 1000 (574~1191); 271.7 (163.7~491.5)

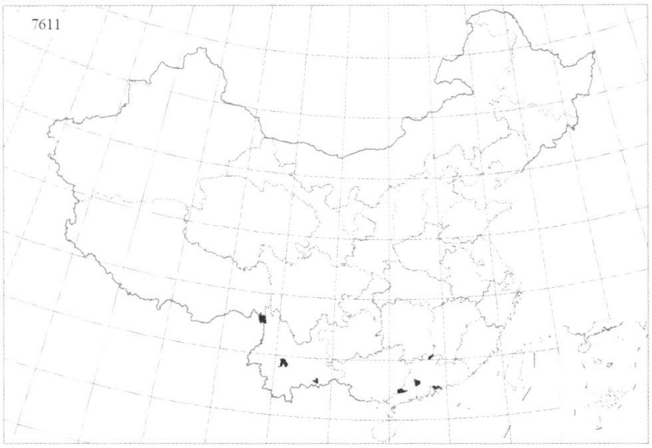

Aucubaceae 桃叶珊瑚科 *Aucuba* 桃叶珊瑚属
Aucuba chlorascens Wang 细齿桃叶珊瑚
Shrubs or small trees, evergreen

17.9 (11.3~21.5); 10.3 (3.8~14.2); 23.5 (17.0~27.4); 17.9 (11.3~21.6); 155.5 (76.6~199.1); 0.1 (0.0~1.5); 892 (644~1131) ❙ 1366 (1028~1776); 729 (544~911); 72 (45~191) ❙ 54.5 (12.5~101.8); 887 (644~1131); 366.7 (225.9~551.6)

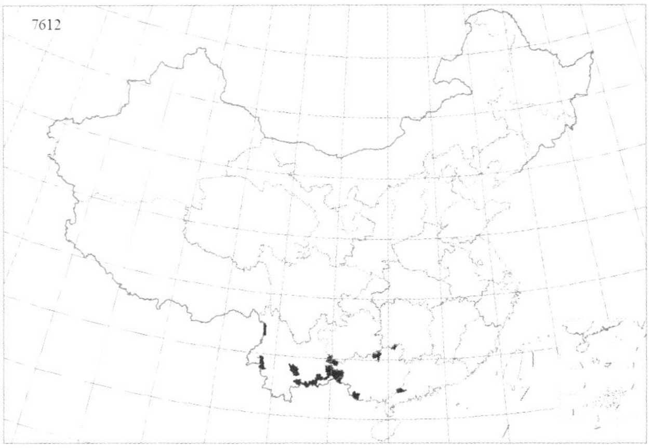

Aucubaceae 桃叶珊瑚科 *Aucuba* 桃叶珊瑚属
Aucuba confertiflora W. P. Fang et Z. P. Song 密花桃叶珊瑚
Small trees, evergreen

17.4; 9.5; 23.2; 17.5; 150.0; 0.0; 853 ❙ 1291; 723; 51 ❙ 51.3; 853; 337.0

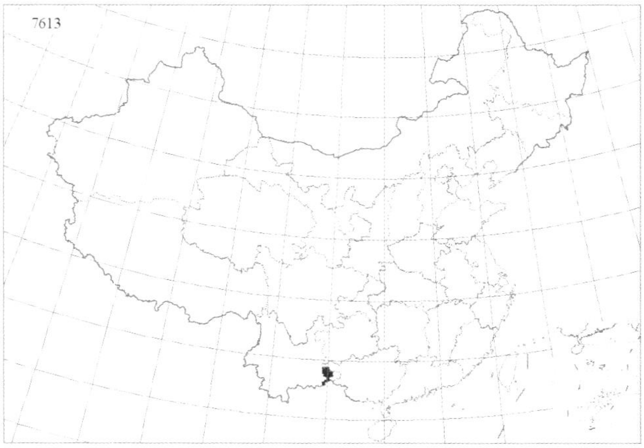

Aucubaceae 桃叶珊瑚科 *Aucuba* 桃叶珊瑚属
Aucuba eriobotryifolia Wang 琵琶叶珊瑚
Trees, evergreen

17.3 (14.9~18.8); 10.7 (8.2~12.7); 21.8 (20.1~23.2); 17.4 (15.0~18.9); 148.6 (119.8~166.2); 0.0 (0.0~0.0); 839 (748~900) ❙ 1229 (956~1530); 712 (562~909); 47 (34~55) ❙ 45.9 (20.8~78.9); 819 (716~867); 375.1 (272.5~482.8)

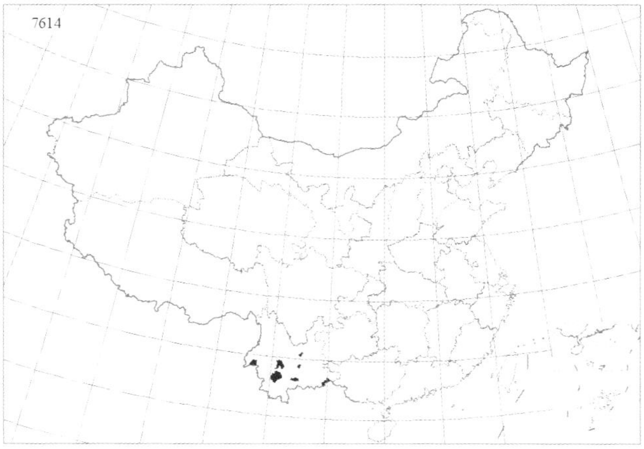

Aucubaceae 桃叶珊瑚科 *Aucuba* 桃叶珊瑚属
Aucuba filicauda Chun et F. C. How 纤尾桃叶珊瑚
Shrubs, evergreen

16.6 (14.3~21.0); 6.7 (3.8~14.2); 25.1 (21.4~27.1); 16.6 (14.3~21.0); 139.9 (112.6~191.8); 0.2 (0.0~1.5); 856 (756~1053) ❙ 1382 (1190~1740); 598 (452~911); 115 (50~191) ❙ 61.7 (44.5~101.8); 856 (756~1053); 271.1 (188.8~551.6)

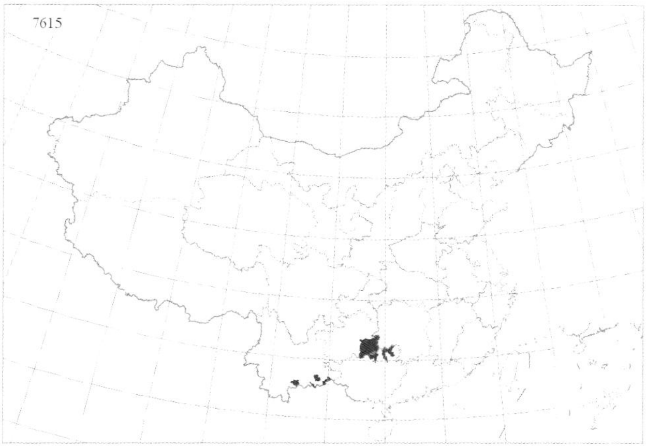

Aucubaceae 桃叶珊瑚科 *Aucuba* 桃叶珊瑚属
Aucuba filicauda var. pauciflora W. P. Fang et Z. P. Song 少花桃叶珊瑚
Shrubs, evergreen

16.3 (15.0~19.1); 6.4 (4.4~10.1); 25.1 (23.4~26.6); 16.3 (15.0~19.1); 136.4 (121.0~169.2); 0.2 (0.0~0.6); 845 (792~969) ❙ 1179 (1072~1334); 560 (477~720); 65 (51~77) ❙ 39.8 (27.5~59.6); 845 (792~969); 204.3 (148.9~283.9)

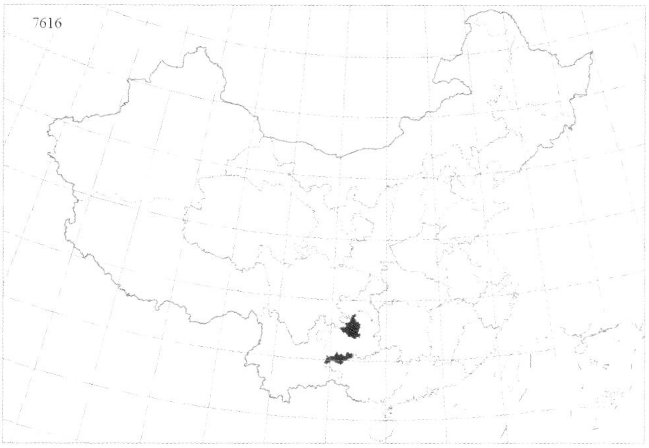

Aucubaceae 桃叶珊瑚科 *Aucuba* 桃叶珊瑚属
Aucuba himalaica Hook. f. et Thomson 喜马拉雅珊瑚
Shrubs or small trees, evergreen

15.6 (2.6~22.3); 5.1 (-7.0~13.5); 25.3 (11.2~29.3); 15.7 (4.3~22.3); 129.5 (20.0~203.1); 1.6 (0.0~48.2); 835 (427~1202) ❙ 1260 (815~1829); 534 (352~1007); 94 (19~230) ❙ 52.1 (7.2~145.8); 833 (427~1202); 212.8 (108.2~487.2)

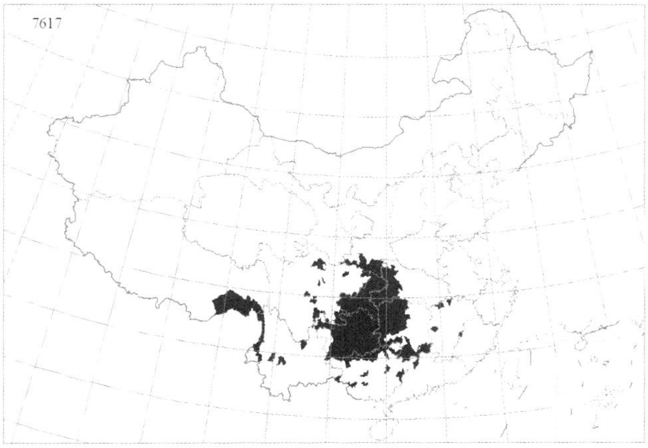

Aucubaceae 桃叶珊瑚科 *Aucuba* 桃叶珊瑚属
Aucuba himalaica var. dolichophylla W. P. Fang et Z. P. Song 长叶珊瑚
Shrubs or small trees, evergreen

15.0 (2.6~19.1); 4.7 (-7.0~10.1); 24.5 (11.2~28.1); 15.1 (4.3~19.1); 124.0 (20.0~169.2); 3.2 (0.0~48.2); 808 (427~969) ❙ 1296 (829~1950); 570 (400~816); 86 (23~226) ❙ 62.4 (11.7~147.0); 808 (427~969); 221.8 (126.1~284.3)

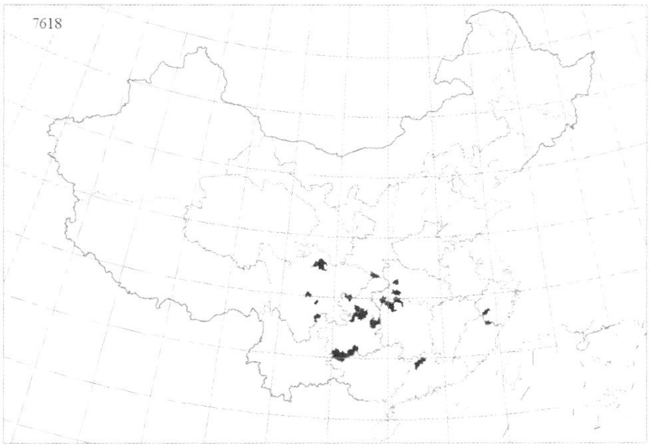

Aucubaceae 桃叶珊瑚科 *Aucuba* 桃叶珊瑚属
Aucuba himalaica var. oblanceolata W. P. Fang et Z. P. Song 倒披针叶珊瑚
Shrubs or small trees, evergreen

17.8 (14.9~21.0); 8.8 (5.3~13.2); 25.3 (22.5~27.7); 17.8 (15.0~21.0); 153.9 (119.6~192.3); 0.0 (0.0~0.0); 911 (782~1083) ❙ 1483 (1076~2005); 689 (513~828); 109 (48~182) ❙ 61.9 (36.4~85.1); 911 (782~1083); 270.6 (197.5~491.5)

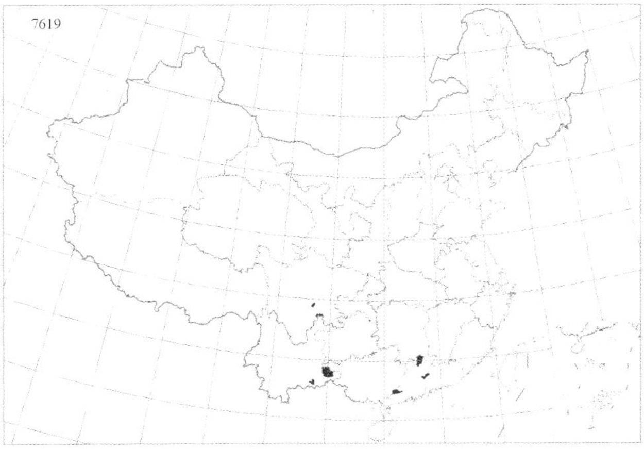

Aucubaceae 桃叶珊瑚科 *Aucuba* 桃叶珊瑚属
Aucuba himalaica var. pilossima W. P. Fang et Z. P. Song 密毛桃叶珊瑚
Shrubs or small trees, evergreen

14.0 (9.3~19.1); 3.2 (-1.7~9.8); 24.3 (20.1~27.5); 14.1 (9.5~19.2); 112.9 (69.8~170.1); 4.1 (0.0~18.0); 769 (603~981) ❙ 1315 (993~1492); 569 (420~816); 84 (46~140) ❙ 72.3 (30.9~107.2); 769 (603~981); 239.1 (187.6~322.1)

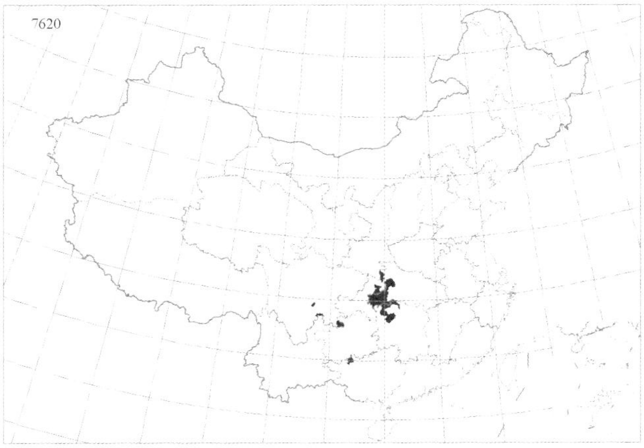

Aucubaceae 桃叶珊瑚科 *Aucuba* 桃叶珊瑚属
Aucuba japonica Thunb. 青木
Shrubs, evergreen

19.0 (15.4~24.1); 12.0 (4.9~18.9); 25.2 (20.1~28.0); 19.0 (15.4~24.1); 168.2 (124.9~229.7); 0.0 (0.0~0.1); 963 (773~1322) ❙ 2380 (1544~3495); 1020 (525~1847); 277 (47~897) ❙ 151.9 (32.5~246.5); 945 (773~1123); 250.4 (154.2~303.6)

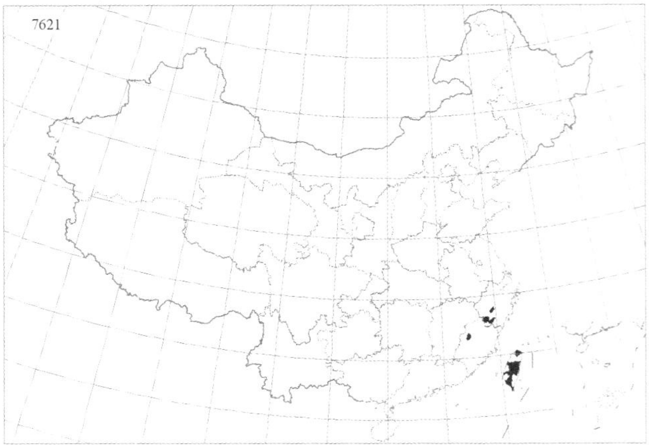

Aucubaceae 桃叶珊瑚科 *Aucuba* 桃叶珊瑚属
Aucuba japonica var. variegata Dombrain 花叶青木
Shrubs, evergreen

16.9; 6.3; 26.6; 16.9; 142.8; 0.0; 892 ❙ 1027; 516; 45 ❙ 15.1; 891; 129.9

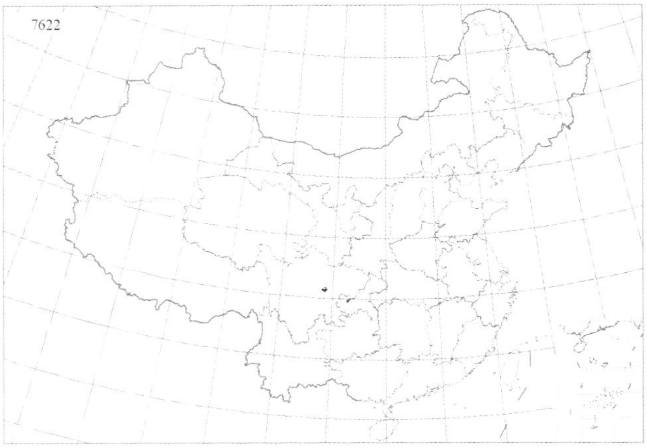

Aucubaceae 桃叶珊瑚科 *Aucuba* 桃叶珊瑚属
Aucuba robusta W. P. Fang et T. P. Song 粗梗桃叶珊瑚
Shrubs, evergreen

20.5; 11.9; 27.1; 20.5; 186.5; 0.0; 1064 ❙ 1642; 881; 111 ❙ 53.5; 1064; 293.0

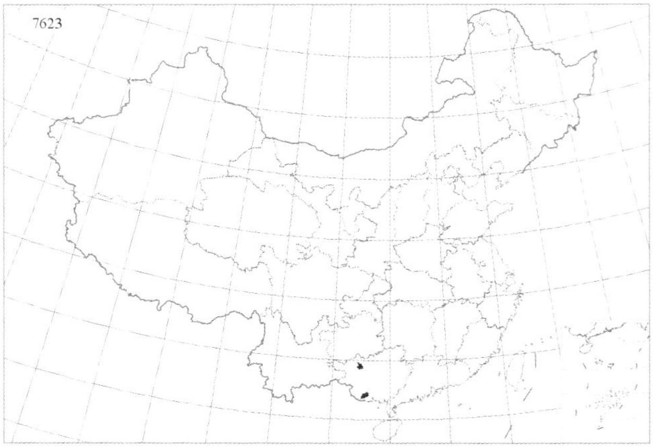

Helwingiaceae 青荚叶科 *Helwingia* 青荚叶属
Helwingia chinensis Batalin 中华青荚叶
Shrubs, evergreen

14.1 (2.9~19.6); 4.2 (−6.5~12.0); 23.1 (11.2~28.8); 14.2 (4.4~19.6); 114.1 (20.5~175.5); 4.2 (0.0~44.6); 766 (428~1038) ❙ 1118 (629~2499); 530 (303~947); 61 (14~303) ❙ 46.5 (−14.2~219.3); 761 (428~1038); 241.5 (120.7~491.5)

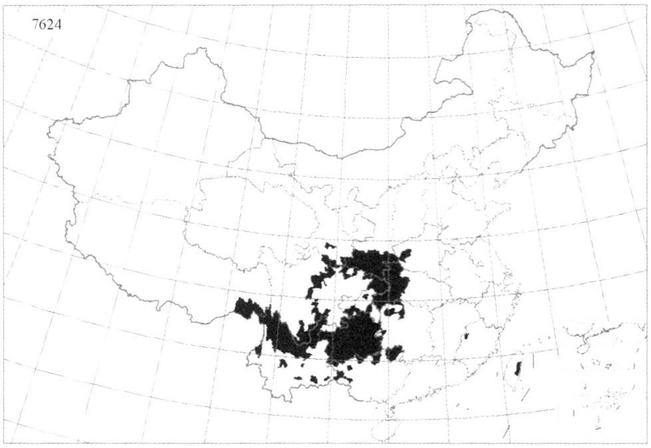

Helwingiaceae 青荚叶科 *Helwingia* 青荚叶属
Helwingia chinensis var. crenata (Lingelsh. ex H. Limpr.) Fang 钝齿青荚叶
Shrubs, evergreen

13.8 (2.9~21.0); 3.5 (−6.3~13.2); 23.2 (11.2~27.0); 13.9 (4.4~21.0); 111.6 (20.5~192.3); 5.7 (0.0~44.6); 769 (428~1081) ❙ 1213 (741~1776); 507 (426~828); 93 (25~146) ❙ 60.0 (41.7~121.7); 769 (428~1081); 236.3 (170.4~352.3)

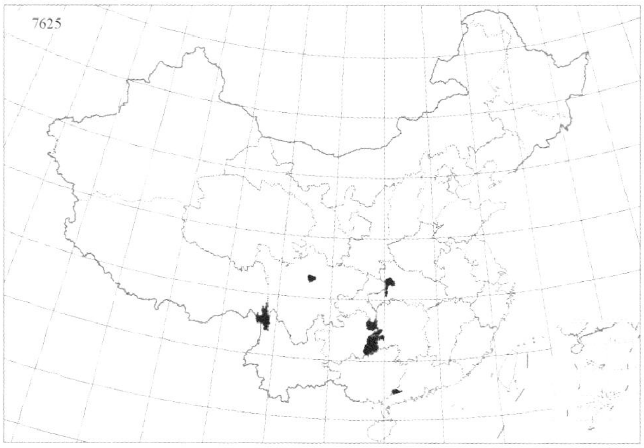

Helwingiaceae 青荚叶科 *Helwingia* 青荚叶属
Helwingia himalaica Hook. f. et Thomson ex C. B. Clarke 西域青荚叶
Shrubs, deciduous

15.2 (3.2~21.0); 5.9 (−6.4~14.2); 23.2 (11.4~29.3); 15.3 (4.7~21.0); 124.4 (22.7~192.3); 1.5 (0.0~43.8); 797 (436~1081) ❙ 1192 (741~1791); 574 (411~1007); 71 (15~198) ❙ 50.2 (−14.2~145.8); 791 (436~1081); 249.4 (130.1~551.6)

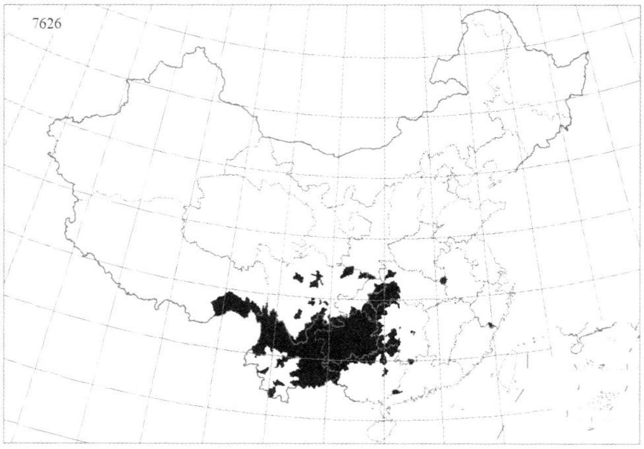

Helwingiaceae 青荚叶科 *Helwingia* 青荚叶属
Helwingia japonica (Thunb. ex Murray) F. Dietr. 青荚叶
Shrubs, deciduous

15.3 (1.8~21.4); 4.7 (-7.7~13.6); 25.2 (9.9~30.1); 15.4 (3.8~21.5); 128.3 (15.4~197.7); 3.7 (0.0~53.3); 835 (404~1128) ❙ 1293 (538~2859); 520 (295~959); 121 (9~402) ❙ 54.8 (-22.2~235.6); 822 (404~1114); 228.7 (123.9~491.5)

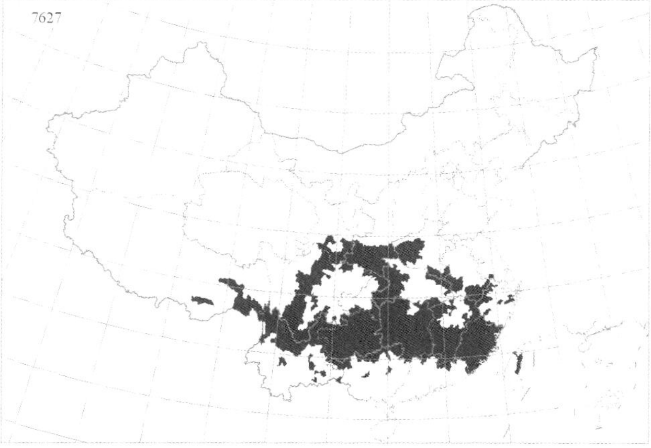

Helwingiaceae 青荚叶科 *Helwingia* 青荚叶属
Helwingia japonica var. hypoleuca Hemsl. ex Rehder 白粉青荚叶
Shrubs, evergreen

12.8 (2.0~16.4); 2.3 (-7.2~5.8); 22.6 (10.1~26.9); 12.9 (3.8~16.4); 100.7 (15.4~136.6); 7.0 (0.0~51.4); 725 (413~871) ❙ 1188 (853~1813); 528 (421~816); 72 (18~226) ❙ 65.9 (18.6~128.6); 725 (413~871); 229.7 (150.8~298.0)

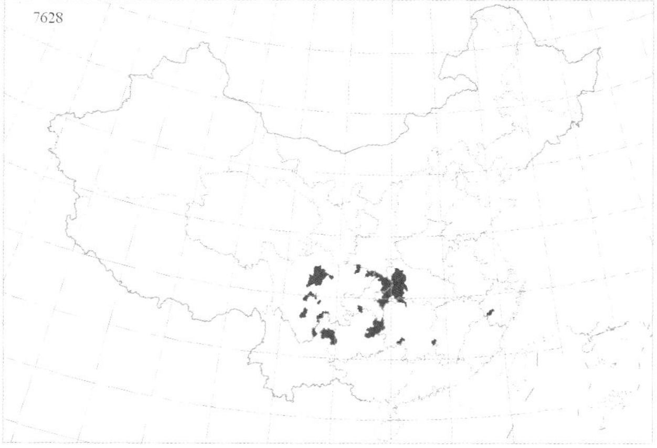

Helwingiaceae 青荚叶科 *Helwingia* 青荚叶属
Helwingia japonica var. papillosa W. P. Fang et Z. P. Song 乳突青荚叶
Shrubs, deciduous

7.3 (2.2~14.9); -2.1 (-7.0~5.3); 15.9 (9.9~23.9); 8.0 (3.8~15.0); 53.3 (15.5~119.6); 24.6 (0.0~49.1); 553 (413~782) ❙ 1018 (731~1414); 536 (346~817); 29 (19~49) ❙ 87.1 (32.5~125.9); 553 (413~782); 255.8 (200.8~293.1)

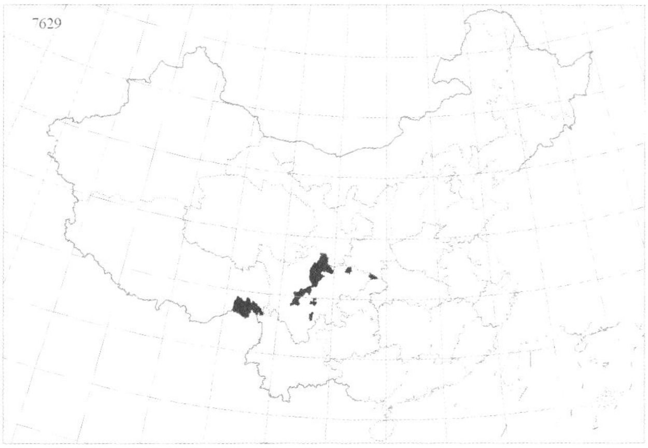

Helwingiaceae 青荚叶科 *Helwingia* 青荚叶属
Helwingia japonica var. zhejiangensis (W. P. Fang et T. P. Soong) M. B. Deng et Yo. Zhang 台湾青荚叶
Shrubs, evergreen

16.2 (15.0~17.6); 5.5 (4.0~7.4); 26.5 (24.7~27.9); 16.2 (15.1~17.6); 134.5 (120.8~151.2); 0.1 (0.0~1.0); 854 (789~920) ❙ 1782 (1573~1950); 637 (555~713); 207 (179~253) ❙ 109.8 (75.7~147.0); 854 (789~920); 241.9 (204.3~283.9)

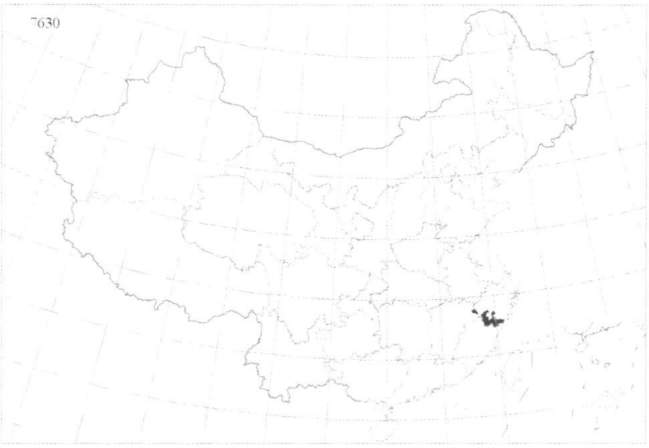

Helwingiaceae 青荚叶科 *Helwingia* 青荚叶属
Helwingia omeiensis (W. P. Fang) H. Hara et S. Kuros. 峨眉青荚叶
Shrubs or small trees, evergreen

14.7 (8.6~19.2); 4.0 (-1.7~9.4); 24.7 (15.0~28.1); 14.8 (8.7~19.2); 119.8 (52.6~170.8); 2.6 (0.0~18.0); 798 (574~1002) ❙ 1234 (822~1577); 533 (390~932); 86 (19~196) ❙ 55.8 (18.5~107.2); 798 (574~1002); 225.3 (135.1~407.1)

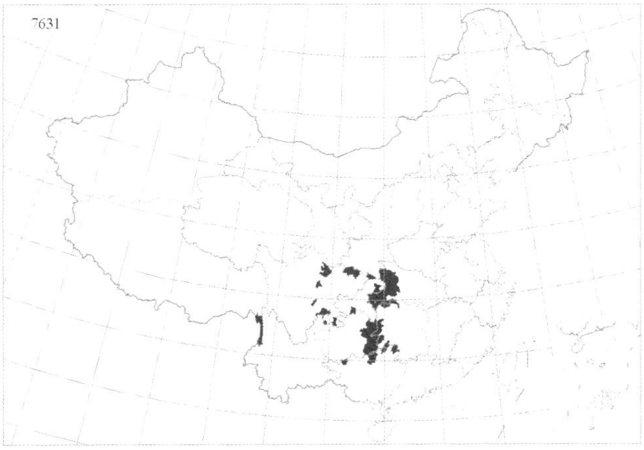

Mastixiaceae 单室茱萸科 *Mastixia* 单室茱萸属
Mastixia pentandra subsp. cambodiana (Pierre) K. M. Matthew 单室茱萸
Trees, evergreen

24.3 (23.9~24.8); 19.3 (18.7~20.1); 27.8 (27.2~28.4); 24.3 (23.9~24.8); 232.2 (226.8~237.6); 0.0 (0.0~0.0); 1339 (1294~1385) ❙ 1421 (1179~1588); 479 (392~539); 47 (41~58) ❙ 6.1 (-11.9~19.1); 1133 (1056~1212); 510.5 (455.2~620.0)

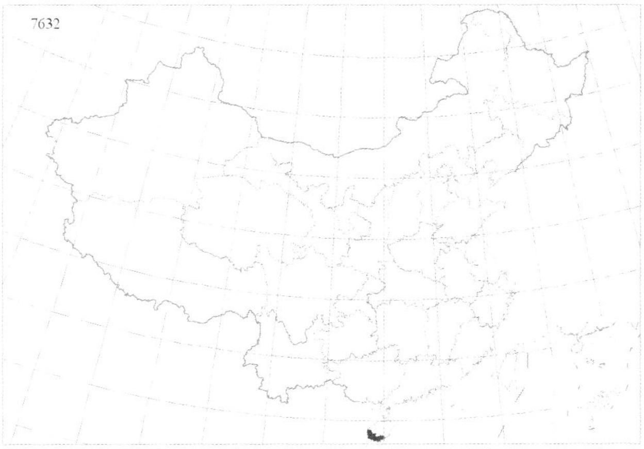

M

Toricelliaceae 鞘柄木科 *Toricellia* 鞘柄木属

Toricellia angulata Oliv. 角叶鞘柄木

Shrubs or small trees, deciduous

15.0 (5.3~20.3); 5.4 (-2.4~13.0); 23.5 (12.2~27.7); 15.0 (5.7~20.4); 121.9 (30.5~184.4); 1.7 (0.0~26.2); 791 (479~1065) ❙ 1175 (711~1736); 560 (357~1007); 68 (15~196) ❙ 49.2 (-14.2~145.8); 784 (479~1065); 239.1 (142.9~506.2)

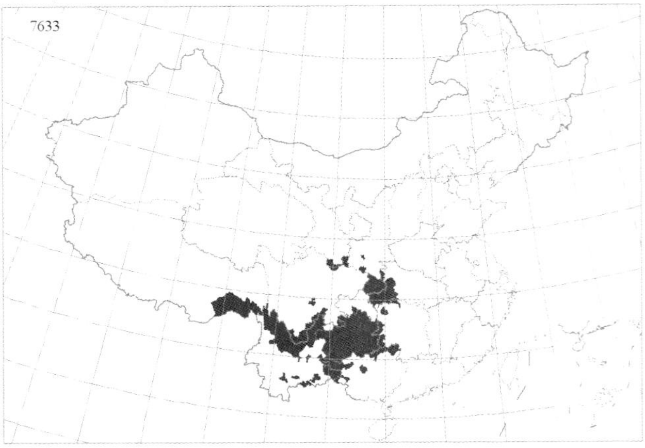

Toricelliaceae 鞘柄木科 *Toricellia* 鞘柄木属

Toricellia tiliifolia DC. 鞘柄木

Small trees, deciduous

16.0 (5.3~19.0); 8.5 (-2.4~13.1); 21.7 (12.2~25.9); 16.1 (5.7~19.1); 134.9 (30.5~168.8); 2.4 (0.0~26.2); 809 (479~909) ❙ 1286 (958~1577); 720 (531~932); 51 (18~107) ❙ 60.3 (13.3~100.7); 797 (479~891); 327.9 (167.5~505.1)

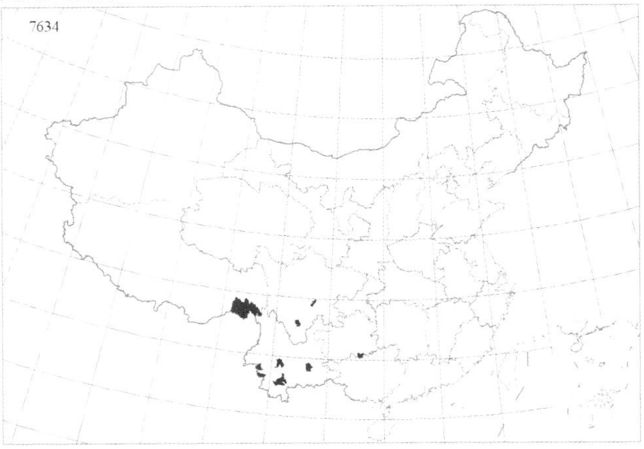

Clethraceae 桤叶树科 *Clethra* 山柳属

Clethra barbinervis Siebold et Zucc. 髭脉桤叶树

Shrubs or trees, deciduous

15.9 (9.3~19.4); 4.3 (-1.7~8.9); 27.0 (20.1~29.3); 15.9 (9.5~19.4); 132.9 (69.8~173.4); 1.8 (0.0~18.0); 864 (603~1029) ❙ 1615 (1202~1997); 587 (474~713); 190 (71~259) ❙ 87.6 (28.8~147.0); 863 (603~1029); 241.8 (169.3~303.4)

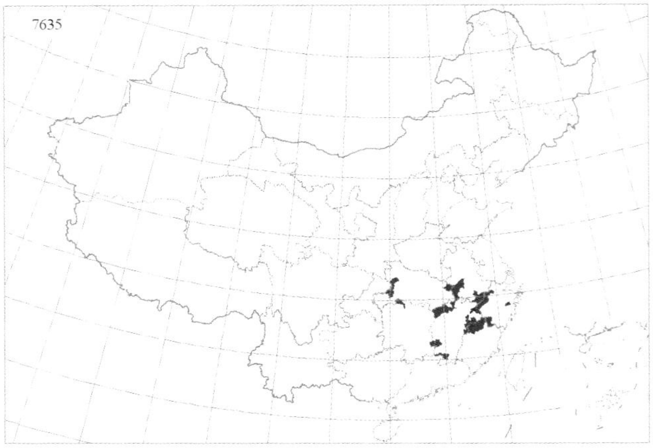

Clethraceae 桤叶树科 *Clethra* 山柳属

Clethra bodinieri H. Lév. 单毛桤叶树

Shrubs, evergreen

18.1 (14.3~24.1); 8.4 (3.8~18.7); 26.5 (22.4~28.7); 18.1 (14.3~24.2); 157.9 (112.6~230.1); 0.1 (0.0~1.5); 948 (756~1326) ❙ 1457 (1159~2435); 629 (444~1428); 121 (43~254) ❙ 54.3 (5.8~130.7); 946 (756~1326); 252.8 (146.0~572.9)

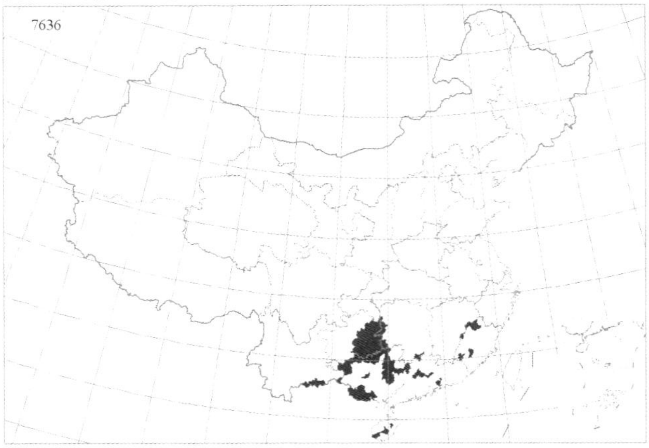

Clethraceae 桤叶树科 *Clethra* 山柳属

Clethra delavayi Franch. 云南桤叶树

Shrubs or small trees, deciduous

16.8 (6.1~22.7); 6.7 (-2.0~15.0); 26.0 (12.7~29.8); 16.9 (6.3~22.8); 143.0 (35.2~213.1); 0.6 (0.0~21.8); 891 (496~1221) ❙ 1449 (801~2435); 583 (392~1428); 140 (18~259) ❙ 63.2 (16.9~147.0); 886 (496~1221); 234.0 (134.6~516.7)

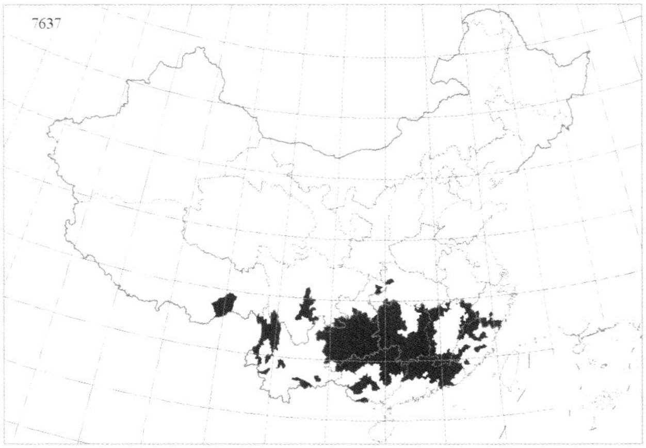

Clethraceae 桤叶树科 *Clethra* 山柳属

Clethra fabri Hance 华南桤叶树

Shrubs or trees, deciduous

18.4 (14.3~23.3); 8.8 (3.3~16.0); 26.5 (21.4~29.8); 18.4 (14.3~23.3); 161.3 (112.6~220.1); 0.2 (0.0~2.3); 960 (756~1264) ❙ 1462 (1018~2435); 647 (403~1428); 123 (43~230) ❙ 53.0 (5.8~106.4); 956 (756~1264); 257.0 (151.0~551.6)

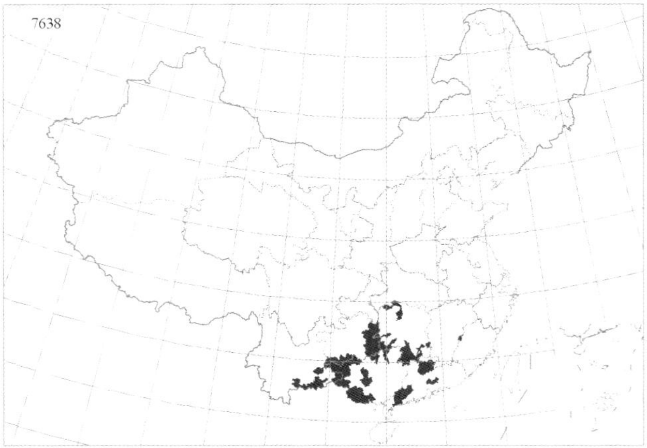

Clethraceae 桤叶树科 *Clethra* 山柳属
Clethra fargesii Franch. 城口桤叶树
Trees, deciduous

15.7 (9.3~19.0); 4.5 (-1.7~9.3); 26.3 (20.1~29.4); 15.8 (9.5~19.1); 130.8 (69.8~168.8); 1.4 (0.0~18.0); 847 (603~989) ❙ 1372 (911~1997); 519 (390~720); 135 (44~259) ❙ 62.2 (16.1~130.7); 841 (603~989); 210.7 (134.6~303.4)

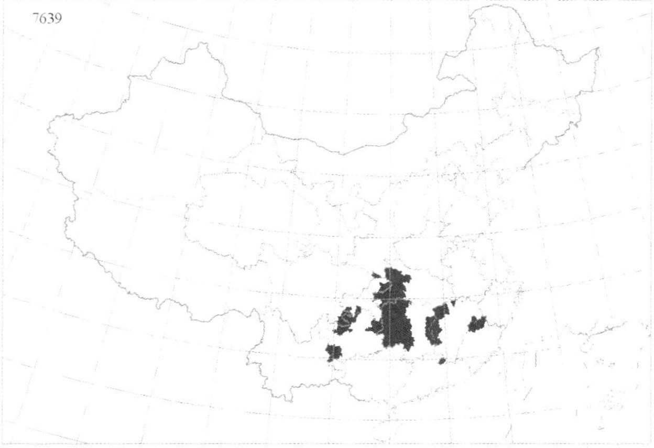

Clethraceae 桤叶树科 *Clethra* 山柳属
Clethra kaipoensis H. Lév. 贵州桤叶树
Shrubs or small trees, deciduous

16.8 (13.5~22.3); 5.9 (2.6~14.9); 26.9 (23.5~29.7); 16.9 (13.6~22.3); 142.9 (106.7~208.2); 0.3 (0.0~4.1); 893 (744~1186) ❙ 1450 (1072~2035); 536 (392~969); 152 (43~259) ❙ 62.3 (16.9~128.9); 887 (744~1186); 210.4 (134.6~354.5)

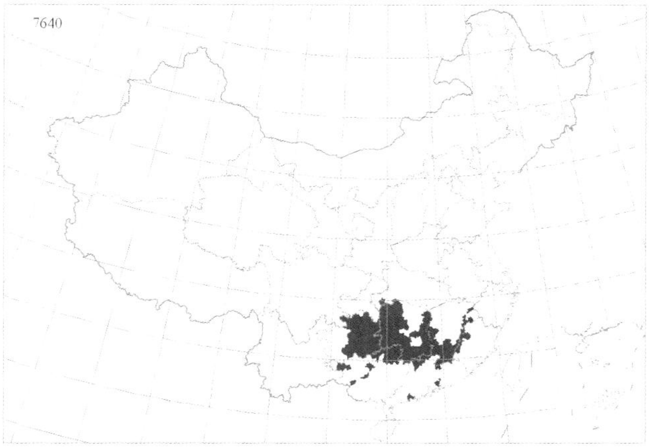

Clethraceae 桤叶树科 *Clethra* 山柳属
Clethra petelotii Dop et Troch.-Marq. 白背桤叶树
Shrubs or small trees, evergreen

19.4; 12.6; 24.1; 19.4; 172.6; 0.0; 954 ❙ 1445; 787; 64 ❙ 52.3; 954; 521.5

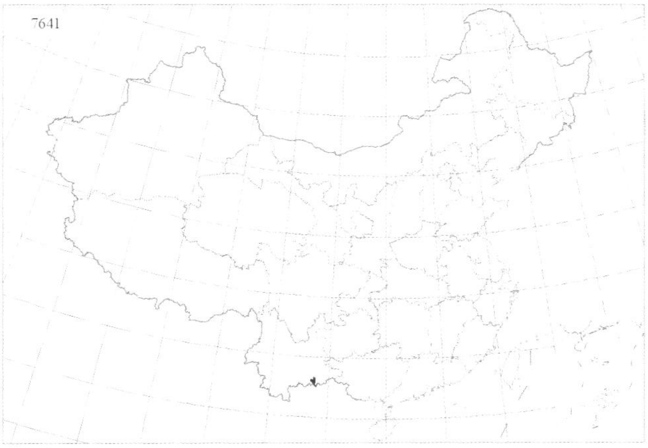

Clethraceae 桤叶树科 *Clethra* 山柳属
Clethra sleumeriana Hao 湖南桤叶树
Shrubs or small trees, deciduous

16.0 (14.2~19.1); 5.4 (3.3~10.1); 25.7 (20.8~28.6); 16.1 (14.2~19.1); 133.6 (110.9~169.2); 0.5 (0.0~2.3); 845 (734~969) ❙ 1303 (1168~1526); 536 (419~720); 107 (51~187) ❙ 54.8 (27.5~99.4); 841 (734~969); 209.0 (134.6~283.9)

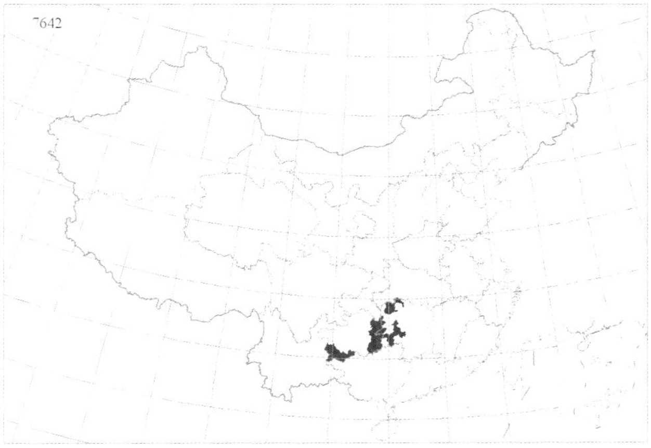

Ericaceae 杜鹃花科 *Agapetes* 树萝卜属
Agapetes aborensis Airy Shaw 阿波树萝卜
Shrubs, evergreen

5.3; -2.4; 12.2; 5.7; 30.5; 26.2; 479 ❙ 961; 558; 31 ❙ 100.7; 479; 234.2

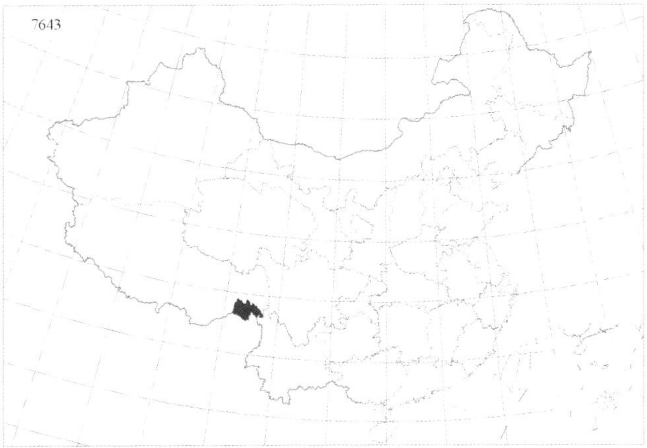

Ericaceae 杜鹃花科 *Agapetes* 树萝卜属
Agapetes angulata (Griff.) Hook. f. 棱枝树萝卜
Shrubs, evergreen

13.3; 5.4; 19.6; 13.4; 100.4; 0.0; 706 ❙ 1736; 1007; 51 ❙ 145.8; 706; 447.4

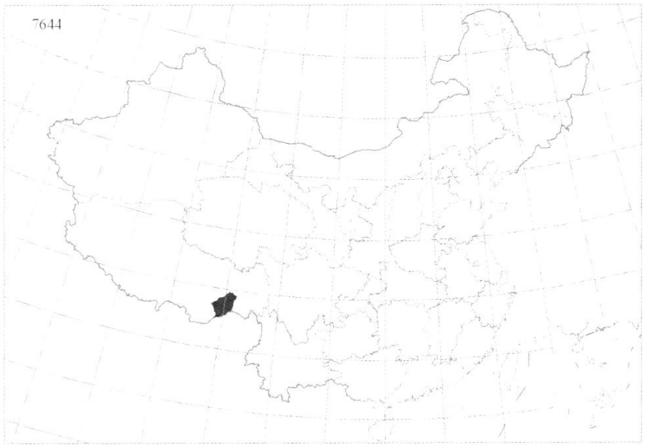

Ericaceae 杜鹃花科 *Agapetes* 树萝卜属
Agapetes anonyma Airy Shaw 锈毛树萝卜
Shrubs, evergreen

1.7 (-1.8~13.3); -7.3 (-10.5~5.4); 9.7 (6.1~19.6); 4.3 (1.8~13.4); 21.8 (1.9~100.4); 60.8 (0.0~82.8); 405 (329~706) ❙ 500 (255~1736); 324 (191~1007); 11 (1~51) ❙ 7.3 (-31.1~145.8); 358 (255~706); 137.2 (60.4~447.4)

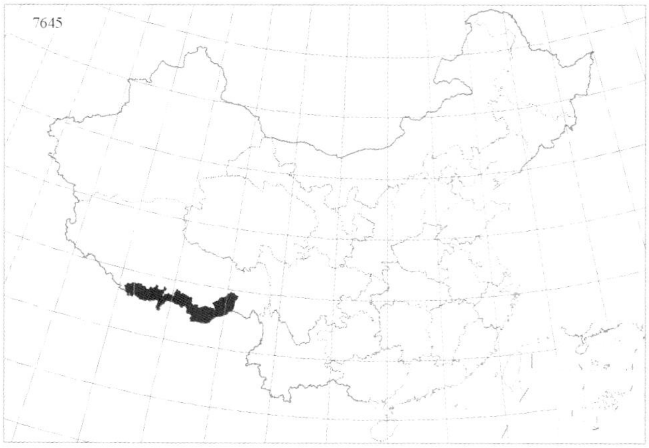

Ericaceae 杜鹃花科 *Agapetes* 树萝卜属
Agapetes brachypoda var. gracilis Airy Shaw 纤细短柄树萝卜
Shrubs, evergreen

13.2; 6.4; 18.4; 13.3; 99.4; 0.0; 695 ❙ 1337; 727; 69 ❙ 92.4; 695; 407.1

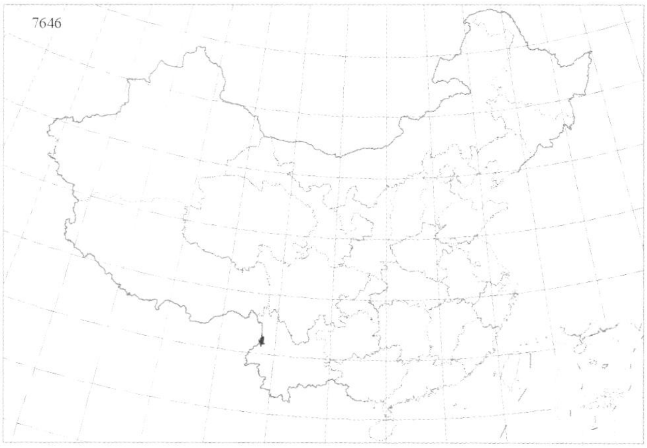

Ericaceae 杜鹃花科 *Agapetes* 树萝卜属
Agapetes brandisiana W. E. Evans 环萼树萝卜
Shrubs, evergreen

18.7 (17.8~20.5); 12.7 (11.1~14.9); 22.3 (21.6~23.7); 18.8 (17.8~20.5); 165.0 (154.1~185.9); 0.0 (0.0~0.0); 896 (854~992) ❙ 1459 (1366~1530); 767 (655~909); 53 (42~66) ❙ 63.4 (47.6~78.9); 875 (833~959); 517.7 (482.8~570.1)

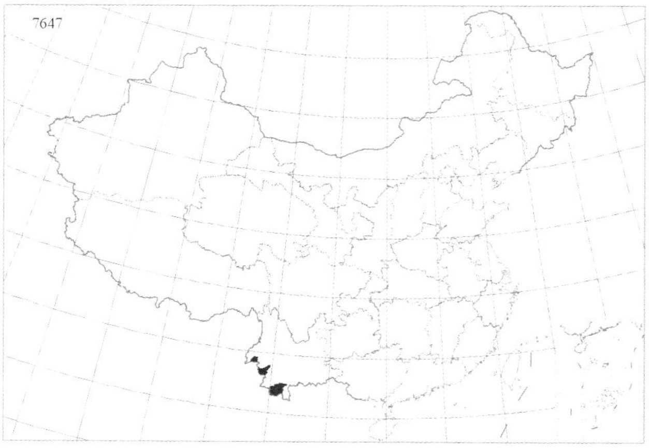

Ericaceae 杜鹃花科 *Agapetes* 树萝卜属
Agapetes burmanica W. E. Evans 缅甸树萝卜
Shrubs, evergreen

15.2 (5.3~20.5); 8.5 (-2.4~15.1); 20.0 (12.2~23.7); 15.3 (5.7~20.5); 126.8 (30.5~185.9); 4.5 (0.0~26.2); 791 (479~992) ❙ 1450 (961~1736); 738 (558~1007); 52 (31~77) ❙ 89.1 (47.6~145.8); 780 (479~986); 464.7 (234.2~659.0)

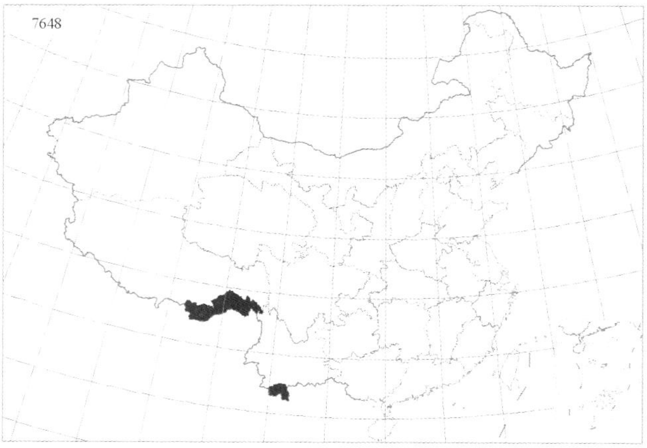

Ericaceae 杜鹃花科 *Agapetes* 树萝卜属
Agapetes buxifolia Nutt. ex Hook. 黄杨叶树萝卜
Shrubs, evergreen

0.7 (-1.8~12.2); -8.3 (-10.5~4.3); 8.9 (6.1~18.5); 3.5 (1.8~12.2); 15.3 (1.9~87.4); 65.9 (0.7~82.8); 380 (329~672) ❙ 397 (255~1431); 267 (191~817); 8 (1~31) ❙ -4.2 (-31.1~113.0); 329 (255~672); 111.4 (60.4~363.7)

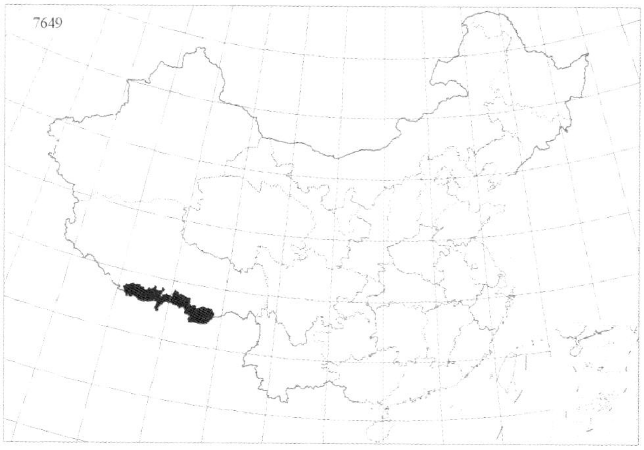

Ericaceae 杜鹃花科 *Agapetes* 树萝卜属
Agapetes camelliifolia S. H. Huang 茶叶树萝卜
Shrubs, evergreen

13.3; 5.4; 19.6; 13.4; 100.4; 0.0; 706 ❙ 1736; 1007; 51 ❙ 145.8; 706; 447.4

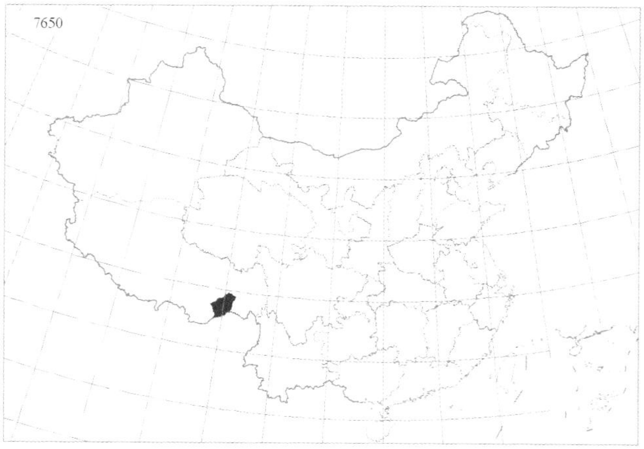

Ericaceae 杜鹃花科 *Agapetes* 树萝卜属
Agapetes ciliata S. H. Huang 纤毛叶树萝卜
Shrubs, evergreen

13.3; 5.4; 19.6; 13.4; 100.4; 0.0; 706 ❙ 1736; 1007; 51 ❙ 145.8; 706; 447.4

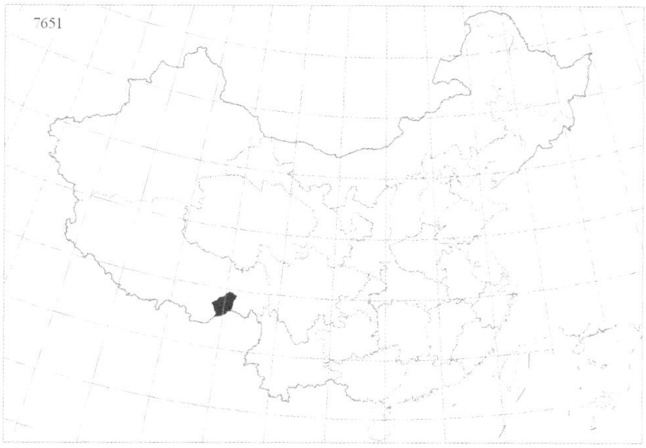

Ericaceae 杜鹃花科 *Agapetes* 树萝卜属
Agapetes discolor C. B. Clarke 异色树萝卜
Shrubs, evergreen

1.7 (-1.8~13.3); -7.3 (-10.5~5.4); 9.7 (6.1~19.6); 4.3 (1.8~13.4); 21.8 (1.9~100.4); 60.8 (0.0~82.8); 405 (329~706) ❙ 500 (255~1736); 324 (191~1007); 11 (1~51) ❙ 7.3 (-31.1~145.8); 358 (255~706); 137.2 (60.4~447.4)

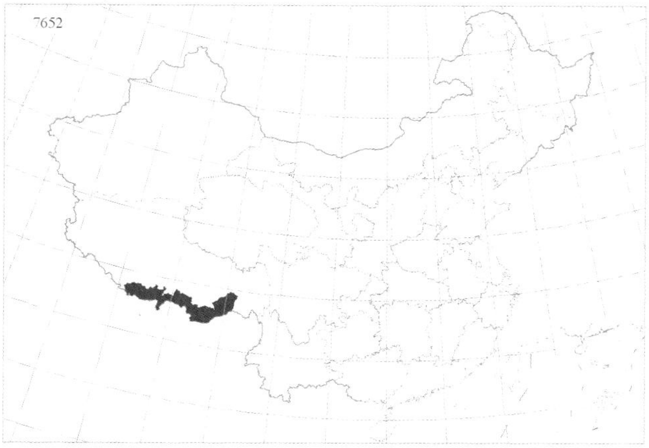

Ericaceae 杜鹃花科 *Agapetes* 树萝卜属
Agapetes epacridea Airy Shaw 尖叶树萝卜
Shrubs, evergreen

13.3; 5.4; 19.6; 13.4; 100.4; 0.0; 706 ❙ 1736; 1007; 51 ❙ 145.8; 706; 447.4

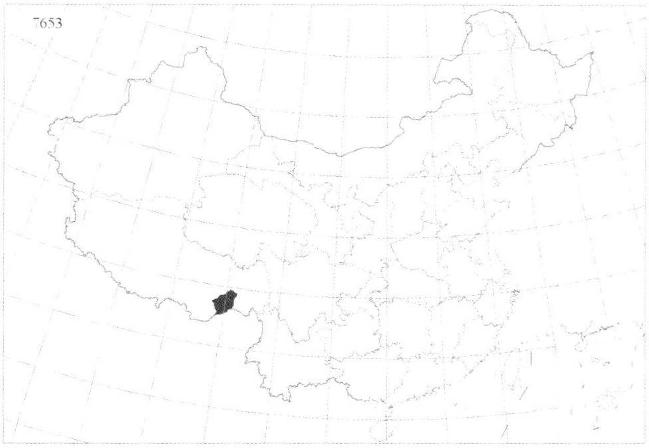

Ericaceae 杜鹃花科 *Agapetes* 树萝卜属
Agapetes flava (Hook. f.) Sleumer 黄花树萝卜
Shrubs, evergreen

10.3; 2.4; 16.7; 10.4; 72.8; 9.0; 619 ❙ 1376; 794; 38 ❙ 119.8; 619; 348.4

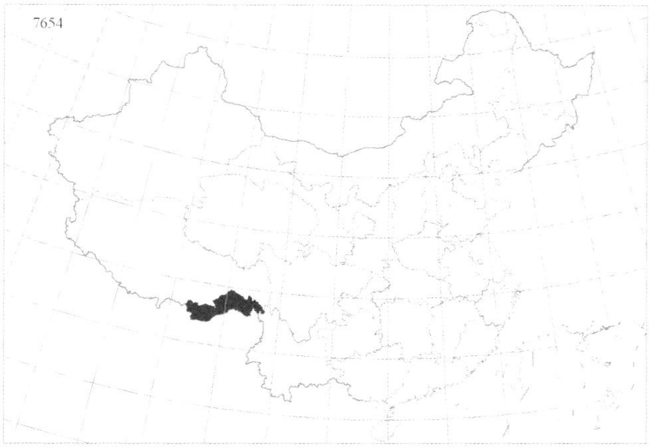

Ericaceae 杜鹃花科 *Agapetes* 树萝卜属
Agapetes forrestii W. E. Evans 伞花树萝卜
Shrubs or small trees, evergreen

11.6 (5.3~14.4); 4.3 (-2.4~7.5); 17.5 (12.2~19.6); 11.7 (5.7~14.4); 84.6 (30.5~113.1); 4.6 (0.0~26.2); 654 (479~727) ❙ 1360 (961~1736); 766 (558~1007); 52 (31~70) ❙ 107.5 (92.4~145.8); 654 (479~727); 387.4 (234.2~487.2)

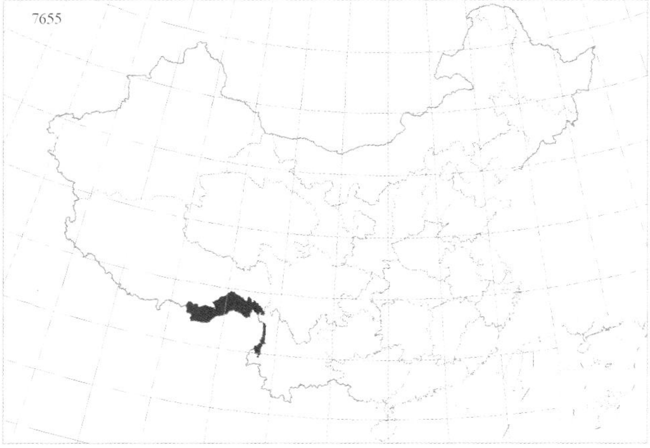

Ericaceae 杜鹃花科 *Agapetes* 树萝卜属
Agapetes griffithii C. B. Clarke 尾叶树萝卜
Climbing shrubs, evergreen

5.3; -2.4; 12.2; 5.7; 30.5; 26.2; 479 ❙ 961; 558; 31 ❙ 100.7; 479; 234.2

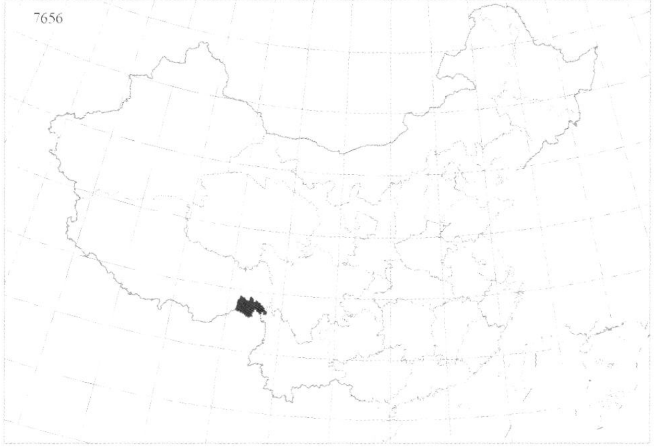

Ericaceae 杜鹃花科 *Agapetes* 树萝卜属
Agapetes hyalocheilos Airy Shaw 透明边树萝卜
Shrubs, evergreen

0.7 (−1.8~12.2); −8.3 (−10.5~4.3); 8.9 (6.1~18.5); 3.5 (1.8~12.2); 15.3 (1.9~87.4); 65.9 (0.7~82.8); 380 (329~672) ❙ 397 (255~1431); 267 (191~817); 8 (1~31) ❙ −4.2 (−31.1~113.0); 329 (255~672); 111.4 (60.4~363.7)

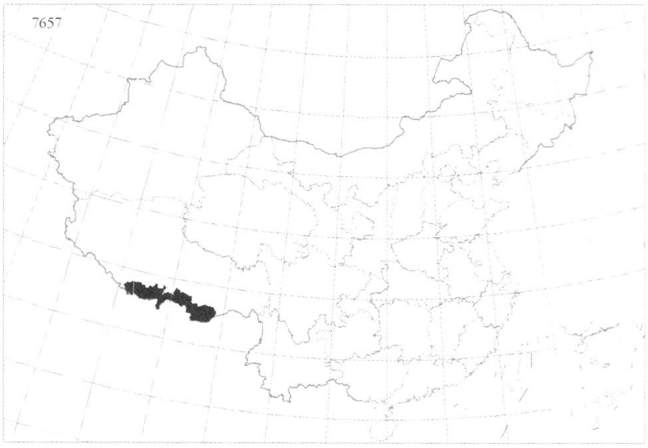

Ericaceae 杜鹃花科 *Agapetes* 树萝卜属
Agapetes incurvata (Griff.) Sleumer 皱叶树萝卜
Shrubs, evergreen

1.7 (−1.8~13.3); −7.3 (−10.5~5.4); 9.7 (6.1~19.6); 4.3 (1.8~13.4); 21.8 (1.9~100.4); 60.8 (0.0~82.8); 405 (329~706) ❙ 500 (255~1736); 324 (191~1007); 11 (1~51) ❙ 7.3 (−31.1~145.8); 358 (255~706); 137.2 (60.4~447.4)

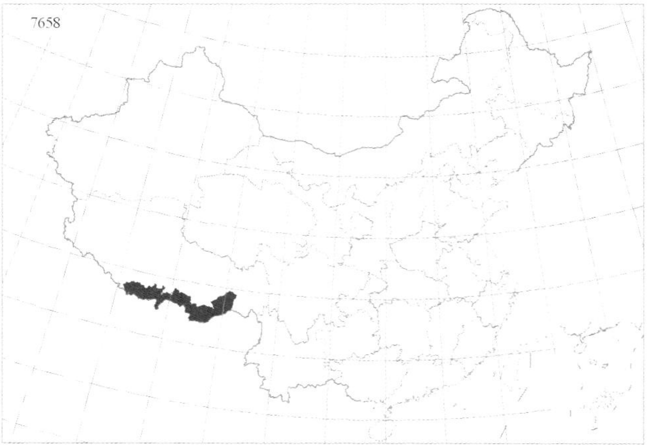

Ericaceae 杜鹃花科 *Agapetes* 树萝卜属
Agapetes inopinata Airy Shaw 沧源树萝卜
Shrubs, evergreen

17.9; 11.9; 21.6; 18.0; 155.8; 0.0; 854 ❙ 1435; 803; 42 ❙ 68.0; 833; 505.1

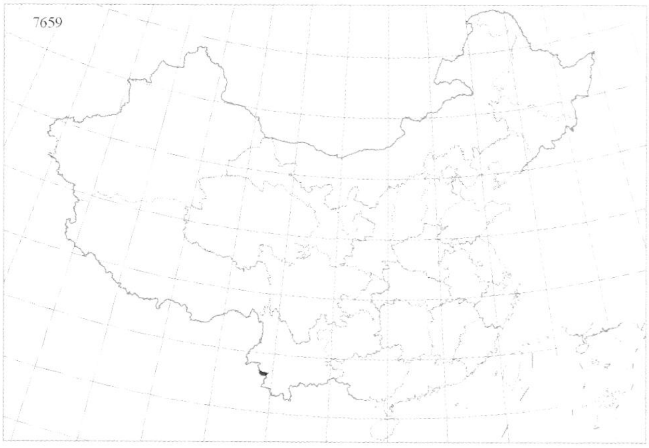

Ericaceae 杜鹃花科 *Agapetes* 树萝卜属
Agapetes interdicta (Hand.-Mazz.) Sleumer 中型树萝卜
Small shrubs, evergreen

10.4 (5.3~13.3); 3.0 (−2.4~6.4); 16.4 (12.2~19.6); 10.5 (5.7~13.4); 71.9 (30.5~100.4); 7.1 (0.0~26.2); 620 (479~706) ❙ 1264 (961~1736); 715 (558~1007); 51 (31~69) ❙ 102.7 (81.5~145.8); 620 (479~706); 365.1 (234.2~447.4)

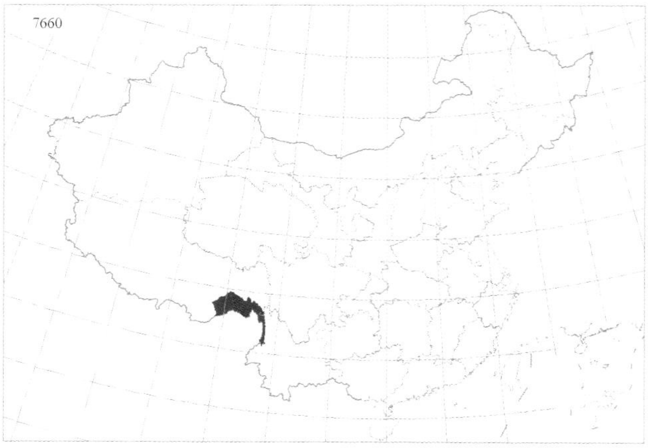

Ericaceae 杜鹃花科 *Agapetes* 树萝卜属
Agapetes lacei Craib 灯笼花
Shrubs, evergreen

10.7 (5.3~13.3); 3.2 (−2.4~6.4); 16.8 (12.2~19.6); 10.8 (5.7~13.4); 74.5 (30.5~100.4); 6.1 (0.0~26.2); 628 (479~706) ❙ 1292 (961~1736); 732 (558~1007); 47 (31~69) ❙ 104.4 (81.5~145.8); 628 (479~706); 364.9 (234.2~447.4)

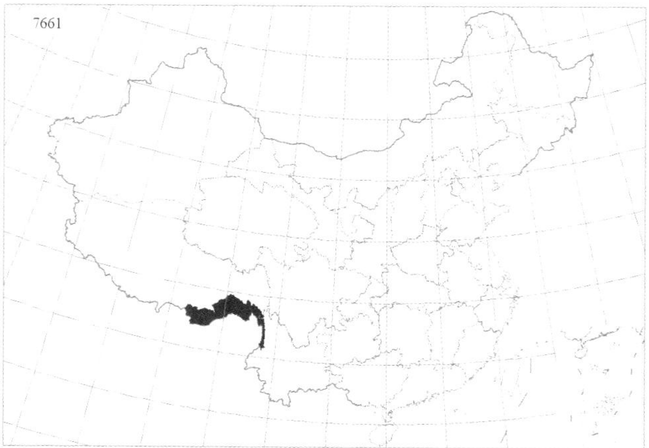

Ericaceae 杜鹃花科 *Agapetes* 树萝卜属
Agapetes lacei var. glaberrima Airy Shaw 无毛灯笼花
Shrubs, evergreen

7.0; −0.5; 13.6; 7.2; 41.5; 17.5; 526 ❙ 1001; 571; 38 ❙ 91.1; 526; 293.2

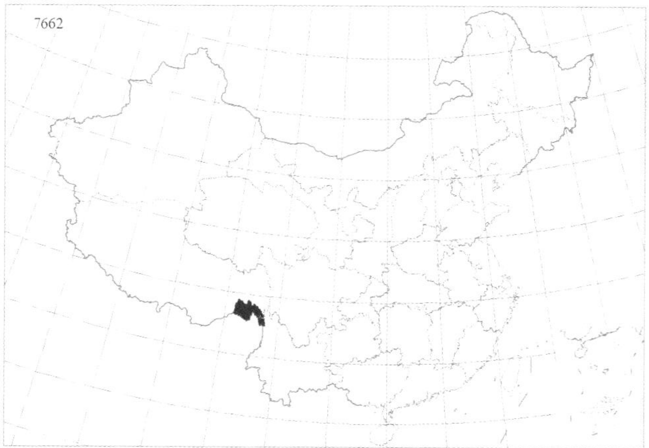

Ericaceae 杜鹃花科 *Agapetes* 树萝卜属
Agapetes lacei var. tomentella Airy Shaw 绒毛灯笼花
Shrubs, evergreen

7.0；-0.5；13.6；7.2；41.5；17.5；526 ❙ 1001；571；38 ❙ 91.1；526；293.2

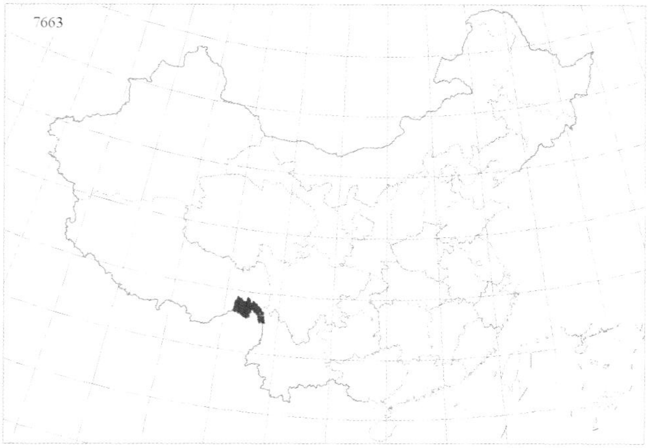

Ericaceae 杜鹃花科 *Agapetes* 树萝卜属
Agapetes leiocarpa S. H. Huang 光果树萝卜
Shrubs, evergreen

13.3；5.4；19.6；13.4；100.4；0.0；706 ❙ 1736；1007；51 ❙ 145.8；706；447.4

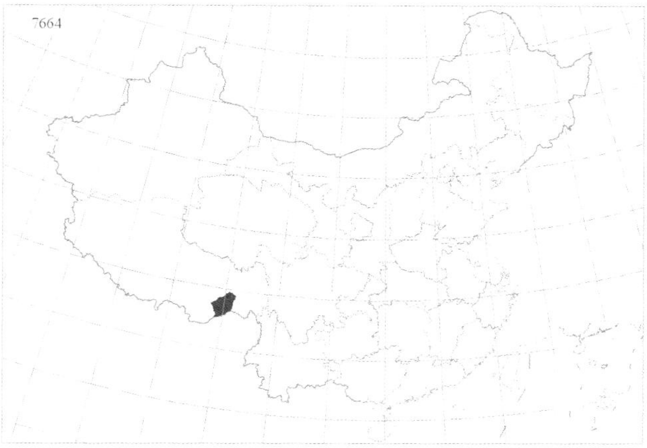

Ericaceae 杜鹃花科 *Agapetes* 树萝卜属
Agapetes leucocarpa S. H. Huang 白果树萝卜
Shrubs, evergreen

7.6；-0.8；14.8；8.6；58.1；26.6；555 ❙ 1159；678；32 ❙ 95.0；555；308.4

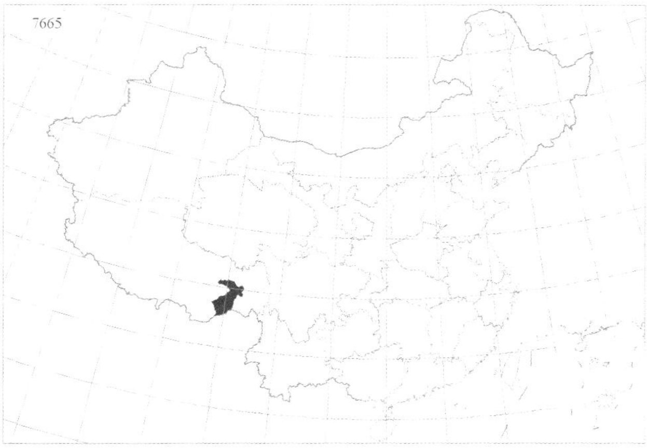

Ericaceae 杜鹃花科 *Agapetes* 树萝卜属
Agapetes linearifolia C. B. Clarke 线叶树萝卜
Shrubs, evergreen

5.3；-2.4；12.2；5.7；30.5；26.2；479 ❙ 961；558；31 ❙ 100.7；479；234.2

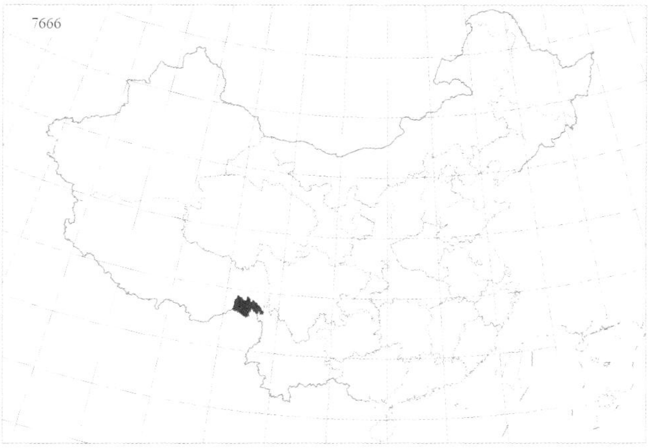

Ericaceae 杜鹃花科 *Agapetes* 树萝卜属
Agapetes listeri (King ex C. B. Clarke) Sleumer 短锥花树萝卜
Shrubs, evergreen

1.7 (-1.8~13.3)；-7.3 (-10.5~5.4)；9.7 (6.1~19.6)；4.3 (1.8~13.4)；21.8 (1.9~100.4)；60.8 (0.0~82.8)；405 (329~706) ❙ 500 (255~1736)；324 (191~1007)；11 (1~51) ❙ 7.3 (-31.1~145.8)；358 (255~706)；137.2 (60.4~447.4)

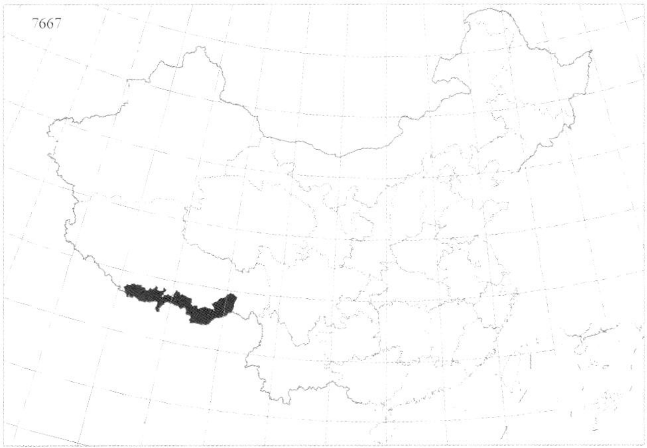

Ericaceae 杜鹃花科 *Agapetes* 树萝卜属
Agapetes lobbii C. B. Clarke 深裂树萝卜
Shrubs, evergreen

17.0 (6.1~20.5)；11.0 (-1.5~15.1)；21.1 (12.7~23.7)；17.1 (6.3~20.5)；148.2 (35.2~185.9)；3.6 (0.0~21.8)；851 (496~992) ❙ 1328 (801~1607)；675 (496~911)；55 (21~77) ❙ 56.2 (26.1~74.2)；832 (496~986)；490.4 (282.9~659.0)

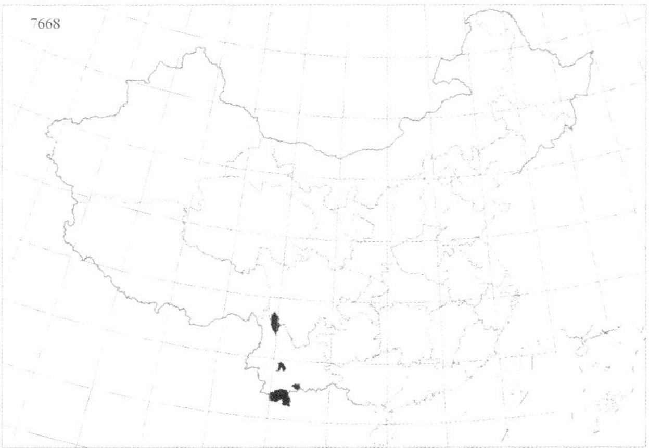

Ericaceae 杜鹃花科 *Agapetes* 树萝卜属

Agapetes macrophylla C. B. Clarke 大叶树萝卜

Shrubs, evergreen

10.3; 2.4; 16.7; 10.4; 72.8; 9.0; 619 ❙ 1376; 794; 38 ❙ 119.8; 619; 348.4

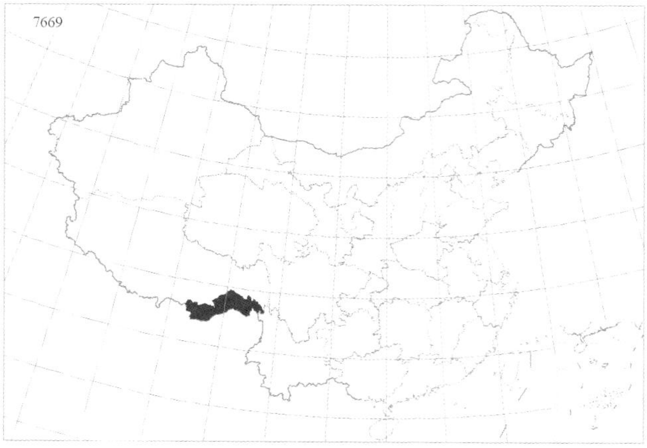

Ericaceae 杜鹃花科 *Agapetes* 树萝卜属

Agapetes malipoensis S. H. Huang 麻栗坡树萝卜

Shrubs, evergreen

18.5 (17.3~21.0); 11.6 (9.9~14.2); 23.4 (22.4~25.6); 18.6 (17.3~21.0); 162.9 (147.1~191.8); 0.0 (0.0~0.0); 903 (833~1053) ❙ 1449 (1368~1522); 817 (756~860); 58 (52~68) ❙ 61.3 (44.5~80.6); 903 (833~1053); 464.5 (359.7~551.6)

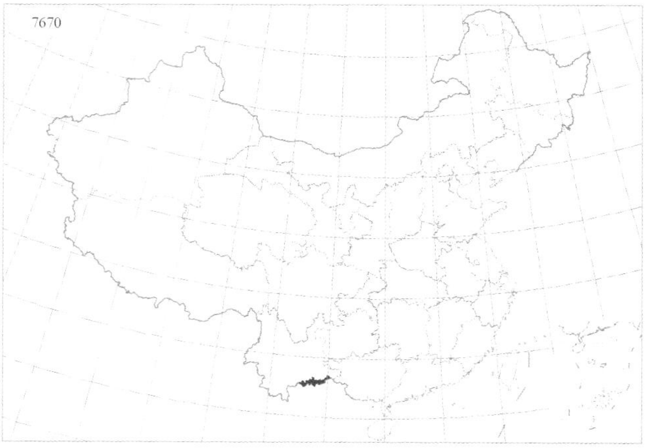

Ericaceae 杜鹃花科 *Agapetes* 树萝卜属

Agapetes mannii Hemsl. 白花树萝卜

Shrubs, evergreen

14.6 (5.4~20.5); 7.4 (-2.1~15.1); 20.4 (12.3~26.6); 14.7 (5.8~20.5); 117.7 (31.5~185.9); 1.4 (0.0~25.6); 755 (481~992) ❙ 1055 (741~1688); 597 (411~959); 37 (15~77) ❙ 40.3 (-14.2~112.2); 729 (481~986); 322.2 (142.9~659.0)

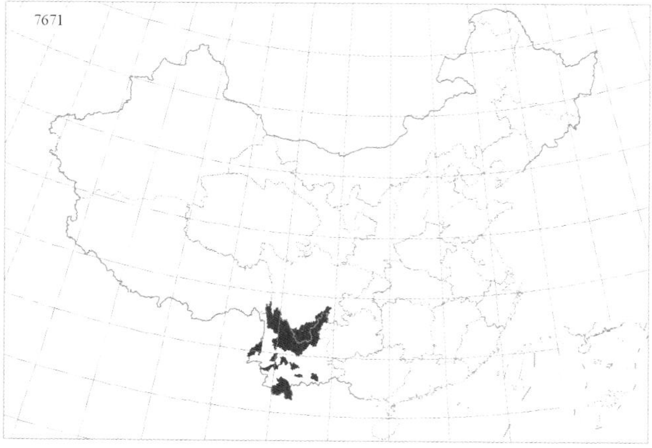

Ericaceae 杜鹃花科 *Agapetes* 树萝卜属

Agapetes marginata Dunn 边脉树萝卜

Shrubs, evergreen

9.3; 1.5; 15.9; 9.6; 65.5; 13.1; 592 ❙ 1348; 783; 41 ❙ 123.3; 592; 340.8

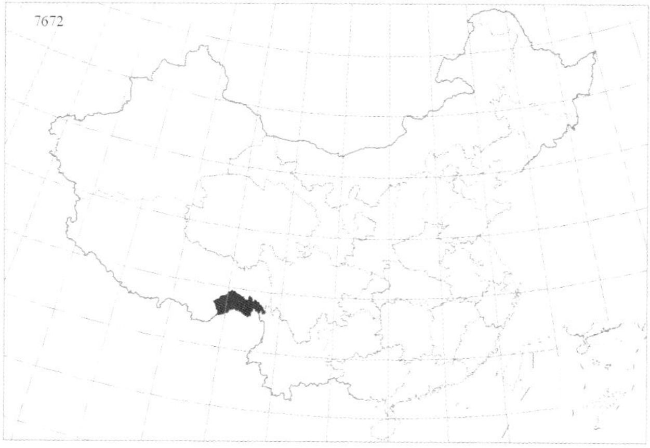

Ericaceae 杜鹃花科 *Agapetes* 树萝卜属

Agapetes medogensis S. H. Huang 墨脱树萝卜

Shrubs, evergreen

13.3; 5.4; 19.6; 13.4; 100.4; 0.0; 706 ❙ 1736; 1007; 51 ❙ 145.8; 706; 447.4

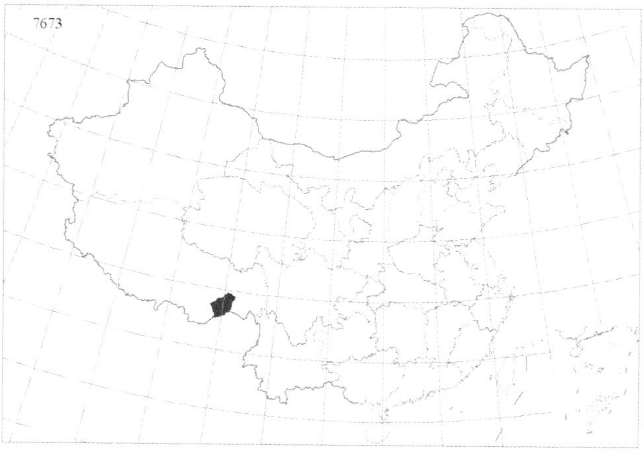

Ericaceae 杜鹃花科 *Agapetes* 树萝卜属

Agapetes megacarpa W. W. Sm. 大果树萝卜

Shrubs, evergreen

14.2 (8.6~19.5); 7.3 (1.3~12.7); 19.2 (15.0~23.3); 14.2 (8.7~19.5); 111.9 (52.6~173.8); 1.2 (0.0~8.8); 736 (574~942) ❙ 1258 (1041~1559); 699 (585~907); 56 (42~69) ❙ 72.2 (50.8~93.0); 733 (574~921); 383.8 (335.1~482.8)

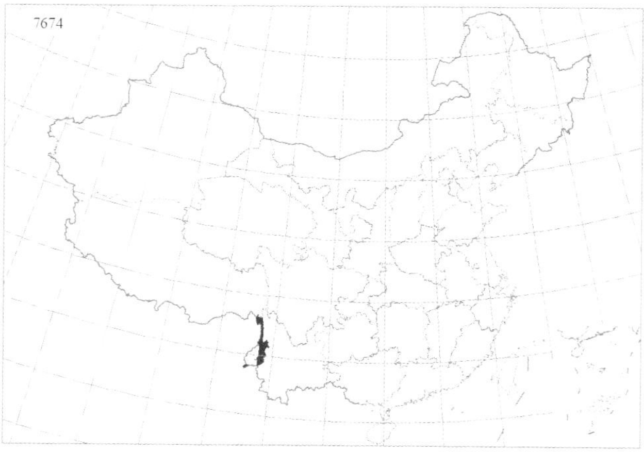

Ericaceae 杜鹃花科 *Agapetes* 树萝卜属
Agapetes miniata (Griff.) Hook. f. 朱红树萝卜
Shrubs, evergreen

5.3；-2.4；12.2；5.7；30.5；26.2；479 ❙ 961；558；31 ❙ 100.7；479；234.2

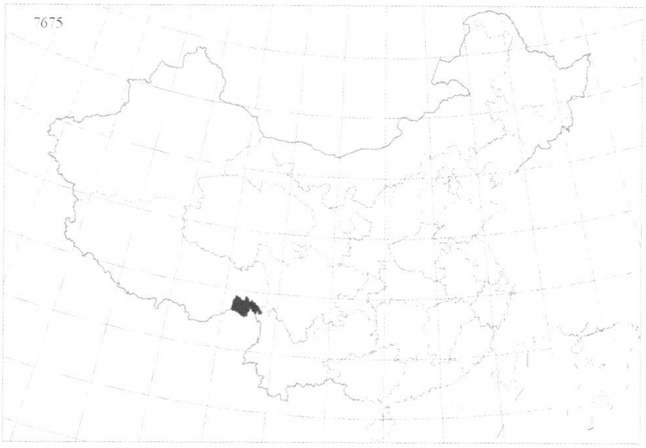

Ericaceae 杜鹃花科 *Agapetes* 树萝卜属
Agapetes miranda Airy Shaw 坛花树萝卜
Shrubs, evergreen

12.2；4.3；18.5；12.2；87.4；0.7；672 ❙ 1431；817；31 ❙ 113.0；672；363.7

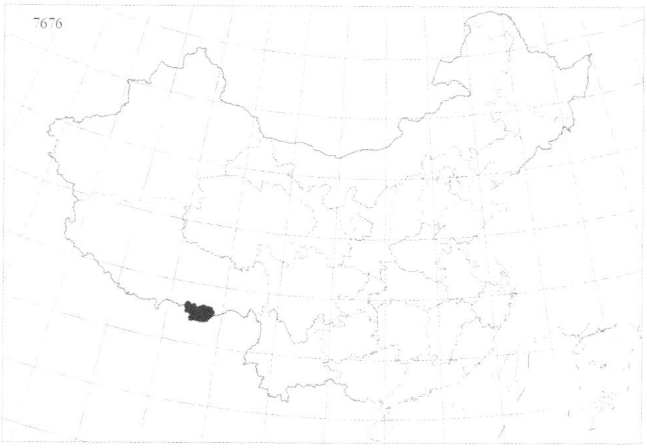

Ericaceae 杜鹃花科 *Agapetes* 树萝卜属
Agapetes mitrarioides Hook. f. ex C. B. Clarke 亮红树萝卜
Small shrubs, evergreen

5.3；-2.4；12.2；5.7；30.5；26.2；479 ❙ 961；558；31 ❙ 100.7；479；234.2

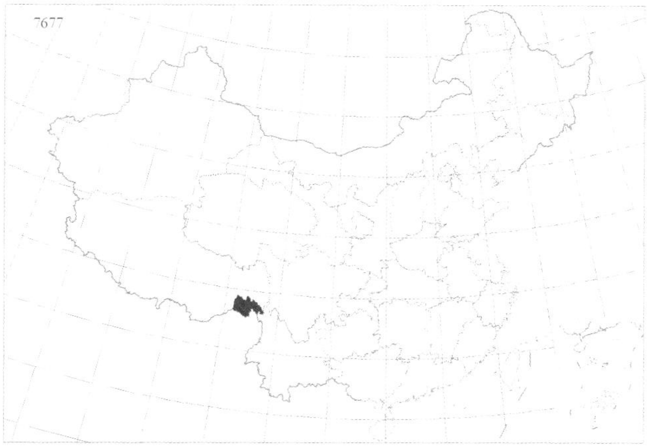

Ericaceae 杜鹃花科 *Agapetes* 树萝卜属
Agapetes neriifolia (King et Prain) Airy Shaw 夹竹桃叶树萝卜
Shrubs, evergreen

17.8；11.1；22.0；17.8；154.1；0.0；855 ❙ 1530；909；51 ❙ 78.9；855；482.8

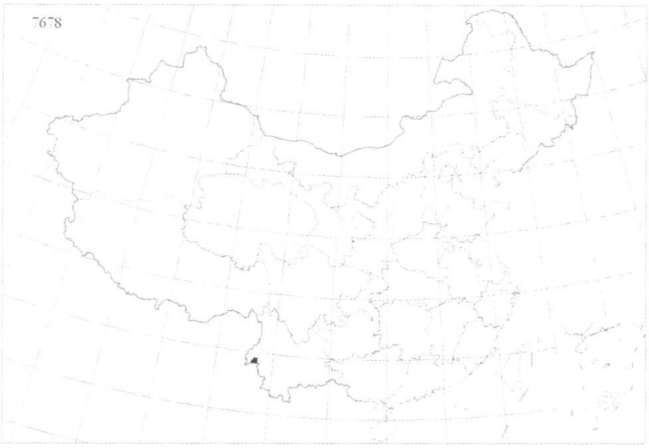

Ericaceae 杜鹃花科 *Agapetes* 树萝卜属
Agapetes nutans Dunn 垂花树萝卜
Shrubs, evergreen

5.3；-2.4；12.2；5.7；30.5；26.2；479 ❙ 961；558；31 ❙ 100.7；479；234.2

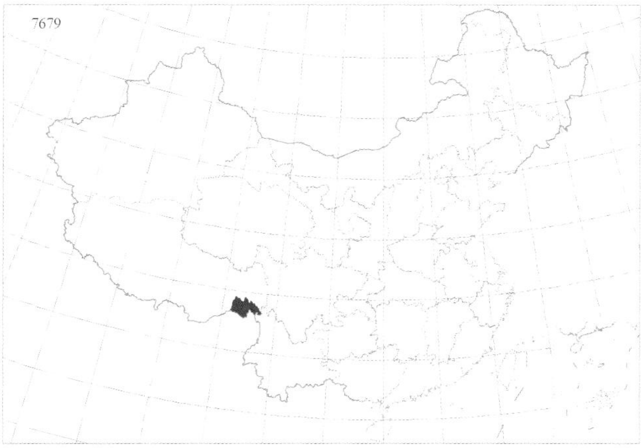

Ericaceae 杜鹃花科 *Agapetes* 树萝卜属
Agapetes oblonga Craib 长圆叶树萝卜
Shrubs, evergreen

14.5 (5.3~19.6)；7.6 (-2.4~12.8)；19.5 (12.2~23.5)；14.6 (5.7~19.6)；117.2 (30.5~175.7)；2.4 (0.0~26.2)；750 (479~953) ❙ 1361 (961~1736)；775 (558~1007)；50 (31~70) ❙ 83.9 (38.0~145.8)；744 (479~932)；424.1 (234.2~627.0)

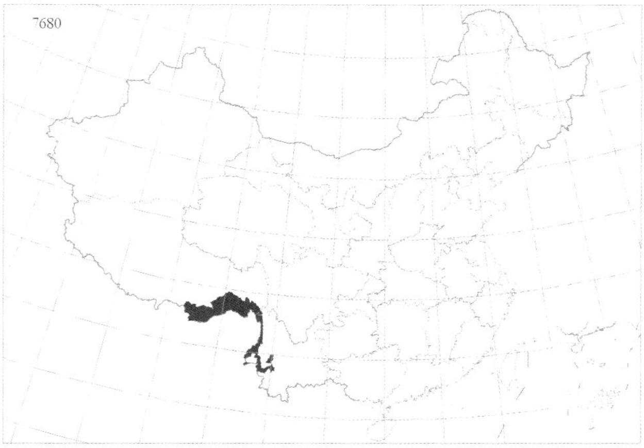

Ericaceae 杜鹃花科 *Agapetes* 树萝卜属
Agapetes obovata (Wight) Hook. f. 倒卵叶树萝卜
Shrubs, evergreen

18.5 (17.7~19.2); 12.4 (10.9~13.3); 22.8 (22.1~23.5); 18.6 (17.8~19.4); 163.4 (153.4~172.6); 0.0 (0.0~0.0); 891 (854~932) ❙ 1391 (1276~1547); 803 (741~911); 57 (51~63) ❙ 56.3 (36.8~74.2); 889 (854~923); 486.8 (432.9~551.3)

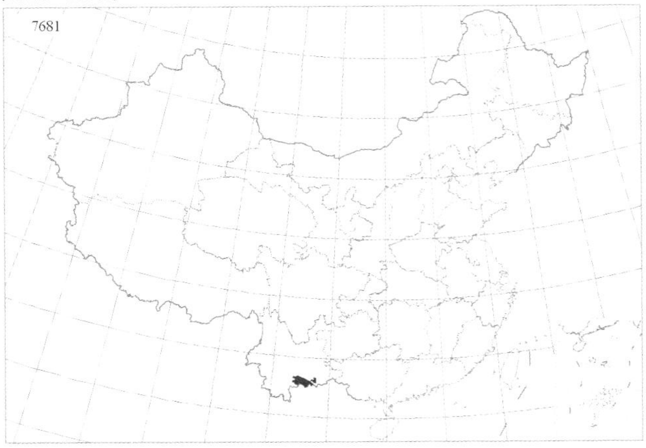

Ericaceae 杜鹃花科 *Agapetes* 树萝卜属
Agapetes pensilis Airy Shaw 倒挂树萝卜
Shrubs, evergreen

9.9; 2.1; 16.3; 10.0; 67.7; 8.9; 608 ❙ 1292; 742; 39 ❙ 110.3; 608; 349.4

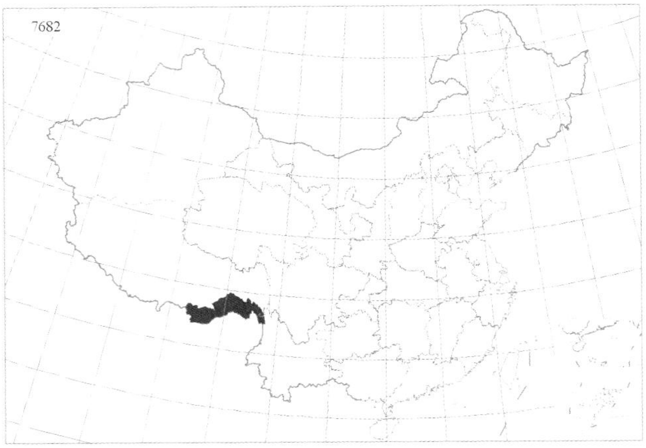

Ericaceae 杜鹃花科 *Agapetes* 树萝卜属
Agapetes pilifera Hook. f. ex C. B. Clarke 钟花树萝卜
Trees, evergreen

5.3; -2.4; 12.2; 5.7; 30.5; 26.2; 479 ❙ 961; 558; 31 ❙ 100.7; 479; 234.2

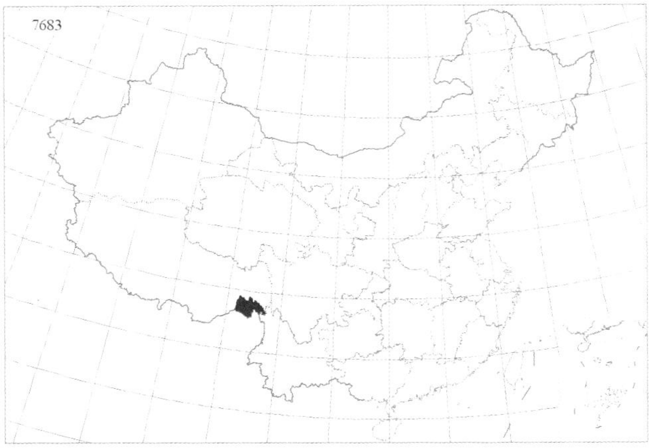

Ericaceae 杜鹃花科 *Agapetes* 树萝卜属
Agapetes praeclara C. Marquand 藏布江树萝卜
Shrubs, evergreen

6.2; -2.3; 13.5; 7.2; 46.1; 31.7; 512 ❙ 930; 547; 24 ❙ 65.0; 512; 282.5

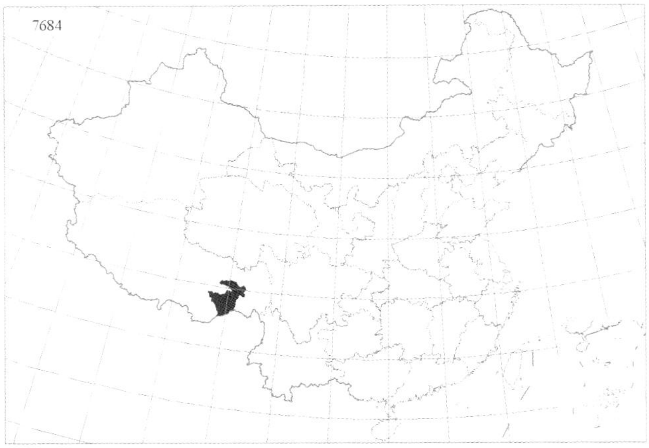

Ericaceae 杜鹃花科 *Agapetes* 树萝卜属
Agapetes praestigiosa Airy Shaw 听邦树萝卜
Shrubs, evergreen

12.2; 4.3; 18.5; 12.2; 87.4; 0.7; 672 ❙ 1431; 817; 31 ❙ 113.0; 672; 363.7

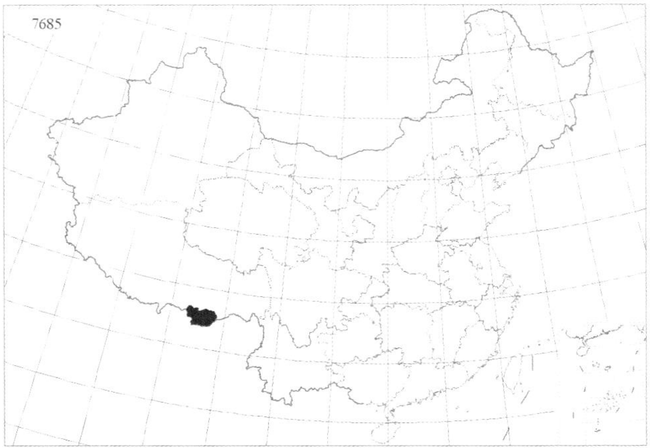

Ericaceae 杜鹃花科 *Agapetes* 树萝卜属
Agapetes pseudogriffithii Airy Shaw 杯梗树萝卜
Shrubs, evergreen

8.6; 1.3; 15.0; 8.7; 52.6; 8.8; 574 ❙ 1041; 585; 45 ❙ 81.5; 574; 352.3

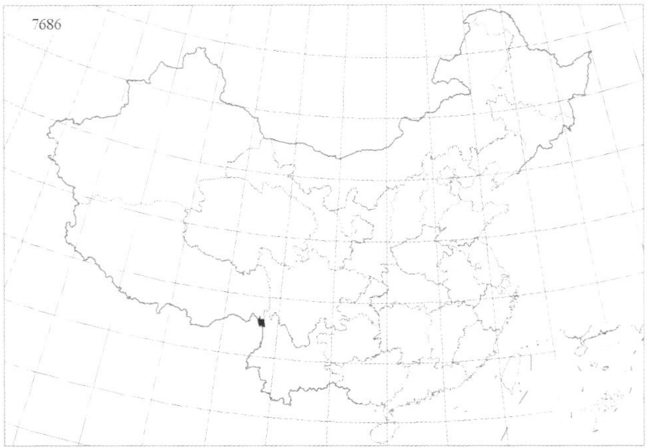

Ericaceae 杜鹃花科 *Agapetes* 树萝卜属
Agapetes pubiflora Airy Shaw 毛花树萝卜
Shrubs, evergreen

9.9; 2.1; 16.3; 10.0; 67.7; 8.9; 608 ▌ 1292; 742; 39 ▌ 110.3; 608; 349.4

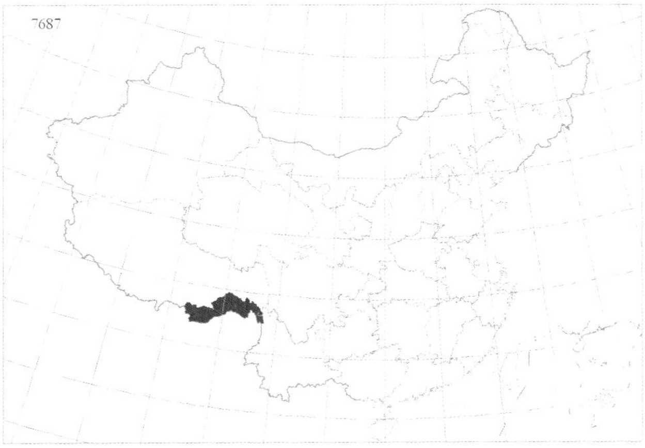

Ericaceae 杜鹃花科 *Agapetes* 树萝卜属
Agapetes pyrolifolia Airy Shaw 鹿蹄草叶树萝卜
Shrubs, evergreen

9.9; 2.1; 16.3; 10.0; 67.7; 8.9; 608 ▌ 1292; 742; 39 ▌ 110.3; 608; 349.4

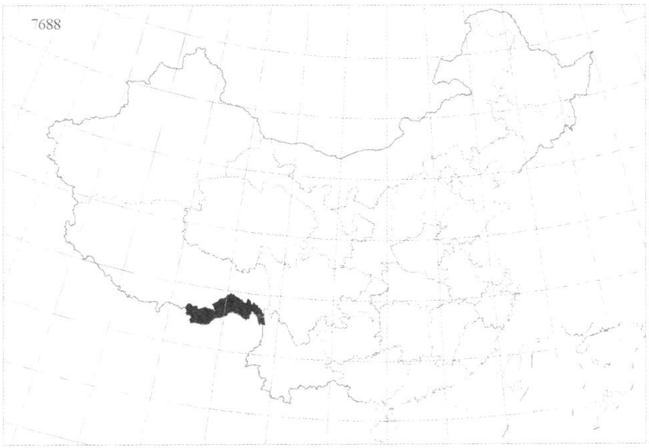

Ericaceae 杜鹃花科 *Agapetes* 树萝卜属
Agapetes refracta Airy Shaw 折瓣树萝卜
Shrubs, evergreen

12.2; 4.3; 18.5; 12.2; 87.4; 0.7; 672 ▌ 1431; 817; 31 ▌ 113.0; 672; 363.7

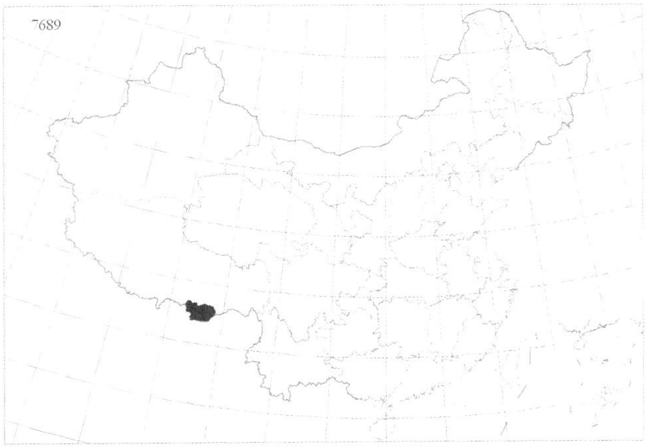

Ericaceae 杜鹃花科 *Agapetes* 树萝卜属
Agapetes rubrobracteata R. C. Fang et S. H. Huang 红苞树萝卜
Shrubs, evergreen

17.8 (14.2~21.0); 10.1 (5.3~14.2); 23.7 (20.1~26.6); 17.9 (14.2~21.0); 154.6 (110.5~191.8); 0.0 (0.0~0.0); 883 (729~1053) ▌ 1306 (1049~1547); 729 (598~911); 55 (44~68) ▌ 48.5 (21.7~80.4); 879 (729~1053); 344.5 (189.1~551.6)

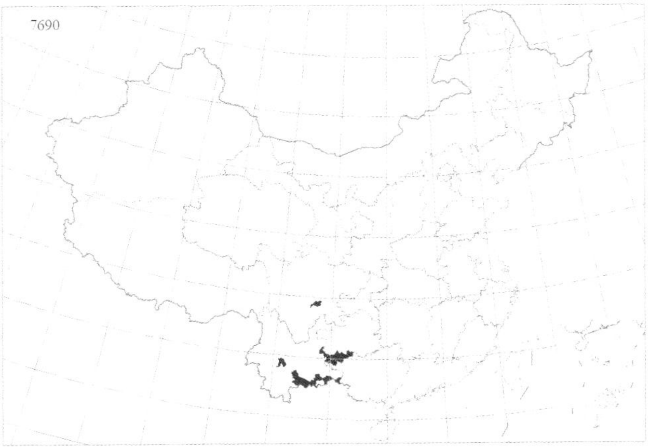

Ericaceae 杜鹃花科 *Agapetes* 树萝卜属
Agapetes salicifolia C. B. Clarke 柳叶树萝卜
Shrubs, evergreen

5.3; -2.4; 12.2; 5.7; 30.5; 26.2; 479 ▌ 961; 558; 31 ▌ 100.7; 479; 234.2

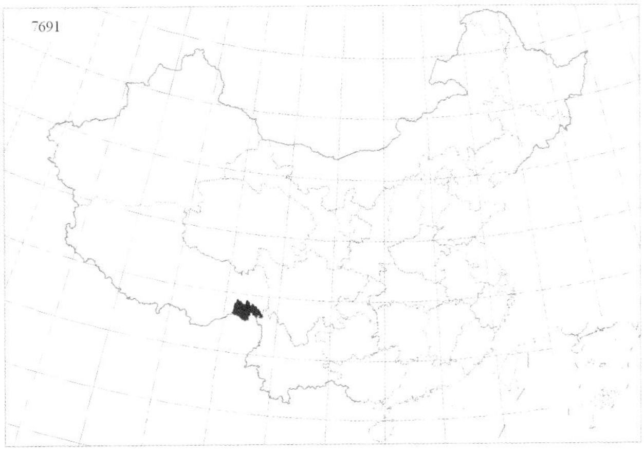

Ericaceae 杜鹃花科 *Agapetes* 树萝卜属
Agapetes serpens (Wight) Sleumer 五翅莓
Shrubs, evergreen

-1.2; -9.9; 6.9; 2.1; 4.4; 78.5; 334 ▌ 345; 240; 15 ▌ 0.9; 329; 66.9

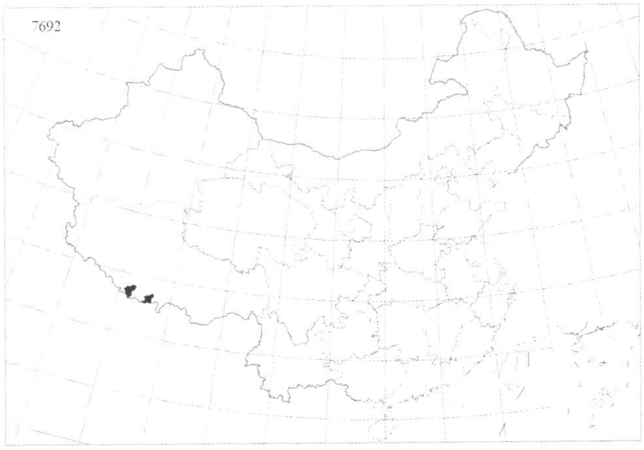

Ericaceae 杜鹃花科 *Arctous* 北极果属

Arctous alpinus (L.) Nied. 北极果

Shrubs, deciduous

2.3 (−5.8~11.3)；−14.7 (−29.8~−1.4)；16.9 (7.8~27.5)；6.5 (2.1~12.1)；46.1 (5.2~107.6)；77.8 (15.2~157.4)；518 (340~813) | 393 (75~856)；214 (39~427)；15 (1~48) | −16.8 (−89.7~94.4)；329 (75~697)；186.7 (7.5~344.4)

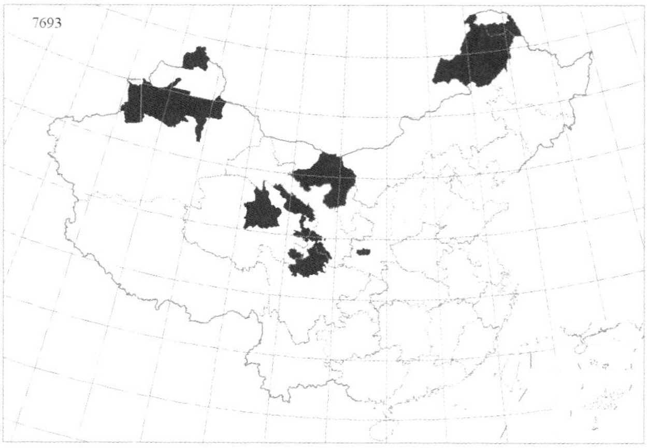

Ericaceae 杜鹃花科 *Arctous* 北极果属

Arctous microphyllus C. Y. Wu 小叶当年枯

Dwarf shrubs, deciduous

6.1；−1.5；12.7；6.3；35.2；21.8；496 | 801；496；21 | 61.4；496；282.9

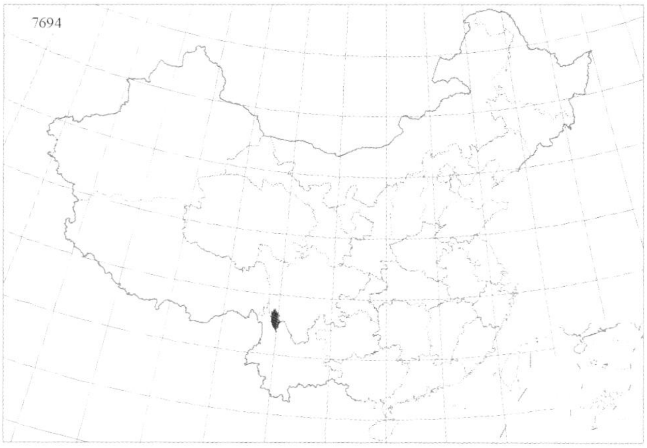

Ericaceae 杜鹃花科 *Arctous* 北极果属

Arctous ruber (Rehder et E. H. Wilson) Nakai 红北极果

Shrubs, deciduous

4.0 (−4.0~11.3)；−10.4 (−28.2~−1.4)；17.0 (10.4~23.8)；6.8 (3.7~11.5)；47.3 (15.1~92.4)；58.2 (15.2~146.2)；522 (409~697) | 641 (132~856)；338 (81~427)；17 (2~25) | 27.5 (−80.1~94.4)；489 (132~697)；283.2 (13.6~346.5)

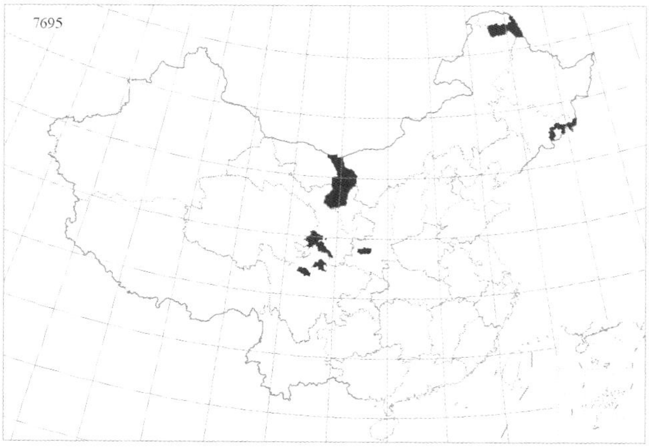

Ericaceae 杜鹃花科 *Cassiope* 锦绦花属

Cassiope abbreviata Hand.-Mazz. 短梗锦绦花

Dwarf shrubs, evergreen

6.1 (2.0~14.8)；−3.5 (−7.2~3.8)；14.7 (10.1~25.3)；7.1 (3.8~14.8)；44.8 (15.4~119.2)；30.9 (1.4~51.4)；521 (413~795) | 975 (791~1358)；493 (400~606)；31 (17~88) | 91.1 (49.4~128.6)；521 (413~795)；253.3 (171.0~357.7)

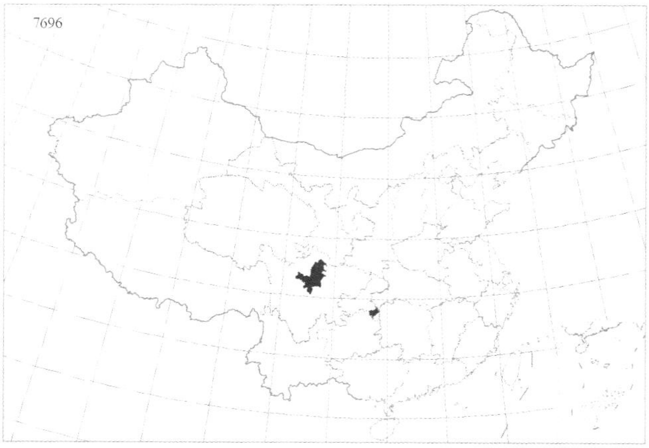

Ericaceae 杜鹃花科 *Cassiope* 锦绦花属

Cassiope argyrotricha T. Z. Hsu 银毛锦绦花

Dwarf shrubs, evergreen

12.3；5.4；17.7；12.3；88.0；0.3；670 | 1290；712；63 | 92.7；670；395.9

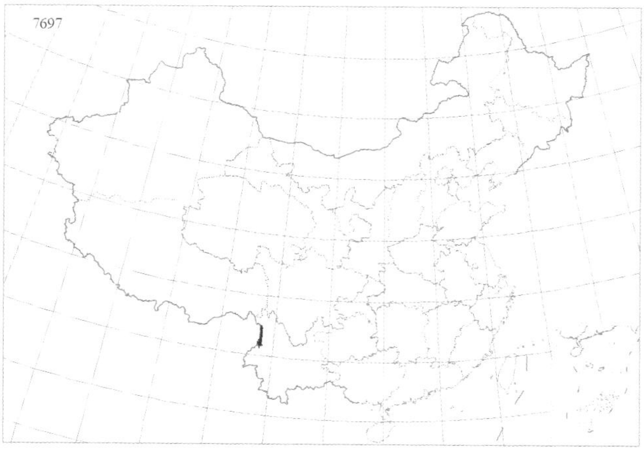

Ericaceae 杜鹃花科 *Cassiope* 锦绦花属

Cassiope fastigiata (Wall.) D. Don 扫帚锦绦花

Dwarf shrubs, evergreen

4.6 (−1.8~13.2)；−3.5 (−10.2~6.4)；11.7 (6.1~18.5)；6.3 (1.8~13.3)；38.4 (1.9~99.4)；42.7 (0.0~82.8)；476 (329~695) | 742 (320~1431)；435 (193~817)；30 (9~69) | 39.8 (−4.0~113.0)；474 (320~695)；223.9 (61.6~407.1)

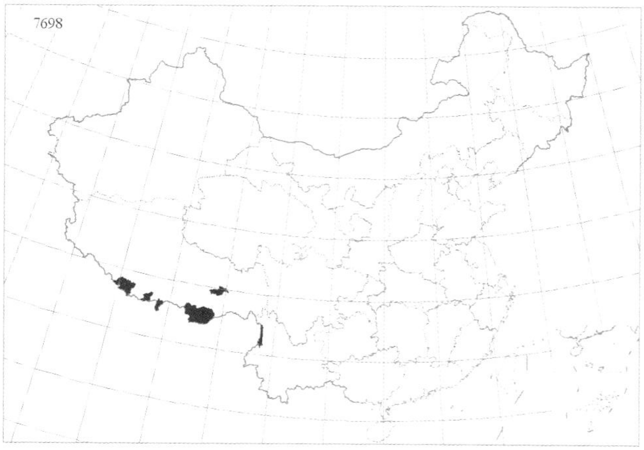

Ericaceae 杜鹃花科 *Cassiope* 锦绦花属
Cassiope fujianensis L. K. Ling et G. S. Hoo 福建锦绦花
Dwarf shrubs, evergreen

18.6; 9.3; 27.6; 18.7; 164.0; 0.0; 965 | 1702; 591; 199 | 76.4; 965; 300.0

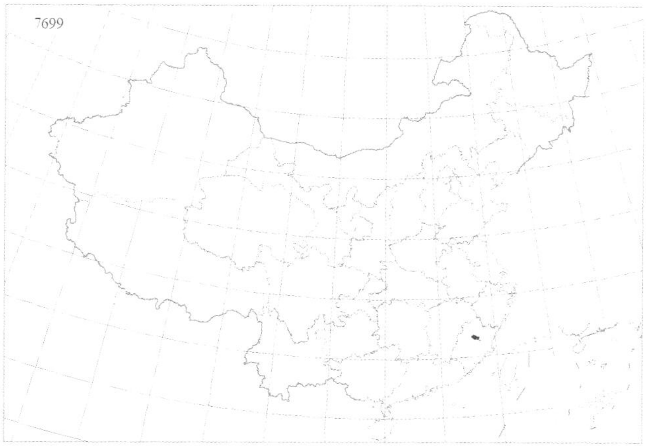

Ericaceae 杜鹃花科 *Cassiope* 锦绦花属
Cassiope membranifolia R. C. Fang 膜叶锦绦花
Dwarf shrubs, evergreen

13.2; 6.4; 18.4; 13.3; 99.4; 0.0; 695 | 1337; 727; 69 | 92.4; 695; 407.1

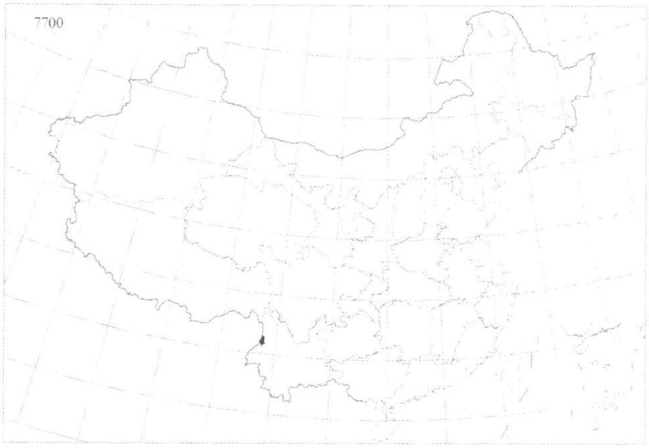

Ericaceae 杜鹃花科 *Cassiope* 锦绦花属
Cassiope myosuroides W. W. Sm. 鼠尾锦绦花
Dwarf shrubs, evergreen

8.3 (2.7~13.2); 0.1 (-7.1~6.4); 15.3 (11.1~18.4); 8.8 (4.4~13.3); 57.7 (20.5~99.4); 17.5 (0.0~48.0); 569 (429~695) | 1049 (800~1337); 573 (427~727); 37 (18~69) | 84.8 (52.8~101.2); 569 (429~695); 319.4 (200.8~407.1)

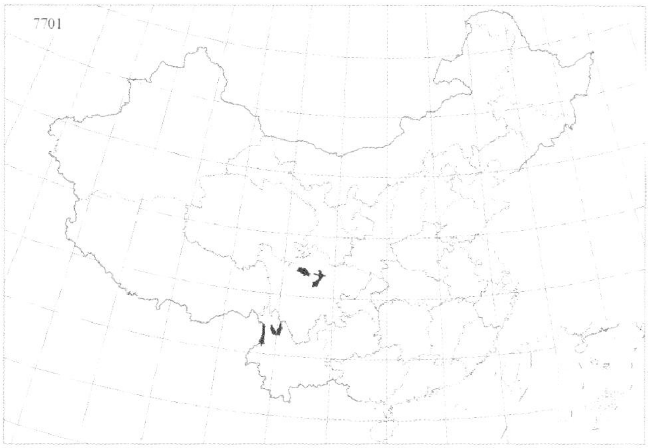

Ericaceae 杜鹃花科 *Cassiope* 锦绦花属
Cassiope nana T. Z. Hsu 矮小锦绦花
Dwarf shrubs, evergreen

5.4; -2.1; 12.3; 5.8; 31.5; 25.6; 481 | 741; 426; 28 | 54.0; 481; 273.5

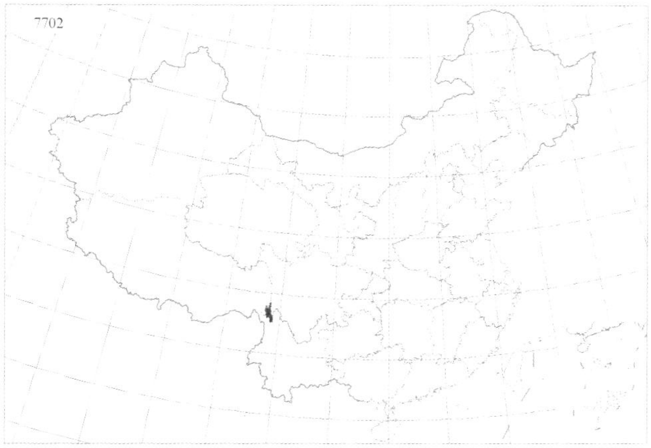

Ericaceae 杜鹃花科 *Cassiope* 锦绦花属
Cassiope palpebrata W. W. Sm. 朝天锦绦花
Dwarf shrubs, evergreen

9.6 (5.4~13.2); 2.5 (-2.1~6.4); 15.7 (12.3~18.4); 9.7 (5.8~13.3); 64.2 (31.5~99.4); 8.1 (0.0~25.6); 599 (481~695) | 1074 (741~1337); 602 (426~727); 48 (28~69) | 77.8 (54.0~93.0); 599 (481~695); 353.9 (273.5~407.1)

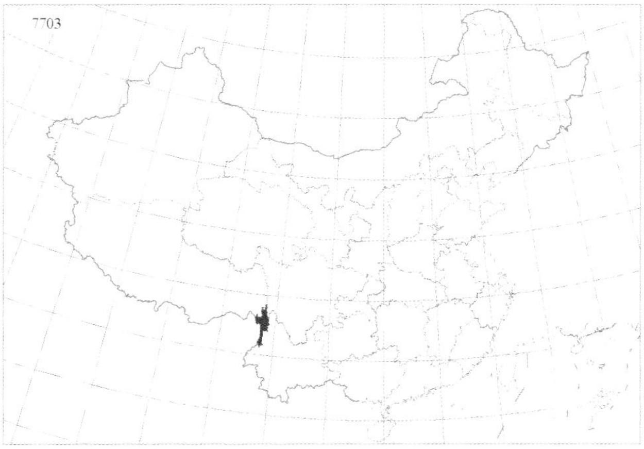

Ericaceae 杜鹃花科 *Cassiope* 锦绦花属
Cassiope pectinata Stapf 篦叶锦绦花
Dwarf shrubs, evergreen

6.1 (2.2~14.1); -2.0 (-7.1~6.5); 13.1 (9.5~20.2); 6.8 (3.9~14.2); 40.5 (15.8~109.8); 27.2 (0.0~48.6); 505 (411~729) | 815 (473~1736); 488 (287~1007); 21 (7~51) | 58.6 (5.0~145.8); 504 (411~707); 282.4 (201.9~447.4)

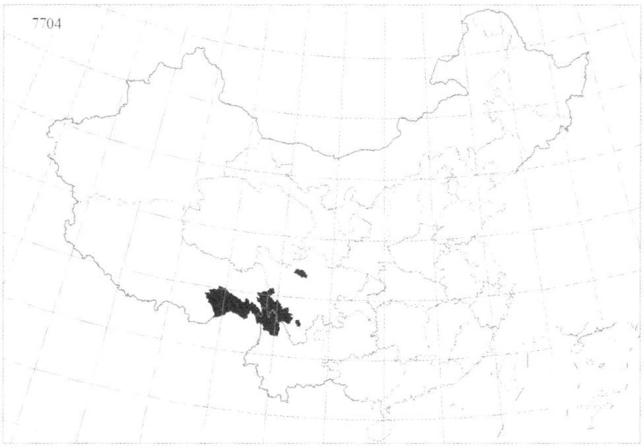

Ericaceae 杜鹃花科 *Cassiope* 锦绦花属

Cassiope selaginoides Hook. f. et Thomson 锦绦花

Dwarf shrubs, evergreen

10.5 (−1.8~18.9); 2.4 (−10.2~12.0); 17.5 (6.1~26.6); 10.9 (1.8~19.0); 80.3 (1.9~167.8); 13.2 (0.0~82.8); 638 (329~929) ❙ 943 (320~1736); 540 (217~1007); 28 (7~58) ❙ 51.7 (−14.2~145.8); 621 (320~912); 266.7 (72.2~447.4)

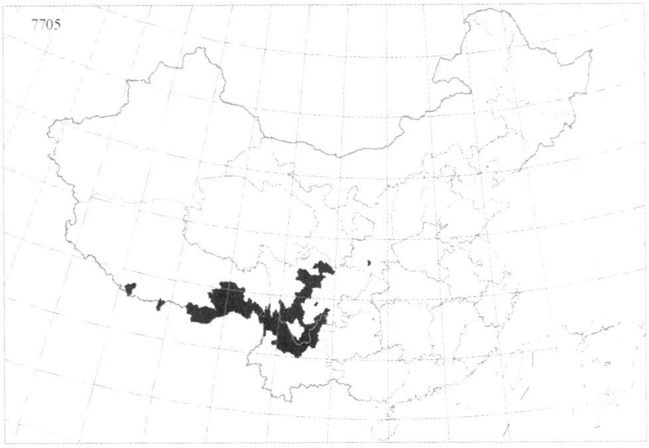

Ericaceae 杜鹃花科 *Cassiope* 锦绦花属

Cassiope wardii C. Marquand et Airy Shaw 长毛锦绦花

Dwarf shrubs, evergreen

7.4 (1.8~13.3); −0.6 (−7.1~6.4); 14.2 (9.9~19.6); 8.1 (3.8~13.4); 52.6 (15.7~100.4); 23.4 (0.0~53.3); 541 (404~706) ❙ 980 (473~1736); 563 (287~1007); 34 (7~69) ❙ 71.9 (5.0~145.8); 541 (404~706); 307.7 (169.5~447.4)

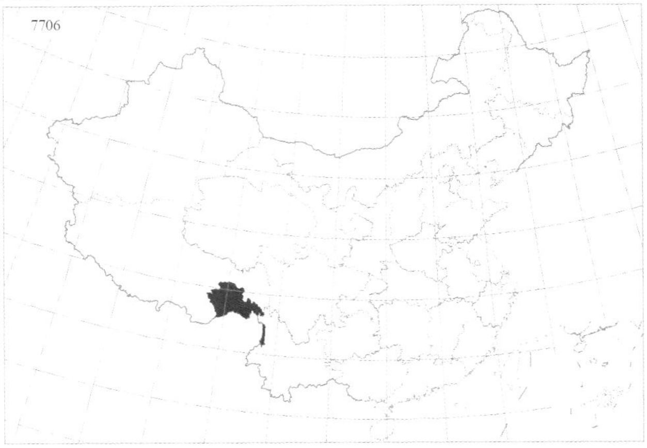

Ericaceae 杜鹃花科 *Chamaedaphne* 地桂属

Chamaedaphne calyculata (L.) Moench 地桂

Shrubs, evergreen

0.1 (−5.8~4.5); −21.9 (−29.8~−14.1); 19.5 (15.3~21.9); 6.7 (4.4~8.3); 50.3 (28.0~64.9); 108.7 (68.3~157.4); 524 (426~597) ❙ 491 (249~808); 322 (179~500); 14 (6~32) ❙ −6.2 (−54.1~45.0); 458 (249~566); 266.1 (115.9~338.4)

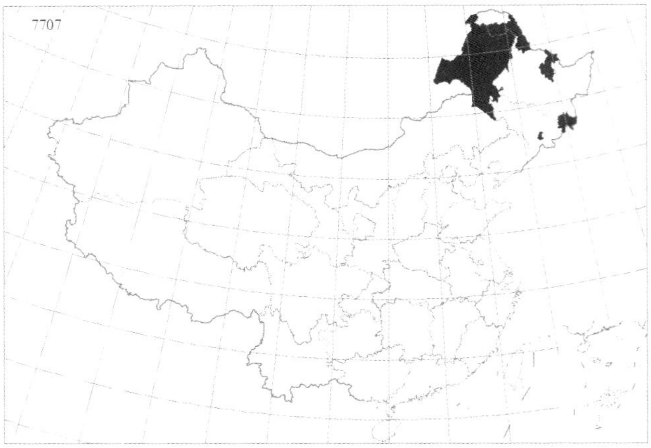

Ericaceae 杜鹃花科 *Craibiodendron* 假木荷属

Craibiodendron henryi W. W. Sm. 柳叶假木荷

Small trees, evergreen

15.0 (10.1~19.0); 7.9 (2.3~13.1); 20.6 (16.9~23.9); 15.1 (10.2~19.1); 121.2 (68.1~168.8); 0.5 (0.0~5.2); 762 (618~929) ❙ 1033 (797~1736); 595 (411~1007); 36 (21~63) ❙ 36.2 (−14.2~145.8); 730 (618~891); 303.7 (190.3~551.3)

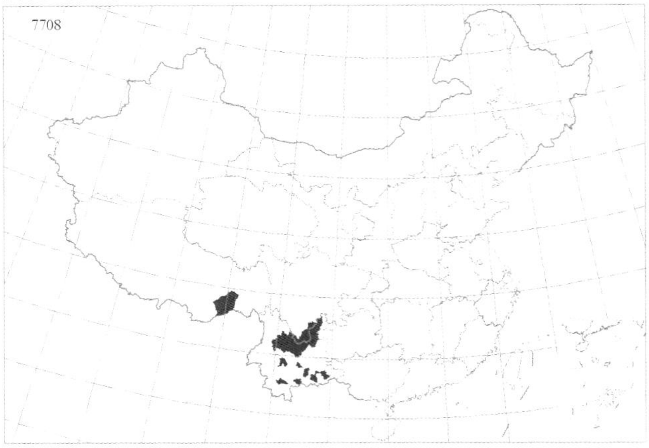

Ericaceae 杜鹃花科 *Craibiodendron* 假木荷属

Craibiodendron scleranthum var. kwangtungense (S. Y. Hu) Judd 广东假木荷

Trees, evergreen

21.4 (17.9~23.4); 13.0 (8.1~16.1); 28.0 (26.3~29.8); 21.4 (17.9~23.5); 196.7 (155.4~221.7); 0.0 (0.0~0.0); 1127 (918~1273) ❙ 1726 (1284~2435); 803 (508~1428); 134 (73~191) ❙ 53.6 (11.9~99.5); 1124 (918~1273); 244.1 (157.7~322.1)

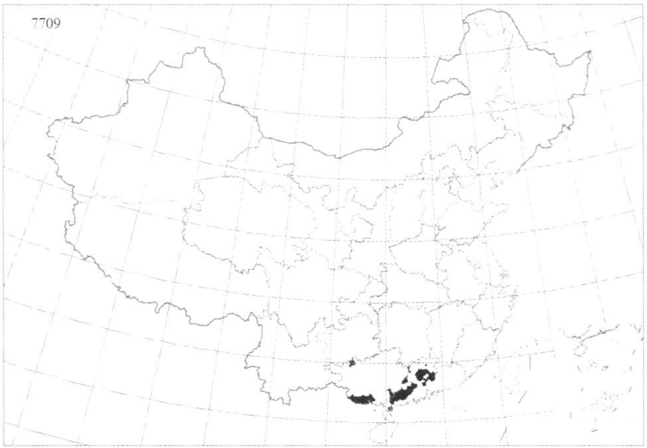

Ericaceae 杜鹃花科 *Craibiodendron* 假木荷属

Craibiodendron stellatum (Pierre) W. W. Sm. 假木荷

Small trees, evergreen

19.2 (14.5~22.3); 11.3 (7.8~15.1); 25.0 (19.6~29.0); 19.2 (14.5~22.3); 170.3 (113.5~208.1); 0.0 (0.0~0.0); 967 (730~1202) ❙ 1350 (958~1784); 695 (531~911); 81 (42~194) ❙ 40.3 (5.8~80.6); 962 (730~1202); 326.1 (157.7~659.0)

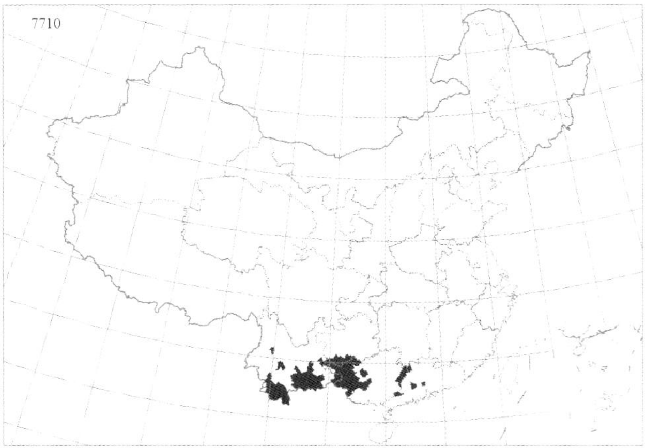

Ericaceae 杜鹃花科 *Craibiodendron* 假木荷属

Craibiodendron yunnanense W. W. Sm. 云南假木荷

Shrubs or small trees, evergreen

14.7 (5.4~22.3); 7.2 (-2.1~13.5); 20.8 (12.3~28.9); 14.8 (5.8~22.3); 118.8 (31.5~208.1); 1.4 (0.0~25.6); 768 (481~1202) ❙ 1048 (741~1736); 596 (411~1007); 41 (15~136) ❙ 38.3 (-14.2~145.8); 744 (481~1202); 293.4 (142.9~447.4)

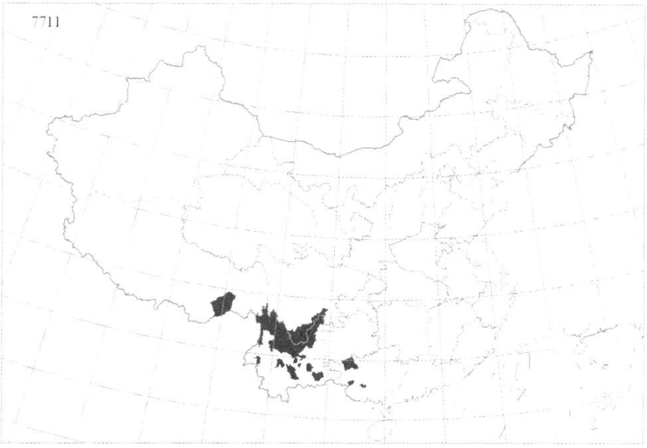

Ericaceae 杜鹃花科 *Diplarche* 杉叶杜鹃属

Diplarche multiflora Hook. f. et Thomson 多花杉叶杜鹃

Small shrubs, evergreen

6.8 (1.8~13.3); -1.1 (-7.1~5.4); 13.7 (9.9~19.6); 7.4 (3.8~13.4); 44.8 (15.7~100.4); 23.0 (0.0~53.3); 524 (404~706) ❙ 935 (473~1736); 541 (287~1007); 31 (7~51) ❙ 71.3 (5.0~145.8); 524 (404~706); 294.2 (169.5~447.4)

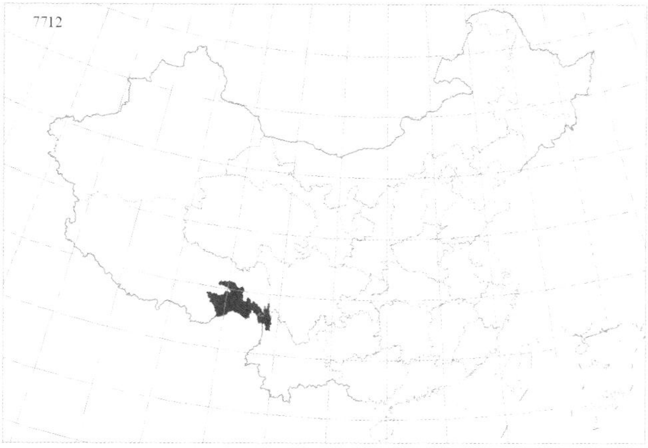

Ericaceae 杜鹃花科 *Diplarche* 杉叶杜鹃属

Diplarche pauciflora Hook. f. et Thomson 少花杉叶杜鹃

Small shrubs, evergreen

6.8; -1.4; 13.9; 7.4; 46.3; 23.9; 525 ❙ 1057; 606; 32 ❙ 95.5; 525; 328.2

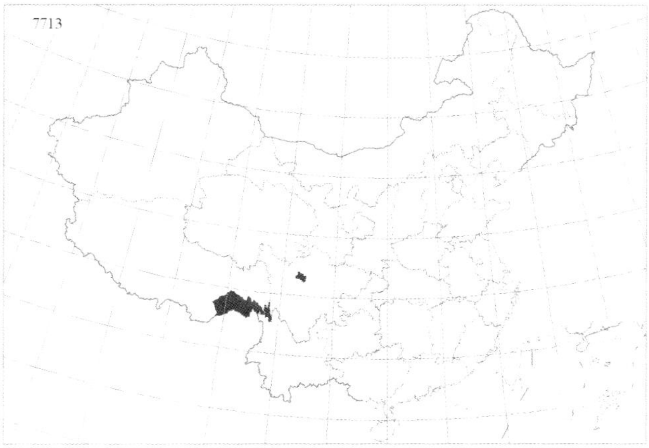

Ericaceae 杜鹃花科 *Empetrum* 岩高兰属

Empetrum nigrum var. japonicum K. Koch 东北岩高兰

Small shrubs, evergreen

0.3 (-5.8~4.5); -21.4 (-29.8~-14.1); 19.1 (15.3~21.6); 6.7 (4.4~8.2); 49.3 (28.0~63.5); 105.5 (68.3~157.4); 522 (426~581) ❙ 554 (187~853); 343 (79~527); 19 (11~35) ❙ 5.6 (-65.1~57.4); 487 (187~569); 281.9 (120.8~346.5)

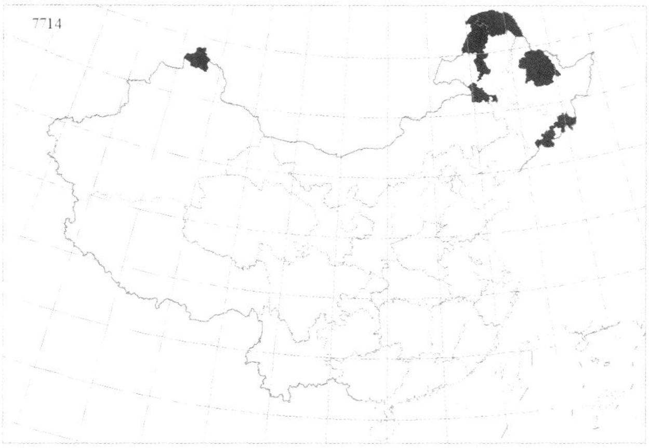

Ericaceae 杜鹃花科 *Enkianthus* 吊钟花属

Enkianthus chinensis Franch. 灯笼吊钟花

Shrubs or small trees, deciduous

15.9 (3.5~22.3); 5.6 (-6.5~14.2); 25.3 (12.2~29.9); 16.0 (5.0~22.3); 133.0 (26.3~208.1); 1.5 (0.0~43.4); 851 (446~1202) ❙ 1365 (634~2014); 557 (290~954); 127 (15~259) ❙ 60.7 (-16.6~147.0); 843 (446~1202); 229.3 (134.6~551.6)

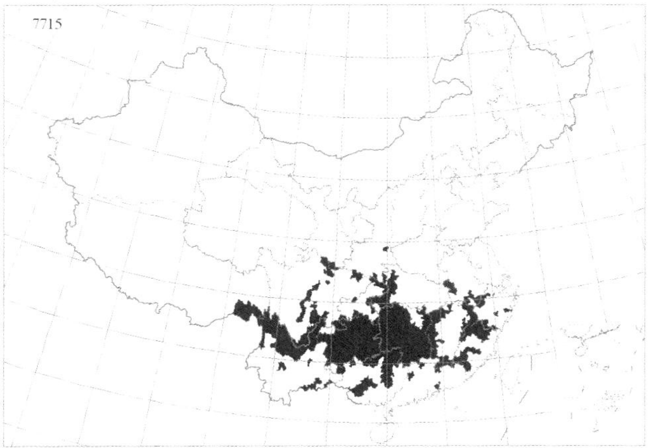

Ericaceae 杜鹃花科 *Enkianthus* 吊钟花属

Enkianthus deflexus (Griff.) C. K. Schneid. 毛叶吊钟花

Shrubs or small trees, deciduous

12.1 (-1.8~18.9); 3.6 (-10.2~12.0); 19.5 (6.1~26.6); 12.3 (1.8~19.0); 93.1 (1.9~167.8); 7.4 (0.0~82.8); 685 (329~929) ❙ 1041 (320~1736); 566 (217~1007); 39 (7~107) ❙ 53.3 (-14.2~145.8); 672 (320~912); 260.4 (72.2~487.2)

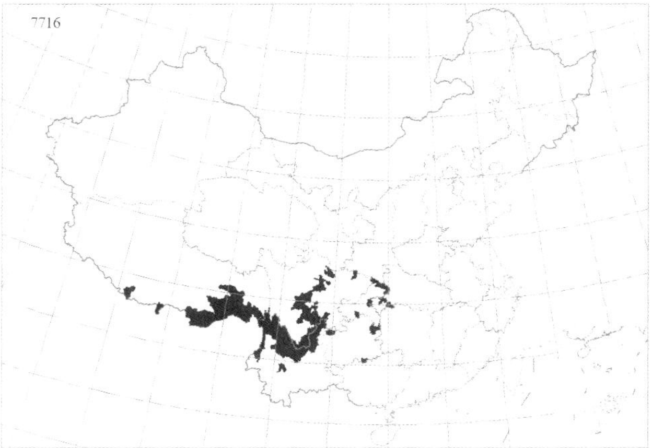

Ericaceae 杜鹃花科 *Enkianthus* 吊钟花属

Enkianthus deflexus var. glabrescens R. C. Fang 腺梗吊钟花

Shrubs or trees, deciduous

10.3; -0.1; 19.8; 10.3; 75.8; 11.7; 634 ❙ 711; 357; 15 ❙ 12.1; 634; 279.5

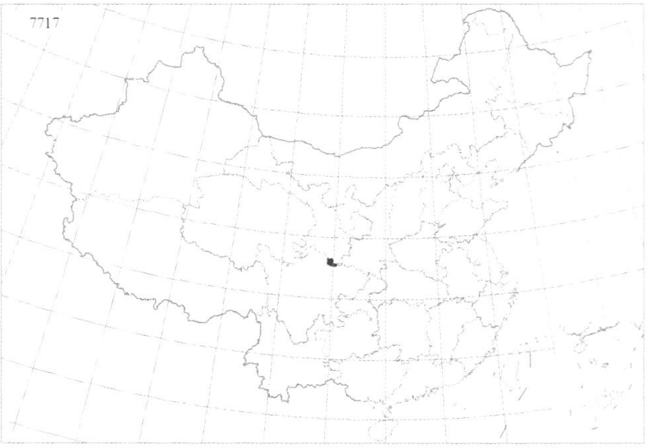

Ericaceae 杜鹃花科 *Enkianthus* 吊钟花属

Enkianthus pauciflorus Wils. 少花吊钟花

Shrubs, deciduous

16.6 (5.4~21.3); 6.6 (-2.1~12.3); 25.4 (12.3~28.7); 16.7 (5.8~21.3); 141.7 (31.5~196.1); 1.7 (0.0~25.6); 891 (481~1130) ❙ 1494 (741~1776); 613 (419~817); 155 (26~198) ❙ 69.5 (22.7~101.8); 889 (481~1130); 238.9 (177.4~352.3)

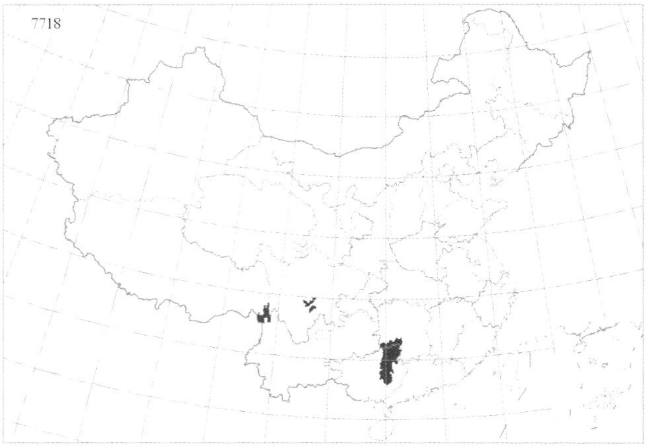

Ericaceae 杜鹃花科 *Enkianthus* 吊钟花属

Enkianthus perulatus C. K. Schneid. 台湾吊钟花

Small shrubs, deciduous

19.7; 13.1; 25.7; 19.8; 177.5; 0.0; 978 ❙ 2208; 750; 341 ❙ 125.7; 978; 260.5

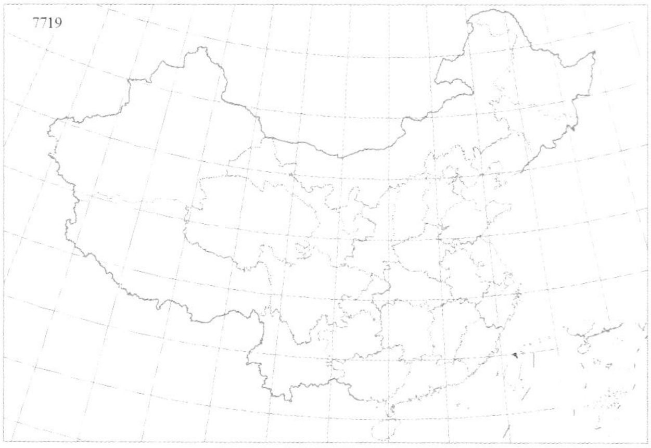

Ericaceae 杜鹃花科 *Enkianthus* 吊钟花属

Enkianthus quinqueflorus Lour. 吊钟花

Shrubs or small trees, deciduous

19.3 (13.4~24.8); 10.0 (2.6~20.1); 27.3 (21.4~29.8); 19.4 (13.4~24.8); 172.5 (104.5~237.6); 0.1 (0.0~4.1); 1020 (735~1385) ❙ 1599 (868~2435); 644 (312~1428); 151 (41~259) ❙ 59.2 (-34.3~130.7); 1009 (735~1360); 251.2 (95.3~620.0)

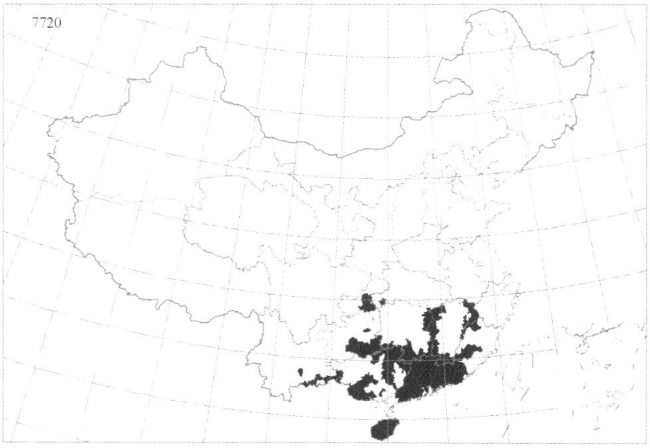

Ericaceae 杜鹃花科 *Enkianthus* 吊钟花属

Enkianthus serotinus Chun et Fang 晚花吊钟花

Shrubs, deciduous

19.7 (15.5~22.3); 10.7 (4.8~14.2); 26.9 (22.5~28.9); 19.7 (15.6~22.3); 176.8 (127.2~208.1); 0.0 (0.0~0.2); 1023 (815~1202) ❙ 1532 (1250~1758); 719 (556~911); 138 (56~195) ❙ 51.6 (20.3~93.2); 1022 (815~1202); 283.8 (185.5~551.6)

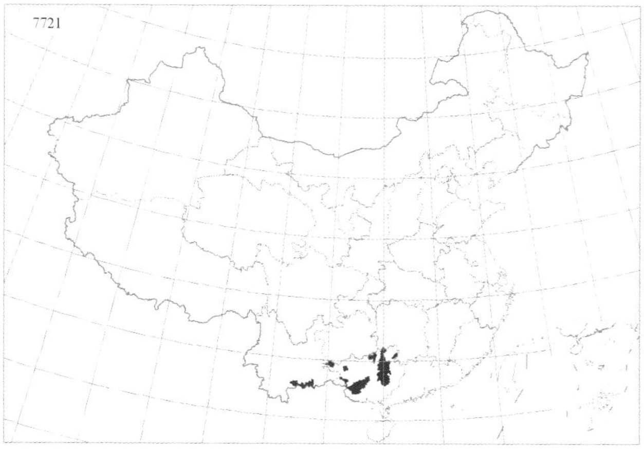

Ericaceae 杜鹃花科 *Enkianthus* 吊钟花属

Enkianthus serrulatus (E. H. Wilson) C. K. Schneid. 齿缘吊钟花

Shrubs or small trees, deciduous

17.7 (11.9~24.8); 7.5 (1.6~20.1); 26.9 (20.0~29.8); 17.8 (12.0~24.8); 153.4 (87.2~237.6); 0.4 (0.0~6.9); 941 (671~1385) ❙ 1478 (868~2024); 561 (312~891); 146 (27~259) ❙ 59.1 (-34.3~130.7); 927 (671~1360); 237.3 (130.1~620.0)

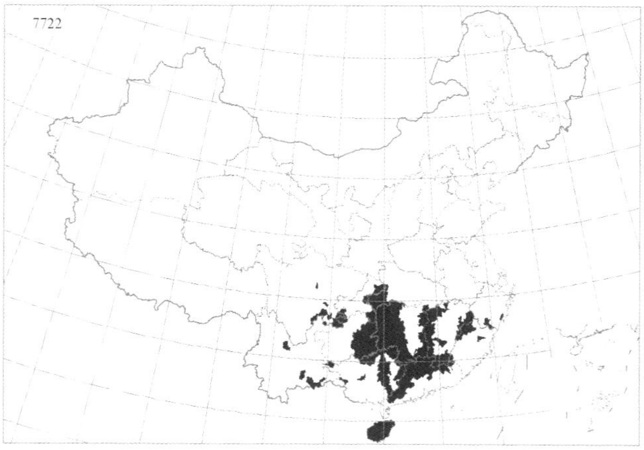

Ericaceae 杜鹃花科 *Gaultheria* 白珠树属
Gaultheria borneensis Stapf 高山白珠
Small shrubs, evergreen

19.3 (16.2~24.1)；14.0 (11.0~18.9)；23.5 (20.1~28.0)；19.3 (16.2~24.1)；171.8 (134.2~229.7)；0.0 (0.0~0.0)；951 (773~1322) ❙ 2473 (1751~2992)；1231 (947~1847)；194 (47~402) ❙ 170.4 (32.5~246.5)；926 (773~1123)；265.1 (154.2~363.7)

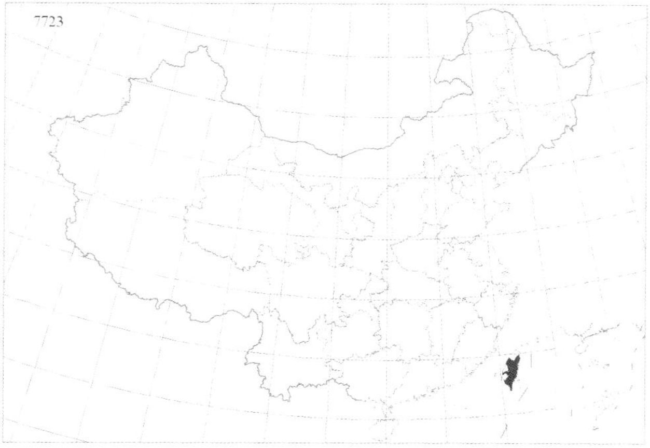

Ericaceae 杜鹃花科 *Gaultheria* 白珠树属
Gaultheria brevistipes (C. Y. Wu et T. Z. Xu) R. C. Fang 短柄白珠
Shrubs, evergreen

9.3；1.5；15.9；9.6；65.5；13.1；592 ❙ 1348；783；41 ❙ 123.3；592；340.8

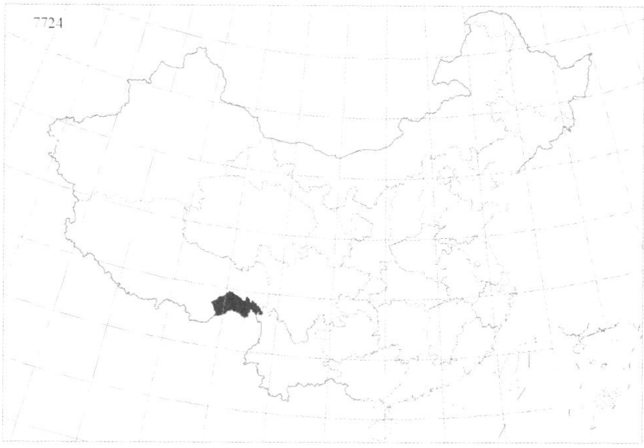

Ericaceae 杜鹃花科 *Gaultheria* 白珠树属
Gaultheria cardiosepala Hand.-Mazz. 苍山白珠
Subshrubs, evergreen

13.6 (5.4~18.9)；6.1 (-2.1~12.0)；19.7 (12.3~26.6)；13.6 (5.8~19.0)；105.1 (31.5~167.8)；1.7 (0.0~25.6)；719 (481~929) ❙ 1001 (741~1348)；578 (411~762)；35 (15~69) ❙ 41.0 (-14.2~93.0)；693 (481~912)；289.7 (142.9~407.1)

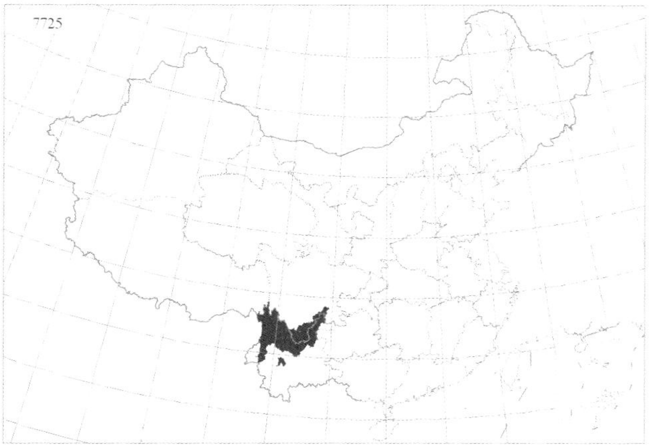

Ericaceae 杜鹃花科 *Gaultheria* 白珠树属
Gaultheria cuneata (Rehder et E. H. Wilson) Bean 四川白珠
Shrubs, evergreen

10.2 (2.9~15.2)；0.9 (-6.4~5.4)；18.6 (11.2~25.9)；10.5 (4.4~15.2)；75.4 (20.5~123.9)；12.6 (0.0~44.6)；634 (428~816) ❙ 1054 (741~1736)；554 (420~1007)；39 (12~94) ❙ 68.8 (18.6~145.8)；634 (428~816)；259.0 (150.8~447.4)

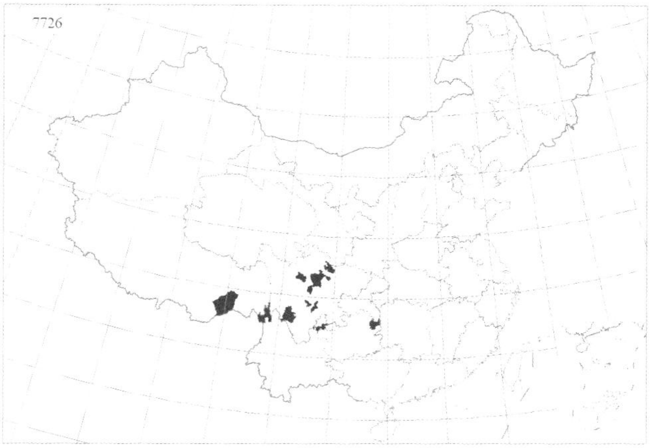

Ericaceae 杜鹃花科 *Gaultheria* 白珠树属
Gaultheria dolichopoda Airy Shaw 长梗白珠
Small shrubs, evergreen

8.3；0.0；15.3；9.0；61.4；20.9；566 ❙ 1104；647；29 ❙ 75.4；566；339.1

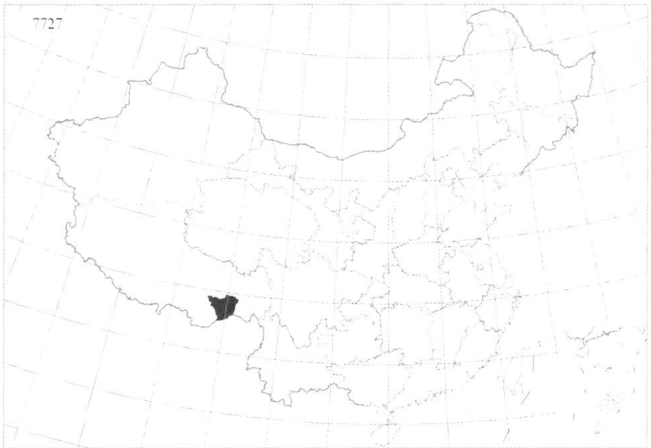

Ericaceae 杜鹃花科 *Gaultheria* 白珠树属
Gaultheria dumicola W. W. Sm. 丛林白珠
Shrubs, evergreen

11.8 (5.3~17.6)；4.7 (-2.4~10.0)；17.5 (12.2~23.2)；11.9 (5.7~17.7)；89.2 (30.5~152.4)；6.7 (0.0~26.2)；665 (479~859) ❙ 1212 (741~1561)；684 (426~897)；51 (28~70) ❙ 81.9 (54.0~100.7)；665 (479~859)；378.5 (234.2~539.7)

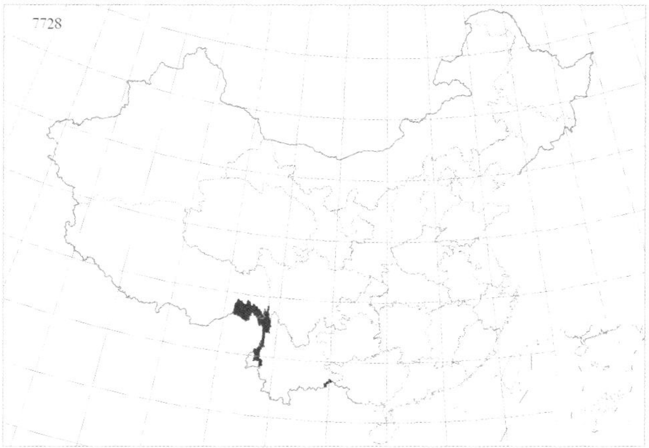

Ericaceae 杜鹃花科 *Gaultheria* 白珠树属
Gaultheria dumicola var. aspera Airy Shaw 粗糙丛林白珠
Shrubs, evergreen

10.4 (5.3~13.3); 3.0 (−2.4~6.4); 16.4 (12.2~19.6); 10.5 (5.7~13.4); 71.9 (30.5~100.4); 7.1 (0.0~26.2); 620 (479~706) ❙ 1264 (961~1736); 715 (558~1007); 51 (31~69) ❙ 102.7 (81.5~145.8); 620 (479~706); 365.1 (234.2~447.4)

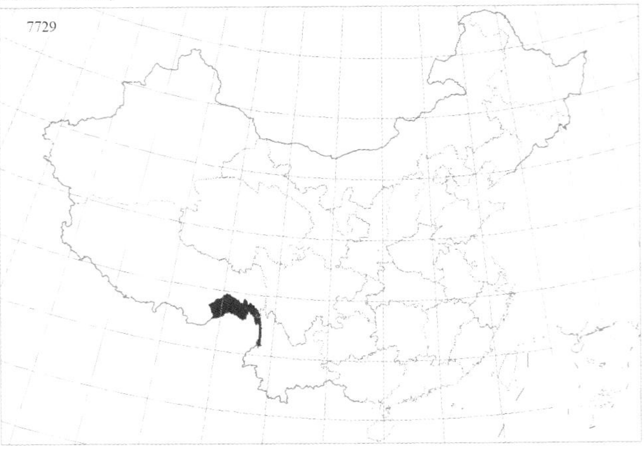

Ericaceae 杜鹃花科 *Gaultheria* 白珠树属
Gaultheria dumicola var. petanoneuron Airy Shaw 高山丛林白珠
Shrubs, evergreen

11.4 (5.3~15.5); 4.3 (−2.4~8.6); 17.0 (12.2~20.4); 11.5 (5.7~15.6); 83.2 (30.5~127.3); 5.9 (0.0~26.2); 648 (479~768) ❙ 1203 (961~1454); 663 (558~792); 56 (31~70) ❙ 86.9 (53.6~100.7); 648 (479~768); 366.8 (234.2~487.2)

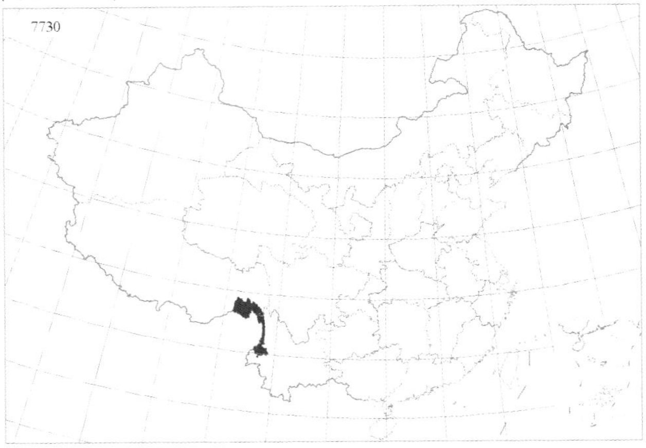

Ericaceae 杜鹃花科 *Gaultheria* 白珠树属
Gaultheria fragrantissima Wall. 芳香白珠
Shrubs, evergreen

14.2 (−1.8~21.0); 6.7 (−10.2~14.2); 20.2 (6.1~26.6); 14.3 (1.8~21.0); 114.2 (1.9~191.8); 3.0 (0.0~82.8); 745 (329~1053) ❙ 1069 (320~1736); 604 (217~1007); 41 (9~94) ❙ 44.4 (−14.2~145.8); 725 (320~1053); 313.6 (72.2~551.6)

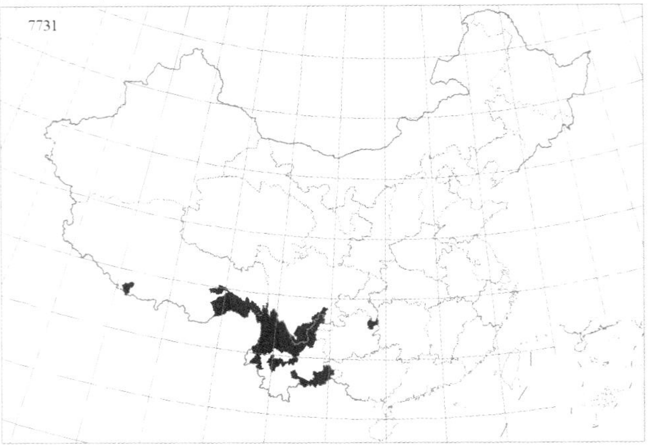

Ericaceae 杜鹃花科 *Gaultheria* 白珠树属
Gaultheria griffithiana Wight 尾叶白珠
Shrubs or small trees, evergreen

13.4 (5.4~18.9); 5.9 (−3.2~12.7); 19.6 (12.3~26.6); 13.5 (5.8~19.0); 103.9 (31.5~167.8); 2.1 (0.0~26.7); 716 (481~929) ❙ 1024 (741~1736); 588 (411~1007); 36 (15~71) ❙ 45.0 (−14.2~145.8); 692 (481~912); 297.1 (142.9~551.3)

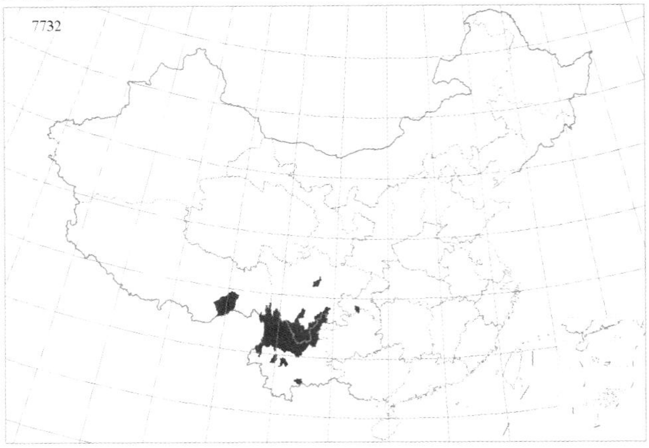

Ericaceae 杜鹃花科 *Gaultheria* 白珠树属
Gaultheria hookeri C. B. Clarke 红粉白珠
Shrubs, evergreen

13.1 (1.8~18.9); 5.2 (−7.1~12.7); 19.7 (9.9~26.6); 13.2 (3.8~19.0); 102.2 (15.5~167.8); 4.2 (0.0~53.3); 710 (404~929) ❙ 1043 (583~1736); 590 (348~1007); 37 (13~71) ❙ 49.3 (−14.2~145.8); 693 (404~912); 273.5 (142.9~551.3)

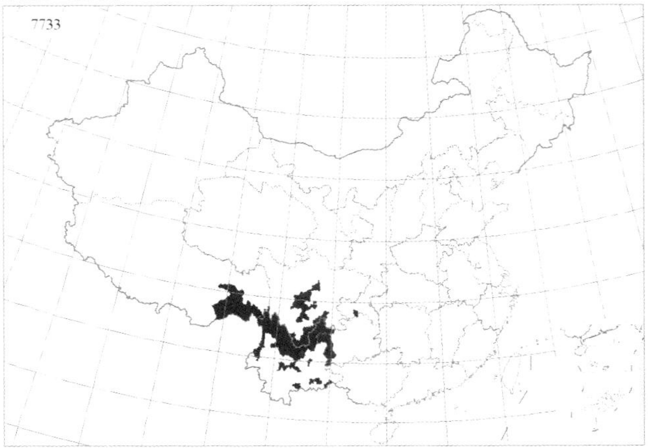

Ericaceae 杜鹃花科 *Gaultheria* 白珠树属
Gaultheria hookeri var. angustifolia C. B. Clarke 狭叶红粉白珠
Shrubs, evergreen

8.4; 1.2; 14.8; 8.6; 53.6; 11.7; 566 ❙ 1009; 569; 43 ❙ 76.2; 566; 336.8

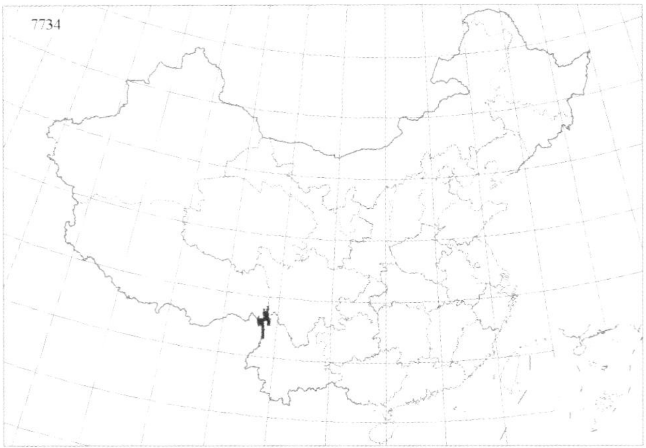

Ericaceae 杜鹃花科 *Gaultheria* 白珠树属
Gaultheria hypochlora Airy Shaw 绿背白珠
Shrubs, evergreen

11.7 (2.2~19.1); 3.8 (-6.6~13.0); 18.2 (9.9~24.3); 11.9 (3.8~19.1); 89.4 (15.5~169.7); 9.2 (0.0~49.1); 674 (413~917) ❙ 1066 (741~1428); 631 (426~835); 31 (12~56) ❙ 62.3 (5.5~125.9); 656 (413~917); 312.8 (214.5~506.2)

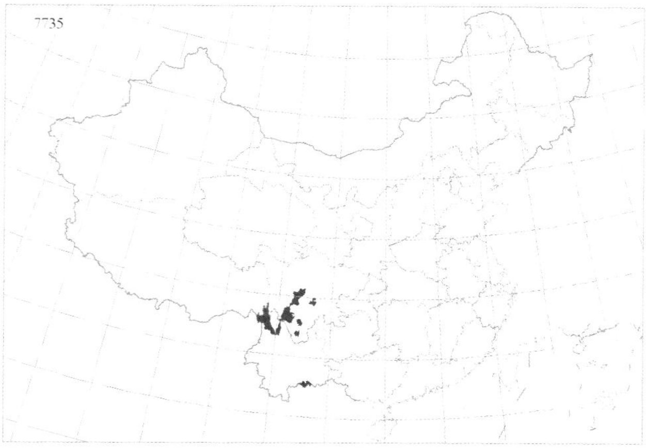

Ericaceae 杜鹃花科 *Gaultheria* 白珠树属
Gaultheria leucocarpa (Franch.) T. Z. Hsu et R. C. Fang 白果白珠
Shrubs, evergreen

17.5 (11.7~24.1); 8.6 (1.9~18.9); 25.3 (17.9~29.8); 17.6 (11.7~24.1); 151.2 (82.6~229.7); 0.2 (0.0~5.9); 909 (658~1322) ❙ 1554 (797~3495); 657 (403~1847); 149 (21~897) ❙ 71.0 (-14.2~246.5); 893 (658~1177); 247.0 (119.4~659.0)

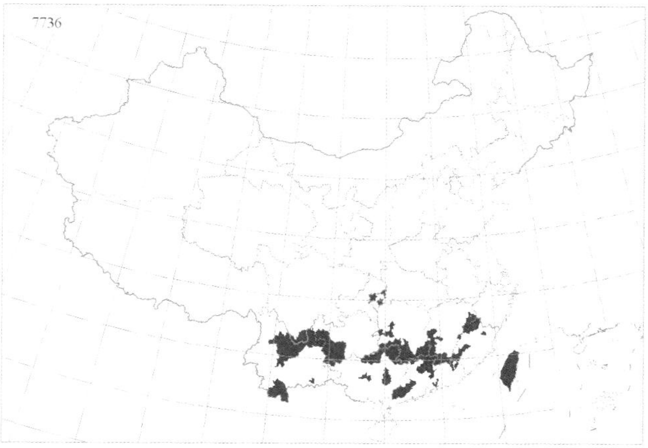

Ericaceae 杜鹃花科 *Gaultheria* 白珠树属
Gaultheria leucocarpa var. crenulata (Kurz) T. Z. Hsu 毛滇白珠
Shrubs, evergreen

16.7 (10.1~21.3); 5.7 (1.8~13.3); 26.8 (17.7~29.9); 16.7 (10.3~21.3); 141.4 (70.1~196.1); 0.5 (0.0~7.0); 890 (621~1130) ❙ 1377 (899~1742); 520 (392~835); 141 (18~221) ❙ 55.2 (8.0~128.9); 876 (621~1130); 210.7 (108.2~506.2)

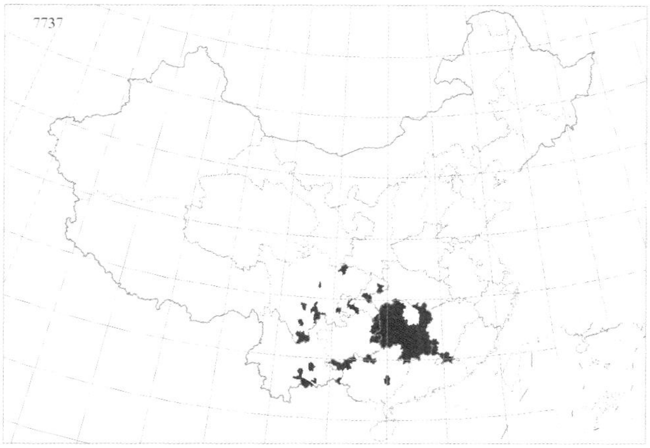

Ericaceae 杜鹃花科 *Gaultheria* 白珠树属
Gaultheria leucocarpa var. yunnanensis (Franch.) T. Z. Hsu et R. C. Fang 滇白珠
Shrubs, evergreen

16.8 (9.6~22.3); 7.1 (1.6~15.3); 25.4 (15.7~29.8); 16.8 (9.6~22.3); 142.6 (60.8~208.1); 0.4 (0.0~6.9); 877 (600~1202) ❙ 1421 (874~3495); 594 (392~1412); 129 (18~897) ❙ 62.2 (6.5~246.5); 873 (600~1202); 233.3 (145.3~551.3)

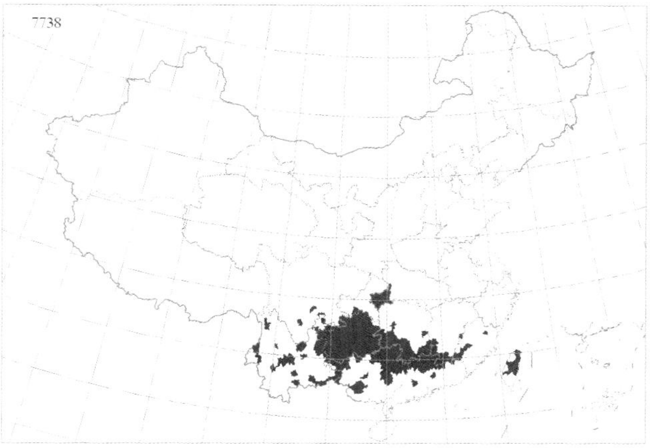

Ericaceae 杜鹃花科 *Gaultheria* 白珠树属
Gaultheria longiracemosa Y. C. Yang 长序白珠
Shrubs, evergreen

7.3; -2.0; 15.6; 7.5; 48.2; 20.4; 537 ❙ 1005; 531; 27 ❙ 87.3; 537; 272.3

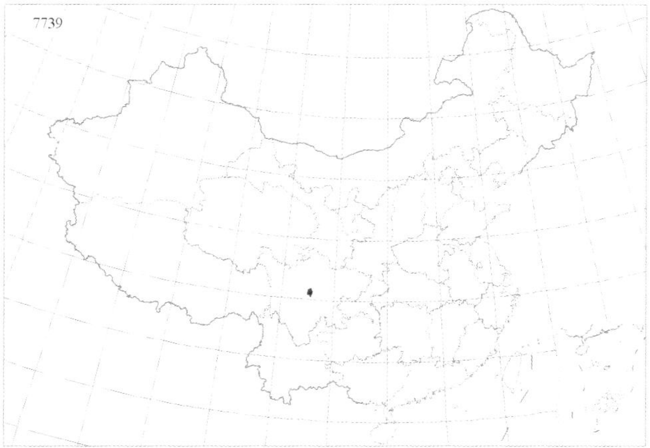

Ericaceae 杜鹃花科 *Gaultheria* 白珠树属
Gaultheria notabilis J. Anthony 短穗白珠
Shrubs, evergreen

11.5; 4.4; 17.1; 11.5; 82.8; 4.4; 651 ❙ 1248; 688; 57 ❙ 90.7; 651; 419.8

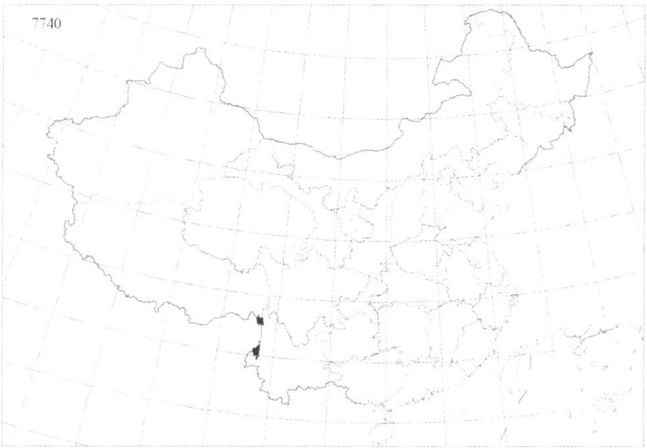

Ericaceae 杜鹃花科 *Gaultheria* 白珠树属

Gaultheria nummularioides D. Don 铜钱叶白珠

Shrubs, evergreen

10.4 (−1.8~19.2); 2.2 (−10.2~13.3); 17.4 (6.1~24.4); 10.9 (1.8~19.4); 80.7 (1.9~172.6); 15.5 (0.0~82.8); 636 (329~932) ▎1040 (320~1736); 588 (217~1007); 36 (7~70) ▎62.5 (−3.0~145.8); 633 (320~923); 292.1 (72.2~506.2)

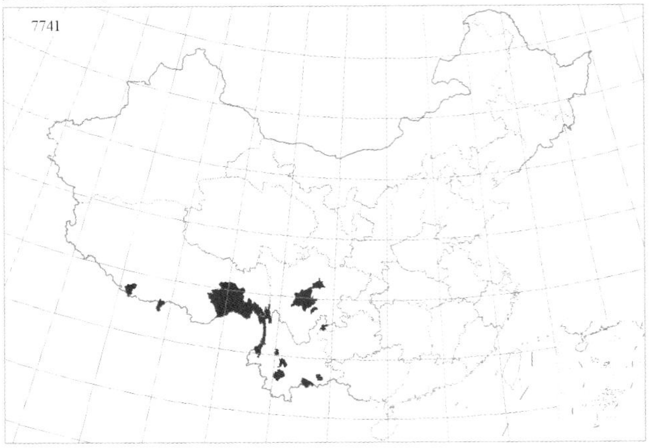

Ericaceae 杜鹃花科 *Gaultheria* 白珠树属

Gaultheria praticola C. Y. Wu et T. A. Hsu 草地白珠

Shrubs, evergreen

7.8 (1.8~13.3); −0.1 (−7.1~5.4); 14.6 (9.9~19.6); 8.3 (3.8~13.4); 52.7 (15.7~100.4); 18.7 (0.0~53.3); 553 (404~706) ▎1054 (583~1736); 600 (348~1007); 36 (13~58) ▎85.3 (44.2~145.8); 553 (404~706); 308.1 (169.5~447.4)

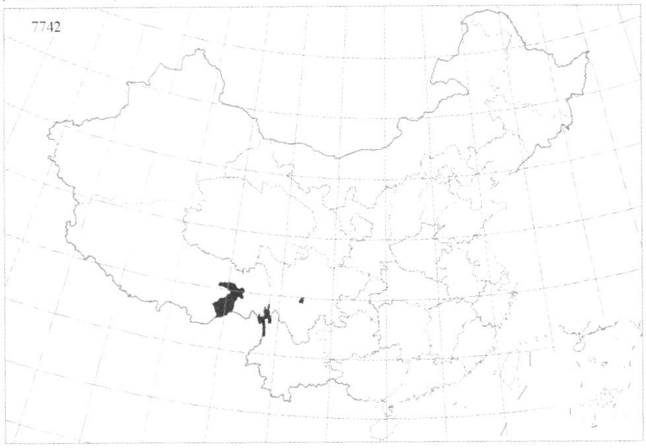

Ericaceae 杜鹃花科 *Gaultheria* 白珠树属

Gaultheria prostrata W. W. Sm. 平卧白珠

Small shrubs, evergreen

8.6; 1.3; 15.0; 8.7; 52.6; 8.8; 574 ▎1041; 585; 45 ▎81.5; 574; 352.3

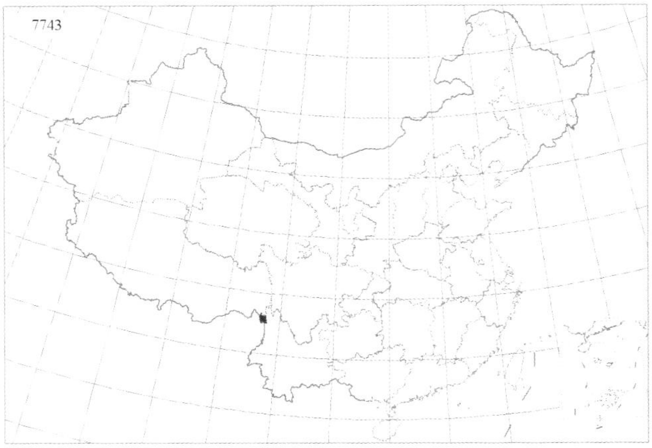

Ericaceae 杜鹃花科 *Gaultheria* 白珠树属

Gaultheria pseudonotabilis H. Li ex R. C. Fang 假短穗白珠

Shrubs, evergreen

8.6; 1.3; 15.0; 8.7; 52.6; 8.8; 574 ▎1041; 585; 45 ▎81.5; 574; 352.3

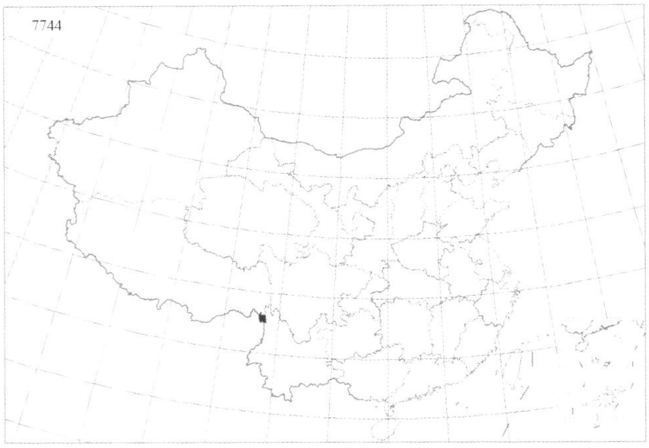

Ericaceae 杜鹃花科 *Gaultheria* 白珠树属

Gaultheria pyrolifolia Hook. f. ex C. B. Clarke 鹿蹄草叶白珠

Small shrubs, evergreen

9.9 (1.8~19.1); 1.8 (−7.1~13.0); 16.9 (9.9~24.0); 10.3 (3.8~19.1); 75.3 (15.7~169.7); 16.1 (0.0~53.3); 626 (404~917) ▎1064 (583~1736); 612 (348~1007); 34 (13~56) ▎70.4 (18.6~145.8); 626 (404~917); 303.0 (150.8~506.2)

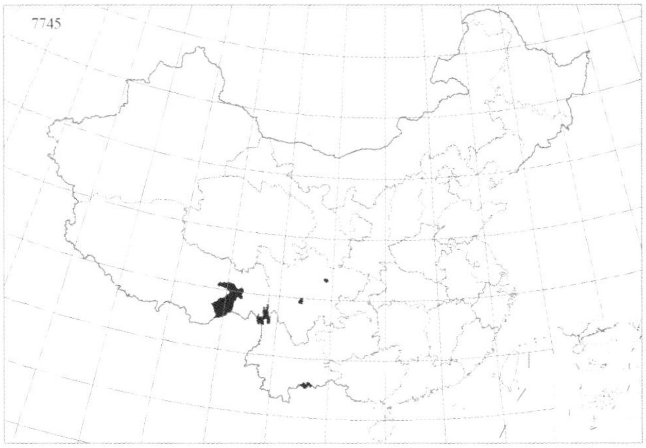

Ericaceae 杜鹃花科 *Gaultheria* 白珠树属

Gaultheria semi-infera (C. B. Clarke) Airy Shaw 五雄白珠

Shrubs, evergreen

12.5 (0.3~19.6); 5.3 (−8.6~13.9); 18.0 (8.3~22.7); 12.7 (2.9~19.6); 98.2 (9.6~175.1); 8.3 (0.0~65.4); 692 (368~933) ▎1138 (389~1736); 637 (246~1007); 42 (11~69) ▎65.3 (0.0~145.8); 678 (368~882); 354.7 (142.5~530.1)

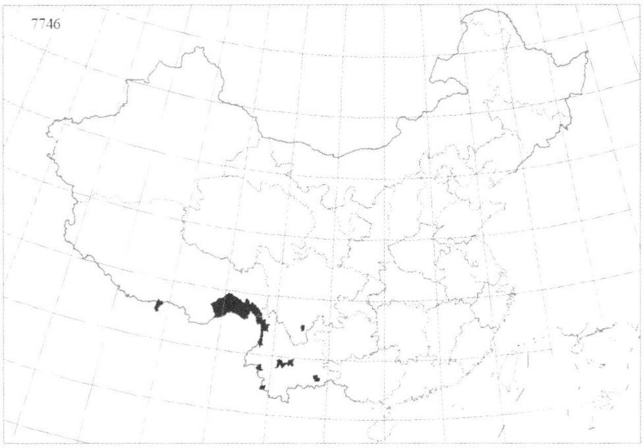

Ericaceae 杜鹃花科 *Gaultheria* 白珠树属
Gaultheria sinensis J. Anthony 华白珠
Subshrubs, evergreen

13.1 (−1.8~20.5); 5.3 (−10.2~14.9); 19.5 (6.1~26.6); 13.2 (1.8~20.5); 101.4 (1.9~185.9); 4.3 (0.0~82.8); 708 (329~992) ❙ 1007 (320~1736); 580 (217~1007); 33 (15~69) ❙ 44.2 (−14.2~145.8); 683 (320~959); 279.4 (72.2~570.1)

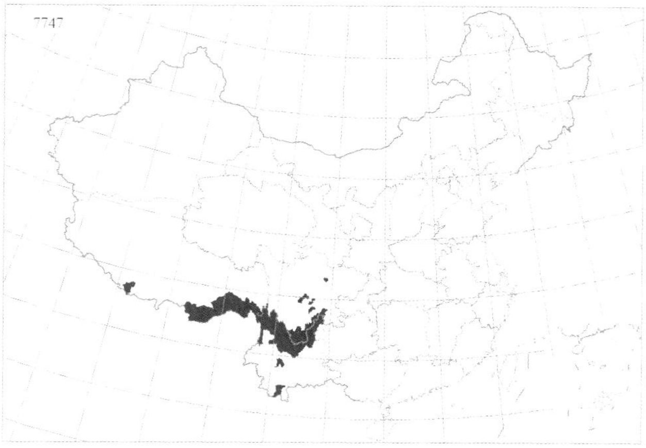

Ericaceae 杜鹃花科 *Gaultheria* 白珠树属
Gaultheria sinensis var. nivea J. Anthony 白果华白珠
Subshrubs, evergreen

8.4; 1.1; 14.7; 8.6; 53.2; 11.9; 566 ❙ 1082; 613; 44 ❙ 91.7; 566; 323.7

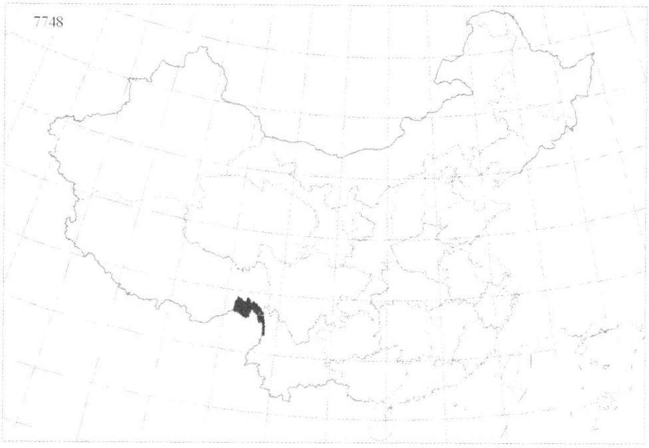

Ericaceae 杜鹃花科 *Gaultheria* 白珠树属
Gaultheria straminea R. C. Fang 草黄白珠
Shrubs, evergreen

9.3; 1.5; 15.9; 9.6; 65.5; 13.1; 592 ❙ 1348; 783; 41 ❙ 123.3; 592; 340.8

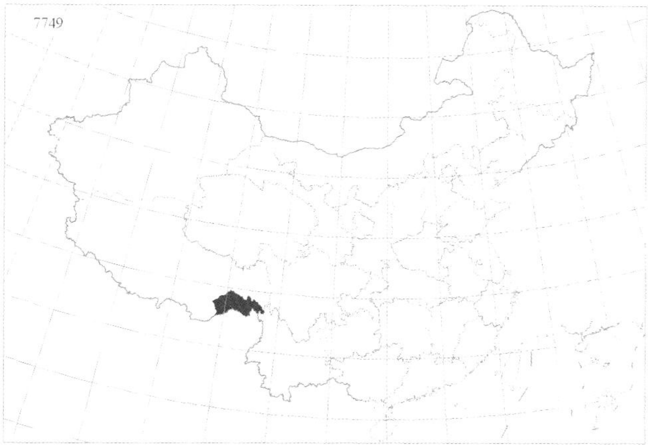

Ericaceae 杜鹃花科 *Gaultheria* 白珠树属
Gaultheria suborbicularis W. W. Sm. 伏地白珠
Dwarf shrubs, evergreen

7.2; −0.2; 13.8; 7.5; 43.8; 16.5; 533 ❙ 938; 537; 36 ❙ 76.0; 533; 302.9

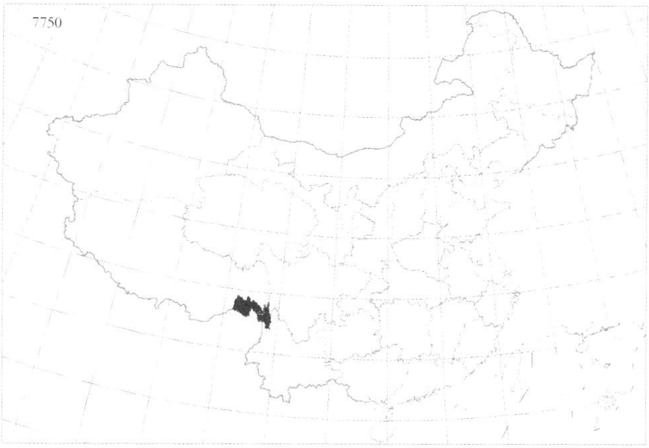

Ericaceae 杜鹃花科 *Gaultheria* 白珠树属
Gaultheria tetramera W. W. Sm. 四裂白珠
Shrubs, evergreen

13.6 (8.6~17.4); 6.2 (1.3~11.0); 19.2 (15.0~21.4); 13.6 (8.7~17.4); 105.3 (52.6~149.3); 1.8 (0.0~8.8); 715 (574~834) ❙ 1378 (1041~1736); 777 (585~1007); 53 (45~70) ❙ 93.3 (68.7~145.8); 715 (574~834); 387.4 (215.5~487.2)

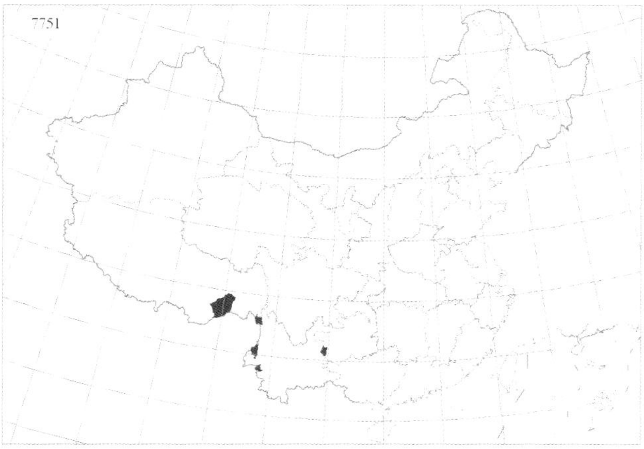

Ericaceae 杜鹃花科 *Gaultheria* 白珠树属
Gaultheria trichophylla Royle 刺毛白珠
Shrubs, evergreen

11.4 (−1.8~18.9); 3.6 (−10.5~12.0); 18.0 (6.1~26.6); 11.8 (1.8~19.0); 89.0 (1.9~167.8); 11.9 (0.0~82.8); 661 (329~929) ❙ 893 (311~1431); 523 (217~817); 28 (2~58) ❙ 35.2 (−25.4~125.9); 634 (311~912); 259.9 (61.6~384.7)

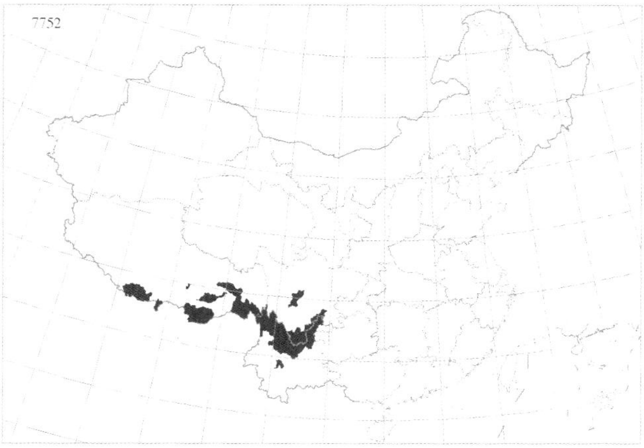

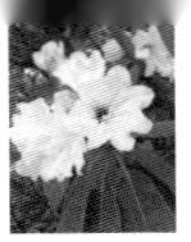

Ericaceae 杜鹃花科 *Gaultheria* 白珠树属
Gaultheria wardii C. Marquand et Airy Shaw 西藏白珠
Shrubs, evergreen

6.5 (1.8~13.3); -1.6 (-7.1~5.4); 13.6 (9.9~19.6); 7.2 (3.8~13.4); 44.5 (15.7~100.4); 26.0 (0.0~53.3); 519 (404~706) ❙ 970 (527~1736); 563 (317~1007); 30 (9~51) ❙ 78.9 (22.4~145.8); 519 (404~706); 296.7 (169.5~447.4)

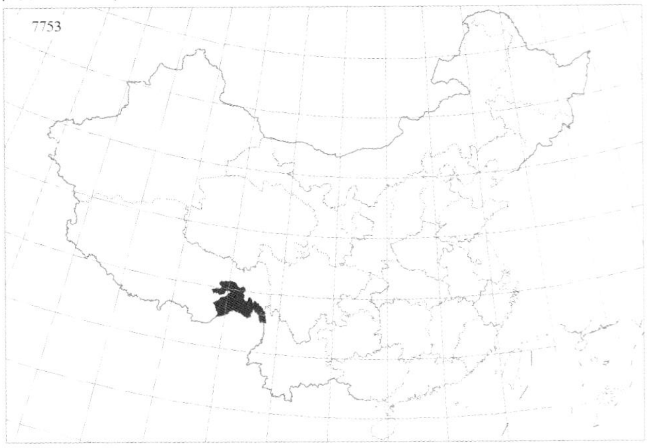

Ericaceae 杜鹃花科 *Ledum* 杜香属
Ledum palustre L. 杜香
Shrubs

-0.1 (-5.8~2.7); -21.6 (-29.8~-16.0); 18.7 (15.3~20.6); 6.4 (4.4~7.7); 46.5 (28.0~57.3); 106.8 (81.0~157.4); 507 (426~557) ❙ 604 (371~853); 389 (260~527); 21 (10~35) ❙ 18.1 (-26.4~57.4); 487 (371~557); 305.3 (217.2~346.5)

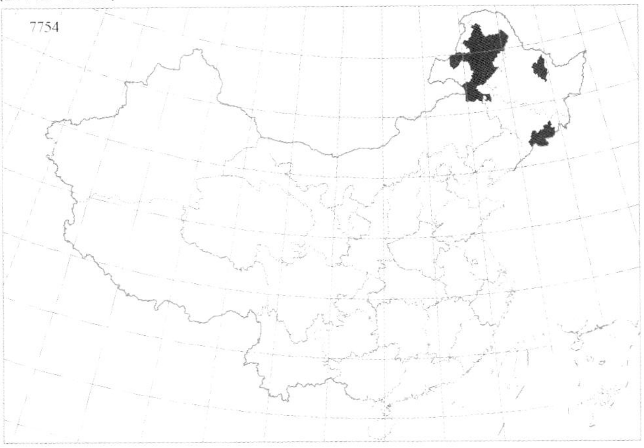

Ericaceae 杜鹃花科 *Ledum* 杜香属
Ledum palustre var. decumbens Aiton 小叶杜香
Shrubs, evergreen

0.7 (-5.8~4.5); -20.9 (-29.8~-14.1); 19.4 (15.3~21.6); 6.8 (4.4~8.2); 50.6 (28.0~63.5); 101.8 (68.3~157.4); 526 (426~584) ❙ 623 (447~854); 387 (300~528); 21 (12~35) ❙ 18.0 (-12.2~57.4); 523 (426~583); 305.4 (263.0~346.5)

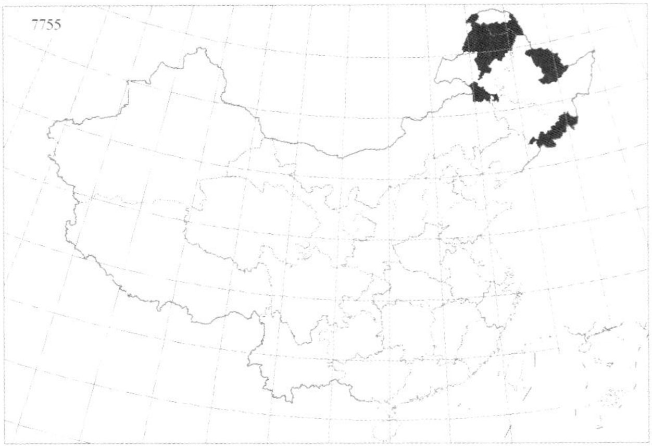

Ericaceae 杜鹃花科 *Ledum* 杜香属
Ledum palustre var. dilatatum Wahlenb. 宽叶杜香
Shrubs, evergreen

1.0 (-5.0~5.2); -20.7 (-29.4~-14.1); 19.8 (17.0~21.8); 7.0 (5.0~8.6); 52.5 (35.2~68.0); 100.4 (65.0~154.8); 535 (450~595) ❙ 630 (468~1005); 387 (300~635); 21 (15~39) ❙ 17.5 (-2.6~68.8); 533 (450~595); 303.5 (263.0~333.3)

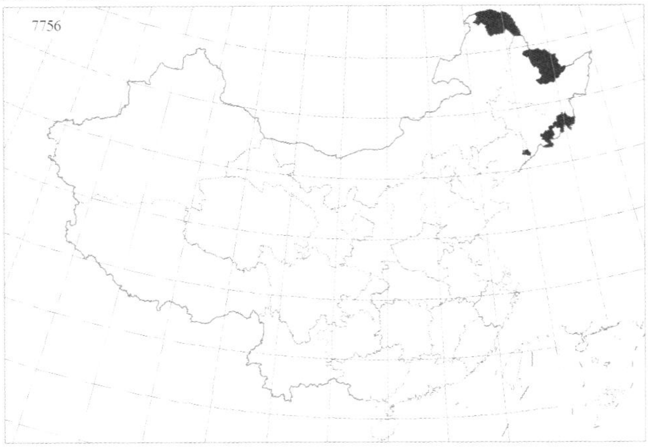

Ericaceae 杜鹃花科 *Leucothoe* 木藜芦属
Leucothoe griffithiana C. B. Clarke 尖基木藜芦
Shrubs, evergreen

13.0 (5.3~16.7); 4.3 (-2.4~6.9); 20.4 (12.2~25.1); 13.1 (5.7~16.7); 99.3 (30.5~140.7); 3.1 (0.0~26.2); 715 (479~851) ❙ 1217 (961~1736); 615 (511~1007); 62 (24~85) ❙ 72.4 (48.8~145.8); 715 (479~851); 290.7 (194.1~447.4)

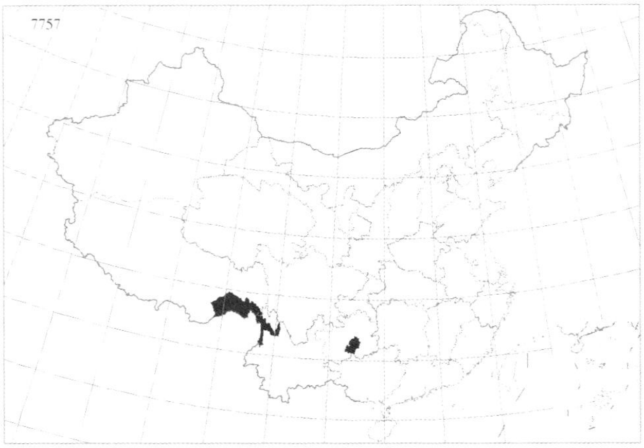

Ericaceae 杜鹃花科 *Leucothoe* 木藜芦属
Leucothoe tonkinensis Dop 圆基木藜芦
Shrubs, evergreen

18.5 (17.2~21.0); 11.4 (9.1~14.2); 23.4 (22.4~25.6); 18.5 (17.3~21.0); 161.9 (147.1~191.8); 0.0 (0.0~0.0); 901 (833~1053) ❙ 1397 (1161~1522); 781 (633~860); 58 (50~68) ❙ 55.6 (37.1~80.6); 901 (833~1053); 455.4 (314.4~551.6)

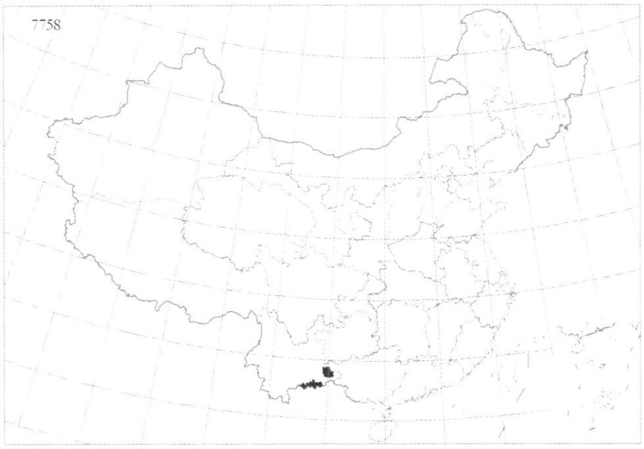

Ericaceae 杜鹃花科 *Rhododendron* 杜鹃花属

Rhododendron agastum var. pennivenium (I. B. Balfour et Forrest) T. L. Ming 光柱迷人杜鹃 Shrubs, evergreen

13.5 (11.3~14.8); 5.4 (3.6~7.5); 20.3 (17.0~24.0); 13.5 (11.3~14.9); 103.0 (76.6~118.8); 0.4 (0.0~1.8); 716 (644~778) ▌ 1228 (1077~1454); 645 (529~792); 62 (58~70) ▌ 72.4 (45.0~99.9); 716 (644~778); 311.8 (153.9~487.2)

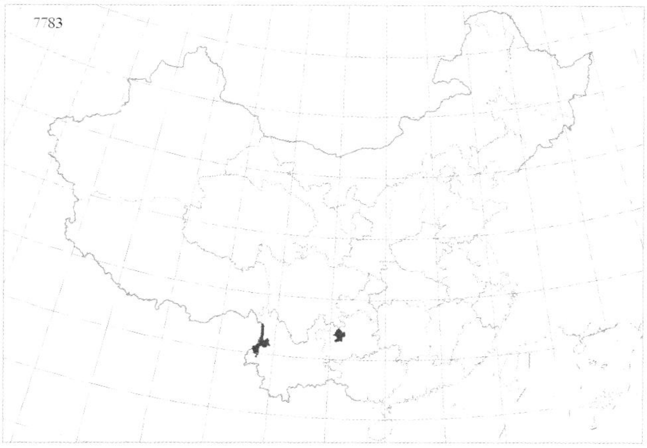

Ericaceae 杜鹃花科 *Rhododendron* 杜鹃花属

Rhododendron albertsenianum Forrest ex I. B. Balfour 亮红杜鹃
Shrubs, evergreen

7.5; 0.2; 14.0; 7.7; 46.1; 15.5; 541 ▌ 874; 502; 34 ▌ 61.0; 541; 312.6

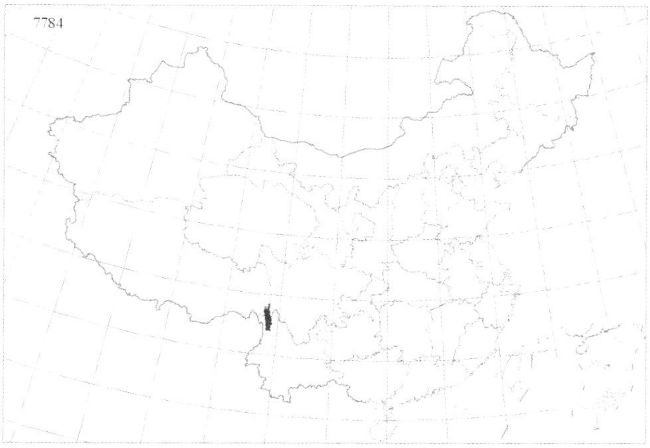

Ericaceae 杜鹃花科 *Rhododendron* 杜鹃花属

Rhododendron alutaceum I. B. Balfour et W. W. Sm. 棕背杜鹃
Shrubs, evergreen

6.8 (2.9~10.8); -1.1 (-6.3~3.9); 13.7 (11.2~16.6); 7.1 (4.4~10.9); 42.3 (20.5~71.8); 20.6 (1.3~44.6); 521 (428~633) ▌ 912 (741~1007); 528 (426~594); 26 (12~40) ▌ 76.7 (52.8~121.7); 521 (428~633); 289.2 (234.2~351.7)

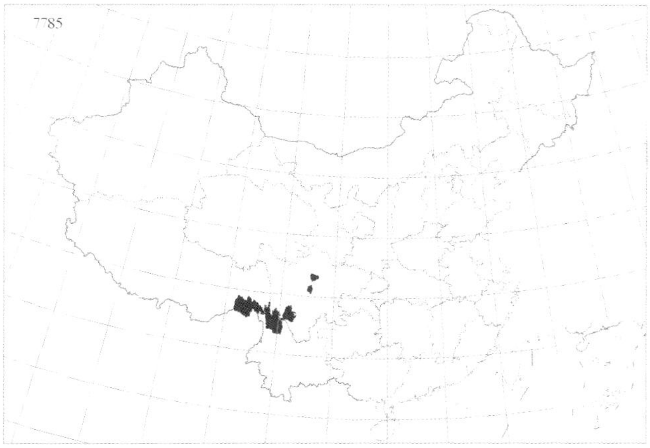

Ericaceae 杜鹃花科 *Rhododendron* 杜鹃花属

Rhododendron alutaceum var. iodes (I. B. Balfour et Forrest) D. F. Chamb. 毛枝棕背杜鹃 Shrubs, evergreen

7.3 (5.3~10.8); -0.2 (-2.4~3.9); 13.8 (12.2~16.6); 7.5 (5.7~10.9); 44.9 (30.5~71.8); 16.7 (1.3~26.2); 534 (479~633) ▌ 890 (741~1007); 534 (426~594); 26 (12~40) ▌ 67.4 (52.8~100.7); 534 (479~633); 295.6 (234.2~351.7)

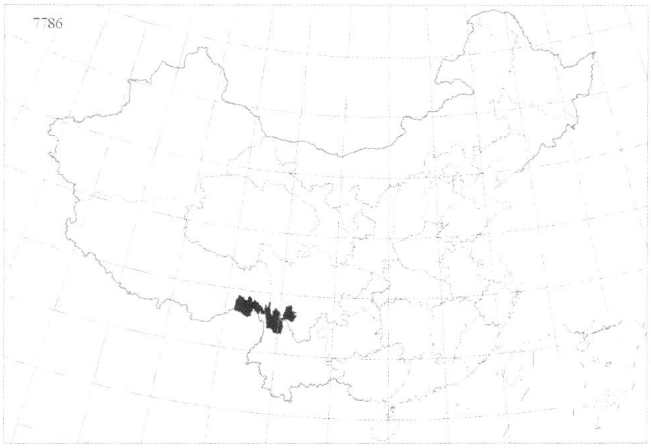

Ericaceae 杜鹃花科 *Rhododendron* 杜鹃花属

Rhododendron alutaceum var. russotinctum (I. B. Balfour et Forrest) D. F. Chamb. 腺房棕背杜鹃 Shrubs, evergreen

6.8; -0.7; 13.4; 7.1; 40.9; 19.1; 520 ▌ 903; 521; 33 ▌ 74.2; 520; 286.5

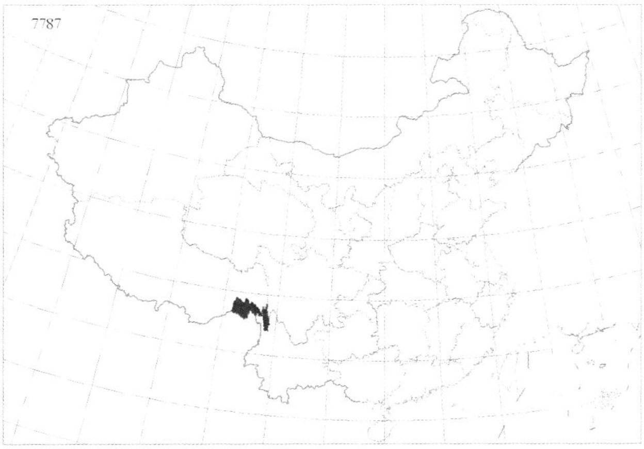

Ericaceae 杜鹃花科 *Rhododendron* 杜鹃花属

Rhododendron amandum Cowan 细枝杜鹃
Shrubs, evergreen

4.4; -4.3; 12.1; 5.3; 28.5; 34.6; 455 ▌ 538; 337; 9 ▌ 18.2; 455; 168.2

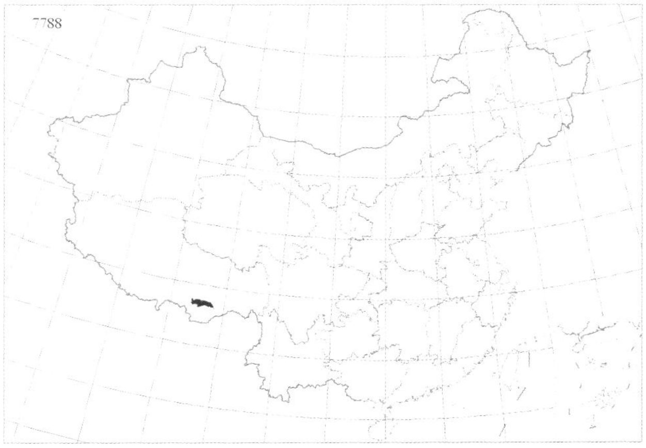

Ericaceae 杜鹃花科 *Rhododendron* 杜鹃花属
Rhododendron ambiguum Hemsl. 问客杜鹃
Shrubs, evergreen

14.5 (2.2~19.1); 4.5 (-6.6~10.1); 23.4 (9.9~27.2); 14.6 (3.8~19.1); 117.7 (15.5~169.2); 3.7 (0.0~49.1); 782 (413~969) ❙ 1215 (890~1494); 566 (444~817); 70 (15~161) ❙ 57.7 (27.5~125.9); 782 (413~969); 221.8 (139.5~298.0)

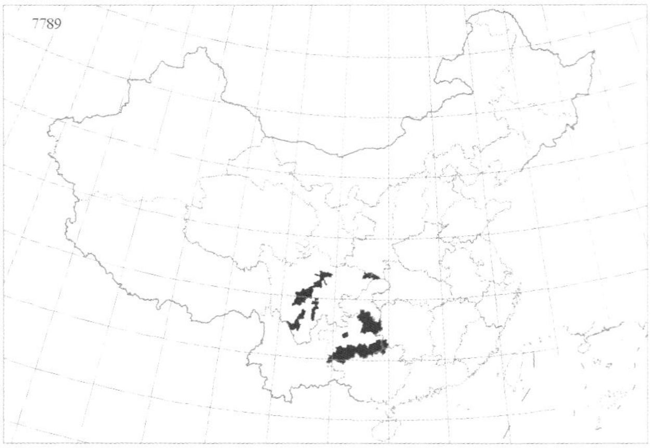

Ericaceae 杜鹃花科 *Rhododendron* 杜鹃花属
Rhododendron amesiae Rehder et E. H. Wilson 紫花杜鹃
Shrubs, evergreen

4.4 (2.0~7.3); -4.8 (-7.2~-2.0); 12.6 (9.9~15.6); 5.5 (3.8~7.5); 29.9 (15.4~48.2); 36.4 (20.4~51.4); 468 (413~537) ❙ 939 (812~1023); 505 (452~546); 23 (17~28) ❙ 102.2 (83.0~128.6); 468 (413~537); 265.3 (200.8~337.3)

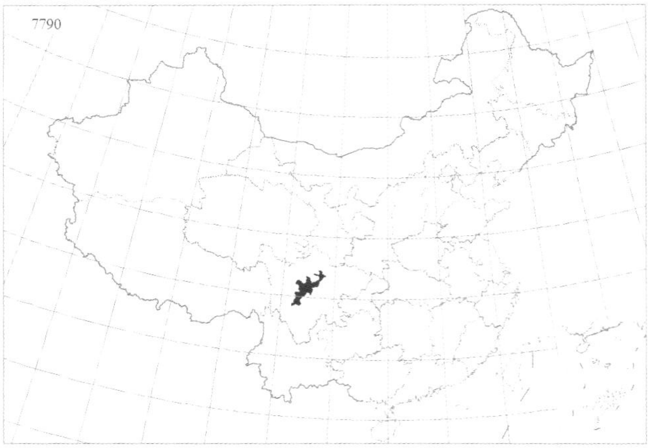

Ericaceae 杜鹃花科 *Rhododendron* 杜鹃花属
Rhododendron amundsenianum Hand.-Mazz. 暗叶杜鹃
Shrubs, evergreen

12.6; 4.9; 18.9; 12.6; 93.4; 2.1; 685 ❙ 1052; 626; 21 ❙ 54.9; 671; 305.6

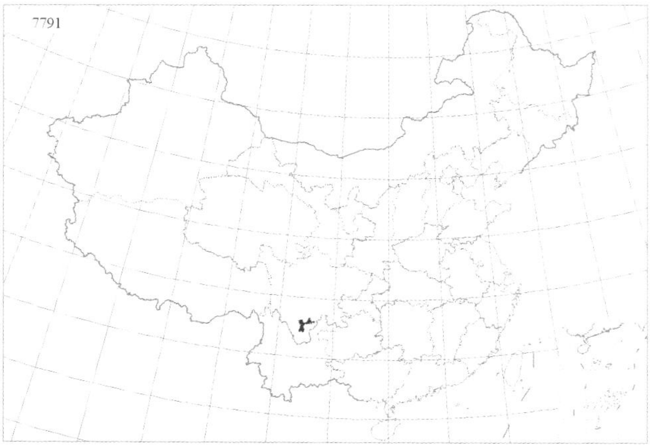

Ericaceae 杜鹃花科 *Rhododendron* 杜鹃花属
Rhododendron annae Franch. 桃叶杜鹃
Shrubs, evergreen

14.9 (11.7~19.6); 6.1 (2.7~11.2); 22.3 (18.8~27.5); 15.0 (11.7~19.6); 120.4 (82.6~175.5); 0.6 (0.0~3.8); 776 (658~1002) ❙ 1175 (916~1495); 616 (489~848); 57 (31~135) ❙ 51.5 (26.1~99.9); 774 (658~1002); 242.1 (153.9~487.2)

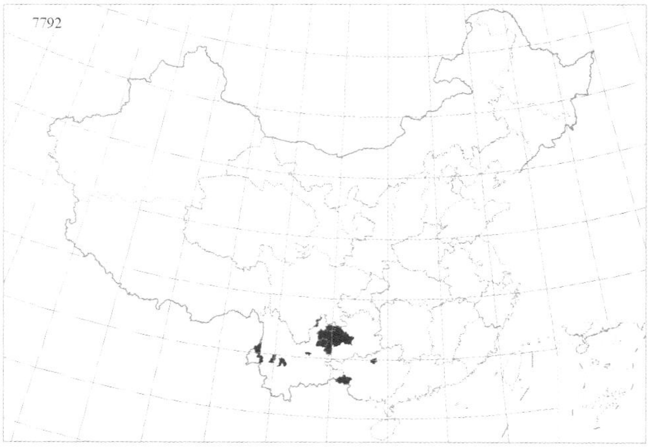

Ericaceae 杜鹃花科 *Rhododendron* 杜鹃花属
Rhododendron annae subsp. laxiflorum (I. B. Balfour et Forrest) T. L. Ming 滇西桃叶杜鹃
Shrubs, evergreen

14.4; 7.5; 19.1; 14.4; 113.1; 0.0; 727 ❙ 1454; 792; 70 ❙ 99.9; 727; 487.2

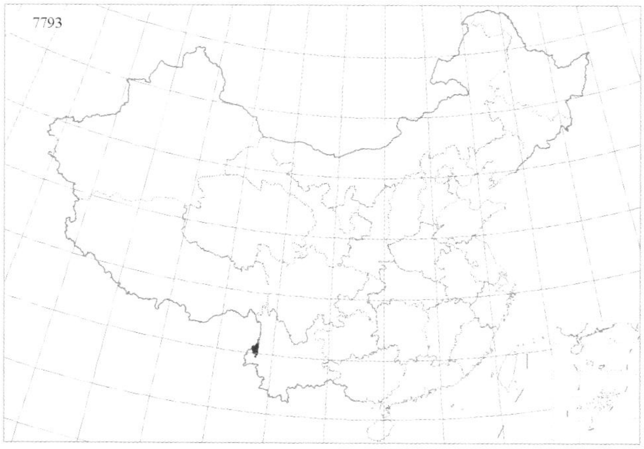

Ericaceae 杜鹃花科 *Rhododendron* 杜鹃花属
Rhododendron anthopogon D. Don 髯花杜鹃
Small shrubs, evergreen

1.7 (-1.8~12.2); -6.9 (-10.5~4.3); 9.5 (6.1~18.5); 4.3 (1.8~12.2); 22.1 (1.9~87.4); 61.1 (0.7~82.8); 408 (329~672) ❙ 564 (311~1431); 354 (217~817); 16 (10~31) ❙ 21.7 (-6.3~113.0); 402 (311~672); 140.5 (61.6~363.7)

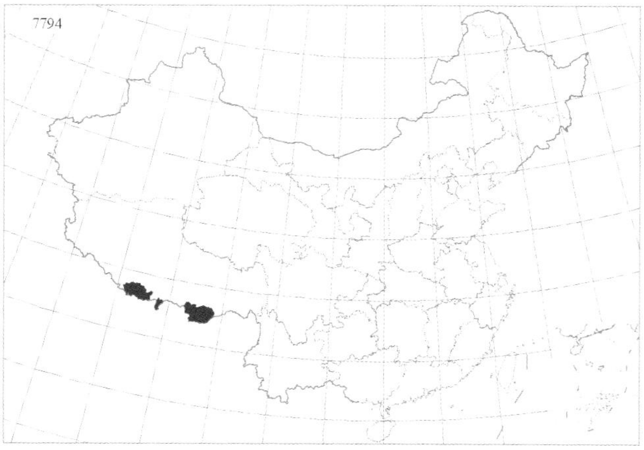

Ericaceae 杜鹃花科 *Rhododendron* 杜鹃花属
Rhododendron anthopogonoides Maxim. 烈香杜鹃
Shrubs, evergreen

2.7 (−1.3~6.4)；−9.0 (−13.5~−2.4)；12.9 (9.2~18.6)；5.0 (2.9~7.9)；27.9 (10.1~58.0)；55.3 (26.2~86.3)；449 (359~554) ❙ 559 (375~961)；305 (203~558)；11 (5~31) ❙ 25.7 (−32.2~125.9)；431 (359~522)；240.3 (95.7~364.1)

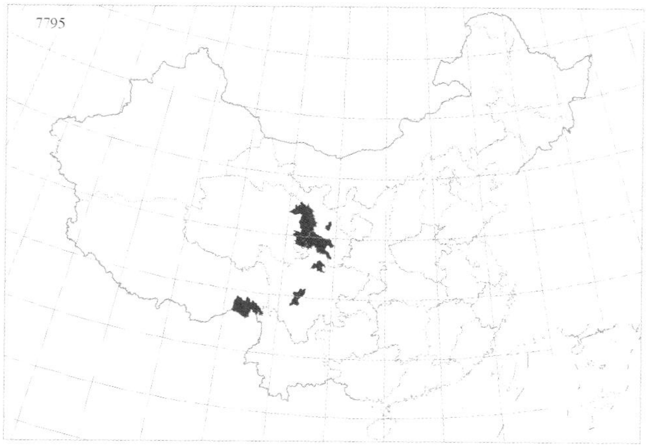

Ericaceae 杜鹃花科 *Rhododendron* 杜鹃花属
Rhododendron anthosphaerum Diels 团花杜鹃
Shrubs or small trees, evergreen

10.7 (5.3~14.5)；3.7 (−2.4~7.8)；16.5 (12.2~19.7)；10.8 (5.7~14.5)；75.2 (30.5~113.5)；6.5 (0.0~26.2)；627 (479~730) ❙ 1053 (741~1454)；604 (426~792)；42 (12~70) ❙ 68.5 (36.1~100.7)；626 (479~730)；344.9 (234.2~487.2)

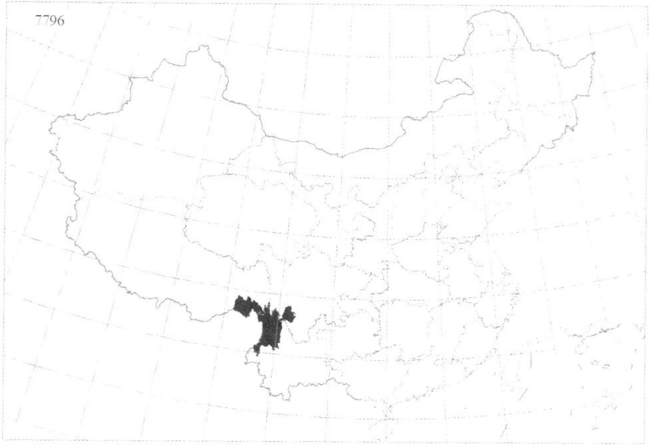

Ericaceae 杜鹃花科 *Rhododendron* 杜鹃花属
Rhododendron aperantum I. B. Balfour et Kingdon-Ward 宿鳞杜鹃
Shrubs, evergreen

11.4；4.4；17.0；11.4；78.9；2.0；646 ❙ 1196；667；56 ❙ 84.4；646；381.2

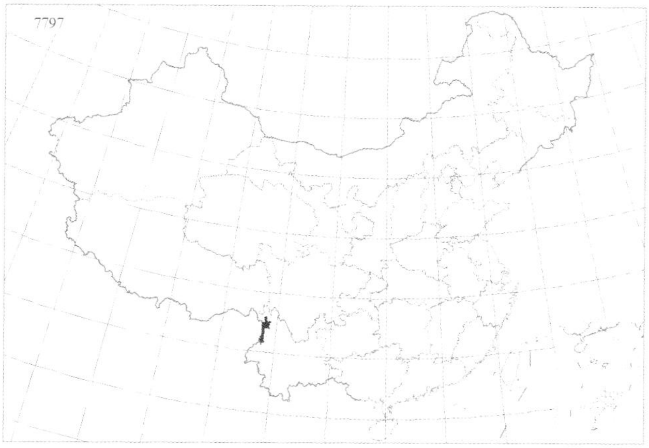

Ericaceae 杜鹃花科 *Rhododendron* 杜鹃花属
Rhododendron apricum Tam 茶绒杜鹃
Small shrubs, evergreen

17.8 (16.6~19.1)；8.4 (5.9~10.4)；26.4 (25.2~27.2)；17.9 (16.6~19.1)；154.2 (138.9~169.1)；0.0 (0.0~0.0)；913 (864~973) ❙ 1762 (1559~1997)；612 (549~669)；208 (163~254) ❙ 93.5 (68.2~130.7)；913 (864~973)；283.9 (254.0~305.7)

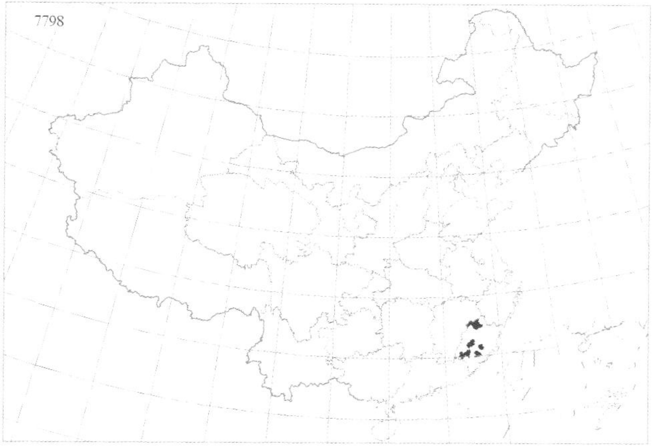

Ericaceae 杜鹃花科 *Rhododendron* 杜鹃花属
Rhododendron araiophyllum I. B. Balfour et W. W. Sm. 窄叶杜鹃
Shrubs, evergreen

14.0 (8.6~18.2)；5.5 (1.3~9.6)；20.9 (15.0~25.5)；14.0 (8.7~18.2)；109.2 (52.6~158.8)；1.1 (0.0~8.8)；736 (574~918) ❙ 1197 (916~1688)；643 (489~959)；55 (31~70) ❙ 63.1 (34.2~112.2)；736 (574~918)；274.3 (183.0~627.0)

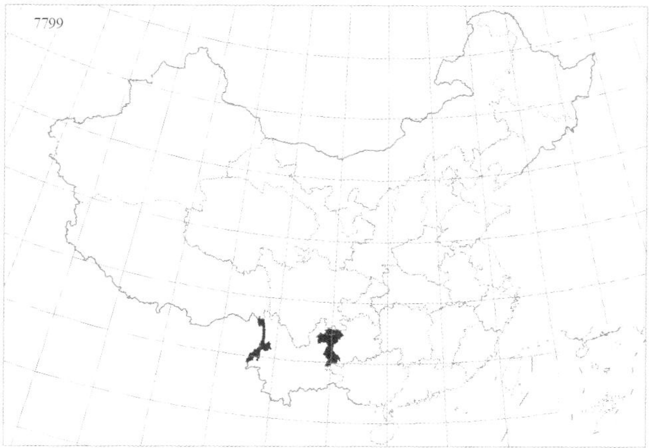

Ericaceae 杜鹃花科 *Rhododendron* 杜鹃花属
Rhododendron arboreum Sm. 树形杜鹃
Trees, evergreen

9.1 (−1.8~15.0)；0.0 (−10.2~5.4)；17.2 (6.1~24.6)；10.1 (1.8~15.1)；73.2 (1.9~121.5)；23.7 (0.0~82.8)；606 (329~792) ❙ 932 (320~1736)；499 (193~1007)；42 (10~65) ❙ 45.1 (−4.0~145.8)；605 (320~792)；188.4 (61.6~447.4)

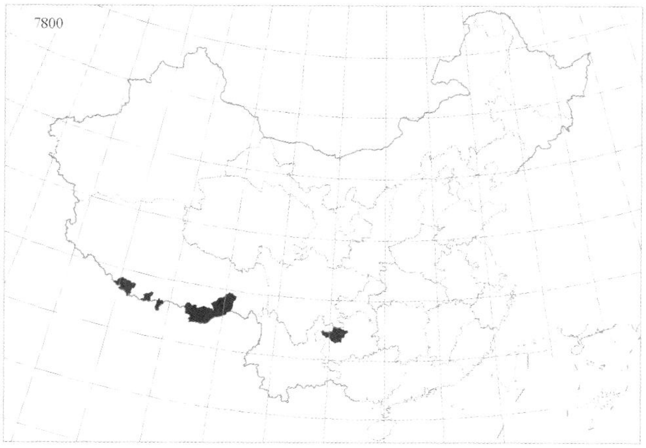

Ericaceae 杜鹃花科 *Rhododendron* 杜鹃花属
Rhododendron arboreum var. roseum Lindley 粉红树形杜鹃
Trees, evergreen

3.6; -4.8; 10.9; 5.6; 33.0; 49.6; 456 ❙ 713; 427; 21 ❙ 36.7; 453; 192.8

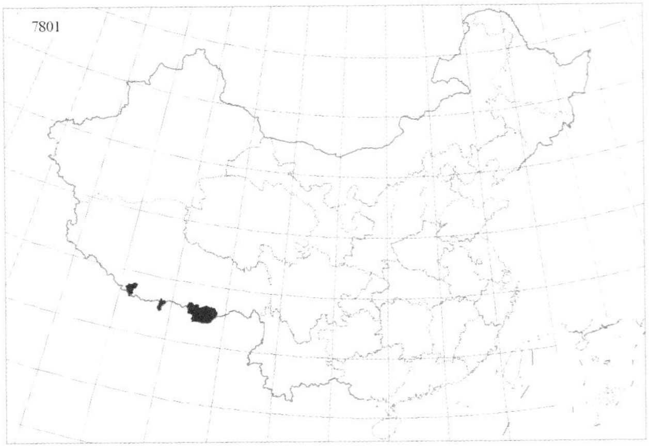

Ericaceae 杜鹃花科 *Rhododendron* 杜鹃花属
Rhododendron argipeplum I. B. Balfour et R. E. Cooper 毛枝杜鹃
Shrubs or small trees, evergreen

5.2; -2.9; 12.3; 7.0; 44.7; 41.8; 500 ❙ 875; 517; 25 ❙ 55.0; 496; 217.9

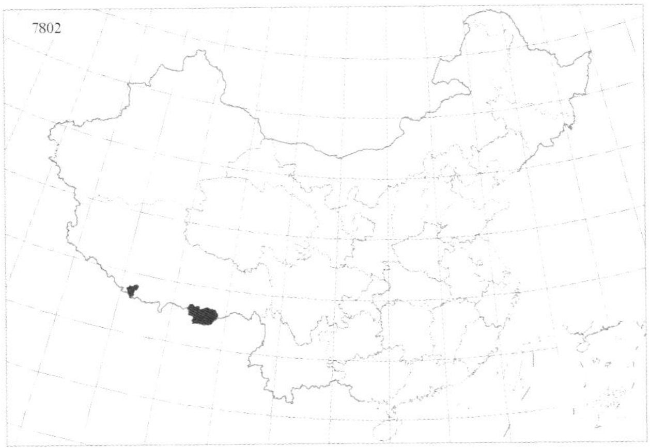

Ericaceae 杜鹃花科 *Rhododendron* 杜鹃花属
Rhododendron argyrophyllum Franch. 银叶杜鹃
Shrubs or small trees, evergreen

13.2 (2.2~17.2); 3.2 (-6.6~9.0); 22.5 (9.9~27.0); 13.3 (3.8~17.3); 104.1 (15.5~147.4); 4.8 (0.0~49.1); 734 (413~905) ❙ 1150 (806~1478); 536 (411~817); 63 (16~129) ❙ 58.6 (8.2~125.9); 734 (413~905); 231.0 (158.7~318.7)

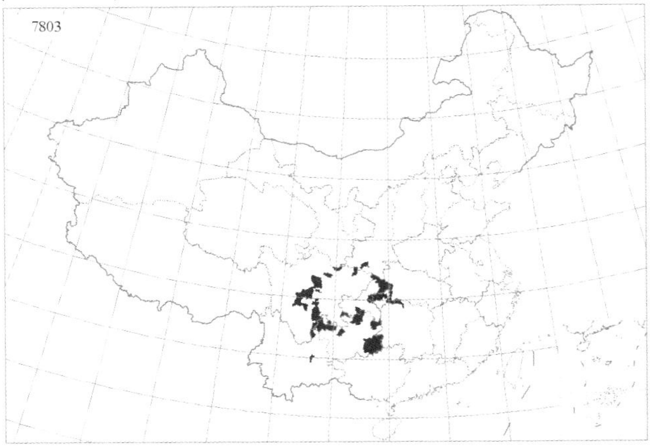

Ericaceae 杜鹃花科 *Rhododendron* 杜鹃花属
Rhododendron argyrophyllum subsp. nankingense (Cowan) D. F. Chamb. 黔东银叶杜鹃
Shrubs or small trees, evergreen

16.4 (9.0~18.7); 6.4 (0.6~8.1); 25.6 (16.5~28.9); 16.4 (9.0~18.8); 137.9 (58.7~166.0); 0.8 (0.0~10.3); 870 (584~1002) ❙ 1115 (985~1410); 561 (454~816); 58 (27~80) ❙ 30.8 (8.0~101.6); 870 (584~1002); 171.7 (108.2~310.8)

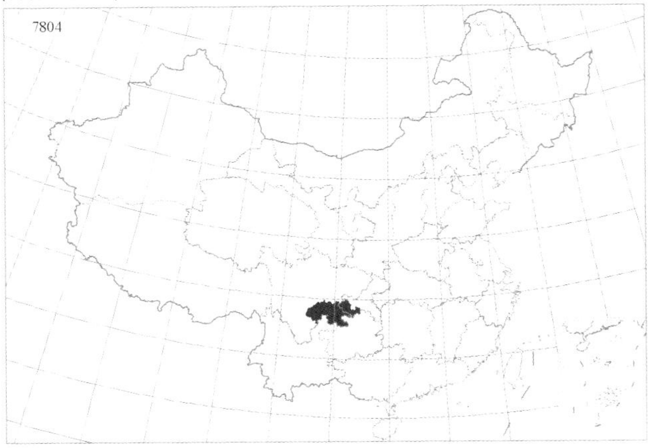

Ericaceae 杜鹃花科 *Rhododendron* 杜鹃花属
Rhododendron argyrophyllum subsp. omeiense (Rehder et E. H. Wilson) D. F. Chamb. 峨眉银叶杜鹃
Shrubs or small trees, evergreen

12.1 (10.1~14.9); 3.1 (1.8~5.3); 20.2 (17.7~23.6); 12.1 (10.3~15.0); 89.2 (70.1~119.6); 3.6 (0.0~7.0); 681 (621~782) ❙ 1140 (1001~1410); 639 (550~816); 34 (22~49) ❙ 67.2 (49.1~87.2); 681 (621~782); 285.4 (214.5~322.4)

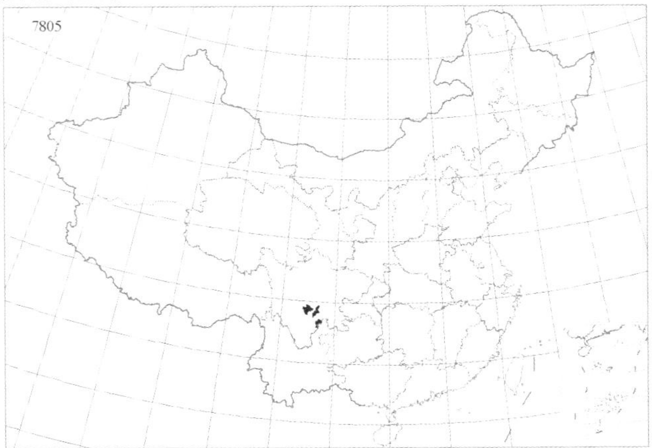

Ericaceae 杜鹃花科 *Rhododendron* 杜鹃花属
Rhododendron arizelum I. B. Balfour et Forrest 夺目杜鹃
Shrubs or small trees, evergreen

9.5 (4.4~14.4); 2.1 (-4.3~7.5); 15.7 (12.1~19.6); 9.7 (5.3~14.4); 65.9 (28.5~113.1); 11.3 (0.0~34.6); 596 (455~727) ❙ 1118 (538~1736); 634 (337~1007); 44 (9~70) ❙ 83.7 (18.2~145.8); 596 (455~727); 345.1 (168.2~487.2)

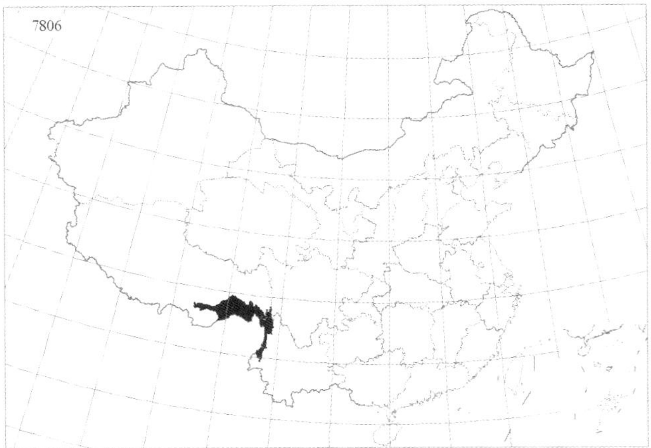

Ericaceae 杜鹃花科 *Rhododendron* 杜鹃花属
Rhododendron asperulum Hutch. et K. Ward 瘤枝杜鹃
Small shrubs, evergreen

9.9; 2.1; 16.3; 10.0; 67.7; 8.9; 608 ❙ 1292; 742; 39 ❙ 110.3; 608; 349.4

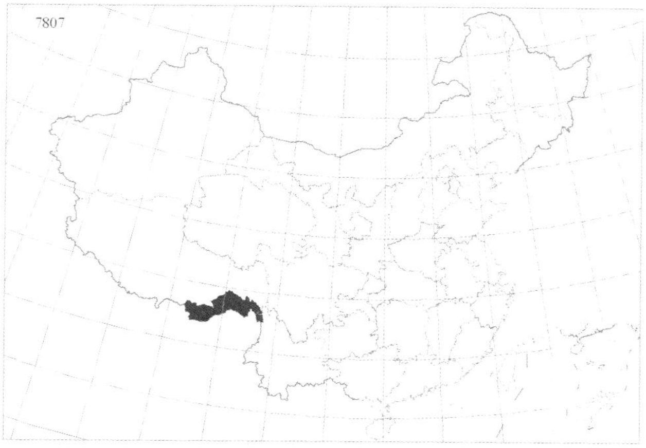

Ericaceae 杜鹃花科 *Rhododendron* 杜鹃花属
Rhododendron asterochnoum Diels 汶川星毛杜鹃
Small trees, evergreen

8.1 (2.2~14.0); -1.0 (-6.6~4.5); 16.4 (9.9~22.7); 8.8 (3.8~14.1); 59.4 (15.5~109.4); 20.9 (0.5~49.1); 572 (413~748) ❙ 1061 (932~1414); 582 (490~817); 30 (19~47) ❙ 90.2 (49.1~125.9); 572 (413~748); 254.8 (200.8~298.1)

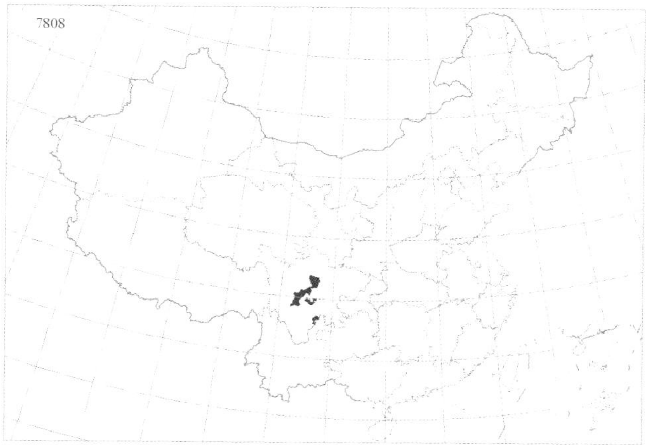

Ericaceae 杜鹃花科 *Rhododendron* 杜鹃花属
Rhododendron atropunicum H. P. Yang 暗紫杜鹃
Shrubs, evergreen

6.6; -1.3; 13.3; 6.8; 39.6; 19.7; 514 ❙ 862; 548; 12 ❙ 67.7; 514; 311.5

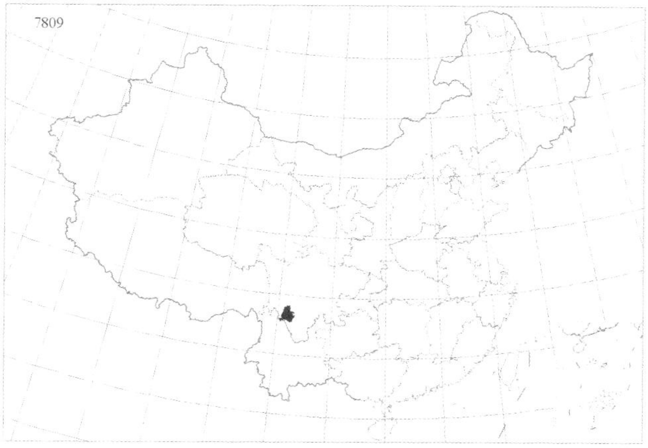

Ericaceae 杜鹃花科 *Rhododendron* 杜鹃花属
Rhododendron atrovirens Franch. 大关杜鹃
Shrubs, evergreen

15.5; 6.6; 23.2; 15.5; 126.5; 0.5; 793 ❙ 1187; 633; 57 ❙ 48.7; 793; 275.1

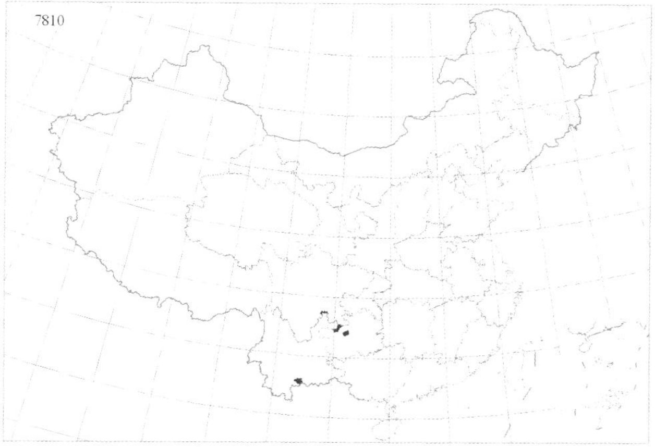

Ericaceae 杜鹃花科 *Rhododendron* 杜鹃花属
Rhododendron augustinii Hemsl. 毛肋杜鹃
Shrubs, evergreen

12.0 (2.2~18.9); 2.6 (-7.0~12.0); 20.6 (9.9~28.1); 12.2 (3.8~19.0); 93.3 (15.5~167.8); 8.4 (0.0~49.1); 691 (413~929) ❙ 1016 (677~1590); 520 (322~946); 38 (14~106) ❙ 49.8 (-14.2~125.9); 680 (413~924); 256.8 (139.5~352.3)

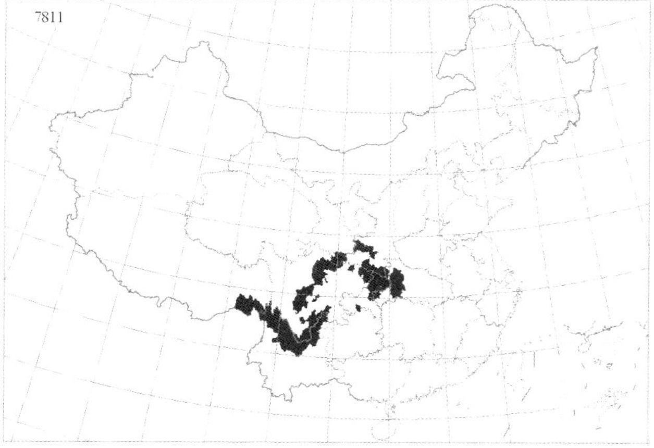

Ericaceae 杜鹃花科 *Rhododendron* 杜鹃花属
Rhododendron augustinii subsp. chasmanthum (Diels) Cullen 张口杜鹃
Shrubs, evergreen

10.0 (3.8~14.1); 1.5 (-4.3~6.5); 17.4 (10.8~21.4); 10.2 (4.7~14.2); 71.1 (21.6~109.8); 10.6 (0.0~35.4); 616 (444~729) ❙ 1051 (981~1220); 600 (535~688); 26 (16~38) ❙ 74.0 (40.1~121.7); 612 (444~708); 276.2 (200.8~331.6)

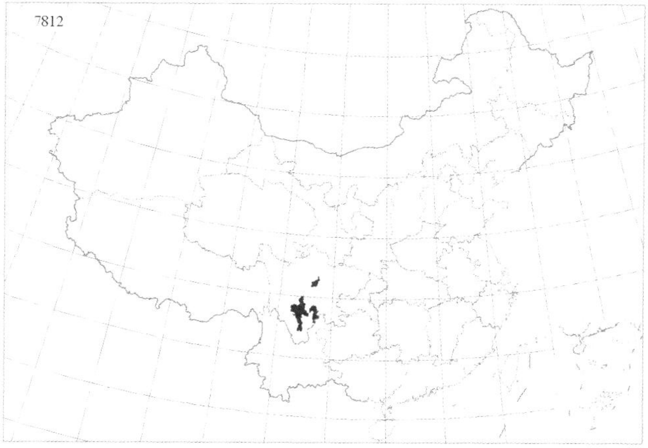

Ericaceae 杜鹃花科 *Rhododendron* 杜鹃花属
Rhododendron aureum Gemerk 牛皮杜鹃
Small shrubs, evergreen

2.5 (1.4~5.2); −17.7 (−20.1~−14.5); 19.9 (18.4~21.8); 7.3 (6.4~8.6); 54.5 (45.8~68.0); 83.5 (65.0~91.9); 542 (501~595) ❙ 774 (657~1005); 482 (415~635); 30 (21~39) ❙ 42.6 (16.5~68.8); 542 (501~595); 324.0 (321.9~329.9)

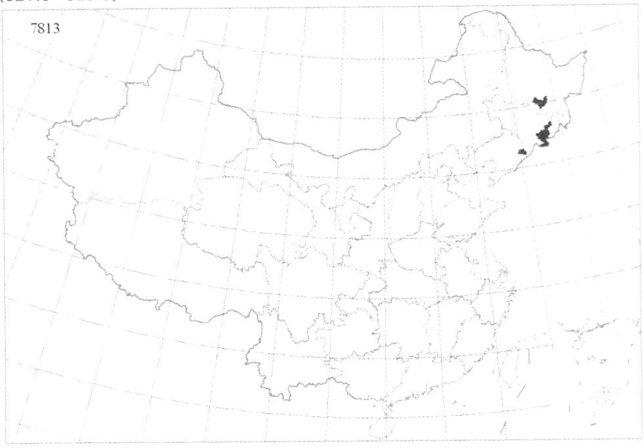

Ericaceae 杜鹃花科 *Rhododendron* 杜鹃花属
Rhododendron auriculatum Hemsl. 耳叶杜鹃
Shrubs or small trees, evergreen

13.9 (9.3~17.0); 2.9 (−1.7~6.1); 24.5 (20.1~27.6); 14.0 (9.5~17.0); 112.0 (69.8~144.3); 4.1 (0.0~18.0); 767 (603~907) ❙ 1291 (963~1660); 527 (421~613); 89 (33~216) ❙ 69.2 (12.0~106.2); 767 (603~907); 231.0 (143.4~305.6)

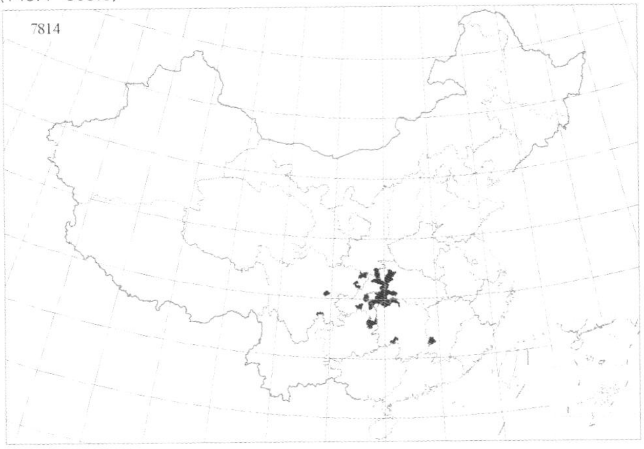

Ericaceae 杜鹃花科 *Rhododendron* 杜鹃花属
Rhododendron auritum Tagg 折萼杜鹃
Shrubs, evergreen

13.3; 5.4; 19.6; 13.4; 100.4; 0.0; 706 ❙ 1736; 1007; 51 ❙ 145.8; 706; 447.4

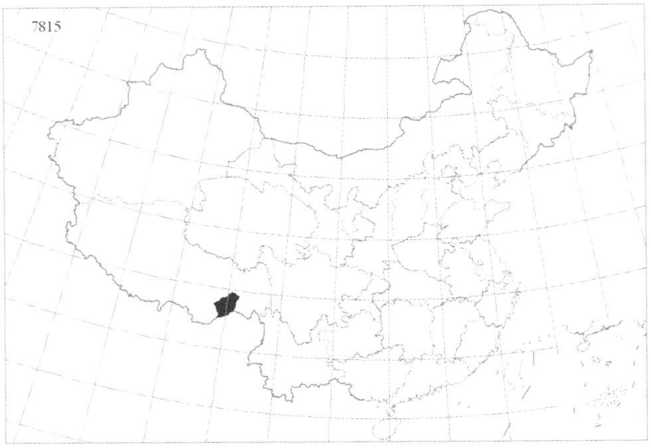

Ericaceae 杜鹃花科 *Rhododendron* 杜鹃花属
Rhododendron bachii H. Lév. 腺萼马银花
Shrubs, evergreen

16.5 (3.5~21.8); 5.6 (−6.5~13.1); 26.6 (12.6~30.1); 16.5 (5.0~21.8); 139.0 (26.3~201.6); 0.7 (0.0~43.4); 879 (446~1160) ❙ 1419 (731~2005); 537 (299~824); 145 (19~259) ❙ 61.7 (−6.7~147.0); 869 (446~1160); 210.0 (126.1~305.7)

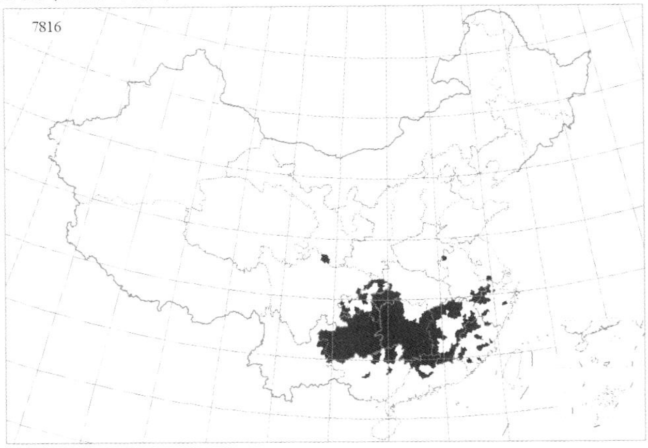

Ericaceae 杜鹃花科 *Rhododendron* 杜鹃花属
Rhododendron baileyi I. B. Balfour 辐花杜鹃
Small shrubs, evergreen

5.9; −2.4; 12.9; 6.9; 42.3; 31.6; 503 ❙ 904; 533; 22 ❙ 67.7; 490; 238.9

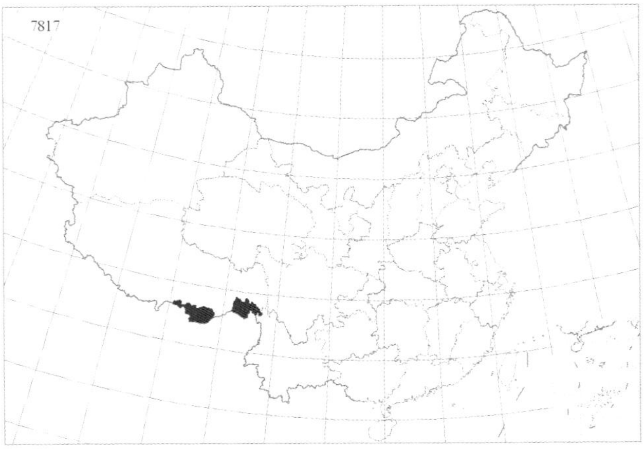

Ericaceae 杜鹃花科 *Rhododendron* 杜鹃花属
Rhododendron bainbridgeanum Tagg et Forrest 毛萼杜鹃
Small shrubs, evergreen

6.2 (2.0~9.6); −1.6 (−7.2~2.4); 13.0 (10.1~15.7); 6.7 (3.8~9.6); 38.1 (15.4~60.8); 23.5 (5.4~51.4); 509 (413~600) ❙ 939 (741~1041); 531 (426~585); 33 (23~45) ❙ 86.5 (54.0~128.6); 509 (413~600); 296.5 (234.2~352.3)

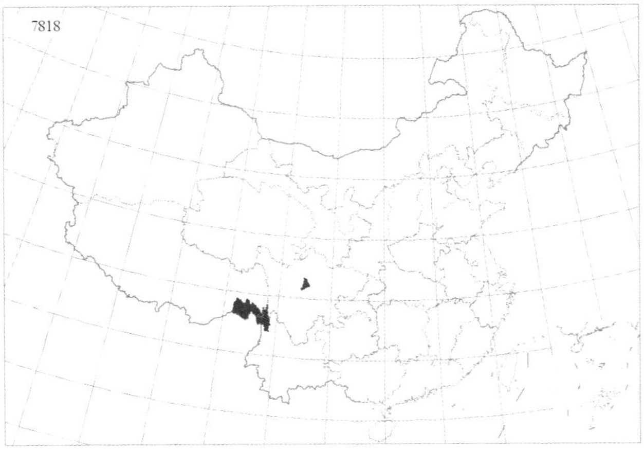

Ericaceae 杜鹃花科 *Rhododendron* 杜鹃花属
Rhododendron balangense Fang 巴朗杜鹃
Shrubs, evergreen

6.0; -3.2; 14.4; 6.6; 39.7; 26.7; 508 | 1023; 535; 28 | 101.2; 508; 200.8

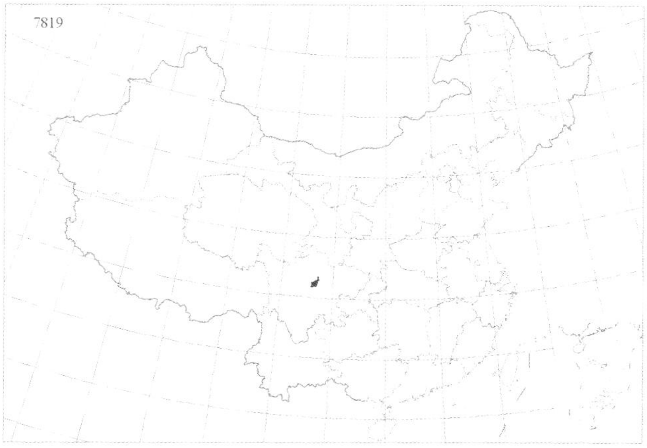

Ericaceae 杜鹃花科 *Rhododendron* 杜鹃花属
Rhododendron balfourianum Diels 粉钟杜鹃
Shrubs, evergreen

8.0 (-1.0~14.3); 0.3 (-11.8~7.6); 14.6 (8.5~19.7); 8.6 (2.7~14.3); 57.5 (8.3~111.9); 21.1 (0.0~79.5); 560 (363~728) | 867 (662~1042); 516 (379~594); 26 (12~46) | 57.5 (36.1~82.4); 557 (363~707); 306.5 (273.5~372.3)

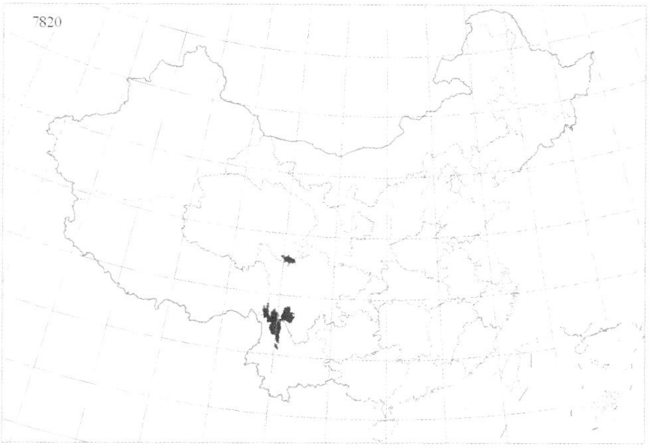

Ericaceae 杜鹃花科 *Rhododendron* 杜鹃花属
Rhododendron balfourianum var. aganniphoides Tagg et Forrest 白毛粉钟杜鹃
Shrubs, evergreen

8.6 (5.4~14.3); 1.3 (-2.1~7.6); 14.9 (12.3~19.7); 8.8 (5.8~14.3); 58.0 (31.5~111.9); 13.7 (0.0~25.6); 570 (481~728) | 873 (741~991); 529 (426~594); 24 (12~35) | 54.4 (36.1~67.7); 566 (481~707); 296.9 (273.5~319.8)

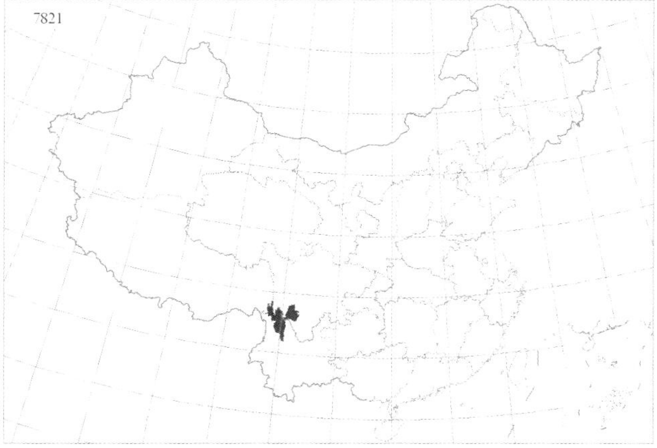

Ericaceae 杜鹃花科 *Rhododendron* 杜鹃花属
Rhododendron barbatum Wall. ex G. Don 硬刺杜鹃
Shrubs or small trees, evergreen

3.1; -5.0; 10.2; 5.3; 30.6; 53.1; 445 | 690; 409; 28 | 35.4; 437; 171.7

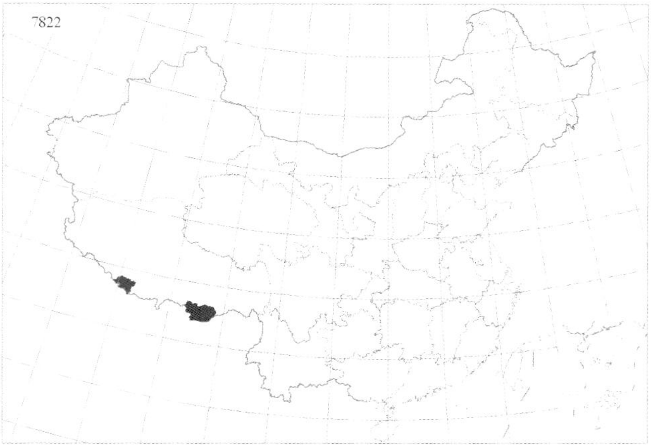

Ericaceae 杜鹃花科 *Rhododendron* 杜鹃花属
Rhododendron barkamense D. F. Chamb. 马尔康杜鹃
Shrubs, evergreen

2.7; -7.1; 11.1; 4.4; 20.5; 48.0; 429 | 800; 427; 18 | 86.5; 429; 341.6

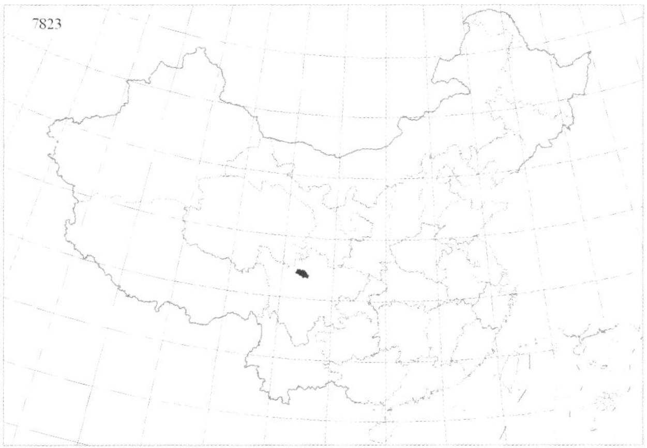

Ericaceae 杜鹃花科 *Rhododendron* 杜鹃花属
Rhododendron basilicum I. B. Balfour et W. W. Sm. 粗枝杜鹃
Shrubs or small trees, evergreen

13.4 (5.4~18.9); 6.6 (-2.1~12.0); 18.9 (12.3~23.9); 13.5 (5.8~19.0); 104.7 (31.5~167.8); 3.0 (0.0~25.6); 714 (481~929) | 997 (741~1454); 578 (411~792); 35 (15~70) | 43.0 (-14.2~99.9); 671 (481~783); 316.6 (218.0~487.2)

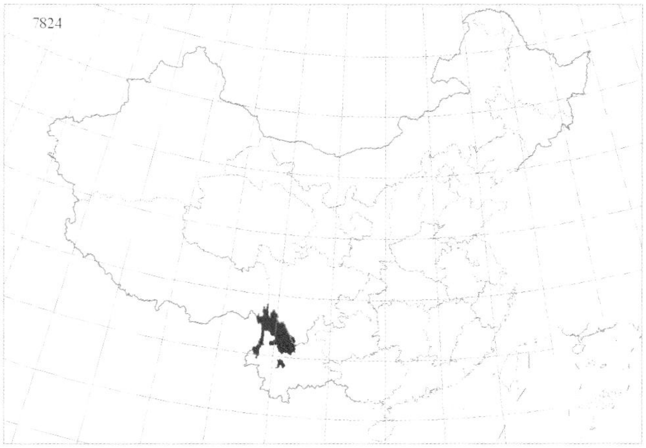

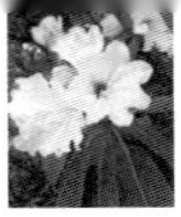

Ericaceae 杜鹃花科 *Rhododendron* 杜鹃花属
Rhododendron beanianum Cowan 刺枝杜鹃
Small shrubs, evergreen

5.3；-2.4；12.2；5.7；30.5；26.2；479 ❙ 961；558；31 ❙ 100.7；479；234.2

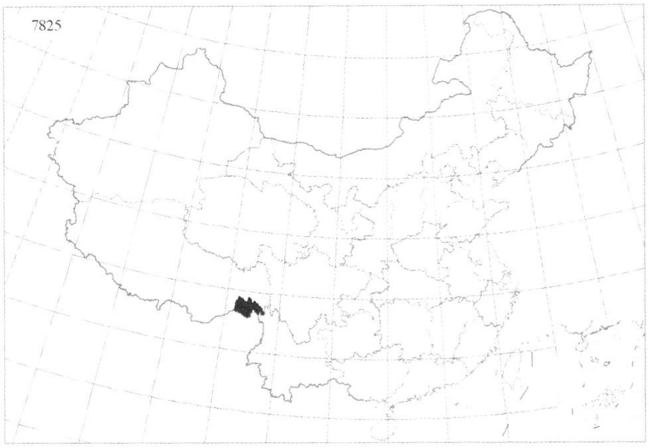

Ericaceae 杜鹃花科 *Rhododendron* 杜鹃花属
Rhododendron beesianum Diels 宽钟杜鹃
Shrubs or small trees, evergreen

7.5 (1.8~14.3)；-0.2 (-7.1~7.6)；14.1 (9.9~19.7)；8.0 (3.8~14.3)；50.5 (15.5~111.9)；20.0 (0.0~53.3)；544 (404~728) ❙ 965 (583~1431)；565 (348~817)；29 (11~69) ❙ 78.5 (36.1~125.9)；542 (404~707)；298.0 (169.5~407.1)

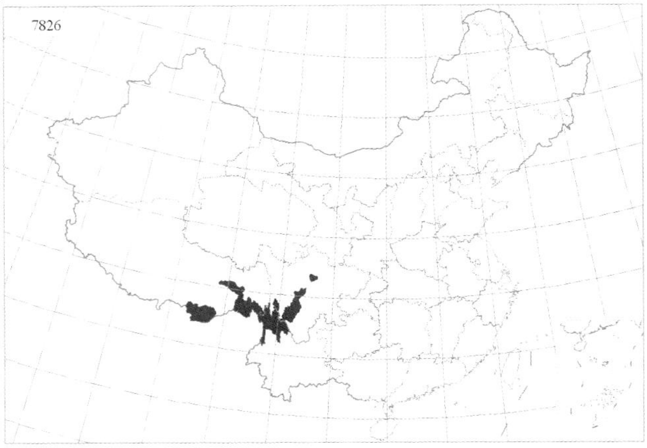

Ericaceae 杜鹃花科 *Rhododendron* 杜鹃花属
Rhododendron bijiangense T. L. Ming 碧江杜鹃
Shrubs, evergreen

12.3；5.4；17.7；12.3；88.0；0.3；670 ❙ 1290；712；63 ❙ 92.7；670；395.9

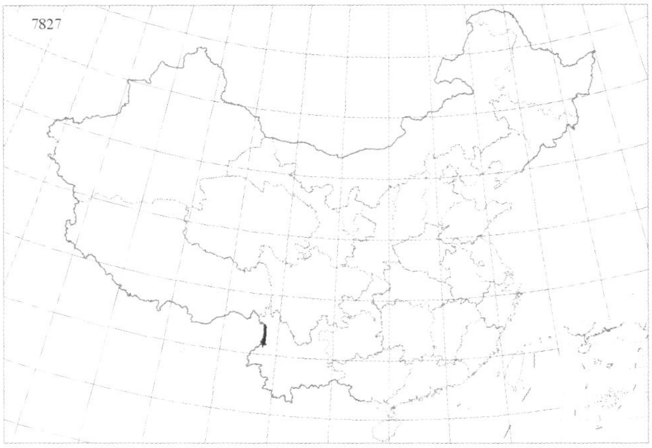

Ericaceae 杜鹃花科 *Rhododendron* 杜鹃花属
Rhododendron bonvalotii Bureau et Franch. 折多杜鹃
Shrubs, evergreen

2.2；-6.6；9.9；3.8；15.5；49.1；413 ❙ 932；546；19 ❙ 125.9；413；248.3

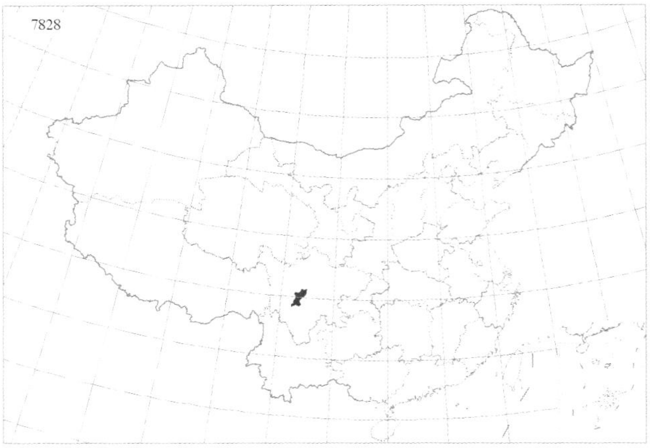

Ericaceae 杜鹃花科 *Rhododendron* 杜鹃花属
Rhododendron boothii Nutt. 柠檬杜鹃
Shrubs, deciduous

6.8；-1.3；13.9；7.6；48.9；26.5；530 ❙ 1093；638；32 ❙ 96.9；530；283.7

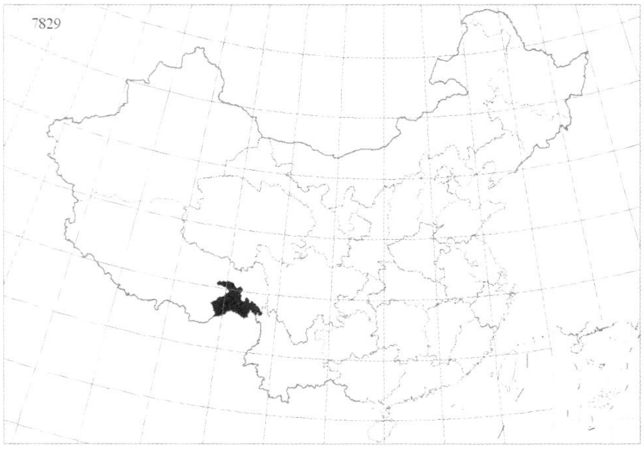

Ericaceae 杜鹃花科 *Rhododendron* 杜鹃花属
Rhododendron brachyanthum Franch. 短花杜鹃
Small shrubs, evergreen

11.2 (5.3~17.3)；4.0 (-2.4~11.0)；16.9 (12.2~21.6)；11.3 (5.7~17.3)；82.9 (30.5~148.0)；8.7 (0.0~26.2)；644 (479~832) ❙ 1146 (741~1736)；651 (426~1007)；45 (28~70) ❙ 79.4 (26.1~145.8)；637 (479~783)；361.7 (234.2~487.2)

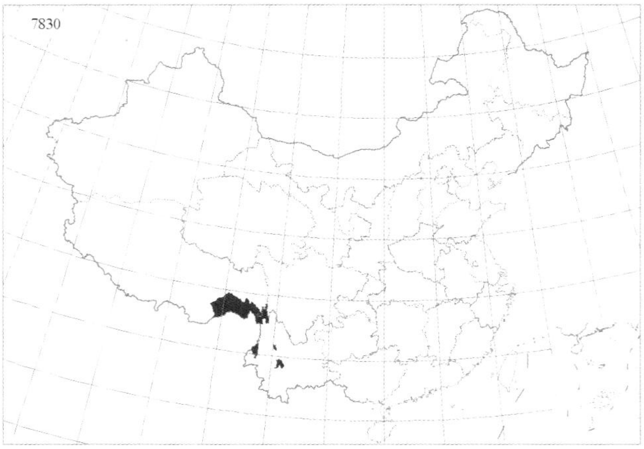

Ericaceae 杜鹃花科 *Rhododendron* 杜鹃花属
Rhododendron brachyanthum subsp. hypolepidotum (Franchet) Cullen 绿柱短花杜鹃
Small shrubs, evergreen

13.3; 5.4; 19.6; 13.4; 100.4; 0.0; 706 ❙ 1736; 1007; 51 ❙ 145.8; 706; 447.4

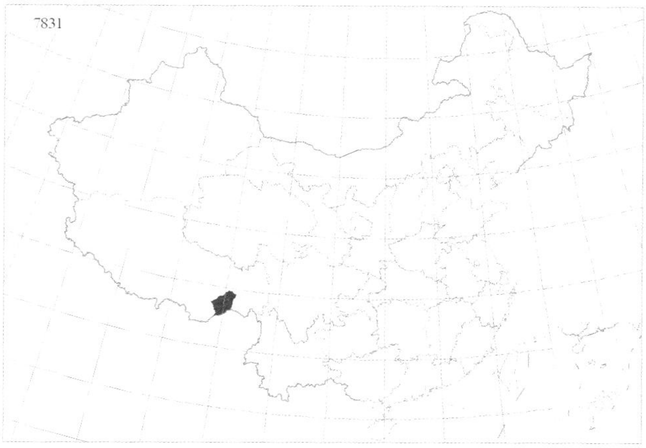

Ericaceae 杜鹃花科 *Rhododendron* 杜鹃花属
Rhododendron brachypodum W. P. Fang et P. S. Liu 短梗杜鹃
Shrubs, evergreen

14.8; 4.3; 24.9; 14.9; 119.2; 0.7; 791 ❙ 1197; 489; 71 ❙ 51.2; 791; 182.9

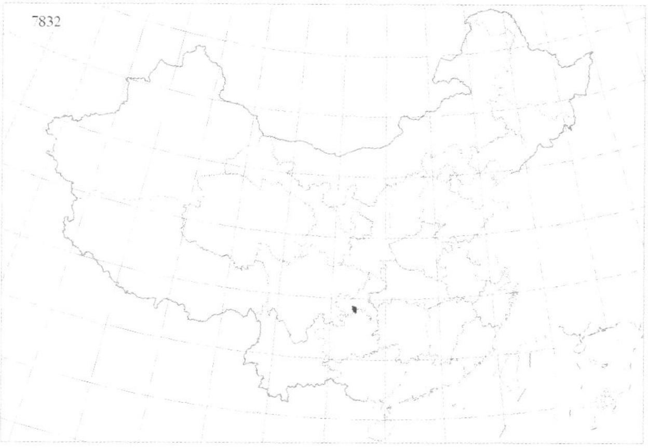

Ericaceae 杜鹃花科 *Rhododendron* 杜鹃花属
Rhododendron bracteatum Rehder et E. H. Wilson 苞叶杜鹃
Shrubs, evergreen

7.6 (2.0~14.7); -1.9 (-7.2~4.3); 16.2 (10.1~24.0); 8.3 (3.8~14.8); 55.6 (15.4~118.0); 23.6 (0.7~51.4); 562 (413~789) ❙ 960 (822~1081); 511 (420~606); 25 (19~32) ❙ 78.5 (18.6~128.6); 562 (413~789); 233.9 (150.8~275.4)

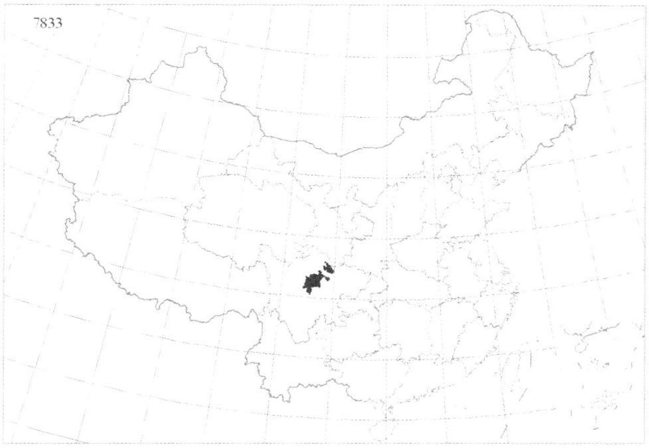

Ericaceae 杜鹃花科 *Rhododendron* 杜鹃花属
Rhododendron brevicaudatum R. C. Fang et S. S. Chang 短尾杜鹃
Shrubs, evergreen

16.2 (14.3~19.1); 6.0 (4.0~10.1); 25.2 (23.0~27.0); 16.3 (14.3~19.1); 135.3 (112.6~169.2); 0.1 (0.0~1.1); 841 (756~969) ❙ 1265 (1160~1425); 556 (448~720); 90 (51~161) ❙ 50.7 (27.5~71.9); 841 (756~969); 220.5 (172.1~283.9)

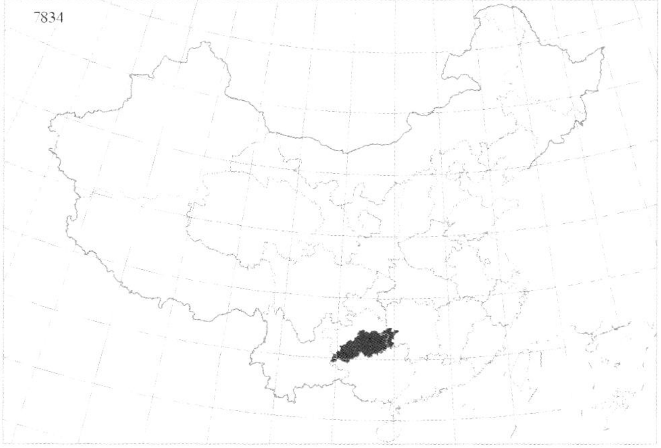

Ericaceae 杜鹃花科 *Rhododendron* 杜鹃花属
Rhododendron brevinerve Chun et Fang 短脉杜鹃
Small trees, evergreen

17.4 (14.3~21.6); 7.0 (3.3~12.6); 26.5 (23.5~29.8); 17.4 (14.3~21.7); 149.1 (112.6~199.8); 0.3 (0.0~2.3); 912 (756~1153) ❙ 1413 (1128~1776); 576 (444~848); 134 (43~194) ❙ 55.5 (16.9~101.8); 911 (756~1153); 224.6 (153.9~371.7)

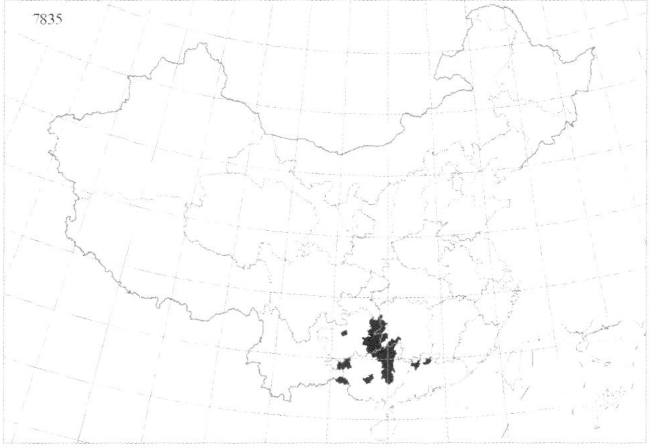

Ericaceae 杜鹃花科 *Rhododendron* 杜鹃花属
Rhododendron breviperulatum Hayata 短鳞芽杜鹃
Small shrubs, semievergreen or green

18.2; 12.4; 22.9; 18.2; 158.4; 0.0; 890 ❙ 2439; 1056; 253 ❙ 184.5; 879; 259.8

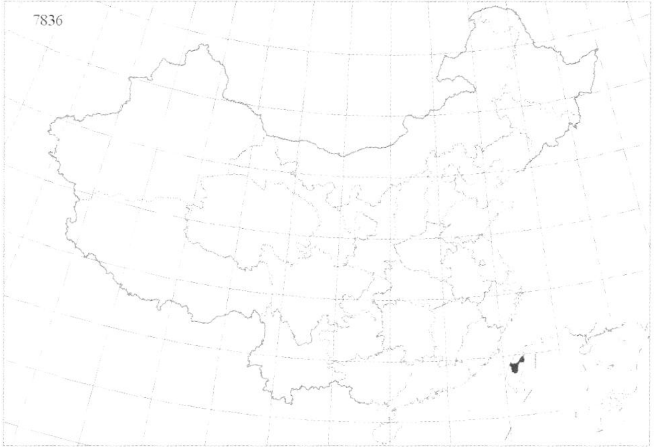

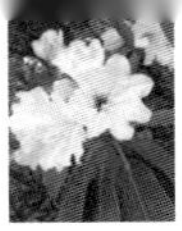

Ericaceae 杜鹃花科 *Rhododendron* 杜鹃花属

Rhododendron brevipetiolatum M. Y. Fang 短柄杜鹃

Shrubs, evergreen

12.7; 3.6; 21.1; 12.8; 95.3; 1.5; 702 ❙ 1102; 605; 30 ❙ 57.1; 702; 298.1

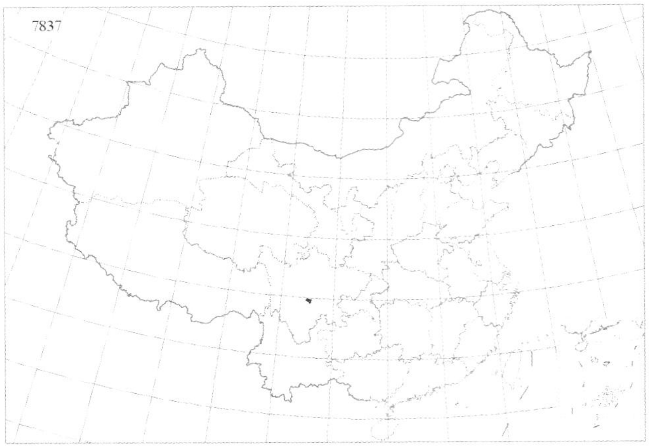

Ericaceae 杜鹃花科 *Rhododendron* 杜鹃花属

Rhododendron bulu Hutch. 蜿蜒杜鹃

Shrubs, evergreen

2.1; −7.0; 10.3; 4.0; 18.3; 52.3; 406 ❙ 456; 282; 7 ❙ 9.2; 406; 238.9

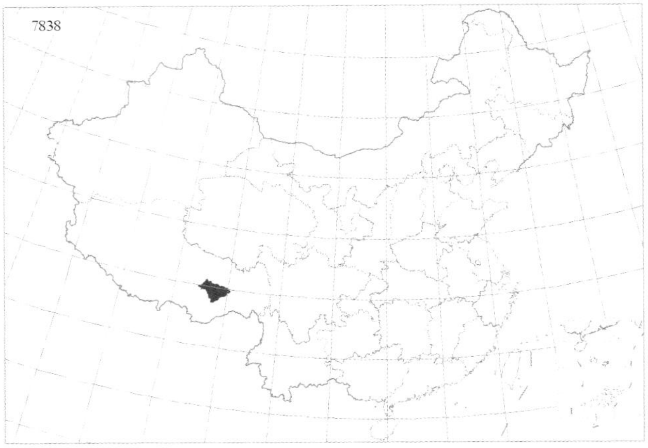

Ericaceae 杜鹃花科 *Rhododendron* 杜鹃花属

Rhododendron bureavii Franch. 锈红杜鹃

Shrubs, evergreen

12.4 (2.0~18.9); 4.5 (−7.2~12.0); 19.0 (9.9~26.6); 12.6 (3.8~19.0); 95.2 (15.4~167.8); 5.9 (0.0~51.4); 689 (413~929) ❙ 980 (741~1414); 571 (411~817); 29 (15~57) ❙ 46.8 (−14.2~128.6); 664 (413~912); 275.5 (142.9~357.7)

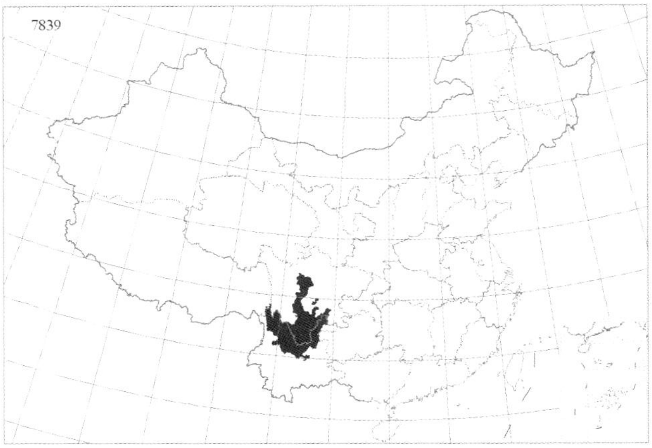

Ericaceae 杜鹃花科 *Rhododendron* 杜鹃花属

Rhododendron callimorphum I. B. Balfour et W. W. Sm. 卵叶杜鹃

Shrubs, evergreen

13.4; 6.5; 18.5; 13.4; 100.6; 0.2; 699 ❙ 1269; 696; 63 ❙ 82.1; 699; 415.4

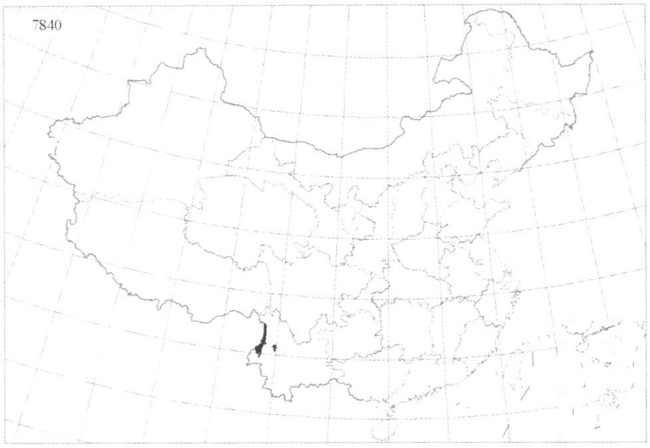

Ericaceae 杜鹃花科 *Rhododendron* 杜鹃花属

Rhododendron callimorphum var. myiagrum (I. B. Balfour et Forrest) D. F. Chamb. 白花卵叶杜鹃

Shrubs, evergreen

13.0; 6.1; 18.2; 13.0; 96.4; 0.2; 689 ❙ 1345; 739; 66 ❙ 95.1; 689; 426.3

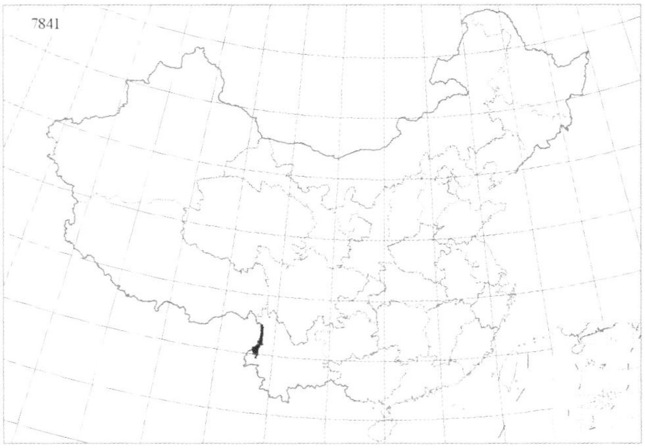

Ericaceae 杜鹃花科 *Rhododendron* 杜鹃花属

Rhododendron calophytum Franch. 美容杜鹃

Shrubs or small trees, evergreen

12.5 (2.2~16.8); 2.4 (−6.6~6.5); 21.9 (9.9~26.5); 12.7 (3.8~16.9); 98.3 (15.5~142.7); 7.5 (0.0~49.1); 713 (413~869) ❙ 1097 (669~1478); 526 (317~817); 53 (15~129) ❙ 55.9 (5.8~125.9); 713 (413~869); 227.9 (139.5~323.3)

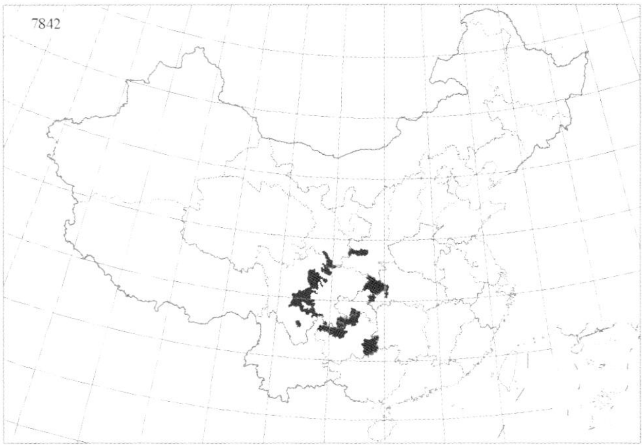

Ericaceae 杜鹃花科 *Rhododendron* 杜鹃花属

Rhododendron calophytum var. jinfuense D. Fang et W. K. Hu 金佛山美容杜鹃 Shrubs or small trees, evergreen

14.8; 4.3; 24.9; 14.9; 119.2; 0.7; 791 ❙ 1197; 489; 71 ❙ 51.2; 791; 182.9

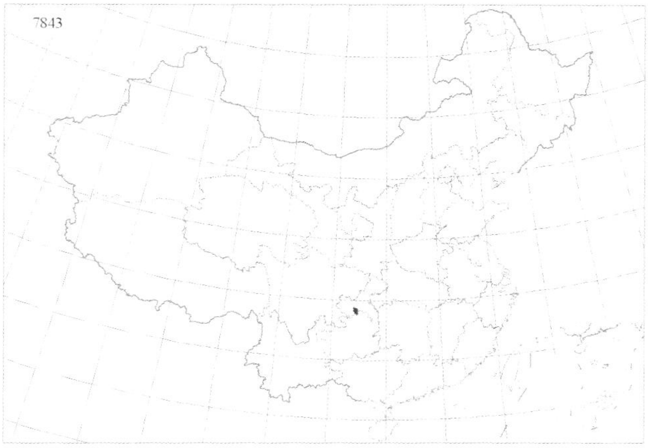

Ericaceae 杜鹃花科 *Rhododendron* 杜鹃花属

Rhododendron calophytum var. openshawianum (Rehder et E. H. Wilson) D. F. Chamb. 尖叶美容杜鹃 Shrubs or small trees, evergreen

13.3 (9.0~16.2); 4.0 (0.6~6.3); 21.5 (16.5~25.4); 13.3 (9.0~16.2); 102.4 (58.7~135.0); 2.4 (0.0~10.3); 724 (584~849) ❙ 1190 (1001~1414); 669 (550~817); 41 (30~50) ❙ 65.3 (36.4~131.6); 724 (584~849); 244.8 (171.5~298.1)

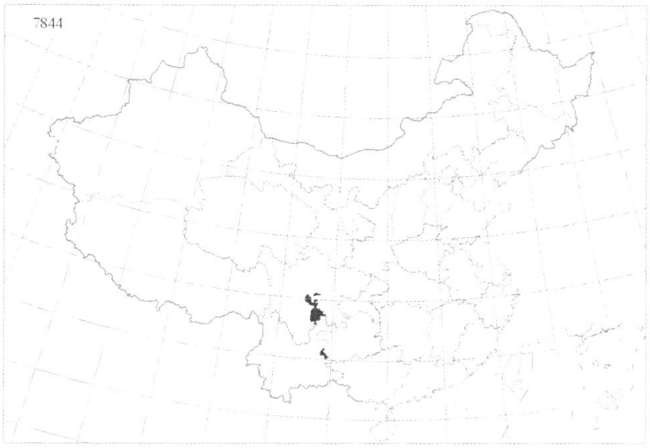

Ericaceae 杜鹃花科 *Rhododendron* 杜鹃花属

Rhododendron calophytum var. pauciflorum W. K. Hu 疏花美容杜鹃 Shrubs or small trees, evergreen

14.0; 3.7; 23.5; 14.0; 110.1; 1.6; 757 ❙ 1139; 547; 51 ❙ 50.3; 757; 176.9

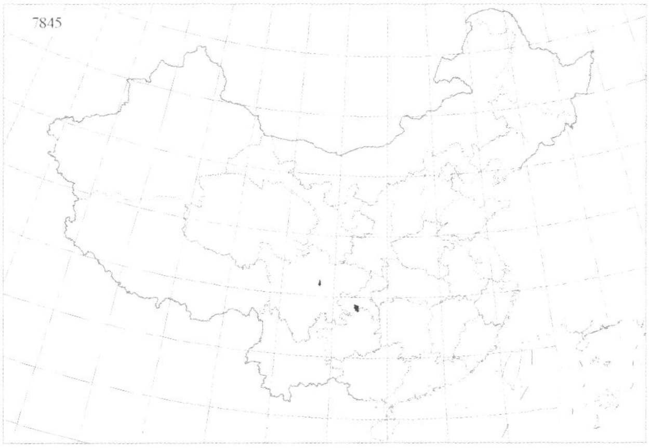

Ericaceae 杜鹃花科 *Rhododendron* 杜鹃花属

Rhododendron calostrotum I. B. Balfour et Kingdon-Ward 美被杜鹃 Small shrubs, evergreen

6.7 (0.0~13.3); -1.2 (-8.7~6.4); 13.6 (8.1~19.6); 7.5 (2.7~13.4); 46.8 (8.0~100.4); 25.8 (0.0~67.9); 525 (363~706) ❙ 911 (341~1736); 525 (226~1007); 33 (4~69) ❙ 65.8 (-15.3~145.8); 519 (341~706); 289.0 (111.0~447.4)

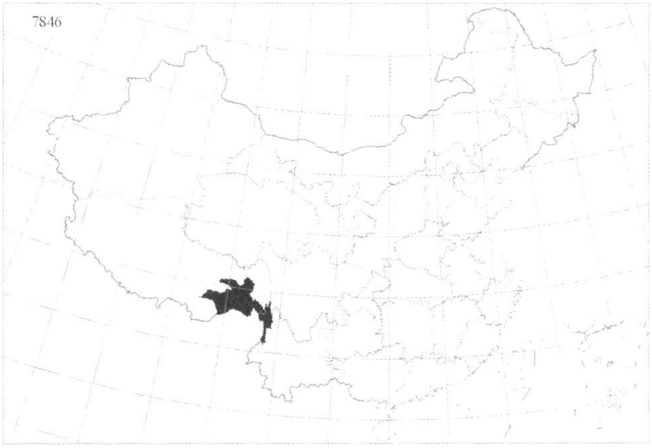

Ericaceae 杜鹃花科 *Rhododendron* 杜鹃花属

Rhododendron calostrotum var. calciphilum (Hutch. et Kingdon-Ward) Davidian 小叶美被杜鹃 Small shrubs, evergreen

5.9 (0.0~13.3); -2.2 (-8.7~6.4); 13.0 (8.1~19.6); 7.0 (2.7~13.4); 43.2 (8.0~100.4); 31.5 (0.0~67.9); 506 (363~706) ❙ 879 (341~1736); 509 (226~1007); 29 (4~69) ❙ 63.6 (-15.3~145.8); 499 (341~706); 271.2 (111.0~447.4)

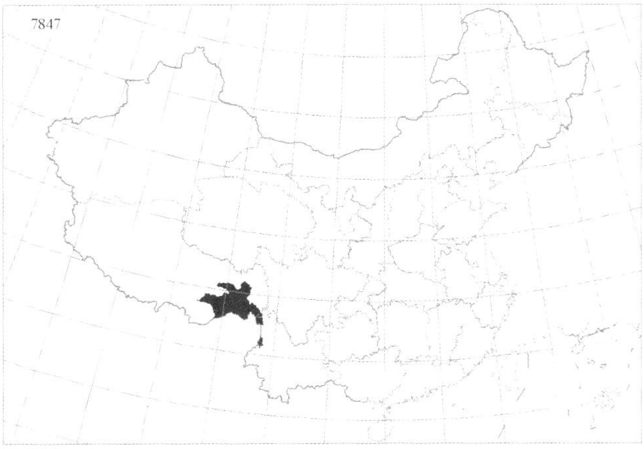

Ericaceae 杜鹃花科 *Rhododendron* 杜鹃花属

Rhododendron calvescens I. B. Balfour et Forrest 变光杜鹃 Small shrubs, evergreen

7.8 (5.3~11.3); 0.4 (-2.4~4.3); 14.2 (12.2~17.0); 8.0 (5.7~11.3); 48.6 (30.5~76.6); 14.4 (0.7~26.2); 549 (479~644) ❙ 976 (741~1243); 565 (426~696); 36 (12~58) ❙ 77.5 (54.0~100.7); 549 (479~644); 318.0 (234.2~384.7)

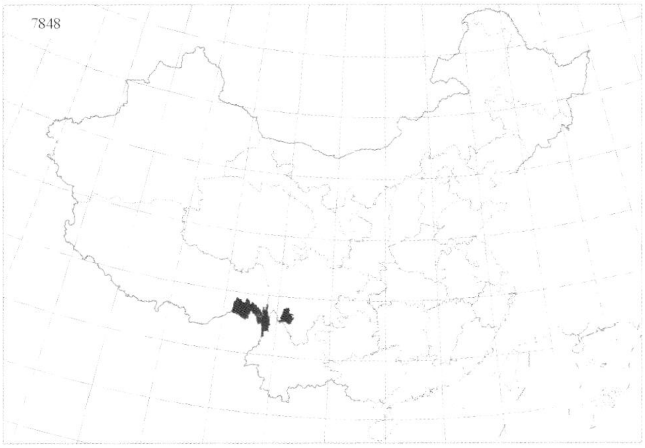

Ericaceae 杜鹃花科 *Rhododendron* 杜鹃花属

Rhododendron calvescens var. duseimatum (I. B. Balfour et Forrest) D. F. Chamb. 长梗变光杜鹃 Small shrubs, evergreen

5.3; -2.4; 12.2; 5.7; 30.5; 26.2; 479 ❙ 961; 558; 31 ❙ 100.7; 479; 234.2

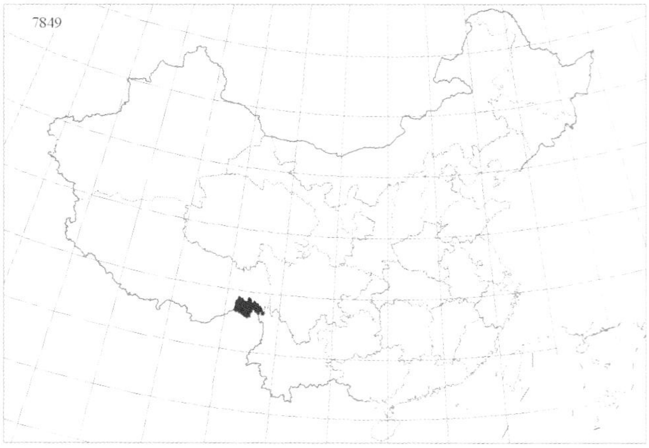

Ericaceae 杜鹃花科 *Rhododendron* 杜鹃花属

Rhododendron camelliiflorum Hook. 茶花杜鹃
Small shrubs, evergreen

1.9; -7.0; 9.9; 3.9; 17.7; 54.4; 396 ❙ 454; 300; 9 ❙ 11.5; 396; 114.9

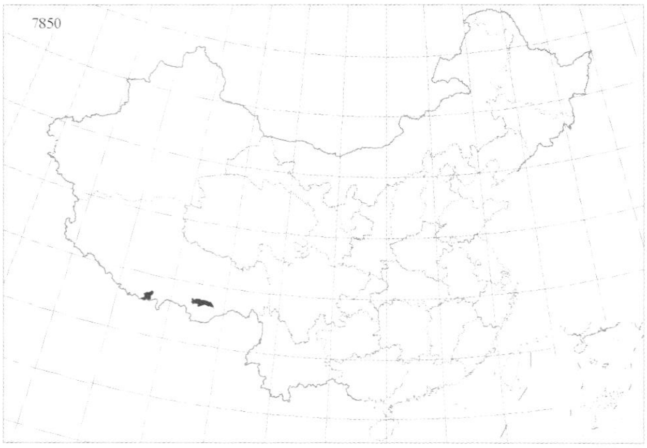

Ericaceae 杜鹃花科 *Rhododendron* 杜鹃花属

Rhododendron campanulatum D. Don 钟花杜鹃
Shrubs, evergreen

1.1 (-1.8~12.2); -7.5 (-10.5~4.3); 8.8 (6.1~18.5); 3.8 (1.8~12.2); 17.4 (1.9~87.4); 64.1 (0.7~82.8); 390 (329~672) ❙ 495 (311~1431); 313 (193~817); 17 (3~33) ❙ 13.4 (-10.6~113.0); 378 (311~672); 128.7 (61.6~363.7)

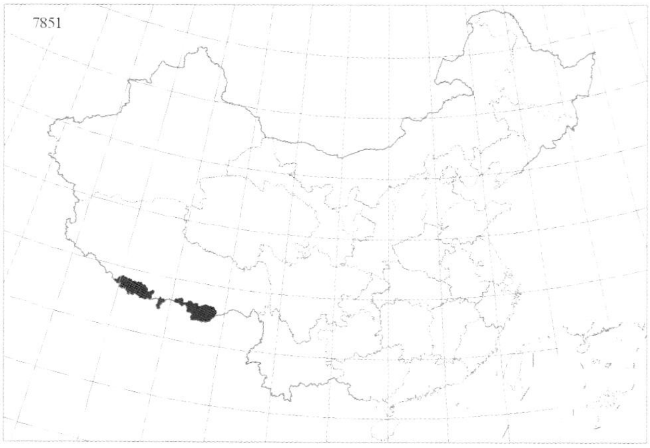

Ericaceae 杜鹃花科 *Rhododendron* 杜鹃花属

Rhododendron campylocarpum Hook. f. 弯果杜鹃
Shrubs, evergreen

5.9 (-0.7~13.2); -2.1 (-10.1~6.4); 12.8 (7.7~18.4); 6.8 (2.5~13.3); 40.8 (6.8~99.4); 30.0 (0.0~74.2); 502 (338~695) ❙ 772 (298~1337); 457 (215~727); 27 (2~69) ❙ 45.6 (-18.0~100.7); 492 (298~695); 262.8 (61.6~407.1)

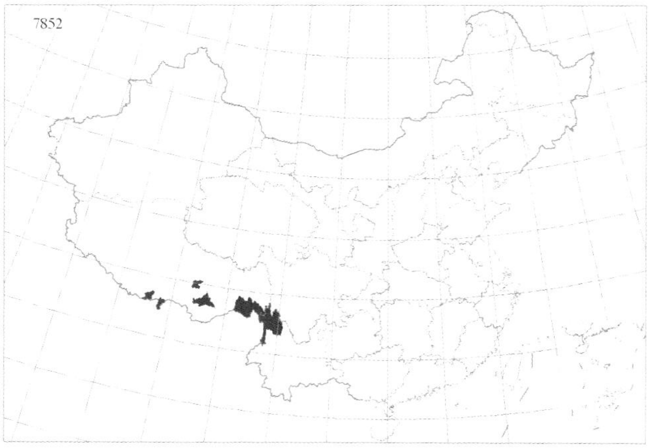

Ericaceae 杜鹃花科 *Rhododendron* 杜鹃花属

Rhododendron campylocarpum subsp. caloxanthum (I. B. Balfour et Farrer) D. F. Chamb. 美丽弯果杜鹃 Shrubs, evergreen

5.3; -2.4; 12.2; 5.7; 30.5; 26.2; 479 ❙ 961; 558; 31 ❙ 100.7; 479; 234.2

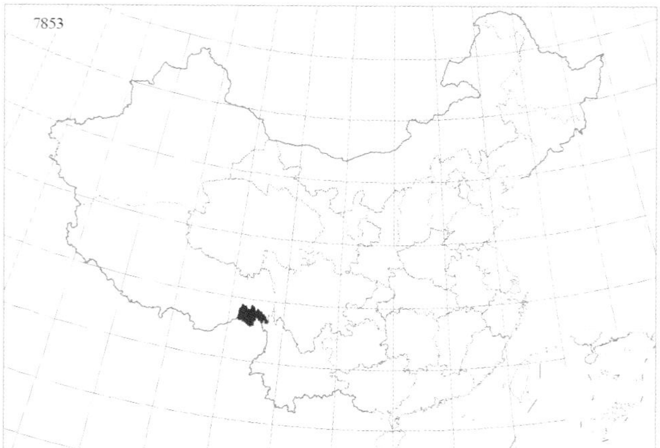

Ericaceae 杜鹃花科 *Rhododendron* 杜鹃花属

Rhododendron campylogynum Franch. 弯柱杜鹃
Small shrubs, evergreen

9.0 (1.8~14.5); 1.5 (-7.1~7.8); 15.4 (9.9~19.6); 9.4 (3.8~14.5); 64.3 (15.7~113.5); 15.7 (0.0~53.3); 584 (404~730) ❙ 977 (473~1736); 560 (287~1007); 37 (7~69) ❙ 64.0 (5.0~145.8); 582 (404~730); 319.4 (169.5~447.4)

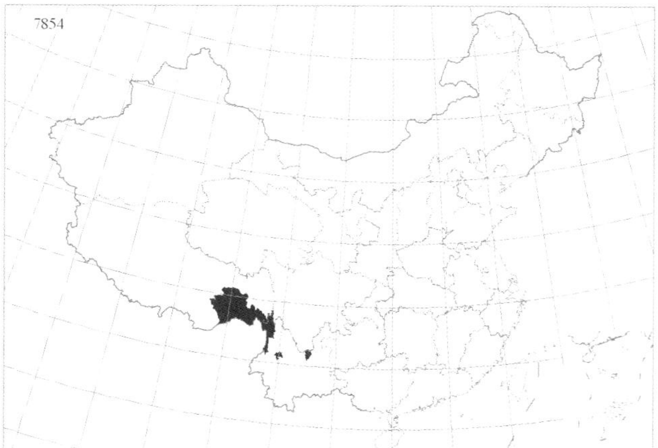

Ericaceae 杜鹃花科 *Rhododendron* 杜鹃花属
Rhododendron capitatum Maxim. 头花杜鹃
Small shrubs, evergreen

6.9 (-3.3~13.5); -5.0 (-15.3~2.3); 17.7 (7.2~24.6); 8.2 (2.0~13.6); 58.3 (3.7~108.0); 35.4 (4.9~103.2); 564 (339~760) ❙ 679 (375~1164); 331 (203~488); 18 (6~51) ❙ 21.4 (-32.2~94.4); 558 (339~760); 260.6 (95.7~364.1)

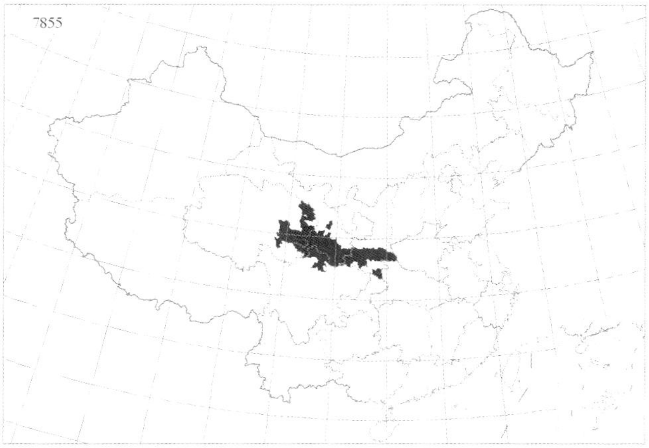

Ericaceae 杜鹃花科 *Rhododendron* 杜鹃花属
Rhododendron catacosmum I. B. Balfour ex Tagg 瓣萼杜鹃
Shrubs, evergreen

7.8; 0.4; 14.3; 8.0; 47.9; 13.5; 551 ❙ 1003; 574; 39 ❙ 83.4; 551; 312.7

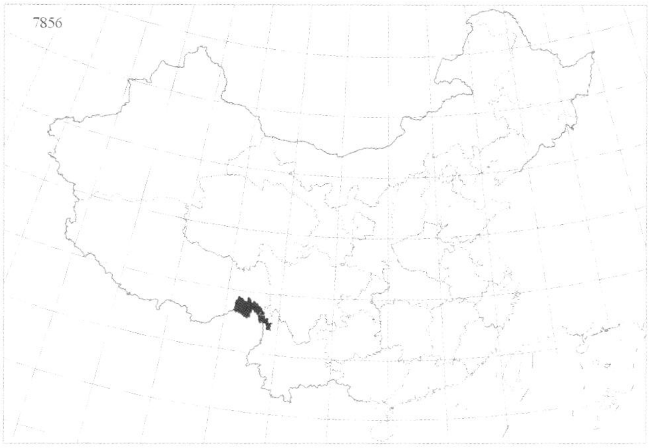

Ericaceae 杜鹃花科 *Rhododendron* 杜鹃花属
Rhododendron cavaleriei H. Lév. 多花杜鹃
Shrubs, evergreen

18.7 (14.3~24.8); 9.5 (3.3~20.1); 26.6 (22.4~29.3); 18.8 (14.3~24.8); 165.3 (112.6~237.6); 0.1 (0.0~2.3); 983 (756~1385) ❙ 1447 (868~1854); 594 (312~911); 122 (41~198) ❙ 49.8 (-34.3~101.8); 965 (756~1360); 275.5 (153.9~620.0)

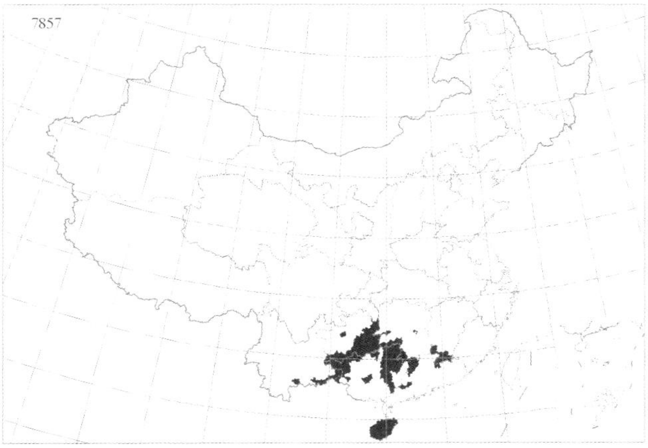

Ericaceae 杜鹃花科 *Rhododendron* 杜鹃花属
Rhododendron cephalanthum Franch. 毛喉杜鹃
Small shrubs, evergreen

7.0 (-1.2~14.3); -1.3 (-11.7~7.6); 14.1 (8.5~20.1); 7.8 (2.6~14.3); 49.9 (8.1~111.9); 25.8 (0.0~82.3); 535 (347~728) ❙ 883 (473~1431); 512 (270~817); 27 (7~69) ❙ 64.0 (0.0~125.9); 534 (347~707); 299.4 (169.5~407.1)

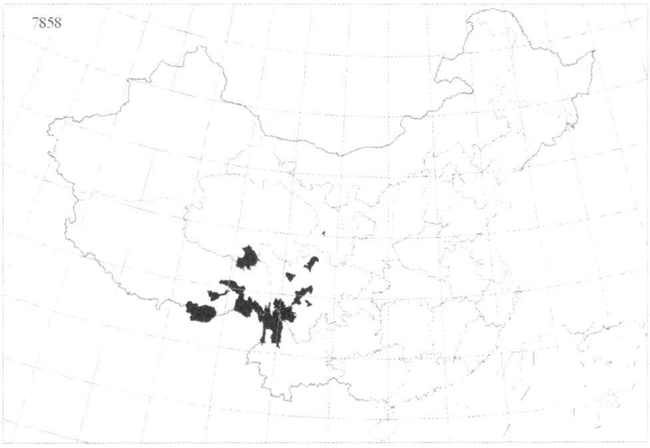

Ericaceae 杜鹃花科 *Rhododendron* 杜鹃花属
Rhododendron cerasinum Tagg 樱花杜鹃
Shrubs, evergreen

4.4; -4.0; 11.8; 5.2; 26.9; 34.0; 455 ❙ 744; 438; 20 ❙ 61.6; 455; 257.1

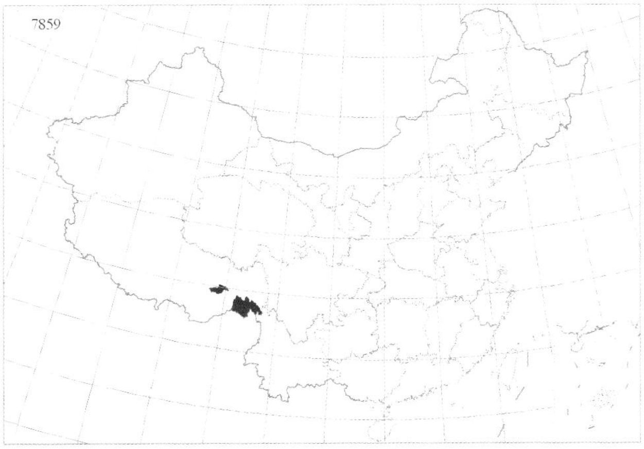

Ericaceae 杜鹃花科 *Rhododendron* 杜鹃花属
Rhododendron chamaethomsonii (Tagg et Forrest) Cowan et Davidian 云雾杜鹃
Shrubs, evergreen

7.2 (3.3~13.3); -0.6 (-5.4~5.4); 14.0 (11.1~19.6); 7.6 (4.6~13.4); 47.5 (22.3~100.4); 20.5 (0.0~41.8); 533 (427~706) ❙ 990 (473~1736); 573 (287~1007); 32 (7~51) ❙ 77.4 (5.0~145.8); 533 (427~706); 307.6 (230.7~447.4)

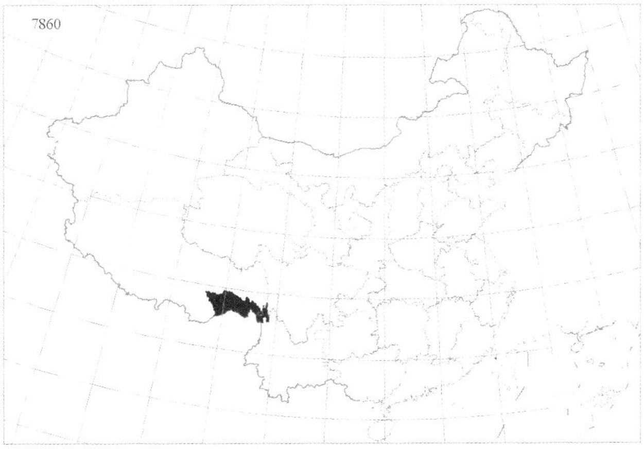

Ericaceae 杜鹃花科 *Rhododendron* 杜鹃花属
Rhododendron chamaethomsonii var. chamaedoron (Tagg et Forrest) D. F. Chamb. 毛背云雾杜鹃 Shrubs, evergreen

6.5; -1.1; 13.2; 6.7; 38.2; 20.2; 511 ❙ 915; 523; 35 ❙ 78.7; 511; 286.6

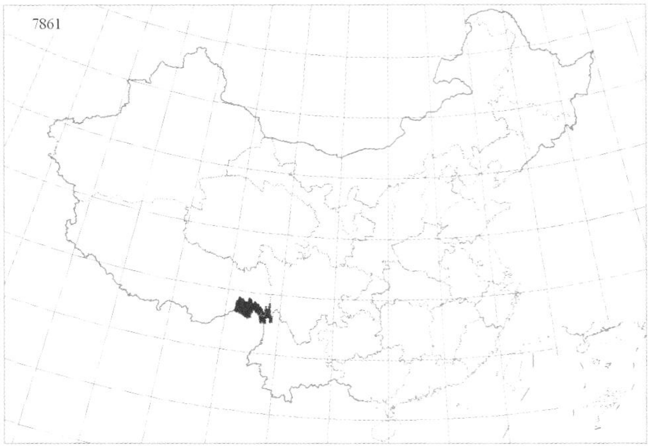

Ericaceae 杜鹃花科 *Rhododendron* 杜鹃花属
Rhododendron chamaethomsonii var. chamaethauma (Tagg) Cowan et Davidian 短萼云雾杜鹃 Shrubs, evergreen

7.3; -0.8; 14.3; 7.9; 51.1; 22.7; 537 ❙ 1057; 617; 30 ❙ 83.8; 537; 304.1

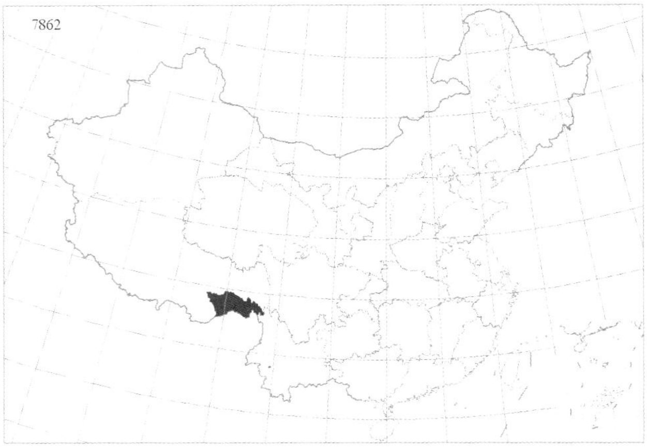

Ericaceae 杜鹃花科 *Rhododendron* 杜鹃花属
Rhododendron championiae Hook. 刺毛杜鹃
Shrubs, evergreen

19.1 (14.3~22.5); 9.8 (3.3~15.2); 27.5 (24.2~29.8); 19.2 (14.3~22.7); 170.5 (113.2~212.7); 0.0 (0.0~2.3); 1002 (761~1216) ❙ 1650 (1070~2037); 621 (356~1033); 181 (82~259) ❙ 66.9 (-5.0~147.0); 996 (761~1216); 233.6 (95.3~332.5)

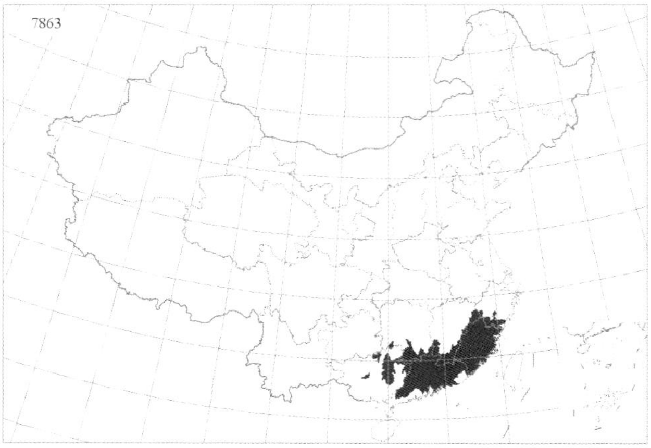

Ericaceae 杜鹃花科 *Rhododendron* 杜鹃花属
Rhododendron changii (Fang) Fang 树枫杜鹃
Shrubs, evergreen

14.8; 4.3; 24.9; 14.9; 119.2; 0.7; 791 ❙ 1197; 489; 71 ❙ 51.2; 791; 182.9

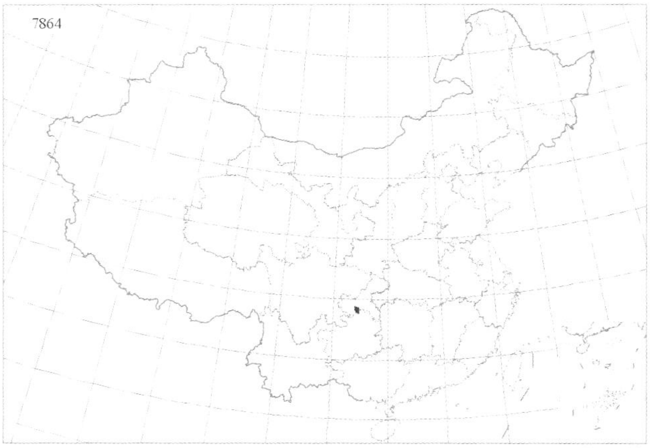

Ericaceae 杜鹃花科 *Rhododendron* 杜鹃花属
Rhododendron charitopes I. B. Balfour et Farrer 雅容杜鹃
Small shrubs, evergreen

6.0; -2.0; 13.0; 6.6; 37.4; 25.3; 500 ❙ 757; 436; 26 ❙ 43.3; 500; 291.5

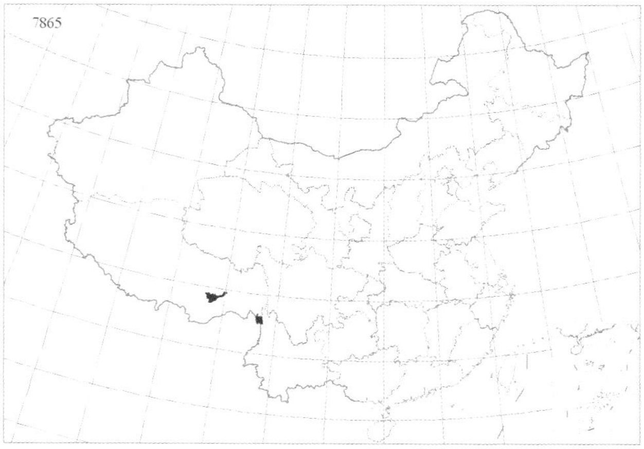

Ericaceae 杜鹃花科 *Rhododendron* 杜鹃花属
Rhododendron charitopes subsp. tsangpoense (Kingdon-Ward) Cullen 藏布雅容杜鹃 Small shrubs, evergreen

3.3; -5.4; 11.1; 4.6; 22.3; 41.8; 427 ❙ 473; 287; 7 ❙ 5.0; 427; 230.7

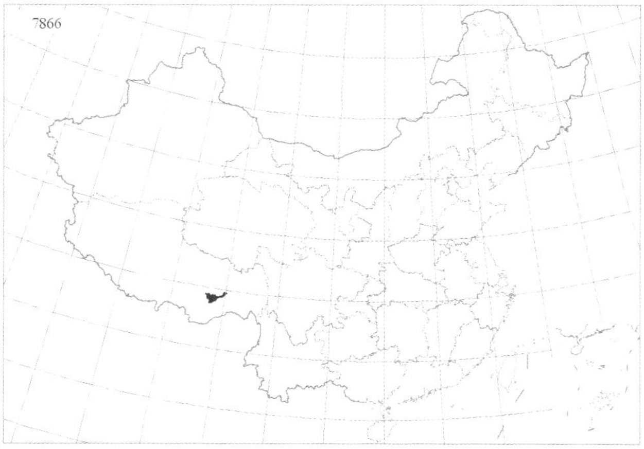

Ericaceae 杜鹃花科 *Rhododendron* 杜鹃花属

Rhododendron chihsinianum Chun et Fang 红滩杜鹃

Small trees, evergreen

16.5 (14.3~19.4); 5.6 (3.3~8.9); 26.4 (24.3~28.7); 16.5 (14.3~19.5); 138.9 (114.1~173.7); 0.5 (0.0~2.3); 873 (765~1038) ❙ 1547 (1415~1776); 558 (419~718); 181 (142~198) ❙ 78.1 (55.1~101.8); 868 (765~1038); 227.9 (177.4~258.0)

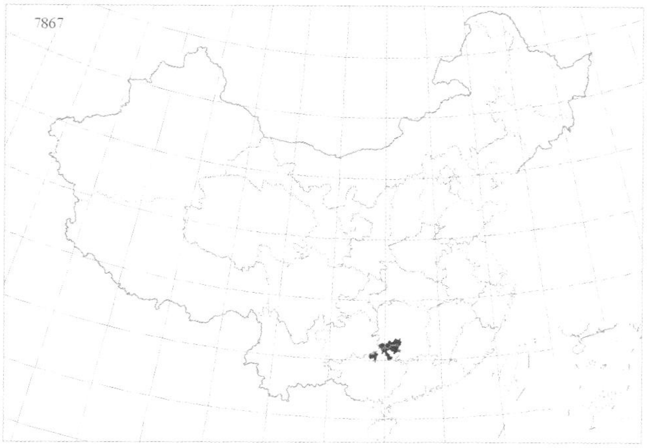

Ericaceae 杜鹃花科 *Rhododendron* 杜鹃花属

Rhododendron chrysocalyx H. Lév. et Vaniot 金萼杜鹃

Shrubs, deciduous

16.8 (12.6~22.7); 6.9 (1.6~15.0); 25.3 (22.2~28.6); 16.8 (12.7~22.8); 142.0 (98.8~213.1); 0.2 (0.0~6.9); 872 (711~1221) ❙ 1329 (1171~2435); 615 (448~1428); 90 (51~161) ❙ 52.8 (12.9~106.2); 872 (711~1221); 230.9 (171.2~327.7)

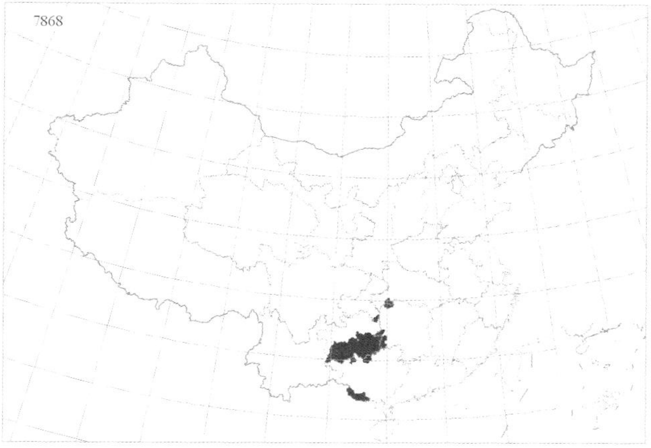

Ericaceae 杜鹃花科 *Rhododendron* 杜鹃花属

Rhododendron chrysocalyx var. xiushanense (Fang) M. Y. He 秀山金萼杜鹃

Shrubs, deciduous

15.7; 4.5; 26.5; 15.7; 129.3; 0.5; 841 ❙ 1299; 511; 92 ❙ 54.4; 841; 182.2

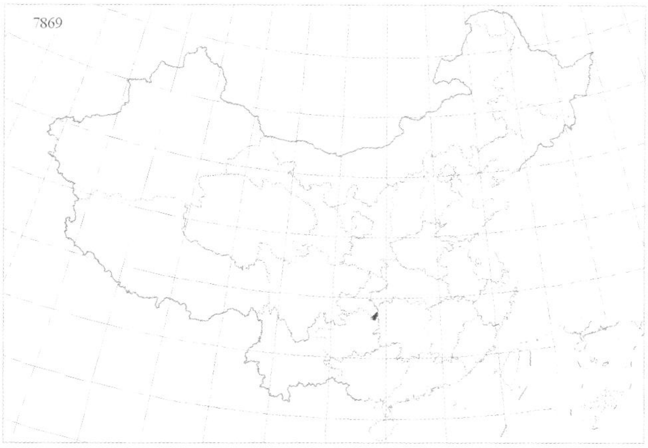

Ericaceae 杜鹃花科 *Rhododendron* 杜鹃花属

Rhododendron chrysodoron Tagg ex Hutch. 纯黄杜鹃

Small shrubs, evergreen

11.4 (5.3~15.5); 4.0 (-2.4~8.6); 17.3 (12.2~20.4); 11.5 (5.7~15.6); 82.0 (30.5~127.3); 5.2 (0.0~26.2); 648 (479~768) ❙ 1276 (961~1736); 716 (558~1007); 50 (31~69) ❙ 97.2 (53.6~145.8); 648 (479~768); 360.7 (234.2~447.4)

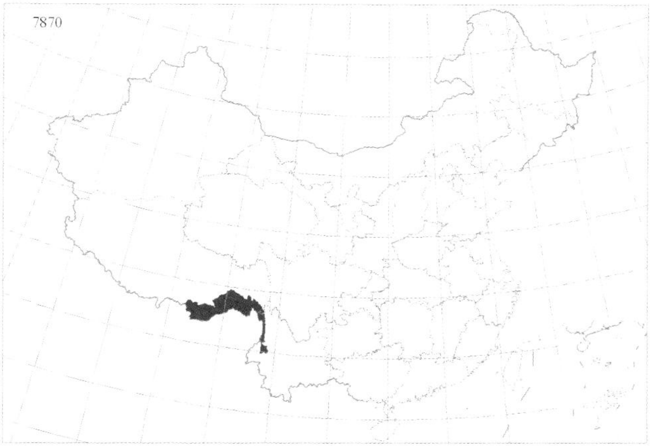

Ericaceae 杜鹃花科 *Rhododendron* 杜鹃花属

Rhododendron chunii W. P. Fang 龙山杜鹃

Shrubs, semievergreen

19.6 (17.7~21.6); 9.8 (6.9~13.2); 28.1 (26.6~29.8); 19.6 (17.7~21.6); 175.3 (152.4~199.2); 0.0 (0.0~0.0); 1030 (916~1139) ❙ 1611 (1486~1953); 599 (505~858); 177 (161~189) ❙ 56.8 (35.6~71.5); 1014 (916~1139); 216.4 (174.9~240.0)

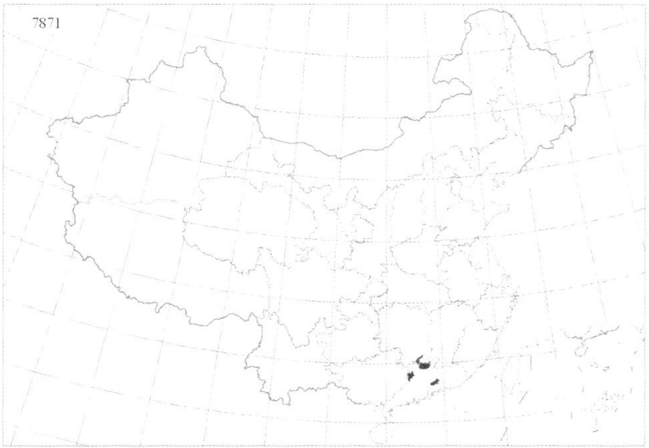

Ericaceae 杜鹃花科 *Rhododendron* 杜鹃花属

Rhododendron ciliatum Hook. 睫毛杜鹃

Shrubs, evergreen

1.9 (-1.8~13.3); -7.1 (-10.5~5.4); 9.9 (6.1~19.6); 4.3 (1.8~13.4); 21.9 (1.9~100.4); 58.6 (0.0~82.8); 408 (329~706) ❙ 492 (255~1736); 319 (191~1007); 11 (1~51) ❙ 6.5 (-31.1~145.8); 363 (255~706); 144.6 (60.4~447.4)

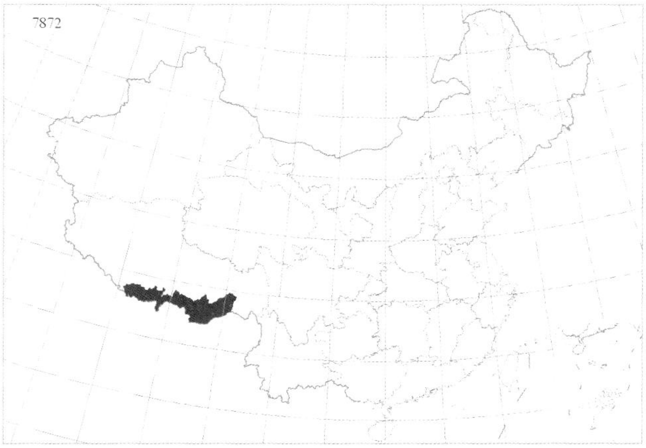

Ericaceae 杜鹃花科 *Rhododendron* 杜鹃花属
Rhododendron ciliicalyx Franch. 睫毛萼杜鹃
Shrubs, evergreen

14.8 (5.4~21.0); 6.7 (-2.1~14.2); 21.3 (12.3~26.6); 14.8 (5.8~21.0); 118.9 (31.5~191.8); 1.3 (0.0~25.6); 766 (481~1053) ❙ 1079 (741~1547); 597 (411~911); 43 (15~94) ❙ 41.4 (-14.2~74.2); 750 (481~1053); 269.1 (142.9~551.6)

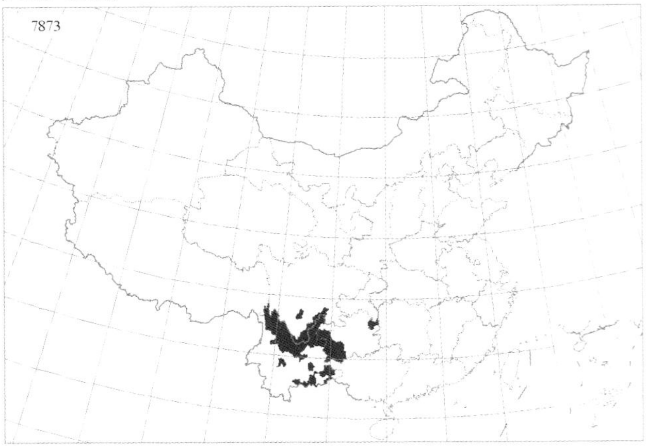

Ericaceae 杜鹃花科 *Rhododendron* 杜鹃花属
Rhododendron ciliicalyx subsp. lyi (H. Lév.) R. C. Fang 长柱睫毛萼杜鹃
Shrubs, evergreen

13.2; 5.1; 19.6; 13.2; 98.5; 0.0; 700 ❙ 989; 554; 37 ❙ 41.2; 700; 200.4

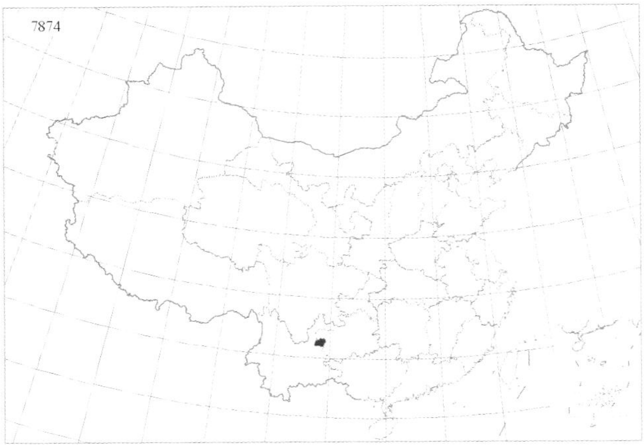

Ericaceae 杜鹃花科 *Rhododendron* 杜鹃花属
Rhododendron ciliipes Hutch. 香花白杜鹃
Shrubs, evergreen

11.9; 4.9; 17.4; 11.9; 85.4; 2.4; 660 ❙ 1269; 700; 60 ❙ 91.7; 660; 407.8

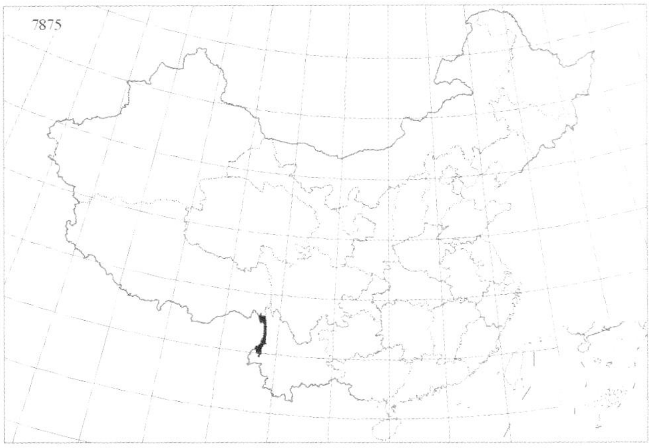

Ericaceae 杜鹃花科 *Rhododendron* 杜鹃花属
Rhododendron cinnabarinum Hook. 朱砂杜鹃
Shrubs, evergreen

4.4; -4.3; 11.9; 6.0; 36.3; 43.5; 467 ❙ 758; 463; 20 ❙ 42.2; 467; 204.9

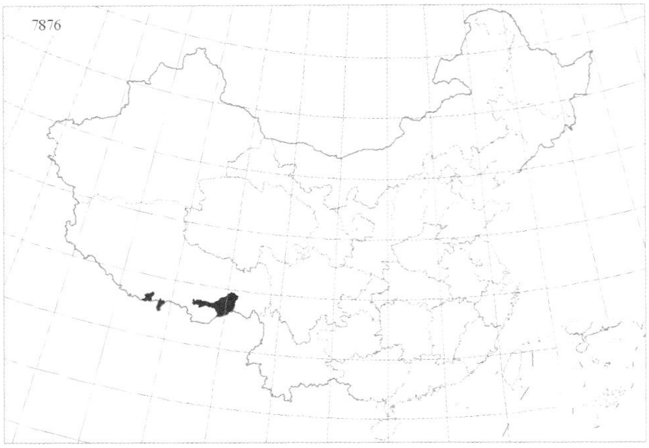

Ericaceae 杜鹃花科 *Rhododendron* 杜鹃花属
Rhododendron circinnatum Cowan et Kingdon-Ward 卷毛杜鹃
Shrubs or small trees

1.9; -7.2; 10.0; 3.8; 16.6; 53.6; 403 ❙ 341; 226; 4 ❙ -15.3; 341; 217.5

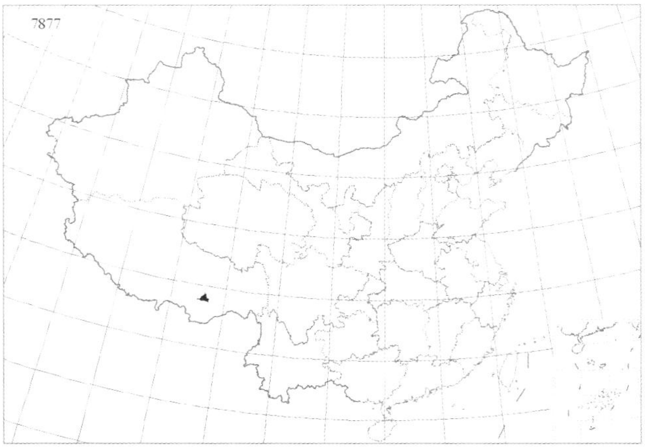

Ericaceae 杜鹃花科 *Rhododendron* 杜鹃花属
Rhododendron citriniflorum I. B. Balfour et Forrest 橙黄杜鹃
Dwarf shrubs, evergreen

8.9 (5.3~13.2); 1.7 (-2.4~6.4); 15.1 (12.2~18.4); 9.1 (5.7~13.3); 58.5 (30.5~99.4); 11.1 (0.0~26.2); 579 (479~695) ❙ 1055 (741~1337); 595 (426~727); 45 (28~69) ❙ 81.6 (54.0~100.7); 579 (479~695); 333.9 (234.2~407.1)

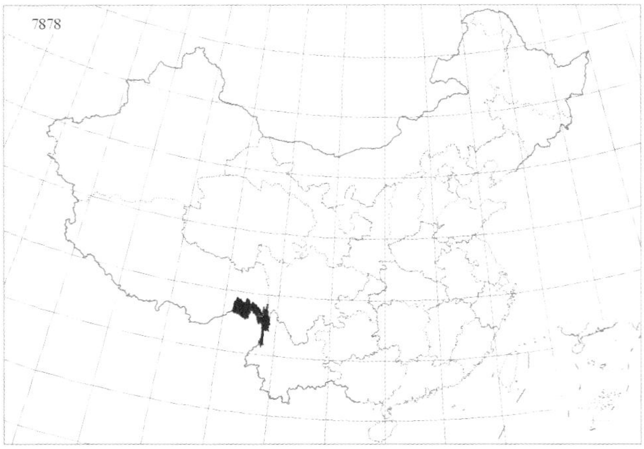

Ericaceae 杜鹃花科 *Rhododendron* 杜鹃花属

Rhododendron citriniflorum var. horaeum (I. B. Balfour et Forrest) D. F. Chamb. 美艳橙黄杜鹃 Dwarf shrubs, evergreen

8.9 (5.3~13.2); 1.7 (-2.4~6.4); 15.1 (12.2~18.4); 9.1 (5.7~13.3); 58.5 (30.5~99.4); 11.1 (0.0~26.2); 579 (479~695) ❙ 1055 (741~1337); 595 (426~727); 45 (28~69) ❙ 81.6 (54.0~100.7); 579 (479~695); 333.9 (234.2~407.1)

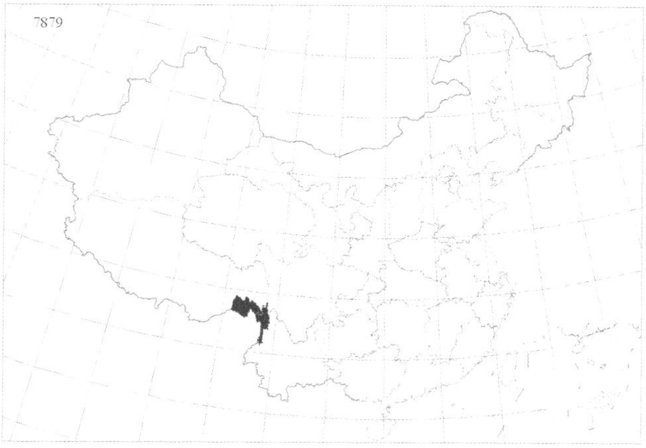

Ericaceae 杜鹃花科 *Rhododendron* 杜鹃花属

Rhododendron clementinae Forrest ex W. W. Smith 麻点杜鹃 Shrubs, evergreen

7.8 (5.4~11.2); 0.4 (-2.1~4.6); 14.1 (12.3~16.7); 8.0 (5.8~11.3); 48.6 (31.5~75.8); 14.6 (0.4~25.6); 547 (481~642) ❙ 893 (741~1054); 535 (426~627); 27 (12~40) ❙ 63.0 (54.0~67.9); 547 (481~642); 314.2 (273.5~351.7)

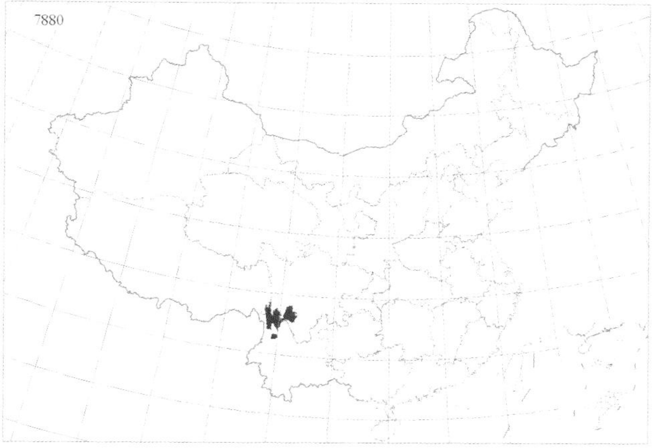

Ericaceae 杜鹃花科 *Rhododendron* 杜鹃花属

Rhododendron clementinae subsp. aureodorsale W. P. Fang 金背杜鹃 Shrubs, evergreen

11.0 (7.4~14.3); -2.1 (-4.2~-0.9); 23.5 (18.7~27.8); 11.3 (8.2~14.4); 90.8 (58.8~123.8); 17.8 (11.9~29.4); 694 (561~854) ❙ 734 (581~882); 328 (257~391); 26 (22~31) ❙ 8.1 (-32.0~52.6); 654 (561~702); 248.7 (182.6~318.2)

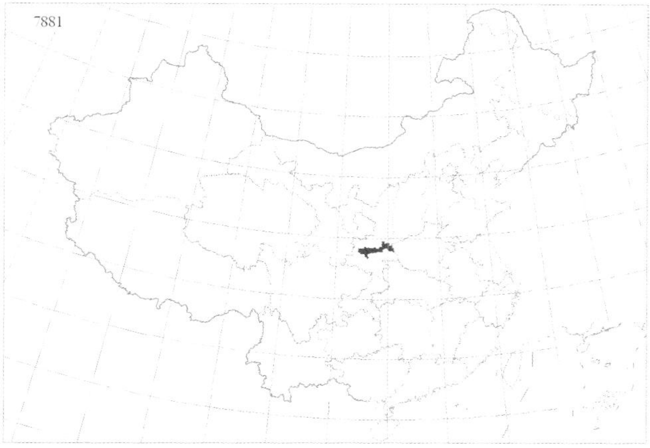

Ericaceae 杜鹃花科 *Rhododendron* 杜鹃花属

Rhododendron codonanthum I. B. Balfour et Forrest 腺蕊杜鹃 Small shrubs, evergreen

9.6; 2.4; 15.7; 9.6; 60.8; 5.4; 600 ❙ 1007; 578; 40 ❙ 67.9; 600; 351.7

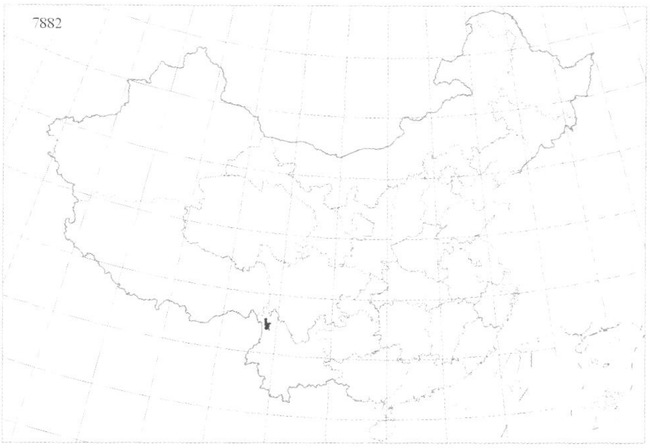

Ericaceae 杜鹃花科 *Rhododendron* 杜鹃花属

Rhododendron coelicum I. B. Balfour et Farrer 滇缅杜鹃 Small shrubs

13.2; 6.4; 18.4; 13.3; 99.4; 0.0; 695 ❙ 1337; 727; 69 ❙ 92.4; 695; 407.1

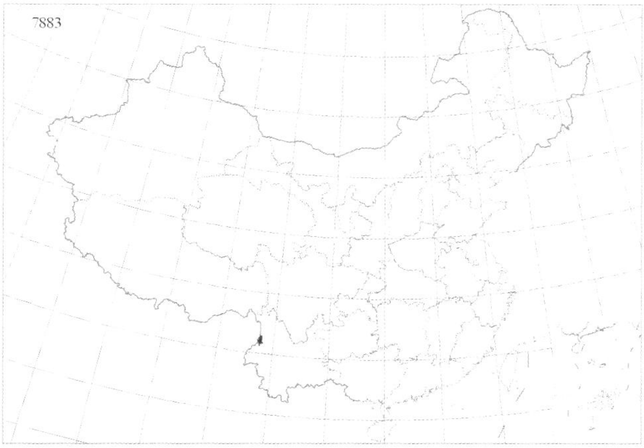

Ericaceae 杜鹃花科 *Rhododendron* 杜鹃花属

Rhododendron coeloneurum Diels 粗脉杜鹃 Trees, evergreen

13.8 (7.3~16.2); 3.7 (-2.0~5.5); 23.4 (15.3~26.8); 13.9 (7.5~16.2); 109.1 (48.2~134.4); 2.4 (0.0~20.4); 758 (537~862) ❙ 1161 (942~1414); 528 (463~817); 62 (18~96) ❙ 54.8 (30.9~101.6); 758 (537~862); 210.4 (146.0~298.0)

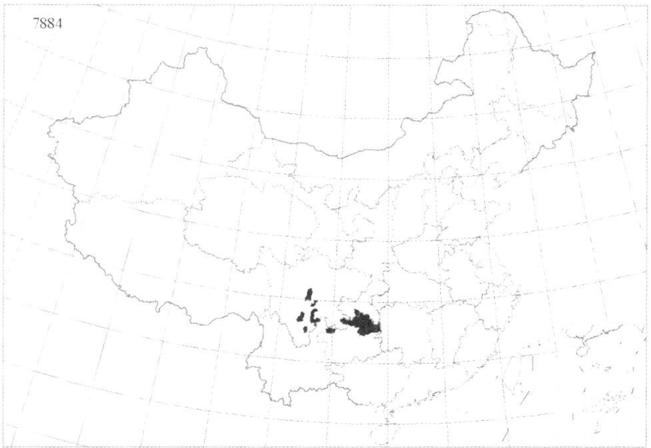

Ericaceae 杜鹃花科 *Rhododendron* 杜鹃花属
Rhododendron comisteum I. B. Balfour et Forrest 砾石杜鹃
Small shrubs

5.4; -2.3; 12.2; 5.8; 31.0; 25.9; 480 ❙ 851; 492; 30 ❙ 77.4; 480; 253.8

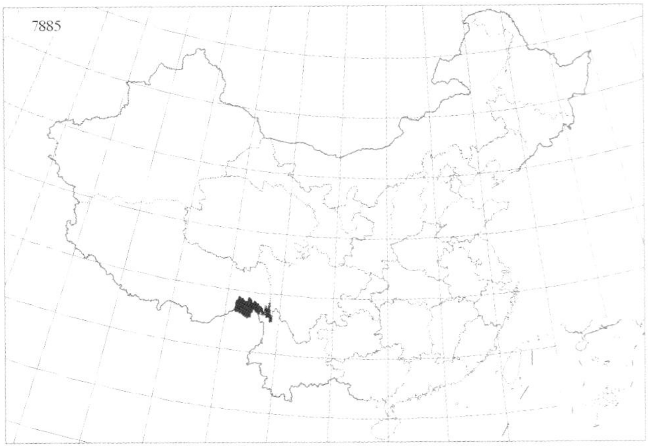

Ericaceae 杜鹃花科 *Rhododendron* 杜鹃花属
Rhododendron complexum I. B. Balfour et W. W. Sm. 环绕杜鹃
Small shrubs, evergreen

8.0; 0.6; 14.4; 8.2; 49.8; 12.9; 554 ❙ 918; 556; 25 ❙ 65.9; 554; 316.6

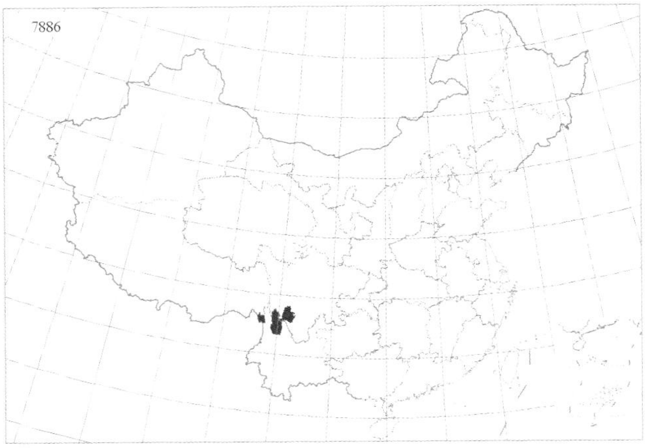

Ericaceae 杜鹃花科 *Rhododendron* 杜鹃花属
Rhododendron concinnum Hemsl. 秀雅杜鹃
Shrubs, evergreen

10.9 (-1.3~17.1); -0.3 (-13.5~7.1); 21.3 (9.7~27.5); 11.2 (2.9~17.1); 86.3 (11.3~145.7); 15.1 (0.0~86.3); 677 (359~891) ❙ 898 (486~1448); 449 (289~817); 35 (8~113) ❙ 36.8 (-22.2~128.6); 659 (359~891); 240.6 (139.1~351.6)

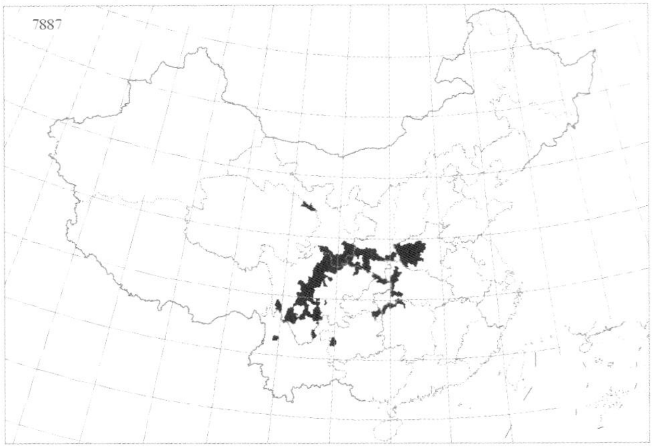

Ericaceae 杜鹃花科 *Rhododendron* 杜鹃花属
Rhododendron coriaceum Franch. 革叶杜鹃
Shrubs or small trees, evergreen

8.8 (5.3~13.2); 1.6 (-2.4~6.4); 15.0 (12.2~18.4); 9.0 (5.7~13.3); 57.3 (30.5~99.4); 11.2 (0.0~26.2); 575 (479~695) ❙ 1013 (741~1337); 583 (426~727); 39 (21~69) ❙ 75.5 (52.8~100.7); 575 (479~695); 325.8 (234.2~407.1)

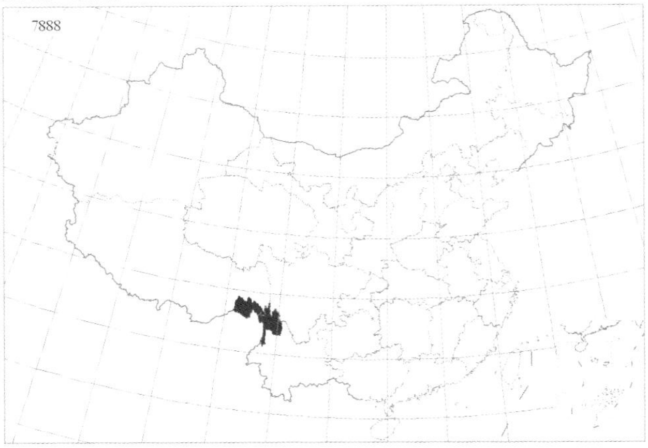

Ericaceae 杜鹃花科 *Rhododendron* 杜鹃花属
Rhododendron coryanum Tagg et Forrest 光蕊杜鹃
Shrubs, evergreen

8.3 (1.8~13.3); 0.6 (-7.1~6.4); 14.9 (9.9~19.6); 8.8 (3.8~13.4); 57.4 (15.7~100.4); 17.0 (0.0~53.3); 567 (404~706) ❙ 1055 (527~1736); 602 (317~1007); 39 (9~69) ❙ 81.0 (22.4~145.8); 567 (404~706); 328.4 (169.5~447.4)

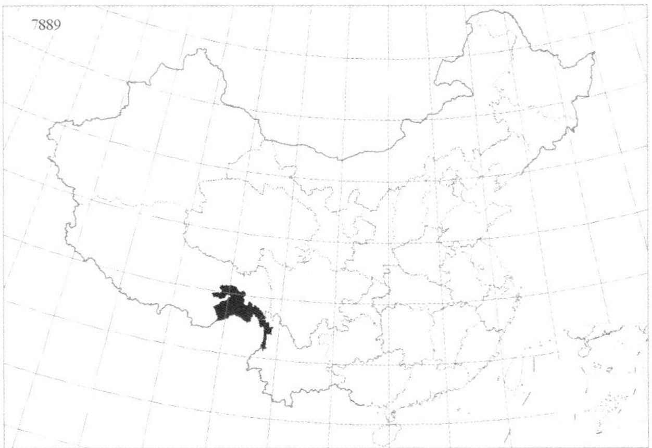

Ericaceae 杜鹃花科 *Rhododendron* 杜鹃花属
Rhododendron crassimedium Tam 棒柱杜鹃
Shrubs, semievergreen or green

16.6; 5.5; 27.1; 16.7; 140.5; 0.3; 883 ❙ 1572; 508; 204 ❙ 79.1; 881; 205.0

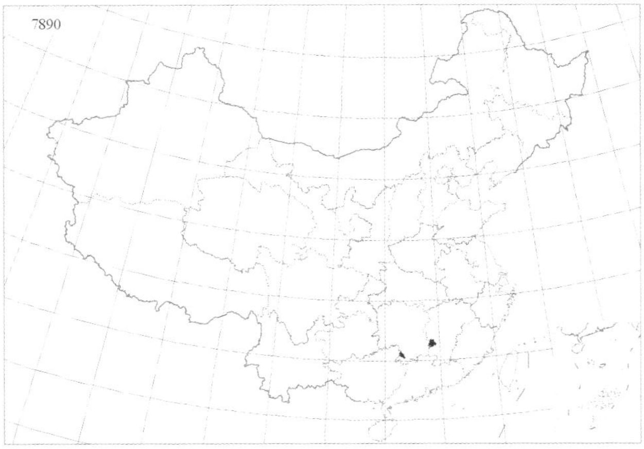

Ericaceae 杜鹃花科 *Rhododendron* 杜鹃花属
Rhododendron crassistylum M. Y. He 粗柱杜鹃
Shrubs, semievergreen or green

17.1 (15.3~18.5); 6.1 (4.2~7.8); 27.3 (26.0~28.1); 17.2 (15.4~18.5); 146.2 (125.8~162.3); 0.2 (0.0~0.8); 906 (818~975) ❙ 1607 (1561~1660); 545 (512~575); 199 (189~216) ❙ 78.1 (65.4~103.1); 906 (818~975); 222.1 (208.1~236.3)

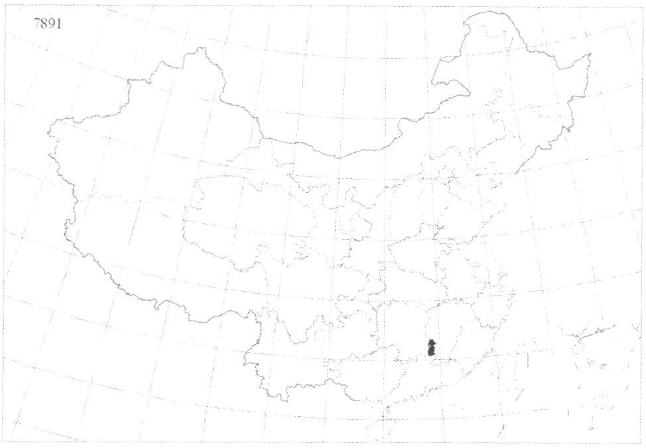

Ericaceae 杜鹃花科 *Rhododendron* 杜鹃花属
Rhododendron cretaceum Tam 白枝杜鹃
Shrubs, semievergreen or green

17.7; 7.7; 26.6; 17.8; 153.6; 0.0; 916 ❙ 1564; 556; 182 ❙ 70.7; 916; 232.5

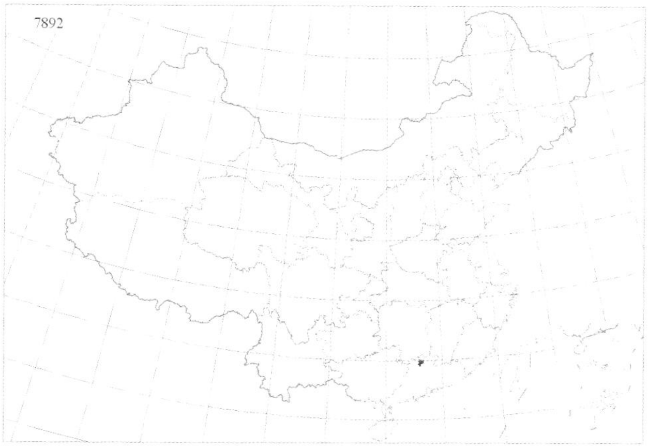

Ericaceae 杜鹃花科 *Rhododendron* 杜鹃花属
Rhododendron crinigerum Franch. 长粗毛杜鹃
Shrubs, evergreen

6.8 (0.3~13.2); -1.2 (-9.9~6.4); 13.8 (9.3~18.4); 7.6 (3.3~13.3); 46.6 (11.9~99.4); 24.0 (0.0~67.5); 532 (391~695) ❙ 953 (591~1337); 538 (362~727); 36 (9~69) ❙ 77.4 (43.8~101.2); 532 (391~695); 311.5 (200.8~407.1)

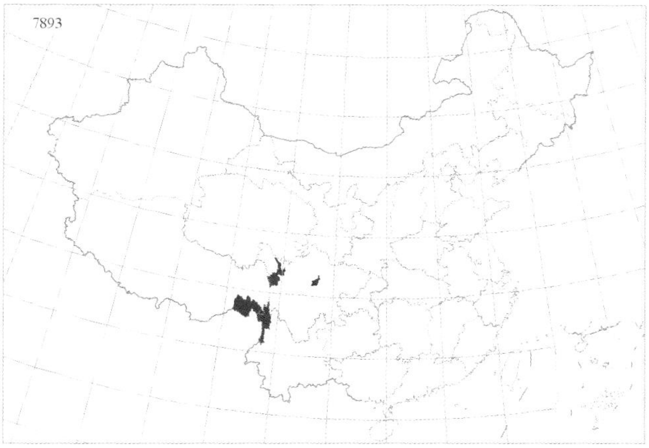

Ericaceae 杜鹃花科 *Rhododendron* 杜鹃花属
Rhododendron crinigerum var. euadenium Tagg et Forrest 腺背长粗毛杜鹃
Shrubs, evergreen

8.6; 1.3; 15.0; 8.7; 52.6; 8.8; 574 ❙ 1041; 585; 45 ❙ 81.5; 574; 352.3

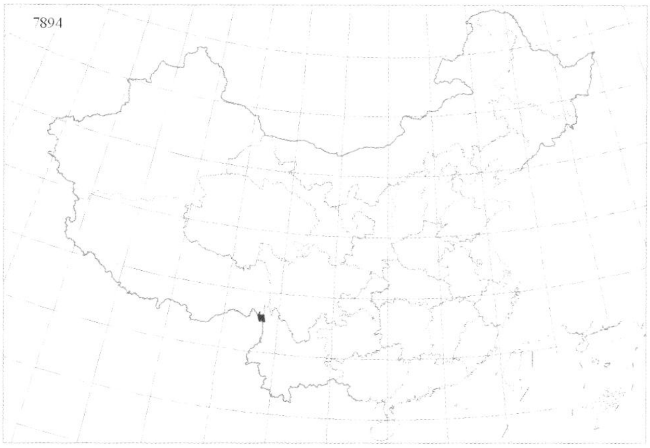

Ericaceae 杜鹃花科 *Rhododendron* 杜鹃花属
Rhododendron cuneatum W. W. Sm. 楔叶杜鹃
Shrubs, evergreen

10.8 (6.1~18.2); 3.5 (-1.5~12.4); 16.7 (12.7~21.6); 10.8 (6.3~18.2); 76.6 (35.2~158.6); 7.0 (0.0~21.8); 634 (496~861) ❙ 1019 (801~1500); 601 (496~785); 28 (12~52) ❙ 60.2 (49.1~74.2); 630 (496~839); 326.3 (278.8~439.5)

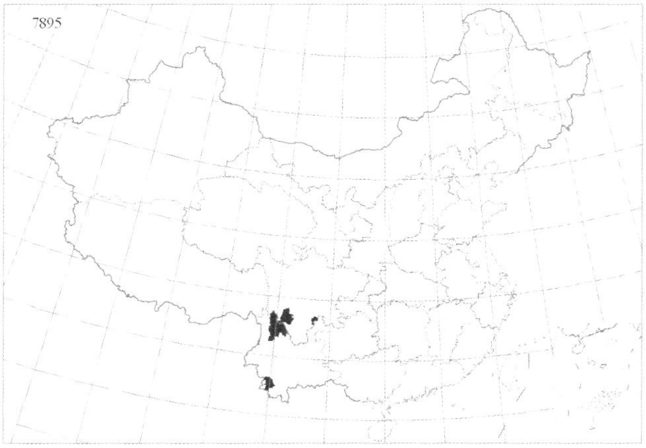

Ericaceae 杜鹃花科 *Rhododendron* 杜鹃花属
Rhododendron cyanocarpum (Franch.) W. W. Sm. 蓝果杜鹃
Shrubs or small trees, evergreen

13.6; 7.0; 18.8; 13.6; 103.1; 0.0; 704 ❙ 1048; 591; 47 ❙ 48.9; 704; 368.2

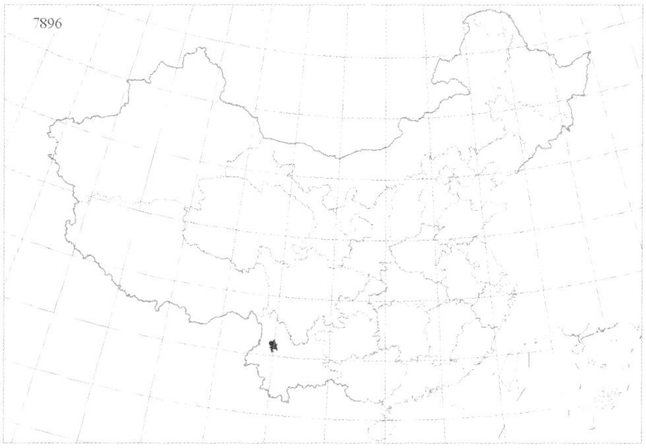

Ericaceae 杜鹃花科 *Rhododendron* 杜鹃花属

Rhododendron dalhousieae (I. B. Balfour et R. E. Cooper) Cullen 长药杜鹃

Shrubs, evergreen

3.2; -5.2; 10.7; 5.5; 32.1; 52.6; 446 | 707; 432; 20 | 38.3; 443; 165.8

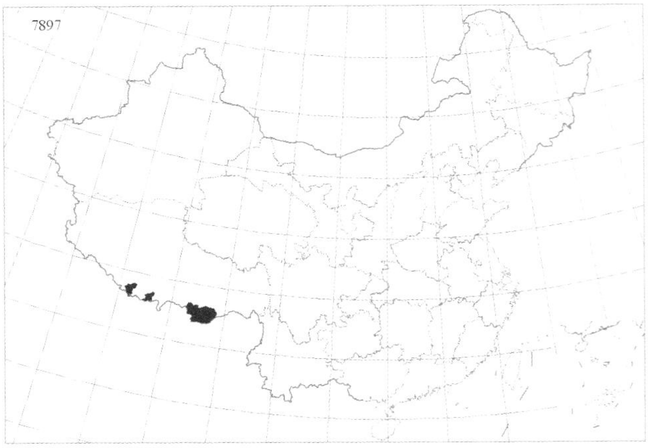

Ericaceae 杜鹃花科 *Rhododendron* 杜鹃花属

Rhododendron dalhousieae var. rhabdotum (I. B. Bal-four et R. E. Cooper) Cullen 红线长药杜鹃

Shrubs

12.2; 4.3; 18.5; 12.2; 87.4; 0.7; 672 | 1431; 817; 31 | 113.0; 672; 363.7

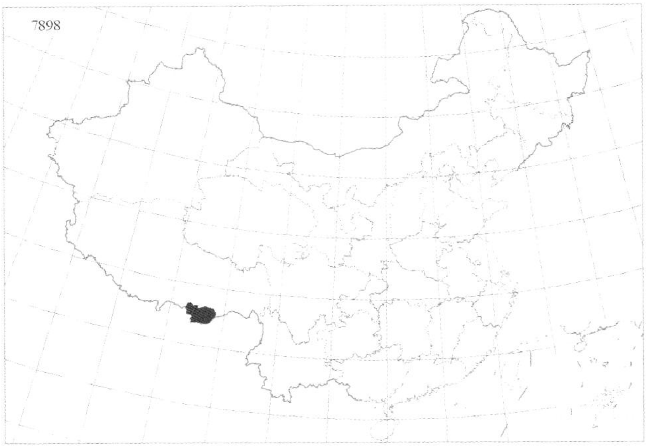

Ericaceae 杜鹃花科 *Rhododendron* 杜鹃花属

Rhododendron danbaense L. C. Hu 丹巴杜鹃

Shrubs, evergreen

3.3; -6.1; 11.4; 4.6; 22.5; 42.7; 434 | 812; 452; 17 | 87.0; 434; 337.3

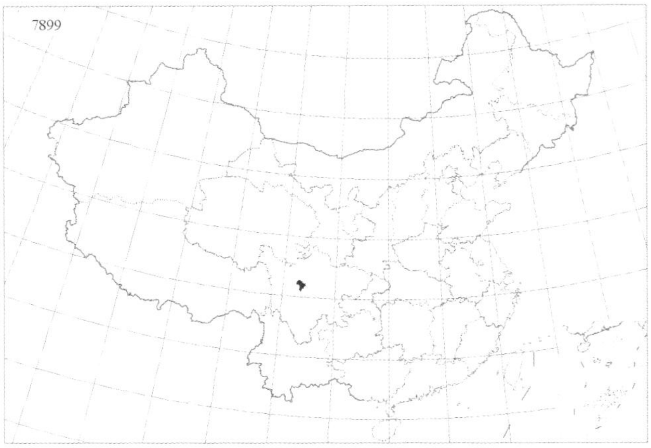

Ericaceae 杜鹃花科 *Rhododendron* 杜鹃花属

Rhododendron dasycladoides Hand.-Mazz. 漏斗杜鹃

Shrubs or small trees, evergreen

6.4; -1.4; 13.0; 6.5; 37.4; 20.7; 505 | 832; 522; 16 | 64.6; 505; 297.2

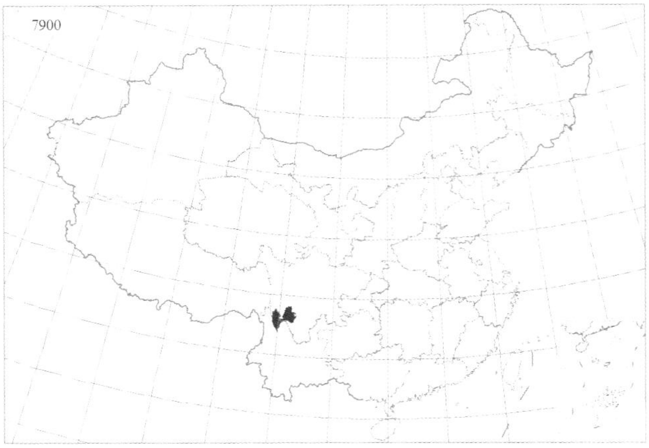

Ericaceae 杜鹃花科 *Rhododendron* 杜鹃花属

Rhododendron dasypetalum I. B. Balfour et Forrest 毛瓣杜鹃

Small shrubs, evergreen

9.6; 2.4; 15.7; 9.6; 60.8; 5.4; 600 | 1007; 578; 40 | 67.9; 600; 351.7

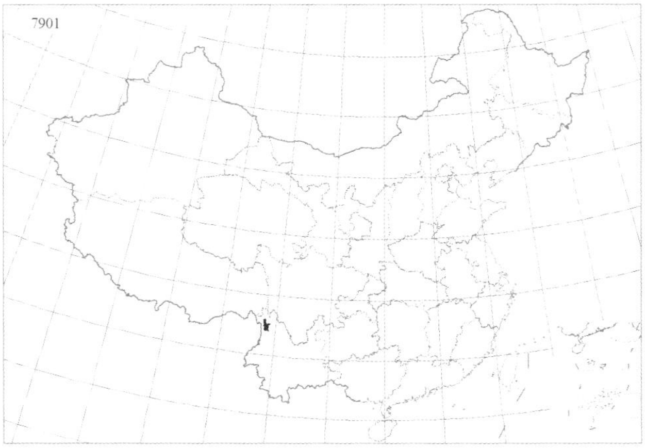

Ericaceae 杜鹃花科 *Rhododendron* 杜鹃花属

Rhododendron dauricum L. 兴安杜鹃

Shrubs, semievergreen

4.2 (-5.8~11.0); -15.3 (-29.8~-4.9); 21.2 (15.3~25.5); 8.3 (4.4~12.0); 65.3 (28.0~101.7); 74.0 (26.8~157.4); 590 (426~757) | 590 (309~1073); 385 (199~691); 18 (7~40) | 0.6 (-42.4~74.1); 533 (309~668); 256.9 (71.4~346.5)

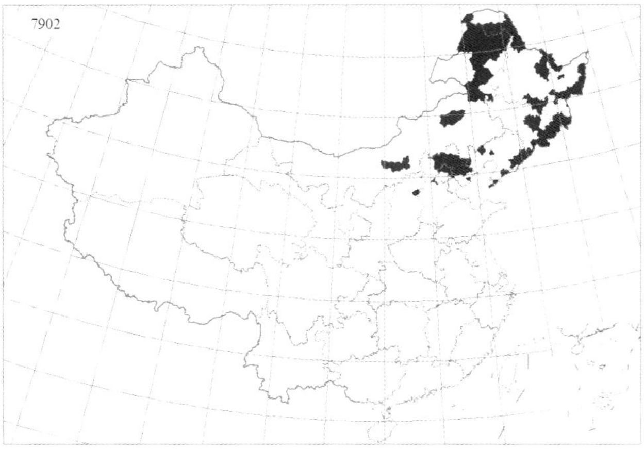

Ericaceae 杜鹃花科 *Rhododendron* 杜鹃花属
Rhododendron davidii Franch. 腺果杜鹃
Shrubs or small trees, evergreen

12.4 (2.2~15.2); 2.8 (-6.6~7.1); 21.2 (9.9~25.9); 12.6 (3.8~15.2); 95.3 (15.5~123.9); 5.6 (0.0~49.1); 701 (413~816) | 1145 (879~1478); 588 (472~817); 48 (19~105) | 65.0 (25.7~125.9); 698 (413~816); 240.5 (160.6~322.4)

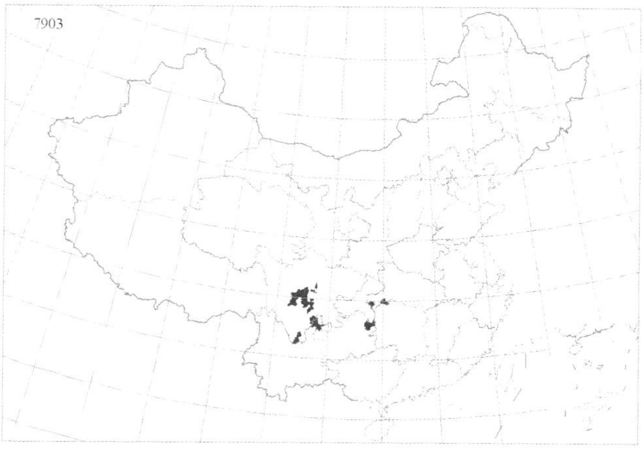

Ericaceae 杜鹃花科 *Rhododendron* 杜鹃花属
Rhododendron davidsonianum Rehder et E. H. Wilson 凹叶杜鹃
Shrubs, deciduous

8.5 (2.0~14.8); -0.2 (-7.2~6.5); 16.1 (9.5~24.0); 9.0 (3.8~14.9); 60.7 (15.4~118.8); 18.1 (0.0~51.4); 578 (411~778) | 970 (628~1220); 551 (410~688); 26 (11~63) | 72.0 (39.1~128.6); 576 (411~778); 257.4 (153.9~322.4)

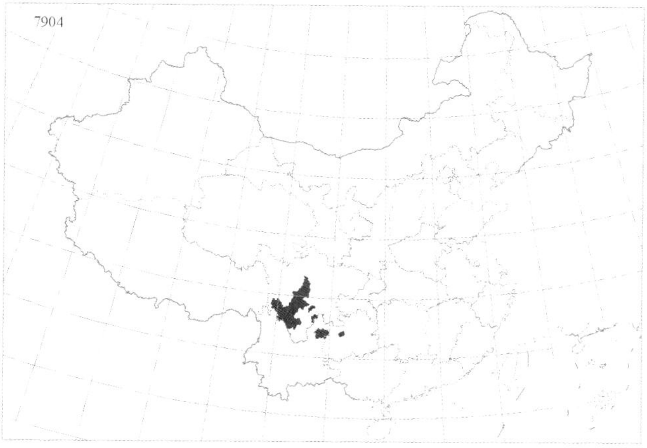

Ericaceae 杜鹃花科 *Rhododendron* 杜鹃花属
Rhododendron dawuense H. P. Yang 道孚杜鹃
Shrubs, evergreen

2.0; -7.4; 10.1; 4.0; 16.9; 51.9; 416 | 777; 457; 15 | 86.8; 416; 271.2

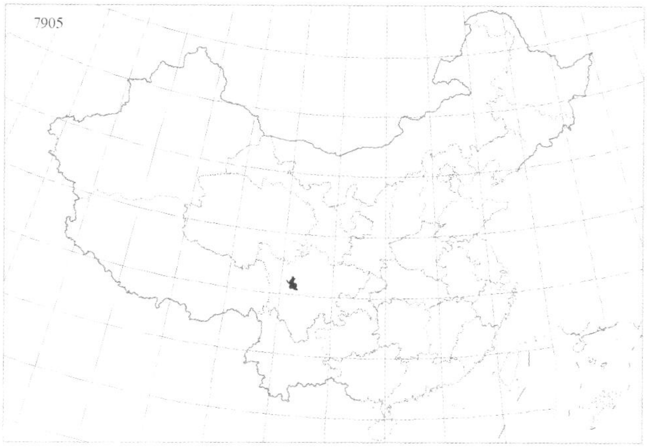

Ericaceae 杜鹃花科 *Rhododendron* 杜鹃花属
Rhododendron declivatum Ching et H. P. Yang 徒生杜鹃
Small shrubs, evergreen

9.5; -2.7; 21.4; 10.0; 76.8; 21.6; 633 | 796; 359; 24 | 27.5; 633; 283.7

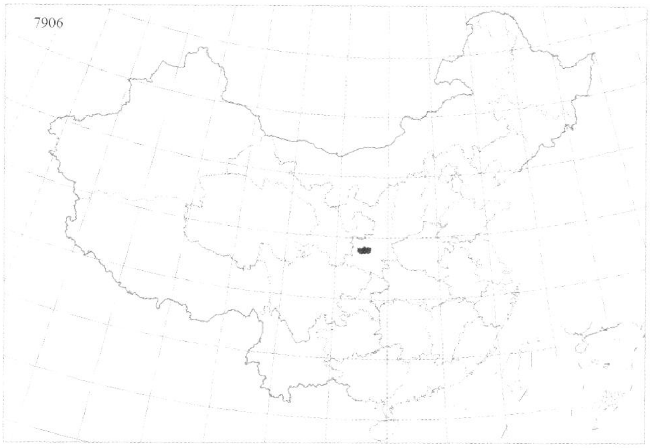

Ericaceae 杜鹃花科 *Rhododendron* 杜鹃花属
Rhododendron decorum Franch. 大白杜鹃
Shrubs or small trees, evergreen

13.0 (0.3~18.9); 5.0 (-9.9~12.0); 19.6 (9.3~26.6); 13.2 (3.3~19.0); 101.7 (11.9~167.8); 5.3 (0.0~67.5); 707 (391~929) | 1040 (628~1454); 575 (372~816); 42 (11~129) | 49.8 (-14.2~128.6); 693 (391~912); 283.5 (142.9~491.5)

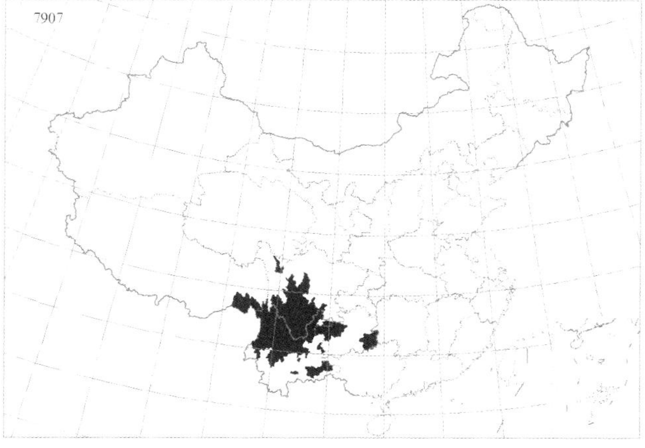

Ericaceae 杜鹃花科 *Rhododendron* 杜鹃花属
Rhododendron decorum subsp. diaprepes (I. B. Balfour et W. W. Sm.) T. L. Ming 高尚大白杜鹃 Shrubs or small trees, evergreen

12.9; 3.8; 20.8; 13.0; 100.0; 5.3; 709 | 1266; 657; 54 | 79.7; 709; 289.2

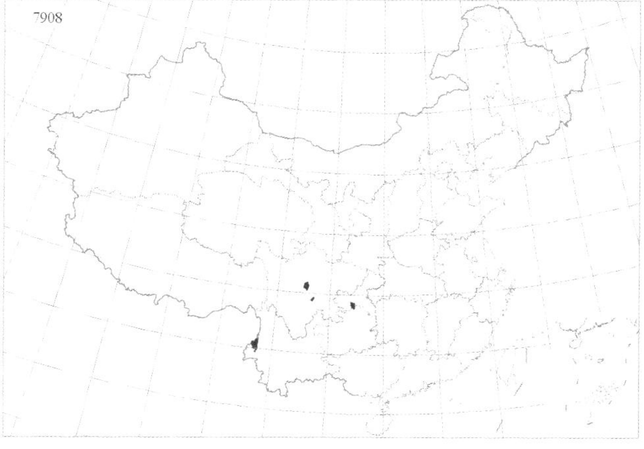

Ericaceae 杜鹃花科 *Rhododendron* 杜鹃花属

Rhododendron decorum subsp. parvistigmatis W. K. Hu 小柱大白杜鹃

Shrubs or small trees, evergreen

11.9; 2.9; 20.0; 12.0; 87.2; 3.4; 671 ❙ 1001; 550; 35 ❙ 49.1; 671; 278.8

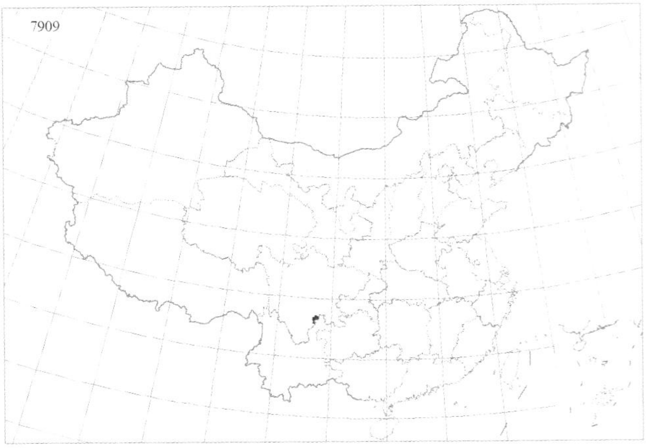

Ericaceae 杜鹃花科 *Rhododendron* 杜鹃花属

Rhododendron dekatanum Cowan 隆子杜鹃

Shrubs, evergreen

4.4; -4.3; 12.1; 5.3; 28.5; 34.6; 455 ❙ 538; 337; 9 ❙ 18.2; 455; 168.2

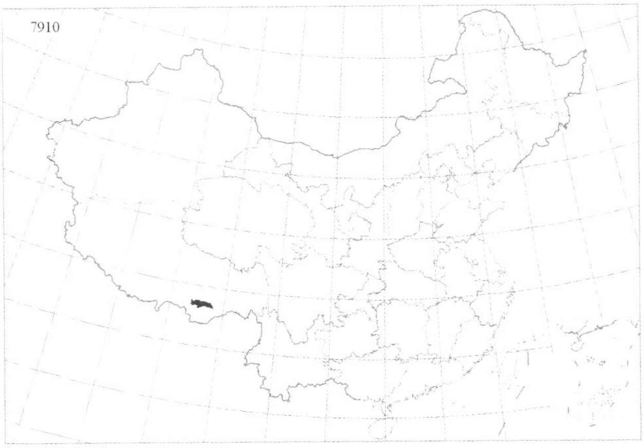

Ericaceae 杜鹃花科 *Rhododendron* 杜鹃花属

Rhododendron delavayi Franch. 马缨杜鹃

Shrubs or small trees, evergreen

14.0 (-1.8~19.2); 6.3 (-10.2~13.3); 20.3 (6.1~26.6); 14.2 (1.8~19.4); 112.2 (1.9~172.6); 3.1 (0.0~82.8); 740 (329~966) ❙ 1059 (320~1736); 589 (217~1007); 43 (11~94) ❙ 44.2 (-14.2~145.8); 724 (320~966); 282.6 (72.2~487.2)

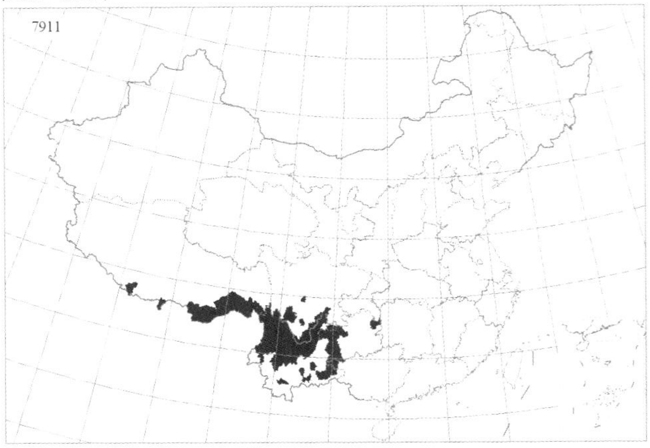

Ericaceae 杜鹃花科 *Rhododendron* 杜鹃花属

Rhododendron delavayi var. peramoenum (I. B. Balfour et Forrest) T. L. Ming 狭叶马缨杜鹃

Shrubs or small trees, evergreen

10.8; 2.8; 17.6; 10.9; 78.7; 8.8; 640 ❙ 1146; 617; 52 ❙ 81.8; 640; 306.9

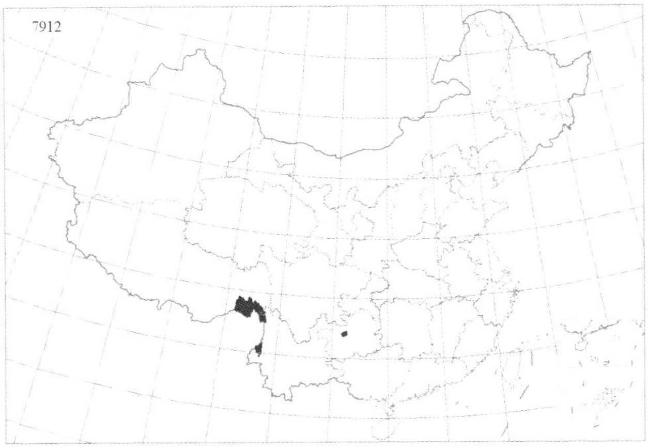

Ericaceae 杜鹃花科 *Rhododendron* 杜鹃花属

Rhododendron delavayi var. pilostylum K. M. Feng 毛柱马缨杜鹃

Shrubs or small trees, evergreen

13.1; 4.1; 20.6; 13.1; 98.4; 0.9; 703 ❙ 1107; 592; 47 ❙ 57.5; 703; 201.7

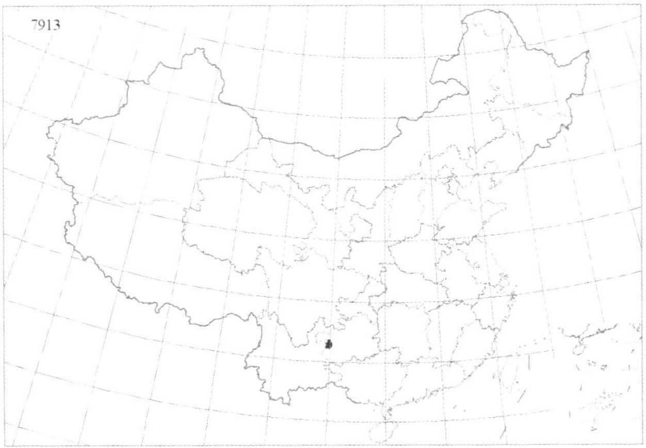

Ericaceae 杜鹃花科 *Rhododendron* 杜鹃花属

Rhododendron dendricola Hutch. 附生杜鹃

Shrubs, evergreen

9.6; 2.5; 15.7; 9.8; 65.0; 8.8; 599 ❙ 1091; 608; 50 ❙ 80.2; 599; 354.4

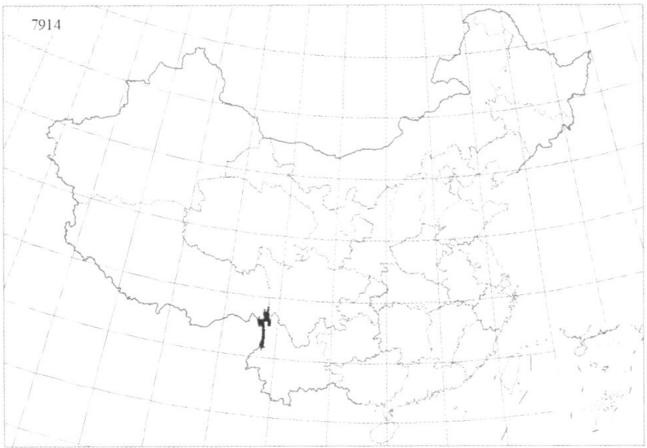

Ericaceae 杜鹃花科 *Rhododendron* 杜鹃花属
Rhododendron dendrocharis Franch. 树生杜鹃
Shrubs, evergreen

9.8 (2.2~14.3); 0.7 (-6.6~5.3); 17.9 (9.9~23.6); 10.1 (3.8~15.0); 71.0 (15.5~119.6) 13.0 (0.0~49.1); 616 (413~782) ❙ 1083 (932~1410); 604 (531~816); 31 (19~49) ❙ 78.7 (49.1~125.9); 616 (413~782); 260.9 (214.5~318.7)

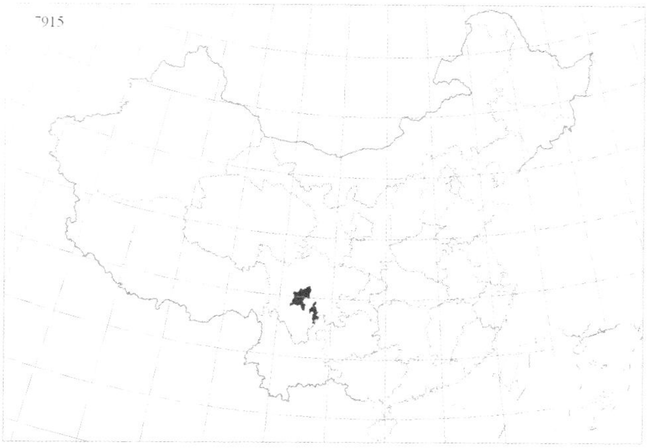

Ericaceae 杜鹃花科 *Rhododendron* 杜鹃花属
Rhododendron densifolium K. M. Feng 密叶杜鹃
Shrubs, evergreen

17.6; 9.9; 23.2; 17.7; 152.4; 0.0; 859 ❙ 1422; 813; 52 ❙ 65.6; 859; 359.7

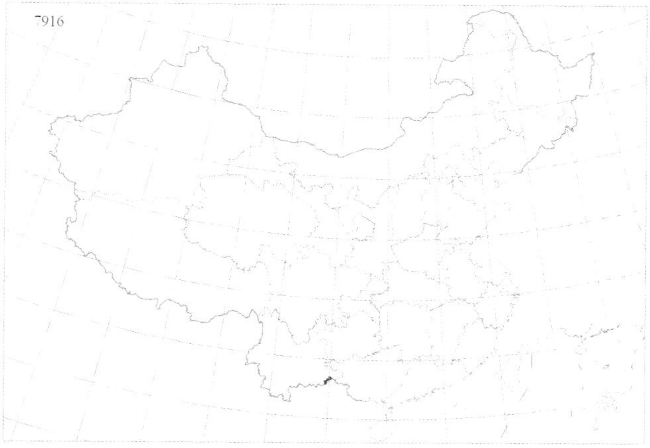

Ericaceae 杜鹃花科 *Rhododendron* 杜鹃花属
Rhododendron denudatum H. Lév. 皱叶杜鹃
Shrubs or small trees, evergreen

12.9 (6.6~16.1); 4.5 (-1.3~8.6); 19.9 (13.3~24.0); 12.9 (6.8~16.1); 97.7 (39.6~133.0); 2.5 (0.0~19.7); 699 (514~797) ❙ 999 (862~1143); 567 (479~630); 33 (12~63) ❙ 43.8 (26.5~67.7); 689 (514~778); 255.3 (153.9~330.0)

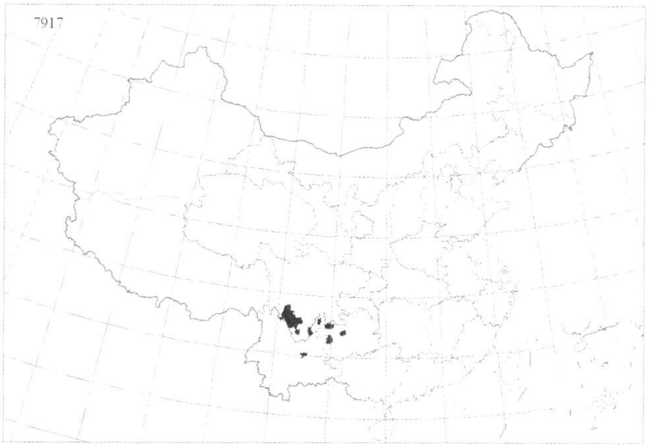

Ericaceae 杜鹃花科 *Rhododendron* 杜鹃花属
Rhododendron detersile Franch. 干净杜鹃
Small shrubs

10.2 (2.9~14.7); -1.0 (-6.3~4.3); 20.8 (11.2~24.0); 10.6 (4.4~14.8); 80.8 (20.5~118.0); 17.6 (0.7~44.6); 650 (428~789) ❙ 933 (738~1207); 443 (332~542); 29 (22~56) ❙ 48.4 (5.8~121.7); 650 (428~789); 252.1 (150.8~318.2)

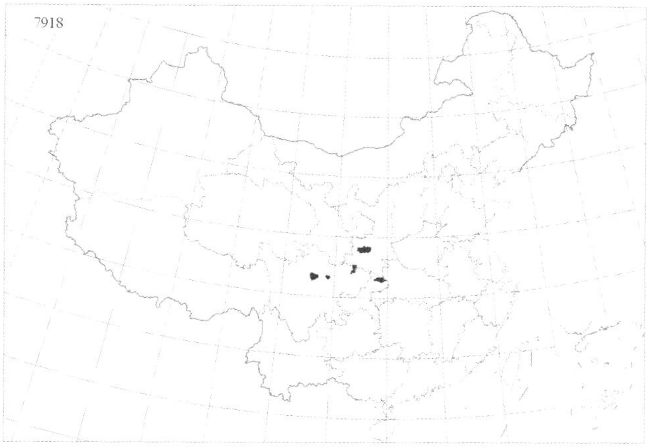

Ericaceae 杜鹃花科 *Rhododendron* 杜鹃花属
Rhododendron dichroanthum Diels 两色杜鹃
Dwarf shrubs, evergreen

11.1; 4.0; 16.8; 11.1; 76.2; 3.2; 638 ❙ 1207; 669; 57 ❙ 89.0; 638; 381.4

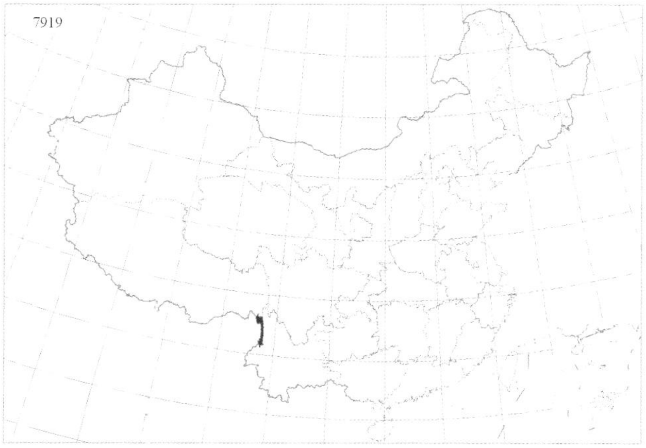

Ericaceae 杜鹃花科 *Rhododendron* 杜鹃花属
Rhododendron dichroanthum subsp. apodectum (I. B. Balfour et W. W. Sm.) Cowan 可喜杜鹃 Dwarf shrubs, evergreen

13.0; 6.1; 18.2; 13.0; 96.4; 0.2; 689 ❙ 1345; 739; 66 ❙ 95.1; 689; 426.3

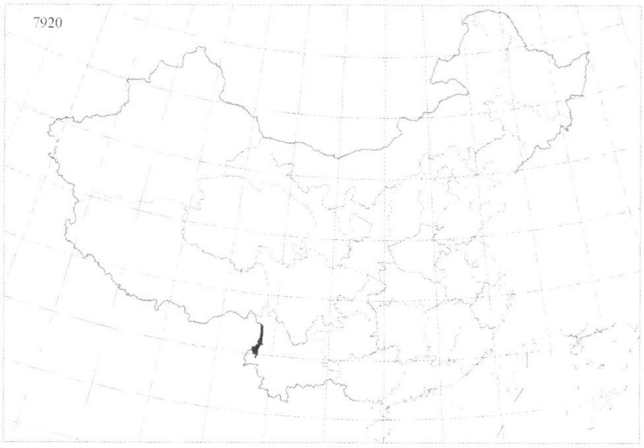

Ericaceae 杜鹃花科 *Rhododendron* 杜鹃花属
Rhododendron dichroanthum subsp. scyphocalyx (I. B. Balfour et Forrest) Cowan 杯萼两色杜鹃
Dwarf shrubs, evergreen

11.9; 4.9; 17.4; 11.9; 85.4; 2.4; 660 ❙ 1269; 700; 60 ❙ 91.7; 660; 407.8

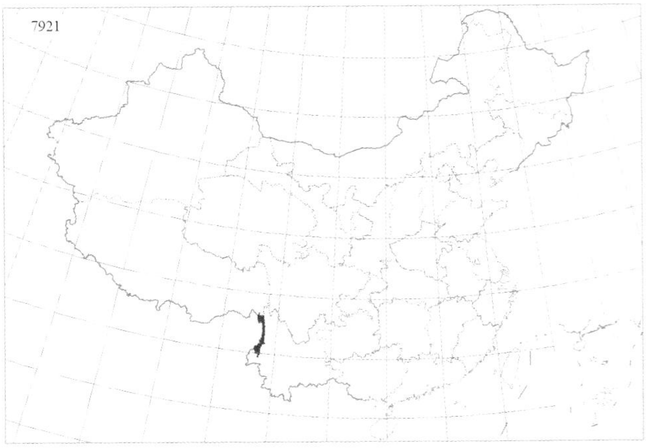

Ericaceae 杜鹃花科 *Rhododendron* 杜鹃花属
Rhododendron dichroanthum subsp. septentrionale Cowan 腺梗两色杜鹃
Dwarf shrubs, evergreen

11.1; 4.0; 16.8; 11.1; 76.2; 3.2; 638 ❙ 1207; 669; 57 ❙ 89.0; 638; 381.4

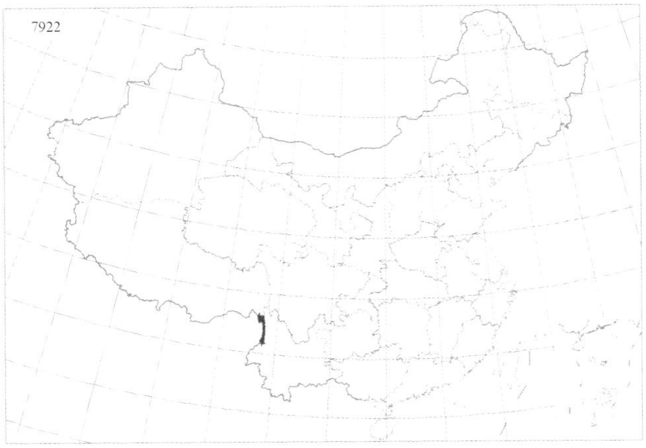

Ericaceae 杜鹃花科 *Rhododendron* 杜鹃花属
Rhododendron dignabile Cowan 疏毛杜鹃
Shrubs or small trees, evergreen

3.2; −5.7; 11.0; 4.6; 22.5; 44.1; 429 ❙ 439; 281; 6 ❙ 1.4; 398; 192.9

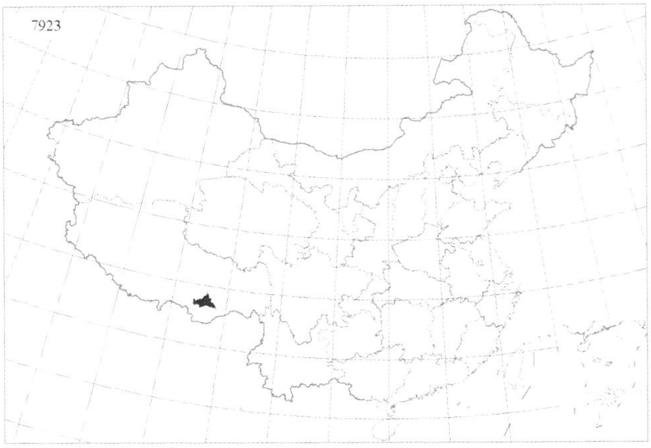

Ericaceae 杜鹃花科 *Rhododendron* 杜鹃花属
Rhododendron dimitrum I. B. Balfour et Forrest 苍山杜鹃
Shrubs, evergreen

13.6; 7.2; 18.8; 13.7; 104.1; 0.0; 706 ❙ 1042; 592; 46 ❙ 47.6; 706; 372.3

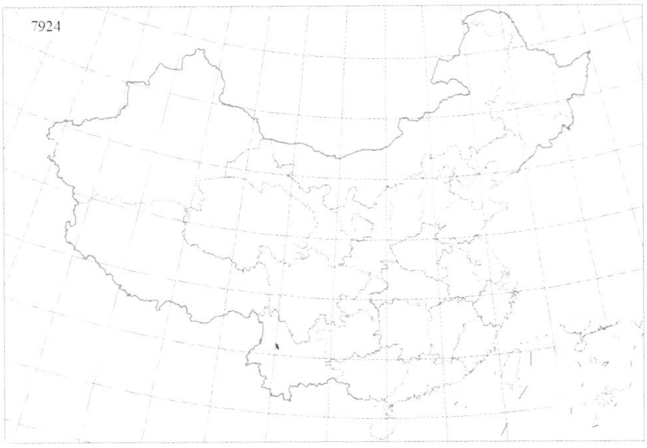

Ericaceae 杜鹃花科 *Rhododendron* 杜鹃花属
Rhododendron diphrocalyx I. B. Balfour 腾冲杜鹃
Shrubs, evergreen

14.4; 7.5; 19.1; 14.4; 113.1; 0.0; 727 ❙ 1454; 792; 70 ❙ 99.9; 727; 487.2

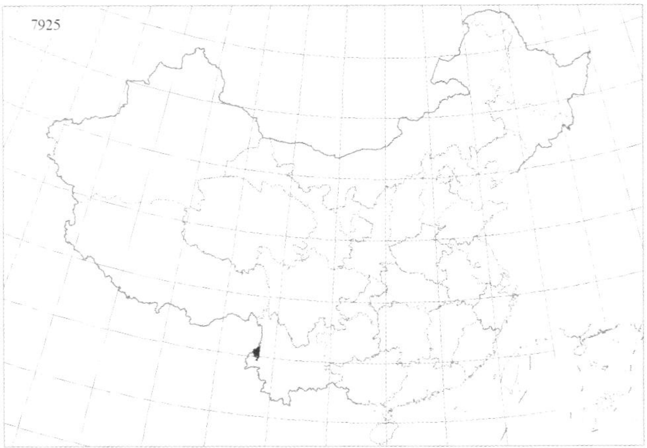

Ericaceae 杜鹃花科 *Rhododendron* 杜鹃花属
Rhododendron discolor Franch. 喇叭杜鹃
Shrubs or small trees, evergreen

15.0 (6.0~19.1); 4.1 (−3.2~9.3); 25.1 (14.4~29.0); 15.0 (6.6~19.1); 123.4 (39.7~169.2); 3.4 (0.0~26.7); 815 (508~1000) ❙ 1372 (899~1829); 523 (403~696); 131 (19~230) ❙ 69.5 (10.0~112.2); 810 (508~988); 226.2 (151.0~305.6)

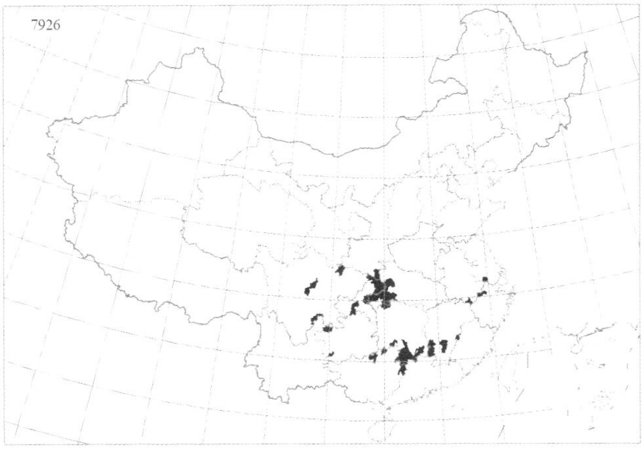

Ericaceae 杜鹃花科 *Rhododendron* 杜鹃花属
Rhododendron dumicola Tagg et Forrest 灌丛杜鹃
Shrubs, evergreen

9.6; 2.4; 15.7; 9.6; 60.8; 5.4; 600 ▎1007; 578; 40 ▎67.9; 600; 351.7

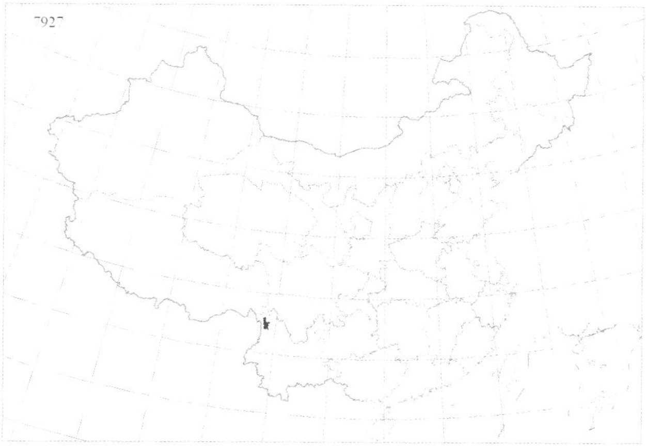

Ericaceae 杜鹃花科 *Rhododendron* 杜鹃花属
Rhododendron ebianense M. Y. Fang 峨边杜鹃
Shrubs or small trees, evergreen

11.1; 2.2; 19.4; 11.3; 80.8; 5.3; 652 ▎1220; 688; 37 ▎87.2; 652; 292.6

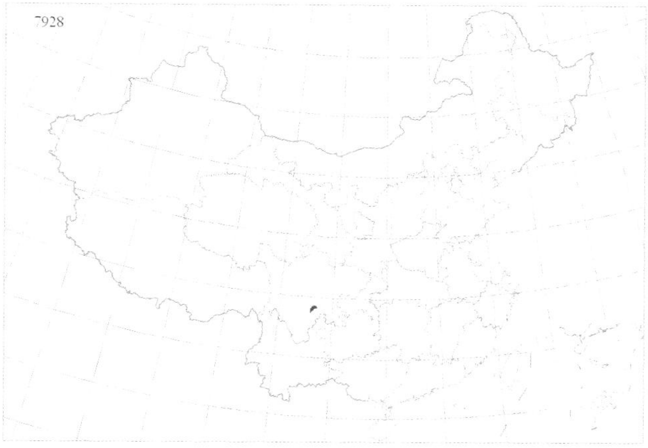

Ericaceae 杜鹃花科 *Rhododendron* 杜鹃花属
Rhododendron eclecteum I. B. Balfour et Forrest 杂色杜鹃
Shrubs, evergreen

5.6 (1.4~11.0); -2.4 (-7.6~3.7); 12.5 (9.2~17.0); 6.3 (3.5~11.1); 35.5 (13.4~74.6); 28.1 (1.8~55.6); 493 (402~640) ▎811 (579~1041); 494 (365~619); 21 (10~45) ▎64.2 (38.4~100.7); 493 (402~638); 267.2 (201.9~352.3)

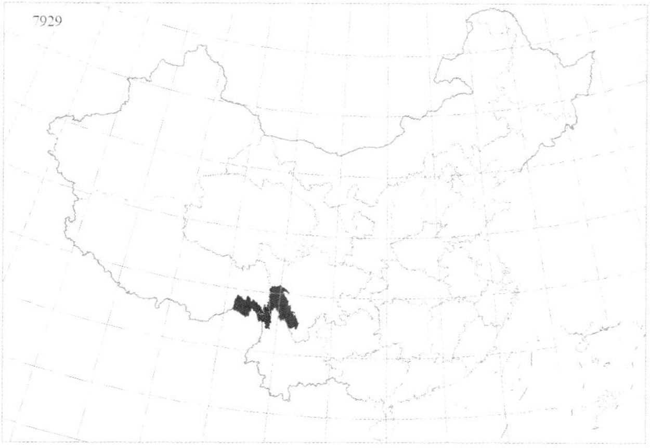

Ericaceae 杜鹃花科 *Rhododendron* 杜鹃花属
Rhododendron eclecteum var. bellatulum I. B. Balfour ex Tagg 长柄杂色杜鹃
Shrubs, evergreen

6.5; -1.1; 13.2; 6.7; 38.2; 20.2; 511 ▎915; 523; 35 ▎78.7; 511; 286.6

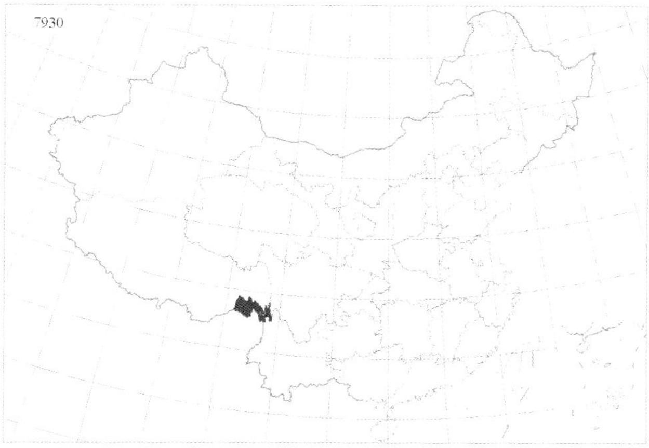

Ericaceae 杜鹃花科 *Rhododendron* 杜鹃花属
Rhododendron edgeworthii Hook. 泡泡叶杜鹃
Shrubs, evergreen

13.1 (4.4~19.2); 5.6 (-4.3~13.3); 19.3 (12.1~26.6); 13.2 (5.3~19.4); 101.1 (28.5~172.6); 3.0 (0.0~34.6); 707 (455~932) ▎1011 (538~1736); 583 (337~1007); 35 (9~70) ▎45.3 (-14.2~145.8); 685 (455~923); 295.8 (142.9~487.2)

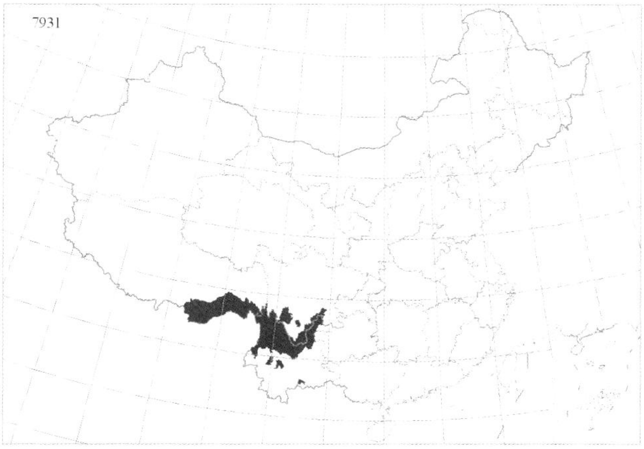

Ericaceae 杜鹃花科 *Rhododendron* 杜鹃花属
Rhododendron elegantulum Tagg et Forrest 金江杜鹃
Small shrubs, evergreen

8.5; 1.0; 14.8; 8.6; 53.8; 11.2; 568 ▎904; 576; 13 ▎59.8; 565; 305.9

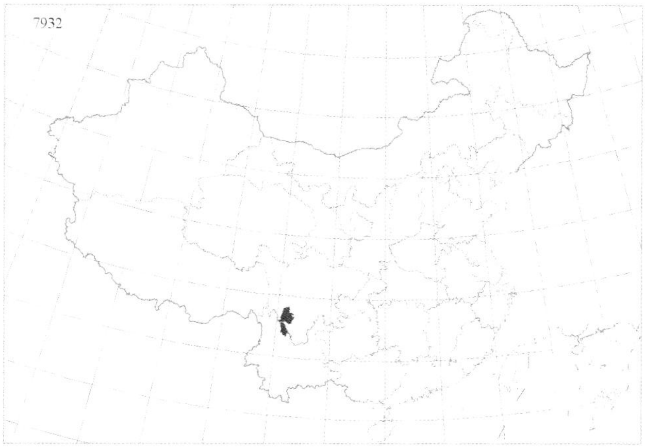

Ericaceae 杜鹃花科 *Rhododendron* 杜鹃花属
Rhododendron emarginatum Hemsl. et E. H. Wilson 缺顶杜鹃
Shrubs, evergreen

18.2 (15.1~21.3); 10.1 (5.0~14.2); 24.3 (21.4~27.3); 18.2 (15.2~21.3); 158.8 (121.9~195.9); 0.0 (0.0~0.0); 907 (788~1111) | 1370 (1178~1900); 744 (545~1056); 63 (48~125) | 51.3 (27.5~80.6); 907 (788~1111); 341.1 (189.5~551.6)

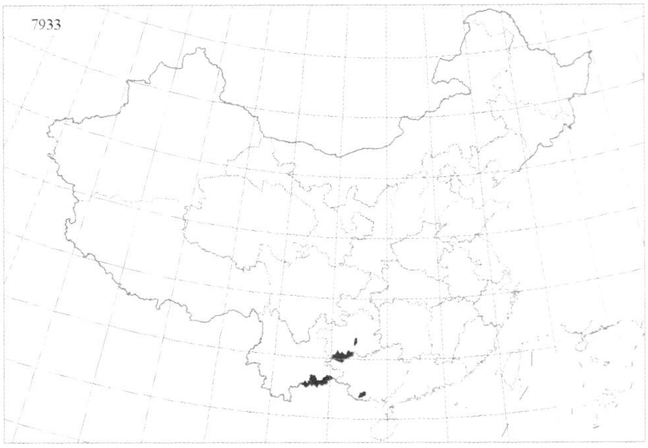

Ericaceae 杜鹃花科 *Rhododendron* 杜鹃花属
Rhododendron erastum I. B. Balfour et Forrest 匍匐杜鹃
Small shrubs, evergreen

5.4; -2.3; 12.2; 5.8; 31.0; 25.9; 480 | 851; 492; 30 | 77.4; 480; 253.8

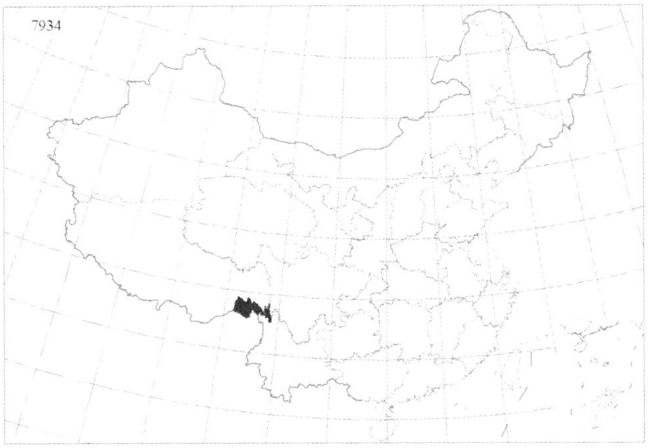

Ericaceae 杜鹃花科 *Rhododendron* 杜鹃花属
Rhododendron erosum Cowan 啮蚀杜鹃
Shrubs or small trees, evergreen

2.1; -6.8; 10.1; 4.0; 18.0; 52.0; 405 | 400; 263; 5 | -2.6; 372; 168.2

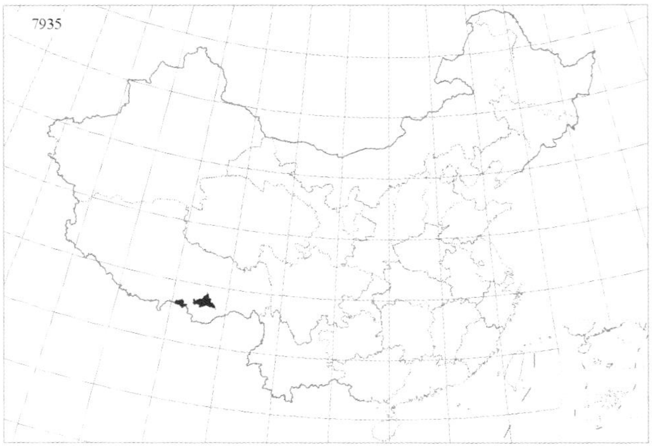

Ericaceae 杜鹃花科 *Rhododendron* 杜鹃花属
Rhododendron esetulosum I. B. Balfour et Forrest 喙尖杜鹃
Small shrubs, evergreen

5.6; -2.0; 12.4; 6.0; 32.4; 24.5; 485 | 834; 494; 27 | 72.1; 485; 263.5

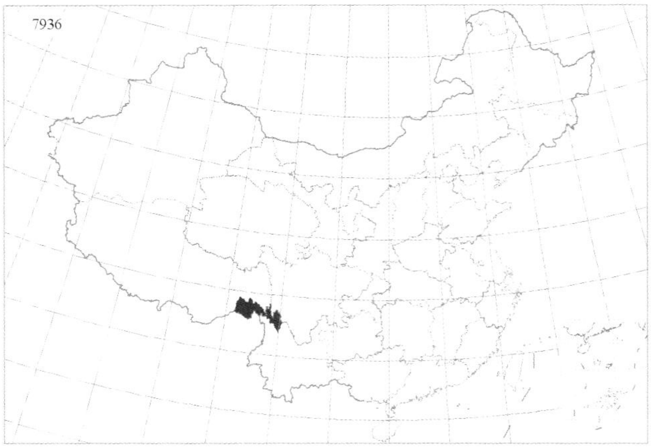

Ericaceae 杜鹃花科 *Rhododendron* 杜鹃花属
Rhododendron euchroum I. B. Balfour et Kingdon-Ward 滇西杜鹃
Small shrubs, evergreen

12.3; 5.4; 17.7; 12.3; 88.0; 0.3; 670 | 1290; 712; 63 | 92.7; 670; 395.9

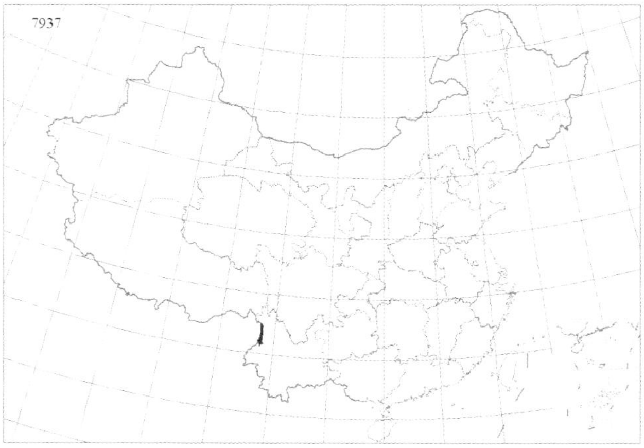

Ericaceae 杜鹃花科 *Rhododendron* 杜鹃花属
Rhododendron eudoxum I. B. Balfour et Forrest 华丽杜鹃
Dwarf shrubs, evergreen

7.2; -0.2; 13.8; 7.5; 43.8; 16.5; 533 | 938; 537; 36 | 76.0; 533; 302.9

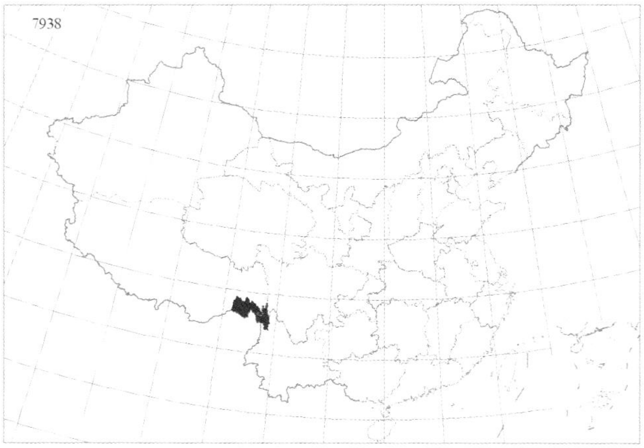

Ericaceae 杜鹃花科 *Rhododendron* 杜鹃花属
Rhododendron eudoxum var. bruneifolium (I. B. Balfour et Forrest) D. F. Chamb. 褐叶华丽杜鹃
Dwarf shrubs, evergreen

5.3; −2.4; 12.2; 5.7; 30.5; 26.2; 479 ❙ 961; 558; 31 ❙ 100.7; 479; 234.2

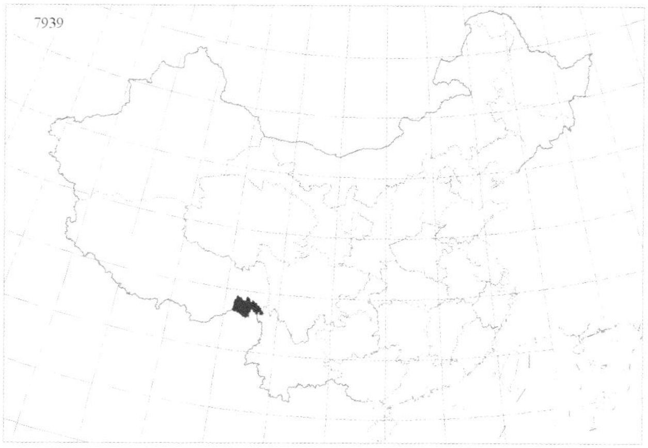

Ericaceae 杜鹃花科 *Rhododendron* 杜鹃花属
Rhododendron eudoxum var. mesopolium (I. B. Balfour et Forrest) D. F. Chamb. 白毛华丽杜鹃
Dwarf shrubs, evergreen

6.5; −1.1; 13.2; 6.7; 38.2; 20.2; 511 ❙ 915; 523; 35 ❙ 78.7; 511; 286.6

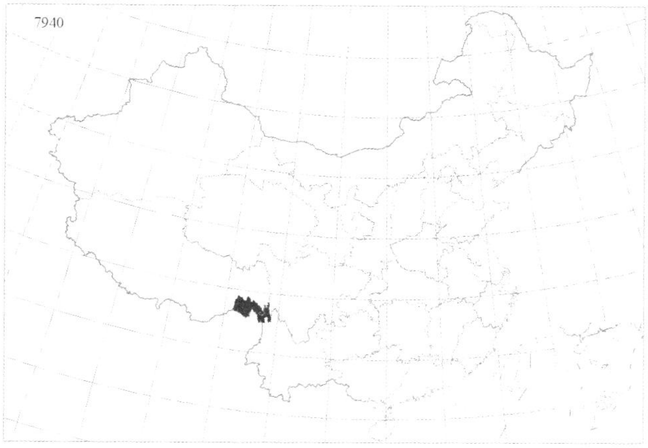

Ericaceae 杜鹃花科 *Rhododendron* 杜鹃花属
Rhododendron eurysiphon Tagg et Forrest 宽筒杜鹃
Shrubs, evergreen

5.3; −2.4; 12.2; 5.7; 30.5; 26.2; 479 ❙ 961; 558; 31 ❙ 100.7; 479; 234.2

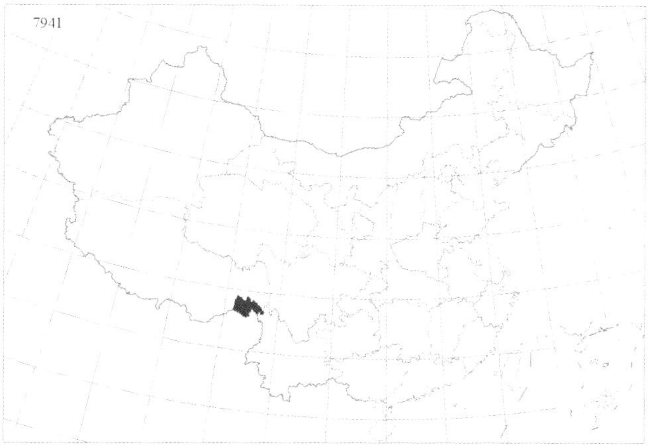

Ericaceae 杜鹃花科 *Rhododendron* 杜鹃花属
Rhododendron exasperatum Tagg 粗糙叶杜鹃
Shrubs or small trees, evergreen

3.5; −4.9; 11.0; 4.7; 22.8; 40.4; 437 ❙ 672; 398; 17 ❙ 50.0; 437; 211.5

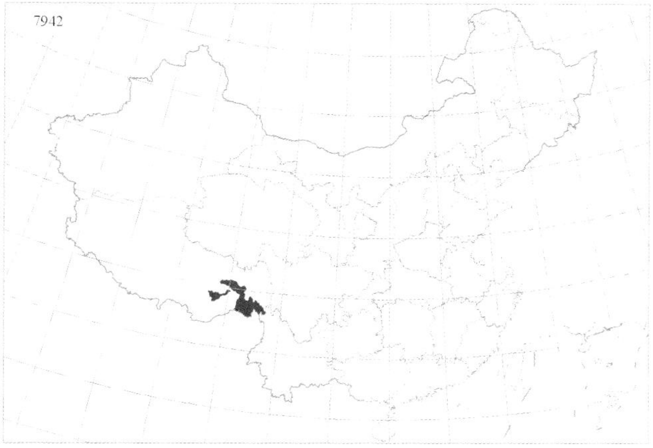

Ericaceae 杜鹃花科 *Rhododendron* 杜鹃花属
Rhododendron excellens Hemsl. et E. H. Wilson 大喇叭杜鹃
Shrubs, evergreen

17.8 (15.1~19.3); 10.1 (5.0~13.1); 23.5 (21.4~26.6); 17.8 (15.2~19.4); 153.9 (121.9~173.2); 0.0 (0.0~0.0); 874 (788~969) ❙ 1294 (1143~1547); 711 (545~911); 56 (48~78) ❙ 48.6 (21.7~80.6); 872 (788~969); 351.4 (189.5~551.3)

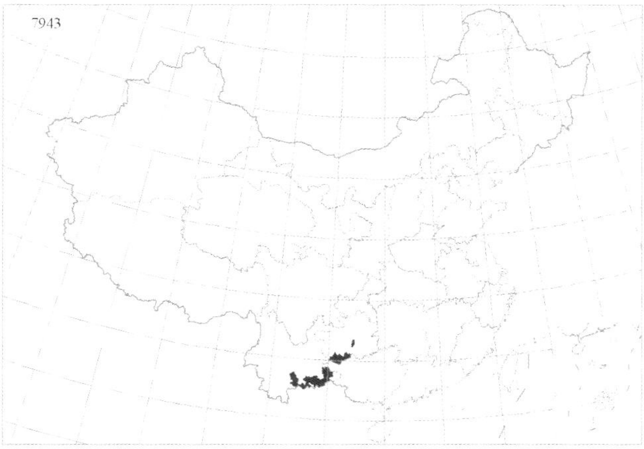

Ericaceae 杜鹃花科 *Rhododendron* 杜鹃花属
Rhododendron faberi Hemsl. 金顶杜鹃
Shrubs, evergreen

7.0 (2.0~14.9); −2.2 (−7.4~5.3); 15.3 (9.9~23.6); 7.8 (3.8~15.0); 51.0 (15.4~119.6); 26.1 (0.0~51.9); 542 (413~782) ❙ 1005 (777~1414); 552 (400~817); 27 (11~49) ❙ 88.3 (38.2~128.6); 542 (413~782); 255.8 (200.8~299.8)

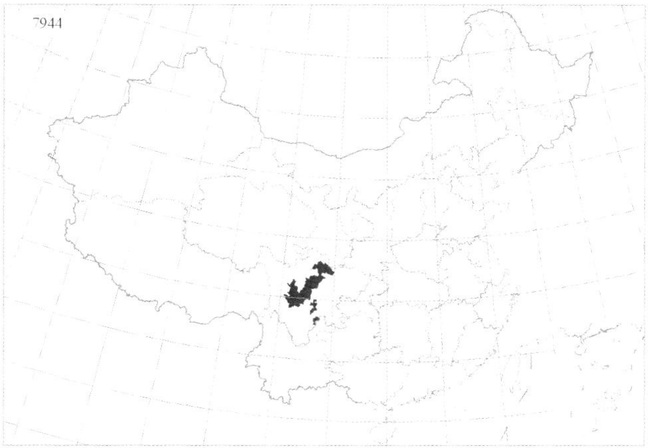

Ericaceae 杜鹃花科 *Rhododendron* 杜鹃花属

Rhododendron faberi subsp. prattii (Franch.) D. F. Chamb. 大叶金顶杜鹃

Shrubs

5.7; −3.5; 14.2; 6.4; 38.1; 28.1; 504 ┃ 922; 460; 25 ┃ 83.0; 504; 262.6

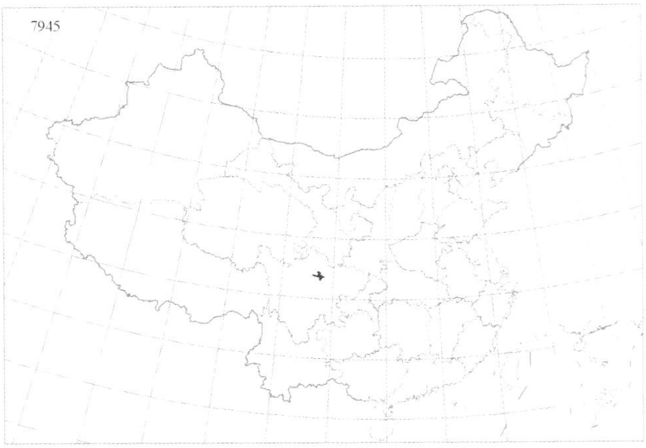

Ericaceae 杜鹃花科 *Rhododendron* 杜鹃花属

Rhododendron facetum I. B. Balfour et Kingdon-Ward 绵毛房杜鹃

Shrubs

13.4 (8.6~17.3); 6.6 (1.3~11.0); 18.6 (15.0~21.6); 13.4 (8.7~17.3); 102.1 (52.6~148.0); 0.9 (0.0~8.8); 704 (574~832) ┃ 1139 (976~1454); 633 (553~792); 53 (38~70) ┃ 63.5 (26.1~99.9); 696 (574~783); 377.4 (309.8~487.2)

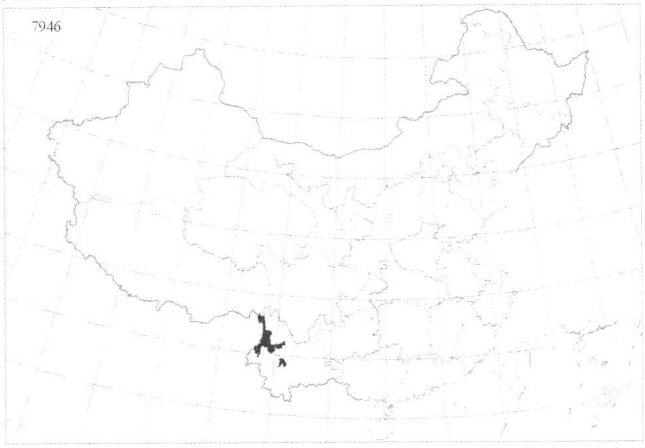

Ericaceae 杜鹃花科 *Rhododendron* 杜鹃花属

Rhododendron faithiae Chun 大云锦杜鹃

Shrubs or small trees, evergreen

18.3 (14.3~22.2); 8.5 (3.8~14.7); 26.6 (23.5~28.6); 18.3 (14.3~22.2); 160.0 (112.6~206.9); 0.2 (0.0~1.5); 958 (756~1177) ┃ 1524 (1175~1876); 628 (448~891); 147 (97~195) ┃ 60.2 (26.8~101.8); 958 (756~1177); 233.4 (197.5~273.3)

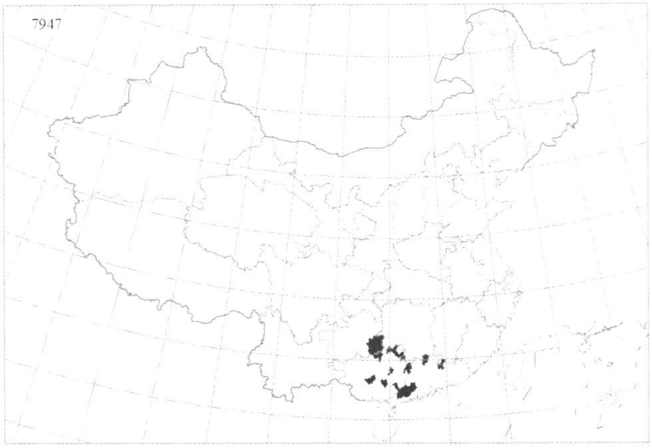

Ericaceae 杜鹃花科 *Rhododendron* 杜鹃花属

Rhododendron fangchengense P. C. Tam 防城杜鹃

Shrubs, evergreen

21.9; 14.1; 27.9; 22.0; 203.7; 0.0; 1161 ┃ 2161; 1240; 126 ┃ 85.7; 1161; 305.5

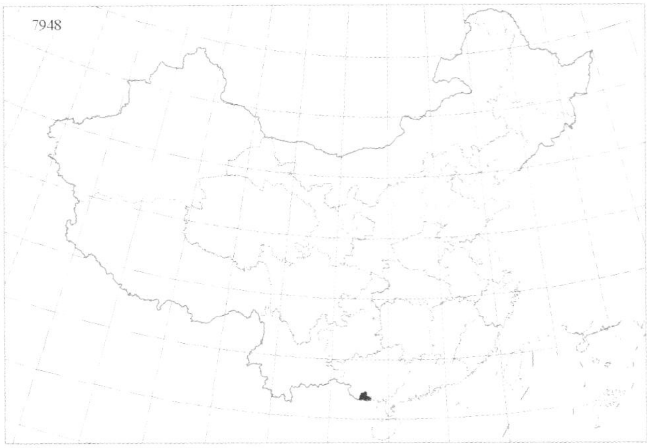

Ericaceae 杜鹃花科 *Rhododendron* 杜鹃花属

Rhododendron farinosum H. Lév. 钝头杜鹃

Shrubs, evergreen

12.7; 4.7; 19.2; 12.7; 93.0; 0.3; 687 ┃ 922; 546; 26 ┃ 34.2; 687; 232.7

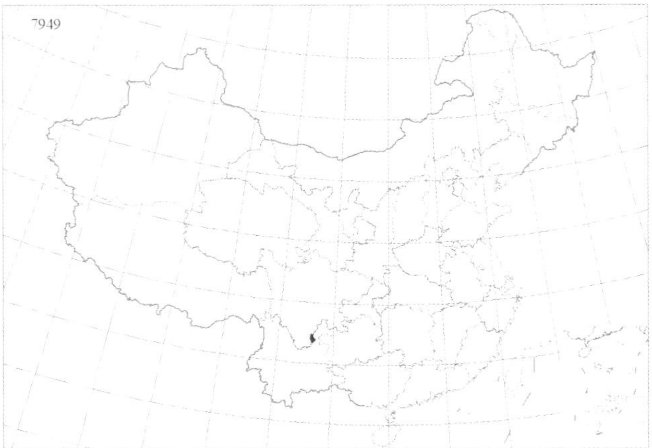

Ericaceae 杜鹃花科 *Rhododendron* 杜鹃花属

Rhododendron farrerae Sweet 丁香杜鹃

Shrubs, deciduous

19.3 (14.3~23.6); 10.0 (3.3~16.7); 27.5 (23.5~29.8); 19.3 (14.3~23.6); 172.1 (112.6~223.5); 0.1 (0.0~2.3); 1013 (756~1281) ┃ 1627 (1039~2037); 628 (356~1033); 170 (75~259) ┃ 63.1 (−6.9~137.5); 1008 (756~1273); 231.7 (95.3~332.5)

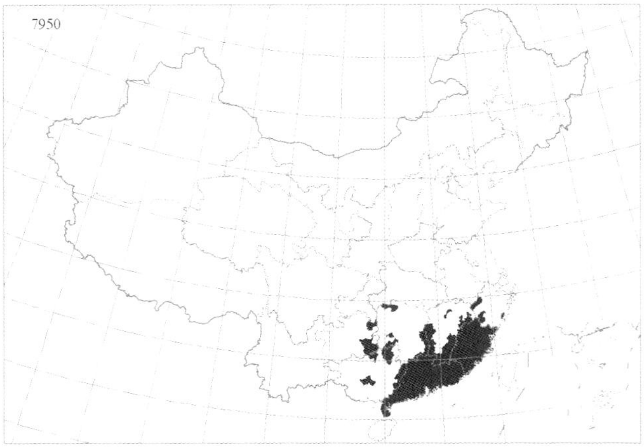

Ericaceae 杜鹃花科 *Rhododendron* 杜鹃花属
Rhododendron fastigiatum Franch. 密枝杜鹃
Shrubs, evergreen

10.8 (-3.3~18.9); 2.6 (-15.2~12.0); 17.6 (7.2~26.6); 11.4 (2.0~19.0); 85.3 (3.7~167.8); 15.5 (0.0~103.2); 650 (339~929) ❙ 907 (488~1141); 526 (284~627); 28 (9~57) ❙ 44.3 (-14.2~125.9); 629 (339~912); 276.9 (142.9~382.6)

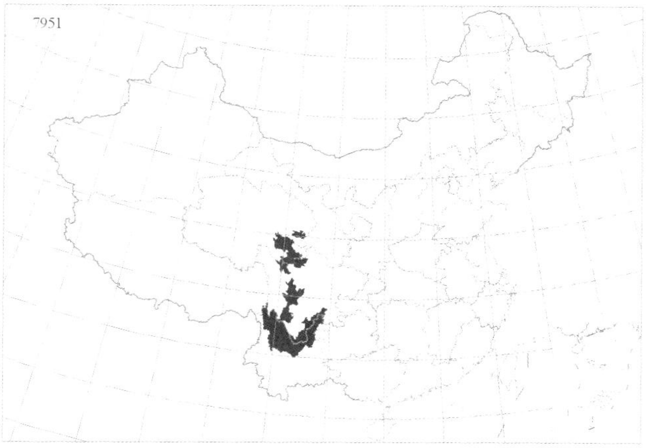

Ericaceae 杜鹃花科 *Rhododendron* 杜鹃花属
Rhododendron faucium D. F. Chamb. 猴斑杜鹃
Shrubs or small trees, evergreen

2.6; -6.3; 10.6; 4.2; 19.5; 47.5; 417 ❙ 555; 333; 11 ❙ 33.3; 417; 224.7

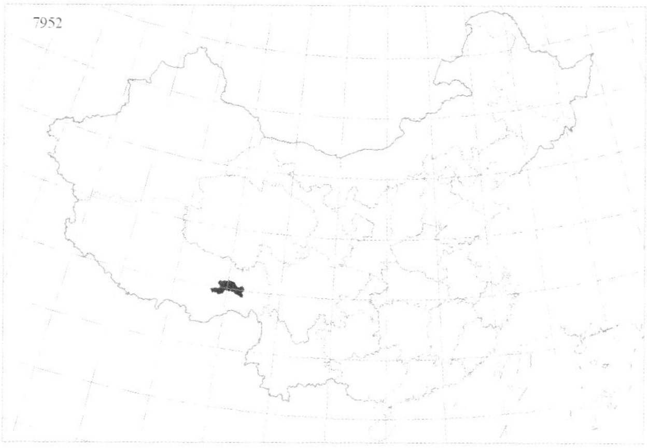

Ericaceae 杜鹃花科 *Rhododendron* 杜鹃花属
Rhododendron feddei H. Lév. 黔中杜鹃
Shrubs or small trees

15.7 (14.9~16.7); 5.8 (4.8~6.9); 24.4 (23.9~24.9); 15.8 (15.0~16.7); 129.4 (119.9~140.7); 0.0 (0.0~0.2); 812 (781~851) ❙ 1261 (1240~1283); 569 (545~600); 76 (68~85) ❙ 55.4 (50.8~58.8); 812 (781~851); 235.0 (221.1~254.8)

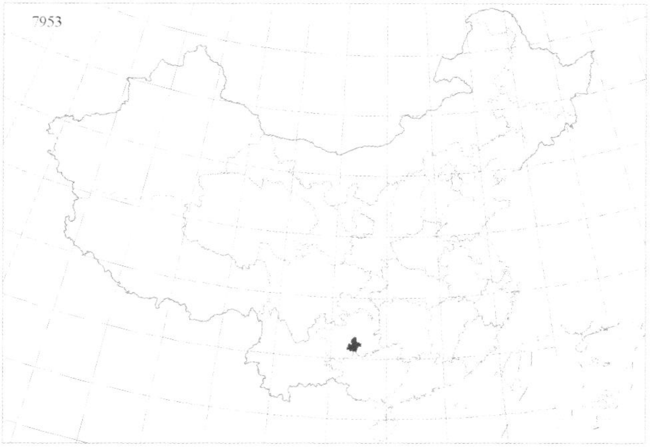

Ericaceae 杜鹃花科 *Rhododendron* 杜鹃花属
Rhododendron flavantherum Hutch. et Kingdon-Ward 黄药杜鹃
Shrubs, evergreen

9.3; 1.5; 15.9; 9.6; 65.5; 13.1; 592 ❙ 1348; 783; 41 ❙ 123.3; 592; 340.8

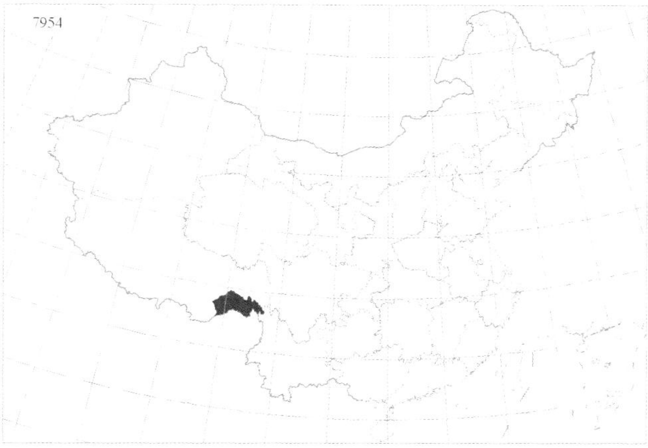

Ericaceae 杜鹃花科 *Rhododendron* 杜鹃花属
Rhododendron flavidum Franch. 淡黄杜鹃
Shrubs, evergreen

6.4; -1.5; 13.0; 7.3; 45.5; 28.2; 517 ❙ 1085; 625; 35 ❙ 113.3; 517; 306.9

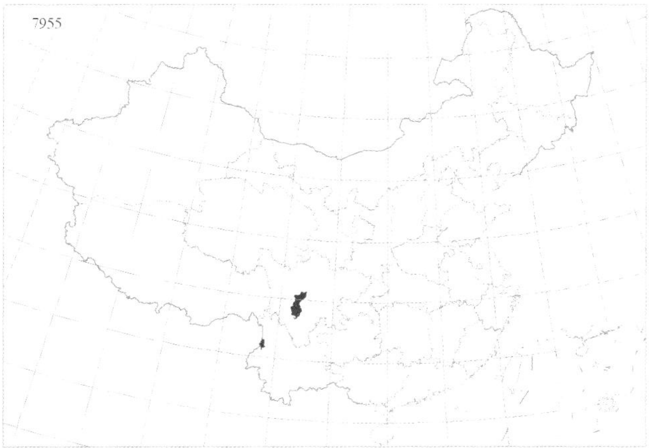

Ericaceae 杜鹃花科 *Rhododendron* 杜鹃花属
Rhododendron flavoflorum T. L. Ming 泸水杜鹃
Small trees

13.2; 6.4; 18.4; 13.3; 99.4; 0.0; 695 ❙ 1337; 727; 69 ❙ 92.4; 695; 407.1

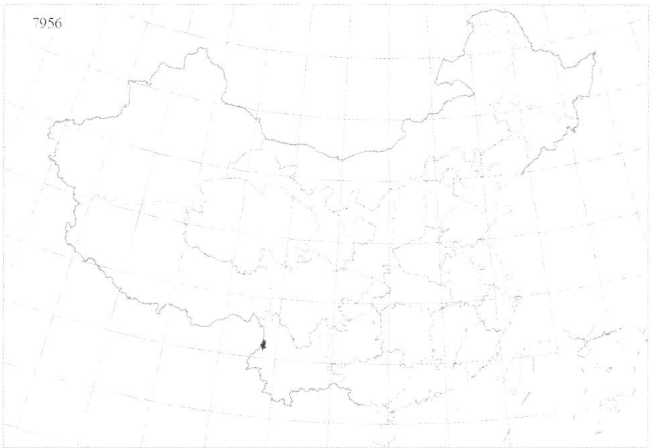

Ericaceae 杜鹃花科 *Rhododendron* 杜鹃花属

Rhododendron fletcherianum Davidian 翅柄杜鹃

Shrubs, evergreen

7.0; −0.5; 13.6; 7.2; 41.5; 17.5; 526 ❙ 1001; 571; 38 ❙ 91.1; 526; 293.2

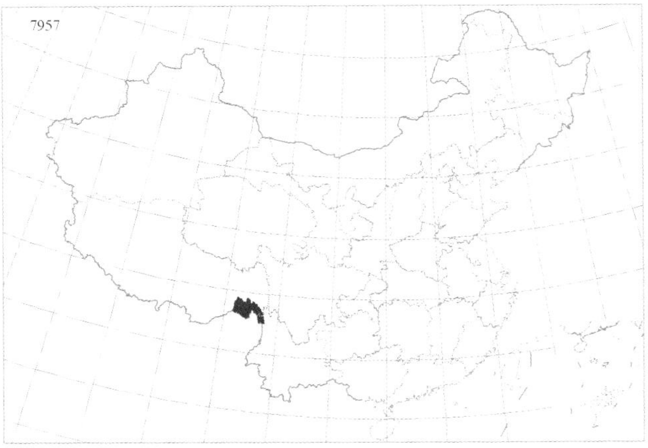

Ericaceae 杜鹃花科 *Rhododendron* 杜鹃花属

Rhododendron floccigerum Franch. 绵毛杜鹃

Shrubs, evergreen

8.5 (5.3~13.2); 1.2 (−2.4~6.4); 14.7 (12.2~18.4); 8.7 (5.7~13.3); 55.2 (30.5~99.4); 12.6 (0.0~26.2); 567 (479~695) ❙ 1019 (741~1337); 581 (426~727); 42 (21~69) ❙ 78.7 (54.0~100.7); 567 (479~695); 326.6 (234.2~407.1)

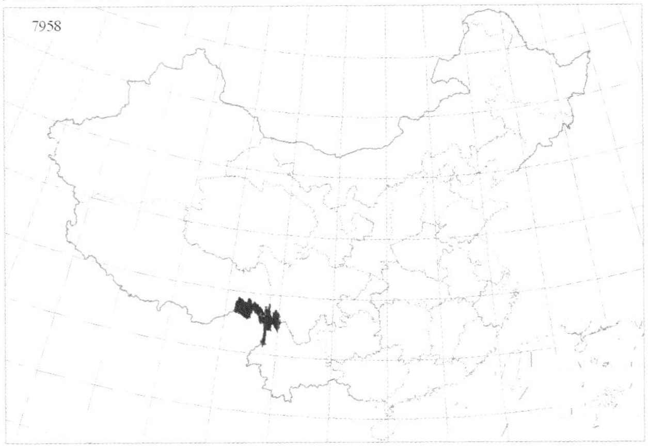

Ericaceae 杜鹃花科 *Rhododendron* 杜鹃花属

Rhododendron floribundum Franch. 繁花杜鹃

Shrubs or small trees, evergreen

11.8 (6.1~15.0); 3.1 (−2.4~6.5); 19.4 (12.7~24.6); 11.9 (6.3~15.1); 87.1 (35.2~121.5); 4.6 (0.0~24.1); 669 (496~792) ❙ 1024 (801~1252); 558 (496~686); 37 (18~65) ❙ 54.2 (27.3~92.9); 668 (496~792); 231.2 (153.9~318.7)

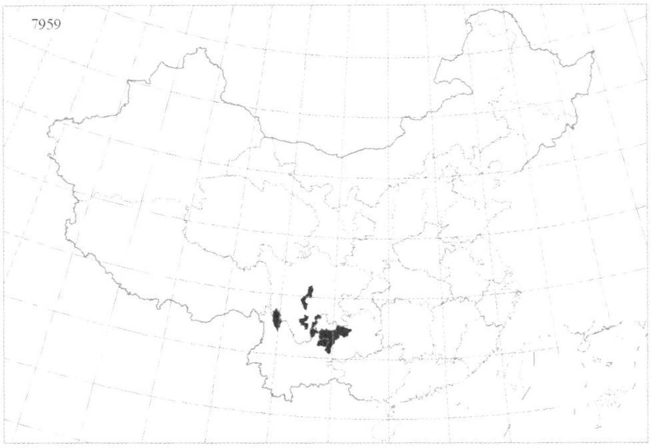

Ericaceae 杜鹃花科 *Rhododendron* 杜鹃花属

Rhododendron florulentum Tam 龙岩杜鹃

Shrubs, semievergreen or green

18.1 (16.6~20.1); 8.6 (5.9~11.5); 26.8 (25.2~28.5); 18.1 (16.6~20.1); 157.7 (138.9~181.7); 0.0 (0.0~0.0); 933 (864~1045) ❙ 1759 (1600~1997); 605 (564~669); 208 (163~254) ❙ 89.4 (56.1~130.7); 931 (864~1045); 282.6 (261.7~305.7)

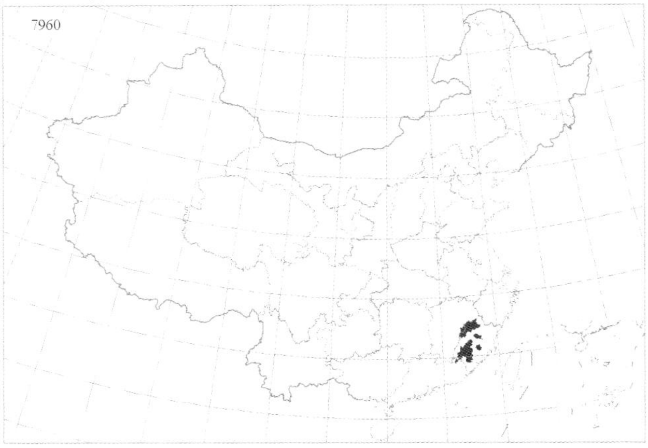

Ericaceae 杜鹃花科 *Rhododendron* 杜鹃花属

Rhododendron flosculum Fang et G. Z. Li 子花杜鹃

Shrubs or small trees, semievergreen or green

15.8; 5.0; 25.7; 15.9; 131.0; 0.8; 835 ❙ 1598; 569; 191 ❙ 92.1; 835; 249.4

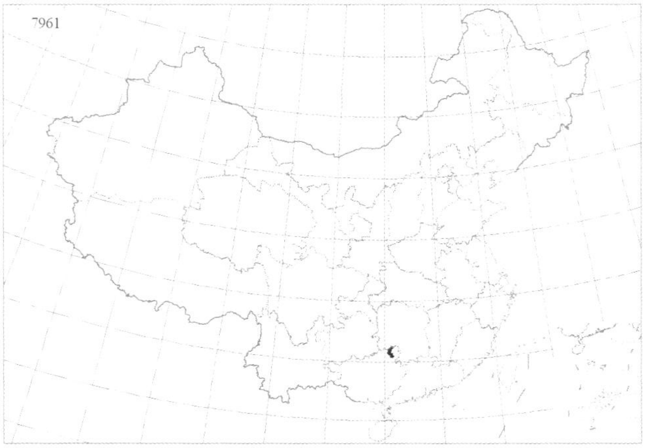

Ericaceae 杜鹃花科 *Rhododendron* 杜鹃花属

Rhododendron flumineum Fang et M. Y. He 河边杜鹃

Shrubs, semievergreen or green

18.3; 11.6; 22.9; 18.3; 159.8; 0.0; 880 ❙ 1441; 829; 58 ❙ 63.9; 880; 477.2

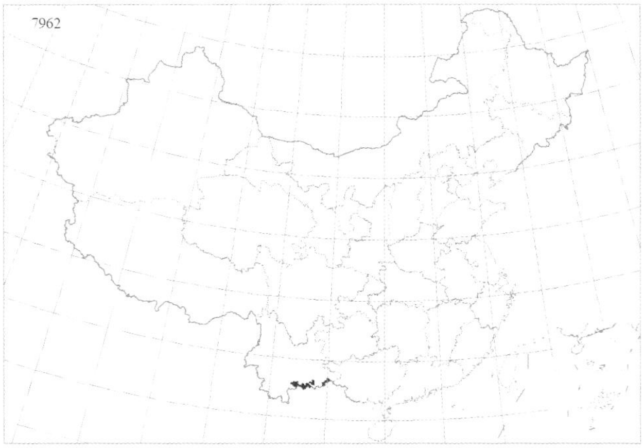

Ericaceae 杜鹃花科 *Rhododendron* 杜鹃花属
Rhododendron formosanum Hemsl. 台湾杜鹃
Shrubs or small trees, evergreen

19.0 (16.2~22.2); 13.4 (11.0~17.7); 23.8 (20.1~27.4); 19.0 (16.2~22.2); 168.4 (134.2~206.7); 0.0 (0.0~0.0); 933 (773~1123) ❙ 2820 (2355~3495); 1091 (821~1847); 394 (107~897) ❙ 204.7 (159.6~246.5); 933 (773~1123); 281.1 (226.8~363.7)

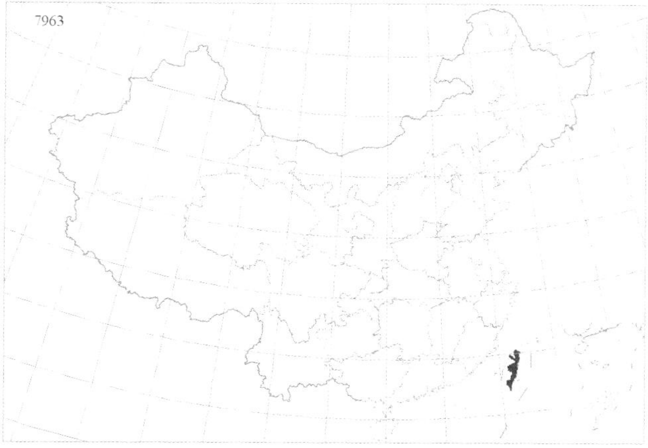

Ericaceae 杜鹃花科 *Rhododendron* 杜鹃花属
Rhododendron forrestii I. B. Balfour ex Diels 紫背杜鹃
Small shrubs, evergreen

7.9 (1.8~14.4); 0.3 (-7.1~7.5); 14.4 (9.9~19.1); 8.4 (3.8~14.4); 55.2 (15.7~113.1); 19.6 (0.0~53.3); 554 (404~727) ❙ 979 (473~1454); 553 (287~792); 40 (7~70) ❙ 71.3 (5.0~100.7); 554 (404~727); 317.4 (169.5~487.2)

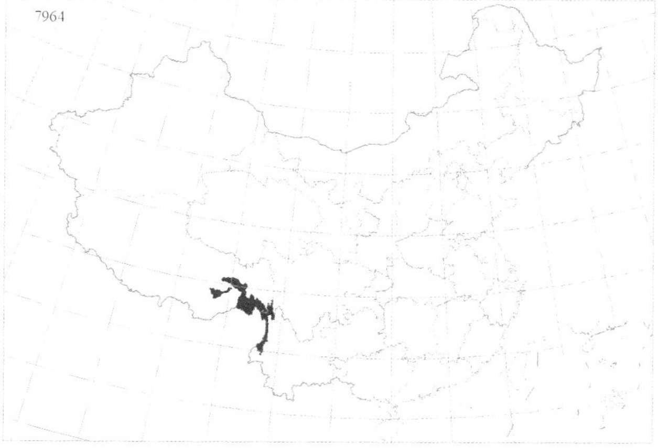

Ericaceae 杜鹃花科 *Rhododendron* 杜鹃花属
Rhododendron forrestii subsp. papillatum D. F. Chamb. 乳突紫背杜鹃
Shrubs, evergreen

2.6; -6.2; 10.5; 4.2; 19.0; 47.5; 415 ❙ 528; 318; 10 ❙ 24.6; 415; 200.1

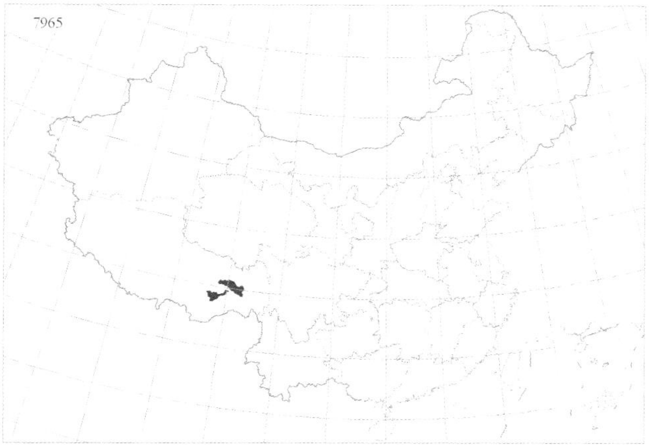

Ericaceae 杜鹃花科 *Rhododendron* 杜鹃花属
Rhododendron fortunei Lindl. 云锦杜鹃
Shrubs or small trees, evergreen

16.3 (2.2~21.6); 5.4 (-6.6~14.1); 26.4 (9.9~29.3); 16.3 (3.8~21.6); 137.1 (15.5~199.4); 1.6 (0.0~49.1); 871 (413~1128) ❙ 1467 (731~1997); 557 (346~902); 155 (19~259) ❙ 68.7 (-0.6~147.0); 868 (413~1128); 231.6 (149.5~384.1)

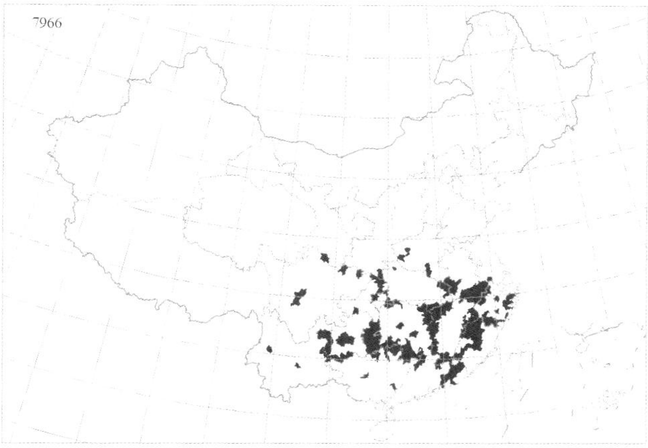

Ericaceae 杜鹃花科 *Rhododendron* 杜鹃花属
Rhododendron fragariiflorum Kingdon-Ward 草莓花杜鹃
Small shrubs, evergreen

5.1 (1.9~12.2); -3.6 (-7.2~4.3); 12.6 (10.0~18.5); 6.1 (3.8~12.2); 35.6 (16.6~87.4); 34.5 (0.7~53.6); 477 (403~672) ❙ 662 (341~1431); 397 (226~817); 12 (4~31) ❙ 28.7 (-15.3~113.0); 465 (341~672); 252.0 (168.2~363.7)

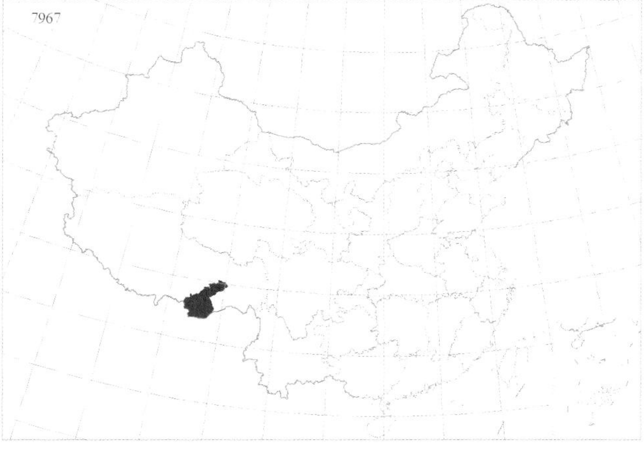

Ericaceae 杜鹃花科 *Rhododendron* 杜鹃花属
Rhododendron fuchsiifolium H. Lév. 贵定杜鹃
Small shrubs, semievergreen or green

14.4; 4.4; 23.5; 14.5; 114.5; 0.7; 763 ❙ 1150; 536; 66 ❙ 50.7; 763; 192.9

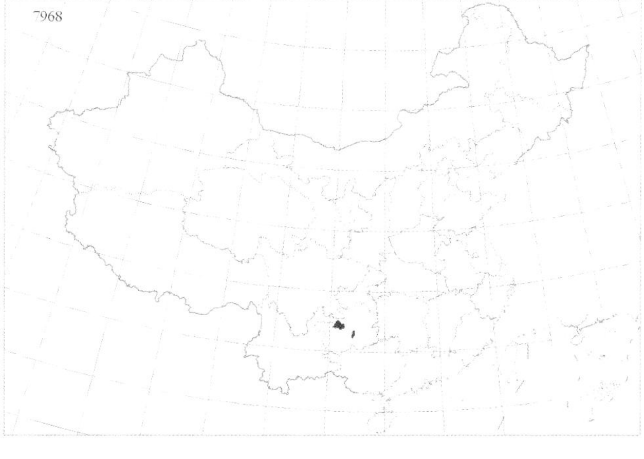

Ericaceae 杜鹃花科 *Rhododendron* 杜鹃花属

Rhododendron fulgens Hook. f. 猩红杜鹃

Shrubs, evergreen

4.7; −3.9; 12.2; 6.1; 35.5; 38.6; 474 ❙ 675; 407; 14 ❙ 29.0; 459; 223.0

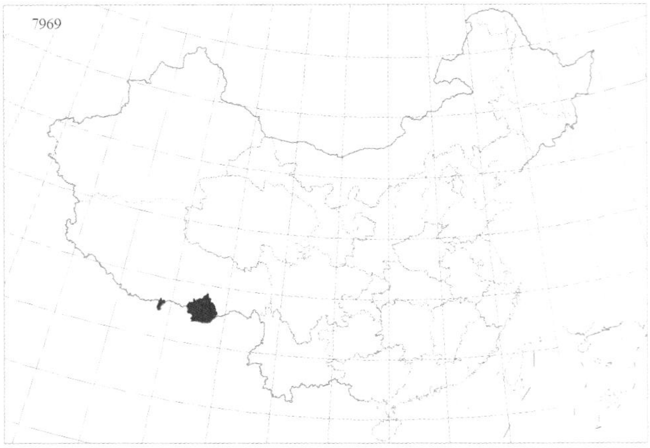

Ericaceae 杜鹃花科 *Rhododendron* 杜鹃花属

Rhododendron fulvum I. B. Balfour et W. W. Sm. 镰果杜鹃

Shrubs or small trees, evergreen

9.3 (2.2~14.4); 2.0 (−6.6~7.6); 15.4 (9.9~19.7); 9.6 (3.8~14.4); 64.4 (15.5~113.1); 12.2 (0.0~49.1); 591 (413~728) ❙ 1039 (741~1454); 597 (426~792); 39 (12~70) ❙ 77.0 (36.1~125.9); 589 (413~727); 332.0 (234.2~487.2)

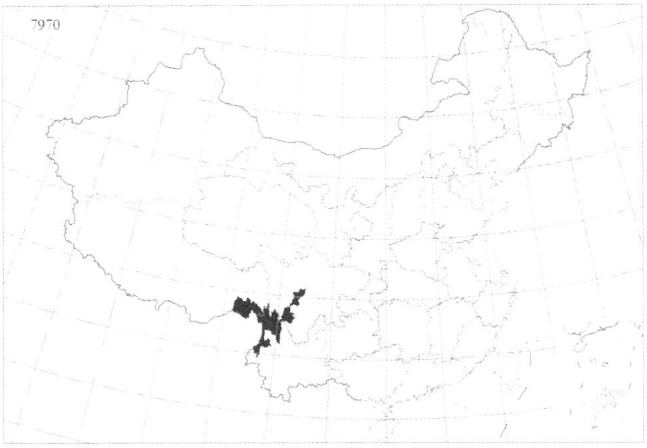

Ericaceae 杜鹃花科 *Rhododendron* 杜鹃花属

Rhododendron fuscipilum M. Y. He 棕毛杜鹃

Shrubs, semievergreen or green

18.0; 8.5; 25.8; 18.0; 156.4; 0.0; 913 ❙ 1551; 672; 169 ❙ 69.8; 913; 243.9

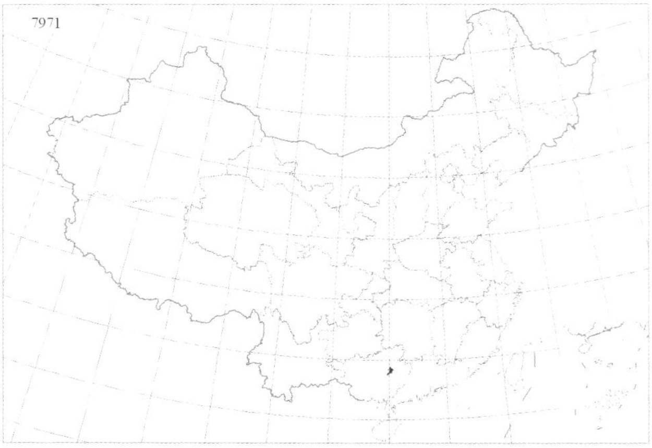

Ericaceae 杜鹃花科 *Rhododendron* 杜鹃花属

Rhododendron galactinum I. B. Balfour ex Tagg 乳黄叶杜鹃

Shrubs or small trees, evergreen

6.3 (2.2~13.3); −3.0 (−6.6~3.2); 14.6 (9.9~22.1); 7.1 (3.8~13.2); 45.0 (15.5~100.9); 28.7 (2.5~49.1); 522 (413~723) ❙ 998 (932~1081); 542 (490~606); 26 (19~32) ❙ 97.1 (49.4~125.9); 522 (413~723); 232.0 (171.0~272.3)

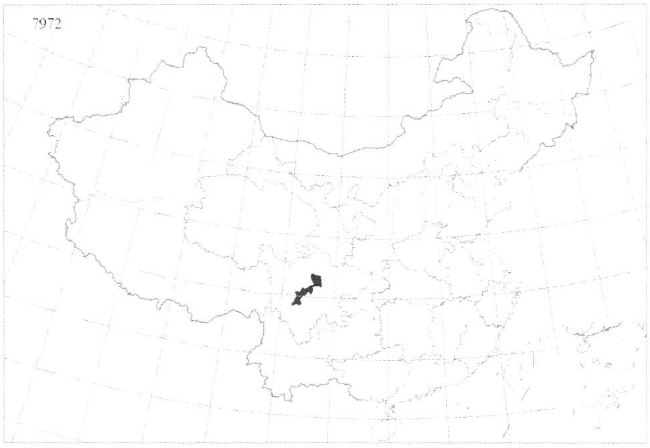

Ericaceae 杜鹃花科 *Rhododendron* 杜鹃花属

Rhododendron gannanense Z. C. Feng et X. G. Sun 甘南杜鹃

Shrubs or trees

4.6; −5.9; 14.1; 5.9; 35.1; 39.6; 482 ❙ 663; 320; 15 ❙ 38.5; 482; 277.3

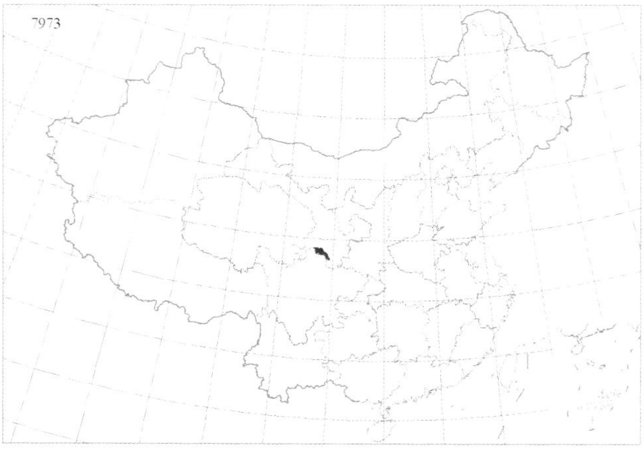

Ericaceae 杜鹃花科 *Rhododendron* 杜鹃花属

Rhododendron gemmiferum Philipson et M. N. Philipson 大芽杜鹃

Shrubs, evergreen

8.7; 1.6; 14.9; 8.9; 56.0; 10.5; 575 ❙ 934; 544; 35 ❙ 62.0; 575; 325.6

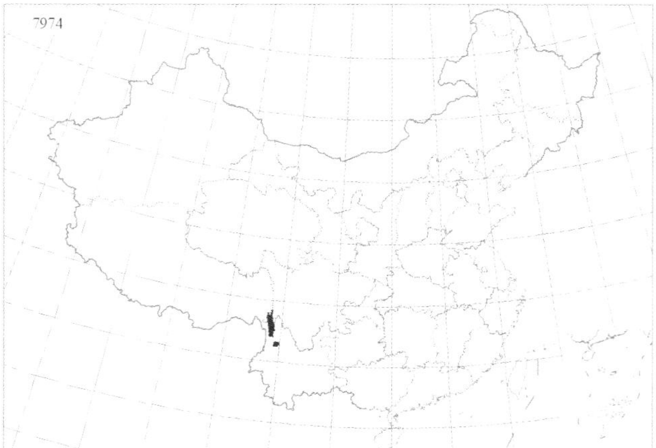

Ericaceae 杜鹃花科 *Rhododendron* 杜鹃花属
Rhododendron genestierianum Forrest 灰白杜鹃
Shrubs, evergreen

9.8 (5.3~13.3); 2.4 (-2.4~6.4); 16.0 (12.2~19.6); 9.9 (5.7~13.4); 66.6 (30.5~100.4); 8.5 (0.0~26.2); 605 (479~706) ❙ 1131 (741~1736); 651 (426~1007); 39 (24~58) ❙ 85.5 (52.8~145.8); 605 (479~706); 338.6 (234.2~447.4)

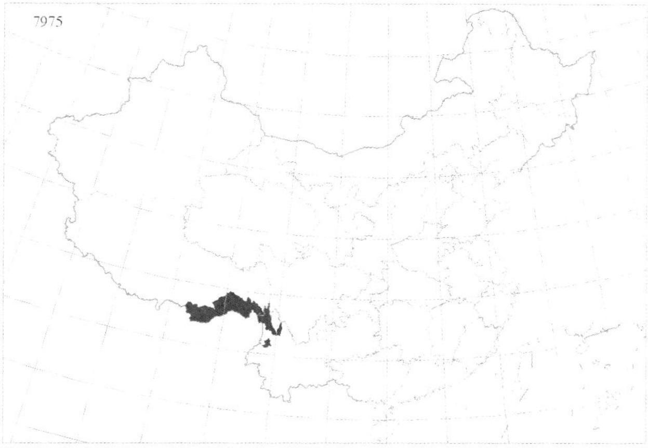

Ericaceae 杜鹃花科 *Rhododendron* 杜鹃花属
Rhododendron glanduliferum Franch. 大果杜鹃
Small shrubs, evergreen

14.2; 4.8; 22.5; 14.3; 112.6; 1.3; 752 ❙ 1135; 584; 55 ❙ 50.8; 752; 215.8

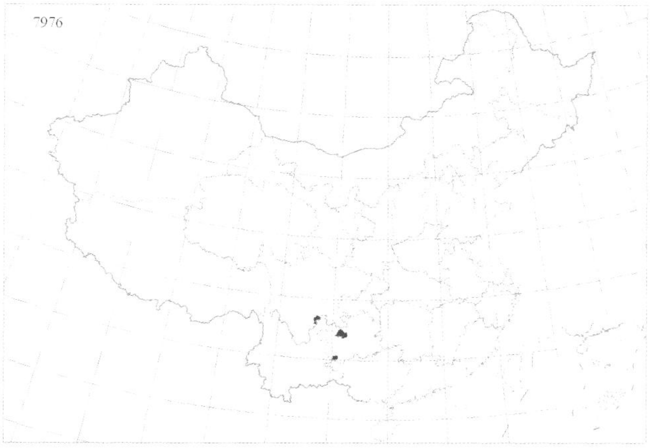

Ericaceae 杜鹃花科 *Rhododendron* 杜鹃花属
Rhododendron glischrum I. B. Balfour et W. W. Sm. 黏毛杜鹃
Shrubs or small trees, evergreen

13.4 (1.8~18.9); 5.7 (-7.1~12.0); 19.7 (9.9~26.6); 13.4 (3.8~19.0); 103.8 (15.7~167.8); 3.1 (0.0~53.3); 716 (404~929) ❙ 964 (583~1337); 560 (348~727); 31 (13~69) ❙ 36.9 (-14.2~93.0); 685 (404~912); 275.6 (142.9~407.1)

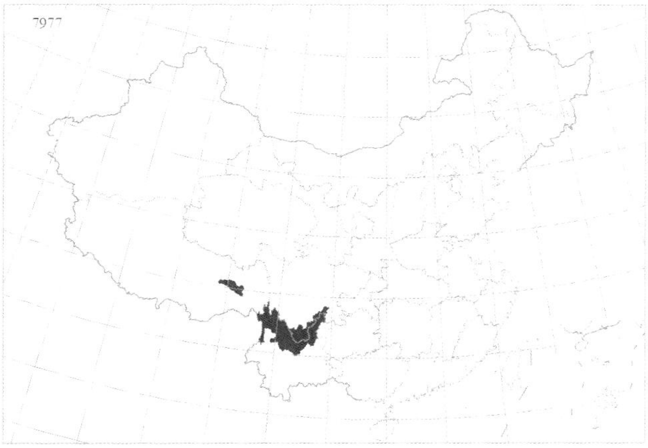

Ericaceae 杜鹃花科 *Rhododendron* 杜鹃花属
Rhododendron glischrum subsp. rude (Tagg et Forrest) D. F. Chamb. 红黏毛杜鹃
Shrubs or small trees, evergreen

11.1; 4.0; 16.8; 11.1; 76.2; 3.2; 638 ❙ 1207; 669; 57 ❙ 89.0; 638; 381.4

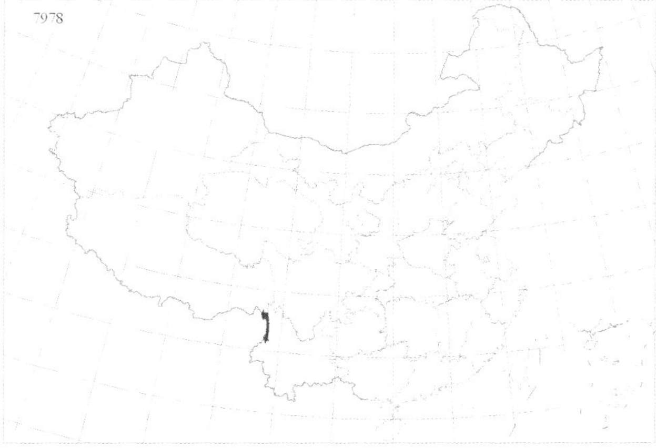

Ericaceae 杜鹃花科 *Rhododendron* 杜鹃花属
Rhododendron gologense C. J. Xu et Z. J. Zhao 果洛杜鹃
Shrubs

-2.4 (-3.6~-1.0); -13.8 (-15.6~-11.8); 7.7 (7.2~8.5); 2.3 (2.0~2.7); 5.3 (3.7~8.3); 93.8 (79.5~107.2); 346 (335~363) ❙ 549 (372~664); 317 (223~379); 13 (11~16) ❙ 56.5 (0.0~87.3); 346 (335~363); 252.5 (135.0~307.6)

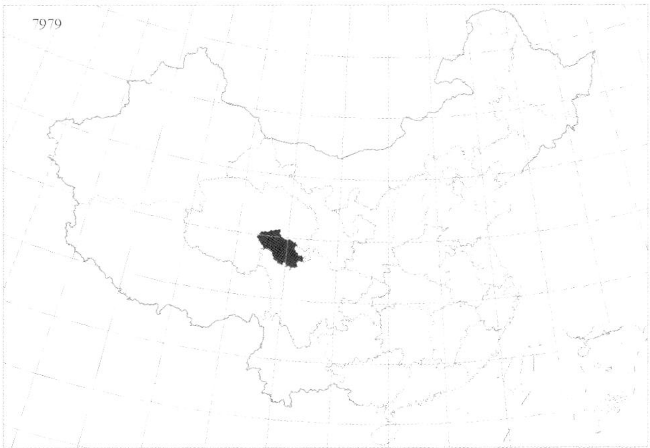

Ericaceae 杜鹃花科 *Rhododendron* 杜鹃花属
Rhododendron gongshanense T. L. Ming 贡山杜鹃
Shrubs, evergreen

8.6; 1.3; 15.0; 8.7; 52.6; 8.8; 574 ❙ 1041; 585; 45 ❙ 81.5; 574; 352.3

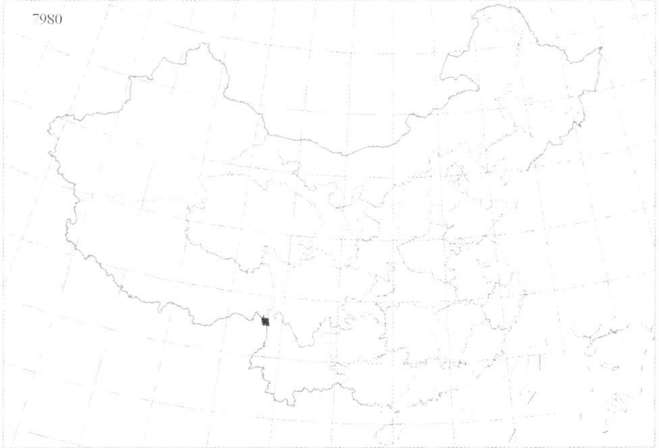

Ericaceae 杜鹃花科 *Rhododendron* 杜鹃花属
Rhododendron grande Wight 大叶杜鹃
Trees, evergreen

8.0; -0.4; 15.0; 8.7; 58.2; 22.2; 559 ❙ 1011; 597; 24 ❙ 65.4; 543; 299.2

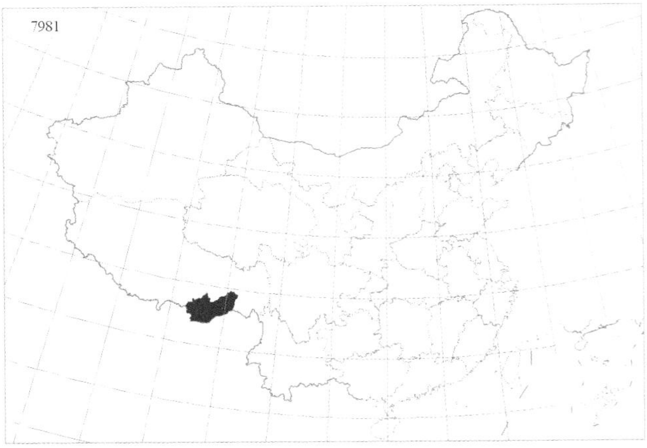

Ericaceae 杜鹃花科 *Rhododendron* 杜鹃花属
Rhododendron griersonianum I. B. Balfour et Forrest 朱红大杜鹃
Shrubs, evergreen

14.4; 7.5; 19.1; 14.4; 113.1; 0.0; 727 ❙ 1454; 792; 70 ❙ 99.9; 727; 487.2

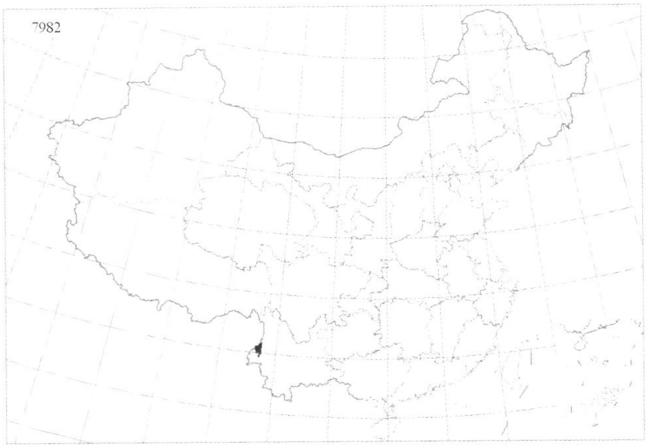

Ericaceae 杜鹃花科 *Rhododendron* 杜鹃花属
Rhododendron griffithianum Wight 不丹杜鹃
Shrubs or small trees, evergreen

8.3; 0.0; 15.3; 8.8; 58.0; 17.7; 563 ❙ 984; 577; 20 ❙ 65.6; 563; 266.0

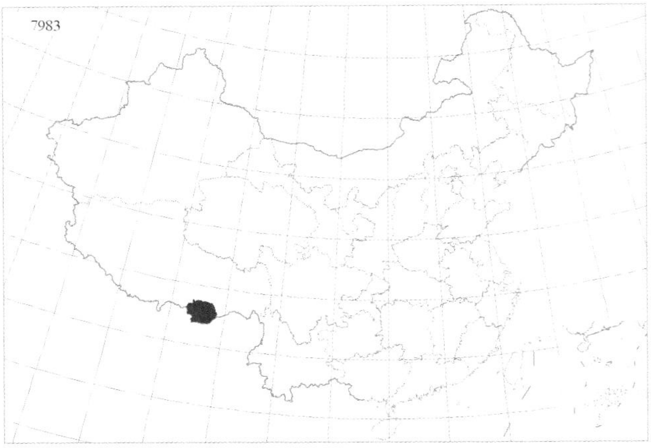

Ericaceae 杜鹃花科 *Rhododendron* 杜鹃花属
Rhododendron guangnanense R. C. Fang 广南杜鹃
Shrubs or small trees, evergreen

16.9; 9.1; 22.5; 17.0; 143.8; 0.0; 829 ❙ 1118; 607; 49 ❙ 34.8; 829; 295.7

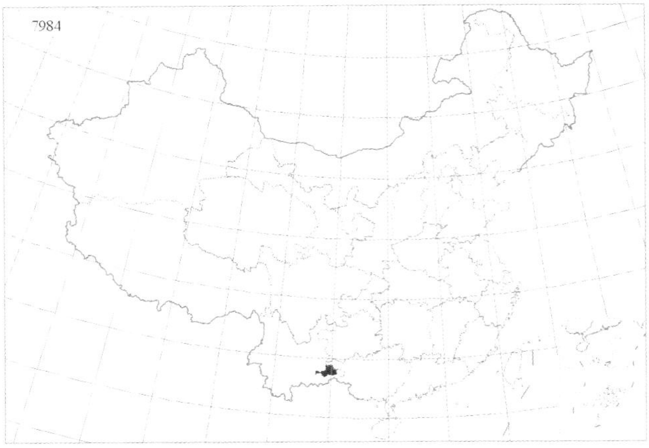

Ericaceae 杜鹃花科 *Rhododendron* 杜鹃花属
Rhododendron guizhongense G. Z. Li 腺柱杜鹃
Small shrubs

18.0; 8.5; 25.8; 18.0; 156.4; 0.0; 913 ❙ 1551; 672; 169 ❙ 69.8; 913; 243.9

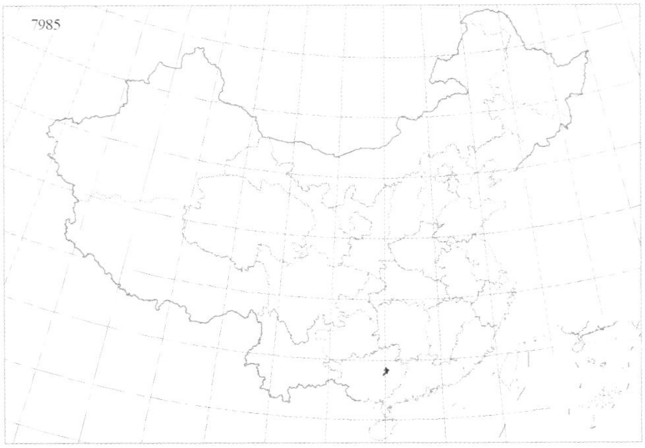

Ericaceae 杜鹃花科 *Rhododendron* 杜鹃花属
Rhododendron guizhouense M. Y. Fang 贵州杜鹃
Shrubs or small trees, evergreen

14.8; 3.8; 25.4; 14.9; 120.1; 1.4; 796 ❙ 1317; 496; 112 ❙ 65.9; 796; 206.7

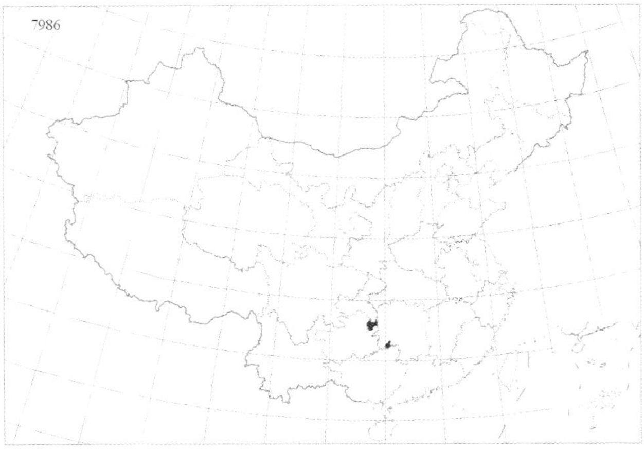

Ericaceae 杜鹃花科 *Rhododendron* 杜鹃花属

Rhododendron habrotrichum I. B. Balfour et W. W. Sm. 粗毛杜鹃

Shrubs, evergreen

14.4；7.5；19.1；14.4；113.1；0.0；727 ❙ 1454；792；70 ❙ 99.9；727；487.2

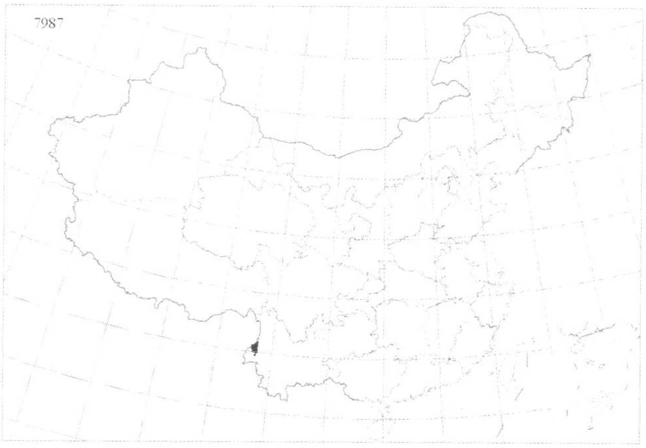

Ericaceae 杜鹃花科 *Rhododendron* 杜鹃花属

Rhododendron haematodes Franch. 似血杜鹃

Small shrubs, evergreen

11.0 (5.3~17.3)；4.0 (-2.4~11.0)；16.7 (12.2~21.6)；11.1 (5.7~17.3)；79.7 (30.5~148.0)；7.4 (0.0~26.2)；638 (479~832) ❙ 1052 (741~1337)；592 (426~727)；46 (28~69) ❙ 67.4 (26.1~100.7)；632 (479~783)；347.1 (234.2~407.1)

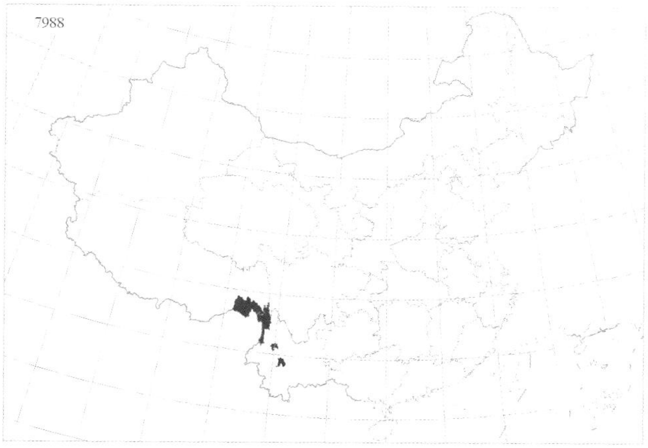

Ericaceae 杜鹃花科 *Rhododendron* 杜鹃花属

Rhododendron haematodes subsp. chaetomallum (I. B. Balfour et Forrest) D. F. Chamb. 绢毛杜鹃

Small shrubs, evergreen

8.9 (5.3~13.2)；1.7 (-2.4~6.4)；15.1 (12.2~18.4)；9.1 (5.7~13.3)；58.5 (30.5~99.4)；11.1 (0.0~26.2)；579 (479~695) ❙ 1055 (741~1337)；595 (426~727)；45 (28~69) ❙ 81.6 (54.0~100.7)；579 (479~695)；333.9 (234.2~407.1)

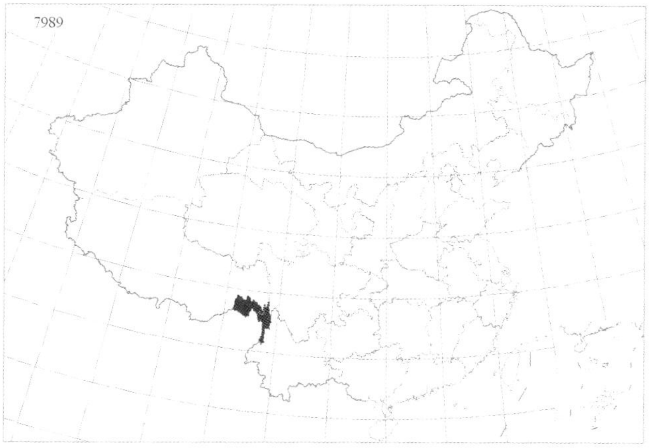

Ericaceae 杜鹃花科 *Rhododendron* 杜鹃花属

Rhododendron hainanense Merr. 海南杜鹃

Small shrubs, semievergreen or green

23.0 (18.0~24.8)；16.3 (8.5~20.1)；27.9 (25.7~28.8)；23.0 (18.0~24.8)；216.6 (156.4~237.6)；0.0 (0.0~0.0)；1242 (913~1385) ❙ 1595 (868~2435)；686 (312~1428)；96 (41~194) ❙ 29.9 (-34.3~99.5)；1172 (868~1360)；372.8 (213.5~620.0)

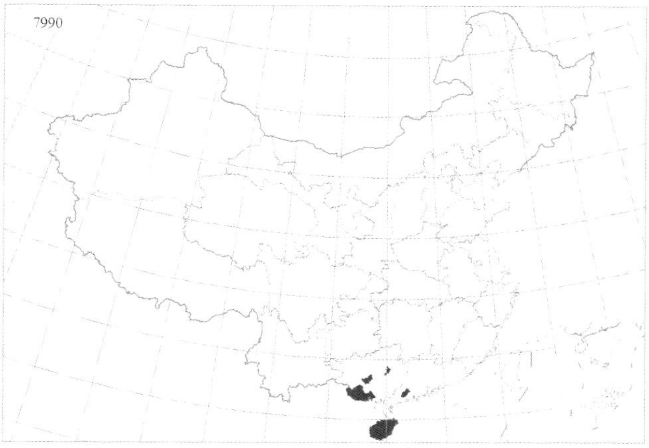

Ericaceae 杜鹃花科 *Rhododendron* 杜鹃花属

Rhododendron hanceanum Hemsl. 疏叶杜鹃

Shrubs, evergreen

11.6 (6.0~15.1)；2.2 (-3.2~5.3)；20.1 (14.4~24.0)；11.8 (6.6~15.1)；87.8 (39.7~121.7)；8.3 (0.0~26.7)；674 (508~789) ❙ 1163 (1005~1414)；643 (531~817)；38 (26~49) ❙ 74.8 (36.4~101.2)；674 (508~789)；239.9 (200.8~322.4)

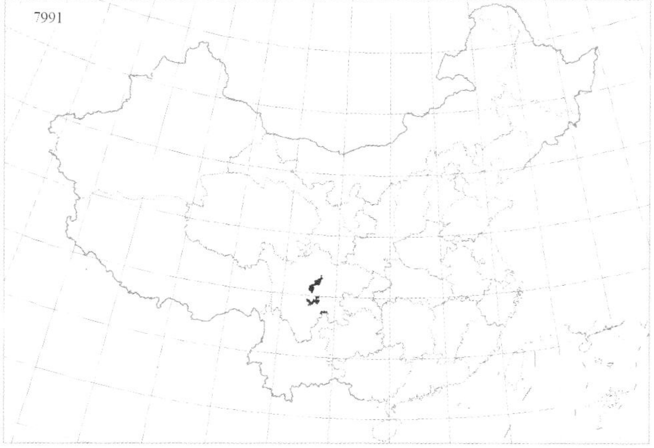

Ericaceae 杜鹃花科 *Rhododendron* 杜鹃花属

Rhododendron hancockii Hemsl. 滇南杜鹃

Shrubs or small trees, evergreen

18.9 (14.6~22.7)；11.5 (7.9~15.1)；24.3 (19.6~28.6)；19.0 (14.6~22.8)；167.4 (115.8~213.1)；0.0 (0.0~0.0)；952 (736~1221) ❙ 1336 (911~2435)；716 (524~1428)；75 (32~197) ❙ 38.9 (6.5~99.5)；938 (725~1221)；328.7 (234.5~659.0)

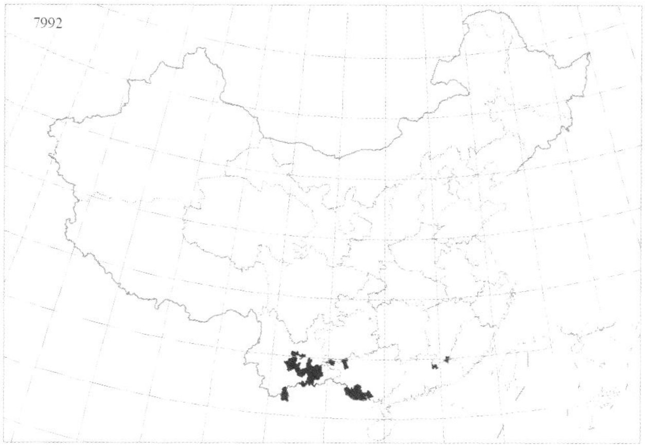

Ericaceae 杜鹃花科 *Rhododendron* 杜鹃花属

Rhododendron haofui Chun et Fang 光枝杜鹃

Shrubs

16.8 (11.1~21.3); 6.1 (1.8~12.3); 26.5 (19.4~29.3); 16.9 (11.1~21.3); 142.9 (79.8~196.1); 0.5 (0.0~6.9); 891 (648~1130) ❙ 1467 (879~1922); 554 (445~813); 158 (30~259) ❙ 65.0 (16.9~128.9); 889 (648~1130); 224.7 (153.9~359.7)

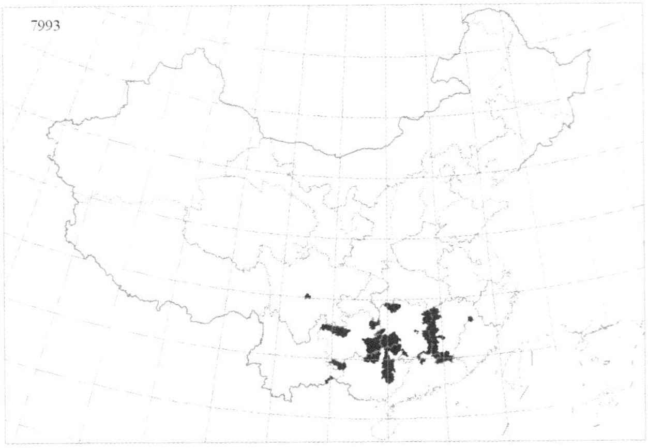

Ericaceae 杜鹃花科 *Rhododendron* 杜鹃花属

Rhododendron hejiangense M. Y. He 合江杜鹃

Small shrubs, deciduous

17.2; 6.8; 27.0; 17.3; 147.4; 0.0; 905 ❙ 1097; 499; 72 ❙ 21.3; 905; 158.7

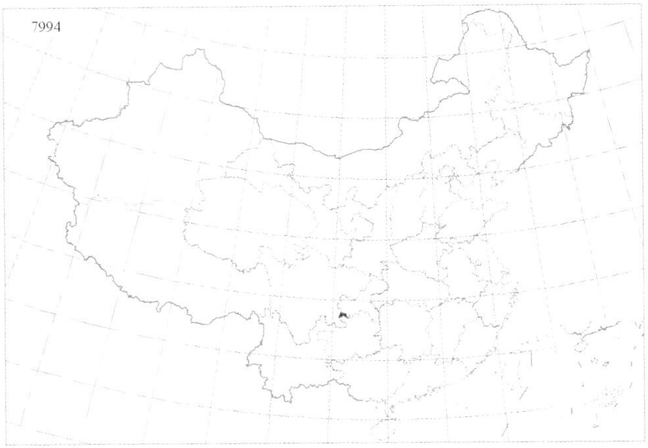

Ericaceae 杜鹃花科 *Rhododendron* 杜鹃花属

Rhododendron heliolepis Franch. 亮鳞杜鹃

Shrubs or small trees, evergreen

10.5 (3.8~14.5); 3.1 (-4.3~7.8); 16.6 (10.8~20.0); 10.6 (4.7~14.5); 74.1 (21.6~113.5); 8.0 (0.0~35.4); 623 (444~730) ❙ 1035 (741~1454); 593 (426~792); 38 (16~70) ❙ 68.1 (34.2~121.7); 620 (444~730); 320.9 (221.3~487.2)

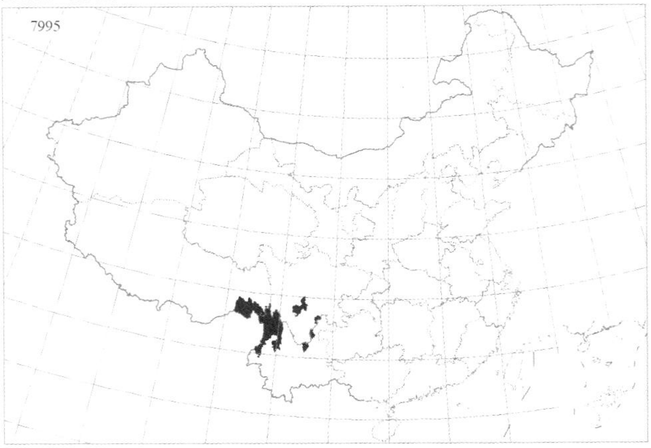

Ericaceae 杜鹃花科 *Rhododendron* 杜鹃花属

Rhododendron heliolepis var. fumidum (I. B. Balfour et W. W. Smith) R. C. Fang 灰褐亮鳞杜鹃

Shrubs or small trees, evergreen

13.2; 5.7; 19.2; 13.3; 99.4; 0.1; 700 ❙ 943; 569; 28 ❙ 34.6; 689; 249.5

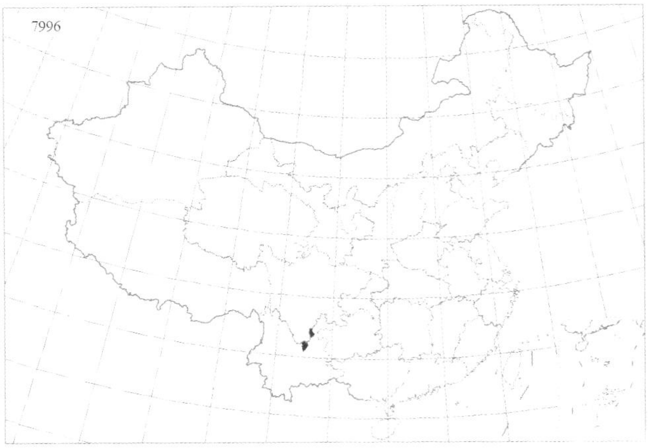

Ericaceae 杜鹃花科 *Rhododendron* 杜鹃花属

Rhododendron heliolepis var. oporinum (I. B. Balfour et Kingdon-Ward) A. L. Chang 毛冠亮鳞杜鹃

Shrubs or small trees, evergreen

11.4 (5.3~14.4); 4.4 (-2.4~7.5); 16.9 (12.2~19.1); 11.5 (5.7~14.4); 82.2 (30.5~113.1); 5.4 (0.0~26.2); 645 (479~727) ❙ 1210 (961~1454); 677 (558~792); 54 (31~70) ❙ 88.4 (56.0~100.7); 645 (479~727); 372.6 (234.2~487.2)

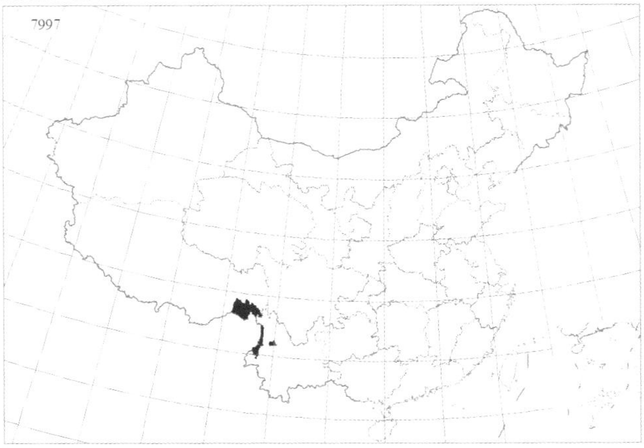

Ericaceae 杜鹃花科 *Rhododendron* 杜鹃花属

Rhododendron hemitrichotum I. B. Balfour et Forrest 粉背碎米花

Small shrubs, evergreen

10.7; 2.5; 17.7; 10.8; 75.5; 6.6; 639 ❙ 1075; 647; 23 ❙ 67.5; 639; 280.0

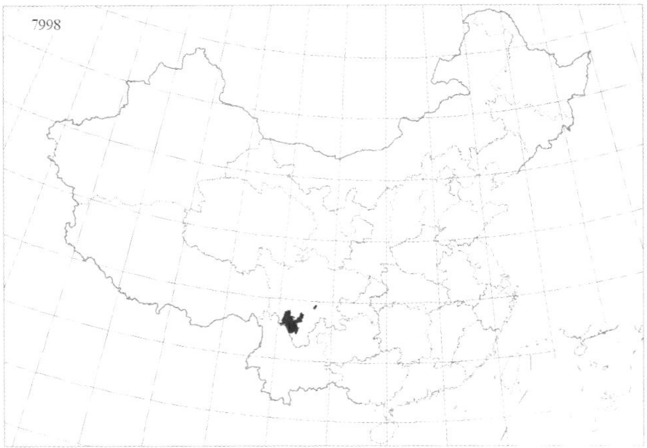

Ericaceae 杜鹃花科 *Rhododendron* 杜鹃花属

Rhododendron hemsleyanum E. H. Wilson 波叶杜鹃

Shrubs or small trees, evergreen

13.9 (11.1~16.3); 4.9 (2.2~9.1); 21.8 (19.4~24.7); 14.0 (11.3~16.4); 109.6 (80.8~136.4); 1.7 (0.0~5.3); 741 (652~801) ❙ 1174 (1001~1410); 644 (507~816); 47 (35~64) ❙ 59.4 (30.8~87.2); 741 (652~801); 263.6 (184.8~347.3)

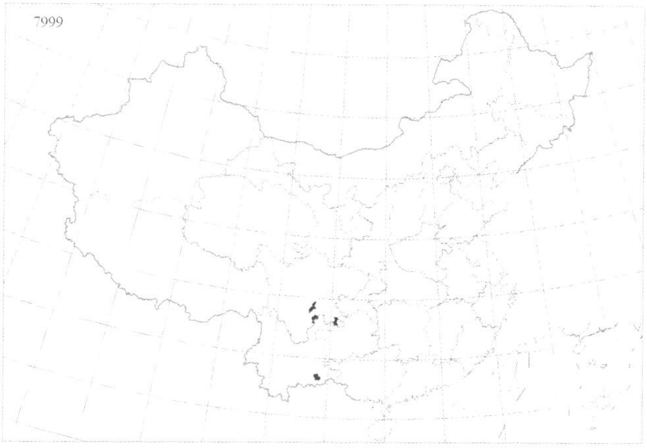

Ericaceae 杜鹃花科 *Rhododendron* 杜鹃花属

Rhododendron hemsleyanum var. chengianum D. Fang ex Ching 无腺杜鹃

Shrubs or trees, evergreen

14.9; 5.3; 23.6; 15.0; 119.6; 0.0; 782 ❙ 1410; 816; 49 ❙ 80.4; 782; 214.5

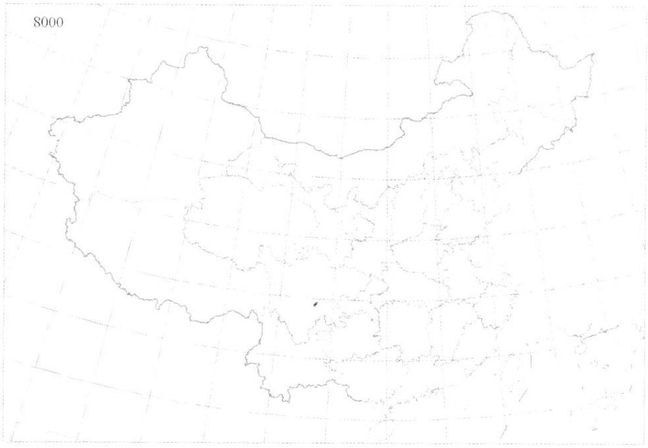

Ericaceae 杜鹃花科 *Rhododendron* 杜鹃花属

Rhododendron henanense Fang 河南杜鹃

Shrubs, evergreen

11.8 (10.2~13.6); -1.4 (-2.9~0.4); 24.2 (22.4~26.1); 12.0 (10.6~13.6); 97.5 (83.1~114.4); 15.3 (10.0~20.5); 717 (656~794) ❙ 757 (622~808); 360 (298~389); 37 (26~49) ❙ 5.5 (-13.4~19.6); 704 (622~786); 230.3 (169.1~274.1)

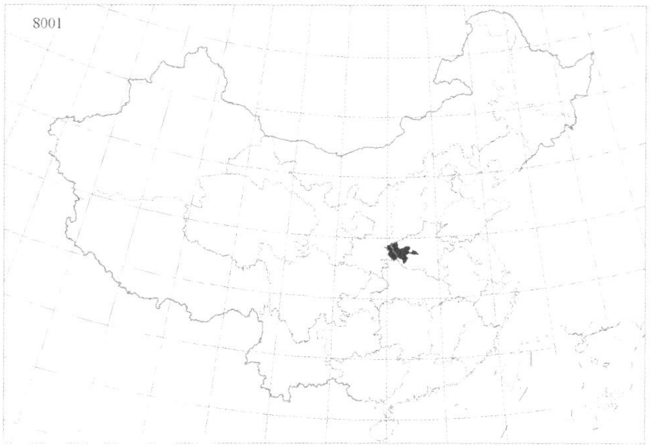

Ericaceae 杜鹃花科 *Rhododendron* 杜鹃花属

Rhododendron henryi Hance 弯蒴杜鹃

Shrubs, evergreen

19.1 (14.3~22.7); 9.9 (3.7~15.0); 27.3 (24.2~29.1); 19.1 (14.3~22.8); 169.4 (113.2~213.1); 0.0 (0.0~[illegible].4); 993 (761~1221) ❙ 1642 (1070~2435); 617 (356~1428); 182 (43~259) ❙ 67.7 (-5.0~147.0); 986 (761~1221); 251.4 (123.9~354.5)

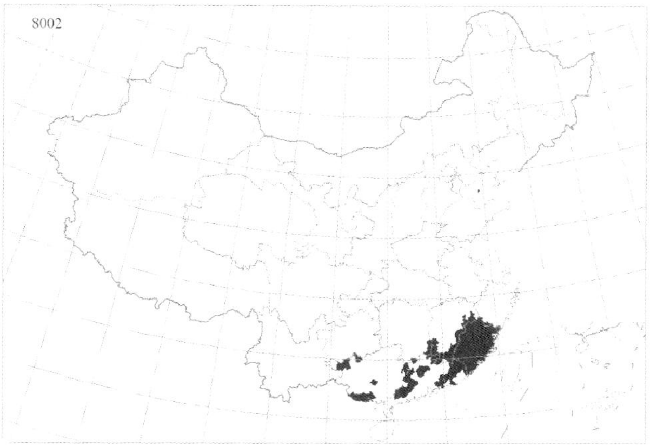

Ericaceae 杜鹃花科 *Rhododendron* 杜鹃花属

Rhododendron henryi var. dunnii (E. H. Wilson) M. Y. He 秃房弯蒴杜鹃

Shrubs, evergreen

18.1 (15.0~19.7); 8.4 (5.0~11.5); 27.0 (24.7~29.1); 18.1 (15.1~19.7); 157.6 (120.8~176.5); 0.0 (0.0~0.0); 936 (789~1001) ❙ 1707 (1427~1950); 583 (495~713); 204 (156~259) ❙ 83.0 (50.3~147.0); 933 (789~1001); 271.5 (183.9~332.5)

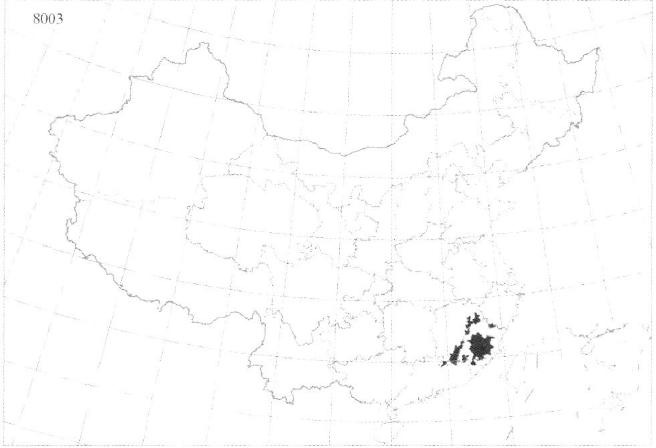

Ericaceae 杜鹃花科 *Rhododendron* 杜鹃花属

Rhododendron hippophaeoides I. B. Balfour et W. W. Sm. 灰背杜鹃

Small shrubs, evergreen

9.5 (2.2~14.6); 1.9 (-6.6~7.8); 15.9 (9.9~20.0); 9.7 (3.8~14.6); 64.6 (15.5~115.2); 10.5 (0.0~49.1); 597 (413~738) ❙ 970 (741~1178); 575 (426~663); 27 (12~46) ❙ 64.8 (30.5~125.9); 593 (413~706); 304.9 (248.3~372.3)

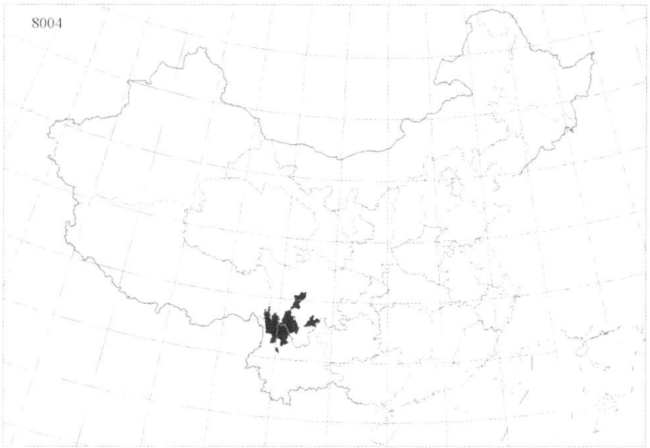

Ericaceae 杜鹃花科 *Rhododendron* 杜鹃花属

Rhododendron hippophaeoides var. occidentale Philipson et M. N. Philipson 长柱灰背杜鹃

Small shrubs, evergreen

9.4；2.4；15.4；9.5；60.9；7.2；593 ❙ 957；574；30 ❙ 61.5；593；326.5

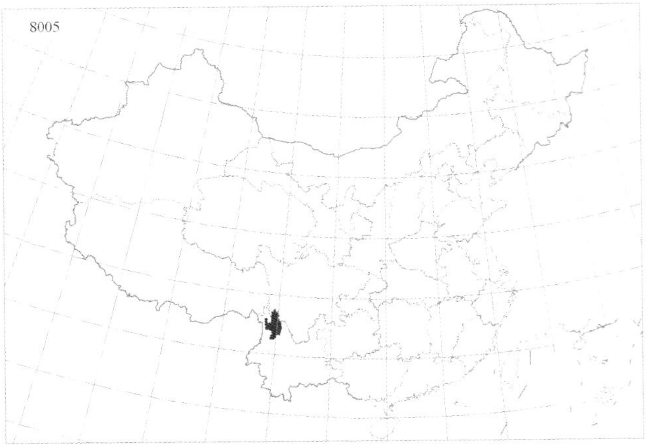

Ericaceae 杜鹃花科 *Rhododendron* 杜鹃花属

Rhododendron hirsutipetiolatum A. L. Chang et R. C. Fang 凸脉杜鹃

Shrubs, evergreen

11.3；4.3；17.0；11.3；76.6；0.7；644 ❙ 1243；696；58 ❙ 93.0；644；384.7

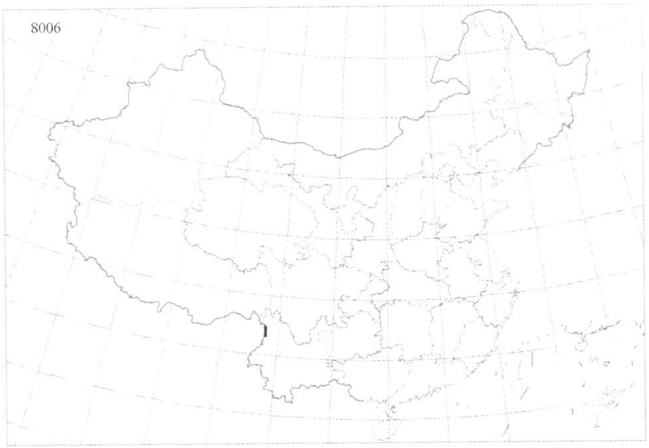

Ericaceae 杜鹃花科 *Rhododendron* 杜鹃花属

Rhododendron hirtipes Tagg 硬毛杜鹃

Shrubs or small trees, evergreen

3.0 (−2.5~13.3)；-6.1 (−12.7~5.4)；11.0 (7.0~19.6)；5.0 (2.1~13.4)；27.4 (4.0~100.4)；51.0 (0.0~93.0)；437 (332~706) ❙ 630 (341~1736)；383 (226~1007)；14 (4~51) ❙ 30.6 (−15.3~145.8)；429 (332~706)；245.2 (165.8~447.4)

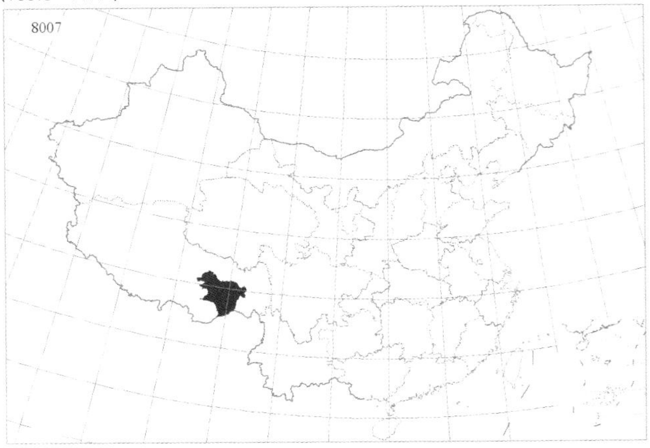

Ericaceae 杜鹃花科 *Rhododendron* 杜鹃花属

Rhododendron hodgsonii Hook. f. 多裂杜鹃

Shrubs or small trees, evergreen

3.6 (−0.7~12.2)；-5.1 (−9.6~4.3)；11.3 (7.7~18.5)；5.3 (2.5~12.2)；29.8 (6.8~87.4)；45.7 (0.7~74.2)；447 (338~672) ❙ 614 (341~1431)；378 (226~817)；13 (4~31) ❙ 24.1 (−15.3~113.0)；435 (338~672)；190.7 (61.6~363.7)

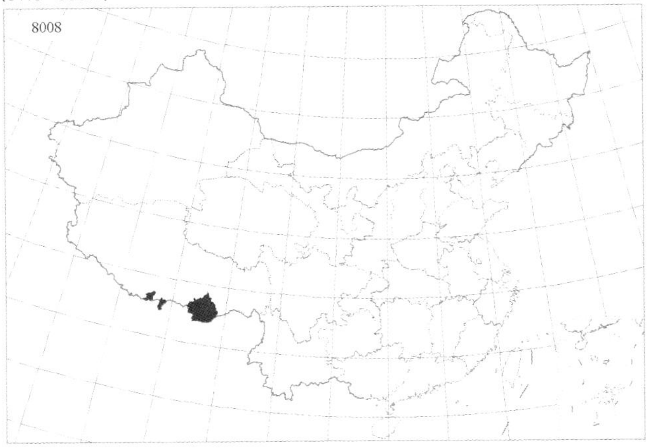

Ericaceae 杜鹃花科 *Rhododendron* 杜鹃花属

Rhododendron hoi Fang 川北杜鹃

Shrubs

2.9；-6.3；11.2；4.4；20.5；44.6；428 ❙ 950；490；25 ❙ 121.7；428；267.5

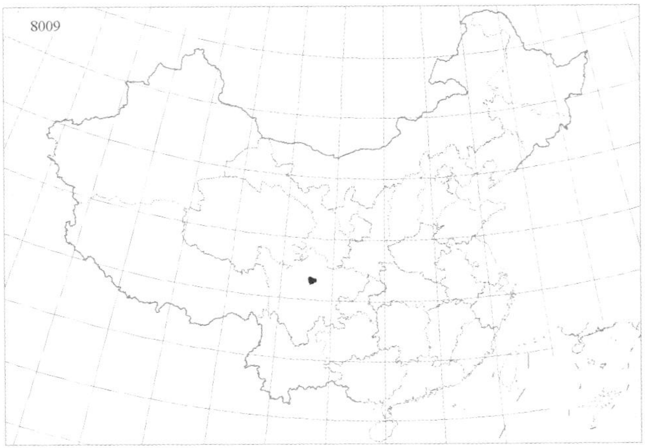

Ericaceae 杜鹃花科 *Rhododendron* 杜鹃花属

Rhododendron hongkongense Hutch. 白马银花

Shrubs, evergreen

19.3 (17.3~22.4)；10.0 (6.4~15.2)；27.8 (26.9~28.7)；19.6 (17.3~22.7)；175.0 (147.9~212.7)；0.0 (0.0~0.0)；1028 (907~1216) ❙ 1736 (1486~2037)；695 (505~1033)；170 (104~197) ❙ 69.0 (51.7~85.1)；1028 (907~1216)；222.5 (163.7~261.3)

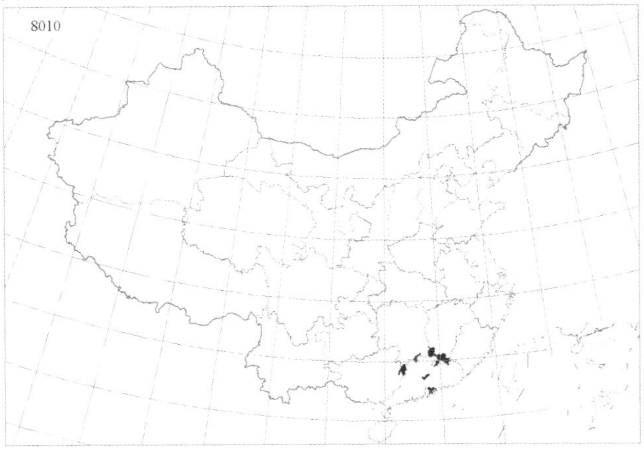

Ericaceae 杜鹃花科 *Rhododendron* 杜鹃花属

Rhododendron hookeri Nutt. 串珠杜鹃

Shrubs or small trees, evergreen

12.2; 4.3; 18.5; 12.2; 87.4; 0.7; 672 ❙ 1431; 817; 31 ❙ 113.0; 672; 363.7

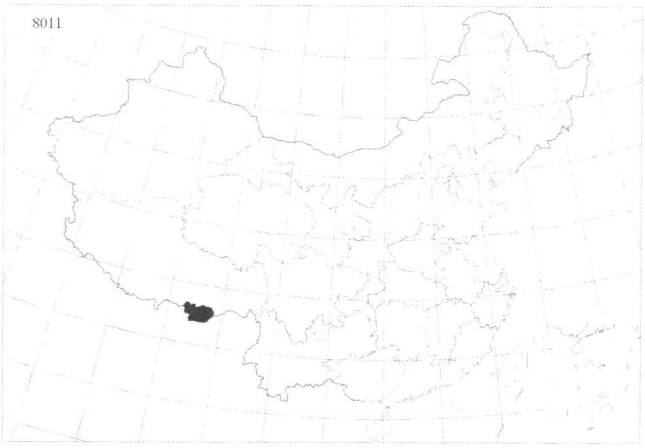

Ericaceae 杜鹃花科 *Rhododendron* 杜鹃花属

Rhododendron huadingense B. Y. Ding et Y. Y. Fang 华顶杜鹃

Shrubs

15.6; 4.4; 27.1; 15.7; 128.8; 0.6; 843 ❙ 1581; 587; 185 ❙ 87.6; 843; 236.5

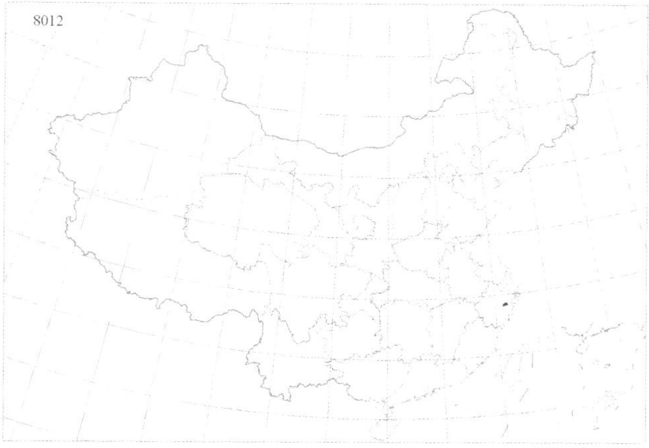

Ericaceae 杜鹃花科 *Rhododendron* 杜鹃花属

Rhododendron huanum W. P. Fang 凉山杜鹃

Shrubs or small trees, evergreen

13.8 (11.9~16.2); 4.0 (2.9~5.5); 22.9 (19.6~26.3); 13.9 (12.0~16.2); 107.6 (87.2~134.6); 1.3 (0.0~3.4); 747 (671~854) ❙ 1079 (879~1280); 509 (472~610); 56 (33~94) ❙ 44.1 (25.7~56.8); 747 (671~854); 219.9 (170.4~278.8)

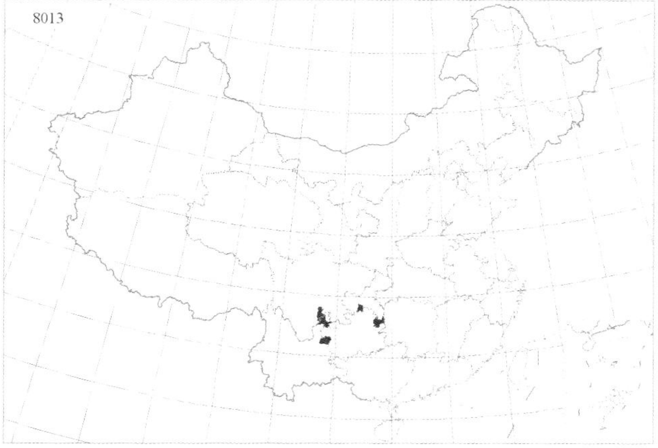

Ericaceae 杜鹃花科 *Rhododendron* 杜鹃花属

Rhododendron huguangense Tam 大鳞杜鹃

Shrubs, evergreen

17.4; 6.1; 27.9; 17.4; 148.9; 0.0; 922 ❙ 1514; 478; 205 ❙ 64.6; 909; 206.0

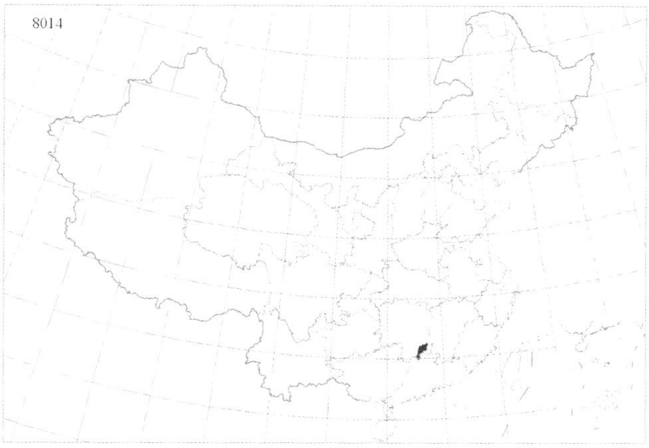

Ericaceae 杜鹃花科 *Rhododendron* 杜鹃花属

Rhododendron huidongense T. L. Ming 会东杜鹃

Shrubs, evergreen

13.7; 6.2; 19.7; 13.7; 104.7; 0.0; 714 ❙ 963; 593; 26 ❙ 34.8; 696; 271.3

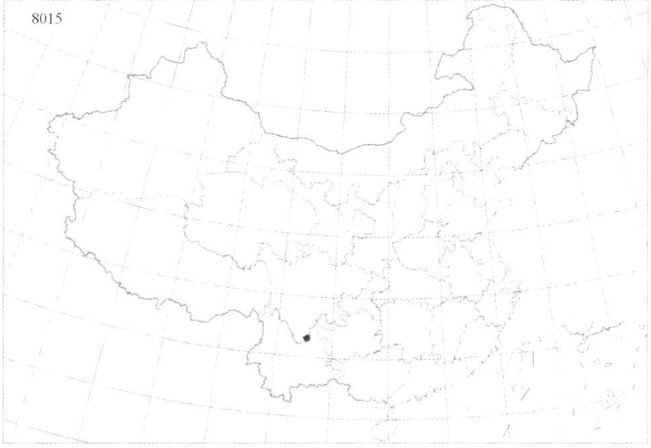

Ericaceae 杜鹃花科 *Rhododendron* 杜鹃花属

Rhododendron hunanense Chun ex Tam 湖南杜鹃

Shrubs, semievergreen or green

17.7 (16.4~20.1); 6.9 (5.2~10.7); 27.4 (26.2~29.1); 17.7 (16.5~20.1); 152.4 (137.7~181.1); 0.0 (0.0~0.0); 929 (864~1048) ❙ 1541 (1410~1922); 509 (403~644); 195 (161~259) ❙ 66.3 (47.4~111.3); 917 (864~1048); 218.2 (151.0~279.5)

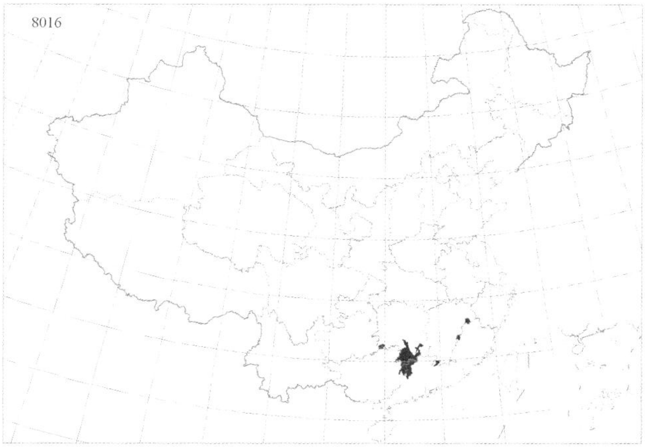

Ericaceae 杜鹃花科 *Rhododendron* 杜鹃花属

Rhododendron hunnewellianum Rehder et E. H. Wilson 岷江杜鹃

Shrubs, evergreen

10.0 (2.6~13.5); -0.4 (-7.0~3.2); 19.4 (11.2~23.5); 10.3 (4.3~13.5); 75.2 (20.0~105.3); 15.4 (2.5~48.2); 633 (427~744) ❙ 842 (707~1081); 439 (357~606); 21 (15~32) ❙ 38.5 (-0.2~101.2); 633 (427~744); 251.8 (171.0~294.3)

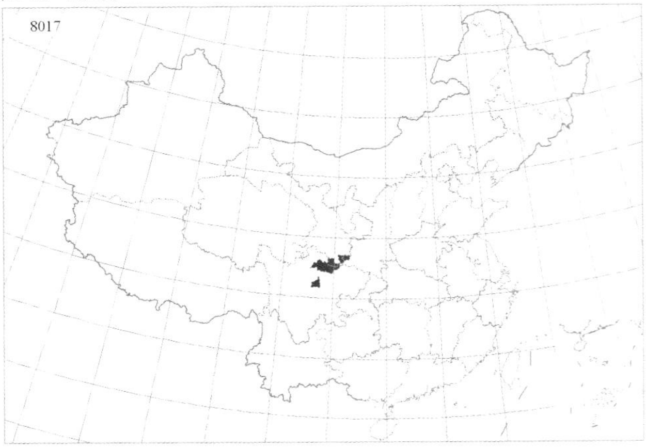

Ericaceae 杜鹃花科 *Rhododendron* 杜鹃花属

Rhododendron hunnewellianum subsp. rockii (E. H. Wilson) D. F. Chamb. 黄毛岷江杜鹃

Shrubs, evergreen

5.9; -3.9; 14.7; 6.8; 42.1; 31.2; 511 ❙ 826; 410; 21 ❙ 66.3; 511; 263.8

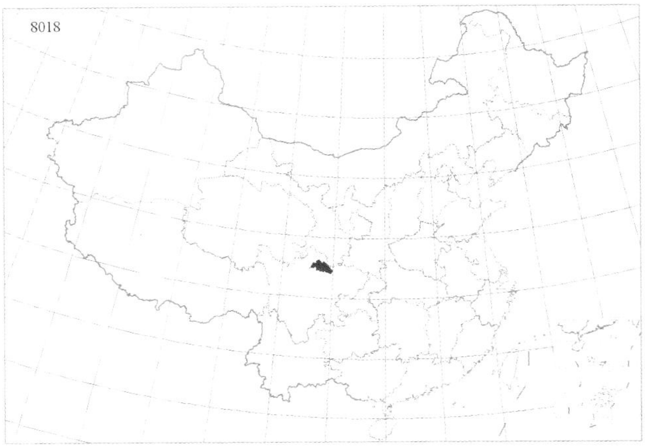

Ericaceae 杜鹃花科 *Rhododendron* 杜鹃花属

Rhododendron hylaeum I. B. Balfour et Farrer 粉果杜鹃

Shrubs or small trees, evergreen

9.1; 1.4; 15.6; 9.3; 61.2; 11.7; 586 ❙ 1246; 717; 42 ❙ 109.3; 586; 344.6

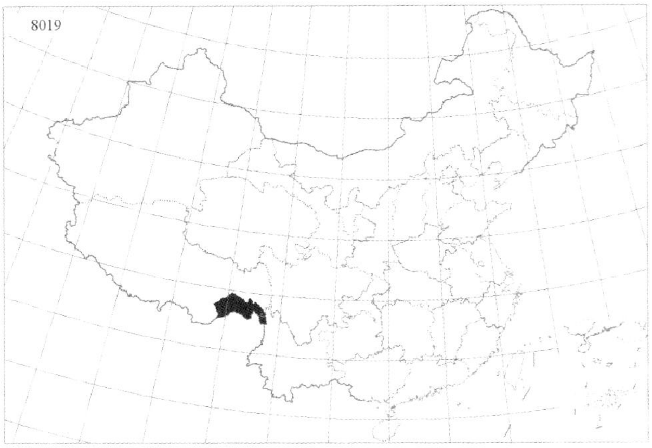

Ericaceae 杜鹃花科 *Rhododendron* 杜鹃花属

Rhododendron hypenanthum I. B. Balfour 毛花杜鹃

Small shrubs, evergreen

-1.0; -9.4; 6.8; 2.2; 4.5; 75.4; 341 ❙ 320; 212; 18 ❙ -5.8; 320; 90.1

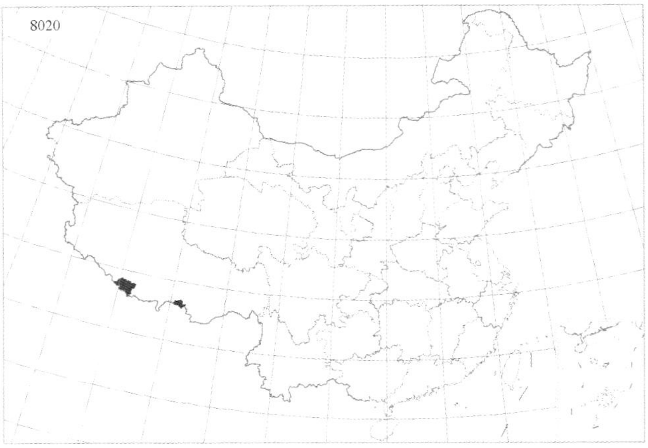

Ericaceae 杜鹃花科 *Rhododendron* 杜鹃花属

Rhododendron hyperythrum Hayata 微笑杜鹃

Shrubs or small trees, evergreen

20.3; 14.4; 25.4; 20.3; 184.2; 0.0; 1003 ❙ 2719; 1168; 329 ❙ 171.7; 1003; 234.5

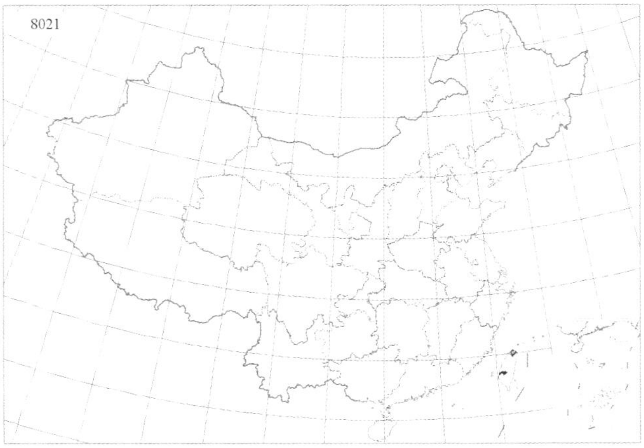

Ericaceae 杜鹃花科 *Rhododendron* 杜鹃花属

Rhododendron hypoblematosum Tam 背绒杜鹃

Shrubs, evergreen

16.8; 5.9; 27.0; 16.9; 142.9; 0.3; 890 ❙ 1578; 519; 199 ❙ 78.5; 890; 226.5

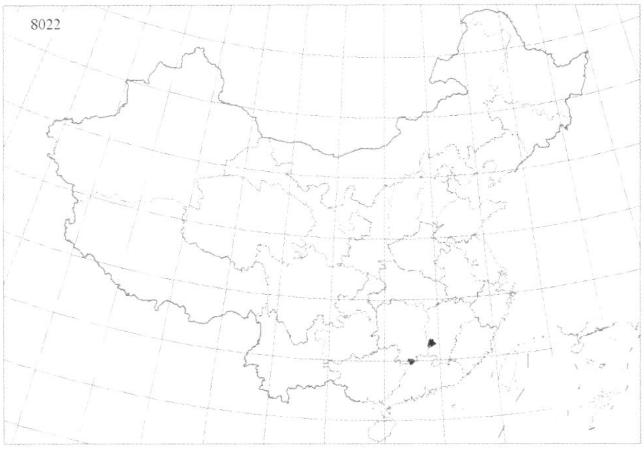

Ericaceae 杜鹃花科 *Rhododendron* 杜鹃花属
Rhododendron hypoglaucum Hemsl. 粉白杜鹃
Shrubs, evergreen

13.0 (7.4~16.9); 1.6 (-4.2~5.5); 24.0 (18.7~28.1); 13.1 (8.2~16.9); 104.3 (58.8~143.4); 7.3 (0.0~29.4); 741 (561~924) ❙ 1084 (738~1466); 464 (332~602); 52 (19~113) ❙ 46.9 (5.8~106.2); 741 (561~924); 248.2 (166.0~318.2)

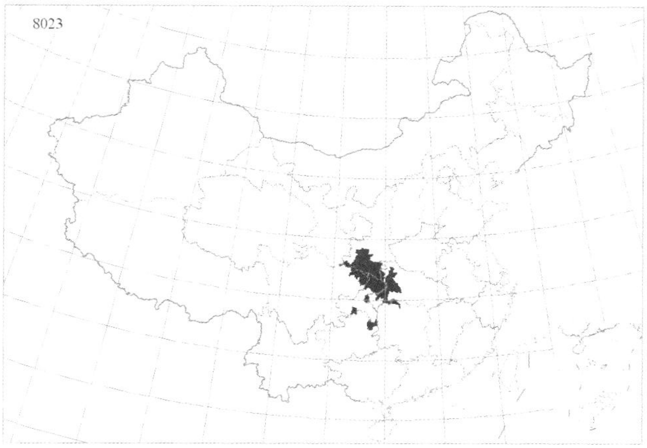

Ericaceae 杜鹃花科 *Rhododendron* 杜鹃花属
Rhododendron igneum Cowan 肉红杜鹃
Shrubs, evergreen

4.4; -4.3; 12.1; 5.3; 28.5; 34.6; 455 ❙ 538; 337; 9 ❙ 18.2; 455; 168.2

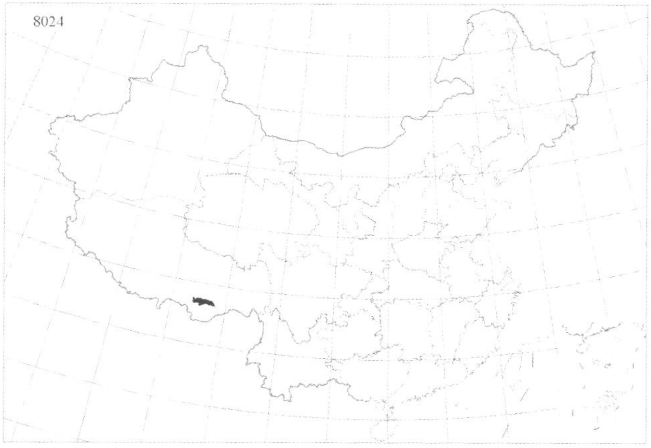

Ericaceae 杜鹃花科 *Rhododendron* 杜鹃花属
Rhododendron impeditum I. B. Balfour et W. W. Sm. 粉紫杜鹃
Shrubs, evergreen

11.4 (-1.7~18.9); 3.4 (-13.5~12.0); 18.1 (8.0~26.6); 11.8 (2.4~19.0); 88.8 (6.4~167.8); 11.4 (0.0~86.3); 663 (344~929) ❙ 912 (486~1141); 534 (289~627); 28 (8~57) ❙ 41.5 (-14.2~125.9); 640 (344~912); 267.9 (142.9~372.3)

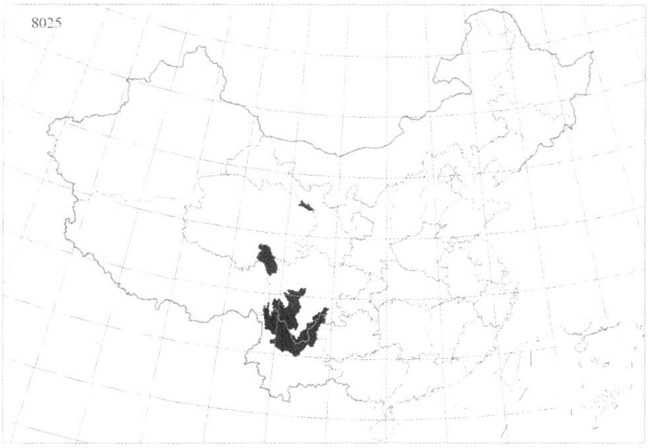

Ericaceae 杜鹃花科 *Rhododendron* 杜鹃花属
Rhododendron insigne Hemsl. et E. H. Wilson 不凡杜鹃
Shrubs, evergreen

14.3 (11.1~18.1); 4.5 (2.2~8.1); 23.1 (19.4~27.3); 14.3 (11.3~18.2); 113.2 (80.8~158.1); 1.3 (0.0~5.3); 765 (652~949) ❙ 1043 (879~1220); 530 (472~688); 52 (33~72) ❙ 37.4 (19.4~87.2); 765 (652~949); 197.6 (135.1~292.6)

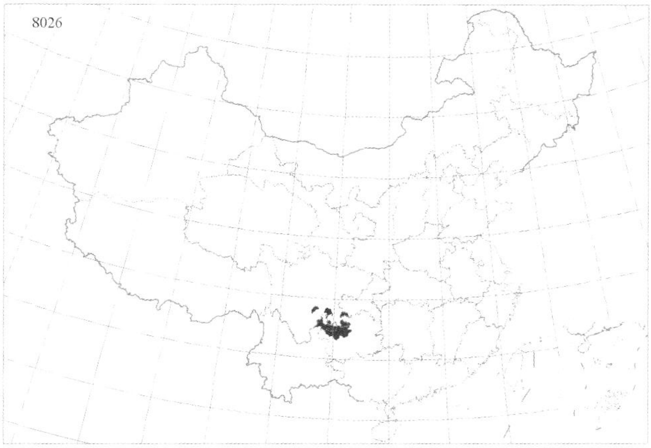

Ericaceae 杜鹃花科 *Rhododendron* 杜鹃花属
Rhododendron intricatum Franch. 隐蕊杜鹃
Shrubs, evergreen

5.7 (0.1~14.6); -2.6 (-10.1~7.8); 13.0 (9.5~20.0); 6.8 (3.2~14.6); 40.6 (12.3~115.2); 31.2 (0.0~71.1); 503 (377~738) ❙ 862 (533~1005); 524 (328~603); 17 (11~27) ❙ 75.0 (30.5~125.9); 498 (377~689); 279.9 (208.8~319.8)

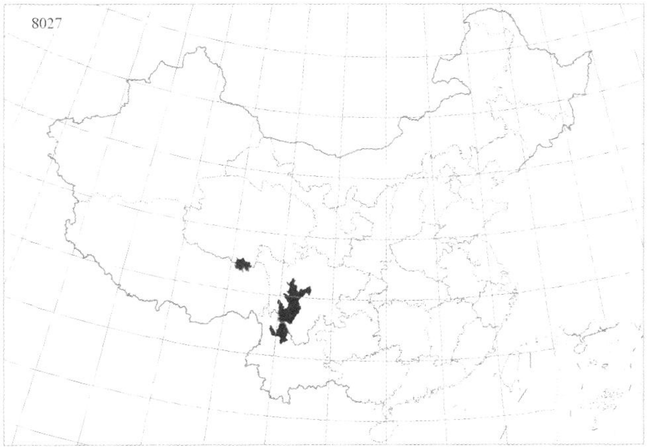

Ericaceae 杜鹃花科 *Rhododendron* 杜鹃花属
Rhododendron invictum I. B. Balfour et Farrer 绝伦杜鹃
Shrubs, evergreen

12.3; 2.1; 21.7; 12.4; 93.8; 5.5; 701 ❙ 848; 462; 18 ❙ 21.8; 701; 260.1

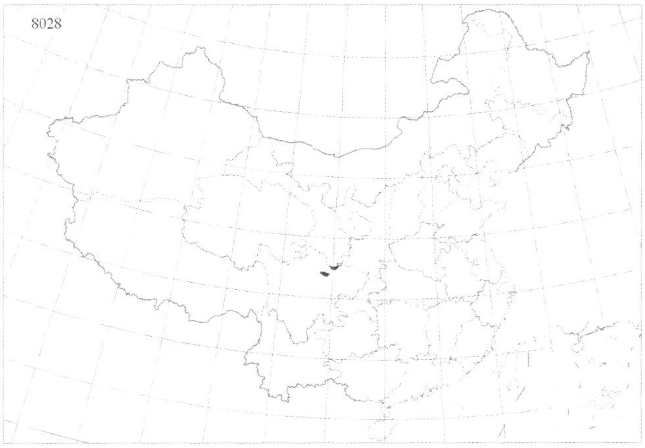

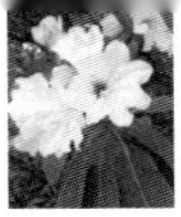

Ericaceae 杜鹃花科 *Rhododendron* 杜鹃花属

Rhododendron irroratum Franch. 露珠杜鹃

Shrubs or small trees, evergreen

14.2 (−0.7~19.8); 6.6 (−9.6~13.1); 20.3 (7.7~26.6); 14.3 (2.5~19.8); 113.5 (6.8~177.9); 2.2 (0.0~74.2); 743 (338~1010) ❙ 1036 (370~1454); 590 (262~835); 38 (10~70) ❙ 40.5 (−14.2~99.9); 724 (338~1010); 288.7 (61.6~506.2)

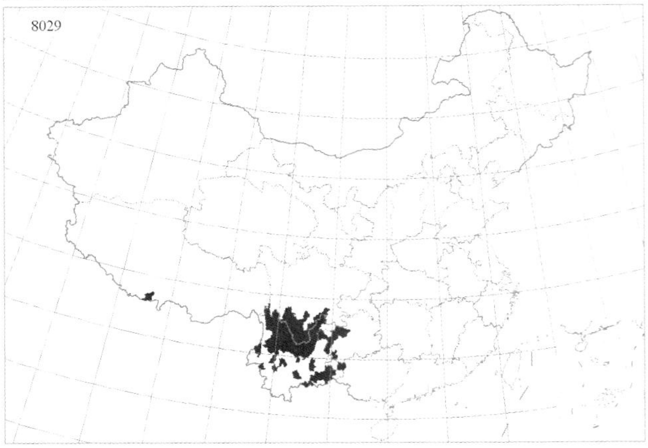

Ericaceae 杜鹃花科 *Rhododendron* 杜鹃花属

Rhododendron irroratum subsp. pogonostylum (I. B. Balfour et W. W. Sm.) D. F. Chamb. 红花露珠杜鹃 Shrubs or small trees, evergreen

17.8; 11.2; 22.4; 17.8; 153.9; 0.0; 858 ❙ 1233; 692; 52 ❙ 43.5; 852; 423.2

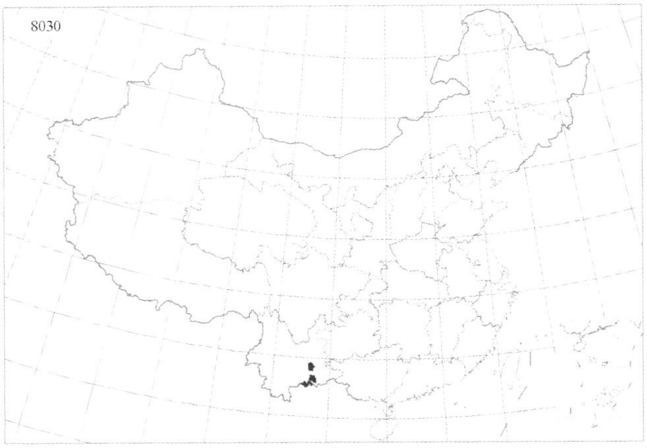

Ericaceae 杜鹃花科 *Rhododendron* 杜鹃花属

Rhododendron jasminoides M. Y. He 素馨杜鹃

Shrubs, semievergreen or green

18.0; 8.5; 25.8; 18.0; 156.4; 0.0; 913 ❙ 1551; 672; 169 ❙ 69.8; 913; 243.9

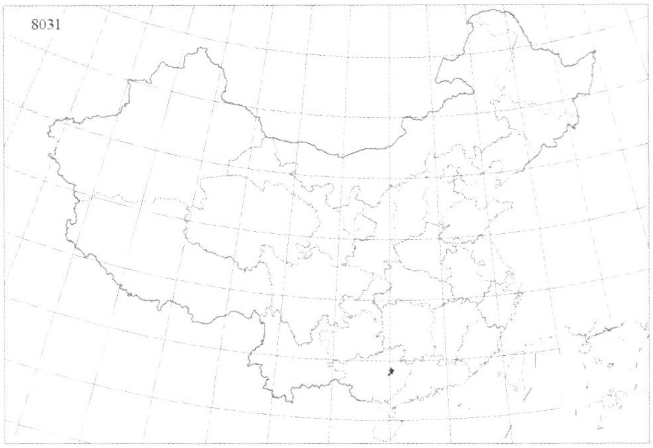

Ericaceae 杜鹃花科 *Rhododendron* 杜鹃花属

Rhododendron jingangshanicum P. C. Tam 井冈山杜鹃

Shrubs, evergreen

16.2; 5.0; 26.8; 16.2; 135.1; 0.4; 860 ❙ 1611; 524; 208 ❙ 88.0; 860; 217.4

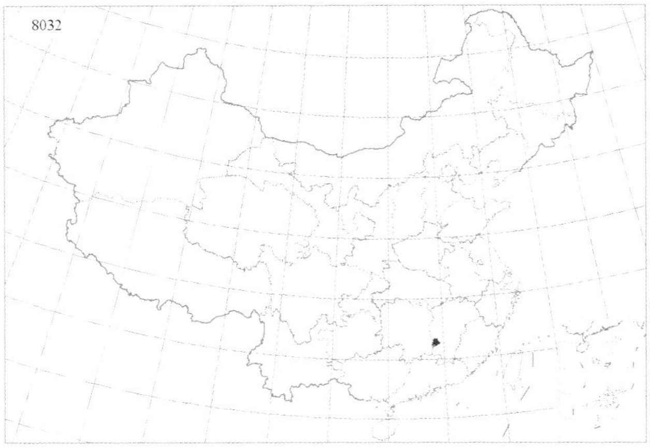

Ericaceae 杜鹃花科 *Rhododendron* 杜鹃花属

Rhododendron jinpingense Fang et M. Y. He 金平杜鹃

Shrubs, semievergreen or green

19.1; 13.0; 23.3; 19.1; 169.7; 0.0; 917 ❙ 1428; 835; 56 ❙ 55.7; 917; 506.2

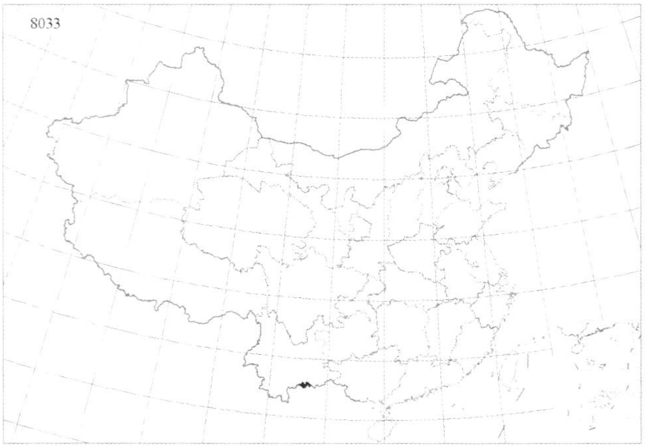

Ericaceae 杜鹃花科 *Rhododendron* 杜鹃花属

Rhododendron jinxiuense Fang et M. Y. He 金秀杜鹃

Shrubs, evergreen

18.0; 8.5; 25.8; 18.0; 156.4; 0.0; 913 ❙ 1551; 672; 169 ❙ 69.8; 913; 243.9

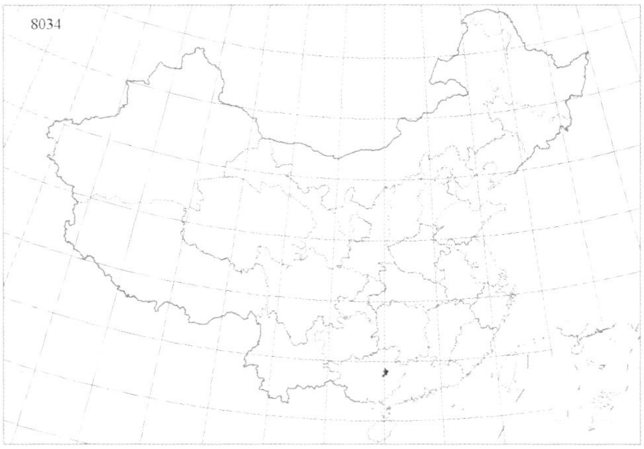

Ericaceae 杜鹃花科 *Rhododendron* 杜鹃花属
Rhododendron joniense Ching et H. P. Yang 卓尼杜鹃
Shrubs, evergreen

1.9; −9.2; 11.7; 4.3; 21.1; 57.8; 427 ❙ 621; 318; 13 ❙ 45.4; 427; 289.4

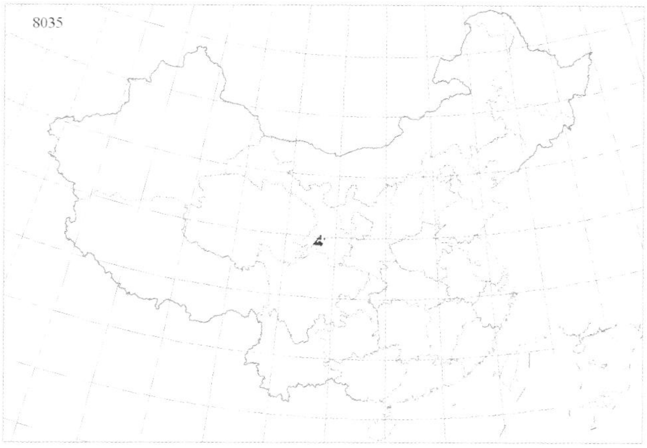

Ericaceae 杜鹃花科 *Rhododendron* 杜鹃花属
Rhododendron kanehirae E. H. Wilson 台北杜鹃
Shrubs, deciduous

20.7; 14.1; 26.9; 20.7; 188.4; 0.0; 1047 ❙ 3095; 863; 665 ❙ 195.8; 1047; 242.2

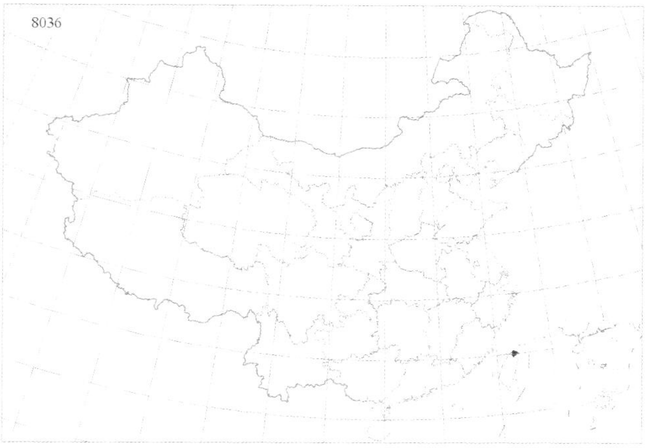

Ericaceae 杜鹃花科 *Rhododendron* 杜鹃花属
Rhododendron kasoense Hutch. et Kingdon-Ward 黄管杜鹃
Shrubs, evergreen

7.6; −0.8; 14.8; 8.6; 58.1; 26.6; 555 ❙ 1159; 678; 32 ❙ 95.0; 555; 308.4

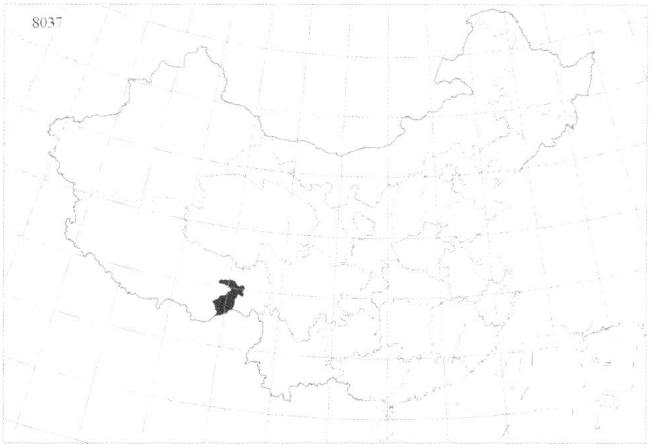

Ericaceae 杜鹃花科 *Rhododendron* 杜鹃花属
Rhododendron kawakamii Hayata 着生杜鹃
Shrubs, evergreen

18.6; 13.0; 23.4; 18.7; 164.2; 0.0; 908 ❙ 2744; 1120; 326 ❙ 206.4; 908; 256.0

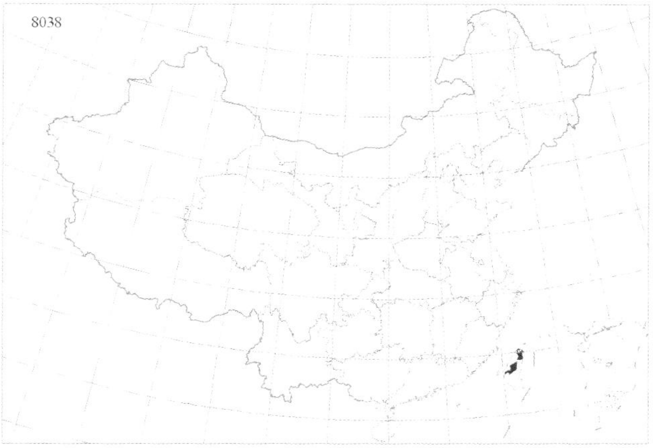

Ericaceae 杜鹃花科 *Rhododendron* 杜鹃花属
Rhododendron keleticum I. B. Balfour et Forrest 独龙杜鹃
Small shrubs, evergreen

8.9 (5.3~13.2); 1.6 (−2.4~6.4); 15.0 (12.2~18.4); 9.1 (5.7~13.3); 58.8 (30.5~99.4); 11.5 (0.0~26.2); 578 (479~695) ❙ 1077 (801~1337); 612 (496~727); 45 (21~69) ❙ 85.8 (61.4~100.7); 578 (479~695); 332.2 (234.2~407.1)

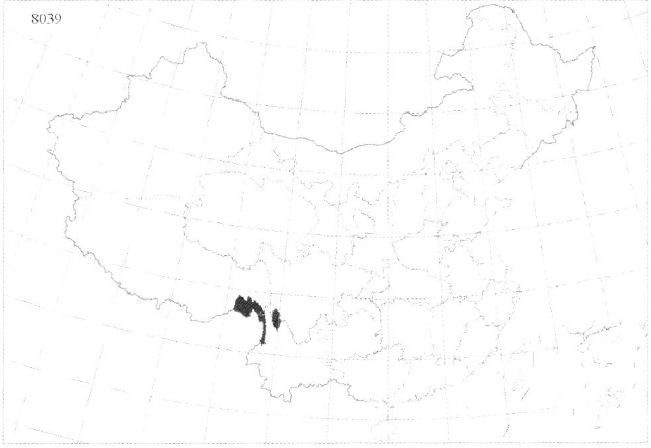

Ericaceae 杜鹃花科 *Rhododendron* 杜鹃花属
Rhododendron kendrickii Nutt. 多斑杜鹃
Shrubs or small trees

8.3; 0.0; 15.3; 8.8; 58.0; 17.7; 563 ❙ 984; 577; 20 ❙ 65.6; 563; 266.0

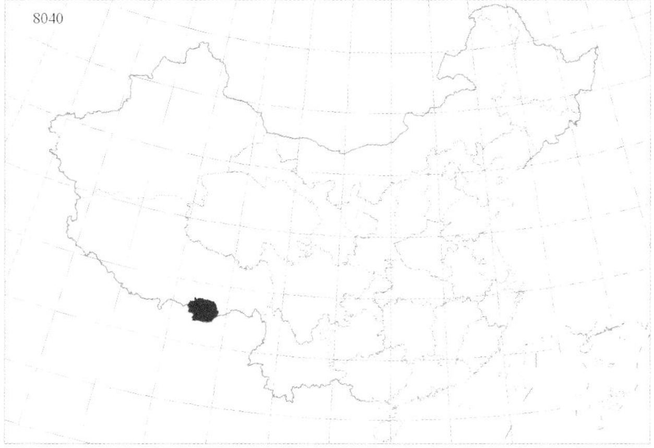

Ericaceae 杜鹃花科 *Rhododendron* 杜鹃花属

Rhododendron keysii Nutt. 管花杜鹃

Shrubs, evergreen

12.8; 4.8; 19.0; 12.8; 93.9; 0.4; 689 ❙ 1583; 912; 41 ❙ 129.4; 689; 405.6

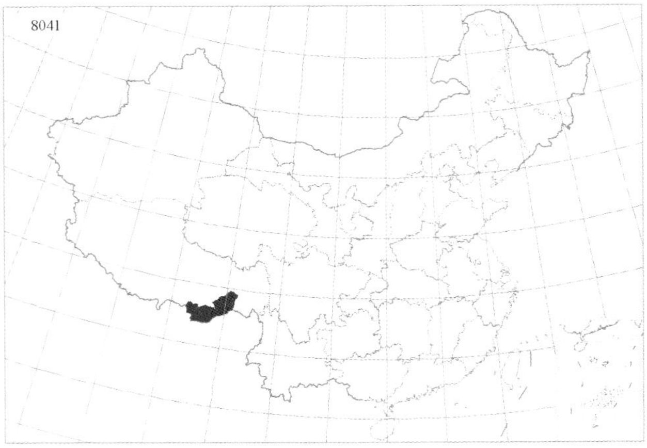

Ericaceae 杜鹃花科 *Rhododendron* 杜鹃花属

Rhododendron kiangsiense Fang 江西杜鹃

Shrubs, evergreen

16.8 (15.1~17.6); 5.5 (4.0~7.2); 27.4 (25.9~28.5); 16.8 (15.2~17.7); 141.9 (122.9~151.9); 0.1 (0.0~1.0); 894 (811~925) ❙ 1749 (1555~1997); 567 (489~669); 232 (200~259) ❙ 96.2 (68.0~130.7); 892 (811~925); 246.8 (212.3~303.4)

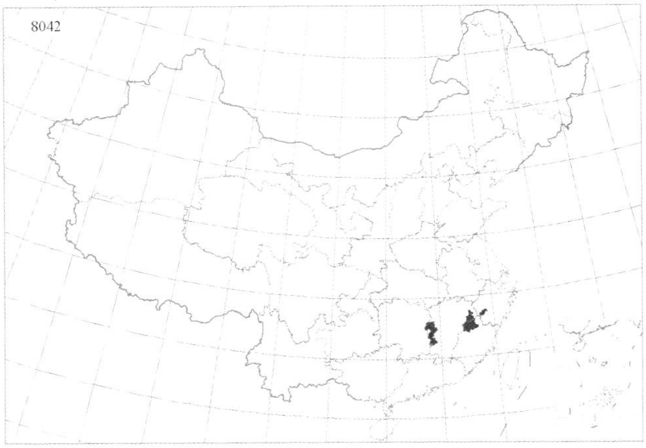

Ericaceae 杜鹃花科 *Rhododendron* 杜鹃花属

Rhododendron kongboense Hutch. 工布杜鹃

Small shrubs, evergreen

0.1; -9.4; 8.7; 3.0; 10.9; 69.1; 365 ❙ 334; 225; 4 ❙ -9.6; 330; 207.7

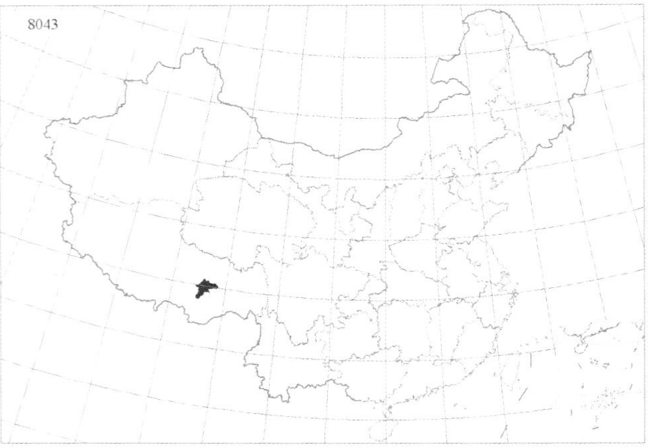

Ericaceae 杜鹃花科 *Rhododendron* 杜鹃花属

Rhododendron kwangsiense Hu ex Tam 广西杜鹃

Shrubs, evergreen

18.2 (14.3~21.8); 8.3 (3.3~13.4); 26.5 (23.0~28.7); 18.2 (14.3~21.8); 158.6 (112.6~201.6); 0.1 (0.0~2.3); 946 (756~1162) ❙ 1451 (1182~1852); 623 (419~828); 134 (43~198) ❙ 54.5 (16.9~101.8); 945 (756~1162); 238.4 (177.4~354.5)

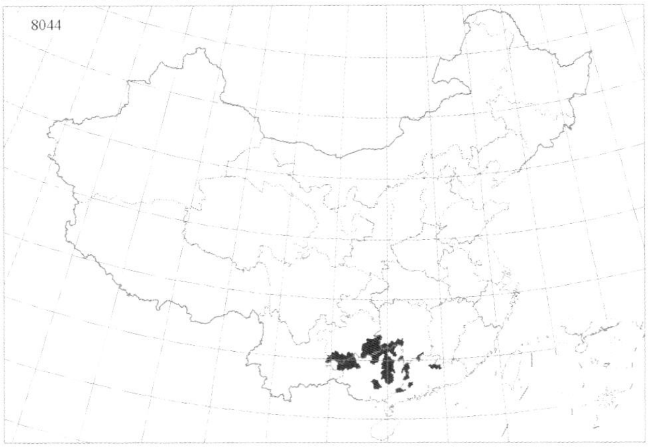

Ericaceae 杜鹃花科 *Rhododendron* 杜鹃花属

Rhododendron kwangtungense Merr. et Chun 广东杜鹃

Shrubs, deciduous

18.4 (14.3~21.5); 8.5 (3.3~12.8); 27.1 (23.6~29.8); 18.4 (14.3~21.5); 161.3 (114.1~198.2); 0.1 (0.0~2.3); 960 (765~1138) ❙ 1568 (1221~2005); 585 (442~824); 170 (64~207) ❙ 63.7 (29.9~99.4); 958 (765~1138); 236.6 (170.4~305.7)

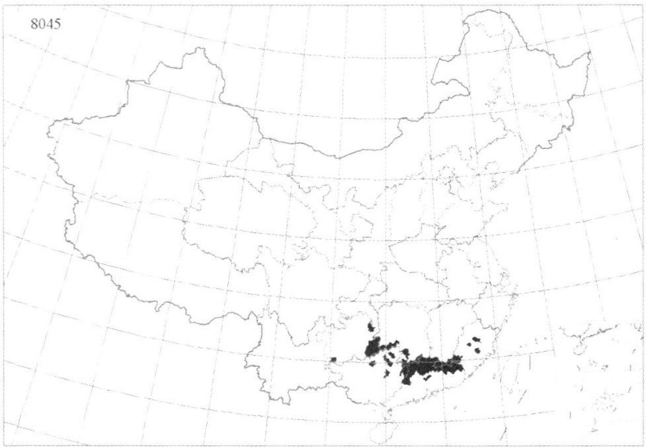

Ericaceae 杜鹃花科 *Rhododendron* 杜鹃花属

Rhododendron kyawii Lace et W. W. Smith 星毛杜鹃

Shrubs, evergreen

11.9; 4.9; 17.4; 11.9; 85.4; 2.4; 660 ❙ 1269; 700; 60 ❙ 91.7; 660; 407.8

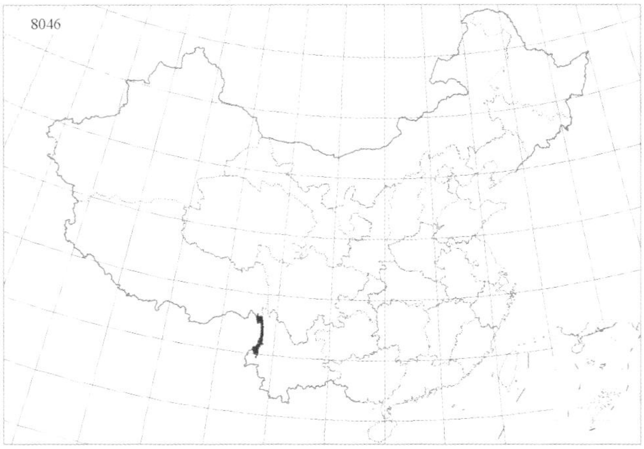

Ericaceae 杜鹃花科 *Rhododendron* 杜鹃花属
Rhododendron litchiifolium T. C. Wu et Tam 荔叶杜鹃
Shrubs, semievergreen or green

21.6; 13.5; 28.1; 21.6; 199.3; 0.0; 1138 | 1906; 832; 148 | 68.0; 1138; 215.5

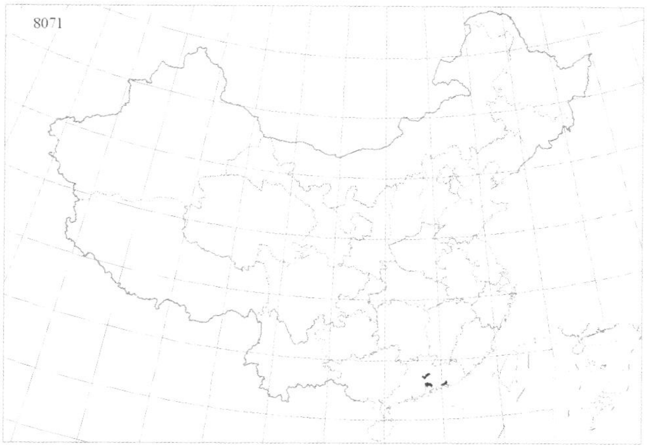

Ericaceae 杜鹃花科 *Rhododendron* 杜鹃花属
Rhododendron longesquamatum C. K. Schneid. 长鳞杜鹃
Shrubs or small trees, evergreen

10.9 (2.2~15.2); 1.0 (-6.6~4.4); 20.1 (9.9~25.9); 11.2 (3.8~15.2); 83.9 (15.5~123.9); 12.0 (0.6~49.1); 666 (413~816) | 1085 (922~1280); 534 (460~688); 46 (19~94) | 68.2 (18.6~125.9); 666 (413~816); 215.0 (147.3~292.6)

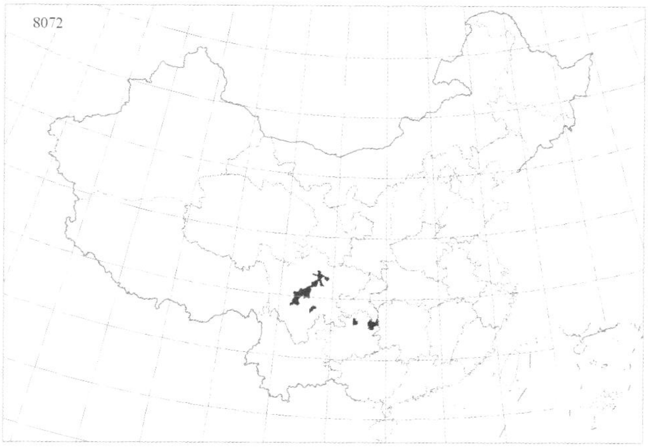

Ericaceae 杜鹃花科 *Rhododendron* 杜鹃花属
Rhododendron longicalyx M. Y. Fang 长萼杜鹃
Shrubs or small trees, evergreen

2.2; -6.6; 9.9; 3.8; 15.5; 49.1; 413 | 932; 546; 19 | 125.9; 413; 248.3

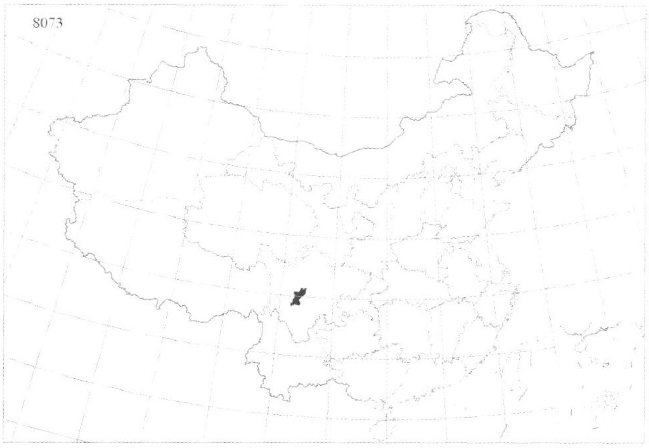

Ericaceae 杜鹃花科 *Rhododendron* 杜鹃花属
Rhododendron longifalcatum P. C. Tam 长尖杜鹃
Small shrubs, semievergreen or green

21.3; 13.2; 27.3; 21.3; 195.9; 0.0; 1111 | 1900; 1056; 125 | 71.0; 1111; 297.2

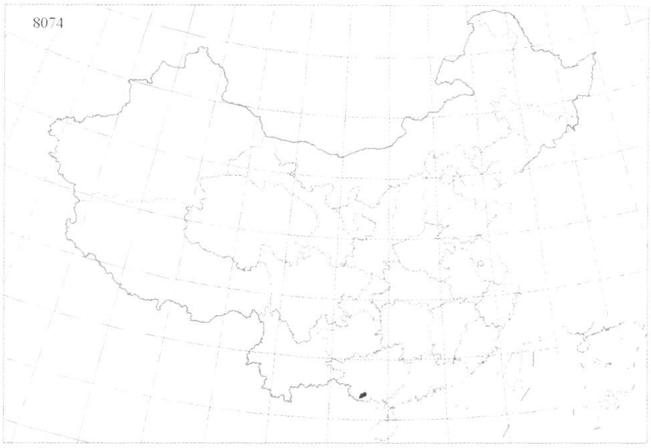

Ericaceae 杜鹃花科 *Rhododendron* 杜鹃花属
Rhododendron longiperulatum Hayata 长鳞芽杜鹃
Small shrubs, semievergreen or green

19.9; 13.5; 25.9; 19.9; 179.1; 0.0; 987 | 2993; 924; 547 | 203.4; 987; 261.4

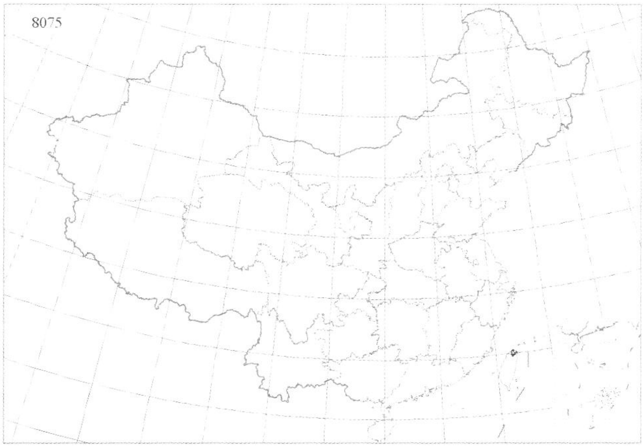

Ericaceae 杜鹃花科 *Rhododendron* 杜鹃花属
Rhododendron longipes Rehder et E. H. Wilson 长柄杜鹃
Shrubs or small trees, evergreen

13.7 (11.1~15.2); 3.6 (1.8~4.8); 23.0 (19.4~25.9); 13.7 (11.1~15.2); 106.4 (79.8~123.9); 2.1 (0.2~6.9); 743 (648~816) | 1127 (999~1280); 522 (473~584); 60 (26~94) | 51.8 (39.0~66.0); 743 (648~816); 214.1 (153.9~322.4)

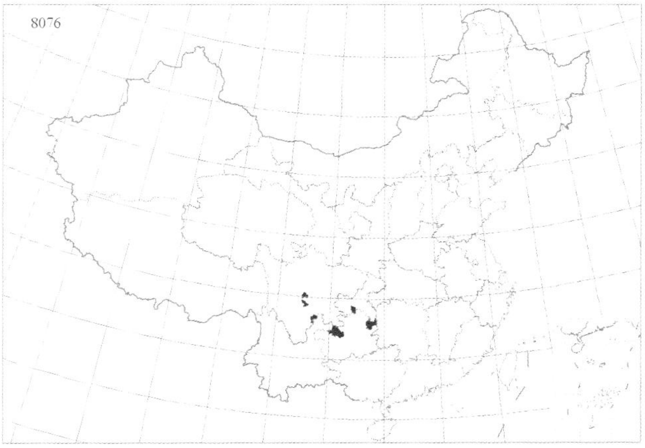

Ericaceae 杜鹃花科 *Rhododendron* 杜鹃花属

Rhododendron longipes var. chienianum (D. Fang) D. F. Chamb. 金山杜鹃

Shrubs or small trees, evergreen

14.4; 4.2; 24.0; 14.5; 114.5; 0.8; 769 ❙ 1092; 494; 63 ❙ 41.7; 769; 198.7

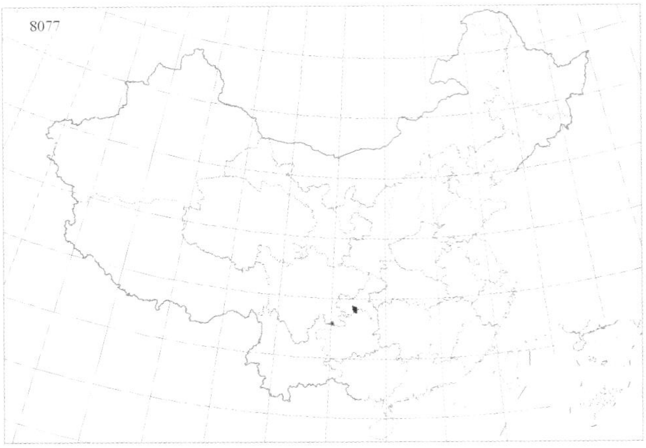

Ericaceae 杜鹃花科 *Rhododendron* 杜鹃花属

Rhododendron longistylum Rehder et A. Wilson 长柱杜鹃

Shrubs, evergreen

11.4 (6.0~15.1); 1.7 (−3.2~4.4); 20.2 (14.4~25.4); 11.5 (6.6~15.2); 84.6 (39.7~122.8); 7.8 (0.6~26.7); 668 (508~817) ❙ 944 (806~1081); 514 (420~606); 25 (16~36) ❙ 44.3 (8.2~101.2); 668 (508~817); 240.0 (171.0~318.7)

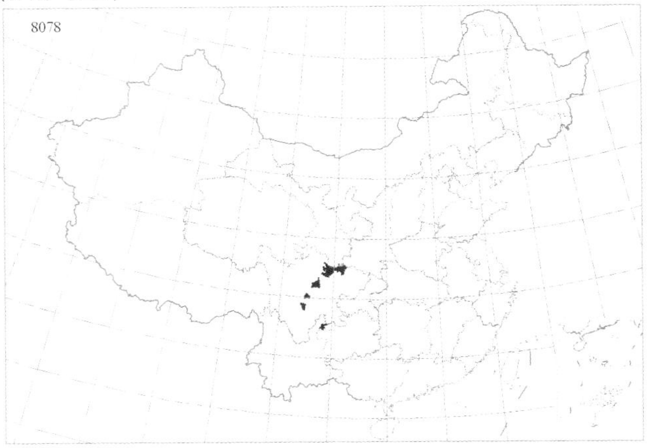

Ericaceae 杜鹃花科 *Rhododendron* 杜鹃花属

Rhododendron loniceriflorum Tam 忍冬杜鹃

Shrubs, semievergreen or green

18.8; 10.2; 26.4; 18.8; 165.7; 0.0; 950 ❙ 1638; 583; 178 ❙ 72.3; 950; 260.8

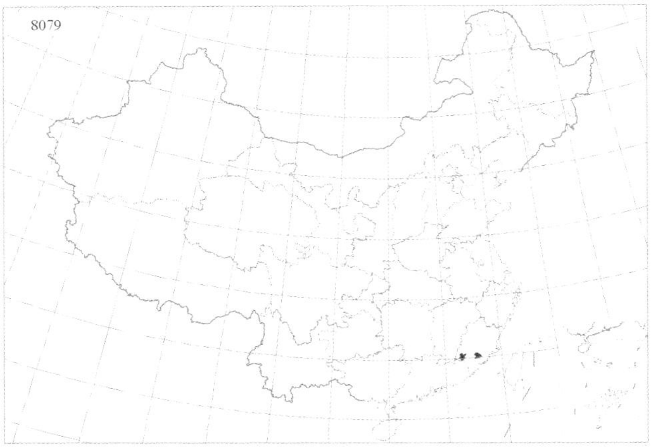

Ericaceae 杜鹃花科 *Rhododendron* 杜鹃花属

Rhododendron ludlowii Cowan 广口杜鹃

Small shrubs, evergreen

13.3; 5.4; 19.6; 13.4; 100.4; 0.0; 706 ❙ 1736; 1007; 51 ❙ 145.8; 706; 447.4

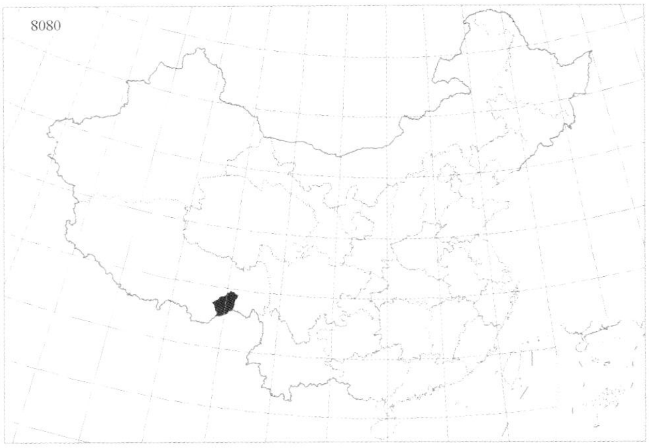

Ericaceae 杜鹃花科 *Rhododendron* 杜鹃花属

Rhododendron luhuoense H. P. Yang 炉霍杜鹃

Shrubs, evergreen

2.2; −7.8; 10.7; 4.2; 19.2; 52.8; 421 ❙ 702; 407; 15 ❙ 66.5; 421; 315.4

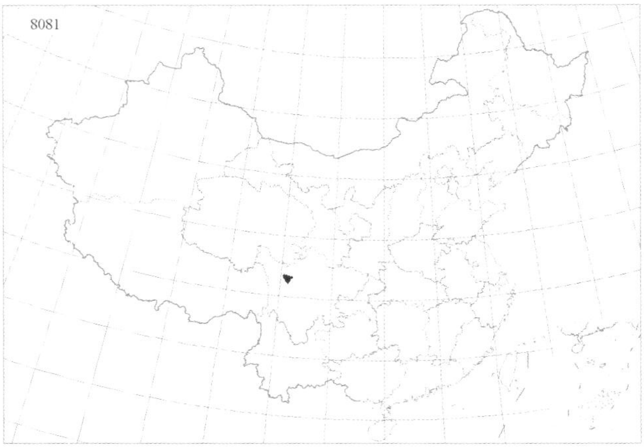

Ericaceae 杜鹃花科 *Rhododendron* 杜鹃花属

Rhododendron lukiangense Franch. 蜡叶杜鹃

Shrubs or small trees, evergreen

10.5 (5.4~18.1); 3.4 (−2.1~12.0); 16.3 (12.3~22.1); 10.6 (5.8~18.1); 73.8 (31.5~157.3); 7.6 (0.0~25.6); 624 (481~863) ❙ 1047 (741~1337); 604 (426~774); 41 (12~69) ❙ 67.9 (43.0~93.0); 624 (481~863); 348.9 (273.5~432.9)

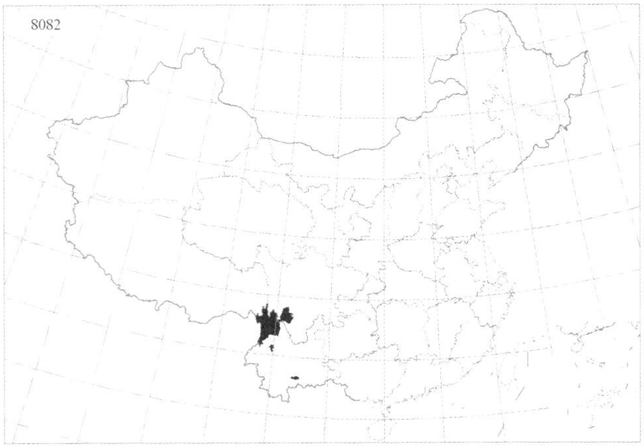

Ericaceae 杜鹃花科 *Rhododendron* 杜鹃花属
Rhododendron lulangense L. C. Hu et Tateishi 鲁浪杜鹃
Shrubs or small trees, evergreen

3.4; -5.5; 11.4; 4.7; 23.3; 41.8; 431 ❙ 527; 317; 9 ❙ 22.4; 431; 280.0

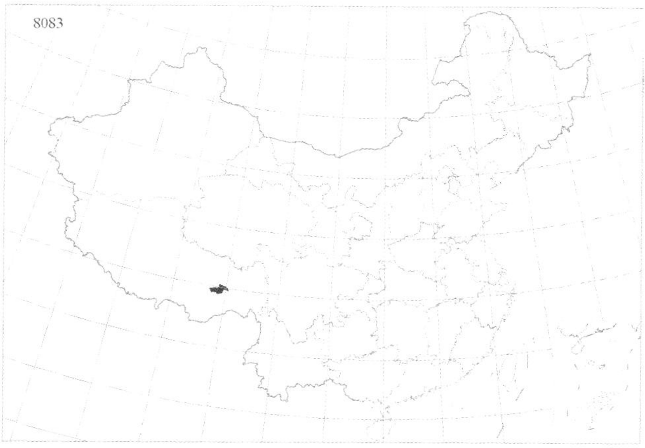

Ericaceae 杜鹃花科 *Rhododendron* 杜鹃花属
Rhododendron lungchiense W. P. Fang 龙溪杜鹃
Shrubs

2.9; -6.3; 11.2; 4.4; 20.5; 44.6; 428 ❙ 950; 490; 25 ❙ 121.7; 428; 267.5

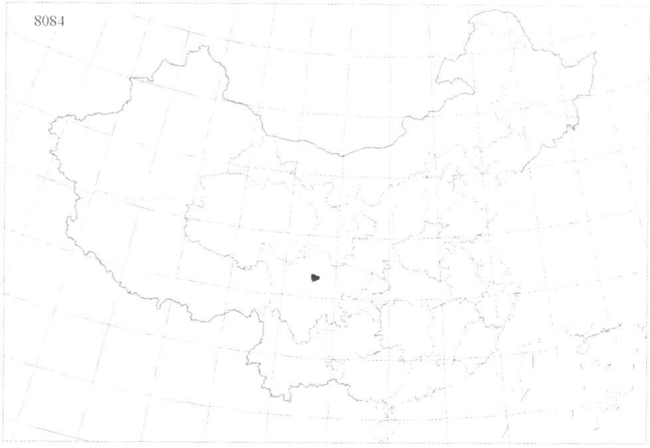

Ericaceae 杜鹃花科 *Rhododendron* 杜鹃花属
Rhododendron lutescens Franch. 黄花杜鹃
Shrubs, deciduous

10.8 (1.4~19.1); 1.7 (-7.6~13.0); 18.8 (9.2~24.6); 11.2 (3.5~19.1); 82.8 (13.4~169.7); 12.9 (0.0~55.6); 653 (402~917) ❙ 1069 (699~1478); 584 (425~835); 34 (10~85) ❙ 68.1 (5.5~128.6); 646 (402~917); 264.3 (171.0~506.2)

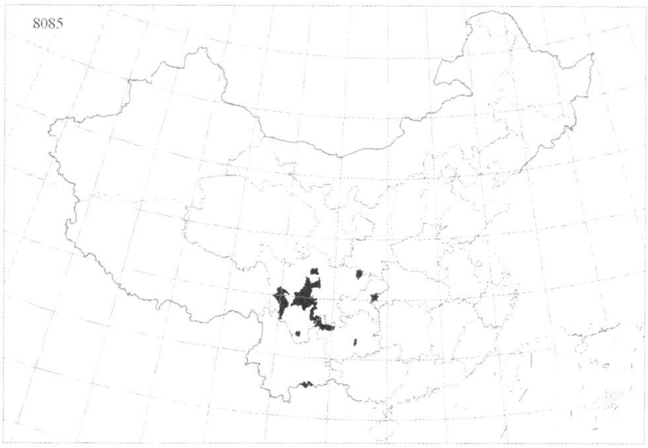

Ericaceae 杜鹃花科 *Rhododendron* 杜鹃花属
Rhododendron maculiferum Franch. 麻花杜鹃
Shrubs, evergreen

13.8 (6.2~18.2); 2.4 (-4.2~7.8); 24.7 (15.8~29.1); 13.9 (7.0~18.2); 112.6 (45.2~158.8); 6.2 (0.0~29.9); 777 (522~984) ❙ 1292 (669~1997); 502 (317~720); 120 (14~259) ❙ 63.8 (-1.2~130.7); 775 (522~927); 251.8 (140.5~334.4)

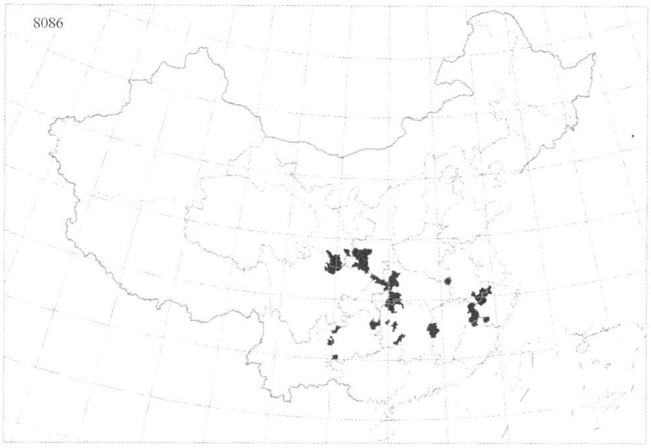

Ericaceae 杜鹃花科 *Rhododendron* 杜鹃花属
Rhododendron maculiferum subsp. anwheiense (E. H. Wilson) D. F. Chamb. 黄山杜鹃
Shrubs, evergreen

15.1 (13.4~16.1); 3.0 (1.5~4.9); 26.9 (25.2~28.0); 15.1 (13.4~16.1); 124.8 (108.3~133.8); 3.1 (0.1~7.3); 834 (755~886) ❙ 1497 (1198~1893); 585 (465~677); 167 (109~227) ❙ 79.9 (39.7~129.9); 834 (755~886); 234.1 (169.3~319.9)

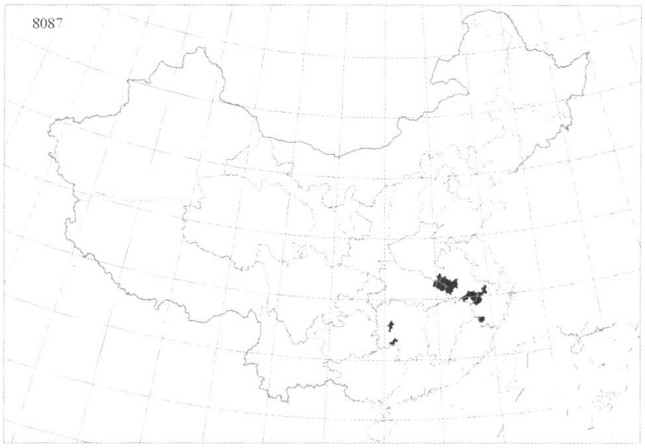

Ericaceae 杜鹃花科 *Rhododendron* 杜鹃花属
Rhododendron maddenii Hook. 隐脉杜鹃
Shrubs or small trees, evergreen

12.6 (2.1~18.6); 5.4 (-7.4~12.7); 18.2 (10.6~23.2); 12.8 (4.1~18.7); 99.1 (19.6~163.9); 7.3 (0.0~53.8); 692 (410~888) ❙ 1155 (286~1547); 659 (211~911); 45 (1~70) ❙ 65.6 (-30.3~113.0); 681 (286~888); 372.2 (174.2~551.3)

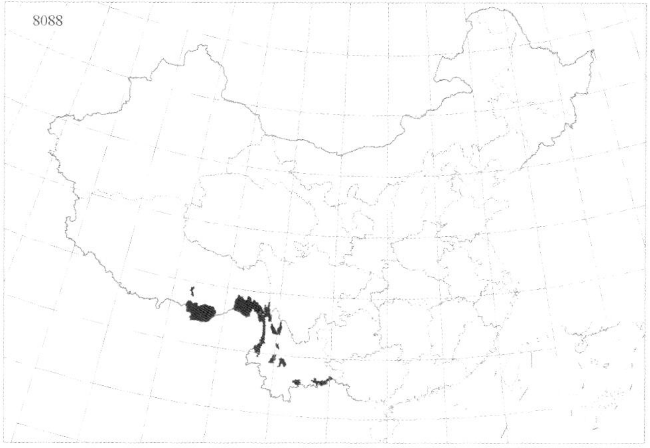

Ericaceae 杜鹃花科 *Rhododendron* 杜鹃花属
Rhododendron maddenii subsp. crassum (Franch.) Cullen 滇隐脉杜鹃
Shrubs or small trees, evergreen

15.7 (2.1~21.0); 8.7 (-7.4~14.2); 20.9 (10.6~25.6); 16.1 (4.1~21.0); 139.7 (19.6~191.8); 10.8 (0.0~53.8); 825 (410~1053) ▌ 1230 (286~1547); 706 (211~911); 50 (1~68) ▌ 40.8 (-30.3~74.2); 800 (286~1053); 454.9 (174.2~551.6)

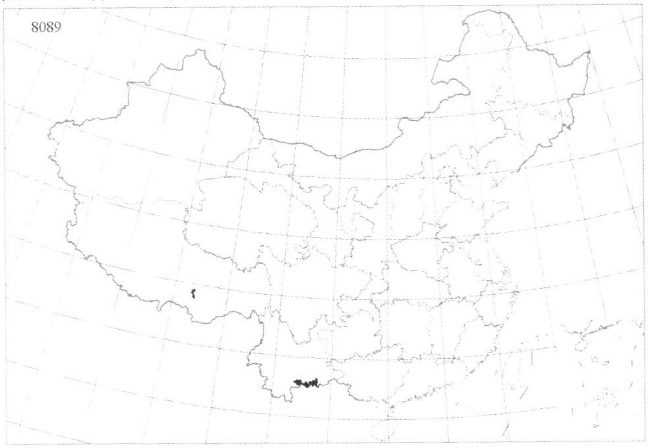

Ericaceae 杜鹃花科 *Rhododendron* 杜鹃花属
Rhododendron maguanense K. M. Feng 马关杜鹃
Shrubs, evergreen

17.4; 9.9; 22.8; 17.5; 149.8; 0.0; 846 ▌ 1463; 837; 54 ▌ 73.1; 846; 386.6

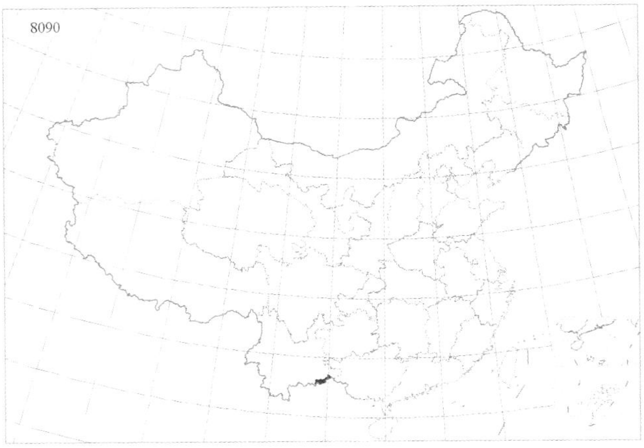

Ericaceae 杜鹃花科 *Rhododendron* 杜鹃花属
Rhododendron mainlingense S. H. Huang et R. C. Fang 米林杜鹃
Small shrubs, evergreen

3.3; -5.4; 11.1; 4.6; 22.3; 41.8; 427 ▌ 473; 287; 7 ▌ 5.0; 427; 230.7

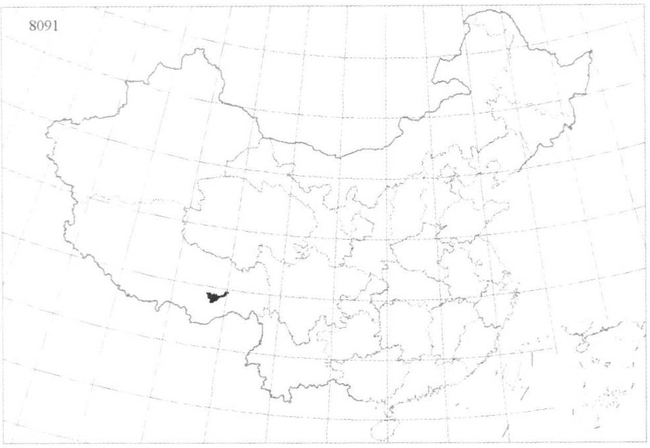

Ericaceae 杜鹃花科 *Rhododendron* 杜鹃花属
Rhododendron malipoense M. Y. He 麻栗坡杜鹃
Shrubs, semievergreen or green

17.6; 9.9; 23.2; 17.7; 152.4; 0.0; 859 ▌ 1422; 813; 52 ▌ 65.6; 859; 359.7

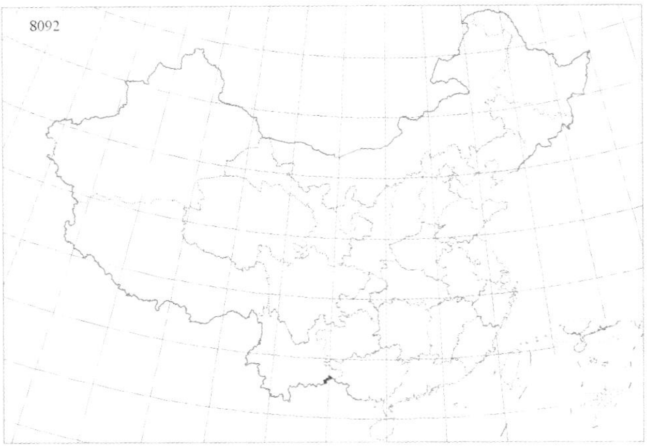

Ericaceae 杜鹃花科 *Rhododendron* 杜鹃花属
Rhododendron mallotum I. B. Balfour et Kingdon 羊毛杜鹃
Shrubs or small trees, evergreen

15.3; 8.7; 20.0; 15.3; 123.7; 0.0; 763 ▌ 1193; 662; 57 ▌ 59.3; 739; 386.2

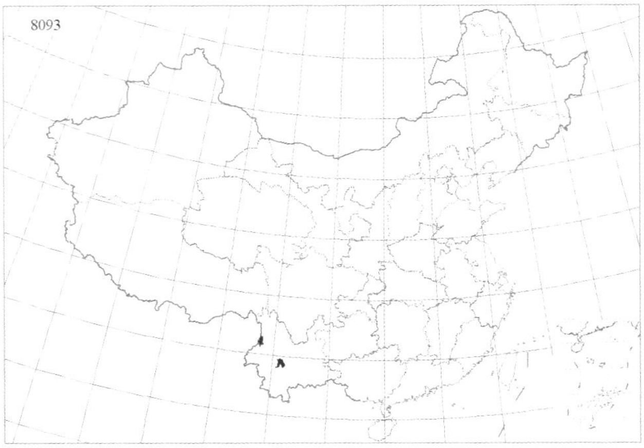

Ericaceae 杜鹃花科 *Rhododendron* 杜鹃花属
Rhododendron maoerense D. Fang et Q. Z. Li 猫儿山杜鹃
Trees, evergreen

15.8; 5.0; 25.7; 15.9; 131.0; 0.8; 835 ▌ 1598; 569; 191 ▌ 92.1; 835; 249.4

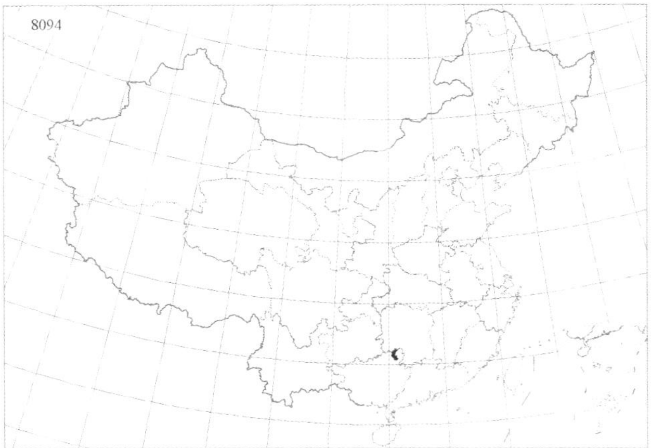

Ericaceae 杜鹃花科 *Rhododendron* 杜鹃花属
Rhododendron maowenense Ching et H. P. Yang 茂汶杜鹃
Shrubs, evergreen

8.3; −1.2; 16.9; 8.8; 59.6; 19.1; 578 ❙ 1008; 534; 29 ❙ 77.9; 578; 211.5

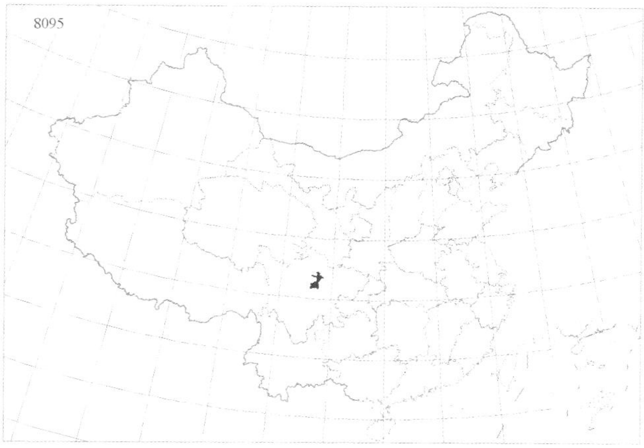

Ericaceae 杜鹃花科 *Rhododendron* 杜鹃花属
Rhododendron mariae Hance 岭南杜鹃
Shrubs, deciduous

19.9 (14.3~23.4); 10.9 (3.3~16.1); 27.4 (24.3~29.8); 19.9 (14.3~23.5); 178.9 (114.1~221.7); 0.0 (0.0~2.3); 1037 (765~1273) ❙ 1630 (1089~2035); 676 (445~1056); 155 (43~254) ❙ 58.4 (5.8~130.7); 1036 (765~1273); 244.6 (139.6~371.7)

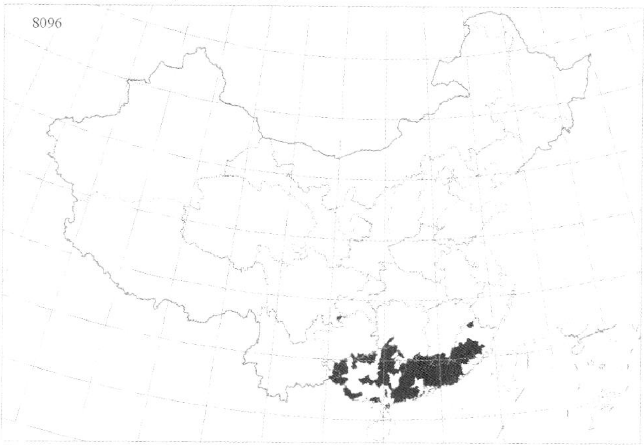

Ericaceae 杜鹃花科 *Rhododendron* 杜鹃花属
Rhododendron mariesii Hemsl. et E. H. Wilson 满山红
Shrubs, deciduous

16.2 (3.9~22.0); 4.9 (−13.9~15.3); 26.8 (16.6~30.1); 16.3 (7.8~22.7); 137.5 (58.3~212.7); 2.6 (0.0~71.1); 879 (560~1216) ❙ 1333 (358~2993); 521 (234~1186); 139 (10~551) ❙ 51.2 (−37.1~246.5); 865 (358~1216); 211.3 (105.6~354.5)

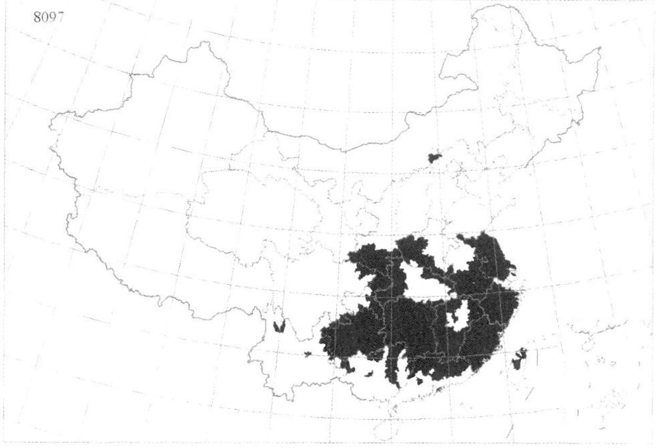

Ericaceae 杜鹃花科 *Rhododendron* 杜鹃花属
Rhododendron martinianum I. B. Balfour et Forrest 少花杜鹃
Shrubs, evergreen

9.4 (5.3~16.0); 1.9 (−2.4~8.0); 15.6 (12.2~21.8); 9.5 (5.7~16.0); 64.0 (30.5~132.2); 11.1 (0.0~26.2); 595 (479~793) ❙ 1017 (741~1243); 575 (426~696); 42 (28~58) ❙ 72.8 (39.7~100.7); 595 (479~793); 315.8 (234.2~384.7)

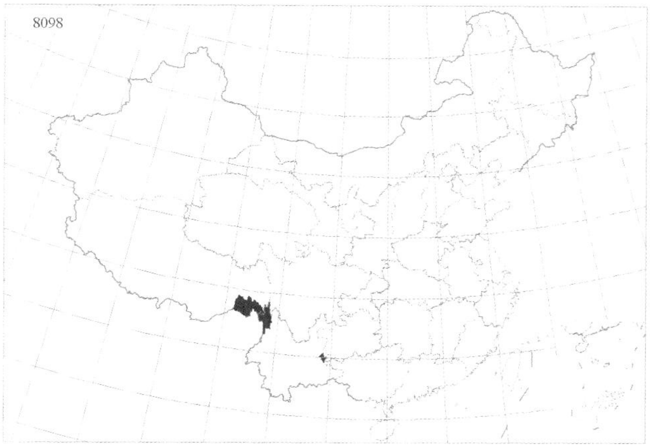

Ericaceae 杜鹃花科 *Rhododendron* 杜鹃花属
Rhododendron meddianum Forrest 红萼杜鹃
Shrubs, evergreen

12.4; 5.5; 17.7; 12.4; 91.1; 1.8; 674 ❙ 1266; 699; 60 ❙ 86.7; 674; 415.4

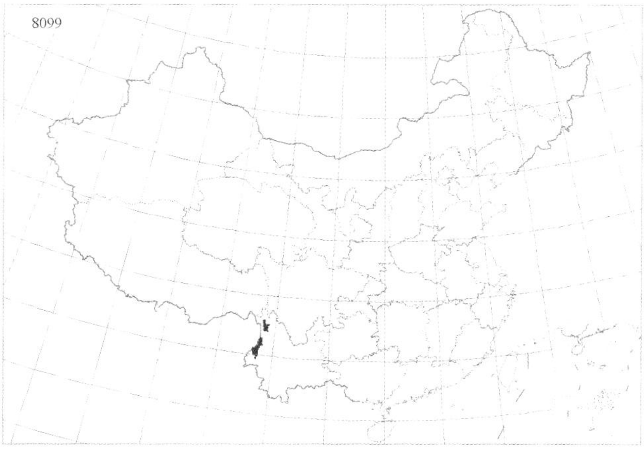

Ericaceae 杜鹃花科 *Rhododendron* 杜鹃花属
Rhododendron meddianum var. atrokermesinum Tagg 腺房红萼杜鹃
Shrubs, evergreen

13.2; 6.4; 18.4; 13.3; 99.4; 0.0; 695 ❙ 1337; 727; 69 ❙ 92.4; 695; 407.1

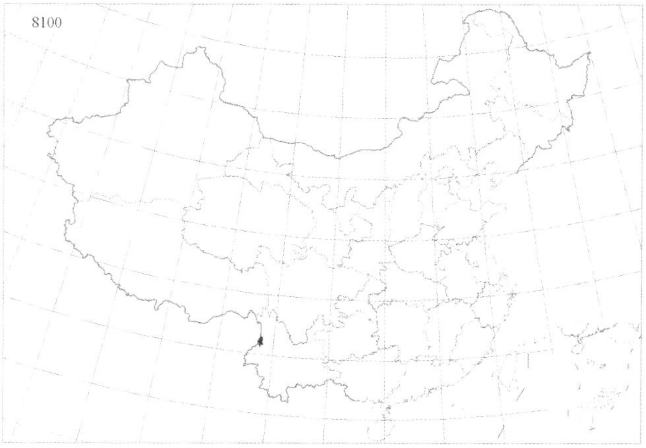

Ericaceae 杜鹃花科 *Rhododendron* 杜鹃花属

Rhododendron medoense Fang et M. Y. He 墨脱马银花

Shrubs, evergreen

13.3; 5.4; 19.6; 13.4; 100.4; 0.0; 706 ❙ 1736; 1007; 51 ❙ 145.8; 706; 447.4

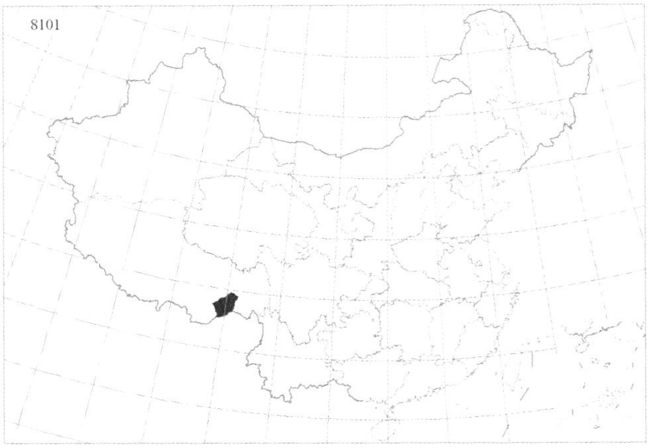

Ericaceae 杜鹃花科 *Rhododendron* 杜鹃花属

Rhododendron megacalyx I. B. Balfour et Kingdon-Ward 大萼杜鹃

Shrubs or small trees, evergreen

9.6; 2.4; 15.6; 9.7; 64.8; 8.9; 598 ❙ 1146; 642; 51 ❙ 91.9; 598; 344.6

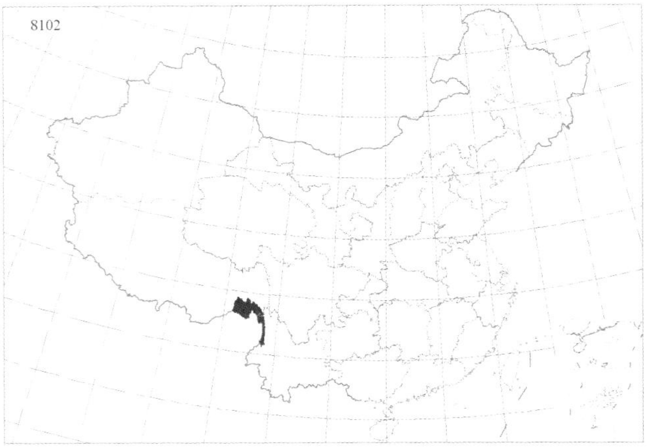

Ericaceae 杜鹃花科 *Rhododendron* 杜鹃花属

Rhododendron megalanthum M. Y. Fang 西藏杜鹃

Shrubs or small trees, evergreen

13.3; 5.4; 19.6; 13.4; 100.4; 0.0; 706 ❙ 1736; 1007; 51 ❙ 145.8; 706; 447.4

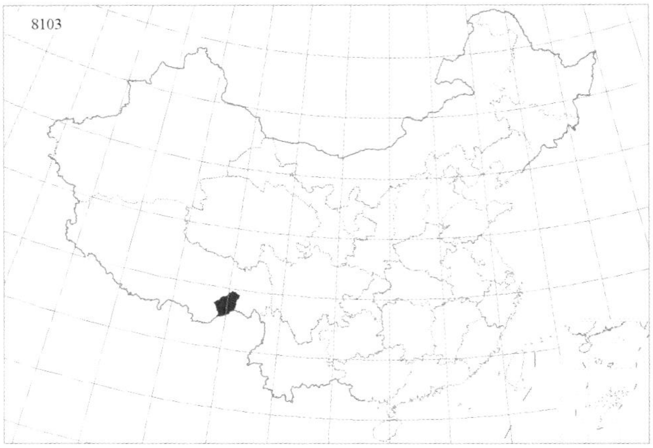

Ericaceae 杜鹃花科 *Rhododendron* 杜鹃花属

Rhododendron megeratum I. B. Balfour et Forrest 招展杜鹃

Small shrubs, evergreen

5.9 (1.9~9.6); -2.0 (-7.2~2.4); 12.9 (10.0~15.7); 6.5 (3.8~9.6); 36.7 (16.6~60.8); 25.7 (5.4~53.6); 499 (403~600) ❙ 772 (341~1041); 452 (226~585); 26 (4~45) ❙ 51.2 (-15.3~100.7); 488 (341~600); 266.2 (168.2~352.3)

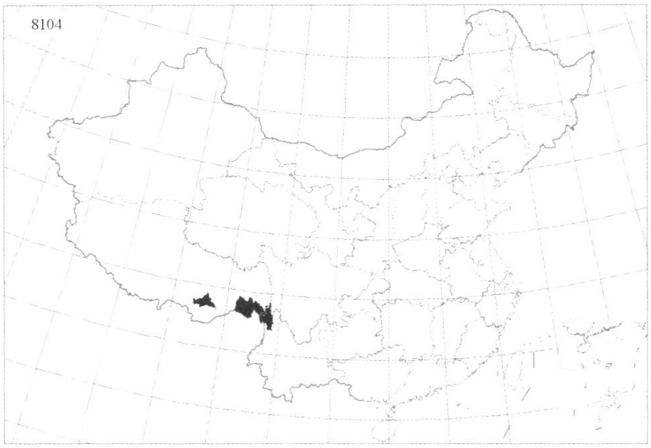

Ericaceae 杜鹃花科 *Rhododendron* 杜鹃花属

Rhododendron mekongense Franch. 弯月杜鹃

Shrubs, deciduous, semievergreen or evergreen

4.1 (-1.4~14.4); -4.4 (-11.4~7.5); 11.7 (7.9~19.1); 5.6 (2.4~14.4); 31.5 (6.6~113.1); 41.9 (0.0~82.3); 463 (341~727) ❙ 729 (510~1454); 425 (311~792); 24 (8~70) ❙ 53.0 (20.7~100.7); 463 (341~727); 259.2 (111.0~487.2)

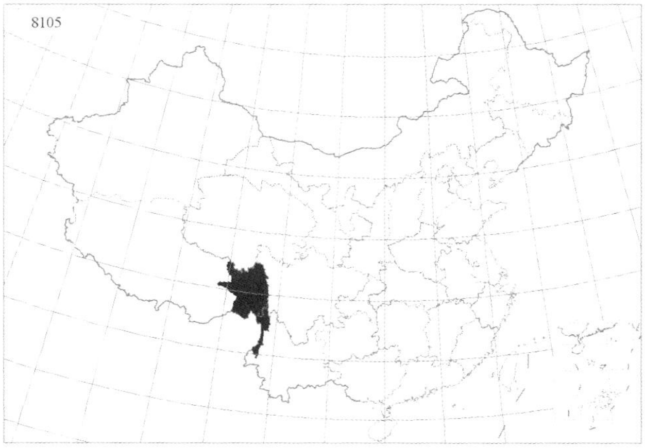

Ericaceae 杜鹃花科 *Rhododendron* 杜鹃花属

Rhododendron mekongense var. longipilosum (Cowan) Cullen 长毛弯月杜鹃

Shrubs, deciduous, semievergreen or evergreen

10.3; 2.4; 16.7; 10.4; 72.8; 9.0; 619 ❙ 1376; 794; 38 ❙ 119.8; 619; 348.4

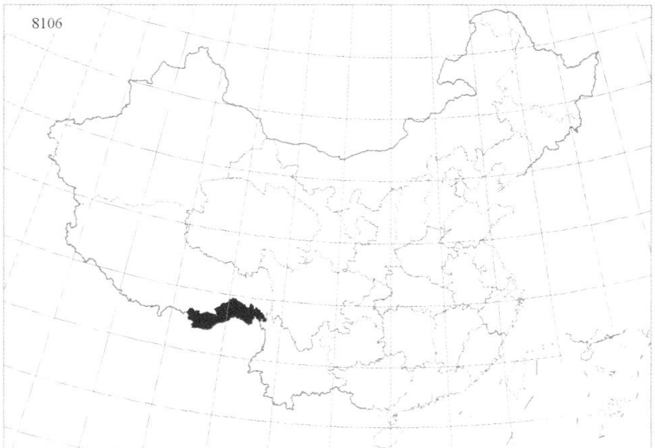

Ericaceae 杜鹃花科 *Rhododendron* 杜鹃花属
Rhododendron mekongense var. melinanthum (I. B. Balfour et Kingdon-Ward) Cullen 密花弯月杜鹃
Shrubs, deciduous, semievergreen or evergreen

6.8; -0.7; 13.4; 7.1; 40.9; 19.1; 520 ❙ 903; 521; 33 ❙ 74.2; 520; 286.5

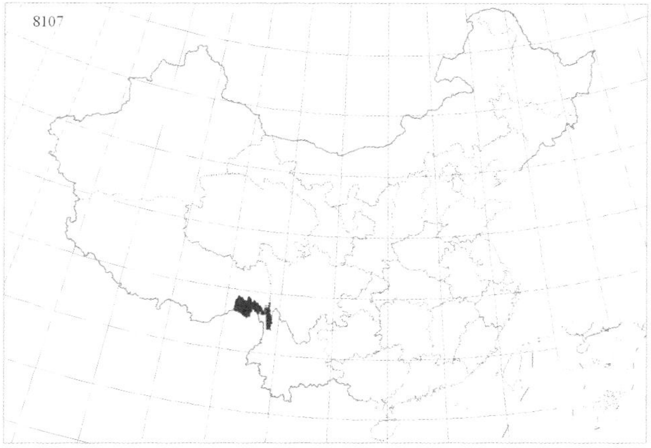

Ericaceae 杜鹃花科 *Rhododendron* 杜鹃花属
Rhododendron mekongense var. rubrolineatum (I. B. Balfour et Forrest) Cullen 红线弯月杜鹃
Shrubs, deciduous, semievergreen or evergreen

6.8; -0.7; 13.4; 7.1; 40.9; 19.1; 520 ❙ 903; 521; 33 ❙ 74.2; 520; 286.5

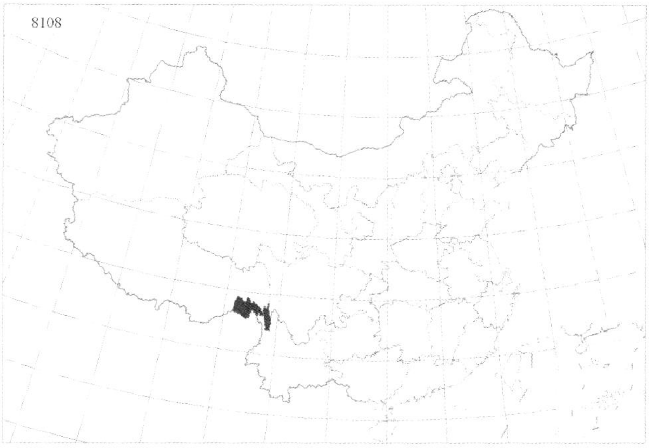

Ericaceae 杜鹃花科 *Rhododendron* 杜鹃花属
Rhododendron mengtszense I. B. Balfour et W. W. Smith 蒙自杜鹃
Shrubs or small trees, evergreen

17.5 (16.5~19.1); 10.3 (9.0~13.0); 22.5 (21.4~23.3); 17.5 (16.6~19.1); 150.2 (139.2~169.7); 0.0 (0.0~0.0); 847 (810~917) ❙ 1260 (1018~1428); 710 (554~835); 51 (48~56) ❙ 48.3 (25.7~65.6); 847 (810~917); 379.4 (311.3~506.2)

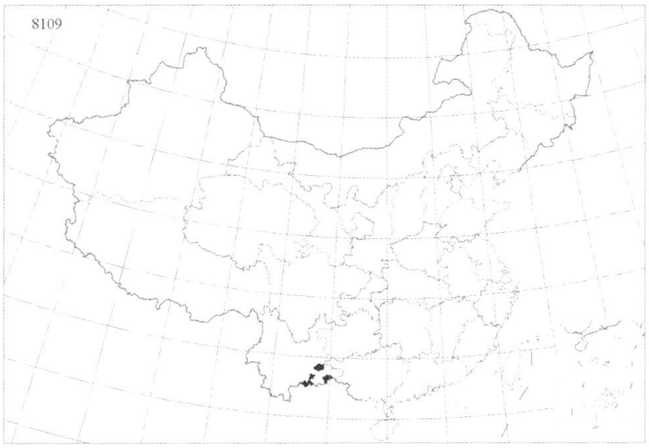

Ericaceae 杜鹃花科 *Rhododendron* 杜鹃花属
Rhododendron meridionale P. C. Tam 南边杜鹃
Shrubs, semievergreen or green

20.6 (17.6~22.8); 11.9 (7.3~15.0); 27.5 (26.3~28.9); 20.6 (17.6~22.8); 187.7 (151.6~213.6); 0.0 (0.0~0.0); 1084 (909~1230) ❙ 1780 (1495~2435); 916 (548~1428); 130 (106~175) ❙ 64.1 (33.9~99.5); 1084 (909~1230); 271.7 (225.9~318.8)

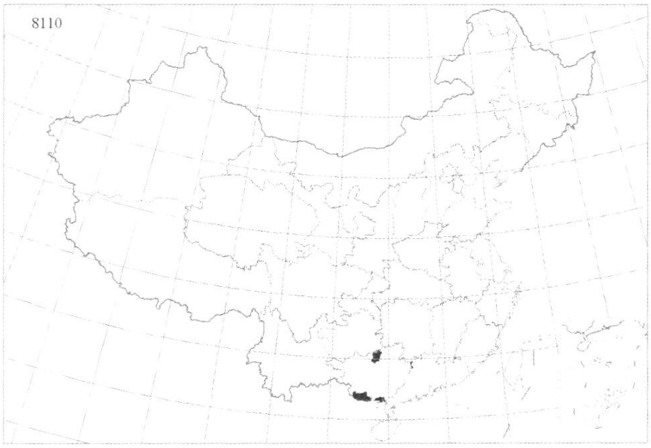

Ericaceae 杜鹃花科 *Rhododendron* 杜鹃花属
Rhododendron meridionale var. minus P. C. Tam 狭叶南边杜鹃
Shrubs

19.2; 9.2; 27.5; 19.2; 170.8; 0.0; 1002 ❙ 1495; 678; 135 ❙ 49.1; 1002; 256.4

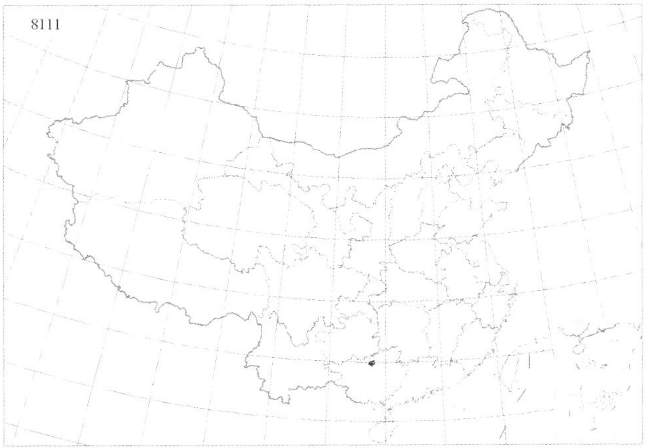

Ericaceae 杜鹃花科 *Rhododendron* 杜鹃花属
Rhododendron meridionale var. setistylum P. C. Tam 糙柱杜鹃
Shrubs, semievergreen or green

20.9; 12.6; 27.5; 21.0; 191.6; 0.0; 1100 ❙ 2006; 1067; 138 ❙ 81.2; 1100; 295.0

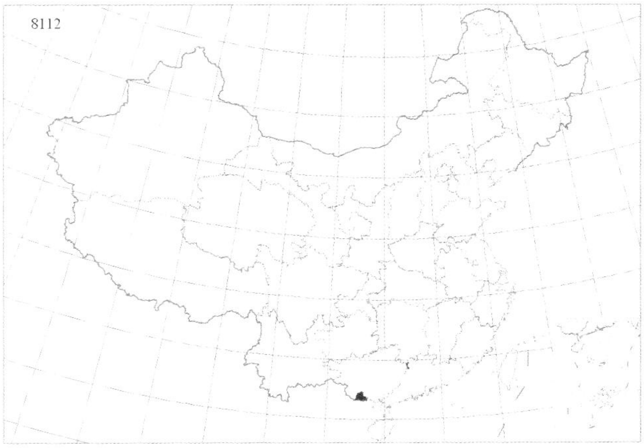

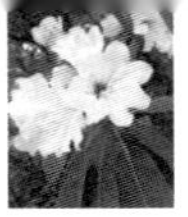

Ericaceae 杜鹃花科 *Rhododendron* 杜鹃花属
Rhododendron mianningense Z. J. Zhao 冕宁杜鹃
Shrubs, evergreen

10.2; 2.4; 16.9; 10.3; 68.2; 5.1; 620 ❙ 1036; 606; 18 ❙ 67.1; 620; 298.0

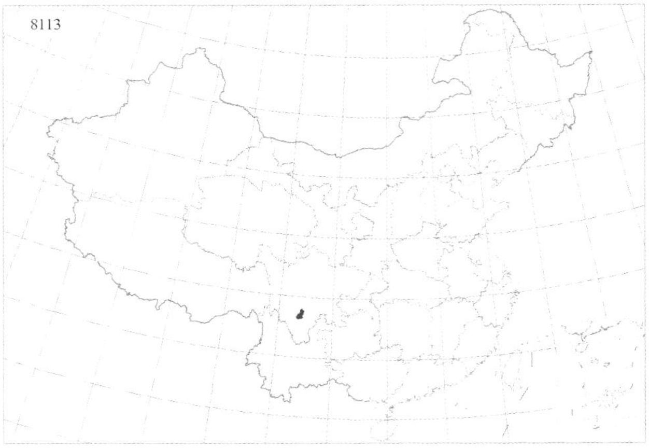

Ericaceae 杜鹃花科 *Rhododendron* 杜鹃花属
Rhododendron micranthum Turcz. 照山白
Shrubs, evergreen

10.7 (−5.8~16.9); −4.3 (−29.8~4.4); 24.2 (11.2~29.2); 11.8 (4.3~16.9); 96.7 (20.5~143.8); 28.0 (0.6~157.4); 724 (426~936) ❙ 682 (249~1621); 380 (179~666); 32 (5~195) ❙ −5.6 (−54.1~121.7); 620 (249~930); 185.9 (70.3~338.4)

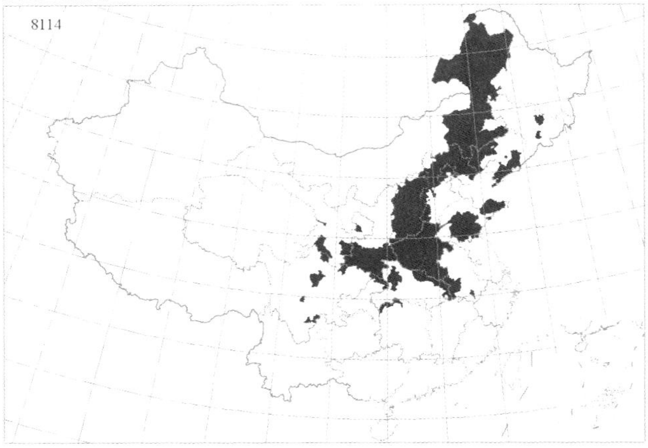

Ericaceae 杜鹃花科 *Rhododendron* 杜鹃花属
Rhododendron microgynum I. B. Balfour et Forrest 短蕊杜鹃
Shrubs, evergreen

5.4; −2.3; 12.2; 5.8; 31.0; 25.9; 480 ❙ 851; 492; 30 ❙ 77.4; 480; 253.8

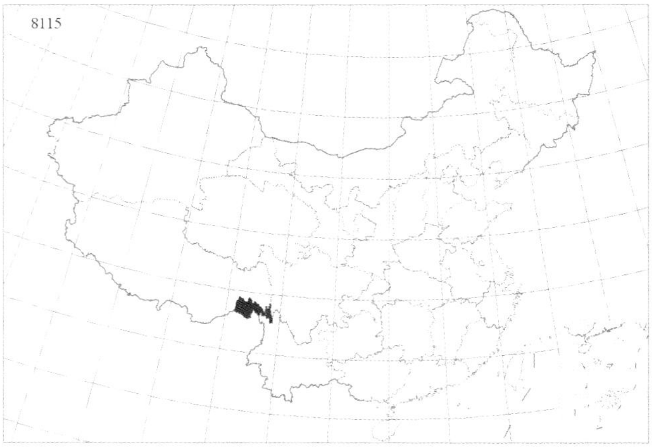

Ericaceae 杜鹃花科 *Rhododendron* 杜鹃花属
Rhododendron microphyton Franch. 亮毛杜鹃
Shrubs, evergreen

15.7 (5.4~21.3); 7.9 (−2.1~14.2); 21.9 (12.3~28.6); 15.7 (5.8~21.3); 129.5 (31.5~196.1); 0.9 (0.0~25.6); 807 (481~1130) ❙ 1128 (741~1758); 610 (411~821); 57 (15~189) ❙ 40.4 (−14.2~99.9); 787 (481~1130); 287.3 (131.4~551.6)

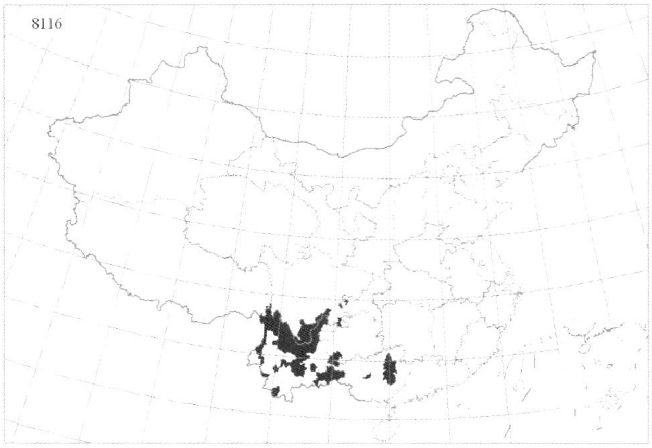

Ericaceae 杜鹃花科 *Rhododendron* 杜鹃花属
Rhododendron microphyton var. trichanthum A. L. Chang ex R. C. Fang 碧江亮毛杜鹃 Shrubs, evergreen

12.3; 5.4; 17.7; 12.3; 88.0; 0.3; 670 ❙ 1290; 712; 63 ❙ 92.7; 670; 395.9

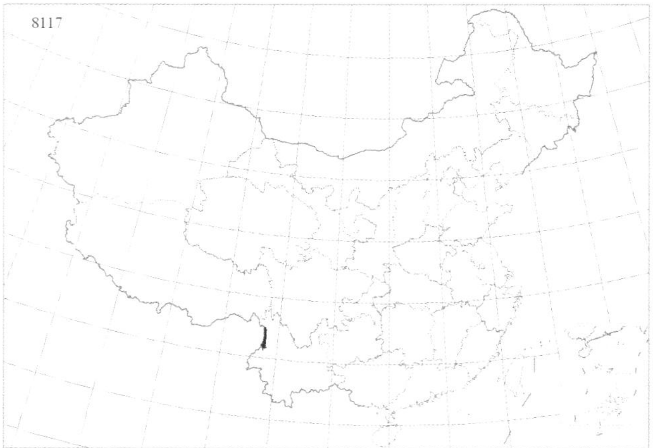

Ericaceae 杜鹃花科 *Rhododendron* 杜鹃花属
Rhododendron mimetes Tagg et Forrest 优异杜鹃
Shrubs, evergreen

11.3; 3.6; 17.6; 11.4; 86.3; 9.9; 655 ❙ 938; 589; 16 ❙ 47.5; 616; 320.7

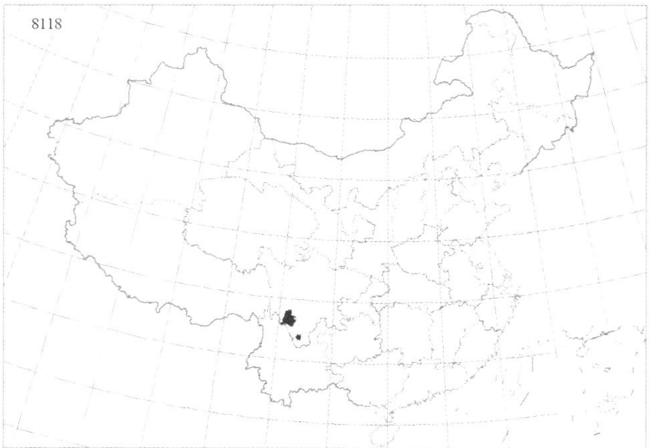

Ericaceae 杜鹃花科 *Rhododendron* 杜鹃花属

Rhododendron miniatum Cowan 焰红杜鹃

Shrubs, evergreen

1.9; -7.2; 10.0; 3.8; 16.6; 53.6; 403 ❙ 341; 226; 4 ❙ -15.3; 341; 217.5

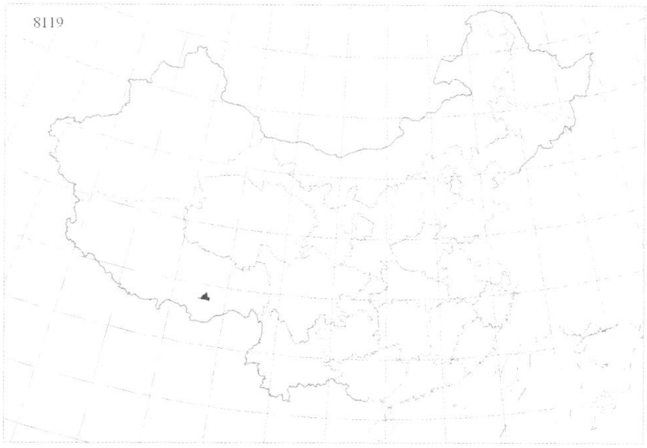

Ericaceae 杜鹃花科 *Rhododendron* 杜鹃花属

Rhododendron minutiflorum Hu 小花杜鹃

Shrubs, semievergreen or green

19.6 (15.5~21.8); 10.5 (4.8~14.1); 27.2 (23.4~29.8); 19.7 (15.6~21.8); 176.1 (127.2~201.6); 0.0 (0.0~0.2); 1022 (815~1162) ❙ 1592 (1232~1991); 676 (508~902); 153 (58~195) ❙ 56.4 (30.5~93.2); 1019 (815~1162); 244.5 (174.9~318.5)

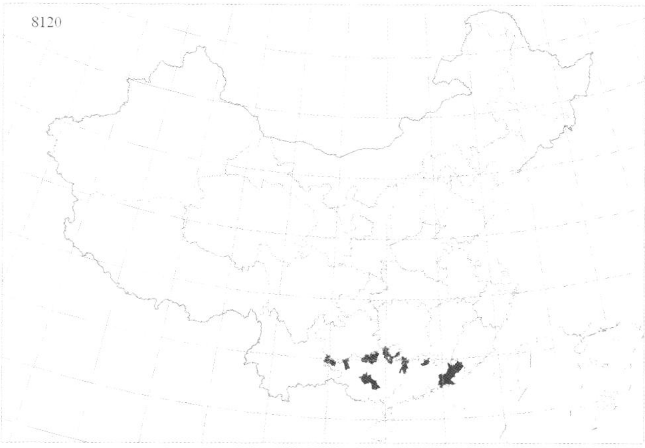

Ericaceae 杜鹃花科 *Rhododendron* 杜鹃花属

Rhododendron minyaense Philipson et M. N. Philipson 黄褐杜鹃

Small shrubs, evergreen

2.0; -7.4; 10.3; 3.9; 16.6; 51.5; 416 ❙ 852; 470; 19 ❙ 105.0; 416; 284.6

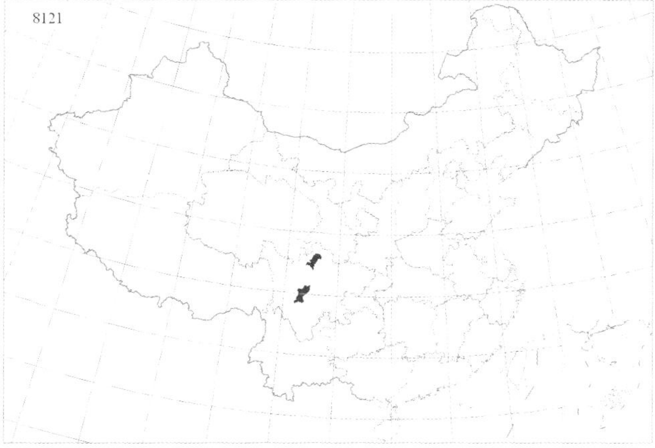

Ericaceae 杜鹃花科 *Rhododendron* 杜鹃花属

Rhododendron mitriforme Tam 头巾马银花

Shrubs or small trees, evergreen

19.2 (17.3~21.6); 9.6 (6.9~12.6); 27.2 (25.8~28.5); 19.2 (17.3~21.7); 170.9 (148.1~199.8); 0.0 (0.0~0.0); 1000 (896~1153) ❙ 1551 (1498~1625); 629 (517~795); 171 (161~185) ❙ 56.2 (40.9~69.8); 1000 (896~1153); 234.3 (214.2~243.9)

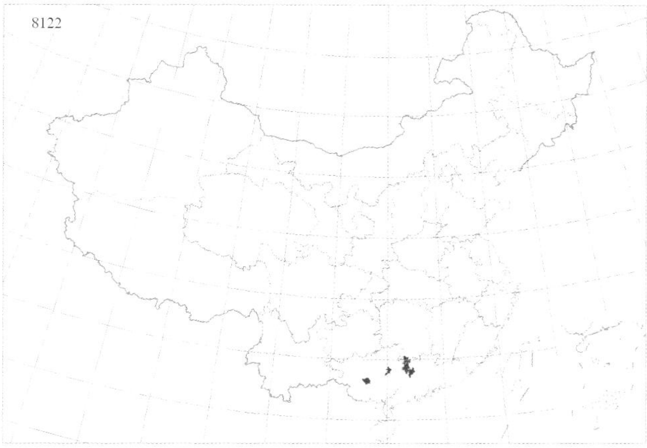

Ericaceae 杜鹃花科 *Rhododendron* 杜鹃花属

Rhododendron mitriforme var. setaceum Tam 腺刺马银花

Shrubs or small trees, evergreen

15.7; 4.9; 25.5; 15.8; 129.8; 0.6; 828 ❙ 1590; 576; 186 ❙ 92.4; 828; 251.9

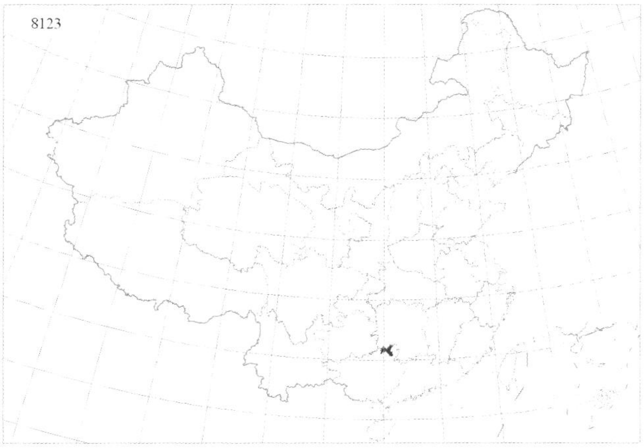

Ericaceae 杜鹃花科 *Rhododendron* 杜鹃花属

Rhododendron miyiense W. K. Hu 米易杜鹃

Shrubs, evergreen

15.7; 8.4; 21.5; 15.8; 129.2; 0.0; 784 ❙ 998; 620; 19 ❙ 27.4; 707; 320.1

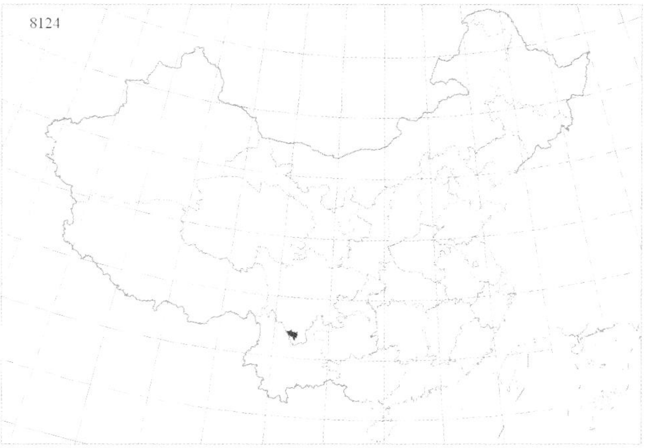

Ericaceae 杜鹃花科 *Rhododendron* 杜鹃花属

Rhododendron molle (Blume) G. Don 羊踯躅

Shrubs, deciduous

16.6 (9.1~22.3); 5.0 (-0.9~14.0); 27.5 (18.1~30.1); 16.6 (9.2~22.4); 140.8 (64.1~208.3); 1.3 (0.0~14.1); 898 (595~1200) ❙ 1405 (750~1997); 528 (323~779); 159 (19~259) ❙ 56.6 (-12.9~137.5); 887 (595~1200); 202.9 (108.2~319.9)

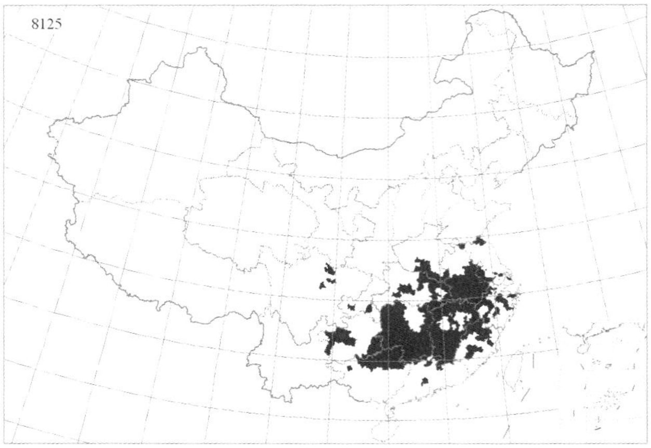

Ericaceae 杜鹃花科 *Rhododendron* 杜鹃花属

Rhododendron mollicomum I. B. Balfour et W. W. Sm. 柔毛碎米花

Shrubs, evergreen

10.3; 2.8; 16.6; 10.4; 71.3; 7.0; 620 ❙ 957; 578; 20 ❙ 55.4; 614; 301.7

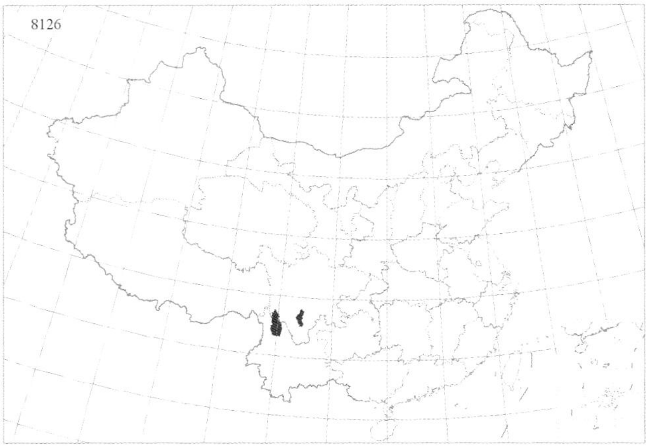

Ericaceae 杜鹃花科 *Rhododendron* 杜鹃花属

Rhododendron monanthum I. B. Balfour et W. W. Sm. 一朵花杜鹃

Small shrubs, evergreen

8.7 (2.9~13.2); 1.3 (-5.7~6.4); 15.0 (10.6~18.4); 9.0 (4.3~13.3); 58.4 (19.9~99.4); 13.7 (0.0~45.2); 574 (421~695) ❙ 1014 (535~1337); 581 (326~727); 39 (10~69) ❙ 74.6 (27.1~100.7); 574 (421~695); 322.2 (234.2~407.1)

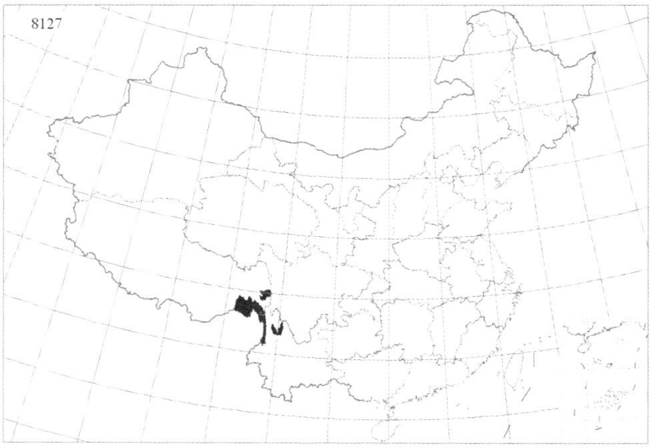

Ericaceae 杜鹃花科 *Rhododendron* 杜鹃花属

Rhododendron montiganum T. L. Ming 山地杜鹃

Shrubs or small trees

13.8; 6.7; 19.3; 13.8; 105.9; 0.0; 714 ❙ 963; 592; 29 ❙ 35.0; 690; 266.2

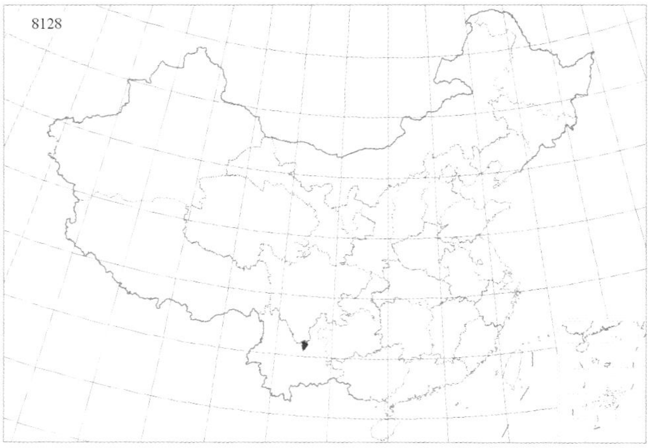

Ericaceae 杜鹃花科 *Rhododendron* 杜鹃花属

Rhododendron montroseanum Davidian 墨脱杜鹃

Trees, evergreen

11.0; 3.4; 17.3; 11.0; 76.5; 4.4; 640 ❙ 1389; 796; 48 ❙ 113.7; 640; 399.9

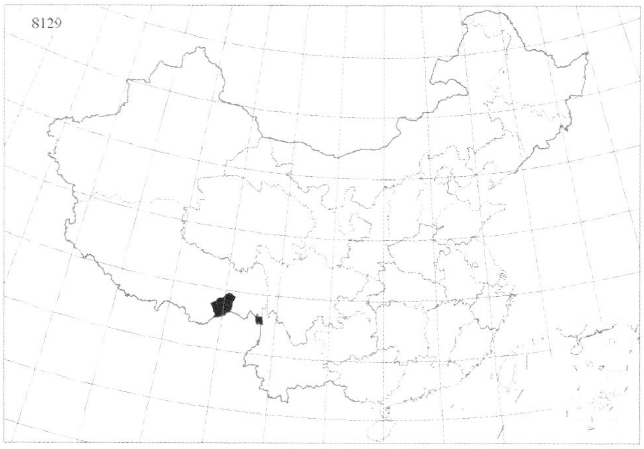

Ericaceae 杜鹃花科 *Rhododendron* 杜鹃花属

Rhododendron morii Hayata 玉山杜鹃

Shrubs or small trees, evergreen

18.5; 13.2; 22.5; 18.5; 161.7; 0.0; 896 ❙ 2563; 1299; 178 ❙ 193.3; 896; 248.4

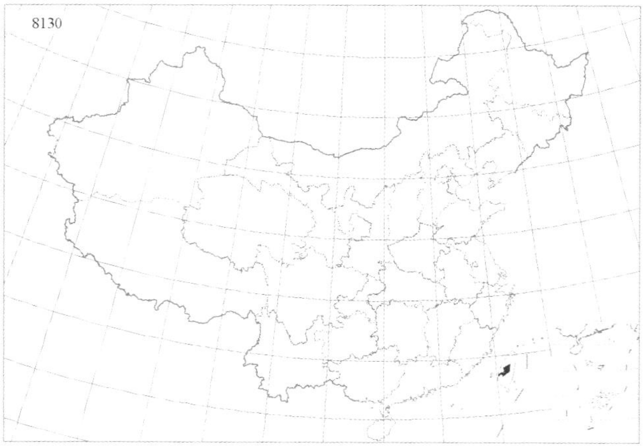

Ericaceae 杜鹃花科 *Rhododendron* 杜鹃花属
Rhododendron moulmainense Hook. 毛棉杜鹃
Shrubs or small trees, evergreen

18.8 (5.3~24.8); 9.6 (-2.4~20.1); 26.5 (12.2~29.8); 18.8 (5.7~24.8); 166.6 (30.5~237.6); 0.6 (0.0~26.2); 991 (479~1385) ❙ 1498 (868~2435); 637 (312~1428); 130 (31~230) ❙ 53.7 (-34.3~128.9); 979 (479~1360); 265.7 (139.6~620.0)

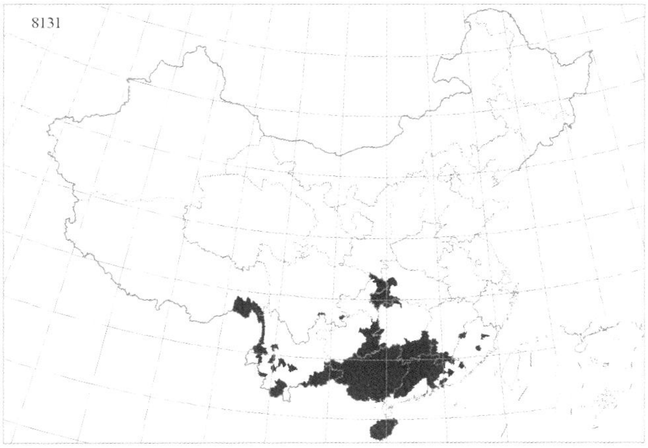

Ericaceae 杜鹃花科 *Rhododendron* 杜鹃花属
Rhododendron moupinense Franch. 宝兴杜鹃
Shrubs, evergreen

12.4 (6.0~15.2); 2.5 (-3.2~5.3); 21.6 (14.4~25.9); 12.5 (6.6~15.2); 95.2 (39.7~123.9); 5.7 (0.0~26.7); 703 (508~816) ❙ 1146 (885~1466); 559 (473~816); 54 (27~105) ❙ 64.5 (26.5~106.2); 703 (508~816); 229.3 (170.4~292.6)

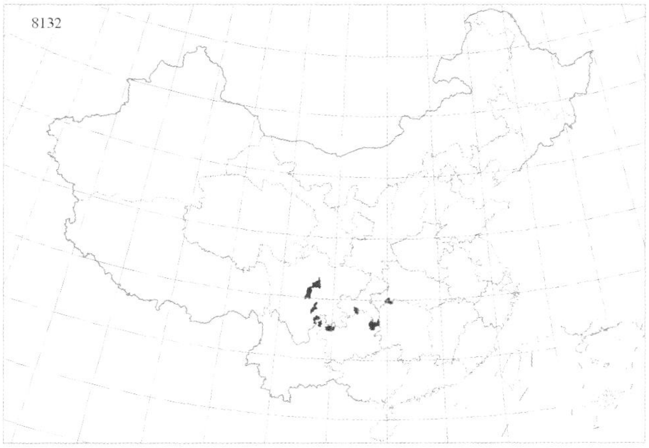

Ericaceae 杜鹃花科 *Rhododendron* 杜鹃花属
Rhododendron mucronatum (Blume) G. Don 白花杜鹃
Shrubs, semievergreen

16.8 (14.3~18.0); 6.5 (2.0~15.2); 27.9 (25.8~29.7); 17.5 (14.3~22.7); 150.3 (117.5~212.7); 0.8 (0.0~5.4); 937 (801~1216) ❙ 1635 (1112~1997); 648 (403~994); 174 (104~254) ❙ 76.1 (13.2~130.7); 921 (801~1216); 237.5 (175.3~303.4)

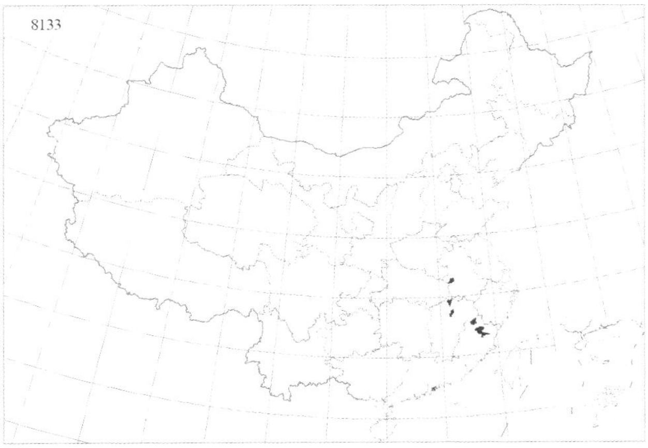

Ericaceae 杜鹃花科 *Rhododendron* 杜鹃花属
Rhododendron mucronulatum Turcz. 迎红杜鹃
Shrubs, deciduous

9.6 (-4.6~15.8); -6.3 (-28.8~2.5); 23.7 (16.6~28.0); 11.0 (5.0~15.8); 90.6 (34.4~133.1); 35.1 (3.4~149.1); 696 (448~891) ❙ 620 (350~1200); 368 (241~691); 24 (5~109) ❙ -10.5 (-47.7~74.1); 591 (350~891); 187.2 (70.3~346.5)

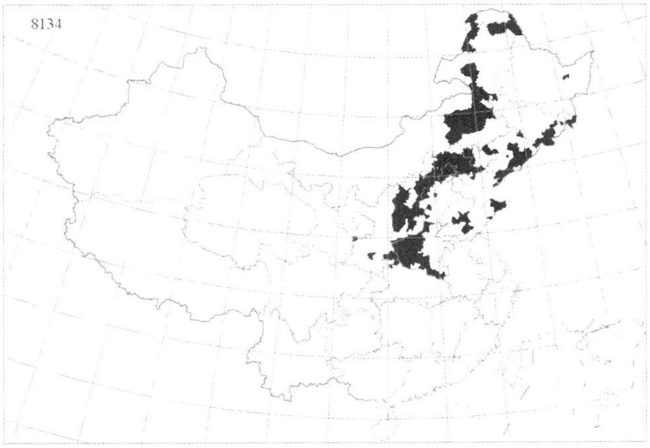

Ericaceae 杜鹃花科 *Rhododendron* 杜鹃花属
Rhododendron myrsinifolium Ching ex Fang et M. Y. He 铁仔叶杜鹃
Small shrubs, evergreen

21.9; 14.1; 27.9; 22.0; 203.7; 0.0; 1161 ❙ 2161; 1240; 126 ❙ 85.7; 1161; 305.5

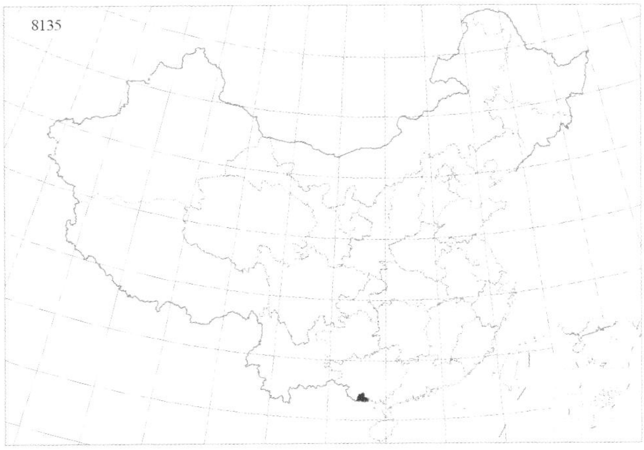

Ericaceae 杜鹃花科 *Rhododendron* 杜鹃花属
Rhododendron naamkwanense Merr. 南昆杜鹃
Small shrubs, semievergreen or green

20.0 (16.6~21.8); 11.2 (6.0~14.1); 27.5 (26.3~28.5); 20.1 (16.6~21.8); 181.0 (138.9~201.9); 0.0 (0.0~0.0); 1043 (866~1158) ❙ 1721 (1494~2005); 666 (539~902); 174 (136~254) ❙ 65.7 (39.6~130.7); 1043 (866~1158); 249.8 (201.9~303.4)

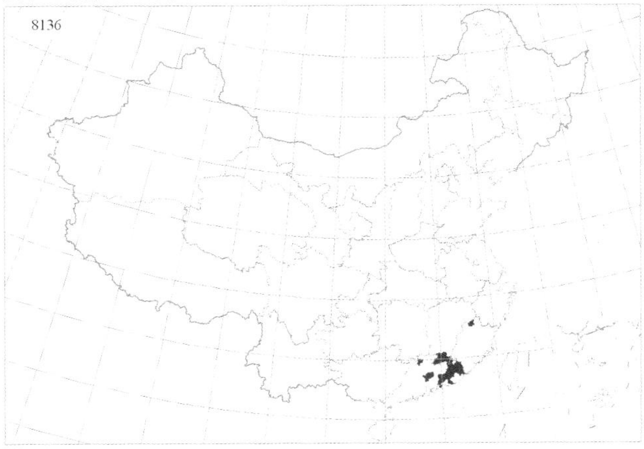

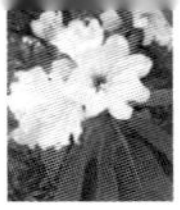

Ericaceae 杜鹃花科 *Rhododendron* 杜鹃花属

Rhododendron naamkwanense var. cryptonerve Tam 紫薇春

Small shrubs, semievergreen or green

18.9 (17.3~20.7); 9.1 (6.4~12.8); 27.5 (26.6~28.5); 19.0 (17.3~20.7); 167.8 (147.9~189.0); 0.0 (0.0~0.0); 989 (907~1083) ❙ 1659 (1494~2005); 600 (539~824); 185 (136~197) ❙ 68.2 (39.6~85.1); 989 (907~1083); 240.4 (208.1~262.4)

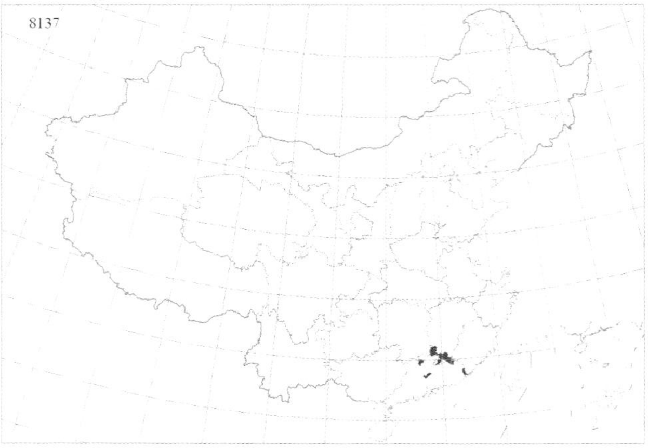

Ericaceae 杜鹃花科 *Rhododendron* 杜鹃花属

Rhododendron nakaharae Hayata 那克哈杜鹃

Shrubs, evergreen

19.9; 13.5; 25.9; 19.9; 179.1; 0.0; 987 ❙ 2993; 924; 547 ❙ 203.4; 987; 261.4

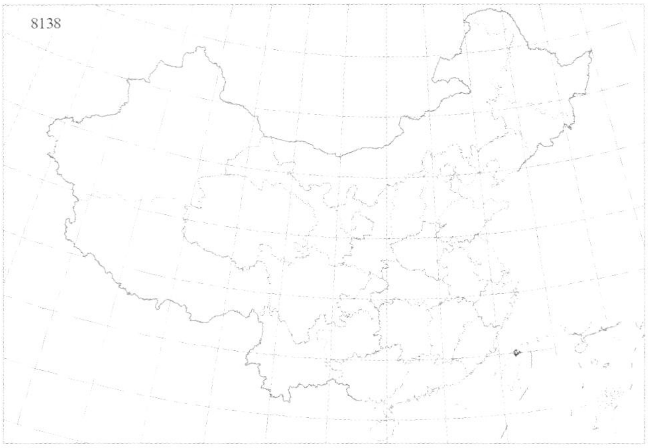

Ericaceae 杜鹃花科 *Rhododendron* 杜鹃花属

Rhododendron nakotiltum I. B. Balfour et Forrest 德钦杜鹃

Shrubs, evergreen

5.4; -2.1; 12.3; 5.8; 31.5; 25.6; 481 ❙ 741; 426; 28 ❙ 54.0; 481; 273.5

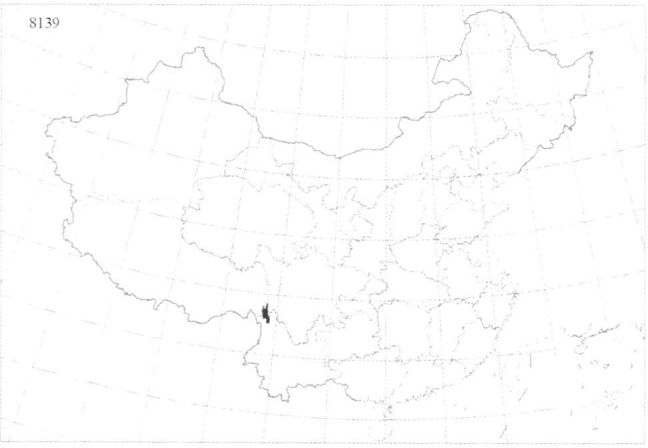

Ericaceae 杜鹃花科 *Rhododendron* 杜鹃花属

Rhododendron nanjianense K. M. Feng et Z. H. Yang 南涧杜鹃

Shrubs, evergreen

15.9; 9.5; 20.6; 16.0; 131.6; 0.0; 779 ❙ 1014; 568; 46 ❙ 30.2; 751; 318.7

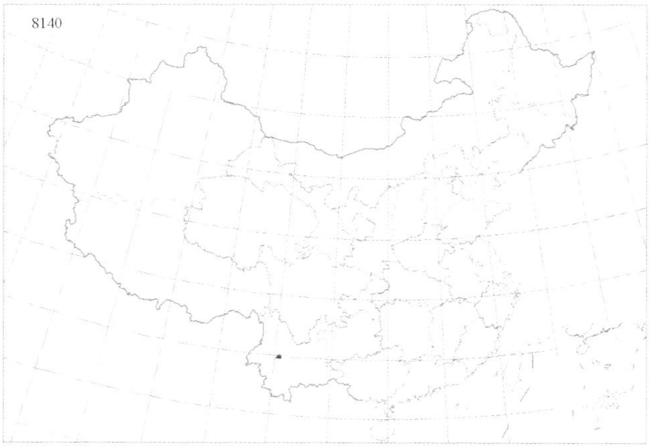

Ericaceae 杜鹃花科 *Rhododendron* 杜鹃花属

Rhododendron nanpingense Tam 南平杜鹃

Shrubs, semievergreen or green

17.6; 7.5; 27.0; 17.6; 151.7; 0.0; 917 ❙ 1845; 628; 227 ❙ 101.9; 917; 299.2

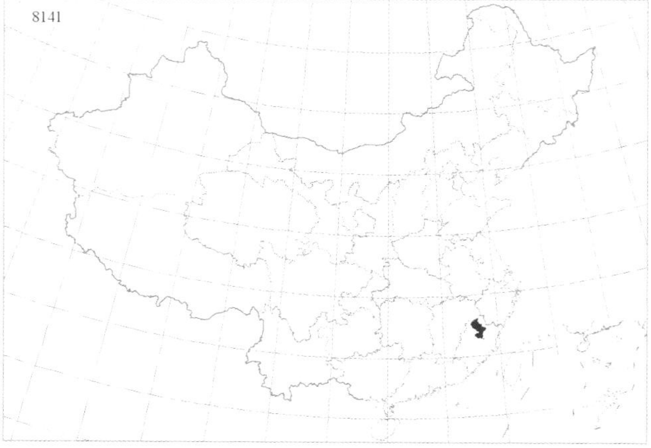

Ericaceae 杜鹃花科 *Rhododendron* 杜鹃花属

Rhododendron neriiflorum Franch. 火红杜鹃

Shrubs, evergreen

10.1 (1.9~17.3); 2.8 (-7.2~11.0); 16.1 (10.0~21.6); 10.4 (3.8~17.3); 73.5 (16.6~148.0); 11.5 (0.0~53.6); 615 (403~832) ❙ 1010 (341~1454); 576 (226~817); 39 (4~70) ❙ 61.0 (-15.3~113.0); 607 (341~783); 333.9 (168.2~487.2)

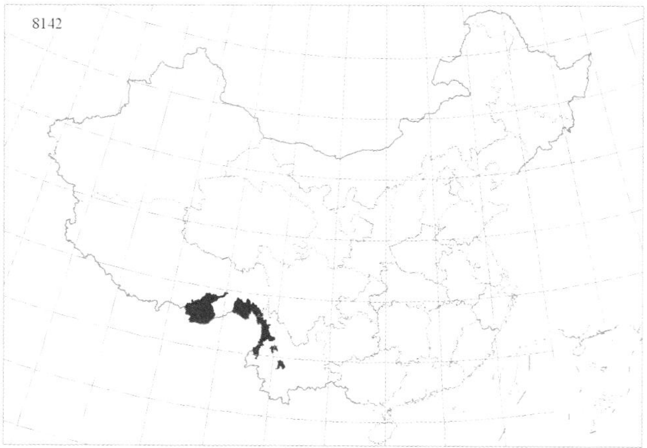

Ericaceae 杜鹃花科 *Rhododendron* 杜鹃花属
Rhododendron pachyphyllum Fang 厚叶杜鹃
Trees, evergreen

15.4; 4.5; 25.2; 15.4; 125.9; 1.0; 812 ❙ 1574; 565; 185 ❙ 94.2; 812; 253.2

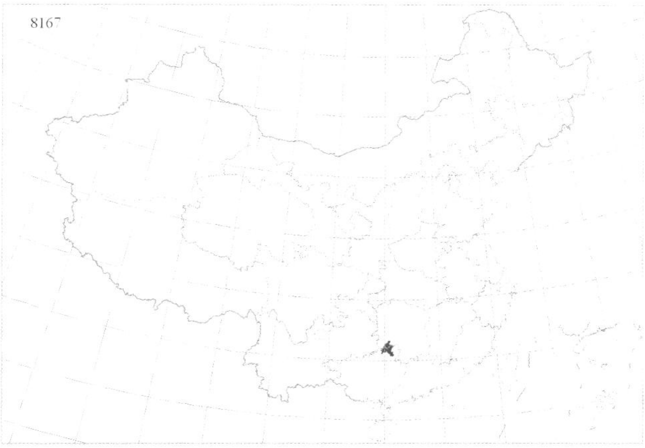

Ericaceae 杜鹃花科 *Rhododendron* 杜鹃花属
Rhododendron pachypodum I. B. Balfour et W. W. Sm. 云上杜鹃
Shrubs, evergreen

16.7 (12.6~21.0); 10.0 (6.0~14.2); 21.4 (17.9~25.6); 16.7 (12.6~21.0); 140.5 (91.6~191.8); 0.0 (0.0~0.0); 818 (678~1053) ❙ 1182 (939~1547); 662 (524~911); 51 (34~70) ❙ 45.0 (6.5~99.9); 803 (678~1053); 375.8 (255.3~551.6)

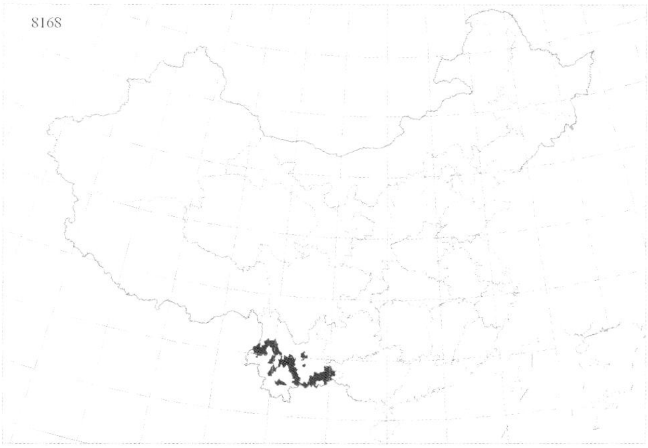

Ericaceae 杜鹃花科 *Rhododendron* 杜鹃花属
Rhododendron pachytrichum Franch. 绒毛杜鹃
Shrubs, evergreen

10.1 (2.0~16.1); 0.7 (-7.4~9.0); 18.6 (9.9~25.4); 10.4 (3.8~16.1); 75.4 (15.5~133.2); 13.8 (0.0~51.9); 631 (413~791) ❙ 985 (760~1414); 535 (363~817); 28 (11~71) ❙ 60.7 (5.0~125.9); 630 (413~791); 257.5 (147.3~334.4)

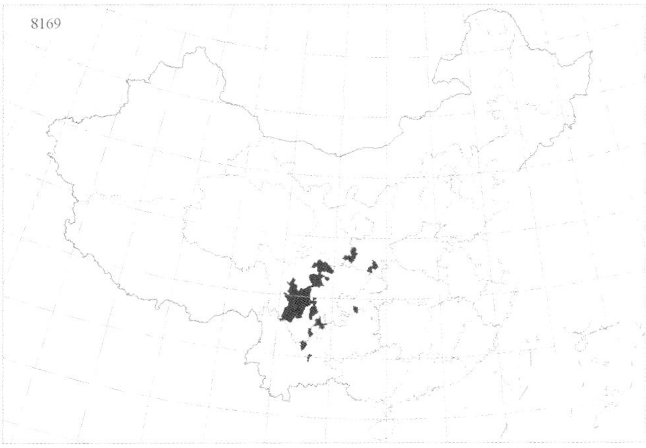

Ericaceae 杜鹃花科 *Rhododendron* 杜鹃花属
Rhododendron papillatum I. B. Balfour et Cooper 乳突杜鹃
Shrubs or small trees, evergreen

12.8; 4.8; 19.0; 12.8; 93.9; 0.4; 689 ❙ 1583; 912; 41 ❙ 129.4; 689; 405.6

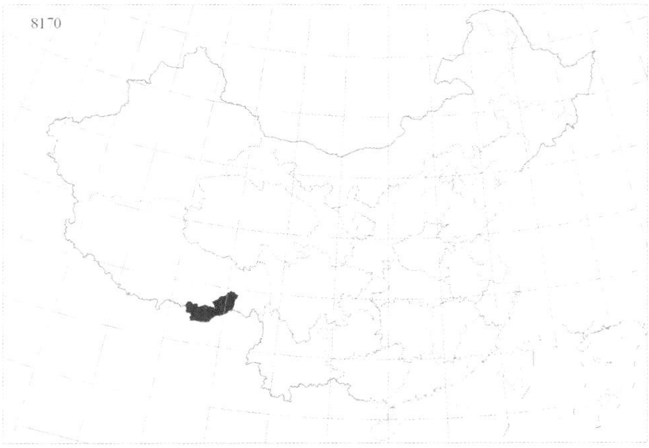

Ericaceae 杜鹃花科 *Rhododendron* 杜鹃花属
Rhododendron parmulatum Cowan 盘萼杜鹃
Small shrubs, evergreen

8.3; 0.0; 15.3; 9.0; 61.4; 20.9; 566 ❙ 1104; 647; 29 ❙ 75.4; 566; 339.1

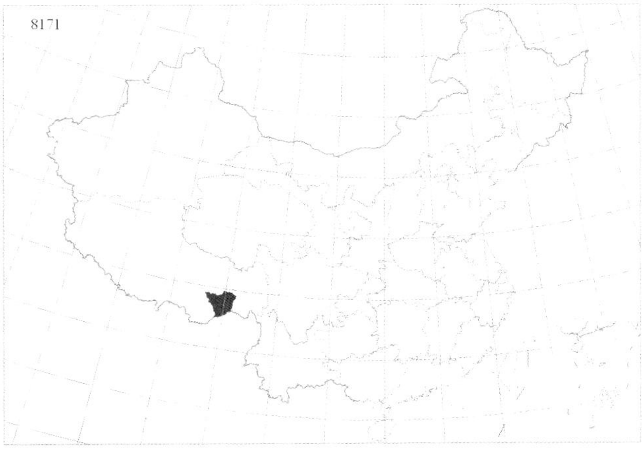

Ericaceae 杜鹃花科 *Rhododendron* 杜鹃花属
Rhododendron pemakoense Kingdon-Ward 假单花杜鹃
Shrubs, evergreen

9.3; 1.5; 15.9; 9.6; 65.5; 13.1; 592 ❙ 1348; 783; 41 ❙ 123.3; 592; 340.8

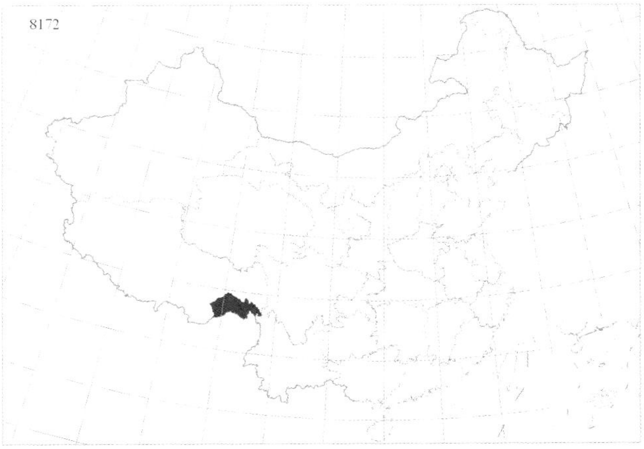

Ericaceae 杜鹃花科 *Rhododendron* 杜鹃花属

Rhododendron pendulum Hook. 凸叶杜鹃

Shrubs, evergreen

6.3; -2.1; 13.4; 7.6; 48.5; 33.1; 520 ❙ 910; 532; 21 ❙ 56.5; 520; 253.1

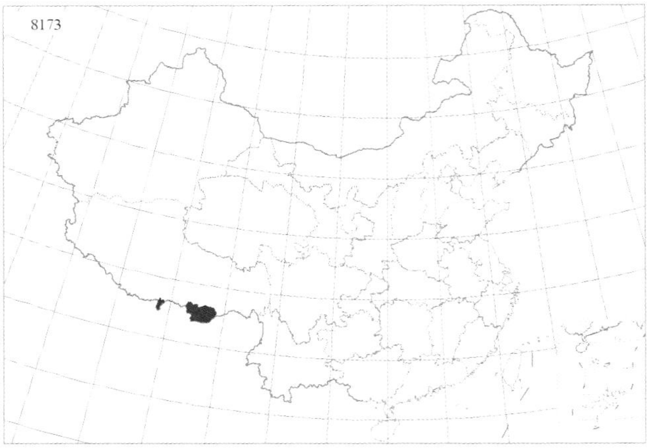

Ericaceae 杜鹃花科 *Rhododendron* 杜鹃花属

Rhododendron petrocharis Diels 饰石杜鹃

Shrubs, evergreen

11.0 (6.0~14.8); 1.7 (-3.2~4.8); 19.4 (14.4~24.0); 11.2 (6.6~14.9); 81.3 (39.7~118.8); 8.5 (0.2~26.7); 652 (508~778) ❙ 1090 (1001~1220); 578 (534~688); 39 (28~63) ❙ 69.7 (45.0~101.2); 652 (508~778); 234.5 (153.9~292.6)

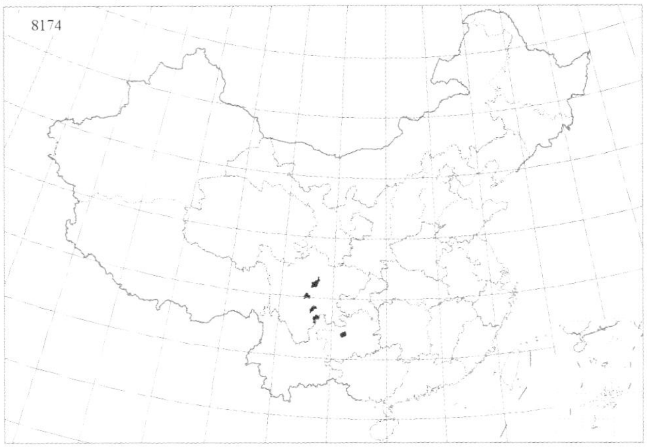

Ericaceae 杜鹃花科 *Rhododendron* 杜鹃花属

Rhododendron phaeochrysum I. B. Balfour et W. W. Sm. 栎叶杜鹃

Shrubs, evergreen

5.6 (1.9~18.5); -2.8 (-8.2~10.5); 12.9 (9.5~24.3); 6.6 (3.8~18.3); 39.2 (15.4~160.0); 31.3 (0.0~53.8); 500 (411~906) ❙ 840 (473~1022); 489 (287~619); 19 (7~40) ❙ 71.9 (5.0~128.6); 494 (411~786); 281.4 (208.8~357.7)

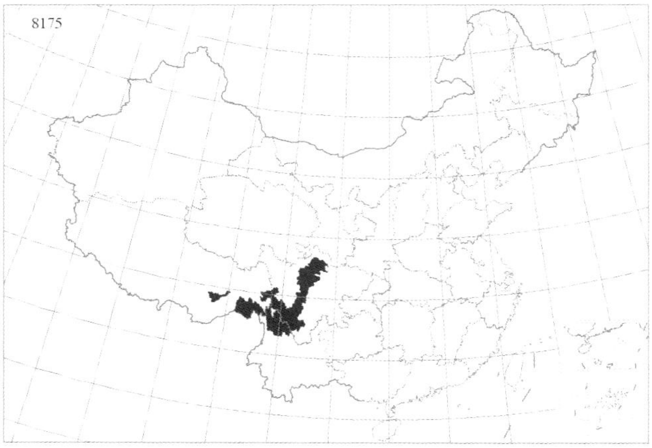

Ericaceae 杜鹃花科 *Rhododendron* 杜鹃花属

Rhododendron phaeochrysum var. agglutinatum (I. B. Balfour et Forrest) D. F. Chamb. 凝毛杜鹃 Shrubs, evergreen

5.7 (2.0~14.7); -2.9 (-7.4~6.1); 13.2 (9.5~24.0); 6.5 (3.8~14.8); 38.5 (15.4~118.0); 29.8 (0.0~51.9); 500 (411~789) ❙ 882 (579~1042); 510 (365~641); 21 (10~40) ❙ 79.9 (18.6~128.6); 499 (411~789); 263.9 (150.8~351.7)

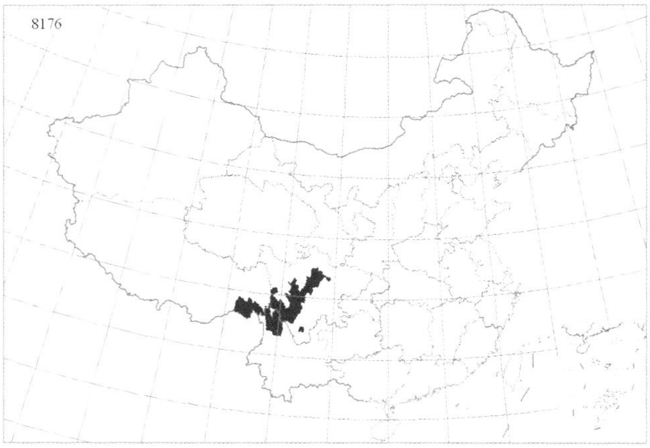

Ericaceae 杜鹃花科 *Rhododendron* 杜鹃花属

Rhododendron phaeochrysum var. levistratum (I. B. Balfour et Forrest) D. F. Chamb. 毡毛栎叶杜鹃 Shrubs, evergreen

7.1 (2.2~18.5); -1.0 (-7.1~10.5); 14.0 (9.9~24.3); 7.7 (3.8~18.3); 49.6 (15.5~160.0); 23.9 (0.0~49.1); 541 (413~906) ❙ 894 (741~990); 530 (426~619); 20 (12~31) ❙ 73.0 (5.5~125.9); 528 (413~786); 287.8 (234.2~341.6)

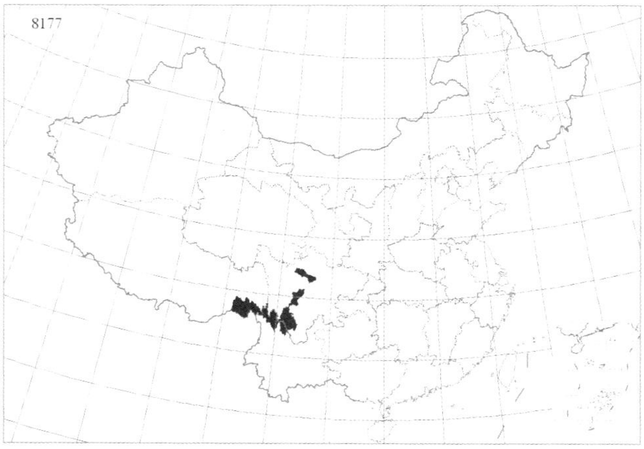

Ericaceae 杜鹃花科 *Rhododendron* 杜鹃花属

Rhododendron piercei Davidian 察隅杜鹃

Shrubs, evergreen

5.3; -2.4; 12.2; 5.7; 30.5; 26.2; 479 ❙ 961; 558; 31 ❙ 100.7; 479; 234.2

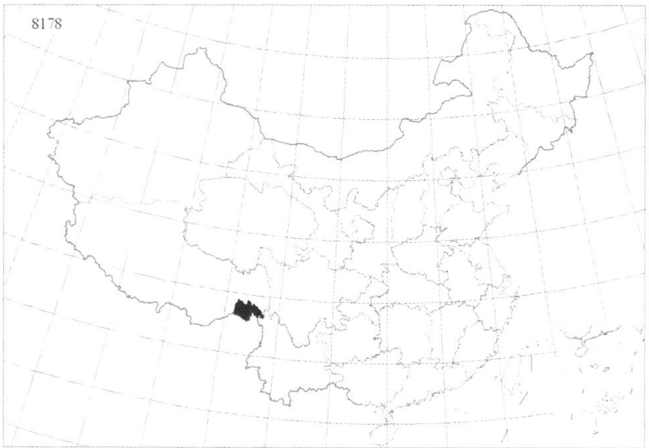

Ericaceae 杜鹃花科 *Rhododendron* 杜鹃花属
Rhododendron pilostylum W. K. Hu 金平毛柱杜鹃
Trees, evergreen

19.1; 13.0; 23.3; 19.1; 169.7; 0.0; 917 ❙ 1428; 835; 56 ❙ 55.7; 917; 506.2

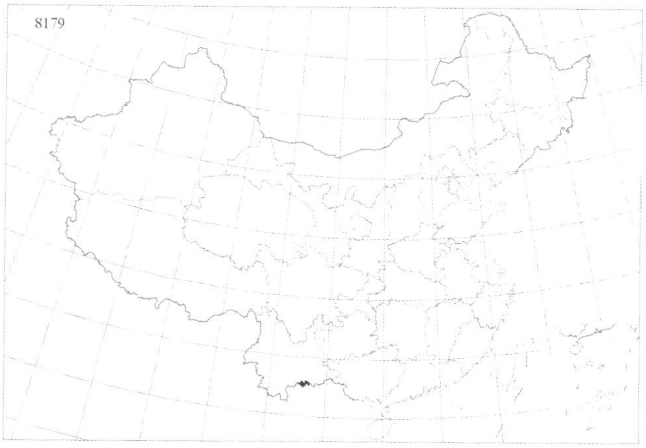

Ericaceae 杜鹃花科 *Rhododendron* 杜鹃花属
Rhododendron pingbianense M. Y. Fang 屏边杜鹃
Shrubs or small trees, evergreen

17.7; 10.9; 22.5; 17.8; 153.4; 0.0; 854 ❙ 1368; 756; 59 ❙ 60.1; 854; 491.5

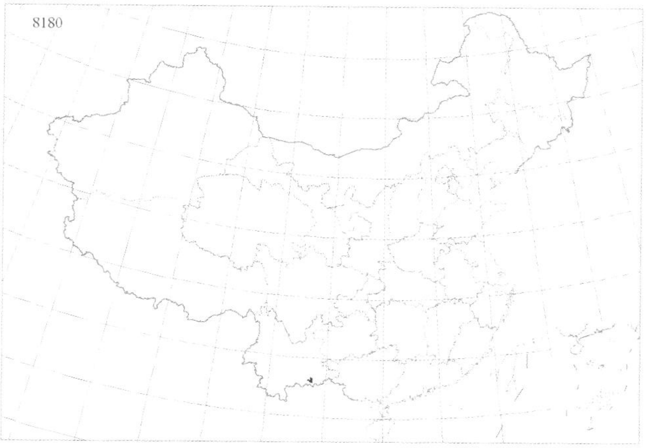

Ericaceae 杜鹃花科 *Rhododendron* 杜鹃花属
Rhododendron pingianum D. Fang 海绵杜鹃
Shrubs or small trees, evergreen

12.3 (6.0~15.1); 3.0 (-3.2~5.3); 20.7 (14.4~24.0); 12.4 (6.6~15.1); 93.1 (39.7~121.7); 5.0 (0.0~26.7); 692 (508~789) ❙ 1148 (944~1414); 640 (520~817); 40 (28~49) ❙ 67.3 (36.4~101.2); 692 (508~789); 240.2 (200.8~292.6)

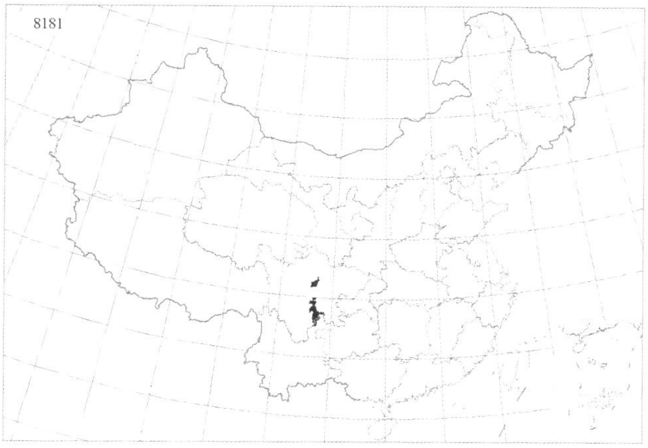

Ericaceae 杜鹃花科 *Rhododendron* 杜鹃花属
Rhododendron platypodum Diels 阔柄杜鹃
Shrubs or small trees, evergreen

14.5 (11.9~16.2); 3.9 (2.9~5.5); 24.5 (20.0~26.3); 14.5 (12.0~16.2); 116.0 (87.2~134.6); 1.5 (0.0~3.4); 783 (671~854) ❙ 1125 (806~1280); 488 (448~550); 65 (16~94) ❙ 43.6 (8.2~56.8); 783 (671~854); 209.8 (170.4~278.8)

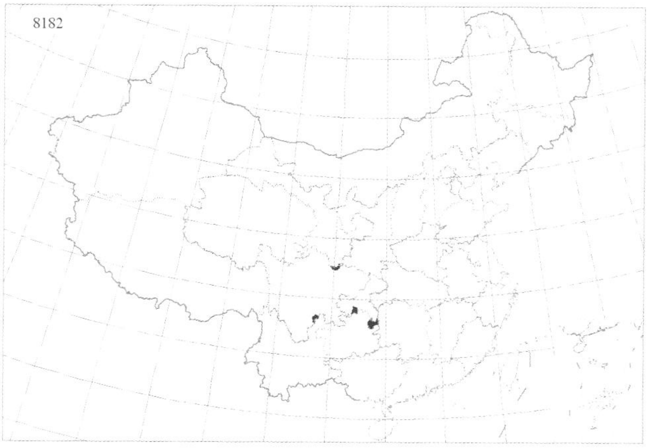

Ericaceae 杜鹃花科 *Rhododendron* 杜鹃花属
Rhododendron pocophorum I. B. Balfour ex Tagg 杯萼杜鹃
Shrubs, evergreen

7.0; -0.5; 13.6; 7.2; 41.5; 17.5; 526 ❙ 1001; 571; 38 ❙ 91.1; 526; 293.2

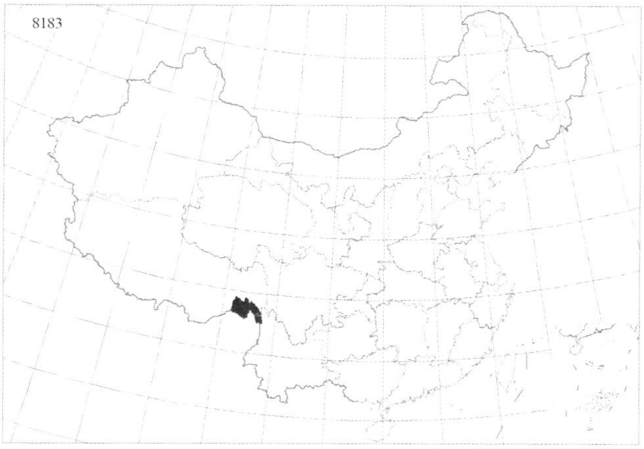

Ericaceae 杜鹃花科 *Rhododendron* 杜鹃花属
Rhododendron pocophorum var. hemidartum (I. B. Balfour ex Tagg) D. F. Chamb. 腺柄杯萼杜鹃
Shrubs, evergreen

7.0; -0.5; 13.6; 7.2; 41.5; 17.5; 526 ❙ 1001; 571; 38 ❙ 91.1; 526; 293.2

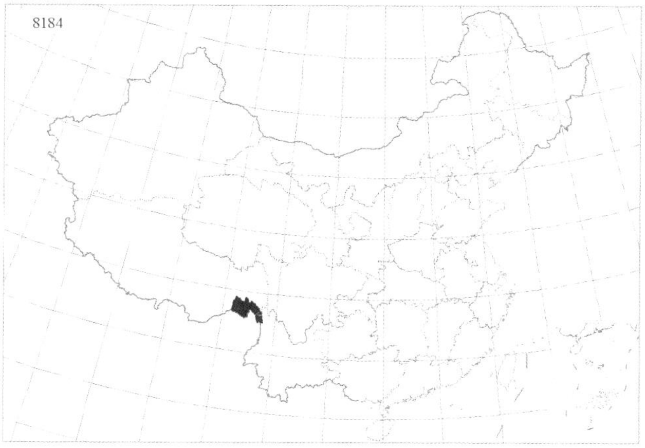

Ericaceae 杜鹃花科 *Rhododendron* 杜鹃花属
Rhododendron poilanei Dop 毛果缺顶杜鹃
Shrubs, evergreen

17.8 (15.1~19.1); 8.8 (5.0~10.9); 24.9 (22.5~26.6); 17.8 (15.2~19.1); 153.8 (121.9~169.2); 0.0 (0.0~0.0); 897 (788~969) ❙ 1281 (1236~1368); 669 (545~756); 61 (51~78) ❙ 43.8 (27.5~60.1); 897 (788~969); 275.3 (189.5~491.5)

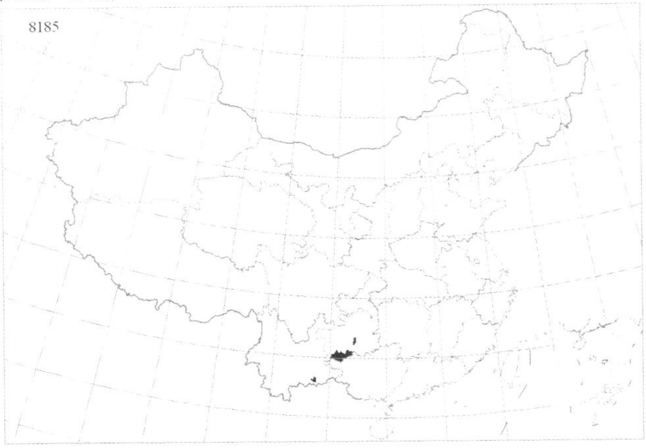

Ericaceae 杜鹃花科 *Rhododendron* 杜鹃花属
Rhododendron polycladum Franch. 多枝杜鹃
Shrubs, evergreen

11.8; 5.0; 17.5; 11.9; 84.0; 1.7; 660 ❙ 1006; 591; 36 ❙ 53.2; 654; 329.6

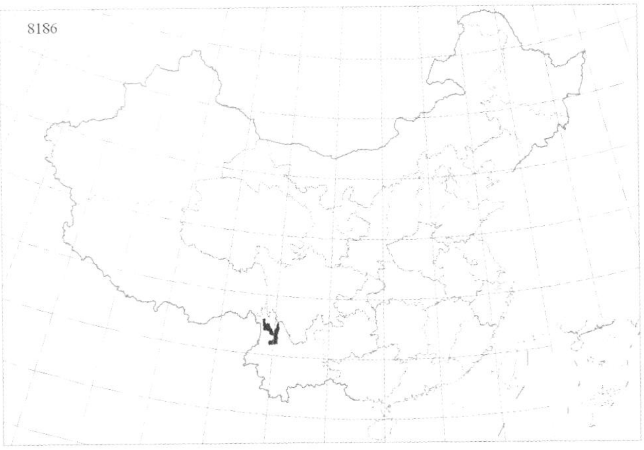

Ericaceae 杜鹃花科 *Rhododendron* 杜鹃花属
Rhododendron polylepis Franch. 多鳞杜鹃
Shrubs or small trees, evergreen

9.6 (2.6~14.9); 0.2 (-7.0~5.3); 18.1 (10.8~24.0); 9.9 (4.3~15.0); 70.1 (20.0~119.6); 14.0 (0.0~48.2); 615 (427~789) ❙ 990 (677~1410); 540 (336~816); 26 (14~49) ❙ 64.8 (-0.2~121.7); 615 (427~789); 246.6 (139.5~318.7)

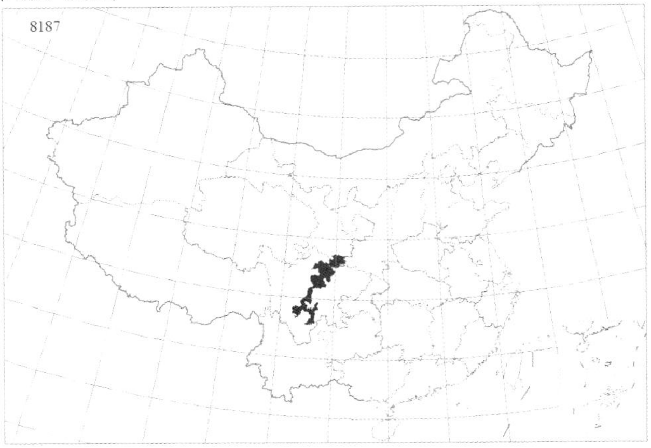

Ericaceae 杜鹃花科 *Rhododendron* 杜鹃花属
Rhododendron polyraphidoideum Tam 千针叶杜鹃
Shrubs, semievergreen or green

18.4 (14.3~21.6); 8.4 (3.7~12.6); 27.3 (24.2~28.5); 18.5 (14.3~21.7); 161.9 (113.2~199.8); 0.1 (0.0~1.4); 972 (761~1153) ❙ 1736 (1483~1997); 652 (481~821); 208 (161~259) ❙ 82.2 (35.6~130.7); 972 (761~1153); 245.3 (188.8~303.4)

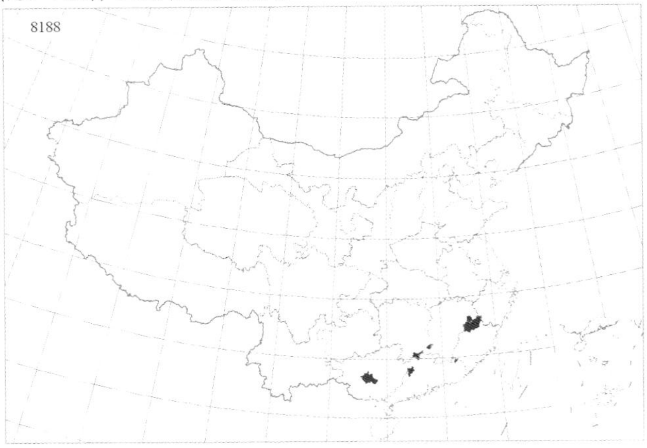

Ericaceae 杜鹃花科 *Rhododendron* 杜鹃花属
Rhododendron polytrichum Fang 多毛杜鹃
Shrubs, evergreen

15.9; 5.1; 25.5; 15.9; 131.7; 0.8; 837 ❙ 1598; 594; 182 ❙ 91.5; 837; 253.0

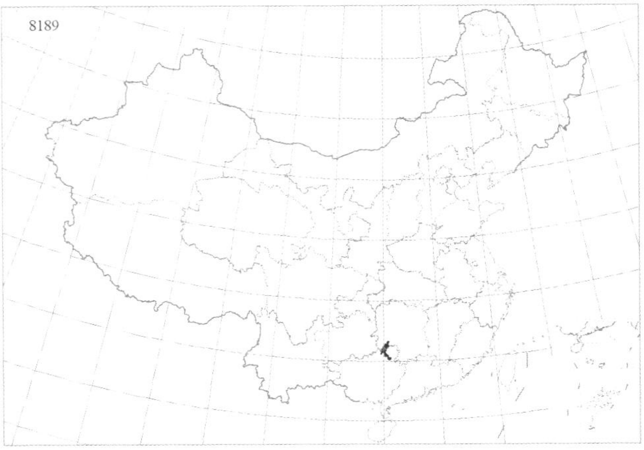

Ericaceae 杜鹃花科 *Rhododendron* 杜鹃花属
Rhododendron pomense Cowan et Davidian 波密杜鹃
Shrubs, evergreen

1.8; -7.1; 9.9; 3.8; 15.7; 53.3; 404 ❙ 583; 348; 13 ❙ 44.2; 404; 169.5

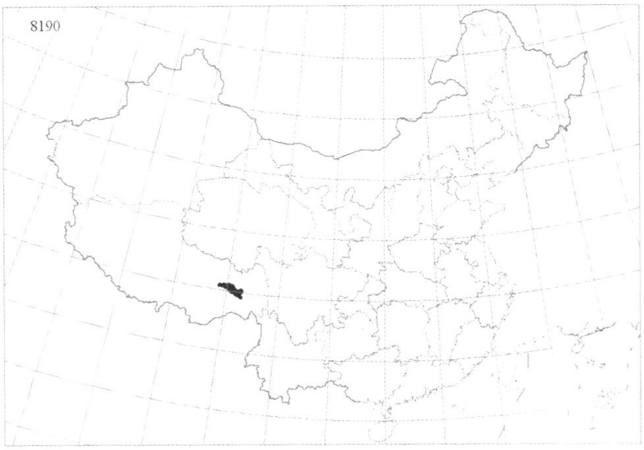

Ericaceae 杜鹃花科 *Rhododendron* 杜鹃花属

Rhododendron populare Cowan 蜜腺杜鹃

Shrubs or small trees, evergreen

4.4; -4.3; 12.1; 5.3; 28.5; 34.6; 455 ❙ 538; 337; 9 ❙ 18.2; 455; 168.2

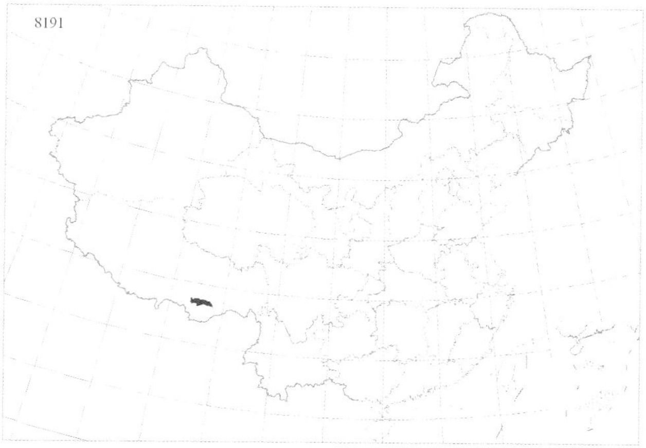

Ericaceae 杜鹃花科 *Rhododendron* 杜鹃花属

Rhododendron potaninii Batalin 甘肃杜鹃

Small trees, evergreen

5.2; -8.0; 16.5; 7.0; 47.4; 44.1; 5'8 ❙ 443; 239; 8 ❙ -14.6; 443; 141.6

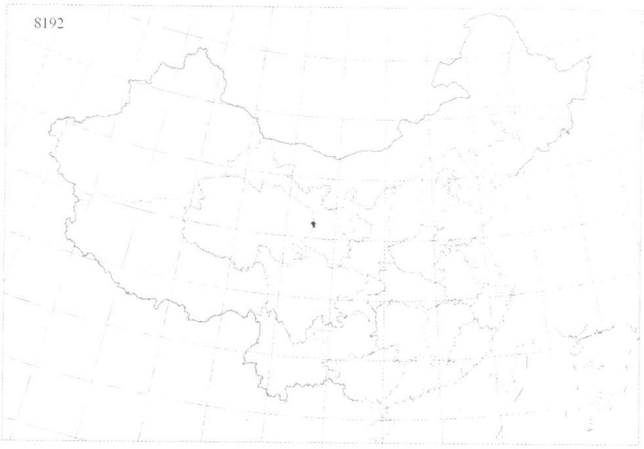

Ericaceae 杜鹃花科 *Rhododendron* 杜鹃花属

Rhododendron praestans I. B. Balfour et W. W. Sm. 优秀杜鹃

Shrubs or small trees, evergreen

9.2 (5.3~13.2); 2.0 (-2.4~6.4); 15.3 (12.2~18.4); 9.3 (5.7~13.3); 60.4 (30.5~99.4); 9.7 (0.0~26.2); 587 (479~695) ❙ 1043 (741~1337); 595 (426~727); 42 (24~69) ❙ 77.5 (52.8~100.7); 587 (479~695); 331.9 (234.2~407.1)

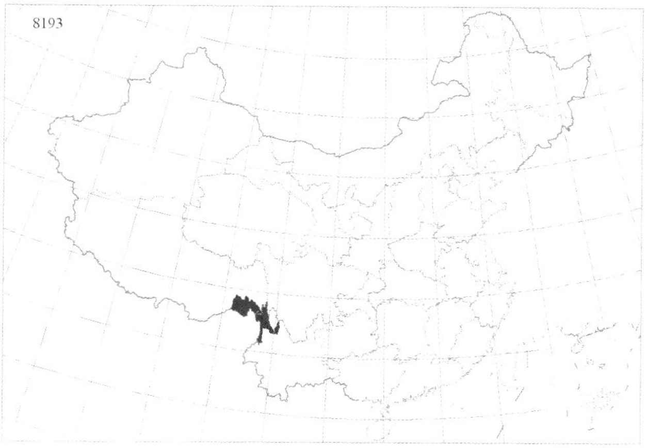

Ericaceae 杜鹃花科 *Rhododendron* 杜鹃花属

Rhododendron praeteritum Hutch. 鄂西杜鹃

Shrubs, evergreen

9.3; -1.7; 20.1; 9.5; 69.8; 18.0; 603 ❙ 1214; 516; 71 ❙ 101.4; 603; 279.5

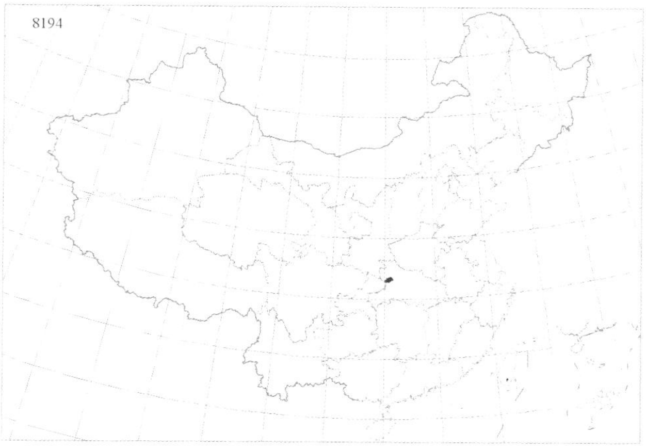

Ericaceae 杜鹃花科 *Rhododendron* 杜鹃花属

Rhododendron praevernum Hutch. 早春杜鹃

Shrubs or small trees, evergreen

12.8 (6.2~17.0); 2.3 (-4.1~6.2); 22.8 (15.8~26.7); 12.9 (7.0~17.0); 100.4 (45.2~144.4); 5.9 (0.0~29.9); 722 (522~891) ❙ 1155 (669~1624); 510 (317~688); 64 (14~191) ❙ 59.7 (2.4~107.2); 722 (522~891); 249.7 (170.4~305.6)

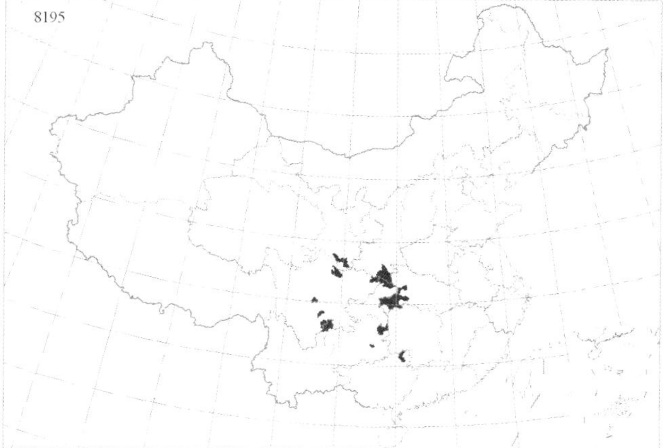

Ericaceae 杜鹃花科 *Rhododendron* 杜鹃花属

Rhododendron preptum I. B. Balfour et Forrest 复毛杜鹃

Shrubs or small trees, evergreen

13.2; 6.4; 18.4; 13.2; 98.8; 0.0; 694 ❙ 1250; 683; 64 ❙ 80.2; 694; 386.7

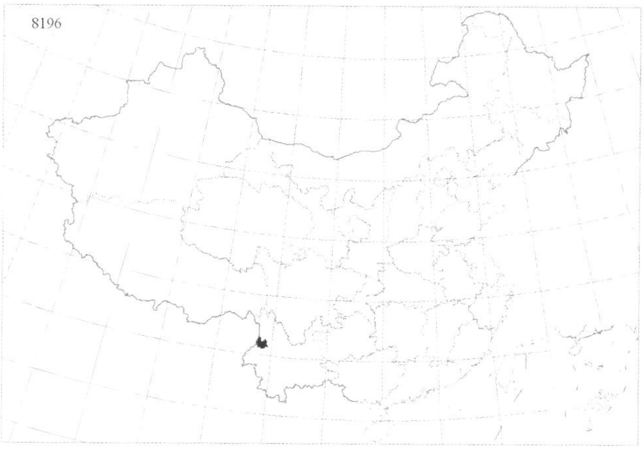

Ericaceae 杜鹃花科 *Rhododendron* 杜鹃花属

Rhododendron primuliflorum Bureau et Franch. 樱草杜鹃

Small shrubs, evergreen

4.2 (-1.2~12.6); -4.5 (-11.7~6.0); 11.9 (8.1~19.8); 5.7 (2.6~12.6); 31.8 (8.0~91.6); 40.8 (0.0~82.3); 465 (345~678) ❙ 647 (250~1243); 393 (190~696); 16 (1~58) ❙ 35.0 (-31.9~128.6); 440 (250~678); 249.0 (111.0~384.7)

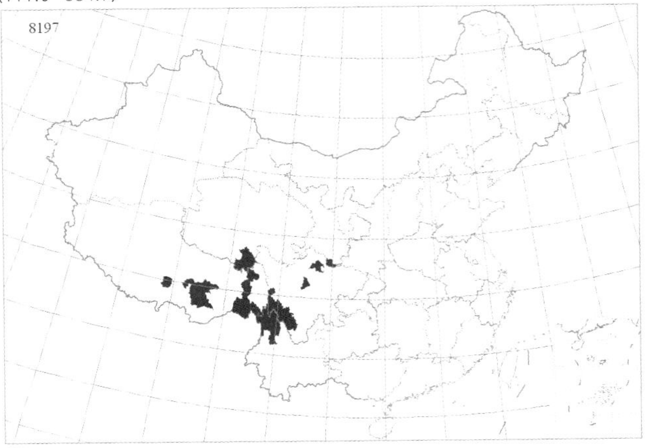

Ericaceae 杜鹃花科 *Rhododendron* 杜鹃花属

Rhododendron primuliflorum var. cephalanthoides (I. B. Balfour et W. W. Sm.) Cowan et Davidian 微毛樱草杜鹃

Small shrubs, evergreen

0.6 (-1.2~3.7); -9.3 (-11.7~-6.1); 9.6 (8.5~12.1); 3.4 (2.6~5.1); 14.6 (8.1~27.8); 66.5 (43.5~82.3); 380 (345~441) ❙ 351 (250~533); 244 (198~328); 4 (1~12) ❙ -6.9 (-30.3~41.8); 313 (250~377); 215.0 (131.4~290.1)

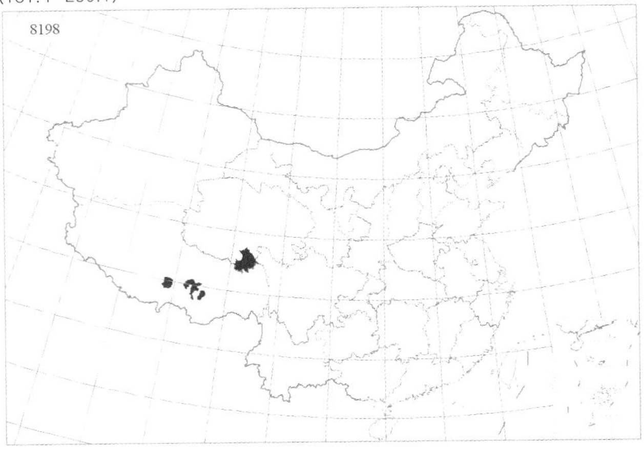

Ericaceae 杜鹃花科 *Rhododendron* 杜鹃花属

Rhododendron principis Bureau et Franch. 藏南杜鹃

Small trees, evergreen

2.5 (-2.5~12.2); -6.7 (-12.7~4.3); 10.8 (7.0~18.5); 4.6 (2.1~12.2); 23.7 (4.0~87.4); 53.0 (0.7~93.0); 427 (332~672) ❙ 658 (382~1431); 393 (254~817); 15 (6~31) ❙ 49.2 (12.1~113.0); 427 (332~672); 235.0 (111.0~363.7)

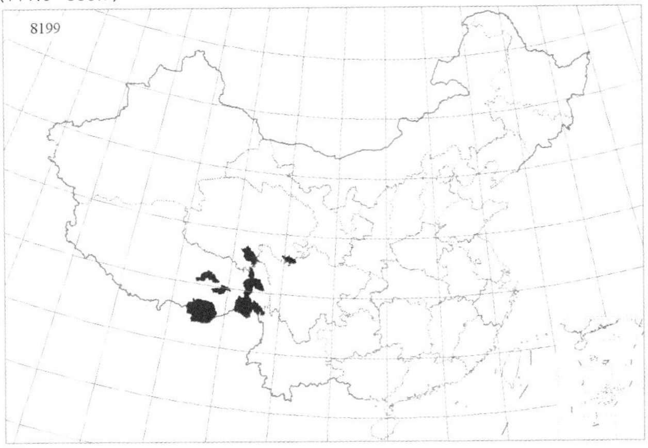

Ericaceae 杜鹃花科 *Rhododendron* 杜鹃花属

Rhododendron pronum Tagg et Forrest 平卧杜鹃

Small shrubs, evergreen

11.1; 4.4; 16.7; 11.2; 74.5; 0.6; 639 ❙ 1094; 639; 41 ❙ 71.2; 639; 345.5

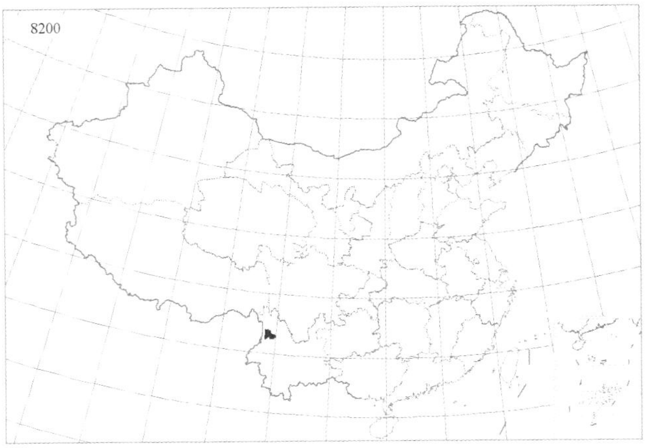

Ericaceae 杜鹃花科 *Rhododendron* 杜鹃花属

Rhododendron proteoides I. B. Balfour et W. W. Sm. 矮生杜鹃

Small shrubs, evergreen

6.6 (5.3~9.6); -1.0 (-2.4~2.4); 13.2 (12.2~15.7); 6.9 (5.7~9.6); 39.5 (30.5~60.8); 19.8 (5.4~26.2); 514 (479~600) ❙ 875 (741~1007); 521 (426~578); 26 (12~40) ❙ 70.4 (54.0~100.7); 514 (479~600); 290.8 (234.2~351.7)

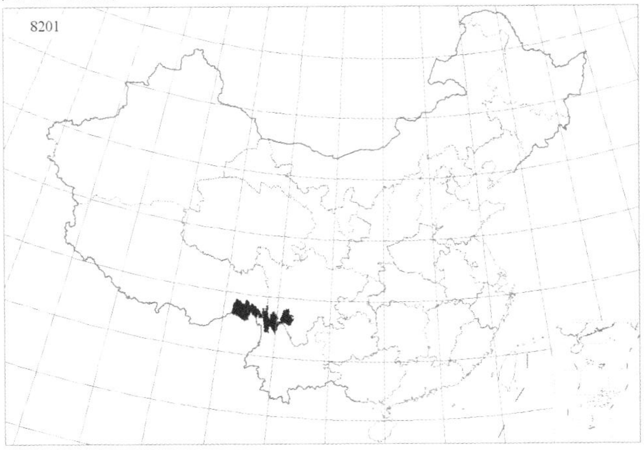

Ericaceae 杜鹃花科 *Rhododendron* 杜鹃花属

Rhododendron protistum I. B. Balfour et Forrest 翘首杜鹃

Trees, evergreen

9.9; 2.7; 15.8; 10.0; 68.2; 8.9; 606 ❙ 1175; 658; 51 ❙ 93.8; 606; 364.6

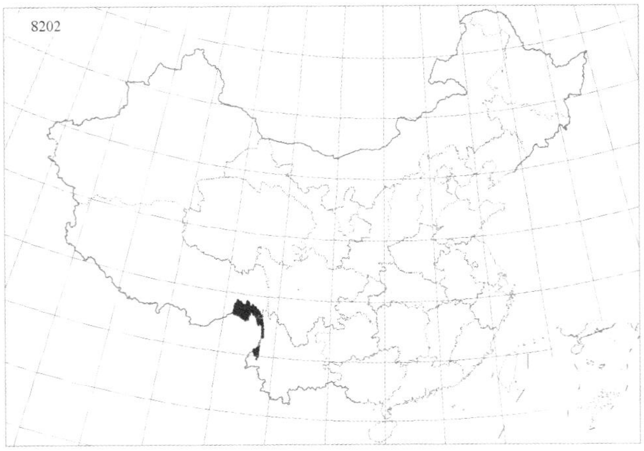

Ericaceae 杜鹃花科 *Rhododendron* 杜鹃花属
Rhododendron protistum var. giganteum (Forrest ex Tagg) D. F. Chamb. 大树杜鹃
Trees, evergreen

11.9; 4.9; 17.4; 11.9; 85.4; 2.4; 660 ❙ 1269; 700; 60 ❙ 91.7; 660; 407.8

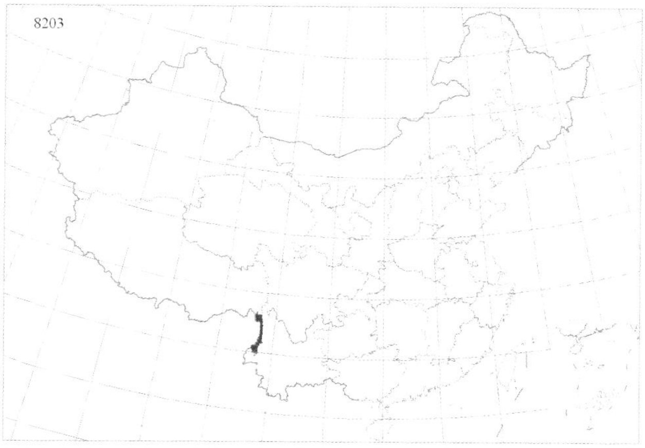

Ericaceae 杜鹃花科 *Rhododendron* 杜鹃花属
Rhododendron pruniflorum Hutch. 桃花杜鹃
Small shrubs, evergreen

3.9; -4.6; 11.4; 4.9; 25.2; 38.2; 445 ❙ 613; 374; 15 ❙ 34.5; 425; 206.6

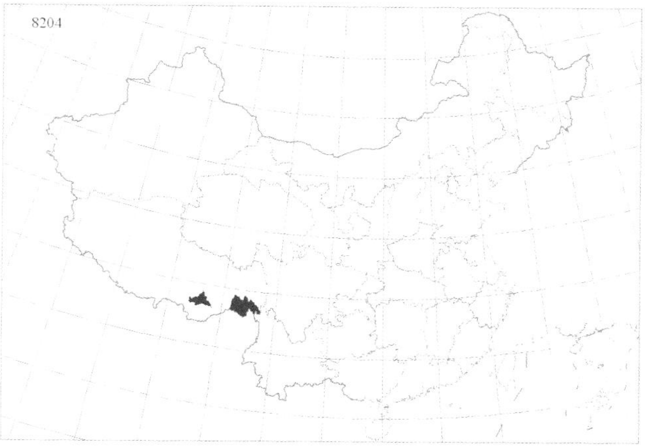

Ericaceae 杜鹃花科 *Rhododendron* 杜鹃花属
Rhododendron przewalski Maxim. 陇蜀杜鹃
Shrubs, evergreen

4.1 (-3.2~14.0); -7.0 (-14.4~4.5); 14.0 (7.3~24.2); 6.0 (2.0~14.1); 36.8 (4.0~109.4); 46.8 (0.5~101.5); 483 (335~748) ❙ 670 (375~1414); 357 (203~817); 15 (5~47) ❙ 40.9 (-32.2~128.6); 473 (335~748); 256.4 (95.7~357.7)

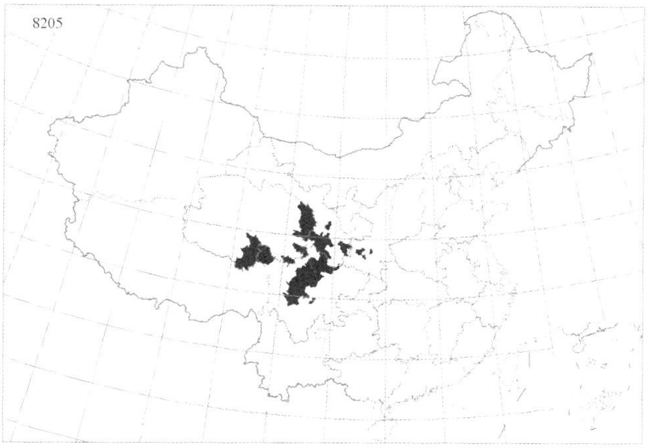

Ericaceae 杜鹃花科 *Rhododendron* 杜鹃花属
Rhododendron pseudochrysanthum Hayata 阿里山杜鹃
Small shrubs, evergreen

18.2; 12.8; 22.6; 18.3; 159.3; 0.0; 881 ❙ 2661; 1186; 252 ❙ 207.4; 881; 254.2

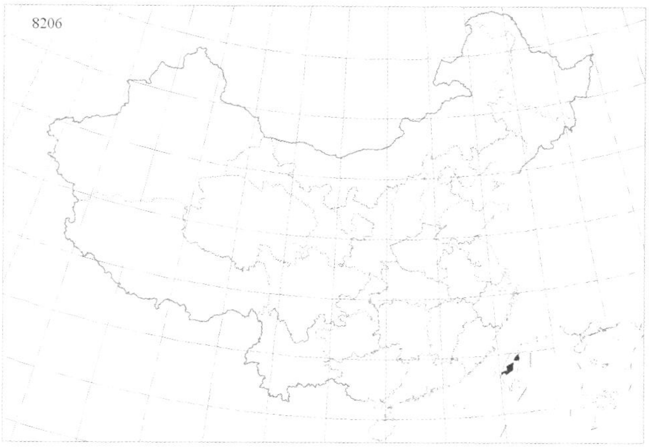

Ericaceae 杜鹃花科 *Rhododendron* 杜鹃花属
Rhododendron pseudociliipes Cullen 褐叶杜鹃
Shrubs, evergreen

11.1; 4.0; 16.8; 11.1; 76.2; 3.2; 638 ❙ 1207; 669; 57 ❙ 89.0; 638; 381.4

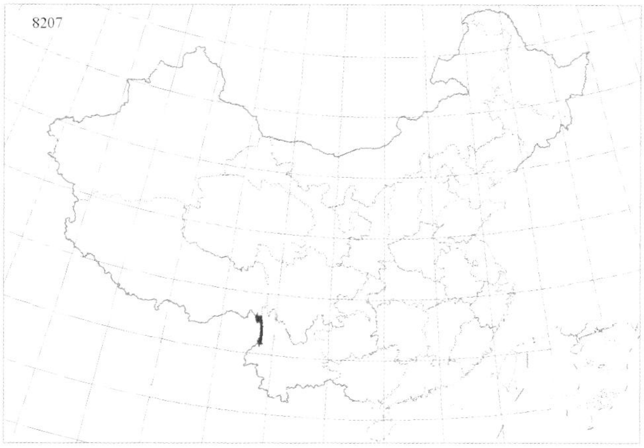

Ericaceae 杜鹃花科 *Rhododendron* 杜鹃花属
Rhododendron pubescens I. B. Balfour et Forrest 柔毛杜鹃
Small shrubs, evergreen

10.4 (6.1~14.6); 2.7 (-1.5~7.8); 16.8 (12.7~20.2); 10.4 (6.3~14.6); 71.9 (35.2~115.2); 7.1 (0.0~21.8); 622 (496~738) ❙ 971 (801~1093); 584 (496~622); 19 (12~27) ❙ 57.4 (30.5~78.3); 613 (496~707); 300.2 (279.1~318.7)

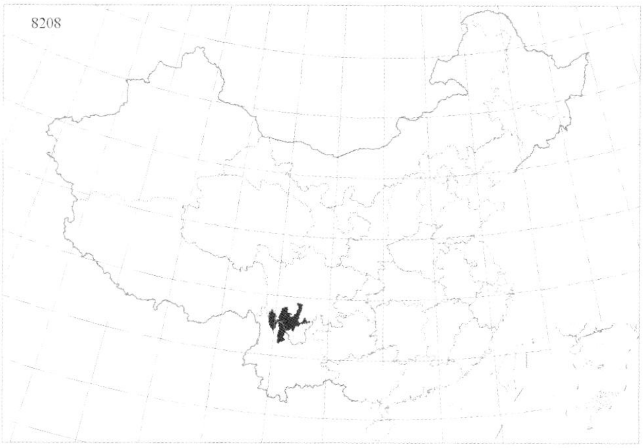

Ericaceae 杜鹃花科 *Rhododendron* 杜鹃花属

Rhododendron pubicostatum T. L. Ming 毛脉杜鹃

Shrubs, evergreen

13.8; 6.7; 19.3; 13.8; 105.9; 0.0; 714 ❙ 963; 592; 29 ❙ 35.0; 690; 266.2

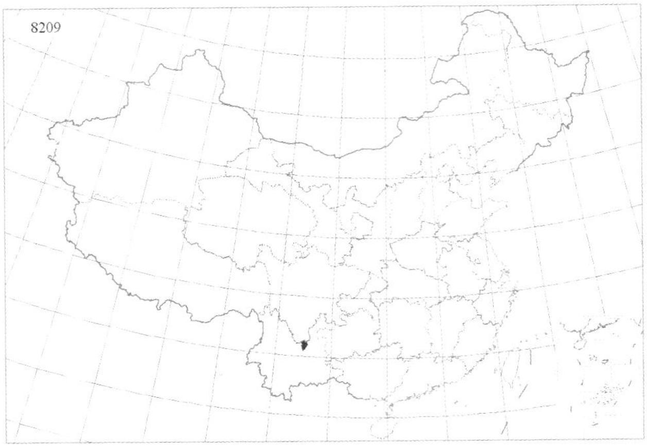

Ericaceae 杜鹃花科 *Rhododendron* 杜鹃花属

Rhododendron pudorosum Cowan 羞怯杜鹃

Trees, evergreen

1.9; -7.0; 9.9; 3.9; 17.3; 54.1; 399 ❙ 416; 275; 8 ❙ 2.5; 378; 149.1

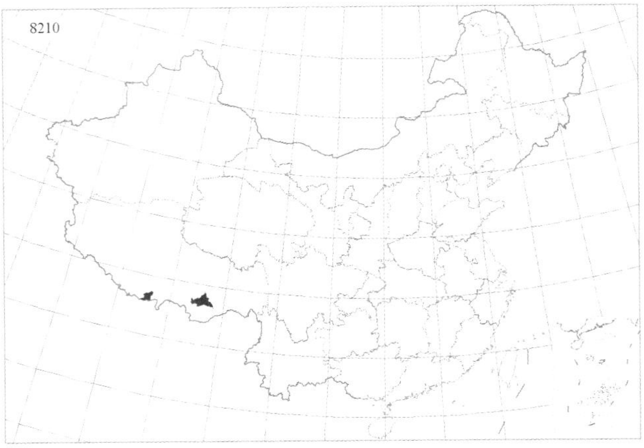

Ericaceae 杜鹃花科 *Rhododendron* 杜鹃花属

Rhododendron pugeense L. C. Hu 普格杜鹃

Shrubs, evergreen

11.3; 3.6; 17.7; 11.4; 78.3; 1.8; 649 ❙ 1058; 632; 23 ❙ 63.1; 649; 297.6

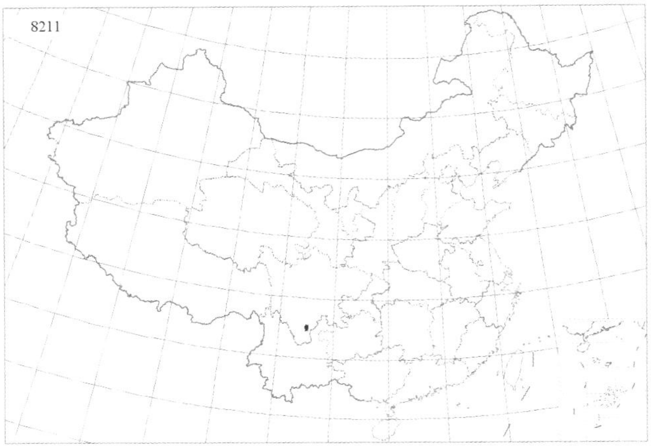

Ericaceae 杜鹃花科 *Rhododendron* 杜鹃花属

Rhododendron pulchroides Chun et Fang 美艳杜鹃

Small shrubs, deciduous

20.2 (15.5~22.3); 11.0 (4.8~13.8); 27.5 (25.1~28.9); 20.2 (15.6~22.3); 183.0 (127.2~208.1); 0.0 (0.0~0.2); 1070 (815~1202) ❙ 1551 (1446~1740); 733 (591~839); 144 (102~187) ❙ 48.5 (20.3~93.2); 1070 (815~1202); 256.0 (201.4~335.1)

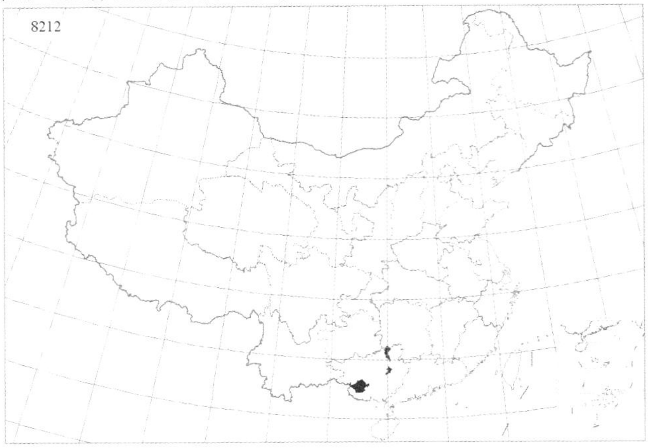

Ericaceae 杜鹃花科 *Rhododendron* 杜鹃花属

Rhododendron pulchrum Sweet 锦绣杜鹃

Shrubs, semievergreen

17.4 (13.4~22.1); 7.2 (3.6~15.2); 27.3 (22.3~28.8); 17.7 (13.4~22.7); 153.1 (102.7~212.7); 0.3 (0.0~1.8); 939 (721~1216) ❙ 1630 (1077~1965); 587 (488~994); 188 (59~222) ❙ 74.8 (49.3~128.9); 939 (721~1216); 223.1 (157.7~254.7)

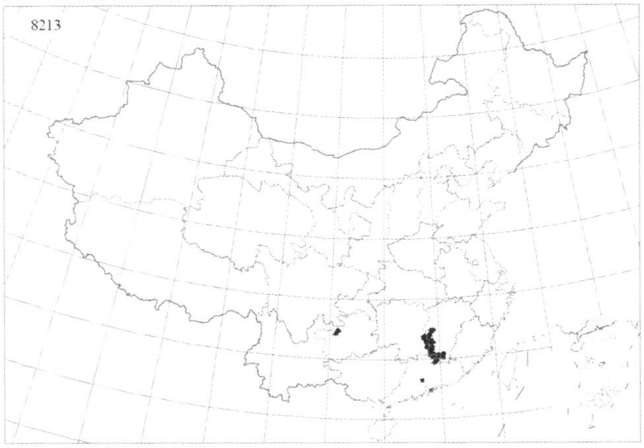

Ericaceae 杜鹃花科 *Rhododendron* 杜鹃花属

Rhododendron pumilum Hook. 矮小杜鹃

Small shrubs, evergreen

10.1 (1.8~16.7); 2.8 (-7.2~9.7); 16.3 (9.9~22.1); 10.5 (3.8~16.7); 73.6 (15.7~140.9); 11.4 (0.0~53.6); 616 (403~819) ❙ 961 (341~1736); 559 (226~1007); 35 (4~69) ❙ 53.8 (-15.3~145.8); 605 (341~748); 316.6 (168.2~447.4)

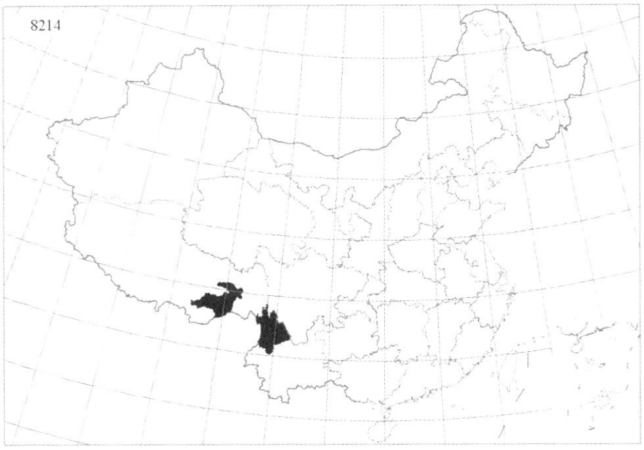

Ericaceae 杜鹃花科 *Rhododendron* 杜鹃花属
Rhododendron punctifolium L. C. Hu 斑叶杜鹃
Small shrubs, evergreen

13.6; 7.2; 18.8; 13.7; 104.1; 0.0; 706 ❙ 1042; 592; 46 ❙ 47.6; 706; 372.3

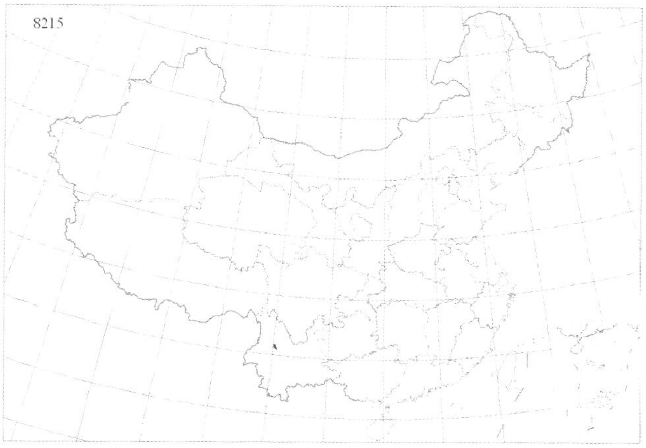

Ericaceae 杜鹃花科 *Rhododendron* 杜鹃花属
Rhododendron purdomii Rehder et E. H. Wilson 太白杜鹃
Shrubs or small trees, evergreen

11.6 (6.2~15.9); -0.5 (-4.2~4.8); 23.2 (15.8~26.5); 11.8 (7.0~16.0); 93.8 (45.2~131.6); 13.4 (0.2~29.9); 701 (522~855) ❙ 893 (669~1265); 402 (315~533); 35 (15~68) ❙ 27.5 (-5.3~93.8); 699 (522~855); 258.1 (167.5~320.6)

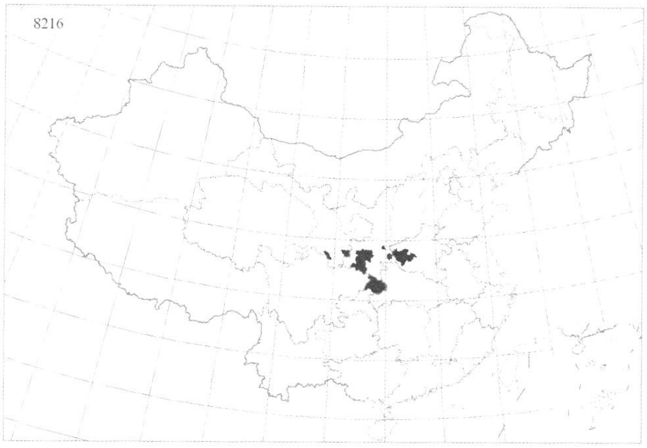

Ericaceae 杜鹃花科 *Rhododendron* 杜鹃花属
Rhododendron qianyangense M. Y. He 黔阳杜鹃
Small shrubs, deciduous

16.2; 4.6; 26.9; 16.2; 135.2; 0.4; 866 ❙ 1348; 450; 153 ❙ 55.6; 866; 206.3

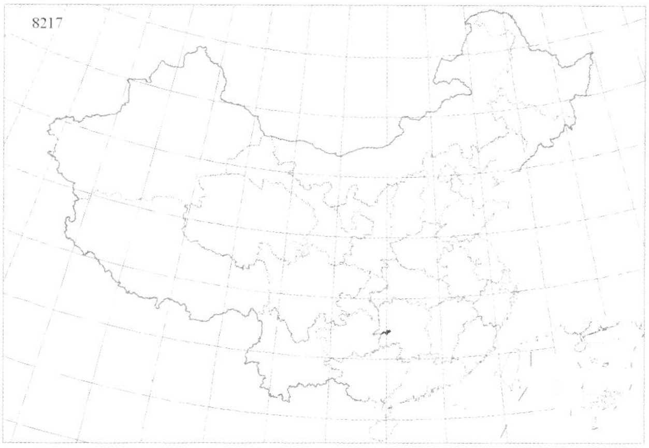

Ericaceae 杜鹃花科 *Rhododendron* 杜鹃花属
Rhododendron qinghaiense Ching ex W. Y. Wang 青海杜鹃
Small shrubs, evergreen

-1.2; -11.7; 8.5; 2.6; 8.1; 82.3; 347 ❙ 491; 301; 11 ❙ 41.8; 347; 290.1

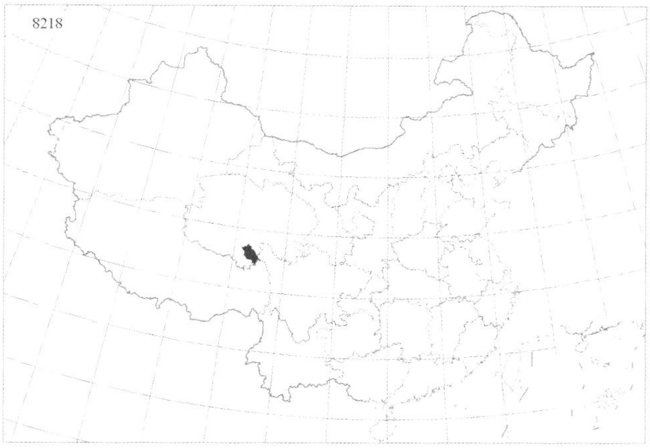

Ericaceae 杜鹃花科 *Rhododendron* 杜鹃花属
Rhododendron racemosum Franch. 腋花杜鹃
Small shrubs, evergreen

12.6 (2.2~18.9); 4.7 (-6.6~12.0); 19.2 (9.5~26.6); 12.7 (3.8~19.0); 96.3 (15.5~167.8); 4.4 (0.0~49.1); 692 (411~929) ❙ 992 (628~1410); 570 (410~816); 33 (11~59) ❙ 46.1 (-14.2~125.9); 674 (411~912); 269.1 (142.9~384.1)

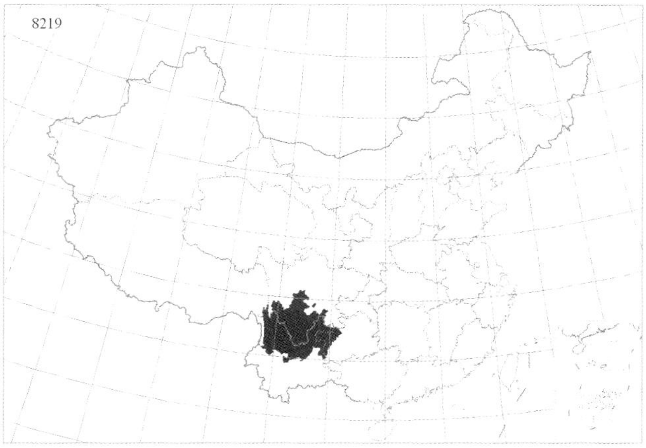

Ericaceae 杜鹃花科 *Rhododendron* 杜鹃花属
Rhododendron radendum Fang 毛叶杜鹃
Small shrubs, evergreen

4.8; -3.4; 11.8; 5.9; 31.9; 33.7; 477 ❙ 909; 562; 15 ❙ 94.9; 476; 254.6

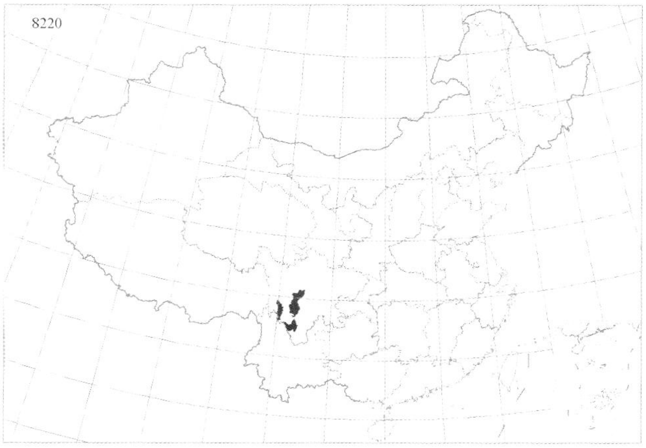

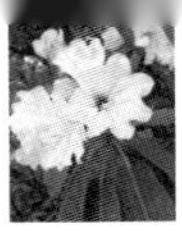

Ericaceae 杜鹃花科 *Rhododendron* 杜鹃花属

Rhododendron ramipilosum T. L. Ming 线裂杜鹃

Shrubs, evergreen

13.3; 5.4; 19.6; 13.4; 100.4; 0.0; 706 ❙ 1736; 1007; 51 ❙ 145.8; 706; 447.4

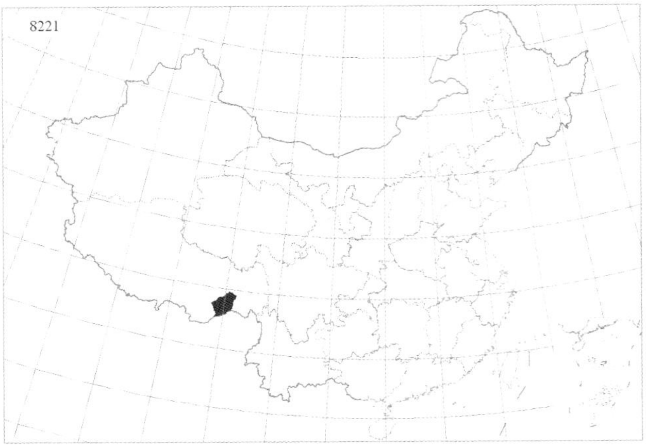

Ericaceae 杜鹃花科 *Rhododendron* 杜鹃花属

Rhododendron ramsdenianum Cowan 长轴杜鹃

Shrubs, evergreen

3.4; -5.4; 11.2; 4.6; 22.8; 41.8; 429 ❙ 500; 302; 8 ❙ 13.7; 429; 255.3

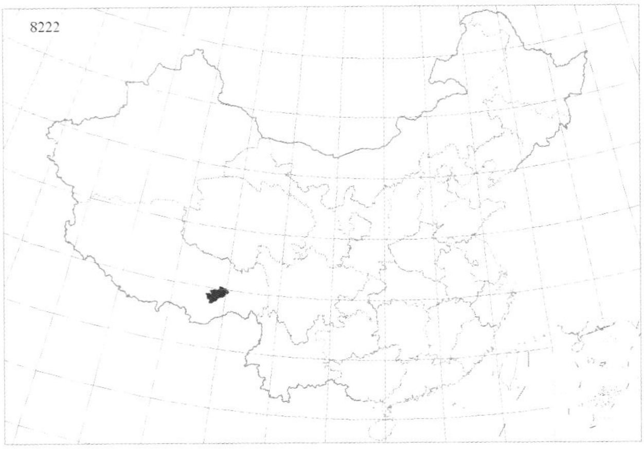

Ericaceae 杜鹃花科 *Rhododendron* 杜鹃花属

Rhododendron redowskianum Maxim. 叶状苞杜鹃

Small shrubs, deciduous

3.2 (1.4~5.2); -16.2 (-18.8~-14.1); 19.9 (18.4~21.8); 7.5 (6.4~8.6); 55.8 (45.8~68.0); 77.5 (65.0~88.8); 548 (501~595) ❙ 706 (561~1005); 428 (333~635); 25 (16~39) ❙ 29.2 (-2.6~68.8); 545 (501~595); 319.6 (289.3~346.5)

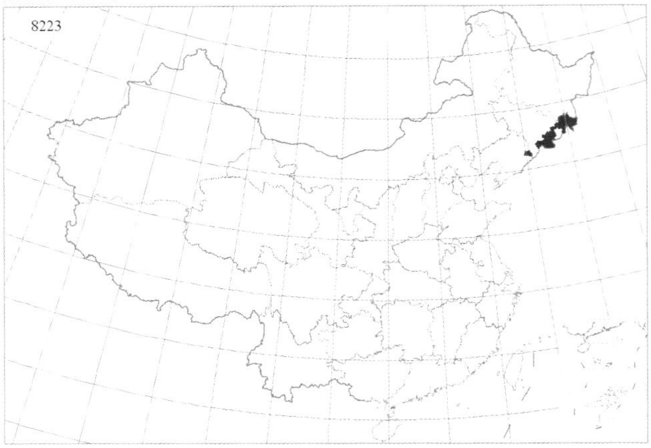

Ericaceae 杜鹃花科 *Rhododendron* 杜鹃花属

Rhododendron rex H. Lév. 大王杜鹃

Small trees, evergreen

12.0 (5.3~18.5); 4.4 (-2.4~11.0); 18.2 (12.2~24.3); 12.0 (5.7~18.3); 88.8 (30.5~160.0); 4.9 (0.0~26.2); 670 (479~906) ❙ 1006 (741~1220); 589 (426~688); 30 (12~58) ❙ 53.0 (5.5~101.6); 652 (479~786); 302.0 (232.7~382.6)

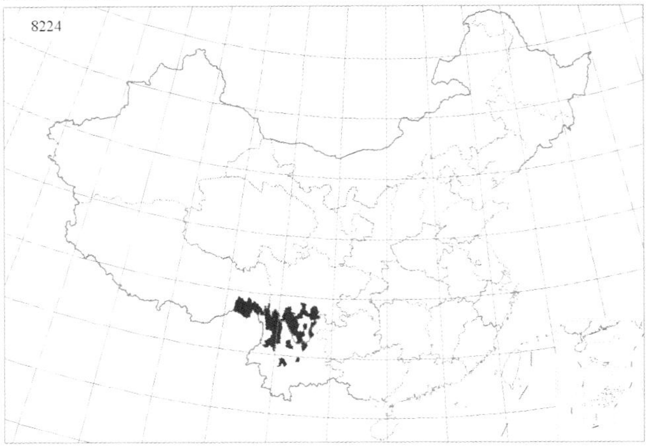

Ericaceae 杜鹃花科 *Rhododendron* 杜鹃花属

Rhododendron rex subsp. fictolacteum (I. B. Balfour) D. F. Chamb. 假乳黄叶杜鹃

Small trees, evergreen

9.7 (5.3~14.5); 2.4 (-2.4~7.8); 16.0 (12.2~19.7); 9.9 (5.7~14.5); 66.7 (30.5~113.5); 9.5 (0.0~26.2); 602 (479~730) ❙ 979 (741~1135); 572 (426~644); 31 (12~54) ❙ 64.0 (36.1~100.7); 600 (479~730); 311.2 (234.2~382.6)

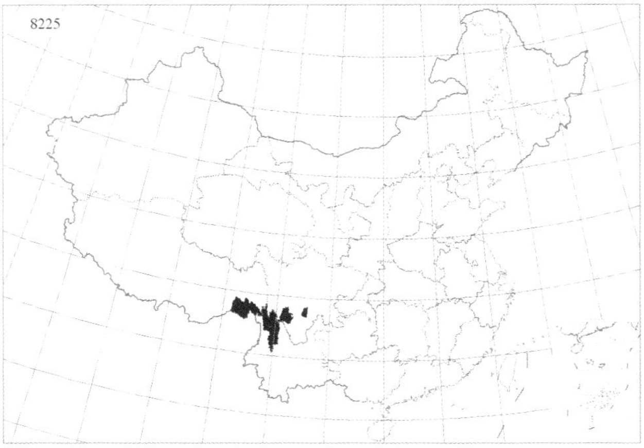

Ericaceae 杜鹃花科 *Rhododendron* 杜鹃花属

Rhododendron rex subsp. gratum (T. L. Ming) M. Y. Fang 可爱杜鹃

Small trees, evergreen

13.1; 6.4; 18.4; 13.2; 98.1; 0.0; 692 ❙ 1164; 639; 58 ❙ 68.1; 692; 366.3

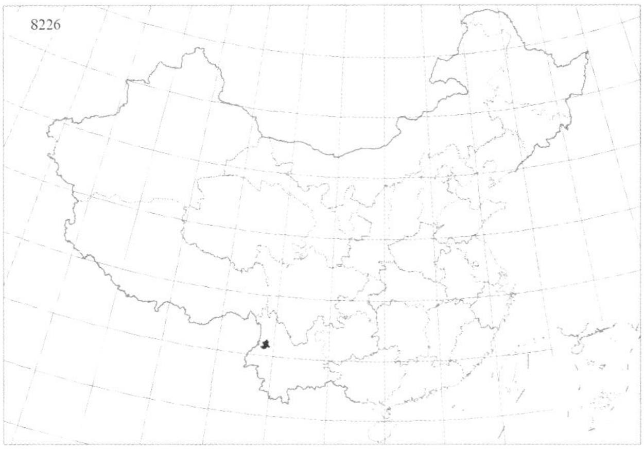

Ericaceae 杜鹃花科 *Rhododendron* 杜鹃花属

Rhododendron rhodanthum M. Y. He 淡红杜鹃

Shrubs, deciduous

16.5; 5.2; 27.1; 16.5; 138.3; 0.0; 875 | 1485; 442; 207 | 69.7; 873; 199.7

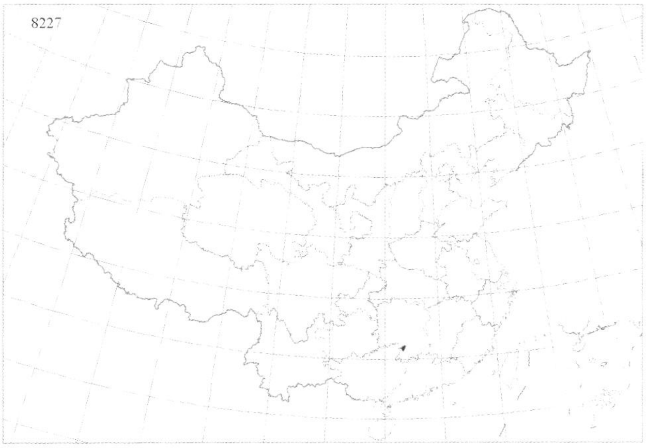

Ericaceae 杜鹃花科 *Rhododendron* 杜鹃花属

Rhododendron rhombifolium R. C. Fang 菱形叶杜鹃

Shrubs, evergreen

8.6; 1.3; 15.0; 8.7; 52.6; 8.8; 574 | 1041; 585; 45 | 81.5; 574; 352.3

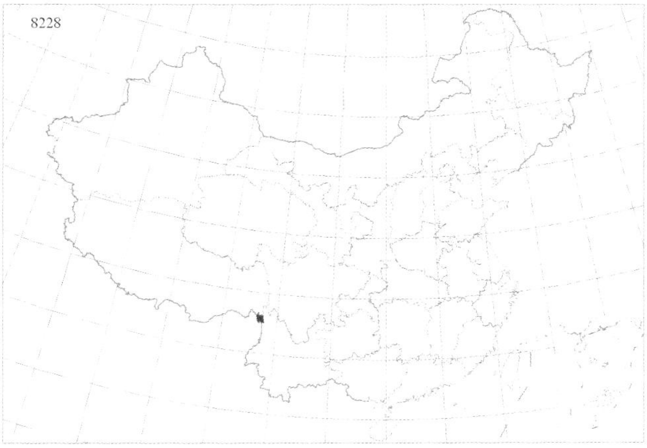

Ericaceae 杜鹃花科 *Rhododendron* 杜鹃花属

Rhododendron rhuyuenense Chun ex Tam 乳源杜鹃

Shrubs, evergreen

18.1 (16.7~19.4); 8.0 (5.9~13.3); 27.1 (22.5~28.7); 18.2 (16.7~19.4); 158.0 (141.0~173.4); 0.0 (0.0~0.0); 943 (854~1029) | 1658 (1276~1950); 599 (481~756); 191 (51~259) | 76.0 (36.8~121.9); 942 (854~1029); 261.5 (188.8~491.5)

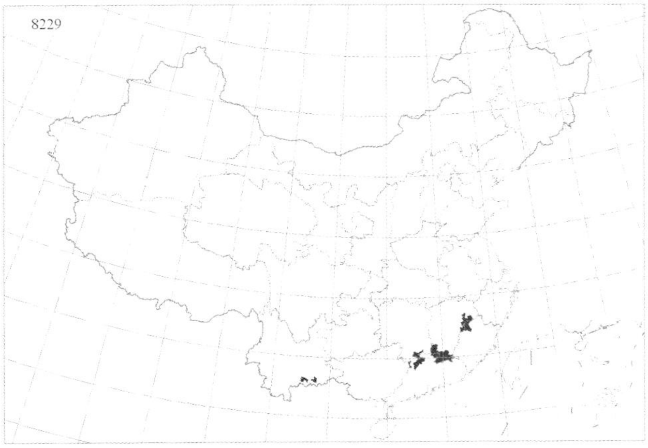

Ericaceae 杜鹃花科 *Rhododendron* 杜鹃花属

Rhododendron riparioides (Cullen) Cubey 雪龙美被杜鹃

Small shrubs, evergreen

10.5; 3.4; 16.3; 10.5; 70.9; 4.7; 623 | 1129; 630; 51 | 80.6; 623; 370.4

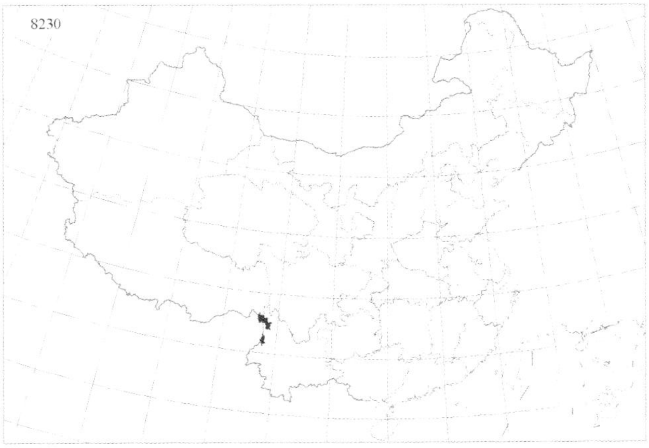

Ericaceae 杜鹃花科 *Rhododendron* 杜鹃花属

Rhododendron ririei Hemsl. et E. H. Wilson 大钟杜鹃

Shrubs or small trees, evergreen

13.0 (9.0~15.2); 3.0 (0.6~5.3); 22.3 (16.5~25.9); 13.0 (9.0~15.2); 100.0 (58.7~123.9); 3.7 (0.0~10.3); 722 (584~816) | 1261 (1001~1492); 596 (473~816); 63 (30~96) | 75.5 (49.1~107.2); 722 (584~816); 242.7 (170.4~298.1)

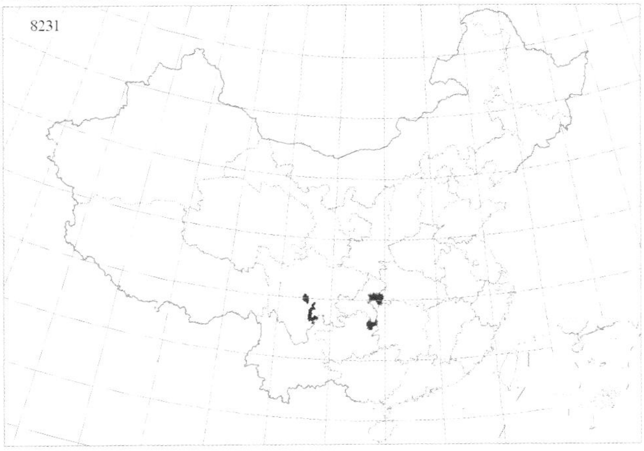

Ericaceae 杜鹃花科 *Rhododendron* 杜鹃花属

Rhododendron roseatum Hutch. 红晕杜鹃

Shrubs, evergreen

12.9; 5.9; 18.1; 12.9; 94.9; 0.3; 686 | 1349; 744; 64 | 96.4; 686; 436.0

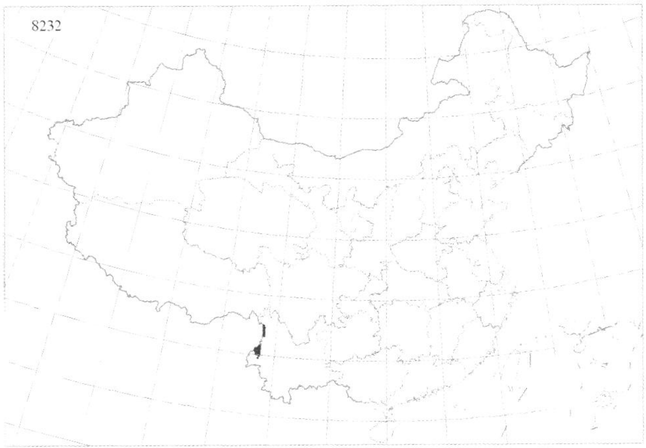

Ericaceae 杜鹃花科 *Rhododendron* 杜鹃花属
Rhododendron rothschildii Davidian 宽柄杜鹃
Shrubs or small trees, evergreen

9.6; 2.4; 15.7; 9.6; 60.8; 5.4; 600 ❙ 1007; 578; 40 ❙ 67.9; 600; 351.7

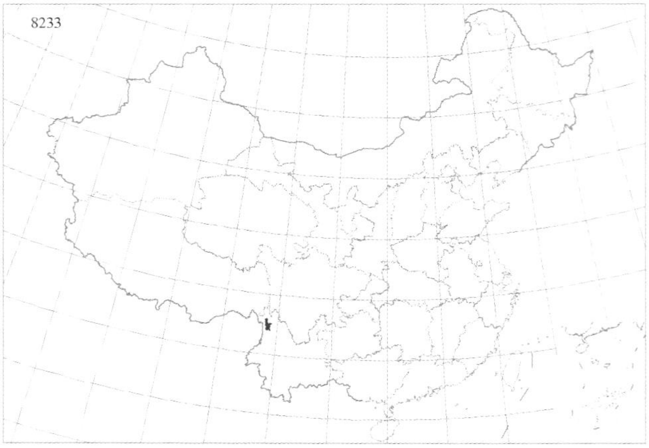

Ericaceae 杜鹃花科 *Rhododendron* 杜鹃花属
Rhododendron roxieanum Forrest ex W. W. Smith 卷叶杜鹃
Shrubs, evergreen

9.8 (5.3~14.1); 0.7 (−2.8~6.5); 17.9 (12.2~22.9); 10.0 (5.7~14.2); 70.4 (30.5~109.8); 12.5 (0.0~26.2); 616 (479~729) ❙ 868 (626~1337); 481 (305~727); 26 (12~69) ❙ 42.7 (−8.1~100.7); 610 (479~707); 300.4 (201.9~407.1)

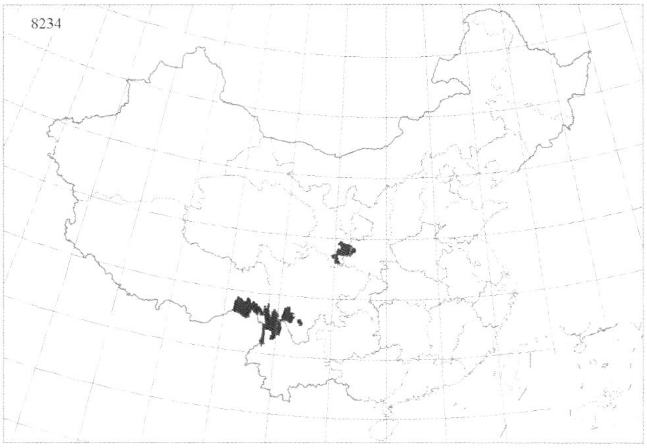

Ericaceae 杜鹃花科 *Rhododendron* 杜鹃花属
Rhododendron roxieanum var. cucullatum (Hand.-Mazz.) D. F. Chamb. 兜尖卷叶杜鹃
Shrubs, evergreen

9.4 (5.3~14.1); 2.0 (−2.4~6.5); 15.5 (12.2~20.2); 9.5 (5.7~14.2); 63.7 (30.5~109.8); 10.7 (0.0~26.2); 591 (479~729) ❙ 1001 (741~1337); 590 (426~727); 32 (12~69) ❙ 69.8 (40.1~100.7); 589 (479~707); 316.4 (234.2~407.1)

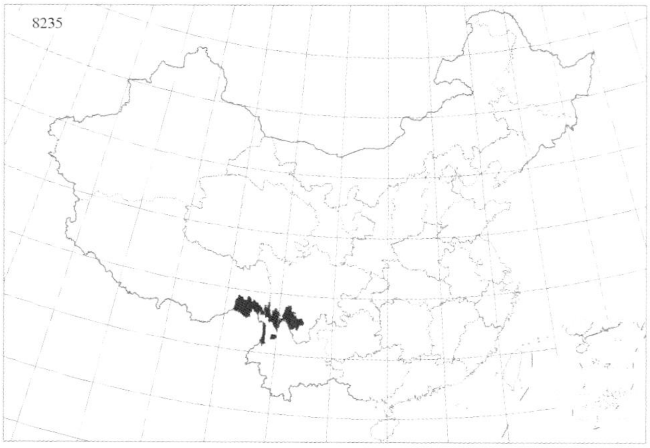

Ericaceae 杜鹃花科 *Rhododendron* 杜鹃花属
Rhododendron roxieanum var. oreonastes (I. B. Balfour et Forrest) T. L. Ming 线形卷叶杜鹃
Shrubs, evergreen

8.6; 1.4; 14.9; 8.8; 54.7; 10.8; 571 ❙ 905; 533; 31 ❙ 58.2; 571; 315.0

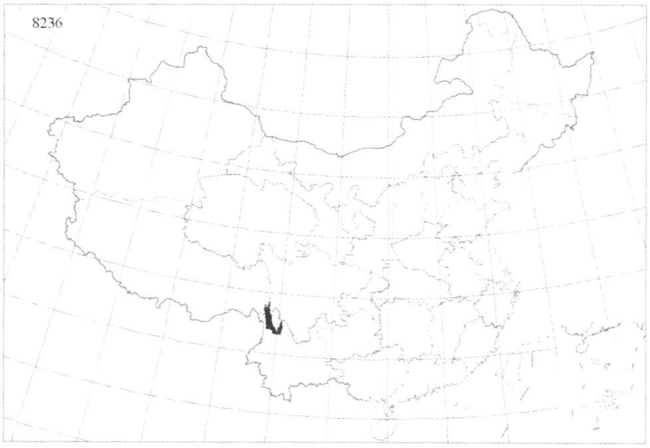

Ericaceae 杜鹃花科 *Rhododendron* 杜鹃花属
Rhododendron roxieoides D. F. Chamb. 巫山杜鹃
Shrubs, evergreen

13.7; 2.6; 24.4; 13.8; 109.1; 4.0; 758 ❙ 1148; 475; 68 ❙ 51.4; 758; 258.0

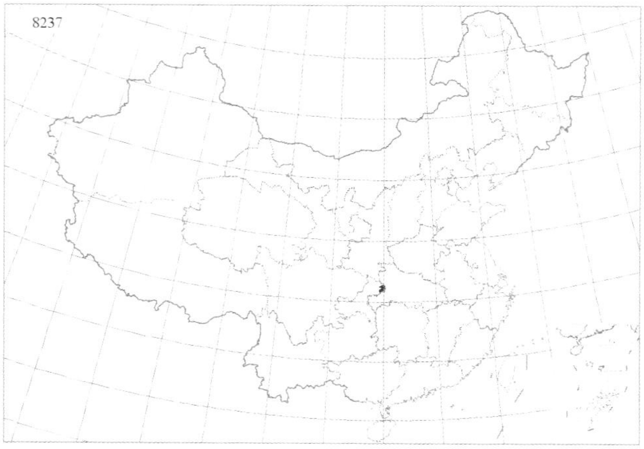

Ericaceae 杜鹃花科 *Rhododendron* 杜鹃花属
Rhododendron rubiginosum Franch. 红棕杜鹃
Shrubs, evergreen

12.3 (2.0~18.9); 4.4 (−7.2~12.7); 18.8 (9.5~26.6); 12.4 (3.8~19.0); 93.7 (15.4~167.8); 6.0 (0.0~51.4); 684 (411~929) ❙ 1003 (729~1547); 579 (411~911); 31 (11~69) ❙ 50.8 (−14.2~128.6); 664 (411~912); 283.9 (142.9~551.3)

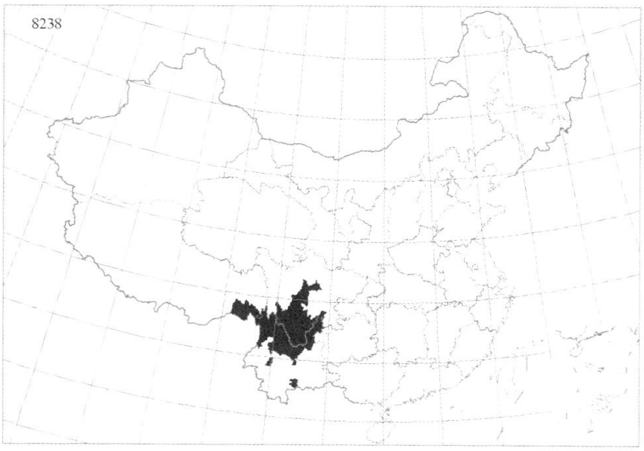

Ericaceae 杜鹃花科 *Rhododendron* 杜鹃花属
Rhododendron rubiginosum var. leclerei (H. Lév.) R. C. Fang 洁净红棕杜鹃
Shrubs, evergreen

12.4; 4.7; 18.6; 12.4; 89.1; 0.3; 675 ❙ 896; 528; 31 ❙ 32.7; 675; 205.2

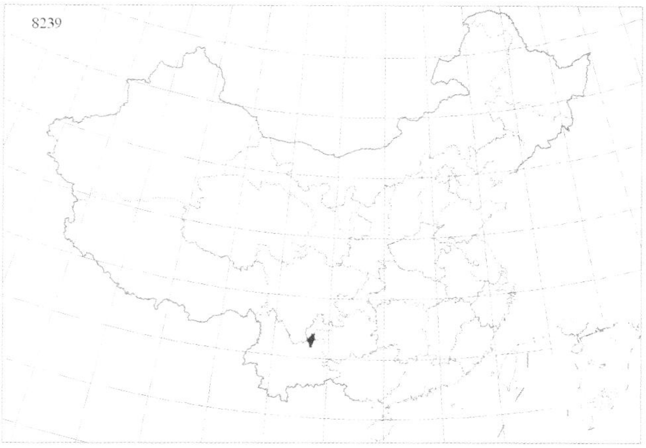

Ericaceae 杜鹃花科 *Rhododendron* 杜鹃花属
Rhododendron rubiginosum var. ptilostylum R. C. Fang 毛柱红棕杜鹃
Shrubs, evergreen

10.8; 3.9; 16.6; 10.9; 71.8; 1.3; 633 ❙ 968; 594; 24 ❙ 52.8; 633; 319.8

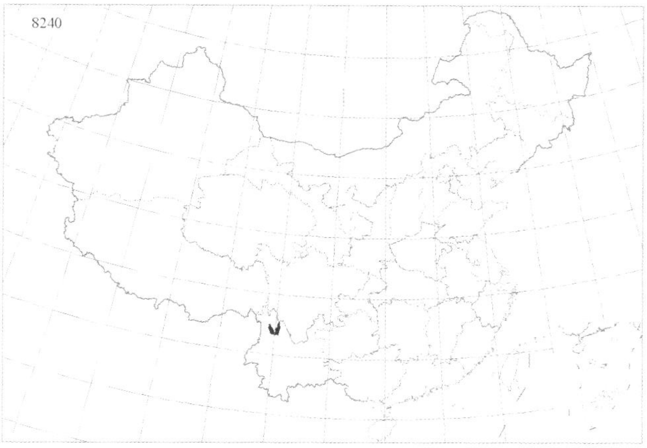

Ericaceae 杜鹃花科 *Rhododendron* 杜鹃花属
Rhododendron rubropilosum Hayata 台红毛杜鹃
Shrubs

20.4 (16.2~23.8); 14.8 (11.0~18.1); 24.8 (20.1~28.2); 20.4 (16.2~23.8); 184.7 (134.2~225.8); 0.0 (0.0~0.0); 1038 (773~1297) ❙ 2371 (1757~2859); 1216 (959~1412); 178 (58~402) ❙ 145.4 (35.5~246.5); 982 (773~1144); 218.0 (119.4~289.1)

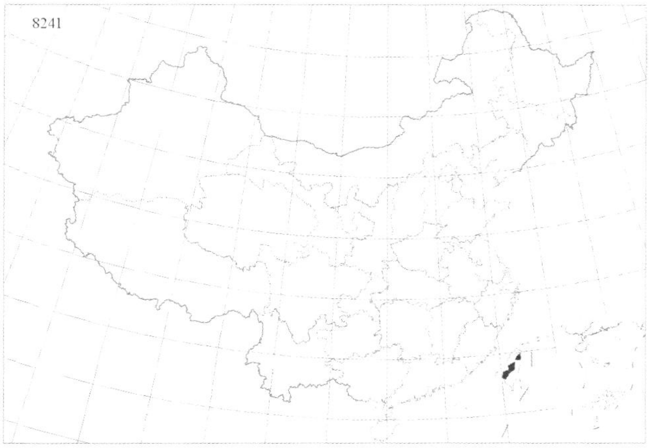

Ericaceae 杜鹃花科 *Rhododendron* 杜鹃花属
Rhododendron rufohirtum Hand.-Mazz. 滇红毛杜鹃
Shrubs, semievergreen or green

14.4 (11.7~17.5); 5.7 (2.7~10.5); 21.5 (18.8~24.0); 14.4 (11.7~17.5); 113.8 (82.6~149.9); 0.7 (0.0~3.8); 748 (658~846) ❙ 1099 (879~1334); 595 (472~720); 49 (29~64) ❙ 46.9 (13.3~70.6); 745 (658~845); 223.6 (183.0~311.9)

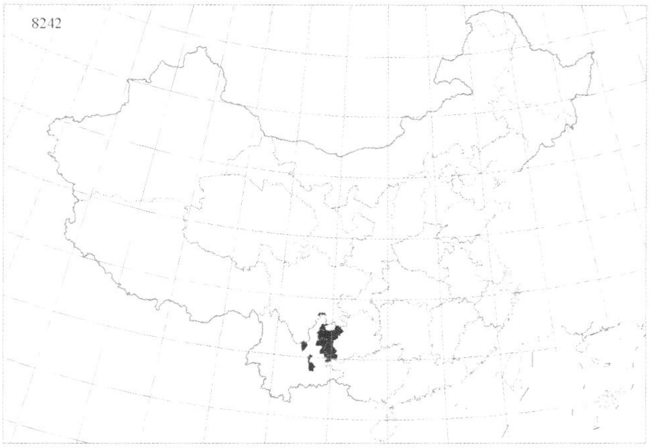

Ericaceae 杜鹃花科 *Rhododendron* 杜鹃花属
Rhododendron rufum Batalin 黄毛杜鹃
Shrubs or small trees, evergreen

4.8 (-1.0~14.8); -6.2 (-12.5~6.2); 14.5 (8.5~25.3); 6.4 (2.7~14.8); 40.0 (8.3~119.2); 42.0 (0.0~79.5); 497 (363~795) ❙ 709 (373~1358); 368 (201~644); 18 (7~88) ❙ 45.3 (-37.6~128.6); 481 (363~795); 257.3 (81.2~357.7)

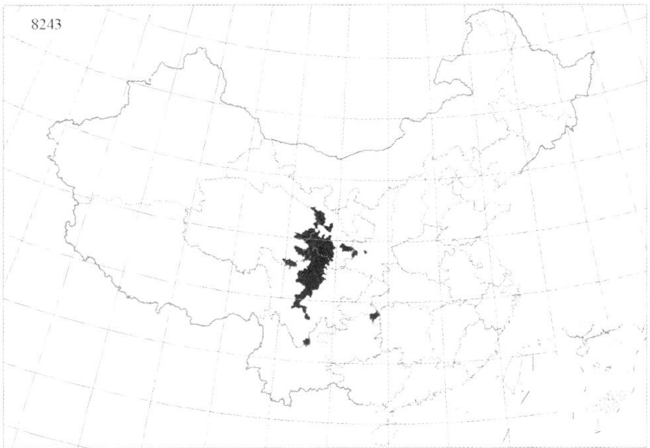

Ericaceae 杜鹃花科 *Rhododendron* 杜鹃花属
Rhododendron rupicola W. W. Sm. 多色杜鹃
Small shrubs

9.0 (5.3~13.2); 1.7 (-2.4~6.4); 15.1 (12.2~18.4); 9.1 (5.7~13.3); 58.3 (30.5~99.4); 10.2 (0.0~26.2); 580 (479~695) ❙ 997 (741~1337); 585 (426~727); 34 (12~69) ❙ 71.6 (51.8~100.7); 579 (479~695); 324.5 (234.2~407.1)

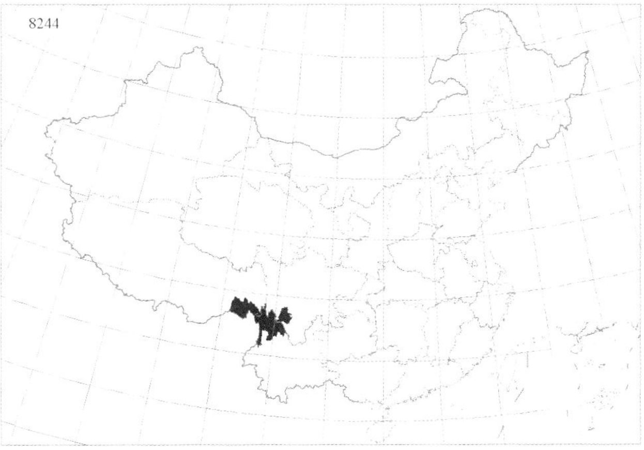

Ericaceae 杜鹃花科　*Rhododendron* 杜鹃花属

Rhododendron rupicola var. chryseum (I. B. Balfour et Kingdon-Ward) Philipson et M. N. Philipson 金黄多色杜鹃　Small shrubs

10.3；2.4；16.7；10.4；72.8；9.0；619 ❙ 1376；794；38 ❙ 119.8；619；348.4

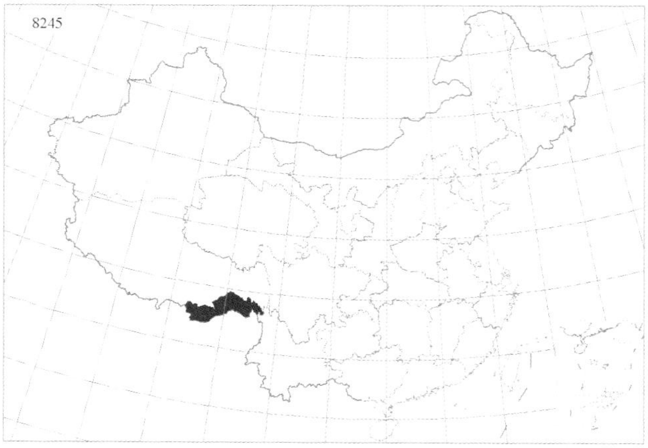

Ericaceae 杜鹃花科　*Rhododendron* 杜鹃花属

Rhododendron rupivalleculatum P. C. Tam 岩谷杜鹃

Small shrubs, evergreen

19.3 (15.5~24.3)；10.2 (4.8~18.7)；26.9 (25.1~28.4)；19.3 (15.6~24.3)；172.1 (127.2~231.9)；0.0 (0.0~0.2)；1021 (815~1338) ❙ 1530 (1179~1900)；638 (392~1056)；133 (42~182) ❙ 55.6 (-11.9~93.2)；965 (815~1111)；286.5 (197.5~455.2)

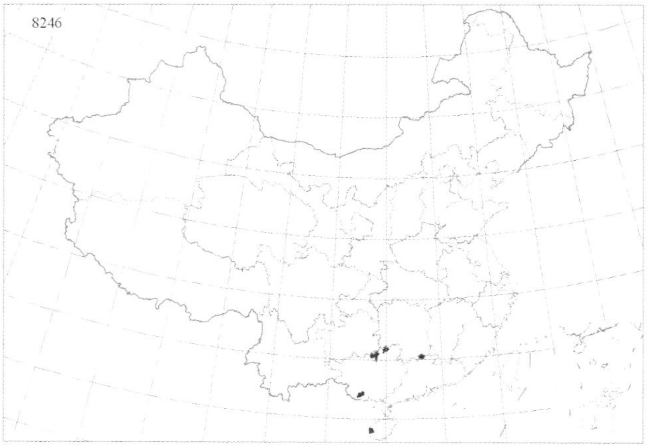

Ericaceae 杜鹃花科　*Rhododendron* 杜鹃花属

Rhododendron russatum I. B. Balfour et Forrest 紫兰杜鹃

Small shrubs, evergreen

7.7 (5.4~10.8)；0.3 (-2.1~3.9)；14.1 (12.3~16.6)；7.9 (5.8~10.9)；47.8 (31.5~71.8)；14.8 (1.3~25.6)；545 (481~633) ❙ 876 (741~1007)；529 (426~594)；25 (12~40) ❙ 60.8 (52.8~67.9)；545 (481~633)；307.9 (273.5~351.7)

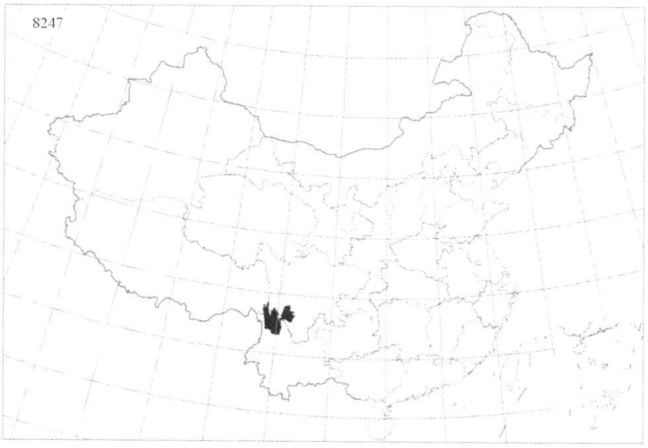

Ericaceae 杜鹃花科　*Rhododendron* 杜鹃花属

Rhododendron saluenense Franch. 怒江杜鹃

Shrubs, evergreen

8.7 (5.3~14.5)；1.1 (-2.4~7.0)；15.2 (12.2~20.2)；8.9 (5.7~14.6)；58.6 (30.5~115.0)；13.3 (0.0~26.2)；575 (479~739) ❙ 936 (741~1041)；549 (426~615)；28 (12~45) ❙ 65.2 (32.4~100.7)；573 (479~739)；288.4 (221.3~352.3)

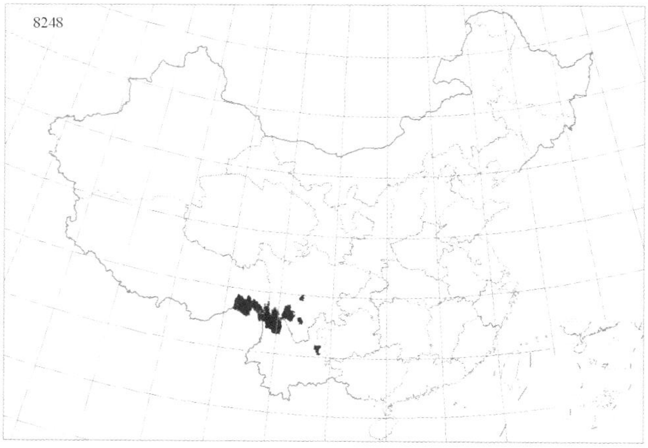

Ericaceae 杜鹃花科　*Rhododendron* 杜鹃花属

Rhododendron saluenense var. prostratum (W. W. Sm.) R. C. Fang 平卧怒江杜鹃　Small shrubs, evergreen

8.1 (5.4~10.8)；0.8 (-2.1~3.9)；14.5 (12.3~16.6)；8.3 (5.8~10.9)；50.4 (31.5~71.8)；12.6 (1.3~25.6)；557 (481~633) ❙ 912 (741~1041)；536 (426~594)；32 (21~45) ❙ 63.5 (52.8~81.5)；557 (481~633)；316.0 (273.5~352.3)

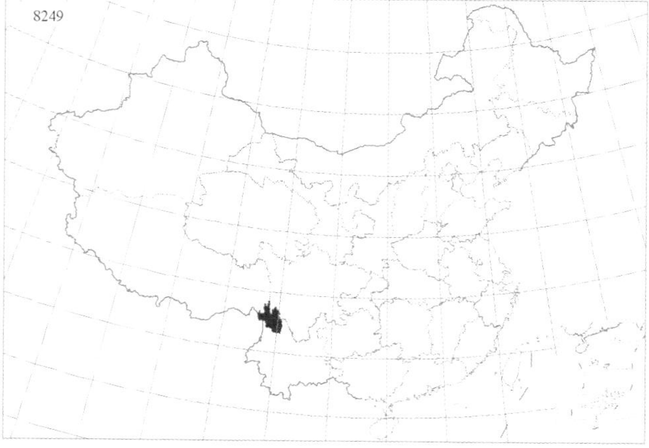

Ericaceae 杜鹃花科　*Rhododendron* 杜鹃花属

Rhododendron sanguineum Franch. 血红杜鹃

Small shrubs, evergreen

6.2 (1.8~9.6)；-1.6 (-7.1~2.4)；13.0 (9.9~15.7)；6.7 (3.8~9.6)；38.2 (15.7~60.8)；23.9 (5.4~53.3)；508 (404~600) ❙ 867 (583~1041)；499 (348~585)；31 (13~45) ❙ 69.7 (44.2~100.7)；508 (404~600)；276.2 (169.5~352.3)

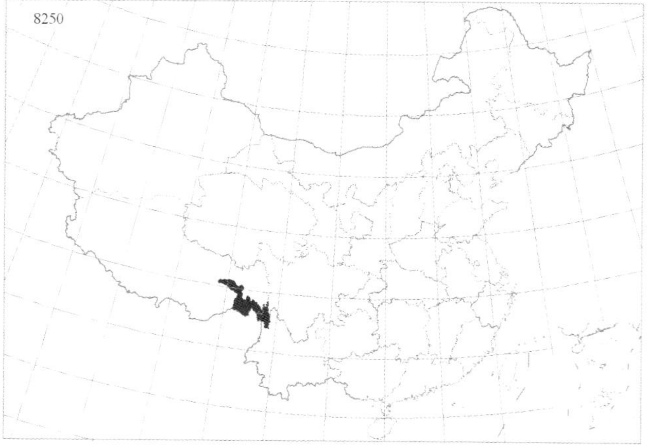

Ericaceae 杜鹃花科 *Rhododendron* 杜鹃花属

Rhododendron sanguineum var. cloiophorum (I. B. Balfour et Forrest) D. F. Chamb. 退色血红杜鹃 Small shrubs, evergreen

6.5; −1.1; 13.2; 6.7; 38.2; 20.2; 511 ❙ 915; 523; 35 ❙ 78.7; 511; 286.6

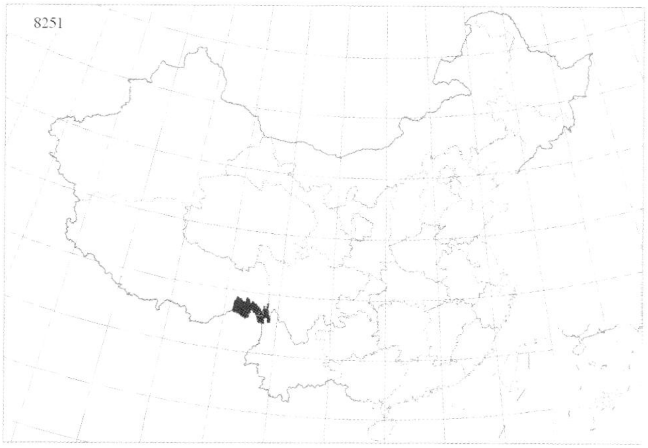

Ericaceae 杜鹃花科 *Rhododendron* 杜鹃花属

Rhododendron sanguineum var. didymoides Tagg et Forrest 变色血红杜鹃 Small shrubs, evergreen

6.5; −1.1; 13.2; 6.7; 38.2; 20.2; 511 ❙ 915; 523; 35 ❙ 78.7; 511; 286.6

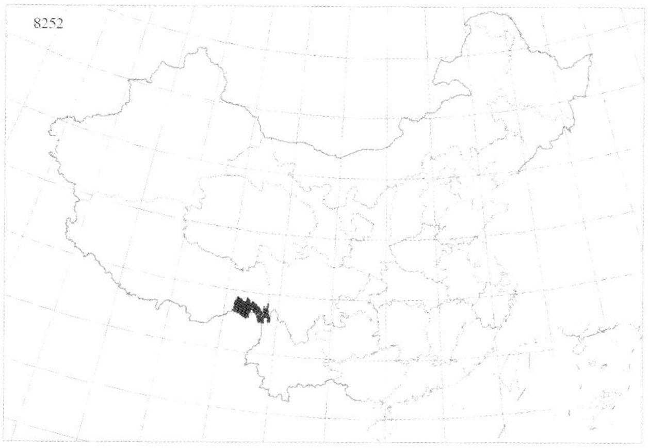

Ericaceae 杜鹃花科 *Rhododendron* 杜鹃花属

Rhododendron sanguineum var. didymum (I. B. Balfour et Forrest) T. L. Ming 黑红血红杜鹃 Small shrubs, evergreen

6.5; −1.1; 13.2; 6.7; 38.2; 20.2; 511 ❙ 915; 523; 35 ❙ 78.7; 511; 286.6

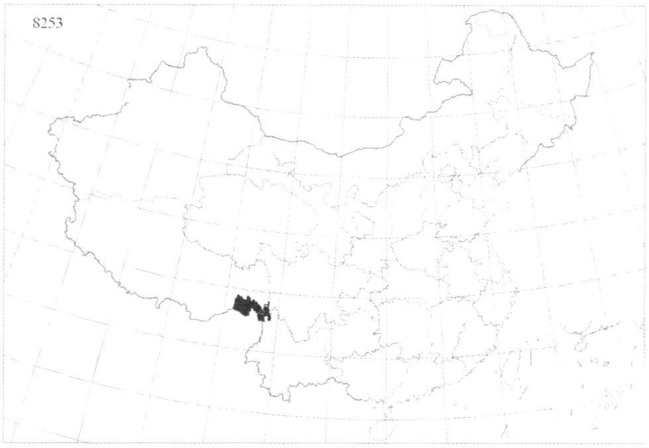

Ericaceae 杜鹃花科 *Rhododendron* 杜鹃花属

Rhododendron sanguineum var. haemaleum (I. B. Balfour et Forrest) D. F. Chamb. 紫血杜鹃 Small shrubs, evergreen

6.5; −1.1; 13.2; 6.7; 38.2; 20.2; 511 ❙ 915; 523; 35 ❙ 78.7; 511; 286.6

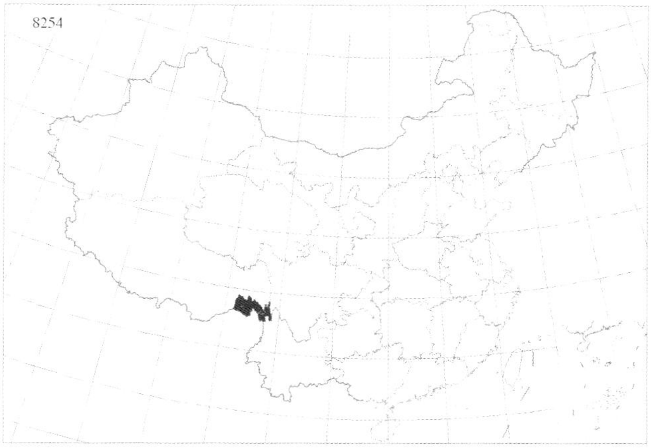

Ericaceae 杜鹃花科 *Rhododendron* 杜鹃花属

Rhododendron sanguineum var. himertum (I. B. Balfour et Forrest) D. F. Chamb. 蜜黄血红杜鹃 Small shrubs, evergreen

5.4; −2.3; 12.2; 5.8; 31.0; 25.9; 480 ❙ 851; 492; 30 ❙ 77.4; 480; 253.8

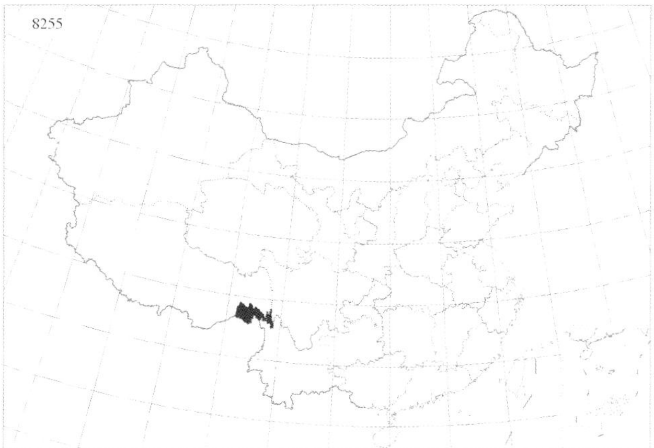

Ericaceae 杜鹃花科 *Rhododendron* 杜鹃花属

Rhododendron sargentianum Rehder et E. H. Wilson 水仙杜鹃
Small shrubs, evergreen

8.2 (2.2~13.3); −1.0 (−6.6~3.2); 16.4 (9.9~22.1); 8.6 (3.8~13.2); 57.8 (15.5~100.9); 19.3 (2.5~49.1); 570 (413~723) ❙ 1045 (932~1220); 576 (531~688); 28 (19~37) ❙ 87.1 (49.4~125.9); 570 (413~723); 236.1 (171.0~292.6)

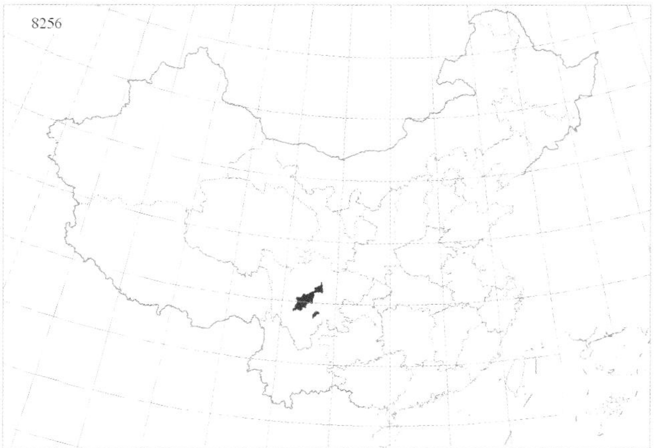

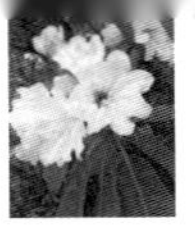

Ericaceae 杜鹃花科 *Rhododendron* 杜鹃花属
Rhododendron saxatile B. Y. Ding et Y. Y. Fang 崖壁杜鹃
Small shrubs, semievergreen or green

17.6; 7.4; 27.6; 17.6; 151.2; 0.0; 919 | 1674; 611; 179 | 82.2; 919; 216.1

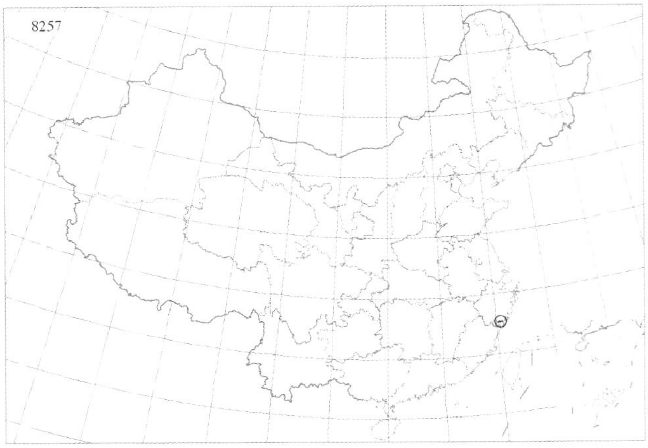

Ericaceae 杜鹃花科 *Rhododendron* 杜鹃花属
Rhododendron scabrifolium Franch. 糙叶杜鹃
Shrubs, evergreen

13.6 (5.4~18.9); 6.0 (-2.1~12.0); 19.8 (12.3~26.6); 13.6 (5.8~19.0); 105.5 (31.5~167.8); 1.9 (0.0~25.6); 720 (481~929) | 973 (741~1454); 564 (411~792); 32 (15~70) | 36.9 (-14.2~99.9); 692 (481~912); 279.8 (142.9~487.2)

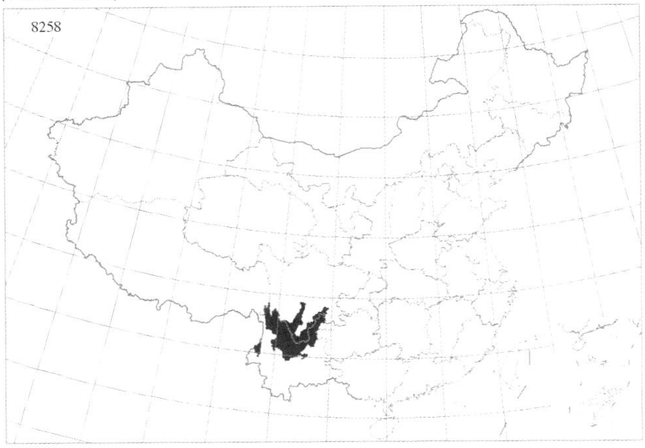

Ericaceae 杜鹃花科 *Rhododendron* 杜鹃花属
Rhododendron schistocalyx I. B. Balfour et Forrest 裂萼杜鹃
Shrubs, evergreen

14.4; 7.5; 19.1; 14.4; 113.1; 0.0; 727 | 1454; 792; 70 | 99.9; 727; 487.2

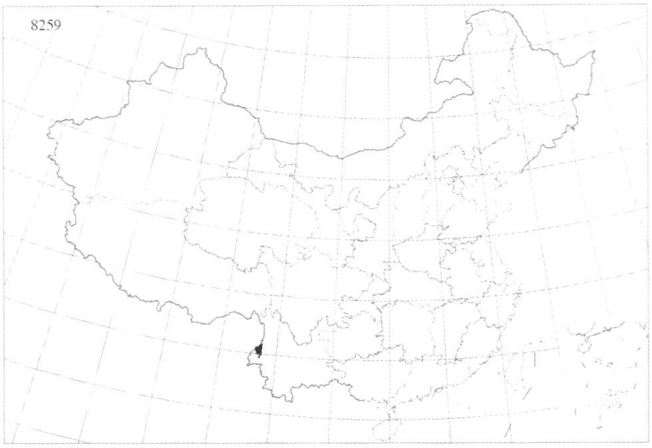

Ericaceae 杜鹃花科 *Rhododendron* 杜鹃花属
Rhododendron schlippenbachii Maxim. 大字杜鹃
Shrubs, deciduous

6.5 (1.4~9.5); -12.0 (-18.8~-7.2); 22.4 (18.4~24.8); 9.2 (6.4~10.8); 74.8 (45.8~91.5); 56.8 (35.9~88.8); 625 (501~701) | 775 (406~1073); 489 (281~691); 29 (7~40) | 25.6 (-38.1~74.1); 593 (406~684); 256.0 (132.8~346.5)

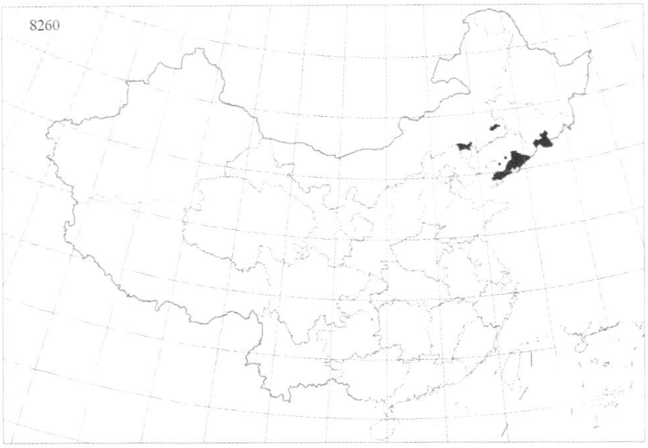

Ericaceae 杜鹃花科 *Rhododendron* 杜鹃花属
Rhododendron scopulorum Hutch. 石峰杜鹃
Shrubs, evergreen

6.7; -1.8; 14.0; 7.5; 48.7; 27.9; 521 | 912; 537; 22 | 57.8; 521; 319.4

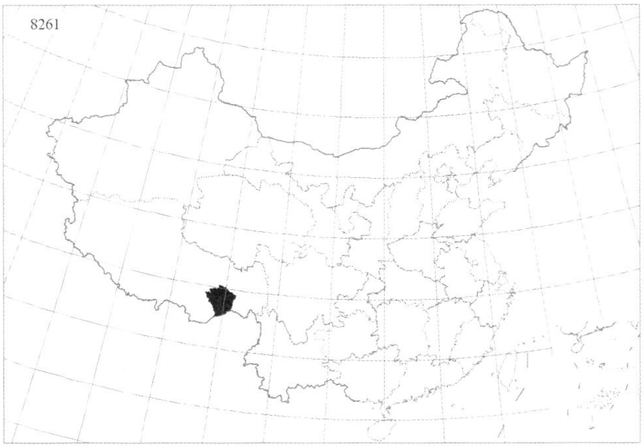

Ericaceae 杜鹃花科 *Rhododendron* 杜鹃花属
Rhododendron searsiae Rehder et E. H. Wilson 绿点杜鹃
Shrubs, evergreen

13.1 (11.1~15.9); 3.9 (2.2~6.1); 21.6 (19.4~24.9); 13.2 (11.3~16.0); 100.8 (80.8~131.7); 2.1 (0.0~5.3); 719 (652~824) | 1144 (985~1414); 634 (509~817); 40 (30~52) | 60.4 (19.4~89.0); 719 (652~824); 257.5 (200.3~298.1)

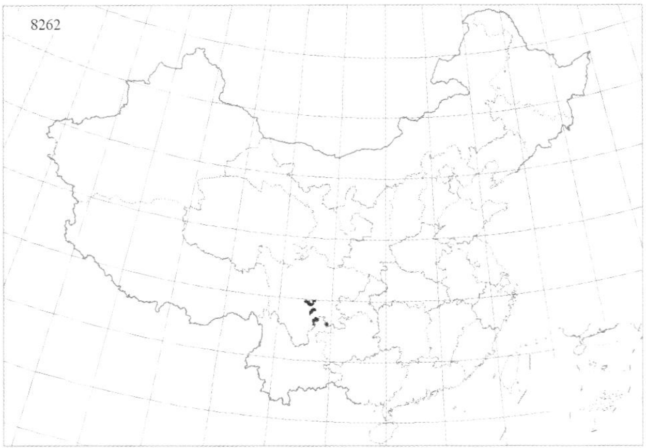

Ericaceae 杜鹃花科 *Rhododendron* 杜鹃花属
Rhododendron seinghkuense Kingdon-Ward ex Hutch. 黄花泡泡叶杜鹃 Shrubs, evergreen

7.0; −0.5; 13.6; 7.2; 41.5; 17.5; 526 ❙ 1001; 571; 38 ❙ 91.1; 526; 293.2

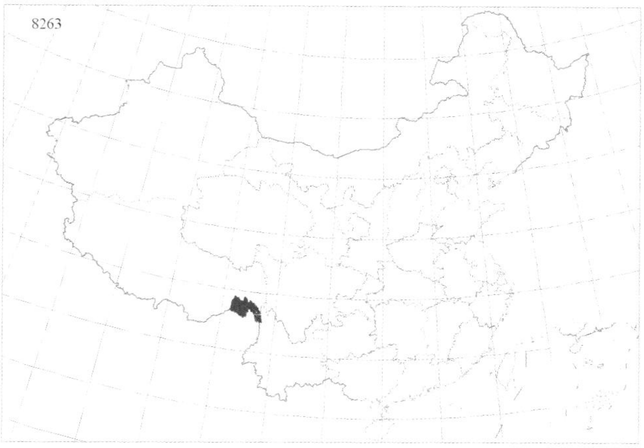

Ericaceae 杜鹃花科 *Rhododendron* 杜鹃花属
Rhododendron selense Franch. 多变杜鹃
Small shrubs, evergreen

8.7 (1.8~14.5); 1.4 (−7.1~7.8); 14.9 (9.9~19.6); 9.0 (3.8~14.5); 58.6 (15.7~113.5); 13.6 (0.0~53.3); 574 (404~730) ❙ 970 (583~1337); 563 (348~727); 36 (12~69) ❙ 68.6 (43.0~100.7); 574 (404~730); 318.4 (169.5~407.1)

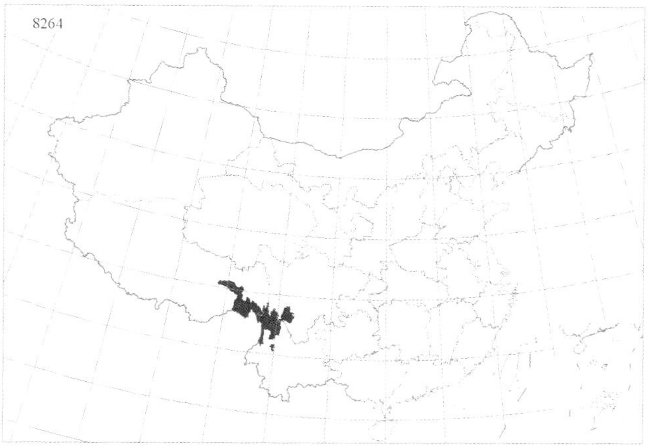

Ericaceae 杜鹃花科 *Rhododendron* 杜鹃花属
Rhododendron selense subsp. dasycladum (I. B. Balfour et W. W. Sm.) D. F. Chamb. 毛枝多变杜鹃 Small shrubs, evergreen

7.9 (5.3~11.2); 0.5 (−2.4~4.6); 14.2 (12.2~16.7); 8.1 (5.7~11.3); 49.3 (30.5~75.8); 14.4 (0.4~26.2); 549 (479~642) ❙ 913 (741~1054); 547 (426~627); 27 (12~40) ❙ 67.0 (52.8~100.7); 549 (479~642); 303.6 (234.2~351.7)

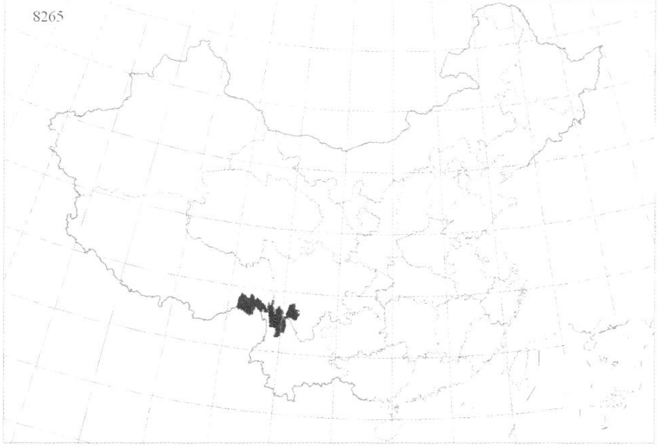

Ericaceae 杜鹃花科 *Rhododendron* 杜鹃花属
Rhododendron selense subsp. jucundum (I. B. Balfour et W. W. Sm.) D. F. Chamb. 粉背多变杜鹃 Small shrubs, evergreen

14.5; 7.8; 19.6; 14.5; 113.5; 0.0; 730 ❙ 1044; 570; 54 ❙ 43.0; 730; 382.6

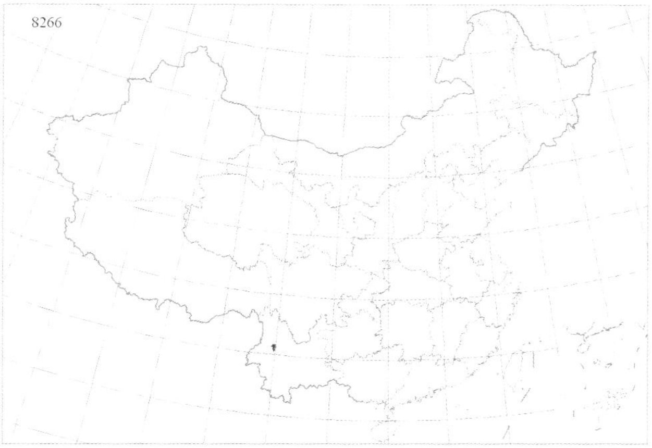

Ericaceae 杜鹃花科 *Rhododendron* 杜鹃花属
Rhododendron semnoides Tagg et Forrest 圆头杜鹃
Shrubs or small trees, evergreen

8.9 (5.3~13.1); 1.7 (−2.4~6.4); 15.0 (12.2~18.4); 9.0 (5.7~13.2); 58.5 (30.5~98.1); 11.7 (0.0~26.2); 577 (479~692) ❙ 968 (741~1164); 559 (426~639); 36 (24~58) ❙ 68.7 (52.8~100.7); 577 (479~692); 309.1 (234.2~366.3)

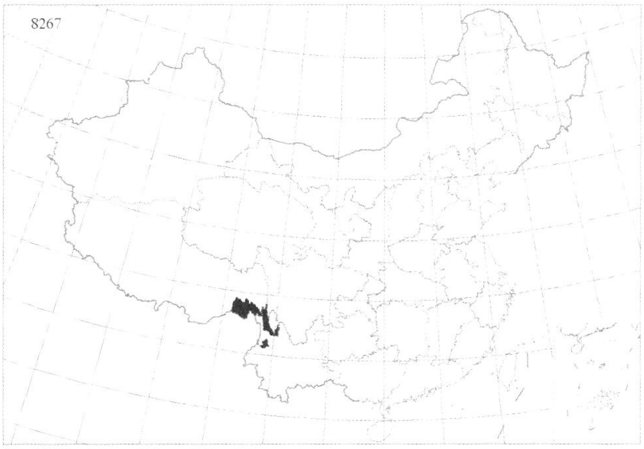

Ericaceae 杜鹃花科 *Rhododendron* 杜鹃花属
Rhododendron seniavinii Maxim. 毛果杜鹃
Shrubs, evergreen

16.7 (14.3~21.6); 6.1 (3.3~12.6); 26.4 (23.5~29.1); 16.7 (14.3~21.7); 141.0 (112.6~199.8); 0.2 (0.0~2.3); 877 (756~1153) ❙ 1507 (1159~1997); 555 (444~795); 161 (64~259) ❙ 71.6 (34.4~147.0); 876 (756~1153); 233.2 (169.7~332.5)

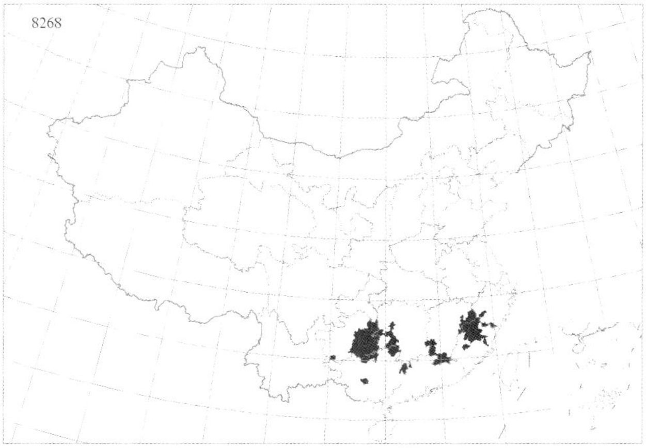

Ericaceae 杜鹃花科 *Rhododendron* 杜鹃花属

Rhododendron serotinum Hucthinson 晚波杜鹃

Shrubs

17.9; 11.1; 22.6; 17.9; 155.0; 0.0; 863 ❙ 1343; 765; 53 ❙ 55.4; 863; 416.4

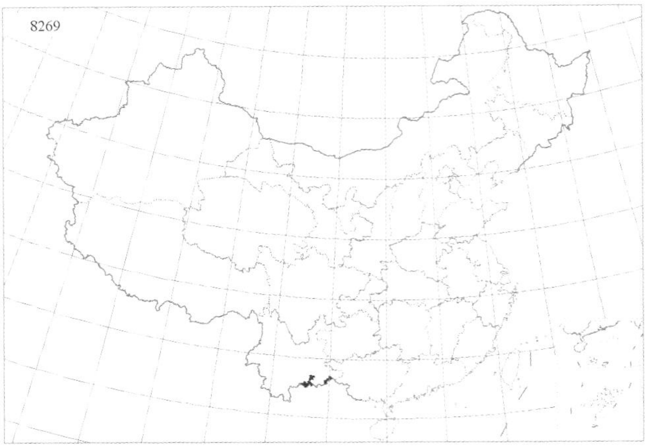

Ericaceae 杜鹃花科 *Rhododendron* 杜鹃花属

Rhododendron setiferum I. B. Balfour et Forrest 刚刺杜鹃

Shrubs, evergreen

7.2; -0.2; 13.8; 7.5; 43.8; 16.5; 533 ❙ 938; 537; 36 ❙ 76.0; 533; 302.9

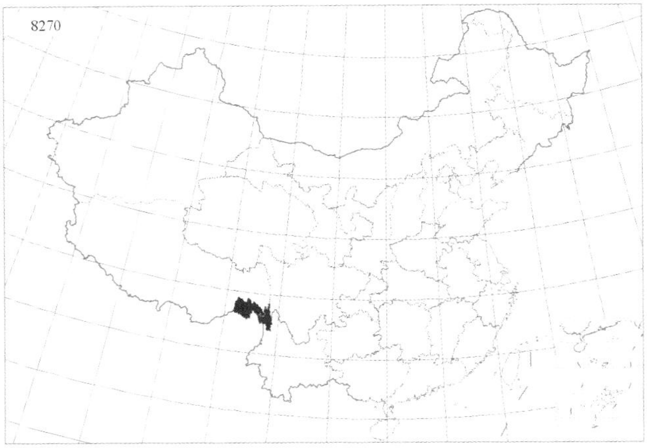

Ericaceae 杜鹃花科 *Rhododendron* 杜鹃花属

Rhododendron setosum D. Don 刚毛杜鹃

Small shrubs, evergreen

1.2 (-1.8~12.2); -7.3 (-10.5~4.3); 9.0 (6.1~18.5); 3.9 (1.8~12.2); 18.8 (1.9~87.4); 63.5 (0.7~82.8); 395 (329~672) ❙ 524 (311~1431); 327 (193~817); 19 (10~33) ❙ 17.4 (-6.3~113.0); 388 (311~672); 130.3 (61.6~363.7)

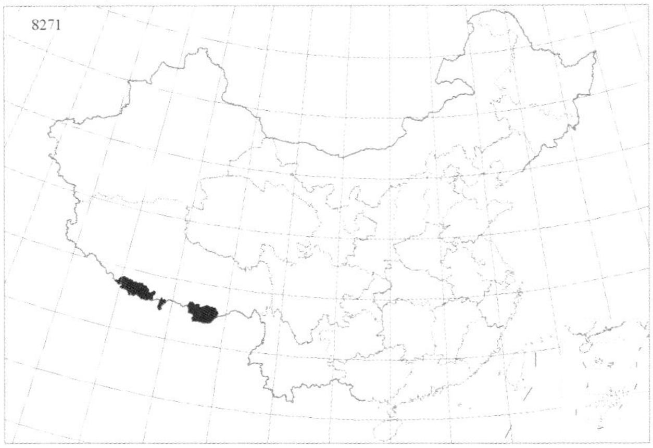

Ericaceae 杜鹃花科 *Rhododendron* 杜鹃花属

Rhododendron shanii Fang 都支杜鹃

Trees, evergreen

15.0 (13.4~16.1); 2.8 (1.5~4.4); 27.1 (25.2~28.0); 15.1 (13.4~16.1); 124.4 (108.3~133.8); 3.5 (0.6~7.3); 838 (755~886) ❙ 1397 (1198~1571); 596 (536~666); 132 (109~153) ❙ 67.4 (39.7~108.0); 838 (755~886); 200.0 (169.3~243.1)

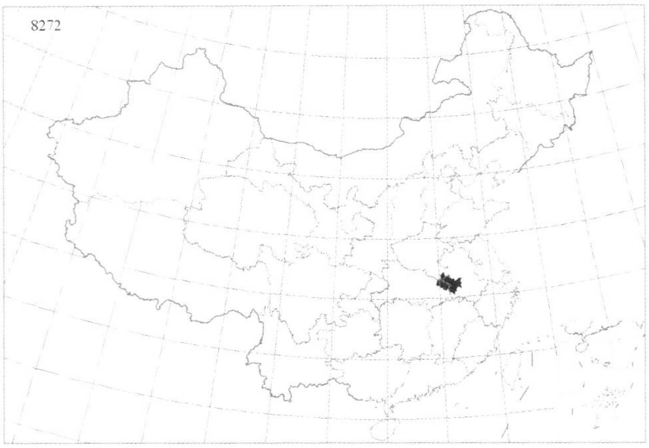

Ericaceae 杜鹃花科 *Rhododendron* 杜鹃花属

Rhododendron sherriffii Cowan 红钟杜鹃

Shrubs, evergreen

4.4; -4.3; 12.1; 5.3; 28.5; 34.6; 455 ❙ 538; 337; 9 ❙ 18.2; 455; 168.2

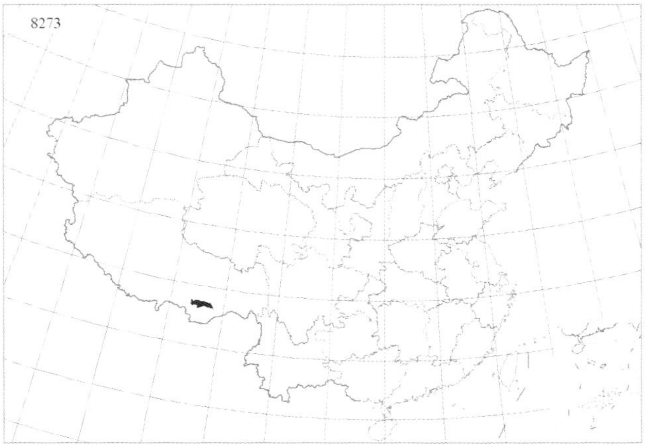

Ericaceae 杜鹃花科 *Rhododendron* 杜鹃花属

Rhododendron shimianense W. P. Fang et P. S. Liu 石棉杜鹃

Shrubs, evergreen

10.1; 1.8; 17.7; 10.3; 70.1; 7.0; 621 ❙ 1023; 568; 22 ❙ 64.7; 621; 318.7

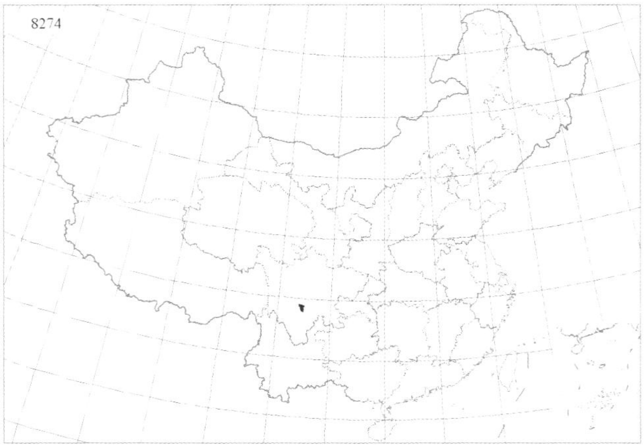

Ericaceae 杜鹃花科 *Rhododendron* 杜鹃花属

Rhododendron tapetiforme I. B. Balfour et Kingdon-Ward 犾萼杜鹃

Shrubs, evergreen

6.0 (0.0~13.3); -1.9 (-8.7~5.4); 13.0 (8.1~19.6); 6.8 (2.7~13.4); 41.1 (8.0~100.4); 28.3 (0.0~67.9); 505 (363~706) ❙ 960 (559~1736); 564 (335~1007); 29 (14~51) ❙ 83.2 (54.0~145.8); 505 (363~706); 269.8 (111.0~447.4)

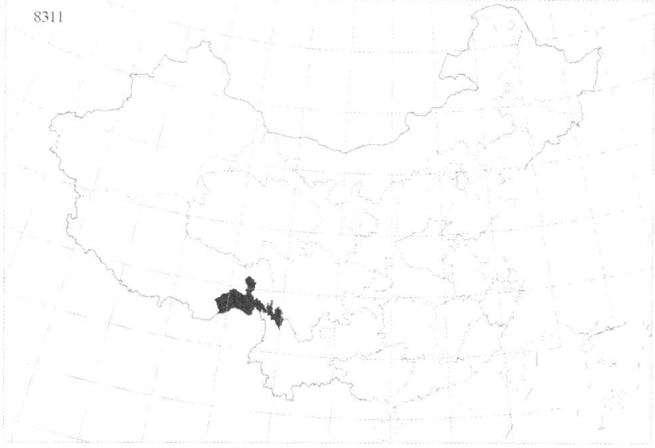

Ericaceae 杜鹃花科 *Rhododendron* 杜鹃花属

Rhododendron taronense Hutch. 薄皮杜鹃

Shrubs, evergreen

8.6; 1.3; 15.0; 8.7; 52.6; 8.8; 574 ❙ 1041; 585; 45 ❙ 81.5; 574; 352.3

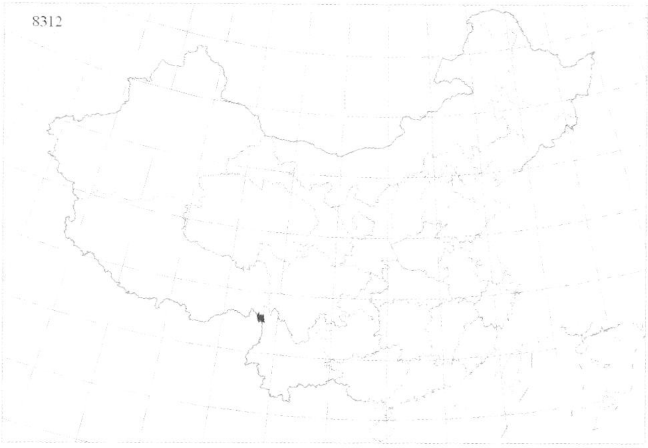

Ericaceae 杜鹃花科 *Rhododendron* 杜鹃花属

Rhododendron tashiroi Maxim. 大武杜鹃

Shrubs, evergreen

18.5; 14.1; 21.7; 18.5; 162.1; 0.0; 863 ❙ 2713; 1388; 188 ❙ 214.4; 863; 363.7

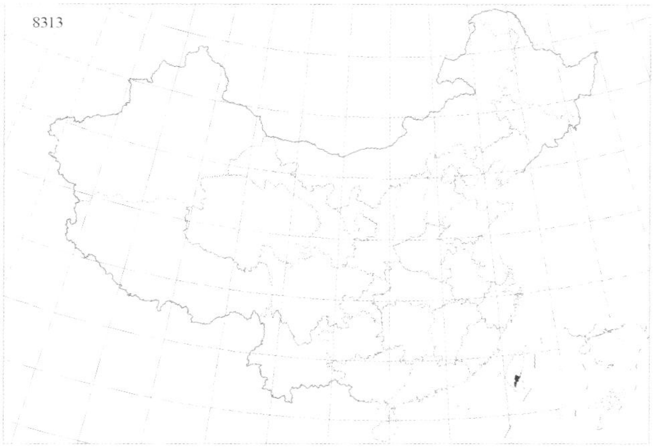

Ericaceae 杜鹃花科 *Rhododendron* 杜鹃花属

Rhododendron tatsienense Franch. 硬叶杜鹃

Shrubs

8.9 (2.0~16.0); 0.4 (-7.2~8.0); 16.2 (9.9~25.9); 9.3 (3.8~16.0); 63.8 (15.4~132.2); 16.9 (0.0~51.4); 590 (413~816) ❙ 991 (628~1280); 543 (346~651); 33 (11~94) ❙ 72.0 (36.1~128.6); 589 (413~816); 277.0 (170.4~372.3)

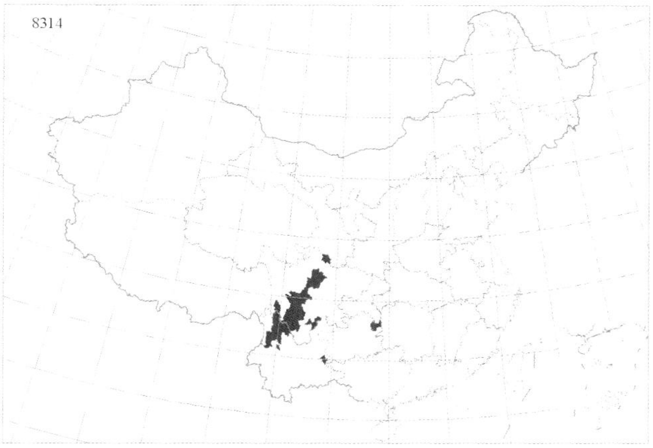

Ericaceae 杜鹃花科 *Rhododendron* 杜鹃花属

Rhododendron tatsienense var. nudatum R. C. Fang 丽江硬叶杜鹃

Shrubs

10.8; 3.9; 16.6; 10.9; 71.8; 1.3; 633 ❙ 968; 594; 24 ❙ 52.8; 633; 319.8

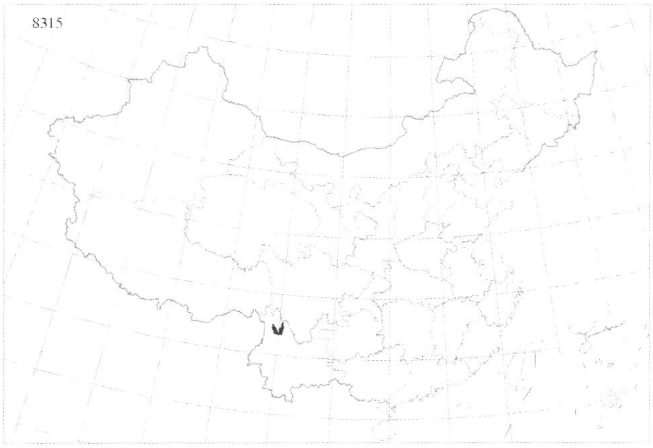

Ericaceae 杜鹃花科 *Rhododendron* 杜鹃花属

Rhododendron telmateium I. B. Balfour et W. W. Sm. 草原杜鹃

Small shrubs

7.6 (2.2~13.6); -0.1 (-6.6~7.2); 14.1 (9.5~18.8); 8.0 (3.8~13.7); 50.2 (15.5~104.1); 18.8 (0.0~49.1); 545 (411~706) ❙ 911 (729~1054); 553 (426~627); 22 (11~46) ❙ 69.8 (47.6~125.9); 544 (411~706); 292.1 (208.8~372.3)

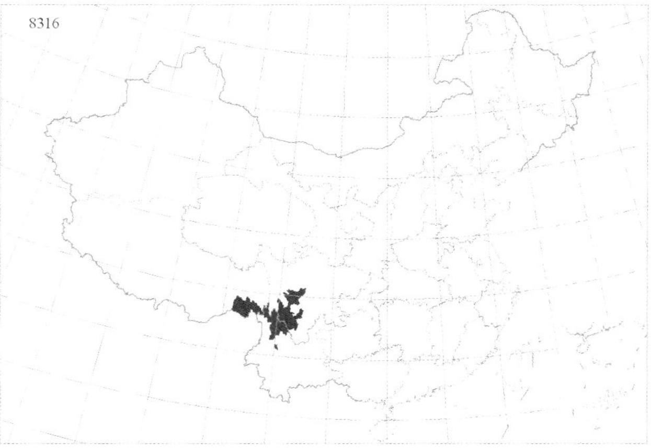

Ericaceae 杜鹃花科 *Rhododendron* 杜鹃花属
Rhododendron temenium I. B. Balfour et Forrest 滇藏杜鹃
Shrubs, evergreen

6.5; -1.1; 13.2; 6.7; 38.2; 20.2; 511 ❙ 915; 523; 35 ❙ 78.7; 511; 286.6

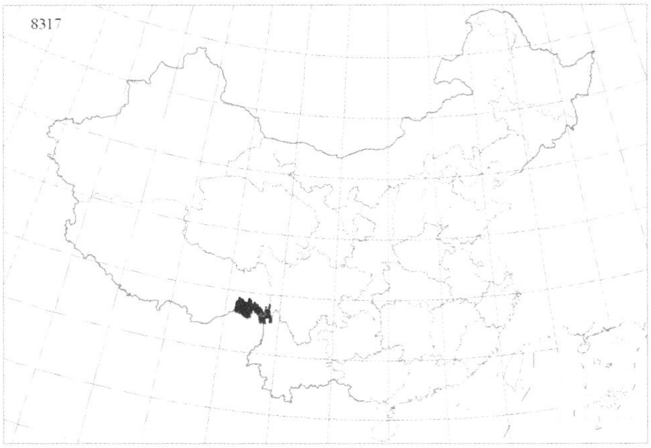

Ericaceae 杜鹃花科 *Rhododendron* 杜鹃花属
Rhododendron temenium var. dealbatum (Cowan) D. F. Chamb. 粉红滇藏杜鹃
Shrubs, evergreen

6.5; -1.1; 13.2; 6.7; 38.2; 20.2; 511 ❙ 915; 523; 35 ❙ 78.7; 511; 286.6

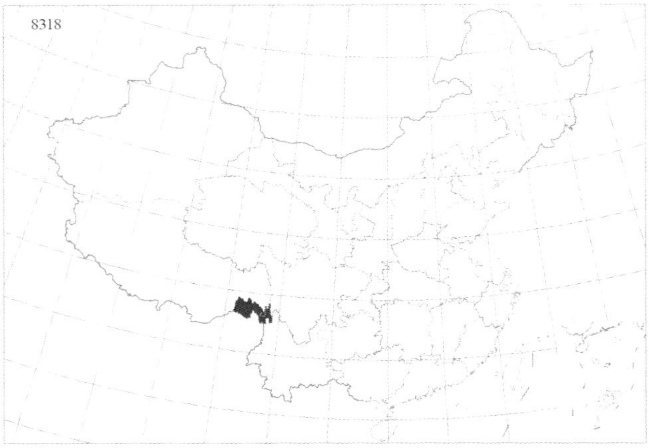

Ericaceae 杜鹃花科 *Rhododendron* 杜鹃花属
Rhododendron temenium var. gilvum (Cowan) D. F. Chamb. 黄花滇藏杜鹃
Shrubs, evergreen

6.5; -1.1; 13.2; 6.7; 38.2; 20.2; 511 ❙ 915; 523; 35 ❙ 78.7; 511; 286.6

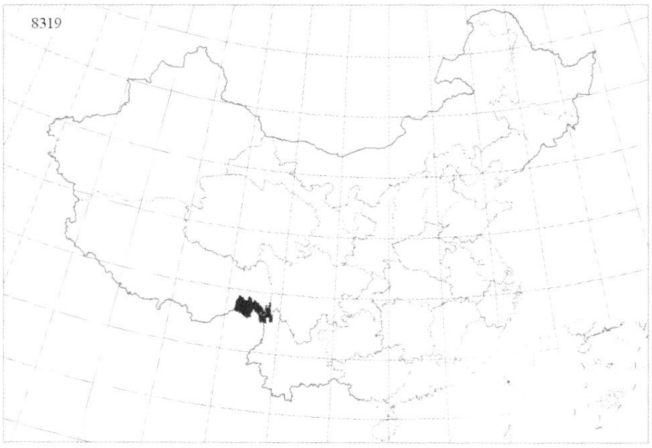

Ericaceae 杜鹃花科 *Rhododendron* 杜鹃花属
Rhododendron tenue Ching ex Fang et M. Y. He 细瘦杜鹃
Shrubs, semievergreen or green

15.8; 5.0; 25.7; 15.9; 131.0; 0.8; 835 ❙ 1598; 569; 191 ❙ 92.1; 835; 249.4

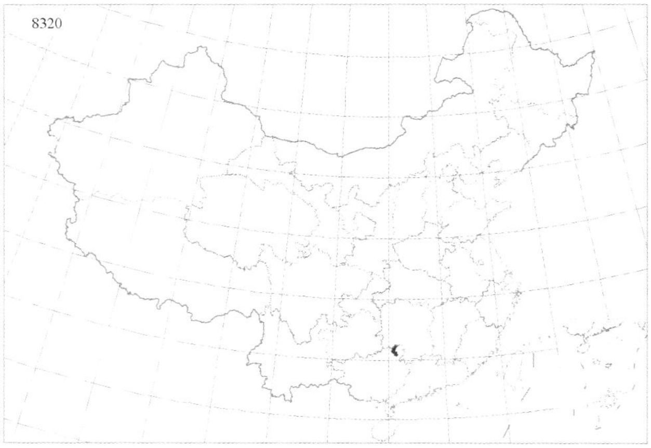

Ericaceae 杜鹃花科 *Rhododendron* 杜鹃花属
Rhododendron tenuifolium R. C. Fang et S. H. Huang 薄叶管花杜鹃
Shrubs, evergreen

13.3; 5.4; 19.6; 13.4; 100.4; 0.0; 706 ❙ 1736; 1007; 51 ❙ 145.8; 706; 447.4

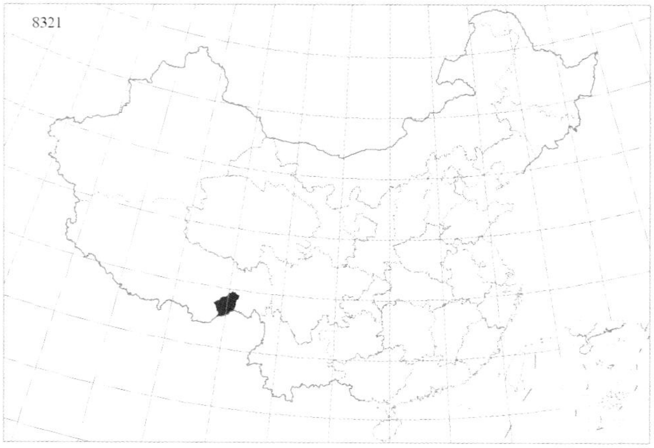

Ericaceae 杜鹃花科 *Rhododendron* 杜鹃花属
Rhododendron tenuilaminare Tam 薄片杜鹃
Shrubs, semievergreen or green

21.2; 13.0; 27.7; 21.2; 194.6; 0.0; 1109 ❙ 1935; 809; 153 ❙ 74.8; 1109; 242.6

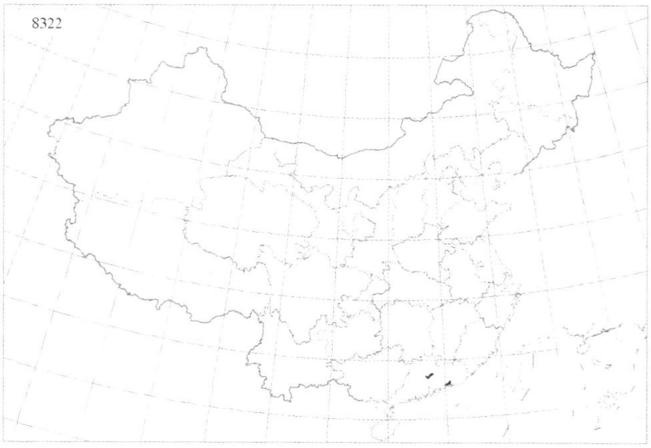

Ericaceae 杜鹃花科 *Rhododendron* 杜鹃花属
Rhododendron tephropeplum I. B. Balfour et Farrer 灰被杜鹃
Small shrubs, evergreen

9.9; 2.1; 16.3; 10.0; 67.7; 8.9; 608 ❙ 1292; 742; 39 ❙ 110.3; 608; 349.4

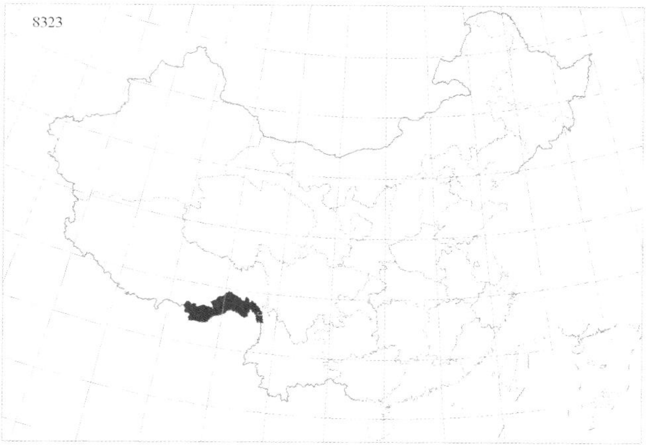

Ericaceae 杜鹃花科 *Rhododendron* 杜鹃花属
Rhododendron thayerianum Rehder et E. H. Wilson 反边杜鹃
Shrubs, evergreen

11.1; 1.7; 19.6; 11.3; 83.9; 10.2; 659 ❙ 1208; 674; 38 ❙ 83.8; 659; 243.4

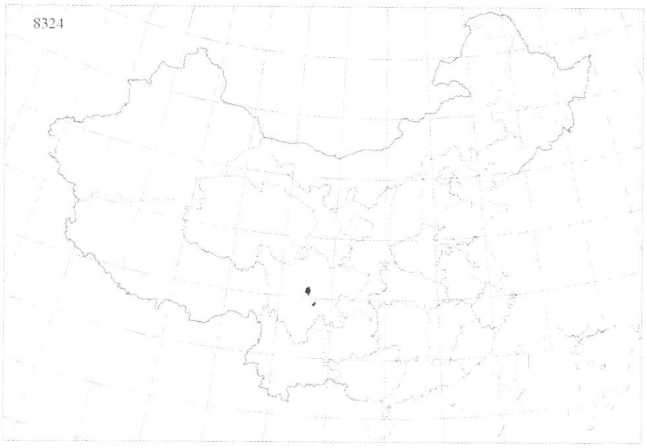

Ericaceae 杜鹃花科 *Rhododendron* 杜鹃花属
Rhododendron thomsonii Hook. f. 半圆叶杜鹃
Shrubs or small trees, evergreen

4.5; −4.2; 12.1; 5.9; 34.8; 40.8; 467 ❙ 670; 411; 14 ❙ 30.2; 451; 202.7

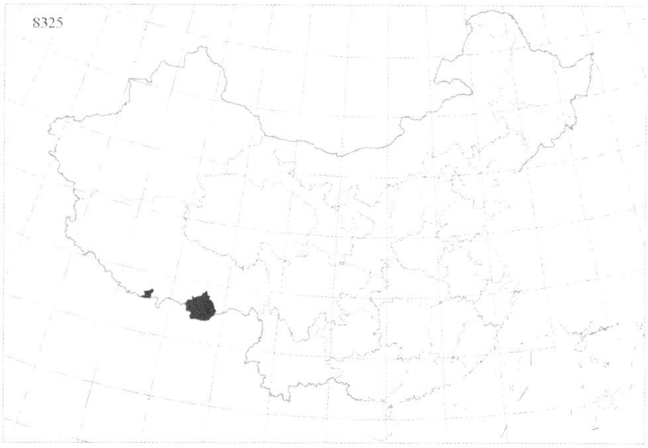

Ericaceae 杜鹃花科 *Rhododendron* 杜鹃花属
Rhododendron thomsonii subsp. lopsangianum (Cowan) D. F. Chamb. 小半圆叶杜鹃
Shrubs or small trees, evergreen

1.9; −7.2; 10.0; 3.8; 16.6; 53.6; 403 ❙ 341; 226; 4 ❙ −15.3; 341; 217.5

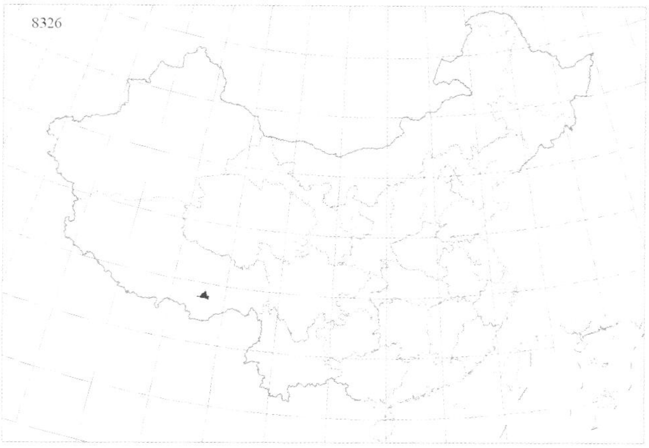

Ericaceae 杜鹃花科 *Rhododendron* 杜鹃花属
Rhododendron thymifolium Maxim. 千里香杜鹃
Small shrubs, evergreen

2.6 (−4.1~14.1); −8.6 (−17.3~6.5); 12.3 (7.8~20.2); 5.0 (2.1~14.2); 27.6 (5.2~109.8); 55.7 (0.0~114.2); 448 (340~729) ❙ 610 (247~1093); 345 (138~622); 12 (5~27) ❙ 36.8 (−55.7~125.9); 429 (247~707); 256.1 (104.0~364.1)

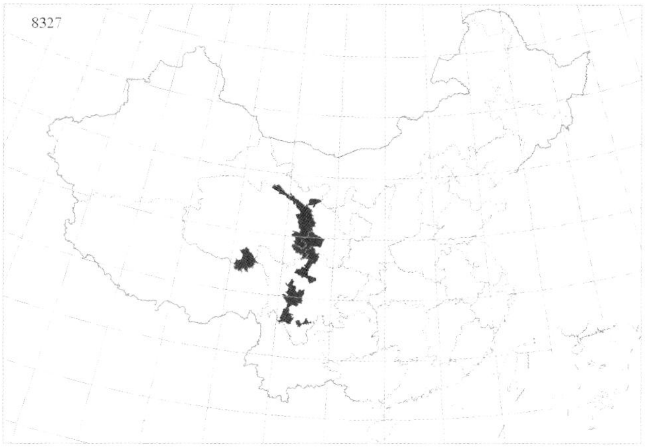

Ericaceae 杜鹃花科 *Rhododendron* 杜鹃花属
Rhododendron tianlinense Tam 田林马银花
Shrubs or small trees, evergreen

19.8; 11.1; 26.4; 19.8; 177.9; 0.0; 1010 ❙ 1182; 673; 43 ❙ 16.9; 1010; 354.5

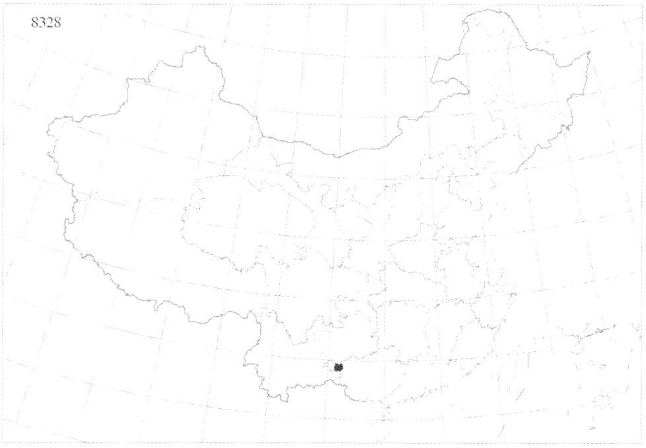

Ericaceae 杜鹃花科 *Rhododendron* 杜鹃花属

Rhododendron tingwuense Tam 鼎湖杜鹃

Shrubs, semievergreen or green

22.0 (21.8~22.2); 13.9 (13.5~14.2); 28.5 (27.8~28.8); 22.0 (21.9~22.3); 204.4 (202.6~207.2); 0.0 (0.0~0.0); 1169 (1150~1186) ❙ 1828 (1636~2014); 841 (722~954); 134 (116~149) ❙ 56.4 (40.1~72.1); 1169 (1150~1186); 217.6 (187.2~273.3)

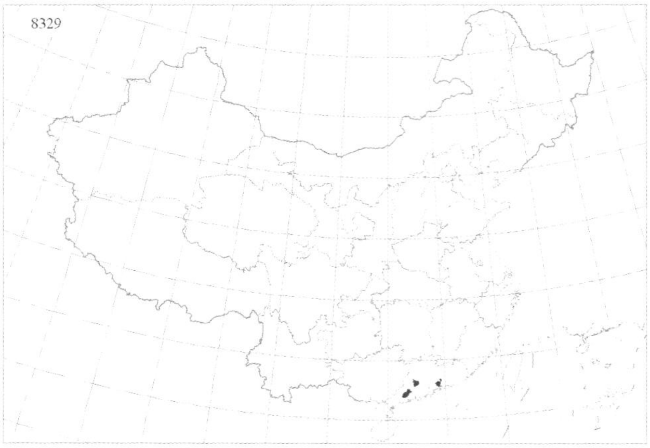

Ericaceae 杜鹃花科 *Rhododendron* 杜鹃花属

Rhododendron traillianum Forrest et W. W. Sm. 川滇杜鹃

Shrubs or small trees, evergreen

5.9 (2.0~10.8); -2.0 (-7.2~3.9); 12.8 (9.9~16.6); 6.6 (3.8~10.9); 37.3 (15.4~71.8); 26.4 (1.3~51.4); 500 (413~633) ❙ 834 (473~1007); 497 (287~603); 21 (7~40) ❙ 67.6 (5.0~128.6); 499 (413~633); 278.0 (230.7~351.7)

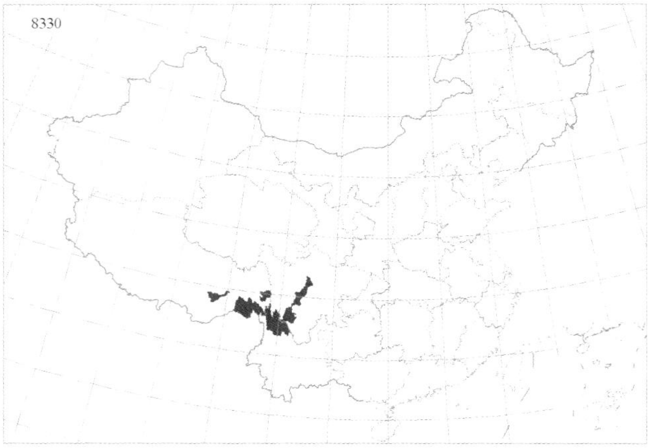

Ericaceae 杜鹃花科 *Rhododendron* 杜鹃花属

Rhododendron traillianum var. dictyotum (I. B. Balfour ex Tagg) D. F. Chamb. 棕背川滇杜鹃

Shrubs or small trees, evergreen

5.6 (2.9~10.8); -2.2 (-5.7~3.9); 12.6 (10.6~16.6); 6.3 (4.3~10.9); 35.2 (19.9~71.8); 27.0 (1.3~45.2); 490 (421~633) ❙ 746 (473~968); 448 (287~594); 20 (7~31) ❙ 50.2 (5.0~100.7); 490 (421~633); 262.7 (230.7~319.8)

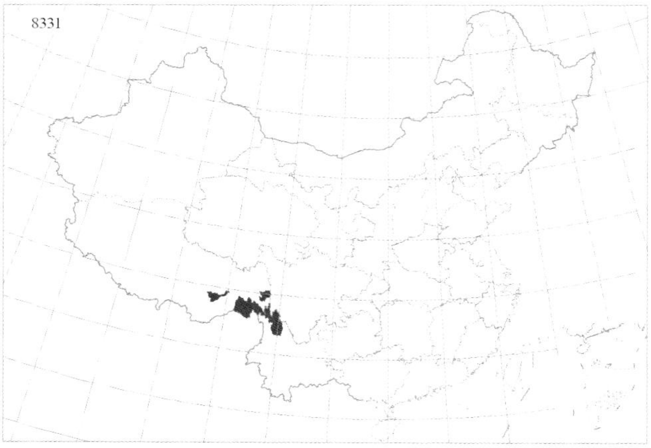

Ericaceae 杜鹃花科 *Rhododendron* 杜鹃花属

Rhododendron trichanthum Rehder 长毛杜鹃

Shrubs, evergreen

7.8 (2.2~11.9); -1.1 (-6.6~2.9); 15.8 (9.9~20.0); 8.2 (3.8~12.0); 53.8 (15.5~87.2); 19.4 (3.4~49.1); 556 (413~671) ❙ 1013 (862~1220); 565 (531~688); 27 (12~37) ❙ 84.7 (49.1~125.9); 556 (413~671); 259.0 (200.8~311.5)

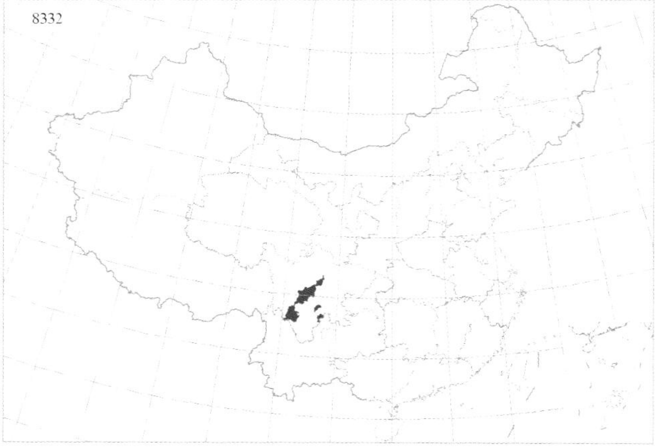

Ericaceae 杜鹃花科 *Rhododendron* 杜鹃花属

Rhododendron trichocladum Franch. 糙毛杜鹃

Shrubs, deciduous

11.3 (5.3~16.1); 4.4 (-2.4~9.6); 17.0 (12.2~20.6); 11.5 (5.7~16.2); 83.4 (30.5~133.9); 6.9 (0.0~26.2); 646 (479~785) ❙ 1075 (741~1454); 605 (426~792); 46 (24~70) ❙ 67.8 (38.0~100.7); 645 (479~779); 359.3 (234.2~487.2)

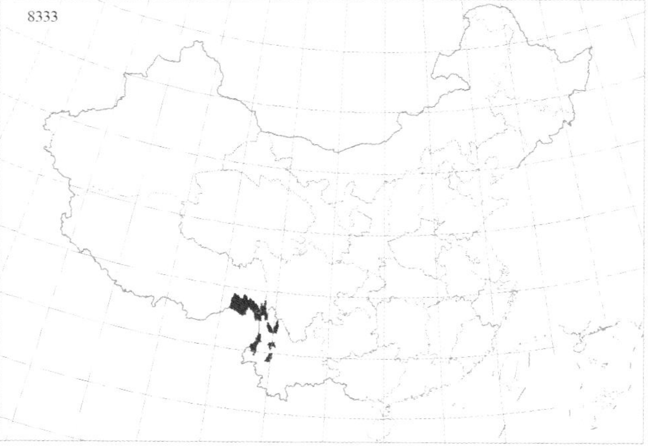

Ericaceae 杜鹃花科 *Rhododendron* 杜鹃花属

Rhododendron trichogynum L. C. Hu 理县杜鹃

Small trees, evergreen

4.5; -4.7; 12.8; 5.5; 30.1; 35.6; 468 ❙ 986; 513; 27 ❙ 111.4; 468; 234.2

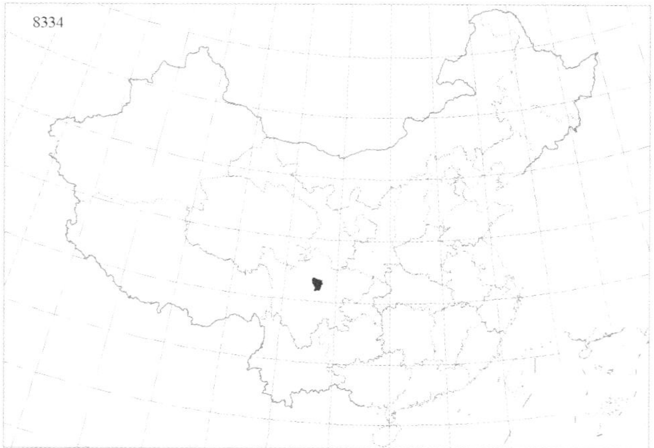

Ericaceae 杜鹃花科 *Rhododendron* 杜鹃花属
Rhododendron trichostomum Franch. 毛嘴杜鹃
Shrubs, evergreen

3.4 (-3.2~12.6); -5.4 (-14.4~6.0); 11.3 (7.3~17.9); 5.1 (2.0~12.6); 26.6 (4.0~91.6); 44.9 (0.0~101.5); 448 (335~678) ❙ 726 (425~1057); 436 (259~619); 16 (9~42) ❙ 61.6 (20.7~125.9); 448 (335~678); 265.3 (130.1~357.7)

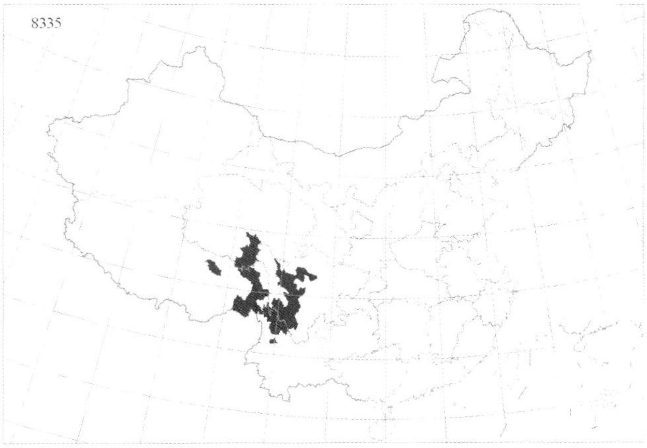

Ericaceae 杜鹃花科 *Rhododendron* 杜鹃花属
Rhododendron trichostomum var. ledoides (I. B. Balfour et W. W. Sm.) Cowan et Davidian 筒花毛嘴杜鹃 Small shrubs, evergreen

8.5; 0.8; 15.0; 8.6; 54.4; 12.1; 567 ❙ 922; 567; 16 ❙ 62.8; 567; 297.1

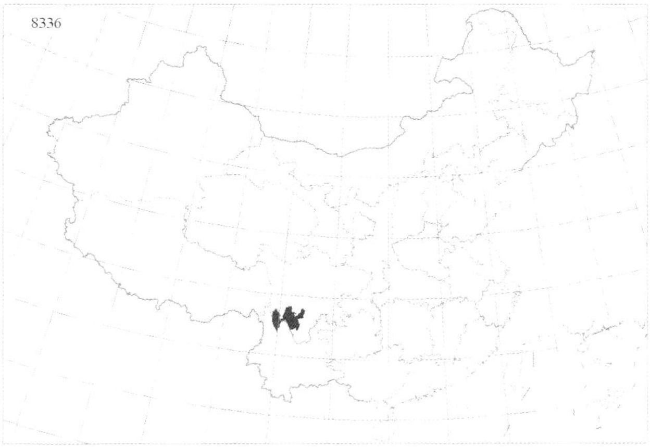

Ericaceae 杜鹃花科 *Rhododendron* 杜鹃花属
Rhododendron trichostomum var. radinum (I. B. Balfour et W. W. Sm.) Cowan et Davidian 鳞斑毛嘴杜鹃 Small shrubs, evergreen

8.5; 1.1; 14.7; 8.6; 53.7; 11.3; 567 ❙ 894; 560; 18 ❙ 58.5; 565; 303.6

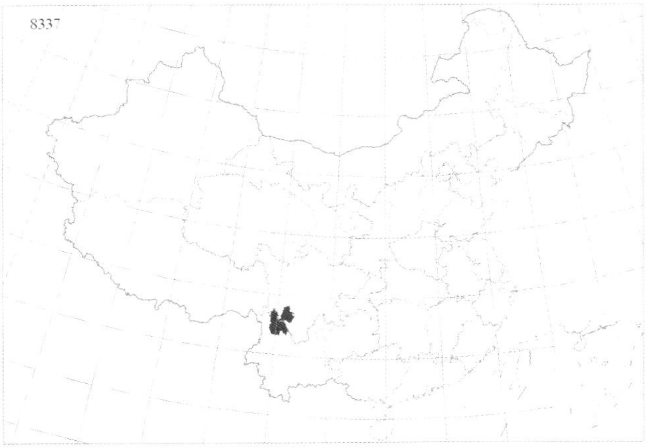

Ericaceae 杜鹃花科 *Rhododendron* 杜鹃花属
Rhododendron triflorum Hook. 三花杜鹃
Shrubs or small trees, evergreen

6.3 (0.3~17.3); -2.0 (-8.6~11.0); 13.5 (8.3~21.6); 7.4 (2.9~17.3); 48.2 (9.6~148.0); 31.7 (0.0~65.4); 518 (368~832) ❙ 803 (341~1736); 474 (226~1007); 21 (4~51) ❙ 46.0 (-15.3~145.8); 506 (341~783); 261.9 (142.5~447.4)

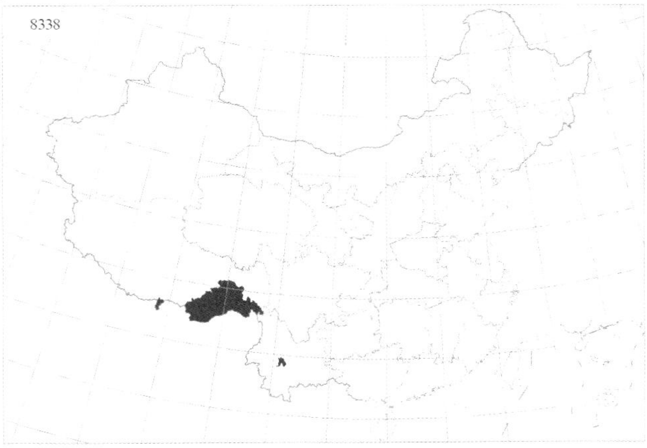

Ericaceae 杜鹃花科 *Rhododendron* 杜鹃花属
Rhododendron trilectorum Cowan 朗贡杜鹃
Small shrubs, evergreen

1.9; -7.2; 10.0; 3.8; 16.6; 53.6; 403 ❙ 341; 226; 4 ❙ -15.3; 341; 217.5

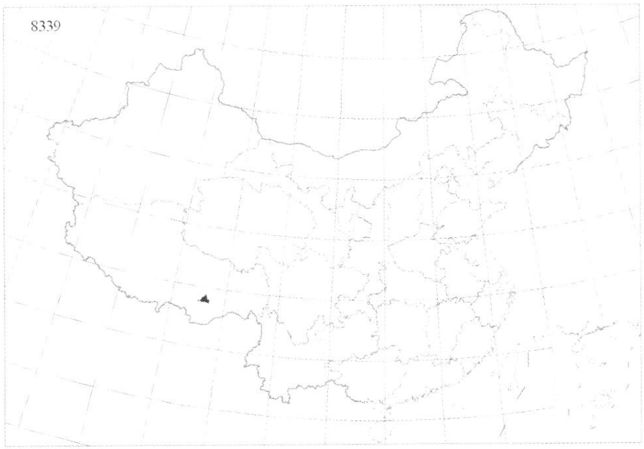

Ericaceae 杜鹃花科 *Rhododendron* 杜鹃花属
Rhododendron tsaii Fang 昭通杜鹃
Small shrubs, evergreen

10.7; 2.3; 17.8; 10.7; 73.6; 4.9; 632 ❙ 983; 558; 26 ❙ 56.2; 632; 243.6

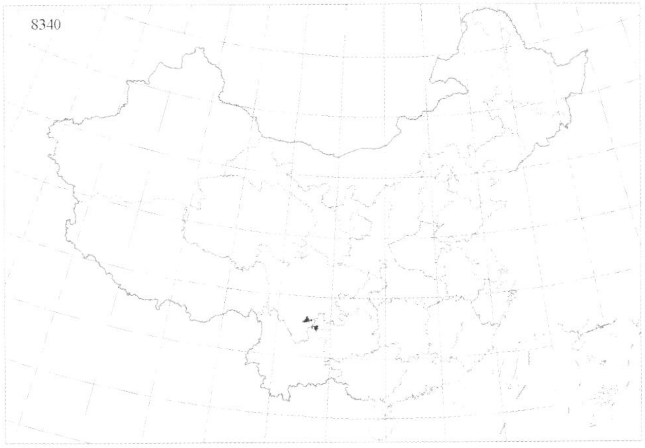

Ericaceae 杜鹃花科 *Rhododendron* 杜鹃花属

Rhododendron tsariense Cowan 白钟杜鹃

Shrubs, evergreen

6.2; −2.4; 13.5; 7.1; 44.2; 29.7; 510 ❙ 770; 460; 15 ❙ 38.6; 489; 249.8

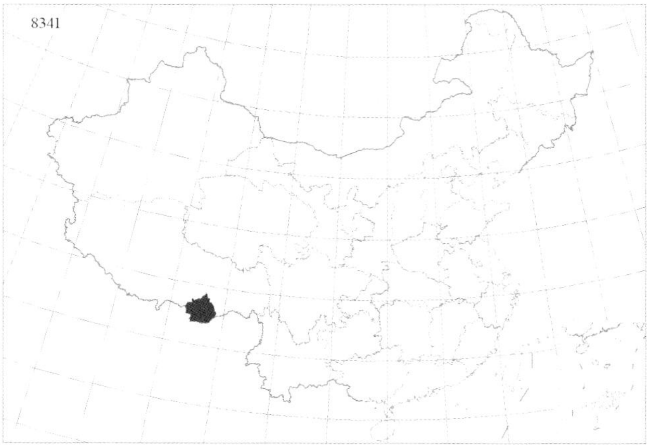

Ericaceae 杜鹃花科 *Rhododendron* 杜鹃花属

Rhododendron tsinlingense W. P. Fang ex J. Q. Fu 秦岭杜鹃

Shrubs

12.6; 1.0; 23.5; 12.6; 99.9; 8.6; 724 ❙ 760; 382; 17 ❙ 5.0; 724; 294.3

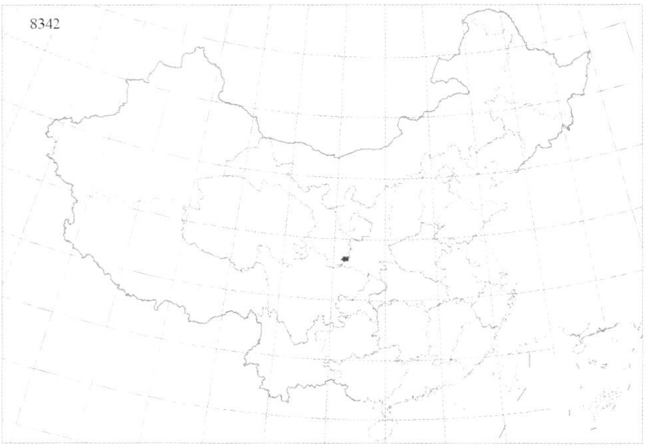

Ericaceae 杜鹃花科 *Rhododendron* 杜鹃花属

Rhododendron tsoi Merr. 两广杜鹃

Shrubs, semievergreen

20.8 (20.2~21.8); 12.3 (10.8~14.1); 27.7 (26.7~28.5); 20.8 (20.2~21.8); 190.1 (183.0~201.9); 0.0 (0.0~0.0); 1085 (1036~1158) ❙ 1684 (1451~1991); 719 (584~902); 153 (99~182) ❙ 55.4 (27.0~83.4); 1085 (1036~1158); 251.9 (201.9~335.1)

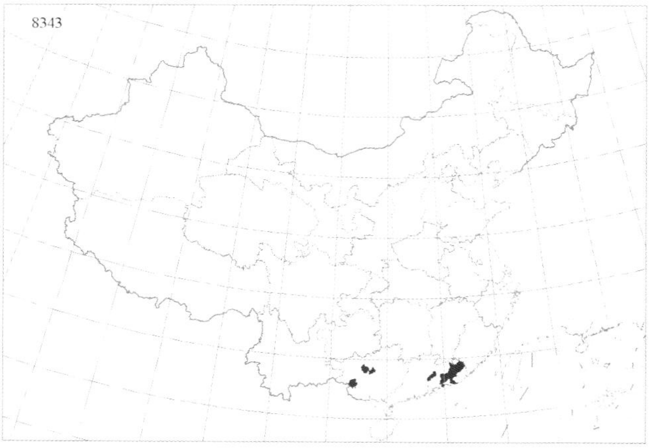

Ericaceae 杜鹃花科 *Rhododendron* 杜鹃花属

Rhododendron tubiforme (Cowan et Davidian) Davidian 苍白杜鹃

Small shrubs, evergreen

−0.7; −9.6; 7.7; 2.5; 6.8; 74.2; 338 ❙ 370; 262; 10 ❙ 4.7; 338; 61.6

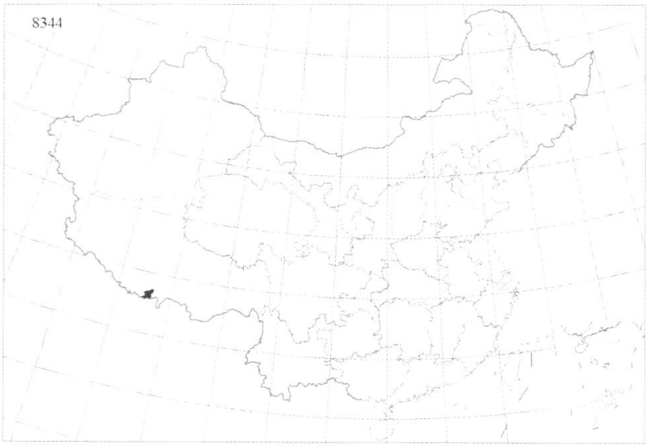

Ericaceae 杜鹃花科 *Rhododendron* 杜鹃花属

Rhododendron tubulosum Ching ex W. Y. Wang 长管杜鹃

Shrubs, evergreen

−1.2; −11.7; 8.5; 2.6; 8.1; 82.3; 347 ❙ 491; 301; 11 ❙ 41.8; 347; 290.1

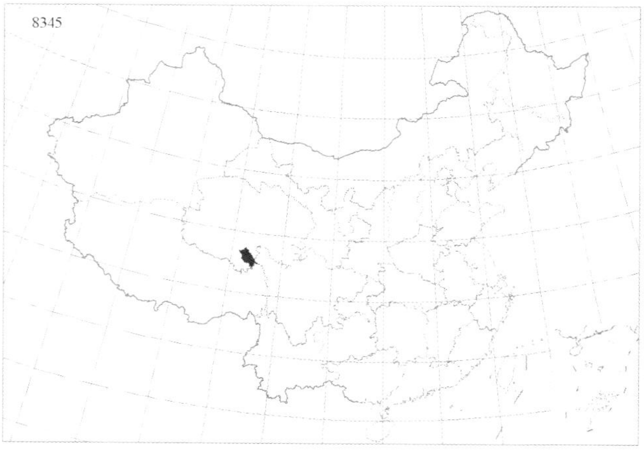

Ericaceae 杜鹃花科 *Rhododendron* 杜鹃花属

Rhododendron tutcherae Hemsl. et E. H. Wilson 香缅树杜鹃

Trees, evergreen

17.9 (16.3~21.0); 10.9 (9.1~14.2); 22.9 (21.4~25.6); 18.0 (16.4~21.0); 155.5 (136.4~191.8); 0.0 (0.0~0.0); 874 (797~1053) ❙ 1300 (1161~1522); 721 (633~835); 55 (48~68) ❙ 48.8 (37.1~60.1); 874 (797~1053); 418.7 (314.4~551.6)

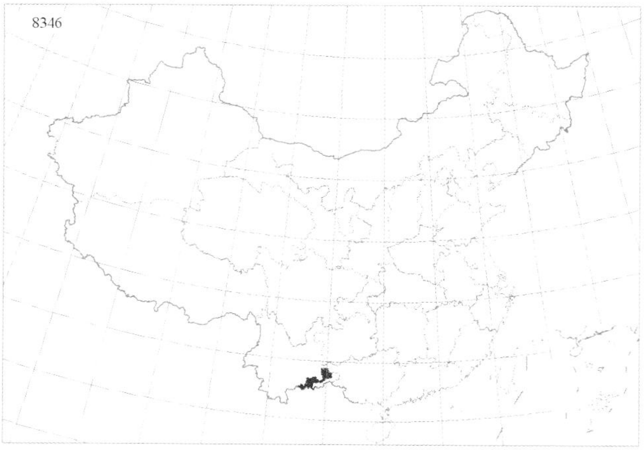

Ericaceae 杜鹃花科 *Rhododendron* 杜鹃花属
Rhododendron unciferum P. C. Tam 垂钩杜鹃
Shrubs, semievergreen or green

21.8; 12.9; 28.5; 21.8; 201.6; 0.0; 1162 ❙ 1519; 735; 145 ❙ 30.7; 1162; 208.5

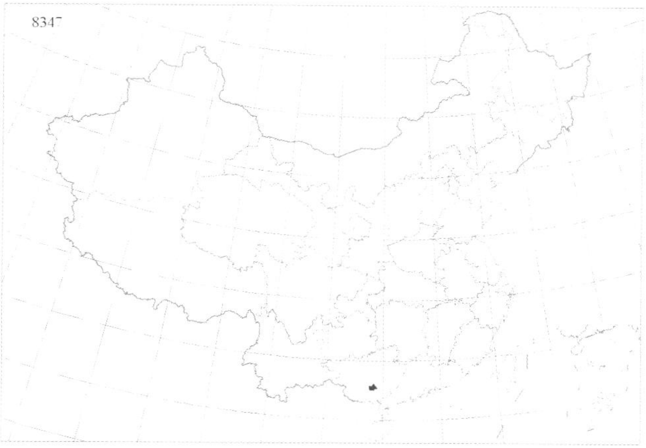

Ericaceae 杜鹃花科 *Rhododendron* 杜鹃花属
Rhododendron uniflorum Hutch. et Kingdon-Ward 单花杜鹃
Small shrubs, evergreen

8.3; 0.0; 15.3; 9.0; 61.4; 20.9; 566 ❙ 1104; 647; 29 ❙ 75.4; 566; 339.1

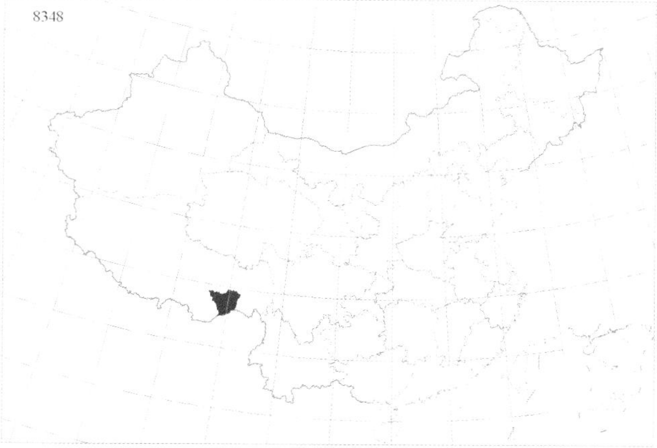

Ericaceae 杜鹃花科 *Rhododendron* 杜鹃花属
Rhododendron urophyllum Fang 尾叶杜鹃
Shrubs, evergreen

11.9; 2.9; 20.0; 12.0; 87.2; 3.4; 671 ❙ 1001; 550; 35 ❙ 49.1; 671; 278.8

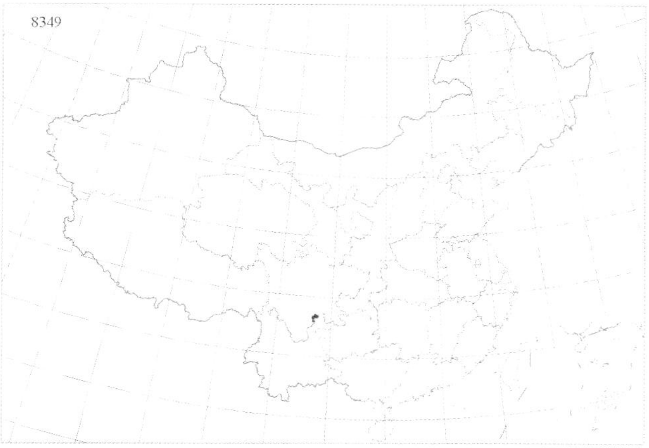

Ericaceae 杜鹃花科 *Rhododendron* 杜鹃花属
Rhododendron uvariifolium Diels 紫玉盘杜鹃
Shrubs or trees, evergreen

8.8 (1.8~14.6); 1.1 (-8.2~7.8); 15.2 (9.9~20.0); 9.2 (3.8~14.6); 62.2 (15.7~115.2); 16.3 (0.0~53.8); 580 (404~738) ❙ 995 (527~1736); 574 (317~1007); 34 (9~70) ❙ 70.6 (22.4~145.8); 576 (404~727); 328.3 (169.5~487.2)

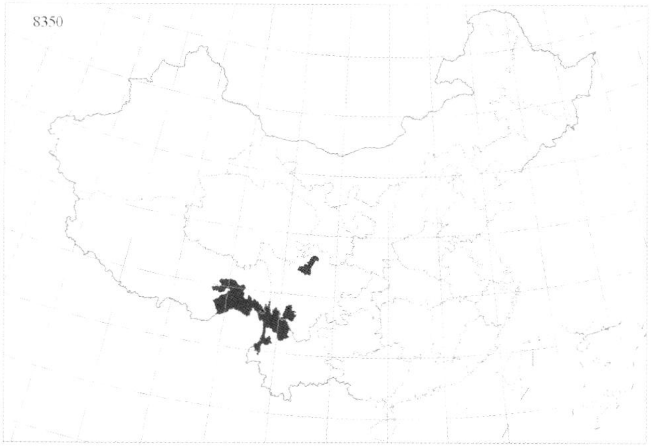

Ericaceae 杜鹃花科 *Rhododendron* 杜鹃花属
Rhododendron vaccinioides Hook. 越橘杜鹃
Small shrubs, evergreen

13.7 (1.8~19.1); 5.9 (-7.1~13.0); 19.9 (9.9~26.0); 13.9 (3.8~19.1); 110.5 (15.7~169.7); 5.7 (0.0~53.3); 738 (404~966) ❙ 1201 (583~1736); 667 (348~1007); 51 (13~69) ❙ 65.3 (25.7~145.8); 738 (404~966); 320.9 (169.5~506.2)

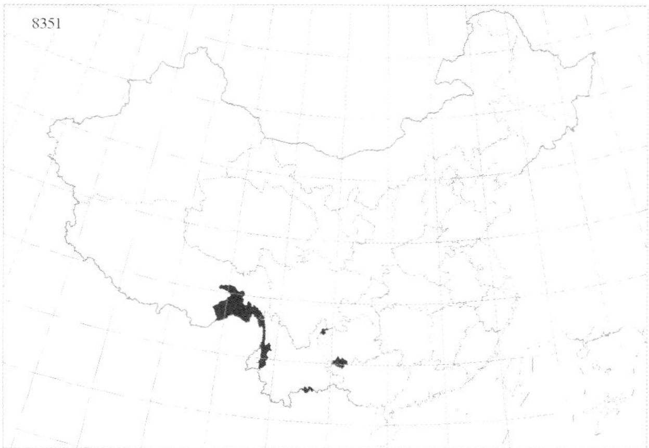

Ericaceae 杜鹃花科 *Rhododendron* 杜鹃花属
Rhododendron valentinianum Forrest ex Hutch. 毛柄杜鹃
Shrubs, evergreen

16.7 (14.4~19.1); 8.0 (4.7~13.0); 23.7 (19.1~26.0); 16.8 (14.4~19.1); 141.1 (113.1~169.7); 0.0 (0.0~0.3); 842 (727~966) ❙ 1275 (1049~1454); 653 (511~835); 64 (44~85) ❙ 52.4 (26.1~99.9); 839 (727~966); 293.3 (189.5~506.2)

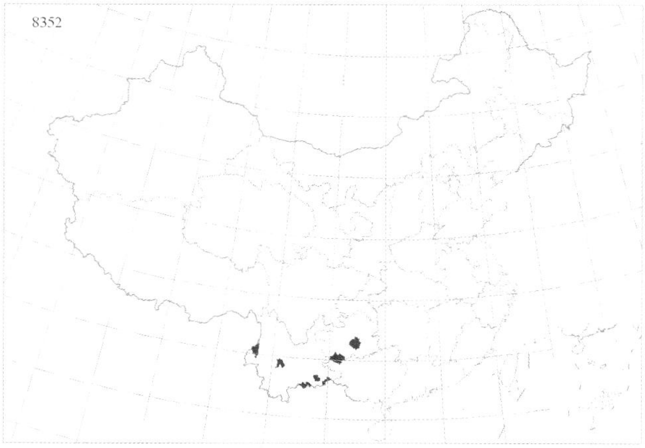

Ericaceae 杜鹃花科 *Rhododendron* 杜鹃花属

Rhododendron valentinianum var. oblongilobatum R. C. Fang 滇南毛柄杜鹃

Shrubs, evergreen

17.7; 10.9; 22.5; 17.8; 153.4; 0.0; 854 ❙ 1368; 756; 59 ❙ 60.1; 854; 491.5

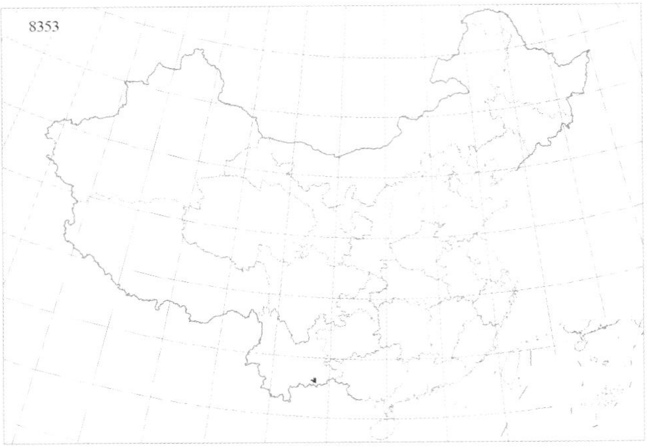

Ericaceae 杜鹃花科 *Rhododendron* 杜鹃花属

Rhododendron vellereum Hutch. ex Tagg 白毛杜鹃

Small trees, evergreen

1.8 (−2.5~4.4); −7.6 (−12.7~−4.3); 10.2 (7.0~12.1); 3.9 (2.1~5.3); 17.6 (4.0~28.5); 55.4 (34.6~93.0); 408 (332~455) ❙ 576 (382~800); 337 (254~427); 12 (6~19) ❙ 40.6 (12.1~86.5); 408 (332~455); 229.4 (111.0~341.6)

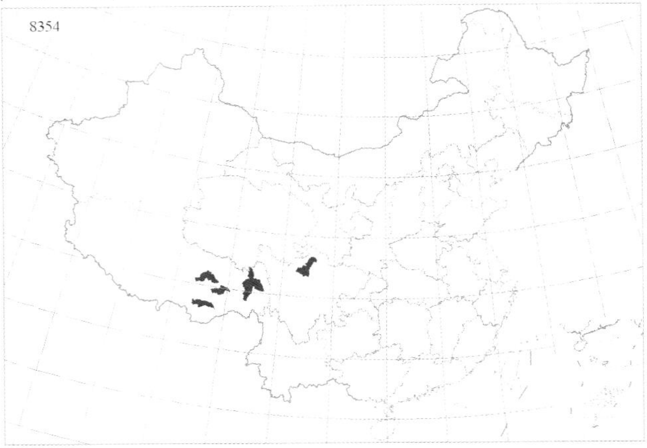

Ericaceae 杜鹃花科 *Rhododendron* 杜鹃花属

Rhododendron venator Tagg ex L. Rothschild 毛柱杜鹃

Shrubs, evergreen

3.3; −5.4; 11.1; 4.6; 22.3; 41.8; 427 ❙ 473; 287; 7 ❙ 5.0; 427; 230.7

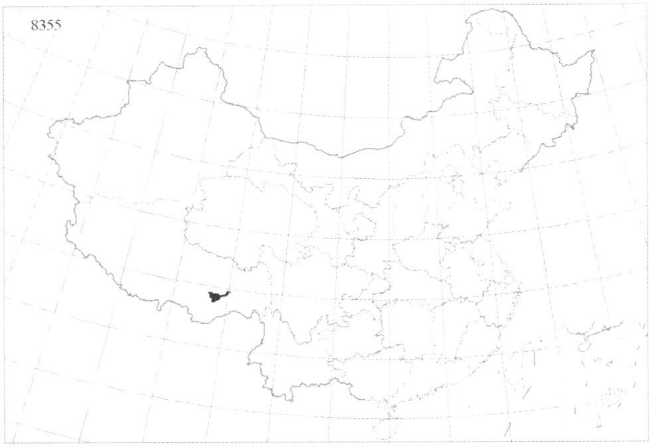

Ericaceae 杜鹃花科 *Rhododendron* 杜鹃花属

Rhododendron vernicosum Franch. 亮叶杜鹃

Shrubs or small trees, evergreen

8.7 (0.3~17.2); −0.3 (−9.9~8.7); 16.7 (9.3~27.0); 9.4 (3.3~17.3); 65.9 (11.9~147.4); 20.8 (0.0~67.5); 594 (391~905) ❙ 936 (535~1414); 511 (326~817); 33 (10~72) ❙ 62.0 (21.3~128.6); 594 (391~905); 250.4 (148.9~381.5)

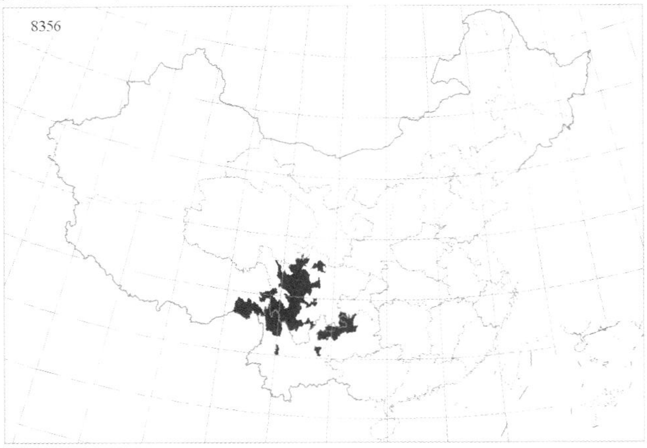

Ericaceae 杜鹃花科 *Rhododendron* 杜鹃花属

Rhododendron vesiculiferum Tagg 泡毛杜鹃

Shrubs or small trees, evergreen

7.0; −0.5; 13.6; 7.2; 41.5; 17.5; 526 ❙ 1001; 571; 38 ❙ 91.1; 526; 293.2

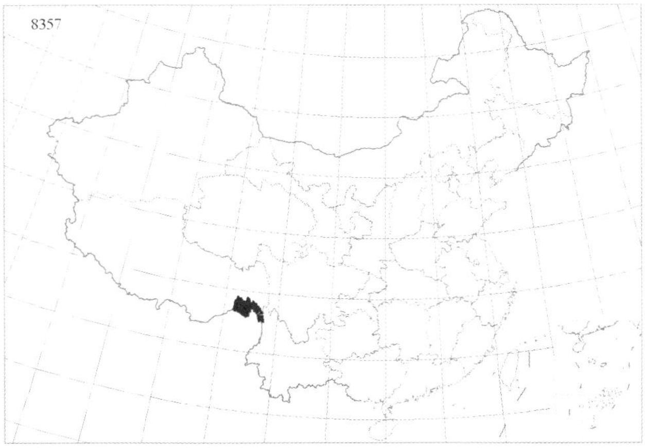

Ericaceae 杜鹃花科 *Rhododendron* 杜鹃花属

Rhododendron vialii Delavay et Franch. 红马银花

Shrubs, evergreen

18.4 (16.8~20.5); 12.1 (9.1~14.9); 22.6 (21.3~23.7); 18.4 (16.9~20.5); 161.2 (142.3~185.9); 0.0 (0.0~0.0); 885 (813~992) ❙ 1257 (984~1547); 689 (562~911); 52 (43~66) ❙ 41.7 (20.8~74.2); 871 (788~959); 413.6 (294.5~570.1)

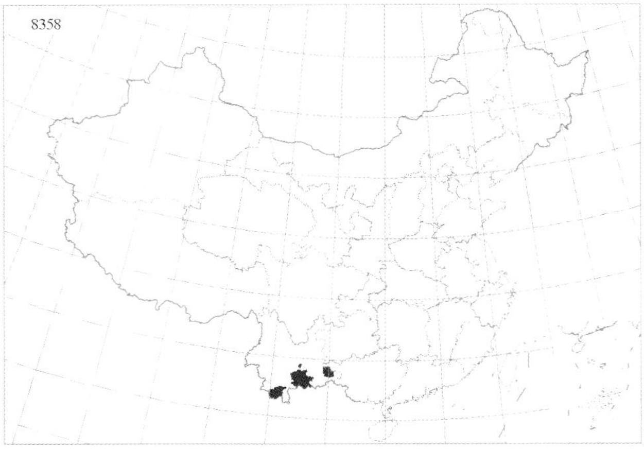

Ericaceae 杜鹃花科 *Rhododendron* 杜鹃花属

Rhododendron virgatum Hook. 柳条杜鹃

Small shrubs, evergreen

11.9 (1.8~18.9); 4.8 (-7.1~12.0); 17.7 (9.9~23.9); 12.2 (3.8~19.0); 91.9 (15.7~167.8); 8.2 (0.0~53.3); 670 (404~929) ❙ 931 (527~1337); 541 (317~727); 32 (9~69) ❙ 41.9 (-14.2~100.7); 637 (404~760); 300.0 (168.2~407.1)

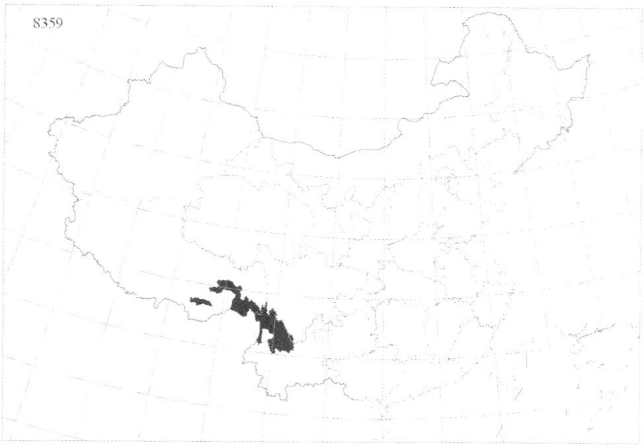

Ericaceae 杜鹃花科 *Rhododendron* 杜鹃花属

Rhododendron viridescens Hutch. 显绿杜鹃

Shrubs, semideciduous

13.3; 5.4; 19.6; 13.4; 100.4; 0.0; 706 ❙ 1736; 1007; 51 ❙ 145.8; 706; 447.4

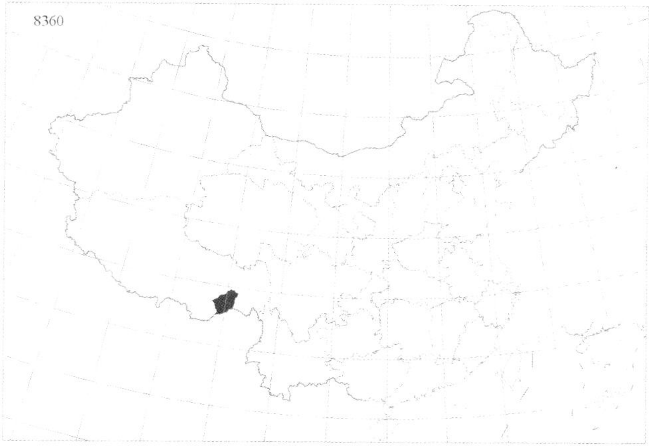

Ericaceae 杜鹃花科 *Rhododendron* 杜鹃花属

Rhododendron viscidifolium Davidian 铜色杜鹃

Shrubs, evergreen

13.3; 5.4; 19.6; 13.4; 100.4; 0.0; 706 ❙ 1736; 1007; 51 ❙ 145.8; 706; 447.4

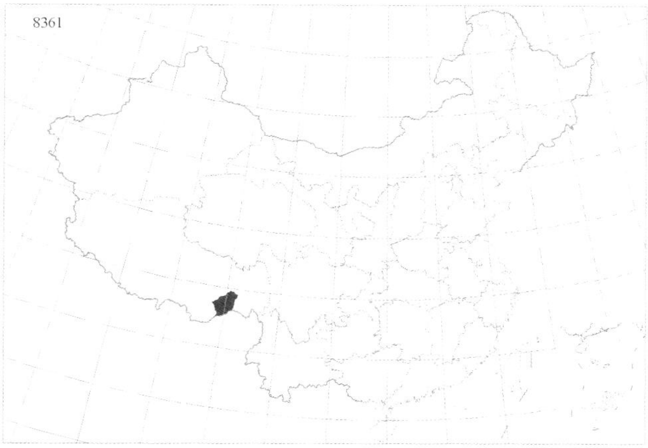

Ericaceae 杜鹃花科 *Rhododendron* 杜鹃花属

Rhododendron viscidum C. Z. Guo et Z. H. Liu 黏质杜鹃

Small shrubs, semievergreen or green

17.6; 6.6; 27.7; 17.6; 151.3; 0.0; 928 ❙ 1494; 476; 197 ❙ 61.1; 922; 180.3

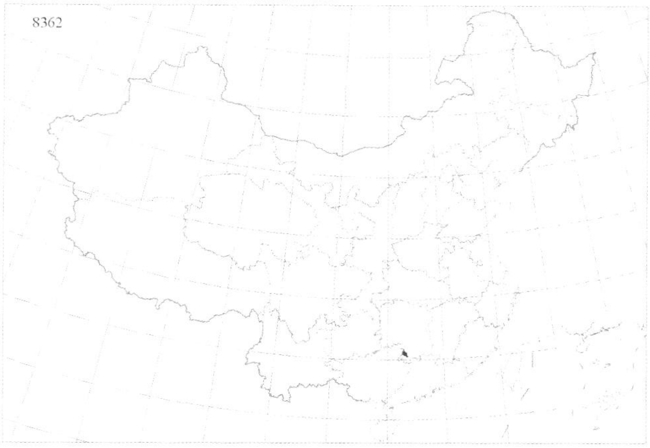

Ericaceae 杜鹃花科 *Rhododendron* 杜鹃花属

Rhododendron wallichii Hook. f. 簇毛杜鹃

Shrubs, evergreen

3.9; -4.6; 11.5; 5.9; 34.6; 46.8; 459 ❙ 730; 442; 17 ❙ 39.3; 459; 189.3

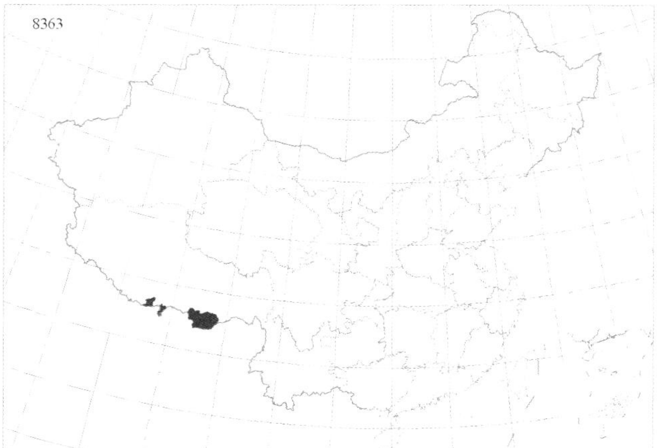

Ericaceae 杜鹃花科 *Rhododendron* 杜鹃花属

Rhododendron wardii W. W. Sm. 黄杯杜鹃

Shrubs, evergreen

7.0 (1.8~13.3); -0.9 (-7.1~5.4); 13.8 (9.5~19.6); 7.5 (3.8~13.4); 45.9 (15.5~100.4); 21.9 (0.0~53.3); 530 (404~706) ❙ 914 (473~1736); 543 (287~1007); 25 (7~58) ❙ 70.1 (5.0~145.8); 529 (404~706); 293.1 (169.5~447.4)

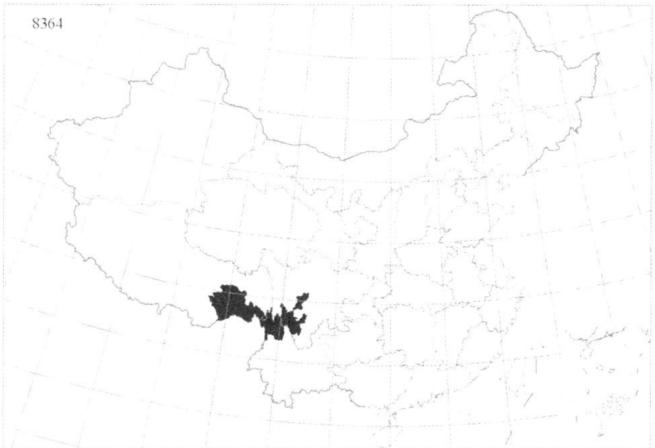

Ericaceae 杜鹃花科 *Rhododendron* 杜鹃花属
Rhododendron wasonii Hemsl. et E. H. Wilson 褐毛杜鹃
Shrubs, evergreen

7.5 (2.0~14.7); -1.7 (-7.2~4.3); 15.8 (9.9~24.0); 8.2 (3.8~14.8); 54.4 (15.4~118.0); 23.5 (0.7~51.4); 558 (413~789) ❙ 990 (922~1081); 541 (460~606); 25 (18~32) ❙ 84.9 (18.6~128.6); 558 (413~789); 238.8 (150.8~298.0)

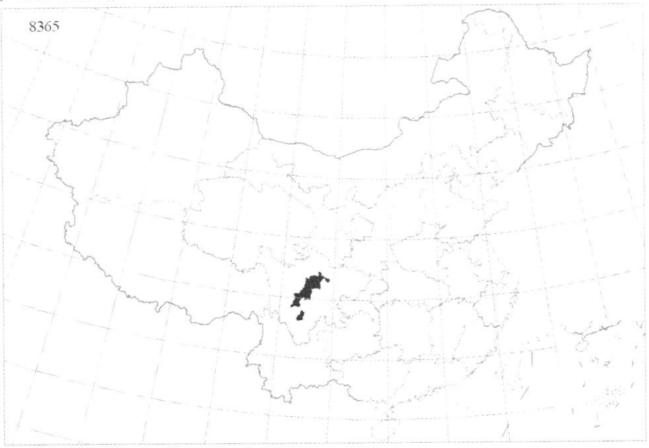

Ericaceae 杜鹃花科 *Rhododendron* 杜鹃花属
Rhododendron wasonii var. wenchuanense L. C. Hu 汶川褐毛杜鹃
Shrubs or small trees, evergreen

6.0; -3.2; 14.4; 6.6; 39.7; 26.7; 508 ❙ 1023; 535; 28 ❙ 101.2; 508; 200.8

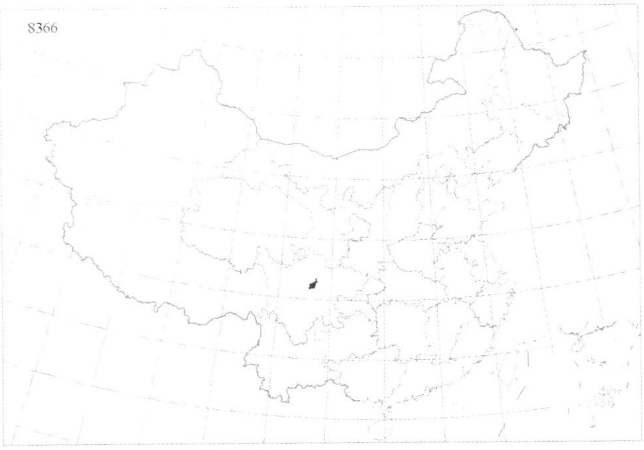

Ericaceae 杜鹃花科 *Rhododendron* 杜鹃花属
Rhododendron watsonii Hemsl. et E. H. Wilson 无柄杜鹃
Shrubs or small trees, evergreen

7.9 (2.2~14.7); -1.7 (-7.0~4.3); 16.5 (9.9~24.0); 8.5 (3.8~14.8); 57.4 (15.5~118.0); 22.2 (0.7~49.1); 567 (413~789) ❙ 914 (637~1081); 489 (306~606); 23 (14~32) ❙ 65.7 (18.6~125.9); 567 (413~789); 239.4 (147.3~298.0)

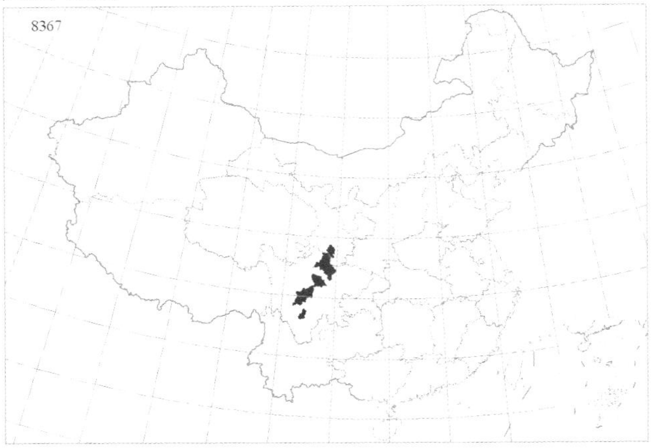

Ericaceae 杜鹃花科 *Rhododendron* 杜鹃花属
Rhododendron websterianum Rehder et E. H. Wilson 毛蕊杜鹃
Small shrubs, evergreen

-1.1 (-3.6~2.2); -11.8 (-15.6~-6.6); 8.4 (7.2~9.9); 2.8 (2.0~3.8); 8.7 (3.7~15.5); 81.3 (49.1~107.2); 368 (335~413) ❙ 617 (372~932); 361 (223~546); 14 (11~19) ❙ 62.8 (0.0~125.9); 368 (335~413); 245.4 (135.0~304.5)

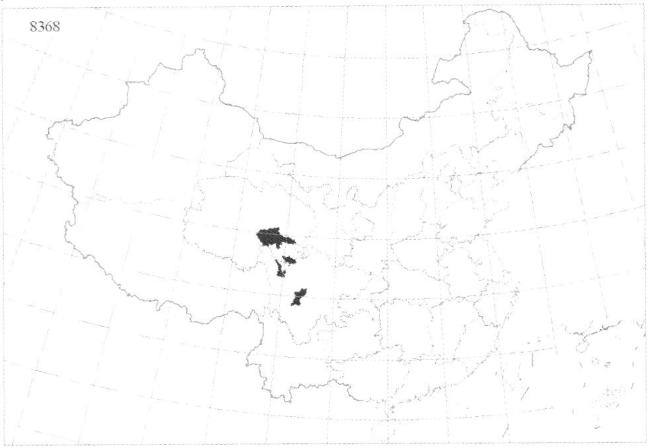

Ericaceae 杜鹃花科 *Rhododendron* 杜鹃花属
Rhododendron westlandii Hemsl. 凯里杜鹃
Shrubs or small trees, evergreen

19.1 (13.0~21.8); 9.7 (1.9~15.2); 27.2 (23.5~29.8); 19.2 (13.1~22.7); 170.3 (102.9~212.7); 0.2 (0.0~6.1); 999 (730~1216) ❙ 1608 (1202~1991); 631 (490~994); 166 (57~233) ❙ 61.7 (30.7~103.1); 997 (730~1216); 242.5 (174.9~371.7)

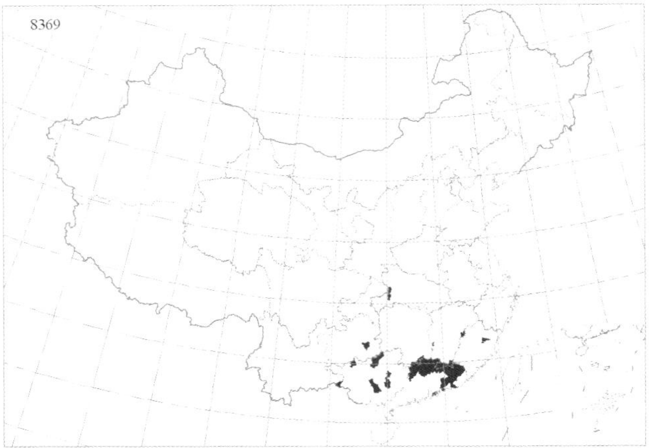

Ericaceae 杜鹃花科 *Rhododendron* 杜鹃花属
Rhododendron wightii Hook. f. 宏钟杜鹃
Shrubs or small trees, evergreen

3.9; -4.6; 11.5; 5.9; 34.6; 46.8; 459 ❙ 730; 442; 17 ❙ 39.3; 459; 189.3

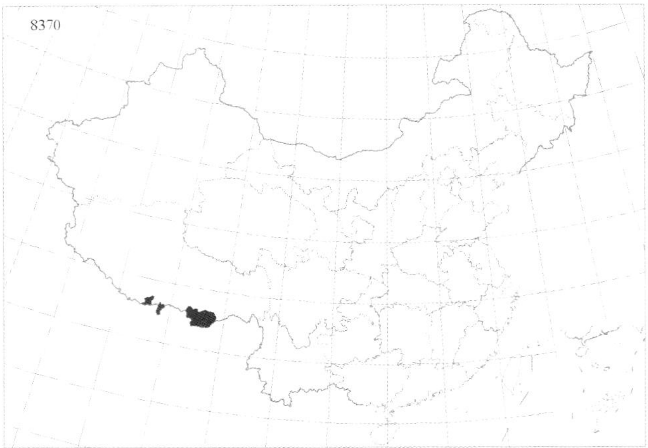

Ericaceae 杜鹃花科 *Rhododendron* 杜鹃花属

Rhododendron williamsianum Rehder et E. H. Wilson 圆叶杜鹃

Shrubs, evergreen

13.8 (11.1~17.1); 4.8 (2.2~8.5); 21.5 (19.4~23.6); 13.9 (11.3~17.1); 107.4 (80.8~145.1); 1.2 (0.0~5.3); 734 (652~845) ❙ 1261 (885~1736); 706 (479~1007); 47 (34~59) ❙ 72.4 (26.5~145.8); 734 (652~845); 271.0 (214.5~447.4)

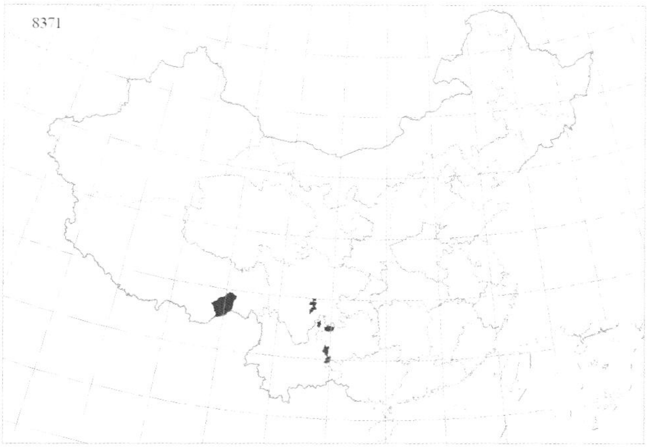

Ericaceae 杜鹃花科 *Rhododendron* 杜鹃花属

Rhododendron wiltonii Hemsl. et E. H. Wilson 皱皮杜鹃

Shrubs, evergreen

12.0 (2.2~15.9); 2.5 (-6.6~5.5); 20.6 (9.9~25.6); 12.1 (3.8~15.9); 91.2 (15.5~131.1); 7.0 (0.0~49.1); 686 (413~833) ❙ 1091 (890~1414); 580 (477~817); 41 (19~71) ❙ 61.8 (30.9~125.9); 686 (413~833); 218.3 (139.5~318.7)

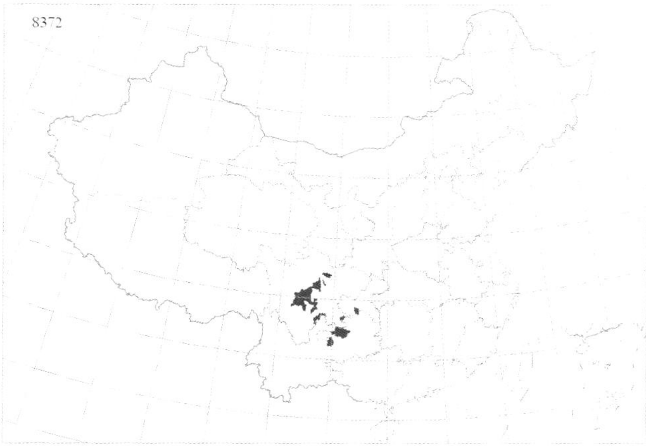

Ericaceae 杜鹃花科 *Rhododendron* 杜鹃花属

Rhododendron wolongense W. K. Hu 卧龙杜鹃

Small trees, evergreen

6.0; -3.2; 14.4; 6.6; 39.7; 26.7; 508 ❙ 1023; 535; 28 ❙ 101.2; 508; 200.8

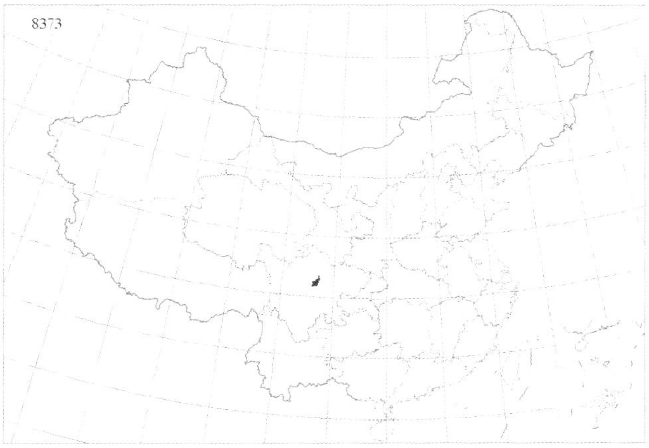

Ericaceae 杜鹃花科 *Rhododendron* 杜鹃花属

Rhododendron wumingense Fang 武鸣杜鹃

Shrubs, evergreen

21.6; 12.6; 28.5; 21.7; 199.8; 0.0; 1153 ❙ 1625; 795; 173 ❙ 40.9; 1153; 237.8

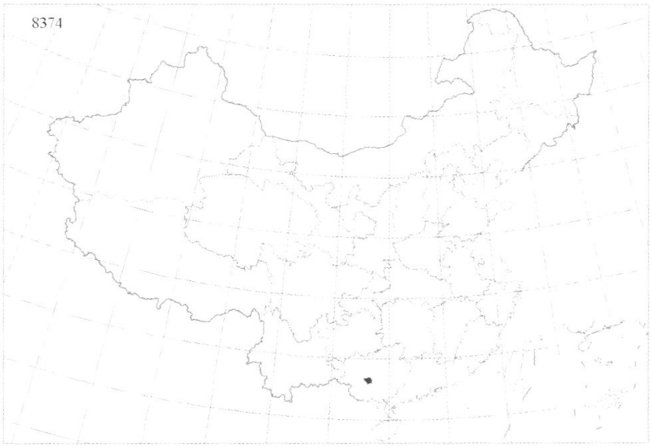

Ericaceae 杜鹃花科 *Rhododendron* 杜鹃花属

Rhododendron xanthocodon Hutch. 黄铃杜鹃

Shrubs or small trees, evergreen

1.7 (-1.8~13.3); -7.3 (-10.5~5.4); 9.7 (6.1~19.6); 4.3 (1.8~13.4); 21.8 (1.9~100.4); 60.8 (0.0~82.8); 405 (329~706) ❙ 500 (255~1736); 324 (191~1007); 11 (1~51) ❙ 7.3 (-31.1~145.8); 358 (255~706); 137.2 (60.4~447.4)

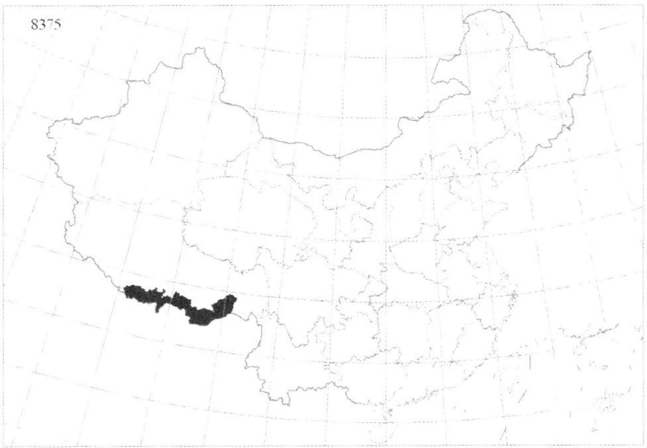

Ericaceae 杜鹃花科 *Rhododendron* 杜鹃花属

Rhododendron xanthostephanum Merr. 鲜黄杜鹃

Shrubs, evergreen

13.6 (5.3~19.1); 6.7 (-2.4~13.0); 19.1 (12.2~23.9); 13.7 (5.7~19.1); 107.5 (30.5~169.7); 3.4 (0.0~26.2); 721 (479~929) ❙ 1024 (741~1504); 596 (411~860); 35 (15~69) ❙ 44.9 (-14.2~100.7); 684 (479~917); 317.4 (218.0~506.2)

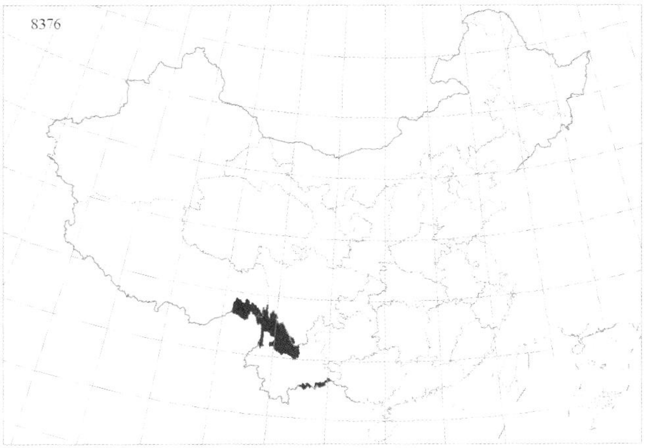

Ericaceae 杜鹃花科 *Rhododendron* 杜鹃花属
Rhododendron xiaoxidongense W. K. Hu 小溪洞杜鹃
Shrubs, evergreen

16.2; 5.0; 26.8; 16.2; 135.1; 0.4; 860 ❙ 1611; 524; 208 ❙ 88.0; 860; 217.4

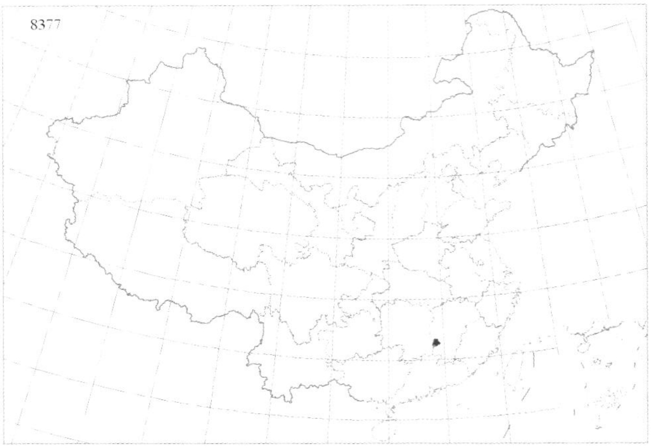

Ericaceae 杜鹃花科 *Rhododendron* 杜鹃花属
Rhododendron xichangense Z. J. Zhao 西昌杜鹃
Shrubs, evergreen

14.5; 5.6; 22.1; 14.5; 114.3; 0.1; 754 ❙ 1075; 575; 40 ❙ 42.6; 743; 230.0

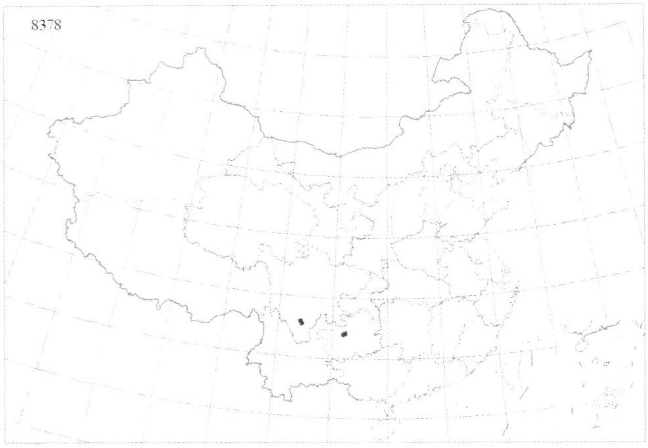

Ericaceae 杜鹃花科 *Rhododendron* 杜鹃花属
Rhododendron xiguense Ching et H. P. Yang 西固杜鹃
Shrubs, evergreen

6.7; -4.0; 16.6; 7.4; 49.5; 29.1; 535 ❙ 637; 306; 14 ❙ 19.1; 535; 263.2

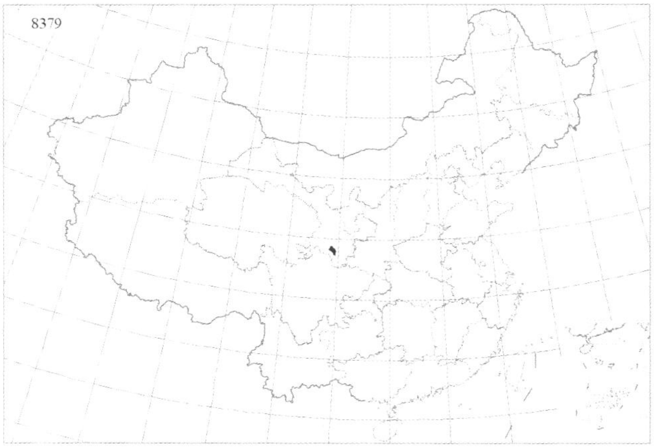

Ericaceae 杜鹃花科 *Rhododendron* 杜鹃花属
Rhododendron yangmingshanense Tam 阳明山杜鹃
Shrubs, deciduous

16.0; 4.8; 26.3; 16.0; 132.9; 0.8; 849 ❙ 1496; 473; 197 ❙ 77.3; 844; 209.2

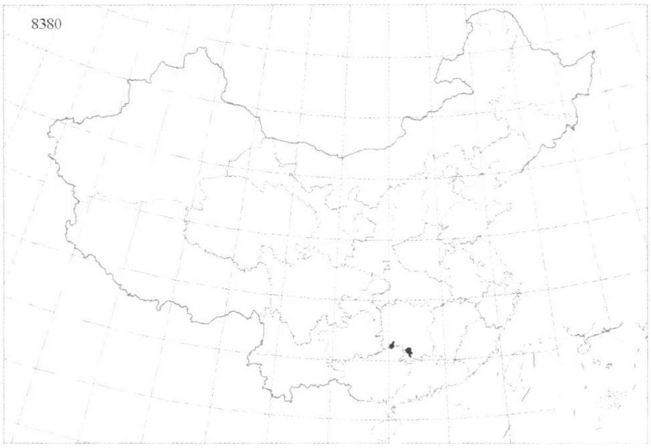

Ericaceae 杜鹃花科 *Rhododendron* 杜鹃花属
Rhododendron yaogangxianense Q. X. Liu 瑶岗仙杜鹃
Small shrubs

17.7; 6.9; 27.5; 17.7; 152.4; 0.0; 927 ❙ 1486; 505; 176 ❙ 60.2; 927; 188.8

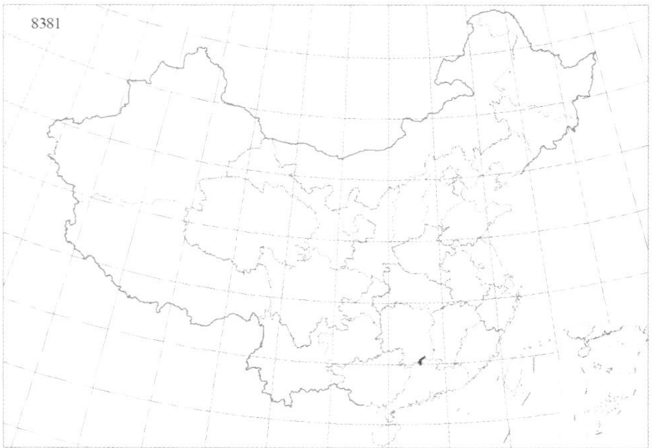

Ericaceae 杜鹃花科 *Rhododendron* 杜鹃花属
Rhododendron yaoshanicum Fang et M. Y. He 瑶山杜鹃
Small shrubs, semievergreen or green

19.2; 9.4; 27.4; 19.2; 170.7; 0.0; 1003 ❙ 1501; 580; 169 ❙ 51.6; 1003; 223.0

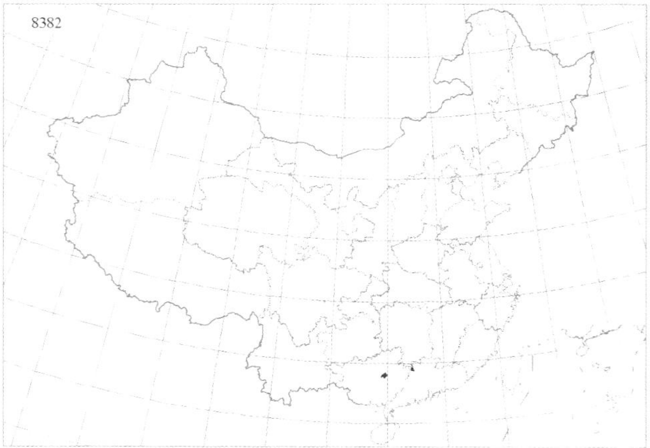

Ericaceae 杜鹃花科 *Rhododendron* 杜鹃花属

Rhododendron yizhangense Q. X. Liu 宜章杜鹃

Small shrubs

17.7；6.9；27.5；17.7；152.4；0.0；927 ▎1486；505；176 ▎60.2；927；188.8

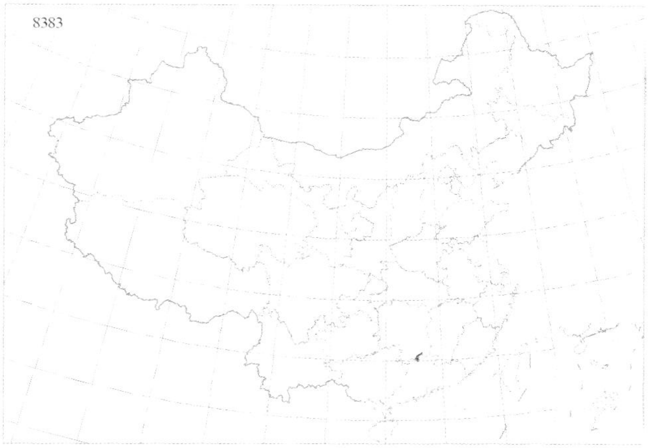

Ericaceae 杜鹃花科 *Rhododendron* 杜鹃花属

Rhododendron yungchangense Cullen 少鳞杜鹃

Shrubs, evergreen

11.0；4.2；16.6；11.0；73.1；0.8；636 ▎1134；651；47 ▎78.3；636；339.5

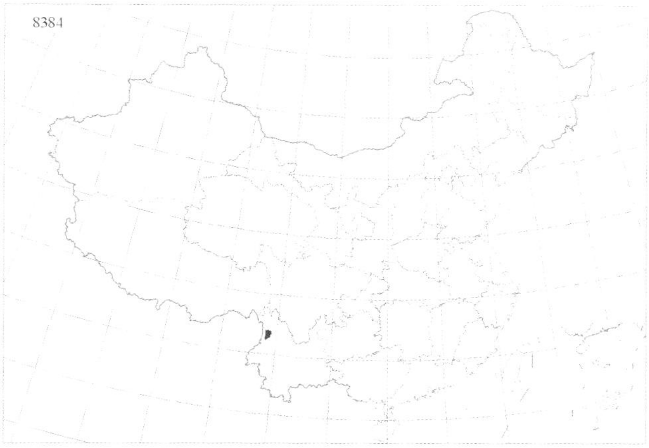

Ericaceae 杜鹃花科 *Rhododendron* 杜鹃花属

Rhododendron yungningense I. B. Balfour ex Hutchinson 永宁杜鹃

Shrubs, evergreen

8.8 (3.8~14.3)；1.5 (-4.3~7.6)；15.0 (10.8~19.7)；9.0 (4.7~14.3)；58.4 (21.6~111.9)；12.3 (0.0~35.4)；577 (444~728) ▎937 (801~1007)；571 (496~603)；23 (12~40) ▎65.6 (36.1~121.7)；573 (444~707)；304.1 (265.2~351.7)

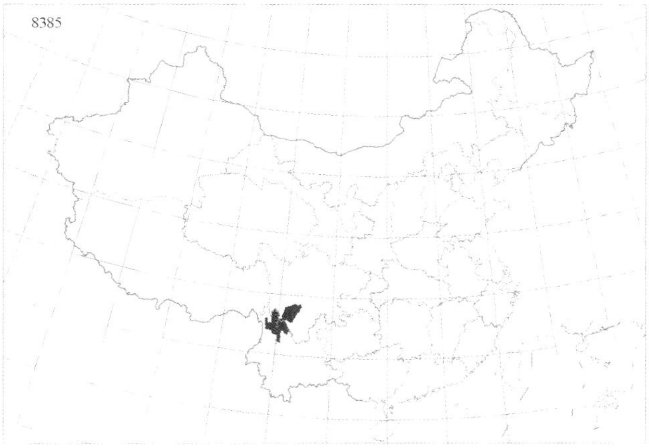

Ericaceae 杜鹃花科 *Rhododendron* 杜鹃花属

Rhododendron yunnanense Franch. 云南杜鹃

Shrubs, deciduous, semievergreen or evergreen

12.9 (0.3~19.1)；4.6 (-9.9~12.0)；20.0 (9.3~26.6)；13.1 (3.3~19.1)；101.3 (11.9~169.2)；5.5 (0.0~67.5)；711 (391~969) ▎1041 (628~1454)；577 (372~817)；38 (11~70) ▎49.2 (-14.2~125.9)；697 (391~969)；255.3 (142.9~487.2)

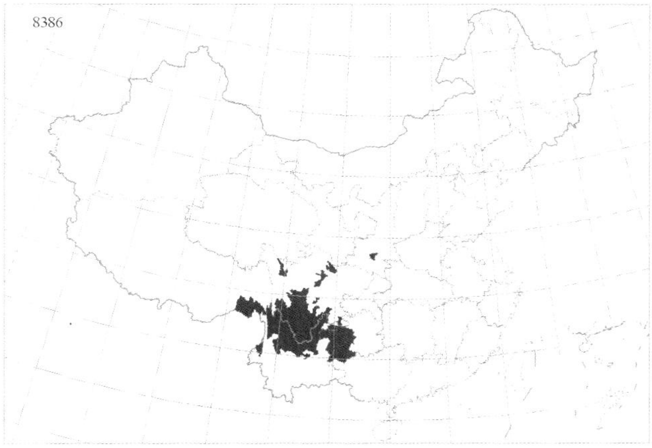

Ericaceae 杜鹃花科 *Rhododendron* 杜鹃花属

Rhododendron zaleucum I. B. Balfour et W. W. Sm. 白面杜鹃

Shrubs or small trees, evergreen

13.4 (8.6~19.5)；6.0 (1.3~12.7)；19.2 (15.0~23.3)；13.4 (8.7~19.5)；102.9 (52.6~173.8)；1.9 (0.0~8.8)；717 (574~942) ▎1272 (999~1559)；701 (501~907)；56 (42~70) ▎78.5 (39.0~99.9)；713 (574~921)；382.9 (183.0~487.2)

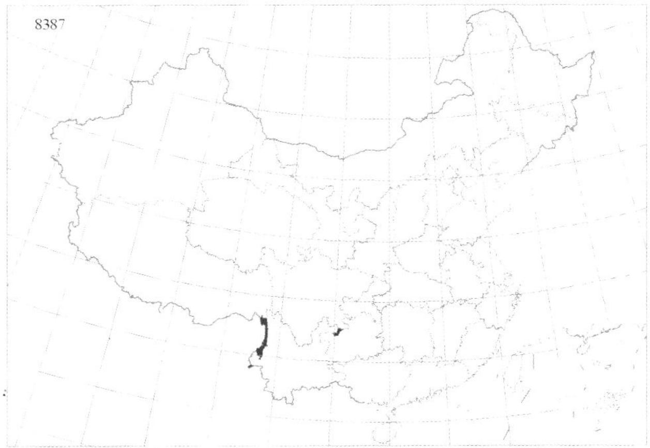

Ericaceae 杜鹃花科 *Rhododendron* 杜鹃花属

Rhododendron zaleucum var. pubifolium R. C. Fang 毛叶白面杜鹃

Shrubs or small trees, evergreen

12.3；5.4；17.7；12.3；88.0；0.3；670 ▎1290；712；63 ▎92.7；670；395.9

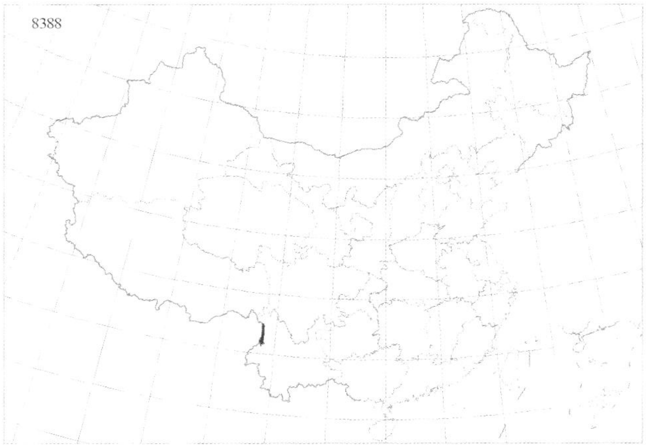

Ericaceae 杜鹃花科 *Rhododendron* 杜鹃花属
Rhododendron zheguense Ching et H. P. Yang 鹧鸪杜鹃
Shrubs, evergreen

2.0; -7.2; 10.1; 3.8; 15.4; 51.4; 413 ❙ 943; 507; 23 ❙ 128.6; 413; 270.7

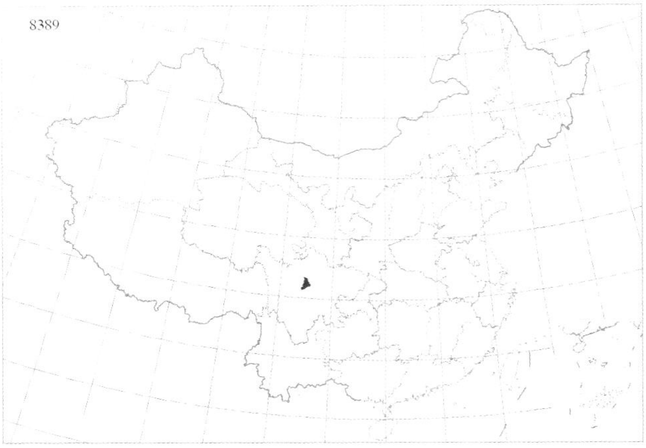

Ericaceae 杜鹃花科 *Rhododendron* 杜鹃花属
Rhododendron zhongdianense L. C. Hu 中甸杜鹃
Shrubs, evergreen

6.1; -1.5; 12.7; 6.3; 35.2; 21.8; 496 ❙ 801; 496; 21 ❙ 61.4; 496; 282.9

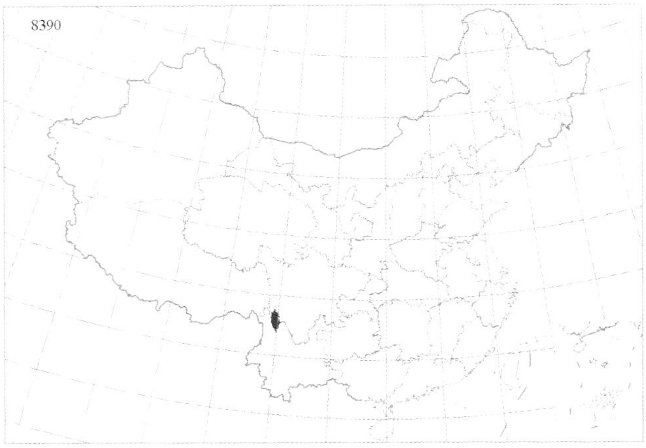

Ericaceae 杜鹃花科 *Rhododendron* 杜鹃花属
Rhododendron ziyuanense P. C. Tam 资源杜鹃
Shrubs or trees, evergreen

16.3; 5.4; 26.3; 16.3; 136.5; 0.4; 860 ❙ 1544; 518; 195 ❙ 80.4; 859; 224.1

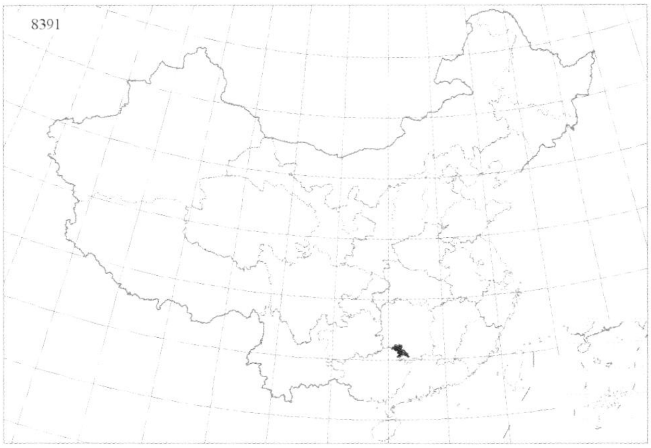

Ericaceae 杜鹃花科 *Rhododendron* 杜鹃花属
Rhododendron × bathyphyllum Balf. f. et Forrest 多叶杜鹃
Small shrubs

5.4; -2.3; 12.2; 5.8; 31.0; 25.9; 480 ❙ 851; 492; 30 ❙ 77.4; 480; 253.8

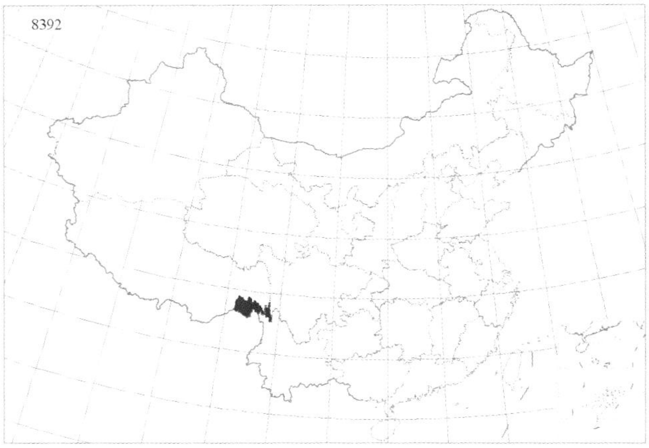

Ericaceae 杜鹃花科 *Rhododendron* 杜鹃花属
Rhododendron × detonsum I. B. Balfour et Forrest 落毛杜鹃
Shrubs, evergreen

9.0 (2.9~14.3); 1.2 (-6.3~7.6); 15.8 (11.2~19.7); 9.5 (4.4~14.3); 63.7 (20.5~111.9); 14.9 (0.0~44.6); 588 (428~728) ❙ 984 (950~1023); 564 (490~619); 26 (15~35) ❙ 73.3 (36.1~121.7); 583 (428~707); 276.2 (200.8~319.8)

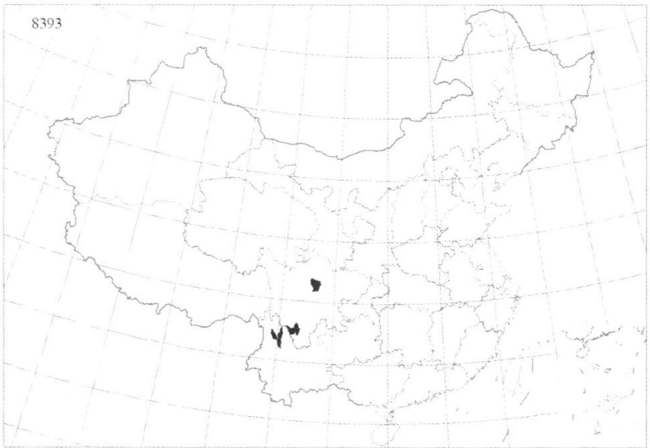

Ericaceae 杜鹃花科 *Rhododendron* 杜鹃花属
Rhododendron × duclouxii H. Lév. 粉红爆杖花
Shrubs

14.6; 7.9; 19.6; 14.6; 115.8; 0.0; 736 ❙ 1009; 596; 39 ❙ 37.1; 732; 278.4

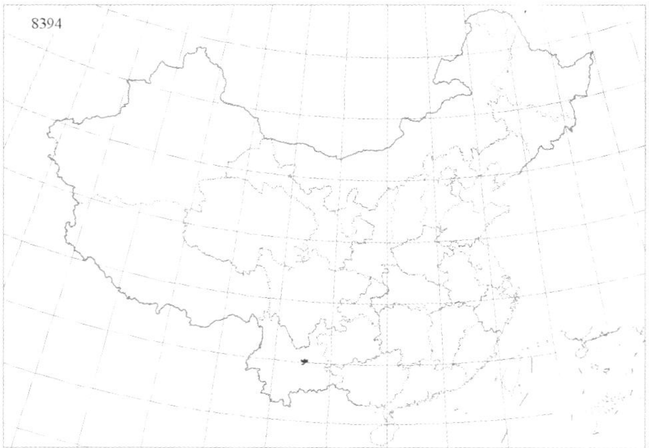

Ericaceae 杜鹃花科 *Vaccinium* 乌饭树属

Vaccinium delavayi subsp. merrillianum (Hayata) R. C. Fang 台湾越橘

Shrubs, evergreen

19.5；13.6；24.6；19.5；174.2；0.0；952 ❙ 2766；1098；353 ❙ 193.0；952；245.0

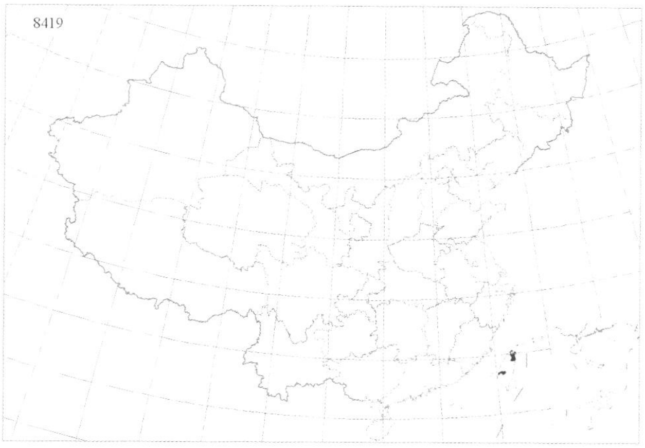

Ericaceae 杜鹃花科 *Vaccinium* 乌饭树属

Vaccinium dendrocharis Hand.-Mazz. 树生越橘

Shrubs, evergreen

10.9 (3.3~19.2)；3.7 (-5.4~13.3)；16.8 (11.?~23.5)；11.2 (4.6~19.4)；81.0 (22.3~172.6)；9.1 (0.0~41.8)；643 (427~932) ❙ 1145 (473~1736)；649 (287~1007)；46 (7~70) ❙ 75.2 (5.0~145.8)；642 (427~923)；376.3 (230.7~487.2)

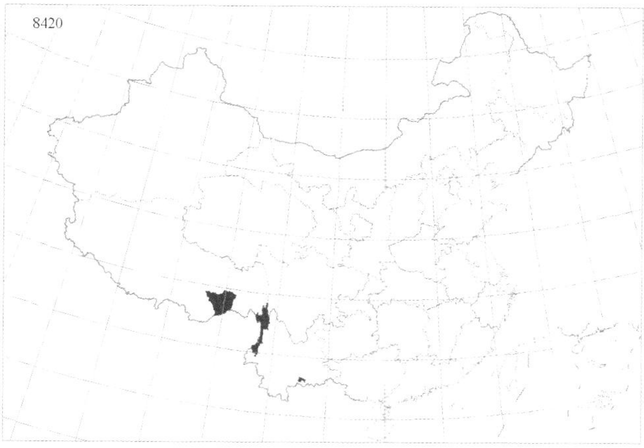

Ericaceae 杜鹃花科 *Vaccinium* 乌饭树属

Vaccinium duclouxii (H. Lév.) Hand.-Mazz. 云南越橘

Shrubs or small trees, evergreen

14.9 (5.4~21.6)；7.6 (-2.1~15.1)；20.5 (12.3~28.5)；14.9 (5.8~21.7)；120.6 (31.5~199.8)；1.4 (0.0~25.6)；765 (481~1153) ❙ 1074 (741~1625)；613 (411~911)．41 (12~173) ❙ 41.3 (-14.2~99.9)；742 (481~1153)；324.4 (142.9~659.0)

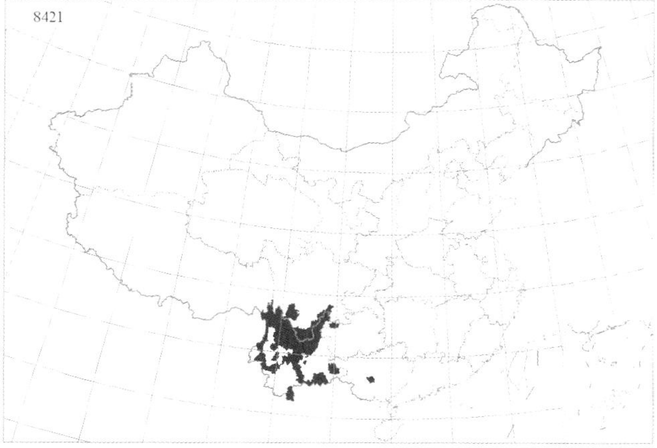

Ericaceae 杜鹃花科 *Vaccinium* 乌饭树属

Vaccinium duclouxii var. hirticaule C. Y. Wu 刚毛云南越橘

Shrubs or small trees, evergreen

16.2；9.1；21.3；16.2；134.9；0.0；798 ❙ 1062；605；42 ❙ 33.1；778；311.9

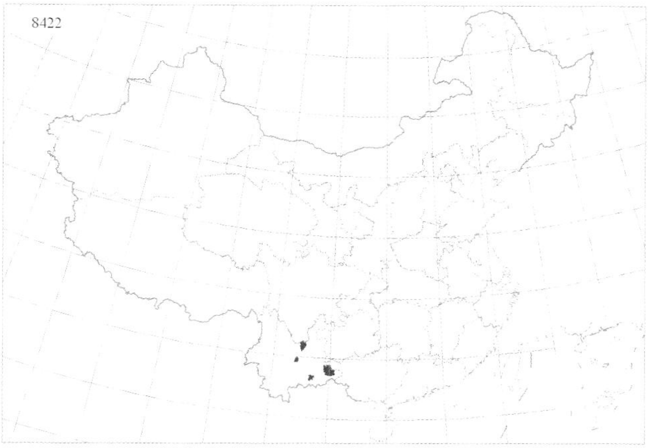

Ericaceae 杜鹃花科 *Vaccinium* 乌饭树属

Vaccinium duclouxii var. pubipes C. Y. Wu 柔毛云南越橘

Shrubs or small trees, evergreen

10.7；3.6；16.5；10.7；72.3；3.7；628 ❙ 1157；647；53 ❙ 83.7；628；374.0

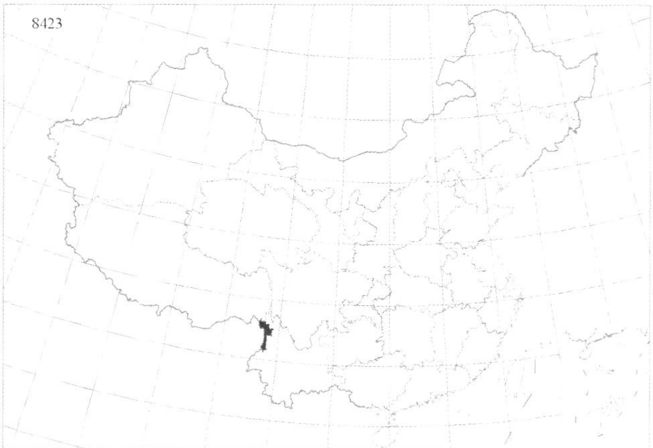

Ericaceae 杜鹃花科 *Vaccinium* 乌饭树属

Vaccinium dunalianum Wight 樟叶越橘

Shrubs, evergreen

15.4 (-0.7~22.2)；7.4 (-9.6~17.7)；21.9 (7.7~26.6)；15.5 (2.5~22.2)；126.5 (6.8~206.7)；1.4 (0.0~74.2)；790 (338~1123) ❙ 1166 (370~2992)；627 (262~1847)；53 (10~402) ❙ 47.3 (-14.2~235.6)；778 (338~1123)；282.1 (61.6~659.0)

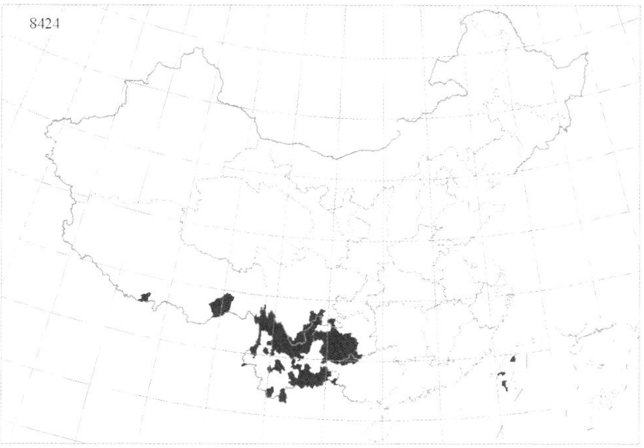

Ericaceae 杜鹃花科 *Vaccinium* 乌饭树属

Vaccinium dunalianum var. megaphyllum Sleumer 大樟叶越橘

Shrubs, evergreen

18.0 (16.3~19.3); 11.5 (9.9~13.3); 22.4 (20.3~23.6); 18.0 (16.4~19.4); 156.0 (136.6~173.2); 0.0 (0.0~0.0); 865 (789~939) ❙ 1298 (1049~1547); 742 (598~911); 52 (44~63) ❙ 50.4 (21.7~80.6); 854 (780~923); 414.2 (295.5~551.3)

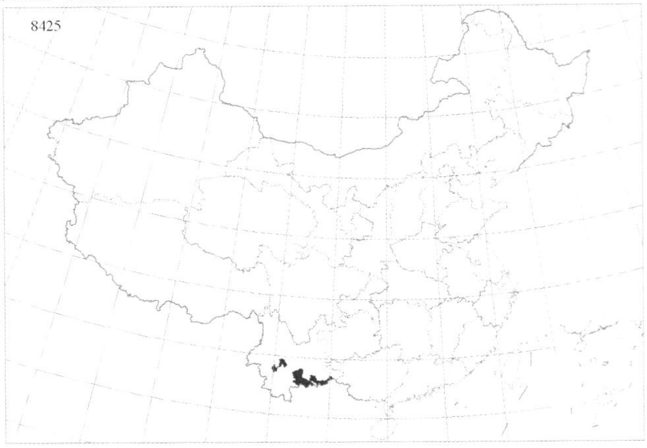

Ericaceae 杜鹃花科 *Vaccinium* 乌饭树属

Vaccinium dunalianum var. urophyllum Rehder et E. H. Wilson 尾叶越橘

Shrubs, evergreen

15.8 (8.6~21.0); 7.2 (1.3~14.2); 22.7 (15.0~26.6); 15.8 (8.7~21.0); 130.1 (52.6~191.8); 0.4 (0.0~8.8); 806 (574~1053) ❙ 1213 (916~1736); 639 (501~1007); 56 (27~74) ❙ 51.7 (12.2~145.8); 802 (574~1053); 268.7 (153.9~551.6)

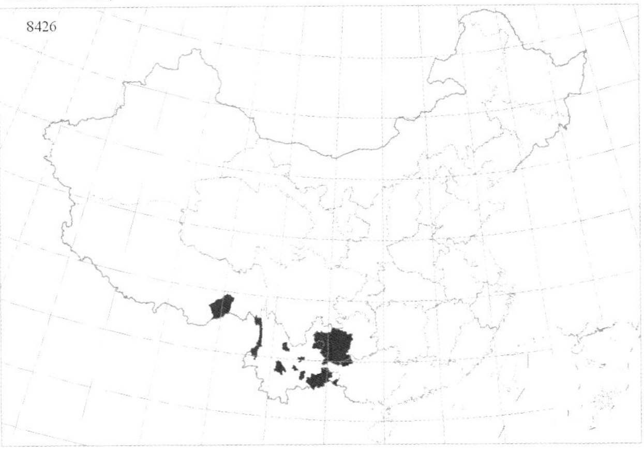

Ericaceae 杜鹃花科 *Vaccinium* 乌饭树属

Vaccinium dunnianum Sleumer 长穗越橘

Shrubs, evergreen

18.7 (16.9~21.0); 11.6 (8.5~14.2); 23.8 (21.4~25.8); 18.7 (16.9~21.0); 164.9 (142.7~191.8); 0.0 (0.0~0.0); 916 (813~1053) ❙ 1398 (1178~1551); 768 (646~848); 71 (51~169) ❙ 52.7 (36.8~69.8); 915 (813~1053); 416.3 (243.9~551.6)

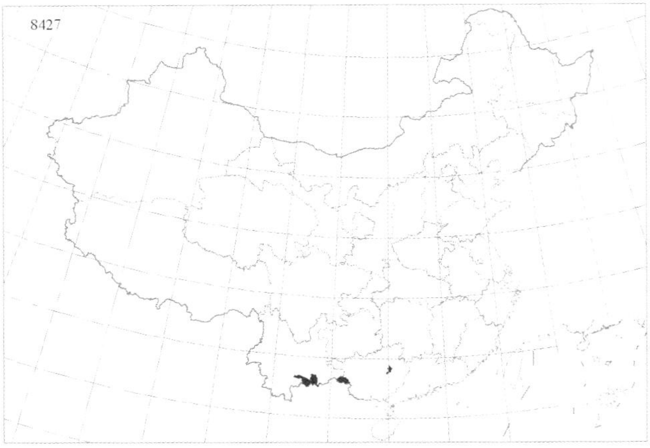

Ericaceae 杜鹃花科 *Vaccinium* 乌饭树属

Vaccinium emarginatum Hayata 凹顶越橘

Shrubs, evergreen

18.8; 13.5; 23.2; 18.8; 166.2; 0.0; 910 ❙ 2708; 1228; 272 ❙ 201.1; 910; 280.5

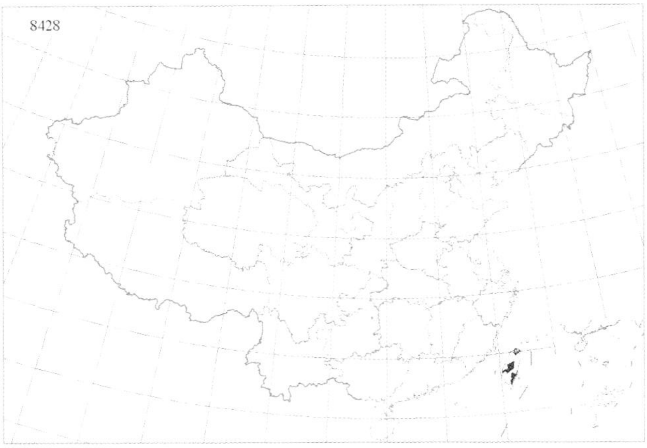

Ericaceae 杜鹃花科 *Vaccinium* 乌饭树属

Vaccinium exaristatum Kurz 隐距越橘

Small trees, evergreen

17.6 (10.8~21.0); 10.2 (3.9~15.1); 23.0 (16.6~26.6); 17.6 (10.9~21.0); 151.8 (71.8~191.8); 0.0 (0.0~1.3); 871 (633~1053) ❙ 1304 (939~1607); 702 (524~911); 56 (24~97) ❙ 50.6 (6.5~99.9); 861 (633~1053); 362.7 (189.5~659.0)

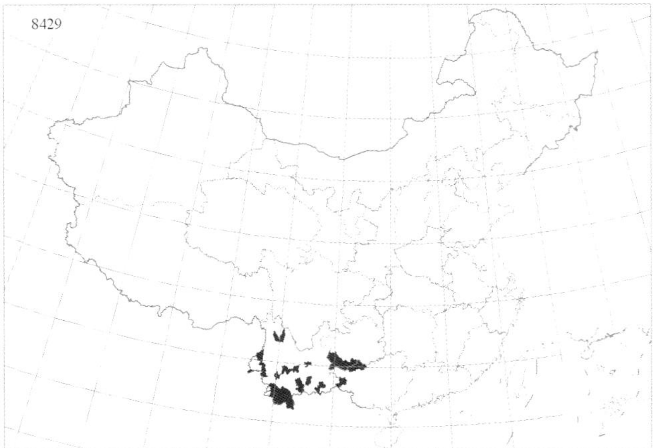

Ericaceae 杜鹃花科 *Vaccinium* 乌饭树属

Vaccinium fimbribracteatum C. Y. Wu 齿苞越橘

Shrubs, evergreen

15.1; 4.3; 25.5; 15.2; 123.2; 1.3; 813 ❙ 1191; 491; 68 ❙ 46.7; 813; 198.4

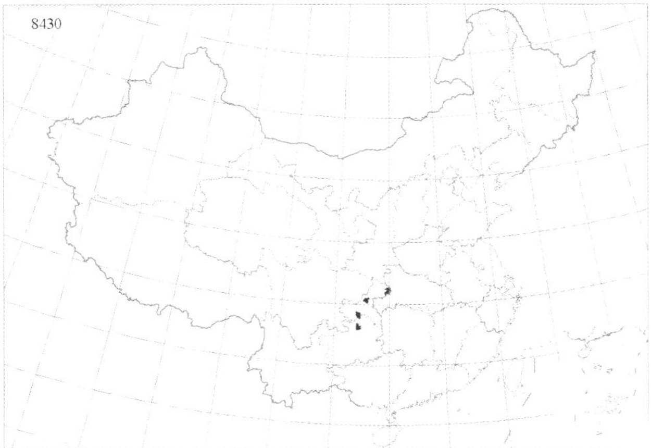

Ericaceae 杜鹃花科 *Vaccinium* 乌饭树属
Vaccinium fimbricalyx Chun et W. P. Fang 流苏萼越橘
Shrubs, evergreen

17.9; 7.9; 26 7; 18.0; 155.9; 0.1; 943 ❙ 1676; 635; 180 ❙ 79.5; 943; 242.7

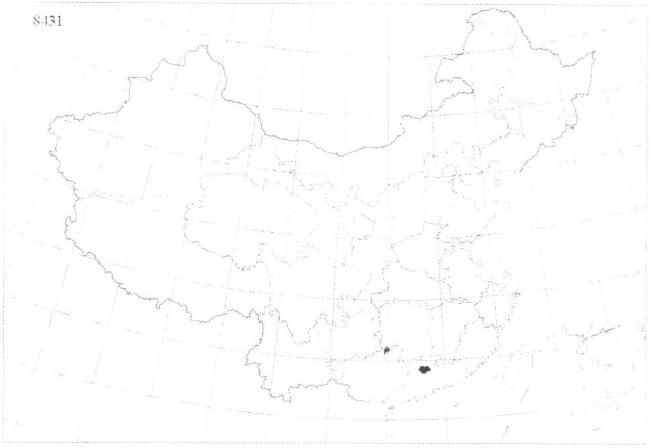

Ericaceae 杜鹃花科 *Vaccinium* 乌饭树属
Vaccinium foetidissimum H. Lév. et Vaniot 臭越橘
Shrubs, evergreen

16.5 (14.3~19.1); 6.5 (4.0~10.1); 25.1 (22.2~27.2); 16.5 (14.3~19.2); 138.5 (112.6~170.1); 0.1 (0.0~1.1); 851 (756~981) ❙ 1278 (1171~1496); 582 (448~720); 86 (51~158) ❙ 50.5 (27.5~71.9); 851 (756~981); 228.1 (188.8~322.1)

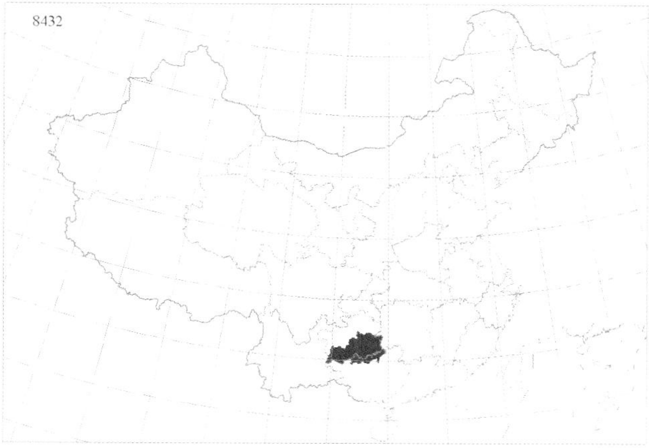

Ericaceae 杜鹃花科 *Vaccinium* 乌饭树属
Vaccinium fragile Franch. 乌鸦果
Small shrubs, evergreen

12.8 (2.2~18.9); 4.9 (-6.6~12.0); 19.4 (9.5~26.6); 12.9 (3.8~19.0); 99.0 (15.5~167.8); 4.6 (0.0~49.1); 700 (411~929) ❙ 1025 (628~1454); 579 (406~792); 37 (11~70) ❙ 49.2 (-14.2~125.9); 683 (411~912); 275 6 (142.9~487.2)

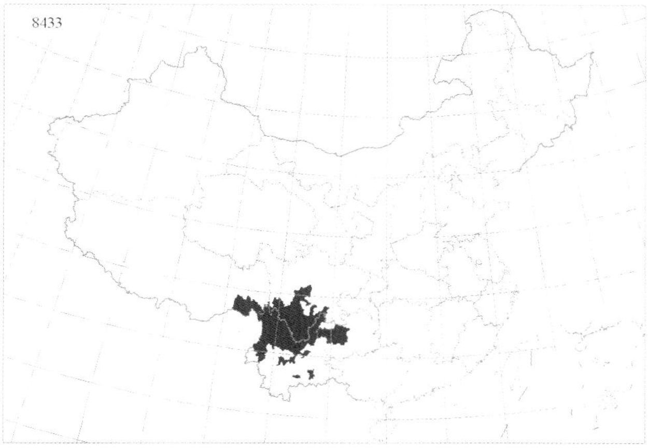

Ericaceae 杜鹃花科 *Vaccinium* 乌饭树属
Vaccinium fragile var. mekongense (W. W. Sm.) Sleumer 大叶乌鸦果
Small shrubs, evergreen

13.7 (11.3~14.6); 6.5 (3.6~7.6); 19.4 (17.7~20.3); 13.8 (11.4~14.6); 105.6 (78.3~114.8); 0.4 (0.0~1.8); 715 (649~741) ❙ 1108 (991~1454); 649 (580~792); 33 (18~70) ❙ 55.2 (36.1~99.9); 696 (649~727); 330.6 (265.0~487.2)

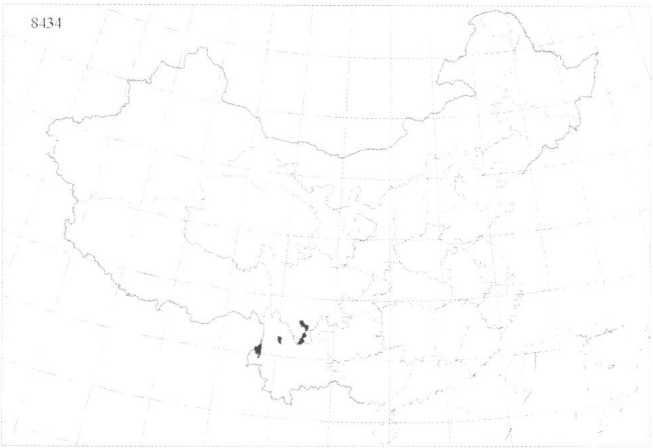

Ericaceae 杜鹃花科 *Vaccinium* 乌饭树属
Vaccinium gaultheriifolium (Griff.) Hook. f. ex C. B. Clarke 软骨边越橘
Shrubs, evergreen

12.4 (5.3~17.6); 4.8 (-2.4~9.9); 18.5 (12.2~23.2); 12.5 (5.7~17.7); 96.6 (30.5~152.4); 7.0 (0.0~26.2); 690 (479~859) ❙ 1333 (961~1736); 765 (558~1007); 47 (31~55) ❙ 94.8 (65.6~145.8); 690 (479~859); 361.4 (234.2~447.4)

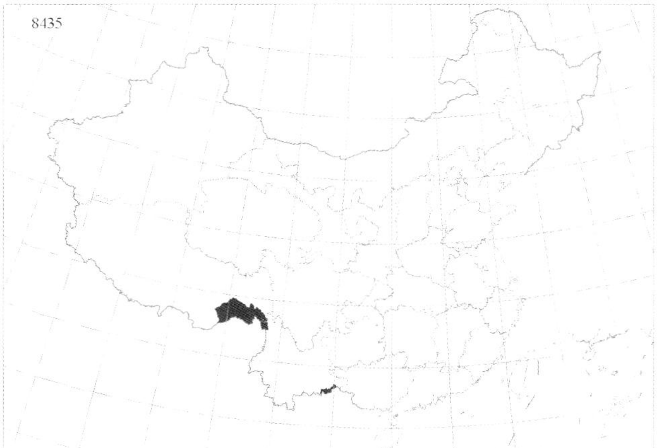

Ericaceae 杜鹃花科 *Vaccinium* 乌饭树属
Vaccinium gaultheriifolium var. glaucorubrum C. Y. Wu 粉花软骨边越橘
Shrubs, evergreen

17.4; 9.9; 22.8; 17.5; 149.8; 0.0; 846 ❙ 1463; 837; 54 ❙ 73.1; 846; 386.6

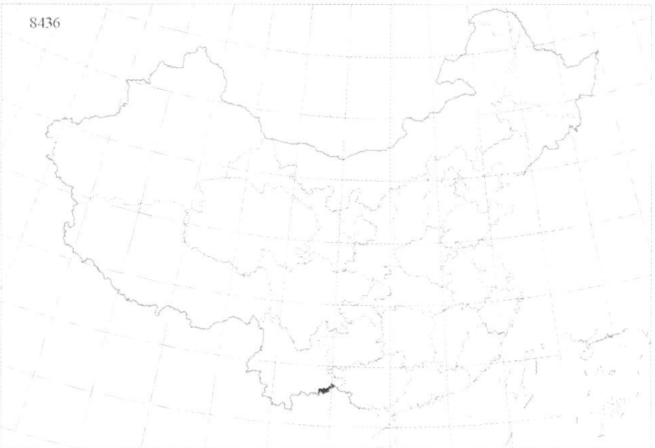

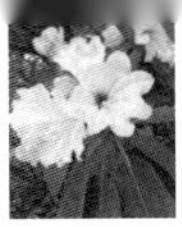

Ericaceae 杜鹃花科 *Vaccinium* 乌饭树属

Vaccinium glaucoalbum Hook. f. ex C. B. Clarke 粉白越橘

Shrubs, evergreen

9.4 (-0.7~17.3); 1.7 (-9.6~11.0); 15.7 (7.7~21.6); 10.0 (2.5~17.3); 71.0 (6.8~148.0); 18.3 (0.0~74.2); 600 (338~832) ❙ 1098 (370~1736); 638 (262~1007); 35 (10~51) ❙ 78.7 (4.7~145.8); 592 (338~783); 304.1 (61.6~447.4)

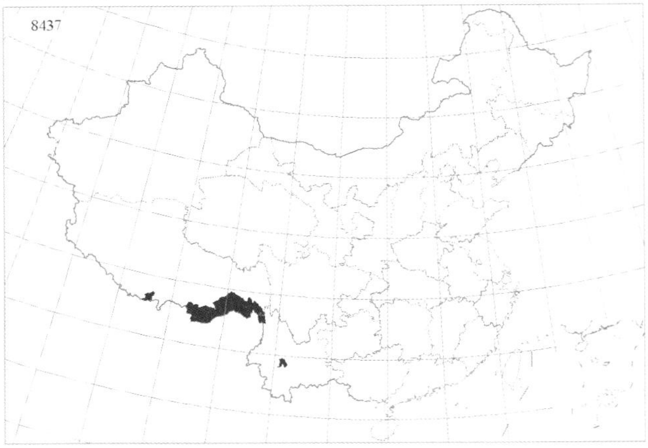

Ericaceae 杜鹃花科 *Vaccinium* 乌饭树属

Vaccinium glaucophyllum C. Y. Wu et R. C. Fang 灰叶乌饭

Shrubs, evergreen

16.8; 7.8; 23.6; 16.8; 141.5; 0.0; 836 ❙ 1334; 720; 64 ❙ 59.6; 836; 226.6

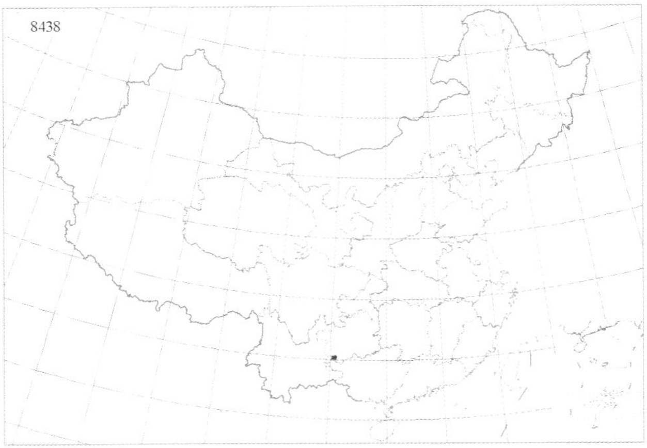

Ericaceae 杜鹃花科 *Vaccinium* 乌饭树属

Vaccinium guangdongense W. P. Fang et Z. H. Pan 广东乌饭

Trees, evergreen

20.5 (17.7~22.3); 11.6 (7.7~14.7); 27.8 (26.6~28.8); 20.5 (17.8~22.4); 186.5 (153.6~208.3); 0.0 (0.0~0.0); 1077 (916~1188) ❙ 1734 (1462~2005); 722 (507~891); 156 (75~191) ❙ 61.9 (29.9~95.5); 1077 (916~1188); 237.2 (203.4~275.2)

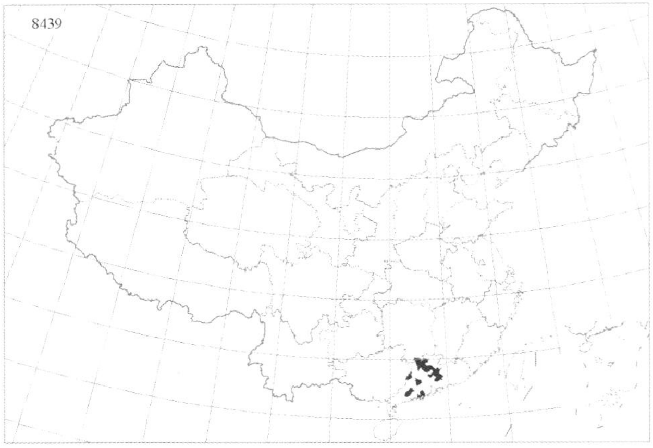

Ericaceae 杜鹃花科 *Vaccinium* 乌饭树属

Vaccinium hainanense Sleumer 海南越橘

Shrubs, evergreen

24.0; 19.1; 27.2; 24.0; 228.1; 0.0; 1302 ❙ 1551; 539; 48 ❙ 19.1; 1147; 620.0

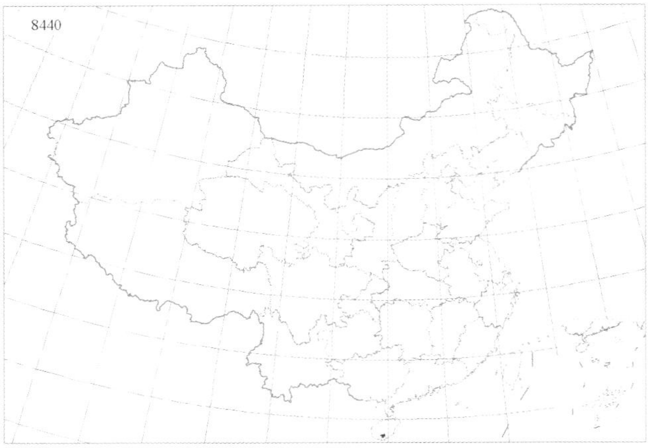

Ericaceae 杜鹃花科 *Vaccinium* 乌饭树属

Vaccinium haitangense Sleumer 海棠越橘

Dwarf shrubs, evergreen

10.8; 2.3; 18.5; 10.9; 76.1; 5.3; 639 ❙ 1124; 629; 28 ❙ 75.8; 639; 310.8

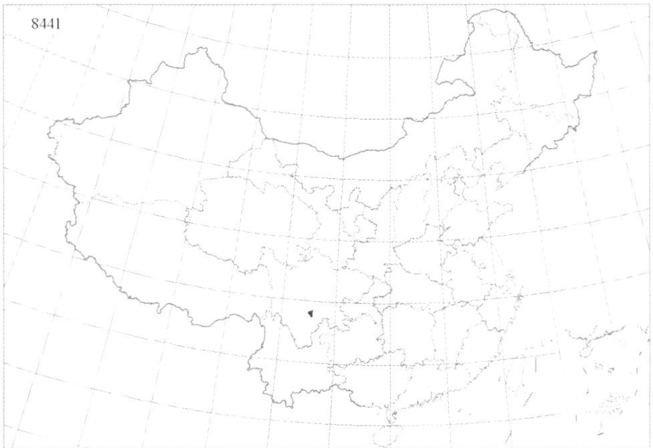

Ericaceae 杜鹃花科 *Vaccinium* 乌饭树属

Vaccinium harmandianum Dop 长冠越橘

Shrubs or trees, evergreen

17.9 (16.3~20.5); 11.6 (9.9~14.9); 22.1 (20.3~23.7); 18.0 (16.4~20.5); 155.7 (136.6~185.9); 0.0 (0.0~0.0); 864 (789~992) ❙ 1340 (1049~1495); 725 (598~813); 54 (44~66) ❙ 55.2 (26.1~65.9); 847 (780~959); 432.3 (359.7~570.1)

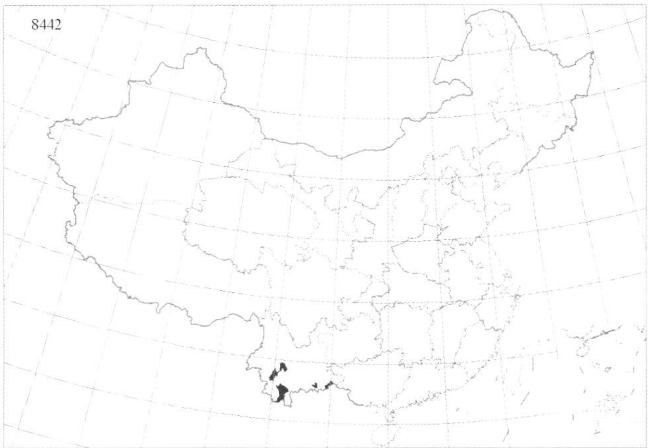

Ericaceae 杜鹃花科 *Vaccinium* 乌饭树属
Vaccinium papillatum P. F. Stevens 粉果越橘
Shrubs, evergreen

17.2; 9.5; 22.6; 17.2; 146.5; 0.0; 836 ❙ 1250; 698; 49 ❙ 49.2; 836; 324.3

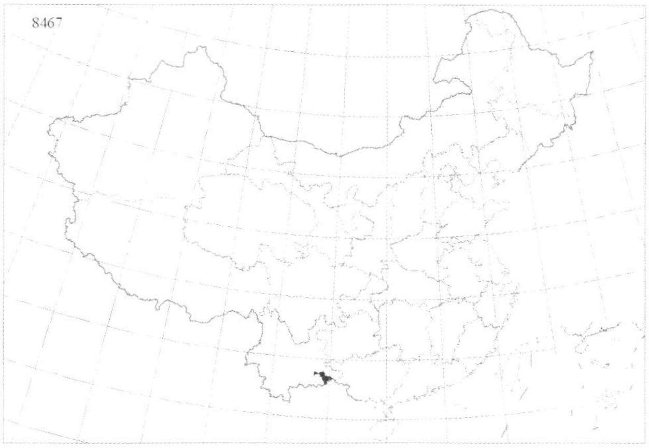

Ericaceae 杜鹃花科 *Vaccinium* 乌饭树属
Vaccinium papulosum C. Y. Wu et R. C. Fang 瘤果越橘
Shrubs, evergreen

13.3; 5.4; 19.6; 13.4; 100.4; 0.0; 706 ❙ 1736; 1007; 51 ❙ 145.8; 706; 447.4

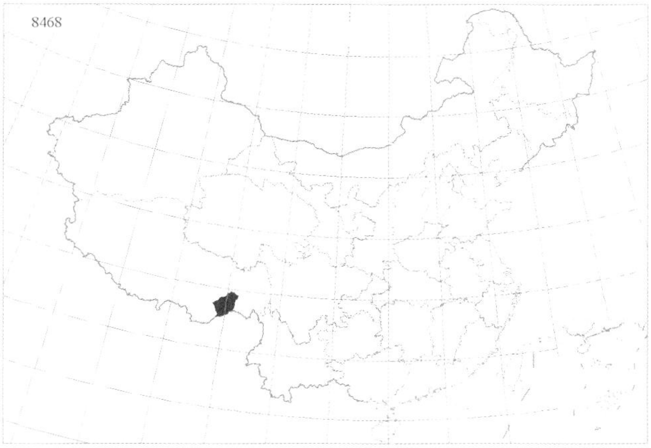

Ericaceae 杜鹃花科 *Vaccinium* 乌饭树属
Vaccinium petelotii Merr. 大叶越橘
Shrubs, evergreen

18.5 (16.9~21.0); 12.0 (9.9~14.2); 23.0 (21.4~25.6); 18.6 (16.9~21.0); 163.0 (142.7~191.8); 0.0 (0.0~0.0); 899 (813~1053) ❙ 1403 (1178~1547); 795 (646~911); 58 (51~68) ❙ 56.7 (36.8~80.6); 897 (813~1053); 478.5 (383.4~551.6)

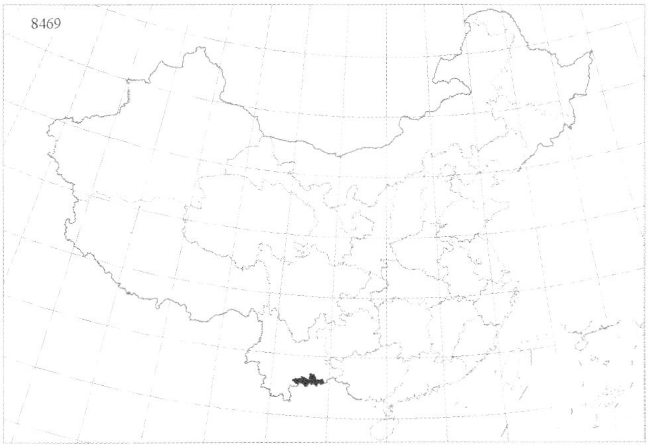

Ericaceae 杜鹃花科 *Vaccinium* 乌饭树属
Vaccinium podocarpoideum Fang et Z. H. Pan 罗汉松叶乌饭
Shrubs, evergreen

15.3 (14.3~15.7); 4.1 (3.3~4.8); 25.6 (24.3~26.6); 15.3 (14.3~15.7); 125.1 (114.1~129.7); 1.0 (0.2~2.3); 814 (765~841) ❙ 1480 (1408~1574); 504 (454~591); 179 (170~192) ❙ 82.1 (67.4~99.4); 814 (765~841); 226.5 (193.4~257.1)

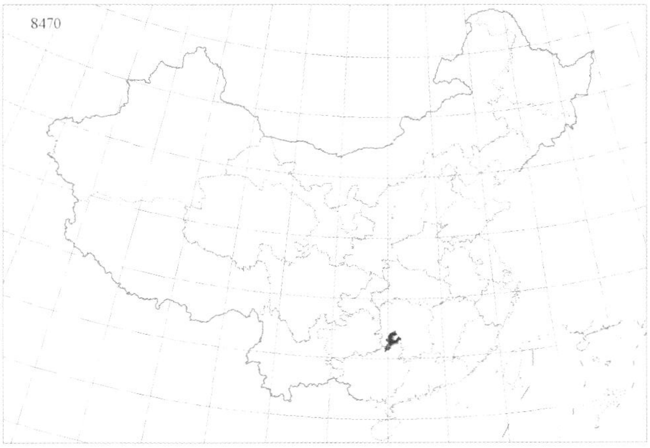

Ericaceae 杜鹃花科 *Vaccinium* 乌饭树属
Vaccinium pratense P. C. Tam ex C. Y. Wu et R. C. Fang 草地越橘
Shrubs, evergreen

22.0; 13.5; 28.8; 22.0; 203.6; 0.0; 1169 ❙ 1641; 724; 132 ❙ 40.5; 1169; 205.0

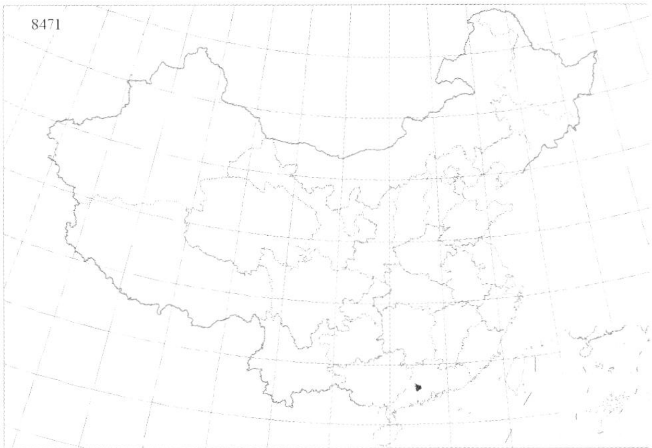

Ericaceae 杜鹃花科 *Vaccinium* 乌饭树属
Vaccinium pseudobullatum Fang et Z. H. Pan 拟泡叶乌饭
Shrubs, evergreen

17.5 (16.3~18.9); 9.6 (9.1~10.6); 23.1 (21.4~25.0); 17.5 (16.4~19.0); 150.2 (136.4~167.5); 0.0 (0.0~0.0); 856 (797~939) ❙ 1264 (1161~1422); 706 (633~813); 50 (48~52) ❙ 47.9 (37.1~65.6); 856 (797~939); 340.0 (314.4~359.7)

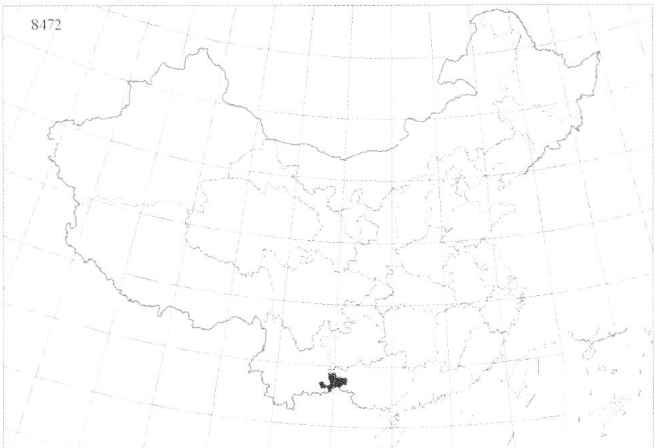

Ericaceae 杜鹃花科 *Vaccinium* 乌饭树属
Vaccinium pseudorobustum Sleumer 椭圆叶越橘
Scandent shrubss or trees, evergreen

19.2 (15.5~21.3); 9.7 (4.8~13.2); 27.0 (25.0~28.6); 19.2 (15.6~21.3); 170.9 (127.2~196.1); 0.0 (0.0~0.2); 998 (815~1130) ❙ 1543 (1250~1776); 682 (562~828); 148 (57~189) ❙ 55.8 (22.7~93.2); 997 (815~1130); 244.6 (185.5~371.7)

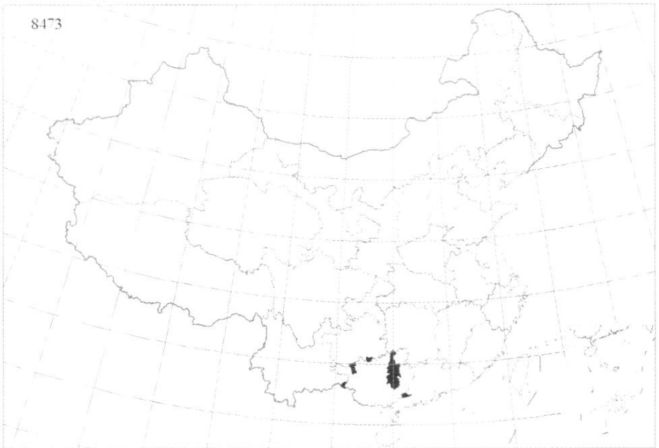

Ericaceae 杜鹃花科 *Vaccinium* 乌饭树属
Vaccinium pseudospadiceum Dop 耳叶越橘
Shrubs, evergreen

17.6; 9.9; 23.2; 17.7; 152.4; 0.0; 859 ❙ 1422; 813; 52 ❙ 65.6; 859; 359.7

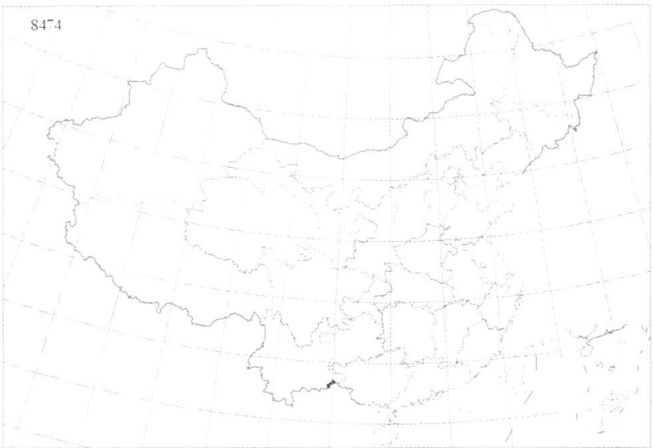

Ericaceae 杜鹃花科 *Vaccinium* 乌饭树属
Vaccinium pseudotonkinense Sleumer 腺萼越橘
Small shrubs, evergreen

18.3 (16.3~21.0); 11.5 (9.1~14.2); 23.0 (21.4~25.6); 18.3 (16.4~21.0); 160.2 (136.4~191.8); 0.0 (0.0~0.0); 891 (797~1053) ❙ 1378 (1190~1522); 775 (656~835); 57 (50~68) ❙ 55.0 (44.5~65.6); 891 (797~1053); 448.2 (347.3~551.6)

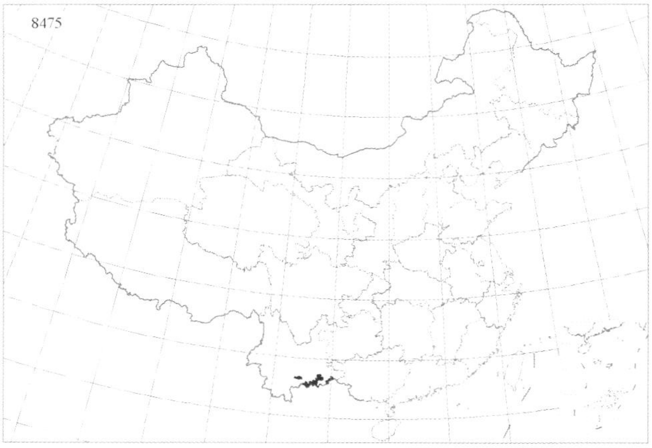

Ericaceae 杜鹃花科 *Vaccinium* 乌饭树属
Vaccinium pubicalyx Franch. 毛萼越橘
Shrubs or trees, evergreen

14.9 (5.4~21.0); 6.3 (-2.1~14.2); 22.2 (12.3~27.0); 15.0 (5.8~21.0); 120.4 (31.5~191.8); 1.0 (0.0~25.6); 779 (481~1053) ❙ 1092 (741~1547); 558 (411~911); 57 (15~161) ❙ 40.7 (-14.2~99.9); 764 (481~1053); 264.4 (142.9~551.6)

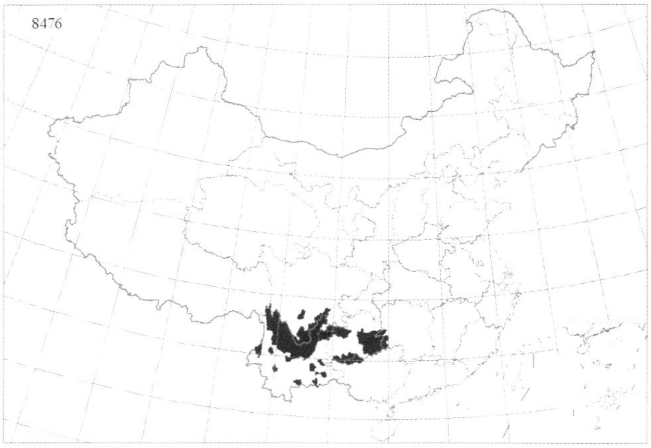

Ericaceae 杜鹃花科 *Vaccinium* 乌饭树属
Vaccinium pubicalyx var. anomalum J. Anthony 少毛毛萼越橘
Shrubs or trees, evergreen

12.5; 5.7; 18.0; 12.6; 90.8; 0.2; 678 ❙ 1008; 609; 32 ❙ 49.5; 666; 308.9

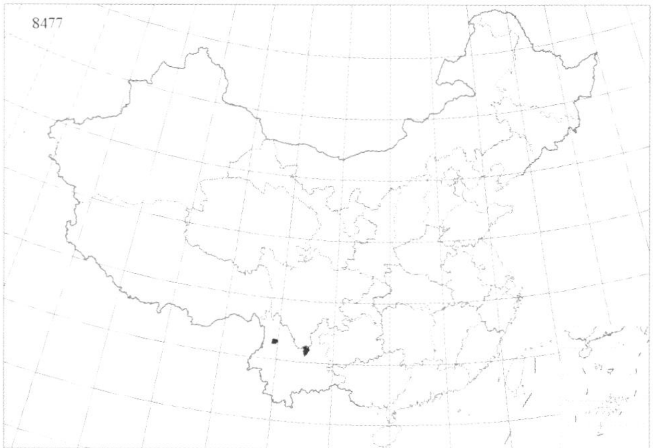

Ericaceae 杜鹃花科 *Vaccinium* 乌饭树属
Vaccinium pubicalyx var. leucocalyx (H. Lév.) Rehder 多毛毛萼越橘
Shrubs or trees, evergreen

17.2 (15.9~19.1); 7.3 (4.7~10.1); 25.5 (23.4~26.6); 17.2 (16.0~19.1); 146.8 (131.7~169.2); 0.0 (0.0~0.3); 882 (836~969) ❙ 1277 (1232~1368); 593 (452~720); 89 (51~129) ❙ 45.4 (27.5~59.6); 882 (836~969); 242.5 (206.2~283.9)

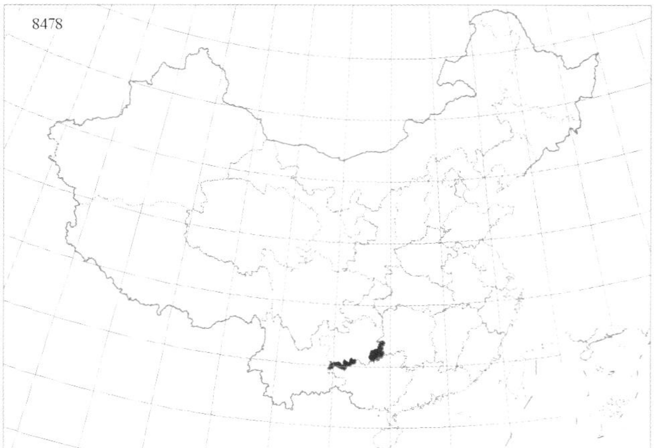

Ericaceae 杜鹃花科 *Vaccinium* 乌饭树属
Vaccinium randaiense Hayata 峦大越橘
Shrubs, evergreen

18.7 (14.3~22.2); 9.8 (3.3~17.7); 26.2 (20.1~28.8); 18.7 (14.3~22.2); 164.8 (114.1~206.7); 0.2 (0.0~2.3); 969 (765~1170) ❙ 1781 (1194~2992); 787 (465~1847); 161 (65~245) ❙ 86.8 (40.1~246.5); 967 (765~1170); 236.6 (160.8~297.2)

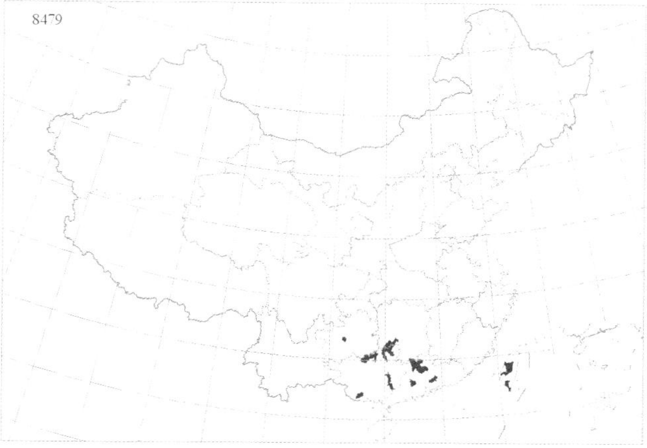

Ericaceae 杜鹃花科 *Vaccinium* 乌饭树属
Vaccinium sikkimense C. B. Clarke 荚蒾叶越橘
Shrubs, evergreen

8.3 (1.8~17.3); 0.0 (-7.1~11.0); 15.6 (9.9~22.7); 8.9 (3.8~17.3); 61.5 (15.5~148.0); 21.0 (0.0~53.3); 574 (404~832) ❙ 1003 (473~1736); 574 (287~1007); 31 (7~51) ❙ 73.1 (5.0~145.8); 569 (404~783); 277.2 (169.5~447.4)

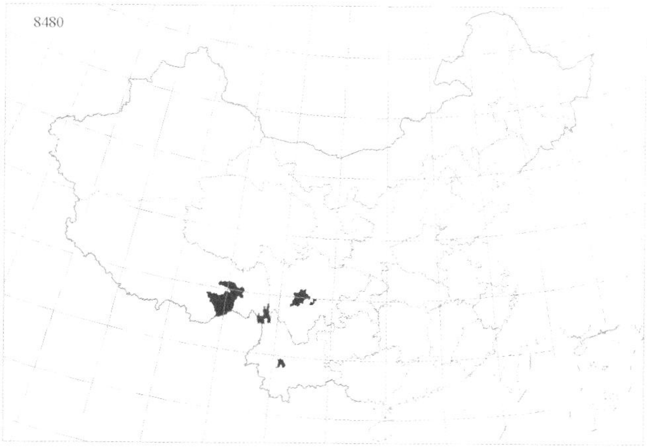

Ericaceae 杜鹃花科 *Vaccinium* 乌饭树属
Vaccinium sinicum Sleumer 广西越橘
Small shrubs, evergreen

18.1 (14.3~21.3); 8.1 (3.3~12.3); 26.9 (24.3~28.7); 18.2 (14.3~21.3); 158.3 (114.1~196.1); 0.2 (0.0~2.3); 947 (765~1130) ❙ 1608 (1371~1893); 618 (465~776); 181 (74~227) ❙ 71.6 (22.7~129.9); 947 (765~1130); 238.0 (180.3~300.5)

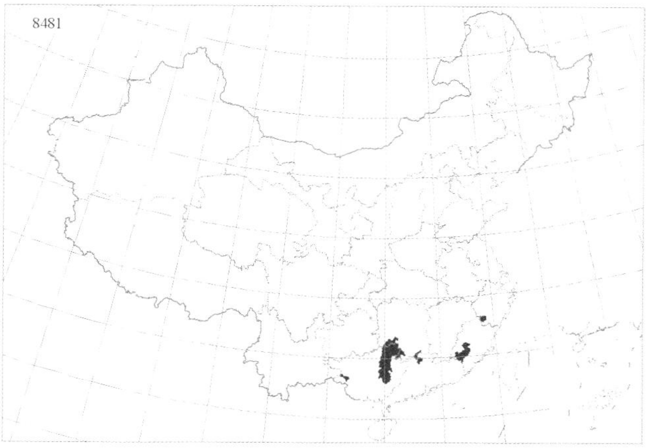

Ericaceae 杜鹃花科 *Vaccinium* 乌饭树属
Vaccinium subfalcatum Merr. ex Sleumer 镰叶越橘
Shrubs or small trees, evergreen

21.3 (17.6~22.7); 12.9 (7.3~15.0); 27.8 (26.3~28.7); 21.4 (17.6~22.8); 196.3 (151.6~213.1); 0.0 (0.0~0.0); 1128 (909~1221) ❙ 1724 (1302~2435); 880 (556~1428); 123 (82~182) ❙ 53.5 (12.9~99.5); 1128 (909~1221); 268.0 (164.9~327.7)

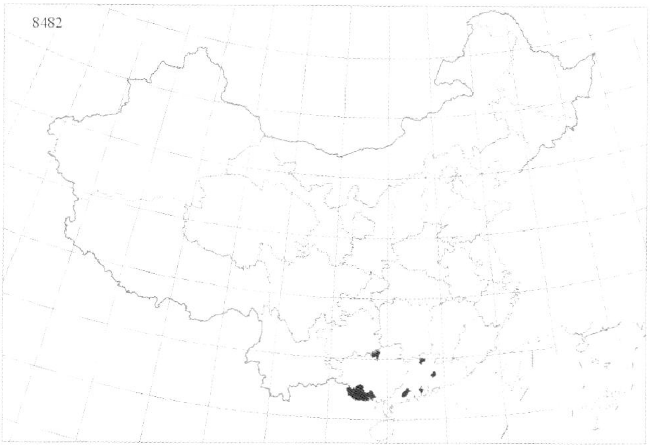

Ericaceae 杜鹃花科 *Vaccinium* 乌饭树属
Vaccinium supracostatum Hand.-Mazz. 凸脉越橘
Shrubs, evergreen

17.5 (15.1~20.4); 8.2 (6.2~10.8); 25.0 (22.2~28.0); 17.5 (15.2~20.4); 150.5 (122.6~184.9); 0.0 (0.0~0.0); 890 (772~1071) ❙ 1307 (1232~1584); 667 (594~750); 72 (51~165) ❙ 47.5 (27.5~69.4); 890 (772~1071); 240.0 (189.5~322.1)

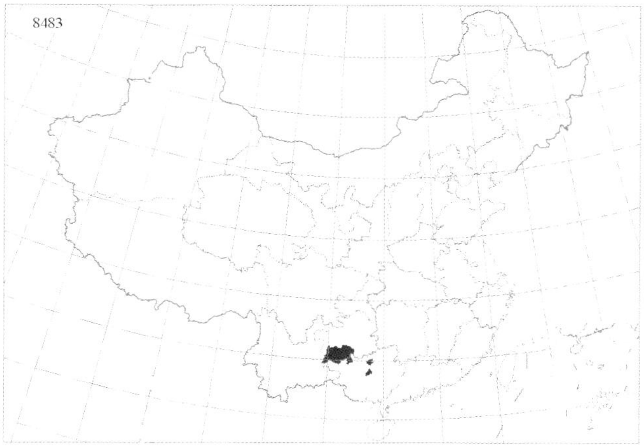

Ericaceae 杜鹃花科 *Vaccinium* 乌饭树属
Vaccinium trichocladum Merr. et F. P. Metcalf 刺毛越橘
Shrubs, evergreen

17.5 (14.4~21.6); 7.3 (2.1~13.2); 26.6 (23.0~29.3); 17.6 (14.5~21.6); 151.4 (118.7~199.2); 0.4 (0.0~4.9); 917 (790~1139) ❙ 1564 (1221~1997); 613 (452~1056); 160 (51~259) ❙ 70.7 (27.5~130.7); 917 (790~1139); 247.6 (171.2~319.9)

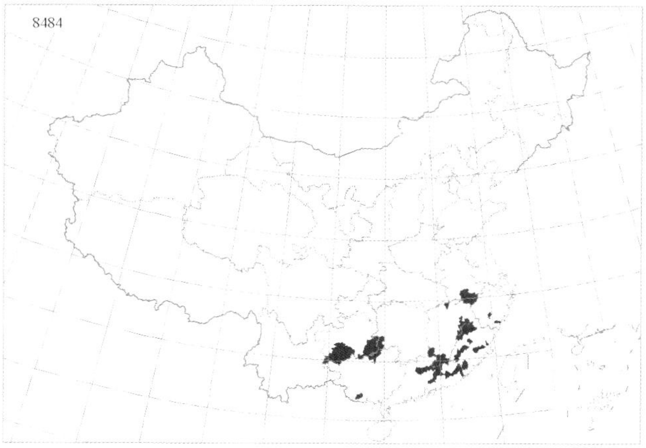

Ericaceae 杜鹃花科 *Vaccinium* 乌饭树属

Vaccinium uliginosum L. 笃斯越橘

Shrubs, deciduous

1.6 (-5.8~5.9); -18.5 (-29.8~-13.8); 19.2 (15.3~23.0); 7.0 (4.4~9.4); 51.8 (28.0~77.9); 91.7 (65.0~157.4); 534 (426~655) ❙ 540 (152~1005); 324 (69~635); 23 (14~39) ❙ 1.7 (-76.8~68.8); 438 (152~595); 257.3 (28.1~346.5)

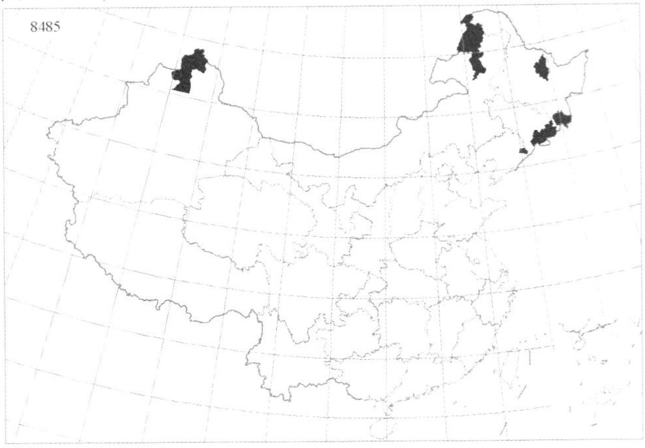

Ericaceae 杜鹃花科 *Vaccinium* 乌饭树属

Vaccinium urceolatum Hemsl. 红花越橘

Shrubs or small trees, evergreen

12.7 (6.0~16.8); 3.5 (-3.2~7.8); 21.0 (14.4~25.6); 12.8 (6.6~16.8); 97.5 (39.7~141.5); 4.2 (0.0~26.7); 706 (508~845) ❙ 1094 (879~1414); 588 (472~817); 44 (27~71) ❙ 56.8 (19.4~101.2); 706 (508~845); 247.5 (154.4~372.3)

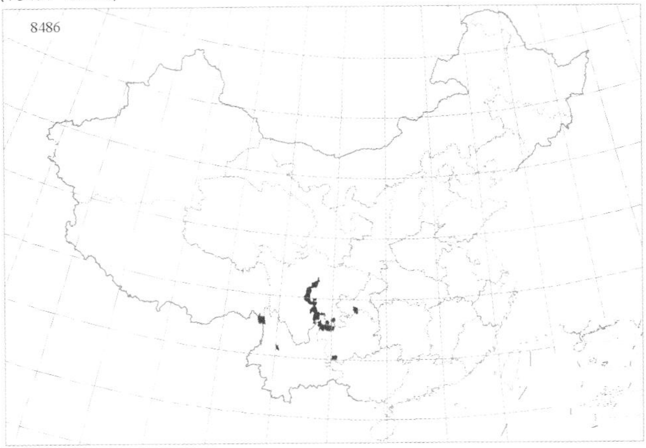

Ericaceae 杜鹃花科 *Vaccinium* 乌饭树属

Vaccinium vitis-idaea L. 越橘

Shrubs, evergreen

1.7 (-5.8~14.4); -18.3 (-29.8~0.5); 19.3 (13.7~27.6); 7.1 (4.4~14.4); 53.2 (27.2~122.6); 92.4 (9.3~157.4); 538 (426~847) ❙ 562 (187~856); 332 (79~527); 21 (6~46) ❙ 4.5 (-65.1~57.4); 492 (187~697); 275.3 (117.1~346.5)

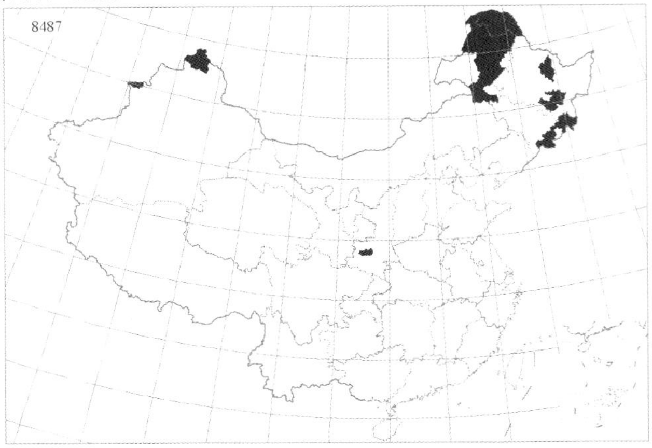

Myrsinaceae 紫金牛科 *Aegiceras* 桐花树属

Aegiceras corniculatum (L.) Blanco 蜡烛果

Shrubs or small trees

22.2 (20.4~24.8); 14.6 (12.6~20.1); 28.3 (27.3~29.2); 22.2 (20.4~24.8); 206.5 (184.9~237.6); 0.0 (0.0~0.0); 1180 (1046~1385) ❙ 1699 (1258~2148); 790 (425~1234); 119 (41~185) ❙ 44.3 (-1.0~86.7); 1169 (1046~1329); 213.5 (95.3~527.5)

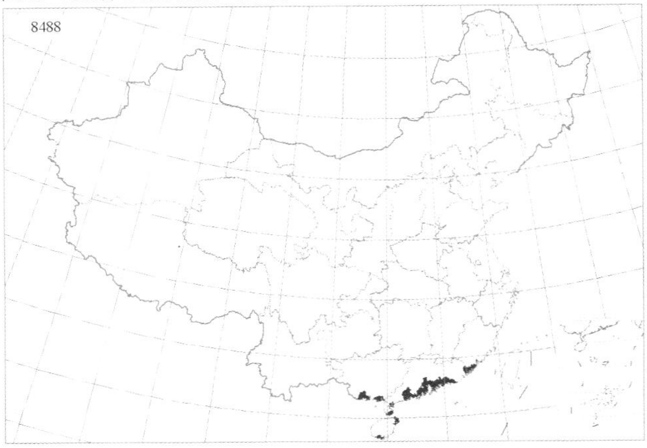

Myrsinaceae 紫金牛科 *Ardisia* 紫金牛属

Ardisia aberrans (E. Walker) C. Y. Wu et C. Chen 狗骨头

Shrubs, evergreen

17.7; 10.9; 22.5; 17.8; 153.4; 0.0; 854 ❙ 1368; 756; 59 ❙ 60.1; 854; 491.5

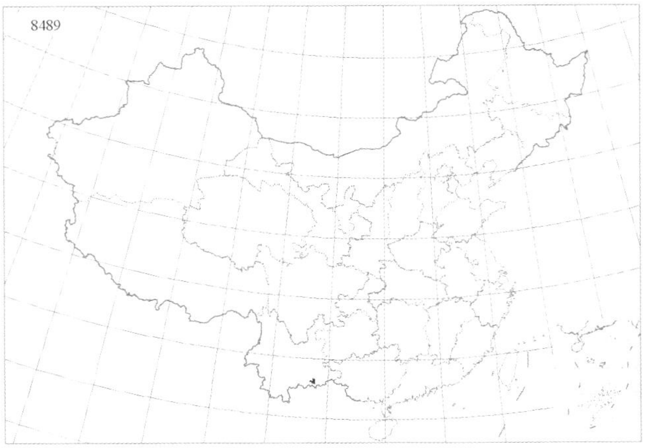

Myrsinaceae 紫金牛科 *Ardisia* 紫金牛属

Ardisia alutacea C. Y. Wu et C. Chen 显脉紫金牛

Shrubs, evergreen

17.4; 9.9; 22.8; 17.5; 149.8; 0.0; 846 ❙ 1463; 837; 54 ❙ 73.1; 846; 386.6

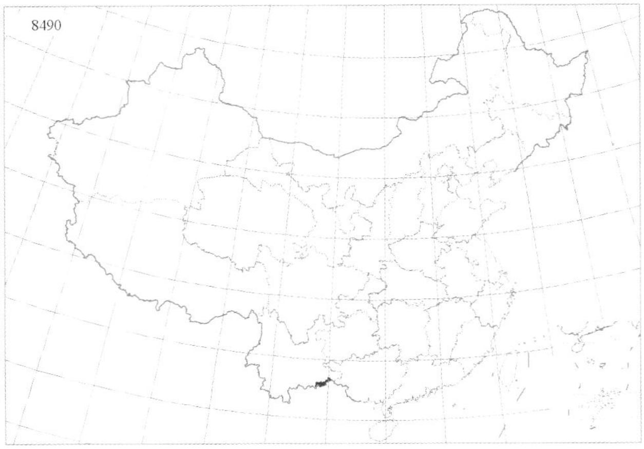

Myrsinaceae 紫金牛科 *Ardisia* 紫金牛属
Ardisia gigantifolia Stapf 走马胎
Shrubs or subshrubs, evergreen

20.5 (16.2~24.3); 12.1 (6.3~20.1); 27.2 (21.4~29.8); 20.5 (16.2~24.8); 186.3 (134.5~237.6); 0.0 (0.0~0.0); 1070 (797~1385) ❙ 1590 (1007~2435); 708 (455~1428); 127 (41~199) ❙ 49.7 (-6.9~99.5); 1063 (797~1326); 264.8 (95.3~659.0)

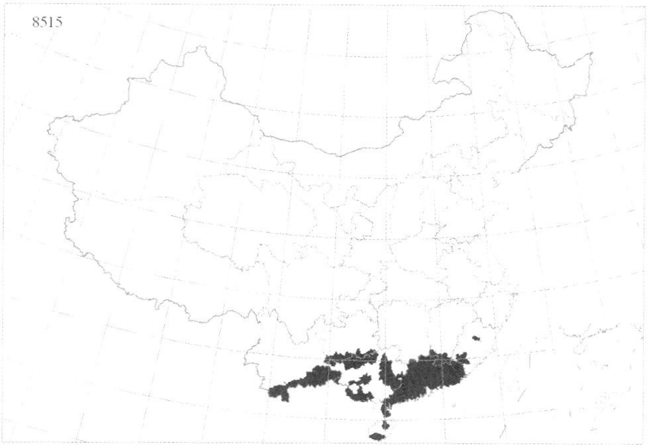

Myrsinaceae 紫金牛科 *Ardisia* 紫金牛属
Ardisia hanceana Mez 大罗伞树
Shrubs, evergreen

19.5 (14.3~24.8); 10.4 (2.5~20.1); 27.6 (24.2~29.8); 19.6 (14.3~24.8); 175.0 (113.2~237.6); 0.1 (0.0~3.9); 1028 (761~1385) ❙ 1618 (1070~2148); 638 (356~1234); 165 (41~259) ❙ 60.5 (-12.0~147.0); 1018 (761~1273); 243.7 (95.3~620.0)

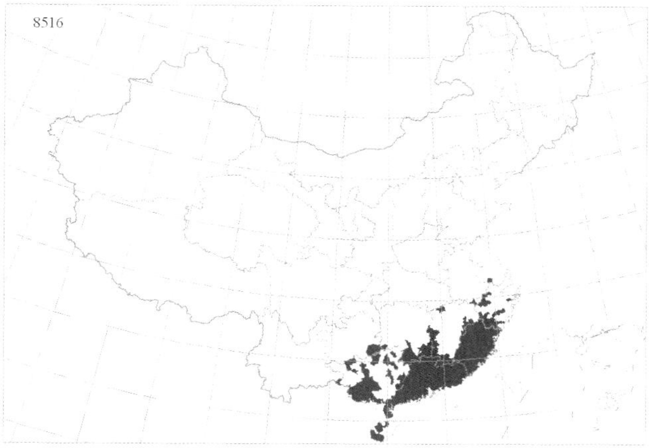

Myrsinaceae 紫金牛科 *Ardisia* 紫金牛属
Ardisia humilis Vahl 矮紫金牛
Shrubs, evergreen

23.9 (22.2~24.8); 18.0 (15.9~20.1); 28.1 (25.7~29.2); 23.9 (22.3~24.8); 227.2 (207.2~237.6); 0.0 (0.0~0.0); 1303 (1127~1385) ❙ 1469 (868~1854); 546 (312~712); 75 (41~115) ❙ 12.9 (-34.3~47.8); 1201 (868~1360); 401.1 (178.7~620.0)

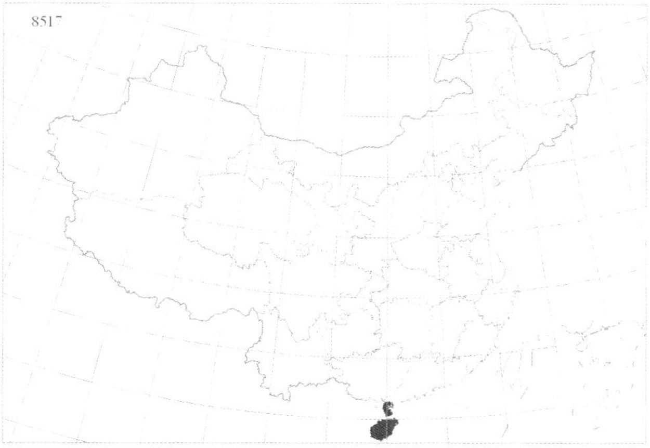

Myrsinaceae 紫金牛科 *Ardisia* 紫金牛属
Ardisia hypargyrea C. Y. Wu et C. Chen 柳叶紫金牛
Shrubs, evergreen

18.8 (15.5~22.7); 10.8 (4.8~15.0); 24.7 (21.4~28.8); 18.8 (15.6~22.8); 166.1 (127.2~213.1); 0.0 (0.0~0.2); 949 (797~1221) ❙ 1487 (1074~2435); 796 (562~1428); 87 (47~177) ❙ 56.2 (22.7~99.5); 949 (797~1221); 338.0 (188.6~551.6)

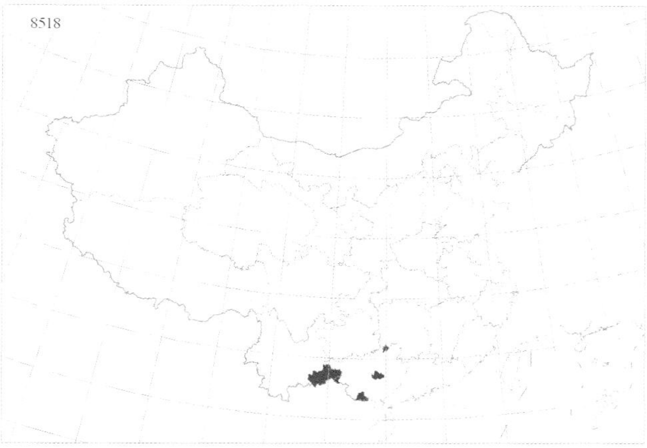

Myrsinaceae 紫金牛科 *Ardisia* 紫金牛属
Ardisia japonica (Thunb.) Blume 紫金牛
Shrubs or subshrubs, evergreen

16.2 (9.3~22.0); 5.4 (-2.9~13.8); 26.2 (18.8~29.9); 16.3 (9.5~22.0); 136.7 (69.8~204.5); 1.5 (0.0~20.5); 868 (603~1179) ❙ 1331 (626~2993); 537 (316~1056); 130 (19~547) ❙ 53.2 (-22.2~203.4); 858 (603~1179); 216.6 (108.2~432.9)

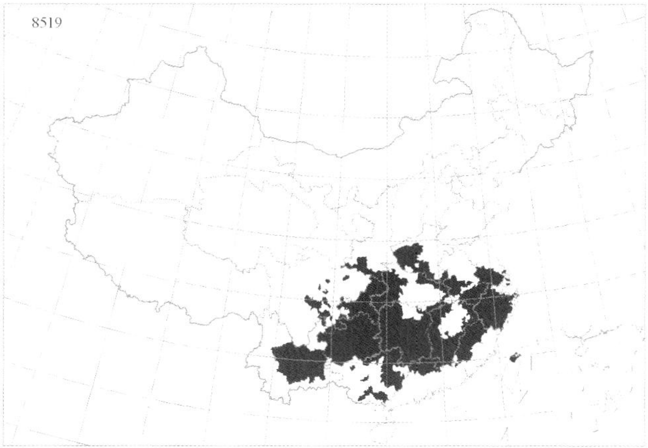

Myrsinaceae 紫金牛科 *Ardisia* 紫金牛属
Ardisia lindleyana D. Dietr. 山血丹
Shrubs, evergreen

18.9 (14.3~23.6); 9.3 (2.1~16.7); 27.6 (24.2~30.1); 19.0 (14.3~23.6); 168.1 (113.2~223.5); 0.1 (0.0~4.9); 998 (761~1281) ❙ 1609 (1039~2435); 614 (356~1428); 173 (73~259) ❙ 63.5 (-6.9~147.0); 990 (761~1273); 228.9 (95.3~332.5)

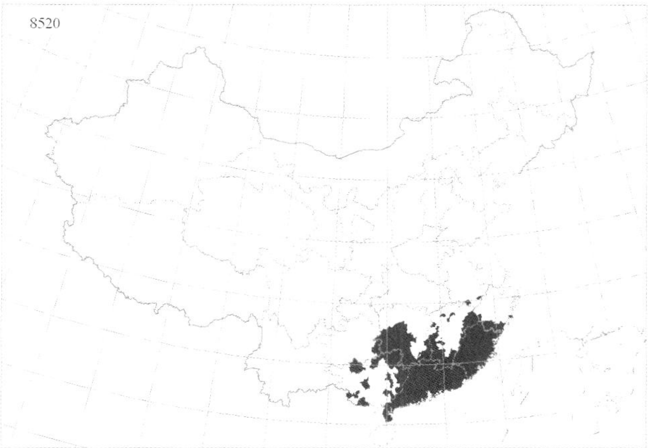

Myrsinaceae 紫金牛科 *Ardisia* 紫金牛属
Ardisia maclurei Merr. 心叶紫金牛
Shrubs or subshrubs, evergreen

20.5 (17.7~24.1); 11.6 (7.4~19.1); 27.8 (25.8~29.8); 20.6 (17.8~24.2); 187.0 (153.6~230.1); 0.0 (0.0~0.0); 1082 (913~1326) ❙ 1593 (1236~2993); 680 (455~924); 154 (48~547) ❙ 48.9 (6.7~203.4); 1071 (913~1326); 262.0 (174.9~620.0)

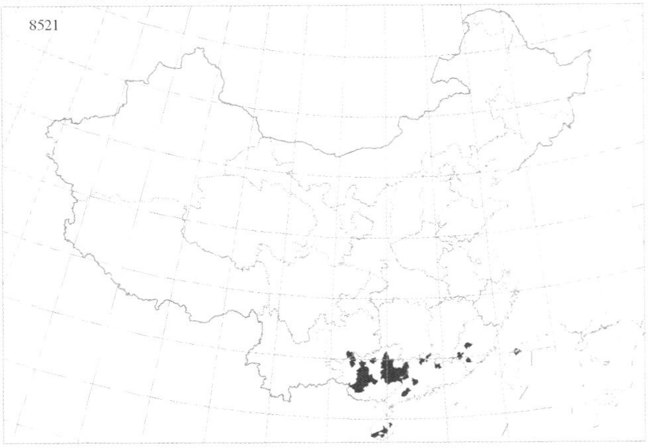

Myrsinaceae 紫金牛科 *Ardisia* 紫金牛属
Ardisia malipoensis C. M. Hu 麻栗坡罗伞
Shrubs, evergreen

17.6; 9.9; 23.2; 17.7; 152.4; 0.0; 859 ❙ 1422; 813; 52 ❙ 65.6; 859; 359.7

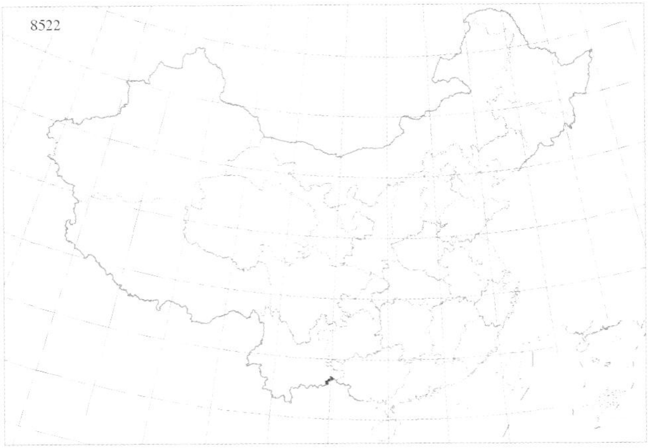

Myrsinaceae 紫金牛科 *Ardisia* 紫金牛属
Ardisia mamillata Hance 虎舌红
Dwarf subshrubs, evergreen

19.4 (11.1~24.3); 10.5 (2.2~18.7); 27.0 (19.1~29.8); 19.4 (11.3~24.3); 173.4 (80.8~231.9); 0.0 (0.0~5.3); 1012 (652~1338) ❙ 1541 (985~2037); 644 (356~1056); 143 (37~257) ❙ 54.3 (−12.0~137.5); 1006 (652~1326); 248.9 (95.3~572.9)

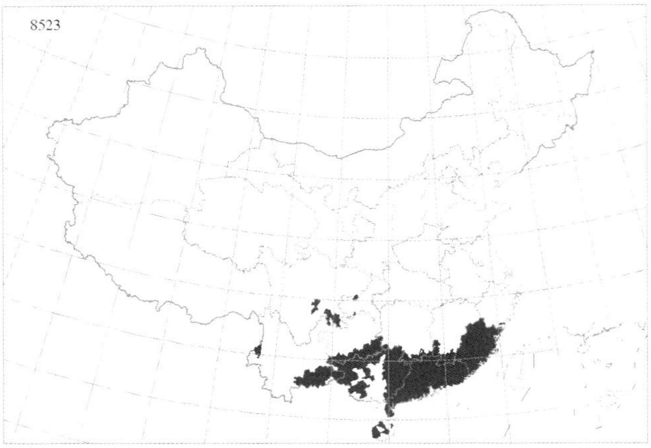

Myrsinaceae 紫金牛科 *Ardisia* 紫金牛属
Ardisia merrillii E. Walker 白花紫金牛
Shrubs, evergreen

22.2; 13.9; 28.3; 22.3; 207.5; 0.0; 1190 ❙ 1374; 713; 87 ❙ 15.4; 1190; 327.7

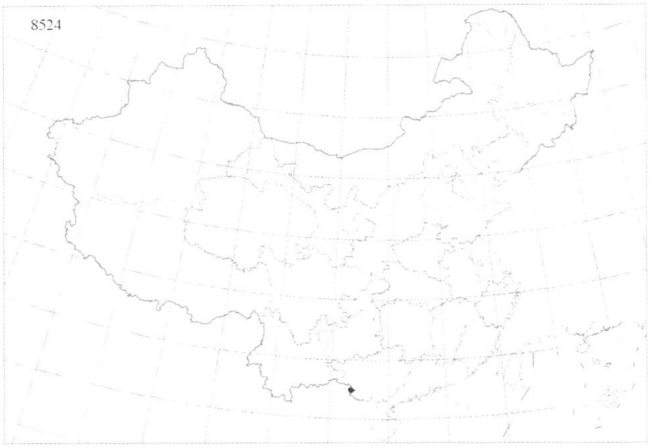

Myrsinaceae 紫金牛科 *Ardisia* 紫金牛属
Ardisia nigropilosa Pit. 星毛紫金牛
Shrubs, evergreen

19.2 (18.1~21.0); 13.1 (12.0~14.2); 23.3 (22.1~25.6); 19.3 (18.1~21.0); 171.1 (157.3~191.8); 0.0 (0.0~0.0); 929 (863~1053) ❙ 1418 (1143~1751); 810 (637~996); 60 (50~77) ❙ 53.0 (21.7~88.8); 921 (863~1053); 463.7 (346.9~551.6)

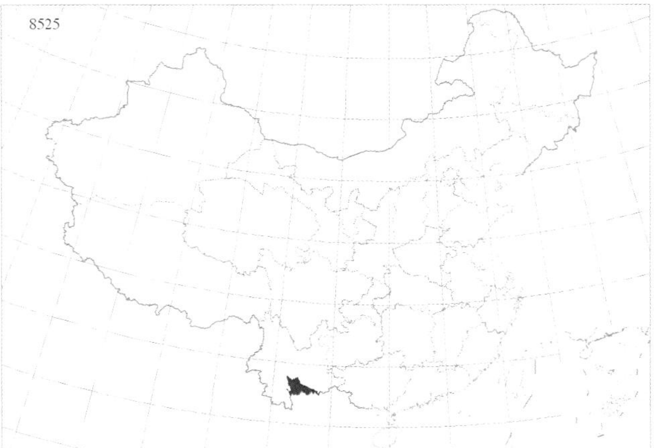

Myrsinaceae 紫金牛科 *Ardisia* 紫金牛属
Ardisia obtusa Mez 铜盆花
Shrubs, evergreen

23.6 (22.2~24.8); 17.5 (13.9~20.1); 28.0 (25.7~29.2); 23.6 (22.3~24.8); 223.0 (207.2~237.6); 0.0 (0.0~0.0); 1275 (1127~1385) ❙ 1504 (1187~1965); 596 (444~994); 71 (41~109) ❙ 18.4 (−6.9~61.6); 1192 (1055~1321); 403.8 (178.7~620.0)

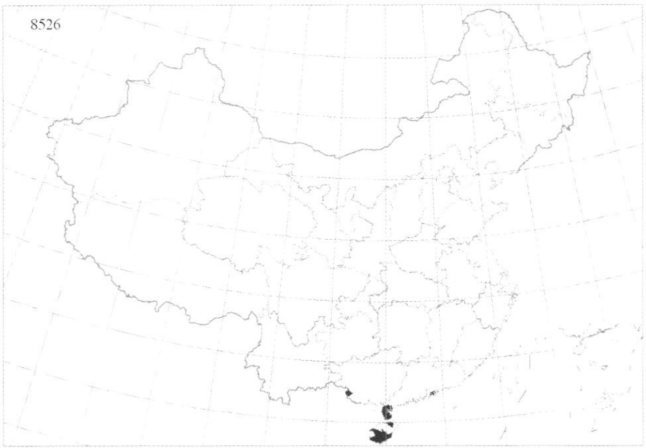

Myrsinaceae 紫金牛科 *Ardisia* 紫金牛属
Ardisia olivacea E. Walker 榄色紫金牛
Shrubs, evergreen

21.2 (19.1~22.7); 13.0 (11.0~15.0); 27.4 (25.0~28.6); 21.3 (19.1~22.8); 195.2 (169.6~213.1); 0.0 (0.0~0.0); 1112 (945~1221) ❙ 1571 (1188~2435); 859 (634~1428); 96 (56~130) ❙ 41.4 (5.8~99.5); 1112 (945~1221); 303.0 (256.6~371.7)

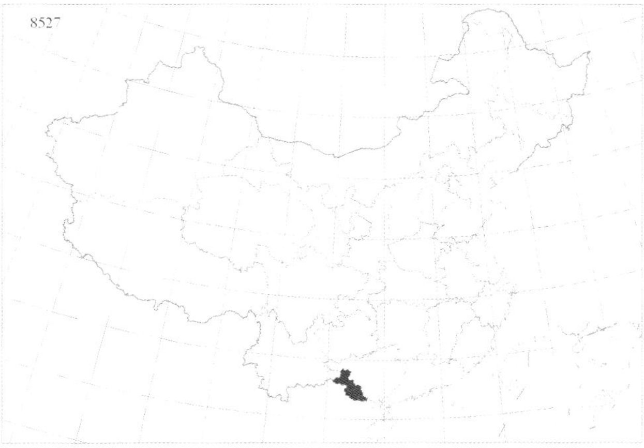

Myrsinaceae 紫金牛科 *Ardisia* 紫金牛属
Ardisia omissa C. M. Hu 光萼紫金牛
Dwarf subshrubs, evergreen

19.5 (17.8~21.0); 10.0 (7.1~13.2); 27.4 (26.3~28.3); 19.5 (17.8~21.0); 173.9 (154.0~192.3); 0.0 (0.0~0.0); 1012 (918~1081) ❙ 1629 (1503~1777); 630 (513~828); 165 (125~182) ❙ 61.2 (49.1~67.8); 1012 (918~1081); 235.0 (197.5~263.5)

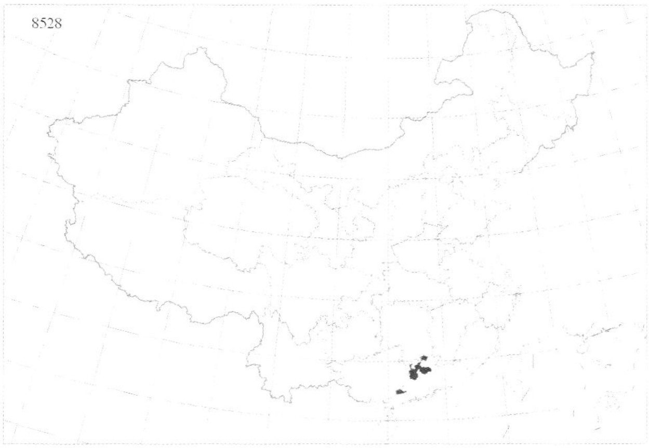

Myrsinaceae 紫金牛科 *Ardisia* 紫金牛属
Ardisia ordinata E. Walker 轮叶紫金牛
Shrubs or subshrubs, evergreen

23.5; 18.0; 27.5; 23.6; 222.8; 0.0; 1266 ❙ 1370; 481; 48 ❙ 8.4; 1115; 477.7

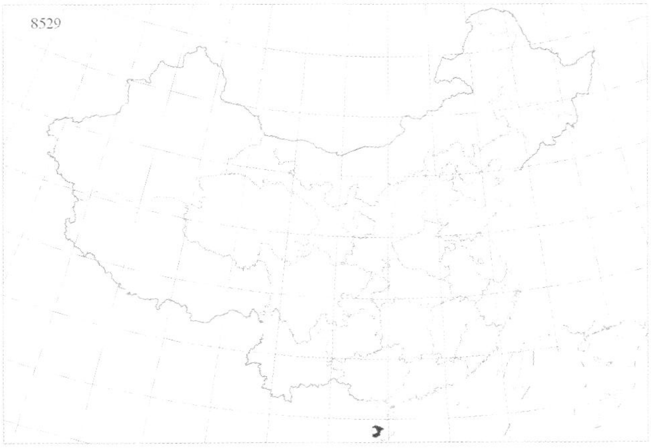

Myrsinaceae 紫金牛科 *Ardisia* 紫金牛属
Ardisia perreticulata C. Chen 花脉紫金牛
Shrubs or subshrubs, evergreen

20.8 (18.2~23.2); 11.8 (8.7~15.8); 27.9 (25.9~29.0); 20.8 (18.3~23.2); 190.0 (159.3~218.9); 0.0 (0.0~0.0); 1095 (928~1257) ❙ 1520 (1279~1776); 701 (561~828); 132 (73~175) ❙ 39.2 (14.8~64.3); 1095 (928~1257); 239.3 (160.7~335.1)

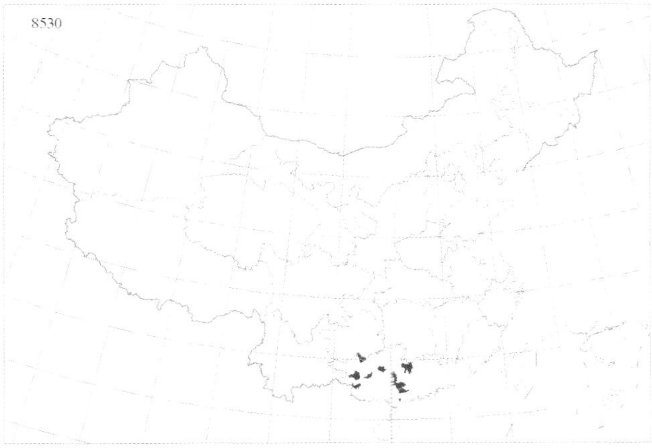

Myrsinaceae 紫金牛科 *Ardisia* 紫金牛属
Ardisia pingbienensis Y. P. Yang 长穗紫金牛
Subshrubs, evergreen

17.7; 10.9; 22.5; 17.8; 153.4; 0.0; 854 ❙ 1368; 756; 59 ❙ 60.1; 854; 491.5

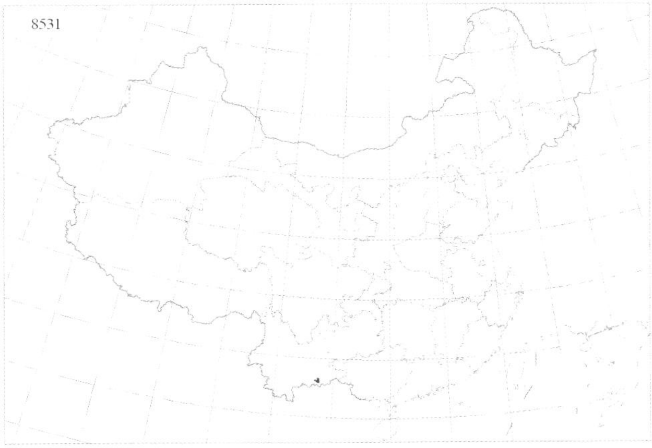

Myrsinaceae 紫金牛科 *Ardisia* 紫金牛属
Ardisia porifera E. Walker 细孔紫金牛
Shrubs or subshrubs, evergreen

23.9; 18.7; 27.5; 23.9; 226.8; 0.0; 1294 ❙ 1380; 455; 48 ❙ 6.7; 1112; 455.2

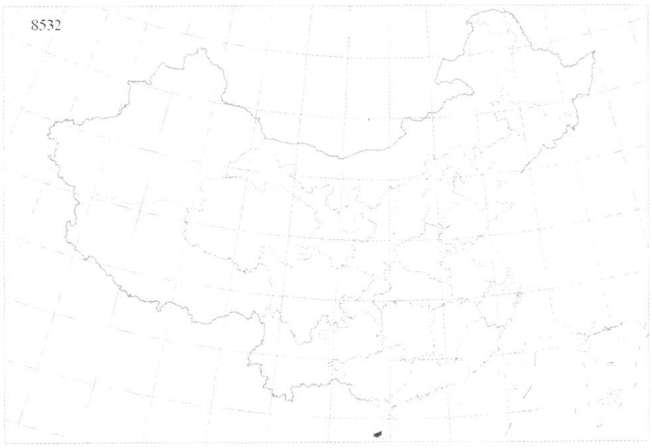

Myrsinaceae 紫金牛科 *Ardisia* 紫金牛属
Ardisia primulifolia Gardner et Champ. 莲座紫金牛
Subshrubs, evergreen

19.8 (14.3~24.0); 10.9 (3.7~19.1); 27.5 (22.7~29.8); 19.8 (14.3~24.0); 178.2 (113.2~228.1); 0.0 (0.0~1.5); 1038 (761~1302) ❙ 1623 (1070~2435); 654 (356~1428); 159 (44~257) ❙ 58.9 (-6.9~147.0); 1031 (761~1273); 244.5 (95.3~620.0)

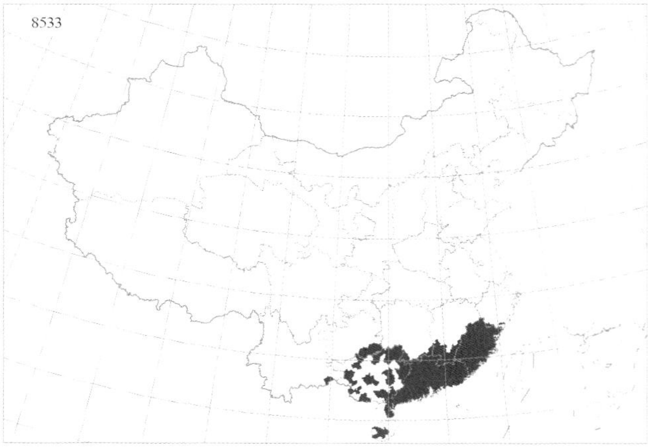

Myrsinaceae 紫金牛科 *Ardisia* 紫金牛属
Ardisia pubivenula E. Walker 毛脉紫金牛
Small shrubs, evergreen

24.0; 19.1; 27.2; 24.0; 228.1; 0.0; 1302 ❙ 1551; 539; 48 ❙ 19.1; 1147; 620.0

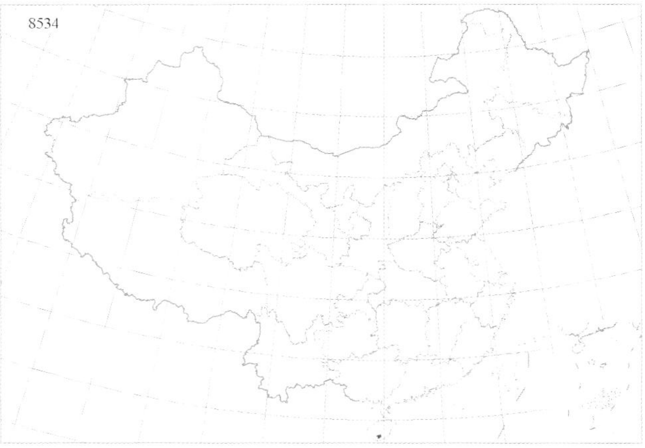

Myrsinaceae 紫金牛科 *Ardisia* 紫金牛属
Ardisia purpureovillosa C. Y. Wu et C. Chen ex C. M. Hu 紫脉紫金牛
Shrubs, evergreen

20.5 (17.2~24.8); 13.5 (9.1~20.1); 25.4 (22.4~28.3); 20.5 (17.3~24.8); 186.1 (147.0~237.6); 0.0 (0.0~0.0); 1050 (833~1385) ❙ 1456 (1161~1725); 716 (473~860); 63 (41~106) ❙ 41.5 (1.5~80.6); 1012 (833~1212); 430.6 (281.7~620.0)

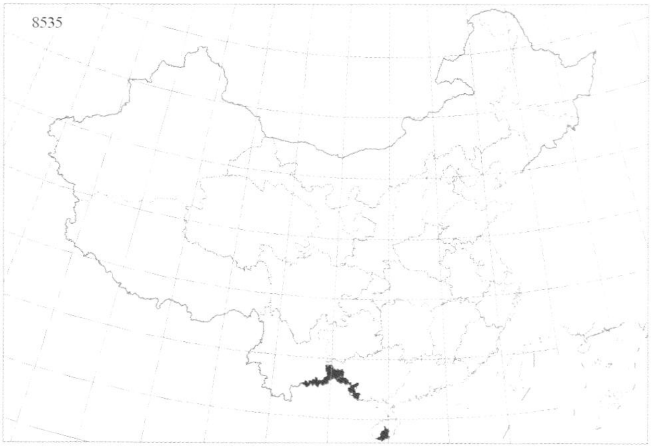

Myrsinaceae 紫金牛科 *Ardisia* 紫金牛属
Ardisia pusilla A. DC. 九节龙
Small shrubs, evergreen

16.9 (2.9~22.4); 6.8 (-6.3~15.3); 26.1 (11.2~29.8); 17.0 (4.4~22.4); 144.4 (20.5~208.9); 0.5 (0.0~44.6); 893 (428~1202) ❙ 1416 (916~3495); 581 (323~1056); 128 (23~897) ❙ 59.0 (-1.6~224.4); 889 (428~1202); 215.8 (108.2~371.7)

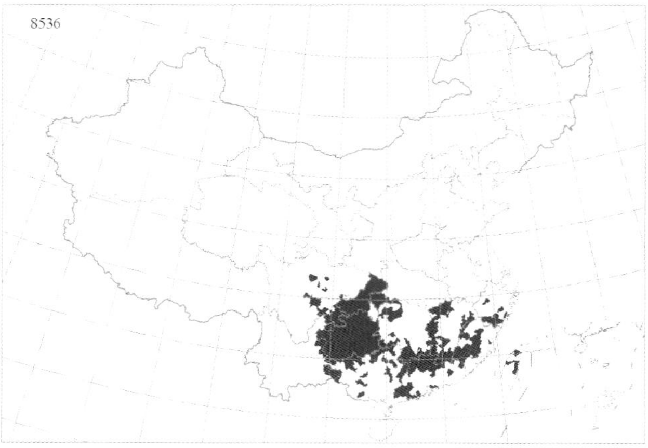

Myrsinaceae 紫金牛科 *Ardisia* 紫金牛属
Ardisia quinquegona Blume 罗伞树
Shrubs, evergreen

19.6 (10.6~24.8); 11.0 (-0.2~20.1); 27.0 (18.6~29.8); 19.7 (10.6~24.8); 176.1 (79.4~237.6); 0.1 (0.0~12.1); 1028 (643~1385) ❙ 1566 (868~3495); 667 (312~1847); 141 (31~897) ❙ 54.8 (-34.3~246.5); 1017 (643~1360); 248.2 (95.3~620.0)

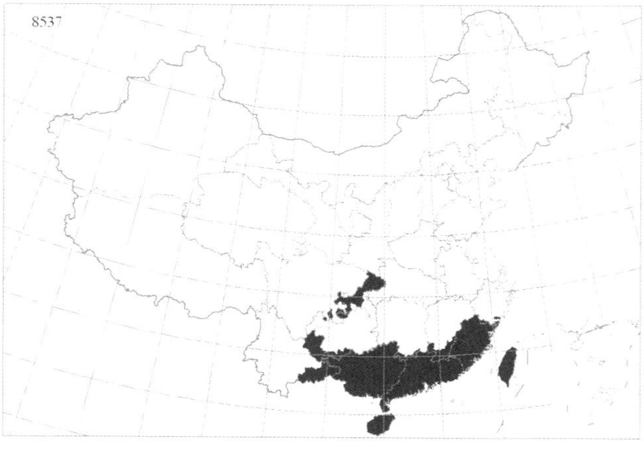

Myrsinaceae 紫金牛科 *Ardisia* 紫金牛属
Ardisia ramondiiformis Pit. 梯脉紫金牛
Small shrubs, evergreen

23.9; 18.9; 27.3; 23.9; 227.2; 0.0; 1298 ❙ 1621; 561; 58 ❙ 25.4; 1174; 562.6

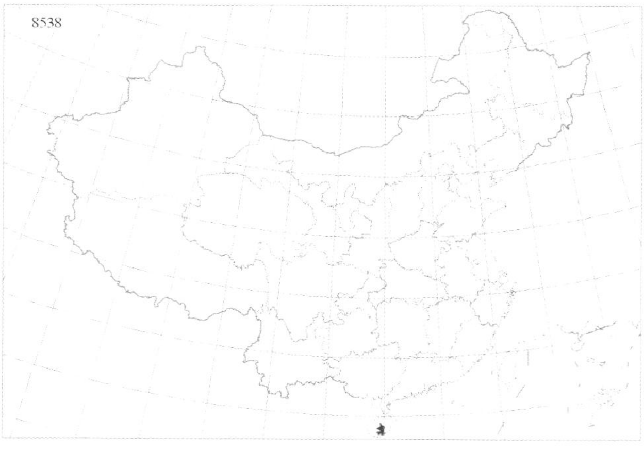

Myrsinaceae 紫金牛科 *Ardisia* 紫金牛属
Ardisia replicata E. Walker 卷边紫金牛
Small shrubs, evergreen

18.0 (16.3~21.0); 10.5 (9.1~14.2); 23.4 (21.4~25.8); 18.1 (16.4~21.0); 156.6 (136.4~191.8); 0.0 (0.0~0.0); 884 (797~1053) ❙ 1307 (1007~1522); 727 (539~860); 56 (44~74) ❙ 47.9 (19.0~80.6); 883 (797~1053); 367.0 (277.0~551.6)

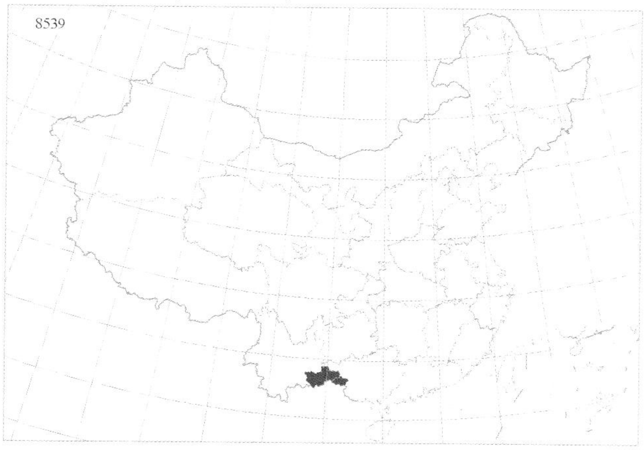

Myrsinaceae 紫金牛科 *Ardisia* 紫金牛属
Ardisia retroflexa E. Walker 弯梗紫金牛
Shrubs, evergreen

23.1; 17.5; 27.2; 23.2; 218.4; 0.0; 1232 ❙ 1487; 544; 53 ❙ 20.7; 1142; 538.2

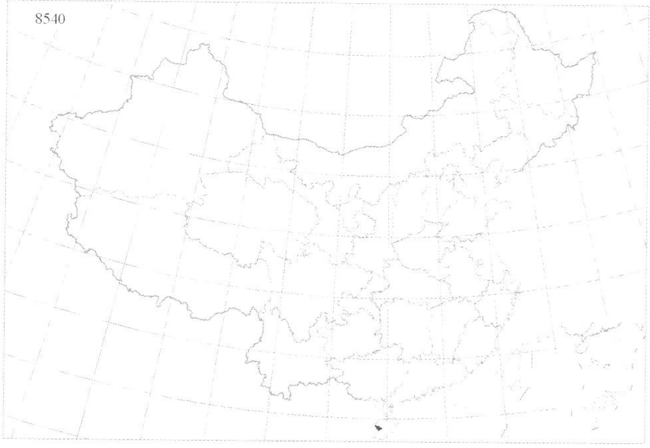

Myrsinaceae 紫金牛科 *Ardisia* 紫金牛属
Ardisia scalarinervis E. Walker 梯脉紫金牛
Shrubs or subshrubs, evergreen

19.3 (18.2~20.5); 13.7 (12.4~15.1); 22.8 (21.6~23.7); 19.4 (18.2~20.5); 172.3 (158.6~185.9); 0.0 (0.0~0.0); 925 (861~992) ❙ 1552 (1464~1751); 787 (655~996); 65 (52~77) ❙ 68.1 (47.6~88.8); 910 (839~986); 525.1 (427.9~659.0)

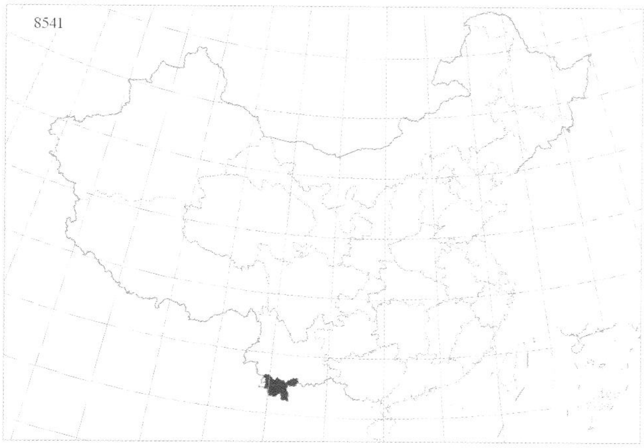

Myrsinaceae 紫金牛科 *Ardisia* 紫金牛属
Ardisia shweliensis W. W. Sm. 瑞丽紫金牛
Shrubs, evergreen

17.8 (16.0~19.5); 11.2 (9.3~13.5); 22.1 (20.3~25.0); 17.8 (16.0~19.5); 153.9 (132.4~173.8); 0.0 (0.0~0.0); 859 (780~945) ❙ 1421 (1143~1751); 813 (637~996); 54 (42~77) ❙ 65.9 (21.7~112.2); 852 (780~945); 451.9 (316.1~627.0)

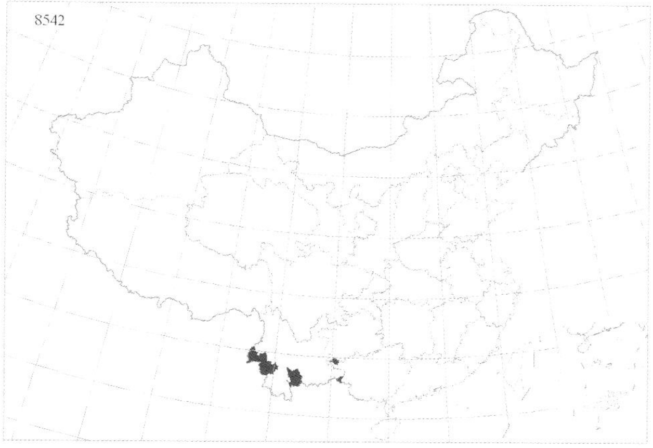

Myrsinaceae 紫金牛科 *Ardisia* 紫金牛属
Ardisia sieboldii Miq. 多枝紫金牛
Shrubs, evergreen

20.1 (16.0~24.1); 12.9 (5.6~18.9); 26.7 (20.1~29.0); 20.2 (16.0~24.1); 182.1 (132.5~229.7); 0.0 (0.0~0.0); 1034 (773~1322) ❙ 1839 (1070~3495); 771 (356~1847); 208 (47~897) ❙ 84.0 (-5.0~246.5); 1001 (773~1206); 224.1 (119.4~363.7)

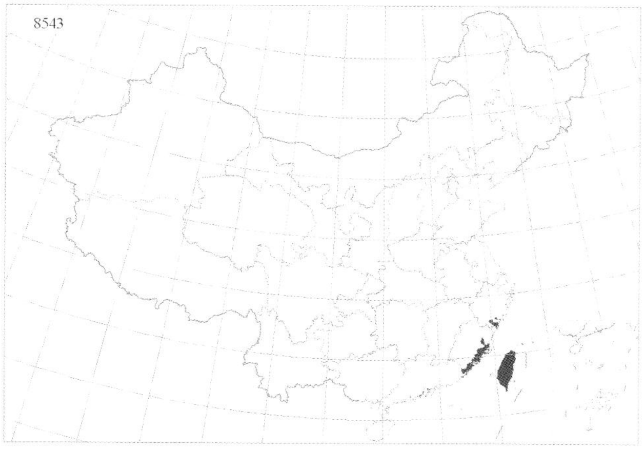

Myrsinaceae 紫金牛科 *Ardisia* 紫金牛属
Ardisia sinoaustralis C. Chen 细罗伞
Small shrubs, evergreen

18.4 (14.3~21.8); 8.4 (3.3~14.2); 27.1 (24.3~28.8); 18.4 (14.3~21.9); 161.1 (114.1~202.6); 0.1 (0.0~2.3); 958 (765~1151) ❙ 1614 (1372~1876); 602 (465~891); 180 (97~230) ❙ 69.4 (27.7~104.6); 958 (765~1151); 241.2 (188.6~301.1)

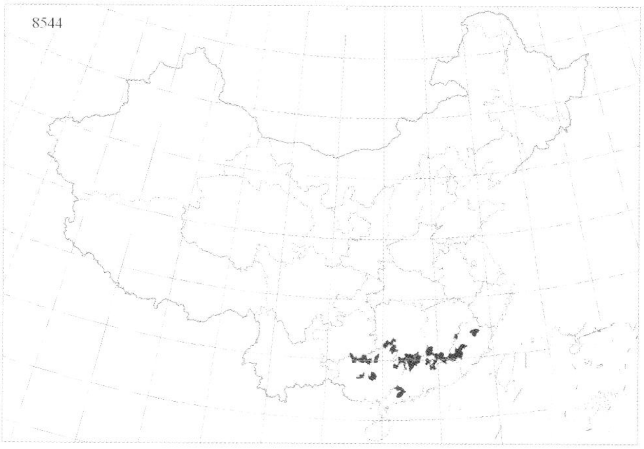

Myrsinaceae 紫金牛科 *Ardisia* 紫金牛属

Ardisia solanacea Roxb. 酸苔菜

Shrubs or trees, evergreen

18.9 (16.0~22.7); 11.4 (7.1~15.1); 24.4 (20.3~28.7); 19.0 (16.0~22.8); 167.8 (132.4~213.1); 0.0 (0.0~0.0); 947 (780~1221) ❙ 1405 (958~2435); 752 (531~1428); 74 (41~166) ❙ 49.2 (5.8~112.2); 941 (780~1221); 368.3 (190.9~659.0)

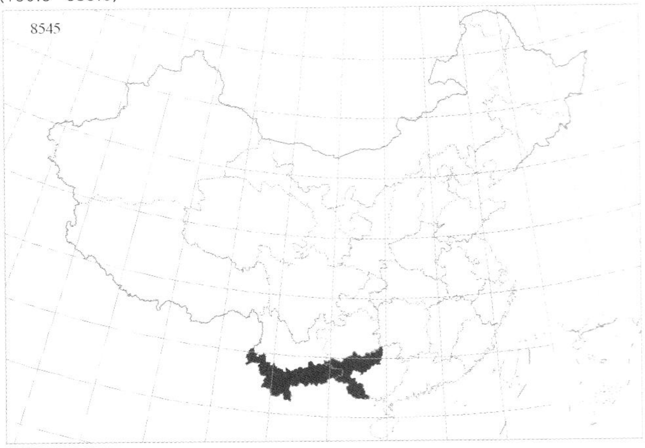

Myrsinaceae 紫金牛科 *Ardisia* 紫金牛属

Ardisia thyrsiflora D. Don 南方紫金牛

Shrubs or small trees, evergreen

18.1 (8.6~24.2); 10.4 (1.3~18.1); 24.0 (15.0~28.9); 18.1 (8.7~24.3); 157.6 (52.6~231.2); 0.1 (0.0~8.8); 916 (574~1335) ❙ 1342 (991~2435); 696 (442~1428); 74 (35~194) ❙ 48.1 (-12.0~145.8); 908 (574~1221); 343.7 (108.2~659.0)

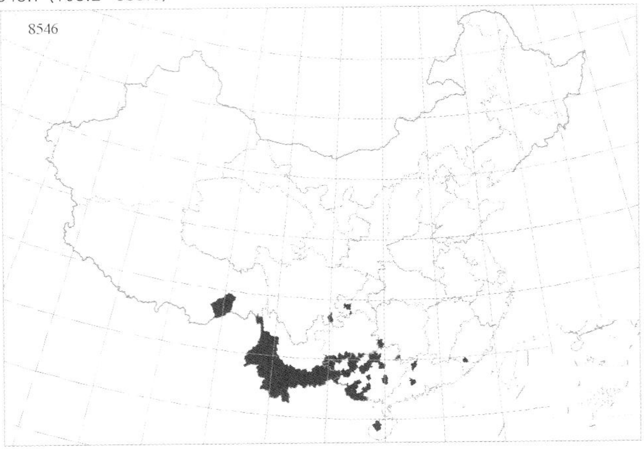

Myrsinaceae 紫金牛科 *Ardisia* 紫金牛属

Ardisia verbascifolia Mez 长毛紫金牛

Small shrubs, evergreen

21.0 (16.3~24.8); 15.0 (9.1~20.1); 25.2 (21.4~28.4); 21.0 (16.4~24.8); 192.4 (136.4~237.6); 0.0 (0.0~0.0); 1096 (797~1385) ❙ 1425 (1163~1725); 629 (392~860); 58 (41~110) ❙ 34.2 (-11.9~80.6); 1021 (797~1326); 468.1 (347.3~620.0)

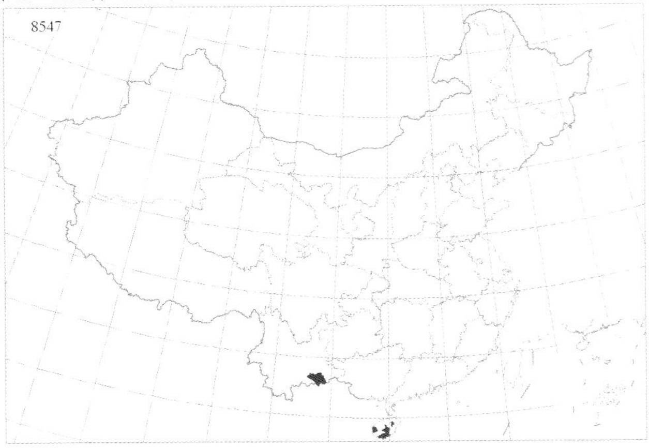

Myrsinaceae 紫金牛科 *Ardisia* 紫金牛属

Ardisia villosa Roxb. 雪下红

Shrubs, evergreen

21.9 (17.3~24.7); 14.3 (10.2~19.8); 27.7 (21.6~29.2); 21.9 (17.3~24.7); 203.4 (148.0~236.5); 0.0 (0.0~0.0); 1158 (832~1376) ❙ 1610 (1049~2435); 771 (455~1428); 109 (44~194) ❙ 39.6 (-6.9~99.5); 1141 (783~1273); 301.7 (157.7~659.0)

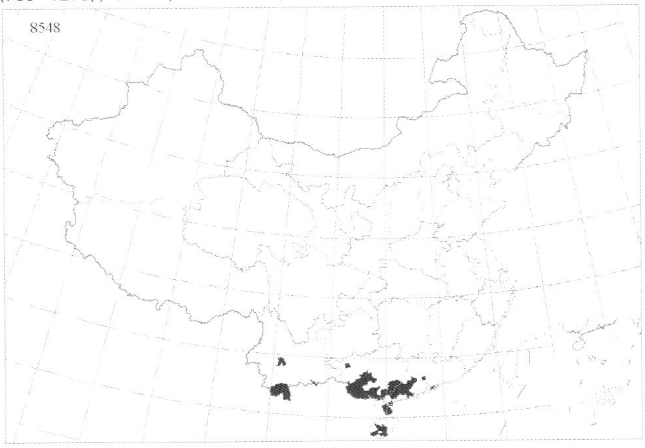

Myrsinaceae 紫金牛科 *Ardisia* 紫金牛属

Ardisia violacea (T. Suzuki) W. Z. Fang et K. Yao 锦花紫金牛

Subshrubs, evergreen

17.4 (16.5~19.9); 6.5 (4.3~13.5); 28.3 (25.9~29.3); 17.5 (16.6~19.9); 150.4 (139.8~179.1); 0.4 (0.0~0.7); 935 (910~987) ❙ 1787 (1377~2993); 653 (493~924); 239 (160~547) ❙ 89.9 (51.2~203.4); 935 (910~987); 213.6 (189.1~261.4)

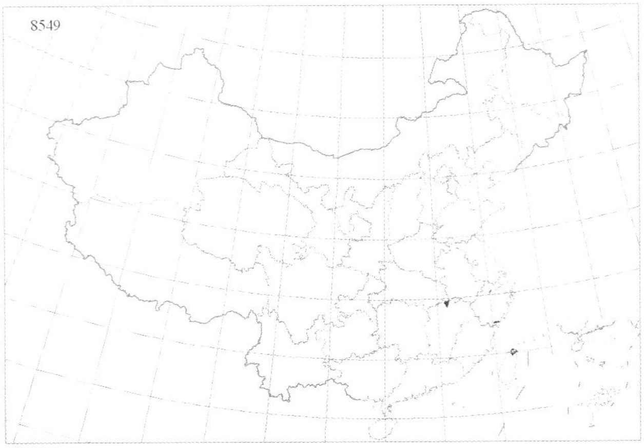

Myrsinaceae 紫金牛科 *Ardisia* 紫金牛属

Ardisia virens Kurz 钮子果

Shrubs or small trees, evergreen

19.2 (8.6~24.7); 12.4 (1.3~19.8); 24.2 (15.0~28.8); 19.2 (8.7~24.7); 170.8 (52.6~236.5); 0.1 (0.0~8.8); 958 (574~1376) ❙ 1639 (958~3495); 830 (531~1847); 114 (41~897) ❙ 72.4 (12.5~246.5); 939 (574~1221); 337.3 (119.4~659.0)

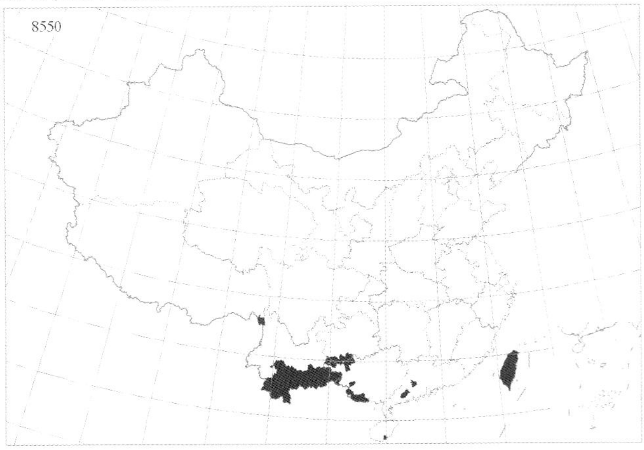

Myrsinaceae 紫金牛科 *Ardisia* 紫金牛属
Ardisia waitakii C. M. Hu 越南紫金牛
Shrubs or small trees, evergreen

22.1 (19.1~24.3); 15.2 (11.0~19.1); 27.4 (25.0~28.9); 22.2 (19.1~24.3); 206.3 (169.6~231.9); 0.0 (0.0~0.0); 1172 (945~1338) ❙ 1639 (1179~1900); 745 (392~1056); 89 (42~125) ❙ 41.7 (-11.9~71.0); 1117 (945~1257); 360.9 (194.5~620.0)

8551

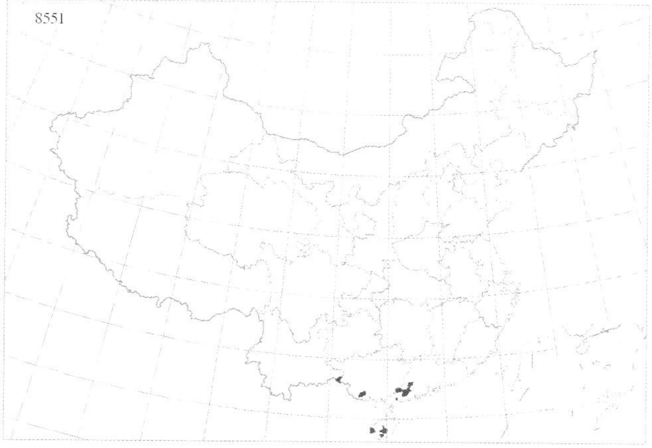

Myrsinaceae 紫金牛科 *Embelia* 酸藤子属
Embelia carnosisperma C. Y. Wu et C. Chen 肉果酸藤子
Scandent shrubs

17.7; 10.9; 22.5; 17.8; 153.4; 0.0; 854 ❙ 1368; 756; 59 ❙ 60.1; 854; 491.5

8552

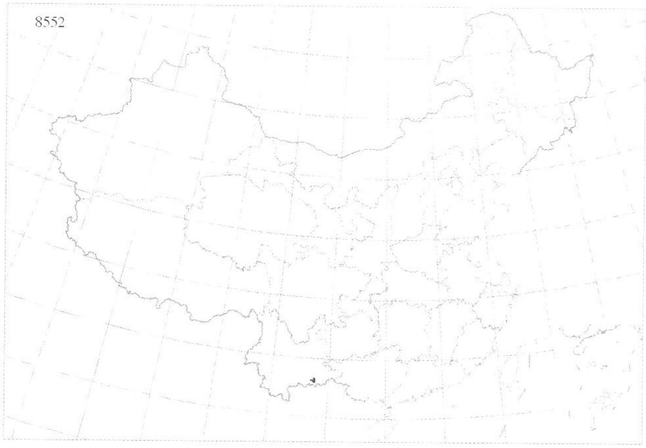

Myrsinaceae 紫金牛科 *Embelia* 酸藤子属
Embelia floribunda Wall. 多花酸藤子
Scandent shrubss or vines

10.7 (5.3~13.3); 3.2 (-2.4~6.4); 16.8 (12.2~19.6); 10.8 (5.7~13.4); 74.5 (30.5~100.4); 6.1 (0.0~26.2); 628 (479~706) ❙ 1292 (961~1736); 732 (558~1007); 47 (31~69) ❙ 104.4 (81.5~145.8); 628 (479~706); 364.9 (234.2~447.4)

8553

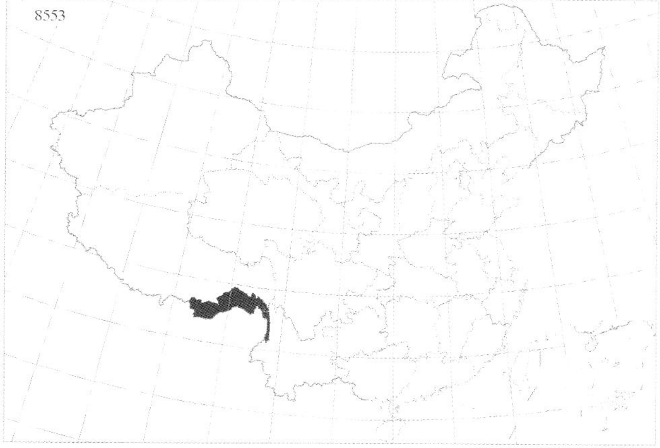

Myrsinaceae 紫金牛科 *Embelia* 酸藤子属
Embelia gamblei Kurz ex C. B. Clarke 皱叶酸藤子
Scandent shrubs

11.2 (5.3~14.4); 3.8 (-2.4~7.5); 17.1 (12.2~19.6); 11.3 (5.7~14.4); 80.0 (30.5~113.1); 5.2 (0.0~26.2); 642 (479~727) ❙ 1315 (961~1736); 740 (558~1007); 51 (31~70) ❙ 103.8 (81.5~145.8); 642 (479~727); 382.4 (234.2~487.2)

8554

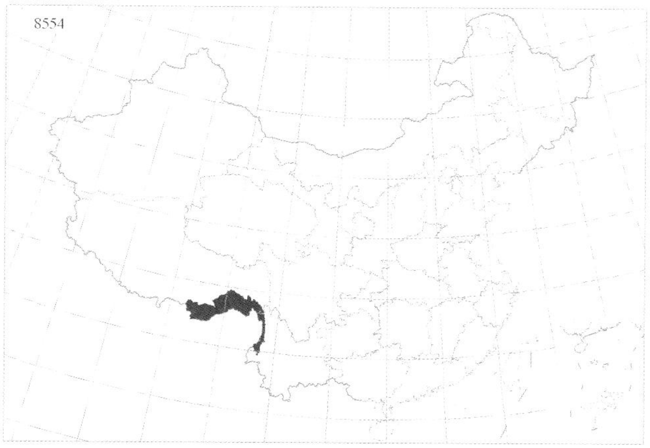

Myrsinaceae 紫金牛科 *Embelia* 酸藤子属
Embelia henryi E. Walker 毛果酸藤子
Scandent shrubs

17.6 (12.6~21.0); 10.0 (1.6~14.2); 23.3 (21.4~25.8); 17.6 (12.7~21.0); 152.5 (98.8~19[illegible].8); 0.7 (0.0~6.9); 871 (711~1053) ❙ 1375 (1074~1751); 743 (582~996); 62 (47~105) ❙ 59.1 (30.5~107.2); 871 (711~1053); 388.7 (234.7~551.6)

8555

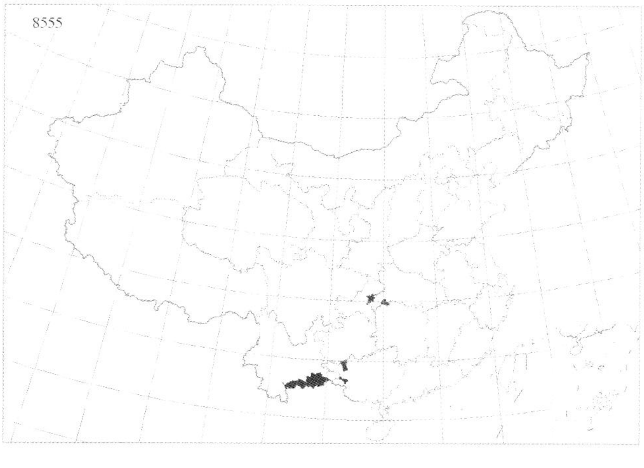

Myrsinaceae 紫金牛科 *Embelia* 酸藤子属
Embelia laeta (L.) Mez 酸藤子
Scandent shrubss or vines

20.4 (15.3~24.8); 12.3 (3.5~20.1); 27.1 (20.1~29.8); 20.5 (15.4~24.8); 185.8 (125.8~237.6); 0.0 (0.0~1.5); 1070 (773~1385) ❙ 1655 (868~3495); 714 (312~1847); 144 (41~897) ❙ 57.6 (-34.3~246.5); 1053 (773~1360); 270.1 (95.3~659.0)

8556

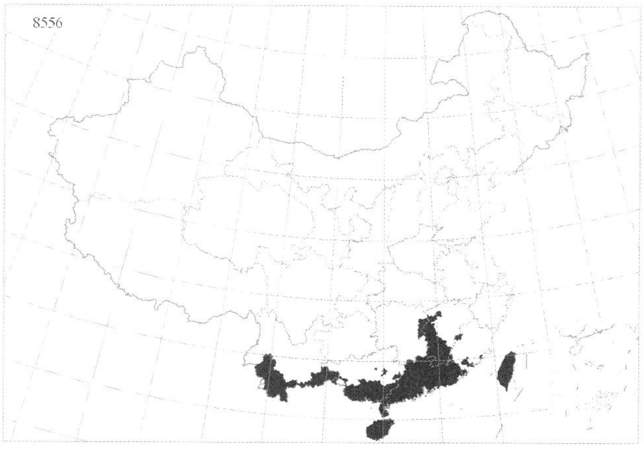

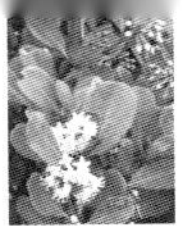

Myrsinaceae 紫金牛科 *Embelia* 酸藤子属
Embelia parviflora Wall. ex A. DC. 当归藤
Scandent shrubss or vines

19.3 (8.6~24.8); 10.8 (1.3~20.1); 26.2 (15.0~29.8); 19.3 (8.7~24.8); 172.2 (52.6~237.6); 0.1 (0.0~8.8); 1003 (574~1385) ▎1480 (896~2435); 676 (444~1428); 115 (31~235) ▎48.5 (-6.9~145.8); 996 (574~1326); 262.5 (95.3~659.0)

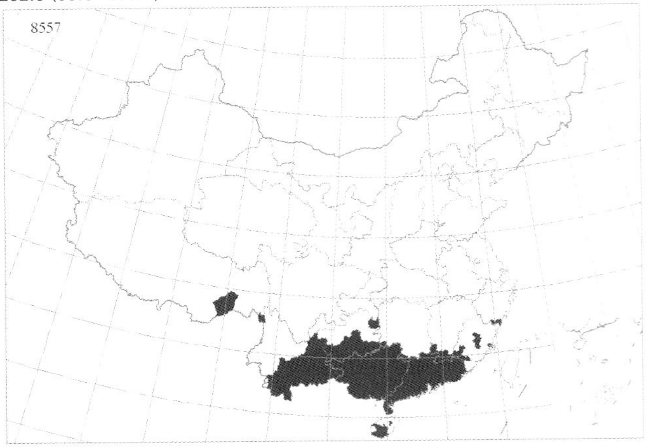

Myrsinaceae 紫金牛科 *Embelia* 酸藤子属
Embelia pauciflora Diels 疏花酸藤子
Vines or shrubs

15.5 (14.3~17.8); 5.3 (3.8~7.2); 24.7 (22.8~27.7); 15.5 (14.3~17.8); 126.3 (112.3~153.7); 0.2 (0.0~1.5); 808 (751~939) ▎1205 (985~1342); 542 (473~648); 72 (52~95) ▎49.4 (16.1~62.6); 808 (751~939); 191.5 (142.5~254.8)

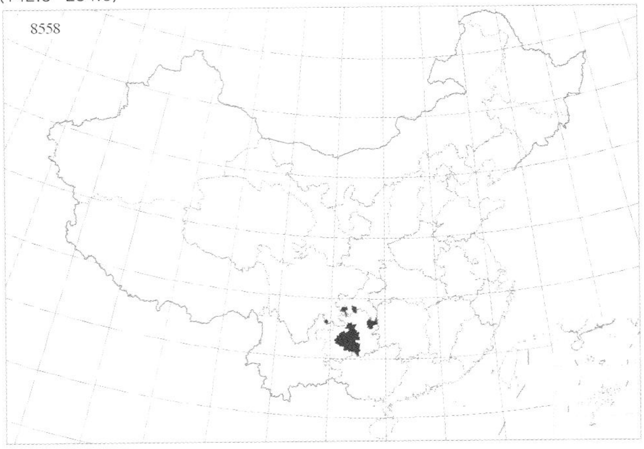

Myrsinaceae 紫金牛科 *Embelia* 酸藤子属
Embelia polypodioides Hemsl. et Mez 龙骨酸藤子
Scandent shrubss or vines

18.0 (15.6~21.0); 11.2 (9.0~14.2); 22.7 (20.3~25.8); 18.0 (15.6~21.0); 156.4 (127.6~191.8); 0.0 (0.0~0.0); 873 (767~1053) ▎1232 (911~1751); 695 (524~996); 51 (32~77) ▎40.9 (6.5~88.8); 856 (717~1053); 373.7 (277.0~551.6)

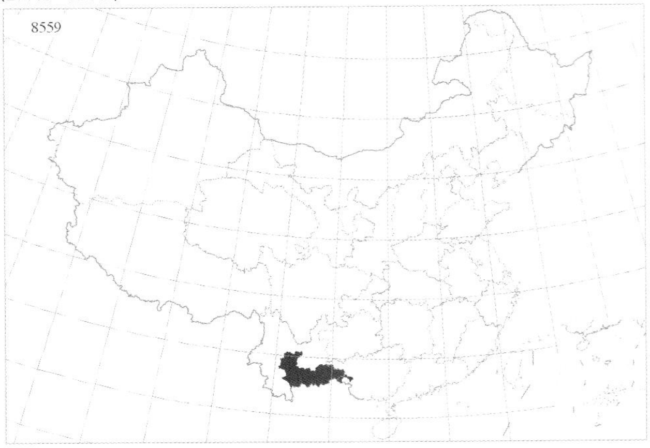

Myrsinaceae 紫金牛科 *Embelia* 酸藤子属
Embelia procumbens Hemsl. 匍匐酸藤子
Vines

16.9 (11.1~19.6); 10.3 (2.2~13.9); 21.4 (19.1~23.6); 16.9 (11.3~19.6); 143.0 (80.8~175.1); 0.1 (0.0~5.3); 821 (652~939) ▎1163 (939~1500); 652 (524~836); 48 (34~70) ▎41.9 (6.5~99.9); 804 (652~923); 356.4 (214.5~530.1)

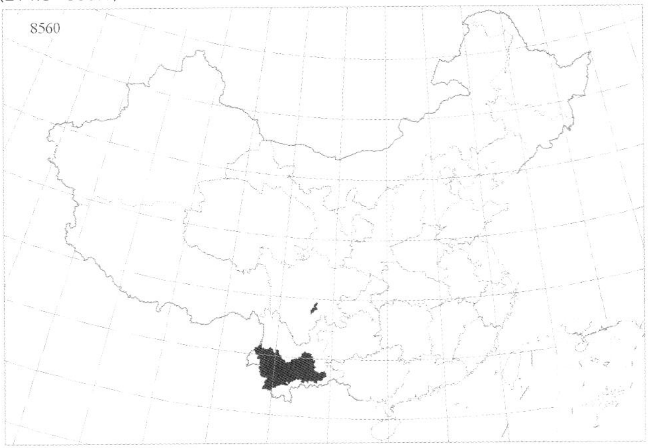

Myrsinaceae 紫金牛科 *Embelia* 酸藤子属
Embelia ribes Burm. f. 白花酸藤果
Scandent shrubss or vines

19.2 (8.6~24.8); 11.2 (1.3~20.1); 25.4 (15.0~29.2); 19.2 (8.7~24.8); 170.6 (52.6~237.6); 0.0 (0.0~8.8); 991 (574~1385) ▎1435 (797~2435); 691 (312~1428); 101 (22~198) ▎45.8 (-34.3~145.8); 978 (574~1326); 286.1 (95.3~659.0)

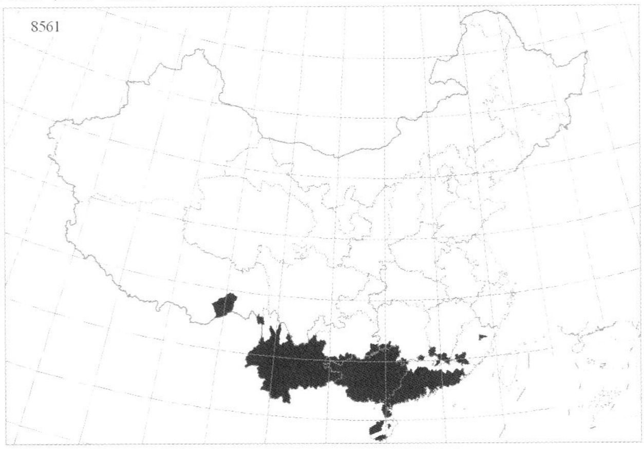

Myrsinaceae 紫金牛科 *Embelia* 酸藤子属
Embelia ribes subsp. pachyphylla (Chun ex C. Y. Wu et C. Chen) Pipoly et C. Chen 厚叶白花酸藤果
Scandent shrubss or vines

19.7 (14.2~22.8); 11.0 (4.8~15.5); 26.6 (20.1~29.8); 19.7 (14.2~22.8); 176.6 (110.5~214.1); 0.0 (0.0~0.2); 1020 (729~1225) ▎1566 (958~2435); 710 (505~1428); 127 (44~203) ▎53.6 (5.8~99.5); 1018 (729~1225); 264.4 (152.9~551.6)

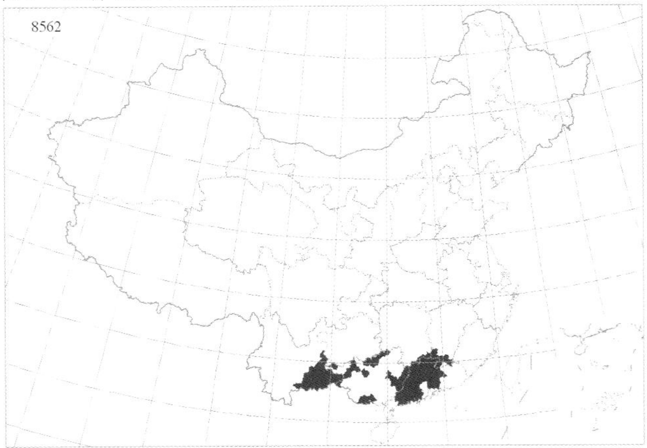

Myrsinaceae 紫金牛科 *Embelia* 酸藤子属
Embelia scandens (Lour.) Mez 瘤皮孔酸藤子
Scandent shrubs

21.0 (16.3~24.8); 13.1 (6.6~20.1); 27.2 (21.4~29.2); 21.1 (16.4~24.8); 192.9 (136.4~237.6); 0.0 (0.0~0.0); 1102 (797~1385) ▎1567 (1007~2037); 742 (473~1033); 110 (41~197) ▎43.1 (-6.9~88.8); 1096 (797~1326); 268.5 (95.3~659.0)

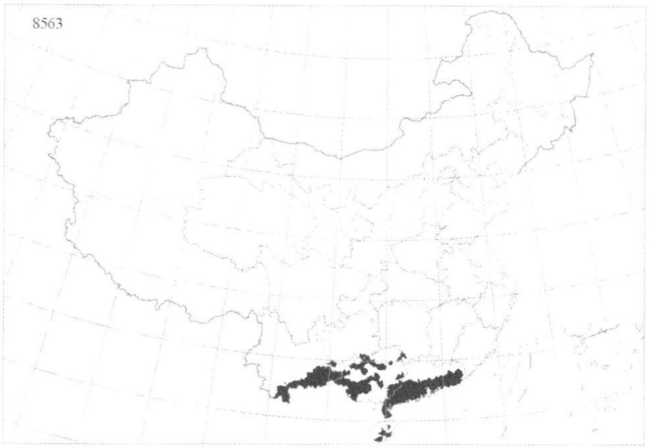

Myrsinaceae 紫金牛科 *Embelia* 酸藤子属
Embelia sessiliflora Kurz 短梗酸藤子
Scandent shrubss or vines

18.1 (16.0~21.0); 11.7 (7.8~15.1); 22.2 (20.3~25.6); 18.1 (16.0~21.0); 157.8 (132.4~191.8); 0.0 (0.0~0.0); 873 (780~1053) ▎1355 (1028~1751); 748 (566~996); 53 (41~77) ▎55.4 (12.5~112.2); 858 (779~1053); 440.4 (226.6~659.0)

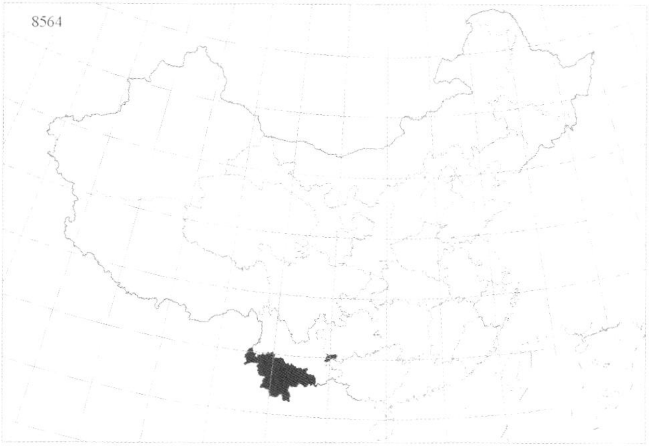

Myrsinaceae 紫金牛科 *Embelia* 酸藤子属
Embelia undulata (Wall.) Mez 平叶酸藤子
Scandent shrubs

17.1 (5.4~24.0); 8.6 (-2.1~19.1); 24.2 (12.3~29.8); 17.2 (5.8~24.0); 146.3 (31.5~228.1); 0.4 (0.0~25.6); 883 (481~1302) ▎1331 (741~2035); 629 (411~996); 90 (15~231) ▎50.6 (-14.2~115.7); 874 (481~1225); 273.6 (142.9~659.0)

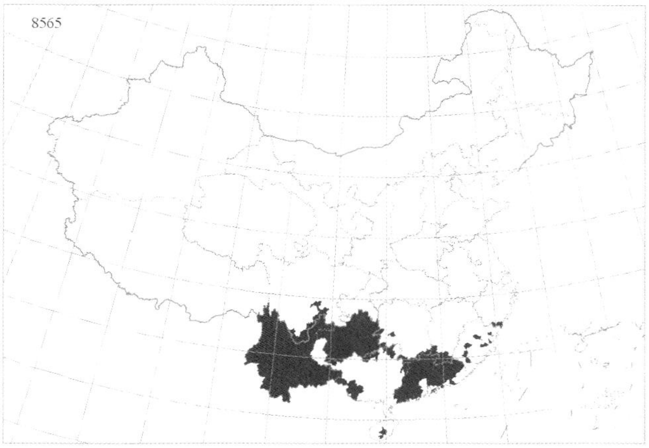

Myrsinaceae 紫金牛科 *Embelia* 酸藤子属
Embelia vestita Roxb. 密齿酸藤子
Shrubs or small trees

18.2 (5.3~24.7); 9.1 (-2.4~19.8); 26.2 (12.2~29.8); 18.2 (5.7~24.7); 159.0 (30.5~236.5); 0.3 (0.0~26.2); 948 (479~1376) ▎1474 (868~2680); 608 (312~1357); 134 (31~258) ▎56.9 (-34.3~246.5); 939 (479~1273); 249.6 (95.3~659.0)

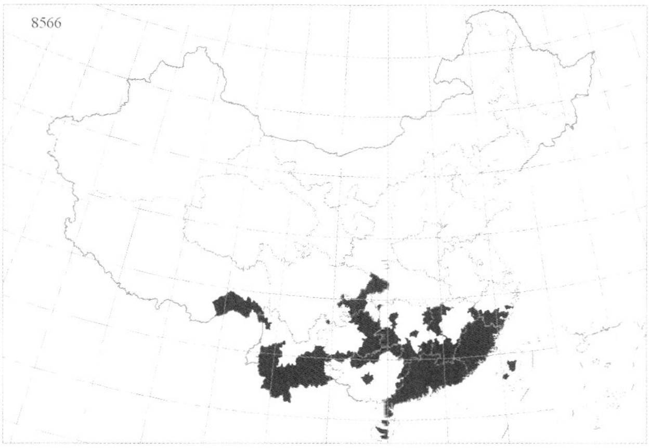

Myrsinaceae 紫金牛科 *Maesa* 杜茎山属
Maesa acuminatissima Merr. 米珍果
Shrubs, evergreen

22.1 (17.7~24.8); 15.7 (10.9~20.1); 26.7 (22.5~28.6); 22.1 (17.8~24.8); 205.2 (153.4~237.6); 0.0 (0.0~0.0); 1159 (854~1385) ▎1613 (1179~2435); 757 (392~1428); 77 (41~171) ▎41.6 (-11.9~99.5); 1085 (854~1221); 450.5 (206.0~620.0)

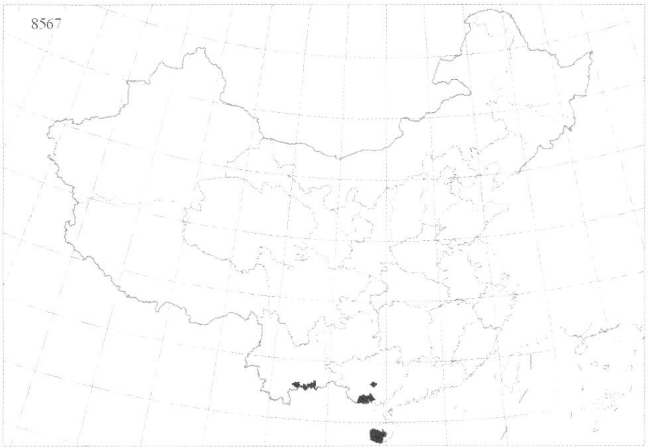

Myrsinaceae 紫金牛科 *Maesa* 杜茎山属
Maesa ambigua C. Y. Wu et C. Chen 坚髓杜茎山
Shrubs, evergreen

18.0 (16.3~21.0); 10.9 (9.1~14.2); 23.0 (21.4~25.6); 18.1 (16.4~21.0); 156.8 (136.4~191.8); 0.0 (0.0~0.0); 879 (797~1053) ▎1384 (1190~1522); 777 (656~860); 56 (48~68) ▎57.9 (44.5~80.6); 879 (797~1053); 429.4 (336.3~551.6)

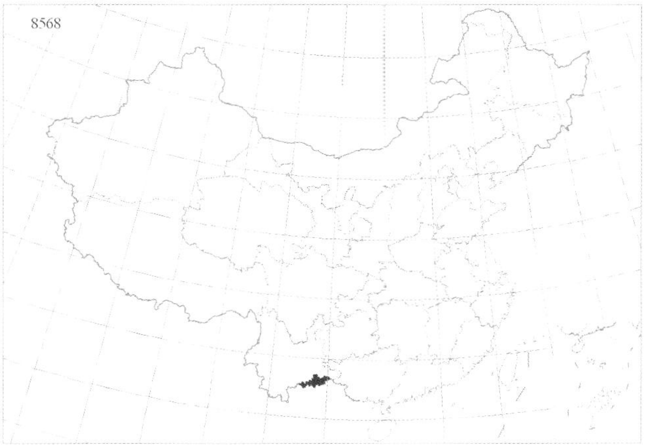

Myrsinaceae 紫金牛科 *Maesa* 杜茎山属
Maesa argentea (Wall.) A. DC. 银叶杜茎山
Shrubs, evergreen

14.9 (3.8~21.0); 7.7 (-4.3~14.2); 20.4 (10.8~26.6); 14.9 (4.7~21.0); 120.5 (21.6~191.8); 1.4 (0.0~35.4); 759 (444~1053) ❙ 1018 (741~1522); 583 (411~818); 39 (15~70) ❙ 35.5 (-14.2~121.7); 735 (444~1053); 303.1 (142.9~551.6)

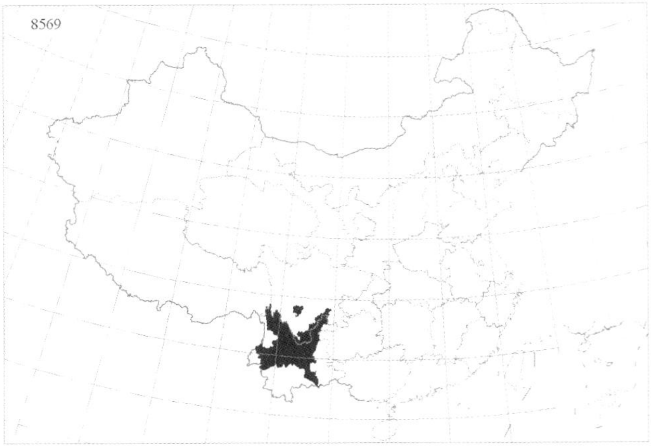

Myrsinaceae 紫金牛科 *Maesa* 杜茎山属
Maesa brevipaniculata (C. Y. Wu et C. Chen) Pipoly et C. Chen 短序杜茎山
Shrubs, evergreen

17.7 (16.1~21.0); 9.2 (5.3~14.2); 24.2 (21.4~26.6); 17.7 (16.1~21.0); 152.4 (133.1~191.8); 0.0 (0.0~0.0); 882 (797~1053) ❙ 1305 (1074~1522); 676 (523~860); 69 (47~142) ❙ 48.3 (27.5~80.6); 882 (797~1053); 313.2 (189.5~551.6)

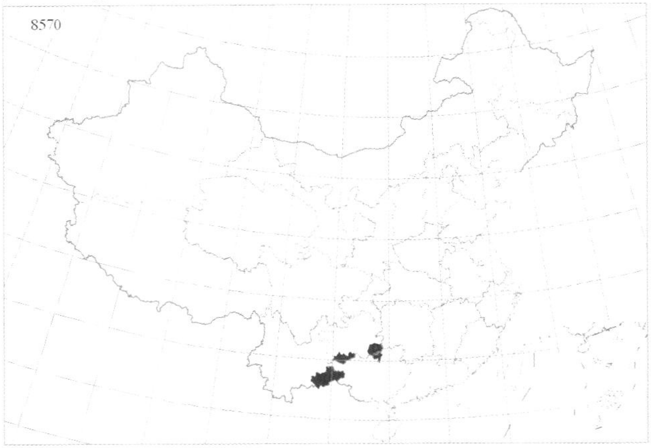

Myrsinaceae 紫金牛科 *Maesa* 杜茎山属
Maesa cavinervis C. Chen 凹脉杜茎山
Shrubs, evergreen

13.3; 5.4; 19.6; 13.4; 100.4; 0.0; 706 ❙ 1736; 1007; 51 ❙ 145.8; 706; 447.4

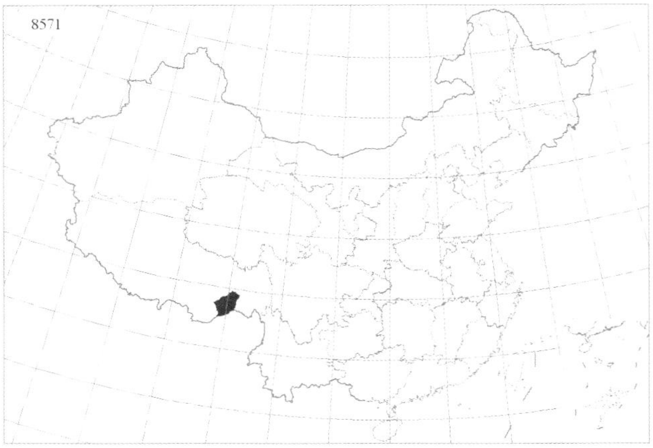

Myrsinaceae 紫金牛科 *Maesa* 杜茎山属
Maesa chisia Buch.-Ham. ex D. Don 密腺杜茎山
Shrubs, evergreen

15.5 (-1.8~19.6); 8.6 (-10.2~13.5); 20.1 (6.1~23.5); 15.8 (1.8~19.6); 132.3 (1.9~175.7); 6.4 (0.0~82.8); 789 (329~953) ❙ 1460 (320~1751); 842 (217~1007); 55 (20~77) ❙ 82.4 (-3.0~145.8); 785 (320~932); 465.2 (72.2~627.0)

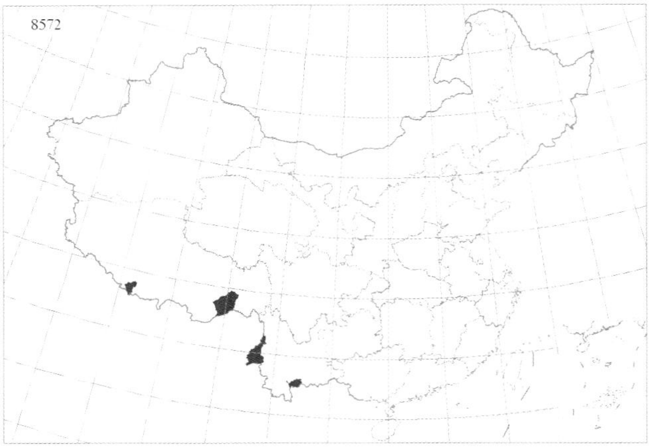

Myrsinaceae 紫金牛科 *Maesa* 杜茎山属
Maesa confusa (C. M. Hu) Pipoly et C. Chen 紊纹杜茎山
Shrubs, evergreen

24.2 (23.0~24.8); 18.9 (17.7~20.1); 28.0 (26.7~28.5); 24.3 (23.1~24.8); 231.3 (217.0~237.6); 0.0 (0.0~0.0); 1334 (1217~1385) ❙ 1579 (1179~1854); 542 (392~622); 72 (41~115) ❙ 18.8 (-11.9~41.8); 1209 (1056~1360); 472.0 (352.9~572.9)

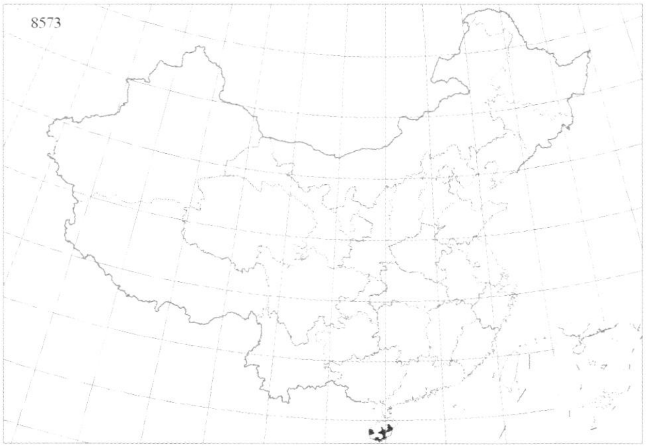

Myrsinaceae 紫金牛科 *Maesa* 杜茎山属
Maesa consanguinea Merr. 拟杜茎山
Scandent shrubs, evergreen

23.5 (22.2~24.3); 18.1 (17.2~18.7); 27.3 (25.7~28.4); 23.6 (22.3~24.3); 222.6 (207.2~231.9); 0.0 (0.0~0.0); 1260 (1127~1338) ❙ 1535 (1179~1725); 535 (392~619); 64 (42~110) ❙ 22.9 (-11.9~47.8); 1143 (1055~1326); 485.4 (352.9~590.7)

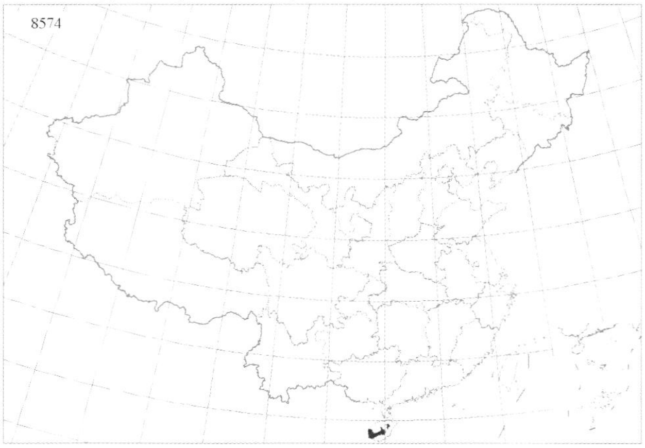

Myrsinaceae 紫金牛科 *Maesa* 杜茎山属
Maesa densistriata C. Chen et C. M. Hu 灰叶杜茎山
Shrubs, evergreen

17.9；11.6；22.1；17.9；155.1；0.0；858 ❙ 1098；634；46 ❙ 27.9；836；295.5

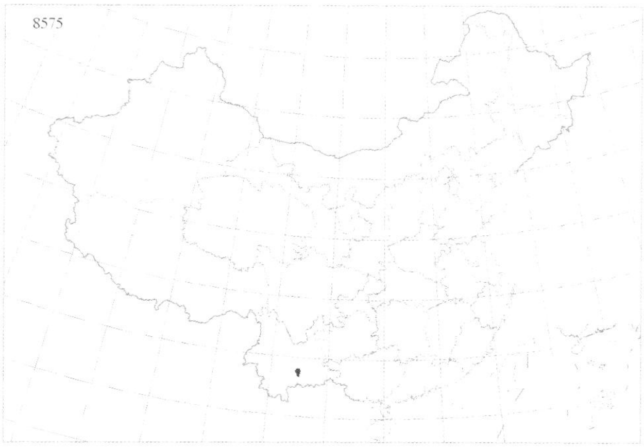

Myrsinaceae 紫金牛科 *Maesa* 杜茎山属
Maesa hupehensis Rehder 湖北杜茎山
Shrubs, evergreen

15.2 (2.2~19.1)；4.8 (-6.6~11.0)；24.8 (9.9~28.9)；15.3 (3.8~19.1)；125.0 (15.5~169.6)；1.5 (0.0~49.1)；811 (413~969) ❙ 1229 (916~1578)；555 (444~932)；80 (19~215) ❙ 52.6 (9.2~125.9)；811 (413~969)；207.5 (130.1~371.7)

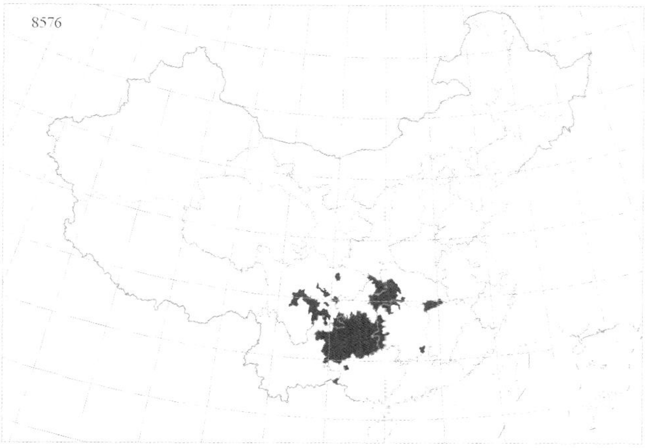

Myrsinaceae 紫金牛科 *Maesa* 杜茎山属
Maesa indica (Roxb.) A. DC. 包疮叶
Tall shrubs, evergreen

18.3 (16.3~22.1)；11.7 (9.0~15.1)；22.7 (20.3~28.8)；18.4 (16.4~22.1)；160.4 (136.4~205.6)；0.0 (0.0~0.0)；889 (789~1182) ❙ 1304 (939~1784)；710 (524~996)；55 (37~144) ❙ 46.6 (6.5~88.8)；872 (744~1182)；401.2 (157.7~659.0)

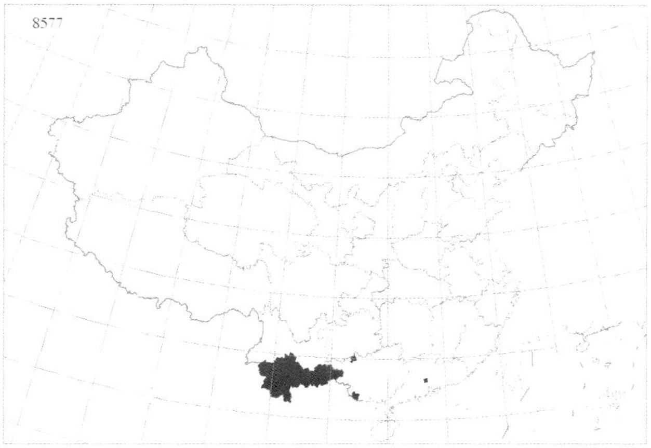

Myrsinaceae 紫金牛科 *Maesa* 杜茎山属
Maesa insignis Chun 毛穗杜茎山
Shrubs, evergreen

17.7 (14.1~21.8)；8.3 (3.5~15.1)；25.8 (22.5~28.5)；17.7 (14.2~21.8)；153.3 (111.5~201.8)；0.3 (0.0~1.7)；920 (762~1151) ❙ 1355 (1136~1776)；616 (465~828)；88 (51~156) ❙ 48.4 (5.8~72.9)；920 (762~1151)；248.2 (146.0~659.0)

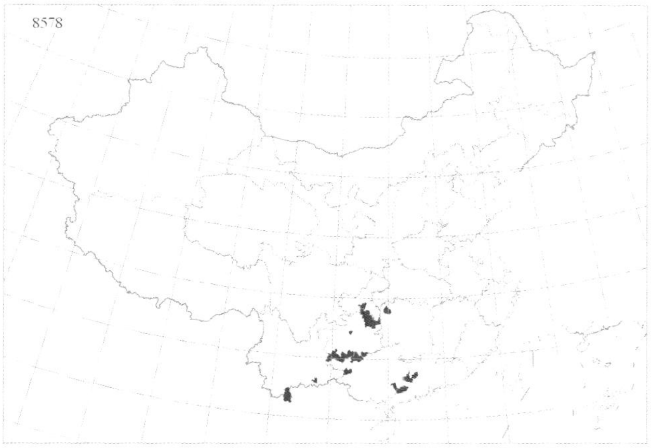

Myrsinaceae 紫金牛科 *Maesa* 杜茎山属
Maesa japonica (Thunb.) Moritzi et Zoll. 杜茎山
Shrubs, evergreen

17.9 (10.6~24.1)；8.1 (-0.2~18.9)；26.8 (18.6~30.1)；18.0 (10.6~24.2)；156.3 (79.4~230.1)；0.4 (0.0~12.1)；947 (643~1326) ❙ 1492 (871~3495)；611 (299~1847)；143 (19~897) ❙ 59.1 (-6.9~246.5)；939 (643~1326)；226.0 (95.3~659.0)

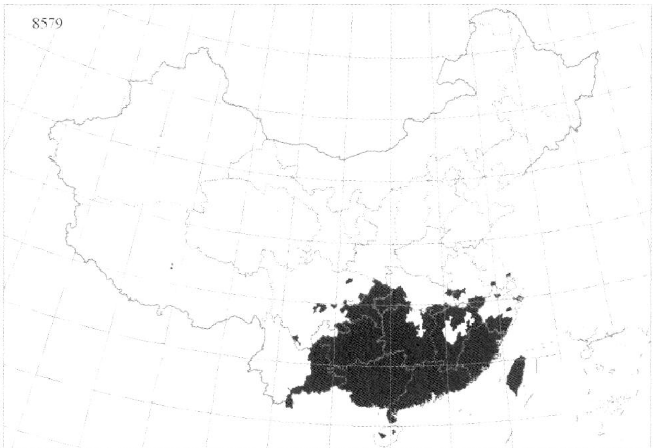

Myrsinaceae 紫金牛科 *Maesa* 杜茎山属
Maesa macilenta E. Walker 细梗杜茎山
Shrubs, evergreen

19.4；12.3；24.3；19.4；172.9；0.0；954 ❙ 1437；772；63 ❙ 50.8；949；431.9

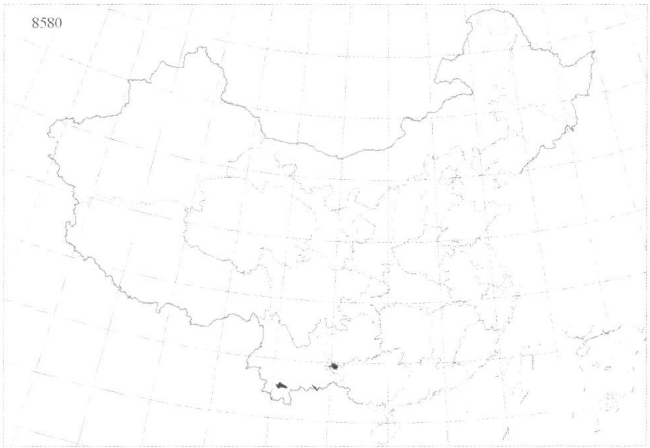

Myrsinaceae 紫金牛科 *Maesa* 杜茎山属

Maesa macilentoides C. Chen 薄叶杜茎山

Shrubs, evergreen

19.3 (18.2~20.5); 13.6 (12.4~15.1); 22.7 (21.6~23.7); 19.3 (18.2~20.5); 172.1 (158.6~185.9); 0.0 (0.0~0.0); 923 (861~992) ❙ 1529 (1434~1751); 768 (655~996); 60 (41~77) ❙ 65.9 (47.6~88.8); 902 (839~986); 520.0 (427.9~659.0)

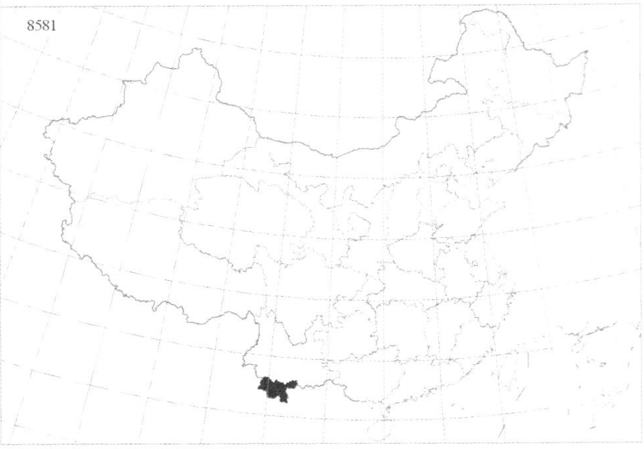

Myrsinaceae 紫金牛科 *Maesa* 杜茎山属

Maesa manipurensis Mez 隐纹杜茎山

Shrubs, evergreen

17.2 (8.6~22.8); 10.3 (1.3~14.7); 21.9 (15.0~28.9); 17.3 (8.7~22.8); 148.3 (52.6~213.6); 1.0 (0.0~8.8); 866 (574~1230) ❙ 1503 (1041~1688); 864 (585~959); 58 (42~106) ❙ 76.5 (33.9~112.2); 861 (574~1230); 461.1 (229.5~627.0)

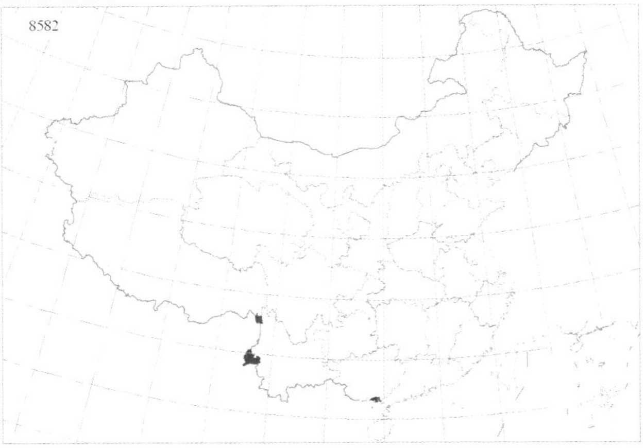

Myrsinaceae 紫金牛科 *Maesa* 杜茎山属

Maesa marioniae Merr. 毛脉杜茎山

Shrubs, evergreen

13.1; 6.3; 18.4; 13.1; 99.6; 2.4; 707 ❙ 1262; 711; 57 ❙ 80.7; 707; 412.6

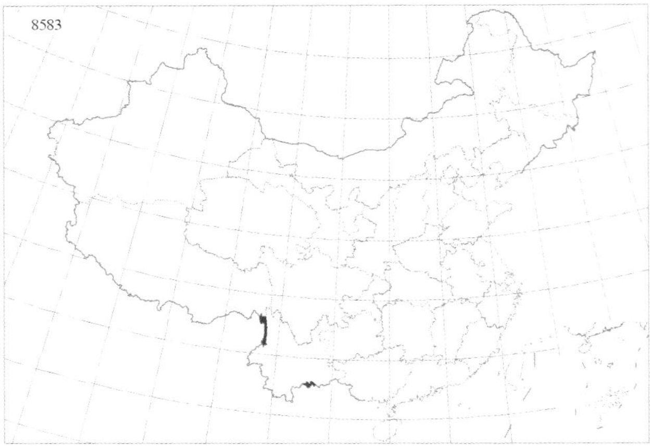

Myrsinaceae 紫金牛科 *Maesa* 杜茎山属

Maesa membranacea A. DC. 腺叶杜茎山

Tall shrubs, evergreen

20.7 (17.2~24.8); 14.6 (9.5~20.1); 25.0 (22.1~28.5); 20.8 (17.3~24.8); 189.4 (147.0~237.6); 0.0 (0.0~0.0); 1059 (833~1385) ❙ 1493 (1143~1900); 705 (392~1056); 66 (41~125) ❙ 43.9 (-11.9~88.8); 1017 (833~1360); 458.0 (248.2~659.0)

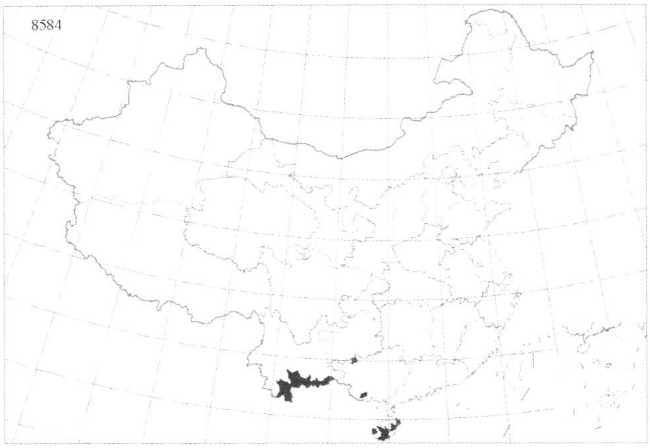

Myrsinaceae 紫金牛科 *Maesa* 杜茎山属

Maesa montana A. DC. 金珠柳

Shrubs or small trees, evergreen

17.4 (5.4~22.8); 9.1 (-2.1~17.7); 24.3 (12.3~29.8); 17.5 (5.8~22.8); 150.3 (31.5~214.1); 0.5 (0.0~25.6); 901 (481~1225) ❙ 1332 (741~2992); 635 (386~1847); 91 (15~257) ❙ 47.9 (-14.2~166.5); 892 (481~1225); 272.1 (108.2~659.0)

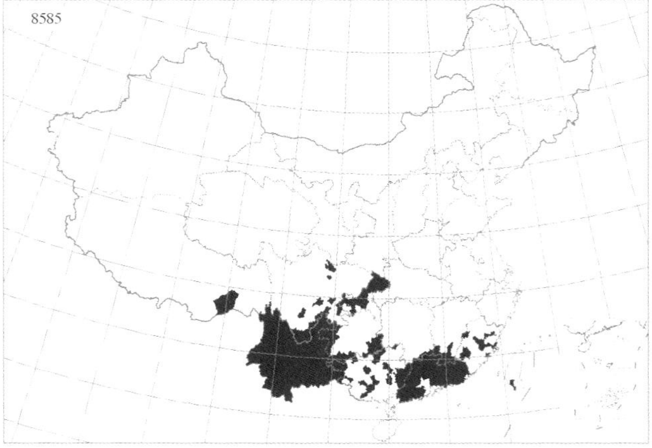

Myrsinaceae 紫金牛科 *Maesa* 杜茎山属

Maesa parvifolia Aug. DC. 小叶杜茎山

Shrubs, evergreen

21.5 (16.3~24.3); 13.7 (7.8~18.7); 27.6 (21.4~29.1); 21.5 (16.4~24.3); 198.5 (136.4~231.9); 0.0 (0.0~0.0); 1134 (797~1338) ❙ 1618 (868~2435); 773 (312~1428); 112 (42~188) ❙ 43.8 (-34.3~99.5); 1120 (797~1326); 258.1 (95.3~572.9)

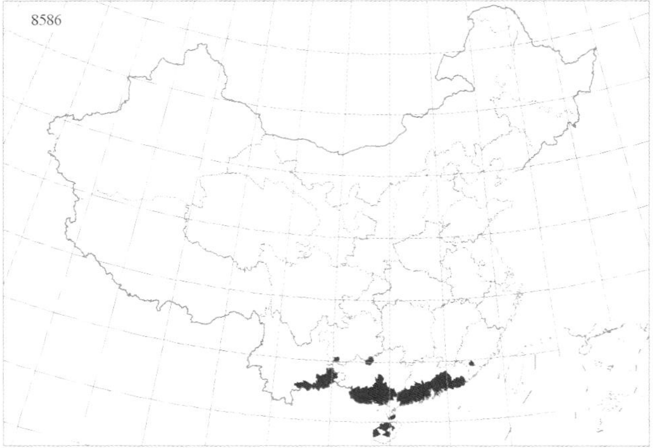

Plumbaginaceae 白花丹科 *Acantholimon* 彩花属
Acantholimon laevigatum (Z. X. Peng) Kamelin 光萼彩花
Shrubs

1.1；-17.3；16.3；6.0；41.0；87.2；494 ❙ 379；156；38 ❙ -20.4；356；107.4

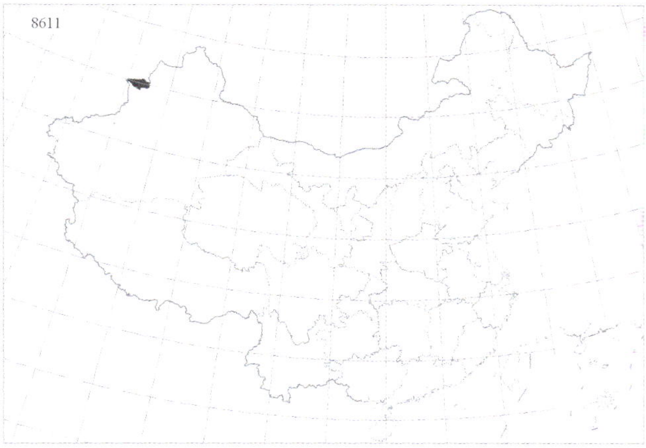

Plumbaginaceae 白花丹科 *Ceratostigma* 蓝雪花属
Ceratostigma griffithii C. B. Clarke 毛蓝雪花
Shrubs, evergreen

7.8 (2.2~14.3)；0.0 (-6.6~7.6)；14.5 (9.9~19.7)；8.2 (3.8~14.3)；52.1 (15.5~111.9)；18.3 (0.0~49.1)；551 (413~728) ❙ 917 (324~1736)；548 (257~1007)；24 (0~51) ❙ 64.8 (-26.0~145.8)；541 (324~707)；286.0 (123.3~447.4)

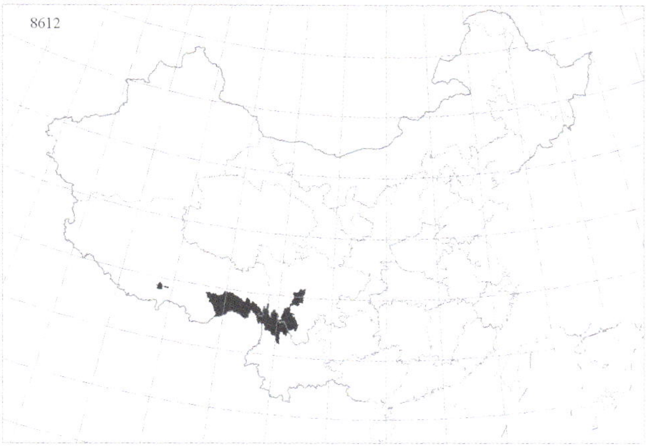

Plumbaginaceae 白花丹科 *Ceratostigma* 蓝雪花属
Ceratostigma minus Stapf ex Prain 小蓝雪花
Shrubs, deciduous

8.6 (0.0~18.9)；0.4 (-9.4~12.0)；15.7 (8.1~26.2)；9.4 (2.7~19.0)；67.1 (8.0~167.8)；22.8 (0.0~67.9)；587 (363~929) ❙ 853 (307~1736)；489 (226~1007)；25 (1~88) ❙ 49.3 (-28.5~145.8)；565 (307~865)；269.6 (111.0~447.4)

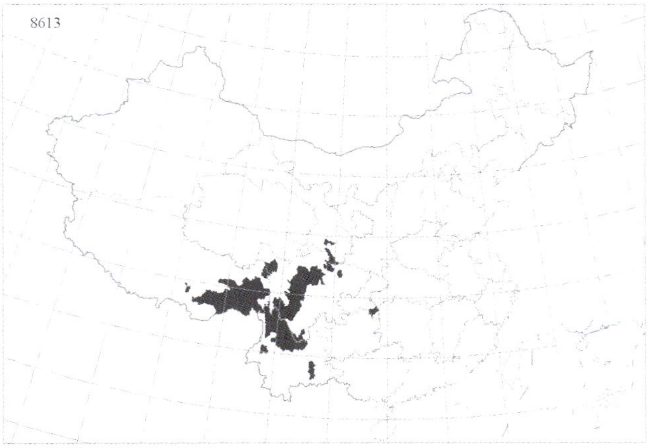

Plumbaginaceae 白花丹科 *Ceratostigma* 蓝雪花属
Ceratostigma ulicinum Prain 刺鳞蓝雪花
Shrubs, deciduous

-0.3 (-2.2~3.3)；-9.6 (-11.8~-6.6)；8.4 (6.1~12.1)；2.9 (1.8~5.0)；10.4 (1.9~27.3)；72.8 (46.3~91.0)；357 (327~438) ❙ 290 (212~370)；213 (156~262)；9 (0~33) ❙ -19.0 (-35.1~4.7)；286 (212~338)；97.1 (61.6~151.1)

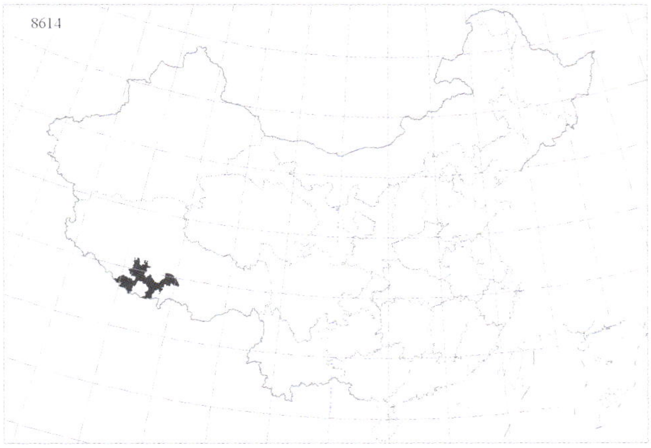

Plumbaginaceae 白花丹科 *Ceratostigma* 蓝雪花属
Ceratostigma willmottianum Stapf 岷江蓝雪花
Subshrubs, deciduous

13.5 (0.6~19.3)；5.9 (-8.8~13.1)；19.6 (9.1~26.6)；13.7 (3.2~19.4)；107.4 (12.4~173.2)；4.8 (0.0~65.0)；721 (371~939) ❙ 988 (300~1736)；561 (208~1007)；37 (3~69) ❙ 38.5 (-19.2~145.8)；703 (300~912)；281.6 (142.9~447.4)

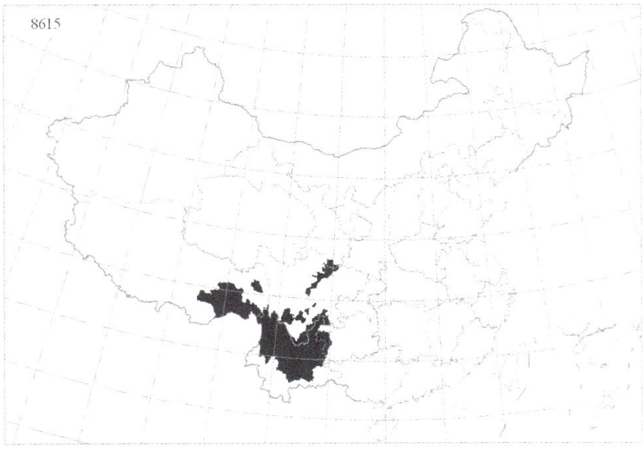

Plumbaginaceae 白花丹科 *Ikonnikovia* 伊犁花属
Ikonnikovia kaufmanniana (Regel) Lincz. 伊犁花
Shrubs, deciduous

4.3 (-2.8~9.3)；-15.2 (-20.9~-8.8)；19.9 (10.9~27.5)；8.1 (3.5~12.1)；63.8 (16.3~107.6)；72.0 (40.4~109.2)；595 (390~813) ❙ 237 (101~450)；108 (51~211)；20 (3~48) ❙ -56.5 (-85.3~3.9)；235 (101~406)；114.4 (21.1~297.3)

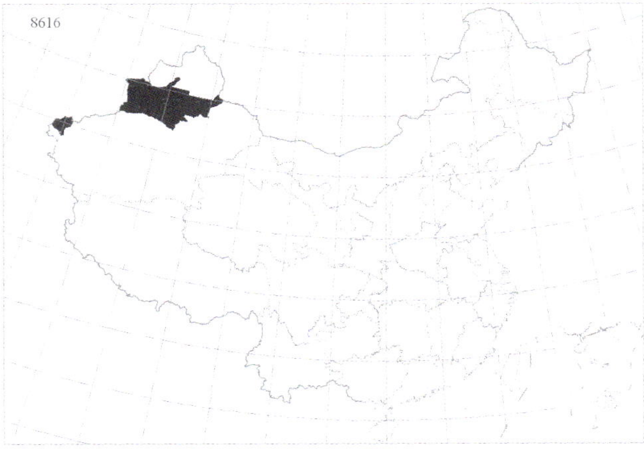

Plumbaginaceae 白花丹科 *Limonium* 补血草属
Limonium suffruticosum (L.) Kuntze 木本补血草
Dwarf shrubs, deciduous

4.4 (-2.8~10.8); -14.7 (-20.9~-8.0); 20.2 (10.9~27.5); 8.3 (3.5~12.2); 64.8 (16.3~107.6); 71.2 (34.1~109.2); 598 (390~813) ❙ 202 (70~450); 94 (39~211); 18 (3~51) ❙ -63.6 (-90.6~3.9); 201 (70~406); 89.6 (10.9~297.3)

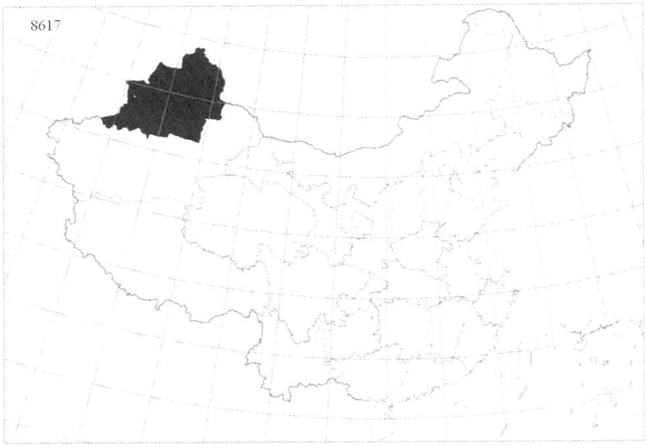

Plumbaginaceae 白花丹科 *Plumbago* 白花丹属
Plumbago zeylanica L. 白花丹
Subshrubs, evergreen

18.8 (9.3~24.8); 10.9 (-1.7~20.1); 25.2 (17.0~29.4); 18.8 (9.5~24.8); 166.2 (69.8~237.6); 0.2 (0.0~18.0); 968 (603~1385) ❙ 1439 (797~3495); 677 (312~1847); 101 (18~897) ❙ 49.9 (-34.3~246.5); 951 (603~1360); 280.7 (108.2~659.0)

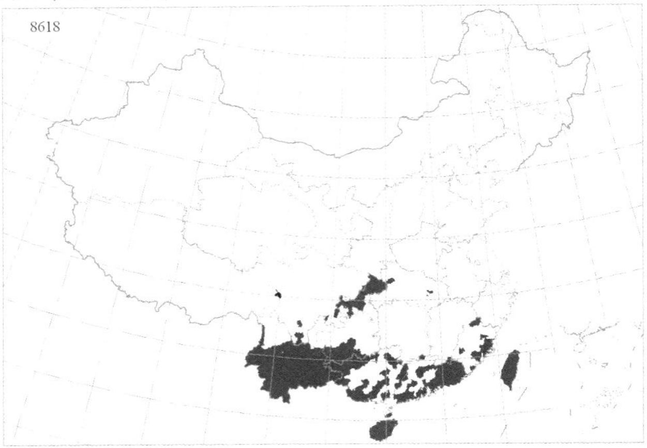

Sapotaceae 山榄科 *Chrysophyllum* 金叶树属
Chrysophyllum lanceolatum var. stellatocarpon P. Royen 金叶树
Trees

22.1 (19.5~24.8); 14.2 (10.6~20.1); 28.2 (26.7~29.1); 22.1 (19.6~24.8); 205.2 (175.1~237.6); 0.0 (0.0~0.0); 1175 (1003~1385) ❙ 1685 (868~2435); 769 (312~1428); 124 (41~194) ❙ 44.6 (-34.3~99.5); 1154 (868~1331); 237.7 (95.3~527.5)

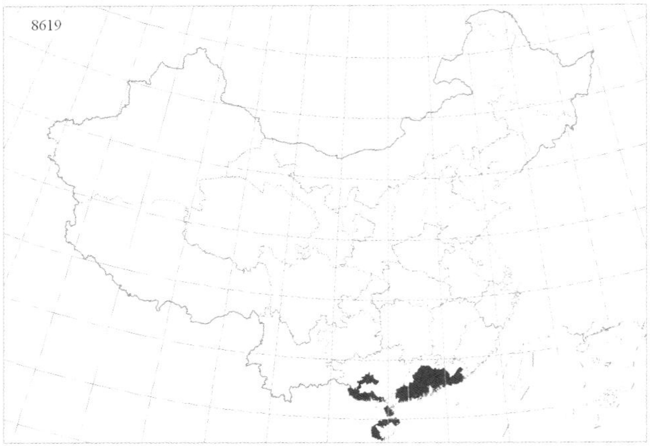

Sapotaceae 山榄科 *Diploknema* 藏榄属
Diploknema butyracea (Roxb.) H. J. Lam 藏榄
Trees

10.3; 2.4; 16.7; 10.4; 72.8; 9.0; 619 ❙ 1376; 794; 38 ❙ 119.8; 619; 348.4

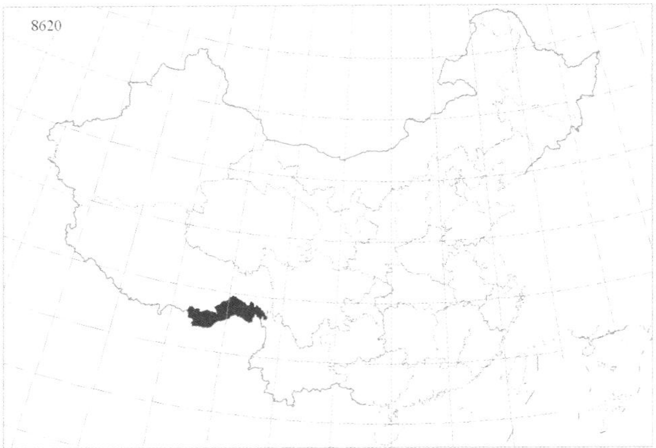

Sapotaceae 山榄科 *Diploknema* 藏榄属
Diploknema yunnanensis D. D. Tao, Z. H. Yang et Q. T. Zhang 云南藏榄
Trees

22.2; 13.9; 28.3; 22.3; 207.5; 0.0; 1190 ❙ 1374; 713; 87 ❙ 15.4; 1190; 327.7

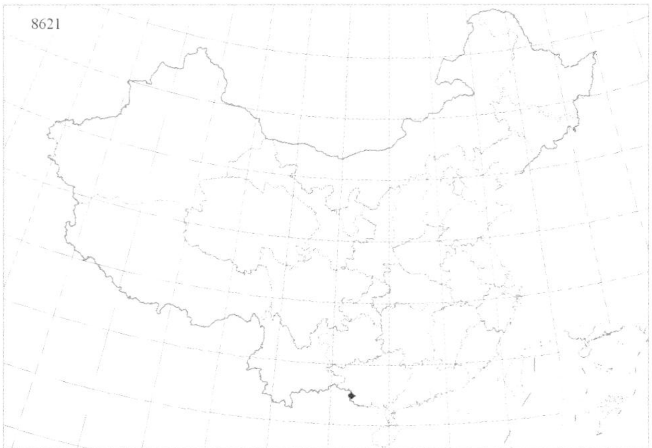

Sapotaceae 山榄科 *Eberhardtia* 梭子果属
Eberhardtia aurata (Pierre ex Dubard) Lecomte 锈毛梭子果
Trees, evergreen

19.7 (16.3~22.7); 11.8 (9.1~15.0); 25.5 (21.4~28.6); 19.8 (16.4~22.8); 177.2 (136.4~213.1); 0.0 (0.0~0.0); 1005 (797~1221) ❙ 1460 (1074~2435); 806 (582~1428); 76 (47~130) ❙ 45.6 (5.8~99.5); 1005 (797~1221); 339.9 (235.1~551.6)

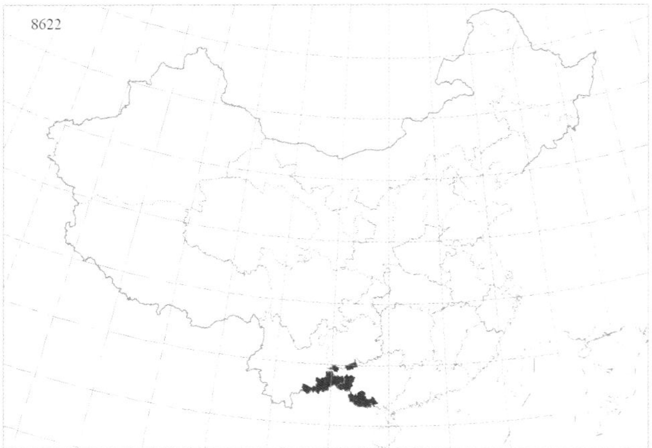

Sapotaceae 山榄科 *Eberhardtia* 梭子果属
Eberhardtia tonkinensis Lecomte 梭子果
Trees, evergreen

18.2 (16.3~21.0); 11.3 (9.1~14.2); 23.0 (21.4~25.6); 18.3 (16.4~21.0); 159.0 (136.4~191.8); 0.0 (0.0~0.0); 884 (797~1053) ❙ 1341 (1074~1751); 756 (582~996); 55 (47~77) ❙ 51.9 (21.7~88.8); 880 (797~1053); 410.3 (277.0~551.6)

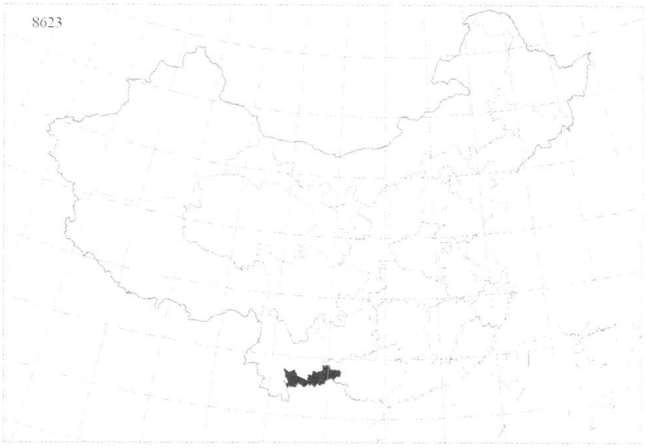

Sapotaceae 山榄科 *Madhuca* 紫荆木属
Madhuca hainanensis Chun et F. C. How 海南紫荆木
Trees, evergreen

23.9 (22.2~24.8); 18.5 (17.2~20.1); 27.8 (25.7~28.7); 24.0 (22.3~24.8); 227.8 (207.2~237.6); 0.0 (0.0~0.0); 1305 (1127~1385) ❙ 1457 (868~1854); 506 (312~622); 64 (41~115) ❙ 12.1 (-34.3~47.8); 1155 (868~1360); 479.8 (288.3~620.0)

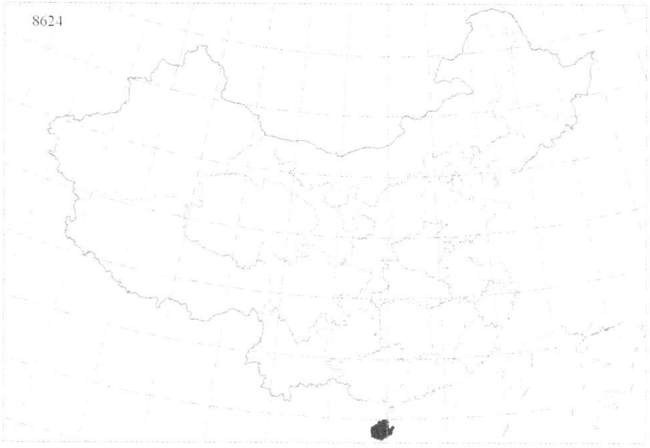

Sapotaceae 山榄科 *Madhuca* 紫荆木属
Madhuca pasquieri (Dubard) H. J. Lam 紫荆木
Trees, evergreen

20.4 (16.3~23.4); 12.5 (7.4~16.1); 26.3 (21.4~29.2); 20.4 (16.4~23.5); 184.8 (136.4~221.7); 0.0 (0.0~0.0); 1049 (797~1273) ❙ 1534 (1074~2435); 796 (582~1428); 98 (47~194) ❙ 47.0 (5.8~99.5); 1048 (797~1273); 308.5 (178.7~551.6)

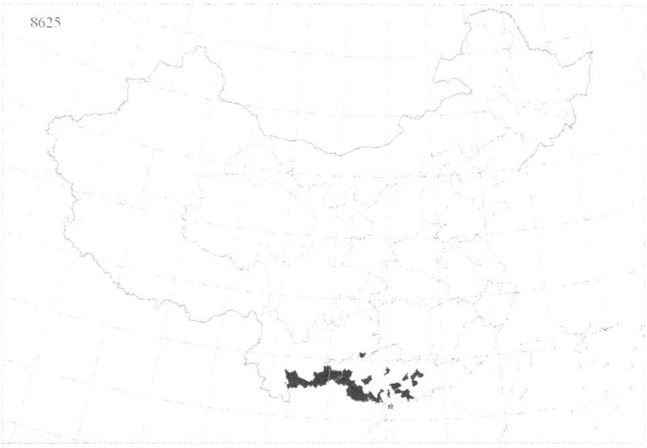

Sapotaceae 山榄科 *Manilkara* 铁线子属
Manilkara hexandra (Roxb.) Dubard 铁线子
Shrubs or trees

23.9 (22.8~24.8); 17.7 (14.7~20.1); 28.5 (27.8~29.0); 24.0 (22.8~24.8); 227.7 (213.6~237.6); 0.0 (0.0~0.0); 1314 (1230~1385) ❙ 1371 (868~1767); 556 (312~997); 71 (41~106) ❙ 4.7 (-34.3~39.9); 1147 (868~1299); 383.4 (199.6~548.0)

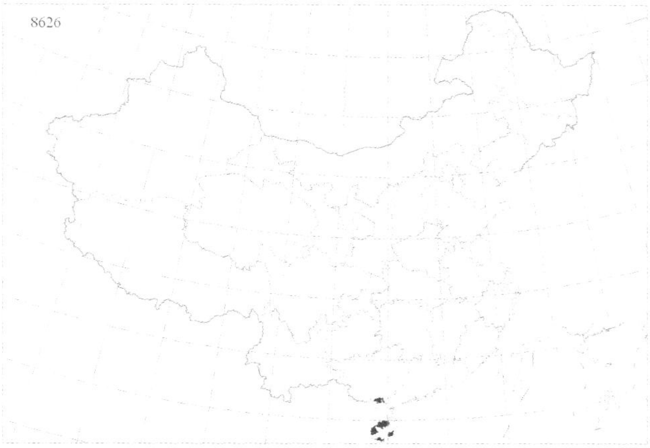

Sapotaceae 山榄科 *Palaquium* 胶木属
Palaquium formosanum Hayata 台湾胶木
Trees

20.3 (18.5~22.2); 14.6 (13.1~17.7); 25.2 (21.7~27.4); 20.3 (18.5~22.2); 183.7 (162.1~206.7); 0.0 (0.0~0.0); 1006 (863~1123) ❙ 2741 (2208~2993); 1151 (750~1847); 347 (107~551) ❙ 173.9 (125.7~214.4); 1006 (863~1123); 284.2 (238.5~363.7)

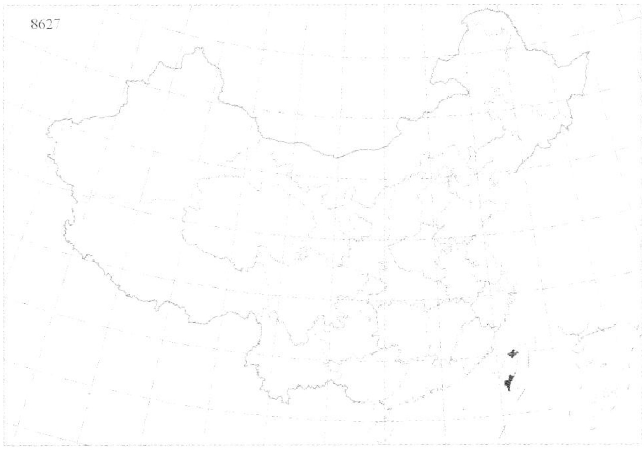

Sapotaceae 山榄科 *Planchonella* 山榄属
Planchonella clemensii (Lecomte) P. Royen 狭叶山榄
Trees, evergreen

24.0 (22.2~24.8); 18.4 (17.2~20.1); 28.0 (25.7~28.8); 24.0 (22.3~24.8); 228.3 (207.2~237.6); 0.0 (0.0~0.0); 1310 (1127~1385) ❙ 1501 (868~1854); 534 (312~712); 74 (41~115) ❙ 15.0 (-34.3~47.8); 1191 (868~1360); 440.3 (248.2~620.0)

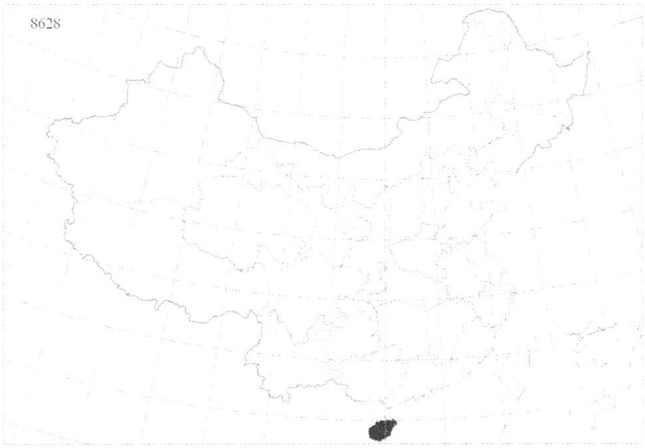

Sapotaceae 山榄科 *Planchonella* 山榄属

Planchonella obovata (R. Br.) Pierre 山榄

Shrubs or trees, evergreen

20.7 (16.2~24.7); 15.0 (11.0~19.8); 25.4 (20.1~29.2); 20.8 (16.2~24.7); 189.3 (134.2~236.5); 0.0 (0.0~0.0); 1061 (773~1376) ❙ 2297 (1425~3495); 1039 (534~1847); 232 (47~897) ❙ 129.6 (11.9~246.5); 1019 (773~1299); 266.1 (119.4~548.0)

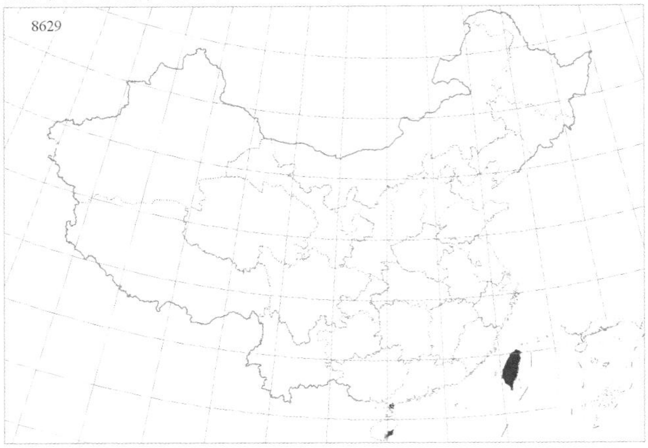

Sapotaceae 山榄科 *Pouteria* 桃榄属

Pouteria annamensis (Pierre) Baehni 桃榄

Trees, evergreen

23.6 (21.5~24.8); 17.5 (13.6~20.1); 28.1 (25.7~29.2); 23.7 (21.6~24.8); 224.1 (199.1~237.6); 0.0 (0.0~0.0); 1285 (1127~1385) ❙ 1495 (868~1854); 584 (312~850); 78 (41~115) ❙ 16.7 (-34.3~47.8); 1195 (868~1360); 394.2 (178.7~620.0)

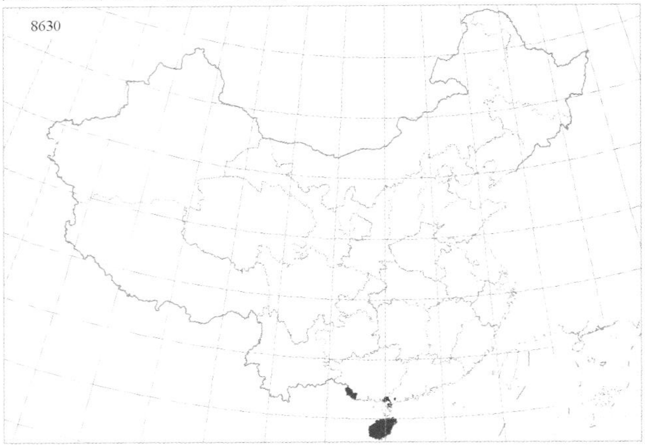

Sapotaceae 山榄科 *Pouteria* 桃榄属

Pouteria grandifolia (Wall.) Baehni 龙果

Trees, evergreen

20.0; 14.6; 23.2; 20.1; 180.8; 0.0; 963 ❙ 1523; 681; 67 ❙ 58.4; 942; 581.0

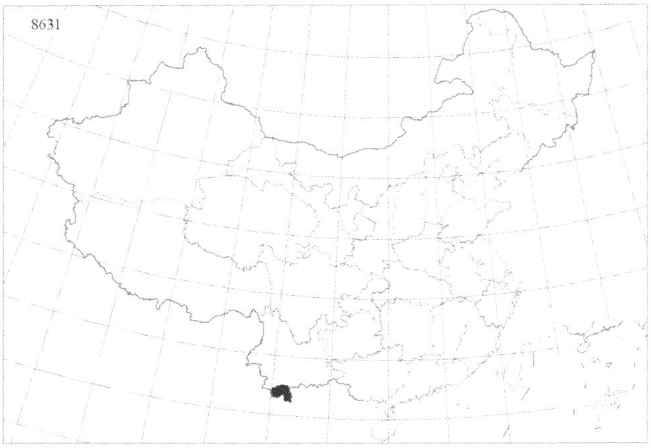

Sapotaceae 山榄科 *Sarcosperma* 肉实树属

Sarcosperma arboreum Buch.-Ham. ex C. B. Clarke 大肉实树

Trees

16.3 (5.4~21.8); 9.1 (-2.1~15.1); 21.7 (12.3~28.4); 16.4 (5.8~21.8); 137.0 (31.5~202.2); 0.5 (0.0~25.6); 820 (481~1162) ❙ 1154 (741~1751); 646 (411~1007); 47 (15~107) ❙ 41.5 (-14.2~145.8); 804 (481~1162); 336.1 (142.9~659.0)

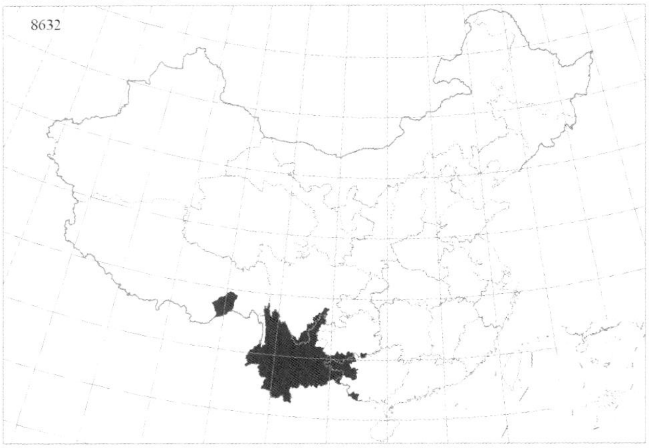

Sapotaceae 山榄科 *Sarcosperma* 肉实树属

Sarcosperma griffithii Hook. f. ex C. B. Clarke 小叶肉实树

Trees

20.0; 14.6; 23.2; 20.1; 180.8; 0.0; 963 ❙ 1523; 681; 67 ❙ 58.4; 942; 581.0

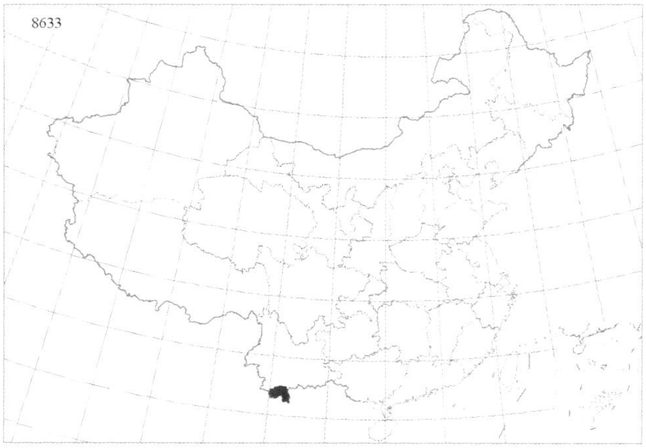

Sapotaceae 山榄科 *Sarcosperma* 肉实树属

Sarcosperma kachinense (King et Prain) Exell 绒毛肉实树

Trees

19.5 (16.3~24.7); 12.7 (9.1~19.8); 24.1 (20.3~28.6); 19.5 (16.4~24.7); 174.2 (136.4~236.5); 0.0 (0.0~0.0); 974 (789~1376) ❙ 1405 (1028~1767); 723 (534~996); 66 (41~194) ❙ 46.0 (5.8~88.8); 954 (780~1326); 410.6 (213.5~659.0)

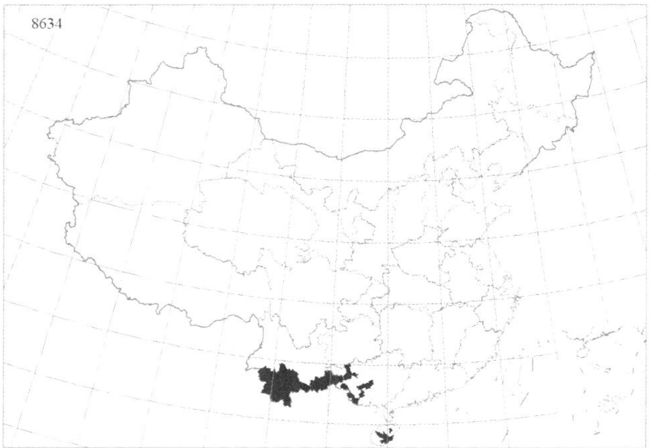

Sapotaceae 山榄科 *Sarcosperma* 肉实树属
Sarcosperma kachinense var. simondii (Gagnep.) H. J. Lam et P. Royen 光序肉实树
Trees

18.4 (16.3~21.0); 11.7 (9.1~15.1); 23.0 (21.4~25.6); 18.4 (16.4~21.0); 161.3 (136.4~191.8); 0.0 (0.0~0.0); 892 (797~1053) ❙ 1372 (1074~1751); 742 (582~996); 57 (47~77) ❙ 53.7 (32.4~88.8); 888 (797~1053); 443.9 (277.0~659.0)

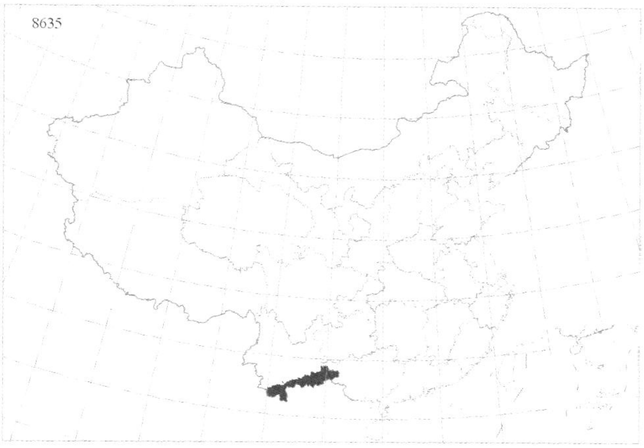

Sapotaceae 山榄科 *Sarcosperma* 肉实树属
Sarcosperma laurinum (Benth.) Hook. f. 肉实树
Trees

21.6 (16.5~24.8); 13.7 (5.3~20.1); 28.0 (25.0~29.2); 21.7 (16.5~24.8); 200.0 (138.0~237.6); 0.0 (0.0~0.0); 1147 (862~1385) ❙ 1643 (868~2435); 707 (312~1428); 129 (41~235) ❙ 44.9 (-34.3~100.8); 1129 (862~1360); 257.6 (95.3~620.0)

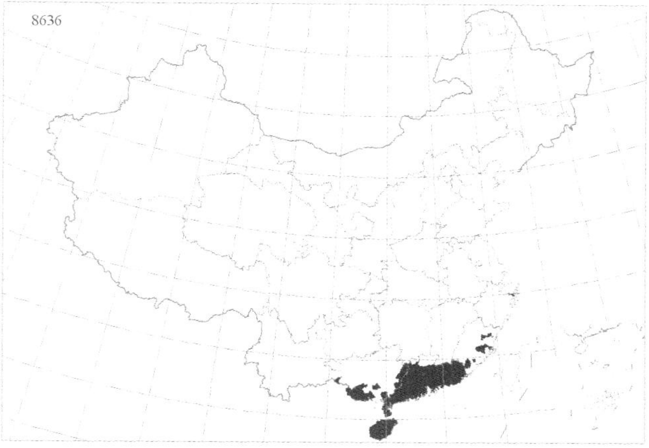

Sapotaceae 山榄科 *Sinosideroxylon* 铁榄属
Sinosideroxylon pedunculatum (Hemsl.) H. Chuang 铁榄
Trees

18.8 (15.5~22.2); 9.6 (4.2~14.2); 26.3 (21.4~29.0); 18.9 (15.6~22.3); 166.5 (123.1~207.5); 0.0 (0.0~0.8); 970 (797~1190) ❙ 1447 (1074~1777); 664 (419~860); 118 (43~198) ❙ 49.9 (15.4~85.5); 969 (797~1190); 270.7 (160.7~551.6)

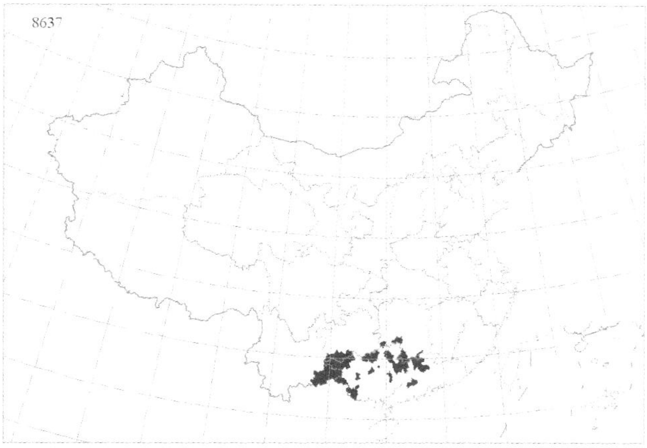

Sapotaceae 山榄科 *Sinosideroxylon* 铁榄属
Sinosideroxylon pedunculatum var. pubifolium H. Chuang 毛叶铁榄
Trees

22.0; 13.8; 28.0; 22.1; 204.8; 0.0; 1172 ❙ 1338; 691; 87 ❙ 14.1; 1172; 315.8

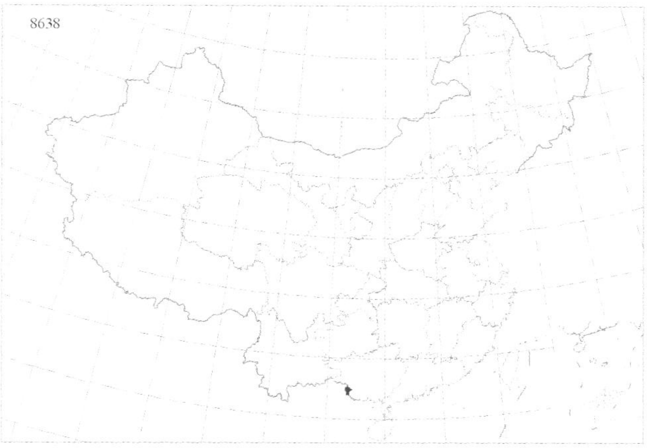

Sapotaceae 山榄科 *Sinosideroxylon* 铁榄属
Sinosideroxylon wightianum (Hook. et Arn.) Aubrév. 革叶铁榄
Trees

20.6 (15.5~24.1); 12.2 (4.8~18.0); 27.5 (22.4~29.1); 20.7 (15.5~24.2); 188.0 (126.5~230.2); 0.0 (0.0~0.2); 1080 (820~1329) ❙ 1630 (1161~2435); 732 (444~1428); 131 (43~230) ❙ 51.7 (-2.2~104.6); 1078 (820~1329); 250.3 (95.3~551.6)

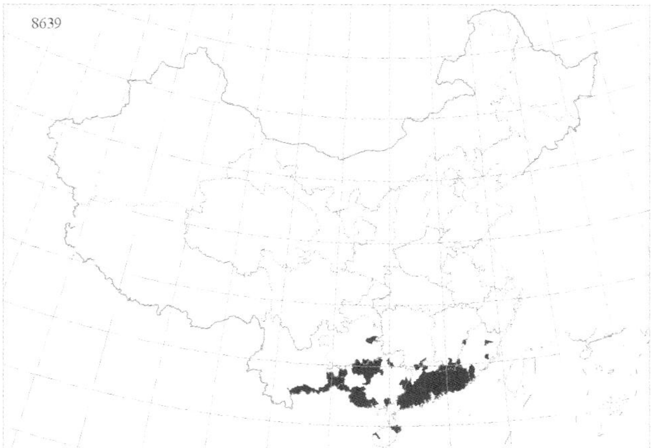

Sapotaceae 山榄科 *Sinosideroxylon* 铁榄属
Sinosideroxylon yunnanense (C. Y. Wu) H. Chuang 滇铁榄
Trees

17.8 (16.3~21.0); 10.1 (9.1~14.2); 23.4 (21.4~25.6); 17.9 (16.4~21.0); 154.5 (136.4~191.8); 0.0 (0.0~0.0); 878 (797~1053) ❙ 1298 (1074~1522); 719 (582~860); 53 (47~68) ❙ 48.1 (32.4~80.6); 878 (797~1053); 362.3 (277.0~551.6)

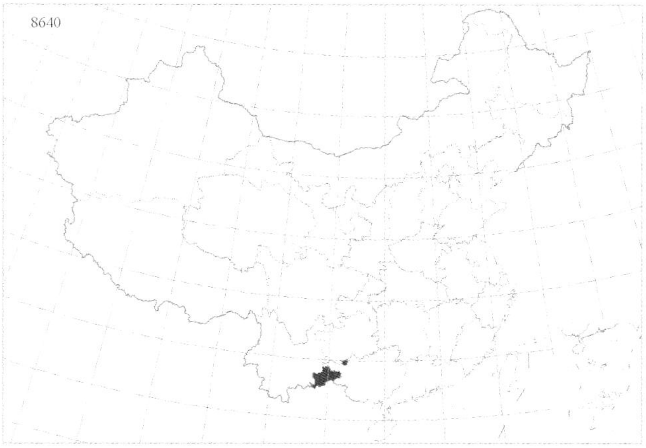

Sapotaceae 山榄科 *Xantolis* 刺榄属
Xantolis boniana var. rostrata (Merr.) P. Royen 喙果刺榄
Trees

21.1; 15.7; 24.7; 21.2; 194.0; 0.0; 1092 ❙ 1538; 637; 62 ❙ 44.0; 1036; 524.8

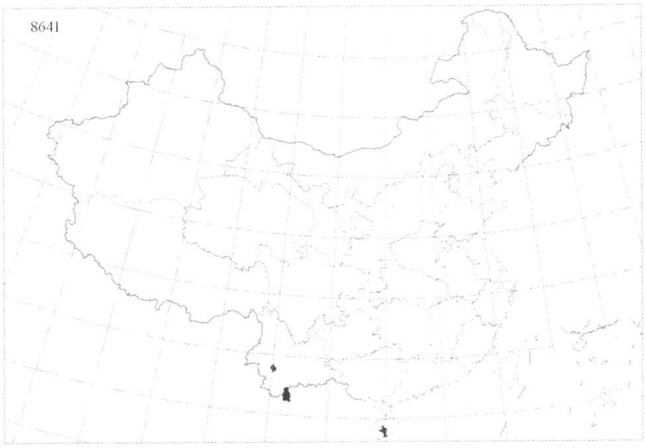

Sapotaceae 山榄科 *Xantolis* 刺榄属
Xantolis longispinosa (Merr.) H. S. Lo 琼刺榄
Shrubs or trees

23.6 (21.8~24.8); 17.6 (13.7~20.1); 28.0 (25.7~29.2); 23.7 (21.8~24.8); 224.2 (202.2~237.6); 0.0 (0.0~0.0); 1285 (1127~1385) ❙ 1484 (868~1854); 573 (312~824); 73 (41~115) ❙ 15.9 (-34.3~47.8); 1178 (868~1360); 409.0 (178.7~620.0)

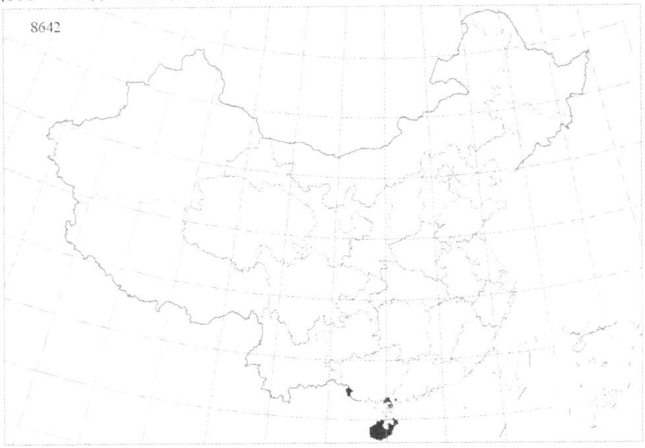

Sapotaceae 山榄科 *Xantolis* 刺榄属
Xantolis shweliensis (W. W. Sm.) P. Royen 瑞丽刺榄
Shrubs

19.5; 12.7; 23.3; 19.5; 173.8; 0.0; 942 ❙ 1559; 907; 42 ❙ 65.4; 921; 482.8

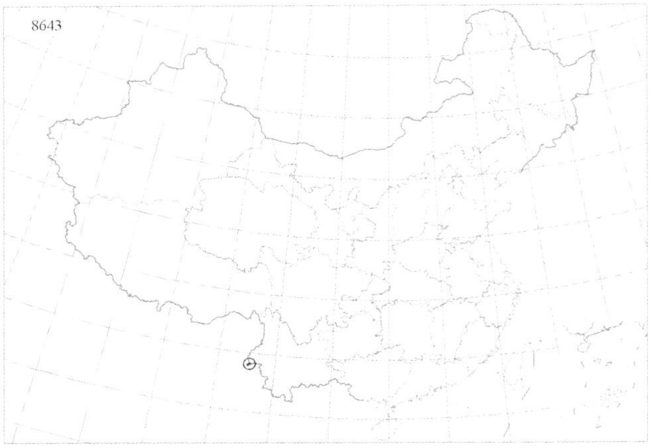

Sapotaceae 山榄科 *Xantolis* 刺榄属
Xantolis stenosepala (Hu) P. Royen 滇刺榄
Trees

20.0; 14.6; 23.2; 20.1; 180.8; 0.0; 963 ❙ 1523; 681; 67 ❙ 58.4; 942; 581.0

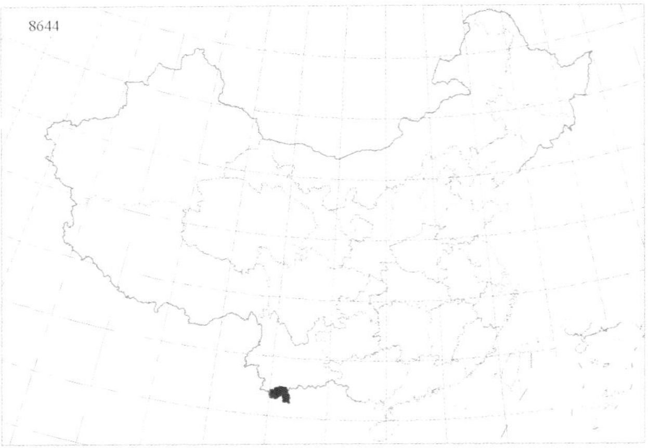

Sapotaceae 山榄科 *Xantolis* 刺榄属
Xantolis stenosepala var. brevistylis C. Y. Wu 短柱滇刺榄
Trees

20.0; 14.6; 23.2; 20.1; 180.8; 0.0; 963 ❙ 1523; 681; 67 ❙ 58.4; 942; 581.0

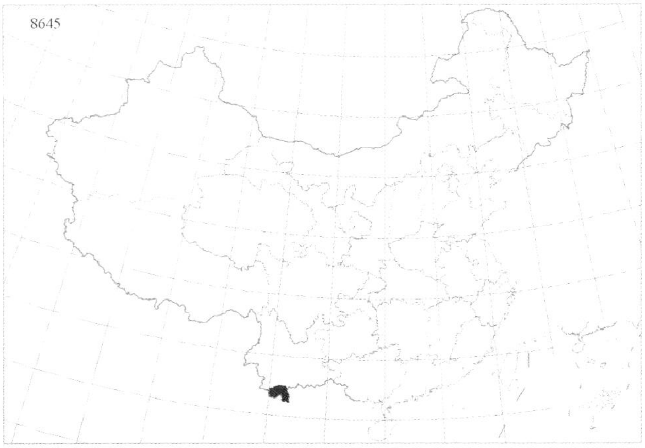

Ebenaceae 柿树科 *Diospyros* 柿属
Diospyros anisocalyx C. Y. Wu 异萼柿
Trees

20.0; 11.6; 26.2; 20.0; 180.3; 0.0; 1020 ❙ 1416; 774; 81 ❙ 39.6; 1020; 328.8

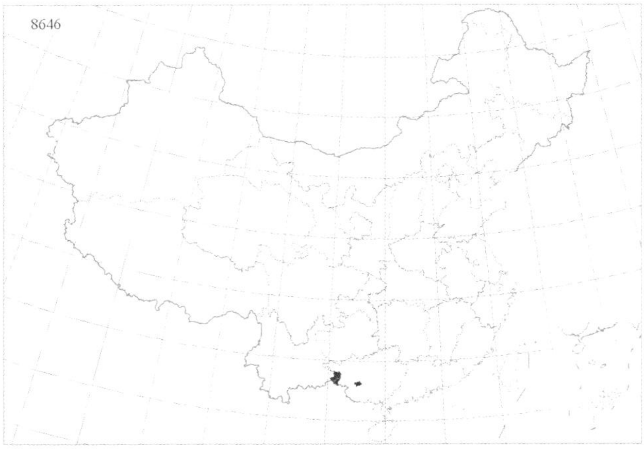

Ebenaceae 柿树科 *Diospyros* 柿属

Diospyros armata Hemsl. 瓶兰花

Trees, semievergreen or deciduous

12.9 (10.3~14.4); 1.4 (-0.1~2.6); 23.8 (19.8~25.8); 13.0 (10.3~14.4); 102.9 (75.8~116.7); 7.4 (3.6~11.7); 736 (634~793) ❙ 990 (711~1335); 444 (357~550); 55 (15~86) ❙ 34.2 (1.9~84.7); 736 (634~793); 258.2 (204.8~297.4)

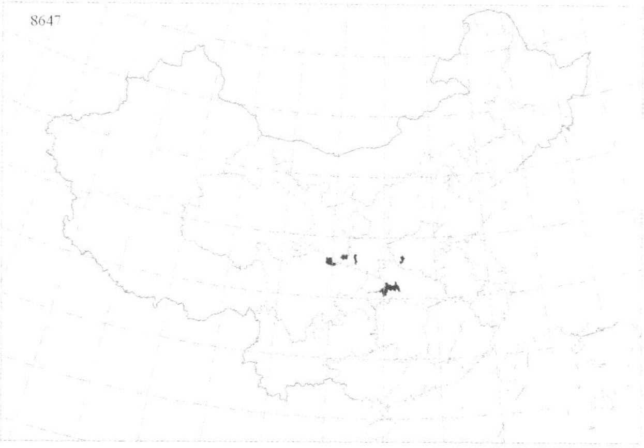

Ebenaceae 柿树科 *Diospyros* 柿属

Diospyros balfouriana Diels 大理柿

Shrubs or trees

13.6; 7.2; 18.8; 13.7; 104.1; 0.0; 706 ❙ 1042; 592; 46 ❙ 47.6; 706; 372.3

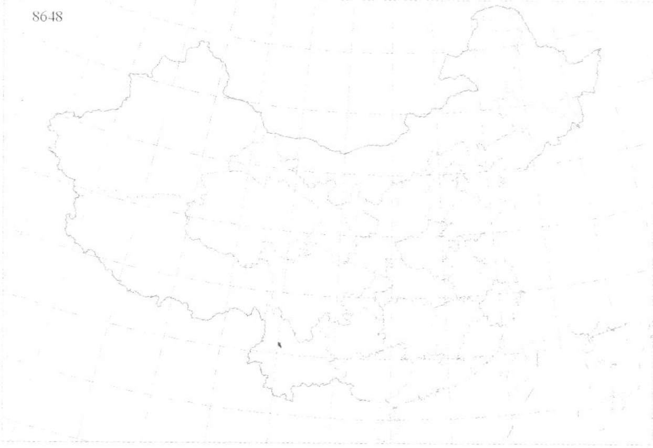

Ebenaceae 柿树科 *Diospyros* 柿属

Diospyros caloneura C. Y. Wu 美脉柿

Small trees

17.3; 11.0; 21.6; 17.3; 148.0; 0.0; 832 ❙ 1049; 598; 44 ❙ 26.1; 783; 365.3

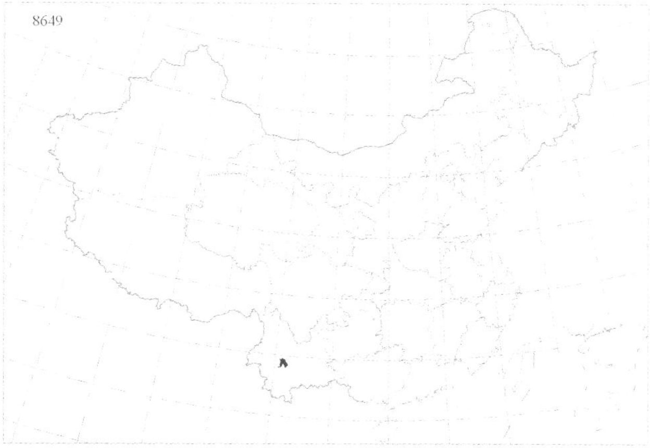

Ebenaceae 柿树科 *Diospyros* 柿属

Diospyros cathayensis Steward 乌柿

Small trees, evergreen or semievergreen

16.4 (5.7~21.6); 5.7 (-3.5~12.6); 26.2 (14.2~29.3); 16.4 (6.4~21.7); 138.2 (38.1~199.8); 1.4 (0.0~28.1); 875 (504~1153) ❙ 1277 (711~1829); 542 (357~821); 103 (15~231) ❙ 47.3 (8.0~112.2); 872 (504~1153); 191.1 (108.2~284.3)

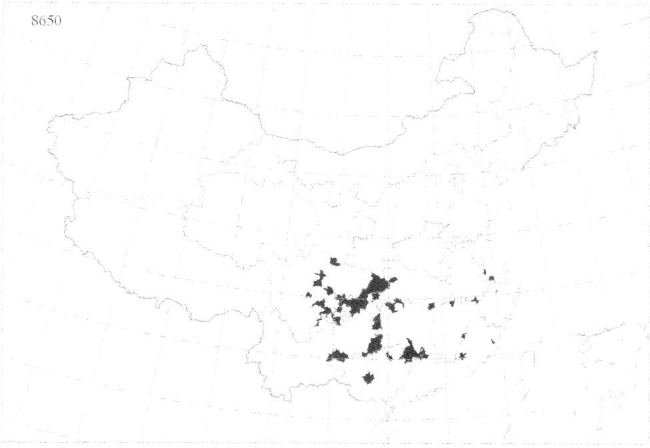

Ebenaceae 柿树科 *Diospyros* 柿属

Diospyros chunii F. P. Metcalf et L. Chen 崖柿

Shrubs or small trees

21.4; 13.3; 28.0; 21.4; 196.8; 0.0; 1149 ❙ 1626; 590; 139 ❙ 43.3; 1066; 331.0

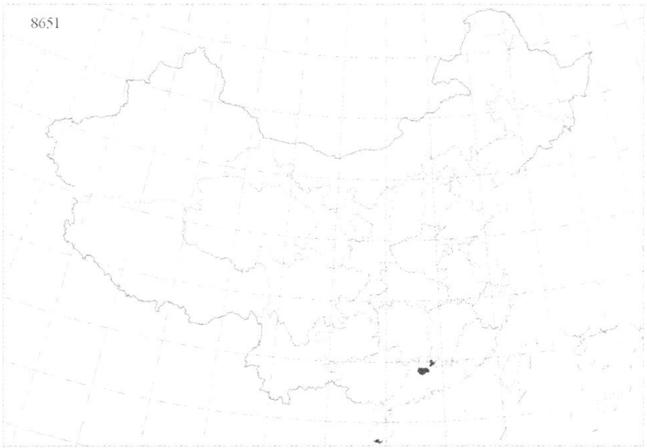

Ebenaceae 柿树科 *Diospyros* 柿属

Diospyros corallina Chun et L. Chen 五蒂柿

Trees

24.8; 20.1; 28.0; 24.8; 237.6; 0.0; 1385 ❙ 1405; 473; 41 ❙ 1.5; 1137; 527.5

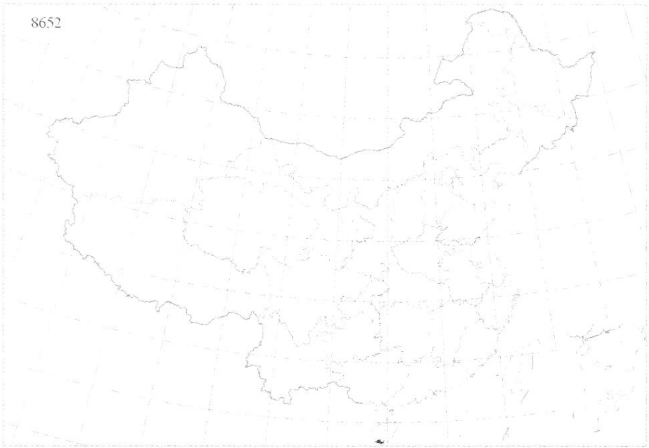

Ebenaceae 柿树科 *Diospyros* 柿属

Diospyros diversilimba Merr. et Chun 光叶柿

Shrubs or trees

22.6 (18.2~24.8); 15.4 (8.7~20.1); 28.1 (25.5~29.0); 22.7 (18.3~24.8); 212.0 (159.3~237.6); 0.0 (0.0~0.0); 1221 (924~1385) ▎1441 (868~1813); 635 (312~997); 90 (41~194) ▎19.7 (-34.3~53.1); 1132 (868~1329); 316.0 (194.5~527.5)

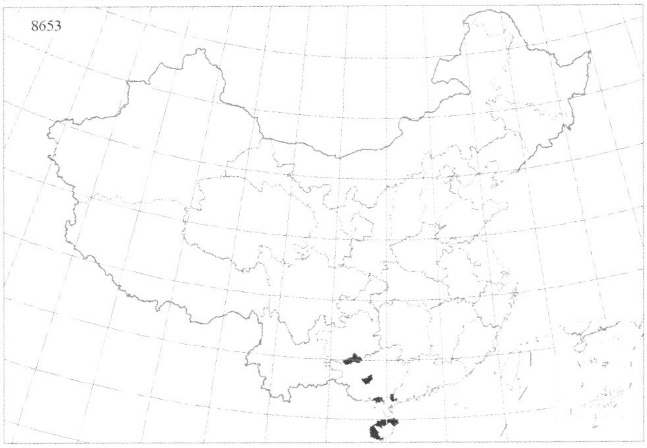

Ebenaceae 柿树科 *Diospyros* 柿属

Diospyros dumetorum W. W. Sm. 岩柿

Trees

16.5 (2.2~22.3); 7.7 (-6.6~14.0); 23.7 (9.9~29.0); 16.6 (3.8~22.3); 140.2 (15.5~208.1); 1.8 (0.0~49.1); 865 (413~1202) ▎1240 (797~2148); 633 (411~1234); 75 (12~194) ▎45.3 (-14.2~125.9); 856 (413~1202); 268.8 (135.1~551.3)

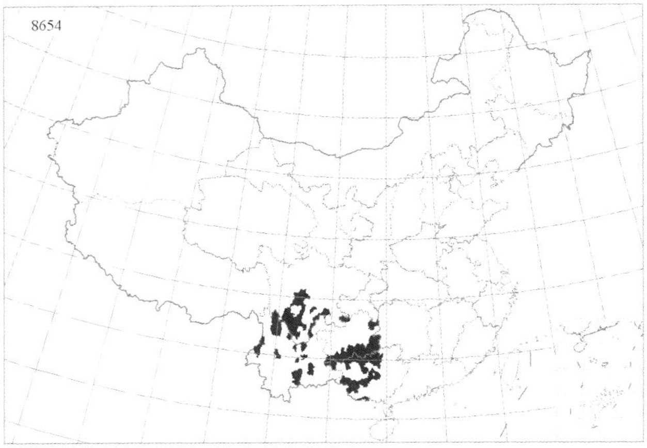

Ebenaceae 柿树科 *Diospyros* 柿属

Diospyros ehretioides Wall. ex A. DC. 红枝柿

Trees

24.0 (23.1~24.8); 18.5 (17.5~20.1); 27.9 (27.2~28.7); 24.0 (23.2~24.8); 228.3 (218.4~237.6); 0.0 (0.0~0.0); 1310 (1232~1385) ▎1286 (868~1551); 451 (312~544); 51 (41~71) ▎-1.5 (-34.3~20.7); 1087 (868~1147); 467.0 (288.3~620.0)

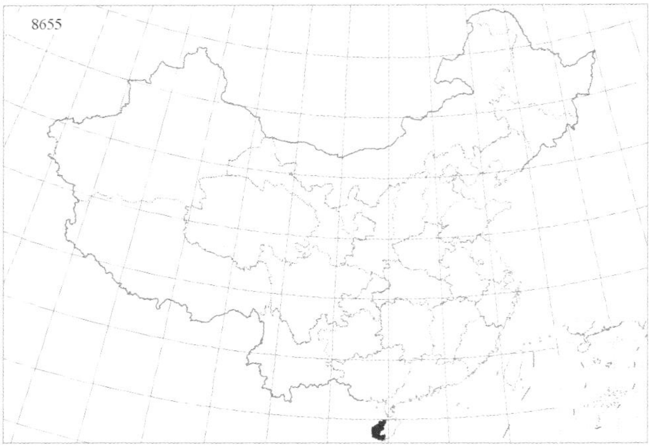

Ebenaceae 柿树科 *Diospyros* 柿属

Diospyros eriantha Champ. ex Benth. 乌材

Shrubs or trees, evergreen

21.3 (16.2~24.8); 13.6 (7.4~20.1); 27.5 (20.1~29.1); 21.3 (16.2~24.8); 195.9 (134.2~237.6); 0.0 (0.0~0.0); 1119 (773~1385) ▎1760 (868~3495); 778 (312~1847); 148 (41~897) ▎61.1 (-34.3~246.5); 1100 (773~1331); 251.6 (95.3~620.0)

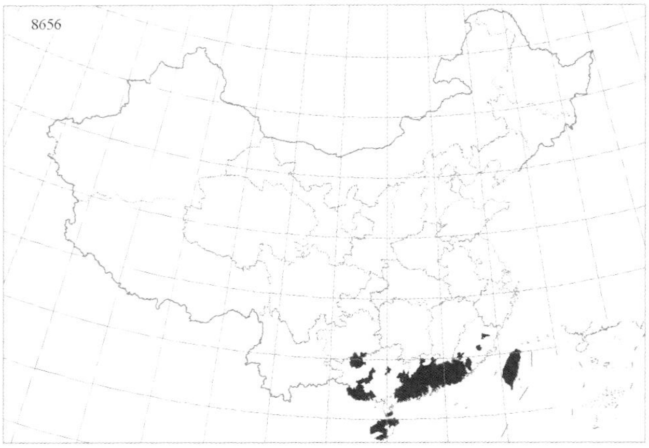

Ebenaceae 柿树科 *Diospyros* 柿属

Diospyros esquirolii H. Lév. 贵阳柿

Trees

15.6; 5.5; 24.6; 15.6; 127.5; 0.0; 808 ▎1206; 546; 69 ▎49.2; 808; 189.4

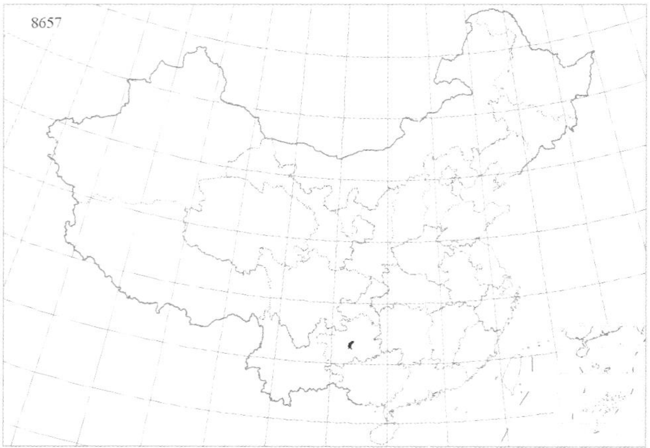

Ebenaceae 柿树科 *Diospyros* 柿属

Diospyros fanjingshanica S. K. Lee 梵净山柿

Shrubs or small trees

15.0; 4.0; 25.7; 15.1; 122.0; 1.1; 806 ▎1247; 484; 89 ▎54.7; 806; 189.9

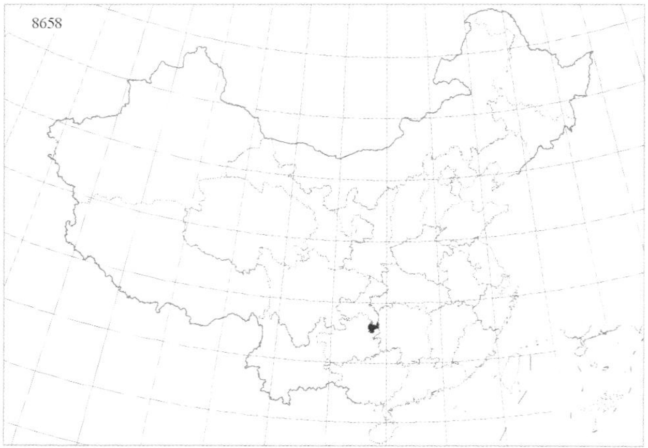

Ebenaceae 柿树科 *Diospyros* 柿属
Diospyros ferrea (Willd.) Bakh. 象牙树
Trees, evergreen

20.2; 15.1; 24.3; 20.2; 182.6; 0.0; 991 | 2899; 1386; 281 | 194.8; 991; 307.3

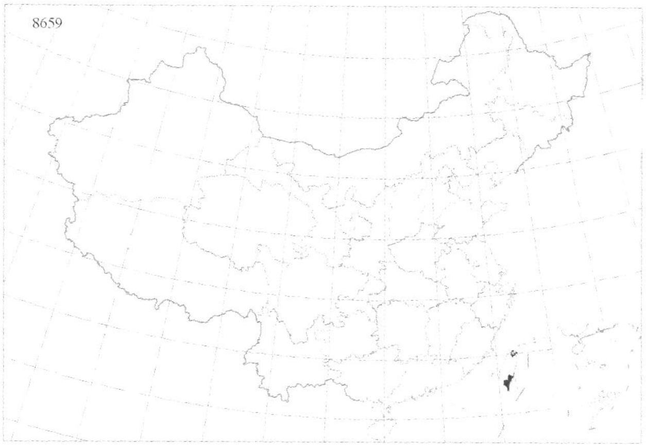

Ebenaceae 柿树科 *Diospyros* 柿属
Diospyros forrestii J. Anthony 腾冲柿
Trees

13.6 (8.6~16.2); 6.7 (1.3~9.4); 18.7 (15.0~20.8); 13.6 (8.7~16.3); 104.3 (52.6~135.1); 1.2 (0.0~8.8); 709 (574~790) | 1245 (1041~1454); 685 (585~792); 60 (45~70) | 76.5 (50.8~99.9); 709 (574~790); 384.3 (335.1~487.2)

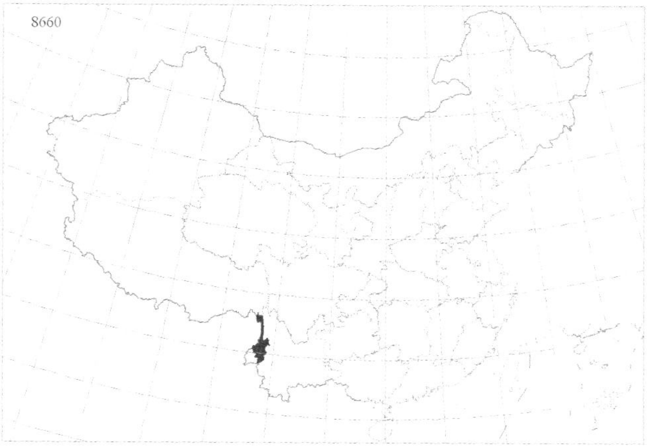

Ebenaceae 柿树科 *Diospyros* 柿属
Diospyros hainanensis Merr. 海南柿
Trees

23.7 (21.5~24.8); 18.1 (13.6~20.1); 27.6 (26.7~28.5); 23.7 (21.6~24.8); 224.4 (199.1~237.6); 0.0 (0.0~0.0); 1286 (1131~1385) | 1599 (1405~1854); 596 (473~850); 70 (41~115) | 25.0 (1.5~41.8); 1185 (1131~1360); 494.8 (281.7~620.0)

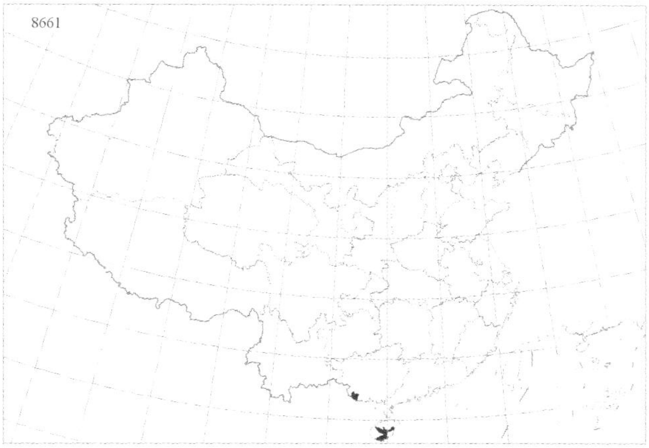

Ebenaceae 柿树科 *Diospyros* 柿属
Diospyros hasseltii Zoll. 黑毛柿
Shrubs or small trees

20.4; 15.1; 23.6; 20.5; 185.8; 0.0; 989 | 1607; 702; 77 | 62.5; 986; 659.0

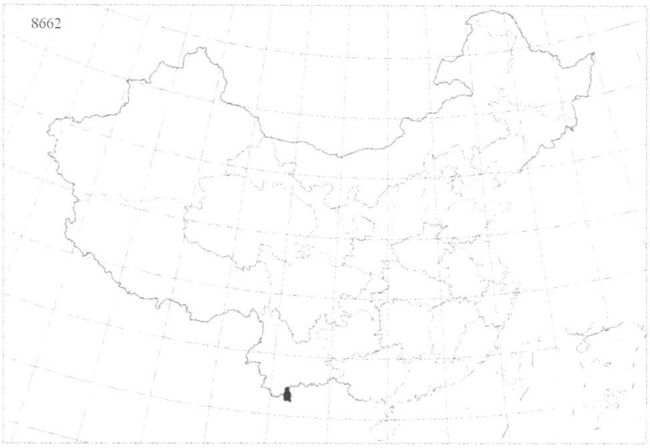

Ebenaceae 柿树科 *Diospyros* 柿属
Diospyros hexamera C. Y. Wu 六花柿
Trees

21.0; 14.2; 25.6; 21.0; 191.8; 0.0; 1053 | 1522; 818; 68 | 44.5; 1053; 551.6

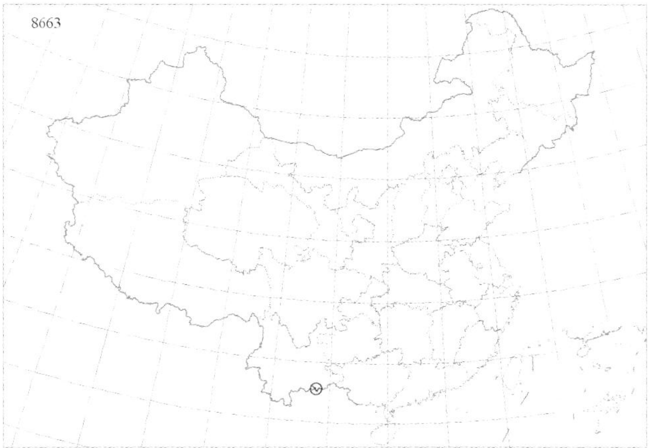

Ebenaceae 柿树科 *Diospyros* 柿属
Diospyros howii Merr. et Chun 琼南柿
Trees

23.6; 18.3; 27.2; 23.6; 223.5; 0.0; 1279 | 1516; 591; 54 | 20.1; 1138; 548.5

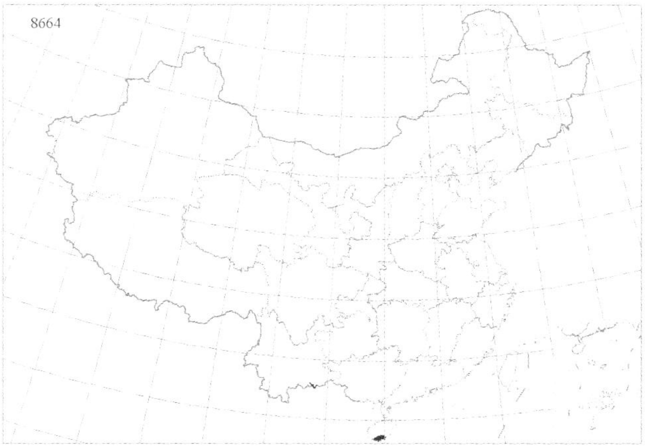

Ebenaceae 柿树科 *Diospyros* 柿属
Diospyros inflata Merr. et Chun 囊萼柿
Trees

23.9；18.8；27.5；24.0；227.6；0.0；1304 ❙ 1424；491；47 ❙ 9.6；1130；507.0

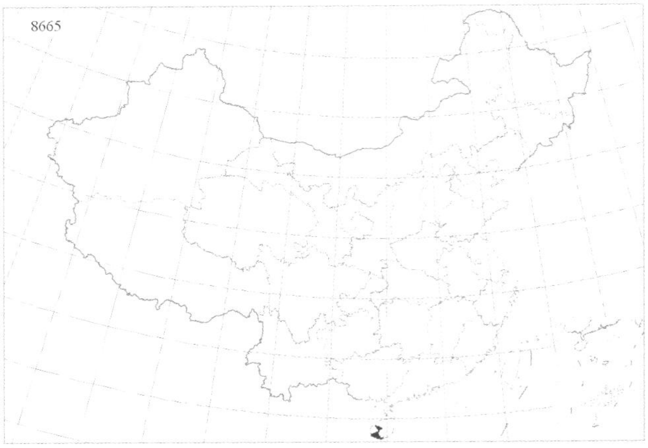

Ebenaceae 柿树科 *Diospyros* 柿属
Diospyros japonica Siebold et Zucc. 山柿
Trees, deciduous

16.6 (12.6~21.0)；5.0 (1.6~11.9)；27.6 (23.2~29.9)；16.7 (12.7~21.1)；140.5 (98.8~193.0)；0.6 (0.0~6.9)；895 (711~1114) ❙ 1478 (992~1922)；517 (392~821)；176 (43~259) ❙ 65.7 (10.9~129.9)；883 (711~1114)；209.9 (131.2~354.5)

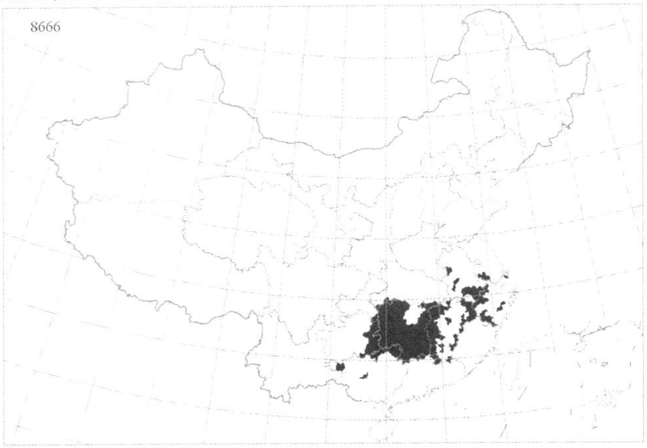

Ebenaceae 柿树科 *Diospyros* 柿属
Diospyros kaki Thunb. 柿
Trees, deciduous

17.0 (2.9~24.8)；6.5 (-5.7~20.1)；26.5 (10.6~30.1)；17.1 (4.3~24.8)；147.4 (19.9~237.6)；2.4 (0.0~45.2)；921 (421~1385) ❙ 1332 (520~2435)；559 (258~1428)；119 (10~259) ❙ 45.0 (-35.9~147.0)；897 (421~1360)；223.8 (95.3~659.0)

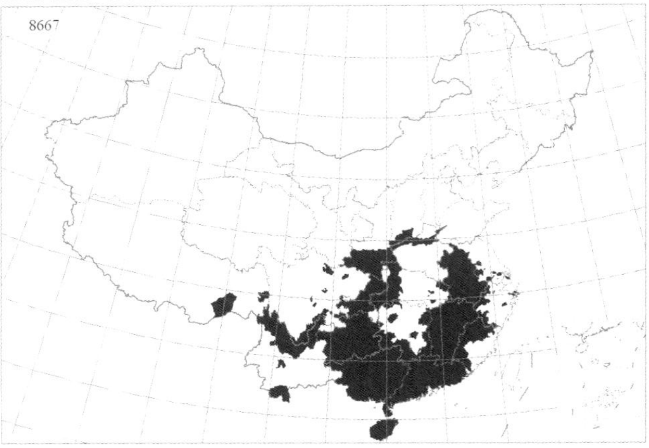

Ebenaceae 柿树科 *Diospyros* 柿属
Diospyros kaki var. silvestris Makino 野柿
Trees, deciduous

15.9 (5.4~21.6)；5.2 (-4.2~15.1)；25.5 (12.3~29.9)；15.9 (5.8~21.6)；133.2 (31.5~199.2)；2.4 (0.0~29.4)；850 (481~1139) ❙ 1235 (604~2005)；543 (332~996)；110 (15~259) ❙ 44.8 (-16.6~128.9)；838 (481~1139)；231.2 (110.1~659.0)

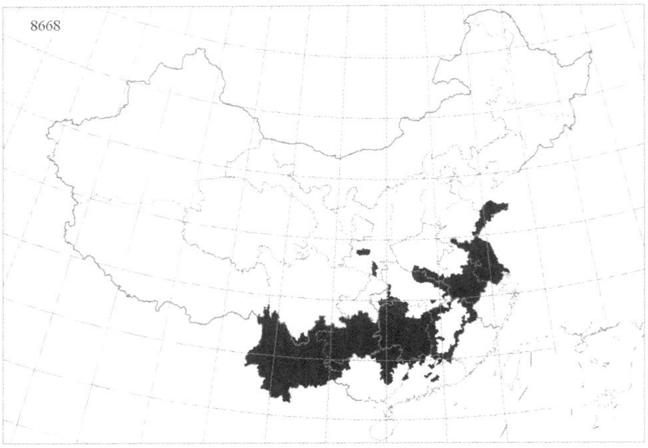

Ebenaceae 柿树科 *Diospyros* 柿属
Diospyros kerrii Craib 傣柿
Trees

19.7；14.1；23.0；19.7；176.5；0.0；944 ❙ 1529；738；66 ❙ 62.3；928；573.6

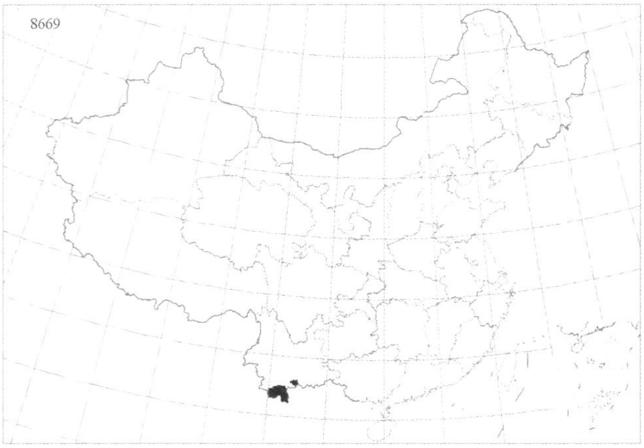

Ebenaceae 柿树科 *Diospyros* 柿属
Diospyros kintungensis C. Y. Wu 景东君迁子
Trees

17.3；11.0；21.6；17.3；148.0；0.0；832 ❙ 1049；598；44 ❙ 26.1；783；365.3

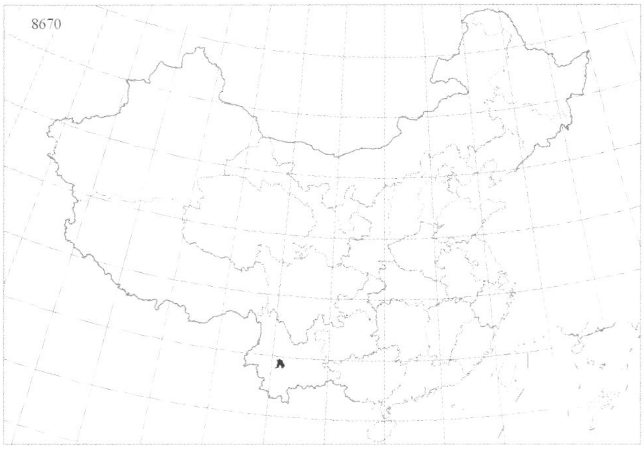

Ebenaceae 柿树科 *Diospyros* 柿属
Diospyros kotcensis T. Yamaz. 兰屿柿
Trees

18.5；14.1；21.7；18.5；162.1；0.0；863 | 2713；1388；188 | 214.4；863；363.7

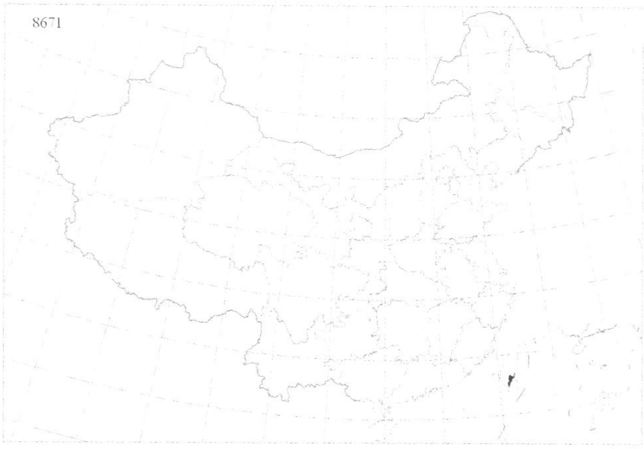

Ebenaceae 柿树科 *Diospyros* 柿属
Diospyros longibracteata Lecomte 长苞柿
Trees

23.3 (18.1~24.8)；17.7 (9.5~20.1)；27.4 (24.6~28.7)；23.4 (18.2~24.8)；220.6 (158.1~237.6)；0.0 (0.0~0.0)；1265 (901~1385) | 1420 (1176~1725)；524 (392~723)；53 (41~68) | 13.9 (-12.0~43.5)；1112 (901~1212)；488.0 (316.1~620.0)

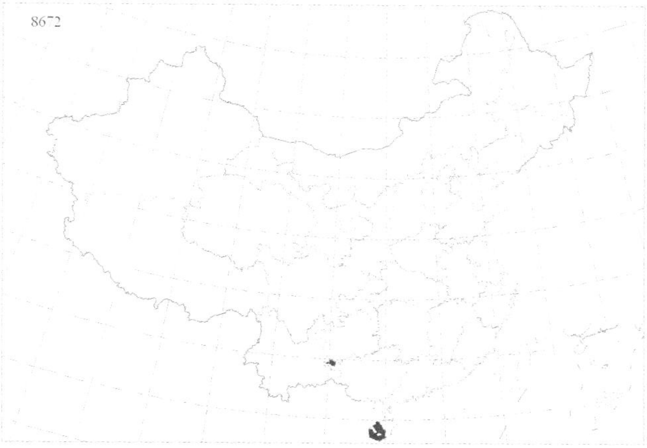

Ebenaceae 柿树科 *Diospyros* 柿属
Diospyros longshengensis S. K. Lee 龙胜柿
Shrubs

16.4；5.7；25.9；16.4；137.4；0.1；857 | 1409；560；136 | 65.1；857；229.4

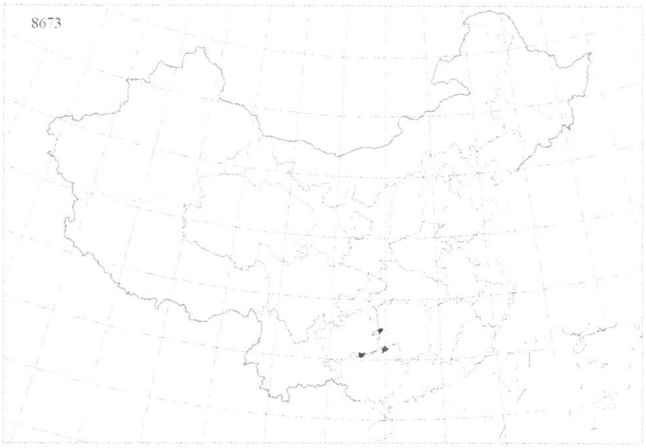

Ebenaceae 柿树科 *Diospyros* 柿属
Diospyros lotus L. 君迁子
Trees, deciduous

14.2 (2.0~21.6)；1.7 (-10.7~14.2)；25.7 (9.9~29.9)；14.4 (3.8~21.7)；120.0 (15.4~199.8)；8.9 (0.0~54.8)；815 (413~1153) | 984 (428~1922)；470 (258~1007)；71 (10~259) | 20.9 (-42.3~145.8)；744 (413~1153)；200.1 (84.7~551.6)

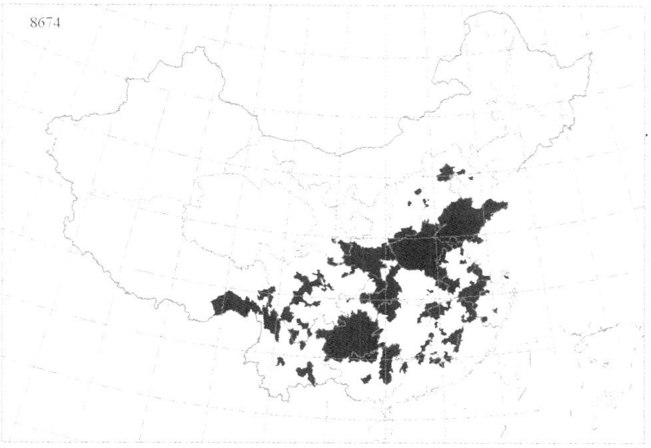

Ebenaceae 柿树科 *Diospyros* 柿属
Diospyros lotus var. mollissima C. Y. Wu 多毛君迁子
Trees, deciduous

13.4 (9.8~21.0)；1.3 (-4.9~14.2)；24.0 (19.3~27.1)；13.6 (10.0~21.0)；113.3 (76.4~191.8)；12.7 (0.0~26.3)；769 (630~1053) | 785 (520~1547)；415 (258~911)；28 (17~68) | 0.8 (-35.9~74.2)；656 (520~1053)；253.3 (128.6~551.6)

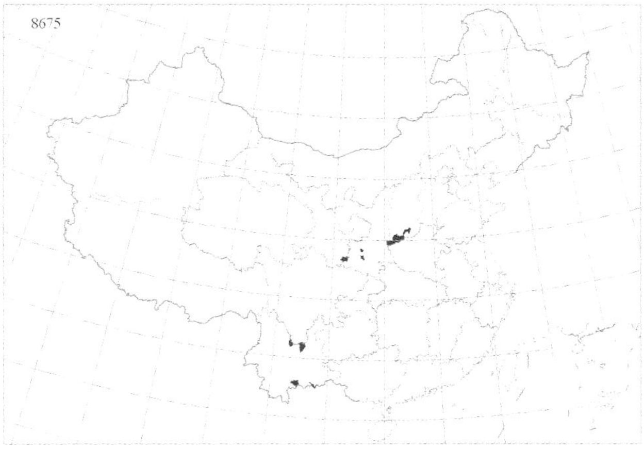

Ebenaceae 柿树科 *Diospyros* 柿属
Diospyros maclurei Merr. 琼岛柿
Trees

24.1；19.2；27.5；24.2；229.8；0.0；1320 | 1567；539；54 | 19.4；1165；553.8

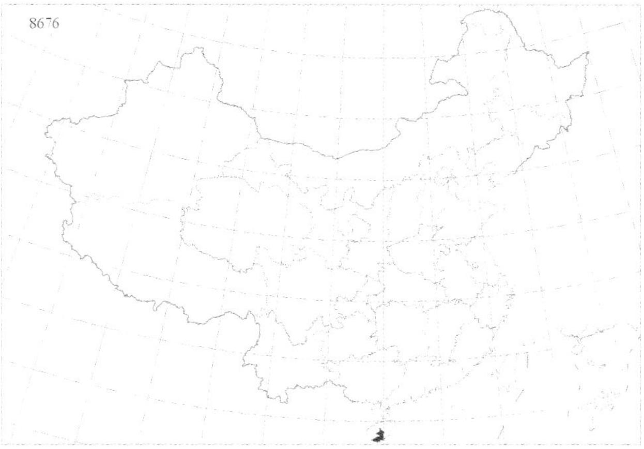

Ebenaceae 柿树科 *Diospyros* 柿属
Diospyros maritima Blume 海边柿
Small trees, evergreen

19.9 (17.8~22.2); 14.3 (12.1~17.7); 24.7 (21.7~27.4); 19.9 (17.9~22.2); 179.0 (154.4~206.7); 0.0 (0.0~0.0); 980 (852~1123) ❙ 2871 (2713~2993); 1192 (844~1847); 359 (107~551) ❙ 195.9 (159.6~235.6); 980 (852~1123); 285.3 (238.5~363.7)

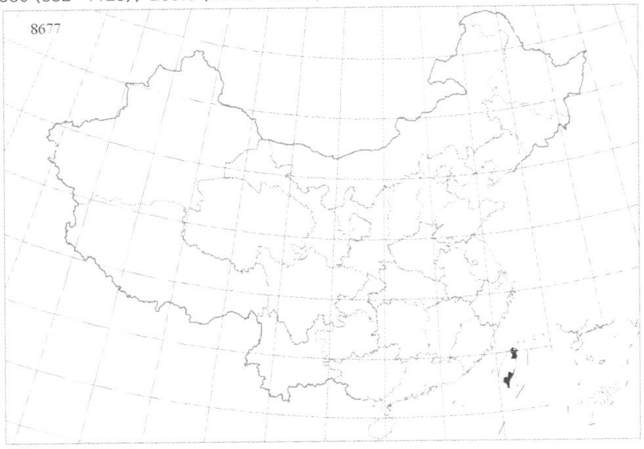

Ebenaceae 柿树科 *Diospyros* 柿属
Diospyros metcalfii Chun et L. Chen 圆萼柿
Trees

24.8; 20.1; 28.0; 24.8; 237.6; 0.0; 1385 ❙ 1405; 473; 41 ❙ 1.5; 1137; 527.5

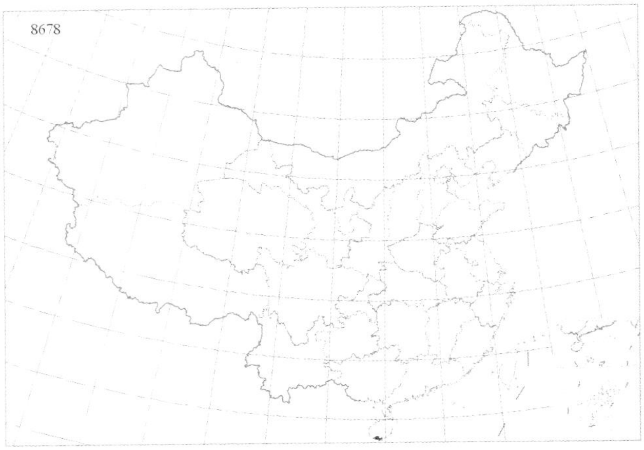

Ebenaceae 柿树科 *Diospyros* 柿属
Diospyros miaoshanica S. K. Lee 苗山柿
Shrubs or small trees

18.3 (15.0~22.7); 8.5 (3.8~15.0); 26.7 (24.6~28.6); 18.3 (15.1~22.8); 160.4 (122.0~213.1); 0.3 (0.0~1.2); 962 (810~1221) ❙ 1543 (1260~2435); 720 (450~1428); 121 (58~185) ❙ 60.7 (12.9~99.5); 962 (810~1221); 253.3 (193.3~322.1)

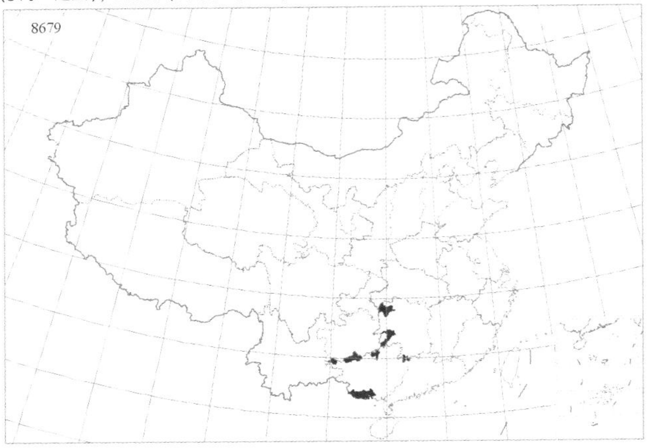

Ebenaceae 柿树科 *Diospyros* 柿属
Diospyros morrisiana Hance 罗浮柿
Trees, evergreen

19.0 (10.6~24.7); 9.7 (-0.2~19.8); 27.2 (20.1~30.1); 19.1 (10.6~24.7); 169.1 (79.4~236.5); 0.2 (0.0~12.1); 999 (643~1376) ❙ 1579 (950~3495); 649 (356~1847); 153 (27~897) ❙ 60.3 (-6.9~246.5); 992 (643~1273); 235.1 (95.3~620.0)

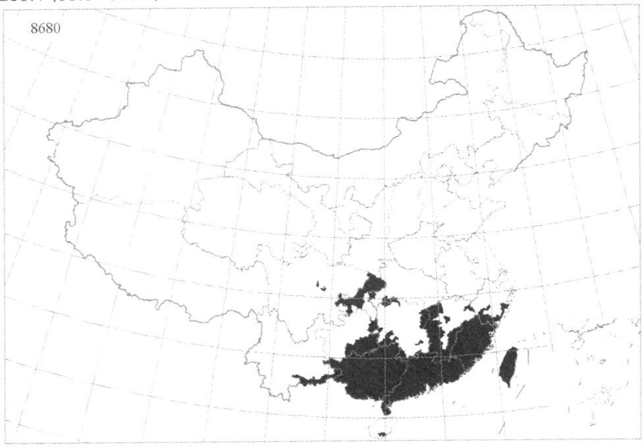

Ebenaceae 柿树科 *Diospyros* 柿属
Diospyros nigricortex C. Y. Wu 黑皮柿
Trees

19.9 (18.6~21.0); 14.0 (12.7~15.1); 23.8 (22.5~25.6); 20.0 (18.7~21.0); 179.4 (163.9~191.8); 0.0 (0.0~0.0); 968 (888~1053) ❙ 1514 (1428~1607); 784 (655~911); 66 (56~77) ❙ 56.9 (44.5~74.2); 961 (888~1053); 567.6 (506.2~659.0)

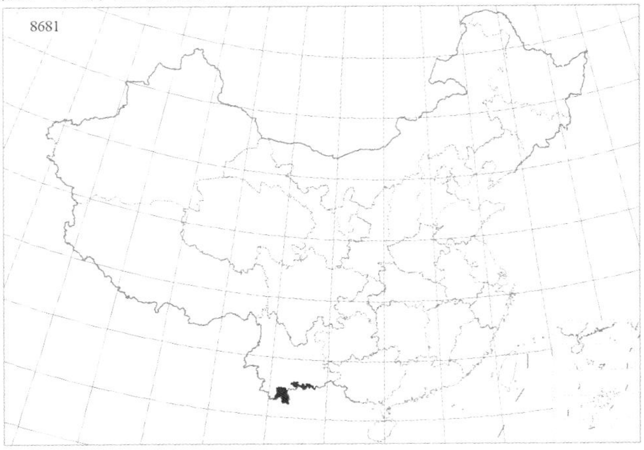

Ebenaceae 柿树科 *Diospyros* 柿属
Diospyros nitida Merr. 黑柿
Trees

24.2 (23.6~24.8); 18.3 (17.7~20.1); 28.4 (27.8~28.7); 24.2 (23.6~24.8); 230.4 (223.3~237.6); 0.0 (0.0~0.0); 1328 (1271~1385) ❙ 1331 (868~1813); 487 (312~712); 71 (41~113) ❙ 0.2 (-34.3~36.4); 1150 (868~1329); 375.7 (248.2~527.5)

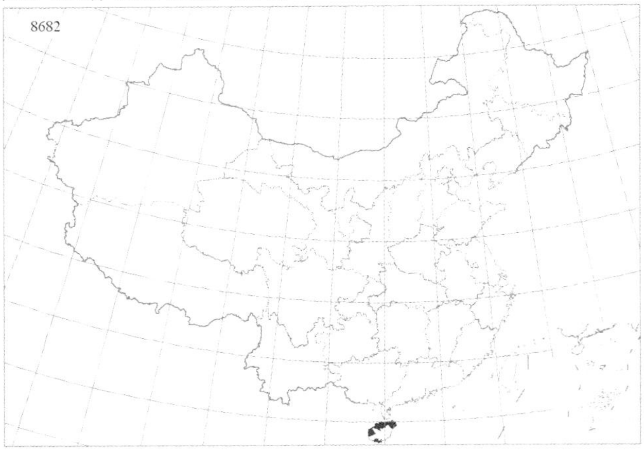

Ebenaceae 柿树科 *Diospyros* 柿属

Diospyros oldhamii Maxim. 红柿

Small trees, deciduous

20.2 (16.2~24.1); 14.5 (11.0~18.9); 25.0 (20.1~28.2); 20.2 (16.2~24.1); 182.9 (134.2~229.7); 0.0 (0.0~0.0); 1019 (773~1322) ❙ 2402 (1498~3495); 1[illegible]06 (750~1847); 255 (47~897) ❙ 146.2 (25.2~246.5); 983 (773~1144); 245.0 (119.4~363.7)

8683

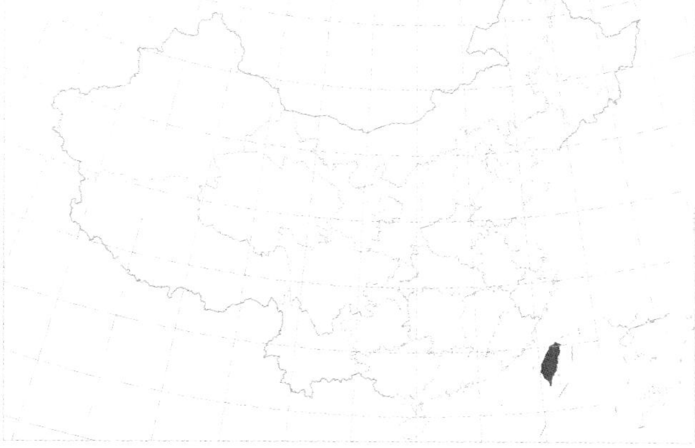

Ebenaceae 柿树科 *Diospyros* 柿属

Diospyros oleifera Cheng 油柿

Trees, deciduous

16.9 (12.1~22.3); 5.9 (1.1~[illegible]3.9); 27.0 (22.9~29.9); 16.9 (12.2~22.3); 143.9 (95.0~208.1); 0.8 (0.0~8.5); 899 (696~1202) ❙ 1517 (1002~1997); 572 (415~858); 167 (58~259) ❙ 69.3 (14.0~137.5); 895 (696~1202); 233.7 (131.2~327.7)

8684

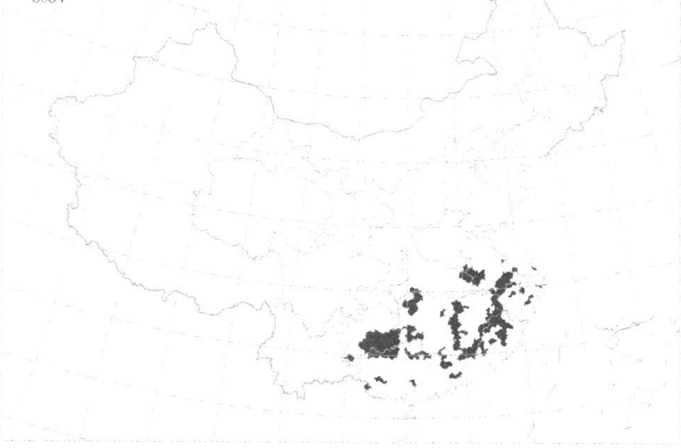

Ebenaceae 柿树科 *Diospyros* 柿属

Diospyros oliviformis R. H. Miao 榄果柿

Trees

24.0 (22.2~24.8); 18.4 (17.2~20.1); 28.0 (25.7~28.8); 24.0 (22.3~24.8); 228.3 (207.2~237.6); 0.0 (0.0~0.0); 1310 (1127~1385) ❙ 1501 (868~1854); 534 (312~712); 74 (41~115) ❙ 15.0 (-34.3~47.8); 1[illegible]91 (868~1360); 440.3 (248.2~620.0)

8685

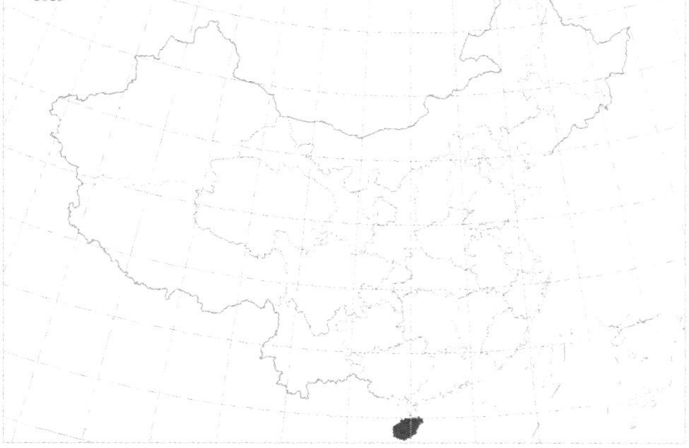

Ebenaceae 柿树科 *Diospyros* 柿属

Diospyros philippensis (Desr.) Gürke 异色柿

Trees, evergreen

20.4 (16.2~24.2); 14.7 (11.0~18.9); 25.2 (20.1~28.7); 20.4 (16.2~24.3); 185.2 (134.2~231.2); 0.0 (0.0~0.0); 1035 (773~1335) ❙ 2344 (1176~3495); 1074 (447~1847); 246 (47~897) ❙ 138.7 (-12.0~246.5); 990 (773~1144); 252.9 (119.4~411.8)

8686

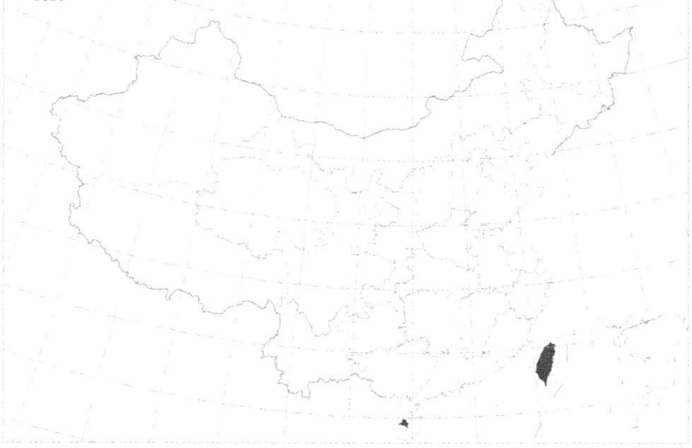

Ebenaceae 柿树科 *Diospyros* 柿属

Diospyros potingensis Merr. et Chun 保亭柿

Trees

21.9; 14.7; 27.1; 21.9; 202.8; 0.0; 1154 ❙ 1627; 815; 72 ❙ 40.9; 1103; 431.1

8687

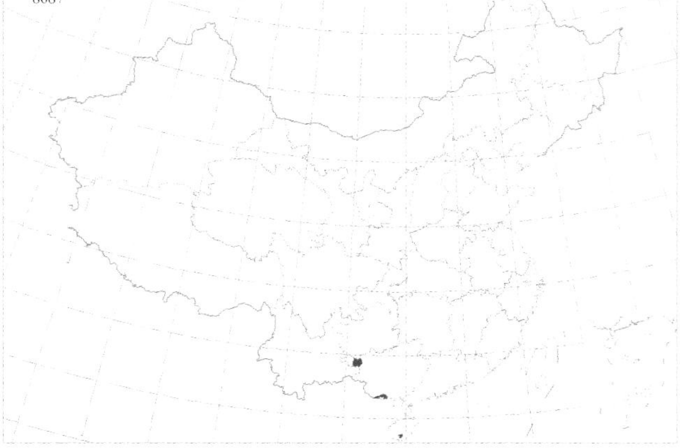

Ebenaceae 柿树科 *Diospyros* 柿属

Diospyros punctilimba C. Y. Wu 点叶柿

Trees

19.3; 12.9; 23.8; 19.4; 172.7; 0.0; 945 ❙ 1327; 738; 57 ❙ 40.8; 934; 460.5

8688

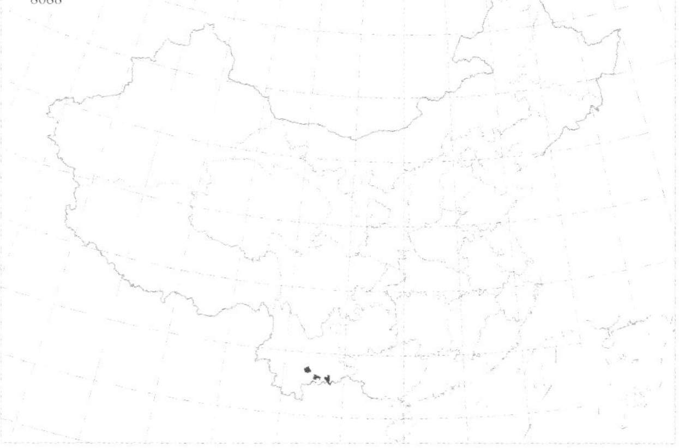

Ebenaceae 柿树科 *Diospyros* 柿属
Diospyros reticulinervis C. Y. Wu 网脉柿
Small trees

17.7 (16.3~21.0); 10.2 (9.1~14.2); 23.0 (20.6~25.6); 17.8 (16.4~21.0); 153.3 (136.4~191.8); 0.0 (0.0~0.0); 872 (795~1053) ❙ 1336 (1074~1688); 741 (582~959); 54 (47~68) ❙ 54.1 (32.4~112.2); 872 (795~1053); 402.8 (277.0~627.0)

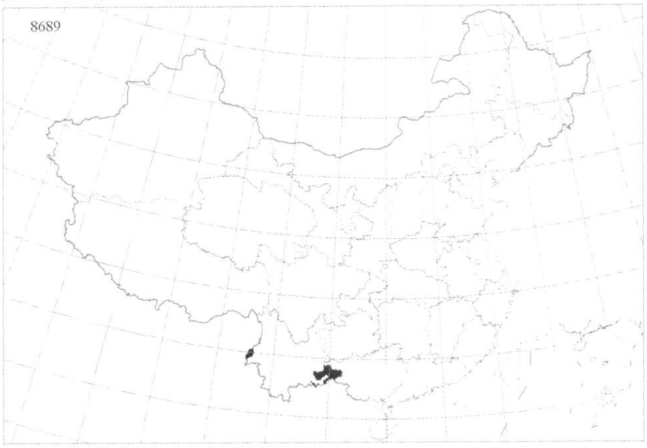

Ebenaceae 柿树科 *Diospyros* 柿属
Diospyros reticulinervis var. glabrescens C. Y. Wu 无毛网脉柿
Small trees

17.3 (16.3~18.9); 9.7 (9.1~10.6); 22.7 (20.6~25.0); 17.4 (16.4~19.0); 148.2 (136.4~167.5); 0.0 (0.0~0.0); 847 (795~939) ❙ 1351 (1161~1688); 758 (633~959); 53 (49~62) ❙ 60.4 (37.1~112.2); 847 (795~939); 398.2 (314.4~627.0)

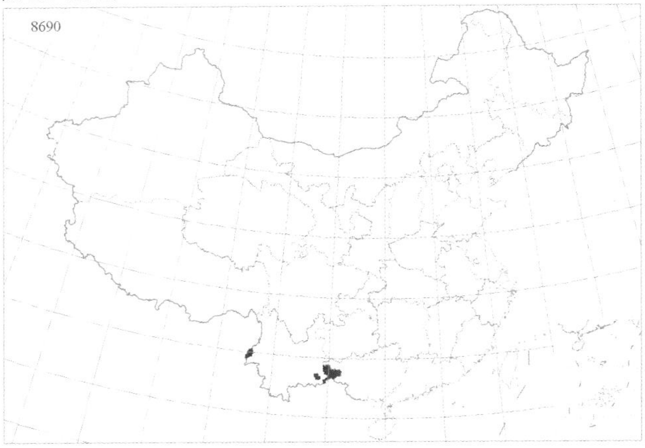

Ebenaceae 柿树科 *Diospyros* 柿属
Diospyros rhombifolia Hemsl. 老鸦柿
Small trees, deciduous

16.2 (13.2~21.7); 4.6 (-0.9~15.3); 27.7 (21.8~29.7); 16.3 (13.3~21.7); 136.8 (112.8~200.2); 1.4 (0.0~13.7); 883 (787~1118) ❙ 1459 (750~2382); 542 (370~1045); 180 (44~259) ❙ 65.5 (-12.9~191.8); 877 (750~1075); 218.6 (111.9~319.9)

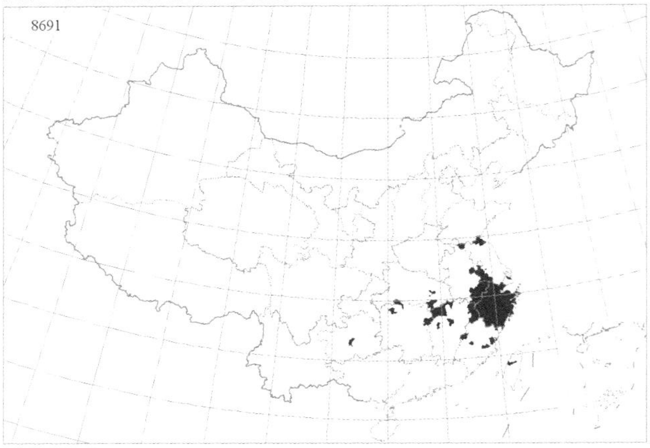

Ebenaceae 柿树科 *Diospyros* 柿属
Diospyros rubra Lecomte 青茶柿
Trees

24.1 (23.6~24.8); 18.5 (17.8~20.1); 28.2 (27.5~28.7); 24.2 (23.6~24.8); 230.1 (223.3~237.6); 0.0 (0.0~0.0); 1324 (1271~1385) ❙ 1208 (868~1405); 420 (312~473); 51 (41~71) ❙ -8.7 (-34.3~6.7); 1067 (868~1139); 429.6 (288.3~527.5)

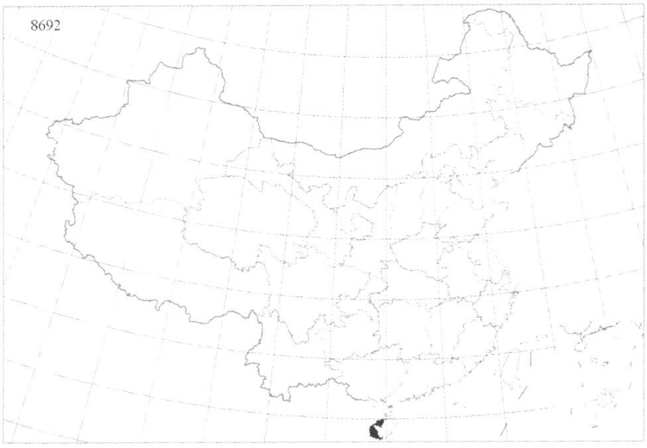

Ebenaceae 柿树科 *Diospyros* 柿属
Diospyros saxicola Chang ex Miau 石生柿
Shrubs

20.3; 10.9; 28.3; 20.4; 184.5; 0.0; 1072 ❙ 1777; 679; 182 ❙ 65.8; 1072; 228.5

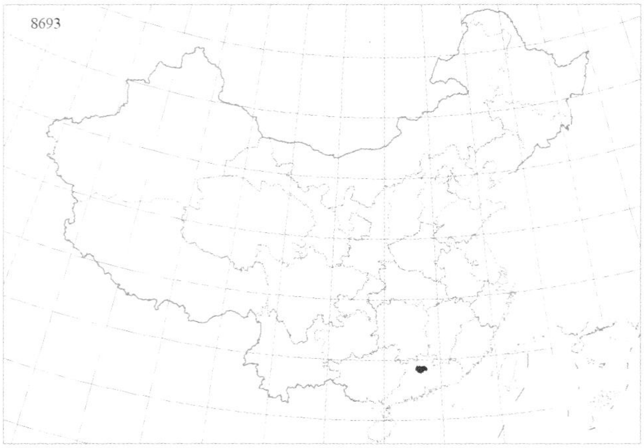

Ebenaceae 柿树科 *Diospyros* 柿属
Diospyros sichourensis C. Y. Wu 西畴君迁子
Trees, evergreen

17.4 (15.0~21.0); 9.2 (4.5~14.2); 23.9 (22.3~25.6); 17.4 (15.0~21.0); 149.0 (120.8~191.8); 0.1 (0.0~0.5); 870 (794~1053) ❙ 1216 (958~1522); 635 (499~818); 59 (44~69) ❙ 39.5 (13.3~60.1); 866 (794~1053); 345.1 (181.2~551.6)

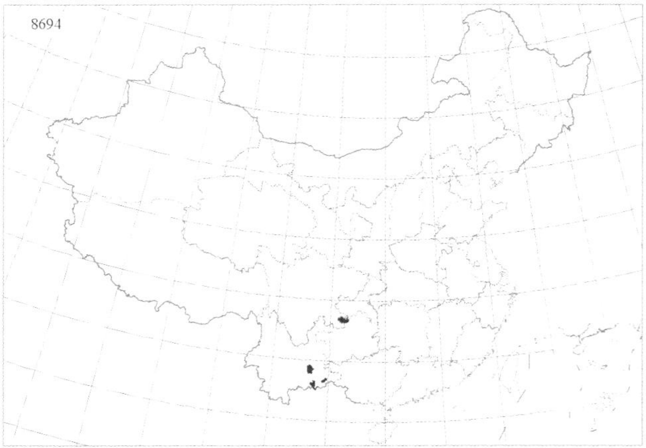

Ebenaceae 柿树科 *Diospyros* 柿属

Diospyros siderophylla H. L. Li 山榄叶柿

Trees

20.6 (18.3~22.4); 12.0 (8.8~14.0); 27.1 (25.0~28.6); 20.7 (18.4~22.4); 187.9 (160.9~209.1); 0.0 (0.0~0.0); 1074 (924~1204) | 1532 (1260~2148); 816 (670~1234); 101 (57~138) | 43.0 (15.4~86.7); 1074 (924~1204); 293.8 (235.1~371.7)

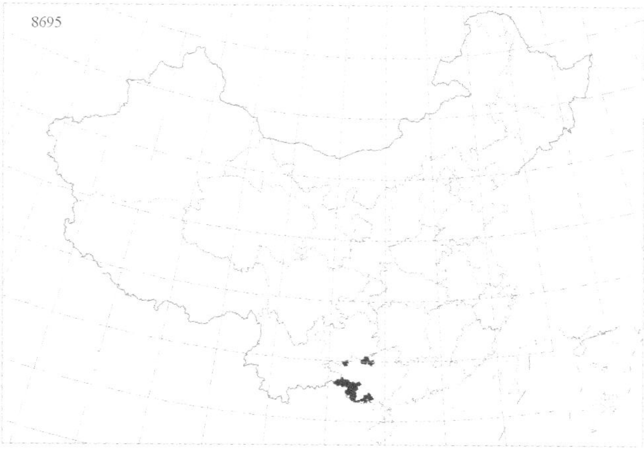

Ebenaceae 柿树科 *Diospyros* 柿属

Diospyros strigosa Hemsl. 毛柿

Shrubs or small trees, evergreen

23.5 (21.6~24.8); 17.2 (13.2~20.1); 28.1 (27.2~28.7); 23.5 (21.6~24.8); 222.0 (199.2~237.6); 0.0 (0.0~0.0); 1273 (1134~1385) | 1500 (1176~1953); 582 (392~858); 84 (41~167) | 19.3 (-12.0~71.5); 1176 (1056~1329); 353.0 (165.9~538.2)

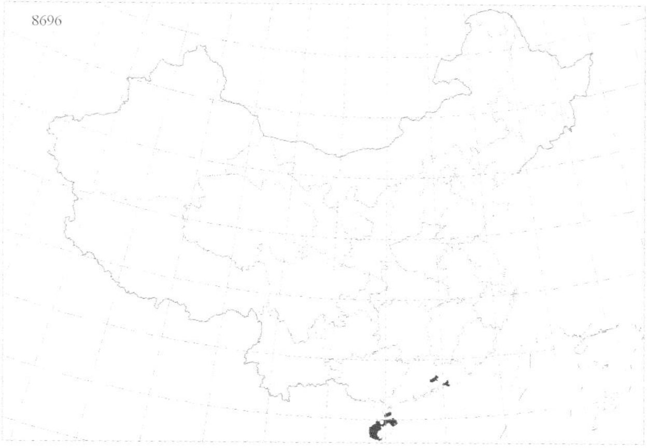

Ebenaceae 柿树科 *Diospyros* 柿属

Diospyros sunyiensis Chun et L. Chen 信宜柿

Shrubs or small trees

21.3 (19.1~23.2); 13.4 (11.0~15.8); 27.4 (25.0~28.9); 21.3 (19.1~23.2); 196.0 (169.6~218.9); 0.0 (0.0~0.0); 1112 (945~1257) | 1561 (1438~1876); 818 (742~891); 105 (57~125) | 49.9 (34.1~64.3); 1112 (945~1257). 256.1 (194.5~371.7)

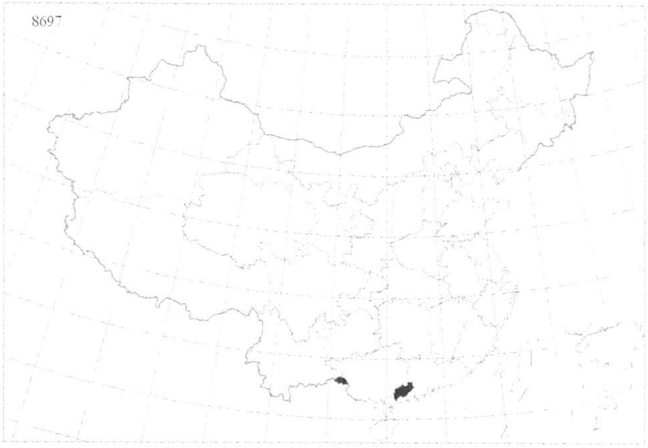

Ebenaceae 柿树科 *Diospyros* 柿属

Diospyros susarticulata Lecomte 过布柿

Trees

23.8 (23.0~24.8); 18.6 (17.5~20.1); 27.5 (26.7~28.4); 23.9 (23.1~24.8); 226.6 (217.0~237.6); 0.0 (0.0~0.0); 1295 (1217~1385) | 1469 (1179~1725); 512 (392~610); 50 (41~68) | 14.2 (-11.9~41.8); 1129 (1056~1163); 542.8 (455.2~620.0)

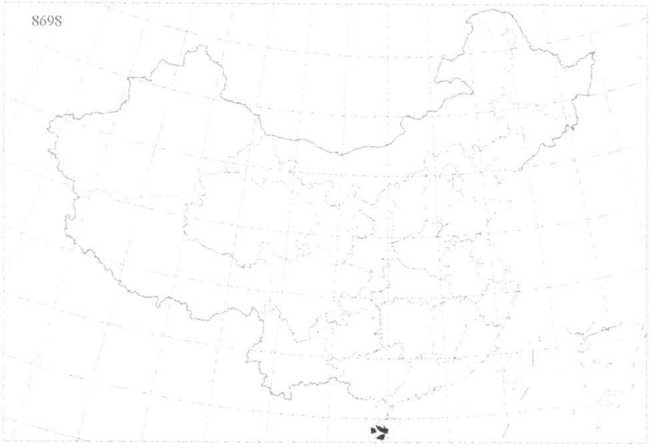

Ebenaceae 柿树科 *Diospyros* 柿属

Diospyros sutchuensis Yang 川柿

Trees

15.8; 4.9; 26.2; 15.9; 131.3; 1.0; 852 | 1174; 486; 61 | 39.9; 852; 175.3

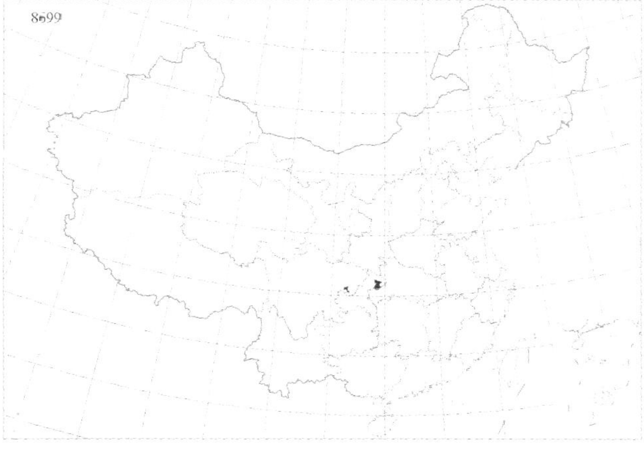

Ebenaceae 柿树科 *Diospyros* 柿属

Diospyros tsangii Merr. 延平柿

Shrubs or small trees

18.1 (15.0~22.4); 7.8 (3.5~14.8); 27.5 (24.3~29.8); 18.1 (15.1~22.4); 157.9 (120.8~208.9); 0.1 (0.0~1.5); 955 (789~1191) | 1686 (1258~2037); 595 (425~1033); 199 (110~259) | 78.1 (14.0~147.0); 949 (789~1191); 240.6 (163.7~303.4)

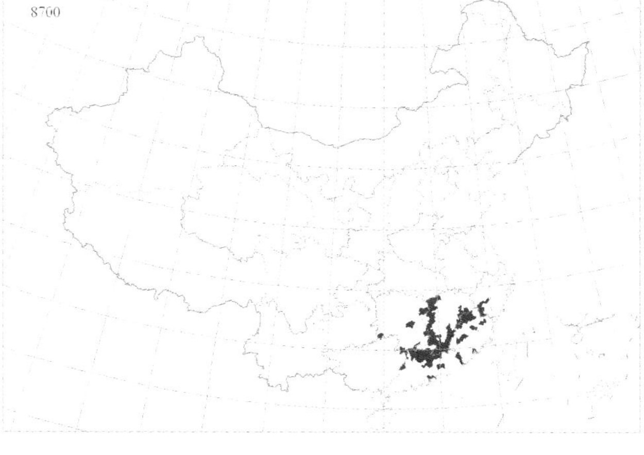

Ebenaceae 柿树科 *Diospyros* 柿属

Diospyros tutcheri Dunn 岭南柿

Small trees

17.3 (12.6~22.4); 6.7 (1.6~15.2); 27.2 (23.0~29.4); 17.5 (12.7~22.7); 150.3 (98.8~212.7); 0.5 (0.0~6.9); 931 (711~1216) ❙ 1491 (1178~2037); 563 (392~1033); 154 (81~209) ❙ 60.9 (27.0~107.2); 920 (711~1216); 204.1 (134.6~285.1)

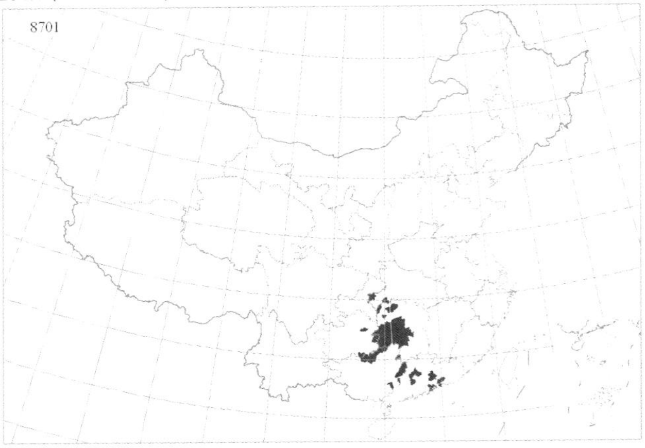

Ebenaceae 柿树科 *Diospyros* 柿属

Diospyros unisemina C. Y. Wu 单子柿

Trees

17.9; 10.3; 23.4; 18.0; 155.6; 0.0; 874 ❙ 1370; 772; 54 ❙ 56.9; 874; 389.8

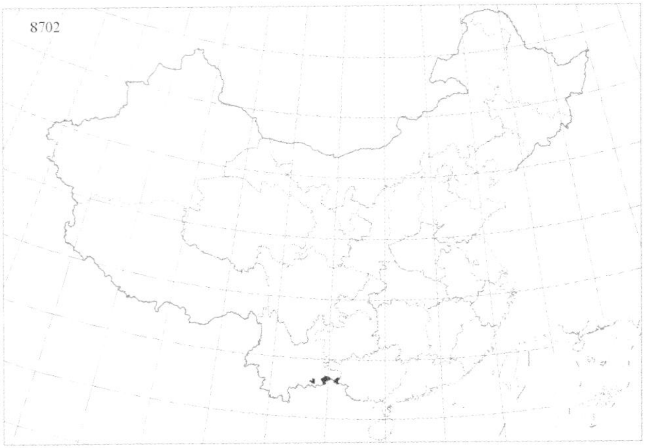

Ebenaceae 柿树科 *Diospyros* 柿属

Diospyros vaccinioides Lindl. 小果柿

Shrubs, evergreen

22.4 (21.8~24.0); 15.2 (14.0~19.1); 28.2 (26.7~28.7); 22.5 (21.9~24.0); 210.1 (202.2~228.1); 0.0 (0.0~0.0); 1198 (1150~1302) ❙ 1897 (1447~2435); 934 (539~1428); 105 (48~149) ❙ 58.8 (19.1~99.5); 1183 (1147~1221); 271.7 (139.6~620.0)

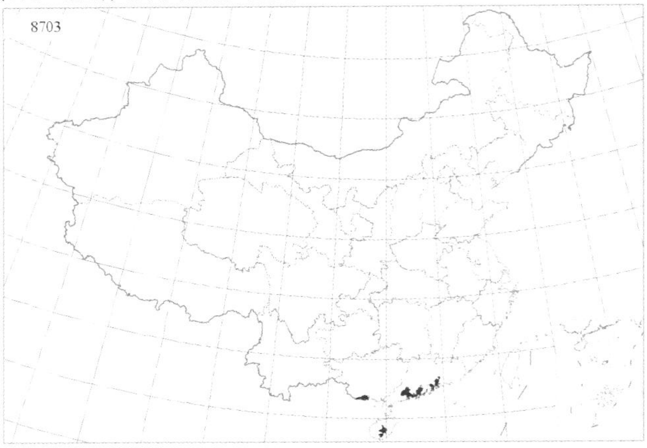

Ebenaceae 柿树科 *Diospyros* 柿属

Diospyros xiangguiensis S. K. Lee 湘桂柿

Shrubs or small trees

18.5 (17.4~19.4); 8.6 (6.9~11.0); 26.9 (25.0~28.7); 18.6 (17.4~19.5); 162.8 (148.9~173.7); 0.0 (0.0~0.0); 959 (886~1038) ❙ 1508 (1284~1776); 667 (505~820); 132 (57~189) ❙ 57.3 (30.9~81.7); 959 (886~1038); 261.9 (188.8~371.7)

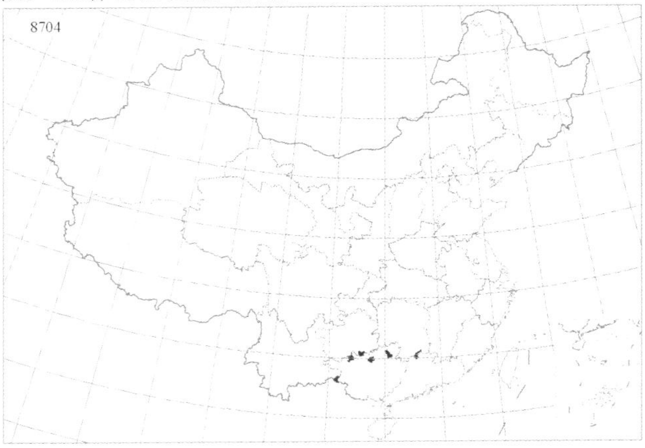

Ebenaceae 柿树科 *Diospyros* 柿属

Diospyros xishuangbannaensis C. Y. Wu et H. Chu 版纳柿

Trees

19.5; 13.9; 23.1; 19.6; 174.8; 0.0; 938 ❙ 1577; 806; 70 ❙ 68.4; 937; 605.2

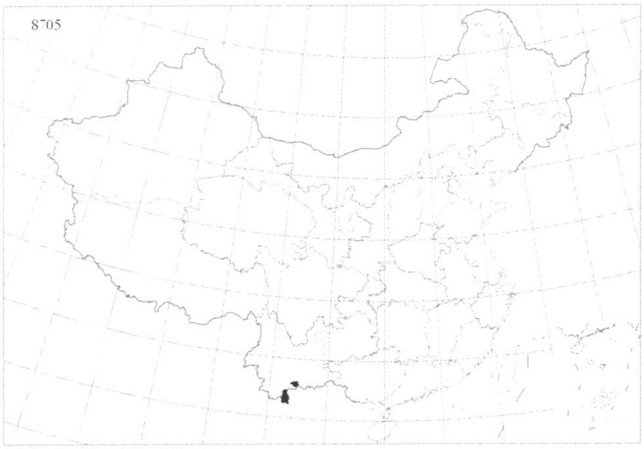

Ebenaceae 柿树科 *Diospyros* 柿属

Diospyros yunnanensis Rehder et E. H. Wilson 云南柿

Trees

19.2 (16.9~22.2); 12.8 (10.3~15.1); 23.5 (21.4~28.3); 19.3 (16.9~22.3); 171.3 (142.7~207.5); 0.0 (0.0~0.0); 938 (813~1190) ❙ 1394 (1143~1607); 727 (637~835); 62 (50~87) ❙ 49.5 (15.4~64.9); 928 (813~1190); 457.4 (327.7~659.0)

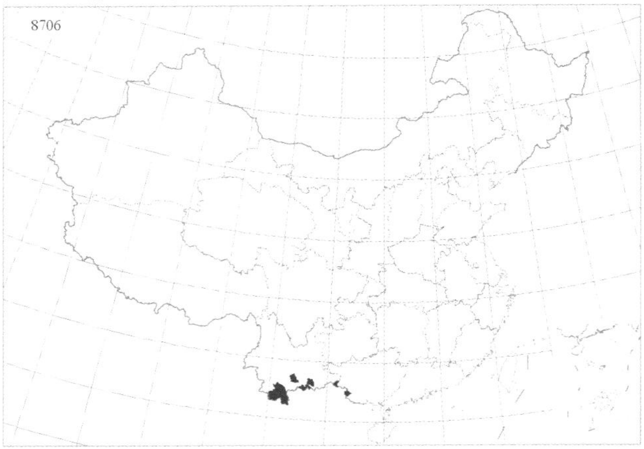

Symplocaceae 山矾科 *Symplocos* 山矾属

Symplocos hookeri var. tomentosa Y. F. Wu 绒毛滇南山矾

Trees, evergreen

19.3; 12.9; 23.7; 19.3; 171.9; 0.0; 939 ❙ 1398; 788; 59 ❙ 49.3; 937; 500.3

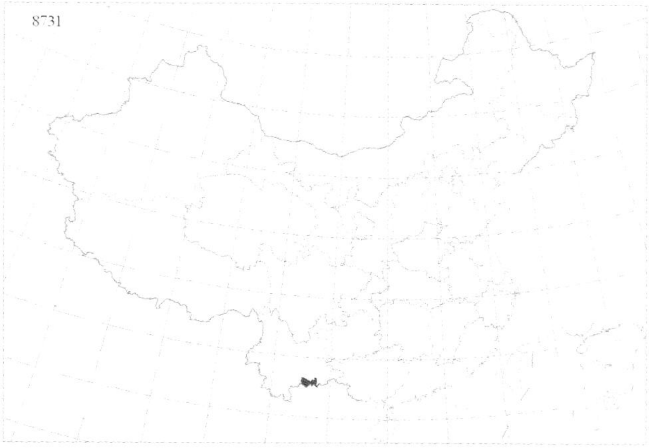

Symplocaceae 山矾科 *Symplocos* 山矾属

Symplocos lancifolia Siebold et Zucc. 光叶山矾

Small trees, evergreen

18.2 (9.3~24.8); 8.6 (-1.7~20.1); 26.8 (20.1~29.8); 18.2 (9.5~24.8); 159.3 (69.8~237.6); 0.3 (0.0~18.0); 956 (603~1385) ❙ 1500 (868~2993); 612 (312~1594); 141 (38~551) ❙ 58.8 (-34.3~246.5); 947 (603~1273); 236.3 (95.3~659.0)

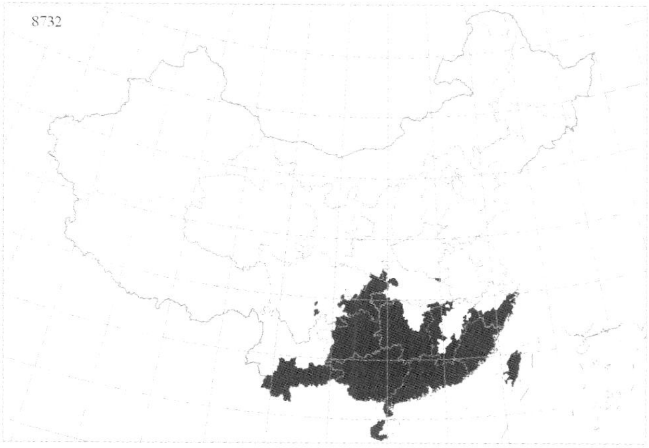

Symplocaceae 山矾科 *Symplocos* 山矾属

Symplocos lucida (Thunb.) Siebold et Zucc. 光亮山矾

Shrubs or trees, evergreen

16.4 (2.6~24.8); 6.6 (-7.0~20.1); 25.1 (11.2~29.9); 16.4 (4.3~24.8); 138.2 (20.0~237.6); 1.0 (0.0~48.2); 865 (427~1385) ❙ 1363 (741~2993); 578 (299~1412); 119 (15~551) ❙ 58.0 (-14.2~246.5); 853 (427~1247); 242.1 (108.2~627.0)

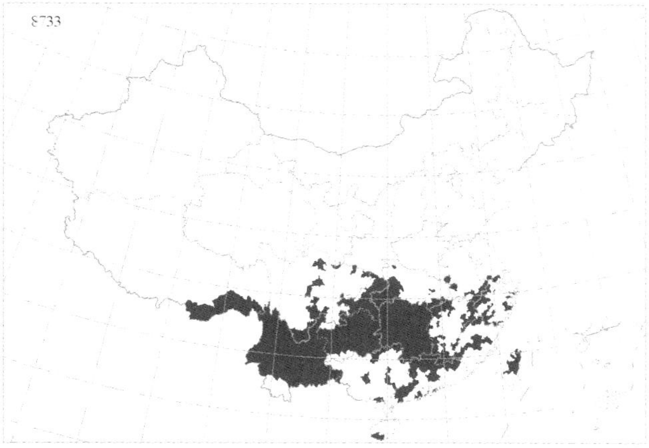

Symplocaceae 山矾科 *Symplocos* 山矾属

Symplocos modesta Brand 长梗山矾

Shrubs, evergreen

19.1 (16.2~22.2); 13.7 (11.0~17.7); 23.4 (20.1~27.0); 19.1 (16.2~22.2); 169.3 (134.2~206.7); 0.0 (0.0~0.0); 935 (773~1123) ❙ 2567 (1836~2993); 1223 (924~1847); 233 (107~547) ❙ 180.7 (64.2~246.5); 930 (773~1123); 275.8 (202.8~363.7)

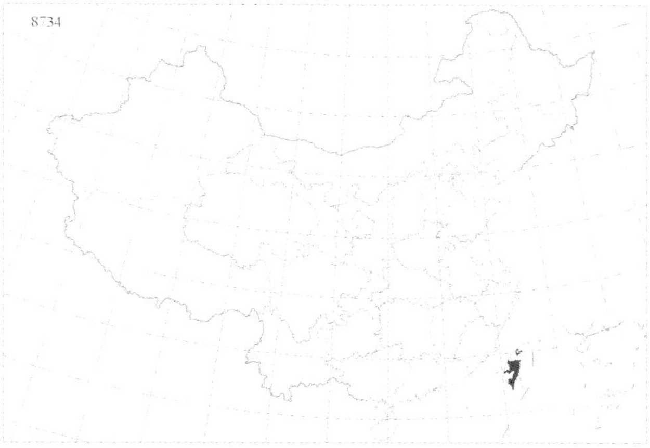

Symplocaceae 山矾科 *Symplocos* 山矾属

Symplocos nokoensis (Hayata) Kaneh. 能高山矾

Shrubs, evergreen

17.0; 11.5; 21.5; 17.0; 144.3; 0.0; 813 ❙ 2770; 1073; 323 ❙ 241.0; 813; 277.5

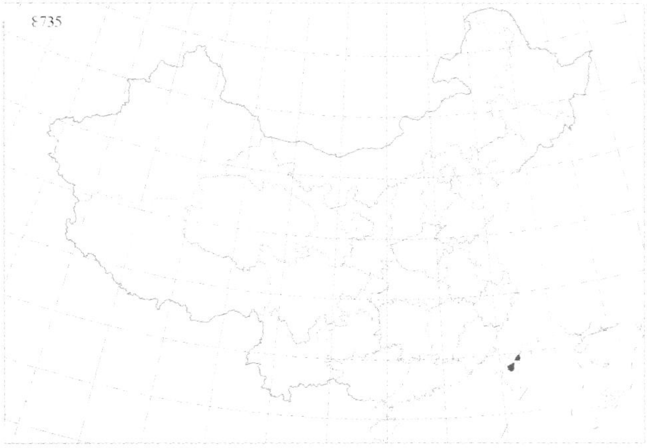

Symplocaceae 山矾科 *Symplocos* 山矾属

Symplocos ovatilobata Noot. 单花山矾

Shrubs or small trees, evergreen

23.9 (22.2~24.8); 18.6 (17.2~20.1); 27.6 (25.7~28.5); 23.9 (22.3~24.8); 227.0 (207.2~237.6); 0.0 (0.0~0.0); 1298 (1127~1385) ❙ 1531 (1179~1854); 525 (392~622); 60 (41~115) ❙ 18.5 (-11.9~47.8); 1161 (1055~1360); 515.5 (428.6~620.0)

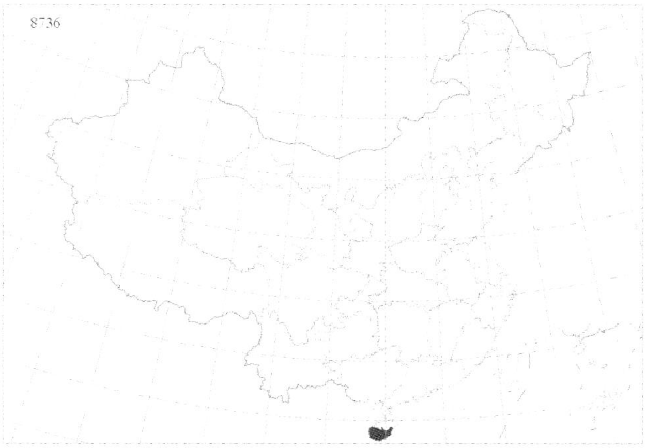

Symplocaceae 山矾科 *Symplocos* 山矾属

Symplocos paniculata (Thunb.) Miq. 白檀

Shrubs or small trees, deciduous

16.1 (-1.8~23.8); 5.4 (-15.2~18.1); 25.9 (6.1~29.9); 16.3 (1.8~23.8); 137.9 (1.9~225.8); 3.6 (0.0~82.8); 876 (329~1297) ❙ 1275 (320~2993); 546 (193~1847); 114 (9~551) ❙ 45.1 (-36.5~246.5); 857 (320~1273); 224.2 (72.2~659.0)

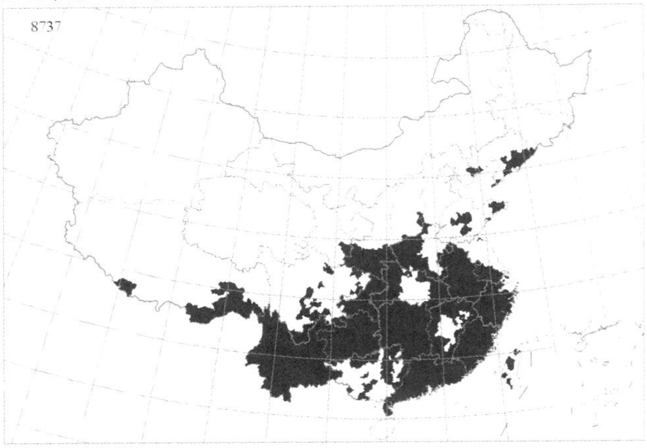

Symplocaceae 山矾科 *Symplocos* 山矾属

Symplocos paucinervia Noot. 少脉山矾

Shrubs, evergreen

19.2; 11.0; 25.2; 19.2; 170.5; 0.0; 952 ❙ 1485; 848; 72 ❙ 55.9; 952; 313.7

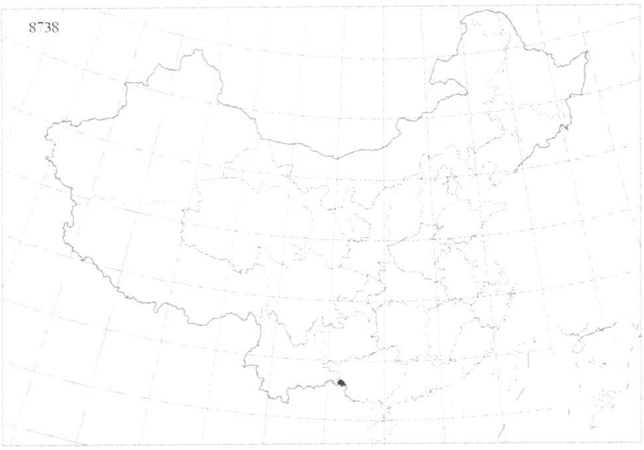

Symplocaceae 山矾科 *Symplocos* 山矾属

Symplocos pendula Wight 吊钟山矾

Shrubs or trees, evergreen

18.6 (14.3~24.8); 8.9 (3.3~20.1); 27.2 (21.4~29.8); 18.6 (14.3~24.8); 163.7 (113.2~237.6); 0.1 (0.0~2.3); 980 (761~1385) ❙ 1637 (1175~2993); 615 (392~1428); 173 (41~547) ❙ 69.8 (-11.9~214.4); 964 (761~1360); 261.7 (151.0~620.0)

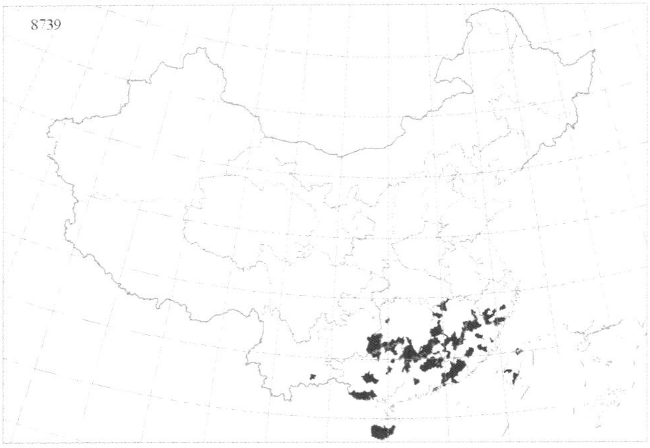

Symplocaceae 山矾科 *Symplocos* 山矾属

Symplocos pendula var. hirtistylis (C. B. Clarke) Noot. 南岭山矾

Shrubs or trees, evergreen

17.8 (14.3~22.2); 7.3 (3.3~14.7); 27.4 (22.1~30.1); 17.8 (14.3~22.2); 154.3 (113.2~206.9); 0.1 (0.0~2.3); 939 (761~1177) ❙ 1598 (1175~3495); 558 (392~924); 194 (55~897) ❙ 70.7 (29.9~224.4); 927 (761~1177); 225.3 (134.6~513.9)

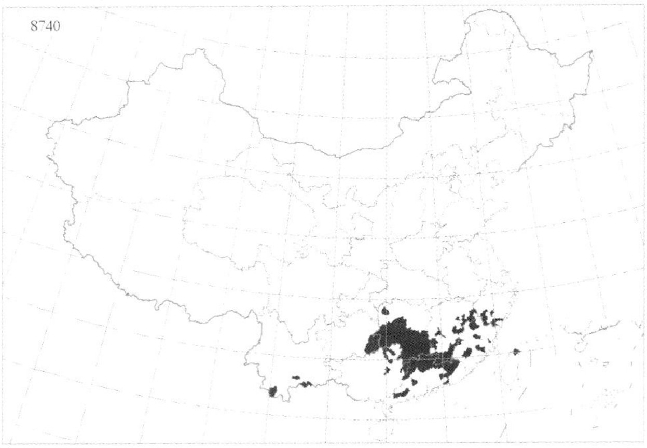

Symplocaceae 山矾科 *Symplocos* 山矾属

Symplocos pilosa Rehder 柔毛山矾

Trees

18.6 (16.9~21.0); 12.3 (10.3~14.2); 22.9 (21.4~25.6); 18.6 (16.9~21.0); 163.5 (142.7~191.8); 0.0 (0.0~0.0); 898 (813~1053) ❙ 1352 (1163~1547); 765 (641~911); 56 (48~68) ❙ 50.7 (34.8~74.2); 897 (813~1053); 472.0 (383.4~551.6)

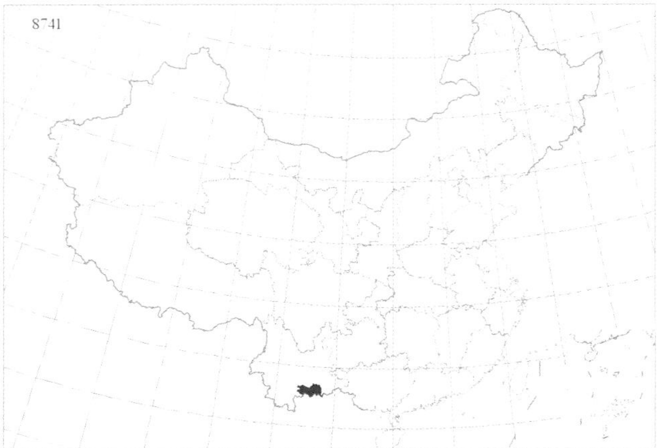

Symplocaceae 山矾科 *Symplocos* 山矾属

Symplocos poilanei Guillaumin 丛花山矾

Shrubs or small trees, evergreen

21.9 (14.3~24.8); 14.2 (4.0~20.1); 27.9 (20.6~29.2); 21.9 (14.3~24.8); 202.9 (112.6~237.6); 0.0 (0.0~1.1); 1171 (756~1385) ❙ 1573 (868~2435); 694 (312~1428); 109 (41~189) ❙ 36.1 (-34.3~99.5); 1139 (756~1360); 282.9 (152.9~620.0)

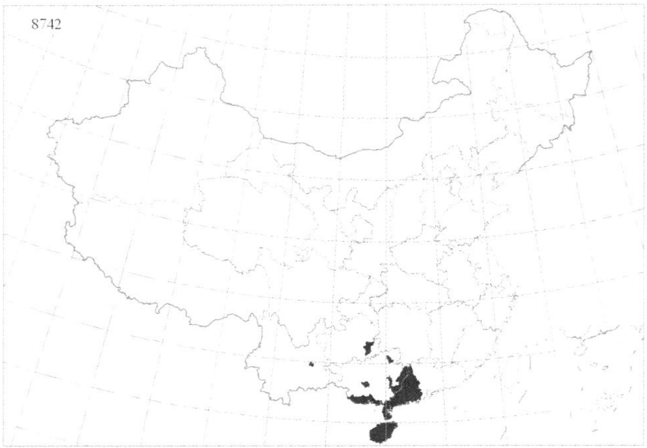

Symplocaceae 山矾科 *Symplocos* 山矾属

Symplocos pseudobarberina Gontsch. 铁山矾

Trees, evergreen

19.3 (11.9~24.7); 10.1 (2.9~19.8); 27.2 (20.0~29.8); 19.4 (12.0~24.7); 172.4 (87.2~236.5); 0.1 (0.0~3.4); 1015 (671~1376) | 1591 (868~2148); 632 (312~1234); 155 (35~231) | 58.8 (-34.3~104.6); 998 (671~1329); 259.4 (120.3~620.0)

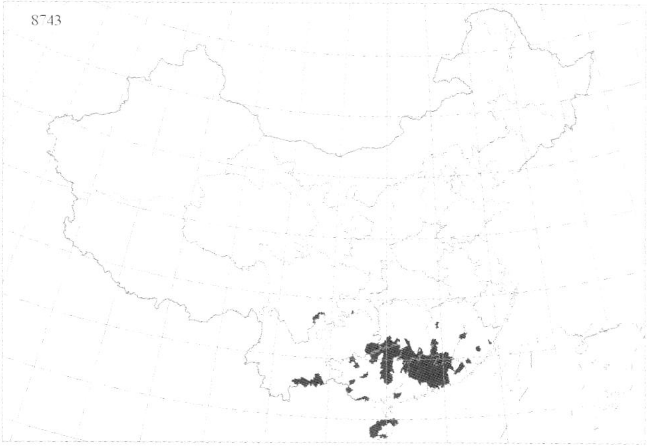

Symplocaceae 山矾科 *Symplocos* 山矾属

Symplocos pyrifolia Wall. ex G. Don 梨叶山矾

Shrubs or small trees, evergreen

0.7 (-1.8~12.2); -8.3 (-10.5~4.3); 8.9 (6.1~18.5); 3.5 (1.8~12.2); 15.3 (1.9~87.4); 65.9 (0.7~82.8); 380 (329~672) | 397 (255~1431); 267 (191~817); 8 (1~31) | -4.2 (-31.[illegible]~113.0); 329 (255~672); 111.4 (60.4~363.7)

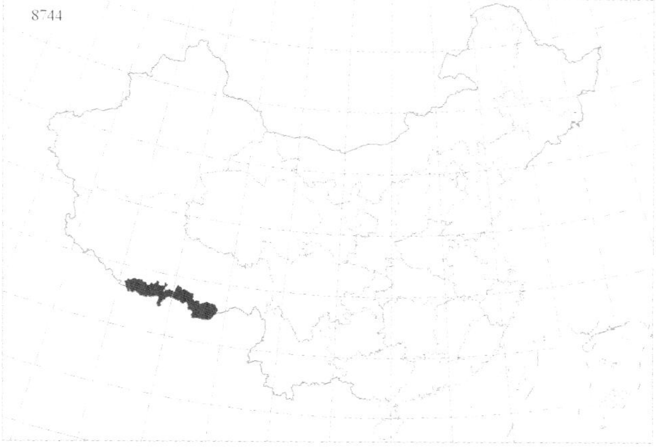

Symplocaceae 山矾科 *Symplocos* 山矾属

Symplocos racemosa Roxb. 珠仔树

Shrubs or small trees, evergreen

17.4 (-1.7~24.8); 9.6 (-12.3~20.1); 23.6 (8.0~29.8); 17.5 (2.4~24.8); 152.6 (6.4~237.6); 3.2 (0.0~86.1); 920 (344~1385) | 1333 (502~2435); 651 (301~1428); 82 (9~200) | 46.5 (-34.3~125.9); 905 (344~1360); 289.6 (95.3~659.0)

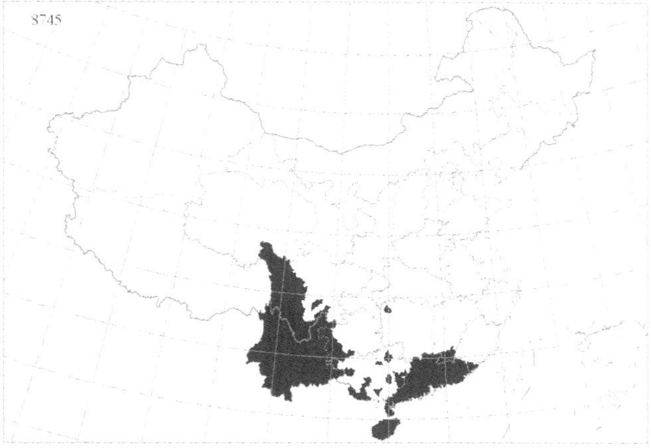

Symplocaceae 山矾科 *Symplocos* 山矾属

Symplocos ramosissima Wall. ex G. Don 多花山矾

Shrubs or small trees, evergreen

15.7 (-1.8~22.4); 5.9 (-10.2~14.0); 24.3 (6.1~29.8); 15.8 (1.8~22.4); 130.5 (1.9~209.1); 1.9 (0.0~82.8); 831 (329~1204) | 1278 (320~2148); 571 (193~1234); 94 (10~230) | 53.7 (-4.0~145.8); 829 (320~1204); 226.3 (61.6~506.2)

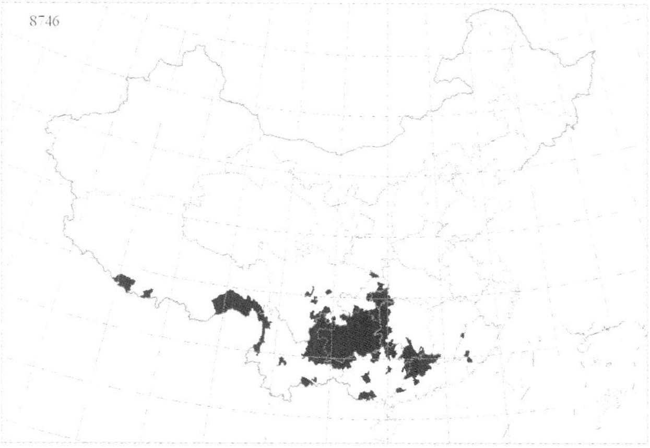

Symplocaceae 山矾科 *Symplocos* 山矾属

Symplocos spectabilis Brand 绿春山矾

Trees, evergreen

17.3 (15.3~21.0); 8.5 (3.8~14.2); 25.6 (21.4~29.3); 17.7 (15.4~21.0); 152.5 (125.8~191.8); 0.3 (0.0~1.2); 903 (813~1053) | 1459 (1163~1751); 676 (474~996); 119 (48~216) | 62.0 (28.8~103.1); 902 (813~1053); 349.3 (198.6~551.6)

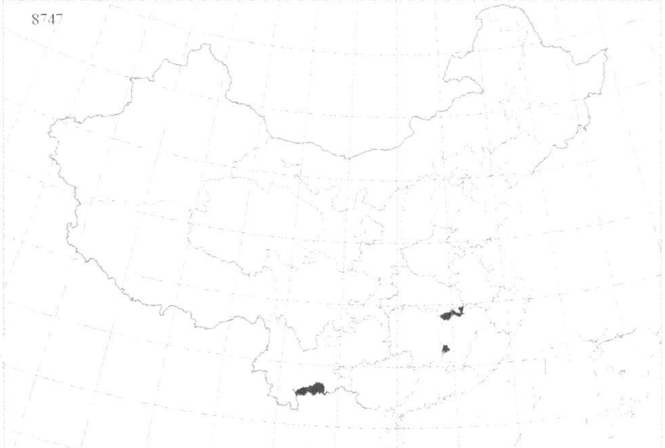

Symplocaceae 山矾科 *Symplocos* 山矾属

Symplocos stellaris Brand 老鼠矢

Shrubs or small trees, evergreen

17.1 (11.1~24.0); 6.6 (1.1~19.1); 27.0 (19.4~30.1); 17.2 (11.1~24.0); 146.8 (79.8~228.1); 0.5 (0.0~8.5); 910 (648~1302) | 1499 (862~2992); 563 (299~1847); 163 (27~551) | 65.6 (-6.7~246.5); 900 (648~1186); 224.7 (123.9~620.0)

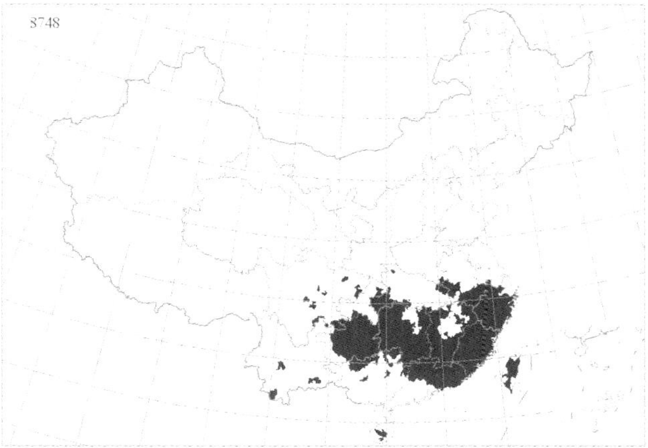

Symplocaceae 山矾科 *Symplocos* 山矾属

Symplocos stellaris var. aenea (Hand.-Mazz.) Noot. 铜绿山矾

Shrubs or small trees, evergreen

13.7 (11.1~16.2); 4.1 (1.8~6.2); 22.4 (18.8~26.3); 13.7 (11.1~16.3); 106.4 (79.8~135.7); 1.7 (0.0~6.9); 739 (648~854) ❙ 1041 (874~1414); 547 (472~817); 45 (25~71) ❙ 41.5 (19.4~89.0); 739 (648~854); 221.9 (154.4~292.6)

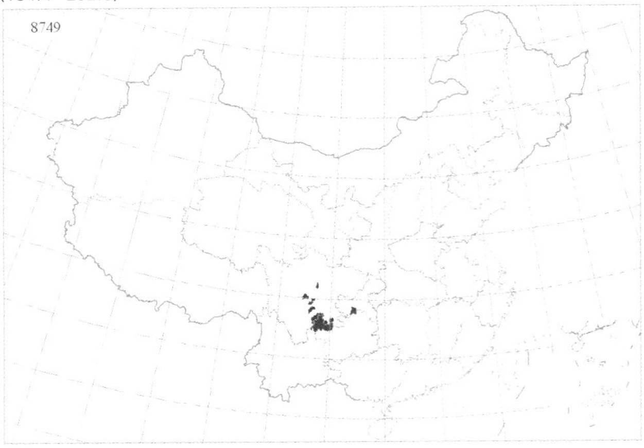

Symplocaceae 山矾科 *Symplocos* 山矾属

Symplocos sulcata Kurz 沟槽山矾

Trees, evergreen

18.3 (8.6~21.0); 12.1 (1.3~15.1); 22.5 (15.0~26.4); 18.4 (8.7~21.0); 160.9 (52.6~191.8); 0.4 (0.0~8.8); 893 (574~1053) ❙ 1363 (939~1751); 737 (524~1007); 55 (39~77) ❙ 54.0 (6.5~145.8); 872 (574~1053); 446.1 (316.6~659.0)

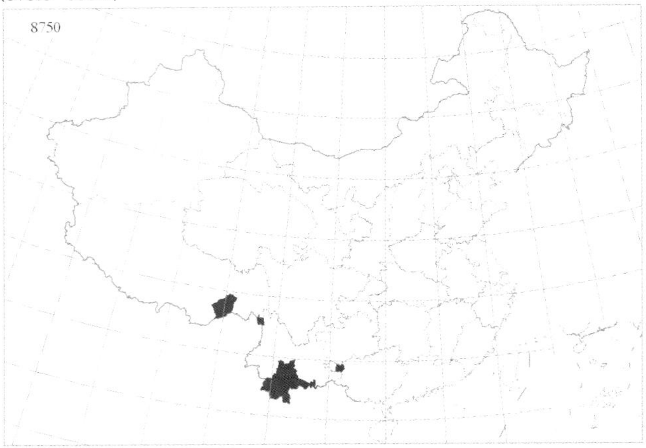

Symplocaceae 山矾科 *Symplocos* 山矾属

Symplocos sumuntia Buch.-Ham. ex D. Don 山矾

Small trees, evergreen

17.2 (5.4~24.3); 7.1 (-3.2~18.7); 26.4 (12.3~30.1); 17.3 (5.8~24.3); 148.0 (31.5~231.9); 0.7 (0.0~26.7); 915 (481~1338) ❙ 1445 (741~2993); 581 (299~1847); 142 (15~547) ❙ 58.8 (-14.2~203.4); 904 (481~1273); 229.6 (95.3~659.0)

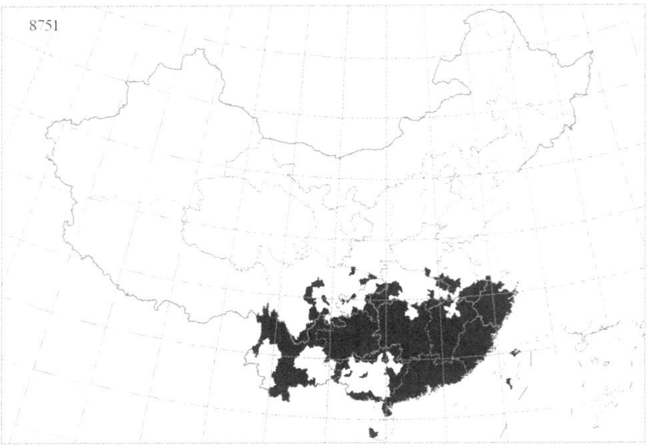

Symplocaceae 山矾科 *Symplocos* 山矾属

Symplocos ulotricha Ling 卷毛山矾

Trees

20.2 (18.2~21.8); 11.6 (8.7~14.2); 27.4 (25.9~28.8); 20.3 (18.3~21.9); 183.3 (159.3~202.6); 0.0 (0.0~0.0); 1049 (921~1150) ❙ 1574 (1284~1953); 676 (539~891); 138 (73~190) ❙ 50.3 (27.7~89.4); 1049 (921~1150); 261.8 (188.6~322.1)

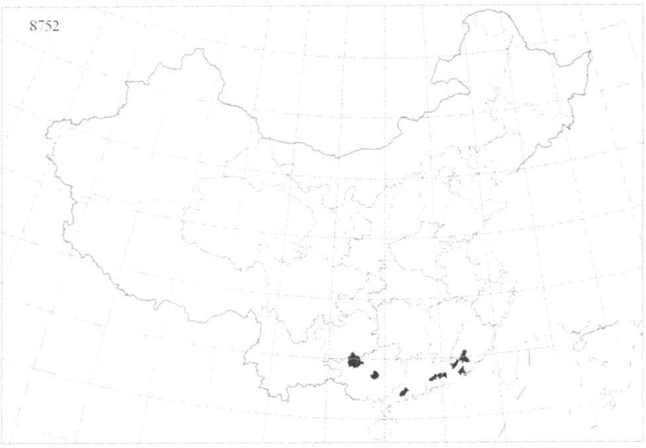

Symplocaceae 山矾科 *Symplocos* 山矾属

Symplocos viridissima Brand 绿枝山矾

Shrubs or small trees

18.5 (3.3~24.8); 10.3 (-5.4~20.1); 25.3 (11.1~28.7); 18.6 (4.6~24.8); 164.0 (22.3~237.6); 1.4 (0.0~41.8); 991 (427~1385) ❙ 1402 (473~2148); 601 (287~1234); 88 (7~197) ❙ 46.6 (-34.3~145.8); 943 (427~1360); 341.2 (185.3~620.0)

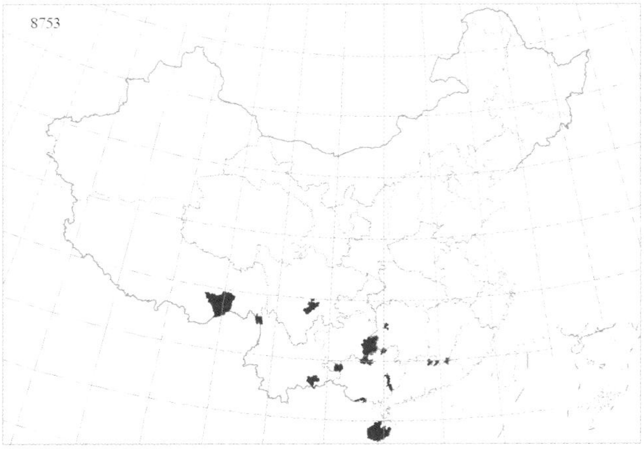

Symplocaceae 山矾科 *Symplocos* 山矾属

Symplocos wikstroemiifolia Hayata 微毛山矾

Shrubs or trees, evergreen

18.8 (14.3~24.3); 9.9 (3.3~19.1); 26.5 (20.1~29.8); 18.9 (14.3~24.3); 166.7 (112.6~231.9); 0.1 (0.0~2.3); 981 (756~1338) ❙ 1657 (868~3495); 681 (312~1847); 156 (42~897) ❙ 71.1 (-34.3~246.5); 966 (756~1200); 256.8 (119.4~620.0)

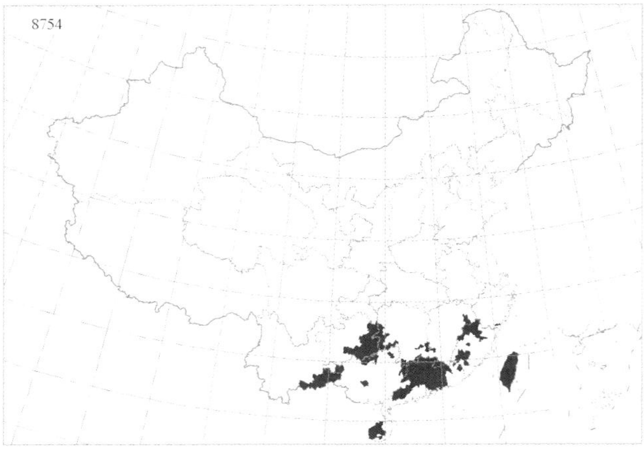

Symplocaceae 山矾科 *Symplocos* 山矾属
Symplocos xylopyrena C. Y. Wu et Y. F. Wu 木核山矾
Trees

11.0; 3.4; 17.3; 11.0; 76.5; 4.4; 640 ❙ 1389; 796; 48 ❙ 113.7; 640; 399.9

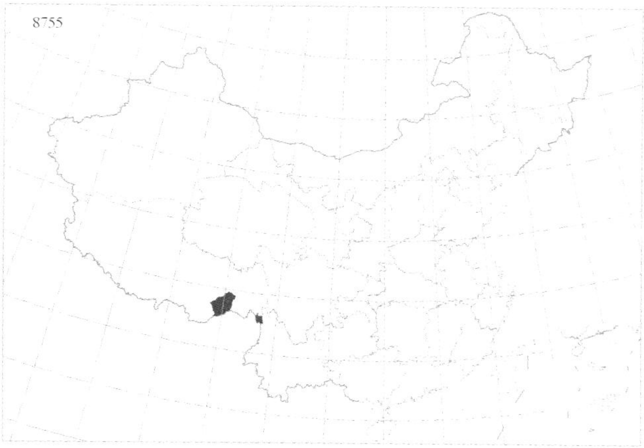

Styracaceae 安息香科 *Alniphyllum* 赤杨叶属
Alniphyllum eberhardtii Guillaumin 滇赤杨叶
Trees

18.6 (16.3~22.2); 11.2 (9.1~14.2); 24.0 (21.4~28.3); 18.7 (16.4~22.3); 164.2 (136.4~207.5); 0.0 (0.0~0.0); 921 (797~1190) ❙ 1311 (1059~1567); 727 (582~911); 61 (43~141) ❙ 42.9 (15.4~80.6); 917 (797~1190); 363.5 (232.6~551.6)

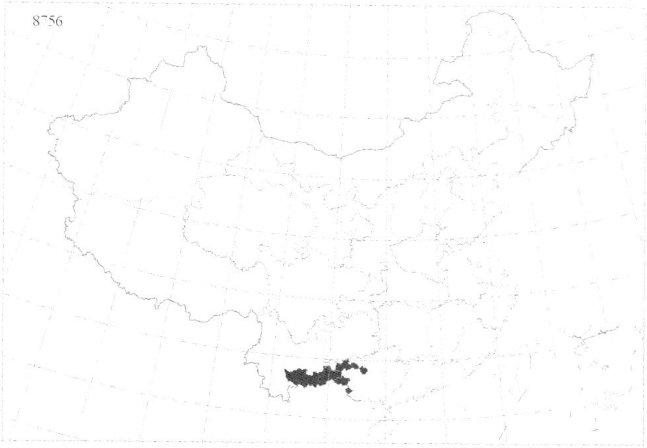

Styracaceae 安息香科 *Alniphyllum* 赤杨叶属
Alniphyllum fortunei (Hemsl.) Makino 赤杨叶
Trees

18.2 (9.3~24.8); 8.2 (−1.7~20.1); 27.2 (20.0~30.1); 18.2 (9.5~24.8); 158.9 (69.8~237.6); 0.4 (0.0~18.0); 960 (603~1385) ❙ 1508 (862~2435); 599 (323~1428); 154 (23~259) ❙ 58.8 (−6.9~147.0); 949 (603~1326); 239.9 (95.3~659.0)

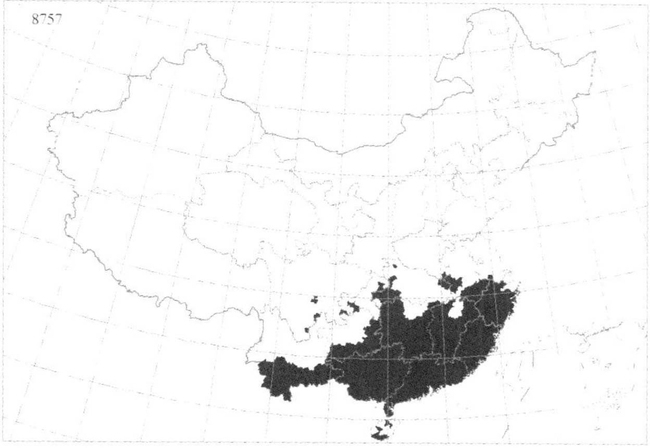

Styracaceae 安息香科 *Bruinsmia* 岐序野茉莉属
Bruinsmia polysperma (C. B. Clarker) Steenis 岐序安息香
Trees

18.8; 12.7; 22.6; 18.9; 166.2; 0.0; 900 ❙ 1323; 754; 51 ❙ 47.1; 867; 403.0

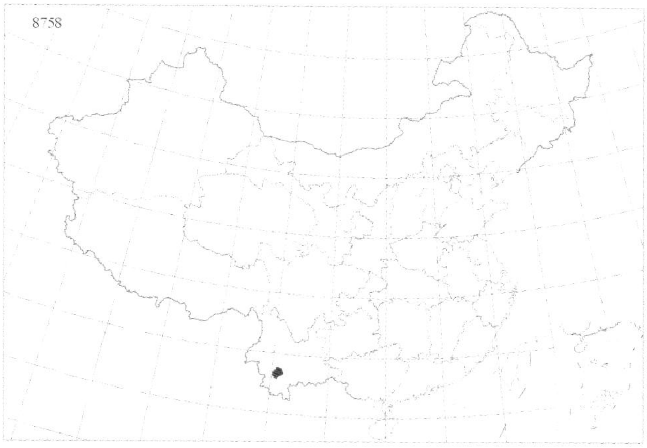

Styracaceae 安息香科 *Halesia* 银钟花属
Halesia macgregorii Chun 银钟花
Trees, deciduous

17.4 (14.3~21.8); 6.7 (3.3~14.2); 27.1 (24.2~29.8); 17.4 (14.3~21.9); 149.3 (113.2~202.6); 0.2 (0.0~2.3); 917 (761~1150) ❙ 1601 (1168~1997); 561 (403~891); 190 (84~259) ❙ 75.3 (34.4~147.0); 913 (761~1150); 232.9 (151.0~303.4)

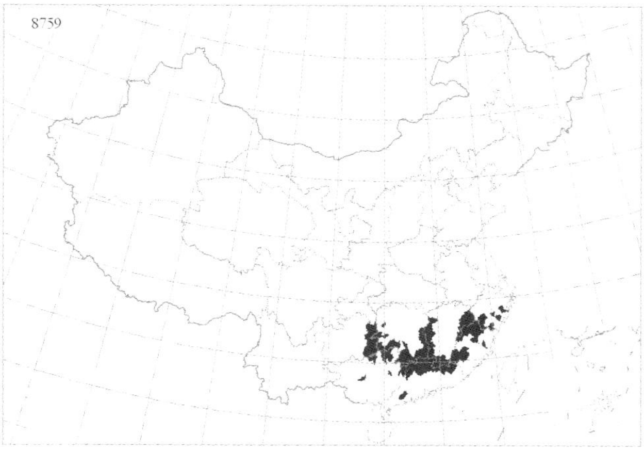

Styracaceae 安息香科 *Huodendron* 山茉莉属
Huodendron biaristatum (W. W. Sm.) Rehder 双齿山茉莉
Shrubs or small trees, deciduous

18.4 (14.4~22.2); 9.3 (4.2~15.1); 25.7 (19.1~29.8); 18.4 (14.4~22.3); 161.0 (113.1~207.5); 0.0 (0.0~0.8); 938 (727~1190) ❙ 1495 (1074~1865); 659 (403~996); 129 (41~258) ❙ 60.2 (15.4~112.2); 931 (727~1190); 303.7 (151.0~659.0)

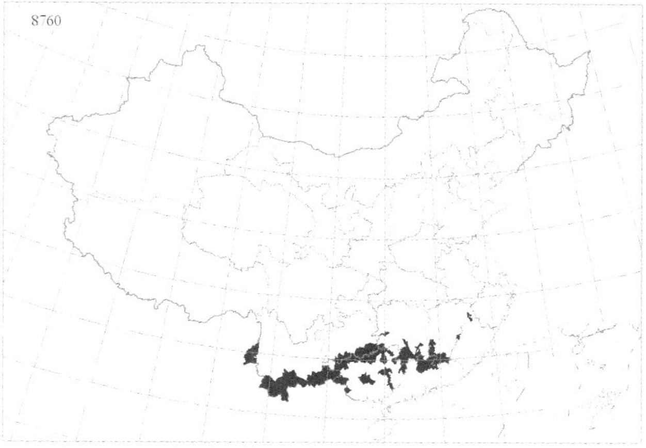

Styracaceae 安息香科 *Huodendron* 山茉莉属

Huodendron biaristatum var. parviflorum (Merr.) Rehder 岭南山茉莉

Shrubs or small trees, deciduous

19.7 (15.5~22.9); 10.3 (4.1~15.2); 27.6 (25.0~29.8); 19.8 (15.6~22.9); 177.3 (127.2~214.8); 0.0 (0.0~0.9); 1036 (815~1232) ❙ 1602 (1182~2024); 659 (445~891); 159 (43~221) ❙ 55.9 (15.4~95.5); 1034 (815~1232); 238.1 (174.9~371.7)

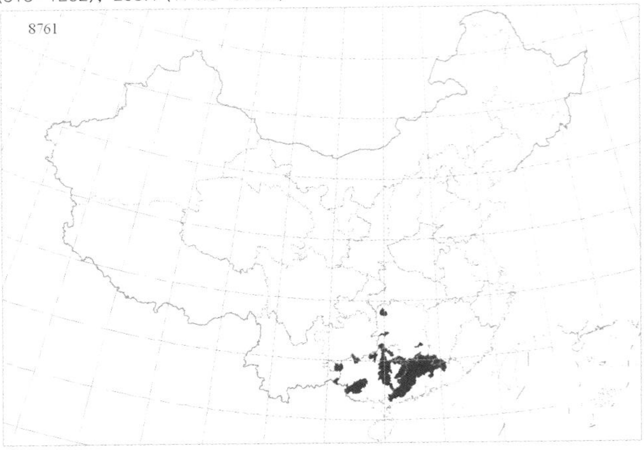

Styracaceae 安息香科 *Huodendron* 山茉莉属

Huodendron tibeticum (J. Anthony) Rehder 西藏山茉莉

Shrubs or trees, deciduous

16.7 (5.3~22.7); 6.8 (-2.4~15.0); 25.3 (12.2~29.1); 16.8 (5.7~22.8); 141.9 (30.5~213.1); 0.8 (0.0~26.2); 878 (479~1221) ❙ 1426 (961~2435); 623 (445~1428); 124 (31~221) ❙ 63.1 (27.5~101.8); 877 (479~1221); 250.3 (169.7~432.9)

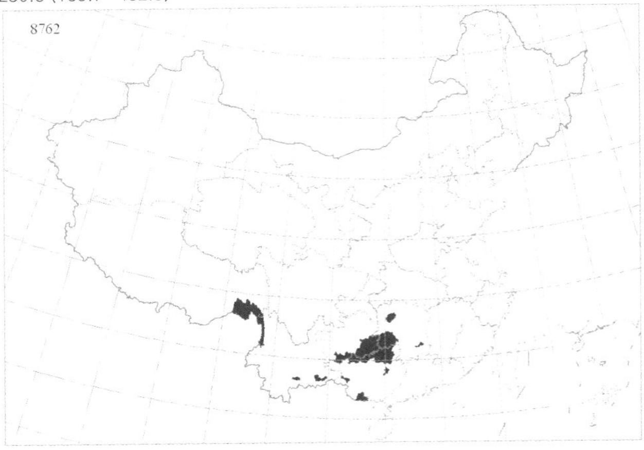

Styracaceae 安息香科 *Huodendron* 山茉莉属

Huodendron tomentosum Y. C. Tang et S. M. Hwang 绒毛山茉莉

Trees, deciduous

17.0 (8.6~19.6); 9.3 (1.3~12.8); 22.5 (15.0~25.8); 17.1 (8.7~19.6); 145.8 (52.6~175.7); 1.1 (0.0~8.8); 854 (574~953) ❙ 1454 (1041~1616); 798 (585~946); 70 (42~169) ❙ 71.1 (55.9~90.4); 849 (574~932); 400.9 (243.9~589.2)

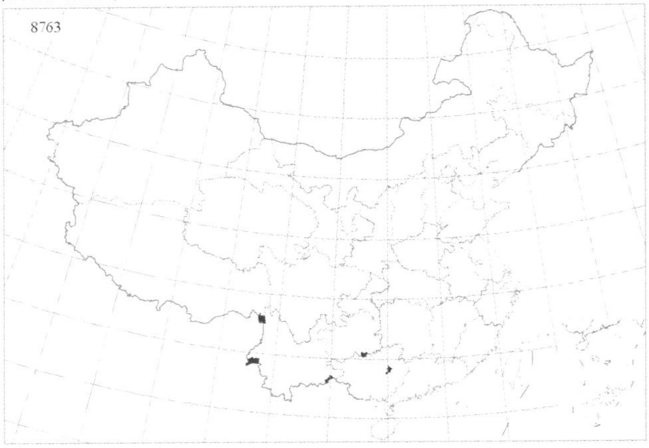

Styracaceae 安息香科 *Huodendron* 山茉莉属

Huodendron tomentosum var. guangxiense S. M. Hwang et C. F. Liang 广西山茉莉

Trees, deciduous

17.8; 7.7; 26.4; 17.8; 154.0; 0.0; 917 ❙ 1555; 610; 180 ❙ 69.6; 917; 245.9

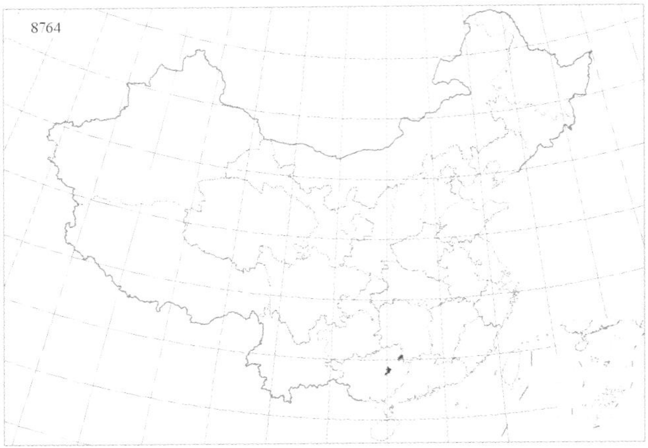

Styracaceae 安息香科 *Melliodendron* 陀罗果属

Melliodendron xylocarpum Hand.-Mazz. 陀螺果

Trees, deciduous

18.3 (10.6~22.2); 8.1 (-0.2~13.9); 27.3 (21.1~29.8); 18.3 (10.6~22.3); 159.9 (79.4~207.5); 0.2 (0.0~12.1); 962 (643~1190) ❙ 1517 (1028~2024); 588 (392~858); 157 (38~259) ❙ 58.8 (5.8~130.7); 955 (643~1190); 225.5 (108.2~371.7)

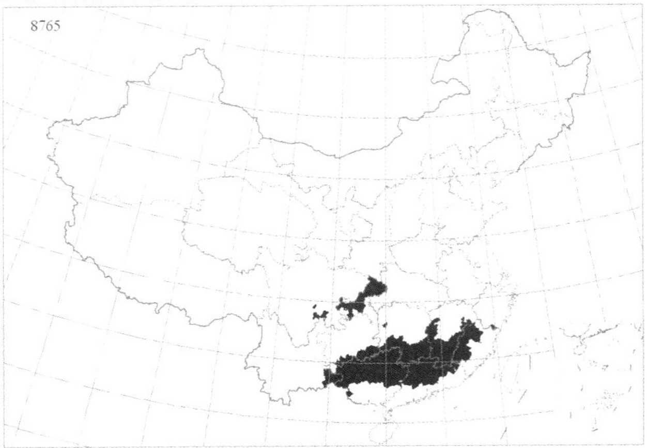

Styracaceae 安息香科 *Parastyrax* 茉莉果属

Parastyrax lacei (W. W. Sm.) W. W. Sm. 茉莉果

Trees

18.2 (16.4~19.6); 11.4 (9.6~12.8); 22.2 (20.6~23.5); 18.2 (16.4~19.6); 158.7 (137.3~175.7); 0.0 (0.0~0.0); 879 (795~953) ❙ 1586 (1530~1688); 927 (907~959); 50 (42~62) ❙ 81.6 (61.2~112.2); 870 (795~932); 526.6 (451.2~627.0)

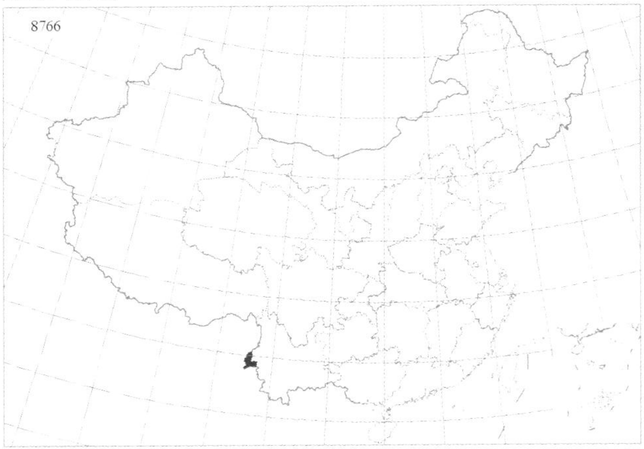

Styracaceae 安息香科 *Parastyrax* 茉莉果属
Parastyrax macrophyllus C. Y. Wu et K. M. Feng 大叶茉莉果
Trees

18.7；11.7；23.5；18.7；164.1；0.0；914 ❙ 1465；812；61 ❙ 61.7；914；485.5

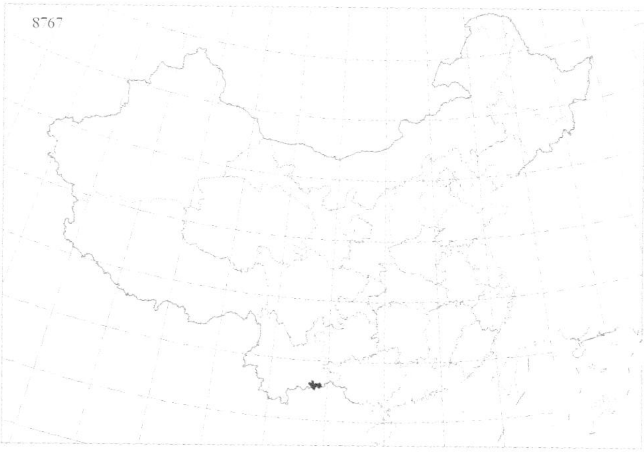

Styracaceae 安息香科 *Pterostyrax* 白辛树属
Pterostyrax corymbosus Siebold et Zucc. 小叶白辛树
Trees, deciduous

16.7 (13.4~21.0)；5.2 (1.5~11.9)；27.7 (24.0~29.9)；16.7 (13.4~21.1)；141.8 (106.7~193.0)；0.8 (0.0~7.3)；900 (744~1114) ❙ 1525 (931~1997)；538 (323~821)；184 (81~259) ❙ 70.1 (-1.6~147.0)；889 (744~1114)；218.8 (111.9~332.5)

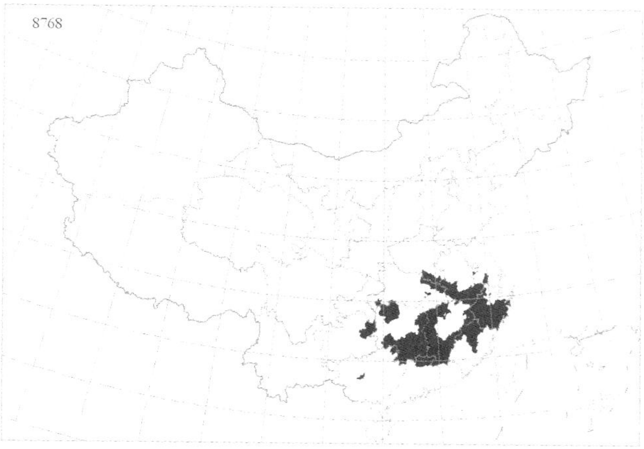

Styracaceae 安息香科 *Pterostyrax* 白辛树属
Pterostyrax psilophyllus Diels ex Perkins 白辛树
Trees, deciduous

15.1 (6.0~21.0)；4.7 (-3.2~12.2)；24.7 (14.4~29.8)；15.2 (6.6~21.0)；125.3 (39.7~192.6)；3.1 (0.0~26.7)；818 (508~1113) ❙ 1267 (760~1852)；537 (352~817)；97 (17~197) ❙ 54.8 (5.0~107.2)；816 (508~1107)；235.2 (134.6~354.5)

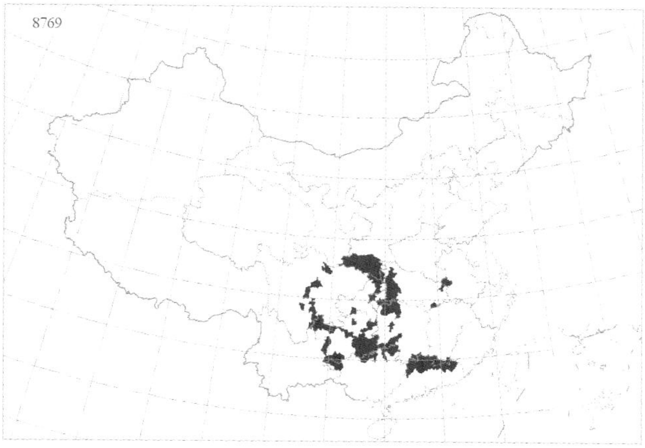

Styracaceae 安息香科 *Rehderodendron* 木瓜红属
Rehderodendron gongshanense Y. C. Tang 贡山木瓜红
Trees, deciduous

8.6；1.3；15.0；8.7；52.6；8.8；574 ❙ 1041；585；45 ❙ 81.5；574；352.3

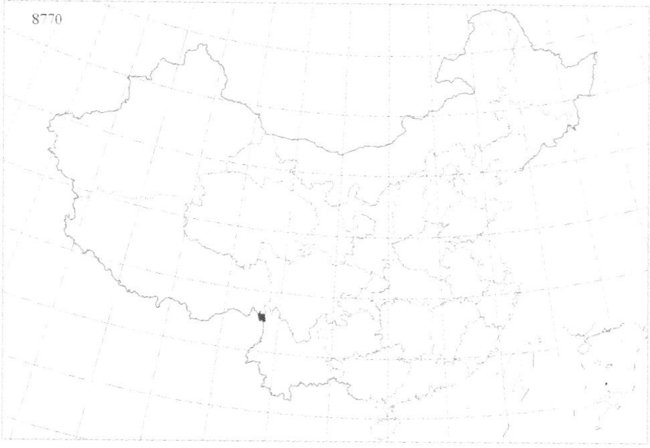

Styracaceae 安息香科 *Rehderodendron* 木瓜红属
Rehderodendron indochinense H. L. Li 越南木瓜红
Trees, deciduous

17.9 (14.9~21.0)；10.9 (8.2~14.2)；22.8 (20.1~25.6)；17.9 (15.0~21.0)；155.1 (119.8~191.8)；0.0 (0.0~0.0)；870 (748~1053) ❙ 1318 (956~1751)；743 (573~996)；54 (34~77) ❙ 51.3 (27.9~88.8)；867 (716~1053)；407.5 (272.5~551.6)

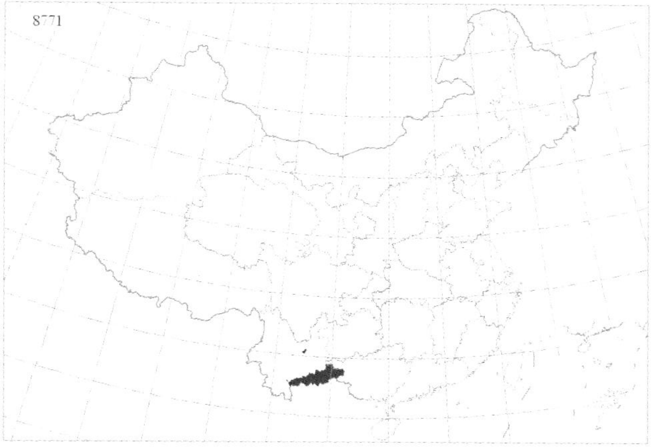

Styracaceae 安息香科 *Rehderodendron* 木瓜红属
Rehderodendron kwangtungense Chun 广东木瓜红
Trees, deciduous

18.6 (14.3~22.4)；8.9 (3.3~15.2)；27.2 (22.5~29.8)；18.7 (14.3~22.7)；164.8 (112.6~212.7)；0.1 (0.0~2.3)；981 (756~1216) ❙ 1591 (1218~2037)；611 (403~1033)；165 (59~230) ❙ 62.9 (17.4~104.6)；976 (756~1216)；230.5 (151.0~551.3)

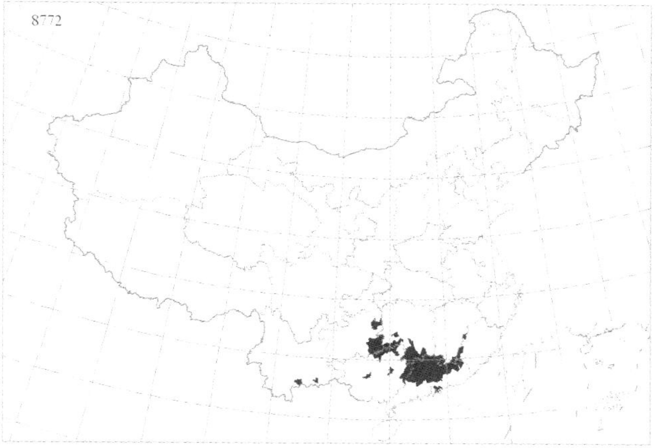

Styracaceae 安息香科 *Rehderodendron* 木瓜红属
Rehderodendron kweichowense Hu 贵州木瓜红
Trees, deciduous

17.3 (13.3~21.7); 8.5 (3.6~14.2); 24.4 (21.4~28.3); 17.3 (13.3~21.7); 148.0 (101.9~200.7); 0.1 (0.0~1.8); 874 (718~1147) ▎1308 (999~1648); 653 (488~860); 80 (47~177) ▎49.5 (30.1~80.6); 874 (718~1147); 294.7 (162.8~551.6)

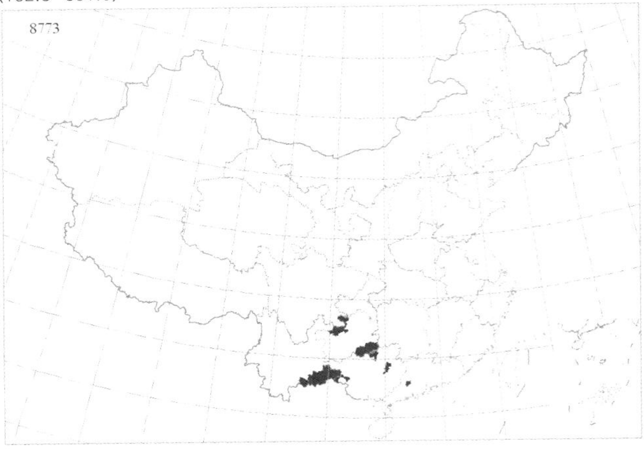

Styracaceae 安息香科 *Rehderodendron* 木瓜红属
Rehderodendron macrocarpum Hu 木瓜红
Small trees, deciduous

15.8 (11.1~21.0); 6.6 (2.2~14.2); 23.7 (19.4~27.0); 15.8 (11.3~21.0); 130.5 (80.8~191.8); 0.5 (0.0~5.3); 816 (652~1053) ▎1222 (879~1547); 614 (452~911); 68 (30~129) ▎50.0 (21.3~89.0); 815 (652~1053); 279.8 (154.4~551.6)

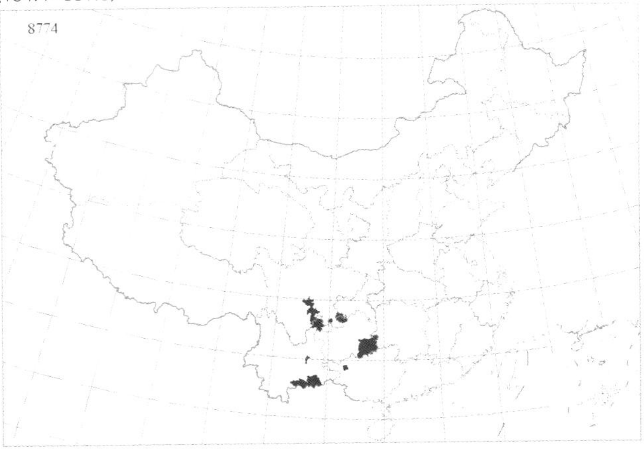

Styracaceae 安息香科 *Sinojackia* 秤锤树属
Sinojackia dolichocarpa C. J. Qi 长果秤锤树
Trees, deciduous

14.3 (11.7~15.5); 3.4 (2.4~4.0); 24.5 (18.8~26.6); 14.3 (11.7~15.6); 113.8 (82.6~128.0); 2.2 (1.0~4.4); 781 (658~839) ▎1254 (916~1451); 544 (478~600); 90 (31~120) ▎59.6 (39.1~79.3); 781 (658~839); 213.2 (185.6~236.0)

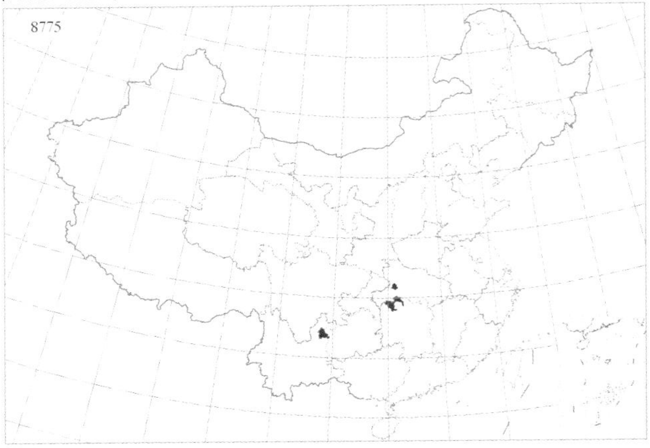

Styracaceae 安息香科 *Sinojackia* 秤锤树属
Sinojackia henryi (Dummer) Merr. 棱果秤锤树
Shrubs or small trees, deciduous

17.8 (0.3~21.0); 7.5 (-9.9~12.2); 26.8 (9.3~29.8); 17.9 (3.3~21.1); 156.4 (11.9~193.0); 2.7 (0.0~67.5); 949 (391~1114) ▎1545 (632~1852); 554 (372~821); 179 (14~230) ▎64.1 (29.7~125.9); 941 (391~1114); 225.4 (111.9~304.5)

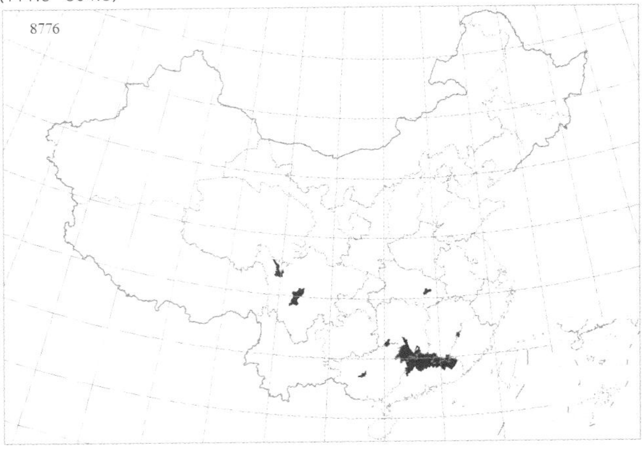

Styracaceae 安息香科 *Sinojackia* 秤锤树属
Sinojackia rehderiana Hu 狭果秤锤树
Shrubs or trees, deciduous

18.6 (14.9~21.0); 8.0 (2.6~12.2); 28.0 (25.7~29.8); 18.6 (15.0~21.0); 163.4 (121.2~192.6); 0.1 (0.0~3.1); 981 (807~1113) ▎1526 (871~1852); 548 (299~711); 177 (105~217) ▎55.6 (-6.7~86.3); 959 (755~1107); 223.2 (111.9~281.2)

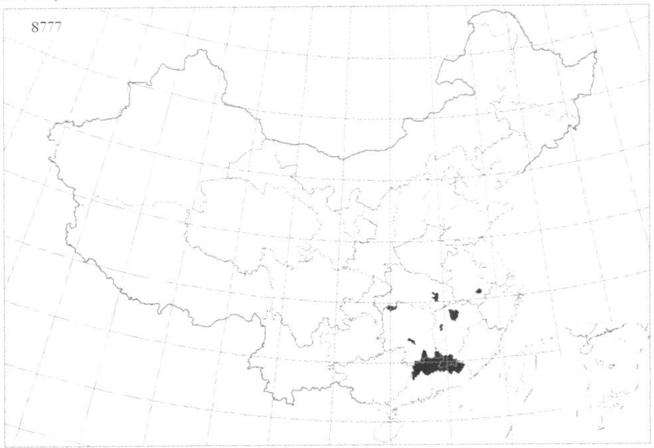

Styracaceae 安息香科 *Sinojackia* 秤锤树属
Sinojackia sarcocarpa L. Q. Luo 肉果秤锤树
Trees, deciduous

15.9; 6.2; 24.6; 15.9; 130.8; 0.0; 822 ▎1318; 753; 50 ▎60.3; 822; 189.1

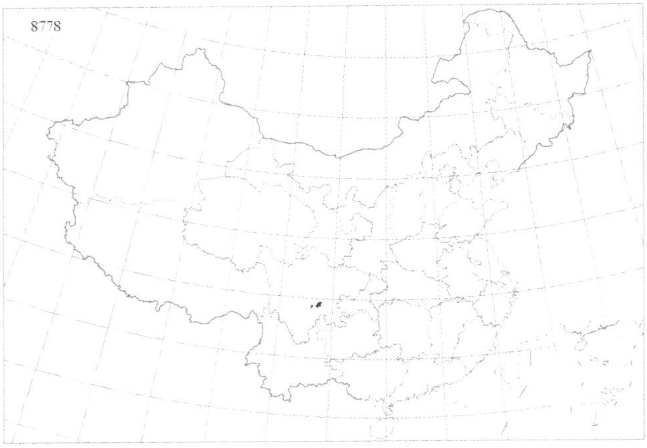

Styracaceae 安息香科 *Sinojackia* 秤锤树属

Sinojackia xylocarpa Hu 秤锤树

Trees, deciduous

15.4 (13.7~17.1); 2.6 (1.7~4.7); 27.5 (24.6~29.4); 15.4 (13.7~17.1); 128.7 (109.4~146.1); 3.5 (0.3~5.7); 861 (758~952) ❙ 1178 (880~1490); 494 (399~565); 121 (58~204) ❙ 37.3 (7.2~82.7); 858 (758~921); 187.4 (111.9~284.0)

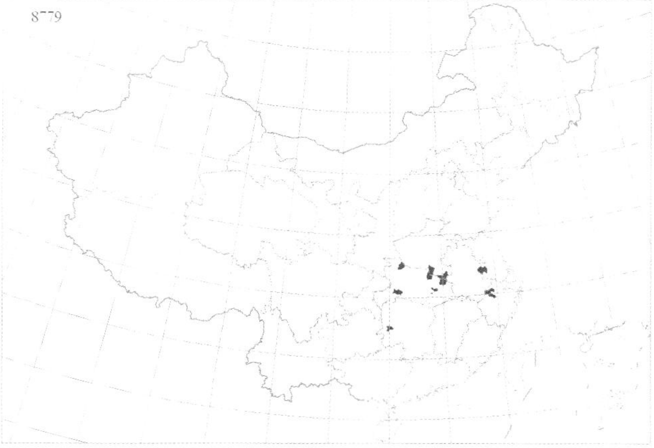

Styracaceae 安息香科 *Styrax* 安息香属

Styrax agrestis (Lour.) G. Don 喙果安息香

Trees

23.7 (19.2~24.8); 18.1 (13.8~20.1); 27.7 (22.3~28.8); 23.8 (19.2~24.8); 225.4 (170.5~237.6); 0.0 (0.0~0.0); 1290 (909~1385) ❙ 1501 (868~1854); 541 (312~712); 73 (41~115) ❙ 17.5 (-34.3~64.9); 1176 (868~1360); 444.0 (248.2~620.0)

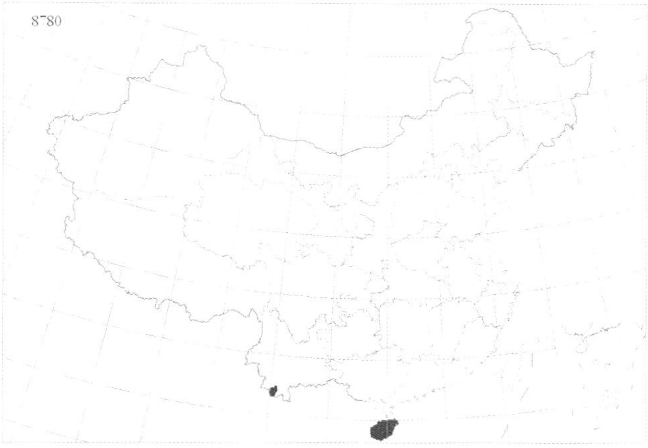

Styracaceae 安息香科 *Styrax* 安息香属

Styrax argentifolius H. L. Li 银叶安息香

Trees

20.0 (15.5~22.2); 11.4 (4.8~14.2); 26.6 (22.5~29.0); 20.0 (15.6~22.3); 179.8 (127.2~207.5); 0.0 (0.0~0.2); 1034 (815~1190) ❙ 1585 (1368~2148); 784 (565~1234); 117 (57~187) ❙ 55.3 (15.4~93.2); 1030 (815~1190); 327.5 (208.7~551.6)

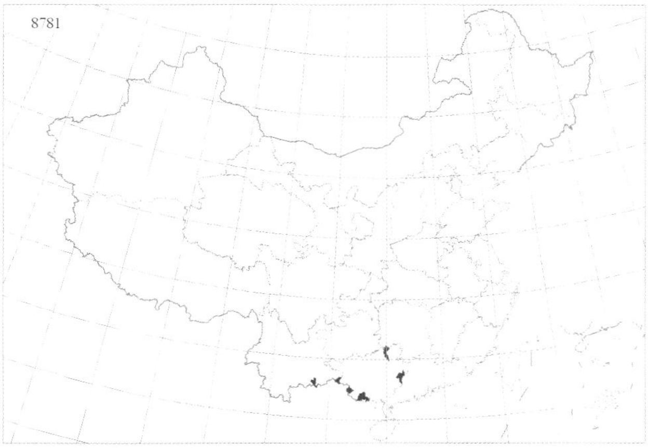

Styracaceae 安息香科 *Styrax* 安息香属

Styrax bashanensis S. Z. Qu et K. Y. Wang 巴山安息香

Trees

10.2; -0.7; 21.0; 10.4; 77.2; 13.6; 634 ❙ 1164; 488; 51 ❙ 83.5; 634; 305.6

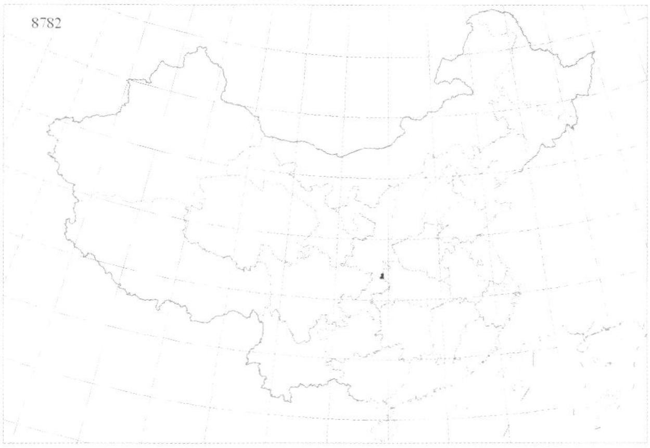

Styracaceae 安息香科 *Styrax* 安息香属

Styrax benzoides Craib 滇南安息香

Trees

20.0; 14.6; 23.2; 20.1; 180.8; 0.0; 963 ❙ 1523; 681; 67 ❙ 58.4; 942; 581.0

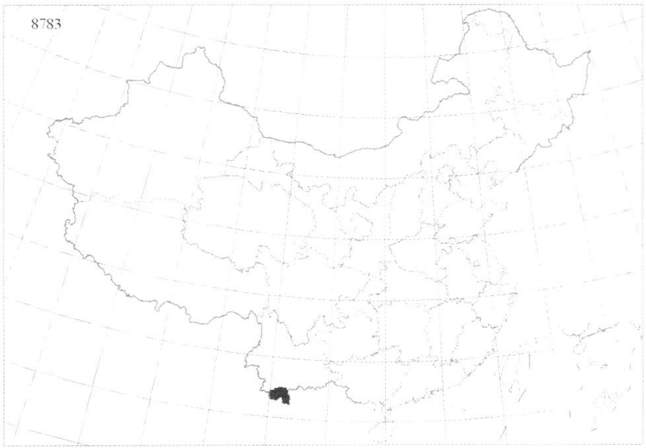

Styracaceae 安息香科 *Styrax* 安息香属

Styrax calvescens Perkins 灰叶安息香

Shrubs or small trees

16.5 (12.0~21.3); 5.1 (0.6~12.3); 27.3 (22.9~29.5); 16.6 (12.1~21.3); 139.9 (94.7~196.1); 1.0 (0.0~9.7); 891 (696~1130) ❙ 1479 (1039~1997); 548 (381~728); 162 (75~259) ❙ 66.5 (7.5~130.7); 886 (696~1130); 215.6 (111.9~303.4)

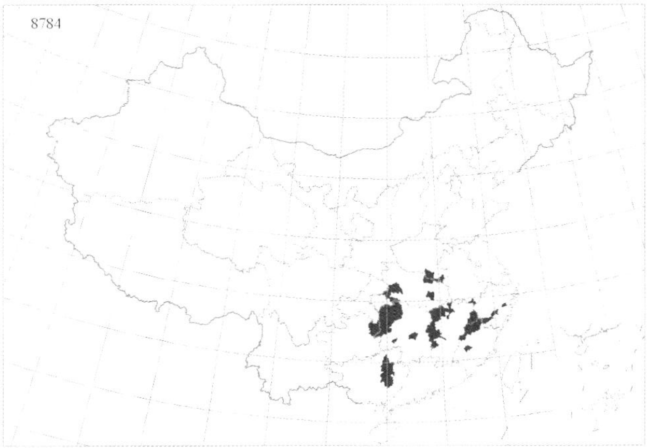

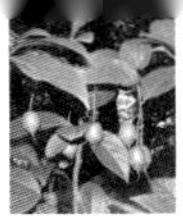

Styracaceae 安息香科 *Styrax* 安息香属
Styrax chinensis Hu et S. Ye Liang 中华安息香
Trees

19.7 (16.7~22.3); 11.2 (5.6~13.9); 26.2 (21.6~28.9); 19.7 (16.8~22.3); 176.4 (141.2~208.1); 0.0 (0.0~0.0); 1014 (832~1202) ❙ 1479 (1049~1900); 732 (497~1056); 119 (44~196) ❙ 47.2 (15.4~71.9); 1008 (783~1202); 297.5 (201.4~506.2)

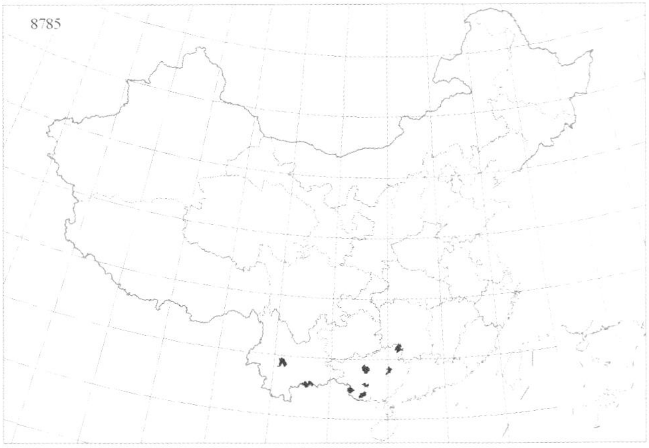

Styracaceae 安息香科 *Styrax* 安息香属
Styrax chrysocarpus H. L. Li 黄果安息香
Trees

17.7; 10.9; 22.5; 17.8; 153.4; 0.0; 854 ❙ 1368; 756; 59 ❙ 60.1; 854; 491.5

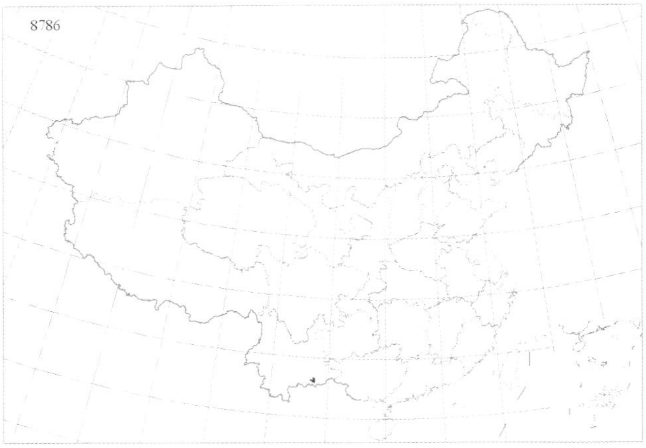

Styracaceae 安息香科 *Styrax* 安息香属
Styrax confusus Hemsl. 赛山梅
Small trees

17.5 (10.5~23.6); 6.6 (−3.8~16.7); 27.3 (19.4~29.9); 17.5 (11.1~23.6); 152.2 (79.8~223.5); 2.1 (0.0~23.0); 941 (648~1281) ❙ 1451 (604~2037); 576 (347~1033); 149 (26~259) ❙ 54.3 (−23.0~130.7); 929 (604~1273); 213.5 (95.3~411.1)

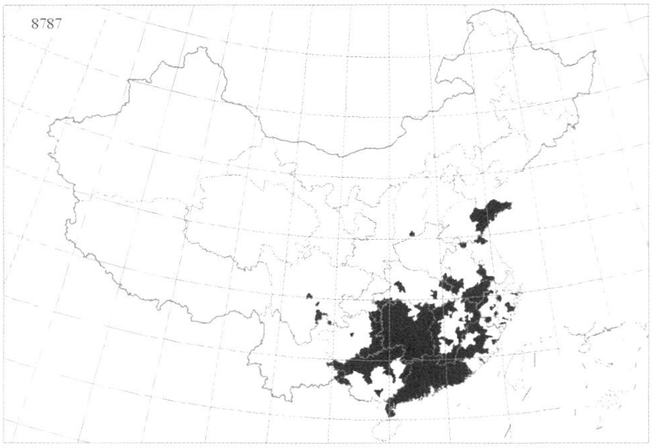

Styracaceae 安息香科 *Styrax* 安息香属
Styrax confusus var. superbus (Chun) S. M. Hwang 华丽赛山梅
Small trees

20.0 (17.6~22.2); 10.7 (7.1~13.7); 27.9 (26.2~29.8); 20.1 (17.7~22.2); 181.0 (152.5~206.4); 0.0 (0.0~0.0); 1051 (906~1190) ❙ 1699 (1462~2024); 652 (507~858); 176 (132~200) ❙ 62.2 (29.9~95.5); 1049 (906~1190); 233.4 (152.9~275.2)

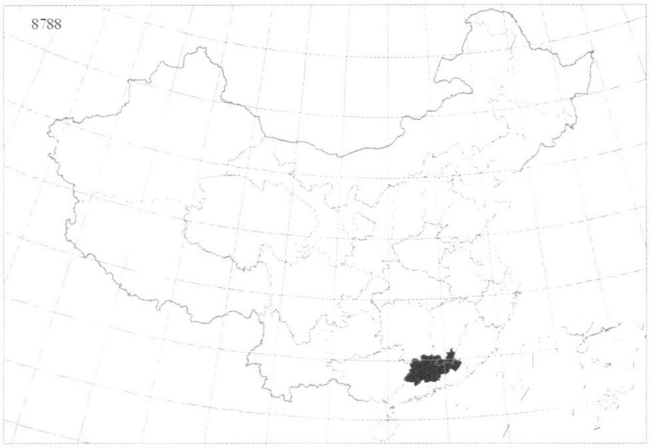

Styracaceae 安息香科 *Styrax* 安息香属
Styrax dasyanthus Perkins 垂珠花
Trees

16.2 (9.3~22.0); 4.8 (−3.7~13.6); 26.8 (20.0~30.1); 16.3 (9.5~22.0); 138.2 (69.8~204.5); 2.9 (0.0~23.0); 880 (603~1179) ❙ 1324 (604~1997); 519 (299~1056); 139 (26~259) ❙ 49.9 (−23.0~147.0); 864 (603~1179); 209.9 (123.9~411.1)

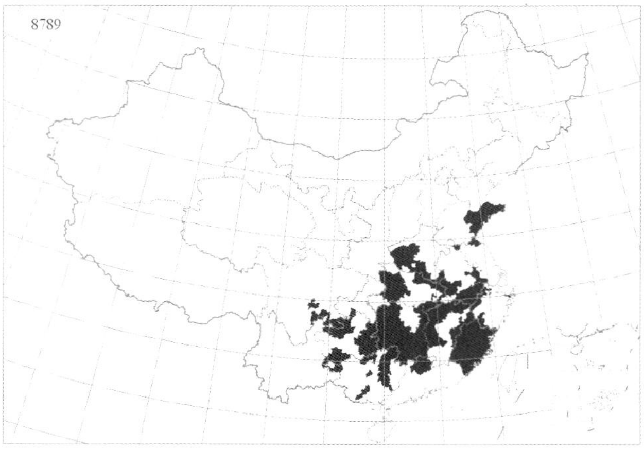

Styracaceae 安息香科 *Styrax* 安息香属
Styrax faberi Perkins 白花龙
Shrubs

18.4 (10.5~23.6); 8.1 (−3.8~16.7); 27.7 (23.2~30.1); 18.5 (11.2~23.6); 162.1 (89.3~223.5); 0.4 (0.0~22.5); 981 (687~1281) ❙ 1529 (640~2435); 602 (347~1428); 158 (26~259) ❙ 57.3 (−6.9~147.0); 973 (640~1273); 219.1 (95.3~411.1)

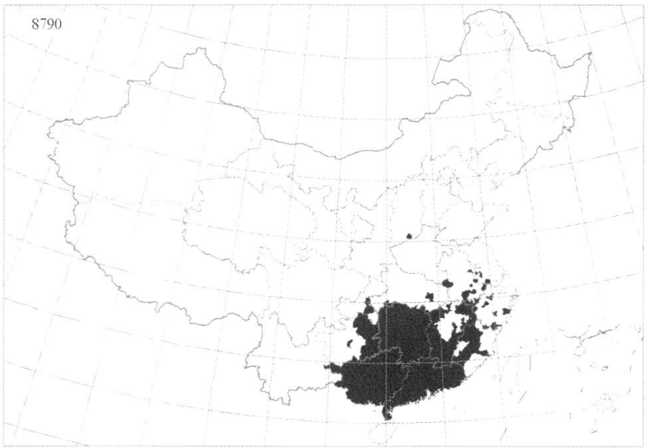

Styracaceae 安息香科 *Styrax* 安息香属
Styrax faberi var. amplexifolius Chun et F. C. How ex S. M. Hwang 抱茎白花龙
Shrubs

15.7; 4.1; 26.6; 15.7; 129.7; 0.9; 841 ❙ 1408; 454; 174 ❙ 67.4; 841; 193.4

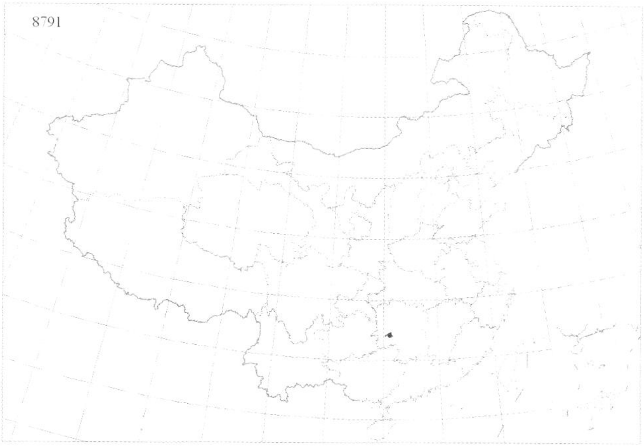

Styracaceae 安息香科 *Styrax* 安息香属
Styrax formosanus Matsum. 台湾安息香
Shrubs

17.6 (12.6~23.2); 7.4 (1.6~17.7); 26.9 (20.1~30.1); 17.7 (12.7~23.2); 152.3 (98.8~218.9); 0.4 (0.0~6.9); 928 (711~1257) ❙ 1628 (994~2993); 610 (411~1847); 180 (67~551) ❙ 76.5 (13.5~246.5); 923 (711~1257); 230.7 (111.9~332.5)

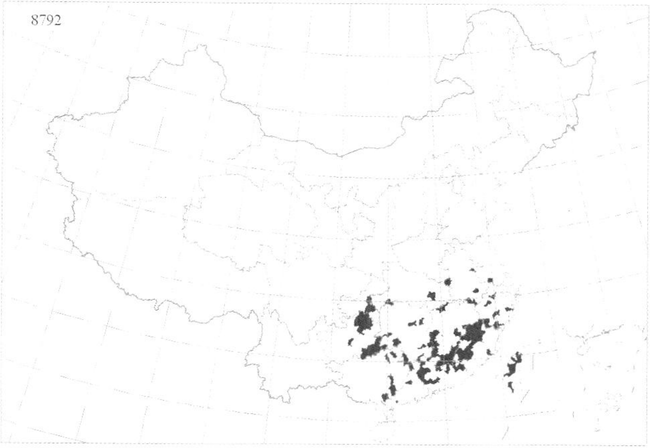

Styracaceae 安息香科 *Styrax* 安息香属
Styrax formosanus var. hirtus S. M. Hwang 长柔毛安息香
Shrubs

16.5 (15.0~17.7); 5.4 (3.3~6.9); 27.0 (24.7~27.9); 16.6 (15.1~17.7); 139.1 (120.8~152.4); 0.4 (0.0~1.7); 880 (789~928) ❙ 1636 (1486~1950); 570 (476~713); 198 (176~223) ❙ 87.9 (60.2~147.0); 878 (789~927); 230.3 (180.3~283.9)

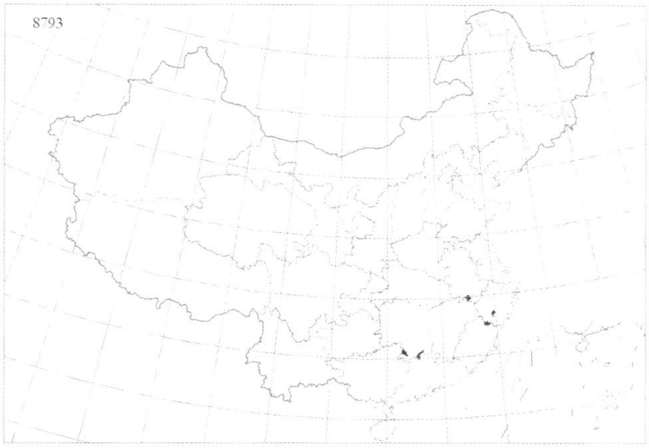

Styracaceae 安息香科 *Styrax* 安息香属
Styrax grandiflorus Griff. 大花野茉莉
Trees

18.8 (11.4~23.6); 10.7 (2.7~17.5); 25.4 (18.4~29.8); 18.9 (11.4~23.6); 166.5 (80.7~223.5); 0.1 (0.0~3.7); 967 (652~1281) ❙ 1453 (874~2713); 674 (444~1388); 104 (25~215) ❙ 50.8 (-6.9~214.4); 961 (652~1273); 277.3 (95.3~659.0)

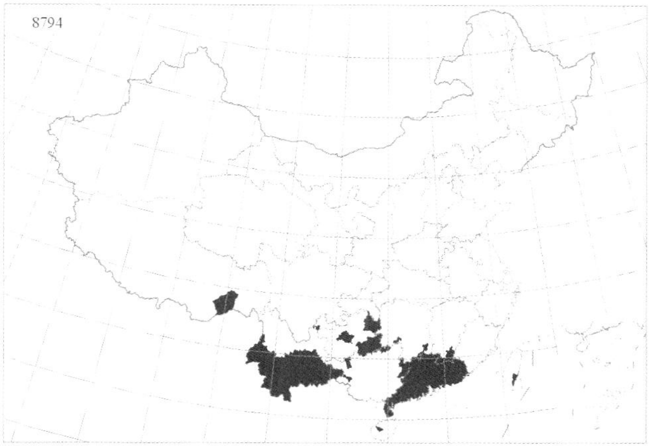

Styracaceae 安息香科 *Styrax* 安息香属
Styrax hainanensis F. C. How 厚叶安息香
Trees

24.8; 20.1; 28.0; 24.8; 237.6; 0.0; 1385 ❙ 1405; 473; 41 ❙ 1.5; 1137; 527.5

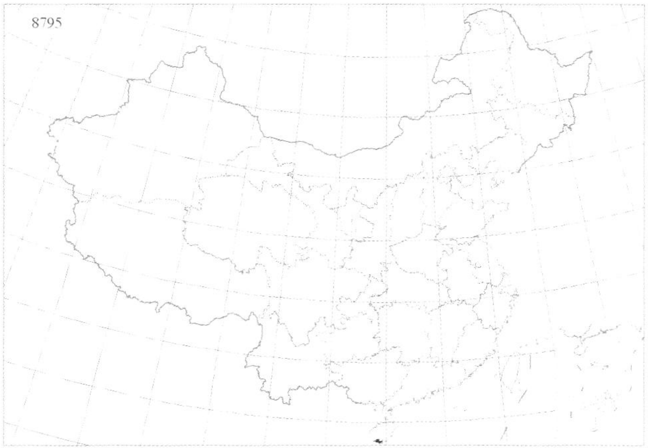

Styracaceae 安息香科 *Styrax* 安息香属
Styrax hemsleyanus Diels 老鸹铃
Trees

13.7 (7.4~21.3); 2.0 (-4.9~12.3); 24.5 (16.5~29.4); 13.8 (8.2~21.3); 112.8 (58.7~195.1); 7.9 (0.0~29.4); 778 (561~1130) ❙ 1012 (512~1758); 461 (258~816); 62 (15~192) ❙ 29.8 (-35.9~107.2); 748 (512~1130); 221.7 (111.9~334.4)

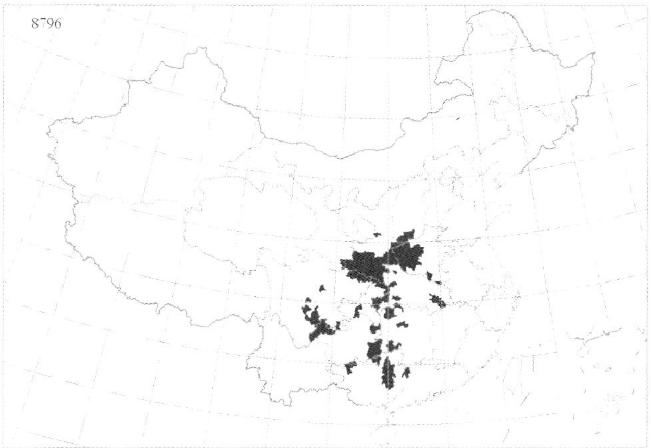

Styracaceae 安息香科 *Styrax* 安息香属

Styrax huanus Rehder 墨泡

Trees

14.3; 3.6; 24.6; 14.4; 115.4; 3.1; 779 ❙ 1178; 488; 64 ❙ 52.9; 779; 205.8

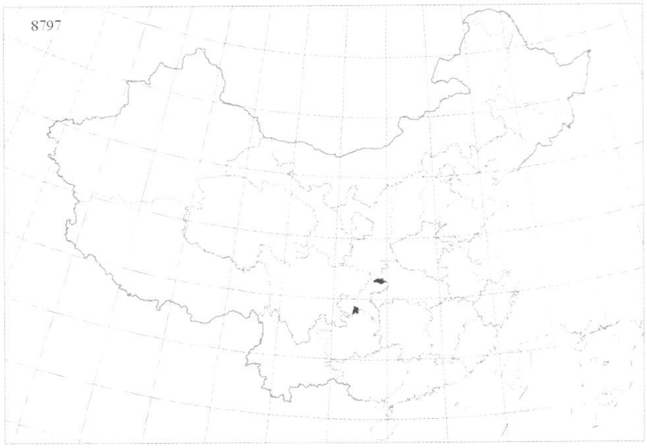

Styracaceae 安息香科 *Styrax* 安息香属

Styrax japonicus Siebold et Zucc. 野茉莉

Shrubs or small trees

15.7 (3.5~21.6); 4.2 (-6.5~12.6); 26.4 (12.6~30.1); 15.8 (5.0~21.7); 131.9 (26.3~199.8); 3.1 (0.0~43.4); 858 (446~1153) ❙ 1237 (520~1997); 502 (258~821); 115 (14~259) ❙ 43.6 (-36.5~147.0); 836 (446~1153); 203.5 (108.2~432.9)

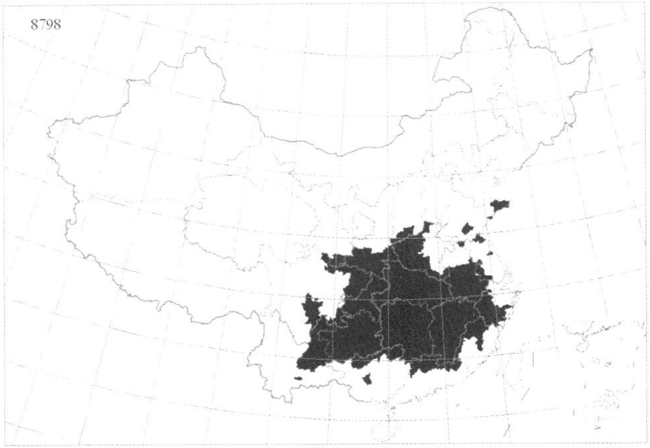

Styracaceae 安息香科 *Styrax* 安息香属

Styrax japonicus var. calycothrix Gilg 毛萼野茉莉

Shrubs or small trees

14.7 (11.1~17.1); 4.3 (-3.3~8.5); 24.0 (18.8~26.8); 14.8 (11.6~17.1); 120.0 (82.6~145.1); 2.7 (0.0~22.0); 785 (658~871) ❙ 1138 (670~1663); 533 (392~720); 72 (31~212) ❙ 44.1 (-9.5~101.3); 782 (658~871); 203.4 (136.4~298.7)

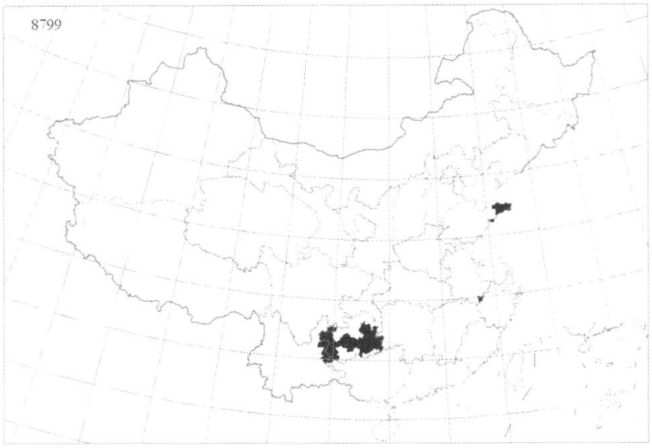

Styracaceae 安息香科 *Styrax* 安息香属

Styrax limprichtii Lingelsh. et Borza 楚雄安息香

Shrubs

14.2 (5.4~18.9); 6.9 (-2.1~12.0); 20.0 (12.3~26.6); 14.3 (5.8~19.0); 112.5 (31.5~167.8); 1.6 (0.0~25.6); 737 (481~929) ❙ 987 (741~1220); 569 (411~688); 36 (12~65) ❙ 35.8 (-14.2~89.0); 712 (481~912); 289.1 (142.9~406.0)

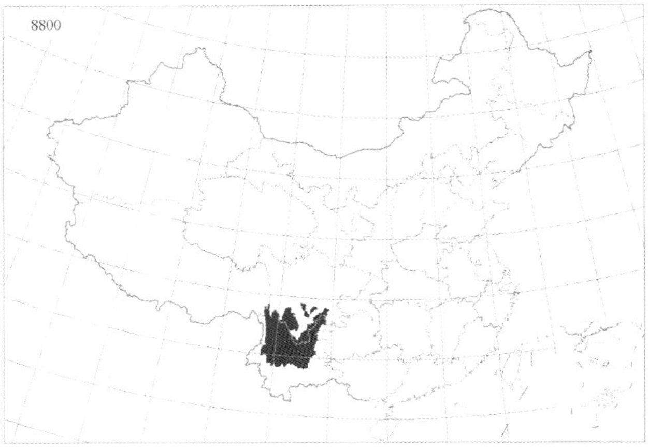

Styracaceae 安息香科 *Styrax* 安息香属

Styrax macranthus Perkins 禄春安息香

Trees

17.2 (14.2~19.6); 10.5 (7.3~13.9); 21.9 (19.4~27.5); 17.3 (14.2~19.6); 147.5 (110.6~175.1); 0.0 (0.0~0.0); 839 (724~1002) ❙ 1161 (939~1547); 648 (524~911); 49 (34~135) ❙ 38.1 (6.5~74.2); 822 (716~1002); 348.5 (234.5~551.3)

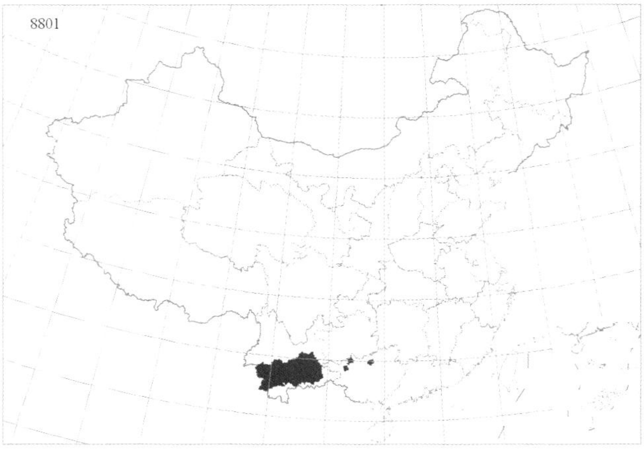

Styracaceae 安息香科 *Styrax* 安息香属

Styrax macrocarpus Cheng 大果安息香

Trees

19.7 (16.5~22.3); 10.4 (5.7~14.7); 27.6 (22.5~29.8); 19.8 (16.5~22.4); 177.3 (138.2~208.7); 0.0 (0.0~0.0); 1033 (854~1200) ❙ 1681 (1368~2037); 666 (435~1026); 168 (56~217) ❙ 63.3 (17.4~95.5); 1029 (854~1200); 243.5 (120.3~551.6)

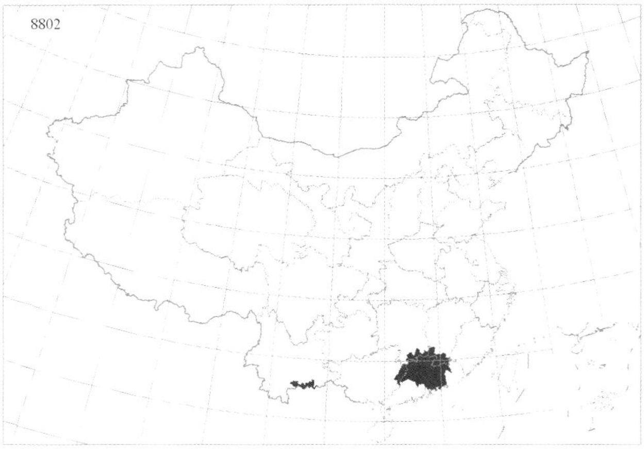

Styracaceae 安息香科 *Styrax* 安息香属
Styrax obassis Siebold et Zucc. 玉铃花
Trees

13.8 (4.8~21.1); 0.4 (−15.5~12.0); 26.2 (21.5~29.3); 14.1 (8.4~21.2); 117.7 (66.0~194.2); 11.4 (0.0~68.8); 809 (588~1121) ❙ 1085 (604~1828); 489 (316~691); 95 (26~253) ❙ 32.7 (−23.0~124.8); 786 (588~1121); 203.0 (135.6~335.3)

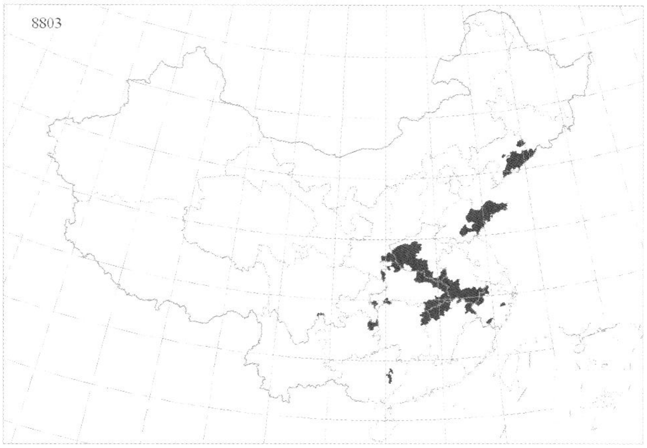

Styracaceae 安息香科 *Styrax* 安息香属
Styrax odoratissimus Champ. ex Benth. 芬芳安息香
Small trees

17.0 (9.3~22.4); 5.8 (−3.8~15.2); 27.4 (20.1~29.9); 17.1 (9.5~22.7); 146.0 (69.8~212.7); 1.2 (0.0~22.5); 913 (603~1216) ❙ 1507 (520~2037); 549 (258~1033); 171 (18~259) ❙ 65.3 (−35.9~147.0); 899 (520~1216); 220.4 (128.6~354.5)

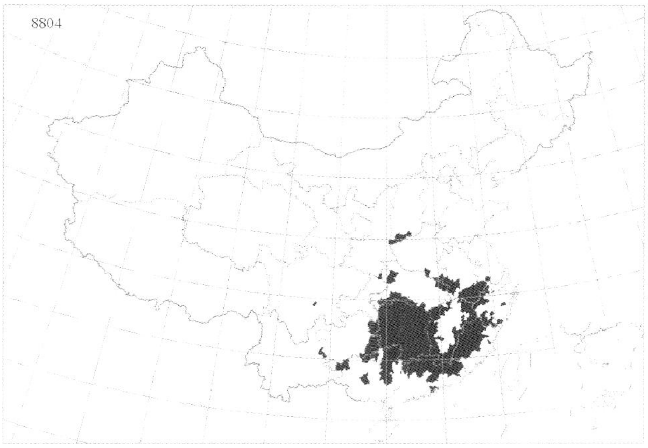

Styracaceae 安息香科 *Styrax* 安息香属
Styrax perkinsiae Rehder 瓦山安息香
Shrubs or small trees

13.5 (6.2~19.6); 6.1 (−2.4~12.8); 19.3 (14.1~23.5); 13.5 (6.7~19.6); 104.8 (39.4~175.7); 2.4 (0.0~24.1); 716 (509~953) ❙ 1207 (879~1688); 680 (472~959); 47 (22~70) ❙ 68.8 (25.7~112.2); 714 (509~932); 370.3 (171.0~627.0)

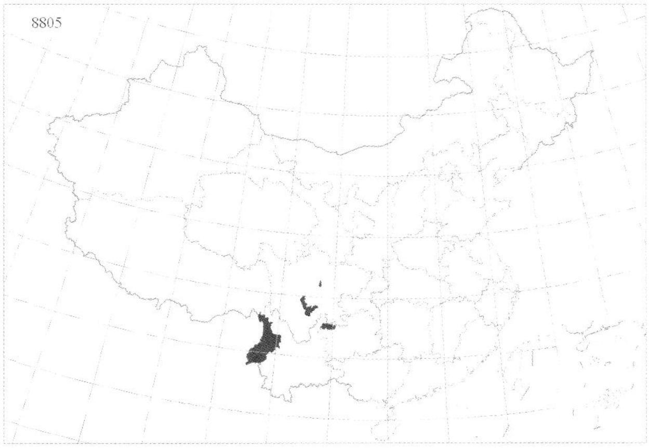

Styracaceae 安息香科 *Styrax* 安息香属
Styrax roseus Dunn 粉花安息香
Small trees

13.9 (6.2~19.1); 4.6 (−2.4~10.1); 22.0 (14.1~26.3); 14.0 (6.7~19.1); 110.0 (39.4~169.2); 2.2 (0.0~24.1); 747 (509~966) ❙ 1138 (863~1736); 585 (420~1007); 54 (22~129) ❙ 53.3 (19.4~145.8); 747 (509~966); 234.7 (153.9~447.4)

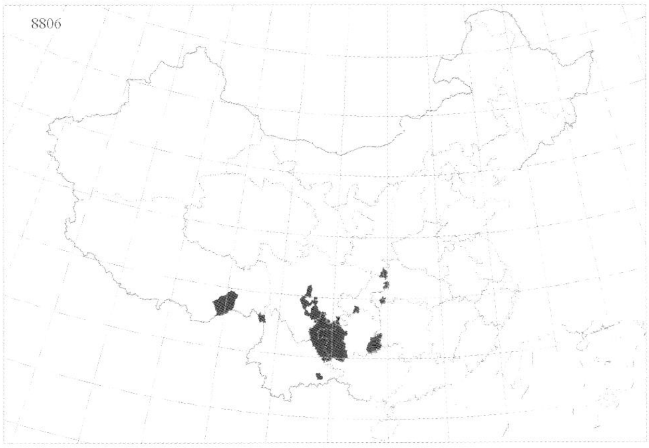

Styracaceae 安息香科 *Styrax* 安息香属
Styrax rugosus Kurz 皱叶安息香
Shrubs or small trees

18.2; 11.2; 23.3; 18.3; 159.4; 0.0; 893 ❙ 1337; 616; 74 ❙ 49.5; 865; 417.8

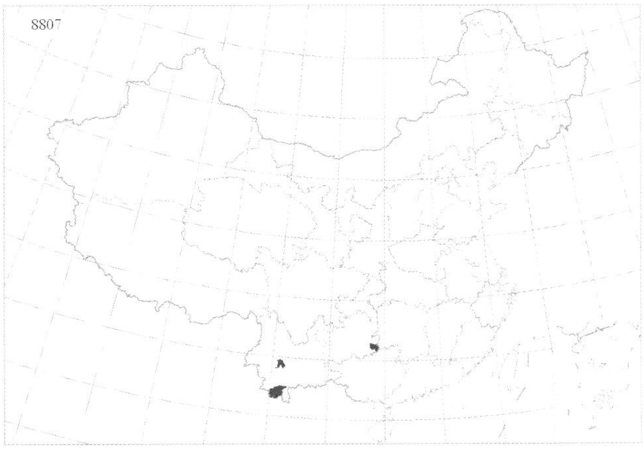

Styracaceae 安息香科 *Styrax* 安息香属
Styrax serrulatus Roxb. 齿叶安息香
Trees

21.8 (13.3~24.8); 14.3 (5.4~20.1); 27.7 (19.6~29.1); 21.9 (13.4~24.8); 202.3 (100.4~237.6); 0.0 (0.0~0.0); 1153 (706~1385) ❙ 1711 (868~2435); 787 (312~1428); 114 (41~188) ❙ 49.9 (−34.3~145.8); 1134 (706~1360); 261.0 (95.3~620.0)

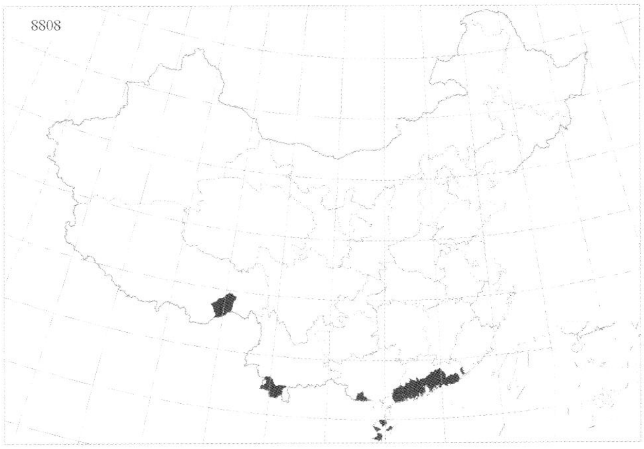

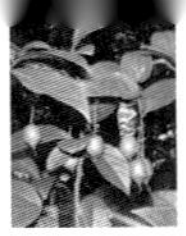

Styracaceae 安息香科 *Styrax* 安息香属
Styrax suberifolius Hook. et Arn. 栓叶安息香
Trees

18.0 (7.3~24.3); 8.2 (-2.0~18.7); 26.8 (15.6~29.8); 18.1 (7.5~24.3); 157.2 (48.2~231.9); 0.5 (0.0~20.4); 950 (537~1338) ❙ 1541 (822~2993); 620 (312~1847); 148 (19~551) ❙ 64.0 (-34.3~246.5); 942 (537~1273); 242.0 (95.3~659.0)

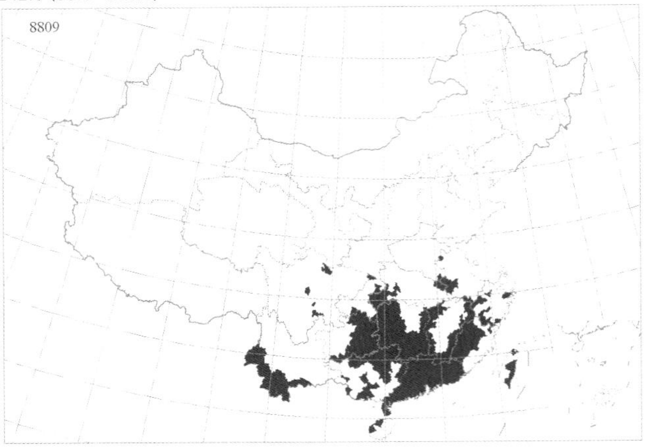

Styracaceae 安息香科 *Styrax* 安息香属
Styrax supaii Chun et F. Chun 裂叶安息香
Shrubs or small trees

19.0 (15.5~21.0); 9.2 (4.8~12.2); 27.5 (25.1~29.8); 19.1 (15.6~21.0); 169.1 (127.2~192.6); 0.0 (0.0~0.2); 994 (815~1113) ❙ 1625 (1425~1852); 589 (505~711); 184 (161~197) ❙ 64.1 (35.6~93.2); 991 (815~1107); 239.9 (174.9~271.3)

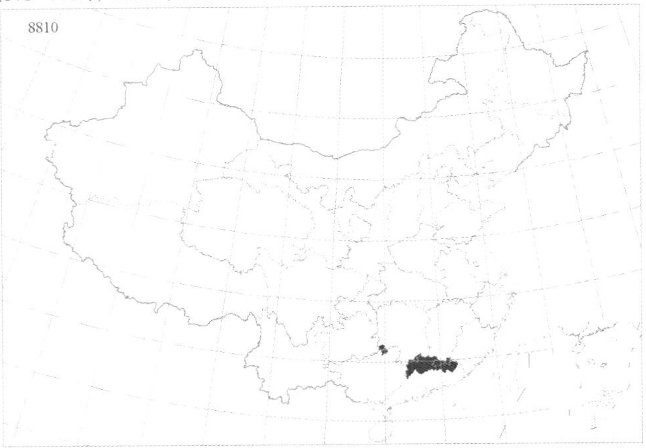

Styracaceae 安息香科 *Styrax* 安息香属
Styrax tonkinensis (Pierre) Craib ex Hartwich 越南安息香
Trees

19.8 (14.3~24.8); 11.3 (3.3~20.1); 26.7 (20.3~29.8); 19.8 (14.3~24.8); 178.0 (114.1~237.6); 0.1 (0.0~2.3); 1031 (765~1385) ❙ 1541 (868~2435); 666 (312~1428); 126 (39~249) ❙ 51.4 (-34.3~121.9); 1018 (765~1360); 276.7 (95.3~659.0)

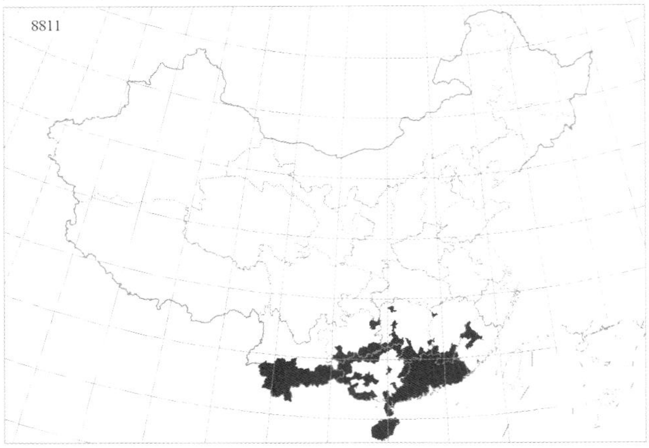

Styracaceae 安息香科 *Styrax* 安息香属
Styrax wilsonii Rehder 小叶安息香
Shrubs

10.3; 0.6; 18.8; 10.4; 74.5; 11.4; 630 ❙ 1043; 569; 30 ❙ 68.4; 630; 221.7

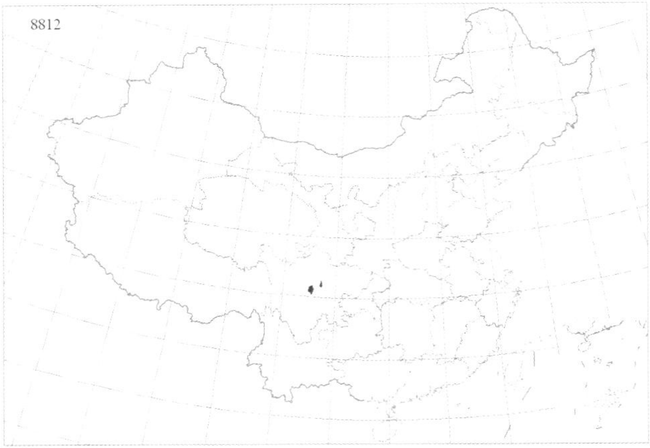

Styracaceae 安息香科 *Styrax* 安息香属
Styrax wuyuanensis S. M. Hwang 婺源安息香
Shrubs

15.5 (13.4~17.0); 3.4 (1.5~5.1); 27.2 (25.2~28.3); 15.5 (13.4~17.0); 128.5 (108.3~144.4); 2.5 (0.0~7.3); 853 (755~921) ❙ 1571 (1317~1820); 600 (552~666); 177 (119~250) ❙ 84.3 (53.3~112.2); 853 (755~921); 224.3 (169.3~264.5)

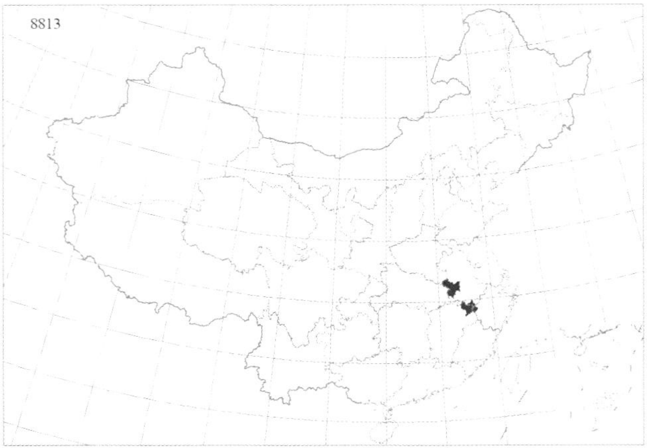

Styracaceae 安息香科 *Styrax* 安息香属
Styrax zhejiangensis S. M. Hwang et L. L. Yu 浙江安息香
Shrubs

16.2; 4.0; 28.1; 16.2; 135.4; 1.0; 884 ❙ 1536; 513; 213 ❙ 73.7; 884; 234.0

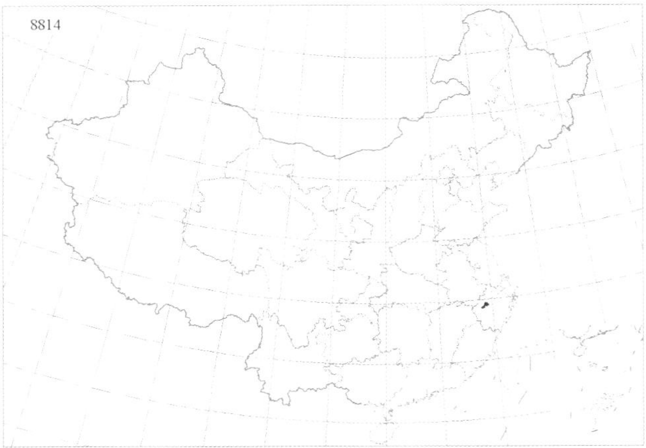

Oleaceae 木犀科 *Chionanthus* 流苏树属
Chionanthus brachythyrsus (Merr.) P. S. Green 白枝流苏树
Shrubs or small trees, deciduous

24.3；19.3；27.7；24.3；231.7；0.0；1335 ❙ 1484；494；53 ❙ 11.1；1162；475.0

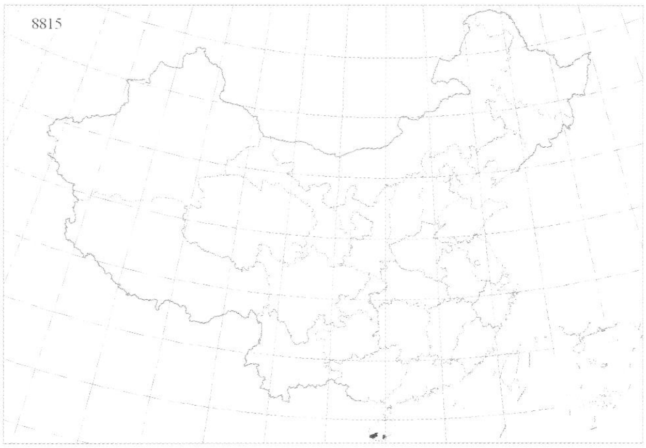

Oleaceae 木犀科 *Chionanthus* 流苏树属
Chionanthus guangxiensis B. M. Miao 广西流苏树
Shrubs or small trees, deciduous

22.2；13.9；28.3；22.3；207.5；0.0；1190 ❙ 1374；713；87 ❙ 15.4；1190；327.7

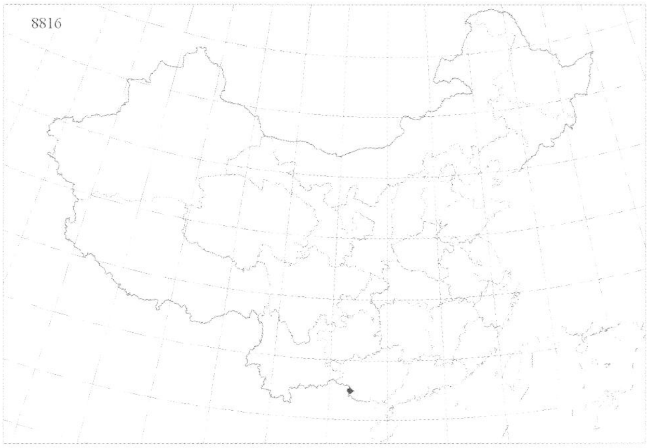

Oleaceae 木犀科 *Chionanthus* 流苏树属
Chionanthus hainanensis (Merr. et Chun) B. M. Miao 海南流苏树
Shrubs or trees, deciduous

22.9 (17.8~24.8)；16.4 (7.1~20.1)；27.8 (26.7~28.4)；22.9 (17.8~24.8)；214.8 (154.0~237.6)；0.0 (0.0~0.0)；1237 (933~1385) ❙ 1462 (1179~1767)；557 (392~850)；78 (41~182) ❙ 20.1 (-11.9~61.1)；1124 (933~1299)；422.8 (197.5~572.9)

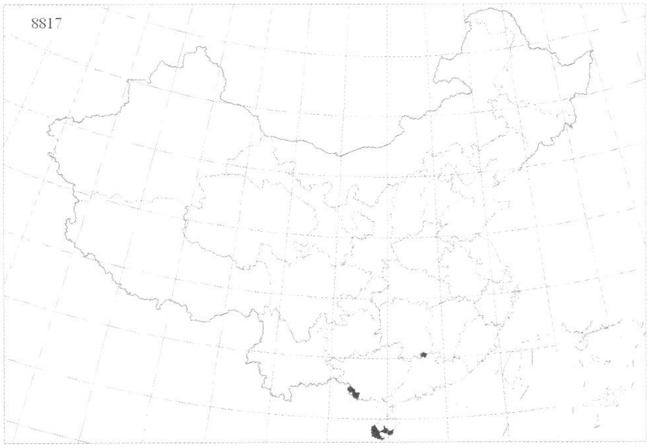

Oleaceae 木犀科 *Chionanthus* 流苏树属
Chionanthus henryanus P. S. Green 李榄
Shrubs or trees, deciduous

18.8 (17.2~21.0)；12.1 (9.1~15.1)；23.3 (22.3~25.6)；18.8 (17.3~21.0)；165.8 (147.0~191.8)；0.0 (0.0~0.0)；911 (833~1053) ❙ 1448 (1161~1751)；776 (633~996)；60 (48~77) ❙ 59.2 (37.1~88.8)；906 (833~1053)；471.1 (314.4~659.0)

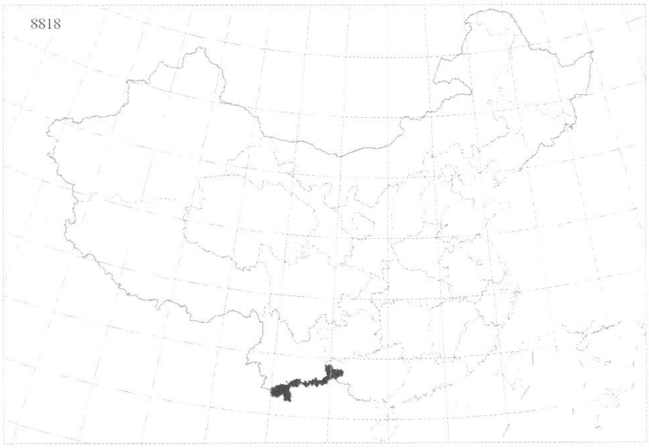

Oleaceae 木犀科 *Chionanthus* 流苏树属
Chionanthus longiflorus (H. L. Li) B. M. Miao 长花流苏树
Trees, deciduous

16.9；9.9；21.5；16.9；142.9；0.0；817 ❙ 1336；759；51 ❙ 63.7；817；388.7

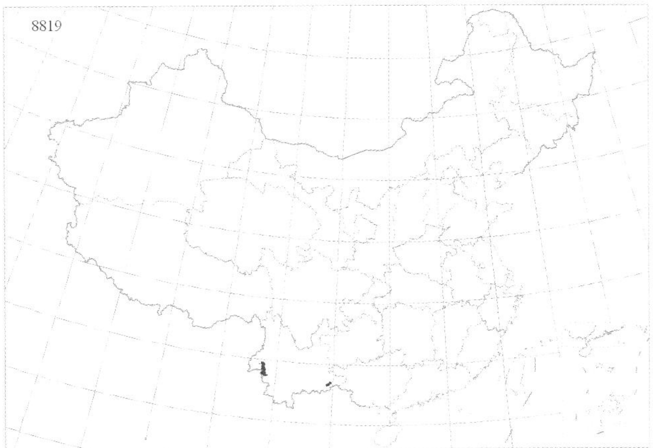

Oleaceae 木犀科 *Chionanthus* 流苏树属
Chionanthus ramiflorus Roxb. 枝花流苏树
Trees, deciduous

20.1 (13.3~24.8)；12.8 (3.8~20.1)；25.7 (19.6~29.0)；20.1 (13.4~24.8)；181.4 (100.4~237.6)；0.0 (0.0~1.2)；1034 (706~1385) ❙ 1668 (868~3495)；783 (312~1847)；126 (41~897) ❙ 66.4 (-34.3~246.5)；1003 (706~1326)；323.9 (119.4~659.0)

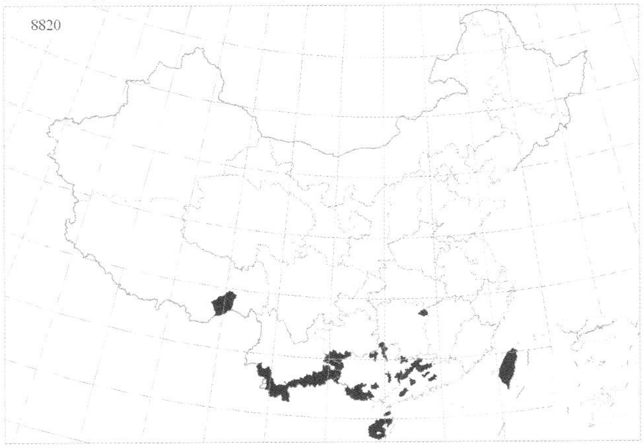

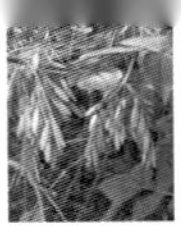

Oleaceae 木犀科 *Chionanthus* 流苏树属
Chionanthus ramiflorus var. grandiflorus B. M. Miao 大花流苏树
Trees

17.8；7.1；27.5；17.8；154.0；0.0；933 ❙ 1503；513；182 ❙ 61.1；933；197.5

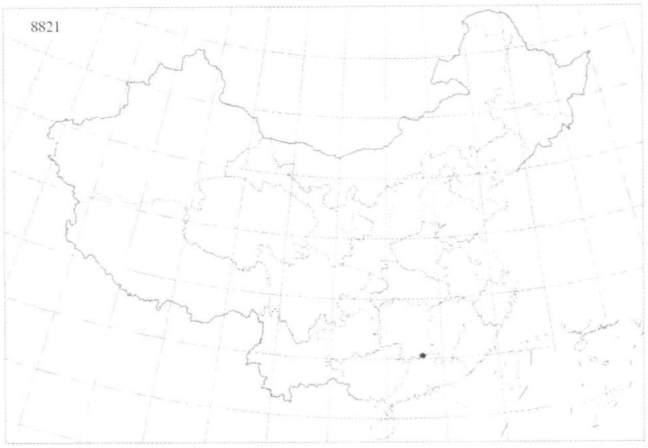

Oleaceae 木犀科 *Chionanthus* 流苏树属
Chionanthus retusus Lindl. et Paxton 流苏树
Shrubs or trees, deciduous

14.1 (2.2~21.7)；2.4 (-12.5~15.2)；24.8 (9.9~29.7)；14.4 (3.8~22.7)；119.7 (15.5~212.7)；9.5 (0.0~62.6)；803 (413~1216) ❙ 1069 (377~3495)；479 (241~994)；89 (8~897) ❙ 30.4 (-37.2~224.4)；759 (377~1216)；219.3 (109.4~491.5)

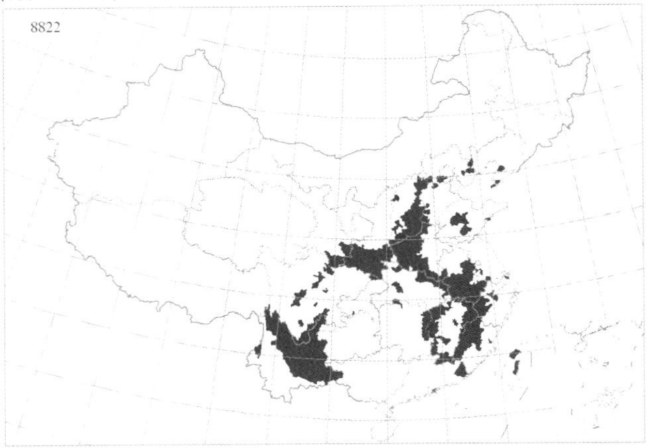

Oleaceae 木犀科 *Fontanesia* 雪柳属
Fontanesia philliraeoides subsp. fortunei (Carrière) Yalt. 雪柳
Shrubs or small trees, deciduous

14.3 (2.7~17.6)；0.4 (-20.0~5.9)；27.1 (21.6~29.4)；14.4 (8.2~17.6)；121.7 (63.1~151.4)；10.2 (0.0~89.7)；836 (591~953) ❙ 893 (432~1893)；441 (258~691)；77 (12~242) ❙ 6.4 (-47.7~129.9)；757 (432~941)；177.5 (110.1~335.3)

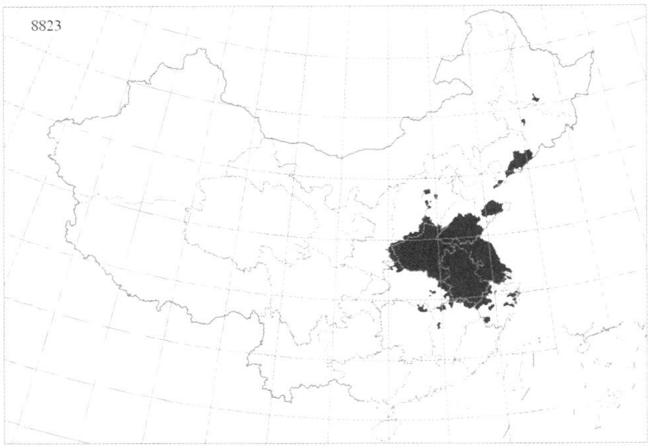

Oleaceae 木犀科 *Forsythia* 连翘属
Forsythia giraldiana Lingelsh. 秦连翘
Shrubs, deciduous

12.6 (2.6~17.1)；0.6 (-7.7~5.4)；24.0 (11.2~29.2)；12.8 (4.3~17.1)；102.6 (20.0~145.8)；11.0 (0.0~49.3)；740 (427~937) ❙ 992 (562~1571)；428 (288~666)；58 (12~161) ❙ 33.6 (-23.0~108.0)；734 (427~927)；233.9 (140.5~334.4)

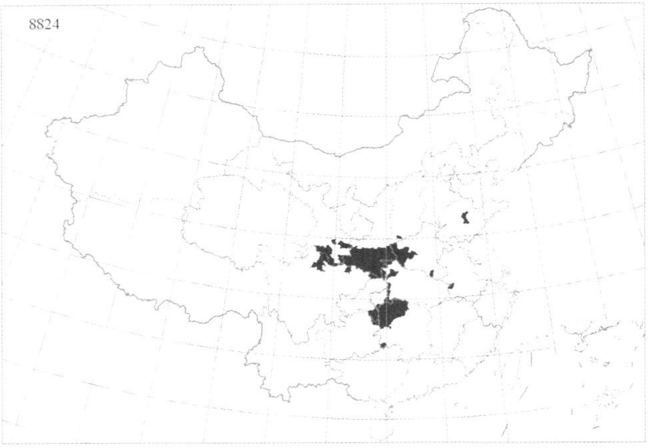

Oleaceae 木犀科 *Forsythia* 连翘属
Forsythia likiangensis Ching et Feng ex P. Y. Bai 丽江连翘
Shrubs, deciduous

7.8；0.4；14.2；8.0；48.9；14.3；548 ❙ 877；546；19 ❙ 60.7；548；304.7

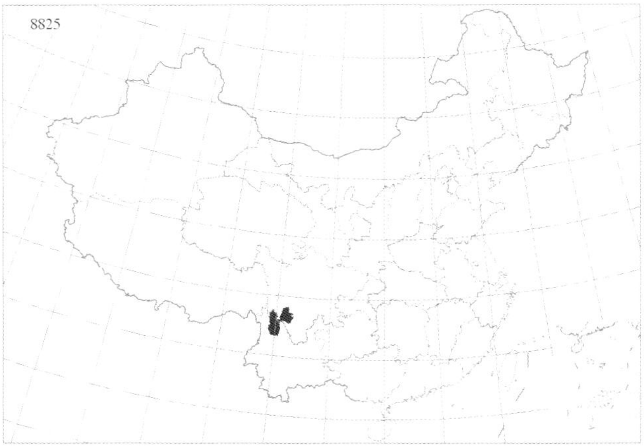

Oleaceae 木犀科 *Forsythia* 连翘属
Forsythia mandschurica Uyeki 东北连翘
Shrubs, deciduous

7.4 (5.6~8.7)；-11.2 (-13.5~-8.3)；23.5 (21.9~25.0)；9.8 (8.7~10.6)；81.5 (69.5~90.9)；51.8 (40.2~61.8)；653 (601~700) ❙ 845 (698~1073)；532 (434~691)；30 (22~40) ❙ 30.1 (-0.3~74.1)；652 (601~698)；264.5 (191.9~335.3)

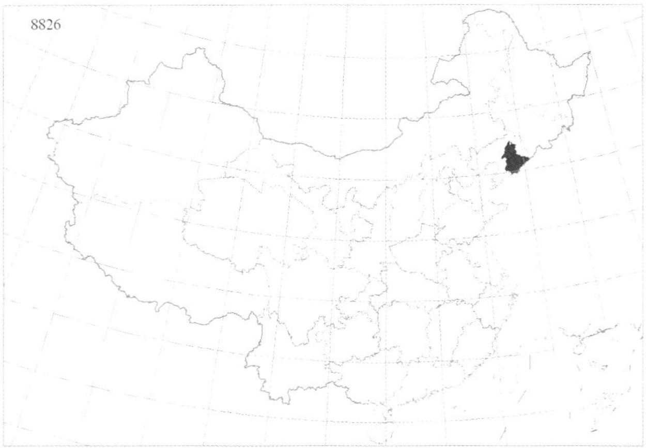

Oleaceae 木犀科 *Forsythia* 连翘属

Forsythia mira M. C. Chang 奇异连翘

Scandent shrubs, deciduous

10.9；-2.3；23.2；11.1；88.4；17.6；679 ▎738；350；33 ▎8.6；679；243.3

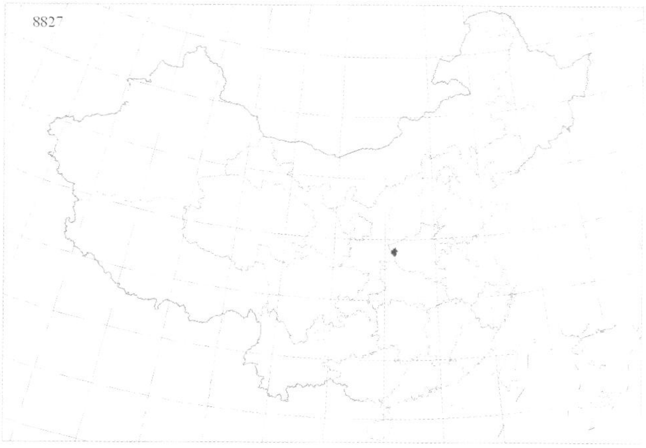

Oleaceae 木犀科 *Forsythia* 连翘属

Forsythia suspensa (Thunb.) Vahl 连翘

Shrubs, deciduous

12.5 (0.0~22.3)；-1.2 (-12.5~13.5)；25.0 (9.7~29.3)；12.9 (3.3~22.3)；106.1 (12.2~208.1)；15.9 (0.0~72.5)；762 (395~1202) ▎821 (367~1619)；409 (219~710)；48 (5~225) ▎7.2 (-43.4~108.0)；692 (367~1202)；188.0 (91.4~364.1)

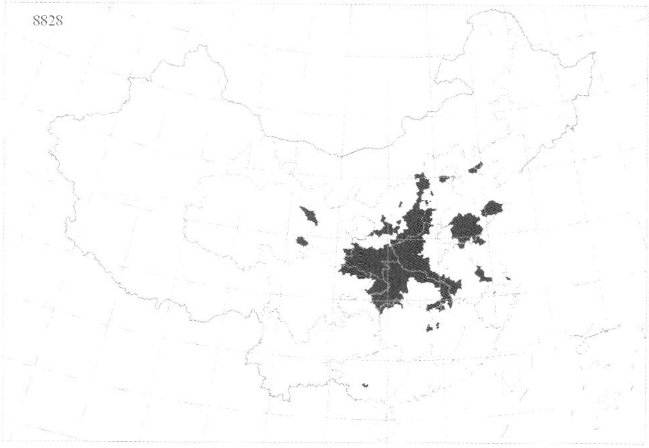

Oleaceae 木犀科 *Forsythia* 连翘属

Forsythia viridissima Lindl. 金钟花

Shrubs, deciduous

15.3 (7.4~19.3)；3.6 (-4.2~10.1)；26.2 (15.7~29.4)；15.3 (8.2~19.3)；127.2 (58.8~172.0)；3.4 (0.0~29.4)；839 (561~1037) ▎1215 (622~1997)；505 (290~723)；111 (16~259) ▎44.5 (-23.1~130.7)；828 (561~1000)；210.5 (110.1~372.3)

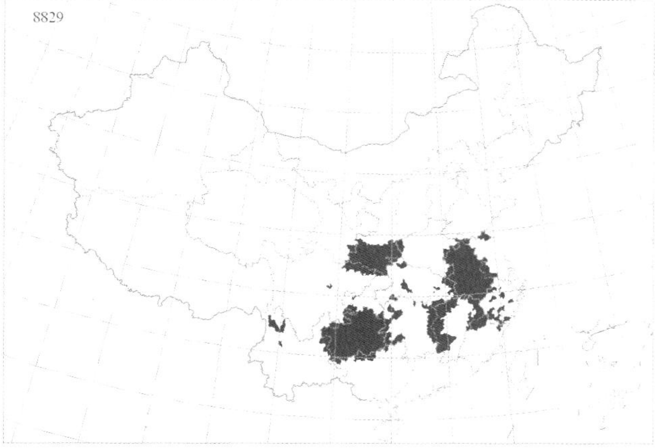

Oleaceae 木犀科 *Fraxinus* 白蜡树属

Fraxinus baroniana Diels 狭叶梣

Small trees, deciduous

11.5 (7.4~13.5)；-0.2 (-4.2~2.3)；22.8 (18.7~25.1)；11.7 (8.2~13.6)；91.5 (58.8~109.4)；12.9 (4.9~29.4)；691 (561~767) ▎868 (675~1118)；398 (331~482)；28 (16~75) ▎26.9 (-1.2~60.4)；690 (561~767)；290.4 (247.3~318.2)

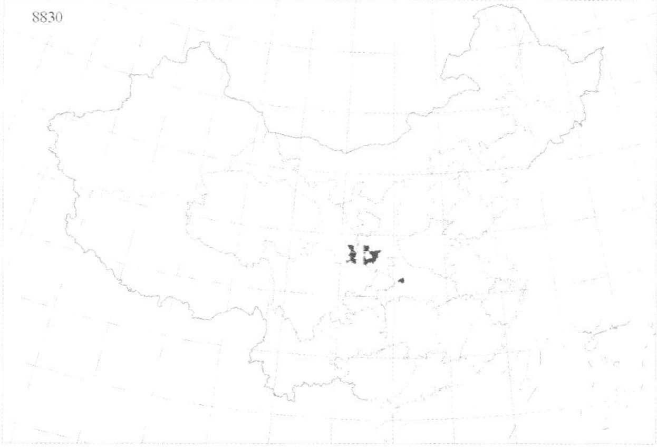

Oleaceae 木犀科 *Fraxinus* 白蜡树属

Fraxinus bungeana A. DC. 小叶梣

Shrubs or small trees, deciduous

11.0 (1.4~17.1)；-4.9 (-18.8~4.1)；25.1 (18.4~29.4)；12.0 (6.4~17.1)；100.8 (45.8~146.1)；28.2 (0.9~88.8)；750 (501~952) ▎581 (349~1663)；367 (227~608)；22 (7~212) ▎-21.3 (-51.1~101.3)；563 (349~930)；173.9 (76.1~346.5)

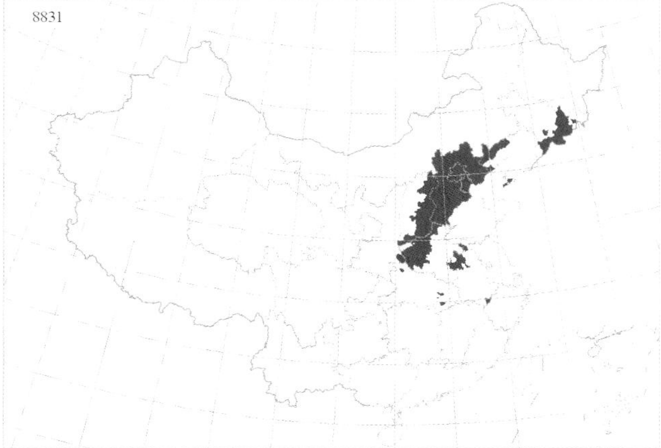

Oleaceae 木犀科 *Fraxinus* 白蜡树属

Fraxinus chinensis Roxb. 白蜡树

Trees, deciduous

14.5 (1.4~22.4)；2.2 (-20.1~14.7)；25.8 (9.9~29.9)；14.9 (3.8~22.5)；124.4 (15.4~209.6)；9.4 (0.0~91.9)；830 (413~1205) ▎1086 (196~2014)；485 (117~1056)；92 (4~259) ▎30.1 (-69.4~137.5)；786 (196~1205)；206.4 (61.2~551.6)

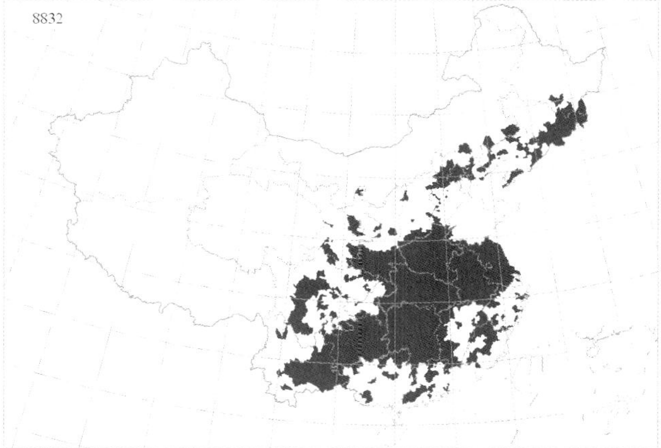

Oleaceae 木犀科 *Fraxinus* 白蜡树属
Fraxinus chinensis subsp. rhynchophylla (Hance) E. Murray 花曲柳
Trees, deciduous

10.7 (0.6~21.0); -4.4 (-20.0~14.2); 24.3 (12.1~29.3); 11.7 (5.1~21.0); 96.9 (27.8~191.8); 27.8 (0.0~101.0); 723 (441~1053) ❙ 679 (337~1663); 382 (227~818); 33 (1~220) ❙ -6.8 (-43.4~106.2); 616 (337~1053); 182.3 (97.2~551.6)

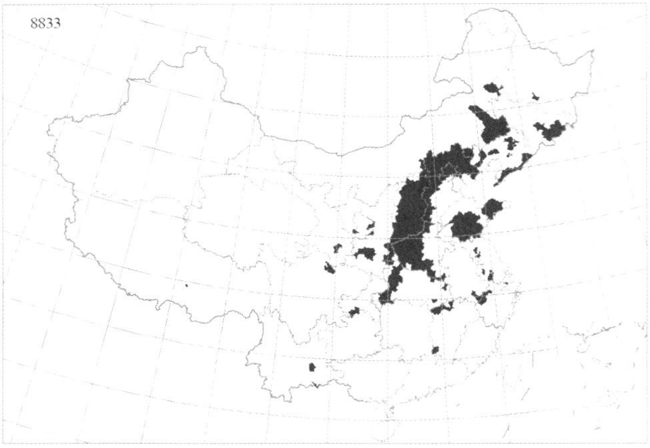

Oleaceae 木犀科 *Fraxinus* 白蜡树属
Fraxinus depauperata (Lingelsh.) Z. Wei 疏花梣
Small trees, deciduous

15.7; 4.1; 26.7; 15.7; 129.3; 0.9; 845 ❙ 1441; 575; 121 ❙ 70.5; 845; 200.9

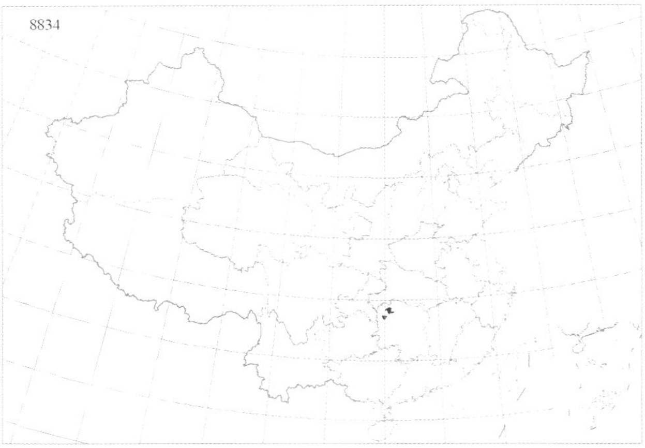

Oleaceae 木犀科 *Fraxinus* 白蜡树属
Fraxinus ferruginea Lingelsh. 锈毛梣
Trees, deciduous

16.2 (5.3~20.7); 6.7 (-2.4~15.1); 24.4 (12.2~29.8); 16.3 (5.7~20.8); 136.3 (30.5~189.3); 1.0 (0.0~26.2); 842 (479~1113) ❙ 1292 (961~1607); 583 (494~775); 85 (31~189) ❙ 54.7 (27.5~100.7); 837 (479~1032); 249.9 (172.1~659.0)

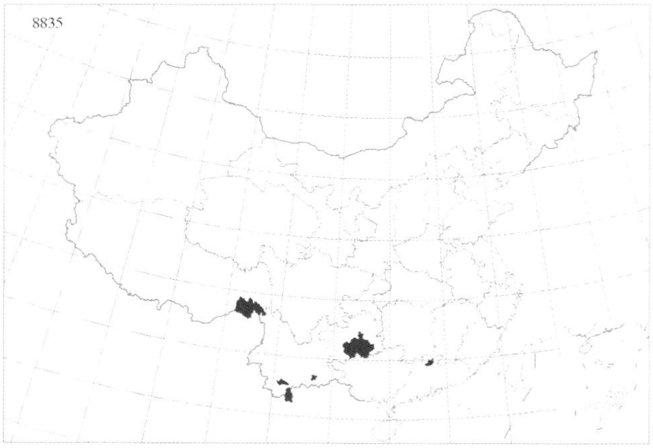

Oleaceae 木犀科 *Fraxinus* 白蜡树属
Fraxinus floribunda Wall. 多花梣
Trees, deciduous

16.3 (8.6~24.1); 7.0 (-1.7~18.9); 24.4 (15.0~29.3); 16.3 (8.7~24.1); 136.5 (52.6~229.7); 0.7 (0.0~18.0); 843 (574~1322) ❙ 1409 (906~3495); 630 (412~1847); 111 (31~897) ❙ 66.9 (9.3~246.5); 839 (574~1123); 272.4 (146.0~659.0)

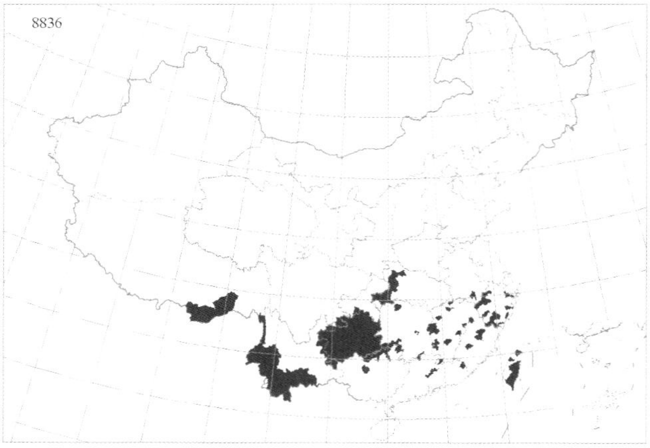

Oleaceae 木犀科 *Fraxinus* 白蜡树属
Fraxinus griffithii C. B. Clarke 光蜡树
Trees, semideciduous

17.8 (5.3~24.8); 8.3 (-2.4~20.1); 26.3 (12.2~29.7); 17.8 (5.7~24.8); 155.3 (30.5~237.6); 1.4 (0.0~26.2); 946 (479~1385) ❙ 1612 (868~3495); 648 (312~1847); 159 (20~897) ❙ 74.0 (-34.3~246.5); 922 (479~1329); 247.9 (111.9~538.2)

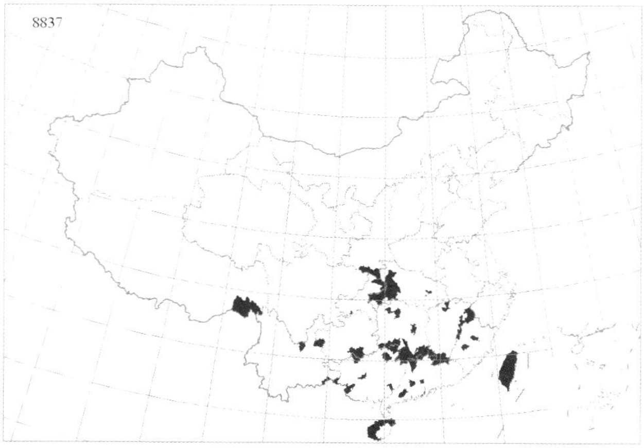

Oleaceae 木犀科 *Fraxinus* 白蜡树属
Fraxinus hupehensis S. T. Chu et C. B. Shang 湖北梣
Trees, deciduous

16.1; 3.3; 28.0; 16.1; 134.9; 1.8; 895 ❙ 986; 431; 89 ❙ 10.2; 895; 172.3

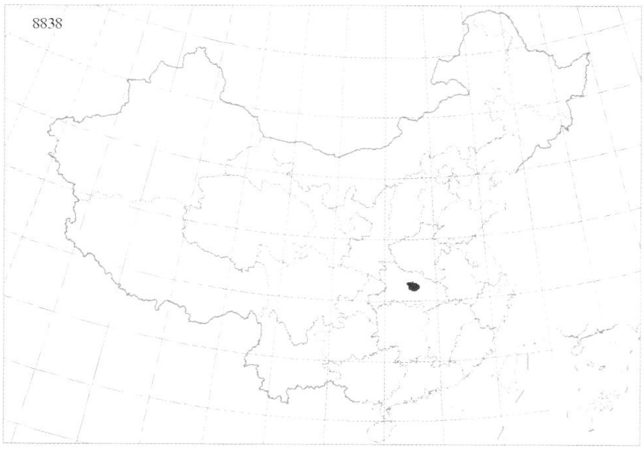

Oleaceae 木犀科 *Fraxinus* 白蜡树属

Fraxinus insularis Hemsl. 苦枥木

Trees, deciduous

16.5 (6.0~22.4); 5.8 (-3.2~17.7); 26.4 (13.3~29.9); 16.5 (6.6~22.7); 139.1 (39.6~212.7); 0.9 (0.0~26.7); 877 (508~1216) ❙ 1464 (815~2993); 562 (352~1847); 149 (12~547) ❙ 66.9 (7.2~235.6); 870 (508~1216); 219.9 (111.9~363.7)

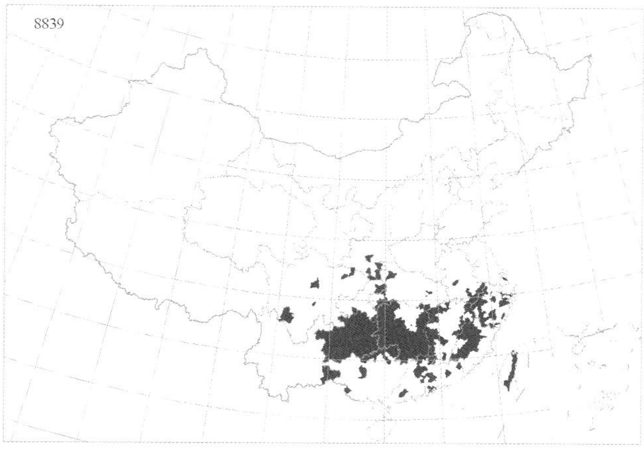

Oleaceae 木犀科 *Fraxinus* 白蜡树属

Fraxinus malacophylla Hemsl. 白枪杆

Trees, deciduous

16.4 (12.4~21.6); 9.3 (3.3~14.2); 21.7 (18.6~27.9); 16.5 (12.4~21.6); 137.7 (89.1~199.6); 0.1 (0.0~2.2); 814 (675~1143) ❙ 1098 (797~1751); 618 (411~996); 46 (22~102) ❙ 34.8 (-14.2~88.8); 794 (675~1143); 308.2 (200.4~551.6)

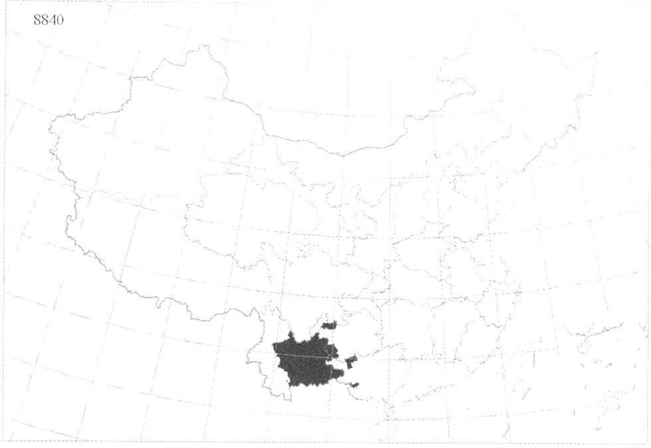

Oleaceae 木犀科 *Fraxinus* 白蜡树属

Fraxinus mandschurica Rupr. 水曲柳

Trees, deciduous

8.0 (-5.0~15.2); -8.6 (-29.4~2.3); 22.8 (12.4~27.5); 10.2 (4.8~15.3); 81.9 (25.1~128.0); 44.9 (4.3~154.8); 658 (441~865) ❙ 692 (341~1200); 390 (190~691); 24 (5~109) ❙ 6.1 (-46.6~83.5); 606 (341~858); 237.3 (61.2~346.5)

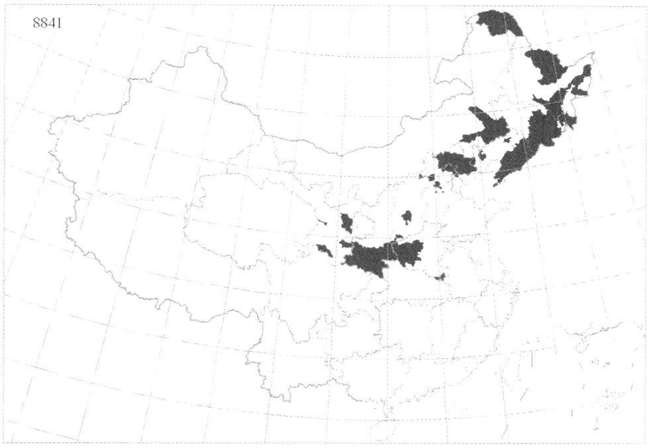

Oleaceae 木犀科 *Fraxinus* 白蜡树属

Fraxinus odontocalyx Hand.-Mazz. ex E. Peter 尖萼梣

Trees, deciduous

15.5 (2.0~21.6); 4.6 (-7.2~13.2); 25.7 (10.1~29.1); 15.6 (3.8~21.6); 129.6 (15.4~199.2); 3.2 (0.0~51.4); 846 (413~1139) ❙ 1545 (943~1997); 565 (473~858); 171 (23~259) ❙ 82.8 (47.0~130.7); 843 (413~1139); 238.7 (170.4~303.4)

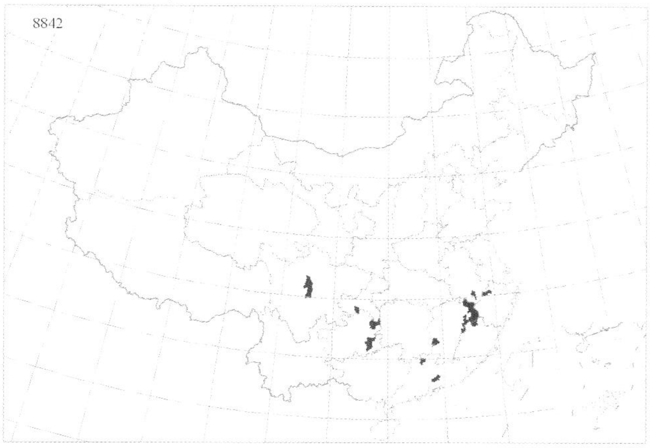

Oleaceae 木犀科 *Fraxinus* 白蜡树属

Fraxinus paxiana Lingelsh. 秦岭梣

Trees, deciduous

10.5 (1.8~17.4); -0.4 (-10.8~11.0); 20.5 (9.9~27.1); 10.9 (3.8~17.4); 83.0 (15.5~149.3); 16.4 (0.0~53.6); 659 (404~834) ❙ 904 (377~1478); 456 (241~817); 34 (11~113) ❙ 40.6 (-37.2~125.9); 636 (377~834); 258.7 (109.4~434.4)

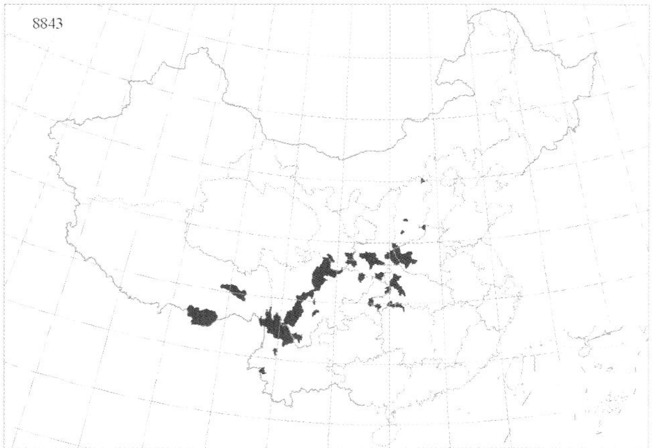

Oleaceae 木犀科 *Fraxinus* 白蜡树属

Fraxinus platypoda Oliv. 象蜡树

Trees, deciduous

8.9 (1.9~14.6); -2.9 (-9.2~3.9); 19.9 (9.9~26.8); 9.6 (3.8~14.7); 71.3 (15.5~122.7); 23.6 (1.3~57.8); 615 (413~834) ❙ 735 (379~1492); 359 (205~613); 24 (9~96) ❙ 21.6 (-34.8~125.9); 589 (379~817); 224.1 (60.5~357.7)

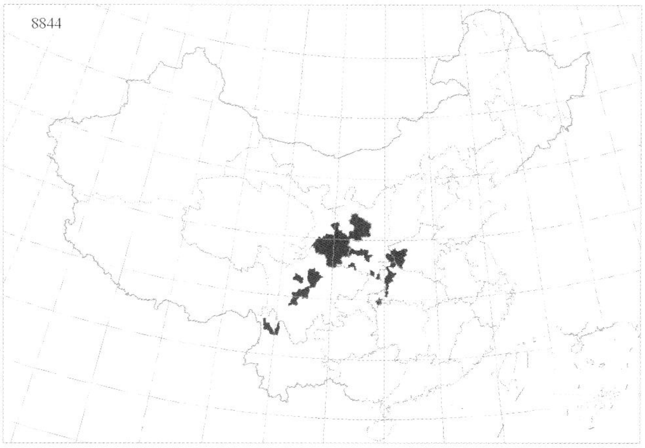

Oleaceae 木犀科 *Fraxinus* 白蜡树属
Fraxinus punctata S. Y. Hu 斑叶梣
Shrubs or small trees, deciduous

12.7；2.1；23.0；12.8；99.0；5.6；713 ▎1478；592；85 ▎107.2；713；234.7

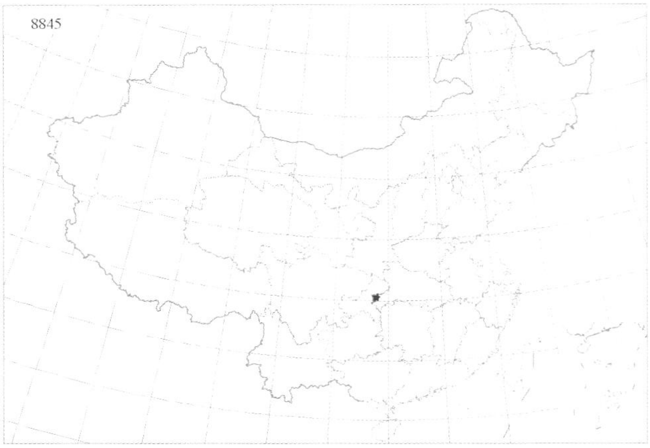

Oleaceae 木犀科 *Fraxinus* 白蜡树属
Fraxinus retusifoliolata Feng ex P. Y. Bai 楷叶梣
Small trees, deciduous

15.6；8.8；20.7；15.7；128.5；0.0；780 ▎985；565；37 ▎27.0；763；303.5

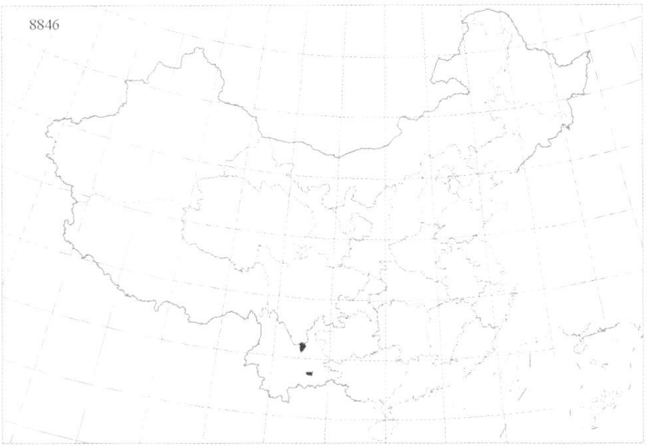

Oleaceae 木犀科 *Fraxinus* 白蜡树属
Fraxinus sieboldiana Blume 庐山梣
Trees, deciduous

16.2 (13.4~19.5)；4.6 (1.5~10.9)；27.5 (25.2~29.3)；16.2 (13.4~19.6)；136.1 (108.3~174.7)；1.2 (0.0~7.3)；878 (755~1007) ▎1589 (1211~1997)；578 (427~685)；192 (134~254) ▎81.9 (28.8~130.7)；876 (755~1007)；235.3 (181.5~303.4)

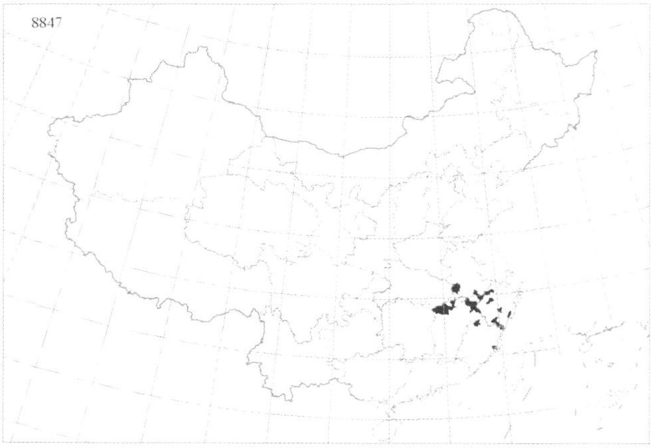

Oleaceae 木犀科 *Fraxinus* 白蜡树属
Fraxinus sikkimensis (Lingelsh.) Hand.-Mazz. 锡金梣
Trees, deciduous

12.9 (1.2~18.9)；5.4 (-7.1~12.0)；19.1 (9.0~26.6)；13.0 (3.3~19.0)；99.3 (11.9~167.8)；3.6 (0.0~56.5)；701 (390~929) ▎995 (578~1454)；573 (345~817)；35 (12~70) ▎44.4 (-14.2~113.0)；678 (390~912)；288.9 (130.1~487.2)

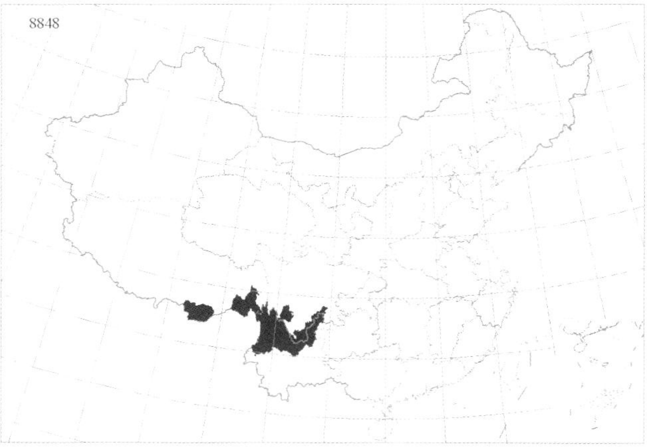

Oleaceae 木犀科 *Fraxinus* 白蜡树属
Fraxinus sogdiana Bunge 天山梣
Trees, deciduous

7.8 (-1.3~12.5)；-10.2 (-20.9~-5.1)；21.9 (12.9~26.5)；10.1 (4.5~13.2)；82.4 (25.5~114.0)；47.8 (23.1~103.9)；673 (433~822) ▎142 (28~406)；70 (17~211)；11 (0~48) ▎-74.6 (-96.6~-7.6)；142 (28~406)；80.2 (1.6~297.3)

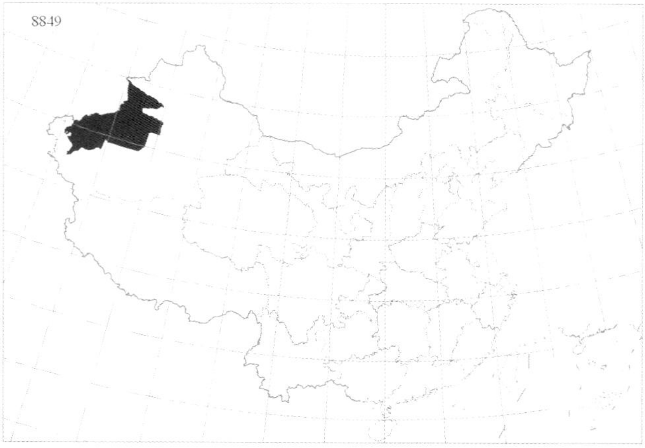

Oleaceae 木犀科 *Fraxinus* 白蜡树属
Fraxinus stylosa Lingelsh. 宿柱梣
Small trees, deciduous

12.0 (2.9~17.2)；0.3 (-6.5~5.3)；23.2 (11.2~29.3)；12.2 (4.4~17.2)；96.2 (20.5~146.8)；11.7 (0.0~44.6)；713 (428~941) ▎981 (622~1590)；436 (298~685)；50 (15~184) ▎38.3 (-13.4~121.7)；712 (428~941)；255.7 (164.0~334.4)

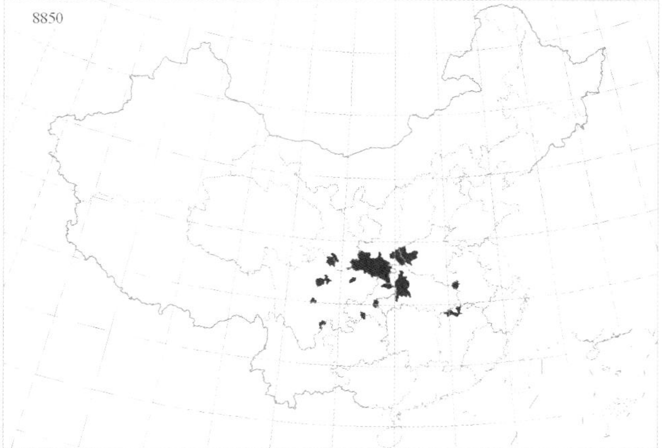

Oleaceae 木犀科 *Fraxinus* 白蜡树属
Fraxinus trifoliolata W. W. Sm. 三叶梣
Shrubs, deciduous

14.8 (5.4~18.9); 7.8 (-2.1~12.5); 20.1 (12.3~24.3); 14.8 (5.8~19.0); 119.3 (31.5~167.8); 1.3 (0.0~25.6); 754 (481~929) ❙ 977 (741~1147); 571 (411~651); 35 (12~52) ❙ 31.3 (-14.2~78.3); 721 (481~851); 288.2 (205.2~365.3)

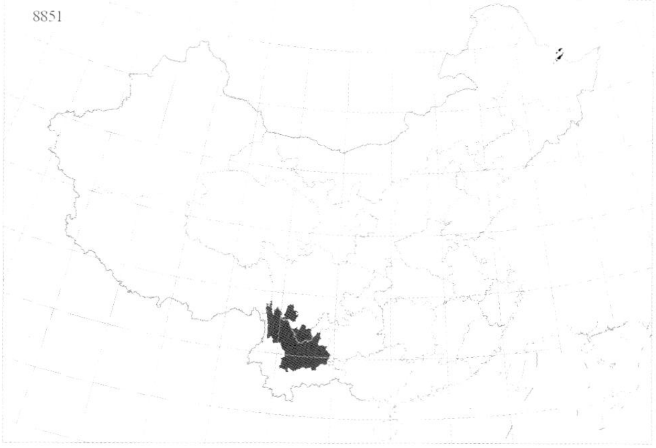

Oleaceae 木犀科 *Fraxinus* 白蜡树属
Fraxinus xanthoxyloides (G. Don) A. DC. 椒叶梣
Shrubs or small trees, deciduous

4.4; -9.3; 16.5; 7.1; 52.9; 59.2; 531 ❙ 276; 126; 51 ❙ -20.5; 188; 36.0

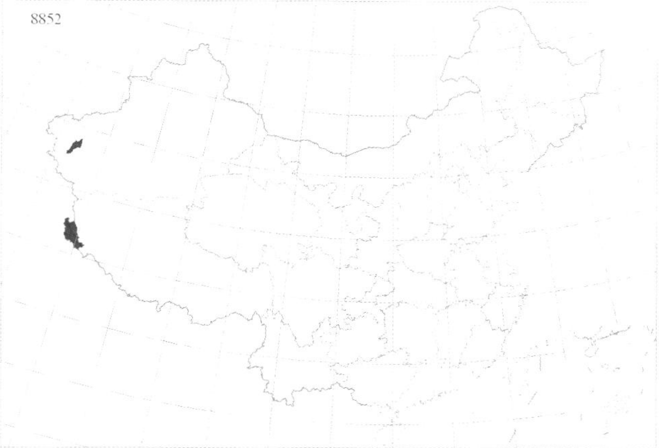

Oleaceae 木犀科 *Jasminum* 素馨属
Jasminum albicalyx Kobuski 白萼素馨
Scandent shrubs

20.0 (17.6~22.2); 10.5 (7.3~13.9); 27.8 (25.5~29.0); 20.1 (17.6~22.3); 181.0 (151.6~207.5); 0.0 (0.0~0.0); 1055 (909~1190) ❙ 1531 (1250~1776); 706 (639~821); 142 (58~189) ❙ 46.0 (15.4~85.5); 1054 (909~1190); 238.7 (160.7~335.1)

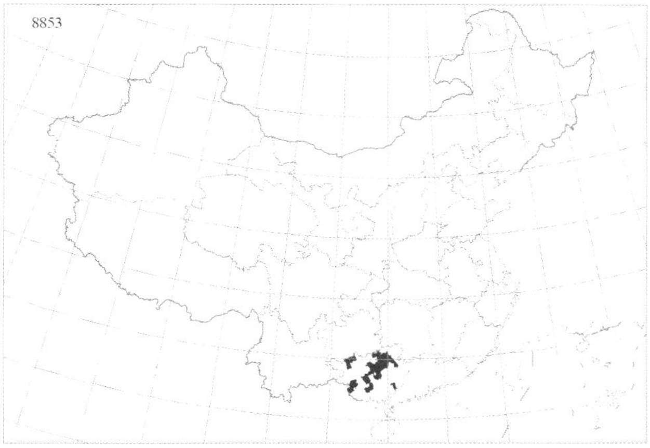

Oleaceae 木犀科 *Jasminum* 素馨属
Jasminum attenuatum Roxb. ex G. Don 大叶素馨
Vines

18.1 (14.4~20.5); 11.8 (7.5~15.1); 21.9 (19.1~23.7); 18.1 (14.4~20.5); 157.5 (113.1~185.9); 0.0 (0.0~0.0); 870 (727~992) ❙ 1471 (1192~1751); 801 (655~996); 55 (41~77) ❙ 69.7 (47.1~112.2); 856 (727~986); 485.1 (346.6~659.0)

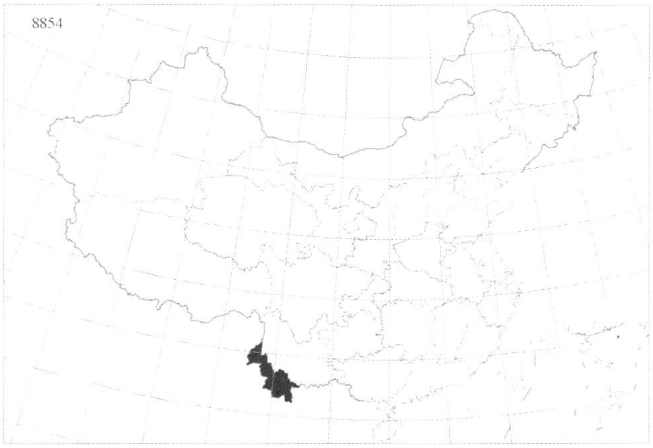

Oleaceae 木犀科 *Jasminum* 素馨属
Jasminum beesianum Forrest et Diels 红素馨
Vines

13.7 (2.0~19.3); 6.2 (-7.4~13.1); 19.7 (9.9~26.6); 13.8 (3.8~19.4); 107.9 (15.5~173.2); 2.8 (0.0~51.9); 724 (413~939) ❙ 1004 (741~1454); 576 (411~817); 37 (11~70) ❙ 41.2 (-14.2~125.9); 705 (413~924); 282.5 (142.9~487.2)

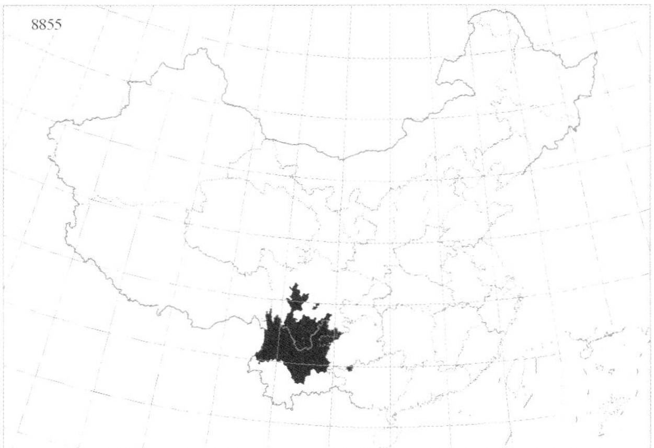

Oleaceae 木犀科 *Jasminum* 素馨属
Jasminum cinnamomifolium Kobuski 樟叶素馨
Scandent shrubs

23.4 (17.4~24.8); 18.1 (11.0~20.1); 27.1 (21.4~28.5); 23.5 (17.4~24.8); 221.5 (149.3~237.6); 0.0 (0.0~0.0); 1272 (834~1385) ❙ 1535 (1179~1813); 566 (392~816); 62 (41~113) ❙ 22.9 (-11.9~68.7); 1143 (834~1329); 484.0 (248.2~620.0)

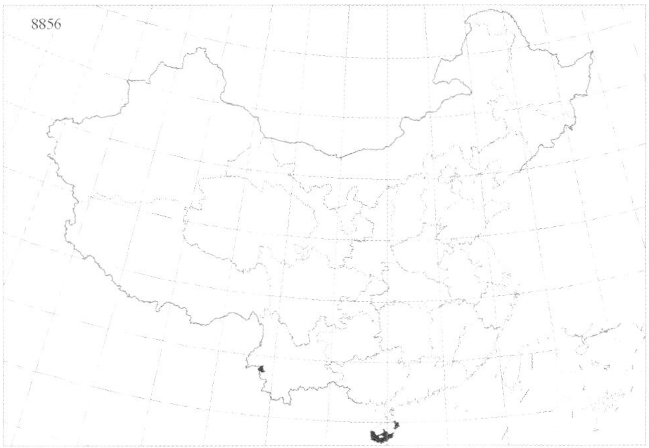

Oleaceae 木犀科 *Jasminum* 素馨属

Jasminum craibianum Kerr 毛萼素馨

Vines

24.0；19.1；27.2；24.0；228.1；0.0；1302 ❙ 1551；539；48 ❙ 19.1；1147；620.0

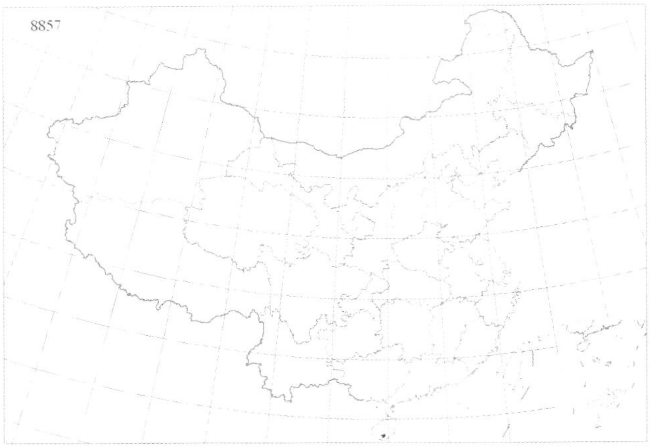

Oleaceae 木犀科 *Jasminum* 素馨属

Jasminum dispermum Wall. 双子素馨

Scandent shrubs

16.7 (9.6~19.6)；10.1 (2.4~13.9)；21.2 (15.7~23.6)；16.7 (9.6~19.6)；140.4 (60.8~175.7)；0.1 (0.0~5.4)；815 (600~953) ❙ 1188 (911~1751)；670 (524~996)；49 (32~77) ❙ 46.1 (6.5~112.2)；799 (600~932)；370.9 (234.5~627.0)

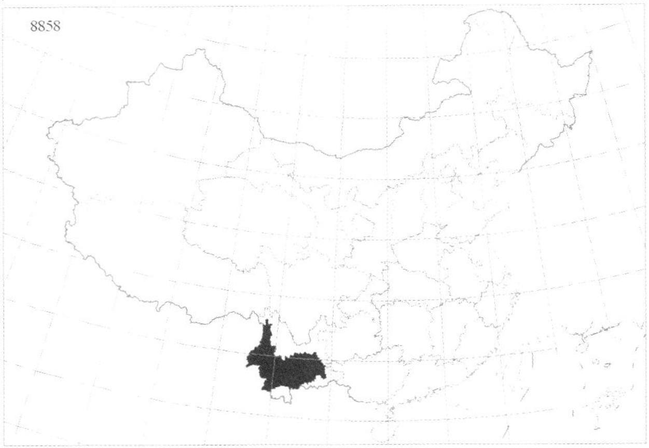

Oleaceae 木犀科 *Jasminum* 素馨属

Jasminum duclouxii (H. Lév.) Rehder 丛林素馨

Scandent shrubs

16.1 (5.4~22.2)；9.2 (−2.1~15.1)；21.0 (12.3~28.3)；16.1 (5.8~22.3)；133.9 (31.5~207.5)；0.6 (0.0~25.6)；802 (481~1190) ❙ 1150 (741~1751)；649 (411~996)；46 (15~106) ❙ 43.8 (−14.2~112.2)；784 (481~1190)；347.9 (185.6~659.0)

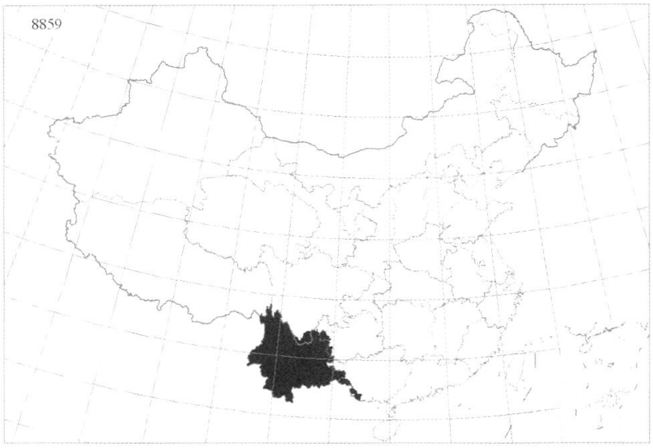

Oleaceae 木犀科 *Jasminum* 素馨属

Jasminum elongatum (Bergius) Willd. 扭肚藤

Scandent shrubs

18.7 (−1.1~24.8)；10.5 (−9.1~20.1)；25.3 (6.2~29.2)；18.8 (2.0~24.8)；167.2 (2.5~237.6)；2.4 (0.0~75.6)；1004 (334~1385) ❙ 1401 (321~2148)；665 (193~1234)；97 (11~194) ❙ 41.4 (−12.0~128.6)；991 (321~1360)；265.1 (79.3~659.0)

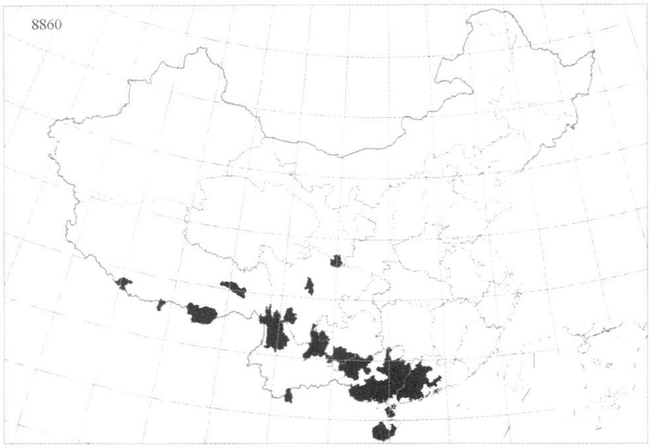

Oleaceae 木犀科 *Jasminum* 素馨属

Jasminum flexile Vahl 盈江素馨

Vines

17.9；11.5；22.1；18.0；155.9；0.0；866 ❙ 1389；790；56 ❙ 62.9；854；468.9

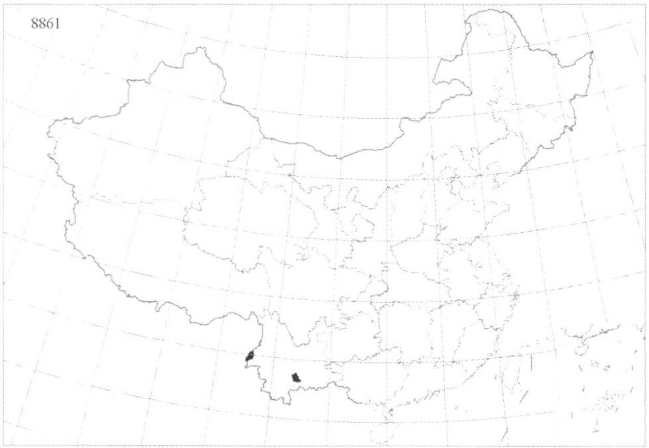

Oleaceae 木犀科 *Jasminum* 素馨属

Jasminum floridum Bunge 探春花

Scandent shrubs

13.1 (2.6~17.8)；0.4 (−7.0~8.0)；24.9 (11.2~29.4)；13.3 (4.3~17.9)；108.8 (20.0~154.7)；11.2 (0.0~48.2)；771 (427~952) ❙ 891 (533~1590)；415 (258~685)；51 (14~197) ❙ 17.1 (−36.5~121.7)；716 (427~942)；197.2 (111.9~334.4)

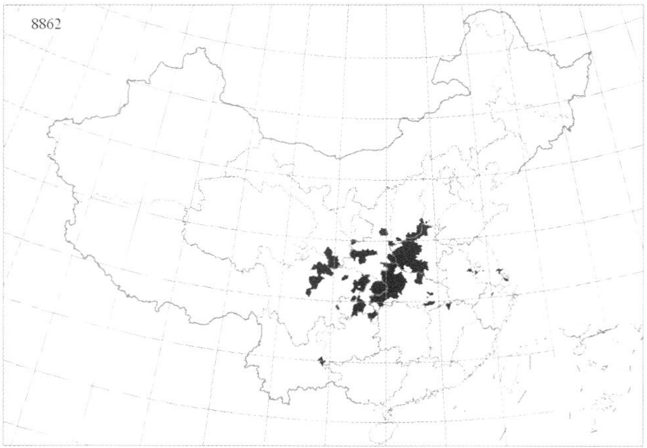

Oleaceae 木犀科 *Jasminum* 素馨属

Jasminum fuchsiifolium Gagnep. 倒吊钟叶素馨

Scandent shrubs

17.3 (14.2~21.0); 9.6 (6.2~14.2); 23.0 (20.1~26.6); 17.4 (14.2~21.0); 148.5 (110.5~191.8); 0.0 (0.0~0.0); 857 (729~1053) ❙ 1186 (958~1522); 655 (531~860); 51 (37~68) ❙ 38.5 (13.3~80.6); 851 (717~1053); 329.2 (189.5~551.6)

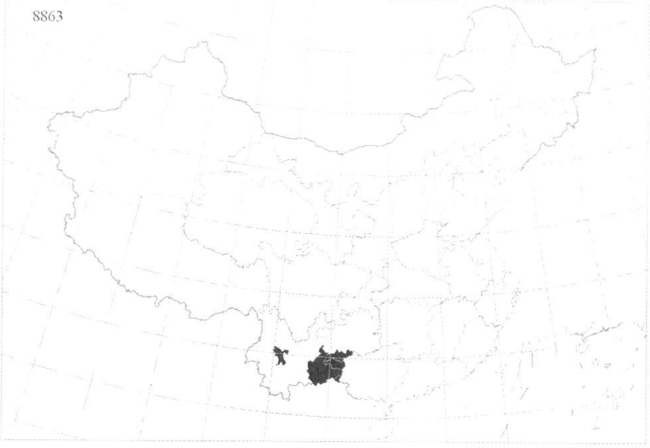

Oleaceae 木犀科 *Jasminum* 素馨属

Jasminum grandiflorum L. 素馨花

Scandent shrubs

15.6 (-1.1~22.4); 8.6 (-9.1~14.3); 20.8 (6.2~29.0); 15.7 (2.0~22.5); 130.4 (2.5~209.6); 3.1 (0.0~75.6); 797 (334~1205) ❙ 1048 (321~1784); 587 (193~836); 47 (7~144) ❙ 31.3 (-14.2~68.1); 772 (321~1205); 287.6 (79.3~427.9)

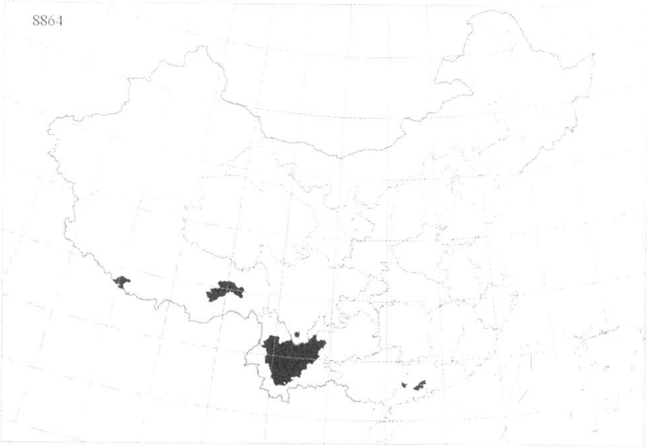

Oleaceae 木犀科 *Jasminum* 素馨属

Jasminum guangxiense B. M. Miao 广西素馨

Vines

22.2; 13.9; 28.3; 22.3; 207.5; 0.0; 1190 ❙ 1374; 713; 87 ❙ 15.4; 1190; 327.7

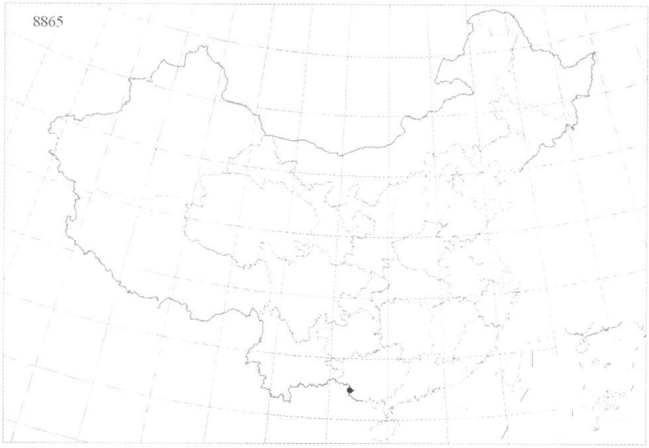

Oleaceae 木犀科 *Jasminum* 素馨属

Jasminum hongshuihoense Z. P. Jien ex B. M. Miao 绒毛素馨

Vines

18.4 (16.8~21.6); 9.8 (7.8~13.1); 24.8 (22.7~27.9); 18.4 (16.8~21.6); 160.8 (141.5~199.6); 0.0 (0.0~0.0); 921 (836~1143) ❙ 1300 (1161~1451); 709 (617~820); 59 (48~102) ❙ 41.9 (27.0~65.6); 921 (836~1143); 296.5 (189.5~371.7)

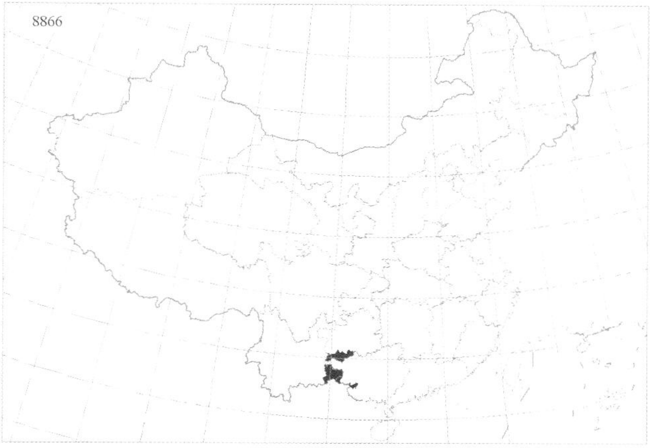

Oleaceae 木犀科 *Jasminum* 素馨属

Jasminum humile L. 矮探春

Shrubs or small trees

12.7 (-1.1~23.0); 4.9 (-9.1~15.1); 19.0 (6.2~29.0); 12.9 (2.0~23.0); 101.0 (2.5~216.2); 8.3 (0.0~75.6); 704 (334~1247) ❙ 978 (321~1771); 550 (193~1007); 38 (9~187) ❙ 41.5 (-14.2~145.8); 687 (321~1247); 276.4 (79.3~447.4)

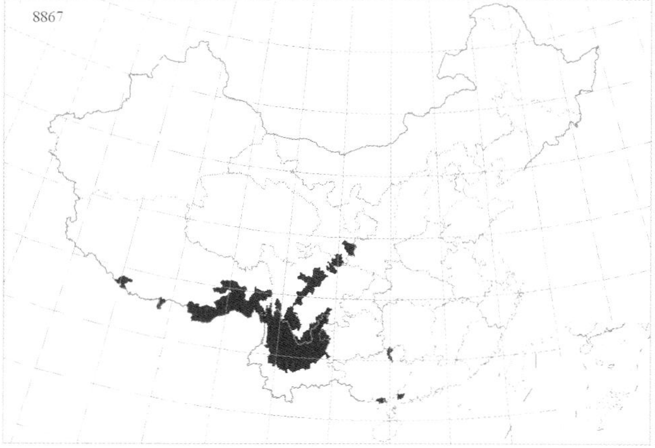

Oleaceae 木犀科 *Jasminum* 素馨属

Jasminum humile var. microphyllum (L. C. Chia) P. S. Green 狭叶矮探春

Shrubs or small trees

6.4 (1.6~15.3); -2.2 (-8.0~8.7); 14.0 (9.8~24.1); 7.2 (3.7~15.4); 45.2 (15.2~124.3); 27.9 (0.0~56.2); 521 (408~761) ❙ 844 (535~1023); 474 (326~594); 22 (10~40) ❙ 65.7 (11.5~128.6); 518 (408~755); 278.6 (200.8~357.7)

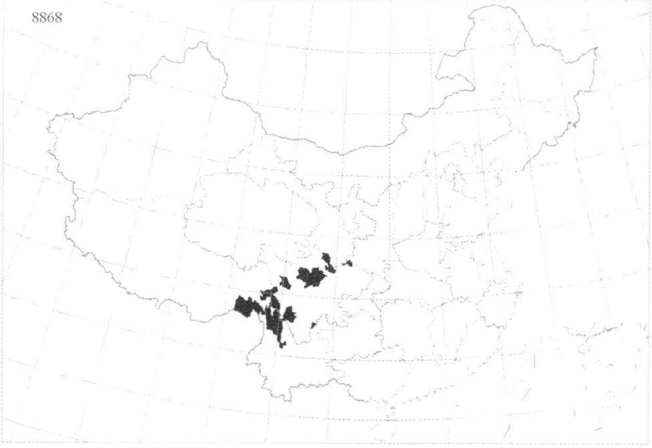

Oleaceae 木犀科 *Jasminum* 素馨属

Jasminum lanceolaria Roxb. 清香藤

Scandent shrubs

17.4 (7.3~24.8)；7.5 (−2.9~20.1)；26.2 (15.0~29.8)；17.5 (7.5~24.8)；151.1 (48.2~237.6)；1.4 (0.0~20.5)；922 (537~1385) ▎1416 (511~2499)；593 (258~1428)；126 (15~303) ▎54.5 (−45.2~219.3)；911 (511~1299)；248.0 (108.2~659.0)

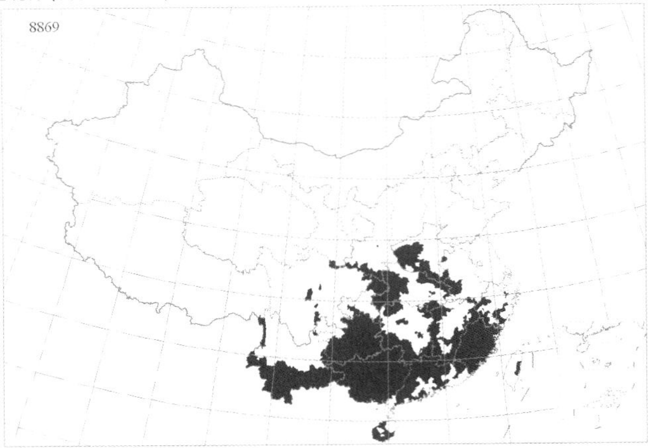

Oleaceae 木犀科 *Jasminum* 素馨属

Jasminum lang Gagnep. 栀花素馨

Vines

20.8 (17.9~22.7)；12.2 (7.4~15.0)；27.4 (25.0~28.6)；20.8 (18.0~22.8)；190.1 (155.6~213.1)；0.0 (0.0~0.0)；1091 (913~1221) ▎1660 (1182~2435)；873 (670~1428)；119 (43~187) ▎52.3 (12.9~99.5)；1091 (913~1221)；283.5 (208.7~354.5)

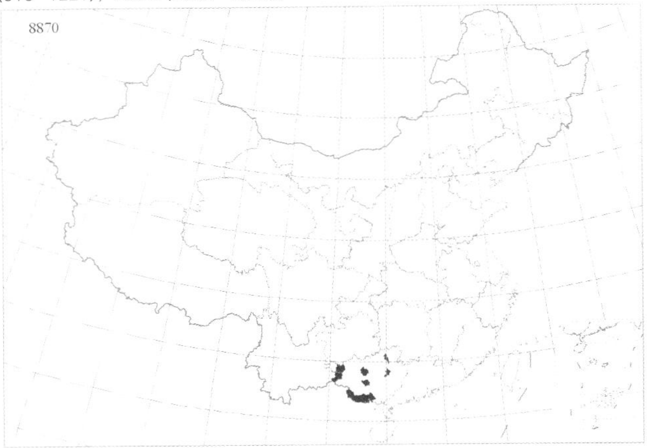

Oleaceae 木犀科 *Jasminum* 素馨属

Jasminum laurifolium Kurz 桂叶素馨

Vines, evergreen

19.8 (13.3~24.7)；11.2 (5.4~19.8)；26.7 (19.6~28.6)；19.9 (13.4~24.7)；178.6 (100.4~236.5)；0.0 (0.0~0.0)；1035 (706~1376) ▎1594 (1182~2435)；764 (497~1428)；122 (42~196) ▎56.0 (12.9~145.8)；1023 (706~1221)；300.7 (185.5~620.0)

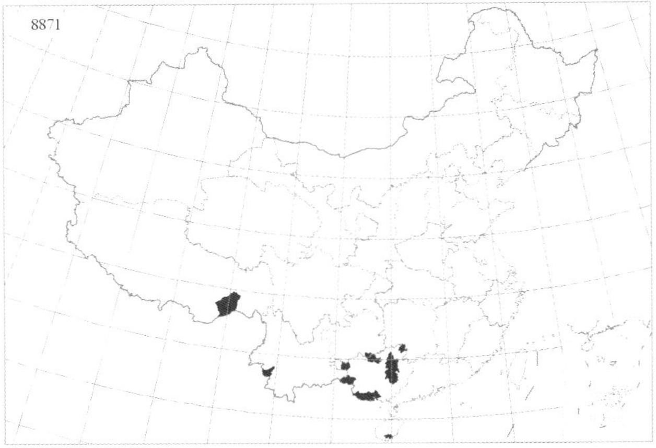

Oleaceae 木犀科 *Jasminum* 素馨属

Jasminum longitubum L. C. Chia ex B. M. Miao 长管素馨

Vines

21.9；13.5；28.1；22.0；203.5；0.0；1167 ▎1413；739；94 ▎21.2；1167；331.4

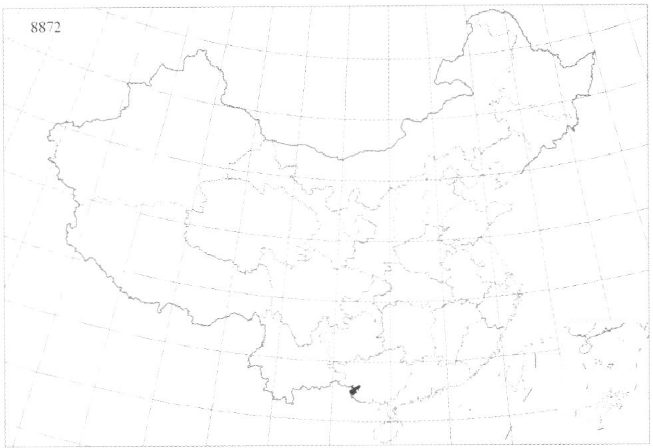

Oleaceae 木犀科 *Jasminum* 素馨属

Jasminum mesnyi Hance 野迎春

Subshrubs, evergreen

15.7 (2.2~24.2)；7.3 (−6.6~17.6)；22.6 (9.9~29.4)；15.7 (3.8~24.2)；129.4 (15.5~230.4)；0.8 (0.0~49.1)；806 (413~1331) ▎1156 (741~1784)；594 (376~996)；58 (15~197) ▎44.1 (−14.2~125.9)；795 (413~1331)；278.4 (111.9~659.0)

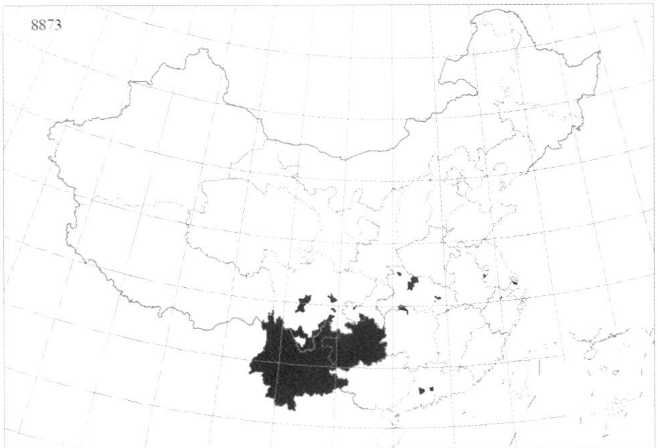

Oleaceae 木犀科 *Jasminum* 素馨属

Jasminum microcalyx Hance 小萼素馨

Scandent shrubs

21.3 (18.1~24.7)；13.7 (9.1~19.8)；27.1 (23.6~28.7)；21.4 (18.2~24.7)；196.5 (158.1~236.5)；0.0 (0.0~0.0)；1125 (901~1376) ▎1440 (868~1900)；646 (312~1056)；92 (48~177) ▎31.0 (−34.3~71.0)；1045 (868~1212)；359.0 (190.9~659.0)

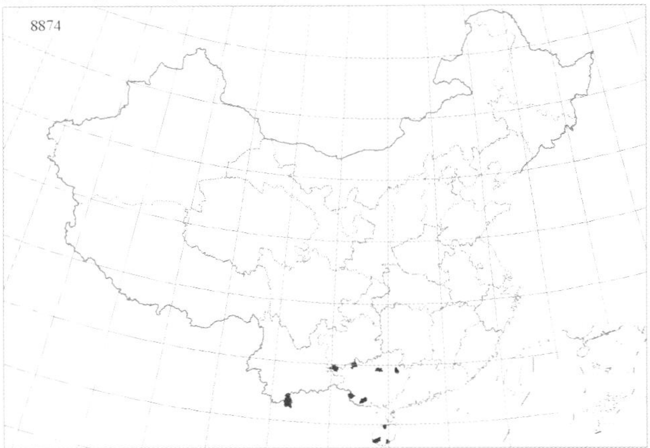

Oleaceae 木犀科 *Jasminum* 素馨属

Jasminum multiflorum (Burm. f.) Andrews 毛茉莉

Scandent shrubs

22.5; 14.6; 28.8; 22.6; 210.8; 0.0; 1213 ❙ 1635; 702; 127 ❙ 35.7; 1213; 213.0

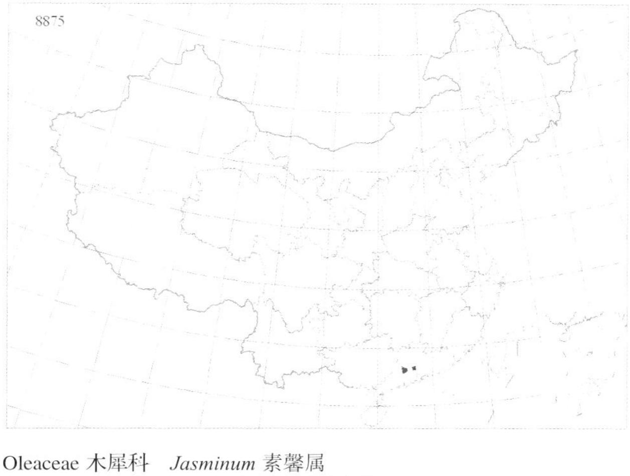

Oleaceae 木犀科 *Jasminum* 素馨属

Jasminum nervosum Lour. 青藤仔

Scandent shrubs

19.5 (15.5~24.8); 12.1 (4.7~20.1); 25.0 (20.1~29.2); 19.5 (15.6~24.8); 174.2 (127.2~237.6); 0.0 (0.0~0.3); 990 (773~1385) ❙ 1506 (868~3495); 747 (312~1847); 100 (38~897) ❙ 55.0 (-34.3~246.5); 966 (763~1360); 327.0 (119.4~659.0)

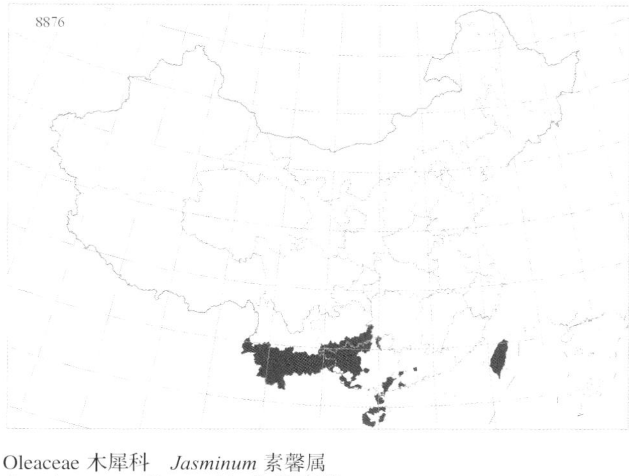

Oleaceae 木犀科 *Jasminum* 素馨属

Jasminum nintooides Rehder 银花素馨

Scandent shrubs

17.6 (16.3~21.0); 10.6 (9.1~14.2); 22.5 (21.4~25.6); 17.7 (16.4~21.0); 151.8 (136.4~191.8); 0.0 (0.0~0.0); 858 (797~1053) ❙ 1263 (1007~1522); 694 (539~860); 53 (44~68) ❙ 47.3 (19.0~80.6); 857 (797~1053); 400.7 (277.0~551.6)

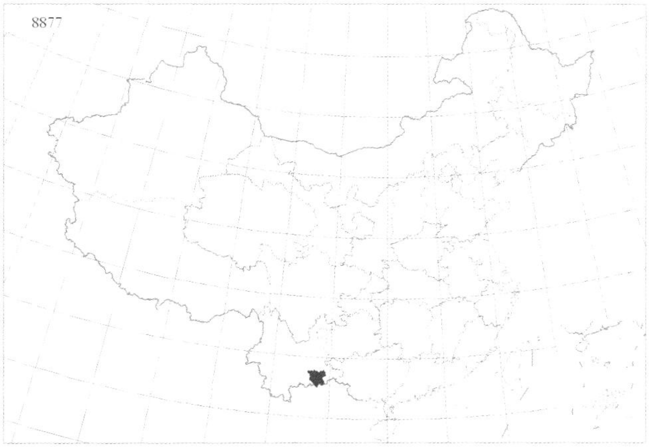

Oleaceae 木犀科 *Jasminum* 素馨属

Jasminum nudiflorum Lindl. 迎春花

Shrubs, deciduous

13.5 (0.0~23.0); 2.1 (-10.6~15.1); 23.9 (8.1~29.5); 13.8 (2.7~23.0); 112.3 (8.0~216.2); 9.3 (0.0~67.9); 771 (363~1247) ❙ 927 (358~1997); 454 (200~997); 50 (8~254) ❙ 20.3 (-45.2~130.7); 713 (358~1247); 208.6 (44.7~411.1)

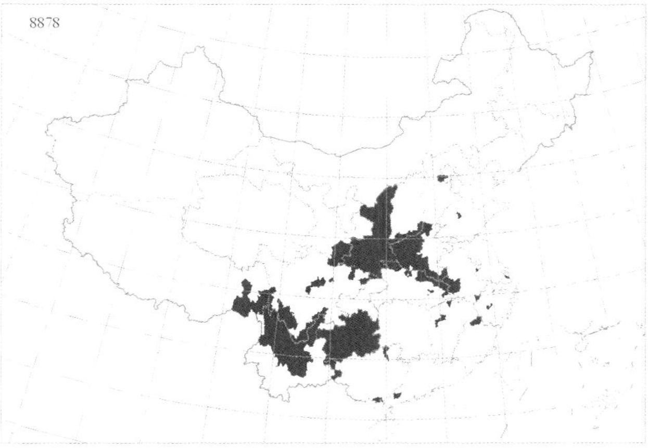

Oleaceae 木犀科 *Jasminum* 素馨属

Jasminum nudiflorum var. pulvinatum (W. W. Sm.) Kobuski 垫状迎春

Small shrubs, deciduous

7.0 (0.0~13.2); -1.1 (-8.7~6.4); 14.2 (8.1~20.4); 7.7 (2.7~13.3); 47.6 (8.0~99.4); 22.6 (0.0~67.9); 534 (363~695) ❙ 881 (535~1337); 495 (323~727); 30 (0~69) ❙ 63.8 (2.4~101.2); 534 (363~695); 280.1 (111.0~407.1)

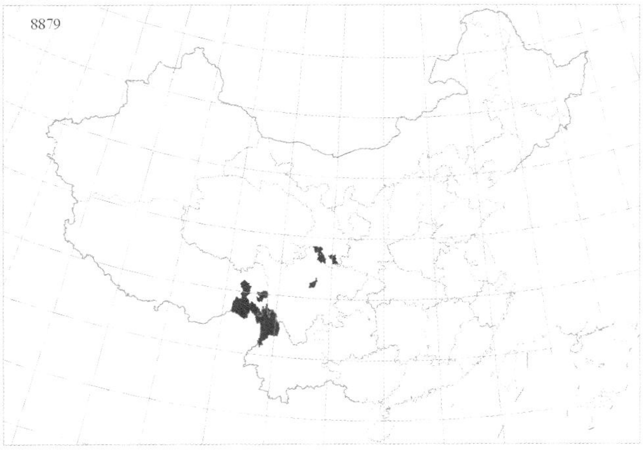

Oleaceae 木犀科 *Jasminum* 素馨属

Jasminum officinale L. 素方花

Scandent shrubs

12.7 (0.0~23.0); 4.9 (-8.7~15.1); 19.0 (8.1~29.0); 13.0 (2.7~23.0); 101.3 (8.0~216.2); 8.9 (0.0~67.9); 705 (363~1247) ❙ 977 (318~1771); 551 (226~997); 39 (1~187) ❙ 41.5 (-25.4~128.6); 687 (318~1247); 273.2 (111.0~407.1)

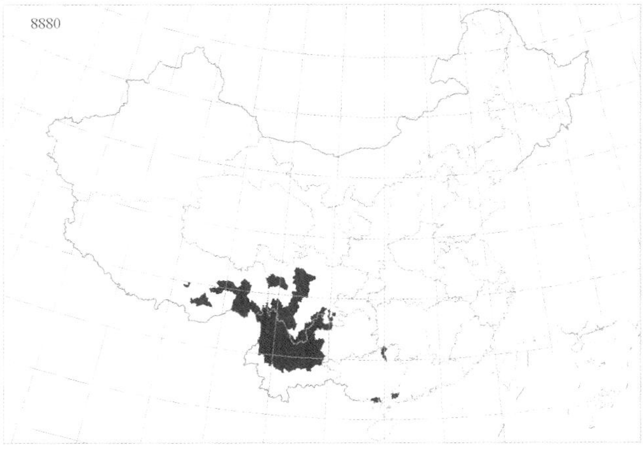

Oleaceae 木犀科 *Jasminum* 素馨属
Jasminum officinale var. piliferum P. Y. Bai 具毛素方花
Scandent shrubs

5.3; -2.4; 12.2; 5.7; 30.5; 26.2; 479 ❙ 961; 558; 31 ❙ 100.7; 479; 234.2

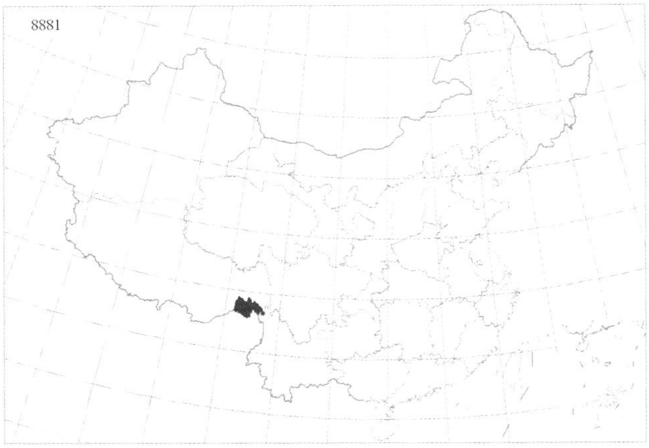

Oleaceae 木犀科 *Jasminum* 素馨属
Jasminum officinale var. tibeticum C. Y. Wu ex P. Y. Bai 西藏素方花
Scandent shrubs

5.5 (2.2~10.8); -2.7 (-7.1~3.9); 12.7 (9.5~16.6); 6.3 (3.9~10.9); 35.9 (15.8~71.8); 29.0 (1.3~48.6); 491 (411~633) ❙ 735 (337~968); 458 (257~603); 15 (1~28) ❙ 49.5 (-23.5~86.5); 479 (337~633); 266.0 (168.2~341.6)

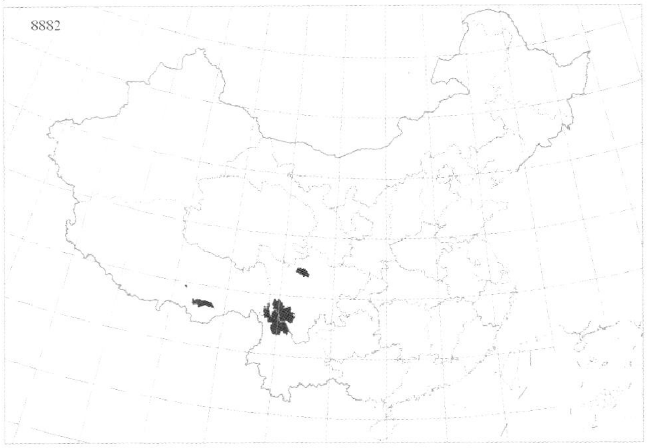

Oleaceae 木犀科 *Jasminum* 素馨属
Jasminum pentaneurum Hand.-Mazz. 厚叶素馨
Scandent shrubs

21.1 (17.7~24.7); 12.5 (7.4~19.8); 28.0 (25.8~29.0); 21.2 (17.8~24.7); 194.2 (153.6~236.5); 0.0 (0.0~0.0); 1120 (913~1376) ❙ 1632 (1188~2435); 729 (455~1428); 140 (48~194) ❙ 46.6 (5.8~99.5); 1111 (913~1360); 251.9 (157.7~620.0)

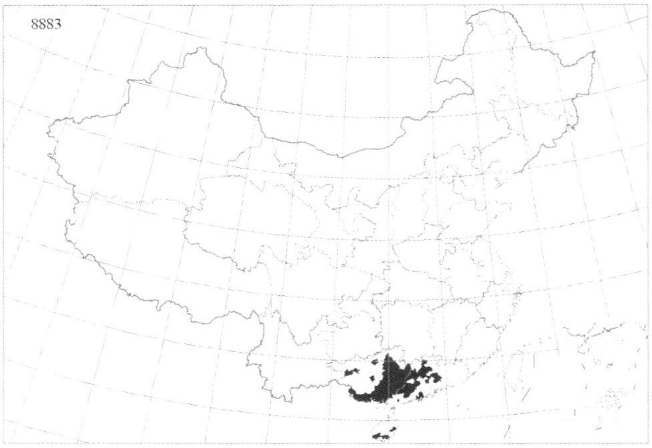

Oleaceae 木犀科 *Jasminum* 素馨属
Jasminum pierreanum Gagnep. 心叶素馨
Scandent shrubs

20.4 (16.9~24.8); 13.5 (10.3~20.1); 25.3 (21.4~28.5); 20.4 (16.9~24.8); 184.8 (142.7~237.6); 0.0 (0.0~0.0); 1044 (813~1385) ❙ 1369 (1049~1600); 701 (392~850); 73 (41~136) ❙ 33.5 (-11.9~60.1); 999 (783~1188); 398.0 (223.8~551.6)

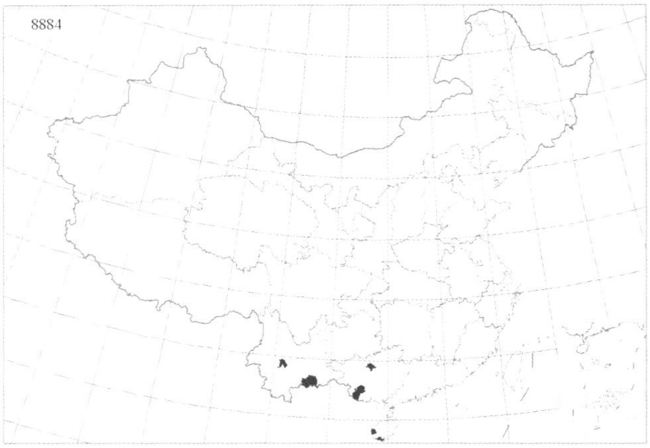

Oleaceae 木犀科 *Jasminum* 素馨属
Jasminum polyanthum Franch. 多花素馨
Vines

15.9 (5.4~23.0); 8.4 (-2.1~15.1); 21.6 (12.3~29.0); 15.9 (5.8~23.0); 131.9 (31.5~216.2); 0.7 (0.0~25.6); 803 (481~1247) ❙ 1134 (741~1771); 625 (411~997); 48 (12~187) ❙ 41.5 (-14.2~93.2); 788 (481~1247); 308.6 (135.3~659.0)

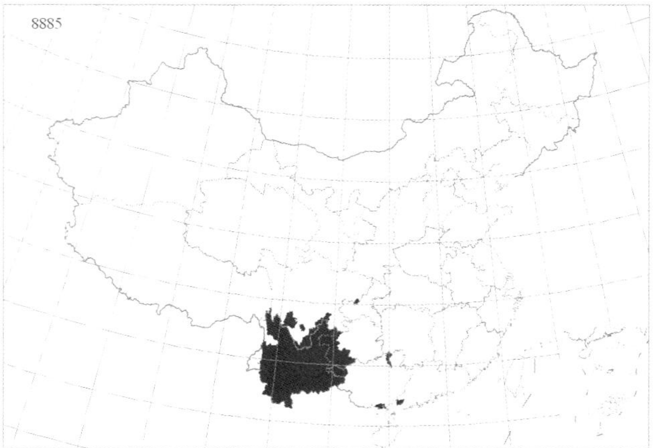

Oleaceae 木犀科 *Jasminum* 素馨属
Jasminum prainii H. Lév. 披针叶素馨
Vines

16.0 (14.3~19.1); 5.3 (3.8~10.1); 25.7 (23.4~27.7); 16.0 (14.3~19.1); 132.3 (112.6~169.2); 0.3 (0.0~1.5); 840 (756~966) ❙ 1248 (1055~1425); 511 (444~720); 96 (51~161) ❙ 48.8 (24.9~71.9); 840 (756~966); 206.6 (146.0~283.9)

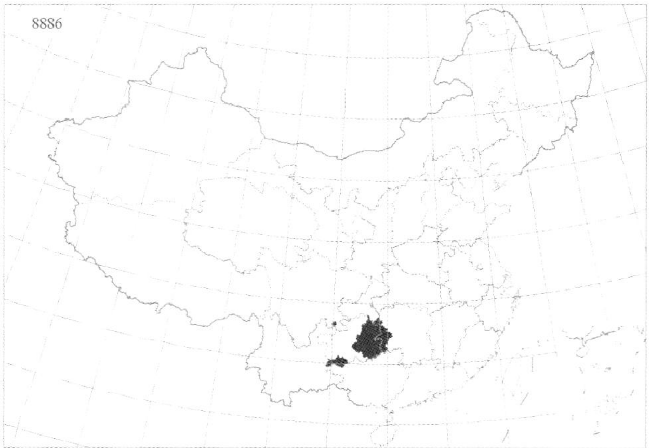

Oleaceae 木犀科 *Jasminum* 素馨属

Jasminum rehderianum Kobuski 白皮素馨

Scandent shrubs

23.7; 17.4; 28.3; 23.8; 225.1; 0.0; 1296 ❙ 1415; 616; 68 ❙ 9.5; 1186; 350.5

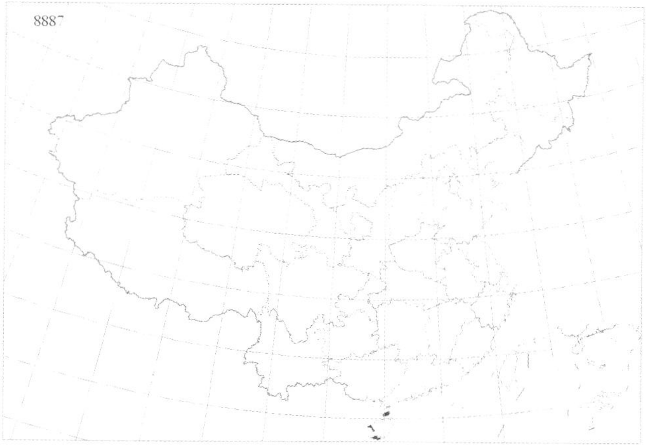

Oleaceae 木犀科 *Jasminum* 素馨属

Jasminum rufohirtum Gagnep. 云南素馨

Vines

18.9; 12.9; 22.9; 18.9; 166.8; 0.0; 902 ❙ 1487; 873; 59 ❙ 65.0; 902; 528.7

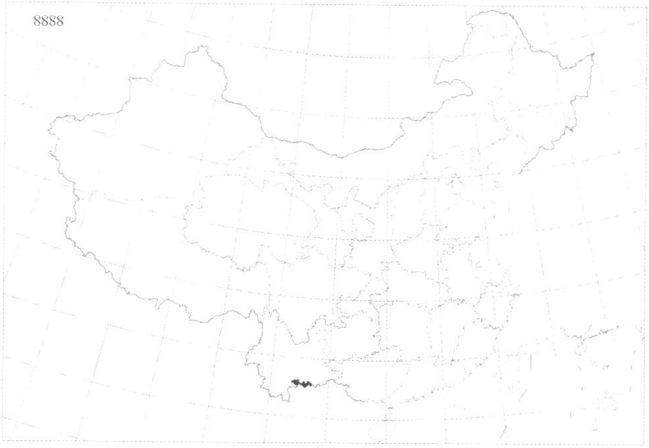

Oleaceae 木犀科 *Jasminum* 素馨属

Jasminum sambac (L.) Aiton 茉莉花

Scandent shrubs

18.0 (11.7~24.8); 8.3 (2.3~20.1); 26.3 (18.8~29.7); 18.0 (11.7~24.8); 156.5 (82.6~237.6); 0.2 (0.0~3.9); 948 (658~1385) ❙ 1372 (916~2435); 624 (403~1428); 108 (20~197) ❙ 46.2 (-12.0~10 .8); 943 (658~1299); 226.7 (95.3~548.0)

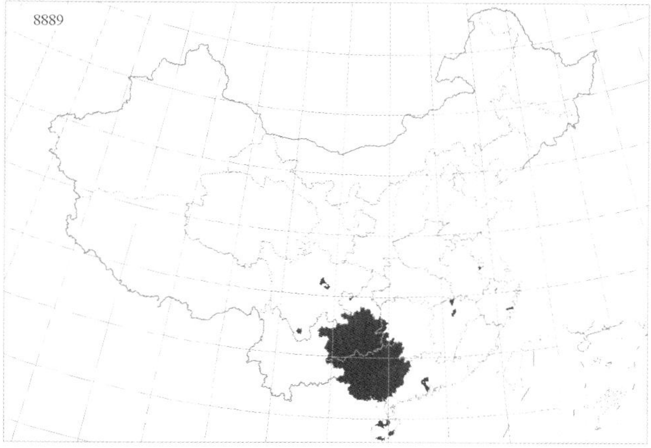

Oleaceae 木犀科 *Jasminum* 素馨属

Jasminum seguinii H. Lév. 亮叶素馨

Vines

17.0 (5.4~24.8); 9.4 (-2.1~20.1); 22.9 (12.3~29.0); 17.1 (5.8~24.8); 145.2 (31.5~237.6); 0.4 (0.0~25.6); 869 (481~1385) ❙ 1233 (741~1813); 641 (312~997); 67 (15~196) ❙ 43.2 (-34.3~112.2); 848 (481~1329); 317.2 (142.9~659.0)

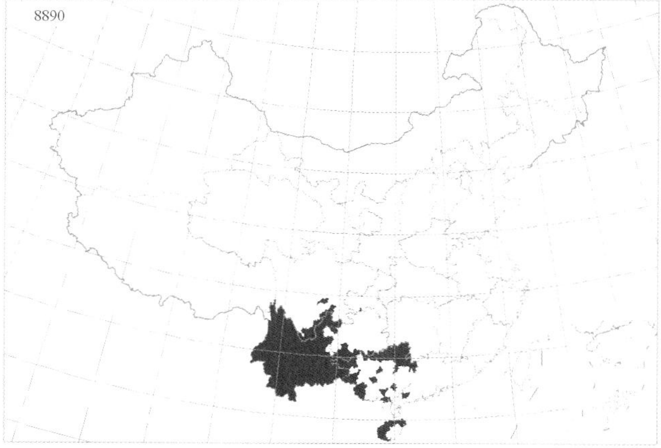

Oleaceae 木犀科 *Jasminum* 素馨属

Jasminum sinense Hemsl. 华素馨

Vines

16.9 (2.6~22.8); 6.7 (-7.0~14.8); 26.0 (11.2~29.3); 16.9 (4.3~22.8); 143.9 (20.0~213.6); 0.9 (0.0~48.2); 889 (427~1230) ❙ 1485 (711~2859); 593 (346~1186); 143 (15~402) ❙ 67.8 (5.8~246.5); 886 (427~1230); 239.2 (126.1~491.5)

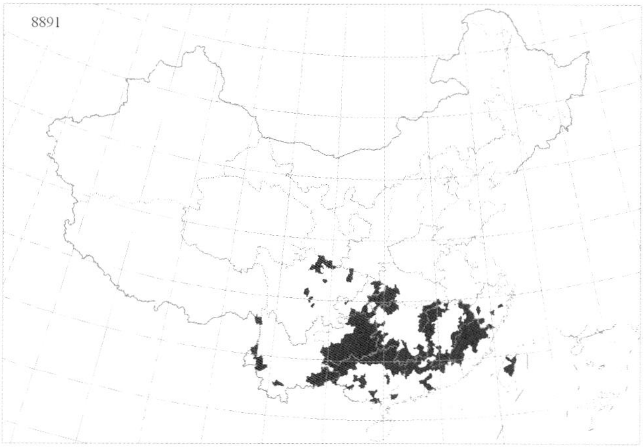

Oleaceae 木犀科 *Jasminum* 素馨属

Jasminum subglandulosum Kurz 腺叶素馨

Scandent shrubs

19.6 (18.6~20.5); 13.9 (12.7~15.1); 23.2 (22.5~23.7); 19.6 (18.7~20.5); 175.6 (163.9~185.9); 0.0 (0.0~0.0); 943 (888~992) ❙ 1559 (1428~1751); 820 (655~996); 68 (56~77) ❙ 65.8 (47.6~88.8); 935 (888~986); 560.1 (506.2~659.0)

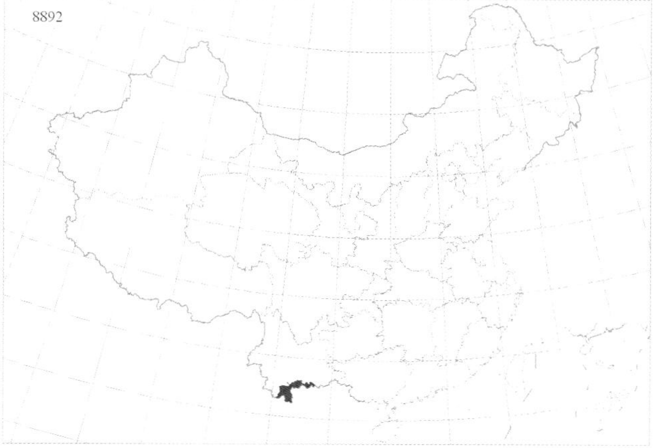

Oleaceae 木犀科 *Jasminum* 素馨属
Jasminum subhumile W. W. Sm. 滇素馨
Shrubs or small trees

14.7 (2.2~21.0); 7.5 (-6.3~15.1); 20.3 (9.5~26.6); 14.8 (3.9~21.0); 119.2 (15.8~191.8); 2.3 (0.0~48.6); 757 (411~1053) ❙ 1079 (628~1751); 616 (406~996); 40 (11~77) ❙ 43.9 (-14.2~121.7); 739 (411~1053); 320.2 (142.9~659.0)

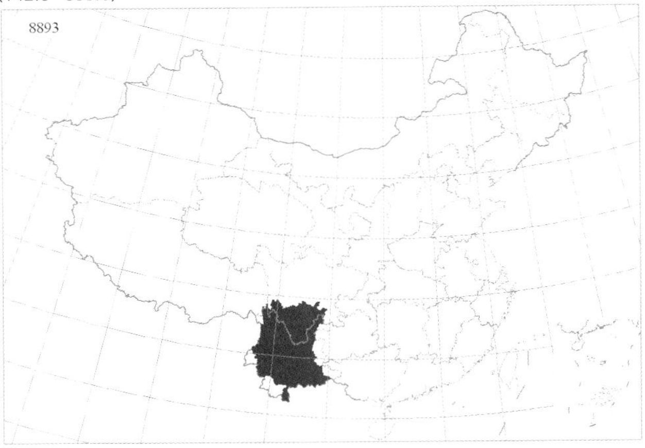

Oleaceae 木犀科 *Jasminum* 素馨属
Jasminum tonkinense Gagnep. 密花素馨
Scandent shrubs

17.2 (5.4~22.2); 10.2 (-2.1~15.1); 22.1 (12.3~28.3); 17.2 (5.8~22.3); 147.3 (31.5~207.5); 0.7 (0.0~25.6); 853 (481~1190) ❙ 1293 (741~1751); 716 (426~996); 53 (21~102) ❙ 52.8 (5.8~112.2); 844 (481~1190); 393.7 (189.5~659.0)

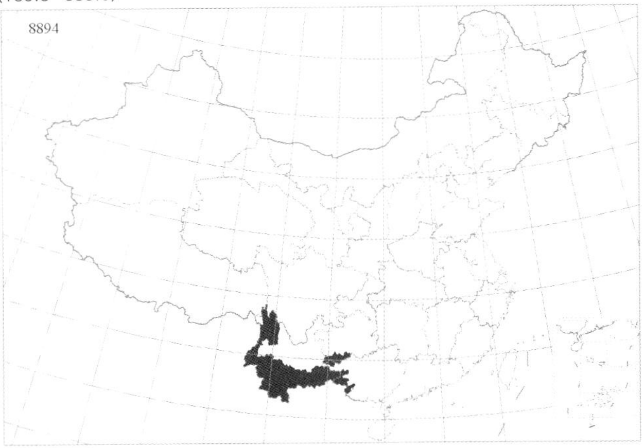

Oleaceae 木犀科 *Jasminum* 素馨属
Jasminum urophyllum Hemsl. 川素馨
Scandent shrubs

16.2 (6.0~22.2); 6.1 (-3.2~14.7); 25.4 (14.4~28.7); 16.3 (6.6~22.3); 136.8 (39.7~207.5); 1.4 (0.0~26.7); 867 (508~1190) ❙ 1339 (945~1876); 601 (444~891); 91 (24~192) ❙ 56.1 (9.2~107.2); 867 (508~1190); 213.6 (130.1~371.7)

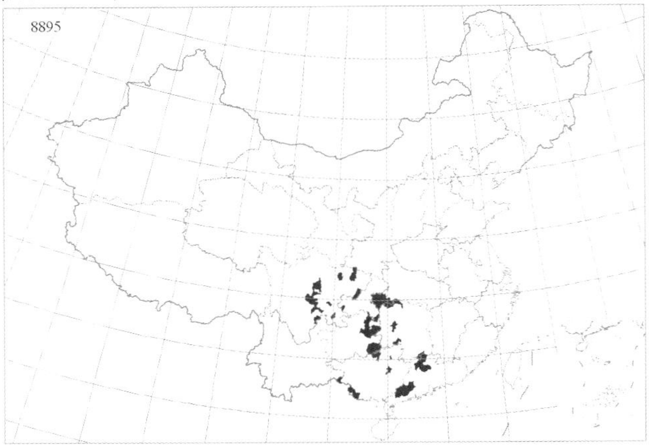

Oleaceae 木犀科 *Jasminum* 素馨属
Jasminum wengeri C. E. C. Fisch. 异叶素馨
Shrubs

19.4; 13.7; 22.9; 19.5; 173.7; 0.0; 938 ❙ 1492; 724; 64 ❙ 59.6; 926; 554.5

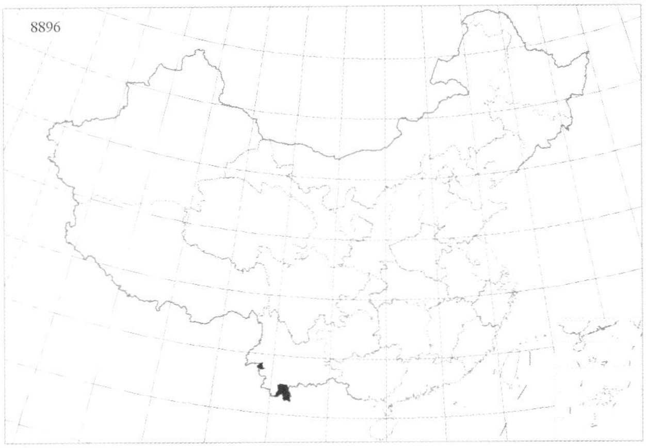

Oleaceae 木犀科 *Jasminum* 素馨属
Jasminum yuanjiangense P. Y. Bai 元江素馨
Scandent shrubs

19.3; 13.2; 23.5; 19.4; 172.9; 0.0; 936 ❙ 1209; 689; 50 ❙ 29.3; 913; 399.5

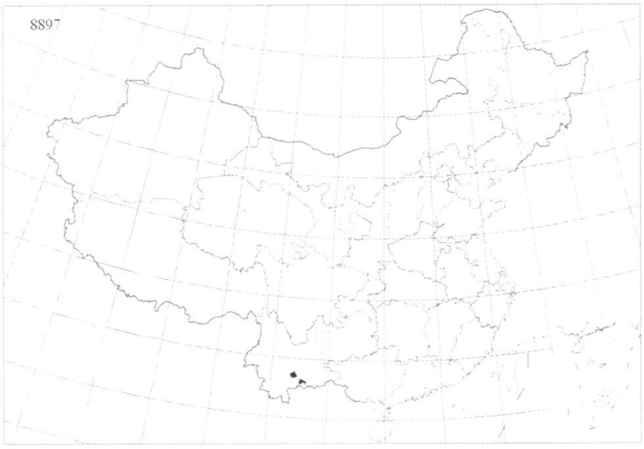

Oleaceae 木犀科 *Jasminum* 素馨属
Jasminum × stephanense Lemoine 淡红素馨
Scandent shrubs

3.7; -6.1; 12.1; 5.1; 27.8; 43.5; 441 ❙ 337; 257; 1 ❙ -23.5; 337; 217.3

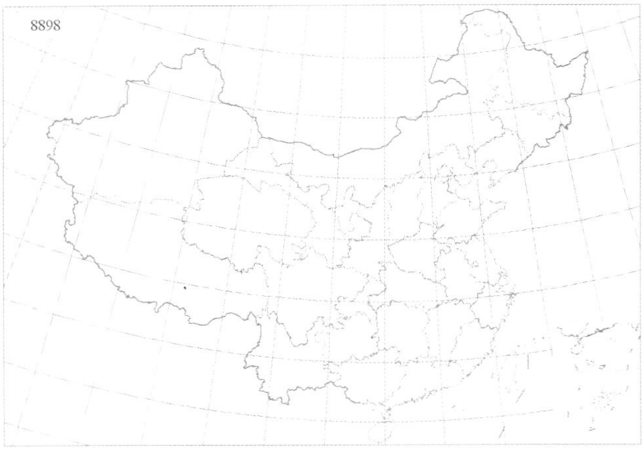

Oleaceae 木犀科 *Ligustrum* 女贞属
Ligustrum amamianum Koidz. 台湾女贞
Shrubs

21.1 (19.7~22.4)；14.3 (13.1~15.2)；27.4 (25.7~28.7)；21.3 (19.8~22.7)；196.0 (177.5~212.7)；0.0 (0.0~0.0)；1102 (978~1216) | 2503 (1965~3495)；913 (750~1033)；381 (104~897) | 131.0 (61.6~224.4)；1102 (978~1216)；220.5 (163.7~261.4)

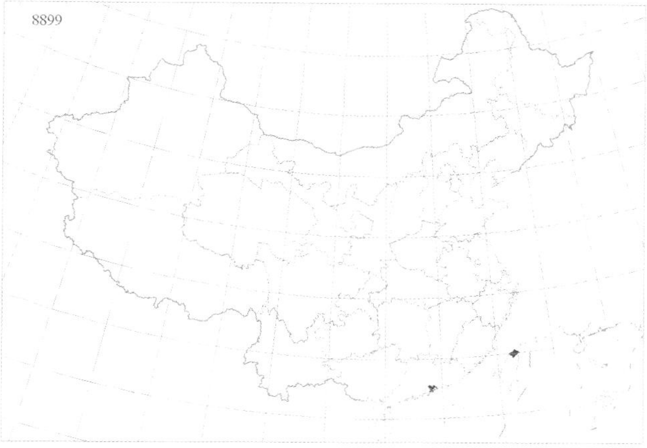

Oleaceae 木犀科 *Ligustrum* 女贞属
Ligustrum angustum B. M. Miao 狭叶女贞
Small shrubs

18.3 (16.8~19.1)；9.2 (7.8~10.1)；25.4 (23.4~26.6)；18.3 (16.8~19.2)；159.5 (141.5~170.1)；0.0 (0.0~0.0)；923 (836~981) | 1267 (1232~1334)；672 (617~720)；60 (51~73) | 37.9 (27.5~59.6)；923 (836~981)；260.1 (189.5~322.1)

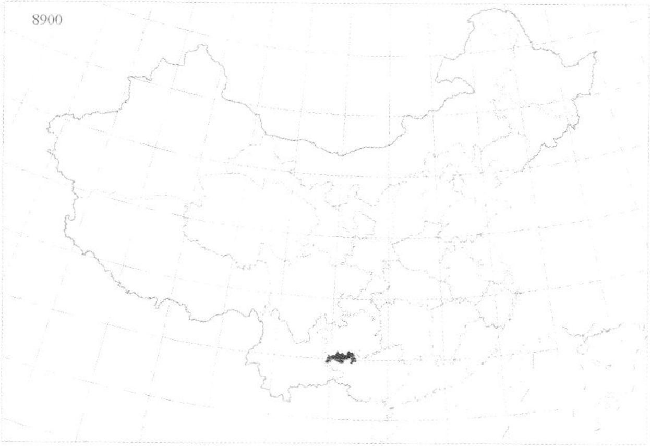

Oleaceae 木犀科 *Ligustrum* 女贞属
Ligustrum compactum (Wall. ex G. Don) Hook. f. et Thomson ex Brandis 长叶女贞
Shrubs or small trees

14.2 (0.7~21.0)；6.6 (-8.8~15.1)；20.1 (9.4~26.6)；14.3 (3.3~21.0)；115.1 (12.7~191.8)；4.7 (0.0~63.9)；744 (389~1053) | 1063 (514~1751)；591 (314~1007)；42 (9~113) | 43.9 (-14.2~145.8)；730 (389~1053)；307.3 (133.8~659.0)

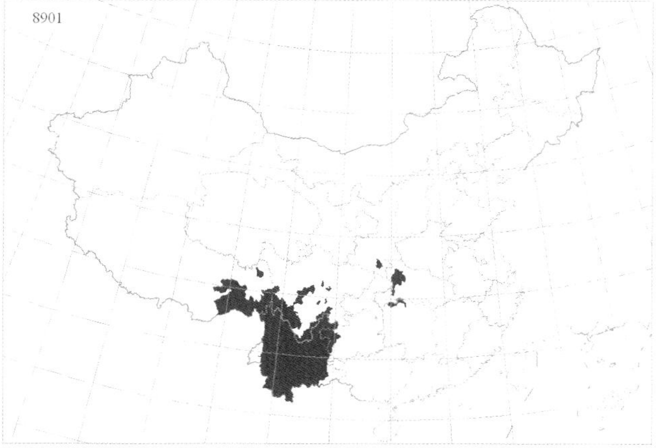

Oleaceae 木犀科 *Ligustrum* 女贞属
Ligustrum compactum var. velutinum P. S. Green 毛长叶女贞
Shrubs or small trees, semievergreen

1.8；-7.1；9.9；3.8；15.7；53.3；404 | 583；348；13 | 44.2；404；169.5

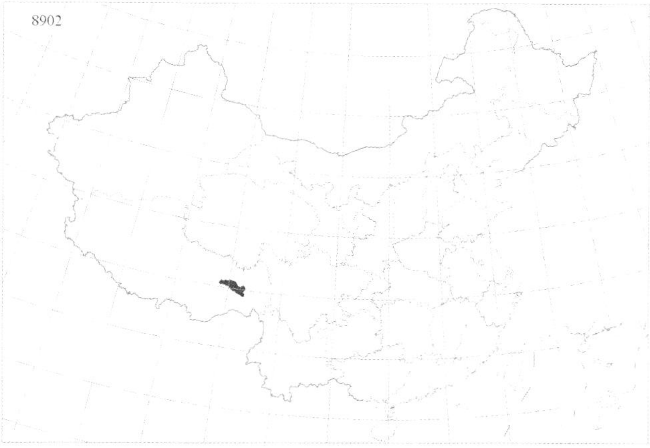

Oleaceae 木犀科 *Ligustrum* 女贞属
Ligustrum confusum Decne. 散生女贞
Shrubs or small trees

15.9 (-1.1~21.0)；9.0 (-9.1~15.1)；21.0 (6.2~25.6)；16.0 (2.0~21.0)；132.8 (2.5~191.8)；1.2 (0.0~75.6)；796 (334~1053) | 1173 (321~1751)；652 (193~1007)；49 (29~77) | 47.8 (-4.0~145.8)；782 (321~1053)；350.3 (79.3~659.0)

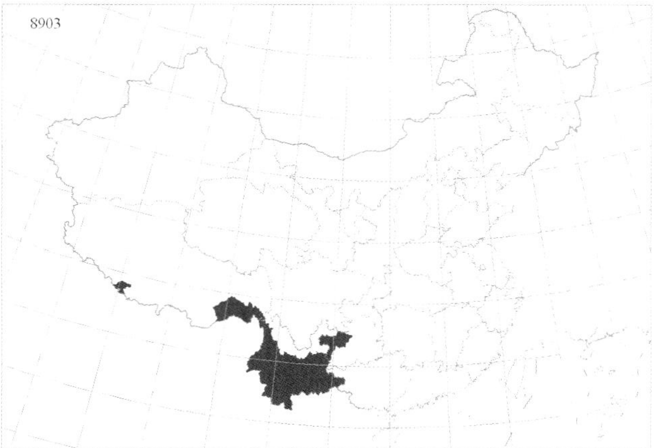

Oleaceae 木犀科 *Ligustrum* 女贞属
Ligustrum delavayanum Har. 紫药女贞
Shrubs

14.2 (2.2~18.9)；5.0 (-6.6~12.5)；21.0 (9.9~27.6)；14.2 (3.8~19.0)；112.8 (15.5~167.8)；2.3 (0.0~49.1)；746 (413~929) | 1087 (741~1577)；582 (411~932)；48 (12~200) | 47.4 (-14.2~125.9)；733 (413~902)；275.8 (155.0~516.7)

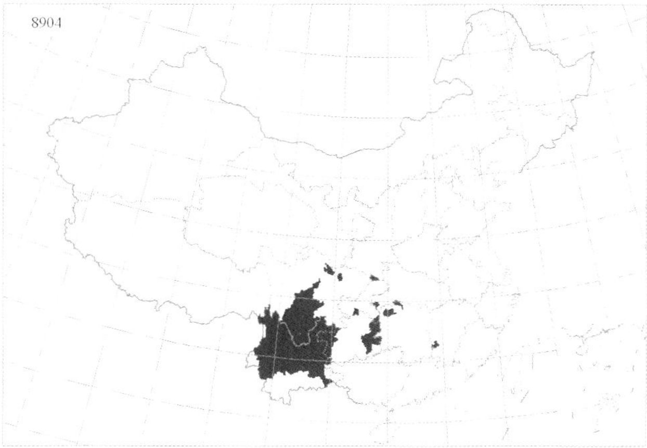

Oleaceae 木犀科 *Ligustrum* 女贞属
Ligustrum expansum Rehder 扩展女贞
Shrubs

14.5 (7.3~18.1); 4.0 (−2.0~7.8); 24.4 (15.6~28.2); 14.6 (7.5~18.2); 118.0 (48.2~158.0); 3.4 (0.0~20.4); 789 (537~962) ❙ 1235 (951~1583); 595 (436~938); 69 (26~173) ❙ 58.8 (9.4~107.2); 789 (537~962); 212.3 (130.1~298.1)

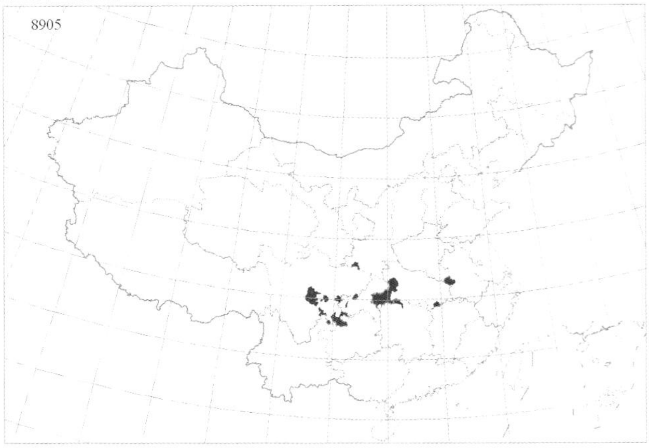

Oleaceae 木犀科 *Ligustrum* 女贞属
Ligustrum gracile Rehder 细女贞
Shrubs

12.1 (1.8~19.1); 4.3 (−7.2~12.0); 18.7 (9.5~26.8); 12.3 (3.8~19.1); 92.9 (15.4~169.1); 7.1 (0.0~53.3); 681 (404~973) ❙ 977 (583~1716); 563 (348~688); 32 (11~193) ❙ 47.2 (−14.2~128.6); 659 (404~973); 273.8 (142.9~366.3)

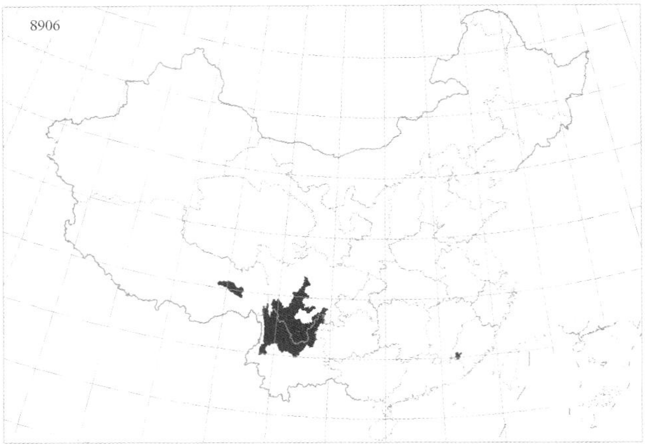

Oleaceae 木犀科 *Ligustrum* 女贞属
Ligustrum henryi Hemsl. 丽叶女贞
Shrubs

14.0 (5.7~18.9); 3.3 (−3.5~10.6); 24.0 (12.7~29.4); 14.1 (6.3~19.0); 113.1 (35.2~167.5); 4.5 (0.0~28.1); 773 (496~1002) ❙ 1130 (711~1828); 503 (332~813); 63 (12~229) ❙ 47.2 (0.0~107.2); 772 (496~1002); 225.5 (108.2~359.7)

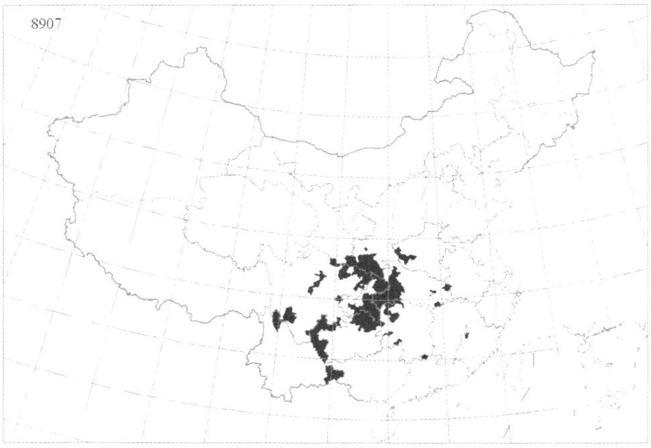

Oleaceae 木犀科 *Ligustrum* 女贞属
Ligustrum leucanthum (S. Moore) P. S. Green 蜡子树
Shrubs or small trees, deciduous

14.9 (1.9~21.0); 2.8 (−9.2~11.9); 26.4 (11.7~29.9); 15.0 (4.3~21.1); 124.8 (21.1~193.0); 5.0 (0.0~57.8); 835 (427~1114) ❙ 1215 (512~1997); 471 (245~821); 119 (10~259) ❙ 44.6 (−36.5~147.0); 803 (427~1114); 209.3 (111.9~334.4)

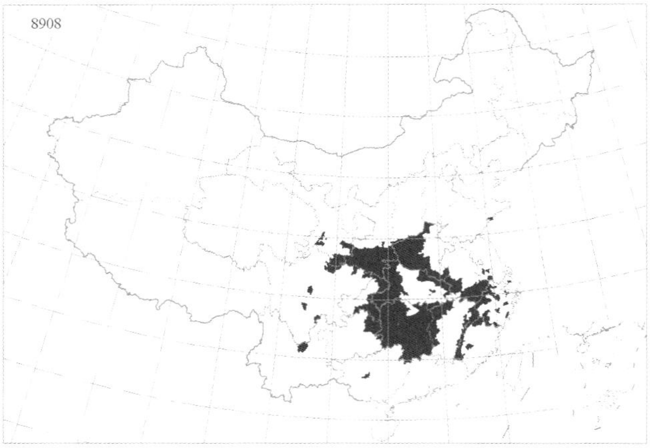

Oleaceae 木犀科 *Ligustrum* 女贞属
Ligustrum lianum P. S. Hsu 华女贞
Shrubs or small trees

18.5 (14.3~24.7); 8.5 (3.3~19.8); 27.4 (24.2~29.8); 18.5 (14.3~24.7); 162.5 (113.2~236.5); 0.1 (0.0~1.7); 974 (761~1376) ❙ 1672 (1175~2148); 614 (448~1234); 184 (48~259) ❙ 73.1 (15.4~130.7); 969 (761~1299); 248.1 (163.7~620.0)

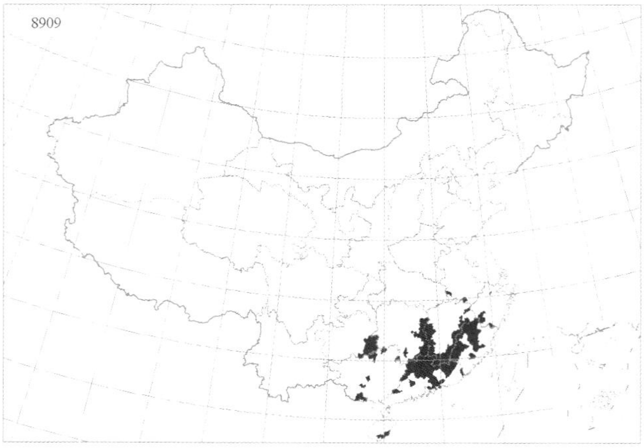

Oleaceae 木犀科 *Ligustrum* 女贞属
Ligustrum longitubum (P. S. Hsu) P. S. Hsu 长筒女贞
Shrubs

16.6 (15.0~18.2); 4.8 (2.6~6.9); 27.8 (26.8~29.1); 16.7 (15.0~18.2); 141.0 (123.7~158.8); 0.8 (0.0~3.5); 901 (826~984) ❙ 1782 (1621~1922); 594 (535~623); 237 (195~259) ❙ 98.1 (78.7~112.2); 895 (826~961); 249.9 (183.9~299.5)

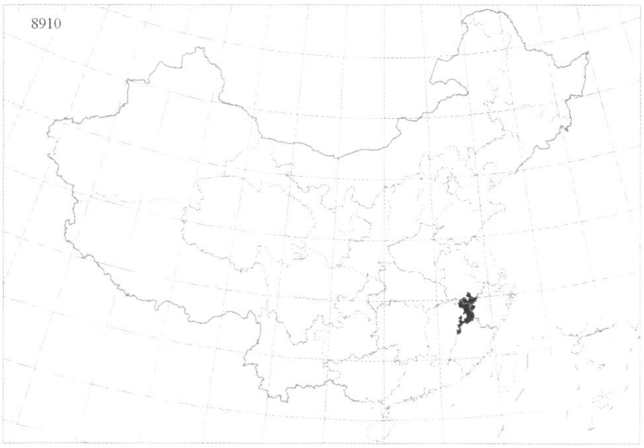

Oleaceae 木犀科 *Ligustrum* 女贞属
Ligustrum lucidum W. T. Aiton 女贞
Shrubs or trees

15.7 (1.5~22.4); 5.0 (-9.0~15.2); 25.5 (9.9~29.9); 15.8 (3.8~22.7); 131.8 (15.4~212.7); 3.2 (0.0~58.4); 849 (404~1216) ❙ 1236 (512~2030); 535 (258~1033); 105 (13~259) ❙ 45.5 (-35.9~147.0); 836 (404~1216); 221.4 (108.2~551.6)

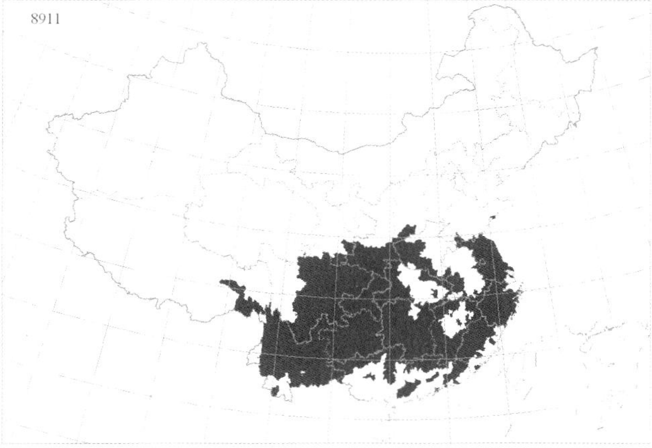

Oleaceae 木犀科 *Ligustrum* 女贞属
Ligustrum morrisonense Kaneh. et Sasaki 玉山女贞
Dwarf shrubs

19.6; 14.7; 23.3; 19.6; 175.6; 0.0; 941 ❙ 2579; 1400; 149 ❙ 177.2; 941; 285.7

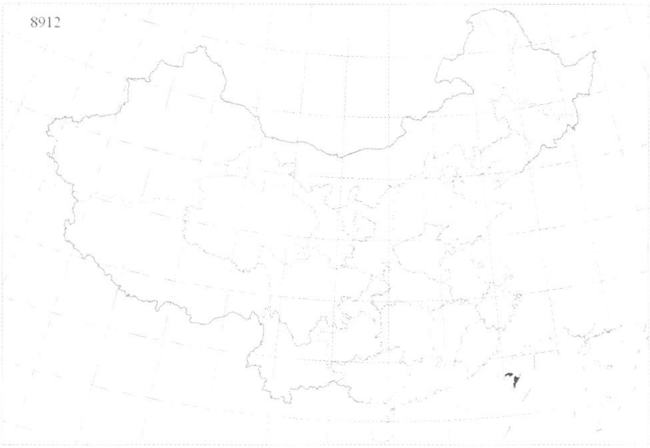

Oleaceae 木犀科 *Ligustrum* 女贞属
Ligustrum obovatilimbum B. M. Miao 倒卵叶女贞
Shrubs

21.6; 13.2; 28.2; 21.6; 199.2; 0.0; 1139 ❙ 1953; 858; 167 ❙ 71.5; 1139; 240.0

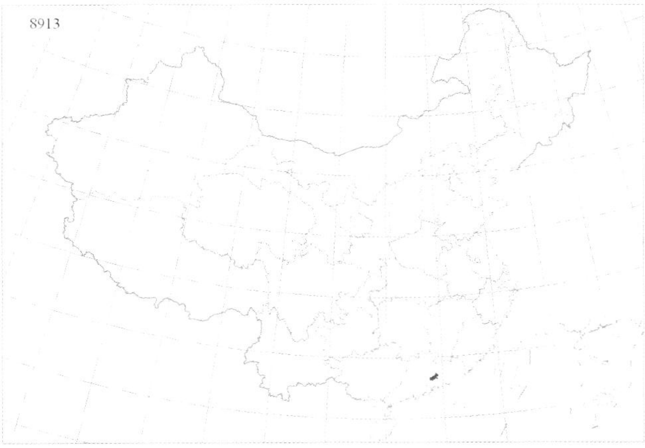

Oleaceae 木犀科 *Ligustrum* 女贞属
Ligustrum obtusifolium Siebold et Zucc. 水蜡树
Shrubs

13.4 (5.2~19.2); 0.2 (-14.5~10.7); 25.8 (18.7~29.4); 13.7 (8.2~19.2); 113.4 (58.8~170.8); 11.6 (0.0~65.0); 792 (561~1002) ❙ 935 (551~1770); 438 (245~705); 67 (15~223) ❙ 18.0 (-36.5~112.2); 742 (551~1002); 206.1 (111.9~411.1)

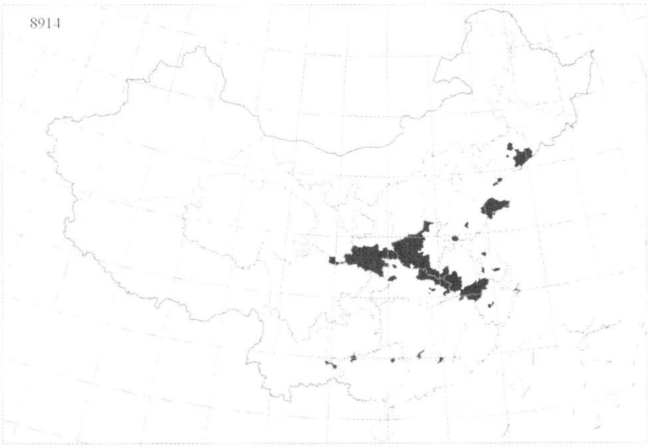

Oleaceae 木犀科 *Ligustrum* 女贞属
Ligustrum obtusifolium subsp. microphyllum (Nakai) P. S. Green 东亚女贞
Shrubs

16.5; 5.3; 27.7; 16.5; 138.0; 0.0; 888 ❙ 1363; 481; 183 ❙ 53.5; 888; 222.6

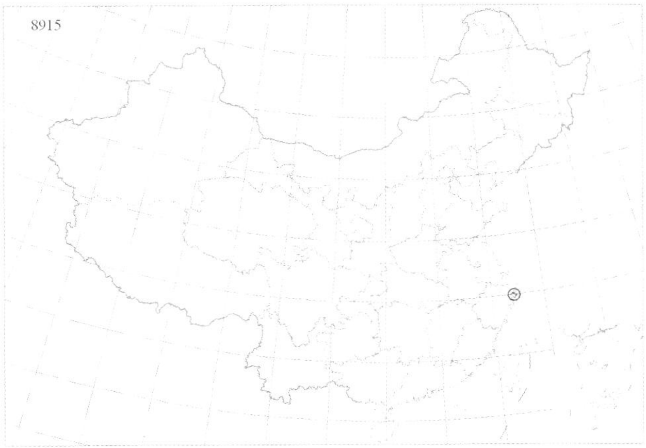

Oleaceae 木犀科 *Ligustrum* 女贞属
Ligustrum obtusifolium subsp. suave (Kitag.) Kitag. 辽东水蜡树
Shrubs, deciduous

9.4; -6.9; 23.4; 10.7; 88.6; 34.8; 683 ❙ 825; 523; 31 ❙ 21.8; 647; 224.9

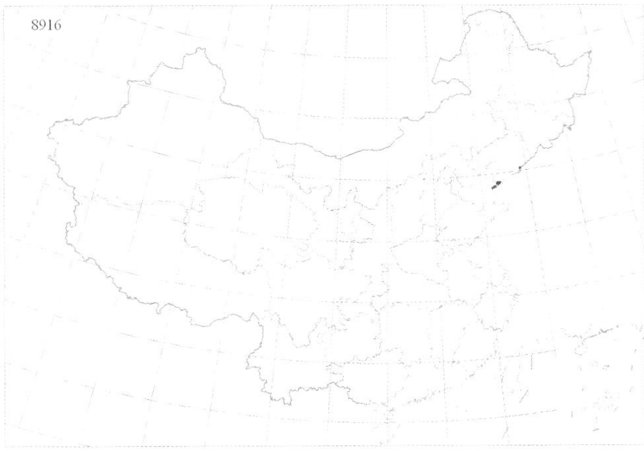

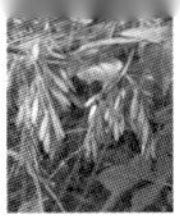

Oleaceae 木犀科 *Ligustrum* 女贞属

Ligustrum pedunculare Rehder 总梗女贞

Shrubs or small trees, evergreen

14.3 (9.3~19.3); 3.7 (−1.7~13.1); 24.1 (20.0~29.3); 14.3 (9.5~19.4); 115.4 (69.8~173.2); 3.6 (0.0~18.0); 776 (603~961) ❙ 1376 (1001~1847); 588 (486~816); 91 (35~242) ❙ 78.4 (21.7~107.2); 775 (603~961); 238.0 (170.4~346.9)

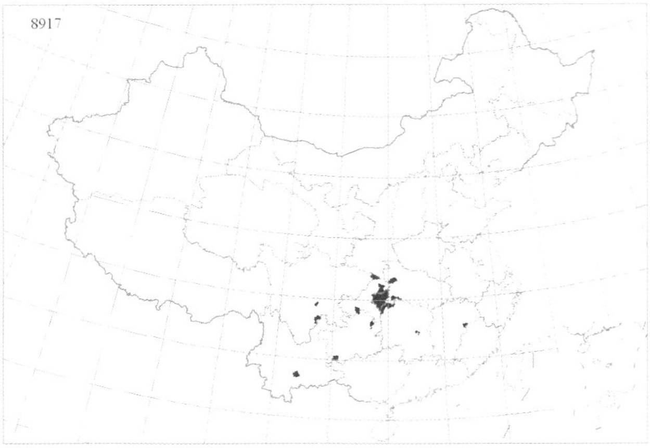

Oleaceae 木犀科 *Ligustrum* 女贞属

Ligustrum pricei Hayata 阿里山女贞

Shrubs or small trees

15.1 (9.3~24.1); 4.9 (−1.7~18.9); 24.6 (20.0~28.0); 15.2 (9.5~24.1); 124.4 (69.8~229.7); 2.2 (0.0~18.0); 810 (603~1322) ❙ 1311 (862~2759); 598 (376~1594); 81 (26~303) ❙ 62.9 (7.2~219.3); 806 (603~1095); 216.3 (139.5~303.6)

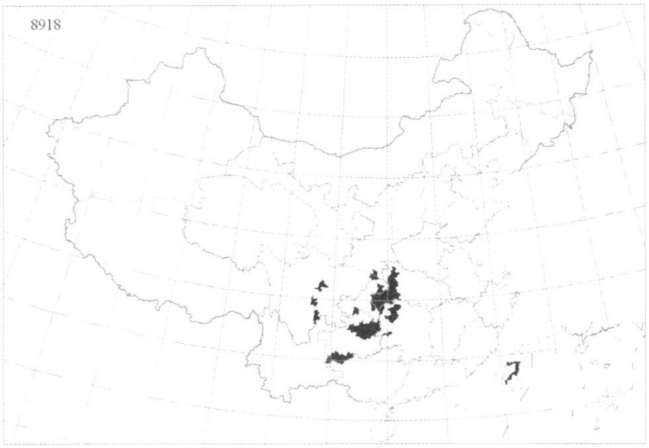

Oleaceae 木犀科 *Ligustrum* 女贞属

Ligustrum punctifolium M. C. Chang 斑叶女贞

Shrubs

22.0 (19.3~23.6); 14.7 (10.5~17.8); 28.3 (27.8~28.7); 22.2 (19.3~23.6); 205.9 (172.0~223.3); 0.0 (0.0~0.0); 1178 (998~1271) ❙ 1733 (1242~2037); 806 (444~1033); 122 (44~257) ❙ 47.3 (−2.2~71.8); 1148 (998~1216); 221.2 (139.6~439.7)

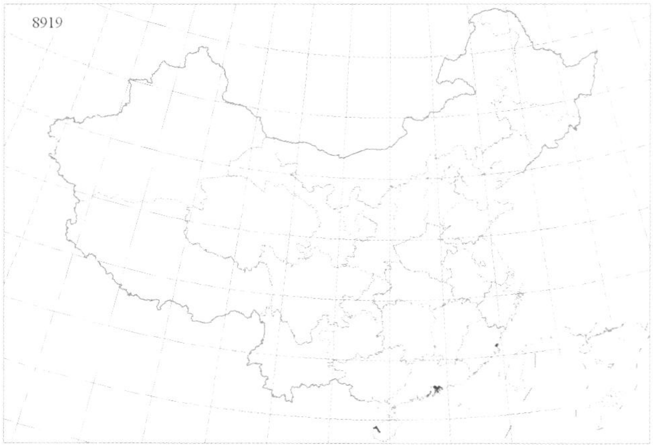

Oleaceae 木犀科 *Ligustrum* 女贞属

Ligustrum quihoui Carrière 小叶女贞

Shrubs, deciduous

14.7 (−1.1~19.4); 3.4 (−9.1~13.1); 25.1 (6.2~29.4); 14.8 (2.0~19.5); 121.7 (2.5~173.7); 4.7 (0.0~75.6); 810 (334~1038) ❙ 1053 (321~1852); 495 (193~1007); 77 (15~253) ❙ 30.3 (−36.5~145.8); 785 (321~1038); 212.6 (79.3~447.4)

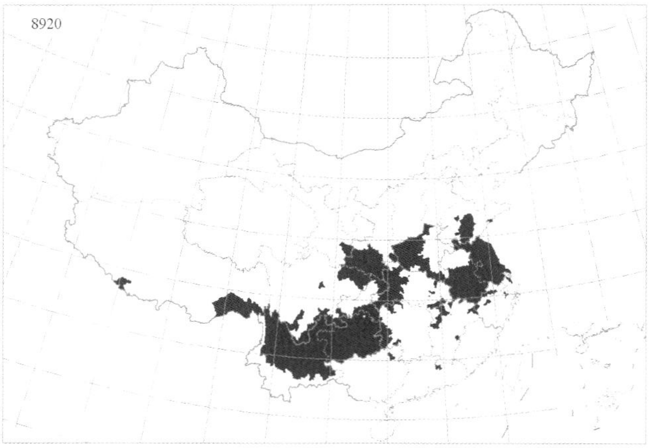

Oleaceae 木犀科 *Ligustrum* 女贞属

Ligustrum retusum Merr. 小叶女贞

Shrubs

19.3 (14.3~24.8); 10.0 (3.3~20.1); 27.4 (24.3~29.4); 19.3 (14.3~24.8); 172.1 (114.1~237.6); 0.1 (0.0~2.3); 1023 (765~1385) ❙ 1561 (868~2148); 599 (312~1234); 158 (41~235) ❙ 56.5 (−34.3~101.8); 980 (765~1186); 253.8 (145.3~527.5)

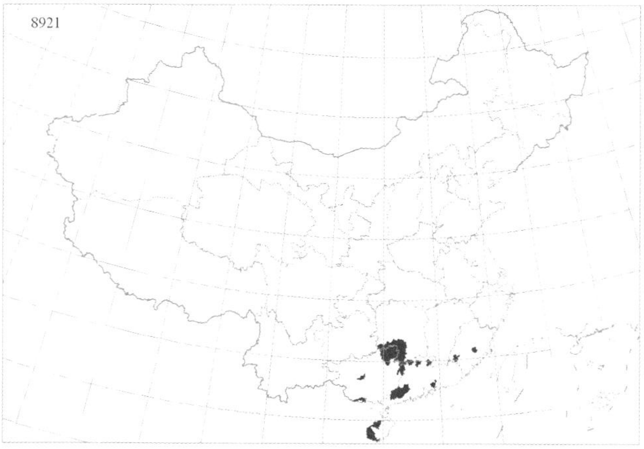

Oleaceae 木犀科 *Ligustrum* 女贞属

Ligustrum robustum subsp. chinense P. S. Green 粗壮女贞

Shrubs or small trees

16.0 (2.2~21.0); 5.8 (−6.6~15.1); 25.1 (9.9~29.7); 16.0 (3.8~21.0); 133.0 (15.5~191.8); 1.0 (0.0~49.1); 843 (413~1053) ❙ 1318 (822~1997); 559 (400~938); 110 (18~259) ❙ 56.3 (7.2~130.7); 838 (413~1053); 226.3 (123.1~659.0)

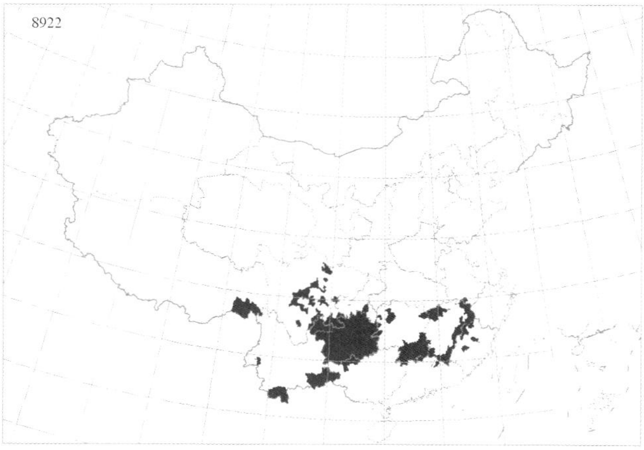

Oleaceae 木犀科 *Ligustrum* 女贞属
Ligustrum sempervirens (Franch.) Lingelsh. 裂果女贞
Shrubs, evergreen

13.7 (5.4~18.9); 6.0 (-2.1~12.0); 20.1 (12.3~26.6); 13.8 (5.8~19.0); 106.9 (31.5~167.8); 1.8 (0.0~25.6); 725 (481~929) ❙ 973 (741~1164); 555 (411~651); 35 (12~68) ❙ 35.8 (-14.2~78.3); 699 (481~912); 274.7 (142.9~382.6)

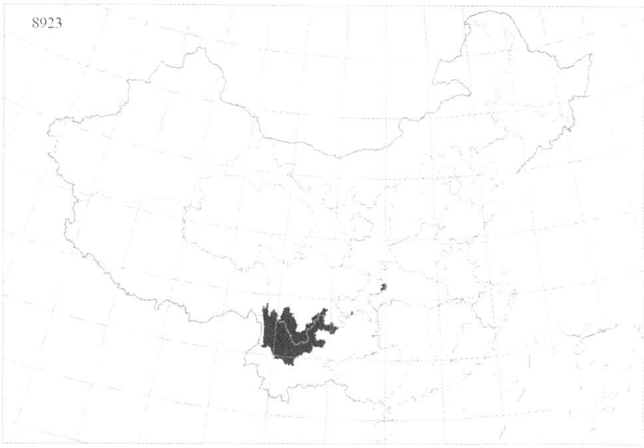

Oleaceae 木犀科 *Ligustrum* 女贞属
Ligustrum sinense Lour. 小蜡
Shrubs or small trees, deciduous

16.9 (5.3~24.7); 6.7 (-4.2~19.8); 26.0 (12.2~29.9); 16.9 (5.7~24.7); 144.3 (30.5~236.5); 1.5 (0.0~29.4); 897 (479~1376) ❙ 1367 (634~3495); 583 (290~1847); 128 (15~897) ❙ 52.6 (-16.6~246.5); 887 (479~1273); 236.0 (111.9~659.0)

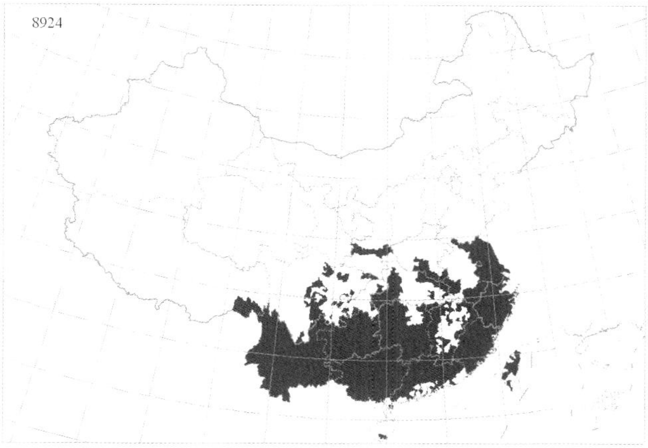

Oleaceae 木犀科 *Ligustrum* 女贞属
Ligustrum sinense var. coryanum (W. W. Sm.) Hand.-Mazz. 多毛小蜡
Shrubs or small trees, deciduous

16.8 (15.0~19.1); 7.4 (4.5~10.5); 24.7 (21.9~26.6); 16.8 (15.0~19.1); 142.1 (120.8~169.2); 0.1 (0.0~0.5); 857 (794~969) ❙ 1197 (958~1342); 594 (481~720); 62 (20~95) ❙ 39.9 (13.3~62.5); 850 (719~969); 239.6 (152.4~330.0)

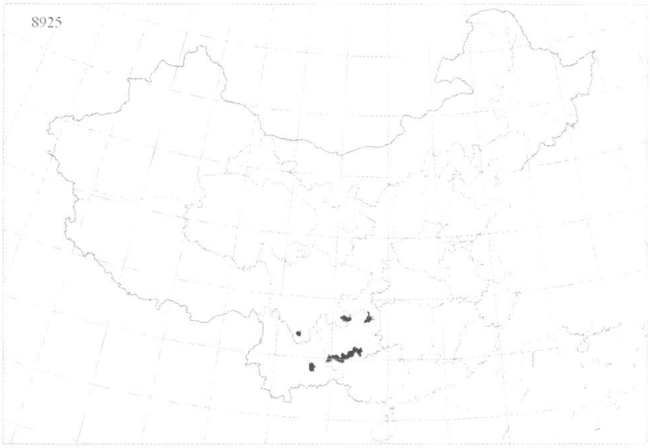

Oleaceae 木犀科 *Ligustrum* 女贞属
Ligustrum sinense var. luodianense M. C. Chang 罗甸小蜡
Shrubs or small trees, deciduous

21.1 (18.9~22.2); 12.8 (9.5~14.1); 27.8 (26.6~28.7); 21.2 (18.9~22.3); 194.0 (167.3~207.2); 0.0 (0.0~0.0); 1108 (969~1186) ❙ 1844 (1236~2014); 842 (617~954); 138 (60~169) ❙ 65.6 (27.5~79.9); 1108 (969~1186); 238.4 (187.2~272.2)

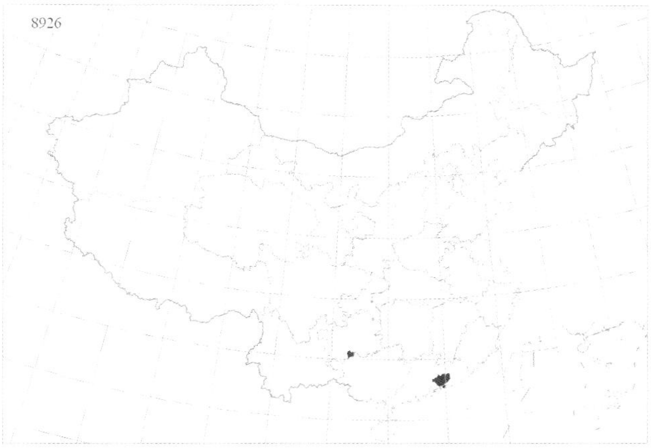

Oleaceae 木犀科 *Ligustrum* 女贞属
Ligustrum sinense var. myrianthum (Diels) Hoefker 光萼小蜡
Shrubs or small trees, deciduous

16.8 (10.2~22.3); 6.8 (-0.7~14.7); 25.8 (18.8~29.8); 16.9 (10.4~22.4); 143.2 (77.2~208.3); 0.6 (0.0~13.6); 885 (634~1202) ❙ 1360 (916~2024); 580 (443~1056); 107 (27~230) ❙ 54.0 (5.8~106.4); 884 (634~1202); 219.0 (118.7~411.1)

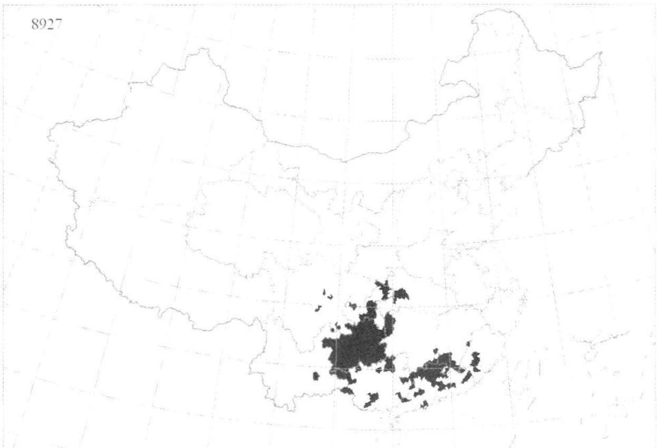

Oleaceae 木犀科 *Ligustrum* 女贞属
Ligustrum sinense var. opienense Y. C. Yang 峨边小蜡
Shrubs or small trees, deciduous

13.0; 3.2; 22.2; 13.1; 100.0; 3.0; 722 ❙ 1208; 588; 54 ❙ 69.2; 722; 237.7

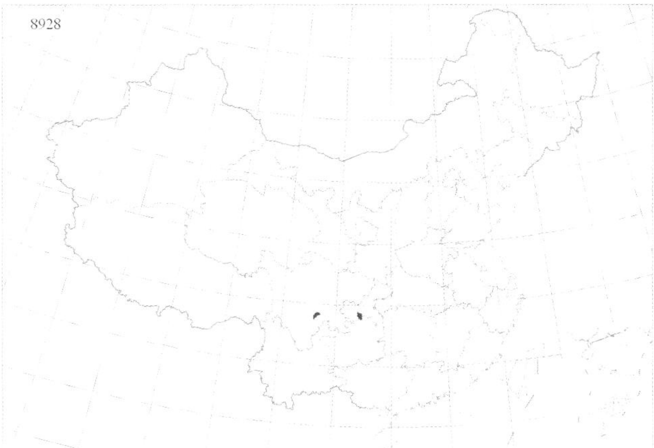

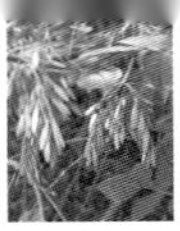

Oleaceae 木犀科 *Ligustrum* 女贞属

Ligustrum sinense var. rugosulum (W. W. Sm.) M. C. Chang 皱叶小蜡

Shrubs or small trees, deciduous

17.0 (5.3~21.0); 10.5 (-2.4~14.2); 21.7 (12.2~25.6); 17.1 (5.7~21.0); 148.6 (30.5~191.8); 3.7 (0.0~26.2); 857 (479~1053) ❙ 1309 (961~1547); 746 (558~911); 53 (31~68) ❙ 56.2 (21.3~100.7); 852 (479~1053); 440.2 (234.2~551.6)

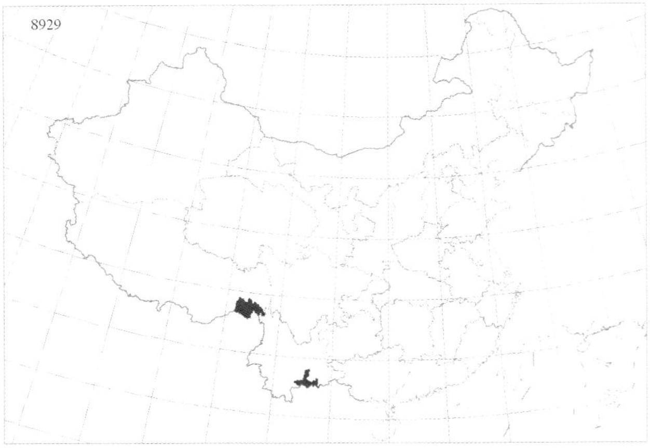

Oleaceae 木犀科 *Ligustrum* 女贞属

Ligustrum strongylophyllum Hemsl. 宜昌女贞

Shrubs

12.8 (2.9~18.7); 1.9 (-6.5~8.1); 23.2 (11.2~28.9); 13.0 (4.4~18.8); 102.9 (20.5~166.0); 8.8 (0.0~44.6); 739 (428~1002) ❙ 1062 (675~1478); 472 (331~606); 49 (15~91) ❙ 47.2 (-1.2~121.7); 739 (428~1002); 233.2 (108.2~320.6)

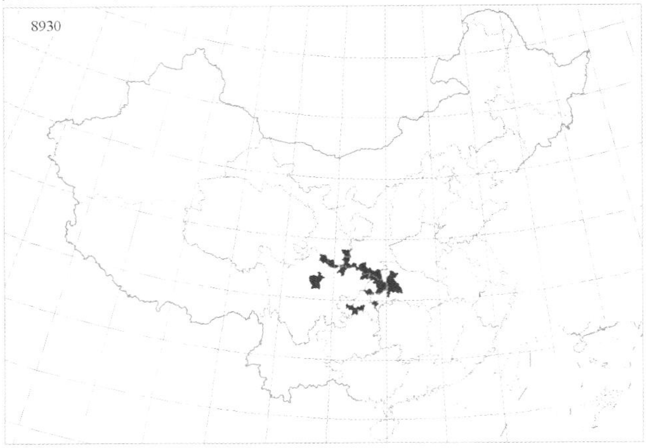

Oleaceae 木犀科 *Ligustrum* 女贞属

Ligustrum tenuipes M. C. Chang 细梗女贞

Shrubs

17.8; 7.2; 27.1; 17.8; 153.6; 0.0; 929 ❙ 1649; 629; 189 ❙ 77.5; 929; 228.4

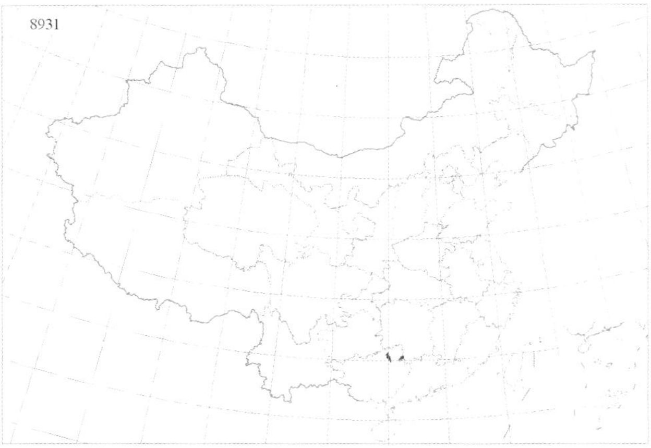

Oleaceae 木犀科 *Ligustrum* 女贞属

Ligustrum xingrenense D. J. Liu 兴仁女贞

Shrubs, evergreen

15.9; 6.9; 23.0; 16.0; 131.6; 0.0; 802 ❙ 1321; 696; 64 ❙ 64.7; 802; 203.5

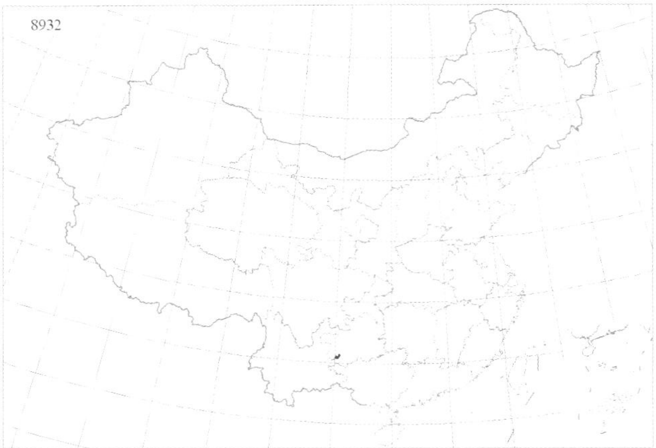

Oleaceae 木犀科 *Ligustrum* 女贞属

Ligustrum yunguiense B. M. Miao 云贵女贞

Shrubs or small trees, evergreen

13.4; 3.6; 22.3; 13.4; 102.7; 1.8; 721 ❙ 1077; 529; 59 ❙ 49.3; 721; 203.8

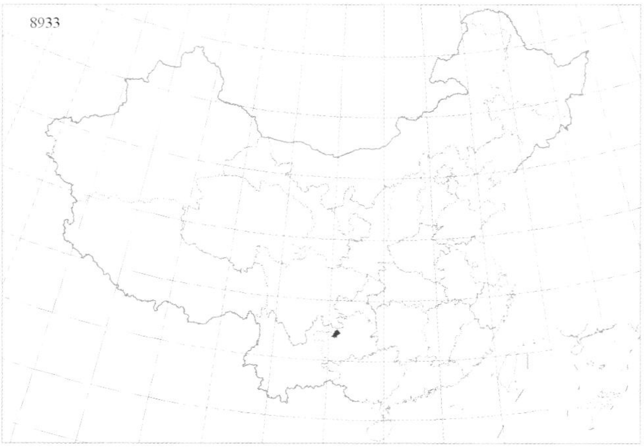

Oleaceae 木犀科 *Myxopyrum* 胶核藤属

Myxopyrum pierrei Gagnep. 海南胶核木

Scandent shrubs

23.8 (22.2~24.8); 18.6 (17.2~20.1); 27.6 (25.7~28.7); 23.9 (22.3~24.8); 226.7 (207.2~237.6); 0.0 (0.0~0.0); 1296 (1127~1385) ❙ 1479 (1176~1767); 515 (392~610); 55 (41~88) ❙ 14.9 (-12.0~47.8); 1144 (1055~1299); 519.9 (411.8~620.0)

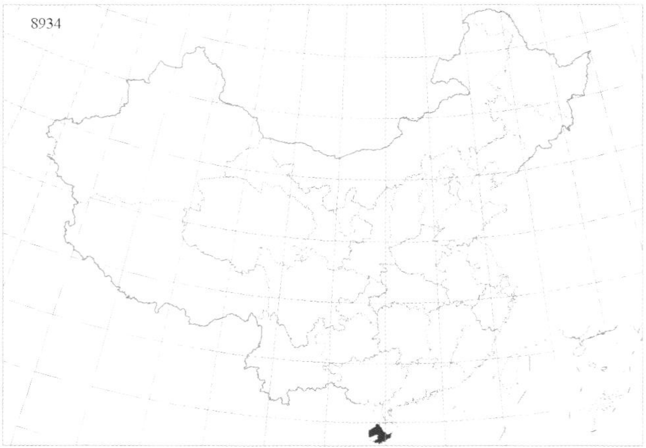

Oleaceae 木犀科 *Myxopyrum* 胶核藤属

Myxopyrum smilacifolium Blume 阔叶胶核木

Scandent shrubs

24.0 (22.2~24.8); 18.4 (17.2~20.1); 28.0 (25.7~28.8); 24.0 (22.3~24.8); 228.3 (207.2~237.6); 0.0 (0.0~0.0); 1310 (1127~1385) ❙ 1501 (868~1854); 534 (312~712); 74 (41~115) ❙ 15.0 (-34.3~47.8); 1191 (868~1360); 440.3 (248.2~620.0)

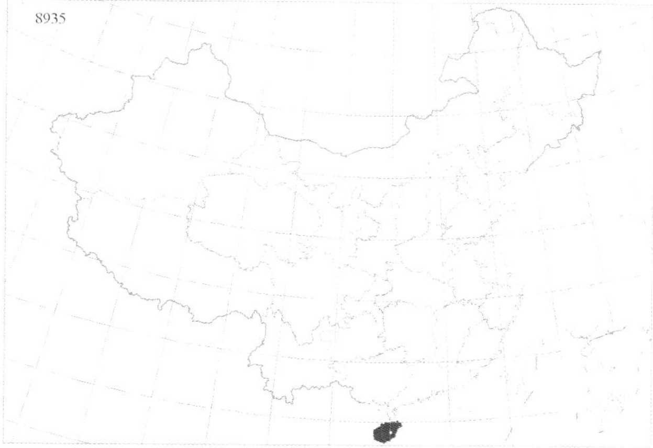

Oleaceae 木犀科 *Olea* 木犀榄属

Olea brachiata (Lour.) Merr. 滨木犀榄

Shrubs, evergreen

23.8 (21.8~24.8); 17.9 (14.0~20.1); 28.0 (25.7~29.0); 23.8 (21.9~24.8); 225.8 (202.2~237.6); 0.0 (0.0~0.0); 1295 (1127~1385) ❙ 1549 (868~2435); 6C5 (312~1428); 78 (41~130) ❙ 20.5 (-34.3~99.5); 1188 (868~1360); 415.4 (225.6~620.0)

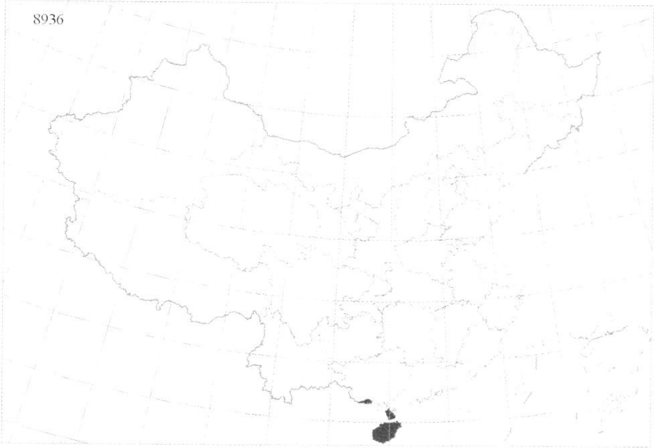

Oleaceae 木犀科 *Olea* 木犀榄属

Olea caudatilimba L. C. Chia 尾叶木犀榄

Small trees, evergreen

17.5 (14.2~21.0); 10.6 (6.2~15.1); 22.3 (20.1~25.6); 17.5 (14.2~21.0); 150.4 (110.5~191.8); 0.0 (0.0~0.0); 853 (729~1053) ❙ 1223 (958~1751); 673 (531~996); 54 (44~77) ❙ 42.8 (13.3~88.8); 850 (729~1053); 383.9 (234.5~659.0)

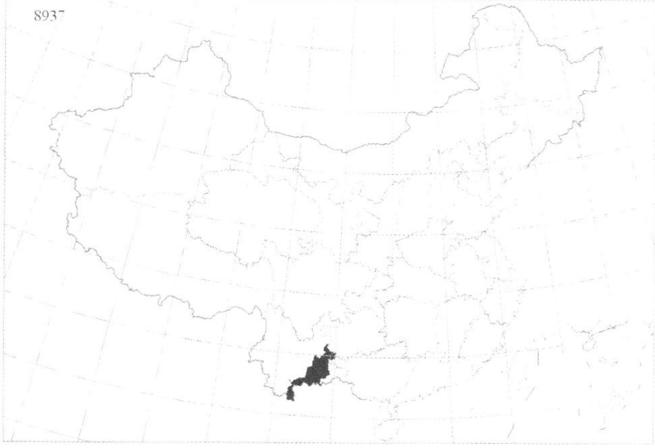

Oleaceae 木犀科 *Olea* 木犀榄属

Olea europaea L. 木犀榄

Small trees, evergreen

16.1 (5.3~24.8); 7.5 (-2.4~20.1); 23.4 (12.2~29.4); 16.1 (5.7~24.8); 134.4 (30.5~237.6); 1.2 (0.0~26.2); 853 (479~1385) ❙ 1140 (741~1854); 553 (312~720); 53 (12~142) ❙ 36.5 (-34.3~100.7); 828 (479~1360); 253.3 (108.2~620.0)

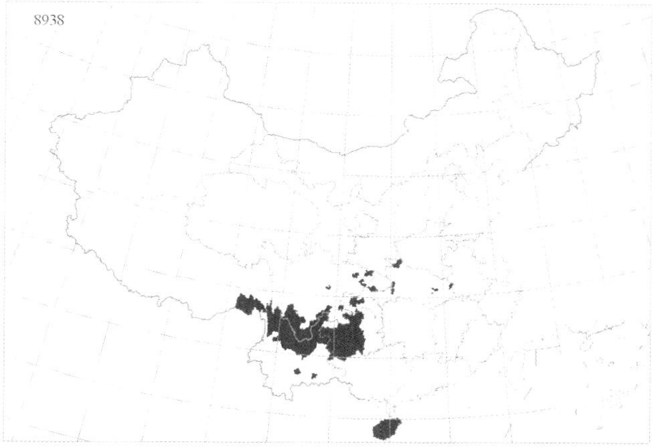

Oleaceae 木犀科 *Olea* 木犀榄属

Olea europaea subsp. cuspidata (Wall. ex G. Don) Cif. 锈鳞木犀榄

Shrubs or small trees, evergreen

13.6 (5.3~18.2); 6.6 (-2.4~11.8); 18.8 (12.2~22.5); 13.6 (5.7~18.2); 108.5 (30.5~158.4); 5.4 (0.0~26.2); 723 (479~873) ❙ 1035 (862~1178); 593 (531~651); 37 (12~51) ❙ 48.3 (13.3~100.7); 717 (479~863); 324.3 (234.2~406.7)

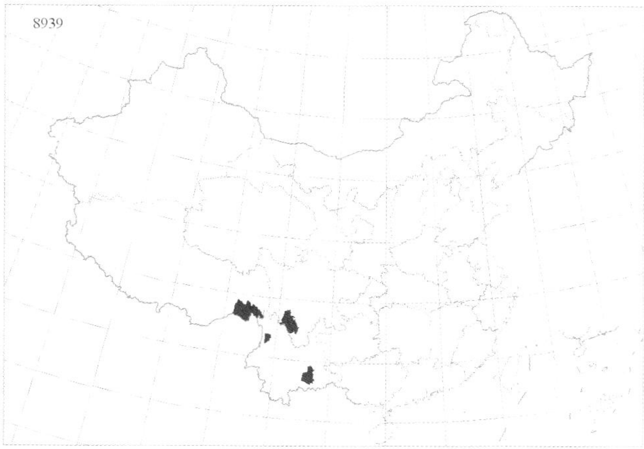

Oleaceae 木犀科 *Olea* 木犀榄属

Olea guangxiensis B. M. Miao 广西木犀榄

Shrubs or small trees, evergreen

19.8 (15.5~22.7); 10.6 (4.8~15.0); 27.2 (25.1~28.6); 19.8 (15.6~22.8); 177.9 (127.2~213.1); 0.0 (0.0~0.2); 1038 (815~1221) ❙ 1681 (1284~2435); 837 (591~1428); 129 (73~189) ❙ 62.4 (12.9~99.5); 1038 (815~1221); 273.0 (208.7~322.1)

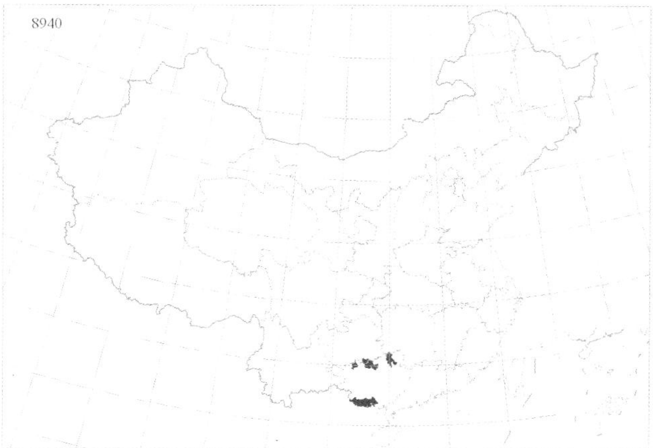

Oleaceae 木犀科 *Olea* 木犀榄属

Olea hainanensis H. L. Li 海南木犀榄

Shrubs or small trees, evergreen

21.3 (14.3~24.7); 13.8 (4.0~19.8); 27.1 (23.5~28.6); 21.3 (14.3~24.7); 195.8 (112.6~236.5); 0.1 (0.0~1.1); 1138 (756~1376) ❙ 1588 (1179~2435); 714 (392~1428); 89 (42~130) ❙ 41.9 (-11.9~99.5); 1083 (756~1326); 359.6 (188.8~620.0)

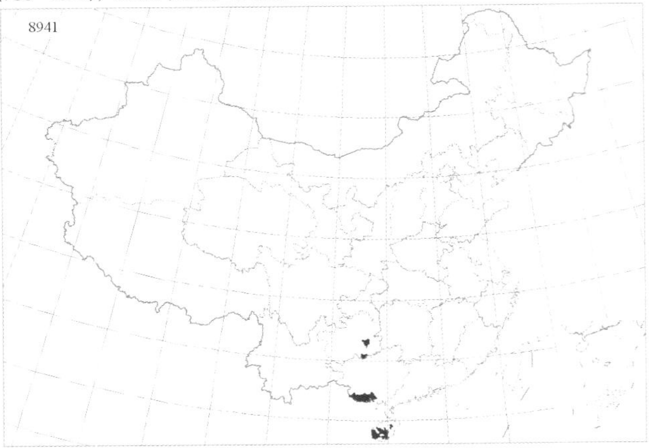

Oleaceae 木犀科 *Olea* 木犀榄属

Olea laxiflora H. L. Li 疏花木犀榄

Shrubs, evergreen

8.6; 1.3; 15.0; 8.7; 52.6; 8.8; 574 ❙ 1041; 585; 45 ❙ 81.5; 574; 352.3

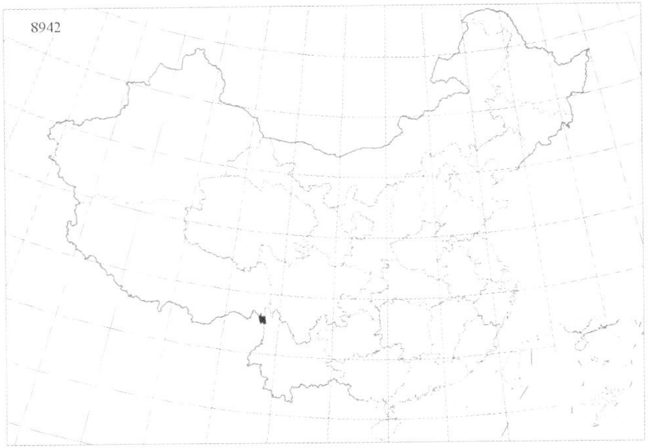

Oleaceae 木犀科 *Olea* 木犀榄属

Olea neriifolia H. L. Li 狭叶木犀榄

Shrubs, evergreen

22.6; 15.7; 27.7; 22.6; 211.1; 0.0; 1221 ❙ 1633; 613; 99 ❙ 34.9; 1170; 435.0

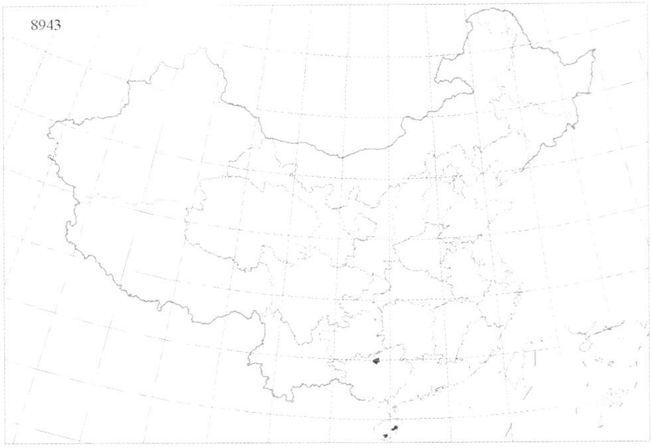

Oleaceae 木犀科 *Olea* 木犀榄属

Olea paniculata R. Br. 腺叶木犀榄

Trees, evergreen

17.7 (16.0~20.5); 11.3 (9.3~15.1); 21.8 (20.4~23.7); 17.8 (16.0~20.5); 153.2 (132.4~185.9); 0.0 (0.0~0.0); 857 (780~992) ❙ 1388 (1049~1688); 767 (594~959); 54 (42~77) ❙ 62.2 (26.1~112.2); 846 (779~986); 470.3 (346.6~659.0)

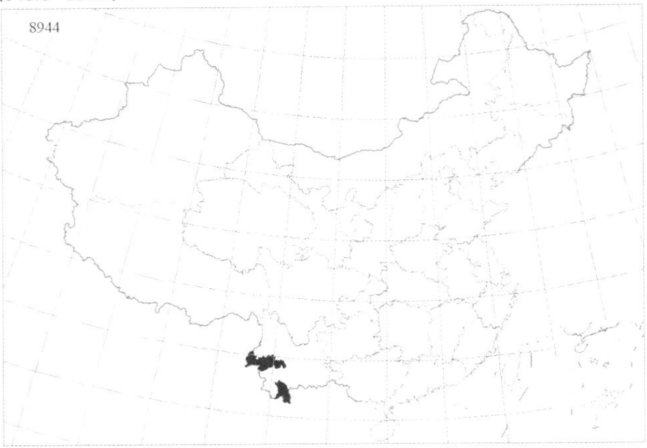

Oleaceae 木犀科 *Olea* 木犀榄属

Olea parvilimba (Merr. et Chun) B. M. Miao 小叶木犀榄

Trees, evergreen

23.7; 18.4; 27.4; 23.8; 225.1; 0.0; 1282 ❙ 1666; 590; 75 ❙ 30.3; 1212; 515.3

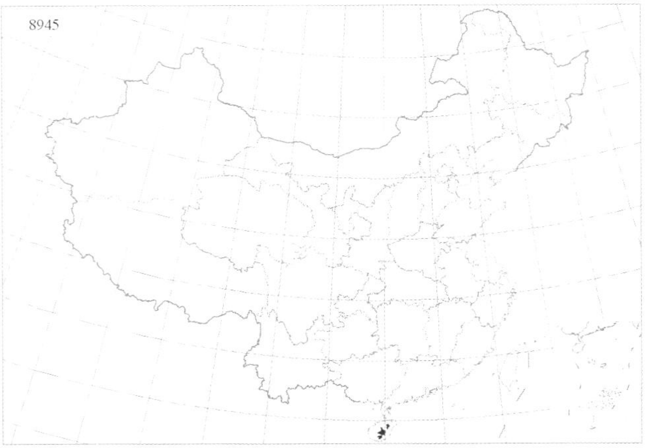

Oleaceae 木犀科 *Olea* 木犀榄属

Olea rosea Craib 红花木犀榄

Shrubs or small trees, evergreen

18.9 (16.9~20.5); 13.0 (10.3~15.1); 22.5 (21.4~23.7); 18.9 (16.9~20.5); 167.0 (142.7~185.9); 0.0 (0.0~0.0); 907 (813~992) ❙ 1382 (1049~1607); 677 (598~775); 60 (44~77) ❙ 51.8 (26.1~64.9); 885 (783~986); 486.6 (365.3~659.0)

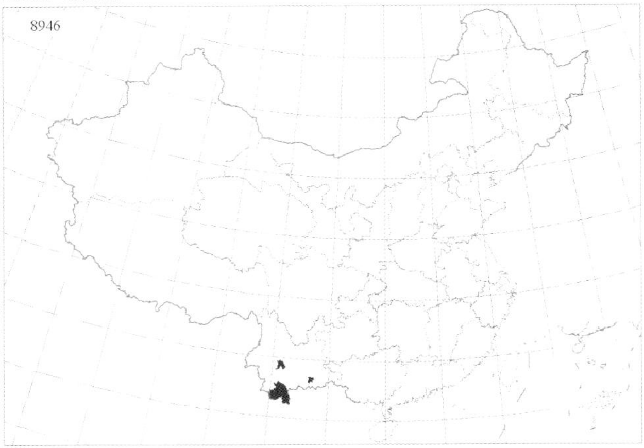

Oleaceae 木犀科 *Olea* 木犀榄属
Olea salicifolia Wall. ex G. Don 喜马木犀榄
Small trees, evergreen

13.3; 5.4; 19.6; 13.4; 100.4; 0.0; 706 ▎1736; 1007; 51 ▎145.8; 706; 447.4

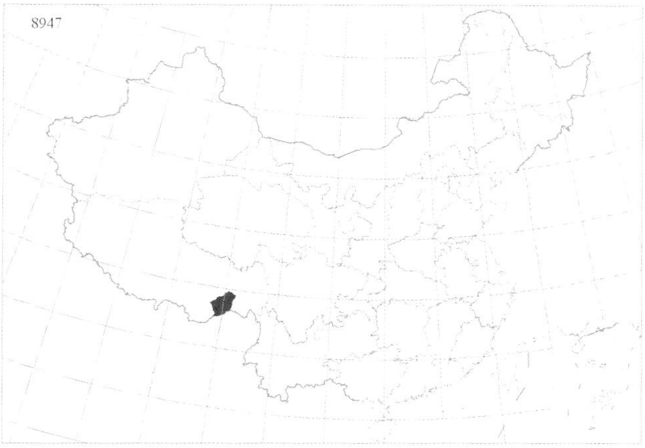

Oleaceae 木犀科 *Olea* 木犀榄属
Olea tetragonoclada L. C. Chia 方枝木犀榄
Shrubs, evergreen

21.6 (18.1~22.7); 13.4 (9.5~15.0); 27.7 (24.6~28.6); 21.7 (18.2~22.8); 200.0 (158.1~213.1); 0.0 (0.0~0.0); 1144 (901~1221) ▎1705 (1293~2435); 936 (670~1428); 108 (58~130) ▎48.9 (12.9~99.5); 1144 (901~1221); 291.1 (235.1~327.7)

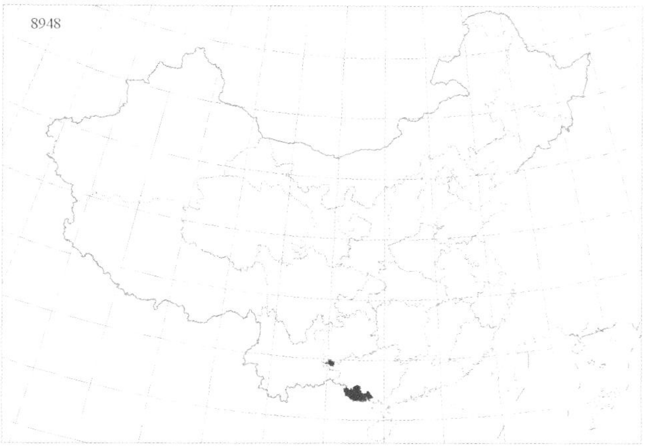

Oleaceae 木犀科 *Olea* 木犀榄属
Olea tsoongii (Merr.) P. S. Green 云南木犀榄
Shrubs or trees, evergreen

17.3 (5.4~24.8); 9.5 (-2.1~20.1); 23.6 (12.3~29.0); 17.4 (5.8~24.8); 148.8 (31.5~237.6); 0.6 (0.0~25.6); 899 (481~1385) ▎1278 (741~2148); 616 (312~1234); 80 (12~235) ▎43.3 (-34.3~95.5); 877 (481~1329); 298.8 (120.3~659.0)

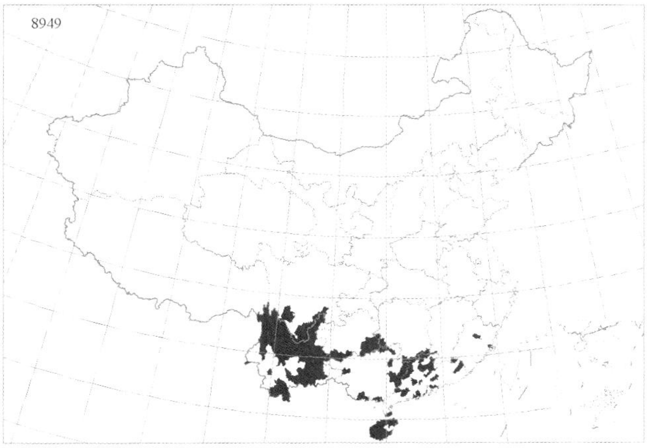

Oleaceae 木犀科 *Osmanthus* 木犀属
Osmanthus armatus Diels 红柄木犀
Shrubs or trees, evergreen

13.9 (9.3~16.7); 2.8 (-1.7~5.9); 24.7 (20.1~28.4); 14.0 (9.5~16.8); 111.9 (69.8~141.5); 4.2 (0.0~18.0); 771 (603~911) ▎1316 (1105~1492); 535 (463~613); 89 (31~175) ▎71.9 (32.3~107.2); 771 (603~911); 232.6 (159.0~285.1)

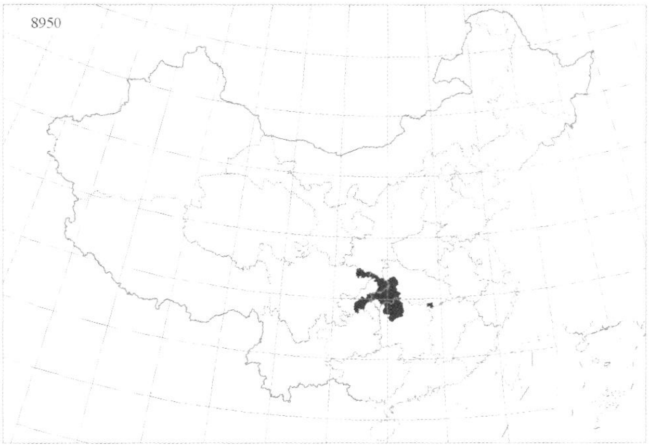

Oleaceae 木犀科 *Osmanthus* 木犀属
Osmanthus attenuatus P. S. Green 狭叶木犀
Shrubs, evergreen

17.9 (14.6~21.9); 8.5 (3.8~13.2); 25.8 (21.6~28.6); 17.9 (14.7~22.0); 155.5 (117.7~203.8); 0.2 (0.0~1.5); 924 (779~1176) ▎1473 (1049~1624); 625 (497~835); 143 (44~196) ▎61.4 (26.1~101.8); 919 (779~1176); 278.2 (188.8~506.2)

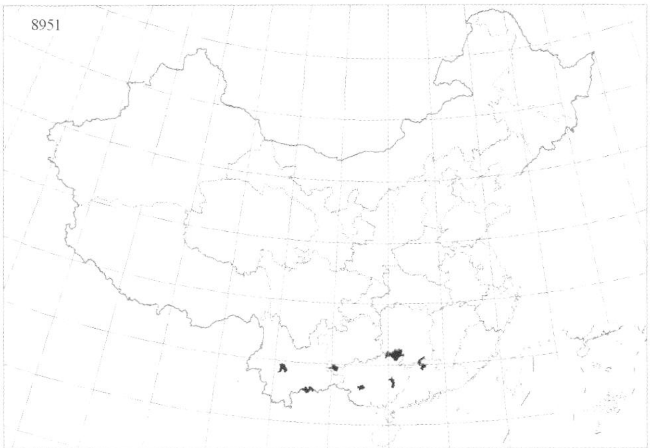

Oleaceae 木犀科 *Osmanthus* 木犀属
Osmanthus cooperi Hemsl. 宁波木犀
Shrubs or small trees, evergreen

16.6 (12.7~19.5); 5.6 (2.1~10.9); 27.3 (23.0~29.4); 16.7 (12.8~19.6); 140.9 (99.0~174.7); 0.7 (0.0~5.6); 888 (713~1007) ▎1635 (1117~1997); 566 (427~713); 205 (85~259) ▎84.7 (28.5~147.0); 883 (713~1007); 250.2 (111.9~332.5)

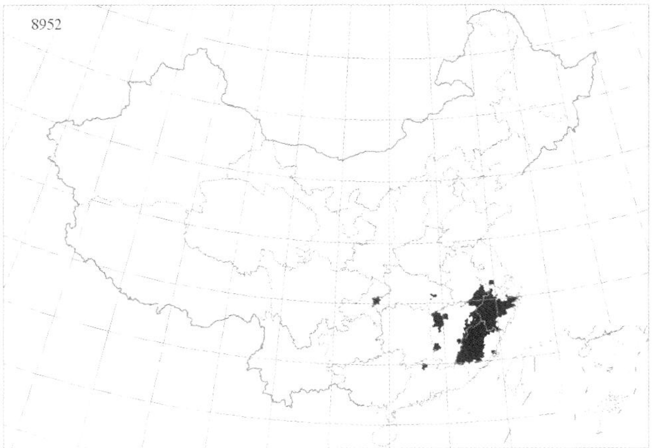

Oleaceae 木犀科 *Osmanthus* 木犀属

Osmanthus delavayi Franch. 山桂花

Shrubs, evergreen

14.4 (2.2~18.9); 6.8 (-6.3~12.5); 20.5 (9.5~26.6); 14.5 (3.9~19.0); 114.8 (15.8~167.8); 1.5 (0.0~48.6); 746 (411~929) ❙ 1000 (729~1252); 565 (411~686); 39 (11~81) ❙ 35.6 (-14.2~78.3); 725 (411~912); 275.5 (142.9~406.0)

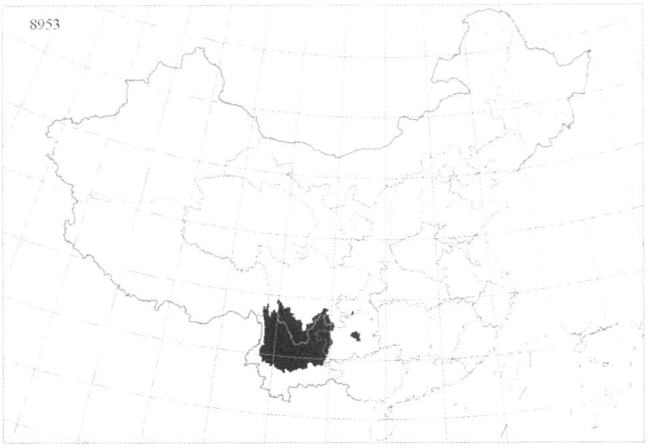

Oleaceae 木犀科 *Osmanthus* 木犀属

Osmanthus didymopetalus P. S. Green 双瓣木犀

Trees, evergreen

23.1 (21.0~24.8); 16.9 (13.2~20.1); 27.7 (25.7~28.7); 23.2 (21.0~24.8); 217.9 (192.3~237.6); 0.0 (0.0~0.0); 1243 (1081~1385) ❙ 1519 (868~1876); 602 (312~891); 78 (41~125) ❙ 23.9 (-34.3~64.3); 1128 (868~1326); 398.5 (209.2~620.0)

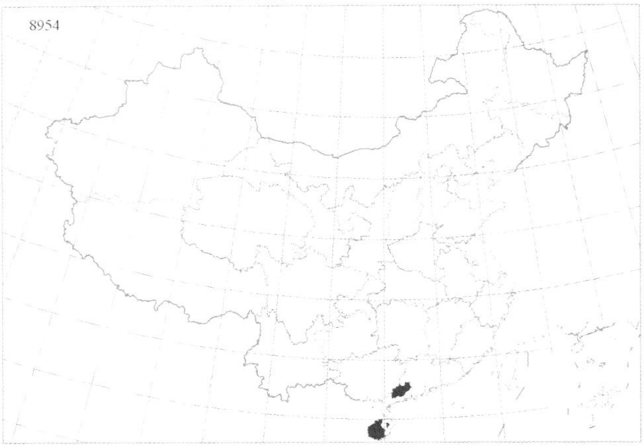

Oleaceae 木犀科 *Osmanthus* 木犀属

Osmanthus enervius Masam. et K. Mori 无脉木犀

Small trees, evergreen

19.4; 13.3; 24.4; 19.4; 172.6; 0.0; 967 ❙ 2109; 1039; 182 ❙ 128.0; 946; 242.1

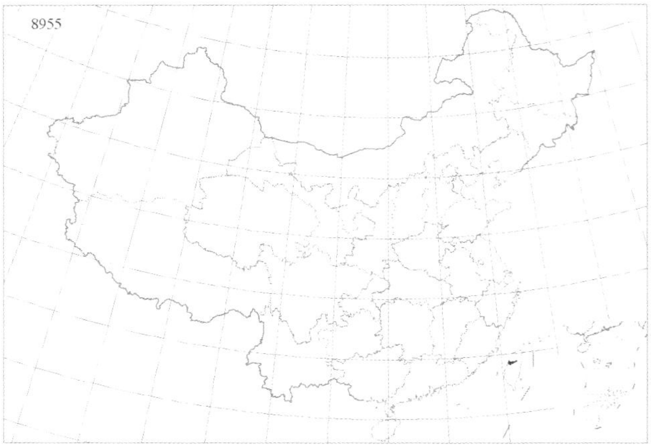

Oleaceae 木犀科 *Osmanthus* 木犀属

Osmanthus fordii Hemsl. 石山桂花

Shrubs, evergreen

18.1 (16.2~19.4); 8.0 (4.6~10.1); 26.7 (25.5~28.7); 18.2 (16.2~19.5); 157.9 (135.2~173.7); 0.0 (0.0~0.4); 939 (850~1038) ❙ 1489 (1236~1776); 621 (450~718); 142 (51~195) ❙ 58.9 (27.5~86.9); 939 (850~1038); 251.3 (203.9~322.1)

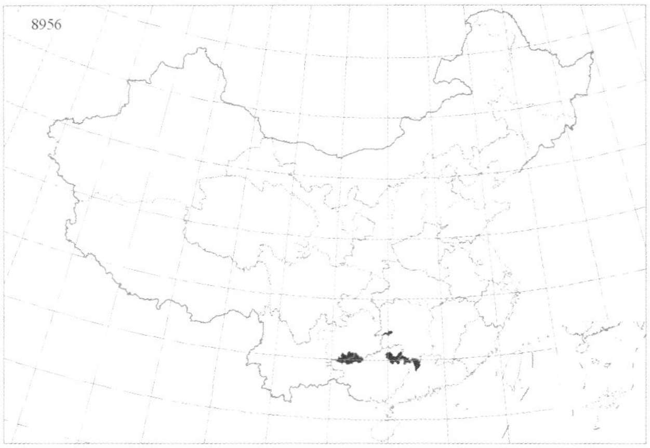

Oleaceae 木犀科 *Osmanthus* 木犀属

Osmanthus fragrans (Thunb.) Lour. 木犀

Shrubs or trees, evergreen

16.5 (5.3~24.8); 7.0 (-7.1~20.1); 24.9 (12.2~29.9); 16.6 (5.7~24.8); 140.4 (30.5~237.6); 1.5 (0.0~40.6); 871 (479~1385) ❙ 1340 (520~2993); 577 (245~1847); 109 (10~547) ❙ 54.0 (-36.2~246.5); 853 (479~1360); 259.5 (111.9~659.0)

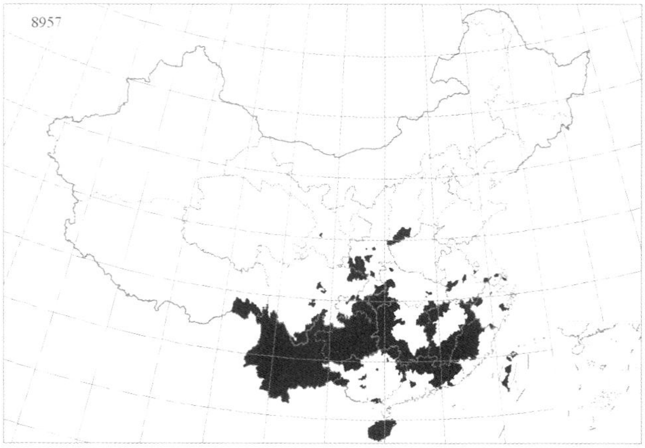

Oleaceae 木犀科 *Osmanthus* 木犀属

Osmanthus gracilinervis L. C. Chia ex R. L. Lu 细脉木犀

Shrubs or small trees, evergreen

17.4 (14.3~19.4); 6.9 (3.7~9.4); 27.1 (23.6~28.8); 17.5 (14.3~19.4); 150.1 (113.2~173.4); 0.1 (0.0~1.4); 919 (761~1029) ❙ 1564 (1090~1846); 577 (505~816); 177 (49~256) ❙ 70.9 (15.8~128.9); 917 (761~1029); 225.9 (130.1~261.3)

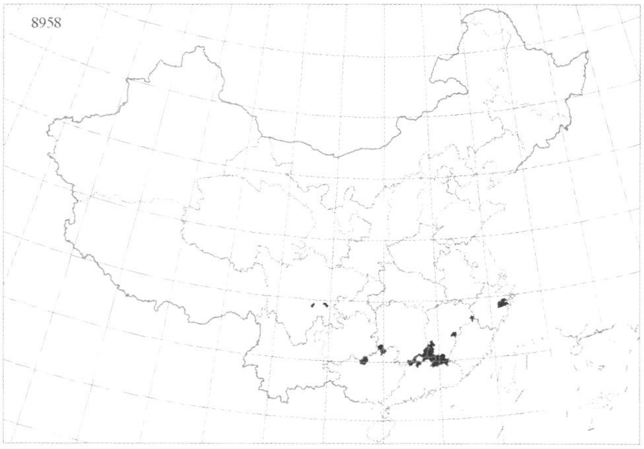

Oleaceae 木犀科 *Osmanthus* 木犀属
Osmanthus hainanensis P. S. Green 显脉木犀
Shrubs or small trees, evergreen

23.9; 19.0; 27.3; 24.0; 227.6; 0.0; 1301 ❙ 1560; 541; 52 ❙ 20.8; 1149; 573.5

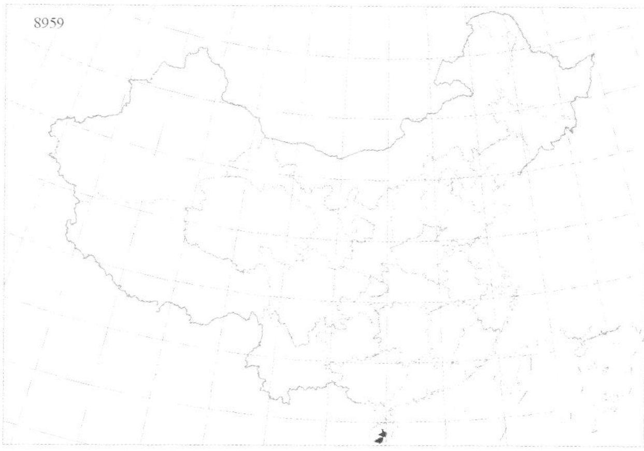

Oleaceae 木犀科 *Osmanthus* 木犀属
Osmanthus henryi P. S. Green 蒙自桂花
Shrubs or small trees, evergreen

16.6 (11.7~21.0); 9.5 (3.2~15.1); 21.9 (18.6~28.4); 16.7 (11.7~21.0); 140.3 (82.6~191.8); 0.1 (0.0~2.3); 822 (658~1053) ❙ 1186 (797~1751); 644 (411~996); 53 (23~197) ❙ 44.0 (-14.2~112.2); 804 (658~1053); 344.0 (170.4~659.0)

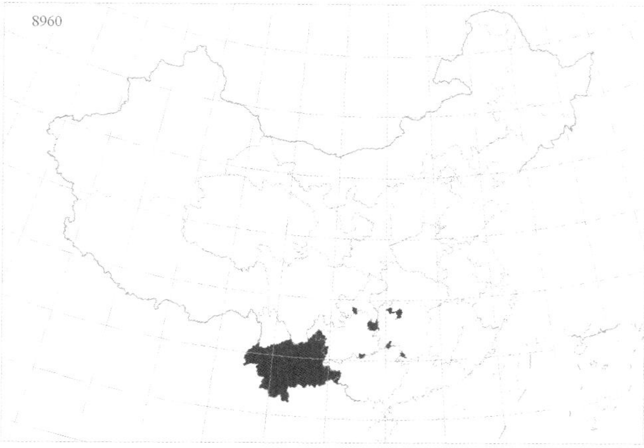

Oleaceae 木犀科 *Osmanthus* 木犀属
Osmanthus heterophyllus (G. Don) P. S. Green 冬树
Shrubs or small trees, evergreen

19.3 (16.4~21.1); 13.6 (11.5~15.3); 24.4 (20.3~27.4); 19.4 (16.5~21.1); 172.7 (137.8~193.1); 0.0 (0.0~0.0); 951 (783~1077) ❙ 2829 (2445~3495); 1042 (821~1412); 428 (110~897) ❙ 199.5 (140.0~235.6); 951 (783~1077); 266.8 (207.7~363.7)

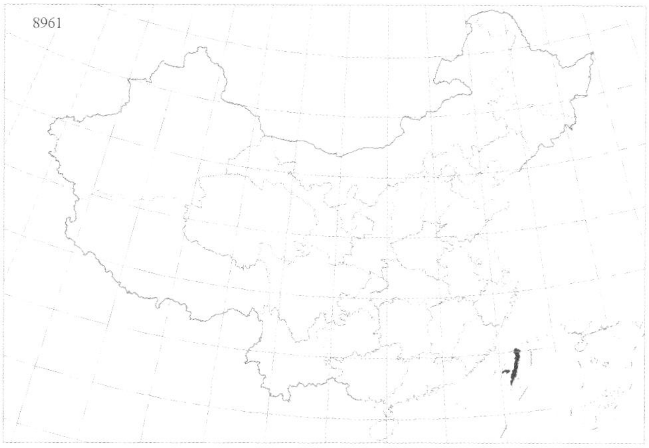

Oleaceae 木犀科 *Osmanthus* 木犀属
Osmanthus lanceolatus Hayata 锐叶木犀
Shrubs or small trees, evergreen

18.5; 13.6; 22.3; 18.6; 163.0; 0.0; 888 ❙ 2552; 1249; 201 ❙ 191.2; 888; 291.7

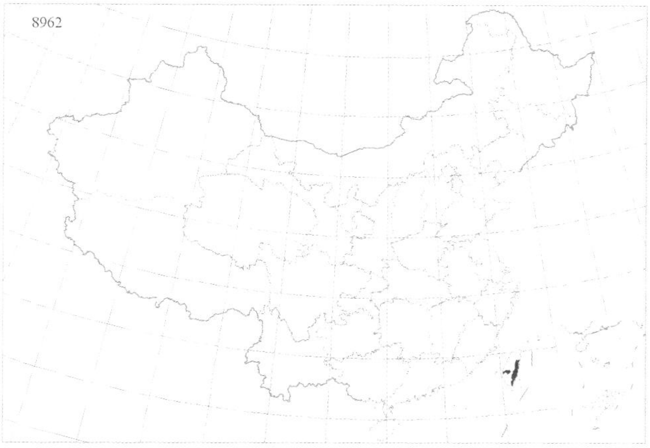

Oleaceae 木犀科 *Osmanthus* 木犀属
Osmanthus marginatus (Champ. ex Benth.) Hemsl. 厚边木犀
Shrubs or trees, evergreen

18.2 (11.9~24.8); 8.8 (2.1~20.1); 26.5 (19.1~29.8); 18.3 (12.0~24.8); 159.9 (87.2~237.6); 0.3 (0.0~5.6); 956 (671~1385) ❙ 1621 (939~3495); 663 (392~1847); 160 (35~897) ❙ 71.7 (-11.9~246.5); 945 (671~1326); 264.2 (119.4~627.0)

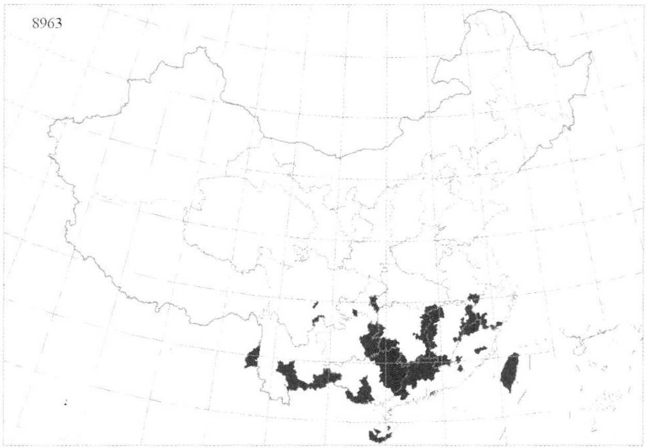

Oleaceae 木犀科 *Osmanthus* 木犀属
Osmanthus marginatus var. longissimus (H. T. Chang) R. L. Lu 长叶木犀
Shrubs or trees, evergreen

17.2 (15.2~19.1); 6.0 (4.1~10.0); 27.6 (25.8~29.1); 17.2 (15.3~19.1); 146.7 (124.1~169.1); 0.1 (0.0~0.9); 914 (810~984) ❙ 1762 (1561~1922); 568 (488~623); 234 (193~259) ❙ 93.1 (69.7~111.3); 908 (810~973); 238.5 (183.9~279.5)

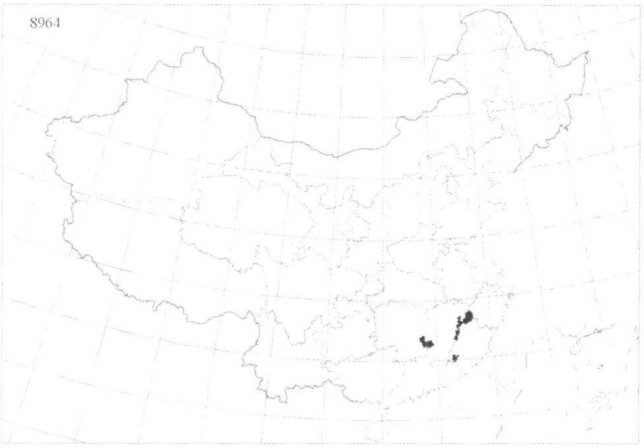

Oleaceae 木犀科 *Osmanthus* 木犀属
Osmanthus matsumuranus Hayata 牛矢果
Shrubs or trees, evergreen

19.3 (14.4~24.8); 10.3 (2.1~20.1); 26.9 (20.1~29.8); 19.3 (14.5~24.8); 171.9 (118.7~237.6); 0.2 (0.0~4.9); 1010 (773~1385) ❙ 1646 (868~3495); 692 (312~1847); 157 (41~897) ❙ 66.3 (-34.3~246.5); 992 (773~1360); 268.8 (119.4~659.0)

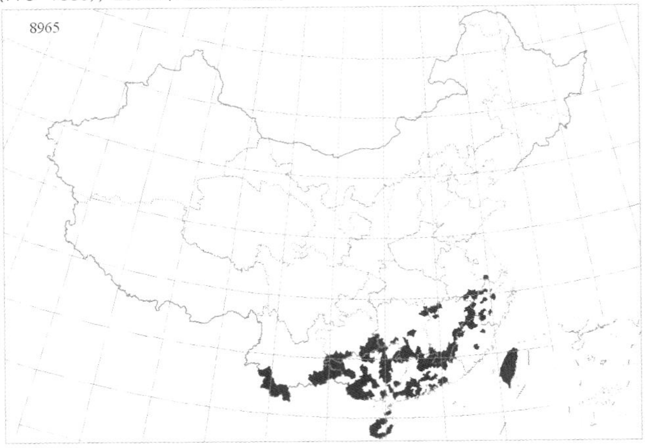

Oleaceae 木犀科 *Osmanthus* 木犀属
Osmanthus minor P. S. Green 小叶月桂
Shrubs or small trees, evergreen

18.5 (15.1~22.8); 8.5 (3.1~15.5); 27.5 (25.8~29.1); 18.6 (15.2~22.8); 162.9 (122.9~214.1); 0.1 (0.0~2.2); 974 (811~1225) ❙ 1755 (1494~1997); 630 (531~891); 202 (107~259) ❙ 82.2 (33.5~130.7); 971 (811~1225); 248.3 (183.9~303.4)

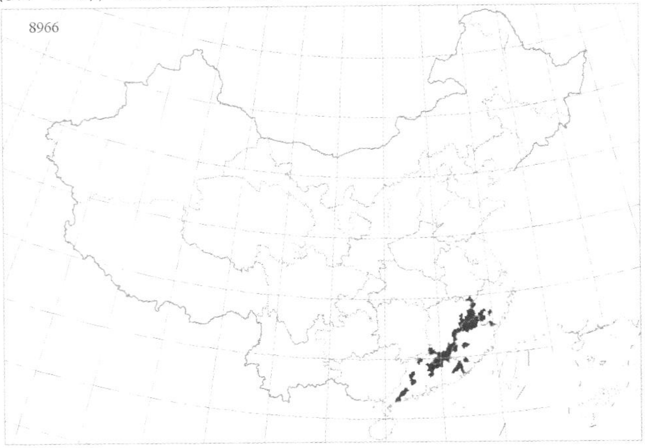

Oleaceae 木犀科 *Osmanthus* 木犀属
Osmanthus pubipedicellatus L. C. Chia ex H. T. Chang 毛柄木犀
Shrubs, evergreen

20.4; 12.0; 27.7; 20.5; 185.5; 0.0; 1057 ❙ 1591; 607; 162 ❙ 50.6; 1057; 266.9

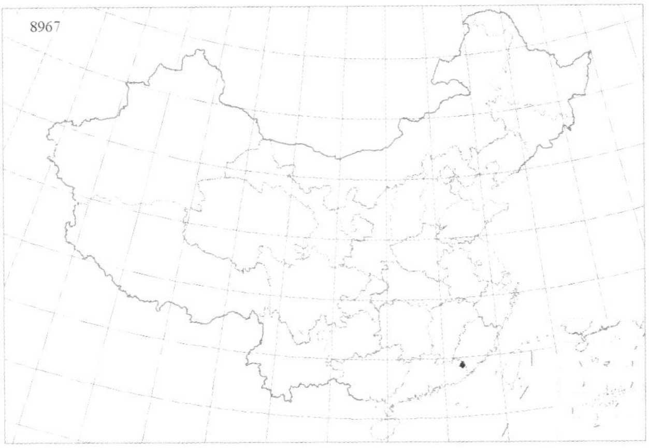

Oleaceae 木犀科 *Osmanthus* 木犀属
Osmanthus reticulatus P. S. Green 网脉木犀
Shrubs or small trees, evergreen

15.9 (12.7~18.2); 5.1 (2.1~7.7); 26.0 (23.0~27.8); 16.0 (12.8~18.2); 132.5 (99.0~158.5); 0.6 (0.0~5.6); 845 (713~949) ❙ 1342 (1132~1624); 512 (444~639); 123 (64~197) ❙ 59.1 (31.5~107.2); 845 (713~949); 209.3 (169.7~258.0)

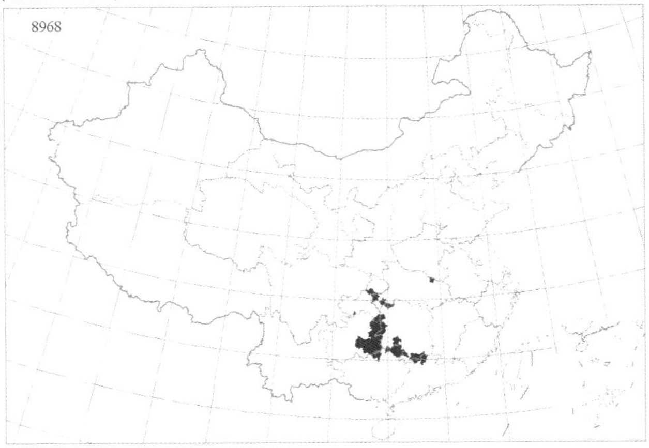

Oleaceae 木犀科 *Osmanthus* 木犀属
Osmanthus serrulatus Rehder 短丝木犀
Shrubs or small trees, evergreen

13.5 (7.3~17.3); 3.7 (-2.0~9.0); 22.5 (15.6~27.5); 13.6 (7.5~17.3); 106.8 (48.2~147.6); 4.4 (0.0~20.4); 744 (537~910) ❙ 1383 (1001~1997); 620 (501~816); 103 (27~259) ❙ 83.9 (49.1~130.7); 744 (537~910); 264.0 (171.0~305.7)

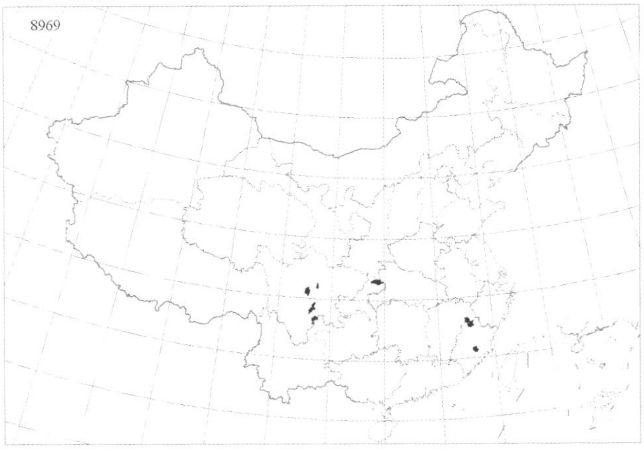

Oleaceae 木犀科 *Osmanthus* 木犀属
Osmanthus suavis King ex C. B. Clarke 香花木犀
Shrubs or small trees, evergreen

15.8 (13.6~18.9); 9.0 (5.6~12.0); 20.8 (19.0~23.9); 15.9 (13.6~19.0); 130.6 (103.6~167.8); 0.0 (0.0~0.0); 782 (707~929) ❙ 1028 (797~1454); 587 (411~792); 42 (20~70) ❙ 32.1 (-14.2~99.9); 754 (690~815); 296.6 (215.5~487.2)

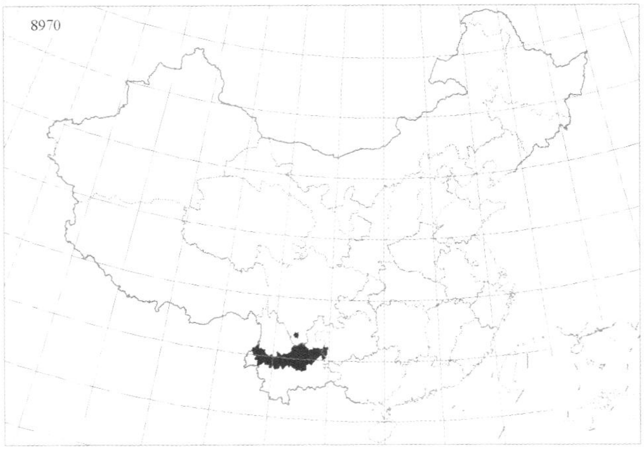

Oleaceae 木犀科 *Osmanthus* 木犀属
Osmanthus urceolatus P. S. Green 坛花木犀
Shrubs, evergreen

13.7; 3.0; 24.0; 13.7; 108.4; 3.4; 750 ❙ 1370; 557; 81 ❙ 83.6; 750; 218.6

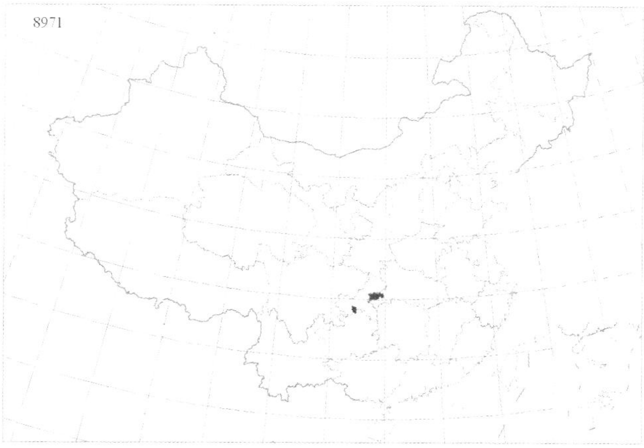

Oleaceae 木犀科 *Osmanthus* 木犀属
Osmanthus venosus Pamp. 毛木犀
Shrubs or small trees, evergreen

13.4 (12.6~14.5); 2.1 (1.6~2.6); 24.1 (23.0~26.1); 13.4 (12.7~14.5); 106.2 (98.8~119.1); 5.3 (4.1~6.9); 746 (711~808) ❙ 1251 (842~1492); 538 (441~613); 76 (19~105) ❙ 69.3 (11.5~107.2); 746 (711~808); 246.5 (204.2~285.1)

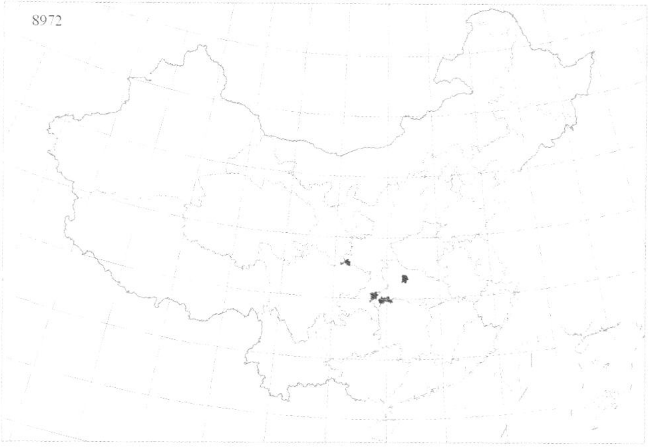

Oleaceae 木犀科 *Osmanthus* 木犀属
Osmanthus yunnanensis (Franch.) P. S. Green 野桂花
Shrubs or trees, evergreen

13.6 (3.8~21.0); 5.5 (-4.3~15.1); 20.4 (10.8~27.5); 13.7 (4.7~21.0); 106.7 (21.6~191.8); 2.8 (0.0~35.4); 730 (444~1053) ❙ 1086 (741~1736); 586 (411~1007); 48 (12~181) ❙ 50.6 (-14.2~145.8); 716 (444~1053); 282.7 (142.9~659.0)

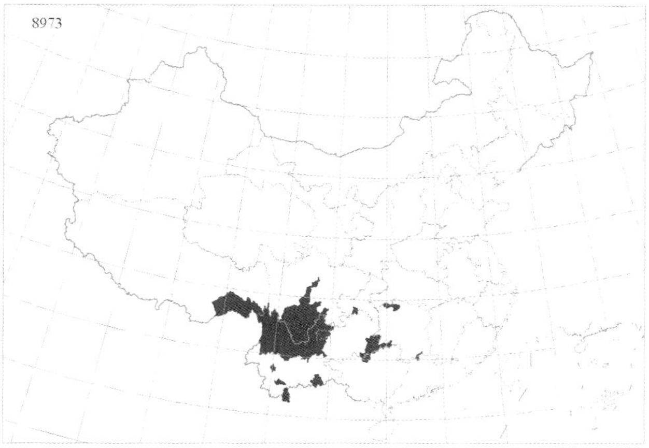

Oleaceae 木犀科 *Syringa* 丁香属
Syringa komarowii C. K. Schneid. 西蜀丁香
Shrubs, deciduous

11.1 (3.5~17.4); 1.0 (-6.5~10.3); 20.4 (12.6~26.4); 11.2 (5.0~17.4); 84.3 (26.3~149.1); 11.2 (0.0~43.4); 663 (446~903) ❙ 981 (632~1410); 500 (306~816); 32 (14~71) ❙ 49.4 (0.0~101.6); 661 (446~903); 268.9 (154.2~351.6)

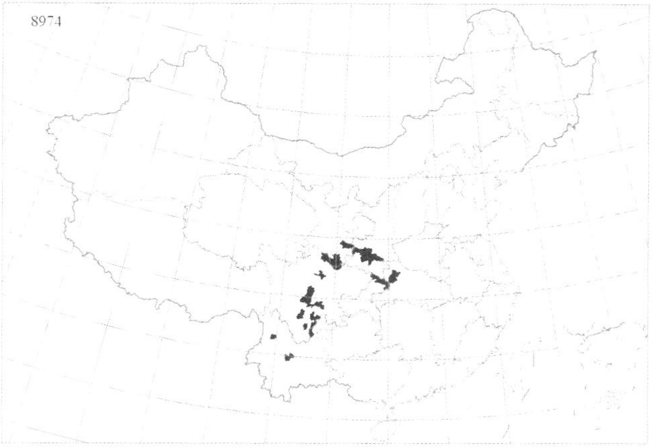

Oleaceae 木犀科 *Syringa* 丁香属
Syringa komarowii subsp. reflexa (C. K. Schneid.) P. S. Green et M. C. Chang 垂丝丁香
Shrubs, deciduous

10.8; -1.0; 22.3; 11.0; 84.8; 14.4; 666 ❙ 951; 410; 37 ❙ 44.6; 666; 278.1

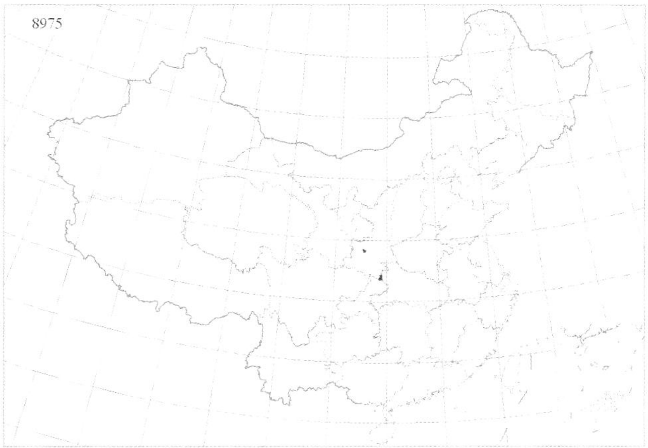

Oleaceae 木犀科 *Syringa* 丁香属
Syringa mairei (H. Lév.) Rehder 皱叶丁香
Shrubs or small trees, deciduous

9.8 (3.2~15.5); 2.4 (-6.4~8.6); 15.9 (11.4~20.4); 10.1 (4.7~15.6); 70.5 (22.7~127.3); 12.7 (0.0~43.8); 605 (436~768) ❙ 1003 (741~1180); 558 (426~651); 43 (17~65) ❙ 68.0 (40.1~100.7); 605 (436~768); 330.3 (234.2~384.1)

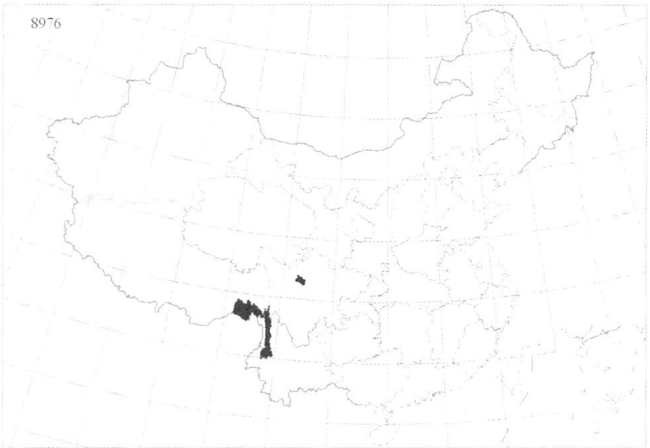

Oleaceae 木犀科 *Syringa* 丁香属
Syringa meyeri C. K. Schneid. 蓝丁香
Dwarf shrubs, deciduous

12.2 (7.4~14.9); -3.2 (-11.0~0.0); 25.9 (20.7~28.6); 12.8 (8.9~14.9); 107.8 (69.1~129.4); 20.9 (10.3~49.4); 779 (598~889) ❙ 609 (584~698); 375 (315~453); 22 (12~31) ❙ -20.3 (-34.2~7.1); 606 (584~698); 162.7 (118.4~192.2)

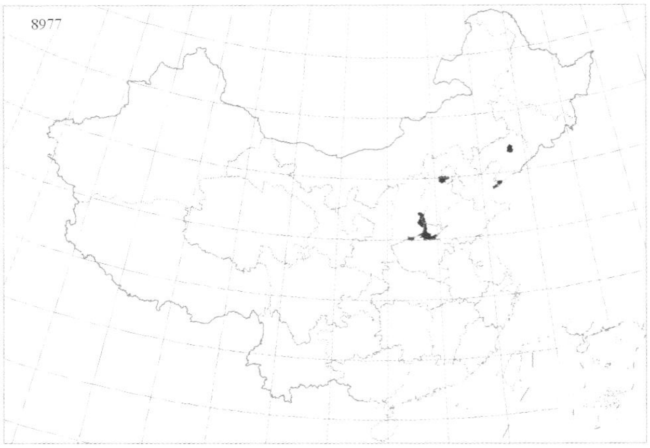

Oleaceae 木犀科 *Syringa* 丁香属
Syringa meyeri var. spontanea M. C. Chang 小叶蓝丁香
Dwarf shrubs

10.5; -4.9; 23.7; 11.3; 93.2; 26.8; 705 ❙ 632; 398; 27 ❙ -10.4; 632; 180.8

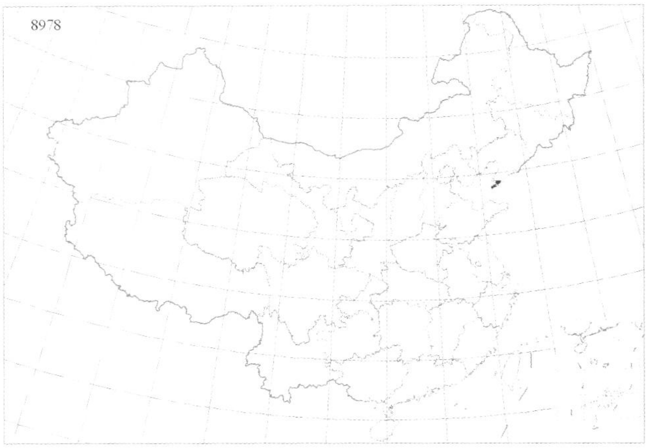

Oleaceae 木犀科 *Syringa* 丁香属
Syringa oblata Lindl. 紫丁香
Shrubs or small trees, deciduous

11.2 (0.8~17.7); -3.9 (-15.2~6.9); 24.6 (10.7~28.3); 11.9 (3.8~17.7); 99.3 (16.2~152.2); 24.3 (0.0~67.8); 741 (411~937) ❙ 598 (132~1214); 357 (81~648); 21 (2~75) ❙ -17.6 (-80.1~101.4); 566 (132~937); 173.5 (13.6~334.4)

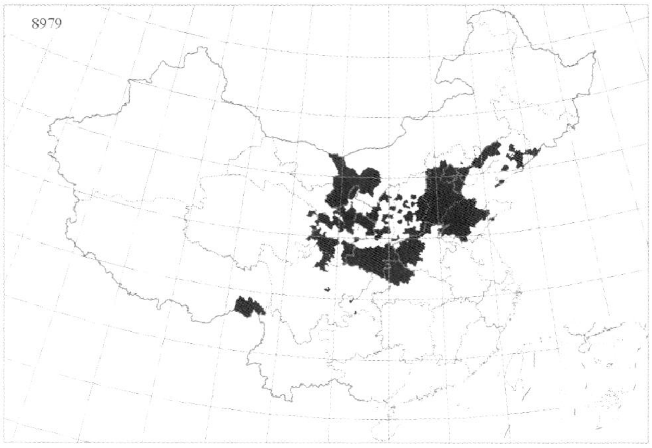

Oleaceae 木犀科 *Syringa* 丁香属
Syringa oblata subsp. dilatata (Nakai) P. S. Green et M. C. Chang 朝阳丁香
Shrubs, deciduous

7.0; -11.6; 22.9; 9.5; 78.1; 53.3; 637 ❙ 848; 532; 32 ❙ 34.5; 637; 275.3

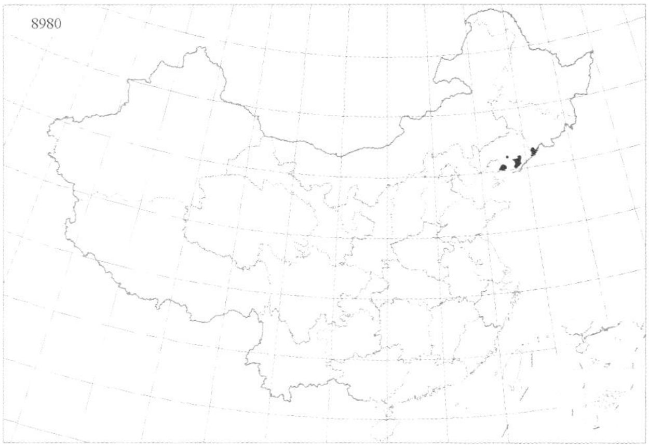

Oleaceae 木犀科 *Syringa* 丁香属
Syringa pinetorum W. W. Sm. 松林丁香
Shrubs, deciduous

4.9 (0.9~10.8); -3.3 (-8.3~3.9); 12.1 (9.5~16.6); 5.8 (3.4~10.9); 32.0 (13.3~71.8); 33.2 (1.3~61.6); 475 (389~633) ❙ 738 (510~968); 448 (311~594); 20 (11~31) ❙ 54.6 (20.7~100.7); 475 (389~633); 249.3 (161.1~319.8)

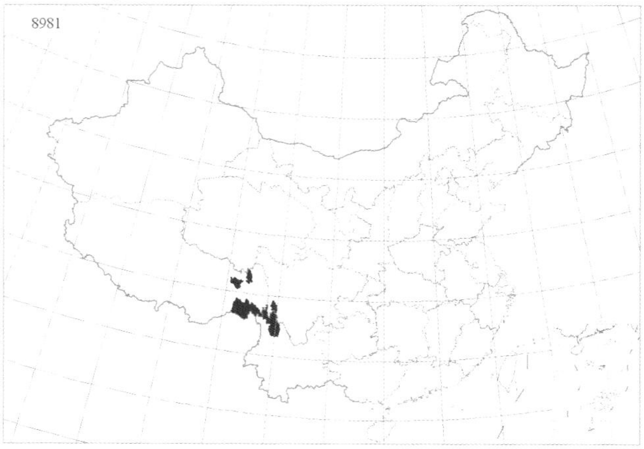

Oleaceae 木犀科 *Syringa* 丁香属
Syringa pinnatifolia Hemsl. 羽叶丁香
Shrubs, deciduous

7.8 (2.2~13.5); -5.1 (-9.8~2.4); 19.5 (9.9~24.2); 8.9 (3.8~13.6); 66.5 (15.5~107.5); 32.4 (4.9~56.4); 597 (413~755) ❙ 588 (132~1036); 303 (81~606); 14 (2~27) ❙ 5.3 (-80.1~125.9); 455 (132~755); 192.3 (13.6~357.7)

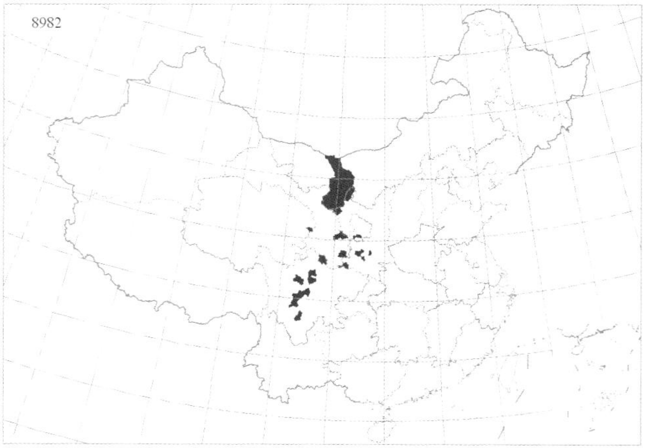

Oleaceae 木犀科 *Syringa* 丁香属

Syringa protolaciniata P. S. Green et M. C. Chang 华丁香

Small shrubs, deciduous

2.5; -9.7; 13.1; 5.0; 27.9; 56.4; 449 ❙ 516; 283; 9 ❙ 14.7; 449; 218.4

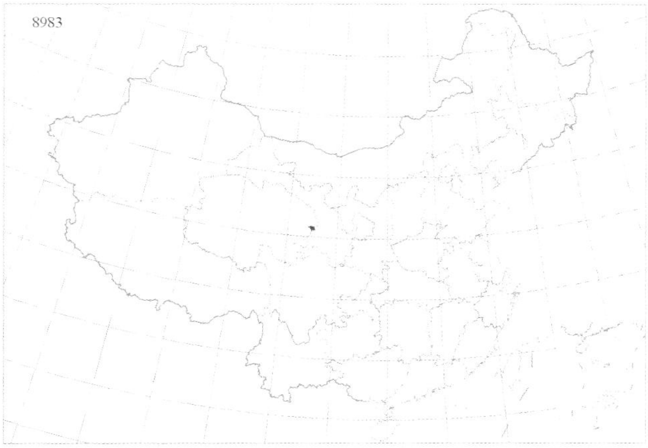

Oleaceae 木犀科 *Syringa* 丁香属

Syringa pubescens Turcz. 巧玲花

Shrubs, deciduous

9.6 (-1.3~15.1); -5.5 (-18.8~1.8); 23.0 (9.7~28.0); 10.7 (2.9~15.2); 87.1 (11.3~128.0); 31.5 (5.6~88.8); 682 (359~880) ❙ 599 (337~1073); 349 (190~691); 20 (1~58) ❙ -10.7 (-47.7~128.6); 568 (337~821); 187.8 (44.7~346.5)

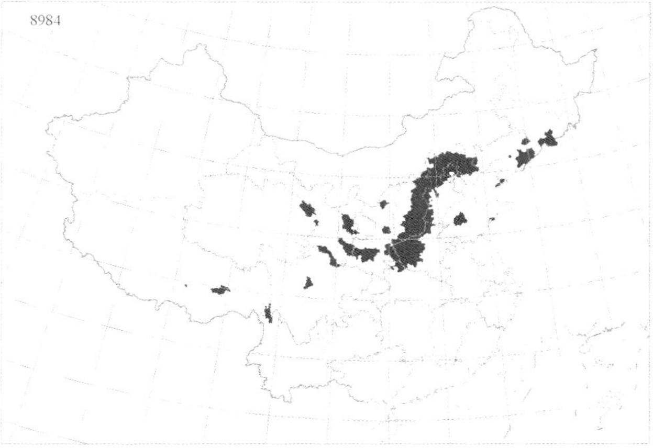

Oleaceae 木犀科 *Syringa* 丁香属

Syringa pubescens subsp. julianae (C. K. Schneid.) M. C. Chang et X. Y. Chen 光萼巧玲花

Shrubs, deciduous

13.1 (8.2~15.8); -0.7 (-6.9~2.5); 25.8 (20.1~28.2); 13.3 (9.5~15.8); 111.4 (69.8~133.1); 13.1 (3.4~34.4); 787 (603~891) ❙ 726 (512~1214); 364 (245~536); 38 (18~109) ❙ -7.0 (-36.5~101.4); 679 (512~891); 174.1 (120.1~311.0)

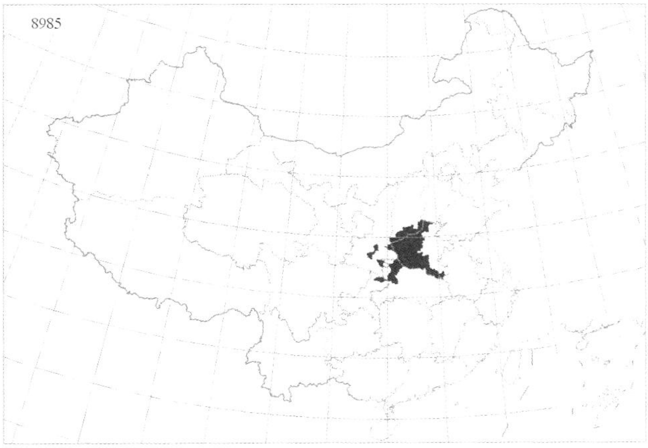

Oleaceae 木犀科 *Syringa* 丁香属

Syringa pubescens subsp. microphylla (Diels) M. C. Chang et X. L. Chen 小叶巧玲花

Shrubs, deciduous

9.2 (0.5~16.0); -7.3 (-23.0~3.1); 23.7 (10.1~28.1); 10.9 (3.8~16.1); 89.4 (15.4~135.0); 38.8 (2.0~107.1); 693 (413~902) ❙ 672 (225~1214); 383 (130~691); 28 (1~109) ❙ -1.5 (-64.1~128.6); 616 (225~891); 202.6 (44.9~346.5)

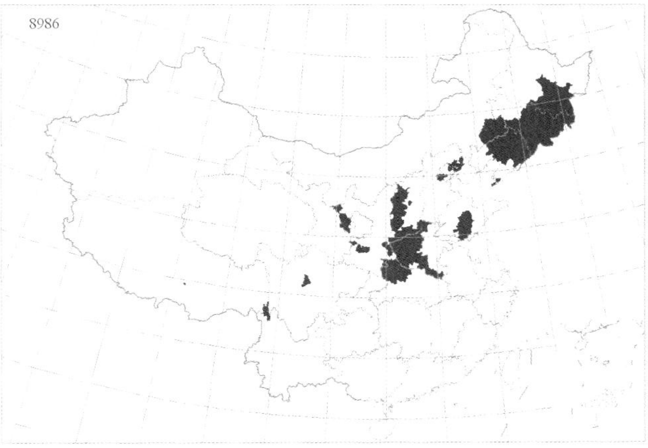

Oleaceae 木犀科 *Syringa* 丁香属

Syringa pubescens subsp. patula (Palib.) M. C. Chang et X. L. Chen 关东巧玲花

Shrubs, deciduous

7.3 (2.5~9.5); -11.1 (-18.8~-7.2); 23.2 (20.0~25.0); 9.7 (7.3~10.9); 80.3 (54.1~91.7); 52.1 (35.9~84.3); 648 (542~704) ❙ 732 (397~1073); 467 (254~691); 25 (10~40) ❙ 13.6 (-36.6~74.1); 614 (397~698); 246.3 (111.2~346.5)

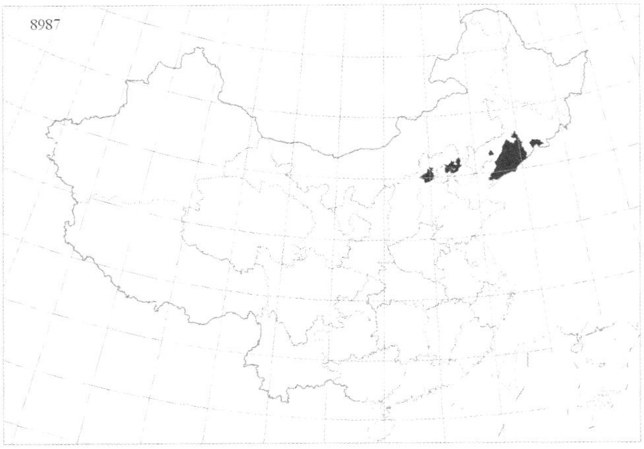

Oleaceae 木犀科 *Syringa* 丁香属

Syringa reticulata subsp. amurensis (Rupr.) P. S. Green et M. C. Chang 暴马丁香

Trees, deciduous

9.7 (5.9~14.5); -4.1 (-9.3~-0.2); 22.5 (17.9~28.2); 10.5 (7.3~14.6); 82.9 (51.7~125.3); 25.4 (10.4~46.8); 660 (529~864) ❙ 614 (341~856); 307 (190~409); 20 (8~31) ❙ -6.3 (-40.9~52.6); 581 (341~702); 194.2 (61.2~320.6)

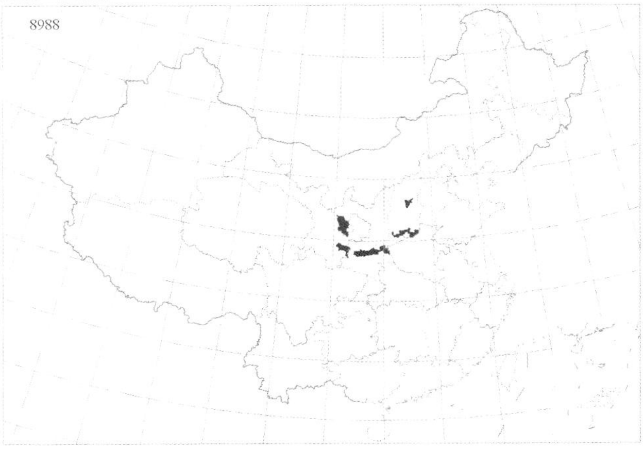

Oleaceae 木犀科　*Syringa* 丁香属

Syringa reticulata subsp. pekinensis (Rupr.) P. S. Green et M. C. Chang 北京丁香

Shrubs or small trees, deciduous

9.5 (−5.0~14.5); −7.0 (−29.4~1.1); 24.0 (10.7~28.3); 11.0 (3.8~14.6); 91.5 (16.2~126.1); 36.6 (8.1~154.8); 708 (411~874) ❙ 572 (341~1073); 367 (190~691); 18 (7~41) ❙ −16.6 (−51.1~74.1); 548 (341~800); 184.4 (61.2~346.5)

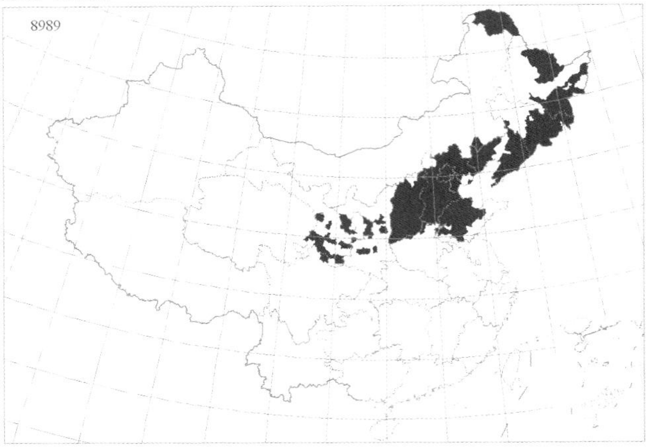

Oleaceae 木犀科　*Syringa* 丁香属

Syringa sweginzowii Koehne et Lingelsh. 四川丁香

Shrubs, deciduous

3.6 (−3.6~14.2); −6.7 (−15.6~3.1); 12.7 (7.2~26.0); 5.6 (2.0~14.2); 32.7 (3.7~116.7); 49.0 (2.8~107.2); 470 (335~800) ❙ 757 (225~1047); 412 (130~606); 19 (5~47) ❙ 66.0 (−64.1~128.6); 457 (225~800); 265.0 (44.9~357.7)

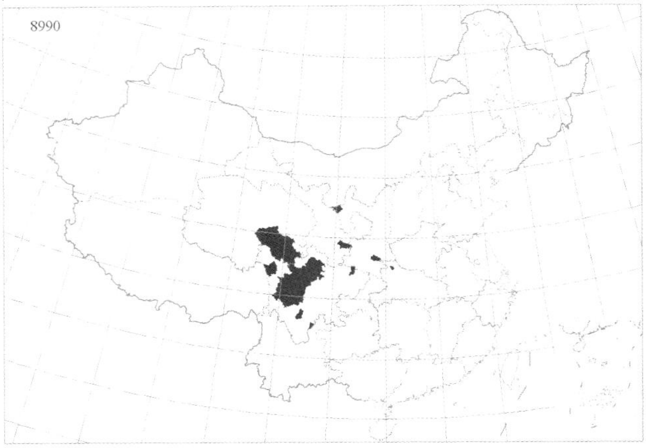

Oleaceae 木犀科　*Syringa* 丁香属

Syringa tibetica P. Y. Bai 藏南丁香

Small trees, deciduous

−1.1; −9.1; 6.2; 2.0; 2.5; 75.6; 334 ❙ 321; 193; 33 ❙ −4.0; 321; 79.3

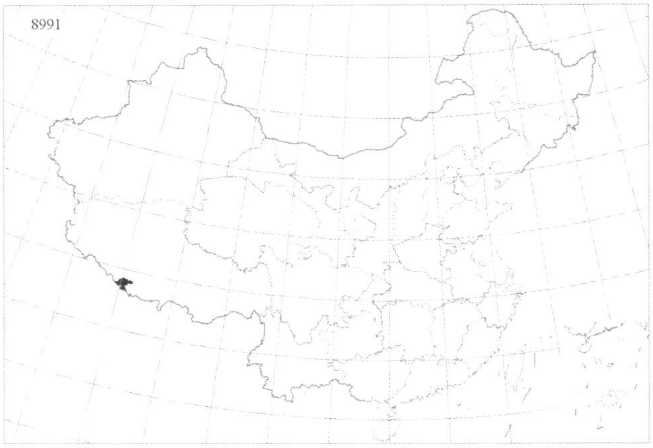

Oleaceae 木犀科　*Syringa* 丁香属

Syringa tomentella Bureau et Franch. 毛丁香

Shrubs, deciduous

6.1 (2.0~11.9); −3.0 (−7.2~2.9); 14.2 (9.9~20.0); 6.9 (3.8~12.0); 42.1 (15.4~87.2); 28.8 (3.4~51.4); 512 (413~671) ❙ 974 (731~1178); 522 (346~663); 26 (19~35) ❙ 93.4 (49.1~128.6); 512 (413~671); 257.4 (200.8~293.1)

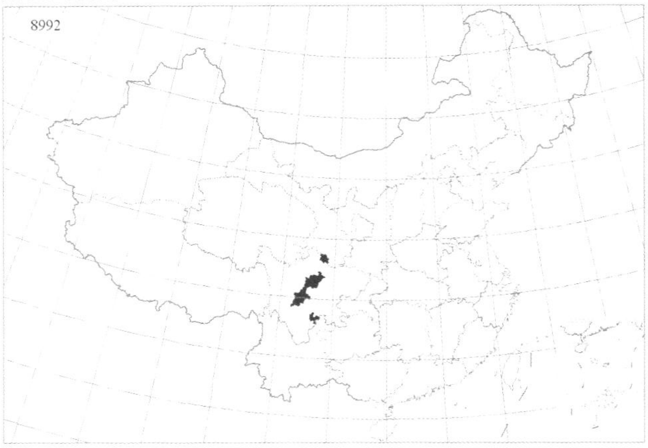

Oleaceae 木犀科　*Syringa* 丁香属

Syringa villosa Vahl 红丁香

Shrubs, deciduous

8.3 (−1.1~13.5); −8.1 (−18.8~−1.3); 22.7 (6.2~27.1); 10.0 (2.0~13.7); 81.5 (2.5~117.5); 41.1 (14.6~84.3); 656 (334~820) ❙ 552 (321~1073); 345 (193~691); 16 (7~40) ❙ −15.1 (−40.9~74.1); 534 (321~684); 180.8 (70.3~346.5)

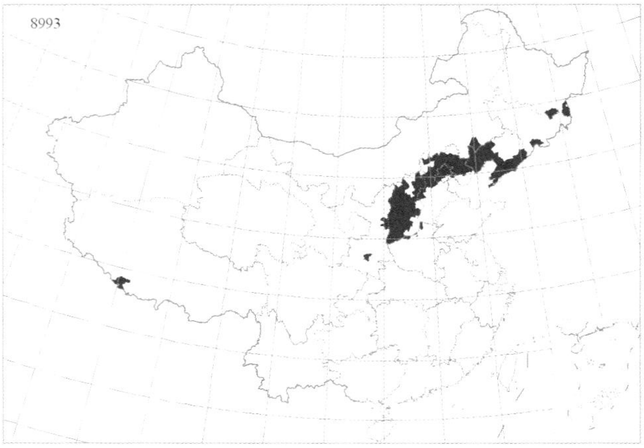

Oleaceae 木犀科　*Syringa* 丁香属

Syringa wardii W. W. Smith 圆叶丁香

Shrubs or small trees

4.0 (0.9~6.1); −4.3 (−8.3~−1.5); 11.6 (9.5~12.7); 5.1 (3.4~6.3); 26.4 (13.3~35.2); 37.4 (21.8~61.6); 454 (389~496) ❙ 715 (510~961); 426 (311~558); 21 (11~31) ❙ 56.1 (20.7~100.7); 454 (389~496); 237.9 (161.1~282.9)

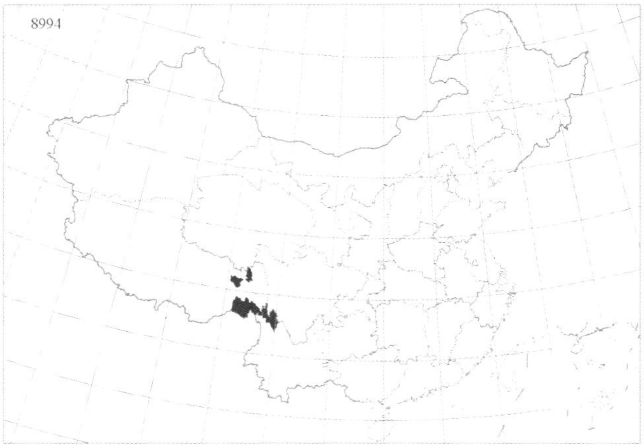

Oleaceae 木犀科 *Syringa* 丁香属

Syringa wolfii C. K. Schneid. 辽东丁香

Shrubs, deciduous

7.6 (1.4~13.4)；-9.1 (-20.1~1.5)；22.3 (18.4~25.6)；9.7 (6.4~13.4)；77.4 (45.8~108.0)；46.0 (7.1~93.6)；637 (501~760) ▎699 (367~1073)；410 (235~691)；22 (8~40) ▎10.4 (-37.7~74.1)；603 (367~760)；257.9 (109.4~346.5)

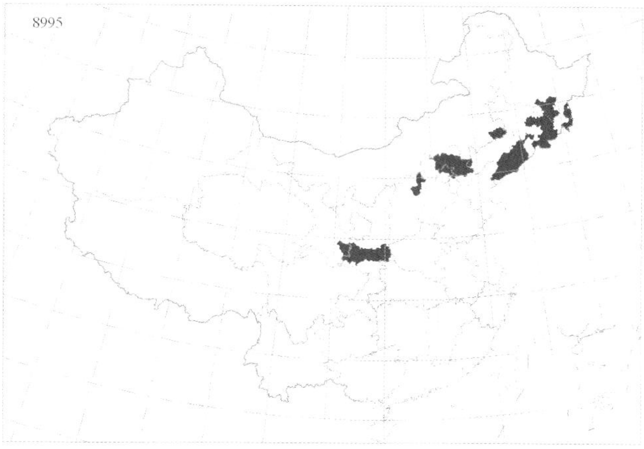

Oleaceae 木犀科 *Syringa* 丁香属

Syringa yunnanensis Franch. 云南丁香

Shrubs, deciduous

12.3 (-0.3~18.9)；4.5 (-10.9~12.0)；18.7 (9.0~26.6)；12.5 (3.0~19.0)；94.3 (10.4~167.8)；6.7 (0.0~73.4)；684 (378~929) ▎947 (656~1141)；555 (382~651)；29 (11~57) ▎42.3 (-14.2~121.7)；659 (378~912)；277.1 (142.9~382.6)

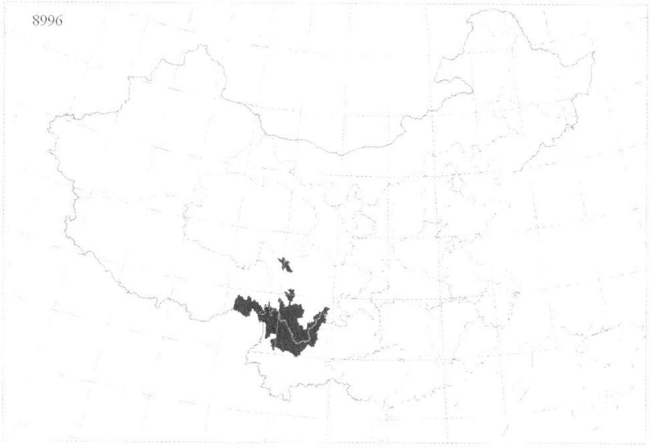

Loganiaceae 马钱科 *Buddleja* 醉鱼草属

Buddleja albiflora Hemsl. 巴东醉鱼草

Shrubs

13.2 (3.0~18.9)；3.7 (-7.7~12.5)；21.7 (12.3~29.3)；13.3 (4.8~19.0)；105.5 (25.1~167.8)；6.5 (0.0~49.3)；728 (441~962) ▎965 (620~1590)；494 (290~685)；41 (13~184) ▎33.2 (-16.6~107.2)；714 (441~962)；269.7 (126.1~406.0)

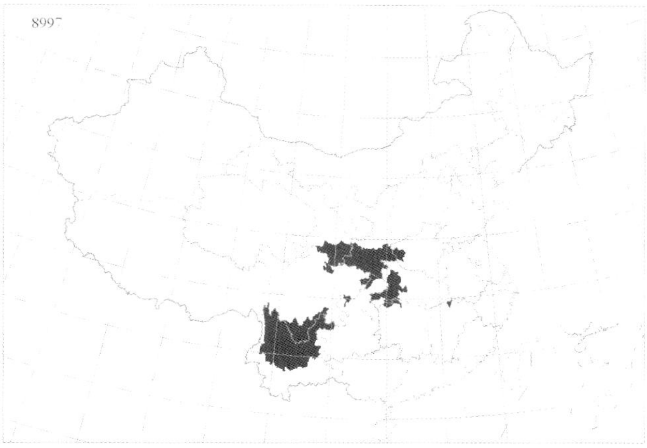

Loganiaceae 马钱科 *Buddleja* 醉鱼草属

Buddleja alternifolia Maxim. 互叶醉鱼草

Shrubs

7.4 (1.2~16.2)；-6.8 (-12.7~5.5)；20.0 (9.2~26.6)；8.9 (3.5~16.2)；68.0 (13.4~134.6)；38.3 (0.0~65.0)；600 (394~854) ▎503 (132~1197)；279 (81~489)；13 (0~71) ▎-13.8 (-80.1~86.8)；464 (132~854)；155.4 (13.6~315.4)

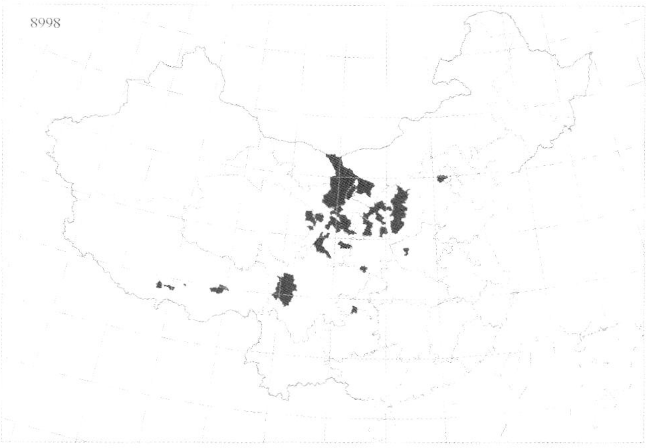

Loganiaceae 马钱科 *Buddleja* 醉鱼草属

Buddleja asiatica Lour. 白背枫

Shrubs or small trees

17.8 (5.3~24.8)；8.9 (-2.4~20.1)；25.6 (12.2~29.8)；17.9 (5.7~24.8)；155.1 (30.5~237.6)；0.3 (0.0~26.2)；931 (479~1385) ▎1431 (797~3495)；613 (312~1847)；122 (20~897) ▎55.0 (-34.3~246.5)；919 (479~1360)；254.4 (95.3~659.0)

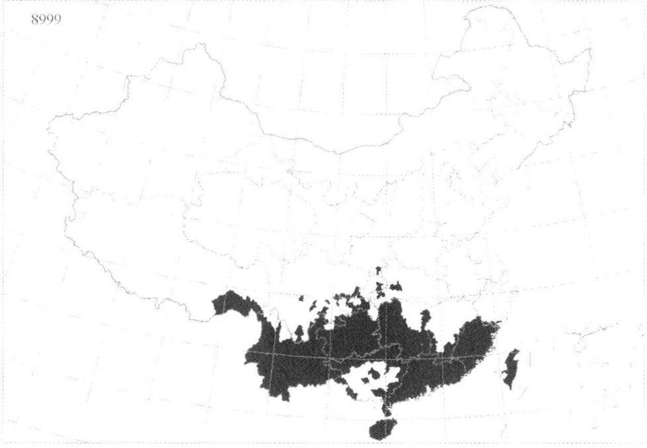

Loganiaceae 马钱科 *Buddleja* 醉鱼草属

Buddleja brachystachya Diels 短序醉鱼草

Shrubs

12.8 (6.2~18.7)；3.9 (-4.1~8.7)；20.7 (14.1~28.9)；13.0 (6.7~18.8)；101.1 (39.4~166.0)；7.2 (0.0~29.9)；727 (509~1002) ▎964 (637~1243)；490 (306~696)；41 (14~66) ▎35.6 (2.4~93.0)；724 (509~1002)；259.0 (108.2~384.7)

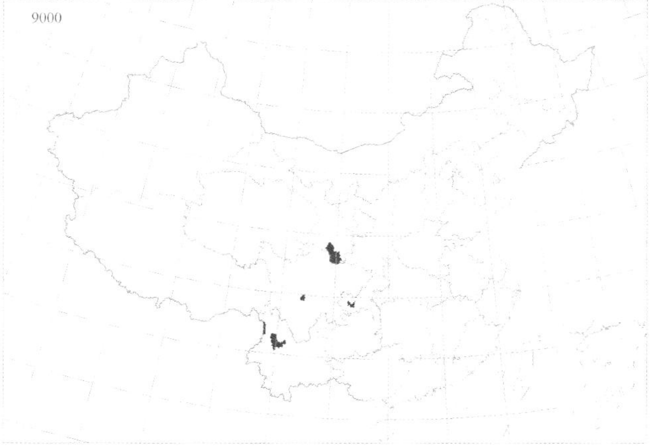

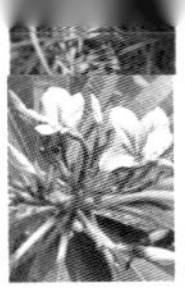

Loganiaceae 马钱科 *Strychnos* 马钱属
Strychnos axillaris Colebr. 腋花马钱
Vines

20.4; 15.1; 23.6; 20.5; 185.8; 0.0; 989 ❙ 1607; 702; 77 ❙ 62.5; 986; 659.0

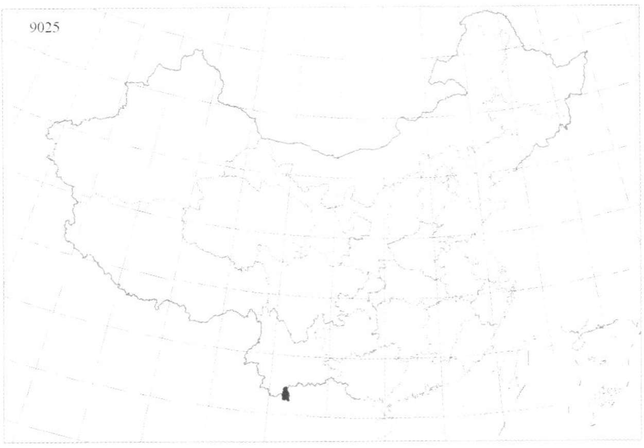

Loganiaceae 马钱科 *Strychnos* 马钱属
Strychnos cathayensis Merr. 华马钱
Vines

20.7 (16.2~24.8); 13.0 (7.1~20.1); 26.9 (20.1~29.8); 20.8 (16.2~24.8); 189.2 (134.2~237.6); 0.0 (0.0~0.0); 1078 (773~1385) ❙ 1679 (868~3495); 747 (312~1847); 139 (41~897) ❙ 58.7 (-34.3~246.5); 1060 (773~1299); 275.0 (95.3~659.0)

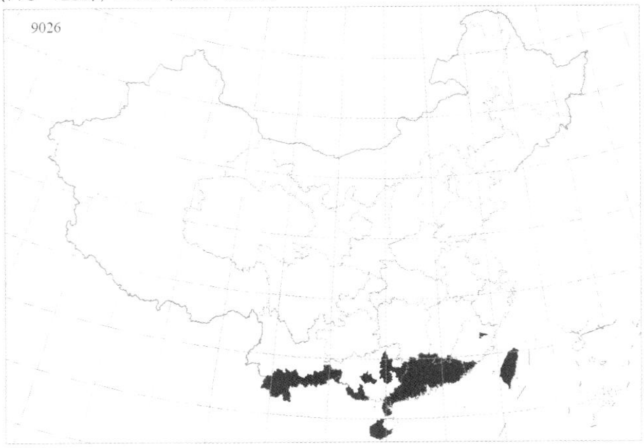

Loganiaceae 马钱科 *Strychnos* 马钱属
Strychnos ignatii P. J. Bergius 吕宋果
Vines

21.2 (16.3~24.8); 14.1 (7.4~20.1); 26.6 (21.4~29.2); 21.3 (16.4~24.8); 195.2 (136.4~237.6); 0.0 (0.0~0.0); 1116 (797~1385) ❙ 1543 (1190~1784); 700 (473~893); 91 (41~188) ❙ 41.4 (-1.4~85.5); 1083 (797~1326); 351.0 (111.7~620.0)

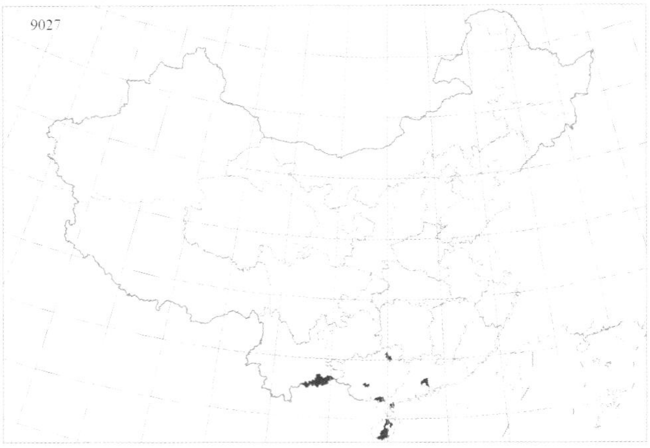

Loganiaceae 马钱科 *Strychnos* 马钱属
Strychnos nitida G. Don 毛柱马钱
Vines

17.9 (14.4~22.2); 11.3 (7.5~15.1); 22.3 (19.1~28.5); 17.9 (14.4~22.2); 155.1 (113.1~206.5); 0.0 (0.0~0.0); 867 (727~1188) ❙ 1270 (958~1751); 700 (531~996); 53 (41~128) ❙ 46.5 (12.2~99.9); 854 (727~1188); 393.7 (255.3~659.0)

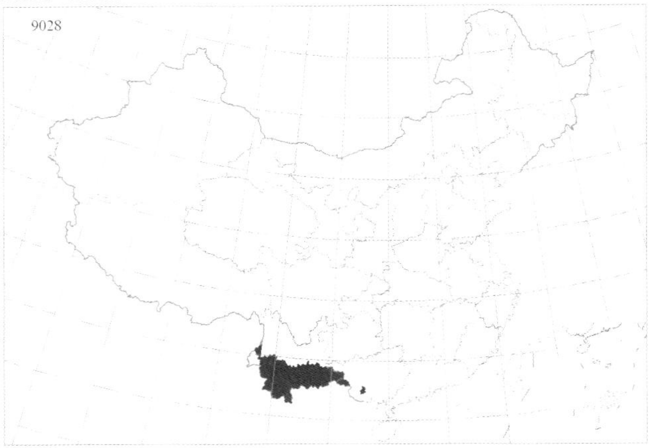

Loganiaceae 马钱科 *Strychnos* 马钱属
Strychnos ovata A. W. Hill 密花马钱
Large climbers

23.7 (22.2~24.8); 18.0 (14.7~20.1); 27.8 (25.7~28.9); 23.7 (22.3~24.8); 224.8 (207.2~237.6); 0.0 (0.0~0.0); 1285 (1127~1385) ❙ 1400 (868~1725); 528 (312~893); 61 (41~106) ❙ 9.9 (-34.3~47.8); 1126 (868~1269); 450.3 (229.5~620.0)

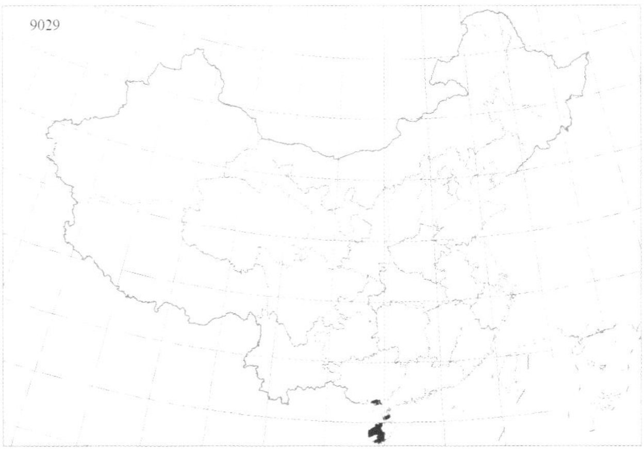

Apocynaceae 夹竹桃科 *Aganosma* 香花藤属
Aganosma cymosa (Roxb.) G. Don 云南香花藤
Vines

18.2 (16.0~20.5); 11.9 (9.3~15.1); 22.2 (20.3~24.6); 18.2 (16.0~20.5); 158.7 (132.4~185.9); 0.0 (0.0~0.0); 875 (780~992) ❙ 1390 (1028~1751); 775 (566~996); 54 (42~77) ❙ 59.2 (12.5~112.2); 860 (780~986); 452.9 (316.1~659.0)

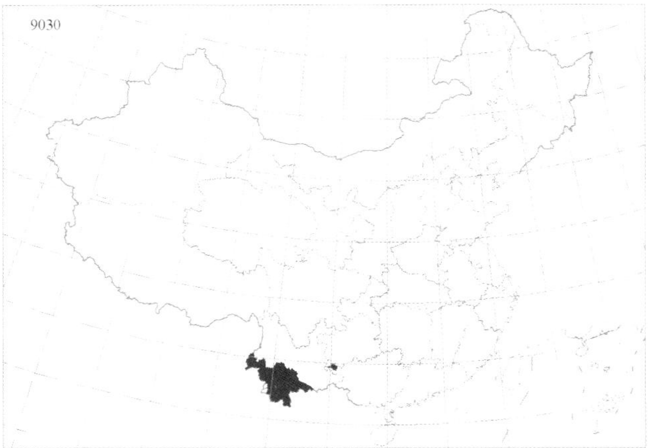

Apocynaceae 夹竹桃科 *Aganosma* 香花藤属
Aganosma marginata (Roxb.) G. Don 香花藤
Vines

23.8 (20.4~24.8); 18.1 (15.1~20.1); 27.8 (23.6~28.8); 23.8 (20.5~24.8); 226.0 (185.8~237.6); 0.0 (0.0~0.0); 1293 (989~1385) ❙ 1495 (868~1854); 542 (312~712); 74 (41~115) ❙ 16.5 (-34.3~62.5); 1185 (868~1360); 441.0 (235.0~659.0)

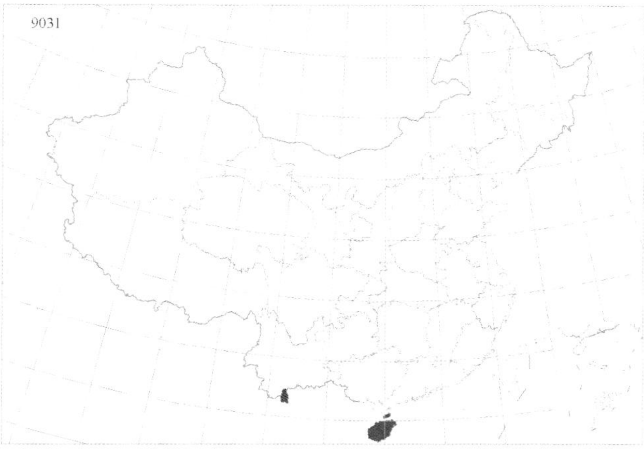

Apocynaceae 夹竹桃科 *Aganosma* 香花藤属
Aganosma schlechteriana H. Lév. 海南香花藤
Scandent shrubs

17.1 (5.4~24.8); 10.0 (-2.1~20.1); 22.3 (12.3~28.7); 17.1 (5.8~24.8); 145.8 (31.5~237.6); 0.4 (0.0~25.6); 857 (481~1385) ❙ 1171 (741~1900); 641 (312~1056); 48 (15~125) ❙ 38.1 (-34.3~88.8); 826 (481~1190); 340.2 (152.1~659.0)

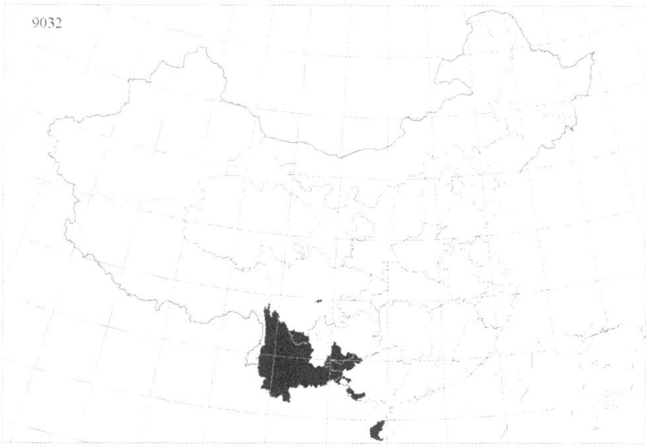

Apocynaceae 夹竹桃科 *Aganosma* 香花藤属
Aganosma siamensis Craib 广西香花藤
Scandent shrubs

18.8 (12.8~22.7); 10.9 (3.5~15.1); 24.7 (20.1~28.6); 18.9 (12.9~22.8); 166.3 (95.9~213.1); 0.0 (0.0~1.5); 948 (699~1221) ❙ 1390 (885~2435); 736 (479~1428); 78 (34~194) ❙ 46.9 (5.8~99.5); 945 (699~1221); 338.4 (189.5~659.0)

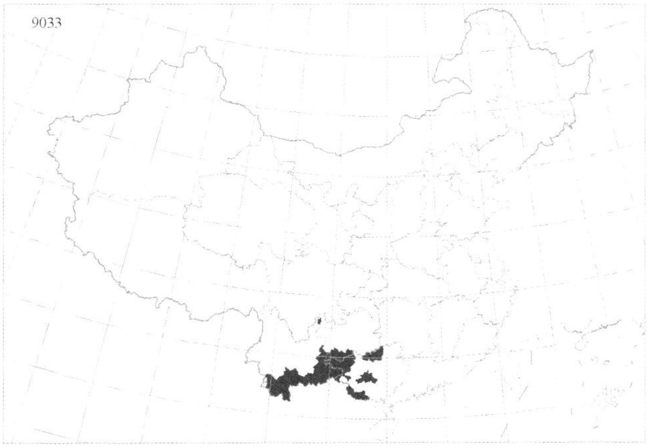

Apocynaceae 夹竹桃科 *Alstonia* 鸡骨常山属
Alstonia henryi Tsiang 黄花羊角棉
Shrubs, evergreen

19.0; 13.1; 22.6; 19.1; 168.8; 0.0; 909 ❙ 1495; 775; 62 ❙ 64.5; 891; 427.9

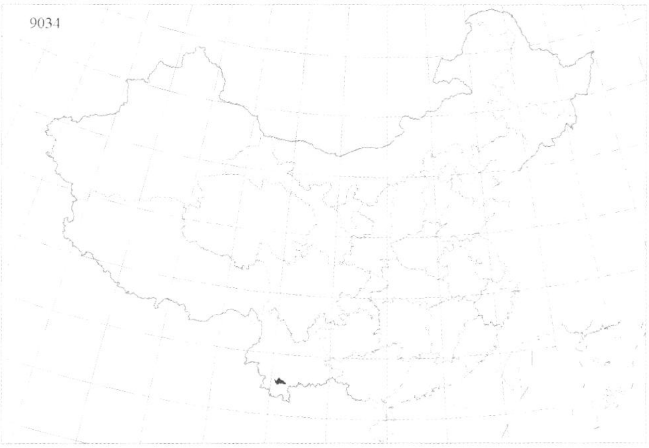

Apocynaceae 夹竹桃科 *Alstonia* 鸡骨常山属
Alstonia macrophylla Wall. ex G. Don 大叶糖胶树
Trees, evergreen

24.0 (22.2~24.8); 18.4 (17.2~20.1); 28.0 (25.7~28.8); 24.0 (22.3~24.8); 228.3 (207.2~237.6); 0.0 (0.0~0.0); 1310 (1127~1385) ❙ 1501 (868~1854); 534 (312~712); 74 (41~115) ❙ 15.0 (-34.3~47.8); 1191 (868~1360); 440.3 (248.2~620.0)

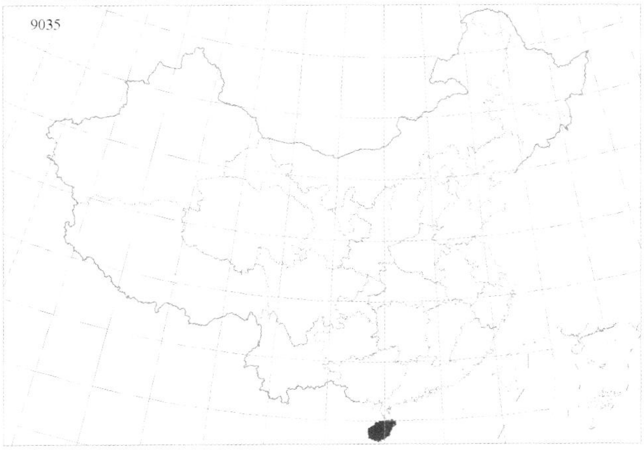

Apocynaceae 夹竹桃科 *Alstonia* 鸡骨常山属
Alstonia mairei H. Lév. 羊角棉
Shrubs, evergreen

15.2 (11.2~18.9); 7.5 (2.7~12.5); 21.2 (16.7~26.9); 15.2 (11.3~19.0); 122.8 (75.8~167.8); 0.3 (0.0~3.8); 769 (642~929) ❙ 1036 (797~1454); 569 (411~792); 45 (18~101) ❙ 35.3 (-14.2~99.9); 752 (642~871); 280.9 (176.4~491.5)

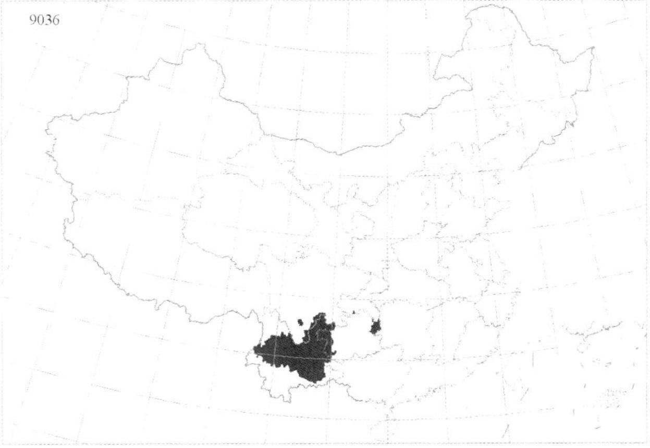

Apocynaceae 夹竹桃科 *Alstonia* 鸡骨常山属
Alstonia neriifolia D. Don 竹叶羊角棉
Shrubs, evergreen

21.1；12.8；27.3；21.2；194.0；0.0；1105 ❙ 1504；794；106 ❙ 37.1；1105；304.7

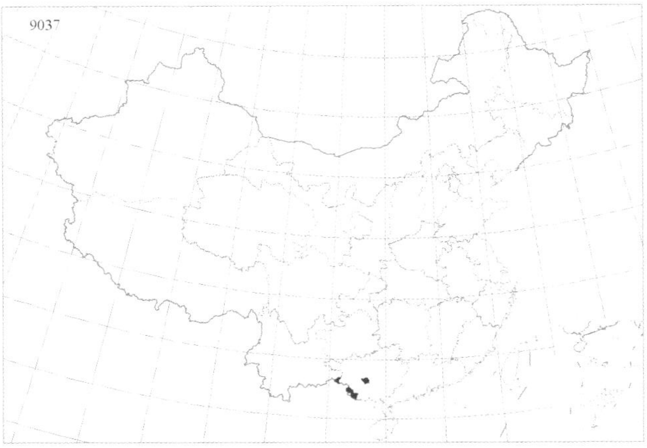

Apocynaceae 夹竹桃科 *Alstonia* 鸡骨常山属
Alstonia rostrata C. E. C. Fisch. 盆架树
Trees, evergreen

20.1 (16.0~24.8)；13.9 (9.0~20.1)；24.2 (20.4~28.8)；20.1 (16.0~24.8)；181.6 (132.4~237.6)；0.0 (0.0~0.0)；1019 (780~1385) ❙ 1436 (868~1854)；696 (312~996)；60 (41~115) ❙ 45.4 (-34.3~112.2)；974 (780~1360)；446.7 (248.2~659.0)

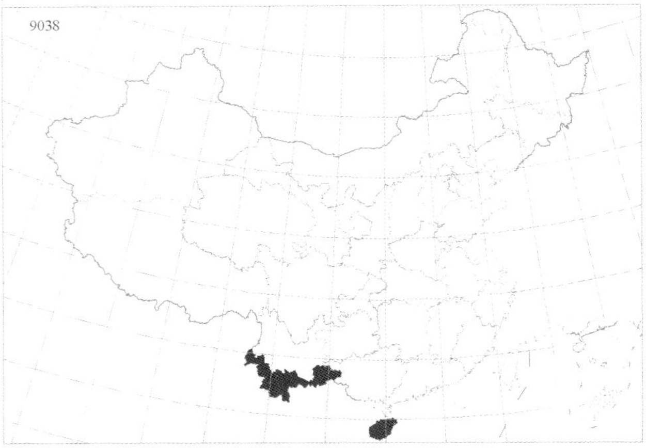

Apocynaceae 夹竹桃科 *Alstonia* 鸡骨常山属
Alstonia rupestris Kerr 岩生羊角棉
Shrubs, evergreen

19.4；11.1；25.5；19.4；173.0；0.0；969 ❙ 1442；812；73 ❙ 49.0；969；304.2

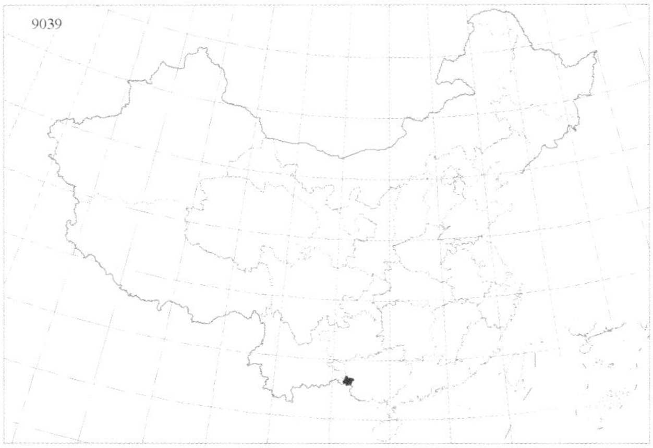

Apocynaceae 夹竹桃科 *Alstonia* 鸡骨常山属
Alstonia scholaris (L.) R. Br. 糖胶树
Trees, evergreen

20.5 (16.2~24.8)；14.0 (9.1~20.1)；25.3 (20.1~28.9)；20.6 (16.2~24.8)；187.1 (134.2~237.6)；0.0 (0.0~0.0)；1054 (773~1385) ❙ 1678 (868~3495)；813 (312~1847)；114 (41~897) ❙ 64.7 (-34.3~246.5)；1021 (773~1360)；366.5 (119.4~659.0)

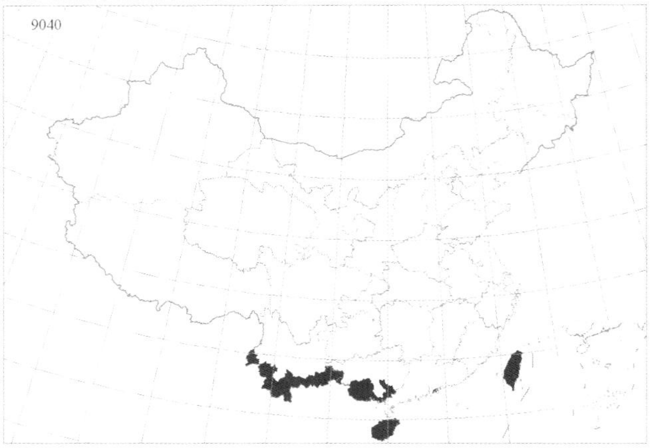

Apocynaceae 夹竹桃科 *Alstonia* 鸡骨常山属
Alstonia yunnanensis Diels 鸡骨常山
Shrubs, evergreen

15.7 (5.4~21.3)；8.3 (-2.1~15.1)；21.4 (12.3~27.9)；15.8 (5.8~21.3)；129.8 (31.5~196.0)；0.7 (0.0~25.6)；796 (481~1123) ❙ 1129 (741~1751)；631 (411~996)；45 (15~97) ❙ 42.6 (-14.2~112.2)；781 (481~1121)；321.5 (133.1~659.0)

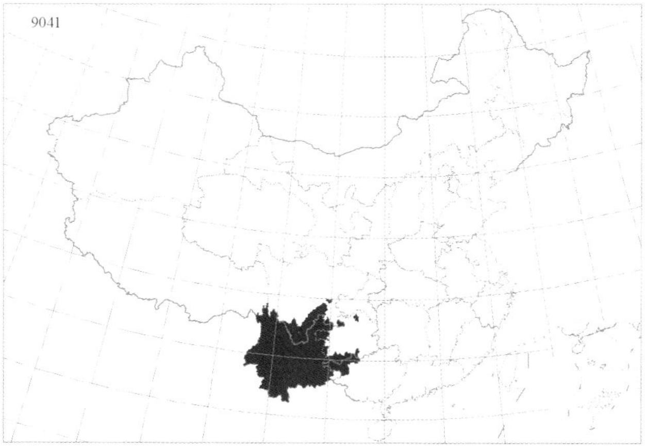

Apocynaceae 夹竹桃科 *Alyxia* 念珠藤属
Alyxia fascicularis Benth. 尾尖链珠藤
Vines, evergreen

13.3；5.4；19.6；13.4；100.4；0.0；706 ❙ 1736；1007；51 ❙ 145.8；706；447.4

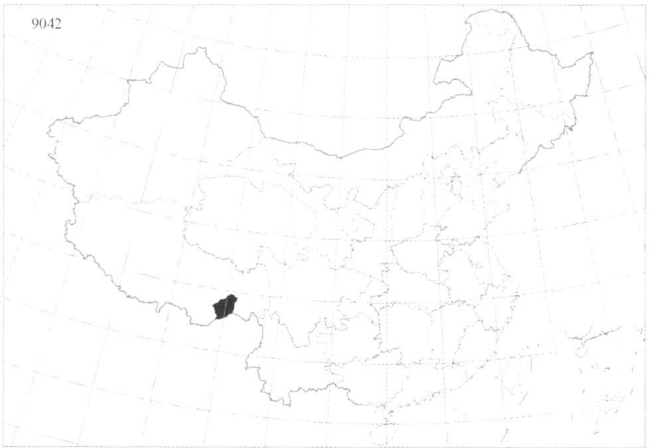

Apocynaceae 夹竹桃科 *Alyxia* 念珠藤属

Alyxia insularis Kaneh. et Sasaki 兰屿链珠藤

Shrubs, evergreen

18.5；14.1；21.7；18.5；162.1；0.0；863 ▎2713；1388；188 ▎214.4；863；363.7

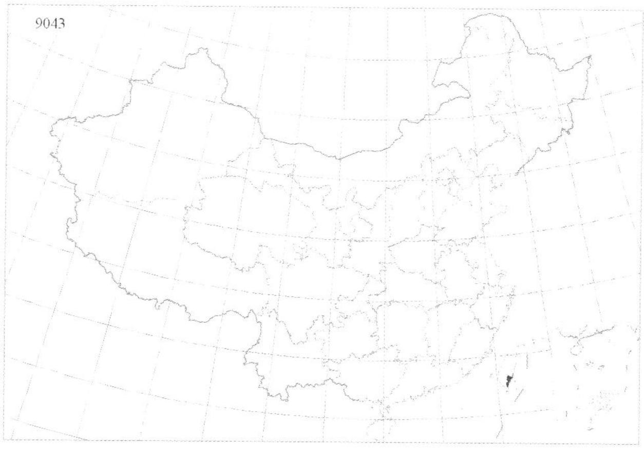

Apocynaceae 夹竹桃科 *Alyxia* 念珠藤属

Alyxia levinei Merr. 筋藤

Scandent shrubs, evergreen

18.8 (14.2~22.7)；9.2 (3.8~15.0)；26.8 (20.1~29.0)；18.8 (14.2~22.8)；166.0 (110.5~213.1)；0.1 (0.0~1.5)；985 (729~1221) ▎1493 (1107~2435)；668 (444~1428)；129 (50~196) ▎52.7 (12.9~101.8)；985 (729~1221)；235.0 (146.0~371.7)

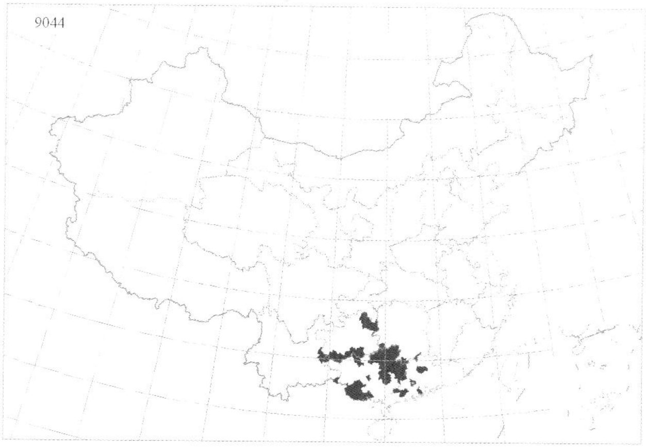

Apocynaceae 夹竹桃科 *Alyxia* 念珠藤属

Alyxia marginata Pit. 陷边链珠藤

Scandent shrubs, evergreen

18.9 (17.2~21.4)；10.8 (7.4~13.8)；24.9 (22.3~28.2)；19.0 (17.3~21.5)；167.6 (147.0~197.8)；0.0 (0.0~0.0)；948 (838~1140) ▎1390 (1254~1499)；730 (637~820)；74 (48~130) ▎47.7 (24.6~65.6)；945 (838~1140)；334.0 (245.3~513.9)

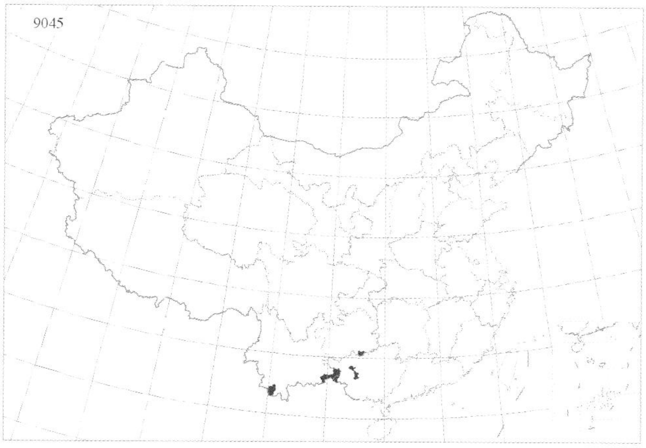

Apocynaceae 夹竹桃科 *Alyxia* 念珠藤属

Alyxia menglungensis Tsiang et P. T. Li 勐龙链珠藤

Vines, evergreen

20.5；14.9；23.7；20.5；185.9；0.0；992 ▎1464；655；66 ▎47.6；959；570.1

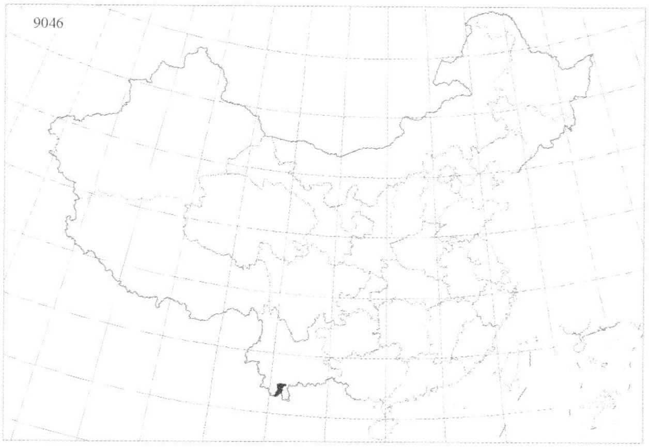

Apocynaceae 夹竹桃科 *Alyxia* 念珠藤属

Alyxia odorata Wall. ex G. Don 海南链珠藤

Scandent shrubs, evergreen

18.5 (7.3~24.8)；8.8 (-2.0~20.1)；27.0 (15.6~29.8)；18.5 (7.5~24.8)；162.7 (48.2~237.6)；0.5 (0.0~20.4)；980 (537~1385) ▎1450 (1005~2435)；593 (392~1428)；122 (27~230) ▎50.0 (-11.9~128.9)；968 (537~1360)；237.0 (108.2~620.0)

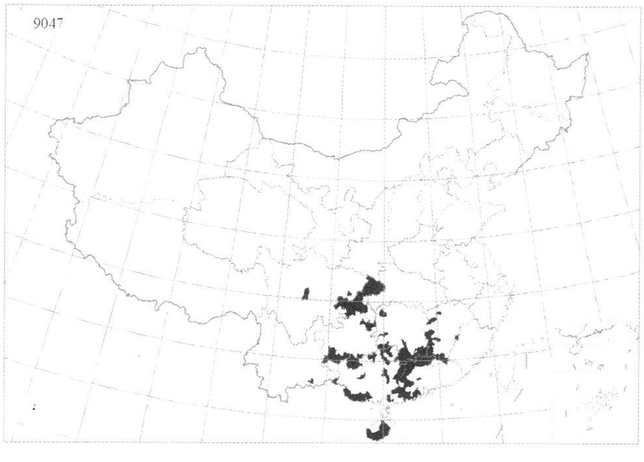

Apocynaceae 夹竹桃科 *Alyxia* 念珠藤属

Alyxia reinwardtii Blume 长花链珠藤

Vines, evergreen

18.3 (17.2~21.0)；10.8 (9.5~14.2)；23.6 (22.4~25.6)；18.3 (17.3~21.0)；159.9 (147.0~191.8)；0.0 (0.0~0.0)；896 (833~1053) ▎1394 (1254~1522)；779 (699~860)；55 (48~58) ▎56.3 (37.6~80.6)；896 (833~1053)；415.8 (336.3~551.6)

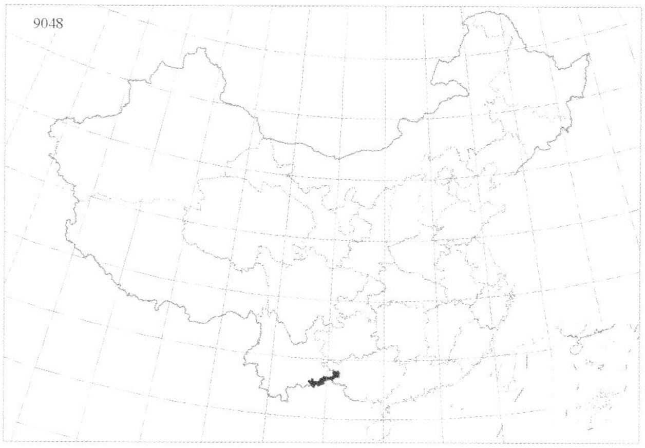

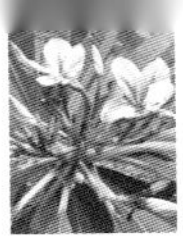

Apocynaceae 夹竹桃科 *Alyxia* 念珠藤属

Alyxia schlechteri H. Lév. 狭叶链珠藤

Vines, evergreen

17.8 (15.1~21.3); 8.7 (4.8~14.2); 25.3 (21.4~29.0); 17.9 (15.2~21.3); 154.5 (122.6~196.1); 0.0 (0.0~0.2); 910 (772~1130) ❙ 1346 (1007~1758); 653 (444~860); 91 (43~189) ❙ 48.4 (16.9~85.5); 909 (772~1130); 264.0 (160.7~551.6)

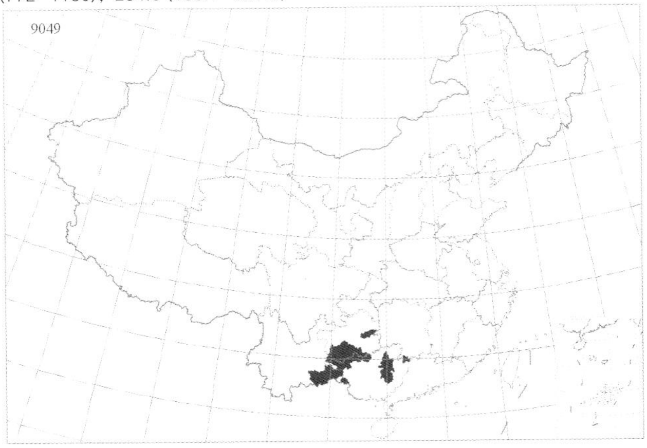

Apocynaceae 夹竹桃科 *Alyxia* 念珠藤属

Alyxia siamensis Craib 长序链珠藤

Vines, evergreen

18.9 (8.6~22.7); 11.4 (1.3~15.1); 24.4 (15.0~28.6); 19.0 (8.7~22.8); 167.9 (52.6~213.1); 0.3 (0.0~8.8); 954 (574~1221) ❙ 1495 (1041~2435); 808 (582~1428); 79 (44~169) ❙ 56.6 (12.9~99.5); 951 (574~1221); 373.0 (235.1~659.0)

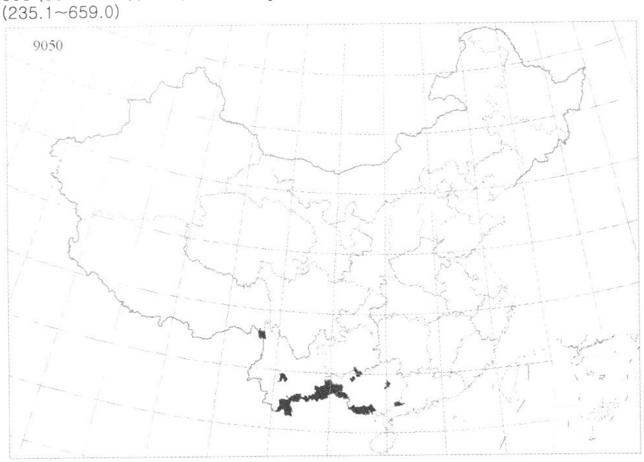

Apocynaceae 夹竹桃科 *Alyxia* 念珠藤属

Alyxia sinensis Champ. ex Benth. 链珠藤

Scandent shrubs, evergreen

19.0 (14.3~24.8); 9.7 (3.3~20.1); 27.3 (23.0~29.8); 19.0 (14.3~24.8); 168.4 (114.1~237.6); 0.0 (0.0~2.3); 992 (765~1385) ❙ 1656 (1070~3495); 631 (356~1428); 180 (41~897) ❙ 69.1 (-5.0~224.4); 983 (765~1273); 247.2 (111.7~620.0)

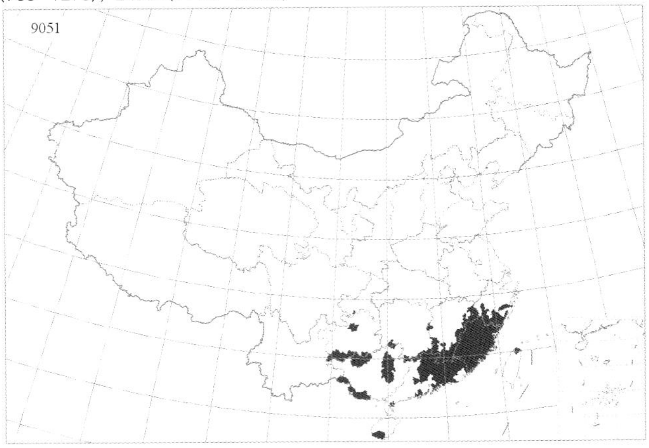

Apocynaceae 夹竹桃科 *Alyxia* 念珠藤属

Alyxia villilimba C. Y. Wu ex Tsiang et P. T. Li 毛叶链珠藤

Shrubs, evergreen

17.8 (16.3~19.6); 9.9 (9.0~11.2); 23.4 (21.4~25.8); 17.8 (16.4~19.6); 153.9 (136.4~175.5); 0.0 (0.0~0.0); 874 (797~986) ❙ 1294 (1018~1504); 725 (554~860); 55 (47~74) ❙ 48.0 (25.7~80.6); 874 (797~986); 334.7 (277.0~413.5)

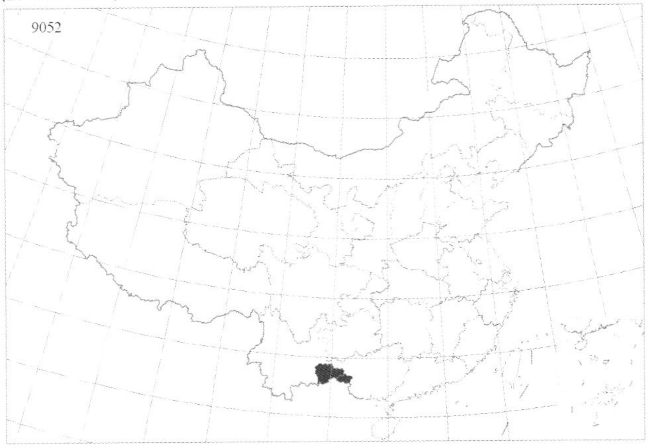

Apocynaceae 夹竹桃科 *Amalocalyx* 毛车藤属

Amalocalyx microlobus Pierre 毛车藤

Vines

18.5 (15.5~21.0); 12.4 (8.4~15.1); 22.5 (20.6~25.6); 18.5 (15.5~21.0); 162.5 (126.3~191.8); 0.0 (0.0~0.0); 893 (768~1053) ❙ 1372 (989~1751); 727 (573~996); 56 (41~77) ❙ 53.1 (25.4~88.8); 881 (768~1053); 457.8 (234.5~659.0)

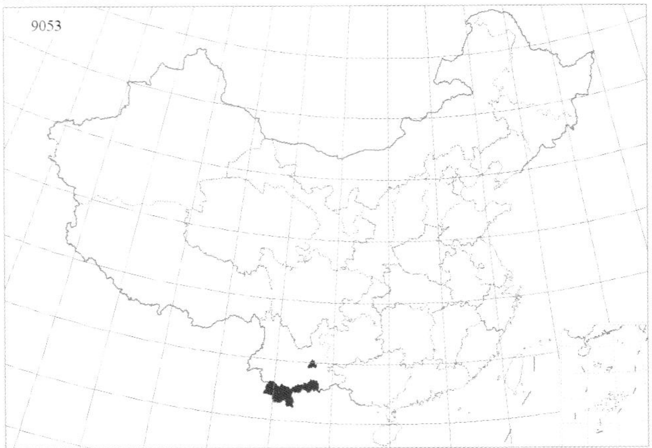

Apocynaceae 夹竹桃科 *Anodendron* 鳝藤属

Anodendron affine (Hook. et Arn.) Druce 鳝藤

Vines

18.5 (9.3~24.8); 9.4 (-1.7~20.1); 26.5 (20.1~29.8); 18.6 (9.5~24.8); 163.4 (69.8~237.6); 0.4 (0.0~18.0); 969 (603~1385) ❙ 1523 (868~3495); 638 (312~1847); 138 (41~897) ❙ 59.2 (-34.3~246.5); 958 (603~1329); 253.0 (95.3~659.0)

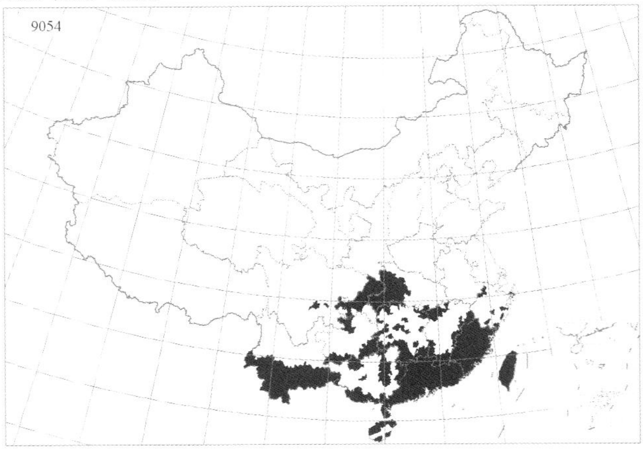

Apocynaceae 夹竹桃科 *Anodendron* 鳝藤属

Anodendron benthamianum Hemsl. 台湾鳝藤

Vines

19.8；14.4；24.[illegible]；19.8；177.3；0.0；975 ❙ 2777；1343；252 ❙ 189.1；975；263.8

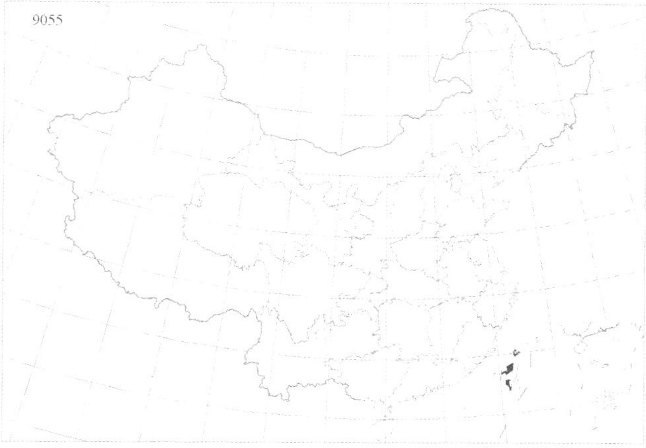

Apocynaceae 夹竹桃科 *Anodendron* 鳝藤属

Anodendron formicinum (Tsiang et P. T. Li) D. J. Middleton 平脉藤

Scandent shrubs

20.0；14.6；23.2；20.1；180.8；0.0；963 ❙ 1523；681；67 ❙ 58.4；942；581.0

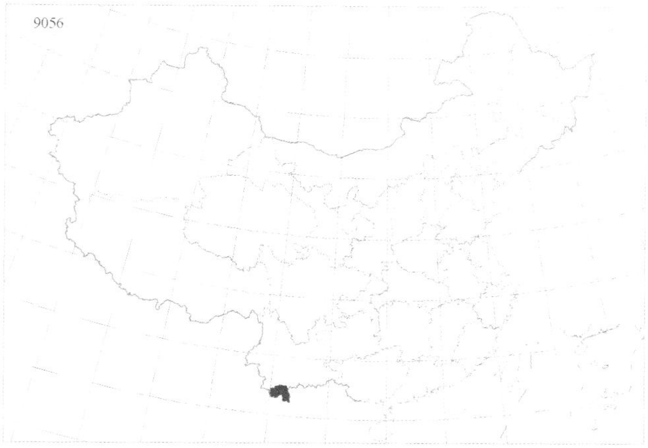

Apocynaceae 夹竹桃科 *Apocynum* 白麻属

Apocynum pictum Schrenk 白麻

Subshrubs

5.9 (-4.8~12.5)；-11.5 (-20.9~-0.1)；20.4 (7.4~27.5)；8.8 (1.9~13.2)；70.0 (4.3~114.1)；58.3 (11.7~121.6)；616 (329~822) ❙ 162 (21~711)；83 (15~357)；10 (0~51) ❙ -69.9 (-97.3~41.8)；160 (21~675)；66.0 (1.6~306.1)

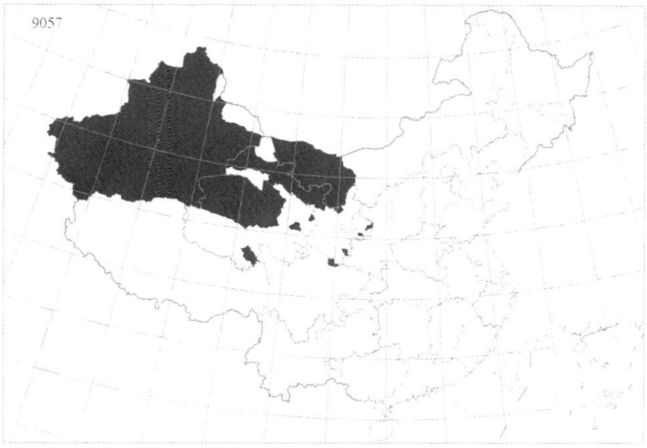

Apocynaceae 夹竹桃科 *Apocynum* 白麻属

Apocynum venetum L. 罗布麻

Subshrubs

10.8 (-4.3~16.8)；-5.0 (-20.9~4.6)；24.7 (7.4~29.5)；12.0 (2.1~16.9)；100.3 (4.6~143.4)；30.2 (0.4~121.1)；748 (329~932) ❙ 508 (21~1483)；290 (15~554)；27 (0~197) ❙ -33.8 (-97.3~67.0)；490 (21~930)；135.0 (1.6~334.4)

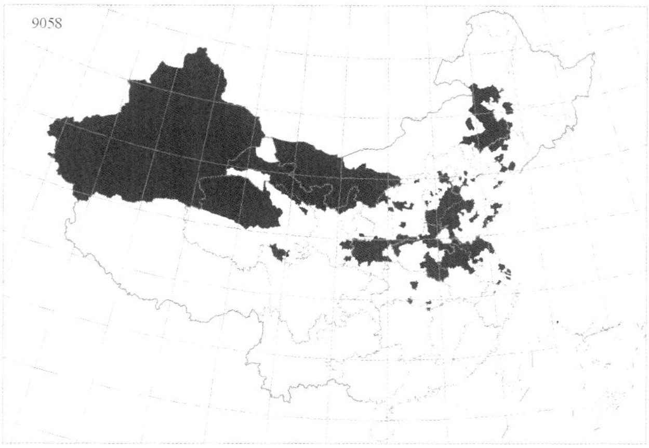

Apocynaceae 夹竹桃科 *Beaumontia* 清明花属

Beaumontia brevituba Oliv. 断肠花

Vines, evergreen

23.5 (21.5~24.3)；17.7 (13.6~19.1)；27.7 (26.7~28.7)；23.6 (21.6~24.3)；222.8 (199.1~231.9)；0.0 (0.0~0.0)；1271 (1131~1338) ❙ 1354 (868~1725)；510 (312~850)；60 (42~106) ❙ 7.5 (-34.3~41.8)；1094 (868~1163)；451.5 (281.7~620.0)

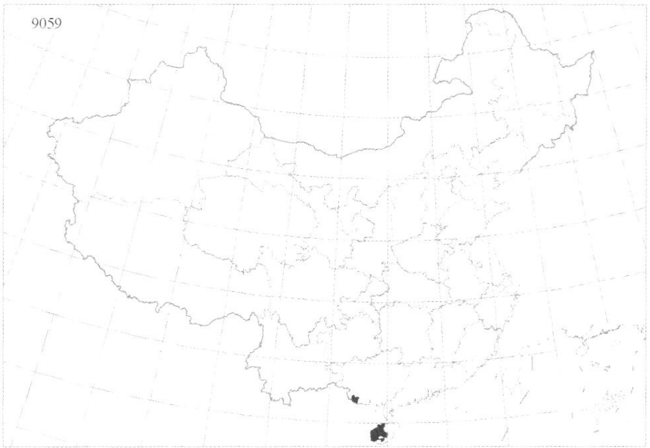

Apocynaceae 夹竹桃科 *Beaumontia* 清明花属

Beaumontia grandiflora Wall. 清明花

Vines, evergreen

19.5 (15.8~24.8)；13.0 (8.8~20.1)；24.0 (20.3~28.8)；19.6 (15.8~24.8)；174.9 (129.7~237.6)；0.0 (0.0~0.0)；985 (776~1385) ❙ 1378 (868~2435)；697 (312~1428)；61 (41~130) ❙ 42.4 (-34.3~112.2)；949 (772~1360)；408.6 (238.3~659.0)

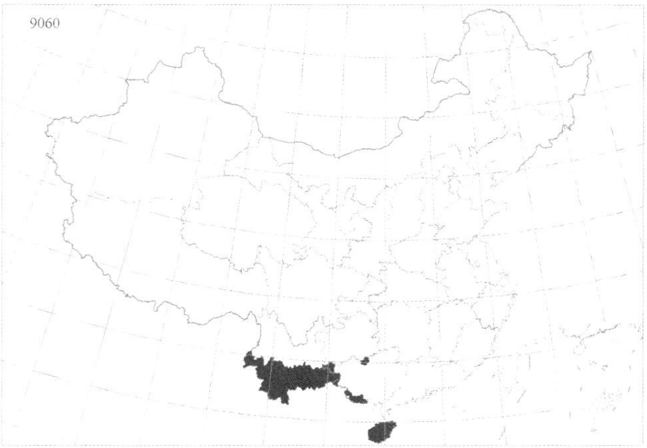

Apocynaceae 夹竹桃科 *Beaumontia* 清明花属
Beaumontia khasiana Hook. f. 云南清明花
Vines, evergreen

18.4 (16.0~20.5); 12.2 (9.3~15.1); 22.1 (20.4~23.7); 18.4 (16.0~20.5); 161.0 (132.4~185.9); 0.0 (0.0~0.0); 883 (780~992) ❙ 1501 (1348~1688); 808 (655~959); 53 (41~77) ❙ 70.8 (47.6~112.2); 867 (780~986); 513.2 (395.4~659.0)

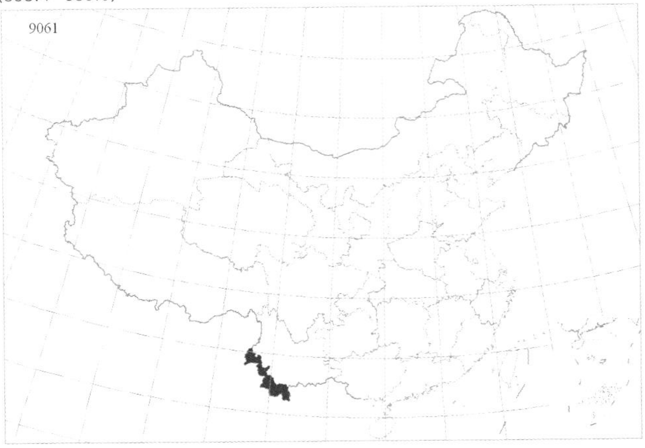

Apocynaceae 夹竹桃科 *Beaumontia* 清明花属
Beaumontia murtonii Craib 思茅清明花
Vines, evergreen

19.8; 14.2; 23.1; 19.8; 177.8; 0.0; 949 ❙ 1516; 704; 66 ❙ 59.9; 929; 542.7

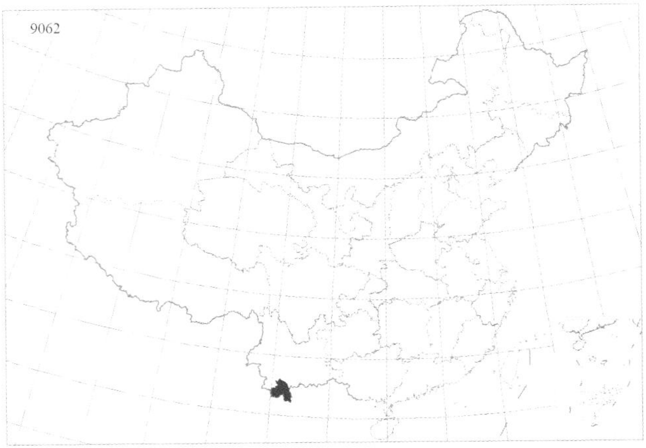

Apocynaceae 夹竹桃科 *Beaumontia* 清明花属
Beaumontia pitardii Tsiang 广西清明花
Vines, evergreen

19.4 (16.3~22.7); 11.7 (9.1~15.0); 25.1 (21.4~28.6); 19.5 (16.4~22.8); 173.9 (136.4~213.1); 0.0 (0.0~0.0); 986 (797~1221) ❙ 1484 (1074~2435); 824 (582~1428); 76 (47~138) ❙ 50.7 (12.9~99.5); 986 (797~1221); 354.2 (272.3~551.6)

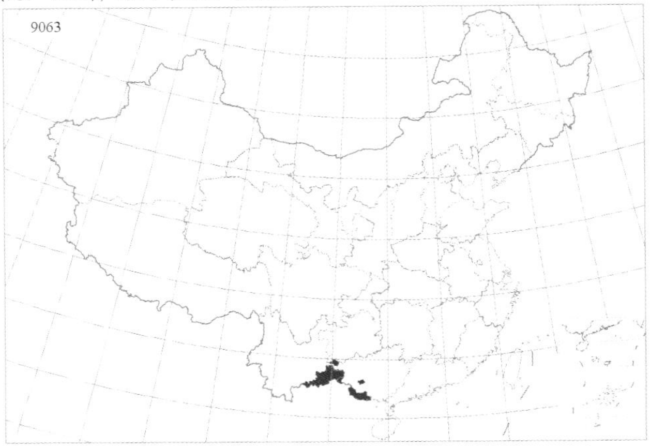

Apocynaceae 夹竹桃科 *Bousigonia* 奶子藤属
Bousigonia angustifolia Pierre 闷奶果
Vines

19.1 (16.9~21.0); 13.0 (10.3~15.1); 23.1 (21.4~25.6); 19.1 (16.9~21.0); 169.6 (142.7~191.8); 0.0 (0.0~0.0); 922 (813~1053) ❙ 1437 (1163~1751); 763 (641~996); 61 (48~77) ❙ 55.9 (34.8~88.8); 915 (813~1053); 509.1 (383.4~659.0)

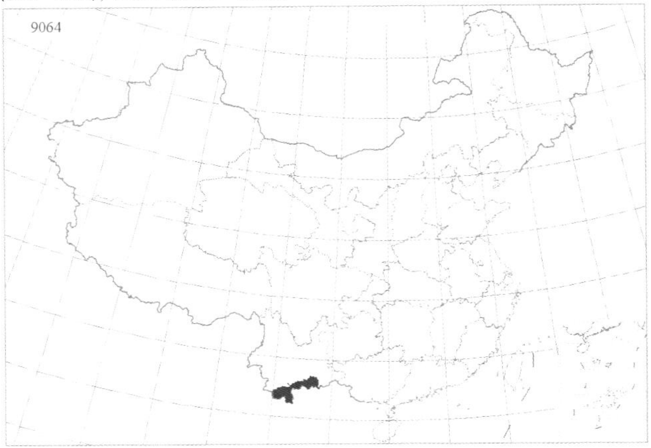

Apocynaceae 夹竹桃科 *Carissa* 假虎刺属
Carissa carandas L. 刺黄果
Shrubs, trees or vines, evergreen

22.0 (16.2~24.8); 16.4 (11.0~20.1); 26.4 (20.1~28.8); 22.1 (16.2~24.8); 205.0 (134.2~237.6); 0.0 (0.0~0.0); 1161 (773~1385) ❙ 1963 (868~3495); 827 (312~1847); 167 (41~897) ❙ 82.3 (-34.3~246.5); 1084 (773~1360); 340.1 (119.4~620.0)

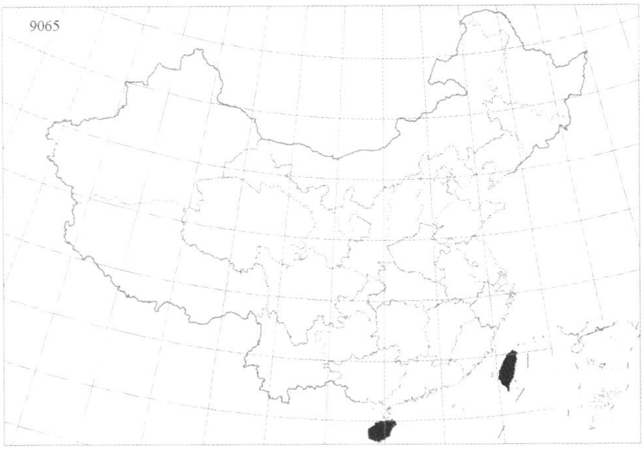

Apocynaceae 夹竹桃科 *Carissa* 假虎刺属
Carissa edulis (Forssk.) Vahl 甜假虎刺
Small trees, evergreen

19.3; 13.1; 23.6; 19.4; 173.2; 0.0; 939 ❙ 1143; 637; 50 ❙ 21.7; 903; 346.9

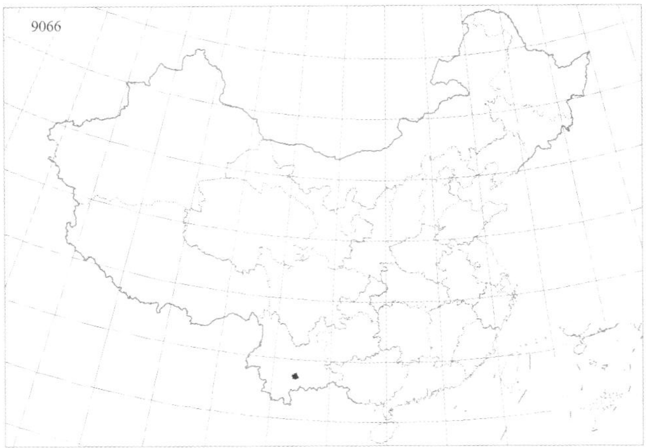

Apocynaceae 夹竹桃科 *Carissa* 假虎刺属

Carissa macrocarpa (Eckl.) A. DC. 大花假虎刺

Shrubs or small trees, evergreen

24.0 (22.2~24.8); 18.4 (17.2~20.1); 28.0 (25.7~28.8); 24.0 (22.3~24.8); 228.3 (207.2~237.6); 0.0 (0.0~0.0); 1310 (1127~1385) ❙ 1501 (868~1854); 534 (312~712); 74 (41~115) ❙ 15.0 (-34.3~47.8); 1191 (868~1360); 440.3 (248.2~620.0)

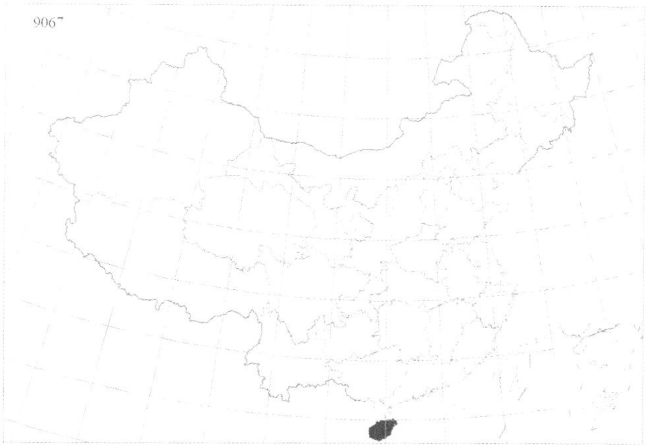

Apocynaceae 夹竹桃科 *Carissa* 假虎刺属

Carissa spinarum L. 假虎刺

Shrubs, evergreen

16.5 (11.2~19.3); 9.6 (3.6~13.5); 21.5 (16.7~26.6); 16.5 (11.3~19.4); 138.1 (75.8~173.2); 0.0 (0.0~1.5); 813 (642~969) ❙ 1104 (797~1751); 625 (411~996); 45 (22~77) ❙ 36.1 (-14.2~89.0); 787 (642~969); 320.2 (189.5~551.3)

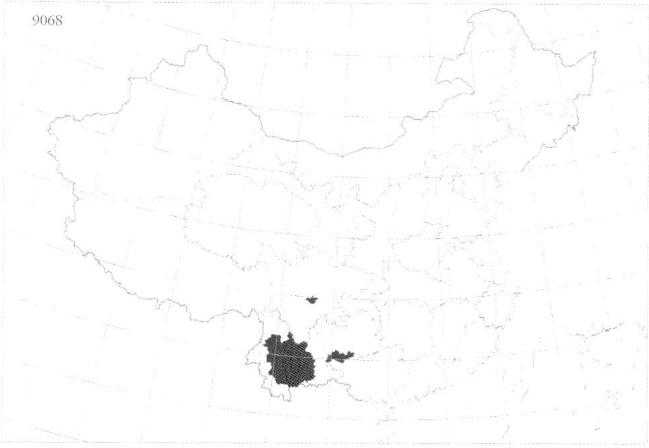

Apocynaceae 夹竹桃科 *Cerbera* 海芒果属

Cerbera manghas L. 海杧果

Trees, evergreen

22.4 (19.9~24.8); 15.0 (12.0~20.1); 28.2 (25.5~29.2); 22.4 (19.9~24.8); 208.9 (179.1~237.6); 0.0 (0.0~0.0); 1193 (987~1385) ❙ 1720 (868~3495); 755 (312~1847); 130 (41~897) ❙ 45.9 (-34.3~224.4); 1168 (868~1360); 257.9 (95.3~620.0)

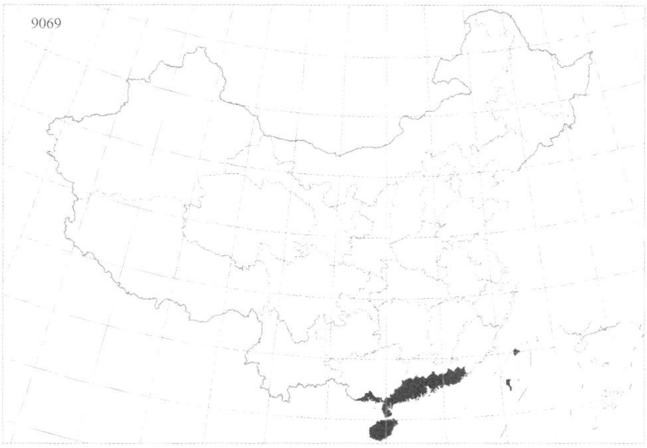

Apocynaceae 夹竹桃科 *Chonemorpha* 鹿角藤属

Chonemorpha eriostylis Pit. 鹿角藤

Large climbers

21.1 (14.4~24.8); 14.2 (7.5~20.1); 25.9 (19.1~29.0); 21.1 (14.4~24.8); 193.4 (113.1~237.6); 0.0 (0.0~0.0); 1096 (727~1385) ❙ 1542 (868~2435); 736 (312~1428); 88 (41~194) ❙ 44.6 (-34.3~112.2); 1057 (727~1360); 400.0 (194.5~659.0)

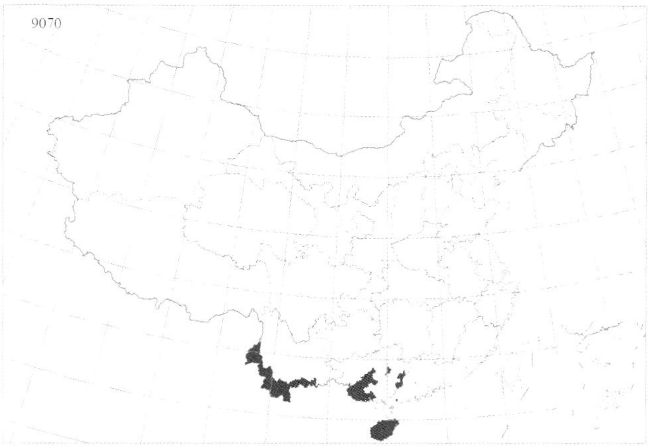

Apocynaceae 夹竹桃科 *Chonemorpha* 鹿角藤属

Chonemorpha floccosa Tsiang et P. T. Li 丛毛鹿角藤

Large climbers

21.3; 13.2; 27.3; 21.3; 195.9; 0.0; 1111 ❙ 1900; 1056; 125 ❙ 71.0; 1111; 297.2

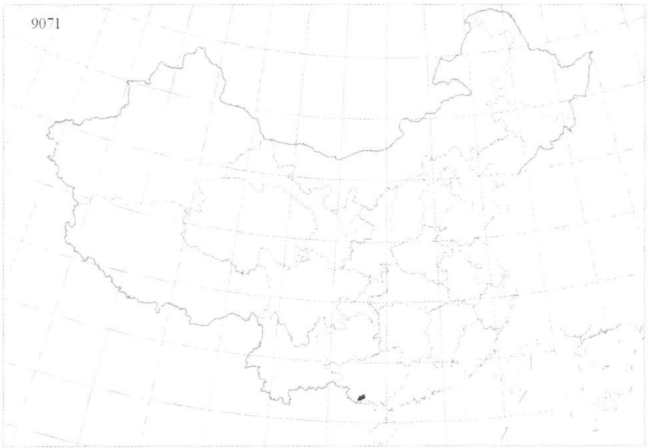

Apocynaceae 夹竹桃科 *Chonemorpha* 鹿角藤属

Chonemorpha fragrans (Moon) Alston 大叶鹿角藤

Large climbers

23.1 (16.9~24.8); 16.9 (10.3~20.1); 27.5 (21.4~28.8); 23.1 (16.9~24.8); 217.5 (142.7~237.6); 0.0 (0.0~0.0); 1244 (813~1385) ❙ 1572 (868~2435); 657 (312~1428); 82 (41~130) ❙ 28.2 (-34.3~99.5); 1159 (813~1360); 407.4 (235.1~620.0)

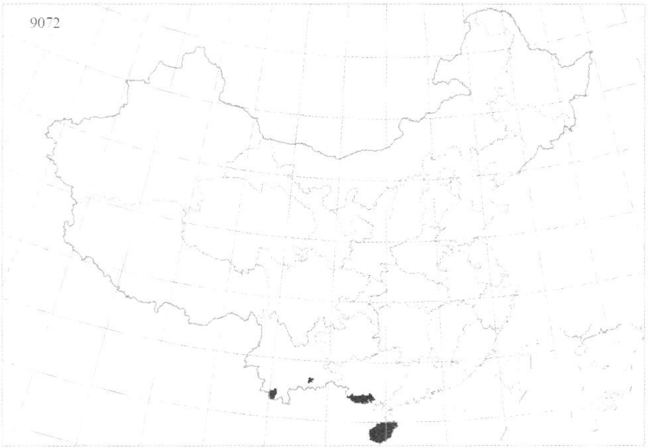

Apocynaceae 夹竹桃科 *Chonemorpha* 鹿角藤属
Chonemorpha griffithii Hook. f. 漾濞鹿角藤
Vines

17.1 (5.3~21.0); 10.8 (−2.4~15.1); 21.4 (12.2~25.6); 17.2 (5.7~21.0); 147.3 (30.5~191.8); 1.7 (0.0~26.2); 847 (479~1053) | 1379 (961~1751); 736 (558~1007); 54 (31~77) | 64.6 (34.8~145.8); 836 (479~1053); 465.0 (234.2~659.0)

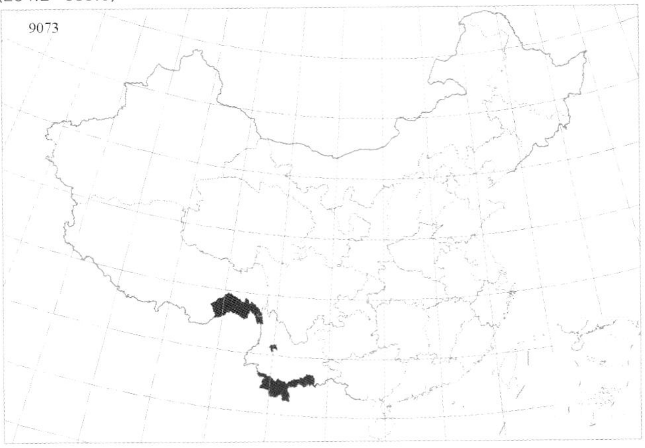

Apocynaceae 夹竹桃科 *Chonemorpha* 鹿角藤属
Chonemorpha megacalyx Pierre 长萼鹿角藤
Vines

22.2 (16.9~24.8); 16.3 (9.9~20.1); 26.1 (21.4~28.8); 22.2 (16.9~24.8); 206.3 (142.7~237.6); 0.0 (0.0~0.0); 1166 (813~1385) | 1496 (868~1854); 626 (312~996); 70 (41~115) | 32.4 (−34.3~88.8); 1088 (813~1360); 465.0 (248.2~659.0)

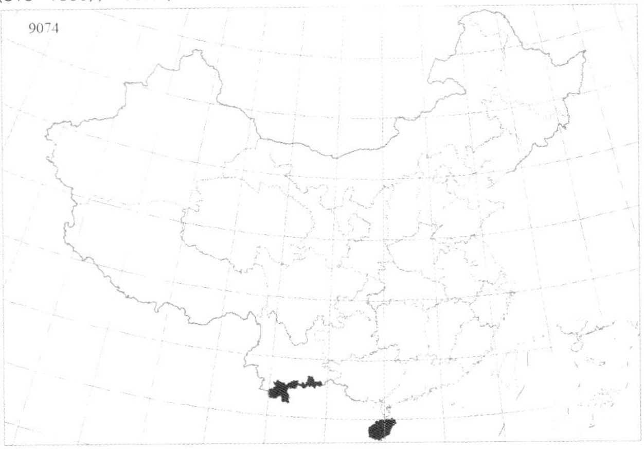

Apocynaceae 夹竹桃科 *Chonemorpha* 鹿角藤属
Chonemorpha parviflora Tsiang et P. T. Li 小花鹿角藤
Vines

21.9 (21.3~22.7); 14.0 (13.2~15.0); 27.9 (27.3~28.6); 22.0 (21.3~22.8); 203.5 (195.9~213.1); 0.0 (0.0~0.0); 1161 (1111~1221) | 1887 (1374~2435); 1056 (713~1428); 114 (87~130) | 62.5 (15.4~99.5); 1161 (1111~1221); 305.2 (281.7~327.7)

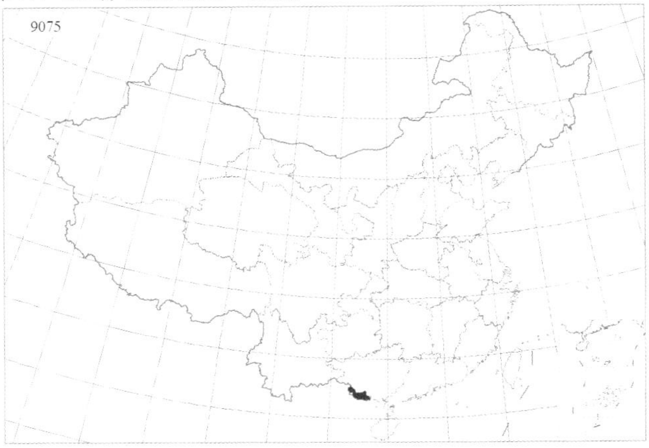

Apocynaceae 夹竹桃科 *Chonemorpha* 鹿角藤属
Chonemorpha splendens Chun et Tsiang 海南鹿角藤
Vines

23.5 (19.3~24.8); 17.8 (13.1~20.1); 27.6 (23.6~28.8); 23.6 (19.4~24.8); 222.9 (173.2~237.6); 0.0 (0.0~0.0); 1273 (939~1385) | 1508 (868~2035); 564 (312~969); 74 (41~115) | 19.3 (−34.3~71.8); 1167 (868~1360); 432.8 (237.9~620.0)

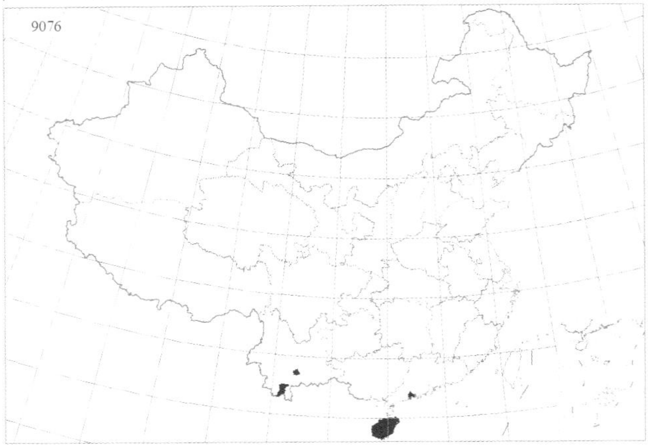

Apocynaceae 夹竹桃科 *Chonemorpha* 鹿角藤属
Chonemorpha verrucosa (Blume) D. J. Middleton 尖子藤
Vines

22.8 (18.2~24.6); 16.7 (12.2~19.5); 27.2 (22.1~29.0); 22.8 (18.3~24.7); 213.9 (160.0~235.9); 0.0 (0.0~0.0); 1216 (872~1372) | 1507 (1179~1854); 645 (392~836); 73 (42~115) | 26.1 (−11.9~66.9); 1150 (866~1360); 414.4 (194.5~620.0)

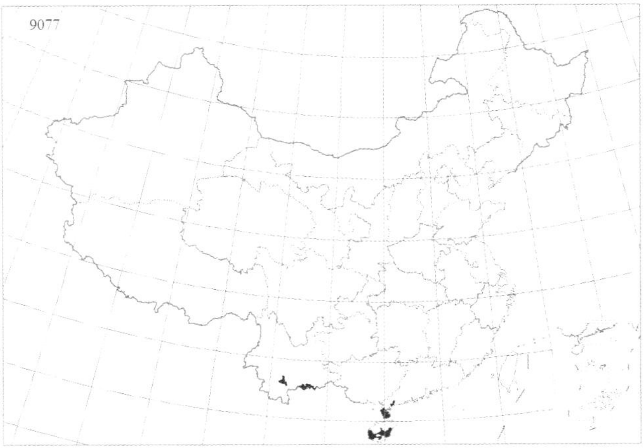

Apocynaceae 夹竹桃科 *Cleghornia* 金平藤属
Cleghornia malaccensis (Hook. f.) King et Gamble 金平藤
Scandent shrubs

18.2 (15.9~21.0); 10.8 (6.3~15.1); 23.5 (21.4~26.6); 18.2 (16.0~21.0); 158.9 (131.6~191.8); 0.0 (0.0~0.0); 893 (797~1053) | 1368 (1074~1751); 729 (582~996); 60 (47~95) | 53.5 (27.5~88.8); 889 (797~1053); 386.3 (189.5~659.0)

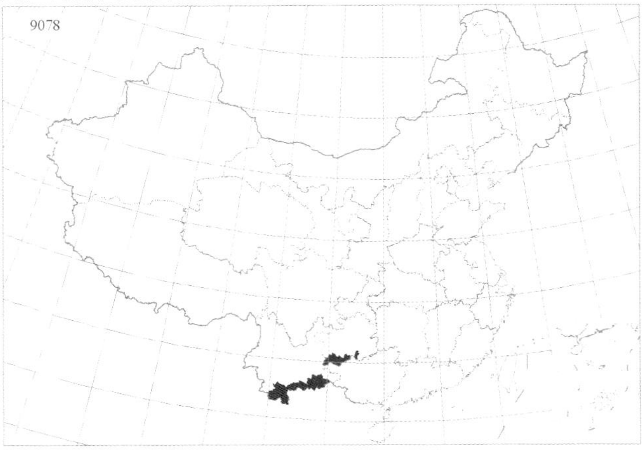

Apocynaceae 夹竹桃科 *Epigynum* 思茅藤属
Epigynum auritum (C. K. Schneid.) Tsiang et P. T. Li 思茅藤
Vines

18.7 (16.9~20.5); 12.7 (10.3~15.1); 22.6 (21.4~23.7); 18.8 (16.9~20.5); 165.4 (142.7~185.9); 0.0 (0.0~0.0); 899 (813~992) ❙ 1354 (1028~1751); 735 (566~996); 57 (44~77) ❙ 50.5 (12.5~88.8); 881 (783~986); 443.4 (339.5~659.0)

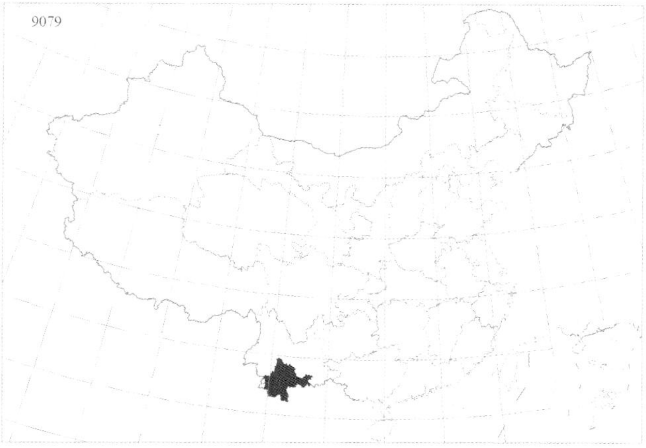

Apocynaceae 夹竹桃科 *Holarrhena* 止泻木属
Holarrhena pubescens Wall. ex G. Don 止泻木
Trees

21.9 (16.8~24.8); 16.0 (10.4~20.1); 26.0 (21.0~29.2); 22.0 (16.9~24.8); 203.4 (142.6~237.6); 0.0 (0.0~0.0); 1148 (811~1385) ❙ 1484 (868~1854); 646 (312~996); 69 (41~115) ❙ 33.5 (-34.3~88.8); 1079 (811~1360); 447.5 (178.7~659.0)

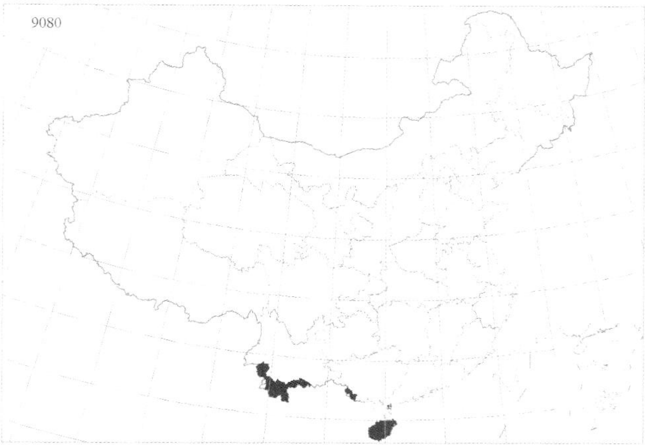

Apocynaceae 夹竹桃科 *Hunteria* 仔榄树属
Hunteria zeylanica (Retz.) Gardner ex Thwaites 仔榄树
Trees

24.0 (22.2~24.8); 18.4 (17.2~20.1); 28.0 (25.7~28.8); 24.0 (22.3~24.8); 228.3 (207.2~237.6); 0.0 (0.0~0.0); 1310 (1127~1385) ❙ 1501 (868~1854); 534 (312~712); 74 (41~115) ❙ 15.0 (-34.3~47.8); 1191 (868~1360); 440.3 (248.2~620.0)

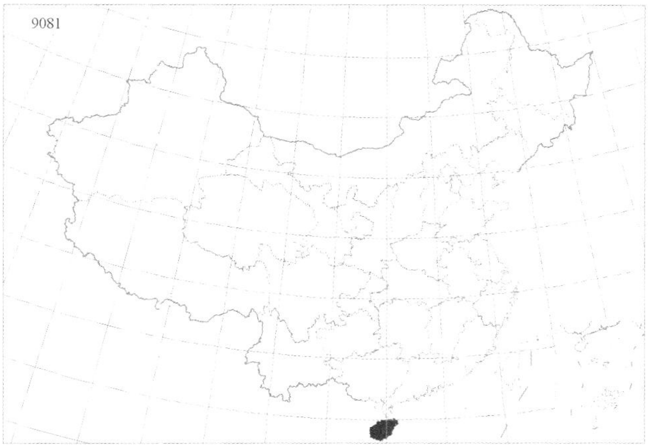

Apocynaceae 夹竹桃科 *Ichnocarpus* 腰骨藤属
Ichnocarpus frutescens (L.) W. T. Aiton 腰骨藤
Vines

19.4 (13.3~24.8); 10.8 (3.6~20.1); 26.4 (21.6~29.8); 19.5 (13.3~24.8); 173.5 (101.4~237.6); 0.1 (0.0~1.8); 1019 (715~1385) ❙ 1417 (868~2435); 655 (312~1428); 99 (41~257) ❙ 41.0 (-34.3~99.5); 1002 (715~1360); 272.3 (111.7~659.0)

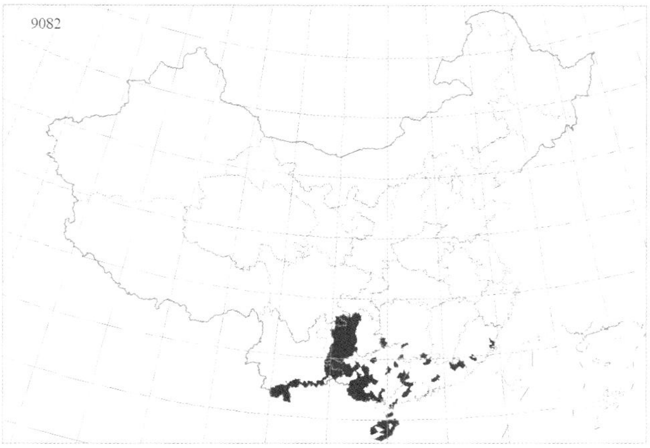

Apocynaceae 夹竹桃科 *Ichnocarpus* 腰骨藤属
Ichnocarpus jacquetii (Pierre) D. J. Middleton 少花腰骨藤
Vines

21.5 (17.8~24.0); 13.4 (7.1~19.1); 28.2 (27.1~28.9); 21.7 (17.8~24.0); 199.8 (154.0~228.1); 0.0 (0.0~0.0); 1149 (933~1302) ❙ 1653 (1446~1965); 732 (513~994); 129 (48~182) ❙ 44.8 (19.1~65.8); 1138 (933~1257); 245.4 (139.6~620.0)

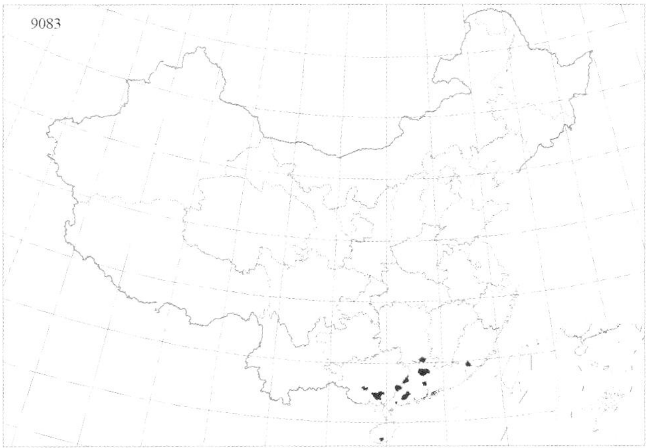

Apocynaceae 夹竹桃科 *Ichnocarpus* 腰骨藤属
Ichnocarpus polyanthus (Blume) P. I. Forst. 小花藤
Vines

21.7 (17.2~24.8); 14.7 (8.8~20.1); 26.8 (21.6~29.0); 21.7 (17.3~24.8); 200.6 (147.0~237.6); 0.0 (0.0~0.0); 1139 (832~1385) ❙ 1572 (868~2435); 740 (312~1428); 89 (41~194) ❙ 40.1 (-34.3~99.5); 1101 (783~1360); 357.8 (139.6~659.0)

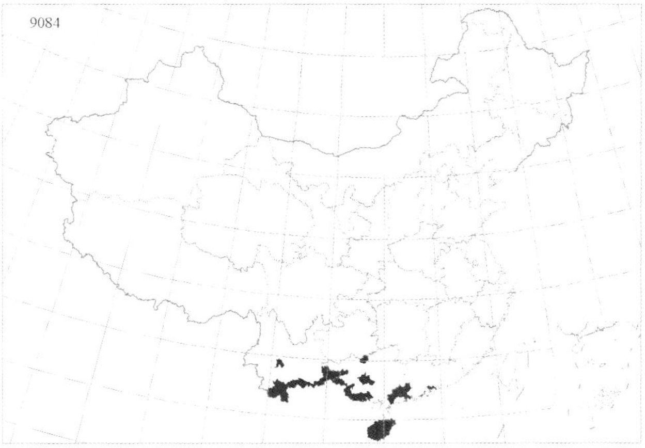

Apocynaceae 夹竹桃科 *Kibatalia* 倒缨木属
Kibatalia macrophylla (Pierre ex Hua) Woodson 倒缨木
Shrubs or small trees

19.0 (15.8~21.0); 13.0 (9.0~15.1); 22.8 (20.6~25.6); 19.0 (15.8~21.0); 168.2 (129.7~191.8); 0.0 (0.0~0.0); 917 (776~1053) ❙ 1460 (1011~1751); 769 (586~996); 62 (44~77) ❙ 59.0 (30.3~88.8); 908 (772~1053); 497.1 (258.7~659.0)

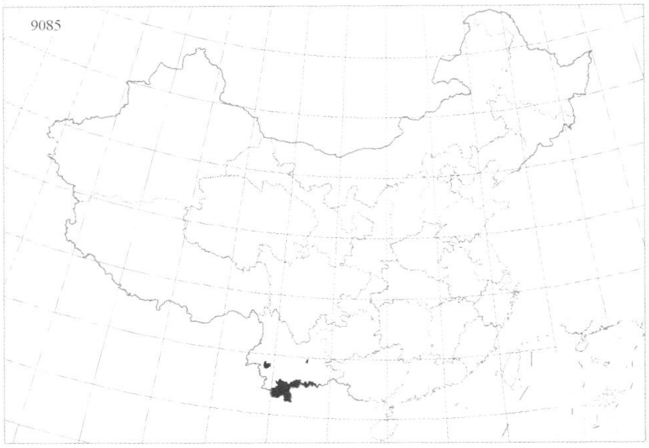

Apocynaceae 夹竹桃科 *Kopsia* 蕊木属
Kopsia arborea Blume 蕊木
Trees

22.4 (18.2~24.8); 15.5 (8.7~20.1); 27.6 (22.3~29.2); 22.5 (18.3~24.8); 209.8 (159.3~237.6); 0.0 (0.0~0.0); 1198 (909~1385) ❙ 1609 (868~2435); 695 (312~1428); 94 (41~183) ❙ 36.0 (-34.3~99.5); 1143 (868~1360); 358.4 (157.7~659.0)

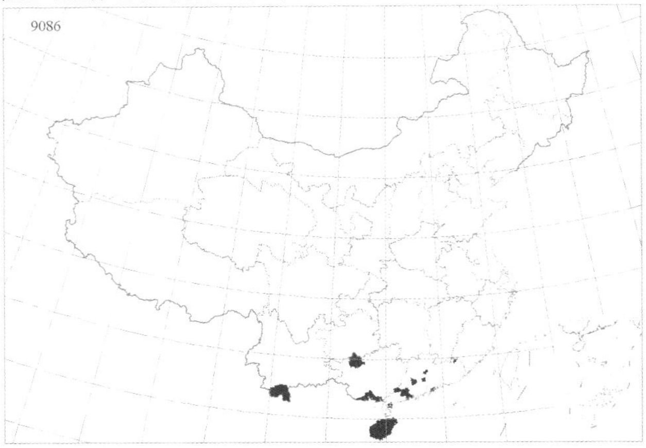

Apocynaceae 夹竹桃科 *Melodinus* 山橙属
Melodinus angustifolius Hayata 台湾山橙
Vines

19.9; 12.9; 26.1; 19.9; 178.9; 0.0; 1013 ❙ 2303; 1191; 158 ❙ 122.6; 1013; 273.9

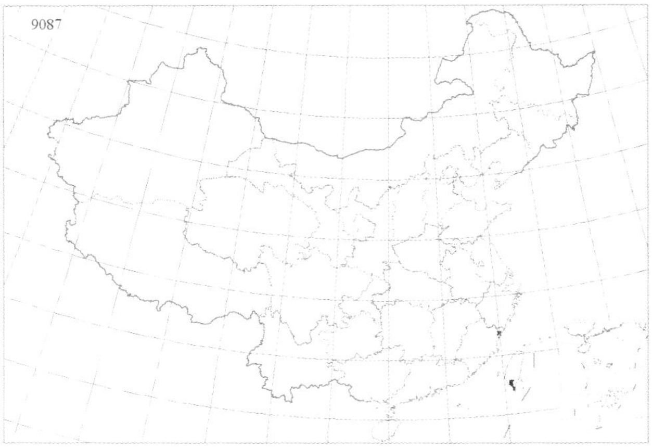

Apocynaceae 夹竹桃科 *Melodinus* 山橙属
Melodinus axillaris W. T. Wang ex Tsiang et P. T. Li 腋花山橙
Vines

19.5 (17.7~21.0); 13.5 (10.9~15.1); 23.3 (22.3~25.6); 19.5 (17.8~21.0); 174.2 (153.4~191.8); 0.0 (0.0~0.0); 941 (854~1053) ❙ 1523 (1368~1751); 795 (655~996); 66 (56~77) ❙ 62.3 (44.5~88.8); 933 (854~1053); 544.7 (491.5~659.0)

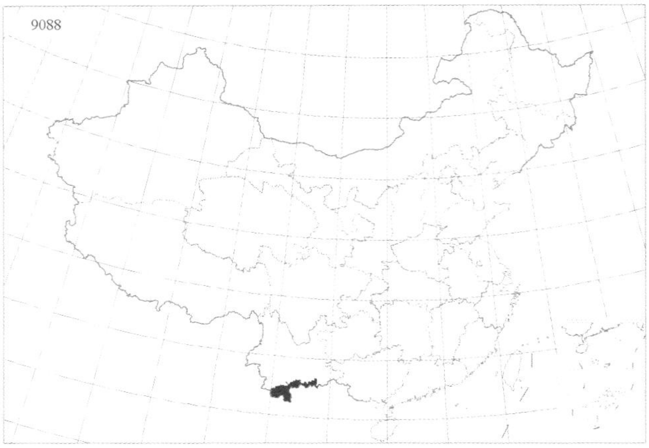

Apocynaceae 夹竹桃科 *Melodinus* 山橙属
Melodinus chinensis P. T. Li et Z. R. Xu 贵州山橙
Shrubs

17.4; 7.4; 25.5; 17.4; 148.9; 0.0; 886 ❙ 1381; 637; 107 ❙ 55.9; 886; 245.3

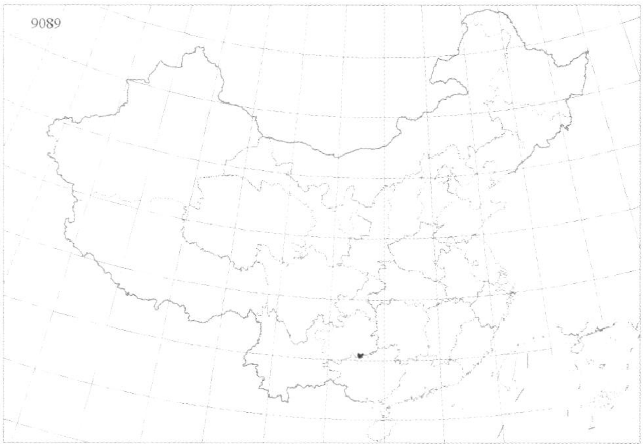

Apocynaceae 夹竹桃科 *Melodinus* 山橙属
Melodinus cochinchinensis (Lour.) Merr. 思茅山橙
Vines

18.6 (16.3~21.0); 12.4 (9.1~15.1); 22.6 (20.8~25.6); 18.7 (16.4~21.0); 164.0 (136.4~191.8); 0.0 (0.0~0.0); 897 (797~1053) ❙ 1457 (1190~1751); 775 (655~996); 56 (41~77) ❙ 62.8 (44.5~88.8); 883 (797~1053); 480.8 (336.3~659.0)

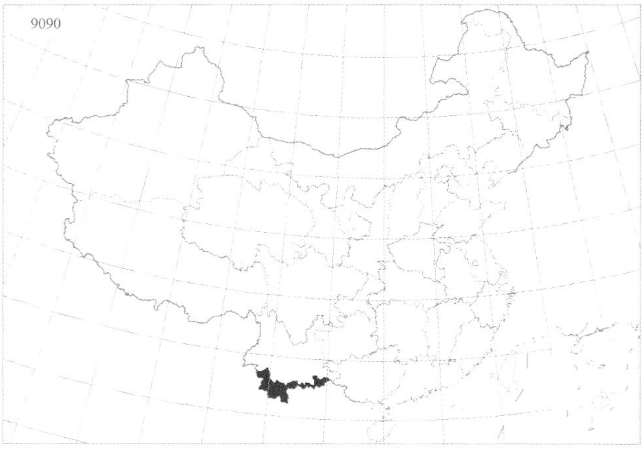

Apocynaceae 夹竹桃科 *Melodinus* 山橙属
Melodinus fusiformis Champ. ex Benth. 尖山橙
Shrubs

18.7 (12.1~23.6); 9.5 (3.3~16.7); 26.7 (20.1~29.8); 18.8 (12.2~23.6); 165.9 (88.5~223.5); 0.1 (0.0~2.4); 988 (679~1281) ❙ 1477 (999~2435); 648 (444~1428); 117 (26~197) ❙ 50.1 (-6.9~99.5); 987 (679~1273); 217.6 (95.3~371.7)

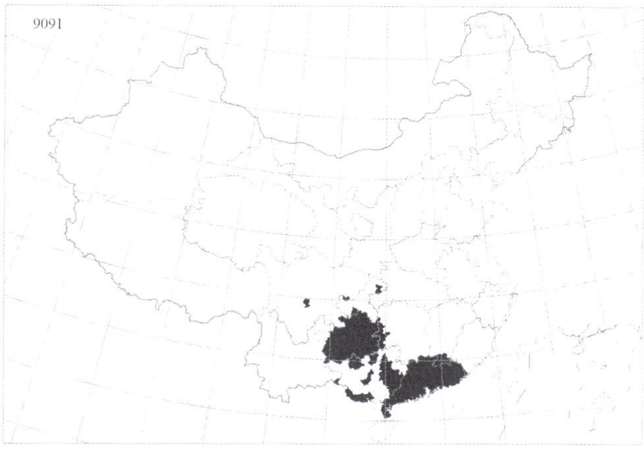

Apocynaceae 夹竹桃科 *Melodinus* 山橙属
Melodinus hemsleyanus Diels 川山橙
Vines

16.3 (10.6~19.2); 5.9 (-0.2~11.6); 25.9 (20.0~28.9); 16.3 (10.6~19.2); 136.3 (79.4~170.8); 0.7 (0.0~12.1); 860 (643~1002) ❙ 1182 (976~1495); 522 (442~820); 68 (35~135) ❙ 39.1 (8.0~107.2); 859 (643~1002); 187.1 (108.2~371.7)

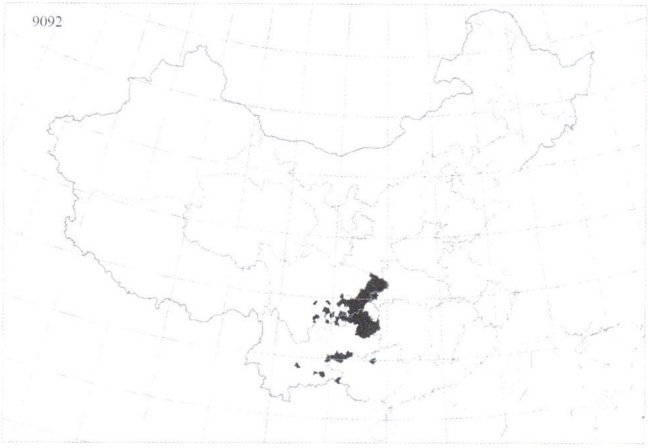

Apocynaceae 夹竹桃科 *Melodinus* 山橙属
Melodinus khasianus Hook. f. 景东山橙
Vines

16.2 (13.6~18.2); 9.7 (7.2~11.9); 20.7 (18.8~22.0); 16.3 (13.7~18.2); 135.2 (104.1~158.7); 0.0 (0.0~0.0); 791 (706~868) ❙ 1184 (986~1454); 663 (553~816); 50 (42~70) ❙ 49.7 (26.1~99.9); 780 (706~847); 392.0 (258.7~516.7)

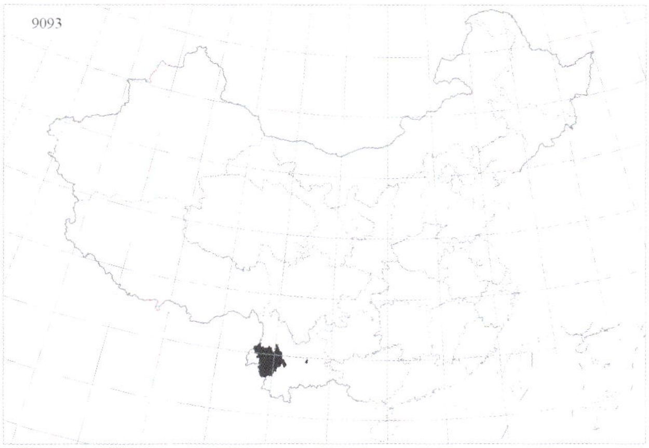

Apocynaceae 夹竹桃科 *Melodinus* 山橙属
Melodinus magnificus Tsiang 茶藤
Vines

20.5 (17.9~22.7); 11.5 (7.4~15.0); 27.6 (25.0~28.6); 20.6 (18.0~22.8); 186.8 (155.6~213.1); 0.0 (0.0~0.0); 1078 (913~1221) ❙ 1622 (1284~2435); 787 (562~1428); 137 (57~189) ❙ 51.2 (12.9~99.5); 1077 (913~1221); 260.6 (185.5~371.7)

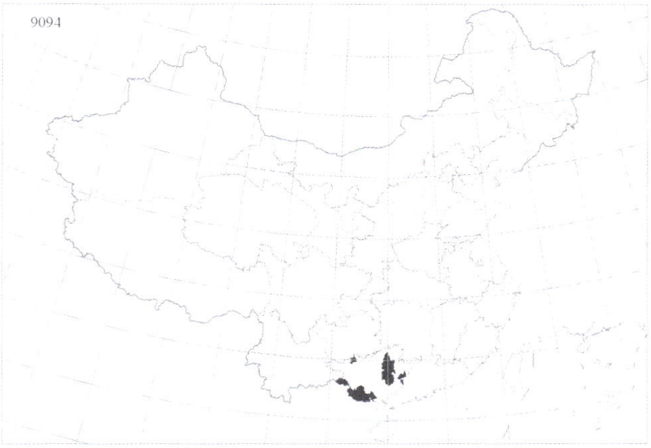

Apocynaceae 夹竹桃科 *Melodinus* 山橙属
Melodinus morsei Tsiang 龙州山橙
Vines

21.2 (18.0~22.7); 13.0 (8.5~15.0); 27.4 (25.0~28.6); 21.2 (18.0~22.8); 195.0 (156.4~213.1); 0.0 (0.0~0.0); 1113 (913~1221) ❙ 1735 (1302~2435); 943 (670~1428); 112 (57~169) ❙ 56.0 (12.9~99.5); 1113 (913~1221); 297.8 (235.1~371.7)

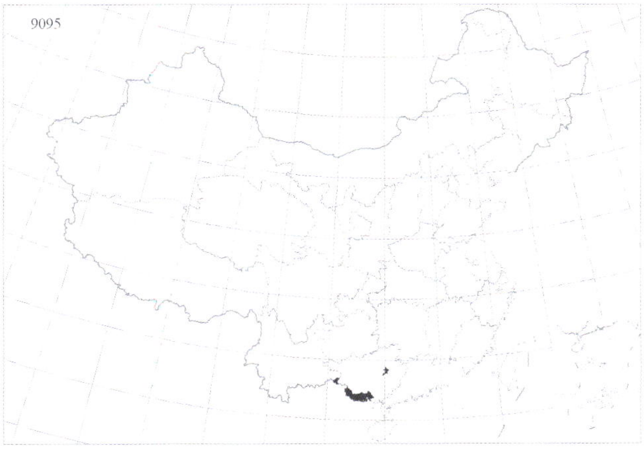

Apocynaceae 夹竹桃科 *Melodinus* 山橙属
Melodinus suaveolens (Hance) Champ. ex Benth. 山橙
Scandent shrubs

21.5 (16.9~24.8); 13.6 (7.1~20.1); 28.0 (22.5~29.8); 21.6 (17.0~24.8); 198.9 (143.6~237.6); 0.0 (0.0~0.0); 1141 (862~1385) ❙ 1645 (868~2435); 705 (312~1428); 131 (41~198) ❙ 45.8 (-34.3~100.8); 1124 (862~1360); 259.1 (95.3~620.0)

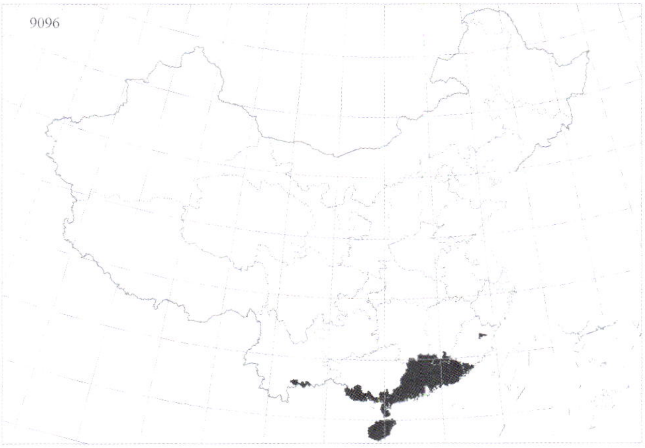

Apocynaceae 夹竹桃科 *Melodinus* 山橙属
Melodinus tenuicaudatus Tsiang et P. T. Li 薄叶山橙
Scandent shrubs

19.0 (16.2~22.7); 11.1 (6.3~15.1); 24.9 (21.4~28.6); 19.0 (16.2~22.8); 168.4 (134.5~213.1); 0.0 (0.0~0.0); 954 (813~1221) ❙ 1400 (1059~2435); 758 (594~1428); 69 (43~130) ❙ 46.8 (5.8~99.5); 952 (813~1221); 351.0 (189.5~659.0)

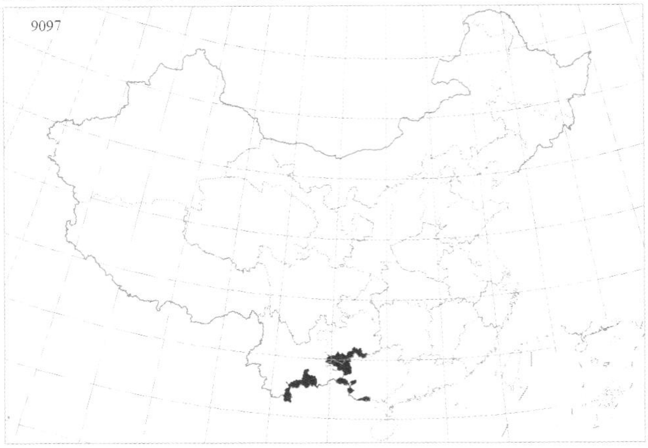

Apocynaceae 夹竹桃科 *Melodinus* 山橙属
Melodinus yunnanensis Tsiang et P. T. Li 雷打果
Scandent shrubs

18.7 (16.9~21.0); 12.3 (10.3~14.2); 23.0 (21.4~25.6); 18.7 (16.9~21.0); 164.6 (142.7~191.8); 0.0 (0.0~0.0); 903 (813~1053) ❙ 1332 (1059~1751); 756 (604~996); 56 (44~77) ❙ 47.5 (21.3~88.8); 895 (813~1053); 425.2 (294.5~551.6)

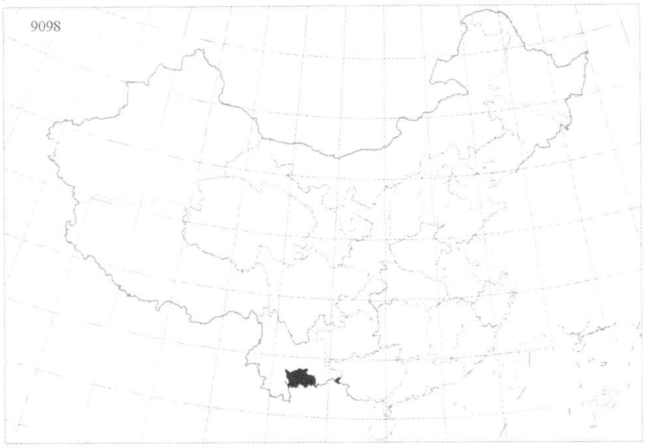

Apocynaceae 夹竹桃科 *Parepigynum* 富宁藤属
Parepigynum funingense Tsiang et P. T. Li 富宁藤
Vines

16.5 (14.2~19.1); 7.3 (4.8~10.6); 24.1 (20.8~26.6); 16.6 (14.2~19.1); 139.1 (110.9~169.2); 0.0 (0.0~0.2); 839 (734~969) ❙ 1282 (1206~1504); 652 (517~860); 64 (48~95) ❙ 53.4 (27.5~80.6); 839 (734~969); 240.6 (171.2~413.5)

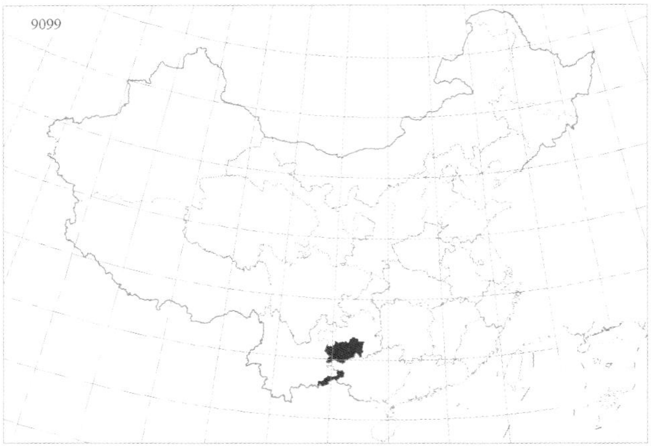

Apocynaceae 夹竹桃科 *Parsonsia* 长节珠属
Parsonsia alboflavescens (Dennst.) Mabb. 海南同心结
Scandent shrubs

20.7 (17.7~24.8); 13.9 (7.7~20.1); 26.2 (22.3~28.3); 20.8 (17.8~24.8); 189.2 (153.6~237.6); 0.0 (0.0~0.0); 1067 (909~1385) ❙ 1796 (1242~3495); 652 (444~924); 204 (41~897) ❙ 72.1 (-2.2~224.4); 1015 (880~1147); 387.7 (223.8~659.0)

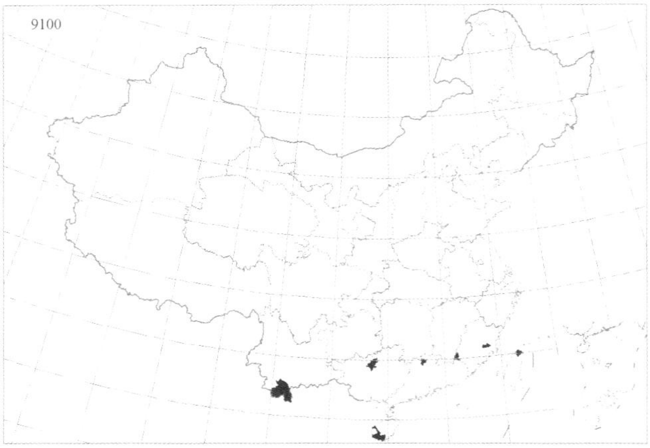

Apocynaceae 夹竹桃科 *Parsonsia* 长节珠属
Parsonsia goniostemon Hand.-Mazz. 广西同心结
Scandent shrubs

20.1 (18.7~22.2); 11.2 (8.8~13.9); 27.1 (25.8~28.3); 20.2 (18.8~22.3); 182.3 (165.2~207.5); 0.0 (0.0~0.0); 1046 (958~1190) ❙ 1431 (1250~1581); 733 (670~850); 99 (60~135) ❙ 37.3 (15.4~49.1); 1046 (958~1190); 276.8 (238.3~327.7)

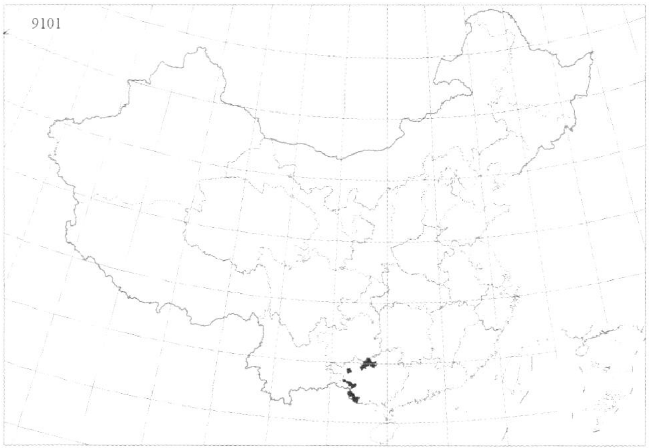

Apocynaceae 夹竹桃科 *Pottsia* 帘子藤属
Pottsia grandiflora Markgr. 大花帘子藤
Scandent shrubs

18.5 (15.3~22.5); 8.3 (4.1~15.2); 27.6 (25.8~29.8); 18.6 (15.4~22.7); 162.8 (125.8~212.7); 0.1 (0.0~0.9); 975 (818~1216) ❙ 1590 (1284~2037); 608 (403~1033); 175 (73~230) ❙ 64.0 (22.7~129.9); 969 (818~1216); 221.7 (139.6~322.1)

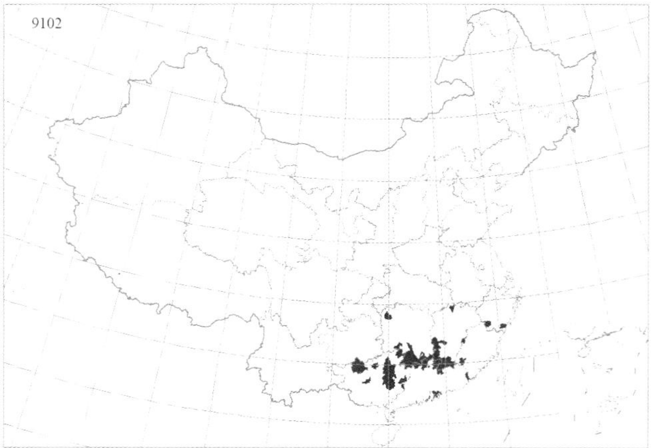

Apocynaceae 夹竹桃科 *Pottsia* 帘子藤属
Pottsia laxiflora (Blume) Kuntze 帘子藤
Scandent shrubs, evergreen

19.8 (14.3~24.8); 10.9 (3.3~20.1); 27.3 (21.4~29.9); 19.8 (14.3~24.8); 178.2 (113.2~237.6); 0.1 (0.0~2.3); 1040 (761~1385) ❙ 1554 (868~2435); 652 (312~1428); 137 (41~259) ❙ 51.2 (-34.3~128.9); 1027 (761~1360); 257.5 (95.3~659.0)

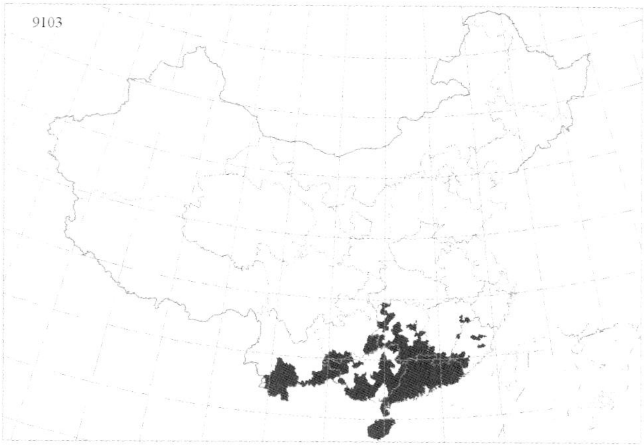

Apocynaceae 夹竹桃科 *Rauvolfia* 萝芙木属
Rauvolfia serpentina (L.) Benth. ex Kurz 蛇根木
Shrubs

18.9 (16.4~24.6); 12.5 (8.9~19.5); 23.0 (20.6~28.9); 18.9 (16.4~24.7); 167.1 (137.3~235.9); 0.0 (0.0~0.0); 920 (795~1372) ❙ 1452 (1059~1776); 775 (567~996); 62 (41~188) ❙ 59.1 (20.3~112.2); 905 (795~1299); 456.9 (201.4~659.0)

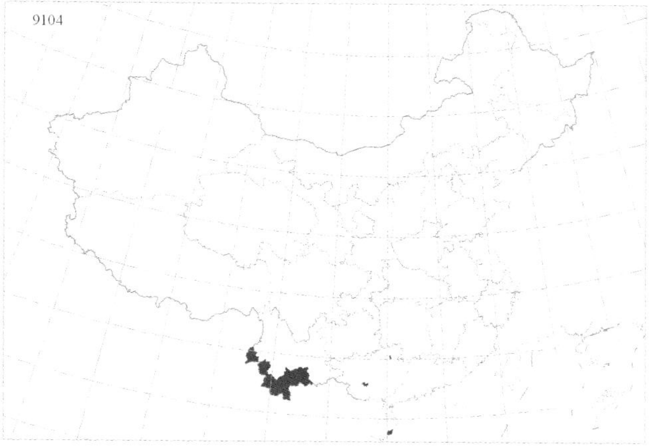

Apocynaceae 夹竹桃科 *Rauvolfia* 萝芙木属
Rauvolfia tiaolushanensis Tsiang 吊罗山萝芙木
Shrubs

24.2; 19.0; 27.9; 24.3; 231.3; 0.0; 1331 ❙ 1702; 580; 81 ❙ 27.7; 1254; 524.3

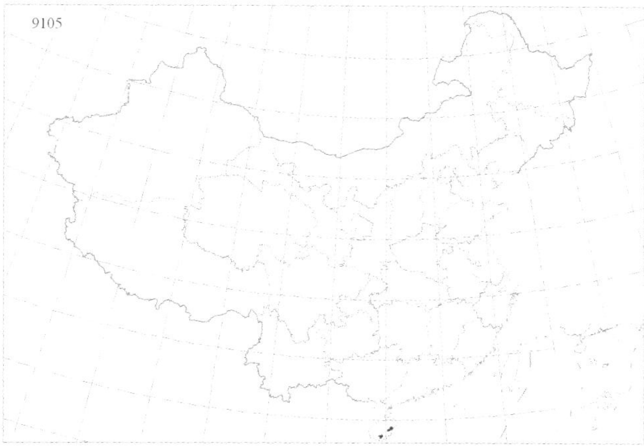

Apocynaceae 夹竹桃科 *Rauvolfia* 萝芙木属
Rauvolfia verticillata (Lour.) Baill. 萝芙木
Shrubs

19.2 (10.1~24.8); 11.3 (2.3~20.1); 25.3 (16.9~29.8); 19.2 (10.2~24.8); 171.0 (68.1~237.6); 0.0 (0.0~5.2); 990 (618~1385) ❙ 1436 (797~3495); 693 (312~1847); 101 (20~897) ❙ 46.1 (-34.3~224.4); 972 (618~1360); 294.1 (111.7~659.0)

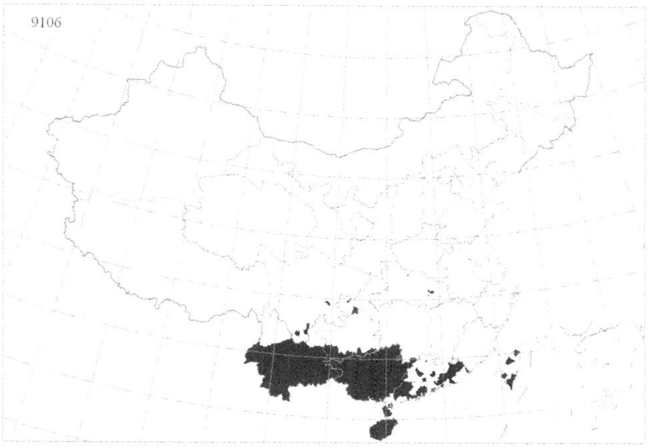

Apocynaceae 夹竹桃科 *Sindechites* 毛药藤属
Sindechites chinensis (Merr.) Markgr. et Tsiang 坭藤
Vines

24.1 (23.1~24.8); 18.9 (17.5~20.1); 27.8 (27.2~28.7); 24.2 (23.2~24.8); 230.1 (218.4~237.6); 0.0 (0.0~0.0); 1323 (1232~1385) ❙ 1395 (1176~1588); 483 (392~544); 50 (41~59) ❙ 5.6 (-12.0~20.7); 1135 (1056~1212); 500.4 (411.8~620.0)

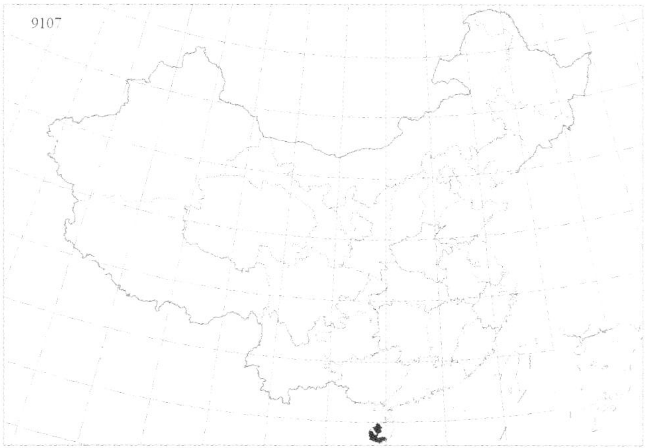

Apocynaceae 夹竹桃科 *Sindechites* 毛药藤属
Sindechites henryi Oliv. 毛药藤
Vines

16.3 (7.3~21.3); 6.0 (-2.0~14.2); 25.6 (15.6~29.3); 16.3 (7.5~21.3); 136.7 (48.2~196.0); 0.9 (0.0~20.4); 855 (537~1123) ❙ 1365 (958~1950); 582 (420~911); 116 (27~253) ❙ 60.3 (5.8~147.0); 853 (537~1121); 251.0 (146.0~551.6)

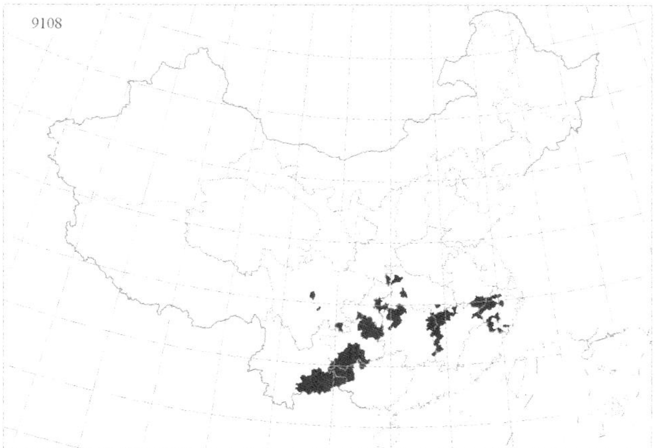

Apocynaceae 夹竹桃科 *Strophanthus* 羊角拗属

Strophanthus caudatus (L.) Kurz 卵萼羊角拗

Shrubs

22.3; 14.5; 28.1; 22.3; 207.7; 0.0; 1186 | 2291; 1331; 127 | 93.1; 1186; 309.7

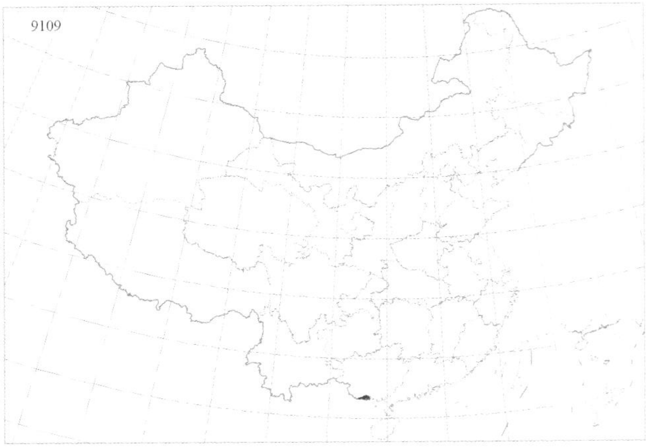

Apocynaceae 夹竹桃科 *Strophanthus* 羊角拗属

Strophanthus divaricatus (Lour.) Hook. et Arn. 羊角拗

Shrubs

19.9 (14.2~24.8); 11.2 (3.6~20.1); 27.1 (19.1~29.8); 19.9 (14.2~24.8); 178.9 (110.5~237.6); 0.0 (0.0~1.7); 1038 (727~1385) | 1562 (1070~2435); 670 (356~1428); 135 (41~249) | 52.0 (-6.9~121.9); 1031 (727~1321); 252.3 (95.3~659.0)

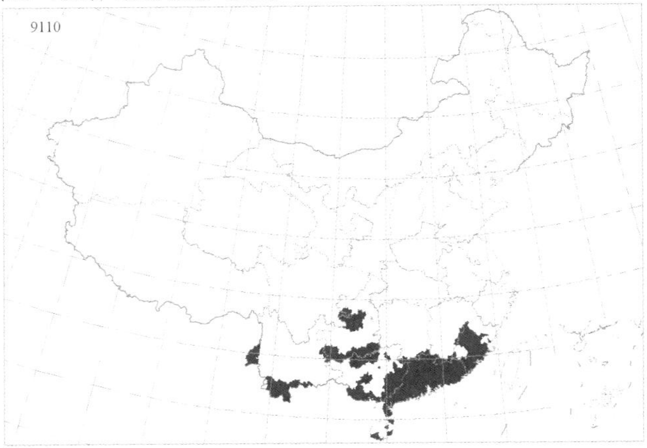

Apocynaceae 夹竹桃科 *Strophanthus* 羊角拗属

Strophanthus wallichii A. DC. 云南羊角拗

Shrubs

20.0; 14.6; 23.2; 20.1; 180.8; 0.0; 963 | 1523; 681; 67 | 58.4; 942; 581.0

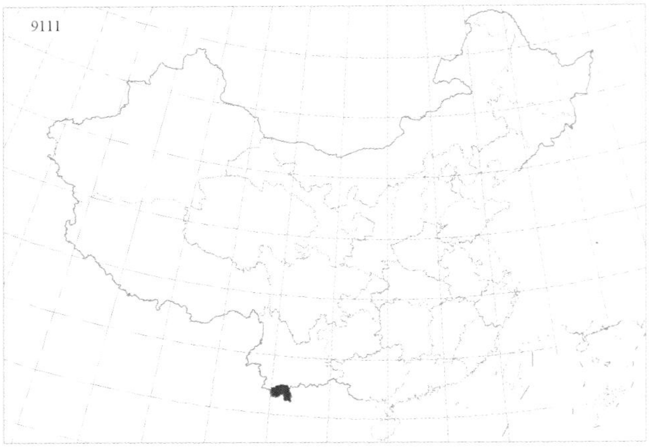

Apocynaceae 夹竹桃科 *Tabernaemontana* 狗牙花属

Tabernaemontana bovina Lour. 药用狗牙花

Shrubs, evergreen

20.4 (16.3~24.8); 14.0 (9.0~20.1); 24.7 (20.6~28.8); 20.5 (16.4~24.8); 185.5 (136.4~237.6); 0.0 (0.0~0.0); 1043 (795~1385) | 1447 (868~1854); 691 (312~996); 64 (41~194) | 43.0 (-34.3~112.2); 997 (795~1360); 441.6 (213.5~659.0)

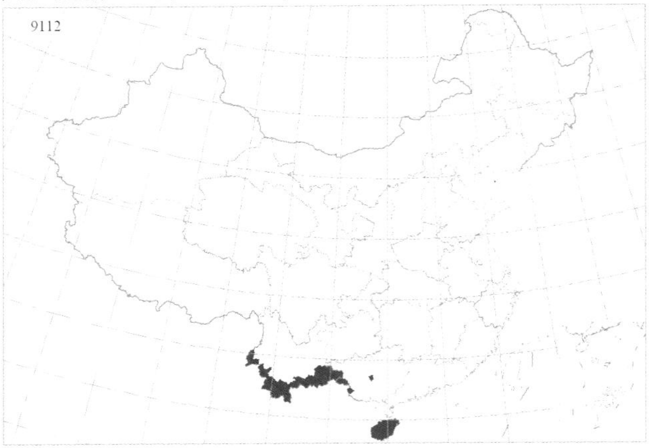

Apocynaceae 夹竹桃科 *Tabernaemontana* 狗牙花属

Tabernaemontana bufalina Lour. 尖蕾狗牙花

Shrubs or small trees, evergreen

22.7 (15.8~24.8); 16.3 (9.0~20.1); 27.3 (20.6~29.0); 22.7 (15.8~24.8); 212.8 (129.7~237.6); 0.0 (0.0~0.0); 1215 (776~1385) | 1594 (868~2435); 702 (312~1428); 85 (41~144) | 32.8 (-34.3~99.5); 1149 (772~1360); 384.7 (157.7~659.0)

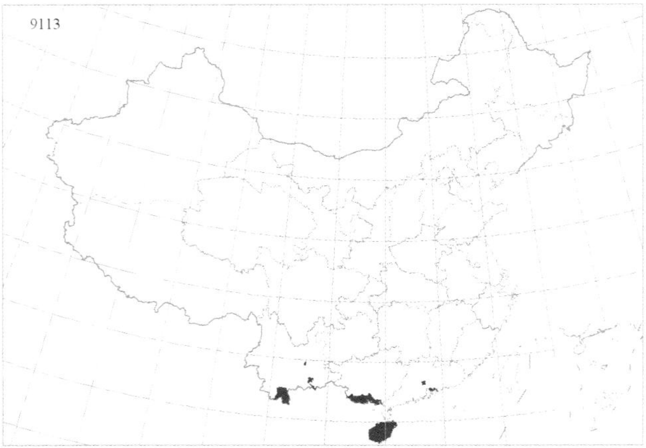

Apocynaceae 夹竹桃科 *Tabernaemontana* 狗牙花属

Tabernaemontana corymbosa Roxb. ex Wall. 伞房狗牙花

Shrubs or small trees, evergreen

19.4 (16.3~22.7); 11.9 (7.8~15.1); 24.8 (21.4~28.6); 19.5 (16.4~22.8); 173.6 (136.4~213.1); 0.0 (0.0~0.0); 974 (797~1221) | 1441 (1074~2435); 779 (582~1428); 69 (41~130) | 48.5 (5.8~99.5); 969 (797~1221); 370.6 (189.5~659.0)

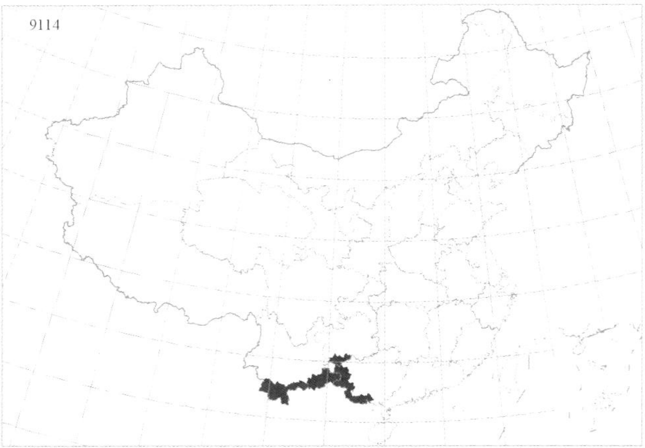

Apocynaceae 夹竹桃科 *Tabernaemontana* 狗牙花属

Tabernaemontana divaricata (L.) R. Br. ex Roem. et Schult. 狗牙花

Shrubs or small trees, evergreen

20.8 (8.6~24.8); 14.5 (1.3~20.1); 25.1 (15.0~28.8); 20.8 (8.7~24.8); 189.9 (52.6~237.6); 0.2 (0.0~8.8); 1076 (574~1385) ❙ 1540 (868~2993); 689 (312~1045); 81 (41~547) ❙ 49.4 (-34.3~203.4); 1018 (574~1360); 439.0 (202.8~659.0)

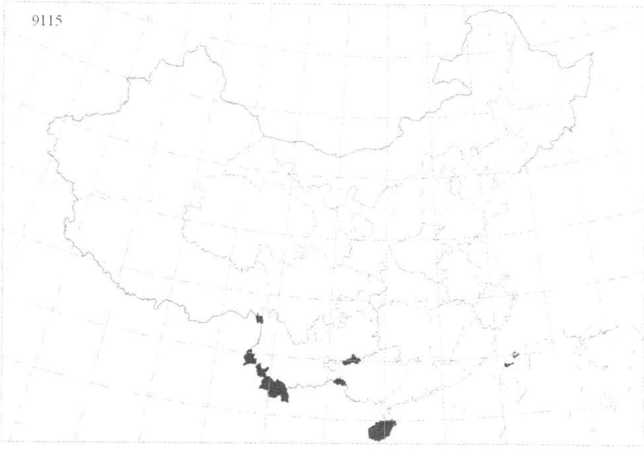

Apocynaceae 夹竹桃科 *Thevetia* 黄花夹竹桃属

Thevetia peruviana (Pers.) K. Schum. 黄花夹竹桃

Trees, evergreen

19.5 (13.3~24.8); 11.3 (3.6~20.1); 26.1 (20.3~29.0); 19.5 (13.3~24.3); 174.0 (101.9~237.6); 0.0 (0.0~1.7); 1030 (718~1385) ❙ 1341 (868~1854); 585 (312~832); 74 (20~188) ❙ 33.6 (-34.3~71.1); 986 (692~1360); 295.7 (126.1~620.0)

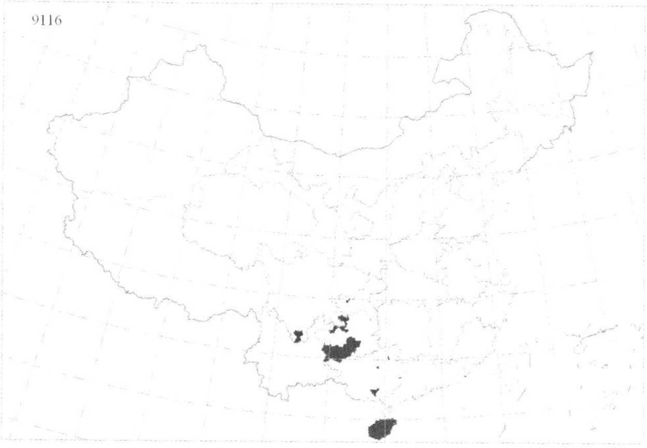

Apocynaceae 夹竹桃科 *Trachelospermum* 络石属

Trachelospermum asiaticum (Siebold et Zucc.) Nakai 亚洲络石

Vines, evergreen

17.0 (5.3~23.9); 6.6 (-2.4~18.7); 26.4 (12.2~29.8); 17.0 (5.7~23.9); 145.6 (30.5~225.8); 1.2 (0.0~26.2); 901 (479~1294) ❙ 1395 (793~2992); 567 (352~1847); 124 (31~242) ❙ 55.6 (-4.0~166.5); 897 (479~1204); 221.1 (108.2~455.2)

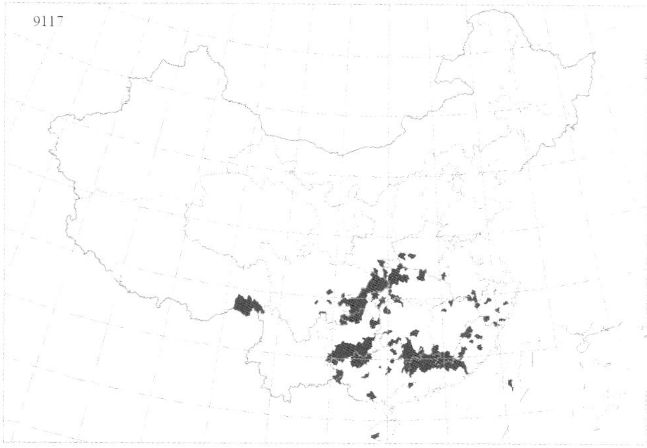

Apocynaceae 夹竹桃科 *Trachelospermum* 络石属

Trachelospermum axillare Hook. f. 紫花络石

Vines, evergreen

16.2 (1.8~22.2); 6.3 (-7.1~15.1); 25.0 (9.9~29.9); 16.3 (3.8~22.3); 136.0 (15.7~207.5); 0.9 (0.0~53.3); 853 (404~1190) ❙ 1340 (583~2024); 570 (348~1056); 116 (13~259) ❙ 57.1 (-14.2~147.0); 842 (404~1190); 247.6 (126.1~659.0)

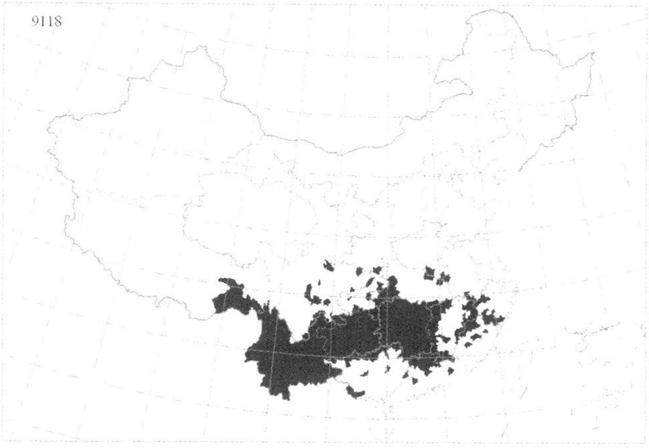

Apocynaceae 夹竹桃科 *Trachelospermum* 络石属

Trachelospermum bodinieri (H. Lév.) Woodson 贵州络石

Vines, evergreen

16.7 (2.2~22.4); 8.2 (-6.6~15.2); 23.8 (9.5~29.4); 16.8 (3.8~22.7); 142.6 (15.5~212.7); 1.2 (0.0~49.1); 868 (411~1216) ❙ 1338 (729~2037); 625 (403~1033); 96 (11~253) ❙ 54.2 (-14.2~145.8); 859 (411~1216); 28[illegible].3 (130.1~659.0)

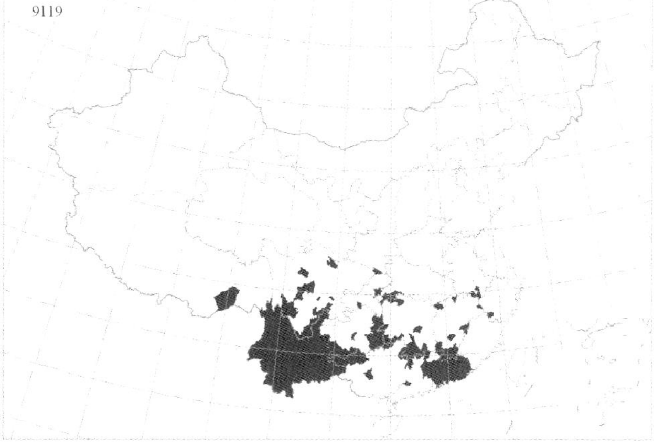

Apocynaceae 夹竹桃科 *Trachelospermum* 络石属

Trachelospermum brevistylum Hand.-Mazz. 短柱络石

Vines, evergreen

17.1 (3.4~22.7); 6.5 (-5.5~15.0); 26.6 (11.4~29.8); 17.1 (4.7~22.8); 146.3 (23.3~213.1); 1.0 (0.0~41.8); 905 (431~1221) ❙ 1492 (527~2435); 588 (317~1428); 153 (9~259) ❙ 65.4 (10.4~130.7); 901 (431~1221); 229.3 (111.9~413.5)

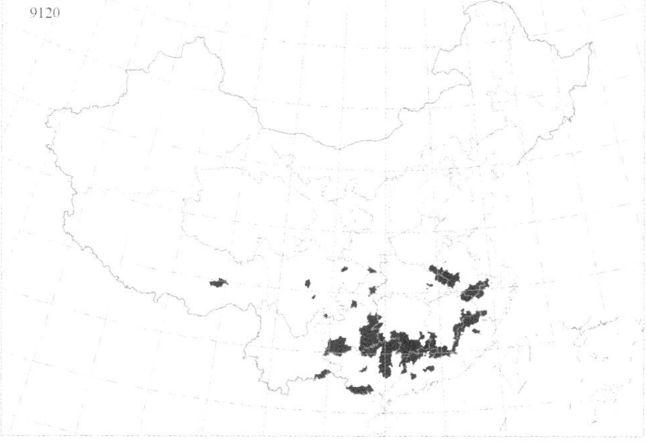

Apocynaceae 夹竹桃科 *Trachelospermum* 络石属
Trachelospermum dunnii (H. Lév.) H. Lév. 锈毛络石
Vines, evergreen

16.7 (14.2~23.0); 7.7 (3.3~15.1); 24.2 (19.4~29.3); 16.7 (14.2~23.0); 140.8 (110.5~216.2); 0.1 (0.0~2.3); 849 (724~1247) ❙ 1232 (946~1751); 597 (419~997); 77 (34~204) ❙ 45.4 (5.8~99.4); 845 (716~1247); 262.6 (148.9~551.6)

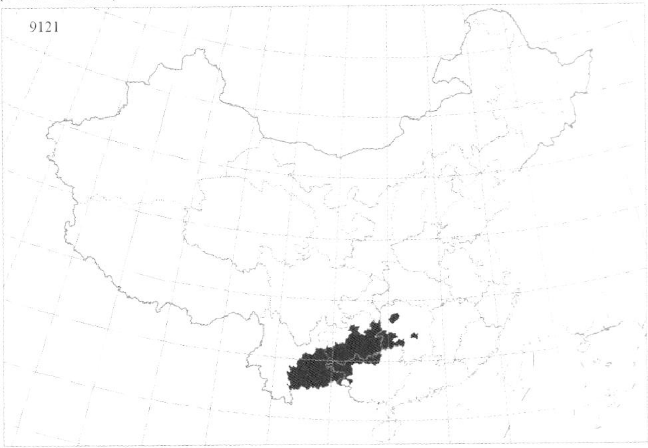

Apocynaceae 夹竹桃科 *Trachelospermum* 络石属
Trachelospermum jasminoides (Lindl.) Lem. 络石
Vines, evergreen

16.4 (2.2~24.3); 5.5 (-12.3~18.7); 26.3 (9.5~30.1); 16.5 (3.9~24.3); 140.8 (15.8~231.9); 3.2 (0.0~61.5); 891 (411~1338) ❙ 1259 (466~3495); 542 (258~1847); 114 (11~897) ❙ 40.9 (-36.5~224.4); 867 (411~1321); 220.4 (95.3~659.0)

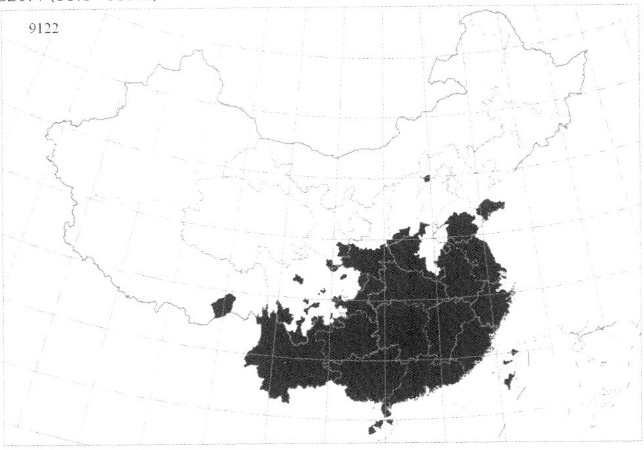

Apocynaceae 夹竹桃科 *Urceola* 水壶藤属
Urceola huaitingii (Chun et Tsiang) D. J. Middleton 毛杜仲藤
Vines

19.3 (16.1~23.6); 9.5 (5.2~17.8); 27.6 (25.5~29.4); 19.3 (16.1~23.6); 171.8 (133.1~223.3); 0.0 (0.0~0.0); 1016 (838~1271) ❙ 1547 (1242~2435); 649 (392~1428); 157 (44~211) ❙ 53.4 (-2.2~99.5); 1005 (838~1221); 231.4 (145.3~439.7)

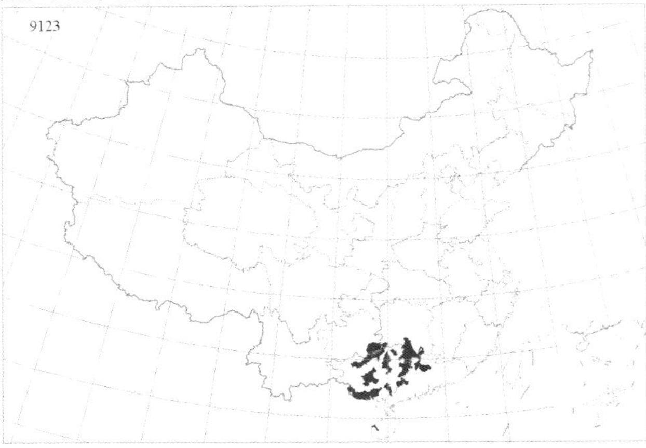

Apocynaceae 夹竹桃科 *Urceola* 水壶藤属
Urceola linearicarpa (Pierre) D. J. Middleton 线果水壶藤
Vines

18.6 (16.3~21.0); 12.3 (9.1~15.1); 22.7 (21.4~25.6); 18.6 (16.4~21.0); 163.2 (136.4~191.8); 0.0 (0.0~0.0); 894 (797~1053) ❙ 1342 (958~1751); 731 (531~996); 56 (44~77) ❙ 50.1 (13.3~88.8); 882 (797~1053); 433.5 (294.5~659.0)

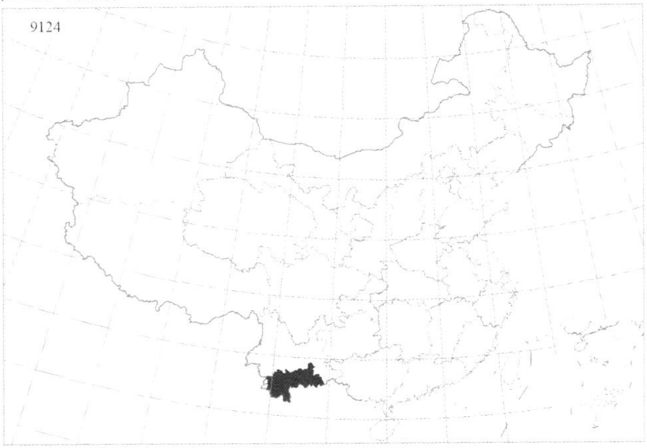

Apocynaceae 夹竹桃科 *Urceola* 水壶藤属
Urceola micrantha (Wall. ex G. Don) D. J. Middleton 杜仲藤
Vines

19.8 (11.4~24.8); 12.1 (2.7~20.1); 25.9 (18.6~29.8); 19.9 (11.4~24.8); 178.4 (80.7~237.6); 0.1 (0.0~3.7); 1032 (652~1385) ❙ 1526 (863~3495); 702 (312~1847); 112 (25~897) ❙ 50.4 (-34.3~246.5); 1013 (652~1360); 298.6 (111.7~659.0)

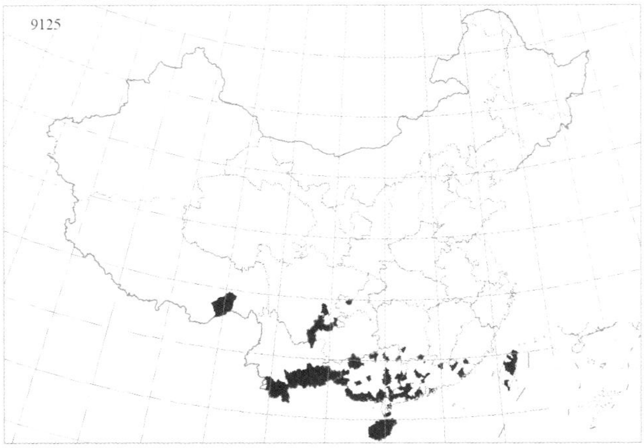

Apocynaceae 夹竹桃科 *Urceola* 水壶藤属
Urceola napeensis (Quintaret) D. J. Middleton 华南水壶藤
Vines

24.0 (22.2~24.8); 18.4 (17.2~20.1); 28.0 (25.7~28.8); 24.0 (22.3~24.8); 228.3 (207.2~237.6); 0.0 (0.0~0.0); 1310 (1127~1385) ❙ 1501 (868~1854); 534 (312~712); 74 (41~115) ❙ 15.0 (-34.3~47.8); 1191 (868~1360); 440.3 (248.2~620.0)

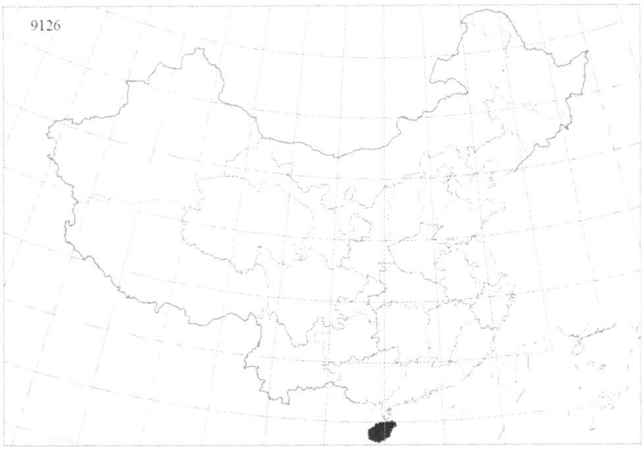

Apocynaceae 夹竹桃科 *Urceola* 水壶藤属
Urceola quintaretii (Pierre) D. J. Middleton 华南杜仲藤
Vines

21.9 (17.6~24.7); 14.2 (7.3~19.8); 27.8 (26.3~28.8); 21.9 (17.6~24.7); 202.9 (151.6~236.5); 0.0 (0.0~0.0); 1165 (909~1376) | 1633 (1179~2435); 732 (392~1428); 110 (42~194) | 41.3 (-11.9~99.5); 1129 (909~1326); 327.6 (203.4~620.0)

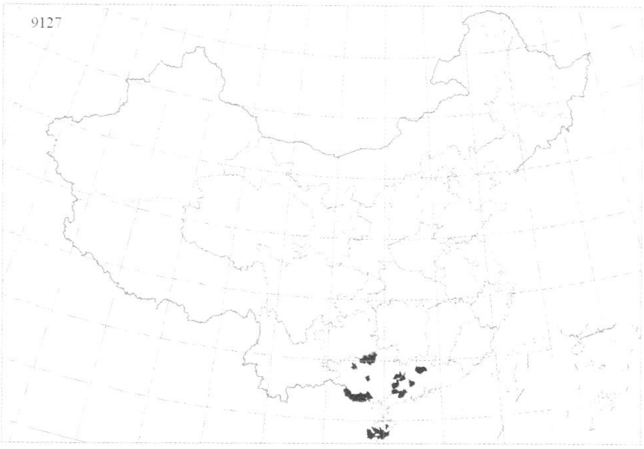

Apocynaceae 夹竹桃科 *Urceola* 水壶藤属
Urceola rosea (Hook. et Arn.) D. J. Middleton 酸叶胶藤
Vines

19.5 (14.1~24.8); 10.9 (3.3~20.1); 26.8 (20.1~29.8); 19.6 (14.2~24.8); 175.0 (111.5~237.6); 0.0 (0.0~2.3); 1019 (762~1385) | 1570 (868~3495); 667 (312~1847); 134 (41~897) | 56.1 (-34.3~246.5); 1003 (762~1360); 270.4 (108.2~659.0)

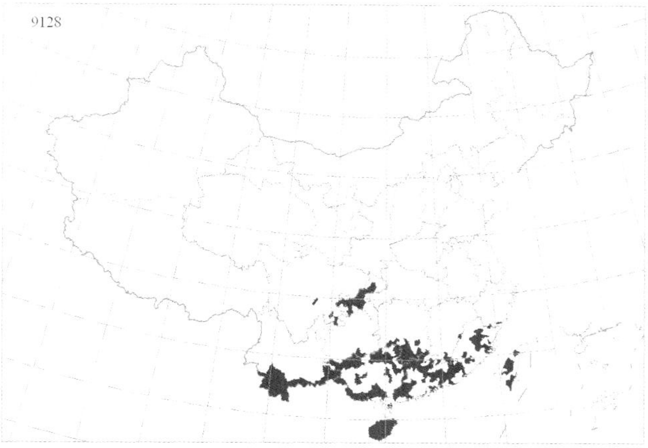

Apocynaceae 夹竹桃科 *Urceola* 水壶藤属
Urceola tournieri (Pierre) D. J. Middleton 云南水壶藤
Vines

19.3 (17.7~21.0); 13.4 (10.9~15.1); 23.0 (21.6~25.6); 19.4 (17.8~21.0); 172.3 (153.4~191.8); 0.0 (0.0~0.0); 929 (854~1053) | 1507 (1368~1751); 777 (655~996); 60 (41~77) | 62.6 (44.5~88.8); 913 (839~1053); 527.4 (439.5~659.0)

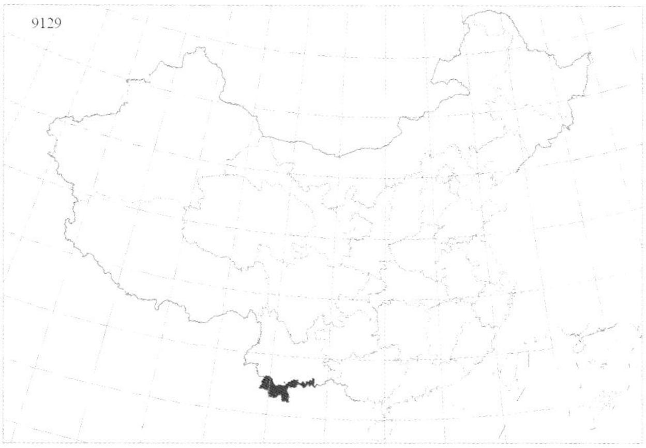

Apocynaceae 夹竹桃科 *Vallaris* 钮子花属
Vallaris indecora (Baill.) Tsiang et P. T. Li 大纽子花
Scandent shrubs

15.6 (5.4~24.8); 7.9 (-2.1~20.1); 21.8 (12.3~29.0); 15.7 (5.8~24.8); 128.8 (31.5~237.6); 0.7 (0.0~25.6); 801 (481~1385) | 1066 (741~1522); 589 (411~860); 45 (15~142) | 34.5 (-14.2~87.2); 780 (481~1137); 279.2 (135.1~551.6)

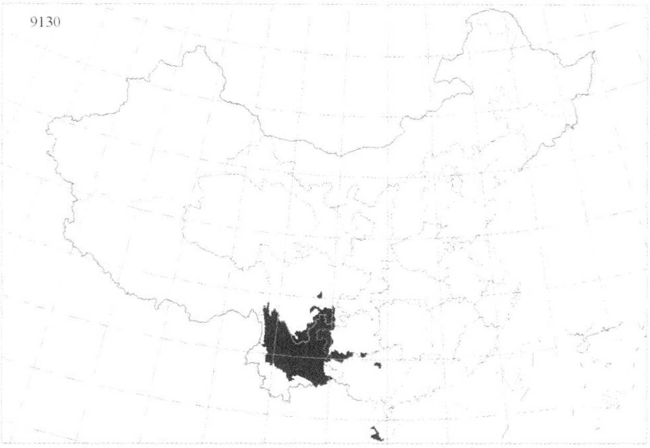

Apocynaceae 夹竹桃科 *Vallaris* 钮子花属
Vallaris solanacea (Roth) Kuntze 纽子花
Climbing shrubs

24.1; 18.8; 27.9; 24.2; 229.9; 0.0; 1322 | 1302; 441; 44 | -1.5; 1099; 469.4

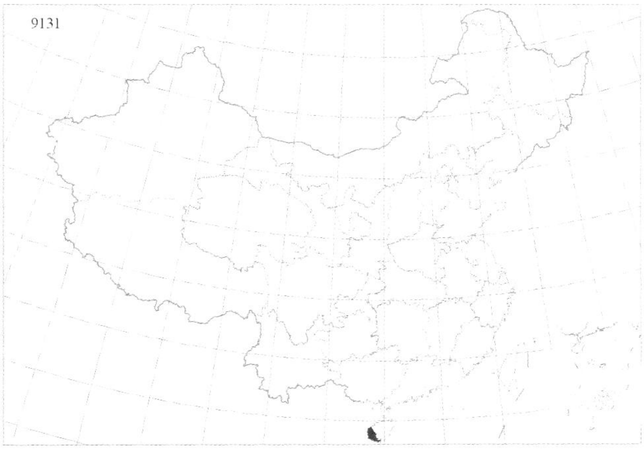

Apocynaceae 夹竹桃科 *Wrightia* 倒吊笔属
Wrightia arborea (Dennst.) Mabb. 胭木
Trees

18.5 (14.3~22.8); 10.3 (4.0~15.1); 24.7 (20.6~28.9); 18.5 (14.3~22.8); 162.1 (112.6~213.6); 0.0 (0.0~1.1); 931 (756~1230) | 1368 (1011~2435); 716 (476~1428); 73 (43~178) | 47.2 (15.4~99.5); 929 (756~1230); 331.2 (188.8~659.0)

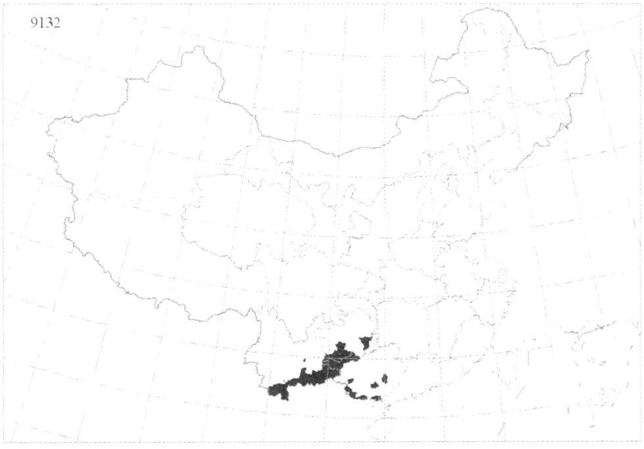

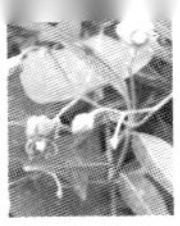

Apocynaceae 夹竹桃科 *Wrightia* 倒吊笔属
Wrightia coccinea (Lodd.) Sims 云南倒吊笔
Small trees

19.2 (16.3~22.2); 12.7 (9.1~15.1); 23.8 (21.4~28.3); 19.3 (16.4~22.3); 171.4 (136.4~207.5); 0.0 (0.0~0.0); 947 (797~1190) ❙ 1445 (1178~1751); 780 (646~996); 67 (50~106) ❙ 53.9 (15.4~88.8); 944 (797~1190); 465.3 (281.7~659.0)

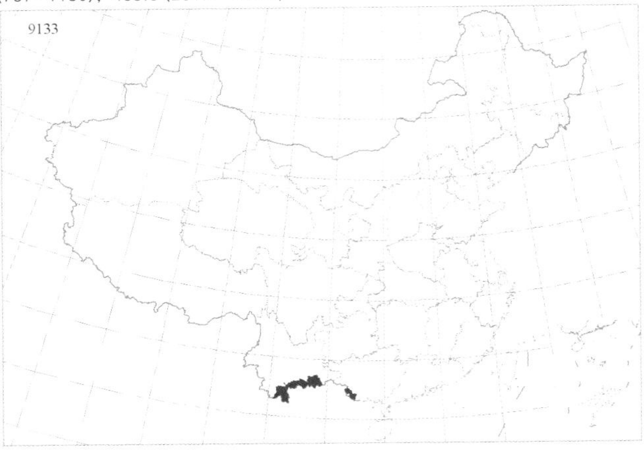

Apocynaceae 夹竹桃科 *Wrightia* 倒吊笔属
Wrightia laevis Hook. f. 蓝树
Trees

21.3 (17.2~24.8); 13.7 (7.4~20.1); 27.1 (22.3~29.8); 21.3 (17.3~24.8); 196.2 (147.0~237.6); 0.0 (0.0~0.0); 1120 (833~1385) ❙ 1577 (868~2435); 710 (312~1428); 107 (41~197) ❙ 43.0 (-34.3~99.5); 1088 (833~1360); 337.7 (157.7~659.0)

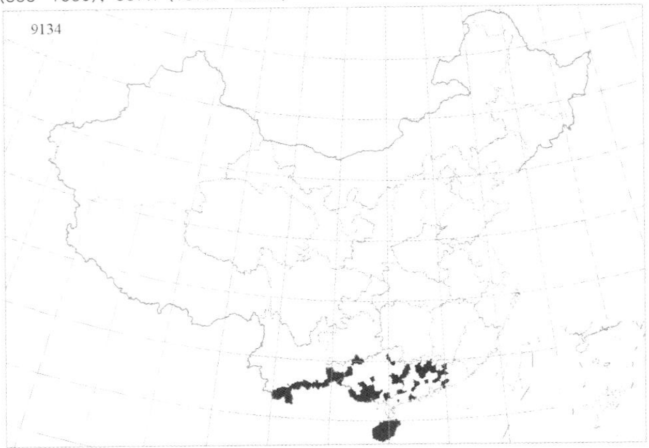

Apocynaceae 夹竹桃科 *Wrightia* 倒吊笔属
Wrightia pubescens R. Br. 倒吊笔
Trees

20.7 (15.1~24.8); 12.6 (5.9~20.1); 27.1 (20.8~29.8); 20.7 (15.2~24.8); 188.4 (122.6~237.6); 0.0 (0.0~0.0); 1082 (772~1385) ❙ 1578 (868~2435); 703 (312~1428); 118 (41~197) ❙ 47.6 (-34.3~99.5); 1068 (772~1360); 273.5 (95.3~659.0)

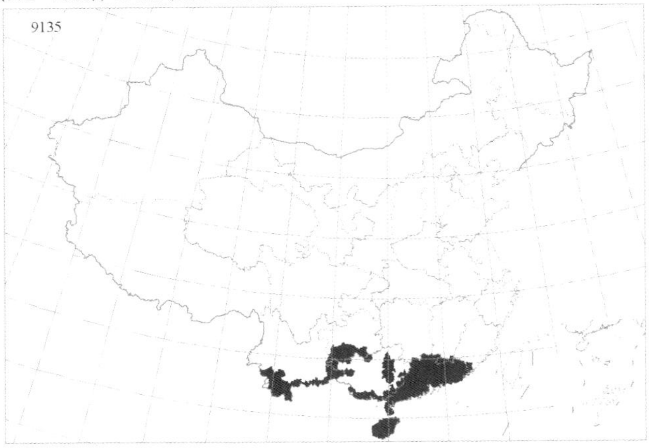

Apocynaceae 夹竹桃科 *Wrightia* 倒吊笔属
Wrightia sikkimensis Gamble 锡金倒吊笔
Trees

19.0 (13.2~22.7); 10.8 (6.4~15.1); 25.3 (18.4~29.0); 19.1 (13.3~22.8); 169.0 (99.4~213.1); 0.0 (0.0~0.0); 966 (695~1221) ❙ 1400 (958~2435); 735 (531~1428); 84 (43~187) ❙ 45.2 (5.8~99.5); 965 (695~1221); 317.8 (160.7~659.0)

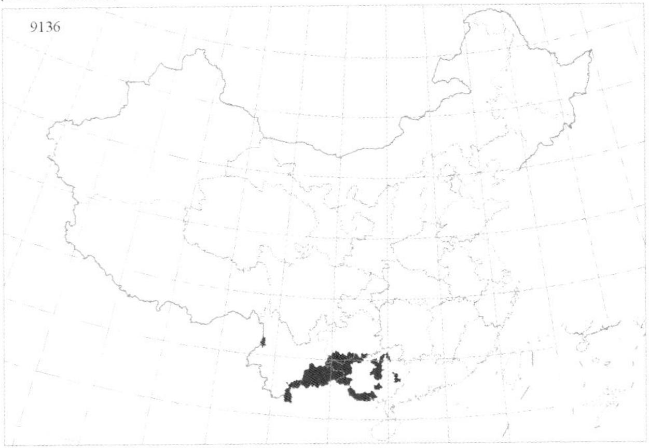

Asclepiadaceae 萝藦科 *Calotropis* 牛角瓜属
Calotropis gigantea (L.) W. T. Aiton 牛角瓜
Shrubs

16.3 (5.4~24.8); 8.8 (-2.1~20.1); 22.3 (12.3~29.2); 16.4 (5.8~24.8); 137.5 (31.5~237.6); 1.0 (0.0~25.6); 854 (481~1385) ❙ 1105 (741~1688); 592 (312~959); 45 (15~137) ❙ 32.9 (-34.3~112.2); 816 (481~1273); 301.8 (108.2~659.0)

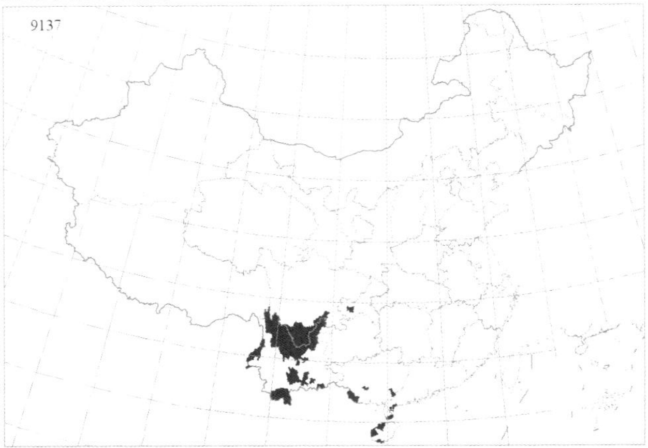

Asclepiadaceae 萝藦科 *Cryptolepis* 白叶藤属
Cryptolepis buchananii Schult. 古钩藤
Vines

17.4 (9.6~22.3); 9.8 (2.4~15.1); 23.2 (15.7~28.9); 17.4 (9.6~22.3); 149.6 (60.8~208.1); 0.2 (0.0~5.4); 878 (600~1202) ❙ 1224 (797~1900); 662 (411~1056); 60 (15~194) ❙ 39.9 (-14.2~99.9); 861 (600~1202); 326.9 (170.4~659.0)

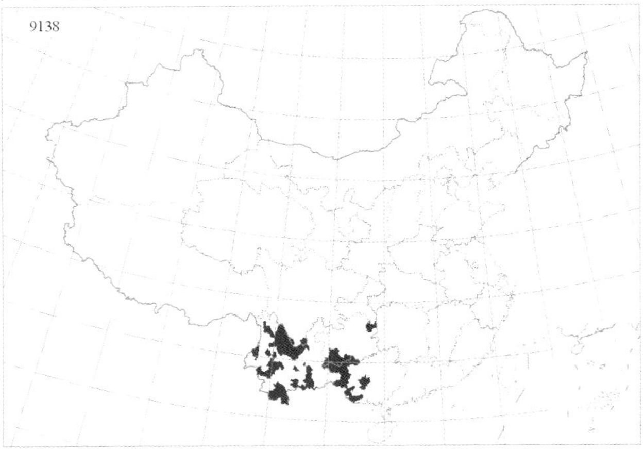

A

Asclepiadaceae 萝藦科 *Cryptolepis* 白叶藤属

Cryptolepis sinensis (Lour.) Merr. 白叶藤

Vines

19.8 (6.1~24.8); 12.1 (-1.5~20.1); 25.9 (12.7~29.8); 19.9 (6.3~24.8); 178.8 (35.2~237.6); 0.2 (0.0~21.8); 1037 (496~1385) ❙ 1521 (797~3495); 733 (392~1594); 111 (15~897) ❙ 47.4 (-14.2~224.4); 1015 (496~1331); 274.2 (111.7~620.0)

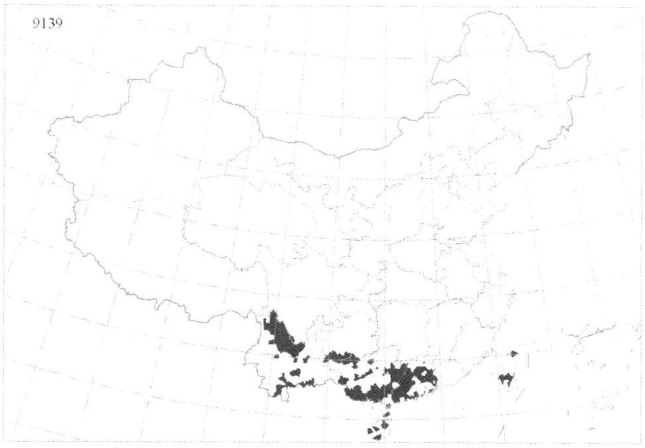

Asclepiadaceae 萝藦科 *Dregea* 南山藤属

Dregea cuneifolia Tsiang et P. T. Li 楔叶南山藤

Vines

18.5; 8.4; 26.9; 18.5; 162.5; 0.0; 959 ❙ 1476; 662; 131 ❙ 54.1; 959; 240.2

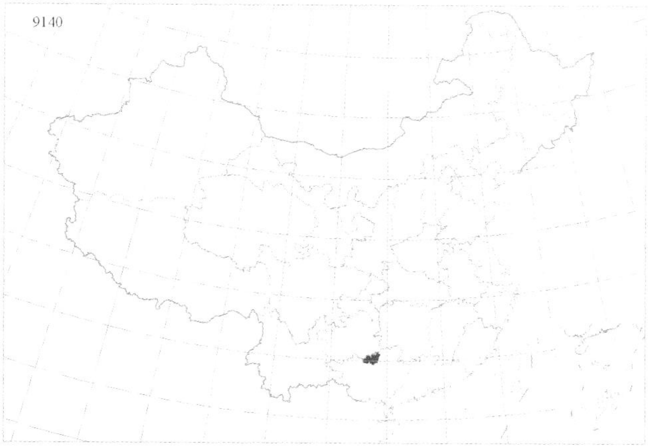

Asclepiadaceae 萝藦科 *Dregea* 南山藤属

Dregea sinensis Hemsl. 苦绳

Vines

12.9 (2.0~21.3); 1.6 (-10.8~12.6); 23.4 (9.5~28.8); 13.0 (3.8~21.3); 104.4 (15.4~196.0); 9.7 (0.0~53.6); 747 (411~1123) ❙ 1010 (377~1571); 471 (241~820); 59 (10~187) ❙ 37.7 (-37.2~128.6); 735 (377~1121); 227.0 (109.4~381.5)

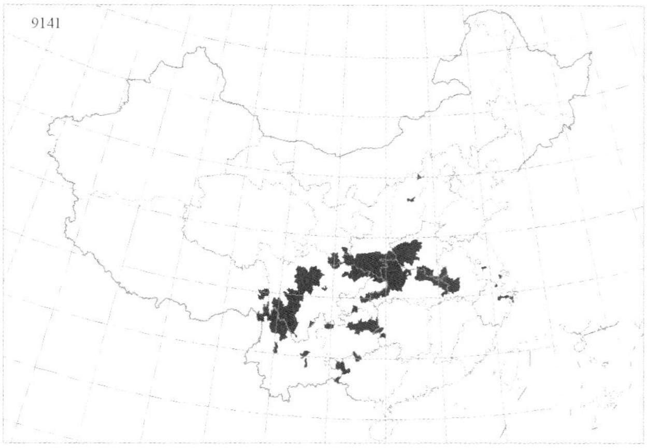

Asclepiadaceae 萝藦科 *Dregea* 南山藤属

Dregea sinensis var. corrugata (C. K. Schneid.) Tsiang et P. T. Li 贯筋藤

Vines

11.1 (2.0~16.4); 0.7 (-7.2~5.3); 20.8 (9.5~26.9); 11.5 (3.8~16.4); 87.7 (15.4~136.6); 14.2 (0.0~51.4); 679 (411~871) ❙ 1036 (579~1280); 474 (357~606); 50 (10~101) ❙ 57.4 (5.0~128.6); 679 (411~871); 221.0 (146.0~298.1)

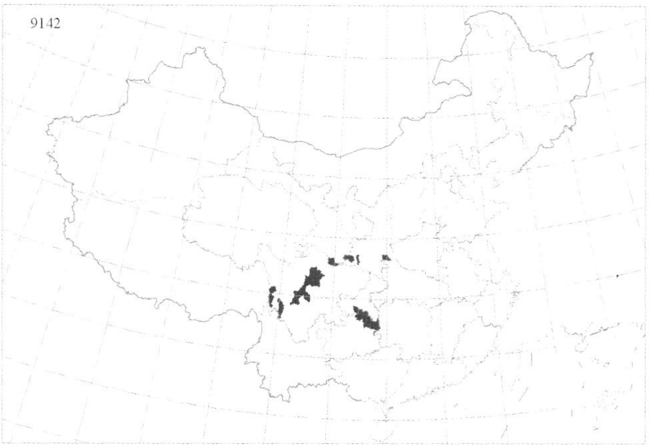

Asclepiadaceae 萝藦科 *Dregea* 南山藤属

Dregea volubilis (L. f.) Benth. ex Hook. f. 南山藤

Vines

20.5 (15.9~24.8); 12.9 (4.8~20.1); 26.4 (20.6~29.4); 20.6 (16.0~24.8); 186.7 (131.6~237.6); 0.0 (0.0~0.2); 1071 (795~1385) ❙ 1555 (868~3495); 7[illegible]6 (312~1847); 115 (41~897) ❙ 48.3 (-34.3~224.4); 1032 (795~1331); 322.3 (111.7~659.0)

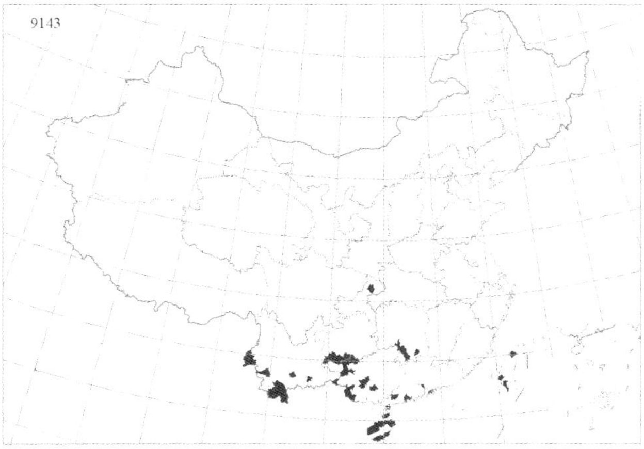

Asclepiadaceae 萝藦科 *Dregea* 南山藤属

Dregea yunnanensis (Tsiang) Tsiang et P. T. Li 丽子藤

Vines

11.6 (1.2~18.9); 3.1 (-7.1~12.0); 18.9 (9.0~26.8); 11.8 (3.3~19.0); 89.6 (11.9~167.8); 9.9 (0.0~56.5); 671 (390~929) ❙ 919 (578~1368); 519 (336~756); 30 (10~76) ❙ 40.6 (-14.2~125.9); 653 (390~912); 263.4 (130.1~491.5)

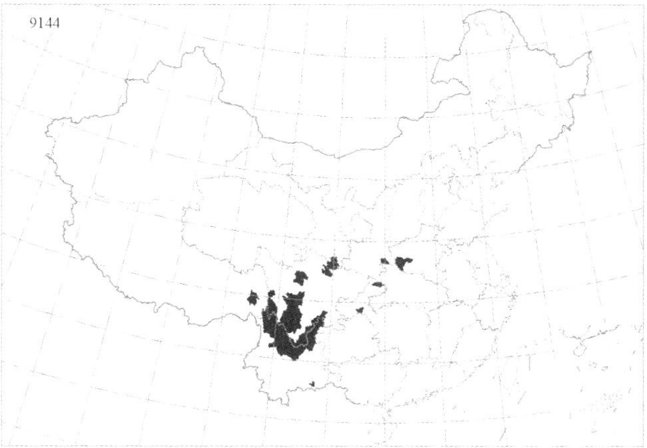

Asclepiadaceae 萝藦科 *Genianthus* 须花藤属

Genianthus bicoronatus Klack. 须花藤

Vines

19.3 (17.3~22.2); 12.9 (9.5~15.1); 23.5 (21.6~28.4); 19.4 (17.3~22.2); 172.3 (148.0~206.6); 0.0 (0.0~0.0); 944 (832~1187) ❙ 1405 (1049~1607); 721 (598~800); 65 (44~116) ❙ 49.2 (26.1~64.9); 926 (783~1187); 444.1 (273.1~659.0)

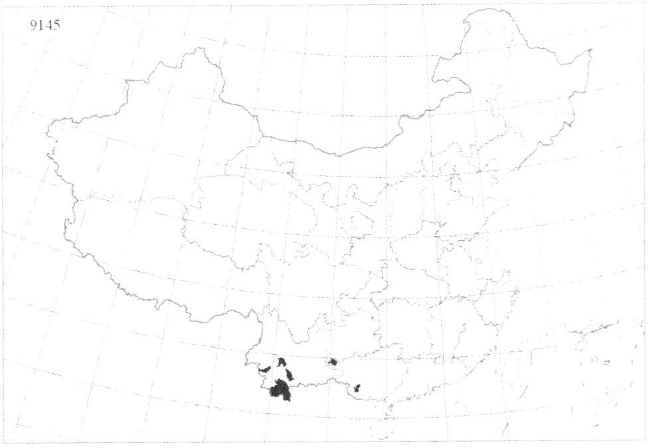

Asclepiadaceae 萝藦科 *Gongronema* 纤冠藤属

Gongronema multibracteolatum P. T. Li et X. M. Wang 多苞纤冠藤

Vines

17.4; 7.4; 25.5; 17.4; 148.9; 0.0; 886 ❙ 1381; 637; 107 ❙ 55.9; 886; 245.3

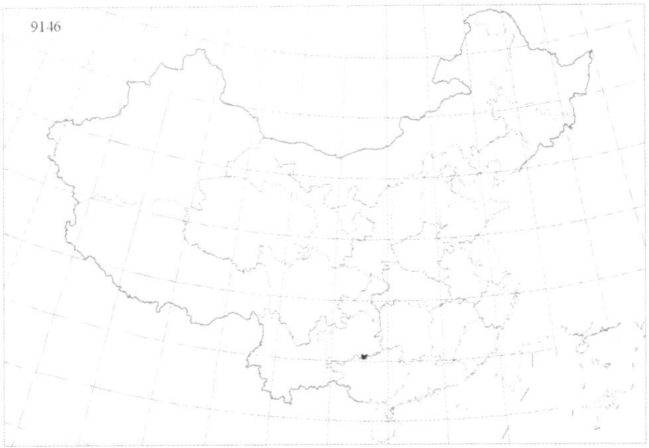

Asclepiadaceae 萝藦科 *Gongronema* 纤冠藤属

Gongronema napalense (Wall.) Decne. 纤冠藤

Vines

20.5 (13.3~24.8); 13.0 (5.3~20.1); 26.3 (19.6~29.2); 20.6 (13.4~24.8); 186.7 (100.4~237.6); 0.0 (0.0~0.0); 1075 (706~1385) ❙ 1439 (868~1876); 657 (312~1007); 85 (41~194) ❙ 36.9 (-34.3~145.8); 1038 (706~1360); 329.6 (142.9~620.0)

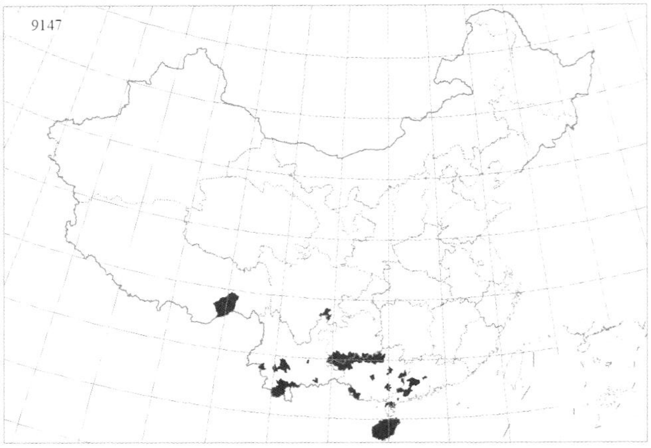

Asclepiadaceae 萝藦科 *Goniostemma* 勐腊藤属

Goniostemma punctatum Tsiang et P. T. Li 勐腊藤

Vines

20.5; 15.0; 23.7; 20.5; 185.9; 0.0; 990 ❙ 1536; 678; 72 ❙ 55.1; 973; 614.6

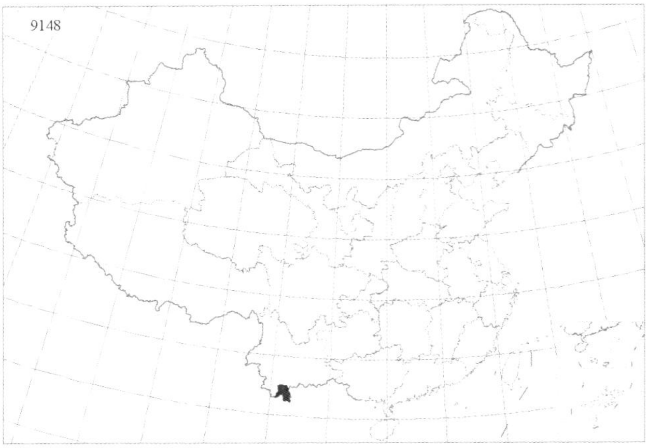

Asclepiadaceae 萝藦科 *Graphistemma* 天星藤属

Graphistemma pictum (Champ. ex Benth.) Benth. et Hook. f. ex Maxim. 天星藤

Vines

22.6 (19.1~24.8); 15.3 (9.8~20.1); 28.1 (25.7~29.2); 22.6 (19.2~24.8); 211.5 (170.1~237.6); 0.0 (0.0~0.0); 1211 (981~1385) ❙ 1586 (868~2435); 709 (312~1428); 101 (41~154) ❙ 31.8 (-34.3~99.5); 1171 (868~1360); 306.7 (139.6~620.0)

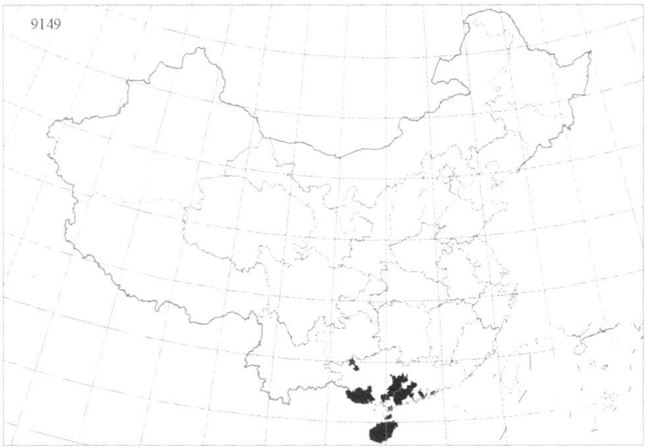

Asclepiadaceae 萝藦科 *Gymnanthera* 海岛藤属

Gymnanthera oblonga (Burm. f.) P. S. Green 海岛藤

Vines

22.7 (20.5~24.8); 15.7 (12.6~20.1); 28.2 (27.0~29.2); 22.8 (20.5~24.8); 213.2 (186.2~237.6); 0.0 (0.0~0.0); 1219 (1051~1385) ❙ 1633 (868~2435); 741 (312~1428); 97 (41~152) ❙ 35.5 (-34.3~99.5); 1156 (868~1273); 270.1 (139.6~538.2)

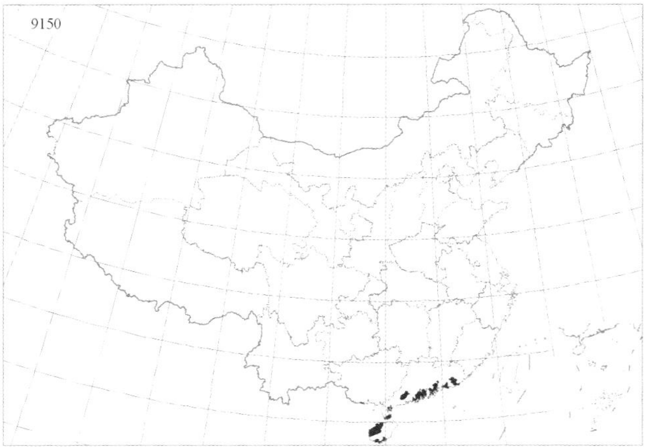

Asclepiadaceae 萝藦科 *Gymnema* 匙羹藤属
Gymnema foetidum Tsiang 华宁藤
Vines

15.3 (10.1~20.4); 7.7 (2.3~15.1); 21.4 (16.9~26.6); 15.4 (10.2~20.5); 125.1 (68.1~185.8); 0.6 (0.0~5.2); 783 (618~989) | 1031 (797~1607); 579 (411~835); 37 (21~77) | 32.1 (-14.2~70.8); 753 (618~986); 291.2 (190.3~659.0)

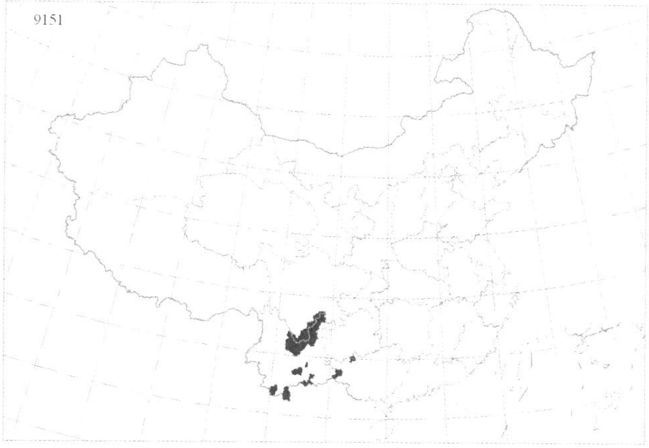

Asclepiadaceae 萝藦科 *Gymnema* 匙羹藤属
Gymnema hainanense Tsiang 海南匙羹藤
Vines

24.3; 18.7; 28.4; 24.3; 231.9; 0.0; 1338 | 1179; 392; 42 | -11.9; 1056; 455.2

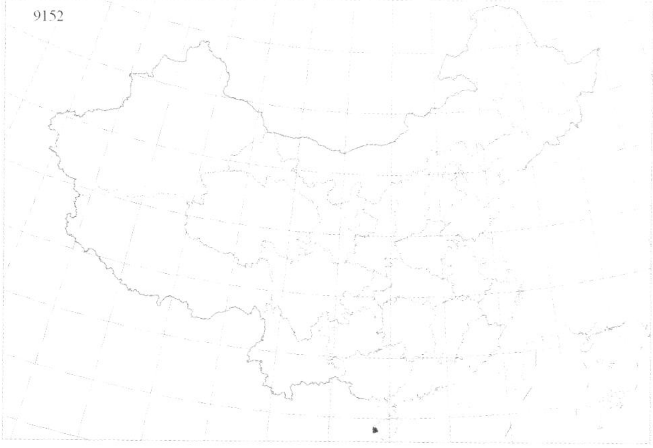

Asclepiadaceae 萝藦科 *Gymnema* 匙羹藤属
Gymnema inodorum (Lour.) Decne. 广东匙羹藤
Vines

21.2 (16.8~24.8); 13.3 (7.4~20.1); 27.4 (22.3~29.8); 21.3 (16.8~24.8); 195.1 (141.5~237.6); 0.0 (0.0~0.0); 1120 (836~1385) | 1555 (868~2435); 699 (312~1428); 110 (41~189) | 40.4 (-34.3~99.5); 1091 (836~1360); 303.7 (95.3~659.0)

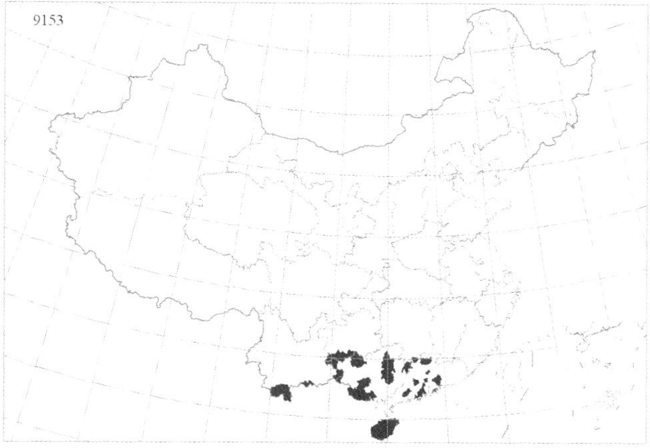

Asclepiadaceae 萝藦科 *Gymnema* 匙羹藤属
Gymnema latifolium Wall. ex Wight 宽叶匙羹藤
Vines

20.1 (16.3~22.2); 12.5 (8.5~15.1); 25.7 (20.3~28.5); 20.2 (16.4~22.3); 182.0 (136.6~207.5); 0.0 (0.0~0.0); 1026 (789~1190) | 1494 (1188~2148); 752 (599~1234); 90 (47~169) | 46.7 (5.8~86.7); 1021 (780~1190); 358.7 (209.2~659.0)

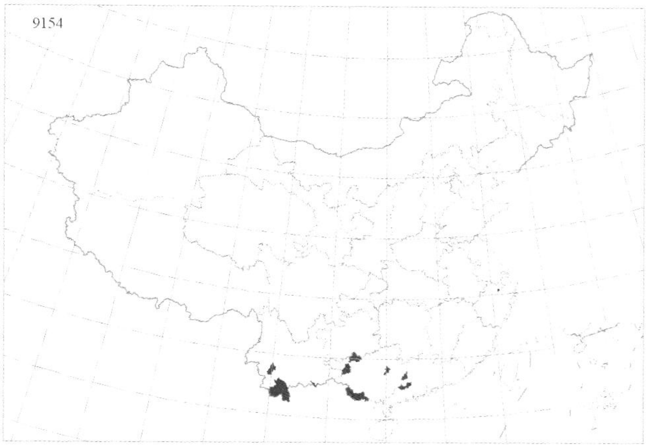

Asclepiadaceae 萝藦科 *Gymnema* 匙羹藤属
Gymnema longiretinaculatum Tsiang 会东藤
Scandent shrubs

16.4 (12.7~19.1); 8.1 (4.7~12.0); 22.8 (19.2~26.0); 16.4 (12.7~19.1); 136.9 (93.0~169.2); 0.0 (0.0~0.3); 826 (687~966) | 1125 (797~1381); 621 (411~720); 46 (18~107) | 36.5 (-14.2~64.7); 801 (687~966); 250.2 (189.5~330.0)

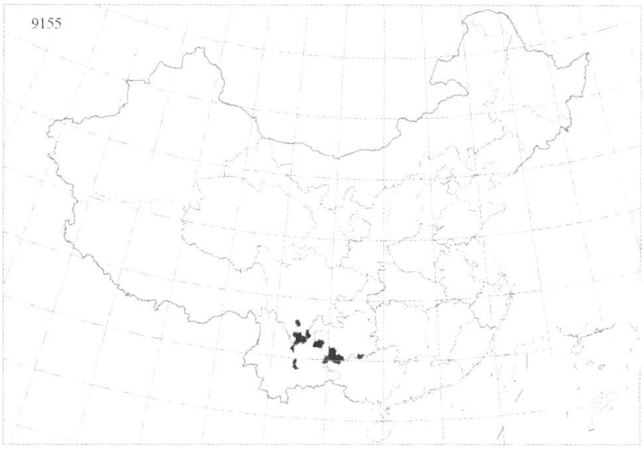

Asclepiadaceae 萝藦科 *Gymnema* 匙羹藤属
Gymnema sylvestre (Retz.) Schult. 匙羹藤
Vines

20.4 (15.3~24.8); 12.0 (5.4~20.1); 27.6 (21.6~29.8); 20.5 (15.4~24.8); 185.7 (124.4~237.6); 0.0 (0.0~0.0); 1073 (791~1385) | 1638 (868~3495); 662 (312~1428); 158 (41~897) | 55.5 (-34.3~224.4); 1059 (783~1360); 259.5 (95.3~659.0)

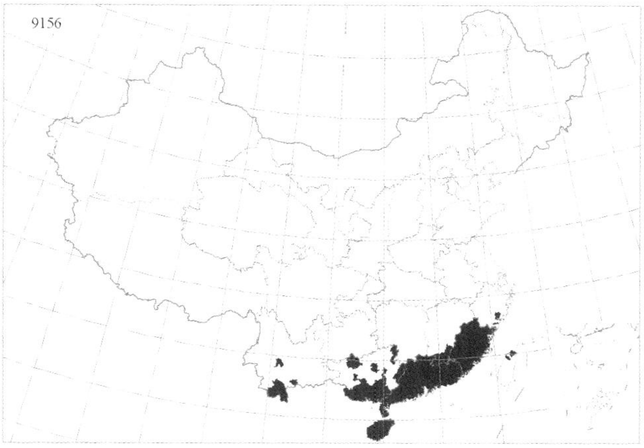

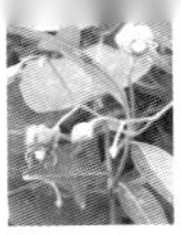

Asclepiadaceae 萝藦科 *Gymnema* 匙羹藤属

Gymnema yunnanense Tsiang 云南匙羹藤

Vines

17.7 (14.2~22.2); 11.3 (7.3~15.1); 22.1 (19.4~28.3); 17.8 (14.2~22.3); 153.3 (110.6~207.5); 0.0 (0.0~0.0); 870 (724~1190) ❙ 1250 (991~1607); 671 (562~816); 55 (38~87) ❙ 43.9 (15.4~68.7); 856 (724~1190); 414.5 (282.6~659.0)

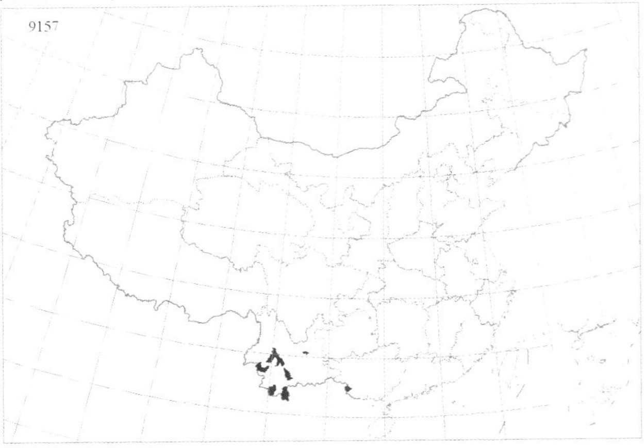

Asclepiadaceae 萝藦科 *Heterostemma* 醉魂藤属

Heterostemma brownii Hayata 台湾醉魂藤

Vines

17.9; 11.6; 23.4; 18.0; 155.6; 0.0; 864 ❙ 2355; 900; 308 ❙ 172.5; 864; 284.4

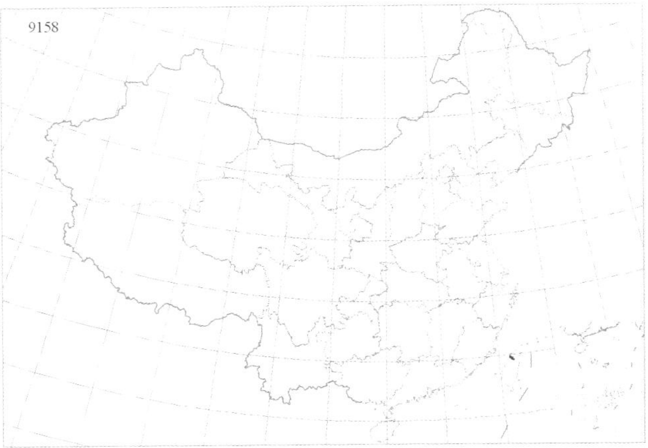

Asclepiadaceae 萝藦科 *Heterostemma* 醉魂藤属

Heterostemma esquirolii (H. Lév.) Tsiang 贵州醉魂藤

Vines

18.5 (15.9~22.2); 9.9 (6.1~15.1); 25.2 (21.4~28.3); 18.5 (16.0~22.3); 162.0 (131.6~207.5); 0.0 (0.0~0.0); 934 (797~1190) ❙ 1302 (985~1607); 661 (465~723); 66 (50~117) ❙ 40.3 (5.8~64.9); 931 (797~1190); 299.6 (135.3~659.0)

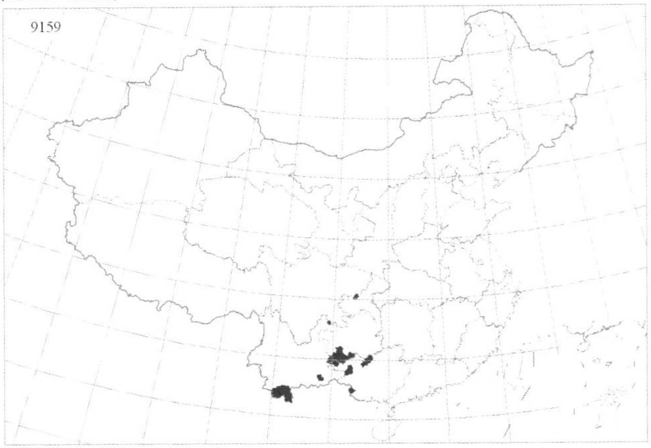

Asclepiadaceae 萝藦科 *Heterostemma* 醉魂藤属

Heterostemma grandiflorum Costantin 大花醉魂藤

Vines

18.4 (7.3~24.7); 10.7 (-2.0~19.8); 24.5 (15.6~29.2); 18.5 (7.5~24.7); 163.0 (48.2~236.5); 1.3 (0.0~20.4); 958 (537~1376) ❙ 1483 (1001~2035); 693 (531~969); 87 (27~191) ❙ 58.3 (11.9~87.3); 938 (537~1273); 363.4 (178.7~659.0)

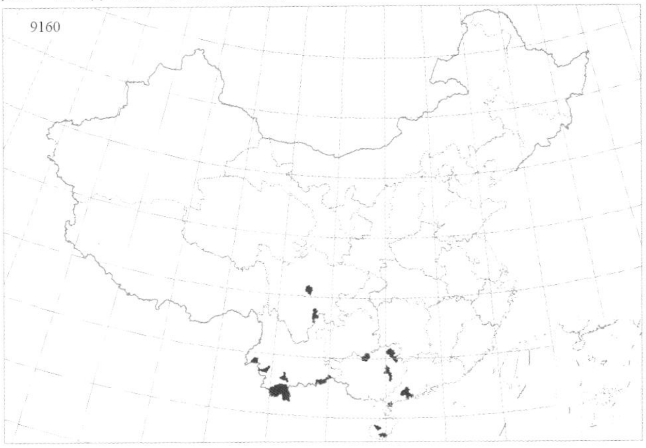

Asclepiadaceae 萝藦科 *Heterostemma* 醉魂藤属

Heterostemma menghaiense (H. Zhu et H. Wang) M. G. Gilbert et P. T. Li 勐海醉魂藤

Vines

18.0 (16.1~20.4); 12.0 (9.6~15.1); 21.9 (20.6~23.6); 18.1 (16.2~20.5); 157.0 (133.9~185.8); 0.0 (0.0~0.0); 867 (785~989) ❙ 1297 (1049~1607); 675 (598~816); 54 (44~77) ❙ 48.9 (26.1~68.7); 846 (779~986); 456.8 (363.2~659.0)

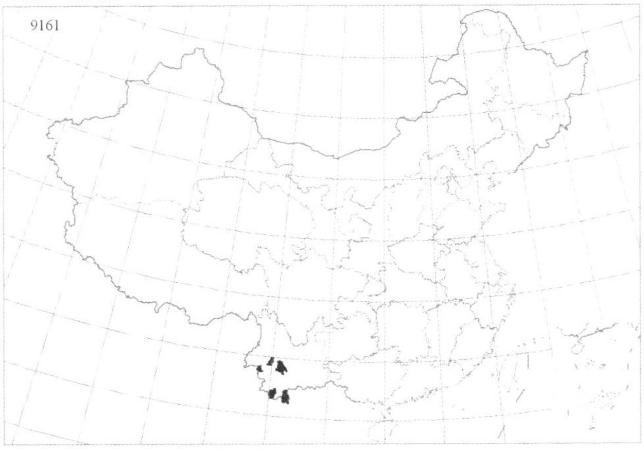

Asclepiadaceae 萝藦科 *Heterostemma* 醉魂藤属

Heterostemma oblongifolium Costantin 催乳藤

Vines

21.6 (17.8~24.8); 14.6 (7.1~20.1); 26.7 (22.0~28.8); 21.6 (17.8~24.8); 199.5 (154.0~237.6); 0.0 (0.0~0.0); 1133 (868~1385) ❙ 1522 (1176~2148); 658 (444~1234); 90 (41~183) ❙ 37.3 (-12.0~86.7); 1072 (847~1212); 400.3 (157.7~659.0)

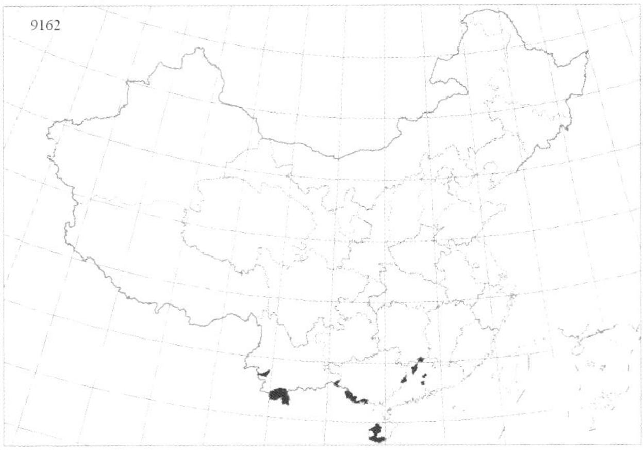

Asclepiadaceae 萝藦科 *Heterostemma* 醉魂藤属
Heterostemma s amicum Craib 心叶醉魂藤
Vines

19.5 (16.1~22.2); 13.2 (9.6~15.1); 23.7 (20.6~28.3); 19.6 (16.2~22.3); 174.6 (133.9~207.5); 0.0 (0.0~0.0); 966 (785~1190) ❙ 1397 (1083~1607); 70 (600~836); 68 (50~87) ❙ 46.1 (15.4~66.9); 956 (779~1190); 471.4 (327.7~659.0)

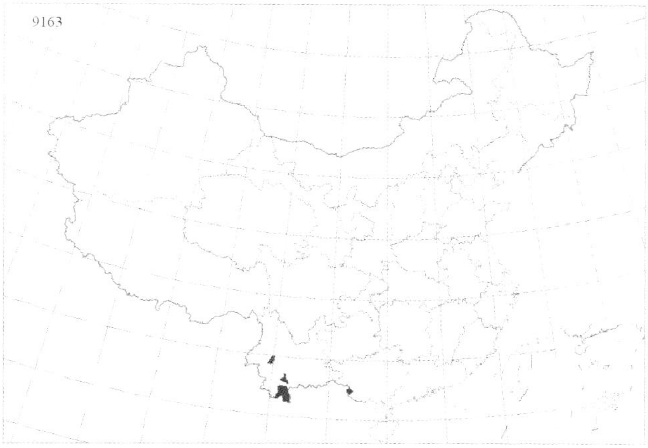

Asclepiadaceae 萝藦科 *Heterostemma* 醉魂藤属
Heterostemma sinicum Tsiang 海南醉魂藤
Vines

24.0; 19.1; 27.2; 24.0; 228.1; 0.0; 1302 ❙ 1551; 539; 48 ❙ 19.1; 1147; 620.0

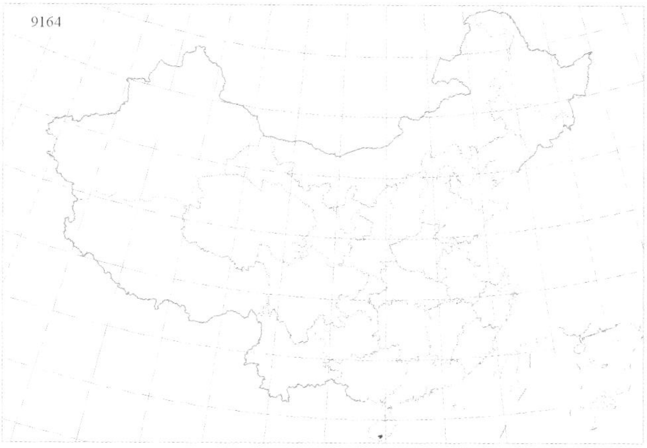

Asclepiadaceae 萝藦科 *Heterostemma* 醉魂藤属
Heterostemma tsoongii Tsiang 广西醉魂藤
Vines

19.4 (17.0~22.2); 10.0 (6.2~13.9); 27.2 (25.0~28.5); 19.5 (17.0~22.3); 173.8 (144.4~207.5); 0.0 (0.0~0.0); 1013 (891~1190) ❙ 1547 (1292~1758); 690 (577~832); 140 (49~191) ❙ 54.0 (15.4~85.5); 1013 (891~1190); 258.1 (208.7~342.5)

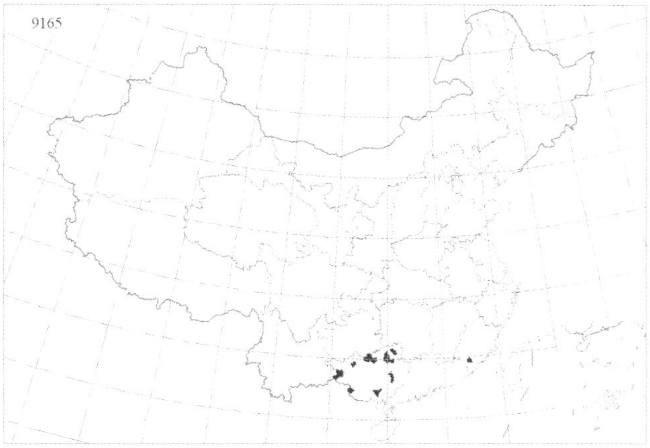

Asclepiadaceae 萝藦科 *Heterostemma* 醉魂藤属
Heterostemma wallichii Wight 云南醉魂藤
Vines

18.4 (13.8~20.5); 12.4 (6.7~15.1); 22.4 (19.3~23.7); 18.5 (13.8~20.5); 161.6 (105.9~185.9); 0.0 (0.0~0.0); 895 (714~992) ❙ 1396 (963~1607); 746 (592~911); 56 (29~77) ❙ 55.4 (35.0~74.2); 881 (690~986); 511.6 (266.2~659.0)

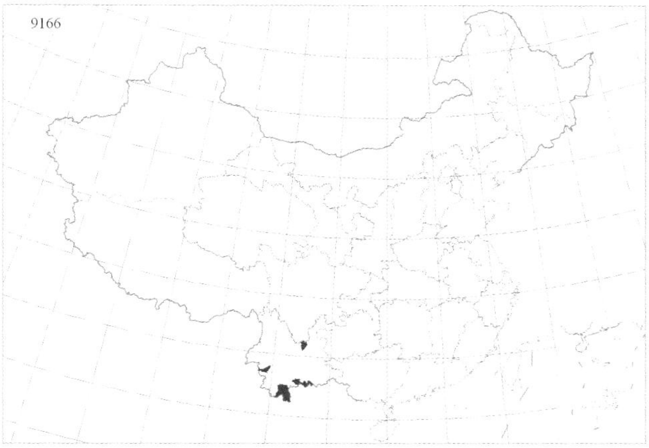

Asclepiadaceae 萝藦科 *Jasminanthes* 黑鳗藤属
Jasminanthes chunii (Tsiang) W. D. Stevens et P. T. Li 假木藤
Vines

19.1 (14.6~22.7); 9.7 (3.8~15.0); 27.1 (24.6~28.8); 19.2 (14.7~22.8); 170.4 (117.7~213.1); 0.2 (0.0~1.5); 1010 (779~1221) ❙ 1595 (1234~2435); 715 (450~1428); 135 (86~191) ❙ 59.0 (12.9~101.8); 1010 (779~1221); 239.9 (186.2~318.8)

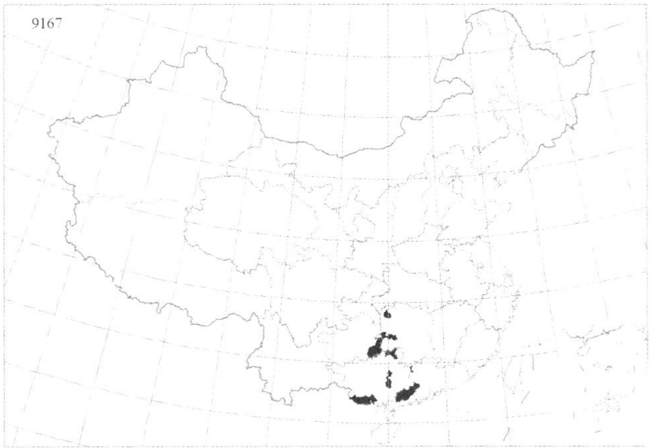

Asclepiadaceae 萝藦科 *Jasminanthes* 黑鳗藤属
Jasminanthes mucronata (Blanco) W. D. Stevens et P. T. Li 黑鳗藤
Scandent shrubs

17.7 (14.3~23.8); 7.3 (3.3~18.1); 27.3 (20.1~29.8); 17.7 (14.3~23.8); 153.1 (113.2~225.8); 0.2 (0.0~2.3); 937 (761~1297) ❙ 1593 (1007~3495); 593 (392~1357); 179 (49~897) ❙ 70.9 (9.0~246.5); 927 (761~1273); 215.5 (111.7~305.7)

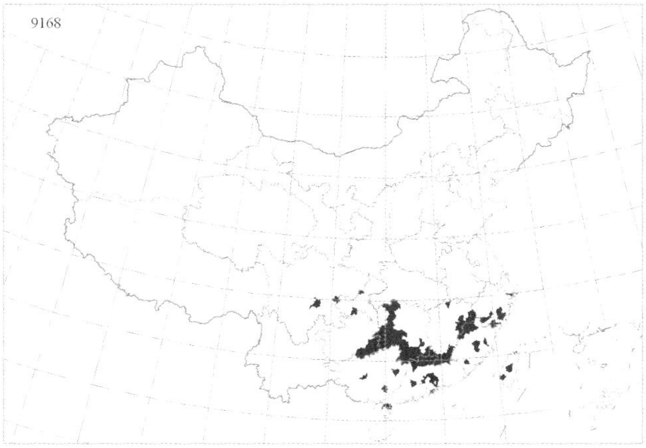

Asclepiadaceae 萝藦科 *Jasminanthes* 黑鳗藤属
Jasminanthes pilosa (Kerr) W. D. Stevens et P. T. Li 茶药藤
Scandent shrubs

21.0 (18.2~22.7); 13.6 (12.4~15.0); 26.4 (21.6~28.6); 21.1 (18.2~22.8); 193.0 (158.6~213.1); 0.0 (0.0~0.0); 1090 (861~1221) ❙ 1788 (1302~2435); 997 (670~1428); 101 (52~130) ❙ 64.4 (12.9~99.5); 1087 (839~1221); 341.0 (235.1~551.3)

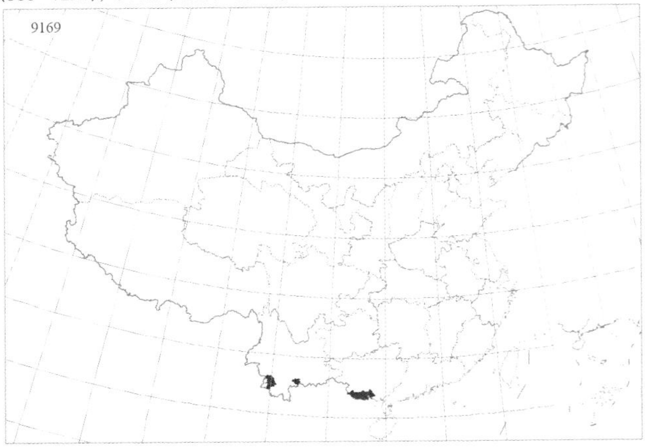

Asclepiadaceae 萝藦科 *Jasminanthes* 黑鳗藤属
Jasminanthes saxatilis (Tsiang et P. T. Li) W. D. Stevens et P. T. Li 云南黑鳗藤
Vines

19.0; 10.8; 25.0; 19.0; 168.6; 0.0; 942 ❙ 1365; 775; 53 ❙ 44.9; 942; 357.1

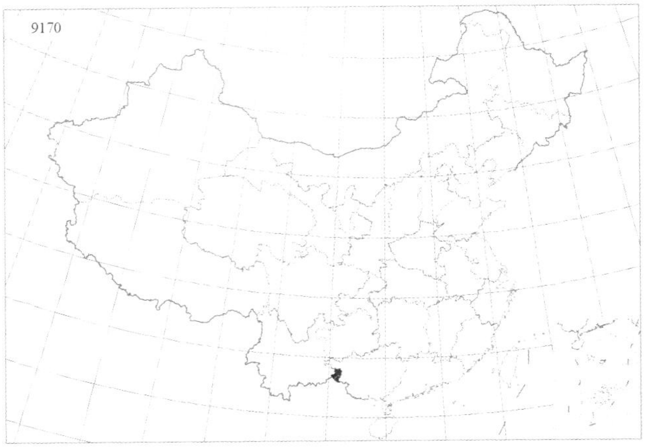

Asclepiadaceae 萝藦科 *Marsdenia* 牛奶菜属
Marsdenia griffithii Hook. f. 白药牛奶菜
Vines

18.4 (16.4~20.5); 11.2 (6.9~14.9); 23.4 (20.6~27.5); 18.5 (16.4~20.5); 161.7 (137.3~185.9); 0.0 (0.0~0.0); 903 (795~992) ❙ 1520 (1292~1688); 782 (505~959); 75 (42~180) ❙ 69.5 (37.6~112.2); 893 (795~959); 455.1 (188.8~627.0)

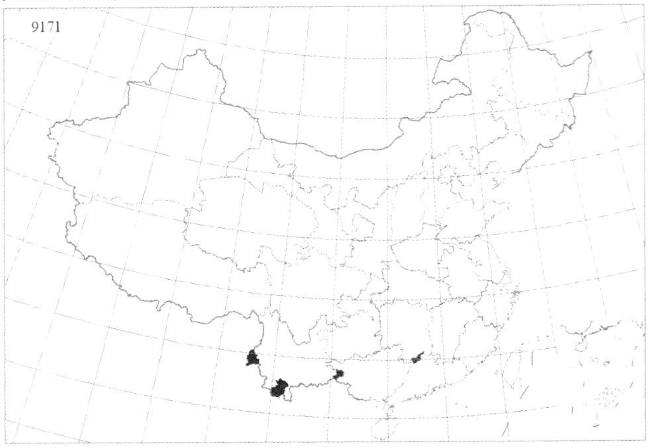

Asclepiadaceae 萝藦科 *Marsdenia* 牛奶菜属
Marsdenia koi Tsiang 大叶牛奶菜
Vines

18.4 (1.8~22.2); 10.6 (−7.1~14.7); 24.4 (9.9~28.5); 18.5 (3.8~22.3); 162.7 (15.7~207.5); 1.6 (0.0~53.3); 949 (404~1190) ❙ 1366 (583~1876); 699 (348~911); 78 (13~138) ❙ 45.5 (6.5~93.0); 944 (404~1190); 290.7 (169.5~551.3)

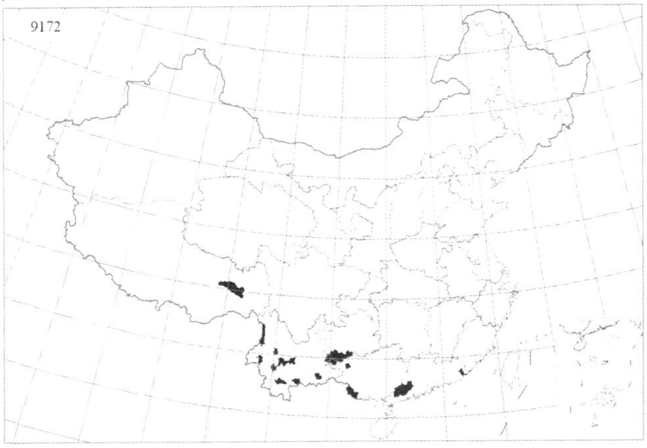

Asclepiadaceae 萝藦科 *Marsdenia* 牛奶菜属
Marsdenia medogensis P. T. Li 墨脱牛奶菜
Vines

13.3; 5.4; 19.6; 13.4; 100.4; 0.0; 706 ❙ 1736; 1007; 51 ❙ 145.8; 706; 447.4

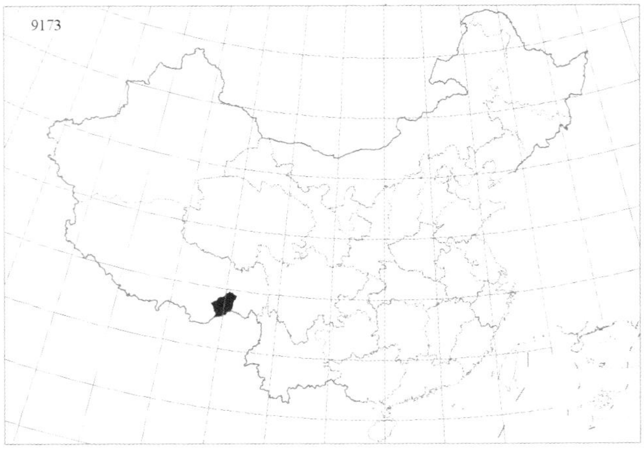

Asclepiadaceae 萝藦科 *Marsdenia* 牛奶菜属
Marsdenia officinalis Tsiang et P. T. Li 海枫藤
Vines

15.0 (7.3~23.0); 5.5 (−2.0~15.1); 23.6 (15.6~29.0); 15.0 (7.5~23.0); 122.7 (48.2~216.2); 2.5 (0.0~20.4); 812 (537~1247) ❙ 1375 (956~1893); 627 (514~997); 104 (26~227) ❙ 70.9 (27.9~129.9); 809 (537~1247); 244.8 (185.4~282.6)

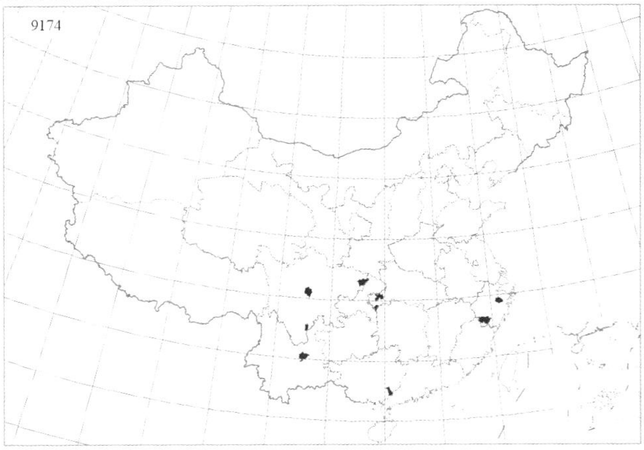

Asclepiadaceae 萝藦科 *Marsdenia* 牛奶菜属
Marsdenia tenacissima (Roxb.) Moon 通光散
Vines

17.9 (13.8~21.3); 10.1 (4.8~15.1); 23.8 (19.3~27.9); 18.0 (13.8~21.3); 155.4 (105.9~196.0); 0.0 (0.0~0.2); 892 (714~1123) | 1298 (956~1607); 707 (520~932); 55 (20~95) | 46.0 (5.5~80.6); 879 (690~1121); 334.7 (152.1~659.0)

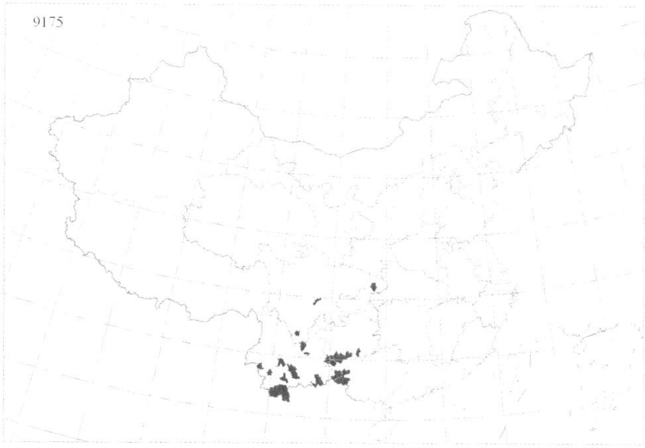

Asclepiadaceae 萝藦科 *Merrillanthus* 驼峰藤属
Merrillanthus hainanensis Chun et Tsiang 驼峰藤
Vines

22.8 (21.0~24.8); 16.2 (12.2~20.1); 27.9 (25.7~28.9); 22.9 (21.0~24.8); 214.6 (192.3~237.6); 0.0 (0.0~0.0); 1226 (1081~1385) | 1568 (1179~1876); 624 (392~891); 90 (41~142) | 29.1 (-11.9~64.3); 1151 (1055~1360); 376.8 (197.7~620.0)

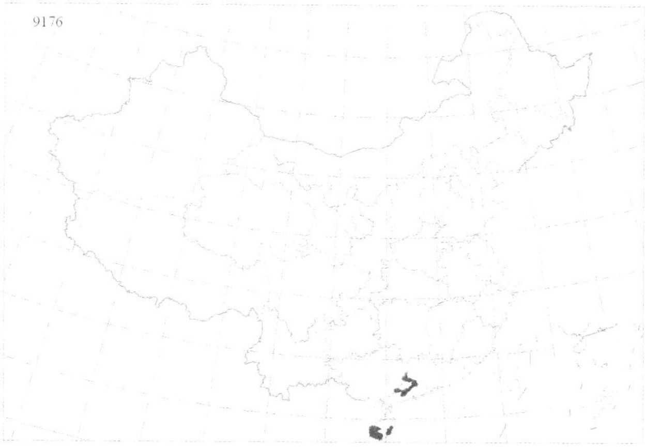

Asclepiadaceae 萝藦科 *Myriopteron* 翅果藤属
Myriopteron extensum (Wight et Arn.) K. Schum. 翅果藤
Vines

18.1 (15.1~21.3); 10.4 (6.2~15.1); 23.8 (20.3~27.9); 18.1 (15.2~21.3); 157.8 (122.6~196.0); 0.0 (0.0~0.0); 898 (767~1123) | 1283 (958~1607); 688 (531~911); 58 (43~107) | 43.6 (5.8~74.2); 892 (767~1121); 352.3 (189.5~659.0)

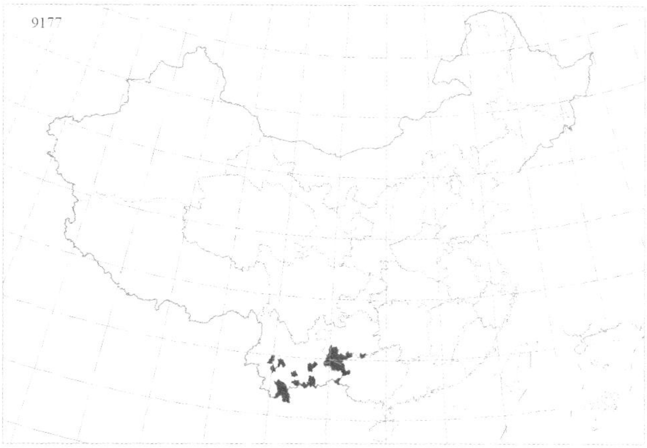

Asclepiadaceae 萝藦科 *Pentastelma* 白水藤属
Pentastelma auritum Tsiang et P. T. Li 白水藤
Vines

23.4; 18.1; 27.0; 23.4; 221.2; 0.0; 1250 | 1588; 564; 56 | 27.2; 1151; 577.1

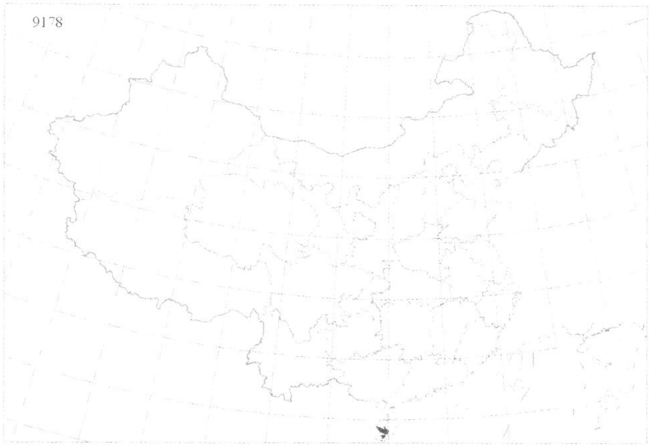

Asclepiadaceae 萝藦科 *Periploca* 杠柳属
Periploca calophylla (Wight) Falc. 青蛇藤
Scandent shrubs

14.2 (1.8~21.0); 4.9 (-7.1~14.2); 22.3 (9.9~28.1); 14.3 (3.8~21.0); 114.8 (15.7~191.8); 4.3 (0.0~53.3); 765 (404~1065) | 1109 (583~1736); 561 (348~1007); 52 (13~141) | 47.0 (-14.2~145.8); 756 (404~1065); 253.1 (118.7~551.6)

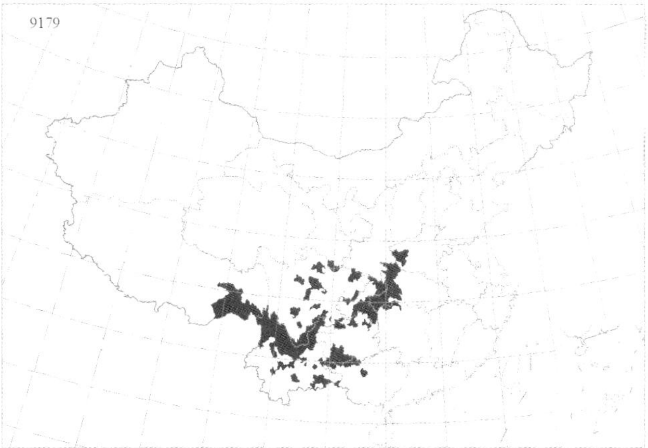

Asclepiadaceae 萝藦科 *Periploca* 杠柳属
Periploca floribunda Tsiang 多花青蛇藤
Scandent shrubs

14.0 (8.6~17.7); 7.2 (1.3~11.0); 19.1 (15.0~22.5); 14.0 (8.7~17.8); 109.2 (52.6~153.4); 1.1 (0.0~8.8); 724 (574~854) | 1178 (1019~1454); 653 (562~792); 55 (44~70) | 64.3 (26.1~99.9); 718 (574~854); 402.7 (352.3~491.5)

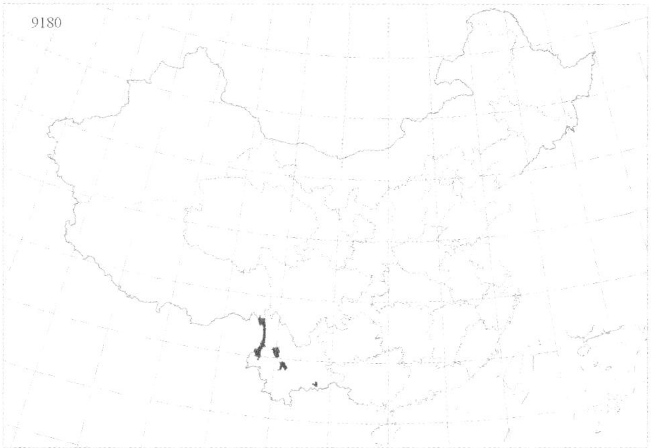

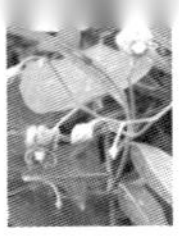

Asclepiadaceae 萝藦科 *Periploca* 杠柳属
Periploca forrestii Schltr. 黑龙骨
Scandent shrubs

14.6 (−4.1~21.3); 5.4 (−17.3~15.1); 22.6 (6.2~27.9); 14.8 (2.0~21.3); 119.8 (2.5~196.0); 3.6 (0.0~114.2); 780 (334~1123) ❙ 1130 (321~1736); 558 (193~1007); 59 (5~153) ❙ 45.7 (−19.1~145.8); 771 (321~1121); 249.3 (79.3~659.0)

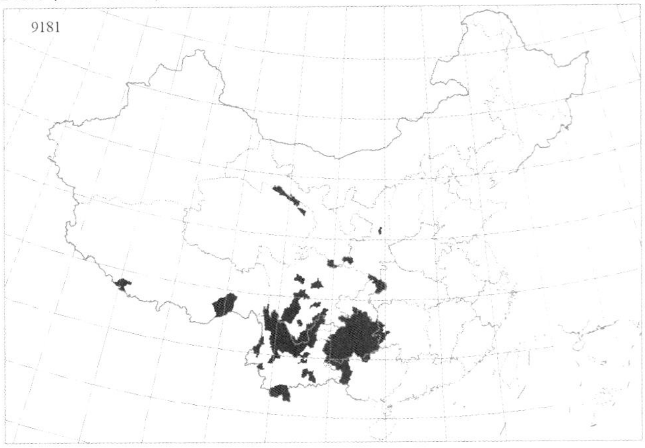

Asclepiadaceae 萝藦科 *Periploca* 杠柳属
Periploca sepium Bunge 杠柳
Shrubs, deciduous

11.5 (−2.6~17.8); −3.4 (−20.7~7.4); 24.9 (8.5~28.9); 12.3 (2.5~17.9); 102.1 (7.7~154.3); 23.6 (0.0~101.4); 752 (347~947) ❙ 647 (127~1444); 372 (81~816); 30 (2~197) ❙ −13.1 (−81.2~121.7); 607 (127~947); 171.6 (13.6~337.1)

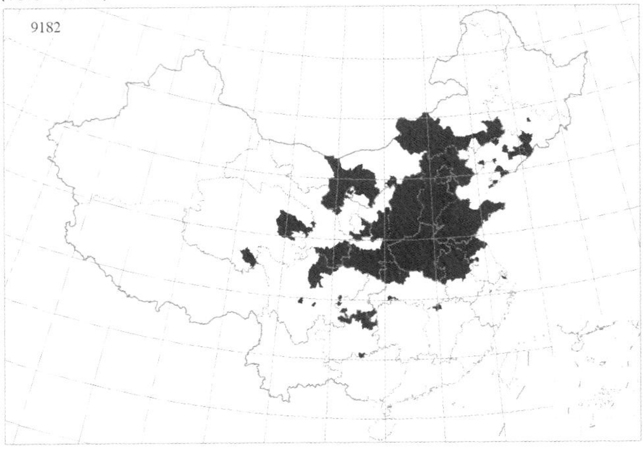

Asclepiadaceae 萝藦科 *Periploca* 杠柳属
Periploca tsangii D. Fang et H. Z. Ling 大花杠柳
Scandent shrubs

18.8; 9.2; 26.6; 18.9; 166.6; 0.0; 975 ❙ 1474; 706; 124 ❙ 51.5; 975; 238.9

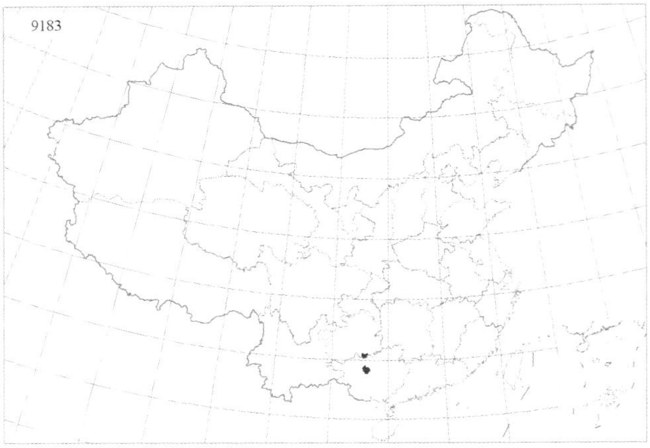

Asclepiadaceae 萝藦科 *Raphistemma* 大花藤属
Raphistemma hooperianum (Blume) Decne. 广西大花藤
Vines

21.9; 13.2; 28.6; 22.0; 203.8; 0.0; 1176 ❙ 1518; 773; 138 ❙ 29.1; 1176; 272.3

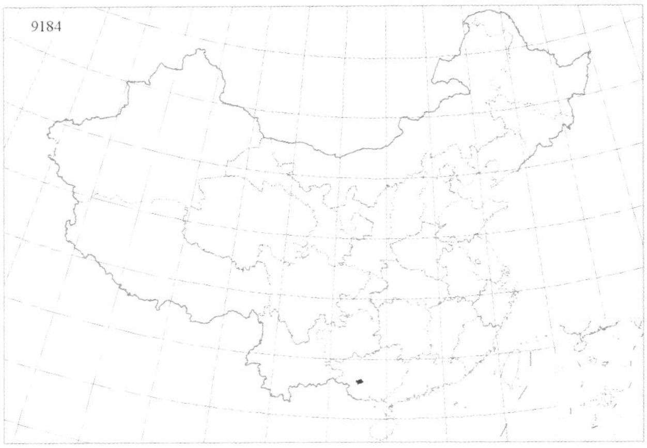

Asclepiadaceae 萝藦科 *Raphistemma* 大花藤属
Raphistemma pulchellum (Roxb.) Wall. 大花藤
Vines

18.6 (15.9~21.4); 10.4 (6.3~14.9); 24.7 (22.3~28.2); 18.6 (16.0~21.5); 163.2 (131.6~197.8); 0.0 (0.0~0.0); 928 (802~1140) ❙ 1330 (1172~1499); 695 (611~835); 64 (43~130) ❙ 44.2 (16.9~64.9); 923 (802~1140); 329.1 (189.5~570.1)

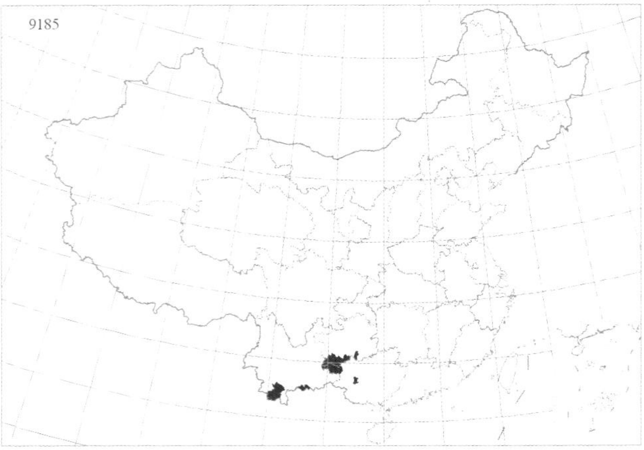

Asclepiadaceae 萝藦科 *Secamone* 鲫鱼藤属
Secamone bonii Costantin 斑皮鲫鱼藤
Scandent shrubs

23.0; 15.7; 28.3; 23.0; 216.0; 0.0; 1249 ❙ 1484; 682; 98 ❙ 19.8; 1167; 357.6

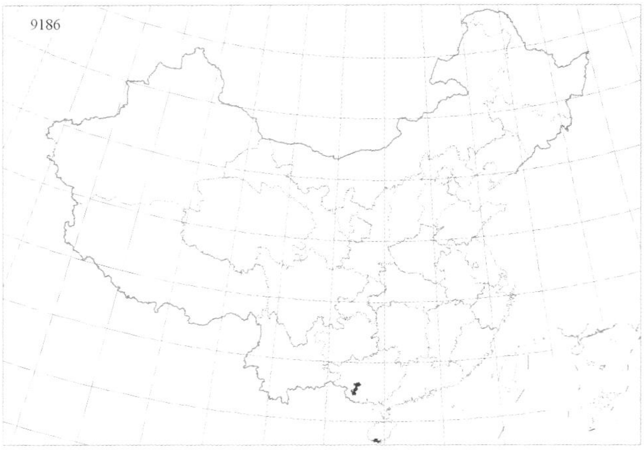

Asclepiadaceae 萝藦科 *Secamone* 鲫鱼藤属
Secamone elliptica R. Br. 鲫鱼藤
Vines

21.9 (13.4~24.8); 14.4 (2.7~20.1); 27.6 (22.1~29.2); 21.9 (13.4~24.8); 202.9 (104.5~237.6); 0.0 (0.0~3.7); 1161 (735~1385) ❙ 1587 (868~2114); 739 (312~1357); 98 (41~181) ❙ 37.9 (-34.3~95.9); 1131 (735~1360); 271.1 (95.3~620.0)

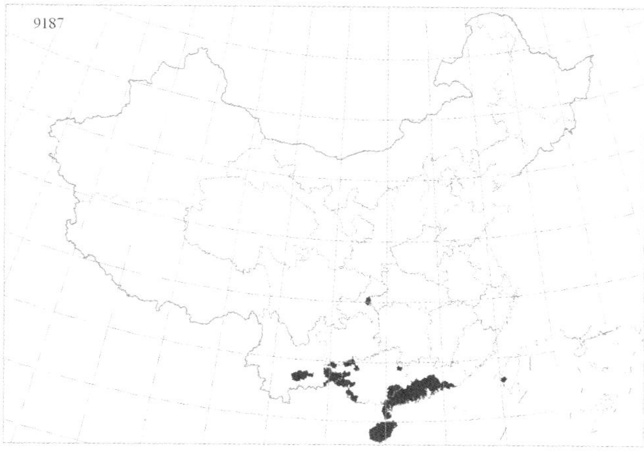

Asclepiadaceae 萝藦科 *Secamone* 鲫鱼藤属
Secamone ferruginea Pierre ex Costantin 锈毛鲫鱼藤
Scandent shrubs

21.9 (19.2~24.8); 14.3 (11.9~20.1); 28.1 (22.3~29.2); 22.0 (19.2~24.8); 203.8 (170.5~237.6); 0.0 (0.0~0.0); 1160 (909~1385) ❙ 1682 (1187~2037); 768 (473~1033); 118 (41~188) ❙ 45.7 (-6.9~83.4); 1154 (880~1273); 226.6 (95.3~659.0)

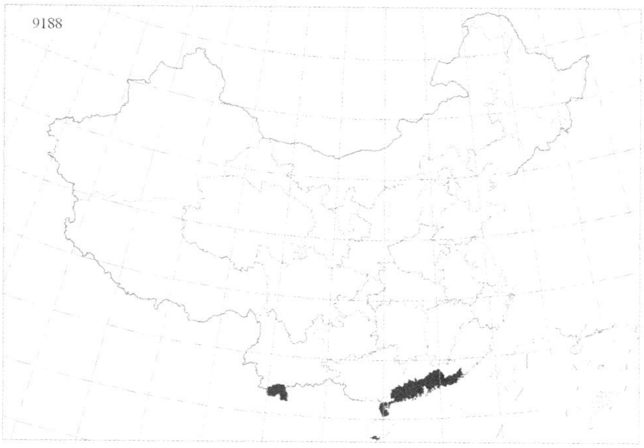

Asclepiadaceae 萝藦科 *Secamone* 鲫鱼藤属
Secamone likiangensis Tsiang 丽江鲫鱼藤
Subshrubs

8.7; 1.3; 15.0; 8.8; 55.7; 10.5; 574 ❙ 915; 571; 18 ❙ 60.3; 574; 315.6

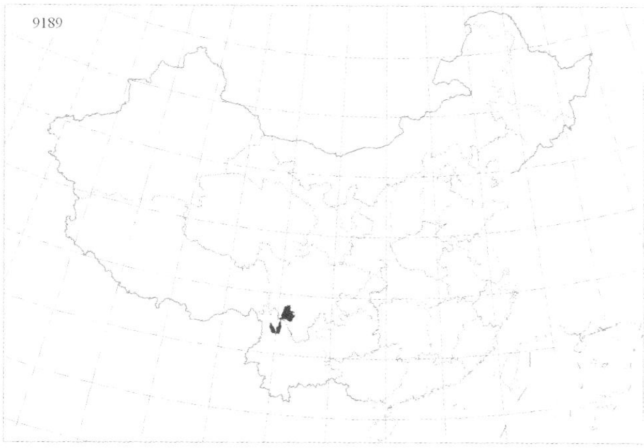

Asclepiadaceae 萝藦科 *Secamone* 鲫鱼藤属
Secamone minutiflora (Woodson) Tsiang 催吐鲫鱼藤
Scandent shrubs

17.6 (11.1~22.2); 8.6 (2.2~13.9); 24.8 (19.4~28.3); 17.6 (11.3~22.3); 151.6 (80.8~207.5); 0.2 (0.0~5.3); 899 (652~1190) ❙ 1296 (963~1584); 677 (570~816); 70 (26~165) ❙ 45.8 (12.9~87.2); 898 (652~1190); 265.3 (186.7~382.6)

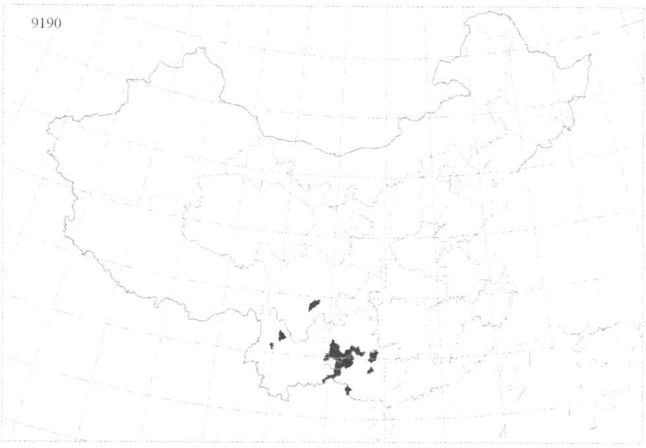

Asclepiadaceae 萝藦科 *Secamone* 鲫鱼藤属
Secamone sinica Hand.-Mazz. 吊山桃
Scandent shrubs

19.7 (15.9~23.6); 10.9 (6.9~16.7); 26.7 (21.4~29.2); 19.8 (16.0~23.6); 177.3 (131.6~223.5); 0.0 (0.0~0.0); 1024 (802~1281) ❙ 1420 (1143~1900); 704 (551~1056); 102 (49~188) ❙ 39.9 (-1.0~85.5); 1023 (802~1273); 266.8 (160.7~452.0)

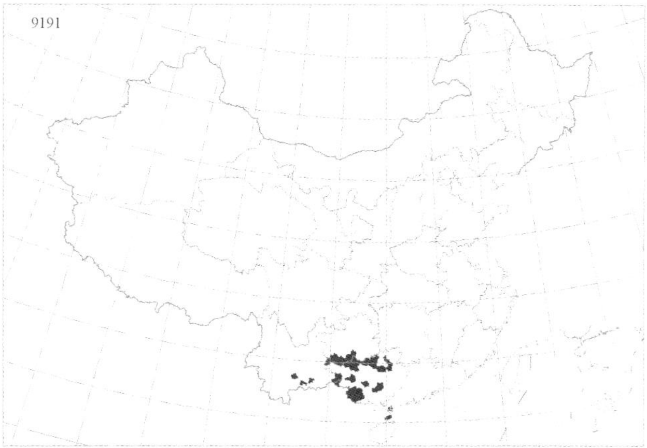

Asclepiadaceae 萝藦科 *Stelmocrypton* 须药藤属
Stelmocrypton khasianum (Kurz) Baill. 须药藤
Vines

18.1 (14.6~21.3); 10.4 (6.9~15.1); 23.7 (19.6~27.9); 18.1 (14.6~21.3); 157.4 (115.8~196.0); 0.0 (0.0~0.0); 895 (736~1123) ❙ 1288 (976~1607); 687 (556~848); 57 (39~77) ❙ 44.4 (5.8~56.9); 885 (732~1121); 331.9 (189.5~659.0)

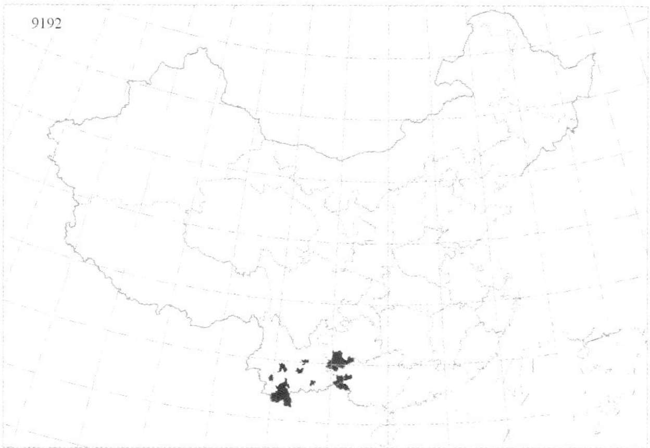

Asclepiadaceae 萝藦科 *Streptocaulon* 马莲鞍属

Streptocaulon juventas (Lour.) Merr. 马莲鞍

Vines, evergreen

19.6 (15.9~22.7); 11.5 (6.3~15.1); 25.8 (20.3~28.9); 19.7 (16.0~22.8); 176.2 (131.6~213.1); 0.0 (0.0~0.0); 1005 (789~1221) ❙ 1465 (1182~2435); 769 (566~1428); 88 (42~194) ❙ 46.5 (5.8~99.5); 1002 (780~1221); 316.4 (189.5~659.0)

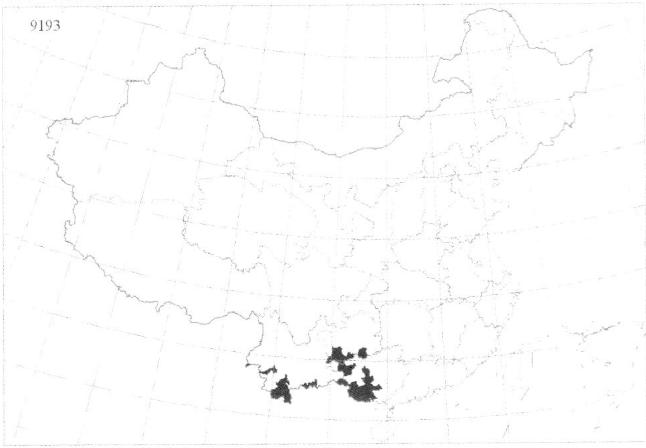

Asclepiadaceae 萝藦科 *Toxocarpus* 弓果藤属

Toxocarpus aurantiacus C. Y. Wu ex Tsiang et P. T. Li 云南弓果藤

Vines

20.0; 14.6; 23.2; 20.1; 180.8; 0.0; 963 ❙ 1523; 681; 67 ❙ 58.4; 942; 581.0

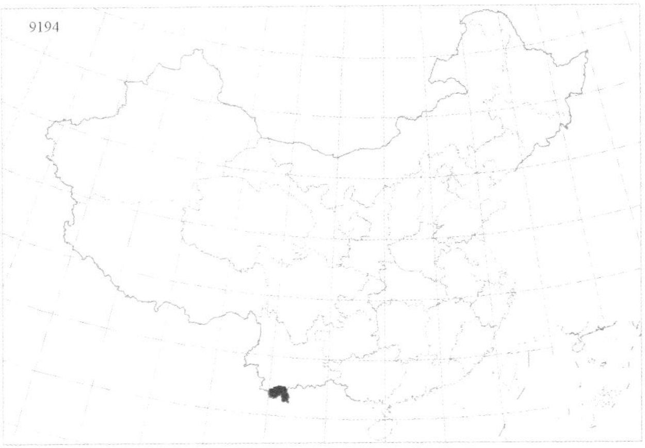

Asclepiadaceae 萝藦科 *Toxocarpus* 弓果藤属

Toxocarpus fuscus Tsiang 锈毛弓果藤

Vines

20.7 (16.9~23.1); 12.5 (9.1~17.5); 26.9 (21.4~28.9); 20.7 (16.9~23.2); 188.6 (142.7~218.4); 0.0 (0.0~0.0); 1079 (813~1232) ❙ 1477 (1178~1703); 717 (544~839); 114 (51~194) ❙ 38.1 (5.8~71.6); 1072 (813~1202); 321.8 (201.4~551.6)

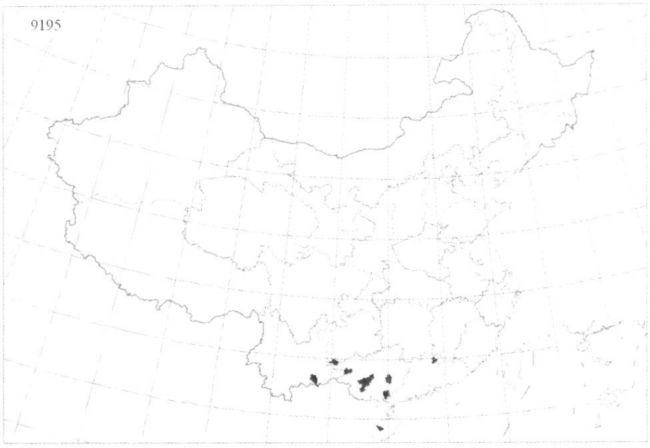

Asclepiadaceae 萝藦科 *Toxocarpus* 弓果藤属

Toxocarpus hainanensis Tsiang 海南弓果藤

Vines

24.4; 19.6; 27.8; 24.5; 233.9; 0.0; 1353 ❙ 1574; 526; 59 ❙ 16.4; 1194; 565.2

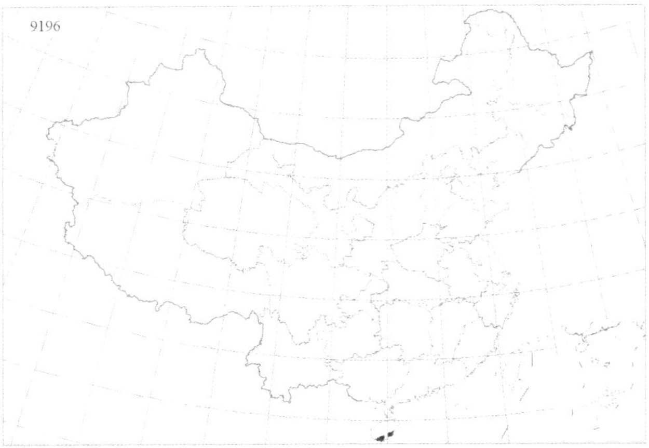

Asclepiadaceae 萝藦科 *Toxocarpus* 弓果藤属

Toxocarpus himalensis Falc. ex Hook. f. 西藏弓果藤

Vines

11.9 (−2.3~21.8); 3.3 (−10.9~15.1); 19.0 (5.4~28.6); 13.0 (1.6~21.8); 109.9 (0.4~201.6); 27.1 (0.0~89.7); 747 (324~1162) ❙ 1063 (255~1758); 509 (158~817); 86 (1~189) ❙ 33.2 (−31.1~113.0); 727 (255~1162); 225.4 (46.9~659.0)

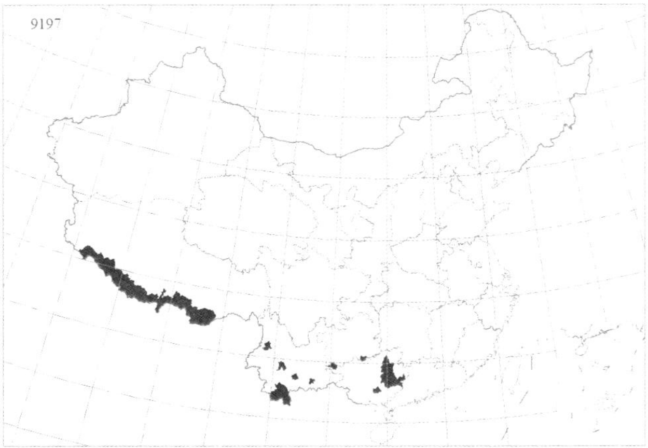

Asclepiadaceae 萝藦科 *Toxocarpus* 弓果藤属

Toxocarpus laevigatus Tsiang 平滑弓果藤

Scandent shrubs

23.1; 17.5; 27.2; 23.2; 218.4; 0.0; 1232 ❙ 1487; 544; 53 ❙ 20.7; 1142; 538.2

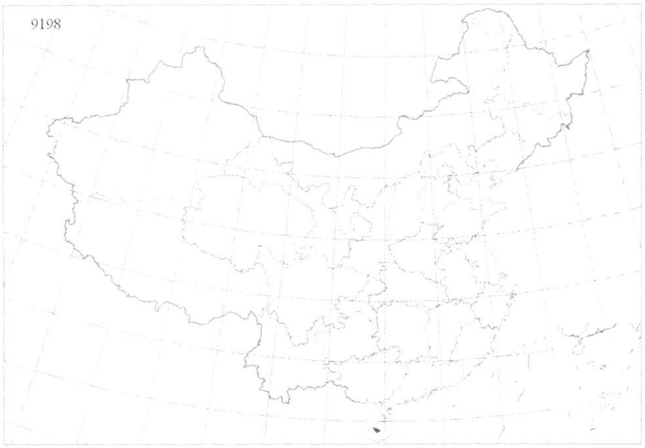

Asclepiadaceae 萝藦科 *Toxocarpus* 弓果藤属
Toxocarpus patens Tsiang 广花弓果藤
Scandent shrubs

23.9; 18.7; 27.5; 23.9; 226.8; 0.0; 1294 ❙ 1380; 455; 48 ❙ 6.7; 1112; 455.2

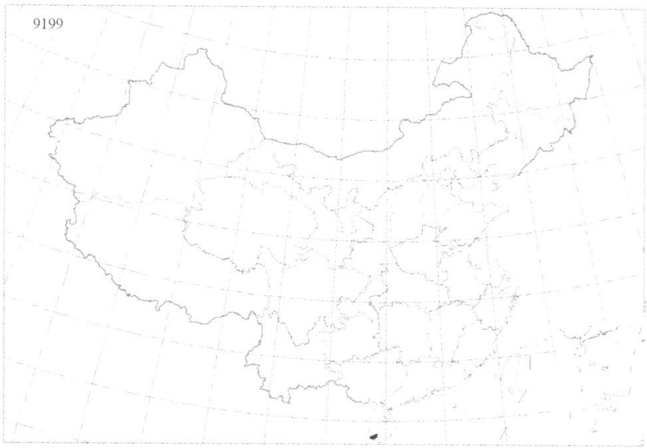

Asclepiadaceae 萝藦科 *Toxocarpus* 弓果藤属
Toxocarpus paucinervius Tsiang 凌云弓果藤
Scandent shrubs

19.9 (19.0~21.2); 10.7 (9.2~12.5); 27.2 (25.0~29.0); 19.9 (19.0~21.2); 179.2 (168.4~194.6); 0.0 (0.0~0.0); 1031 (945~1114) ❙ 1367 (1182~1500); 690 (639~820); 92 (43~142) ❙ 33.1 (14.8~52.2); 1031 (945~1114); 272.0 (160.7~371.7)

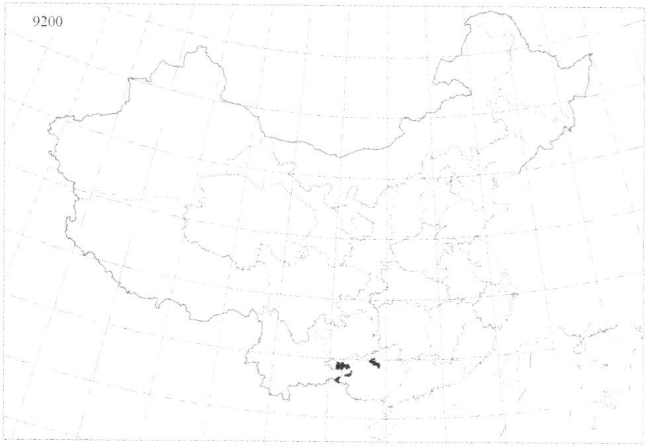

Asclepiadaceae 萝藦科 *Toxocarpus* 弓果藤属
Toxocarpus villosus (Blume) Decne. 毛弓果藤
Vines

18.6 (11.1~22.7); 10.1 (0.6~15.1); 25.7 (19.4~29.0); 18.7 (11.3~22.8) 164.2 (80.8~213.1); 0.4 (0.0~9.7); 955 (652~1221) ❙ 1416 (958~2435) 657 (419~1428); 104 (37~205) ❙ 49.2 (12.9~99.5); 949 (652~1221) 274.0 (152.1~659.0)

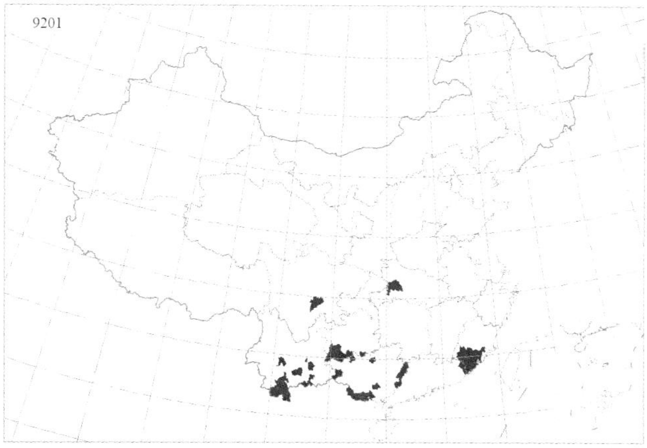

Asclepiadaceae 萝藦科 *Toxocarpus* 弓果藤属
Toxocarpus villosus var. thorelii Costantin 小叶弓果藤
Vines

16.9; 10.3; 21.4; 16.9; 142.7; 0.0; 813 ❙ 1178; 646; 51 ❙ 44.9; 813; 383.4

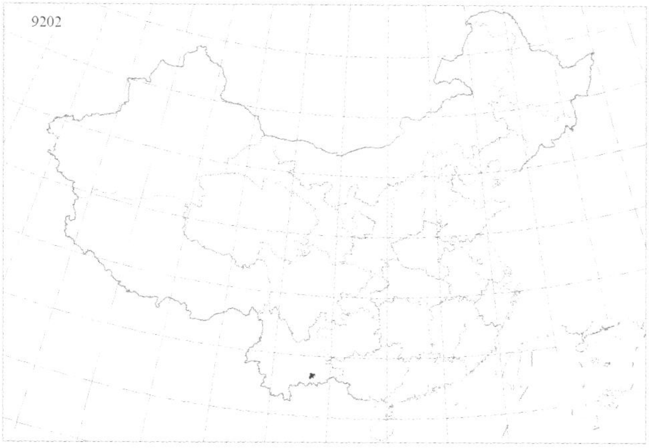

Asclepiadaceae 萝藦科 *Toxocarpus* 弓果藤属
Toxocarpus wangianus Tsiang 澜沧弓果藤
Vines

23.0 (17.7~24.8); 16.9 (10.9~20.1); 27.3 (21.6~28.8); 23.0 (17.8~24.8); 216.0 (153.4~237.6); 0.0 (0.0~0.0); 1233 (854~1385) ❙ 1510 (868~1888); 608 (312~1044); 76 (41~138) ❙ 24.8 (-34.3~74.2); 1145 (839~1360); 416.5 (215.3~620.0)

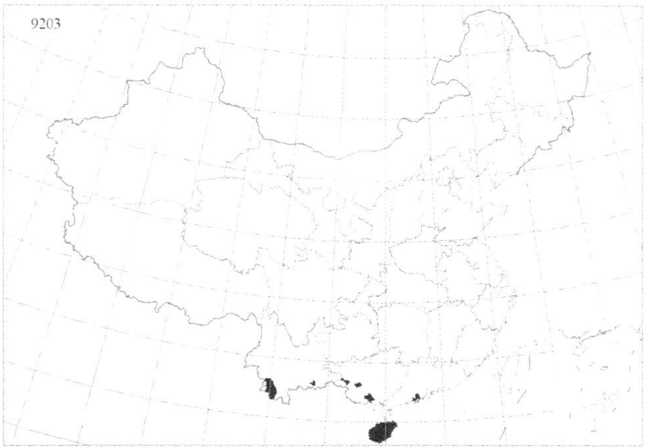

Asclepiadaceae 萝藦科 *Toxocarpus* 弓果藤属
Toxocarpus wightianus Hook. et Arn. 弓果藤
Vines

21.8 (15.9~24.8); 14.4 (6.9~20.1); 27.5 (22.3~29.2); 21.9 (16.0~24.8); 202.4 (131.6~237.6); 0.0 (0.0~0.0); 1159 (802~1385) ❙ 1574 (868~2435); 716 (312~1428); 95 (41~168) ❙ 37.1 (-34.3~99.5); 1123 (802~1360); 301.8 (111.7~620.0)

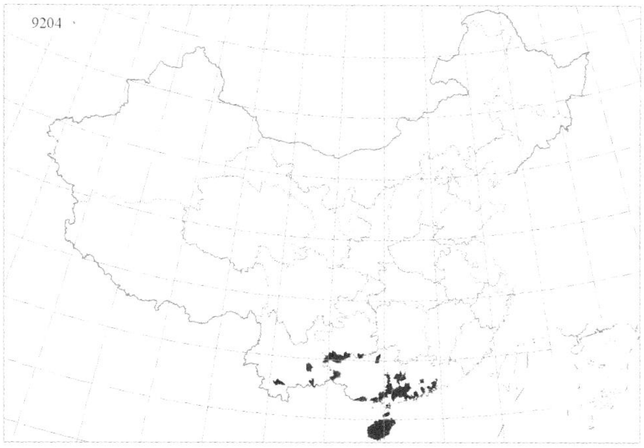

Asclepiadaceae 萝藦科 *Tylophora* 娃儿藤属
Tylophora anthopotamica (Hand.-Mazz.) Tsiang et H. D. Zhang 花溪娃儿藤
Scandent shrubs

15.2 (14.9~15.6); 4.9 (4.6~5.5); 24.5 (23.9~24.9); 15.2 (15.0~15.6); 122.7 (119.9~127.5); 0.1 (0.0~0.4); 795 (781~808) ❙ 1207 (1160~1244); 525 (494~553); 75 (69~81) ❙ 51.9 (44.2~58.8); 795 (781~808); 202.2 (185.3~233.1)

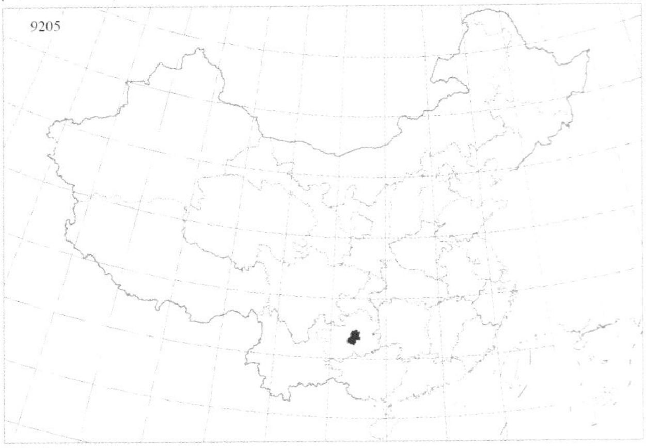

Asclepiadaceae 萝藦科 *Tylophora* 娃儿藤属
Tylophora arenicola Merr. 虎须娃儿藤
Vines

22.1 (20.2~24.8); 14.4 (11.9~20.1); 28.1 (25.7~29.1); 22.1 (20.2~24.8); 205.5 (183.0~237.6); 0.0 (0.0~0.0); 1173 (1036~1385) ❙ 1703 (1374~2037); 766 (455~1033); 120 (41~188) ❙ 45.9 (1.5~83.4); 1158 (1036~1299); 247.6 (95.3~620.0)

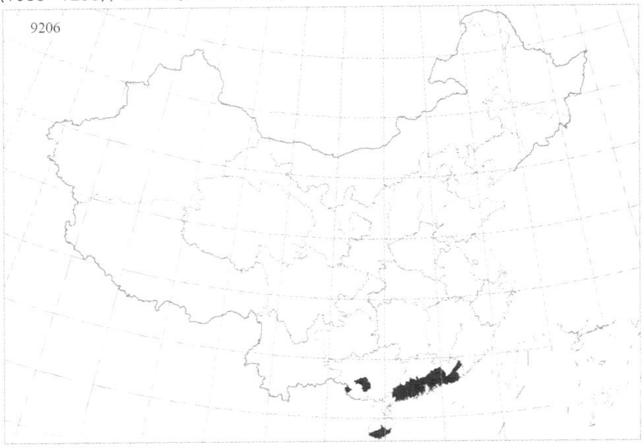

Asclepiadaceae 萝藦科 *Tylophora* 娃儿藤属
Tylophora augustiniana (Hemsl.) Craib 宜昌娃儿藤
Vines

17.2; 6.6; 26.6; 17.3; 149.1; 1.8; 922 ❙ 1296; 616; 96 ❙ 40.8; 922; 232.3

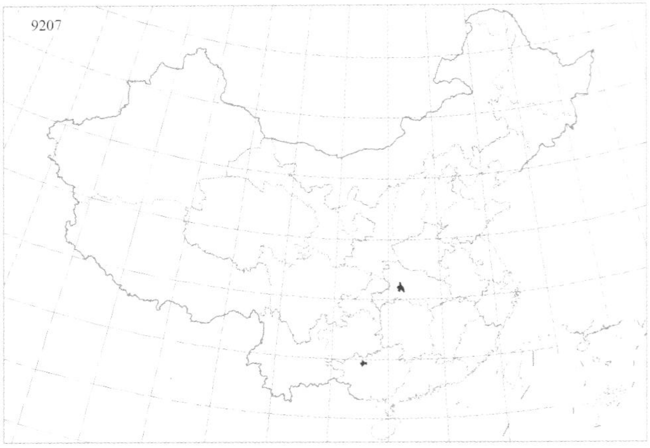

Asclepiadaceae 萝藦科 *Tylophora* 娃儿藤属
Tylophora brownii Hayata 光叶娃儿藤
Vines

23.9 (22.2~24.8); 18.2 (15.8~20.1); 28.0 (25.7~28.9); 24.0 (22.3~24.8); 227.6 (207.2~237.6); 0.0 (0.0~0.0); 1306 (1127~1385) ❙ 1499 (868~1854); 548 (312~824); 76 (41~115) ❙ 15.1 (-34.3~47.8); 1198 (868~1360); 418.8 (194.5~620.0)

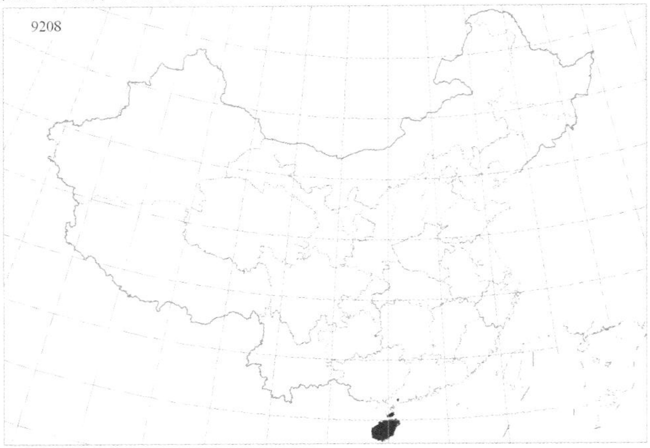

Asclepiadaceae 萝藦科 *Tylophora* 娃儿藤属
Tylophora cycleoides Tsiang 轮环娃儿藤
Vines

22.2 (17.6~24.8); 15.4 (7.3~20.1); 27.3 (25.7~28.5); 22.2 (17.6~24.8); 206.5 (151.6~237.6); 0.0 (0.0~0.0); 1189 (909~1385) ❙ 1543 (1380~1767); 578 (455~678); 85 (41~142) ❙ 32.1 (1.5~64.7); 1103 (909~1299); 432.4 (225.9~620.0)

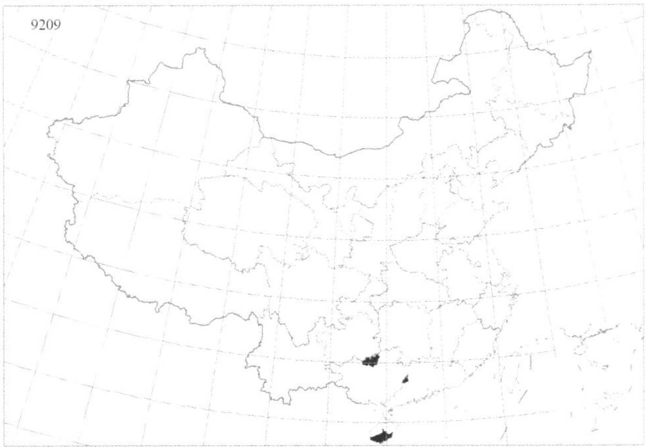

Asclepiadaceae 萝藦科 *Tylophora* 娃儿藤属
Tylophora glabra Costantin 长梗娃儿藤
Vines

22.2 (15.2~24.8); 14.6 (4.1~20.1); 28.0 (25.7~29.2); 22.2 (15.2~24.8); 206.7 (123.4~237.6); 0.0 (0.0~0.9); 1193 (812~1385) ❙ 1632 (1221~2435); 751 (463~1428); 104 (41~178) ❙ 37.7 (1.5~99.5); 1166 (812~1360); 308.1 (178.7~620.0)

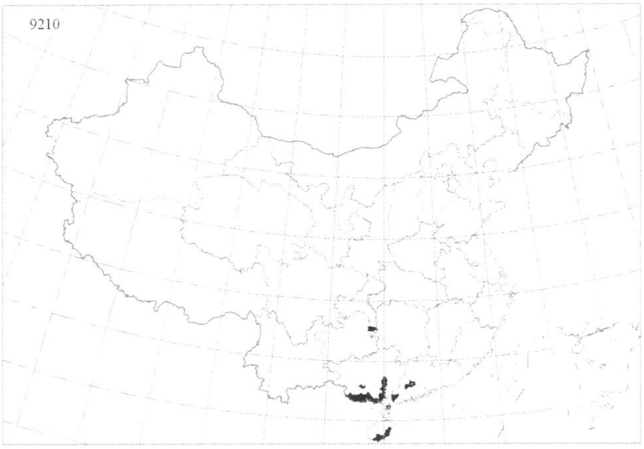

Asclepiadaceae 萝藦科 *Tylophora* 娃儿藤属
Tylophora gracilenta Tsiang et P. T. Li 天峨娃儿藤
Vines

19.1; 10.4; 25.8; 19.2; 169.9; 0.0; 963 | 1361; 736; 65 | 41.5; 963; 346.9

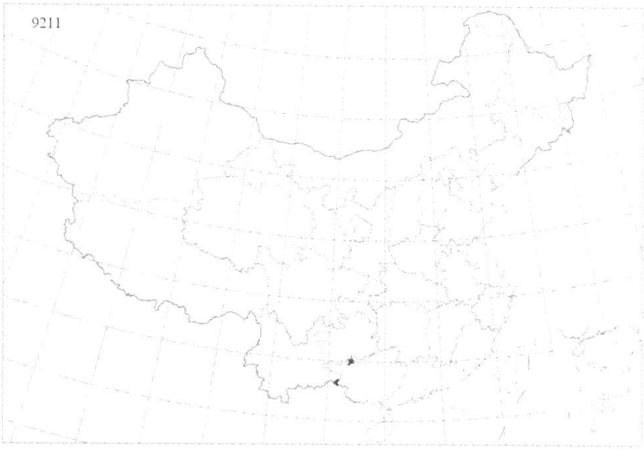

Asclepiadaceae 萝藦科 *Tylophora* 娃儿藤属
Tylophora hui Tsiang 建水娃儿藤
Vines

16.7 (14.9~18.2); 7.8 (3.6~11.8); 24.1 (21.3~25.7); 16.8 (15.0~18.2); 141.5 (121.2~158.4); 0.2 (0.0~1.6); 845 (807~873) | 1262 (991~1448); 595 (523~756); 87 (44~129) | 49.5 (21.3~79.3); 839 (794~869); 279.7 (222.1~491.5)

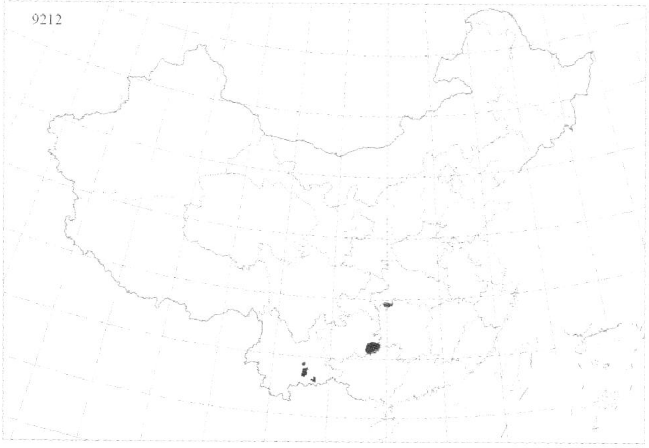

Asclepiadaceae 萝藦科 *Tylophora* 娃儿藤属
Tylophora insulana Tsiang et P. T. Li 台湾娃儿藤
Vines

20.2 (16.2~24.1); 14.5 (11.0~18.9); 25.0 (20.1~28.2); 20.2 (16.2~24.1); 182.9 (134.2~229.7); 0.0 (0.0~0.0); 1019 (773~1322) | 2402 (1498~3495); 1106 (750~1847); 255 (47~897) | 146.2 (25.2~246.5); 983 (773~1144); 245.0 (119.4~363.7)

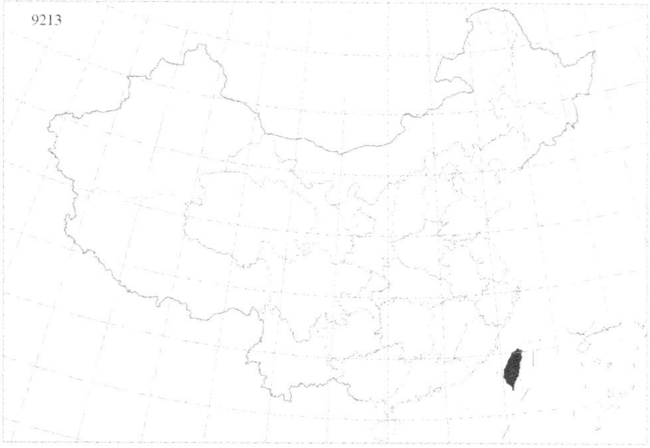

Asclepiadaceae 萝藦科 *Tylophora* 娃儿藤属
Tylophora kerrii Craib 人参娃儿藤
Vines

20.2 (14.8~23.4); 11.7 (4.3~16.1); 26.9 (20.3~29.2); 20.2 (14.9~23.5); 182.9 (119.2~221.7); 0.0 (0.0~0.7); 1053 (789~1273) | 1482 (1049~2435); 725 (475~1428); 102 (43~177) | 41.2 (5.8~99.5); 1052 (780~1273); 261.5 (152.9~372.4)

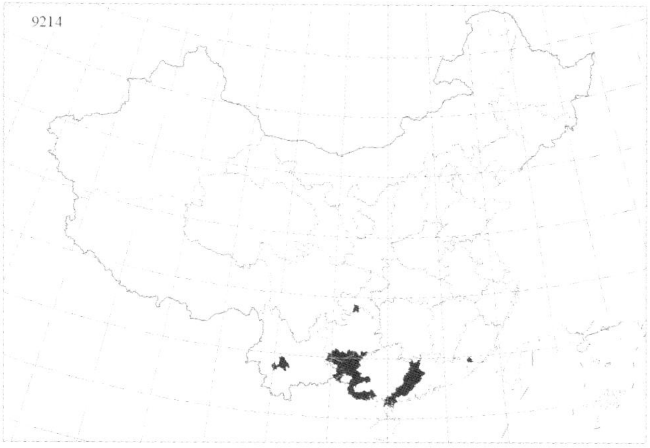

Asclepiadaceae 萝藦科 *Tylophora* 娃儿藤属
Tylophora koi Merr. 通天连
Vines

20.2 (5.3~24.8); 11.9 (-2.4~20.1); 27.3 (12.2~29.8); 20.2 (5.7~24.8); 183.1 (30.5~237.6); 0.3 (0.0~26.2); 1064 (479~1385) | 1625 (868~3495); 631 (312~1428); 160 (31~897) | 56.0 (-34.3~224.4); 1036 (479~1350); 278.2 (119.4~620.0)

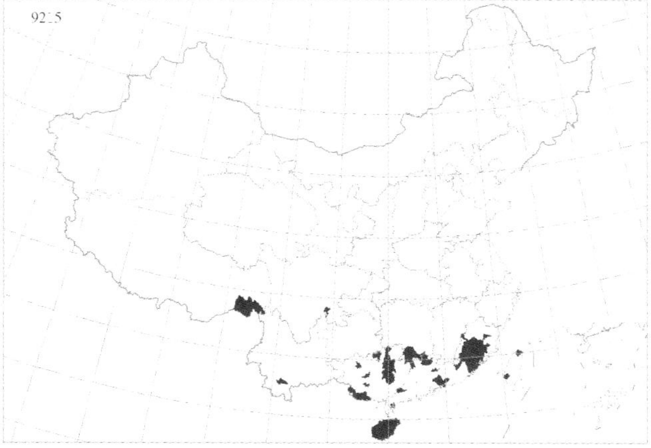

Asclepiadaceae 萝藦科 *Tylophora* 娃儿藤属
Tylophora leptantha Tsiang 广花娃儿藤
Scandent shrubs

22.2 (21.2~24.0); 14.6 (12.2~19.1); 28.0 (25.7~29.0); 22.3 (21.3~24.0); 207.5 (195.2~228.1); 0.0 (0.0~0.0); 1184 (1111~1302) | 1658 (1302~2435); 826 (539~1428); 101 (48~142) | 40.2 (12.9~99.5); 1168 (1055~1221); 297.8 (111.7~620.0)

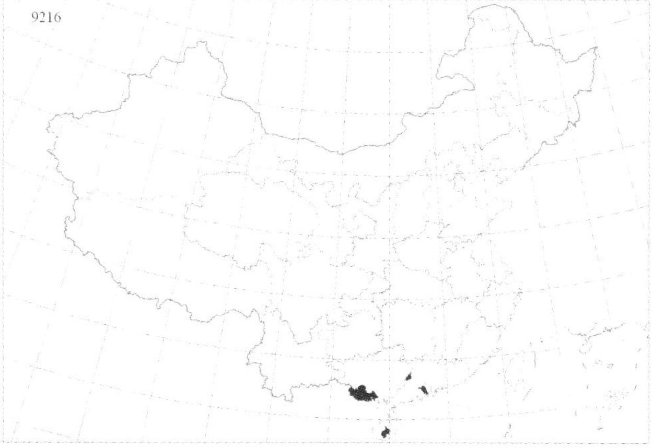

Asclepiadaceae 萝藦科 *Tylophora* 娃儿藤属
Tylophora nana C. K. Schneid. 汶川娃儿藤
Small shrubs

6.4 (1.6~13.3); -4.0 (-9.3~3.2); 15.8 (10.9~22.7); 7.4 (4.0~13.2); 48.4 (17.9~100.9); 31.5 (2.5~58.4); 532 (419~723) ▎759 (629~1081); 382 (306~606); 19 (14~32) ▎46.3 (-0.2~101.2); 531 (419~723); 263.4 (171.0~344.4)

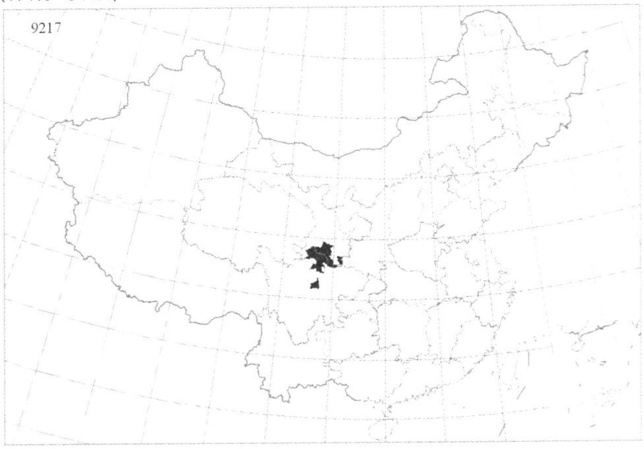

Asclepiadaceae 萝藦科 *Tylophora* 娃儿藤属
Tylophora ovata (Lindl.) Hook. ex Steud. 娃儿藤
Vines

19.7 (11.6~24.8); 10.8 (3.1~20.1); 27.4 (18.6~29.8); 19.8 (11.5~24.8); 177.7 (80.9~237.6); 0.1 (0.0~2.8); 1043 (654~1385) ▎1582 (868~2993); 658 (312~1847); 146 (23~547) ▎53.9 (-34.3~235.6); 1027 (654~1360); 244.0 (95.3~620.0)

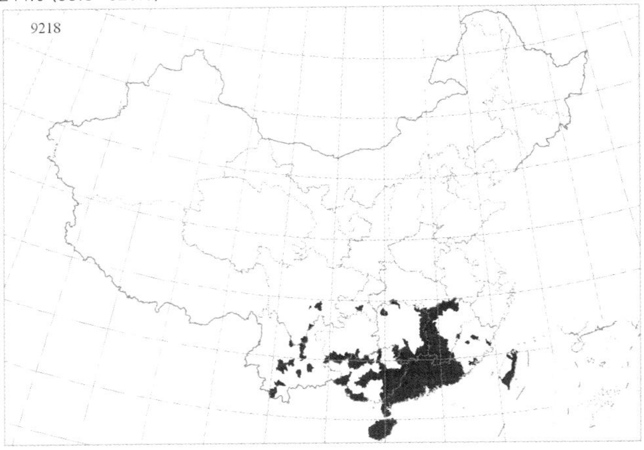

Asclepiadaceae 萝藦科 *Tylophora* 娃儿藤属
Tylophora picta Tsiang 紫叶娃儿藤
Scandent shrubs

24.0; 19.1; 27.2; 24.0; 228.1; 0.0; 1302 ▎1551; 539; 48 ▎19.1; 1147; 620.0

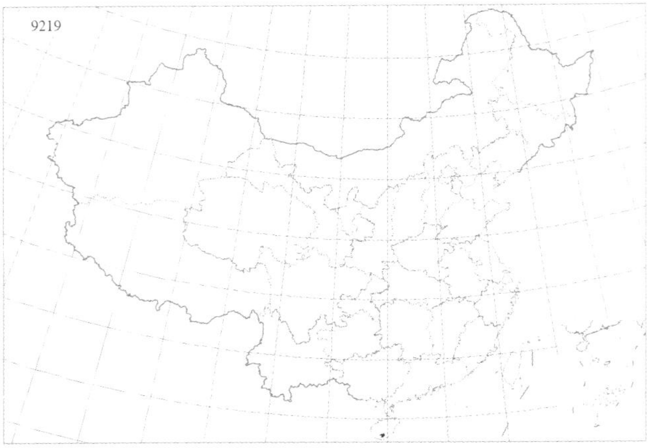

Asclepiadaceae 萝藦科 *Tylophora* 娃儿藤属
Tylophora secamonoides Tsiang 蛇胆草
Dwarf shrubs

24.1; 17.8; 28.7; 24.1; 229.4; 0.0; 1320 ▎868; 312; 71 ▎-34.3; 868; 288.3

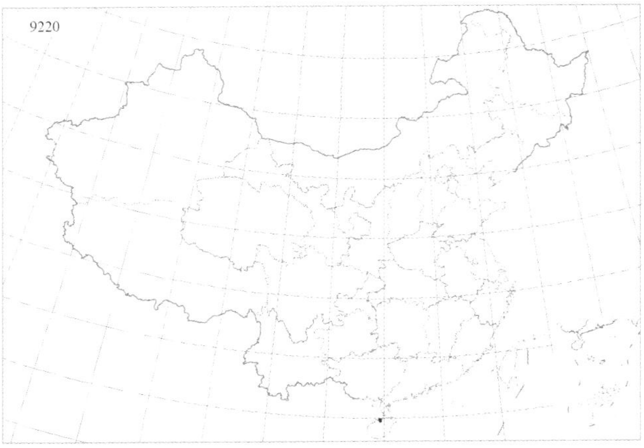

Convolvulaceae 旋花科 *Argyreia* 银背藤属
Argyreia acuta Lour. 白鹤藤
Scandent shrubs, evergreen

21.9 (17.6~24.8); 14.0 (7.3~20.1); 28.1 (23.6~29.0); 22.0 (17.6~24.8); 203.8 (151.6~237.6); 0.0 (0.0~0.0); 1171 (909~1385) ▎1612 (1143~2435); 722 (392~1428); 122 (41~182) ▎38.7 (-11.9~99.5); 1152 (903~1360); 268.3 (120.3~548.0)

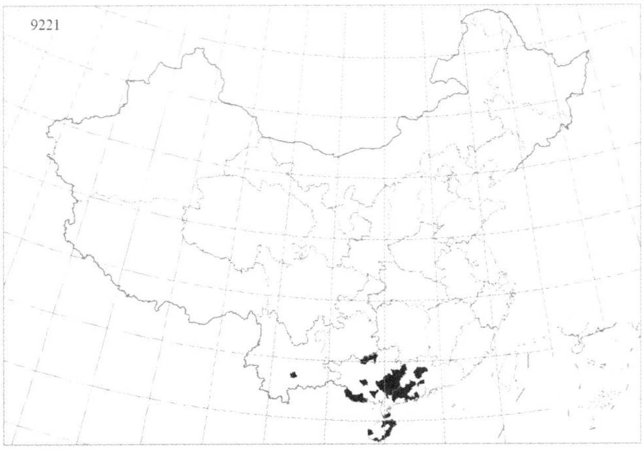

Convolvulaceae 旋花科 *Argyreia* 银背藤属
Argyreia baoshanensis S. H. Huang 保山银背藤
Scandent shrubs

15.5; 8.6; 20.4; 15.6; 127.3; 0.0; 768 ▎1180; 622; 65 ▎53.6; 768; 335.1

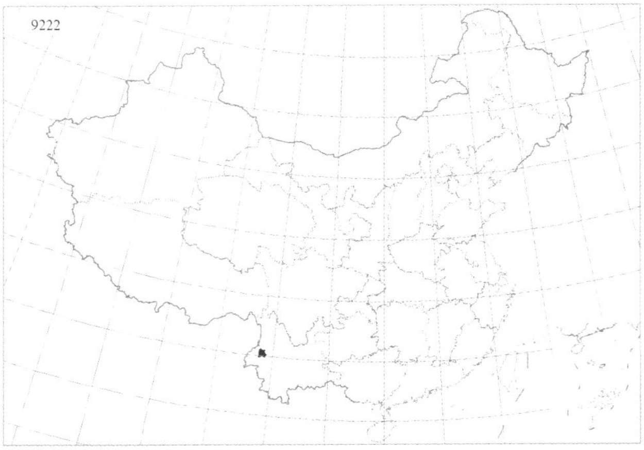

Convolvulaceae 旋花科 *Argyreia* 银背藤属
Argyreia capitiformis (Poir.) Ooststr. 头花银背藤
Scandent shrubs

20.5 (15.9~24.8); 12.9 (6.9~20.1); 26.3 (21.4~28.8); 20.6 (16.0~24.8); 187.1 (131.6~237.6); 0.0 (0.0~0.0); 1066 (802~1385) ❙ 1497 (868~2435); 741 (312~1428); 86 (41~194) ❙ 42.2 (-34.3~99.5); 1035 (783~1299); 330.7 (157.7~659.0)

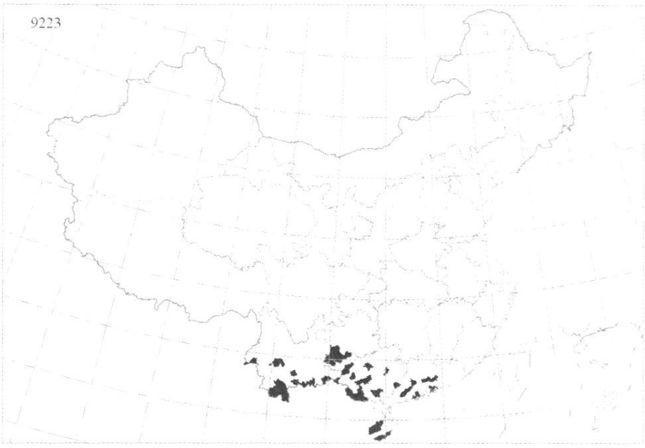

Convolvulaceae 旋花科 *Argyreia* 银背藤属
Argyreia cheliensis C. Y. Wu et Li 车里银背藤
Scandent shrubs

20.5; 14.9; 23.7; 20.5; 185.9; 0.0; 992 ❙ 1464; 655; 66 ❙ 47.6; 959; 570.1

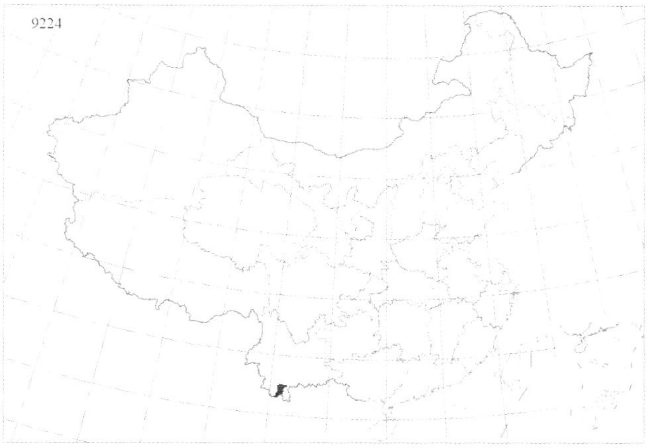

Convolvulaceae 旋花科 *Argyreia* 银背藤属
Argyreia eriocephala C. Y. Wu et Li 毛头银背藤
Scandent shrubs

15.1; 8.3; 19.9; 15.1; 121.1; 0.0; 754 ❙ 1258; 687; 60 ❙ 68.6; 754; 395.3

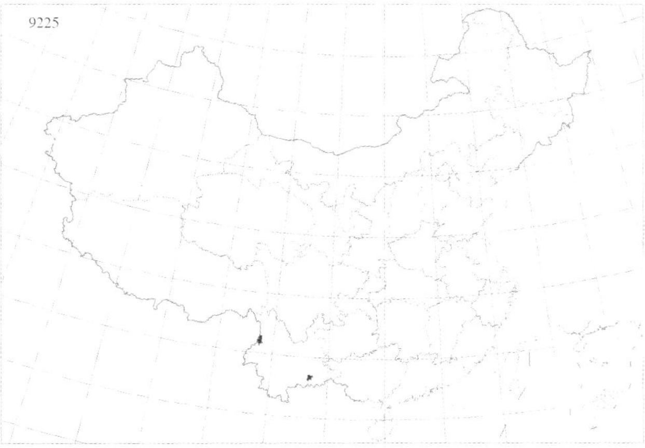

Convolvulaceae 旋花科 *Argyreia* 银背藤属
Argyreia formosana Ishigami ex T. Yamazaki 台湾银背藤
Scandent shrubs

21.9; 17.0; 25.4; 21.9; 202.8; 0.0; 1119 ❙ 2501; 1537; 94 ❙ 133.8; 1043; 234.5

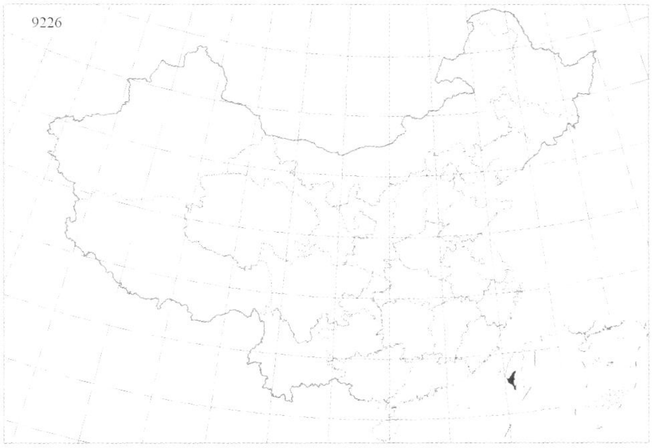

Convolvulaceae 旋花科 *Argyreia* 银背藤属
Argyreia fulvocymosa C. Y. Wu et Li 黄伞白鹤藤
Scandent shrubs

19.7; 13.4; 23.9; 19.8; 177.0; 0.0; 965 ❙ 1499; 759; 68 ❙ 55.7; 965; 567.4

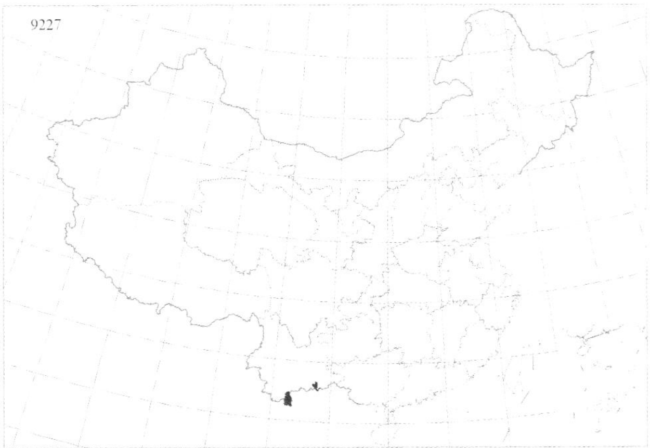

Convolvulaceae 旋花科 *Argyreia* 银背藤属
Argyreia henryi (Craib) Craib 长叶银背藤
Scandent shrubs

18.7 (14.4~20.5); 12.9 (7.5~15.1); 22.3 (19.1~23.7); 18.7 (14.4~20.5); 164.8 (113.1~185.9); 0.0 (0.0~0.0); 905 (727~992) ❙ 1504 (1454~1607); 722 (655~792); 67 (58~77) ❙ 67.9 (47.6~99.9); 889 (727~986); 531.6 (427.9~659.0)

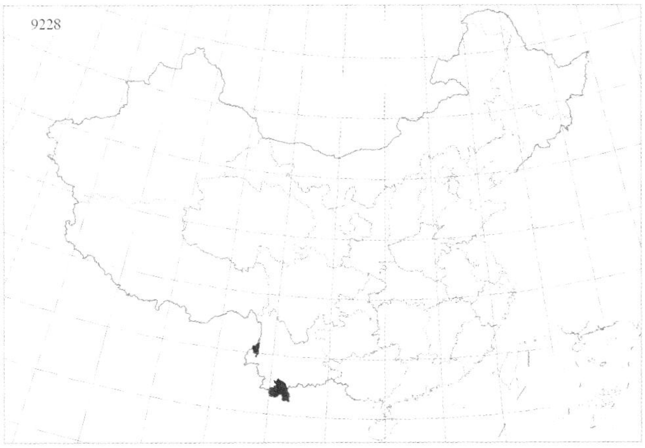

Convolvulaceae 旋花科 *Argyreia* 银背藤属
Argyreia henryi var. hypochrysa C. Y. Wu et Li 金背长叶藤
Scandent shrubs

4.4; -3.8; 11.4; 5.3; 27.7; 34.2; 462 ❙ 795; 515; 11 ❙ 72.6; 462; 260.2

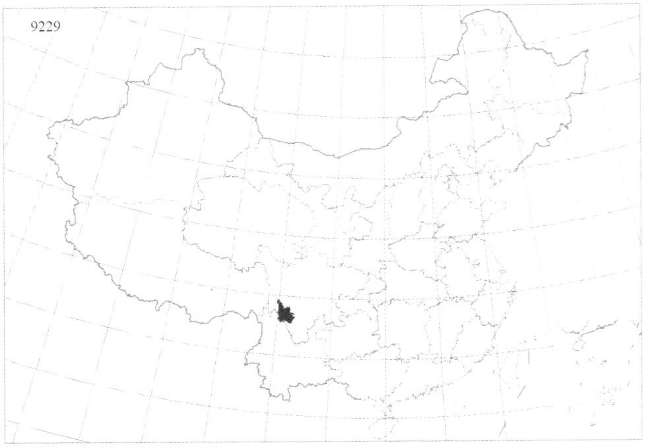

Convolvulaceae 旋花科 *Argyreia* 银背藤属
Argyreia lineariloba C. Y. Wu 线叶银背藤
Scandent shrubs

16.7; 10.3; 21.3; 16.8; 141.5; 0.0; 811 ❙ 951; 547; 37 ❙ 17.2; 744; 301.1

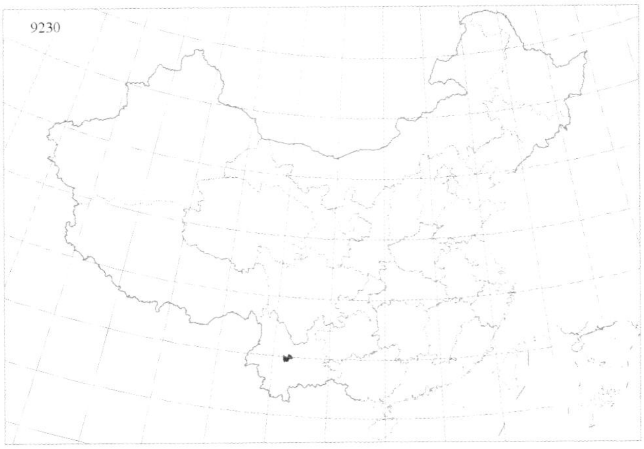

Convolvulaceae 旋花科 *Argyreia* 银背藤属
Argyreia marlipoensis C. Y. Wu et S. H. Huang 麻栗坡银背藤
Vines

18.3; 11.6; 23.0; 18.4; 160.5; 0.0; 885 ❙ 1287; 732; 50 ❙ 45.7; 882; 406.1

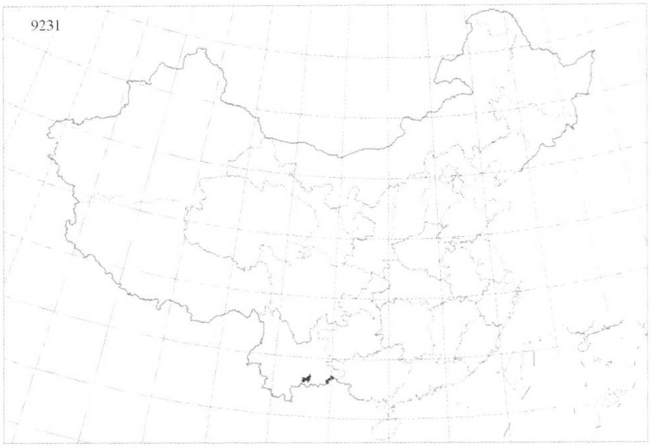

Convolvulaceae 旋花科 *Argyreia* 银背藤属
Argyreia mastersii (Prain) Raizada 叶苞银背藤
Scandent shrubs

18.3 (14.4~20.5); 12.5 (7.5~15.1); 21.9 (19.1~23.7); 18.3 (14.4~20.5); 160.1 (113.1~185.9); 0.0 (0.0~0.0); 886 (727~992) ❙ 1475 (1352~1607); 721 (655~792); 64 (47~77) ❙ 68.2 (47.6~99.9); 871 (727~986); 533.8 (438.9~659.0)

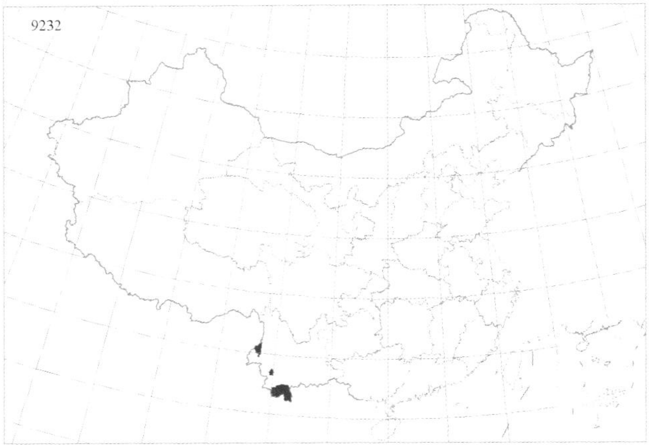

Convolvulaceae 旋花科 *Argyreia* 银背藤属
Argyreia maymyo (W. W. Sm.) Raizada 思茅银背藤
Scandent shrubs

19.0; 13.1; 22.6; 19.1; 168.8; 0.0; 909 ❙ 1495; 775; 62 ❙ 64.5; 891; 427.9

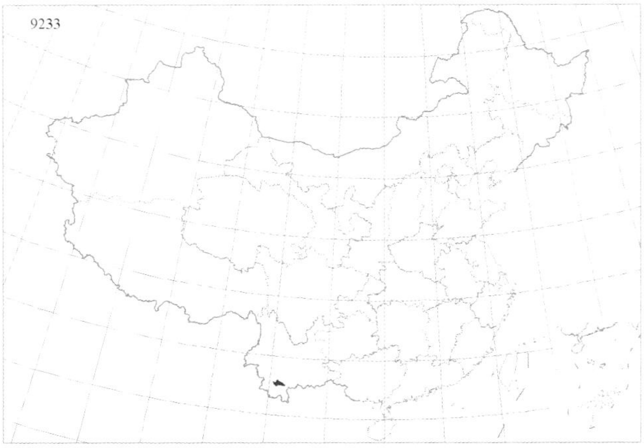

Convolvulaceae 旋花科 *Argyreia* 银背藤属
Argyreia mollis (Burm. f.) Choisy 银背藤
Vines

23.2 (19.3~24.8); 16.8 (13.2~20.1); 28.0 (22.8~28.9); 23.2 (19.3~24.8); 218.6 (171.9~237.6); 0.0 (0.0~0.0); 1250 (912~1385) ❙ 1690 (868~2759); 723 (312~1594); 93 (41~167) ❙ 37.9 (-34.3~202.5); 1177 (868~1360); 329.6 (116.6~620.0)

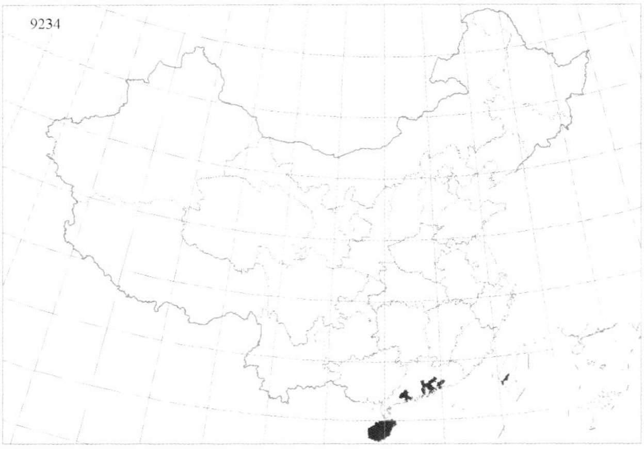

Convolvulaceae 旋花科 *Argyreia* 银背藤属
Argyreia monosperma C. Y. Wu et Li 单籽银背藤
Scandent shrubs

18.7; 12.3; 23.0; 18.7; 164.6; 0.0; 907 ❙ 1392; 727; 59 ❙ 53.6; 900; 485.1

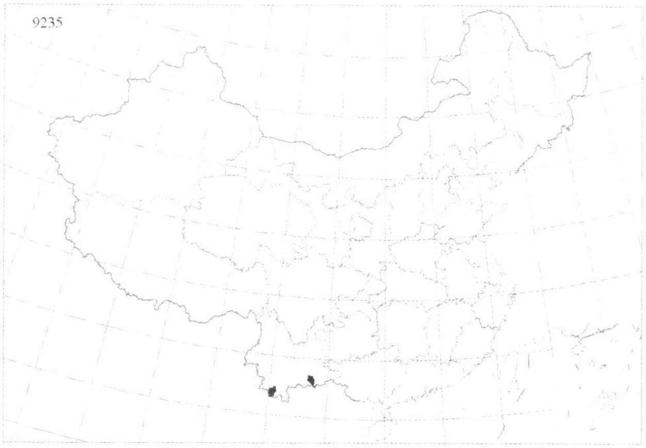

Convolvulaceae 旋花科 *Argyreia* 银背藤属
Argyreia osyrensis (Roth) Choisy 聚花白鹤藤
Scandent shrubs

20.5 (17.4~24.8); 14.6 (11.0~20.1); 24.4 (21.4~28.4); 20.5 (17.4~24.8); 186.5 (149.3~237.6); 0.0 (0.0~0.0); 1032 (834~1385) ❙ 1368 (976~1607); 668 (392~911); 55 (41~77) ❙ 36.6 (-11.9~74.2); 961 (815~1137); 480.3 (293.0~659.0)

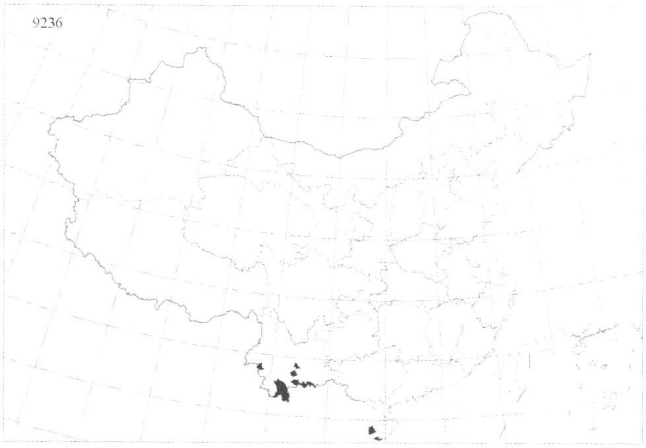

Convolvulaceae 旋花科 *Argyreia* 银背藤属
Argyreia osyrensis var. cinerea Hand.-Mazz. 灰毛白鹤藤
Scandent shrubs

18.4 (16.9~21.0); 12.0 (10.3~14.2); 22.9 (21.4~25.6); 18.5 (16.9~21.0); 161.9 (142.7~191.8); 0.0 (0.0~0.0); 894 (813~1053) ❙ 1285 (958~1547); 722 (531~911); 54 (44~68) ❙ 43.6 (13.3~74.2); 888 (813~1053); 448.6 (311.9~551.6)

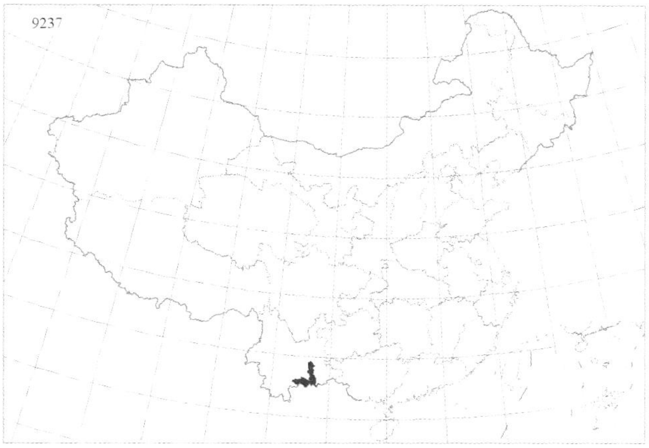

Convolvulaceae 旋花科 *Argyreia* 银背藤属
Argyreia pierreana Bois 东京银背藤
Vines

18.9; 10.6; 25.0; 19.0; 167.5; 0.0; 939 ❙ 1292; 729; 49 ❙ 37.6; 939; 342.5

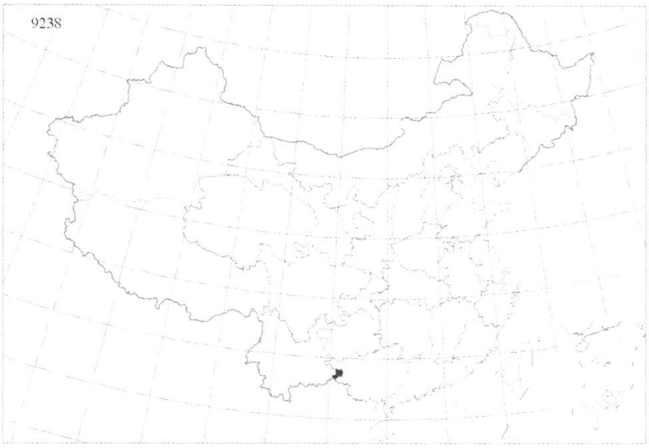

Convolvulaceae 旋花科 *Argyreia* 银背藤属
Argyreia splendens (Hornem.) Sweet 亮叶银背藤
Scandent shrubs

16.4; 8.9; 22.0; 16.5; 137.4; 0.0; 859 ❙ 1472; 771; 104 ❙ 77.7; 859; 377.4

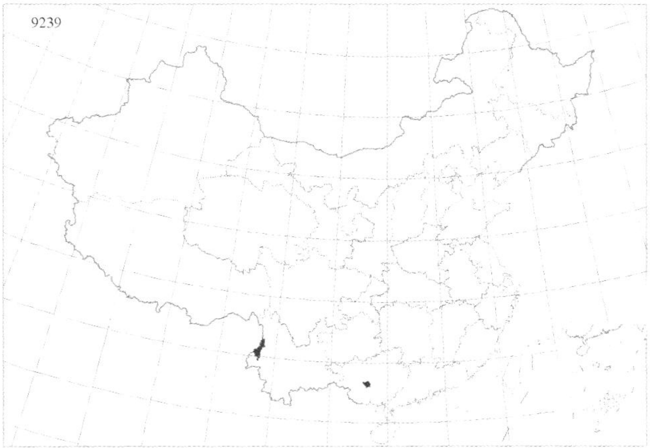

Convolvulaceae 旋花科 *Argyreia* 银背藤属
Argyreia strigillosa C. Y. Wu et Li 细毛银背藤
Scandent shrubs

18.5; 12.3; 22.6; 18.6; 163.0; 0.0; 890 ❙ 1480; 866; 55 ❙ 67.2; 887; 530.8

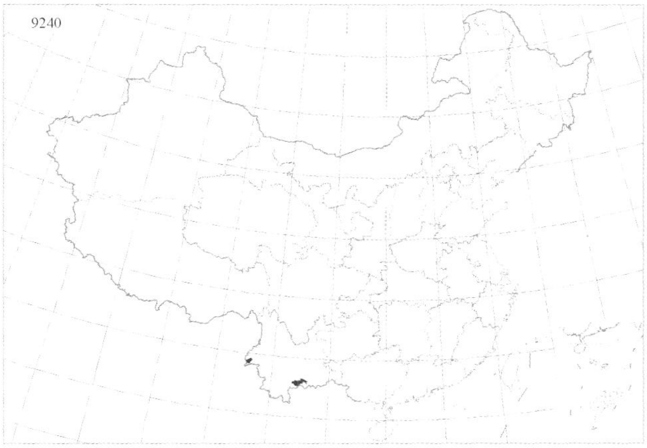

Convolvulaceae 旋花科 *Argyreia* 银背藤属

Argyreia velutina C. Y. Wu et Li 黄毛银背藤

Scandent shrubs

19.2 (17.5~21.0); 12.9 (10.5~15.1); 23.3 (22.3~25.6); 19.2 (17.5~21.0); 170.3 (149.9~191.8); 0.0 (0.0~0.0); 930 (846~1053) ❙ 1391 (958~1607); 699 (531~818); 61 (44~77) ❙ 49.1 (13.3~64.9); 920 (823~1053); 505.6 (311.9~659.0)

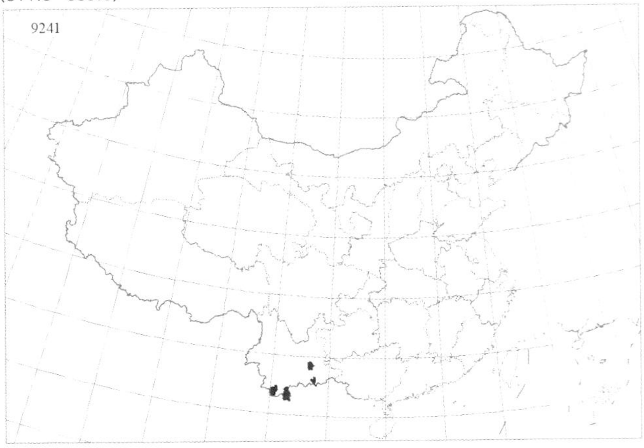

Convolvulaceae 旋花科 *Argyreia* 银背藤属

Argyreia wallichii Choisy 大叶银背藤

Vines

18.4 (14.4~22.2); 11.1 (5.5~15.1); 23.5 (19.1~28.6); 18.4 (14.4~22.3); 160.7 (113.1~207.5); 0.0 (0.0~0.0); 906 (727~1190) ❙ 1303 (956~1751); 702 (546~996); 62 (20~194) ❙ 44.2 (5.5~99.9); 890 (719~1190); 360.7 (189.4~659.0)

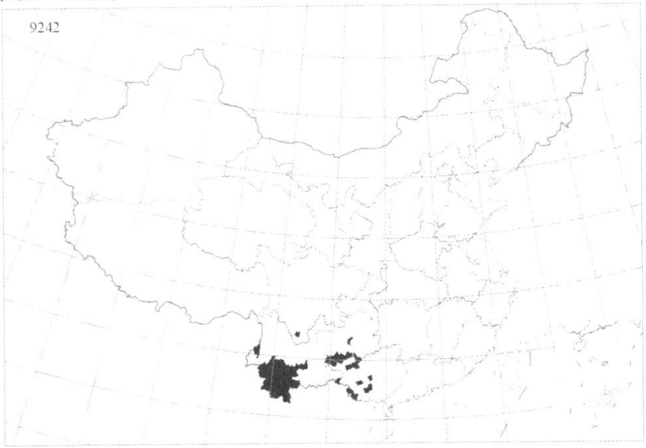

Convolvulaceae 旋花科 *Blinkworthia* 苞叶藤属

Blinkworthia convolvuloides Prain 苞叶藤

Scandent shrubs

18.8 (13.1~21.7); 11.1 (6.4~13.3); 24.6 (18.4~28.4); 18.9 (13.2~21.8); 166.3 (98.1~201.3); 0.0 (0.0~0.0); 945 (692~1162) ❙ 1227 (1059~1372); 671 (604~774); 60 (43~97) ❙ 31.3 (5.8~68.1); 939 (692~1162); 341.6 (273.6~452.0)

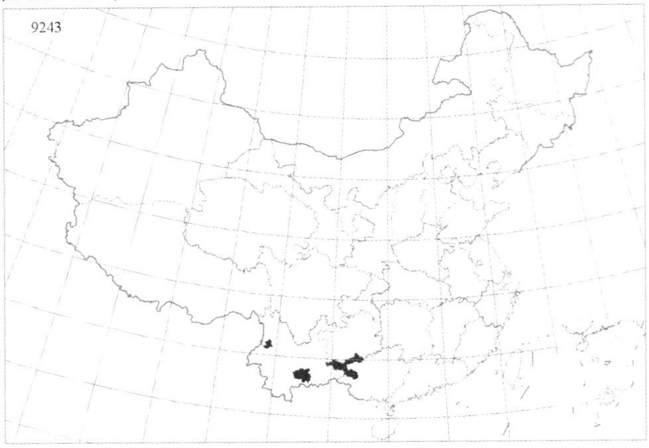

Convolvulaceae 旋花科 *Convolvulus* 旋花属

Convolvulus fruticosus Pall. 灌木旋花

Shrubs or subshrubs, deciduous

4.7 (-4.3~11.9); -13.4 (-20.9~-5.6); 19.6 (9.6~27.5); 8.1 (2.5~12.7); 62.5 (9.7~107.9); 65.7 (25.2~121.1); 583 (329~813) ❙ 193 (53~450); 96 (23~211); 14 (2~48) ❙ -64.1 (-92.4~3.9); 193 (53~406); 80.7 (10.3~297.3)

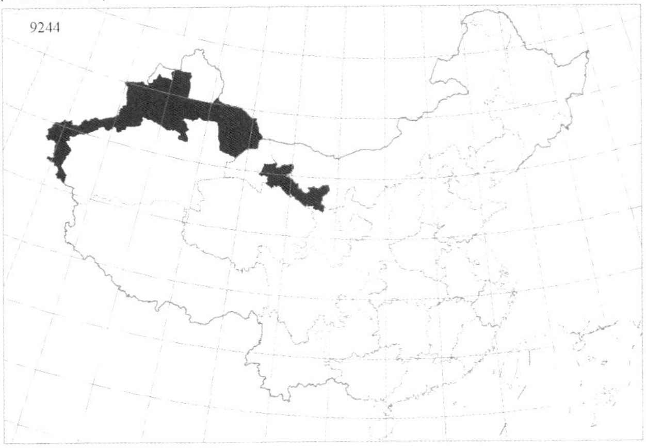

Convolvulaceae 旋花科 *Convolvulus* 旋花属

Convolvulus gortschakovii Schrenk 鹰爪柴

Shrubs or subshrubs, deciduous

5.4 (-4.3~12.4); -12.9 (-20.1~-5.3); 21.2 (9.6~27.5); 8.7 (2.5~13.2); 69.6 (9.7~114.1); 63.8 (24.7~121.1); 611 (329~813) ❙ 181 (38~450); 93 (19~265); 13 (1~51) ❙ -69.3 (-95.3~3.9); 180 (38~416); 58.2 (2.8~173.1)

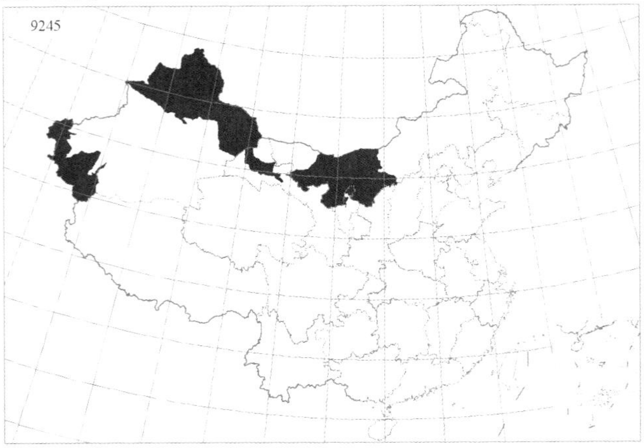

Convolvulaceae 旋花科 *Convolvulus* 旋花属

Convolvulus tragacanthoides Turcz. 刺旋花

Subshrubs, deciduous

5.9 (-2.8~13.5); -10.8 (-21.5~1.4); 20.5 (10.1~27.5); 8.6 (2.9~13.5); 67.1 (12.0~113.3); 55.5 (7.1~109.2); 600 (358~813) ❙ 337 (32~950); 178 (21~507); 13 (0~51) ❙ -41.5 (-95.9~128.6); 304 (32~767); 122.1 (1.6~334.4)

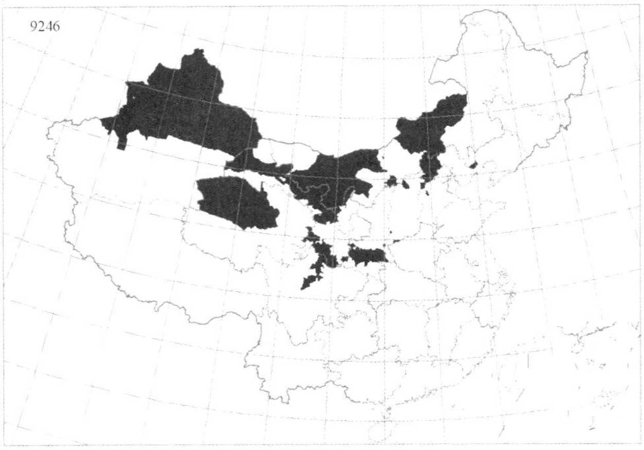

Convolvulaceae 旋花科 *Erycibe* 丁公藤属
Erycibe elliptilimba Merr. et Chun 九来龙
Scandent shrubs

24.0 (22.2~24.8); 18.4 (17.2~20.1); 28.0 (25.7~28.8); 24.0 (22.3~24.8); 228.3 (207.2~237.6); 0.0 (0.0~0.0); 1310 (1127~1385) ❙ 1501 (868~1854); 534 (312~712); 74 (41~115) ❙ 15.0 (-34.3~47.8); 1191 (868~1360); 440.3 (248.2~620.0)

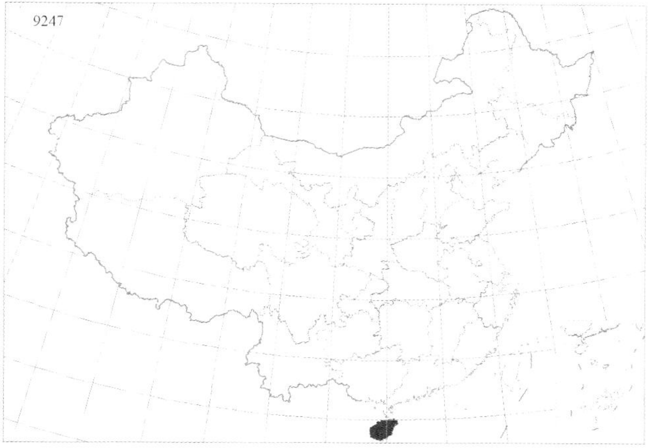

Convolvulaceae 旋花科 *Erycibe* 丁公藤属
Erycibe expansa Wall. ex G. Don 锈毛丁公藤
Scandent shrubs

17.4; 9.7; 22.9; 17.5; 149.7; 0.0; 848 ❙ 1338; 756; 50 ❙ 57.6; 848; 348.0

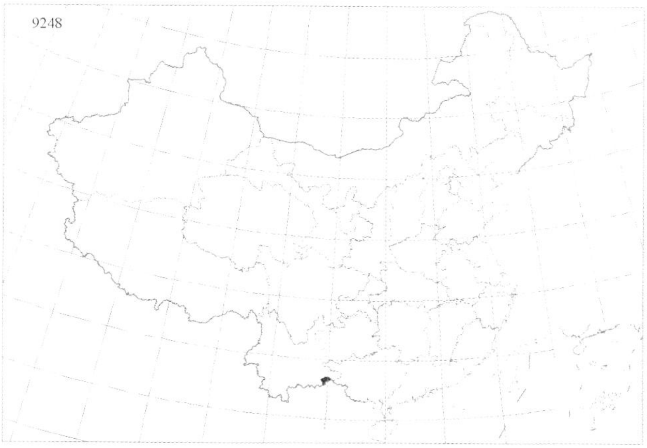

Convolvulaceae 旋花科 *Erycibe* 丁公藤属
Erycibe hainanensis Merr. 毛叶丁公藤
Scandent shrubs

23.5 (21.8~24.8); 17.5 (13.9~20.1); 27.8 (25.7~28.6); 23.5 (21.9~24.8); 222.5 (202.2~237.6); 0.0 (0.0~0.0); 1273 (1127~1385) ❙ 1651 (1179~2435); 676 (392~1428); 77 (41~130) ❙ 30.8 (-11.9~99.5); 1179 (1055~1360); 450.4 (235.1~620.0)

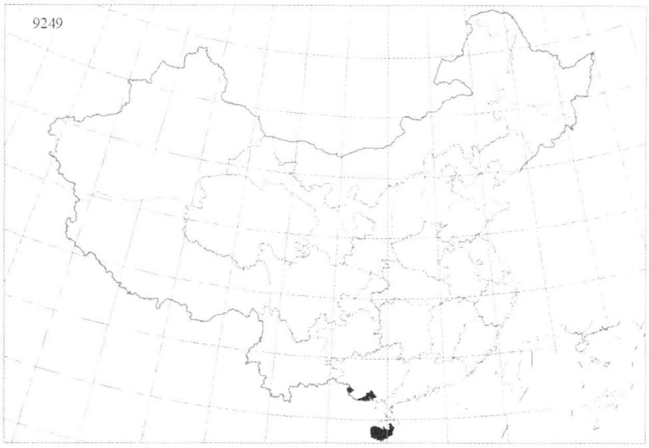

Convolvulaceae 旋花科 *Erycibe* 丁公藤属
Erycibe henryi Prain 台湾丁公藤
Scandent shrubs

19.6 (16.2~22.2); 13.9 (11.0~17.7); 24.5 (20.1~27.4); 19.6 (16.2~22.2); 175.3 (134.2~206.7); 0.0 (0.0~0.0); 968 (773~1123) ❙ 2851 (2288~3495); 1139 (821~1847); 397 (107~897) ❙ 196.6 (159.6~246.5); 968 (773~1123); 282.2 (226.8~363.7)

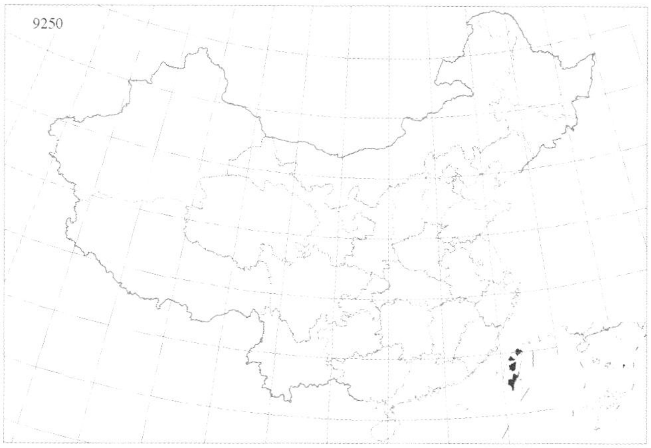

Convolvulaceae 旋花科 *Erycibe* 丁公藤属
Erycibe myriantha Merr. 多花丁公藤
Scandent shrubs

23.7 (22.2~24.8); 18.3 (14.9~20.1); 27.6 (25.7~28.5); 23.8 (22.3~24.8); 225.5 (207.2~237.6); 0.0 (0.0~0.0); 1288 (1127~1385) ❙ 1573 (1179~2035); 562 (392~969); 64 (41~115) ❙ 23.0 (-11.9~71.8); 1163 (1055~1360); 492.4 (237.9~620.0)

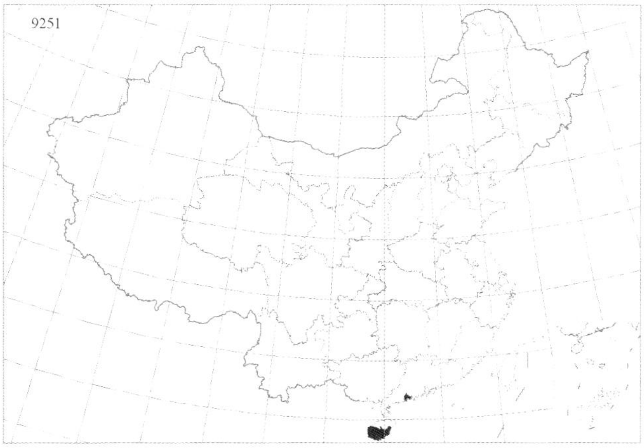

Convolvulaceae 旋花科 *Erycibe* 丁公藤属
Erycibe oligantha Merr. et Chun 疏花丁公藤
Scandent shrubs

24.2; 19.0; 27.9; 24.3; 231.3; 0.0; 1331 ❙ 1702; 580; 81 ❙ 27.7; 1254; 524.3

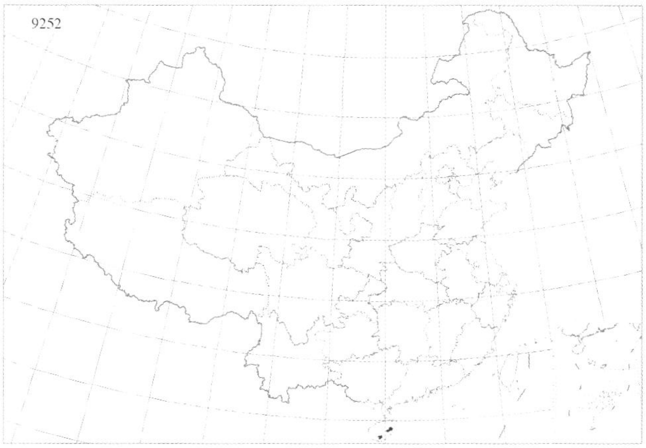

Convolvulaceae 旋花科 *Erycibe* 丁公藤属
Erycibe schmidtii Craib 光叶丁公藤
Vines

21.6 (16.9~24.8); 14.0 (7.3~20.1); 27.5 (21.4~29.2); 21.7 (16.9~24.8); 199.9 (142.7~237.6); 0.0 (0.0~0.0); 1142 (813~1385) ❙ 1602 (868~2435); 719 (312~1428); 111 (41~194) ❙ 42.0 (-34.3~99.5); 1115 (813~1360); 306.3 (139.6~659.0)

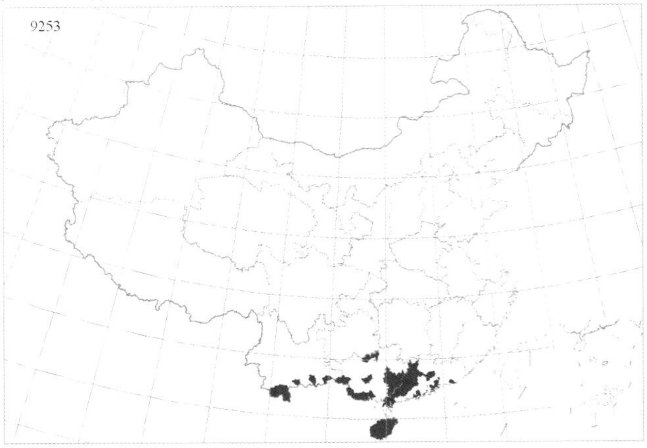

Convolvulaceae 旋花科 *Erycibe* 丁公藤属
Erycibe sinii F. C. How 瑶山丁公藤
Scandent shrubs

19.5 (17.6~21.3); 9.7 (7.4~12.3); 27.6 (25.8~28.6); 19.5 (17.7~21.3); 174.2 (152.5~196.1); 0.0 (0.0~0.0); 1021 (906~1130) ❙ 1593 (1371~1758); 659 (547~728); 175 (161~189) ❙ 57.3 (22.7~85.5); 1020 (906~1130); 223.4 (185.5~260.5)

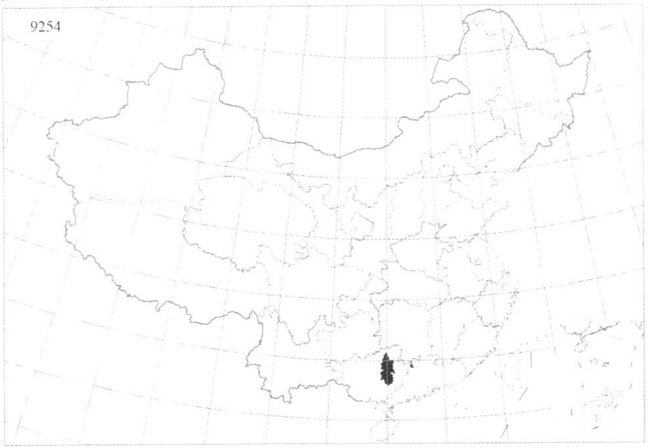

Convolvulaceae 旋花科 *Erycibe* 丁公藤属
Erycibe subspicata Wall. ex G. Don 锥序丁公藤
Scandent shrubs

18.9 (15.9~22.2); 12.2 (6.9~15.1); 23.4 (21.6~28.3); 18.9 (16.0~22.3); 167.1 (131.6~207.5); 0.0 (0.0~0.0); 926 (802~1190) ❙ 1400 (1049~1607); 711 (598~813); 63 (44~87) ❙ 52.4 (15.4~65.6); 912 (783~1190); 435.4 (203.5~659.0)

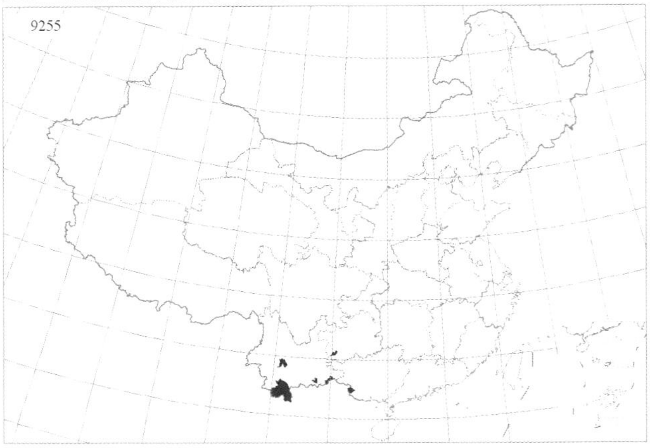

Convolvulaceae 旋花科 *Neuropeltis* 盾苞藤属
Neuropeltis racemosa Wall. 盾苞藤
Vines

23.5 (20.4~24.8); 18.1 (14.9~20.1); 27.2 (23.6~28.7); 23.5 (20.5~24.8); 222.0 (185.8~237.6); 0.0 (0.0~0.0); 1261 (989~1385) ❙ 1464 (868~1854); 526 (312~702); 62 (41~115) ❙ 17.9 (-34.3~62.5); 1115 (868~1360); 506.7 (288.3~659.0)

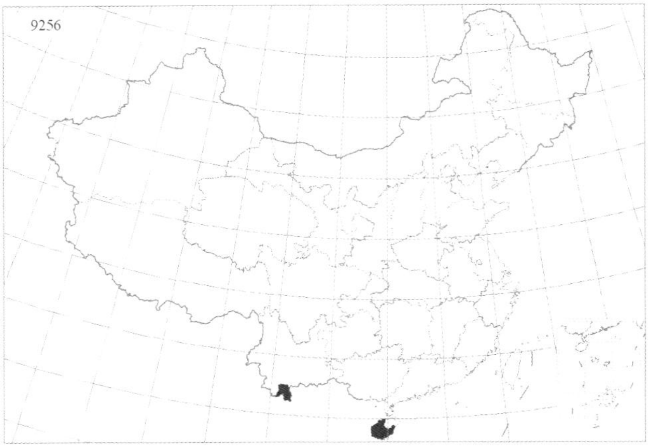

Convolvulaceae 旋花科 *Poranopsis* 白花叶属
Poranopsis discifera (C. K. Schneid.) Staples 搭棚藤
Vines

18.0 (13.2~20.5); 11.9 (6.4~15.1); 22.0 (18.4~23.7); 18.1 (13.3~20.5); 156.7 (99.4~185.9); 0.0 (0.0~0.0); 872 (695~992) ❙ 1326 (976~1607); 690 (556~835); 57 (42~77) ❙ 52.7 (12.2~92.4); 852 (695~986); 436.4 (293.0~659.0)

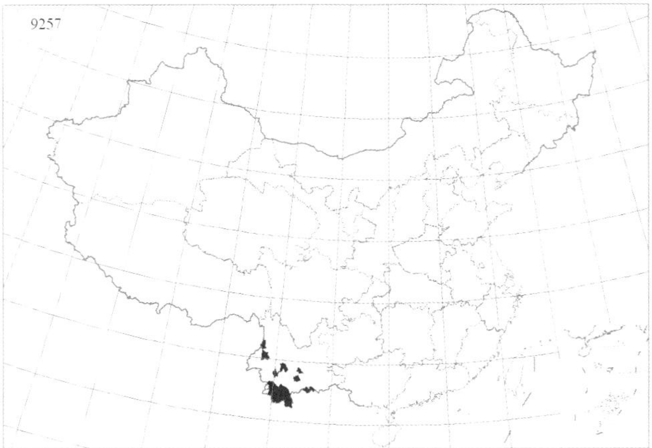

Convolvulaceae 旋花科 *Poranopsis* 白花叶属
Poranopsis paniculata (Roxb.) Roberty 圆锥白花叶
Vines

16.4; 9.6; 20.6; 16.4; 137.3; 0.0; 795 ❙ 1688; 959; 62 ❙ 112.2; 795; 627.0

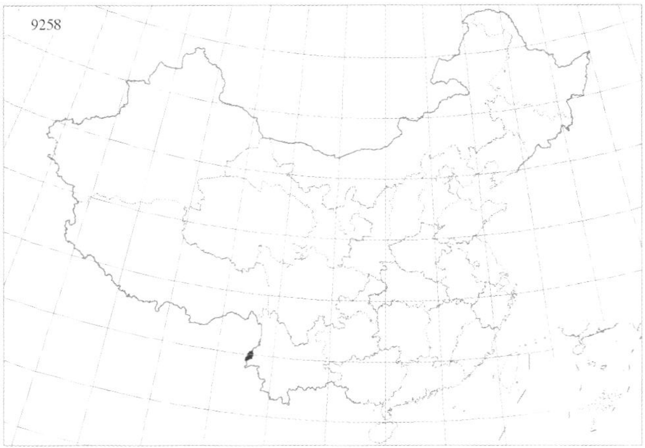

Convolvulaceae 旋花科 *Poranopsis* 白花叶属
Poranopsis sinensis (Hand.-Mazz.) Staples 白花叶
Vines

15.1 (11.6~18.9); 7.6 (2.9~12.7); 21.1 (18.6~24.0); 15.1 (11.5~19.0); 122.3 (80.9~167.8); 0.7 (0.0~3.4); 770 (654~929) ❙ 1064 (797~1547); 609 (411~911); 37 (22~63) ❙ 38.9 (-14.2~74.2); 743 (654~888); 300.9 (212.0~551.3)

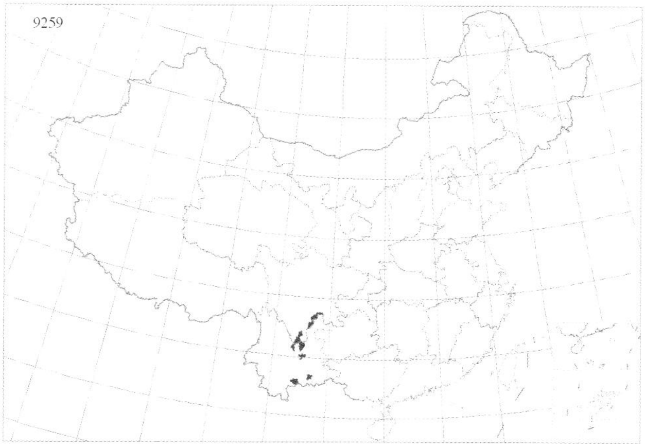

Convolvulaceae 旋花科 *Stictocardia* 腺叶藤属
Stictocardia tiliifolia (Desr.) Hallier f. 腺叶藤
Vines

22.3 (18.5~24.8); 15.8 (13.2~20.1); 27.5 (21.7~29.2); 22.4 (18.5~24.8); 208.3 (162.1~237.6); 0.0 (0.0~0.0); 1182 (863~1385) ❙ 2005 (1269~3495); 925 (473~1847); 171 (41~897) ❙ 76.6 (-1.0~224.4); 1146 (863~1331); 258.6 (119.4~548.0)

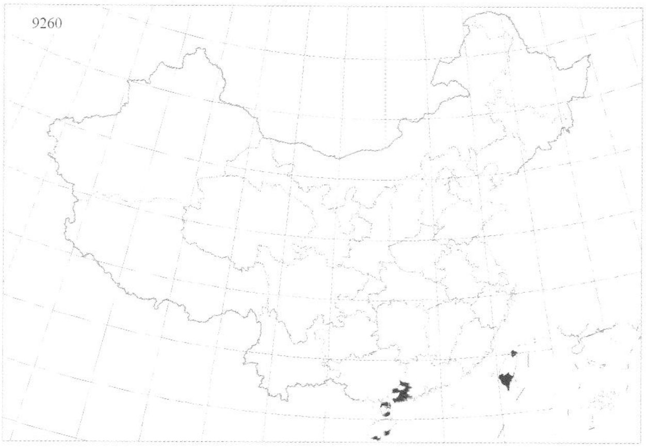

Convolvulaceae 旋花科 *Tridynamia* 三翅藤属
Tridynamia megalantha (Merr.) Staples 大花三翅藤
Vines

20.8 (17.6~24.8); 12.5 (7.3~20.1); 27.2 (23.6~28.8); 20.9 (17.6~24.8); 190.5 (151.6~237.6); 0.0 (0.0~0.0); 1091 (901~1385) ❙ 1457 (1284~1625); 702 (455~850); 102 (41~173) ❙ 35.0 (1.5~64.7); 1070 (901~1190); 323.0 (188.6~659.0)

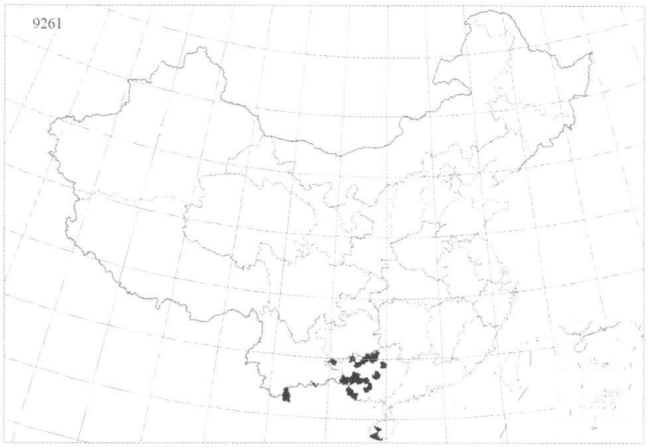

Convolvulaceae 旋花科 *Tridynamia* 三翅藤属
Tridynamia sinensis (Hemsl.) Staples 大果三翅藤
Vines

16.4 (3.2~21.8); 5.9 (-6.4~15.1); 26.0 (11.4~29.8); 16.5 (4.7~21.9); 138.9 (22.7~202.6); 1.3 (0.0~43.8); 873 (436~1150) ❙ 1364 (791~1876); 546 (392~891); 125 (16~230) ❙ 56.9 (5.8~128.9); 862 (436~1150); 225.8 (130.1~659.0)

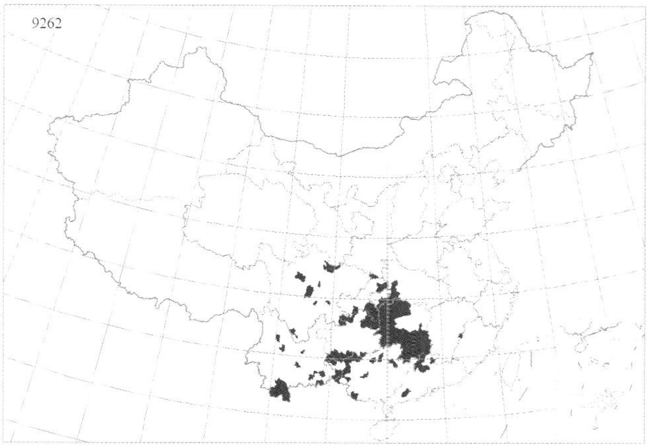

Convolvulaceae 旋花科 *Tridynamia* 三翅藤属
Tridynamia sinensis var. delavayi (Gagnep. et Courchet) Staples 近无毛三翅藤 Vines

13.3 (2.2~19.1); 3.5 (-6.6~10.1); 22.0 (9.9~26.9); 13.5 (3.8~19.1); 107.0 (15.5~169.2); 7.0 (0.0~49.1); 742 (413~969) ❙ 1142 (711~1418); 575 (357~816); 49 (15~153) ❙ 56.9 (8.2~125.9); 739 (413~969); 243.7 (130.1~357.7)

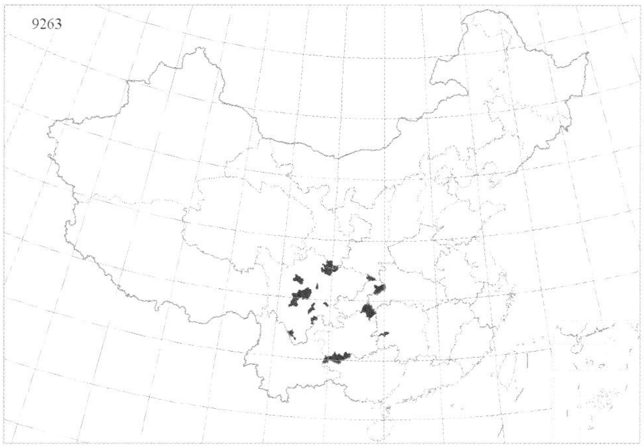

Boraginaceae 紫草科 *Carmona* 基及树属
Carmona microphylla (Lam.) G. Don 基及树
Shrubs, evergreen

23.3 (19.3~24.8); 17.1 (13.5~20.1); 27.8 (22.8~29.0); 23.3 (19.3~24.8); 219.6 (171.9~237.6); 0.0 (0.0~0.0); 1254 (912~1385) ❙ 1708 (868~2993); 773 (312~1847); 100 (41~547) ❙ 40.8 (-34.3~203.4); 1158 (868~1331); 327.1 (119.4~572.9)

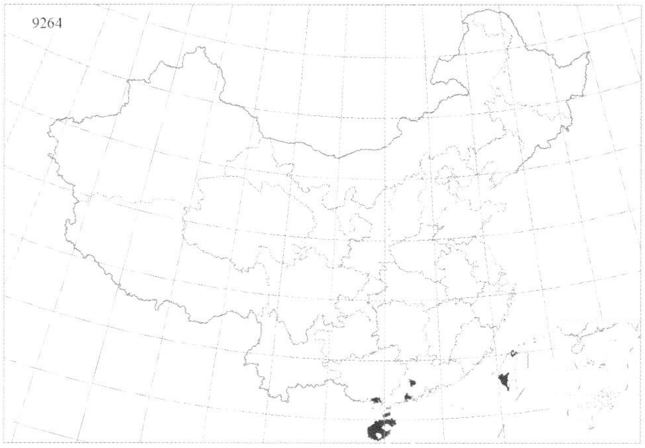

Boraginaceae 紫草科 *Cordia* 破布木属

Cordia cochinchinensis Gagnep. 越南破布木

Shrubs or small trees

24.3; 19.4; 27.7; 24.4; 232.2; 0.0; 1339 ❙ 1393; 464; 45 ❙ 4.1; 1124; 491.4

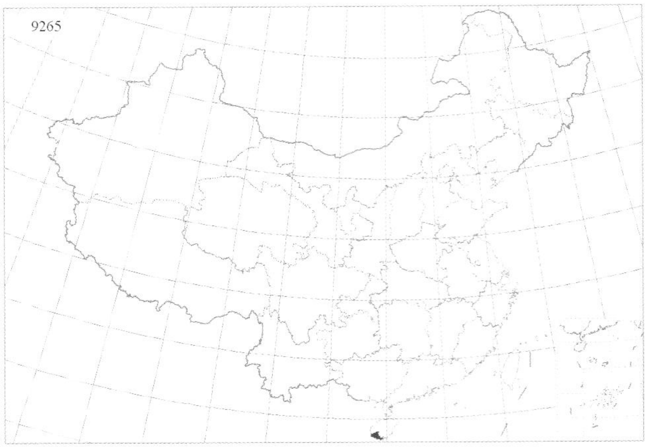

Boraginaceae 紫草科 *Cordia* 破布木属

Cordia dichotoma G. Forst. 破布木

Trees, deciduous

19.7 (−1.8~24.8); 11.8 (−10.2~20.1); 26.5 (6.1~29.8); 19.9 (1.8~24.8); 179.0 (1.9~237.6); 1.2 (0.0~82.8); 1031 (329~1385) ❙ 1534 (320~3495); 632 (193~1594); 144 (20~897) ❙ 50.8 (−34.3~224.4); 1004 (320~1331); 290.1 (72.2~659.0)

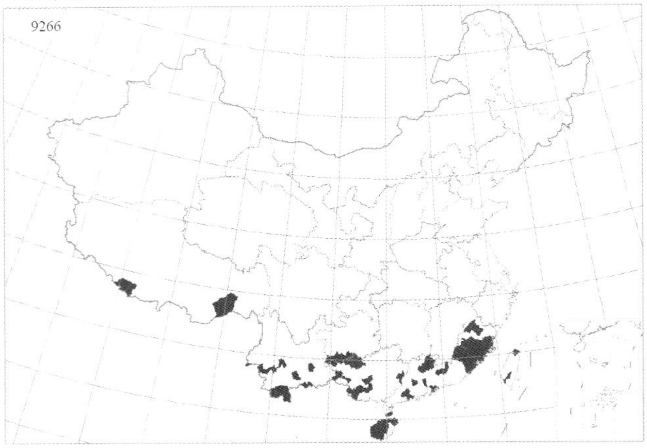

Boraginaceae 紫草科 *Cordia* 破布木属

Cordia furcans I. M. Johnst. 二叉破布木

Trees

21.1 (13.8~24.8); 13.9 (6.7~20.1); 26.2 (19.1~28.8); 21.1 (13.8~24.8); 193.2 (105.9~237.6); 0.0 (0.0~0.0); 1100 (714~1385) ❙ 1459 (868~2435); 715 (312~1428); 80 (29~173) ❙ 35.3 (−34.3~99.9); 1058 (690~1360); 367.1 (237.8~659.0)

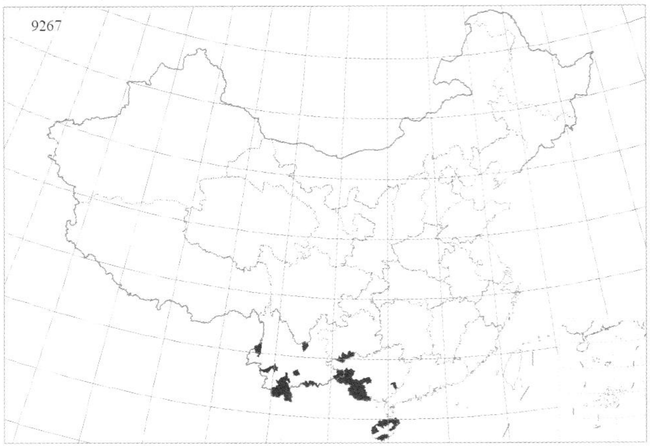

Boraginaceae 紫草科 *Ehretia* 厚壳树属

Ehretia acuminata (DC.) R. Br. 厚壳树

Trees, deciduous

16.4 (−1.8~24.8); 5.6 (−10.7~20.1); 26.1 (6.1~29.9); 16.4 (1.8~24.8); 140.3 (1.9~237.6); 3.5 (0.0~82.8); 886 (329~1385) ❙ 1297 (320~2993); 538 (217~1847); 123 (11~551) ❙ 45.9 (−36.5~246.5); 862 (320~1360); 227.9 (72.2~659.0)

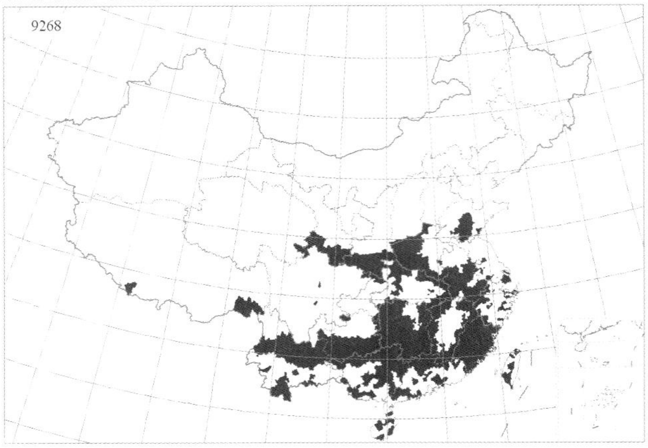

Boraginaceae 紫草科 *Ehretia* 厚壳树属

Ehretia asperula Zoll. et Moritzi 宿苞厚壳树

Scandent shrubs

22.0 (18.2~24.8); 14.8 (8.7~20.1); 27.5 (25.5~29.0); 22.1 (18.3~24.8); 205.0 (159.3~237.6); 0.0 (0.0~0.0); 1177 (924~1385) ❙ 1513 (1260~2035); 687 (473~997); 81 (41~113) ❙ 29.6 (−1.0~71.8); 1135 (924~1299); 367.4 (199.6~620.0)

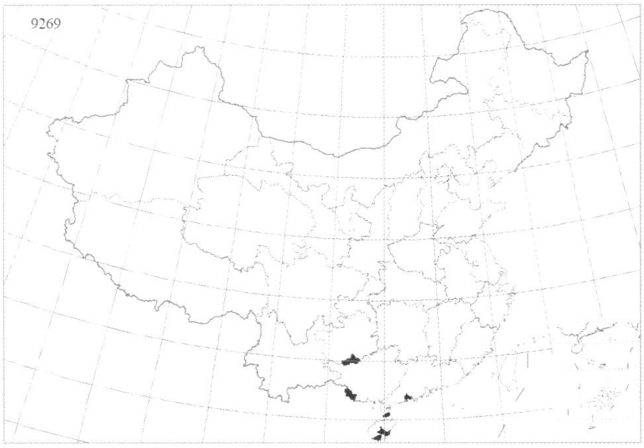

Boraginaceae 紫草科 *Ehretia* 厚壳树属

Ehretia changjiangensis F. W. Xing et Z. X. Li 昌江厚壳树

Shrubs

23.6; 17.8; 27.8; 23.6; 223.3; 0.0; 1271 ❙ 1242; 444; 44 ❙ −2.2; 1091; 439.7

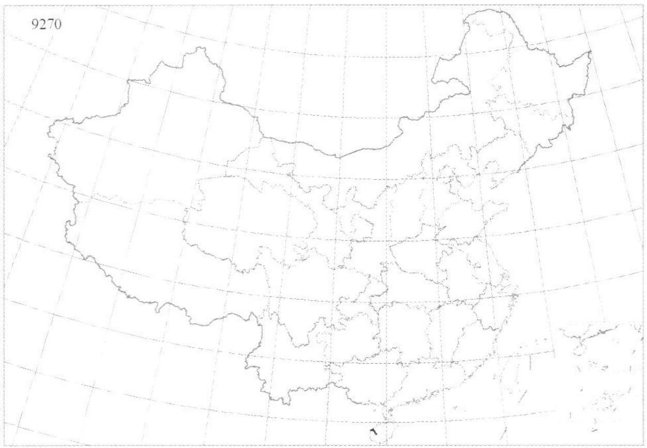

Boraginaceae 紫草科 *Ehretia* 厚壳树属
Ehretia confinis I. M. Johnst. 云南厚壳树
Trees

14.5; 7.3; 19.9; 14.5; 116.8; 2.2; 756 | 1218; 677; 53 | 64.2; 755; 396.7

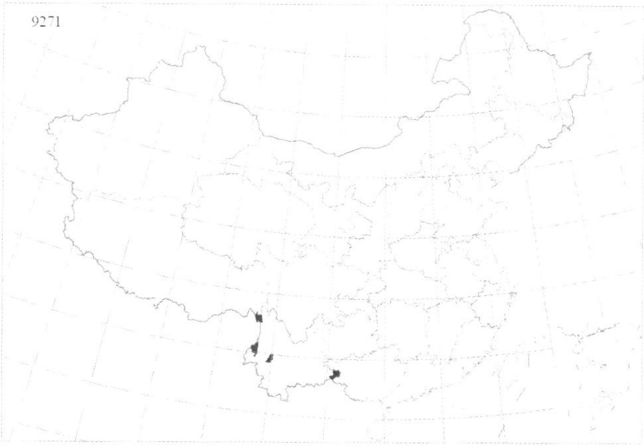

Boraginaceae 紫草科 *Ehretia* 厚壳树属
Ehretia corylifolia C. H. Wright 西南粗糠树
Trees

13.4 (2.2~19.3); 6.0 (-6.6~13.1); 19.4 (9.5~26.6); 13.6 (3.8~19.4); 105.9 (15.5~173.2); 4.2 (0.0~49.1); 718 (411~939) | 1016 (729~1495); 586 (411~835); 35 (11~70) | 44.5 (-14.2~125.9); 696 (411~917); 296.8 (142.9~506.2)

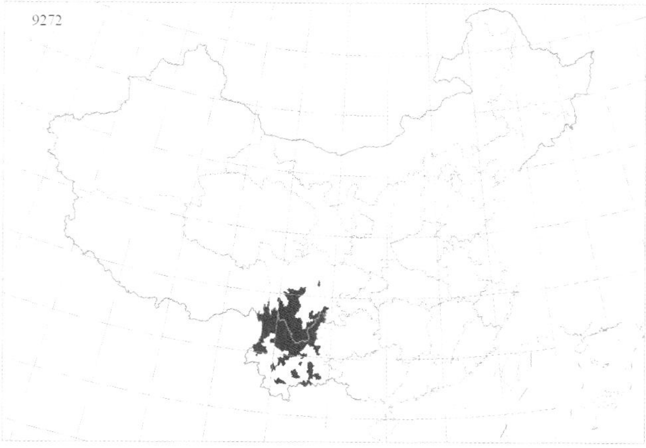

Boraginaceae 紫草科 *Ehretia* 厚壳树属
Ehretia dicksonii Hance 粗糠树
Trees

15.8 (-1.8~24.2); 4.9 (-10.7~18.0); 25.8 (6.1~29.9); 15.9 (1.8~24.2); 133.5 (1.9~230.4); 3.9 (0.0~82.8); 858 (329~1331) | 1304 (320~2993); 519 (217~1847); 123 (11~551) | 51.9 (-22.2~246.5); 847 (320~1331); 222.3 (72.2~659.0)

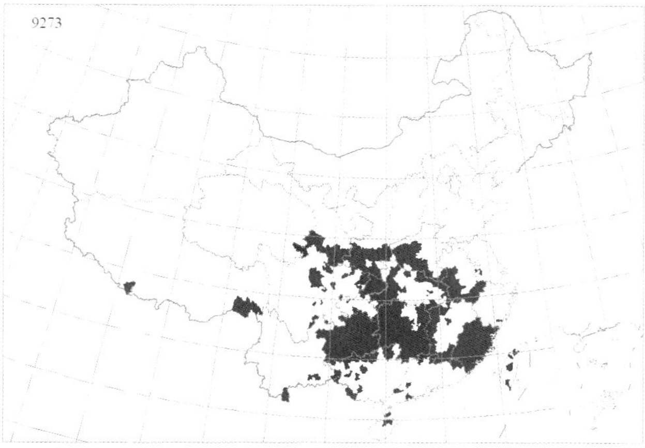

Boraginaceae 紫草科 *Ehretia* 厚壳树属
Ehretia dunniana H. Lév. 云贵厚壳树
Trees

18.5 (16.8~21.0); 11.0 (7.8~14.2); 23.9 (21.6~26.6); 18.5 (16.8~21.0); 161.7 (141.5~191.8); 0.0 (0.0~0.0); 908 (832~1053) | 1319 (1049~1547); 722 (598~911); 57 (44~68) | 45.6 (26.1~74.2); 903 (783~1053); 370.2 (189.5~551.6)

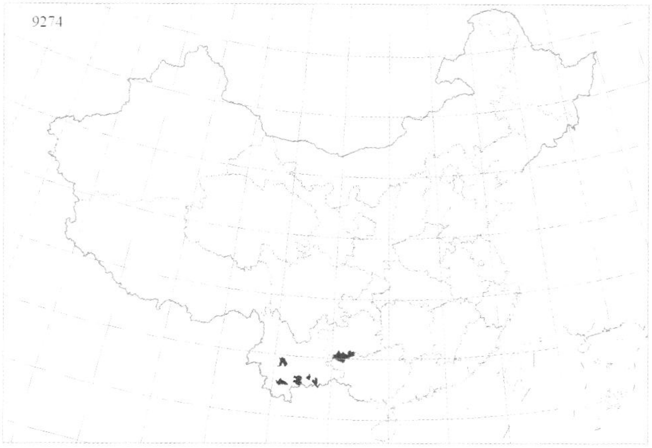

Boraginaceae 紫草科 *Ehretia* 厚壳树属
Ehretia hainanensis I. M. Johnst. 海南厚壳树
Trees

24.8; 20.1; 28.0; 24.8; 237.6; 0.0; 1385 | 1405; 473; 41 | 1.5; 1137; 527.5

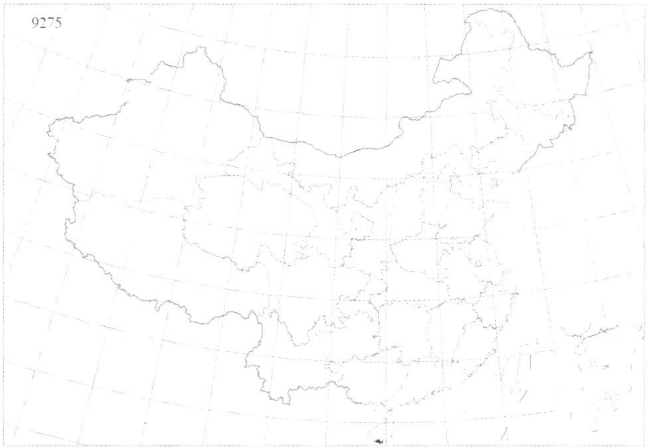

Boraginaceae 紫草科 *Ehretia* 厚壳树属
Ehretia laevis Roxb. 毛萼厚壳树
Trees

23.9 (23.1~24.8); 18.7 (17.5~20.1); 27.7 (27.2~28.4); 24.0 (23.2~24.8); 227.7 (218.4~237.6); 0.0 (0.0~0.0); 1304 (1232~1385) | 1374 (1179~1551); 474 (392~544); 46 (41~53) | 5.6 (-11.9~20.7); 1114 (1056~1147); 506.0 (439.7~620.0)

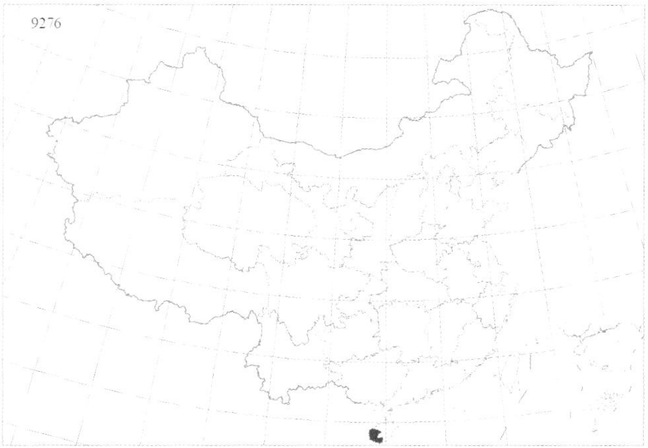

Boraginaceae 紫草科 *Ehretia* 厚壳树属
Ehretia longiflora Champ. ex Benth. 长花厚壳树
Trees

18.6 (14.3~24.8); 9.0 (3.3~20.1); 27.0 (20.1~29.8); 18.6 (14.3~24.8); 163.9 (112.6~237.6); 0.1 (0.0~2.3); 977 (756~1385) ❙ 1510 (868~2993); 605 (312~1186); 146 (41~547) ❙ 56.4 (-34.3~246.5); 964 (756~1331); 243.6 (134.6~659.0)

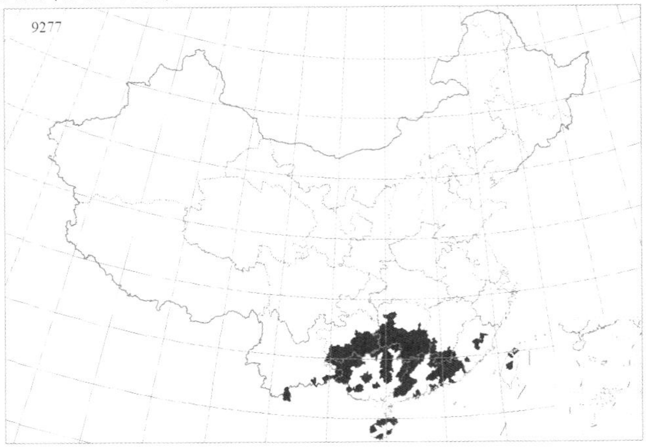

Boraginaceae 紫草科 *Ehretia* 厚壳树属
Ehretia pingbianensis Y. L. Liu 屏边厚壳树
Trees

18.7 (14.4~21.0); 12.4 (7.5~15.1); 22.8 (19.1~25.6); 18.7 (14.4~21.0); 164.6 (113.1~191.8); 0.0 (0.0~0.0); 907 (727~1053) ❙ 1441 (1292~1607); 755 (655~835); 60 (45~77) ❙ 60.1 (37.6~99.9); 896 (727~1053); 500.5 (342.5~659.0)

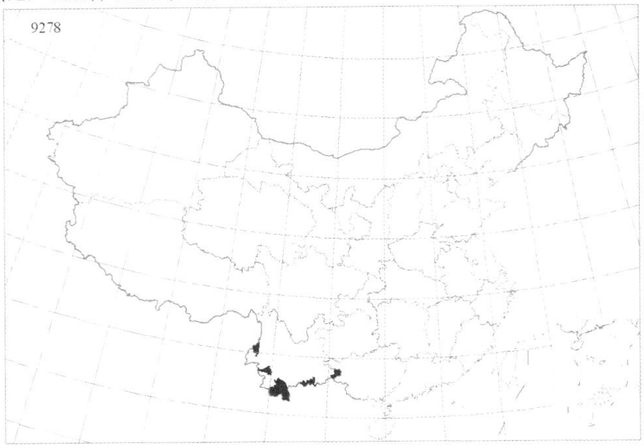

Boraginaceae 紫草科 *Ehretia* 厚壳树属
Ehretia resinosa Hance 台湾厚壳树
Shrubs or trees

19.3; 14.6; 22.9; 19.4; 172.3; 0.0; 953 ❙ 2746; 1397; 205 ❙ 192.9; 953; 300.3

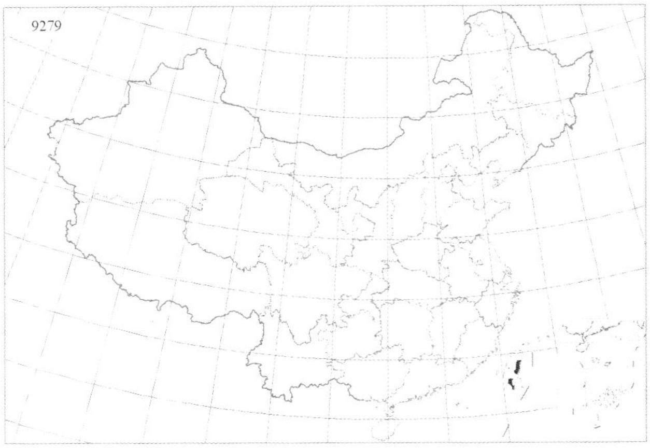

Boraginaceae 紫草科 *Ehretia* 厚壳树属
Ehretia tsangii I. M. Johnst. 上思厚壳树
Small trees

20.0 (14.4~22.7); 12.1 (7.5~15.1); 25.9 (19.1~28.9); 20.0 (14.4~22.8); 180.5 (113.1~213.1); 0.0 (0.0~0.0); 1024 (727~1221) ❙ 1483 (1182~2435); 793 (617~1428); 82 (43~173) ❙ 45.7 (12.9~99.9); 1021 (727~1221); 346.2 (189.5~659.0)

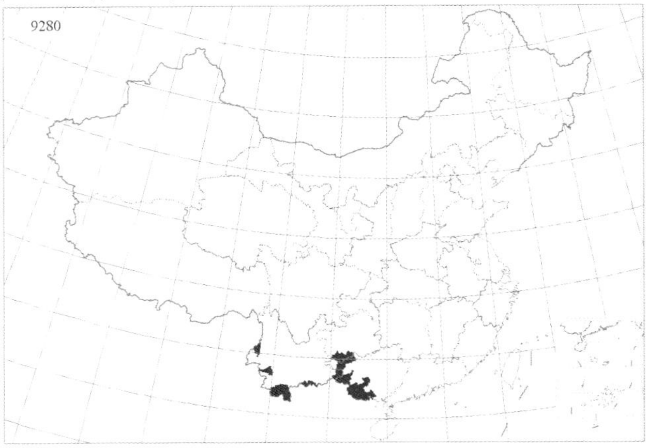

Boraginaceae 紫草科 *Rotula* 轮冠木属
Rotula aquatica Lour. 轮冠木
Shrubs, deciduous

18.8 (15.9~22.3); 10.4 (6.9~15.1); 25.2 (23.0~28.9); 18.8 (16.0~22.3); 165.8 (131.6~208.1); 0.0 (0.0~0.0); 947 (802~1202) ❙ 1325 (1172~1607); 701 (617~835); 64 (43~136) ❙ 40.8 (16.9~64.7); 947 (802~1202); 322.9 (189.5~659.0)

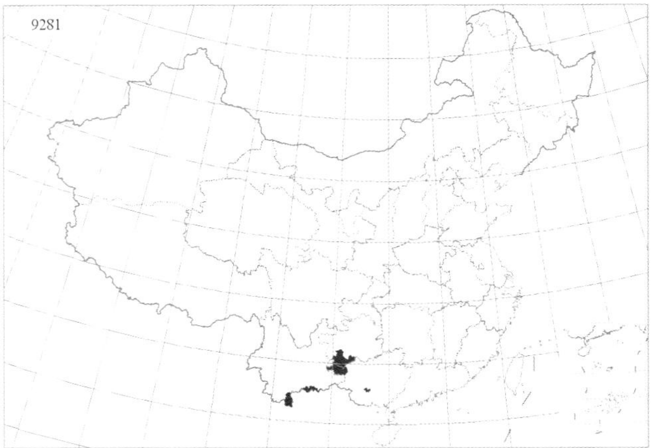

Boraginaceae 紫草科 *Tournefortia* 紫丹属
Tournefortia argentea L. f. 银毛树
Shrubs or small trees

21.3; 16.3; 25.2; 21.4; 196.4; 0.0; 1089 ❙ 2526; 1158; 221 ❙ 146.4; 1027; 362.4

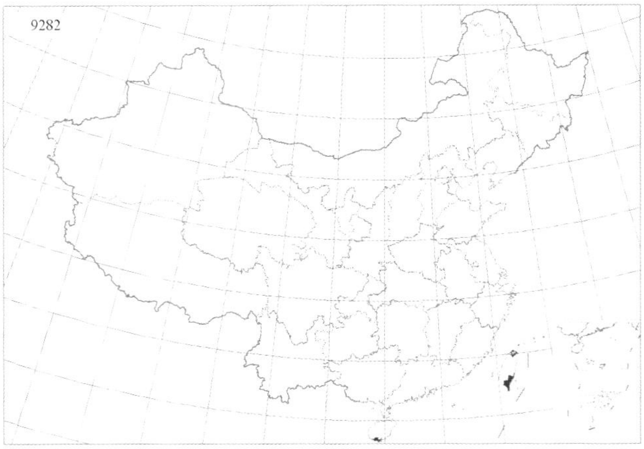

Boraginaceae 紫草科 *Tournefortia* 紫丹属
Tournefortia montana Lour. 紫丹
Scandent shrubs

21.6 (16.3~24.8); 14.9 (9.1~20.1); 26.3 (21.4~28.8); 21.6 (16.4~24.8); 199.7 (136.4~237.6); 0.0 (0.0~0.0); 1127 (797~1385) ❙ 1484 (1188~1813); 680 (473~850); 79 (41~144) ❙ 34.1 (-1.4~64.9); 1094 (797~1331); 408.5 (157.7~659.0)

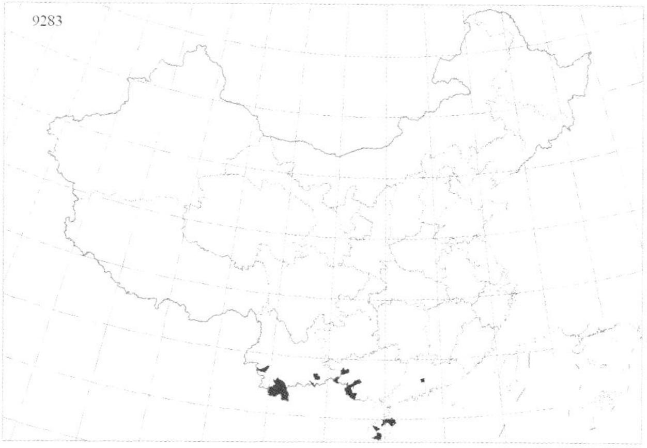

Boraginaceae 紫草科 *Tournefortia* 紫丹属
Tournefortia sarmentosa Lam. 台湾紫丹
Scandent shrubs

22.4; 17.2; 26.3; 22.5; 209.5; 0.0; 1164 ❙ 2118; 1268; 79 ❙ 83.5; 1093; 320.9

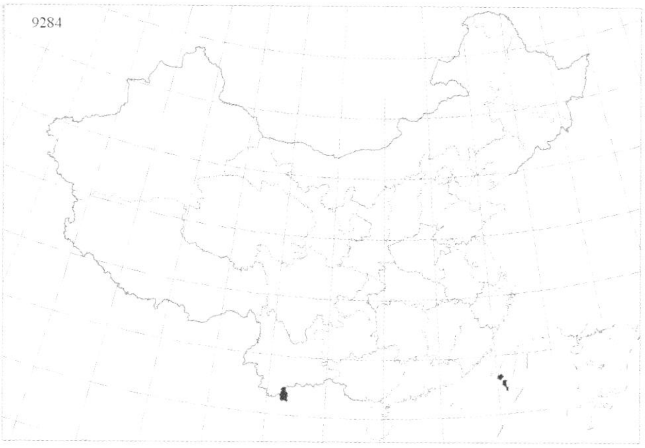

Verbenaceae 马鞭草科 *Avicennia* 海榄雌属
Avicennia marina (Forssk.) Vierh. 海榄雌
Shrubs

21.9 (17.9~24.8); 14.5 (10.5~20.1); 28.2 (23.4~29.2); 21.9 (18.0~24.8); 203.3 (155.6~237.6); 0.0 (0.0~0.0); 1161 (864~1385) ❙ 1627 (868~3495); 701 (312~1428); 147 (41~897) ❙ 41.6 (-34.3~224.4); 1129 (864~1331); 227.9 (116.6~548.0)

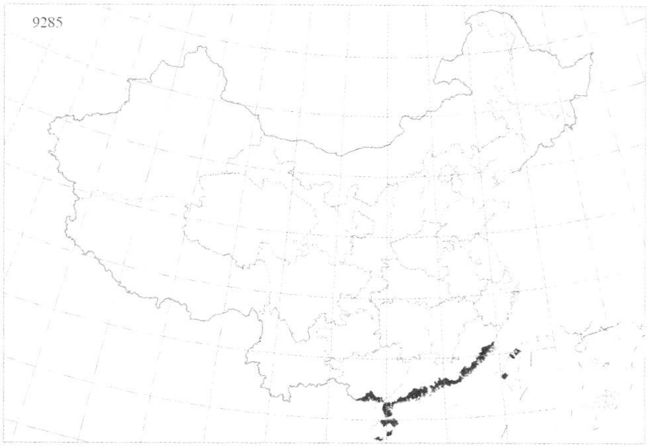

Verbenaceae 马鞭草科 *Callicarpa* 紫珠属
Callicarpa acutifolia H. T. Chang 尖叶紫珠
Shrubs

22.3; 14.5; 28.4; 22.3; 208.1; 0.0; 1190 ❙ 1637; 741; 117 ❙ 38.5; 1187; 218.8

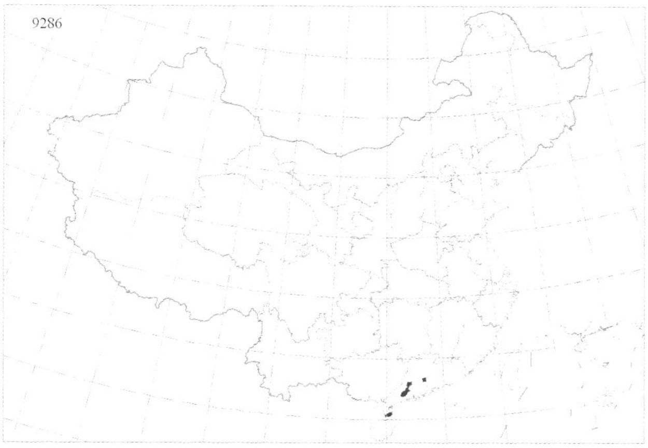

Verbenaceae 马鞭草科 *Callicarpa* 紫珠属
Callicarpa anisophylla C. Y. Wu ex W. Z. Fang 异叶紫珠
Shrubs or subshrubs

17.6 (16.1~19.2); 7.4 (5.3~10.1); 26.1 (25.5~27.5); 17.6 (16.1~19.2); 151.3 (133.1~170.8); 0.0 (0.0~0.0); 908 (838~1002) ❙ 1370 (1256~1496); 609 (523~701); 114 (51~142) ❙ 51.2 (30.0~64.7); 908 (838~1002); 243.0 (222.1~282.2)

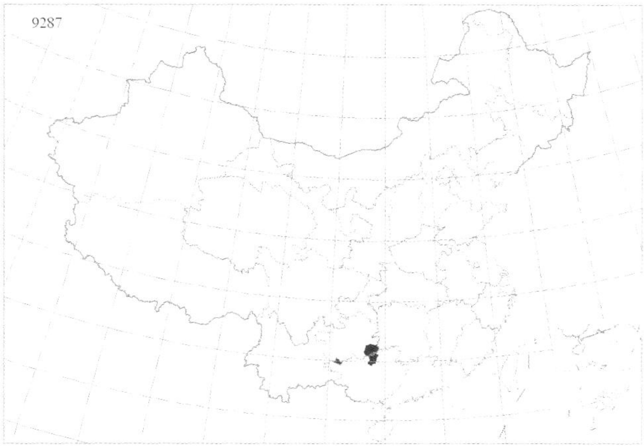

Verbenaceae 马鞭草科 *Callicarpa* 紫珠属
Callicarpa arborea Roxb. 木紫珠
Trees

17.5 (8.6~22.2); 10.7 (1.3~15.1); 22.1 (15.0~28.4); 17.5 (8.7~22.3); 150.5 (52.6~207.5); 0.1 (0.0~8.8); 859 (574~1190) ❙ 1287 (939~1751); 711 (524~1007); 54 (37~106) ❙ 51.1 (6.5~145.8); 846 (574~1190); 403.0 (273.6~659.0)

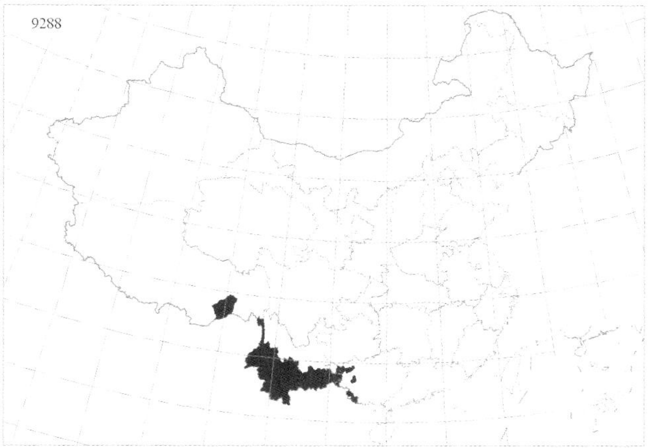

Verbenaceae 马鞭草科 *Callicarpa* 紫珠属
Callicarpa basitruncata Merr. et Moldenke 平基紫珠
Shrubs

24.0 (22.2~24.8); 18.4 (17.2~20.1); 28.0 (25.7~28.8); 24.0 (22.3~24.8); 228.3 (207.2~237.6); 0.0 (0.0~0.0); 1310 (1127~1385) ❙ 1501 (868~1854); 534 (312~712); 74 (41~115) ❙ 15.0 (-34.3~47.8); 1191 (868~1360); 440.3 (248.2~620.0)

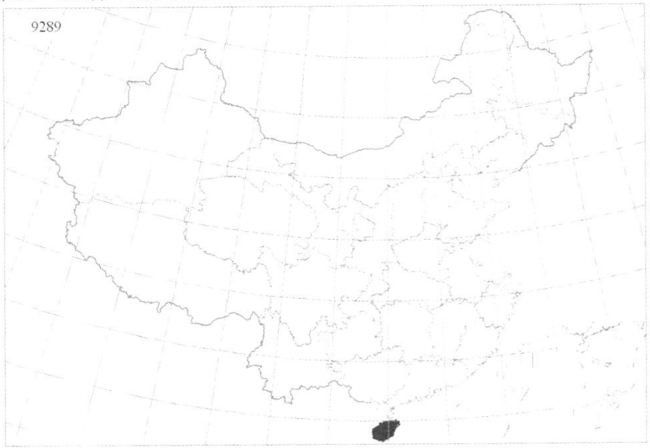

Verbenaceae 马鞭草科 *Callicarpa* 紫珠属
Callicarpa bodinieri H. Lév. 紫珠
Shrubs

16.3 (5.4~22.2); 6.1 (-4.2~15.1); 25.6 (12.3~30.1); 16.4 (5.8~22.3); 137.9 (31.5~207.5); 1.6 (0.0~29.4); 867 (481~1190) ❙ 1341 (738~2024); 559 (332~996); 122 (17~259) ❙ 54.4 (-14.2~130.7); 858 (481~1190); 242.9 (108.2~659.0)

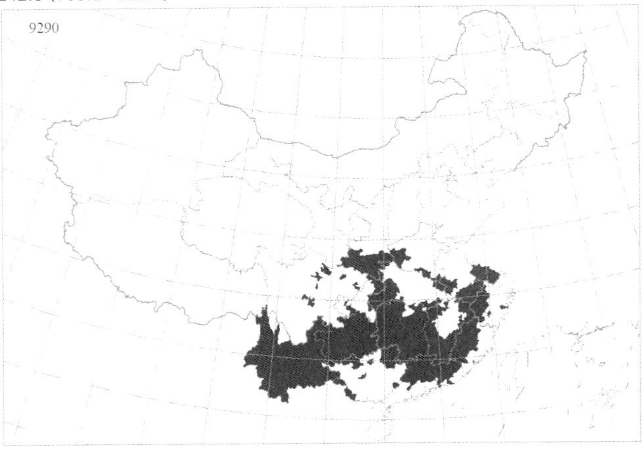

Verbenaceae 马鞭草科 *Callicarpa* 紫珠属
Callicarpa bodinieri var. iteophylla C. Y. Wu 柳叶紫珠
Shrubs

19.3 (17.2~21.0); 13.0 (9.1~15.1); 23.3 (21.6~25.6); 19.3 (17.3~21.0); 171.6 (147.5~191.8); 0.0 (0.0~0.0); 937 (832~1053) ❙ 1384 (1049~1607); 682 (598~818); 61 (44~77) ❙ 47.1 (26.1~64.9); 918 (783~1053); 495.7 (314.4~659.0)

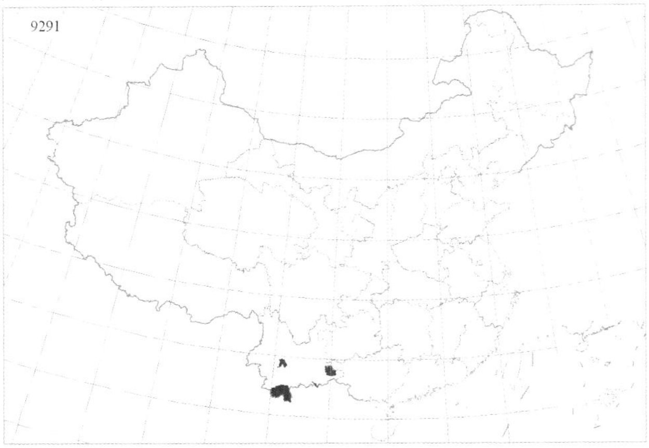

Verbenaceae 马鞭草科 *Callicarpa* 紫珠属
Callicarpa bodinieri var. rosthornii (Diels) Rehder 南川紫珠
Shrubs

12.8 (9.4~22.2); 1.2 (-2.5~13.9); 23.7 (21.0~28.3); 13.0 (9.8~22.3); 105.2 (73.9~207.5); 10.7 (0.0~20.6); 744 (619~1190) ❙ 913 (760~1374); 421 (352~723); 36 (17~87) ❙ 23.0 (5.0~51.2); 744 (619~1190); 271.5 (182.9~334.4)

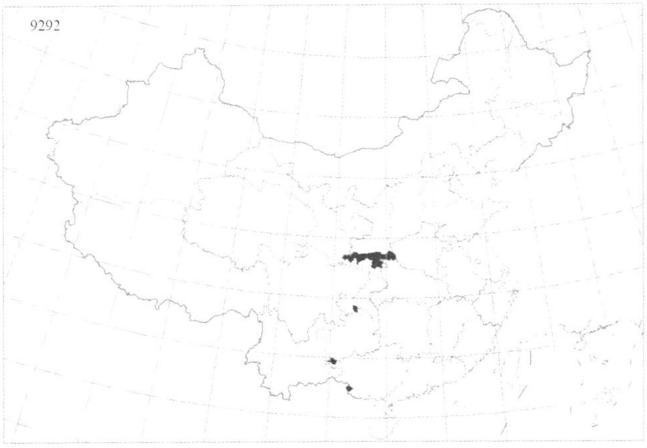

Verbenaceae 马鞭草科 *Callicarpa* 紫珠属
Callicarpa brevipes (Benth.) Hance 短柄紫珠
Shrubs

20.4 (14.3~24.8); 11.7 (3.3~20.1); 27.8 (24.3~29.8); 20.5 (14.3~24.8); 185.5 (114.1~237.6); 0.1 (0.0~2.3); 1082 (765~1385) ❙ 1604 (868~2037); 647 (312~1033); 142 (41~233) ❙ 50.7 (-34.3~147.0); 1065 (765~1360); 245.1 (95.3~620.0)

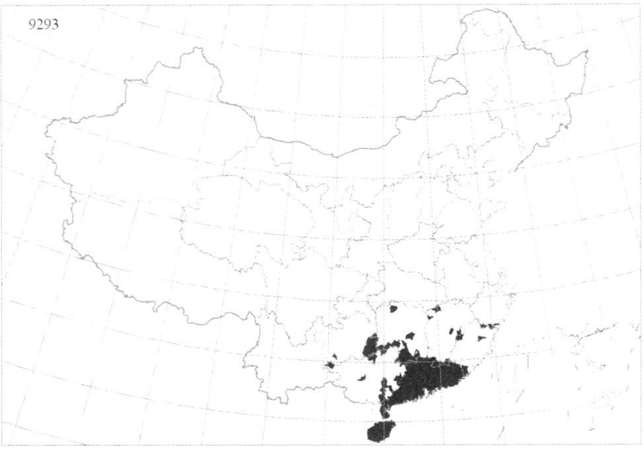

Verbenaceae 马鞭草科 *Callicarpa* 紫珠属
Callicarpa brevipes var. obovata H. T. Chang 倒卵叶短柄紫珠
Shrubs

22.5 (20.4~24.8); 15.2 (11.9~20.1); 28.2 (25.7~29.1); 22.5 (20.4~24.8); 210.4 (184.9~237.6); 0.0 (0.0~0.0); 1203 (1039~1385) ❙ 1676 (868~2037); 729 (312~1033); 111 (41~188) ❙ 40.6 (-34.3~83.4); 1169 (868~1360); 270.6 (95.3~620.0)

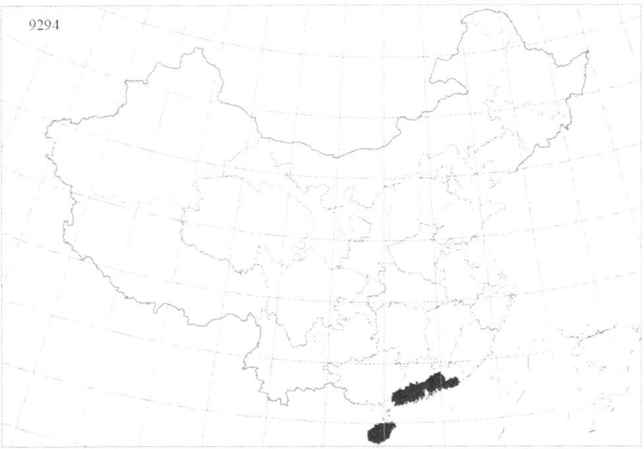

Verbenaceae 马鞭草科 *Callicarpa* 紫珠属

Callicarpa longibracteata H. T. Chang 长苞紫珠

Shrubs

22.4；14.9；28.5；22.5；210.1；0.0；1200 ❙ 1947；979；107 ❙ 62.4；1200；174.1

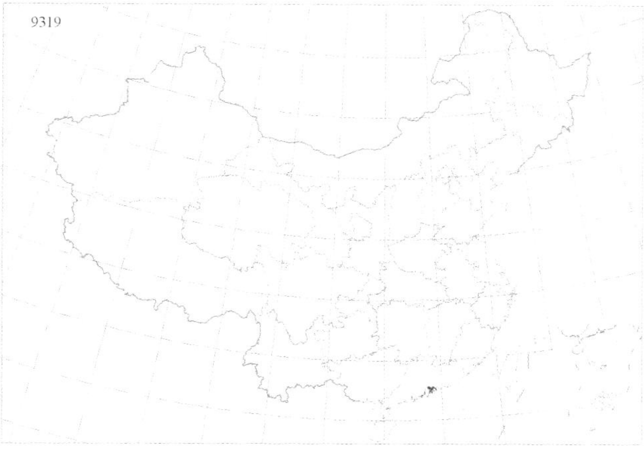

Verbenaceae 马鞭草科 *Callicarpa* 紫珠属

Callicarpa longifolia Lam. 长叶紫珠

Shrubs

20.3 (12.0~24.8)；12.6 (1.1~20.1)；26.4 (19.1~29.1)；20.4 (12.0~24.8)；184.5 (92.3~237.6)；0.1 (0.0~8.5)；1057 (690~1385) ❙ 1559 (1176~2037)；737 (392~1033)；96 (41~188) ❙ 49.7 (-12.0~112.2)；1040 (690~1326)；306.0 (95.3~659.0)

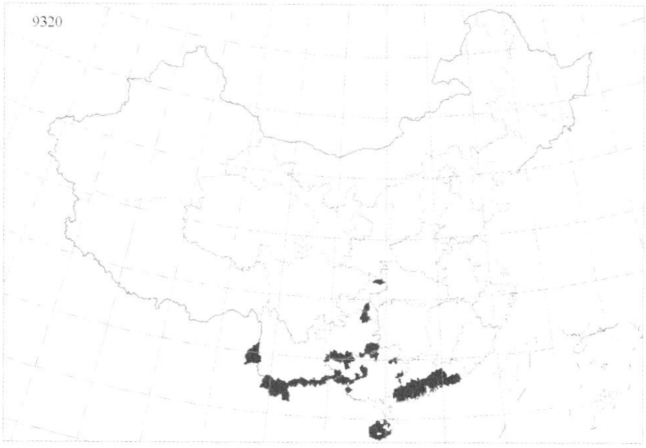

Verbenaceae 马鞭草科 *Callicarpa* 紫珠属

Callicarpa longifolia var. floccosa Schauer 白毛长叶紫珠

Shrubs

18.5 (14.4~22.2)；9.1 (3.8~13.9)；26.2 (23.4~28.3)；18.5 (14.5~22.3)；162.3 (115.2~207.5)；0.1 (0.0~1.5)；954 (776~1190) ❙ 1369 (1206~1710)；675 (481~848)；90 (51~186) ❙ 44.6 (15.4~79.6)；954 (776~1190)；256.4 (152.4~335.1)

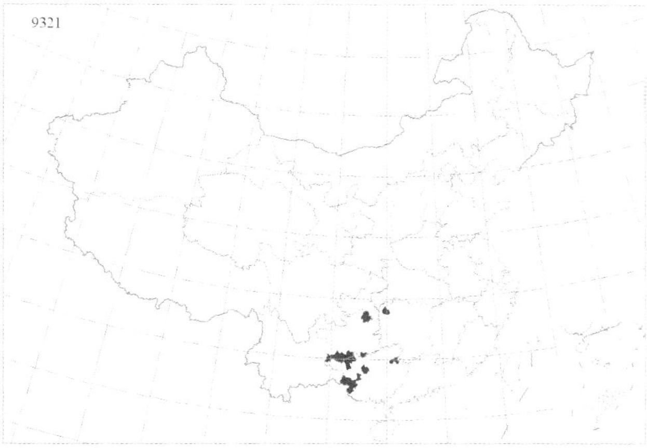

Verbenaceae 马鞭草科 *Callicarpa* 紫珠属

Callicarpa longifolia var. lanceolaria (Roxb.) C. B. Clarke 披针叶紫珠

Shrubs

23.2 (17.6~24.8)；17.1 (7.3~20.1)；27.7 (22.5~28.8)；23.3 (17.6~24.8)；219.1 (151.6~237.6)；0.0 (0.0~0.0)；1257 (888~1385) ❙ 1506 (868~1854)；569 (312~911)；84 (41~173) ❙ 21.7 (-34.3~74.2)；1159 (868~1360)；417.2 (220.1~620.0)

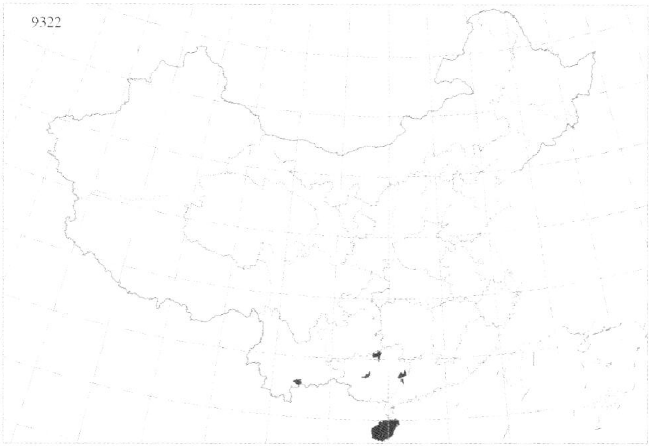

Verbenaceae 马鞭草科 *Callicarpa* 紫珠属

Callicarpa longipes Dunn 长柄紫珠

Shrubs

18.6 (14.4~21.8)；8.8 (2.1~14.3)；27.3 (24.6~29.1)；18.6 (14.5~21.8)；163.9 (117.7~202.0)；0.3 (0.0~4.9)；975 (779~1168) ❙ 1676 (1188~2005)；629 (497~902)；185 (56~259) ❙ 74.0 (5.8~147.0)；974 (779~1153)；254.3 (183.9~322.1)

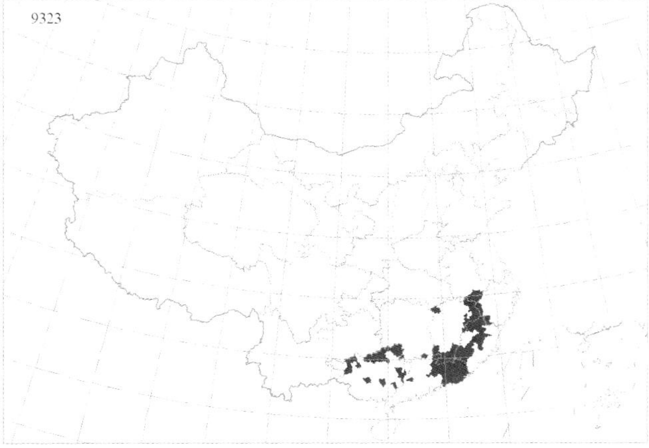

Verbenaceae 马鞭草科 *Callicarpa* 紫珠属

Callicarpa longissima (Hemsl.) Merr. 尖尾枫

Shrubs or small trees

19.4 (12.4~24.8)；10.7 (3.0~20.1)；26.7 (18.6~29.8)；19.5 (12.4~24.8)；174.1 (89.1~237.6)；0.1 (0.0~3.0)；1021 (675~1385) ❙ 1545 (868~2993)；659 (312~1428)；136 (29~551) ❙ 52.9 (-34.3~246.5)；1011 (675~1360)；250.7 (95.3~620.0)

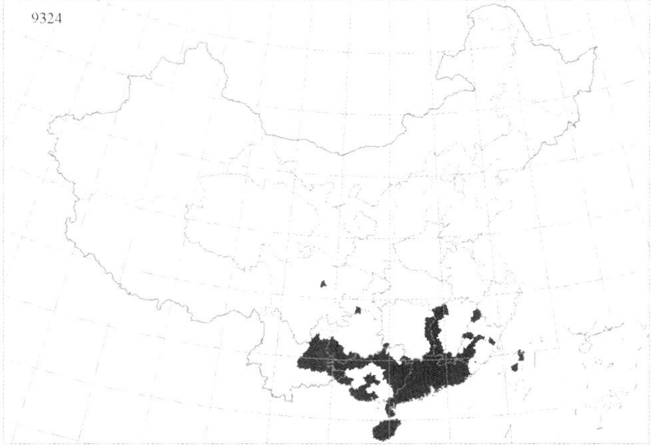

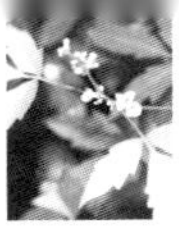

Verbenaceae 马鞭草科 *Callicarpa* 紫珠属

Callicarpa luteopunctata H. T. Chang 黄腺紫珠

Shrubs

15.4 (11.6~21.0); 6.3 (2.9~15.1); 23.4 (18.6~26.6); 15.5 (11.5~21.0); 126.3 (80.9~191.8); 0.9 (0.0~3.4); 808 (654~1053) ❙ 1195 (944~1607); 639 (507~911); 49 (24~77) ❙ 48.3 (9.2~89.0); 808 (654~1053); 277.1 (142.9~659.0)

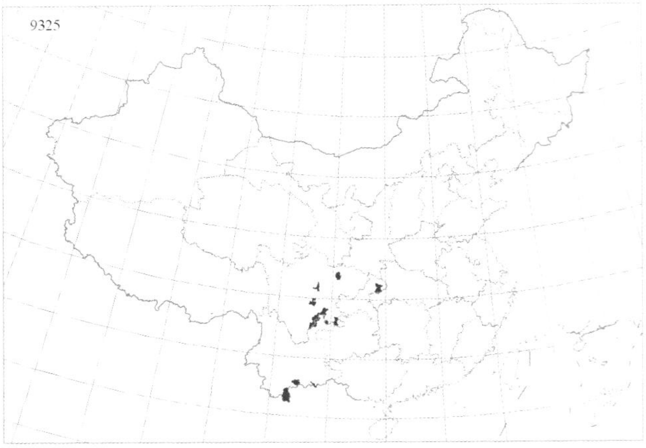

Verbenaceae 马鞭草科 *Callicarpa* 紫珠属

Callicarpa macrophylla Vahl 大叶紫珠

Shrubs or small trees

19.3 (12.4~24.8); 10.9 (3.3~20.1); 25.9 (18.4~29.8); 19.3 (12.4~24.8); 171.7 (89.1~237.6); 0.0 (0.0~2.3); 1000 (675~1385) ❙ 1455 (868~2435); 674 (312~1428); 107 (31~197) ❙ 46.7 (-34.3~112.2); 989 (675~1360); 275.5 (95.3~659.0)

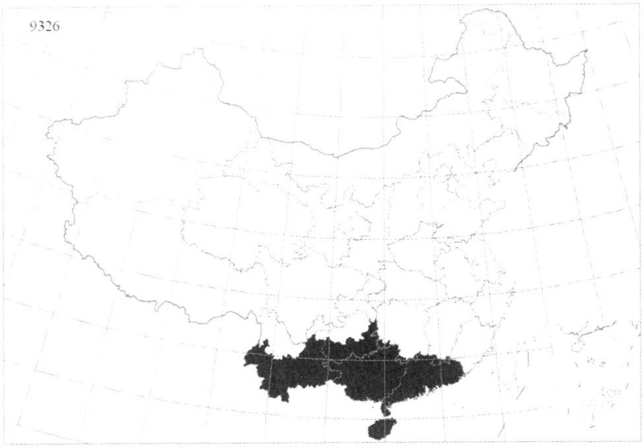

Verbenaceae 马鞭草科 *Callicarpa* 紫珠属

Callicarpa membranacea H. T. Chang 窄叶紫珠

Shrubs

15.0 (7.4~21.8); 3.4 (-4.2~14.2); 25.9 (18.1~29.3); 15.0 (8.2~21.9); 124.5 (58.8~202.6); 4.4 (0.0~29.4); 827 (561~1150) ❙ 1307 (520~1893); 505 (258~891); 129 (14~253) ❙ 57.2 (-35.9~129.9); 811 (520~1150); 225.6 (128.6~334.4)

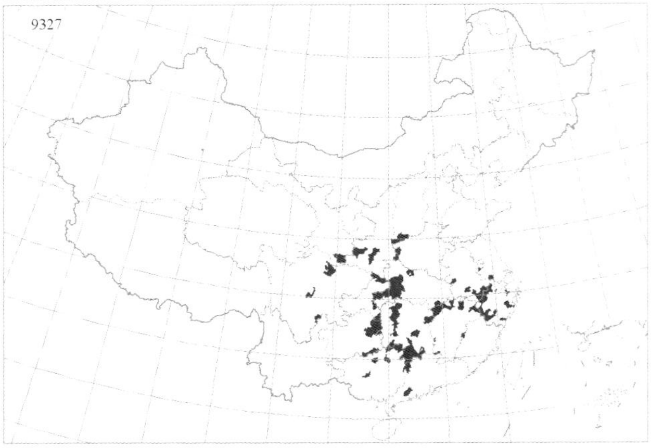

Verbenaceae 马鞭草科 *Callicarpa* 紫珠属

Callicarpa minutiflora Y. Y. Qian 细花紫珠

Shrubs

19.3; 13.5; 23.1; 19.4; 172.8; 0.0; 927 ❙ 1751; 996; 77 ❙ 88.8; 927; 514.1

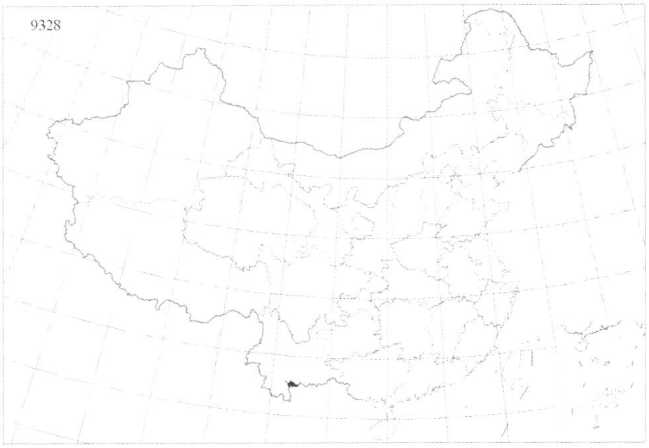

Verbenaceae 马鞭草科 *Callicarpa* 紫珠属

Callicarpa nudiflora Hook. et Arn. 裸花紫珠

Shrubs or small trees

21.5 (17.6~24.8); 13.4 (7.1~20.1); 28.0 (23.6~29.8); 21.5 (17.7~24.8); 198.2 (152.5~237.6); 0.0 (0.0~0.0); 1139 (906~1385) ❙ 1639 (868~2148); 701 (312~1234); 131 (41~197) ❙ 45.4 (-34.3~95.5); 1121 (868~1360); 254.8 (95.3~620.0)

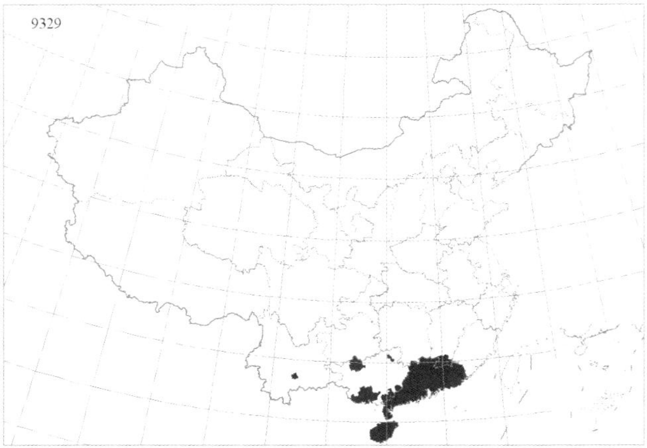

Verbenaceae 马鞭草科 *Callicarpa* 紫珠属

Callicarpa oligantha Merr. 罗浮紫珠

Shrubs

21.6; 13.2; 28.2; 21.6; 199.2; 0.0; 1139 ❙ 1953; 858; 167 ❙ 71.5; 1139; 240.0

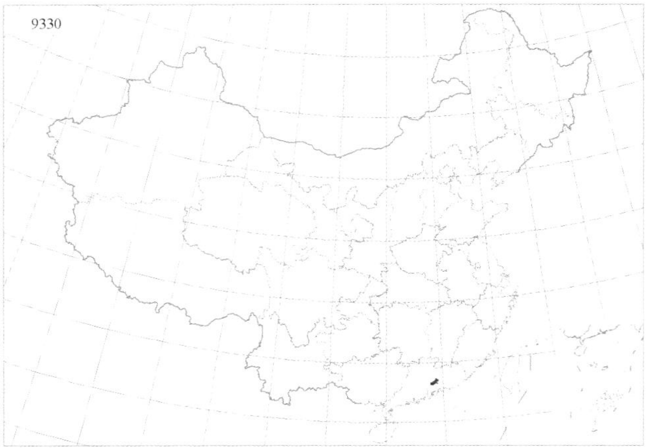

Verbenaceae 马鞭草科 *Callicarpa* 紫珠属

Callicarpa pauciflora Chun ex H. T. Chang 少花紫珠

Shrubs or subshrubs

18.5 (17.3~21.6)；8.4 (6.4~13.2)；27.4 (26.6~28.2)；18.6 (17.3~21.6)；162.6 (147.9~199.2)；0.0 (0.0~0.0)；967 (907~1139) | 1657 (1486~1953)；610 (505~858)；184 (167~197) | 71.3 (60.2~86.3)；967 (907~1139)；229.7 (188.8~271.3)

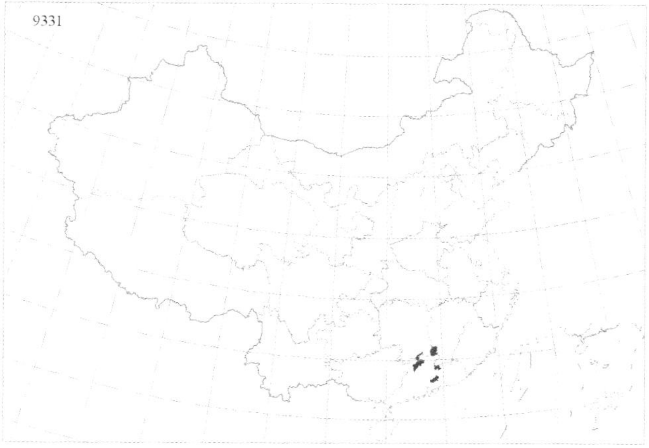

Verbenaceae 马鞭草科 *Callicarpa* 紫珠属

Callicarpa peichieniana Chun et S. L. Chen 钩毛紫珠

Shrubs

19.1 (14.2~22.7)；9.2 (3.3~15.0)；27.7 (23.6~29.8)；19.1 (14.3~22.8)；169.6 (113.3~213.1)；0.1 (0.0~2.1)；1009 (776~1221) | 1617 (1152~2435)；637 (383~1428)；168 (42~250) | 61.3 (12.9~129.9)；1003 (776~1221)；222.3 (95.3~318.8)

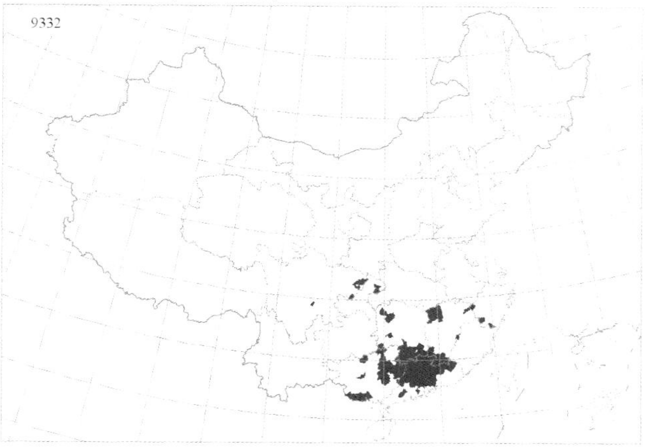

Verbenaceae 马鞭草科 *Callicarpa* 紫珠属

Callicarpa pilosissima Maxim. 长毛紫珠

Shrubs

19.5 (16.4~22.2)；14.4 (11.5~17.7)；23.7 (20.3~25.9)；19.6 (16.5~22.2)；175.0 (137.8~206.7)；0.0 (0.0~0.0)；955 (783~1123) | 2728 (2445~2993)；1304 (924~1847)；251 (107~547) | 188.7 (140.0~219.3)；955 (783~1123)；286.7 (207.7~363.7)

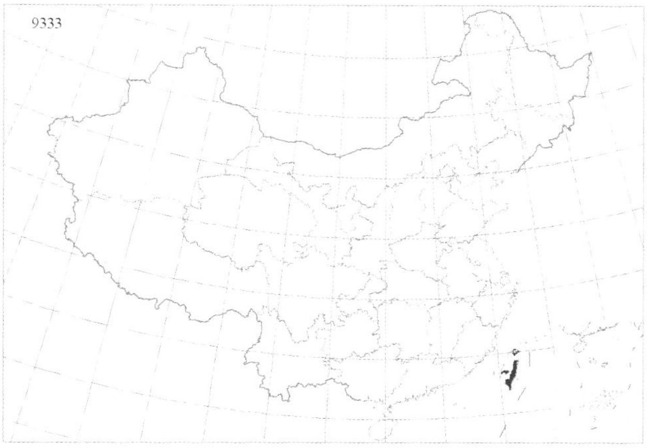

Verbenaceae 马鞭草科 *Callicarpa* 紫珠属

Callicarpa pingshanensis C. Y. Wu ex W. Z. Fang 屏山紫珠

Shrubs

14.4；6.0；21.6；14.5；113.8；0.0；751 | 1020；589；39 | 35.7；740；239.1

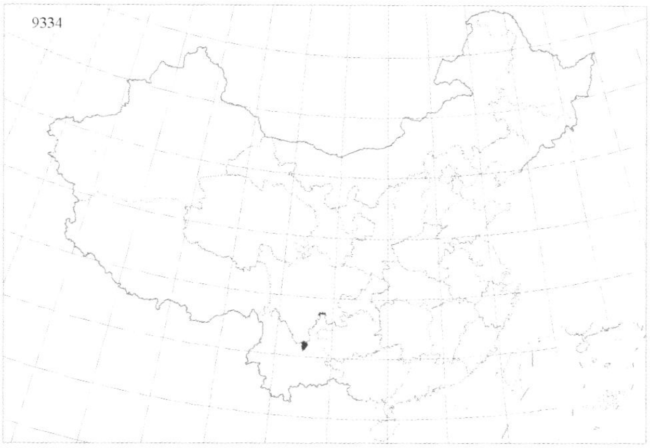

Verbenaceae 马鞭草科 *Callicarpa* 紫珠属

Callicarpa poilanei Dop 白背紫珠

Small trees

21.0；14.2；25.6；21.0；191.8；0.0；1053 | 1522；818；68 | 44.5；1053；551.6

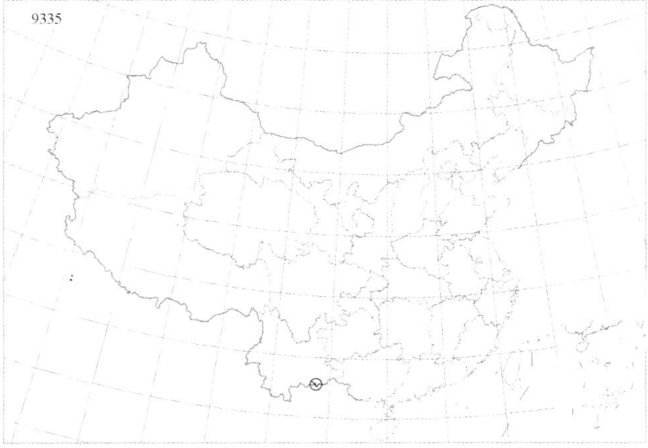

Verbenaceae 马鞭草科 *Callicarpa* 紫珠属

Callicarpa prolifera C. Y. Wu 拙芽紫珠

Shrubs

18.3；10.2；24.3；18.3；160.1；0.0；910 | 1297；727；52 | 44.4；910；348.9

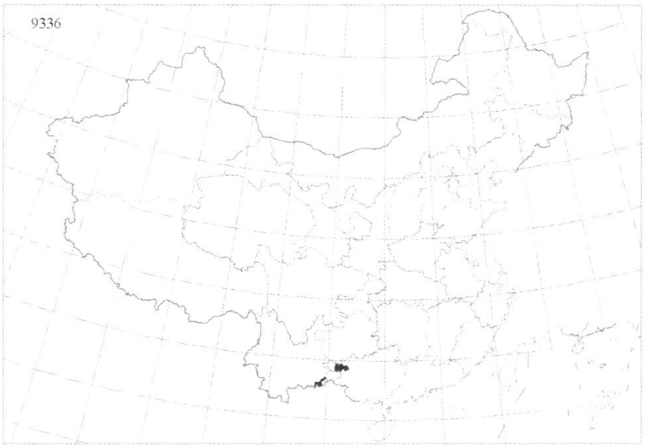

Verbenaceae 马鞭草科 *Callicarpa* 紫珠属
Callicarpa prolifera var. rubroglandulosa S. L. Chen 红腺抽芽紫珠
Shrubs

17.6; 7.3; 26.3; 17.6; 151.6; 0.0; 909 ❙ 1496; 639; 142 ❙ 64.7; 909; 225.9

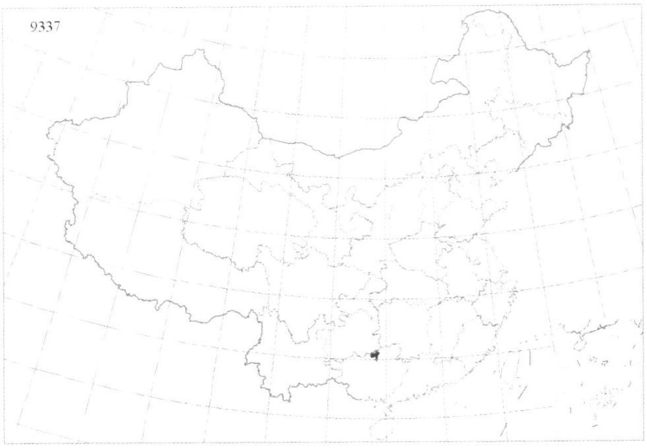

Verbenaceae 马鞭草科 *Callicarpa* 紫珠属
Callicarpa pseudorubella H. T. Chang 拟红紫珠
Shrubs or subshrubs

22.0; 13.5; 28.8; 22.0; 203.6; 0.0; 1169 ❙ 1641; 724; 132 ❙ 40.5; 1169; 205.0

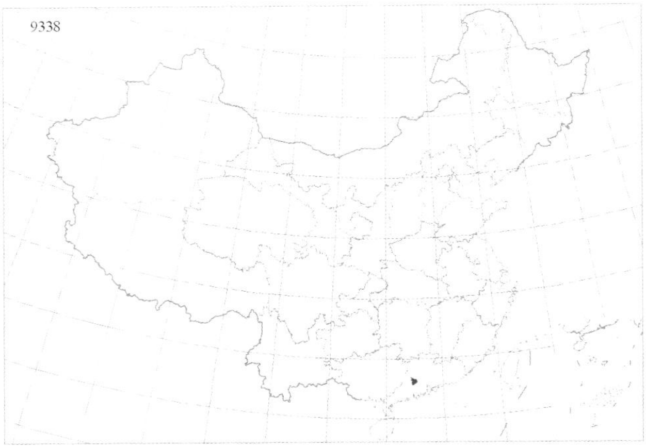

Verbenaceae 马鞭草科 *Callicarpa* 紫珠属
Callicarpa randaiensis Hayata 峦大紫珠
Small shrubs

19.0 (16.2~21.7); 13.1 (11.0~15.3); 24.3 (20.1~27.4); 19.1 (16.2~21.7); 168.9 (134.2~200.2); 0.0 (0.0~0.0); 947 (773~1118) ❙ 2669 (1836~3495); 972 (821~1186); 415 (120~897) ❙ 187.0 (64.2~246.5); 941 (773~1077); 257.7 (202.8~303.6)

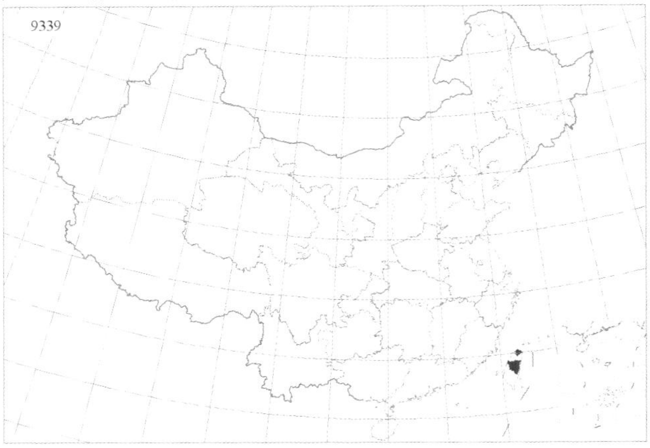

Verbenaceae 马鞭草科 *Callicarpa* 紫珠属
Callicarpa remotiserrulata Hayata 疏齿紫珠
Shrubs

22.2; 17.7; 25.5; 22.2; 206.7; 0.0; 1123 ❙ 2992; 1847; 107 ❙ 166.5; 1123; 296.9

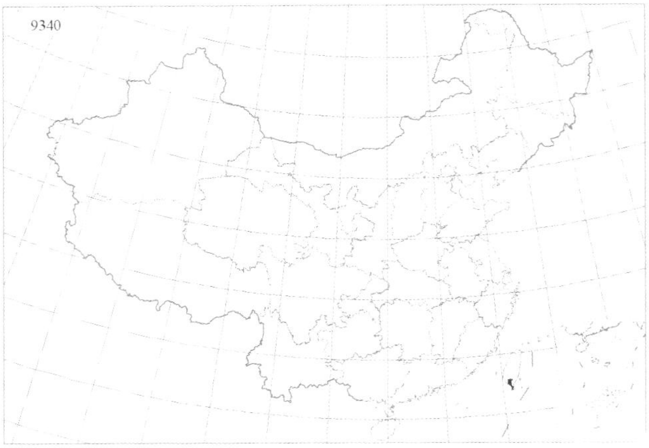

Verbenaceae 马鞭草科 *Callicarpa* 紫珠属
Callicarpa rubella Lindl. 红紫珠
Shrubs

17.5 (6.0~23.6); 7.9 (-3.2~17.5); 25.9 (14.4~30.1); 17.5 (6.6~23.6); 150.7 (39.7~223.5); 0.4 (0.0~26.7); 918 (508~1281) ❙ 1424 (797~2435); 608 (392~1428); 127 (15~259) ❙ 55.9 (-14.2~147.0); 909 (508~1273); 246.1 (95.3~659.0)

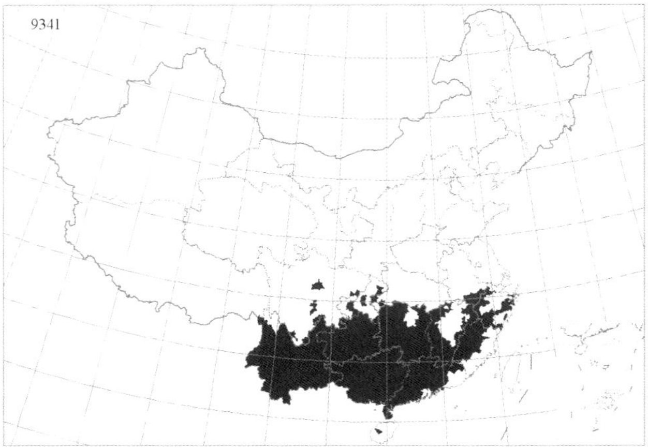

Verbenaceae 马鞭草科 *Callicarpa* 紫珠属
Callicarpa rubella var. subglabra (C. P'ei) H. T. Chang 秃红紫珠
Shrubs

18.4 (14.3~22.4); 8.3 (2.5~14.3); 27.3 (23.5~29.8); 18.4 (14.3~22.5); 161.5 (112.6~209.7); 0.2 (0.0~3.9); 972 (756~1205) ❙ 1561 (1168~2024); 608 (419~954); 159 (57~227) ❙ 61.4 (22.7~147.0); 970 (756~1205); 220.2 (95.3~371.7)

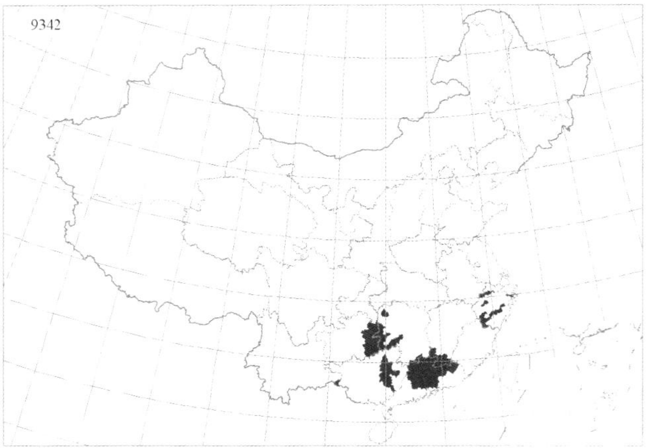

Verbenaceae 马鞭草科 *Callicarpa* 紫珠属
Callicarpa salicifolia C. P'ei et W. Z. Fang 水金花
Shrubs

15.5 (12.1~17.8); 5.7 (2.1~7.8); 24.4 (21.1~27.0); 15.5 (12.2~17.9); 127.2 (91.8~154.2); 0.8 (0.0~5.5); 817 (690~929) ❙ 1135 (985~1410); 618 (509~816); 49 (28~67) ❙ 40.3 (18.3~80.4); 817 (690~929); 188.8 (139.5~298.1)

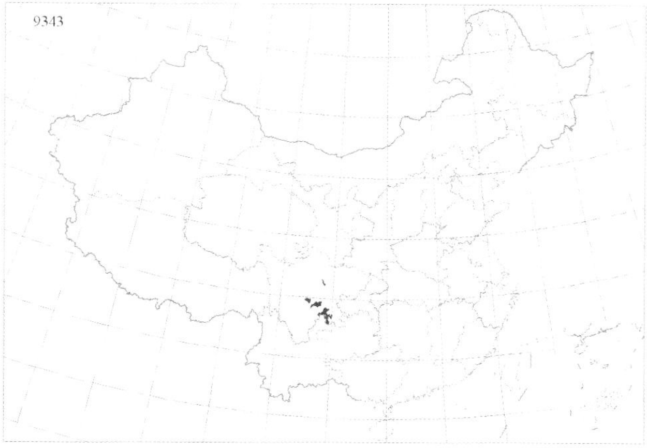

Verbenaceae 马鞭草科 *Callicarpa* 紫珠属
Callicarpa siongsaiensis F. P. Metcalf 上狮紫珠
Shrubs

17.6; 7.4; 27.6; 17.6; 151.2; 0.0; 919 ❙ 1674; 611; 179 ❙ 82.2; 919; 216.1

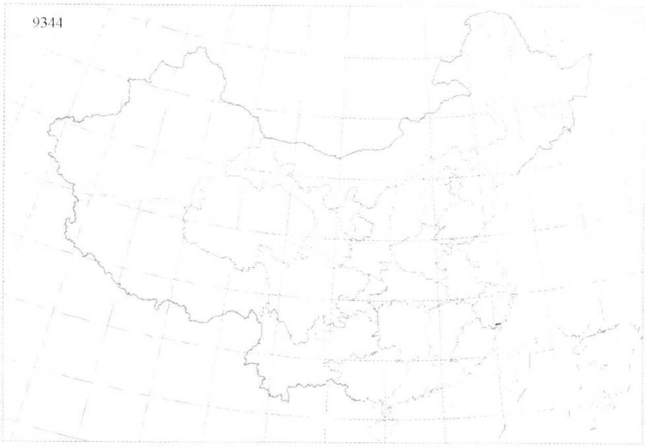

Verbenaceae 马鞭草科 *Callicarpa* 紫珠属
Callicarpa tingwuensis H. T. Chang 鼎湖紫珠
Shrubs

21.0 (19.5~22.0); 12.3 (10.6~13.5); 28.0 (26.9~28.8); 21.0 (19.6~22.0); 191.7 (175.1~203.8); 0.0 (0.0~0.0); 1099 (1003~1170) ❙ 1725 (1556~1990); 715 (601~827); 156 (132~191) ❙ 57.9 (38.6~95.5); 1099 (1003~1170); 234.7 (203.4~275.2)

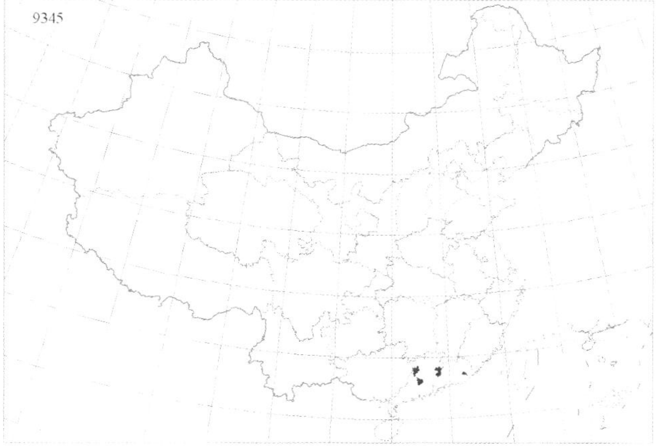

Verbenaceae 马鞭草科 *Callicarpa* 紫珠属
Callicarpa yunnanensis W. Z. Fang 云南紫珠
Trees

20.4; 15.1; 23.6; 20.5; 185.8; 0.0; 989 ❙ 1607; 702; 77 ❙ 62.5; 986; 659.0

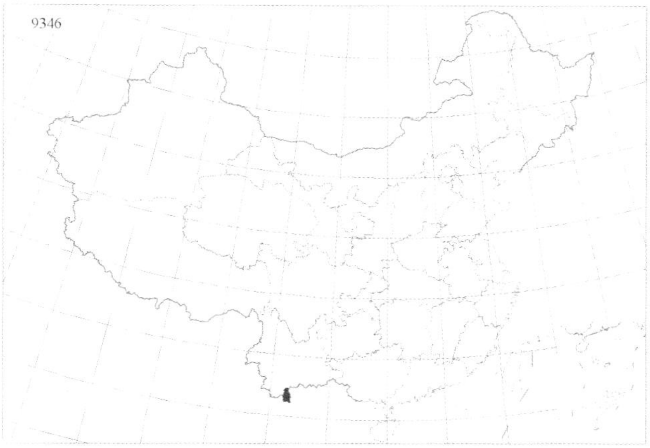

Verbenaceae 马鞭草科 *Caryopteris* 莸属
Caryopteris aureoglandulosa (Vaniot) C. Y. Wu 金腺莸
Subshrubs

16.6 (9.3~20.5); 7.3 (-1.7~15.1); 24.6 (20.1~26.9); 16.7 (9.5~20.5); 141.0 (69.8~185.9); 1.3 (0.0~18.0); 855 (603~1010) ❙ 1253 (1118~1607); 603 (444~729); 66 (43~95) ❙ 47.9 (16.9~101.4); 853 (603~1010); 275.3 (178.6~659.0)

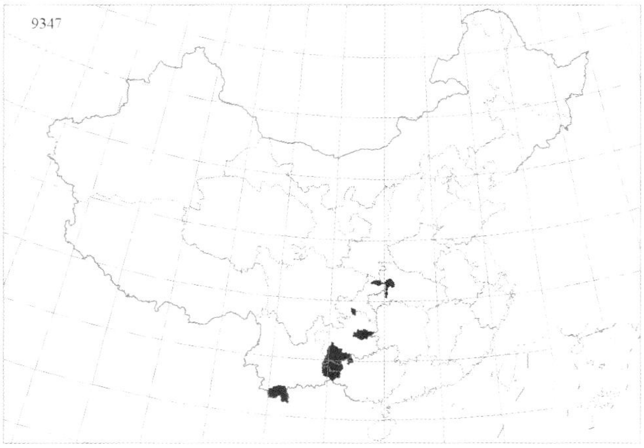

Verbenaceae 马鞭草科 *Caryopteris* 莸属
Caryopteris bicolor (Roxb. ex Hardw.) Mabb. 香莸
Shrubs

13.6 (5.4~18.9); 5.9 (-2.1~12.0); 19.9 (12.3~26.6); 13.6 (5.8~19.0); 105.4 (31.5~167.8); 1.9 (0.0~25.6); 721 (481~929) ❙ 970 (741~1454); 566 (411~792); 31 (15~70) ❙ 36.4 (-14.2~99.9); 691 (481~912); 275.5 (142.9~487.2)

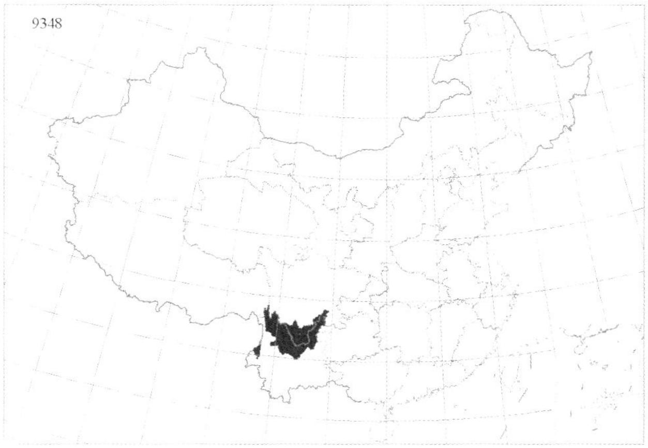

Verbenaceae 马鞭草科 *Caryopteris* 莸属
Caryopteris forrestii Diels 灰毛莸
Small shrubs, deciduous

14.3 (0.0~19.1); 6.3 (-8.7~13.1); 20.8 (8.1~26.6); 14.4 (2.7~19.1); 115.1 (8.0~169.2); 3.1 (0.0~67.9); 750 (363~969) ❙ 1055 (535~1736); 575 (326~1007); 45 (10~107) ❙ 41.8 (-14.2~145.8); 736 (363~969); 261.2 (111.0~447.4)

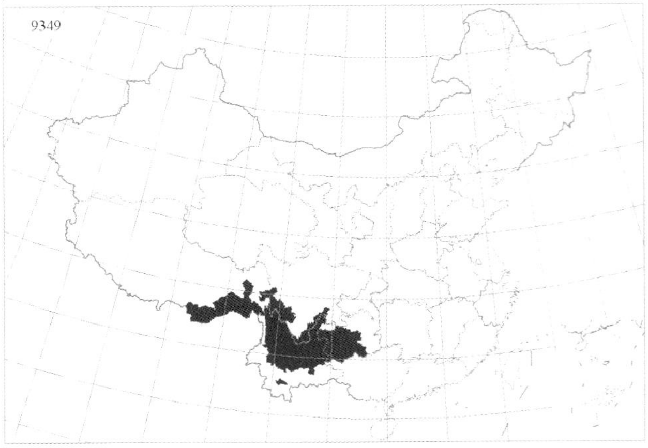

Verbenaceae 马鞭草科 *Caryopteris* 莸属
Caryopteris forrestii var. minor C. P'ei et S. L. Chen ex C. Y. Wu 小叶灰毛莸
Small shrubs, deciduous

4.1 (2.2~6.6); -4.2 (-6.3~-1.3); 11.3 (9.5~13.3); 5.1 (3.9~6.8); 25.6 (15.8~39.6); 36.1 (19.7~48.6); 451 (411~514) ❙ 722 (535~961); 446 (326~558); 16 (10~31) ❙ 59.7 (27.1~100.7); 451 (411~514); 248.4 (201.9~311.5)

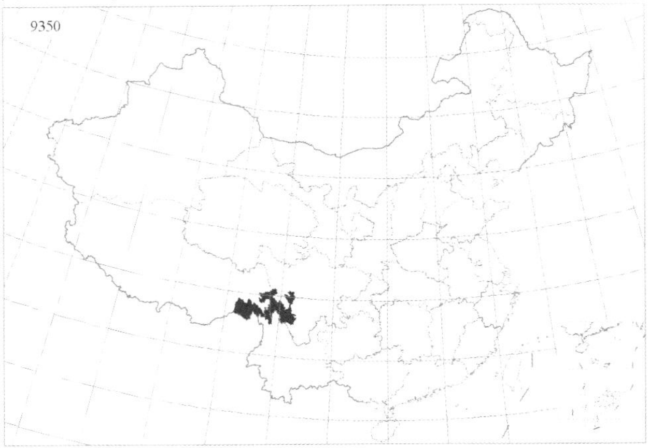

Verbenaceae 马鞭草科 *Caryopteris* 莸属
Caryopteris glutinosa Rehder 黏叶莸
Shrubs, deciduous

4.9; -4.3; 13.3; 5.8; 32.7; 33.1; 480 ❙ 965; 495; 26 ❙ 102.0; 480; 243.6

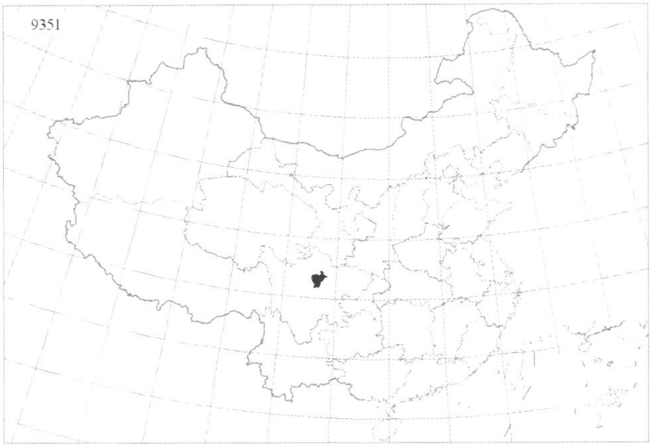

Verbenaceae 马鞭草科 *Caryopteris* 莸属
Caryopteris jinshajiangensis Y. K. Yang et X. D. Cong 金沙江莸
Subshrubs

13.6 (5.4~18.9); 5.9 (-2.1~12.0); 20.0 (12.3~26.6); 13.7 (5.8~19.0); 105.9 (31.5~167.8); 1.9 (0.0~25.6); 722 (481~929) ❙ 958 (741~1141); 559 (411~641); 30 (15~57) ❙ 34.4 (-14.2~70.8); 691 (481~912); 270.0 (142.9~351.7)

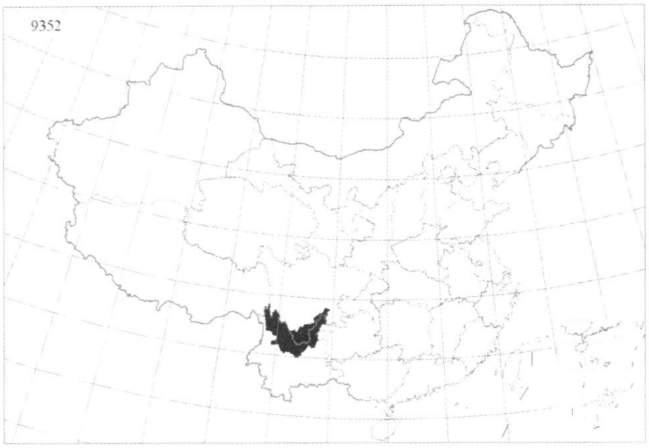

Verbenaceae 马鞭草科 *Caryopteris* 莸属
Caryopteris mongholica Bunge 蒙古莸
Small shrubs, deciduous

7.5 (-4.8~15.1); -8.0 (-22.1~4.4); 21.5 (7.4~26.8); 9.5 (1.9~15.2); 74.8 (4.3~122.8); 43.8 (0.6~121.6); 625 (332~834) ❙ 491 (49~1246); 264 (34~528); 14 (1~53) ❙ -22.7 (-92.8~93.8); 444 (49~817); 149.1 (5.1~334.4)

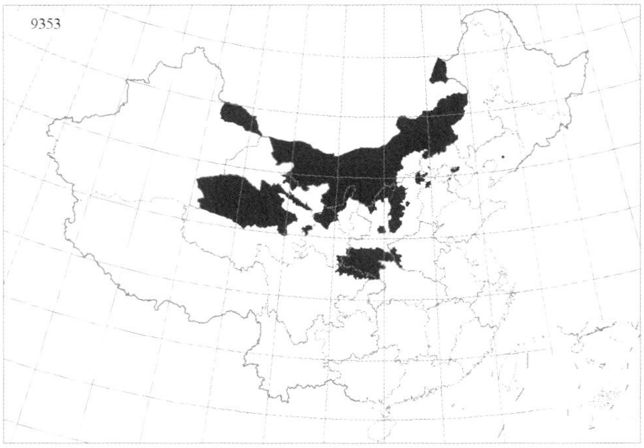

Verbenaceae 马鞭草科 *Caryopteris* 莸属
Caryopteris paniculata C. B. Clarke 锥花莸
Shrubs

17.6 (7.3~21.0); 10.5 (-2.0~15.1); 22.6 (15.6~27.7); 17.6 (7.5~21.0); 151.9 (48.2~191.8); 0.3 (0.0~20.4); 864 (537~1065) ❙ 1278 (939~1751); 700 (489~996); 55 (27~141) ❙ 48.3 (6.5~112.2); 853 (537~1065); 382.6 (189.5~659.0)

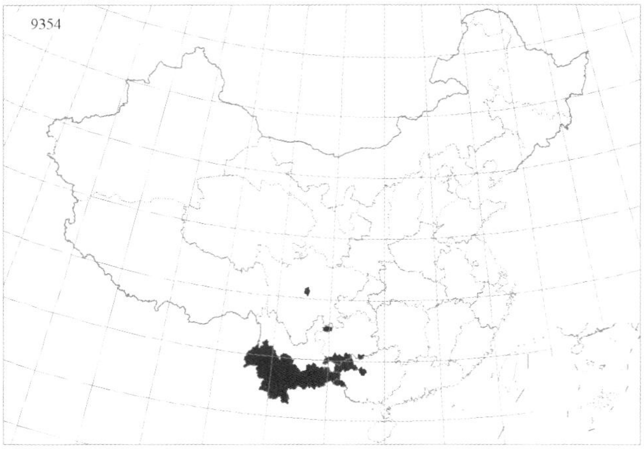

Verbenaceae 马鞭草科 *Caryopteris* 莸属
Caryopteris tangutica Maxim. 光果莸
Shrubs

10.3 (−1.7~17.1)；−2.2 (−13.5~4.4)；21.9 (8.0~29.4)；11.0 (2.4~17.1)；86.5 (6.4~146.1)；21.8 (0.6~86.3)；681 (344~952) | 772 (417~1358)；380 (239~546)；30 (6~123) | 18.1 (−37.6~128.6)；632 (344~921)；225.2 (111.9~357.7)

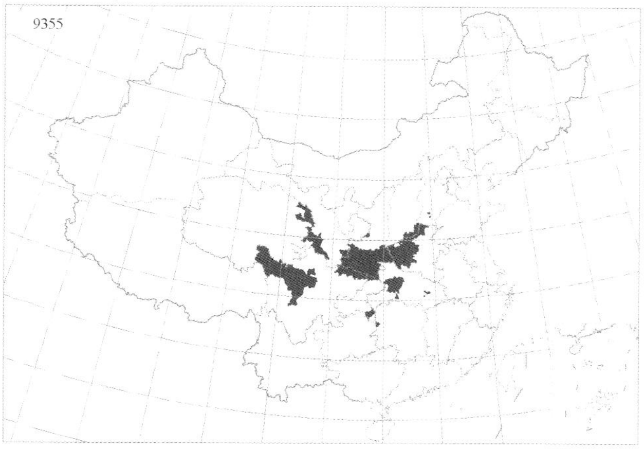

Verbenaceae 马鞭草科 *Caryopteris* 莸属
Caryopteris terniflora Maxim. 三花莸
Subshrubs

13.2 (2.2~18.9)；0.7 (−12.5~9.5)；24.8 (9.9~29.8)；13.6 (3.8~18.9)；111.3 (15.5~167.3)；11.7 (0.0~62.6)；778 (413~1002) | 922 (377~1922)；429 (241~740)；57 (8~259) | 18.3 (−47.7~125.9)；722 (377~1002)；199.3 (108.2~351.7)

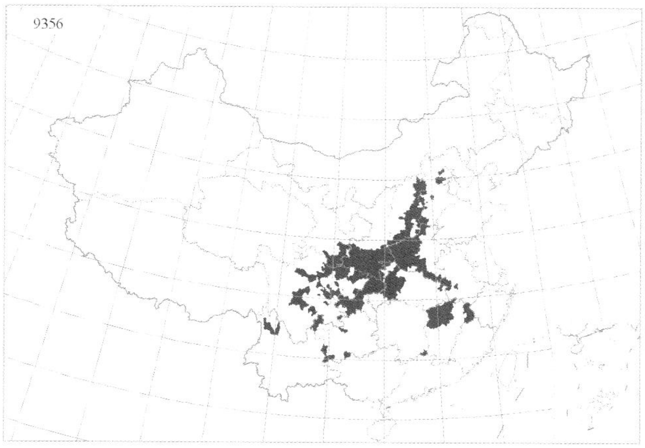

Verbenaceae 马鞭草科 *Caryopteris* 莸属
Caryopteris trichosphaera W. W. Sm. 毛球莸
Shrubs

3.9 (−3.2~13.2)；−4.9 (−14.4~6.4)；11.6 (7.3~18.4)；5.5 (2.0~13.3)；29.7 (4.0~99.4)；42.8 (0.0~101.5)；460 (335~695) | 741 (425~1337)；436 (259~727)；21 (8~69) | 58.8 (20.7~128.6)；460 (335~695)；271.3 (130.1~407.1)

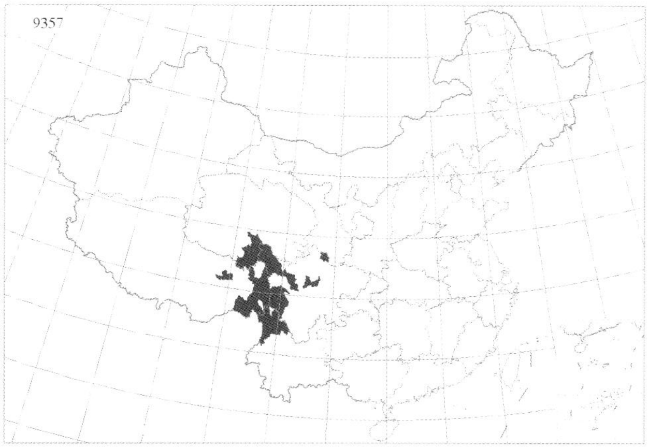

Verbenaceae 马鞭草科 *Clerodendrum* 大青属
Clerodendrum brachystemon C. Y. Wu et R. C. Fang 短蕊大青
Shrubs, deciduous

18.4；12.3；22.3；18.4；160.7；0.0；899 | 1576；762；63 | 80.2；883；547.6

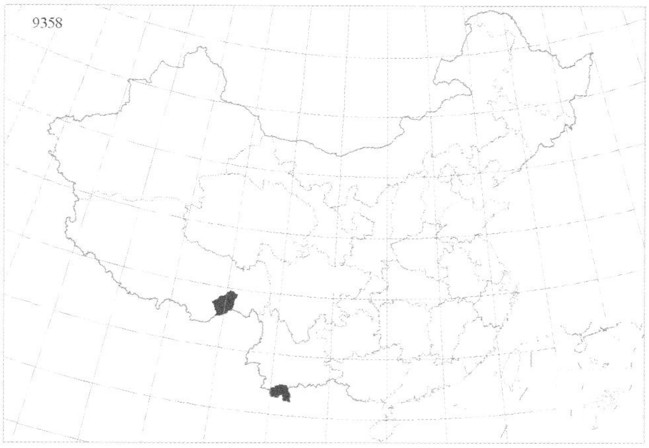

Verbenaceae 马鞭草科 *Clerodendrum* 大青属
Clerodendrum bracteatum Wall. ex Walp. 苞花大青
Shrubs or small trees

9.1；1.4；15.6；9.3；61.2；11.7；586 | 1246；717；42 | 109.3；586；344.6

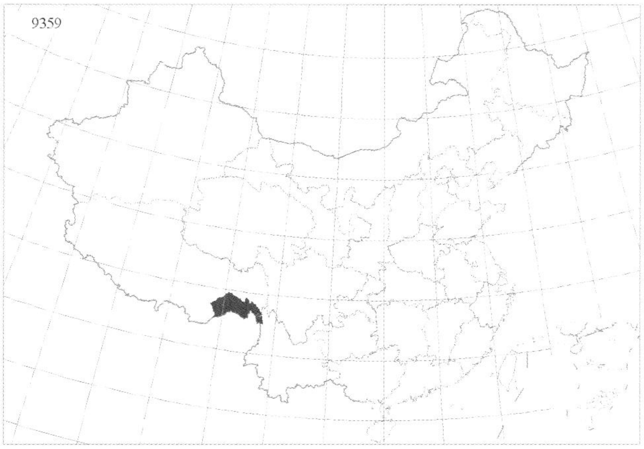

Verbenaceae 马鞭草科 *Clerodendrum* 大青属
Clerodendrum bungei Steud. 臭牡丹
Shrubs

15.6 (2.2~23.6)；4.9 (−7.0~16.7)；25.2 (9.9~30.1)；15.6 (3.8~23.6)；130.9 (15.5~223.5)；3.4 (0.0~49.1)；842 (413~1281) | 1218 (512~1997)；531 (258~996)；102 (14~259) | 44.4 (−36.5~130.7)；820 (413~1269)；235.3 (110.1~659.0)

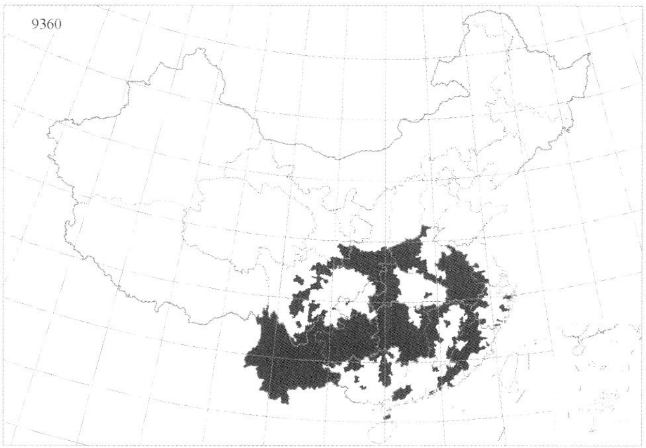

Verbenaceae 马鞭草科 *Clerodendrum* 大青属
Clerodendrum bungei var. megacalyx C. Y. Wu ex S. L. Chen 大萼臭牡丹
Shrubs

14.8; 4.3; 24.9; 14.9; 119.2; 0.7; 791 ❙ 1197; 489; 71 ❙ 51.2; 791; 182.9

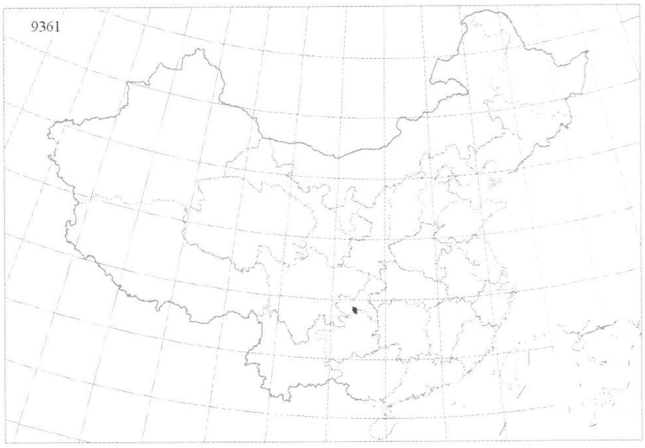

Verbenaceae 马鞭草科 *Clerodendrum* 大青属
Clerodendrum canescens Wall. ex Walp. 灰毛大青
Shrubs

19.1 (12.4~24.8); 9.8 (3.2~20.1); 27.1 (18.6~30.1); 19.2 (12.4~24.8); 170.4 (89.1~237.6); 0.1 (0.0~2.5); 1010 (675~1385) ❙ 1549 (868~3495); 636 (312~1428); 149 (18~897) ❙ 55.2 (-34.3~246.5); 998 (675~1360); 243.0 (95.3~620.0)

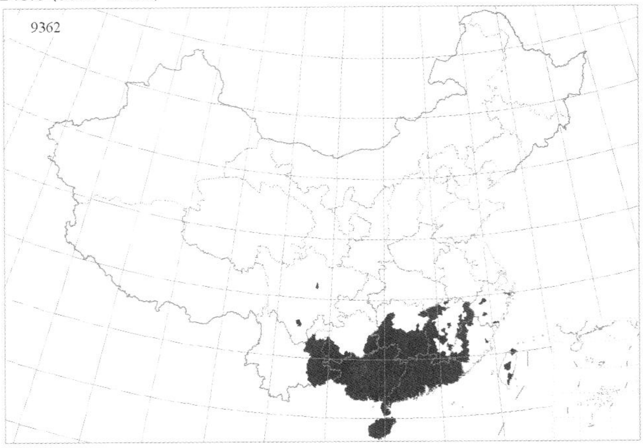

Verbenaceae 马鞭草科 *Clerodendrum* 大青属
Clerodendrum chinense var. simplex (Moldenke) S. L. Chen 臭茉莉
Shrubs

18.1 (13.2~21.0); 11.6 (6.4~14.2); 22.7 (18.4~25.6); 18.2 (13.3~21.0); 158.1 (99.4~191.8); 0.0 (0.0~0.0); 882 (695~1053) ❙ 1303 (1007~1751); 725 (539~996); 59 (44~99) ❙ 48.3 (19.0~92.4); 874 (695~1053); 410.4 (236.8~551.6)

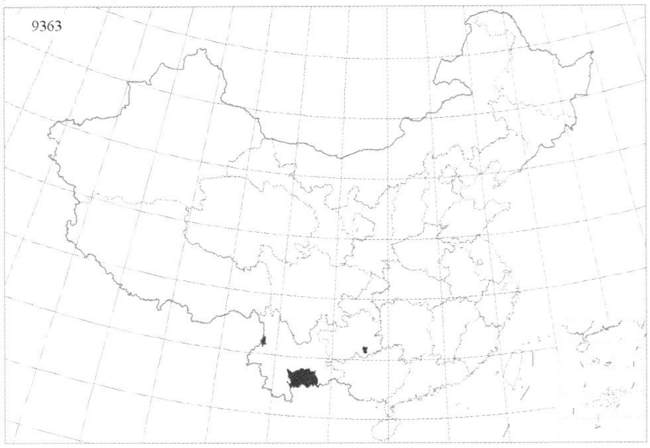

Verbenaceae 马鞭草科 *Clerodendrum* 大青属
Clerodendrum colebrookianum Walp. 腺茉莉
Shrubs or small trees

19.2 (7.3~23.6); 11.7 (-2.0~16.7); 24.9 (15.0~29.2); 19.3 (7.5~23.6); 171.6 (48.2~223.5); 0.2 (0.0~20.4); 986 (537~1281) ❙ 1464 (939~2435); 725 (507~1428); 91 (27~191) ❙ 49.3 (-6.9~145.8); 979 (537~1273); 297.8 (95.3~659.0)

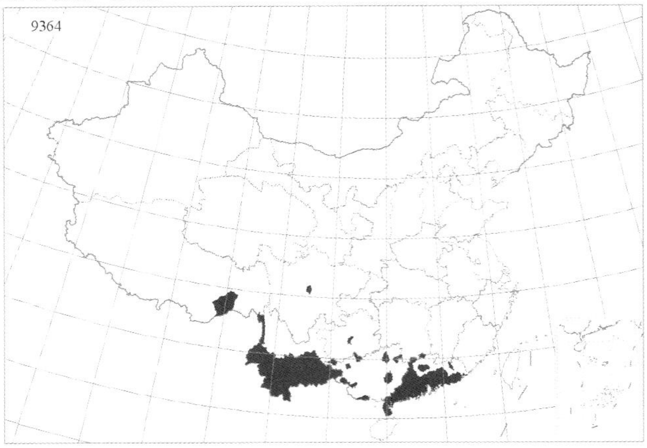

Verbenaceae 马鞭草科 *Clerodendrum* 大青属
Clerodendrum confine S. L. Chen et T. D. Zhuang 川黔大青
Shrubs

11.4; 2.3; 19.5; 11.5; 84.3; 6.7; 659 ❙ 1211; 619; 42 ❙ 84.5; 659; 256.5

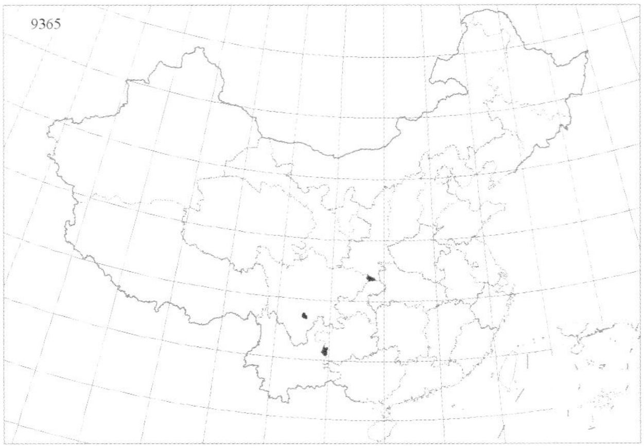

Verbenaceae 马鞭草科 *Clerodendrum* 大青属
Clerodendrum cyrtophyllum Turcz. 大青
Shrubs or small trees

18.1 (13.4~24.8); 7.8 (1.5~20.1); 27.4 (21.4~29.9); 18.2 (13.4~24.8); 158.4 (108.3~237.6); 0.6 (0.0~7.3); 964 (755~1385) ❙ 1495 (868~3495); 593 (312~1847); 156 (41~897) ❙ 56.4 (-34.3~224.4); 954 (755~1360); 233.4 (95.3~659.0)

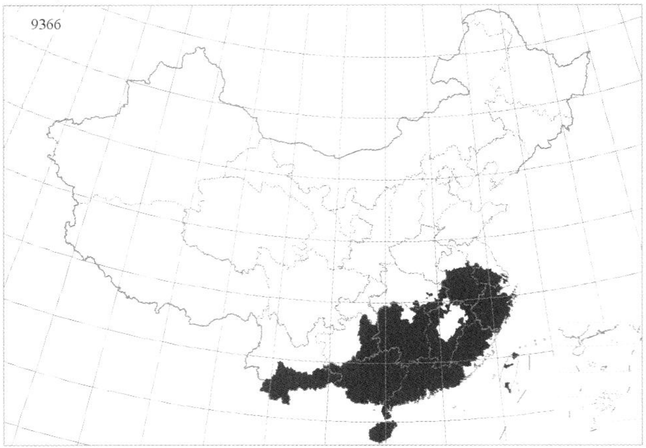

Verbenaceae 马鞭草科 *Clerodendrum* 大青属
Clerodendrum ervatamioides C. Y. Wu 狗牙大青
Shrubs

20.0; 13.6; 24.4; 20.1; 180.7; 0.0; 985 ❙ 1475; 827; 62 ❙ 50.1; 985; 528.9

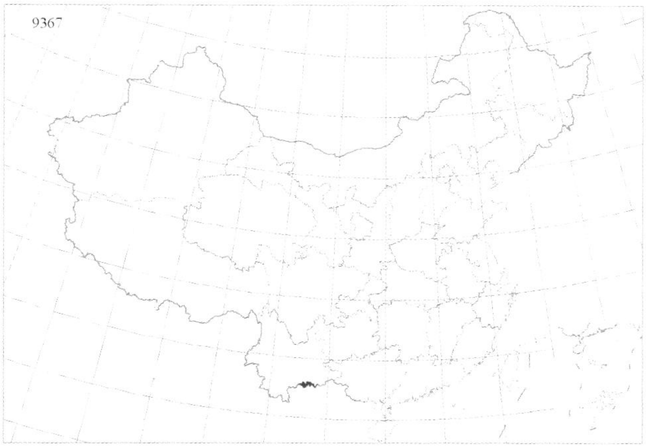

Verbenaceae 马鞭草科 *Clerodendrum* 大青属
Clerodendrum fortunatum L. 白花灯笼
Shrubs

20.5 (14.6~24.5); 11.8 (3.8~19.1); 27.8 (24.6~29.8); 20.6 (14.7~24.5); 186.7 (117.7~234.4); 0.0 (0.0~1.5); 1078 (779~1360) ❙ 1635 (1174~2435); 695 (386~1428); 148 (48~230) ❙ 53.3 (-6.9~104.6); 1075 (779~1360); 238.9 (95.3~620.0)

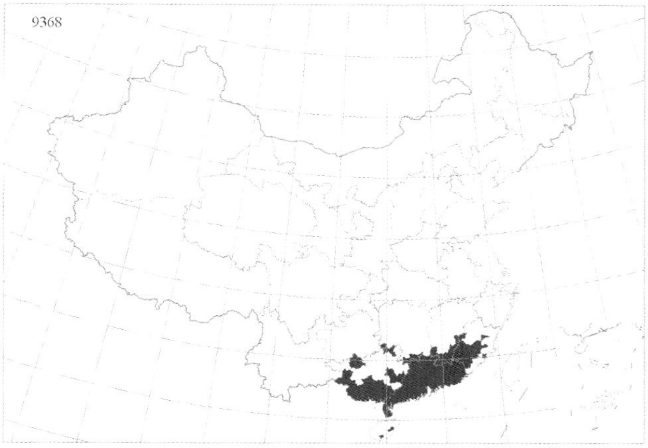

Verbenaceae 马鞭草科 *Clerodendrum* 大青属
Clerodendrum garrettianum Craib 泰国垂茉莉
Shrubs

20.0; 14.6; 23.2; 20.1; 180.8; 0.0; 963 ❙ 1523; 681; 67 ❙ 58.4; 942; 581.0

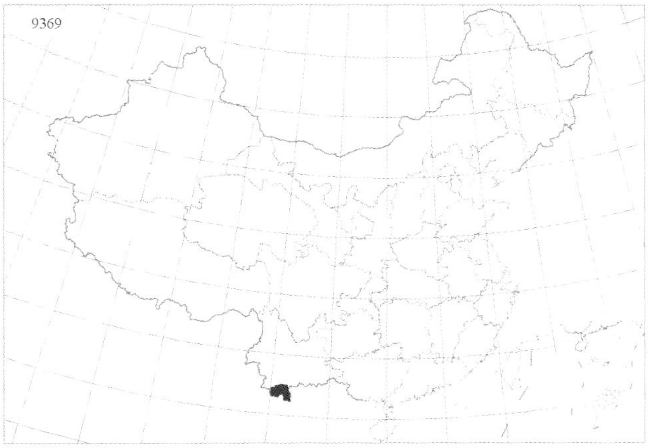

Verbenaceae 马鞭草科 *Clerodendrum* 大青属
Clerodendrum griffithianum C. B. Clarke 西垂茉莉
Shrubs

18.3 (14.4~21.0); 12.0 (7.5~15.1); 22.4 (19.1~25.6); 18.3 (14.4~21.0); 160.2 (113.1~191.8); 0.0 (0.0~0.0); 885 (727~1053) ❙ 1476 (1163~1751); 812 (641~996); 59 (42~77) ❙ 67.8 (34.8~112.2); 880 (727~1053); 508.2 (383.4~659.0)

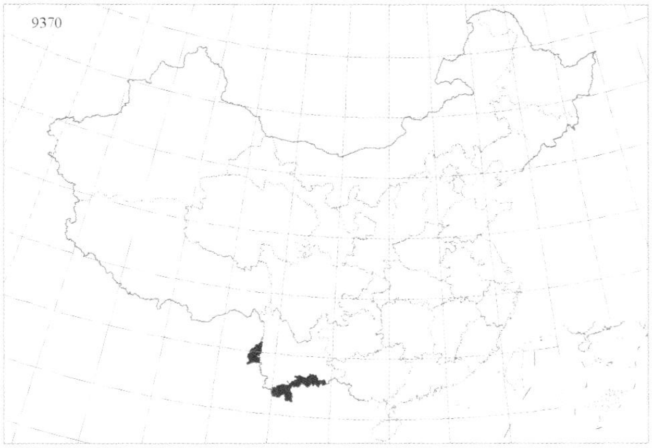

Verbenaceae 马鞭草科 *Clerodendrum* 大青属
Clerodendrum hainanense Hand.-Mazz. 海南赪桐
Shrubs

21.8 (14.8~24.8); 14.6 (4.3~20.1); 27.5 (24.7~28.8); 21.9 (14.9~24.8); 202.3 (119.2~237.6); 0.0 (0.0~0.7); 1175 (791~1385) ❙ 1425 (868~1900); 595 (312~1056); 78 (41~130) ❙ 22.6 (-34.3~71.0); 1107 (791~1360); 348.7 (130.1~620.0)

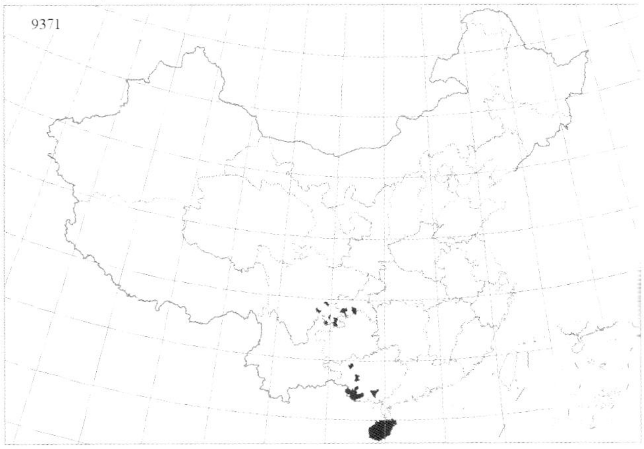

Verbenaceae 马鞭草科 *Clerodendrum* 大青属
Clerodendrum henryi C. P'ei 南垂茉莉
Shrubs

18.8 (17.2~21.3); 12.0 (9.1~15.1); 23.5 (22.1~27.9); 18.9 (17.3~21.3); 166.5 (147.0~196.0); 0.0 (0.0~0.0); 918 (833~1123) ❙ 1403 (1161~1751); 763 (633~996); 59 (48~77) ❙ 53.6 (5.8~88.8); 913 (833~1121); 435.1 (314.4~659.0)

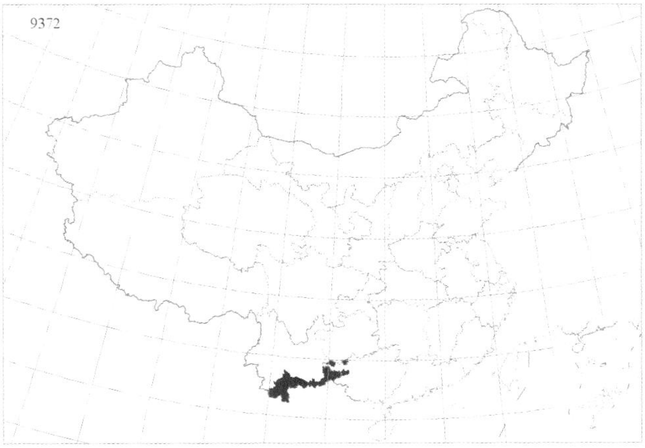

Verbenaceae 马鞭草科 *Clerodendrum* 大青属
Clerodendrum indicum (L.) Kuntze 长管大青
Shrubs

20.7 (14.4~23.6); 13.2 (7.5~16.7); 26.3 (19.1~29.2); 20.7 (14.4~23.6); 189.0 (113.1~223.5); 0.0 (0.0~0.0); 1073 (727~1281) ❙ 1572 (1187~2435); 770 (545~1428); 98 (41~177) ❙ 49.1 (-6.9~112.2); 1067 (727~1273); 313.6 (152.9~659.0)

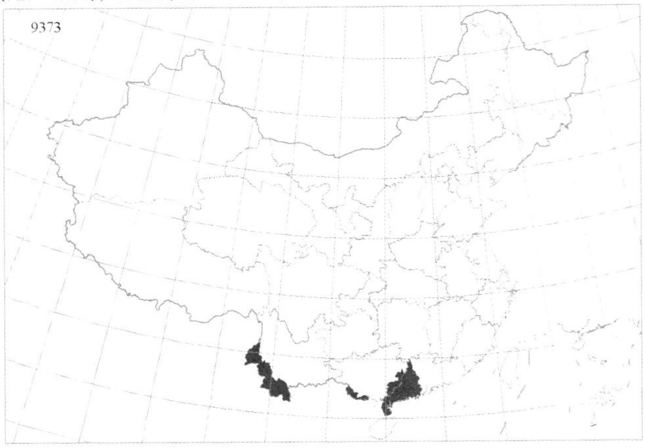

Verbenaceae 马鞭草科 *Clerodendrum* 大青属
Clerodendrum inerme (L.) Gaertn. 苦郎树
Scandent shrubs

21.8 (17.6~24.8); 14.1 (7.4~20.1); 28.0 (22.8~29.2); 21.8 (17.6~24.8); 201.7 (151.2~237.6); 0.0 (0.0~0.0); 1155 (874~1385) ❙ 1690 (868~3495); 755 (312~1594); 139 (41~897) ❙ 48.7 (-34.3~224.4); 1126 (868~1331); 242.5 (111.7~548.0)

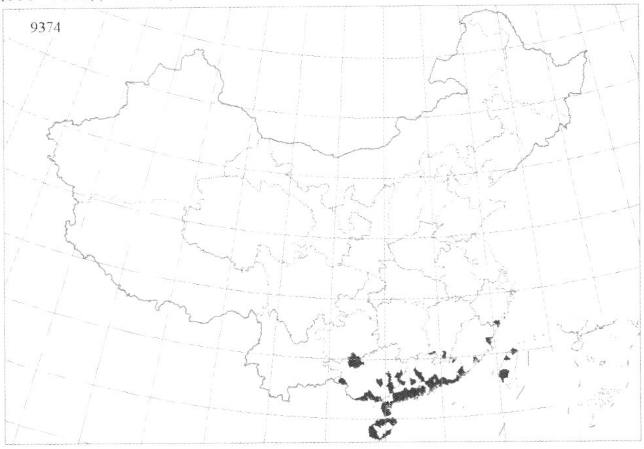

Verbenaceae 马鞭草科 *Clerodendrum* 大青属
Clerodendrum intermedium Cham. 垦丁苦林盘
Shrubs

20.3; 15.9; 23.6; 20.4; 184.4; 0.0; 993 ❙ 2853; 1618; 147 ❙ 190.4; 993; 330.3

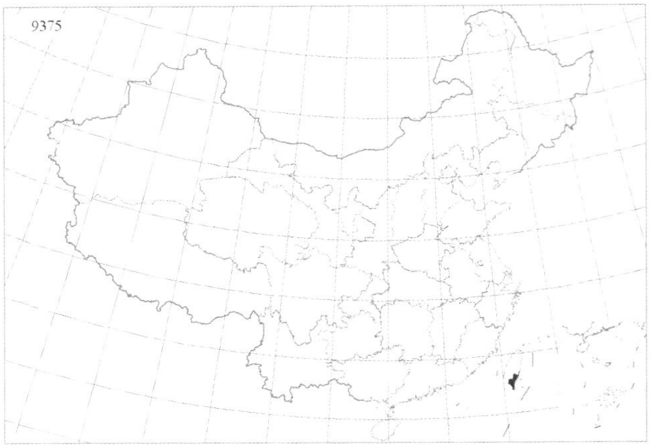

Verbenaceae 马鞭草科 *Clerodendrum* 大青属
Clerodendrum japonicum (Thunb.) Sweet 赪桐
Shrubs

18.5 (5.3~24.8); 9.1 (-2.4~20.1); 26.6 (12.2~29.9); 18.5 (5.7~24.8); 162.5 (30.5~237.6); 0.3 (0.0~26.2); 970 (479~1385) ❙ 1471 (868~2037); 615 (312~1033); 128 (27~259) ❙ 53.4 (-34.3~145.8); 959 (479~1360); 249.9 (95.3~659.0)

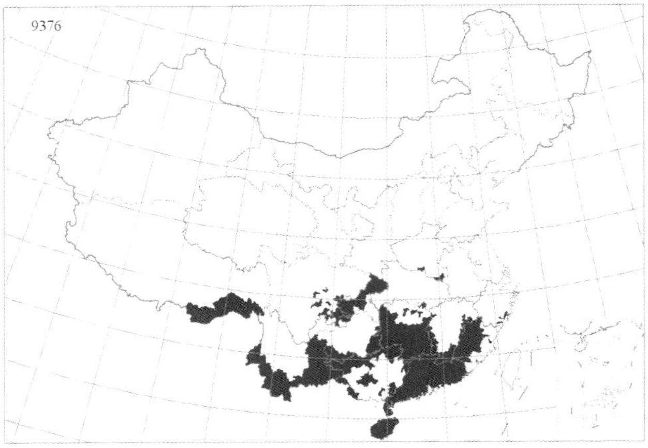

Verbenaceae 马鞭草科 *Clerodendrum* 大青属
Clerodendrum kaichianum Hsu 浙江大青
Shrubs or small trees, deciduous

16.1 (14.1~23.6); 4.3 (1.6~17.8); 27.4 (24.7~29.1); 16.1 (14.2~23.6); 135.2 (116.7~223.3); 1.8 (0.0~6.5); 878 (789~1271) ❙ 1609 (1206~1950); 578 (444~713); 198 (44~259) ❙ 84.7 (-2.2~147.0); 871 (789~1091); 252.3 (149.5~439.7)

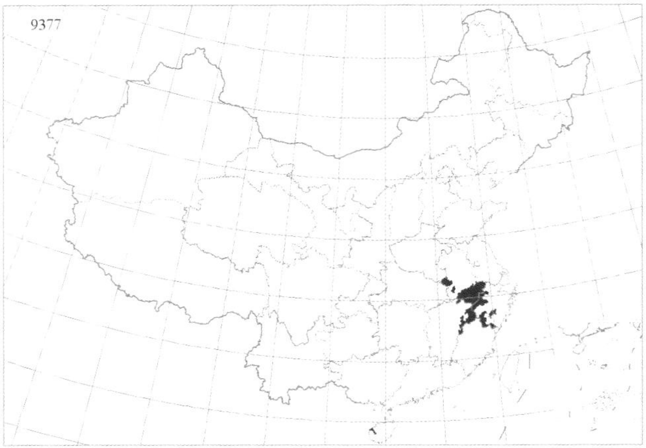

Verbenaceae 马鞭草科 *Clerodendrum* 大青属
Clerodendrum kiangsiense Merr. ex H. L. Li 江西大青
Shrubs

17.5 (16.7~18.2); 6.1 (5.0~7.4); 28.1 (27.5~29.1); 17.5 (16.8~18.2); 150.0 (141.2~158.8); 0.0 (0.0~0.0); 933 (903~984) ❙ 1815 (1674~1922); 588 (535~623); 243 (179~259) ❙ 94.7 (78.7~111.3); 923 (903~961); 234.5 (183.9~279.5)

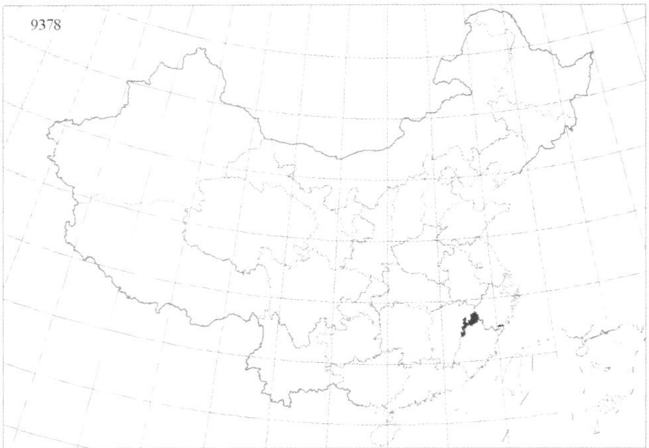

Verbenaceae 马鞭草科 *Clerodendrum* 大青属
Clerodendrum kwangtungense Hand.-Mazz. 广东大青
Shrubs

19.4 (14.3~24.3); 9.9 (3.3~18.7); 27.6 (21.4~29.8); 19.4 (14.3~24.3); 173.4 (112.6~231.9); 0.1 (0.0~2.3); 1024 (756~1338) ❙ 1571 (1074~2435); 635 (392~1428); 152 (42~233) ❙ 54.8 (-11.9~128.9); 1015 (756~1273); 223.3 (95.3~551.6)

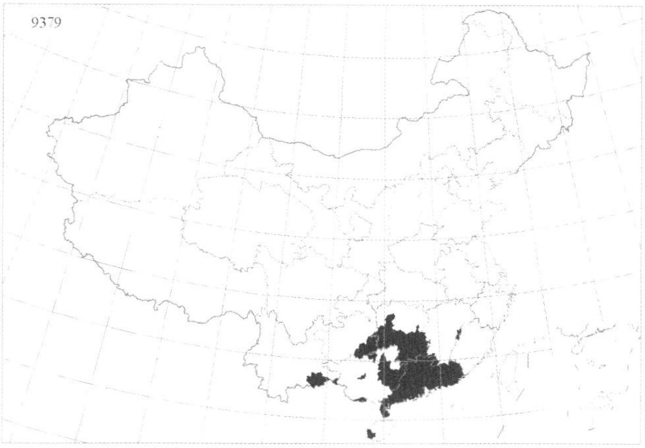

Verbenaceae 马鞭草科 *Clerodendrum* 大青属
Clerodendrum lindleyi Decne. ex Planch. 尖齿臭茉莉
Shrubs

18.6 (5.4~24.8); 9.0 (-2.1~20.1); 27.0 (12.3~29.9); 18.6 (5.8~24.8); 164.2 (31.5~237.6); 0.7 (0.0~25.6); 988 (481~1385) ❙ 1499 (741~2435); 626 (312~1428); 139 (12~256) ❙ 53.7 (-34.3~147.0); 980 (481~1360); 242.8 (95.3~620.0)

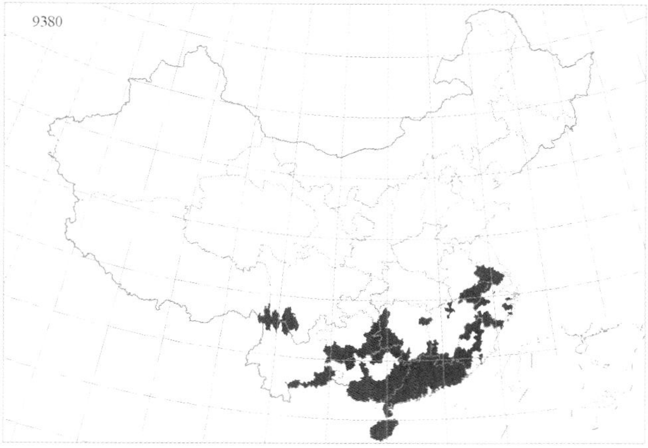

Verbenaceae 马鞭草科 *Clerodendrum* 大青属
Clerodendrum longilimbum C. P'ei 长叶大青
Shrubs

18.0 (13.1~21.9); 11.2 (6.4~15.1); 22.6 (18.4~28.6); 18.1 (13.2~22.0); 157.0 (98.1~203.8); 0.0 (0.0~0.0); 883 (692~1176) ❙ 1389 (1048~1751); 759 (553~996); 59 (41~138) ❙ 58.5 (5.8~112.2); 875 (692~1176); 434.3 (272.3~659.0)

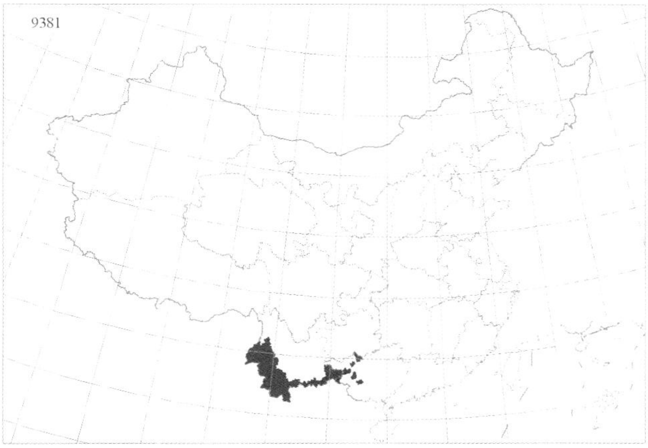

Verbenaceae 马鞭草科 *Clerodendrum* 大青属
Clerodendrum luteopunctatum C. P'ei et S. L. Chen 黄腺大青
Shrubs

15.7 (10.6~18.9); 4.7 (-0.2~9.5); 26.0 (21.1~29.4); 15.7 (10.6~18.9); 130.0 (79.4~167.3); 1.1 (0.0~12.1); 840 (643~969) ❙ 1290 (985~1492); 530 (450~637); 100 (46~164) ❙ 54.9 (18.1~107.2); 838 (643~969); 200.6 (140.5~285.1)

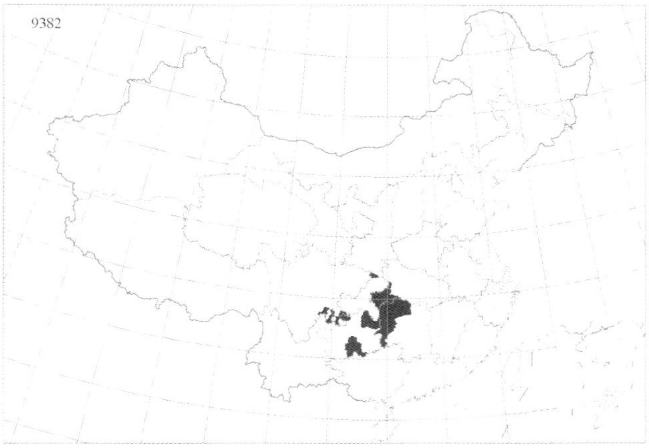

Verbenaceae 马鞭草科 *Clerodendrum* 大青属
Clerodendrum mandarinorum Diels 海通
Shrubs or trees

17.1 (11.7~23.4); 6.8 (0.6~16.1); 26.4 (18.8~29.8); 17.2 (11.7~23.5); 146.7 (82.6~221.7); 0.5 (0.0~9.7); 904 (658~1273) ❙ 1398 (916~1876); 567 (392~996); 129 (31~253) ❙ 55.4 (5.8~128.9); 898 (658~1273); 225.0 (126.1~551.6)

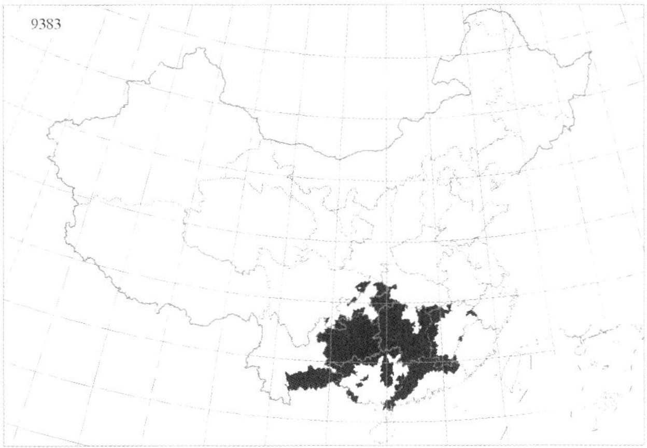

Verbenaceae 马鞭草科 *Clerodendrum* 大青属
Clerodendrum paniculatum L. 圆齿大青
Shrubs

21.6 (17.8~24.1); 13.7 (7.1~18.9); 28.1 (23.4~29.2); 21.6 (17.8~24.1); 199.7 (154.0~229.7); 0.0 (0.0~0.0); 1140 (864~1322) ❙ 1741 (1187~3495); 770 (425~1847); 150 (47~897) ❙ 54.1 (-6.9~224.4); 1132 (864~1273); 217.0 (95.3~304.0)

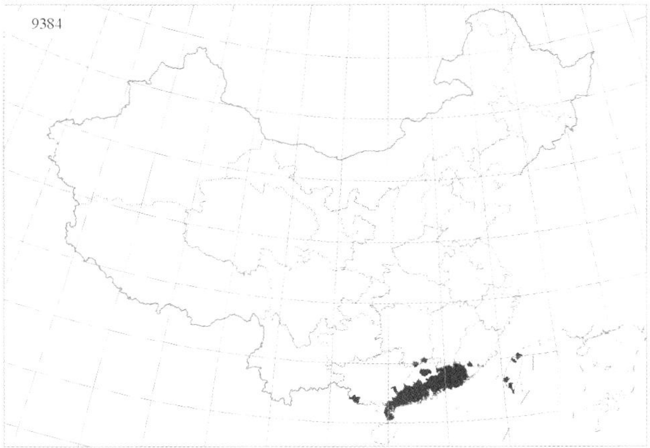

Verbenaceae 马鞭草科 *Clerodendrum* 大青属
Clerodendrum peii Moldenke 长梗大青
Shrubs

18.1 (16.3~21.0); 11.4 (9.0~14.2); 22.6 (21.4~25.6); 18.1 (16.4~21.0); 157.1 (136.4~191.8); 0.0 (0.0~0.0); 873 (797~1053) ❙ 1269 (958~1751); 714 (531~996); 53 (44~77) ❙ 45.2 (13.3~88.8); 866 (797~1053); 394.2 (277.0~551.6)

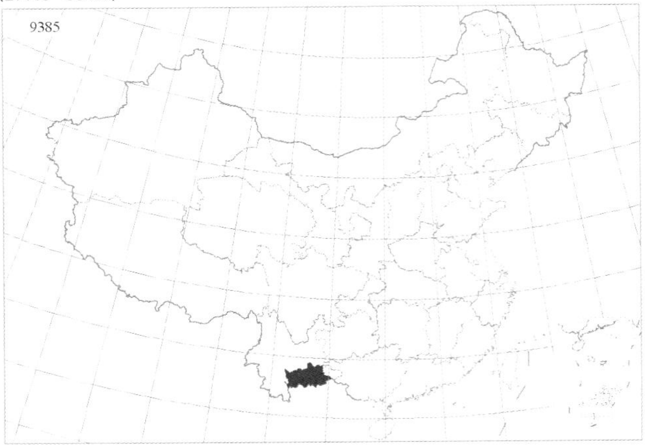

Verbenaceae 马鞭草科 *Clerodendrum* 大青属
Clerodendrum serratum (L.) Moon 三对节
Shrubs

18.6 (13.3~22.3); 11.1 (5.4~15.1); 23.9 (19.1~28.9); 18.6 (13.4~22.3); 163.6 (100.4~208.1); 0.0 (0.0~0.0); 926 (706~1202) ❙ 1350 (958~1751); 729 (531~1007); 68 (41~194) ❙ 47.1 (5.8~145.8); 919 (706~1202); 364.3 (189.5~659.0)

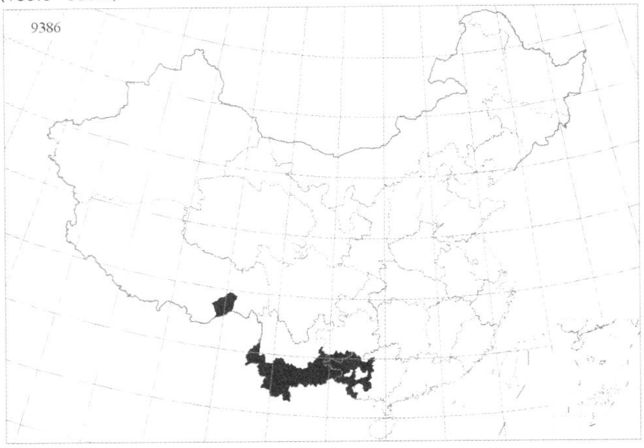

Verbenaceae 马鞭草科 *Clerodendrum* 大青属
Clerodendrum serratum var. amplexifolium Moldenke 三台花
Shrubs

18.7 (15.9~22.2); 11.2 (6.9~15.1); 24.1 (21.4~28.3); 18.7 (16.0~22.3); 164.6 (131.6~207.5); 0.0 (0.0~0.0); 924 (797~1190) ❙ 1338 (958~1751); 714 (531~996); 65 (41~142) ❙ 45.2 (13.3~88.8); 918 (797~1190); 362.0 (203.5~659.0)

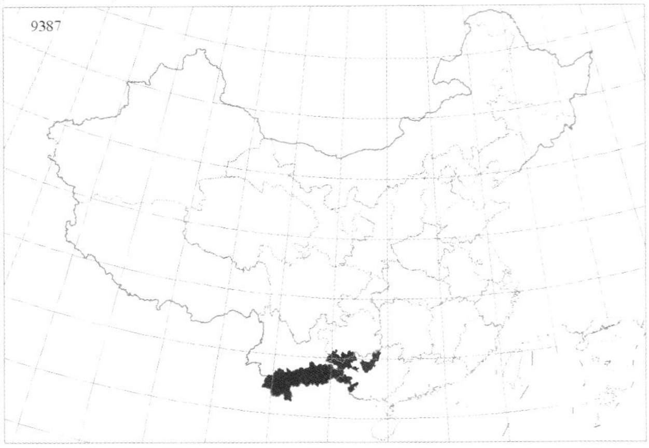

Verbenaceae 马鞭草科 *Clerodendrum* 大青属
Clerodendrum serratum var. herbaceum (Roxb. ex Schauer) C. Y. Wu 草本三对节
Shrubs

18.3 (16.3~21.0); 11.7 (8.5~15.1); 22.7 (20.8~25.6); 18.3 (16.4~21.0); 160.0 (136.4~191.8); 0.0 (0.0~0.0); 883 (797~1053) ❙ 1343 (1007~1751); 733 (539~996); 54 (41~77) ❙ 52.1 (19.0~88.8); 872 (797~1053); 422.7 (277.0~659.0)

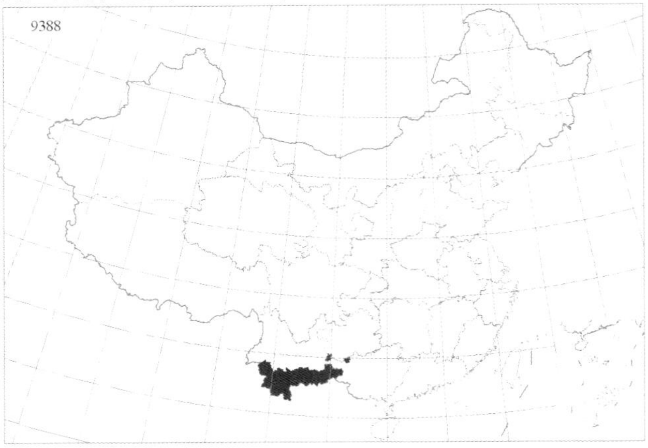

Verbenaceae 马鞭草科 *Clerodendrum* 大青属
Clerodendrum serratum var. wallichii C. B. Clarke 大序三对节
Shrubs

18.3 (13.3~21.0); 11.9 (5.4~15.1); 22.4 (19.6~25.6); 18.3 (13.4~21.0); 159.5 (100.4~191.8); 0.0 (0.0~0.0); 880 (706~1053) ❙ 1358 (1028~1751); 743 (566~1007); 53 (41~77) ❙ 54.9 (12.5~145.8); 864 (706~1053); 435.3 (294.5~659.0)

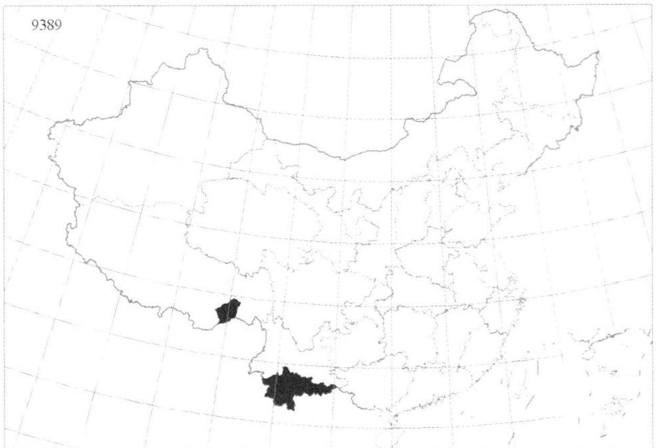

Verbenaceae 马鞭草科 *Clerodendrum* 大青属
Clerodendrum subscaposum Hemsl. 抽葶大青
Subshrubs

17.3; 10.2; 22.2; 17.3; 147.7; 0.0; 835 ❙ 1267; 701; 53 ❙ 51.6; 835; 403.7

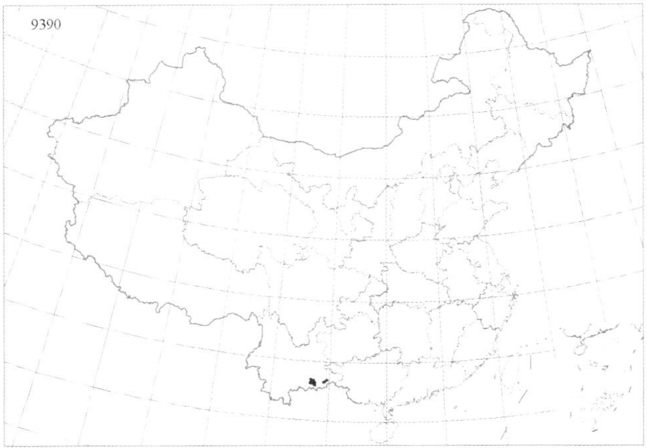

Verbenaceae 马鞭草科 *Premna* 豆腐柴属
Premna crassa var. yui Moldenke 凤庆豆腐柴
Shrubs

16.1 (15.5~16.7); 9.4 (8.6~10.0); 20.7 (20.4~21.3); 16.2 (15.6~16.7); 133.9 (127.3~140.8); 0.0 (0.0~0.0); 787 (768~809) ❙ 1179 (1083~1348); 646 (594~762); 56 (50~65) ❙ 50.0 (34.7~72.8); 785 (768~809); 377.6 (335.1~406.0)

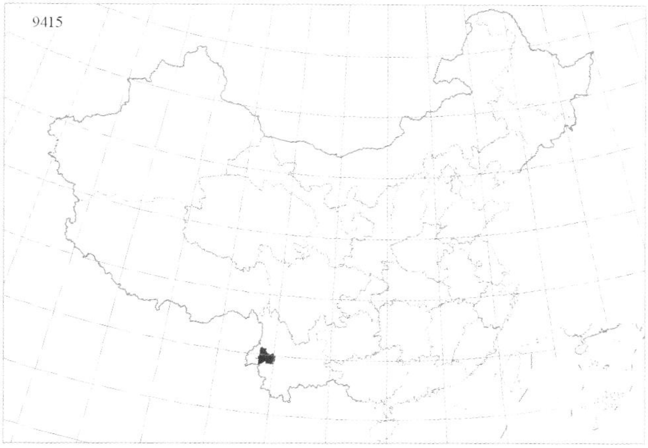

Verbenaceae 马鞭草科 *Premna* 豆腐柴属
Premna flavescens Buch.-Ham. ex C. B. Clarke 淡黄豆腐柴
Shrubs

21.0 (16.3~24.5); 13.1 (5.8~18.9); 27.4 (21.4~29.8); 21.1 (16.4~24.5); 193.2 (136.4~234.4); 0.0 (0.0~0.0); 1105 (797~1360) ❙ 1583 (868~2435); 726 (312~1428); 114 (42~203) ❙ 44.6 (-34.3~99.5); 1094 (797~1360); 270.3 (95.3~659.0)

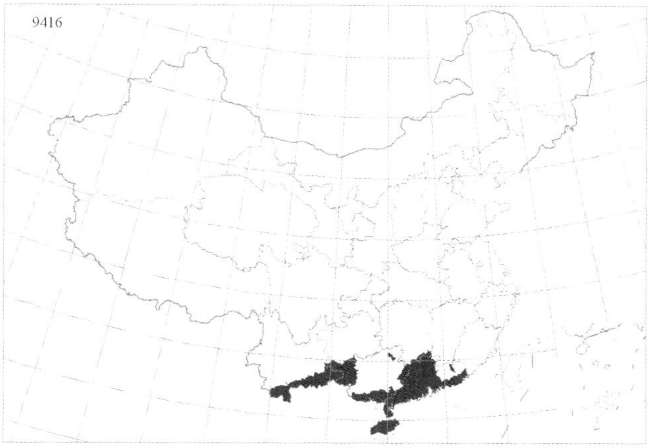

Verbenaceae 马鞭草科 *Premna* 豆腐柴属
Premna fohaiensis C. P'ei et S. L. Chen ex C. Y. Wu 勐海豆腐柴
Shrubs or trees

19.2 (17.3~20.5); 13.4 (11.0~15.1); 22.9 (21.6~23.7); 19.3 (17.3~20.5); 171.2 (148.0~185.9); 0.0 (0.0~0.0); 923 (832~992) ❙ 1453 (1049~1751); 765 (598~996); 61 (44~77) ❙ 57.1 (26.1~88.8); 908 (783~986); 516.5 (365.3~659.0)

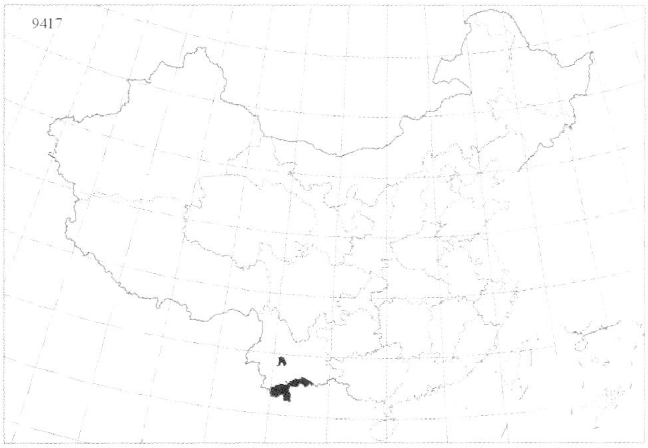

Verbenaceae 马鞭草科 *Premna* 豆腐柴属
Premna fordii Dunn 长序臭黄荆
Shrubs

20.2 (14.2~24.8); 11.2 (3.8~20.1); 27.6 (20.1~29.0); 20.2 (14.2~24.8); 183.0 (110.5~237.6); 0.0 (0.0~1.5); 1070 (729~1385) ❙ 1602 (1147~2435); 684 (411~1428); 146 (41~229) ❙ 51.8 (1.5~101.8); 1058 (729~1360); 252.4 (160.7~620.0)

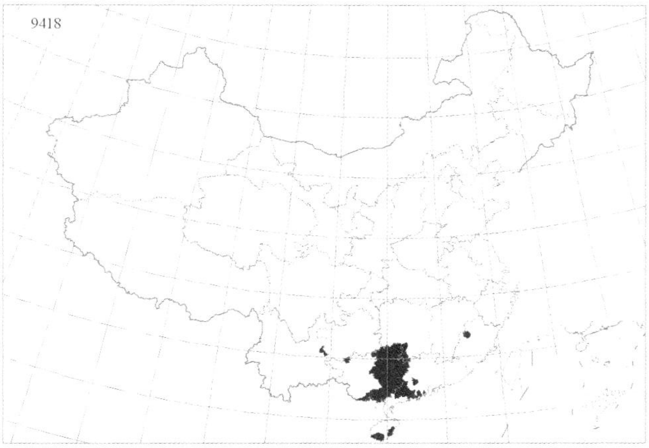

Verbenaceae 马鞭草科 *Premna* 豆腐柴属
Premna fordii var. glabra S. L. Chen 无毛臭黄荆
Shrubs

21.9; 14.1; 27.9; 22.0; 203.7; 0.0; 1161 ❙ 2161; 1240; 126 ❙ 85.7; 1161; 305.5

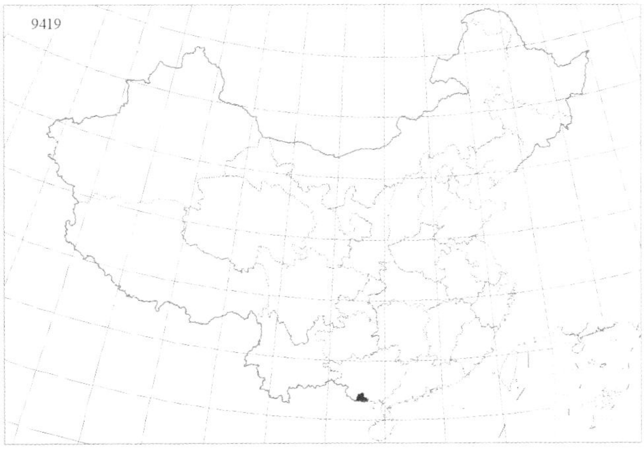

Verbenaceae 马鞭草科 *Premna* 豆腐柴属
Premna fulva Craib 黄毛豆腐柴
Shrubs or trees

18.6 (6.1~22.2); 11.4 (-1.5~15.1); 23.7 (12.7~28.5); 18.7 (6.3~22.3); 164.5 (35.2~207.5); 0.5 (0.0~21.8); 930 (496~1190) ❙ 1334 (801~1751); 722 (496~996); 62 (21~130) ❙ 44.7 (12.5~88.8); 920 (496~1190); 378.3 (189.5~659.0)

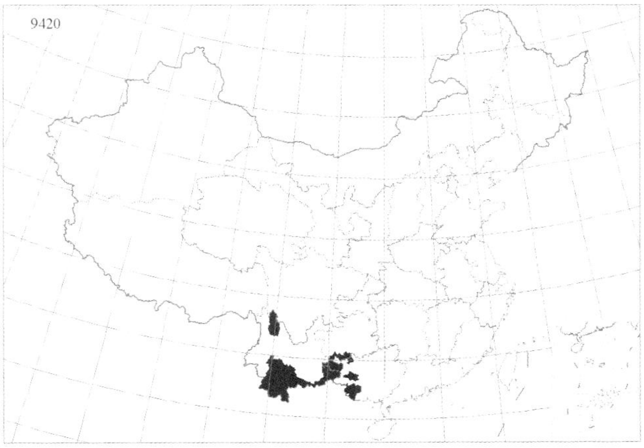

Verbenaceae 马鞭草科 *Premna* 豆腐柴属
Premna glandulosa Hand.-Mazz. 腺叶豆腐柴
Shrubs

16.6 (14.9~18.9); 10.1 (8.4~12.5); 21.3 (19.9~23.2); 16.7 (15.0~19.0); 139.9 (119.9~167.5); 0.0 (0.0~0.0); 811 (747~914) ❙ 981 (907~1139); 564 (521~651); 36 (23~47) ❙ 21.2 (6.5~33.4); 746 (683~851); 310.0 (260.2~365.3)

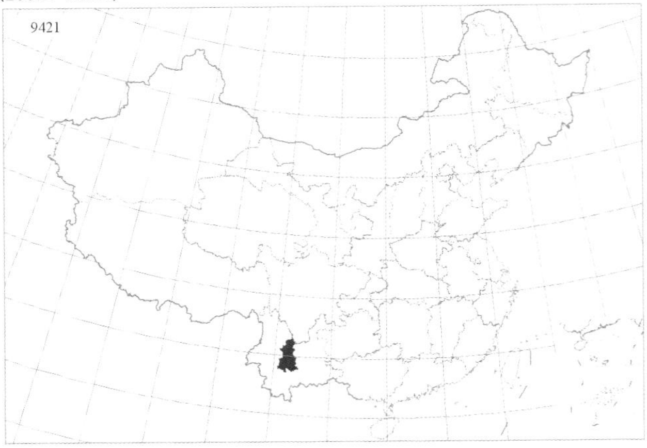

Verbenaceae 马鞭草科 *Premna* 豆腐柴属
Premna hainanensis Chun et F. C. How 海南臭黄荆
Shrubs

23.5 (20.8~24.8); 17.3 (12.3~20.1); 28.1 (27.1~28.7); 23.6 (20.8~24.8); 223.0 (190.0~237.6); 0.0 (0.0~0.0); 1284 (1080~1385) ❙ 1324 (868~1625); 514 (312~795); 71 (41~173) ❙ 4.3 (-34.3~40.9); 1094 (868~1212); 398.5 (237.8~527.5)

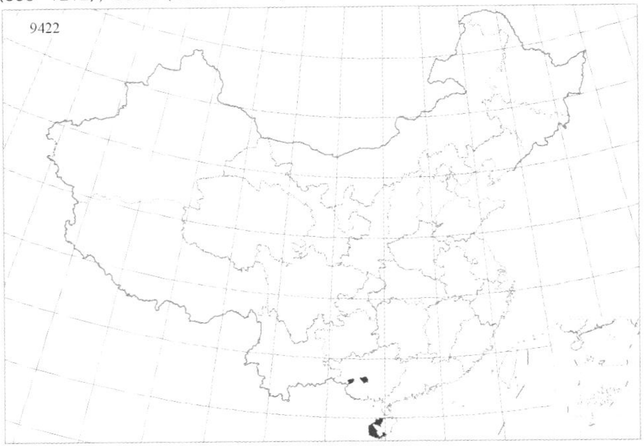

Verbenaceae 马鞭草科 *Premna* 豆腐柴属
Premna henryana (Hand.-Mazz.) C. Y. Wu 蒙自豆腐柴
Shrubs

17.9 (16.3~21.0); 11.2 (9.1~14.2); 22.6 (21.4~25.6); 18.0 (16.4~21.0); 155.9 (136.4~191.8); 0.0 (0.0~0.0); 869 (797~1053) ❙ 1305 (1007~1547); 734 (539~911); 53 (44~68) ❙ 50.2 (19.0~80.6); 868 (797~1053); 417.9 (277.0~551.6)

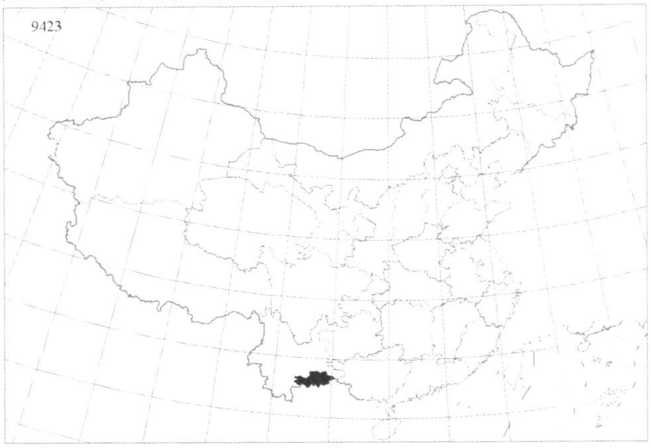

Verbenaceae 马鞭草科 *Premna* 豆腐柴属
Premna herbacea Roxb. 千解草
Dwarf subshrubs

16.8 (5.4~23.6); 9.9 (-2.1~17.8); 21.8 (12.3~29.2); 16.8 (5.8~23.6); 142.7 (31.5~223.5); 0.8 (0.0~25.6); 850 (481~1281) ❙ 1197 (741~1725); 664 (411~959); 51 (15~145) ❙ 43.3 (-14.2~112.2); 826 (481~1273); 358.4 (178.7~627.0)

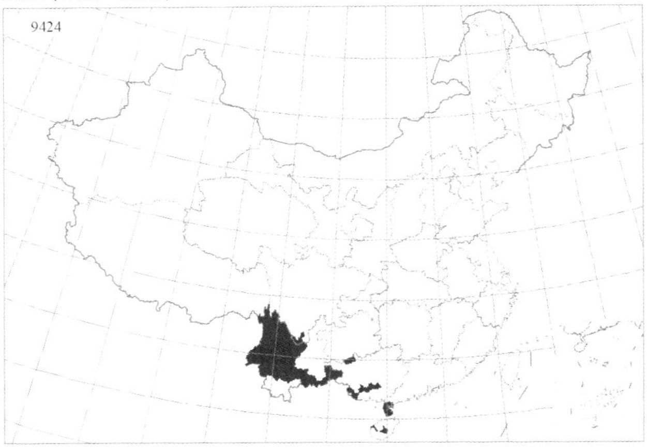

Verbenaceae 马鞭草科 *Premna* 豆腐柴属
Premna interrupta Wall. ex Schauer 间序豆腐柴
Shrubs

16.3 (-0.4~22.7); 8.0 (-10.0~15.0); 23.0 (8.4~28.9); 16.4 (2.8~22.8); 139.3 (9.4~213.1); 2.9 (0.0~73.3); 857 (360~1221) ❙ 1196 (341~2435); 648 (226~1428); 62 (4~194) ❙ 40.4 (-15.3~145.8); 851 (341~1221); 284.2 (135.1~551.6)

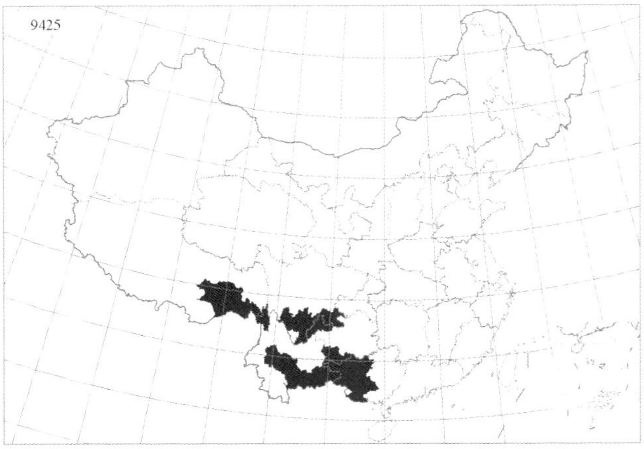

Verbenaceae 马鞭草科 *Premna* 豆腐柴属
Premna laevigata C. Y. Wu 平滑豆腐柴
Shrubs

20.4; 15.1; 23.6; 20.5; 185.8; 0.0; 989 ❙ 1607; 702; 77 ❙ 62.5; 986; 659.0

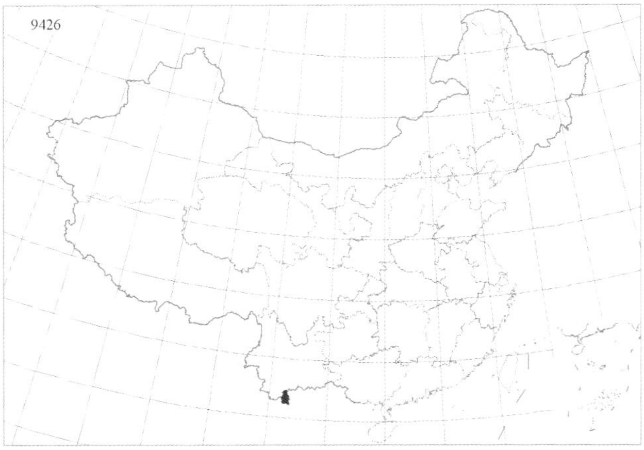

Verbenaceae 马鞭草科 *Premna* 豆腐柴属

Premna latifolia Roxb. 大叶豆腐柴

Shrubs or small trees

20.0; 14.6; 23.2; 20.1; 180.8; 0.0; 963 ❙ 1523; 681; 67 ❙ 58.4; 942; 581.0

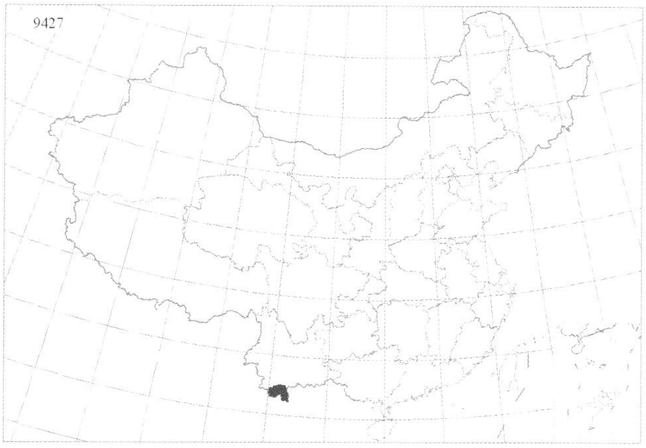

Verbenaceae 马鞭草科 *Premna* 豆腐柴属

Premna latifolia var. cuneata C. B. Clarke 楔叶豆腐柴

Shrubs or small trees

20.5; 14.9; 23.7; 20.5; 185.9; 0.0; 992 ❙ 1464; 655; 66 ❙ 47.6; 959; 570.1

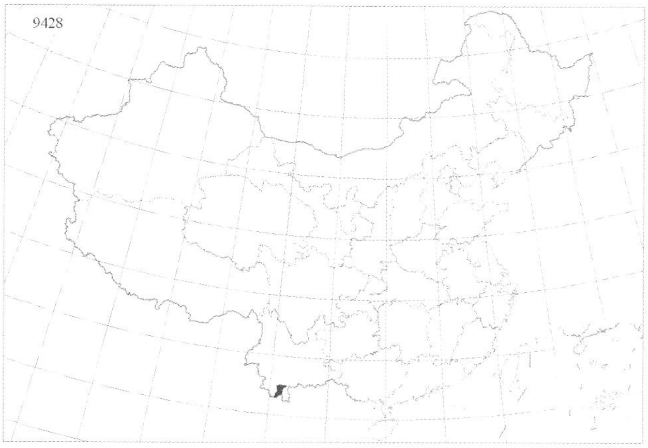

Verbenaceae 马鞭草科 *Premna* 豆腐柴属

Premna ligustroides Hemsl. 臭黄荆

Shrubs

16.4 (9.1~21.7); 5.3 (-1.7~13.4); 26.8 (17.7~29.8); 16.5 (9.2~21.7); 139.1 (64.1~200.7); 1.5 (0.0~18.0); 886 (595~1147) ❙ 1328 (807~1990); 537 (299~827); 121 (19~259) ❙ 50.2 (-6.7~111.3); 878 (595~1147); 208.9 (108.2~322.1)

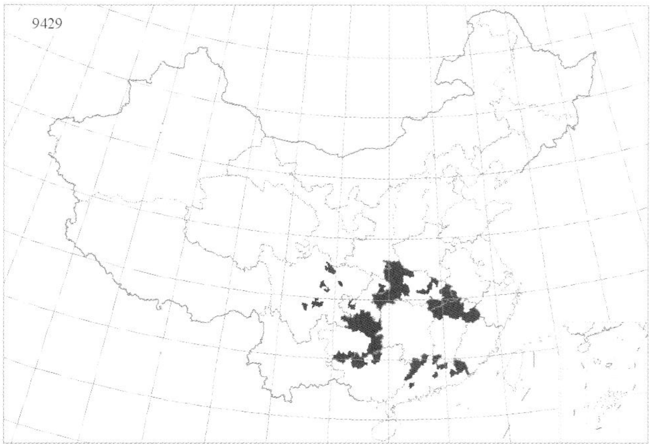

Verbenaceae 马鞭草科 *Premna* 豆腐柴属

Premna maclurei Merr. 弯毛臭黄荆

Scandent shrubs

20.5 (14.3~24.8); 12.1 (3.3~20.1); 27.4 (24.3~28.9); 20.6 (14.3~24.8); 186.9 (114.1~237.6); 0.1 (0.0~2.3); 1096 (765~1385) ❙ 1524 (868~1876); 606 (312~891); 125 (41~198) ❙ 43.3 (-34.3~101.8); 1053 (765~1360); 308.9 (177.4~620.0)

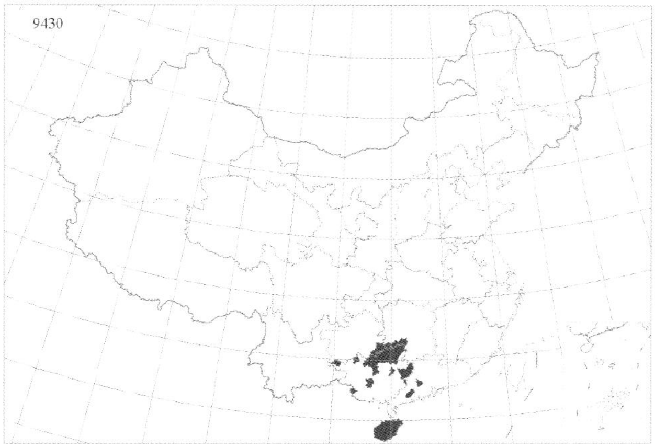

Verbenaceae 马鞭草科 *Premna* 豆腐柴属

Premna mekongensis W. W. Sm. 澜沧豆腐柴

Small shrubs

13.0 (5.4~17.9); 6.1 (-2.1~10.7); 18.5 (12.3~23.4); 13.0 (5.8~18.0); 98.8 (31.5~155.7); 2.9 (0.0~25.6); 696 (481~878) ❙ 984 (741~1164); 572 (426~651); 36 (15~59) ❙ 43.6 (4.4~78.3); 669 (481~749); 323.1 (242.3~384.1)

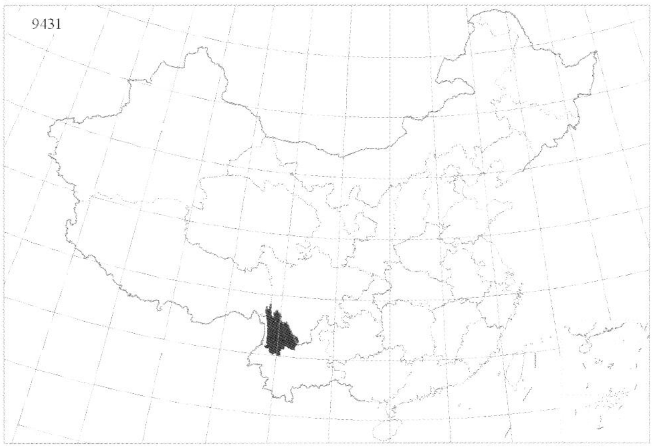

Verbenaceae 马鞭草科 *Premna* 豆腐柴属

Premna microphylla Turcz. 豆腐柴

Shrubs

16.8 (10.2~24.8); 5.8 (-2.9~20.1); 27.0 (18.8~29.8); 16.9 (10.6~24.8); 143.6 (79.4~237.6); 1.3 (0.0~20.5); 903 (643~1385) ❙ 1405 (626~3495); 550 (316~1388); 145 (20~897) ❙ 55.6 (-22.2~224.4); 896 (626~1205); 209.9 (95.3~527.5)

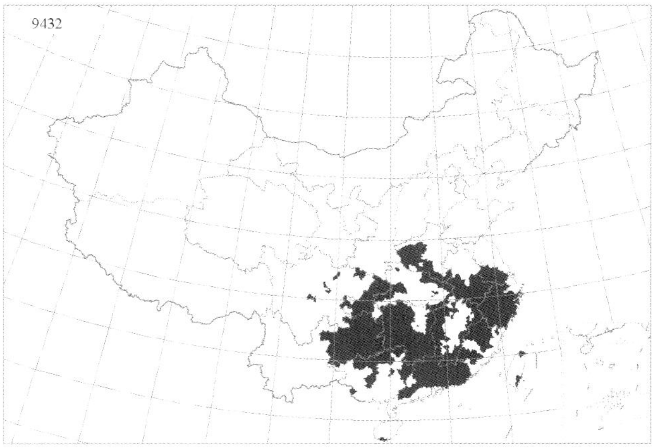

Verbenaceae 马鞭草科 *Premna* 豆腐柴属
Premna octonervia Merr. et F. P. Metcalf 八脉臭黄荆
Shrubs or trees

24.0 (22.8~24.8); 18.0 (14.7~20.1); 28.4 (27.5~29.2); 24.0 (22.8~24.8); 228.5 (213.6~237.6); 0.0 (0.0~0.0); 1318 (1230~1385) ▎1413 (868~1813); 556 (312~893); 70 (41~113) ▎7.5 (-34.3~36.4); 1152 (868~1329); 359.7 (178.7~527.5)

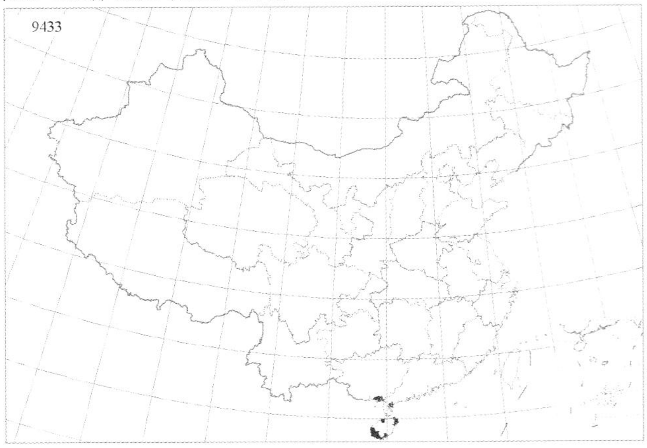

Verbenaceae 马鞭草科 *Premna* 豆腐柴属
Premna odorata Blanco 毛鱼臭目
Small trees

21.9; 17.0; 25.4; 21.9; 202.8; 0.0; 1119 ▎2501; 1537; 94 ▎133.8; 1043; 234.5

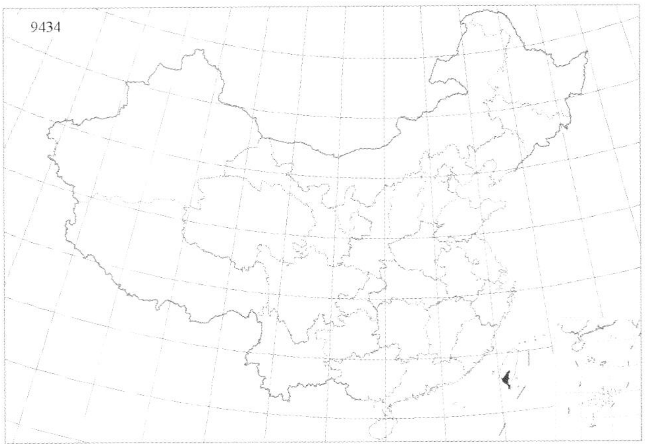

Verbenaceae 马鞭草科 *Premna* 豆腐柴属
Premna oligantha C. Y. Wu 少花豆腐柴
Small shrubs

5.4; -2.3; 12.3; 5.8; 31.5; 26.0; 481 ▎796; 464; 27 ▎65.5; 481; 236.5

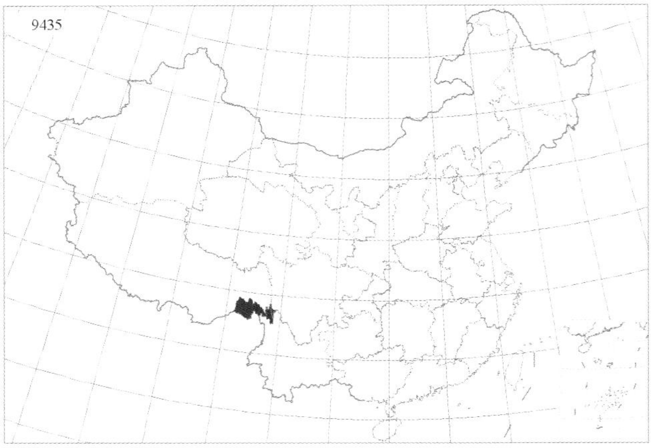

Verbenaceae 马鞭草科 *Premna* 豆腐柴属
Premna paisehensis C. P'ei et S. L. Chen 百色豆腐柴
Trees

21.3; 12.6; 27.9; 21.3; 196.0; 0.0; 1123 ▎1188; 634; 56 ▎5.8; 1121; 315.6

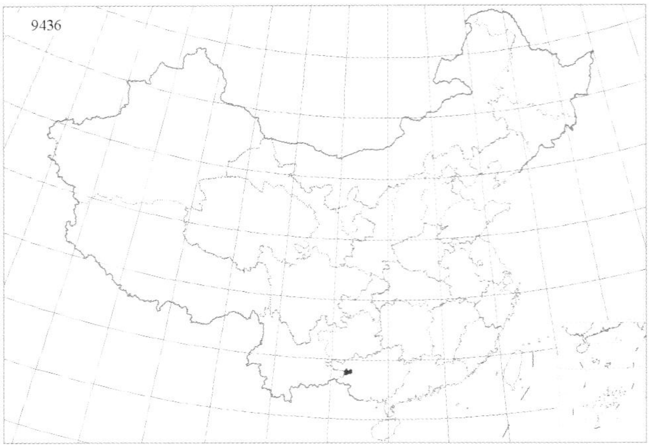

Verbenaceae 马鞭草科 *Premna* 豆腐柴属
Premna parvilimba C. P'ei 小叶豆腐柴
Shrubs

8.5 (5.4~10.8); 1.2 (-2.1~3.9); 14.7 (12.3~16.6); 8.6 (5.8~10.9); 53.5 (31.5~71.8); 11.3 (1.3~25.6); 567 (481~633) ▎893 (741~1007); 539 (426~603); 26 (15~40) ▎57.6 (51.8~67.9); 565 (481~633); 305.6 (273.5~351.7)

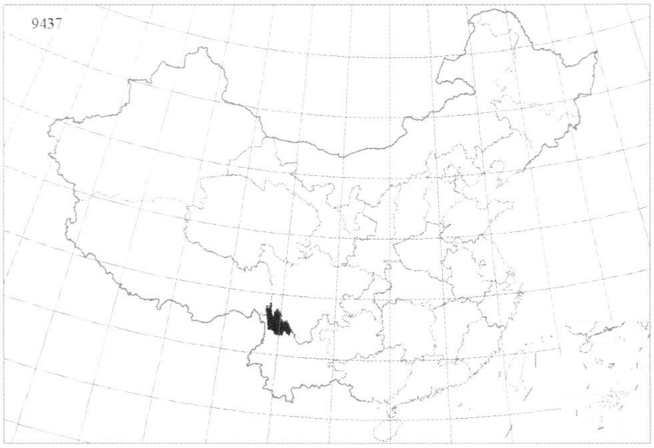

Verbenaceae 马鞭草科 *Premna* 豆腐柴属
Premna puberula Pamp. 狐臭柴
Shrubs or small trees

16.3 (9.3~22.7); 6.2 (-1.7~15.0); 25.3 (18.8~29.8); 16.3 (9.5~22.8); 137.1 (69.8~213.1); 1.0 (0.0~18.0); 859 (603~1221) ▎1304 (711~2435); 577 (357~1428); 96 (15~259) ▎52.2 (5.4~111.3); 858 (603~1221); 228.4 (108.2~413.5)

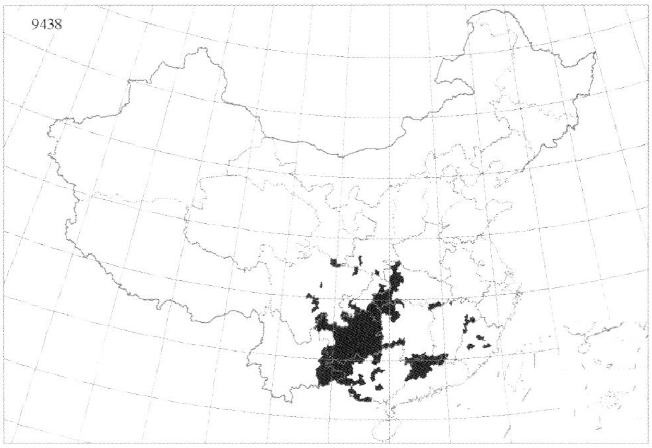

Verbenaceae 马鞭草科 *Premna* 豆腐柴属
Premna puberula var. bodinieri (H. Lév.) C. Y. Wu et S. Y. Pao 毛狐臭柴
Shrubs or small trees

16.5 (13.5~19.8); 6.7 (2.6~11.1); 25.0 (20.1~27.7); 16.6 (13.6~19.8); 139.4 (106.7~177.9); 0.3 (0.0~4.1); 853 (729~1010) ❙ 1271 (1018~1504); 583 (444~860); 86 (43~161) ❙ 49.6 (16.9~100.5); 853 (729~1010); 248.8 (175.2~413.5)

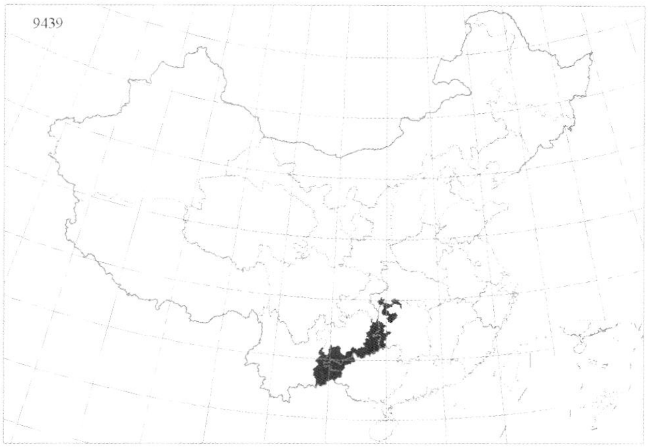

Verbenaceae 马鞭草科 *Premna* 豆腐柴属
Premna puerensis Y. Y. Qian 普洱豆腐柴
Climbing shrubs

18.2; 12.2; 22.1; 18.3; 160.0; 0.0; 872 ❙ 1456; 836; 62 ❙ 66.9; 866; 395.0

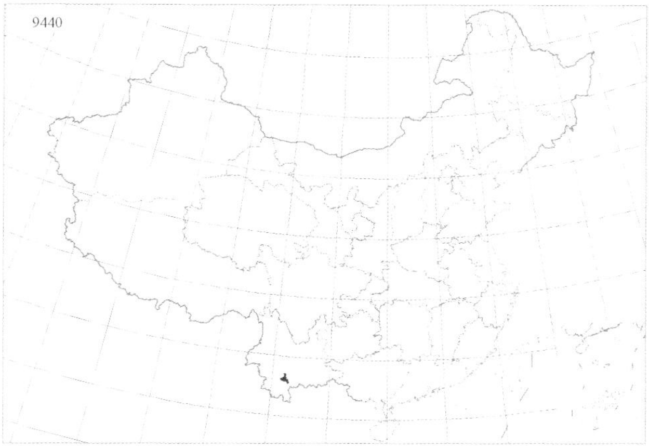

Verbenaceae 马鞭草科 *Premna* 豆腐柴属
Premna punicea C. Y. Wu 玫花豆腐柴
Shrubs

14.2; 7.2; 19.8; 14.3; 112.0; 0.4; 740 ❙ 1127; 662; 37 ❙ 51.5; 733; 325.5

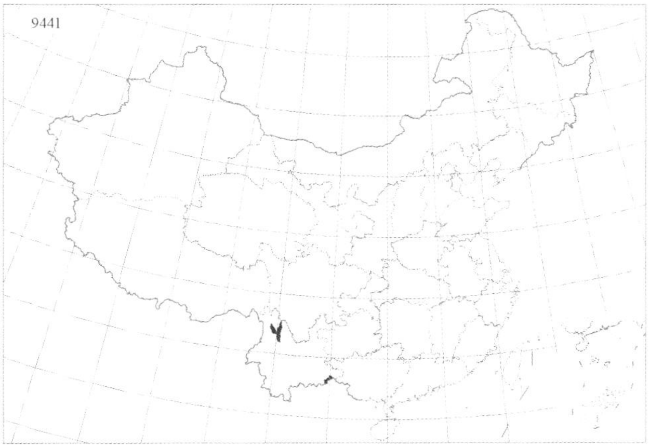

Verbenaceae 马鞭草科 *Premna* 豆腐柴属
Premna pyramidata Wall. ex Schauer 塔序豆腐柴
Shrubs or small trees

22.0; 13.5; 28.8; 22.0; 203.6; 0.0; 1169 ❙ 1641; 724; 132 ❙ 40.5; 1169; 205.0

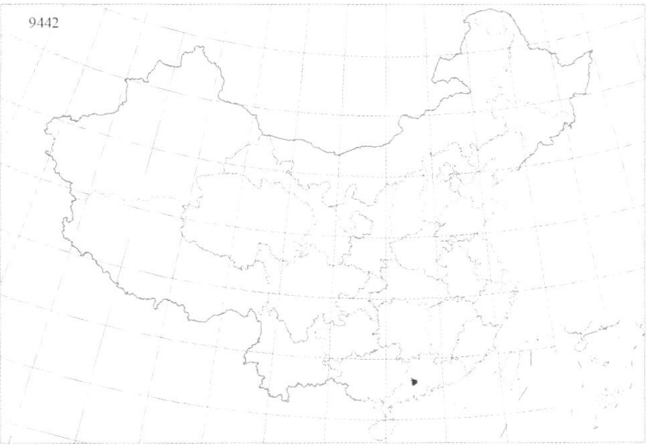

Verbenaceae 马鞭草科 *Premna* 豆腐柴属
Premna racemosa Wall. ex Schauer 总序豆腐柴
Vines or shrubs

17.0 (−1.8~21.0); 10.4 (−10.2~14.2); 21.7 (6.1~25.6); 17.2 (1.8~21.0); 148.6 (1.9~191.8); 3.7 (0.0~82.8); 845 (329~1053) ❙ 1254 (320~1751); 706 (217~996); 52 (20~77) ❙ 47.6 (−3.0~88.8); 834 (320~1053); 391.4 (72.2~551.6)

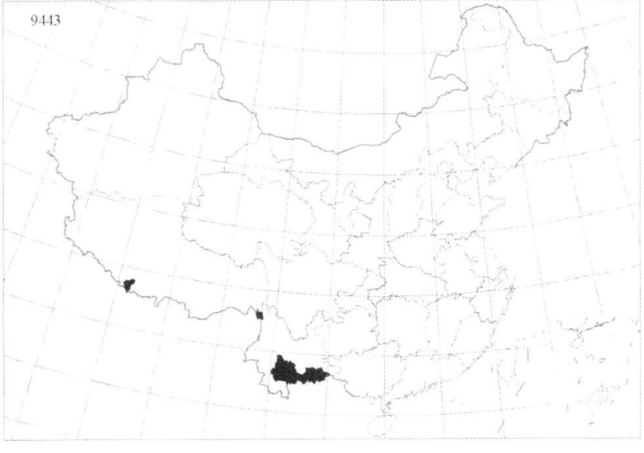

Verbenaceae 马鞭草科 *Premna* 豆腐柴属
Premna rubroglandulosa C. Y. Wu 红腺豆腐柴
Shrubs

17.1; 11.0; 20.8; 17.1; 145.0; 0.0; 815 ❙ 1352; 771; 47 ❙ 65.9; 802; 438.9

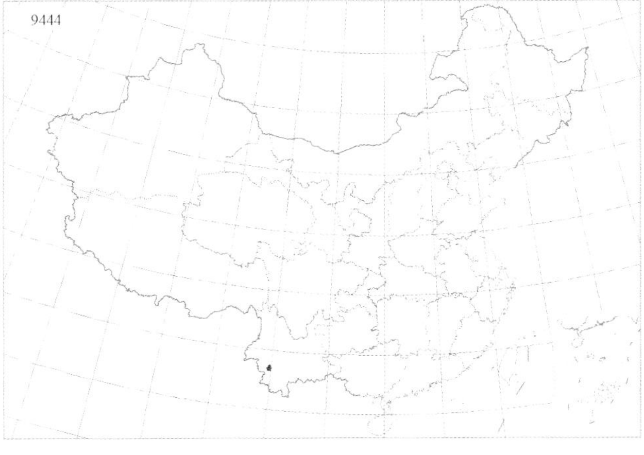

Verbenaceae 马鞭草科 *Premna* 豆腐柴属

Premna scandens Roxb. 藤豆腐柴

Large climbers

18.5 (16.3~21.0); 12.2 (9.1~15.1); 22.9 (21.4~25.6); 18.6 (16.4~21.0); 163.0 (136.4~191.8); 0.0 (0.0~0.0); 897 (797~1053) ❙ 1412 (1163~1751); 778 (641~996); 59 (48~77) ❙ 57.5 (34.8~88.8); 896 (797~1053); 479.7 (347.3~659.0)

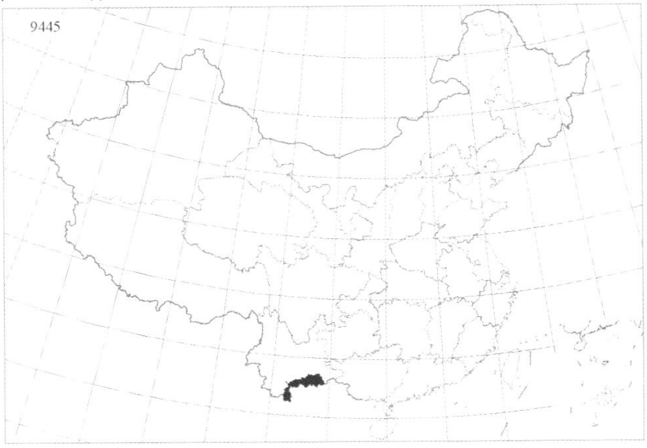

Verbenaceae 马鞭草科 *Premna* 豆腐柴属

Premna scoriarum W. W. Sm. 腾冲豆腐柴

Shrubs or trees

17.5 (14.4~19.6); 10.7 (7.5~12.8); 21.6 (19.1~23.5); 17.5 (14.4~19.6); 149.9 (113.1~175.7); 0.0 (0.0~0.0); 848 (727~953) ❙ 1563 (1454~1688); 903 (792~959); 55 (42~70) ❙ 85.7 (61.2~112.2); 842 (727~932); 522.8 (451.2~627.0)

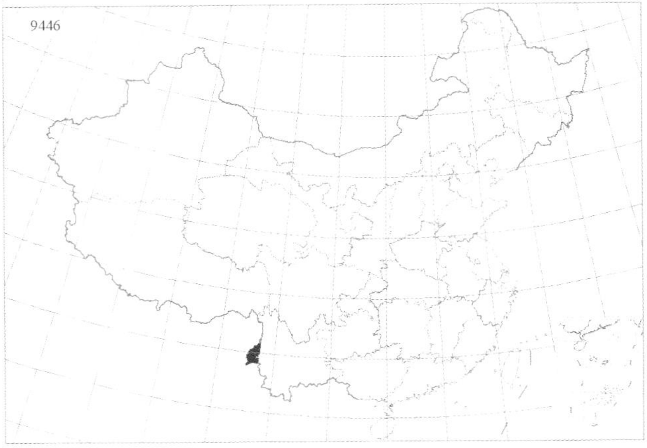

Verbenaceae 马鞭草科 *Premna* 豆腐柴属

Premna serratifolia L. 伞序臭黄荆

Shrubs or trees

22.4 (19.3~24.8); 15.3 (11.9~20.1); 28.0 (22.8~29.1); 22.5 (19.3~24.8); 209.6 (171.9~237.6); 0.0 (0.0~0.0); 1196 (912~1385) ❙ 1727 (868~2993); 773 (312~1847); 116 (41~547) ❙ 46.5 (-34.3~203.4); 1164 (868~1360); 267.1 (95.3~620.0)

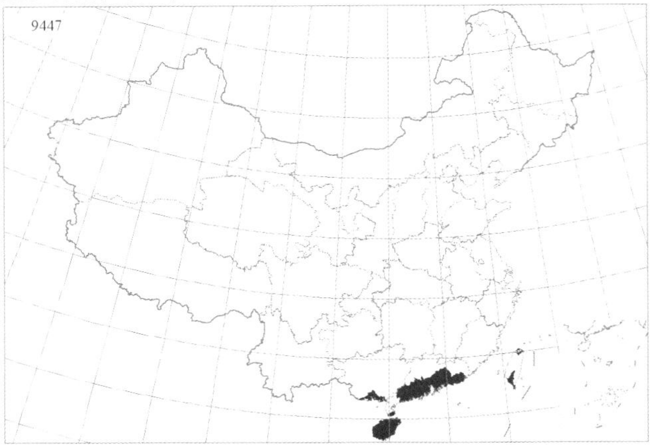

Verbenaceae 马鞭草科 *Premna* 豆腐柴属

Premna steppicola Hand.-Mazz. 草坡豆腐柴

Shrubs

12.3 (5.4~17.9); 5.3 (-2.1~10.7); 18.0 (12.3~23.4); 12.4 (5.8~18.0); 93.5 (31.5~155.7); 5.5 (0.0~25.6); 680 (481~878) ❙ 931 (741~1057); 556 (426~610); 27 (12~42) ❙ 40.2 (4.4~67.9); 642 (481~749); 299.6 (242.3~351.7)

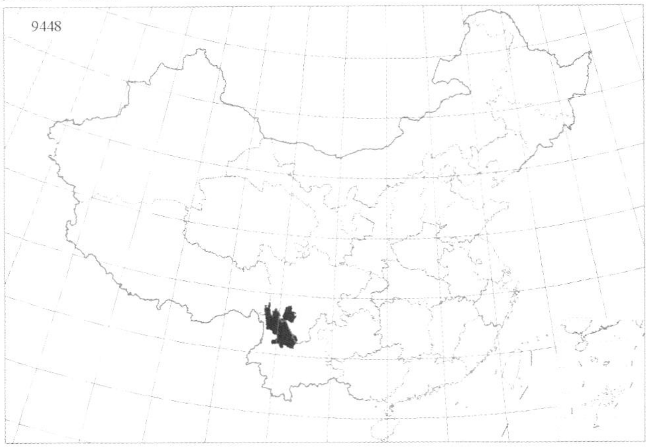

Verbenaceae 马鞭草科 *Premna* 豆腐柴属

Premna straminicaulis C. Y. Wu 草黄枝豆腐柴

Shrubs

19.5; 12.7; 23.3; 19.5; 173.8; 0.0; 942 ❙ 1559; 907; 42 ❙ 65.4; 921; 482.8

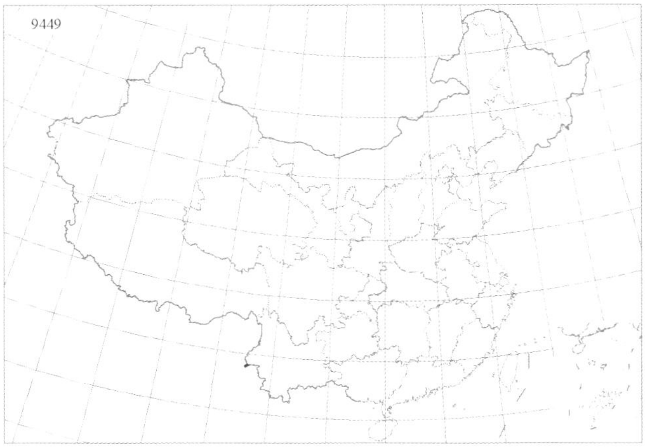

Verbenaceae 马鞭草科 *Premna* 豆腐柴属

Premna subcapitata Rehder 近头状豆腐柴

Shrubs

13.9 (5.4~18.9); 6.0 (-2.4~12.0); 20.4 (12.3~27.4); 13.9 (5.8~19.0); 108.9 (31.5~167.8); 2.0 (0.0~25.6); 734 (481~951) ❙ 978 (741~1410); 561 (411~816); 35 (15~68) ❙ 35.2 (-14.2~92.9); 711 (481~951); 255.3 (135.1~351.7)

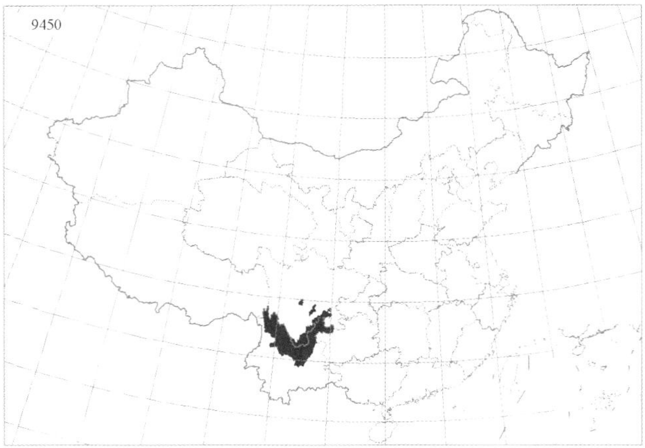

Verbenaceae 马鞭草科 *Premna* 豆腐柴属

Premna subscandens Merr. 攀援臭黄荆

Scandent shrubs

24.2；18.5；28.3；24.3；231.0；0.0；1335 ❙ 1284；472；56 ❙ -3.8；1154；405.9

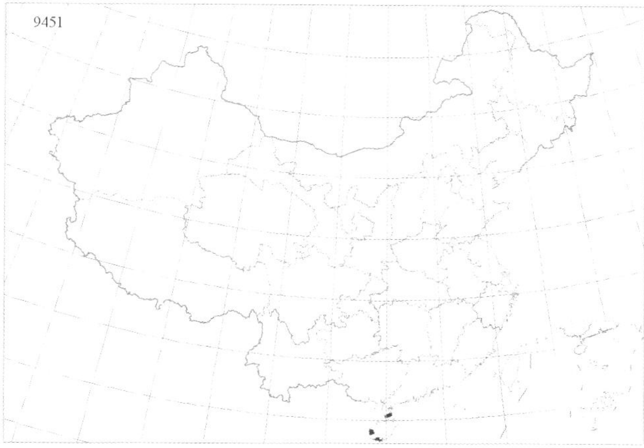

Verbenaceae 马鞭草科 *Premna* 豆腐柴属

Premna sunyiensis C. P'ei 塘虱角

Scandent shrubs

21.1 (17.7~22.3)；12.4 (7.7~13.9)；27.9 (26.3~28.9)；21.2 (17.8~22.3)；194.2 (153.6~208.1)；0.0 (0.0~0.0)；1119 (916~1202) ❙ 1569 (1374~1776)；751 (548~850)；143 (87~194) ❙ 41.4 (15.4~70.7)；1119 (916~1202)；254.8 (201.4~335.1)

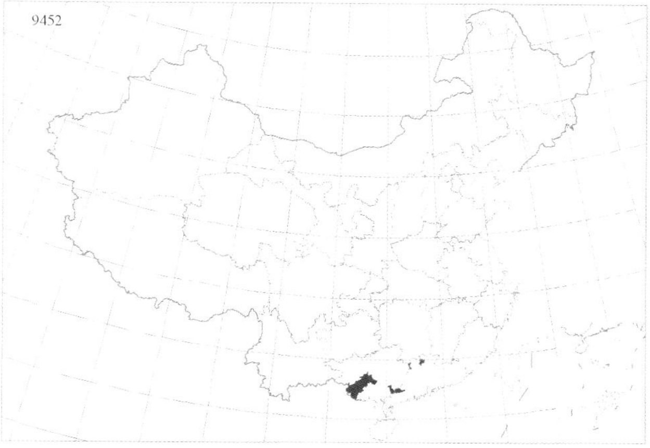

Verbenaceae 马鞭草科 *Premna* 豆腐柴属

Premna szemaoensis C. P'ei 思茅豆腐柴

Trees

18.2 (16.0~21.0)；11.9 (6.5~15.1)；22.4 (20.3~25.7)；18.3 (16.0~21.0)；159.5 (132.4~191.8)；0.0 (0.0~0.0)；880 (780~1053) ❙ 1393 (1007~1751)；764 (539~996)；55 (41~122) ❙ 58.7 (19.0~112.2)；867 (780~1053)；451.6 (225.1~659.0)

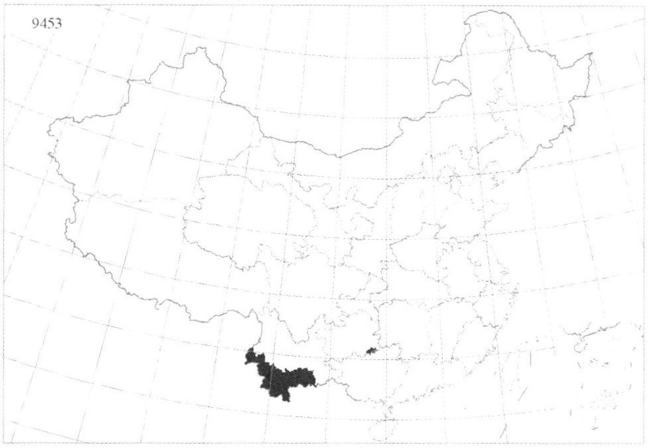

Verbenaceae 马鞭草科 *Premna* 豆腐柴属

Premna tapintzeana Dop 大坪子豆腐柴

Shrubs

13.8 (5.4~18.9)；6.3 (-2.1~12.0)；19.8 (12.3~26.6)；13.8 (5.8~19.0)；107.3 (31.5~167.8)；1.6 (0.0~25.6)；724 (481~929) ❙ 977 (741~1348)；563 (411~762)；35 (15~58) ❙ 36.6 (-14.2~78.3)；697 (481~912)；283.1 (142.9~395.4)

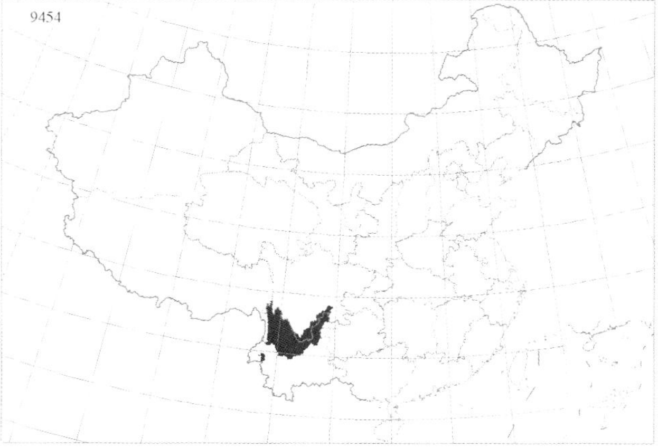

Verbenaceae 马鞭草科 *Premna* 豆腐柴属

Premna tenii C. P'ei 圆叶豆腐柴

Shrubs

15.0；8.4；20.2；15.1；121.2；0.0；753 ❙ 950；570；27 ❙ 26.3；683；307.7

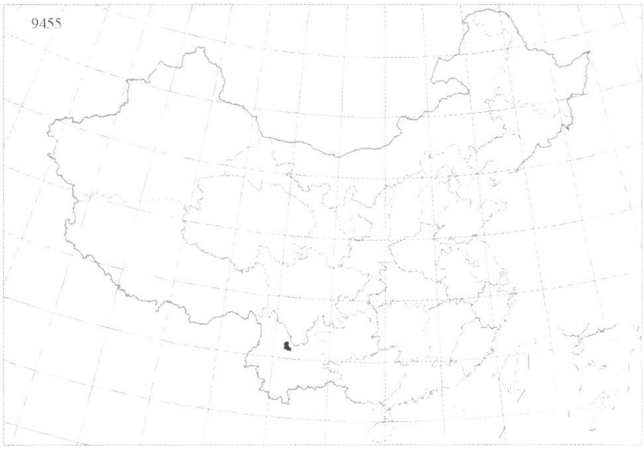

Verbenaceae 马鞭草科 *Premna* 豆腐柴属

Premna urticifolia Rehder 麻叶豆腐柴

Shrubs

17.5；11.2；21.8；17.5；150.5；0.0；845 ❙ 1172；627；51 ❙ 37.9；826；362.4

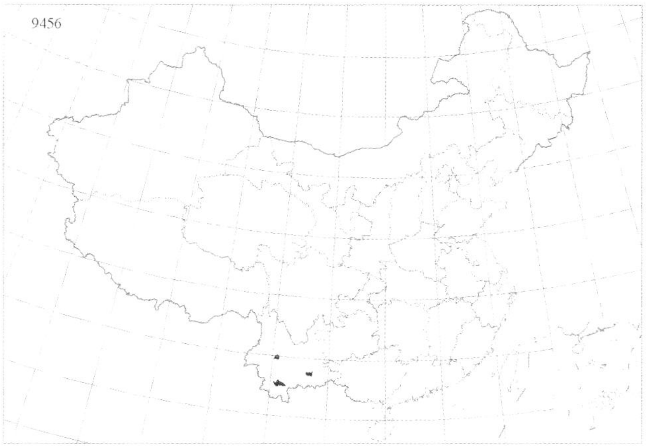

Verbenaceae 马鞭草科 *Premna* 豆腐柴属
Premna velutina C. Y. Wu 黄绒豆腐柴
Shrubs

16.0; 9.3; 20.4; 16.0; 132.4; 0.0; 780 ❙ 1348; 762; 56 ❙ 72.8; 780; 395.4

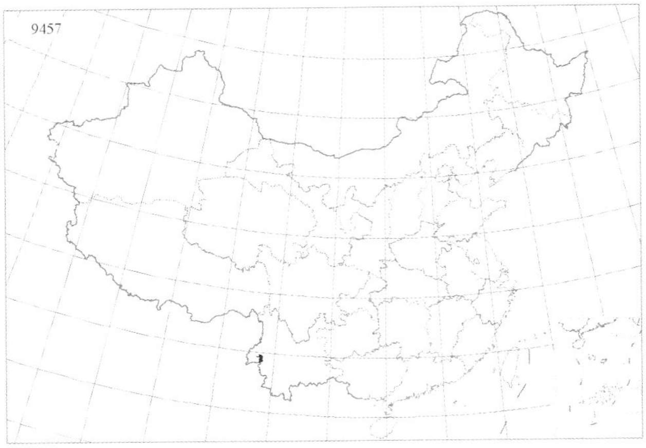

Verbenaceae 马鞭草科 *Premna* 豆腐柴属
Premna yunnanensis W. W. Sm. 云南豆腐柴
Small shrubs

13.3 (2.2~18.9); 5.8 (-6.6~12.0); 19.4 (9.5~26.6); 13.4 (3.8~19.0); 103.4 (15.5~167.8); 3.6 (0.0~49.1); 711 (411~929) ❙ 966 (729~1164); 558 (411~651); 34 (11~59) ❙ 38.7 (-14.2~125.9); 685 (411~912); 284.5 (142.9~405.0)

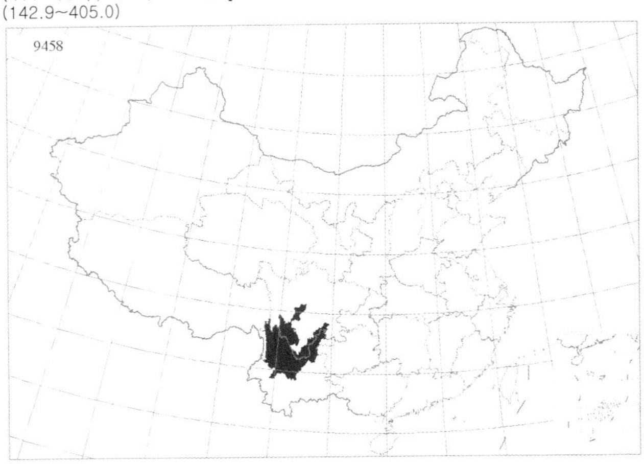

Verbenaceae 马鞭草科 *Sphenodesme* 楔翅藤属
Sphenodesme floribunda Chun et F. C. How 多花楔翅藤
Scandent shrubs

23.8 (22.2~24.8); 18.3 (17.2~20.1); 27.7 (25.7~28.7); 23.8 (22.3~24.8); 225.9 (207.2~237.6); 0.0 (0.0~0.0); 1289 (1127~1385) ❙ 1387 (868~1725); 489 (312~610); 59 (41~92) ❙ 8.5 (-34.3~47.8); 1122 (868~1294); 477.1 (288.3~620.0)

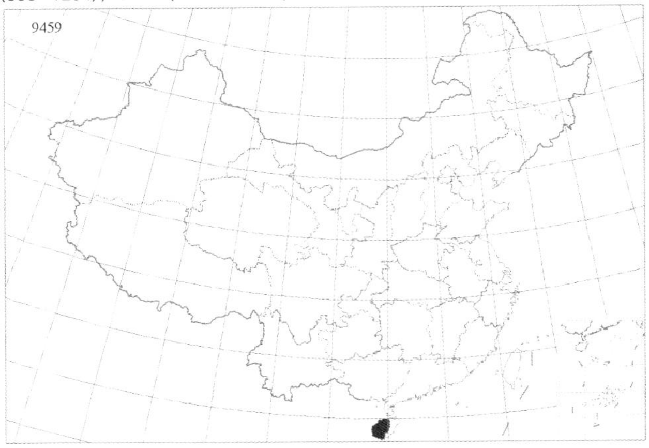

Verbenaceae 马鞭草科 *Sphenodesme* 楔翅藤属
Sphenodesme involucrata (Presl) B. L. Rob. 爪楔翅藤
Scandent shrubs

23.4 (17.1~24.8); 17.8 (11.2~20.1); 27.5 (21.8~28.8); 23.5 (17.1~24.8); 221.4 (145.0~237.6); 0.0 (0.0~0.0); 1265 (816~1385) ❙ 1599 (868~2445); 619 (312~1412); 85 (41~244) ❙ 30.9 (-34.3~191.8); 1161 (816~1360); 411.7 (202.8~620.0)

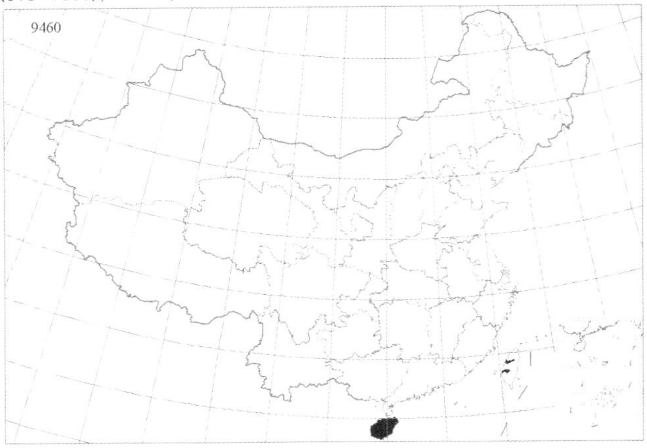

Verbenaceae 马鞭草科 *Sphenodesme* 楔翅藤属
Sphenodesme mollis Craib 毛楔翅藤
Scandent shrubs

18.8 (17.8~20.4); 12.7 (11.6~15.1); 22.7 (21.9~23.6); 18.8 (17.9~20.5); 166.1 (154.5~185.8); 0.0 (0.0~0.0); 902 (854~989) ❙ 1356 (1028~1751); 751 (566~996); 58 (45~77) ❙ 50.2 (12.5~88.8); 883 (816~986); 423.6 (295.5~659.0)

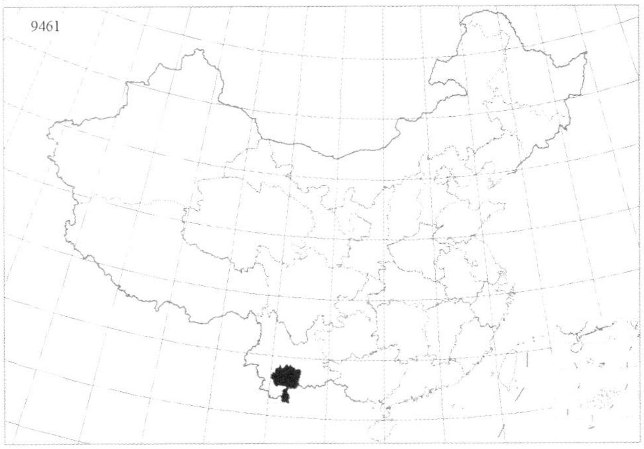

Verbenaceae 马鞭草科 *Symphorema* 六苞藤属
Symphorema involucratum Roxb. 六苞藤
Climbing shrubs

19.2 (17.9~20.5); 13.6 (11.9~15.1); 22.5 (21.6~23.7); 19.3 (18.0~20.5); 171.3 (155.8~185.9); 0.0 (0.0~0.0); 920 (854~992) ❙ 1486 (1434~1607); 719 (655~803); 54 (41~77) ❙ 61.9 (47.6~74.2); 892 (833~986); 527.4 (439.5~659.0)

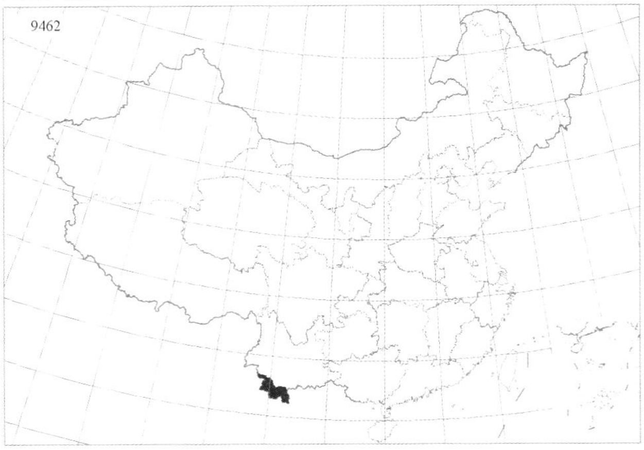

Verbenaceae 马鞭草科 *Tsoongia* 假紫珠属

Tsoongia axillariflora Merr. 假紫珠

Shrubs

22.2 (17.7~24.8); 14.8 (7.4~20.1); 27.8 (22.5~28.9); 22.2 (17.8~24.8); 206.3 (153.4~237.6); 0.0 (0.0~0.0); 1187 (854~1385) ❙ 1596 (868~2435); 668 (312~1428); 112 (41~189) ❙ 37.0 (-34.3~99.5); 1131 (854~1360); 347.5 (185.5~620.0)

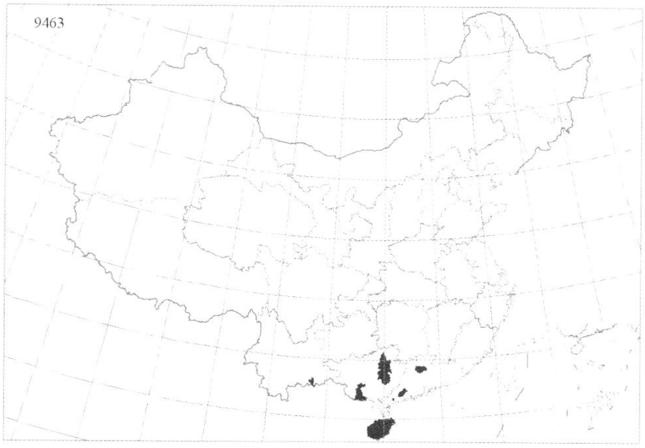

Verbenaceae 马鞭草科 *Vitex* 牡荆属

Vitex burmensis Moldenke 长叶荆

Shrubs or trees

18.3 (13.3~21.0); 11.4 (5.4~15.1); 23.0 (19.6~26.6); 18.3 (13.4~21.0); 159.6 (100.4~191.8); 0.0 (0.0~0.0); 889 (706~1053) ❙ 1377 (1143~1751); 747 (617~1007); 56 (41~77) ❙ 55.7 (21.7~145.8); 879 (706~1053); 413.2 (189.5~659.0)

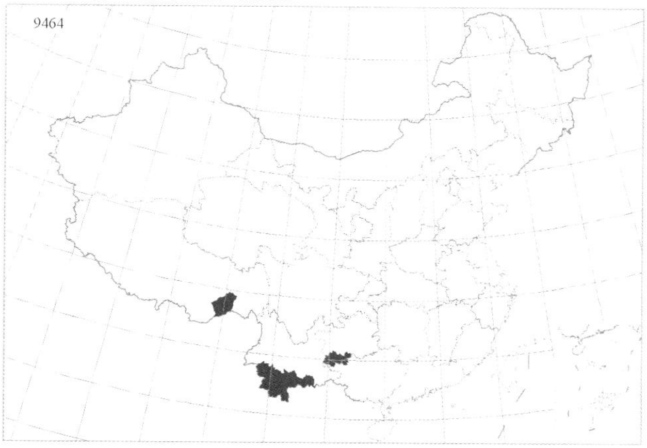

Verbenaceae 马鞭草科 *Vitex* 牡荆属

Vitex canescens Kurz 灰毛牡荆

Trees

19.3 (14.2~24.8); 10.6 (2.6~20.1); 26.5 (20.8~29.8); 19.3 (14.2~24.8); 172.1 (110.9~237.6); 0.1 (0.0~3.6); 1008 (734~1385) ❙ 1485 (868~2435); 651 (312~1428); 112 (41~221) ❙ 49.0 (-34.3~101.8); 997 (734~1360); 262.2 (95.3~659.0)

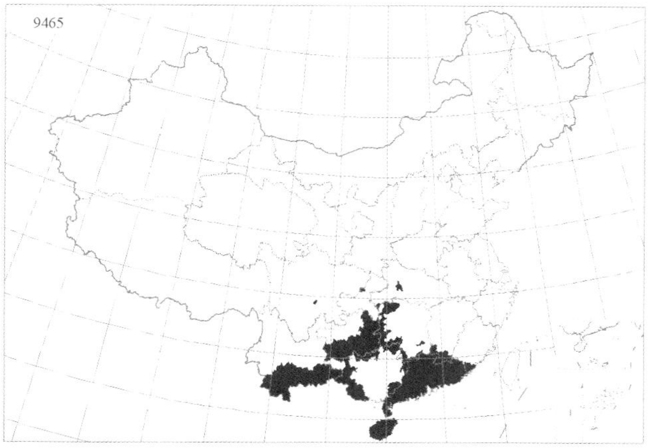

Verbenaceae 马鞭草科 *Vitex* 牡荆属

Vitex duclouxii P. Dop 金沙荆

Shrubs or trees

12.8 (11.4~14.8); 4.1 (2.7~6.1); 20.3 (18.6~23.8); 12.9 (11.4~14.9); 96.0 (80.7~119.4); 1.3 (0.0~3.7); 699 (652~781) ❙ 937 (863~1044); 527 (479~559); 34 (25~50) ❙ 34.2 (26.5~49.1); 699 (652~781); 229.9 (190.3~278.8)

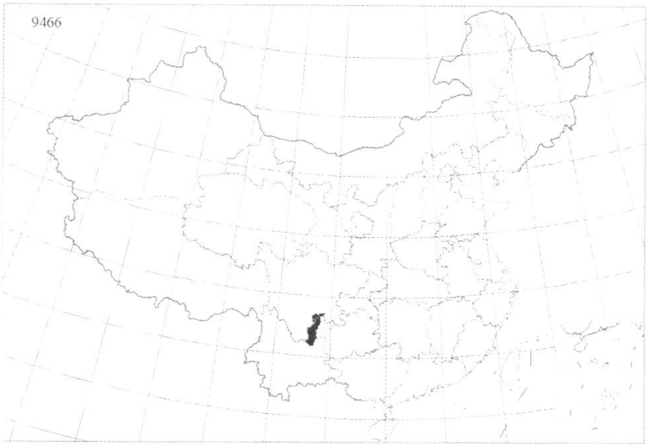

Verbenaceae 马鞭草科 *Vitex* 牡荆属

Vitex kwangsiensis C. P'ei 广西牡荆

Trees

21.8; 13.6; 27.9; 21.8; 202.0; 0.0; 1155 ❙ 1469; 776; 98 ❙ 27.4; 1155; 314.8

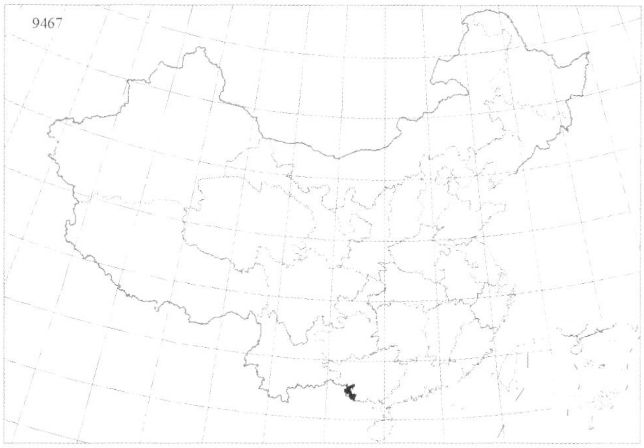

Verbenaceae 马鞭草科 *Vitex* 牡荆属

Vitex negundo L. 黄荆

Shrubs or small trees

15.2 (2.6~24.8); 3.4 (-19.6~20.1); 25.8 (10.4~29.9); 15.5 (4.2~24.8); 132.6 (18.9~237.6); 9.0 (0.0~87.4); 861 (419~1385) ❙ 1103 (189~3495); 505 (114~1847); 90 (4~897) ❙ 25.8 (-70.2~224.4); 798 (189~1360); 212.4 (33.5~659.0)

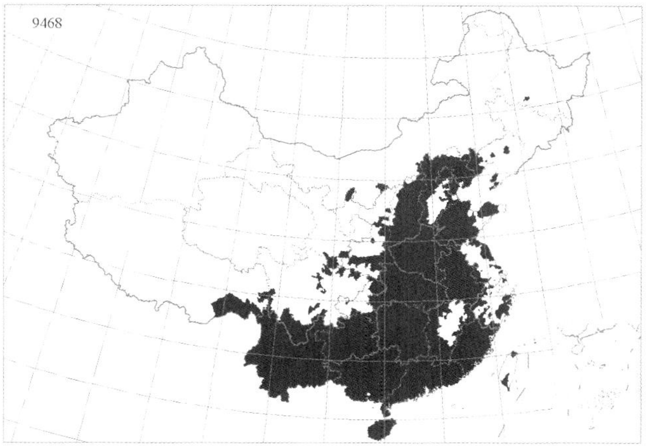

Verbenaceae 马鞭草科 *Vitex* 牡荆属
Vitex negundo var. cannabifolia (Siebold et Zucc.) Hand.-Mazz. 牡荆
Shrubs or small trees, deciduous

15.8 (3.3~23.2); 3.9 (−12.7~15.8); 26.9 (10.8~30.1); 16.0 (4.7~23.2); 136.0 (21.6~218.9); 5.6 (0.0~65.0); 879 (444~1257) ❙ 1299 (377~2037); 508 (241~1033); 138 (8~259) ❙ 45.7 (−37.2~147.0); 850 (377~1257); 201.8 (70.3~506.2)

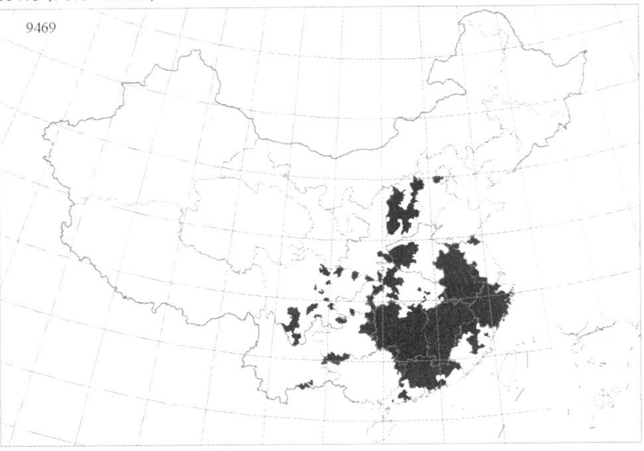

Verbenaceae 马鞭草科 *Vitex* 牡荆属
Vitex negundo var. heterophylla (Franch.) Rehder 荆条
Shrubs or small trees

12.6 (2.2~18.2); −2.0 (−15.9~8.6); 25.6 (9.9~29.9); 13.1 (3.8~18.3); 110.1 (15.5~159.4); 18.9 (0.0~80.4); 783 (413~986) ❙ 781 (357~1742); 397 (235~753); 54 (7~225) ❙ −1.8 (−51.1~128.9); 649 (357~977); 175.3 (57.3~334.4)

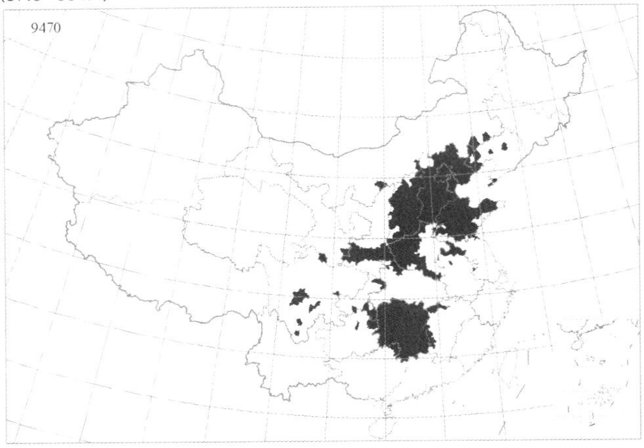

Verbenaceae 马鞭草科 *Vitex* 牡荆属
Vitex negundo var. microphylla Hand.-Mazz. 小叶荆
Shrubs or small trees

6.5 (2.7~13.3); −1.3 (−6.0~5.4); 13.4 (10.4~19.6); 7.1 (4.2~13.4); 42.6 (18.9~100.4); 24.2 (0.0~46.9); 515 (419~706) ❙ 876 (535~1736); 522 (326~1007); 25 (10~51) ❙ 65.3 (27.1~145.8); 515 (419~706); 278.6 (201.9~447.4)

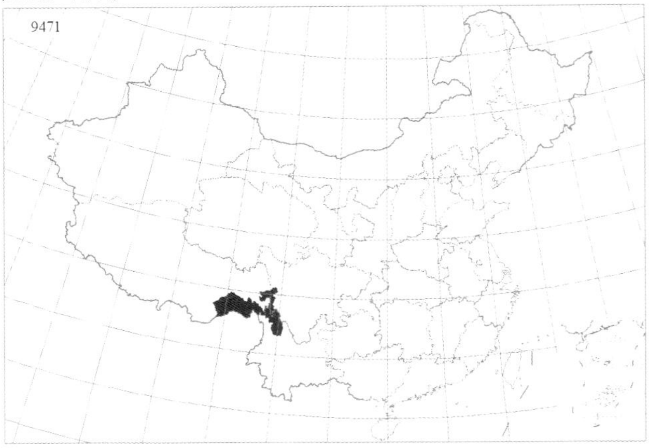

Verbenaceae 马鞭草科 *Vitex* 牡荆属
Vitex negundo var. thyrsoides C. P'ei et S. L. Liou 拟黄荆
Shrubs or small trees

19.3 (10.6~22.4); 9.8 (−0.2~14.8); 27.5 (21.1~29.8); 19.4 (10.6~22.4); 172.7 (79.4~208.9); 0.4 (0.0~12.1); 1013 (643~1191) ❙ 1651 (1246~2037); 624 (505~1033); 174 (46~197) ❙ 63.7 (35.6~93.8); 1010 (643~1191); 235.0 (157.7~284.3)

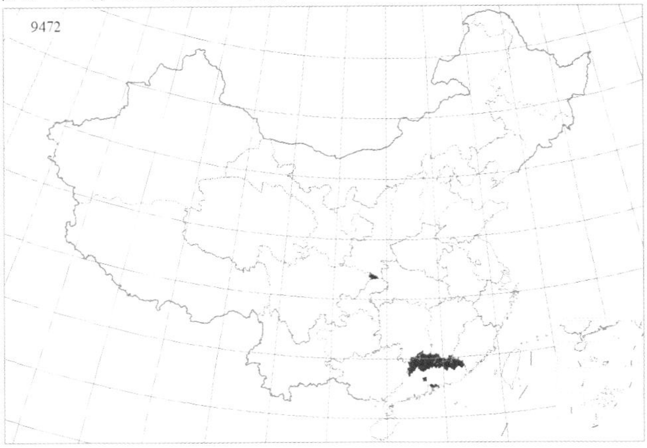

Verbenaceae 马鞭草科 *Vitex* 牡荆属
Vitex peduncularis Wall. ex Schauer 长序荆
Trees

19.5 (18.6~20.5); 13.8 (12.7~15.1); 23.1 (22.3~23.7); 19.5 (18.7~20.5); 174.5 (163.9~185.9); 0.0 (0.0~0.0); 936 (888~992) ❙ 1510 (1276~1751); 789 (655~996); 64 (51~77) ❙ 61.5 (36.8~88.8); 926 (880~986); 538.1 (452.0~659.0)

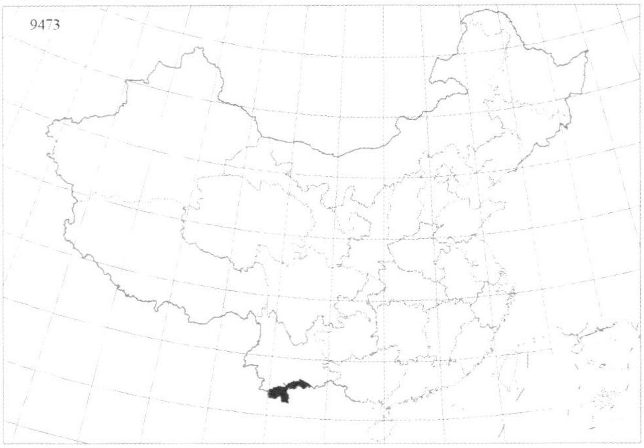

Verbenaceae 马鞭草科 *Vitex* 牡荆属
Vitex pierreana P. Dop 莺哥木
Trees

23.9 (22.2~24.8); 18.5 (17.2~20.1); 27.8 (25.7~28.7); 24.0 (22.3~24.8); 227.8 (207.2~237.6); 0.0 (0.0~0.0); 1305 (1127~1385) ❙ 1457 (868~1854); 506 (312~622); 64 (41~115) ❙ 12.1 (−34.3~47.8); 1155 (868~1360); 479.8 (288.3~620.0)

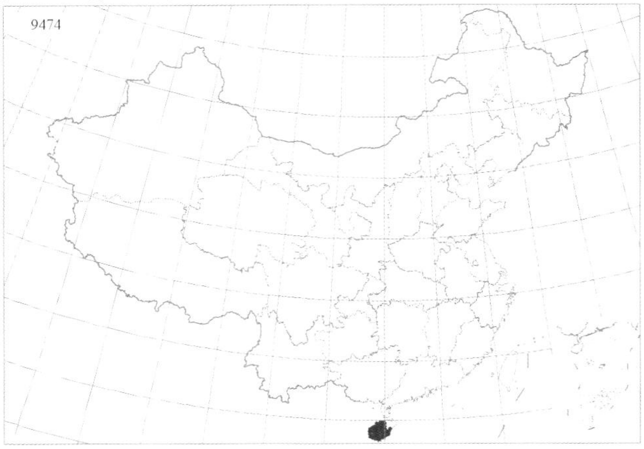

Verbenaceae 马鞭草科 *Vitex* 牡荆属
Vitex quinata (Lour.) F. N. Williams 山牡荆
Trees, evergreen

19.1 (13.3~24.8); 9.6 (3.3~20.1); 27.6 (19.6~29.9); 19.2 (13.4~24.8); 170.3 (100.4~237.6); 0.1 (0.0~2.3); 1013 (706~1385) ❙ 1595 (868~2993); 622 (312~1594); 159 (41~547) ❙ 60.2 (-34.3~246.5); 999 (706~1360); 239.3 (95.3~659.0)

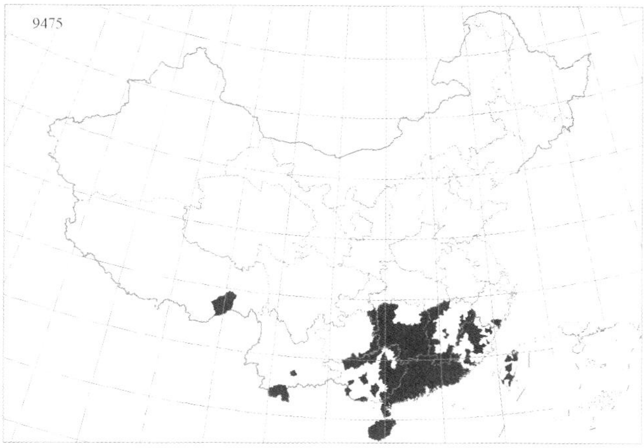

Verbenaceae 马鞭草科 *Vitex* 牡荆属
Vitex quinata var. puberula (H. J. Lam) Moldenke 微毛布惊
Trees, evergreen

17.8 (13.3~22.2); 10.8 (2.6~15.1); 22.7 (19.1~28.3); 17.8 (13.4~22.3); 153.8 (100.4~207.5); 0.1 (0.0~4.0); 871 (706~1190) ❙ 1280 (958~1751); 702 (458~1007); 54 (41~102) ❙ 47.8 (5.8~145.8); 861 (706~1190); 387.4 (126.1~659.0)

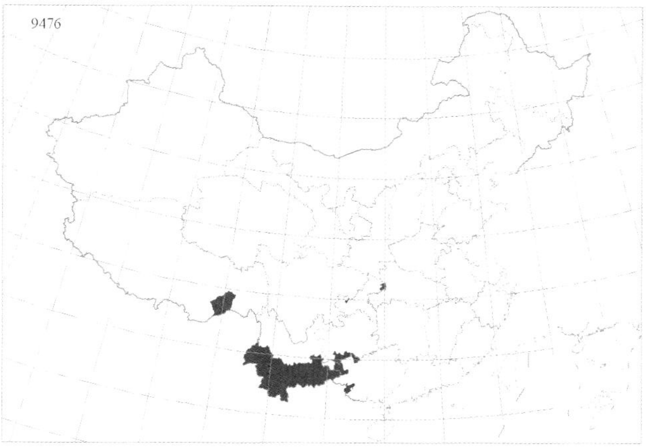

Verbenaceae 马鞭草科 *Vitex* 牡荆属
Vitex rotundifolia L. f. 单叶蔓荆
Shrubs

16.4 (8.3~24.8); 4.2 (-8.6~20.1); 27.2 (22.8~29.7); 16.6 (10.0~24.8); 146.4 (83.0~237.6); 9.2 (0.0~42.9); 935 (659~1385) ❙ 1101 (425~3495); 525 (312~1594); 91 (10~897) ❙ 14.6 (-49.3~224.4); 831 (425~1360); 204.2 (76.1~620.0)

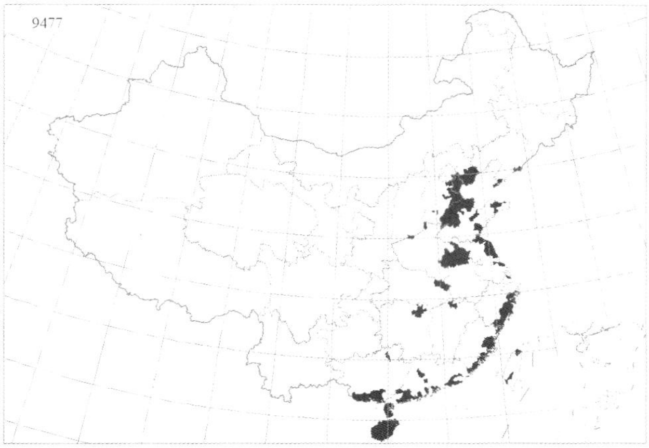

Verbenaceae 马鞭草科 *Vitex* 牡荆属
Vitex sampsonii Hance 广东牡荆
Shrubs

19.4 (14.3~23.4); 9.6 (3.7~16.1); 27.9 (24.2~29.8); 19.4 (14.3~23.5); 173.1 (113.2~221.7); 0.0 (0.0~1.4); 1024 (761~1273) ❙ 1585 (1371~1876); 592 (400~891); 174 (82~221) ❙ 56.5 (9.0~128.9); 1015 (761~1273); 221.3 (151.0~273.3)

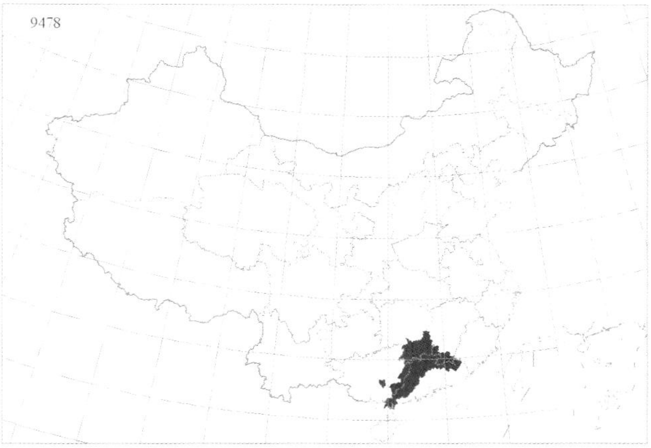

Verbenaceae 马鞭草科 *Vitex* 牡荆属
Vitex trifolia L. 蔓荆
Shrubs, deciduous

18.3 (11.1~24.8); 9.6 (-3.7~20.1); 25.3 (19.1~29.2); 18.4 (11.6~24.8); 162.9 (93.8~237.6); 3.1 (0.0~23.0); 958 (697~1385) ❙ 1332 (576~3495); 657 (356~1594); 92 (17~897) ❙ 37.9 (-27.4~224.4); 932 (576~1360); 277.9 (84.7~659.0)

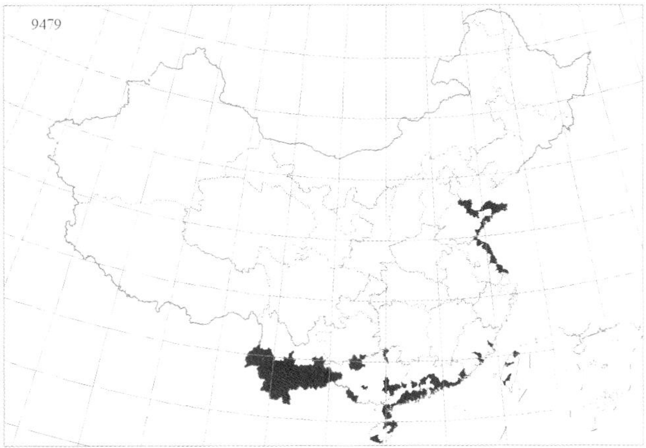

Verbenaceae 马鞭草科 *Vitex* 牡荆属
Vitex trifolia var. subtrisecta (Kuntze) Moldenke 异叶蔓荆
Shrubs or small trees

19.4 (14.4~22.8); 12.3 (7.5~15.5); 24.7 (19.1~29.1); 19.5 (14.4~22.8); 174.0 (113.1~214.1); 0.0 (0.0~0.0); 985 (727~1225) ❙ 1478 (958~2037); 750 (531~1033); 84 (41~188) ❙ 50.1 (12.2~112.2); 978 (727~1225); 319.4 (95.3~659.0)

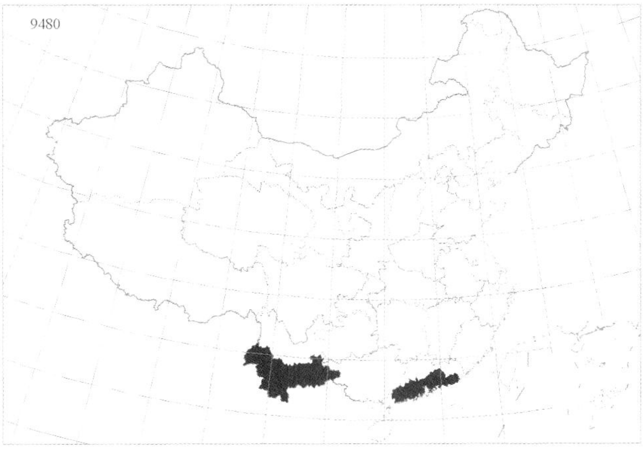

Verbenaceae 马鞭草科 *Vitex* 牡荆属

Vitex tripinnata (Lour.) Merr. 越南牡荆

Shrubs or trees

22.5 (17.7~24.6); 15.9 (10.9~19.5); 27.3 (22.5~29.2); 22.5 (17.8~24.7); 209.9 (153.4~235.9); 0.0 (0.0~0.0); 1194 (854~1372) ❙ 1520 (1187~2148); 697 (539~1234); 79 (48~124) ❙ 29.3 (-6.9~86.7); 1162 (854~1360); 399.1 (178.7~620.0)

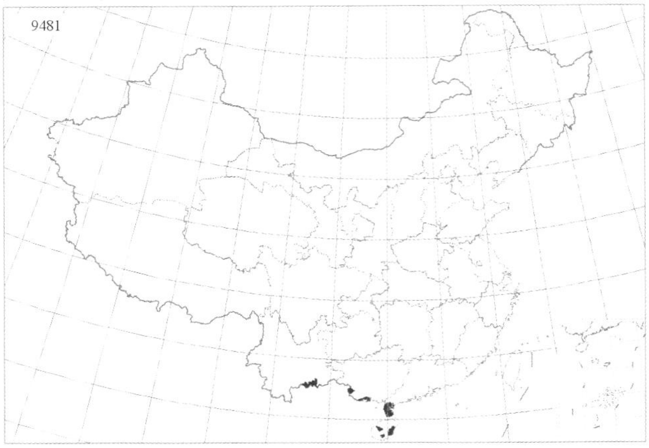

Verbenaceae 马鞭草科 *Vitex* 牡荆属

Vitex vestita Wall. ex Schauer 黄毛牡荆

Shrubs or small trees

18.5 (16.3~21.3); 11.8 (9.0~15.1); 23.2 (21.4~27.9); 18.6 (16.4~21.3); 163.0 (136.4~196.0); 0.0 (0.0~0.0); 901 (797~1123) ❙ 1313 (1018~1751); 715 (554~996); 54 (41~77) ❙ 46.0 (5.8~88.8); 888 (783~1121); 405.7 (277.0~659.0)

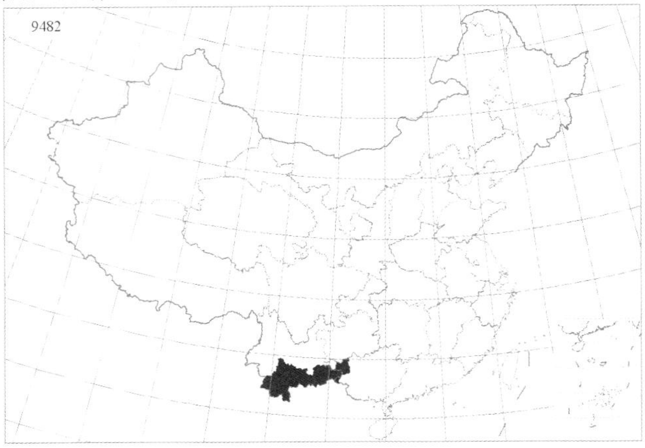

Verbenaceae 马鞭草科 *Vitex* 牡荆属

Vitex yunnanensis W. W. Sm. 滇牡荆

Shrubs or small trees

12.7 (2.2~19.0); 4.9 (-6.3~12.0); 19.3 (9.5~26.6); 12.8 (3.9~19.0); 97.9 (15.8~168.4); 4.7 (0.0~48.6); 698 (411~958) ❙ 963 (729~1250); 560 (411~688); 30 (11~60) ❙ 41.3 (-14.2~121.7); 675 (411~958); 265.3 (142.9~351.7)

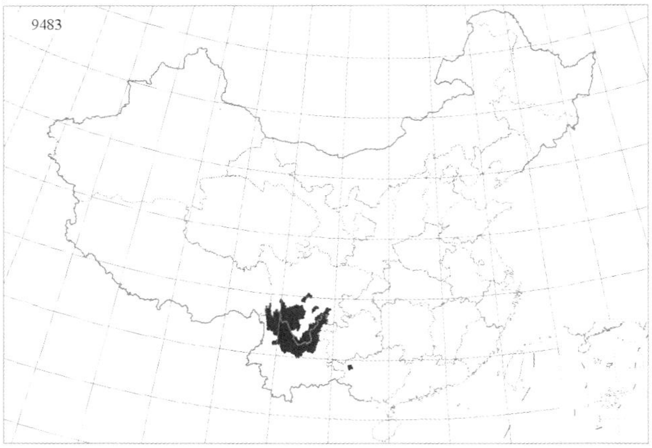

Lamiaceae 唇形科 *Chelonopsis* 铃子香属

Chelonopsis abbreviata C. Y. Wu et H. W. Li 缩序铃子香

Shrubs

13.4 (8.6~21.6); 6.5 (1.3~12.6); 18.9 (15.0~28.5); 13.5 (8.7~21.7); 102.6 (52.6~199.8); 1.0 (0.0~8.8); 717 (574~1153) ❙ 1146 (968~1625); 637 (553~795); 57 (24~173) ❙ 61.1 (36.1~93.0); 716 (574~1153); 351.4 (237.8~407.1)

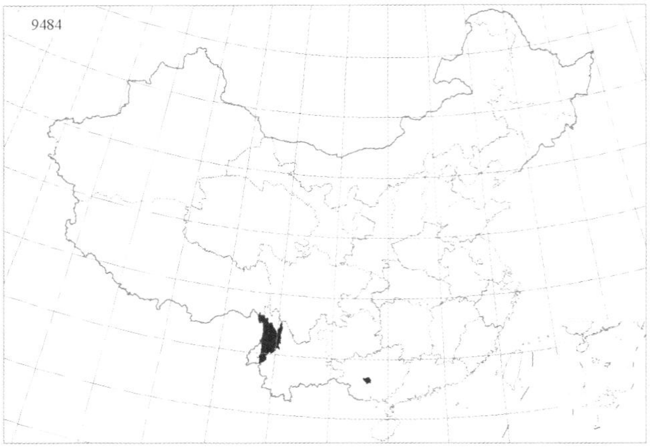

Lamiaceae 唇形科 *Chelonopsis* 铃子香属

Chelonopsis albiflora Pax et K. Hoffm. ex Limpr. 白花铃子香

Shrubs

3.0 (0.6~6.6); -5.8 (-8.8~-1.3); 10.6 (9.1~13.3); 4.5 (3.2~6.8); 21.2 (12.4~39.6); 45.0 (19.7~65.0); 428 (371~514) ❙ 648 (300~932); 410 (208~548); 10 (3~19) ❙ 50.5 (-19.2~125.9); 418 (300~514); 235.2 (168.2~311.5)

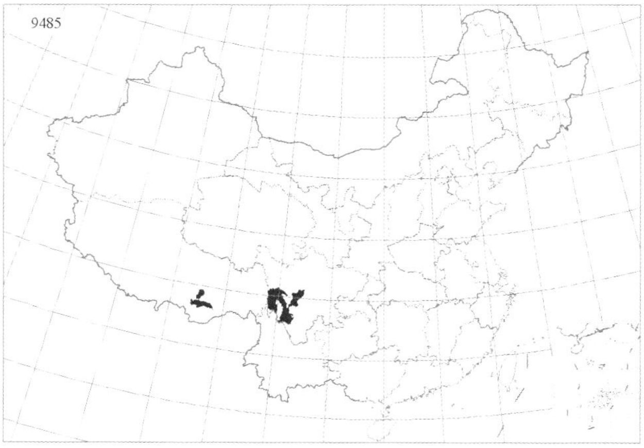

Lamiaceae 唇形科 *Chelonopsis* 铃子香属

Chelonopsis bracteata W. W. Sm. 具苞铃子香

Shrubs

7.8; 0.4; 14.2; 8.0; 48.9; 14.3; 548 ❙ 877; 546; 19 ❙ 60.7; 548; 304.7

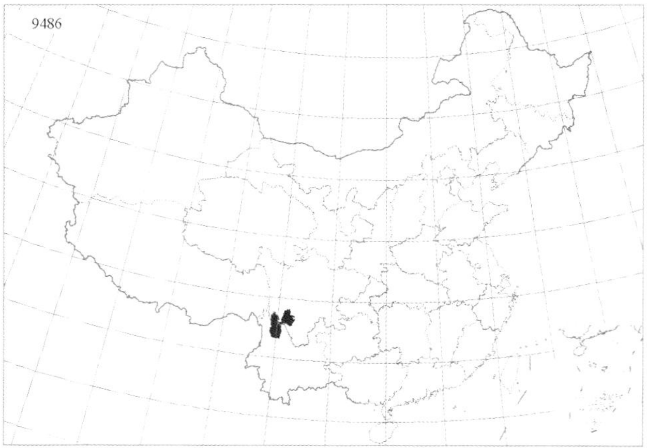

Lamiaceae 唇形科 *Elsholtzia* 香薷属

Elsholtzia fruticosa (D. Don) Rehder 鸡骨柴

Shrubs

13.7 (-3.3~21.9); 4.8 (-15.2~15.1); 21.3 (6.1~28.6); 14.0 (1.8~22.0); 113.9 (1.9~203.8); 8.8 (0.0~103.2); 753 (329~1176) ▎1059 (300~1751); 553 (193~1007); 48 (3~141) ▎41.0 (-28.9~145.8); 743 (300~1176); 282.8 (72.2~659.0)

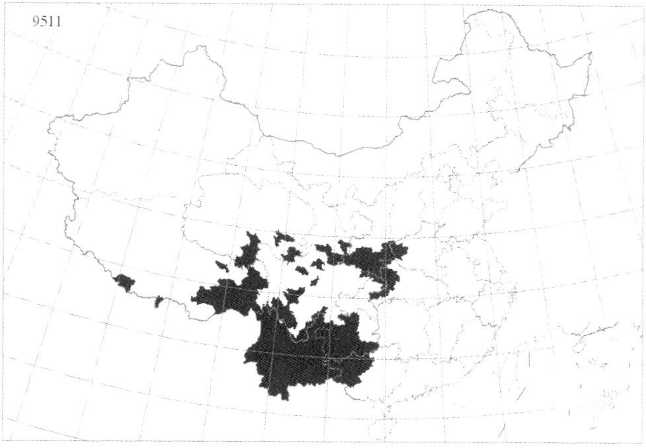

Lamiaceae 唇形科 *Elsholtzia* 香薷属

Elsholtzia fruticosa var. glabrifolia C. Y. Wu et S. C. Huang 光叶鸡骨柴

Shrubs

1.6 (-0.9~2.6); -7.7 (-11.3~-6.8); 10.2 (8.7~11.3); 3.8 (2.7~4.4); 16.4 (9.4~21.4); 56.5 (48.2~79.1); 403 (354~427) ▎582 (510~829); 339 (311~400); 13 (8~23) ▎44.3 (20.7~94.4); 403 (354~427); 221.9 (130.1~254.4)

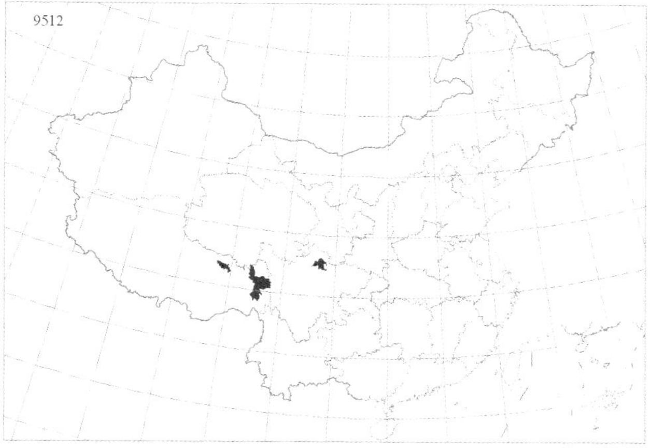

Lamiaceae 唇形科 *Elsholtzia* 香薷属

Elsholtzia glabra C. Y. Wu et S. C. Huang 光香薷

Shrubs

12.8 (2.2~18.9); 5.5 (-6.3~12.0); 18.8 (9.5~23.9); 12.9 (3.9~19.0); 99.6 (15.8~167.8); 5.4 (0.0~48.6); 696 (411~929) ▎950 (729~1214); 552 (411~627); 31 (11~75) ▎39.9 (-14.2~101.4); 659 (411~760); 287.5 (207.5~372.3)

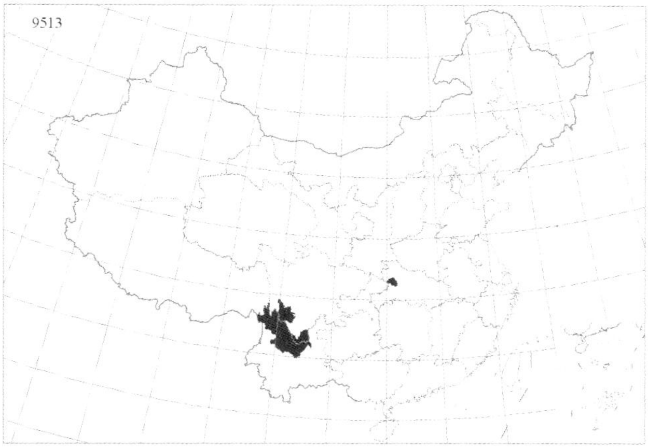

Lamiaceae 唇形科 *Elsholtzia* 香薷属

Elsholtzia myosurus Dunn 鼠尾香薷

Shrubs

13.7 (5.4~18.9); 5.9 (-2.1~12.0); 20.1 (12.3~26.6); 13.7 (5.8~19.0); 106.5 (31.5~167.8); 1.8 (0.0~25.6); 724 (481~929) ▎979 (741~1406); 565 (411~816); 33 (15~57) ▎36.8 (-14.2~87.2); 697 (481~912); 274.7 (142.9~434.4)

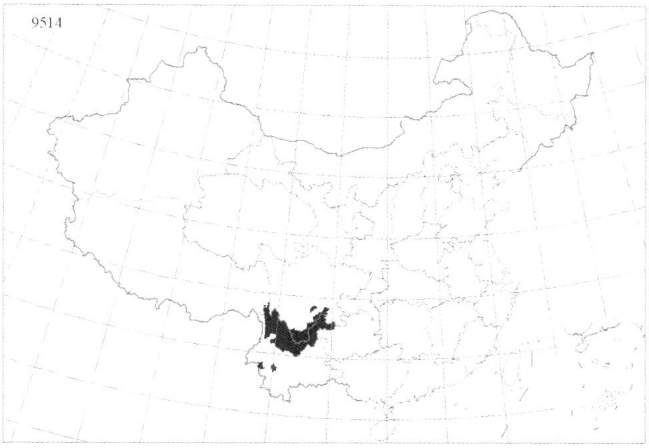

Lamiaceae 唇形科 *Elsholtzia* 香薷属

Elsholtzia ochroleuca Dunn 黄白香薷

Shrubs

12.6 (2.7~16.9); 5.2 (-7.1~10.3); 18.3 (11.1~21.4); 12.9 (4.4~16.9); 99.2 (20.5~142.7); 8.0 (0.0~48.0); 686 (429~813) ▎976 (800~1178); 562 (427~646); 36 (18~51) ▎45.1 (30.0~86.5); 685 (429~813); 288.0 (207.5~383.4)

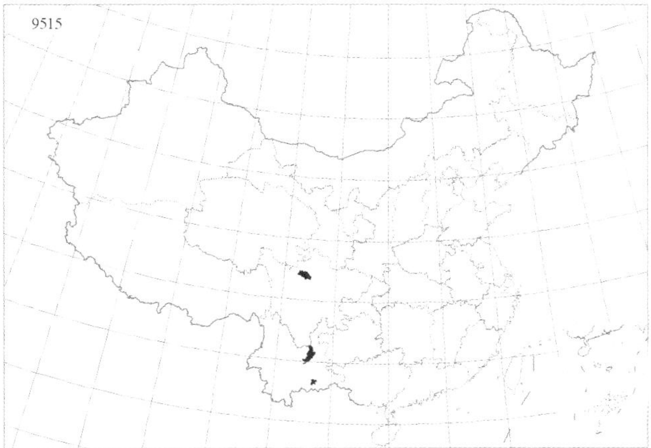

Lamiaceae 唇形科 *Elsholtzia* 香薷属

Elsholtzia penduliflora W. W. Sm. 大黄药

Subshrubs

17.3 (13.6~21.0); 10.4 (7.2~14.2); 22.1 (18.8~25.6); 17.3 (13.7~21.0); 147.4 (104.1~191.8); 0.0 (0.0~0.0); 842 (706~1053) ▎1264 (1014~1522); 694 (562~860); 56 (44~107) ▎49.9 (26.1~80.6); 835 (706~1053); 394.9 (245.3~551.6)

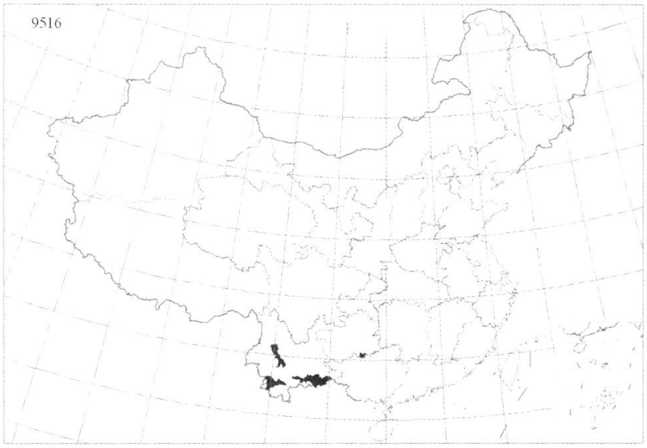

Lamiaceae 唇形科 *Elsholtzia* 香薷属
Elsholtzia stauntonii Benth. 木香薷
Shrubs, deciduous

10.5 (2.2~15.1); -4.2 (-16.1~2.0); 23.8 (18.6~28.0); 11.4 (6.8~15.2); 93.0 (48.4~128.0); 26.2 (5.6~81.0); 706 (516~880) ❙ 630 (332~1164); 349 (215~488); 22 (8~53) ❙ -10.2 (-47.7~83.5); 595 (332~817); 195.3 (105.8~334.4)

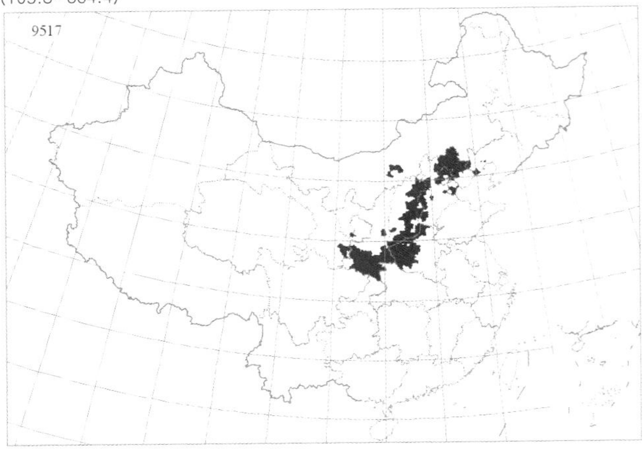

Lamiaceae 唇形科 *Eurysolen* 宽管花属
Eurysolen gracilis Prain 宽管花
Shrubs

18.4 (15.4~21.0); 11.7 (8.1~15.1); 22.9 (20.7~25.6); 18.4 (15.5~21.0); 160.8 (125.7~191.8); 0.0 (0.0~0.0); 894 (767~1053) ❙ 1330 (1007~1607); 691 (554~835); 57 (47~77) ❙ 48.0 (25.7~65.6); 888 (767~1053); 436.0 (274.9~659.0)

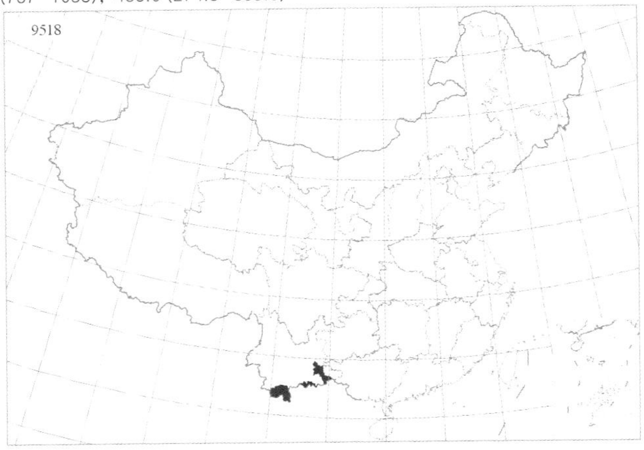

Lamiaceae 唇形科 *Geniosporum* 网萼木属
Geniosporum coloratum (D. Don) Kuntze 网萼木
Shrubs

17.7 (12.8~20.5); 11.2 (3.3~15.1); 21.8 (20.3~23.7); 17.7 (12.9~20.5); 152.6 (97.1~185.9); 0.1 (0.0~2.2); 854 (702~992) ❙ 1353 (942~1688); 746 (489~959); 53 (42~77) ❙ 58.5 (12.5~112.2); 840 (702~986); 446.5 (219.7~659.0)

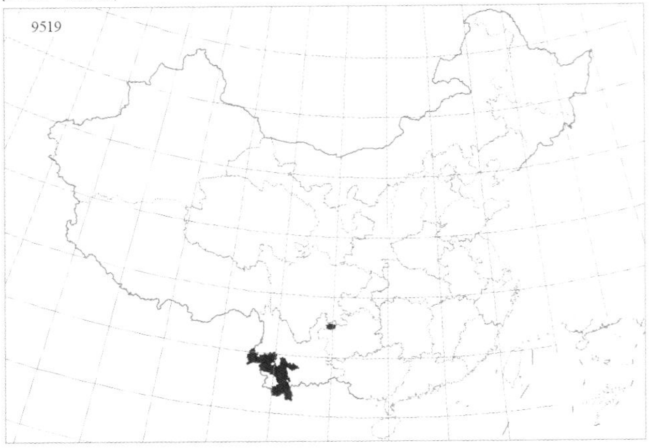

Lamiaceae 唇形科 *Gomphostemma* 锥花属
Gomphostemma arbusculum C. Y. Wu 木锥花
Shrubs

18.5 (16.3~21.0); 12.3 (9.6~15.1); 22.4 (20.3~25.6); 18.5 (16.4~21.0); 162.4 (136.6~191.8); 0.0 (0.0~0.0); 892 (789~1053) ❙ 1401 (1028~1688); 765 (566~959); 54 (42~77) ❙ 57.8 (12.5~112.2); 875 (780~1053); 481.5 (339.5~659.0)

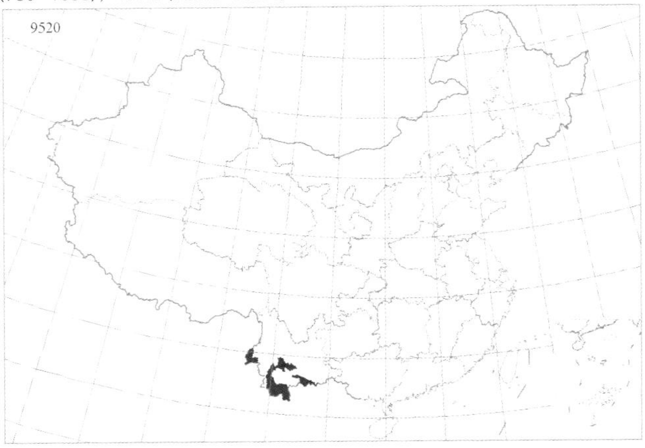

Lamiaceae 唇形科 *Gomphostemma* 锥花属
Gomphostemma callicarpoides (Yamam.) Masam. 紫珠状锥花
Shrubs

20.7; 15.7; 24.4; 20.7; 188.2; 0.0; 1029 ❙ 2417; 1391; 118 ❙ 147.3; 972; 244.5

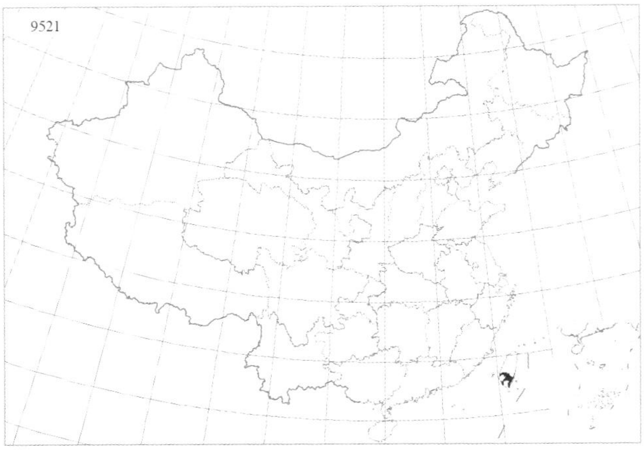

Lamiaceae 唇形科 *Gomphostemma* 锥花属
Gomphostemma latifolium C. Y. Wu 宽叶锥花
Shrubs

21.7 (17.6~23.6); 13.8 (7.8~16.7); 28.0 (22.3~29.2); 21.8 (17.7~23.6); 201.4 (152.5~223.5); 0.0 (0.0~0.0); 1151 (906~1281) ❙ 1617 (1187~2435); 765 (545~1428); 115 (58~177) ❙ 41.4 (-6.9~99.5); 1147 (880~1273); 254.3 (152.9~659.0)

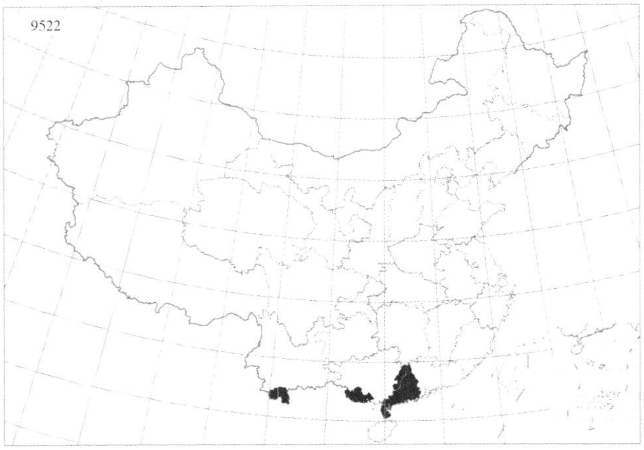

Lamiaceae 唇形科 *Gomphostemma* 锥花属
Gomphostemma leptodon Dunn 细齿锥花
Shrubs

20.7 (18.6~22.2); 13.6 (12.7~14.2); 25.8 (22.5~28.4); 20.8 (18.7~22.3); 189.2 (163.9~207.5); 0.0 (0.0~0.0); 1056 (888~1190) ❙ 1610 (1374~2148); 895 (713~1234); 87 (56~124) ❙ 54.3 (15.4~88.8); 1056 (888~1190); 415.5 (273.1~551.6)

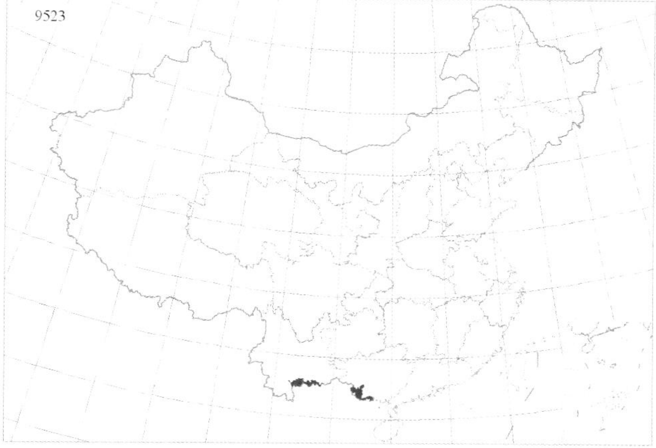

Lamiaceae 唇形科 *Gomphostemma* 锥花属
Gomphostemma lucidum Wall. ex Benth. 光泽锥花
Shrubs

21.4 (16.9~24.8); 13.8 (7.3~20.1); 27.3 (21.4~29.1); 21.4 (16.9~24.8); 197.4 (142.7~237.6); 0.0 (0.0~0.0); 1124 (813~1385) ❙ 1623 (868~2037); 752 (312~1033); 107 (41~188) ❙ 45.8 (-34.3~83.4); 1103 (813~1326); 274.4 (95.3~659.0)

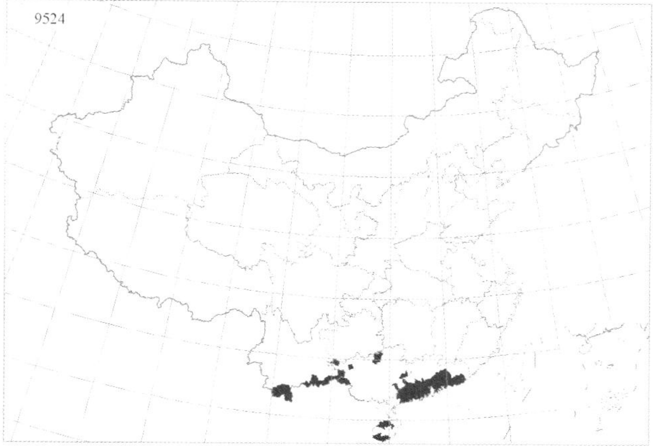

Lamiaceae 唇形科 *Gomphostemma* 锥花属
Gomphostemma lucidum var. intermedium (Craib) C. Y. Wu 中间光泽锥花
Shrubs

19.3; 12.9; 23.7; 19.3; 171.9; 0.0; 939 ❙ 1398; 788; 59 ❙ 49.3; 937; 500.3

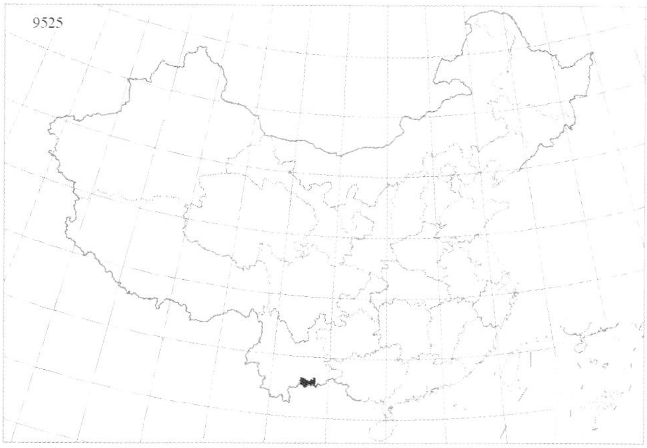

Lamiaceae 唇形科 *Gomphostemma* 锥花属
Gomphostemma pseudocrinitum C. Y. Wu 拟长毛锥花
Shrubs

19.0; 10.2; 25.8; 19.0; 168.4; 0.0; 958 ❙ 1250; 675; 60 ❙ 30.5; 958; 291.2

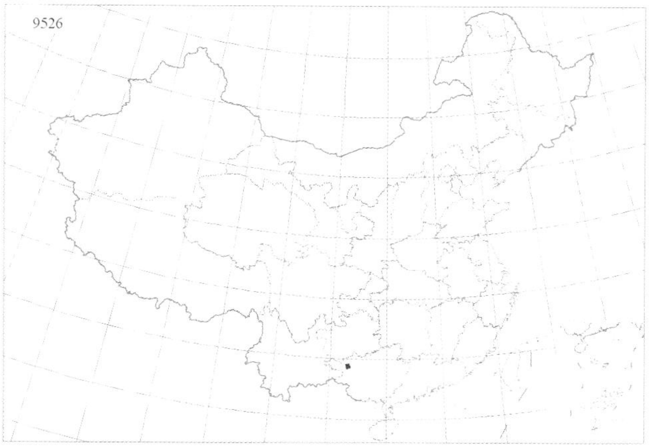

Lamiaceae 唇形科 *Isodon* 香茶菜属
Isodon adenolomus (Hand.-Mazz.) H. Hara 腺叶香茶菜
Shrubs

13.4 (6.1~18.2); 6.4 (-1.5~11.8); 18.8 (12.7~22.5); 13.5 (6.3~18.2); 106.9 (35.2~158.4); 5.5 (0.0~21.8); 717 (496~873) ❙ 989 (801~1178); 583 (496~646); 32 (12~51) ❙ 41.1 (19.0~67.7); 707 (496~847); 310.0 (271.3~383.4)

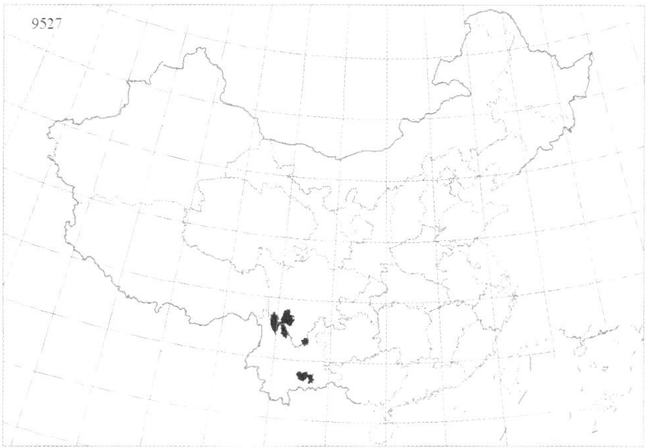

Lamiaceae 唇形科 *Isodon* 香茶菜属
Isodon barbeyanus (H. Lév.) H. W. Li 线齿香茶菜
Shrubs

2.2 (2.0~2.7); -7.2 (-7.8~-6.6); 10.4 (9.9~11.1); 4.0 (3.8~4.4); 17.5 (15.4~20.5); 50.7 (48.0~52.8); 418 (413~429) ❙ 831 (702~943); 469 (407~546); 18 (15~23) ❙ 98.9 (66.5~128.6); 418 (413~429); 289.4 (248.3~341.6)

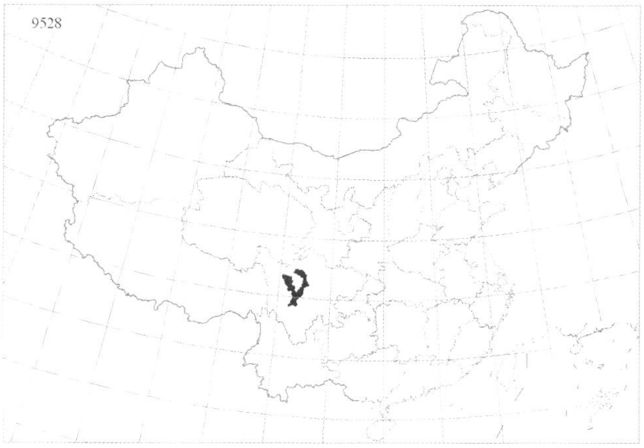

Lamiaceae 唇形科 *Isodon* 香茶菜属
Isodon brevifolius (Hand.-Mazz.) H. W. Li 短叶香茶菜
Shrubs

15.0；8.4；20.2；15.1；121.2；0.0；753 ❙ 950；570；27 ❙ 26.3；683；307.7

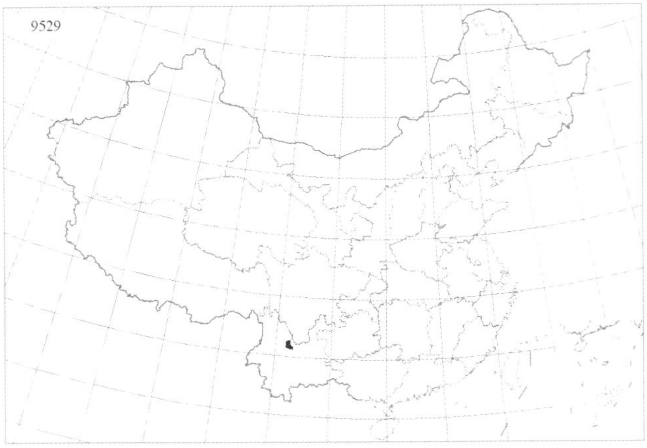

Lamiaceae 唇形科 *Isodon* 香茶菜属
Isodon bulleyanus (Diels) Kudo 苍山香茶菜
Shrubs

13.9；7.2；19.2；14.0；107.8；0.0；718 ❙ 1000；588；36 ❙ 39.7；695；313.4

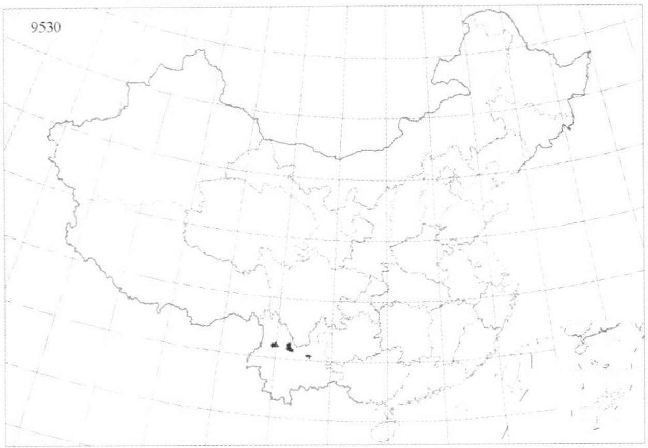

Lamiaceae 唇形科 *Isodon* 香茶菜属
Isodon dawoensis (Hand.-Mazz.) H. Hara 道孚香茶菜
Small shrubs

2.0；-7.4；10.1；4.0；16.9；51.9；416 ❙ 777；457；15 ❙ 86.8；416；271.2

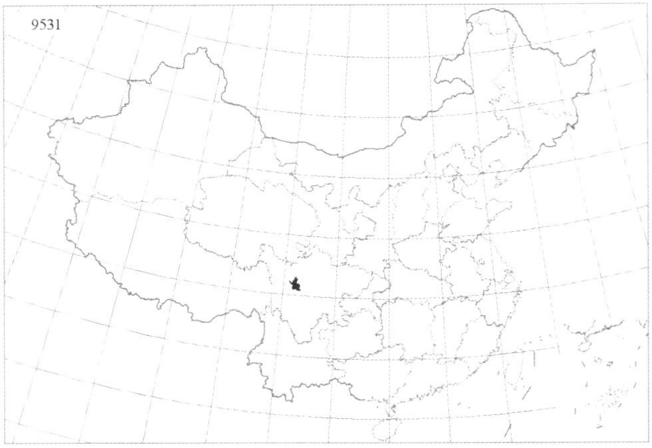

Lamiaceae 唇形科 *Isodon* 香茶菜属
Isodon enanderianus (Hand.-Mazz.) H. W. Li 紫毛香茶菜
Shrubs

17.2 (9.1~19.3)；10.5 (-0.7~13.3)；22.0 (18.1~23.6)；17.2 (9.2~19.4)；148.0 (64.1~173.2)；1.4 (0.0~14.1)；844 (595~939) ❙ 1135 (822~1428)；639 (420~835)；46 (19~56) ❙ 34.6 (19.0~55.7)；833 (595~923)；373.2 (275.4~506.2)

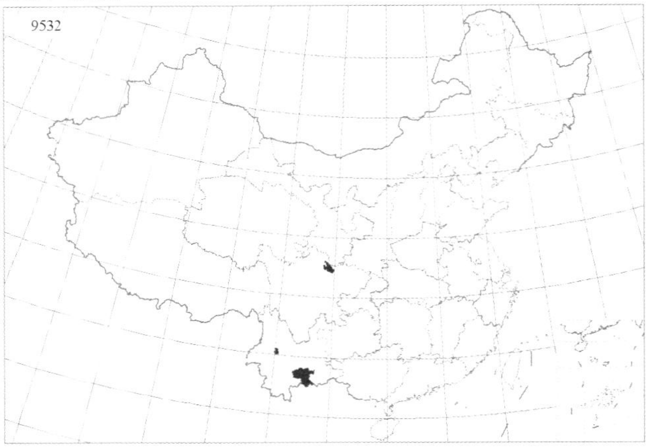

Lamiaceae 唇形科 *Isodon* 香茶菜属
Isodon flexicaulis (C. Y. Wu et H. W. Li) H. Hara 柔茎香茶菜
Shrubs

9.0 (5.4~13.2)；1.7 (-2.1~6.4)；15.1 (12.3~18.4)；9.1 (5.8~13.3)；58.6 (31.5~99.4)；10.3 (0.0~25.6)；580 (481~695) ❙ 1021 (741~1337)；588 (426~727)；40 (12~69) ❙ 74.5 (54.0~93.0)；580 (481~695)；337.9 (273.5~407.1)

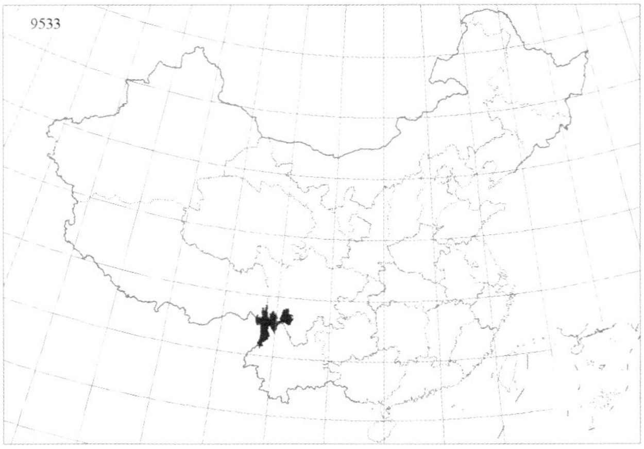

Lamiaceae 唇形科 *Isodon* 香茶菜属
Isodon glutinosus (C. Y. Wu et H. W. Li) H. Hara 胶粘香茶菜
Small shrubs

6.0；-1.6；12.8；6.3；35.5；22.4；497 ❙ 801；490；20 ❙ 61.1；497；289.3

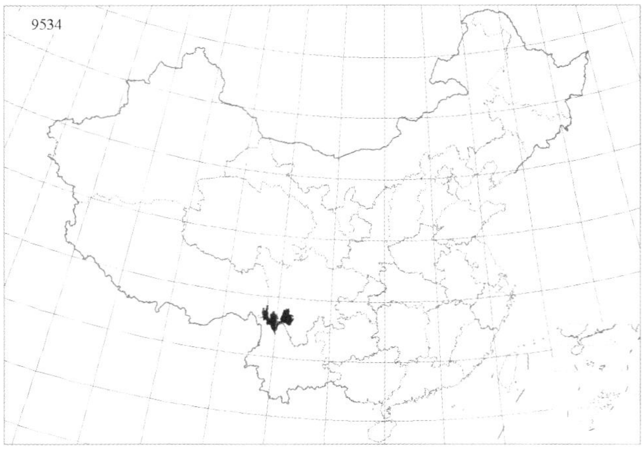

Lamiaceae 唇形科 *Isodon* 香茶菜属

Isodon grandifolius var. atuntzeensis (C. Y. Wu) H. W. Li 大叶香茶菜德钦变种

Shrubs

2.7 (0.6~5.4); -6.3 (-9.4~-2.1); 10.8 (9.5~12.3); 4.4 (3.4~5.8); 20.7 (12.7~31.5); 47.5 (25.6~65.4); 428 (394~481) ❙ 633 (520~780); 382 (317~487); 14 (9~28) ❙ 47.7 (26.6~81.6); 428 (394~481); 275.4 (210.1~307.8)

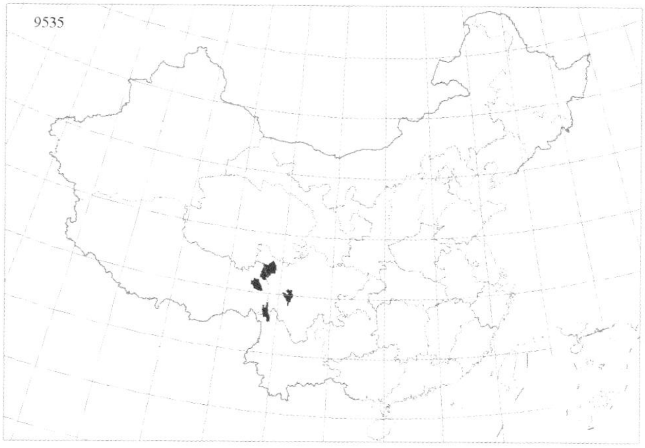

Lamiaceae 唇形科 *Isodon* 香茶菜属

Isodon hirtellus (Hand.-Mazz.) H. Hara 细毛香茶菜

Shrubs or subshrubs

13.7; 6.2; 19.7; 13.7; 104.7; 0.0; 714 ❙ 963; 593; 26 ❙ 34.8; 696; 271.3

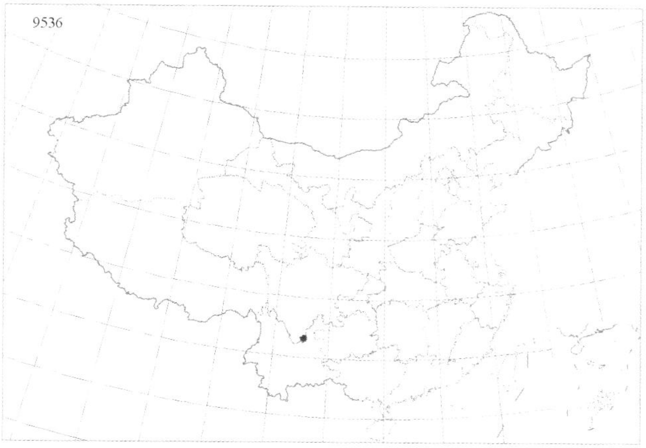

Lamiaceae 唇形科 *Isodon* 香茶菜属

Isodon interruptus (C. Y. Wu et H. W. Li) H. Hara 间断香茶菜

Shrubs

14.2 (13.6~14.6); 5.7 (3.1~7.9); 21.4 (19.0~24.9); 14.2 (13.6~14.6); 111.3 (103.6~115.9); 0.8 (0.0~2.6); 742 (707~781) ❙ 1171 (957~1478); 584 (570~599); 58 (36~91) ❙ 57.1 (35.3~94.5); 741 (707~781); 237.7 (180.4~282.6)

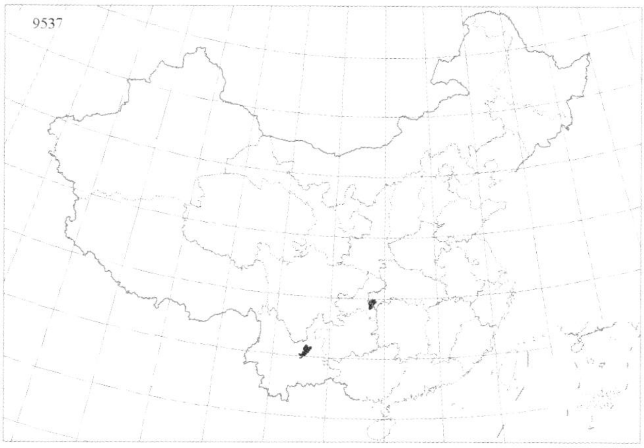

Lamiaceae 唇形科 *Isodon* 香茶菜属

Isodon irroratus (Forrest ex Diels) Kudo 露珠香茶菜

Shrubs

6.8 (-1.5~10.8); -0.7 (-10.5~3.9); 13.4 (7.1~16.6); 7.5 (2.2~10.9); 45.3 (4.7~71.8); 23.1 (1.3~82.2); 528 (332~633) ❙ 796 (311~1007); 487 (227~603); 23 (11~40) ❙ 47.0 (-6.3~67.9); 523 (311~633); 265.2 (62.8~351.7)

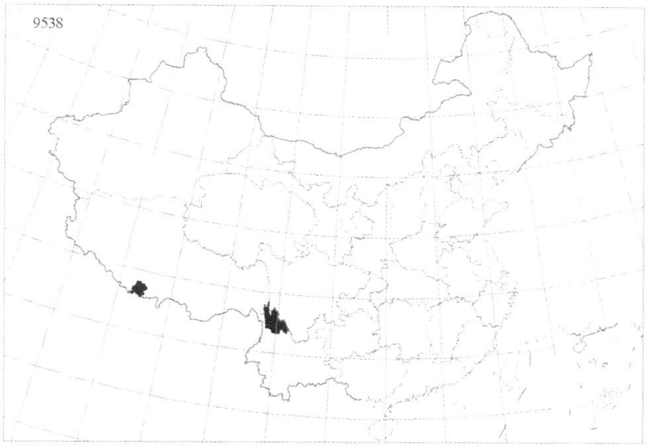

Lamiaceae 唇形科 *Isodon* 香茶菜属

Isodon leucophyllus (Dunn) Kudo 白叶香茶菜

Small shrubs

8.3 (2.2~16.7); 0.0 (-7.1~9.7); 15.4 (9.9~24.6); 8.8 (3.8~16.7); 59.9 (15.5~140.9); 20.4 (0.0~49.1); 572 (413~822) ❙ 921 (741~1318); 540 (426~753); 22 (11~50) ❙ 66.0 (14.4~125.9); 562 (413~822); 291.9 (189.1~357.7)

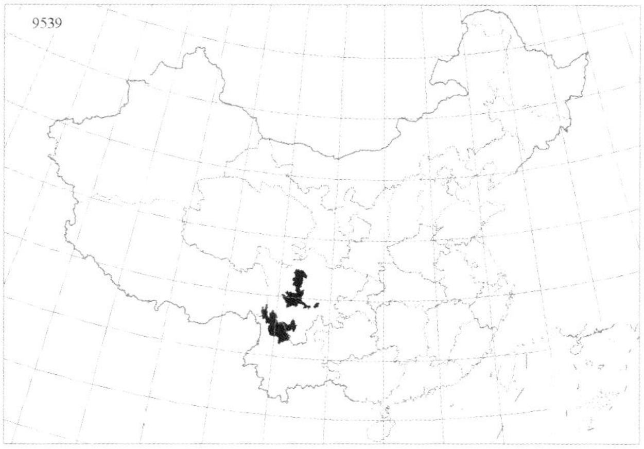

Lamiaceae 唇形科 *Isodon* 香茶菜属

Isodon lihsienensis (C. Y. Wu et H. W. Li) H. Hara 理县香茶菜

Shrubs

2.9; -6.3; 11.2; 4.4; 20.5; 44.6; 428 ❙ 950; 490; 25 ❙ 121.7; 428; 267.5

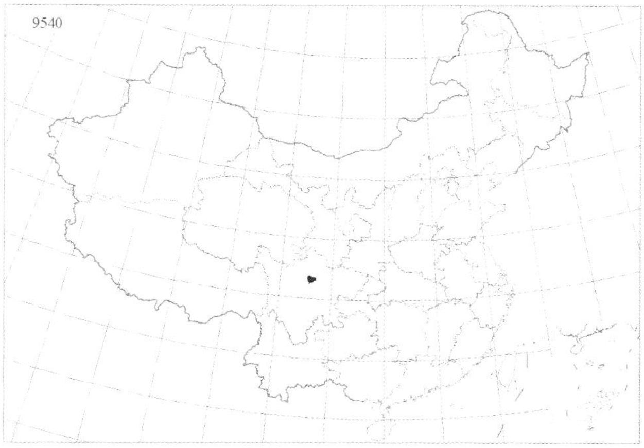

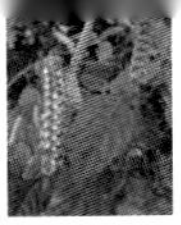

Lamiaceae 唇形科 *Isodon* 香茶菜属

Isodon loxothyrsus (Hand.-Mazz.) H. Hara 弯锥香茶菜

Shrubs

12.7 (2.0~18.9); 5.7 (-7.0~12.0); 18.3 (10.3~23.9); 12.9 (3.9~19.0); 99.9 (17.3~167.8); 6.6 (0.0~52.8); 695 (410~929) ❙ 916 (520~1041); 540 (317~603); 29 (10~45) ❙ 35.9 (-14.2~100.7); 650 (410~799); 289.2 (210.1~352.3)

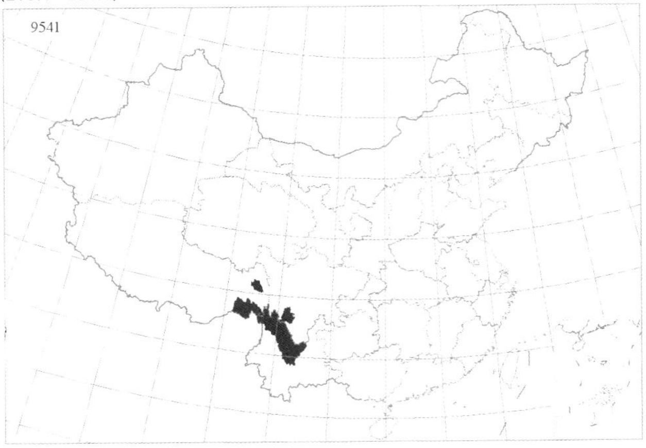

Lamiaceae 唇形科 *Isodon* 香茶菜属

Isodon medilungensis (C. Y. Wu et H. W. Li) H. Hara 麦地龙香茶菜

Shrubs

6.6; -1.3; 13.3; 6.8; 39.6; 19.7; 514 ❙ 862; 548; 12 ❙ 67.7; 514; 311.5

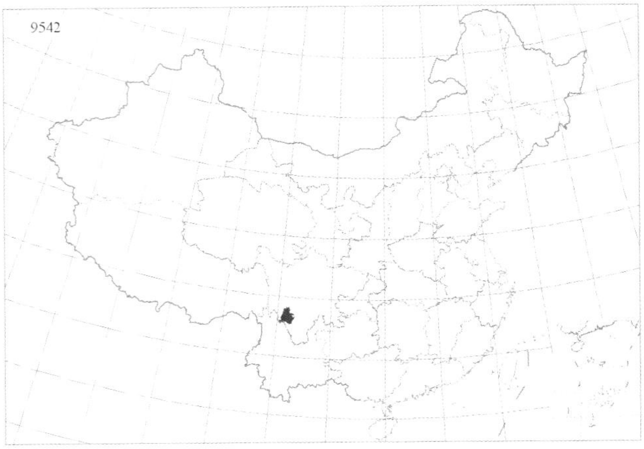

Lamiaceae 唇形科 *Isodon* 香茶菜属

Isodon muliensis (W. W. Sm.) Kudo 木里香茶菜

Subshrubs

6.6; -1.3; 13.3; 6.8; 39.6; 19.7; 514 ❙ 862; 548; 12 ❙ 67.7; 514; 311.5

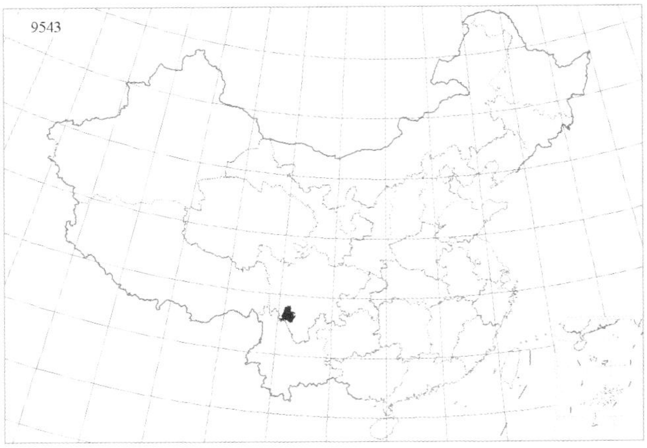

Lamiaceae 唇形科 *Isodon* 香茶菜属

Isodon oresbius (W. W. Sm.) Kudo 山地香茶菜

Small shrubs

4.0 (0.3~10.8); -4.8 (-9.9~3.9); 11.7 (9.1~16.6); 5.4 (3.2~10.9); 28.9 (11.9~71.8); 40.0 (1.3~67.5); 459 (371~633) ❙ 678 (300~1007); 401 (208~594); 16 (3~40) ❙ 45.7 (-19.2~86.8); 447 (300~633); 280.4 (168.2~357.7)

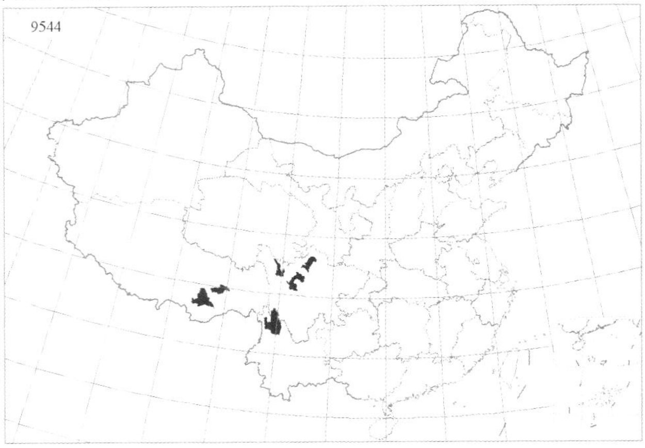

Lamiaceae 唇形科 *Isodon* 香茶菜属

Isodon parvifolius (Batalin) H. Hara 小叶香茶菜

Small shrubs

4.9 (2.0~13.3); -4.5 (-7.7~5.4); 13.3 (9.9~20.4); 6.0 (3.8~13.4); 35.5 (15.5~100.4); 36.4 (0.0~52.8); 482 (410~706) ❙ 756 (473~1736); 413 (287~1007); 18 (7~51) ❙ 56.5 (2.4~145.8); 482 (410~706); 276.4 (201.9~447.4)

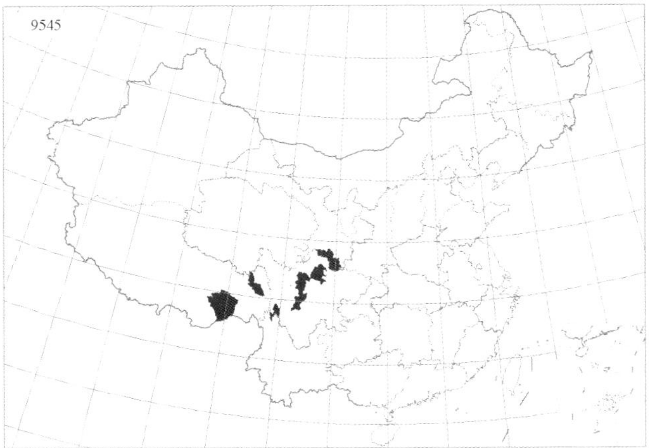

Lamiaceae 唇形科 *Isodon* 香茶菜属

Isodon pharicus (Prain) Murata 川藏香茶菜

Small shrubs

1.2 (-2.2~5.3); -8.2 (-11.8~-2.4); 9.8 (7.2~12.2); 3.6 (2.1~5.7); 15.7 (4.8~30.5); 60.7 (26.2~91.0); 392 (327~479) ❙ 440 (212~961); 287 (156~558); 8 (1~31) ❙ 9.7 (-35.1~100.7); 351 (212~479); 202.9 (77.4~290.1)

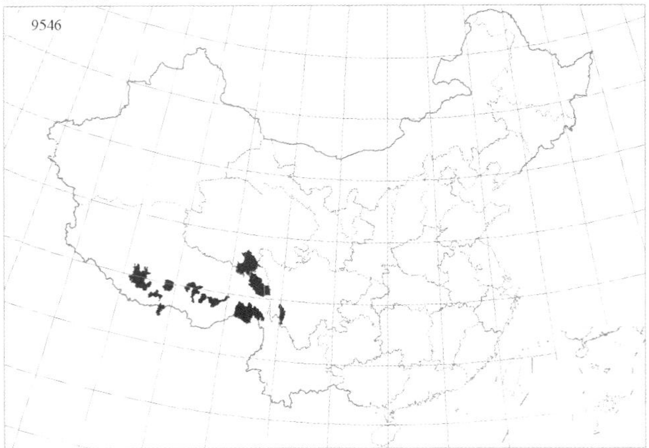

Lamiaceae 唇形科 *Isodon* 香茶菜属
Isodon phyllostachys (Diels) Kudo 叶穗香茶菜
Shrubs or subshrubs

10.5 (5.4~15.3); 3.4 (-2.1~8.7); 16.5 (12.3~20.5); 10.6 (5.8~15.4); 74.7 (31.5~124.3); 8.1 (0.0~25.6); 624 (481~761) ❙ 915 (741~991); 557 (426~619); 24 (12~38) ❙ 48.6 (28.3~67.7); 610 (481~717); 300.0 (273.5~319.8)

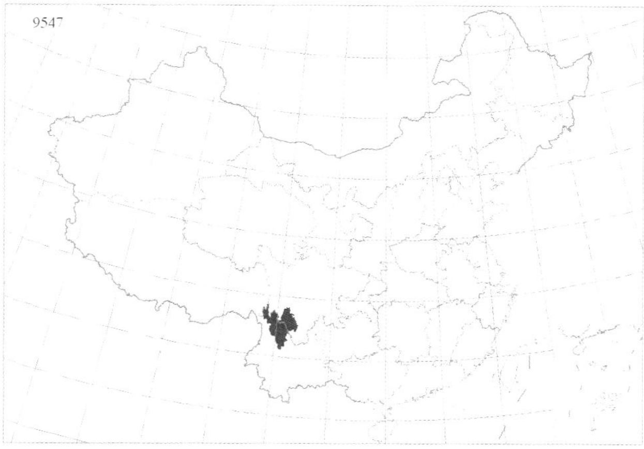

Lamiaceae 唇形科 *Isodon* 香茶菜属
Isodon pleiophyllus (Diels) Kudo 多叶香茶菜
Shrubs

10.7 (5.4~14.6); 3.5 (-2.1~7.8); 16.7 (12.3~20.0); 10.8 (5.8~14.6); 76.3 (31.5~115.2); 7.3 (0.0~25.6); 629 (481~738) ❙ 1021 (741~1736); 612 (426~1007); 29 (15~51) ❙ 61.8 (30.5~145.8); 618 (481~707); 318.6 (273.5~447.4)

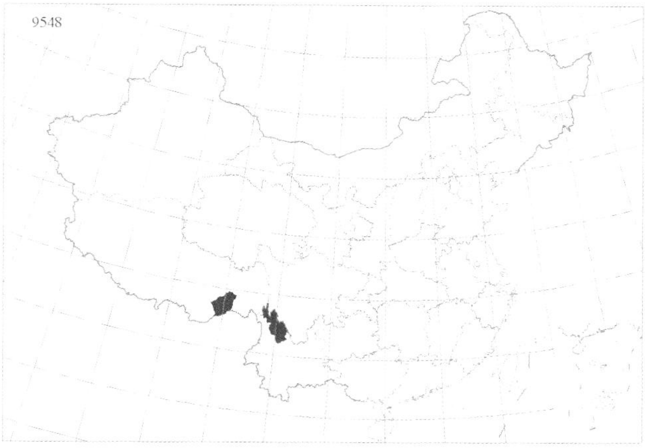

Lamiaceae 唇形科 *Isodon* 香茶菜属
Isodon rubescens (Hemsl.) H. Hara 碎米桠
Small shrubs

13.4 (0.3~19.4); 1.6 (-12.3~8.9); 24.5 (9.3~29.6); 13.6 (3.3~19.5); 110.1 (11.9~173.7); 8.9 (0.0~67.5); 767 (391~1038) ❙ 1028 (431~1776); 449 (258~718); 62 (10~225) ❙ 34.3 (-47.7~125.9); 748 (391~1038); 216.6 (108.2~337.3)

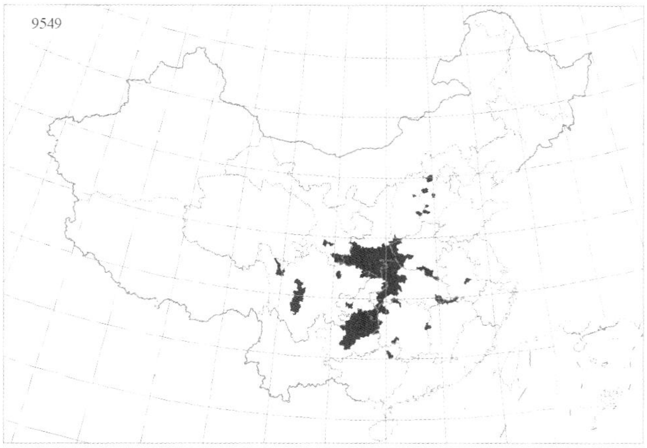

Lamiaceae 唇形科 *Isodon* 香茶菜属
Isodon rugosiformis (Hand.-Mazz.) H. Hara 类皱叶香茶菜
Subshrubs

6.6; -1.2; 13.2; 7.4; 44.3; 25.0; 524 ❙ 1033; 599; 33 ❙ 102.1; 524; 293.9

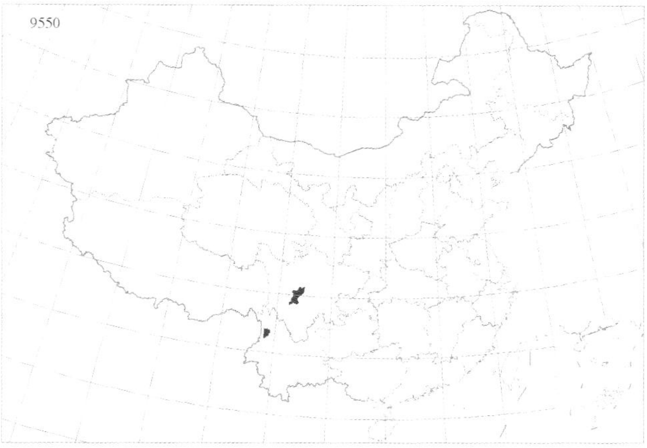

Lamiaceae 唇形科 *Isodon* 香茶菜属
Isodon rugosus (Wall. ex Benth.) Murata 皱叶香茶菜
Shrubs

4.0; -4.5; 11.5; 5.8; 34.4; 45.6; 460 ❙ 707; 424; 23 ❙ 31.9; 438; 241.7

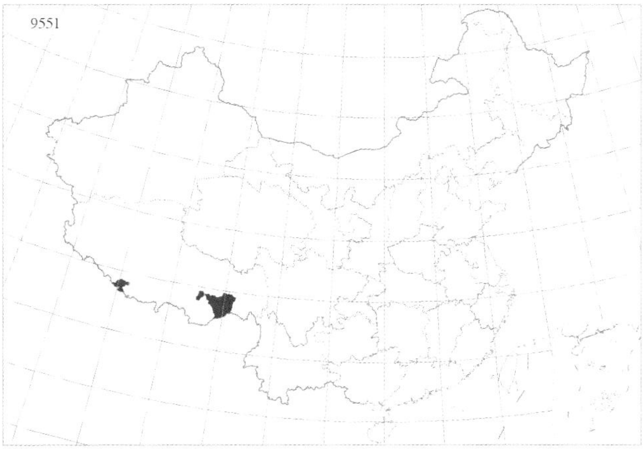

Lamiaceae 唇形科 *Isodon* 香茶菜属
Isodon scoparius (C. Y. Wu et H. W. Li) H. Hara 帚状香茶菜
Small shrubs

5.8; -1.8; 12.5; 6.1; 33.4; 23.7; 489 ❙ 771; 461; 25 ❙ 57.7; 489; 278.2

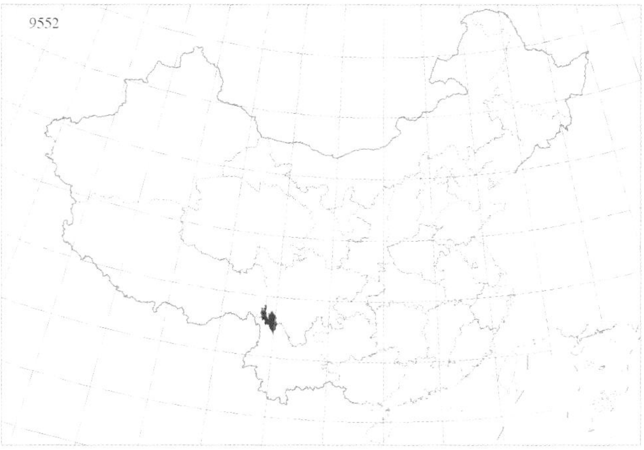

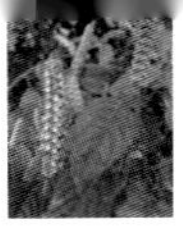

Lamiaceae 唇形科 *Isodon* 香茶菜属
Isodon secundiflorus (C. Y. Wu) H. Hara 侧花香茶菜
Shrubs

6.9; -0.6; 13.5; 7.1; 41.8; 18.1; 523 ❙ 853; 512; 25 ❙ 62.8; 523; 304.9

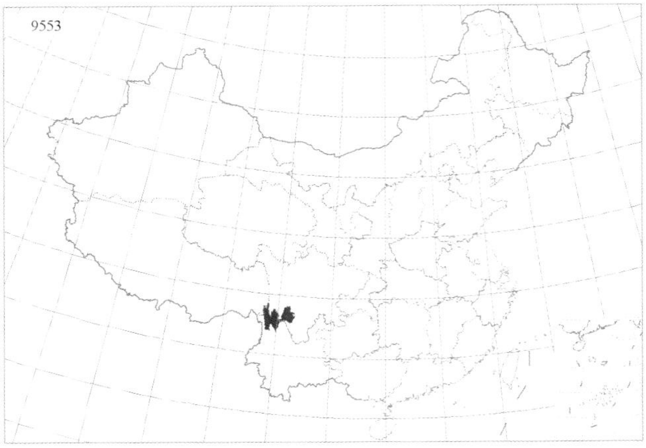

Lamiaceae 唇形科 *Isodon* 香茶菜属
Isodon setschwanensis (Hand.-Mazz.) H. Hara 四川香茶菜
Small shrubs

11.7 (6.6~14.6); 4.8 (-1.3~7.8); 17.5 (13.3~20.0); 11.8 (6.8~14.6); 84.6 (39.6~115.2); 3.2 (0.0~19.7); 657 (514~738) ❙ 977 (862~1057); 591 (548~619); 27 (12~46) ❙ 49.7 (30.5~67.7); 648 (514~707); 319.5 (296.0~372.3)

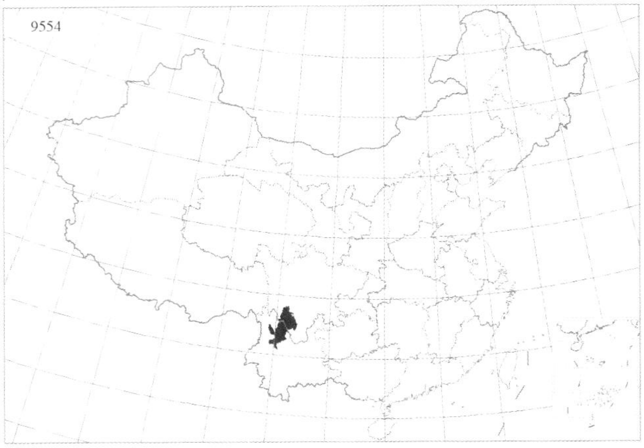

Lamiaceae 唇形科 *Isodon* 香茶菜属
Isodon silvaticus (C. Y. Wu et H. W. Li) H. W. Li 林生香茶菜
Subshrubs

2.4; -7.2; 11.3; 4.4; 21.4; 51.6; 423 ❙ 510; 311; 11 ❙ 20.7; 423; 237.6

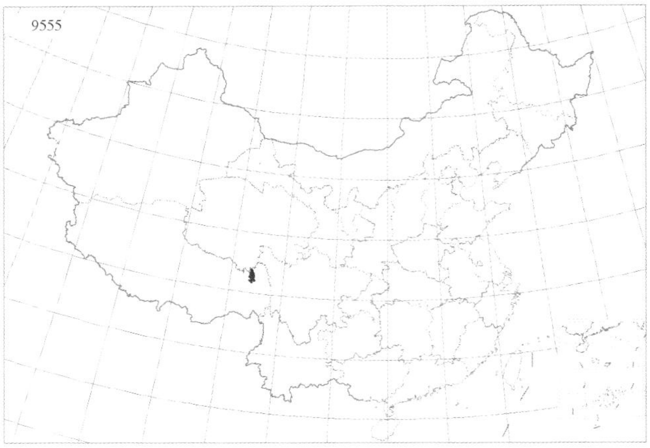

Lamiaceae 唇形科 *Isodon* 香茶菜属
Isodon smithianus (Hand.-Mazz.) H. Hara 马尔康香茶菜
Shrubs

0.9; -9.5; 9.8; 3.6; 14.4; 63.7; 396 ❙ 731; 403; 17 ❙ 84.5; 396; 315.2

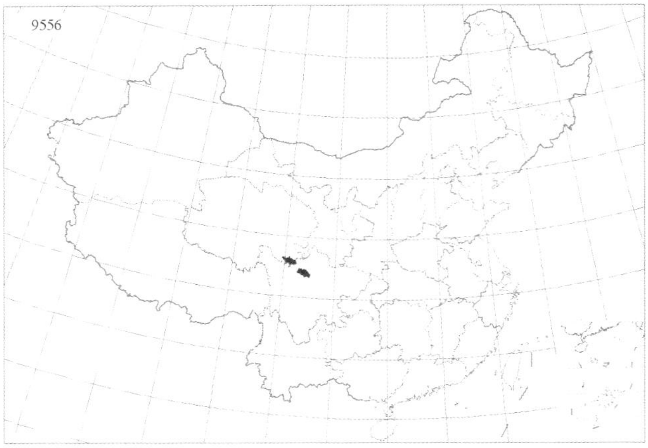

Lamiaceae 唇形科 *Isodon* 香茶菜属
Isodon tenuifolius (W. W. Sm.) Kudo 细叶香茶菜
Small shrubs

6.3 (3.8~9.6); -1.4 (-4.3~2.4); 13.0 (10.8~15.7); 6.6 (4.7~9.6); 37.7 (21.6~60.8); 21.6 (5.4~35.4); 507 (444~600) ❙ 879 (741~1007); 530 (426~601); 23 (12~40) ❙ 74.6 (54.0~121.7); 507 (444~600); 297.0 (265.2~351.7)

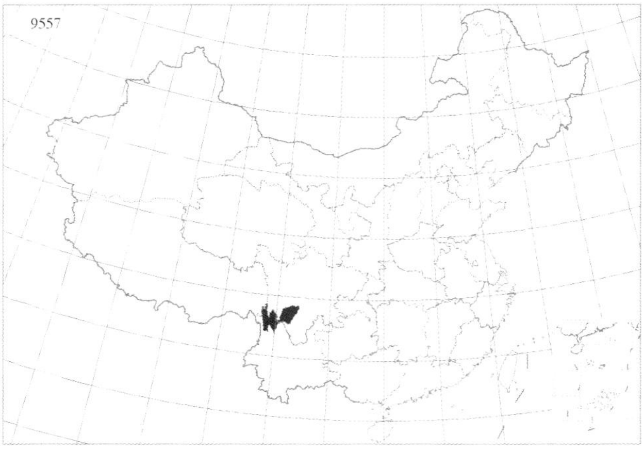

Lamiaceae 唇形科 *Isodon* 香茶菜属
Isodon wardii (C. Marquand et Airy Shaw) H. Hara 西藏香茶菜
Shrubs

7.8; -0.6; 14.8; 8.4; 54.9; 21.3; 549 ❙ 952; 552; 19 ❙ 59.0; 549; 297.2

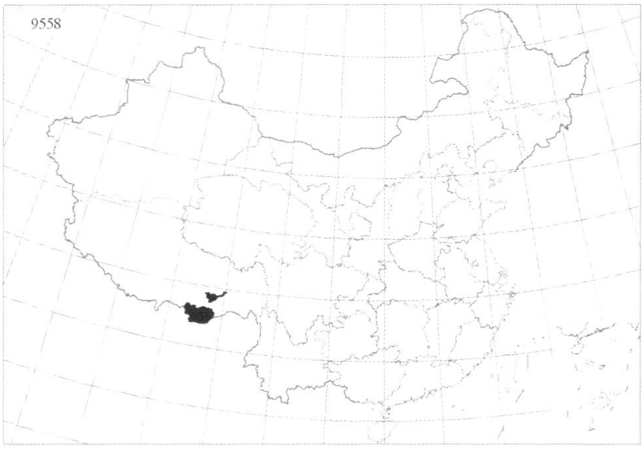

Lamiaceae 唇形科 *Isodon* 香茶菜属
Isodon wikstrcemioides (Hand.-Mazz.) H. Hara 荛花香茶菜
Small shrubs

5.2 (1.2~11.3); -2.9 (-7.1~4.3); 12.4 (9.0~17.0); 6.1 (3.3~11.3); 33.9 (11.9~76.6); 30.7 (0.7~56.5); 485 (390~644) ❙ 824 (578~1243); 480 (345~696); 25 (10~58) ❙ 68.8 (38.4~94.4); 485 (390~644); 275.7 (130.1~384.7)

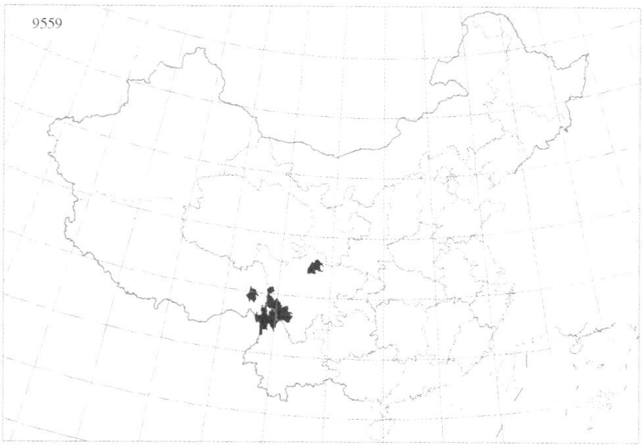

Lamiaceae 唇形科 *Isodon* 香茶菜属
Isodon xerophilus (C. Y. Wu et H. W. Li) H. Hara 旱生香茶菜
Small shrubs

17.9; 11.6; 22.1; 17.9; 155.1; 0.0; 858 ❙ 1098; 634; 46 ❙ 27.9; 836; 295.5

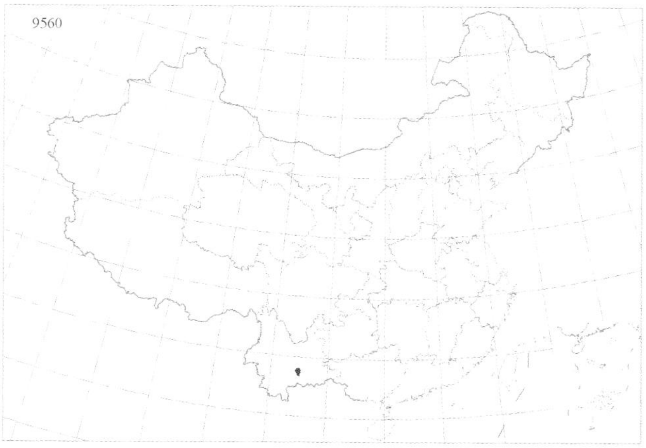

Lamiaceae 唇形科 *Leucosceptrum* 米团花属
Leucosceptrum canum Sm. 米团花
Shrubs or small trees

15.5 (-1.8~21 0); 8.9 (-10.2~15.1); 20.4 (6.1~25.6); 15.6 (1.8~21.0); 128.9 (1.9~191.8); 2.2 (0.0~82.8); 783 (329~1053) ❙ 1140 (320~1751); 643 (217~1007); 44 (12~77) ❙ 46.4 (-14.2~145.8); 759 (320~1053); 360.4 (72.2~659.0)

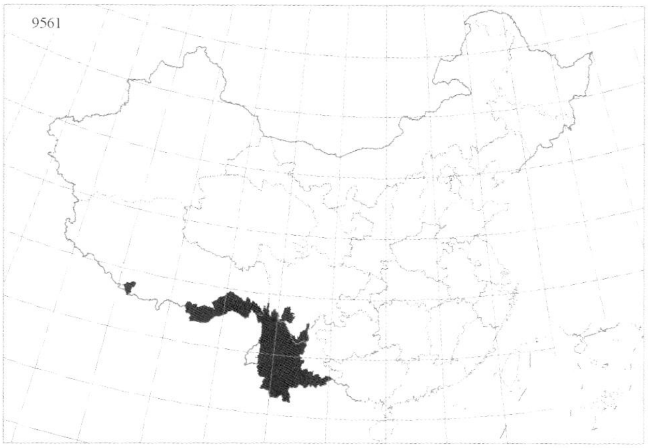

Lamiaceae 唇形科 *Ocimum* 罗勒属
Ocimum sanctum L. 圣罗勒
Subshrubs

20.2 (14.3~24.8); 11.9 (4.0~20.1); 27.2 (23.5~28.8); 20.3 (14.3~24.8); 183.5 (112.6~237.6); 0.1 (0.0~1.1); 1080 (756~1385) ❙ 1565 (868~3495); 609 (312~1847); 149 (41~897) ❙ 48.7 (-34.3~224.4); 1032 (756~1360); 284.3 (185.5~548.0)

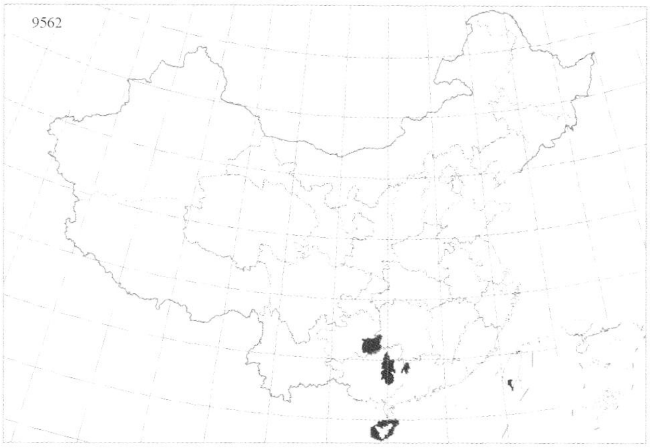

Lamiaceae 唇形科 *Rostrinucula* 钩子木属
Rostrinucula dependens (Rehder) Kudo 钩子木
Shrubs

11.8 (2.2~15 7); 1.7 (-6.6~5.8); 21.2 (9.9~25.5); 11.9 (3.8~15.7); 90.6 (15.5~128.9); 8.4 (0.0~49.1); 686 (413~824) ❙ 1007 (707~1414); 517 (357~817); 34 (15~65) ❙ 49.3 (-0.2~125.9); 686 (413~824); 254.2 (162.8~322.4)

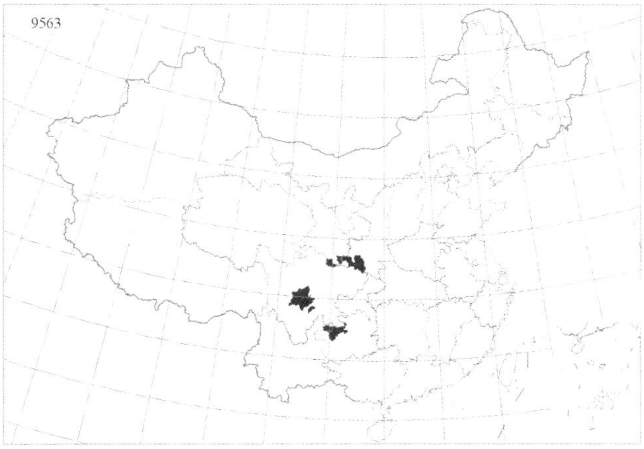

Lamiaceae 唇形科 *Rostrinucula* 钩子木属
Rostrinucula sinensis (Hemsl.) C. Y. Wu 长叶钩子木
Shrubs

14.9 (12.6~17.4); 4.1 (1.6~7.4); 25.0 (23.0~26.7); 14.9 (12.7~17.4); 121.5 (98.8~148.9); 2.1 (0.0~6.9); 800 (711~891) ❙ 1389 (1206~1624); 572 (481~637); 100 (75~191) ❙ 74.7 (47.4~107.2); 800 (711~891); 218.6 (152.4~285.1)

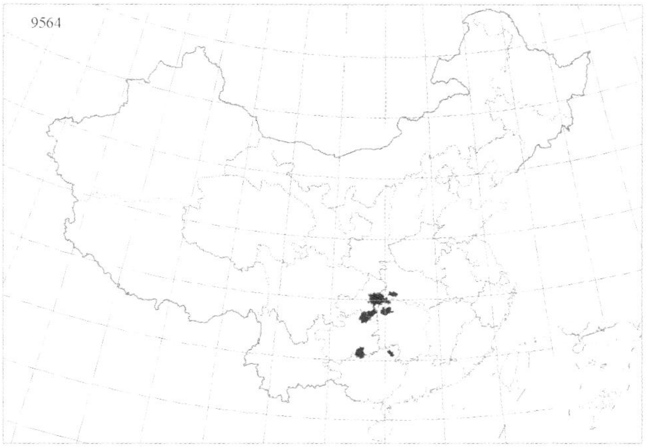

Lamiaceae 唇形科 *Wenchengia* 保亭花属
Wenchengia alternifolia C. Y. Wu et S. Chow 保亭花
Subshrubs

23.6; 18.7; 26.9; 23.7; 223.9; 0.0; 1269 ❙ 1602; 557; 53 ❙ 27.4; 1138; 568.5

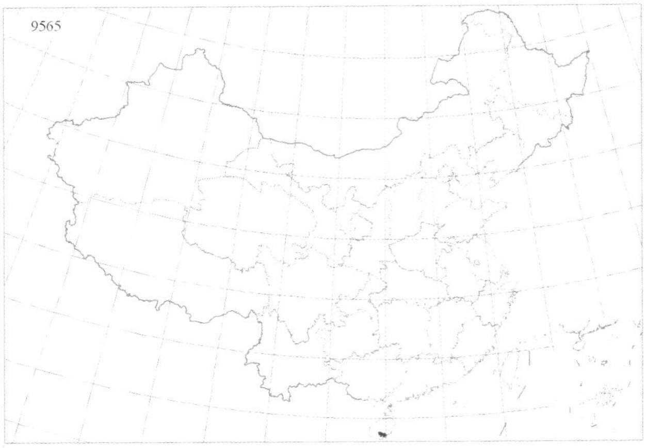

Solanaceae 茄科 *Capsicum* 辣椒属
Capsicum annuum L. 辣椒
Shrubs or herbs, deciduous

16.6 (5.4~24.8); 8.9 (-2.1~20.1); 22.7 (12.3~29.7); 16.7 (5.8~24.8); 140.5 (31.5~237.6); 0.7 (0.0~25.6); 855 (481~1385) ❙ 1215 (741~1997); 606 (312~996); 67 (15~254) ❙ 43.7 (-34.3~130.7); 828 (481~1360); 326.2 (110.1~659.0)

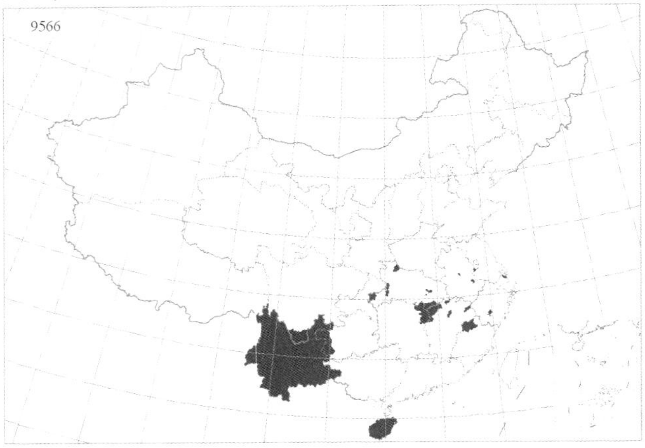

Solanaceae 茄科 *Lycianthes* 红丝线属
Lycianthes neesiana (Wall. ex Nees) D'Arcy et Z. Y. Zhang 截齿红丝线
Shrubs

16.9 (13.6~21.0); 9.9 (4.3~15.1); 22.1 (18.8~29.6); 17.0 (13.6~21.0); 143.4 (103.6~191.8); 0.0 (0.0~0.7); 833 (706~1053) ❙ 1187 (797~1751); 635 (411~996); 60 (23~211) ❙ 42.0 (-14.2~99.9); 813 (683~1053); 344.5 (188.8~659.0)

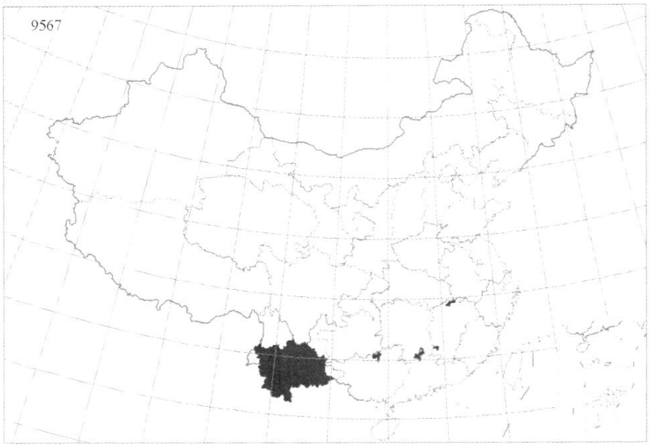

Solanaceae 茄科 *Lycianthes* 红丝线属
Lycianthes yunnanensis (Bitter) C. Y. Wu et S. C. Huang 滇红丝线
Shrubs

20.8 (8.6~24.8); 14.5 (1.3~20.1); 25.4 (15.0~28.8); 20.9 (8.7~24.8); 190.6 (52.6~237.6); 0.2 (0.0~8.8); 1094 (574~1385) ❙ 1411 (868~1854); 617 (312~911); 70 (41~182) ❙ 33.1 (-34.3~81.5); 1033 (574~1360); 416.6 (228.5~620.0)

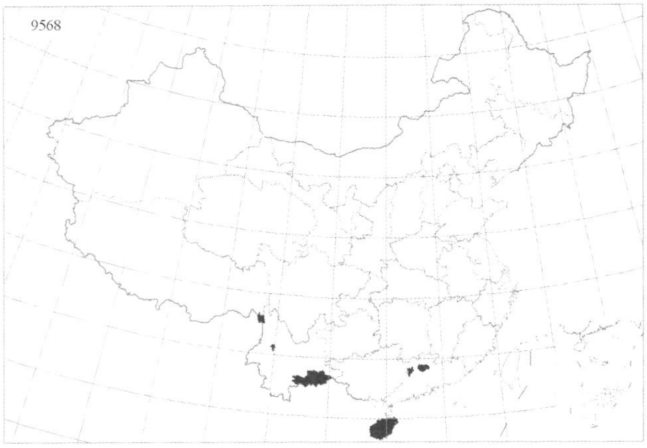

Solanaceae 茄科 *Lycium* 枸杞属
Lycium barbarum L. 宁夏枸杞
Shrubs

6.3 (-4.3~17.1); -10.3 (-25.0~4.3); 20.4 (8.0~29.4); 8.8 (2.4~17.1); 69.3 (6.4~146.5); 53.4 (0.7~124.5); 611 (329~952) ❙ 281 (21~1483); 151 (15~619); 13 (0~183) ❙ -50.9 (-97.3~128.6); 258 (21~929); 92.3 (1.6~313.2)

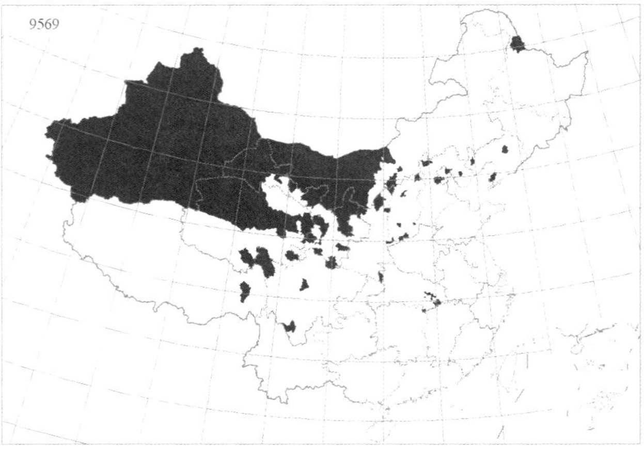

Solanaceae 茄科 *Lycium* 枸杞属
Lycium barbarum var. auranticarpum K. F. Ching 黄果枸杞
Shrubs

8.0; -7.5; 21.9; 9.6; 76.2; 39.0; 627 ❙ 225; 130; 5 ❙ -64.1; 225; 44.9

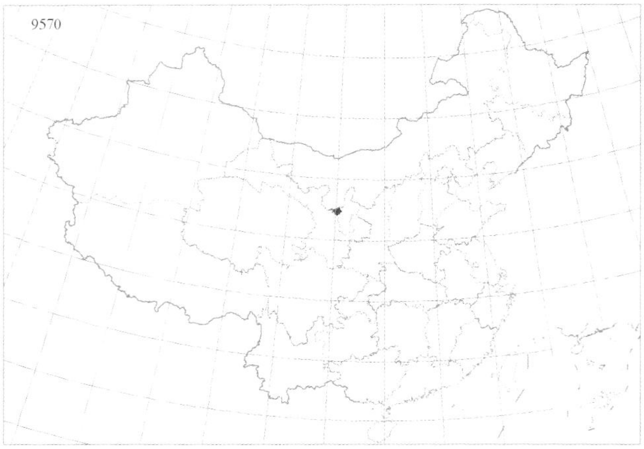

Solanaceae 茄科 *Lycium* 枸杞属
Lycium chinense Mill. 枸杞
Shrubs

13.9 (-1.4~24.8); 0.8 (-20.9~20.1); 25.8 (8.3~30.1); 14.4 (2.7~24.8); 121.8 (9.3~237.5); 14.6 (0.0~106.1); 827 (344~1385) ❙ 977 (35~2993); 437 (23~1388); 83 (0~547) ❙ 14.6 (-94.9~214.4); 729 (35~1360); 187.5 (7.4~620.0)

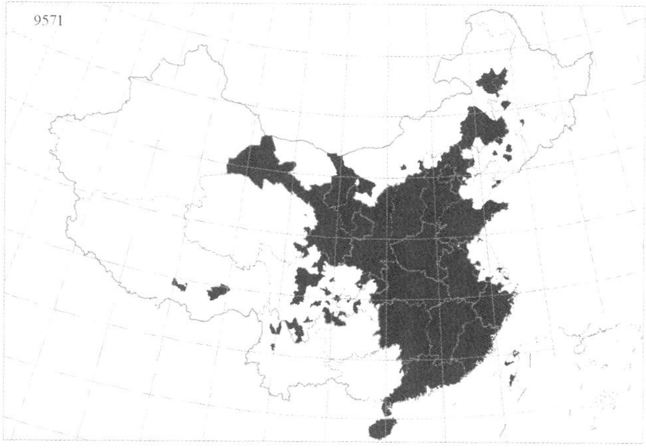

Solanaceae 茄科 *Lycium* 枸杞属
Lycium chinense var. potaninii (Pojark.) A. M. Lu 北方枸杞
Shrubs

6.0 (0.4~10.5); -11.0 (-15.9~-6.5); 21.0 (10.5~25.2); 8.6 (3.6~11.7); 68.0 (15.5~99.1); 55.2 (32.3~80.4); 598 (406~743) ❙ 467 (244~698); 311 (152~499); 11 (5~22) ❙ -21.6 (-60.8~35.3); 463 (244~698); 177.9 (38.8~287.6)

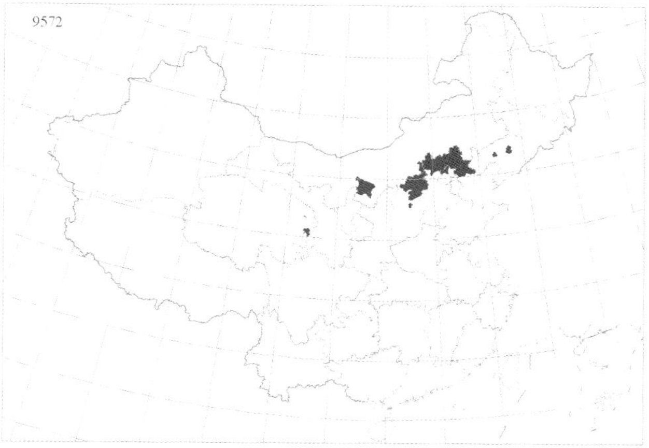

Solanaceae 茄科 *Lycium* 枸杞属
Lycium cylindricum Kuang et A. M. Lu 柱筒枸杞
Shrubs

1.3 (-2.5~5.4); -14.3 (-16.1~-13.1); 15.2 (10.1~22.7); 5.5 (2.9~9.3); 36.9 (12.0~76.8); 81.1 (71.4~101.3); 469 (358~643) ❙ 176 (63~252); 99 (45~153); 9 (2~17) ❙ -60.6 (-84.9~-37.4); 176 (63~252); 43.3 (8.4~77.1)

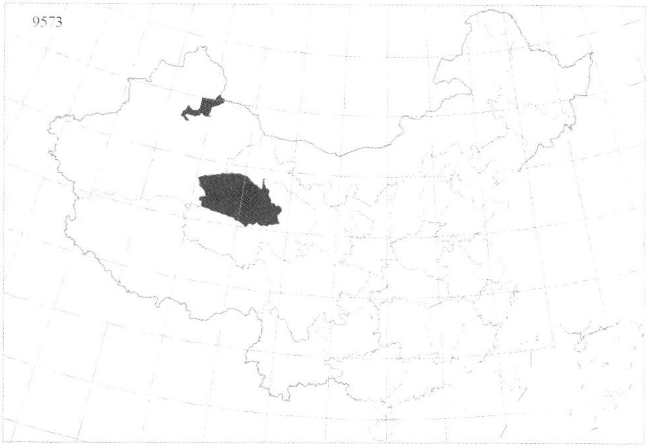

Solanaceae 茄科 *Lycium* 枸杞属
Lycium dasystemum Pojark. 新疆枸杞
Shrubs

3.2 (-2.5~9.0); -13.8 (-20.9~-9.1); 17.5 (10.1~23.2); 6.9 (2.9~10.7); 50.1 (12.0~88.9); 71.2 (40.7~103.9); 529 (358~703) ❙ 224 (77~457); 113 (48~259); 16 (2~48) ❙ -55.0 (-87.2~12.5); 221 (77~406); 104.9 (13.3~297.3)

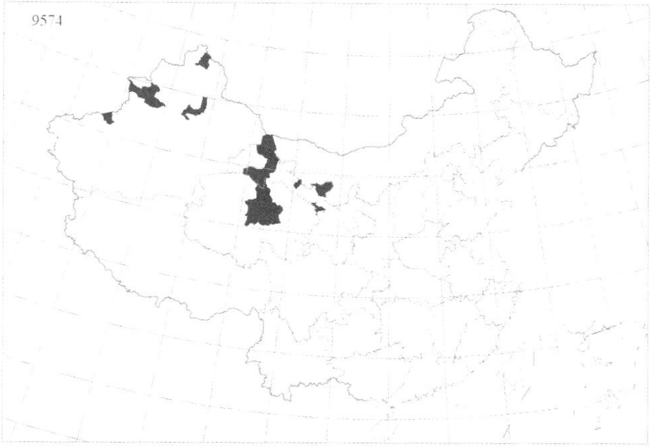

Solanaceae 茄科 *Lycium* 枸杞属
Lycium ruthenicum Murray 黑果枸杞
Shrubs

6.5 (-4.3~12.5); -11.0 (-20.3~-5.1); 21.3 (9.6~27.5); 9.1 (2.5~13.2); 73.4 (9.7~114.1); 55.1 (23.1~121.1); 627 (329~822) ❙ 160 (21~580); 87 (15~298); 8 (0~51) ❙ -72.3 (-97.3~7.5); 159 (21~539); 47.9 (1.6~229.4)

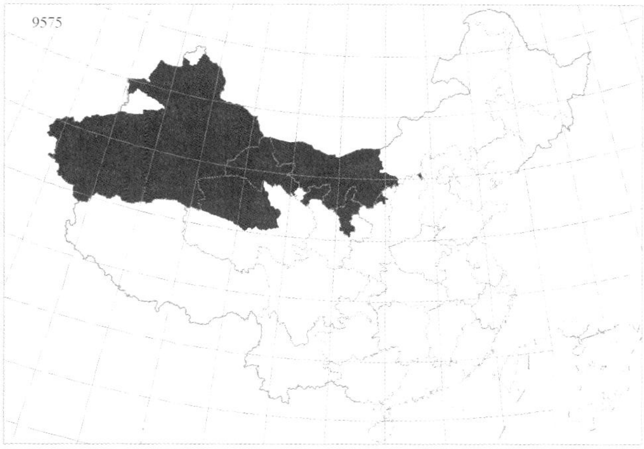

Solanaceae 茄科 *Lycium* 枸杞属
Lycium truncatum Y. C. Wang 截萼枸杞
Shrubs

6.4 (0.1~13.0); -11.0 (-21.5~-2.4); 21.6 (17.5~26.6); 9.0 (6.6~13.3); 71.2 (45.4~113.3); 54.4 (17.1~108.6); 611 (507~800) ❙ 382 (53~646); 234 (33~478); 10 (2~17) ❙ -37.3 (-92.4~-2.4); 382 (53~645); 111.1 (10.3~225.2)

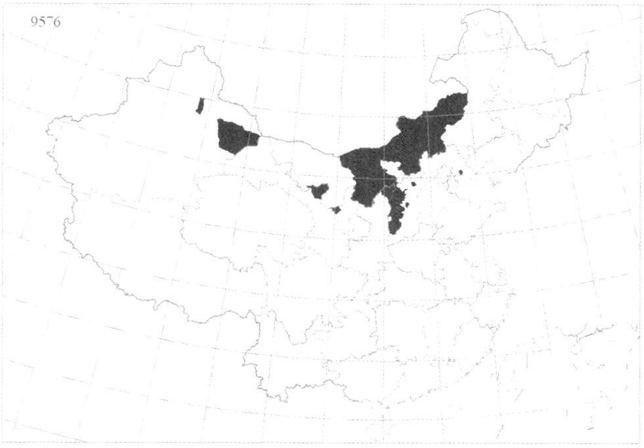

Solanaceae 茄科 *Lycium* 枸杞属
Lycium yunnanense Kuang et A. M. Lu 云南枸杞
Shrubs

15.7 (5.4~21.0); 8.6 (-2.1~15.1); 20.9 (12.3~25.6); 15.7 (5.8~21.0); 129.2 (31.5~191.8); 0.7 (0.0~25.6); 786 (481~1053) ❙ 1123 (741~1751); 632 (411~996); 44 (15~77) ❙ 43.0 (-14.2~112.2); 768 (481~1053); 337.8 (185.6~659.0)

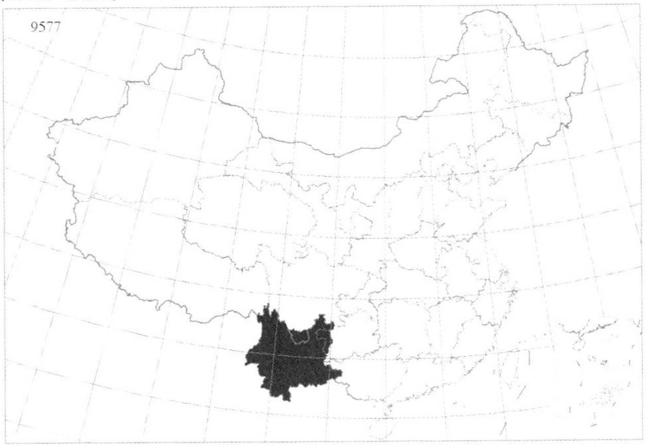

Solanaceae 茄科 *Solanum* 茄属
Solanum chrysotrichum Schltdl. 多裂水茄
Shrubs

20.2 (16.2~24.1); 14.5 (11.0~18.9); 25.0 (20.1~28.2); 20.2 (16.2~24.1); 182.9 (134.2~229.7); 0.0 (0.0~0.0); 1019 (773~1322) ❙ 2402 (1498~3495); 1106 (750~1847); 255 (47~897) ❙ 146.2 (25.2~246.5); 983 (773~1144); 245.0 (119.4~363.7)

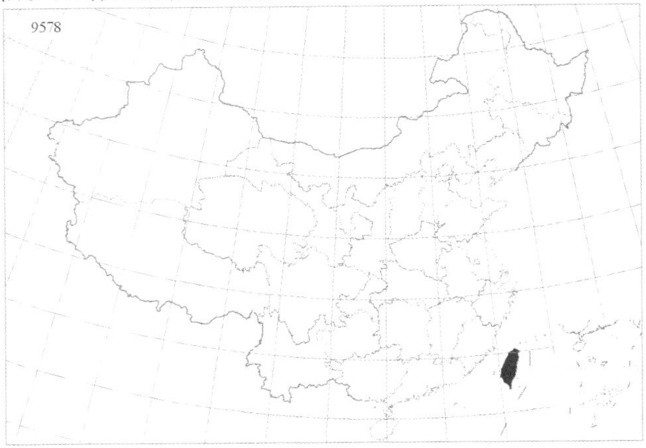

Solanaceae 茄科 *Solanum* 茄属
Solanum deflexicarpum C. Y. Wu et S. C. Huang 苦刺
Shrubs

18.0 (14.4~21.0); 11.5 (7.5~15.1); 22.2 (19.1~25.6); 18.0 (14.4~21.0); 156.6 (113.1~191.8); 0.0 (0.0~0.0); 870 (727~1053) ❙ 1376 (1028~1751); 760 (566~996); 54 (41~77) ❙ 58.6 (12.5~112.2); 857 (727~1053); 442.0 (277.0~659.0)

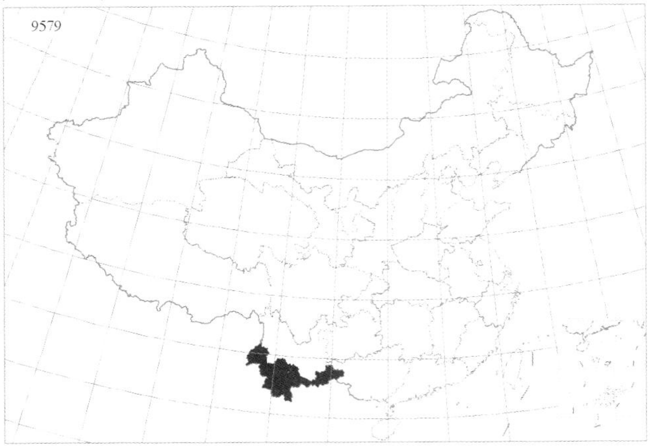

Solanaceae 茄科 *Solanum* 茄属
Solanum diphyllum L. 黄果龙葵
Shrubs

20.2 (16.2~24.1); 14.5 (11.0~18.9); 25.0 (20.1~28.2); 20.2 (16.2~24.1); 182.9 (134.2~229.7); 0.0 (0.0~0.0); 1019 (773~1322) ❙ 2402 (1498~3495); 1106 (750~1847); 255 (47~897) ❙ 146.2 (25.2~246.5); 983 (773~1144); 245.0 (119.4~363.7)

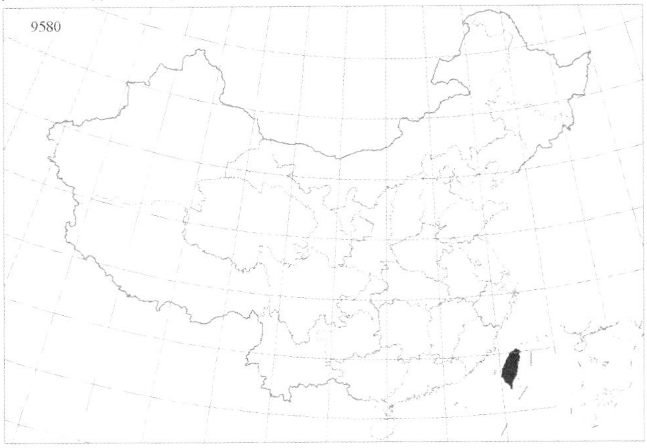

Solanaceae 茄科 *Solanum* 茄属
Solanum erianthum D. Don 假烟叶树
Shrubs or small trees

18.0 (5.3~24.8); 9.9 (-2.4~20.1); 24.7 (12.2~29.0); 18.1 (5.7~24.8); 157.1 (30.5~237.6); 0.4 (0.0~26.2); 932 (479~1385) ❙ 1404 (741~3495); 645 (312~1428); 111 (15~897) ❙ 52.1 (-34.3~246.5); 918 (479~1360); 288.5 (111.7~659.0)

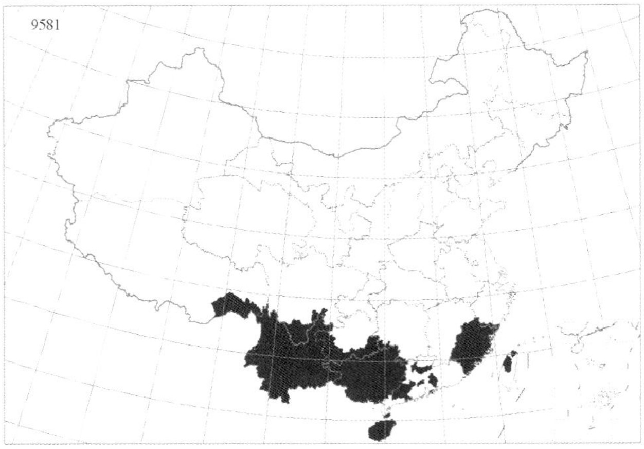

Solanaceae 茄科 *Solanum* 茄属
Solanum laciniatum Aiton 澳洲茄
Shrubs

21.6 (17.0~24.8); 14.3 (7.6~20.1); 27.9 (25.7~29.0); 21.6 (17.0~24.8); 199.4 (144.5~237.6); 0.0 (0.0~0.0); 1147 (870~1385) ❙ 1424 (868~1854); 495 (312~712); 140 (41~257) ❙ 26.7 (-34.3~97.5); 1078 (868~1360); 315.7 (123.9~620.0)

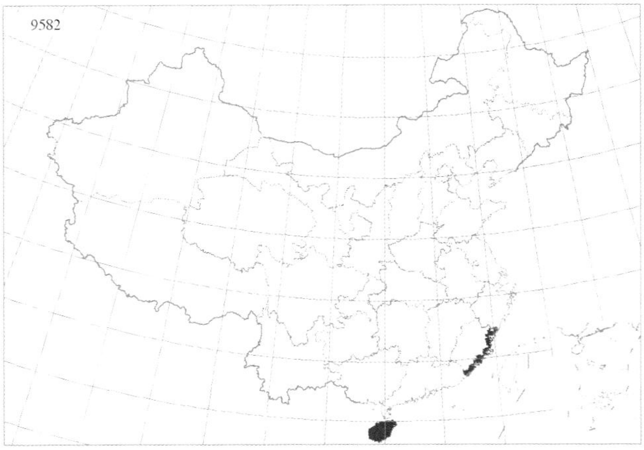

Solanaceae 茄科 *Solanum* 茄属
Solanum luzoniense Merr. 吕宋茄
Shrubs or subshrubs

18.5; 14.1; 21.7; 18.5; 162.1; 0.0; 863 ❙ 2713; 1388; 188 ❙ 214.4; 863; 363.7

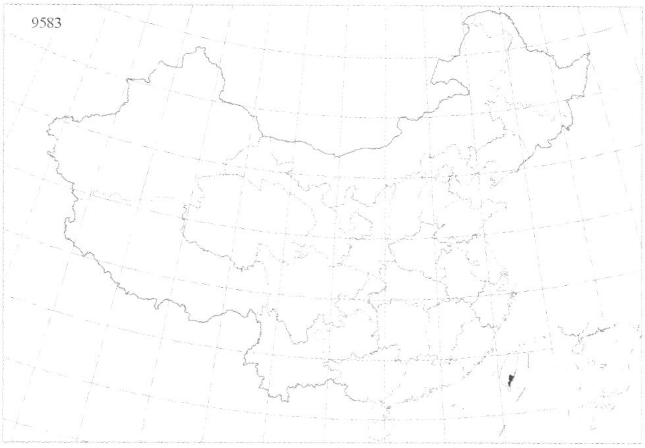

Solanaceae 茄科 *Solanum* 茄属
Solanum macaonense Dunal 山茄
Shrubs

21.8 (17.5~24.8); 14.0 (7.8~20.1); 28.1 (22.3~29.2); 21.8 (17.5~24.8); 201.5 (149.9~237.6); 0.0 (0.0~0.0); 1153 (846~1385) ❙ 1620 (958~2759); 717 (356~1594); 126 (41~193) ❙ 41.6 (-12.0~202.5); 1125 (823~1273); 241.3 (95.3~620.0)

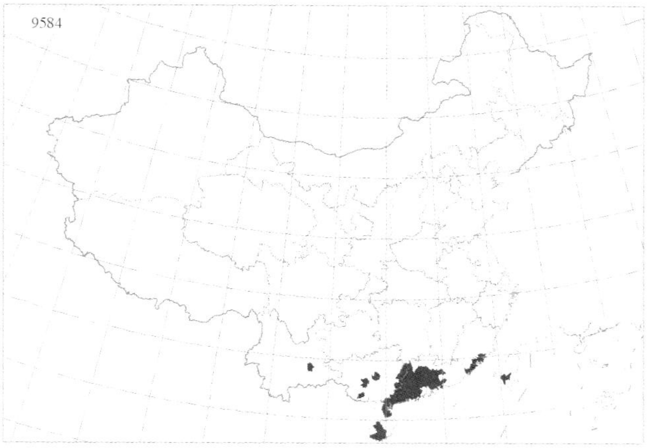

Solanaceae 茄科 *Solanum* 茄属
Solanum nienkui Merr. et Chun 疏刺茄
Shrubs

22.7 (17.8~24.8); 16.0 (7.1~20.1); 28.0 (27.2~28.7); 22.8 (17.8~24.8); 213.2 (154.0~237.6); 0.0 (0.0~0.0); 1233 (933~1385) ❙ 1311 (868~1561); 458 (312~544); 81 (41~193) ❙ 9.7 (-34.3~61.1); 1052 (868~1142); 392.6 (197.5~538.2)

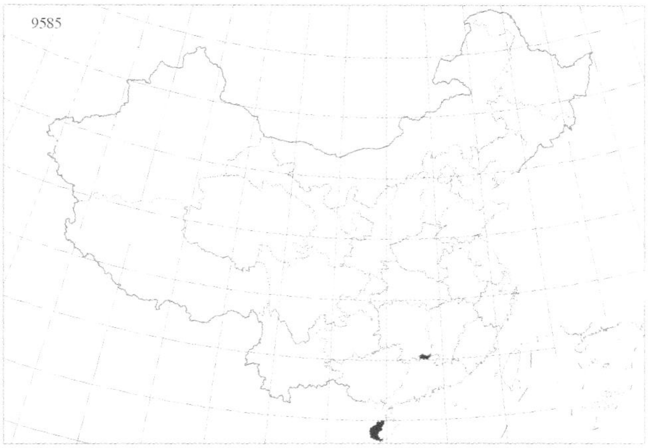

Solanaceae 茄科 *Solanum* 茄属
Solanum pittosporifolium Hemsl. 海桐叶白英
Shrubs

17.8 (2.4~24.1); 7.6 (-8.6~18.9); 27.0 (11.0~30.1); 17.8 (4.3~24.1); 155.0 (20.4~229.7); 1.0 (0.0~51.0); 941 (423~1322) ❙ 1502 (466~3495); 593 (299~1847); 159 (9~897) ❙ 60.5 (-34.7~246.5); 925 (423~1247); 235.4 (119.4~659.0)

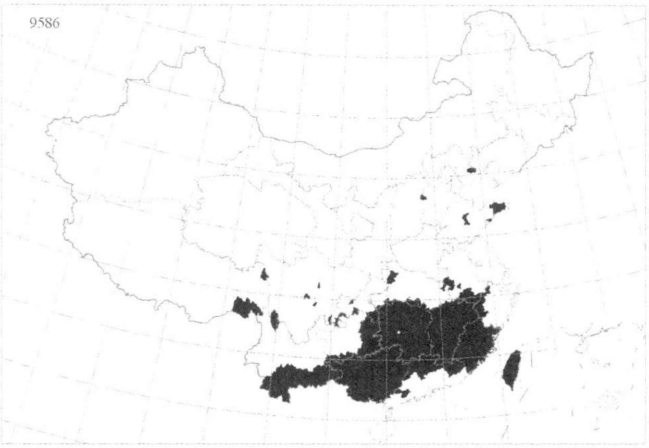

Solanaceae 茄科 *Solanum* 茄属
Solanum procumbens Lour. 海南茄
Shrubs

22.4 (19.5~24.8); 14.9 (10.6~20.1); 28.2 (25.7~29.2); 22.4 (19.6~24.8); 208.8 (175.1~237.6); 0.0 (0.0~0.0); 1194 (1003~1385) ❙ 1675 (868~2037); 731 (312~1033); 116 (41~191) ❙ 41.6 (-34.3~95.5); 1167 (868~1360); 259.1 (95.3~620.0)

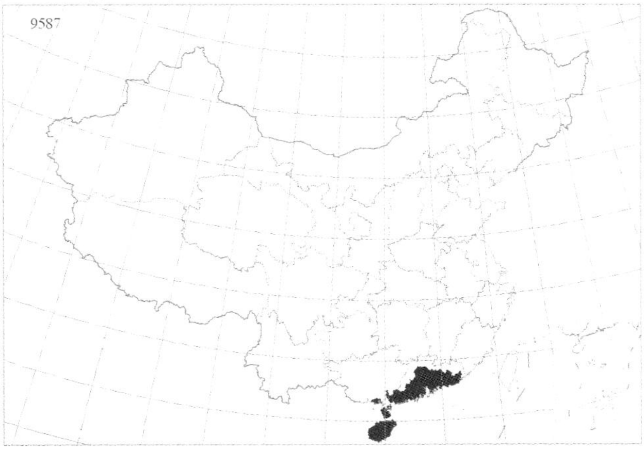

Solanaceae 茄科 *Solanum* 茄属
Solanum pseudocapsicum L. 珊瑚樱
Shrubs

17.9 (5.4~24.8); 9.7 (-2.4~20.1); 24.8 (12.3~29.8); 18.0 (5.8~24.8); 156.6 (31.5~237.6); 0.6 (0.0~25.6); 938 (481~1385) ❙ 1327 (741~2037); 632 (312~1033); 83 (15~226) ❙ 42.3 (-34.3~123.6); 923 (481~1360); 270.8 (95.3~659.0)

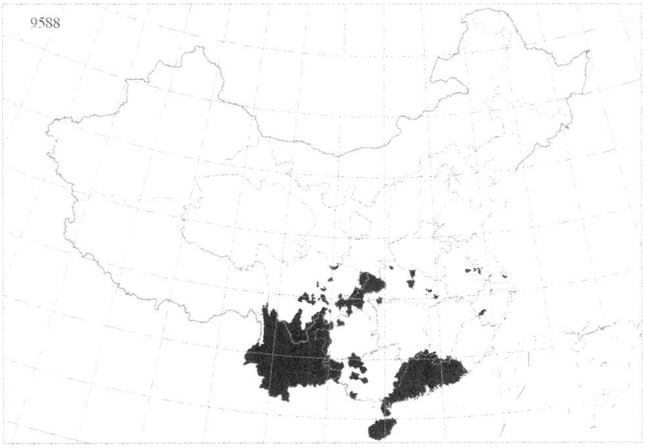

Solanaceae 茄科 *Solanum* 茄属
Solanum pseudocapsicum var. diflorum (Vell.) Bitter 珊瑚豆
Small shrubs

18.5 (11.7~23.6); 9.3 (2.7~16.7); 26.5 (18.8~29.8); 18.6 (11.7~23.6); 163.0 (82.6~223.5); 0.2 (0.0~3.8); 974 (658~1281) ❙ 1452 (916~2037); 631 (444~1033); 113 (31~197) ❙ 49.7 (-6.9~95.5); 972 (658~1273); 213.2 (95.3~340.7)

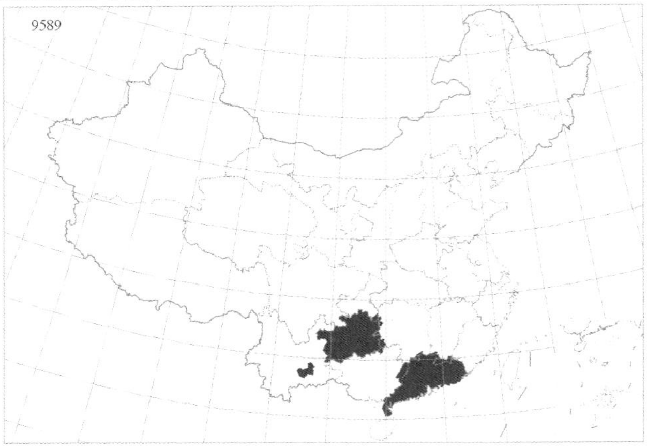

Solanaceae 茄科 *Solanum* 茄属
Solanum spirale Roxb. 旋花茄
Shrubs

16.7 (9.6~21.7); 8.5 (2.4~15.1); 23.2 (15.7~29.4); 16.7 (9.6~21.8); 141.1 (60.8~201.3); 0.1 (0.0~5.4); 846 (600~1162) ❙ 1246 (797~1751); 618 (392~1007); 80 (15~209) ❙ 47.6 (-14.2~145.8); 830 (600~1162); 304.7 (134.6~659.0)

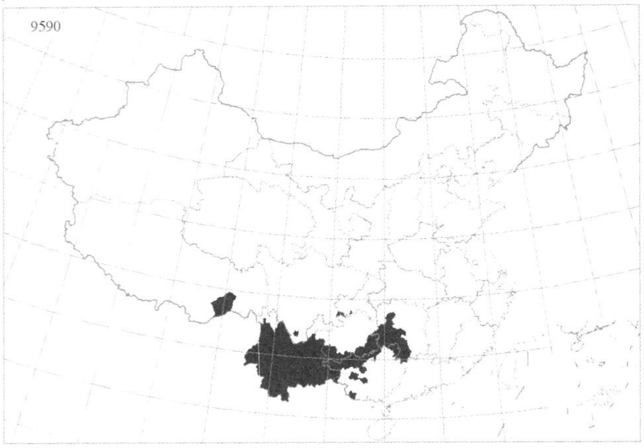

Solanaceae 茄科 *Solanum* 茄属
Solanum torvum Sw. 水茄
Shrubs

18.5 (5.4~24.8); 10.6 (-2.1~20.1); 24.9 (12.3~29.2); 18.5 (5.8~24.8); 162.8 (31.5~237.6); 0.3 (0.0~25.6); 958 (481~1385) ❙ 1429 (741~2759); 659 (312~1594); 107 (15~259) ❙ 50.7 (-34.3~246.5); 942 (481~1360); 285.5 (95.3~659.0)

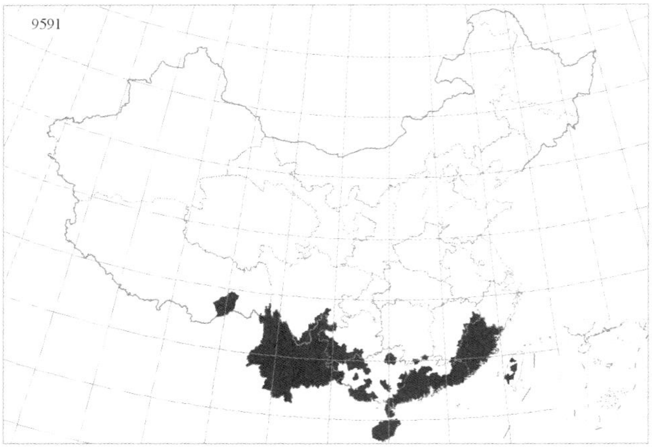

Solanaceae 茄科 *Solanum* 茄属
Solanum violaceum Ortega 刺天茄
Small shrubs

17.6 (5.4~24.0); 9.3 (-2.1~17.7); 24.6 (12.3~29.8); 17.7 (5.8~24.1); 152.3 (31.5~229.0); 0.4 (0.0~25.6); 911 (481~1321) ❙ 1362 (741~3495); 633 (386~1847); 99 (15~897) ❙ 49.8 (-14.2~246.5); 902 (481~1321); 264.4 (95.3~659.0)

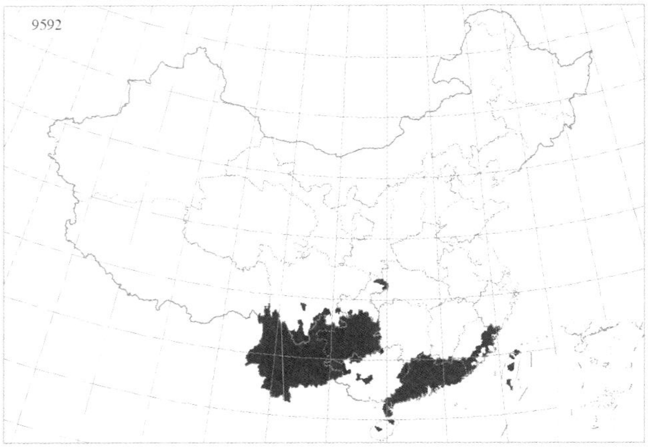

Scrophulariaceae 玄参科 *Brandisia* 来江藤属
Brandisia cauliflora Tsoong et L. T. Lu 茎花来江藤
Scandent shrubs

20.7; 12.4; 26.9; 20.8; 189.3; 0.0; 1074 ❙ 1437; 779; 86 ❙ 35.2; 1074; 305.7

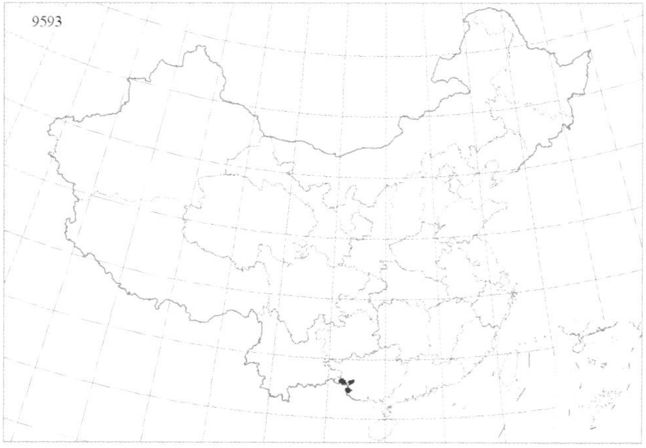

Scrophulariaceae 玄参科 *Brandisia* 来江藤属
Brandisia discolor Hook. f. et Thomson 异色来江藤
Scandent shrubs

18.4 (16.3~21.0); 12.0 (9.1~15.1); 22.6 (21.3~25.6); 18.4 (16.4~21.0); 161.1 (136.4~191.8); 0.0 (0.0~0.0); 887 (797~1053) ❙ 1320 (939~1751); 722 (524~996); 55 (37~77) ❙ 48.8 (6.5~88.8); 869 (744~1053); 422.1 (277.0~659.0)

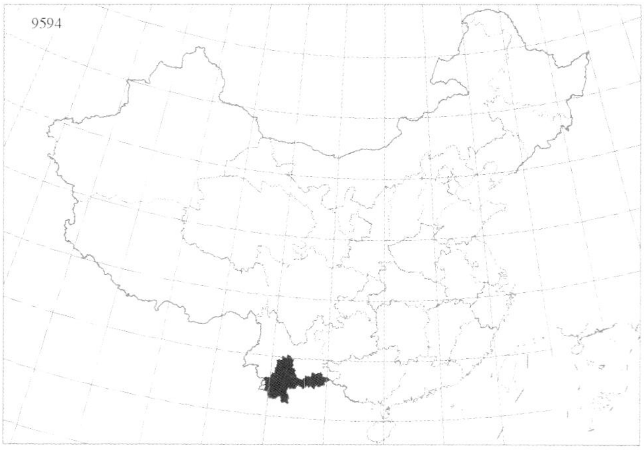

S

Scrophulariaceae 玄参科 *Brandisia* 来江藤属
Brandisia glabrescens Rehder 退毛来江藤
Shrubs

17.3; 10.6; 22.0; 17.3; 148.1; 0.0; 834 ❙ 1273; 701; 55 ❙ 52.5; 834; 437.5

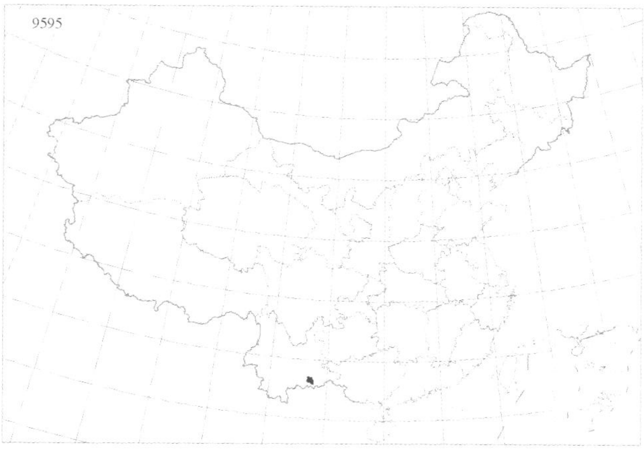

Scrophulariaceae 玄参科 *Brandisia* 来江藤属
Brandisia glabrescens var. hypochrysa P. C. Tsoong 退毛来江藤黄背变种
Shrubs

16.9; 10.3; 21.4; 16.9; 142.7; 0.0; 813 ❙ 1178; 646; 51 ❙ 44.9; 813; 383.4

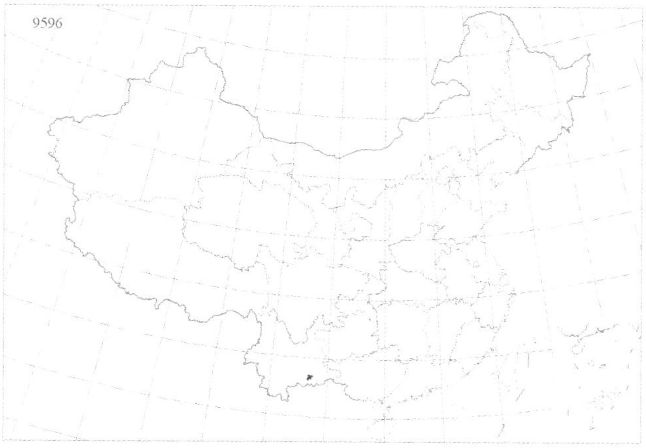

Scrophulariaceae 玄参科 *Brandisia* 来江藤属
Brandisia hancei Hook. f. 来江藤
Shrubs

15.7 (5.4~22.2); 6.6 (-2.5~15.1); 23.5 (12.3~29.8); 15.7 (5.8~22.2); 130.4 (31.5~206.9); 1.7 (0.0~25.6); 819 (481~1177) ❙ 1207 (711~1990); 577 (352~996); 73 (12~218) ❙ 47.6 (-14.2~128.9); 811 (481~1177); 266.6 (108.2~659.0)

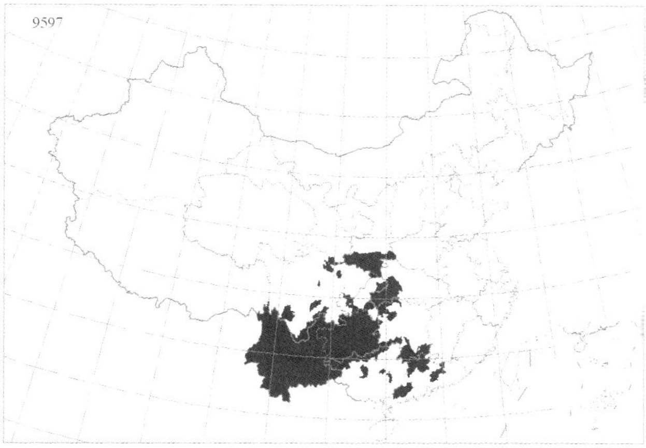

Scrophulariaceae 玄参科 *Brandisia* 来江藤属
Brandisia kwangsiensis H. L. Li 广西来江藤
Scandent shrubs

18.5 (16.3~21.3); 9.9 (7.3~14.2); 25.2 (21.4~28.6); 18.5 (16.4~21.3); 162.4 (136.4~196.1); 0.0 (0.0~0.0); 934 (797~1130) ❙ 1387 (1018~1758); 691 (554~860); 96 (47~189) ❙ 48.7 (22.7~85.5); 934 (797~1130); 301.3 (185.5~551.6)

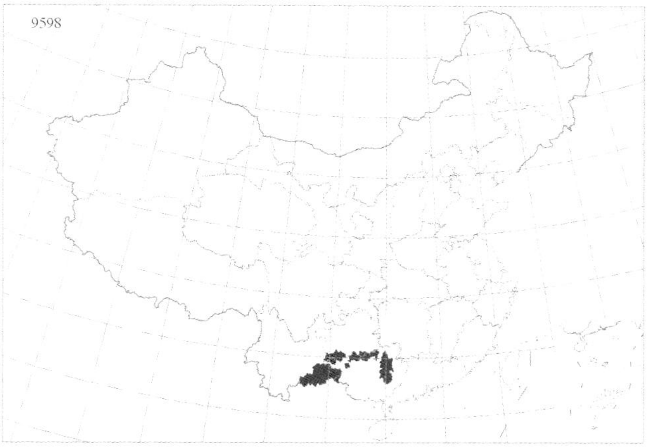

Scrophulariaceae 玄参科 *Brandisia* 来江藤属
Brandisia racemosa Hemsl. 总花来江藤
Scandent shrubs

16.3 (12.6~19.3); 9.7 (6.0~13.1); 21.2 (17.9~23.9); 16.4 (12.6~19.4) 136.6 (91.6~173.2); 0.0 (0.0~0.0); 803 (678~939) ❙ 1038 (797~1495) 594 (411~836); 41 (21~62) ❙ 29.6 (-14.2~66.9); 771 (678~903); 311.6 (218.0~432.9)

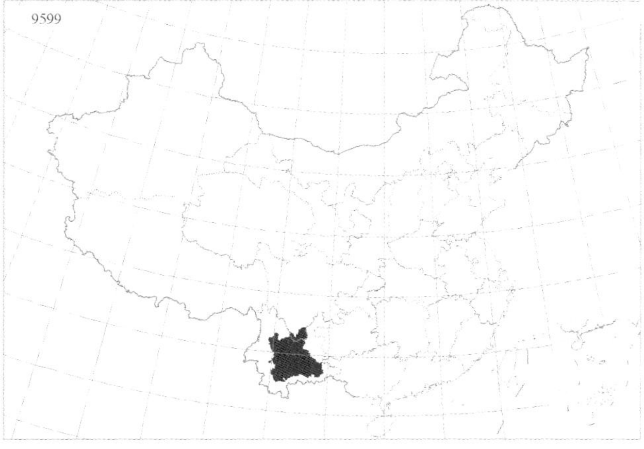

Scrophulariaceae 玄参科 *Brandisia* 来江藤属
Brandisia rosea W. W. Sm. 红花来江藤
Shrubs

12.9 (1.4~19.3); 5.4 (-7.6~13.1); 18.8 (9.2~24.6); 13.1 (3.5~19.4); 102.8 (13.4~173.2); 7.6 (0.0~55.6); 705 (402~939) ❙ 1092 (579~1547); 629 (365~911); 36 (10~70) ❙ 58.0 (4.4~125.9); 689 (402~903); 327.3 (189.1~551.3)

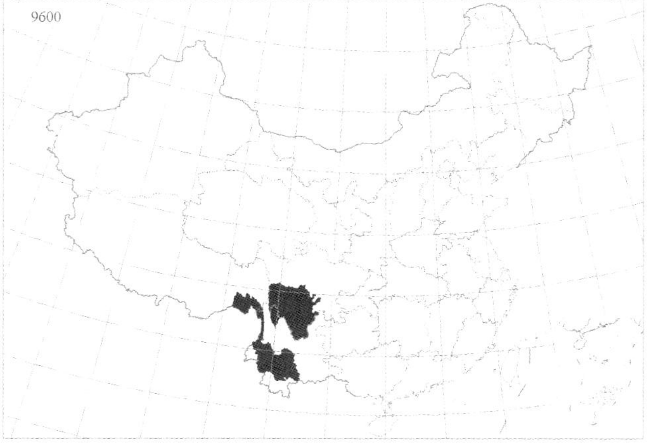

Scrophulariaceae 玄参科 *Brandisia* 来江藤属
Brandisia rosea var. flava C. E. C. Fisch. 红花来江藤黄花变种
Shrubs

5.3；−2.4；12.2；5.7；30.5；26.2；479 ❙ 961；558；31 ❙ 100.7；479；234.2

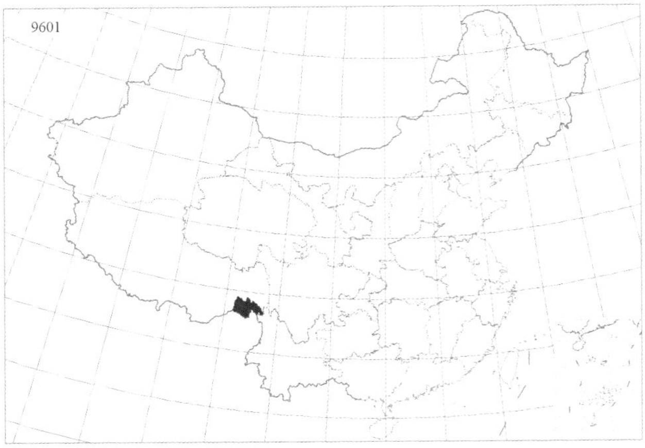

Scrophulariaceae 玄参科 *Brandisia* 来江藤属
Brandisia swinglei Merr. 岭南来江藤
Shrubs

19.8 (16.8~22.4)；10.3 (5.7~14.3)；27.8 (24.6~29.8)；19.9 (16.9~22.5)；178.2 (142.5~209.7)；0.0 (0.0~0.0)；1041 (880~1205) ❙ 1659 (1293~2024)；658 (435~954)；171 (57~211) ❙ 60.1 (22.7~95.5)；1039 (880~1205)；222.9 (95.3~371.7)

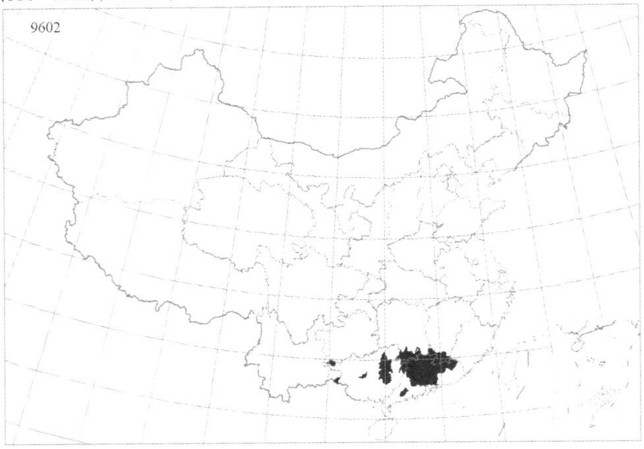

Scrophulariaceae 玄参科 *Paulownia* 泡桐属
Paulownia catalpifolia T. Gong ex D. Y. Hong 楸叶泡桐
Trees, deciduous

9.9 (2.5~14.5)；−5.7 (−15.9~0.2)；23.8 (18.6~27.5)；11.1 (6.9~14.5)；91.1 (48.9~124.7)；31.2 (10.4~80.4)；699 (523~856) ❙ 566 (349~877)；355 (227~542)；19 (8~65) ❙ −18.7 (−43.4~17.0)；561 (349~750)；168.2 (101.0~287.6)

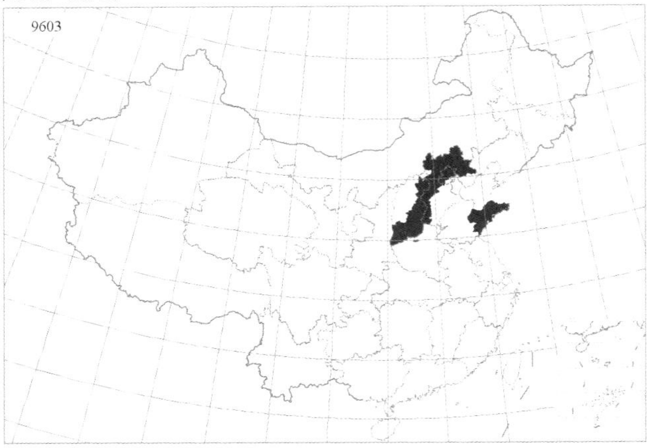

Scrophulariaceae 玄参科 *Paulownia* 泡桐属
Paulownia elongata S. Y. Hu 兰考泡桐
Trees, deciduous

14.4 (12.4~16.1)；0.6 (−4.2~4.2)；27.0 (24.0~28.1)；14.6 (13.0~16.1)；122.5 (107.0~135.3)；8.8 (0.8~22.0)；841 (746~902) ❙ 844 (481~1439)；430 (343~592)；53 (11~94) ❙ 1.4 (−42.0~92.3)；723 (481~884)；169.0 (116.6~252.6)

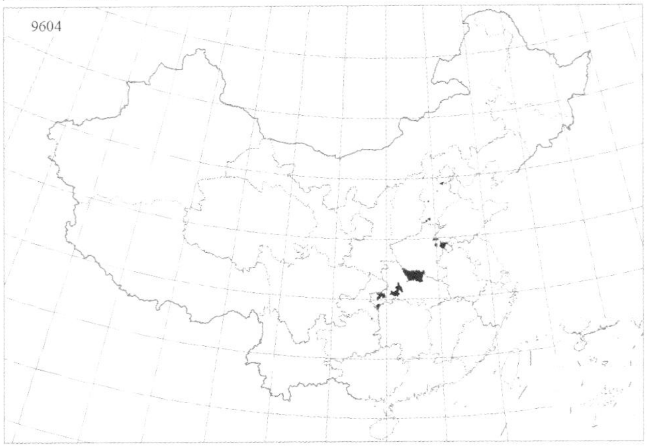

Scrophulariaceae 玄参科 *Paulownia* 泡桐属
Paulownia fargesii Franch. 川泡桐
Trees, deciduous

15.1 (2.2~21.0)；4.6 (−7.0~14.2)；24.9 (9.9~28.9)；15.2 (3.8~21.0)；124.8 (15.5~191.8)；2.9 (0.0~49.1)；818 (413~1053) ❙ 1200 (806~1893)；541 (389~860)；75 (16~227) ❙ 49.1 (−5.3~129.9)；818 (413~1053)；212.0 (108.2~551.6)

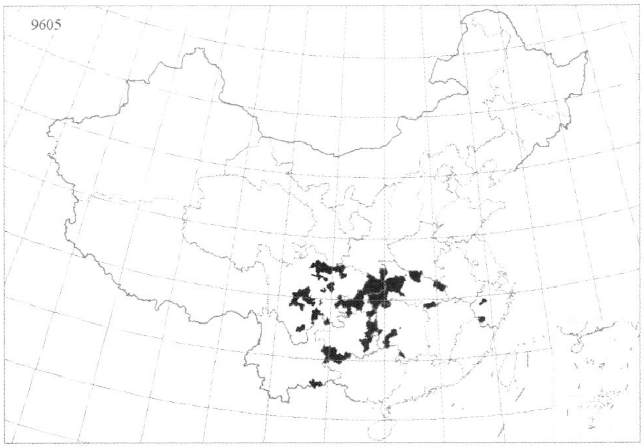

Scrophulariaceae 玄参科 *Paulownia* 泡桐属
Paulownia fortunei (Seem.) Hemsl. 白花泡桐
Trees, deciduous

17.2 (10.6~24.2)；6.6 (−0.2~17.6)；26.9 (18.8~30.1)；17.3 (10.6~24.2)；148.4 (79.4~230.4)；0.9 (0.0~12.1)；918 (643~1331) ❙ 1410 (687~2680)；575 (363~1428)；140 (31~308) ❙ 54.4 (−23.1~246.5)；909 (643~1331)；223.7 (108.2~659.0)

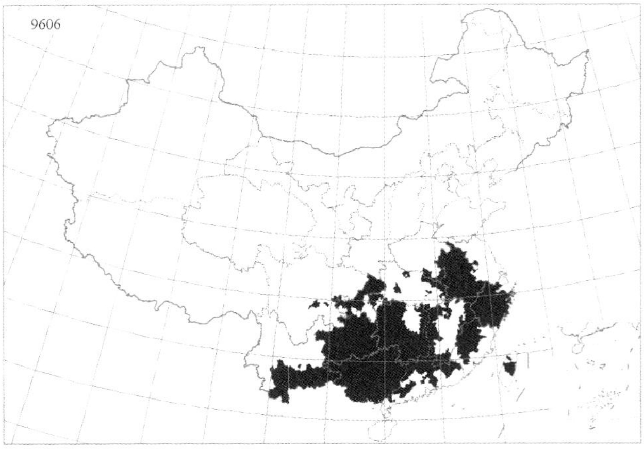

Scrophulariaceae 玄参科 *Paulownia* 泡桐属
Paulownia kawakamii T. Ito 台湾泡桐
Trees, deciduous

16.7 (11.7~24.1); 6.0 (0.6~18.9); 26.6 (18.8~29.8); 16.7 (11.7~24.1); 141.4 (82.6~229.7); 0.6 (0.0~9.7); 885 (658~1322) ❙ 1481 (916~2859); 560 (392~1594); 156 (29~402) ❙ 67.6 (19.6~246.5); 879 (658~1182); 221.5 (111.9~336.6)

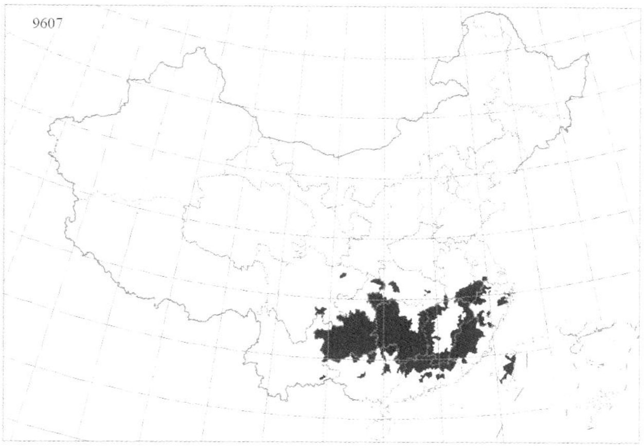

Scrophulariaceae 玄参科 *Paulownia* 泡桐属
Paulownia tomentosa (Thunb.) Steud. 毛泡桐
Trees, deciduous

13.9 (2.5~21.6); 0.4 (-15.9~14.2); 26.0 (15.6~29.7); 14.3 (6.6~21.6); 121.1 (43.0~199.4); 13.7 (0.0~80.4); 823 (505~1128) ❙ 915 (349~1991); 445 (227~911); 70 (8~259) ❙ 9.1 (-51.1~128.9); 710 (349~1128); 197.1 (76.1~551.6)

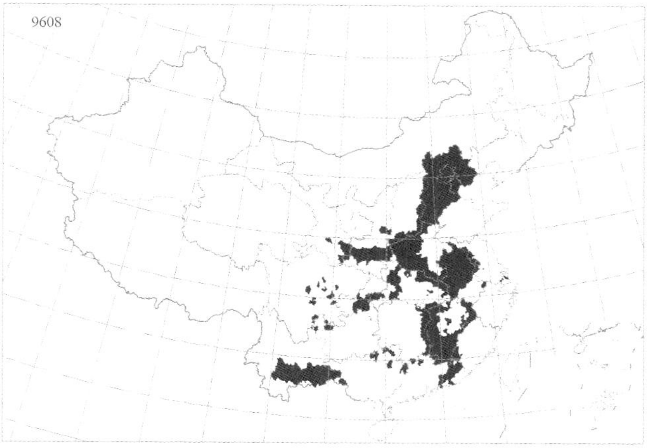

Scrophulariaceae 玄参科 *Paulownia* 泡桐属
Paulownia tomentosa var. tsinlingensis (Pai) T. Gong 光泡桐
Trees, deciduous

13.9 (9.3~16.8); 0.4 (-2.9~6.2); 26.3 (20.1~28.1); 14.0 (9.5~16.9); 117.5 (69.8~142.6); 10.1 (0.0~20.5); 813 (603~902) ❙ 776 (558~1418); 391 (290~816); 44 (19~109) ❙ -3.9 (-36.5~101.4); 715 (558~893); 169.0 (120.1~279.5)

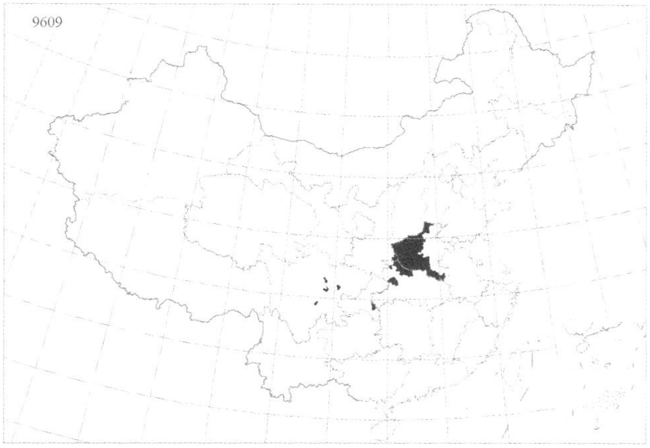

Scrophulariaceae 玄参科 *Paulownia* 泡桐属
Paulownia × taiwaniana T. W. Hu et H. J. Chang 南方泡桐
Trees, deciduous

17.4 (15.2~21.7); 8.0 (3.6~15.3); 26.0 (20.1~28.9); 17.4 (15.3~21.7); 149.4 (124.1~200.2); 0.3 (0.0~1.4); 898 (773~1118) ❙ 1752 (1337~2713); 687 (445~1388); 189 (113~245) ❙ 97.9 (51.2~246.5); 895 (773~1075); 236.3 (169.7~363.7)

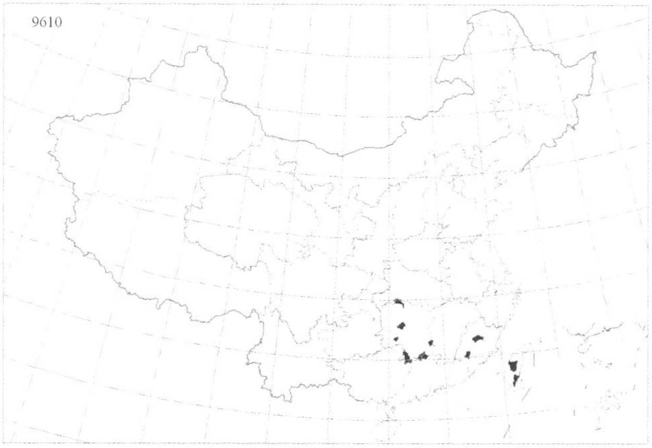

Scrophulariaceae 玄参科 *Wightia* 美丽桐属
Wightia speciosissima (D. Don) Merr. 美丽桐
Trees, deciduous

17.5 (8.6~21.0); 11.0 (1.3~15.1); 21.9 (15.0~25.6); 17.5 (8.7~21.0); 150.6 (52.6~197.8); 0.2 (0.0~8.8); 850 (574~1053) ❙ 1308 (939~1751); 723 (524~996); 53 (39~77) ❙ 54.2 (6.5~112.2); 837 (574~1053); 423.8 (277.0~659.0)

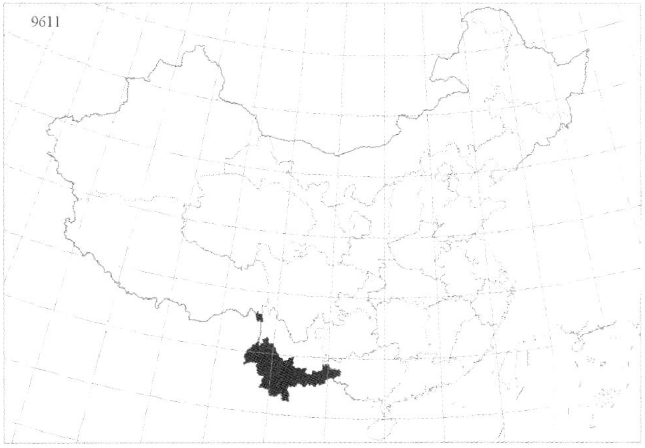

Bignoniaceae 紫葳科 *Campsis* 紫葳属
Campsis grandiflora (Thunb.) K. Schumanan 凌霄
Vines, deciduous

16.5 (2.6~24.8); 5.7 (-7.0~20.1); 26.7 (11.2~29.9); 16.6 (4.3~24.8); 140.4 (20.0~237.6); 1.3 (0.0~48.2); 889 (427~1385) ❙ 1388 (555~1997); 536 (254~845); 139 (14~259) ❙ 56.8 (-37.3~147.0); 876 (427~1360); 220.8 (108.2~620.0)

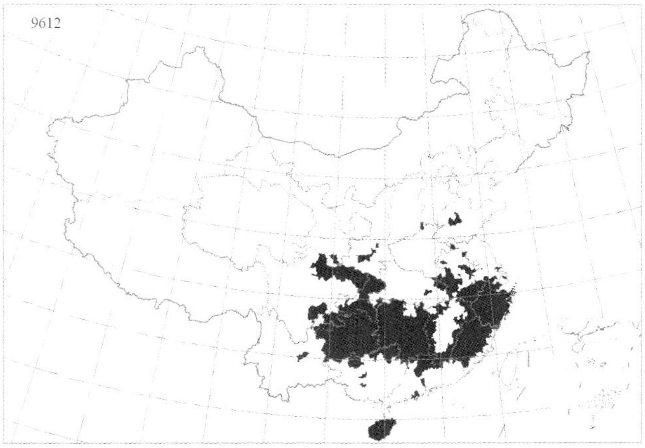

B

Bignoniaceae 紫葳科 *Catalpa* 梓属
Catalpa bungei C. A. Mey. 楸
Small trees, deciduous

14.6 (5.4~22.3); 2.2 (-11.4~13.5); 26.1 (18.1~29.5); 14.7 (8.1~22.3); 122.5 (62.7~208.1); 6.6 (0.0~57.0); 823 (570~1202) ❙ 992 (428~1776); 459 (245~817); 72 (11~234) ❙ 20.3 (-37.6~108.0); 765 (428~1202); 186.3 (110.1~316.6)

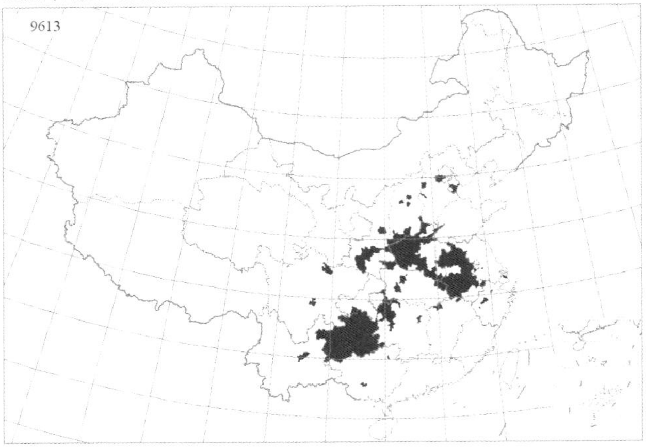

Bignoniaceae 紫葳科 *Catalpa* 梓属
Catalpa fargesii Bureau 灰楸
Trees, deciduous

13.7 (2.9~22.7); 2.5 (-9.5~15.0); 23.9 (11.2~29.2); 13.9 (4.4~22.8); 112.7 (20.5~213.1); 7.5 (0.0~47.8); 770 (428~1221) ❙ 1013 (409~2435); 484 (235~1428); 56 (8~212) ❙ 31.5 (-36.7~121.7); 734 (409~1221); 217.2 (53.3~487.2)

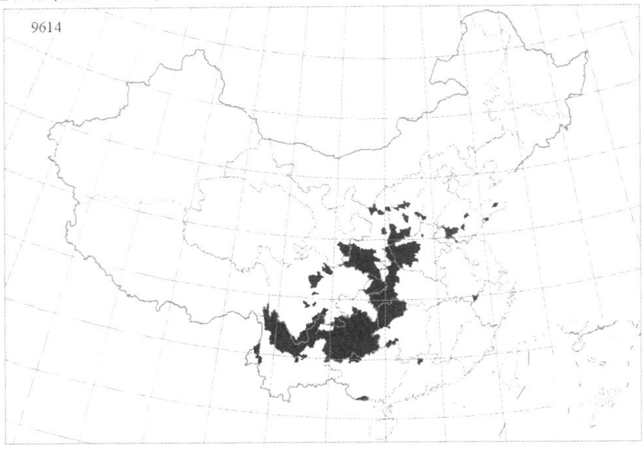

Bignoniaceae 紫葳科 *Catalpa* 梓属
Catalpa ovata G. Don 梓
Trees, deciduous

13.1 (2.2~22.1); -0.9 (-17.2~14.2); 25.7 (16.3~29.9); 13.6 (6.6~22.1); 113.5 (45.4~205.6); 16.1 (0.0~83.8); 793 (507~1182) ❙ 807 (367~1997); 423 (201~891); 49 (7~259) ❙ 1.0 (-51.1~130.7); 666 (367~1182); 179.2 (60.2~384.7)

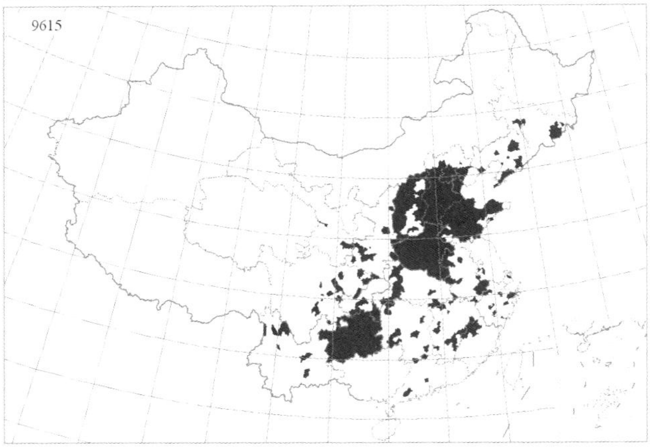

Bignoniaceae 紫葳科 *Catalpa* 梓属
Catalpa tibetica Forrest 藏楸
Shrubs or small trees, deciduous

10.1 (5.3~13.3); 2.5 (-2.4~5.4); 16.4 (12.2~19.6); 10.2 (5.7~13.4); 68.5 (30.5~100.4); 7.4 (0.0~26.2); 613 (479~706) ❙ 1227 (961~1736); 712 (558~1007); 36 (24~51) ❙ 98.8 (52.8~145.8); 613 (479~706); 343.5 (234.2~447.4)

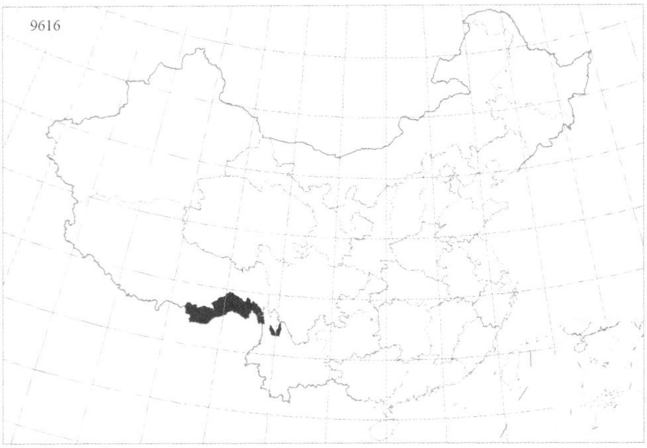

Bignoniaceae 紫葳科 *Dolichandrone* 海岸猫尾木属
Dolichandrone spathacea (L. f.) K. Schumann 海滨猫尾树
Trees

24.5; 19.2; 28.2; 24.5; 234.6; 0.0; 1362 ❙ 1661; 584; 81 ❙ 22.2; 1255; 441.2

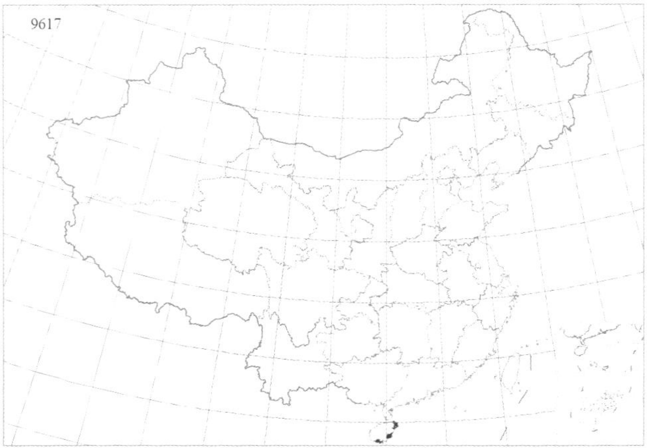

Bignoniaceae 紫葳科 *Fernandoa* 厚膜树属
Fernandoa guangxiensis D. D. Tao 广西厚膜树
Trees

21.3; 14.5; 26.0; 21.4; 196.7; 0.0; 1089 ❙ 1490; 707; 82 ❙ 39.0; 1088; 493.3

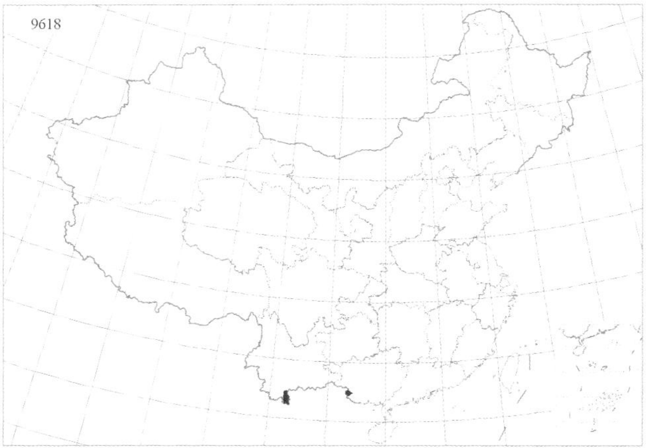

Bignoniaceae 紫葳科 *Markhamia* 猫尾树属
Markhamia stipulata (Wall.) Seem. ex K. Schum. 西南猫尾木
Trees

20.4 (16.3~24.8); 13.7 (7.4~20.1); 25.3 (20.3~28.8); 20.5 (16.4~24.8); 185.8 (136.6~237.6); 0.0 (0.0~0.0); 1048 (789~1385) | 1443 (1028~2035); 682 (392~969); 72 (41~187) | 40.5 (−11.9~85.5); 1009 (780~1326); 404.1 (203.4~659.0)

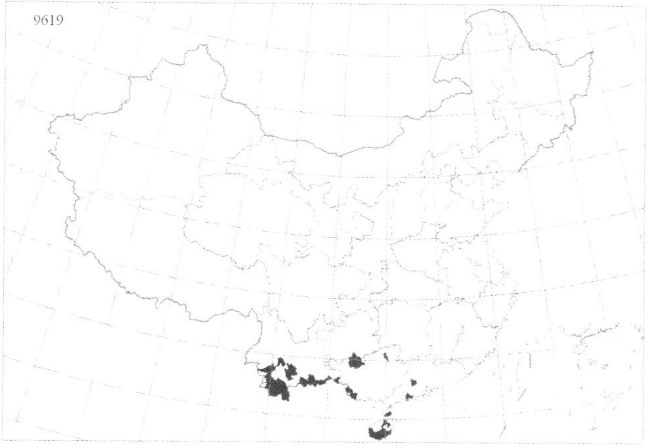

Bignoniaceae 紫葳科 *Mayodendron* 火烧花属
Mayodendron igneum (Kurz) Kurz 火烧花
Trees, evergreen

20.3 (16.9~24.8); 13.1 (8.9~20.1); 25.7 (21.4~28.9); 20.3 (16.9~24.8); 184.2 (142.7~237.6); 0.0 (0.0~0.0); 1037 (813~1385) | 1581 (939~3495); 730 (473~969); 123 (39~897) | 53.6 (1.5~224.4); 1017 (783~1212); 349.3 (145.3~659.0)

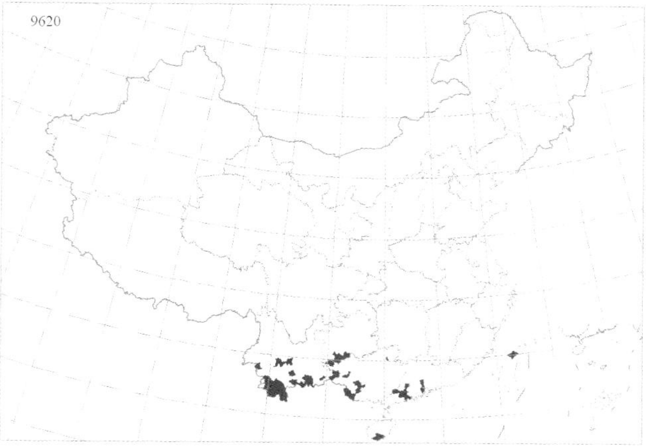

Bignoniaceae 紫葳科 *Millingtonia* 烟筒花属
Millingtonia hortensis L. 老鸦烟筒花
Trees

18.2 (15.5~20.5); 11.9 (5.7~15.1); 22.4 (20.3~24.2); 18.3 (15.6~20.5); 159.2 (126.7~185.9); 0.0 (0.0~0.0); 879 (789~992) | 1413 (1232~1607); 741 (600~835); 56 (42~77) | 60.8 (47.6~74.2); 866 (780~986); 460.8 (172.1~659.0)

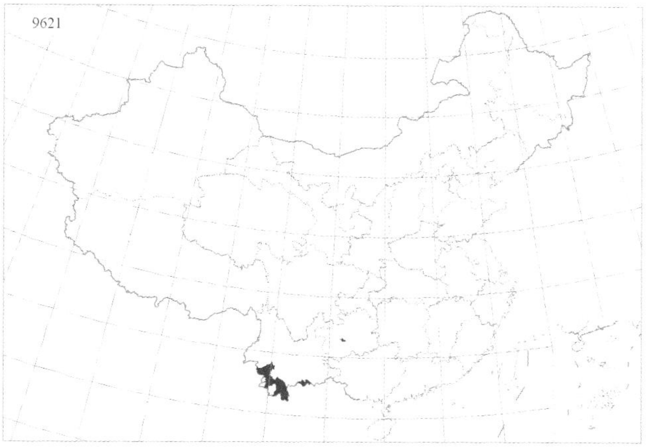

Bignoniaceae 紫葳科 *Nyctocalos* 照夜白属
Nyctocalos brunfelsiiflorum Teijsmann et Binn. 照夜白
Vines

19.1 (16.3~21.0); 12.9 (9.1~15.1); 23.1 (21.4~25.6); 19.1 (16.4~21.0); 169.4 (136.4~191.8); 0.0 (0.0~0.0); 925 (797~1053) | 1369 (1143~1607); 718 (637~835); 57 (42~77) | 48.0 (21.7~68.0); 914 (797~1053); 486.6 (346.9~659.0)

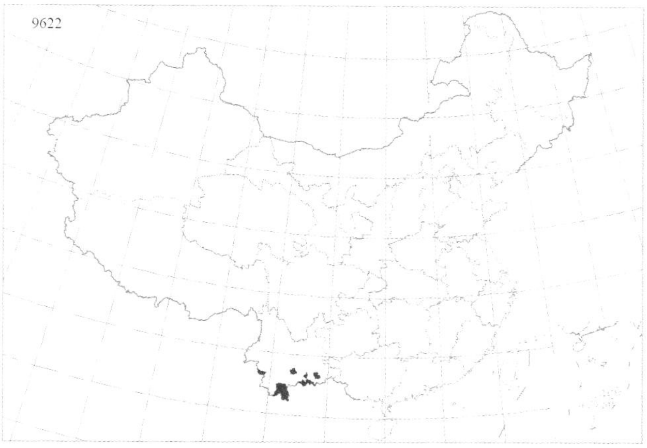

Bignoniaceae 紫葳科 *Nyctocalos* 照夜白属
Nyctocalos pinnatum Steenis 羽叶照夜白
Vines

20.2; 13.7; 24.6; 20.2; 182.5; 0.0; 996 | 1333; 727; 59 | 33.1; 978; 449.3

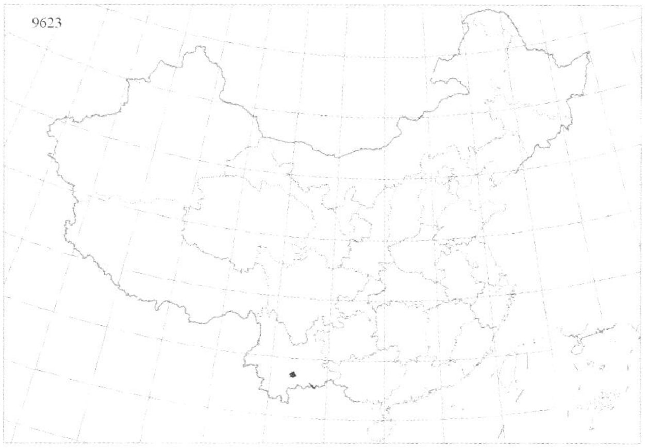

Bignoniaceae 紫葳科 *Oroxylum* 千张纸属
Oroxylum indicum (L.) Benth. ex Kurz 木蝴蝶
Small trees

17.8 (5.4~24.8); 10.0 (−2.1~20.1); 23.9 (12.3~29.0); 17.9 (5.8~24.8); 155.0 (31.5~237.6); 0.6 (0.0~25.6); 928 (481~1385) | 1312 (741~3495); 660 (411~997); 85 (15~897) | 42.4 (−14.2~224.4); 909 (481~1257); 299.4 (142.9~659.0)

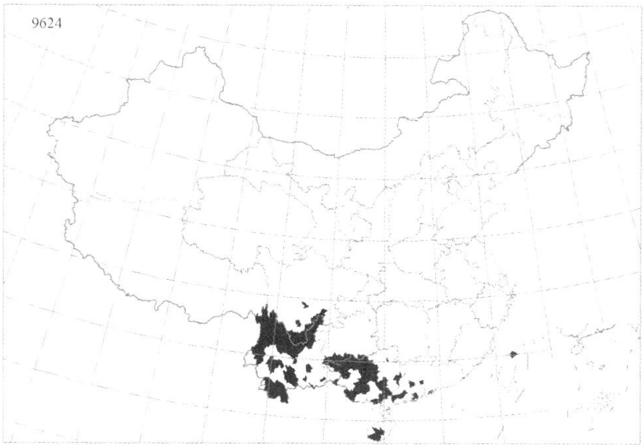

Bignoniaceae 紫葳科 *Pauldopia* 翅叶木属
Pauldopia ghorta (Buch.-Ham. ex G. Don) Steenis 翅叶木
Shrubs

18.4 (16.9~19.2)；12.4 (10.3~13.8)；22.4 (21.4~23.5)；18.5 (16.9~19.4)；161.9 (142.7~172.6)；0.0 (0.0~0.0)；883 (813~932) ❙ 1355 (1163~1499)；739 (641~835)；52 (42~62) ❙ 53.4 (34.8~68.0)；871 (813~923)；464.0 (383.4~516.7)

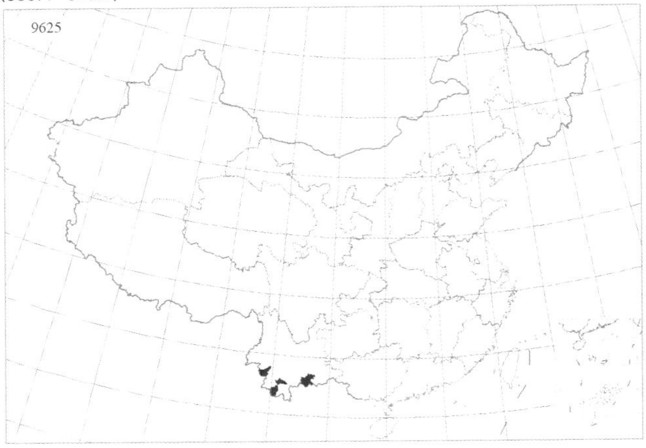

Bignoniaceae 紫葳科 *Radermachera* 菜豆树属
Radermachera frondosa Chun et F. C. How 美叶菜豆树
Trees, evergreen

22.5 (21.3~24.8)；14.9 (12.4~20.1)；28.2 (26.7~29.0)；22.5 (21.3~24.8)；210.4 (195.7~237.6)；0.0 (0.0~0.0)；1207 (1111~1385) ❙ 1612 (1269~2435)；760 (455~1428)；110 (41~171) ❙ 34.1 (-1.0~99.5)；1179 (1111~1360)；308.4 (199.4~620.0)

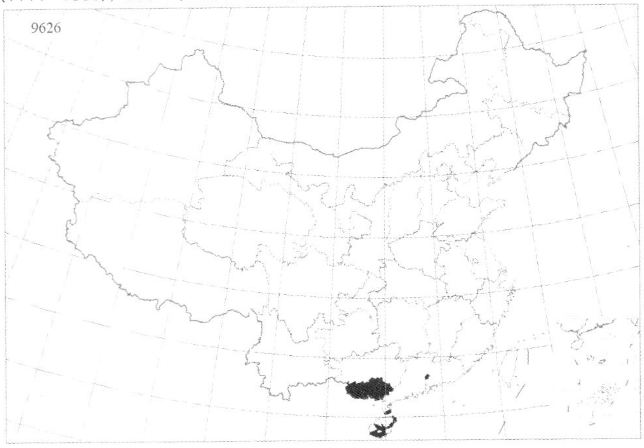

Bignoniaceae 紫葳科 *Radermachera* 菜豆树属
Radermachera glandulosa (Blume) Miq. 广西菜豆树
Small trees, evergreen

22.2；13.9；28.3；22.3；207.5；0.0；1190 ❙ 1374；713；87 ❙ 15.4；1190；327.7

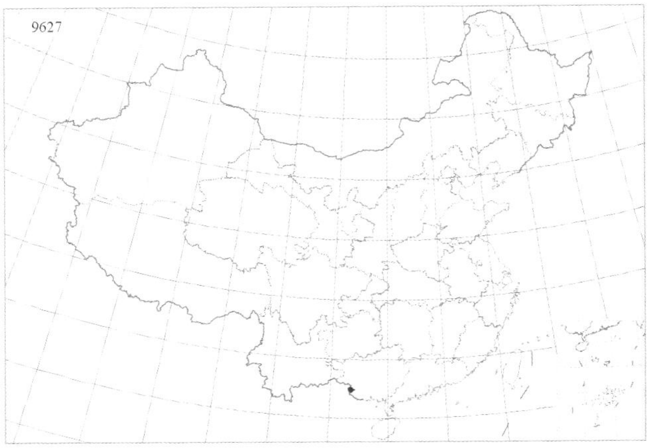

Bignoniaceae 紫葳科 *Radermachera* 菜豆树属
Radermachera hainanensis Merr. 海南菜豆树
Trees, evergreen

22.7 (18.0~24.8)；16.4 (8.5~20.1)；27.3 (23.7~28.7)；22.8 (18.0~24.8)；213.1 (156.4~237.6)；0.0 (0.0~0.0)；1215 (913~1385) ❙ 1450 (868~2035)；593 (312~969)；72 (41~169) ❙ 21.8 (-34.3~71.8)；1097 (868~1299)；424.9 (237.9~620.0)

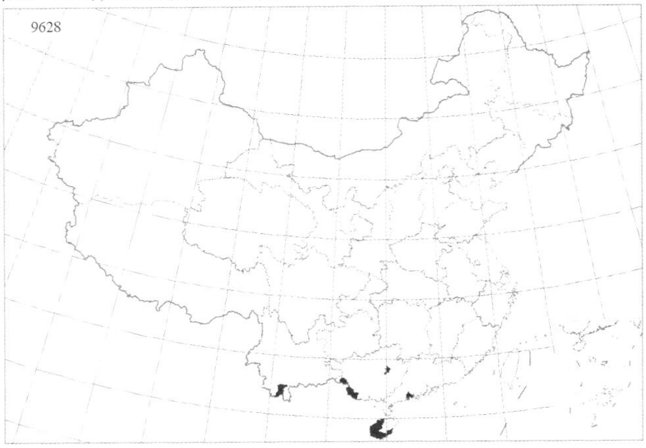

Bignoniaceae 紫葳科 *Radermachera* 菜豆树属
Radermachera microcalyx C. Y. Wu 小萼菜豆树
Trees, evergreen

19.6 (16.9~21.6)；12.6 (10.3~15.1)；24.6 (21.4~28.5)；19.7 (16.9~21.7)；176.2 (142.7~199.8)；0.0 (0.0~0.0)；978 (813~1153) ❙ 1484 (1178~1703)；760 (646~911)；89 (51~194) ❙ 52.7 (14.8~74.2)；971 (813~1153)；417.3 (213.5~659.0)

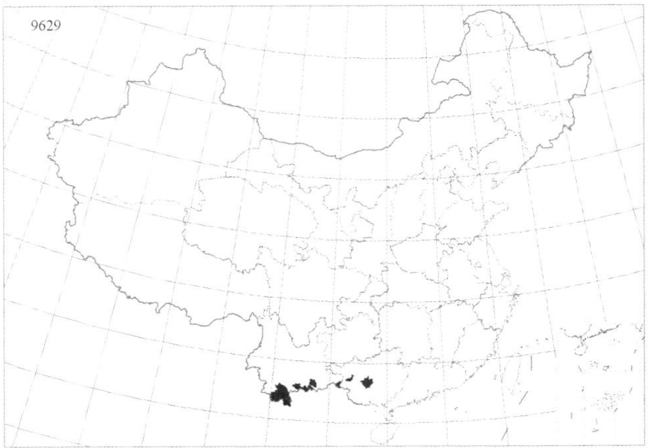

Bignoniaceae 紫葳科 *Radermachera* 菜豆树属
Radermachera pentandra Hemsl. 豇豆树
Trees, evergreen

17.7 (16.9~19.1)；10.7 (9.5~13.0)；22.6 (21.4~23.3)；17.8 (16.9~19.1)；153.1 (142.7~169.7)；0.0 (0.0~0.0)；856 (813~917) ❙ 1330 (1178~1428)；750 (646~835)；53 (48~59) ❙ 55.2 (44.9~65.6)；856 (813~917)；415.4 (336.3~506.2)

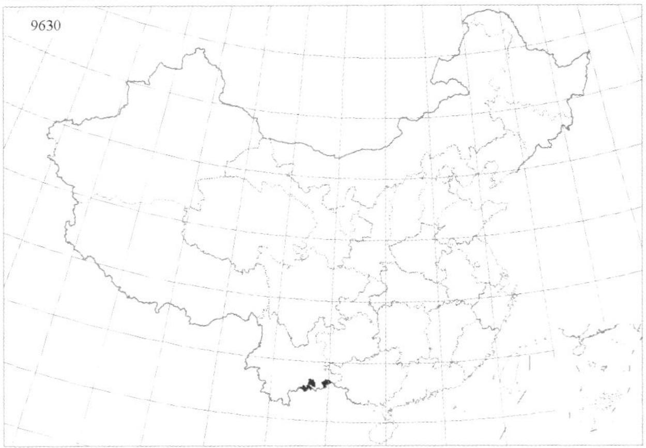

Bignoniaceae 紫葳科 *Radermachera* 菜豆树属
Radermachera sinica (Hance) Hemsl. 菜豆树
Small trees, evergreen

20.3 (13.2~24.8); 11.8 (6.1~20.1); 27.1 (18.4~29.0); 20.3 (13.3~24.8); 184.1 (99.4~237.6); 0.0 (0.0~0.0); 1064 (695~1385) ❙ 1622 (1049~3495); 699 (411~1847); 145 (41~897) ❙ 54.0 (-1.4~224.4); 1051 (695~1360); 281.8 (157.7~620.0)

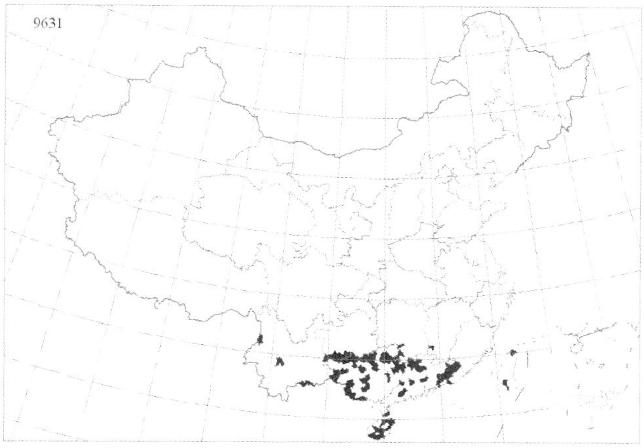

Bignoniaceae 紫葳科 *Radermachera* 菜豆树属
Radermachera yunnanensis C. Y. Wu et W. C. Yin 滇菜豆树
Small trees, evergreen

15.9 (11.3~18.9); 8.8 (4.3~11.0); 21.1 (17.0~25.0); 16.0 (11.3~19.0); 131.5 (76.6~167.5); 0.1 (0.0~0.7); 794 (644~939) ❙ 1302 (1049~1736); 733 (598~1007); 53 (44~69) ❙ 66.5 (26.1~145.8); 788 (644~939); 391.2 (336.3~491.5)

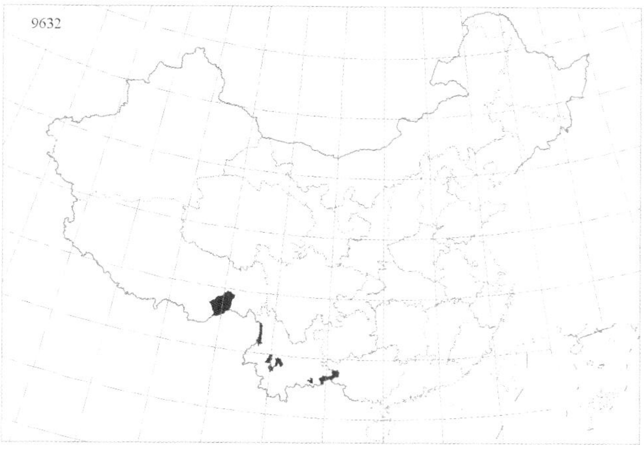

Bignoniaceae 紫葳科 *Stereospermum* 羽叶楸属
Stereospermum colais (Buch.-Ham. ex Dillwyn) Mabb. 羽叶楸
Trees, deciduous

20.5 (14.4~24.8); 13.4 (5.5~20.1); 25.6 (19.1~28.9); 20.6 (14.4~24.8); 186.7 (113.1~237.6); 0.0 (0.0~0.0); 1063 (727~1385) ❙ 1406 (868~1854); 663 (312~911); 72 (39~194) ❙ 35.5 (-34.3~99.9); 1021 (727~1360); 383.3 (189.4~659.0)

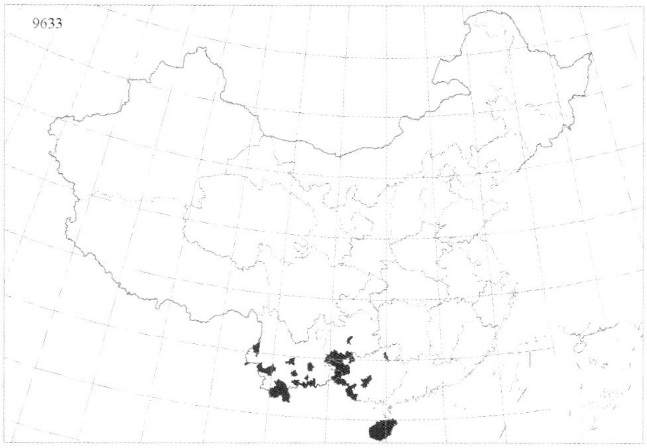

Bignoniaceae 紫葳科 *Stereospermum* 羽叶楸属
Stereospermum neuranthum Kurz 毛叶羽叶楸
Small trees, deciduous

18.1; 12.1; 21.8; 18.1; 157.6; 0.0; 870 ❙ 1403; 751; 54 ❙ 61.2; 862; 495.6

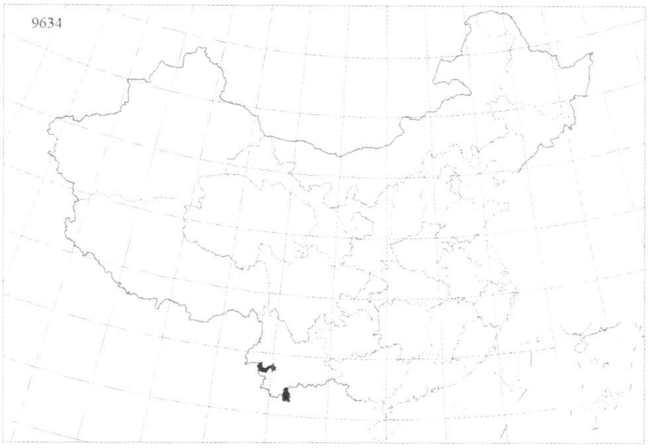

Bignoniaceae 紫葳科 *Stereospermum* 羽叶楸属
Stereospermum strigillosum C. Y. Wu et W. C. Yin 伏毛萼羽叶楸
Small trees, deciduous

19.4; 12.6; 24.1; 19.4; 172.6; 0.0; 954 ❙ 1445; 787; 64 ❙ 52.3; 954; 521.5

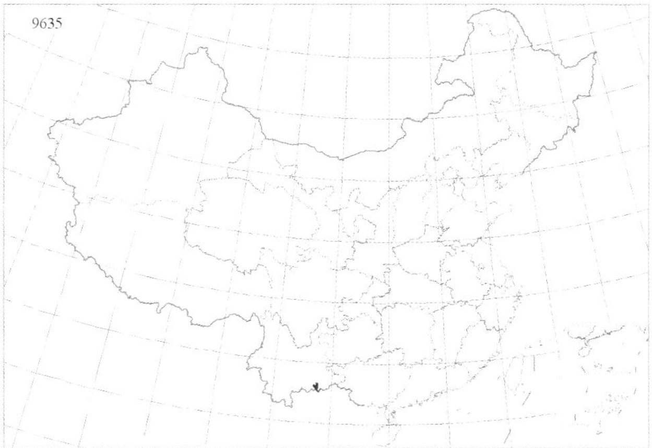

Gesneriaceae 苦苣苔科 *Aeschynanthus* 芒毛苣苔属
Aeschynanthus acuminatissimus W. T. Wang 长尖芒毛苣苔
Scandent shrubs

18.4 (17.2~21.0); 11.5 (9.5~14.2); 23.2 (22.4~25.6); 18.4 (17.3~21.0); 160.8 (147.0~191.8); 0.0 (0.0~0.0); 892 (833~1053) ❙ 1435 (1254~1547); 813 (699~911); 57 (48~68) ❙ 61.5 (44.5~80.6); 892 (833~1053); 458.6 (336.3~551.6)

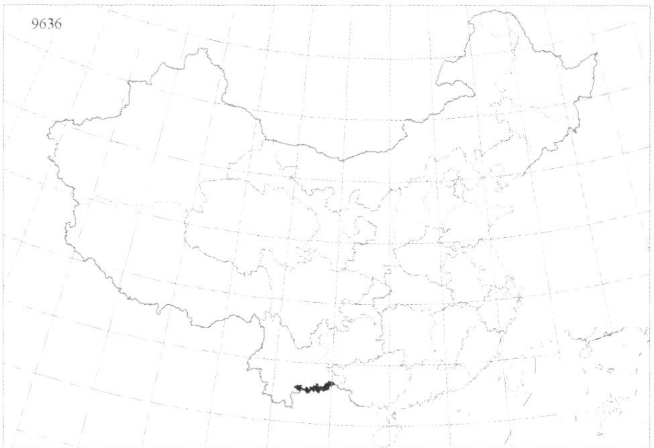

Gesneriaceae 苦苣苔科 *Aeschynanthus* 芒毛苣苔属
Aeschynanthus acuminatus Wall. ex A. DC. 芒毛苣苔
Small shrubs

19.8 (13.3~24.1); 12.3 (5.3~18.9); 25.8 (19.6~29.2); 19.9 (13.4~24.1); 178.8 (100.4~229.7); 0.0 (0.0~0.0); 1014 (706~1322) ❙ 1787 (1097~2993); 851 (479~1847); 141 (47~551) ❙ 79.8 (11.9~246.5); 1008 (706~1273); 295.3 (147.7~659.0)

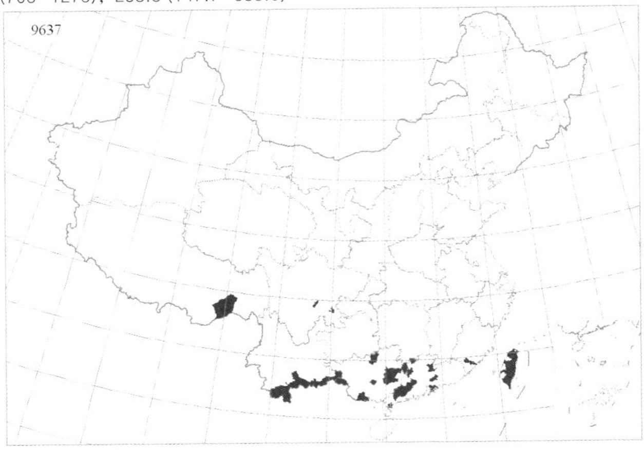

Gesneriaceae 苦苣苔科 *Aeschynanthus* 芒毛苣苔属
Aeschynanthus andersonii C. B. Clarke 轮叶芒毛苣苔
Subshrubs

17.9; 11.9; 21.6; 18.0; 155.8; 0.0; 854 ❙ 1435; 803; 42 ❙ 68.0; 833; 505.1

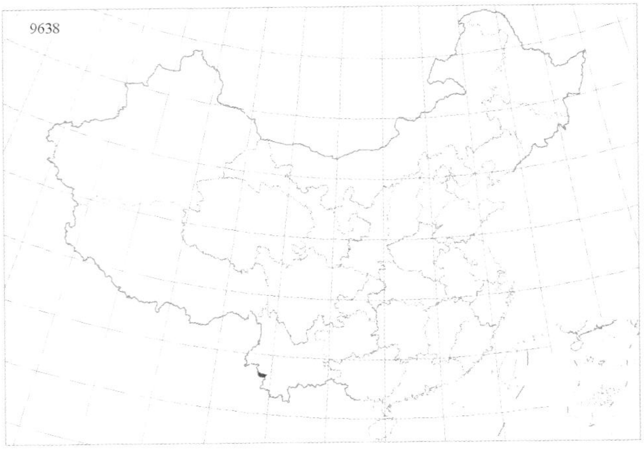

Gesneriaceae 苦苣苔科 *Aeschynanthus* 芒毛苣苔属
Aeschynanthus angustioblongus W. T. Wang 狭矩芒毛苣苔
Small shrubs

8.6; 1.3; 15.0; 8.7; 52.6; 8.8; 574 ❙ 1041; 585; 45 ❙ 81.5; 574; 352.3

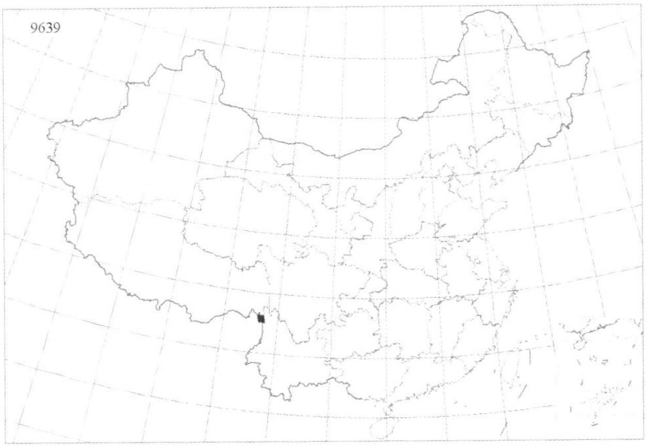

Gesneriaceae 苦苣苔科 *Aeschynanthus* 芒毛苣苔属
Aeschynanthus angustissimus W. T. Wang 狭叶芒毛苣苔
Small shrubs

13.3; 5.4; 19.6; 13.4; 100.4; 0.0; 706 ❙ 1736; 1007; 51 ❙ 145.8; 706; 447.4

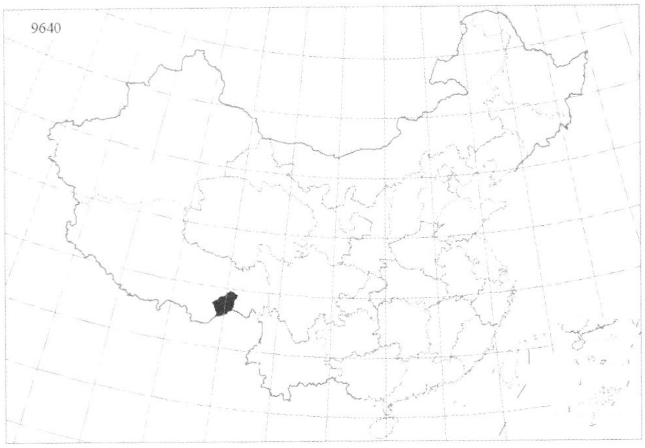

Gesneriaceae 苦苣苔科 *Aeschynanthus* 芒毛苣苔属
Aeschynanthus austroyunnanensis W. T. Wang 滇南芒毛苣苔
Small shrubs

19.0 (16.8~22.2); 11.7 (7.8~15.1); 24.3 (21.6~28.3); 19.1 (16.8~22.3); 168.6 (141.5~207.5); 0.0 (0.0~0.0); 940 (832~1190) ❙ 1373 (1028~1751); 741 (566~996); 67 (43~194) ❙ 46.8 (12.5~88.8); 930 (783~1190); 378.7 (189.5~659.0)

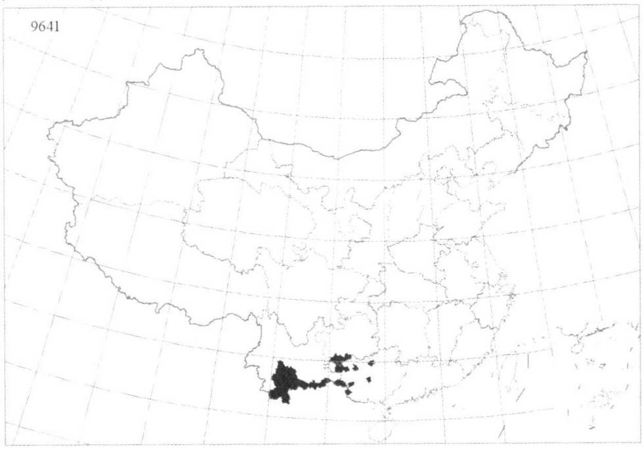

Gesneriaceae 苦苣苔科 *Aeschynanthus* 芒毛苣苔属
Aeschynanthus austroyunnanensis var. guangxiensis (W. Y. Chun ex W. T. Wang) W. T. Wang 广西芒毛苣苔
Small shrubs

19.8 (18.7~22.2); 10.9 (8.8~13.9); 26.7 (25.0~28.3); 19.9 (18.7~22.3); 178.3 (164.2~207.5); 0.0 (0.0~0.0); 1020 (945~1190) ❙ 1402 (1250~1495); 732 (667~848); 88 (54~135) ❙ 38.1 (15.4~55.9); 1020 (945~1190); 296.1 (238.3~371.7)

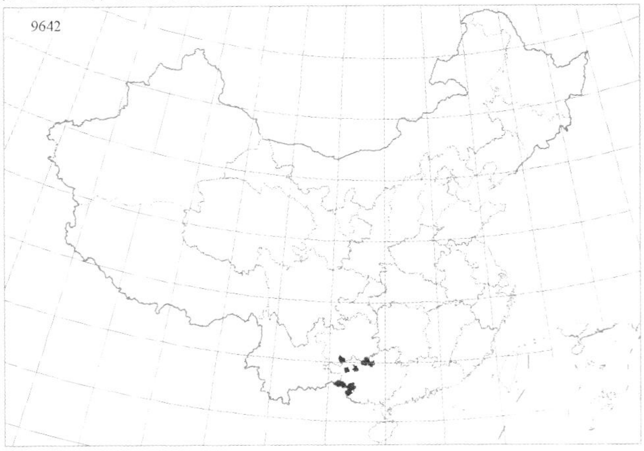

Gesneriaceae 苦苣苔科 *Aeschynanthus* 芒毛苣苔属
Aeschynanthus bracteatus Wall. ex A. DC. 显苞芒毛苣苔
Small shrubs

15.7 (5.3~21.0); 9.0 (-2.4~15.1); 20.5 (12.2~25.8); 15.7 (5.7~21.0); 130.0 (30.5~191.8); 1.5 (0.0~26.2); 790 (479~1053) ▎1224 (741~1751); 683 (426~1007); 53 (15~77) ▎56.0 (12.5~145.8); 778 (479~1053); 391.3 (234.2~659.0)

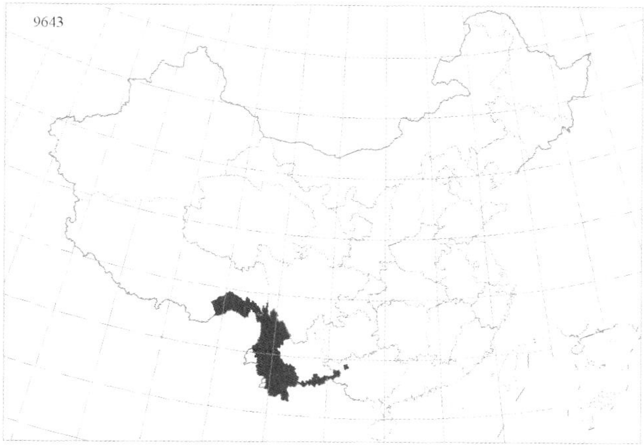

Gesneriaceae 苦苣苔科 *Aeschynanthus* 芒毛苣苔属
Aeschynanthus buxifolius Hemsl. 黄杨叶芒毛苣苔
Small shrubs

18.3 (14.6~21.3); 9.7 (3.8~14.2); 25.2 (21.4~28.6); 18.4 (14.7~21.3); 160.6 (117.7~196.1); 0.0 (0.0~1.5); 930 (779~1130) ▎1412 (1049~1758); 694 (544~911); 100 (44~191) ▎52.6 (22.7~101.8); 928 (779~1130); 295.4 (185.5~551.6)

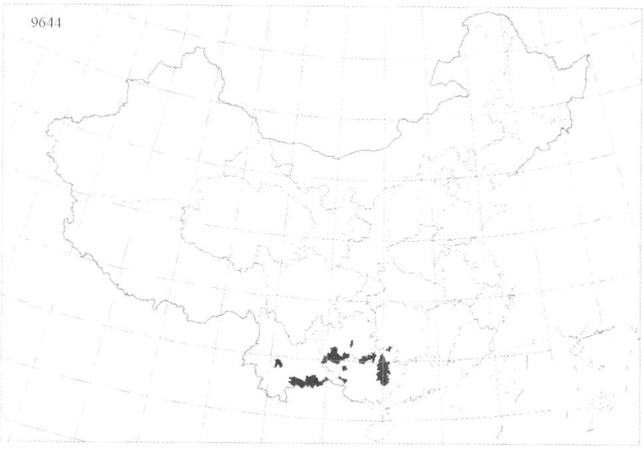

Gesneriaceae 苦苣苔科 *Aeschynanthus* 芒毛苣苔属
Aeschynanthus denticuliger W. T. Wang 小齿芒毛苣苔
Subshrubs

18.3; 10.9; 23.5; 18.3; 159.6; 0.0; 896 ▎1425; 798; 56 ▎60.1; 896; 415.3

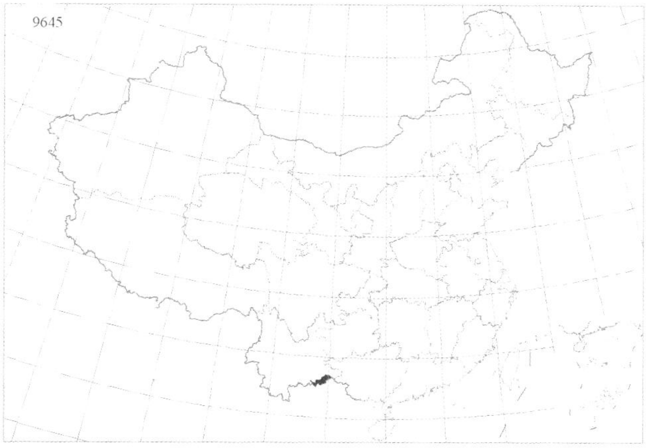

Gesneriaceae 苦苣苔科 *Aeschynanthus* 芒毛苣苔属
Aeschynanthus dolichanthus W. T. Wang 长花芒毛苣苔
Small shrubs

13.3; 5.4; 19.6; 13.4; 100.4; 0.0; 706 ▎1736; 1007; 51 ▎145.8; 706; 447.4

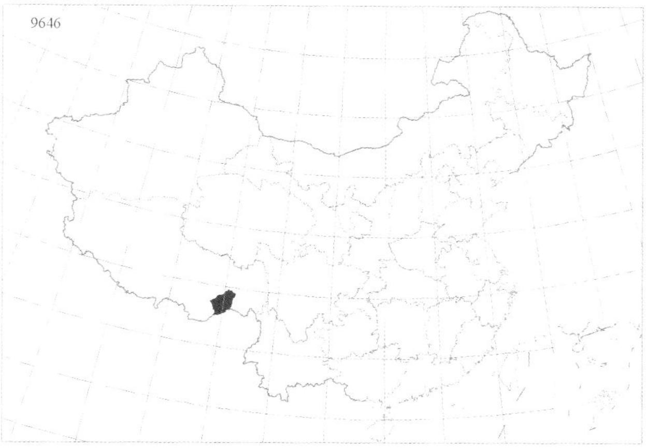

Gesneriaceae 苦苣苔科 *Aeschynanthus* 芒毛苣苔属
Aeschynanthus hookeri C. B. Clarke 束花芒毛苣苔
Small shrubs

16.5 (8.6~20.5); 10.1 (1.3~15.1); 20.9 (15.0~23.7); 16.6 (8.7~20.5); 139.1 (52.6~185.9); 0.3 (0.0~8.8); 814 (574~992) ▎1299 (986~1751); 717 (553~996); 53 (35~77) ▎59.9 (12.5~112.2); 801 (574~986); 429.6 (303.3~659.0)

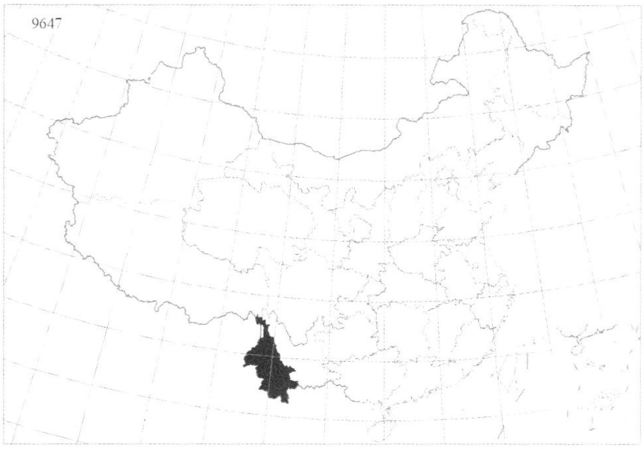

Gesneriaceae 苦苣苔科 *Aeschynanthus* 芒毛苣苔属
Aeschynanthus humilis Hemsl. 矮芒毛苣苔
Small shrubs

18.9 (17.3~21.0); 12.6 (10.9~14.2); 23.2 (21.6~25.6); 18.9 (17.3~21.0); 167.4 (148.0~191.8); 0.0 (0.0~0.0); 916 (832~1053) ▎1356 (1049~1522); 754 (598~835); 57 (44~68) ▎48.0 (26.1~64.5); 904 (783~1053); 465.7 (365.3~551.6)

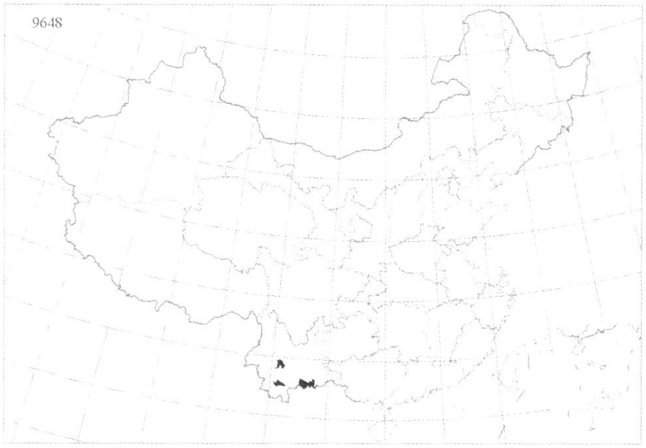

Gesneriaceae 苦苣苔科 *Aeschynanthus* 芒毛苣苔属
Aeschynanthus lancilimbus W. T. Wang 披针芒毛苣苔
Small shrubs

17.6 (16.3~21.0); 10.4 (9.1~14.2); 22.6 (21.4~25.6); 17.6 (16.4~21.0); 151.4 (136.4~191.8); 0.0 (0.0~0.0); 857 (797~1053) ❙ 1314 (1074~1522); 729 (582~860); 54 (47~68) ❙ 53.4 (32.4~80.6); 857 (797~1053); 395.0 (277.0~551.6)

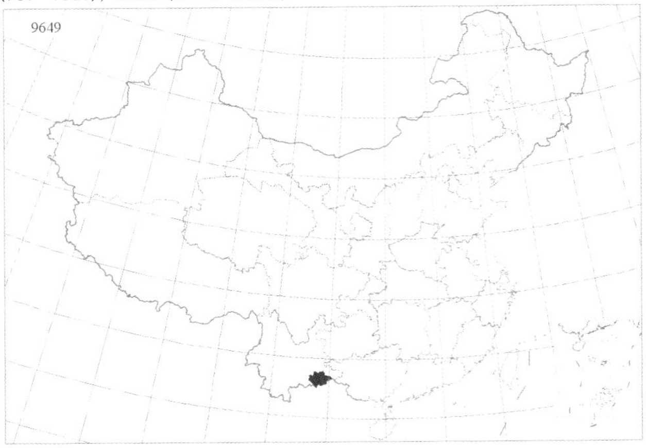

Gesneriaceae 苦苣苔科 *Aeschynanthus* 芒毛苣苔属
Aeschynanthus lasianthus W. T. Wang 毛花芒毛苣苔
Small shrubs

10.0; 2.8; 16.0; 10.0; 64.6; 4.7; 609 ❙ 1142; 640; 51 ❙ 87.3; 609; 368.5

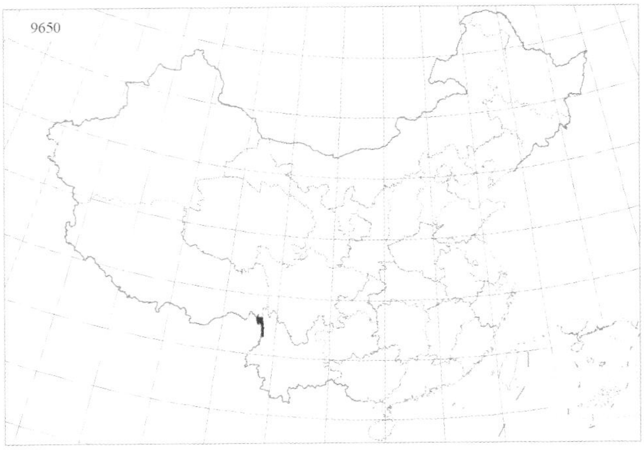

Gesneriaceae 苦苣苔科 *Aeschynanthus* 芒毛苣苔属
Aeschynanthus lasiocalyx W. T. Wang 毛萼芒毛苣苔
Small shrubs

13.3; 5.4; 19.6; 13.4; 100.4; 0.0; 706 ❙ 1736; 1007; 51 ❙ 145.8; 706; 447.4

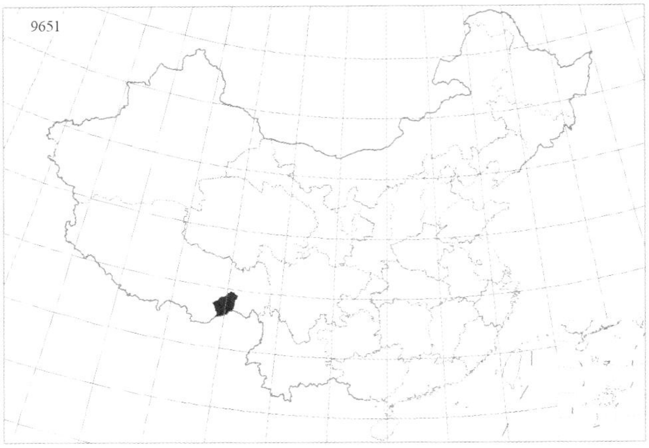

Gesneriaceae 苦苣苔科 *Aeschynanthus* 芒毛苣苔属
Aeschynanthus linearifolius C. E. C. Fisch. 华丽芒毛苣苔
Small shrubs

13.3 (5.3~16.7); 6.3 (−2.4~10.0); 18.6 (12.2~21.3); 13.4 (5.7~16.7); 103.8 (30.5~140.8); 3.2 (0.0~26.2); 705 (479~809) ❙ 1242 (961~1736); 691 (558~1007); 55 (31~70) ❙ 78.5 (34.7~145.8); 705 (479~809); 381.9 (234.2~487.2)

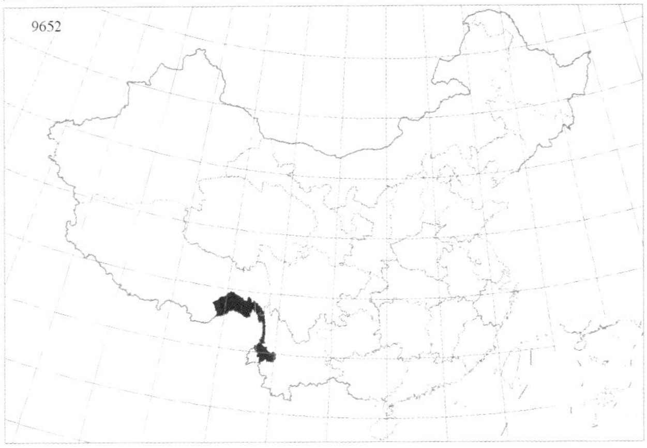

Gesneriaceae 苦苣苔科 *Aeschynanthus* 芒毛苣苔属
Aeschynanthus lineatus Craib 线条芒毛苣苔
Small shrubs

16.9 (8.6~19.5); 10.5 (1.3~13.5); 21.1 (15.0~23.3); 16.9 (8.7~19.5); 143.8 (52.6~173.8); 0.4 (0.0~8.8); 826 (574~942) ❙ 1429 (1041~1751); 810 (585~996); 55 (42~77) ❙ 73.8 (47.1~112.2); 819 (574~927); 459.4 (352.3~627.0)

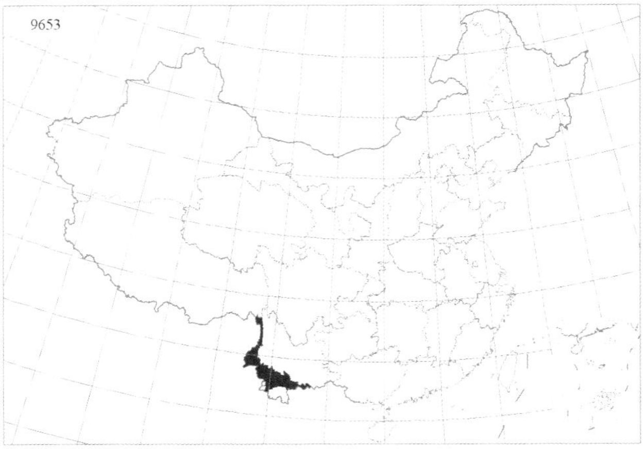

Gesneriaceae 苦苣苔科 *Aeschynanthus* 芒毛苣苔属
Aeschynanthus longicaulis Wall. ex R. Br. 长茎芒毛苣苔
Shrubs

19.4 (18.1~20.5); 13.7 (12.0~15.1); 22.9 (22.1~23.7); 19.4 (18.1~20.5); 172.7 (157.3~185.9); 0.0 (0.0~0.0); 928 (863~992) ❙ 1534 (1336~1751); 787 (655~996); 66 (55~77) ❙ 65.5 (47.6~88.8); 917 (863~986); 540.2 (432.9~659.0)

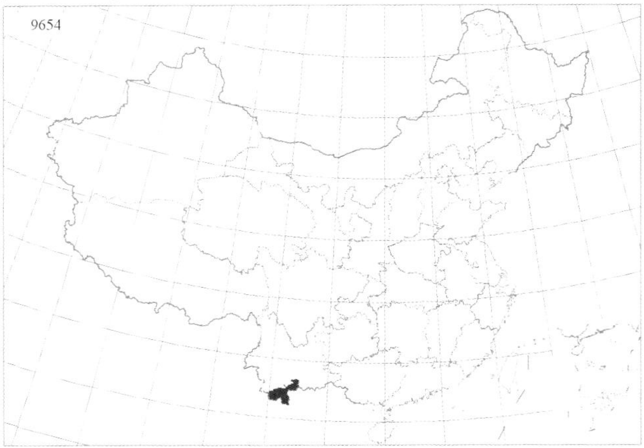

Gesneriaceae 苦苣苔科 *Aeschynanthus* 芒毛苣苔属
Aeschynanthus macranthus (Merr.) Pellegr. 伞花芒毛苣苔
Small shrubs

17.4; 9.9; 22.8; 17.5; 149.8; 0.0; 846 | 1463; 837; 54 | 73.1; 846; 386.6

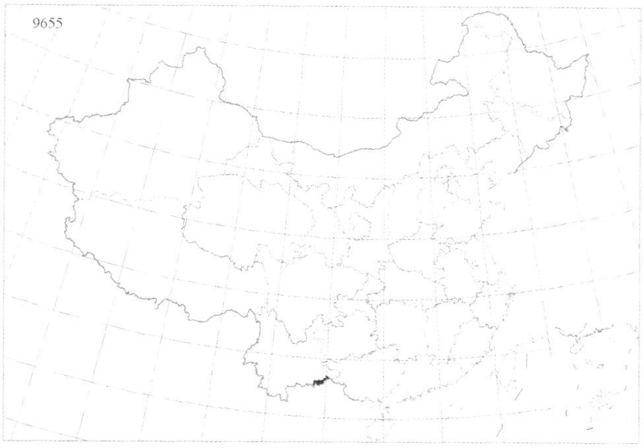

Gesneriaceae 苦苣苔科 *Aeschynanthus* 芒毛苣苔属
Aeschynanthus maculatus Lindl. 具斑芒毛苣苔
Small shrubs

7.5 (-1.8~14.4); 0.0 (-10.2~7.5); 13.9 (6.1~19.1); 8.7 (1.8~14.4); 58.4 (1.9~113.1); 27.7 (0.0~82.8); 551 (329~727) | 961 (320~1454); 547 (217~792); 45 (10~70) | 61.4 (-3.0~99.9); 550 (320~727); 294.2 (61.6~487.2)

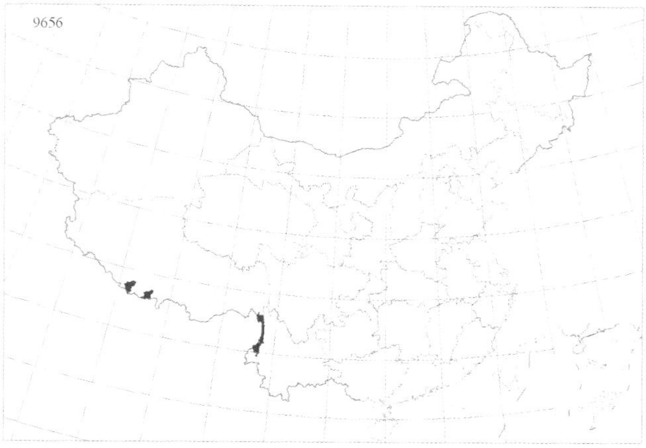

Gesneriaceae 苦苣苔科 *Aeschynanthus* 芒毛苣苔属
Aeschynanthus medogensis W. T. Wang 墨脱芒毛苣苔
Small shrubs

13.3; 5.4; 19.6; 13.4; 100.4; 0.0; 706 | 1736; 1007; 51 | 145.8; 706; 447.4

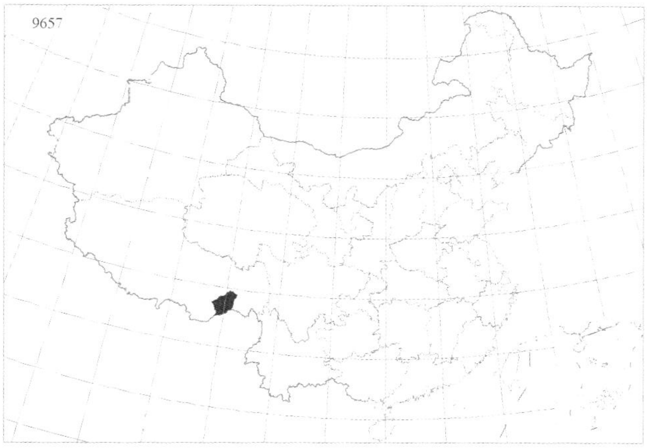

Gesneriaceae 苦苣苔科 *Aeschynanthus* 芒毛苣苔属
Aeschynanthus mengxingensis W. T. Wang 勐醒芒毛苣苔
Small shrubs

20.0; 14.6; 23.2; 20.1; 180.8; 0.0; 963 | 1523; 681; 67 | 58.4; 942; 581.0

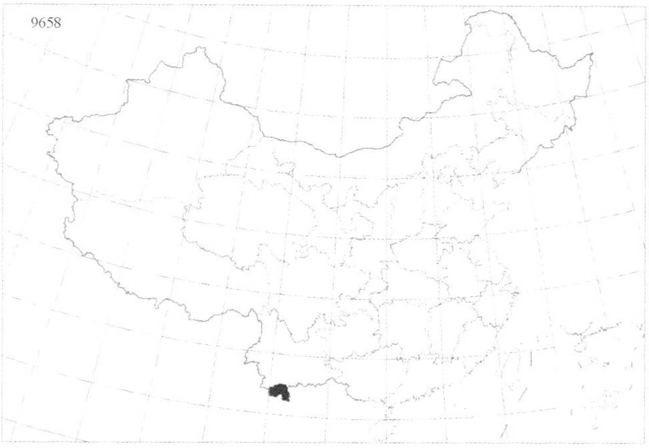

Gesneriaceae 苦苣苔科 *Aeschynanthus* 芒毛苣苔属
Aeschynanthus mimetes B. L. Burtt 大花芒毛苣苔
Small shrubs

17.3 (13.3~20.5); 11.0 (5.4~15.1); 21.4 (19.1~23.7); 17.3 (13.4~20.5); 147.8 (100.4~185.9); 0.0 (0.0~0.0); 837 (706~992) | 1332 (1049~1736); 714 (594~1007); 56 (44~77) | 59.9 (26.1~145.8); 822 (706~986); 429.8 (335.1~659.0)

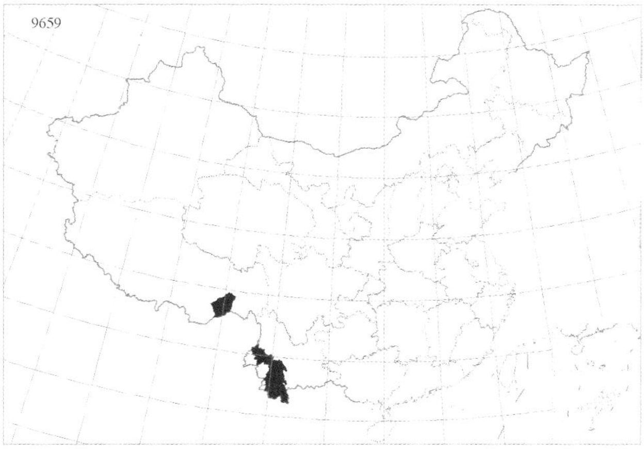

Gesneriaceae 苦苣苔科 *Aeschynanthus* 芒毛苣苔属
Aeschynanthus moningeriae (Merr.) Chun 红花芒毛苣苔
Small shrubs

23.9 (23.0~24.8); 18.6 (17.5~20.1); 27.7 (26.7~28.4); 24.0 (23.1~24.8); 227.9 (217.0~237.6); 0.0 (0.0~0.0); 1306 (1217~1385) | 1488 (1179~1725); 519 (392~619); 58 (41~110) | 14.3 (-11.9~41.8); 1160 (1056~1326); 500.2 (352.9~620.0)

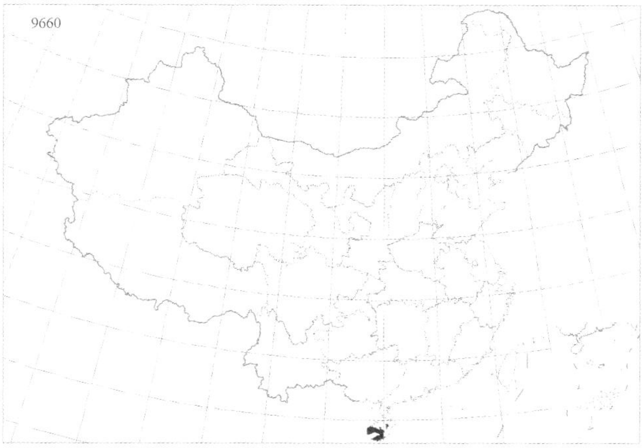

Gesneriaceae 苦苣苔科 *Aeschynanthus* 芒毛苣苔属
Aeschynanthus pachytrichus W. T. Wang 粗毛芒毛苣苔
Vines

16.4；9.6；20.6；16.4；137.3；0.0；795 ❙ 1688；959；62 ❙ 112.2；795；627.0

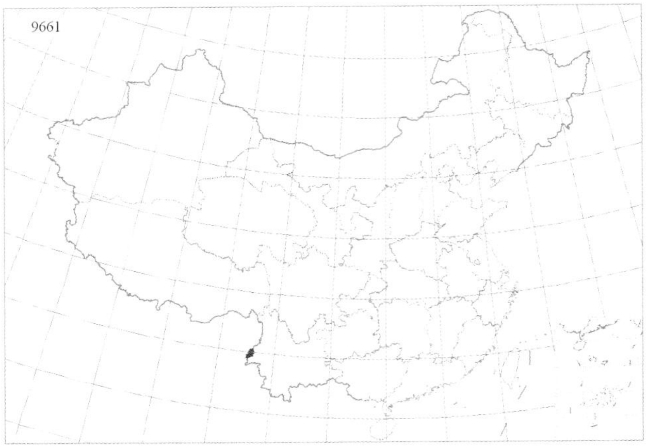

Gesneriaceae 苦苣苔科 *Aeschynanthus* 芒毛苣苔属
Aeschynanthus planipetiolatus H. W. Li 扁柄芒毛苣苔
Small shrubs

20.0；14.6；23.2；20.1；180.8；0.0；963 ❙ 1523；681；67 ❙ 58.4；942；581.0

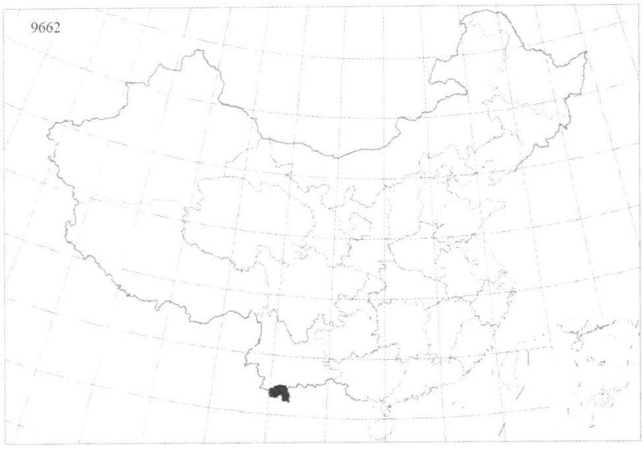

Gesneriaceae 苦苣苔科 *Aeschynanthus* 芒毛苣苔属
Aeschynanthus poilanei Pellegr. 药用芒毛苣苔
Small shrubs

20.0；13.6；24.4；20.1；180.7；0.0；985 ❙ 1475；827；62 ❙ 50.1；985；528.9

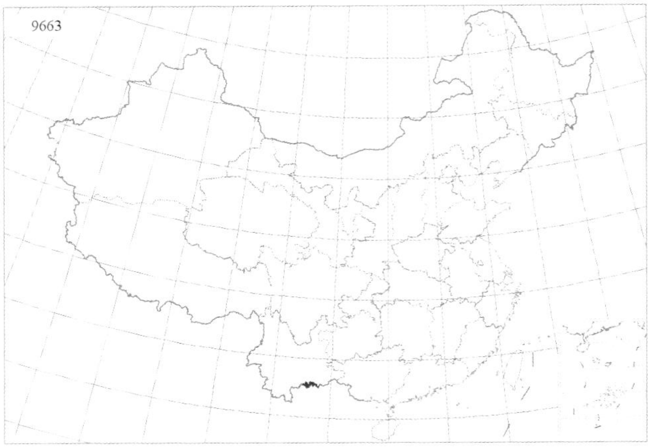

Gesneriaceae 苦苣苔科 *Aeschynanthus* 芒毛苣苔属
Aeschynanthus sinolongicalyx W. T. Wang 长萼芒毛苣苔
Small shrubs

19.4；12.6；24.1；19.4；172.6；0.0；954 ❙ 1445；787；64 ❙ 52.3；954；521.5

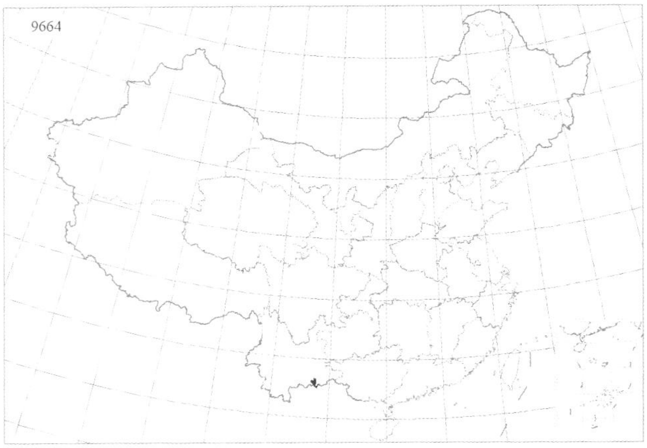

Gesneriaceae 苦苣苔科 *Aeschynanthus* 芒毛苣苔属
Aeschynanthus stenosepalus Anthony 尾叶芒毛苣苔
Small shrubs

12.8；5.7；18.3；12.8；96.3；2.4；694 ❙ 1364；773；49 ❙ 97.1；689；422.4

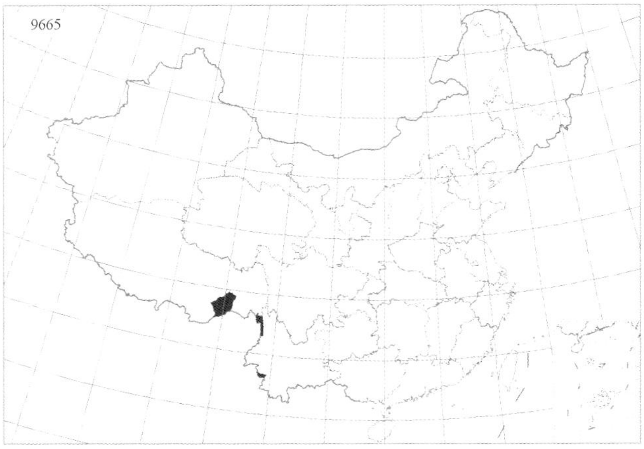

Gesneriaceae 苦苣苔科 *Aeschynanthus* 芒毛苣苔属
Aeschynanthus superbus C. B. Clarke 华丽芒毛苣苔
Small shrubs

16.9 (1.8~21.0)；10.0 (−7.1~15.1)；21.8 (9.9~25.6)；17.0 (3.8~21.0)；145.8 (15.7~191.8)；2.4 (0.0~53.3)；841 (404~1053) ❙ 1293 (583~1751)；704 (348~1007)；53 (13~77) ❙ 54.2 (19.0~145.8)；837 (404~1053)；409.1 (169.5~659.0)

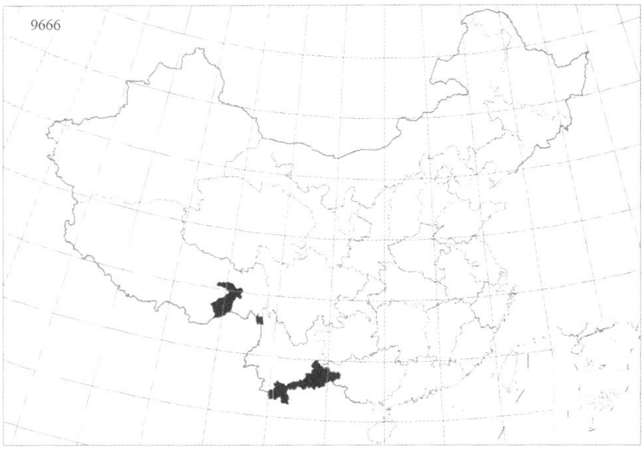

Gesneriaceae 苦苣苔科 *Aeschynanthus* 芒毛苣苔属
Aeschynanthus tengchungensis W. T. Wang 腾冲芒毛苣苔
Small shrubs

11.9; 4.9; 17.4; 11.9; 85.4; 2.4; 660 | 1269; 700; 60 | 91.7; 660; 407.8

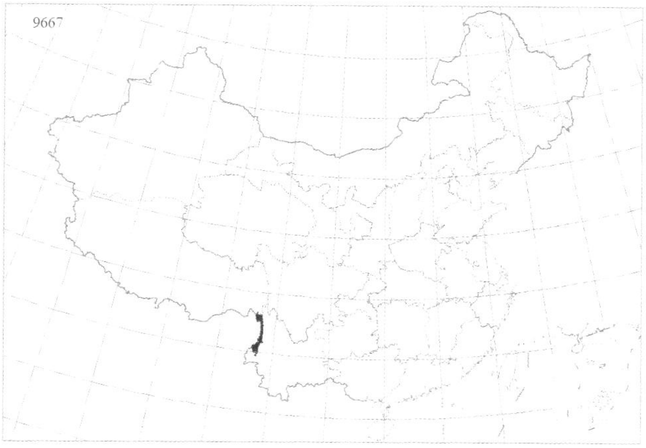

Gesneriaceae 苦苣苔科 *Aeschynanthus* 芒毛苣苔属
Aeschynanthus tubulosus J. Anthony 筒花芒毛苣苔
Small shrubs

15.4; 8.9; 19.7; 15.4; 124.9; 0.0; 758 | 1343; 748; 58 | 78.0; 754; 429.8

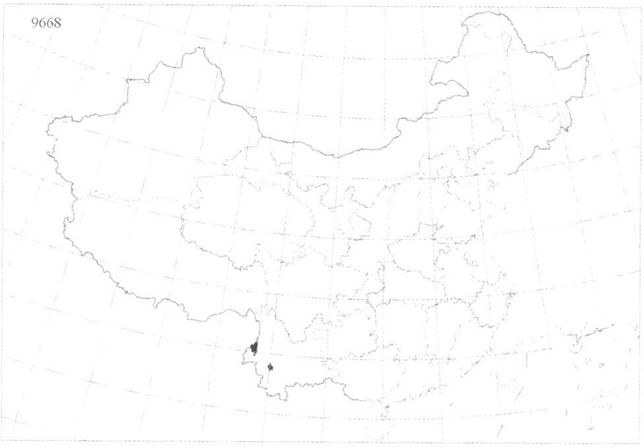

Gesneriaceae 苦苣苔科 *Aeschynanthus* 芒毛苣苔属
Aeschynanthus wardii Merr. 狭花芒毛苣苔
Small shrubs

12.4; 5.5; 17.9; 12.4; 92.4; 3.2; 683 | 1111; 626; 49 | 66.9; 667; 367.4

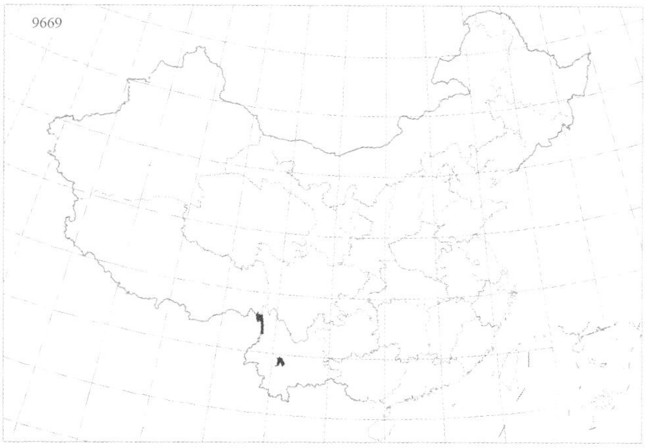

Gesneriaceae 苦苣苔科 *Ancylostemon* 直瓣苣苔属
Ancylostemon rhombifolius K. Y. Pan 菱叶直瓣苣苔
Small shrubs

9.0; 0.6; 16.5; 9.0; 58.7; 10.3; 584 | 1178; 663; 34 | 101.6; 584; 293.1

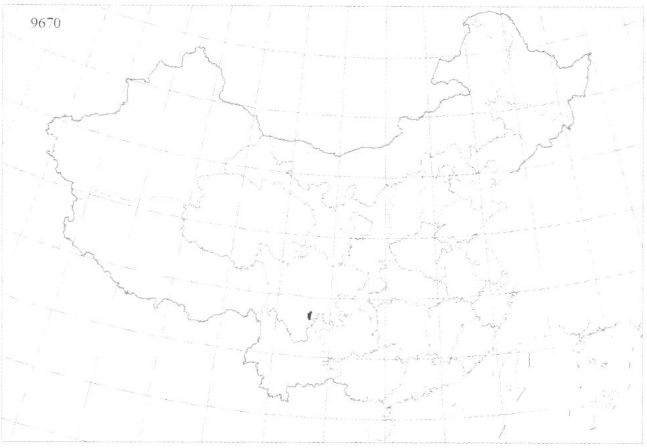

Gesneriaceae 苦苣苔科 *Ancylostemon* 直瓣苣苔属
Ancylostemon ronganensis K. Y. Pan 融安直瓣苣苔
Small shrubs

18.9; 8.5; 27.7; 18.9; 166.9; 0.0; 992 | 1620; 676; 166 | 63.3; 992; 232.4

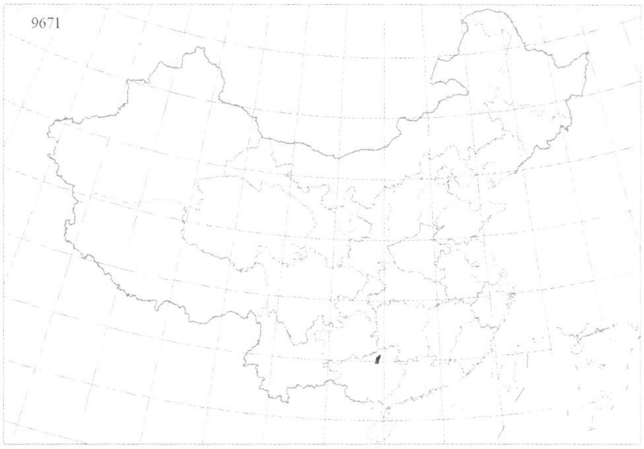

Gesneriaceae 苦苣苔科 *Anna* 大苞苣苔属
Anna mollifolia (W. T. Wang) W. T. Wang et K. Y. Pan 软叶大苞苣苔
Subshrubs

18.0; 10.1; 23.6; 18.0; 156.3; 0.0; 880 | 1371; 777; 53 | 55.8; 880; 355.9

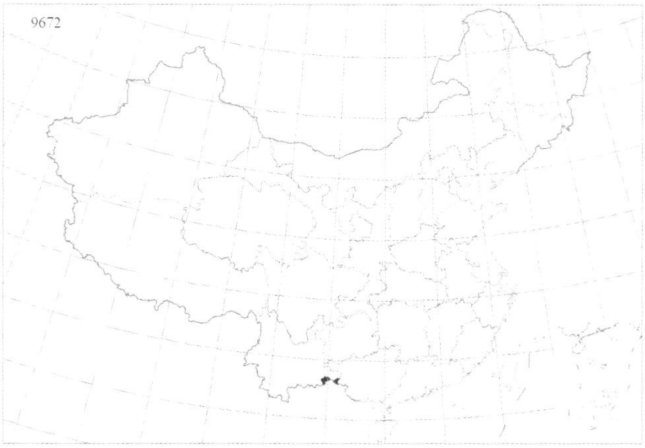

Gesneriaceae 苦苣苔科 *Anna* 大苞苣苔属
Anna ophiorrhizoides (Hemsl.) B. L. Burtt et R. Davidson 白花大苞苣苔
Shrubs or subshrubs

16.6 (13.3~19.1); 6.7 (3.2~10.1); 25.3 (22.1~27.4); 16.6 (13.2~19.1); 139.4 (100.9~169.2); 0.2 (0.0~2.5); 861 (721~969) ❙ 1192 (985~1583); 624 (499~938); 60 (32~107) ❙ 39.3 (18.1~89.0); 861 (721~969); 193.6 (135.1~283.9)

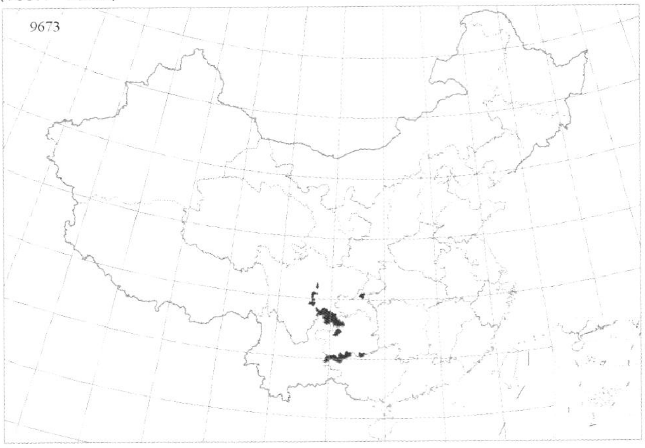

Gesneriaceae 苦苣苔科 *Anna* 大苞苣苔属
Anna submontana Pellegr. 大苞苣苔
Small shrubs

18.4 (17.2~21.0); 10.7 (9.1~14.2); 23.9 (22.4~25.6); 18.4 (17.3~21.0); 160.8 (147.0~191.8); 0.0 (0.0~0.0); 902 (833~1053) ❙ 1383 (1161~1522); 775 (633~860); 57 (48~72) ❙ 53.7 (37.1~80.6); 902 (833~1053); 388.3 (313.7~551.6)

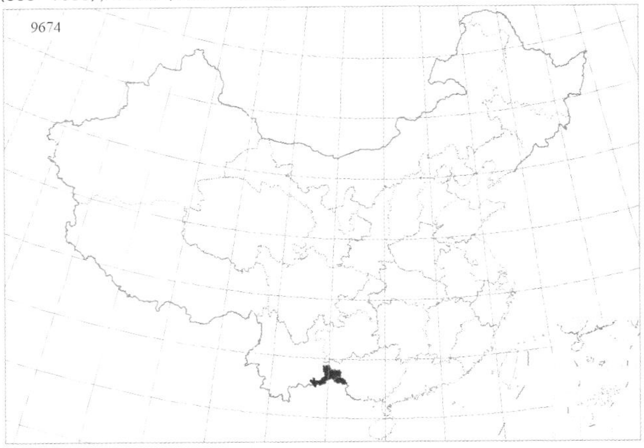

Gesneriaceae 苦苣苔科 *Beccarinda* 横蒴苣苔属
Beccarinda erythrotricha W. T. Wang 红毛横蒴苣苔
Small shrubs

18.7 (16.9~21.0); 12.0 (9.5~15.1); 23.1 (21.4~25.6); 18.7 (16.9~21.0); 164.4 (142.7~191.8); 0.0 (0.0~0.0); 905 (813~1053) ❙ 1409 (1178~1607); 749 (646~860); 57 (45~77) ❙ 55.9 (37.6~80.6); 898 (813~1053); 470.4 (336.3~659.0)

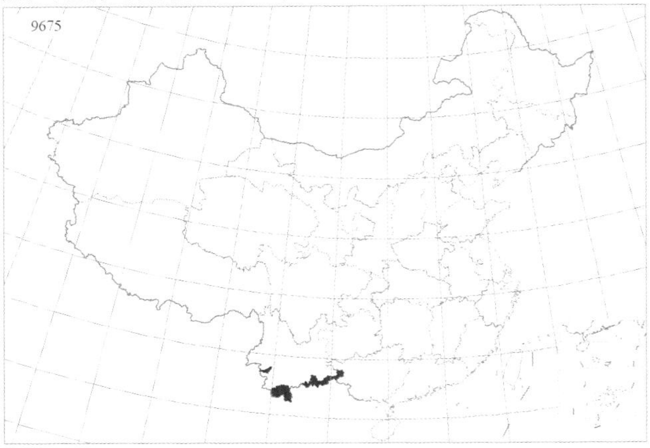

Gesneriaceae 苦苣苔科 *Boeica* 短筒苣苔属
Boeica porosa C. B. Clarke 孔药短筒苣苔
Subshrubs

20.2 (17.7~21.8); 12.8 (10.9~14.2); 25.5 (22.5~28.4); 20.2 (17.8~21.9); 182.3 (153.4~202.2); 0.0 (0.0~0.0); 1019 (854~1150) ❙ 1609 (1368~2148); 861 (662~1234); 95 (56~168) ❙ 57.6 (40.8~86.7); 1019 (854~1150); 417.2 (218.0~551.6)

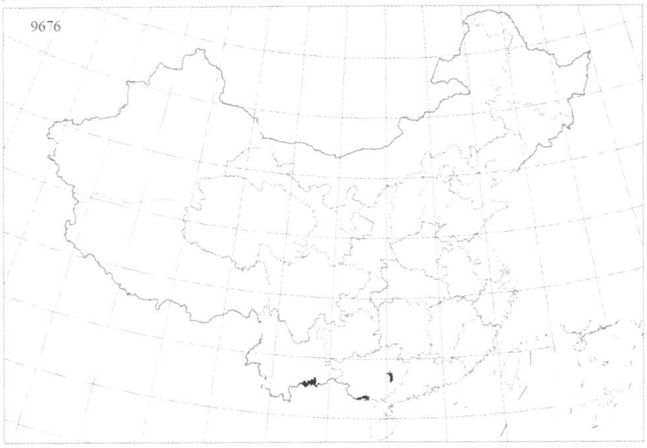

Gesneriaceae 苦苣苔科 *Boeica* 短筒苣苔属
Boeica yunnanensis (H. W. Li) K. Y. Pan 翼柱短筒苣苔
Subshrubs

17.3; 9.9; 22.4; 17.3; 147.1; 0.0; 833 ❙ 1504; 860; 55 ❙ 80.6; 833; 413.5

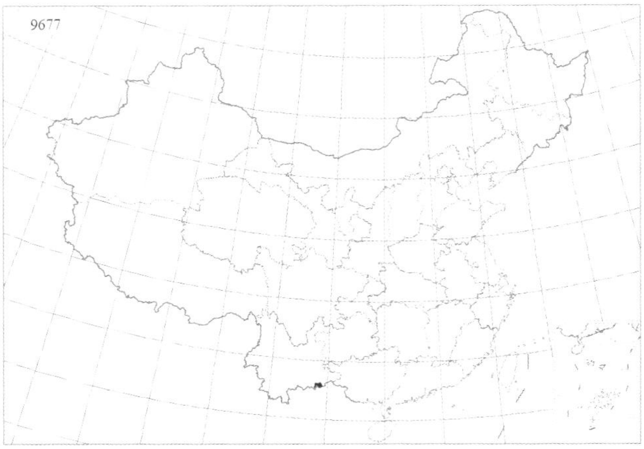

Gesneriaceae 苦苣苔科 *Cyrtandra* 浆果苣苔属
Cyrtandra umbellifera Merr. 浆果苣苔
Small shrubs

18.5; 14.1; 21.7; 18.5; 162.1; 0.0; 863 ❙ 2713; 1388; 188 ❙ 214.4; 863; 363.7

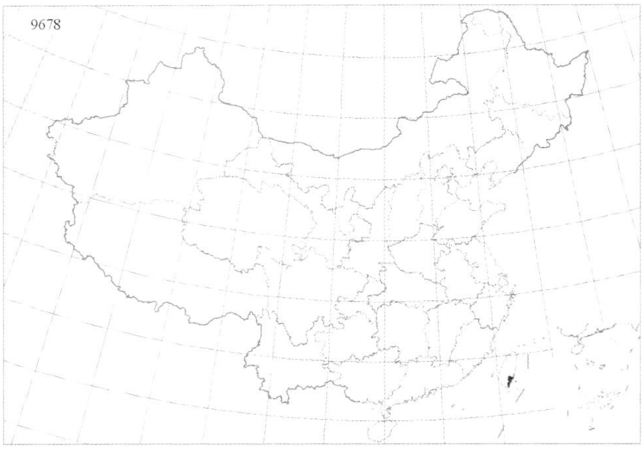

Gesneriaceae 苦苣苔科 *Paraboea* 蛛毛苣苔属
Paraboea sinensis (Oliv.) B. L. Burtt 蛛毛苣苔
Small shrubs

15.7 (8.6~22.2); 5.9 (−1.7~14.9); 24.4 (15.0~28.4); 15.8 (8.7~22.3); 130.6 (52.6~207.5); 1.4 (0.0~18.0); 824 (574~1190) ❙ 1256 (964~1581); 584 (421~911); 73 (33~135) ❙ 53.6 (12.9~107.2); 823 (574~1190); 246.3 (146.0~570.1)

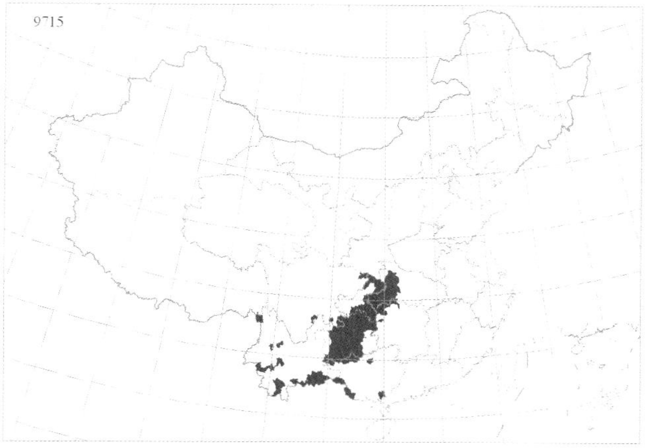

Gesneriaceae 苦苣苔科 *Paraboea* 蛛毛苣苔属
Paraboea swinhoei (Hance) B. L. Burtt 锥序蛛毛苣苔
Small shrubs

19.4 (16.2~22.6); 11.7 (6.3~17.7); 25.4 (20.1~28.3); 19.4 (16.2~22.6); 172.7 (134.2~211.5); 0.0 (0.0~0.0); 979 (773~1197) ❙ 1896 (1250~2993); 910 (611~1847); 164 (58~547) ❙ 99.4 (15.4~246.5); 974 (773~1190); 270.3 (202.8~363.7)

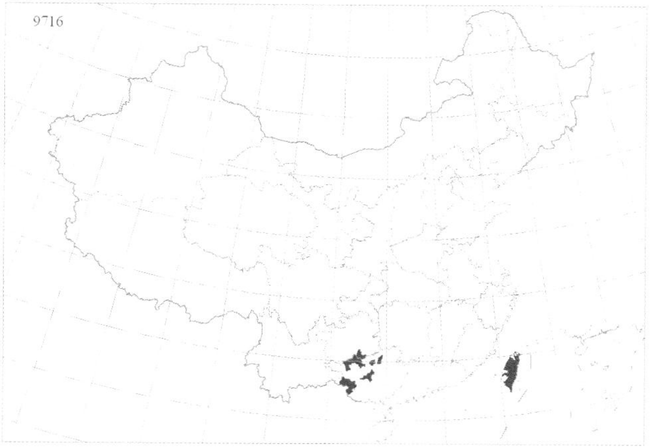

Gesneriaceae 苦苣苔科 *Rhabdothamnopsis* 长冠苣苔属
Rhabdothamnopsis sinensis Hemsl. 长冠苣苔
Small shrubs

13.1 (2.2~18 9); 5.0 (−6.6~12.0); 20.0 (9.5~26.6); 13.2 (3.8~19.0); 102.0 (15.5~167.8); 3.9 (0.0~49.1); 714 (411~969) ❙ 1032 (729~1418); 576 (411~667); 37 (11~95) ❙ 47.2 (−14.2~125.9); 696 (411~969); 265.9 (142.9~372.3)

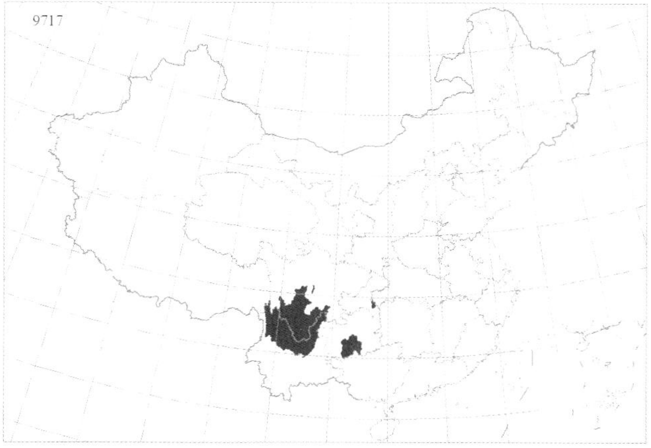

Gesneriaceae 苦苣苔科 *Rhynchotechum* 线柱苣苔属
Rhynchotechum discolor (Maxim.) B. L. Burtt 异色线柱苣苔
Subshrubs

20.3 (16.2~24.1); 13.7 (9.0~19.1); 25.9 (20.1~28.9); 20.4 (16.2~24.1); 184.6 (134.2~229.7); 0.0 (0.0~0.0); 1039 (773~1322) ❙ 2143 (1258~3495); 885 (425~1847); 228 (47~897) ❙ 114.8 (6.7~246.5); 1015 (773~1205); 282.4 (145.3~620.0)

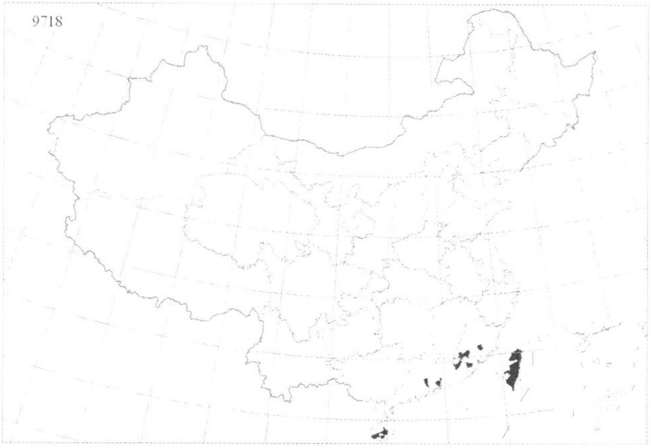

Gesneriaceae 苦苣苔科 *Rhynchotechum* 线柱苣苔属
Rhynchotechum ellipticum (Wall. ex D. Dietr.) A. DC. 线柱苣苔
Small shrubs

20.3 (13.3~24.8); 12.3 (5.2~20.1); 26.6 (19.6~29.8); 20.4 (13.4~24.8); 184.3 (100.4~237.6); 0.0 (0.0~0.0); 1057 (706~1385) ❙ 1541 (868~2435); 694 (312~1428); 111 (41~202) ❙ 47.7 (−34.3~145.8); 1031 (706~1326); 312.6 (158.7~659.0)

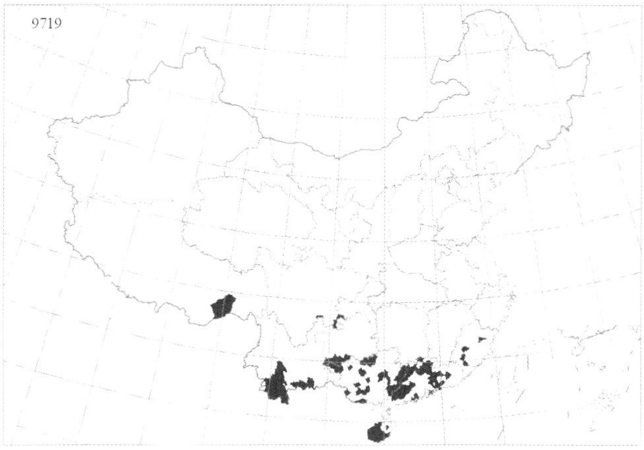

Gesneriaceae 苦苣苔科 *Rhynchotechum* 线柱苣苔属
Rhynchotechum formosanum Hatus. 冠萼线柱苣苔
Subshrubs

21.6 (17.2~24.8); 14.1 (9.5~20.1); 27.3 (22.7~28.7); 21.6 (17.3~24.8); 199.2 (147.0~237.6); 0.0 (0.0~0.0); 1136 (838~1385) ❙ 1712 (868~3495); 769 (312~1428); 149 (41~897) ❙ 53.8 (−34.3~224.4); 1089 (838~1299); 317.6 (163.7~620.0)

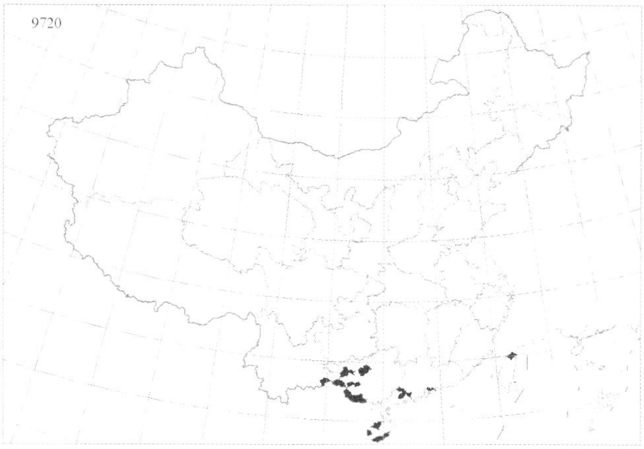

Gesneriaceae 苦苣苔科 *Rhynchotechum* 线柱苣苔属
Rhynchotechum longipes W. T. Wang 长梗线柱苣苔
Subshrubs

21.6 (21.0~22.2); 13.3 (11.8~13.9); 27.8 (27.3~28.3); 21.6 (21.1~22.3); 199.4 (192.6~207.5); 0.0 (0.0~0.0); 1140 (1111~1190) ❙ 1572 (1302~1900); 819 (670~1056); 120 (86~194) ❙ 38.5 (12.9~71.0); 1140 (1111~1190); 284.8 (213.5~327.7)

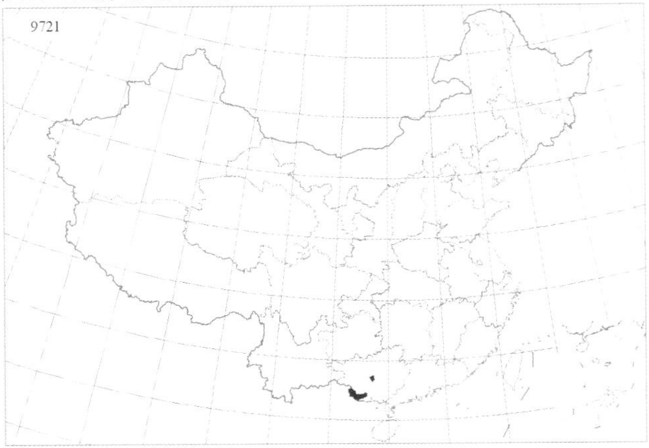

Gesneriaceae 苦苣苔科 *Rhynchotechum* 线柱苣苔属
Rhynchotechum vestitum Wall. ex C. B. Clarke 毛线柱苣苔
Subshrubs

17.1 (8.6~21.0); 10.6 (1.3~15.1); 21.5 (15.0~25.6); 17.2 (8.7~21.0); 146.2 (52.6~191.8); 0.2 (0.0~8.8); 836 (574~1053) ❙ 1389 (1041~1751); 770 (585~1007); 56 (42~77) ❙ 67.1 (26.1~145.8); 827 (574~1053); 452.1 (335.1~659.0)

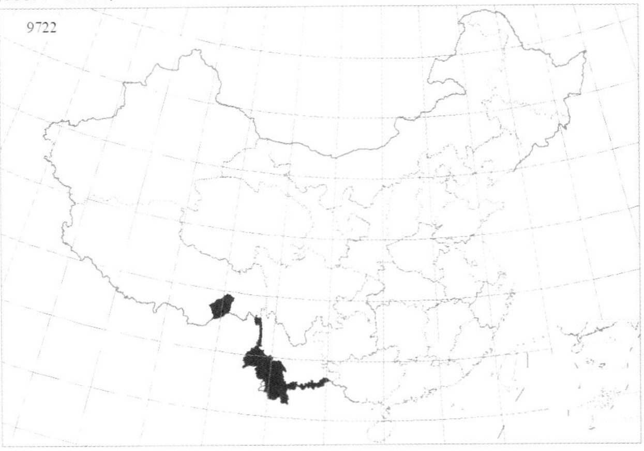

Berberidaceae 小檗科 *Berberis* 小檗属
Berberis acuminata Franchet 渐尖小檗
Shrubs, evergreen

14.8 (12.1~17.1); 4.6 (1.1~8.5); 24.2 (21.5~25.9); 14.9 (12.2~17.1); 119.9 (95.0~145.1); 1.5 (0.0~8.5); 786 (696~863) ❙ 1265 (942~1492); 577 (473~817); 78 (46~106) ❙ 61.8 (19.4~106.2); 786 (696~863); 219.9 (154.4~285.1)

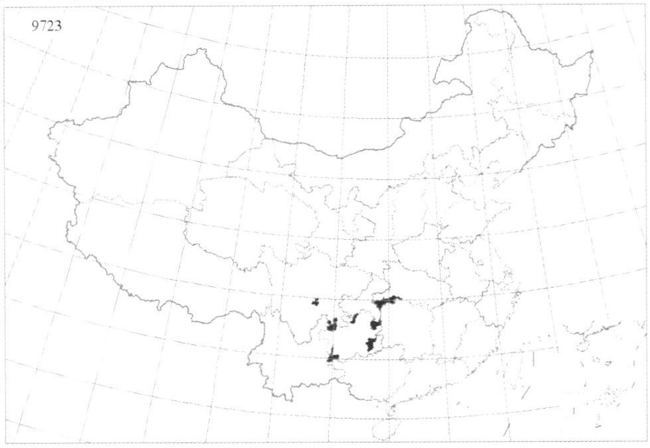

Berberidaceae 小檗科 *Berberis* 小檗属
Berberis aemulans C. K. Schneid. 峨眉小檗
Shrubs, deciduous

7.8; −1.4; 16.0; 8.7; 59.6; 25.0; 570 ❙ 1047; 612; 28 ❙ 84.0; 570; 269.6

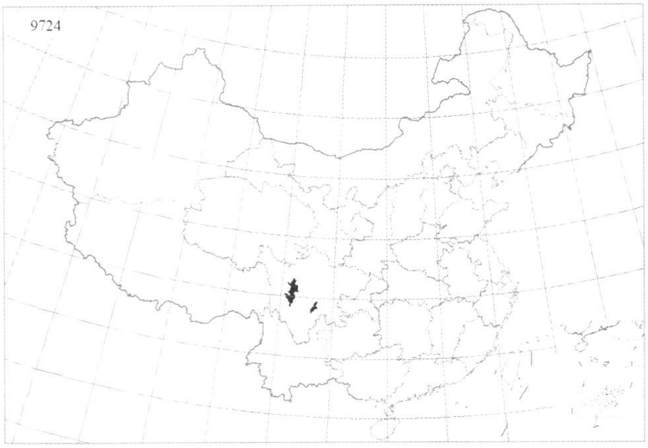

Berberidaceae 小檗科 *Berberis* 小檗属
Berberis aggregata C. K. Schneid. 堆花小檗
Shrubs, deciduous

10.2 (−1.3~15.9); −1.2 (−13.5~8.2); 20.5 (9.7~28.0); 10.7 (2.9~16.0); 82.0 (11.3~134.0); 19.6 (0.0~86.3); 657 (359~897) ❙ 834 (460~1358); 409 (249~610); 36 (8~88) ❙ 30.1 (−35.9~121.7); 623 (359~849); 226.4 (128.6~341.6)

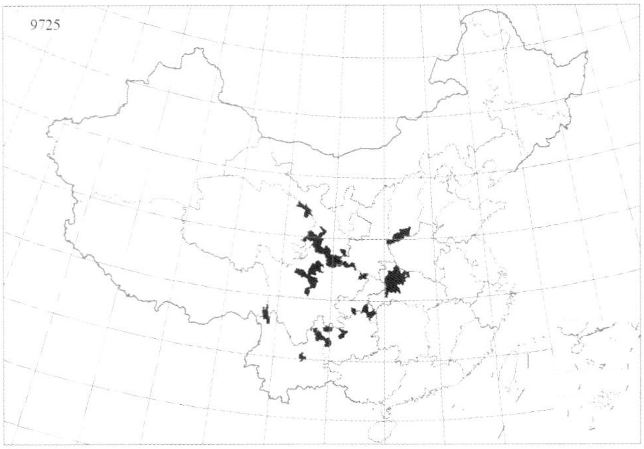

Berberidaceae 小檗科 *Berberis* 小檗属
Berberis agricola Ahrendt 暗红小檗
Shrubs, deciduous

2.8; −6.0; 10.6; 4.4; 21.1; 47.1; 423 ❙ 619; 373; 15 ❙ 41.0; 423; 240.0

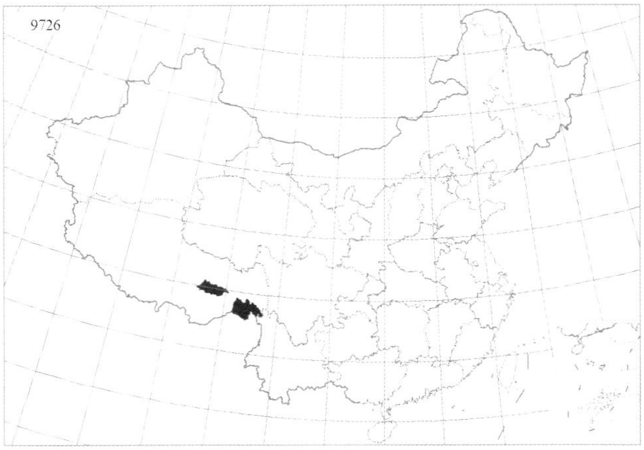

Berberidaceae 小檗科 *Berberis* 小檗属
Berberis alpicola C. K. Schneid. 高山小檗
Shrubs, evergreen

20.8; 15.3; 25.0; 20.8; 189.2; 0.0; 1019 ❙ 2445; 1412; 110 ❙ 140.0; 1019; 207.7

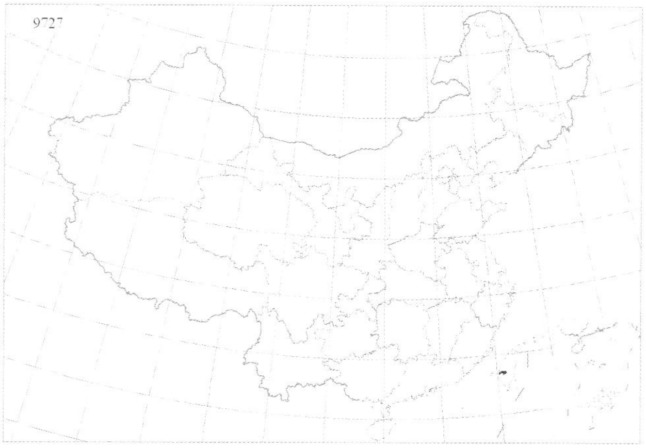

Berberidaceae 小檗科 *Berberis* 小檗属
Berberis amoena Dunn 美丽小檗
Shrubs, deciduous

8.9 (3.4~14.7); 0.3 (−5.5~6.0); 16.4 (11.4~24.0); 9.3 (4.7~14.8); 63.8 (23.3~118.0); 16.4 (0.0~41.8); 590 (431~789) ❙ 905 (527~1057); 510 (317~610); 25 (9~42) ❙ 55.7 (18.6~101.2); 590 (431~789); 260.6 (150.8~349.8)

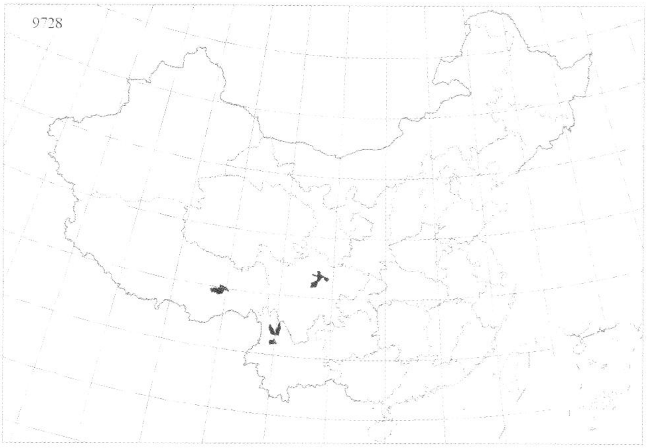

Berberidaceae 小檗科 *Berberis* 小檗属
Berberis amurensis Rupr. 黄芦木
Shrubs, deciduous

9.8 (−0.8~16.2); −6.4 (−25.0~6.1); 24.0 (10.7~28.3); 11.2 (3.8~16.2); 92.6 (16.2~135.0); 34.6 (0.0~118.4); 712 (411~880) ❙ 583 (183~1379); 363 (65~812); 20 (4~65) ❙ −16.1 (−71.0~87.3); 560 (183~849); 181.5 (26.5~346.5)

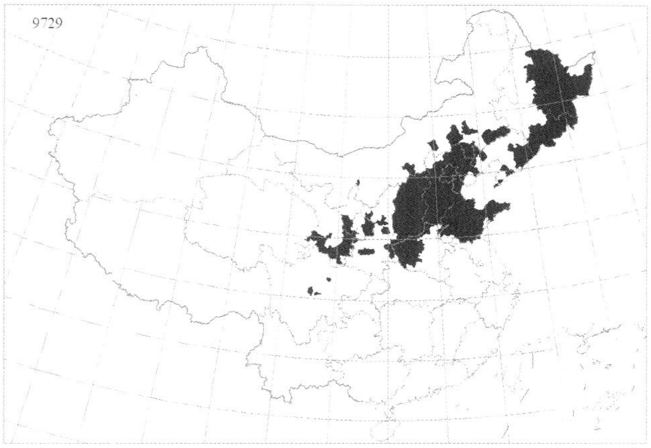

Berberidaceae 小檗科 *Berberis* 小檗属
Berberis angulosa Wall. 有棱小檗
Shrubs, deciduous

−1.5; −10.4; 6.7; 2.1; 3.8; 81.6; 334 ❙ 309; 208; 18 ❙ −7.4; 309; 110.1

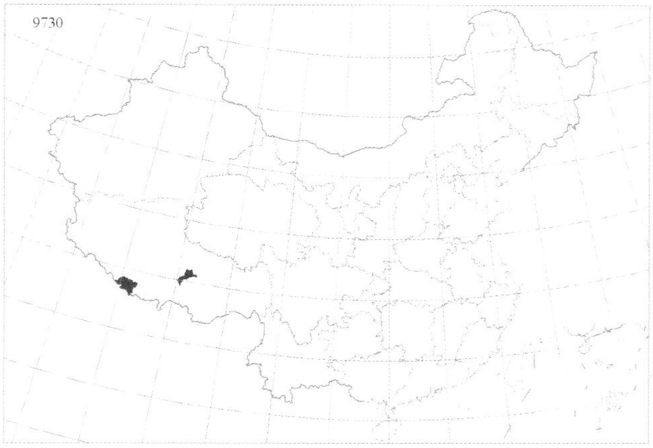

Berberidaceae 小檗科 *Berberis* 小檗属
Berberis aridocalida Ahrendt 陇南小檗
Shrubs, deciduous

7.7 (0.9~16.4); −2.0 (−9.5~6.4); 16.5 (9.8~25.6); 8.5 (3.6~16.4); 58.7 (14.0~137.2); 25.8 (0.0~63.1); 570 (403~857) ❙ 952 (620~1583); 518 (301~938); 26 (13~57) ❙ 70.2 (−1.2~128.6); 570 (403~857); 264.4 (171.5~357.7)

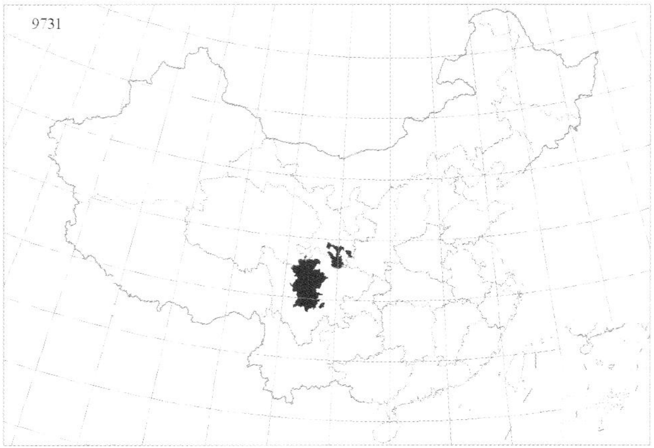

Berberidaceae 小檗科 *Berberis* 小檗属
Berberis aristatoserrulata Hayata 密齿小檗
Shrubs, evergreen

16.4; 11.5; 20.3; 16.5; 137.8; 0.0; 783 ❙ 2499; 947; 303 ❙ 219.3; 783; 303.6

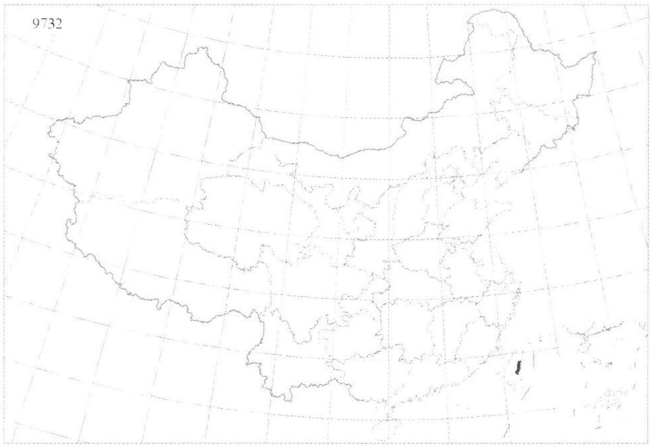

Berberidaceae 小檗科 *Berberis* 小檗属
Berberis asiatica Roxburgh ex Candolle 喜马拉雅小檗
Shrubs, evergreen

-1.5; -9.6; 6.1; 1.9; 2.2; 79.2; 332 ❙ 320; 205; 26 ❙ -3.5; 320; 75.7

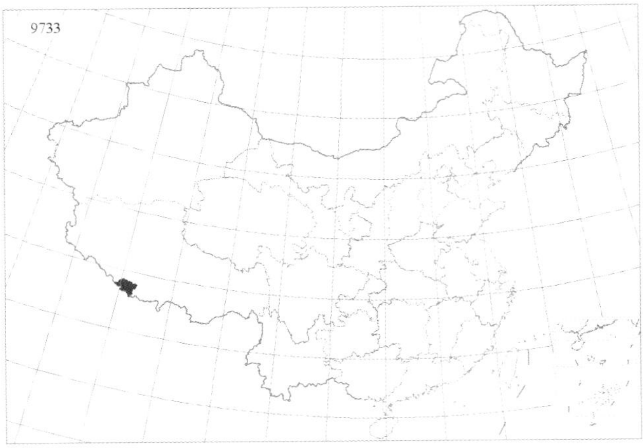

Berberidaceae 小檗科 *Berberis* 小檗属
Berberis asmyana C. K. Schneid. 直梗小檗
Shrubs, evergreen

5.6; -3.1; 13.2; 6.1; 34.9; 27.9; 490 ❙ 995; 566; 22 ❙ 104.5; 490; 268.8

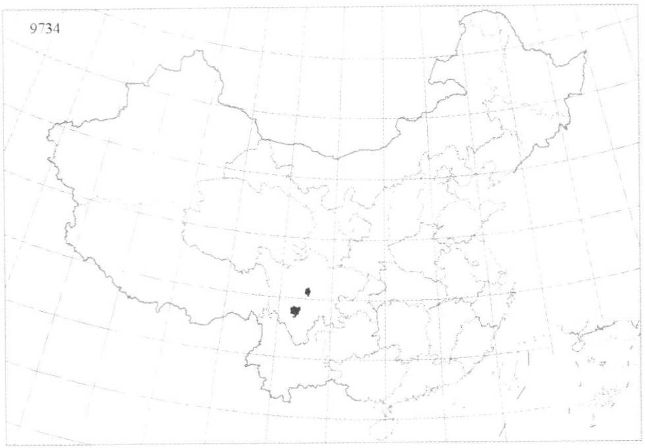

Berberidaceae 小檗科 *Berberis* 小檗属
Berberis atrocarpa C. K. Schneid. 黑果小檗
Shrubs, evergreen

13.8 (7.3~16.7); 4.0 (-2.0~8.6); 22.6 (15.6~28.4); 13.8 (7.5~16.8); 108.4 (48.2~141.5); 2.6 (0.0~20.4); 748 (537~911) ❙ 1114 (864~1379); 549 (393~812); 56 (19~146) ❙ 49.1 (13.7~87.3); 741 (537~911); 221.6 (171.0~330.0)

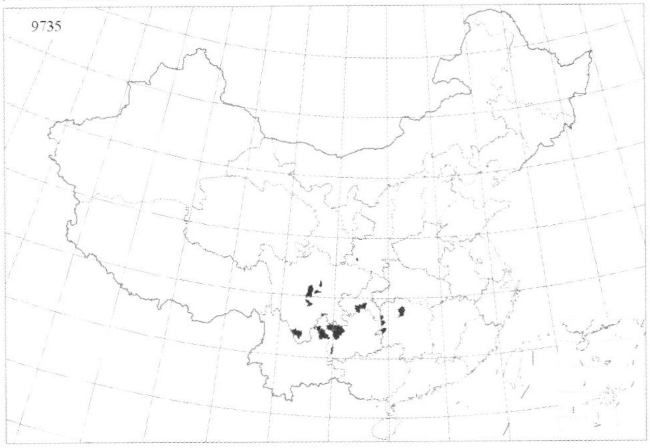

Berberidaceae 小檗科 *Berberis* 小檗属
Berberis atroviridiana T. S. Ying 那觉小檗
Shrubs, deciduous

5.3; -2.4; 12.2; 5.7; 30.5; 26.2; 479 ❙ 961; 558; 31 ❙ 100.7; 479; 234.2

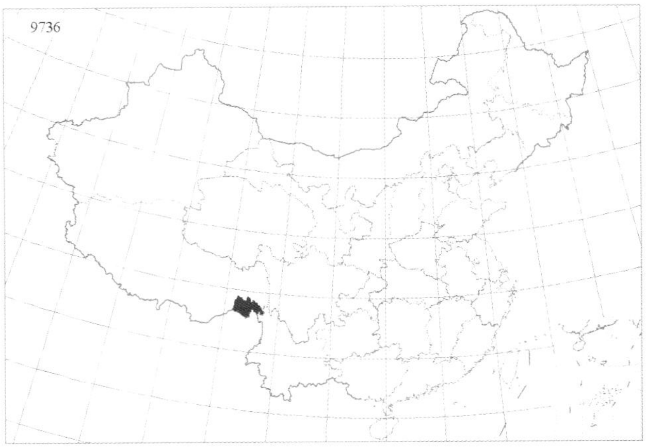

Berberidaceae 小檗科 *Berberis* 小檗属
Berberis batangensis T. S. Ying 巴塘小檗
Shrubs, deciduous

2.7; -6.0; 10.4; 4.2; 18.9; 46.9; 419 ❙ 579; 365; 10 ❙ 38.4; 419; 234.5

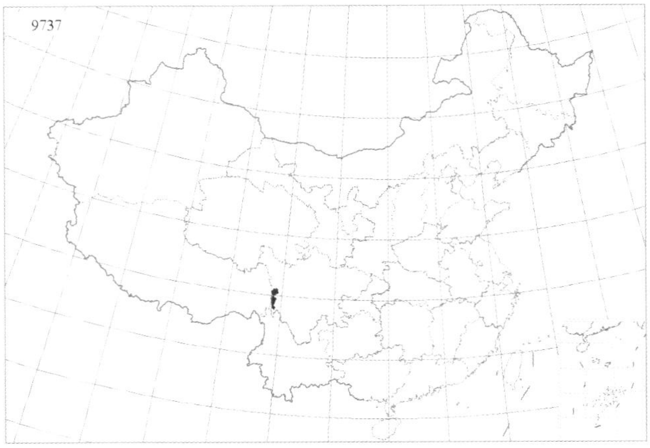

Berberidaceae 小檗科 *Berberis* 小檗属
Berberis bergmanniae C. K. Schneid. 汉源小檗
Shrubs, evergreen

14.6 (12.0~16.9); 2.7 (0.6~4.3); 25.9 (22.9~29.0); 14.6 (12.1~16.9); 119.3 (94.7~143.8); 4.0 (0.7~9.7); 810 (696~933) ❙ 1250 (862~1495); 505 (376~613); 105 (47~183) ❙ 55.5 (7.2~106.2); 807 (696~931); 218.0 (157.1~285.1)

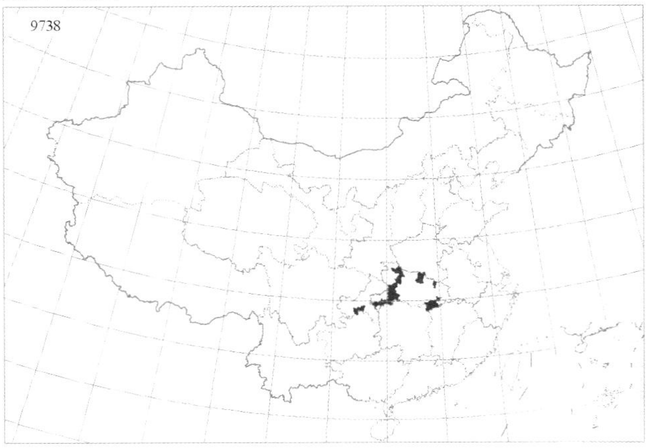

Berberidaceae 小檗科 *Berberis* 小檗属
Berberis bicolor H. Lév. 二色小檗
Shrubs, evergreen

14.7 (10.1~18.5); 6.5 (2.3~14.1); 21.7 (16.9~25.7); 14.8 (10.2~18.5); 117.9 (68.1~162.1); 0.5 (0.0~5.2); 763 (618~878) ❙ 1192 (863~2713); 610 (476~1388); 64 (19~303) ❙ 55.6 (4.4~219.3); 739 (618~863); 260.5 (189.4~363.7)

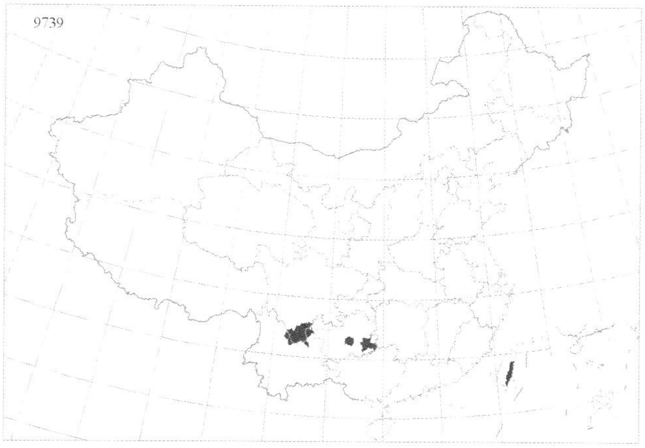

Berberidaceae 小檗科 *Berberis* 小檗属
Berberis bimbilaica (Ahrendt) Harber 比巴小檗
Shrubs, deciduous

1.9; -7.2; 10.0; 3.8; 16.6; 53.6; 403 ❙ 341; 226; 4 ❙ -15.3; 341; 217.5

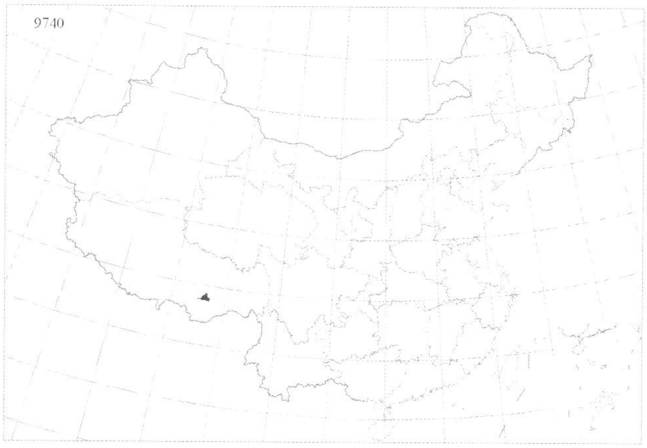

Berberidaceae 小檗科 *Berberis* 小檗属
Berberis brachypoda Maxim. 短柄小檗
Shrubs, deciduous

9.8 (0.8~15.1); -4.0 (-12.7~2.6); 22.3 (10.7~27.5); 10.7 (3.8~15.2); 84.7 (16.2~128.0); 26.4 (3.9~66.0); 669 (411~865) ❙ 643 (315~1478); 325 (174~592); 25 (6~88) ❙ -3.0 (-47.5~121.7); 579 (315~817); 183.7 (44.5~318.2)

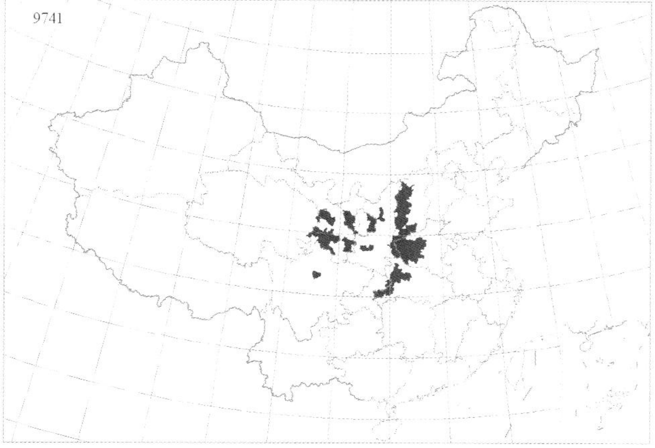

Berberidaceae 小檗科 *Berberis* 小檗属
Berberis calcipratorum Ahrendt 钙原小檗
Shrubs, deciduous

9.9; 2.9; 15.7; 10.0; 69.7; 10.9; 601 ❙ 922; 544; 33 ❙ 54.5; 601; 327.6

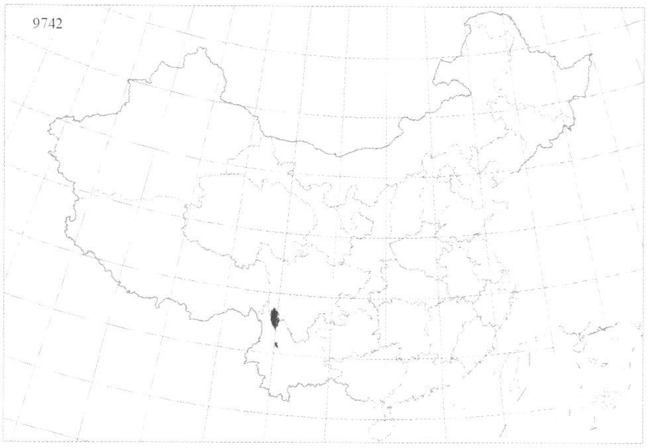

Berberidaceae 小檗科 *Berberis* 小檗属
Berberis calliantha Mulligan 美花小檗
Shrubs, evergreen

13.3; 5.4; 19.6; 13.4; 100.4; 0.0; 706 ❙ 1736; 1007; 51 ❙ 145.8; 706; 447.4

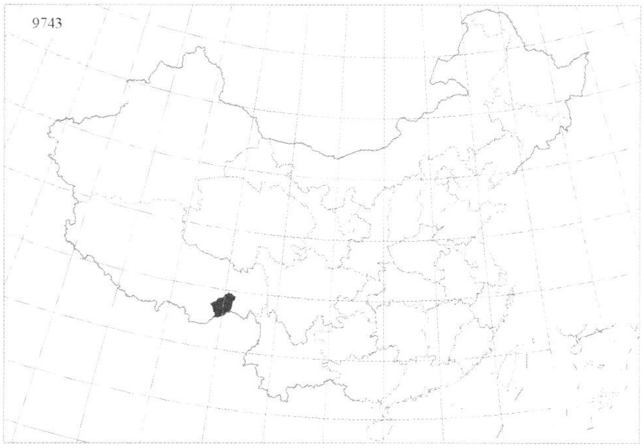

Berberidaceae 小檗科 *Berberis* 小檗属
Berberis campylogyna (Ahrendt) Harber 弯雌蕊小檗
Shrubs, deciduous

9.7 (5.4~15.0); 2.5 (-2.1~8.4); 15.8 (12.3~20.2); 9.9 (5.8~15.1); 68.5 (31.5~121.2); 11.4 (0.0~25.6); 601 (481~753) ❙ 886 (741~991); 536 (426~594); 25 (12~35) ❙ 49.7 (26.3~67.7); 586 (481~707); 298.7 (273.5~319.8)

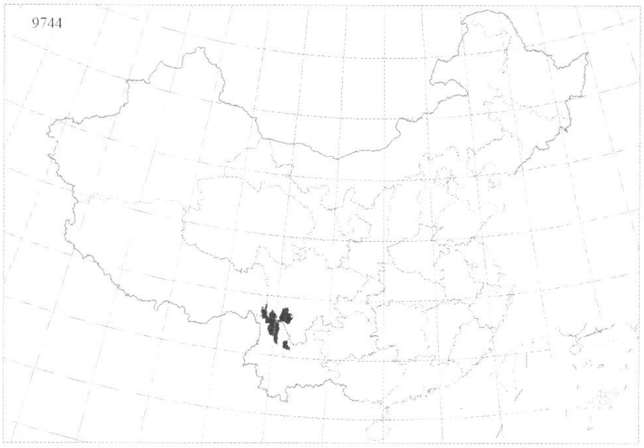

Berberidaceae 小檗科 *Berberis* 小檗属
Berberis campylotropa T. S. Ying 弯果小檗
Shrubs, deciduous

0.7；-9.0；9.6；3.3；13.3；64.8；388 ┃ 536；328；12 ┃ 38.3；388；236.1

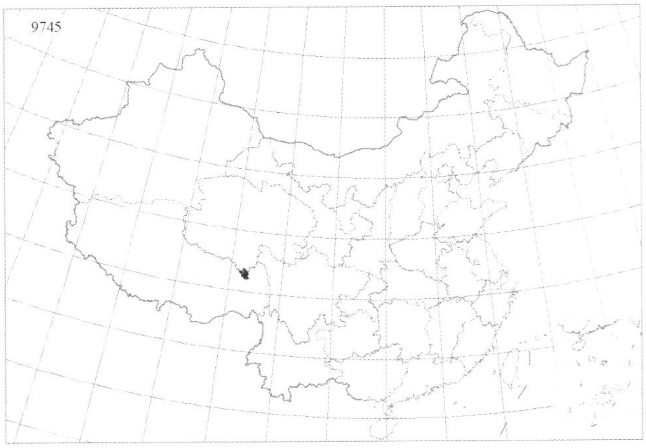

Berberidaceae 小檗科 *Berberis* 小檗属
Berberis candidula C. K. Schneid. 单花小檗
Shrubs, evergreen

12.0 (9.3~13.0)；0.7 (-1.7~2.1)；22.8 (20.1~24.2)；12.0 (9.5~13.1)；93.8 (69.8~102.9)；9.6 (5.6~18.0)；695 (603~732) ┃ 1195 (964~1478)；502 (421~592)；73 (52~85) ┃ 73.1 (31.7~107.2)；695 (603~732)；260.6 (234.7~279.5)

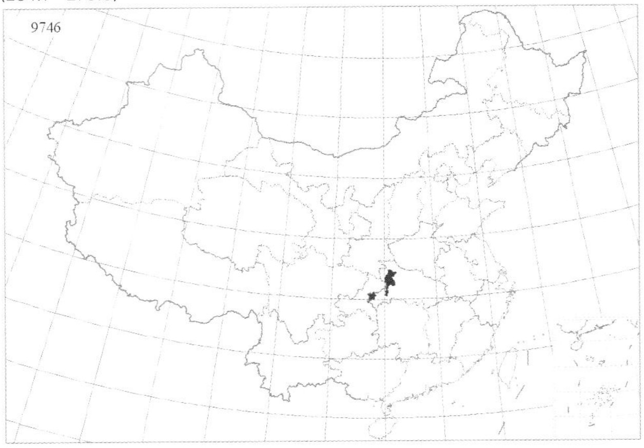

Berberidaceae 小檗科 *Berberis* 小檗属
Berberis capillaris Cox 中缅小檗
Shrubs, deciduous

7.6；-0.3；14.3；8.0；50.9；19.9；547 ┃ 1128；658；34 ┃ 103.7；547；332.7

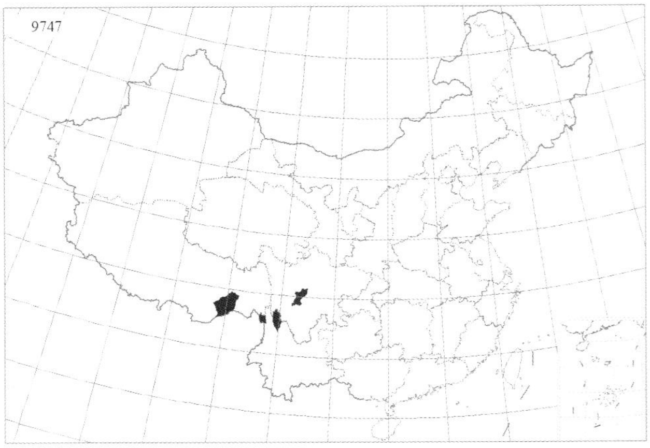

Berberidaceae 小檗科 *Berberis* 小檗属
Berberis cavaleriei H. Lév. 贵州小檗
Shrubs, evergreen

14.9 (11.4~18.3)；5.7 (2.7~11.8)；22.8 (18.8~26.4)；15.0 (11.4~18.3)；119.9 (80.7~159.8)；0.4 (0.0~3.7)；776 (652~882) ┃ 1128 (874~1307)；557 (451~700)；58 (25~89) ┃ 45.4 (6.5~69.4)；769 (652~843)；213.2 (153.9~316.6)

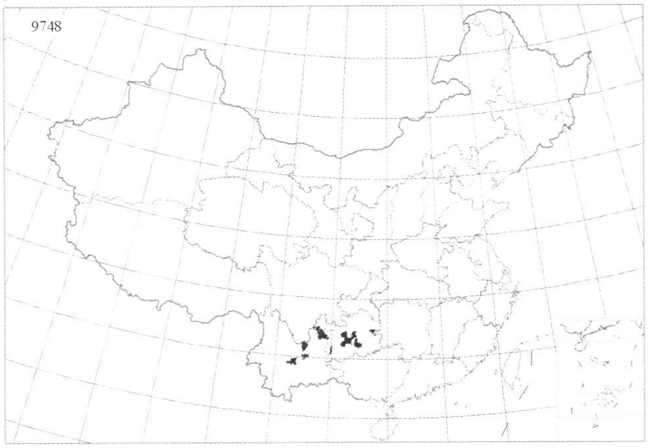

Berberidaceae 小檗科 *Berberis* 小檗属
Berberis centiflora Diels 多花大黄连刺
Shrubs, evergreen

12.9 (8.6~16.9)；6.1 (1.3~10.4)；18.3 (15.0~21.5)；13.0 (8.7~17.0)；97.5 (52.6~143.4)；2.0 (0.0~8.8)；694 (574~819) ┃ 1001 (946~1042)；583 (548~596)；38 (24~46) ┃ 46.9 (15.5~81.5)；682 (574~763)；321.2 (278.4~372.3)

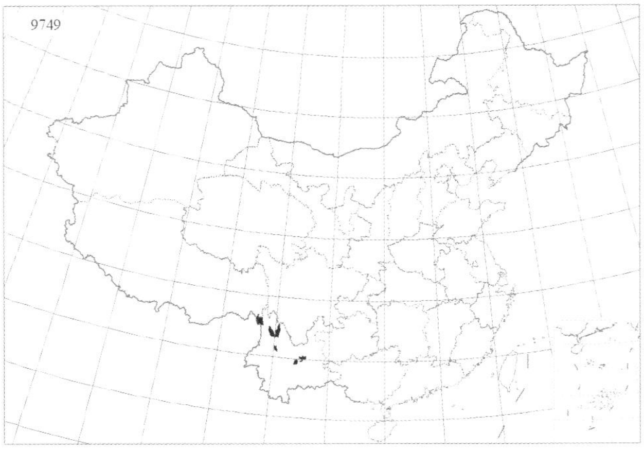

Berberidaceae 小檗科 *Berberis* 小檗属
Berberis chingii Cheng 华东小檗
Shrubs, evergreen

16.3 (14.3~18.8)；4.8 (2.6~7.7)；27.3 (24.2~30.1)；16.4 (14.3~18.9)；137.3 (113.2~166.5)；0.6 (0.0~3.5)；880 (761~1024) ┃ 1530 (1197~1865)；521 (427~607)；189 (71~258) ┃ 74.7 (40.8~128.9)；869 (761~933)；223.4 (148.9~299.5)

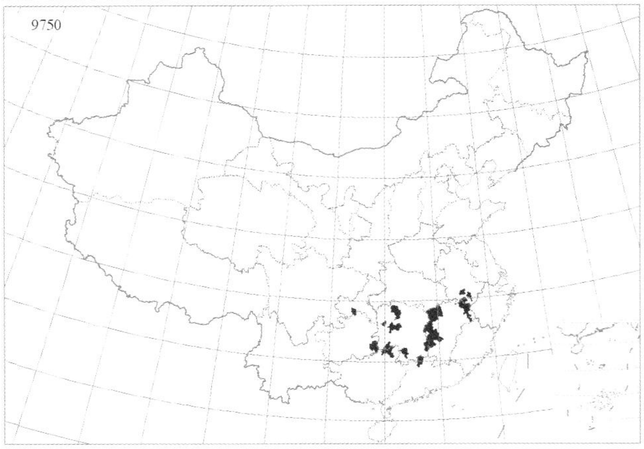

Berberidaceae 小檗科 *Berberis* 小檗属

Berberis chrysosphaera Mulligan 黄球小檗

Shrubs, evergreen

5.3; -2.4; 12.2; 5.7; 30.5; 26.2; 479 | 961; 558; 31 | 100.7; 479; 234.2

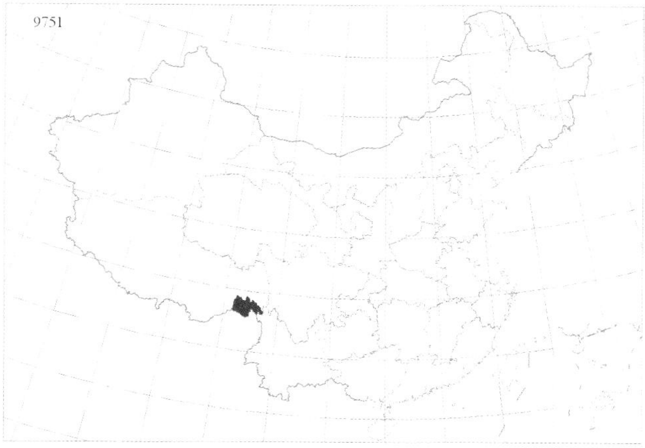

Berberidaceae 小檗科 *Berberis* 小檗属

Berberis circumserrata (C. K. Schneid.) C. K. Schneid. 秦岭小檗

Shrubs, deciduous

8.8 (-4.1~15.1); -4.4 (-17.3~1.8); 20.9 (7.8~27.5); 9.8 (2.1~15.2); 76.4 (5.2~128.0); 30.5 (5.6~114.2); 638 (340~865) | 625 (39~1214); 310 (29~516); 24 (1~75) | -0.3 (-92.3~101.4); 555 (39~817); 195.3 (10.8~364.1)

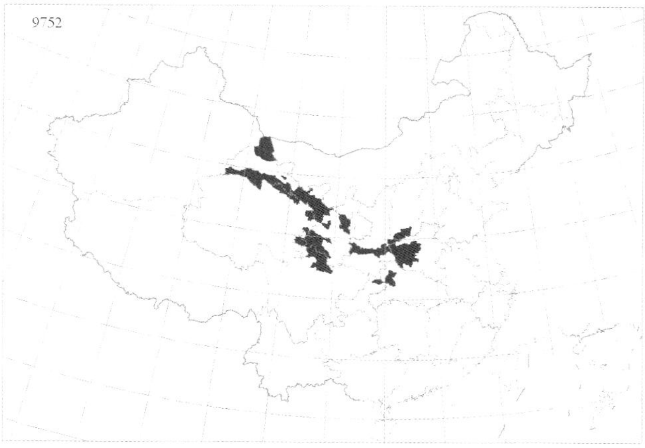

Berberidaceae 小檗科 *Berberis* 小檗属

Berberis concolor W. W. Sm. 同色小檗

Shrubs, deciduous

3.8; -4.2; 10.9; 4.9; 23.7; 37.1; 446 | 735; 454; 20 | 65.7; 446; 241.1

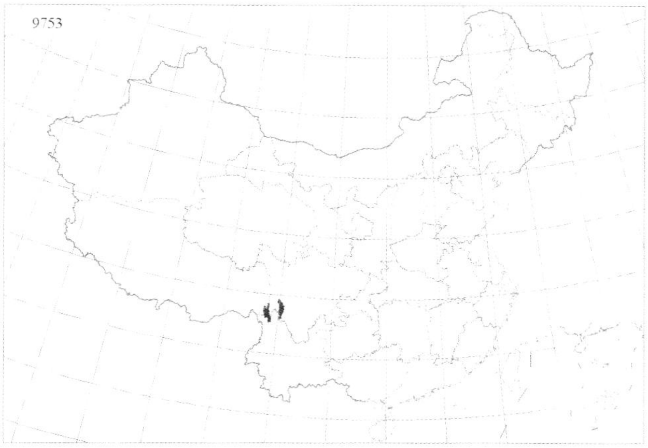

Berberidaceae 小檗科 *Berberis* 小檗属

Berberis coxii C. K. Schneid. 柯氏小檗

Shrubs, evergreen

13.2; 6.4; 18.4; 13.3; 99.4; 0.0; 695 | 1337; 727; 69 | 92.4; 695; 407.1

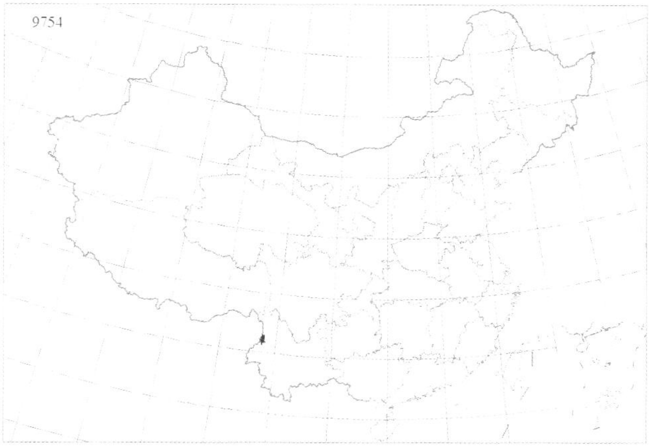

Berberidaceae 小檗科 *Berberis* 小檗属

Berberis crassilimba C. C. Wu 厚檐小檗

Shrubs, deciduous

9.9; 2.6; 16.0; 10.1; 70.5; 10.9; 605 | 882; 544; 25 | 48.2; 593; 274.6

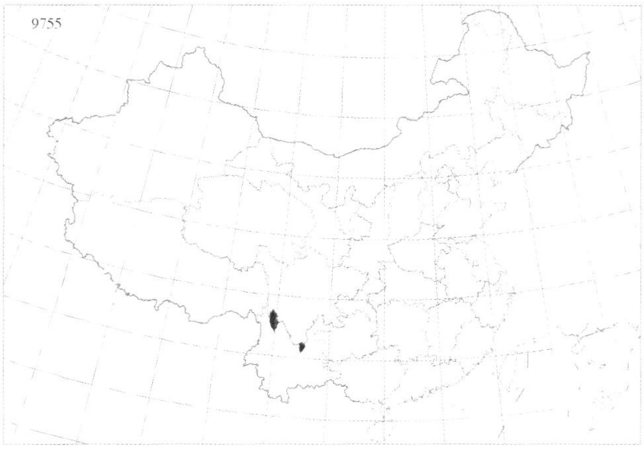

Berberidaceae 小檗科 *Berberis* 小檗属

Berberis daiana T. S. Ying 城口小檗

Shrubs, deciduous

10.6; -0.2; 21.1; 10.6; 79.4; 12.1; 643 | 1246; 525; 46 | 93.8; 643; 284.3

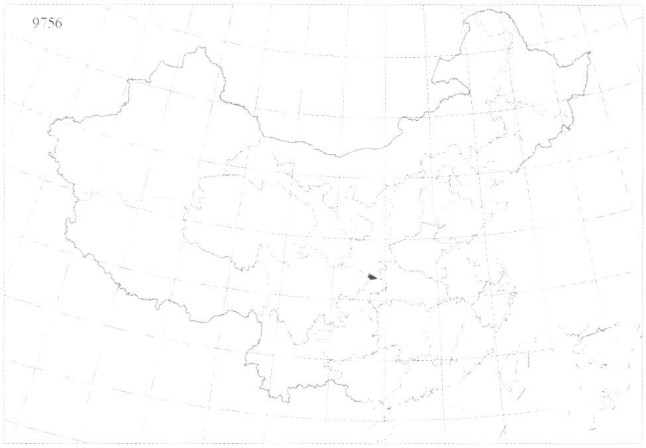

Berberidaceae 小檗科 *Berberis* 小檗属

Berberis dasyclada Ahrendt 粗枝小檗

Shrubs, evergreen

9.3; 1.5; 15.9; 9.6; 65.5; 13.1; 592 ❙ 1348; 783; 41 ❙ 123.3; 592; 340.8

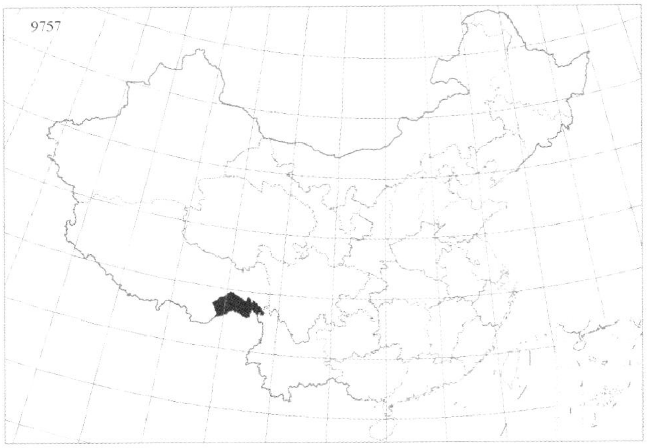

Berberidaceae 小檗科 *Berberis* 小檗属

Berberis dasystachya Maxim. 直穗小檗

Shrubs, deciduous

8.7 (-4.1~16.2); -5.4 (-17.3~4.1); 21.4 (7.8~28.1); 10.0 (2.1~16.3); 78.5 (5.2~136.5); 33.3 (0.9~114.2); 646 (340~903) ❙ 598 (39~1254); 327 (29~533); 20 (1~83) ❙ -5.1 (-92.3~128.6); 543 (39~903); 188.8 (10.8~364.1)

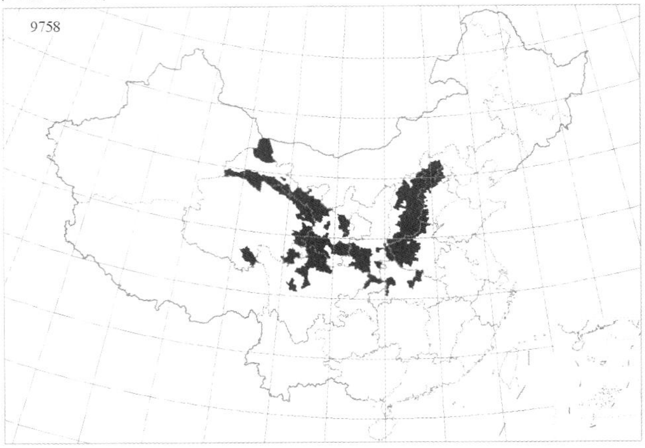

Berberidaceae 小檗科 *Berberis* 小檗属

Berberis davidii Ahrendt 密叶小檗

Shrubs, evergreen

12.0 (6.6~15.8); 5.1 (-1.3~9.0); 17.6 (13.3~20.6); 12.0 (6.8~15.8); 88.0 (39.6~129.7); 3.6 (0.0~19.7); 665 (514~776) ❙ 998 (862~1054); 583 (548~627); 37 (12~54) ❙ 51.9 (27.9~81.5); 659 (514~772); 327.0 (258.7~382.6)

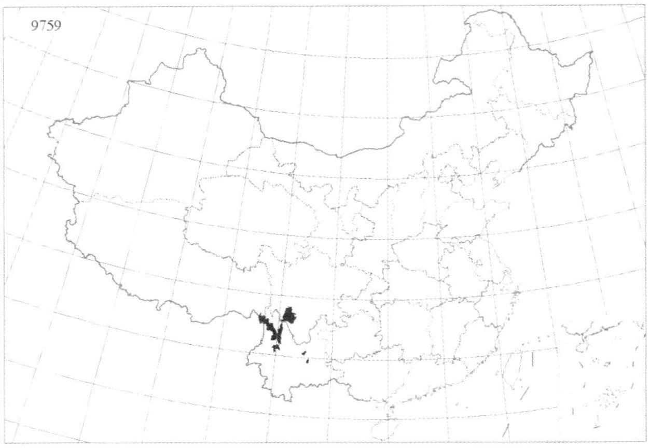

Berberidaceae 小檗科 *Berberis* 小檗属

Berberis deinacantha C. K. Schneid. 壮刺小檗

Shrubs, evergreen

13.7 (9.6~19.1); 6.3 (2.4~13.0); 19.5 (15.7~23.3); 13.7 (9.6~19.1); 105.9 (60.8~169.7); 1.2 (0.0~5.4); 722 (600~917) ❙ 1058 (874~1428); 616 (494~835); 35 (18~56) ❙ 47.0 (26.1~70.6); 708 (600~917); 302.1 (185.6~506.2)

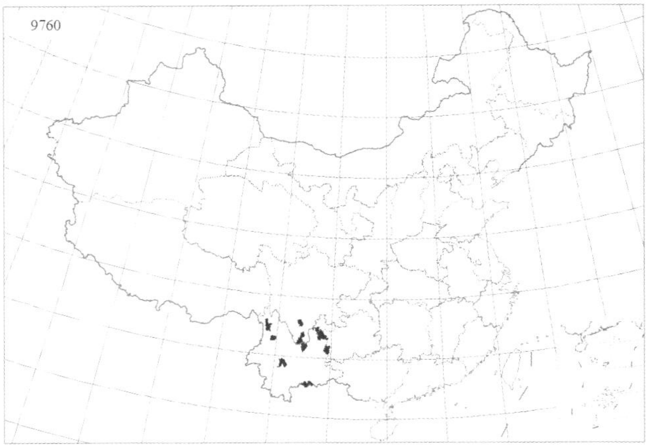

Berberidaceae 小檗科 *Berberis* 小檗属

Berberis delavayi C. K. Schneid. 洱源小檗

Shrubs, evergreen

8.8 (1.8~15.3); 1.5 (-7.1~8.7); 15.1 (9.9~20.5); 9.2 (3.8~15.4); 61.1 (15.7~124.3); 14.4 (0.0~53.3); 579 (404~761) ❙ 897 (527~1134); 533 (317~651); 28 (9~47) ❙ 54.3 (22.4~78.3); 575 (404~717); 306.6 (169.5~351.7)

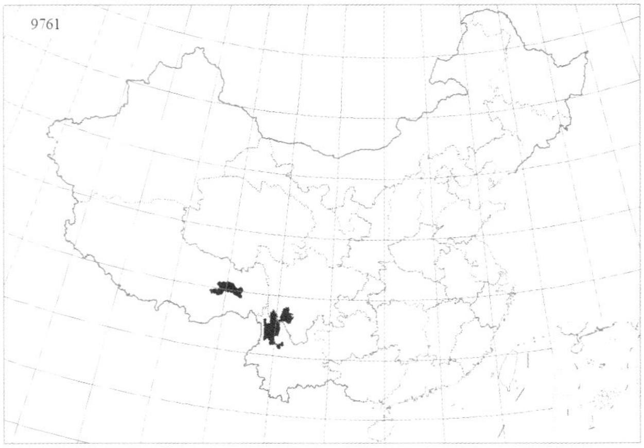

Berberidaceae 小檗科 *Berberis* 小檗属

Berberis derongensis T. S. Ying 得荣小檗

Shrubs, deciduous

4.1; -4.2; 11.5; 5.1; 25.8; 36.6; 451 ❙ 632; 386; 16 ❙ 40.1; 451; 218.2

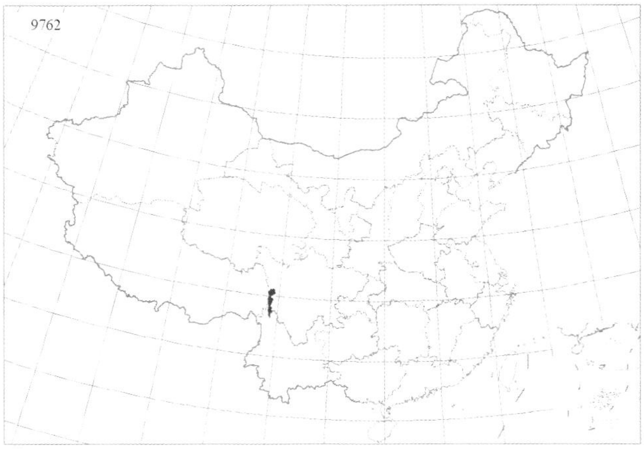

Berberidaceae 小檗科 *Berberis* 小檗属
Berberis diaphana Maxim. 鲜黄小檗
Shrubs, deciduous

6.0 (−4.1~15.1); −6.7 (−17.3~2.2); 17.4 (7.0~27.5); 7.7 (1.9~15.2); 56.0 (3.6~128.0); 43.8 (5.3~114.2); 556 (328~865) ❙ 584 (39~1220); 313 (29~688); 18 (1~53) ❙ 9.6 (−92.3~125.9); 497 (39~817); 204.4 (10.8~364.1)

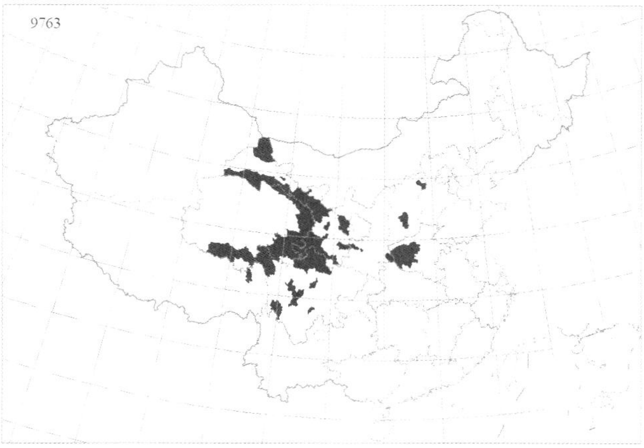

Berberidaceae 小檗科 *Berberis* 小檗属
Berberis dictyoneura C. K. Schneid. 松潘小檗
Shrubs, deciduous

9.2 (0.0~15.1); −2.0 (−9.2~4.4); 19.8 (8.1~25.4); 9.9 (2.7~15.2); 73.7 (8.0~122.8); 22.8 (0.6~67.9); 624 (363~817) ❙ 808 (341~1208); 392 (190~546); 22 (8~51) ❙ 31.8 (−40.9~125.9); 615 (341~817); 251.1 (61.2~334.4)

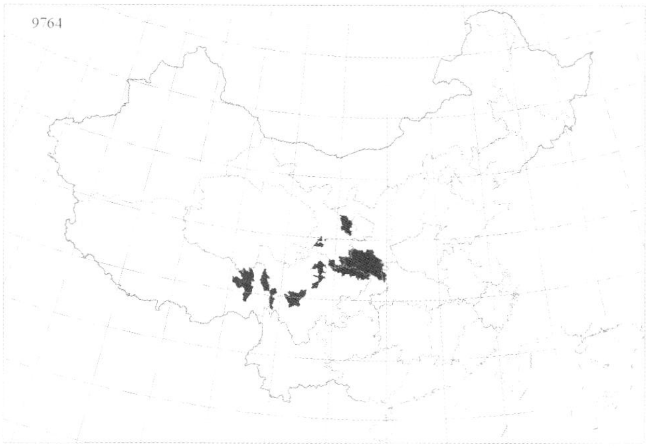

Berberidaceae 小檗科 *Berberis* 小檗属
Berberis dictyophylla Franch. 刺红珠
Shrubs, deciduous

2.3 (−2.2~11.0); −6.8 (−12.8~3.9); 10.4 (7.6~17.0); 4.3 (2.2~11.1); 20.4 (5.5~74.6); 52.8 (1.3~91.3); 420 (336~640) ❙ 622 (298~990); 379 (215~619); 14 (2~31) ❙ 47.5 (−18.0~125.9); 418 (298~638); 237.7 (111.0~319.8)

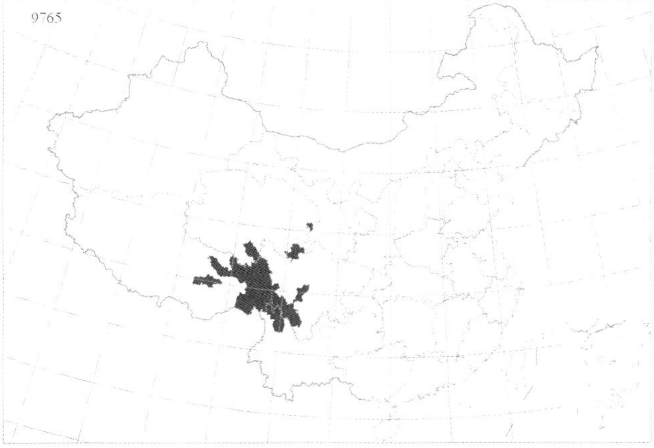

Berberidaceae 小檗科 *Berberis* 小檗属
Berberis dictyophylla var. epruinosa C. K. Schneid. 无粉刺红珠
Shrubs, deciduous

2.4 (−2.2~11.9); −6.8 (−12.8~2.9); 10.6 (7.6~20.0); 4.4 (2.2~12.0); 21.1 (5.5~87.2); 51.4 (3.4~91.3); 424 (336~671) ❙ 690 (298~1001); 410 (215~550); 15 (2~35) ❙ 61.3 (−18.0~125.9); 420 (298~671); 254.3 (169.5~341.6)

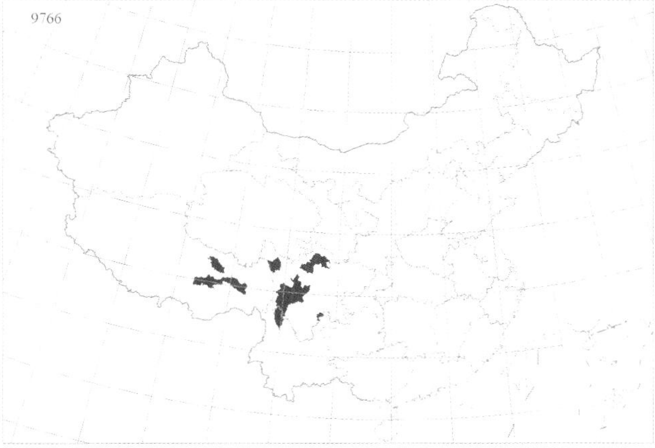

Berberidaceae 小檗科 *Berberis* 小檗属
Berberis dielsiana Fedde 首阳小檗
Shrubs, deciduous

11.4 (5.9~15.1); −1.7 (−8.7~2.3); 23.6 (17.9~28.2); 11.7 (7.3~15.2); 94.1 (51.7~128.0); 17.2 (4.9~45.4); 707 (529~865) ❙ 718 (341~1220); 346 (190~688); 26 (8~56) ❙ 2.2 (−40.9~87.2); 648 (341~817); 213.2 (61.2~334.4)

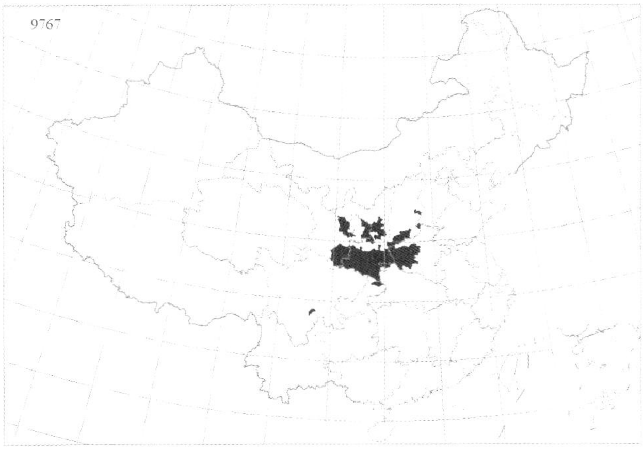

Berberidaceae 小檗科 *Berberis* 小檗属
Berberis dubia C. K. Schneid. 置疑小檗
Shrubs, deciduous

3.7 (−5.5~13.5); −8.9 (−17.3~1.4); 15.1 (5.8~25.1); 6.2 (1.3~13.5); 40.8 (1.2~109.4); 55.5 (7.1~127.1); 492 (299~767) ❙ 512 (39~931); 272 (29~420); 12 (1~29) ❙ 5.1 (−92.3~94.4); 424 (39~767); 204.7 (10.8~364.1)

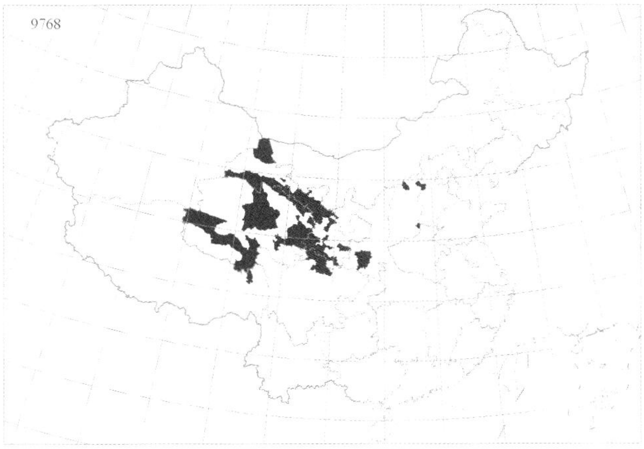

Berberidaceae 小檗科 *Berberis* 小檗属
Berberis dumicola C. K. Schneid. 丛林小檗
Shrubs, evergreen

9.9; 2.8; 15.8; 10.0; 67.1; 7.9; 605 ❙ 1082; 607; 49 ❙ 76.8; 605; 354.3

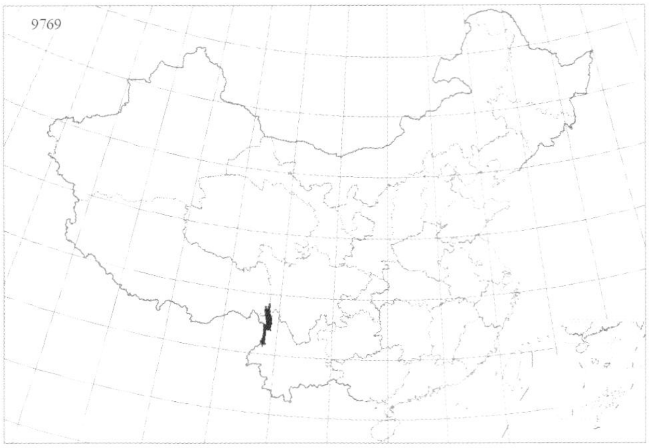

Berberidaceae 小檗科 *Berberis* 小檗属
Berberis elliotii Ahrendt 埃氏小檗
Shrubs, deciduous

3.3; −5.4; 11.1; 4.6; 22.3; 41.8; 427 ❙ 473; 287; 7 ❙ 5.0; 427; 230.7

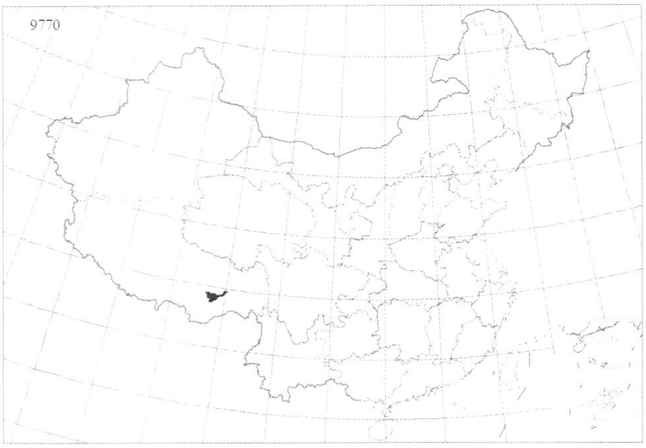

Berberidaceae 小檗科 *Berberis* 小檗属
Berberis erythroclada Ahrendt 红枝小檗
Shrubs, deciduous

3.5; −4.9; 11.0; 4.7; 22.8; 40.4; 437 ❙ 672; 398; 17 ❙ 50.0; 437; 211.5

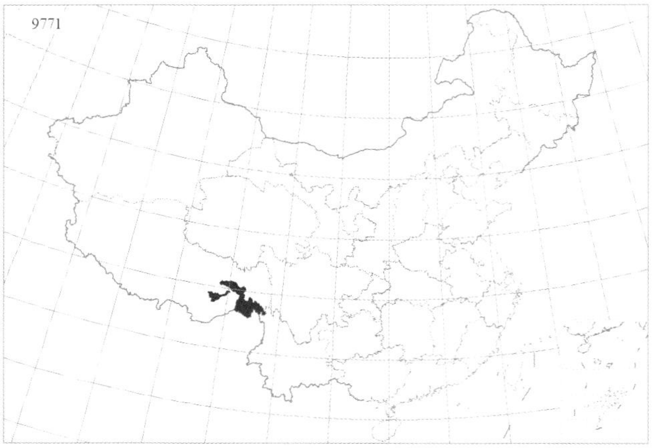

Berberidaceae 小檗科 *Berberis* 小檗属
Berberis fallaciosa C. K. Schneid. 南川小檗
Shrubs, evergreen

13.3 (10.6~17.8); 3.3 (−0.2~7.2); 22.6 (18.6~27.7); 13.4 (10.6~17.8); 103.9 (79.4~153.7); 3.7 (0.0~12.1); 735 (643~939) ❙ 1136 (988~1246); 527 (475~610); 54 (28~71) ❙ 56.7 (16.1~93.8); 735 (643~939); 223.1 (142.5~284.3)

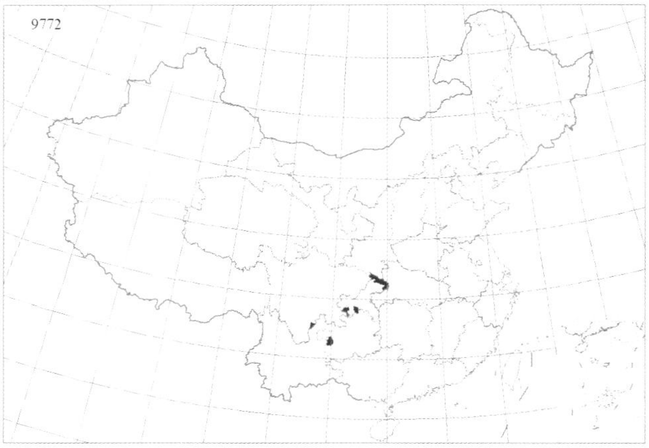

Berberidaceae 小檗科 *Berberis* 小檗属
Berberis fallax C. K. Schneid. 假小檗
Shrubs, evergreen

13.7 (5.4~18.9); 6.1 (−2.1~12.0); 20.1 (12.3~28.5); 13.8 (5.8~19.0); 107.3 (31.5~167.8); 1.8 (0.0~25.6); 727 (481~929) ❙ 980 (741~1337); 565 (411~727); 34 (15~113) ❙ 36.5 (−14.2~93.0); 699 (481~912); 279.8 (142.9~407.1)

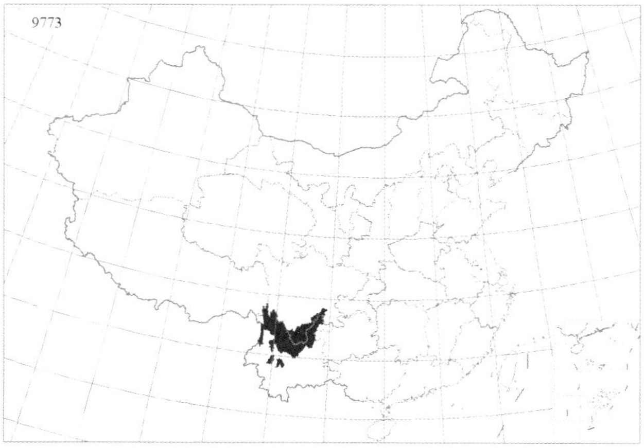

Berberidaceae 小檗科 *Berberis* 小檗属
Berberis fengqingensis Harber 凤庆小檗
Shrubs, evergreen

16.1; 9.6; 20.6; 16.2; 133.9; 0.0; 785 ❙ 1083; 600; 50 ❙ 38.0; 779; 405.0

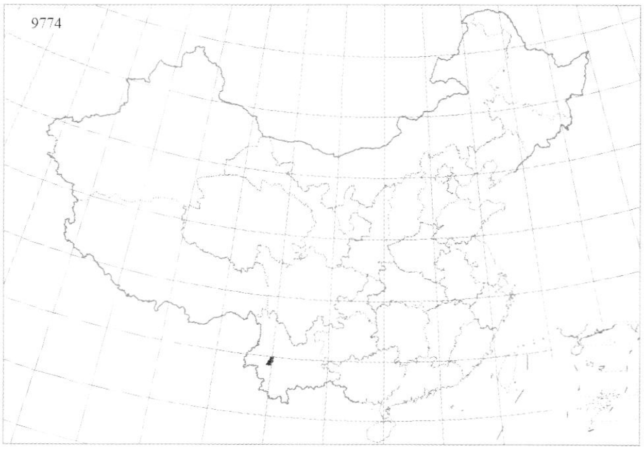

Berberidaceae 小檗科 *Berberis* 小檗属

Berberis ferdinandi-coburgii C. K. Schneid. 大叶小檗

Shrubs, evergreen

16.3 (12.7~19.1); 7.8 (1.9~13.1); 23.3 (19.6~26.9); 16.4 (12.8~19.1); 136.6 (95.3~169.7); 0.4 (0.0~6.1); 825 (702~939) ❙ 1186 (939~1495); 620 (463~835); 54 (30~83) ❙ 44.3 (6.5~89.0); 820 (702~939); 283.6 (142.9~506.2)

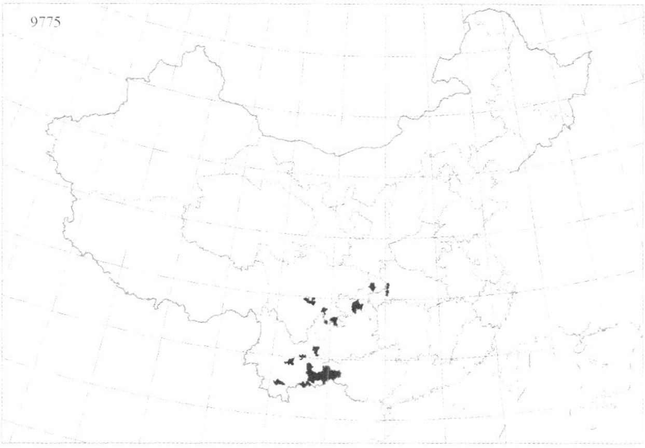

Berberidaceae 小檗科 *Berberis* 小檗属

Berberis francisci-ferdinandi C. K. Schneid. 大黄檗

Shrubs, deciduous

7.4 (2.2~13.5); -2.2 (-7.2~3.7); 16.1 (9.9~23.3); 8.1 (3.8~13.5); 54.5 (15.5~105.3); 25.1 (1.8~51.6); 556 (413~744) ❙ 804 (510~1049); 437 (311~619); 18 (9~28) ❙ 48.6 (-0.2~125.9); 556 (413~744); 269.6 (200.8~341.6)

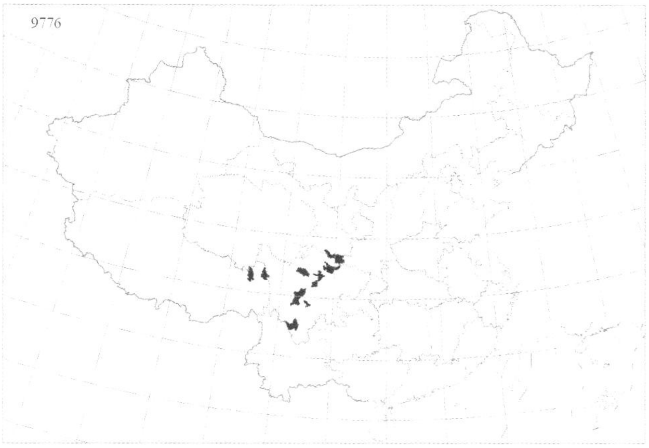

Berberidaceae 小檗科 *Berberis* 小檗属

Berberis fujianensis C. M. Hu 武夷小檗

Shrubs, evergreen

16.6 (14.7~18.2); 5.5 (3.8~6.9); 27.1 (25.3~29.1); 16.7 (14.8~18.2); 140.2 (118.7~158.8); 0.3 (0.0~1.5); 886 (791~984) ❙ 1655 (1185~1997); 564 (473~669); 196 (84~259) ❙ 85.8 (44.5~130.7); 879 (791~961); 234.4 (170.4~303.4)

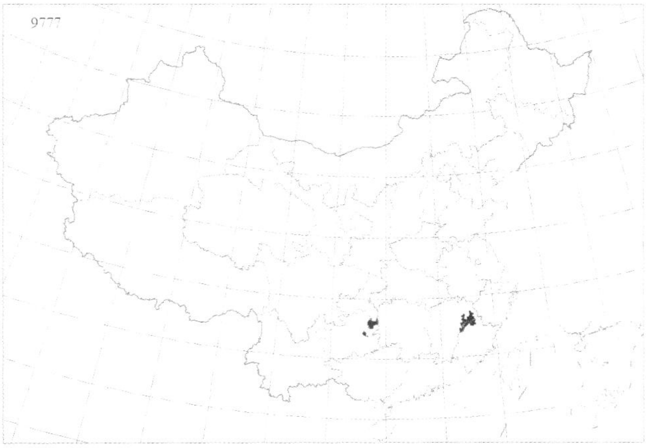

Berberidaceae 小檗科 *Berberis* 小檗属

Berberis gagnepainii C. K. Schneid. 湖北小檗

Shrubs, evergreen

14.0 (5.3~18.2); 3.8 (-2.4~9.0); 23.5 (12.2~27.1); 14.1 (5.7~18.2); 112.5 (30.5~158.8); 4.2 (0.0~26.2); 768 (479~921) ❙ 1178 (711~1478); 559 (357~817); 59 (15~94) ❙ 56.1 (9.4~107.2); 768 (479~921); 202.6 (130.1~292.6)

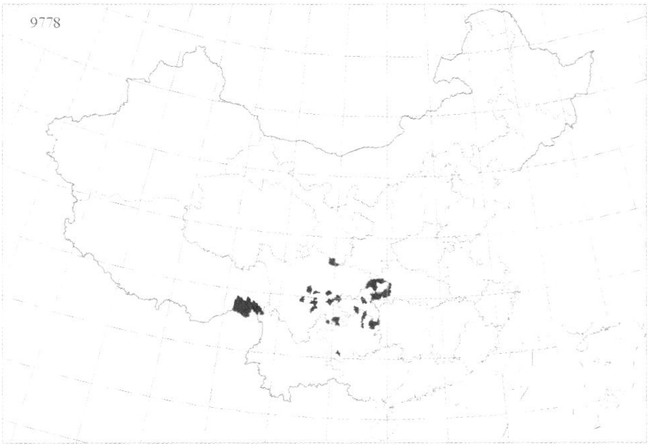

Berberidaceae 小檗科 *Berberis* 小檗属

Berberis gilgiana Fedde 涝峪小檗

Shrubs, deciduous

10.1 (7.4~12.8); -3.6 (-9.8~3.3); 22.8 (18.7~25.6); 10.8 (8.2~12.8); 86.1 (58.8~106.5); 23.8 (2.4~47.7); 670 (561~760) ❙ 653 (189~1214); 306 (114~574): 26 (4~75) ❙ -2.4 (-70.2~101.4); 532 (189~733); 209.9 (33.5~334.4)

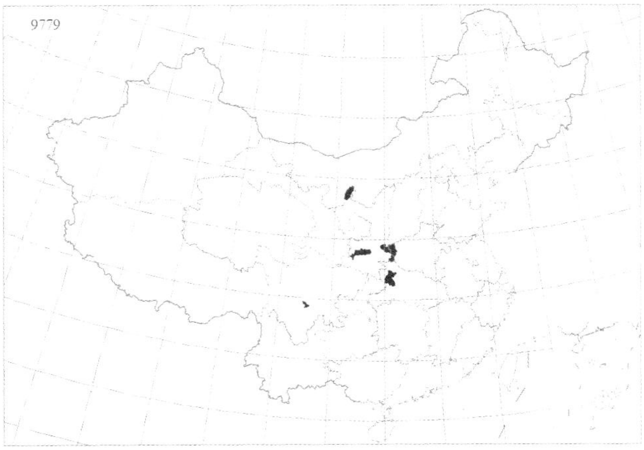

Berberidaceae 小檗科 *Berberis* 小檗属

Berberis gilungensis T. S. Ying 吉隆小檗

Shrubs, deciduous

-1.1; -9.1; 6.2; 2.0; 2.5; 75.6; 334 ❙ 321; 193; 33 ❙ -4.0; 321; 79.3

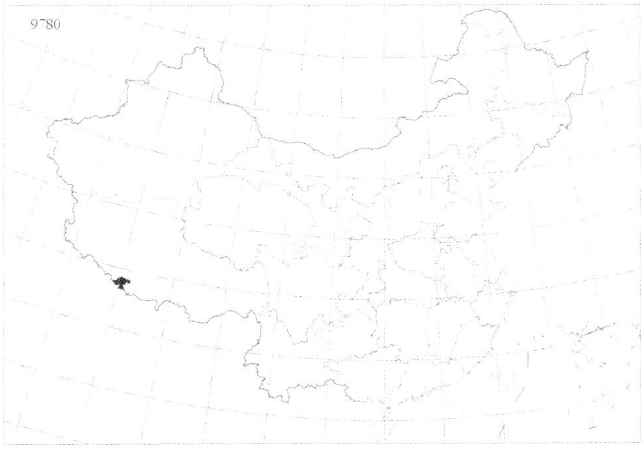

Berberidaceae 小檗科 *Berberis* 小檗属
Berberis griffithiana C. K. Schneid. 错那小檗
Shrubs, evergreen

12.8; 4.8; 19.0; 12.8; 93.9; 0.4; 689 ❙ 1583; 912; 41 ❙ 129.4; 689; 405.6

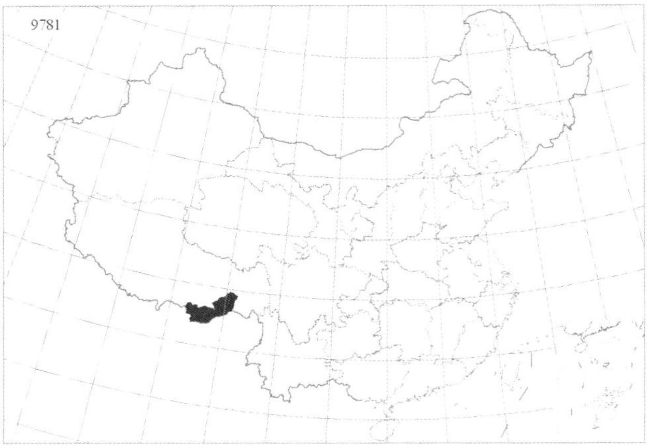

Berberidaceae 小檗科 *Berberis* 小檗属
Berberis griffithiana var. pallida (Hook. f. et Thomson) D. F. Chamb. et C. M. Hu 灰叶小檗
Shrubs, evergreen

13.3; 5.4; 19.6; 13.4; 100.4; 0.0; 706 ❙ 1736; 1007; 51 ❙ 145.8; 706; 447.4

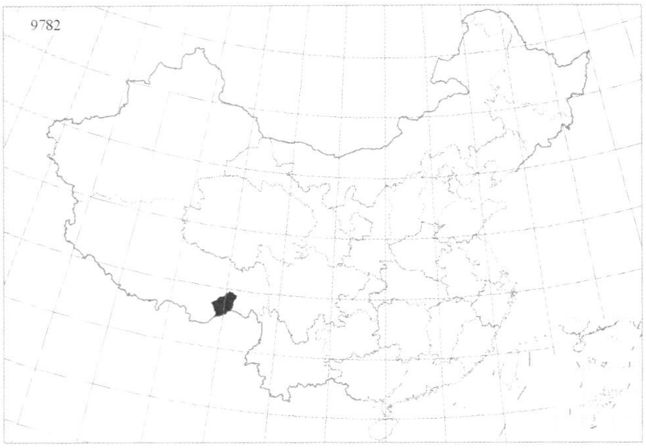

Berberidaceae 小檗科 *Berberis* 小檗属
Berberis grodtmanniana C. K. Schneid. 安宁小檗
Shrubs, evergreen

10.7 (6.6~13.6); 2.9 (-1.3~6.1); 17.3 (13.3~20.0); 10.8 (6.8~13.7); 74.7 (39.6~103.9); 5.5 (0.0~19.7); 632 (514~713) ❙ 968 (862~1042); 592 (548~641); 19 (12~35) ❙ 53.9 (46.1~67.7); 626 (514~691); 303.6 (278.8~331.6)

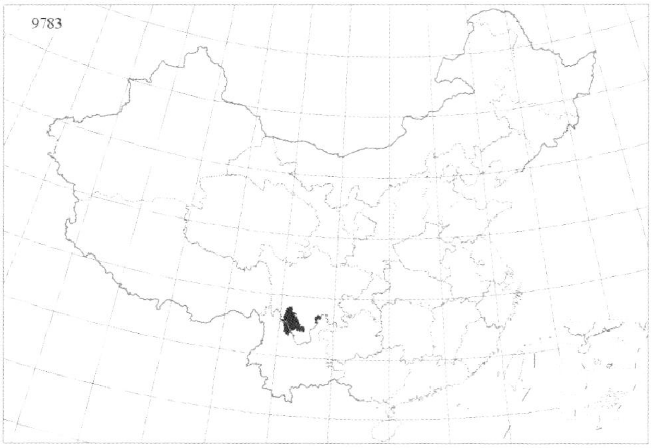

Berberidaceae 小檗科 *Berberis* 小檗属
Berberis guizhouensis T. S. Ying 毕节小檗
Shrubs, evergreen

12.2; 3.1; 20.2; 12.3; 89.7; 2.6; 679 ❙ 959; 509; 41 ❙ 41.3; 679; 189.3

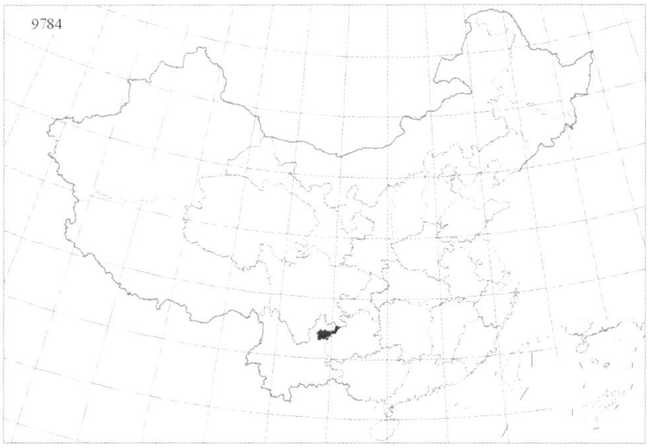

Berberidaceae 小檗科 *Berberis* 小檗属
Berberis gyalaica Ahrendt 波密小檗
Shrubs, deciduous

3.3; -5.6; 11.1; 4.6; 22.5; 42.9; 429 ❙ 530; 323; 10 ❙ 22.5; 429; 212.1

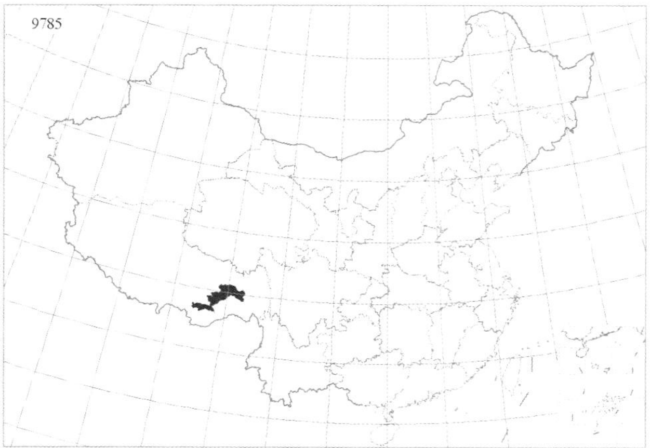

Berberidaceae 小檗科 *Berberis* 小檗属
Berberis hayatana M. Mizush. 南湖小檗
Shrubs, evergreen

17.8; 12.1; 22.8; 17.9; 154.4; 0.0; 852 ❙ 2859; 959; 402 ❙ 235.6; 852; 265.8

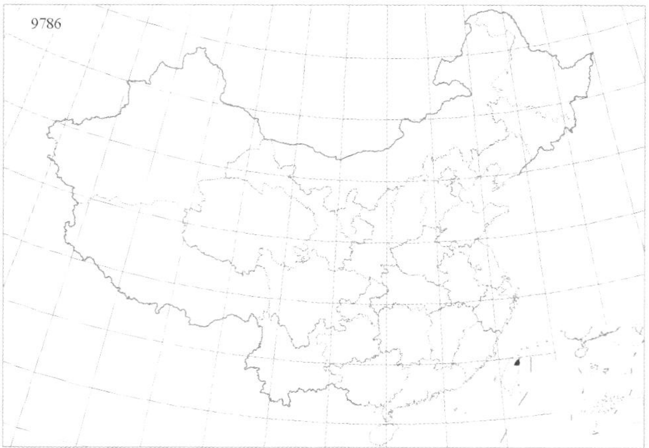

Berberidaceae 小檗科 *Berberis* 小檗属
Berberis hemsleyana Ahrendt 拉萨小檗
Shrubs, deciduous

1.3 (-1.7~4.0)；-8.4 (-11.9~-5.5)；10.0 (7.8~12.3)；3.8 (2.4~5.3)；17.1 (7.0~29.1)；60.5 (39.8~86.3)；393 (339~447) ❙ 312 (246~579)；231 (191~365)；2 (0~10) ❙ -20.9 (-35.9~38.4)；300 (246~419)；162.4 (93.6~234.5)

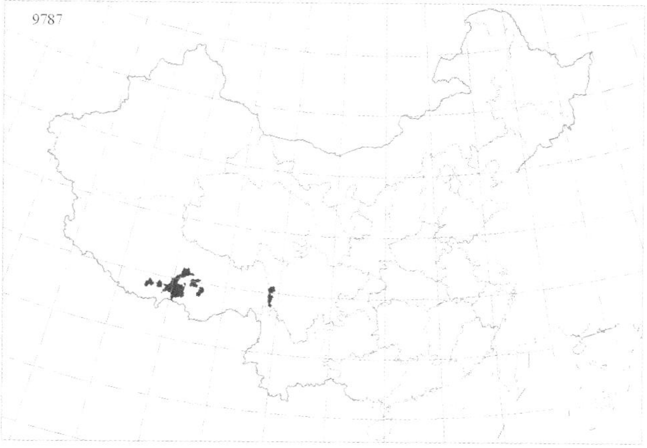

Berberidaceae 小檗科 *Berberis* 小檗属
Berberis henryana C. K. Schneid. 川鄂小檗
Shrubs, deciduous

13.3 (1.6~17.2)；0.8 (-9.3~5.1)；25.2 (10.9~29.3)；13.5 (4.0~17.2)；110.3 (17.9~146.8)；9.9 (0.0~58.4)；775 (419~941) ❙ 1077 (341~1828)；455 (190~685)；88 (8~253) ❙ 37.9 (-40.9~112.2)；746 (341~941)；217.4 (61.2~344.4)

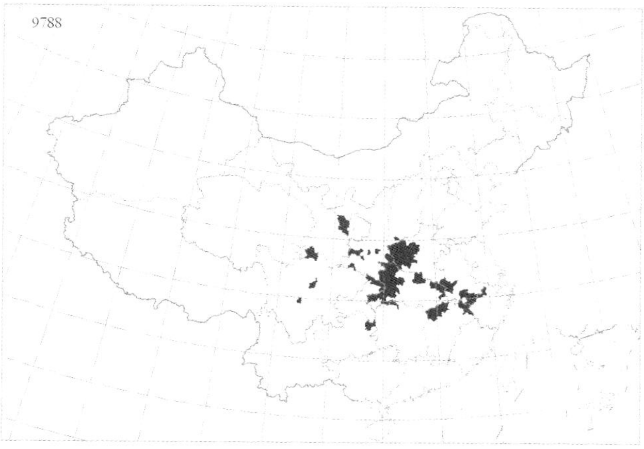

Berberidaceae 小檗科 *Berberis* 小檗属
Berberis heteropoda Schrenk 异果小檗
Shrubs, deciduous

4.5 (-4.3~12.0)；-14.2 (-20.9~-5.2)；19.9 (9.6~27.5)；8.2 (2.5~12.8)；63.8 (9.7~109.4)；69.6 (24.3~121.1)；592 (329~813) ❙ 182 (39~450)；84 (17~211)；16 (1~51) ❙ -66.7 (-94.7~3.9)；181 (39~406)；81.0 (7.6~297.3)

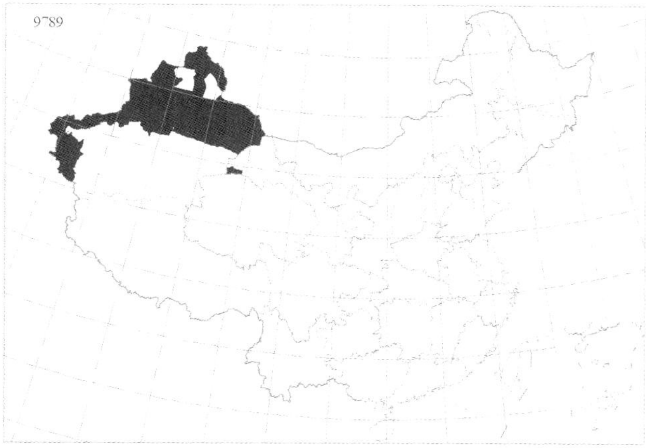

Berberidaceae 小檗科 *Berberis* 小檗属
Berberis heteropsis Ahrendt 龙里小檗
Shrubs, evergreen

11.7；3.2；18.8；11.7；82.6；2.3；658 ❙ 916；510；31 ❙ 39.1；658；185.6

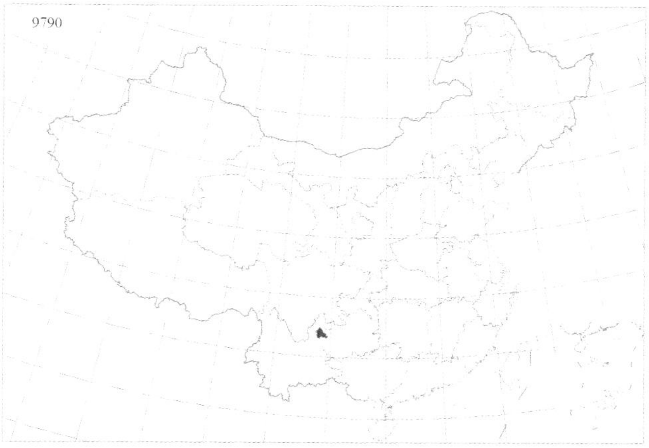

Berberidaceae 小檗科 *Berberis* 小檗属
Berberis hobsonii Ahrendt 毛梗小檗
Shrubs, deciduous

4.2；-4.4；11.6；6.0；35.3；44.6；466 ❙ 713；429；15 ❙ 34.2；453；208.3

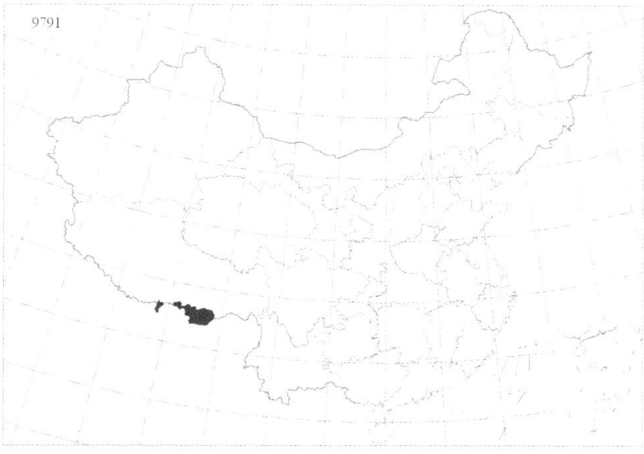

Berberidaceae 小檗科 *Berberis* 小檗属
Berberis holocraspedon Ahrendt 风庆小檗
Shrubs, evergreen

14.1；7.4；19.2；14.1；112.0；2.2；730 ❙ 1054；588；48 ❙ 47.2；716；376.3

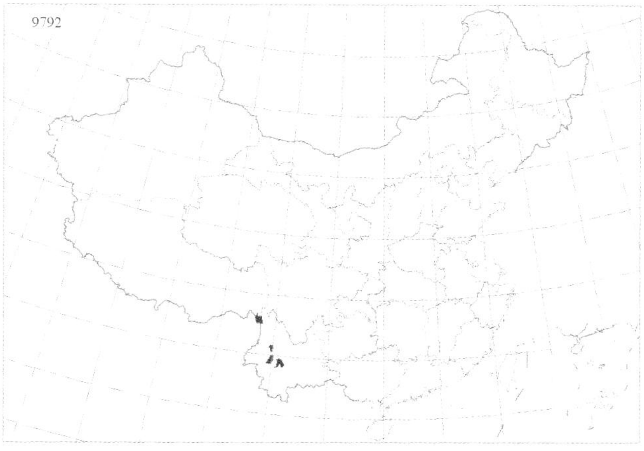

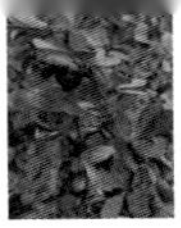

Berberidaceae 小檗科 *Berberis* 小檗属
Berberis hookeri D. F. Chamb. et C. H. Hu 虎克小檗
Shrubs, evergreen

2.1; -6.5; 9.8; 4.7; 25.2; 60.0; 418 | 608; 381; 18 | 27.1; 410; 140.0

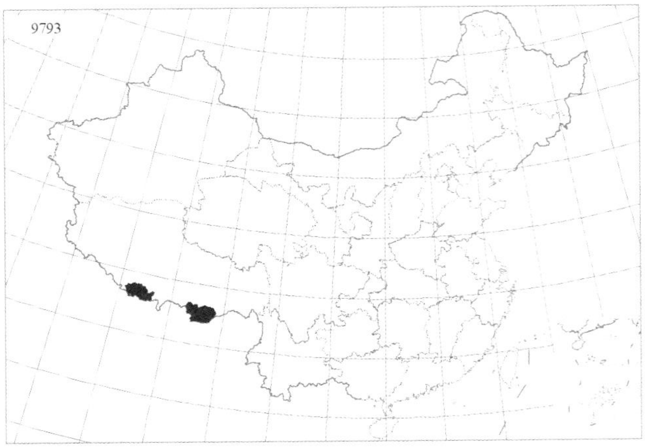

Berberidaceae 小檗科 *Berberis* 小檗属
Berberis hsuyunensis P. G. Xiao et W. C. Sung 叙永小檗
Shrubs, evergreen

15.0; 4.8; 24.6; 15.1; 121.0; 0.3; 790 | 1056; 503; 64 | 33.7; 790; 186.2

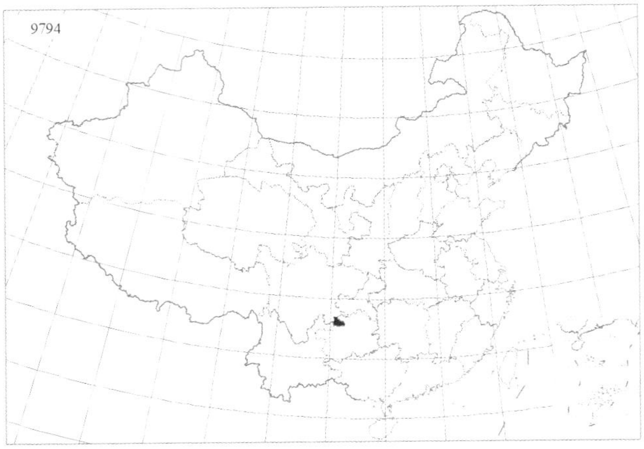

Berberidaceae 小檗科 *Berberis* 小檗属
Berberis humidoumbrosa Ahrendt 阴湿小檗
Shrubs, deciduous

8.8; 0.7; 15.6; 9.2; 61.7; 15.4; 578 | 1166; 680; 30 | 94.4; 578; 303.4

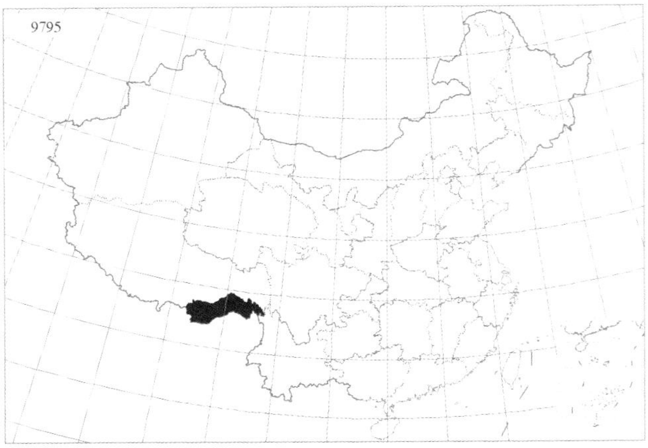

Berberidaceae 小檗科 *Berberis* 小檗属
Berberis impedita C. K. Schneid. 南岭小檗
Shrubs, evergreen

17.1 (11.0~20.3); 6.7 (1.6~10.9); 26.5 (17.0~29.2); 17.2 (11.1~20.4); 146.6 (74.6~184.5); 0.4 (0.0~6.9); 904 (640~1072) | 1545 (983~1865); 571 (411~679); 174 (15~258) | 71.6 (27.6~106.2); 899 (638~1072); 236.2 (157.2~310.2)

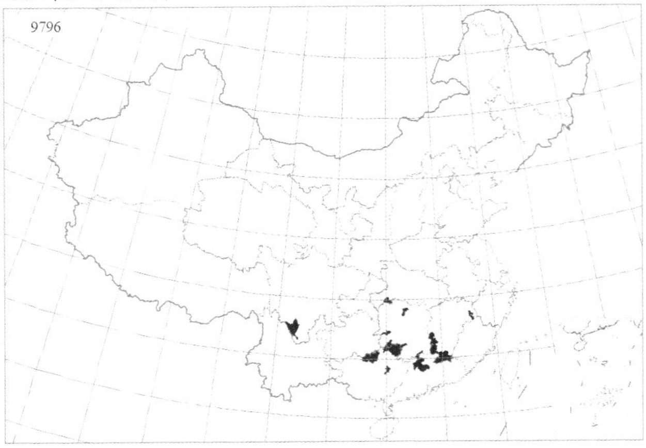

Berberidaceae 小檗科 *Berberis* 小檗属
Berberis insignis subsp. incrassata (Ahrendt) D. F. Chamb. et C. M. Hu 球果小檗
Shrubs, evergreen

10.2 (5.3~13.3); 2.6 (-2.4~5.4); 16.4 (12.2~19.6); 10.3 (5.7~13.4); 69.5 (30.5~100.4); 7.3 (0.0~26.2); 615 (479~706) | 1282 (961~1736); 733 (558~1007); 43 (31~58) | 106.8 (81.5~145.8); 615 (479~706); 356.5 (234.2~447.4)

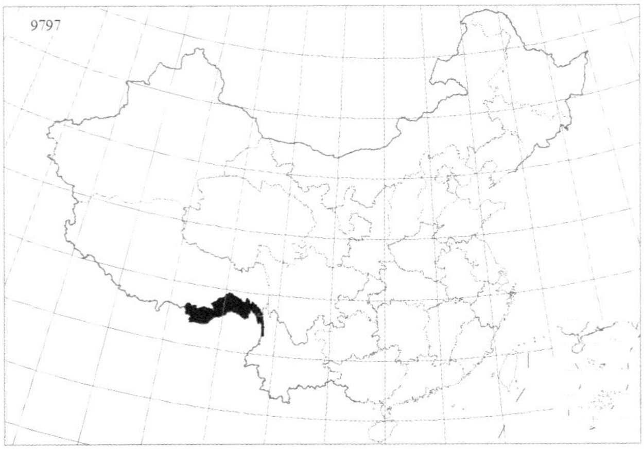

Berberidaceae 小檗科 *Berberis* 小檗属
Berberis insolita C. K. Schneid. 西昌小檗
Shrubs, evergreen

15.8; 8.2; 21.5; 15.8; 130.2; 0.0; 799 | 1256; 712; 46 | 59.2; 791; 378.6

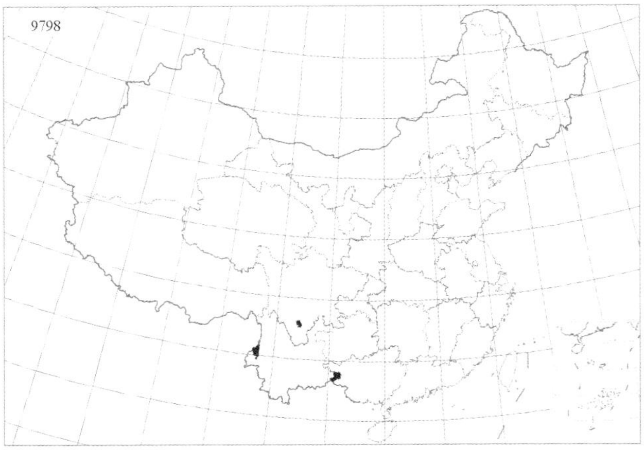

Berberidaceae 小檗科 *Berberis* 小檗属
Berberis integerrima Bunge 全缘小檗
Shrubs, deciduous

3.4; −5.5; 11.4; 4.7; 23.3; 41.8; 431 ❙ 527; 317; 9 ❙ 22.4; 431; 280.0

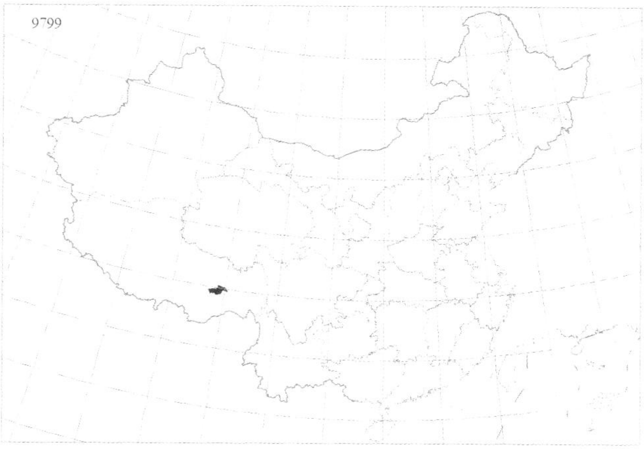

Berberidaceae 小檗科 *Berberis* 小檗属
Berberis integripetala T. S. Ying 甘南小檗
Shrubs, deciduous

11.9; 0.2; 22.9; 12.0; 94.3; 10.9; 701 ❙ 644; 321; 14 ❙ −8.1; 644; 268.5

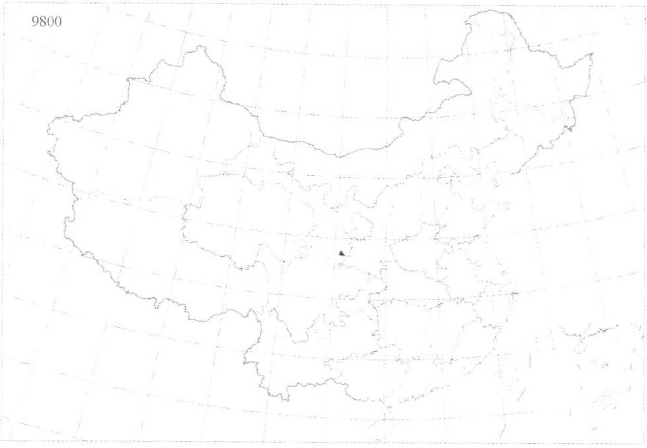

Berberidaceae 小檗科 *Berberis* 小檗属
Berberis jamesiana Forrest et W. W. Sm. 川滇小檗
Shrubs, deciduous

11.6 (−3.3~18.9); 3.6 (−15.3~12.0); 18.4 (7.2~26.6); 12.1 (2.0~19.0); 91.1 (3.7~167.8); 10.9 (0.0~103.2); 671 (339~929) ❙ 915 (380~1197); 527 (225~627); 30 (8~71) ❙ 38.7 (−14.2~121.7); 649 (339~912); 263.0 (142.9~352.3)

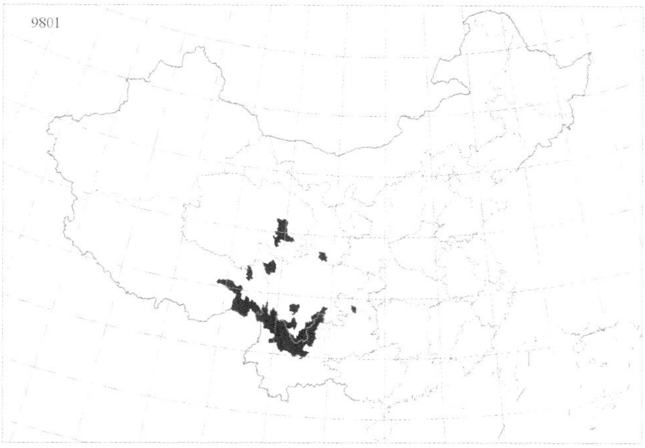

Berberidaceae 小檗科 *Berberis* 小檗属
Berberis jiangxiensis C. M. Hu 江西小檗
Shrubs, evergreen

16.8 (15.2~18.2); 5.1 (4.1~6.6); 27.9 (25.8~28.8); 16.8 (15.3~18.2); 141.9 (124.1~158.6); 0.2 (0.0~0.9); 902 (810~975) ❙ 1611 (1555~1715); 507 (488~537); 219 (200~228) ❙ 79.1 (68.0~106.4); 896 (810~936); 225.9 (206.0~255.1)

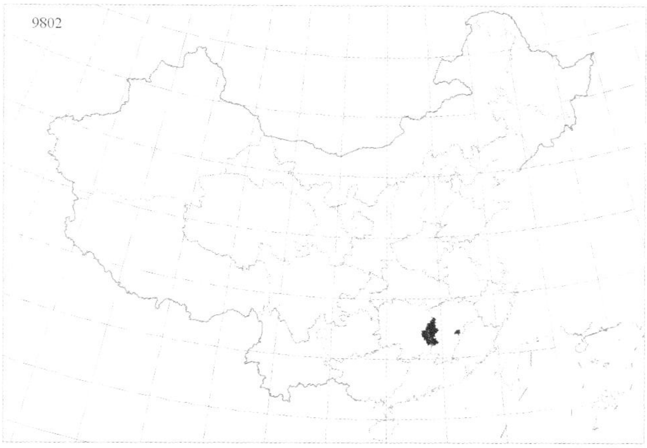

Berberidaceae 小檗科 *Berberis* 小檗属
Berberis jinfoshanensis T. S. Ying 南川县小檗
Shrubs, evergreen

14.8; 4.3; 24.9; 14.9; 119.2; 0.7; 791 ❙ 1197; 489; 71 ❙ 51.2; 791; 182.9

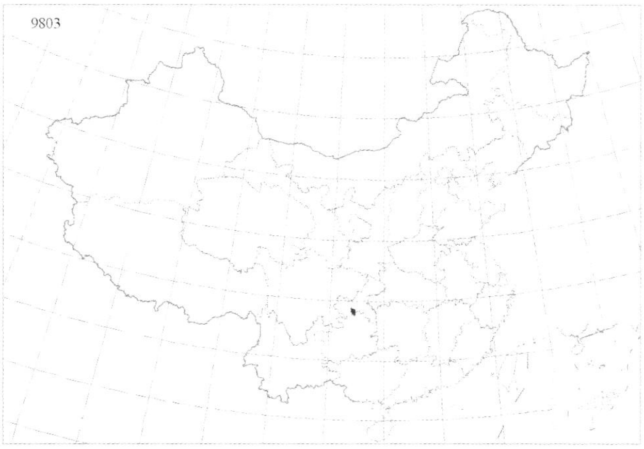

Berberidaceae 小檗科 *Berberis* 小檗属
Berberis jiulongensis T. S. Ying 九龙小檗
Shrubs, semievergreen

16.6; 6.6; 25.7; 16.6; 139.7; 0.0; 866 ❙ 1590; 946; 59 ❙ 83.5; 866; 166.3

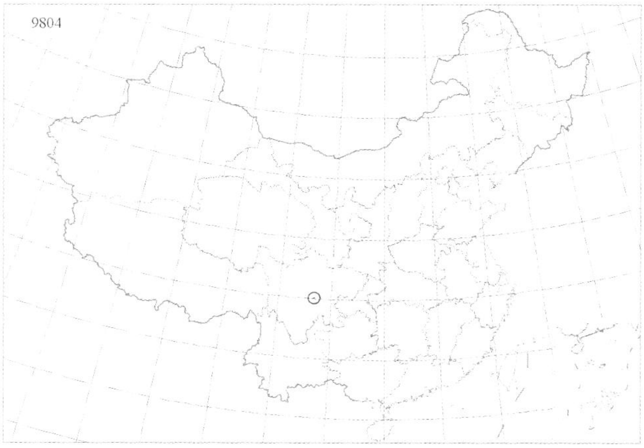

Berberidaceae 小檗科 *Berberis* 小檗属
Berberis johannis Ahrendt 腰果小檗
Shrubs, deciduous

2.9; -6.0; 10.8; 4.3; 20.4; 45.6; 421 ▎528; 318; 10 ▎23.9; 421; 226.7

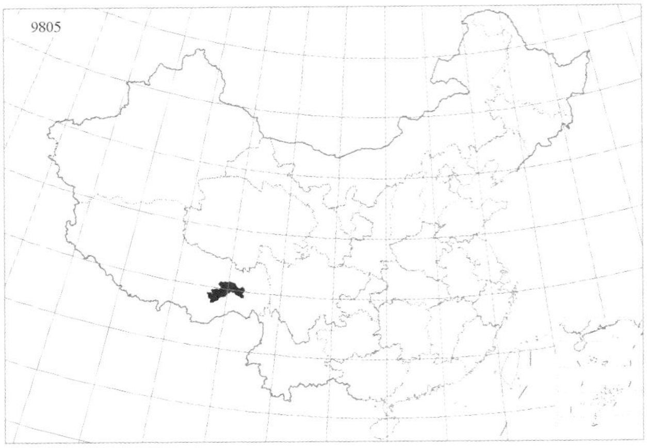

Berberidaceae 小檗科 *Berberis* 小檗属
Berberis julianae C. K. Schneid. 豪猪刺
Shrubs, evergreen

15.6 (9.3~21.3); 4.7 (-2.5~12.8); 25.8 (18.8~29.3); 15.7 (9.5~21.3); 131.1 (69.8~196.1); 3.0 (0.0~20.6); 845 (603~1130) ▎1354 (760~1997); 539 (345~728); 124 (17~254) ▎59.9 (5.0~130.7); 842 (603~1130); 236.2 (140.5~334.4)

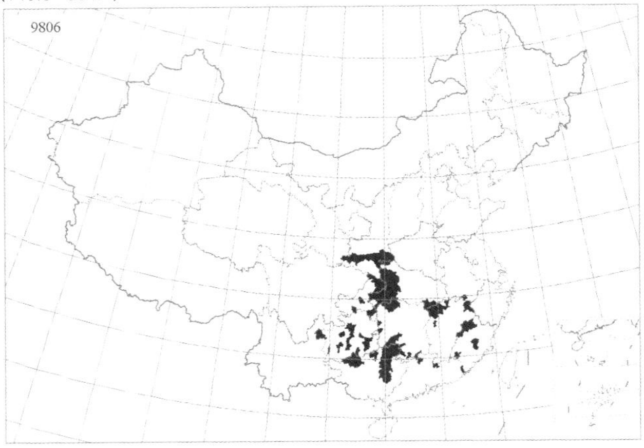

Berberidaceae 小檗科 *Berberis* 小檗属
Berberis kartanica Ahrendt 卡达小檗
Shrubs, deciduous

2.4; -6.4; 10.2; 4.1; 19.0; 50.0; 411 ▎463; 292; 10 ▎9.1; 411; 155.4

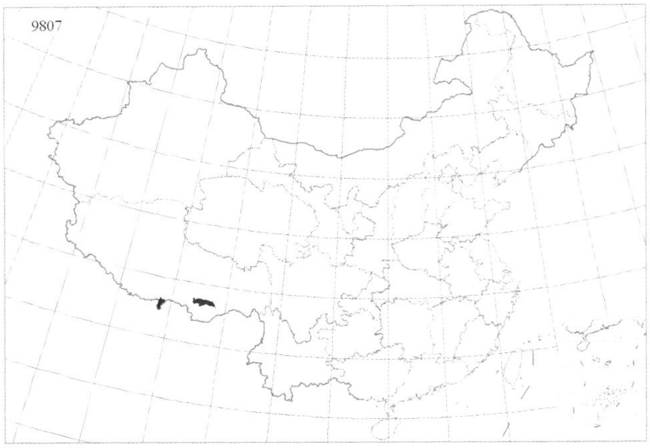

Berberidaceae 小檗科 *Berberis* 小檗属
Berberis kawakamii Hayata 台湾小檗
Shrubs, evergreen

18.7 (16.2~22.2); 13.2 (11.0~17.7); 23.2 (20.1~27.0); 18.8 (16.2~22.2); 165.4 (134.2~206.7); 0.0 (0.0~0.0); 919 (773~1123) ▎2506 (1836~2992); 1166 (900~1847); 230 (107~402) ▎179.5 (64.2~246.5); 913 (773~1123); 266.5 (202.8~303.6)

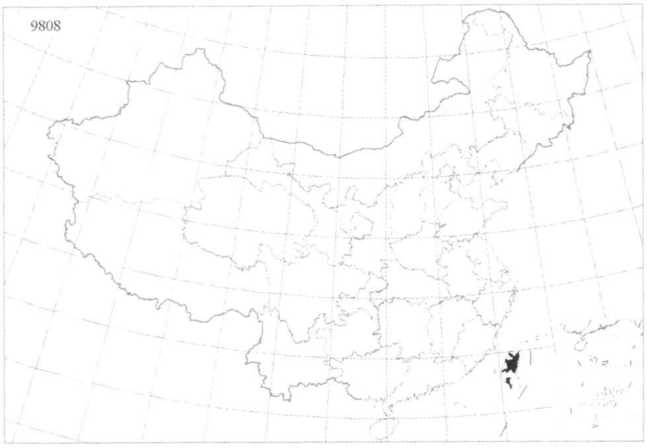

Berberidaceae 小檗科 *Berberis* 小檗属
Berberis kongboensis Ahrendt 工布小檗
Shrubs, deciduous

1.8; -7.0; 9.7; 3.7; 15.9; 53.6; 397 ▎431; 267; 9 ▎2.5; 397; 186.6

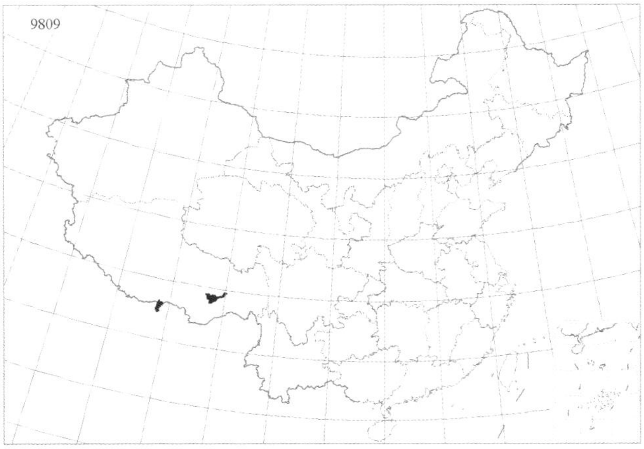

Berberidaceae 小檗科 *Berberis* 小檗属
Berberis lecomtei C. K. Schneid. 光叶小檗
Shrubs, deciduous

5.2 (2.0~11.0); -3.2 (-7.4~3.7); 12.3 (9.9~17.0); 6.1 (3.8~11.1); 33.9 (15.5~74.6); 31.2 (1.8~51.9); 485 (413~640) ▎857 (777~990); 526 (457~619); 15 (11~21) ▎79.7 (54.7~125.9); 484 (413~638); 285.0 (248.3~311.5)

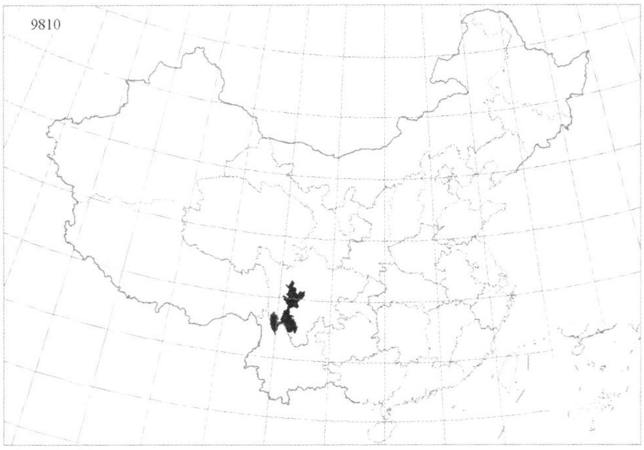

Berberidaceae 小檗科 *Berberis* 小檗属

Berberis lempergiana Ahrendt 天台小檗

Shrubs, evergreen

16.3 (15.4~17.7)；5.4 (3.3~7.8)；27.2 (25.6~28.9)；16.4 (15.4~17.7)；136.9 (124.9~152.4)；0.3 (0.0~1.9)；871 (810~920) ❙ 1637 (1360~1893)；592 (481~677)；194 (176~242) ❙ 88.5 (51.2~129.9)；871 (810~920)；230.5 (189.1~273.1)

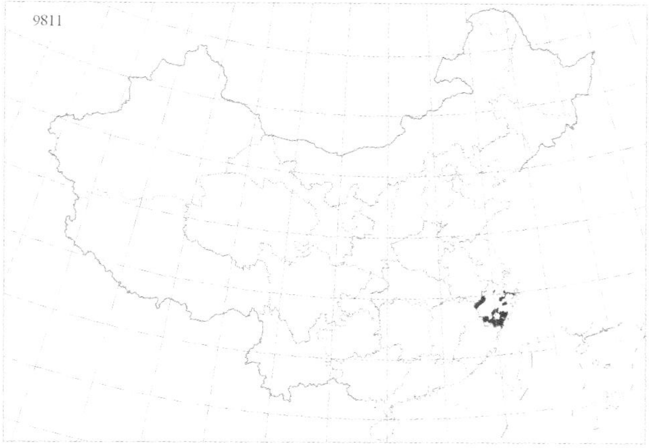

Berberidaceae 小檗科 *Berberis* 小檗属

Berberis lepidifolia Ahrendt 鳞叶小檗

Shrubs, deciduous

12.5 (6.1~17.9)；5.2 (-1.5~10.7)；18.4 (12.7~23.4)；12.6 (6.3~18.0)；95.0 (35.2~155.7)；4.4 (0.0~21.8)；687 (496~878) ❙ 973 (801~1057)；595 (496~641)；23 (12~42) ❙ 44.6 (4.4~67.7)；652 (496~749)；307.1 (242.3~351.6)

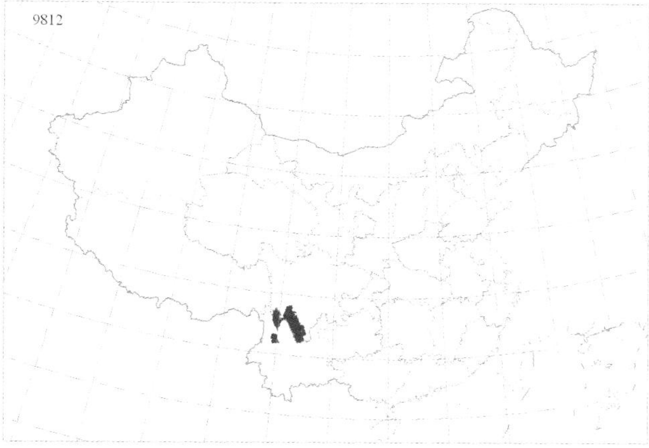

Berberidaceae 小檗科 *Berberis* 小檗属

Berberis leptoclada Diels 细枝小檗

Shrubs, deciduous

5.8；-1.8；12.5；6.1；33.4；23.7；489 ❙ 771；461；25 ❙ 57.7；489；278.2

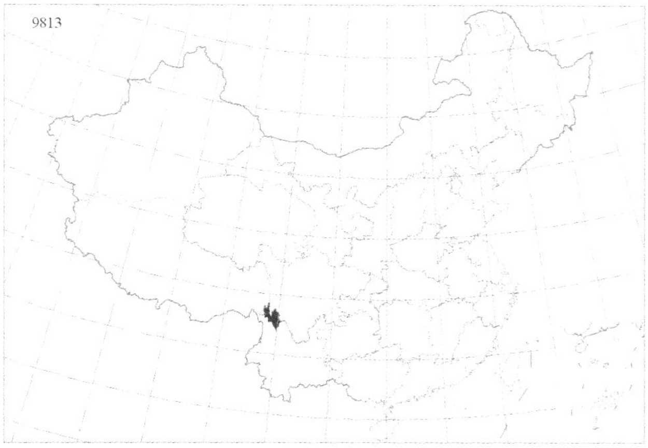

Berberidaceae 小檗科 *Berberis* 小檗属

Berberis levis Franch. 平滑小檗

Shrubs, evergreen

13.3 (5.4~18.9)；5.6 (-3.2~12.0)；19.8 (12.3~26.6)；13.4 (5.8~19.0)；103.2 (31.5~167.8)；2.7 (0.0~26.7)；715 (481~929) ❙ 973 (741~1299)；560 (411~641)；31 (15~92) ❙ 38.3 (-14.2~101.2)；687 (481~912)；271.4 (142.9~372.3)

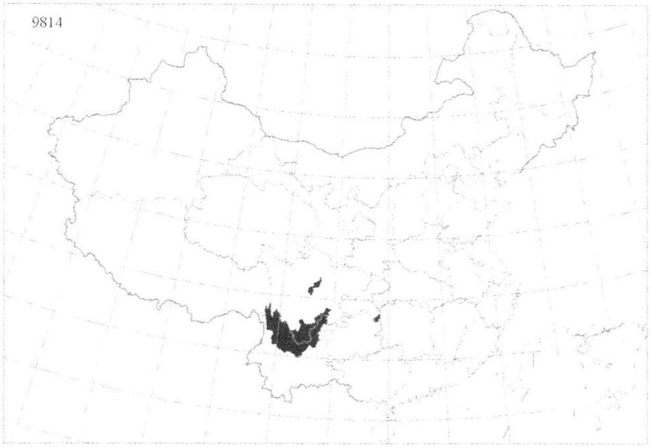

Berberidaceae 小檗科 *Berberis* 小檗属

Berberis lhapsoensis Harber 藏南小檗

Shrubs, deciduous

2.1；-7.2；10.6；4.1；19.0；52.8；408 ❙ 446；279；7 ❙ 8.0；384；242.4

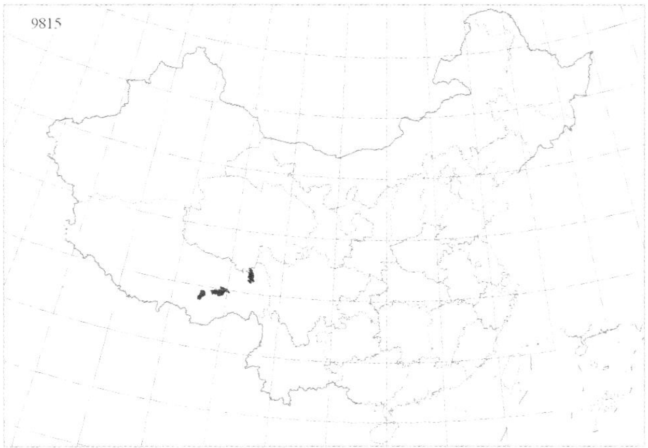

Berberidaceae 小檗科 *Berberis* 小檗属

Berberis lijiangensis C. Y. Wu ex S. Y. Bao 丽江小檗

Shrubs, evergreen

10.8；3.9；16.6；10.9；71.8；1.3；633 ❙ 968；594；24 ❙ 52.8；633；319.8

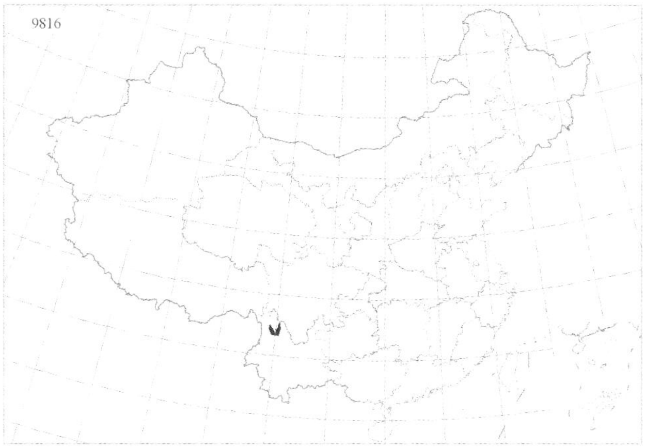

Berberidaceae 小檗科 *Berberis* 小檗属
Berberis lubrica C. K. Schneid. 亮叶小檗
Shrubs, evergreen

11.0; 3.7; 17.0; 11.1; 74.6; 1.8; 640 ❙ 990; 619; 15 ❙ 54.7; 638; 296.0

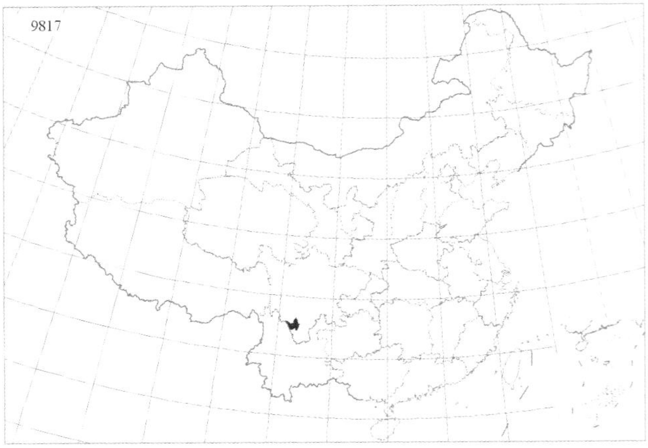

Berberidaceae 小檗科 *Berberis* 小檗属
Berberis ludlowii Ahrendt 卢氏小檗
Shrubs, deciduous

4.6; −3.7; 12.0; 5.7; 30.5; 34.7; 468 ❙ 618; 366; 19 ❙ 23.7; 447; 266.8

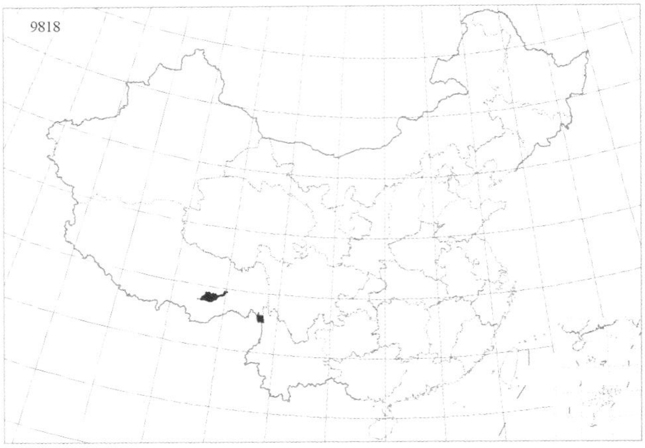

Berberidaceae 小檗科 *Berberis* 小檗属
Berberis malipoensis C. Y. Wu et S. Y. Bao 麻栗坡小檗
Shrubs, evergreen

17.4; 9.7; 22.9; 17.5; 149.7; 0.0; 848 ❙ 1338; 756; 50 ❙ 57.6; 848; 348.0

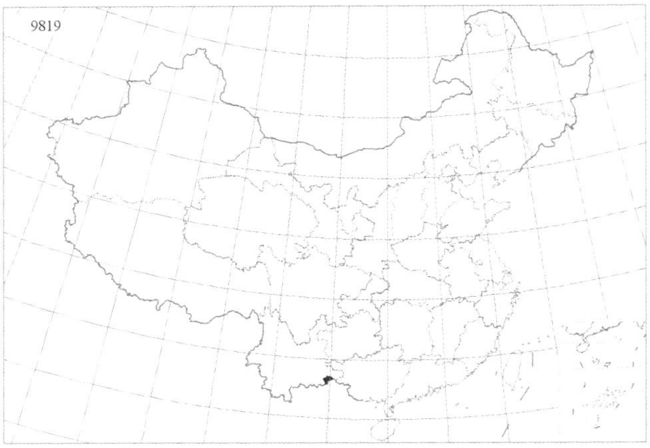

Berberidaceae 小檗科 *Berberis* 小檗属
Berberis mekongensis W. W. Sm. 湄公小檗
Shrubs, deciduous

5.7; −2.1; 12.6; 6.4; 35.9; 26.7; 498 ❙ 894; 517; 29 ❙ 82.6; 498; 291.2

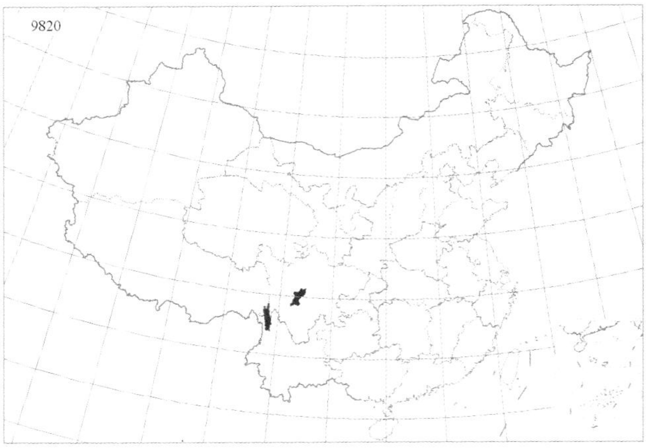

Berberidaceae 小檗科 *Berberis* 小檗属
Berberis mianningensis T. S. Ying 冕宁小檗
Shrubs, deciduous

10.2; 2.4; 16.9; 10.3; 68.2; 5.1; 620 ❙ 1036; 606; 18 ❙ 67.1; 620; 298.0

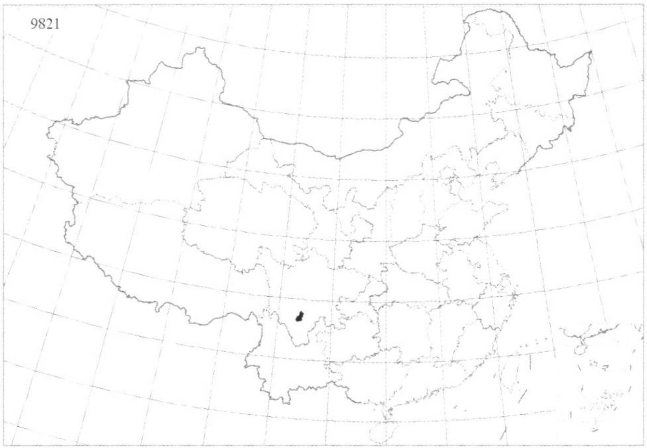

Berberidaceae 小檗科 *Berberis* 小檗属
Berberis minutiflora C. K. Schneid. 小花小檗
Shrubs, deciduous

9.8; 2.4; 16.0; 9.9; 65.4; 7.8; 602 ❙ 888; 539; 25 ❙ 49.0; 602; 269.3

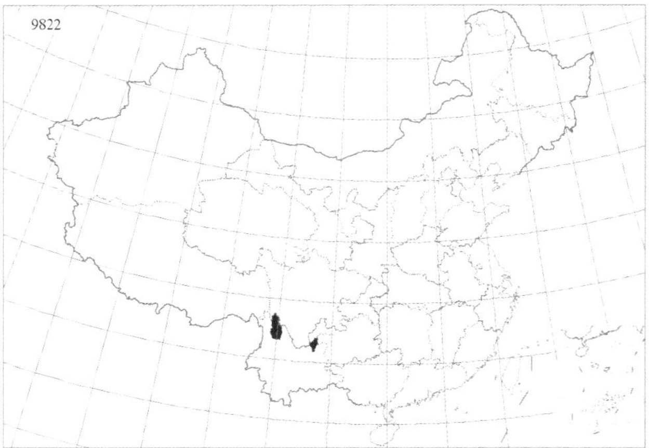

Berberidaceae 小檗科 *Berberis* 小檗属
Berberis moloensis (Ahrendt) Harber 莫洛小檗
Shrubs, deciduous

3.4; −5.5; 11.4; 4.7; 23.3; 41.8; 431 ❙ 527; 317; 9 ❙ 22.4; 431; 280.0

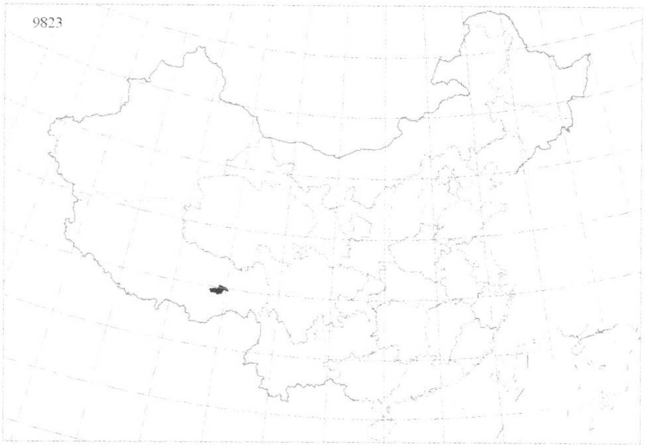

Berberidaceae 小檗科 *Berberis* 小檗属
Berberis morrisonensis Hayata 玉山小檗
Shrubs, deciduous

18.2; 12.3; 23.1; 18.2; 158.8; 0.0; 893 ❙ 2313; 1041; 229 ❙ 168.7; 882; 264.4

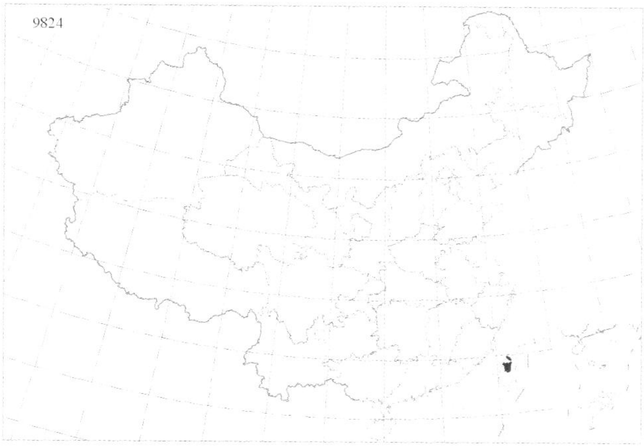

Berberidaceae 小檗科 *Berberis* 小檗属
Berberis mouillacana C. K. Schneid. 变刺小檗
Shrubs, deciduous

3.8 (−1.7~14.8); −5.9 (−12.3~3.8); 12.4 (8.0~25.3); 5.4 (2.4~14.8); 30.5 (6.4~119.2); 44.3 (1.4~86.1); 464 (344~795) ❙ 862 (502~1358); 451 (301~546); 26 (12~88) ❙ 85.8 (45.7~125.9); 464 (344~795); 273.7 (187.9~357.7)

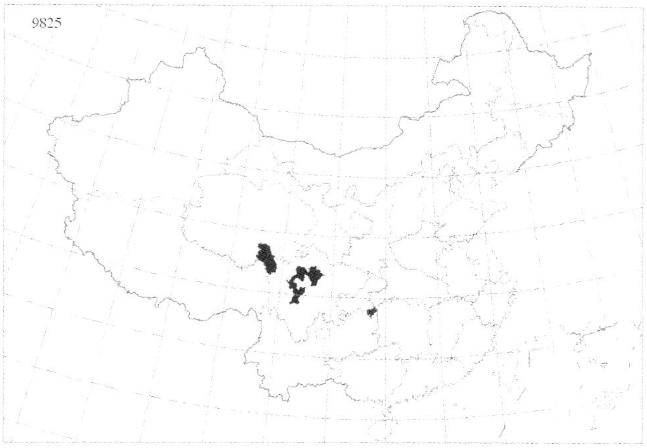

Berberidaceae 小檗科 *Berberis* 小檗属
Berberis muliensis Ahrendt 木里小檗
Shrubs, deciduous

5.3 (1.6~12.6); −2.8 (−8.0~6.0); 12.3 (9.8~17.9); 6.3 (3.7~12.6); 36.2 (15.2~91.6); 31.8 (0.0~56.2); 489 (408~678) ❙ 828 (628~1057); 505 (410~610); 19 (11~42) ❙ 71.3 (48.6~125.9); 489 (408~678); 280.1 (236.3~349.8)

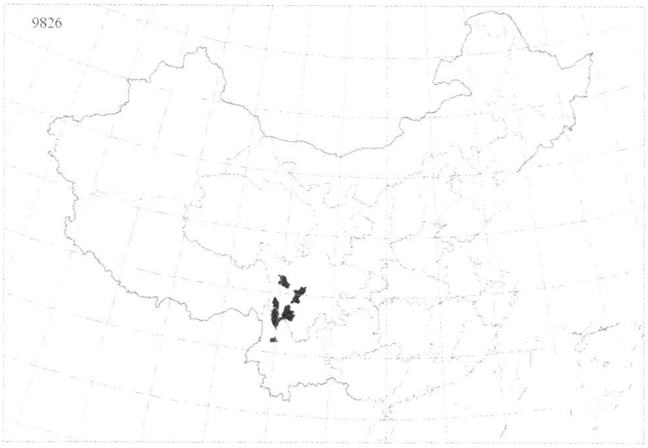

Berberidaceae 小檗科 *Berberis* 小檗属
Berberis multicaulis T. S. Ying 多枝小檗
Shrubs, deciduous

−1.3; −9.8; 6.6; 2.1; 3.6; 78.9; 333 ❙ 316; 210; 22 ❙ −5.1; 316; 71.0

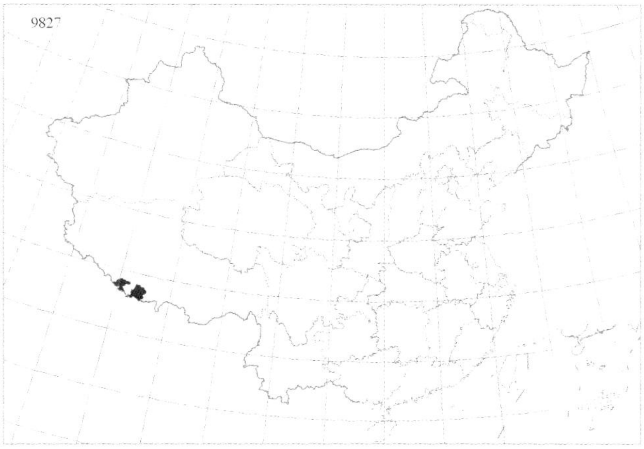

Berberidaceae 小檗科 *Berberis* 小檗属
Berberis multiserrata T. S. Ying 粗齿小檗
Shrubs, deciduous

5.3; −2.4; 12.2; 5.7; 30.5; 26.2; 479 ❙ 961; 558; 31 ❙ 100.7; 479; 234.2

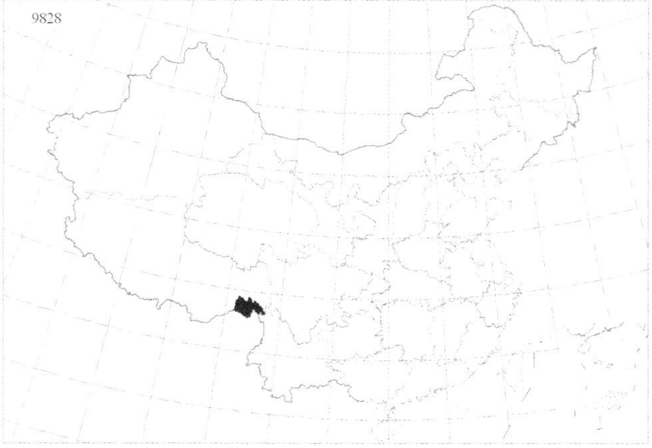

Berberidaceae 小檗科 *Berberis* 小檗属
Berberis nantoensis C. K. Schneid. 南投小檗
Shrubs, evergreen

16.2；11.0；20.1；16.2；134.2；0.0；773 ▎2680；1186；245 ▎246.5；773；289.1

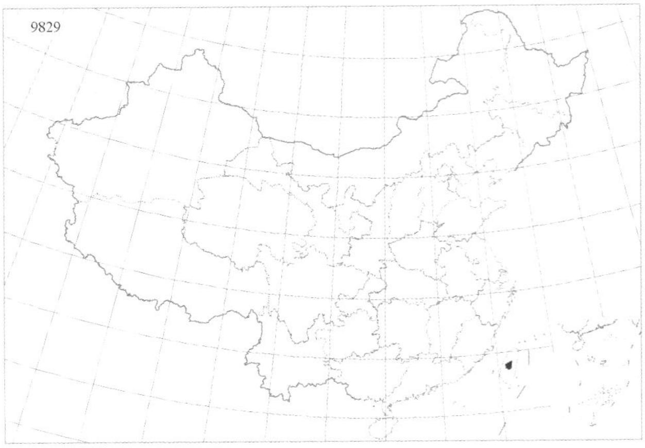

Berberidaceae 小檗科 *Berberis* 小檗属
Berberis nemorosa C. K. Schneid. 林地小檗
Shrubs, evergreen

18.4；8.2；26.9；18.4；161.2；0.0；955 ▎1495；658；138 ▎56.9；955；241.2

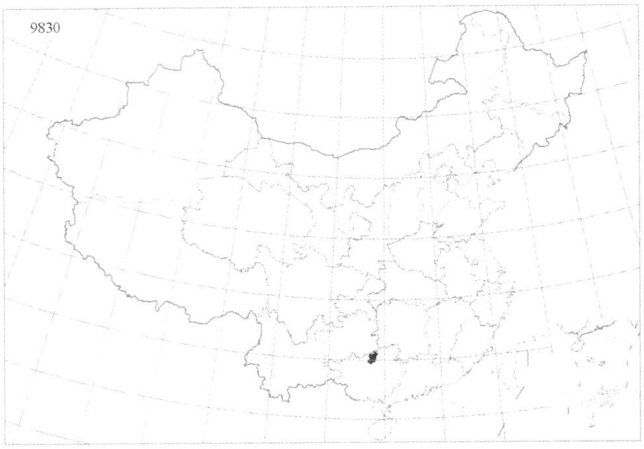

Berberidaceae 小檗科 *Berberis* 小檗属
Berberis nullinervis T. S. Ying 无脉小檗
Shrubs, deciduous

-0.7；-10.8；8.6；2.8；10.0；77.9；345 ▎250；198；1 ▎-27.5；250；131.4

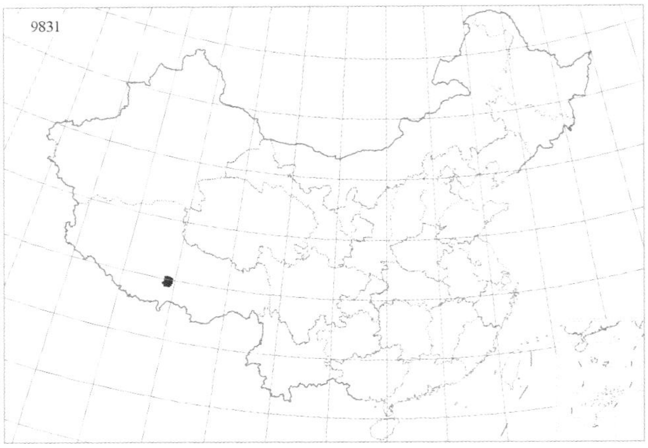

Berberidaceae 小檗科 *Berberis* 小檗属
Berberis nummularia Bunge 新疆小檗
Shrubs, deciduous

5.1 (-4.3~12.5)；-13.2 (-20.9~-5.1)；19.9 (9.6~27.1)；8.4 (2.5~13.2)；66.1 (9.7~114.0)；64.5 (23.1~121.1)；600 (329~817) ▎168 (32~450)；79 (16~211)；14 (0~48) ▎-68.9 (-95.9~3.9)；167 (32~406)；76.7 (1.6~297.3)

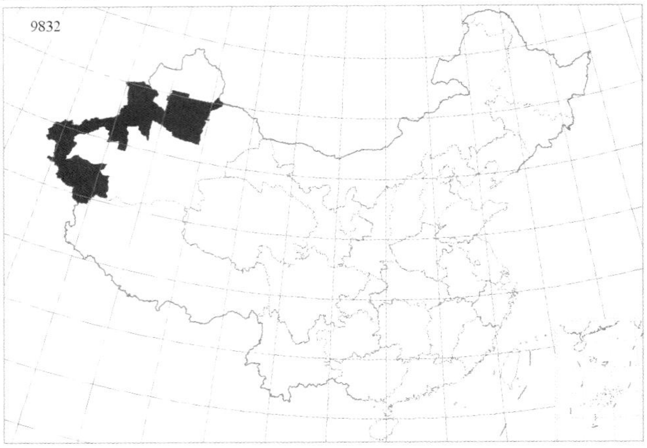

Berberidaceae 小檗科 *Berberis* 小檗属
Berberis oblanceifolia C. M. Hu 石门小檗
Shrubs, deciduous

13.5；3.4；22.6；13.5；104.2；1.9；745 ▎1126；528；72 ▎49.9；745；199.5

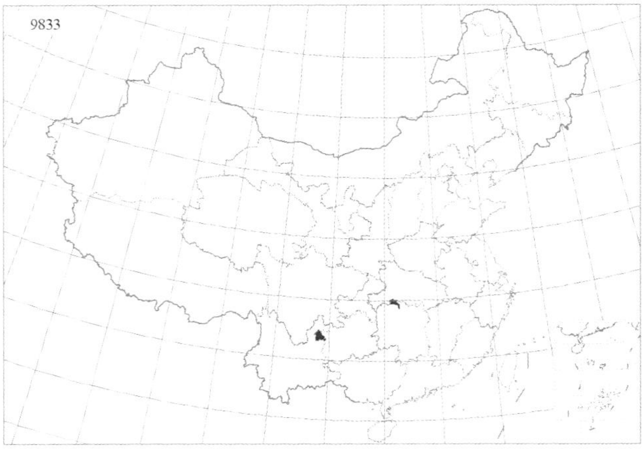

Berberidaceae 小檗科 *Berberis* 小檗属
Berberis obovatifolia T. S. Ying 裂瓣小檗
Shrubs, evergreen

3.7；-6.1；12.1；5.1；27.8；43.5；441 ▎337；257；1 ▎-23.5；337；217.3

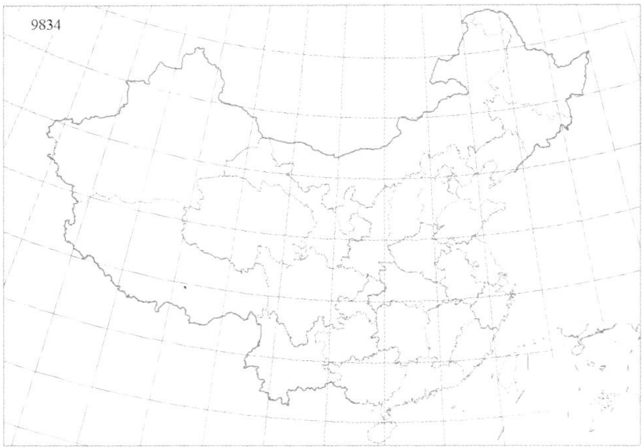

Annonaceae 番荔枝科 *Alphonsea* 藤春属

Alphonsea monogyna Merr. et Chun 藤春

Trees, evergreen

21.7 (17.2~24.8); 15.2 (9.1~20.1); 26.4 (22.3~28.8); 21.8 (17.3~24.8); 201.5 (147.0~237.6); 0.0 (0.0~0.0); 1142 (833~1385) ❙ 1483 (868~1900); 661 (312~1056); 75 (41~168) ❙ 33.2 (-34.3~88.8); 1084 (833~1360); 421.0 (218.0~659.0)

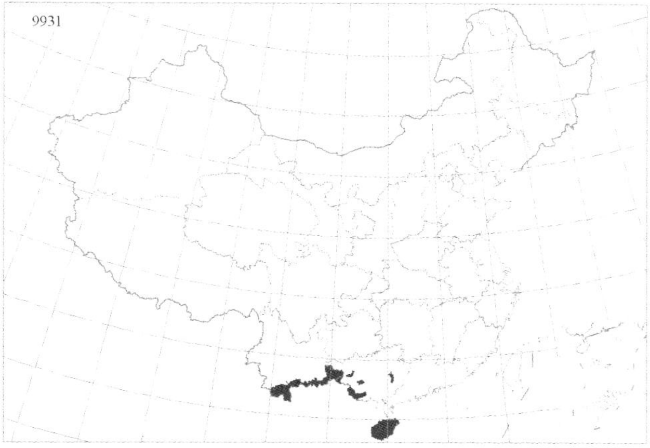

Annonaceae 番荔枝科 *Alphonsea* 藤春属

Alphonsea squamosa Finet et Gagnep. 多苞藤春

Small trees, evergreen

19.6; 12.9; 24.2; 19.6; 175.7; 0.0; 988 ❙ 1355; 671; 71 ❙ 38.6; 985; 463.9

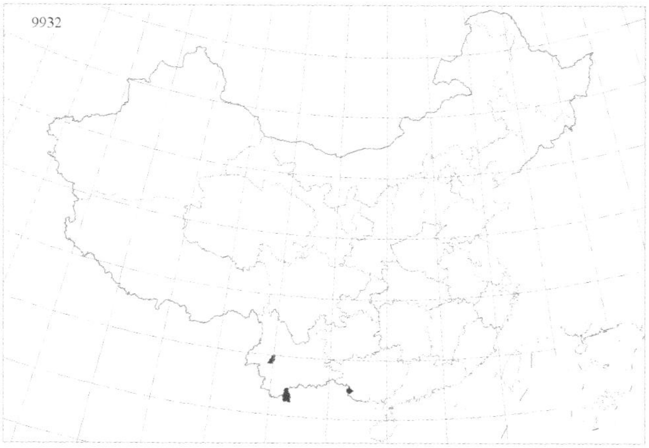

Annonaceae 番荔枝科 *Alphonsea* 藤春属

Alphonsea tsangyuanensis P. T. Li 多脉藤春

Trees, evergreen

18.9 (16.4~21.0); 13.0 (9.6~15.1); 22.6 (20.6~25.6); 18.9 (16.4~21.0); 167.1 (137.3~191.8); 0.0 (0.0~0.0); 907 (795~1053) ❙ 1503 (1352~1751); 792 (655~996); 57 (41~77) ❙ 66.4 (44.5~112.2); 891 (795~1053); 521.7 (427.9~659.0)

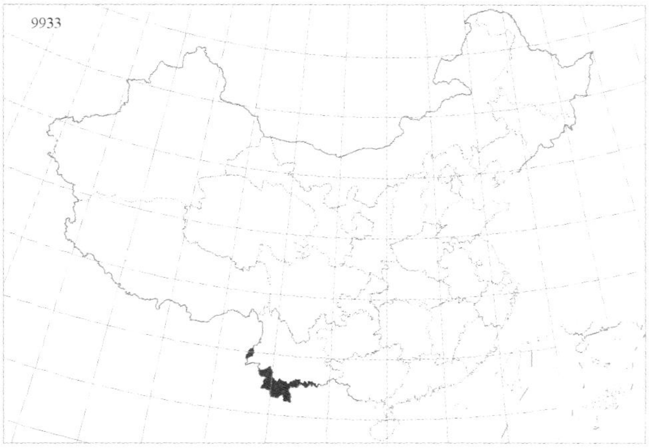

Annonaceae 番荔枝科 *Anaxagorea* 蒙蒿子属

Anaxagorea luzonensis A. Gray 蒙蒿子

Shrubs, evergreen

23.3 (21.3~24.8); 16.8 (13.2~20.1); 28.1 (27.2~29.0); 23.4 (21.3~24.8); 220.2 (195.9~237.6); 0.0 (0.0~0.0); 1266 (1111~1385) ❙ 1774 (1405~2435); 812 (473~1428); 91 (41~130) ❙ 41.6 (1.5~99.5); 1205 (1111~1360); 386.9 (139.6~620.0)

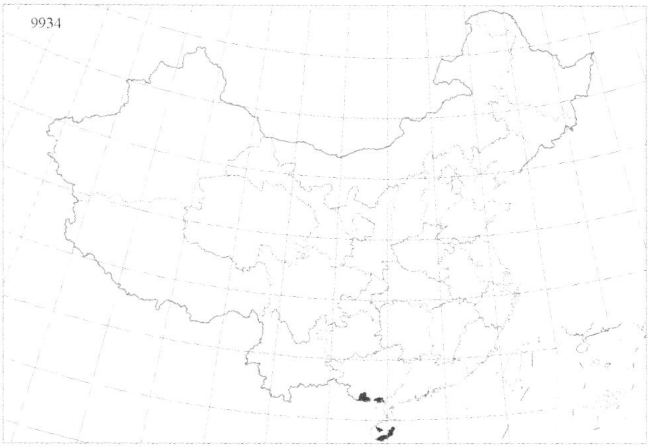

Annonaceae 番荔枝科 *Artabotrys* 鹰爪属

Artabotrys fragrans Ast ex Jovet-Ast 香鹰爪花

Vines, evergreen

19.7; 13.3; 24.0; 19.8; 177.0; 0.0; 967 ❙ 1451; 743; 65 ❙ 50.7; 956; 537.7

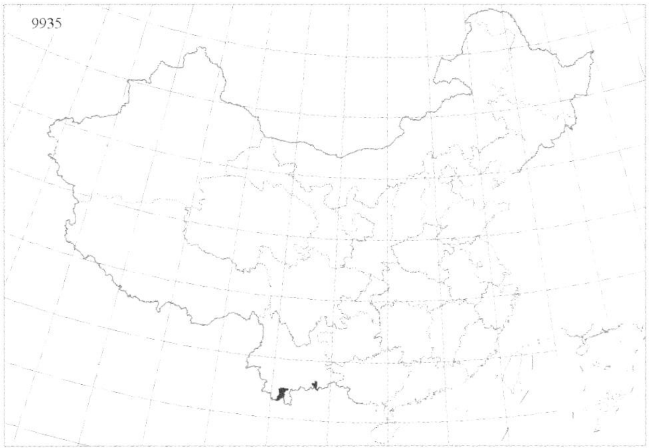

Annonaceae 番荔枝科 *Artabotrys* 鹰爪属

Artabotrys hainanensis R. E. Fries 海南鹰爪花

Scandent shrubs, evergreen

22.2 (19.6~24.8); 15.1 (11.2~20.1); 27.4 (25.8~28.6); 22.3 (19.6~24.8); 207.3 (175.5~237.6); 0.0 (0.0~0.0); 1185 (986~1385) ❙ 1592 (1380~1900); 776 (455~1056); 87 (41~125) ❙ 36.3 (1.5~71.0); 1113 (986~1204); 348.6 (235.1~527.5)

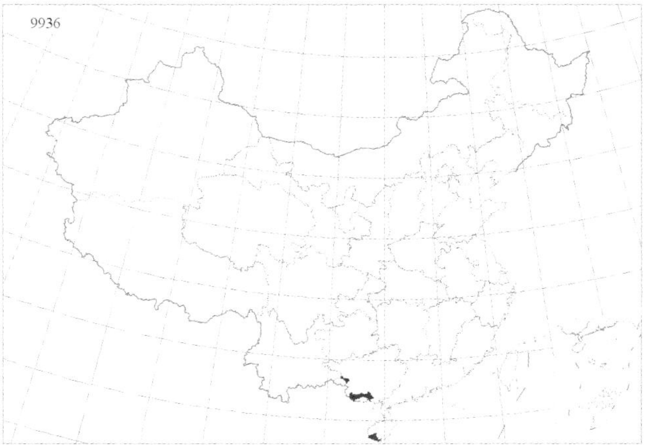

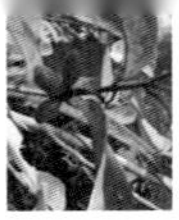

Annonaceae 番荔枝科 *Artabotrys* 鹰爪属

Artabotrys hexapetalus (L. f.) Bhandari 鹰爪花

Scandent shrubs, evergreen

19.8 (14.6~24.8); 11.4 (3.8~20.1); 26.8 (21.4~29.3); 19.9 (14.7~24.8); 178.4 (117.7~237.6); 0.0 (0.0~1.5); 1034 (779~1385) ❙ 1569 (868~3495); 661 (312~994); 142 (41~897) ❙ 54.1 (-34.3~224.4); 1009 (779~1360); 293.7 (139.6~620.0)

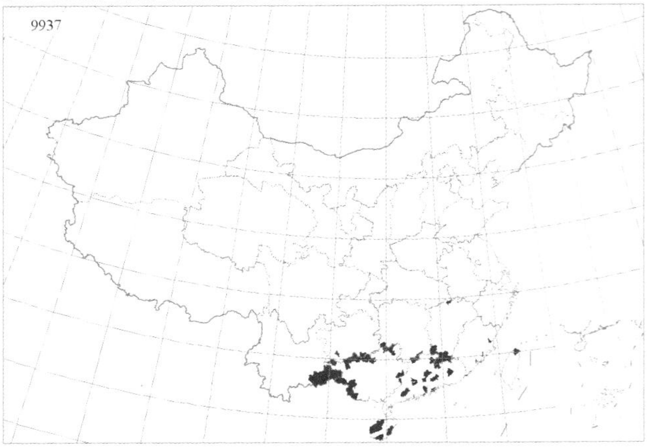

Annonaceae 番荔枝科 *Artabotrys* 鹰爪属

Artabotrys hongkongensis Hance 香港鹰爪花

Scandent shrubs, evergreen

20.0 (14.3~24.8); 11.5 (3.3~20.1); 26.9 (20.3~29.8); 20.0 (14.3~24.8); 180.3 (114.1~237.6); 0.0 (0.0~2.3); 1042 (765~1385) ❙ 1556 (868~2435); 681 (312~1428); 128 (41~217) ❙ 51.0 (-34.3~112.2); 1028 (765~1299); 266.5 (95.3~659.0)

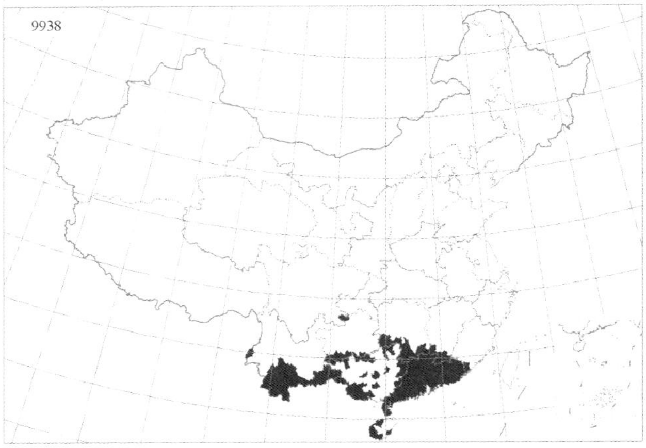

Annonaceae 番荔枝科 *Artabotrys* 鹰爪属

Artabotrys multiflorus C. E. C. Fisch. 多花鹰爪

Vines, evergreen

18.2; 9.2; 25.3; 18.3; 159.2; 0.0; 915 ❙ 1409; 728; 82 ❙ 54.0; 915; 308.5

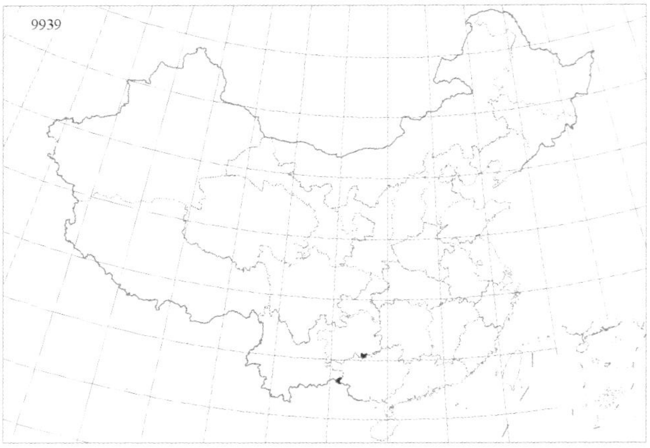

Annonaceae 番荔枝科 *Artabotrys* 鹰爪属

Artabotrys pilosus Merr. et Chun 毛叶鹰爪花

Scandent shrubs, evergreen

23.4 (21.0~24.8); 17.3 (13.2~20.1); 27.9 (26.7~29.2); 23.5 (21.0~24.8); 221.6 (192.3~237.6); 0.0 (0.0~0.0); 1269 (1081~1385) ❙ 1574 (1176~1876); 629 (447~891); 78 (41~125) ❙ 25.3 (-12.0~64.3); 1186 (1081~1299); 407.8 (178.7~620.0)

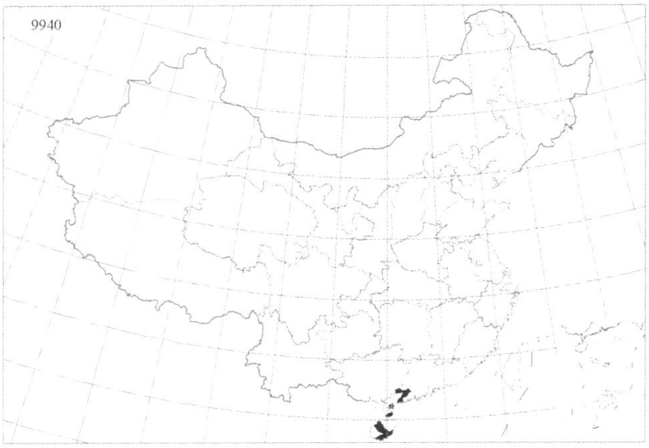

Annonaceae 番荔枝科 *Artabotrys* 鹰爪属

Artabotrys punctulatus C. Y. Wu ex S. H. Yuan 点叶鹰爪花

Scandent shrubs, evergreen

20.5; 14.9; 23.7; 20.5; 185.9; 0.0; 992 ❙ 1464; 655; 66 ❙ 47.6; 959; 570.1

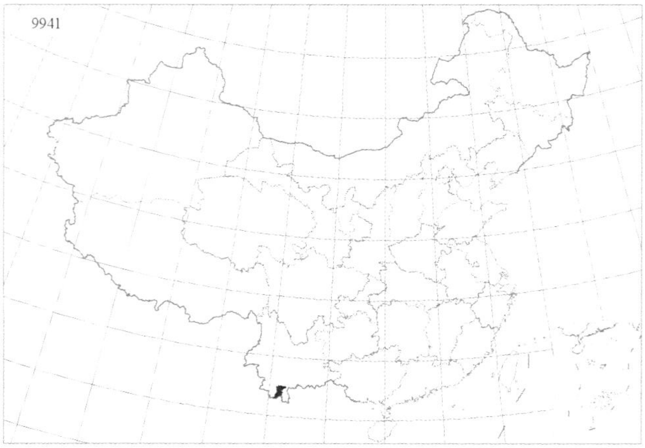

Annonaceae 番荔枝科 *Artabotrys* 鹰爪属

Artabotrys rhynchocarpus C. Y. Wu ex S. H. Yuan 喙果鹰爪花

Scandent shrubs, evergreen

19.4; 12.6; 24.1; 19.4; 172.6; 0.0; 954 ❙ 1445; 787; 64 ❙ 52.3; 954; 521.5

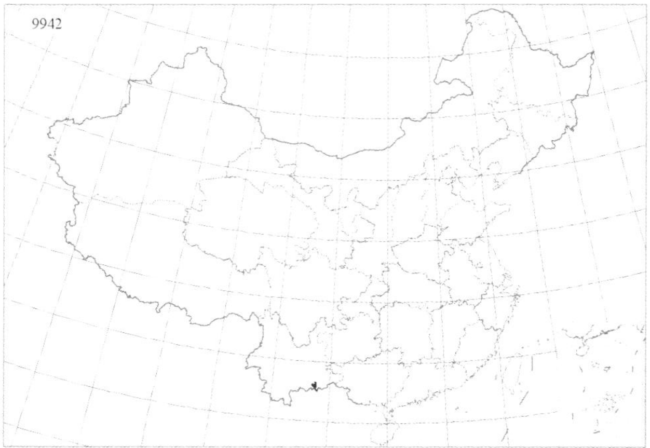

Annonaceae 番荔枝科 *Chieniodendron* 蕉木属
Chieniodendron hainanense (Merr.) Tsiang et P. T. Li 蕉木
Trees, evergreen

21.4 (16.3~24.8); 14.7 (9.1~20.1); 26.2 (21.4~29.0); 21.5 (16.4~24.8); 197.5 (136.4~237.6); 0.0 (0.0~0.0); 1122 (797~1385) ❙ 1508 (868~2435); 712 (312~1428); 79 (41~168) ❙ 37.3 (-34.3~99.5); 1074 (797~1360); 393.2 (199.6~659.0)

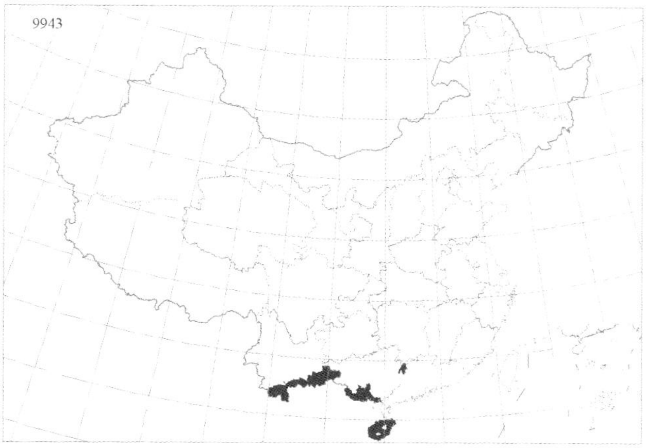

Annonaceae 番荔枝科 *Cyathostemma* 杯冠木属
Cyathostemma yunnanense Hu 杯冠木
Scandent shrubs, evergreen

19.3 (16.4~21.0); 13.4 (9.6~15.1); 23.1 (20.6~25.6); 19.4 (16.4~21.0); 172.3 (137.3~191.8); 0.0 (0.0~0.0); 934 (795~1053) ❙ 1531 (1276~1751); 811 (655~996); 64 (51~77) ❙ 65.3 (36.8~112.2); 925 (795~1053); 549.5 (452.0~659.0)

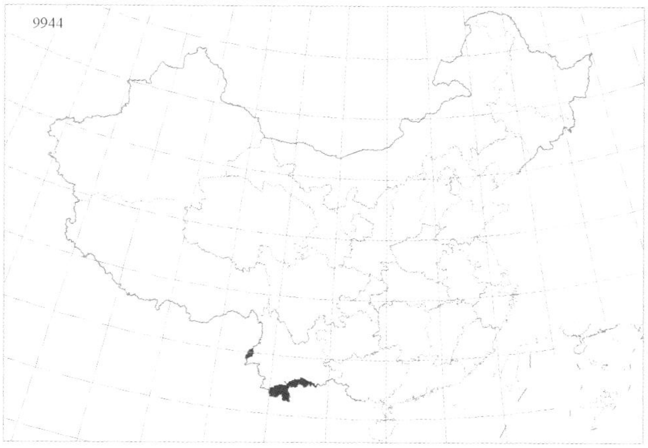

Annonaceae 番荔枝科 *Desmos* 假鹰爪属
Desmos chinensis Lour. 假鹰爪
Shrubs, evergreen

20.2 (12.4~24.8); 12.2 (4.7~20.1); 26.5 (18.6~29.8); 20.3 (12.4~24.8); 183.1 (89.1~237.6); 0.0 (0.0~0.3); 1055 (675~1385) ❙ 1495 (868~2435); 689 (312~1428); 111 (31~198) ❙ 42.4 (-34.3~100.8); 1043 (675~1360); 270.2 (95.3~620.0)

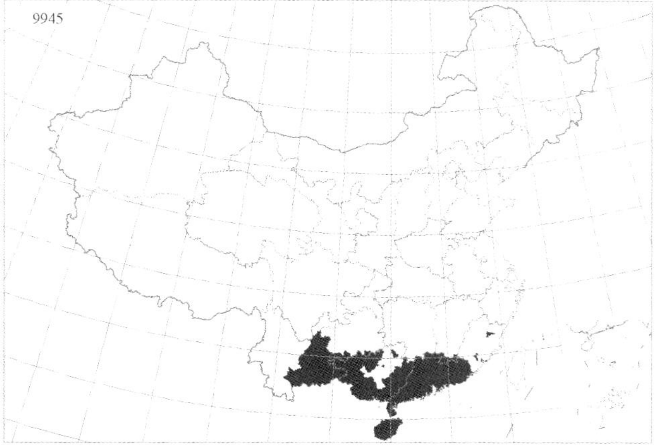

Annonaceae 番荔枝科 *Desmos* 假鹰爪属
Desmos dumosus (Roxb.) Saff. 毛叶假鹰爪
Shrubs, evergreen

18.9 (15.3~22.2); 10.6 (4.5~15.1); 25.4 (21.4~28.6); 19.0 (15.4~22.3); 167.8 (125.0~207.5); 0.0 (0.0~0.5); 960 (797~1190) ❙ 1470 (1028~1758); 712 (481~996); 107 (45~189) ❙ 54.2 (12.5~93.2); 956 (797~1190); 332.6 (152.4~659.0)

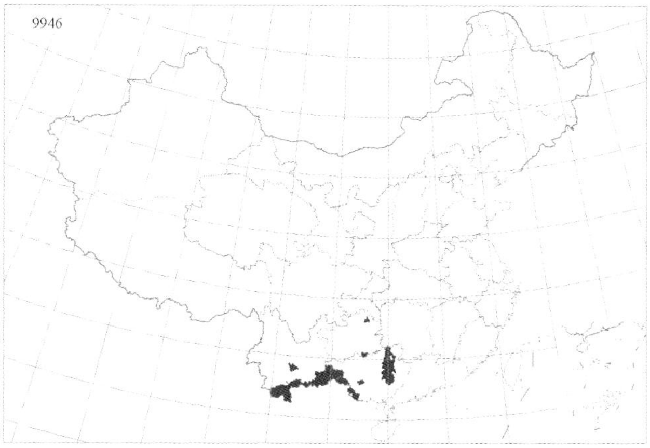

Annonaceae 番荔枝科 *Desmos* 假鹰爪属
Desmos grandifolius (Finet et Gagnep.) C. Y. Wu ex P. T. Li 大叶假鹰爪
Shrubs, evergreen

21.2; 13.4; 26.7; 21.2; 193.9; 0.0; 1088 ❙ 1355; 726; 62 ❙ 25.1; 1087; 433.6

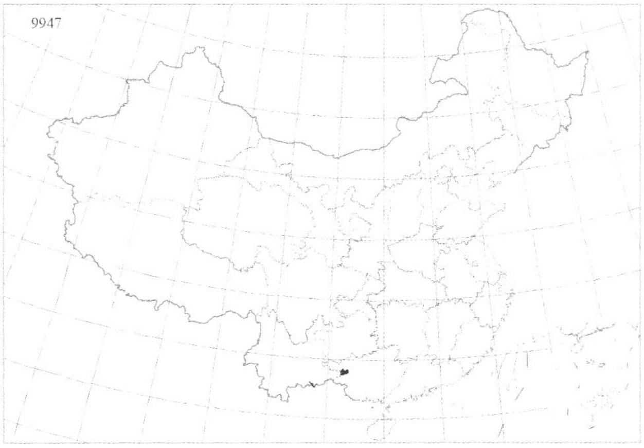

Annonaceae 番荔枝科 *Desmos* 假鹰爪属
Desmos macrocalyx (Finet et Gagnepain) P. T. Li 大萼假鹰爪
Shrubs

21.6 (17.3~24.8); 13.9 (7.4~20.1); 27.5 (22.4~29.2); 21.6 (17.3~24.8); 199.4 (147.1~237.6); 0.0 (0.0~0.0); 1147 (833~1385) ❙ 1509 (1176~1771); 663 (392~997); 109 (41~189) ❙ 34.9 (-12.0~85.5); 1099 (833~1273); 323.4 (178.7~620.0)

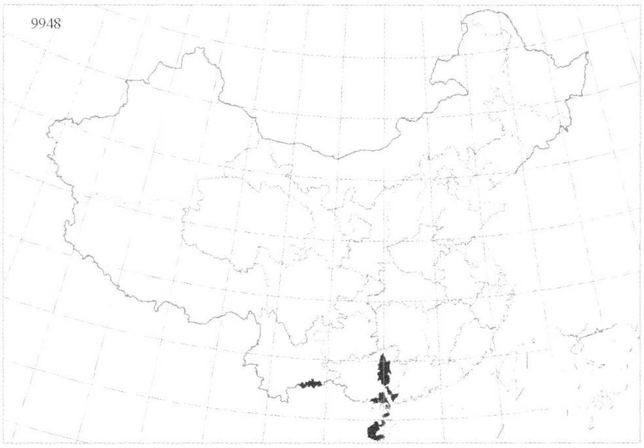

Annonaceae 番荔枝科 *Desmos* 假鹰爪属

Desmos robinsonii (Jovet-Ast) P. T. Li 钝叶假鹰爪

Shrubs

17.4; 7.4; 25.5; 17.4; 148.9; 0.0; 886 ❙ 1381; 637; 107 ❙ 55.9; 886; 245.3

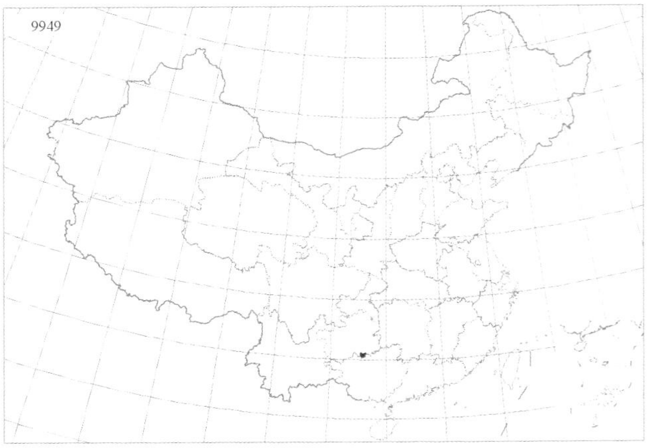

Annonaceae 番荔枝科 *Desmos* 假鹰爪属

Desmos rostratus (Merr. et Chun) P. T. Li 喙果假鹰爪

Shrubs, evergreen

20.1 (13.3~24.8); 12.6 (4.8~20.1); 25.7 (19.6~28.7); 20.1 (13.4~24.8); 181.6 (100.4~237.6); 0.0 (0.0~0.2); 1037 (706~1385) ❙ 1510 (868~2435); 718 (312~1428); 95 (41~196) ❙ 49.5 (-34.3~145.8); 1000 (706~1269); 370.1 (185.5~659.0)

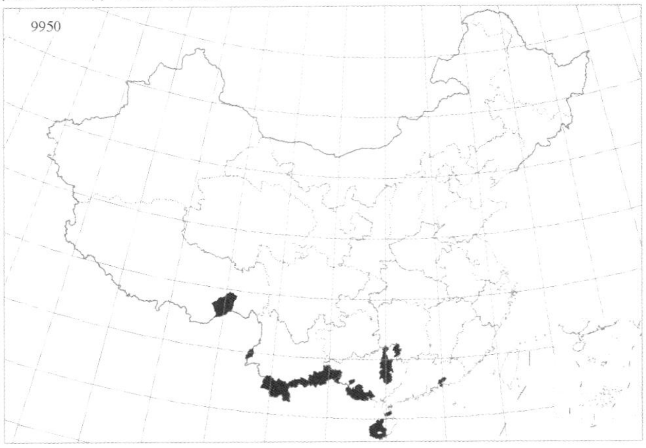

Annonaceae 番荔枝科 *Desmos* 假鹰爪属

Desmos saccopetaloides (W. T. Wang) P. T. Li 亮花假鹰爪

Trees, evergreen

18.9; 13.0; 22.5; 19.0; 167.5; 0.0; 911 ❙ 1507; 759; 62 ❙ 65.6; 910; 546.7

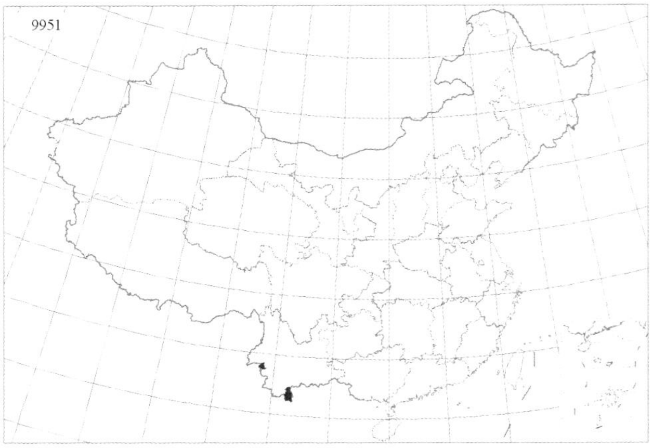

Annonaceae 番荔枝科 *Desmos* 假鹰爪属

Desmos sootepensis (Craib) J. F. Maxwell 黄花假鹰爪

Shrubs, evergreen

19.5; 13.7; 23.0; 19.5; 173.9; 0.0; 936 ❙ 1485; 700; 65 ❙ 58.8; 920; 558.6

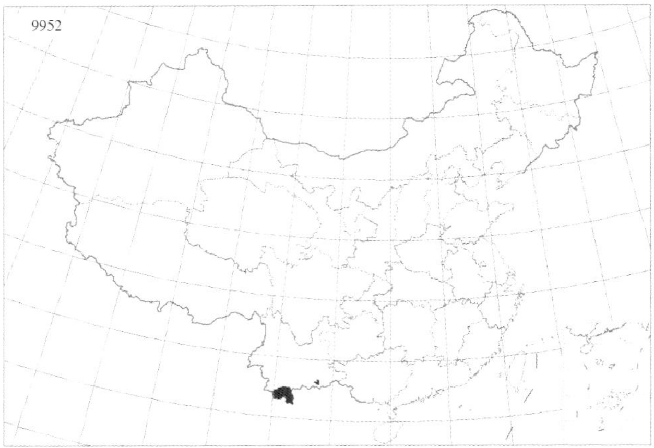

Annonaceae 番荔枝科 *Desmos* 假鹰爪属

Desmos yunnanensis (Hu) P. T. Li 云南假鹰爪

Shrubs or small trees, evergreen

19.1; 13.2; 22.7; 19.1; 169.7; 0.0; 925 ❙ 1586; 772; 68 ❙ 74.1; 913; 618.7

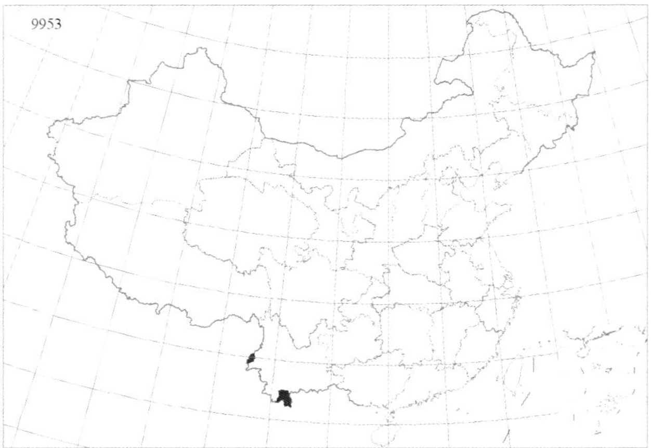

Annonaceae 番荔枝科 *Fissistigma* 瓜馥木属

Fissistigma acuminatissimum Merr. 尖叶瓜馥木

Scandent shrubs, evergreen

18.4 (15.9~21.0); 11.9 (6.9~15.1); 22.7 (21.4~25.6); 18.5 (16.0~21.0); 161.4 (131.6~191.8); 0.0 (0.0~0.0); 890 (797~1053) ❙ 1403 (1074~1751); 745 (582~996); 57 (41~77) ❙ 57.7 (32.4~88.8); 880 (797~1053); 448.0 (203.5~659.0)

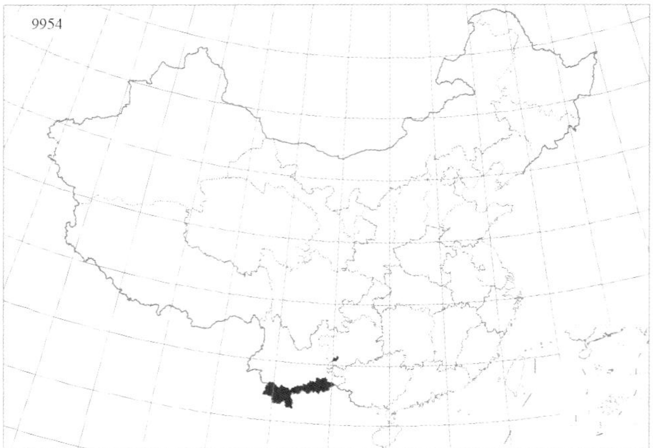

Annonaceae 番荔枝科 *Fissistigma* 瓜馥木属

Fissistigma balansae (Aug. DC.) Merr. 多脉瓜馥木

Scandent shrubs, evergreen

19.0 (17.3~21.0); 11.7 (9.9~14.2); 24.3 (22.4~27.7); 19.1 (17.3~21.0); 168.7 (147.1~191.8); 0.0 (0.0~0.0); 942 (833~1053) ❙ 1456 (1368~1522); 799 (723~860); 70 (55~111) ❙ 55.9 (38.7~80.6); 942 (833~1053); 442.2 (248.4~551.6)

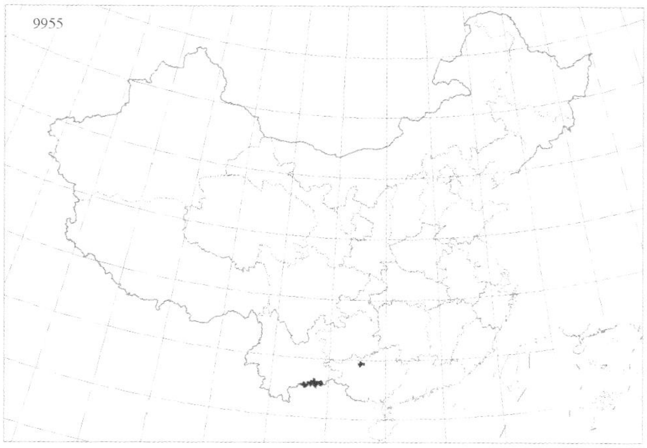

Annonaceae 番荔枝科 *Fissistigma* 瓜馥木属

Fissistigma bracteolatum Chatterjee 多苞瓜馥木

Scandent shrubs, evergreen

18.8 (16.0~22.7); 12.1 (9.1~15.1); 23.2 (20.4~28.6); 18.8 (16.0~22.8); 166.0 (132.4~213.1); 0.0 (0.0~0.0); 921 (780~1221) ❙ 1475 (1074~2435); 814 (582~1428); 62 (41~130) ❙ 60.7 (12.9~112.2); 913 (780~1221); 436.0 (277.0~659.0)

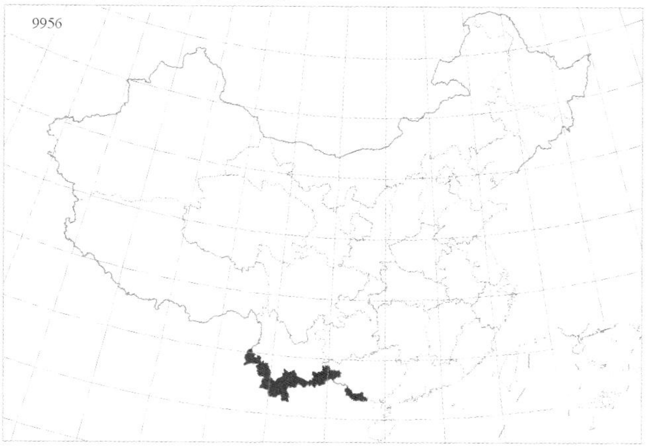

Annonaceae 番荔枝科 *Fissistigma* 瓜馥木属

Fissistigma cavaleriei (H. Lév.) Rehder 独山瓜馥木

Scandent shrubs, evergreen

17.2 (14.9~21.3); 8.1 (4.8~14.2); 24.7 (21.4~27.9); 17.3 (15.0~21.3); 147.3 (119.9~196.0); 0.0 (0.0~0.2); 874 (781~1123) ❙ 1277 (1074~1522); 636 (461~860); 70 (43~129) ❙ 47.1 (5.8~80.6); 874 (781~1121); 279.2 (189.5~551.6)

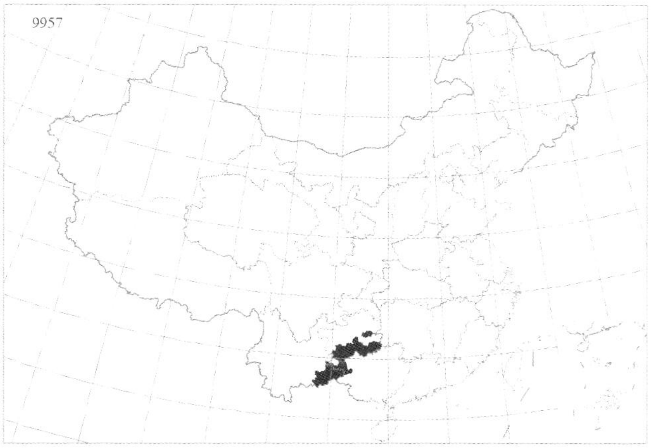

Annonaceae 番荔枝科 *Fissistigma* 瓜馥木属

Fissistigma chloroneurum (Hand.-Mazz.) Tsiang 阔叶瓜馥木

Scandent shrubs, evergreen

19.0 (14.6~22.7); 10.9 (3.8~15.0); 25.2 (21.4~28.6); 19.0 (14.7~22.8); 168.2 (117.7~213.1); 0.1 (0.0~1.5); 964 (779~1221) ❙ 1526 (1163~2435); 788 (518~1428); 99 (48~191) ❙ 59.2 (12.9~101.8); 964 (779~1221); 356.6 (239.6~551.6)

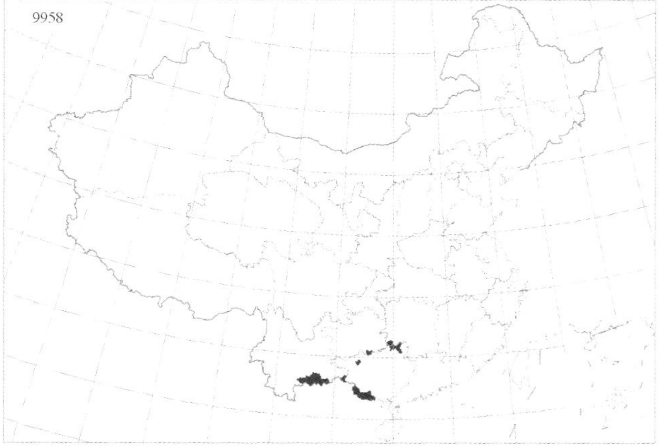

Annonaceae 番荔枝科 *Fissistigma* 瓜馥木属

Fissistigma cupreonitens Merr. et Chun 金果瓜馥木

Scandent shrubs, evergreen

21.7 (19.1~22.7); 13.4 (9.3~15.0); 28.0 (27.2~28.6); 21.8 (19.1~22.8); 201.1 (169.2~213.1); 0.0 (0.0~0.0); 1152 (988~1221) ❙ 1726 (1302~2435); 919 (561~1428); 119 (86~168) ❙ 49.7 (12.9~99.5); 1152 (988~1221); 283.5 (235.1~327.7)

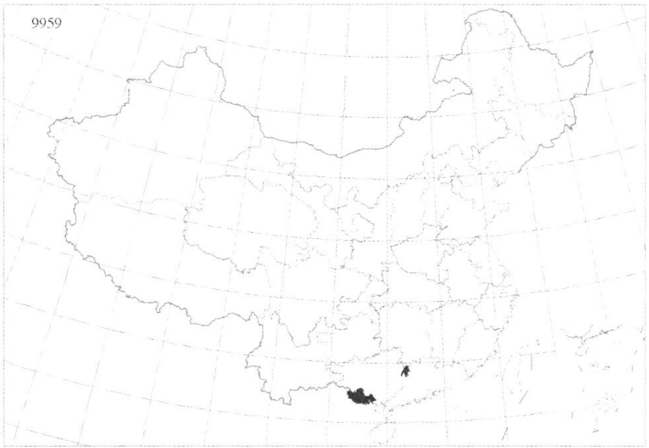

Annonaceae 番荔枝科 *Fissistigma* 瓜馥木属

Fissistigma glaucescens (Hance) Merr. 白叶瓜馥木

Scandent shrubs, evergreen

21.0 (14.6~24.8); 12.6 (3.8~20.1); 27.9 (24.6~29.8); 21.0 (14.7~24.8); 192.2 (117.7~237.6); 0.0 (0.0~1.5); 1109 (779~1385) ❙ 1667 (868~3495); 691 (312~1428); 149 (41~897) ❙ 52.5 (-34.3~224.4); 1094 (779~1360); 252.9 (95.3~620.0)

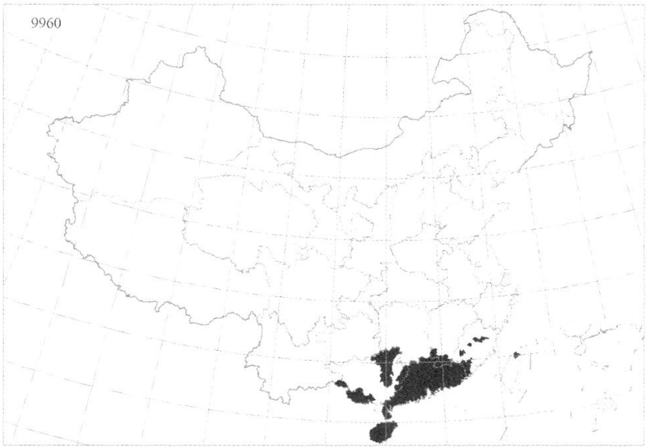

Annonaceae 番荔枝科 *Fissistigma* 瓜馥木属

Fissistigma kwangsiense Tsiang et P. T. Li 广西瓜馥木

Scandent shrubs, evergreen

19.8 (16.3~22.7); 12.2 (8.7~15.0); 25.3 (21.4~28.6); 19.8 (16.4~22.8); 177.6 (136.4~213.1); 0.0 (0.0~0.0); 1004 (797~1221) ❙ 1515 (1074~2435); 837 (582~1428); 81 (47~130) ❙ 51.4 (12.9~99.5); 1004 (797~1221); 361.2 (235.1~551.6)

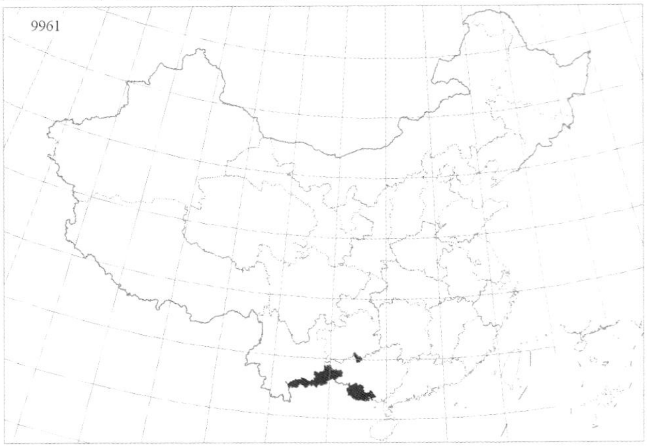

Annonaceae 番荔枝科 *Fissistigma* 瓜馥木属

Fissistigma latifolium (Dunal) Merr. 大叶瓜馥木

Scandent shrubs, evergreen

18.4 (16.3~21.0); 11.9 (9.1~15.1); 22.8 (20.6~25.6); 18.4 (16.4~21.0); 161.4 (136.4~191.8); 0.0 (0.0~0.0); 891 (795~1053) ❙ 1431 (1074~1751); 778 (582~996); 58 (42~77) ❙ 60.9 (32.4~112.2); 886 (795~1053); 472.7 (277.0~659.0)

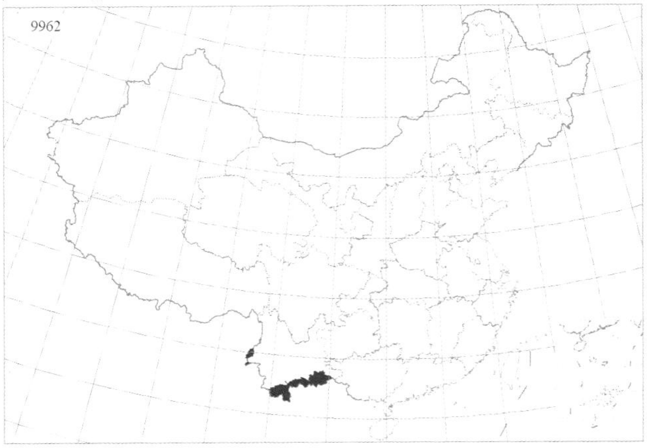

Annonaceae 番荔枝科 *Fissistigma* 瓜馥木属

Fissistigma oldhamii (Hemsl.) Merr. 瓜馥木

Scandent shrubs, evergreen

19.3 (14.3~24.8); 9.9 (3.1~20.1); 27.5 (20.1~30.1); 19.3 (14.3~24.8); 172.0 (113.2~237.6); 0.1 (0.0~2.3); 1014 (761~1385) ❙ 1593 (871~3495); 639 (299~1847); 163 (41~897) ❙ 59.4 (-12.0~246.5); 1002 (755~1360); 247.8 (95.3~659.0)

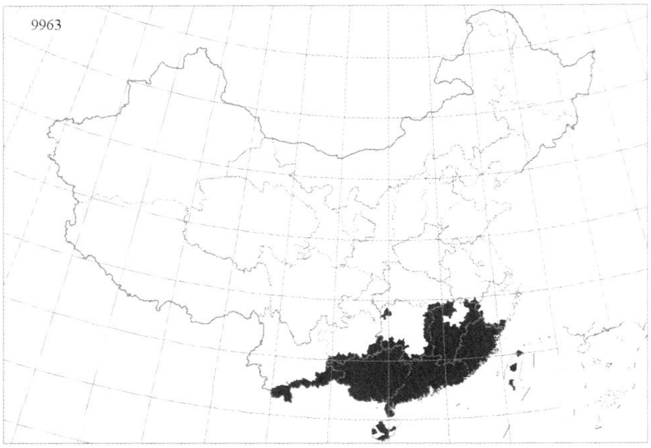

Annonaceae 番荔枝科 *Fissistigma* 瓜馥木属

Fissistigma poilanei (Ast) Tsiang et P. T. Li 火绳藤

Scandent shrubs, evergreen

18.6 (16.3~21.0); 12.1 (9.1~15.1); 23.0 (21.4~25.6); 18.6 (16.4~21.0); 163.5 (136.4~191.8); 0.0 (0.0~0.0); 900 (797~1053) ❙ 1422 (1074~1751); 765 (582~996); 59 (47~77) ❙ 58.1 (32.4~88.8); 895 (797~1053); 467.4 (277.0~659.0)

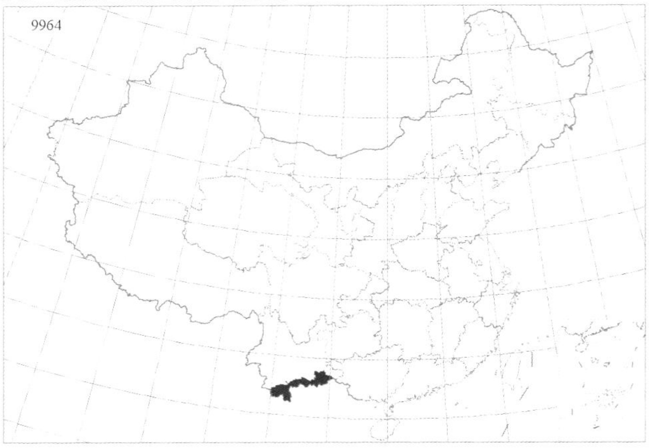

Annonaceae 番荔枝科 *Fissistigma* 瓜馥木属

Fissistigma polyanthoides (A. DC.) Merr. 小萼瓜馥木

Scandent shrubs, evergreen

16.8 (14.6~18.9); 6.9 (3.9~9.5); 25.5 (24.6~26.6); 16.9 (14.6~18.9); 142.8 (116.9~167.3); 0.2 (0.0~1.1); 870 (784~969) ❙ 1274 (1236~1349); 582 (473~667); 77 (54~123) ❙ 47.5 (27.5~72.0); 870 (784~969); 237.5 (200.5~262.1)

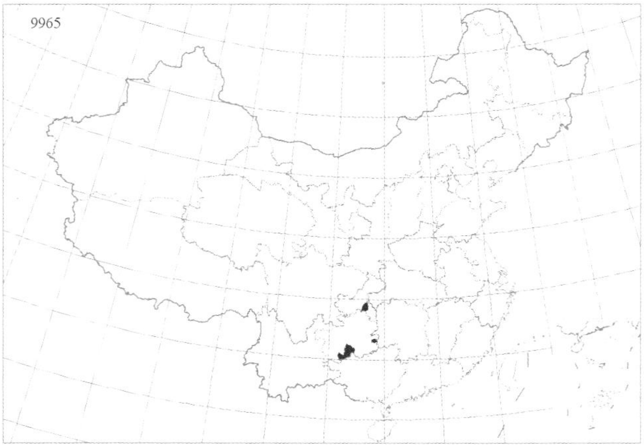

Annonaceae 番荔枝科 *Fissistigma* 瓜馥木属

Fissistigma polyanthum (Hook. f. et Thomson) Merr. 多花瓜馥木

Scandent shrubs, evergreen

19.4 (12.5~24.8); 11.0 (3.6~20.1); 26.0 (19.6~29.2); 19.4 (12.5~24.8); 173.0 (92.0~237.6); 0.0 (0.0~1.5); 1000 (685~1385) ❙ 1480 (868~2435); 687 (312~1428); 109 (41~197) ❙ 50.3 (-34.3~145.8); 986 (685~1331); 305.6 (170.4~659.0)

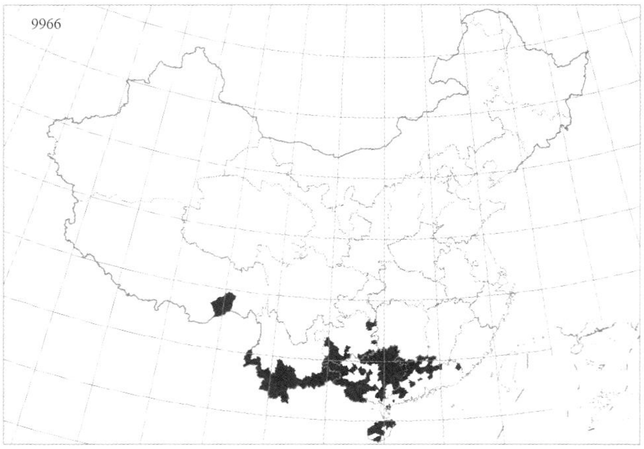

Annonaceae 番荔枝科 *Fissistigma* 瓜馥木属

Fissistigma retusum (H. Lév.) Rehder 凹叶瓜馥木

Scandent shrubs, evergreen

18.3 (13.3~24.8); 10.3 (3.3~20.1); 24.4 (19.6~29.0); 18.3 (13.4~24.8); 159.8 (100.4~237.6); 0.1 (0.0~2.3); 918 (706~1385) ❙ 1379 (1007~1751); 699 (454~1007); 80 (41~194) ❙ 52.3 (-6.9~145.8); 909 (706~1239); 351.5 (167.9~659.0)

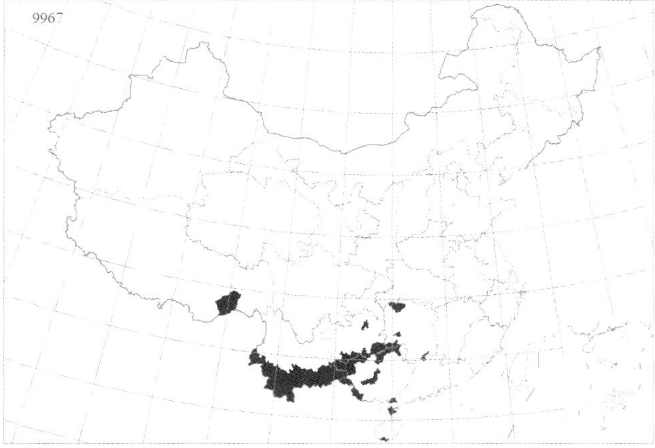

Annonaceae 番荔枝科 *Fissistigma* 瓜馥木属

Fissistigma shangtzeense Tsiang et P. T. Li 上思瓜馥木

Scandent shrubs, evergreen

21.3 (17.4~22.7); 13.0 (7.4~15.0); 27.5 (25.0~28.6); 21.4 (17.4~22.8); 196.4 (148.9~213.1); 0.0 (0.0~0.0); 1124 (886~1221) ❙ 1689 (1302~2435); 917 (637~1428); 108 (57~130) ❙ 50.3 (12.9~99.5); 1124 (886~1221); 292.0 (235.1~371.7)

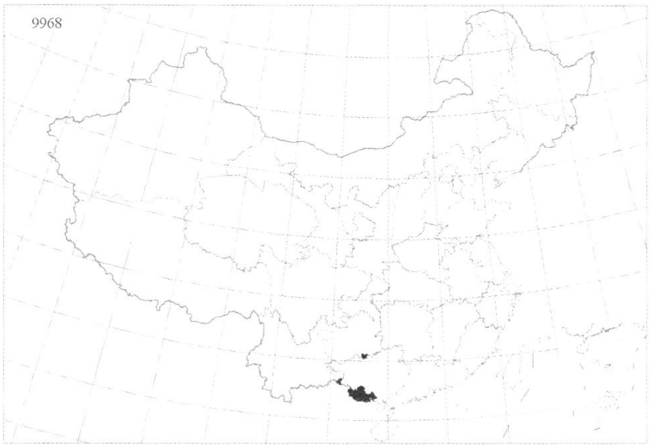

Annonaceae 番荔枝科 *Fissistigma* 瓜馥木属

Fissistigma tientangense Tsiang et P. T. Li 天堂瓜馥木

Scandent shrubs, evergreen

21.9 (19.1~24.0); 14.4 (9.3~19.1); 27.6 (27.2~28.1); 21.9 (19.1~24.0); 202.9 (169.2~228.1); 0.0 (0.0~0.0); 1160 (988~1302) ❙ 1599 (1380~1876); 663 (455~891); 113 (48~168) ❙ 39.1 (6.7~63.2); 1112 (988~1151); 325.2 (221.6~620.0)

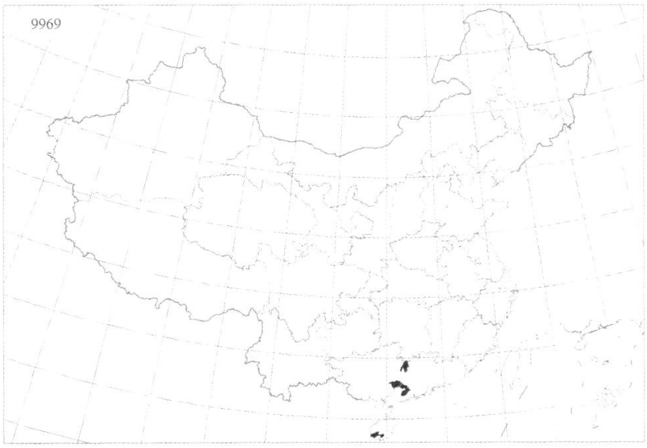

Annonaceae 番荔枝科 *Fissistigma* 瓜馥木属

Fissistigma tonkinense Finet et Gagn. 麻栗坡瓜馥木

Scandent shrubs, evergreen

17.7 (16.3~21.0); 10.4 (9.1~14.2); 22.8 (21.4~25.6); 17.7 (16.4~21.0); 152.6 (136.4~191.8); 0.0 (0.0~0.0); 864 (797~1053) ❙ 1333 (1074~1522); 741 (582~860); 54 (47~68) ❙ 54.6 (32.4~80.6); 864 (797~1053); 396.7 (277.0~551.6)

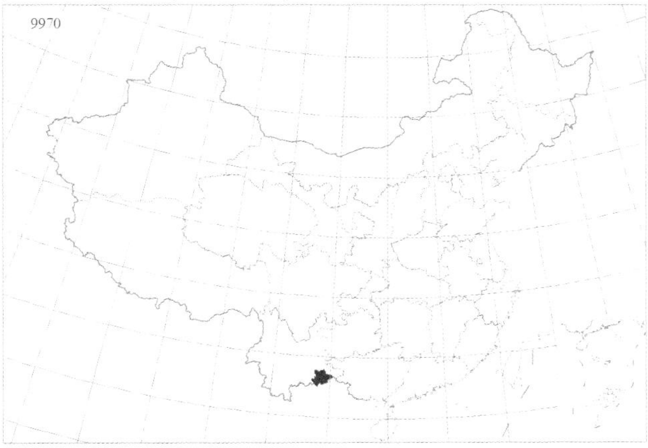

Annonaceae 番荔枝科 *Fissistigma* 瓜馥木属

Fissistigma tungfangense Tsiang et P. T. Li 东方瓜馥木

Scandent shrubs, evergreen

24.3; 18.7; 28.4; 24.3; 231.9; 0.0; 1338 ❙ 1179; 392; 42 ❙ -11.9; 1056; 455.2

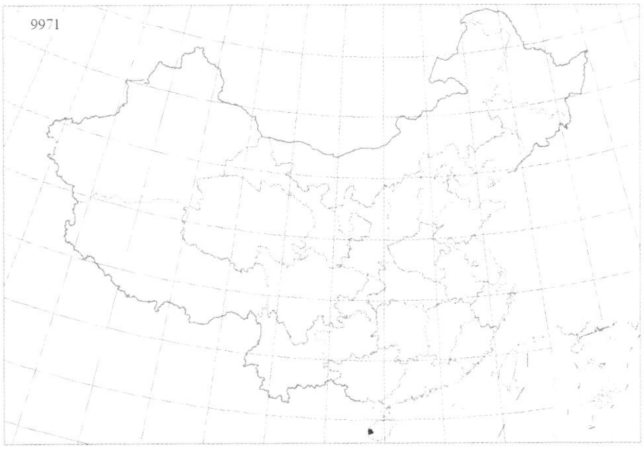

Annonaceae 番荔枝科 *Fissistigma* 瓜馥木属

Fissistigma uonicum (Dunn) Merr. 香港瓜馥木

Scandent shrubs, evergreen

20.7 (14.3~24.3); 12.0 (3.3~19.1); 27.9 (24.3~29.8); 20.7 (14.3~24.3); 188.5 (114.1~231.9); 0.0 (0.0~2.3); 1091 (765~1338) ❙ 1632 (868~2435); 691 (312~1428); 145 (42~221) ❙ 51.5 (-34.3~106.4); 1081 (765~1326); 238.3 (95.3~620.0)

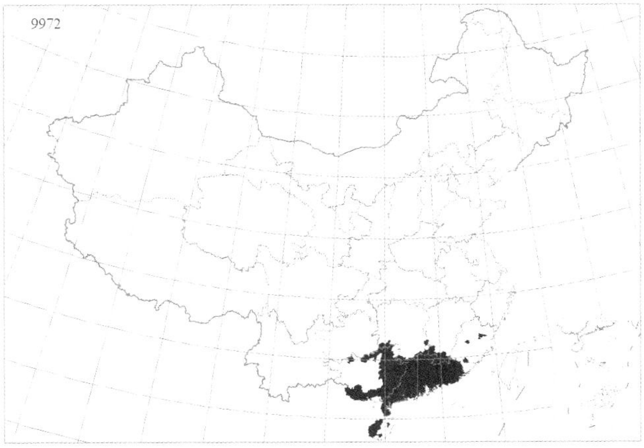

Annonaceae 番荔枝科 *Fissistigma* 瓜馥木属
Fissistigma wallichii (Hook. f. et Thomson) Merr. 贵州瓜馥木
Scandent shrubs, evergreen

18.7 (14.6~21.9); 10.9 (3.8~15.1); 24.4 (21.4~28.6); 18.7 (14.7~22.0); 164.9 (117.7~203.8); 0.0 (0.0~1.5); 932 (779~1176) ❙ 1346 (1007~1751); 707 (518~996); 71 (41~191) ❙ 45.5 (5.8~101.8); 925 (779~1176); 351.7 (189.5~659.0)

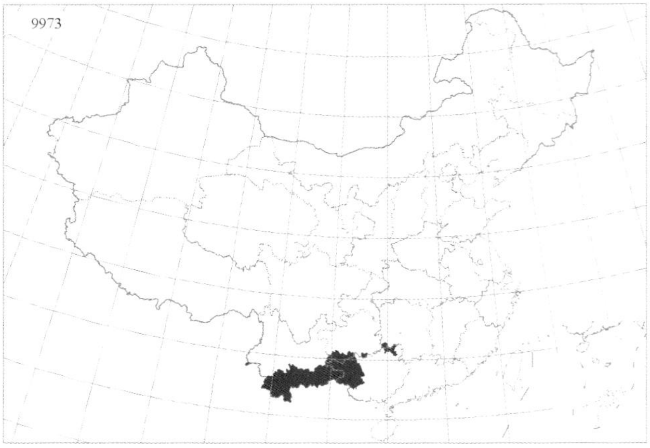

Annonaceae 番荔枝科 *Fissistigma* 瓜馥木属
Fissistigma xylopetalum Tsiang et P. T. Li 木瓣瓜馥木
Scandent shrubs, evergreen

20.1 (17.7~24.0); 14.0 (10.9~19.1); 24.3 (22.5~27.4); 20.2 (17.8~24.0); 181.9 (153.4~228.1); 0.0 (0.0~0.0); 999 (854~1302) ❙ 1514 (1276~1751); 794 (539~996); 67 (48~106) ❙ 53.5 (19.1~88.8); 981 (854~1147); 514.2 (281.7~659.0)

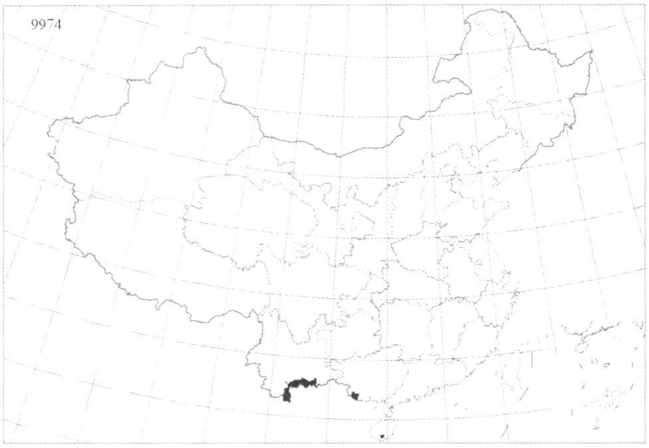

Annonaceae 番荔枝科 *Goniothalamus* 哥纳香属
Goniothalamus amuyon (Blanco) Merr. 台湾哥纳香
Shrubs or small trees, evergreen

22.2; 17.7; 25.5; 22.2; 206.7; 0.0; 1123 ❙ 2992; 1847; 107 ❙ 166.5; 1123; 296.9

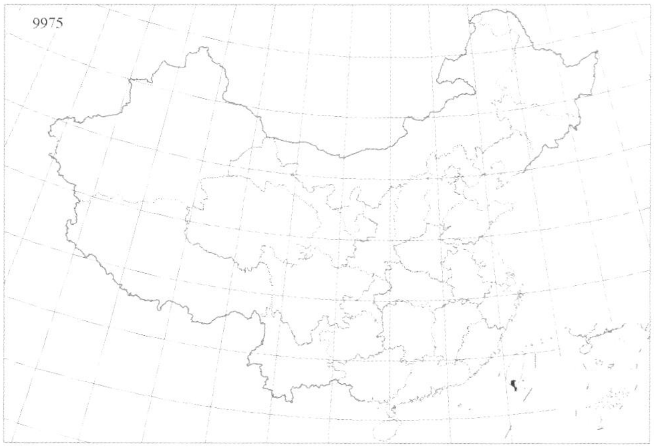

Annonaceae 番荔枝科 *Goniothalamus* 哥纳香属
Goniothalamus cheliensis Hu 景洪哥纳香
Small trees, evergreen

20.5; 14.9; 23.7; 20.5; 185.9; 0.0; 992 ❙ 1464; 655; 66 ❙ 47.6; 959; 570.1

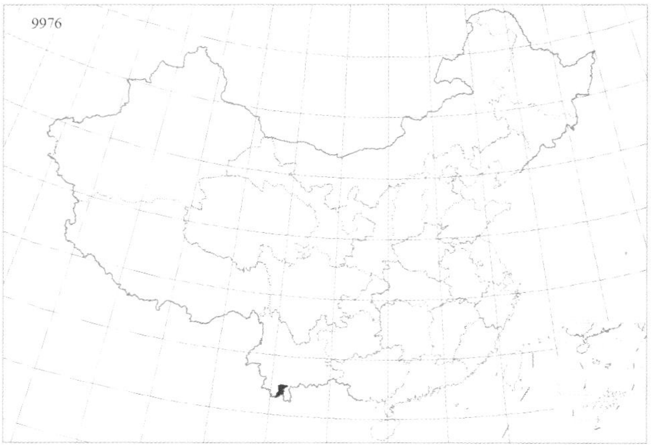

Annonaceae 番荔枝科 *Goniothalamus* 哥纳香属
Goniothalamus chinensis Merr. et Chun 哥纳香
Shrubs, evergreen

22.3 (18.0~24.8); 15.1 (8.5~20.1); 27.6 (23.7~28.6); 22.3 (18.0~24.8); 208.2 (156.4~237.6); 0.0 (0.0~0.0); 1192 (913~1385) ❙ 1672 (1302~2435); 805 (473~1428); 100 (41~169) ❙ 41.9 (1.5~99.5); 1147 (913~1299); 373.3 (235.1~620.0)

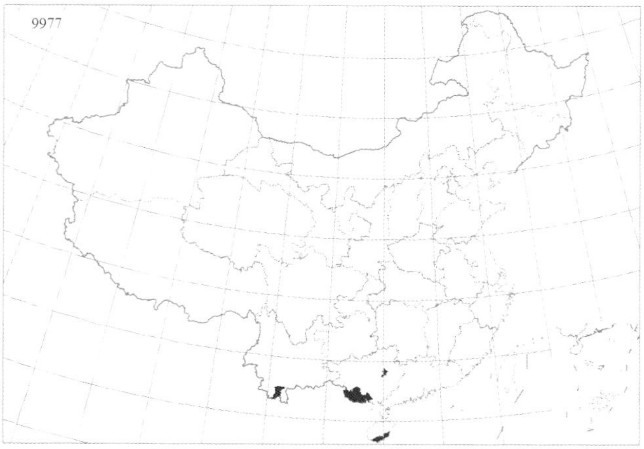

Annonaceae 番荔枝科 *Goniothalamus* 哥纳香属
Goniothalamus donnaiensis Finet et Gagnep. 田方骨
Shrubs or small trees, evergreen

20.3 (17.2~22.7); 12.2 (7.4~15.0); 26.3 (22.4~28.6); 20.4 (17.3~22.8); 184.4 (147.0~213.1); 0.0 (0.0~0.0); 1048 (833~1221) ❙ 1514 (1188~2435); 825 (634~1428); 88 (48~138) ❙ 45.4 (5.8~99.5); 1048 (833~1221); 328.6 (245.3~551.6)

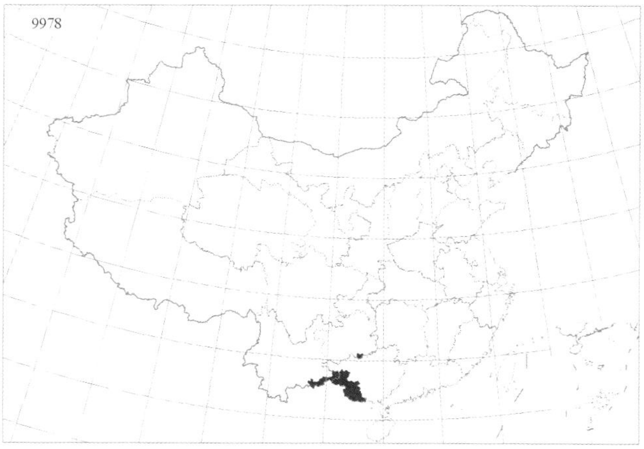

Annonaceae 番荔枝科 *Goniothalamus* 哥纳香属

Goniothalamus gabriacianus (Baill.) Ast 保亭哥纳香

Shrubs or small trees, evergreen

24.0；19.1；27.2；24.0；228.1；0.0；1302 ❙ 1551；539；48 ❙ 19.1；1147；620.0

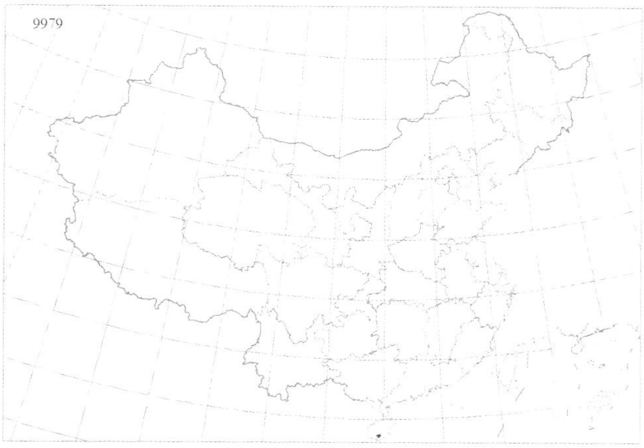

Annonaceae 番荔枝科 *Goniothalamus* 哥纳香属

Goniothalamus gardneri Hook. f. et Thomson 长叶哥纳香

Shrubs or small trees, evergreen

23.9 (23.0~24.8)；18.9 (17.5~20.1)；27.4 (26.7~28.0)；24.0 (23.1~24.8)；227.5 (217.0~237.6)；0.0 (0.0~0.0)；1302 (1217~1385) ❙ 1551 (1405~1725)；540 (473~610)；54 (41~68) ❙ 19.7 (1.5~41.8)；1161 (1137~1212)；550.7 (494.7~620.0)

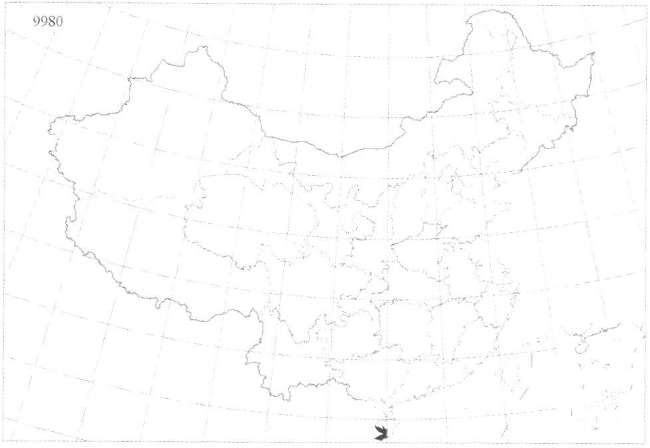

Annonaceae 番荔枝科 *Goniothalamus* 哥纳香属

Goniothalamus griffithii Hook. f. et Thomson 大花哥纳香

Trees, evergreen

18.8 (16.4~20.5)；12.6 (9.6~15.1)；22.5 (20.6~23.7)；18.8 (16.4~20.5)；165.7 (137.3~185.9)；0.0 (0.0~0.0)；901 (795~992) ❙ 1550 (1434~1751)；829 (655~996)；57 (41~77) ❙ 72.9 (47.6~112.2)；885 (795~986)；523.5 (427.9~659.0)

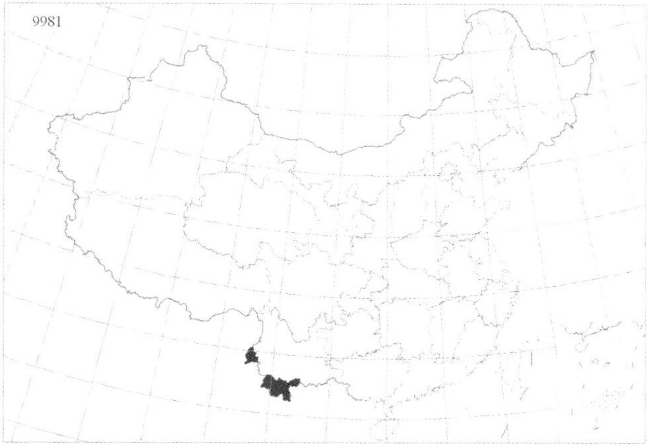

Annonaceae 番荔枝科 *Goniothalamus* 哥纳香属

Goniothalamus howii Merr. et Chun 海南哥纳香

Trees, evergreen

22.0 (17.7~24.8)；16.4 (10.9~20.1)；25.9 (22.5~28.0)；22.1 (17.8~24.8)；204.9 (153.4~237.6)；0.0 (0.0~0.0)；1158 (854~1385) ❙ 1512 (1368~1725)；652 (473~835)；57 (41~68) ❙ 34.0 (1.5~60.1)；1069 (854~1212)；537.8 (491.5~620.0)

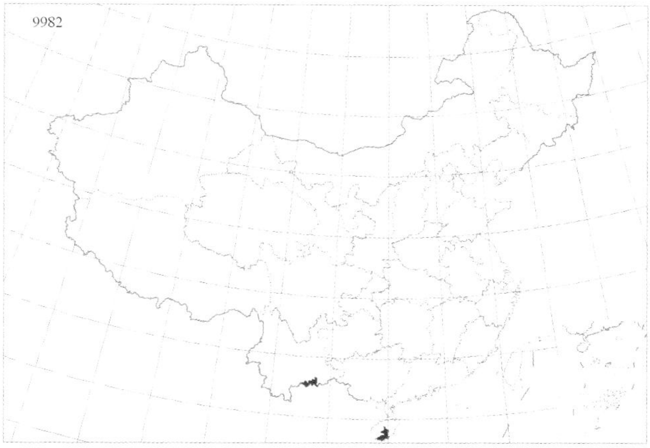

Annonaceae 番荔枝科 *Goniothalamus* 哥纳香属

Goniothalamus leiocarpus (W. T. Wang) P. T. Li 金平哥纳香

Trees, evergreen

19.0；12.5；23.4；19.0；167.9；0.0；922 ❙ 1370；763；58 ❙ 48.8；922；489.0

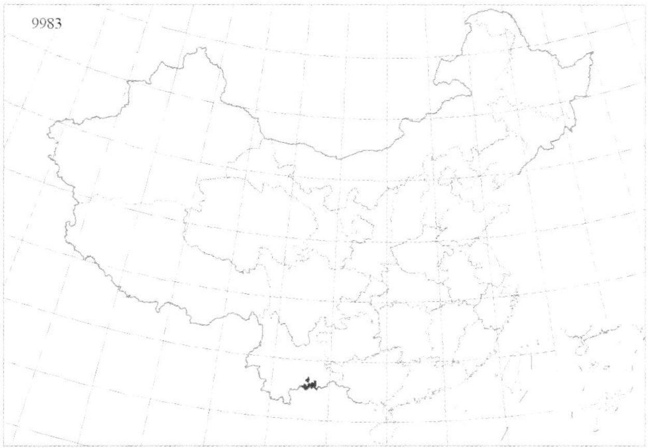

Annonaceae 番荔枝科 *Goniothalamus* 哥纳香属

Goniothalamus yunnanensis W. T. Wang 云南哥纳香

Shrubs or small trees, evergreen

19.3；12.7；23.8；19.3；171.6；0.0；942 ❙ 1439；803；61 ❙ 53.4；942；516.4

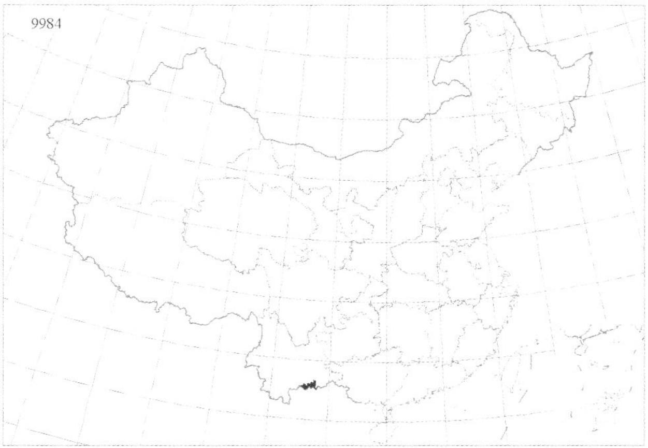

Annonaceae 番荔枝科 *Meiogyne* 鹿茸木属

Meiogyne kwangtungensis P. T. Li 鹿茸木

Shrubs, evergreen

24.2; 19.3; 27.6; 24.2; 230.9; 0.0; 1327 ❙ 1445; 489; 46 ❙ 9.1; 1132; 534.2

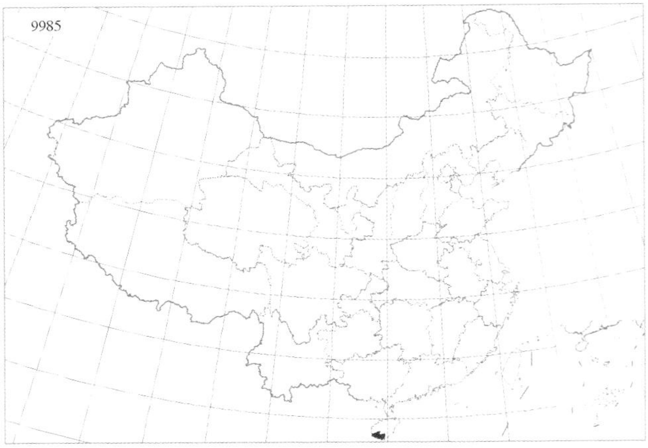

Annonaceae 番荔枝科 *Mezzettiopsis* 蚁花属

Mezzettiopsis creaghii Ridl. 蚁花

Trees, evergreen

23.5 (20.4~24.8); 18.0 (14.9~20.1); 27.2 (23.6~28.7); 23.5 (20.5~24.8); 222.1 (185.8~237.6); 0.0 (0.0~0.0); 1264 (989~1385) ❙ 1379 (868~1767); 502 (312~702); 59 (41~88) ❙ 11.2 (-34.3~62.5); 1091 (868~1299); 489.8 (288.3~659.0)

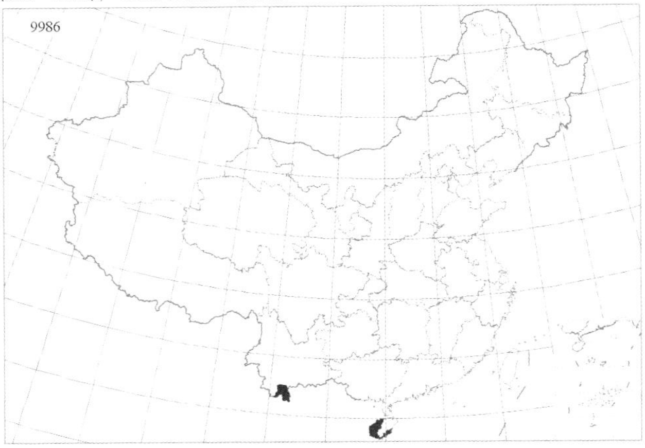

Annonaceae 番荔枝科 *Miliusa* 野独活属

Miliusa balansae Finet et Gagnep. 野独活

Shrubs, evergreen

18.8 (14.3~24.8); 9.8 (3.3~20.1); 26.0 (20.6~29.2); 18.8 (14.3~24.8); 165.7 (114.1~237.6); 0.1 (0.0~2.3); 970 (765~1385) ❙ 1468 (1074~2435); 674 (411~1428); 112 (41~204) ❙ 53.0 (-1.0~112.2); 961 (765~1273); 300.5 (178.7~659.0)

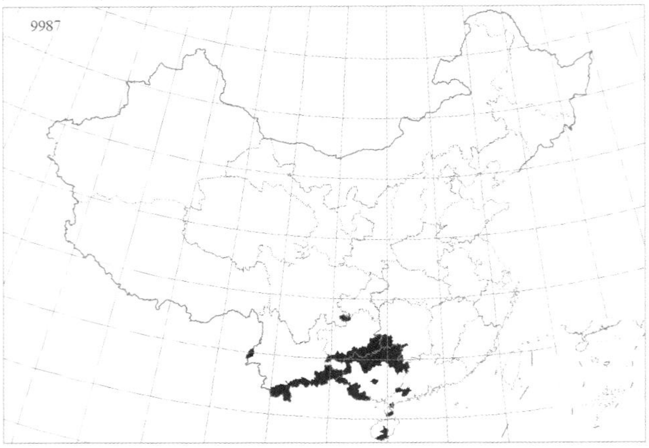

Annonaceae 番荔枝科 *Miliusa* 野独活属

Miliusa cuneata Craib 楔叶野独活

Small trees, evergreen

19.1 (16.4~20.5); 13.3 (9.6~15.1); 22.6 (20.6~23.7); 19.1 (16.4~20.5); 169.7 (137.3~185.9); 0.0 (0.0~0.0); 919 (795~992) ❙ 1550 (1464~1688); 755 (655~959); 65 (58~77) ❙ 70.4 (47.6~112.2); 902 (795~986); 559.6 (427.9~659.0)

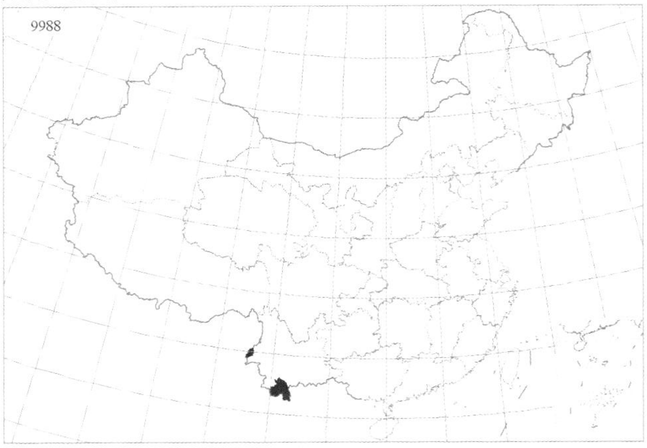

Annonaceae 番荔枝科 *Miliusa* 野独活属

Miliusa prolifica (Chun et F. C. How) P. T. Li 囊瓣木

Trees, evergreen

23.8 (21.8~24.8); 17.9 (13.7~20.1); 28.0 (25.7~28.8); 23.8 (21.9~24.8); 226.0 (202.6~237.6); 0.0 (0.0~0.0); 1296 (1127~1385) ❙ 1533 (868~1876); 562 (312~891); 79 (41~144) ❙ 19.0 (-34.3~63.2); 1189 (868~1360); 418.9 (157.7~620.0)

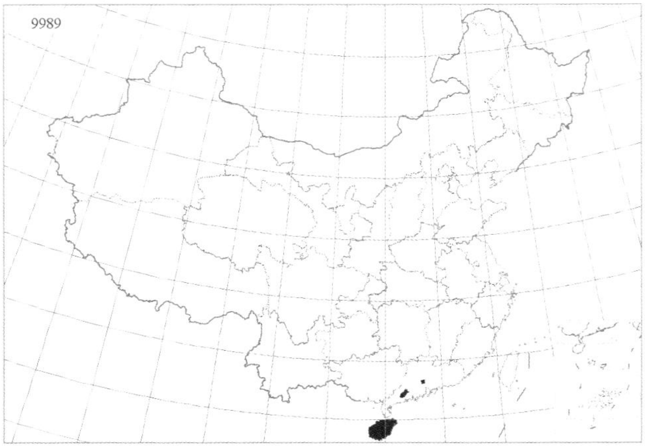

Annonaceae 番荔枝科 *Miliusa* 野独活属

Miliusa sinensis Finet et Gagnep. 中华野独活

Trees, evergreen

19.4 (15.0~24.0); 11.1 (4.5~17.7); 25.7 (22.3~29.2); 19.4 (15.0~24.1); 172.8 (120.8~229.0); 0.0 (0.0~0.5); 988 (794~1321) ❙ 1405 (1089~2435); 731 (472~1428); 75 (43~138) ❙ 43.1 (-1.0~99.5); 987 (794~1321); 317.9 (156.3~659.0)

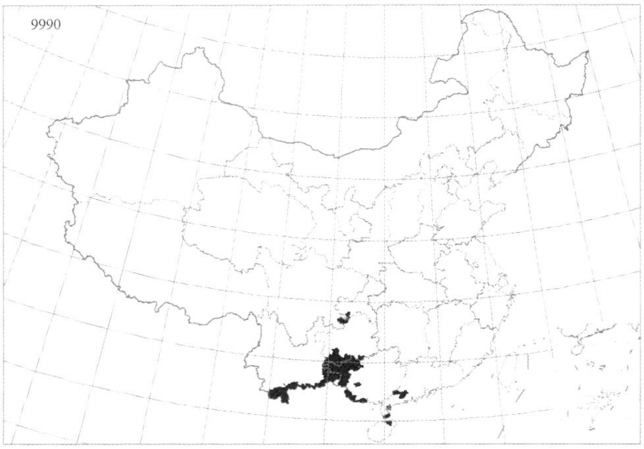

Annonaceae 番荔枝科 *Miliusa* 野独活属

Miliusa tenuistipitata W. T. Wang 云南野独活

Trees, evergreen

19.1 (17.3~21.0); 13.1 (9.9~15.1); 23.0 (21.6~25.6); 19.2 (17.3~21.0); 170.2 (147.1~191.8); 0.0 (0.0~0.0); 921 (833~1053) ❙ 1490 (1276~1751); 730 (655~996); 59 (41~77) ❙ 62.2 (36.8~88.8); 907 (833~1053); 506.8 (413.5~659.0)

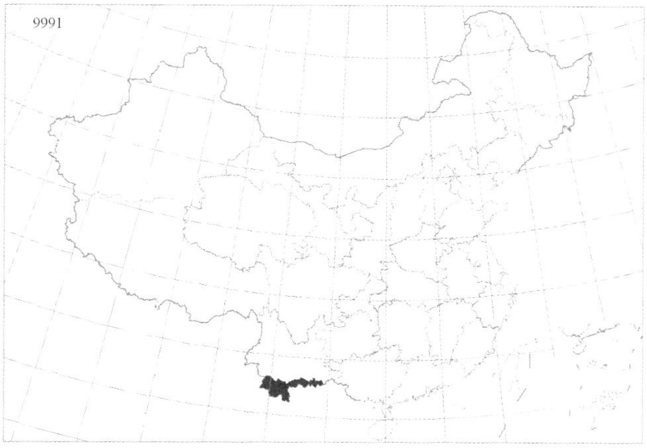

Annonaceae 番荔枝科 *Mitrephora* 银钩花属

Mitrephora maingayi Hook. f. et Thomson 山蕉

Trees, evergreen

19.5 (15.9~24.1); 11.9 (6.9~19.1); 25.1 (21.4~28.6); 19.6 (16.0~24.2); 174.6 (131.6~230.1); 0.0 (0.0~0.0); 987 (797~1326) ❙ 1442 (1074~2435); 747 (455~1428); 72 (41~141) ❙ 47.2 (5.8~99.5); 975 (797~1326); 363.2 (189.5~659.0)

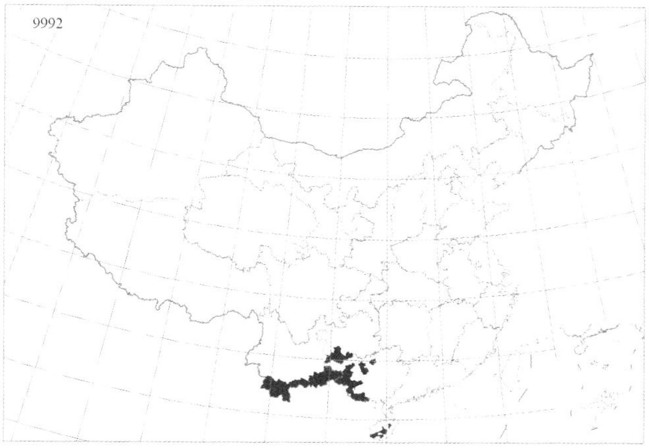

Annonaceae 番荔枝科 *Mitrephora* 银钩花属

Mitrephora thorelii Pierre 银钩花

Trees, evergreen

20.4 (15.9~24.8); 13.4 (6.9~20.1); 25.4 (20.6~28.7); 20.4 (16.0~24.8); 184.9 (131.6~237.6); 0.0 (0.0~0.0); 1046 (795~1385) ❙ 1429 (868~1854); 688 (312~996); 68 (41~141) ❙ 40.9 (-34.3~112.2); 993 (795~1360); 409.7 (189.5~659.0)

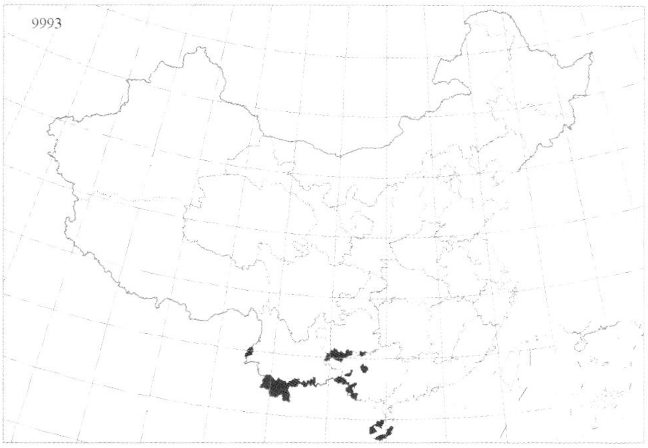

Annonaceae 番荔枝科 *Mitrephora* 银钩花属

Mitrephora wangii Hu 云南银钩花

Trees, evergreen

19.4 (18.2~21.0); 13.6 (12.4~15.1); 23.0 (21.6~25.6); 19.5 (18.2~21.0); 173.7 (158.6~191.8); 0.0 (0.0~0.0); 934 (861~1053) ❙ 1519 (1428~1751); 778 (655~996); 60 (41~77) ❙ 63.0 (44.5~88.8); 917 (839~1053); 521.6 (427.9~659.0)

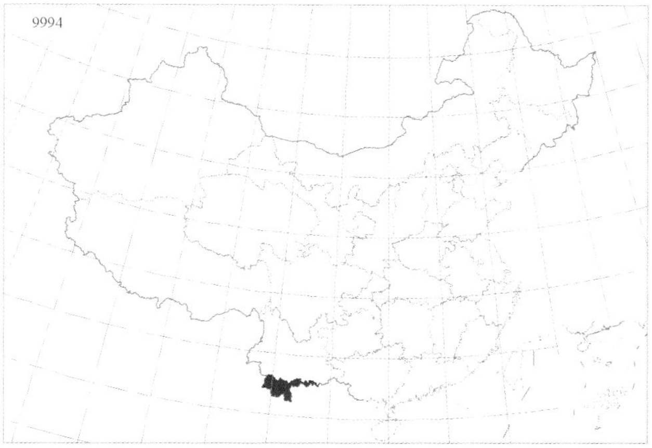

Annonaceae 番荔枝科 *Orophea* 澄广花属

Orophea hainanensis Merr. 澄广花

Trees, evergreen

22.7 (20.4~24.8); 16.1 (13.2~20.1); 27.4 (23.6~29.0); 22.7 (20.5~24.8); 212.8 (185.8~237.6); 0.0 (0.0~0.0); 1209 (989~1385) ❙ 1554 (1242~1900); 703 (444~1056); 78 (41~125) ❙ 30.1 (-2.2~71.0); 1140 (986~1247); 411.7 (199.6~659.0)

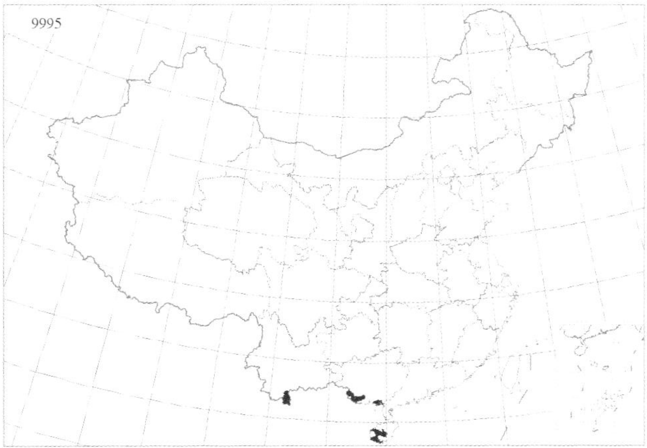

Annonaceae 番荔枝科 *Orophea* 澄广花属

Orophea hirsuta King 毛澄广花

Shrubs, evergreen

22.2 (19.2~24.8); 16.0 (13.1~20.1); 26.4 (22.3~28.8); 22.2 (19.2~24.8); 206.3 (170.5~237.6); 0.0 (0.0~0.0); 1163 (909~1385) ❙ 1525 (1374~1784); 650 (473~779); 76 (41~144) ❙ 33.8 (1.5~64.9); 1093 (880~1212); 467.3 (157.7~659.0)

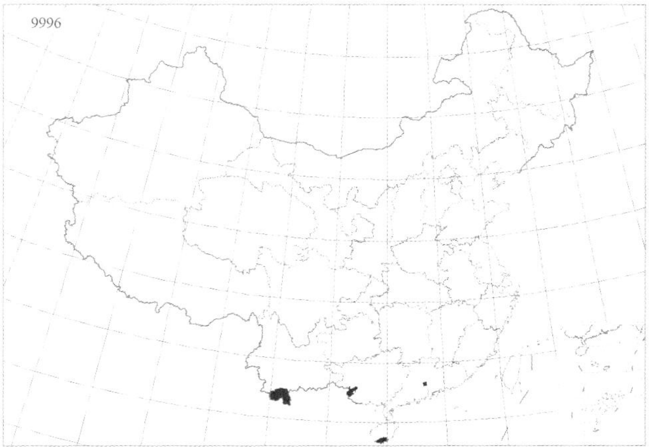

Annonaceae 番荔枝科 *Orophea* 澄广花属

Orophea multiflora Ast ex Jovet-Ast 多花澄广花

Shrubs, evergreen

18.5 (16.4~21.5); 11.5 (9.6~13.6); 22.8 (20.6~27.4); 18.5 (16.4~21.6); 162.2 (137.3~199.1); 0.0 (0.0~0.0); 906 (795~1131) ❙ 1582 (1530~1688); 912 (850~959); 60 (42~106) ❙ 77.1 (39.9~112.2); 900 (795~1131); 493.5 (281.7~627.0)

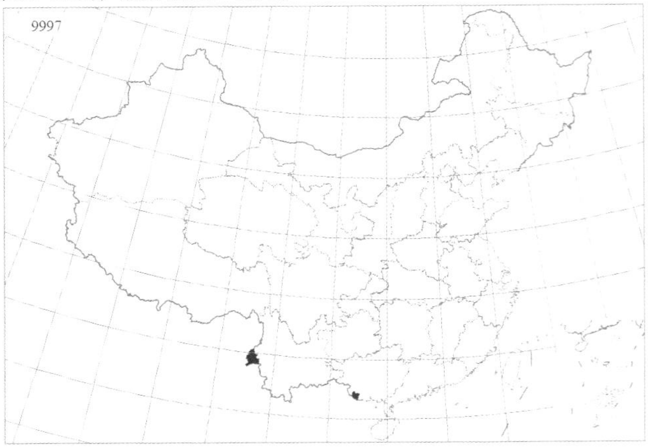

Annonaceae 番荔枝科 *Orophea* 澄广花属

Orophea polycarpa A. DC. 广西澄广花

Trees, evergreen

22.1 (18.9~24.8); 14.9 (8.7~20.1); 27.6 (25.0~28.8); 22.2 (19.0~24.8); 206.1 (167.5~237.6); 0.0 (0.0~0.0); 1181 (939~1385) ❙ 1487 (868~2148); 657 (312~1234); 88 (41~186) ❙ 27.6 (-34.3~86.7); 1125 (868~1360); 361.1 (204.5~620.0)

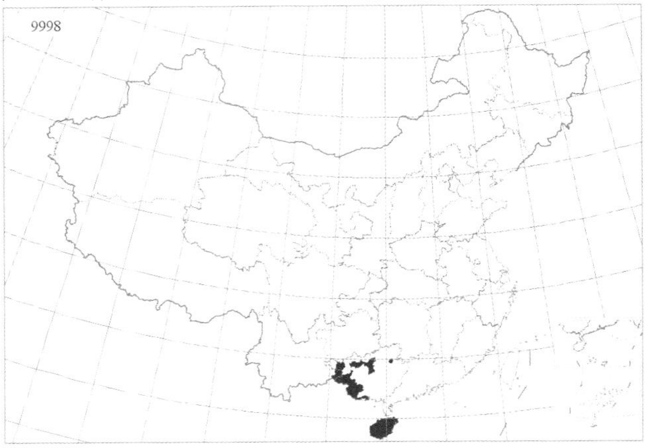

Annonaceae 番荔枝科 *Orophea* 澄广花属

Orophea yunnanensis P. T. Li 云南澄广花

Shrubs, evergreen

14.0; 7.4; 19.0; 14.0; 108.6; 0.2; 727 ❙ 1022; 597; 39 ❙ 43.0; 718; 303.4

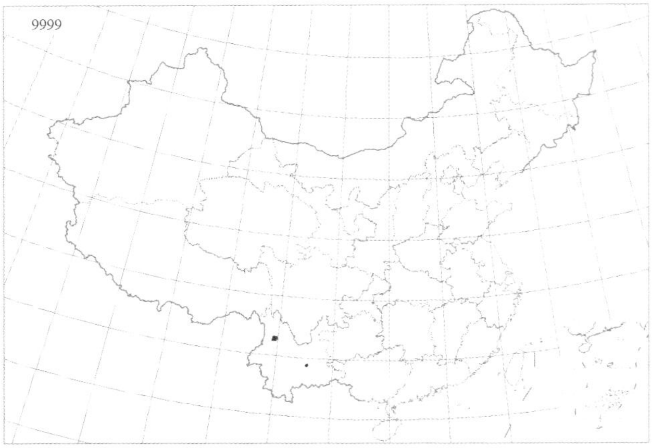

Annonaceae 番荔枝科 *Polyalthia* 暗罗属

Polyalthia cerasoides (Roxb.) Benth. et Hook. f. ex Bedd. 细基丸

Trees, evergreen

21.2 (16.9~24.8); 14.9 (9.5~20.1); 25.5 (21.4~29.2); 21.2 (16.9~24.8); 194.8 (142.7~237.6); 0.0 (0.0~0.0); 1096 (813~1385) ❙ 1427 (868~1854); 666 (312~997); 67 (41~115) ❙ 33.7 (-34.3~88.8); 1044 (813~1360); 407.7 (178.7~659.0)

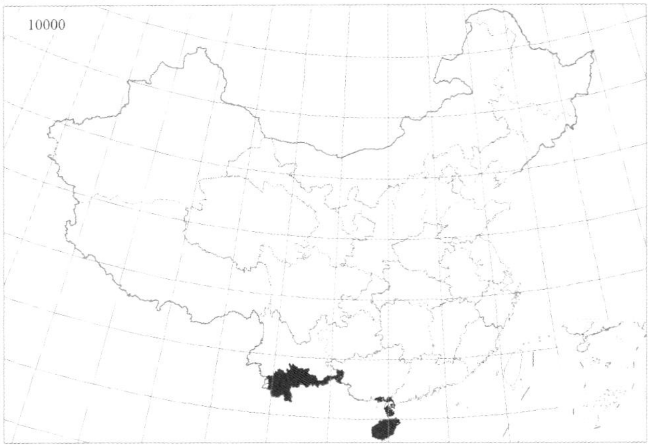

Annonaceae 番荔枝科 *Polyalthia* 暗罗属

Polyalthia florulenta C. Y. Wu ex P. T. Li 小花暗罗

Shrubs, evergreen

18.2; 11.9; 22.0; 18.2; 158.7; 0.0; 868 ❙ 1366; 783; 45 ❙ 57.4; 847; 516.7

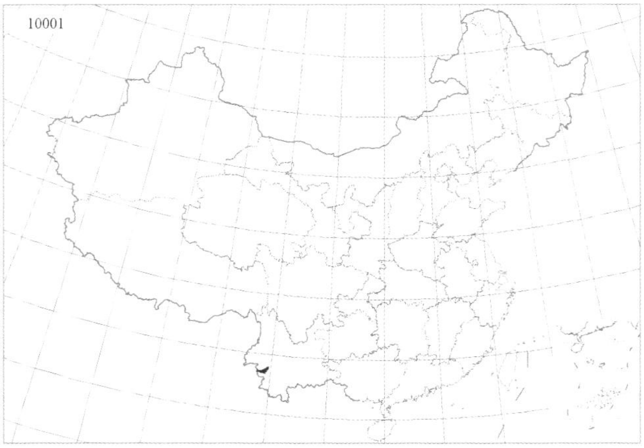

Annonaceae 番荔枝科 *Polyalthia* 暗罗属

Polyalthia lancilimba C. Y. Wu ex P. T. Li 剑叶暗罗

Trees, evergreen

19.4; 12.6; 24.1; 19.4; 172.6; 0.0; 954 ❙ 1445; 787; 64 ❙ 52.3; 954; 521.5

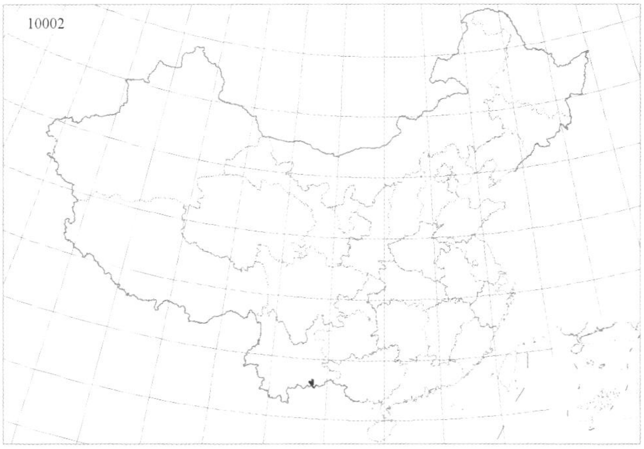

Annonaceae 番荔枝科 *Polyalthia* 暗罗属

Polyalthia laui Merr. 海南暗罗

Trees, evergreen

23.1 (20.5~24.8); 16.9 (13.6~20.1); 27.4 (23.7~28.6); 23.1 (20.5~24.8); 217.0 (185.9~237.6); 0.0 (0.0~0.0); 1236 (992~1385) ❙ 1580 (1179~2435); 689 (392~1428); 72 (41~130) ❙ 29.4 (-11.9~99.5); 1136 (959~1299); 459.2 (281.7~620.0)

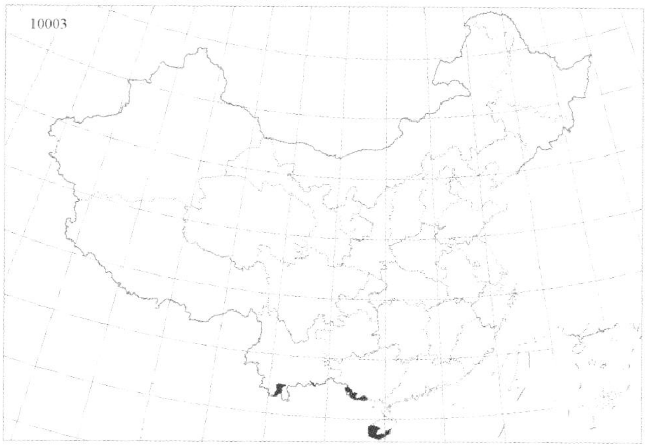

Annonaceae 番荔枝科 *Polyalthia* 暗罗属

Polyalthia litseifolia C. Y. Wu ex P. T. Li 木姜叶暗罗

Trees, semievergreen

20.0; 14.6; 23.2; 20.1; 180.8; 0.0; 963 ❙ 1523; 681; 67 ❙ 58.4; 942; 581.0

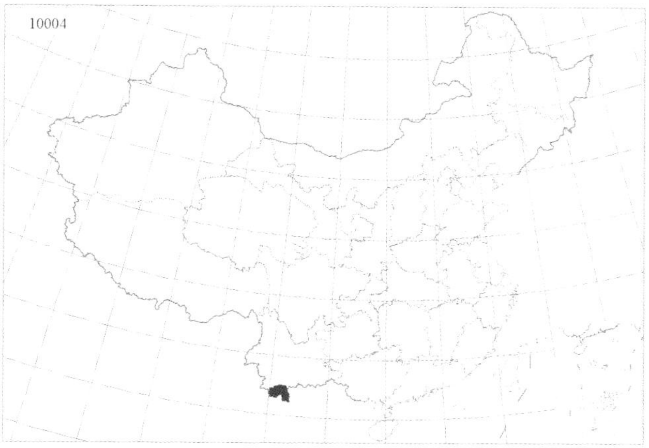

Annonaceae 番荔枝科 *Polyalthia* 暗罗属

Polyalthia nemoralis A. DC. 陵水暗罗

Shrubs or small trees, evergreen

23.5 (19.1~24.8); 17.4 (13.0~20.1); 27.9 (23.3~29.0); 23.5 (19.1~24.8); 222.5 (169.7~237.6); 0.0 (0.0~0.0); 1274 (917~1385) ❙ 1475 (868~1854); 576 (312~850); 78 (41~138) ❙ 16.7 (-34.3~55.7); 1180 (868~1360); 408.6 (225.6~620.0)

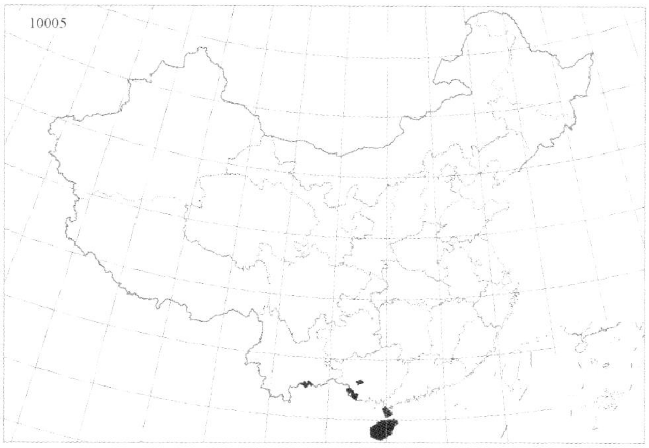

Annonaceae 番荔枝科 *Polyalthia* 暗罗属

Polyalthia obliqua Hook. f. et Thomson 沙煲暗罗

Trees, evergreen

24.0 (22.2~24.8); 18.4 (17.2~20.1); 28.0 (25.7~28.8); 24.0 (22.3~24.8); 228.3 (207.2~237.6); 0.0 (0.0~0.0); 1310 (1127~1385) ❙ 1501 (868~1854); 534 (312~712); 74 (41~115) ❙ 15.0 (-34.3~47.8); 1191 (868~1360); 440.3 (248.2~620.0)

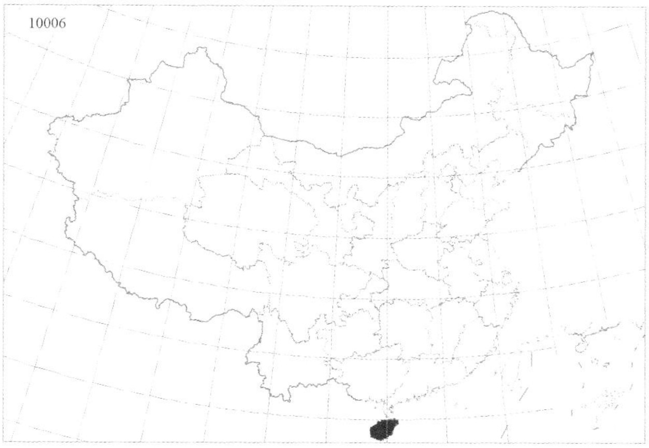

Annonaceae 番荔枝科 *Polyalthia* 暗罗属

Polyalthia petelotii Merr. 云桂暗罗

Shrubs or small trees, evergreen

19.0 (16.4~22.2); 11.6 (7.3~15.1); 24.3 (20.6~28.3); 19.0 (16.4~22.3); 168.4 (137.3~207.5); 0.0 (0.0~0.0); 944 (795~1190) ❙ 1452 (1161~1751); 771 (633~996); 73 (48~142) ❙ 55.8 (12.9~112.2); 942 (795~1190); 423.4 (225.9~659.0)

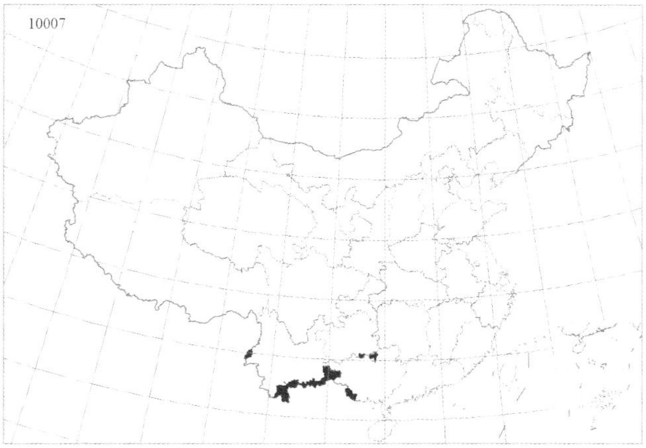

Annonaceae 番荔枝科 *Polyalthia* 暗罗属

Polyalthia pingpinensis P. T. Li 多脉暗罗

Trees, evergreen

18.3; 11.1; 23.3; 18.3; 159.8; 0.0; 895 ❙ 1412; 784; 58 ❙ 58.7; 895; 448.2

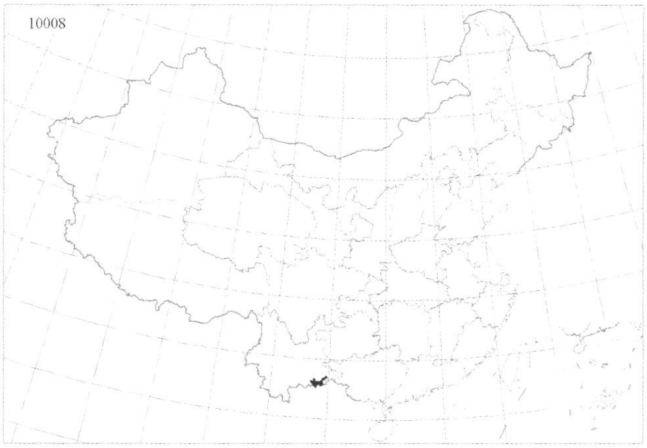

Annonaceae 番荔枝科 *Polyalthia* 暗罗属

Polyalthia plagioneura Diels 斜脉暗罗

Trees, evergreen

21.5 (15.9~24.8); 13.6 (6.9~20.1); 27.7 (22.5~29.8); 21.5 (16.0~24.8); 198.4 (131.6~237.6); 0.0 (0.0~0.0); 1143 (802~1385) ❙ 1586 (868~2435); 679 (312~1428); 120 (41~194) ❙ 41.1 (-34.3~99.5); 1105 (802~1360); 308.8 (174.9~620.0)

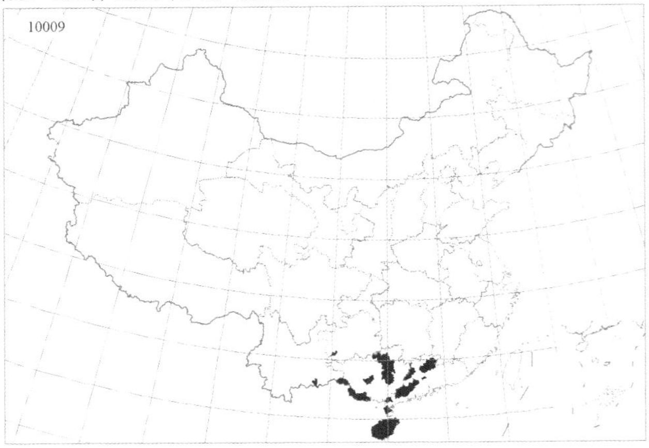

Annonaceae 番荔枝科 *Polyalthia* 暗罗属

Polyalthia rumphii (Blume ex Hensch.) Merr. 香花暗罗

Trees, evergreen

23.7 (21.5~24.8); 17.8 (13.6~20.1); 27.9 (26.7~28.7); 23.7 (21.6~24.8); 224.7 (199.1~237.6); 0.0 (0.0~0.0); 1289 (1131~1385) ❙ 1412 (868~1767); 535 (312~850); 64 (41~106) ❙ 10.2 (-34.3~41.8); 1131 (868~1299); 458.8 (281.7~620.0)

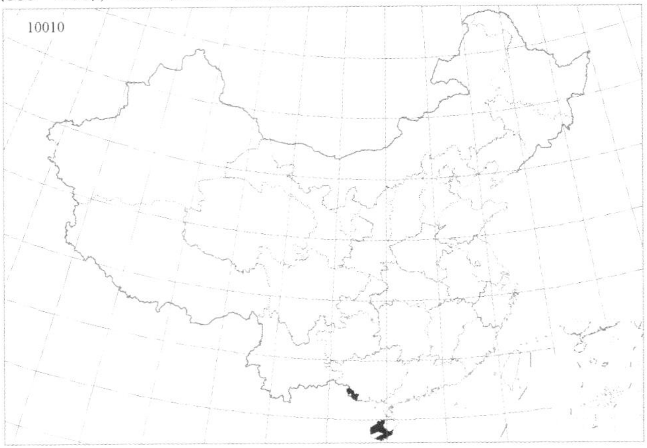

Annonaceae 番荔枝科 *Polyalthia* 暗罗属

Polyalthia simiarum (Buch.-Ham. ex Hook. f. et Thomson) Hook. f. et Thomson 腺叶暗罗

Trees, evergreen

19.7 (16.4~23.6); 13.2 (9.6~16.7); 23.9 (20.6~29.2); 19.7 (16.4~23.6); 176.5 (137.3~223.5); 0.0 (0.0~0.0); 979 (795~1281) ❙ 1501 (1187~1751); 791 (545~996); 64 (42~86) ❙ 58.9 (-6.9~112.2); 964 (795~1269); 471.5 (219.0~659.0)

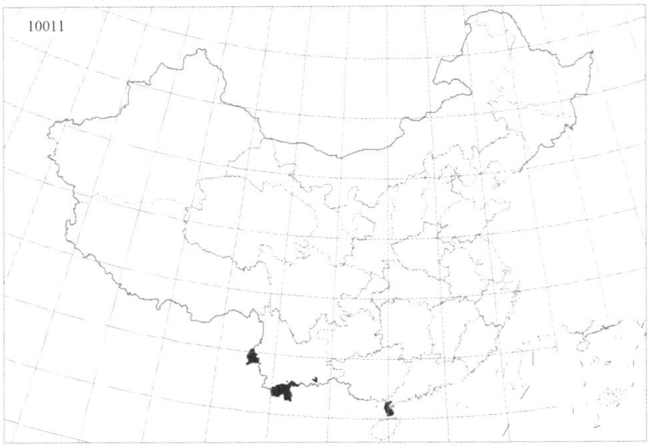

Annonaceae 番荔枝科 *Polyalthia* 暗罗属

Polyalthia suberosa (Roxb.) Thwaites 暗罗

Small trees, evergreen

23.7 (22.2~24.8); 17.6 (14.7~20.1); 28.2 (25.7~29.2); 23.8 (22.3~24.8); 225.4 (207.2~237.6); 0.0 (0.0~0.0); 1294 (1127~1385) ❙ 1497 (868~1854); 590 (312~997); 79 (41~115) ❙ 16.1 (-34.3~47.8); 1203 (868~1360); 376.3 (146.8~620.0)

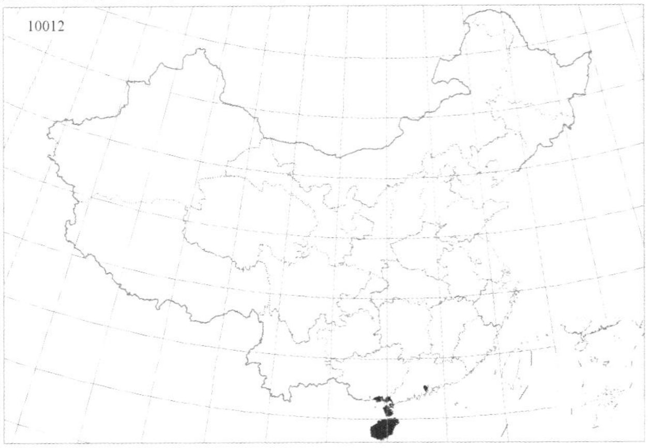

Annonaceae 番荔枝科 *Polyalthia* 暗罗属

Polyalthia verrucipes C. Y. Wu ex P. T. Li 疣叶暗罗

Trees, evergreen

18.2 (13.3~20.5); 12.0 (5.4~15.1); 22.4 (19.6~23.7); 18.3 (13.4~20.5); 159.2 (100.4~185.9); 0.0 (0.0~0.0); 890 (706~992) ❙ 1535 (1368~1736); 761 (655~1007); 62 (51~77) ❙ 76.2 (47.6~145.8); 877 (706~986); 536.4 (447.4~659.0)

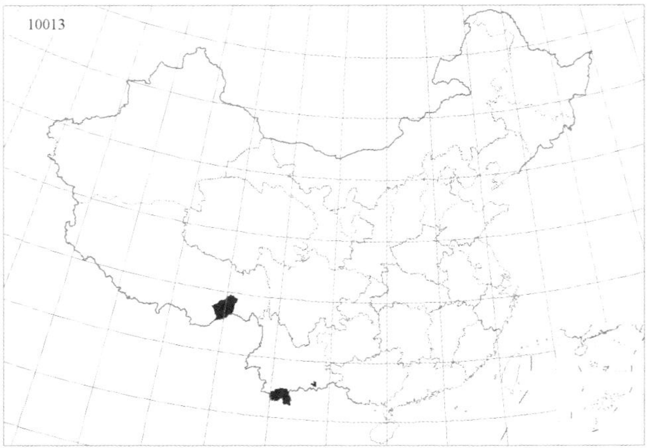

Annonaceae 番荔枝科 *Polyalthia* 暗罗属

Polyalthia viridis Craib 毛脉暗罗

Trees, evergreen

19.5; 13.7; 23.0; 19.5; 173.9; 0.0; 936 ❙ 1485; 700; 65 ❙ 58.8; 920; 558.6

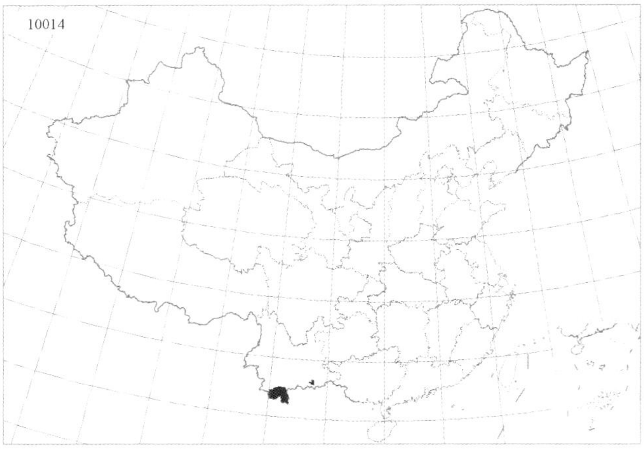

Annonaceae 番荔枝科 *Popowia* 嘉陵花属
Popowia pisocarpa (Blume) Endl. 嘉陵花
Shrubs or small trees, evergreen

23.7 (21.8~24.8); 17.7 (14.2~20.1); 28.0 (25.7~28.8); 23.7 (21.9~24.8); 224.5 (202.6~237.6); 0.0 (0.0~0.0); 1287 (1127~1385) ❙ 1554 (868~2035); 601 (312~1033); 79 (41~116) ❙ 21.7 (-34.3~71.8); 1193 (868~1360); 395.5 (163.7~620.0)

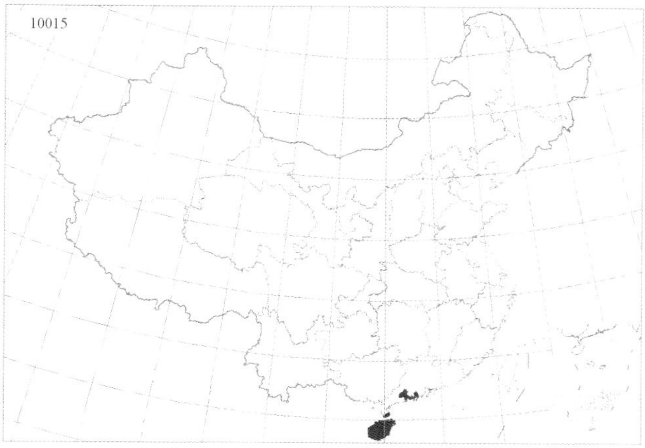

Annonaceae 番荔枝科 *Pseuduvaria* 金钩花属
Pseuduvaria indochinensis Merr. 金钩花
Trees, evergreen

19.9 (16.4~22.2); 13.1 (9.6~15.1); 24.5 (20.6~28.3); 19.9 (16.4~22.3); 179.0 (137.3~207.5); 0.0 (0.0~0.0); 993 (795~1190) ❙ 1579 (1302~2148); 847 (655~1234); 78 (55~125) ❙ 61.1 (12.9~112.2); 989 (795~1190); 461.8 (281.7~659.0)

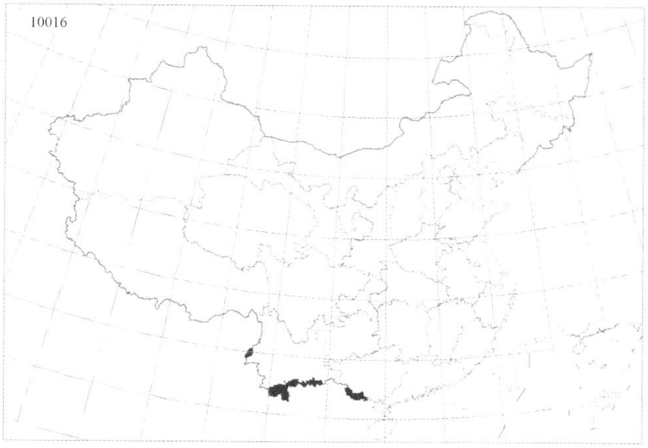

Annonaceae 番荔枝科 *Richella* 尖花藤属
Richella hainanensis (Tsiang et P. T. Li) Tsiang et P. T. Li 尖花藤
Vines, evergreen

24.0; 19.1; 27.2; 24.0; 228.1; 0.0; 1302 ❙ 1551; 539; 48 ❙ 19.1; 1147; 620.0

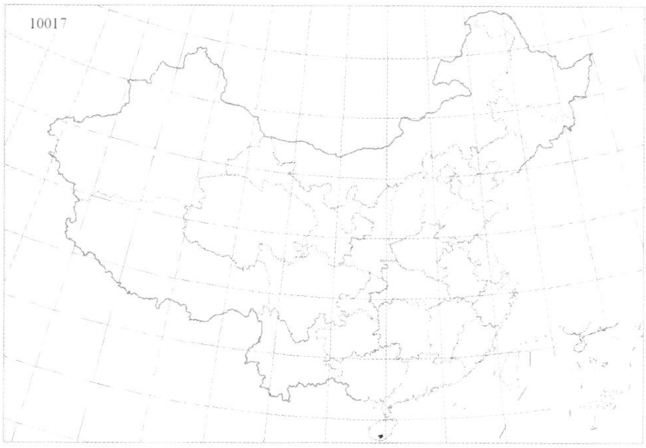

Annonaceae 番荔枝科 *Uvaria* 紫玉盘属
Uvaria boniana Finet et Gagnep. 光叶紫玉盘
Scandent shrubs, evergreen

21.3 (14.6~24.8); 13.1 (3.8~20.1); 28.0 (24.6~29.8); 21.3 (14.7~24.8); 195.9 (117.7~237.6); 0.0 (0.0~1.5); 1127 (779~1385) ❙ 1634 (868~2435); 698 (312~1428); 135 (41~197) ❙ 46.7 (-34.3~101.8); 1110 (779~1299); 252.0 (95.3~620.0)

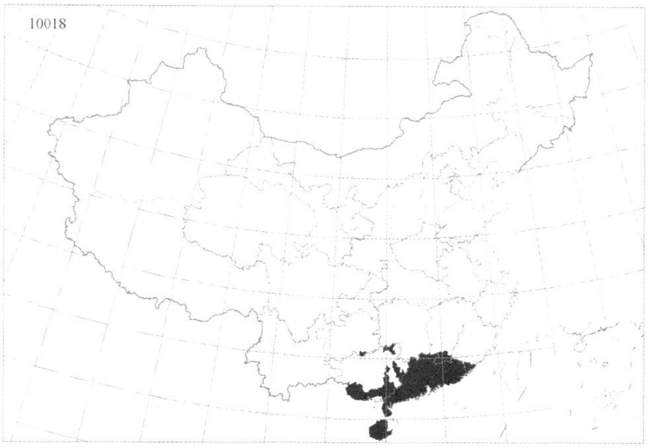

Annonaceae 番荔枝科 *Uvaria* 紫玉盘属
Uvaria calamistrata Hance 刺果紫玉盘
Scandent shrubs, evergreen

23.2 (20.8~24.8); 16.4 (12.2~20.1); 28.2 (27.1~28.8); 23.2 (20.8~24.8); 218.4 (189.7~237.6); 0.0 (0.0~0.0); 1254 (1080~1385) ❙ 1536 (868~2035); 647 (312~994); 90 (41~181) ❙ 23.9 (-34.3~83.4); 1173 (868~1331); 340.9 (139.6~620.0)

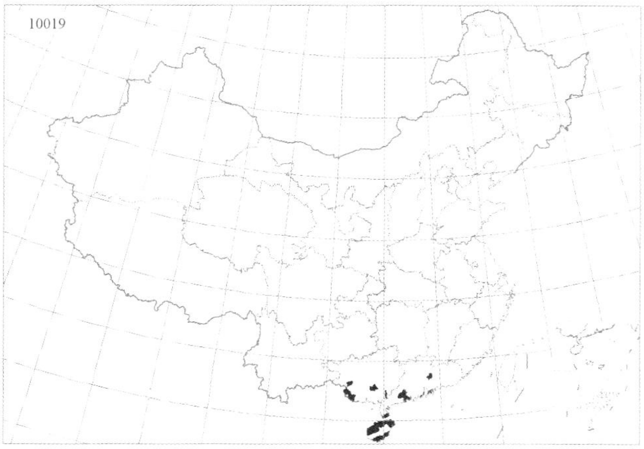

Annonaceae 番荔枝科 *Uvaria* 紫玉盘属
Uvaria grandiflora Roxb. 大花紫玉盘
Scandent shrubs, evergreen

22.0 (16.9~24.8); 14.3 (7.8~20.1); 28.1 (25.7~29.2); 22.0 (17.0~24.8); 204.0 (143.6~237.6); 0.0 (0.0~0.0); 1166 (862~1385) ❙ 1664 (868~2037); 726 (312~1033); 122 (41~198) ❙ 44.0 (-34.3~100.8); 1143 (862~1326); 248.8 (95.3~620.0)

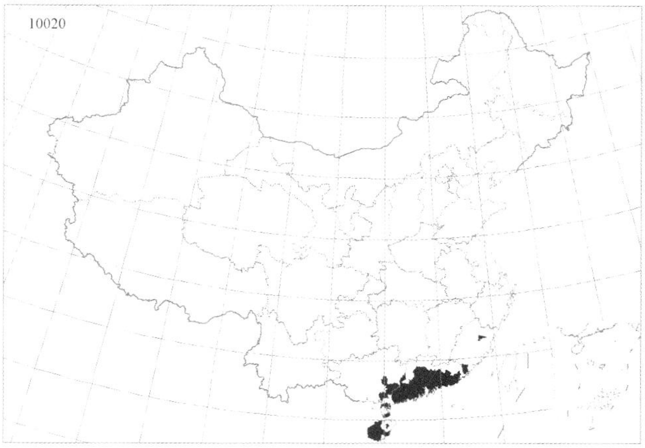

Annonaceae 番荔枝科 *Uvaria* 紫玉盘属

Uvaria kurzii (King) P. T. Li 黄花紫玉盘

Scandent shrubs, evergreen

20.8 (17.3~22.7); 13.2 (10.9~15.2); 26.5 (21.6~28.9); 20.9 (17.3~22.8); 191.3 (148.0~213.1); 0.0 (0.0~0.0); 1079 (832~1221) ❙ 1597 (1049~2435); 851 (598~1428); 102 (44~194) ❙ 48.5 (12.9~99.5); 1076 (783~1221); 349.7 (201.4~659.0)

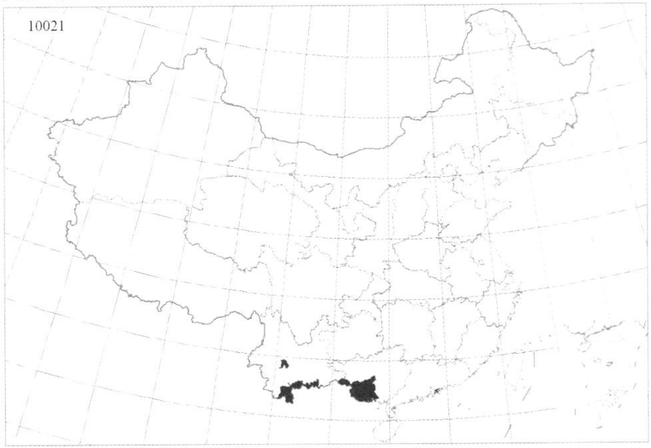

Annonaceae 番荔枝科 *Uvaria* 紫玉盘属

Uvaria kweichowensis P. T. Li 贵州紫玉盘

Scandent shrubs, evergreen

18.4 (16.8~20.4); 9.7 (7.4~15.1); 25.2 (23.4~26.6); 18.5 (16.8~20.5); 161.6 (141.5~185.8); 0.0 (0.0~0.0); 927 (836~1010) ❙ 1301 (1182~1607); 678 (617~723); 63 (43~107) ❙ 40.8 (16.9~62.5); 926 (836~1010); 308.4 (189.5~659.0)

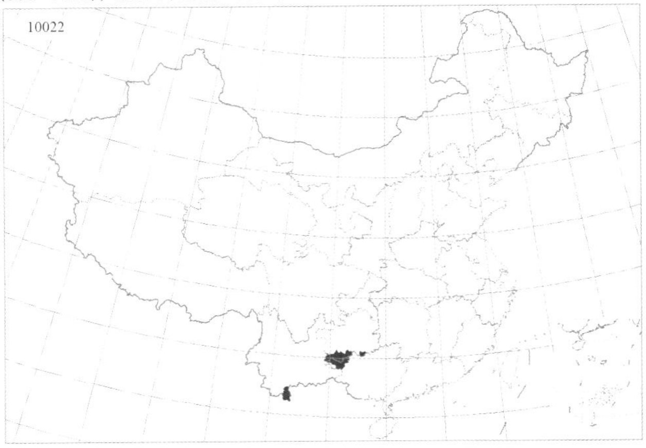

Annonaceae 番荔枝科 *Uvaria* 紫玉盘属

Uvaria macrophylla Roxb. 紫玉盘

Scandent shrubs

21.1 (16.3~24.8); 13.1 (7.1~20.1); 27.3 (21.4~29.8); 21.1 (16.4~24.8); 193.3 (136.4~237.6); 0.0 (0.0~0.0); 1104 (797~1385) ❙ 1611 (868~3495); 716 (312~1428); 128 (41~897) ❙ 47.4 (-34.3~224.4); 1090 (797~1360); 278.9 (95.3~659.0)

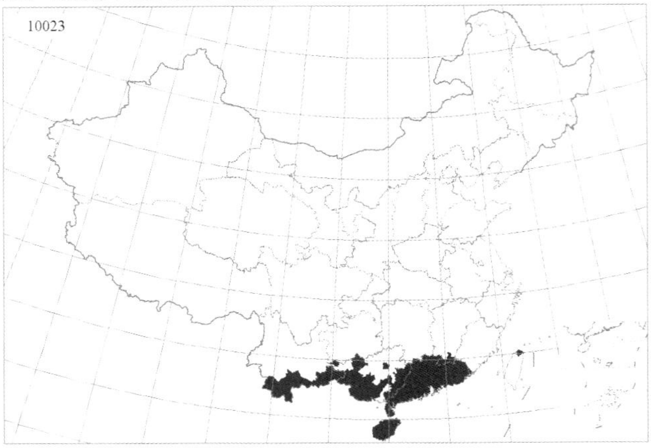

Annonaceae 番荔枝科 *Uvaria* 紫玉盘属

Uvaria rufa Blume 小花紫玉盘

Scandent shrubs, evergreen

19.1 (16.4~24.0); 12.9 (9.5~19.1); 23.2 (20.6~27.2); 19.2 (16.4~24.0); 170.0 (137.3~228.1); 0.0 (0.0~0.0); 934 (795~1302) ❙ 1516 (1254~1751); 788 (539~996); 60 (48~77) ❙ 64.6 (19.1~112.2); 915 (795~1147); 506.4 (336.3~659.0)

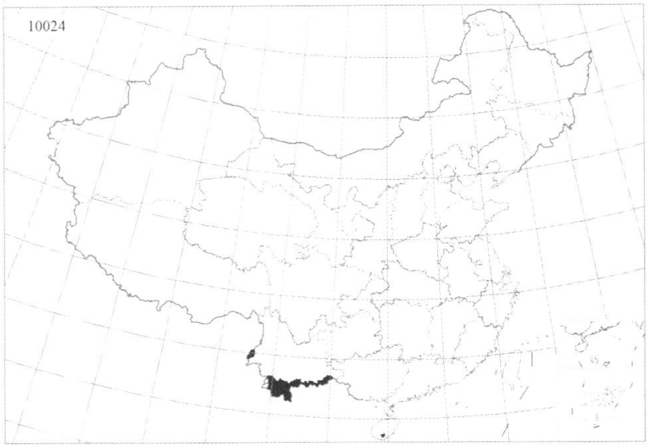

Annonaceae 番荔枝科 *Uvaria* 紫玉盘属

Uvaria tonkinensis Finet et Gagnep. 东京紫玉盘

Scandent shrubs, evergreen

22.7 (19.0~24.8); 16.1 (9.9~20.1); 27.4 (23.6~28.8); 22.7 (19.0~24.8); 212.5 (168.6~237.6); 0.0 (0.0~0.0); 1212 (939~1385) ❙ 1489 (868~2024); 613 (312~850); 82 (41~183) ❙ 24.5 (-34.3~86.5); 1139 (868~1360); 398.2 (226.9~659.0)

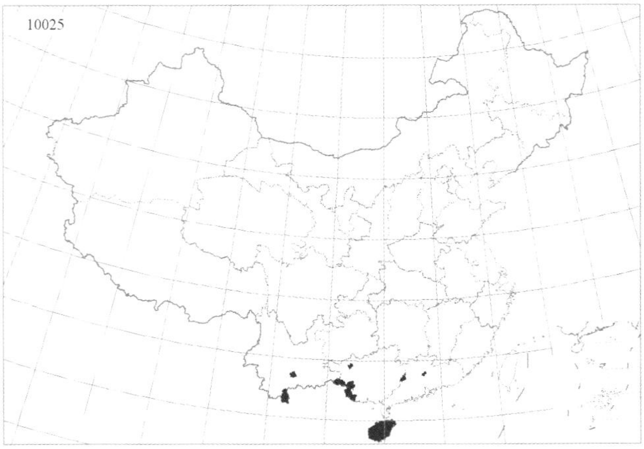

Annonaceae 番荔枝科 *Xylopia* 木瓣树属

Xylopia vielana Pierre 木瓣树

Trees, evergreen

22.3; 14.5; 28.1; 22.3; 207.7; 0.0; 1186 ❙ 2291; 1331; 127 ❙ 93.1; 1186; 309.7

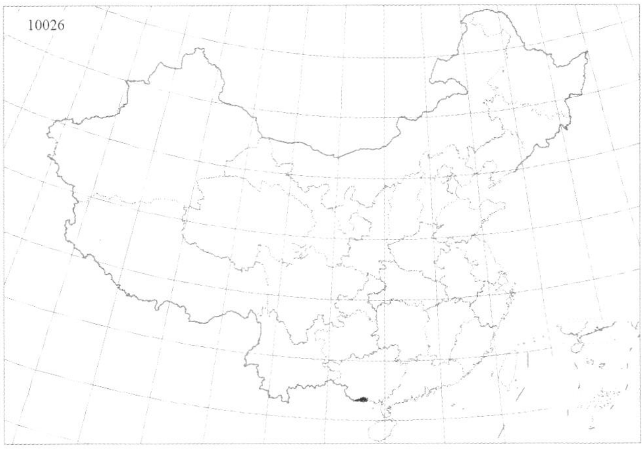

Acanthaceae 爵床科 *Acanthus* 老鼠簕属

Acanthus ebracteatus Vahl 小花老鼠簕

Shrubs

23.6 (21.8~24.8); 17.1 (14.0~20.1); 28.4 (27.7~28.9); 23.6 (21.9~24.8); 223.3 (202.2~237.6); 0.0 (0.0~0.0); 1286 (1150~1385) ❙ 1565 (868~2148); 685 (312~1234); 87 (41~124) ❙ 23.3 (-34.3~86.7); 1183 (868~1329); 327.9 (215.3~527.5)

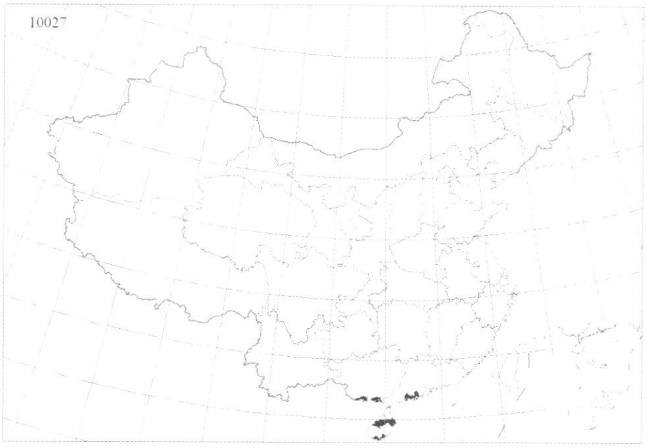

Acanthaceae 爵床科 *Acanthus* 老鼠簕属

Acanthus ebracteatus var. xiamenensis (R. T. Zhang) C. Y. Wu et C. C. Hu 厦门老鼠簕

Shrubs

20.8; 12.9; 28.2; 20.9; 190.5; 0.0; 1084 ❙ 1298; 446; 168 ❙ 19.7; 1059; 214.3

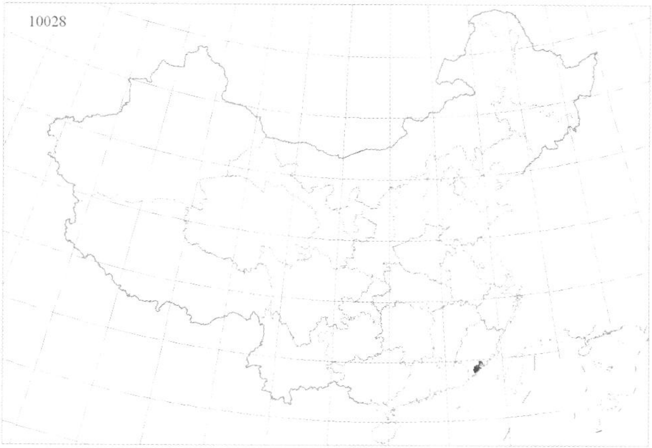

Acanthaceae 爵床科 *Acanthus* 老鼠簕属

Acanthus ilicifolius L. 老鼠簕

Shrubs

22.5 (20.4~24.2); 15.2 (12.6~18.0); 28.3 (25.7~29.2); 22.5 (20.4~24.2); 210.3 (184.9~230.4); 0.0 (0.0~0.0); 1200 (1046~1331) ❙ 1710 (1269~2435); 791 (455~1428); 112 (52~163) ❙ 43.0 (-1.4~99.5); 1193 (1046~1331); 251.8 (139.6~590.7)

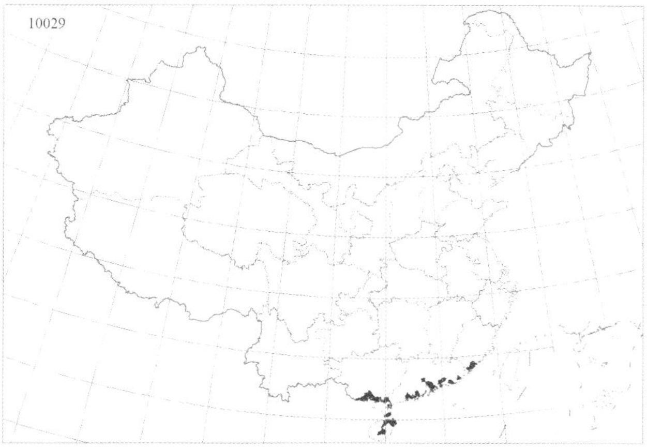

Acanthaceae 爵床科 *Adhatoda* 鸭嘴花属

Adhatoda vasica Nees 鸭嘴花

Tall shrubs

18.2 (8.6~24.2); 11.5 (1.3~18.1); 22.8 (15.0~28.8); 18.2 (8.7~24.3); 158.7 (52.6~231.2); 0.2 (0.0~8.8); 894 (574~1335) ❙ 1393 (1028~1776); 736 (447~996); 66 (41~188) ❙ 57.5 (-12.0~112.2); 878 (574~1331); 426.2 (197.5~659.0)

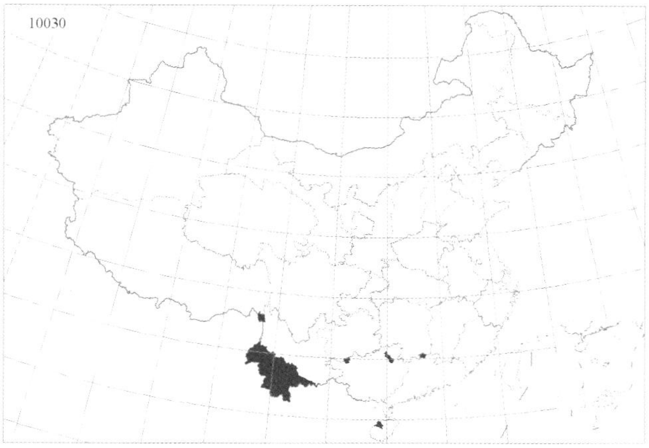

Acanthaceae 爵床科 *Barleria* 假杜鹃属

Barleria cristata L. 假杜鹃

Small shrubs

17.1 (2.2~24.8); 9.7 (-6.6~20.1); 22.8 (9.9~29.0); 17.1 (3.8~24.8); 146.4 (15.5~237.6); 0.9 (0.0~49.1); 881 (413~1385) ❙ 1229 (741~2037); 635 (312~1044); 61 (12~231) ❙ 41.7 (-34.3~125.9); 858 (413~1360); 319.4 (120.3~659.0)

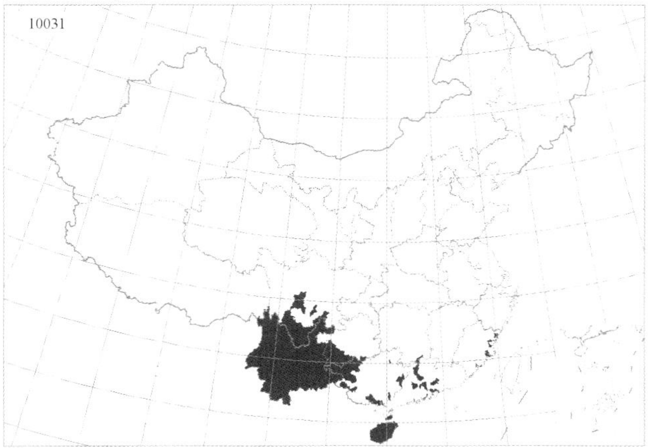

Acanthaceae 爵床科 *Barleria* 假杜鹃属

Barleria integrisepala H. P. Tsui 全缘萼假杜鹃

Shrubs, evergreen

6.6; -1.3; 13.3; 6.8; 39.6; 19.7; 514 ❙ 862; 548; 12 ❙ 67.7; 514; 311.5

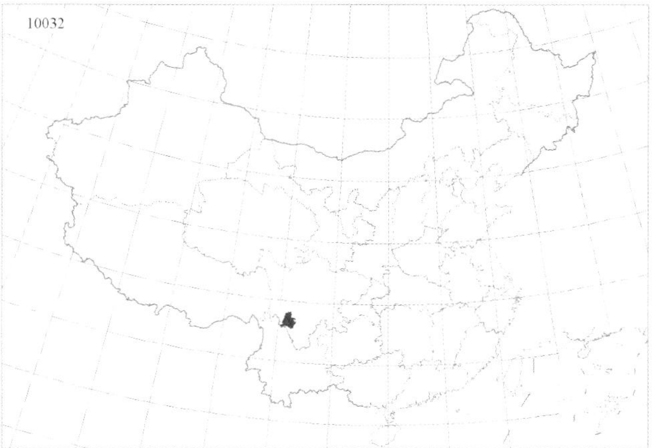

Acanthaceae 爵床科 *Barleria* 假杜鹃属
Barleria prionitis L. 黄花假杜鹃
Small shrubs

20.0; 14.6; 23.2; 20.1; 180.8; 0.0; 963 ❙ 1523; 681; 67 ❙ 58.4; 942; 581.0

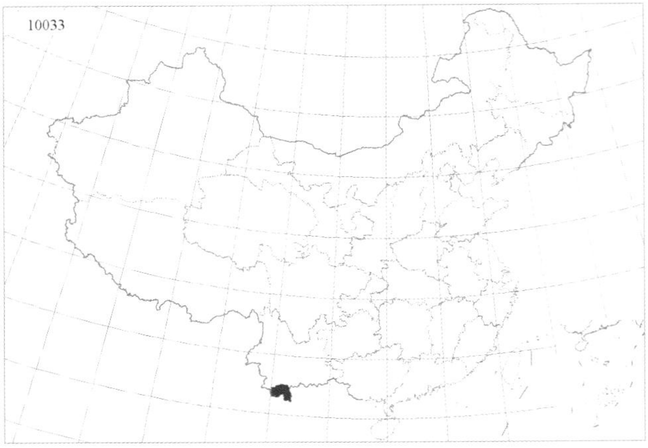

Acanthaceae 爵床科 *Barleria* 假杜鹃属
Barleria purpureosepala H. P. Tsui 紫萼假杜鹃
Small shrubs

20.5; 14.9; 23.7; 20.5; 185.9; 0.0; 992 ❙ 1464; 655; 66 ❙ 47.6; 959; 570.1

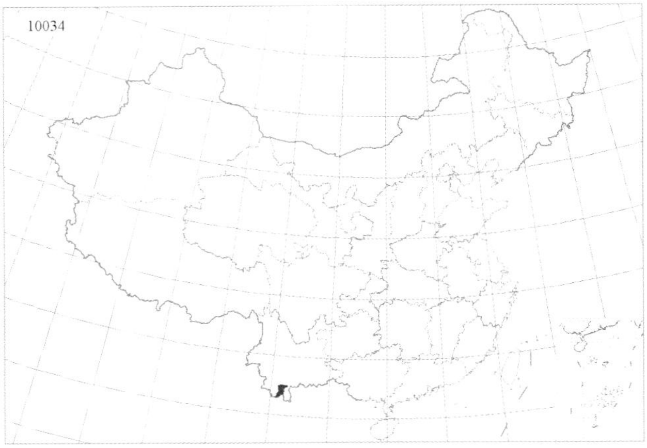

Acanthaceae 爵床科 *Calophanoides* 杜根藤属
Calophanoides kwangsiensis H. S. Lo 广西赛爵床
Shrubs

20.1 (19.0~22.2); 11.1 (9.1~14.2); 27.3 (25.8~28.4); 20.1 (19.0~22.2); 181.3 (168.4~206.6); 0.0 (0.0~0.0); 1045 (958~1187) ❙ 1494 (1250~1876); 694 (561~891); 118 (60~175) ❙ 42.8 (28.8~63.2); 1045 (958~1187); 273.1 (239.0~322.1)

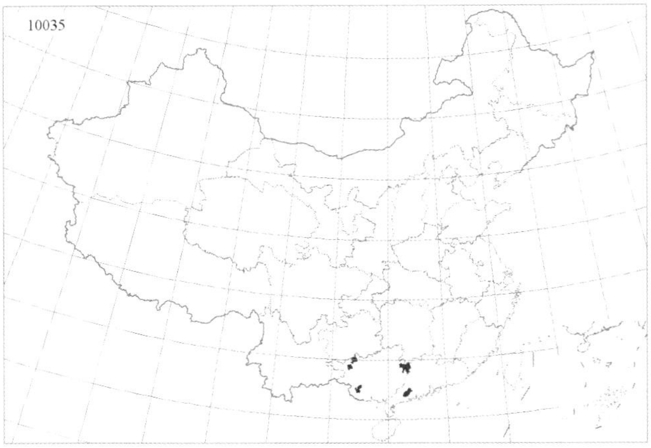

Acanthaceae 爵床科 *Calophanoides* 杜根藤属
Calophanoides siccanea (W. W. Smith) C. Y. Wu ex C. C. Hu 旱杜根藤
Shrubs

9.3; 2.0; 15.4; 9.4; 59.8; 7.9; 590 ❙ 925; 582; 17 ❙ 57.5; 588; 310.5

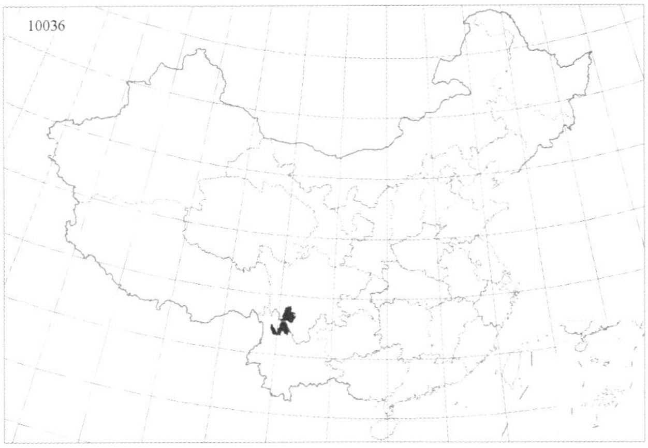

Acanthaceae 爵床科 *Calophanoides* 杜根藤属
Calophanoides wardii (W. W. Smith) C. Y. Wu ex C. C. Hu 高山杜根藤
Shrubs

10.6 (5.4~15.0); 3.6 (-2.1~8.4); 16.5 (12.3~20.2); 10.7 (5.8~15.1); 75.0 (31.5~121.2); 7.5 (0.0~25.6); 625 (481~753) ❙ 932 (741~1042); 556 (426~603); 29 (15~46) ❙ 50.8 (26.3~81.5); 610 (481~706); 314.8 (273.5~372.3)

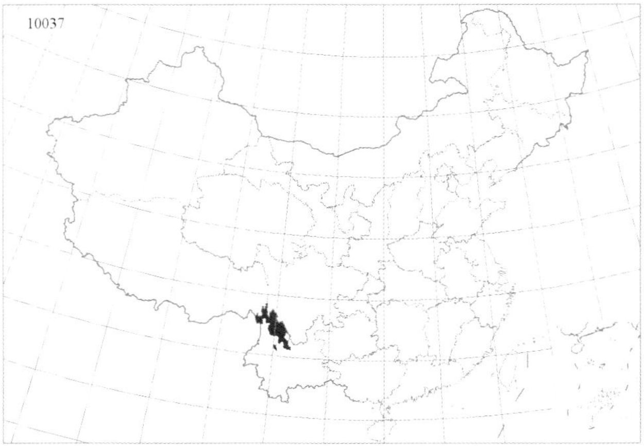

Acanthaceae 爵床科 *Calophanoides* 杜根藤属
Calophanoides xantholeuca (W. W. Smith) C. Y. Wu ex C. C. Hu 黄白杜根藤
Shrubs

14.6; 7.9; 19.6; 14.6; 115.8; 0.0; 736 ❙ 1009; 596; 39 ❙ 37.1; 732; 278.4

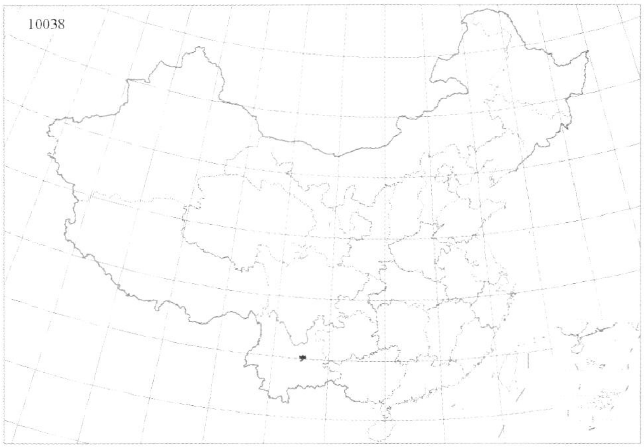

Acanthaceae 爵床科 *Calophanoides* 杜根藤属

Calophanoides xylopoda (W. W. Smith) C. Y. Wu ex C. C. Hu 木柄杜根藤 Shrubs

8.7; 1.3; 15.0; 8.8; 55.7; 10.5; 574 ❙ 915; 571; 18 ❙ 60.3; 574; 315.6

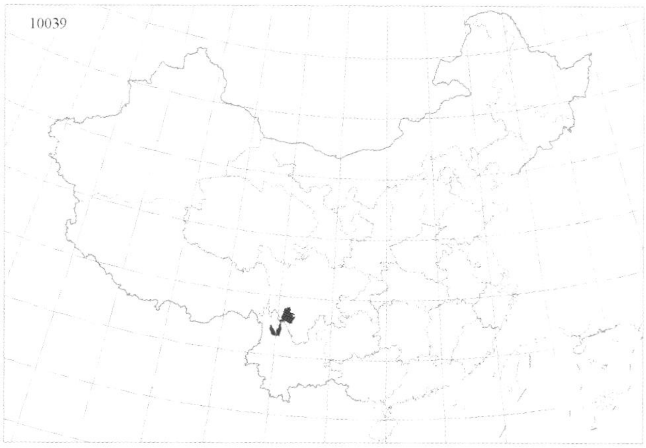

Acanthaceae 爵床科 *Calophanoides* 杜根藤属

Calophanoides yunnanensis (W. W. Smith) C. Y. Wu ex H. P. Tsui 滇杜根藤 Shrubs

14.6; 7.8; 20.0; 14.6; 115.2; 0.0; 738 ❙ 963; 581; 26 ❙ 30.5; 689; 309.7

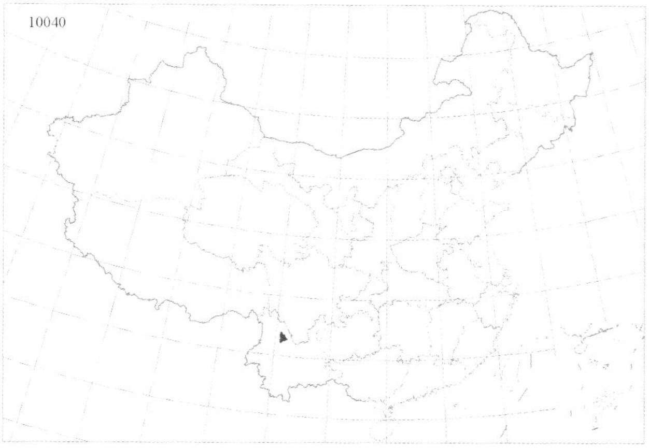

Acanthaceae 爵床科 *Chroesthes* 色萼花属

Chroesthes lanceolata (T. Anders.) B. Hansen 色萼花 Shrubs

19.0 (16.3~22.0); 12.0 (9.1~15.1); 23.8 (21.4~28.6); 19.1 (16.4~22.0); 168.6 (136.4~204.5); 0.0 (0.0~0.0); 936 (797~1179) ❙ 1412 (1074~1751); 747 (582~996); 69 (41~171) ❙ 51.6 (30.4~88.8); 929 (797~1179); 407.2 (206.0~659.0)

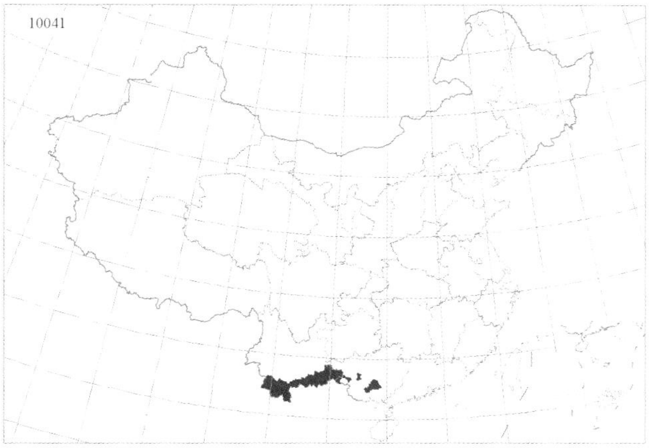

Acanthaceae 爵床科 *Cosmianthemum* 秋英爵床属

Cosmianthemum knoxifolium (C. B. Clarke) B. Hansen 节叶秋英爵床 Shrubs

23.9; 18.8; 27.4; 23.9; 227.0; 0.0; 1297 ❙ 1681; 572; 68 ❙ 29.9; 1203; 580.3

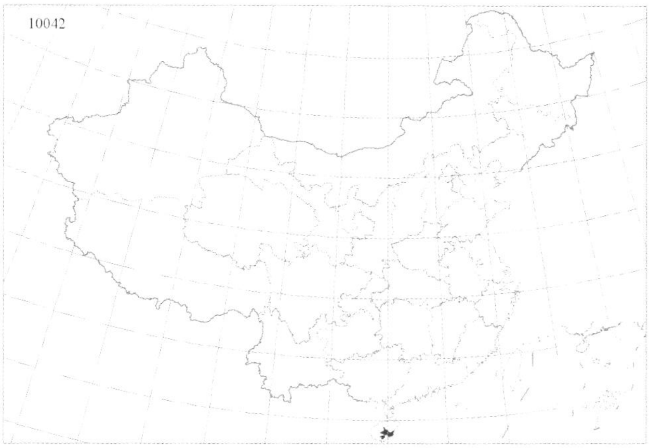

Acanthaceae 爵床科 *Cystacanthus* 鳔冠花属

Cystacanthus affinis W. W. Sm. 丽江鳔冠花 Shrubs

6.4 (-1.8~10.8); -1.1 (-10.2~3.9); 13.0 (6.1~16.6); 7.2 (1.8~10.9); 43.3 (1.9~71.8); 25.6 (1.3~82.8); 519 (329~633) ❙ 779 (320~968); 492 (217~603); 18 (12~24) ❙ 46.2 (-3.0~67.7); 516 (320~633); 257.3 (72.2~319.8)

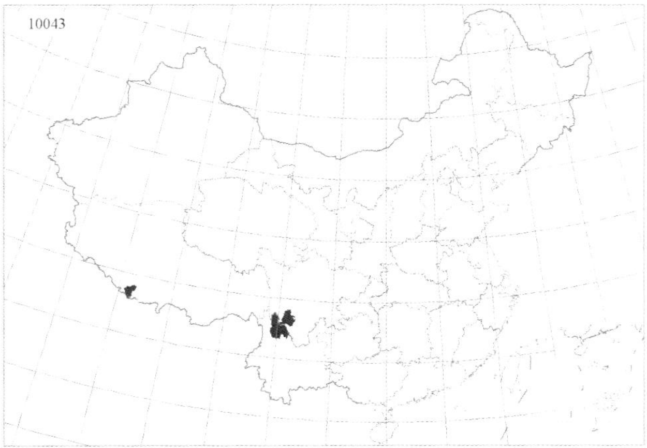

Acanthaceae 爵床科 *Cystacanthus* 鳔冠花属

Cystacanthus paniculatus T. Anderson 鳔冠花 Shrubs

15.9 (5.4~21.0); 8.9 (-2.1~15.1); 21.0 (12.3~26.6); 15.9 (5.8~21.0); 131.6 (31.5~191.8); 0.7 (0.0~25.6); 793 (481~1053) ❙ 1126 (741~1751); 635 (411~996); 44 (15~77) ❙ 42.1 (-14.2~112.2); 774 (481~1053); 342.2 (142.9~659.0)

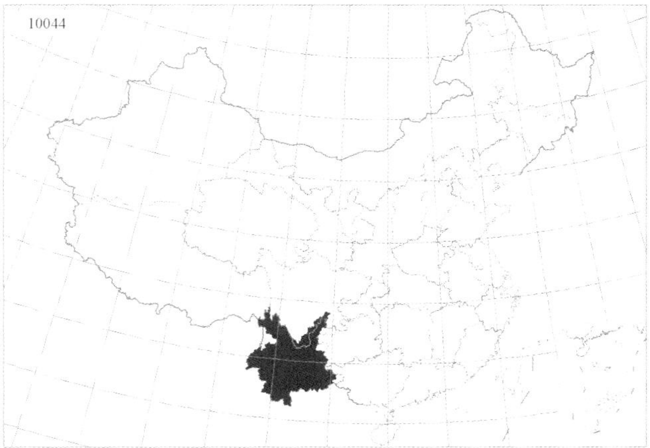

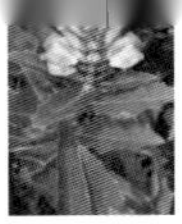

Acanthaceae 爵床科 *Cystacanthus* 鳔冠花属

Cystacanthus yangtsekiangensis (Lév.) Rehder 金江鳔冠花

Shrubs

13.6 (5.4~18.9); 5.9 (−2.1~12.0); 20.0 (12.3~26.6); 13.6 (5.8~19.0); 105.5 (31.5~167.8); 2.1 (0.0~25.6); 721 (481~929) ❙ 952 (741~1141); 552 (411~627); 31 (15~57) ❙ 33.7 (−14.2~70.8); 692 (481~912); 265.0 (142.9~351.7)

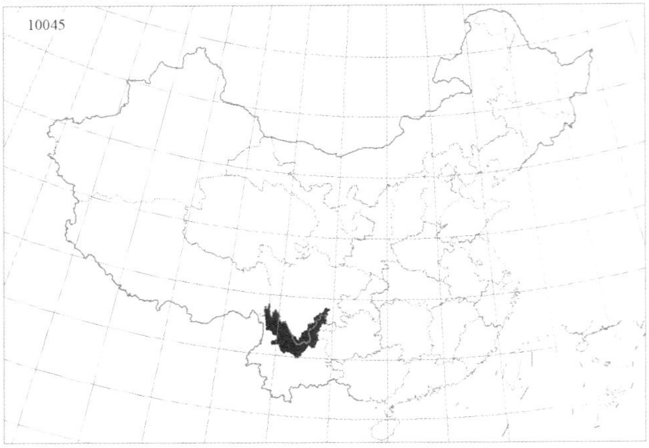

Acanthaceae 爵床科 *Cystacanthus* 鳔冠花属

Cystacanthus yunnanensis W. W. Sm. 滇鳔冠花

Shrubs

14.3 (5.4~18.9); 7.1 (−2.1~12.0); 20.1 (12.3~26.6); 14.4 (5.8~19.0); 113.5 (31.5~167.8); 1.2 (0.0~25.6); 740 (481~929) ❙ 975 (741~1164); 563 (411~651); 36 (15~58) ❙ 33.3 (−14.2~78.3); 715 (481~912); 278.3 (142.9~382.6)

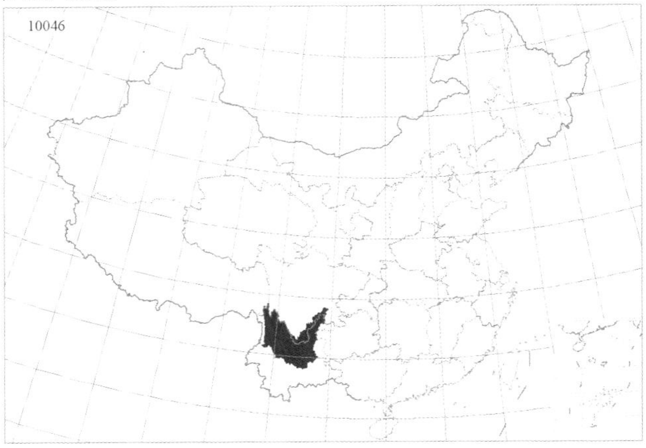

Acanthaceae 爵床科 *Echinacanthus* 恋岩花属

Echinacanthus lofouensis (Lév.) J. R. I. Wood 黄花恋岩花

Shrubs

18.4 (16.2~20.3); 8.9 (6.3~10.9); 26.0 (24.4~27.7); 18.4 (16.2~20.4); 161.1 (134.5~184.4); 0.0 (0.0~0.0); 941 (826~1065) ❙ 1344 (1236~1567); 668 (611~775); 86 (51~141) ❙ 43.3 (27.5~62.5); 941 (826~1065); 253.3 (215.4~282.2)

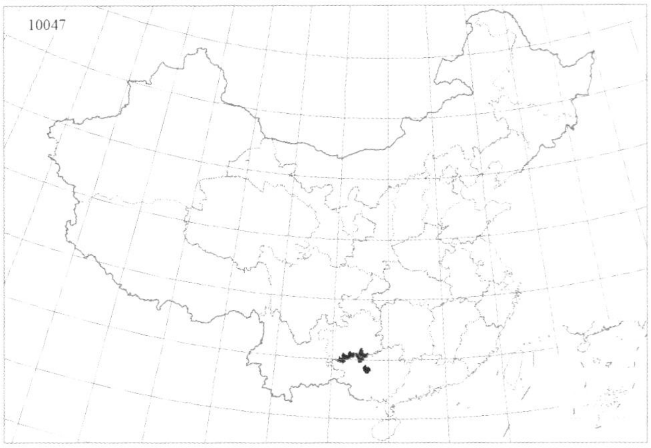

Acanthaceae 爵床科 *Echinacanthus* 恋岩花属

Echinacanthus longzhouensis H. S. Lo 龙州恋岩花

Shrubs

22.2; 13.9; 28.3; 22.3; 207.5; 0.0; 1190 ❙ 1374; 713; 87 ❙ 15.4; 1190; 327.7

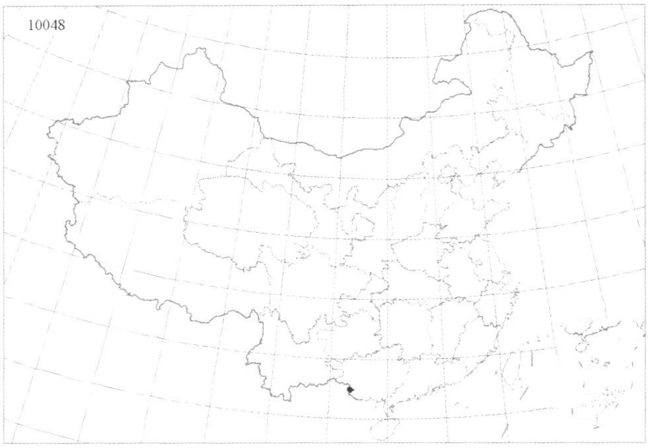

Acanthaceae 爵床科 *Mananthes* 野靛棵属

Mananthes amblyosepala (D. Fang et H. S. Lo) C. Y. Wu et C. C. Hu in C. C. Hu 钝萼野靛棵

Shrubs

19.1; 11.0; 25.0; 19.1; 169.6; 0.0; 945 ❙ 1438; 820; 57 ❙ 52.2; 945; 371.7

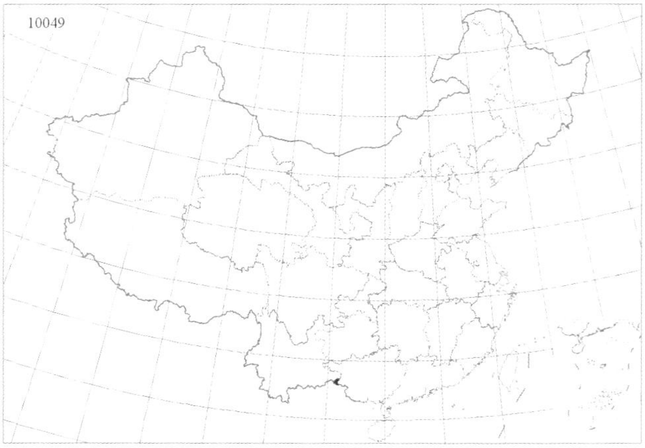

Acanthaceae 爵床科 *Mananthes* 野靛棵属

Mananthes kampotiana (Benoist) C. Y. Wu et C. C. Hu 那坡野靛棵

Shrubs

19.1; 11.0; 25.0; 19.1; 169.6; 0.0; 945 ❙ 1438; 820; 57 ❙ 52.2; 945; 371.7

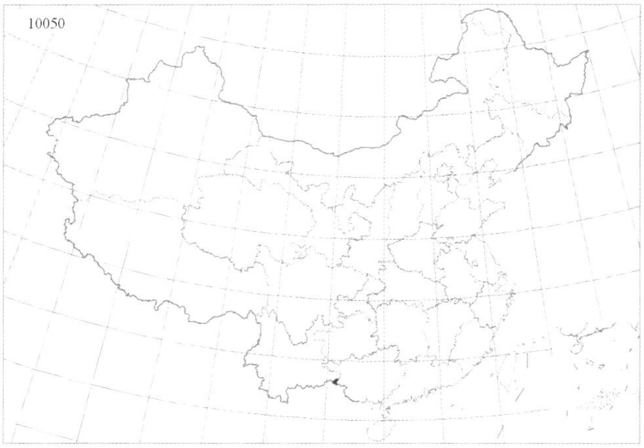

Acanthaceae 爵床科 *Mananthes* 野靛棵属
Mananthes latiflora (Hemsl.) C. Y. Wu et C. C. Hu 紫苞野靛棵
Shrubs

17.7; 10.9; 22.5; 17.8; 153.4; 0.0; 854 | 1368; 756; 59 | 60.1; 854; 491.5

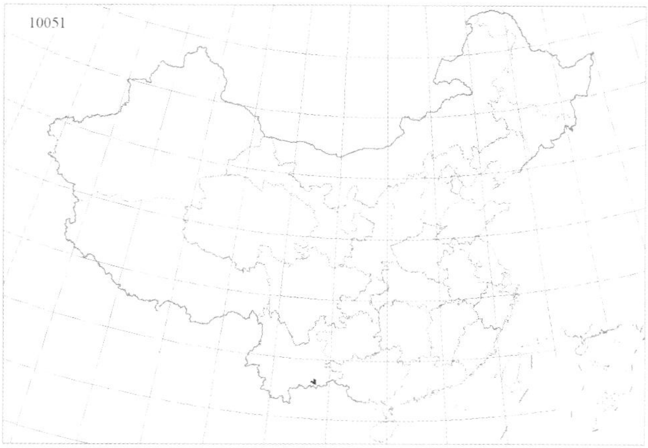

Acanthaceae 爵床科 *Mananthes* 野靛棵属
Mananthes microdonta (W. W. Sm.) C. Y. Wu et C. C. Hu 小齿野靛棵
Shrubs

9.2; 1.6; 15.7; 9.4; 63.4; 12.5; 586 | 873; 519; 27 | 50.4; 580; 269.5

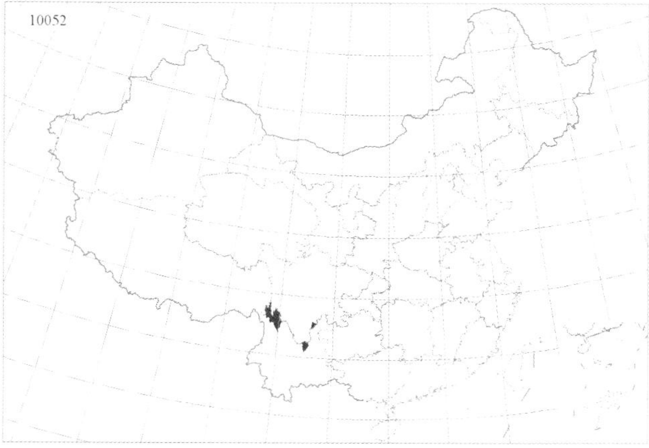

Acanthaceae 爵床科 *Mananthes* 野靛棵属
Mananthes panduriformis (R. Ben.) C. Y. Wu et C. C. Hu 琴叶野靛棵
Shrubs

19.7; 11.9; 25.4; 19.8; 177.3; 0.0; 1012 | 1439; 787; 71 | 48.0; 1012; 370.6

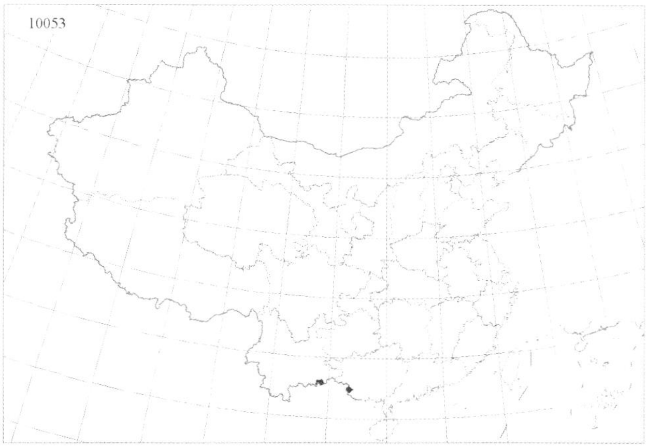

Acanthaceae 爵床科 *Mananthes* 野靛棵属
Mananthes vasculosa (Nees) Bremekamp 滇野靛棵
Shrubs

19.2 (17.2~22.2); 11.7 (9.5~14.2); 24.6 (22.4~28.6); 19.3 (17.3~22.3); 171.1 (147.0~207.5); 0.0 (0.0~0.0); 962 (833~1190) | 1409 (1254~1522); 778 (699~860); 68 (48~138) | 48.7 (15.4~80.6); 962 (833~1190); 400.1 (272.3~551.6)

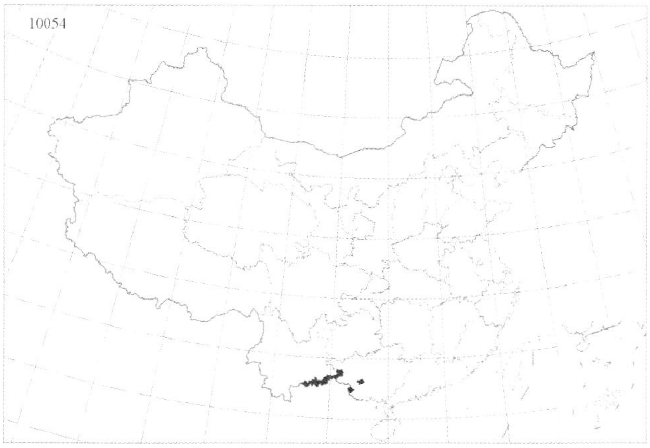

Acanthaceae 爵床科 *Perilepta* 耳叶爵床属
Perilepta ferruginea (D. Fang et H. S. Lo) C. Y. Wu et C. C. Hu 锈背耳叶马蓝
Shrubs

22.2; 13.9; 28.3; 22.3; 207.5; 0.0; 1190 | 1374; 713; 87 | 15.4; 1190; 327.7

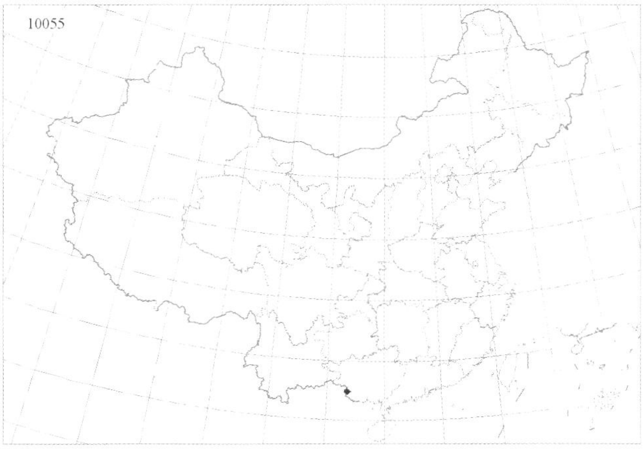

Acanthaceae 爵床科 *Perilepta* 耳叶爵床属
Perilepta longzhouensis (H. S. Lo et D. Fang) C. Y. Wu et C. C. Hu 龙州耳叶马蓝
Shrubs

22.2; 13.9; 28.3; 22.3; 207.5; 0.0; 1190 | 1374; 713; 87 | 15.4; 1190; 327.7

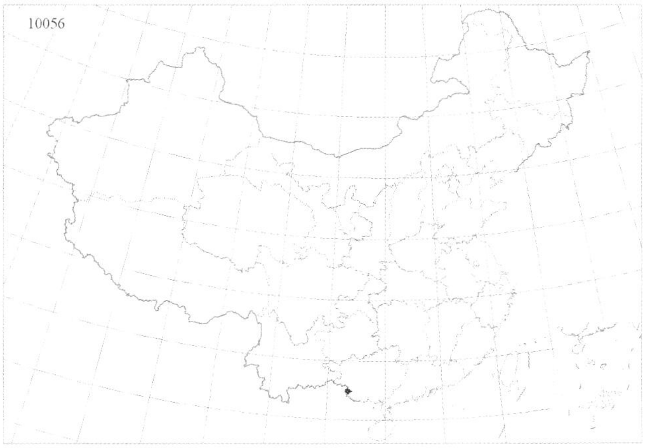

Acanthaceae 爵床科 *Phlogacanthus* 火焰花属

Phlogacanthus curviflorus (Wall.) Nees 火焰花

Shrubs

18.3 (13.3~21.0); 11.9 (5.4~15.1); 22.4 (19.6~26.6); 18.3 (13.4~21.0); 159.9 (100.4~191.8); 0.0 (0.0~0.0); 884 (706~1053) | 1467 (1178~1751); 805 (646~1007); 55 (41~77) | 67.2 (30.9~145.8); 872 (706~1053); 483.8 (322.1~659.0)

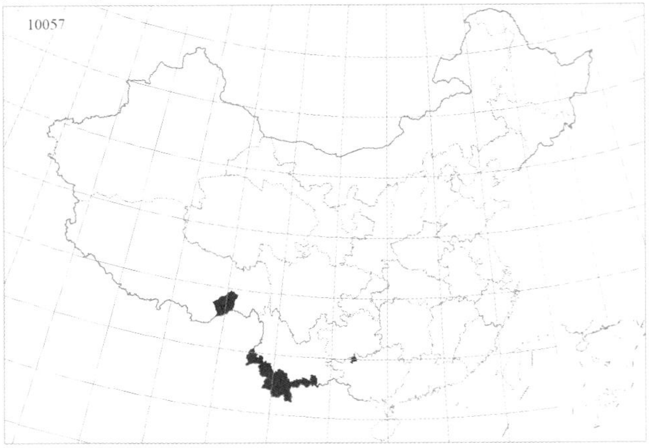

Acanthaceae 爵床科 *Phlogacanthus* 火焰花属

Phlogacanthus pubinervius T. Anderson 毛脉火焰花

Shrubs or small trees

17.1 (6.6~21.3); 9.2 (-1.3~15.1); 23.2 (13.3~27.9); 17.1 (6.8~21.3); 145.7 (39.6~196.0); 0.2 (0.0~19.7); 853 (514~1123) | 1282 (862~1751); 661 (448~996); 66 (12~161) | 50.8 (5.8~112.2); 847 (514~1121); 349.9 (188.8~659.0)

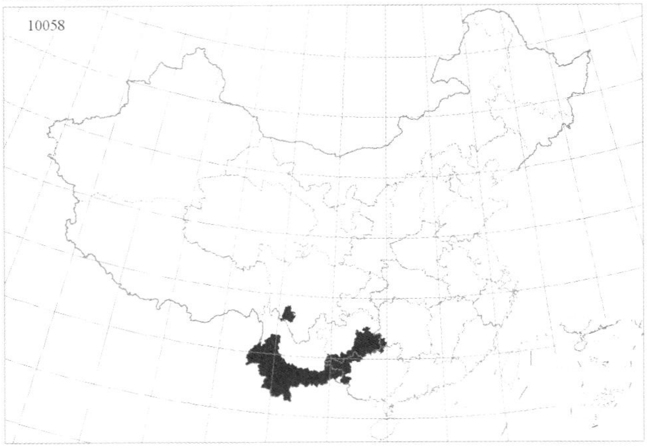

Acanthaceae 爵床科 *Pseuderanthemum* 钩粉草属

Pseuderanthemum graciliflorum (Nees) Ridl. 云南山壳骨

Shrubs or subshrubs

18.1 (14.4~20.5); 11.9 (7.5~15.1); 22.0 (19.1~23.7); 18.2 (14.4~20.5); 158.2 (113.1~185.9); 0.0 (0.0~0.0); 873 (727~992) | 1449 (1192~1751); 799 (655~996); 55 (41~77) | 66.6 (36.8~112.2); 861 (727~986); 477.0 (346.6~659.0)

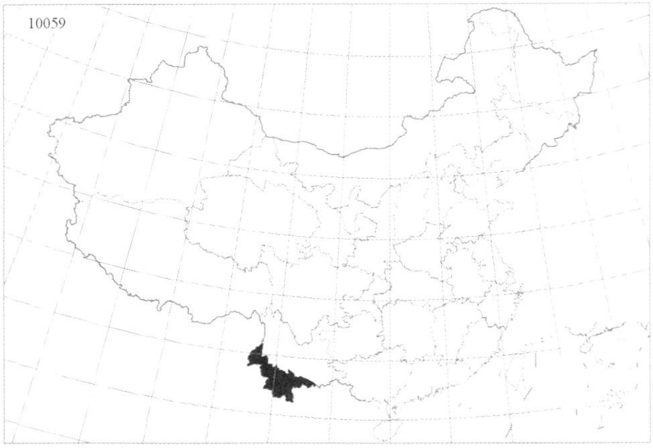

Acanthaceae 爵床科 *Semnostachya* 糯米香属

Semnostachya longespicata (Hayata) Hsieh et Huang 长穗糯米香

Shrubs

19.8 (16.3~24.7); 13.6 (10.2~19.8); 23.8 (20.3~28.3); 19.8 (16.4~24.7); 177.6 (136.6~236.5); 0.0 (0.0~0.0); 989 (789~1376) | 1527 (1188~2992); 797 (534~1847); 61 (41~107) | 57.0 (5.8~166.5); 960 (780~1299); 443.1 (281.7~620.0)

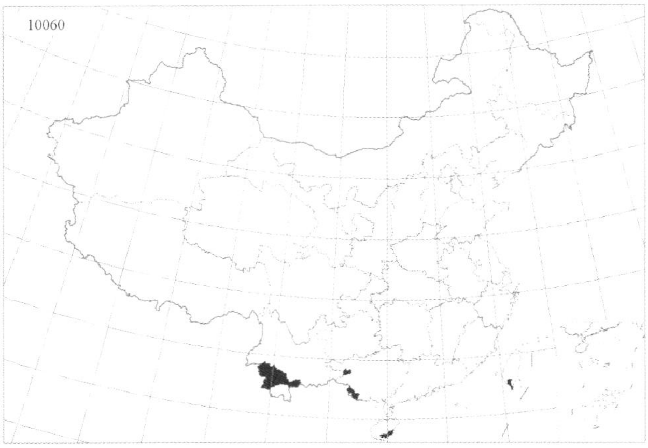

Acanthaceae 爵床科 *Thunbergia* 山牵牛属

Thunbergia coccinea Wall. 红花山牵牛

Scandent shrubs

16.9 (6.1~21.0); 10.4 (-1.5~14.2); 21.4 (12.7~25.6); 17.0 (6.3~21.0); 144.3 (35.2~191.8); 0.7 (0.0~21.8); 828 (496~1053) | 1261 (801~1736); 715 (496~1007); 51 (21~70) | 53.3 (19.0~145.8); 818 (496~1053); 405.8 (282.9~551.6)

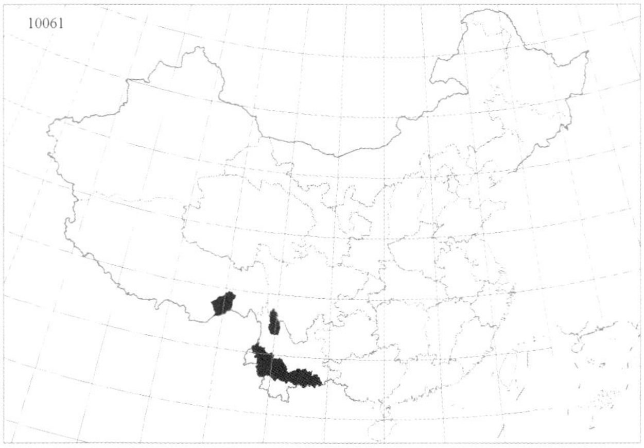

Acanthaceae 爵床科 *Thunbergia* 山牵牛属

Thunbergia grandiflora Roxb. 山牵牛

Scandent shrubs

20.3 (15.1~24.7); 12.8 (7.1~19.8); 26.1 (20.1~29.0); 20.4 (15.2~24.7); 184.9 (121.9~236.5); 0.0 (0.0~0.0); 1055 (753~1376) | 1534 (986~2037); 719 (513~1033); 103 (41~192) | 46.8 (-6.9~88.8); 1045 (739~1360); 316.7 (139.6~659.0)

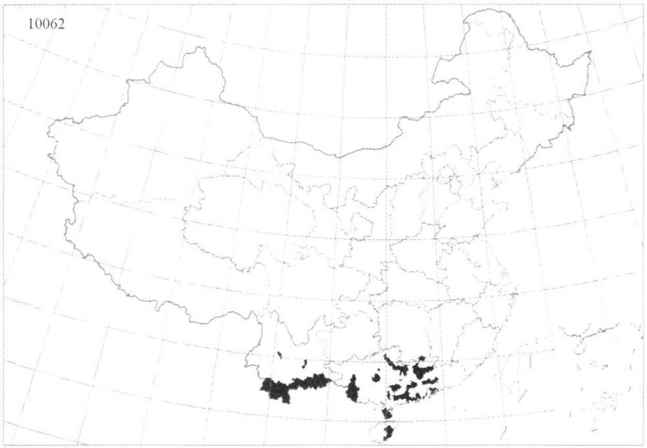

Myoporaceae 苦槛蓝科 *Myoporum* 苦槛蓝属
Myoporum bontioides (Siebold et Zucc.) A. Gray 苦槛蓝
Shrubs, evergreen

21.4 (17.1~24.7)；14.0 (6.9~19.8)；27.8 (21.8~29.2)；21.4 (17.1~24.7)；197.4 (145.0~236.5)；0.0 (0.0~0.0)；1132 (816~1376) | 1715 (1190~2992)；809 (411~1847)；131 (47~257) | 54.6 (7.3~202.5)；1105 (816~1360)；239.2 (119.4~548.0)

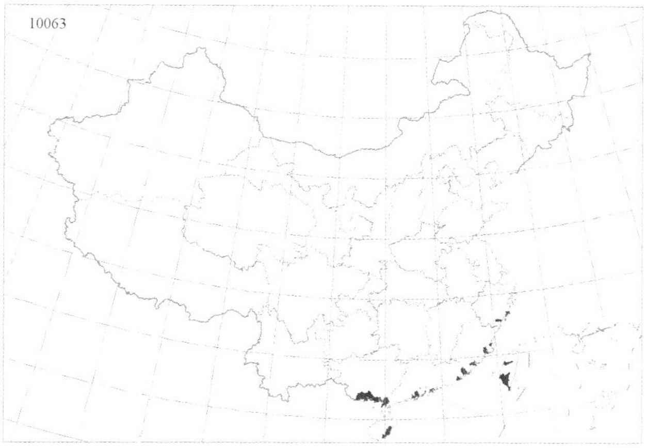

Rubiaceae 茜草科 *Adina* 水团花属
Adina pilulifera (Lam.) Franch. ex Drake 水团花
Shrubs or small trees, evergreen

18.6 (11.7~24.8)；8.9 (2.3~20.1)；27.2 (18.8~30.1)；18.6 (11.7~24.8)；163.9 (82.6~237.6)；0.2 (0.0~4.0)；983 (658~1385) | 1514 (868~2435)；603 (312~1428)；149 (31~259) | 56.1 (-34.3~147.0)；971 (658~1360)；233.6 (95.3~620.0)

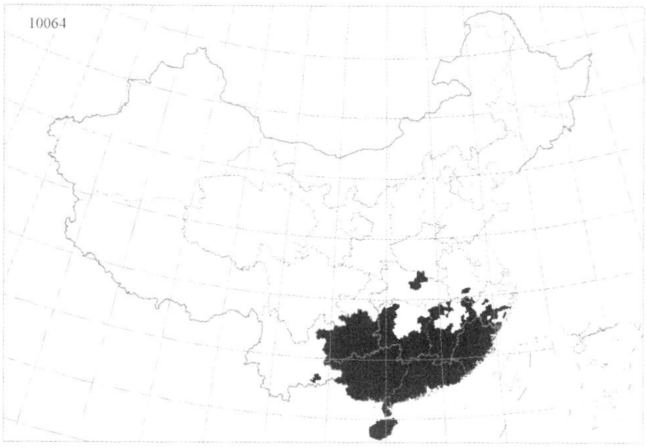

Rubiaceae 茜草科 *Adina* 水团花属
Adina rubella Hance 细叶水团花
Small shrubs, deciduous

16.4 (10.2~22.3)；4.9 (-2.9~13.9)；27.3 (16.6~29.8)；16.5 (10.6~22.3)；139.9 (71.8~208.1)；2.3 (0.0~20.5)；893 (633~1202) | 1338 (626~1997)；512 (316~770)；149 (24~259) | 49.7 (-22.2~147.0)；880 (626~1202)；207.9 (110.1~332.5)

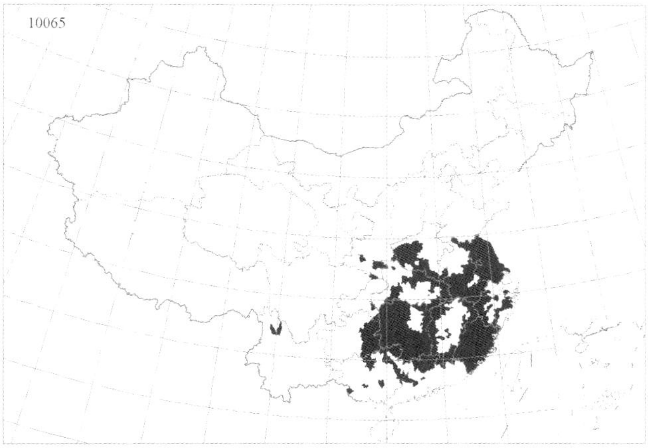

Rubiaceae 茜草科 *Aidia* 茜木属
Aidia canthioides (Champ. ex Benth.) Masam. 香楠
Shrubs or trees

19.7 (13.3~24.8)；10.5 (3.2~20.1)；27.7 (22.1~29.8)；19.7 (13.2~24.8)；176.8 (100.9~237.6)；0.1 (0.0~2.5)；1043 (723~1385) | 1608 (868~3495)；634 (312~1428)；159 (32~897) | 56.6 (-34.3~224.4)；1030 (723~1360)；241.0 (95.3~620.0)

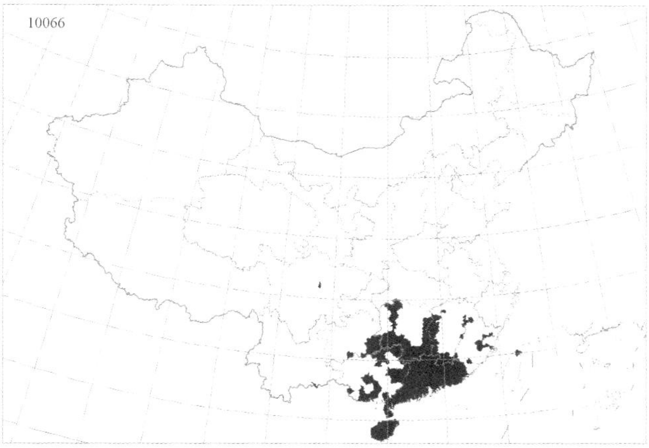

Rubiaceae 茜草科 *Aidia* 茜木属
Aidia cochinchinensis Lour. 茜树
Shrubs or trees

18.6 (8.6~24.8)；9.5 (0.6~20.1)；26.7 (15.0~29.9)；18.7 (8.7~24.8)；164.4 (52.6~237.6)；0.2 (0.0~9.7)；975 (574~1385) | 1530 (868~3495)；620 (312~1234)；145 (37~897) | 59.2 (-34.3~246.5)；963 (574~1360)；255.8 (95.3~659.0)

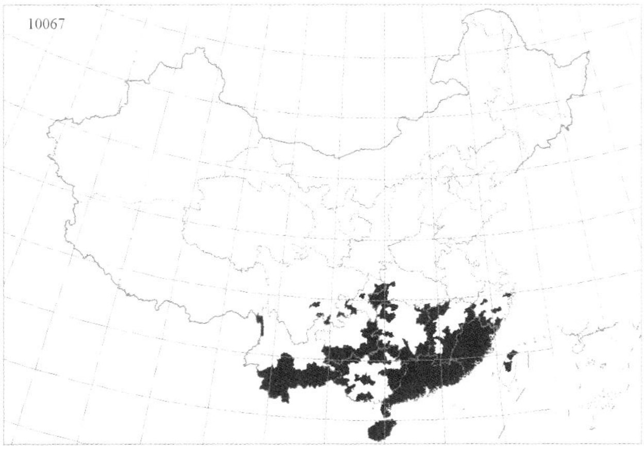

Rubiaceae 茜草科 *Aidia* 茜木属
Aidia oxyodonta (Drake) T. Yamaz. 尖萼茜树
Shrubs or trees

22.3 (19.7~24.8)；14.4 (10.1~20.1)；28.4 (27.7~29.2)；22.3 (19.8~24.8)；207.9 (177.4~237.6)；0.0 (0.0~0.0)；1198 (1031~1385) | 1582 (1377~2148)；709 (473~1234)；127 (41~178) | 33.2 (1.5~86.7)；1173 (1031~1294)；282.5 (199.4~527.5)

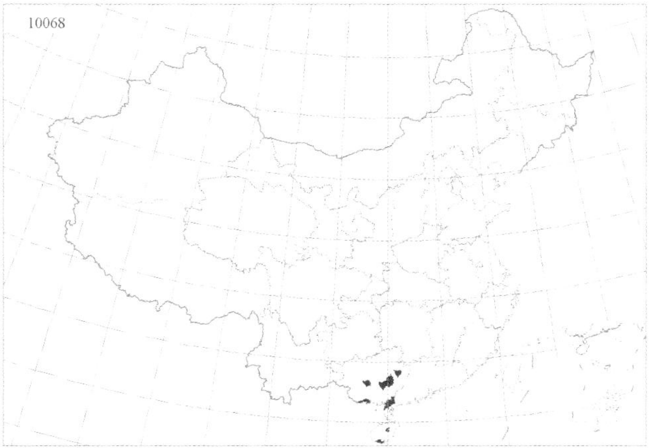

Rubiaceae 茜草科 *Aidia* 茜木属
Aidia pycnantha (Drake) Tirveng. 多毛茜草树
Shrubs or trees

20.8 (14.6~24.8); 12.5 (3.8~20.1); 27.6 (21.4~29.8); 20.8 (14.7~24.8); 190.1 (117.7~237.6); 0.0 (0.0~1.5); 1094 (779~1385) ❙ 1581 (868~2148); 666 (312~1234); 135 (41~197) ❙ 46.3 (-34.3~101.8); 1078 (779~1360); 257.2 (95.3~620.0)

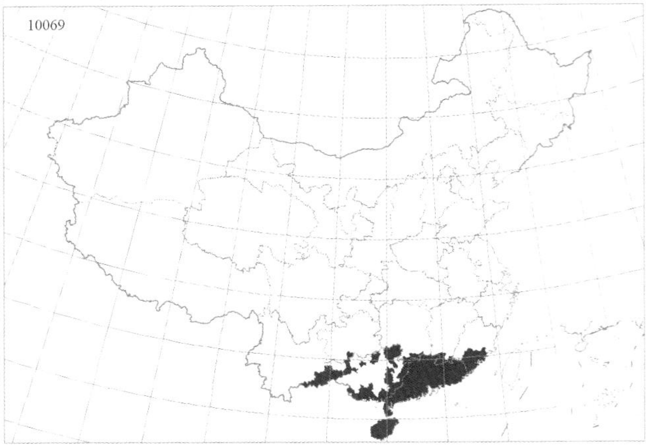

Rubiaceae 茜草科 *Aidia* 茜木属
Aidia salicifolia (H. L. Li) T. Yamaz. 柳叶香楠
Shrubs

20.4 (18.0~22.0); 11.3 (8.5~13.8); 27.7 (25.8~28.6); 20.5 (18.0~22.0); 185.5 (156.4~204.5); 0.0 (0.0~0.0); 1072 (913~1172) ❙ 1541 (1447~1621); 674 (613~720); 155 (115~169) ❙ 44.9 (24.9~69.8); 1072 (913~1172); 231.8 (185.5~264.4)

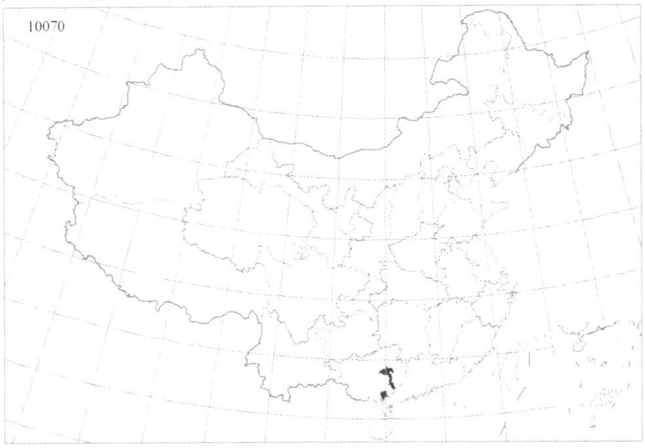

Rubiaceae 茜草科 *Aidia* 茜木属
Aidia yunnanensis (Hutch.) T. Yamaz. 滇茜树
Shrubs or trees

18.4 (16.3~21.0); 11.9 (9.0~15.1); 22.7 (21.4~25.6); 18.4 (16.4~21.0); 160.7 (136.4~191.8); 0.0 (0.0~0.0); 886 (797~1053) ❙ 1362 (1007~1751); 725 (539~996); 55 (41~77) ❙ 53.4 (19.0~88.8); 877 (797~1053); 442.2 (277.0~659.0)

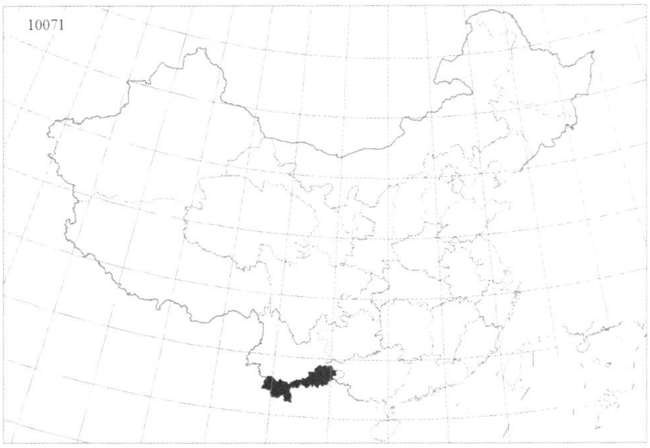

Rubiaceae 茜草科 *Alleizettella* 白香楠属
Alleizettella leucocarpa (Champ. ex Benth.) Tirveng. 白果香楠
Shrubs

21.9 (16.9~24.8); 14.8 (7.8~20.1); 27.6 (25.2~28.8); 21.9 (17.0~24.8); 203.4 (143.6~237.6); 0.0 (0.0~0.0); 1166 (862~1385) ❙ 1635 (868~2005); 636 (312~994); 120 (41~202) ❙ 44.0 (-34.3~100.8); 1113 (862~1360); 340.4 (164.9~620.0)

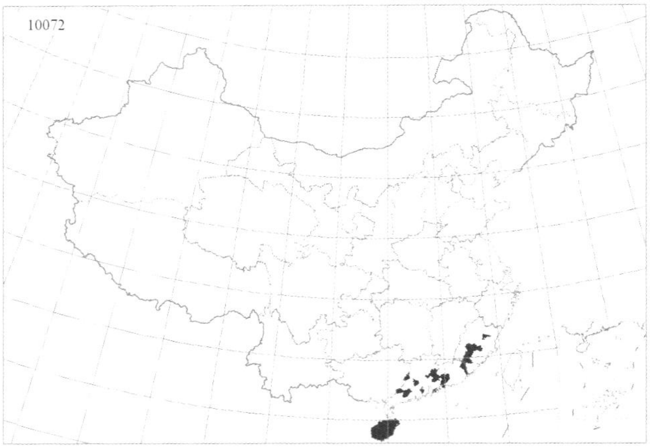

Rubiaceae 茜草科 *Antirhea* 毛茶属
Antirhea chinensis (Champ. ex Benth.) F. B. Forbes et Hemsl. 毛茶
Shrubs

22.5 (16.6~24.8); 15.4 (5.9~20.1); 28.2 (26.3~29.0); 22.6 (16.6~24.8); 210.7 (138.9~237.6); 0.0 (0.0~0.0); 1212 (866~1385) ❙ 1574 (868~2037); 639 (312~1033); 108 (41~254) ❙ 32.6 (-34.3~130.7); 1147 (866~1360); 301.2 (122.7~620.0)

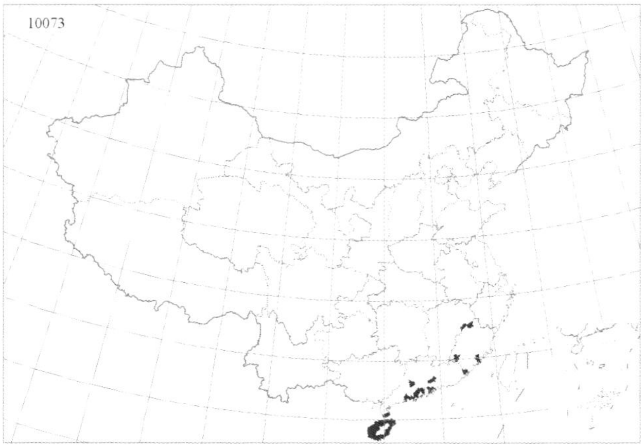

Rubiaceae 茜草科 *Brachytome* 短萼齿木属
Brachytome hainanensis C. Y. Wu ex W. C. Chen 海南短萼齿木
Shrubs

19.3 (8.6~23.1); 13.1 (1.3~17.7); 23.5 (15.0~27.9); 19.3 (8.7~23.2); 172.6 (52.6~218.4); 0.8 (0.0~8.8); 971 (574~1232) ❙ 1481 (1041~1725); 684 (544~848); 65 (41~137) ❙ 55.2 (20.7~81.5); 942 (574~1163); 471.5 (223.4~659.0)

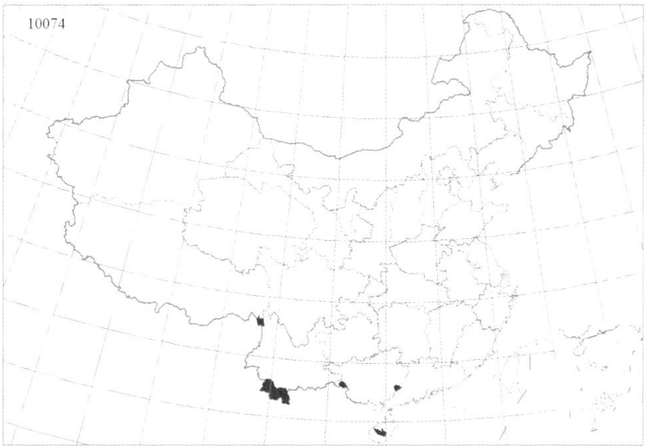

Rubiaceae 茜草科 *Brachytome* 短萼齿木属
Brachytome hirtellata Hu 滇短萼齿木
Shrubs

18.0 (8.6~21.0); 11.5 (1.3~15.1); 22.4 (15.0~25.6); 18.1 (8.7~21.0); 157.0 (52.6~191.8); 0.2 (0.0~8.8); 875 (574~1053) ❙ 1377 (1007~1751); 757 (539~996); 54 (41~77) ❙ 58.2 (19.0~112.2); 864 (574~1053); 440.5 (277.0~659.0)

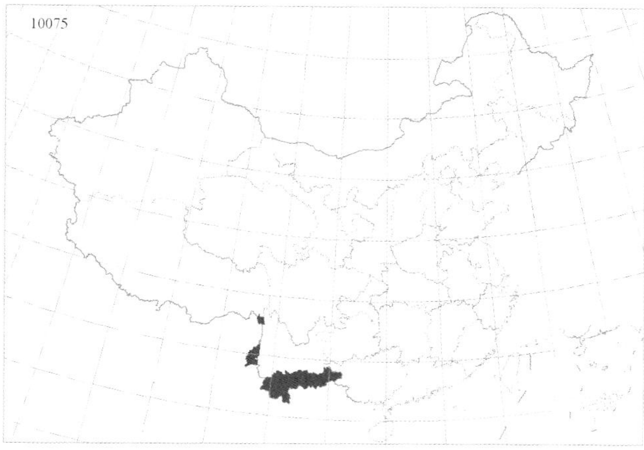

Rubiaceae 茜草科 *Brachytome* 短萼齿木属
Brachytome hirtellata var. glabrescens W. C. Chen 疏毛短萼齿木
Shrubs

18.0 (8.6~21.0); 11.4 (1.3~15.1); 22.5 (15.0~25.6); 18.1 (8.7~21.0); 157.0 (52.6~191.8); 0.3 (0.0~8.8); 876 (574~1053) ❙ 1308 (958~1751); 706 (531~996); 53 (41~77) ❙ 49.6 (13.3~88.8); 865 (574~1053); 418.3 (277.0~659.0)

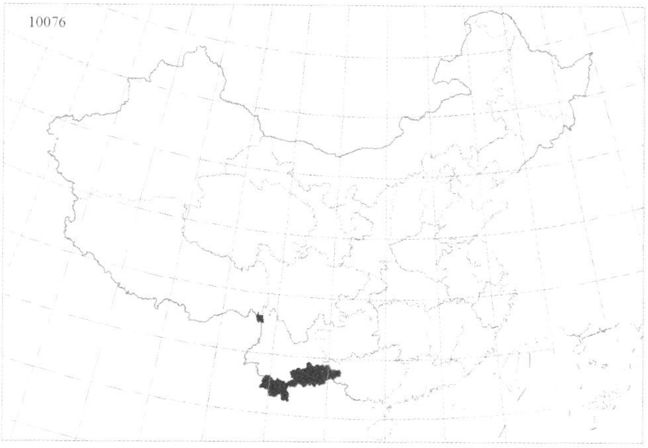

Rubiaceae 茜草科 *Canthium* 鱼骨木属
Canthium dicoccum (Gaertn.) Merr. 鱼骨木
Shrubs or trees

20.4 (13.3~24.8); 13.0 (3.1~20.1); 25.9 (19.1~28.8); 20.4 (13.4~24.8); 185.0 (100.4~237.6); 0.0 (0.0~2.2); 1055 (706~1385) ❙ 1553 (868~2435); 752 (312~1428); 90 (41~223) ❙ 50.4 (-34.3~145.8); 1028 (706~1360); 361.5 (157.7~659.0)

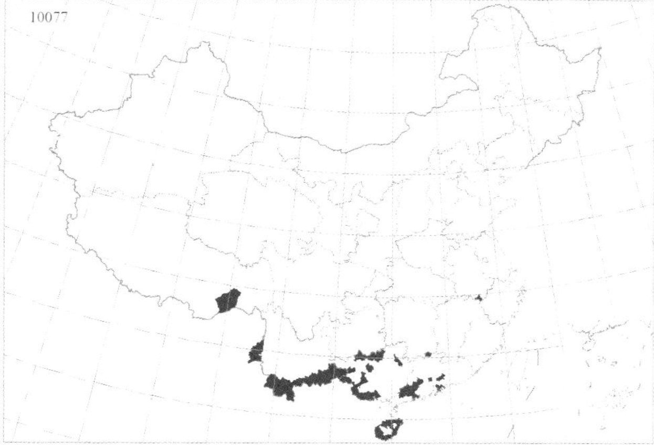

Rubiaceae 茜草科 *Canthium* 鱼骨木属
Canthium dicoccum var. obovatifolium G. A. Fu 倒卵叶鱼骨木
Shrubs or trees

24.1; 18.3; 28.2; 24.1; 229.4; 0.0; 1321 ❙ 1660; 572; 92 ❙ 25.6; 1294; 474.7

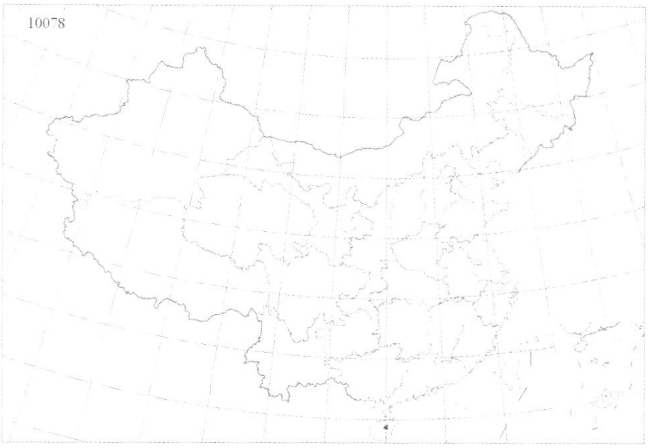

Rubiaceae 茜草科 *Canthium* 鱼骨木属
Canthium horridum Blume 猪肚木
Shrubs

20.3 (14.4~24.8); 12.7 (6.9~20.1); 26.1 (19.1~29.8); 20.4 (14.4~24.8); 184.8 (113.1~237.6); 0.0 (0.0~0.0); 1053 (727~1385) ❙ 1510 (1059~2148); 724 (392~1234); 96 (41~194) ❙ 45.7 (-12.0~112.2); 1034 (727~1360); 330.4 (139.6~659.0)

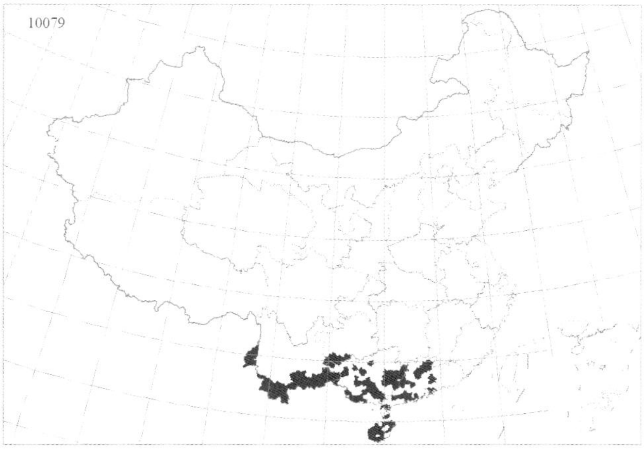

Rubiaceae 茜草科 *Canthium* 鱼骨木属
Canthium simile Merr. et Chun 大叶鱼骨木
Shrubs or small trees

20.0 (16.0~24.8); 13.1 (7.1~20.1); 24.9 (20.4~29.2); 20.1 (16.0~24.8); 180.8 (132.4~237.6); 0.0 (0.0~0.0); 1018 (780~1385) ❙ 1523 (1182~1854); 752 (473~996); 78 (41~189) ❙ 52.9 (1.5~112.2); 997 (780~1360); 419.1 (178.7~659.0)

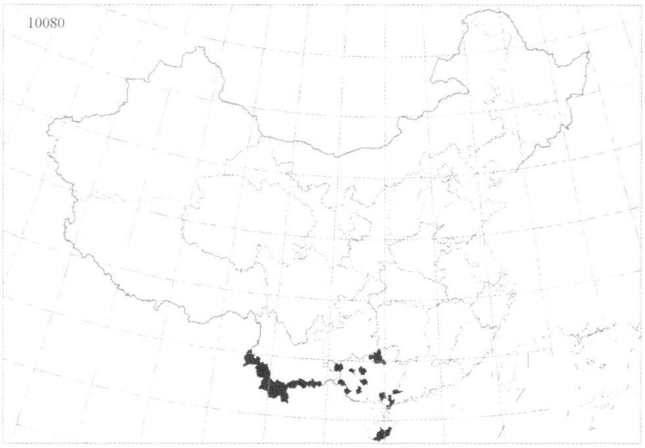

Rubiaceae 茜草科 *Catunaregam* 山石榴属

Catunaregam spinosa (Thunb.) Tirveng. 山石榴

Shrubs or small trees

20.6 (14.4~24.8); 12.8 (5.8~20.1); 26.7 (19.1~29.8); 20.6 (14.4~24.8); 187.5 (113.1~237.6); 0.0 (0.0~0.0); 1070 (727~1385) ❙ 1581 (868~2993); 716 (312~1428); 118 (41~547) ❙ 49.7 (-34.3~203.4); 1054 (727~1360); 293.7 (95.3~659.0)

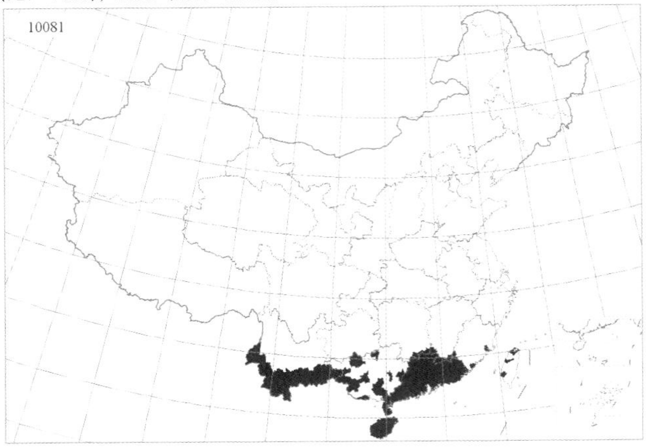

Rubiaceae 茜草科 *Cephaelis* 头九节属

Cephaelis laui (Merr. et F. P. Metcalf) F. C. How et W. C. Ko 头九节

Shrubs

24.0 (23.6~24.3); 18.1 (17.7~18.7); 28.3 (27.5~28.7); 24.1 (23.6~24.3); 229.0 (223.3~231.9); 0.0 (0.0~0.0); 1317 (1271~1338) ❙ 1328 (868~1813); 485 (312~712); 72 (42~113) ❙ 0.8 (-34.3~36.4); 1146 (868~1329); 366.7 (248.2~455.2)

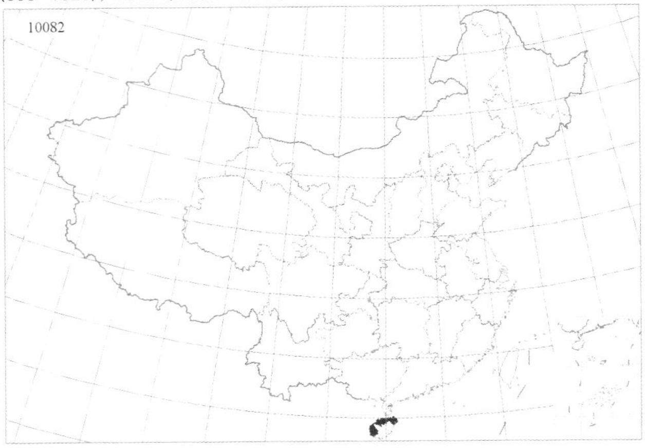

Rubiaceae 茜草科 *Cephalanthus* 风箱树属

Cephalanthus tetrandrus (Roxb.) Ridsdale et Bakh. f. 风箱树

Shrubs or small trees, deciduous

18.7 (14.3~24.3); 9.2 (3.3~18.7); 27.2 (21.6~29.8); 18.8 (14.3~24.3); 165.1 (112.6~231.9); 0.1 (0.0~2.3); 980 (756~1338) ❙ 1603 (868~3495); 591 (312~994); 181 (41~897) ❙ 66.0 (-34.3~224.4); 965 (756~1321); 249.8 (123.9~659.0)

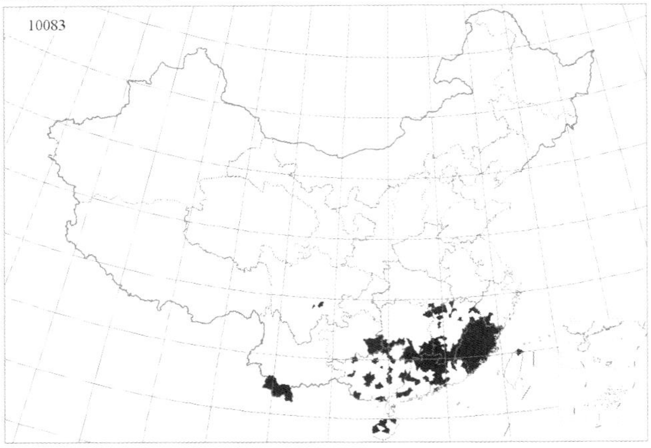

Rubiaceae 茜草科 *Ceriscoides* 木瓜榄属

Ceriscoides howii H. S. Lo 木瓜榄

Shrubs

24.5 (24.0~24.8); 19.5 (18.9~20.1); 28.0 (27.2~28.5); 24.5 (24.0~24.8); 234.5 (228.1~237.6); 0.0 (0.0~0.0); 1359 (1302~1385) ❙ 1633 (1405~1854); 547 (473~622); 70 (41~115) ❙ 20.2 (1.5~36.4); 1231 (1137~1360); 523.8 (428.6~620.0)

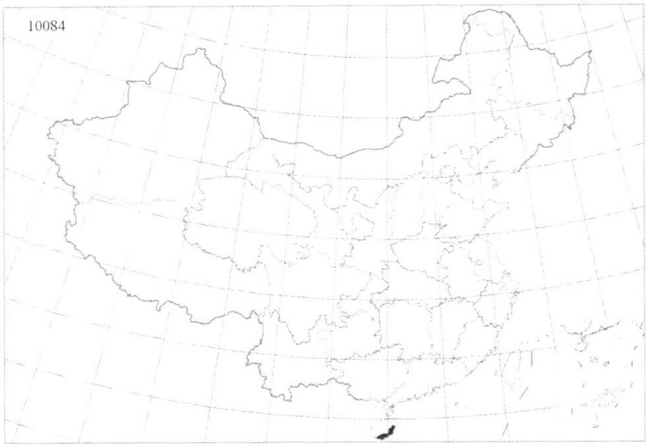

Rubiaceae 茜草科 *Chassalia* 弯管花属

Chassalia curviflora Thwaites 弯管花

Shrubs

19.4 (2.9~24.8); 12.1 (-5.7~20.1); 24.7 (10.6~29.2); 19.4 (4.3~24.8); 173.8 (19.9~237.6); 0.6 (0.0~45.2); 992 (421~1385) ❙ 1415 (535~2435); 714 (312~1428); 77 (10~194) ❙ 45.1 (-34.3~145.8); 967 (421~1331); 359.3 (178.7~659.0)

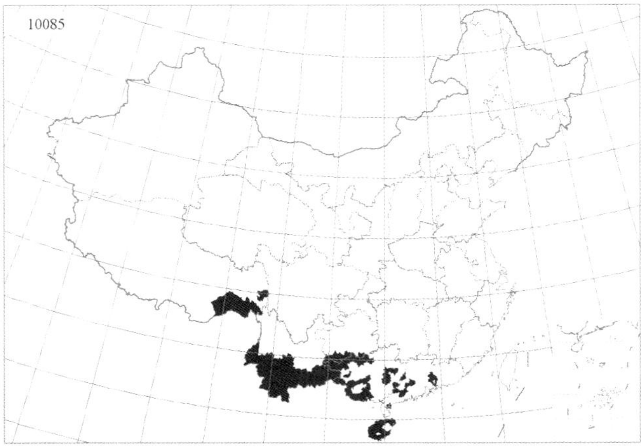

Rubiaceae 茜草科 *Chassalia* 弯管花属

Chassalia curviflora var. longifolia Hook. f. 尖叶弯管花

Shrubs

24.1 (23.1~24.8); 18.9 (17.5~20.1); 27.7 (27.2~28.6); 24.1 (23.2~24.8); 229.6 (218.4~237.6); 0.0 (0.0~0.0); 1319 (1232~1385) ❙ 1453 (1309~1588); 504 (455~544); 56 (41~88) ❙ 10.3 (-1.4~20.7); 1168 (1112~1254); 497.9 (351.4~620.0)

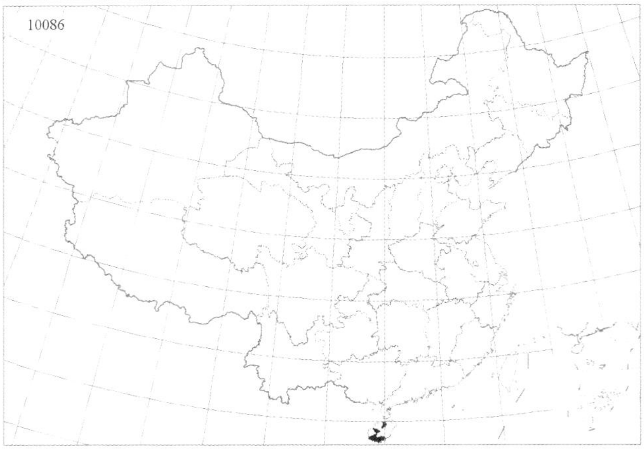

Rubiaceae 茜草科 *Coelospermum* 穴果木属

Coelospermum kanehirae Merr. 穴果木

Vines

23.3 (19.6~24.8); 17.1 (11.2~20.1); 27.8 (23.6~29.0); 23.4 (19.6~24.8); 220.2 (175.5~237.6); 0.0 (0.0~0.0); 1259 (986~1385) | 1581 (868~2035); 634 (312~969); 85 (41~194) | 27.0 (-34.3~71.8); 1179 (868~1360); 390.4 (194.5~659.0)

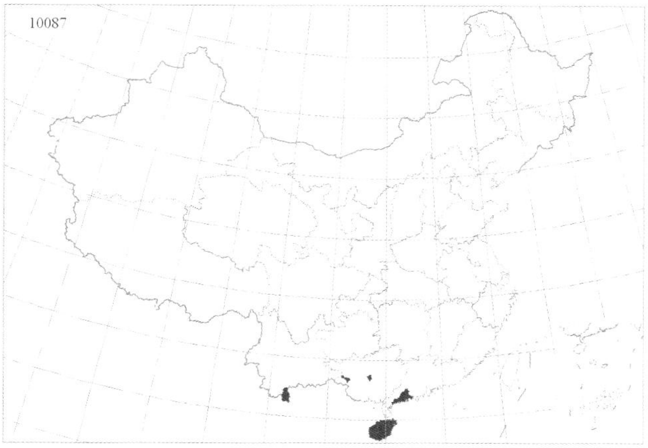

Rubiaceae 茜草科 *Coptosapelta* 流苏子属

Coptosapelta diffusa (Champ. ex Benth.) Steenis 流苏子

Vines or shrubs

18.4 (14.3~23.6); 8.5 (0.5~17.7); 27.4 (20.1~29.8); 18.5 (14.3~23.6); 161.9 (112.6~223.5); 0.2 (0.0~8.9); 973 (756~1281) | 1591 (706~3495); 606 (356~1847); 172 (42~897) | 65.5 (-17.2~246.5); 964 (706~1273); 228.5 (95.3~551.6)

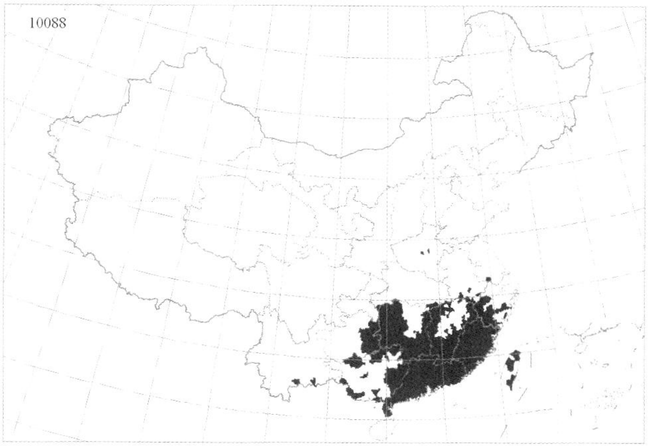

Rubiaceae 茜草科 *Damnacanthus* 虎刺属

Damnacanthus giganteus (Mak.) Nakai 短刺虎刺

Shrubs

17.2 (12.7~21.7); 6.6 (2.1~15.3); 26.8 (20.1~29.9); 17.2 (12.8~21.7); 146.7 (99.0~200.2); 0.3 (0.0~5.6); 904 (713~1118) | 1631 (1171~2993); 599 (403~1412); 186 (49~547) | 81.3 (34.4~246.5); 898 (713~1114); 231.6 (134.6~371.7)

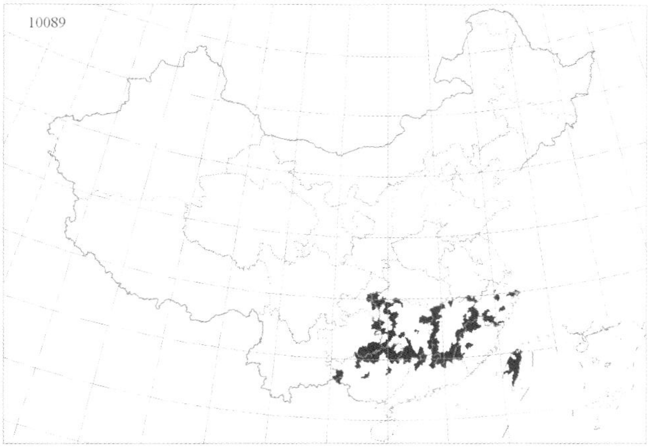

Rubiaceae 茜草科 *Damnacanthus* 虎刺属

Damnacanthus hainanensis (Lo) Lo ex Y. Z. Ruan 海南虎刺

Shrubs

23.6; 17.8; 27.8; 23.7; 224.2; 0.0; 1279 | 1605; 582; 82 | 25.3; 1234; 445.6

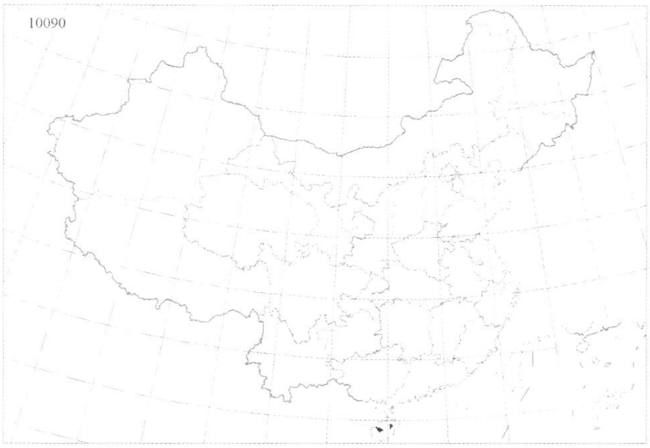

Rubiaceae 茜草科 *Damnacanthus* 虎刺属

Damnacanthus henryi (H. Lév.) H. S. Lo 云桂虎刺

Shrubs

19.8 (16.9~24.1); 13.6 (10.3~18.2); 24.1 (21.4~28.4); 19.9 (16.9~24.2); 178.4 (142.7~230.1); 0.0 (0.0~0.0); 993 (813~1326) | 1449 (1178~1723); 738 (544~911); 64 (51~110) | 48.1 (20.7~74.2); 982 (813~1326); 476.0 (352.9~551.6)

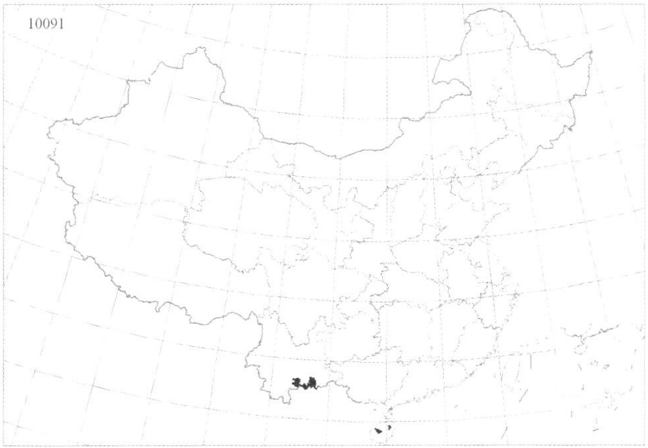

Rubiaceae 茜草科 *Damnacanthus* 虎刺属

Damnacanthus indicus Gaertn. 虎刺

Shrubs

17.1 (1.8~22.4); 6.8 (-7.1~17.7); 26.5 (9.9~29.8); 17.1 (3.8~22.4); 146.7 (15.7~209.1); 1.1 (0.0~53.3); 904 (404~1204) | 1502 (550~3495); 586 (313~1847); 160 (13~897) | 67.0 (-36.1~246.5); 888 (404~1204); 247.3 (123.9~659.0)

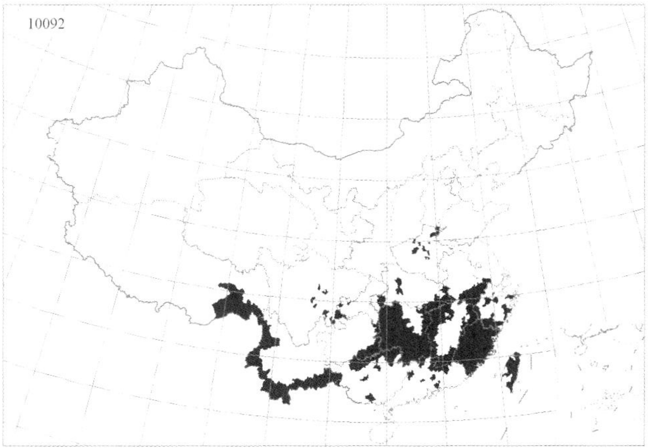

Rubiaceae 茜草科 *Damnacanthus* 虎刺属
Damnacanthus labordei (H. Lév.) H. S. Lo 柳叶虎刺
Shrubs

17.4 (14.3~21.3); 7.5 (3.3~14.2); 26.0 (21.4~29.0); 17.5 (14.3~21.3); 149.8 (112.6~195.9); 0.2 (0.0~2.3); 902 (756~1115) ❙ 1426 (1035~1900); 597 (403~1056); 129 (48~230) ❙ 58.4 (20.8~106.4); 900 (756~1115); 251.7 (142.9~551.6)

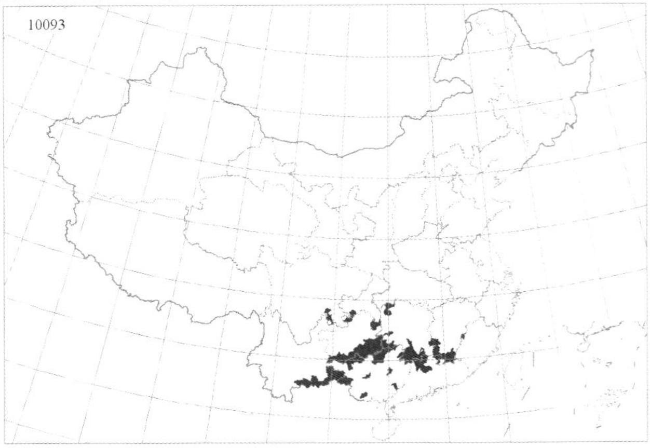

Rubiaceae 茜草科 *Damnacanthus* 虎刺属
Damnacanthus macrophyllus Sieb. ex Miq. 浙皖虎刺
Shrubs

17.2 (14.7~22.4); 6.1 (2.5~14.0); 27.5 (25.0~29.3); 17.2 (14.8~22.4); 147.2 (121.1~209.1); 0.4 (0.0~3.9); 917 (810~1204) ❙ 1637 (1211~1997); 593 (427~1044); 195 (57~259) ❙ 79.5 (28.8~130.7); 914 (810~1204); 244.4 (189.1~371.7)

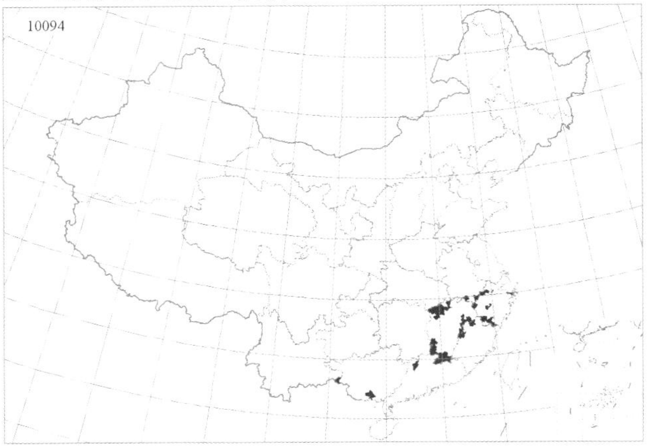

Rubiaceae 茜草科 *Damnacanthus* 虎刺属
Damnacanthus major Sieb. et Zucc. 大卵叶虎刺
Shrubs

20.0; 10.3; 28.2; 20.1; 181.0; 0.0; 1067 ❙ 1474; 612; 159 ❙ 40.7; 1067; 199.5

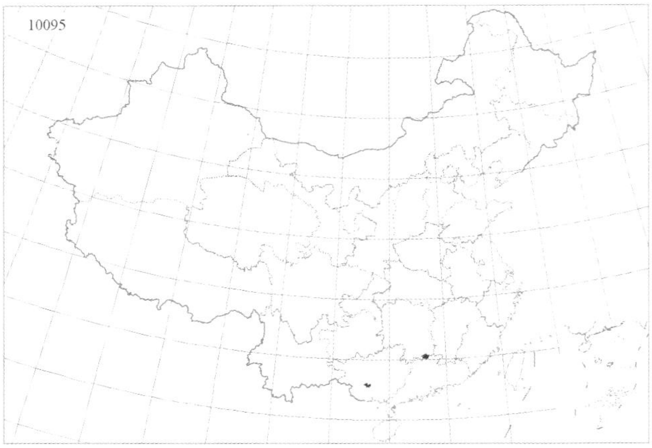

Rubiaceae 茜草科 *Damnacanthus* 虎刺属
Damnacanthus officinarum C. C. Huang 四川虎刺
Shrubs

15.0 (4.0~17.1); 3.7 (-8.4~5.3); 25.8 (14.7~29.2); 15.1 (6.0~17.1); 123.7 (36.4~145.8); 2.7 (0.0~46.8); 824 (480~937) ❙ 1330 (541~1492); 540 (282~816); 104 (11~154) ❙ 62.0 (12.7~107.2); 823 (480~927); 199.9 (152.4~292.6)

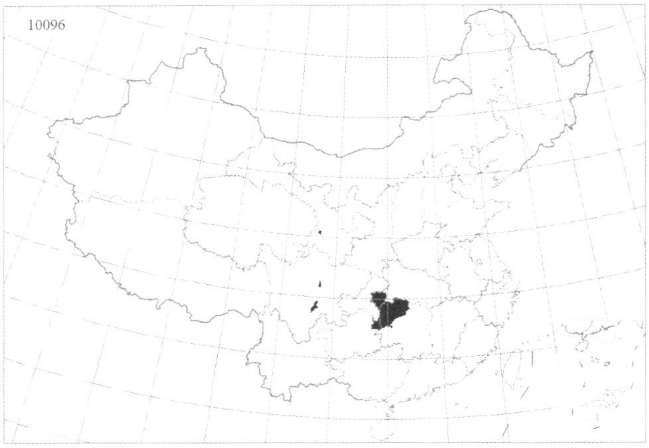

Rubiaceae 茜草科 *Damnacanthus* 虎刺属
Damnacanthus tsaii Hu 西南虎刺
Shrubs

16.7 (11.1~18.2); 9.4 (1.8~11.9); 22.0 (19.4~23.2); 16.8 (11.1~18.2); 142.0 (79.8~158.7); 0.6 (0.0~6.9); 821 (648~868) ❙ 1251 (1074~1504); 696 (582~860); 49 (30~59) ❙ 52.5 (32.4~80.6); 819 (648~863); 372.1 (246.3~516.7)

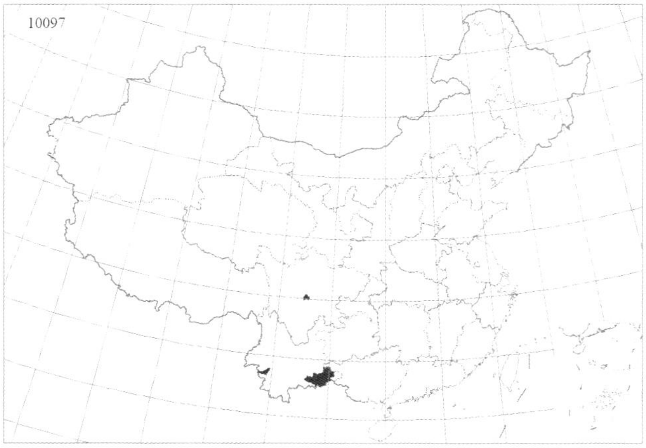

Rubiaceae 茜草科 *Diplospora* 狗骨柴属
Diplospora dubia (Lindl.) Masam. 狗骨柴
Shrubs or trees

18.6 (13.3~24.8); 9.1 (2.6~20.1); 27.3 (20.1~29.9); 18.7 (13.2~24.8); 164.6 (100.9~237.6); 0.2 (0.0~3.5); 985 (723~1385) ❙ 1605 (868~3495); 618 (312~1847); 168 (32~897) ❙ 65.9 (-34.3~246.5); 972 (723~1331); 238.6 (95.3~620.0)

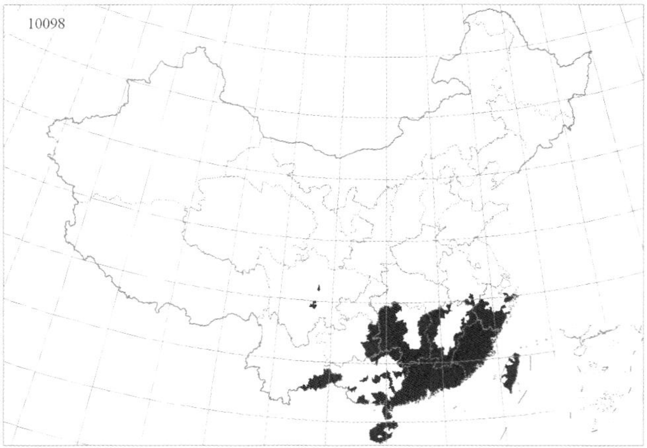

Rubiaceae 茜草科 *Diplospora* 狗骨柴属

Diplospora fruticosa Hemsl. 毛狗骨柴

Shrubs or trees

17.7 (10.2~22.2); 8.7 (-2.9~15.1); 25.1 (19.1~29.8); 17.7 (10.6~22.3); 153.1 (83.1~207.5); 0.5 (0.0~20.5); 904 (656~1190) ❙ 1392 (596~1876); 630 (305~996); 109 (27~230) ❙ 54.2 (-24.7~128.9); 894 (596~1190); 293.6 (126.1~659.0)

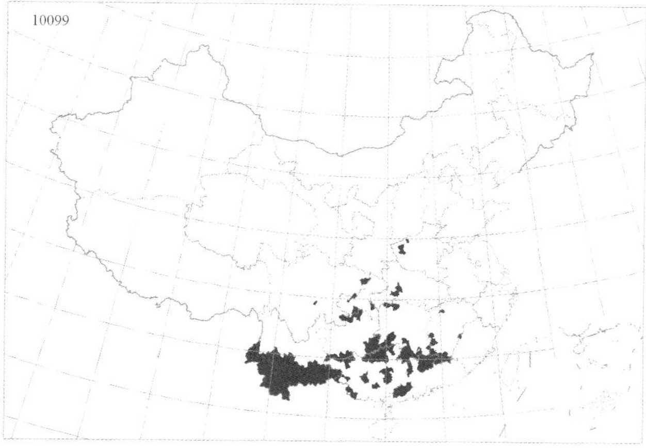

Rubiaceae 茜草科 *Diplospora* 狗骨柴属

Diplospora mollissima Hutch. 云南狗骨柴

Shrubs or trees

18.8 (17.1~20.5); 12.8 (10.8~15.1); 22.7 (21.4~23.7); 18.9 (17.2~20.5); 166.5 (145.8~185.9); 0.0 (0.0~0.0); 904 (823~992) ❙ 1352 (1028~1751); 736 (566~996); 57 (44~77) ❙ 49.4 (12.5~88.8); 883 (783~986); 446.5 (339.5~659.0)

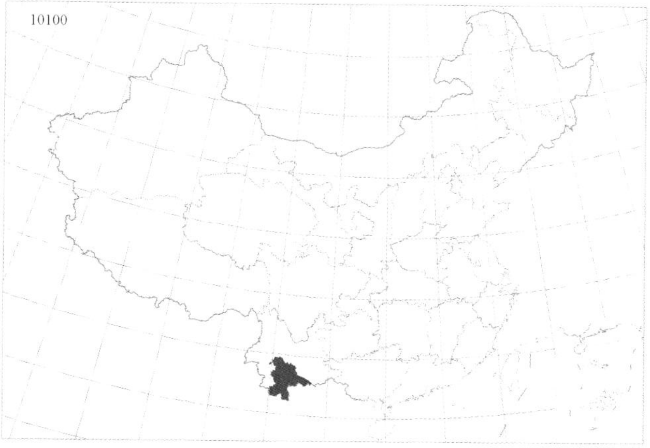

Rubiaceae 茜草科 *Dunnia* 绣球茜属

Dunnia sinensis Tutcher 绣球茜草

Shrubs

22.2 (20.8~22.8); 14.4 (12.2~15.5); 28.3 (27.7~28.8); 22.2 (20.8~22.8); 206.7 (189.7~214.1); 0.0 (0.0~0.0); 1181 (1085~1225) ❙ 1814 (1447~2035); 866 (777~969); 115 (75~181) ❙ 54.0 (20.8~83.4); 1181 (1085~1225); 215.9 (139.6~273.3)

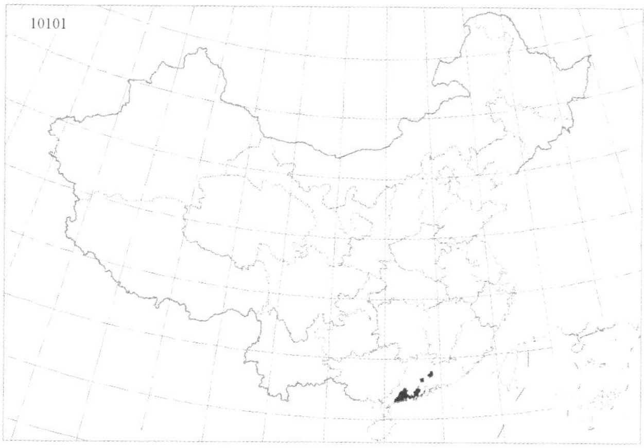

Rubiaceae 茜草科 *Duperrea* 长柱山丹属

Duperrea pavettaefolia (Kurz) Pit. 长柱山丹

Shrubs or small trees

21.0 (16.3~24.8); 14.8 (9.1~20.1); 25.3 (21.4~28.8); 21.1 (16.4~24.8); 192.9 (136.4~237.6); 0.0 (0.0~0.0); 1086 (797~1385) ❙ 1450 (868~1854); 659 (312~996); 65 (41~115) ❙ 37.4 (-34.3~88.8); 1030 (797~1360); 438.8 (248.2~659.0)

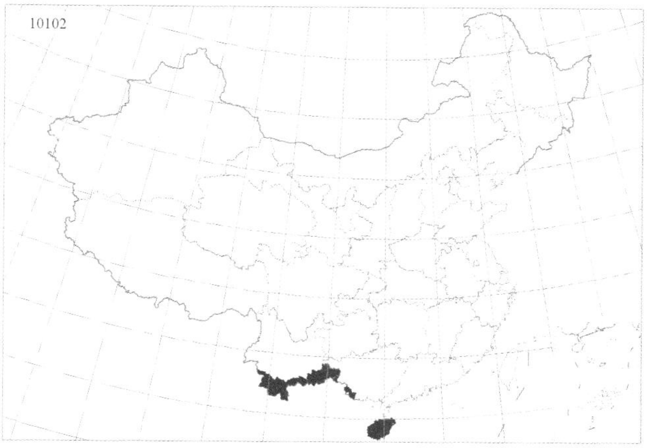

Rubiaceae 茜草科 *Emmenopterys* 香果树属

Emmenopterys henryi Oliv. 香果树

Trees, deciduous

15.6 (5.4~21.6); 5.1 (-2.9~14.2); 25.3 (12.3~29.4); 15.7 (5.8~21.6); 130.3 (31.5~199.6); 2.1 (0.0~25.6); 837 (481~1143) ❙ 1276 (626~1997); 539 (316~860); 111 (15~259) ❙ 52.3 (-22.2~147.0); 826 (481~1143); 230.6 (126.1~551.6)

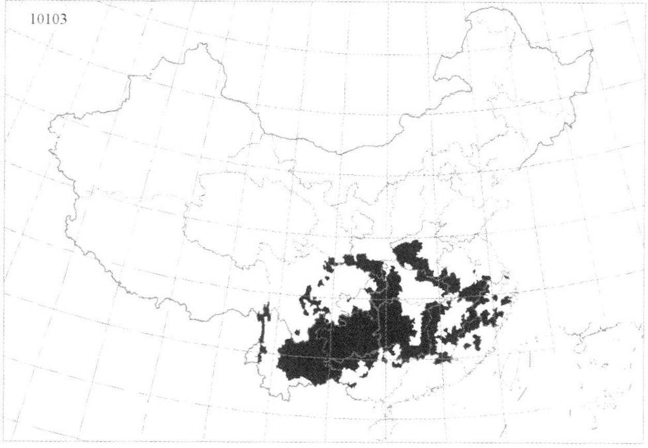

Rubiaceae 茜草科 *Fagerlindia* 浓子茉莉属

Fagerlindia depauperata (Drake) Tirveng. 多刺山黄皮

Shrubs

23.3 (20.4~24.8); 17.1 (12.7~20.1); 27.9 (26.7~29.0); 23.4 (20.5~24.8); 220.3 (185.7~237.6); 0.0 (0.0~0.0); 1265 (1050~1385) ❙ 1627 (1405~1888); 649 (473~1044); 86 (41~130) ❙ 29.2 (1.5~56.8); 1194 (1050~1360); 424.0 (199.6~620.0)

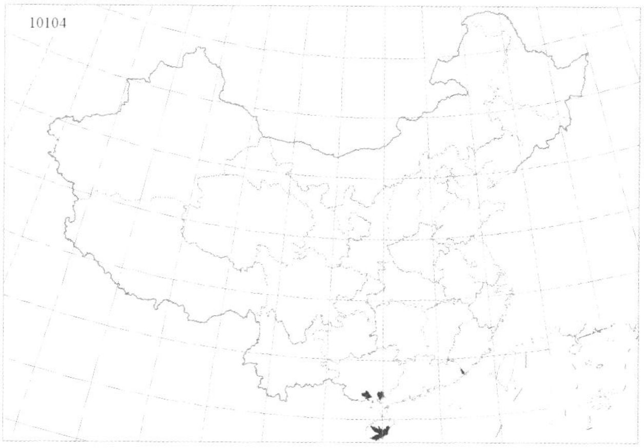

Rubiaceae 茜草科　*Fagerlindia* 浓子茉莉属

Fagerlindia scandens (Thunb.) Tirveng. 浓子茉莉

Shrubs

21.5 (16.3~24.6); 13.9 (9.1~19.5); 27.4 (21.4~29.0); 21.6 (16.4~24.7); 199.1 (136.4~235.9); 0.0 (0.0~0.0); 1141 (797~1372) ❙ 1602 (1074~2435); 797 (544~1428); 96 (47~173) ❙ 41.2 (-6.9~99.5); 1134 (797~1360); 258.3 (116.6~548.0)

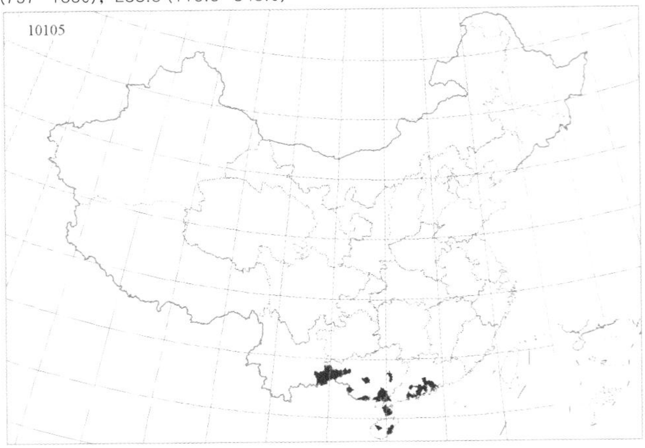

Rubiaceae 茜草科　*Fosbergia* 大果茜属

Fosbergia shweliensis (Anth.) Tirveng. 瑞丽茜

Shrubs

19.5; 12.7; 23.3; 19.5; 173.8; 0.0; 942 ❙ 1559; 907; 42 ❙ 65.4; 921; 482.8

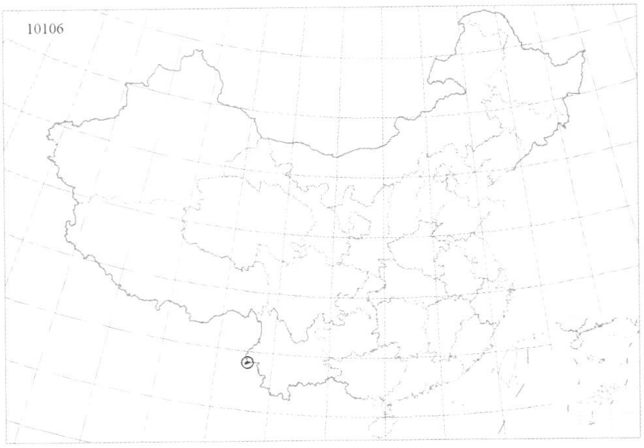

Rubiaceae 茜草科　*Gardenia* 栀子属

Gardenia angkorensis Pit. 匙叶栀子

Shrubs, evergreen

24.2; 18.9; 28.0; 24.2; 230.9; 0.0; 1331 ❙ 1275; 436; 42 ❙ -4.2; 1095; 474.2

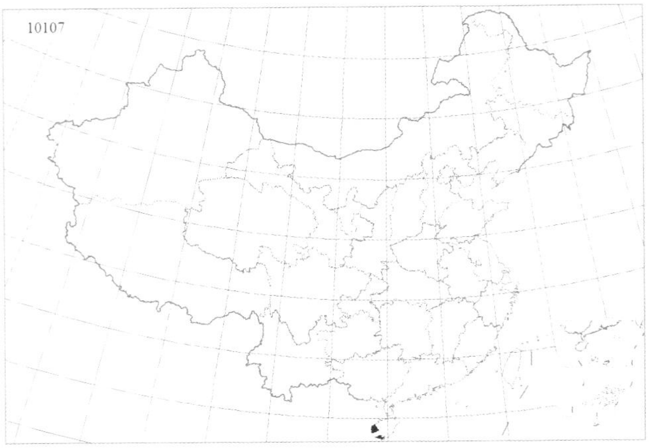

Rubiaceae 茜草科　*Gardenia* 栀子属

Gardenia augusta (L.) Merr. 栀子

Shrubs, evergreen

17.6 (5.4~24.6); 7.8 (-2.1~19.5); 26.3 (12.3~30.1); 17.6 (5.8~24.7); 152.1 (31.5~235.9); 0.5 (0.0~25.6); 927 (481~1372) ❙ 1440 (741~3495); 599 (356~1847); 137 (15~897) ❙ 56.2 (-14.2~246.5); 917 (481~1329); 238.4 (95.3~659.0)

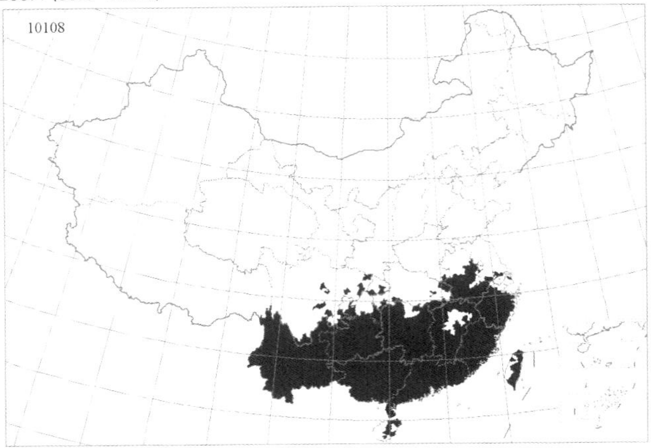

Rubiaceae 茜草科　*Gardenia* 栀子属

Gardenia hainanensis Merr. 海南栀子

Trees, evergreen

23.1 (17.6~24.8); 16.2 (7.3~20.1); 28.1 (26.3~28.9); 23.1 (17.6~24.8); 217.3 (151.6~237.6); 0.0 (0.0~0.0); 1252 (909~1385) ❙ 1574 (868~2435); 704 (312~1428); 94 (41~142) ❙ 27.3 (-34.3~99.5); 1176 (868~1360); 351.3 (225.9~620.0)

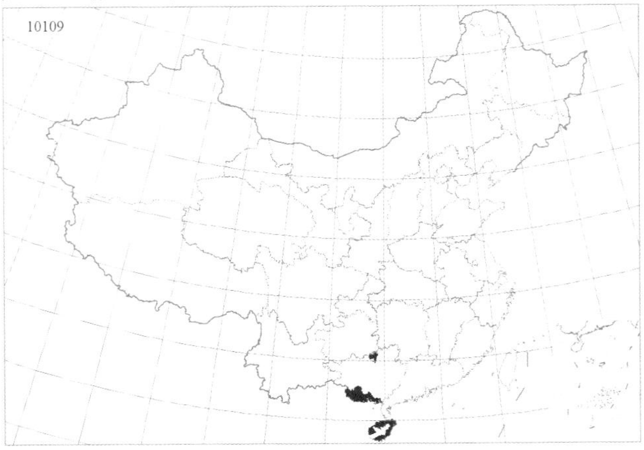

Rubiaceae 茜草科　*Gardenia* 栀子属

Gardenia sootepensis Hutch. 大黄栀子

Trees

19.4 (18.2~20.5); 13.8 (12.4~15.1); 22.7 (21.6~23.7); 19.4 (18.2~20.5); 173.2 (158.6~185.9); 0.0 (0.0~0.0); 928 (861~992) ❙ 1494 (1434~1607); 715 (655~785); 57 (41~77) ❙ 61.4 (47.6~74.2); 900 (839~986); 516.3 (427.9~659.0)

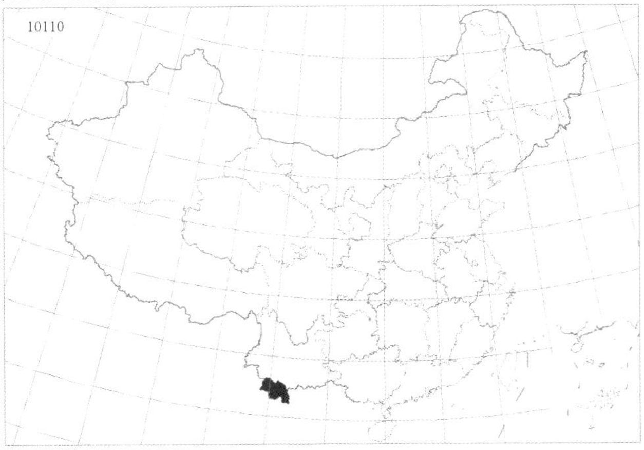

Rubiaceae 茜草科 *Gardenia* 栀子属

Gardenia stenophylla Merr. 狭叶栀子

Shrubs, evergreen

20.3 (15.0~24.8); 11.8 (2.7~20.1); 27.5 (25.8~28.8); 20.4 (15.0~24.8); 184.5 (123.7~237.6); 0.1 (0.0~3.2); 1081 (818~1385) ❙ 1628 (868~2148); 647 (312~1234); 138 (41~227) ❙ 55.2 (-34.3~129.9); 1037 (818~1360); 309.7 (157.7~620.0)

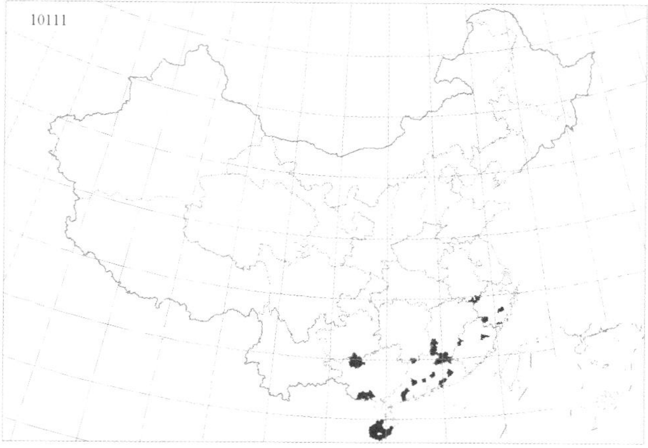

Rubiaceae 茜草科 *Guettarda* 海岸桐属

Guettarda speciosa L. 海岸桐

Small trees, evergreen

21.9 (16.4~24.8); 17.1 (11.5~20.1); 25.3 (20.3~28.2); 21.9 (16.5~24.8); 202.8 (137.8~237.6); 0.0 (0.0~0.0); 1150 (783~1385) ❙ 2161 (1405~2992); 959 (473~1847); 131 (41~303) ❙ 107.6 (1.5~219.3); 1069 (783~1299); 422.4 (296.9~548.0)

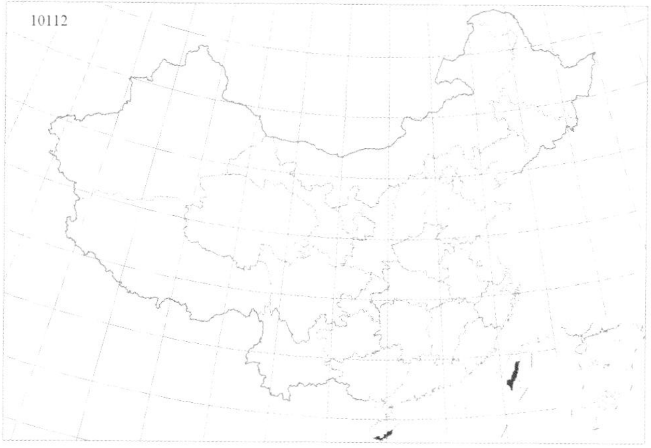

Rubiaceae 茜草科 *Guihaiothamnus* 桂海木属

Guihaiothamnus acaulis H. S. Lo 桂海木

Dwarf shrubs

17.6; 7.3; 26.3; 17.6; 151.6; 0.0; 909 ❙ 1496; 639; 142 ❙ 64.7; 909; 225.9

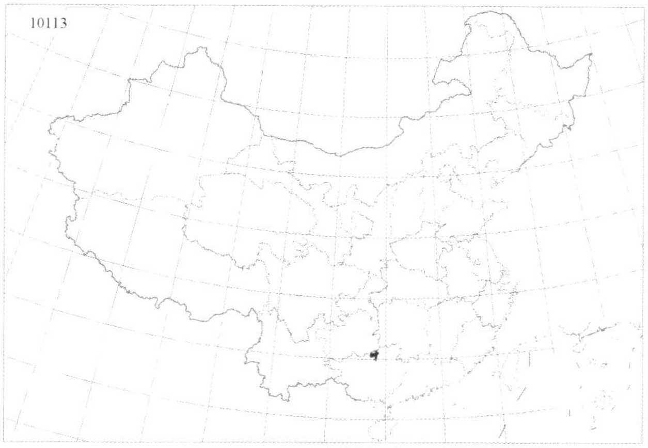

Rubiaceae 茜草科 *Haldina* 心叶树属

Haldina cordifolia (Roxb.) Ridsdale 心叶树

Trees, deciduous

18.2 (12.7~21.0); 11.8 (4.7~14.2); 22.5 (19.2~25.6); 18.2 (12.7~21.0); 158.3 (93.0~191.8); 0.0 (0.0~0.3); 878 (687~1053) ❙ 1227 (922~1751); 697 (524~996); 51 (26~77) ❙ 39.5 (6.5~88.8); 855 (687~1053); 390.5 (232.7~551.6)

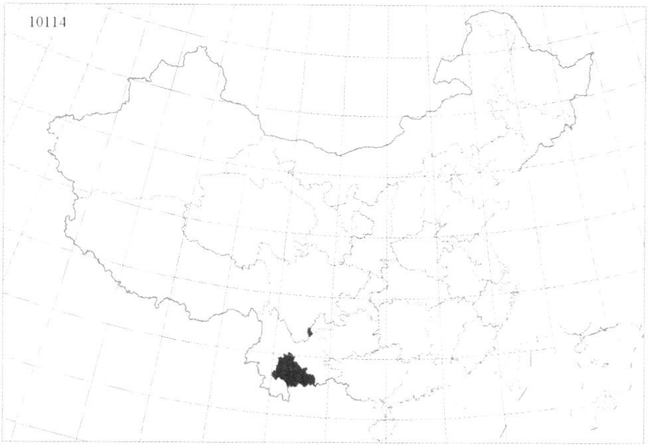

Rubiaceae 茜草科 *Hedyotis* 耳草属

Hedyotis ampliflora Hance 广花耳草

Scandent shrubs

21.4 (16.8~24.7); 13.8 (7.8~19.8); 27.2 (23.6~29.0); 21.4 (16.8~24.7); 196.8 (141.5~236.5); 0.0 (0.0~0.0); 1134 (836~1376) ❙ 1436 (868~1854); 631 (312~997); 82 (48~162) ❙ 28.5 (-34.3~59.6); 1079 (836~1360); 336.9 (189.5~620.0)

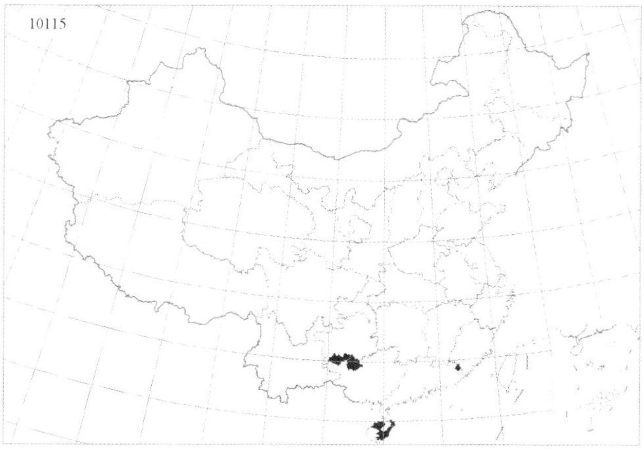

Rubiaceae 茜草科 *Hedyotis* 耳草属

Hedyotis dianxiensis W. C. Ko 滇西耳草

Vines

20.0; 14.6; 23.2; 20.1; 180.8; 0.0; 963 ❙ 1523; 681; 67 ❙ 58.4; 942; 581.0

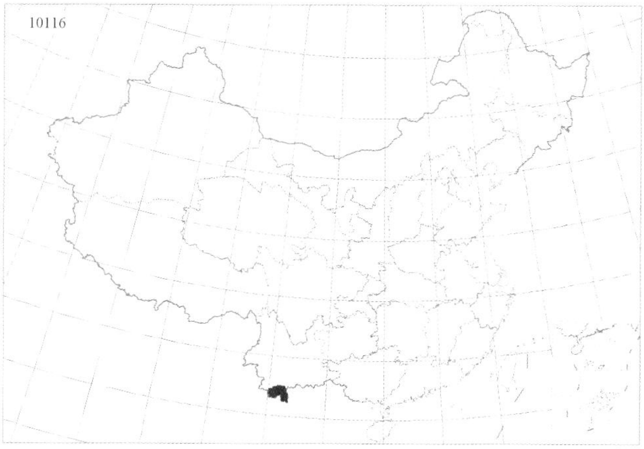

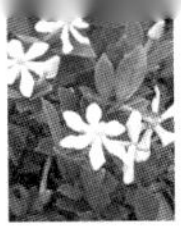

Rubiaceae 茜草科 *Hedyotis* 耳草属
Hedyotis hainanensis (Chun) W. C. Ko 海南耳草
Shrubs

24.5; 19.0; 28.3; 24.5; 234.0; 0.0; 1357 ❙ 1661; 571; 89 ❙ 22.6; 1274; 436.3

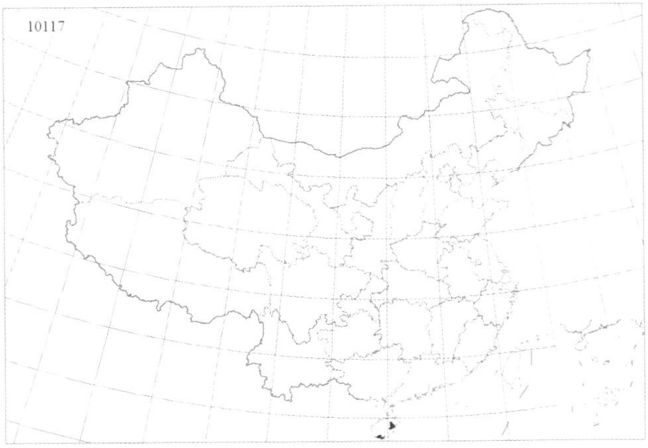

Rubiaceae 茜草科 *Hedyotis* 耳草属
Hedyotis hedyotidea (DC.) Merr. 牛白藤
Scandent shrubs

19.7 (12.8~23.5); 11.4 (3.3~16.4); 26.3 (20.1~29.8); 19.8 (12.9~23.5); 177.1 (97.1~222.3); 0.0 (0.0~2.2); 1014 (702~1276) ❙ 1534 (939~2680); 703 (425~1412); 123 (37~245) ❙ 52.0 (-6.9~246.5); 1008 (702~1247); 278.3 (139.6~659.0)

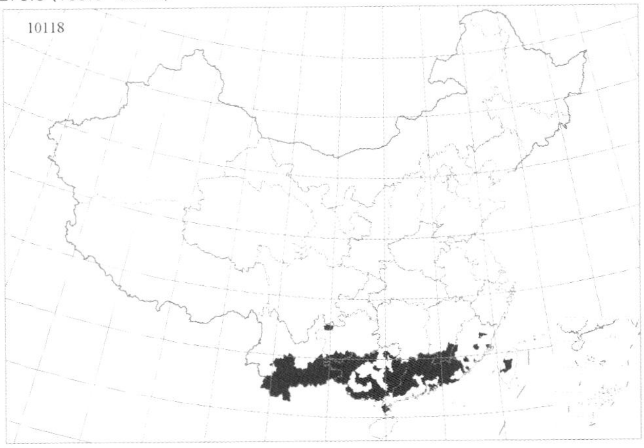

Rubiaceae 茜草科 *Hedyotis* 耳草属
Hedyotis obliquinervis Merr. 偏脉耳草
Scandent shrubs

23.7 (23.0~24.8); 18.5 (17.5~20.1); 27.4 (26.7~28.0); 23.8 (23.1~24.8); 225.2 (217.0~237.6); 0.0 (0.0~0.0); 1283 (1217~1385) ❙ 1465 (1242~1725); 511 (444~610); 50 (41~68) ❙ 14.6 (-2.2~41.8); 1132 (1091~1163); 525.6 (439.7~620.0)

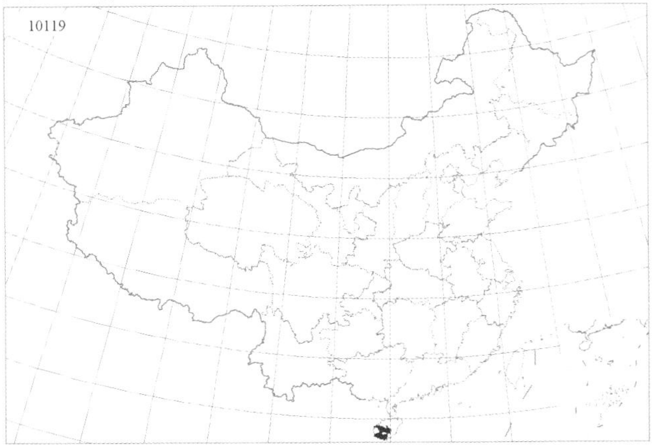

Rubiaceae 茜草科 *Hedyotis* 耳草属
Hedyotis scandens Roxb. 攀茎耳草
Scandent shrubs

16.5 (8.6~21.0); 9.9 (1.3~15.1); 21.0 (15.0~25.6); 16.5 (8.7~21.0); 138.1 (52.6~191.8); 0.2 (0.0~8.8); 810 (574~1053) ❙ 1193 (907~1751); 669 (521~1007); 48 (24~77) ❙ 47.7 (6.5~145.8); 792 (574~1053); 381.9 (234.5~659.0)

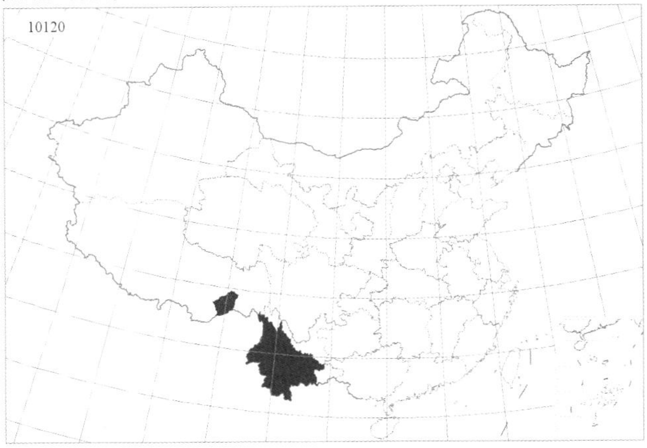

Rubiaceae 茜草科 *Himalrandia* 须弥茜树属
Himalrandia lichiangensis (W. W. Sm.) Tirveng. 须弥茜树
Shrubs

13.0 (2.2~18.9); 5.3 (-6.6~12.0); 19.3 (9.5~26.6); 13.1 (3.8~19.0); 101.1 (15.5~167.8); 4.7 (0.0~49.1); 705 (411~929) ❙ 952 (729~1178); 553 (411~646); 31 (11~57) ❙ 38.3 (-14.2~125.9); 679 (411~912); 266.5 (142.9~383.4)

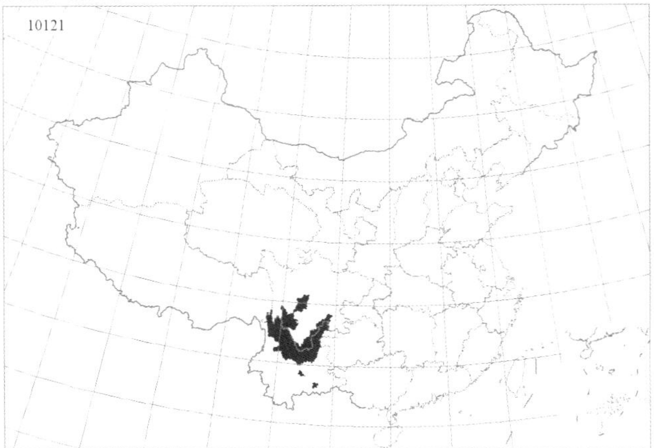

Rubiaceae 茜草科 *Hymenodictyon* 土连翘属
Hymenodictyon flaccidum Wall. 土连翘
Trees, deciduous

15.4 (5.4~22.2); 8.1 (-2.1~15.1); 20.9 (12.3~28.3); 15.4 (5.8~22.3); 125.8 (31.5~207.5); 1.0 (0.0~25.6); 784 (481~1190) ❙ 1056 (741~1607); 593 (411~848); 41 (15~107) ❙ 35.4 (-14.2~74.2); 760 (481~1190); 307.5 (142.9~659.0)

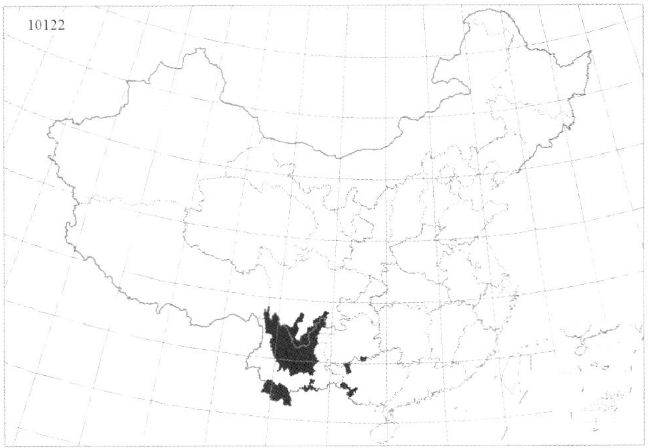

Rubiaceae 茜草科 *Hymenodictyon* 土连翘属

Hymenodictyon orixense (Roxb.) Mabb. 毛土连翘

Trees, deciduous

18.3 (10.2~21.0); 12.1 (2.4~15.1); 22.5 (16.9~25.6); 18.4 (10.3~21.0); 160.9 (68.2~191.8); 0.4 (0.0~5.1); 894 (620~1053) ❙ 1398 (1036~1751); 751 (606~996); 57 (18~77) ❙ 56.7 (34.8~88.8); 888 (620~1053); 487.0 (298.0~659.0)

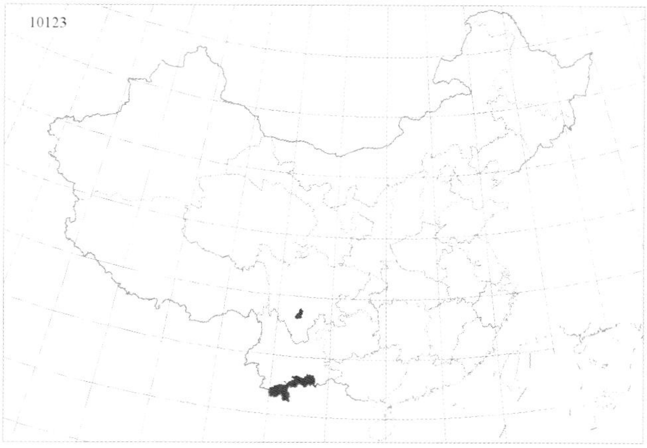

Rubiaceae 茜草科 *Hyptianthera* 藏药木属

Hyptianthera stricta (Roxb.) Wight et Arn. 藏药木

Shrubs or small trees

18.4 (13.3~21.0); 12.1 (5.4~15.1); 22.5 (19.6~25.6); 18.5 (13.4~21.0); 161.7 (100.4~191.8); 0.0 (0.0~0.0); 891 (706~1053) ❙ 1497 (1163~1751); 815 (641~1007); 56 (41~77) ❙ 69.4 (34.8~145.8); 880 (706~1053); 500.2 (383.4~659.0)

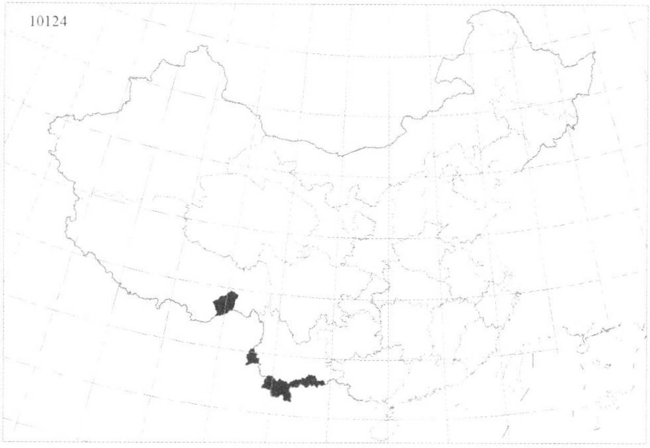

Rubiaceae 茜草科 *Ixora* 龙船花属

Ixora amplexicaulis C. Y. Wu ex Ko 抱茎龙船花

Small trees, evergreen

20.5; 15.0; 23.7; 20.5; 185.9; 0.0; 990 ❙ 1536; 678; 72 ❙ 55.1; 973; 614.6

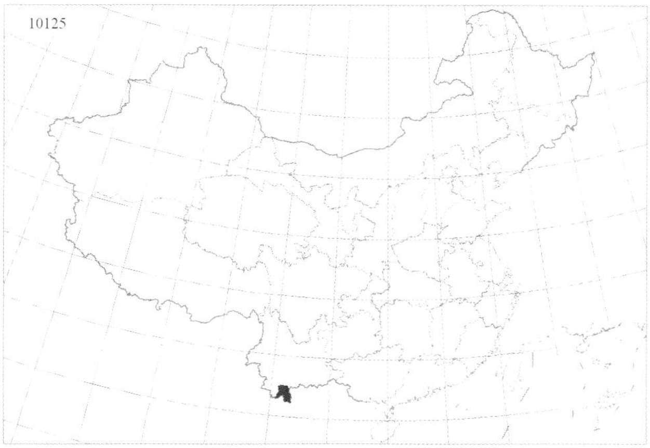

Rubiaceae 茜草科 *Ixora* 龙船花属

Ixora auricularis How ex Ko 耳叶龙船花

Shrubs or small trees, evergreen

20.2 (18.2~22.2); 13.8 (11.9~15.1); 24.3 (22.0~28.3); 20.2 (18.2~22.3); 182.3 (158.7~207.5); 0.0 (0.0~0.0); 997 (868~1190) ❙ 1480 (1366~1607); 764 (655~911); 68 (45~87) ❙ 50.3 (15.4~74.2); 987 (847~1190); 529.4 (327.7~659.0)

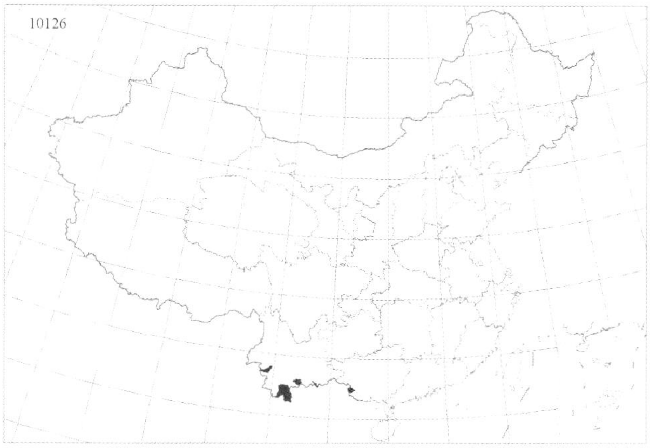

Rubiaceae 茜草科 *Ixora* 龙船花属

Ixora cephalophora Merr. 团花龙船花

Shrubs, evergreen

23.6 (20.4~24.8); 18.2 (15.1~20.1); 27.4 (23.6~28.6); 23.7 (20.5~24.8); 223.9 (185.8~237.6); 0.0 (0.0~0.0); 1277 (989~1385) ❙ 1524 (1242~1813); 557 (444~712); 68 (41~113) ❙ 20.5 (-2.2~62.5); 1172 (986~1329); 471.9 (248.2~659.0)

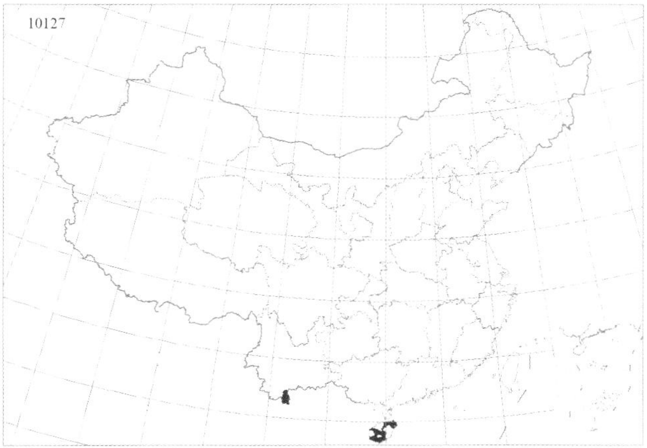

Rubiaceae 茜草科 *Ixora* 龙船花属

Ixora chinensis Lam. 龙船花

Shrubs, evergreen

21.2 (16.4~24.6); 12.8 (5.1~19.5); 28.1 (23.6~29.8); 21.3 (16.5~24.7); 195.6 (137.7~235.9); 0.0 (0.0~0.0); 1126 (864~1372) ❙ 1620 (871~2435); 750 (299~1428); 121 (58~194) ❙ 44.1 (-6.7~99.5); 1117 (755~1331); 239.4 (95.3~659.0)

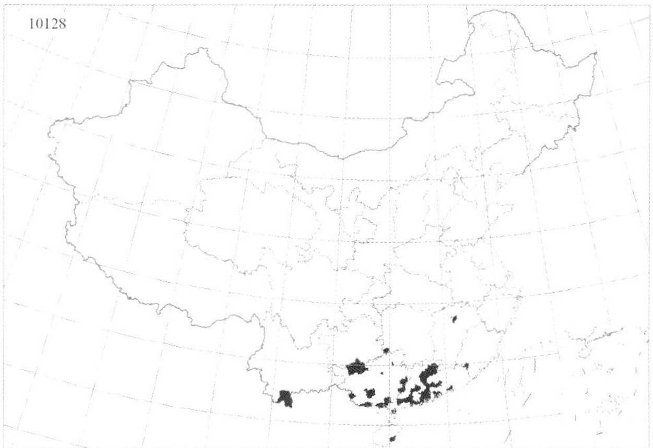

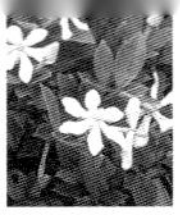

Rubiaceae 茜草科 *Ixora* 龙船花属

Ixora effusa Chun et F. C. How ex W. C. Ko 散花龙船花

Shrubs, evergreen

22.9 (19.1~24.7); 16.7 (9.8~19.8); 27.4 (25.7~28.4); 22.9 (19.2~24.7); 215.0 (170.1~236.5); 0.0 (0.0~0.0); 1228 (981~1376) ❙ 1607 (1179~2148); 669 (392~1234); 73 (42~125) ❙ 32.7 (-11.9~86.7); 1125 (981~1299); 455.8 (297.2~620.0)

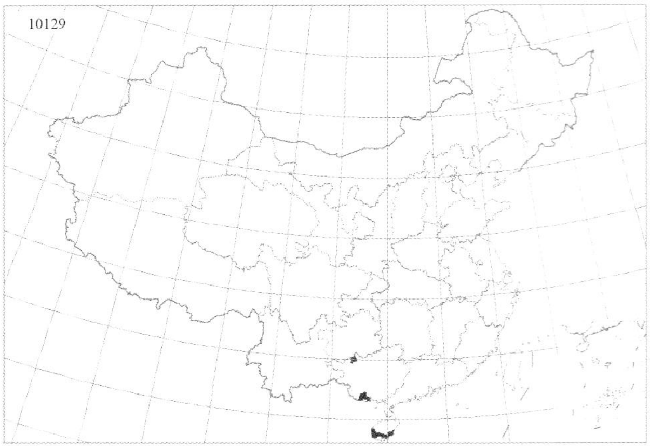

Rubiaceae 茜草科 *Ixora* 龙船花属

Ixora finlaysoniana Wall. ex G. Don 薄叶龙船花

Shrubs or small trees, evergreen

24.1 (23.6~24.8); 18.5 (17.8~20.1); 28.1 (27.5~28.7); 24.1 (23.6~24.8); 229.7 (223.3~237.6); 0.0 (0.0~0.0); 1321 (1271~1385) ❙ 1214 (868~1405); 426 (312~473); 53 (41~71) ❙ -8.1 (-34.3~6.7); 1069 (868~1139); 424.5 (288.3~527.5)

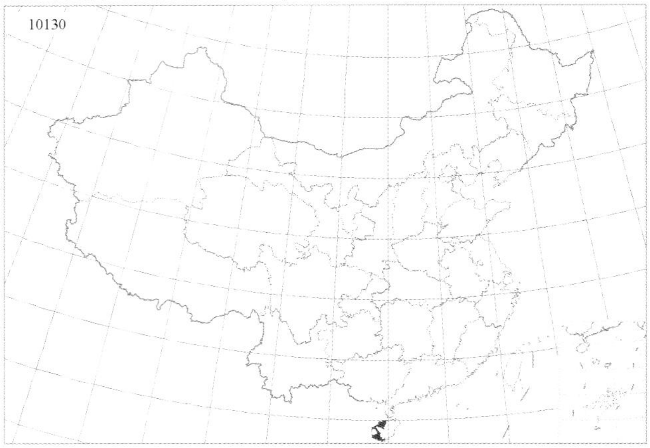

Rubiaceae 茜草科 *Ixora* 龙船花属

Ixora foonchowii Ko 宽昭龙船花

Small trees, evergreen

19.1; 13.0; 23.3; 19.1; 169.7; 0.0; 917 ❙ 1428; 835; 56 ❙ 55.7; 917; 506.2

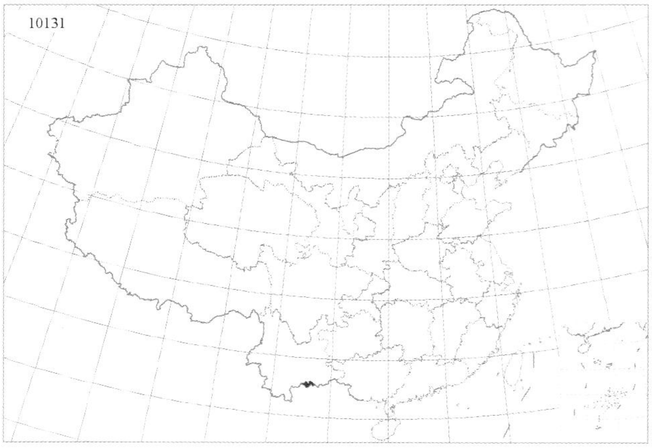

Rubiaceae 茜草科 *Ixora* 龙船花属

Ixora fulgens Roxb. 亮叶龙船花

Shrubs, evergreen

19.3 (17.7~21.0); 13.4 (10.9~15.1); 23.0 (21.6~25.6); 19.3 (17.8~21.0); 172.0 (153.4~191.8); 0.0 (0.0~0.0); 928 (854~1053) ❙ 1489 (1276~1751); 774 (655~996); 60 (41~77) ❙ 60.8 (36.8~88.8); 912 (839~1053); 513.9 (427.9~659.0)

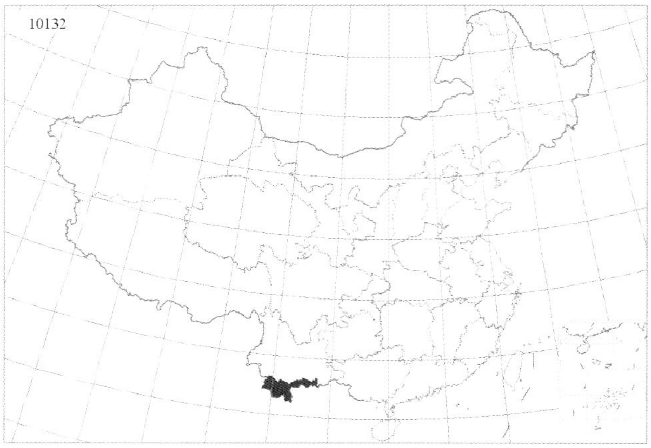

Rubiaceae 茜草科 *Ixora* 龙船花属

Ixora gracilis Ko 纤花龙船花

Climbing shrubs, evergreen

21.0; 14.2; 25.6; 21.0; 191.8; 0.0; 1053 ❙ 1522; 818; 68 ❙ 44.5; 1053; 551.6

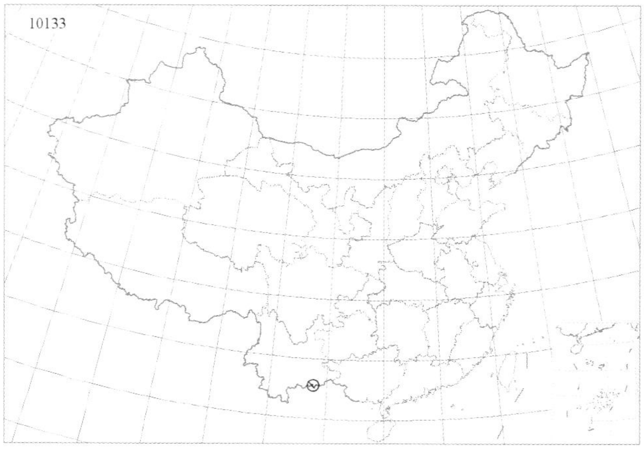

Rubiaceae 茜草科 *Ixora* 龙船花属

Ixora hainanensis Merr. 海南龙船花

Shrubs, evergreen

23.4 (19.3~24.8); 17.2 (13.1~20.1); 28.0 (23.6~29.2); 23.4 (19.4~24.8); 221.2 (173.2~237.6); 0.0 (0.0~0.0); 1266 (939~1385) ❙ 1533 (868~1864); 607 (312~893); 83 (41~144) ❙ 21.9 (-34.3~64.4); 1181 (868~1360); 370.2 (139.6~620.0)

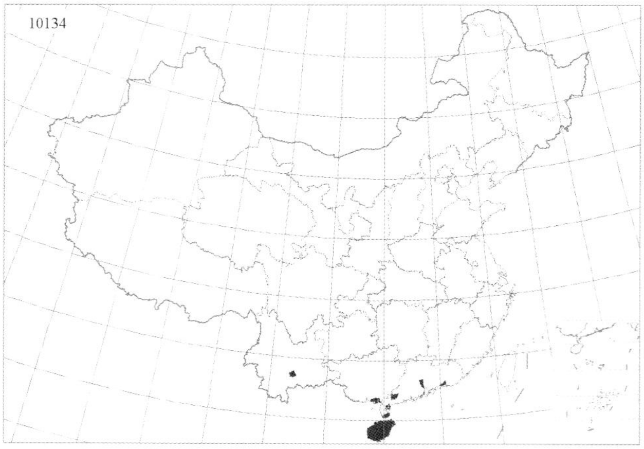

Rubiaceae 茜草科 *Ixora* 龙船花属
Ixora henryi H. Lév. 白花龙船花
Shrubs, evergreen

18.8 (8.6~24.7); 10.7 (1.3~19.8); 25.0 (15.0~29.2); 18.8 (8.7~24.7); 166.2 (52.6~236.5); 0.1 (0.0~8.8); 964 (574~1376) ▎ 1370 (868~2435); 672 (312~1428); 96 (32~197) ▎ 43.6 (-34.3~112.2); 950 (574~1329); 304.4 (158.7~659.0)

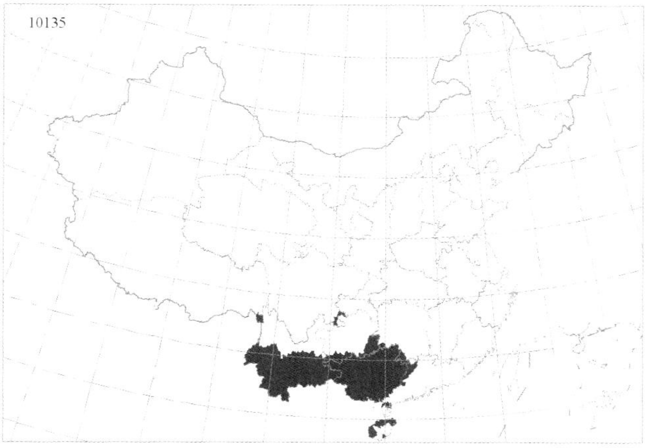

Rubiaceae 茜草科 *Ixora* 龙船花属
Ixora insignis Chun et How ex Ko 长序龙船花
Small trees, evergreen

21.0; 14.2; 25.6; 21.0; 191.8; 0.0; 1053 ▎ 1522; 818; 68 ▎ 44.5; 1053; 551.6

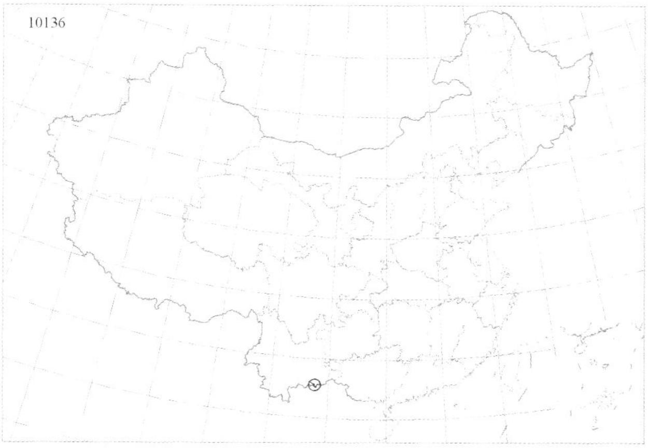

Rubiaceae 茜草科 *Ixora* 龙船花属
Ixora nienkui Merr. et Chun 泡叶龙船花
Shrubs, evergreen

22.4 (15.5~24.8); 15.9 (4.8~20.1); 27.1 (22.5~28.9); 22.5 (15.6~24.8); 209.5 (127.2~237.6); 0.0 (0.0~0.2); 1192 (815~1385) ▎ 1579 (868~2435); 697 (312~1428); 86 (41~177) ▎ 35.4 (-34.3~99.5); 1132 (815~1360); 406.0 (145.3~659.0)

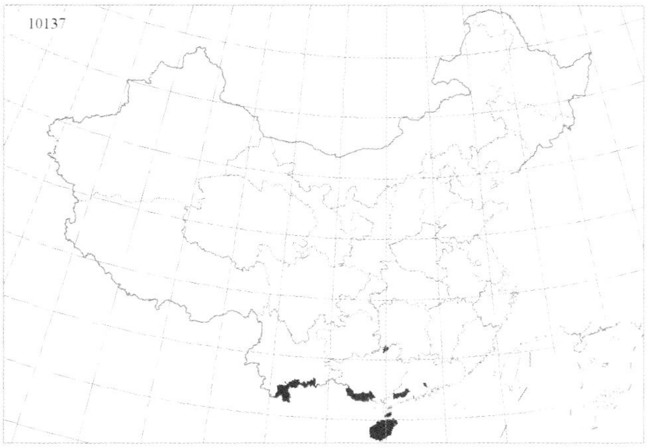

Rubiaceae 茜草科 *Ixora* 龙船花属
Ixora paraopaca Ko 版纳龙船花
Shrubs, evergreen

20.0; 14.6; 23.2; 20.1; 180.8; 0.0; 963 ▎ 1523; 681; 67 ▎ 58.4; 942; 581.0

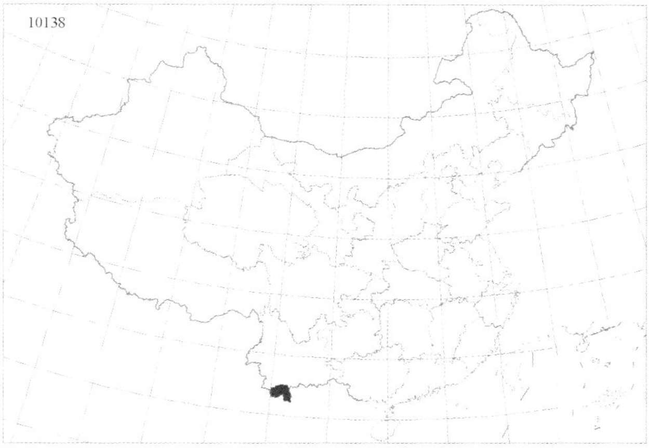

Rubiaceae 茜草科 *Ixora* 龙船花属
Ixora philippinensis Merr. 小仙龙船花
Shrubs, evergreen

18.5; 14.1; 21.7; 18.5; 162.1; 0.0; 863 ▎ 2713; 1388; 188 ▎ 214.4; 863; 363.7

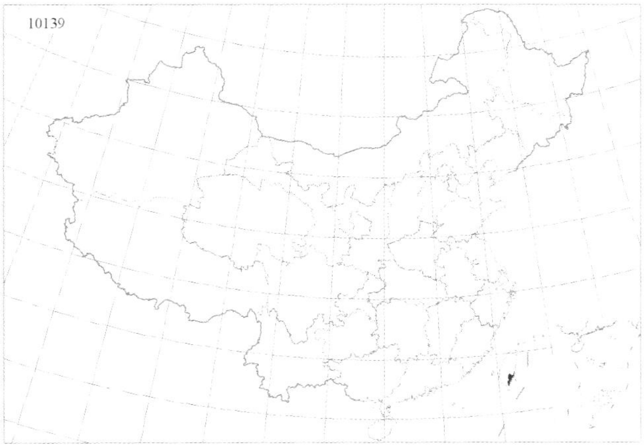

Rubiaceae 茜草科 *Ixora* 龙船花属
Ixora subsessilis Wall. ex Don. 囊果龙船花
Shrubs or small trees, evergreen

13.3; 5.4; 19.6; 13.4; 100.4; 0.0; 706 ▎ 1736; 1007; 51 ▎ 145.8; 706; 447.4

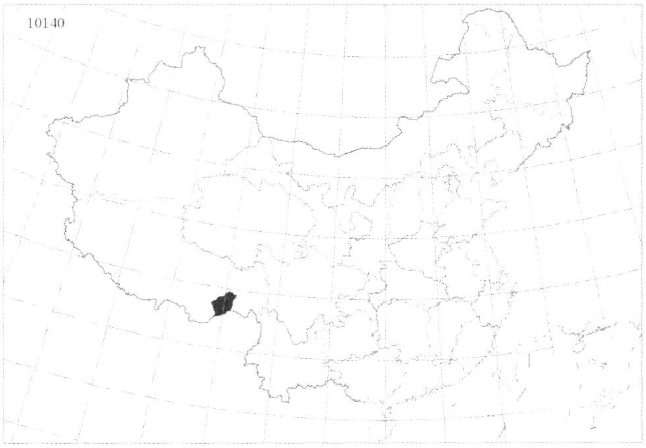

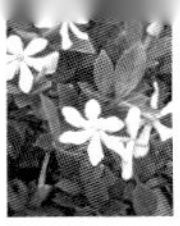

Rubiaceae 茜草科 *Ixora* 龙船花属

Ixora tsangii Merr. ex H. L. Li 上思龙船花

Small shrubs, evergreen

21.3; 13.2; 27.3; 21.3; 195.9; 0.0; 1111 ❙ 1900; 1056; 125 ❙ 71.0; 1111; 297.2

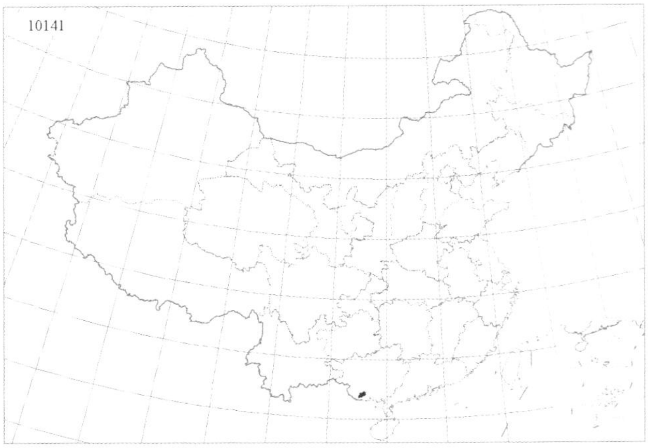

Rubiaceae 茜草科 *Ixora* 龙船花属

Ixora yunnanensis Hutch. 云南龙船花

Shrubs, evergreen

19.0 (16.9~21.0); 13.0 (10.3~15.1); 23.0 (21.4~25.6); 19.0 (16.9~21.0); 168.6 (142.7~191.8); 0.0 (0.0~0.0); 917 (813~1053) ❙ 1428 (1163~1751); 763 (641~996); 61 (48~77) ❙ 55.8 (34.8~88.8); 911 (813~1053); 502.7 (383.4~659.0)

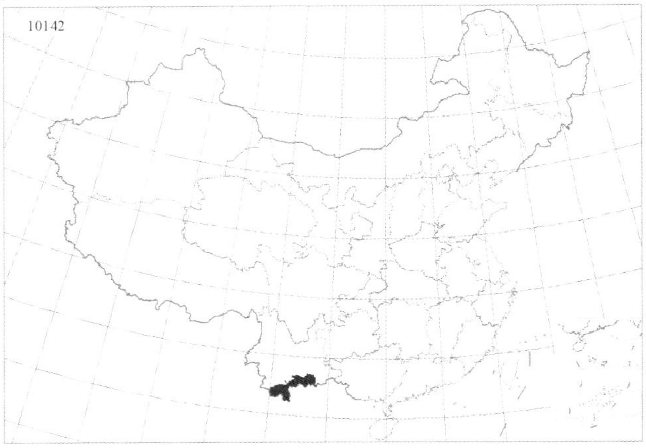

Rubiaceae 茜草科 *Keenania* 溪楠属

Keenania flava H. S. Lo 黄溪楠

Herbs or subshrubs

19.1; 11.0; 25.0; 19.1; 169.6; 0.0; 945 ❙ 1438; 820; 57 ❙ 52.2; 945; 371.7

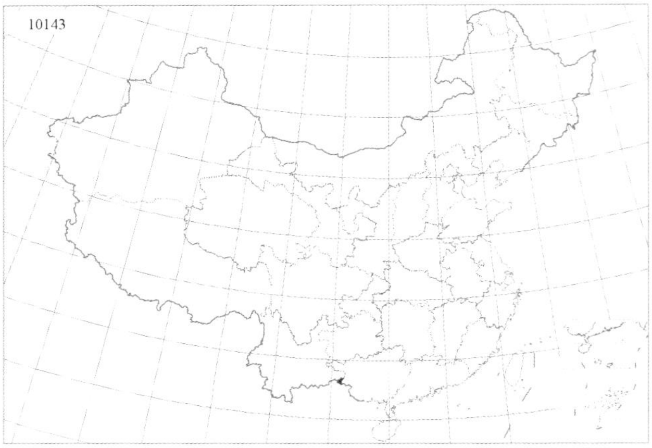

Rubiaceae 茜草科 *Keenania* 溪楠属

Keenania tonkinensis Drake 溪楠

Herbs or subshrubs

22.2; 13.9; 28.3; 22.3; 207.5; 0.0; 1190 ❙ 1374; 713; 87 ❙ 15.4; 1190; 327.7

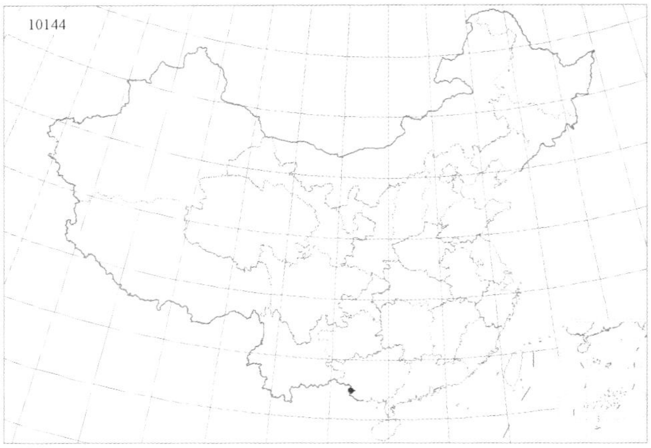

Rubiaceae 茜草科 *Lasianthus* 粗叶木属

Lasianthus attenuatus Jack 斜基粗叶木

Shrubs

20.6 (13.3~24.8); 13.6 (5.4~20.1); 25.9 (19.6~29.0); 20.6 (13.4~24.8); 187.6 (100.4~237.6); 0.0 (0.0~0.0); 1065 (706~1385) ❙ 1709 (1007~3495); 787 (392~1847); 129 (41~897) ❙ 65.4 (−12.0~246.5); 1041 (706~1329); 336.3 (154.2~659.0)

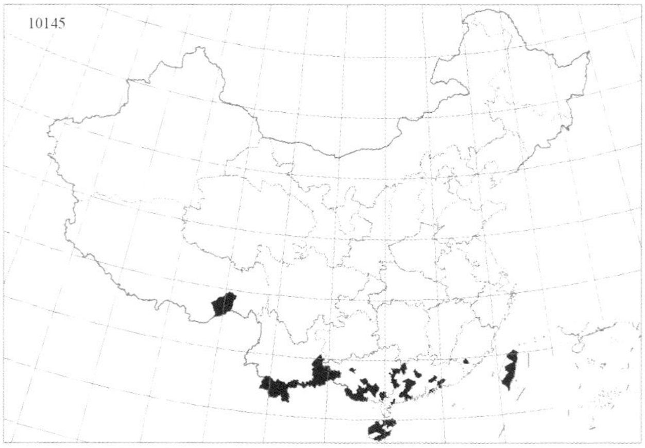

Rubiaceae 茜草科 *Lasianthus* 粗叶木属

Lasianthus austrosinensis H. S. Lo 华南粗叶木

Shrubs

21.5 (21.0~22.0); 13.2 (12.0~13.6); 28.1 (26.9~28.8); 21.6 (21.0~22.0); 198.7 (192.3~203.8); 0.0 (0.0~0.0); 1136 (1081~1170) ❙ 1620 (1502~1776); 756 (662~850); 131 (106~168) ❙ 43.1 (32.9~64.3); 1136 (1081~1170); 236.4 (203.4~281.7)

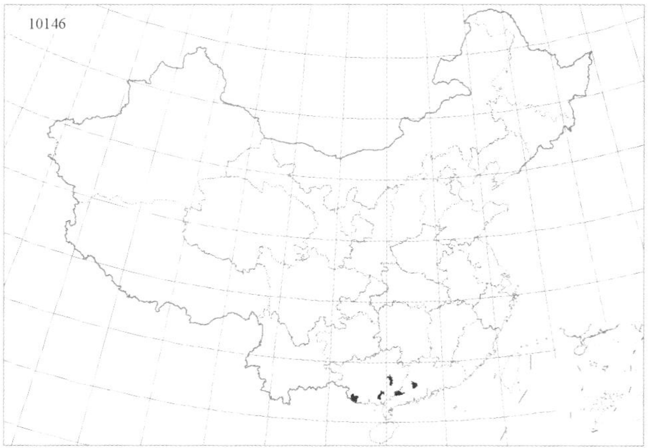

Rubiaceae 茜草科 *Lasianthus* 粗叶木属
Lasianthus austroyunnanensis H. Zhu 滇南粗叶木
Shrubs

19.0 (16.2~24.1); 12.9 (9.5~18.9); 23.8 (20.1~28.0); 19.0 (16.2~24.1); 168.3 (134.2~229.7); 0.0 (0.0~0.0); 937 (773~1322) ❙ 2192 (1254~3495); 982 (699~1594); 236 (47~897) ❙ 138.2 (32.5~246.5); 921 (773~1095); 323.4 (154.2~551.6)

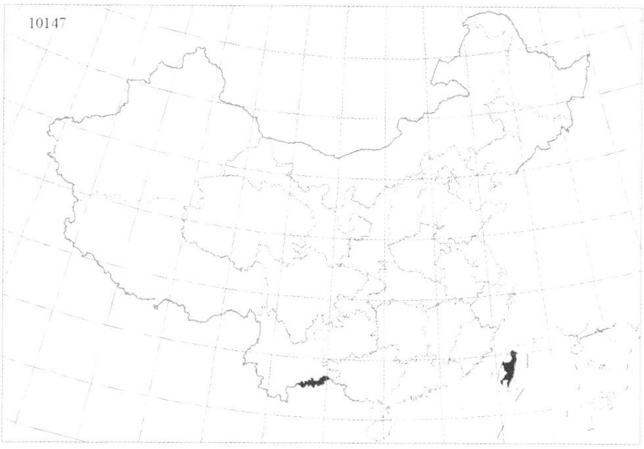

Rubiaceae 茜草科 *Lasianthus* 粗叶木属
Lasianthus biermannii King ex Hook. f. 梗花粗叶木
Shrubs

17.6 (8.6~22.7); 10.0 (1.3~15.0); 23.2 (15.0~28.6); 17.6 (8.7~22.8); 151.6 (52.6~213.1); 0.2 (0.0~8.8); 877 (574~1221) ❙ 1353 (939~2435); 724 (452~1428); 69 (37~194) ❙ 54.9 (6.5~145.8); 869 (574~1221); 355.3 (189.5~627.0)

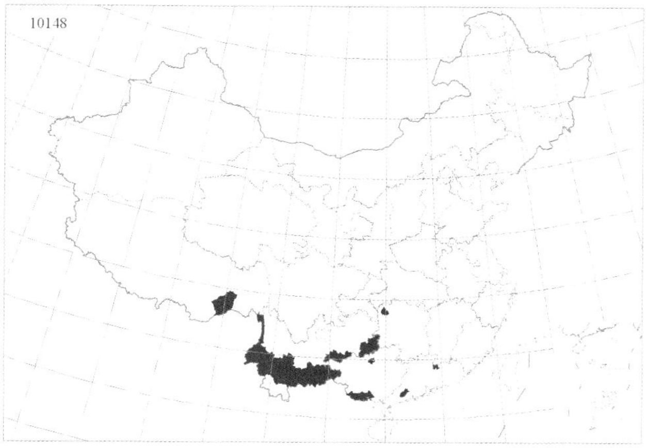

Rubiaceae 茜草科 *Lasianthus* 粗叶木属
Lasianthus biermannii subsp. crassipedunculatus C. Y. Wu et H. Zhu 粗梗粗叶木
Shrubs

18.6 (16.3~23.0); 11.7 (9.1~17.7); 23.4 (21.4~27.8); 18.6 (16.4~23.1); 163.6 (136.4~217.0); 0.0 (0.0~0.0); 914 (797~1217) ❙ 1396 (1074~1876); 754 (582~996); 59 (47~116) ❙ 52.8 (32.4~88.8); 910 (797~1163); 417.3 (273.3~572.9)

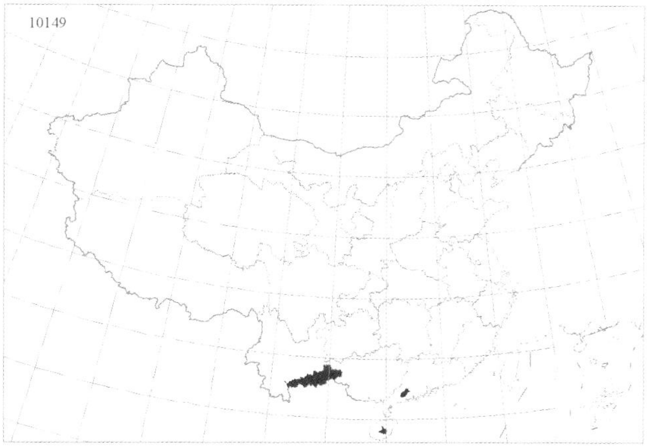

Rubiaceae 茜草科 *Lasianthus* 粗叶木属
Lasianthus biflorus (Bl.) Gangop. et Chakrab. 石核木
Small shrubs

22.8 (16.2~24.8); 16.8 (11.0~20.1); 27.2 (20.1~28.8); 22.9 (16.2~24.8); 214.5 (134.2~237.6); 0.0 (0.0~0.0); 1223 (773~1385) ❙ 1715 (868~2992); 760 (312~1847); 94 (41~245) ❙ 47.4 (-34.3~246.5); 1140 (773~1331); 393.0 (235.1~659.0)

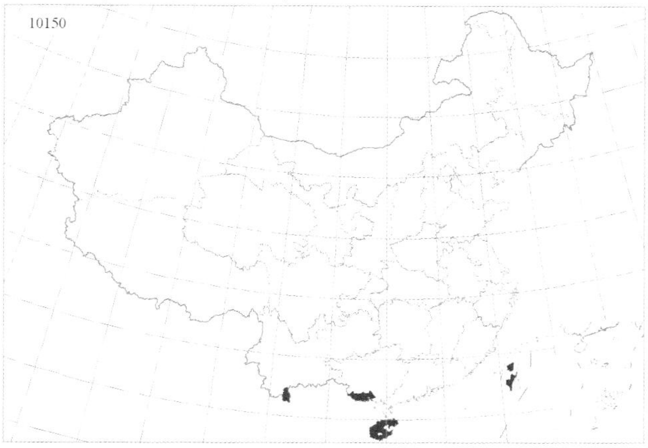

Rubiaceae 茜草科 *Lasianthus* 粗叶木属
Lasianthus calycinus Dunn 黄果粗叶木
Shrubs

23.9 (23.0~24.7); 18.5 (17.5~19.8); 27.8 (26.7~28.7); 24.0 (23.1~24.7); 227.8 (217.0~236.5); 0.0 (0.0~0.0); 1305 (1217~1376) ❙ 1410 (868~1767); 488 (312~610); 61 (42~88) ❙ 8.3 (-34.3~41.8); 1135 (868~1299); 476.4 (288.3~620.0)

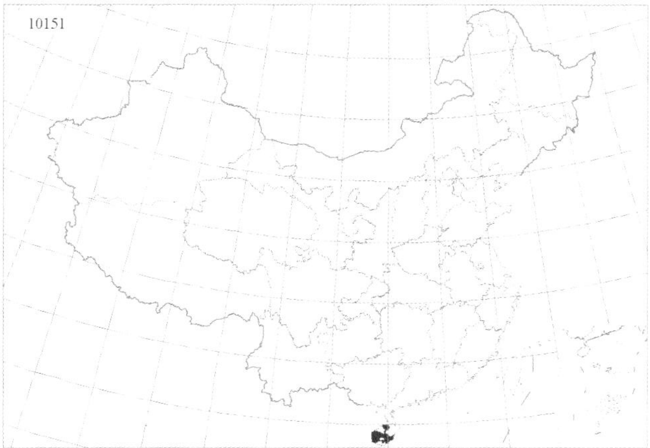

Rubiaceae 茜草科 *Lasianthus* 粗叶木属
Lasianthus chevalieri Pitard 长萼粗叶木
Shrubs

23.9; 18.7; 27.5; 23.9; 226.8; 0.0; 1294 ❙ 1380; 455; 48 ❙ 6.7; 1112; 455.2

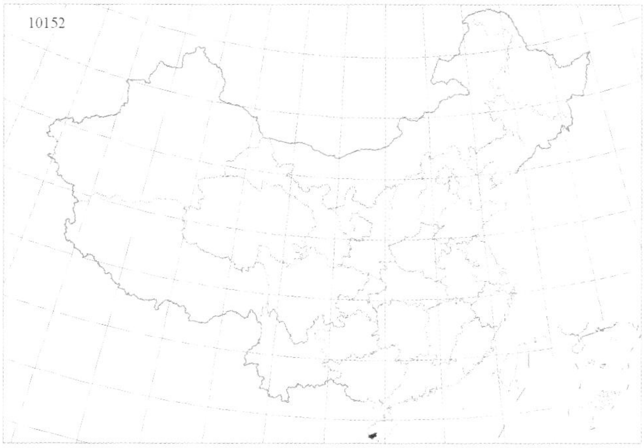

Rubiaceae 茜草科 *Lasianthus* 粗叶木属
Lasianthus chinensis Benth. 粗叶木
Shrubs

19.6 (14.3~24.8); 11.2 (2.1~20.1); 26.7 (20.8~29.8); 19.7 (14.3~24.8); 176.8 (114.1~237.6); 0.2 (0.0~4.9); 1031 (765~1385) ❙ 1598 (868~3495); 670 (312~1428); 138 (41~897) ❙ 58.1 (-34.3~224.4); 1011 (765~1360); 297.2 (122.7~659.0)

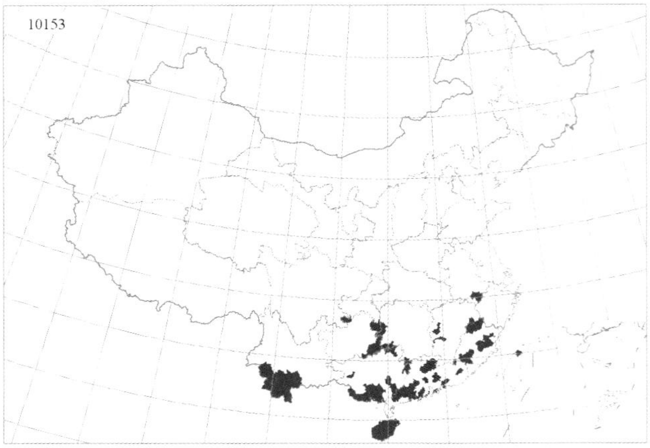

Rubiaceae 茜草科 *Lasianthus* 粗叶木属
Lasianthus chrysoneurus (Korth.) Miq. 库兹粗叶木
Shrubs

18.6 (16.8~20.5); 12.7 (10.4~15.1); 22.1 (20.8~23.7); 18.6 (16.9~20.5); 163.6 (142.6~185.9); 0.0 (0.0~0.0); 889 (811~992) ❙ 1441 (1275~1607); 742 (655~816); 53 (41~77) ❙ 62.3 (47.6~74.2); 869 (802~986); 494.0 (418.5~659.0)

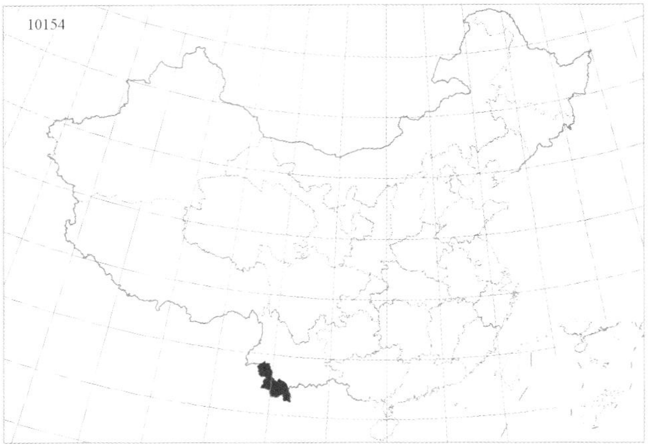

Rubiaceae 茜草科 *Lasianthus* 粗叶木属
Lasianthus chunii H. S. Lo 焕镛粗叶木
Shrubs

19.5 (17.3~21.8); 10.4 (6.9~14.2); 27.3 (25.2~28.5); 19.6 (17.3~21.9); 174.7 (147.6~202.6); 0.0 (0.0~0.0); 1013 (864~1150) ❙ 1704 (1462~2005); 664 (505~902); 175 (107~231) ❙ 68.7 (29.9~89.4); 1013 (864~1150); 248.3 (188.8~305.7)

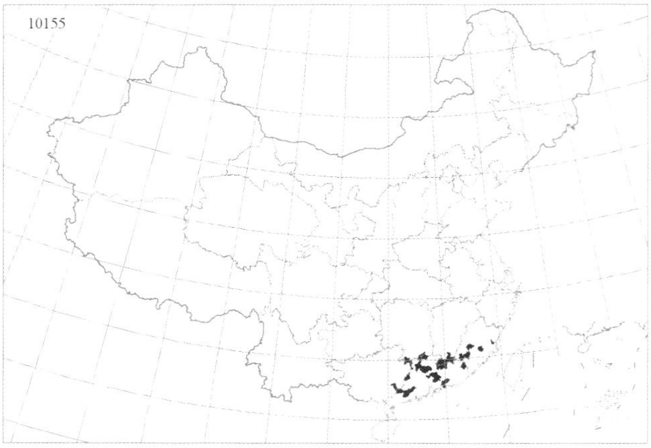

Rubiaceae 茜草科 *Lasianthus* 粗叶木属
Lasianthus curtisii King et Gamble 广东粗叶木
Shrubs or small trees

21.8 (16.2~24.7); 14.8 (10.7~19.8); 27.4 (20.1~28.8); 21.9 (16.2~24.7); 202.8 (134.2~236.5); 0.0 (0.0~0.0); 1155 (773~1376) ❙ 1820 (868~3495); 755 (312~1847); 160 (42~897) ❙ 63.6 (-34.3~246.5); 1115 (773~1331); 312.1 (201.9~620.0)

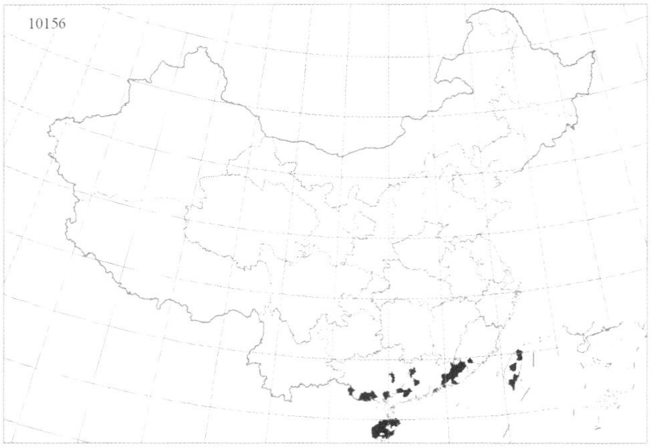

Rubiaceae 茜草科 *Lasianthus* 粗叶木属
Lasianthus filipes Chun ex H. S. Lo 长梗粗叶木
Shrubs

19.4 (16.3~23.3); 11.3 (7.3~16.0); 25.5 (21.4~29.0); 19.4 (16.4~23.3); 172.9 (136.4~220.1); 0.0 (0.0~0.0); 988 (797~1264) ❙ 1521 (1074~2148); 757 (577~1234); 99 (47~190) ❙ 54.5 (25.4~89.4); 988 (797~1264); 305.7 (194.5~551.6)

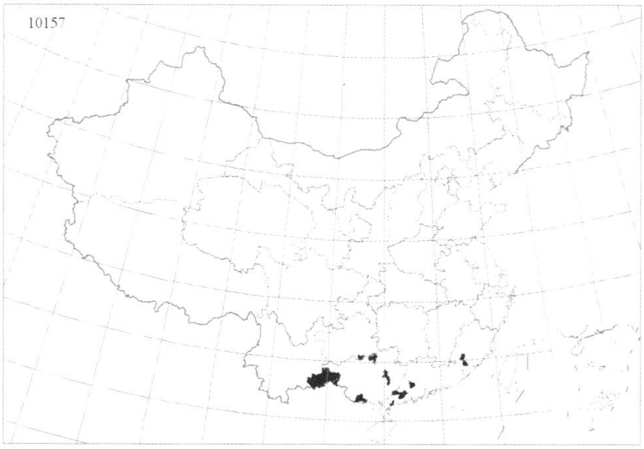

Rubiaceae 茜草科 *Lasianthus* 粗叶木属
Lasianthus fordii Hance 罗浮粗叶木
Shrubs

19.2 (11.4~24.8); 11.7 (2.7~20.1); 25.1 (18.8~29.8); 19.3 (11.4~24.8); 171.4 (80.7~237.6); 0.1 (0.0~3.7); 983 (652~1385) ❙ 1549 (868~3495); 726 (312~1847); 113 (25~897) ❙ 59.9 (-34.3~246.5); 962 (652~1331); 316.4 (119.4~659.0)

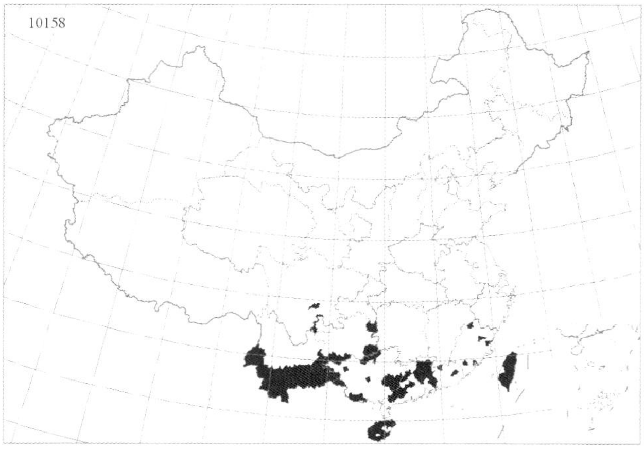

Rubiaceae 茜草科 *Lasianthus* 粗叶木属
Lasianthus formosensis Matsum. 台湾粗叶木
Shrubs

20.8 (17.7~23.2); 13.6 (10.6~17.7); 26.4 (21.7~28.9); 20.8 (17.8~23.2); 189.6 (153.4~218.9); 0.0 (0.0~0.0); 1064 (852~1257) ❙ 2072 (1292~3495); 900 (702~1847); 217 (49~897) ❙ 99.0 (15.4~235.6); 1064 (852~1257); 296.4 (194.5~659.0)

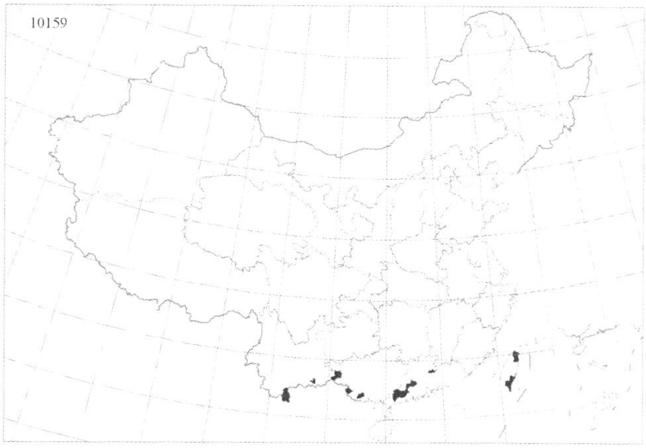

Rubiaceae 茜草科 *Lasianthus* 粗叶木属
Lasianthus henryi Hutch. 西南粗叶木
Shrubs

18.6 (11.1~22.8); 10.2 (2.2~15.2); 25.4 (17.0~28.9); 18.6 (11.3~22.8); 163.8 (76.6~213.6); 0.1 (0.0~5.3); 948 (644~1230) ❙ 1439 (950~2993); 689 (448~996); 99 (27~547) ❙ 52.0 (12.1~203.4); 945 (644~1230); 294.8 (130.1~659.0)

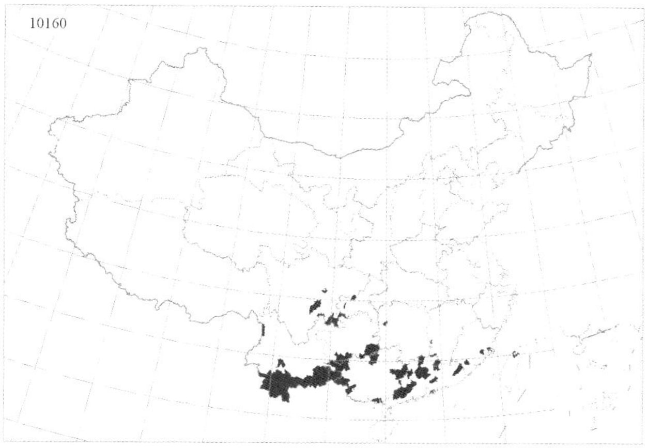

Rubiaceae 茜草科 *Lasianthus* 粗叶木属
Lasianthus hirsutus (Roxb.) Merr. 鸡屎树
Shrubs

23.2 (18.5~24.8); 17.3 (13.2~20.1); 27.6 (21.7~28.8); 23.3 (18.5~24.8); 219.1 (162.1~237.6); 0.0 (0.0~0.0); 1249 (863~1385) ❙ 1732 (868~2993); 706 (312~1847); 104 (41~547) ❙ 43.8 (-34.3~214.4); 1166 (863~1360); 386.7 (139.6~620.0)

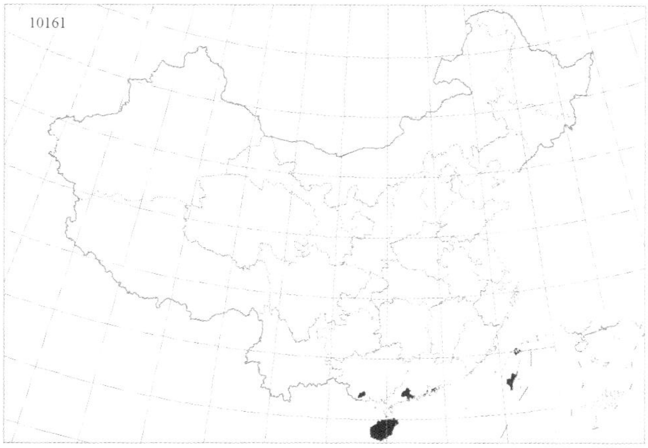

Rubiaceae 茜草科 *Lasianthus* 粗叶木属
Lasianthus hispidulus (Drake) Pitard 文山粗叶木
Shrubs

20.5 (16.2~24.7); 15.1 (11.0~19.8); 24.9 (20.1~28.2); 20.5 (16.2~24.7); 186.3 (134.2~236.5); 0.0 (0.0~0.0); 1045 (773~1376) ❙ 2555 (1588~3495); 966 (534~1847); 314 (58~897) ❙ 159.6 (15.4~246.5); 1019 (773~1299); 351.0 (226.8~572.9)

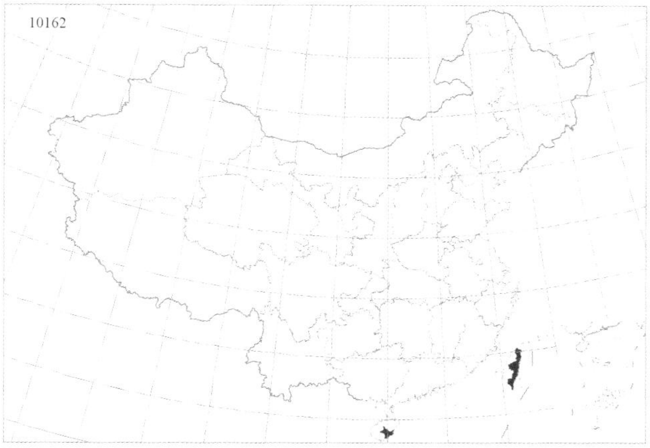

Rubiaceae 茜草科 *Lasianthus* 粗叶木属
Lasianthus hookeri C. B. Clarke ex Hook. f. 虎克粗叶木
Shrubs

18.2 (13.3~21.0); 11.5 (5.4~15.1); 22.7 (19.6~27.0); 18.2 (13.4~21.0); 158.8 (100.4~191.8); 0.0 (0.0~0.0); 882 (706~1053) ❙ 1348 (939~1751); 749 (524~1007); 55 (37~96) ❙ 53.5 (6.5~145.8); 867 (706~1053); 424.0 (277.0~659.0)

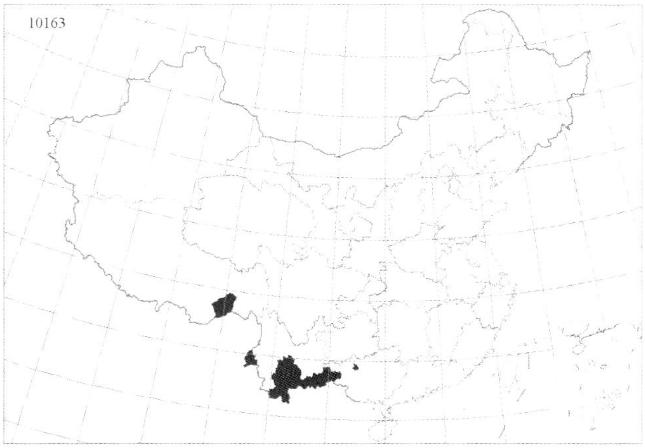

Rubiaceae 茜草科 *Lasianthus* 粗叶木属
Lasianthus hookeri var. dunniana (Lév.) H. Zhu 睫毛虎克粗叶木
Shrubs

18.8 (17.2~21.0); 12.2 (9.5~15.1); 23.2 (21.6~25.6); 18.8 (17.3~21.0); 165.8 (147.0~191.8); 0.0 (0.0~0.0); 910 (832~1053) ❙ 1439 (1049~1751); 774 (598~996); 60 (44~77) ❙ 58.3 (26.1~88.8); 901 (783~1053); 475.0 (336.3~659.0)

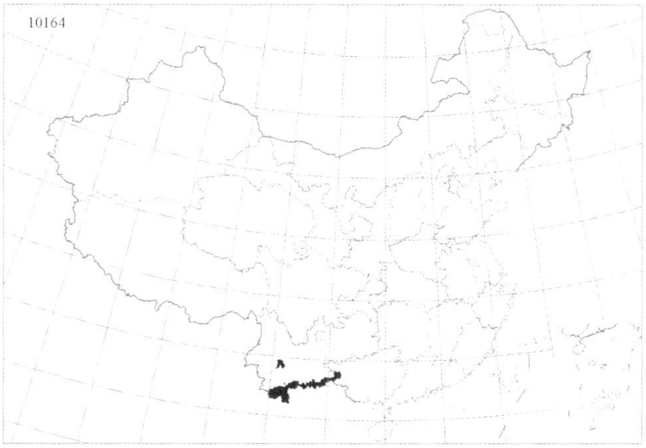

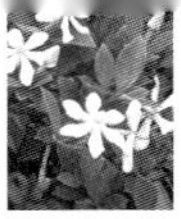

Rubiaceae 茜草科 *Lasianthus* 粗叶木属

Lasianthus inodorus Blume 革叶粗叶木

Shrubs

18.7 (16.4~20.5); 12.6 (9.6~15.1); 22.4 (20.6~23.7); 18.8 (16.4~20.5); 165.3 (137.3~185.9); 0.0 (0.0~0.0); 900 (795~992) ❙ 1535 (1434~1688); 810 (655~959); 55 (41~77) ❙ 71.5 (47.6~112.2); 882 (795~986); 522.1 (427.9~659.0)

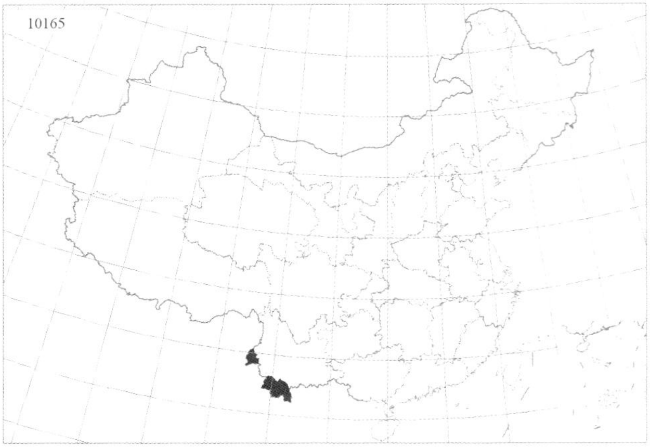

Rubiaceae 茜草科 *Lasianthus* 粗叶木属

Lasianthus japonicus Miq. 日本粗叶木

Shrubs

17.7 (5.3~22.7); 7.6 (−2.4~15.0); 26.9 (12.2~30.1); 17.8 (5.7~22.8); 153.8 (30.5~213.1); 0.3 (0.0~26.2); 933 (479~1221) ❙ 1544 (961~3495); 588 (356~1428); 167 (31~897) ❙ 66.8 (−5.0~246.5); 925 (479~1221); 235.3 (108.2~551.6)

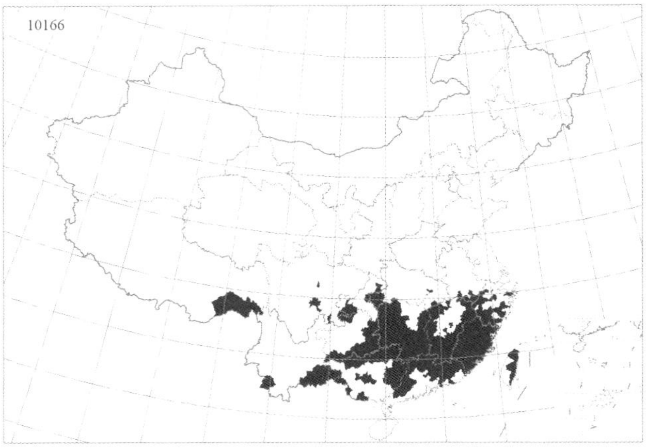

Rubiaceae 茜草科 *Lasianthus* 粗叶木属

Lasianthus japonicus subsp. longicaudus (Hook. f.) C. Y. Wu et H. Zhu 云广粗叶木

Shrubs

18.1 (5.3~22.2); 9.1 (−2.4~14.1); 26.0 (12.2~29.0); 18.1 (5.7~22.3); 157.7 (30.5~207.2); 0.3 (0.0~26.2); 931 (479~1186) ❙ 1511 (961~2014); 606 (356~1007); 151 (31~257) ❙ 64.1 (−5.0~145.8); 925 (479~1186); 272.2 (123.9~530.1)

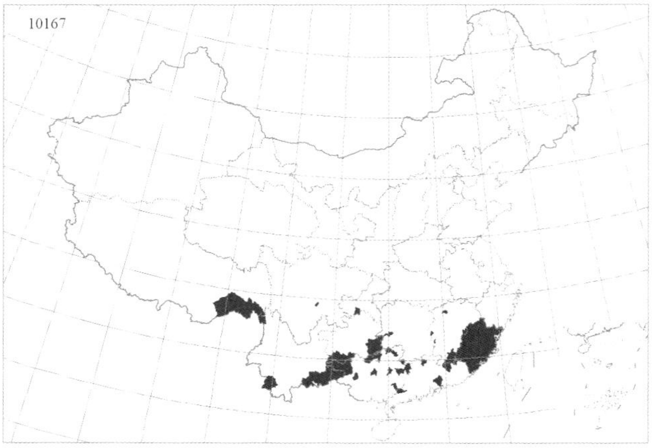

Rubiaceae 茜草科 *Lasianthus* 粗叶木属

Lasianthus japonicus Miq. var. latifolius H. Zhu 宽叶日本粗叶木

Shrubs

17.6 (16.3~20.3); 9.7 (9.1~10.9); 23.3 (21.4~28.3); 17.6 (16.4~20.4); 151.3 (136.4~184.5); 0.0 (0.0~0.0); 868 (797~1072) ❙ 1370 (1074~1777); 715 (582~860); 73 (47~182) ❙ 57.2 (32.4~80.6); 868 (797~1072); 327.0 (228.5~413.5)

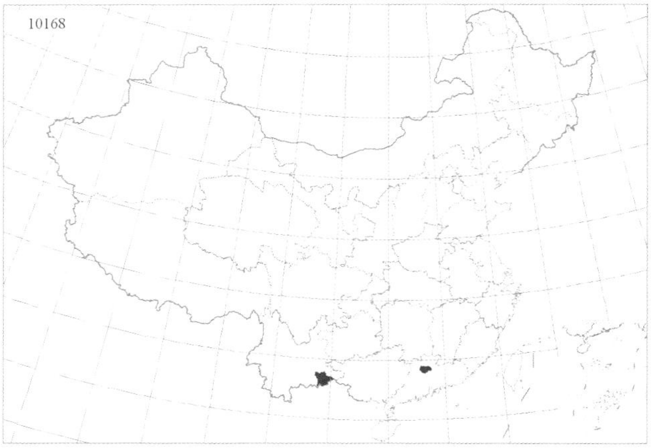

Rubiaceae 茜草科 *Lasianthus* 粗叶木属

Lasianthus lancifolius Hook. f. 美脉粗叶木

Shrubs

20.2 (17.2~24.8); 13.1 (9.1~20.1); 25.5 (22.4~28.8); 20.3 (17.3~24.8); 183.2 (147.0~237.6); 0.0 (0.0~0.0); 1039 (833~1385) ❙ 1429 (1161~1725); 690 (392~860); 71 (41~142) ❙ 40.9 (−11.9~80.6); 997 (833~1170); 393.4 (203.4~572.9)

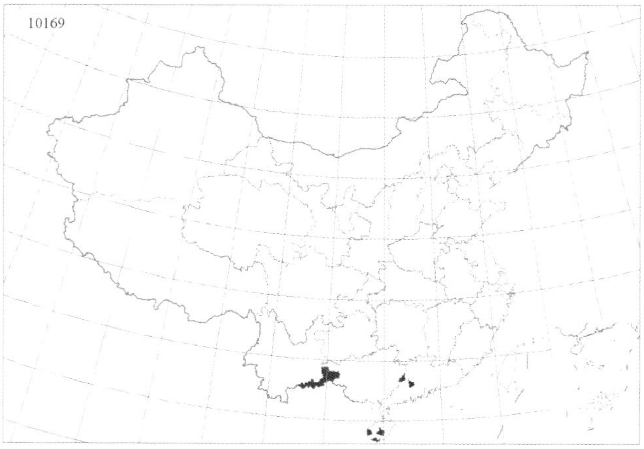

Rubiaceae 茜草科 *Lasianthus* 粗叶木属

Lasianthus linearisepalus C.Y.Wu et H.Zhu 线萼粗叶木

Shrubs

14.4; 7.5; 19.1; 14.4; 113.1; 0.0; 727 ❙ 1454; 792; 70 ❙ 99.9; 727; 487.2

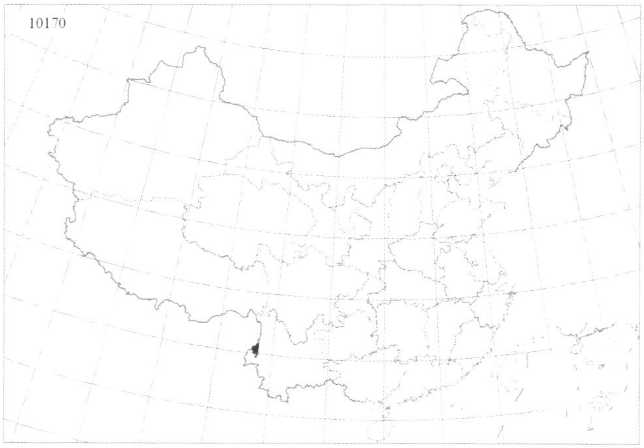

Rubiaceae 茜草科 *Lasianthus* 粗叶木属
Lasianthus lucidus Blume 无苞粗叶木
Shrubs

19.1 (16.4~24.1); 13.0 (9.6~18.2); 23.0 (20.6~28.4); 19.2 (16.4~24.2); 170.0 (137.3~230.1); 0.0 (0.0~0.0); 933 (795~1326) ❙ 1501 (1049~1751); 785 (598~996); 59 (41~110) ❙ 62.3 (26.1~112.2); 918 (783~1326); 501.2 (352.9~659.0)

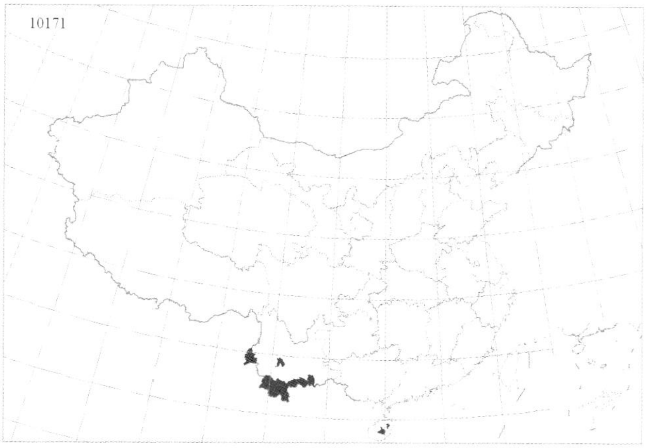

Rubiaceae 茜草科 *Lasianthus* 粗叶木属
Lasianthus micranthus Hook. f. 小花粗叶木
Shrubs

19.4 (5.3~24.3); 12.0 (-2.4~18.7); 25.1 (12.2~29.0); 19.4 (5.7~24.3); 173.5 (30.5~231.9); 0.5 (0.0~26.2); 994 (479~1338) ❙ 1641 (868~2993); 770 (312~1847); 121 (31~547) ❙ 70.3 (-34.3~246.5); 971 (479~1264); 334.4 (185.5~659.0)

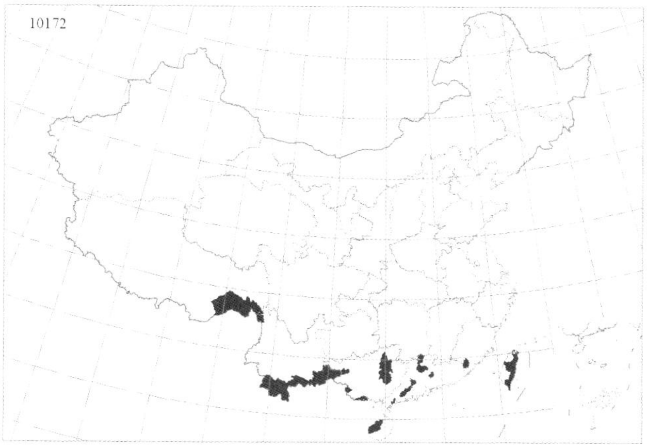

Rubiaceae 茜草科 *Lasianthus* 粗叶木属
Lasianthus obscurus (Blume ex DC.) Miq. 林生粗叶木
Shrubs

20.0; 14.6; 23.2; 20.1; 180.8; 0.0; 963 ❙ 1523; 681; 67 ❙ 58.4; 942; 581.0

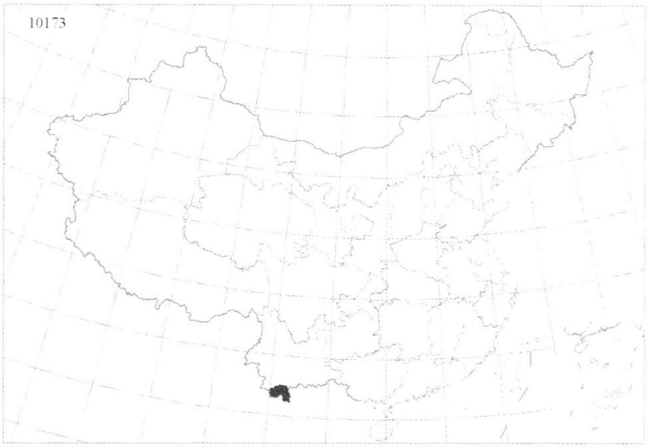

Rubiaceae 茜草科 *Lasianthus* 粗叶木属
Lasianthus rhinocerotis subsp. pedunculatus (Pitard) H. Zhu 黄毛粗叶木
Shrubs

22.1 (17.6~24.3); 14.7 (7.3~19.1); 27.6 (25.0~29.0); 22.1 (17.6~24.3); 205.4 (151.6~231.9); 0.0 (0.0~0.0); 1179 (909~1338) ❙ 1469 (868~2435); 674 (312~1428); 84 (42~142) ❙ 26.4 (-34.3~99.5); 1117 (868~1331); 339.8 (199.6~620.0)

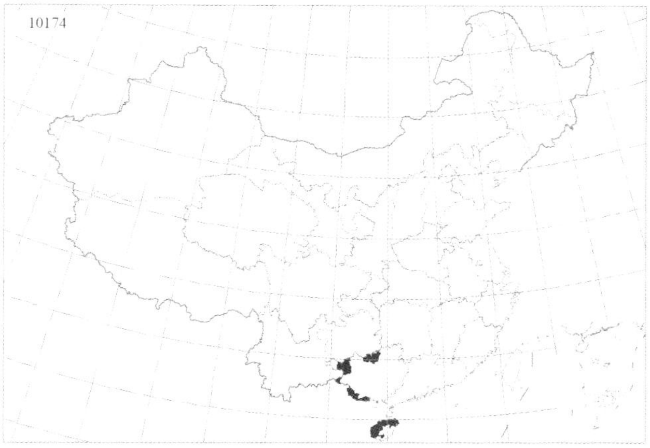

Rubiaceae 茜草科 *Lasianthus* 粗叶木属
Lasianthus rhinocerotis subsp. xishuangbannaensis H. Zhu et H. Wang 版纳粗叶木
Shrubs

20.0; 14.6; 23.2; 20.1; 180.8; 0.0; 963 ❙ 1523; 681; 67 ❙ 58.4; 942; 581.0

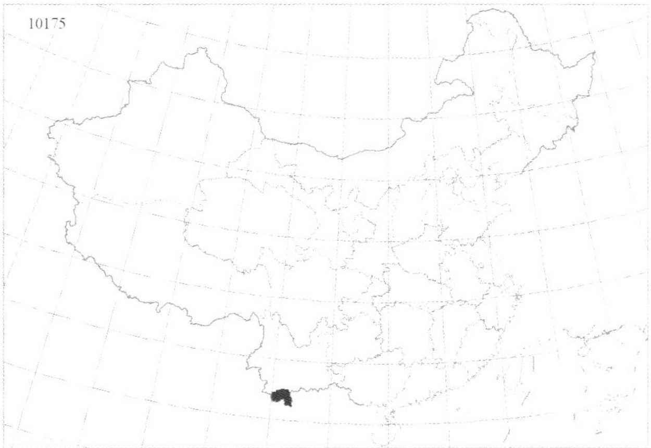

Rubiaceae 茜草科 *Lasianthus* 粗叶木属
Lasianthus rigidus Miq. 大叶粗叶木
Shrubs

19.2; 13.5; 22.6; 19.2; 170.8; 0.0; 921 ❙ 1521; 752; 60 ❙ 65.3; 910; 582.1

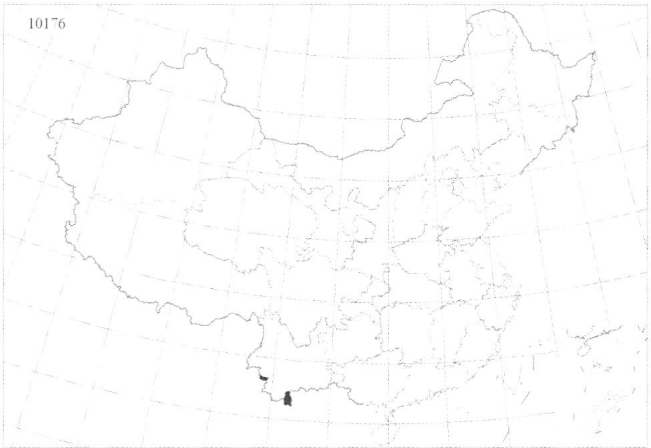

Rubiaceae 茜草科 *Lasianthus* 粗叶木属
Lasianthus schmidtii K. Schum. 泰北粗叶木
Shrubs

19.8；14.2；23.1；19.8；177.8；0.0；949 ▎1516；704；66 ▎59.9；929；542.7

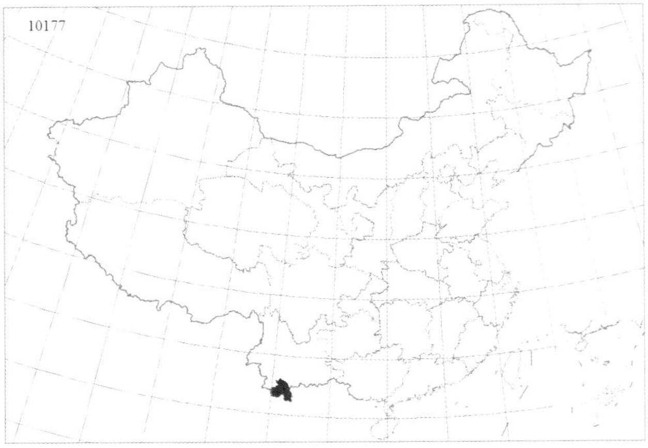

Rubiaceae 茜草科 *Lasianthus* 粗叶木属
Lasianthus shimizui (Liu et Chao) H. Zhu 清水氏鸡屎树
Shrubs

19.4；13.3；24.4；19.4；172.6；0.0；967 ▎2109；1039；182 ▎128.0；946；242.1

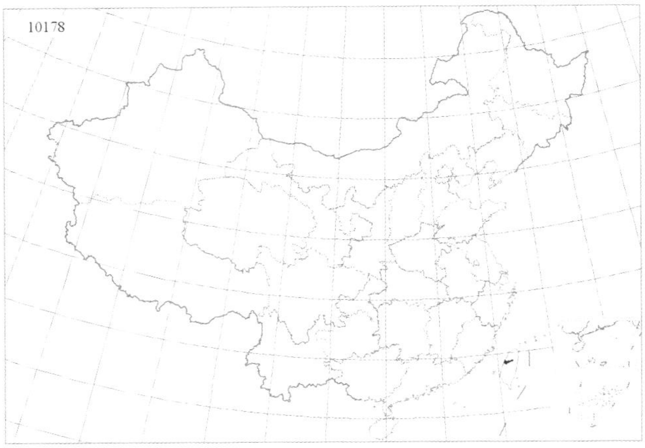

Rubiaceae 茜草科 *Lasianthus* 粗叶木属
Lasianthus sikkimensis Hook. f. 锡金粗叶木
Shrubs

18.0 (8.6~24.3)；10.5 (1.3~18.7)；23.6 (15.0~28.8)；18.0 (8.7~24.3)；156.5 (52.6~231.9)；0.1 (0.0~8.8)；903 (574~1338) ▎1266 (911~1900)；662 (392~1056)；69 (28~194) ▎40.6 (−11.9~145.8)；888 (574~1176)；324.3 (188.8~659.0)

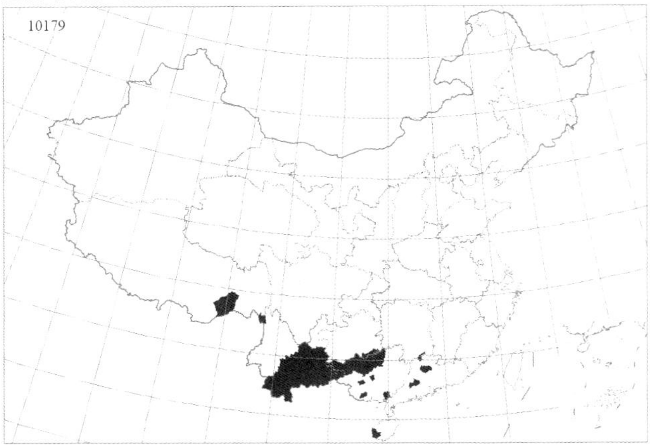

Rubiaceae 茜草科 *Lasianthus* 粗叶木属
Lasianthus trichophlebus Hemsl. 钟萼粗叶木
Shrubs

22.7 (18.5~24.8)；16.1 (13.0~20.1)；27.6 (21.7~28.8)；22.7 (18.5~24.8)；212.3 (162.1~237.6)；0.0 (0.0~0.0)；1210 (863~1385) ▎1638 (1176~2713)；678 (392~1388)；94 (41~188) ▎39.0 (−12.0~214.4)；1139 (863~1326)；371.0 (203.4~620.0)

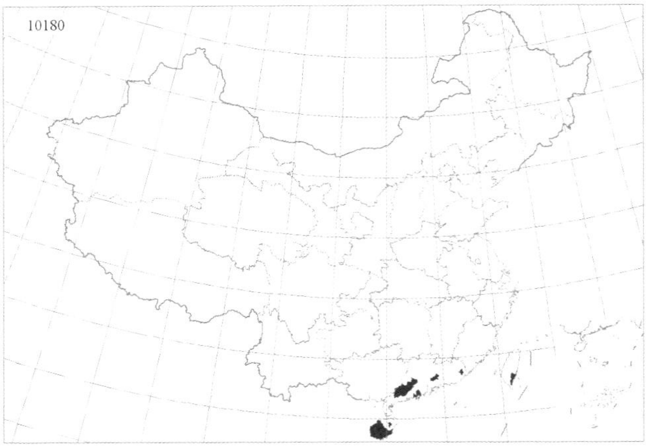

Rubiaceae 茜草科 *Lasianthus* 粗叶木属
Lasianthus trichophlebus var. latifolius (Miq.) H. Zhu 栖兰钟萼粗叶木
Shrubs

19.0；14.4；22.5；19.1；168.9；0.0；923 ▎2735；1394；199 ▎200.0；923；321.4

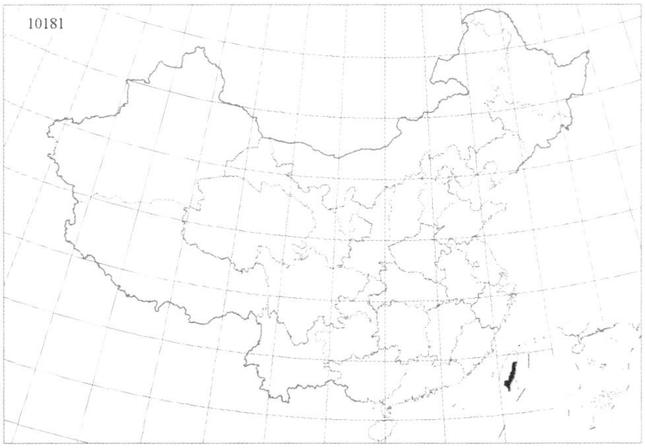

Rubiaceae 茜草科 *Lasianthus* 粗叶木属
Lasianthus verticillatus (Lour.) Merr. 斜脉粗叶木
Shrubs

22.0 (16.2~24.8)；16.1 (11.0~20.1)；26.4 (20.1~28.7)；22.0 (16.2~24.8)；204.3 (134.2~237.6)；0.0 (0.0~0.0)；1155 (773~1385) ▎2026 (1176~3495)；830 (392~1847)；174 (41~897) ▎87.1 (−12.0~246.5)；1091 (773~1360)；378.6 (154.2~620.0)

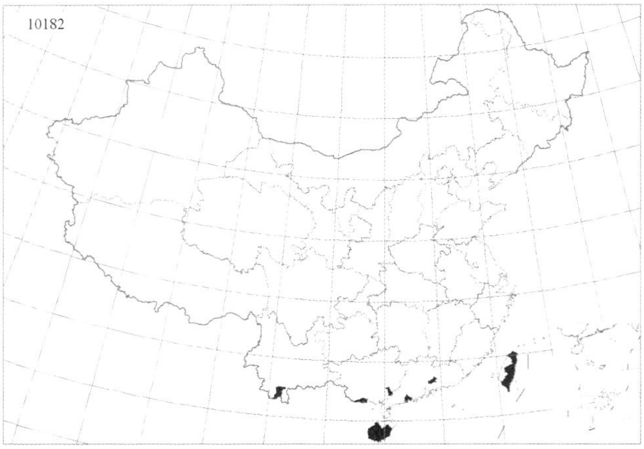

Rubiaceae 茜草科 *Lasianthus* 粗叶木属
Lasianthus warcii Fischer et Kaul 滇西粗叶木
Shrubs

16.9 (16.1~18.2); 10.7 (9.6~11.9); 21.0 (20.3~22.0); 17.0 (16.2~18.2); 144.0 (133.9~158.7); 0.0 (0.0~0.0); 816 (785~868) ❙ 1227 (1083~1366); 696 (600~783); 47 (45~50) ❙ 50.3 (33.9~65.9); 800 (779~847); 425.5 (372.4~516.7)

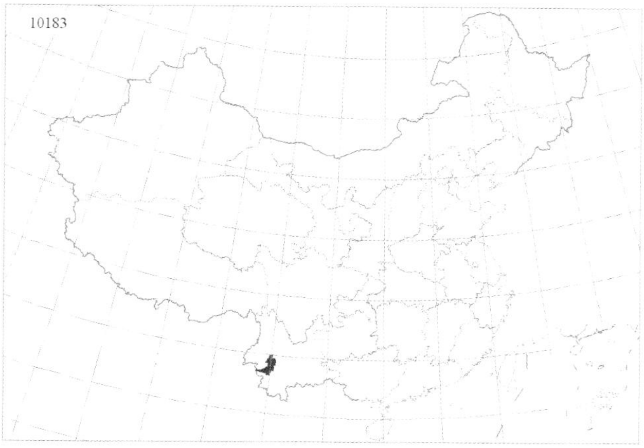

Rubiaceae 茜草科 *Leptodermis* 野丁香属
Leptodermis bechuanensis Lo 北川野丁香
Shrubs

11.1; 1.3; 20.1; 11.2; 82.3; 8.0; 658 ❙ 890; 477; 21 ❙ 35.4; 658; 259.3

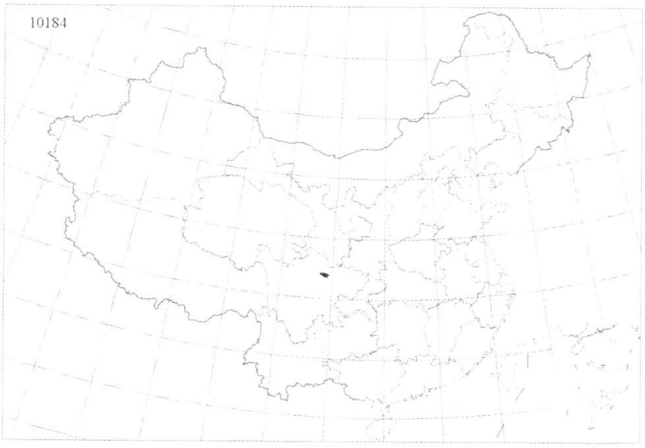

Rubiaceae 茜草科 *Leptodermis* 野丁香属
Leptodermis brevisepala Lo 短萼野丁香
Shrubs

14.6; 7.1; 20.3; 14.6; 114.8; 0.0; 741 ❙ 1014; 627; 22 ❙ 37.0; 692; 265.0

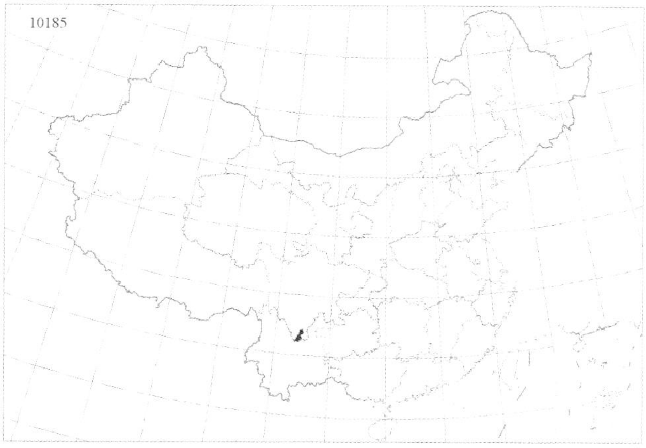

Rubiaceae 茜草科 *Leptodermis* 野丁香属
Leptodermis buxifolia H. S. Lo 黄杨叶野丁香
Shrubs

5.5 (2.0~13.5); -4.3 (-7.2~2.3); 14.4 (10.1~24.1); 6.6 (3.8~13.6); 41.3 (15.4~107.5); 34.1 (4.9~51.4); 507 (413~755) ❙ 835 (711~1023); 434 (346~535); 20 (15~28) ❙ 72.5 (11.5~128.6); 507 (413~755); 280.6 (200.8~357.7)

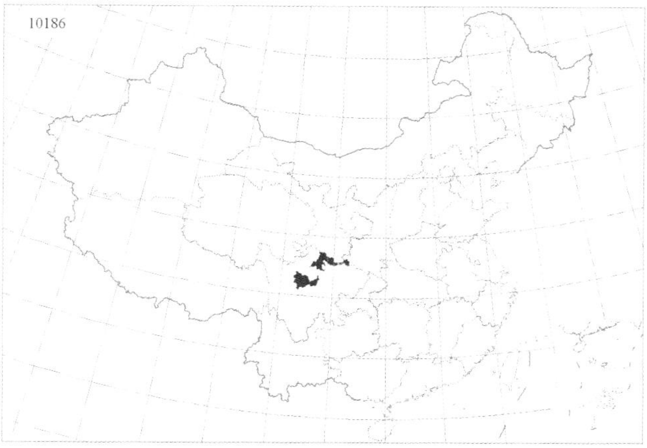

Rubiaceae 茜草科 *Leptodermis* 野丁香属
Leptodermis dielsiana H. Winkl. 丽江野丁香
Small shrubs

10.8; 3.9; 16.6; 10.9; 71.8; 1.3; 633 ❙ 968; 594; 24 ❙ 52.8; 633; 319.8

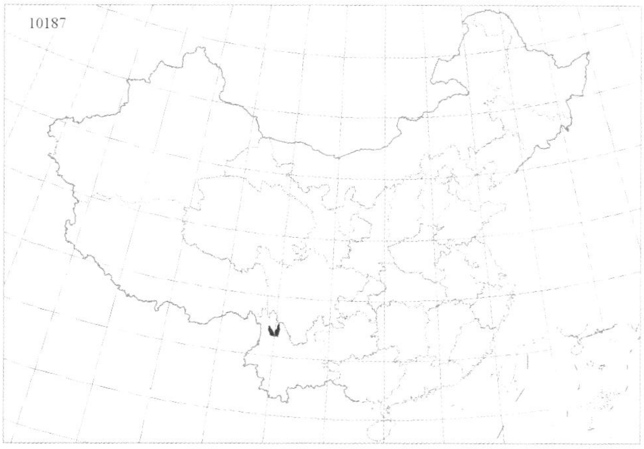

Rubiaceae 茜草科 *Leptodermis* 野丁香属
Leptodermis diffusa Batalin 文水野丁香
Shrubs

12.4 (2.6~17.0); 2.0 (-7.0~6.3); 22.0 (11.2~27.0); 12.7 (4.3~17.1); 99.6 (20.0~145.1); 10.4 (0.0~48.2); 728 (427~908) ❙ 928 (669~1135); 475 (317~554); 26 (14~42) ❙ 32.6 (2.4~121.7); 726 (427~908); 207.0 (130.0~290.3)

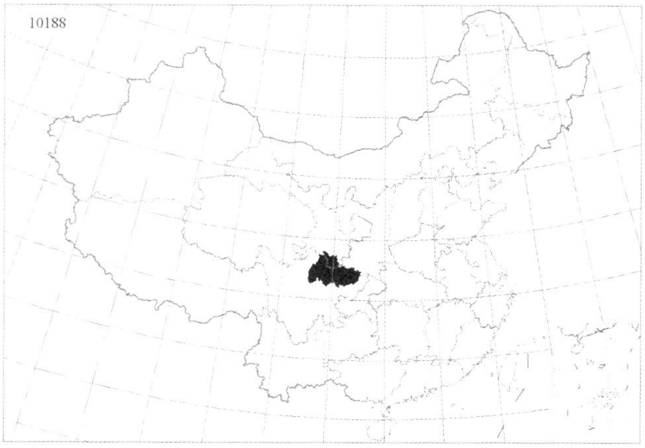

Rubiaceae 茜草科 *Leptodermis* 野丁香属
Leptodermis forrestii Diels 高山野丁香
Shrubs

13.2 (-1.8~19.3); 6.2 (-10.2~13.1); 18.7 (6.1~23.9); 13.5 (1.8~19.4); 107.0 (1.9~173.2); 8.0 (0.0~82.8); 711 (329~939) ❙ 962 (320~1736); 556 (193~1007); 37 (4~69) ❙ 36.5 (-15.3~145.8); 688 (320~903); 298.5 (72.2~447.4)

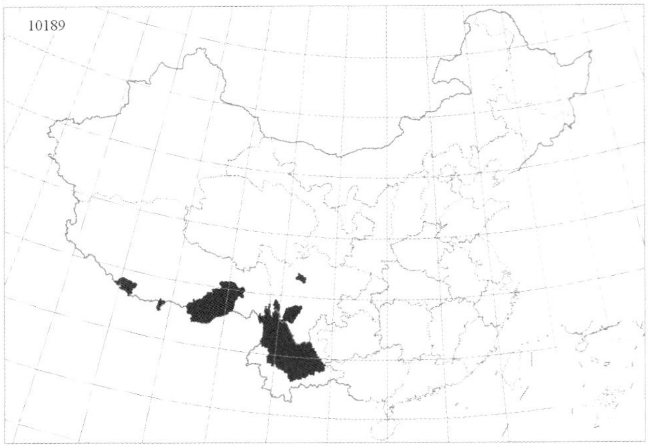

Rubiaceae 茜草科 *Leptodermis* 野丁香属
Leptodermis glomerata Hutch. 聚花野丁香
Shrubs

11.0 (3.0~18.5); 3.5 (-6.4~10.9); 17.1 (10.5~24.3); 11.3 (4.3~18.3); 85.1 (20.1~160.0); 12.3 (0.0~43.9); 652 (423~906) ❙ 917 (628~1178); 546 (410~646); 25 (11~51) ❙ 45.6 (5.5~81.6); 638 (423~836); 314.8 (236.3~383.4)

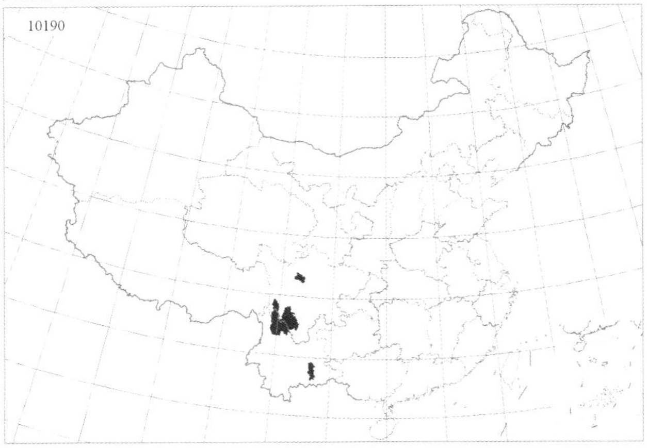

Rubiaceae 茜草科 *Leptodermis* 野丁香属
Leptodermis gracilis C. E. C. Fischer 柔枝野丁香
Shrubs

6.8; -1.3; 13.9; 7.6; 48.9; 26.5; 530 ❙ 1093; 638; 32 ❙ 96.9; 530; 283.7

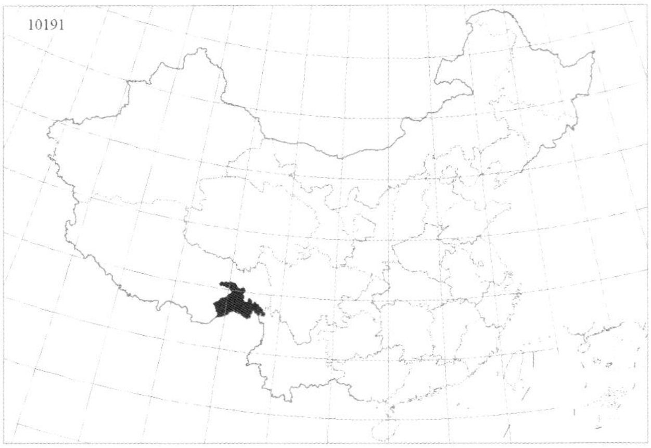

Rubiaceae 茜草科 *Leptodermis* 野丁香属
Leptodermis gracilis var. longiflora Lo 长花野丁香
Shrubs

2.7; -7.1; 11.1; 4.4; 20.5; 48.0; 429 ❙ 800; 427; 18 ❙ 86.5; 429; 341.6

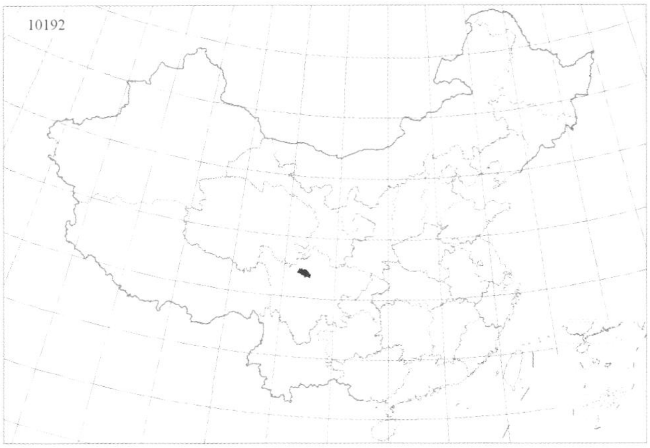

Rubiaceae 茜草科 *Leptodermis* 野丁香属
Leptodermis handeliana H. Winkl. ex Hand-Mazz. 川南野丁香
Shrubs

11.0; 3.7; 17.0; 11.1; 74.6; 1.8; 640 ❙ 990; 619; 15 ❙ 54.7; 638; 296.0

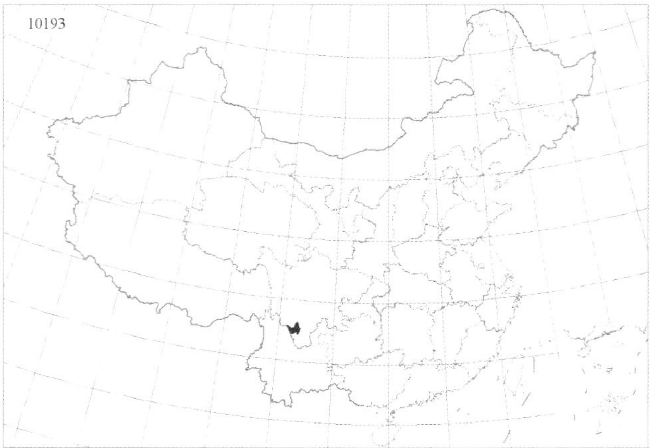

Rubiaceae 茜草科 *Leptodermis* 野丁香属
Leptodermis hirsutiflora Lo 拉萨野丁香
Shrubs

2.5 (0.6~4.4); -6.8 (-8.8~-4.3); 10.8 (9.1~12.1); 4.3 (3.2~5.3); 21.0 (12.4~28.5); 50.1 (34.6~65.0); 416 (371~455) ❙ 360 (286~538); 248 (208~337); 4 (1~9) ❙ -14.0 (-30.3~18.2); 344 (286~455); 197.3 (168.2~217.5)

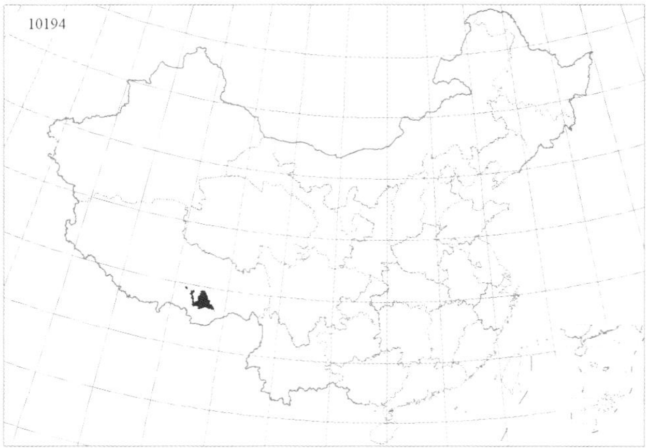

Rubiaceae 茜草科 *Leptodermis* 野丁香属
Leptodermis hirsutiflora var. ciliata Lo 光萼野丁香
Shrubs

0.6; -8.8; 9.1; 3.2; 12.4; 65.0; 371 ❙ 300; 208; 3 ❙ -19.2; 300; 209.5

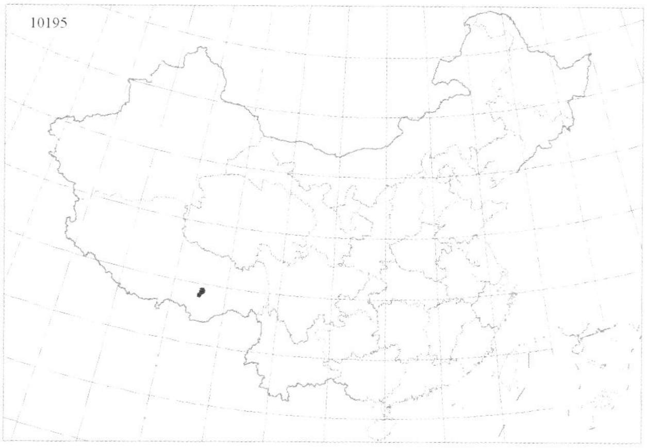

Rubiaceae 茜草科 *Leptodermis* 野丁香属
Leptodermis huashanica Lo 华山野丁香
Shrubs

12.5; -1.8; 25.6; 12.7; 106.5; 15.4; 760 ❙ 634; 290; 25 ❙ -16.6; 634; 212.0

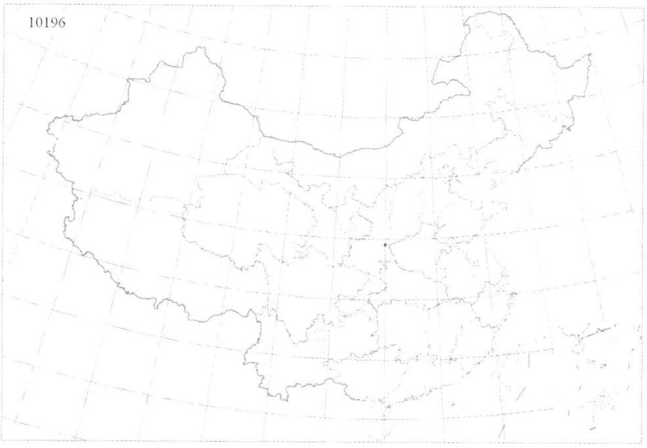

Rubiaceae 茜草科 *Leptodermis* 野丁香属
Leptodermis kumaonensis R. Parker 吉隆野丁香
Shrubs

-1.1; -9.1; 6.2; 2.0; 2.5; 75.6; 334 ❙ 321; 193; 33 ❙ -4.0; 321; 79.3

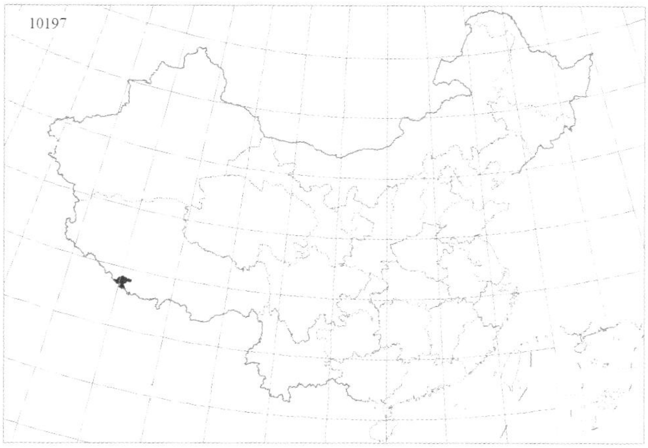

Rubiaceae 茜草科 *Leptodermis* 野丁香属
Leptodermis lanata Lo 绵毛野丁香
Shrubs

10.4; 3.4; 16.3; 10.5; 68.0; 2.6; 623 ❙ 946; 603; 15 ❙ 51.8; 615; 300.3

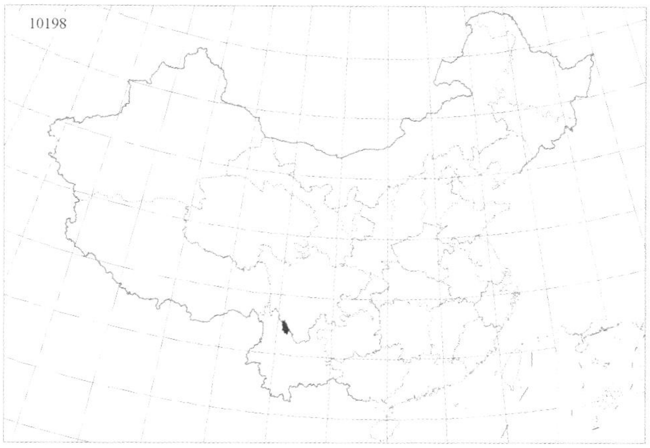

Rubiaceae 茜草科 *Leptodermis* 野丁香属
Leptodermis limprichtii H. Winkl. 川南野丁香
Small shrubs

11.1; 1.8; 19.4; 11.1; 79.8; 6.9; 648 ❙ 1076; 584; 30 ❙ 66.0; 648; 246.3

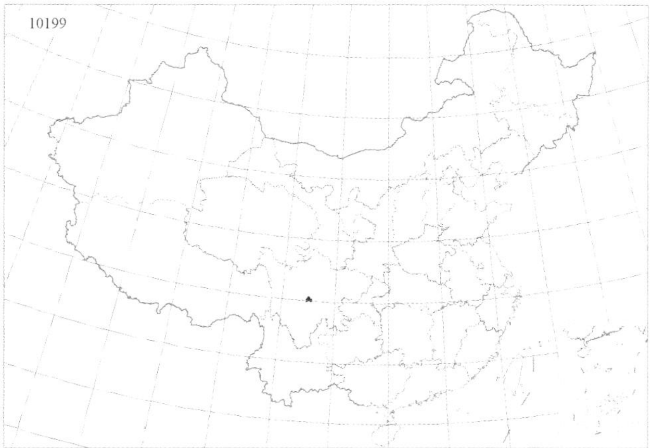

Rubiaceae 茜草科 *Leptodermis* 野丁香属
Leptodermis oblonga Bange 薄皮木
Shrubs

11.3 (1.8~19.3); -1.8 (-15.9~13.1); 22.8 (9.9~28.1); 12.1 (3.8~19.4); 98.5 (15.4~173.2); 22.4 (0.0~80.4); 712 (404~939) ❙ 698 (205~1454); 398 (119~816); 25 (5~83) ❙ -2.2 (-68.3~128.6); 610 (205~924); 209.9 (46.6~516.7)

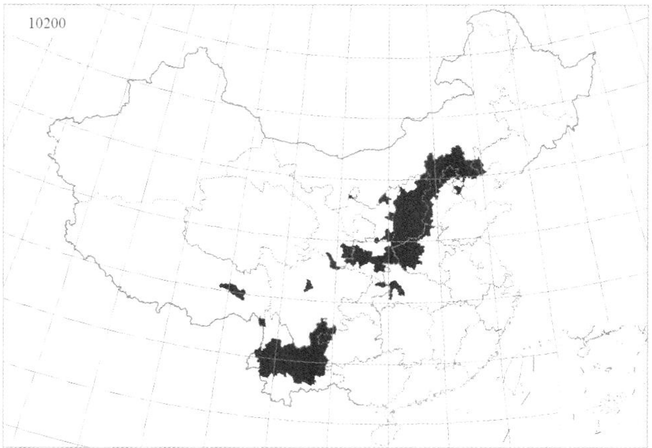

Rubiaceae 茜草科 *Leptodermis* 野丁香属
Leptodermis ordosica H. C. Fu et E. W. Ma 内蒙野丁香
Shrubs

7.9 (7.1~8.7); -9.1 (-10.1~-7.5); 22.8 (21.9~23.8); 9.7 (9.3~10.3); 79.5 (75.2~84.8); 44.5 (39.0~49.3); 642 (622~667) ❙ 205 (132~282); 122 (81~166); 4 (2~6) ❙ -68.0 (-80.1~-55.2); 205 (132~282); 57.5 (13.6~116.9)

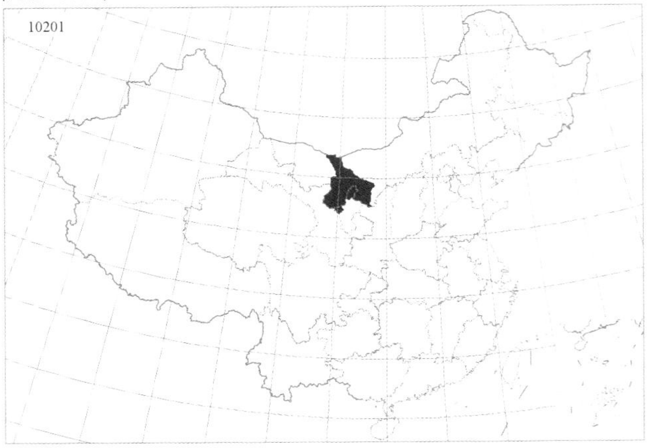

Rubiaceae 茜草科 *Leptodermis* 野丁香属
Leptodermis ovata H. Winkl. 卵叶野丁香
Shrubs

17.7 (10.0~22.2); 9.2 (1.8~14.2); 24.4 (16.9~29.0); 17.7 (10.0~22.3); 153.5 (66.5~207.5); 0.6 (0.0~7.0); 904 (613~1190) ❙ 1384 (1023~1740); 692 (507~860); 89 (22~187) ❙ 54.4 (15.4~85.5); 904 (613~1190); 315.2 (160.7~551.6)

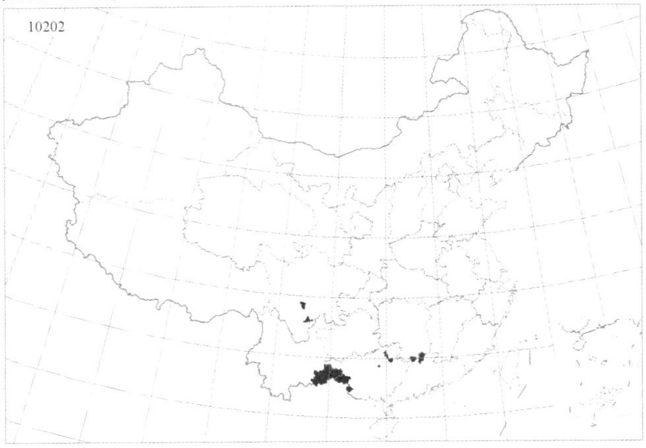

Rubiaceae 茜草科 *Leptodermis* 野丁香属
Leptodermis parkeri Dunn 大叶野丁香
Shrubs

-1.1; -9.1; 6.2; 2.0; 2.5; 75.6; 334 ❙ 321; 193; 33 ❙ -4.0; 321; 79.3

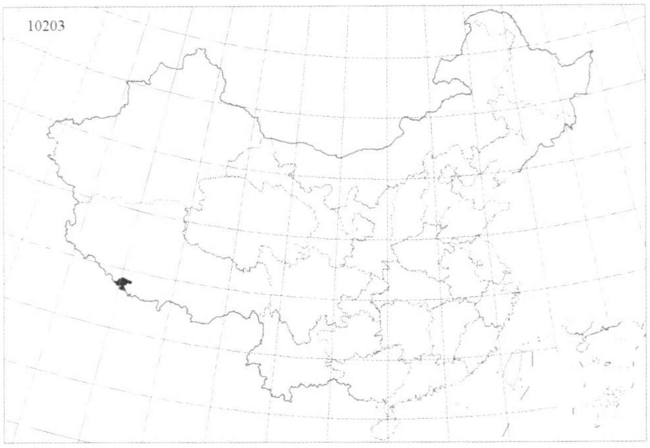

Rubiaceae 茜草科 *Leptodermis* 野丁香属
Leptodermis parvifolia Hutch. 瓦山野丁香
Shrubs

9.1 (1.6~14.2); 0.5 (-8.0~6.2); 16.7 (9.8~24.7); 9.5 (3.7~14.3); 64.3 (15.2~113.3); 14.7 (0.0~56.2); 595 (408~776) ❙ 1029 (687~1265); 576 (419~688); 28 (12~65) ❙ 75.4 (34.8~125.9); 594 (408~776); 268.7 (214.7~318.7)

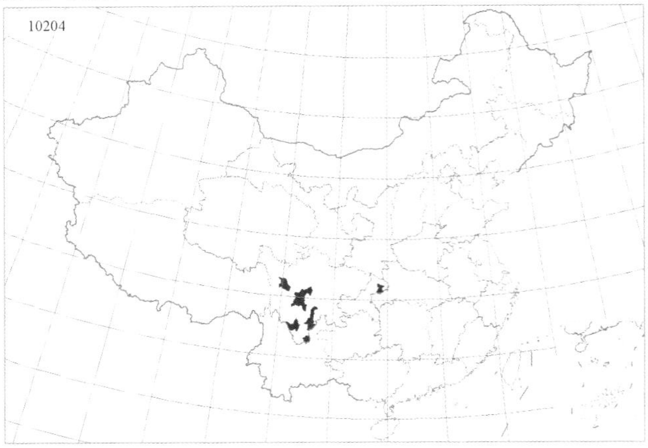

Rubiaceae 茜草科 *Leptodermis* 野丁香属
Leptodermis pilosa Franch. ex Diels 川滇野丁香
Shrubs

10.6 (-1.8~21.0); 2.4 (-11.4~14.2); 17.4 (6.1~26.6); 11.3 (1.8~21.0); 86.2 (1.9~191.8); 18.3 (0.0~82.8); 646 (329~1053) ❙ 886 (255~1736); 507 (191~1007); 30 (1~68) ❙ 38.6 (-31.1~145.8); 626 (255~1053); 267.3 (60.4~551.6)

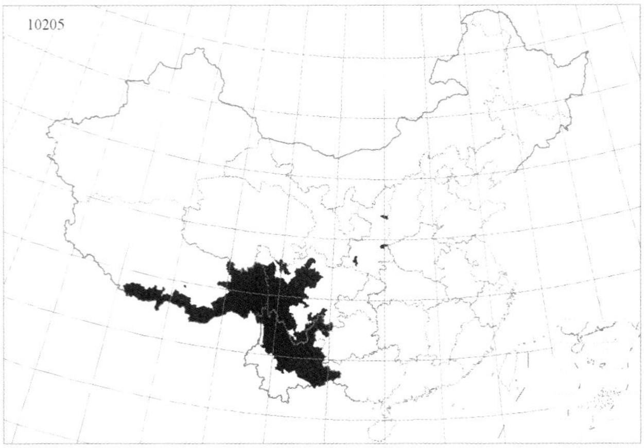

Rubiaceae 茜草科 *Leptodermis* 野丁香属
Leptodermis pilosa var. acanthoclada H. S. Lo ex X. Y. Wen et Q. Lin 刺枝野丁香
Small shrubs

2.2 (-0.7~5.3); -6.7 (-10.8~-2.4); 10.2 (8.1~12.2); 4.0 (2.7~5.7); 18.3 (8.0~30.5); 51.4 (26.2~77.9); 409 (345~479) ❙ 561 (250~961); 344 (198~558); 12 (1~31) ❙ 34.5 (-27.5~128.6); 384 (250~479); 206.5 (111.0~280.0)

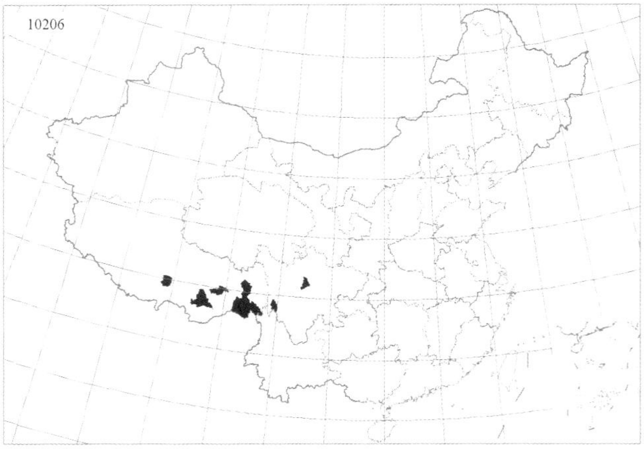

Rubiaceae 茜草科 *Leptodermis* 野丁香属

Leptodermis pilosa var. glabrescens H. Winkl. 光叶野丁香

Shrubs

15.4 (10.4~18.9); 8.8 (3.4~12.5); 20.4 (16.3~23.9); 15.5 (10.5~19.0); 125.9 (68.0~167.8); 0.1 (0.0~2.6); 771 (623~929) ❙ 983 (797~1139); 570 (411~651); 38 (5~54) ❙ 28.6 (-14.2~64.0); 736 (615~851); 295.6 (218.0~382.6)

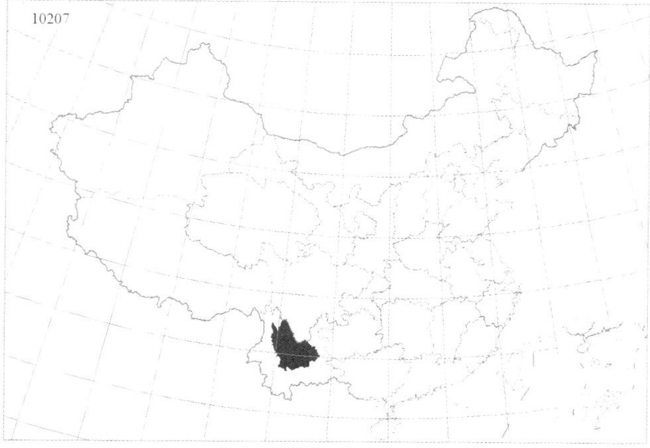

Rubiaceae 茜草科 *Leptodermis* 野丁香属

Leptodermis pilosa var. spicatiformis H. S. Lo ex X. Y. Wen et Q. Lin 穗花野丁香

Shrubs

10.6 (9.2~12.6); -1.5 (-2.8~1.0); 22.1 (20.7~23.5); 10.8 (9.6~12.6); 83.9 (72.1~99.9); 16.3 (8.6~21.0); 662 (615~724) ❙ 711 (632~773); 343 (306~382); 18 (15~20) ❙ 7.6 (-1.2~25.6); 660 (615~724); 290.8 (241.9~323.3)

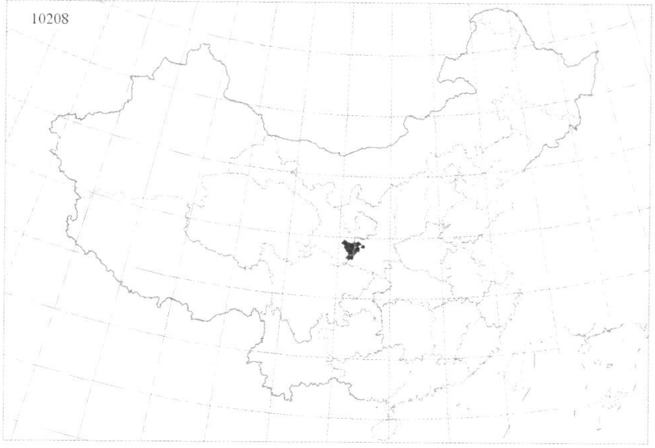

Rubiaceae 茜草科 *Leptodermis* 野丁香属

Leptodermis potanina Batalin 野丁香

Shrubs

13.0 (-1.8~19.3); 5.2 (-10.2~13.1); 19.4 (6.1~26.6); 13.2 (1.8~19.4); 103.7 (1.9~173.2); 7.2 (0.0~82.8); 708 (329~939) ❙ 1018 (320~1736); 568 (217~1007); 39 (7~113) ❙ 46.5 (-14.2~145.8); 693 (320~912); 284.0 (72.2~491.5)

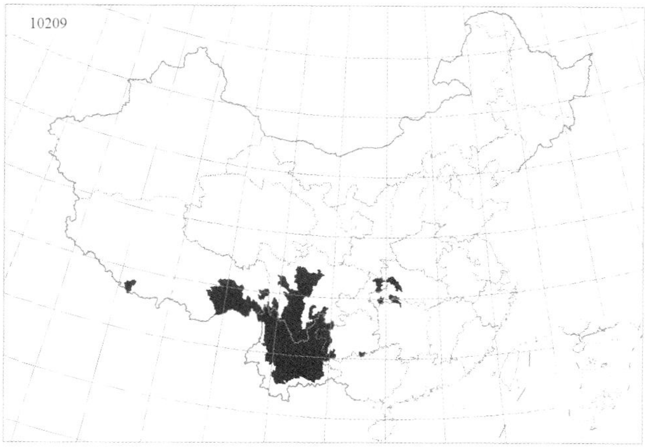

Rubiaceae 茜草科 *Leptodermis* 野丁香属

Leptodermis potanina var. glauca (Diels) H. Winkl. 粉绿野丁香

Shrubs

14.0 (1.8~19.3); 6.7 (-7.1~13.1); 19.7 (9.9~23.9); 14.1 (3.8~19.4); 112.3 (15.7~173.2); 3.8 (0.0~53.3); 732 (404~939) ❙ 1018 (527~1561); 580 (317~897); 39 (9~70) ❙ 40.8 (-14.2~121.7); 714 (404~903); 300.4 (169.5~539.7)

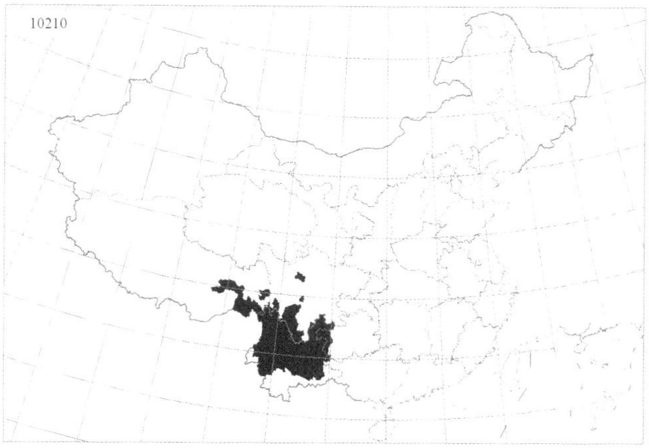

Rubiaceae 茜草科 *Leptodermis* 野丁香属

Leptodermis potanina var. tomentosa H. Winkl. 绒毛野丁香

Shrubs

15.0 (2.2~19.2); 8.2 (-6.6~13.3); 20.1 (9.9~23.9); 15.1 (3.8~19.4); 122.5 (15.5~172.6); 1.5 (0.0~49.1); 761 (413~932) ❙ 1061 (741~1561); 607 (411~897); 42 (12~70) ❙ 41.1 (-14.2~125.9); 741 (413~923); 327.8 (207.5~539.7)

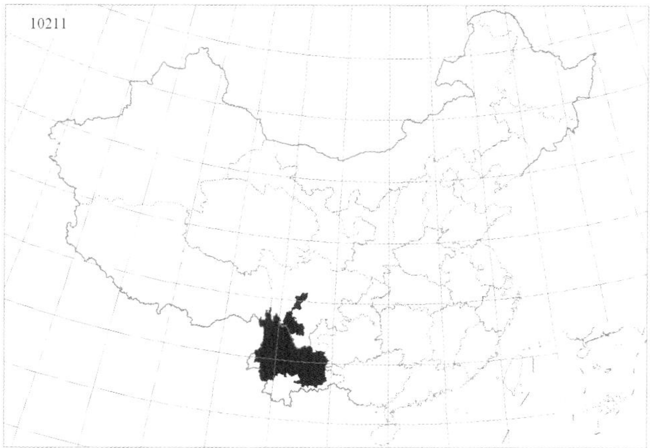

Rubiaceae 茜草科 *Leptodermis* 野丁香属

Leptodermis pumila Lo 矮小野丁香

Dwarf shrubs

10.8; 3.9; 16.6; 10.9; 71.8; 1.3; 633 ❙ 968; 594; 24 ❙ 52.8; 633; 319.8

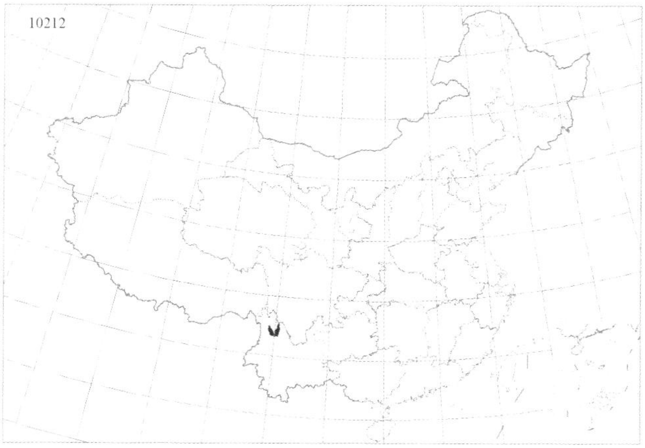

Rubiaceae 茜草科 *Leptodermis* 野丁香属
Leptodermis purdomii Hutch. 甘肃野丁香
Shrubs

5.6 (1.9~13.3); -3.9 (-7.7~5.4); 14.1 (9.9~22.6); 6.6 (3.8~13.4); 40.7 (15.5~100.4); 32.9 (0.0~53.6); 502 (403~706) ❙ 736 (341~1736); 411 (226~1007); 17 (4~51) ❙ 47.3 (-15.3~145.8); 499 (341~706); 278.6 (201.9~447.4)

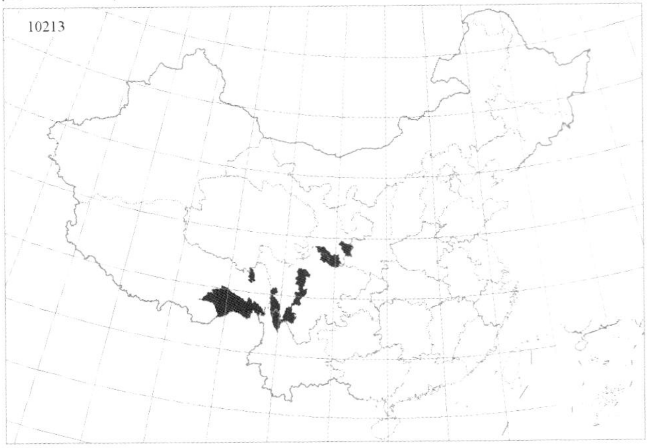

Rubiaceae 茜草科 *Leptodermis* 野丁香属
Leptodermis rehderiana H. Winkl. 白毛野丁香
Shrubs

6.1; -1.5; 12.7; 6.3; 35.2; 21.8; 496 ❙ 801; 496; 21 ❙ 61.4; 496; 282.9

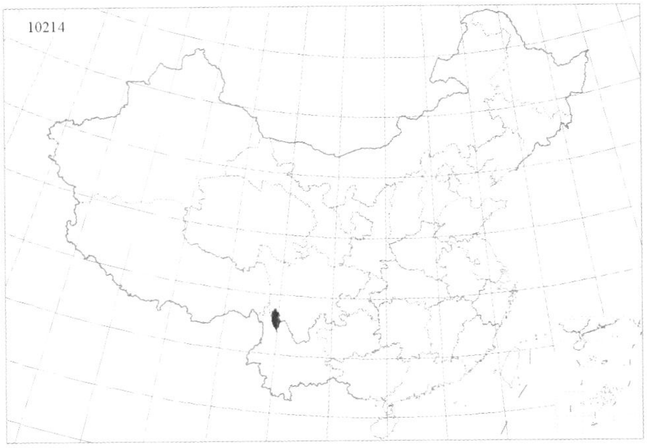

Rubiaceae 茜草科 *Leptodermis* 野丁香属
Leptodermis scabrida Hook. f. 糙叶野丁香
Shrubs

4.7; -3.5; 11.9; 6.2; 37.1; 40.6; 480 ❙ 900; 533; 29 ❙ 71.9; 477; 230.8

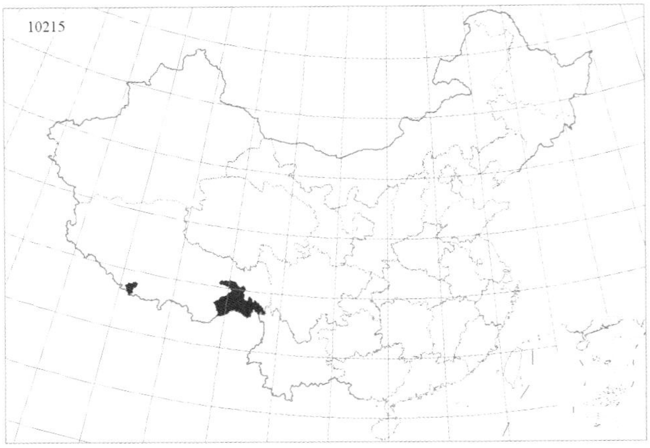

Rubiaceae 茜草科 *Leptodermis* 野丁香属
Leptodermis schneideri H. Winkl. 纤枝野丁香
Shrubs

6.6 (1.8~13.6); -2.1 (-7.2~7.2); 14.2 (9.9~19.8); 7.3 (3.8~13.7); 45.5 (15.4~104.1); 26.3 (0.0~53.3); 524 (404~706) ❙ 891 (583~1220); 501 (348~688); 25 (13~46) ❙ 72.5 (12.1~128.6); 524 (404~706); 284.9 (169.5~372.3)

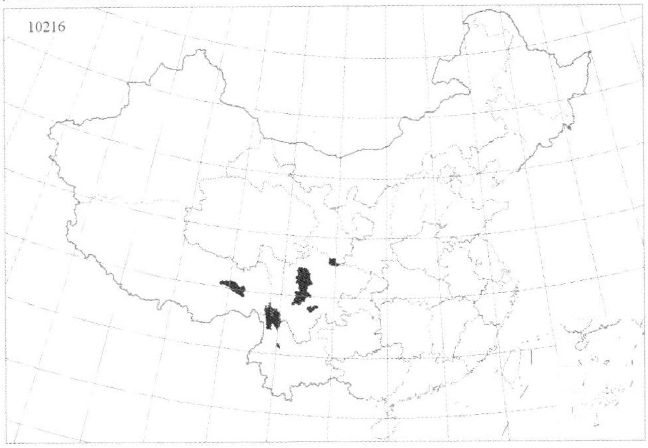

Rubiaceae 茜草科 *Leptodermis* 野丁香属
Leptodermis scissa H. Winkl. 撕裂野丁香
Shrubs

11.1 (5.4~16.1); 3.7 (-2.1~8.6); 17.2 (12.3~21.9); 11.2 (5.8~16.1); 80.4 (31.5~133.0); 6.8 (0.0~25.6); 643 (481~797) ❙ 969 (741~1134); 587 (426~651); 24 (12~47) ❙ 52.3 (27.3~78.3); 626 (481~719); 316.0 (273.5~339.5)

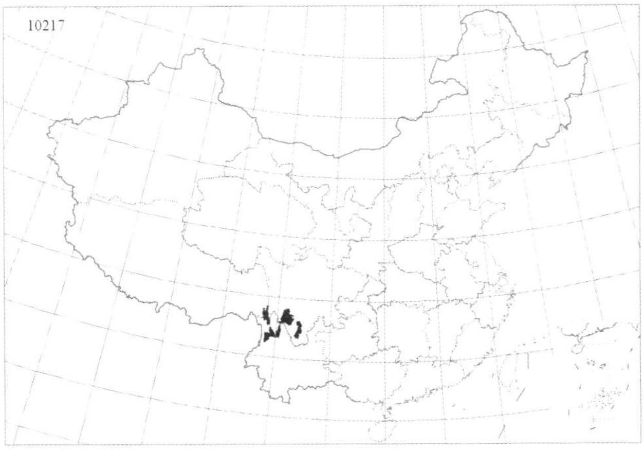

Rubiaceae 茜草科 *Leptodermis* 野丁香属
Leptodermis tomentella (Franch.) H. Winkl. ex Lo 蒙自野丁香
Shrubs

14.1 (2.2~19.3); 6.8 (-6.6~13.1); 19.8 (9.5~26.6); 14.2 (3.8~19.4); 113.2 (15.5~173.2); 3.7 (0.0~49.1); 736 (411~939) ❙ 981 (579~1178); 563 (365~651); 37 (10~59) ❙ 36.2 (-14.2~125.9); 714 (411~912); 285.5 (142.9~406.7)

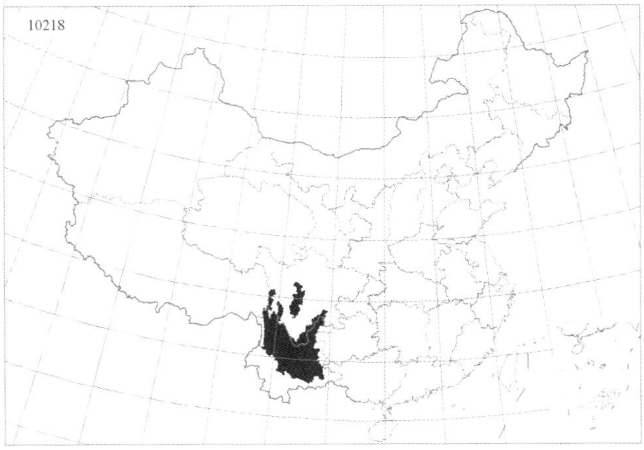

Rubiaceae 茜草科 *Leptodermis* 野丁香属
Leptodermis tubicalyx Lo 管萼野丁香
Shrubs

5.9; -2.2; 13.0; 6.9; 42.5; 30.8; 506 ❙ 927; 541; 24 ❙ 71.2; 506; 246.8

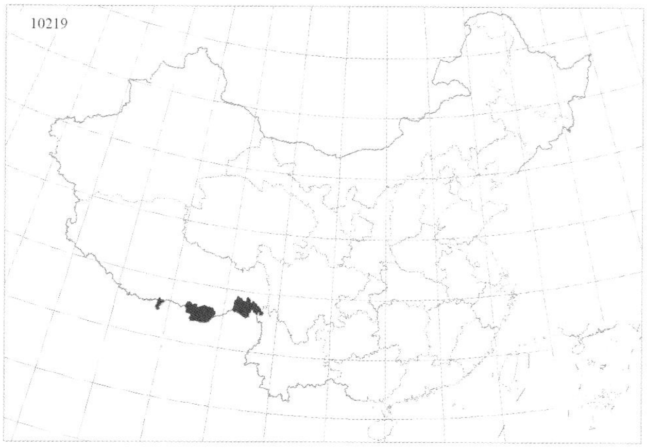

Rubiaceae 茜草科 *Leptodermis* 野丁香属
Leptodermis umbellata Batalin 伞花野丁香
Shrubs

11.9; 1.4; 21.5; 11.9; 90.5; 7.3; 689 ❙ 758; 403; 16 ❙ 10.2; 689; 270.2

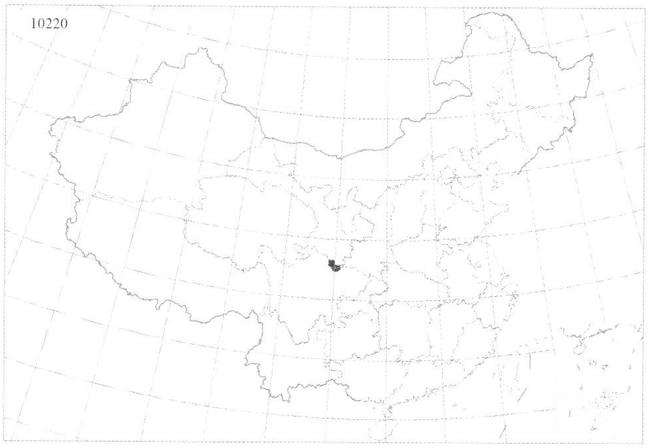

Rubiaceae 茜草科 *Leptodermis* 野丁香属
Leptodermis velutiniflora Lo 毛花野丁香
Shrubs

11.4; 3.2; 18.4; -1.5; 82.2; 4.7; 654 ❙ 1057; 611; 27 ❙ 63.7; 642; 283.7

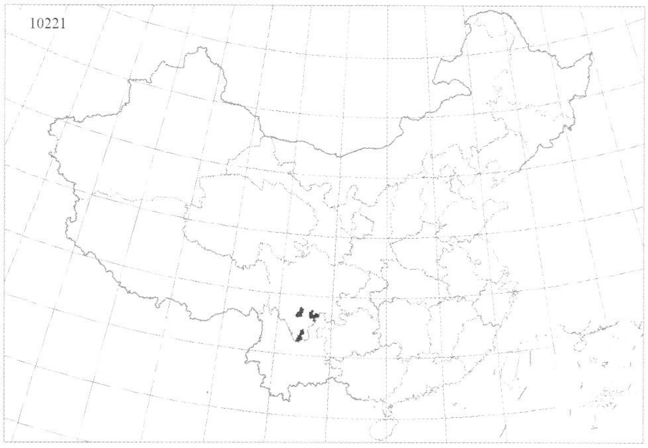

Rubiaceae 茜草科 *Leptodermis* 野丁香属
Leptodermis velutiniflora var. tenera Lo 薄叶野丁香
Shrubs

7.3 (1.8~15.3); -0.6 (-7.1~8.7); 14.1 (9.9~20.5); 8.0 (3.8~15.4); 51.5 (15.7~124.3); 23.4 (0.0~53.3); 539 (404~761) ❙ 774 (473~1007); 466 (287~619); 19 (7~40) ❙ 41.5 (5.0~67.9); 533 (404~717); 278.4 (169.5~351.7)

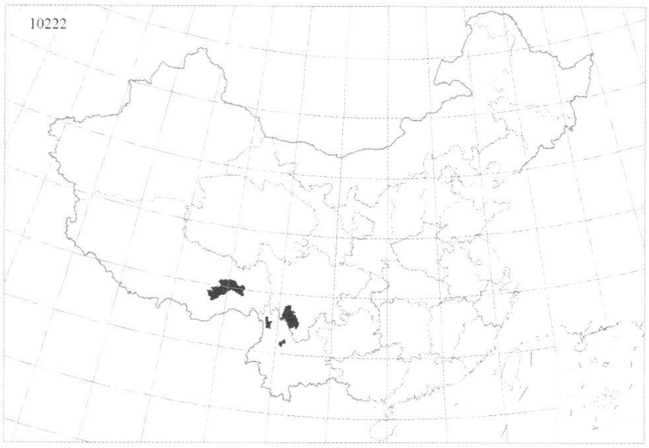

Rubiaceae 茜草科 *Leptodermis* 野丁香属
Leptodermis vestita Hemsl. 广东野丁香
Small shrubs

17.7; 7.7; 26.6; 17.8; 153.6; 0.0; 916 ❙ 1564; 556; 182 ❙ 70.7; 916; 232.5

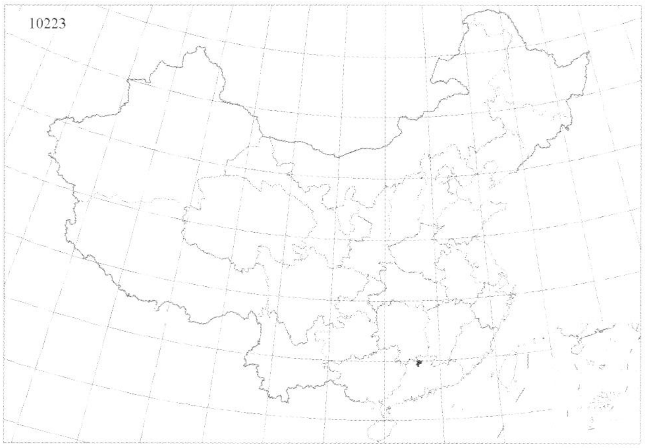

Rubiaceae 茜草科 *Leptodermis* 野丁香属
Leptodermis wilsoni Hort. et Diels 大果野丁香
Shrubs

9.1 (2.0~14.3); 1.6 (-7.2~7.6); 15.6 (10.1~19.7); 9.5 (3.8~14.3); 63.1 (15.4~111.9); 12.7 (0.0~51.4); 589 (413~728) ❙ 982 (741~1134); 562 (426~651); 32 (21~47) ❙ 69.7 (36.1~128.6); 587 (413~707); 303.8 (200.8~372.3)

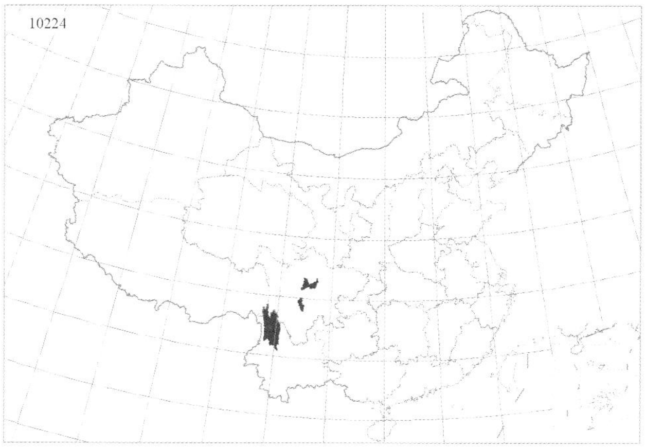

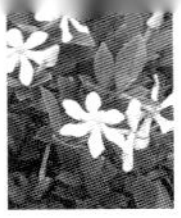

Rubiaceae 茜草科 *Leptodermis* 野丁香属

Leptodermis xizanensis Lo 西藏野丁香

Shrubs

-1.1；-9.1；6.2；2.0；2.5；75.6；334 ▎321；193；33 ▎-4.0；321；79.3

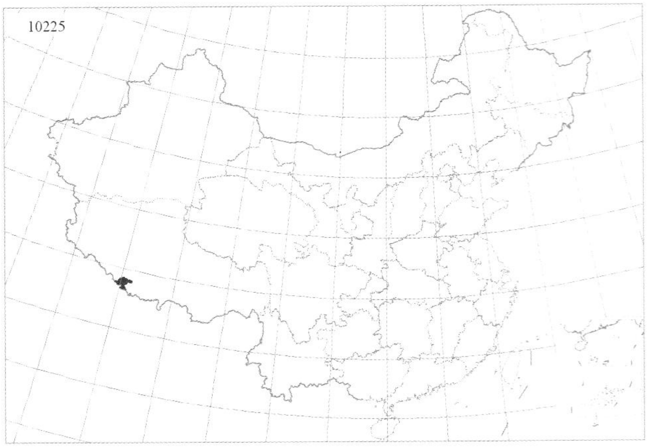

Rubiaceae 茜草科 *Leptodermis* 野丁香属

Leptodermis yui Lo 德浚野丁香

Shrubs

6.6；-1.3；13.3；6.8；39.6；19.7；514 ▎862；548；12 ▎67.7；514；311.5

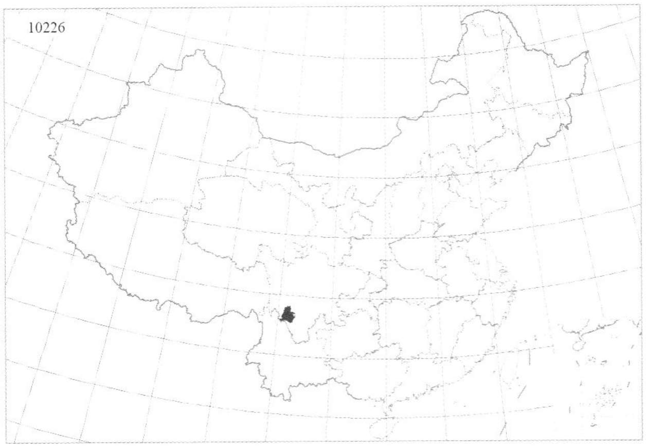

Rubiaceae 茜草科 *Luculia* 滇丁香属

Luculia pinceana Hook. 滇丁香

Shrubs or trees

16.4 (5.3~21.7)；9.5 (-2.4~15.1)；21.5 (12.2~28.4)；16.5 (5.7~21.8)；138.4 (30.5~201.3)；0.7 (0.0~26.2)；821 (479~1162) ▎1187 (741~1751)；661 (411~1007)；51 (15~130) ▎45.3 (-14.2~145.8)；805 (479~1162)；352.5 (218.0~659.0)

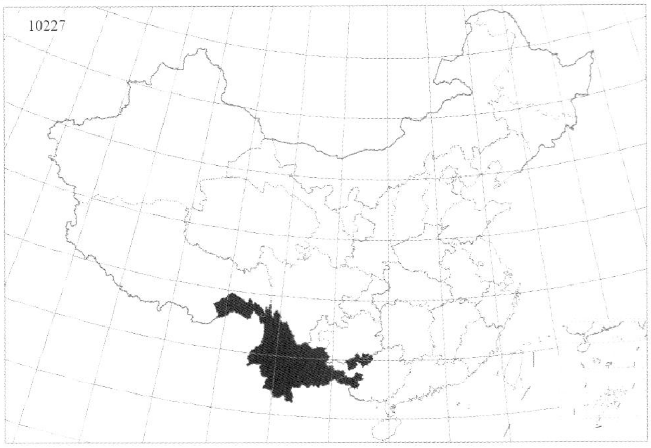

Rubiaceae 茜草科 *Metadina* 黄棉木属

Metadina trichotoma (Zoll. et Moritzi) Bakh. f. 黄棉木

Trees

18.0 (14.3~22.8)；8.1 (3.3~15.1)；26.6 (21.4~29.8)；18.0 (14.3~22.8)；156.7 (112.6~213.6)；0.2 (0.0~2.3)；941 (756~1230) ▎1479 (1007~1900)；598 (392~1056)；144 (43~250) ▎58.1 (12.9~128.9)；937 (756~1230)；246.5 (145.3~659.0)

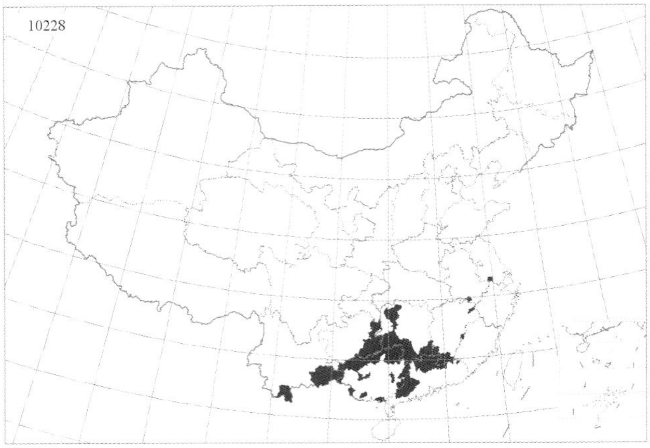

Rubiaceae 茜草科 *Meyna* 琼梅属

Meyna hainanensis Merr. 琼梅

Trees

23.5；17.6；27.7；23.5；222.2；0.0；1276 ▎1644；721；71 ▎31.6；1133；433.8

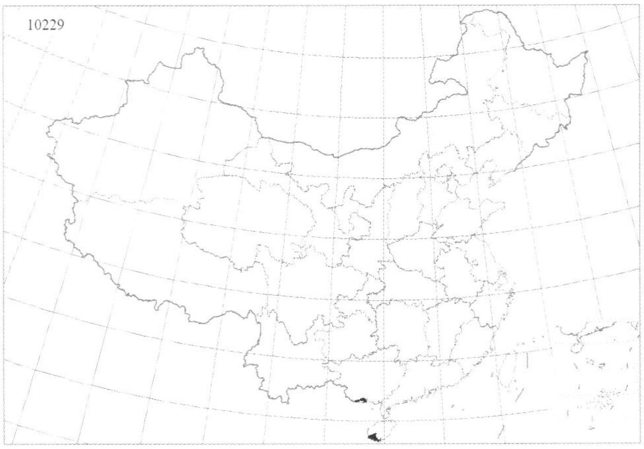

Rubiaceae 茜草科 *Mitragyna* 帽柱木属

Mitragyna rotundifolia (Roxb.) Kuntze 帽蕊木

Trees

18.8 (16.8~21.0)；12.8 (10.4~15.1)；22.4 (20.8~25.6)；18.8 (16.9~21.0)；165.8 (142.6~191.8)；0.0 (0.0~0.0)；902 (811~1053) ▎1447 (1275~1607)；748 (655~818)；54 (41~77) ▎60.9 (44.5~74.2)；883 (802~1053)；498.4 (418.5~659.0)

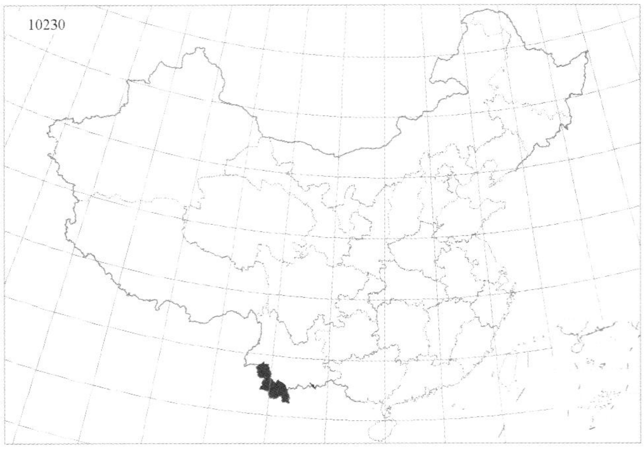

Rubiaceae 茜草科 *Morinda* 巴戟天属

Morinda angustifolia Roxb. 黄木巴戟

Shrubs or small trees

18.5 (16.3~21.0); 12.0 (9.1~15.1); 22.9 (21.4~25.6); 18.6 (16.4~21.0); 162.7 (136.4~191.8); 0.0 (0.0~0.0); 895 (797~1053) ▎1401 (1074~1751); 747 (582~996); 56 (41~77) ▎56.5 (32.4~88.8); 886 (797~1053); 450.2 (277.0~659.0)

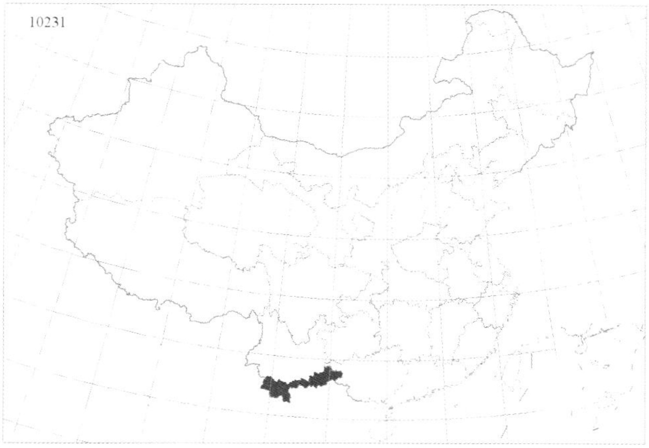

Rubiaceae 茜草科 *Morinda* 巴戟天属

Morinda brevipes S. Y. Hu 短柄鸡眼藤

Vines

23.3 (21.8~24.8); 17.0 (13.5~20.1); 27.9 (27.2~28.8); 23.3 (21.9~24.8); 219.5 (202.6~237.6); 0.0 (0.0~0.0); 1257 (1150~1385) ▎1489 (1179~1876); 576 (392~891); 73 (41~132) ▎19.8 (-11.9~63.2); 1130 (1056~1170); 413.2 (203.4~620.0)

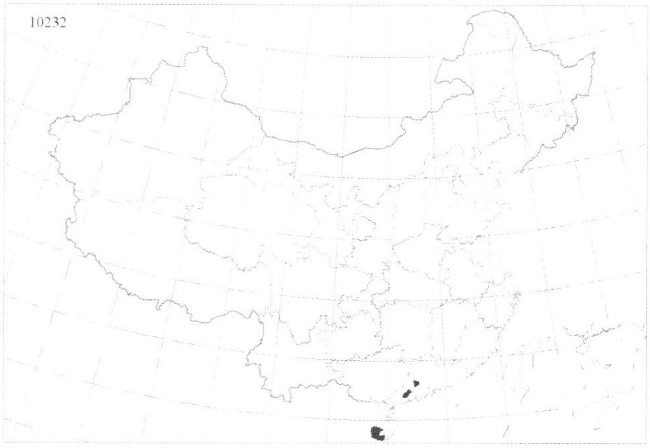

Rubiaceae 茜草科 *Morinda* 巴戟天属

Morinda brevipes var. stenophylla Chun et How ex Ko 狭叶鸡眼藤

Vines

23.9; 18.7; 27.5; 23.9; 226.8; 0.0; 1294 ▎1380; 455; 48 ▎6.7; 1112; 455.2

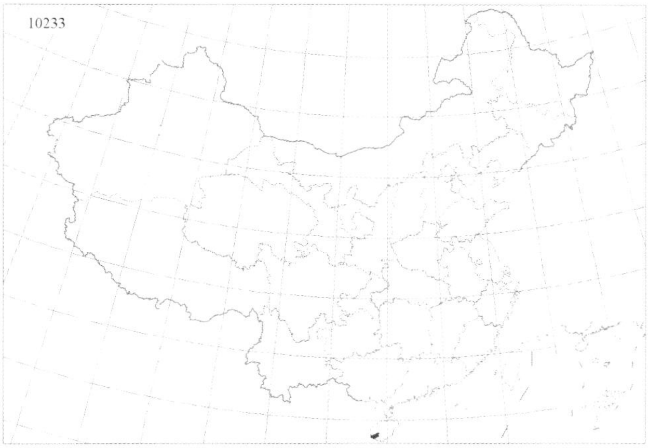

Rubiaceae 茜草科 *Morinda* 巴戟天属

Morinda callicarpaefolia Y. Z. Ruan 紫珠叶巴戟

Vines

17.1; 8.2; 24.5; 17.1; 145.7; 0.0; 864 ▎1265; 637; 62 ▎47.6; 864; 273.7

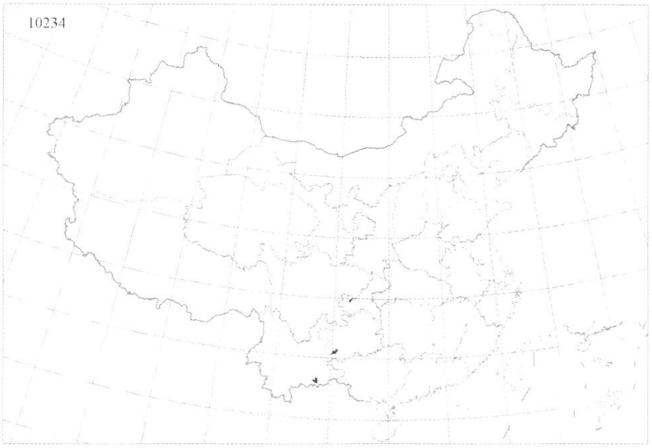

Rubiaceae 茜草科 *Morinda* 巴戟天属

Morinda cinnamomifoliata Y. Z. Ruan 樟叶巴戟

Vines

21.6; 12.6; 28.6; 21.6; 199.3; 0.0; 1150 ▎1459; 650; 164 ▎26.8; 1148; 199.4

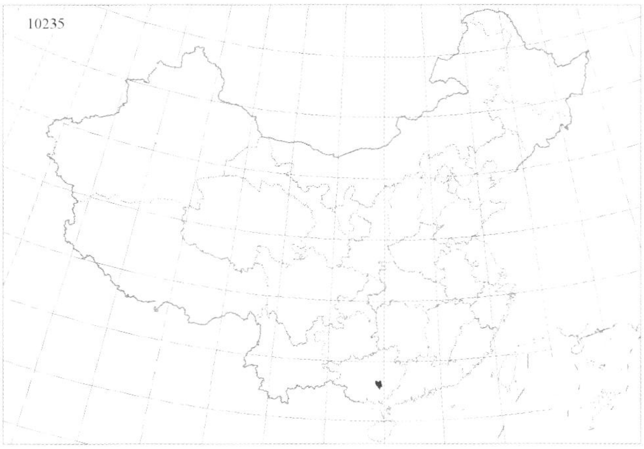

Rubiaceae 茜草科 *Morinda* 巴戟天属

Morinda citrifolia L. 海滨木巴戟

Shrubs or small trees, evergreen

23.5 (18.5~24.8); 18.2 (14.1~20.1); 27.3 (21.7~28.7); 23.5 (18.5~24.8); 222.6 (162.1~237.6); 0.0 (0.0~0.0); 1272 (863~1385) ▎1766 (868~2992); 749 (312~1847); 91 (41~188) ▎47.2 (-34.3~214.4); 1155 (863~1329); 402.3 (248.2~620.0)

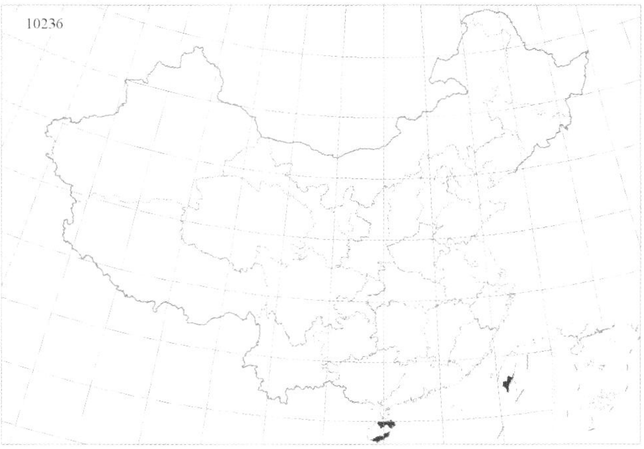

Rubiaceae 茜草科 *Morinda* 巴戟天属
Morinda citrina Y. Z. Ruan 金时巴戟
Vines

15.6；3.3；27.3；15.6；129.5；1.7；857 ❙ 1675；600；204 ❙ 95.4；857；273.6

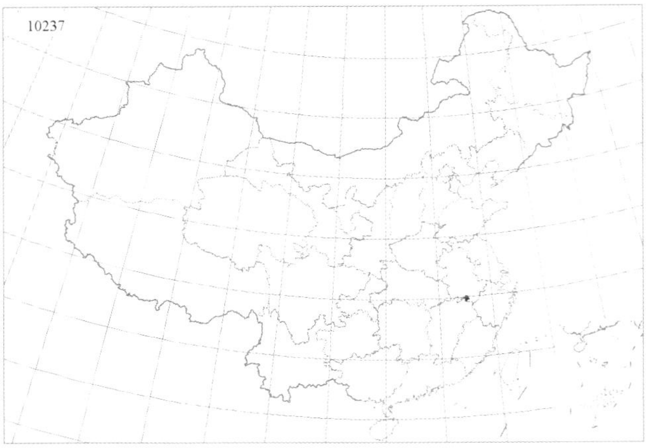

Rubiaceae 茜草科 *Morinda* 巴戟天属
Morinda cochinchinensis DC. 大果巴戟
Vines

21.5 (17.6~24.8)；13.7 (7.8~20.1)；27.5 (21.6~29.2)；21.5 (17.7~24.8)；198.3 (152.5~237.6)；0.0 (0.0~0.0)；1133 (861~1385) ❙ 1600 (868~2435)；710 (312~1428)；115 (41~194) ❙ 43.0 (-34.3~99.5)；1109 (839~1360)；299.7 (95.3~659.0)

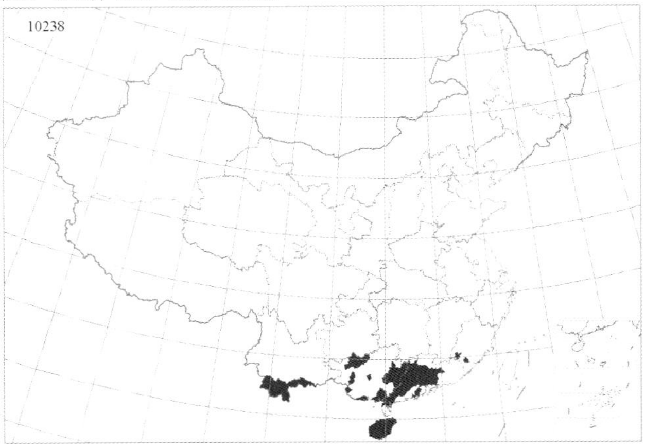

Rubiaceae 茜草科 *Morinda* 巴戟天属
Morinda hainanensis Merr. et How 海南巴戟
Vines

23.5；18.3；27.2；23.6；223.2；0.0；1267 ❙ 1519；541；50 ❙ 19.9；1145；579.1

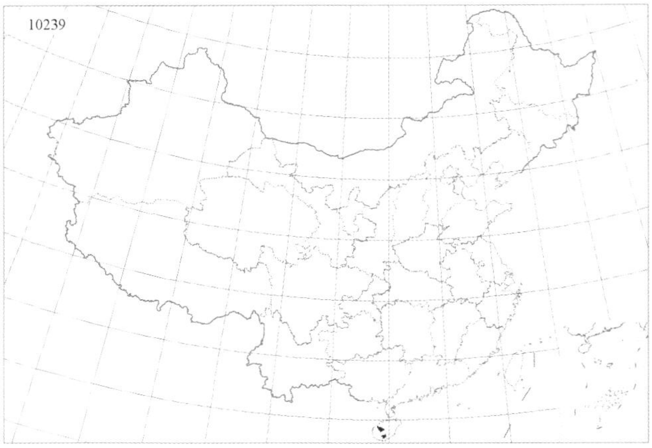

Rubiaceae 茜草科 *Morinda* 巴戟天属
Morinda howlana S. Y. Hu 糠藤
Vines

22.3；14.9；28.1；22.3；208.2；0.0；1185 ❙ 2035；969；113 ❙ 71.8；1185；237.9

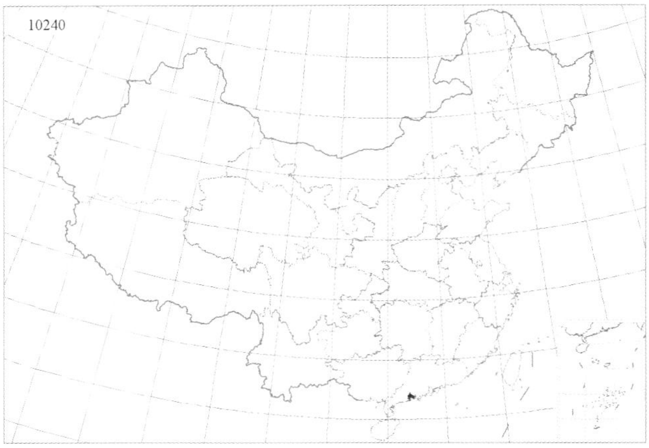

Rubiaceae 茜草科 *Morinda* 巴戟天属
Morinda hupehensis S. Y. Hu 湖北巴戟
Vines

13.6；2.6；24.0；13.6；107.0；3.9；746 ❙ 1435；591；88 ❙ 92.3；746；238.3

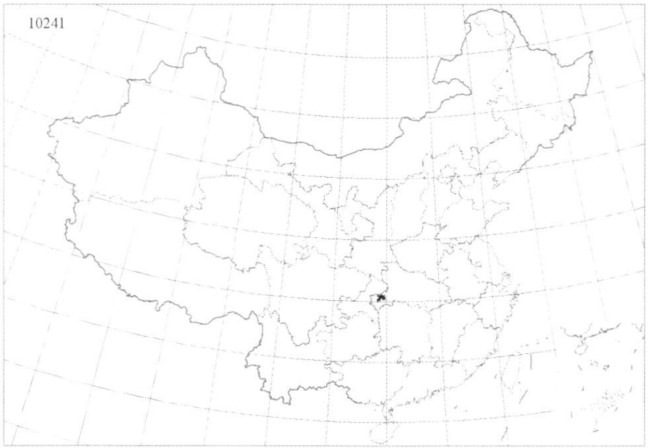

Rubiaceae 茜草科 *Morinda* 巴戟天属
Morinda leiantha Kurz 顶花木巴戟
Shrubs or small trees

20.5；14.9；23.7；20.5；185.9；0.0；992 ❙ 1464；655；66 ❙ 47.6；959；570.1

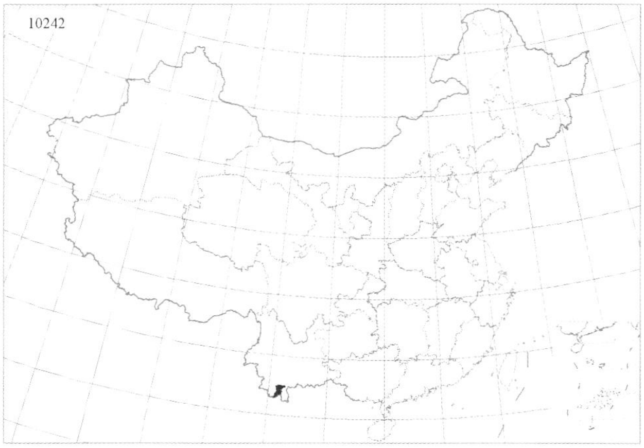

Rubiaceae 茜草科 *Morinda* 巴戟天属
Morinda longissima Y. Z. Ruan 大花木巴戟
Shrubs

20.4; 15.1; 23.6; 20.5; 185.8; 0.0; 989 ▎1607; 702; 77 ▎62.5; 986; 659.0

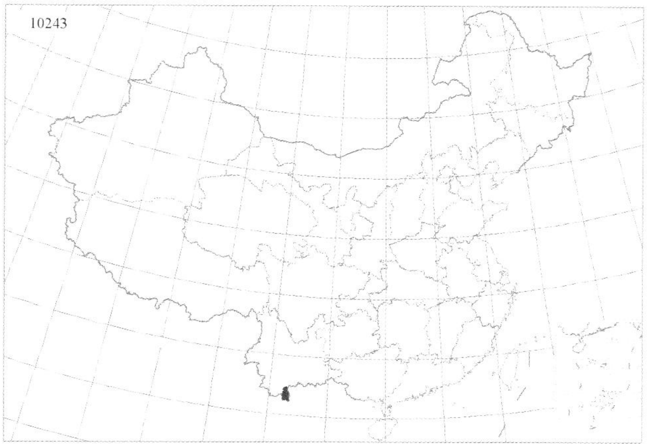

Rubiaceae 茜草科 *Morinda* 巴戟天属
Morinda nanlingensis Y. Z. Ruan 南岭鸡眼藤
Vines

17.7; 6.9; 27.5; 17.7; 152.4; 0.0; 927 ▎1486; 505; 176 ▎60.2; 927; 188.8

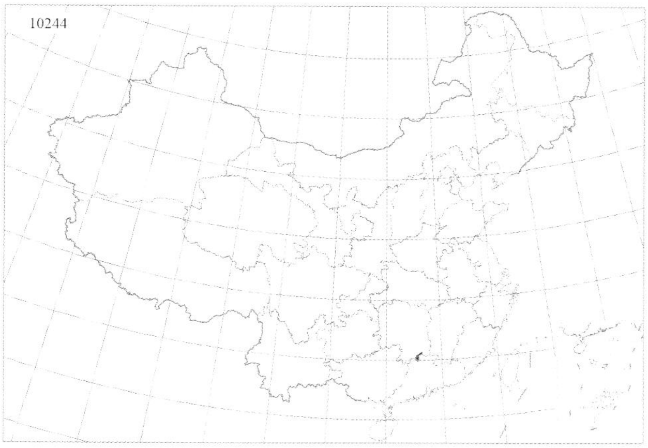

Rubiaceae 茜草科 *Morinda* 巴戟天属
Morinda nanlingensis var. pauciflora Y. Z. Ruan 少花鸡眼藤
Vines

16.5; 4.4; 28.9; 16.6; 139.8; 0.6; 910 ▎1377; 493; 197 ▎51.2; 910; 189.1

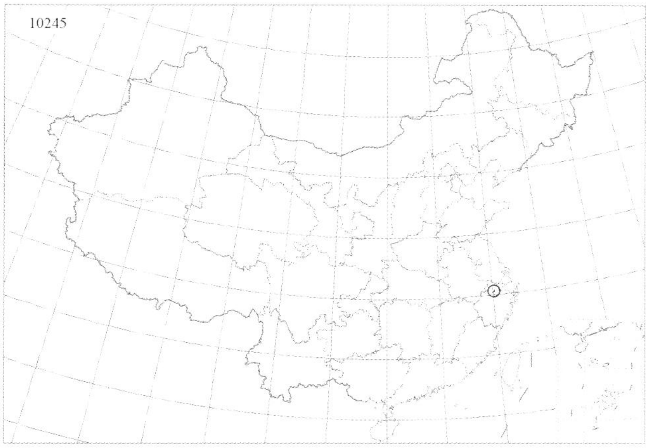

Rubiaceae 茜草科 *Morinda* 巴戟天属
Morinda officinalis F. C. How 巴戟天
Vines

20.6 (14.6~24.7); 12.2 (3.8~19.8); 27.5 (22.5~29.2); 20.7 (14.7~24.7); 188.2 (117.7~236.5); 0.0 (0.0~1.5); 1084 (779~1376) ▎1608 (868~2435); 689 (312~1428); 138 (42~235) ▎50.2 (-34.3~103.1); 1069 (779~1326); 267.1 (122.7~620.0)

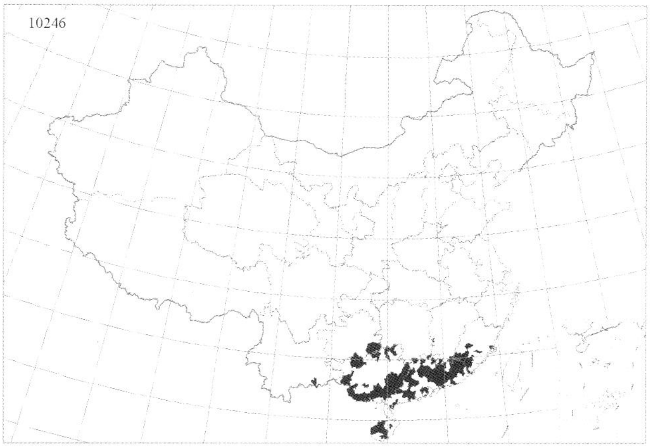

Rubiaceae 茜草科 *Morinda* 巴戟天属
Morinda officinalis var. hirsute F. C. How 毛巴戟天
Vines

24.3; 18.7; 28.4; 24.3; 231.9; 0.0; 1338 ▎1179; 392; 42 ▎-11.9; 1056; 455.2

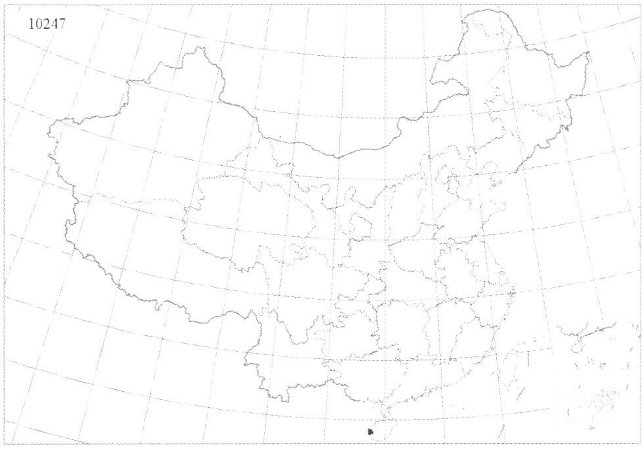

Rubiaceae 茜草科 *Morinda* 巴戟天属
Morinda parvifolia Bartl. ex DC. 鸡眼藤
Vines

21.4 (16.2~24.8); 13.9 (6.4~20.1); 27.6 (20.1~29.0); 21.5 (16.2~24.8); 197.8 (134.2~237.6); 0.0 (0.0~0.0); 1136 (773~1385) ▎1615 (868~2992); 697 (312~1847); 124 (41~245) ▎45.9 (-34.3~246.5); 1089 (773~1360); 284.7 (119.4~620.0)

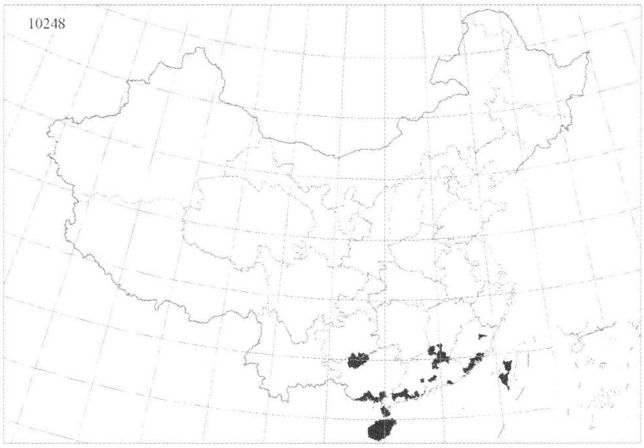

Rubiaceae 茜草科 *Morinda* 巴戟天属
Morinda rugulosa Y. Z. Ruan 皱面鸡眼藤
Vines

17.5; 7.2; 26.4; 17.5; 150.1; 0.0; 904 ❙ 1625; 626; 186 ❙ 79.6; 904; 263.3

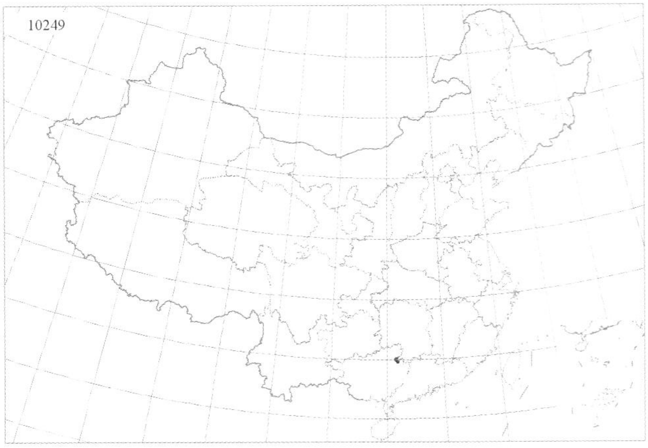

Rubiaceae 茜草科 *Morinda* 巴戟天属
Morinda shuanghuaensis C. Y. Chen et M. S. Huang 假巴戟
Vines

20.0 (17.1~22.3); 11.4 (6.5~14.7); 27.6 (25.7~29.0); 20.0 (17.1~22.4); 180.2 (145.5~208.3); 0.0 (0.0~0.0); 1036 (894~1188) ❙ 1517 (1070~1829); 544 (356~845); 172 (75~230) ❙ 48.4 (-5.0~104.6); 1015 (894~1188); 238.1 (123.9~290.3)

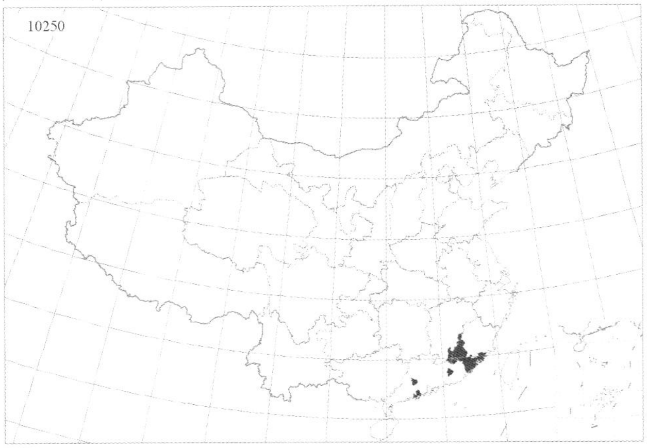

Rubiaceae 茜草科 *Morinda* 巴戟天属
Morinda umbellata Y. Z. Ruan 印度羊角藤
Vines

18.1 (8.6~24.8); 8.2 (1.3~20.1); 27.1 (15.0~29.8); 18.2 (8.7~24.8); 158.6 (52.6~237.6); 0.3 (0.0~8.8); 956 (574~1385) ❙ 1525 (985~2993); 597 (356~1428); 157 (35~547) ❙ 61.2 (-12.0~246.5); 948 (574~1299); 243.8 (108.2~659.0)

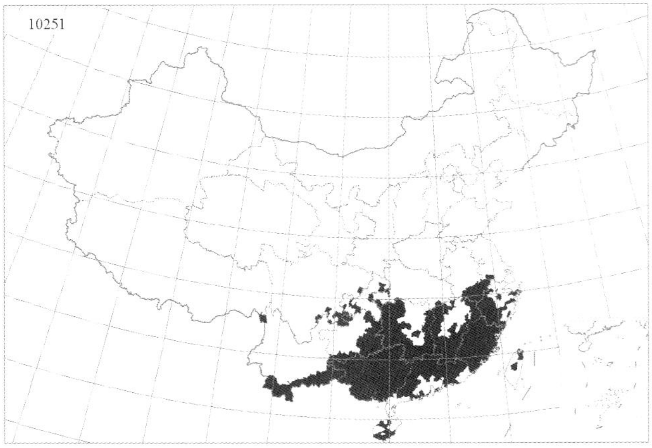

Rubiaceae 茜草科 *Morinda* 巴戟天属
Morinda undulata Y. Z. Ruan 波叶木巴戟
Small trees

20.4; 15.1; 23.6; 20.5; 185.8; 0.0; 989 ❙ 1607; 702; 77 ❙ 62.5; 986; 659.0

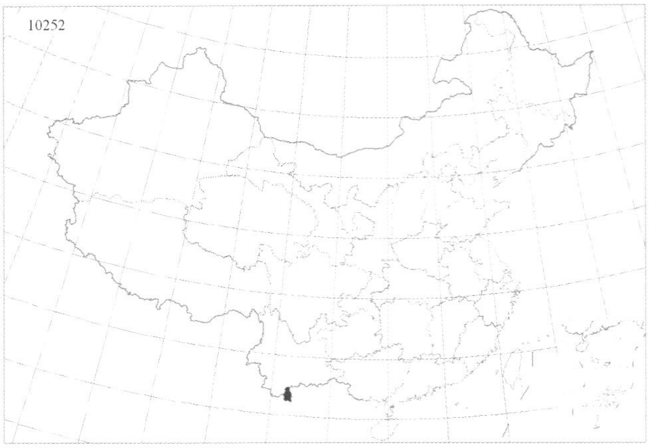

Rubiaceae 茜草科 *Morinda* 巴戟天属
Morinda villosa Hook. f. 须弥巴戟
Vines

20.5; 15.0; 23.7; 20.5; 185.9; 0.0; 990 ❙ 1536; 678; 72 ❙ 55.1; 973; 614.6

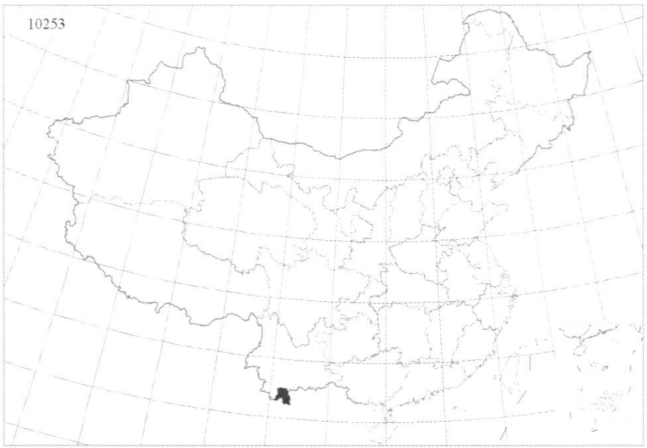

Rubiaceae 茜草科 *Mussaenda* 玉叶金花属
Mussaenda antiloga Y. H. Chun et W. C. Ko 壮丽玉叶金花
Climbing shrubs

24.1; 19.2; 27.5; 24.2; 229.8; 0.0; 1320 ❙ 1567; 539; 54 ❙ 19.4; 1165; 553.8

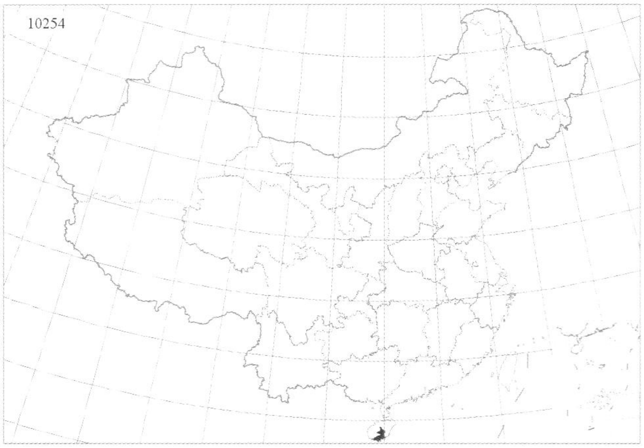

Rubiaceae 茜草科 *Mussaenda* 玉叶金花属
Mussaenda breviloba S. Moore 短裂玉叶金花
Shrubs

18.9 (17.3~21.0); 12.9 (10.9~15.1); 22.9 (21.6~25.6); 19.0 (17.3~21.0); 167.8 (148.0~191.8); 0.0 (0.0~0.0); 911 (832~1053) ❙ 1383 (1028~1751); 750 (566~996); 58 (44~77) ❙ 51.8 (12.5~88.8); 893 (783~1053); 456.7 (339.5~659.0)

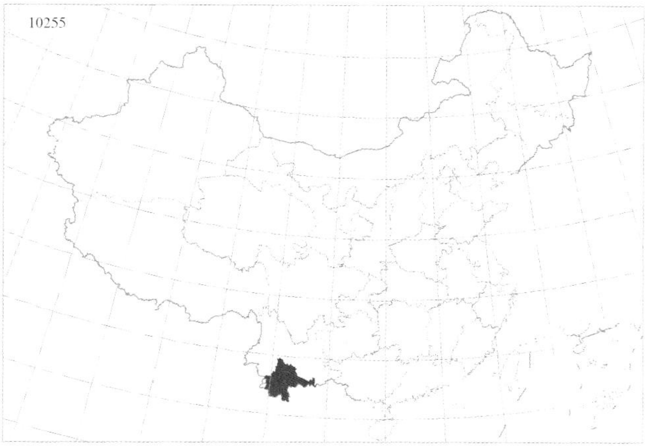

Rubiaceae 茜草科 *Mussaenda* 玉叶金花属
Mussaenda chingii C. Y. Wu ex Hsue et H. Wu 仁昌玉叶金花
Shrubs

19.2; 9.2; 27.5; 19.2; 170.8; 0.0; 1002 ❙ 1495; 678; 135 ❙ 49.1; 1002; 256.4

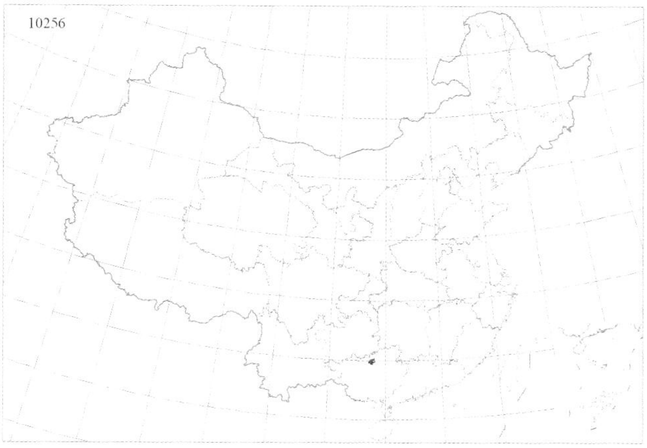

Rubiaceae 茜草科 *Mussaenda* 玉叶金花属
Mussaenda decipiens H. Li 墨脱玉叶金花
Shrubs

11.0; 3.4; 17.3; 11.0; 76.5; 4.4; 640 ❙ 1389; 796; 48 ❙ 113.7; 640; 399.9

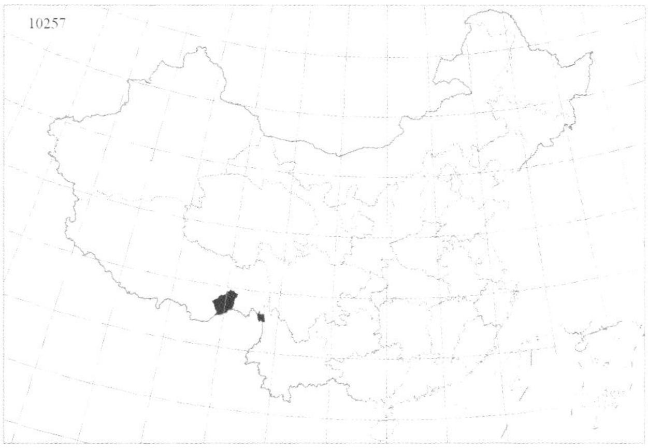

Rubiaceae 茜草科 *Mussaenda* 玉叶金花属
Mussaenda densiflora H. L. Li 密花玉叶金花
Scandent shrubs

22.2; 13.9; 28.3; 22.3; 207.5; 0.0; 1190 ❙ 1374; 713; 87 ❙ 15.4; 1190; 327.7

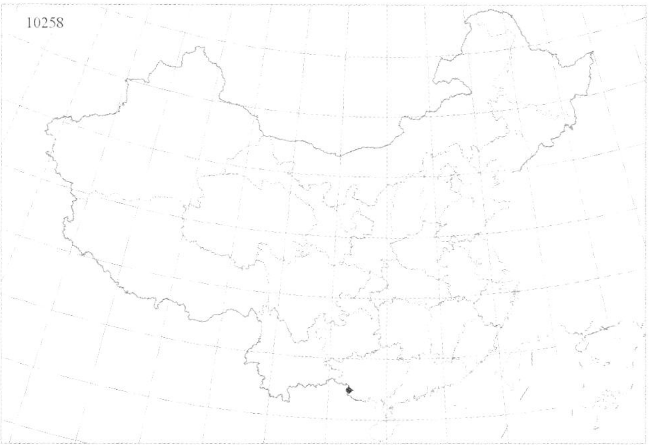

Rubiaceae 茜草科 *Mussaenda* 玉叶金花属
Mussaenda divaricata Hutch. 展枝玉叶金花
Climbing shrubs

16.0 (5.4~22.3); 7.3 (-2.1~14.2); 23.2 (12.3~28.9); 16.0 (5.8~22.3); 132.9 (31.5~208.1); 0.8 (0.0~25.6); 826 (481~1202) ❙ 1187 (741~1893); 594 (411~996); 73 (15~227) ❙ 44.1 (-14.2~129.9); 817 (481~1202); 268.4 (126.1~551.6)

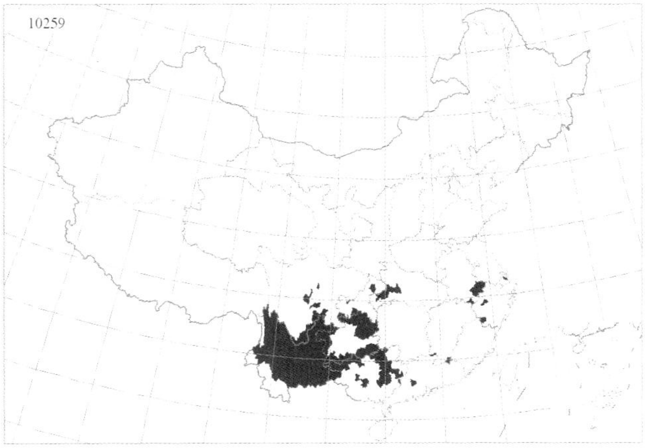

Rubiaceae 茜草科 *Mussaenda* 玉叶金花属
Mussaenda divaricata var. mollis Hutch. 柔毛玉叶金花
Climbing shrubs

17.6 (16.3~21.0); 10.4 (9.1~14.2); 22.6 (21.4~25.6); 17.6 (16.4~21.0); 151.3 (136.4~191.8); 0.0 (0.0~0.0); 856 (797~1053) ❙ 1280 (1007~1522); 708 (539~860); 53 (44~68) ❙ 49.6 (19.0~80.6); 855 (797~1053); 389.0 (277.0~551.6)

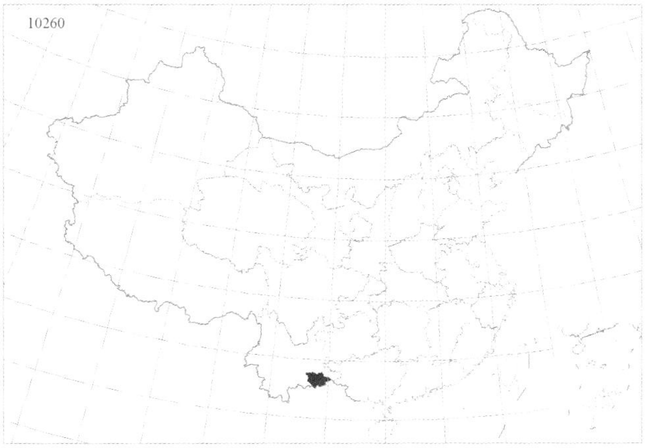

Rubiaceae 茜草科 *Mussaenda* 玉叶金花属
Mussaenda elliptica Hutch. 椭圆玉叶金花
Shrubs

18.2 (16.8~19.1); 9.3 (6.0~13.0); 25.5 (23.3~27.4); 18.3 (16.8~19.1); 159.2 (141.5~169.7); 0.0 (0.0~0.0); 923 (836~969) ❙ 1277 (1154~1428); 673 (488~835); 56 (48~64) ❙ 38.8 (27.3~59.6); 923 (836~969); 269.3 (164.1~506.2)

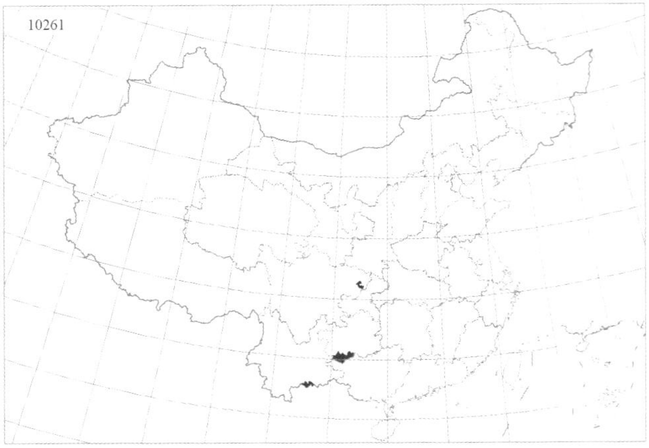

Rubiaceae 茜草科 *Mussaenda* 玉叶金花属
Mussaenda elongata Hutch. 长玉叶金花
Shrubs

18.6 (16.3~21.0); 11.9 (9.1~15.1); 23.1 (21.4~25.6); 18.6 (16.4~21.0); 163.2 (136.4~191.8); 0.0 (0.0~0.0); 900 (797~1053) ❙ 1410 (1074~1751); 761 (582~996); 58 (47~77) ❙ 56.8 (32.4~88.8); 894 (797~1053); 444.2 (277.0~659.0)

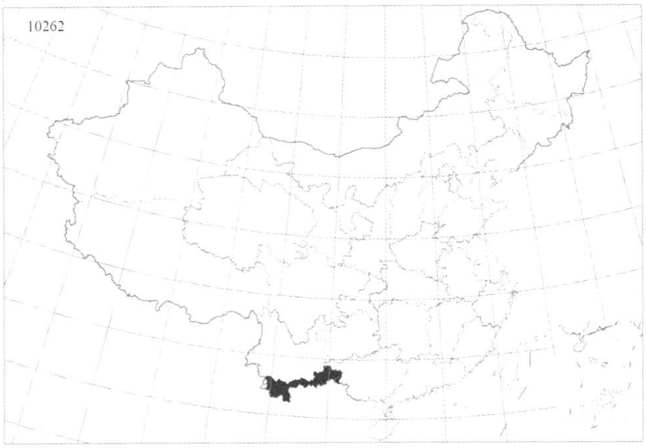

Rubiaceae 茜草科 *Mussaenda* 玉叶金花属
Mussaenda erosa Champ. ex Benth. 楠藤
Scandent shrubs

20.3 (8.6~24.8); 12.5 (1.3~20.1); 26.3 (15.0~28.9); 20.4 (8.7~24.8); 184.3 (52.6~237.6); 0.1 (0.0~8.8); 1054 (574~1385) ❙ 1524 (868~2435); 719 (312~1428); 105 (41~198) ❙ 46.2 (-34.3~112.2); 1036 (574~1360); 318.8 (185.5~659.0)

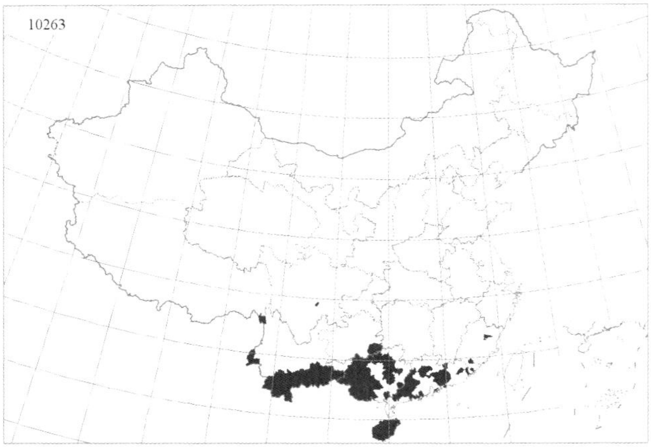

Rubiaceae 茜草科 *Mussaenda* 玉叶金花属
Mussaenda esquirolii H. Levl 黐花
Shrubs

17.3 (11.1~22.9); 6.9 (0.6~15.2); 26.7 (19.4~30.1); 17.3 (11.1~22.9); 148.4 (79.8~214.8); 0.5 (0.0~9.7); 912 (648~1232) ❙ 1508 (890~2014); 575 (381~996); 158 (21~259) ❙ 66.2 (5.8~147.0); 905 (648~1232); 243.0 (130.1~659.0)

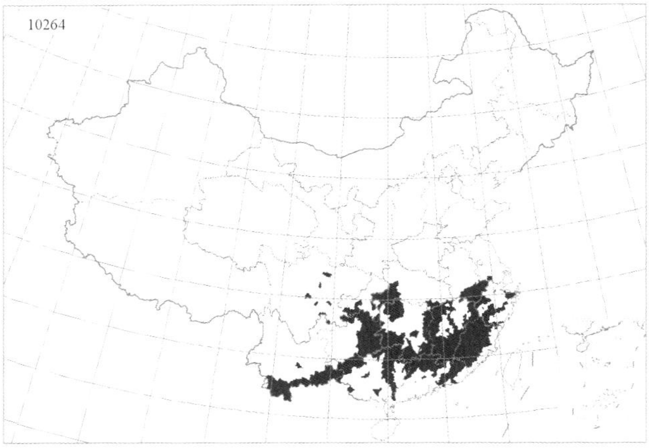

Rubiaceae 茜草科 *Mussaenda* 玉叶金花属
Mussaenda hainanensis Merr. 海南玉叶金花
Climbing shrubs

23.5 (20.8~24.8); 17.7 (12.2~20.1); 27.8 (25.7~28.7); 23.6 (20.8~24.8); 223.0 (189.7~237.6); 0.0 (0.0~0.0); 1277 (1085~1385) ❙ 1470 (868~1990); 527 (312~827); 75 (41~181) ❙ 16.5 (-34.3~83.4); 1138 (868~1326); 445.9 (226.9~620.0)

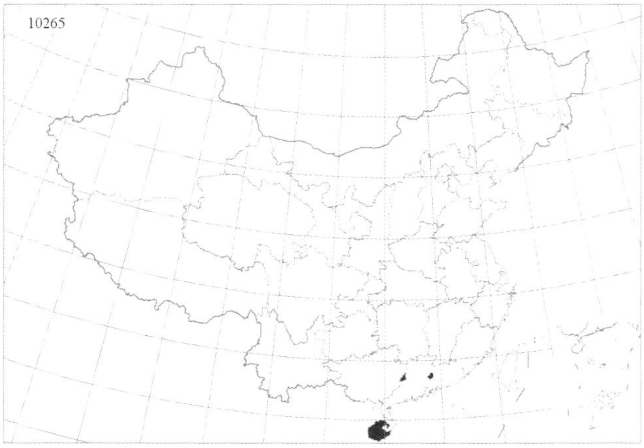

Rubiaceae 茜草科 *Mussaenda* 玉叶金花属
Mussaenda henryi Hutch. 南玉叶金花
Trees

18.6 (16.3~22.2); 11.4 (9.1~14.2); 23.6 (21.4~28.5); 18.6 (16.4~22.3); 163.2 (136.4~207.5); 0.0 (0.0~0.0); 914 (797~1190) ❙ 1311 (1007~1703); 724 (539~911); 65 (44~194) ❙ 43.9 (15.4~80.6); 910 (797~1190); 373.7 (213.5~551.6)

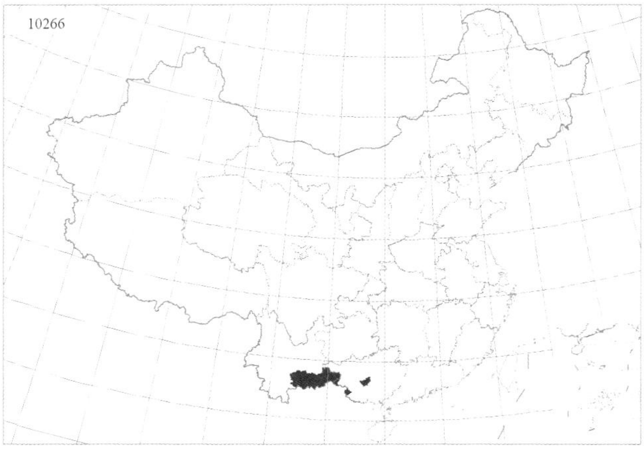

Rubiaceae 茜草科 *Mussaenda* 玉叶金花属
Mussaenda hirsutula Miq. 粗毛玉叶金花
Climbing shrubs

18.7 (13.4~24.8); 10.0 (2.7~20.1); 26.0 (21.4~28.9); 18.8 (13.4~24.8); 165.4 (104.5~237.6); 0.2 (0.0~3.7); 982 (735~1385) ❙ 1349 (868~1854); 599 (312~860); 89 (41~191) ❙ 40.4 (-34.3~99.4); 959 (735~1360); 287.0 (134.1~620.0)

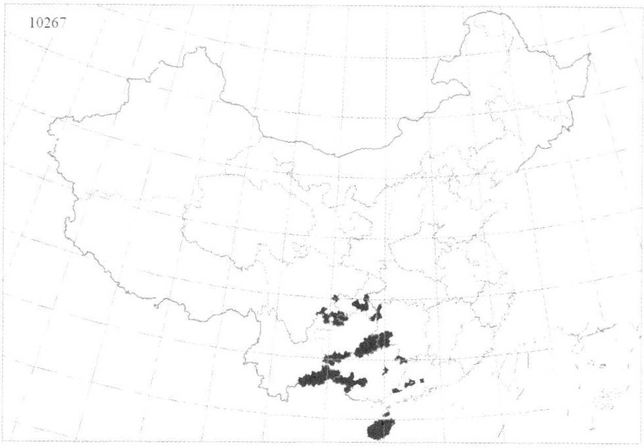

Rubiaceae 茜草科 *Mussaenda* 玉叶金花属
Mussaenda hossei Craib 红毛玉叶金花
Subshrubs

18.3 (16.3~21.0); 11.4 (9.1~15.1); 23.1 (21.4~25.6); 18.3 (16.4~21.0); 160.1 (136.4~191.8); 0.0 (0.0~0.0); 891 (797~1053) ❙ 1345 (1049~1607); 722 (582~860); 56 (44~77) ❙ 51.0 (26.1~80.6); 883 (783~1053); 425.9 (277.0~659.0)

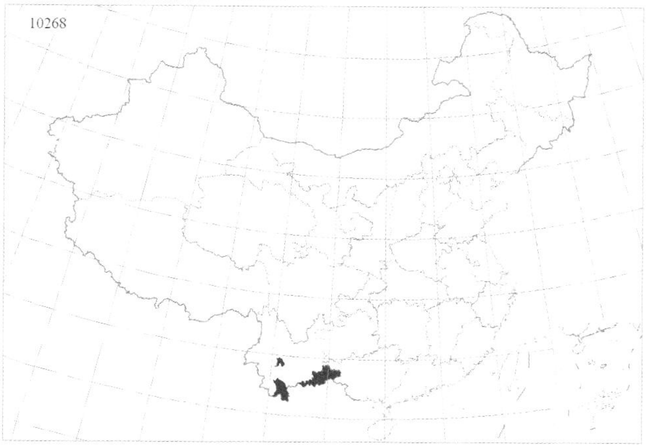

Rubiaceae 茜草科 *Mussaenda* 玉叶金花属
Mussaenda inflata Hsue et H. Wu 胀管玉叶金花
Climbing shrubs

21.8; 14.2; 27.8; 21.9; 202.6; 0.0; 1150 ❙ 1876; 891; 116 ❙ 63.2; 1150; 273.3

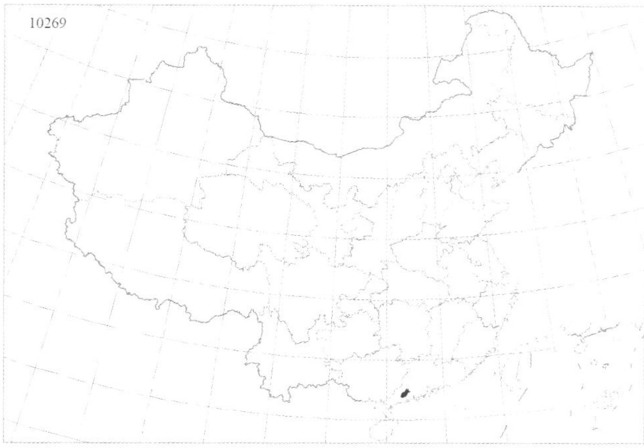

Rubiaceae 茜草科 *Mussaenda* 玉叶金花属
Mussaenda kwangsiensis H. L. Li 广西玉叶金花
Climbing shrubs

20.0 (17.7~22.7); 10.6 (7.4~15.0); 27.8 (25.8~28.7); 20.1 (17.8~22.8); 180.9 (153.6~213.1); 0.0 (0.0~0.0); 1054 (913~1221) ❙ 1658 (1371~2435); 734 (556~1428); 163 (82~189) ❙ 57.9 (20.8~99.5); 1054 (913~1221); 229.0 (164.9~300.5)

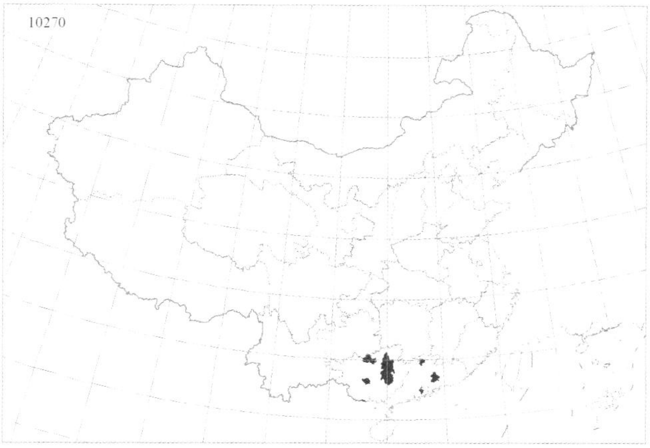

Rubiaceae 茜草科 *Mussaenda* 玉叶金花属
Mussaenda kwangtungensis H. L. Li 广东玉叶金花
Climbing shrubs

20.5 (17.7~22.4); 11.9 (7.7~14.5); 27.4 (25.8~28.7); 20.5 (17.8~22.4); 186.0 (153.6~209.2); 0.0 (0.0~0.0); 1069 (913~1198) ❙ 1811 (1447~2148); 824 (556~1234); 151 (82~191) ❙ 70.2 (20.8~95.5); 1069 (913~1198); 251.2 (164.9~318.8)

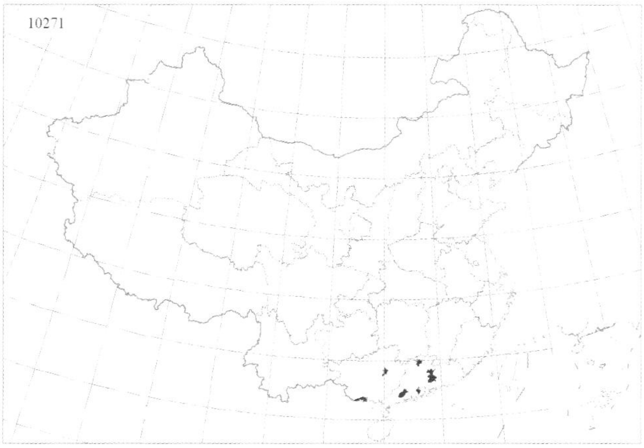

Rubiaceae 茜草科 *Mussaenda* 玉叶金花属
Mussaenda laxiflora Hutch. 疏花玉叶金花
Shrubs

18.1 (15.4~21.0); 11.3 (8.1~15.1); 22.7 (20.7~25.6); 18.1 (15.5~21.0); 157.7 (125.7~191.8); 0.0 (0.0~0.0); 878 (767~1053) ❙ 1293 (958~1751); 702 (531~996); 55 (44~77) ❙ 46.9 (13.3~88.8); 873 (767~1053); 414.1 (274.9~659.0)

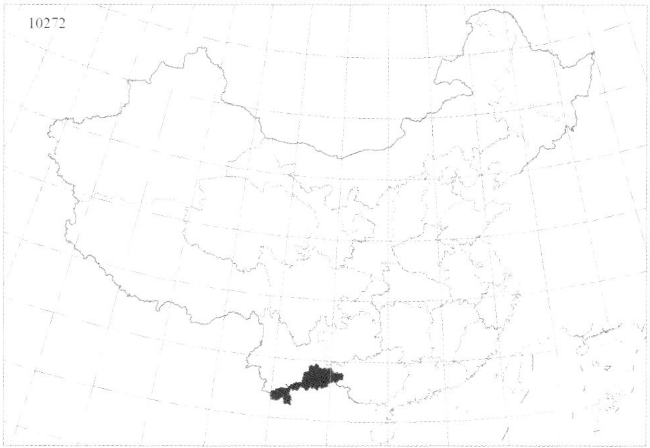

Rubiaceae 茜草科 *Mussaenda* 玉叶金花属
Mussaenda lotungensis Y. H. Chun et W. C. Ko 乐东玉叶金花
Climbing shrubs

23.9；18.7；27.5；23.9；226.8；0.0；1294 ❙ 1380；455；48 ❙ 6.7；1112；455.2

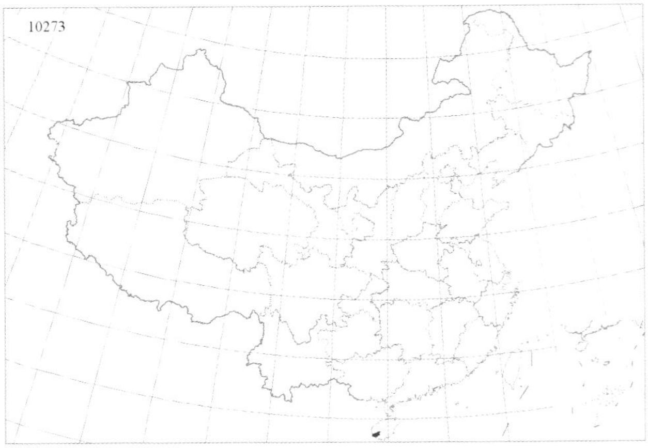

Rubiaceae 茜草科 *Mussaenda* 玉叶金花属
Mussaenda macrophylla Wall. 大叶玉叶金花
Shrubs

18.2 (8.6~22.3)；10.7 (1.3~15.1)；23.7 (15.0~28.9)；18.3 (8.7~22.3)；159.4 (52.6~208.1)；0.3 (0.0~8.8)；914 (574~1202) ❙ 1465 (1009~2713)；725 (562~1388)；99 (18~189) ❙ 61.4 (20.3~214.4)；904 (574~1202)；355.5 (185.5~659.0)

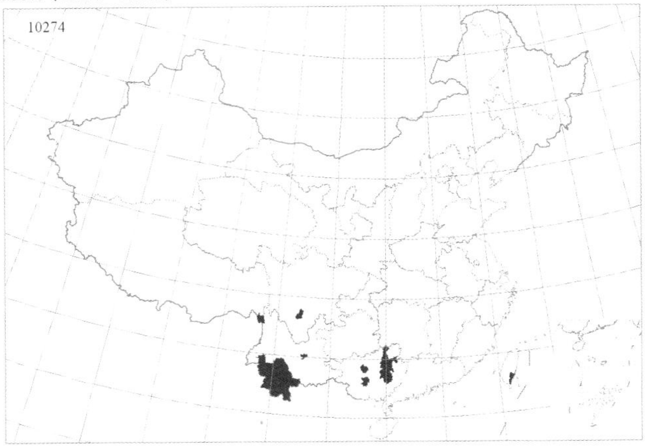

Rubiaceae 茜草科 *Mussaenda* 玉叶金花属
Mussaenda membranifolia Merr. 膜叶玉叶金花
Climbing shrubs

24.0；18.7；27.6；24.0；227.9；0.0；1305 ❙ 1558；539；67 ❙ 20.0；1185；477.1

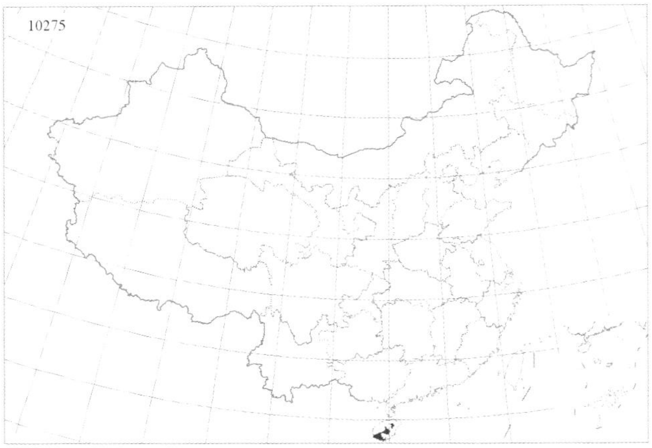

Rubiaceae 茜草科 *Mussaenda* 玉叶金花属
Mussaenda mollissima C. Y. Wu ex Hsue et H. Wu 多毛玉叶金花
Shrubs

18.9 (16.9~21.0)；12.8 (10.3~15.1)；22.8 (21.4~25.6)；18.9 (16.9~21.0)；167.0 (142.7~191.8)；0.0 (0.0~0.0)；907 (813~1053) ❙ 1384 (1059~1751)；744 (604~996)；56 (41~77) ❙ 52.6 (21.3~88.8)；892 (813~1053)；454.3 (294.5~659.0)

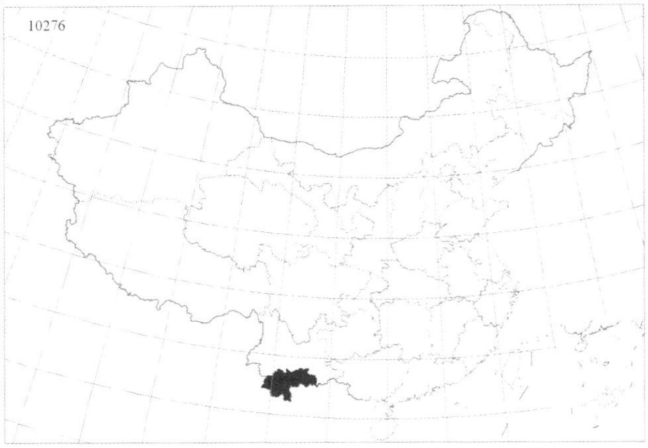

Rubiaceae 茜草科 *Mussaenda* 玉叶金花属
Mussaenda multinervis C. Y. Wu ex Hsue et H. Wu 多脉玉叶金花
Shrubs

19.5 (18.2~21.0)；13.6 (12.4~15.1)；23.2 (21.6~25.6)；19.5 (18.2~21.0)；174.0 (158.6~191.8)；0.0 (0.0~0.0)；938 (861~1053) ❙ 1509 (1276~1751)；790 (655~996)；63 (51~77) ❙ 61.4 (36.8~88.8)；926 (839~1053)；518.6 (427.9~659.0)

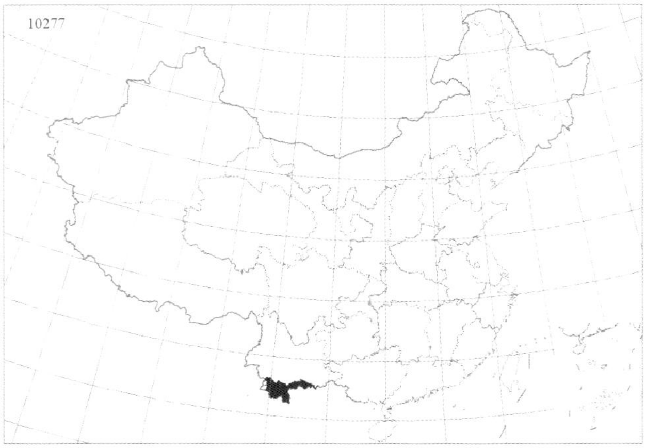

Rubiaceae 茜草科 *Mussaenda* 玉叶金花属
Mussaenda parviflora Miq. 小玉叶金花
Scandent shrubss or vines

17.9 (2.6~24.1)；11.1 (-7.0~18.9)；23.6 (11.2~28.8)；18.1 (4.3~24.1)；159.0 (20.0~229.7)；3.4 (0.0~48.2)；930 (427~1322) ❙ 2081 (829~2993)；985 (400~1847)；179 (23~547) ❙ 129.2 (32.5~246.5)；913 (427~1170)；248.1 (154.2~363.7)

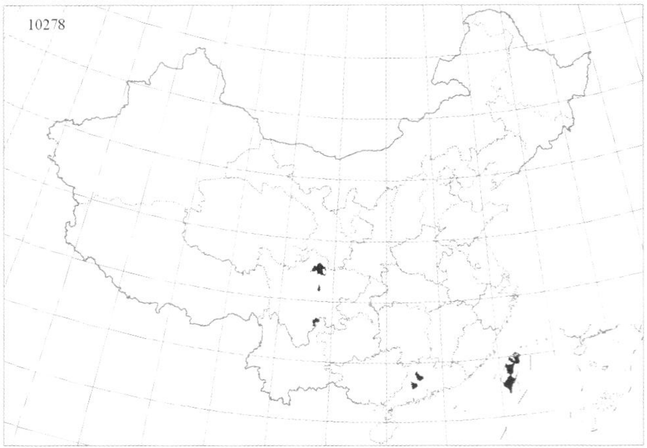

Rubiaceae 茜草科 *Mussaenda* 玉叶金花属
Mussaenda pingpienensis C. Y. Wu ex Hsue et H. Wu 屏边玉叶金花
Shrubs

18.7 (16.3~21.0); 12.2 (9.1~15.1); 23.1 (21.4~25.6); 18.7 (16.4~21.0); 164.8 (136.4~191.8); 0.0 (0.0~0.0); 906 (797~1053) ❙ 1455 (1190~1751); 771 (655~996); 61 (48~77) ❙ 60.9 (44.5~88.8); 900 (797~1053); 478.5 (336.3~659.0)

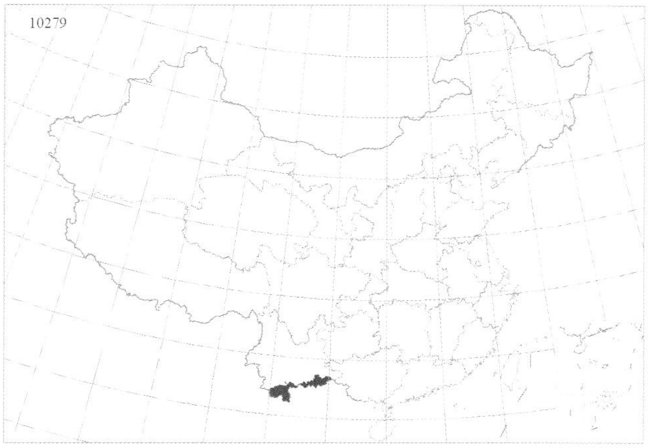

Rubiaceae 茜草科 *Mussaenda* 玉叶金花属
Mussaenda pubescens W. T. Aiton 玉叶金花
Climbing shrubs

18.9 (11.4~24.8); 10.1 (1.9~20.1); 26.3 (18.6~29.8); 18.9 (11.4~24.8); 167.2 (80.7~237.6); 0.1 (0.0~6.1); 983 (652~1385) ❙ 1529 (863~3495); 659 (312~1847); 137 (25~897) ❙ 57.4 (-34.3~246.5); 973 (652~1360); 260.9 (95.3~659.0)

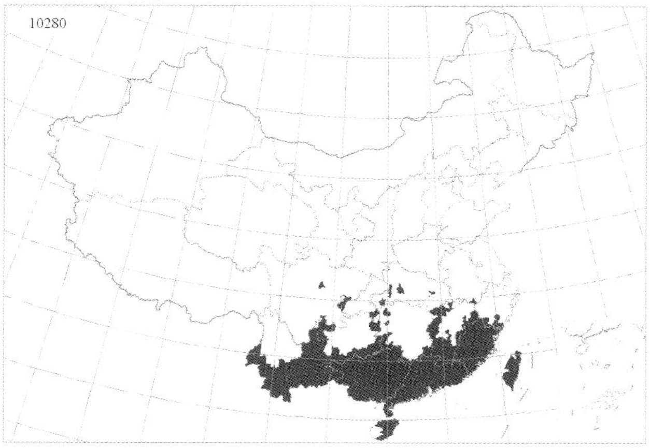

Rubiaceae 茜草科 *Mussaenda* 玉叶金花属
Mussaenda sessilifolia Hutch. 无柄玉叶金花
Climbing shrubs

20.0; 14.4; 23.3; 20.0; 180.2; 0.0; 963 ❙ 1522; 711; 68 ❙ 58.2; 945; 552.4

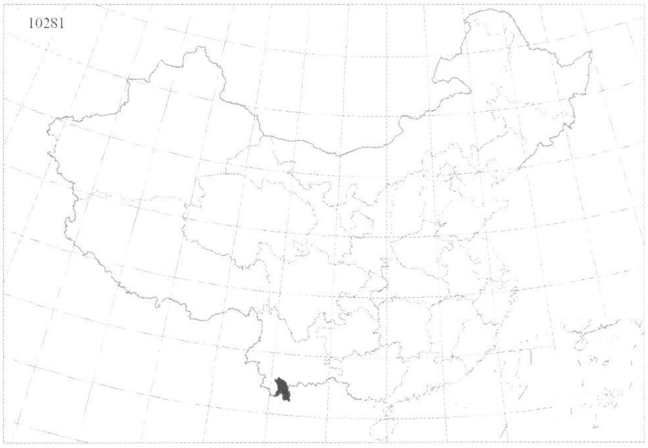

Rubiaceae 茜草科 *Mussaenda* 玉叶金花属
Mussaenda simpliciloba Hand.-Mazz. 单裂玉叶金花
Climbing shrubs

14.0 (6.1~18.9); 6.5 (-1.5~10.5); 20.0 (12.7~26.5); 14.0 (6.3~18.9); 109.3 (35.2~167.3); 1.5 (0.0~21.8); 736 (496~969) ❙ 1042 (801~1342); 593 (496~667); 38 (15~95) ❙ 43.5 (5.5~78.3); 710 (496~969); 311.9 (215.4~384.1)

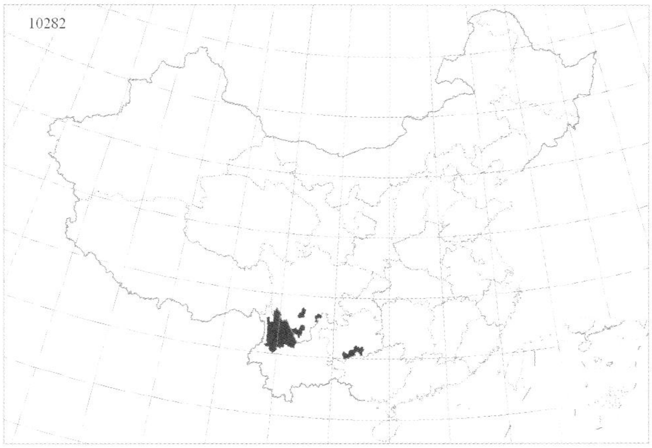

Rubiaceae 茜草科 *Mussaenda* 玉叶金花属
Mussaenda treutleri Stapf 贡山玉叶金花
Shrubs

14.1 (8.6~21.0); 7.2 (1.3~14.2); 19.3 (15.0~25.6); 14.1 (8.7~21.0); 110.8 (52.6~191.8); 1.9 (0.0~8.8); 750 (574~1053) ❙ 1245 (1041~1522); 685 (585~818); 58 (45~69) ❙ 69.9 (38.0~93.0); 749 (574~1053); 420.1 (352.3~551.6)

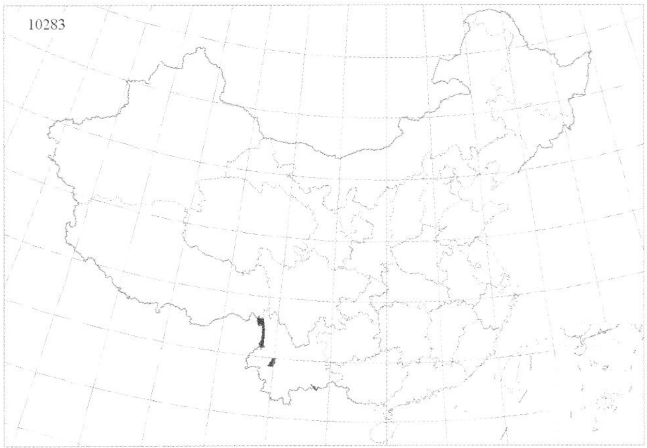

Rubiaceae 茜草科 *Mycetia* 腺萼木属
Mycetia anlongensis H. S. Lo 安龙腺萼木
Shrubs

16.8; 7.8; 23.6; 16.8; 141.5; 0.0; 836 ❙ 1334; 720; 64 ❙ 59.6; 836; 226.6

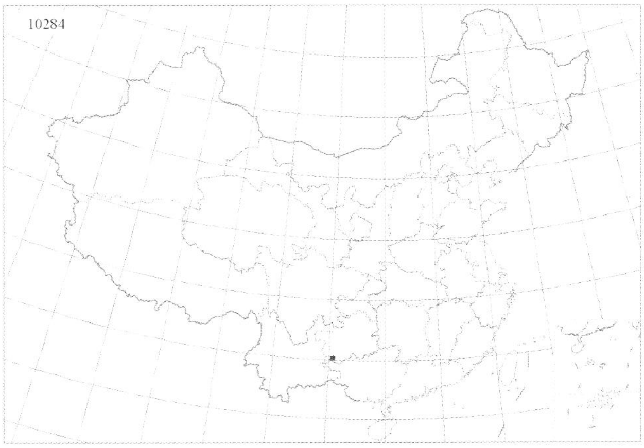

Rubiaceae 茜草科 *Mycetia* 腺萼木属

Mycetia bracteata Hutch. 长苞腺萼木

Small shrubs

20.1 (19.0~21.3); 13.9 (12.6~15.1); 24.0 (22.3~27.9); 20.1 (19.1~21.3); 181.4 (168.8~196.0); 0.0 (0.0~0.0); 984 (909~1123) | 1450 (1188~1607); 690 (634~775); 64 (56~77) | 49.1 (5.8~64.9); 968 (880~1121); 497.3 (315.6~659.0)

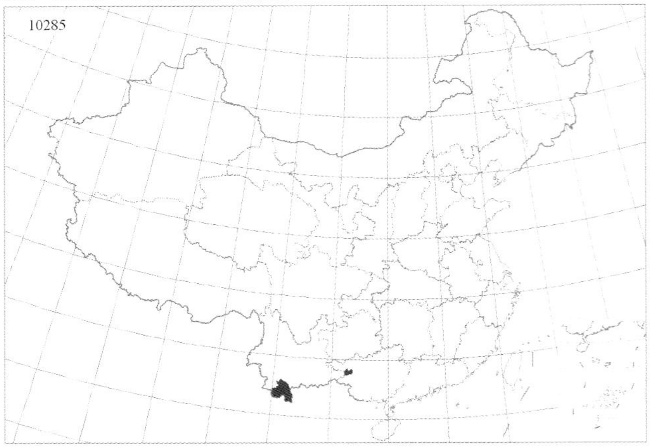

Rubiaceae 茜草科 *Mycetia* 腺萼木属

Mycetia brevipes F. C. How ex H. S. Lo 短柄腺萼木

Shrubs

8.6; 1.3; 15.0; 8.7; 52.6; 8.8; 574 | 1041; 585; 45 | 81.5; 574; 352.3

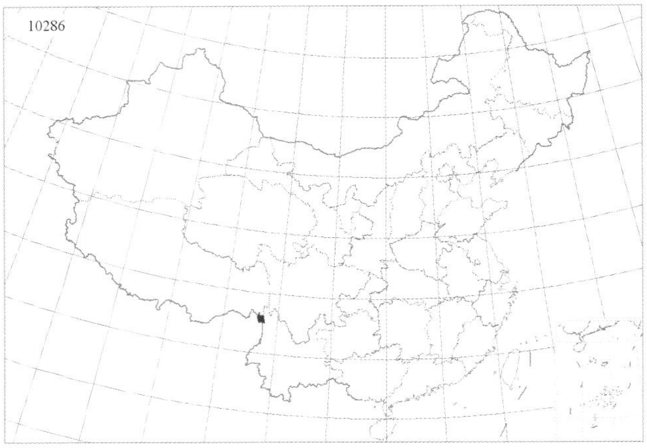

Rubiaceae 茜草科 *Mycetia* 腺萼木属

Mycetia brevisepala H. S. Lo 短萼腺萼木

Shrubs

18.9 (17.2~21.0); 12.6 (9.5~15.1); 23.0 (21.6~25.6); 18.9 (17.3~21.0); 167.3 (147.0~191.8); 0.0 (0.0~0.0); 911 (833~1053) | 1483 (1254~1751); 780 (655~996); 58 (41~77) | 63.2 (44.5~88.8); 898 (833~1053); 493.6 (336.3~659.0)

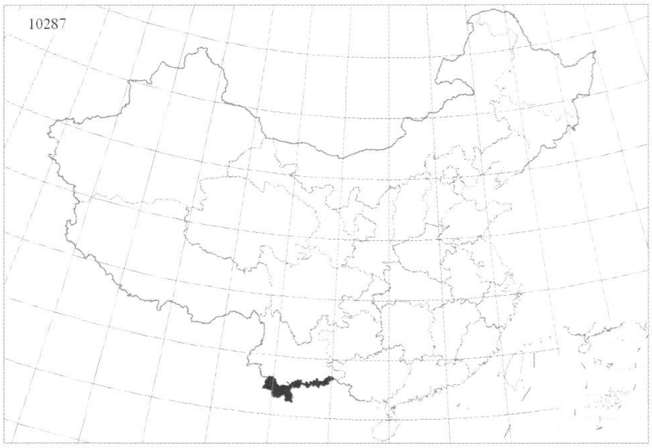

Rubiaceae 茜草科 *Mycetia* 腺萼木属

Mycetia coriacea (Dunn) Merr. 革叶腺萼木

Shrubs

19.9 (16.9~21.7); 11.4 (7.8~14.0); 27.3 (25.7~28.5); 19.9 (17.0~21.7); 179.1 (143.6~200.5); 0.0 (0.0~0.0); 1028 (862~1140) | 1652 (1375~1991); 631 (476~902); 166 (125~209) | 62.0 (30.2~100.8); 1028 (862~1140); 256.5 (197.7~302.2)

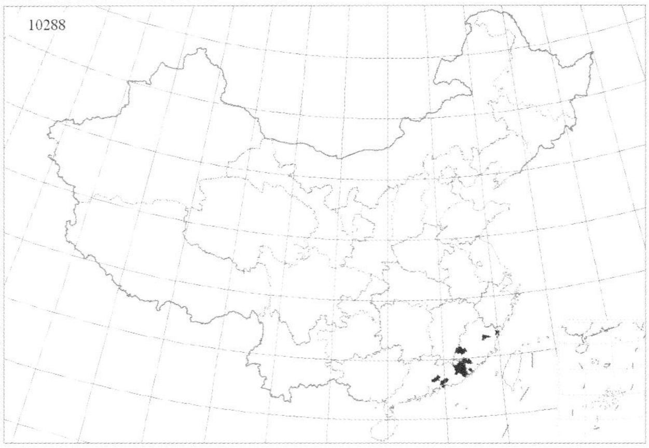

Rubiaceae 茜草科 *Mycetia* 腺萼木属

Mycetia glandulosa Craib 腺萼木

Shrubs

18.4 (8.6~21.0); 12.4 (1.3~15.1); 22.4 (15.0~25.6); 18.5 (8.7~21.0); 162.1 (52.6~191.8); 0.6 (0.0~8.8); 897 (574~1053) | 1438 (1041~1751); 753 (585~996); 58 (41~77) | 61.1 (36.8~88.8); 883 (574~1053); 494.5 (352.3~659.0)

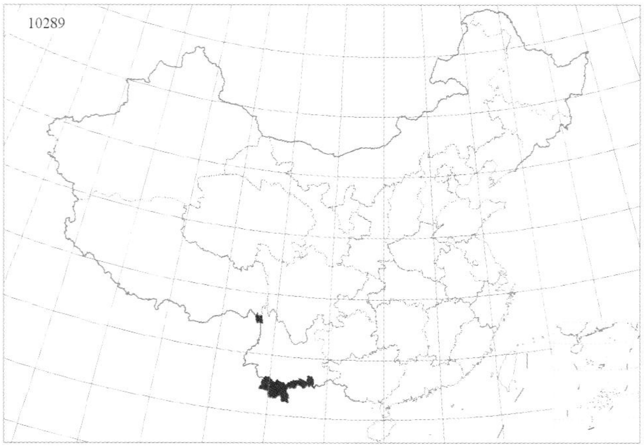

Rubiaceae 茜草科 *Mycetia* 腺萼木属

Mycetia gracilis Craib 纤梗腺萼木

Shrubs

19.2 (17.7~21.0); 13.3 (10.9~15.1); 23.0 (21.6~25.6); 19.3 (17.8~21.0); 171.1 (153.4~191.8); 0.0 (0.0~0.0); 924 (854~1053) | 1455 (1143~1751); 765 (637~996); 59 (41~77) | 57.8 (21.7~88.8); 909 (839~1053); 497.4 (346.9~659.0)

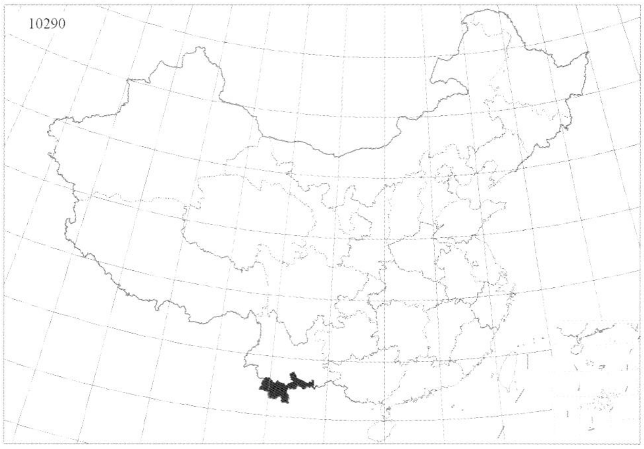

Rubiaceae 茜草科 *Mycetia* 腺萼木属
Mycetia hainanensis H. S. Lo 海南腺萼木
Subshrubs

24.3; 19.2; 28.0; 24.3; 232.1; 0.0; 1339 ❙ 1321; 440; 44 ❙ -1.2; 1102; 479.3

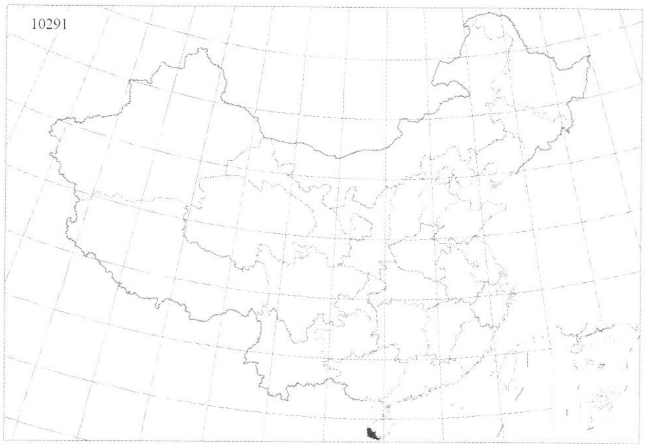

Rubiaceae 茜草科 *Mycetia* 腺萼木属
Mycetia hirta Hutch. 毛腺萼木
Shrubs

18.6 (13.3~24.7); 11.9 (5.4~19.8); 23.2 (19.1~29.8); 18.7 (13.4~24.7); 163.9 (100.4~236.5); 0.0 (0.0~0.0); 920 (706~1376) ❙ 1432 (939~1803); 738 (508~1007); 67 (39~194) ❙ 57.5 (6.5~145.8); 898 (706~1299); 430.3 (174.9~659.0)

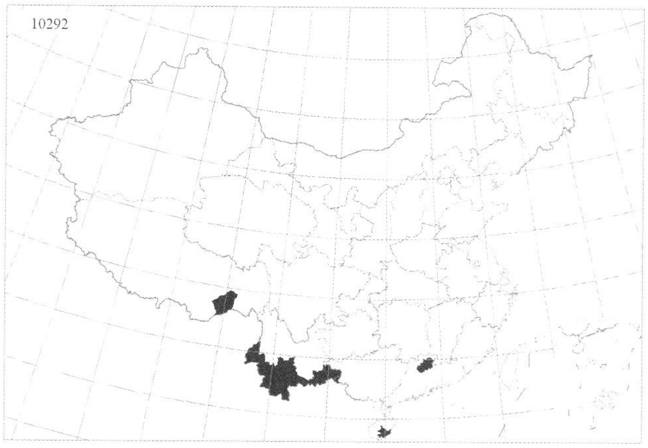

Rubiaceae 茜草科 *Mycetia* 腺萼木属
Mycetia longiflora F. C. How 长花腺萼木
Shrubs

18.3 (13.2~21.0); 12.1 (5.4~15.1); 22.5 (18.4~25.6); 18.3 (13.3~21.0); 160.0 (99.4~191.8); 0.0 (0.0~0.0); 897 (695~1053) ❙ 1543 (1337~1751); 815 (655~1007); 65 (51~77) ❙ 75.2 (44.5~145.8); 890 (695~1053); 524.5 (407.1~659.0)

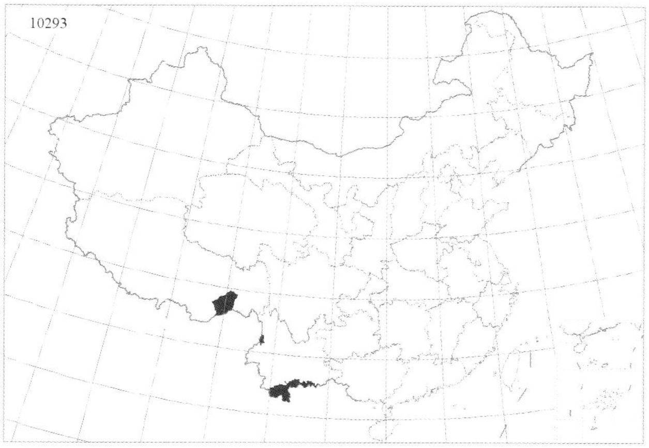

Rubiaceae 茜草科 *Mycetia* 腺萼木属
Mycetia longifolia (Wall.) Kuntze 长叶腺萼木
Shrubs

16.0; 9.1; 21.1; 16.0; 132.1; 0.0; 797 ❙ 1641; 959; 57 ❙ 110.0; 797; 499.4

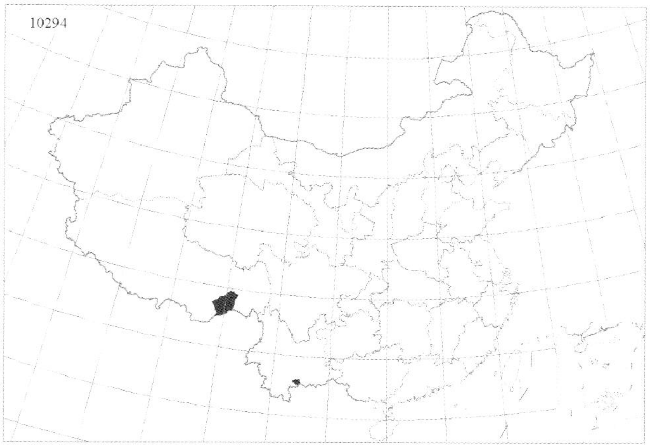

Rubiaceae 茜草科 *Mycetia* 腺萼木属
Mycetia macrocarpa F. C. How ex H. S. Lo 大果腺萼木
Shrubs

17.6; 9.9; 23.2; 17.7; 152.4; 0.0; 859 ❙ 1422; 813; 52 ❙ 65.6; 859; 359.7

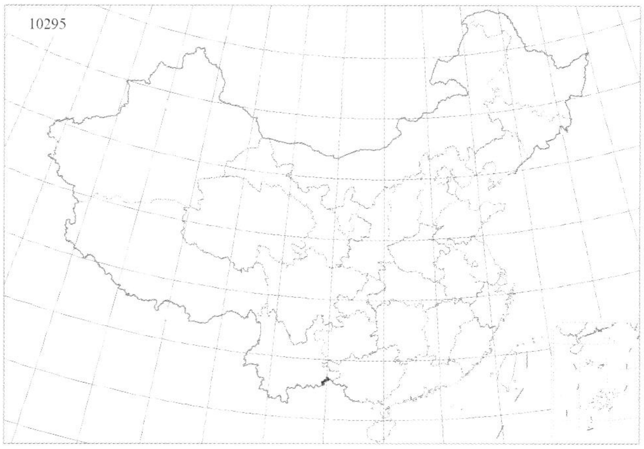

Rubiaceae 茜草科 *Mycetia* 腺萼木属
Mycetia nepalensis (Wall.) Hara 垂花腺萼木
Shrubs

13.3; 5.4; 19.6; 13.4; 100.4; 0.0; 706 ❙ 1736; 1007; 51 ❙ 145.8; 706; 447.4

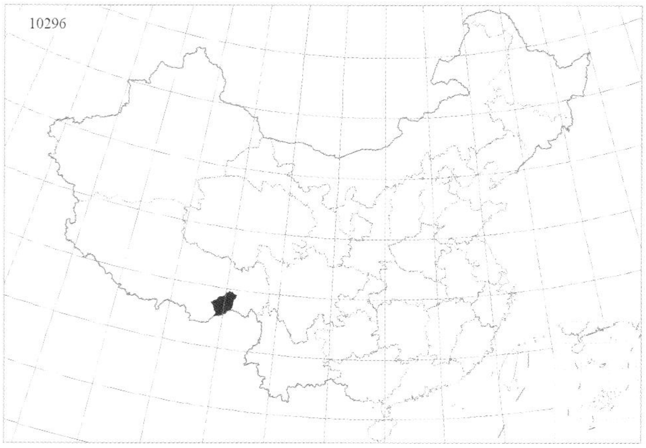

Rubiaceae 茜草科 *Mycetia* 腺萼木属
Mycetia sinensis (Hemsl.) Craib 华腺萼木
Shrubs

19.7 (14.4~24.8); 11.8 (6.3~20.1); 25.8 (19.1~29.8); 19.8 (14.4~24.8); 177.1 (113.1~237.6); 0.0 (0.0~0.0); 1015 (727~1385) ❙ 1515 (868~2435); 681 (312~1428); 111 (39~197) ❙ 51.3 (-34.3~112.2); 996 (727~1360); 328.3 (174.9~659.0)

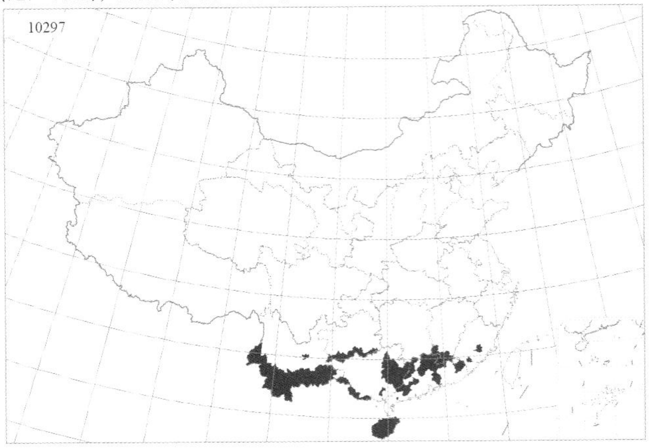

Rubiaceae 茜草科 *Mycetia* 腺萼木属
Mycetia yunnanica Lo 云南腺萼木
Shrubs

17.2 (15.4~19.5); 10.3 (8.1~12.7); 21.5 (20.6~23.3); 17.2 (15.5~19.5); 146.4 (125.7~173.8); 0.0 (0.0~0.0); 833 (767~942) ❙ 1486 (1007~1688); 854 (561~959); 53 (42~62) ❙ 78.2 (31.4~112.2); 829 (767~921); 502.7 (274.9~627.0)

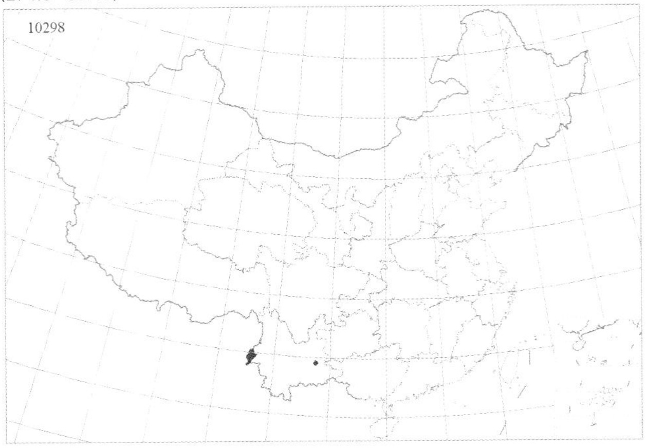

Rubiaceae 茜草科 *Myrioneuron* 密脉木属
Myrioneuron effusum (Pit.) H. L. Li 大叶密脉木
Shrubs

17.7; 10.9; 22.5; 17.8; 153.4; 0.0; 854 ❙ 1368; 756; 59 ❙ 60.1; 854; 491.5

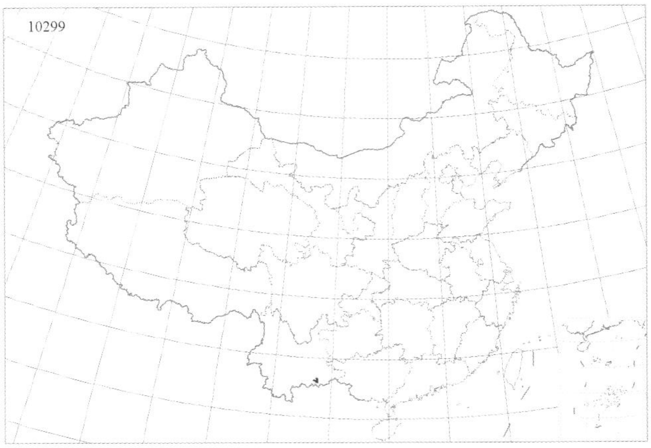

Rubiaceae 茜草科 *Myrioneuron* 密脉木属
Myrioneuron fabri Hemsl. 密脉木
Herbs or subshrubs

17.4 (12.9~22.2); 8.0 (1.9~14.2); 25.5 (21.4~28.3); 17.5 (12.9~22.3); 149.9 (100.9~207.5); 0.4 (0.0~6.1); 899 (723~1190) ❙ 1313 (985~1758); 635 (458~911); 84 (32~221) ❙ 47.2 (5.8~106.4); 898 (723~1190); 258.2 (126.1~551.6)

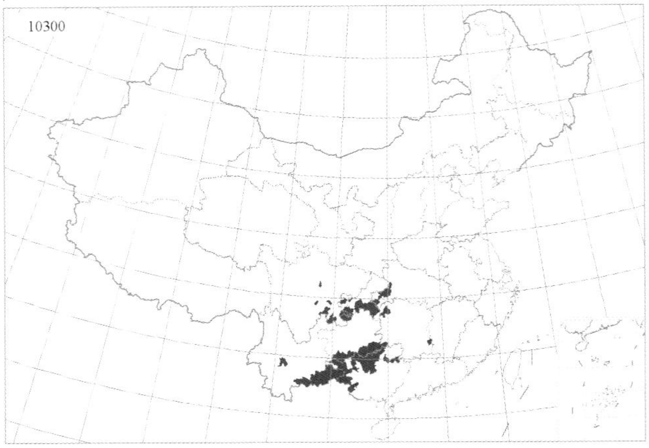

Rubiaceae 茜草科 *Myrioneuron* 密脉木属
Myrioneuron nutans Wall. ex Kurz 垂花密脉木
Small shrubs

19.2 (16.9~21.0); 13.2 (10.3~15.1); 23.2 (21.4~25.6); 19.2 (16.9~21.0); 170.9 (142.7~191.8); 0.0 (0.0~0.0); 927 (813~1053) ❙ 1464 (1178~1751); 775 (646~996); 63 (51~77) ❙ 58.0 (36.8~88.8); 920 (813~1053); 519.3 (383.4~659.0)

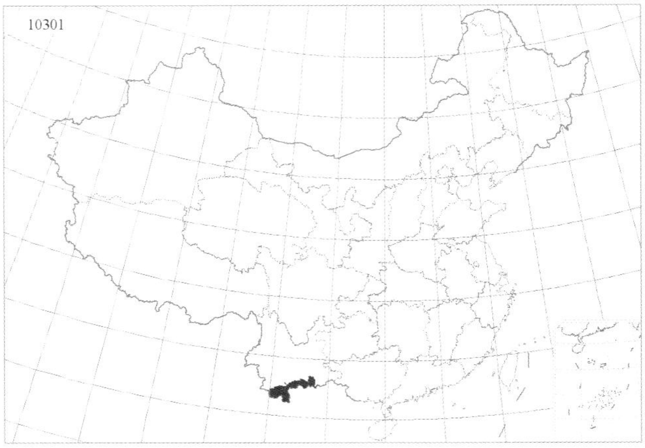

Rubiaceae 茜草科 *Myrioneuron* 密脉木属
Myrioneuron tonkinensis Pit. 越南密脉木
Herbs or shrubs

19.6 (16.3~24.7); 12.4 (9.1~19.8); 24.8 (21.4~28.6); 19.6 (16.4~24.7); 175.6 (136.4~236.5); 0.0 (0.0~0.0); 988 (797~1376) ❙ 1426 (1049~1776); 725 (480~996); 77 (44~177) ❙ 46.2 (-1.4~88.8); 972 (783~1254); 392.7 (223.8~659.0)

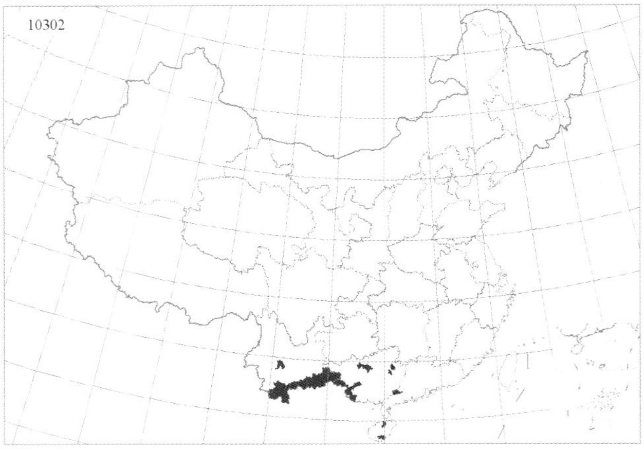

Rubiaceae 茜草科 *Nauclea* 乌檀属
Nauclea officinalis (Pierre ex Pit.) Merr. et Chun 乌檀
Trees

21.2 (17.3~24.8); 13.1 (6.4~20.1); 27.7 (21.6~29.2); 21.3 (17.3~24.8); 195.3 (147.9~237.6); 0.0 (0.0~0.0); 1131 (832~1385) ❙ 1590 (1049~2148); 654 (392~1234); 130 (41~197) ❙ 43.5 (−11.9~86.7); 1095 (783~1331); 295.6 (178.7~620.0)

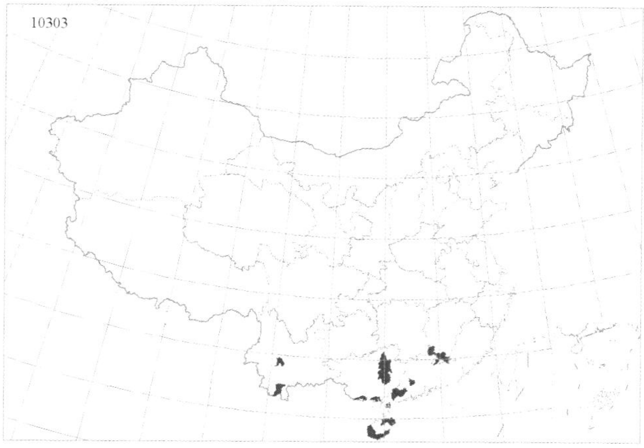

Rubiaceae 茜草科 *Neohymenopogon* 石丁香属
Neohymenopogon oligocarpus (H. L. Li) Bennet 疏果石丁香
Shrubs

18.7 (16.9~21.0); 12.3 (10.3~14.2); 23.0 (21.4~25.6); 18.7 (16.9~21.0); 164.4 (142.7~191.8); 0.0 (0.0~0.0); 903 (813~1053) ❙ 1354 (1163~1547); 764 (641~911); 57 (48~68) ❙ 50.1 (34.8~74.2); 902 (813~1053); 477.5 (383.4~551.6)

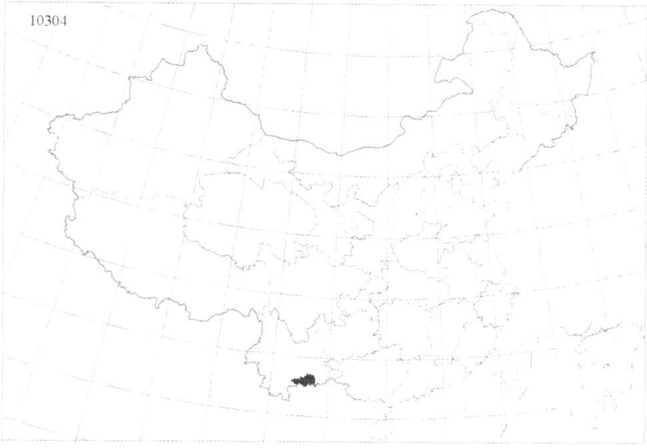

Rubiaceae 茜草科 *Neohymenopogon* 石丁香属
Neohymenopogon parasiticus (Wall.) Bennet 石丁香
Small shrubs

16.6 (−1.8~21.0); 9.9 (−10.2~15.1); 21.3 (6.1~25.6); 16.7 (1.8~21.0); 140.9 (1.9~191.8); 1.0 (0.0~82.8); 817 (329~1053) ❙ 1179 (320~1751); 660 (217~996); 48 (20~77) ❙ 43.7 (−3.0~112.2); 800 (320~1053); 365.7 (72.2~659.0)

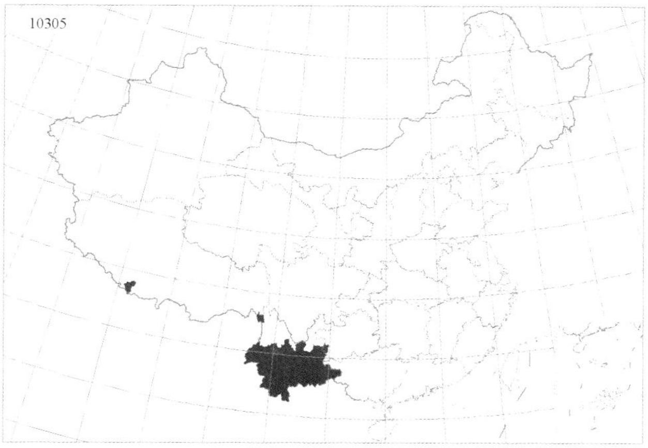

Rubiaceae 茜草科 *Neolamarckia* 团花属
Neolamarckia cadamba (Roxb.) Bosser 团花
Trees, deciduous

18.9 (14.4~24.2); 12.2 (7.5~17.6); 23.3 (19.1~28.8); 18.9 (14.4~24.2); 167.2 (113.1~230.4); 0.0 (0.0~0.0); 936 (727~1331) ❙ 1477 (1083~2435); 794 (581~1428); 67 (41~144) ❙ 59.3 (5.8~112.2); 927 (727~1331); 428.9 (157.7~659.0)

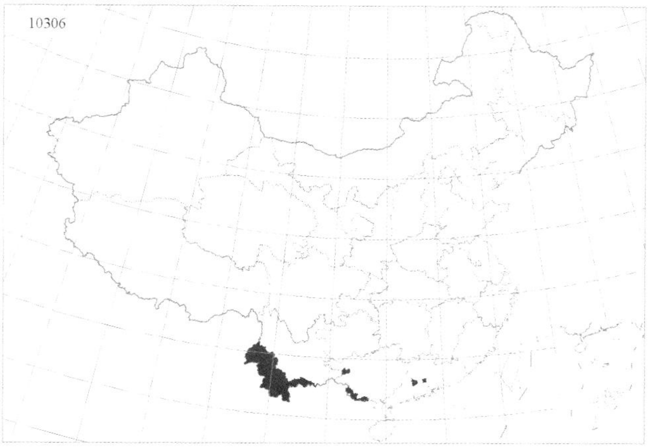

Rubiaceae 茜草科 *Neonauclea* 新乌檀属
Neonauclea griffithii (Hook. f.) Merr. 新乌檀
Trees, evergreen

18.9 (16.3~22.2); 12.6 (9.9~15.1); 22.9 (20.3~28.3); 18.9 (16.4~22.3); 167.0 (136.6~207.5); 0.0 (0.0~0.0); 914 (789~1190) ❙ 1428 (1139~1751); 767 (651~996); 56 (41~87) ❙ 57.1 (15.4~88.8); 899 (780~1190); 465.1 (327.7~659.0)

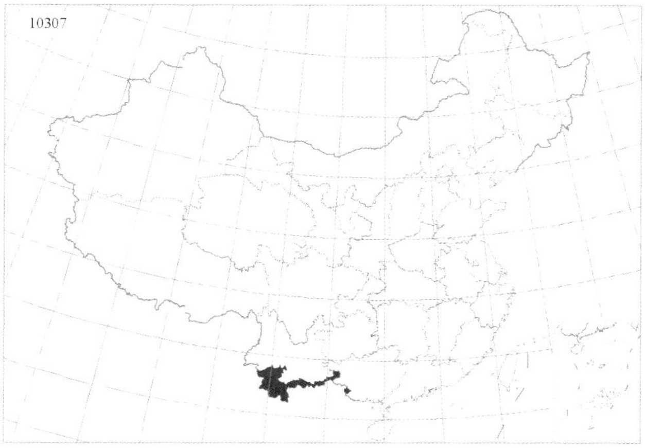

Rubiaceae 茜草科 *Neonauclea* 新乌檀属
Neonauclea sessilifolia (Roxb.) Merr. 无柄新乌檀
Trees

20.0; 14.6; 23.2; 20.1; 180.8; 0.0; 963 ❙ 1523; 681; 67 ❙ 58.4; 942; 581.0

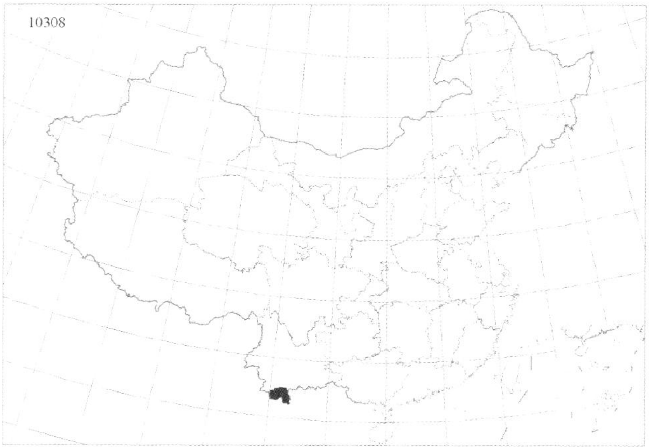

Rubiaceae 茜草科 *Neonauclea* 新乌檀属

Neonauclea truncata (Hayata) Yamam. 台湾新乌檀

Trees, evergreen

21.0；16.3；24.5；21.1；192.6；0.0；1055 ❙ 2554；1499；117 ❙ 154.0；998；266.8

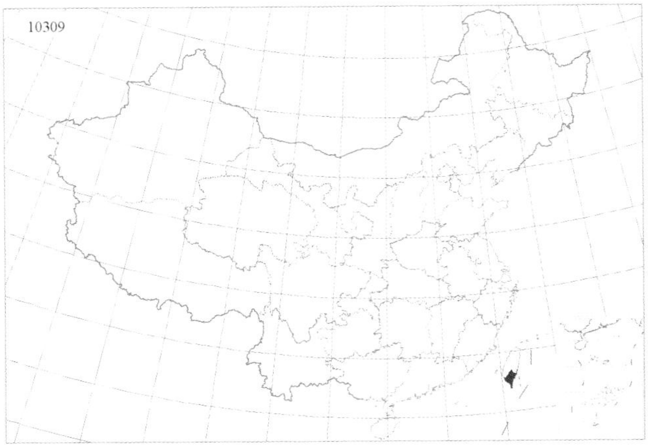

Rubiaceae 茜草科 *Neonauclea* 新乌檀属

Neonauclea tsaiana S. Q. Zou 滇南新乌檀

Trees, evergreen

20.1；14.4；23.6；20.1；181.6；0.0；973 ❙ 1405；664；64 ❙ 44.0；950；525.4

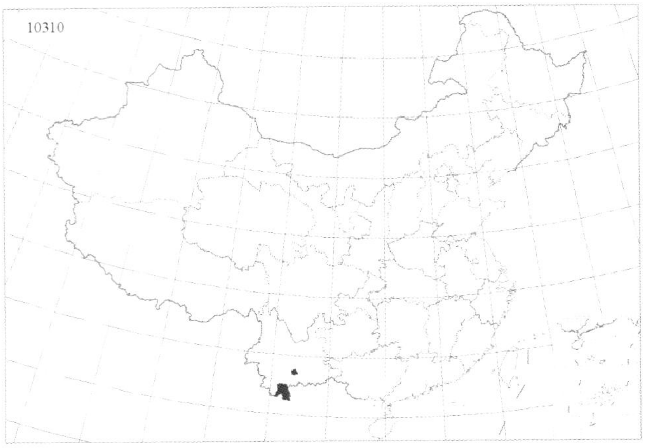

Rubiaceae 茜草科 *Ophiorrhiza* 蛇根草属

Ophiorrhiza lignosa Merr. 木茎蛇根草

Small shrubs

18.2 (16.8~19.1)；9.2 (7.8~10.1)；25.3 (23.2~26.6)；18.2 (16.8~19.1)；159.0 (141.5~169.2)；0.0 (0.0~0.0)；918 (836~969) ❙ 1307 (1236~1422)；695 (617~813)；62 (51~97) ❙ 43.1 (27.5~65.6)；918 (836~969)；265.0 (189.5~359.7)

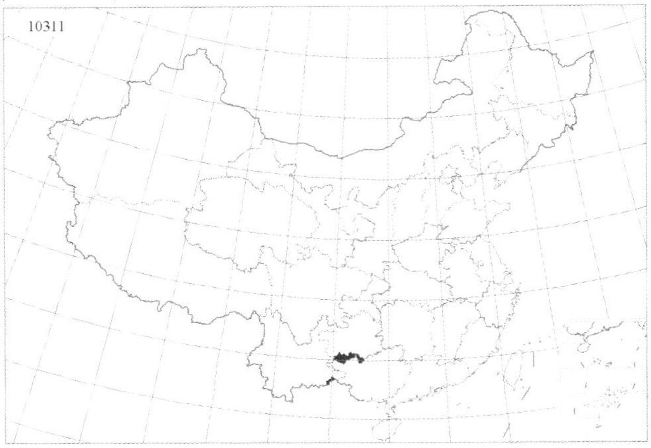

Rubiaceae 茜草科 *Oxyceros* 鸡爪簕属

Oxyceros evenosa (Hutch.) T. Yamaz. 无脉鸡爪簕

Shrubs

18.4 (16.3~21.0)；11.9 (9.1~14.9)；22.8 (21.4~25.6)；18.4 (16.4~21.0)；161.1 (136.4~191.8)；0.0 (0.0~0.0)；893 (797~1053) ❙ 1389 (1178~1522)；726 (646~860)；58 (50~68) ❙ 56.0 (44.5~80.6)；884 (797~1053)；467.3 (347.3~570.1)

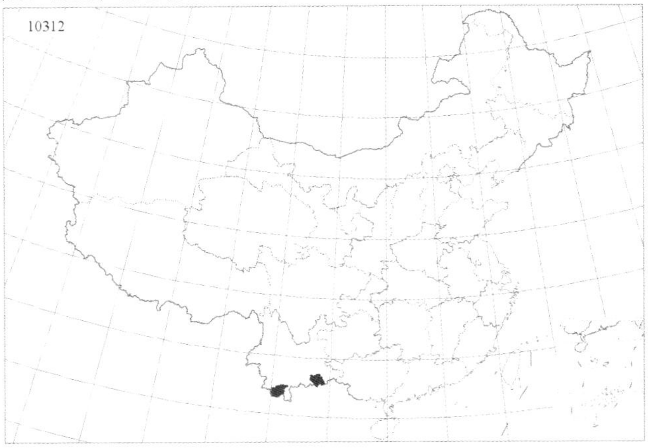

Rubiaceae 茜草科 *Oxyceros* 鸡爪簕属

Oxyceros griffithii (Hook. f.) W. C. Chen 琼滇鸡爪簕

Shrubs or trees

18.2 (12.5~24.8)；11.4 (3.6~20.1)；23.0 (18.4~28.7)；18.3 (12.5~24.8)；159.2 (92.0~237.6)；0.0 (0.0~1.4)；912 (685~1385) ❙ 1303 (868~1813)；680 (312~996)；56 (39~113) ❙ 46.1 (-34.3~112.2)；882 (685~1329)；390.8 (189.5~659.0)

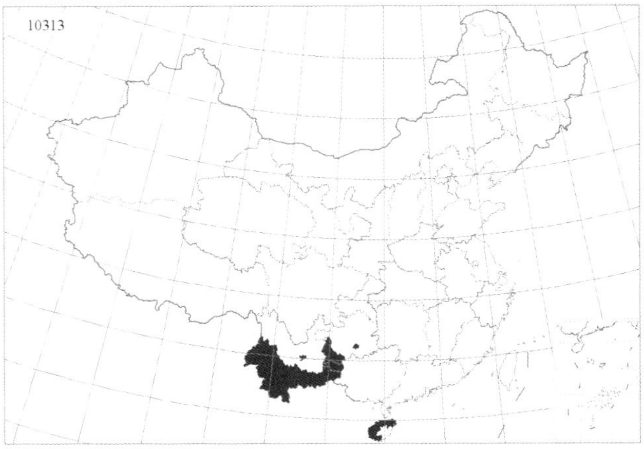

Rubiaceae 茜草科 *Oxyceros* 鸡爪簕属

Oxyceros rectispina (Merr.) T. Yamaz. 直刺鸡爪簕

Shrubs

24.2；18.9；28.0；24.2；230.9；0.0；1331 ❙ 1275；436；42 ❙ -4.2；1095；474.2

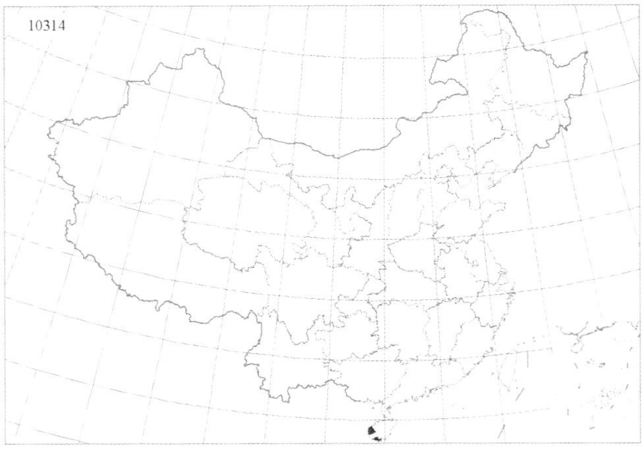

Rubiaceae 茜草科 *Oxyceros* 鸡爪簕属
Oxyceros sinensis Lour. 鸡爪簕
Shrubs or small trees

21.1 (13.6~24.8); 13.7 (6.1~20.1); 27.0 (19.6~29.0); 21.2 (13.7~24.8); 194.3 (103.9~237.6); 0.0 (0.0~0.0); 1108 (713~1385) ❙ 1580 (868~2992); 746 (312~1847); 109 (20~245) ❙ 45.7 (-34.3~246.5); 1071 (691~1329); 278.6 (95.3~659.0)

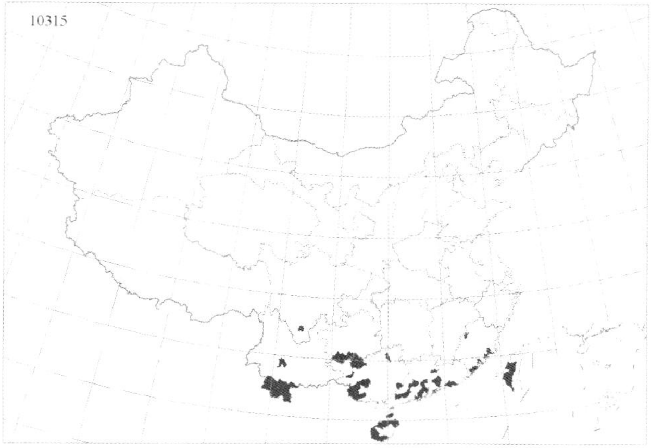

Rubiaceae 茜草科 *Paederia* 鸡矢藤属
Paederia yunnanensis (H. Lév.) Rehder 云南鸡矢藤
Shrubs

16.0 (5.4~22.2); 8.5 (-2.1~15.1); 21.6 (12.3~28.4); 16.0 (5.8~22.3); 132.8 (31.5~207.5); 0.8 (0.0~25.6); 808 (481~1190) ❙ 1149 (741~1751); 635 (411~996); 49 (12~161) ❙ 43.1 (-14.2~112.2); 794 (481~1190); 325.2 (179.0~659.0)

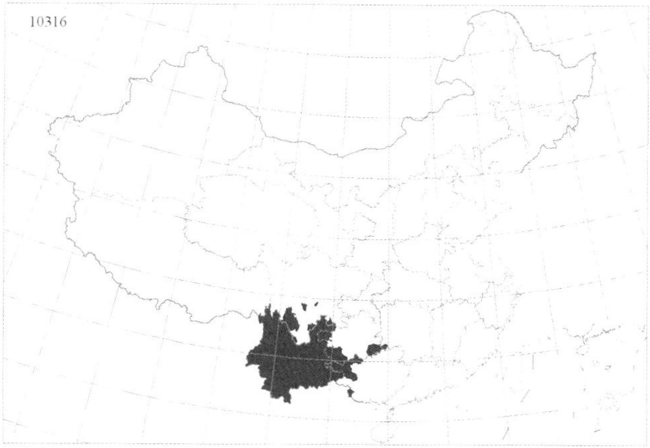

Rubiaceae 茜草科 *Pavetta* 大沙叶属
Pavetta arenosa Lour. 大沙叶
Shrubs

21.4 (16.3~24.8); 13.7 (7.3~20.1); 27.4 (21.4~29.8); 21.5 (16.4~24.8); 197.5 (136.4~237.6); 0.0 (0.0~0.0); 1132 (797~1385) ❙ 1560 (868~2148); 708 (312~1234); 110 (41~194) ❙ 39.9 (-34.3~86.7); 1092 (783~1326); 303.4 (157.7~659.0)

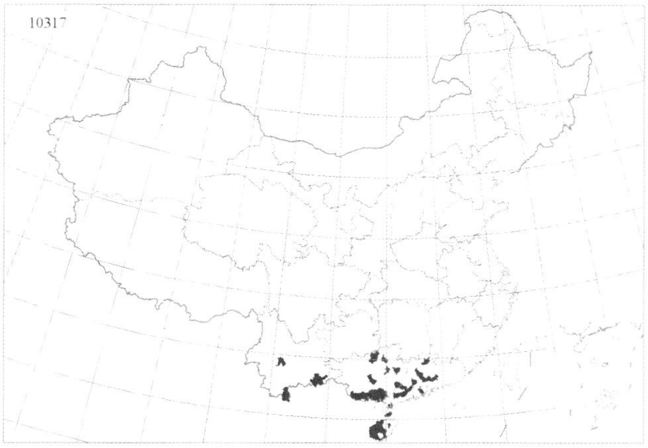

Rubiaceae 茜草科 *Pavetta* 大沙叶属
Pavetta hongkongensis Bremek. 香港大沙叶
Shrubs or small trees

20.7 (16.3~24.8); 12.7 (6.2~20.1); 27.1 (20.8~29.8); 20.8 (16.4~24.8); 189.1 (136.4~237.6); 0.0 (0.0~0.0); 1082 (797~1385) ❙ 1574 (868~2435); 702 (312~1428); 120 (41~197) ❙ 46.8 (-34.3~99.5); 1068 (797~1360); 278.8 (95.3~659.0)

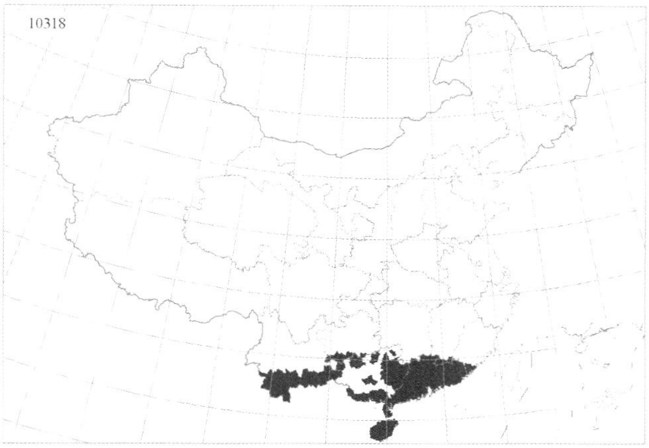

Rubiaceae 茜草科 *Pavetta* 大沙叶属
Pavetta polyantha R. Br ex Bremek. 多花大沙叶
Shrubs

19.3 (15.9~22.7); 11.1 (6.9~15.1); 25.6 (21.4~28.7); 19.4 (16.0~22.8); 172.4 (131.6~213.1); 0.0 (0.0~0.0); 983 (797~1221) ❙ 1501 (1049~2435); 740 (562~1428); 106 (41~194) ❙ 53.0 (21.9~99.5); 979 (783~1221); 316.5 (185.5~659.0)

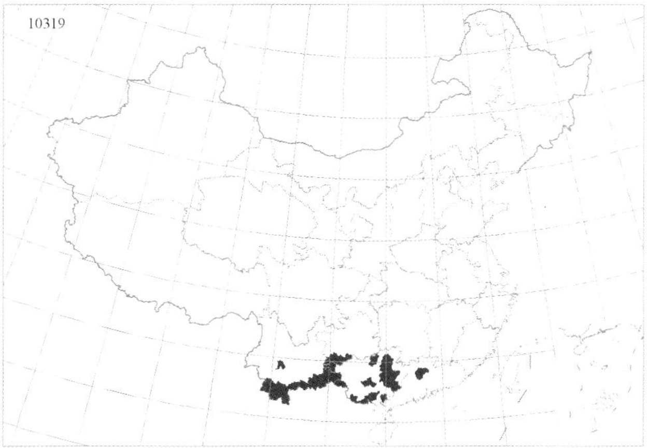

Rubiaceae 茜草科 *Pavetta* 大沙叶属
Pavetta scabrifolia Bremek. 糙叶大沙叶
Shrubs

19.2 (16.9~21.0); 13.2 (10.3~15.1); 23.1 (21.4~25.6); 19.2 (16.9~21.0); 170.7 (142.7~191.8); 0.0 (0.0~0.0); 926 (813~1053) ❙ 1467 (1178~1751); 775 (646~996); 63 (51~77) ❙ 58.6 (36.8~88.8); 918 (813~1053); 511.0 (383.4~659.0)

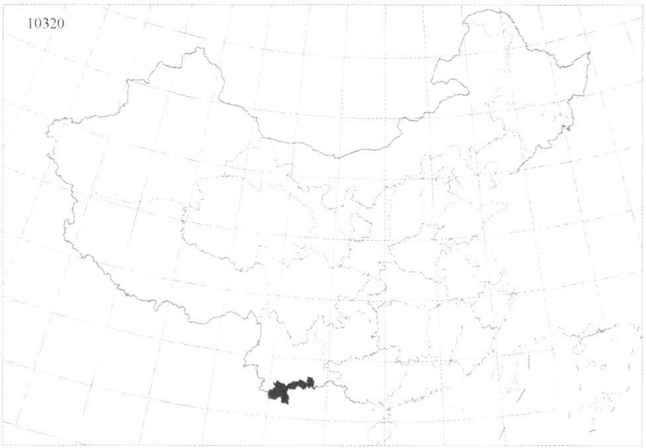

Rubiaceae 茜草科 *Pavetta* 大沙叶属

Pavetta swatouica Bremekamp 汕头大沙叶

Shrubs

21.5; 13.9; 28.2; 21.6; 198.8; 0.0; 1130 ❙ 1537; 657; 131 ❙ 36.1; 1130; 207.9

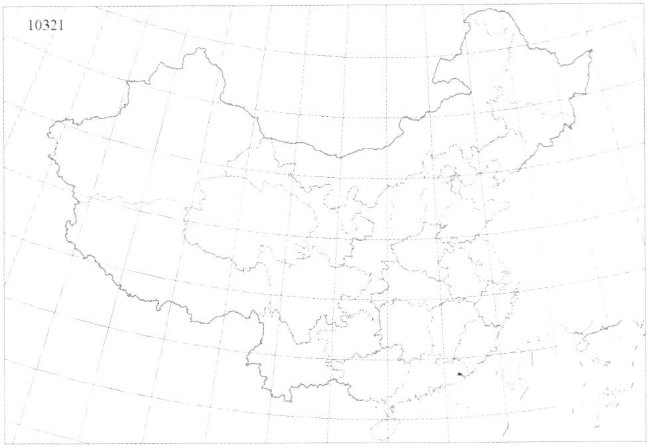

Rubiaceae 茜草科 *Pavetta* 大沙叶属

Pavetta tomentosa Roxb. ex Sm. 绒毛大沙叶

Shrubs or small trees

20.2 (16.2~24.1); 14.5 (11.0~18.9); 25.0 (20.1~28.2); 20.2 (16.2~24.1); 182.9 (134.2~229.7); 0.0 (0.0~0.0); 1019 (773~1322) ❙ 2402 (1498~3495); 1106 (750~1847); 255 (47~897) ❙ 146.2 (25.2~246.5); 983 (773~1144); 245.0 (119.4~363.7)

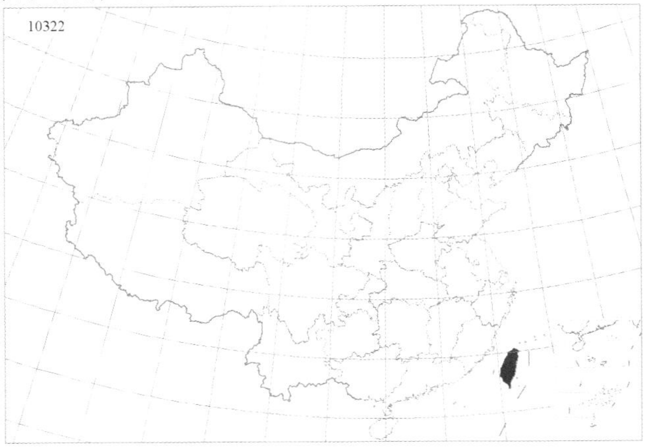

Rubiaceae 茜草科 *Pertusadina* 槽裂木属

Pertusadina hainanensis (F. C. How) Ridsdale 海南槽裂木

Shrubs or trees

18.9 (14.7~24.6); 9.5 (3.8~19.5); 27.2 (24.7~29.4); 18.9 (14.8~24.7); 167.3 (118.7~235.9); 0.1 (0.0~1.5); 1005 (789~1372) ❙ 1517 (868~2037); 577 (312~1033); 137 (42~253) ❙ 54.7 (-34.3~147.0); 973 (789~1329); 259.8 (145.3~620.0)

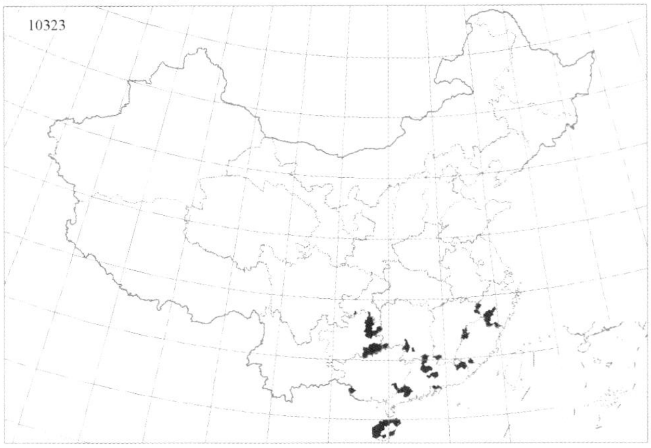

Rubiaceae 茜草科 *Porterandia* 绢冠茜属

Porterandia sericantha (W. C. Chen) W. C. Chen 绢冠茜

Shrubs or trees

18.9 (16.6~22.2); 10.9 (9.1~14.2); 24.7 (21.9~28.3); 18.9 (16.7~22.3); 167.1 (140.0~207.5); 0.0 (0.0~0.0); 946 (811~1190) ❙ 1349 (1074~1522); 741 (582~860); 63 (47~111) ❙ 44.3 (5.8~80.6); 946 (811~1190); 358.7 (248.4~551.6)

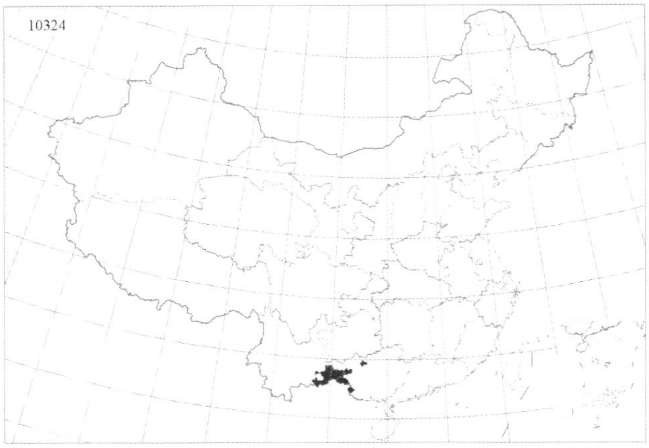

Rubiaceae 茜草科 *Prismatomeris* 南山花属

Prismatomeris connata Y. Z. Ruan 南山花

Shrubs or small trees

22.2 (21.3~23.2); 14.0 (12.9~15.8); 28.4 (27.3~29.0); 22.2 (21.3~23.2); 206.3 (195.9~218.9); 0.0 (0.0~0.0); 1183 (1111~1257) ❙ 1748 (1463~2435); 901 (706~1428); 123 (97~145) ❙ 48.1 (19.2~99.5); 1183 (1111~1257); 239.4 (157.7~318.8)

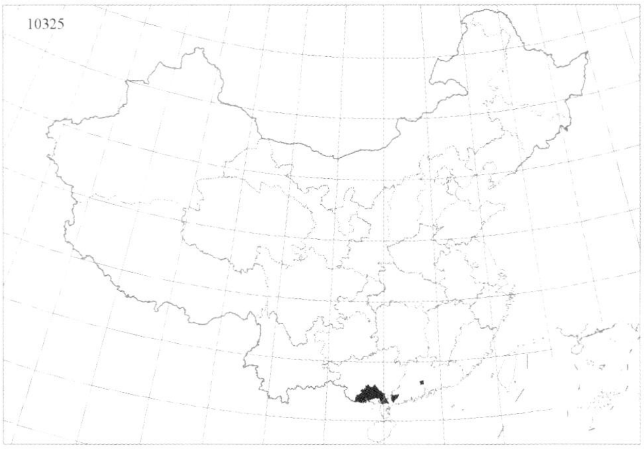

Rubiaceae 茜草科 *Prismatomeris* 南山花属

Prismatomeris connata subsp. hainanensis Y. Z. Ruan 海南三角瓣花

Shrubs or small trees

24.0 (22.2~24.8); 18.4 (17.2~20.1); 27.9 (25.7~28.7); 24.0 (22.3~24.8); 228.2 (207.2~237.6); 0.0 (0.0~0.0); 1308 (1127~1385) ❙ 1503 (868~1854); 531 (312~712); 72 (41~115) ❙ 15.2 (-34.3~47.8); 1184 (868~1360); 449.0 (248.2~620.0)

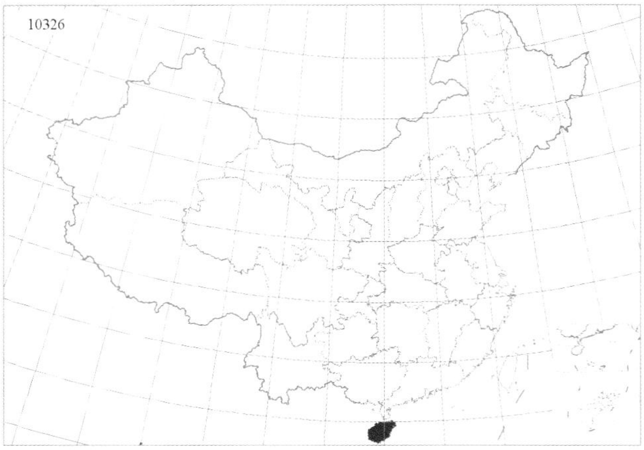

Rubiaceae 茜草科 *Prismatomeris* 南山花属
Prismatomeris tetrantra (Roxb.) K. Schum. 四蕊三角瓣花
Shrubs or small trees

20.0; 14.6; 23.2; 20.1; 180.8; 0.0; 963 ❙ 1523; 681; 67 ❙ 58.4; 942; 581.0

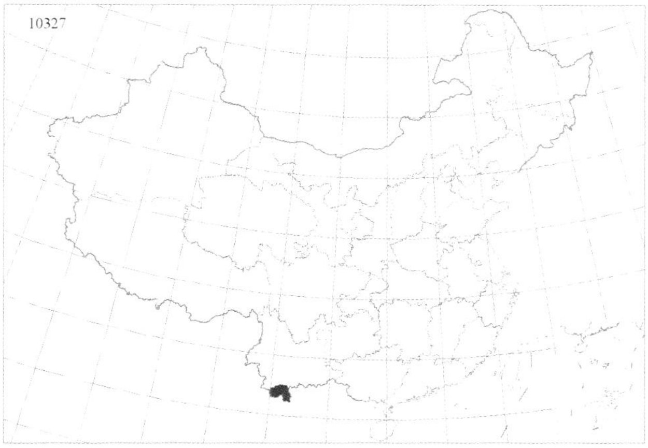

Rubiaceae 茜草科 *Psychotria* 九节属
Psychotria asiatica Wall. 九节
Shrubs or small trees

20.1 (14.6~24.8); 11.7 (3.8~20.1); 27.0 (20.1~29.8); 20.1 (14.7~24.8); 181.4 (117.7~237.6); 0.0 (0.0~1.5); 1045 (773~1385) ❙ 1601 (868~3495); 687 (312~1847); 147 (41~897) ❙ 56.1 (-34.3~246.5); 1032 (773~1360); 268.5 (95.3~620.0)

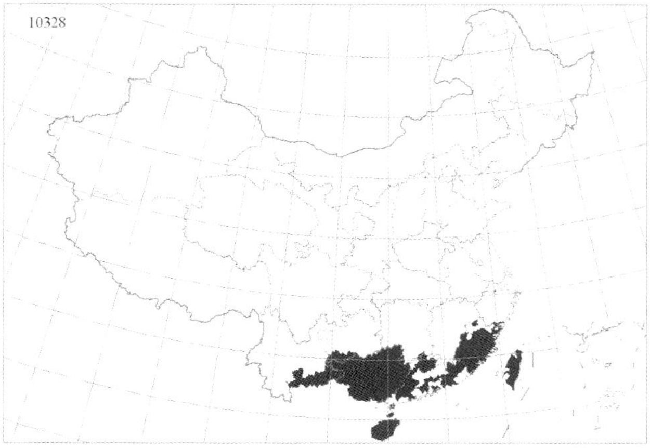

Rubiaceae 茜草科 *Psychotria* 九节属
Psychotria calocarpa Kurz 美果九节
Subshrubs

17.7 (8.6~21.0); 11.3 (1.3~15.1); 22.1 (15.0~25.6); 17.7 (8.7~21.0); 153.6 (52.6~191.8); 0.8 (0.0~8.8); 878 (574~1053) ❙ 1443 (1007~1751); 780 (561~1007); 60 (45~77) ❙ 66.7 (31.4~145.8); 872 (574~1053); 482.4 (274.9~659.0)

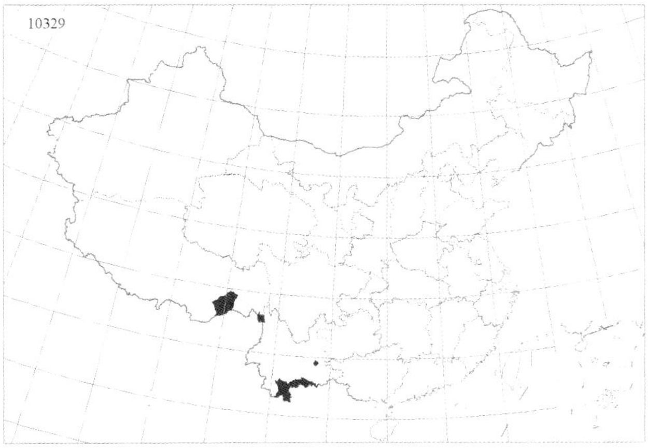

Rubiaceae 茜草科 *Psychotria* 九节属
Psychotria cephalophora Merr. 兰屿九节木
Shrubs

18.5; 14.1; 21.7; 18.5; 162.1; 0.0; 863 ❙ 2713; 1388; 188 ❙ 214.4; 863; 363.7

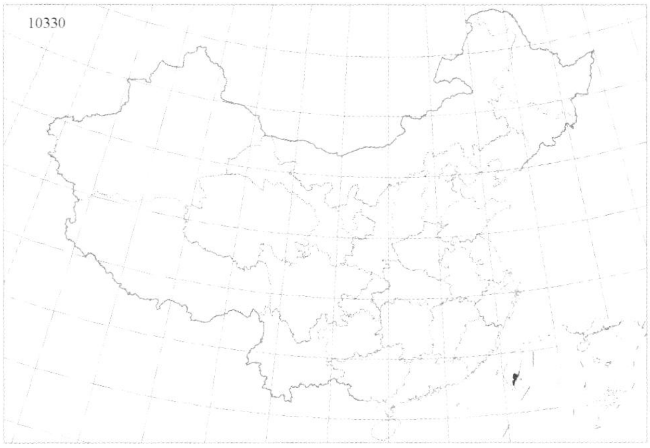

Rubiaceae 茜草科 *Psychotria* 九节属
Psychotria densa W. C. Chen 密脉九节
Shrubs

18.7; 11.7; 23.5; 18.7; 164.1; 0.0; 914 ❙ 1465; 812; 61 ❙ 61.7; 914; 485.5

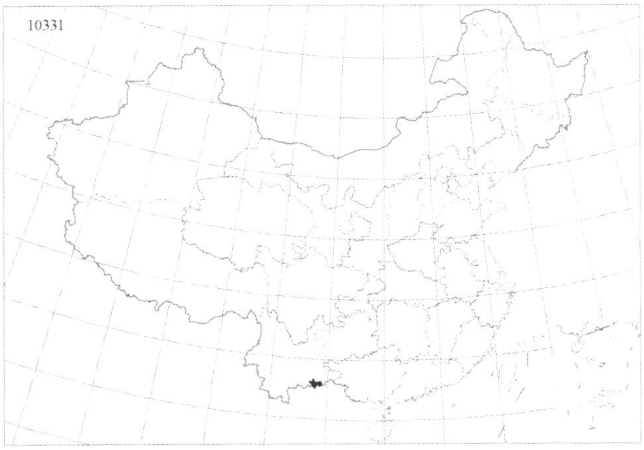

Rubiaceae 茜草科 *Psychotria* 九节属
Psychotria erratica Hook f. 西藏九节
Shrubs

13.3; 5.4; 19.6; 13.4; 100.4; 0.0; 706 ❙ 1736; 1007; 51 ❙ 145.8; 706; 447.4

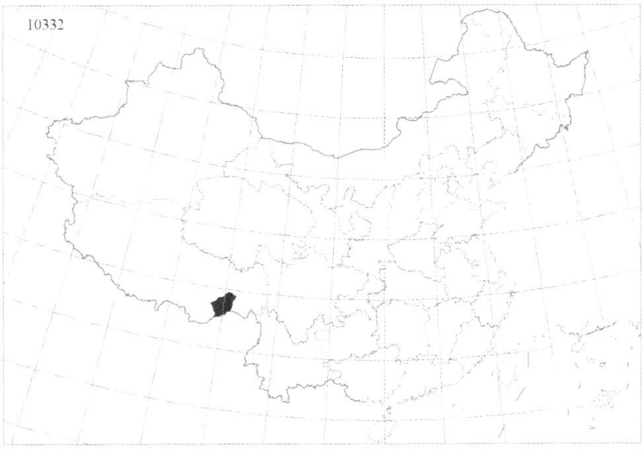

Rubiaceae 茜草科 *Psychotria* 九节属
Psychotria fluviatilis Chun ex W. C. Chen 溪边九节
Shrubs

19.9 (17.7~21.8); 10.8 (7.1~15.2); 27.8 (26.3~29.8); 20.1 (17.8~22.7); 180.6 (153.6~212.7); 0.0 (0.0~0.0); 1049 (916~1216) ❙ 1675 (1462~1990); 675 (508~994); 165 (104~194) ❙ 60.0 (29.9~83.4); 1044 (916~1216); 237.6 (174.9~281.7)

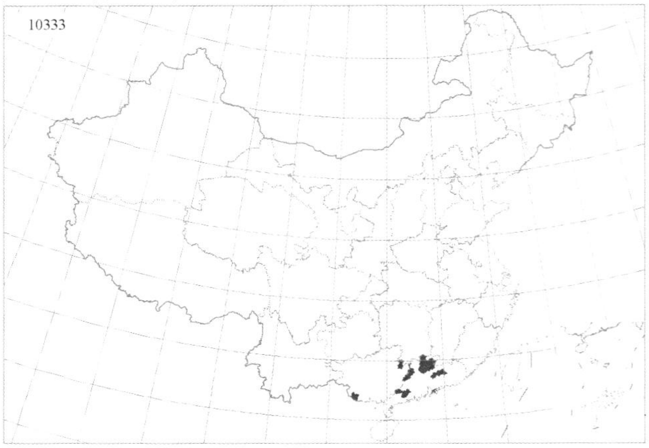

Rubiaceae 茜草科 *Psychotria* 九节属
Psychotria hainanensis H. L. Li 海南九节
Shrubs

23.9 (23.0~24.7); 18.6 (17.5~19.8); 27.6 (26.7~28.4); 24.0 (23.1~24.7); 227.7 (217.0~236.5); 0.0 (0.0~0.0); 1304 (1217~1376) ❙ 1640 (1487~1767); 569 (534~619); 71 (48~110) ❙ 25.9 (15.4~41.8); 1215 (1142~1326); 521.1 (352.9~620.0)

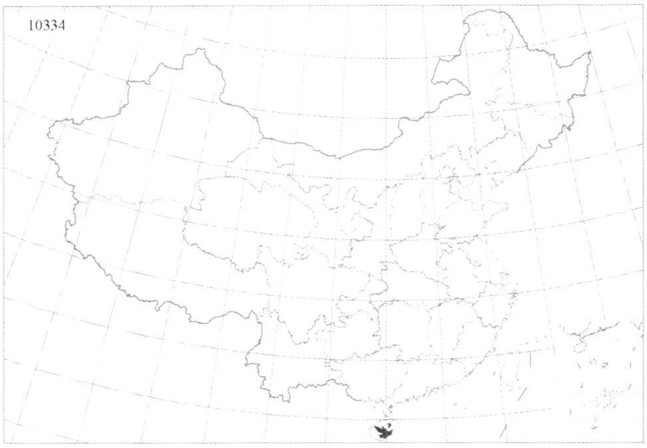

Rubiaceae 茜草科 *Psychotria* 九节属
Psychotria henryi Levl. 滇南九节
Shrubs

19.4 (17.7~21.0); 13.5 (10.9~15.1); 23.3 (22.3~25.6); 19.5 (17.8~21.0); 173.5 (153.4~191.8); 0.0 (0.0~0.0); 937 (854~1053) ❙ 1496 (1276~1751); 787 (655~996); 64 (51~77) ❙ 60.0 (36.8~88.8); 928 (854~1053); 523.8 (427.9~659.0)

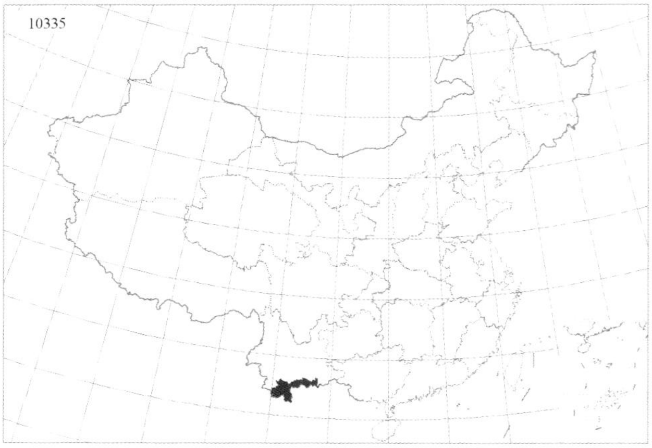

Rubiaceae 茜草科 *Psychotria* 九节属
Psychotria manillensis Bartl. ex DC. 琉球九节木
Shrubs

18.5; 14.1; 21.7; 18.5; 162.1; 0.0; 863 ❙ 2713; 1388; 188 ❙ 214.4; 863; 363.7

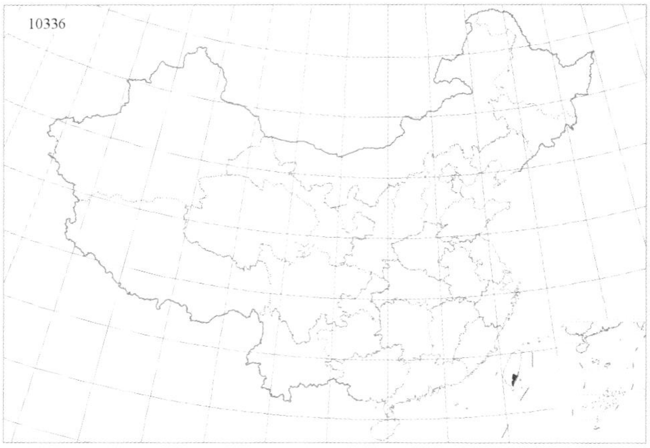

Rubiaceae 茜草科 *Psychotria* 九节属
Psychotria morindoides Hutch. 聚果九节
Shrubs

18.3 (8.6~21.0); 12.1 (1.3~15.1); 22.4 (15.0~25.6); 18.3 (8.7~21.0); 160.4 (52.6~191.8); 0.6 (0.0~8.8); 892 (574~1053) ❙ 1385 (1041~1751); 739 (585~996); 59 (44~77) ❙ 55.9 (21.3~88.8); 884 (574~1053); 476.8 (294.5~659.0)

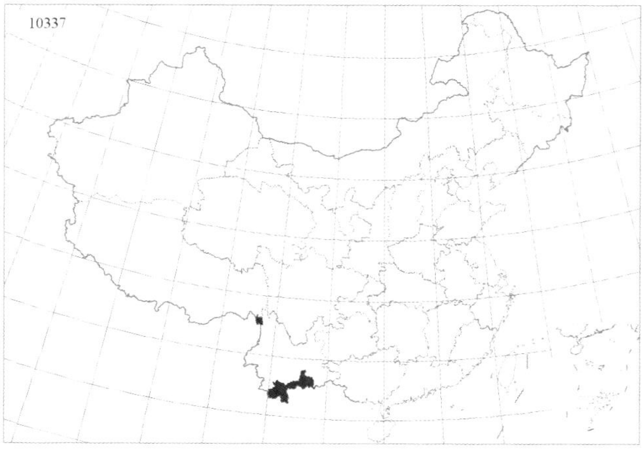

Rubiaceae 茜草科 *Psychotria* 九节属
Psychotria pilifera Hutch. 毛九节
Shrubs

19.0 (16.9~21.0); 12.9 (9.9~15.1); 22.8 (21.4~25.6); 19.0 (16.9~21.0); 168.1 (142.7~191.8); 0.0 (0.0~0.0); 913 (813~1053) ❙ 1482 (1178~1751); 773 (646~996); 59 (41~77) ❙ 62.6 (44.5~88.8); 900 (813~1053); 507.5 (383.4~659.0)

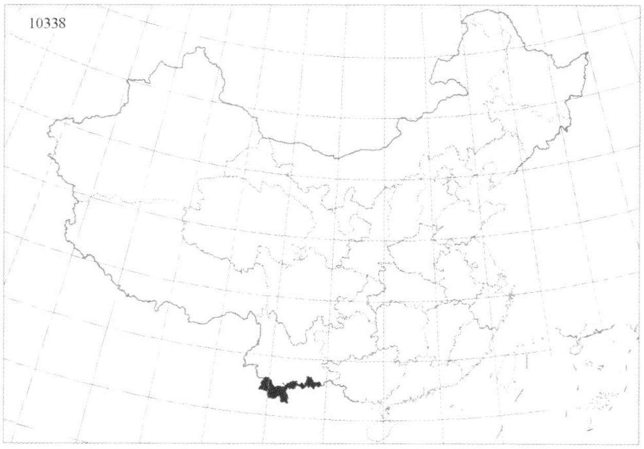

Rubiaceae 茜草科 *Psychotria* 九节属
Psychotria prainii H. Lév. 驳骨九节
Shrubs

18.8 (13.7~22.7); 10.7 (4.7~15.1); 24.8 (20.7~28.8); 18.8 (13.7~22.8); 165.6 (104.8~213.1); 0.0 (0.0~0.3); 947 (722~1221) ❙ 1386 (958~2435); 721 (531~1428); 80 (41~186) ❙ 46.7 (12.9~99.5); 943 (722~1221); 324.6 (188.6~659.0)

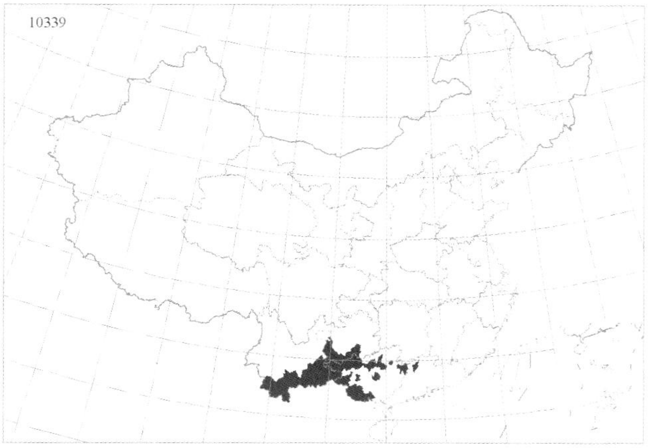

Rubiaceae 茜草科 *Psychotria* 九节属
Psychotria serpens L. 蔓九节
Vines

20.7 (15.9~24.8); 12.7 (5.3~20.1); 27.5 (20.1~29.0); 20.8 (16.0~24.8); 189.0 (131.9~237.6); 0.0 (0.0~0.0); 1086 (773~1385) ❙ 1684 (868~3495); 698 (312~1847); 155 (41~897) ❙ 58.5 (-34.3~246.5); 1065 (773~1360); 266.5 (122.7~620.0)

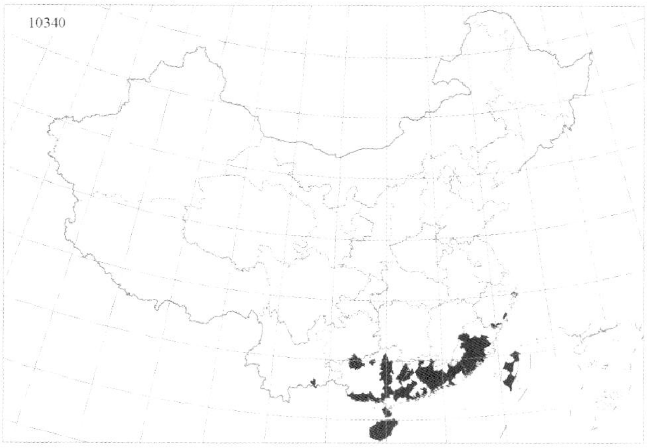

Rubiaceae 茜草科 *Psychotria* 九节属
Psychotria straminea Hutch. 黄脉九节
Shrubs

20.4 (16.0~24.8); 13.7 (9.1~20.1); 25.1 (20.3~29.0); 20.5 (16.0~24.8); 185.5 (132.4~237.6); 0.0 (0.0~0.0); 1047 (780~1385) ❙ 1500 (868~2435); 731 (312~1428); 77 (41~188) ❙ 46.6 (-34.3~112.2); 1013 (780~1360); 404.6 (194.5~659.0)

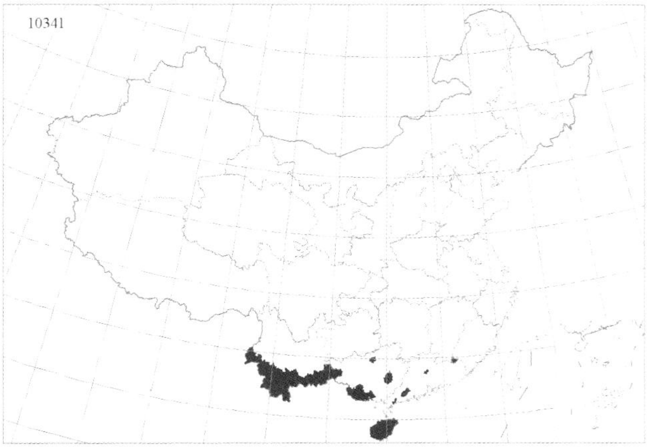

Rubiaceae 茜草科 *Psychotria* 九节属
Psychotria symplocifolia Kurz 山矾叶九节
Shrubs or small trees

19.3 (17.3~20.5); 13.6 (11.0~15.1); 22.8 (21.6~23.7); 19.3 (17.3~20.5); 171.8 (148.0~185.9); 0.0 (0.0~0.0); 926 (832~992) ❙ 1423 (1049~1607); 683 (598~775); 61 (44~77) ❙ 53.1 (26.1~64.9); 900 (783~986); 507.2 (365.3~659.0)

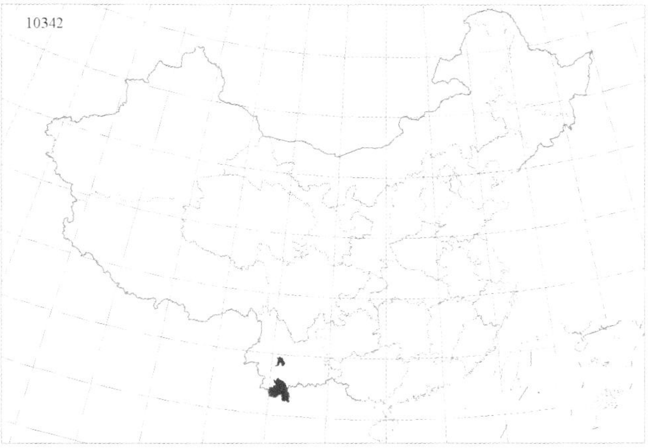

Rubiaceae 茜草科 *Psychotria* 九节属
Psychotria tutcheri Dunn 假九节
Shrubs

20.7 (17.6~24.7); 12.1 (7.1~19.8); 27.8 (25.8~28.8); 20.8 (17.6~24.7); 189.9 (151.6~236.5); 0.0 (0.0~0.0); 1097 (909~1376) ❙ 1642 (1269~2435); 713 (455~1428); 141 (48~197) ❙ 51.0 (-1.0~99.5); 1086 (909~1360); 269.0 (185.5~620.0)

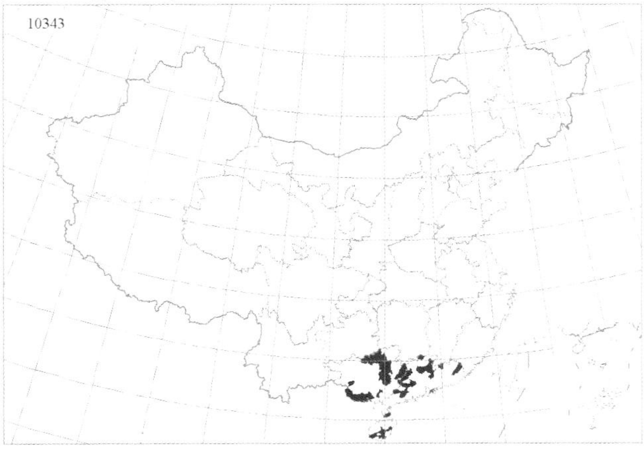

Rubiaceae 茜草科 *Psychotria* 九节属
Psychotria yunnanensis Hutch. 云南九节
Shrubs

18.2 (13.3~21.4); 11.4 (5.2~15.1); 23.0 (19.6~28.2); 18.3 (13.4~21.5); 159.2 (100.4~197.8); 0.0 (0.0~0.0); 888 (706~1140) ❙ 1326 (1007~1751); 719 (507~1007); 56 (41~130) ❙ 49.8 (18.3~145.8); 876 (706~1140); 399.1 (149.6~659.0)

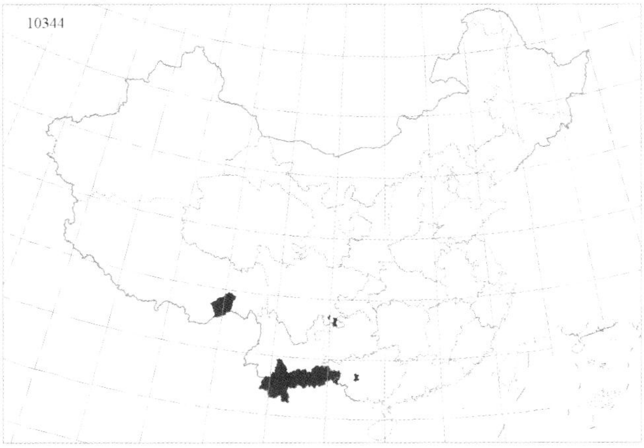

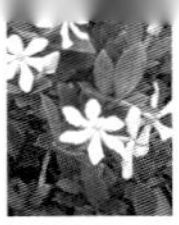

Rubiaceae 茜草科 *Saprosma* 染木树属
Saprosma crassipes Lo 厚梗染木树
Shrubs

23.1 (19.2~24.8); 17.7 (13.8~20.1); 26.8 (22.3~28.7); 23.1 (19.2~24.8); 217.7 (170.5~237.6); 0.0 (0.0~0.0); 1231 (909~1385) ❙ 1440 (868~1767); 530 (312~702); 58 (41~88) ❙ 19.8 (-34.3~64.9); 1083 (868~1299); 512.4 (288.3~659.0)

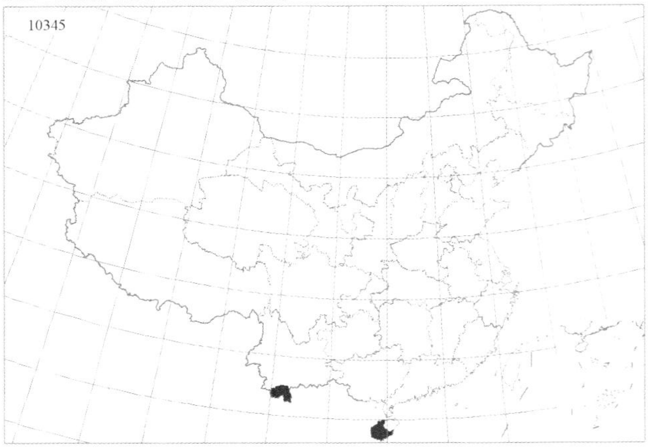

Rubiaceae 茜草科 *Saprosma* 染木树属
Saprosma hainanense Merr. 海南染木树
Shrubs

24.2 (23.6~24.8); 19.0 (17.8~20.1); 28.0 (27.2~28.7); 24.3 (23.6~24.8); 231.4 (223.3~237.6); 0.0 (0.0~0.0); 1334 (1271~1385) ❙ 1411 (1176~1767); 481 (392~567); 54 (41~88) ❙ 5.7 (-12.0~28.8); 1149 (1056~1299); 494.0 (411.8~620.0)

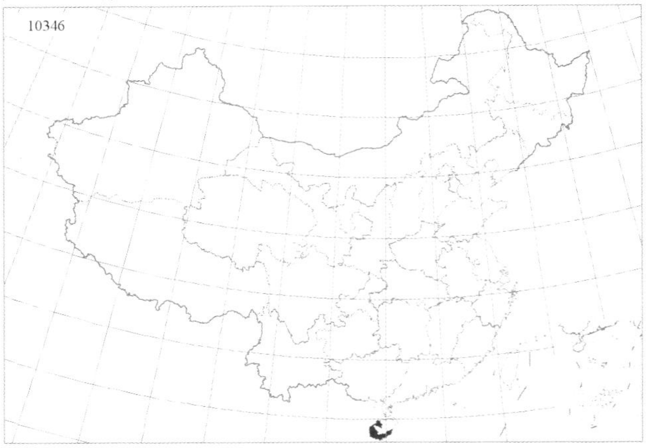

Rubiaceae 茜草科 *Saprosma* 染木树属
Saprosma henryi Hutch. 云南染木树
Shrubs

19.0; 13.1; 22.6; 19.1; 168.8; 0.0; 909 ❙ 1495; 775; 62 ❙ 64.5; 891; 427.9

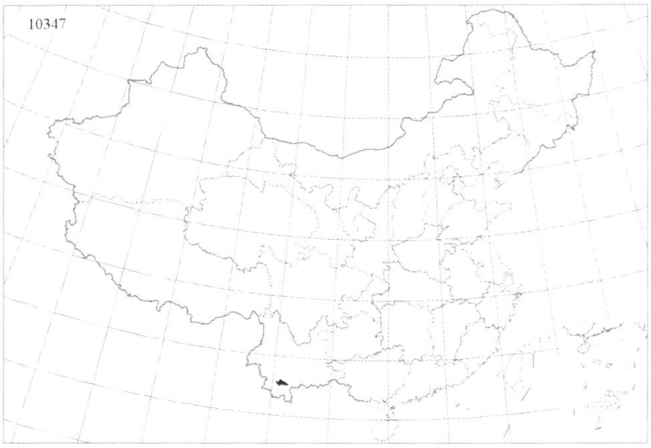

Rubiaceae 茜草科 *Saprosma* 染木树属
Saprosma merrillii Lo 琼岛染木树
Shrubs

24.0 (23.0~24.8); 18.7 (17.5~20.1); 27.8 (26.7~28.6); 24.1 (23.1~24.8); 228.8 (217.0~237.6); 0.0 (0.0~0.0); 1313 (1217~1385) ❙ 1531 (1242~1854); 527 (444~622); 65 (41~115) ❙ 16.7 (-2.2~41.8); 1192 (1091~1360); 497.6 (351.4~620.0)

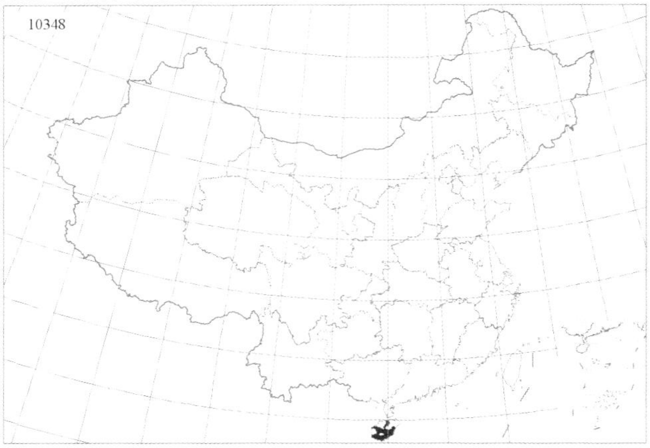

Rubiaceae 茜草科 *Saprosma* 染木树属
Saprosma ternatum Hook. f. 染木树
Shrubs

19.1 (5.3~24.7); 12.8 (-2.4~19.8); 23.3 (12.2~28.4); 19.1 (5.7~24.7); 170.5 (30.5~236.5); 1.3 (0.0~26.2); 967 (479~1376) ❙ 1454 (961~1767); 709 (534~996); 58 (31~110) ❙ 54.3 (15.4~100.7); 939 (479~1326); 469.1 (234.2~659.0)

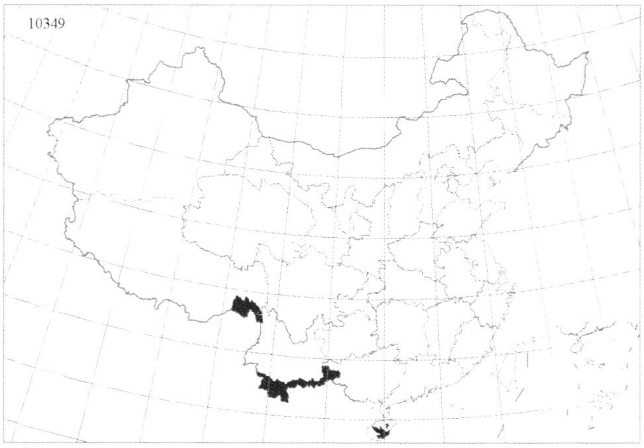

Rubiaceae 茜草科 *Schizomussaenda* 裂果金花属
Schizomussaenda dehiscens (Craib) H. L. Li 裂果金花
Tall shrubs

19.3 (16.3~22.7); 12.2 (9.0~15.1); 24.2 (21.4~28.9); 19.3 (16.4~22.8); 171.6 (136.4~213.1); 0.0 (0.0~0.0); 956 (797~1221) ❙ 1422 (1007~2435); 770 (539~1428); 70 (41~136) ❙ 48.8 (15.4~99.5); 948 (797~1221); 381.7 (201.4~659.0)

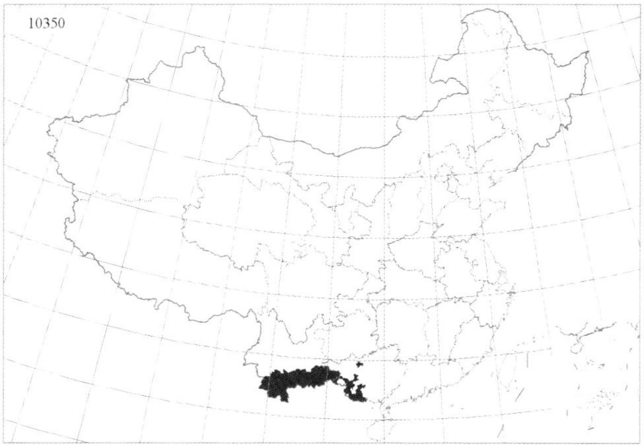

Rubiaceae 茜草科 *Scyphiphora* 瓶花木属
Scyphiphora hydrophyllacea C. F. Gaertn. 瓶花木
Shrubs or small trees

24.2 (23.0~24.8); 18.5 (17.6~20.1); 28.2 (26.7~28.8); 24.2 (23.1~24.8); 230.9 (217.0~237.6); 0.0 (0.0~0.0); 1332 (1217~1385) ❙ 1592 (1309~1813); 574 (473~712); 83 (41~113) ❙ 19.8 (-1.4~41.8); 1256 (1137~1331); 413.9 (248.2~572.9)

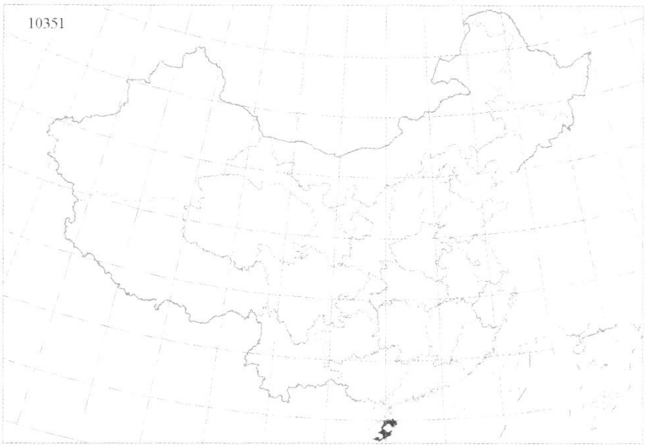

Rubiaceae 茜草科 *Serissa* 白马骨属
Serissa japonica (Thunb.) Thunb. 六月雪
Small shrubs

16.1 (2.2~22.4); 5.7 (-6.6~15.1); 25.4 (9.9~29.7); 16.1 (3.8~22.5); 135.4 (15.5~209.7); 1.7 (0.0~49.1); 855 (413~1205) ❙ 1259 (626~2014); 549 (316~996); 108 (19~259) ❙ 47.0 (-22.2~130.7); 844 (413~1205); 232.4 (110.1~659.0)

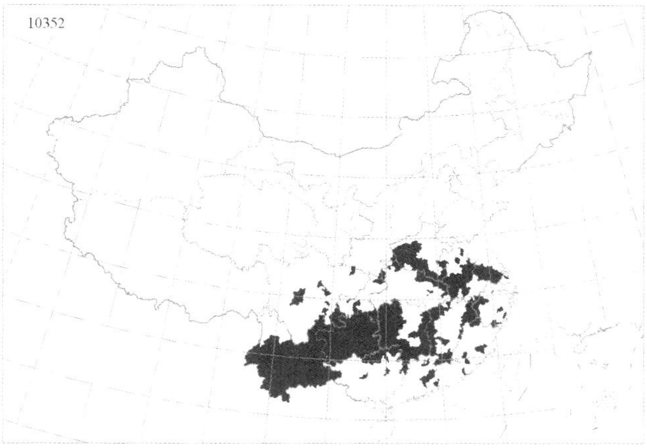

Rubiaceae 茜草科 *Serissa* 白马骨属
Serissa serissoides (DC.) Druce 白马骨
Small shrubs

16.5 (1.9~23.6); 5.0 (-9.2~17.8); 27.1 (11.7~30.1); 16.5 (4.3~23.6); 140.4 (21.1~223.5); 2.4 (0.0~57.8); 893 (427~1281) ❙ 1313 (429~2992); 521 (228~1847); 136 (8~303) ❙ 46.9 (-23.1~219.3); 875 (427~1269); 203.3 (85.5~439.7)

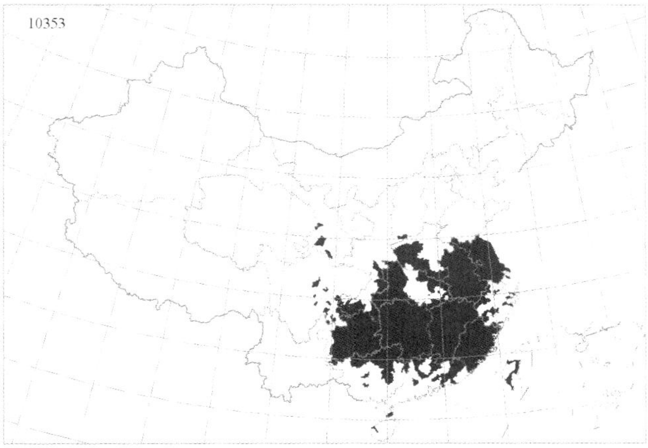

Rubiaceae 茜草科 *Sinoadina* 鸡仔木属
Sinoadina racemosa (Siebold et Zucc.) Ridsdale 鸡仔木
Trees, semievergreen or deciduous

16.4 (5.4~24.1); 6.7 (-2.9~18.2); 24.8 (12.3~29.9); 16.4 (5.8~24.2); 138.5 (31.5~230.2); 1.7 (0.0~25.6); 862 (481~1329) ❙ 1310 (626~3495); 584 (316~1847); 110 (15~897) ❙ 51.5 (-22.2~246.5); 847 (481~1329); 262.6 (134.1~659.0)

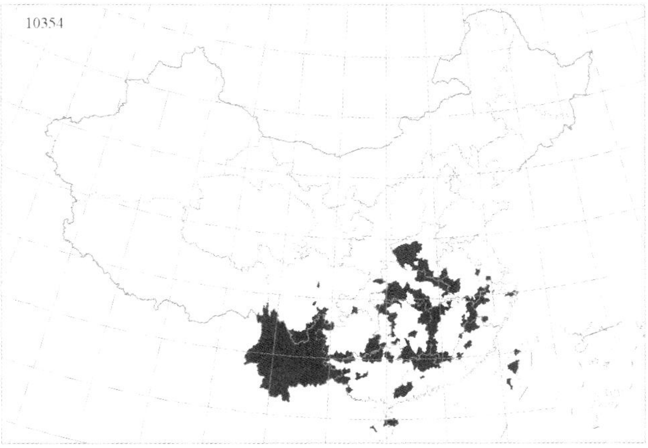

Rubiaceae 茜草科 *Spermadictyon* 香叶木属
Spermadictyon suaveolens Roxb. 香叶木
Subshrubs

0.6 (-1.8~12.2); -8.4 (-10.5~4.3); 8.7 (6.1~18.5); 3.4 (1.8~12.2); 14.3 (1.9~87.4); 66.3 (0.7~82.8); 376 (329~672) ❙ 391 (255~1431); 262 (191~817); 10 (1~33) ❙ -4.2 (-31.1~113.0); 328 (255~672); 108.9 (60.4~363.7)

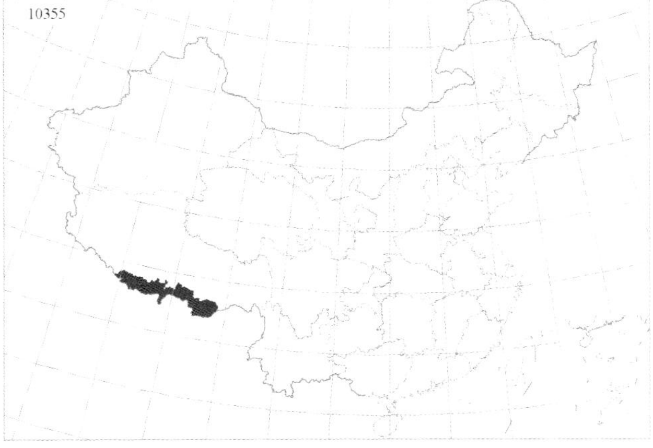

Rubiaceae 茜草科 *Tarenna* 乌口树属
Tarenna acutisepala How ex W. C. Chen 尖萼乌口树
Shrubs

18.2 (14.3~22.3); 8.4 (3.6~14.1); 27.2 (23.5~29.8); 18.2 (14.3~22.4); 158.9 (112.6~208.2); 0.1 (0.0~1.4); 951 (756~1198) ❙ 1611 (1070~2024); 566 (356~954); 191 (66~259) ❙ 71.4 (-5.0~147.0); 943 (756~1198); 241.0 (111.7~332.5)

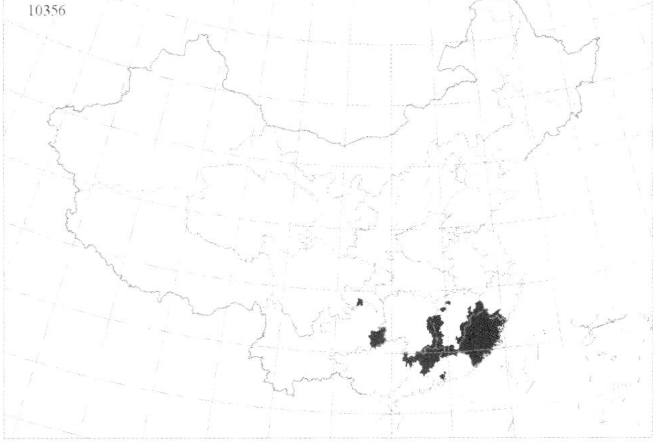

Rubiaceae 茜草科 *Tarenna* 乌口树属
Tarenna tsangii Merr. 海南乌口树
Shrubs or trees

23.1 (17.8~24.8); 16.8 (7.1~20.1); 27.7 (22.5~29.0); 23.1 (17.8~24.8); 217.2 (154.0~237.6); 0.0 (0.0~0.0); 1249 (888~1385) ❙ 1623 (1380~1900); 649 (455~1056); 90 (41~182) ❙ 32.1 (1.5~74.2); 1191 (888~1360); 400.5 (194.5~620.0)

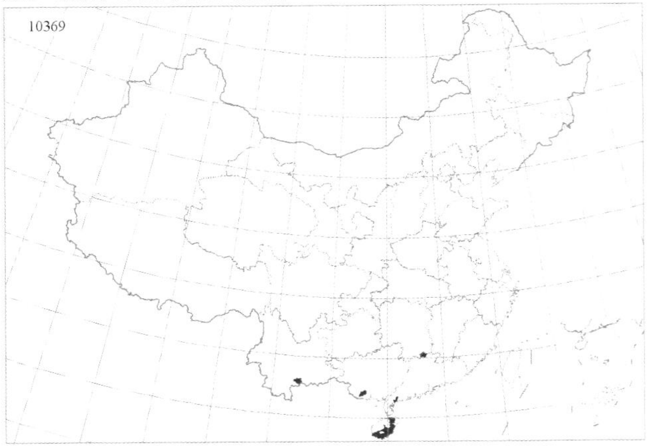

Rubiaceae 茜草科 *Tarenna* 乌口树属
Tarenna tsangii f. elliptica W. Y. Chun et F. C. How ex W. C. Ko 椭圆叶乌口树
Shrubs or trees

24.3; 19.5; 27.6; 24.4; 232.3; 0.0; 1339 ❙ 1569; 536; 53 ❙ 17.2; 1180; 557.4

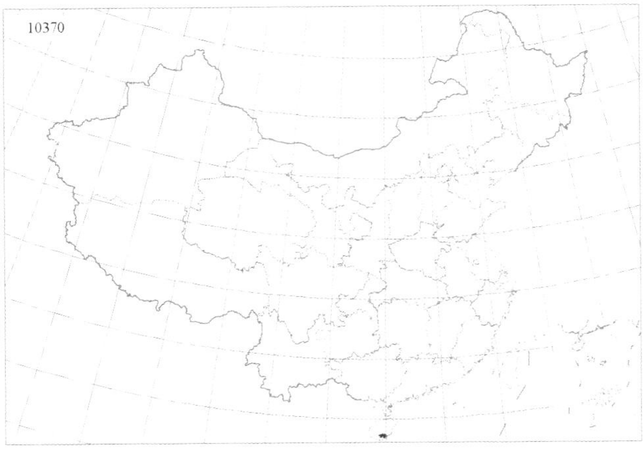

Rubiaceae 茜草科 *Tarenna* 乌口树属
Tarenna wangii Chun et How ex W. C. Chen 长叶乌口树
Shrubs

19.5 (18.6~20.5); 13.8 (12.7~15.1); 23.1 (22.3~23.7); 19.6 (18.7~20.5); 174.8 (163.9~185.9); 0.0 (0.0~0.0); 937 (888~992) ❙ 1549 (1428~1751); 797 (655~996); 66 (56~77) ❙ 65.7 (47.6~88.8); 926 (880~986); 552.4 (506.2~659.0)

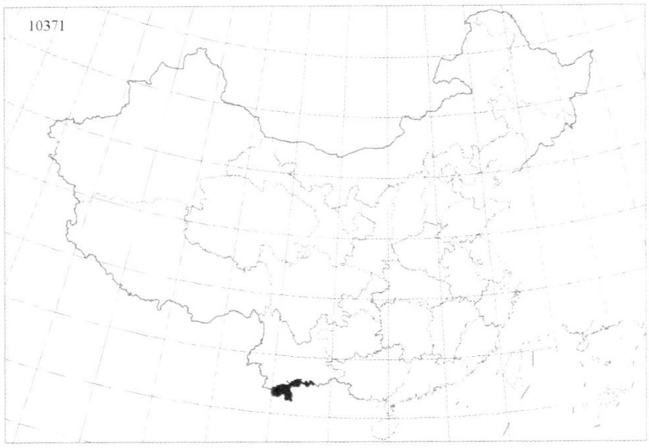

Rubiaceae 茜草科 *Tarenna* 乌口树属
Tarenna yunnanensis How ex W. C. Chen 云南乌口树
Shrubs or trees

20.0; 13.6; 24.4; 20.1; 180.7; 0.0; 985 ❙ 1475; 827; 62 ❙ 50.1; 985; 528.9

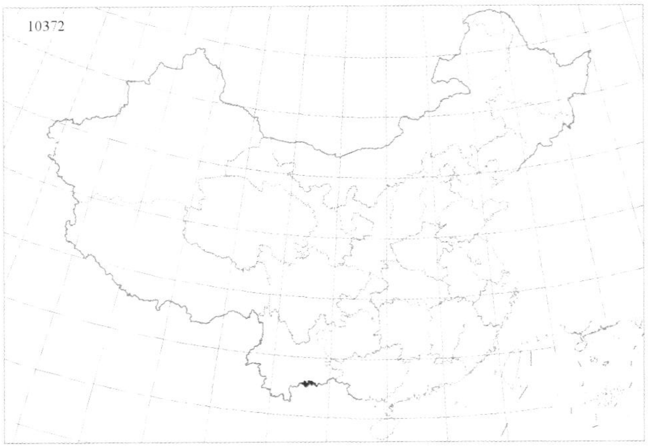

Rubiaceae 茜草科 *Tarenna* 乌口树属
Tarenna zeylanica Gaertn. 锡兰玉心花
Shrubs, evergreen

22.2; 17.7; 25.5; 22.2; 206.7; 0.0; 1123 ❙ 2992; 1847; 107 ❙ 166.5; 1123; 296.9

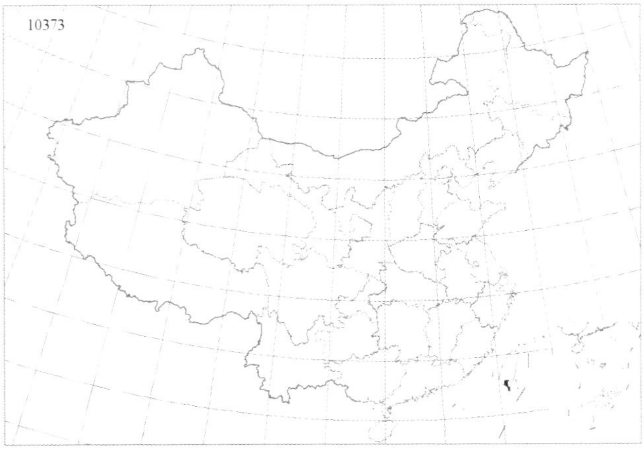

Rubiaceae 茜草科 *Tarennoidea* 岭罗麦属
Tarennoidea wallichii (Hook. f.) Tirveng. et Sastre 岭罗麦
Trees

16.8 (5.4~24.8); 9.5 (-2.1~20.1); 22.3 (12.3~28.8); 16.8 (5.8~24.8); 142.6 (31.5~237.6); 0.5 (0.0~25.6); 856 (481~1385) ❙ 1196 (741~2035); 628 (312~996); 51 (15~115) ❙ 41.5 (-34.3~112.2); 831 (481~1360); 333.3 (142.9~659.0)

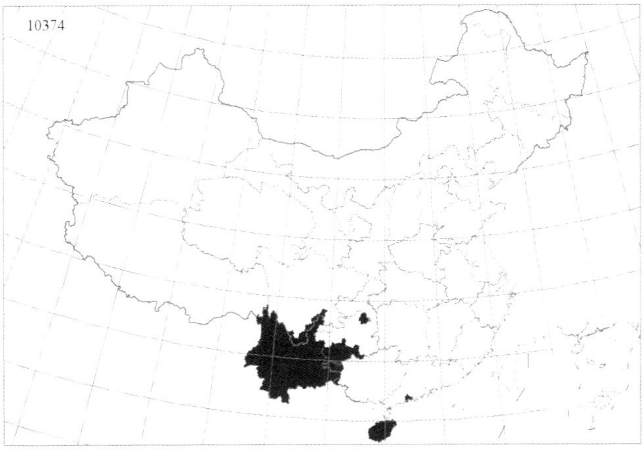

Rubiaceae 茜草科 *Timonius* 海茜树属
Timonius arboreus Elmer 海茜树
Trees, evergreen

18.5；14.1；21.7；18.5；162.1；0.0；863 ❙ 2713；1388；188 ❙ 214.4；863；363.7

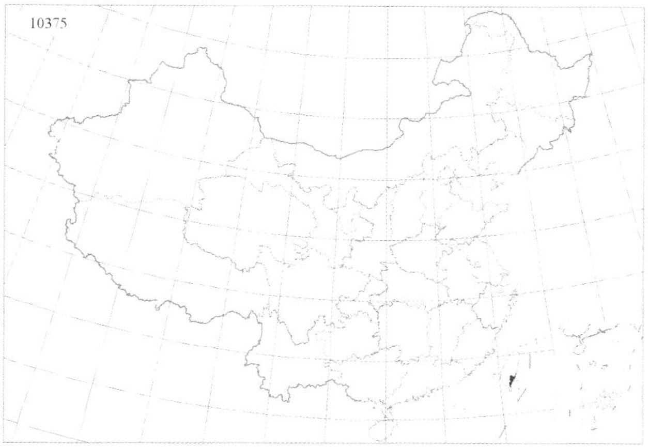

Rubiaceae 茜草科 *Traillaedoxa* 丁茜属
Trailliaedoxa gracilis W. W. Sm. et Forrest 丁茜
Subshrubs

15.4 (12.6~18.9)；8.6 (6.0~12.0)；20.7 (17.9~23.9)；15.5 (12.6~19.0)；125.9 (91.6~167.8)；0.0 (0.0~0.0)；773 (678~929) ❙ 954 (797~1057)；559 (411~627)；31 (21~44) ❙ 24.5 (-14.2~56 0)；715 (678~836)；282.6 (207.5~349.8)

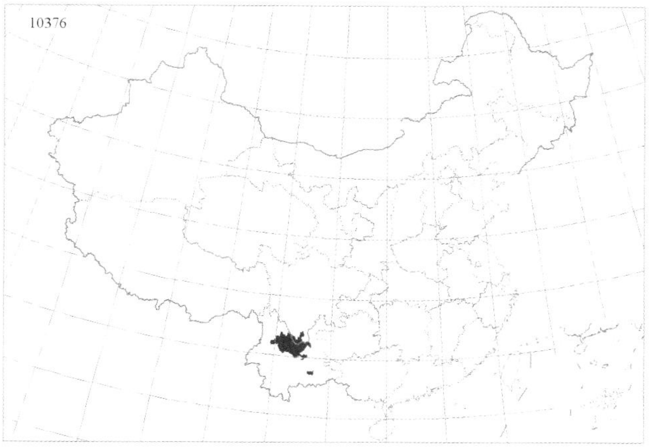

Rubiaceae 茜草科 *Tsiangia* 蒋英木属
Tsiangia hongkongensis (Seem.) But, H. H. Hsue et P. T. Li 蒋英木
Shrubs or small trees

15.2；28.7；22.7；212.7；0.0；1216 ❙ 1965；994；104 ❙ 61.6；1216；204.4

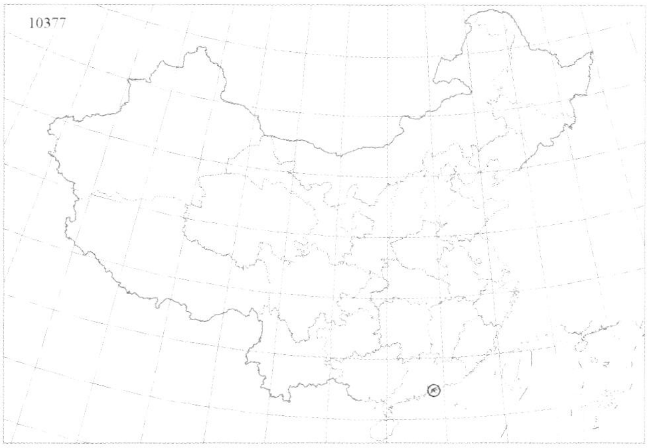

Rubiaceae 茜草科 *Uncaria* 钩藤属
Uncaria hirsuta Havil. 毛钩藤
Vines

19.3 (14.3~22.2)；10.3 (4.0~15.3)；26.9 (20.1~28.8)；19.4 (14.3~22.7)；173.0 (112.6~212.7)；0.0 (0.0~1.1)；1008 (756~1216) ❙ 1604 (983~2993)；703 (452~1186)；148 (19~547) ❙ 60.7 (15.4~246.5)；1007 (696~1216)；247.1 (157.7~411.1)

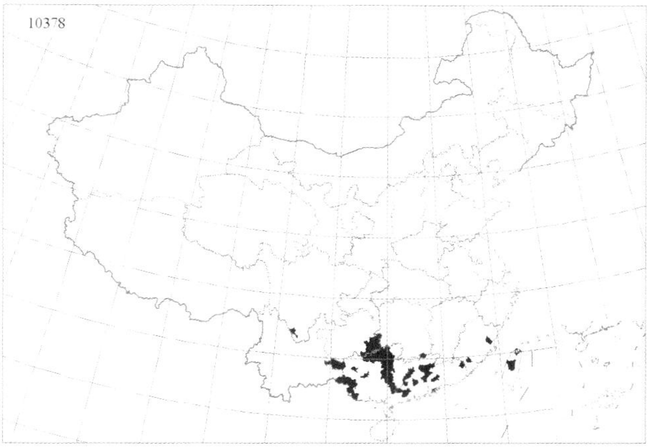

Rubiaceae 茜草科 *Uncaria* 钩藤属
Uncaria homomalla Miq. 北越钩藤
Vines

19.8 (17.2~22.5)；12.1 (9.1~15.0)；25.5 (22.4~28.7)；19.8 (17.3~22.6)；177.9 (147.0~210.9)；0.0 (0.0~0.0)；1003 (833~1206) ❙ 1478 (1161~2037)；801 (633~1033)；77 (48~138) ❙ 48.2 (12.9~80.6)；1002 (833~1206)；341.7 (139.6~551.6)

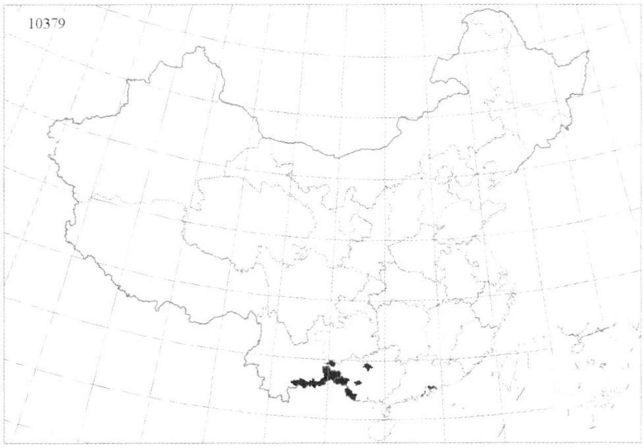

Rubiaceae 茜草科 *Uncaria* 钩藤属
Uncaria laevigata Wall. ex G. Don 平滑钩藤
Vines

19.0 (17.2~22.2)；12.4 (9.1~15.1)；23.4 (21.6~28.3)；19.0 (17.3~22.3)；168.5 (147.0~207.5)；0.0 (0.0~0.0)；925 (832~1190) ❙ 1424 (1049~1751)；758 (598~996)；58 (41~87) ❙ 54.7 (15.4~88.8)；912 (783~1190)；451.0 (314.4~659.0)

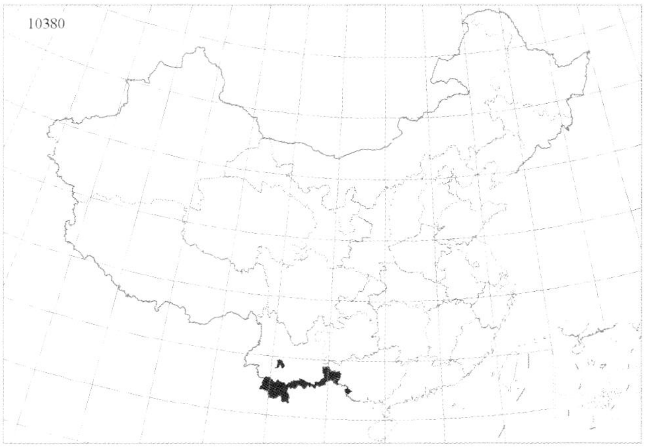

Rubiaceae 茜草科 *Uncaria* 钩藤属

Uncaria lancifolia Hutch. 倒挂金钩

Vines

18.6 (8.6~21.6); 11.2 (1.3~15.1); 24.0 (15.0~28.5); 18.6 (8.7~21.7); 163.9 (52.6~199.8); 0.4 (0.0~8.8); 930 (574~1153) ▎1432 (1041~1751); 746 (585~996); 86 (44~194) ▎54.8 (26.1~88.8); 925 (574~1153); 387.7 (206.0~659.0)

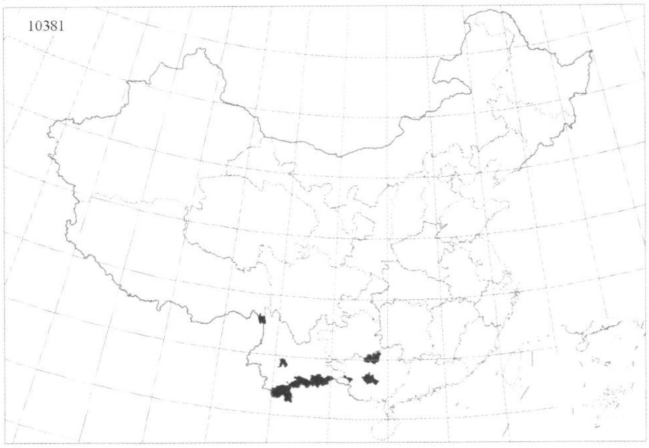

Rubiaceae 茜草科 *Uncaria* 钩藤属

Uncaria lanosa f. setiloba (Benth.) Ridsdale 恒春钩藤

Vines

20.3; 15.9; 23.6; 20.4; 184.4; 0.0; 993 ▎2853; 1618; 147 ▎190.4; 993; 330.3

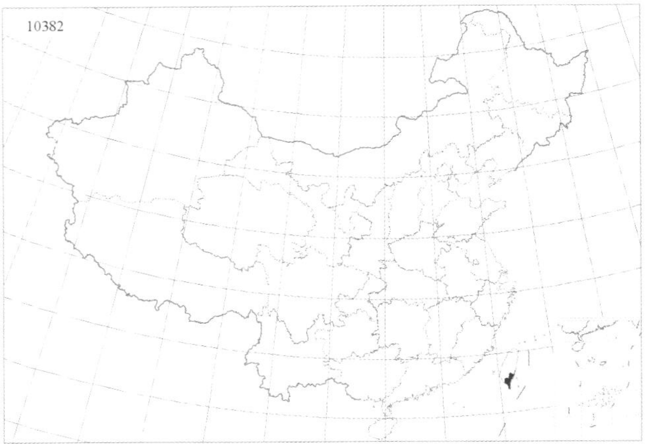

Rubiaceae 茜草科 *Uncaria* 钩藤属

Uncaria macrophylla Wall. 大叶钩藤

Vines

21.2 (17.2~24.8); 13.7 (9.1~20.1); 26.7 (21.6~29.2); 21.2 (17.3~24.8); 194.7 (147.0~237.6); 0.0 (0.0~0.0); 1106 (833~1385) ▎1558 (1161~2435); 756 (392~1428); 94 (41~194) ▎42.5 (-11.9~99.5); 1081 (833~1299); 347.0 (178.7~659.0)

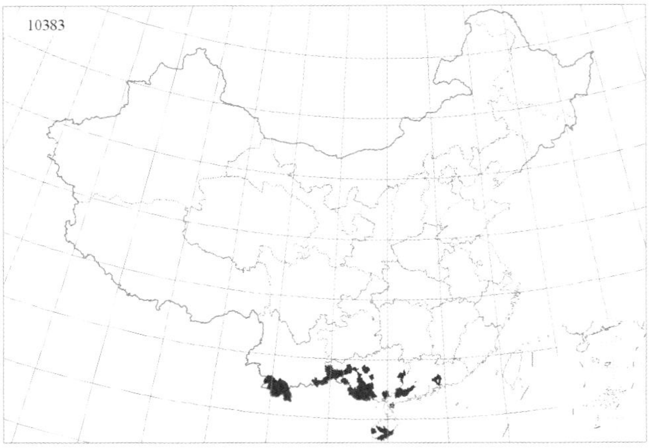

Rubiaceae 茜草科 *Uncaria* 钩藤属

Uncaria rhynchophylla (Miq.) Miq. ex Havil. 钩藤

Vines

17.5 (9.3~23.0); 7.1 (-1.7~15.1); 27.0 (18.8~30.1); 17.6 (9.5~23.0); 151.3 (69.8~216.2); 0.4 (0.0~18.0); 929 (603~1247) ▎1476 (871~2435); 576 (299~1428); 153 (18~259) ▎60.2 (-6.7~147.0); 920 (603~1247); 225.0 (123.9~551.6)

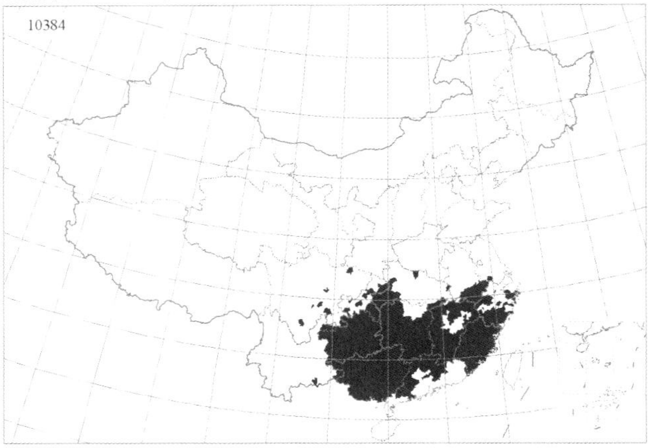

Rubiaceae 茜草科 *Uncaria* 钩藤属

Uncaria rhynchophylloides F. C. How 侯钩藤

Vines

20.5 (17.9~22.4); 11.5 (7.4~14.2); 27.9 (25.0~28.8); 20.6 (18.0~22.4); 186.8 (155.6~209.1); 0.0 (0.0~0.0); 1080 (913~1204) ▎1644 (1371~1900); 726 (562~1056); 150 (57~189) ▎52.9 (22.7~85.5); 1080 (913~1204); 228.4 (157.7~371.7)

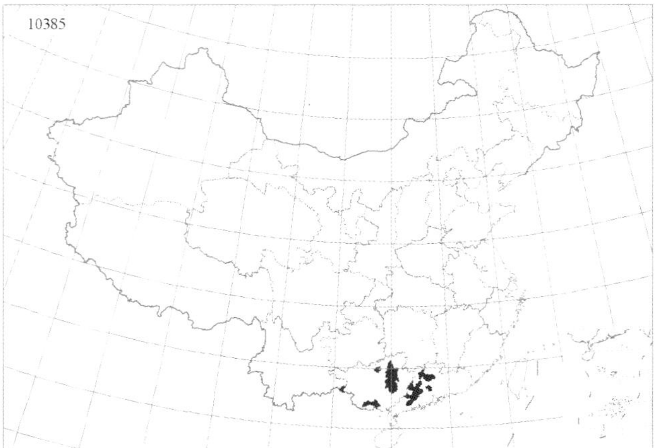

Rubiaceae 茜草科 *Uncaria* 钩藤属

Uncaria scandens (Sm.) Hutch. 攀茎钩藤

Vines

19.2 (11.3~24.8); 11.4 (2.9~20.1); 25.1 (17.0~28.6); 19.2 (11.3~24.8); 170.4 (76.6~237.6); 0.1 (0.0~3.4); 986 (644~1385) ▎1411 (1001~1900); 683 (392~1056); 74 (35~178) ▎46.5 (-11.9~145.8); 961 (644~1326); 353.6 (134.1~659.0)

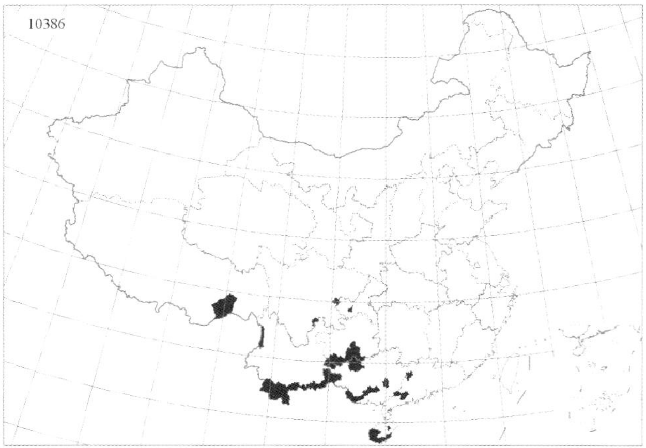

Rubiaceae 茜草科 *Uncaria* 钩藤属
Uncaria sessilifructus Roxb. 白钩藤
Vines

18.8 (14.4~22.4); 11.2 (6.5~15.1); 24.2 (19.1~28.9); 18.8 (14.4~22.5); 165.7 (113.1~209.7); 0.0 (0.0~0.0); 937 (727~1205) ❙ 1383 (1007~2035); 733 (539~1044); 74 (41~173) ❙ 48.5 (5.8~112.2); 930 (727~1205); 361.8 (145.3~659.0)

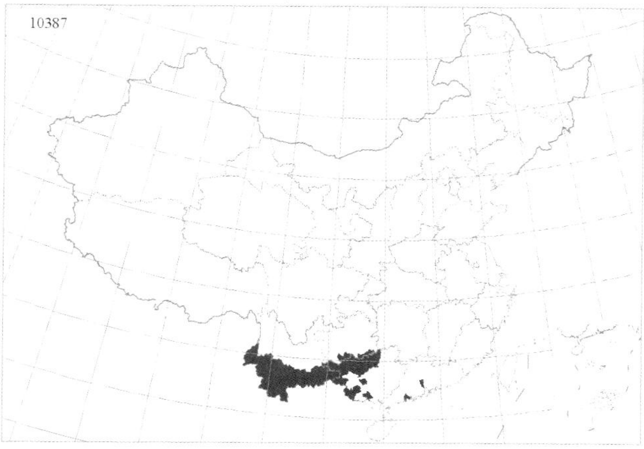

Rubiaceae 茜草科 *Uncaria* 钩藤属
Uncaria sinensis (Oliv.) Havil. 华钩藤
Vines

15.4 (2.6~21.6); 4.7 (-7.0~13.6); 25.3 (11.2~29.3); 15.5 (4.3~21.6); 127.8 (20.0~199.3); 2.5 (0.0~48.2); 833 (427~1150) ❙ 1309 (711~1776); 550 (357~1007); 109 (15~211) ❙ 58.0 (8.2~145.8); 831 (427~1148); 223.7 (133.8~447.4)

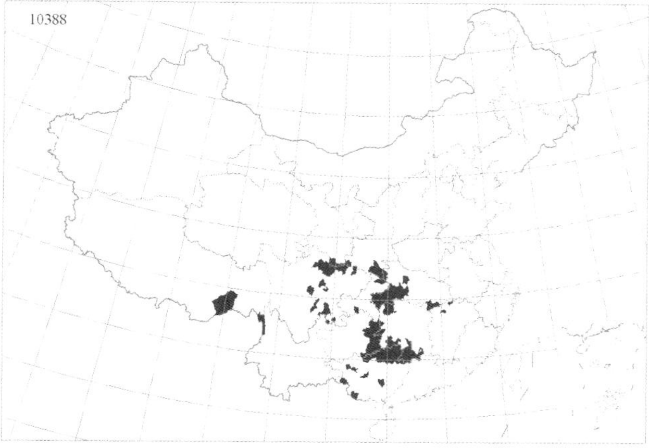

Rubiaceae 茜草科 *Uncaria* 钩藤属
Uncaria yunnanensis K. C. Hsia 云南钩藤
Vines

20.5; 15.0; 23.7; 20.5; 185.9; 0.0; 990 ❙ 1536; 678; 72 ❙ 55.1; 973; 614.6

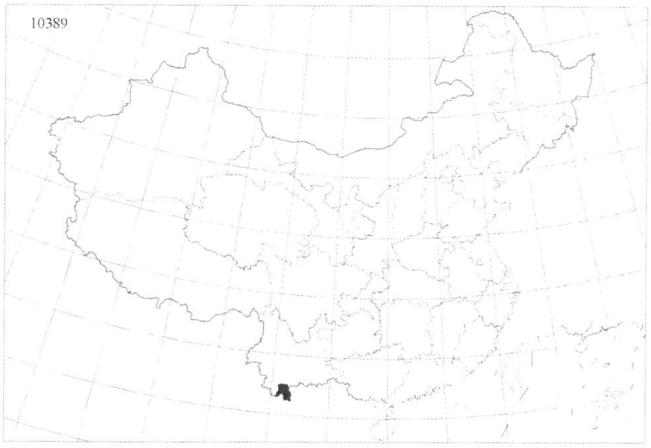

Rubiaceae 茜草科 *Urophyllum* 尖叶木属
Urophyllum chinense Merr. et Chun 尖叶木
Shrubs or small trees

20.3 (16.1~23.2); 11.9 (5.3~15.8); 26.8 (22.3~29.0); 20.3 (16.1~23.2); 183.9 (133.1~218.9); 0.0 (0.0~0.0); 1052 (833~1257) ❙ 1594 (1254~2435); 757 (523~1428); 123 (48~189) ❙ 52.5 (12.9~99.5); 1050 (833~1257); 297.1 (185.5~659.0)

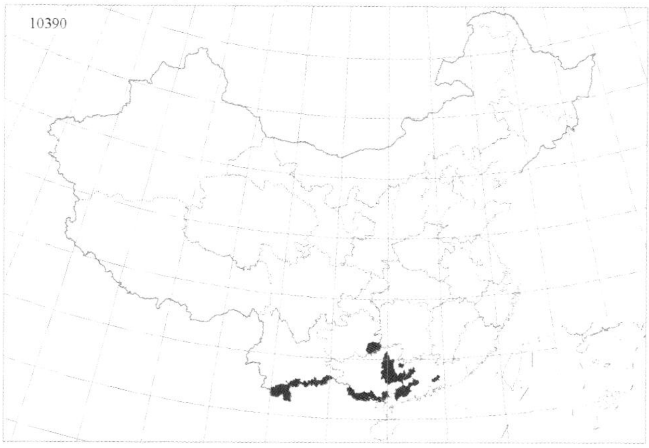

Rubiaceae 茜草科 *Urophyllum* 尖叶木属
Urophyllum parviflorum How ex Lo 小花尖叶木
Shrubs or small trees

19.1; 13.0; 23.3; 19.1; 169.7; 0.0; 917 ❙ 1428; 835; 56 ❙ 55.7; 917; 506.2

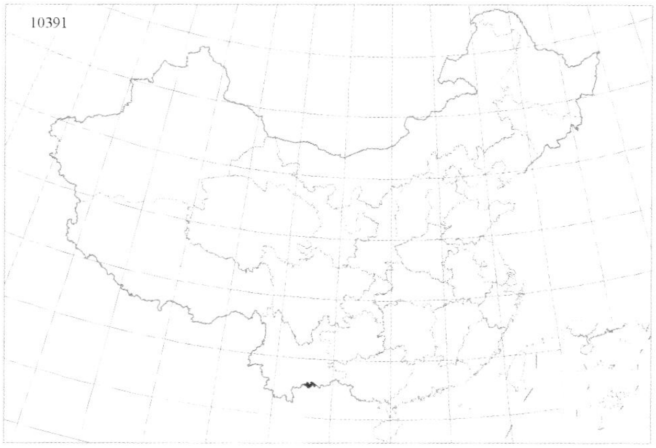

Rubiaceae 茜草科 *Urophyllum* 尖叶木属
Urophyllum tsaianum How ex Lo 滇南尖叶木
Shrubs or small trees

19.2; 12.5; 23.8; 19.2; 170.9; 0.0; 939 ❙ 1480; 772; 64 ❙ 58.2; 938; 515.4

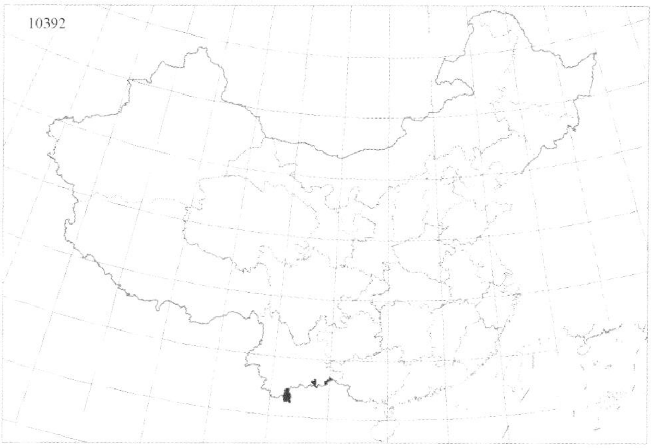

Rubiaceae 茜草科 *Wendlandia* 水锦树属
Wendlandia aberrans F. C. How 广西水锦树
Shrubs

17.7 (15.2~19.2); 8.7 (5.8~11.0); 24.8 (23.0~26.6); 17.7 (15.2~19.2); 153.0 (122.5~170.5); 0.0 (0.0~0.0); 893 (779~981) ❙ 1298 (1232~1485); 693 (617~848); 61 (51~73) ❙ 46.0 (27.5~64.7); 893 (779~981); 252.2 (186.7~371.7)

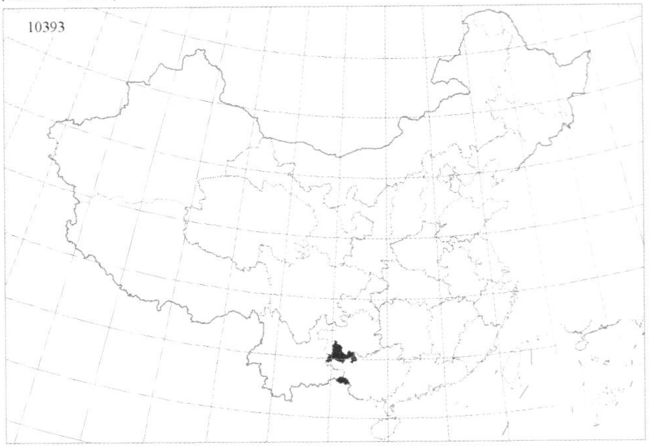

Rubiaceae 茜草科 *Wendlandia* 水锦树属
Wendlandia augustinii Cowan 思茅水锦树
Shrubs

18.8 (18.2~19.3); 12.9 (12.2~13.5); 22.6 (22.1~23.1); 18.9 (18.3~19.4); 166.3 (160.0~172.8); 0.0 (0.0~0.0); 899 (872~927) ❙ 1514 (1323~1751); 854 (754~996); 63 (51~77) ❙ 68.3 (47.1~88.8); 888 (866~927); 458.3 (395.0~551.3)

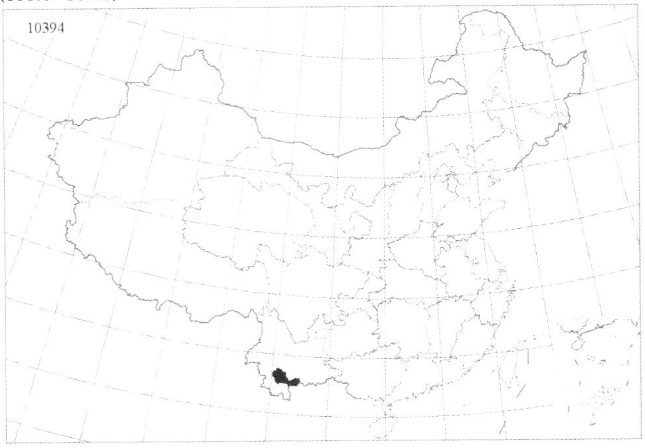

Rubiaceae 茜草科 *Wendlandia* 水锦树属
Wendlandia bouvardioides Hutch. 薄叶水锦树
Shrubs or small trees

19.2 (16.9~21.0); 13.2 (10.3~15.1); 23.2 (21.4~25.6); 19.2 (16.9~21.0); 170.9 (142.7~191.8); 0.0 (0.0~0.0); 927 (813~1053) ❙ 1464 (1178~1751); 775 (646~996); 63 (51~77) ❙ 58.0 (36.8~88.8); 920 (813~1053); 519.3 (383.4~659.0)

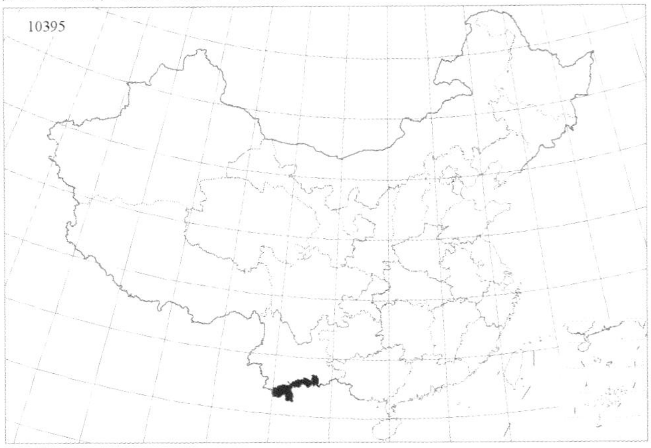

Rubiaceae 茜草科 *Wendlandia* 水锦树属
Wendlandia brevipaniculata W. C. Chen 吹树
Trees

17.3; 11.0; 21.6; 17.3; 148.0; 0.0; 832 ❙ 1049; 598; 44 ❙ 26.1; 783; 365.3

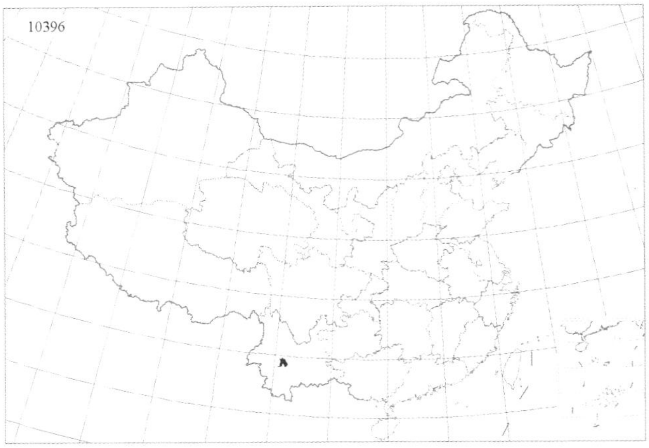

Rubiaceae 茜草科 *Wendlandia* 水锦树属
Wendlandia brevituba Chun et F. C. How ex W. C. Chen 短筒水锦树
Shrubs

20.8 (16.9~22.2); 13.0 (10.3~14.3); 26.7 (21.4~28.5); 20.9 (16.9~22.3); 190.3 (142.7~207.5); 0.0 (0.0~0.0); 1079 (813~1190) ❙ 1551 (1143~1900); 766 (599~1056); 103 (50~142) ❙ 43.6 (15.4~71.0); 1075 (813~1190); 283.3 (216.5~383.4)

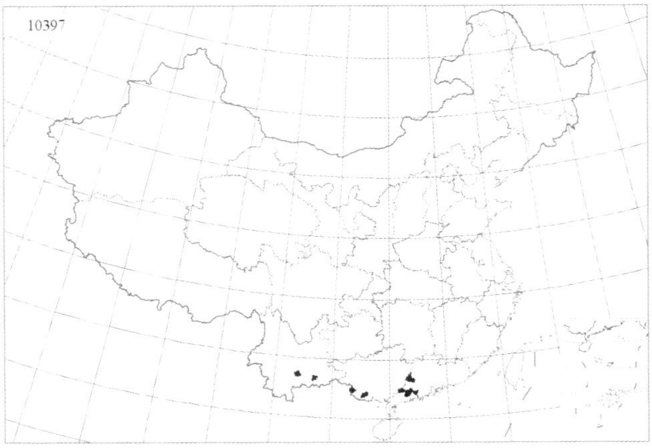

Rubiaceae 茜草科 *Wendlandia* 水锦树属
Wendlandia cavaleriei H. Lév. 贵州水锦树
Shrubs or small trees

19.1 (16.8~21.2); 10.5 (7.8~14.2); 25.6 (23.6~27.7); 19.1 (16.8~21.2); 169.5 (141.5~194.6); 0.0 (0.0~0.0); 964 (836~1114) ❙ 1337 (1236~1522); 728 (617~848); 61 (49~73) ❙ 39.3 (14.8~59.6); 964 (836~1114); 307.3 (189.5~551.6)

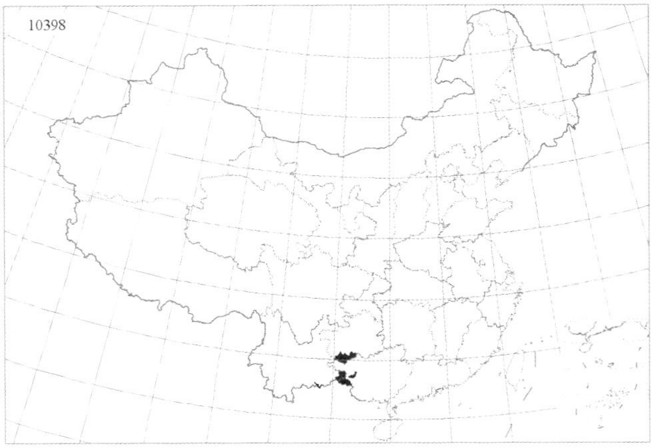

Rubiaceae 茜草科 *Wendlandia* 水锦树属
Wendlandia erythroxylon Cowan 红木水锦树
Trees

20.2 (16.2~24.1); 14.5 (11.0~18.9); 25.0 (20.1~28.2); 20.2 (16.2~24.1); 182.9 (134.2~229.7); 0.0 (0.0~0.0); 1019 (773~1322) ❙ 2402 (1498~3495); 1105 (750~1847); 255 (47~897) ❙ 146.2 (25.2~246.5); 983 (773~1144); 245.0 (119.4~363.7)

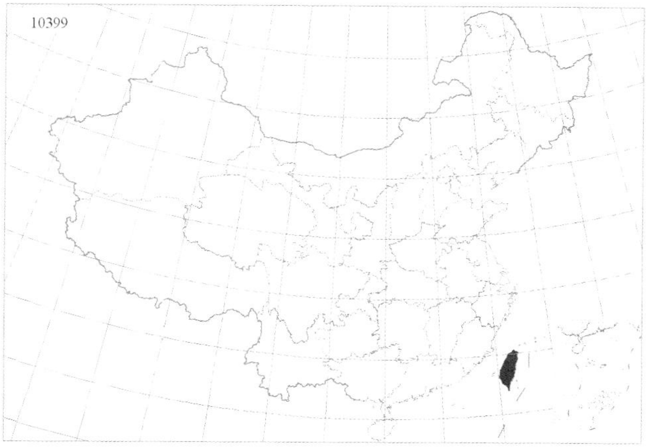

Rubiaceae 茜草科 *Wendlandia* 水锦树属
Wendlandia formosana Cowan 水金京
Shrubs or trees

19.4 (16.2~22.6); 13.6 (11.0~17.7); 24.4 (20.1~28.0); 19.4 (16.2~22.6); 173.3 (134.2~211.5); 0.0 (0.0~0.0); 963 (773~1197) ❙ 2557 (1498~3495); 1043 (750~1847); 330 (102~897) ❙ 172.2 (25.2~246.5); 952 (773~1123); 268.8 (202.8~363.7)

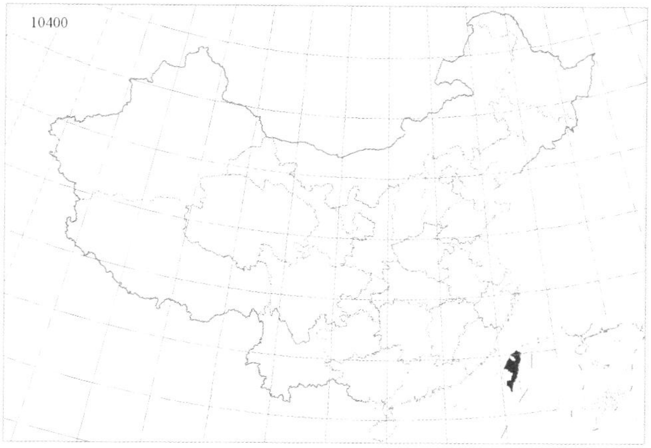

Rubiaceae 茜草科 *Wendlandia* 水锦树属
Wendlandia formosana subsp. breviflora F. C. How 短花水金京
Shrubs or trees

20.2 (17.2~22.7); 12.2 (9.1~15.0); 26.1 (22.4~28.6); 20.2 (17.3~22.8); 182.8 (147.0~213.1); 0.0 (0.0~0.0); 1040 (833~1221) ❙ 1614 (1161~2435); 886 (633~1428); 95 (48~173) ❙ 55.0 (12.9~99.5); 1040 (833~1221); 323.3 (232.6~551.6)

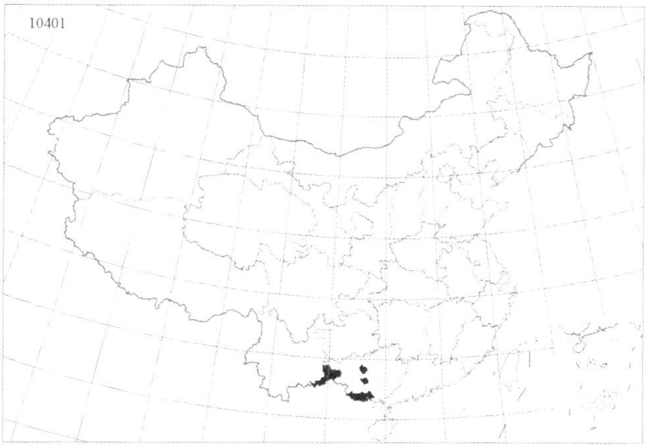

Rubiaceae 茜草科 *Wendlandia* 水锦树属
Wendlandia grandis (Hook. f.) Cowan 西藏水锦树
Trees

8.6; 1.3; 15.0; 8.7; 52.6; 8.8; 574 ❙ 1041; 585; 45 ❙ 81.5; 574; 352.3

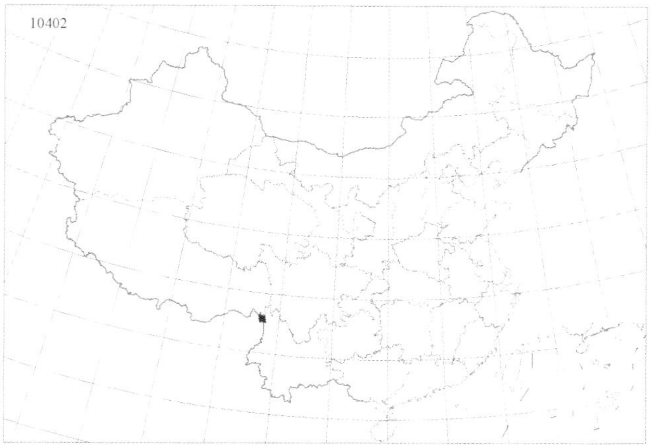

Rubiaceae 茜草科 *Wendlandia* 水锦树属
Wendlandia guangdongensis W. C. Chen 广东水锦树
Shrubs or trees

23.4 (21.0~24.8); 17.4 (13.2~20.1); 27.6 (25.7~29.0); 23.4 (21.0~24.8); 220.7 (192.3~237.6); 0.0 (0.0~0.0); 1259 (1081~1385) ❙ 1570 (1187~1776); 609 (473~842); 76 (41~125) ❙ 26.1 (-6.9~64.3); 1173 (1055~1299); 431.3 (225.6~620.0)

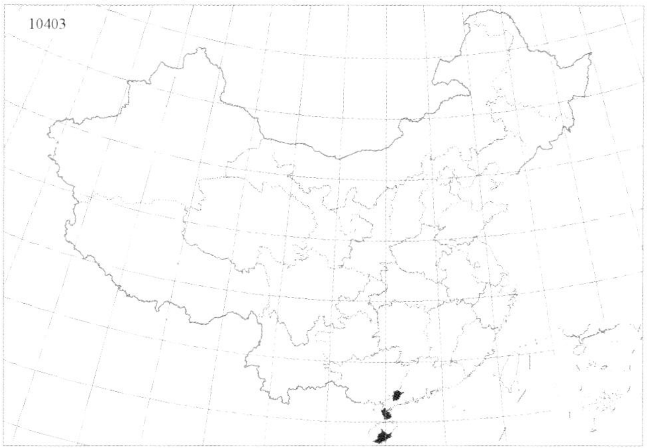

Rubiaceae 茜草科 *Wendlandia* 水锦树属
Wendlandia jingdongensis W. C. Chen 景东水锦树
Shrubs

17.3; 11.0; 21.6; 17.3; 148.0; 0.0; 832 ❙ 1049; 598; 44 ❙ 26.1; 783; 365.3

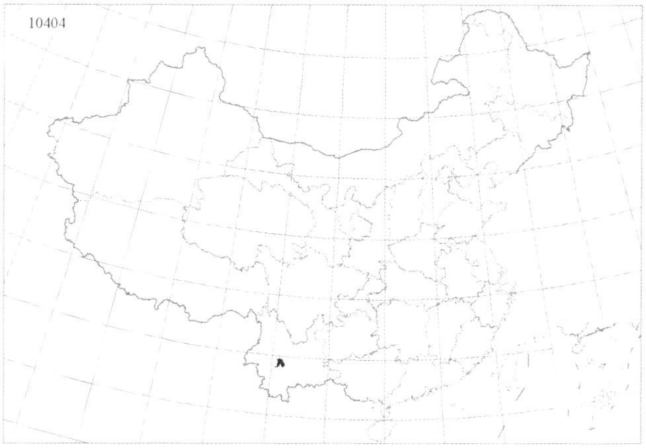

Rubiaceae 茜草科　*Wendlandia* 水锦树属
Wendlandia laxa S. K. Wu 疏花水锦树
Small trees

17.3；11.0；21.6；17.3；148.0；0.0；832 ❙ 1049；598；44 ❙ 26.1；783；365.3

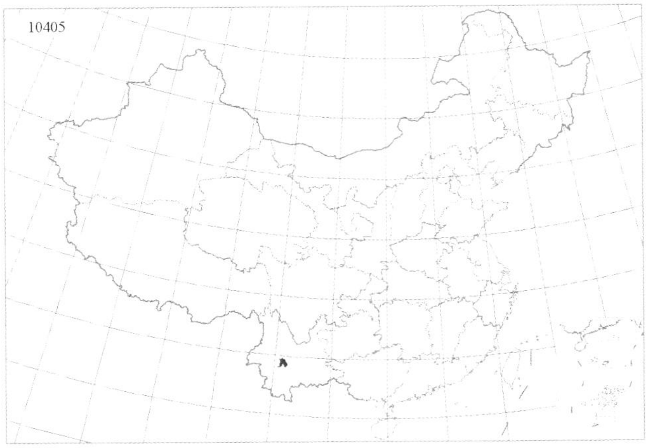

Rubiaceae 茜草科　*Wendlandia* 水锦树属
Wendlandia ligustrina Wall. ex G. Don 小叶水锦树
Shrubs

16.0 (8.6~21.0)；9.5 (1.3~14.9)；20.7 (15.0~25.6)；16.1 (8.7~21.0)；133.6 (52.6~191.8)；0.7 (0.0~8.8)；807 (574~1053) ❙ 1313 (958~1751)；720 (531~1007)；53 (24~77) ❙ 64.4 (13.3~145.8)；799 (574~1053)；431.1 (274.9~570.1)

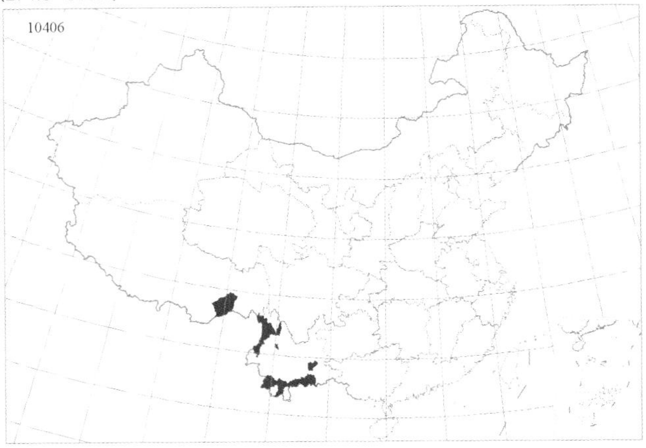

Rubiaceae 茜草科　*Wendlandia* 水锦树属
Wendlandia litseifolia F. C. How 木姜子叶水锦树
Trees

19.1 (16.8~22.2)；10.2 (7.8~13.9)；26.0 (23.6~28.3)；19.1 (16.8~22.3)；169.6 (141.5~207.5)；0.0 (0.0~0.0)；974 (836~1190) ❙ 1299 (1182~1485)；702 (617~848)；61 (43~87) ❙ 34.6 (15.4~59.6)；974 (836~1190)；276.3 (189.5~354.5)

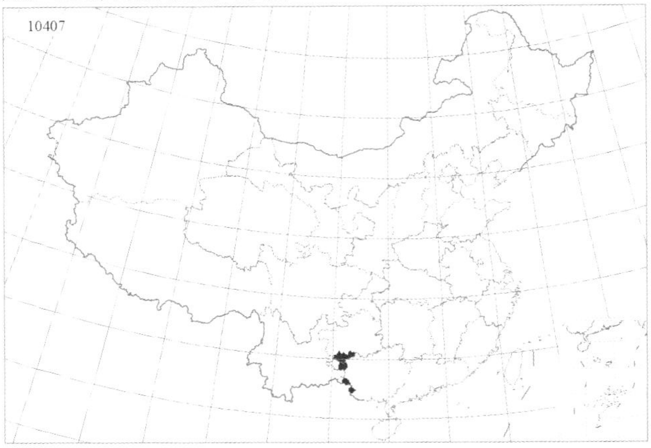

Rubiaceae 茜草科　*Wendlandia* 水锦树属
Wendlandia longidens (Hance) Hutch. 水晶棵子
Small shrubs

14.1 (10.1~17.1)；4.7 (2.3~10.2)；22.6 (16.9~26.1)；14.2 (10.2~17.1)；111.5 (68.1~145.2)；1.4 (0.0~5.2)；754 (618~889) ❙ 1093 (885~1441)；568 (475~809)；50 (22~94) ❙ 45.2 (22.1~95.9)；749 (618~889)；237.2 (152.1~302.7)

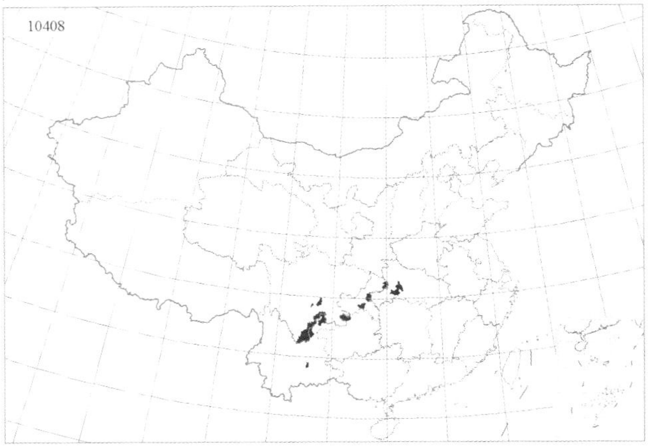

Rubiaceae 茜草科　*Wendlandia* 水锦树属
Wendlandia longipedicellata F. C. How 长梗水锦树
Shrubs

17.6 (16.0~19.6)；10.9 (9.3~12.8)；21.8 (20.4~23.5)；17.7 (16.0~19.6)；152.1 (132.4~175.7)；0.0 (0.0~0.0)；853 (780~953) ❙ 1486 (1049~1688)；862 (598~959)；52 (42~62) ❙ 74.9 (26.1~112.2)；841 (780~932)；491.7 (365.3~627.0)

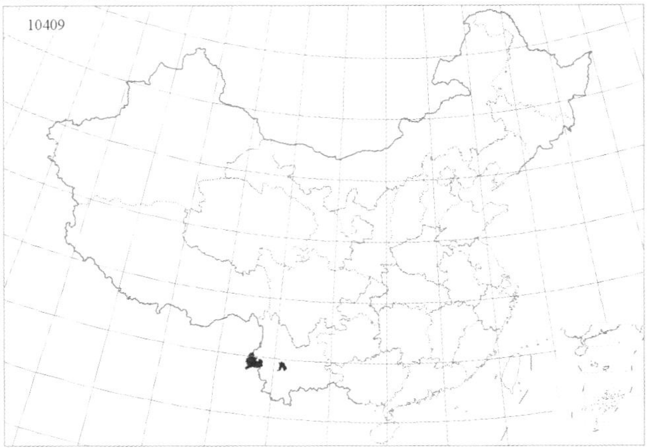

Rubiaceae 茜草科　*Wendlandia* 水锦树属
Wendlandia luzoniensis DC. 吕宋水锦树
Shrubs or small trees

18.5；14.1；21.7；18.5；162.1；0.0；863 ❙ 2713；1388；188 ❙ 214.4；863；363.7

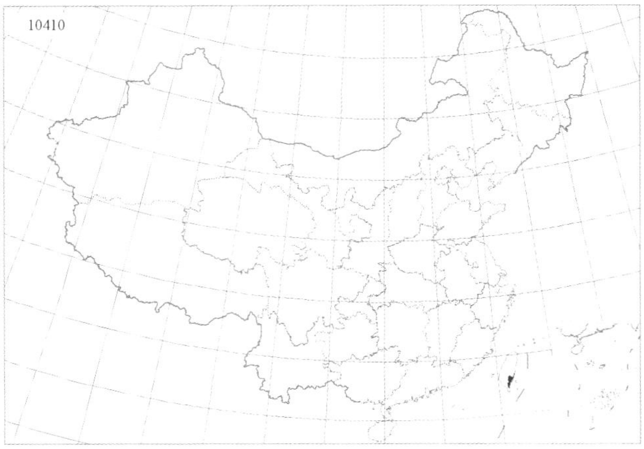

Rubiaceae 茜草科 *Wendlandia* 水锦树属
Wendlandia merrilliana Cowan 海南水锦树
Shrubs

23.9 (23.0~24.7)；18.6 (17.7~19.8)；27.7 (26.7~28.4)；24.0 (23.1~24.7)；227.7 (217.0~236.5)；0.0 (0.0~0.0)；1303 (1217~1376) ❙ 1484 (1179~1725)；513 (392~619)；60 (42~110) ❙ 14.1 (-11.9~41.8)；1158 (1056~1326)；484.4 (352.9~620.0)

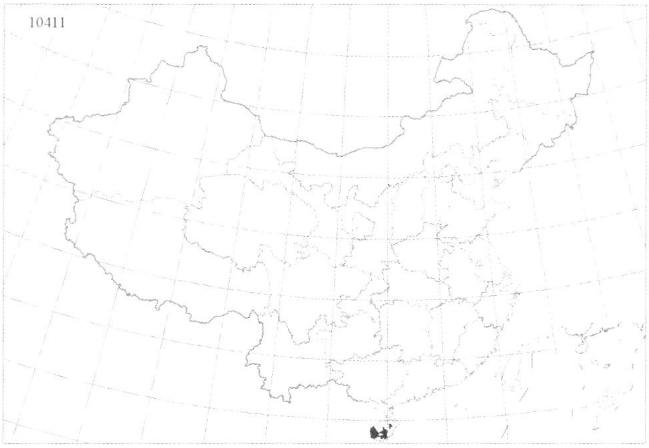

Rubiaceae 茜草科 *Wendlandia* 水锦树属
Wendlandia merrilliana var. parvifolia F. C. How 细叶海南水锦树
Shrubs

23.9；18.3；28.1；24.0；227.6；0.0；1305 ❙ 1211；418；43 ❙ -7.1；1074；447.5

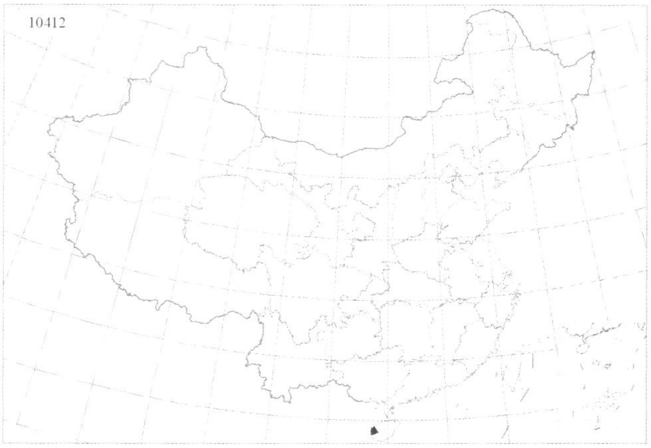

Rubiaceae 茜草科 *Wendlandia* 水锦树属
Wendlandia myriantha F. C. How 密花水锦树
Shrubs

22.0 (21.3~22.7)；13.9 (13.2~15.0)；27.9 (27.3~28.6)；22.0 (21.3~22.8)；204.1 (195.9~213.1)；0.0 (0.0~0.0)；1166 (1111~1221) ❙ 1804 (1302~2435)；999 (670~1428)；112 (86~130) ❙ 54.6 (12.9~99.5)；1166 (1111~1221)；295.0 (235.1~327.7)

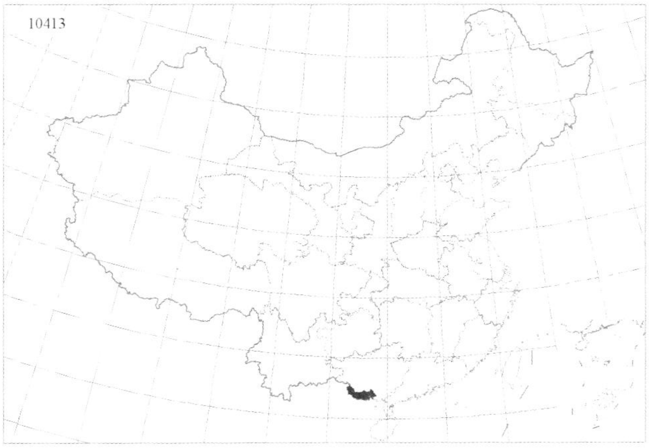

Rubiaceae 茜草科 *Wendlandia* 水锦树属
Wendlandia oligantha W. C. Chen 龙州水锦树
Shrubs or trees

20.2；12.0；26.2；20.2；182.5；0.0；1029 ❙ 1432；794；72 ❙ 41.2；1029；337.7

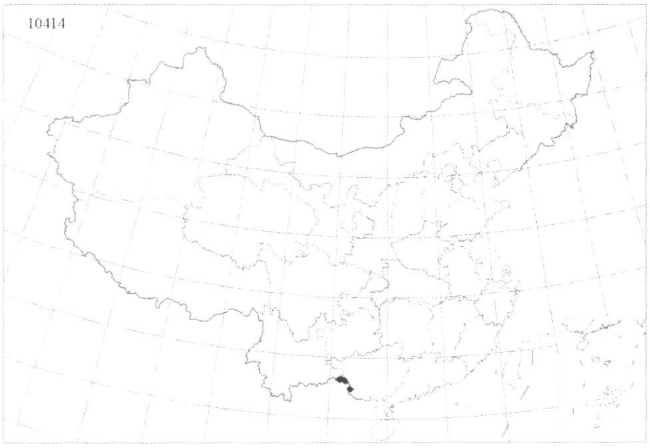

Rubiaceae 茜草科 *Wendlandia* 水锦树属
Wendlandia parviflora W. C. Chen 小花水锦树
Small trees

20.0；14.6；23.2；20.1；180.8；0.0；963 ❙ 1523；681；67 ❙ 58.4；942；581.0

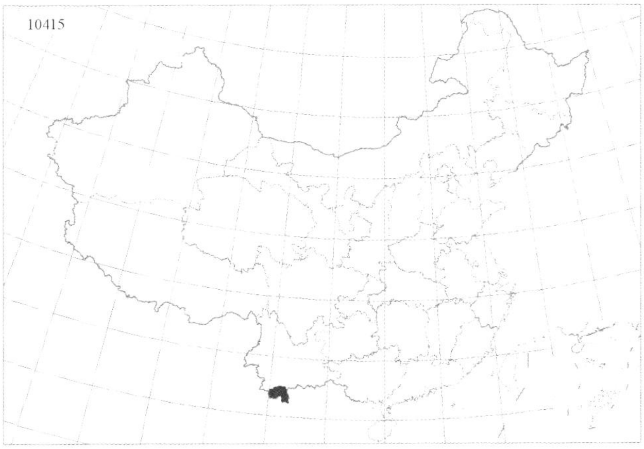

Rubiaceae 茜草科 *Wendlandia* 水锦树属
Wendlandia pendula (Wall.) DC. 垂枝水锦树
Shrubs

18.6 (16.3~20.5)；12.6 (10.2~15.1)；22.3 (20.3~23.7)；18.6 (16.4~20.5)；163.5 (136.6~185.9)；0.0 (0.0~0.0)；890 (789~992) ❙ 1383 (1028~1751)；746 (566~996)；54 (41~77) ❙ 55.5 (12.5~88.8)；869 (780~986)；449.0 (295.5~659.0)

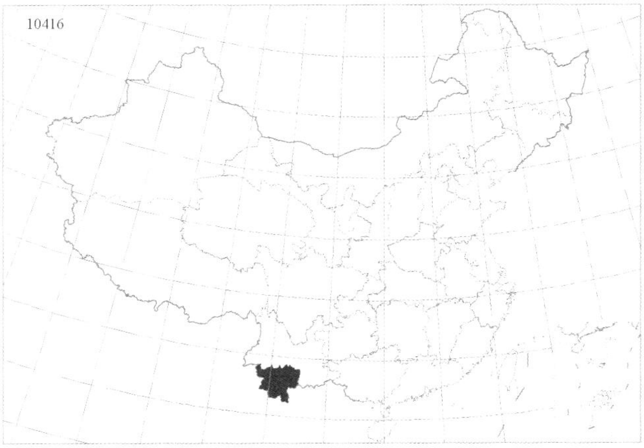

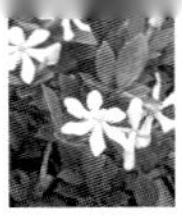

Rubiaceae 茜草科 *Wendlandia* 水锦树属

Wendlandia pingpienensis F. C. How 屏边水锦树

Shrubs or trees

18.5 (17.2~21.0); 11.5 (9.5~14.2); 23.3 (22.4~25.6); 18.5 (17.3~21.0); 162.0 (147.0~191.8); 0.0 (0.0~0.0); 898 (833~1053) ❙ 1396 (1254~1522); 789 (699~860); 56 (48~68) ❙ 56.1 (36.8~80.6); 897 (833~1053); 444.4 (336.3~551.6)

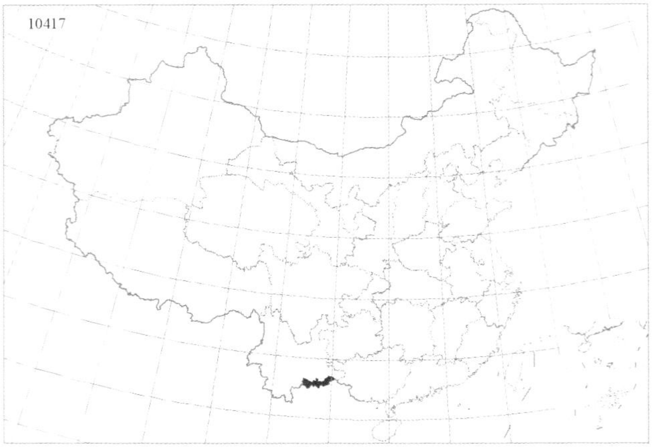

Rubiaceae 茜草科 *Wendlandia* 水锦树属

Wendlandia pubigera W. C. Chen 大叶木莲红

Shrubs

21.9 (21.3~22.7); 13.9 (13.2~15.0); 27.9 (27.3~28.6); 22.0 (21.3~22.8); 203.6 (195.9~213.1); 0.0 (0.0~0.0); 1162 (1111~1221) ❙ 1876 (1302~2435); 1047 (670~1428); 116 (86~130) ❙ 61.1 (12.9~99.5); 1162 (1111~1221); 289.6 (235.1~318.8)

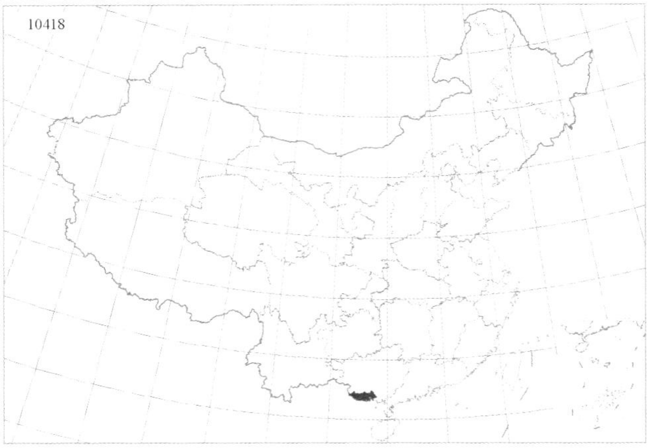

Rubiaceae 茜草科 *Wendlandia* 水锦树属

Wendlandia salicifolia Franch. ex Drake 柳叶水锦树

Shrubs

18.6 (15.9~21.0); 11.3 (6.9~15.1); 23.9 (21.4~27.0); 18.6 (16.0~21.0); 163.6 (131.6~191.8); 0.0 (0.0~0.0); 914 (802~1053) ❙ 1384 (1178~1751); 727 (617~996); 61 (41~96) ❙ 51.9 (27.5~88.8); 906 (802~1053); 393.7 (189.5~659.0)

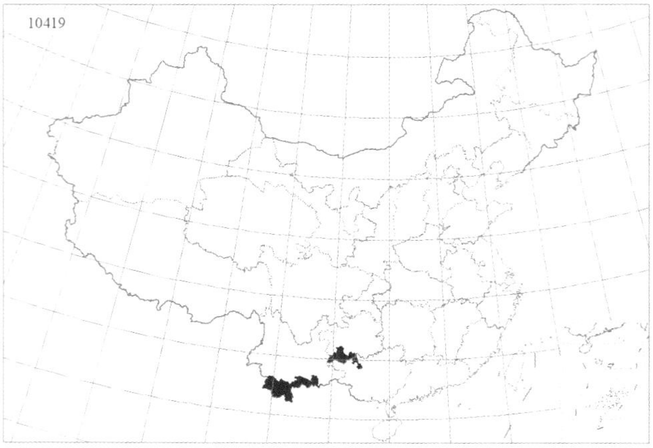

Rubiaceae 茜草科 *Wendlandia* 水锦树属

Wendlandia scabra Kurz 粗叶水锦树

Shrubs or trees

17.1 (8.6~22.7); 10.0 (1.3~15.1); 22.2 (15.0~28.6); 17.1 (8.7~22.8); 145.8 (52.6~213.1); 0.1 (0.0~8.8); 852 (574~1221) ❙ 1227 (797~2435); 681 (411~1428); 55 (23~173) ❙ 44.3 (−14.2~112.2); 837 (574~1221); 348.4 (189.5~659.0)

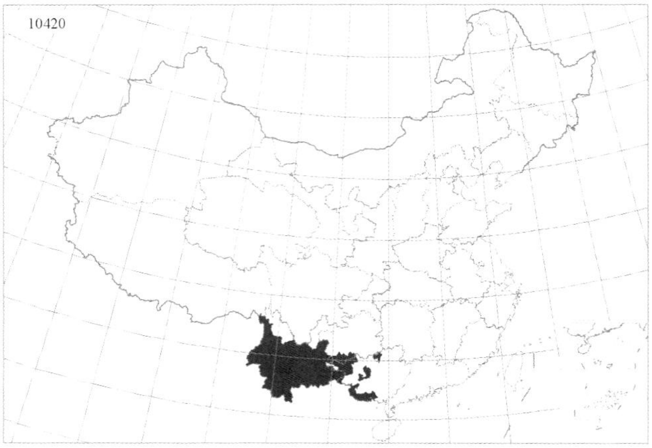

Rubiaceae 茜草科 *Wendlandia* 水锦树属

Wendlandia scabra var. pilifera K. C. How 毛粗叶水锦树

Shrubs or trees

21.9; 14.1; 27.9; 22.0; 203.7; 0.0; 1161 ❙ 2161; 1240; 126 ❙ 85.7; 1161; 305.5

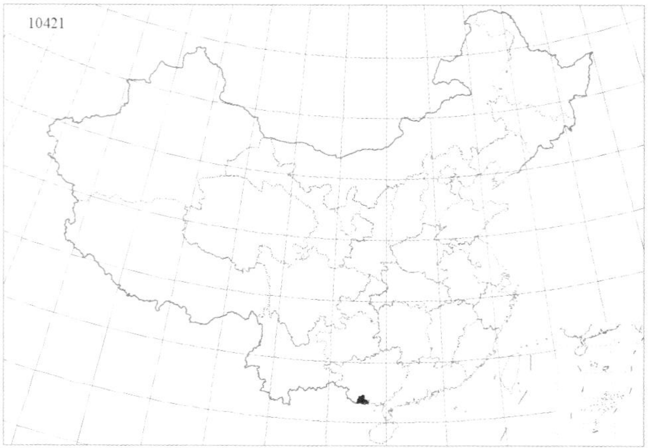

Rubiaceae 茜草科 *Wendlandia* 水锦树属

Wendlandia speciosa Cowan 美丽水锦树

Shrubs or trees

12.8 (3.4~16.4); 5.5 (−5.5~9.6); 18.2 (11.4~20.8); 12.9 (4.7~16.4); 98.5 (23.3~137.3); 5.1 (0.0~41.8); 688 (431~795) ❙ 1245 (527~1736); 691 (317~1007); 54 (9~70) ❙ 79.2 (22.4~145.8); 688 (431~795); 405.1 (280.0~627.0)

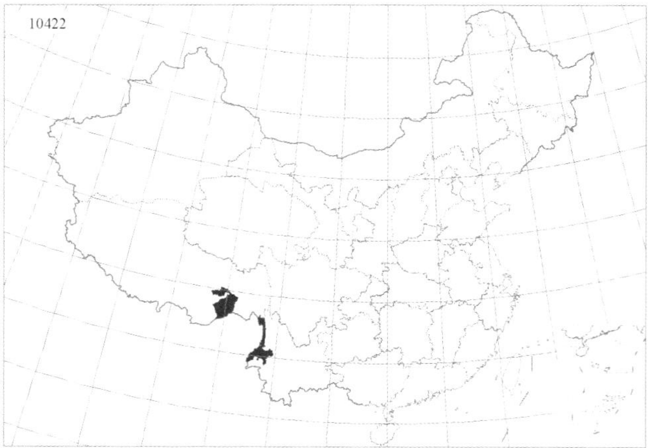

Rubiaceae 茜草科 *Wendlandia* 水锦树属
Wendlandia subalpina W. W. Sm. 高山水锦树
Shrubs

10.2；3.1；16.2；10.3；71.0；7.7；614 ❙ 911；561；25 ❙ 49.7；607；289.6

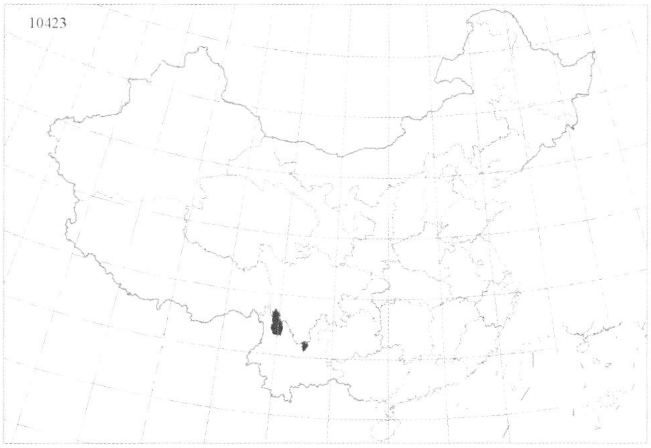

Rubiaceae 茜草科 *Wendlandia* 水锦树属
Wendlandia tinctoria (Roxb.) DC. 染色水锦树
Shrubs or small trees

18.3 (8.6~22.7)；11.1 (1.3~15.1)；23.3 (15.0~28.6)；18.3 (8.7~22.8)；160.1 (52.6~213.1)；0.1 (0.0~8.8)；908 (574~1221) ❙ 1347 (939~2435)；736 (524~1428)；66 (38~194) ❙ 49.0 (6.5~145.8)；897 (574~1221)；371.4 (189.5~659.0)

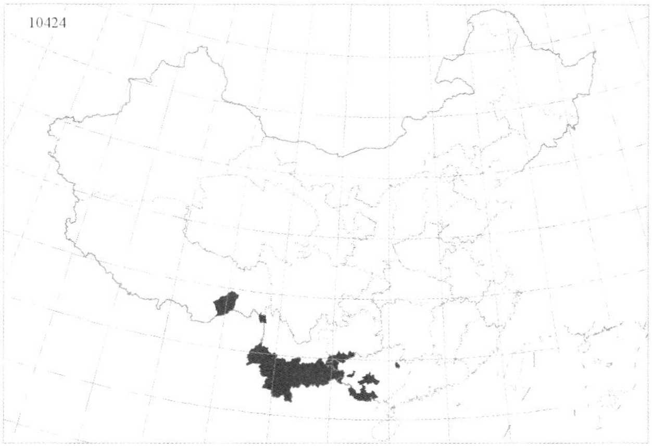

Rubiaceae 茜草科 *Wendlandia* 水锦树属
Wendlandia tinctoria subsp. affinis K. C. How 毛冠水锦树
Shrubs or small trees

18.0 (16.0~21.0)；11.5 (8.4~15.1)；22.3 (20.3~25.6)；18.1 (16.0~21.0)；157.1 (132.4~191.8)；0.0 (0.0~0.0)；871 (780~1053) ❙ 1324 (939~1751)；733 (524~996)；52 (37~77) ❙ 52.2 (6.5~112.2)；855 (744~1053)；425.3 (277.0~659.0)

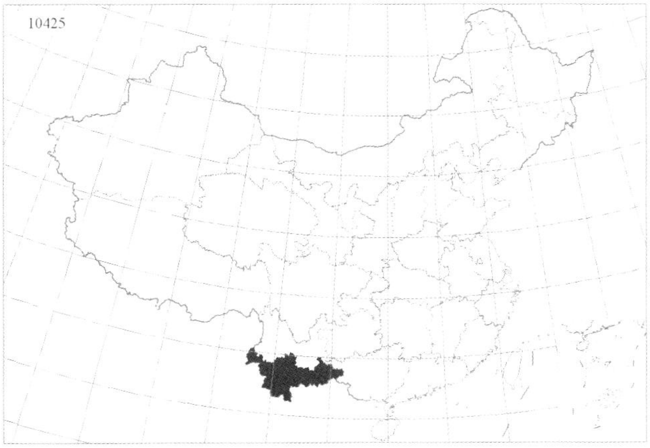

Rubiaceae 茜草科 *Wendlandia* 水锦树属
Wendlandia tinctoria subsp. barbata Cowan 粗毛水锦树
Shrubs or small trees

18.2 (15.4~21.6)；11.6 (8.1~15.1)；22.5 (20.3~28.5)；18.2 (15.5~21.7)；158.5 (125.7~199.8)；0.0 (0.0~0.0)；881 (767~1153) ❙ 1336 (958~1751)；737 (531~996)；55 (41~173) ❙ 52.1 (12.5~112.2)；868 (767~1153)；420.3 (237.8~659.0)

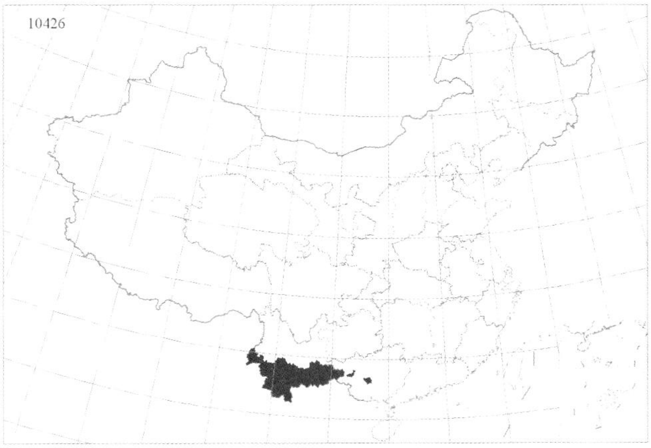

Rubiaceae 茜草科 *Wendlandia* 水锦树属
Wendlandia tinctoria subsp. callitricha (Cowan) W. C. Chen 厚毛水锦树
Shrubs or small trees

17.5 (8.6~21.0)；10.9 (1.3~15.1)；21.9 (15.0~25.6)；17.5 (8.7~21.0)；150.7 (52.6~191.8)；0.1 (0.0~8.8)；849 (574~1053) ❙ 1261 (939~1751)；701 (524~996)；51 (38~77) ❙ 48.6 (6.5~112.2)；835 (574~1053)；399.3 (234.5~659.0)

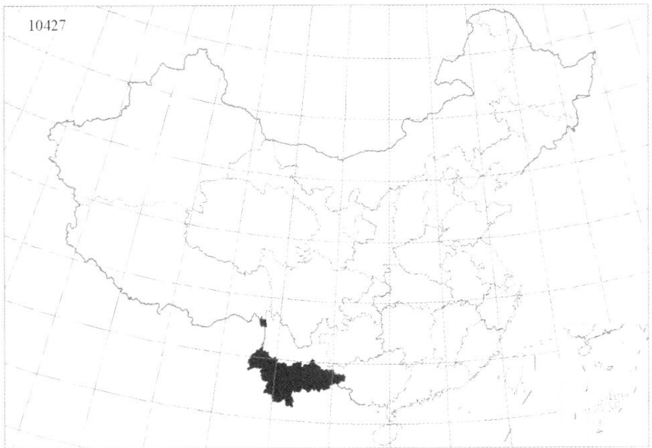

Rubiaceae 茜草科 *Wendlandia* 水锦树属
Wendlandia tinctoria subsp. floribunda (Craib) Cowan 多花水锦树
Shrubs or small trees

19.5 (18.2~20.5)；13.9 (12.4~15.1)；22.7 (21.6~23.7)；19.5 (18.2~20.5)；173.9 (158.6~185.9)；0.0 (0.0~0.0)；931 (861~992) ❙ 1494 (1434~1607)；705 (655~785)；56 (41~77) ❙ 60.9 (47.6~74.2)；901 (839~986)；531.1 (439.5~659.0)

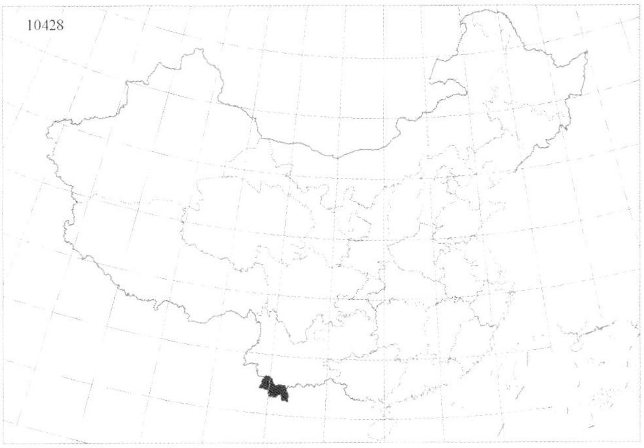

Rubiaceae 茜草科 *Wendlandia* 水锦树属

Wendlandia tinctoria subsp. handelii Cowan 麻栗水锦树

Shrubs or small trees

18.6 (16.9~21.6); 12.2 (10.3~14.2); 22.9 (21.4~28.5); 18.6 (16.9~21.7); 163.7 (142.7~199.8); 0.0 (0.0~0.0); 901 (813~1153) ❙ 1286 (958~1751); 719 (531~996); 59 (44~173) ❙ 42.6 (12.5~88.8); 885 (783~1153); 398.4 (237.8~551.6)

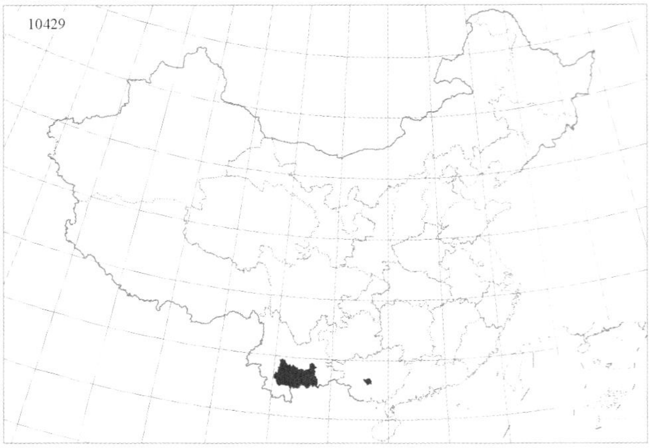

Rubiaceae 茜草科 *Wendlandia* 水锦树属

Wendlandia tinctoria subsp. intermedia (F. C. How) W. C. Chen 红皮水锦树

Shrubs or small trees

18.2 (16.0~21.0); 11.9 (9.3~15.1); 22.2 (20.3~25.6); 18.2 (16.0~21.0); 158.3 (132.4~191.8); 0.0 (0.0~0.0); 873 (780~1053) ❙ 1366 (1049~1751); 756 (594~996); 53 (41~77) ❙ 56.6 (21.3~112.2); 859 (779~1053); 451.9 (294.5~659.0)

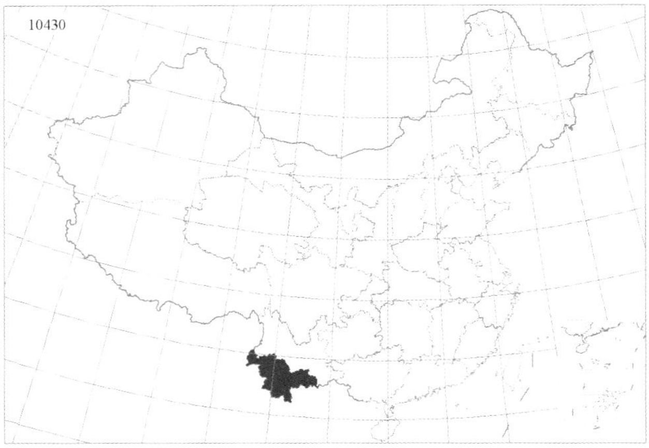

Rubiaceae 茜草科 *Wendlandia* 水锦树属

Wendlandia tinctoria subsp. orientalis Cowan 东方水锦树

Shrubs or small trees

18.5 (14.4~22.4); 11.7 (7.5~15.1); 23.0 (19.1~28.6); 18.5 (14.4~22.4); 162.4 (113.1~209.1); 0.0 (0.0~0.0); 908 (727~1204) ❙ 1378 (958~2148); 756 (531~1234); 64 (41~173) ❙ 52.3 (12.5~112.2); 897 (727~1204); 404.8 (206.0~659.0)

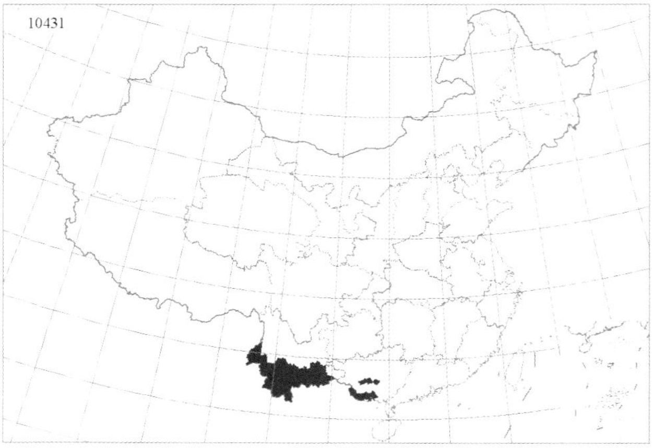

Rubiaceae 茜草科 *Wendlandia* 水锦树属

Wendlandia uvariifolia Hance 水锦树

Shrubs or trees

20.9 (15.5~24.8); 13.1 (4.8~20.1); 26.8 (20.1~29.2); 20.9 (15.6~24.8); 190.8 (127.2~237.6); 0.0 (0.0~0.2); 1089 (773~1385) ❙ 1606 (868~3495); 757 (312~1847); 118 (41~897) ❙ 50.1 (-34.3~246.5); 1067 (773~1360); 295.0 (95.3~659.0)

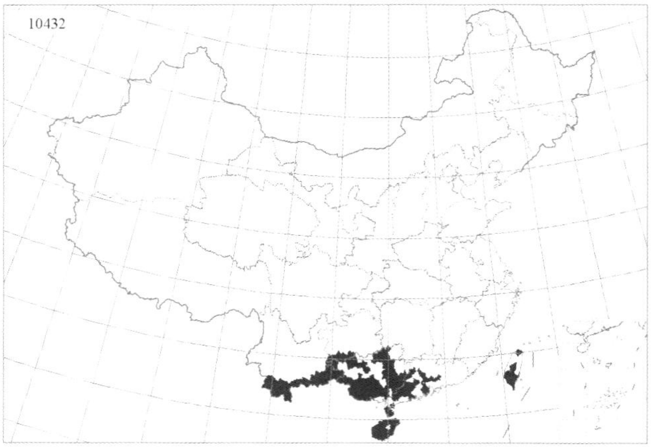

Rubiaceae 茜草科 *Wendlandia* 水锦树属

Wendlandia uvariifolia subsp. pilosa W. C. Chen 疏毛水锦树

Shrubs or trees

20.4; 15.1; 23.6; 20.5; 185.8; 0.0; 989 ❙ 1607; 702; 77 ❙ 62.5; 986; 659.0

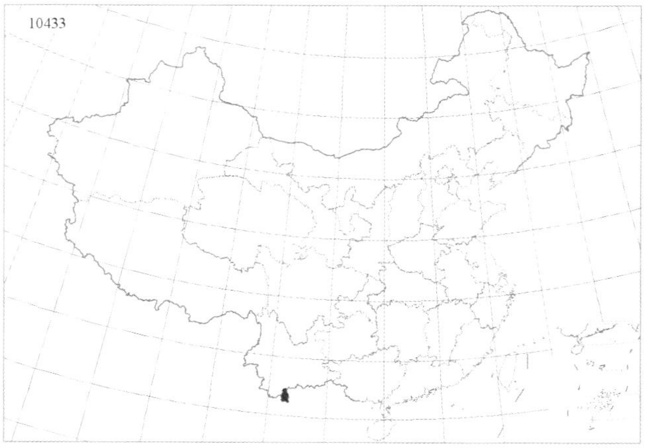

Rubiaceae 茜草科 *Wendlandia* 水锦树属

Wendlandia villosa W. C. Chen 毛叶水锦树

Shrubs or trees

17.9; 11.6; 22.1; 17.9; 155.1; 0.0; 858 ❙ 1098; 634; 46 ❙ 27.9; 836; 295.5

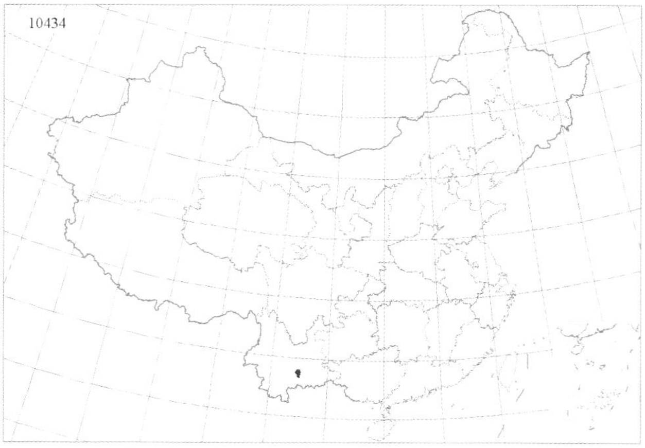

Rubiaceae 茜草科 *Xanthophytum* 岩黄树属
Xanthophytum attopevense (Pierre ex Pit.) H. S. Lo 琼岛岩黄树
Herbs or subshrubs

21.7 (16.3~24.8); 15.6 (9.1~20.1); 26.0 (21.4~28.7); 21.7 (16.4~24.8); 200.6 (136.4~237.6); 0.0 (0.0~0.0); 1142 (797~1385) ❙ 1429 (868~1854); 607 (312~860); 65 (41~115) ❙ 28.6 (-34.3~80.6); 1064 (797~1360); 434.7 (248.2~620.0)

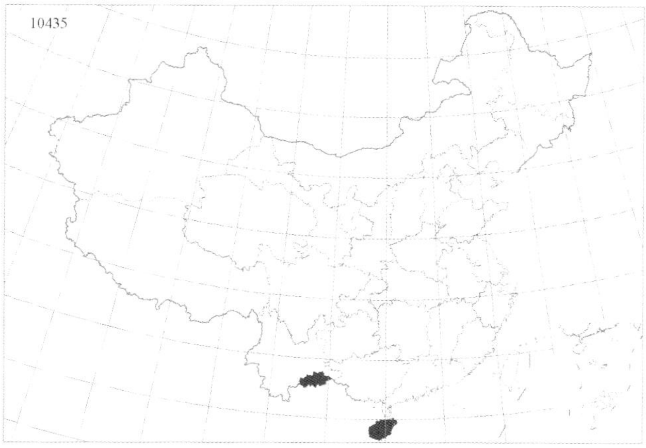

Rubiaceae 茜草科 *Xanthophytum* 岩黄树属
Xanthophytum balansae (Pit.) H. S. Lo 长梗岩黄树
Shrubs

21.9 (21.3~22.7); 13.9 (13.2~15.0); 27.9 (27.3~28.6); 22.0 (21.3~22.8); 203.6 (195.9~213.1); 0.0 (0.0~0.0); 1162 (1111~1221) ❙ 1876 (1302~2435); 1047 (670~1428); 116 (86~130) ❙ 61.1 (12.9~99.5); 1162 (1111~1221); 289.6 (235.1~318.8)

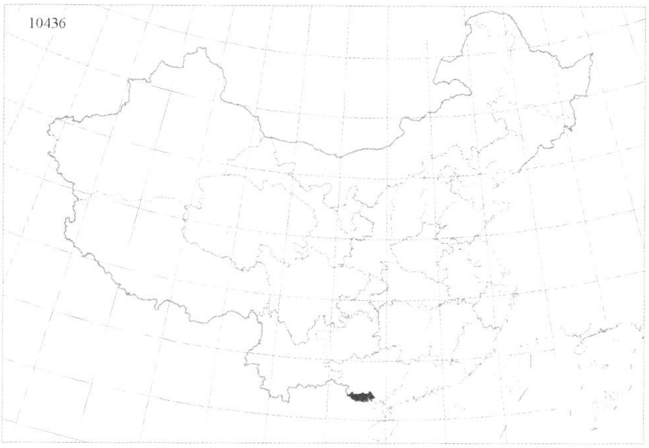

Rubiaceae 茜草科 *Xanthophytum* 岩黄树属
Xanthophytum kwangtungense (Chun et F. C. How) H. S. Lo 岩黄树
Shrubs

21.8 (21.0~22.7); 14.0 (13.2~15.0); 27.6 (25.6~28.6); 21.8 (21.0~22.8); 201.9 (191.8~213.1); 0.0 (0.0~0.0); 1146 (1053~1221) ❙ 1825 (1302~2435); 1014 (670~1428); 109 (68~130) ❙ 58.7 (12.9~99.5); 1146 (1053~1221); 327.0 (235.1~551.6)

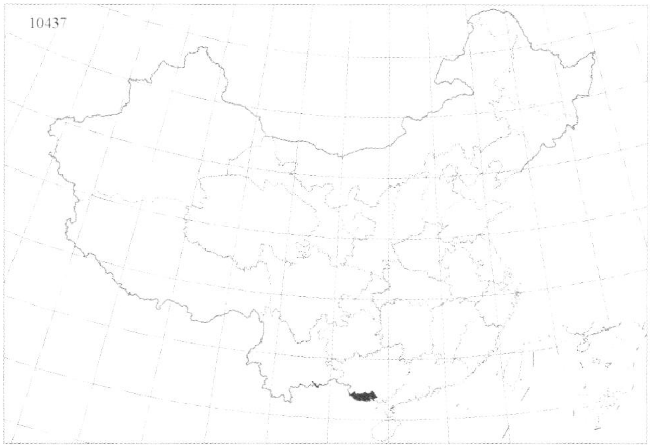

Carlemanniaceae 香茜科 *Silvianthus* 蜘蛛花属
Silvianthus tonkinensis (Gagnep.) Ridsdale 线萼蜘蛛花
Shrubs

18.1 (14.4~21.0); 11.4 (7.5~15.1); 22.5 (19.1~25.6); 18.1 (14.4~21.0); 157.6 (113.1~191.8); 0.0 (0.0~0.0); 877 (727~1053) ❙ 1414 (1074~1751); 784 (582~996); 57 (42~77) ❙ 61.8 (32.4~112.2); 872 (727~1053); 449.2 (277.0~659.0)

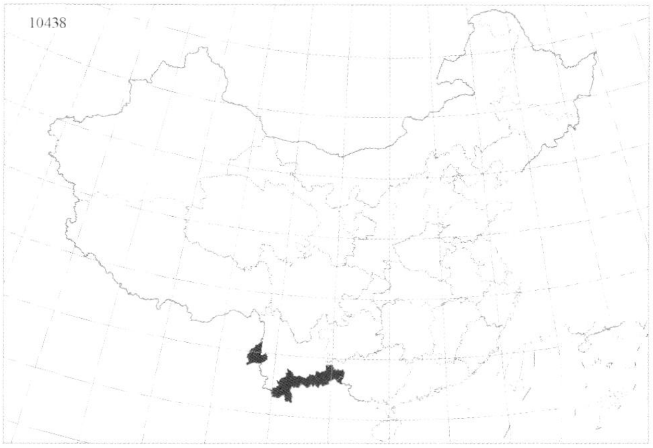

Adoxaceae 五福花科 *Viburnum* 荚蒾属
Viburnum amplifolium Rehder 广叶荚蒾
Shrubs, deciduous

18.0 (16.9~21.0); 10.8 (9.5~14.2); 23.0 (21.4~25.6); 18.0 (16.9~21.0); 155.7 (142.7~191.8); 0.0 (0.0~0.0); 875 (813~1053) ❙ 1375 (1178~1522); 766 (646~860); 56 (48~68) ❙ 57.6 (44.5~80.6); 875 (813~1053); 422.7 (336.3~551.6)

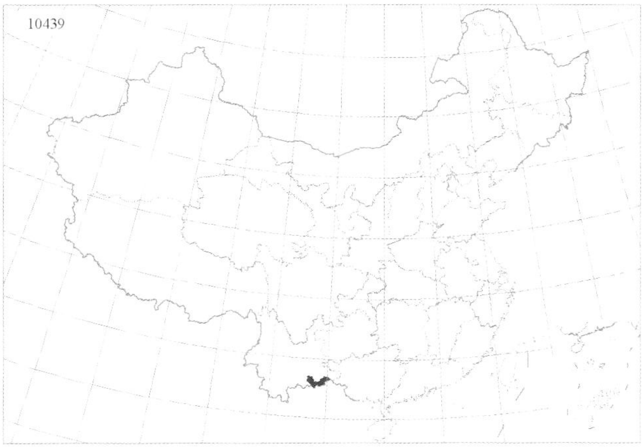

Adoxaceae 五福花科 *Viburnum* 荚蒾属
Viburnum atrocyaneum C. B. Clarke 蓝黑果荚蒾
Shrubs, evergreen

13.7 (1.8~19.1); 5.5 (-7.1~12.0); 20.6 (9.9~26.6); 13.8 (3.8~19.1); 108.9 (15.7~169.2); 3.5 (0.0~53.3); 735 (404~966) ❙ 1057 (473~1736); 588 (287~1007); 40 (7~71) ❙ 44.6 (-14.2~145.8); 719 (404~966); 255.5 (142.9~491.5)

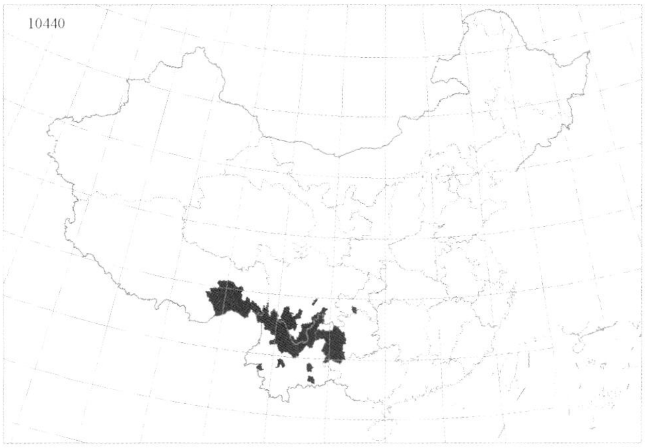

Adoxaceae 五福花科 *Viburnum* 荚蒾属

Viburnum betulifolium Batalin 桦叶荚蒾

Shrubs or small trees, deciduous

12.6 (1.9~24.1); 1.1 (-12.7~18.9); 23.1 (9.5~29.3); 13.0 (3.8~24.1); 104.8 (15.4~229.7); 12.7 (0.0~65.0); 741 (411~1322) ▌ 996 (341~3495); 480 (190~1847); 60 (7~897) ▌ 32.9 (-40.9~246.5); 695 (341~1144); 216.1 (61.2~447.4)

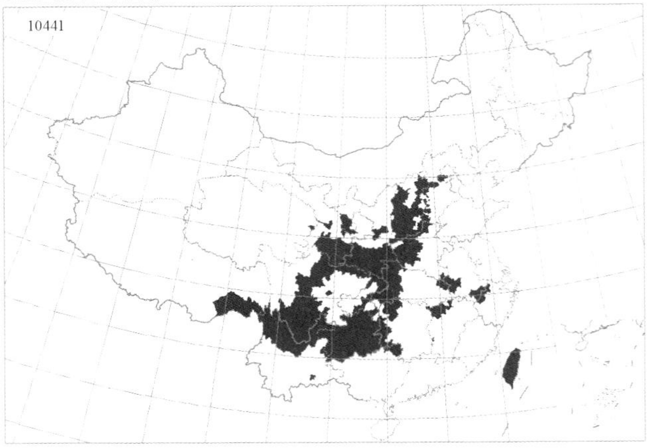

Adoxaceae 五福花科 *Viburnum* 荚蒾属

Viburnum brachybotryum Hemsl. 短序荚蒾

Shrubs or small trees, evergreen

16.8 (11.1~22.2); 6.8 (0.6~15.1); 25.7 (18.6~29.8); 16.9 (11.1~22.3); 143.0 (79.8~207.5); 0.6 (0.0~9.7); 885 (648~1190) ▌ 1360 (896~2024); 578 (392~891); 109 (30~209) ▌ 54.1 (12.9~107.2); 881 (648~1190); 226.0 (134.6~659.0)

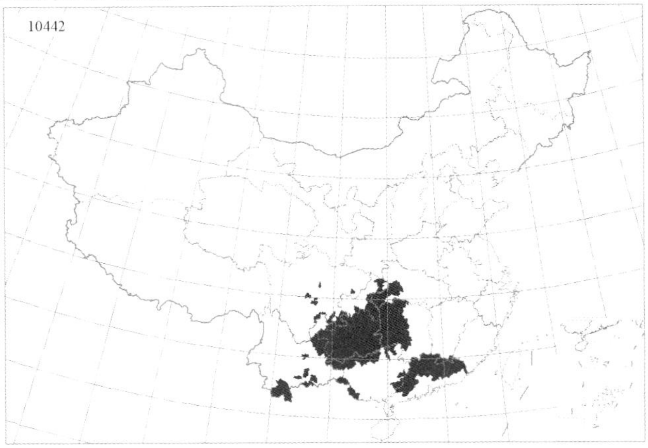

Adoxaceae 五福花科 *Viburnum* 荚蒾属

Viburnum brevitubum (P. S. Hsu) P. S. Hsu 短筒荚蒾

Shrubs, deciduous

14.4 (9.3~17.6); 3.1 (-1.7~5.7); 25.3 (19.4~29.1); 14.5 (9.5~17.7); 117.8 (69.8~152.5); 4.0 (0.0~18.0); 799 (603~955) ▌ 1326 (964~1630); 498 (421~584); 120 (30~228) ▌ 66.2 (31.7~101.4); 798 (603~946); 232.7 (170.4~284.3)

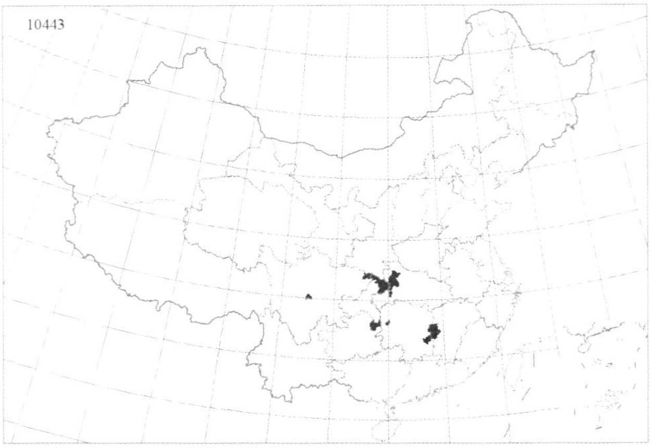

Adoxaceae 五福花科 *Viburnum* 荚蒾属

Viburnum burejaeticum Regel et Herd. 修枝荚蒾

Shrubs, deciduous

5.7 (-0.9~13.3); -12.8 (-25.0~-1.3); 21.7 (17.5~26.6); 8.9 (6.4~13.5); 70.5 (45.4~115.0); 61.5 (14.6~120.7); 611 (501~803) ▌ 594 (309~1073); 372 (199~691); 18 (8~40) ▌ -1.6 (-42.4~74.1); 538 (309~698); 229.4 (70.3~346.5)

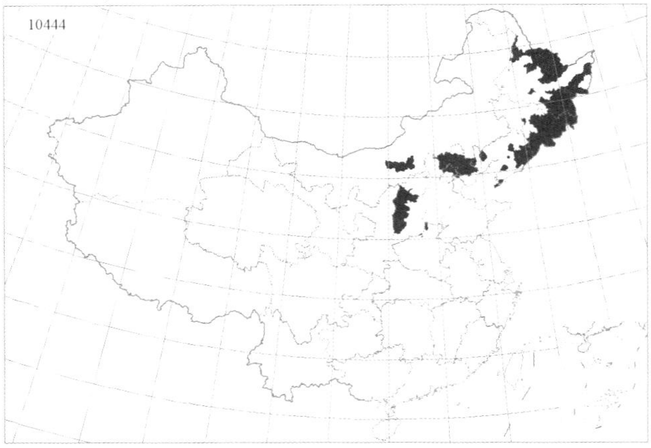

Adoxaceae 五福花科 *Viburnum* 荚蒾属

Viburnum chingii P. S. Hsu 漾濞荚蒾

Shrubs or small trees, evergreen

14.6 (5.4~21.0); 7.3 (-2.1~15.1); 20.3 (12.3~26.6); 14.6 (5.8~21.0); 116.6 (31.5~191.8); 1.3 (0.0~25.6); 753 (481~1053) ▌ 1077 (741~1607); 607 (411~907); 41 (15~77) ▌ 43.7 (-14.2~99.9); 734 (481~1053); 323.7 (142.9~659.0)

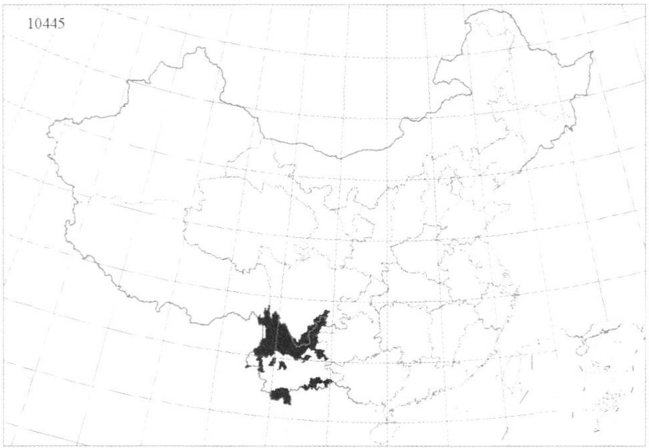

Adoxaceae 五福花科 *Viburnum* 荚蒾属

Viburnum chinshanense Graebn. 金佛山荚蒾

Shrubs

15.2 (2.9~19.2); 4.9 (-6.3~11.0); 24.6 (11.2~28.9); 15.2 (4.4~19.2); 124.9 (20.5~170.8); 2.3 (0.0~44.6); 809 (428~1002) ▌ 1182 (711~1577); 541 (349~932); 68 (15~140) ▌ 47.4 (-0.9~121.7); 809 (428~1002); 212.2 (108.2~413.5)

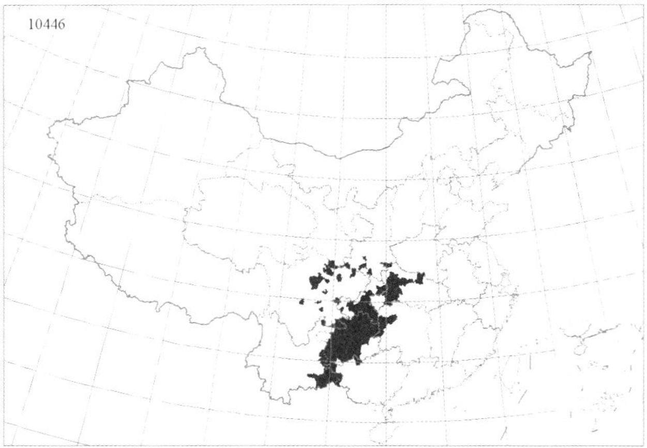

Adoxaceae 五福花科 *Viburnum* 荚蒾属

Viburnum chunii Hsu 金腺荚蒾

Shrubs, evergreen

17.9 (14.3~22.3); 7.4 (2.1~14.7); 27.4 (24.3~29.0); 17.9 (14.3~22.3); 155.4 (114.1~208.1); 0.5 (0.0~4.9); 948 (765~1202) ▎1560 (1082~1950); 597 (403~891); 174 (63~242) ▎66.6 (8.0~147.0); 946 (765~1202); 232.0 (108.2~335.1)

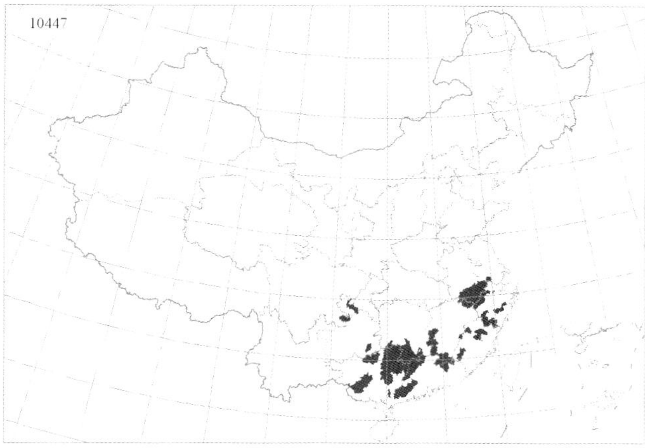

Adoxaceae 五福花科 *Viburnum* 荚蒾属

Viburnum cinnamomifolium Rehder 樟叶荚蒾

Shrubs or small trees, evergreen

15.6 (2.6~22.2); 6.6 (-7.0~13.9); 23.2 (11.2~28.7); 15.7 (4.3~22.3); 131.3 (20.0~207.5); 3.5 (0.0~48.2); 824 (427~1190) ▎1283 (829~1776); 652 (400~946); 72 (19~188) ▎58.0 (12.1~94.4); 822 (427~1190); 270.8 (133.8~491.5)

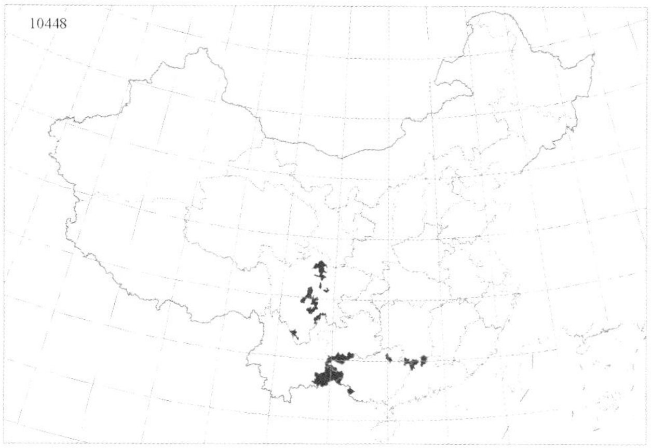

Adoxaceae 五福花科 *Viburnum* 荚蒾属

Viburnum congestum Rehder 密花荚蒾

Shrubs, evergreen

13.5 (2.2~18.9); 4.9 (-6.6~12.0); 20.9 (9.9~26.9); 13.6 (3.8~19.0); 106.3 (15.5~167.8); 3.9 (0.0~49.1); 731 (413~929) ▎1038 (637~1406); 543 (306~816); 45 (12~111) ▎44.0 (-14.2~125.9); 715 (413~912); 256.4 (142.9~434.4)

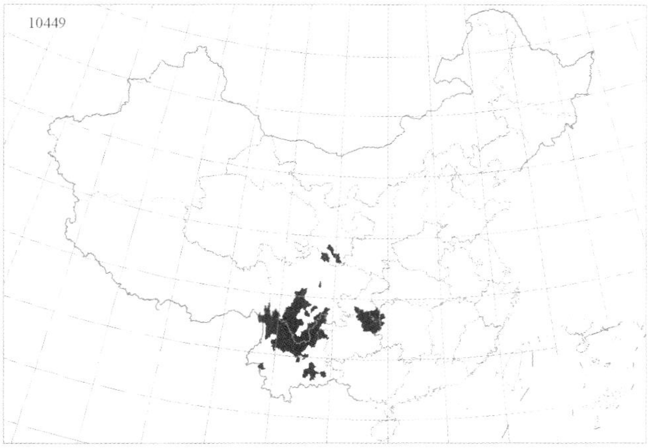

Adoxaceae 五福花科 *Viburnum* 荚蒾属

Viburnum corymbiflorum P. S. Hsu et S. C. Hsu 伞房荚蒾

Shrubs or small trees

16.2 (11.7~21.8); 6.2 (2.6~13.7); 25.2 (18.8~29.1); 16.2 (11.7~21.9); 135.2 (82.6~202.7); 0.4 (0.0~4.1); 845 (658~1159) ▎1357 (916~1997); 573 (444~860); 111 (31~259) ▎60.5 (6.5~130.7); 843 (658~1159); 229.6 (146.0~413.5)

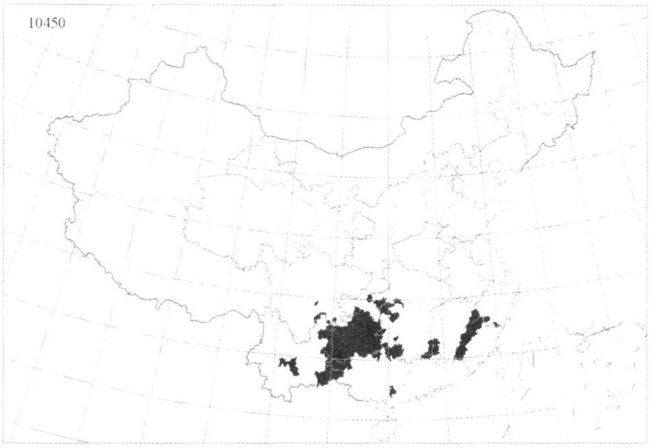

Adoxaceae 五福花科 *Viburnum* 荚蒾属

Viburnum corymbiflorum subsp. malifolium P. S. Hsu 苹果叶荚蒾

Shrubs or small trees

17.7; 11.2; 22.0; 17.8; 153.2; 0.0; 856 ▎1176; 662; 48 ▎39.4; 807; 412.1

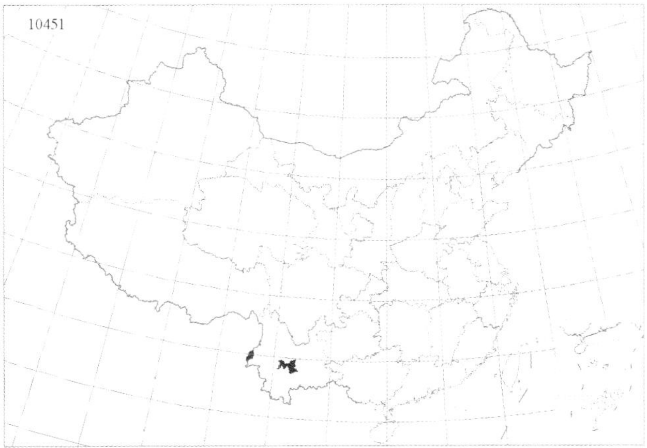

Adoxaceae 五福花科 *Viburnum* 荚蒾属

Viburnum cotinifolium D. Don 黄栌叶荚蒾

Shrubs, deciduous

-0.6; -9.2; 7.3; 2.4; 6.2; 73.5; 350 ▎384; 247; 19 ▎7.8; 339; 95.9

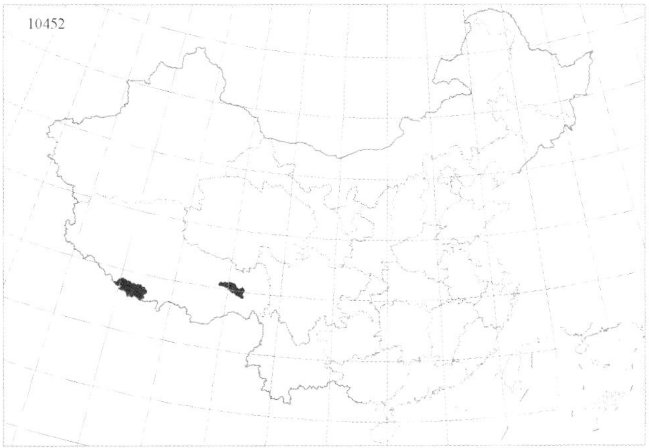

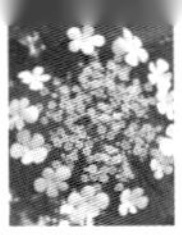

Adoxaceae 五福花科 *Viburnum* 荚蒾属

Viburnum cylindricum Buch.-Ham. ex D. Don 水红木

Shrubs or small trees, evergreen

15.9 (−1.1~23.0); 6.9 (−9.1~15.1); 23.7 (6.2~29.8); 16.0 (2.0~23.0); 133.8 (2.5~216.2); 2.2 (0.0~75.6); 841 (334~1247) ❙ 1253 (321~2435); 605 (193~1428); 81 (9~214) ❙ 50.3 (−14.2~145.8); 834 (321~1247); 261.7 (79.3~659.0)

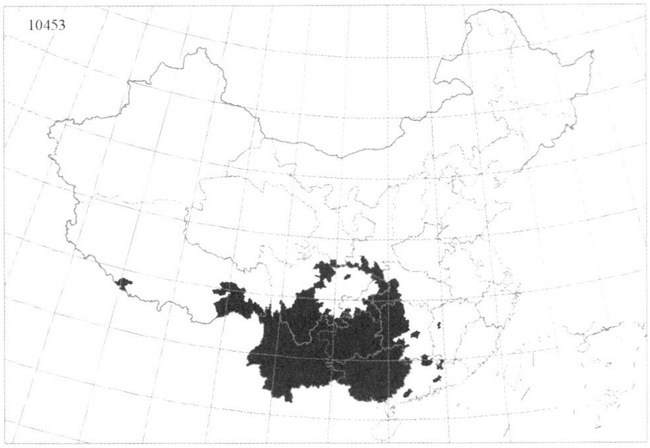

Adoxaceae 五福花科 *Viburnum* 荚蒾属

Viburnum dalzielii W. W. Sm. 粤赣荚蒾

Shrubs

18.3 (15.2~21.8); 7.8 (4.1~13.4); 27.7 (25.8~29.1); 18.3 (15.3~21.8); 159.7 (124.1~201.9); 0.1 (0.0~0.9); 966 (810~1158) ❙ 1661 (1368~1990); 594 (445~858); 193 (131~228) ❙ 72.3 (52.0~106.4); 960 (810~1158); 231.1 (179.5~275.2)

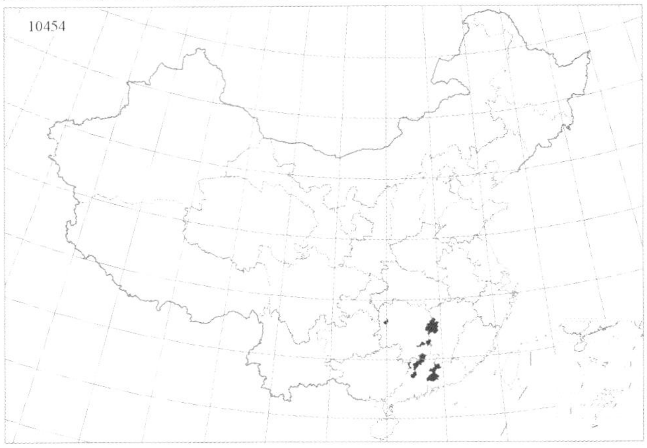

Adoxaceae 五福花科 *Viburnum* 荚蒾属

Viburnum davidii Franch. 川西荚蒾

Shrubs, evergreen

11.1 (2.2~17.4); 1.8 (−6.6~7.4); 19.5 (9.9~25.5); 11.5 (3.8~17.4); 86.0 (15.5~148.9); 11.7 (0.0~49.1); 666 (413~886) ❙ 1206 (932~1414); 660 (531~817); 45 (19~107) ❙ 84.5 (55.9~125.9); 666 (413~886); 248.1 (214.5~292.6)

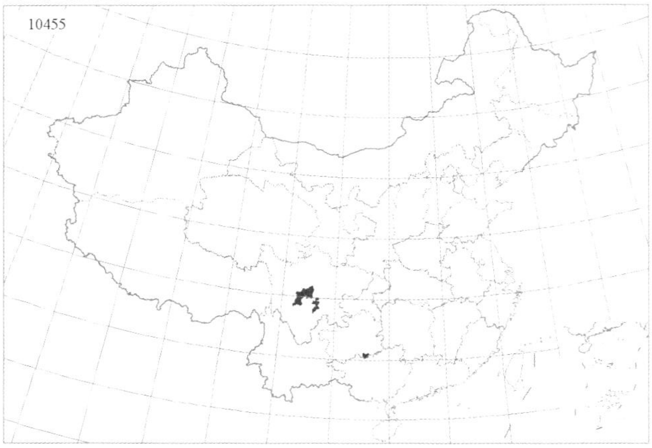

Adoxaceae 五福花科 *Viburnum* 荚蒾属

Viburnum dilatatum Thunb. 荚蒾

Shrubs, deciduous

15.1 (0.0~22.3); 3.1 (−13.7~13.6); 26.2 (9.7~30.1); 15.4 (3.3~22.3); 129.1 (12.2~208.1); 7.1 (0.0~72.2); 846 (395~1202) ❙ 1196 (341~1997); 487 (190~860); 118 (8~259) ❙ 39.1 (−47.7~147.0); 807 (341~1202); 204.3 (61.2~413.5)

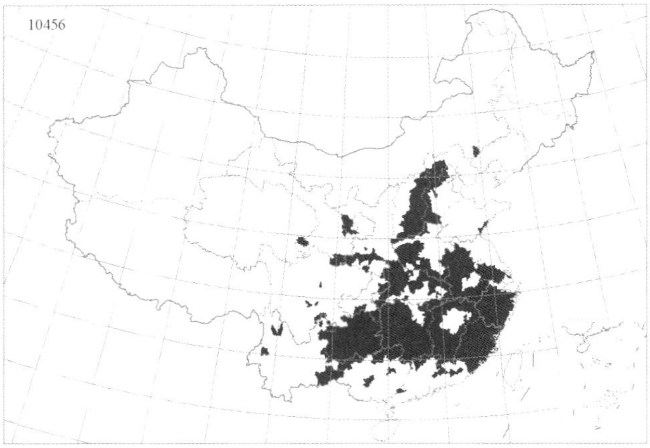

Adoxaceae 五福花科 *Viburnum* 荚蒾属

Viburnum erosum Thunb. 宜昌荚蒾

Shrubs, deciduous

15.4 (0.0~24.1); 4.1 (−11.4~18.9); 26.0 (9.7~30.1); 15.5 (3.3~24.1); 129.1 (12.2~229.7); 3.4 (0.0~72.2); 842 (395~1322) ❙ 1310 (611~3495); 522 (316~1594); 126 (11~897) ❙ 54.6 (−22.2~235.6); 832 (395~1160); 213.7 (123.9~364.1)

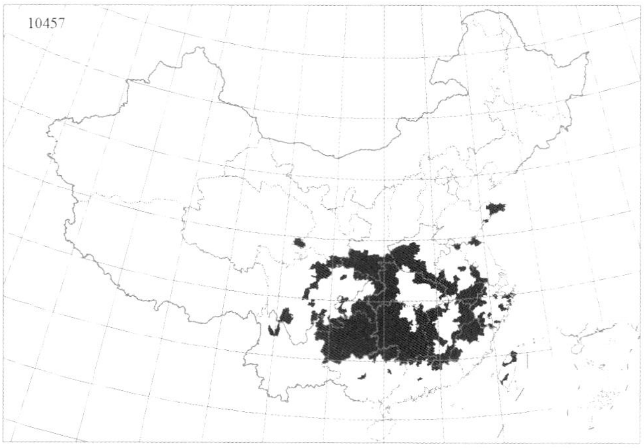

Adoxaceae 五福花科 *Viburnum* 荚蒾属

Viburnum erubescens Wall. 淡红荚蒾

Shrubs or small trees, deciduous

13.2 (−1.8~19.1); 3.4 (−10.5~12.0); 22.1 (6.1~27.7); 13.4 (1.8~19.1); 105.8 (1.9~169.2); 6.7 (0.0~82.8); 736 (329~969) ❙ 1083 (311~1740); 519 (217~1007); 56 (9~191) ❙ 47.7 (−14.2~145.8); 730 (311~969); 238.4 (61.6~447.4)

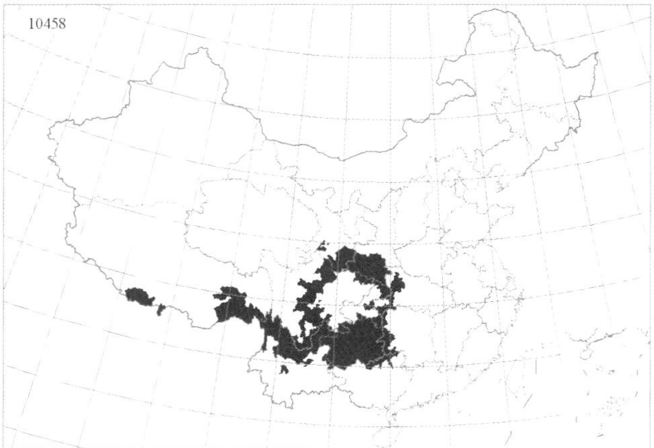

Adoxaceae 五福花科 *Viburnum* 荚蒾属
Viburnum farreri Stearn 香荚蒾
Shrubs, deciduous

9.1 (−2.8~16.6); −5.9 (−20.9~4.4); 22.2 (9.7~28.8); 10.6 (2.9~16.7); 85.0 (11.3~140.6); 35.7 (0.6~109.2); 677 (359~920) ❙ 610 (30~1571); 291 (21~666); 39 (1~157) ❙ −12.9 (−96.3~108.0); 518 (30~910); 156.3 (2.1~313.2)

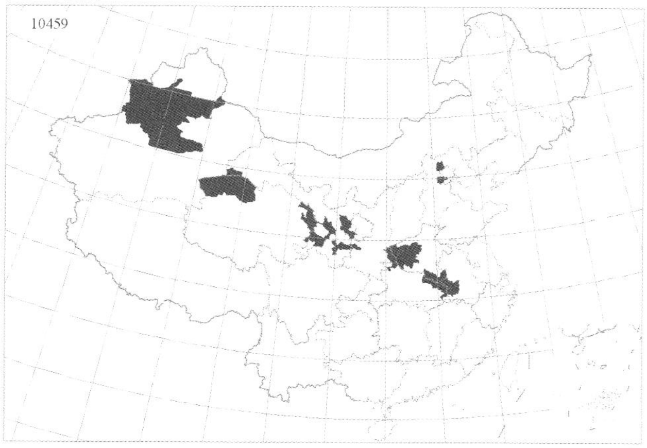

Adoxaceae 五福花科 *Viburnum* 荚蒾属
Viburnum foetidum (C. H. Wright) Hand.-Mazz. 臭荚蒾
Shrubs, deciduous

14.9 (−1.1~24.1); 5.4 (−9.6~18.9); 23.5 (6.2~29.0); 15.0 (2.0~24.1); 122.7 (2.5~229.7); 3.0 (0.0~75.6); 797 (334~1322) ❙ 1225 (321~2992); 574 (193~1847); 76 (10~402) ❙ 54.0 (−14.2~235.6); 789 (321~1123); 233.4 (61.6~513.9)

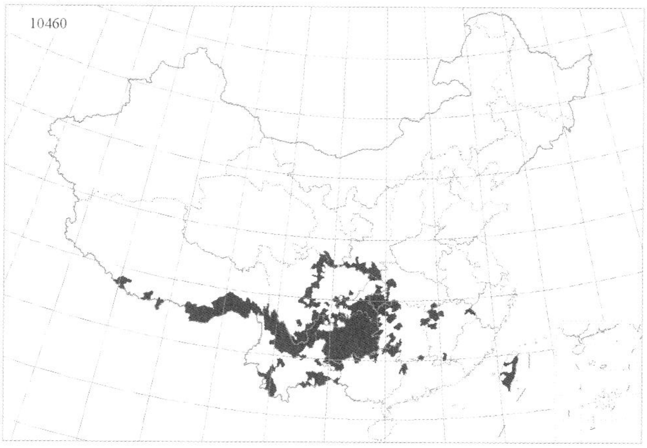

Adoxaceae 五福花科 *Viburnum* 荚蒾属
Viburnum foetidum var. ceanothoides (C. H. Wright) Hand.-Mazz. 珍珠荚蒾
Shrubs, deciduous

15.4 (2.2~21.0); 6.2 (−6.6~14.2); 23.5 (9.9~27.7); 15.5 (3.8~21.0); 126.9 (15.5~191.8); 1.1 (0.0~49.1); 804 (413~1053) ❙ 1179 (907~1522); 568 (444~820); 67 (18~161) ❙ 47.7 (6.5~125.9); 796 (413~1053); 241.8 (146.0~551.6)

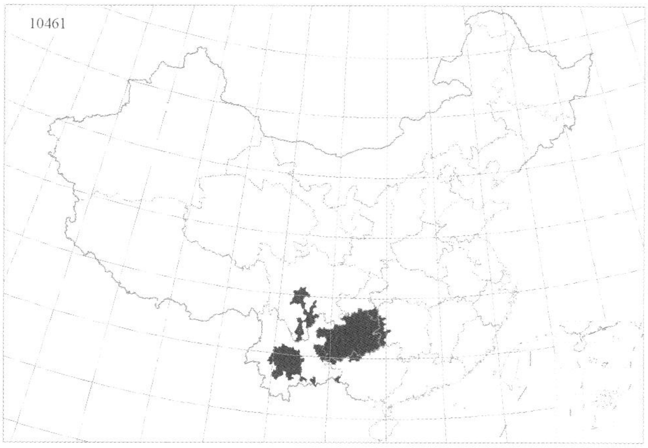

Adoxaceae 五福花科 *Viburnum* 荚蒾属
Viburnum foetidum var. rectangulatum (Graebn.) Rehder 直角荚蒾
Shrubs, deciduous

15.9 (3.5~20.7); 5.0 (−6.5~11.1); 26.0 (12.6~29.9); 16.0 (5.0~20.8); 133.0 (26.3~189.3); 1.2 (0.0~43.4); 854 (446~1113) ❙ 1338 (711~1803); 522 (346~817); 126 (15~228) ❙ 57.1 (11.7~128.9); 844 (446~1072); 204.5 (128.7~354.5)

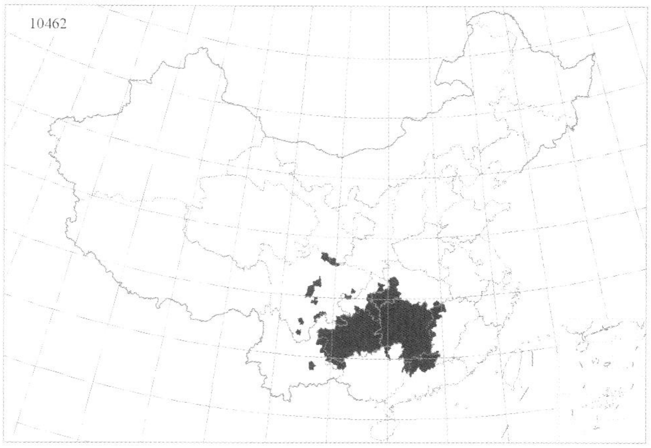

Adoxaceae 五福花科 *Viburnum* 荚蒾属
Viburnum fordiae Hance 南方荚蒾
Shrubs or small trees

18.4 (11.7~23.6); 3.7 (2.7~16.7); 27.0 (18.6~29.8); 18.5 (11.7~23.6); 161.8 (82.6~223.5); 0.1 (0.0~3.8); 970 (658~1281) ❙ 1516 (896~2435); 606 (356~1428); 151 (31~259) ❙ 57.7 (−6.9~147.0); 963 (658~1273); 226.4 (95.3~551.6)

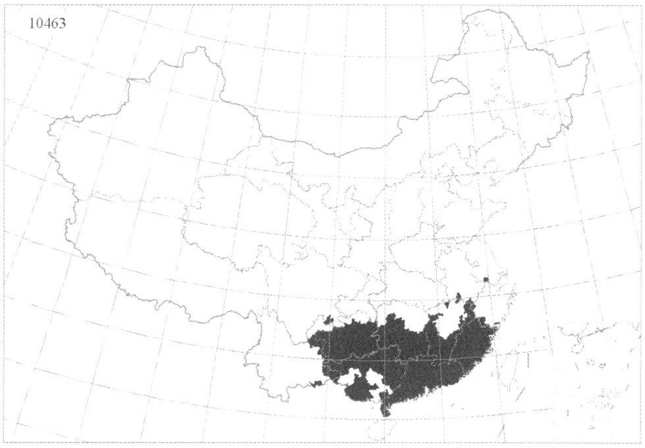

Adoxaceae 五福花科 *Viburnum* 荚蒾属
Viburnum formosanum Hayata 台中荚蒾
Shrubs

17.8 (14.3~22.2); 7.8 (3.3~17.7); 26.9 (20.1~29.4); 17.9 (14.3~22.2); 154.4 (114.1~206.7); 0.1 (0.0~2.3); 931 (765~1123) ❙ 1738 (1197~2993); 640 (403~1847); 202 (60~547) ❙ 88.2 (30.5~246.5); 925 (765~1123); 233.9 (151.0~303.4)

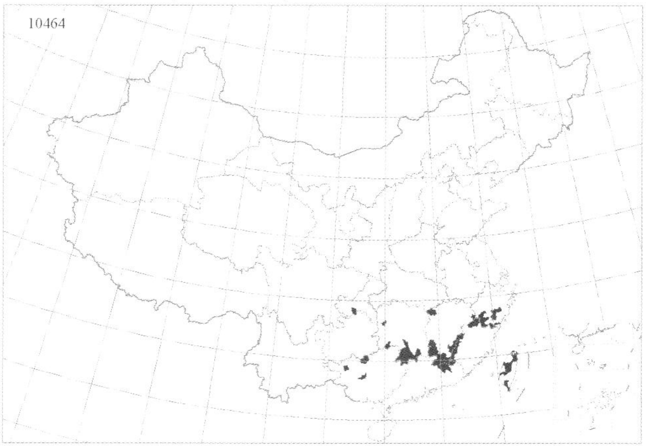

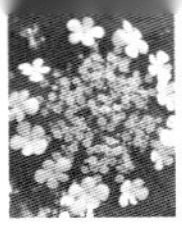

Adoxaceae 五福花科 *Viburnum* 荚蒾属
Viburnum formosanum var. pubigerum P. S. Hsu 毛枝台中荚蒾
Shrubs

18.9 (17.3~20.6); 8.9 (6.4~11.6); 27.5 (26.6~28.5); 18.9 (17.3~20.7); 167.0 (147.9~187.9); 0.0 (0.0~0.0); 984 (907~1081) ❙ 1656 (1503~1852); 594 (507~711); 190 (176~197) ❙ 68.5 (50.5~86.3); 984 (907~1081); 239.7 (197.5~271.3)

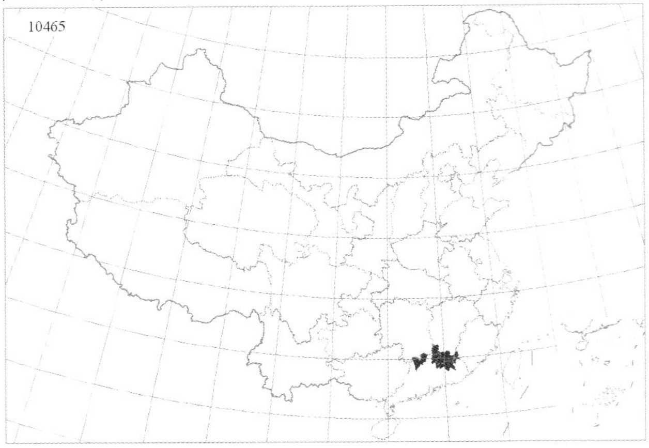

Adoxaceae 五福花科 *Viburnum* 荚蒾属
Viburnum glomeratum Maxim. 聚花荚蒾
Shrubs or small trees, deciduous

10.8 (0.8~19.1); -1.2 (-10.6~13.0); 21.9 (9.9~29.3); 11.2 (3.8~19.1); 87.3 (15.4~169.7); 17.7 (0.0~66.0); 681 (411~941) ❙ 856 (315~1663); 401 (174~835); 43 (6~220) ❙ 26.5 (-47.5~128.6); 653 (315~941); 235.2 (44.5~506.2)

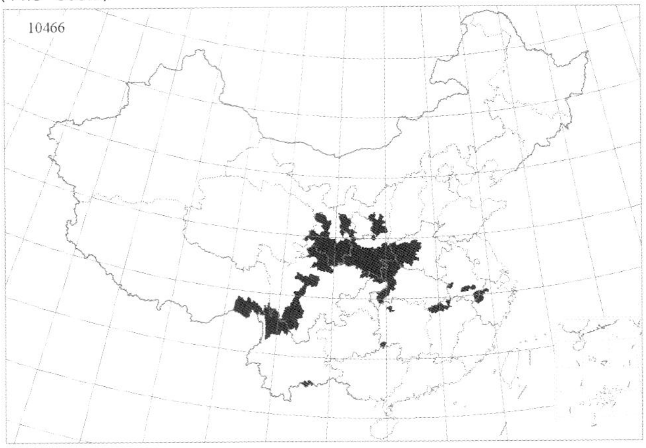

Adoxaceae 五福花科 *Viburnum* 荚蒾属
Viburnum grandiflorum Wall. ex DC. 大花荚蒾
Shrubs or small trees, deciduous

3.1 (-1.8~9.6); -4.8 (-10.2~2.4); 10.2 (6.1~15.7); 5.0 (1.8~9.6); 25.5 (1.9~60.8); 47.6 (5.4~82.8); 441 (329~600) ❙ 616 (320~1041); 364 (193~585); 30 (11~45) ❙ 28.5 (-4.0~81.5); 437 (320~600); 199.6 (72.2~352.3)

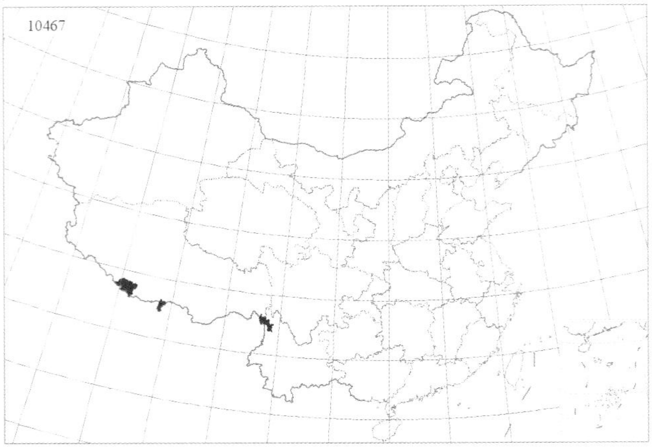

Adoxaceae 五福花科 *Viburnum* 荚蒾属
Viburnum hainanense Merr. et Chun 海南荚蒾
Shrubs, evergreen

22.1 (15.5~24.7); 14.5 (4.8~19.8); 27.9 (25.1~29.0); 22.2 (15.6~24.7); 205.8 (127.2~236.5); 0.0 (0.0~0.2); 1187 (815~1376) ❙ 1664 (1269~2435); 763 (513~1428); 111 (48~191) ❙ 42.1 (-1.0~99.5); 1167 (815~1360); 324.9 (197.5~620.0)

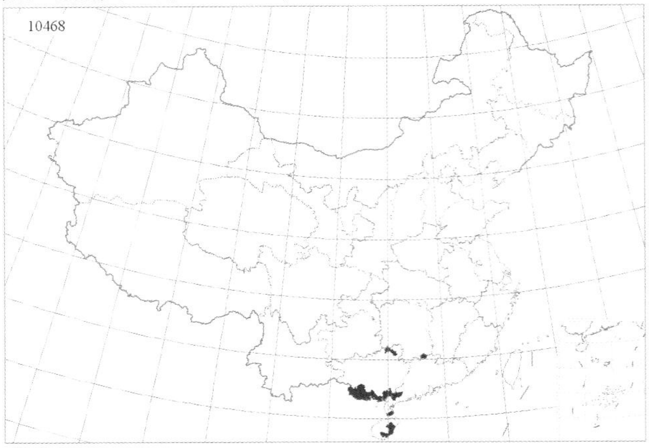

Adoxaceae 五福花科 *Viburnum* 荚蒾属
Viburnum hanceanum Maxim. 蝶花荚蒾
Shrubs

18.4 (14.2~23.2); 8.5 (3.3~15.8); 27.2 (24.2~29.8); 18.5 (14.2~23.2); 162.4 (112.5~218.9); 0.1 (0.0~2.3); 971 (761~1257) ❙ 1603 (1141~2024); 606 (411~994); 169 (67~259) ❙ 66.0 (28.9~128.9); 969 (761~1257); 230.8 (146.0~322.1)

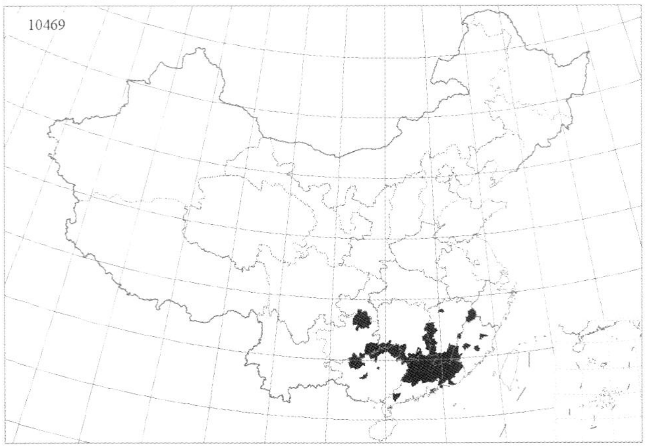

Adoxaceae 五福花科 *Viburnum* 荚蒾属
Viburnum hengshanicum Tsiang 衡山荚蒾
Shrubs, deciduous

15.6 (11.7~19.0); 3.8 (1.5~8.7); 26.8 (18.8~29.3); 15.6 (11.7~19.1); 129.2 (82.6~168.7); 1.8 (0.0~7.3); 851 (658~1002) ❙ 1489 (916~1828); 559 (473~705); 169 (31~253) ❙ 74.9 (28.8~112.2); 850 (658~1002); 234.1 (166.9~319.9)

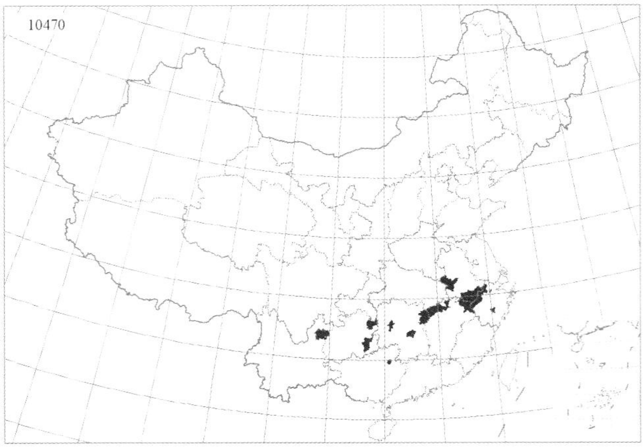

Adoxaceae 五福花科 *Viburnum* 荚蒾属

Viburnum henryi Hemsl. 巴东荚蒾

Shrubs or small trees, evergreen or semievergreen

15.2 (9.3~19.1); 4.6 (−1.7~10.1); 25.1 (16.7~29.2); 15.3 (9.5~19.1); 125.1 (62.7~169.2); 1.6 (0.0~18.0); 816 (600~984) ❙ 1305 (842~1997); 540 (375~932); 103 (19~259) ❙ 59.9 (10.5~147.0); 815 (600~969); 216.6 (146.0~305.6)

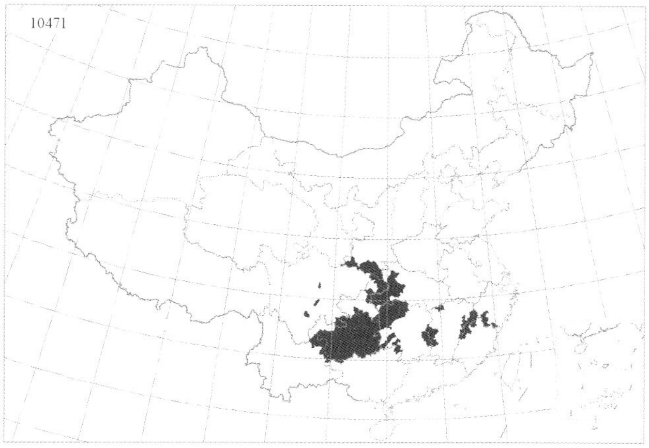

Adoxaceae 五福花科 *Viburnum* 荚蒾属

Viburnum inopinatum Craib 厚绒荚蒾

Shrubs or small trees, evergreen

19.3 (17.2~22.2); 11.9 (9.4~15.1); 24.5 (22.1~28.3); 19.3 (17.3~22.3); 171.6 (147.0~207.5); 0.0 (0.0~0.0); 954 (838~1190) ❙ 1379 (1250~1607); 725 (651~835); 63 (48~87) ❙ 45.2 (15.4~64.9); 950 (838~1190); 421.7 (289.7~659.0)

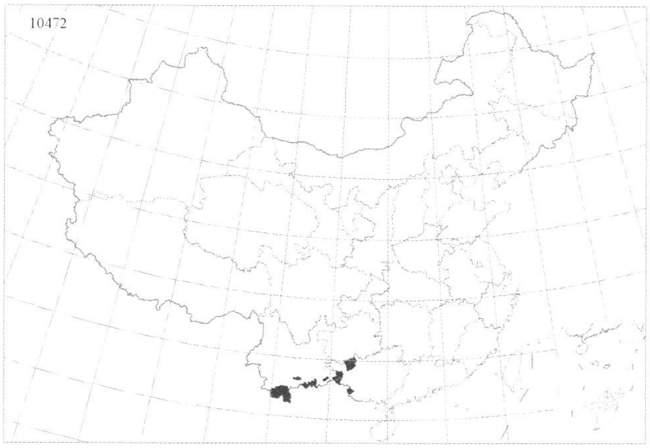

Adoxaceae 五福花科 *Viburnum* 荚蒾属

Viburnum integrifolium Hayata 全叶荚蒾

Shrubs

19.2; 14.0; 23.4; 19.3; 171.1; 0.0; 942 ❙ 2744; 1351; 216 ❙ 197.2; 942; 264.9

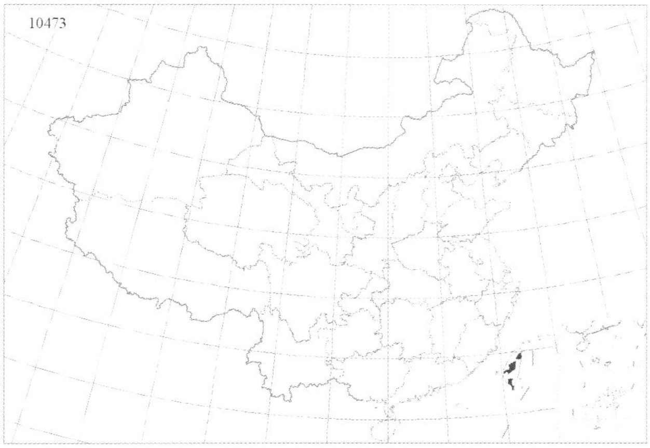

Adoxaceae 五福花科 *Viburnum* 荚蒾属

Viburnum kansuense Batalin 甘肃荚蒾

Shrubs, deciduous

7.8 (0.7~14.9); −1.8 (−8.8~7.6); 16.5 (9.5~24.2); 8.4 (3.3~15.0); 57.5 (12.7~119.6); 23.2 (0.0~63.9); 566 (389~782) ❙ 842 (373~1414); 460 (201~817); 22 (7~49) ❙ 51.0 (−37.6~128.6); 561 (373~782); 258.4 (88.1~322.4)

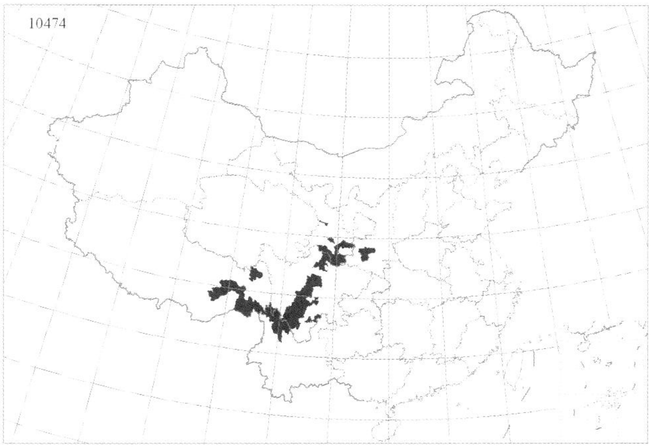

Adoxaceae 五福花科 *Viburnum* 荚蒾属

Viburnum koreanum Nakai 朝鲜荚蒾

Shrubs, deciduous

2.5 (1.4~4.9); −18.1 (−20.1~−16.5); 20.2 (18.4~23.3); 7.4 (6.4~9.1); 55.8 (45.8~73.8); 85.3 (73.9~91.9); 551 (501~629) ❙ 658 (573~777); 409 (350~473); 22 (15~35) ❙ 20.2 (−1.0~55.1); 550 (501~623); 297.3 (186.7~329.9)

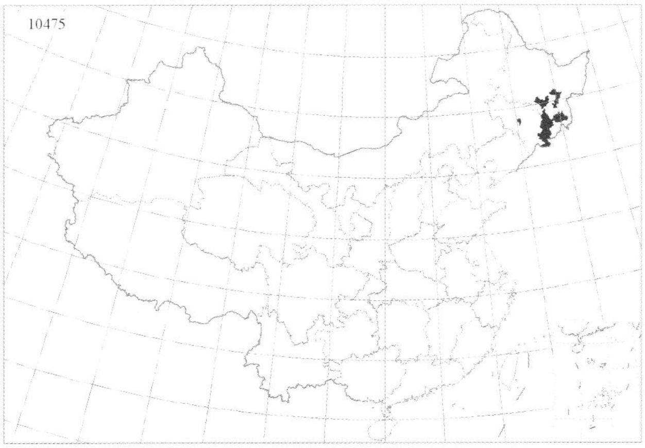

Adoxaceae 五福花科 *Viburnum* 荚蒾属

Viburnum lancifolium P. S. Hsu 披针叶荚蒾

Shrubs, evergreen

17.5 (15.0~20.7); 6.8 (3.9~10.7); 27.5 (24.7~30.1); 17.6 (15.1~20.8); 150.9 (120.8~189.3); 0.1 (0.0~1.1); 926 (789~1108) ❙ 1696 (1370~1997); 591 (453~718); 205 (117~254) ❙ 84.5 (28.9~147.0); 918 (789~1108); 242.5 (148.9~332.5)

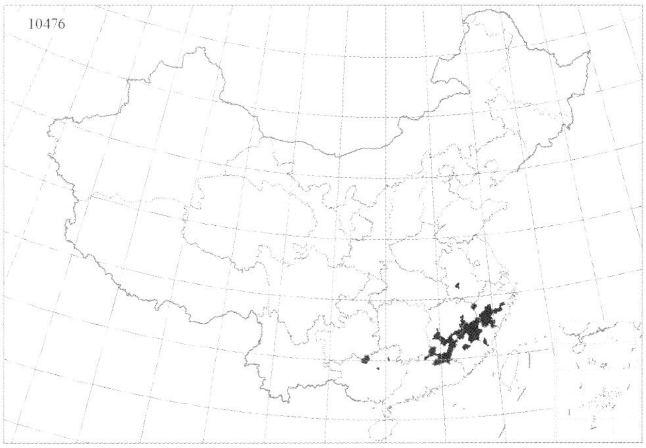

Adoxaceae 五福花科 *Viburnum* 荚蒾属

Viburnum laterale Rehder 侧花荚蒾

Shrubs, deciduous

17.5；7.1；27.1；17.6；150.8；0.0；916 ❙ 1889；627；238 ❙ 106.2；916；274.6

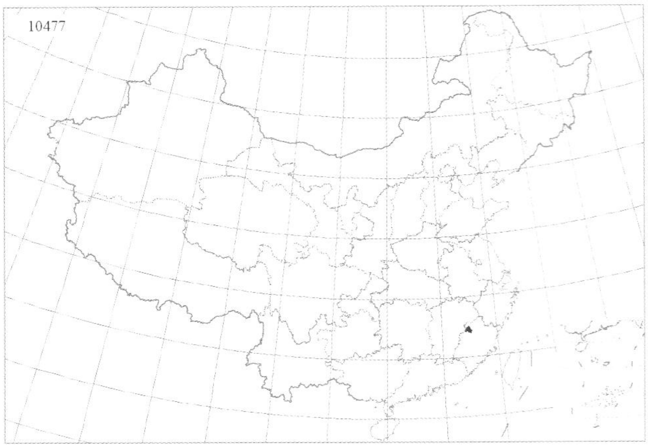

Adoxaceae 五福花科 *Viburnum* 荚蒾属

Viburnum leiocarpum P. S. Hsu 光果荚蒾

Shrubs or small trees

18.7 (16.9~21.0)；12.2 (9.5~15.1)；23.2 (21.4~25.6)；18.8 (16.9~21.0)；165.4 (142.7~191.8)；0.0 (0.0~0.0)；909 (813~1053) ❙ 1412 (1178~1607)；758 (646~911)；58 (48~77) ❙ 55.8 (36.8~80.6)；903 (813~1053)；471.6 (336.3~659.0)

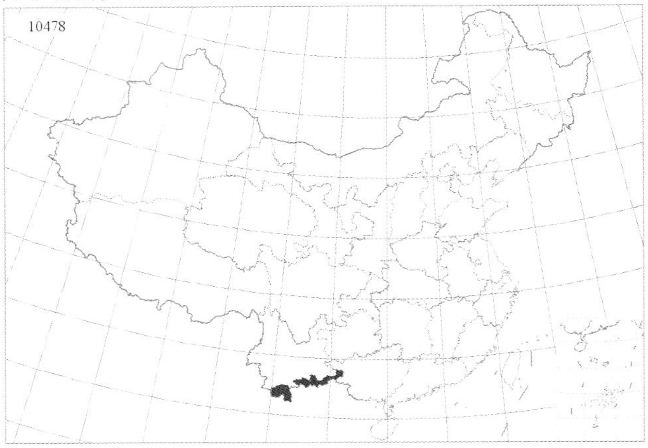

Adoxaceae 五福花科 *Viburnum* 荚蒾属

Viburnum leiocarpum var. punctatum P. S. Hsu 斑点光果荚蒾

Shrubs or small trees

18.4 (17.2~21.0)；11.5 (9.5~14.2)；23.2 (22.4~25.6)；18.4 (17.3~21.0)；160.8 (147.0~191.8)；0.0 (0.0~0.0)；892 (833~1053) ❙ 1435 (1254~1547)；813 (699~911)；57 (48~68) ❙ 61.5 (44.5~80.6)；892 (833~1053)；458.6 (336.3~551.6)

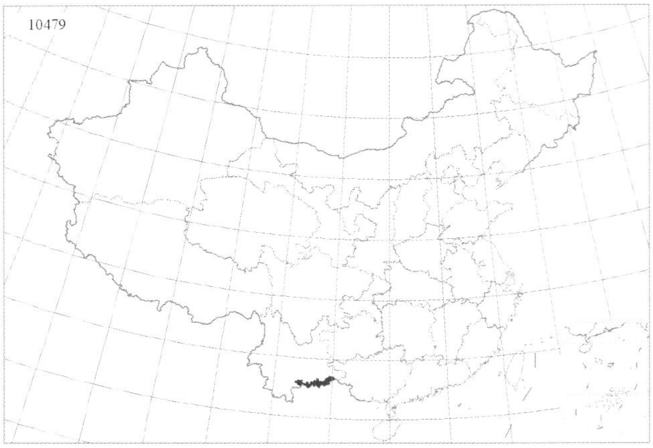

Adoxaceae 五福花科 *Viburnum* 荚蒾属

Viburnum longipedunculatum (P. S. Hsu) P. S. Hsu 长梗荚蒾

Shrubs

19.0 (17.2~20.5)；12.1 (7.3~15.1)；23.7 (22.3~26.3)；19.0 (17.3~20.5)；168.2 (147.0~185.9)；0.0 (0.0~0.0)；927 (838~992) ❙ 1464 (1254~1607)；676 (639~702)；78 (48~142) ❙ 57.9 (47.6~64.9)；914 (838~986)；461.0 (225.9~659.0)

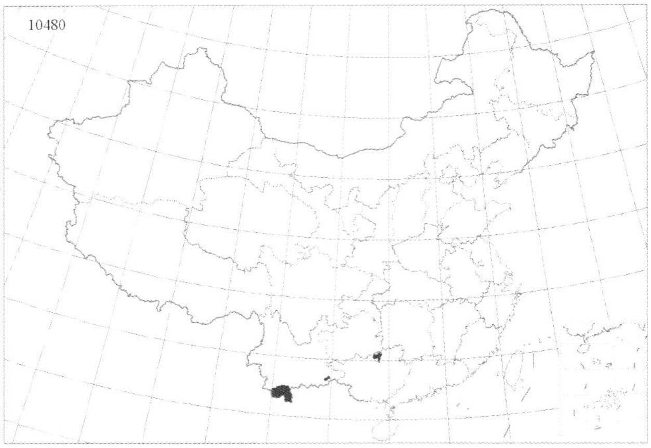

Adoxaceae 五福花科 *Viburnum* 荚蒾属

Viburnum longiradiatum P. S. Hsu et S. W. Fan 长伞梗荚蒾

Shrubs or small trees, deciduous

13.0 (10.6~15.1)；3.3 (-0.2~5.3)；21.8 (19.4~24.0)；13.0 (10.6~15.1)；99.3 (79.4~121.7)；3.3 (0.0~12.1)；714 (643~789) ❙ 1169 (944~1414)；632 (520~817)；39 (30~49) ❙ 64.0 (36.4~93.8)；714 (643~789)；236.7 (171.0~298.1)

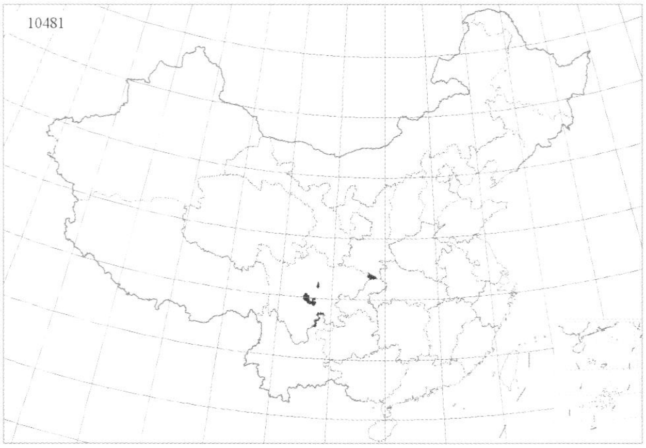

Adoxaceae 五福花科 *Viburnum* 荚蒾属

Viburnum lutescens Blume 淡黄荚蒾

Shrubs, evergreen

20.9 (15.5~24.8)；12.4 (4.8~20.1)；27.8 (25.1~29.8)；20.9 (15.6~24.8)；190.8 (127.2~237.6)；0.0 (0.0~0.2)；1107 (815~1385) ❙ 1593 (868~2148)；652 (312~1234)；138 (41~198) ❙ 46.6 (-34.3~100.8)；1076 (815~1360)；279.9 (174.9~620.0)

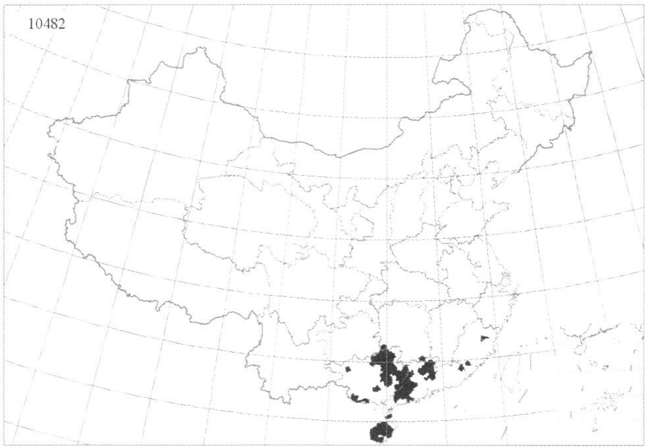

Adoxaceae 五福花科 *Viburnum* 荚蒾属
Viburnum luzonicum Rolfe 吕宋荚蒾
Shrubs

18.9 (14.3~24.1); 9.7 (3.1~18.9); 27.2 (20.1~29.8); 19.0 (14.3~24.1); 167.7 (113.2~229.7); 0.0 (0.0~2.2); 985 (761~1322) ❙ 1676 (1070~3495); 627 (356~1594); 191 (47~897) ❙ 72.8 (-5.0~246.5); 977 (761~1230); 251.3 (123.9~659.0)

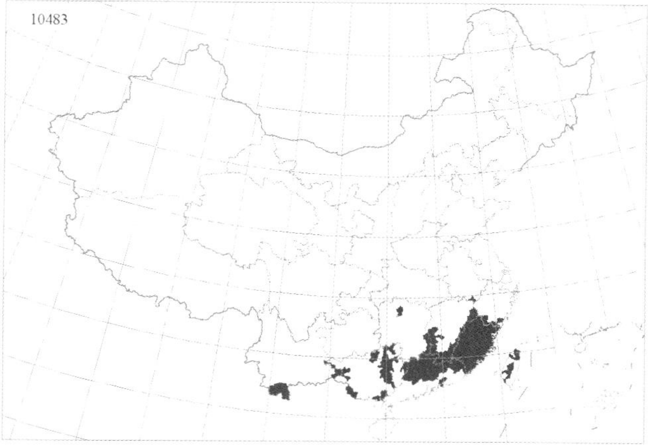

Adoxaceae 五福花科 *Viburnum* 荚蒾属
Viburnum macrocephalum Fort. 绣球荚蒾
Shrubs, deciduous or semievergreen

15.6 (10.9~19.1); 4.5 (-2.3~10.1); 25.9 (18.8~29.9); 15.7 (11.1~19.1); 129.5 (82.6~169.2); 1.4 (0.0~17.6); 839 (658~976) ❙ 1227 (584~1638); 519 (299~720); 102 (21~225) ❙ 46.4 (-34.1~108.0); 829 (584~969); 198.2 (111.9~283.9)

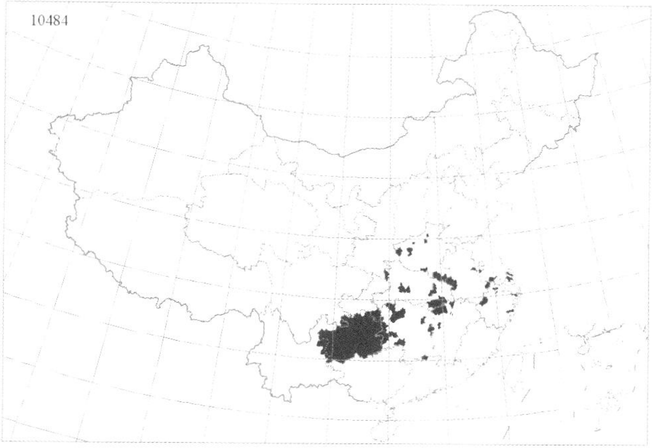

Adoxaceae 五福花科 *Viburnum* 荚蒾属
Viburnum macrocephalum Fort. f. keteleeri (Carr.) Rehd. 八仙花
Shrubs, deciduous or semievergreen

15.1 (12.4~17.2); 2.3 (-1.8~6.3); 27.4 (23.5~29.4); 15.1 (12.6~17.2); 126.6 (101.8~146.8); 5.0 (0.0~15.9); 851 (729~952) ❙ 1057 (714~1660); 485 (417~666); 1[illegible]1 (38~225) ❙ 24.0 (-16.6~108.0); 838 (714~925); 182.7 (111.9~264.9)

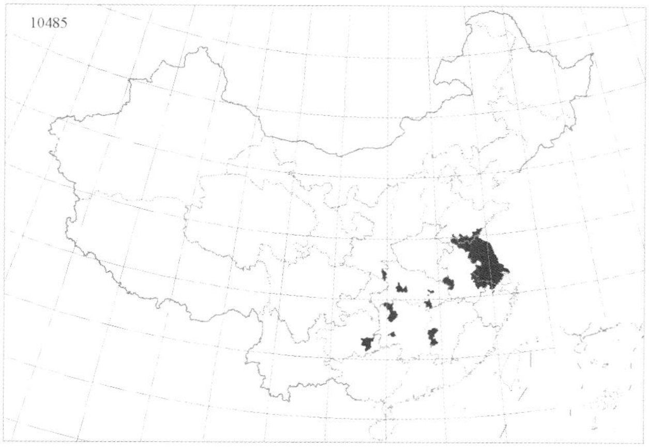

Adoxaceae 五福花科 *Viburnum* 荚蒾属
Viburnum melanocarpum Hsu 黑果荚蒾
Shrubs, deciduous

15.7 (12.9~18.0); 3.7 (1.5~7.0); 27.4 (23.4~29.3); 15.7 (12.9~18.0); 130.6 (101.2~156.2); 2.1 (0.0~7.3); 860 (723~956) ❙ 1411 (1002~1893); 537 (415~685); 165 (83~242) ❙ 64.7 (14.0~129.9); 859 (723~941); 226.6 (131.2~319.9)

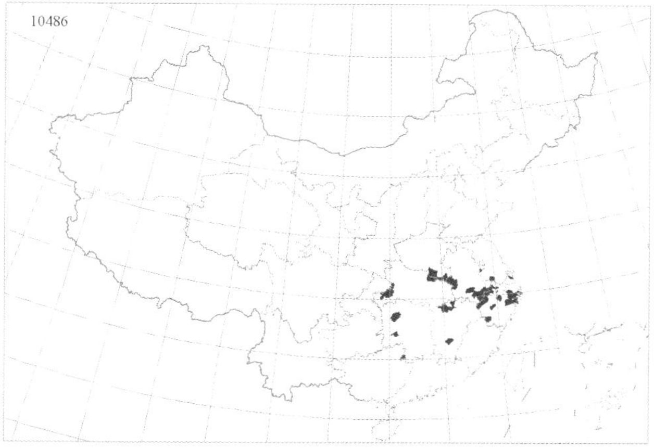

Adoxaceae 五福花科 *Viburnum* 荚蒾属
Viburnum mongolicum (Pall.) Rehder 蒙古荚蒾
Shrubs, deciduous

7.9 (-5.8~15.1); -8.0 (-29.8~1.8); 22.0 (7.8~28.5); 9.8 (2.1~15.2); 77.9 (5.2~128.7); 42.9 (5.6~157.4); 642 (340~886) ❙ 511 (49~856); 299 (34~453); 16 (1~53) ❙ -19.7 (-92.8~52.6); 502 (49~817); 169.0 (5.1~318.2)

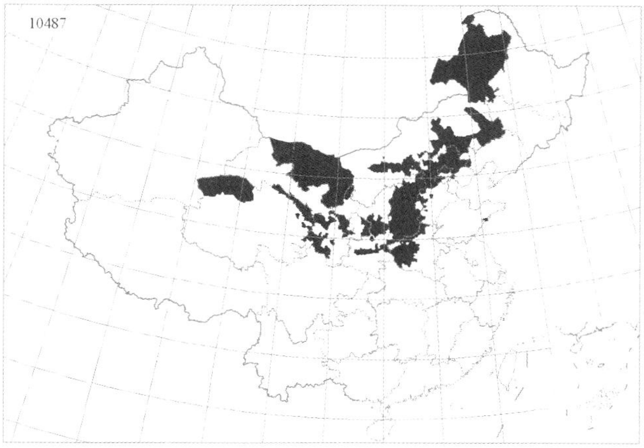

Adoxaceae 五福花科 *Viburnum* 荚蒾属
Viburnum mullaha (C. B. Clarke) Kitam. 西域荚蒾
Shrubs or small trees, deciduous

4.7 (-1.8~13.3); -3.5 (-10.2~5.4); 11.9 (6.1~19.6); 6.1 (1.8~13.4); 36.0 (1.9~100.4); 39.4 (0.0~82.8); 477 (329~706) ❙ 812 (320~1736); 477 (193~1007); 27 (9~51) ❙ 55.6 (-4.0~145.8); 475 (320~706); 237.9 (72.2~447.4)

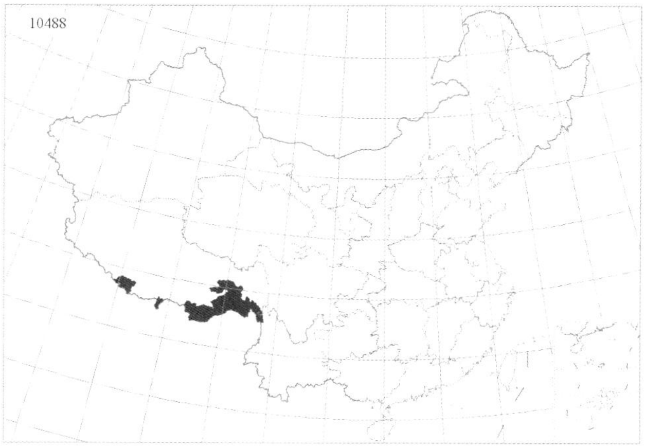

Adoxaceae 五福花科 *Viburnum* 荚蒾属

Viburnum mullaha var. glabrescens (C. B. Clarke) Kitam. 少毛西域荚蒾

Shrubs or small trees, deciduous

6.1 (0.3~13.3); -2.3 (-8.6~5.4); 13.3 (8.3~19.6); 7.1 (2.9~13.4); 44.5 (9.6~100.4); 31.2 (0.0~65.4); 510 (368~706) ❙ 938 (389~1736); 549 (246~1007); 24 (9~51) ❙ 71.0 (0.0~145.8); 510 (368~706); 272.9 (142.5~447.4)

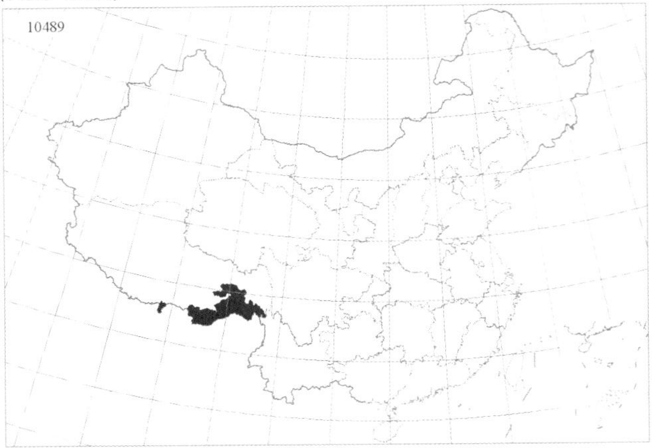

Adoxaceae 五福花科 *Viburnum* 荚蒾属

Viburnum nervosum D. Don 显脉荚蒾

Shrubs or small trees, deciduous

15.1 (-1.8~21.7); 4.3 (-10.2~15.3); 25.1 (6.1~29.8); 15.2 (1.8~21.7); 125.3 (1.9~200.2); 4.0 (0.0~82.8); 825 (329~1121) ❙ 1307 (320~3495); 528 (193~1412); 123 (7~897) ❙ 57.5 (-37.6~246.5); 813 (320~1121); 225.1 (61.6~506.2)

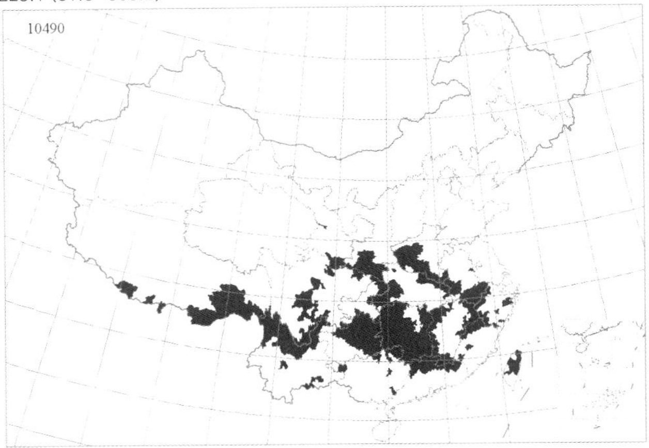

Adoxaceae 五福花科 *Viburnum* 荚蒾属

Viburnum odoratissimum Ker-Gawler 珊瑚树

Shrubs or small trees, evergreen

19.2 (12.7~24.8); 9.8 (1.9~20.1); 27.4 (21.7~29.8); 19.3 (12.8~24.8); 171.2 (99.0~237.6); 0.2 (0.0~6.1); 1016 (713~1385) ❙ 1563 (868~2993); 631 (312~1847); 147 (27~547) ❙ 56.0 (-34.3~214.4); 1003 (713~1360); 238.5 (95.3~659.0)

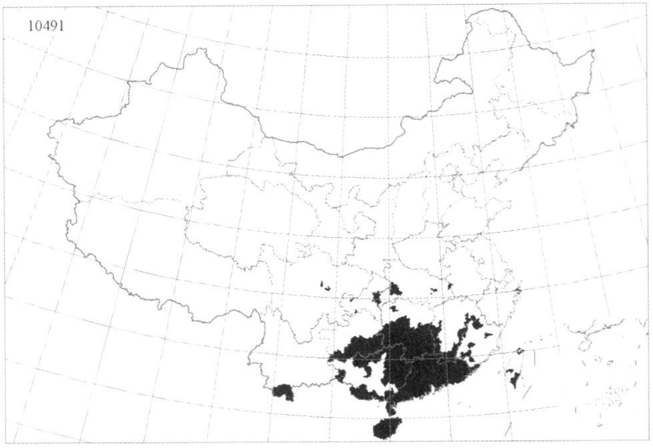

Adoxaceae 五福花科 *Viburnum* 荚蒾属

Viburnum oliganthum Batalin 少花荚蒾

Shrubs or small trees, evergreen

14.4 (-0.7~19.1); 4.3 (-9.6~10.1); 23.4 (7.7~27.7); 14.5 (2.5~19.1); 116.0 (6.8~169.2); 3.3 (0.0~74.2); 775 (338~969) ❙ 1183 (370~1740); 548 (262~871); 71 (10~187) ❙ 53.5 (4.7~125.9); 775 (338~969); 216.3 (61.6~322.4)

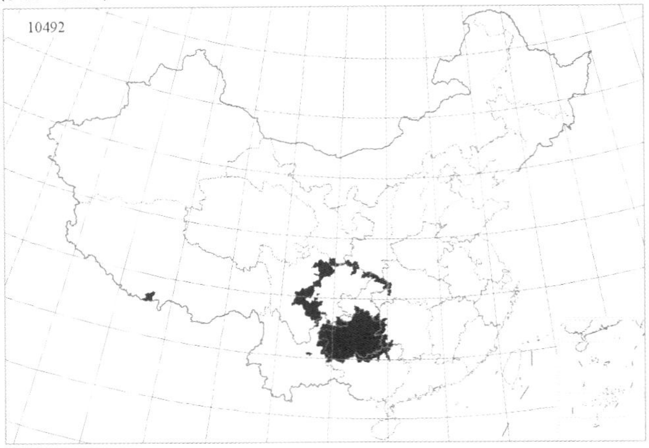

Adoxaceae 五福花科 *Viburnum* 荚蒾属

Viburnum omeiense P. S. Hsu 峨眉荚蒾

Shrubs, deciduous

14.9; 5.3; 23.6; 15.0; 119.6; 0.0; 782 ❙ 1410; 816; 49 ❙ 80.4; 782; 214.5

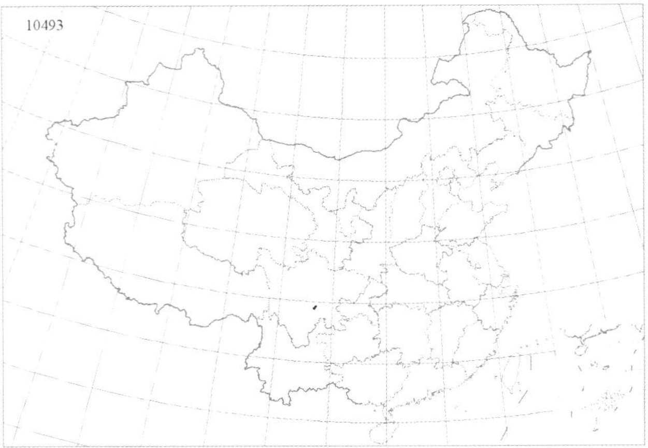

Adoxaceae 五福花科 *Viburnum* 荚蒾属

Viburnum opulus L. 欧洲荚蒾

Shrubs, deciduous

7.7 (-5.8~17.2); -8.8 (-29.8~5.7); 22.2 (11.7~29.3); 9.9 (4.3~17.2); 79.1 (21.1~146.8); 46.4 (0.0~157.4); 647 (426~941) ❙ 605 (120~1663); 335 (51~691); 30 (6~220) ❙ -8.1 (-85.3~103.1); 529 (120~941); 188.1 (16.6~338.4)

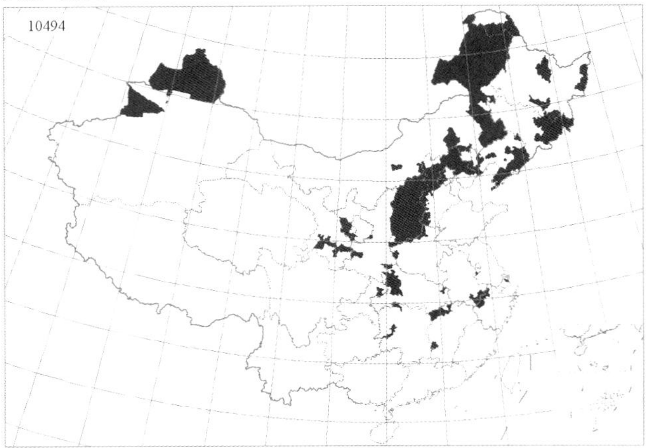

Adoxaceae 五福花科 *Viburnum* 荚蒾属
Viburnum opulus var. sargentii (Koehne) Takeda 鸡树条
Shrubs, deciduous

9.7 (-3.1~16.9); -5.6 (-26.4~4.8); 23.5 (11.7~28.1); 11.0 (4.3~16.9); 90.0 (21.1~143.4); 32.5 (0.2~134.6); 693 (427~924) ❙ 675 (309~1689); 380 (190~691); 31 (6~220) ❙ -3.1 (-42.4~107.2); 599 (309~924); 189.0 (61.2~335.3)

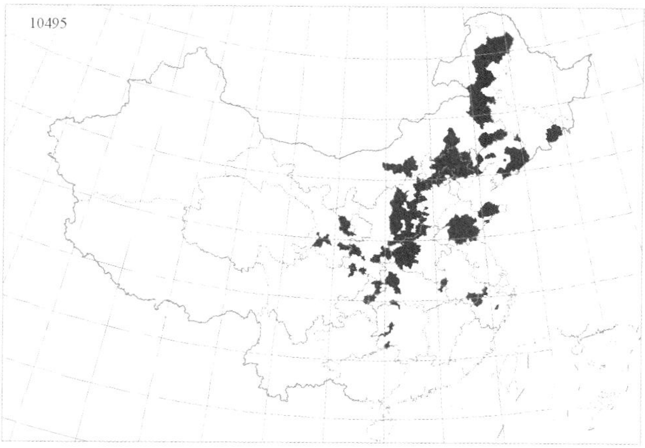

Adoxaceae 五福花科 *Viburnum* 荚蒾属
Viburnum plicatum Thunb. 粉团
Shrubs, deciduous

15.7 (7.3~21.0); 4.6 (-2.0~15.1); 26.2 (15.6~29.4); 15.8 (7.5~21.1); 130.9 (48.2~193.0); 1.8 (0.0~20.4); 846 (537~1114) ❙ 1395 (815~1997); 531 (352~821); 140 (19~259) ❙ 64.7 (7.2~147.0); 843 (537~1114); 228.8 (111.9~659.0)

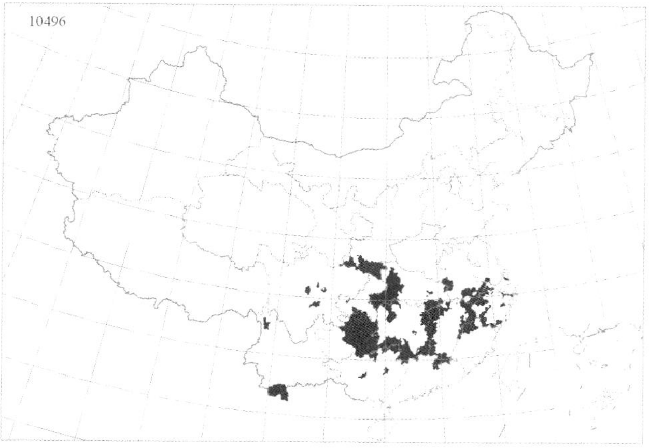

Adoxaceae 五福花科 *Viburnum* 荚蒾属
Viburnum plicatum var. tomentosum Miq. 蝴蝶戏珠花
Shrubs, deciduous

15.4 (7.3~19.1); 3.7 (-2.9~10.5); 26.3 (15.6~29.8); 15.4 (7.5~19.1); 128.1 (48.2~169.2); 3.2 (0.0~20.5); 840 (537~992) ❙ 1228 (626~1997); 503 (316~720); 117 (19~255) ❙ 45.9 (-22.2~130.7); 829 (537~992); 203.9 (134.1~319.9)

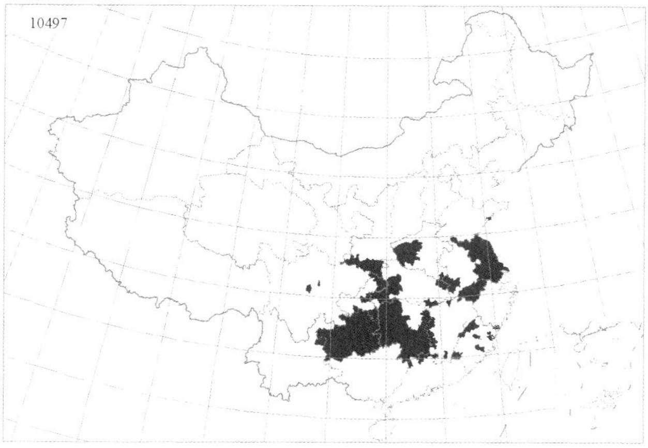

Adoxaceae 五福花科 *Viburnum* 荚蒾属
Viburnum propinquum Hemsl. 球核荚蒾
Shrubs, evergreen

16.0 (9.3~21.8); 5.2 (-1.7~15.3); 25.9 (16.9~29.4); 16.0 (9.5~21.8); 133.4 (68.1~201.6); 1.4 (0.0~18.0); 852 (603~1162) ❙ 1360 (707~2859); 543 (332~1186); 122 (15~402) ❙ 60.0 (-0.2~246.5); 848 (603~1162); 216.8 (108.2~371.7)

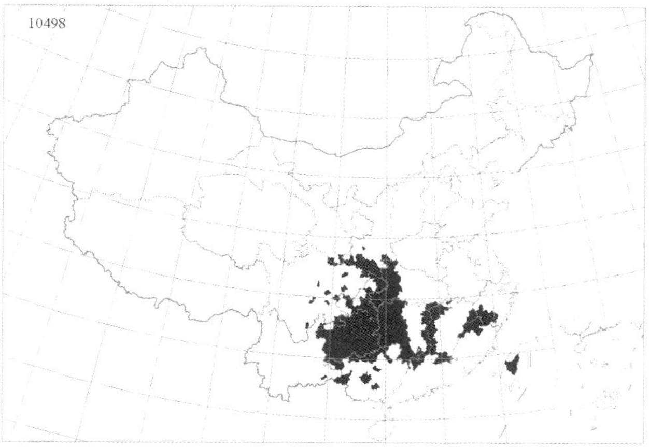

Adoxaceae 五福花科 *Viburnum* 荚蒾属
Viburnum propinquum var. mairei W. W. Sm. 狭叶球核荚蒾
Shrubs, evergreen

13.3 (9.3~16.9); 2.3 (-1.7~5.5); 23.8 (20.1~28.1); 13.4 (9.5~16.9); 105.8 (69.8~143.4); 5.6 (0.0~18.0); 745 (603~924) ❙ 1132 (879~1466); 505 (421~602); 63 (31~105) ❙ 54.1 (23.3~106.2); 745 (603~924); 243.1 (166.0~285.1)

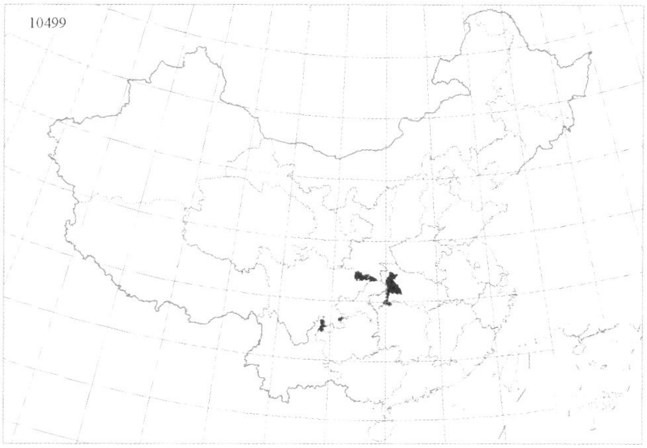

Adoxaceae 五福花科 *Viburnum* 荚蒾属
Viburnum punctatum Buch.-Ham. ex D. Don 鳞斑荚蒾
Shrubs or small trees, evergreen

18.9 (10.2~24.8); 10.8 (2.4~20.1); 25.3 (16.9~29.0); 18.9 (10.3~24.8); 167.1 (68.2~237.6); 0.1 (0.0~5.1); 987 (620~1385) ❙ 1357 (868~2435); 634 (312~1428); 78 (18~184) ❙ 39.9 (-34.3~99.5); 950 (620~1360); 312.6 (160.1~620.0)

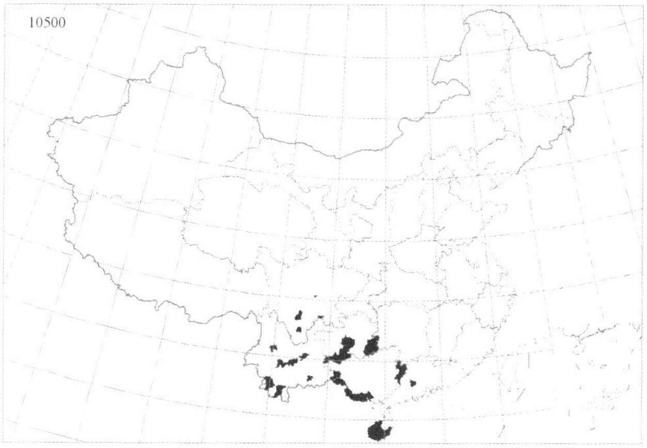

Adoxaceae 五福花科 *Viburnum* 荚蒾属

Viburnum punctatum var. lepidotulum (Merr. et Chun) P. S. Hsu 大果鳞斑荚蒾

Shrubs or small trees, evergreen

21.7 (17.9~24.7); 14.2 (8.1~19.8); 27.6 (25.0~28.8); 21.8 (17.9~24.7); 201.3 (155.4~236.5); 0.0 (0.0~0.0); 1155 (918~1376) ❙ 1574 (1179~1990); 646 (392~828); 109 (42~181) ❙ 38.8 (-11.9~83.4); 1094 (918~1212); 326.6 (203.4~620.0)

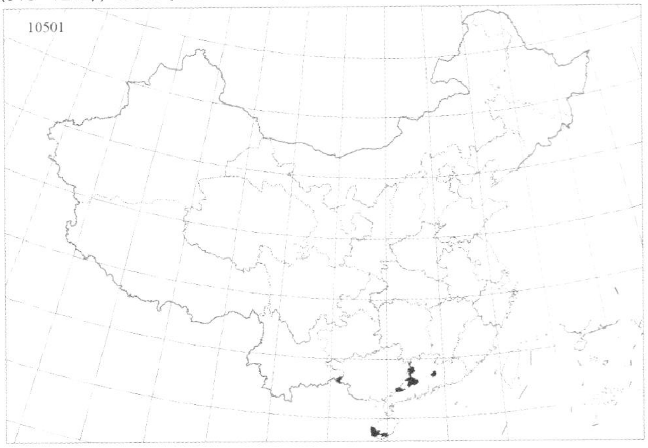

Adoxaceae 五福花科 *Viburnum* 荚蒾属

Viburnum pyramidatum Rehder 锥序荚蒾

Shrubs or small trees

19.4 (16.9~22.2); 11.8 (9.2~14.2); 24.9 (21.4~28.5); 19.4 (16.9~22.3); 172.7 (142.7~207.5); 0.0 (0.0~0.0); 973 (813~1190) ❙ 1461 (1178~2148); 794 (646~1234); 80 (48~171) ❙ 50.9 (15.4~86.7); 972 (813~1190); 384.6 (206.0~551.6)

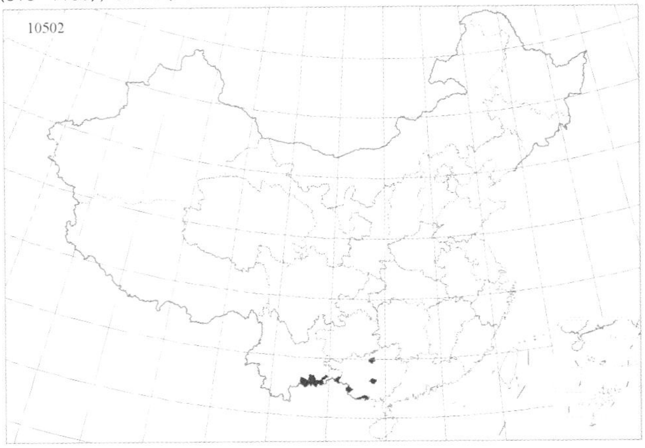

Adoxaceae 五福花科 *Viburnum* 荚蒾属

Viburnum rhytidophyllum Hemsl. 皱叶荚蒾

Shrubs or small trees, evergreen

14.3 (2.6~19.4); 2.9 (-7.0~9.0); 25.0 (11.2~29.4); 14.3 (4.3~19.5); 116.8 (20.0~173.7); 5.0 (0.0~48.2); 791 (427~1038) ❙ 1136 (626~1776); 493 (316~718); 77 (14~220) ❙ 43.8 (-22.2~108.0); 783 (427~1038); 213.3 (111.9~334.4)

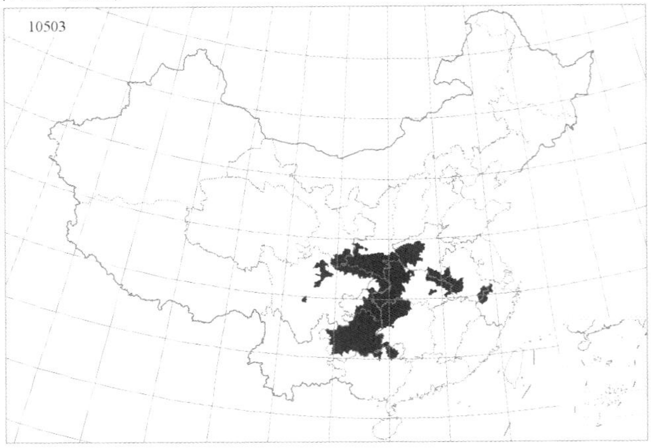

Adoxaceae 五福花科 *Viburnum* 荚蒾属

Viburnum schensianum Maxim. 陕西荚蒾

Shrubs, deciduous

10.6 (2.5~17.0); -3.4 (-12.7~10.3); 23.4 (11.2~28.9); 11.4 (4.3~17.0); 91.9 (20.0~144.0); 23.7 (0.0~65.0); 700 (427~917) ❙ 673 (252~1581); 347 (144~646); 30 (6~213) ❙ -3.9 (-59.3~94.4); 603 (252~917); 185.9 (50.4~383.4)

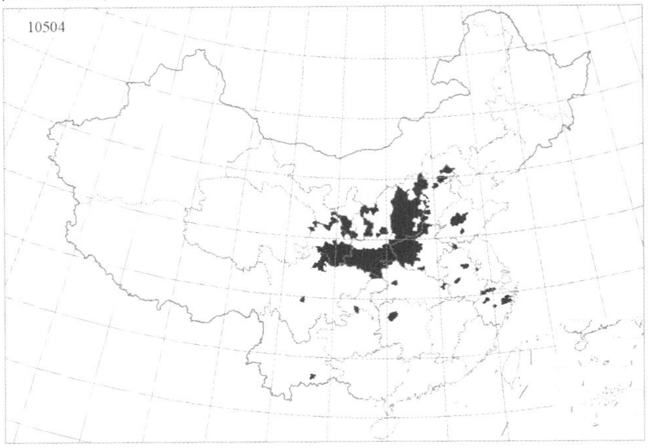

Adoxaceae 五福花科 *Viburnum* 荚蒾属

Viburnum sempervirens K. Koch 常绿荚蒾

Shrubs, evergreen

18.8 (14.3~24.8); 9.2 (2.5~20.1); 27.4 (20.8~29.8); 18.9 (14.3~24.8); 166.6 (112.6~237.6); 0.1 (0.0~3.9); 994 (756~1385) ❙ 1568 (868~2435); 601 (312~1428); 161 (41~259) ❙ 60.2 (-34.3~147.0); 982 (756~1360); 240.3 (95.3~620.0)

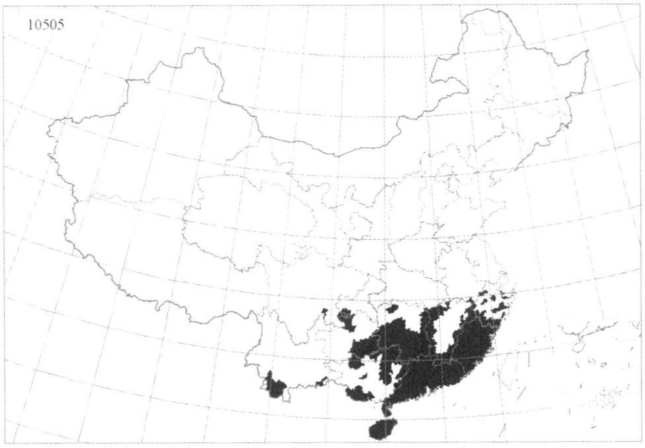

Adoxaceae 五福花科 *Viburnum* 荚蒾属

Viburnum setigerum Hance 茶荚蒾

Shrubs, deciduous

16.1 (3.3~21.0); 4.7 (-6.1~11.9); 26.8 (11.4~30.1); 16.1 (4.6~21.1); 135.2 (22.5~193.0); 1.7 (0.0~42.7); 870 (434~1114) ❙ 1359 (626~1997); 519 (316~821); 138 (17~259) ❙ 56.4 (-22.2~147.0); 859 (434~1114); 205.3 (108.2~337.3)

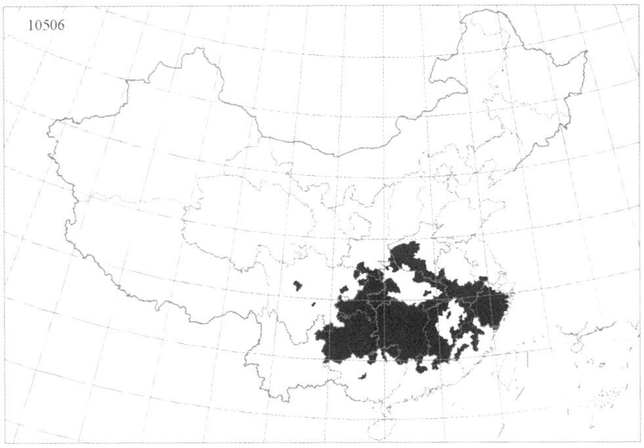

Adoxaceae 五福花科 *Viburnum* 荚蒾属
Viburnum shweliense W. W. Sm. 瑞丽荚蒾
Shrubs or small trees, deciduous

19.5；12.7；23.3；19.5；173.8；0.0；942 ❙ 1559；907；42 ❙ 65.4；921；482.8

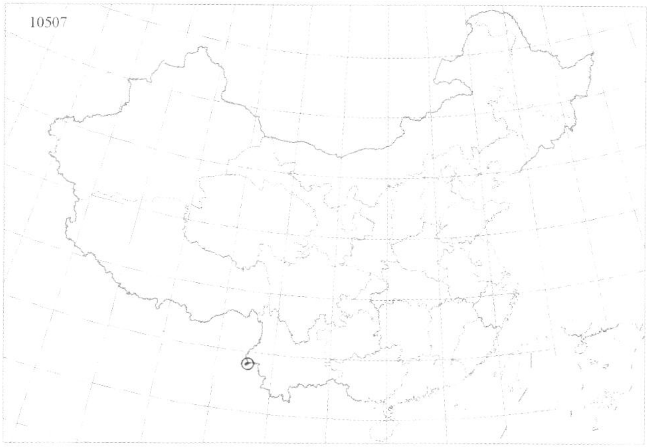

Adoxaceae 五福花科 *Viburnum* 荚蒾属
Viburnum squamulosum P. S. Hsu 瑶山荚蒾
Shrubs

17.7 (17.1~18.7)；7.0 (5.9~8.8)；27.4 (26.2~28.7)；17.8 (17.1~18.8)；153.4 (145.7~165.2)；0.0 (0.0~0.0)；933 (906~968) ❙ 1693 (1437~1922)；607 (547~670)；203 (117~259) ❙ 81.6 (48.5~111.3)；930 (906~968)；251.8 (225.9~279.5)

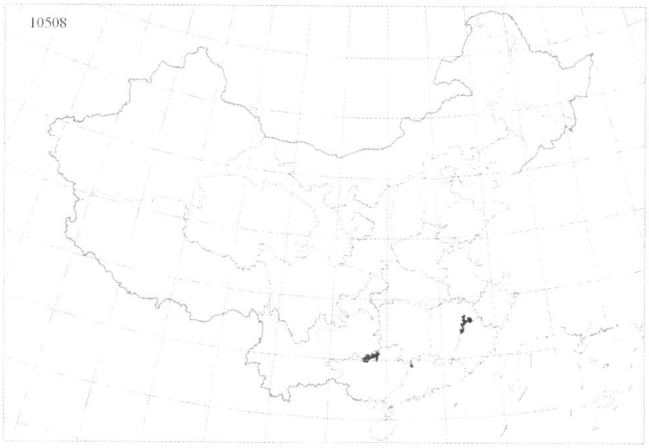

Adoxaceae 五福花科 *Viburnum* 荚蒾属
Viburnum subalpinum Hand.-Mazz. 亚高山荚蒾
Shrubs, deciduous

14.1 (8.6~17.3)；6.9 (1.3~11.0)；19.5 (15.0~21.6)；14.1 (8.7~17.3)；110.9 (52.6~148.0)；1.6 (0.0~8.8)；728 (574~832) ❙ 1179 (942~1454)；650 (489~792)；54 (44~70) ❙ 63.5 (26.1~99.9)；720 (574~783)；376.0 (219.7~487.2)

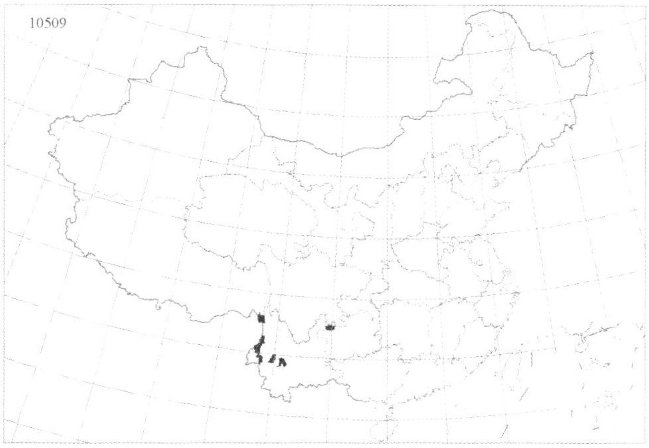

Adoxaceae 五福花科 *Viburnum* 荚蒾属
Viburnum taitoense Hayata 台东荚蒾
Shrubs

17.7 (14.3~22.2)；7.8 (3.3~17.7)；26.5 (20.1~29.4)；17.7 (14.3~22.2)；152.7 (114.1~206.7)；0.1 (0.0~2.3)；919 (765~1123) ❙ 1566 (1250~2992)；656 (392~1847)；181 (58~308) ❙ 82.8 (30.5~246.5)；907 (765~1123)；230.3 (145.3~363.7)

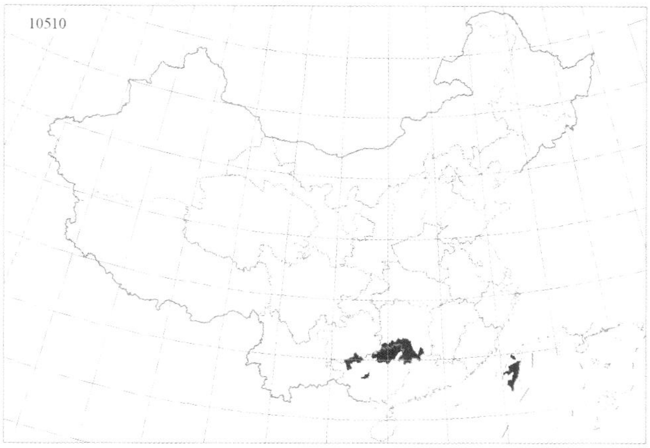

Adoxaceae 五福花科 *Viburnum* 荚蒾属
Viburnum tengyuehense (W. W. Sm.) P. S. Hsu 腾冲荚蒾
Shrubs, deciduous

14.8 (9.6~21.0)；7.8 (2.4~14.2)；20.0 (15.7~25.6)；14.8 (9.6~21.0)；118.6 (60.8~191.8)；0.8 (0.0~5.4)；760 (600~1053) ❙ 1224 (916~1522)；681 (510~818)；53 (31~70) ❙ 62.3 (26.9~99.9)；755 (600~1053)；392.2 (185.6~55[illegible].6)

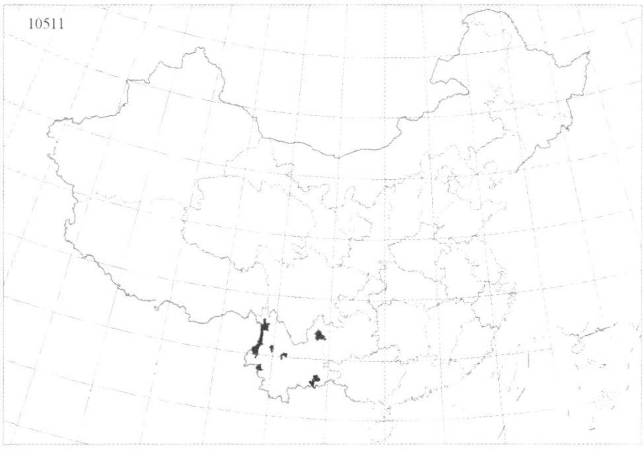

Adoxaceae 五福花科 *Viburnum* 荚蒾属
Viburnum tengyuehense var. polyneurum (P. S. Hsu) P. S. Hsu 多脉腾越荚蒾
Shrubs, deciduous

14.0；6.1；20.1；14.0；109.5；1.2；727 ❙ 1053；583；41 ❙ 44.3；727；266.5

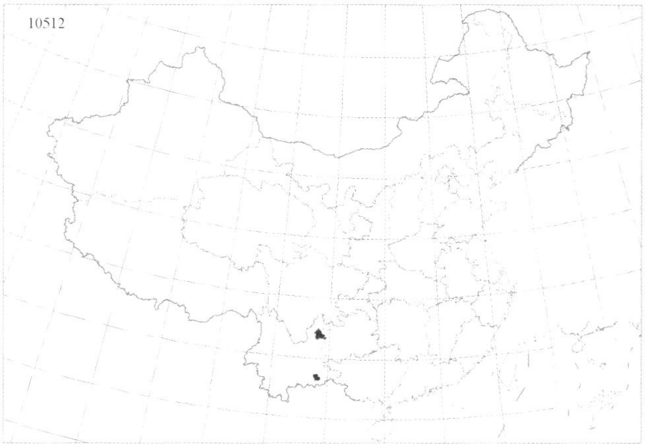

Adoxaceae 五福花科 *Viburnum* 荚蒾属
Viburnum ternatum Rehder 三叶荚蒾
Shrubs or small trees, deciduous

15.3 (11.1~19.1); 4.9 (1.6~11.0); 25.0 (18.6~28.9); 15.3 (11.1~19.1); 125.2 (79.8~169.6); 1.1 (0.0~6.9); 816 (648~1002) ❙ 1205 (942~1492); 537 (444~820); 71 (28~112) ❙ 49.2 (8.0~107.2); 816 (648~1002); 200.0 (108.2~371.7)

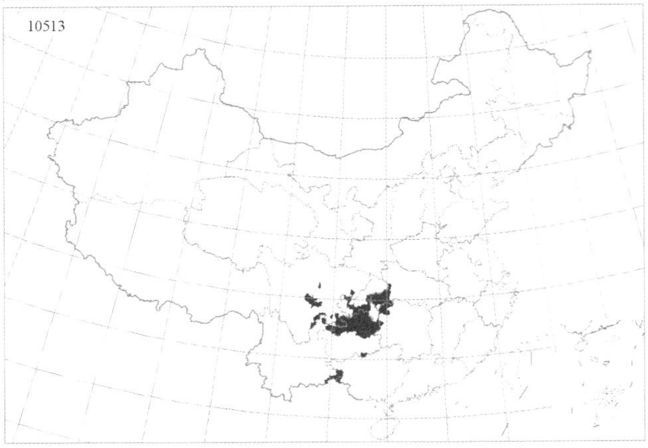

Adoxaceae 五福花科 *Viburnum* 荚蒾属
Viburnum trabeculosum C. Y. Wu ex Hsu 横脉荚蒾
Trees, deciduous

18.9; 12.9; 22.9; 18.9; 166.8; 0.0; 902 ❙ 1487; 873; 59 ❙ 65.0; 902; 528.7

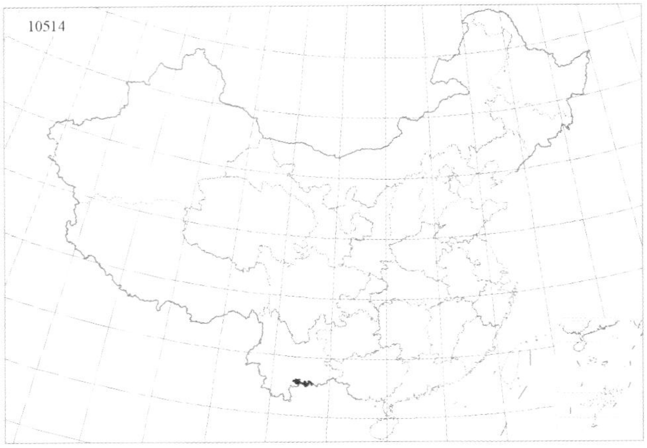

Adoxaceae 五福花科 *Viburnum* 荚蒾属
Viburnum urceolatum Siebold et Zucc. 壶花荚蒾
Shrubs, deciduous

17.4 (14.1~24.1); 7.8 (1.6~18.9); 26.3 (20.1~29.1); 17.5 (14.2~24.1); 149.9 (114.1~229.7); 0.3 (0.0~6.5); 909 (765~1322) ❙ 1931 (1187~3495); 722 (469~1847); 236 (47~897) ❙ 113.3 (32.5~246.5); 900 (765~1123); 243.8 (154.2~303.6)

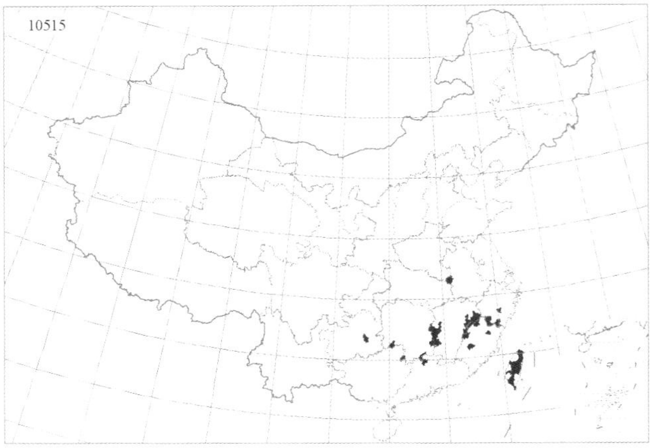

Adoxaceae 五福花科 *Viburnum* 荚蒾属
Viburnum utile Hesml. 烟管荚蒾
Shrubs, evergreen

14.5 (2.9~19.1); 3.5 (-6.3~10.4); 24.7 (10.8~29.2); 14.6 (4.4~19.1); 118.5 (20.5~169.2); 4.0 (0.0~44.6); 792 (428~969) ❙ 1130 (626~1492); 501 (316~720); 70 (16~161) ❙ 43.4 (-22.2~121.7); 783 (428~969); 210.4 (127.1~318.7)

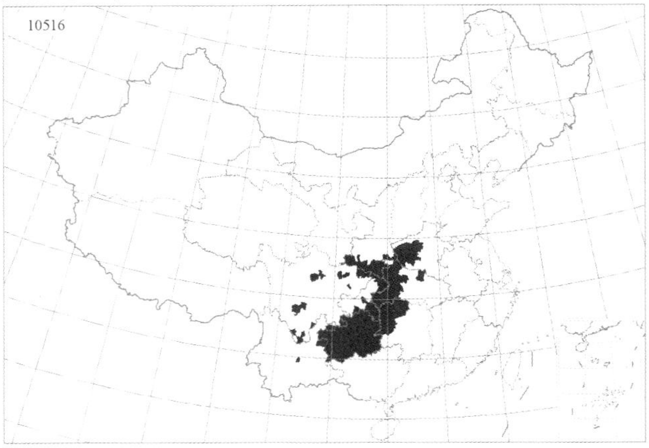

Diervillaceae 黄锦带科 *Weigela* 锦带花属
Weigela florida (Bunge) A. DC. 锦带花
Shrubs, deciduous

11.0 (1.2~17.6); -4.2 (-17.8~5.6); 24.7 (18.2~29.4); 12.0 (6.4~17.6); 99.8 (45.9~151.4); 26.8 (0.0~90.7); 737 (506~953) ❙ 729 (377~1590); 418 (241~691); 39 (7~214) ❙ -2.1 (-47.7~108.0); 658 (377~941); 185.6 (106.8~337.1)

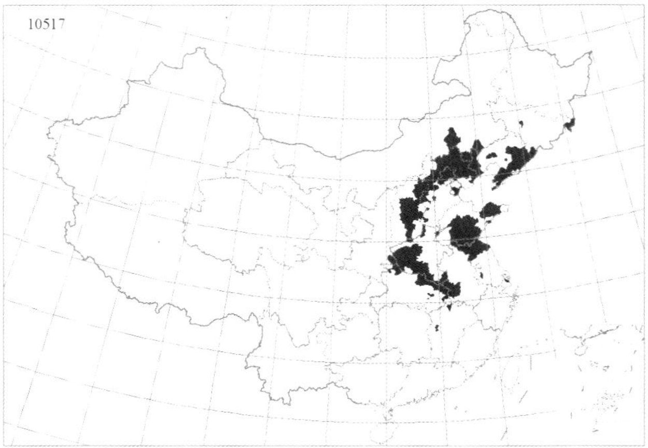

Diervillaceae 黄锦带科 *Weigela* 锦带花属
Weigela japonica var. sinica Rehder 半边月
Shrubs, deciduous

16.5 (9.3~24.8); 5.5 (-1.7~20.1); 26.8 (20.1~29.9); 16.6 (9.5~24.8); 139.9 (69.8~237.6); 1.0 (0.0~18.0); 889 (603~1385) ❙ 1415 (868~1997); 534 (312~821); 143 (32~259) ❙ 60.6 (-34.3~147.0); 877 (603~1360); 224.1 (111.9~620.0)

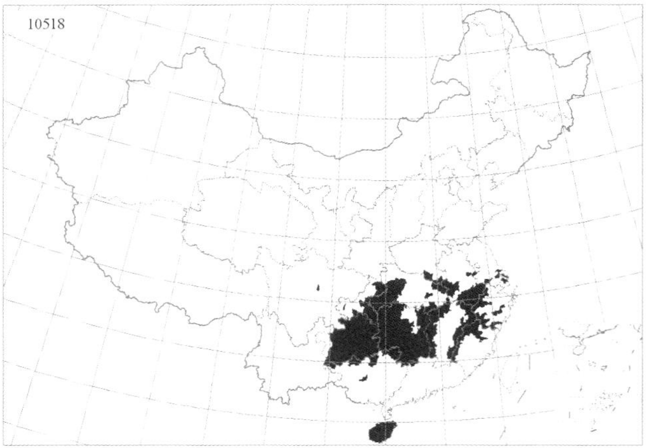

Caprifoliaceae 忍冬科 *Heptacodium* 七子花属
Heptacodium miconioides Rehder 七子花
Shrubs or small trees, deciduous

15.3 (12.0~16.8); 3.3 (0.6~5.7); 27.1 (22.9~28.9); 15.3 (12.1~16.8); 126.6 (94.7~141.7); 2.6 (0.0~9.7); 842 (696~912) ❙ 1421 (993~1648); 536 (420~613); 170 (46~213) ❙ 69.2 (35.6~105.3); 842 (696~912); 238.4 (149.5~284.3)

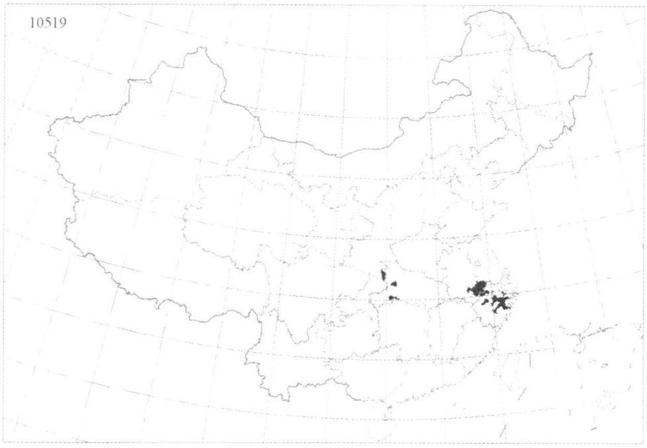

Caprifoliaceae 忍冬科 *Leycesteria* 鬼吹箫属
Leycesteria formosa Wall. 鬼吹箫
Shrubs, deciduous

10.4 (-1.8~19.1); 2.3 (-10.5~13.0); 17.2 (6.1~24.6); 10.9 (1.8~19.1); 80.0 (1.9~169.7); 14.3 (0.0~82.8); 634 (329~917) ❙ 1000 (311~1736); 572 (193~1007); 33 (9~69) ❙ 57.8 (-6.3~145.8); 628 (311~917); 276.1 (61.6~506.2)

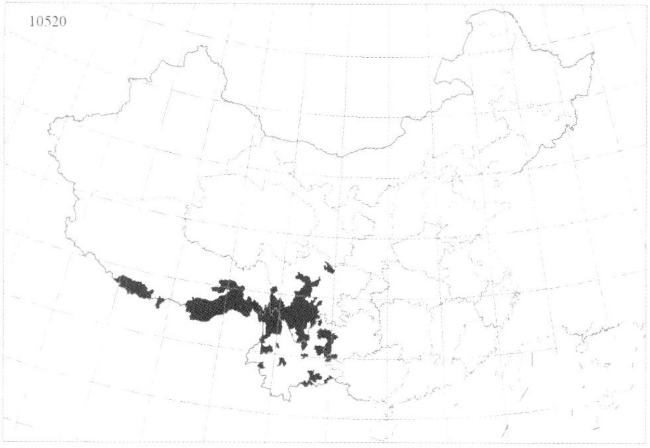

Caprifoliaceae 忍冬科 *Leycesteria* 鬼吹箫属
Leycesteria glaucophylla (Hook. f. et Thomson) C. B. Clarke 西域鬼吹箫
Shrubs, deciduous

13.3; 5.4; 19.6; 13.4; 100.4; 0.0; 706 ❙ 1736; 1007; 51 ❙ 145.8; 706; 447.4

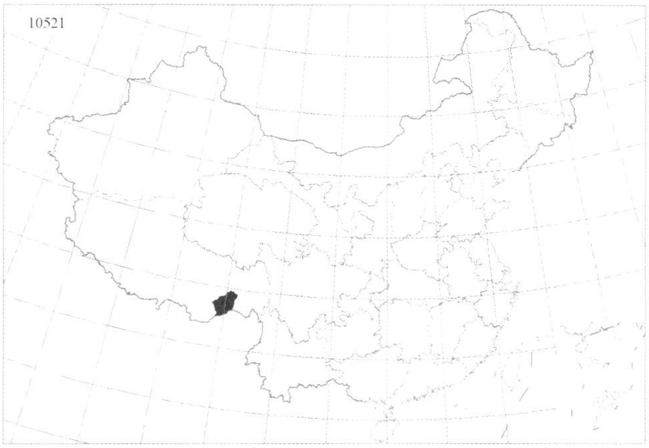

Caprifoliaceae 忍冬科 *Leycesteria* 鬼吹箫属
Leycesteria gracilis (Kurz) Airy Shaw 纤细鬼吹箫
Shrubs, deciduous

15.4 (5.3~19.1); 8.5 (-2.4~13.0); 20.3 (12.2~23.3); 15.4 (5.7~19.1); 126.9 (30.5~169.7); 2.0 (0.0~26.2); 776 (479~917) ❙ 1313 (961~1736); 744 (558~1007); 52 (31~70) ❙ 71.5 (26.1~145.8); 771 (479~917); 416.3 (234.2~551.3)

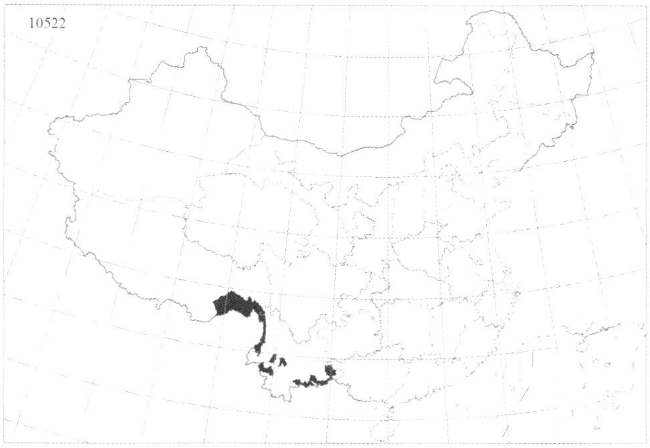

Caprifoliaceae 忍冬科 *Leycesteria* 鬼吹箫属
Leycesteria sinensis Hemsl. 华鬼吹箫
Shrubs, deciduous

17.2 (16.3~17.7); 9.9 (9.1~11.0); 22.3 (21.4~23.2); 17.2 (16.4~17.8); 146.8 (136.4~153.4); 0.0 (0.0~0.0); 834 (797~859) ❙ 1232 (1049~1422); 686 (598~813); 51 (44~59) ❙ 47.6 (26.1~65.6); 827 (783~859); 371.1 (314.4~491.5)

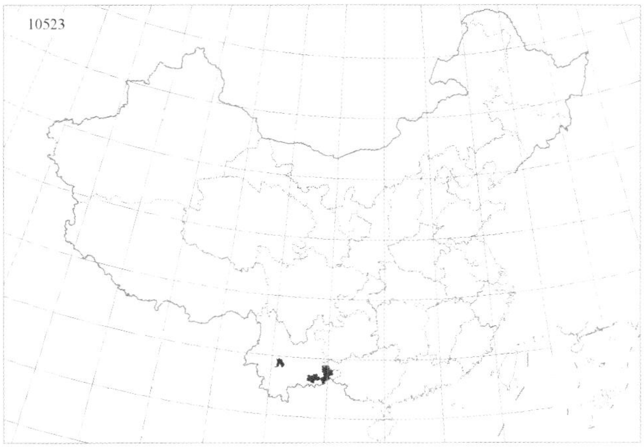

Caprifoliaceae 忍冬科 *Leycesteria* 鬼吹箫属
Leycesteria stipulata (Hook. f. et Thomson) Fritsch 绵毛鬼吹箫
Shrubs, deciduous

8.6; 1.3; 15.0; 8.7; 52.6; 8.8; 574 ❙ 1041; 585; 45 ❙ 81.5; 574; 352.3

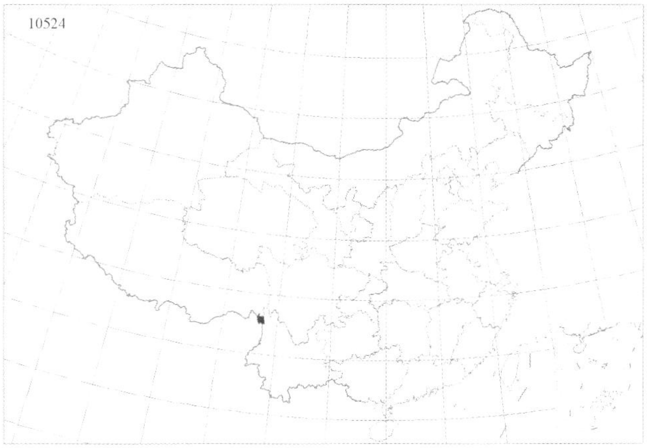

Caprifoliaceae 忍冬科 *Lonicera* 忍冬属

Lonicera acuminata Wall. 淡红忍冬

Vines, deciduous or semievergreen

14.4 (−1.8~24.1); 4.2 (−10.2~18.9); 23.6 (6.1~29.4); 14.5 (1.8~24.1); 117.5 (1.9~229.7); 4.7 (0.0~82.8); 786 (329~1322) ❙ 1227 (320~2992); 548 (217~1847); 90 (9~402) ❙ 55.6 (−22.2~246.5); 775 (320~1123); 244.1 (72.2~627.0)

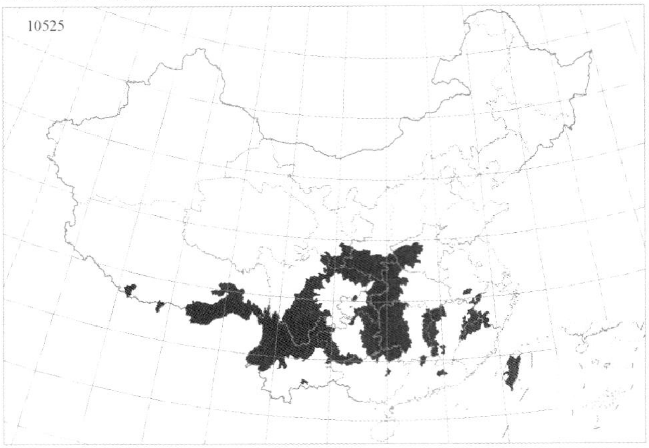

Caprifoliaceae 忍冬科 *Lonicera* 忍冬属

Lonicera acuminata var. depilata P. S. Hsu et H. J. Wang 无毛淡红忍冬

Vines, deciduous or semievergreen

13.0 (2.2~19.3); 3.7 (−6.6~8.6); 21.3 (9.9~29.3); 13.1 (3.8~19.3); 101.4 (15.5~172.0); 5.4 (0.0~49.1); 728 (413~1037) ❙ 1255 (538~1997); 597 (337~817); 81 (9~254) ❙ 73.0 (18.2~147.0); 721 (413~1000); 261.1 (167.5~331.6)

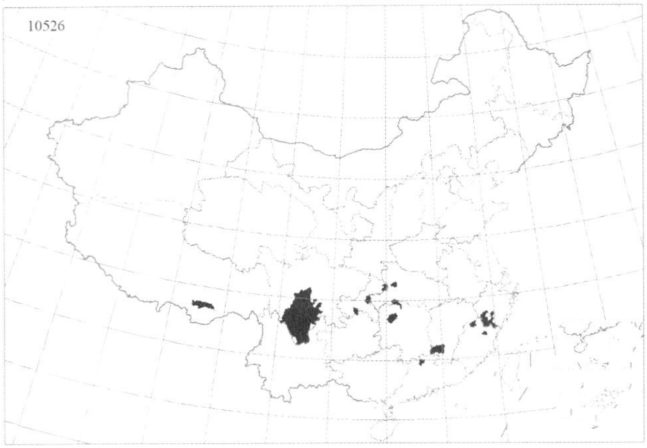

Caprifoliaceae 忍冬科 *Lonicera* 忍冬属

Lonicera albertii Regel 沼生忍冬

Dwarf shrubs, deciduous

2.9 (−1.3~9.0); −15.7 (−20.9~−9.3); 17.3 (13.2~23.2); 7.0 (4.6~10.7); 50.4 (26.7~88.9); 74.7 (40.7~103.9); 537 (440~703) ❙ 310 (225~406); 143 (65~211); 27 (12~48) ❙ −40.1 (−65.1~−7.6); 310 (225~406); 216.4 (165.7~297.3)

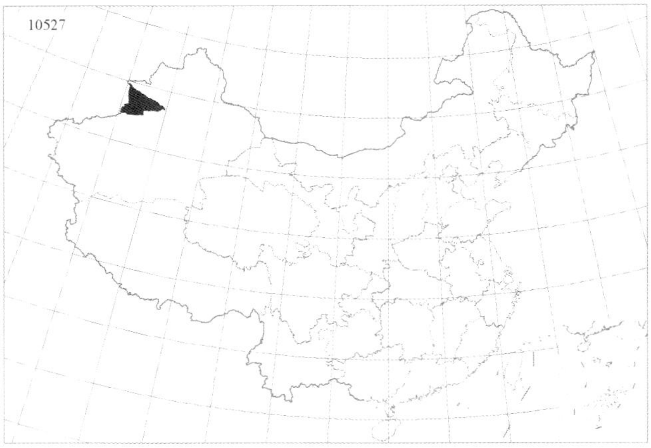

Caprifoliaceae 忍冬科 *Lonicera* 忍冬属

Lonicera altmannii Regel et Schmalh. 截萼忍冬

Shrubs, deciduous

4.2 (−1.3~9.3); −15.3 (−20.9~−11.6); 20.3 (13.2~27.5); 8.2 (4.5~12.1); 64.5 (26.7~107.6); 73.2 (51.6~103.9); 598 (433~813) ❙ 238 (107~450); 106 (51~211); 22 (8~51) ❙ −57.3 (−85.3~3.9); 237 (107~406); 112.5 (21.1~297.3)

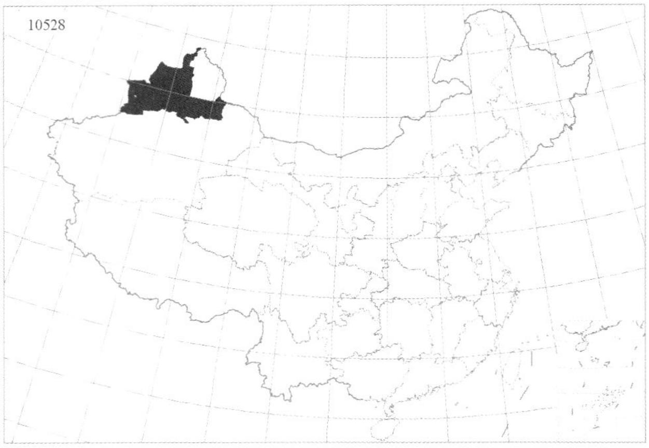

Caprifoliaceae 忍冬科 *Lonicera* 忍冬属

Lonicera angustifolia Wall. ex DC. 狭叶忍冬

Shrubs, deciduous

0.6; −8.5; 8.8; 3.2; 12.8; 64.7; 374 ❙ 321; 221; 11 ❙ −14.7; 316; 128.5

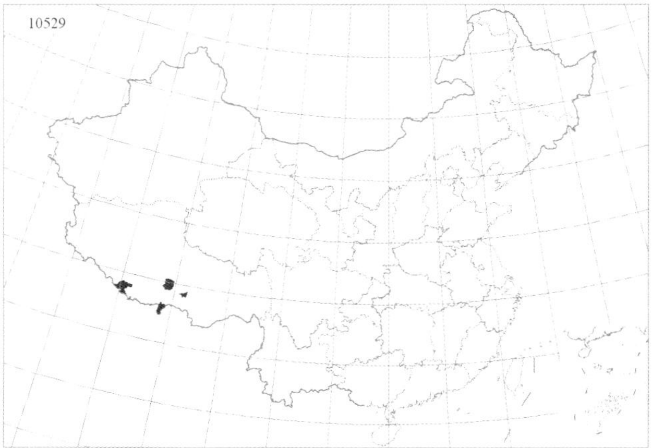

Caprifoliaceae 忍冬科 *Lonicera* 忍冬属

Lonicera anisocalyx Rehder 异萼忍冬

Shrubs, deciduous

2.3 (−2.7~10.3); −8.0 (−14.2~−0.1); 11.4 (7.4~19.8); 4.5 (2.1~10.3); 23.0 (4.3~75.8); 55.3 (11.7~96.5); 434 (341~634) ❙ 718 (562~943); 377 (324~507); 17 (11~23) ❙ 68.7 (12.1~128.6); 434 (341~634); 297.3 (252.2~364.1)

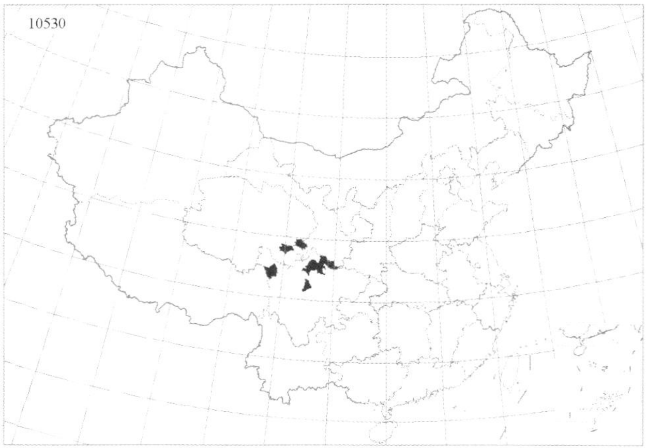

Caprifoliaceae 忍冬科 *Lonicera* 忍冬属
Lonicera bournei Hemsl. 西南忍冬
Vines

17.1 (8.6~20.4)；9.7 (1.3~15.1)；22.7 (15.0~26.7)；17.1 (8.7~20.5)；146.3 (52.6~185.8)；0.7 (0.0~8.8)；850 (574~989) ❙ 1316 (939~1624)；642 (524~816)；81 (39~191) ❙ 55.9 (6.5~101.8)；835 (574~986)；368.1 (240.7~659.0)

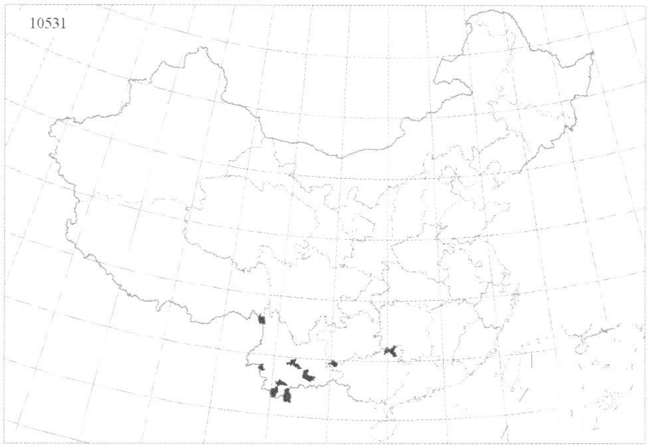

Caprifoliaceae 忍冬科 *Lonicera* 忍冬属
Lonicera brevisepala P. S. Hsu et H. J. Wang 短萼忍冬
Shrubs, deciduous

9.1；-2.7；20.2；9.6；71.0；21.0；613 ❙ 626；304；14 ❙ 0.0；613；219.0

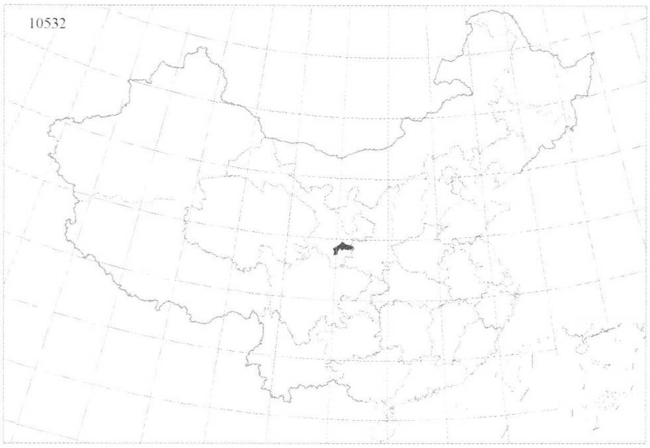

Caprifoliaceae 忍冬科 *Lonicera* 忍冬属
Lonicera buchananii Lace 滇西忍冬
Vines

16.4；9.6；20.6；16.4；137.3；0.0；795 ❙ 1688；959；62 ❙ 112.2；795；627.0

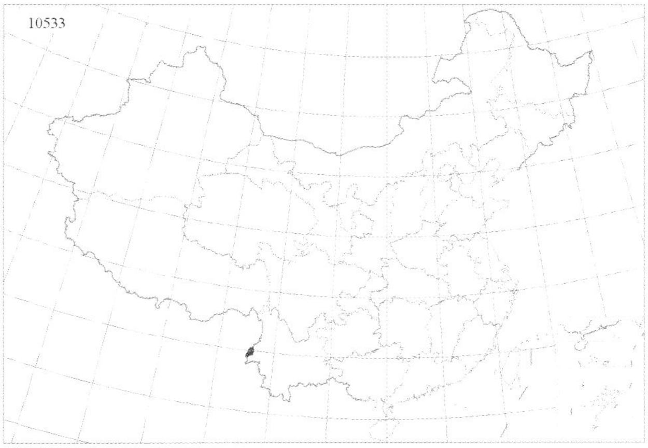

Caprifoliaceae 忍冬科 *Lonicera* 忍冬属
Lonicera buddleioides P. S. Hsu et S. C. Cheng 醉鱼草状忍冬
Vines

22.2；13.9；28.3；22.3；207.5；0.0；1190 ❙ 1374；713；87 ❙ 15.4；1190；327.7

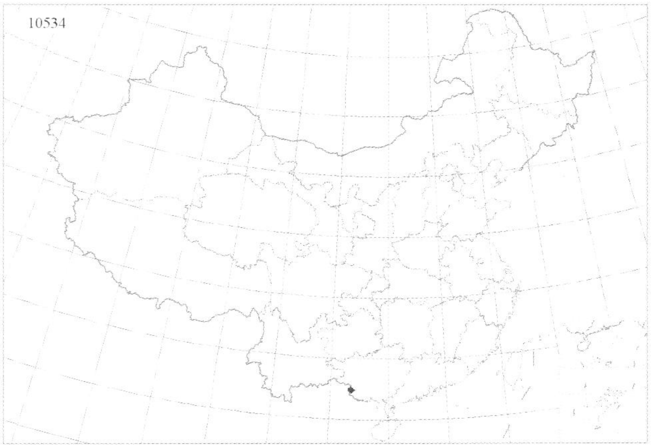

Caprifoliaceae 忍冬科 *Lonicera* 忍冬属
Lonicera caerulea var. altaica Pall. 阿尔泰忍冬
Shrubs, deciduous

1.8 (0.4~4.1)；-16.6 (-18.4~-14.4)；18.2 (16.6~20.4)；6.7 (5.8~8.1)；49.0 (40.7~61.7)；87.7 (72.2~97.3)；533 (494~589) ❙ 231 (167~315)；92 (74~106)；31 (19~51) ❙ -56.9 (-66.8~-46.4)；231 (167~312)；138.1 (120.7~173.1)

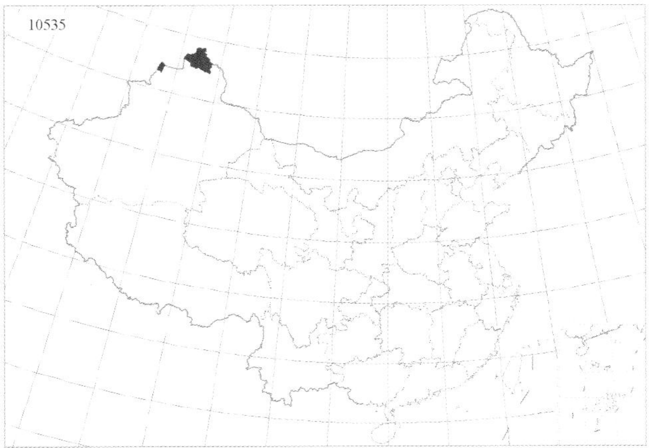

Caprifoliaceae 忍冬科 *Lonicera* 忍冬属
Lonicera caerulea var. edulis Turcz. ex Herder 蓝靛果
Shrubs, deciduous

7.5 (-5.8~15.1)；-8.0 (-29.8~1.8)；21.4 (9.7~28.0)；9.5 (2.9~15.2)；74.9 (11.3~128.0)；44.0 (5.6~157.4)；632 (359~880) ❙ 544 (137~943)；319 (92~507)；17 (3~58) ❙ -12.1 (-77.5~128.6)；516 (137~821)；187.3 (25.4~346.5)

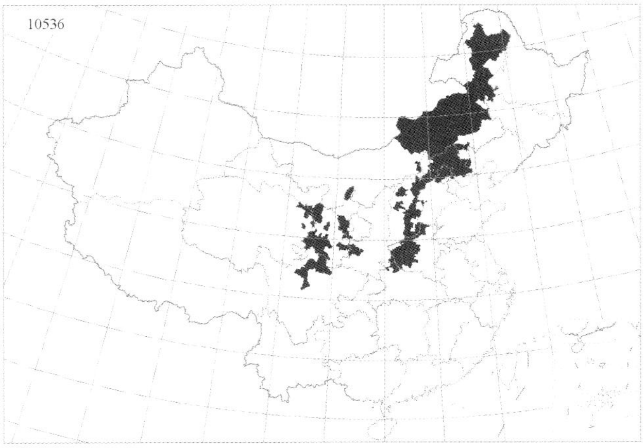

Caprifoliaceae 忍冬科 *Lonicera* 忍冬属

Lonicera calcarata Hemsl. 长距忍冬

Vines

15.3 (11.1~19.2); 5.8 (0.6~11.0); 23.7 (18.8~26.6); 15.4 (11.1~19.2); 125.4 (79.8~170.8); 1.2 (0.0~9.7); 799 (648~969) ❙ 1236 (879~1504); 603 (472~860); 63 (20~113) ❙ 55.6 (5.5~107.2); 795 (648~969); 244.8 (142.9~491.5)

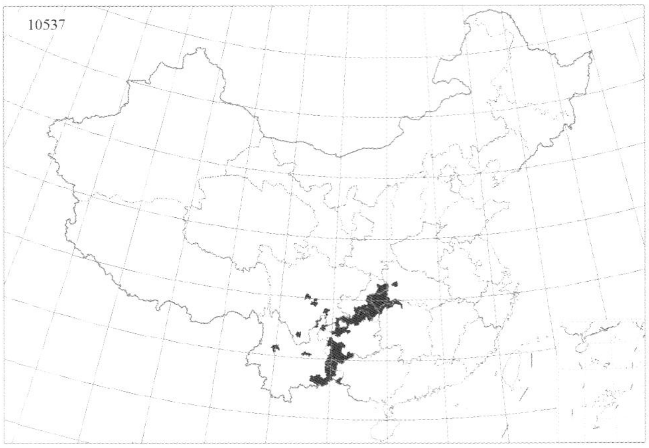

Caprifoliaceae 忍冬科 *Lonicera* 忍冬属

Lonicera calvescens (Chun et F. C. How) P. S. Hsu et H. J. Wang 海南忍冬

Vines, semievergreen

23.5 (21.2~24.8); 17.5 (12.2~20.1); 27.9 (27.2~28.5); 23.5 (21.3~24.8); 222.2 (195.2~237.6); 0.0 (0.0~0.0); 1274 (1126~1385) ❙ 1359 (1179~1487); 484 (392~599); 62 (41~142) ❙ 7.4 (−11.9~29.9); 1111 (1056~1142); 440.5 (226.9~538.2)

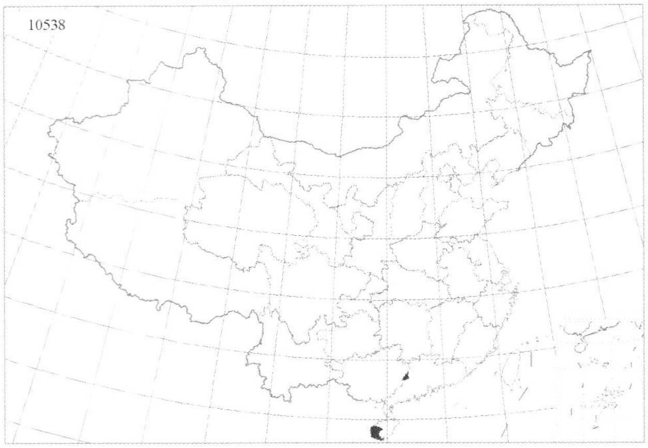

Caprifoliaceae 忍冬科 *Lonicera* 忍冬属

Lonicera carnosifolia C. Y. Wu 肉叶忍冬

Vines

16.8 (14.8~18.0); 7.3 (4.3~9.9); 24.9 (21.8~28.1); 16.8 (14.9~18.0); 142.3 (119.2~156.3); 0.1 (0.0~0.7); 862 (791~958) ❙ 1258 (1070~1496); 600 (453~813); 75 (52~142) ❙ 46.6 (11.7~65.6); 862 (791~958); 239.2 (128.7~359.7)

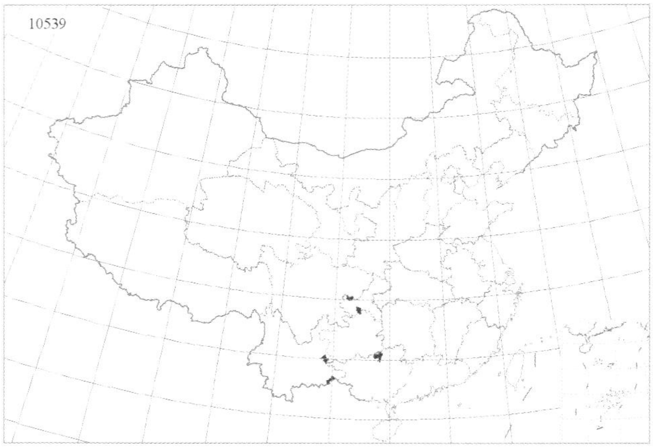

Caprifoliaceae 忍冬科 *Lonicera* 忍冬属

Lonicera chrysantha Turcz. ex Ledeb. 金花忍冬

Shrubs, deciduous

9.0 (−4.0~18.9); −6.1 (−28.2~12.0); 22.5 (8.5~29.3); 10.6 (2.7~19.0); 85.0 (8.3~167.8); 36.1 (0.0~146.2); 671 (359~941) ❙ 657 (132~1813); 369 (81~691); 26 (2~226) ❙ −1.6 (−80.1~123.6); 580 (132~941); 201.0 (13.6~346.5)

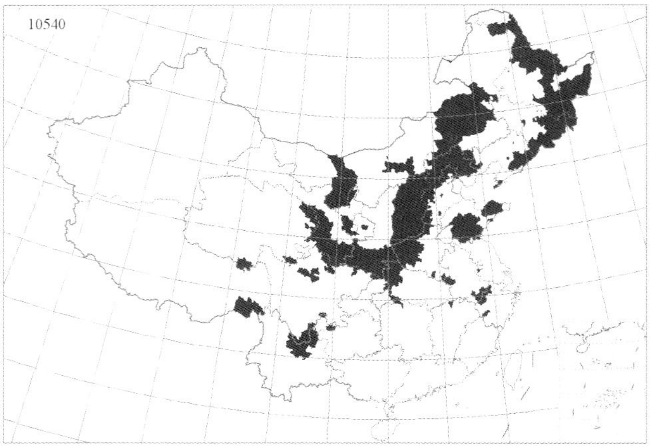

Caprifoliaceae 忍冬科 *Lonicera* 忍冬属

Lonicera chrysantha subsp. koehneana (Rehder) P. S. Hsu et H. J. Wang 须蕊忍冬

Shrubs, deciduous

11.9 (2.2~18.5); 0.3 (−6.6~10.5); 22.7 (9.5~27.6); 12.1 (3.8~18.3); 94.9 (15.5~160.0); 11.9 (0.0~49.1); 706 (411~906) ❙ 948 (517~1813); 440 (257~644); 52 (11~226) ❙ 35.4 (−36.7~125.9); 679 (411~850); 248.1 (111.8~334.4)

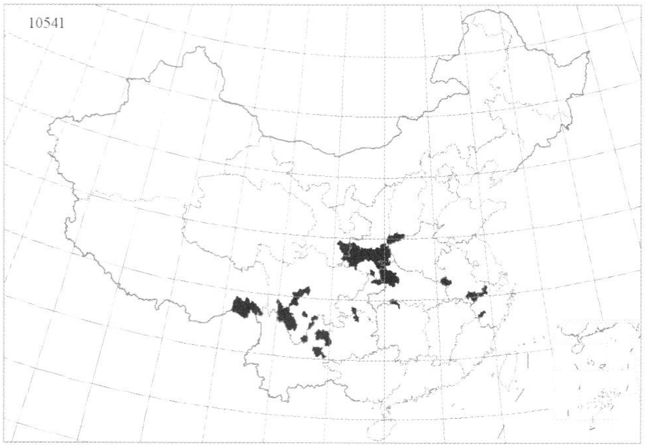

Caprifoliaceae 忍冬科 *Lonicera* 忍冬属

Lonicera ciliosissima C. Y. Wu ex P. S. Hsu et H. J. Wang 长睫毛忍冬

Vines

14.1; 6.5; 20.2; 14.2; 109.8; 0.0; 729 ❙ 1022; 615; 18 ❙ 40.1; 707; 306.0

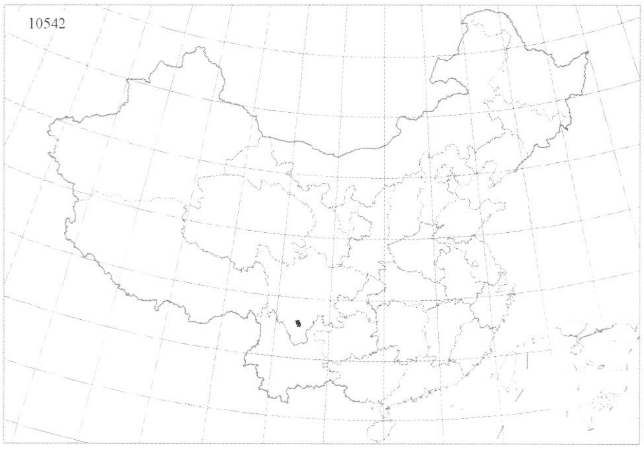

Caprifoliaceae 忍冬科 *Lonicera* 忍冬属
Lonicera cinerea Pojark. 灰毛忍冬
Dwarf shrubs, deciduous

−2.8；−18.9；11.1；3.2；15.3；108.6；360 ❙ 104；35；16 ❙ −71.8；104；19.3

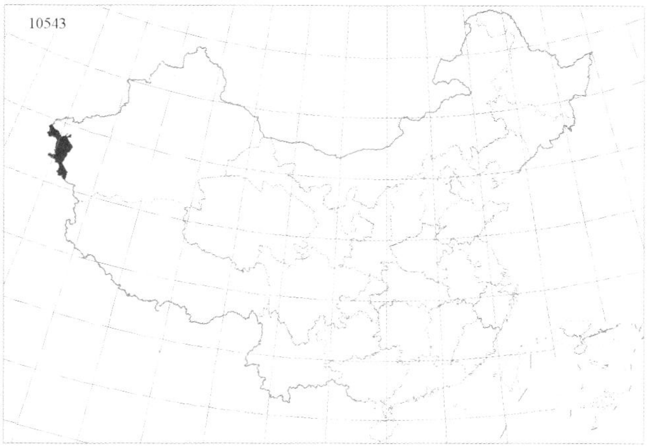

Caprifoliaceae 忍冬科 *Lonicera* 忍冬属
Lonicera codonantha Rehder 钟花忍冬
Shrubs, deciduous

17.4；11.0；21.4；17.4；149.3；0.0；834 ❙ 1406；816；48 ❙ 68.7；834；434.4

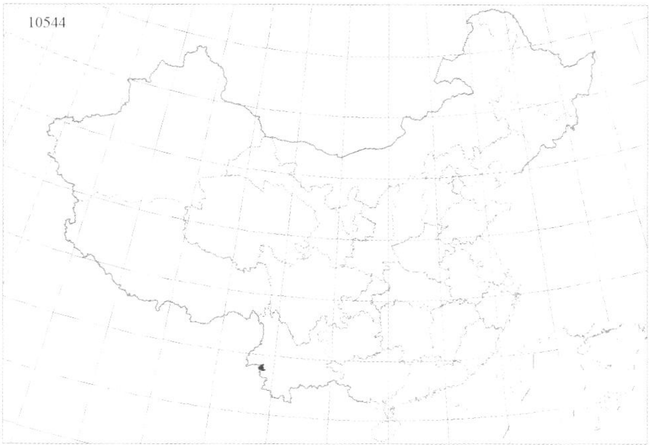

Caprifoliaceae 忍冬科 *Lonicera* 忍冬属
Lonicera confusa (Sweet) DC. 华南忍冬
Vines, semievergreen

21.1 (14.6~24.8)；12.8 (3.8~20.1)；27.9 (24.4~29.8)；21.2 (14.7~24.8)；194.0 (117.7~237.6)；0.0 (0.0~1.5)；1120 (779~1385) ❙ 1621 (868~2435)；696 (312~1428)；130 (41~204) ❙ 46.5 (−34.3~101.8)；1103 (779~1360)；253.3 (95.3~620.0)

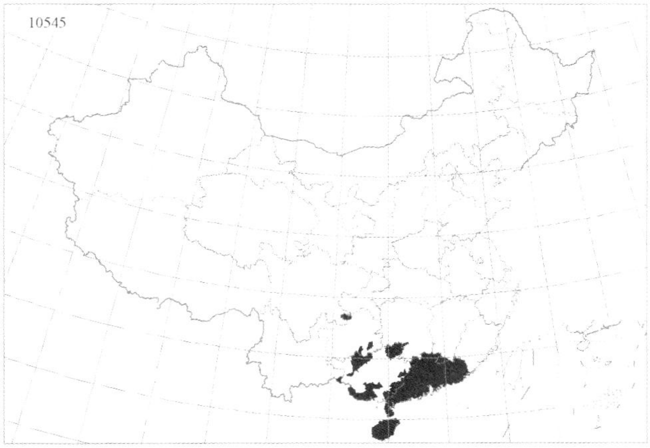

Caprifoliaceae 忍冬科 *Lonicera* 忍冬属
Lonicera crassifolia Batalin 匍匐忍冬
Shrubs, evergreen

13.9 (2.2~17.6)；3.7 (−6.6~9.9)；23.3 (9.9~26.5)；13.9 (3.8~17.7)；110.3 (15.5~152.4)；3.4 (0.0~49.1)；753 (413~859) ❙ 1250 (879~1504)；572 (472~860)；69 (19~113) ❙ 66.7 (24.9~125.9)；753 (413~859)；228.2 (154.4~413.5)

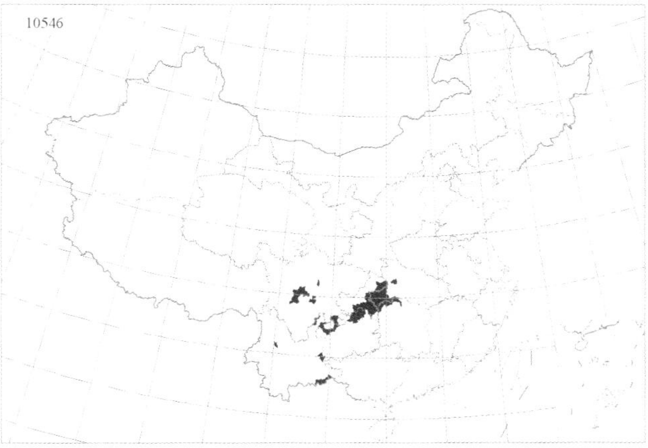

Caprifoliaceae 忍冬科 *Lonicera* 忍冬属
Lonicera cyanocarpa Franch. 微毛忍冬
Shrubs, deciduous

6.1 (−0.9~13.3)；−1.9 (−11.3~6.0)；13.1 (8.1~19.6)；7.0 (2.7~13.4)；42.2 (8.0~100.4)；28.2 (0.0~79.1)；511 (354~706) ❙ 884 (473~1736)；527 (287~1007)；25 (7~51) ❙ 70.7 (5.0~145.8)；511 (354~706)；285.8 (111.0~447.4)

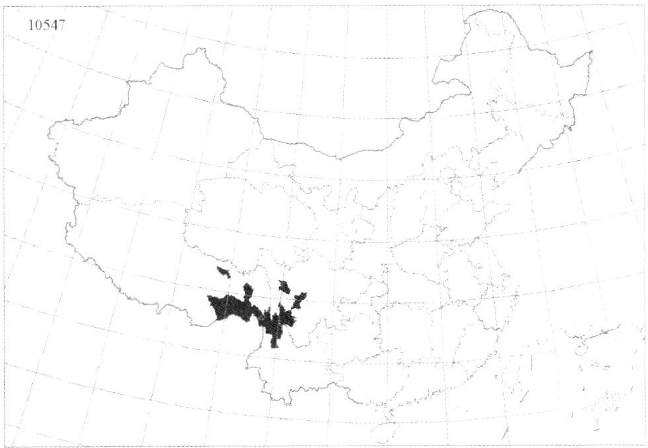

Caprifoliaceae 忍冬科 *Lonicera* 忍冬属
Lonicera dasystyla Rehder 水忍冬
Vines

20.7 (17.8~22.2)；11.4 (7.1~13.9)；28.2 (27.1~29.0)；20.7 (17.8~22.3)；188.6 (153.8~207.5)；0.0 (0.0~0.0)；1096 (932~1190) ❙ 1548 (1302~1776)；713 (617~850)；143 (86~189) ❙ 42.5 (12.9~85.5)；1096 (932~1190)；227.5 (160.7~327.7)

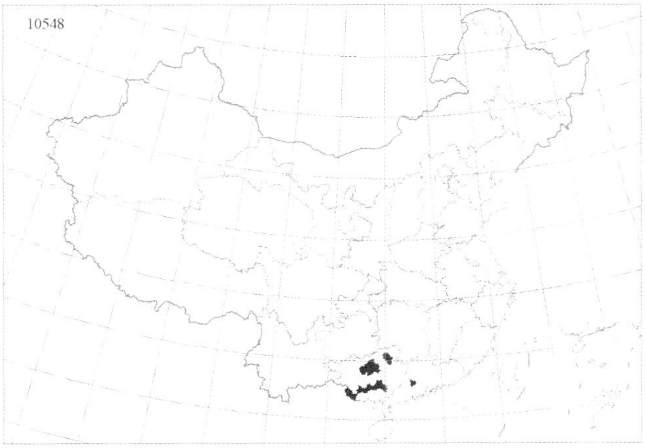

Caprifoliaceae 忍冬科 *Lonicera* 忍冬属
Lonicera elisae Franch. 北京忍冬
Shrubs, deciduous

11.3 (3.0~15.5); -2.5 (-12.3~3.4); 24.0 (12.4~28.0); 11.8 (4.8~15.6); 96.3 (25.1~128.6); 19.9 (1.7~61.5); 719 (441~880) ❙ 701 (341~1638); 354 (190~666); 33 (8~220) ❙ -2.3 (-40.9~108.0); 628 (341~851); 190.2 (61.2~323.3)

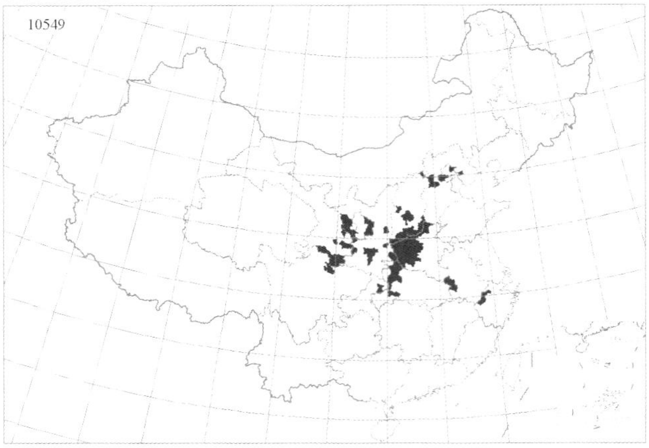

Caprifoliaceae 忍冬科 *Lonicera* 忍冬属
Lonicera fargesii Franch. 黏毛忍冬
Shrubs, deciduous

9.8 (3.3~18.9); -4.3 (-12.7~10.6); 22.7 (17.5~28.2); 10.7 (6.6~19.0); 85.9 (45.4~167.5); 27.4 (0.0~65.0); 671 (507~939) ❙ 645 (374~1292); 329 (205~729); 22 (8~49) ❙ -3.5 (-38.3~93.8); 589 (374~939); 204.6 (81.2~342.5)

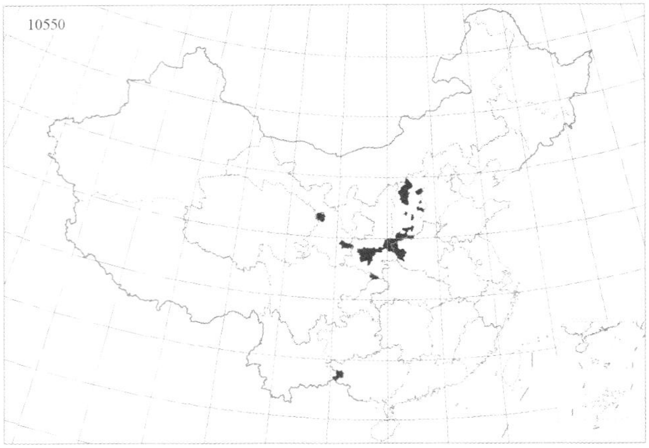

Caprifoliaceae 忍冬科 *Lonicera* 忍冬属
Lonicera ferdinandi Franch. 葱皮忍冬
Shrubs, deciduous

9.2 (0.8~15.1); -4.7 (-14.5~7.6); 21.6 (9.9~28.0); 10.2 (3.8~15.2); 80.5 (15.4~128.0); 29.5 (0.0~66.0); 653 (404~880) ❙ 634 (132~1414); 348 (81~817); 20 (1~88) ❙ -0.4 (-80.1~128.6); 553 (132~817); 186.7 (13.6~372.3)

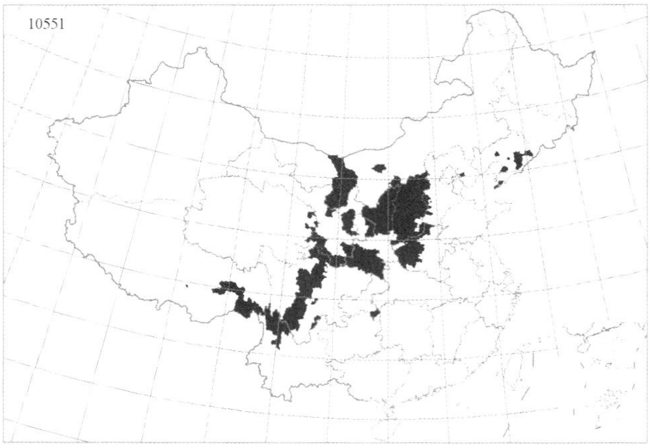

Caprifoliaceae 忍冬科 *Lonicera* 忍冬属
Lonicera ferruginea Rehder 锈毛忍冬
Vines

17.7 (11.1~21.3); 8.9 (2.2~15.1); 25.0 (18.4~29.1); 17.8 (11.3~21.3); 153.2 (80.8~196.0); 0.2 (0.0~5.3); 902 (652~1123) ❙ 1501 (939~2005); 658 (458~911); 125 (35~259) ❙ 66.9 (5.8~130.7); 897 (652~1121); 316.0 (126.1~659.0)

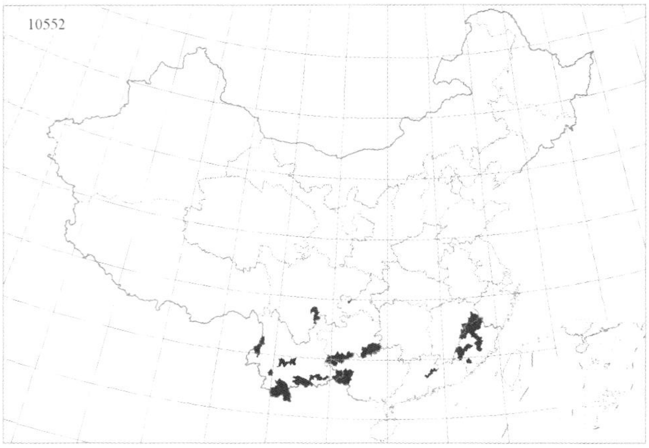

Caprifoliaceae 忍冬科 *Lonicera* 忍冬属
Lonicera fragilis H. Lév. 短柱忍冬
Shrubs, deciduous

12.9; 5.4; 18.9; 13.0; 95.6; 0.2; 691 ❙ 907; 537; 31 ❙ 31.3; 691; 206.4

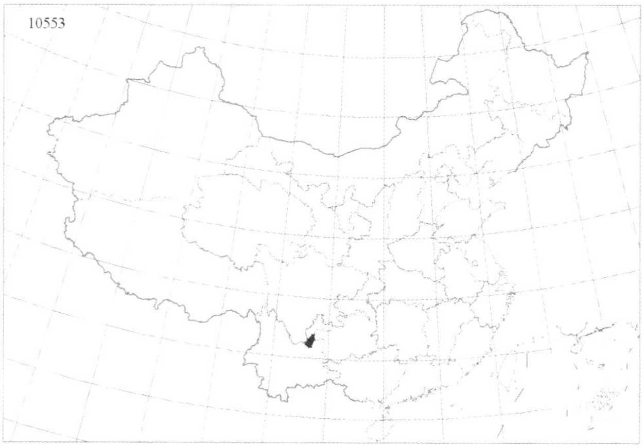

Caprifoliaceae 忍冬科 *Lonicera* 忍冬属
Lonicera fragrantissima Lindley et Paxton 郁香忍冬
Shrubs, semievergreen or deciduous

13.5 (2.0~17.7); 0.7 (-7.7~7.4); 25.5 (9.9~29.6); 13.7 (3.8~17.7); 112.2 (15.4~152.8); 10.2 (0.0~51.4); 789 (413~966) ❙ 981 (503~1828); 448 (257~685); 89 (10~253) ❙ 24.0 (-38.8~128.6); 759 (413~959); 201.5 (111.8~323.3)

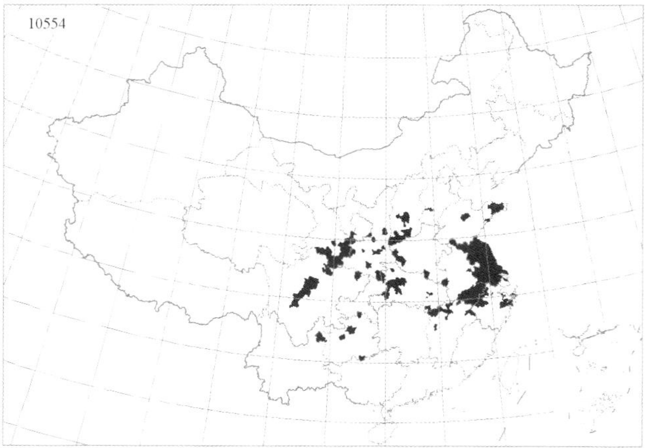

Caprifoliaceae 忍冬科 *Lonicera* 忍冬属

Lonicera fragrantissima subsp. phyllocarpa (Maxim.) P. S. Hsu et H. J. Wang 樱桃忍冬 Shrubs, semievergreen or deciduous

12.8 (9.4~15.4); -0.8 (-3.4~1.8); 25.4 (21.0~28.0); 12.9 (9.8~15.4); 107.6 (73.9~130.0); 13.5 (5.0~20.6); 768 (619~879) ❙ 746 (626~994); 379 (303~485); 36 (18~102) ❙ -2.1 (-23.9~44.0); 710 (619~875); 192.6 (139.1~311.0)

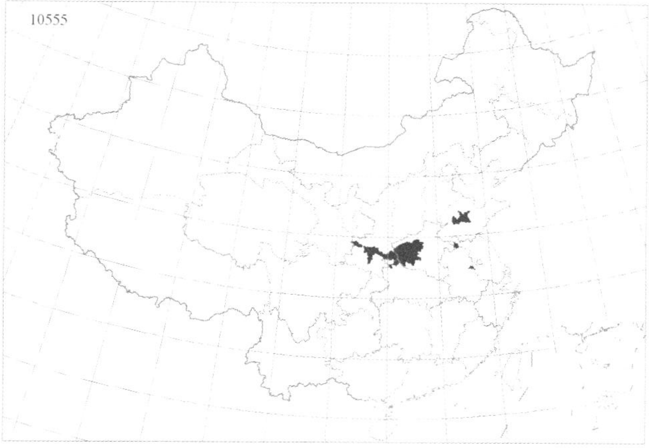

Caprifoliaceae 忍冬科 *Lonicera* 忍冬属

Lonicera fragrantissima subsp. standishii (Carrière) P. S. Hsu et H. J. Wang 苦糖果 Shrubs, semievergreen or deciduous

13.0 (2.0~16.9); -0.3 (-7.7~5.3); 25.2 (10.1~28.9); 13.2 (3.8~16.9); 109.0 (15.4~143.4); 12.7 (0.0~51.4); 776 (413~924) ❙ 887 (557~1663); 442 (286~666); 57 (10~212) ❙ 15.7 (-36.5~128.6); 718 (413~924); 190.8 (120.1~341.6)

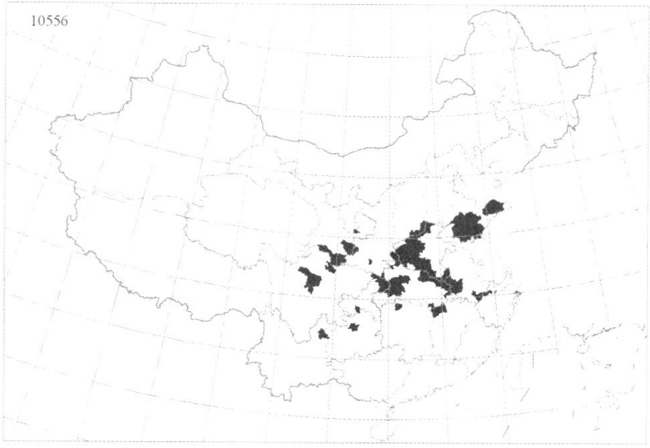

Caprifoliaceae 忍冬科 *Lonicera* 忍冬属

Lonicera fulvotomentosa P. S. Hsu et S. C. Cheng 黄褐毛忍冬
Vines

16.4 (14.2~19.1); 6.1 (3.8~10.2); 25.6 (20.8~27.0); 16.4 (14.2~19.2); 137.2 (110.9~170.1); 0.3 (0.0~1.5); 854 (734~981) ❙ 1256 (1107~1451); 549 (444~723); 88 (51~146) ❙ 47.5 (27.5~72.9); 854 (734~981); 222.9 (146.0~322.1)

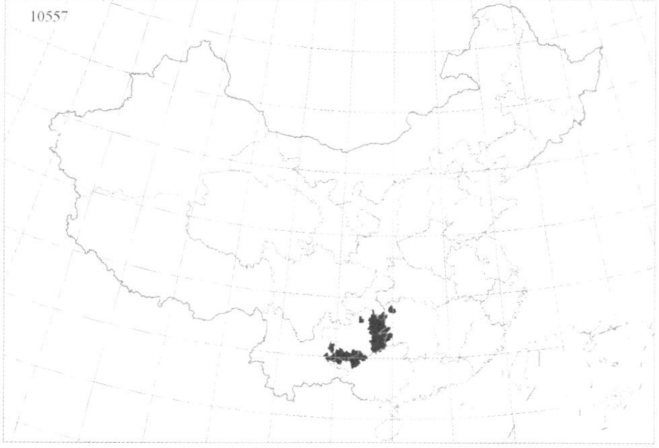

Caprifoliaceae 忍冬科 *Lonicera* 忍冬属

Lonicera graebneri Rehder 短梗忍冬
Shrubs, deciduous

11.2 (3.0~15.1); -2.1 (-7.7~1.8); 23.5 (12.4~27.5); 11.6 (4.8~15.2); 93.6 (25.1~128.0); 19.0 (5.6~49.3); 707 (441~865) ❙ 691 (341~856); 338 (190~408); 30 (8~53) ❙ -1.3 (-40.9~52.6); 655 (341~817); 194.2 (61.2~318.2)

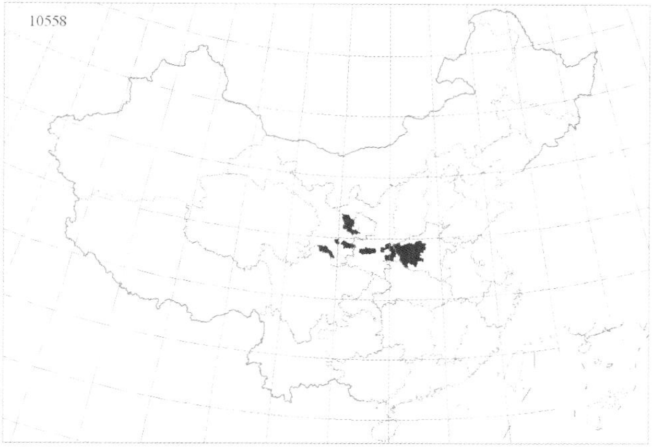

Caprifoliaceae 忍冬科 *Lonicera* 忍冬属

Lonicera gynochlamydea Hemsl. 蕊被忍冬
Shrubs, deciduous

13.3 (5.7~18.0); 1.5 (-4.2~7.2); 24.5 (14.2~28.8); 13.4 (6.4~18.0); 108.0 (38.1~156.3); 7.7 (0.0~29.4); 762 (504~958) ❙ 1048 (626~1648); 465 (306~666); 63 (14~192) ❙ 37.9 (-22.2~108.0); 751 (504~958); 221.5 (128.7~334.4)

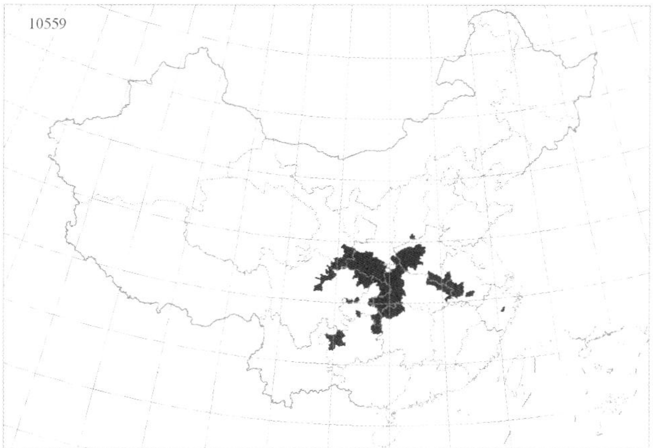

Caprifoliaceae 忍冬科 *Lonicera* 忍冬属

Lonicera hemsleyana (Kuntze) Rehder 倒卵叶忍冬
Shrubs or small trees, deciduous

15.3 (9.3~17.1); 3.0 (-1.7~4.7); 27.3 (20.1~29.3); 15.3 (9.5~17.1); 126.8 (69.8~145.7); 3.1 (0.3~18.0); 849 (603~941) ❙ 1436 (1118~1770); 567 (482~685); 162 (71~223) ❙ 70.0 (31.0~112.2); 849 (603~941); 231.7 (149.5~319.9)

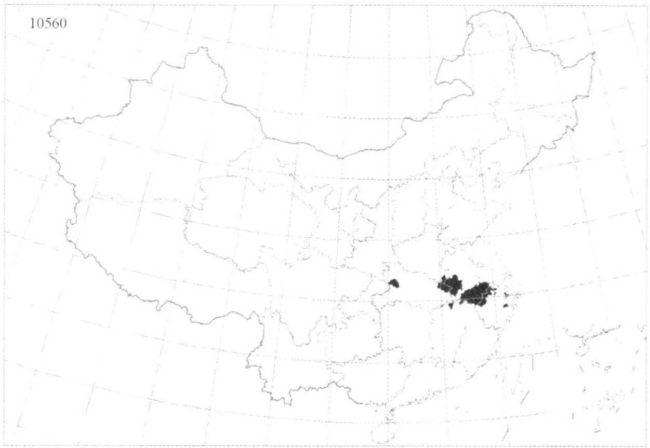

Caprifoliaceae 忍冬科 *Lonicera* 忍冬属
Lonicera heterophylla Decne. 异叶忍冬
Shrubs, deciduous

4.2 (-4.3~12.3); -14.2 (-20.9~-5.2); 19.3 (9.6~26.5); 7.9 (2.5~13.0); 61.2 (9.7~112.3); 69.8 (24.3~121.1); 580 (329~801) ❙ 202 (42~450); 92 (17~211); 18 (2~51) ❙ -62.2 (-94.7~3.9); 201 (42~406); 92.4 (9.3~297.3)

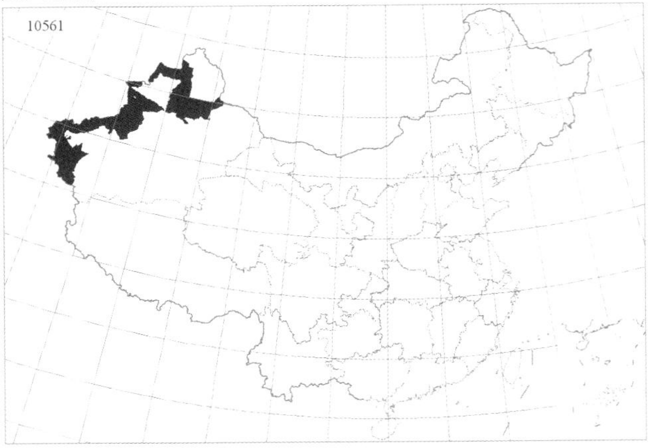

Caprifoliaceae 忍冬科 *Lonicera* 忍冬属
Lonicera hildebrandiana Collett et Hemsl. 大果忍冬
Vines, evergreen

18.1 (16.0~21.0); 11.4 (9.3~14.2); 22.5 (20.3~25.6); 18.1 (16.0~21.0); 157.7 (132.4~191.8); 0.0 (0.0~0.0); 878 (780~1053) ❙ 1380 (1049~1559); 773 (598~916); 52 (42~68) ❙ 57.4 (26.1~80.6); 869 (780~1053); 433.5 (336.3~551.6)

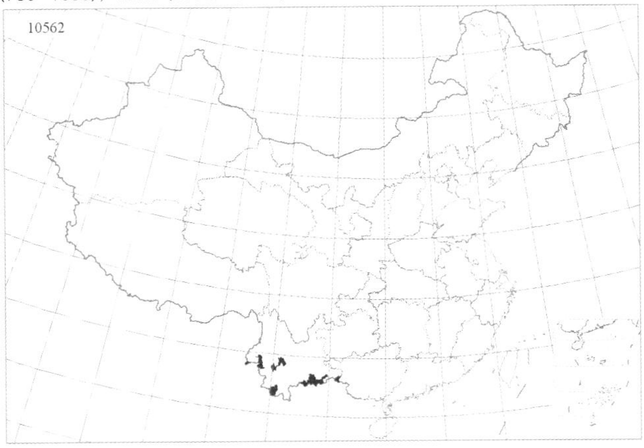

Caprifoliaceae 忍冬科 *Lonicera* 忍冬属
Lonicera hispida Pall. ex Roem. et Schult. 刚毛忍冬
Shrubs, deciduous

6.8 (-5.5~15.5); -7.1 (-20.9~4.5); 19.2 (5.8~28.0); 8.6 (1.3~15.5); 65.5 (1.2~128.0); 42.9 (0.5~127.1); 595 (299~880) ❙ 551 (32~1414); 309 (21~817); 17 (0~122) ❙ -3.5 (-95.9~128.6); 481 (32~859); 187.3 (1.6~364.1)

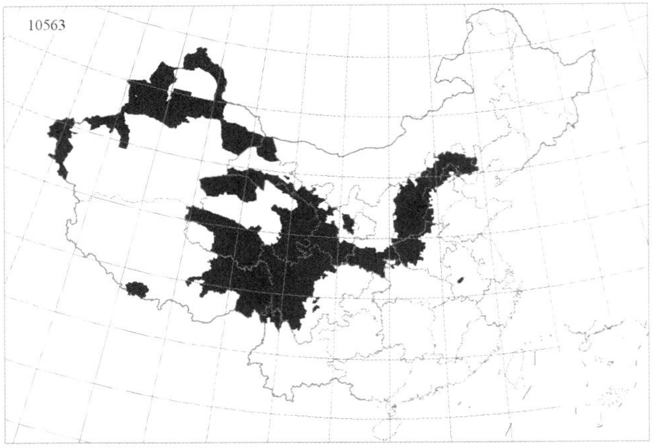

Caprifoliaceae 忍冬科 *Lonicera* 忍冬属
Lonicera humilis Kar. et Kir. 矮小忍冬
Dwarf shrubs, deciduous

2.4 (-4.3~9.0); -16.0 (-20.9~-9.3); 17.4 (9.6~23.2); 6.7 (2.5~10.7); 48.6 (9.7~88.9); 79.9 (40.7~121.1); 527 (329~703) ❙ 277 (71~450); 119 (24~211); 28 (10~51) ❙ -46.3 (-78.5~3.9); 274 (71~406); 153.1 (15.7~297.3)

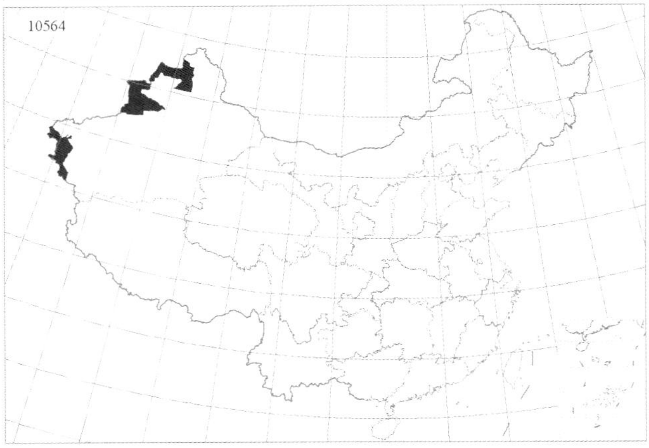

Caprifoliaceae 忍冬科 *Lonicera* 忍冬属
Lonicera hypoglauca Miq. 菰腺忍冬
Vines, deciduous

17.9 (12.6~24.1); 7.8 (1.9~18.9); 27.0 (17.9~30.1); 18.0 (12.6~24.1); 156.0 (91.6~229.7); 0.3 (0.0~6.1); 944 (678~1322) ❙ 1529 (871~3495); 606 (299~1594); 160 (29~897) ❙ 63.2 (-6.7~246.5); 933 (678~1247); 239.7 (108.2~659.0)

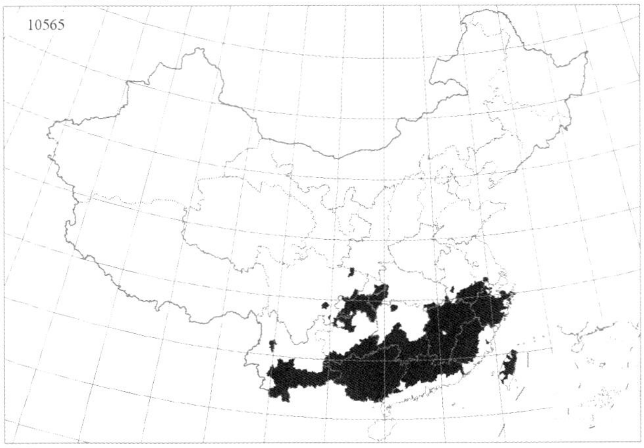

Caprifoliaceae 忍冬科 *Lonicera* 忍冬属
Lonicera hypoglauca subsp. nudiflora P. S. Hsu et H. J. Wang 净花菰腺忍冬
Vines, deciduous

17.0 (14.3~21.2); 8.0 (4.0~13.0); 24.5 (19.6~27.7); 17.1 (14.3~21.2); 144.9 (112.6~194.6); 0.1 (0.0~1.1); 865 (730~1114) ❙ 1268 (939~1624); 615 (477~835); 78 (39~191) ❙ 47.2 (6.5~82.3); 861 (730~1114); 269.8 (188.8~506.2)

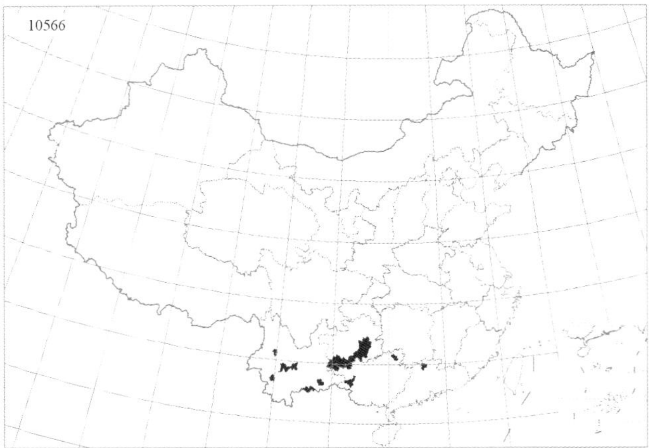

Caprifoliaceae 忍冬科 *Lonicera* 忍冬属
Lonicera hypoleuca Decaisne 白背忍冬
Shrubs, deciduous

-1.8; -12.0; 8.6; 2.4; 7.9; 88.8; 333 ❙ 509; 233; 98 ❙ 52.9; 333; 32.2

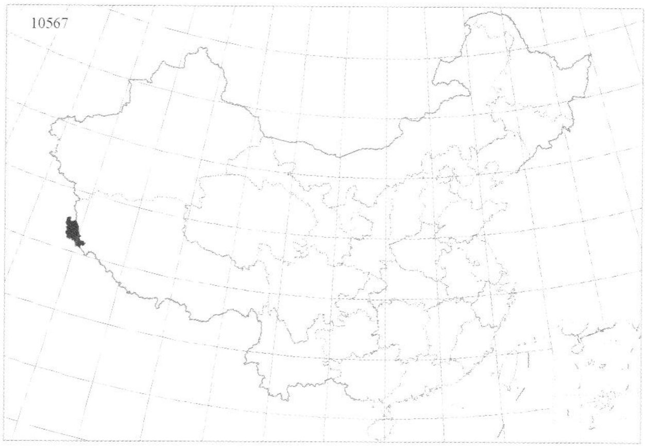

Caprifoliaceae 忍冬科 *Lonicera* 忍冬属
Lonicera inodora W. W. Sm. 卵叶忍冬
Vines

13.9; 6.5; 19.4; 13.9; 106.8; 0.0; 717 ❙ 1595; 900; 60 ❙ 122.8; 717; 467.3

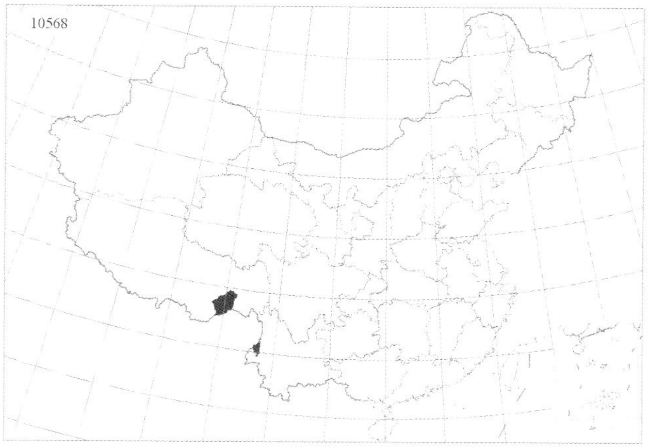

Caprifoliaceae 忍冬科 *Lonicera* 忍冬属
Lonicera japonica Thunb. 忍冬
Vines, semievergreen

14.5 (2.0~22.7); 2.3 (-19.2~15.3); 25.7 (9.9~30.1); 14.8 (3.8~22.8); 123.9 (15.4~213.1); 9.3 (0.0~84.3); 826 (413~1221) ❙ 1095 (349~3495); 483 (223~1428); 94 (7~897) ❙ 30.9 (-43.4~224.4); 782 (349~1221); 200.1 (60.5~551.6)

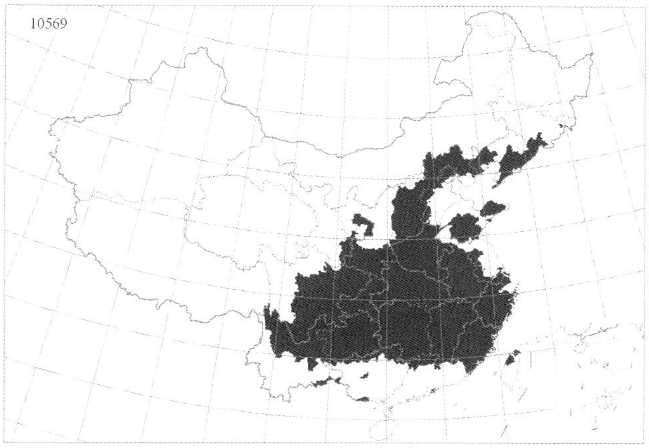

Caprifoliaceae 忍冬科 *Lonicera* 忍冬属
Lonicera japonica var. chinensis (P. Watson) Baker 红白忍冬
Vines, semievergreen

16.4 (13.4~18.1); 4.8 (1.5~6.9); 27.4 (25.2~28.7); 16.4 (13.4~18.1); 139.0 (108.3~157.7); 1.9 (0.0~7.3); 887 (755~965) ❙ 1737 (1489~1922); 604 (564~666); 222 (152~259) ❙ 96.2 (82.7~111.3); 883 (755~961); 256.8 (219.0~284.0)

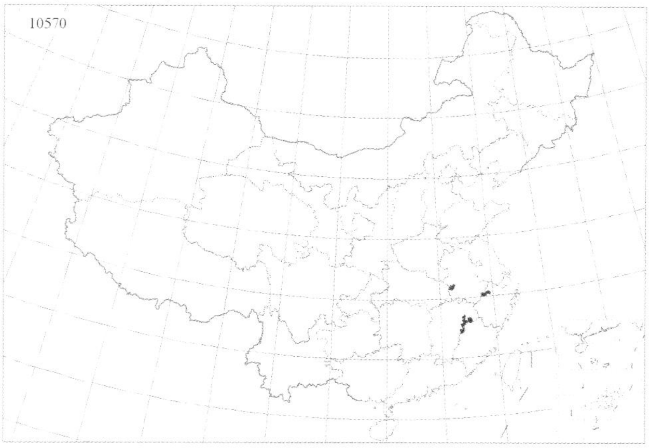

Caprifoliaceae 忍冬科 *Lonicera* 忍冬属
Lonicera jilongensis P. S. Hsu et H. J. Wang 吉隆忍冬
Shrubs, deciduous

-1.1; -9.1; 6.2; 2.0; 2.5; 75.6; 334 ❙ 321; 193; 33 ❙ -4.0; 321; 79.3

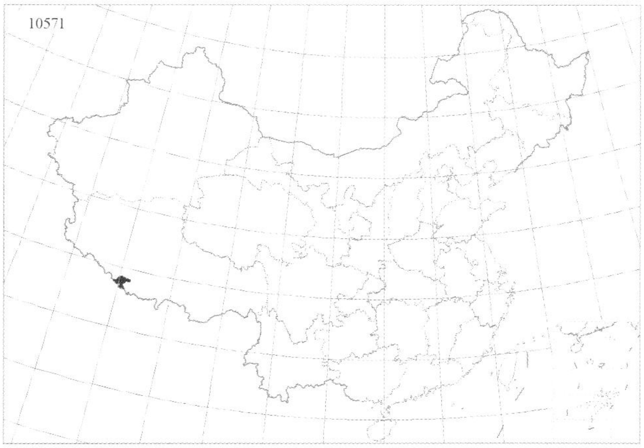

Caprifoliaceae 忍冬科 *Lonicera* 忍冬属
Lonicera kansuensis (Batalin ex Rehder) Pojark. 甘肃忍冬
Shrubs, deciduous

11.0 (2.6~15.1); -0.9 (-7.7~4.4); 22.3 (11.2~25.6); 11.3 (4.3~15.2); 87.7 (20.0~122.8); 15.4 (0.6~49.3); 678 (427~817) ❙ 841 (587~1246); 381 (290~528); 27 (12~56) ❙ 25.6 (-16.6~94.4); 673 (427~817); 263.1 (192.1~334.4)

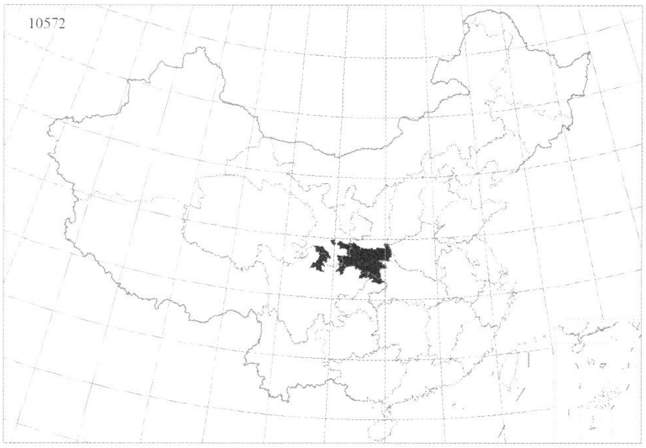

Caprifoliaceae 忍冬科 *Lonicera* 忍冬属
Lonicera kawakamii (Hayata) Masam. 玉山忍冬
Small shrubs, deciduous

19.6 (16.4~24.1); 14.1 (11.2~18.9); 24.0 (20.3~28.0); 19.6 (16.5~24.1); 175.5 (137.8~229.7); 0.0 (0.0~0.0); 975 (783~1322) ❙ 2362 (1751~2859); 1166 (947~1594); 193 (47~402) ❙ 155.1 (32.5~235.6); 936 (783~1095); 238.3 (154.2~303.6)

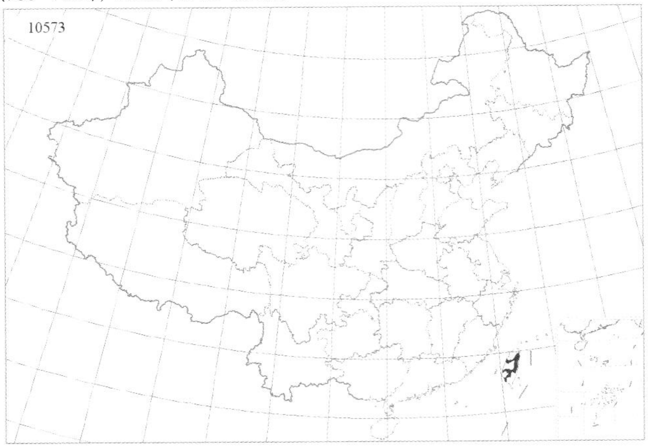

Caprifoliaceae 忍冬科 *Lonicera* 忍冬属
Lonicera lanceolata Wall. 柳叶忍冬
Shrubs, deciduous

7.9 (−1.8~17.4); −1.0 (−10.5~11.0); 15.8 (6.1~26.3); 8.7 (1.8~17.4); 59.1 (1.9~149.3); 24.3 (0.0~82.8); 567 (329~834) ❙ 911 (311~1736); 507 (193~1007); 32 (3~140) ❙ 58.7 (−10.6~145.8); 565 (311~834); 253.6 (61.6~447.4)

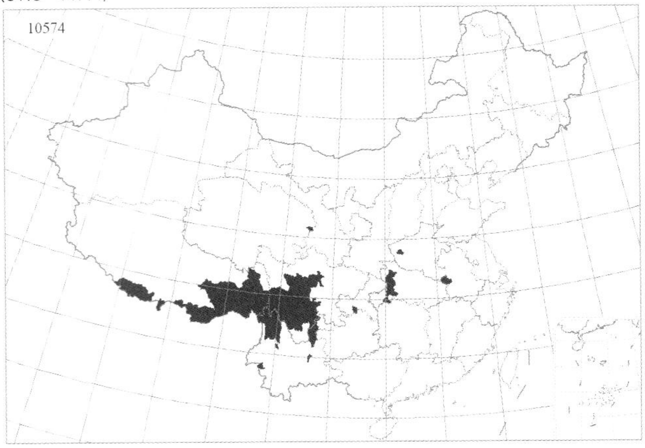

Caprifoliaceae 忍冬科 *Lonicera* 忍冬属
Lonicera lanceolata var. glabra S. S. Chien ex P. S. Hsu et H. J. Wang 光枝柳叶忍冬
Shrubs, deciduous

13.3 (3.4~15.2); 3.0 (−6.0~5.3); 22.9 (11.7~26.3); 13.4 (4.7~15.2); 104.2 (23.2~123.9); 4.5 (0.0~41.9); 736 (442~816) ❙ 1148 (853~1466); 525 (425~688); 68 (22~140) ❙ 56.8 (25.7~106.2); 736 (442~816); 223.8 (170.4~292.6)

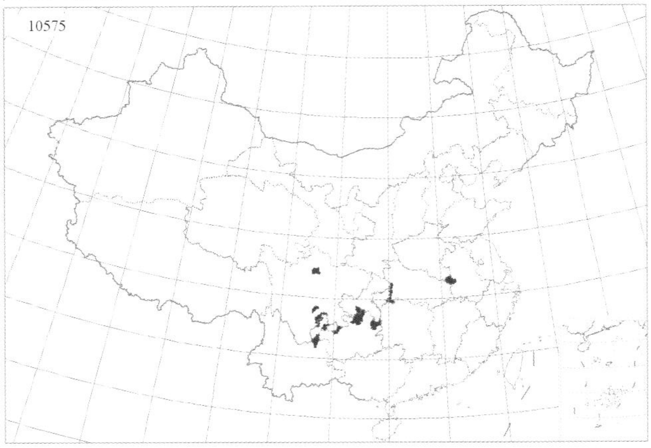

Caprifoliaceae 忍冬科 *Lonicera* 忍冬属
Lonicera ligustrina Wall. 女贞叶忍冬
Shrubs, evergreen or semievergreen

14.4 (5.4~20.4); 5.4 (−3.2~15.1); 22.4 (12.3~28.1); 14.5 (5.8~20.5); 115.8 (31.5~185.8); 2.1 (0.0~26.7); 769 (481~1010) ❙ 1126 (741~1607); 570 (382~860); 54 (15~123) ❙ 47.5 (−14.2~107.2); 758 (481~1010); 251.9 (128.7~659.0)

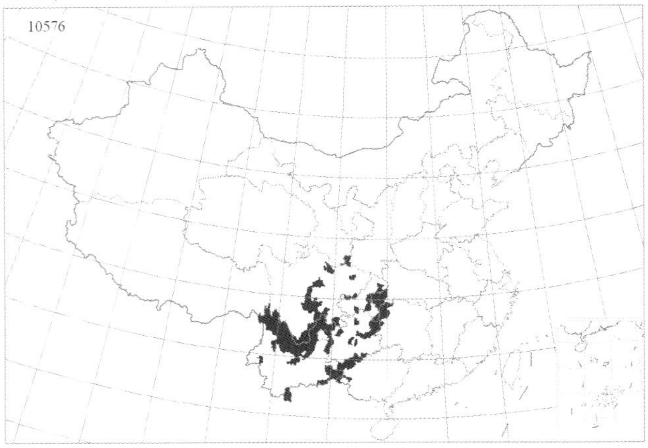

Caprifoliaceae 忍冬科 *Lonicera* 忍冬属
Lonicera ligustrina var. yunnanensis Franch. 亮叶忍冬
Shrubs, evergreen or semievergreen

10.9 (2.2~19.0); 2.5 (−7.0~13.1); 18.3 (9.9~25.4); 11.2 (3.8~19.1); 82.2 (15.5~168.8); 10.5 (0.0~49.1); 650 (413~939) ❙ 980 (677~1495); 552 (336~817); 30 (12~62) ❙ 54.5 (−0.2~125.9); 646 (413~939); 281.0 (171.0~427.9)

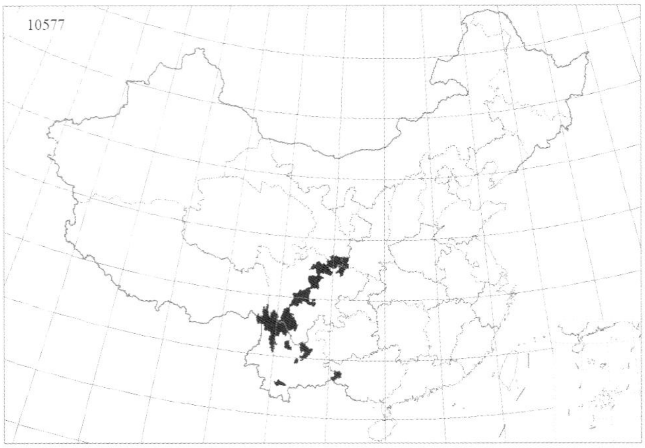

Caprifoliaceae 忍冬科 *Lonicera* 忍冬属
Lonicera litangensis Batalin 理塘忍冬
Dwarf shrubs, deciduous

6.2 (−1.8~16.2); −2.8 (−10.5~7.2); 14.3 (6.1~26.3); 7.3 (1.8~16.3); 47.6 (1.9~135.7); 32.5 (0.0~82.8); 526 (329~854) ❙ 808 (252~1448); 453 (196~816); 27 (0~105) ❙ 52.0 (−35.9~125.9); 521 (252~854); 233.2 (62.8~372.3)

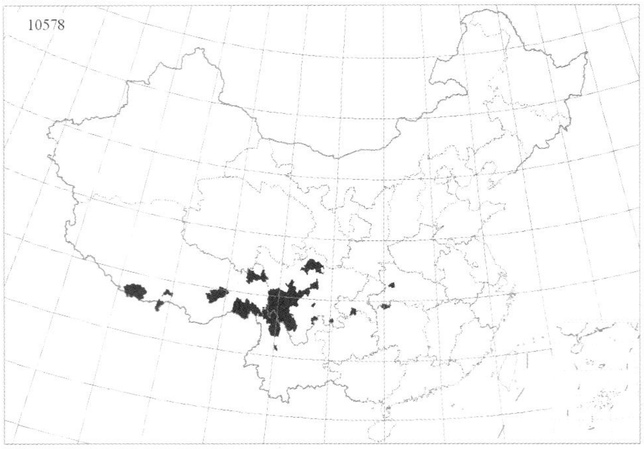

Caprifoliaceae 忍冬科 *Lonicera* 忍冬属
Lonicera longiflora (Lindl.) DC. 长花忍冬
Vines

21.2 (17.3~22.7); 13.2 (8.5~15.2); 27.4 (22.4~28.7); 21.3 (17.3~22.8); 195.9 (147.1~213.1); 0.0 (0.0~0.0); 1119 (833~1221) ❙ 1766 (1302~2435); 912 (670~1428); 121 (52~194) ❙ 58.3 (12.9~99.5); 1119 (833~1221); 277.9 (163.7~551.6)

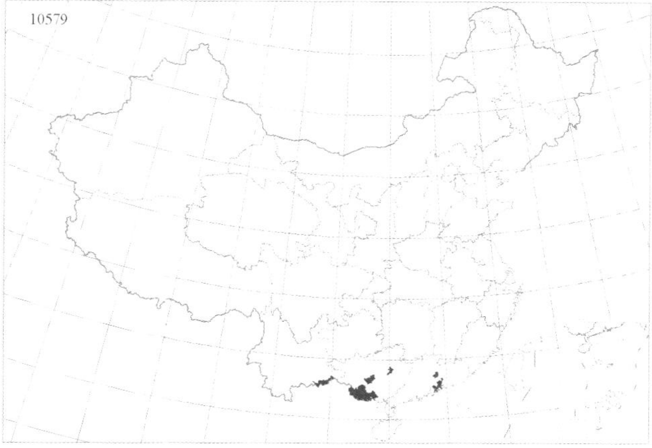

Caprifoliaceae 忍冬科 *Lonicera* 忍冬属
Lonicera longituba H. T. Chang ex P. S. Hsu et H. J. Wang 卷瓣忍冬
Vines

20.8 (18.0~22.0); 12.3 (8.5~14.1); 27.4 (25.8~28.5); 20.8 (18.0~22.1); 189.9 (156.4~205.1); 0.0 (0.0~0.0); 1086 (913~1170) ❙ 1773 (1551~2014); 861 (672~1056); 147 (125~173) ❙ 63.6 (40.9~72.1); 1086 (913~1170); 250.1 (217.4~297.2)

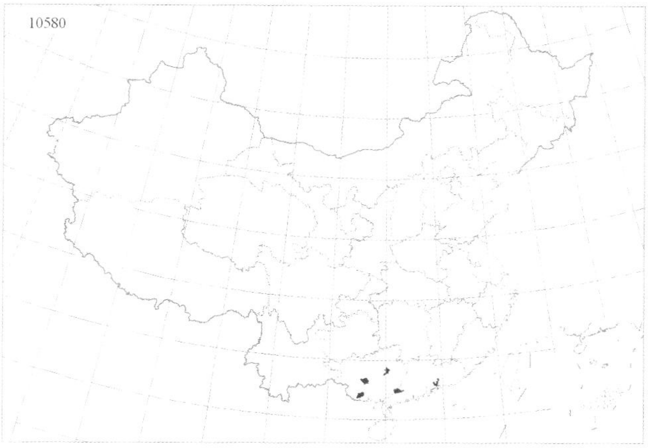

Caprifoliaceae 忍冬科 *Lonicera* 忍冬属
Lonicera maackii (Rupr.) Maxim. 金银忍冬
Shrubs, deciduous

11.6 (−1.1~18.9); −2.3 (−24.7~12.0); 24.2 (6.2~29.4); 12.5 (2.0~19.0); 102.1 (2.5~167.8); 22.3 (0.0~120.7); 739 (334~984) ❙ 869 (321~1865); 414 (190~756); 60 (6~258) ❙ 15.3 (−40.9~112.2); 682 (321~941); 205.0 (61.2~491.5)

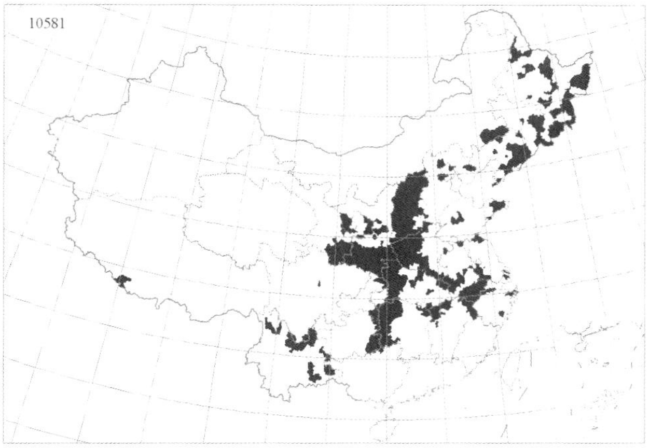

Caprifoliaceae 忍冬科 *Lonicera* 忍冬属
Lonicera maackii var. erubescens Rehder 红花金银忍冬
Shrubs, deciduous

14.9 (10.8~15.6); 2.4 (1.7~3.9); 26.6 (16.6~28.3); 15.0 (10.9~15.7); 123.8 (71.8~131.4); 3.9 (1.3~5.4); 849 (633~882) ❙ 1047 (968~1156); 488 (448~594); 96 (24~122) ❙ 24.2 (13.5~52.8); 849 (633~882); 175.3 (134.1~319.8)

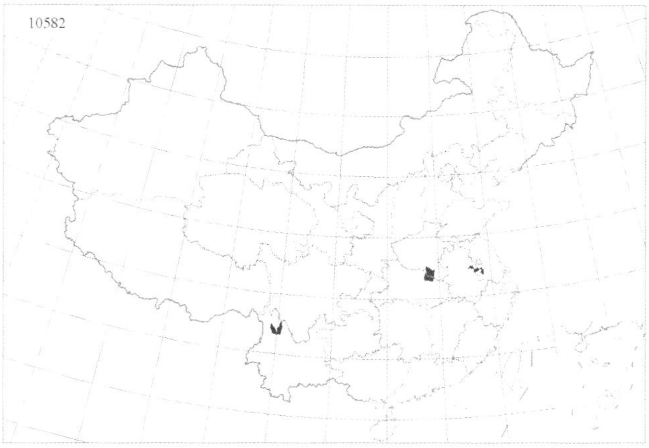

Caprifoliaceae 忍冬科 *Lonicera* 忍冬属
Lonicera macrantha (D. Don) Spreng. 大花忍冬
Vines, semievergreen

18.2 (13.3~24.8); 8.5 (2.3~20.1); 26.8 (19.6~29.4); 18.3 (13.3~24.8); 159.9 (100.4~237.6); 0.2 (0.0~4.9); 962 (706~1385) ❙ 1503 (868~2680); 623 (312~1428); 140 (29~259) ❙ 58.7 (−34.3~246.5); 952 (706~1331); 245.8 (123.1~620.0)

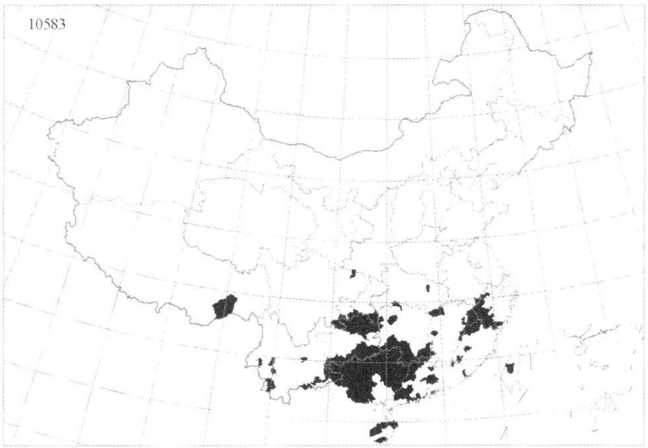

Caprifoliaceae 忍冬科 *Lonicera* 忍冬属
Lonicera macrantha var. heterotricha P. S. Hsu et H. J. Wang 异毛忍冬
Vines, semievergreen

17.9 (13.3~24.8); 8.1 (2.3~20.1); 26.4 (20.3~28.7); 18.0 (13.3~24.8); 155.7 (104.4~237.6); 0.2 (0.0~4.9); 937 (741~1385) ❙ 1460 (1047~1997); 615 (411~848); 131 (29~254) ❙ 57.8 (−2.2~130.7); 927 (741~1162); 259.3 (146.0~538.2)

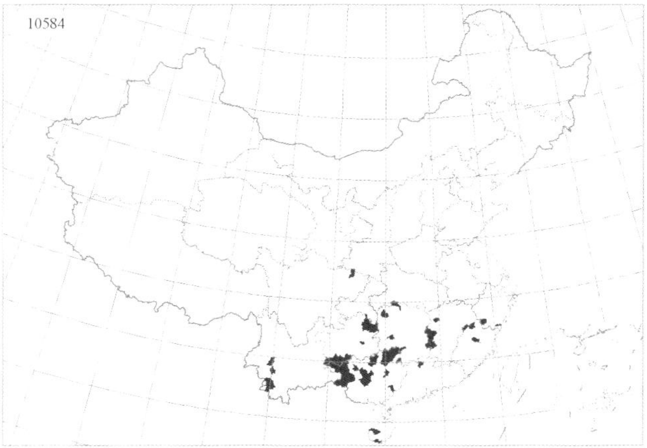

Caprifoliaceae 忍冬科 *Lonicera* 忍冬属

Lonicera macranthoides Hand.-Mazz. 灰毡毛忍冬

Vines

16.2 (8.6~22.2); 5.4 (0.7~13.9); 26.4 (15.0~29.2); 16.3 (8.7~22.3); 136.0 (52.6~207.5); 0.8 (0.0~9.4); 864 (574~1190) ❙ 1444 (842~1997); 557 (411~858); 146 (19~256) ❙ 67.5 (10.0~147.0); 863 (574~1190); 229.9 (146.0~411.1)

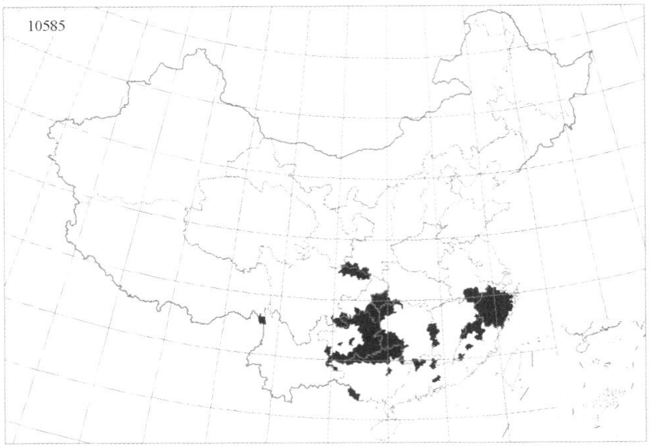

Caprifoliaceae 忍冬科 *Lonicera* 忍冬属

Lonicera maximowiczii (Rupr.) Regel 紫花忍冬

Shrubs, deciduous

3.7 (−0.9~11.9); −16.2 (−24.7~−1.2); 20.8 (18.5~24.3); 8.1 (6.5~12.1); 62.1 (47.6~98.7); 77.4 (14.2~120.7); 578 (513~720) ❙ 632 (470~1073); 387 (309~691); 23 (8~65) ❙ 9.6 (−23.1~74.1); 563 (470~712); 282.7 (173.0~346.5)

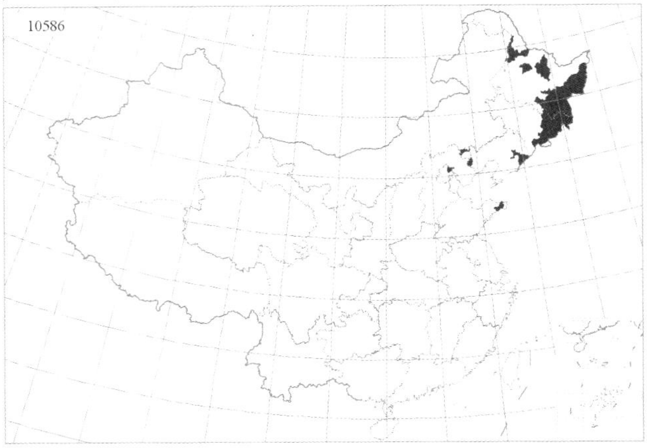

Caprifoliaceae 忍冬科 *Lonicera* 忍冬属

Lonicera microphylla Willd. ex Roem. et Schult. 小叶忍冬

Shrubs, deciduous

6.3 (−4.3~14.5); −10.1 (−21.5~0.1); 20.6 (8.5~28.0); 8.8 (2.4~14.6); 69.2 (7.9~127.2); 52.6 (10.7~121.1); 608 (329~880) ❙ 376 (53~741); 216 (23~453); 14 (2~98) ❙ −36.4 (−92.4~82.4); 366 (53~711); 123.4 (10.3~313.2)

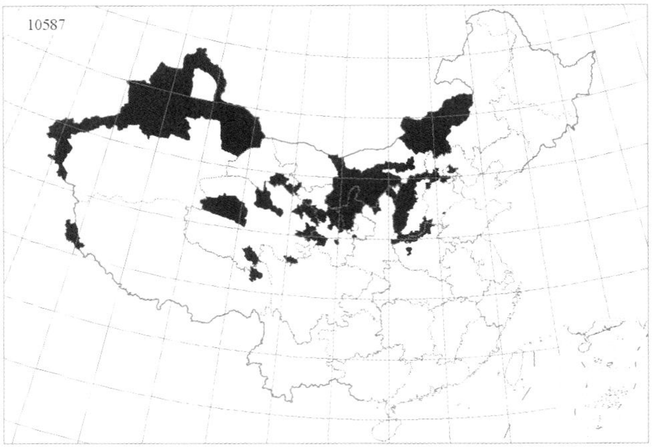

Caprifoliaceae 忍冬科 *Lonicera* 忍冬属

Lonicera minuta Batalin 矮生忍冬

Dwarf shrubs, deciduous

−0.8 (−5.5~5.4); −12.8 (−18.0~−7.1); 10.0 (5.8~16.3); 3.3 (1.3~6.9); 14.1 (1.2~46.4); 82.9 (40.6~127.1); 383 (299~515) ❙ 457 (213~678); 265 (135~379); 10 (4~16) ❙ 19.3 (−44.9~87.3); 354 (213~515); 222.5 (32.7~364.1)

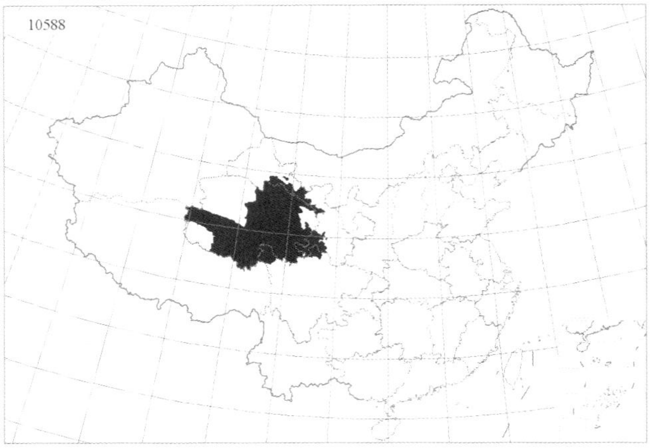

Caprifoliaceae 忍冬科 *Lonicera* 忍冬属

Lonicera minutifolia Kitam. 细叶忍冬

Small shrubs, deciduous

−1.1; −9.1; 6.2; 2.0; 2.5; 75.6; 334 ❙ 321; 193; 33 ❙ −4.0; 321; 79.3

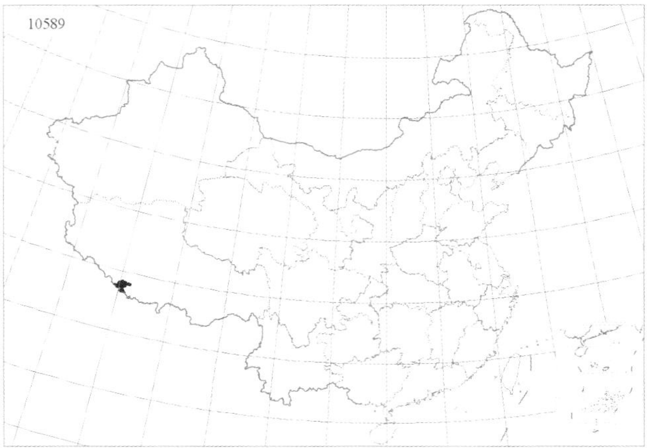

Caprifoliaceae 忍冬科 *Lonicera* 忍冬属

Lonicera modesta Rehder 下江忍冬

Shrubs, deciduous

15.9 (13.4~18.2); 3.8 (0.5~9.0); 27.5 (24.6~29.3); 16.0 (13.4~18.2); 133.6 (108.3~158.6); 2.0 (0.0~9.2); 873 (755~975) ❙ 1469 (710~1997); 554 (410~713); 170 (49~259) ❙ 68.7 (−19.8~147.0); 866 (710~961); 225.9 (134.1~319.9)

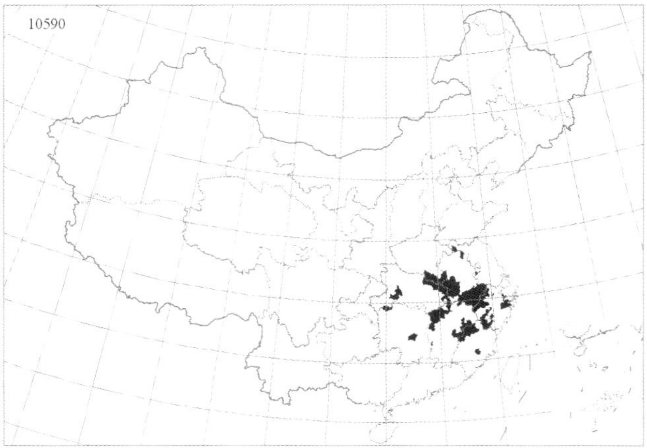

Caprifoliaceae 忍冬科 *Lonicera* 忍冬属
Lonicera modesta var. lushanensis Rehder 庐山忍冬
Shrubs, deciduous

15.3 (14.1~16.6); 3.3 (1.6~6.0); 26.8 (25.9~28.1); 15.3 (14.2~16.6); 126.4 (116.7~138.9); 2.5 (0.0~6.5); 835 (796~894) ❙ 1646 (1364~1997); 597 (485~677); 202 (129~254) ❙ 97.2 (62.3~130.7); 834 (796~876); 266.6 (213.0~319.9)

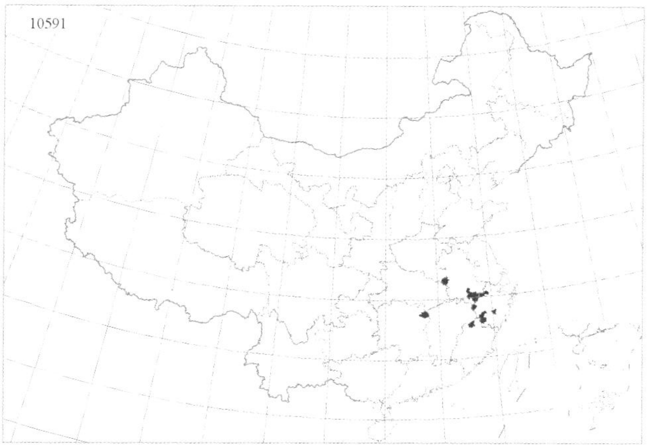

Caprifoliaceae 忍冬科 *Lonicera* 忍冬属
Lonicera mucronata Rehder 短尖忍冬
Shrubs

13.3; 2.1; 24.0; 13.3; 105.0; 5.5; 740 ❙ 1183; 494; 73 ❙ 59.9; 740; 245.6

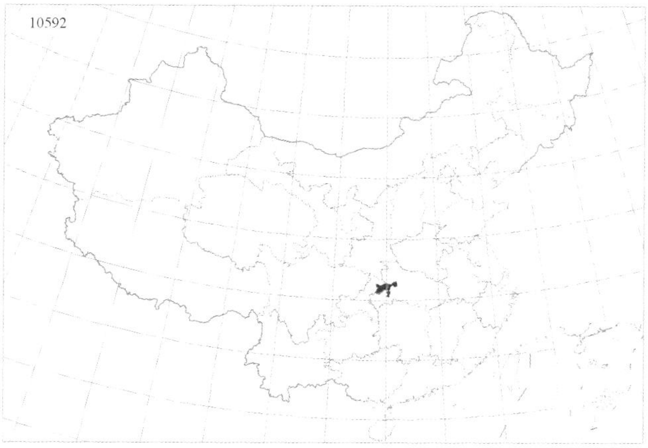

Caprifoliaceae 忍冬科 *Lonicera* 忍冬属
Lonicera myrtillus Hook. f. et Thomson 越橘叶忍冬
Shrubs, deciduous

4.8 (−2.2~17.4); −3.5 (−11.8~11.0); 12.2 (6.1~21.4); 6.2 (1.8~17.4); 36.9 (1.9~149.3); 38.3 (0.0~91.0); 479 (327~834) ❙ 697 (212~1736); 419 (156~1007); 20 (0~51) ❙ 38.6 (−35.1~145.8); 463 (212~834); 232.6 (62.8~447.4)

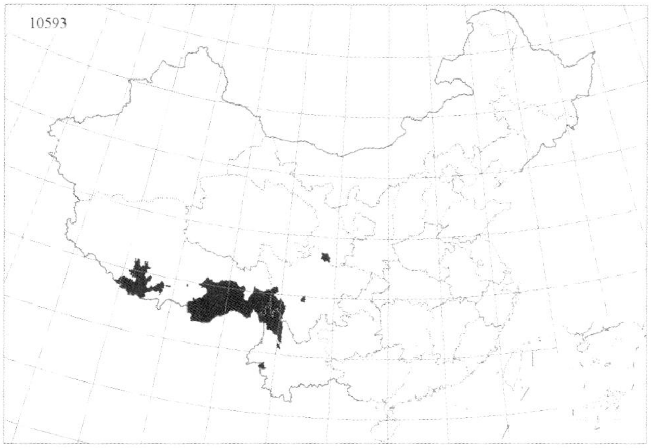

Caprifoliaceae 忍冬科 *Lonicera* 忍冬属
Lonicera myrtillus var. cyclophylla Rehder 圆叶忍冬
Shrubs, deciduous

5.8; −2.2; 12.7; 6.5; 36.4; 26.4; 498 ❙ 835; 497; 28 ❙ 65.1; 498; 294.3

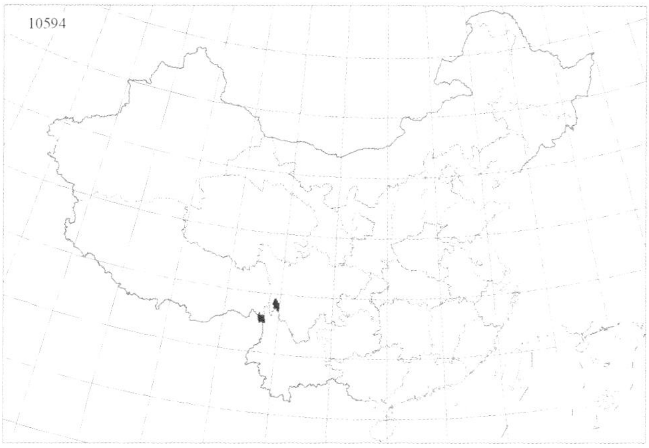

Caprifoliaceae 忍冬科 *Lonicera* 忍冬属
Lonicera nervosa Maxim. 红脉忍冬
Shrubs, deciduous

8.8 (−4.1~17.4); −3.6 (−17.3~11.0); 20.2 (7.8~27.5); 9.8 (2.1~17.4); 74.5 (5.2~149.3); 27.7 (0.0~114.2); 630 (340~865) ❙ 696 (271~1406); 345 (165~816); 24 (4~83) ❙ 13.2 (−44.6~128.6); 584 (271~834); 219.1 (61.2~434.4)

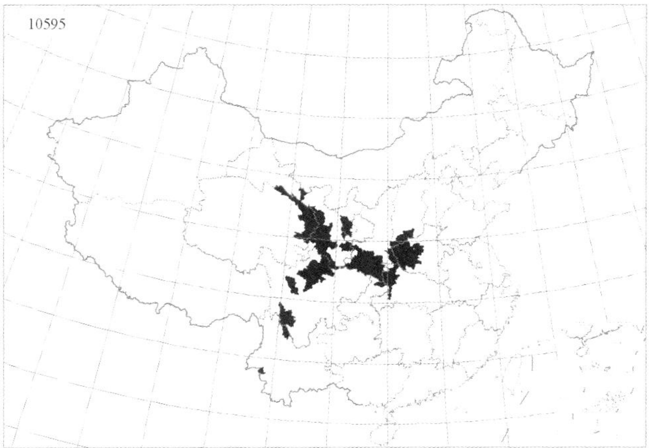

Caprifoliaceae 忍冬科 *Lonicera* 忍冬属
Lonicera nigra L. 黑果忍冬
Shrubs, deciduous

5.0 (2.2~10.2); −13.3 (−17.2~−6.2); 21.1 (19.2~24.7); 8.5 (6.9~11.4); 65.9 (50.3~95.3); 65.0 (32.8~83.8); 593 (525~722) ❙ 600 (541~668); 379 (333~453); 17 (10~24) ❙ 3.1 (−23.1~27.4); 552 (525~590); 289.9 (190.5~322.3)

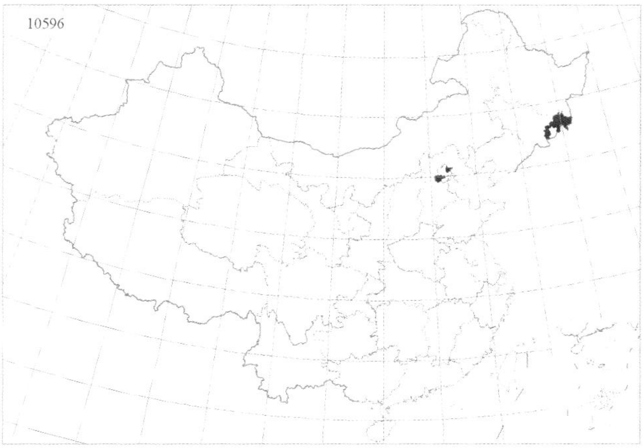

Caprifoliaceae 忍冬科 *Lonicera* 忍冬属
Lonicera nubium (Hand.-Mazz.) Hand.-Mazz. 云雾忍冬
Vines

16.6 (6.2~20.2); 5.4 (-6.7~11.1); 27.0 (15.8~29.8); 16.6 (7.0~20.3); 140.2 (45.2~183.3); 0.9 (0.0~35.7); 886 (522~1056) ❙ 1436 (373~1852); 497 (201~711); 166 (7~233) ❙ 61.8 (-37.6~128.9); 873 (373~1056); 206.8 (88.1~278.8)

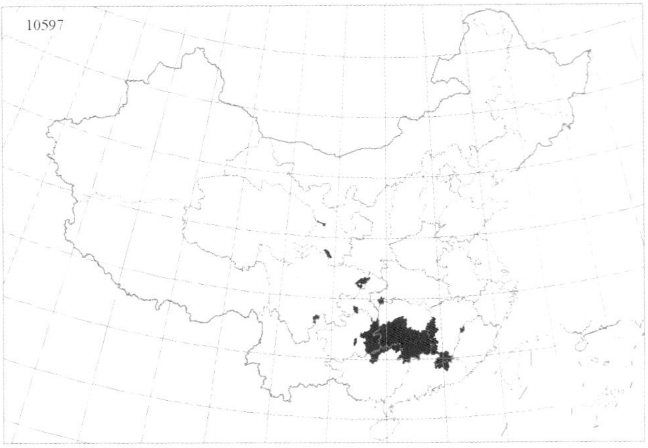

Caprifoliaceae 忍冬科 *Lonicera* 忍冬属
Lonicera oblata K. S. Hao ex P. S. Hsu et H. J. Wang 丁香叶忍冬
Shrubs, deciduous

10.2 (6.3~13.0); -5.8 (-10.8~-1.3); 24.3 (21.4~25.7); 11.4 (8.8~13.1); 94.9 (71.0~110.3); 31.9 (14.7~54.9); 716 (604~779) ❙ 553 (471~600); 370 (313~453); 16 (12~26) ❙ -22.7 (-27.9~-18.2); 553 (471~600); 185.9 (143.7~214.0)

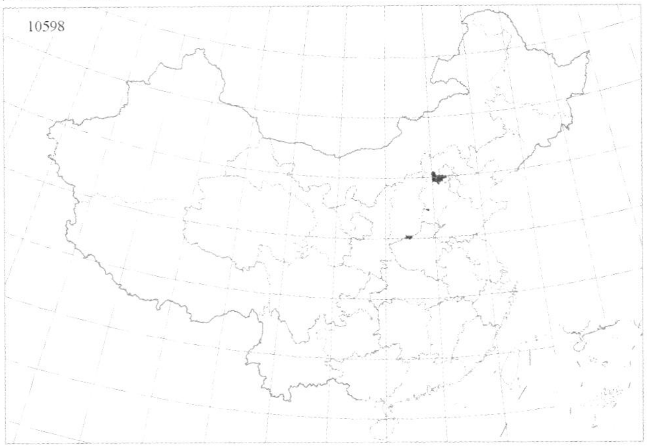

Caprifoliaceae 忍冬科 *Lonicera* 忍冬属
Lonicera oiwakensis Hayata 瘤基忍冬
Shrubs

18.5; 13.2; 22.5; 18.5; 161.7; 0.0; 896 ❙ 2563; 1299; 178 ❙ 193.3; 896; 248.4

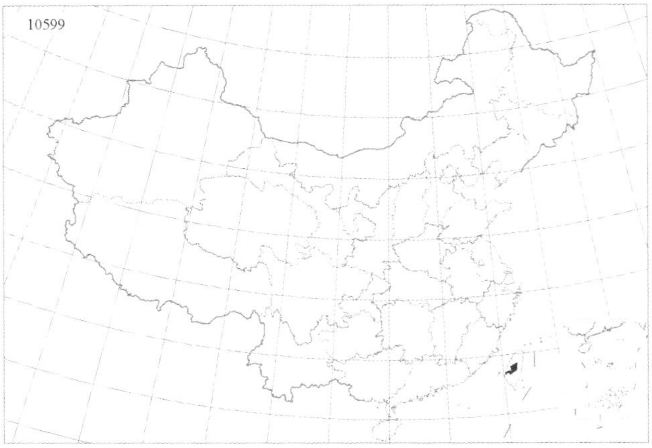

Caprifoliaceae 忍冬科 *Lonicera* 忍冬属
Lonicera oreodoxa Harry Sm. ex Rehder 垫状忍冬
Dwarf shrubs, deciduous

2.6; -7.0; 11.2; 4.3; 20.0; 48.2; 427 ❙ 829; 400; 23 ❙ 94.4; 427; 252.2

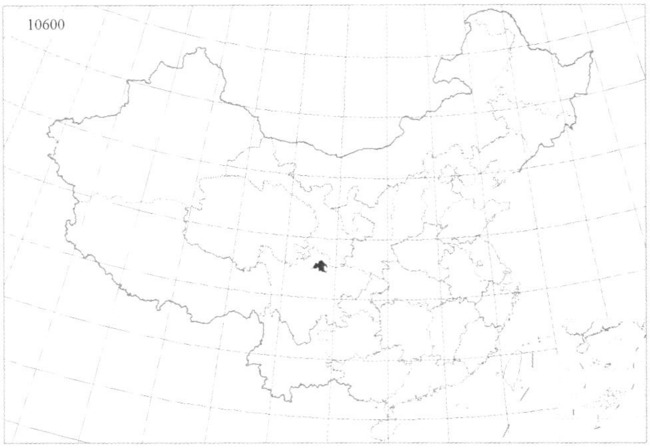

Caprifoliaceae 忍冬科 *Lonicera* 忍冬属
Lonicera pampaninii H. Lév. 短柄忍冬
Vines

16.2 (3.5~21.8); 5.4 (-6.5~13.7); 26.2 (12.6~29.8); 16.2 (5.0~21.9); 135.5 (26.3~202.7); 0.7 (0.0~43.4); 860 (446~1159) ❙ 1387 (731~1997); 543 (299~720); 134 (19~259) ❙ 61.2 (-6.7~130.7); 854 (446~1159); 215.0 (146.0~332.5)

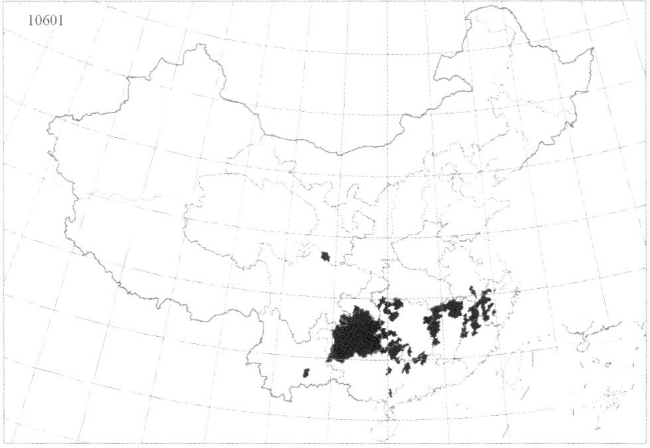

Caprifoliaceae 忍冬科 *Lonicera* 忍冬属
Lonicera pileata Oliv. 蕊帽忍冬
Shrubs, evergreen or semievergreen

15.5 (5.7~20.5); 5.7 (-3.5~15.1); 24.2 (14.2~28.8); 15.5 (6.4~20.5); 128.0 (38.1~185.9); 1.8 (0.0~28.1); 813 (504~994) ❙ 1213 (760~1751); 561 (349~996); 72 (17~182) ❙ 50.2 (5.0~107.2); 811 (504~992); 240.9 (120.7~659.0)

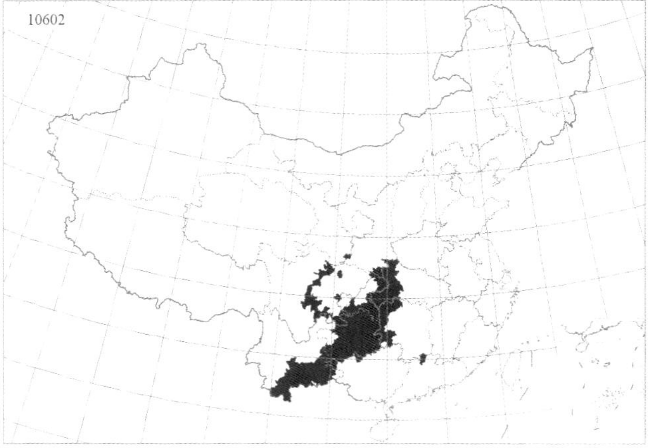

Caprifoliaceae 忍冬科 *Lonicera* 忍冬属

Lonicera pileata var. linearis Rehder 条叶忍冬

Shrubs, evergreen or semievergreen

19.2; 12.5; 23.7; 19.2; 170.9; 0.0; 934 ❙ 1417; 718; 62 ❙ 51.9; 917; 438.1

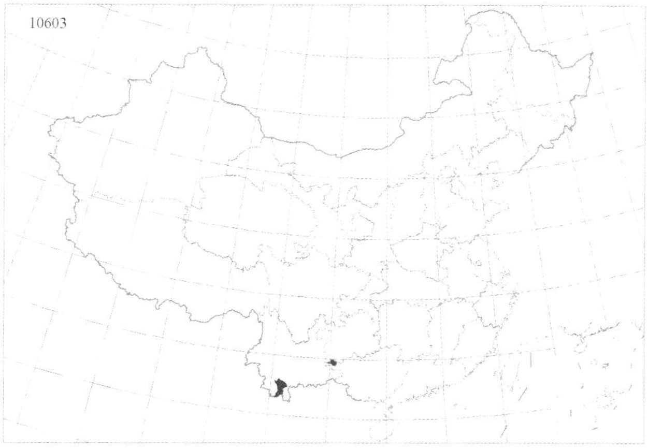

Caprifoliaceae 忍冬科 *Lonicera* 忍冬属

Lonicera praeflorens Batalin 早花忍冬

Shrubs, deciduous

3.2 (−0.9~10.5); −17.5 (−25.0~−4.9); 21.0 (18.4~25.0); 7.9 (6.4~11.3); 61.4 (45.8~93.2); 82.3 (26.8~120.7); 576 (501~705) ❙ 655 (513~1073); 404 (309~691); 22 (9~40) ❙ 13.8 (−10.4~74.1); 570 (501~698); 286.7 (180.8~346.5)

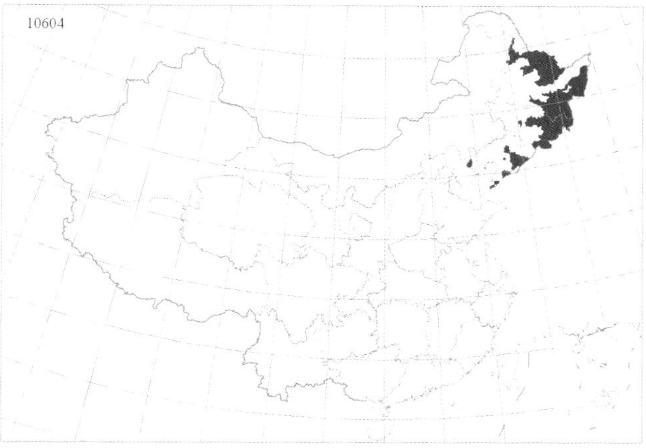

Caprifoliaceae 忍冬科 *Lonicera* 忍冬属

Lonicera prostrata Rehder 平卧忍冬

Shrubs

4.9; −4.5; 13.4; 5.9; 34.1; 34.3; 482 ❙ 917; 466; 25 ❙ 90.8; 482; 262.3

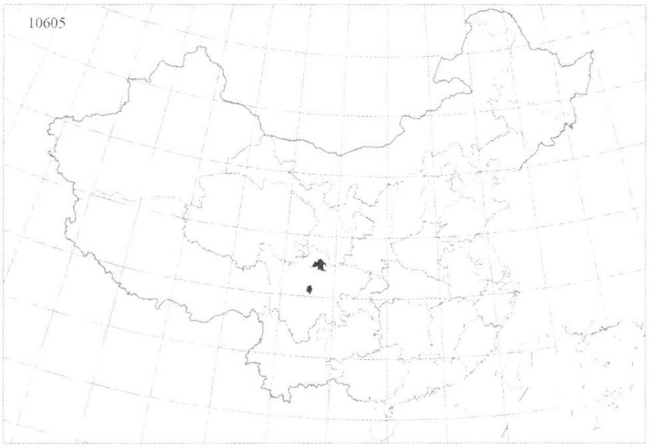

Caprifoliaceae 忍冬科 *Lonicera* 忍冬属

Lonicera retusa Franch. 凹叶忍冬

Shrubs, deciduous

7.3 (2.0~14.5); −3.3 (−10.7~3.9); 17.0 (9.9~28.2); 8.2 (3.8~14.6); 56.7 (15.4~125.3); 28.1 (1.3~54.8); 568 (413~864) ❙ 795 (478~1246); 416 (245~619); 21 (11~46) ❙ 50.1 (−36.2~128.6); 535 (413~692); 263.5 (128.6~357.7)

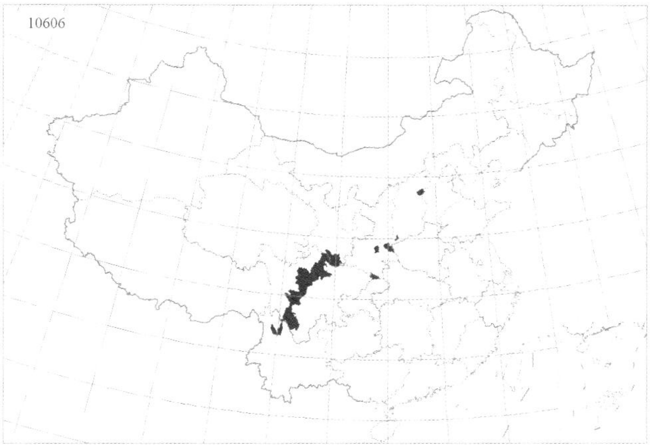

Caprifoliaceae 忍冬科 *Lonicera* 忍冬属

Lonicera rhytidophylla Hand.-Mazz. 皱叶忍冬

Vines, evergreen

18.3 (14.3~22.4); 8 4 (3.3~15.2); 27.2 (23.4~29.4); 18.4 (14.3~22.7); 161.1 (113.2~212.7); 0.1 (0.0~2.3); 963 (761~1216) ❙ 1580 (1175~1997); 599 (403~994); 172 (51~254) ❙ 65.2 (17.4~130.7); 959 (761~1216); 234.2 (51.0~332.5)

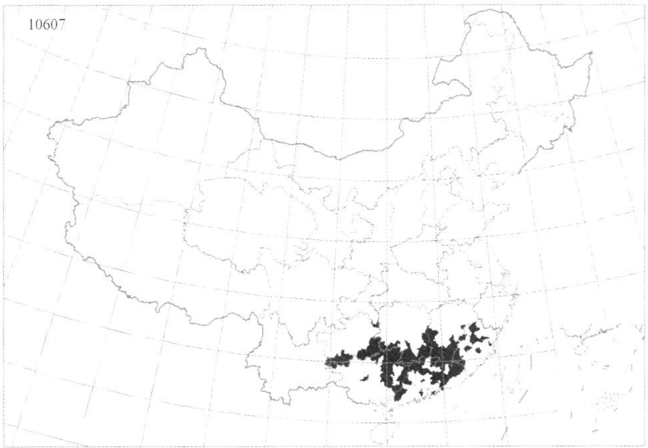

Caprifoliaceae 忍冬科 *Lonicera* 忍冬属

Lonicera rupicola Hook. f. et Thomson 岩生忍冬

Shrubs, deciduous

2.7 (−5.5~12.6); −8.1 (−18.0~6.0); 12.3 (5.8~21.2); 5.0 (1.3~12.6); 28.3 (1.2~91.6); 55.5 (0.0~127.1); 446 (299~678) ❙ 564 (213~1431); 323 (135~817); 12 (1~42) ❙ 27.0 (−40.9~128.6); 423 (213~678); 221.0 (32.7~364.1)

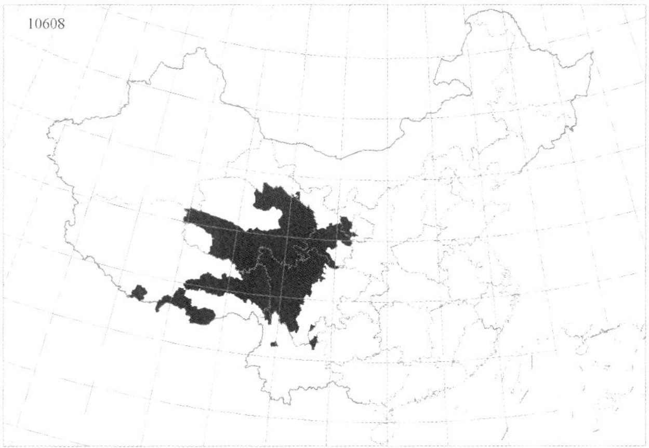

Caprifoliaceae 忍冬科 *Lonicera* 忍冬属

Lonicera rupicola var. syringantha (Maxim.) Zabel 红花岩生忍冬

Shrubs, deciduous

3.2 (-4.1~12.2); -8.9 (-17.3~4.3); 13.7 (7.2~21.4); 5.5 (2.0~12.2); 33.5 (3.7~87.4); 54.9 (0.7~114.2); 467 (339~672) ❙ 546 (101~1431); 302 (61~817); 11 (4~31) ❙ 19.5 (-83.8~128.6); 426 (101~672); 219.5 (25.5~364.1)

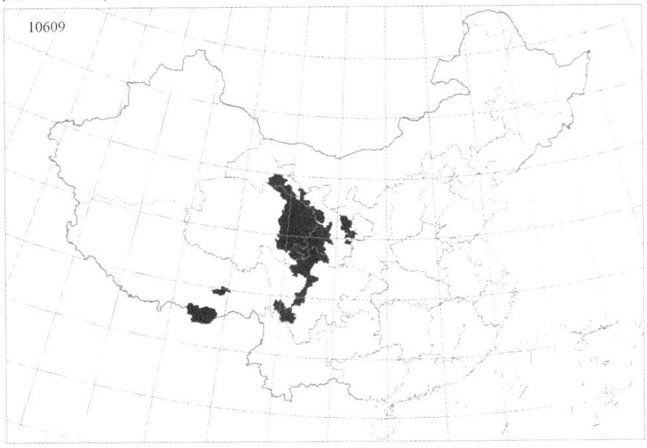

Caprifoliaceae 忍冬科 *Lonicera* 忍冬属

Lonicera ruprechtiana Regel 长白忍冬

Shrubs, deciduous

3.5 (-5.0~8.6); -17.3 (-29.4~-8.9); 21.2 (17.0~25.0); 8.1 (5.0~10.7); 62.8 (35.2~91.6); 80.7 (42.8~154.8); 581 (450~704) ❙ 662 (468~1018); 408 (300~648); 22 (12~38) ❙ 13.6 (-18.5~64.7); 574 (450~698); 283.1 (177.1~346.5)

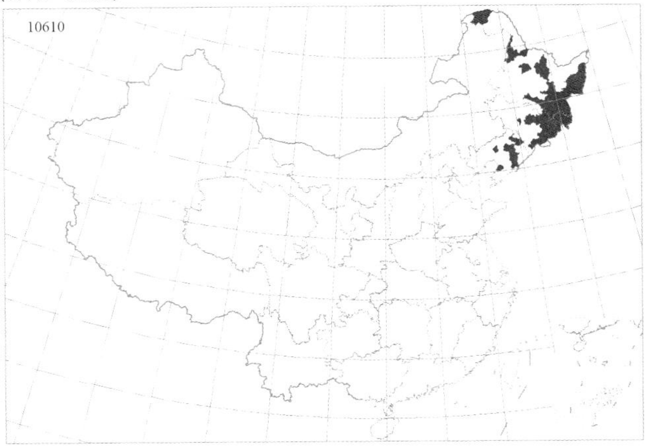

Caprifoliaceae 忍冬科 *Lonicera* 忍冬属

Lonicera ruprechtiana var. calvescens Rehder 光枝长白忍冬

Shrubs, deciduous

-0.2; -23.8; 19.8; 6.8; 50.9; 112.6; 531 ❙ 657; 415; 20 ❙ 23.8; 531; 316.0

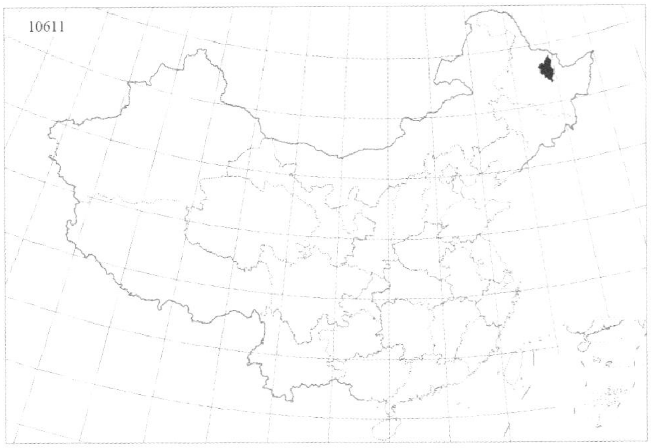

Caprifoliaceae 忍冬科 *Lonicera* 忍冬属

Lonicera semenovii Regel 藏西忍冬

Dwarf shrubs, deciduous

-0.1 (-4.3~5.7); -13.3 (-20.3~-7.4); 11.3 (5.4~19.5); 3.9 (1.6~8.2); 21.4 (0.4~63.7); 82.3 (53.3~121.1); 403 (300~576) ❙ 270 (71~620); 152 (24~306); 27 (1~103) ❙ -27.2 (-78.5~91.4); 236 (71~333); 58.1 (15.7~178.0)

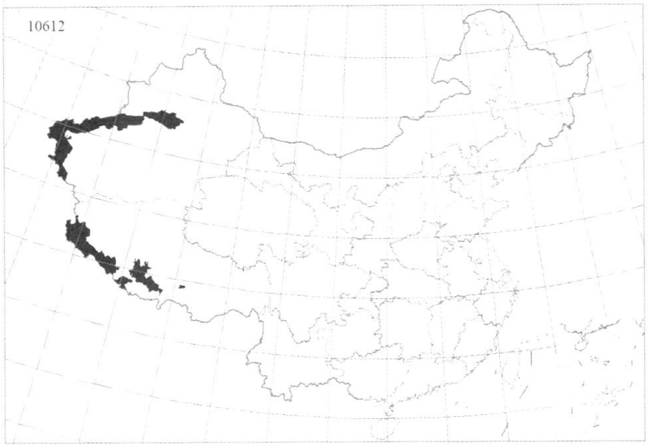

Caprifoliaceae 忍冬科 *Lonicera* 忍冬属

Lonicera serreana Hand.-Mazz. 毛药忍冬

Shrubs, deciduous

9.9 (-1.2~15.1); -4.2 (-12.7~1.8); 22.7 (8.5~28.0); 10.8 (2.6~15.2); 86.4 (8.1~128.0); 27.0 (5.6~82.3); 678 (347~880) ❙ 604 (252~951); 322 (144~436); 21 (6~53) ❙ -9.5 (-59.3~63.8); 579 (252~817); 186.7 (50.4~323.3)

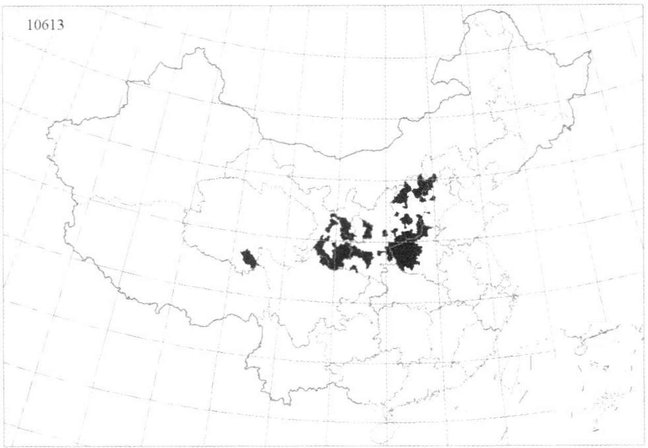

Caprifoliaceae 忍冬科 *Lonicera* 忍冬属

Lonicera setifera Franch. 齿叶忍冬

Shrubs or small trees, deciduous

8.6 (1.8~14.3); 0.4 (-7.1~7.6); 15.7 (9.9~22.7); 8.9 (3.8~14.3); 59.5 (15.5~111.9); 16.4 (0.0~53.3); 575 (404~748) ❙ 937 (337~1414); 545 (257~817); 25 (1~47) ❙ 63.8 (-23.5~125.9); 570 (337~748); 271.1 (169.5~372.3)

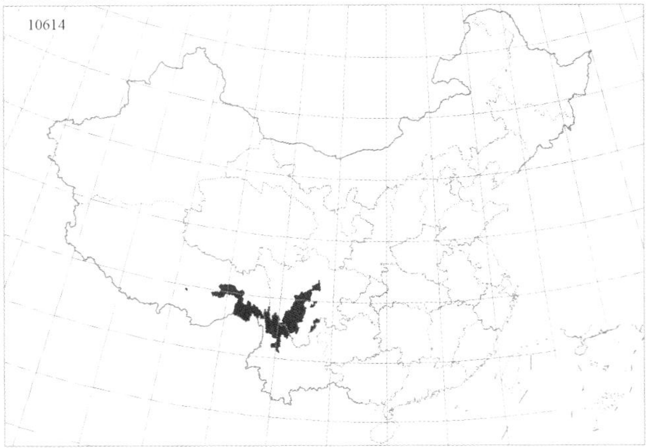

Caprifoliaceae 忍冬科 *Lonicera* 忍冬属
Lonicera similis Hemsl. 细毡毛忍冬
Vines, deciduous

14.7 (5.4~21.0); 4.7 (-3.5~14.2); 23.8 (12.3~29.1); 14.8 (5.8~21.0); 119.8 (31.5~191.8); 2.5 (0.0~28.1); 790 (481~1065) ▎1159 (632~1997); 545 (306~938); 67 (14~259) ▎47.1 (-14.2~147.0); 784 (481~1065); 232.8 (126.1~551.6)

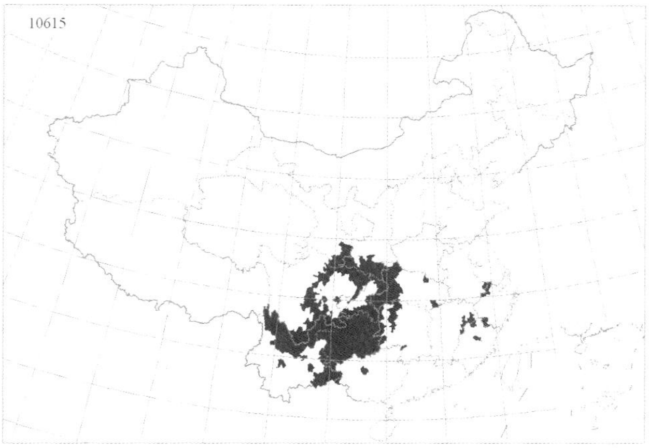

Caprifoliaceae 忍冬科 *Lonicera* 忍冬属
Lonicera similis var. omeiensis P. S. Hsu et H. J. Wang 峨眉忍冬
Vines, deciduous

16.0 (13.6~17.4); 5.4 (2.7~6.6); 25.9 (23.6~27.7); 16.0 (13.6~17.5); 132.8 (107.0~149.6); 0.5 (0.0~3.8); 853 (752~930) ▎1048 (899~1410); 536 (457~816); 38 (19~58) ▎23.5 (8.3~80.4); 853 (752~930); 169.9 (127.1~239.9)

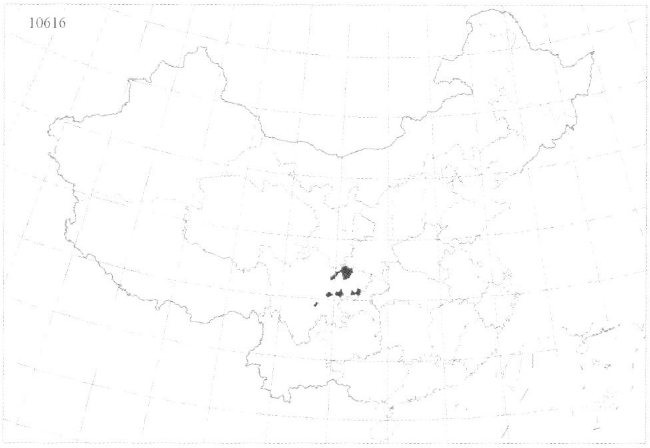

Caprifoliaceae 忍冬科 *Lonicera* 忍冬属
Lonicera spinosa Jacquem. ex Walp. 棘枝忍冬
Dwarf shrubs, deciduous

-1.4 (-5.7~3.3); -11.6 (-20.1~-6.6); 8.4 (6.1~12.1); 2.7 (1.7~5.0); 9.3 (1.9~27.3); 85.3 (46.3~133.0); 348 (294~438) ▎327 (71~620); 197 (24~324); 30 (0~103) ▎-6.4 (-78.5~91.4); 274 (71~417); 79.4 (15.7~254.4)

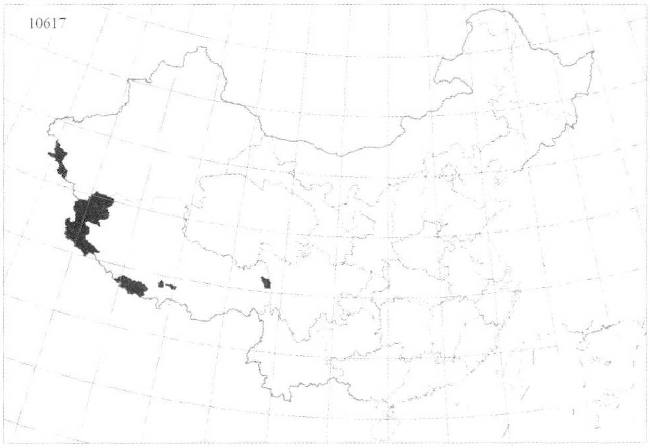

Caprifoliaceae 忍冬科 *Lonicera* 忍冬属
Lonicera stephanocarpa Franch. 冠果忍冬
Shrubs, deciduous

8.7 (2.6~12.5); -3.6 (-8.3~-0.1); 20.2 (11.2~25.6); 9.4 (4.3~12.7); 70.6 (20.0~106.5); 25.7 (11.7~49.3); 611 (427~760) ▎689 (341~1246); 326 (190~525); 20 (7~46) ▎14.2 (-40.9~94.4); 581 (341~702); 235.8 (61.2~323.3)

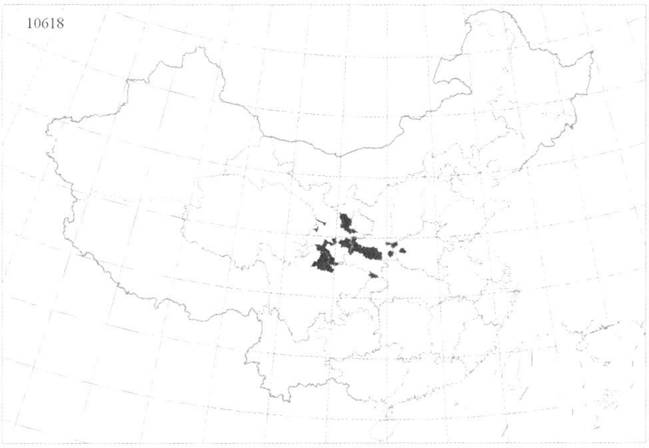

Caprifoliaceae 忍冬科 *Lonicera* 忍冬属
Lonicera subaequalis Rehder 川黔忍冬
Vines

13.6 (10.0~15.9); 4.1 (1.8~6.2); 22.1 (16.9~25.6); 13.7 (10.0~16.0); 105.6 (66.5~131.7); 1.8 (0.0~6.9); 734 (613~833) ▎1136 (964~1414); 589 (488~817); 51 (27~69) ▎55.6 (19.4~89.0); 734 (613~833); 214.0 (162.8~292.6)

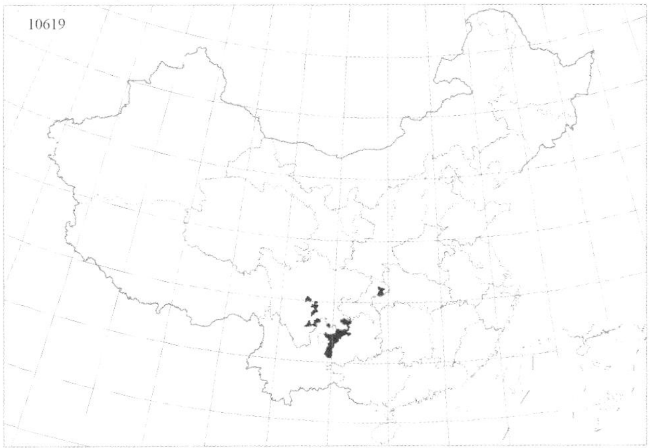

Caprifoliaceae 忍冬科 *Lonicera* 忍冬属
Lonicera subhispida Nakai 单花忍冬
Shrubs, deciduous

5.2 (2.5~6.5); -14.2 (-18.8~-12.0); 21.8 (20.0~22.8); 8.6 (7.3~9.2); 68.1 (54.1~75.2); 64.7 (55.1~84.3); 595 (542~624) ▎948 (853~1073); 600 (527~691); 35 (30~40) ▎59.2 (38.7~74.1); 595 (542~624); 322.9 (287.1~346.5)

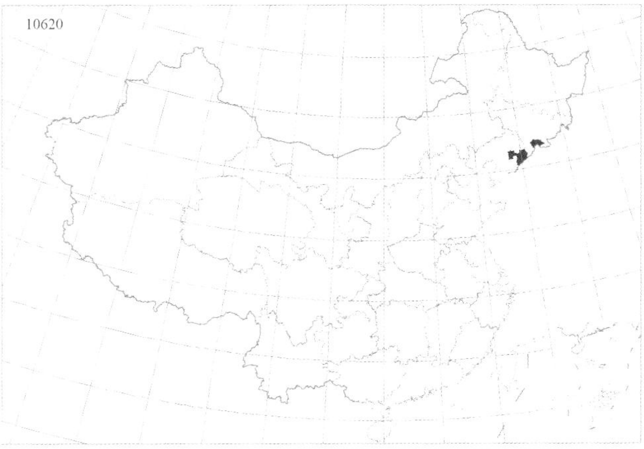

Caprifoliaceae 忍冬科 *Lonicera* 忍冬属
Lonicera sublabiata P. S. Hsu et H. J. Wang 唇花忍冬
Shrubs, deciduous

6.9 (−1.7~16.1); −3.4 (−12.3~3.9); 16.3 (8.0~27.8); 8.2 (2.4~16.1); 57.8 (6.4~134.4); 34.9 (1.1~86.1); 571 (344~878) ❙ 1008 (502~1613); 461 (301~537); 76 (10~209) ❙ 76.9 (45.7~121.7); 571 (344~878); 251.4 (208.8~275.0)

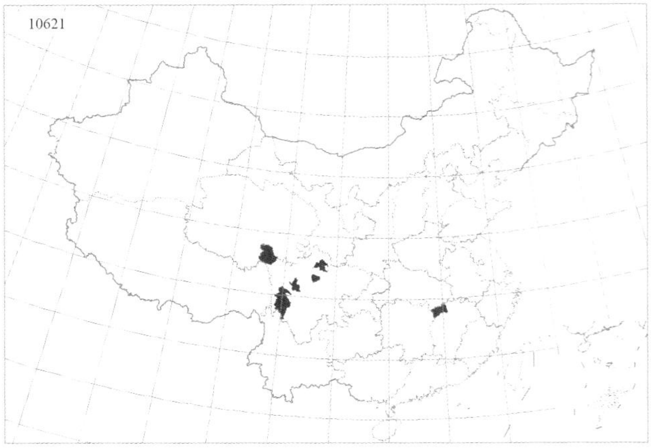

Caprifoliaceae 忍冬科 *Lonicera* 忍冬属
Lonicera tangutica Maxim. 唐古特忍冬
Shrubs, deciduous

9.5 (−4.1~18.9); −2.6 (−17.3~12.0); 20.4 (6.2~28.8); 10.4 (2.0~19.0); 80.3 (2.5~167.8); 25.6 (0.0~114.2); 649 (334~929) ❙ 757 (39~1571); 392 (29~817); 30 (1~157) ❙ 17.4 (−92.3~128.6); 589 (39~924); 208.5 (8.4~382.6)

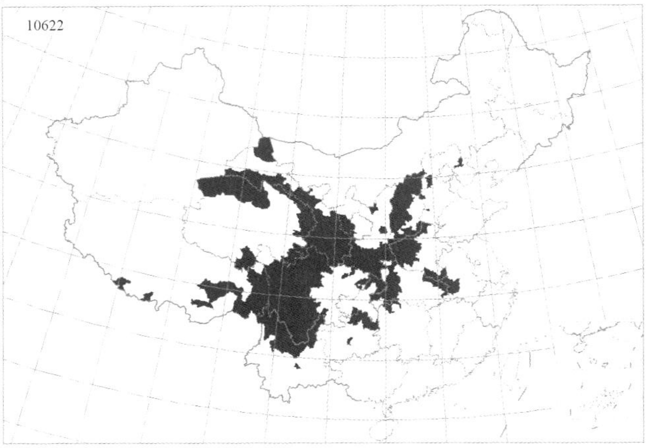

Caprifoliaceae 忍冬科 *Lonicera* 忍冬属
Lonicera tatarica L. 新疆忍冬
Shrubs, deciduous

2.6 (−1.3~9.0); −16.2 (−20.9~−9.3); 18.3 (13.2~23.3); 7.1 (4.5~10.7); 52.2 (26.7~88.9); 80.7 (40.7~104.2); 545 (433~703) ❙ 264 (152~450); 114 (56~211); 27 (12~51) ❙ −50.0 (−76.8~3.9); 262 (152~406); 146.5 (27.7~297.3)

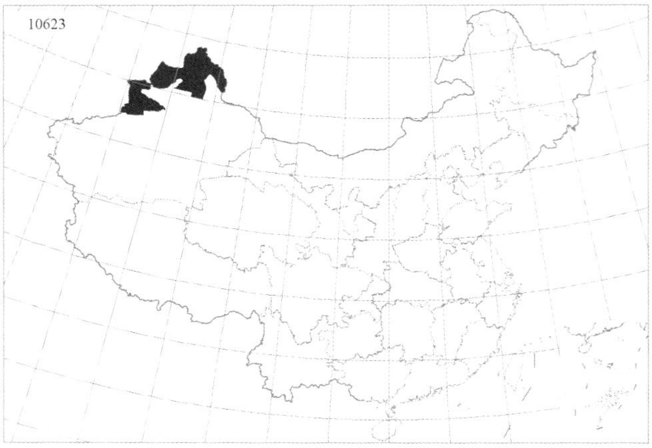

Caprifoliaceae 忍冬科 *Lonicera* 忍冬属
Lonicera tatarica var. micrantha Trautv. 小花忍冬
Shrubs, deciduous

2.3 (−1.3~6.1); −16.4 (−20.9~−11.6); 17.3 (13.2~20.4); 6.7 (4.5~8.7); 47.8 (26.7~68.8); 79.7 (53.6~103.9); 526 (433~608) ❙ 322 (225~450); 143 (102~211); 30 (12~51) ❙ −37.0 (−56.7~3.9); 319 (225~406); 189.8 (97.7~297.3)

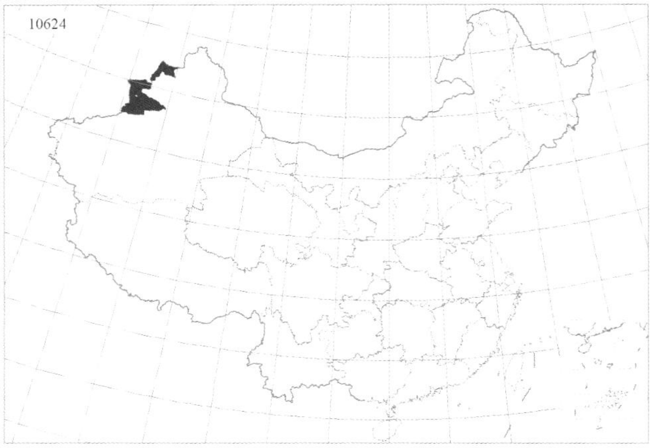

Caprifoliaceae 忍冬科 *Lonicera* 忍冬属
Lonicera tatarinowii Maxim. 华北忍冬
Shrubs, deciduous

8.9 (2.2~14.5); −7.0 (−17.4~0.1); 23.0 (17.5~28.0); 10.5 (6.6~14.6); 84.9 (45.4~127.2); 37.2 (10.7~83.8); 671 (507~880) ❙ 571 (397~1005); 349 (254~635); 19 (8~65) ❙ −13.7 (−47.7~68.8); 549 (397~729); 181.8 (70.3~337.1)

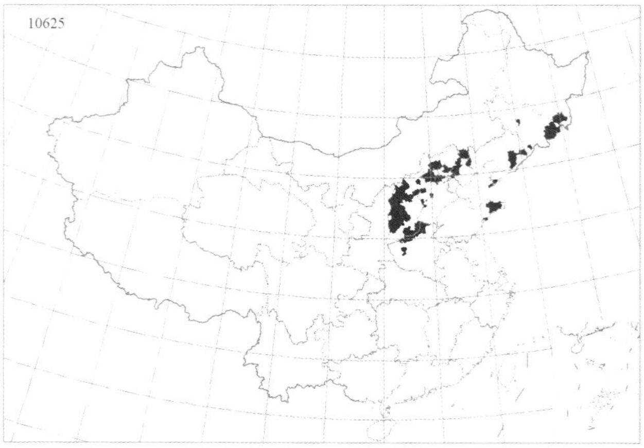

Caprifoliaceae 忍冬科 *Lonicera* 忍冬属
Lonicera tomentella Hook. f. et Thomson 毛冠忍冬
Shrubs, deciduous

5.7 (0.3~12.2); −2.5 (−8.6~4.3); 12.8 (8.3~18.5); 6.6 (2.9~12.2); 39.1 (9.6~87.4); 30.9 (0.7~65.4); 499 (368~672) ❙ 881 (389~1431); 511 (246~817); 26 (11~45) ❙ 67.9 (0.0~113.0); 499 (368~672); 252.4 (142.5~363.7)

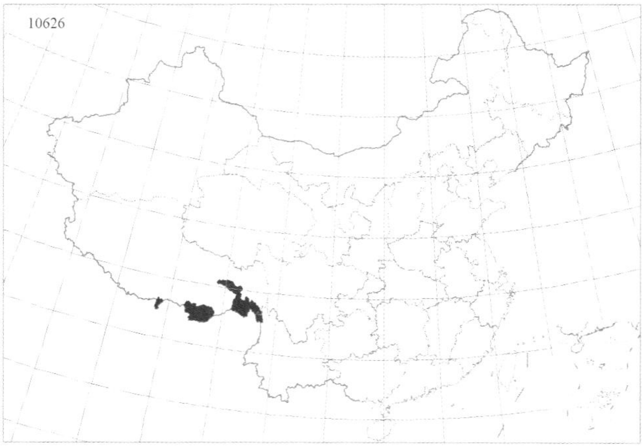

Caprifoliaceae 忍冬科 *Lonicera* 忍冬属
Lonicera tomentella var. tsarongensis W. W. Sm. 察瓦龙忍冬
Shrubs, deciduous

5.3; −2.7; 12.3; 6.0; 32.9; 29.4; 486 ❙ 862; 497; 29 ❙ 75.5; 486; 252.0

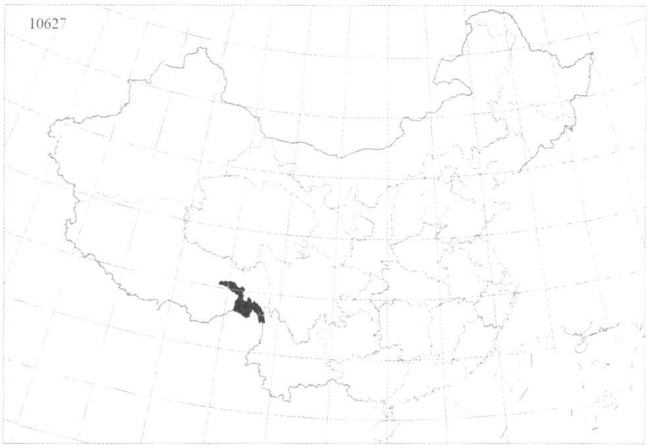

Caprifoliaceae 忍冬科 *Lonicera* 忍冬属
Lonicera tragophylla Hemsl. 盘叶忍冬
Vines, deciduous

12.6 (1.5~18.2); −0.4 (−9.7~8.6); 24.6 (9.9~29.1); 12.9 (3.8~18.2); 105.2 (15.5~158.8); 13.5 (0.0~58.4); 759 (413~984) ❙ 845 (315~1922); 403 (174~817); 50 (6~259) ❙ 12.8 (−47.5~129.9); 677 (315~961); 202.4 (44.5~344.4)

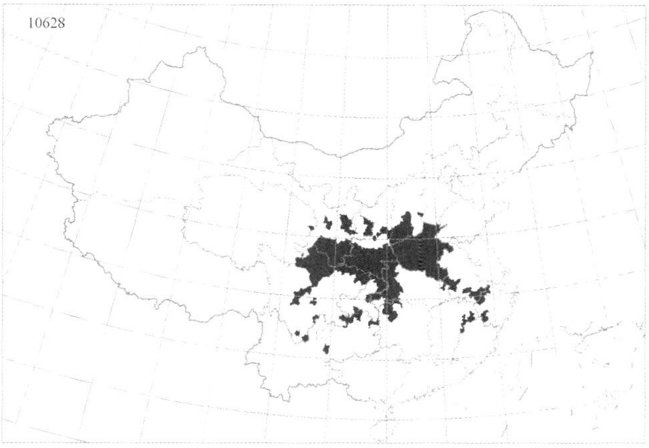

Caprifoliaceae 忍冬科 *Lonicera* 忍冬属
Lonicera tricalysioides C. Y. Wu 赤水忍冬
Shrubs, evergreen or semievergreen

15.9; 5.5; 25.6; 15.9; 131.1; 0.0; 833 ❙ 1089; 503; 69 ❙ 30.9; 833; 181.2

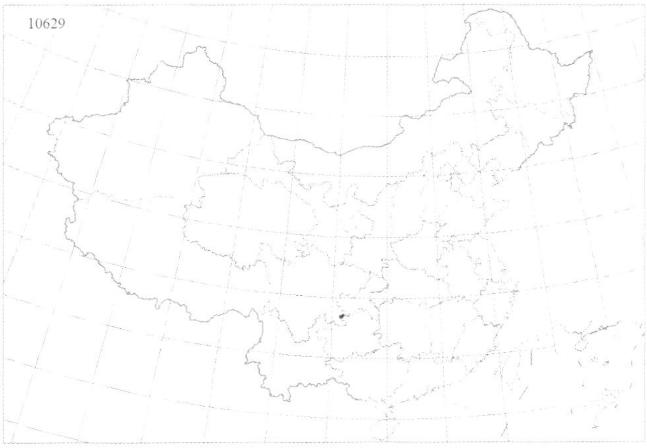

Caprifoliaceae 忍冬科 *Lonicera* 忍冬属
Lonicera trichogyne Rehder 毛果忍冬
Shrubs, deciduous

11.1 (2.9~15.0); 0.2 (−6.3~2.7); 21.4 (11.2~26.9); 11.3 (4.4~15.0); 86.8 (20.5~123.9); 12.8 (3.2~44.6); 682 (428~829) ❙ 1191 (677~1689); 512 (336~617); 96 (12~212) ❙ 74.0 (−0.2~121.7); 682 (428~829); 273.5 (200.8~319.9)

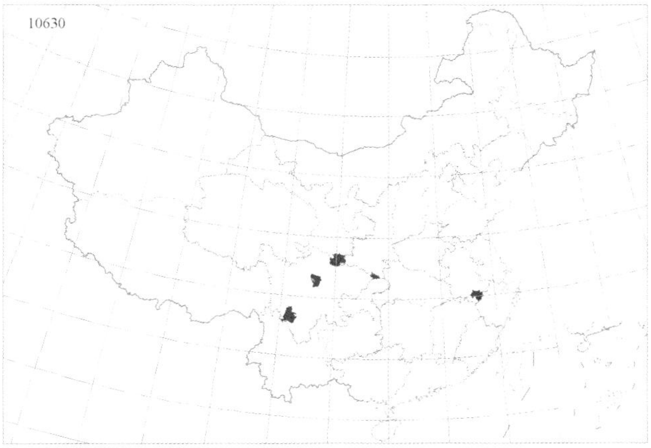

Caprifoliaceae 忍冬科 *Lonicera* 忍冬属
Lonicera trichosantha Bureau et Franch. 毛花忍冬
Dwarf shrubs, deciduous

4.0 (−2.2~17.1); −6.0 (−12.8~9.0); 12.9 (7.6~29.3); 5.7 (2.2~17.1); 33.3 (5.5~145.7); 45.1 (0.0~91.3); 472 (336~941) ❙ 710 (300~1660); 389 (201~619); 21 (3~216) ❙ 51.1 (−37.6~128.6); 467 (300~941); 255.2 (88.1~357.7)

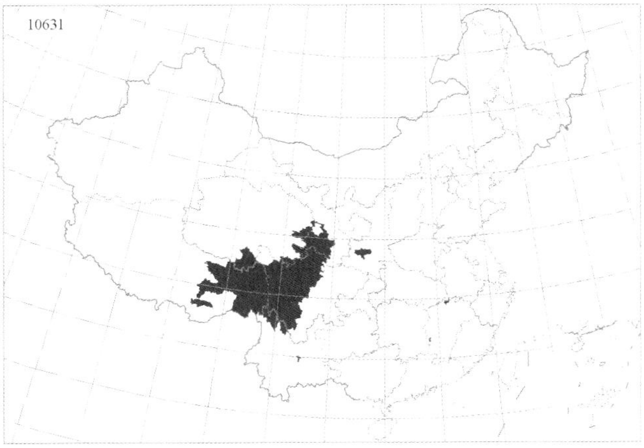

Caprifoliaceae 忍冬科 *Lonicera* 忍冬属
Lonicera trichosantha var. deflexicalyx (Diels) P. S. Hsu et H. J. Wang 长叶毛花忍冬
Shrubs, deciduous

5.9 (1.4~14.3); −3.3 (−9.2~7.6); 14.0 (9.2~22.7); 6.9 (3.5~14.3); 42.4 (13.4~111.9); 30.8 (0.0~57.8); 511 (402~748) ❙ 843 (528~1414); 471 (270~817); 20 (10~47) ❙ 67.1 (0.0~125.9); 510 (402~748); 270.5 (192.9~341.6)

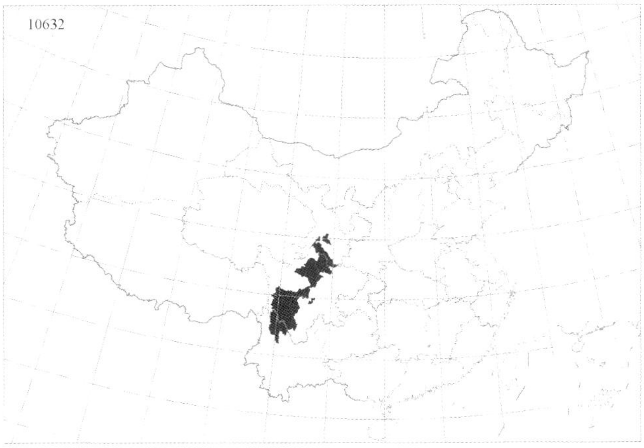

Caprifoliaceae 忍冬科 *Lonicera* 忍冬属
Lonicera trichosepala (Rehder) P. S. Hsu 毛萼忍冬
Vines

16.2 (14.4~17.7); 4.1 (2.1~6.5); 27.8 (25.7~29.3); 16.2 (14.5~17.7); 135.9 (118.7~152.4); 1.3 (0.0~4.9); 883 (801~964) ❙ 1523 (1211~1813); 540 (427~685); 191 (122~242) ❙ 73.1 (28.8~123.6); 877 (801~961); 232.8 (166.9~299.5)

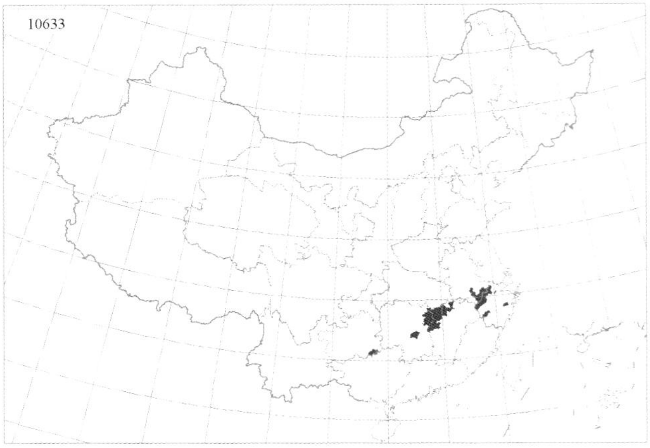

Caprifoliaceae 忍冬科 *Lonicera* 忍冬属
Lonicera tubuliflora Rehder 管花忍冬
Shrubs, deciduous

2.0; −7.2; 10.1; 3.8; 15.4; 51.4; 413 ❙ 943; 507; 23 ❙ 128.6; 413; 270.7

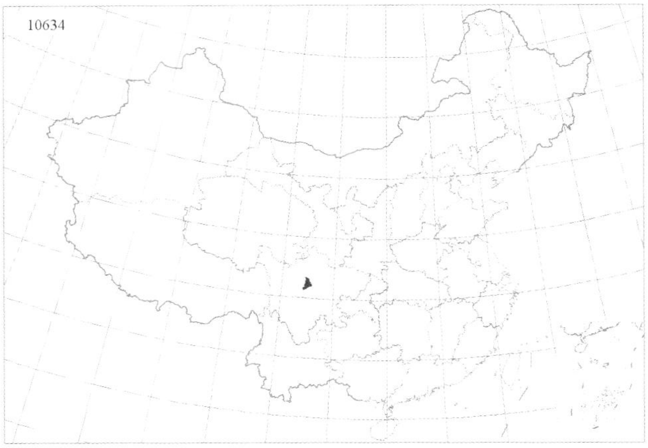

Caprifoliaceae 忍冬科 *Lonicera* 忍冬属
Lonicera virgultorum W. W. Sm. 绢柳林忍冬
Shrubs, deciduous

17.2 (14.4~19.6); 10.4 (7.5~12.8); 21.4 (19.1~23.5); 17.2 (14.4~19.6); 146.2 (113.1~175.7); 0.0 (0.0~0.0); 833 (727~953) ❙ 1486 (1083~1688); 854 (600~959); 54 (42~70) ❙ 79.0 (38.0~112.2); 828 (727~932); 495.6 (395.4~627.0)

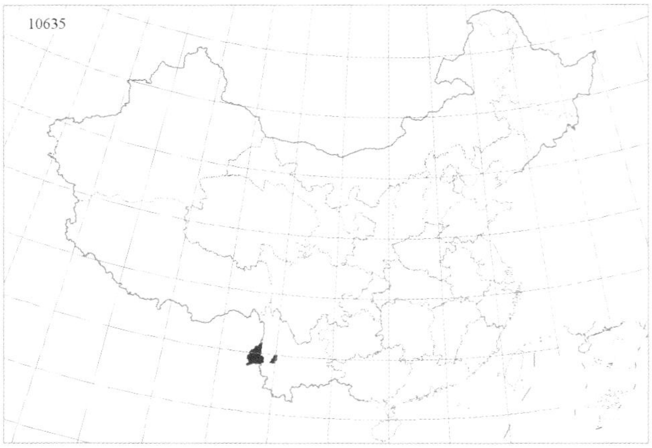

Caprifoliaceae 忍冬科 *Lonicera* 忍冬属
Lonicera webbiana Wall. ex DC. 华西忍冬
Shrubs, deciduous

7.0 (−4.1~17.8); −4.6 (−17.3~7.6); 17.6 (6.2~28.7); 8.3 (2.0~17.9); 59.5 (2.5~154.6); 34.5 (0.0~114.2); 571 (334~961) ❙ 668 (39~1820); 351 (29~816); 22 (1~248) ❙ 20.8 (−92.3~128.6); 523 (39~961); 219.1 (10.8~364.1)

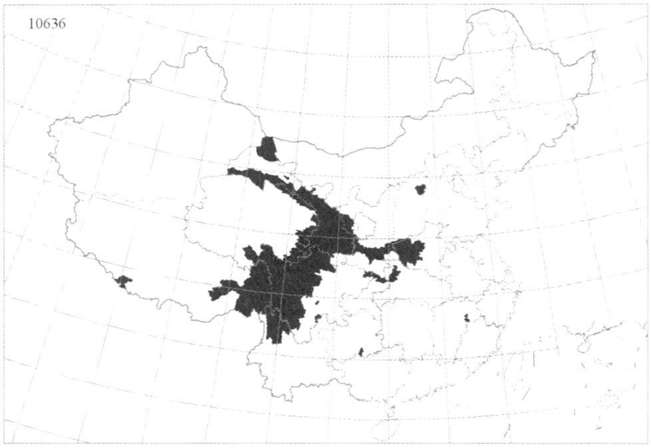

Caprifoliaceae 忍冬科 *Lonicera* 忍冬属
Lonicera webbiana var. mupinensis (Rehder) P. S. Hsu et H. J. Wang 川西忍冬
Shrubs, deciduous

10.9 (6.6~14.9); 1.9 (−2.0~5.3); 18.9 (13.3~23.6); 11.0 (6.8~15.0); 79.6 (39.6~119.6); 8.4 (0.0~20.4); 647 (514~782) ❙ 1126 (862~1414); 643 (531~817); 32 (12~49) ❙ 73.6 (51.0~89.0); 647 (514~782); 252.9 (214.5~311.5)

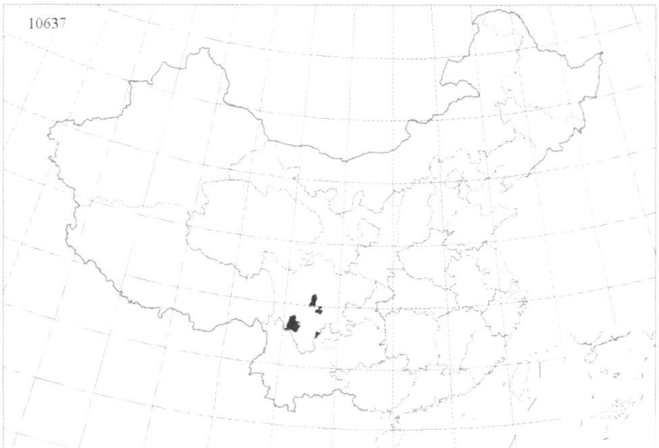

Caprifoliaceae 忍冬科 *Lonicera* 忍冬属
Lonicera yunnanensis Franch. 云南忍冬
Vines

13.5 (3.8~18.9); 5.8 (−4.3~12.0); 19.8 (10.8~26.6); 13.6 (4.7~19.0); 104.9 (21.6~167.8); 2.7 (0.0~35.4); 718 (444~929) ❙ 960 (741~1178); 557 (411~646); 31 (15~57) ❙ 36.1 (−14.2~121.7); 692 (444~912); 269.9 (142.9~383.4)

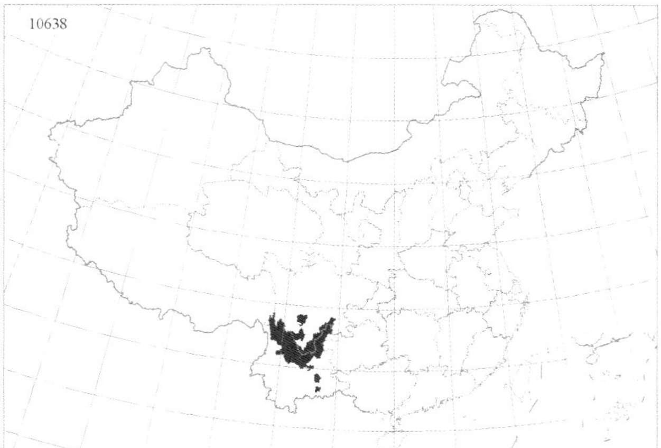

Caprifoliaceae 忍冬科 *Sambucus* 接骨木属
Sambucus williamsii Hance 接骨木
Shrubs or small trees, deciduous

13.4 (−5.8~22.7); 0.0 (−29.8~15.0); 25.6 (11.2~29.9); 14.0 (4.3~22.8); 116.6 (20.0~213.1); 14.7 (0.0~157.4); 803 (426~1221) ❙ 955 (372~2435); 451 (201~1428); 73 (5~259) ❙ 17.6 (−51.1~130.7); 716 (372~1221); 196.2 (46.6~551.6)

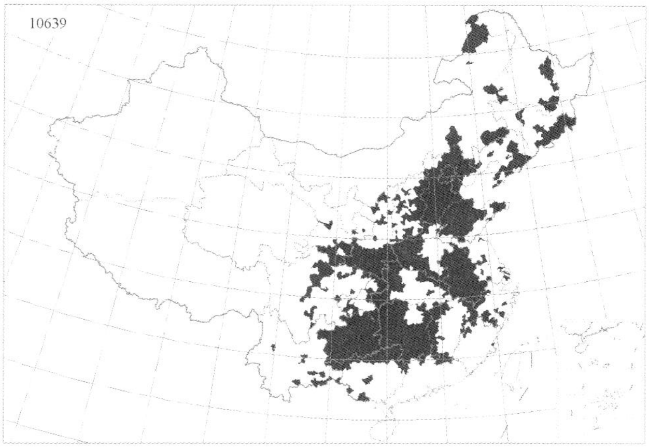

Caprifoliaceae 忍冬科 *Symphoricarpos* 毛核木属
Symphoricarpos sinensis Rehder 毛核木
Shrubs, deciduous

13.2 (5.4~18.9); 4.9 (−2.1~12.0); 20.4 (12.3~26.6); 13.3 (5.8~19.0); 102.3 (31.5~167.8); 3.5 (0.0~25.6); 714 (481~929) ❙ 980 (711~1335); 539 (357~651); 37 (15~86) ❙ 39.1 (−14.2~101.4); 691 (481~912); 267.3 (142.9~351.7)

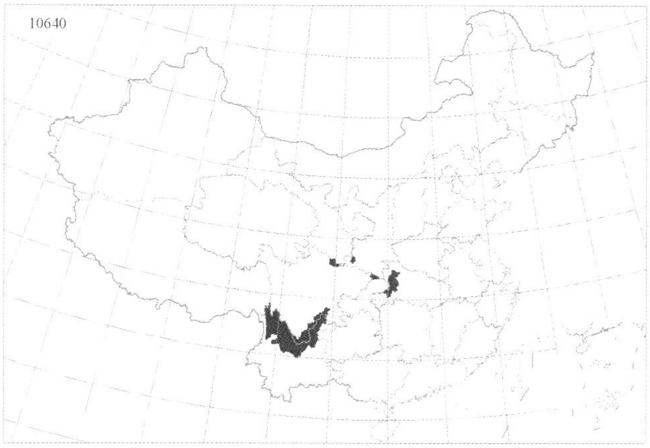

Linnaeaceae 北极花科 *Abelia* 糯米条属
Abelia chinensis R. Br. 糯米条
Shrubs, deciduous

17.0 (11.1~22.3); 5.8 (0.6~17.7); 27.4 (18.8~29.9); 17.1 (11.3~22.4); 145.2 (80.7~208.3); 0.5 (0.0~9.7); 911 (652~1202) ❙ 1518 (863~3495); 540 (299~1847); 178 (25~897) ❙ 66.8 (−6.7~235.6); 896 (652~1202); 212.5 (131.2~551.6)

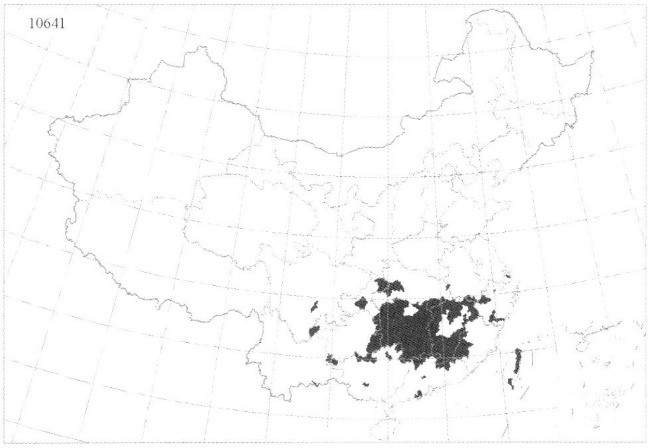

Linnaeaceae 北极花科 *Abelia* 糯米条属
Abelia forrestii (Diels) W. W. Sm. 细瘦六道木
Shrubs, deciduous

11.5 (6.1~17.4); 3.3 (−1.5~7.4); 18.5 (12.7~25.7); 11.6 (6.3~17.4); 84.7 (35.2~148.9); 6.4 (0.0~21.8); 668 (496~886) ❙ 1149 (801~1448); 614 (496~727); 57 (12~107) ❙ 71.8 (52.8~93.0); 668 (496~886); 310.6 (223.0~407.1)

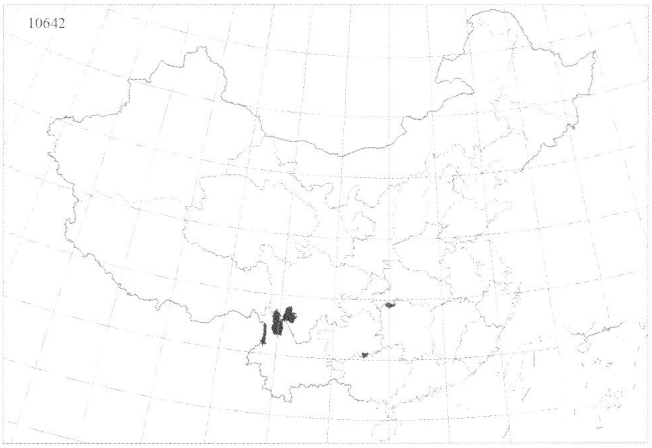

Linnaeaceae 北极花科 *Abelia* 糯米条属
Abelia serrata Siebold et Zuccarini 黄花六道木
Shrubs, deciduous

16.1; 5.4; 26.8; 16.1; 133.4; 0.0; 850 ❙ 1743; 654; 191 ❙ 105.0; 850; 235.1

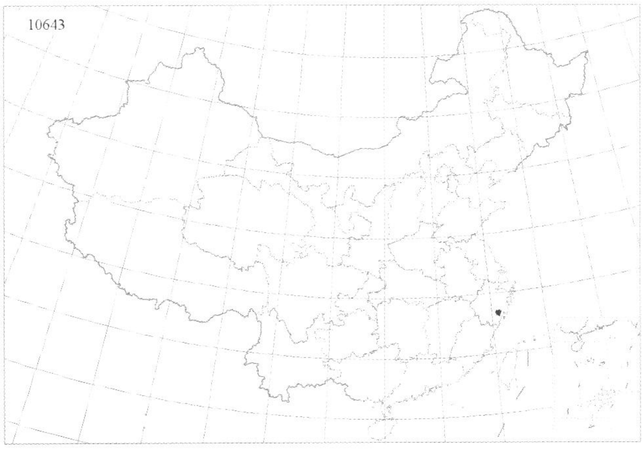

Linnaeaceae 北极花科 *Abelia* 糯米条属
Abelia spathulata Siebold et Zuccarini 温州六道木
Shrubs, deciduous

18.2; 7.8; 28.7; 18.3; 159.6; 0.0; 965 ❙ 1706; 645; 182 ❙ 76.7; 965; 189.8

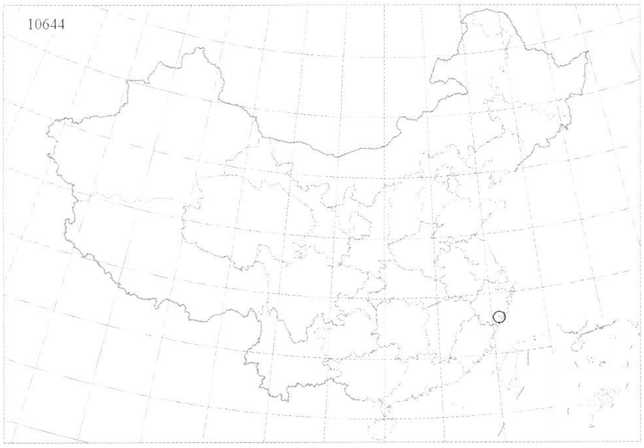

Linnaeaceae 北极花科 *Abelia* 糯米条属

Abelia uniflora R. Br. 二翅六道木

Shrubs, deciduous

13.9 (2.0~19.1); 3.5 (-7.2~12.0); 23.4 (9.9~28.9); 14.0 (3.8~19.2); 112.4 (15.4~170.1); 5.3 (0.0~51.4); 765 (413~1002) ▎1078 (626~1997); 509 (290~817); 58 (12~254) ▎41.8 (-22.2~130.7); 754 (413~1002); 227.9 (108.2~351.7)

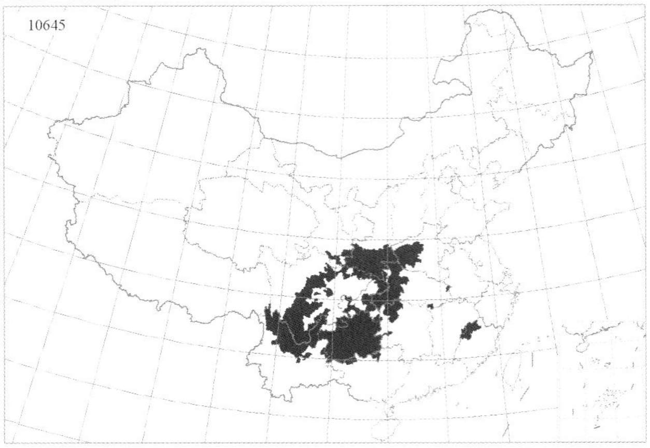

Linnaeaceae 北极花科 *Dipelta* 双盾木属

Dipelta elegans Batalin 优美双盾木

Shrubs, deciduous

6.2 (1.9~10.3); -4.4 (-9.2~-0.1); 16.0 (11.7~19.8); 7.2 (4.3~10.3); 47.5 (21.1~75.8); 32.1 (11.7~57.8); 527 (427~634) ▎665 (620~731); 324 (301~357); 15 (13~19) ▎28.1 (0.0~63.8); 527 (427~634); 264.0 (196.0~289.4)

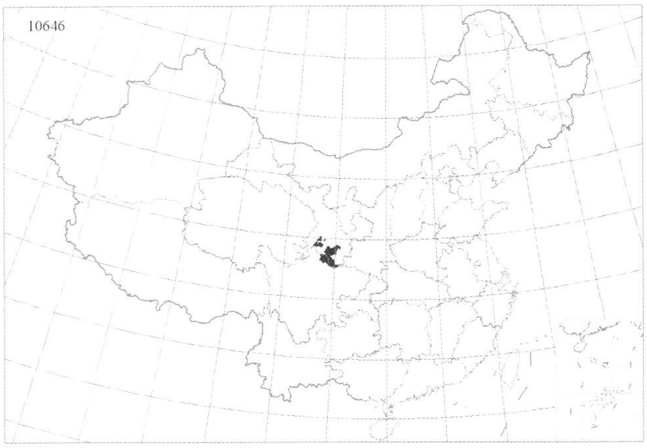

Linnaeaceae 北极花科 *Dipelta* 双盾木属

Dipelta floribunda Maxim. 双盾木

Shrubs or small trees, deciduous

12.2 (3.5~16.9); 0.7 (-6.5~5.3); 23.1 (12.6~28.1); 12.3 (5.0~16.9); 97.2 (26.3~143.4); 10.4 (0.0~43.4); 715 (446~924) ▎953 (620~1466); 425 (301~606); 43 (13~118) ▎33.1 (-8.1~106.2); 714 (446~924); 248.8 (164.0~334.4)

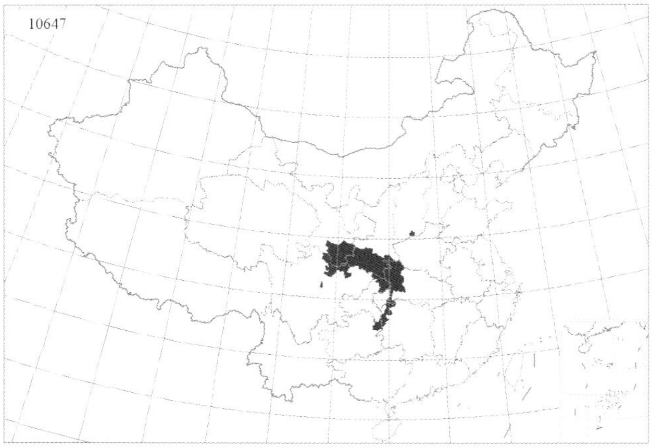

Linnaeaceae 北极花科 *Dipelta* 双盾木属

Dipelta yunnanensis Franch. 云南双盾木

Shrubs, deciduous

12.5 (2.2~21.0); 3.9 (-7.0~14.2); 20.0 (9.5~26.6); 12.7 (3.8~21.0); 97.1 (15.5~191.8); 6.2 (0.0~49.1); 701 (411~1053) ▎1033 (632~1522); 558 (306~818); 38 (11~105) ▎50.5 (-14.2~125.9); 689 (411~1053); 261.3 (142.9~551.6)

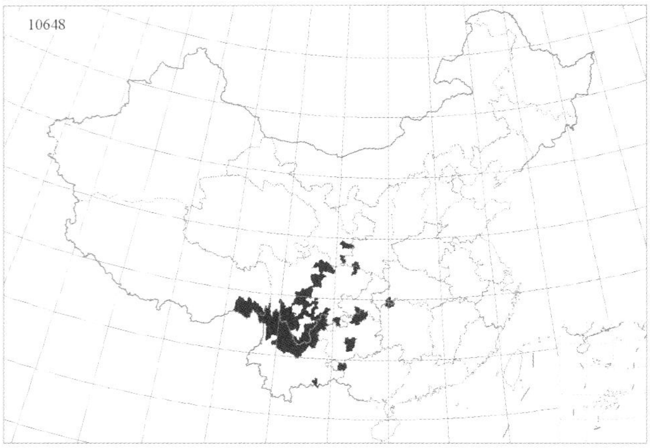

Linnaeaceae 北极花科 *Kolkwitzia* 蝟实属

Kolkwitzia amabilis Graebn. 蝟实

Shrubs, deciduous

12.3 (0.0~16.2); -1.5 (-11.4~3.5); 24.9 (9.7~28.3); 12.6 (3.3~16.2); 104.1 (12.2~136.1); 16.1 (1.5~72.2); 752 (395~898) ▎740 (483~1689); 363 (258~617); 41 (11~212) ▎-1.2 (-36.5~105.7); 658 (395~898); 190.1 (120.1~364.1)

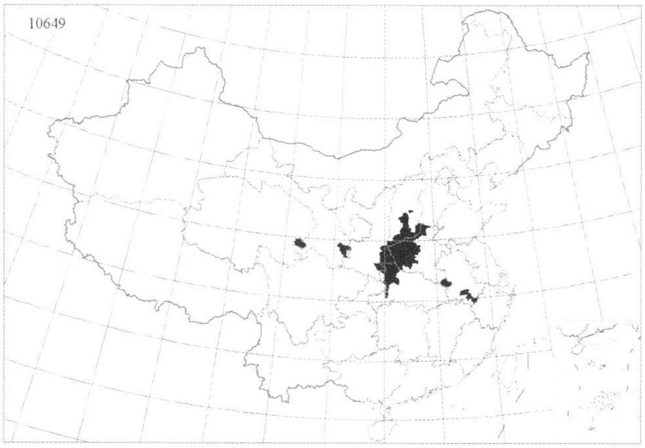

Linnaeaceae 北极花科 *Linnaea* 北极花属

Linnaea borealis L. 北极花

Small shrubs, evergreen

0.3 (-5.8~5.2); -20.8 (-29.8~-14.1); 18.8 (15.3~21.8); 6.6 (4.4~8.6); 48.4 (28.0~68.0); 103.9 (65.0~157.4); 518 (426~595) ▎514 (161~1005); 319 (74~635); 19 (12~39) ▎-2.2 (-68.1~68.8); 455 (161~595); 267.2 (74.6~323.4)

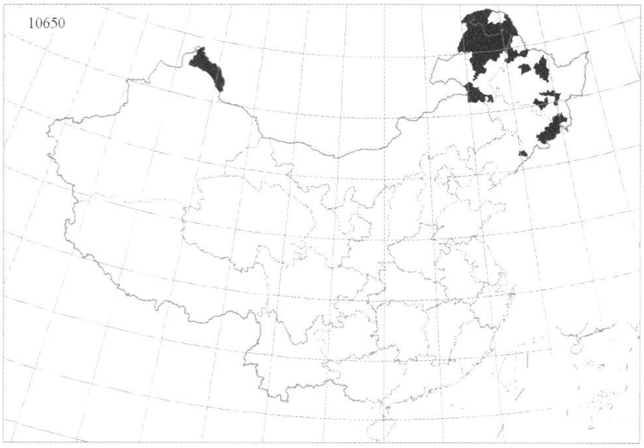

Linnaeaceae 北极花科 *Zabelia* 六道木属
Zabelia biflora (Turczaninow) Makino 六道木
Shrubs, deciduous

11.1 (1.6~18.9); -2.2 (-20.1~12.0); 23.1 (9.5~30.1); 11.8 (3.8~19.0); 95.1 (15.4~167.8); 21.8 (0.0~91.9); 711 (404~1024) ❙ 816 (341~1997); 412 (190~816); 46 (6~254) ❙ 14.5 (-47.7~147.0); 639 (341~941); 207.9 (61.2~372.3)

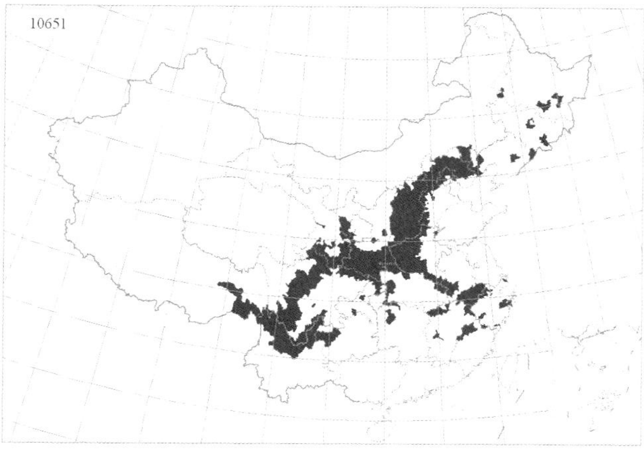

Linnaeaceae 北极花科 *Zabelia* 六道木属
Zabelia triflora (R. Br. ex. Wallich) Makino 醉鱼草状六道木
Shrubs, deciduous

5.8 (2.9~10.8); -2.1 (-5.7~3.9); 12.7 (10.5~16.6); 6.3 (4.3~10.9); 35.6 (19.9~71.8); 26.1 (1.3~45.2); 493 (421~633) ❙ 770 (473~1041); 464 (287~594); 21 (7~45) ❙ 54.1 (5.0~100.7); 493 (421~633); 267.8 (201.9~352.3)

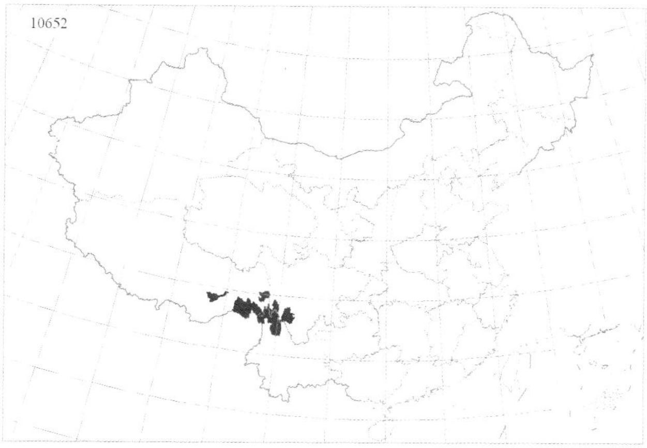

Cucurbitaceae 葫芦科 *Hodgsonia* 油渣果属
Hodgsonia heteroclita (Roxb.) Hook. f. et Thomson 油渣果
Vines, evergreen

19.1 (8.6~24.2); 11.9 (1.3~18.1); 24.1 (15.0~28.9); 19.1 (8.7~24.3); 169.3 (52.6~231.2); 0.1 (0.0~8.8); 959 (574~1335) ❙ 1413 (958~1900); 755 (447~1056); 74 (41~173) ❙ 49.6 (-12.0~145.8); 948 (574~1202); 380.1 (201.4~659.0)

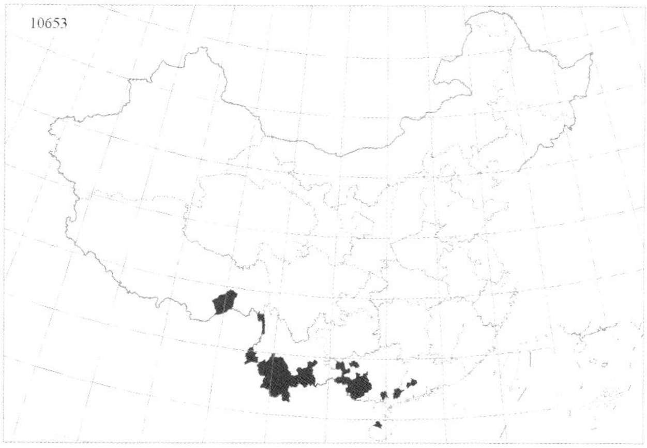

Cucurbitaceae 葫芦科 *Indofevillea* 藏瓜属
Indofevillea khasiana Chatterjee 藏瓜
Vines

9.3; 1.5; 15.9; 9.6; 65.5; 13.1; 592 ❙ 1348; 783; 41 ❙ 123.3; 592; 340.8

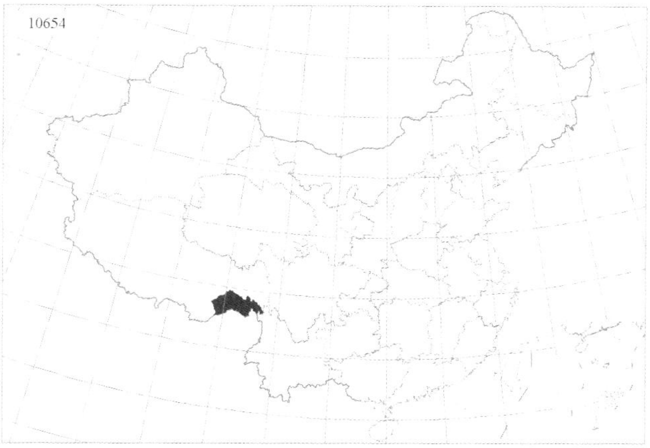

Cucurbitaceae 葫芦科 *Zanonia* 翅子瓜属
Zanonia indica L. 翅子瓜
Vines

19.9; 12.8; 24.9; 20.0; 179.4; 0.0; 986 ❙ 1477; 785; 66 ❙ 50.0; 977; 451.8

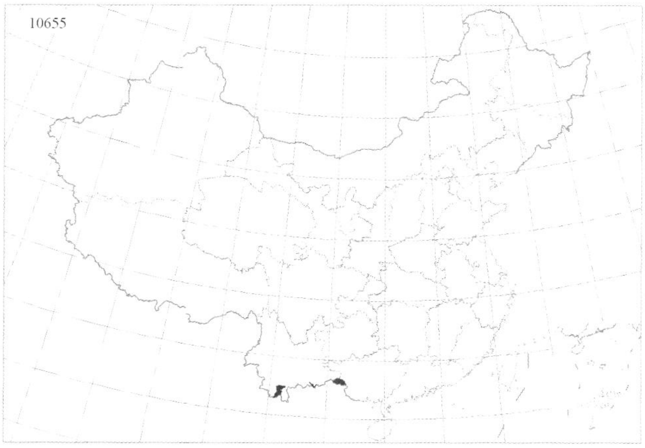

Cucurbitaceae 葫芦科 *Zanonia* 翅子瓜属
Zanonia indica var. pubescens Cogn. 滇南翅子瓜
Vines

20.4; 15.1; 23.6; 20.5; 185.8; 0.0; 989 ❙ 1607; 702; 77 ❙ 62.5; 986; 659.0

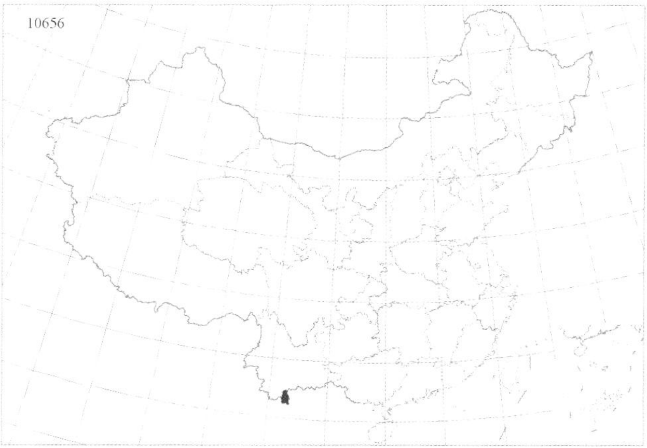

Goodeniaceae 草海桐科 *Scaevola* 草海桐属

Scaevola hainanensis Hance 小草海桐

Small shrubs

22.6 (18.7~24.6); 15.7 (10.5~19.5); 28.1 (25.0~29.0); 22.7 (18.8~24.7); 212.2 (165.4~235.9); 0.0 (0.0~0.0); 1212 (950~1372) ❙ 1719 (1176~2445); 829 (425~1428); 107 (44~185) ❙ 43.4 (-12.0~140.0); 1171 (950~1329); 261.5 (119.4~548.0)

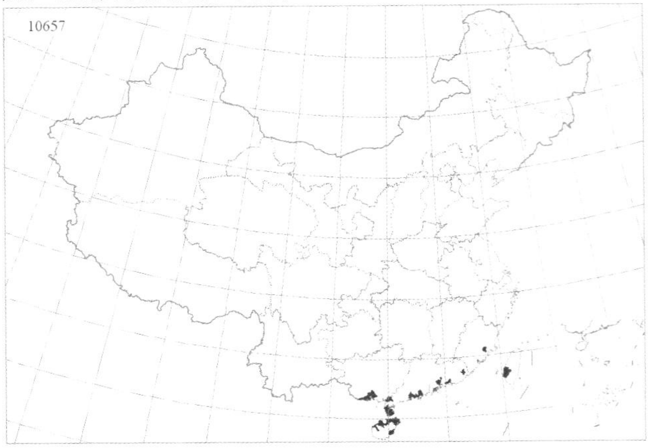

Goodeniaceae 草海桐科 *Scaevola* 草海桐属

Scaevola sericea Vahl 草海桐

Shrubs or small trees

22.1 (16.4~24.6); 15.4 (11.5~19.5); 27.5 (20.3~29.1); 22.2 (16.5~24.7); 206.2 (137.8~235.9); 0.0 (0.0~0.0); 1171 (783~1372) ❙ 1842 (868~3495); 794 (312~1847); 164 (42~897) ❙ 64.1 (-34.3~224.4); 1121 (783~1329); 270.1 (139.6~572.9)

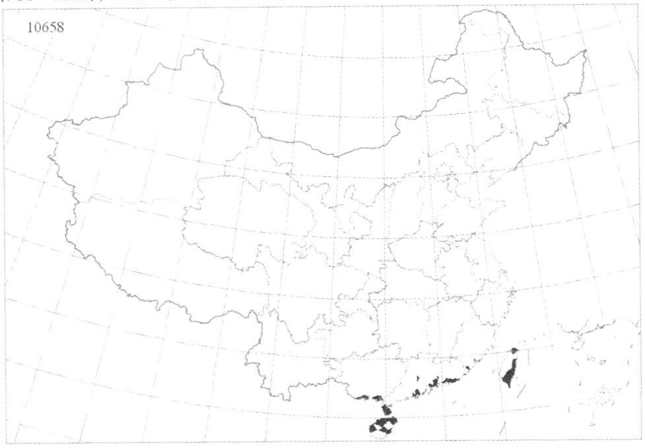

Asteraceae 菊科 *Ajania* 亚菊属

Ajania fruticulosa (Ledeb.) Poljakov 灌木亚菊

Subshrubs

5.0 (-5.7~12.5); -12.0 (-28.8~-5.1); 19.6 (7.0~27.5); 8.1 (1.7~13.2); 62.0 (3.5~114.0); 61.4 (23.1~149.1); 577 (294~822) ❙ 239 (34~657); 137 (16~415); 9 (1~98) ❙ -55.0 (-95.6~52.9); 233 (34~531); 83.8 (5.1~316.0)

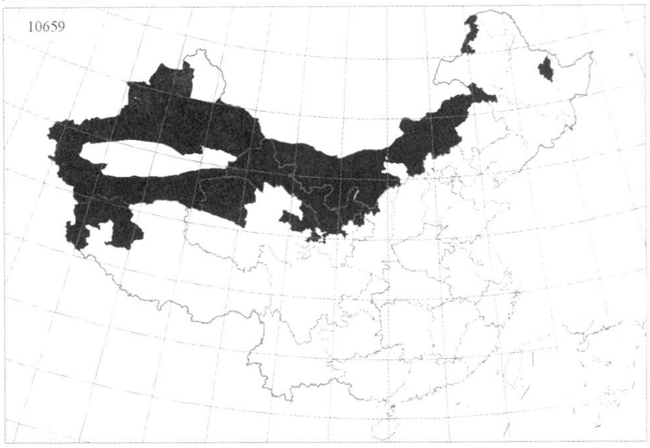

Asteraceae 菊科 *Ajania* 亚菊属

Ajania nematoloba (Hand.-Mazz.) Ling et C. Shih 丝裂亚菊

Subshrubs

4.0 (-4.6~7.9); -10.2 (-28.8~-6.5); 16.5 (11.0~24.3); 6.6 (3.5~10.2); 45.5 (17.2~85.9); 57.0 (35.7~149.1); 509 (375~679) ❙ 361 (49~516); 207 (34~300); 7 (1~15) ❙ -26.6 (-92.8~14.7); 353 (49~463); 137.4 (5.1~289.8)

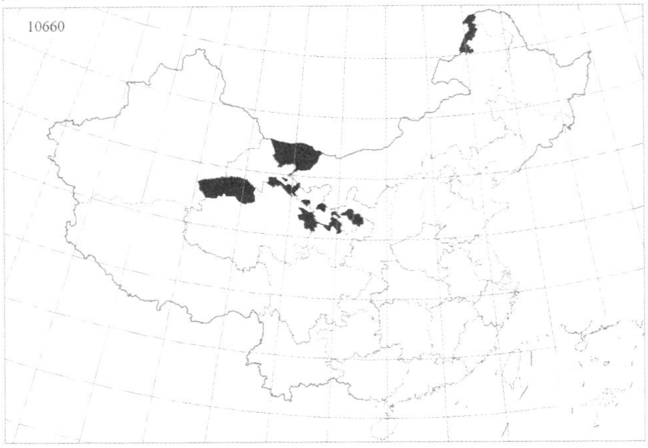

Asteraceae 菊科 *Ajania* 亚菊属

Ajania nitida C. Shih 光苞亚菊

Small shrubs

5.6; -2.4; 12.6; 5.9; 32.6; 26.2; 483 ❙ 685; 406; 23 ❙ 41.8; 483; 201.9

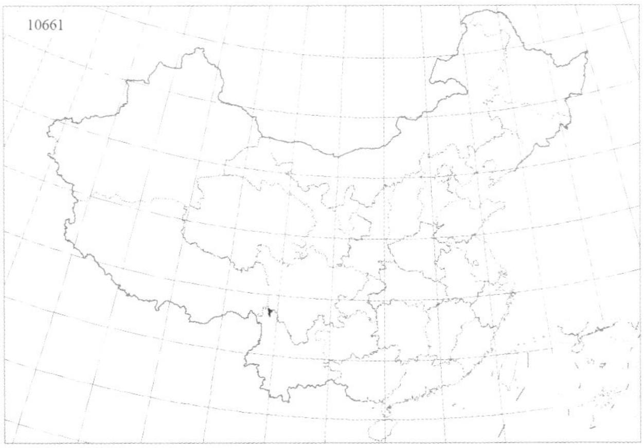

Asteraceae 菊科 *Ajania* 亚菊属

Ajania nubigena (Wall.) C. Shih 黄花亚菊

Subshrubs

-1.1; -9.1; 6.2; 2.0; 2.5; 75.6; 334 ❙ 321; 193; 33 ❙ -4.0; 321; 79.3

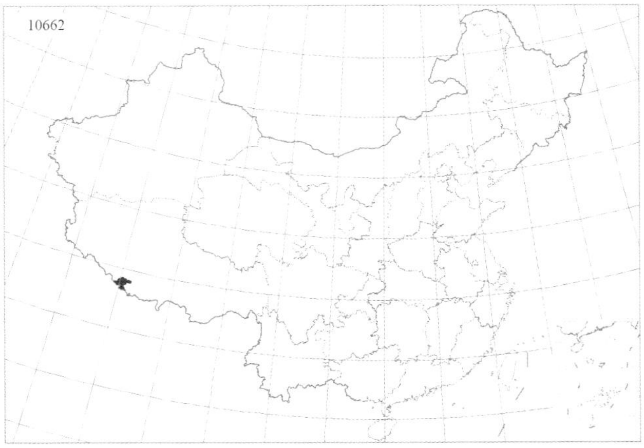

Asteraceae 菊科 *Ajania* 亚菊属
Ajania potaninii (Krasch.) Poljakov 川甘亚菊
Small shrubs

6.5 (2.0~13.4); -3.7 (-7.7~1.5); 15.8 (9.9~24.6); 7.4 (3.8~13.4); 49.1 (15.4~108.0); 30.9 (7.1~51.4); 536 (413~760) ❙ 811 (656~1023); 411 (317~546); 20 (14~28) ❙ 59.9 (-1.2~128.6); 535 (413~760); 275.1 (200.8~357.7)

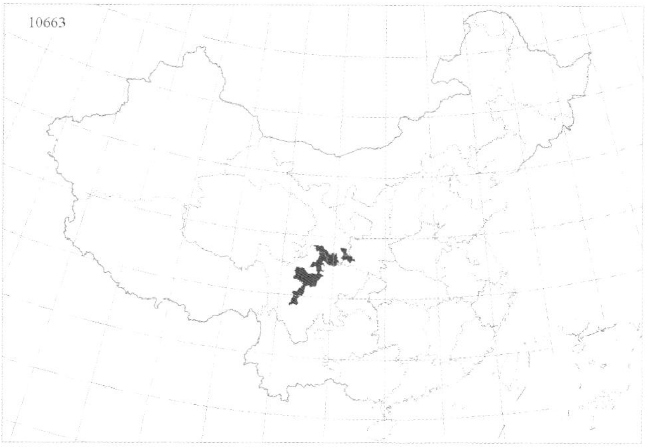

Asteraceae 菊科 *Ajania* 亚菊属
Ajania purpurea C. Shih 紫花亚菊
Subshrubs

2.3 (-0.7~4.4); -7.1 (-10.8~-4.3); 10.6 (8.6~12.1); 4.2 (2.8~5.3); 20.4 (10.0~28.5); 52.5 (34.6~77.9); 408 (345~455) ❙ 390 (250~538); 263 (198~337); 4 (1~9) ❙ -5.9 (-27.5~22.4); 354 (250~455); 201.3 (131.4~280.0)

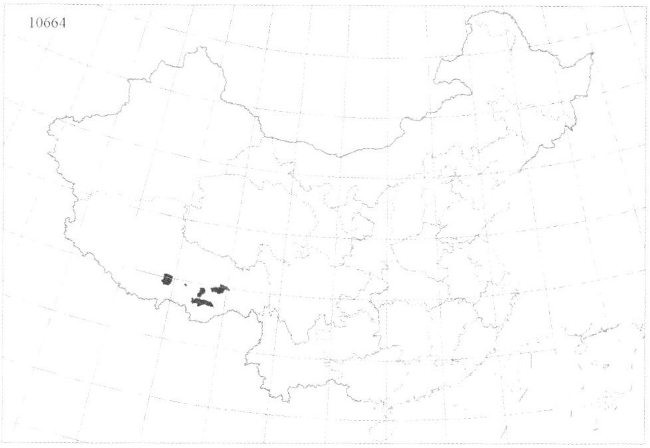

Asteraceae 菊科 *Ajania* 亚菊属
Ajania quercifolia (W. W. Sm.) Ling et C. Shih 栎叶亚菊
Subshrubs

8.1; 0.7; 14.4; 8.5; 54.7; 16.7; 557 ❙ 863; 527; 25 ❙ 54.7; 557; 297.2

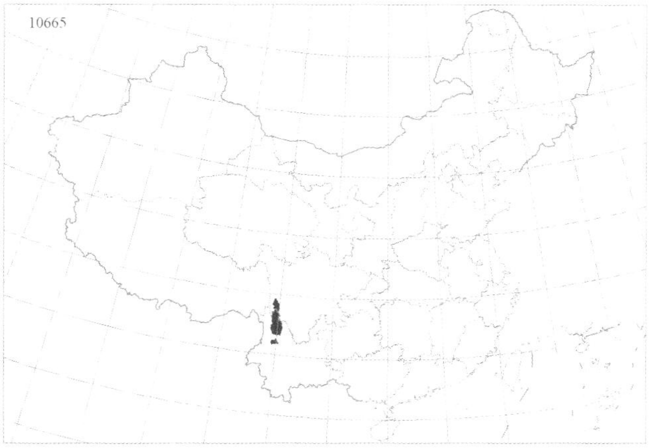

Asteraceae 菊科 *Ajania* 亚菊属
Ajania ramosa (C. C. Chang) C. Shih 分枝亚菊
Shrubs

-0.4 (-4.4~2.4); -10.6 (-16.2~-6.6); 9.0 (6.9~11.3); 3.1 (1.7~4.4); 11.5 (3.2~21.4); 75.4 (49.1~115.9); 376 (323~423) ❙ 575 (312~932); 345 (198~546); 12 (8~19) ❙ 49.8 (-3.2~125.9); 374 (312~423); 233.9 (120.4~290.1)

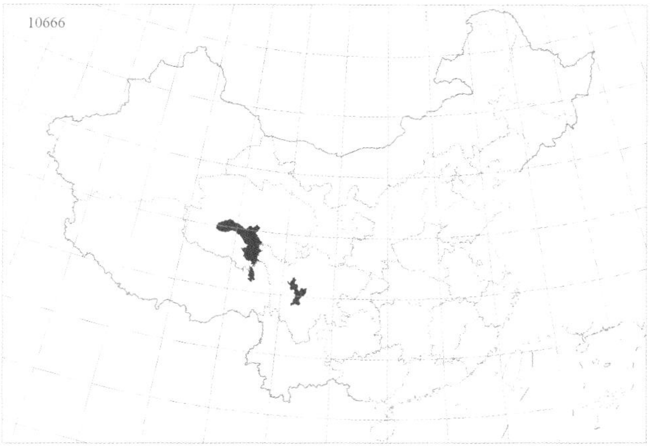

Asteraceae 菊科 *Ajania* 亚菊属
Ajania salicifolia (Mattf. ex Rehder) Poljakov 柳叶亚菊
Subshrubs

5.5 (-1.3~13.6); -6.5 (-13.5~7.2); 16.3 (9.2~23.7); 7.1 (2.9~13.7); 48.5 (10.1~104.1); 41.7 (0.0~86.3); 525 (359~706) ❙ 592 (317~1042); 311 (190~592); 14 (5~46) ❙ 14.1 (-40.9~121.7); 490 (317~706); 223.1 (61.2~372.3)

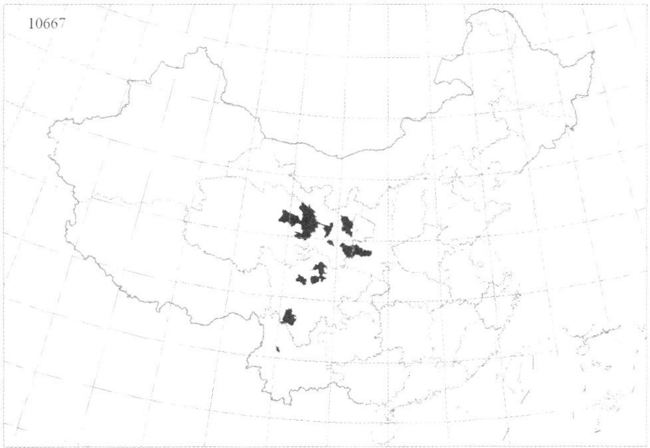

Asteraceae 菊科 *Ajania* 亚菊属
Ajania scharnhorstii (Regel et Schmalh.) Tzvelev 单头亚菊
Subshrubs

3.9 (-5.7~12.4); -11.7 (-20.1~-5.3); 17.3 (7.5~26.8); 7.3 (1.7~13.2); 55.3 (4.7~114.1); 67.5 (24.7~133.0); 544 (294~811) ❙ 93 (21~509); 53 (15~233); 8 (0~98) ❙ -76.4 (-97.3~52.9); 87 (21~333); 19.1 (2.6~86.2)

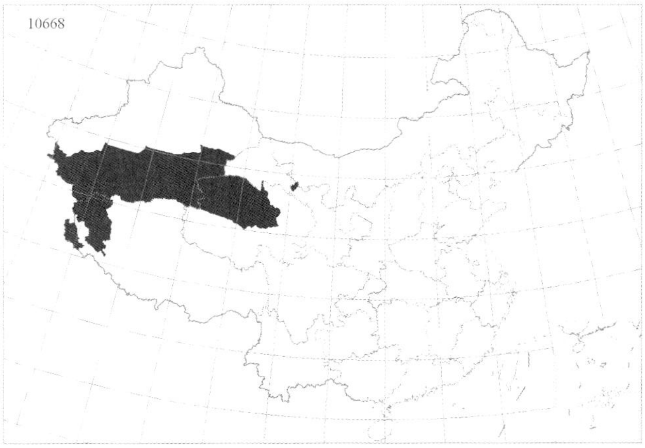

Asteraceae 菊科 *Ajania* 亚菊属

Ajania tibetica (Hook. f. et Thomson) Tzvelev 西藏亚菊

Subshrubs

3.6 (−4.3~12.4); −12.4 (−20.1~−5.3); 17.0 (6.4~26.8); 7.0 (1.8~13.2); 52.3 (2.5~114.1); 68.3 (24.7~121.1); 535 (319~811) ❙ 106 (21~628); 60 (15~410); 7 (0~27) ❙ −75.4 (−97.3~48.6); 97 (21~423); 25.9 (2.8~236.3)

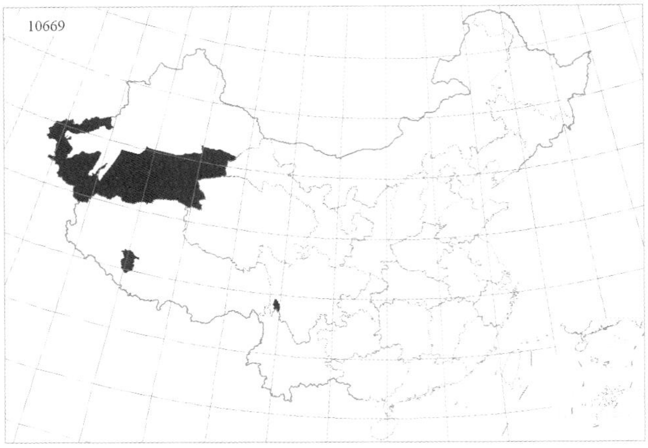

Asteraceae 菊科 *Ajania* 亚菊属

Ajania variifolia (C. C. Chang) Tzvelev 异叶亚菊

Subshrubs

11.6 (7.4~13.8); 0.0 (−4.2~2.6); 22.8 (18.7~24.8); 11.8 (8.2~13.9); 91.9 (58.8~111.2); 12.2 (3.9~29.4); 690 (561~767) ❙ 1049 (724~1435); 449 (316~591); 55 (22~94) ❙ 51.9 (2.5~101.4); 690 (561~767); 270.1 (236.0~318.2)

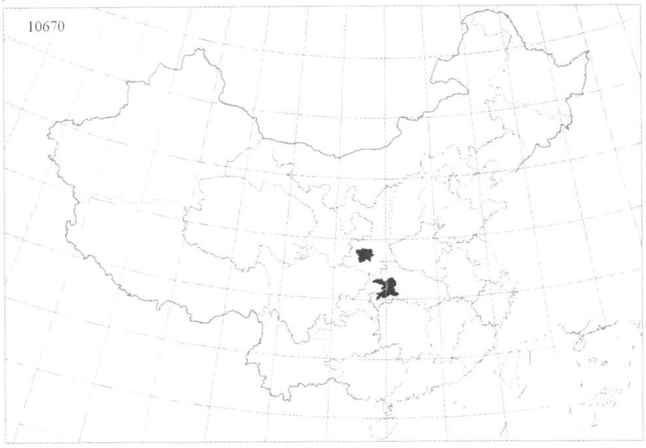

Asteraceae 菊科 *Artemisia* 艾蒿属

Artemisia campestris L. 荒野蒿

Small shrubs

2.4 (−1.3~9.0); −16.4 (−20.9~−9.3); 18.0 (13.2~23.3); 7.0 (4.6~10.7); 51.4 (26.7~88.9); 81.7 (40.7~104.2); 541 (440~703) ❙ 247 (156~406); 110 (56~211); 25 (12~48) ❙ −53.0 (−76.7~−7.6); 247 (156~406); 151.3 (16.6~297.3)

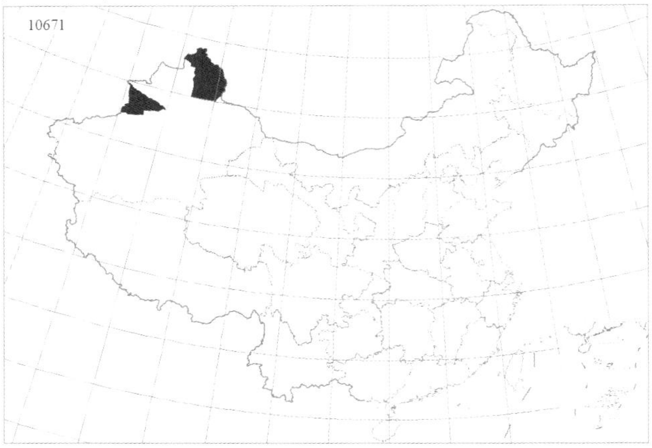

Asteraceae 菊科 *Artemisia* 艾蒿属

Artemisia halodendron Turcz. ex Besser 盐蒿

Small shrubs

5.9 (−3.1~12.7); −12.5 (−25.5~−3.6); 22.2 (13.8~26.9); 9.1 (4.4~13.2); 73.1 (26.2~113.0); 61.2 (20.4~132.1); 622 (413~819) ❙ 423 (44~658); 281 (19~465); 9 (2~15) ❙ −31.8 (−94.5~2.9); 422 (44~658); 126.6 (9.3~308.8)

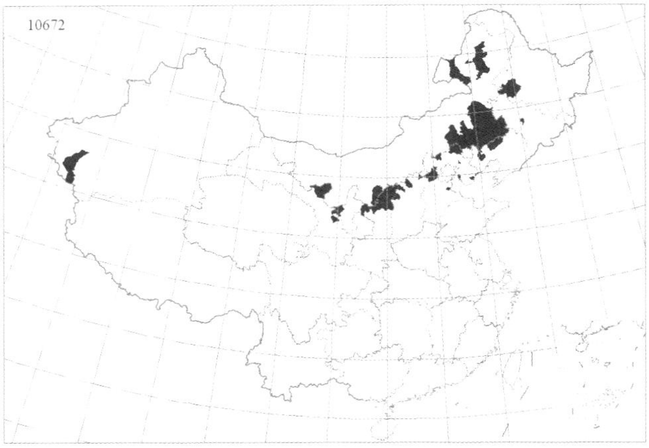

Asteraceae 菊科 *Artemisia* 艾蒿属

Artemisia klementzae Krasch. 蒙古沙地蒿

Small shrubs

2.3; −17.9; 20.2; 7.4; 55.4; 87.0; 548 ❙ 293; 200; 7 ❙ −46.4; 293; 103.5

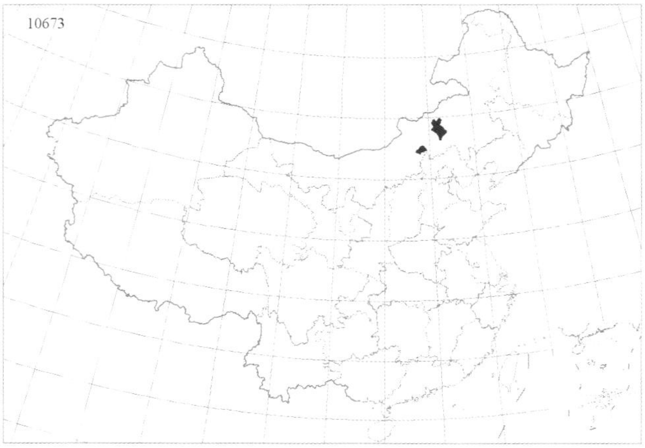

Asteraceae 菊科 *Artemisia* 艾蒿属

Artemisia ordosica Krasch. 黑沙蒿

Small shrubs

5.9 (−2.6~11.6); −10.4 (−17.0~−3.0); 20.3 (8.7~24.8); 8.4 (2.5~12.0); 65.2 (7.7~99.3); 54.2 (19.2~101.3); 586 (349~731) ❙ 275 (49~671); 166 (34~471); 7 (1~33) ❙ −51.5 (−92.8~28.2); 272 (49~668); 80.5 (5.1~284.0)

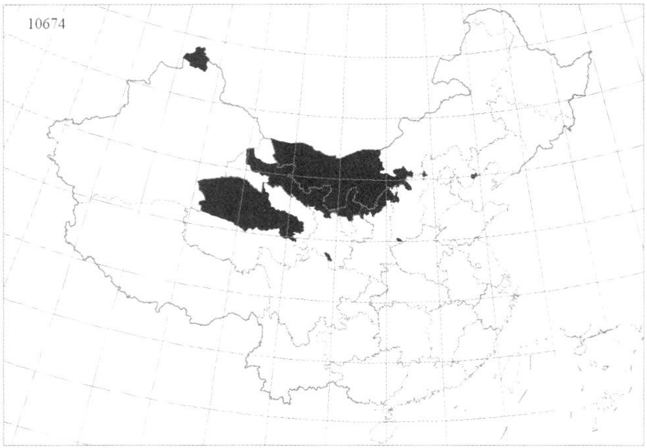

Asteraceae 菊科 *Artemisia* 艾蒿属

Artemisia prattii (Pamp.) Ling et Y. R. Ling 藏岩蒿

Small shrubs

0.2 (-2.6~3.2); -11.8 (-16.1~-6.4); 11.2 (8.7~13.6); 3.8 (2.5~4.7); 18.5 (7.7~27.1); 74.9 (43.8~101.3); 399 (349~436) ❙ 362 (63~932); 217 (45~546); 8 (2~19) ❙ -11.2 (-84.9~125.9); 272 (63~436); 133.7 (8.4~357.7)

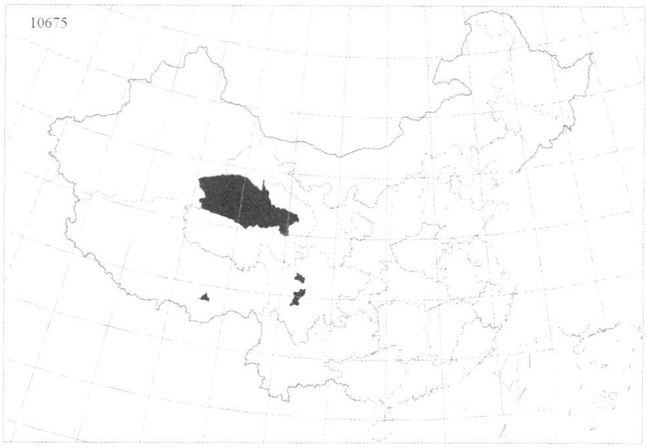

Asteraceae 菊科 *Artemisia* 艾蒿属

Artemisia songarica Schrenk 准噶尔沙蒿

Small shrubs

5.3 (-0.6~11.7); -14.3 (-19.5~-7.6); 21.9 (15.9~27.5); 9.0 (5.4~13.0); 73.2 (37.5~113.4); 69.4 (32.4~104.2); 634 (475~820) ❙ 160 (30~251); 73 (21~117); 15 (1~33) ❙ -73.5 (-96.3~-52.0); 160 (30~251); 55.8 (2.1~139.1)

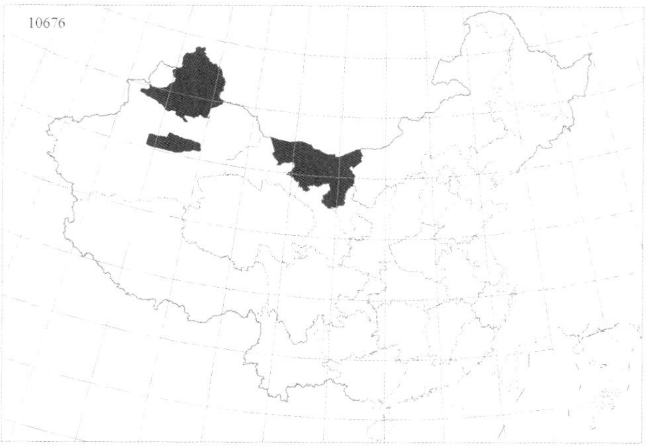

Asteraceae 菊科 *Artemisia* 艾蒿属

Artemisia sphaerocephala Krasch. 圆头蒿

Small shrubs

6.8 (-2.8~13.0); -10.6 (-20.3~-2.4); 21.9 (10.9~27.5); 9.2 (3.5~13.3); 74.4 (16.3~113.3); 52.4 (17.1~109.2); 628 (384~813) ❙ 233 (35~551); 135 (23~306); 8 (1~21) ❙ -61.5 (-94.9~-2.8); 233 (35~551); 60.0 (5.1~189.8)

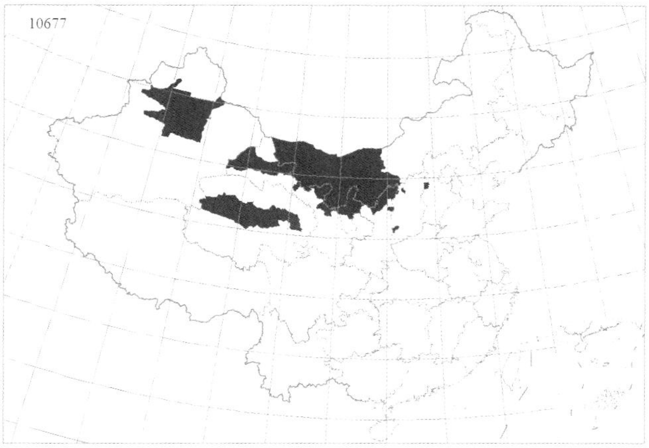

Asteraceae 菊科 *Artemisia* 艾蒿属

Artemisia waltonii J. R. Drumm. ex Pamp. 藏龙蒿

Shrubs or subshrubs

3.5 (-5.5~15.9); -6.0 (-17.1~6.2); 11.9 (5.8~24.6); 5.4 (1.3~15.9); 31.0 (1.2~130.8); 48.9 (0.0~127.1); 458 (299~822) ❙ 712 (212~1414); 419 (156~817); 17 (0~50) ❙ 50.5 (-35.9~128.6); 444 (212~822); 239.3 (60.4~357.7)

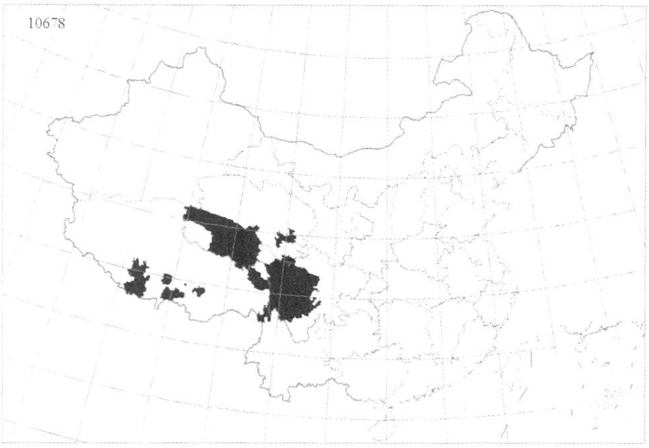

Asteraceae 菊科 *Artemisia* 艾蒿属

Artemisia wudanica Liou et W. Wang 乌丹蒿

Small shrubs

4.6 (1.2~6.7); -13.7 (-17.8~-11.7); 21.3 (18.2~23.4); 8.3 (6.4~9.5); 64.8 (45.9~79.6); 68.6 (58.2~90.7); 589 (506~651) ❙ 415 (383~512); 295 (275~355); 7 (5~9) ❙ -29.3 (-37.4~-15.0); 415 (383~512); 153.9 (97.2~200.5)

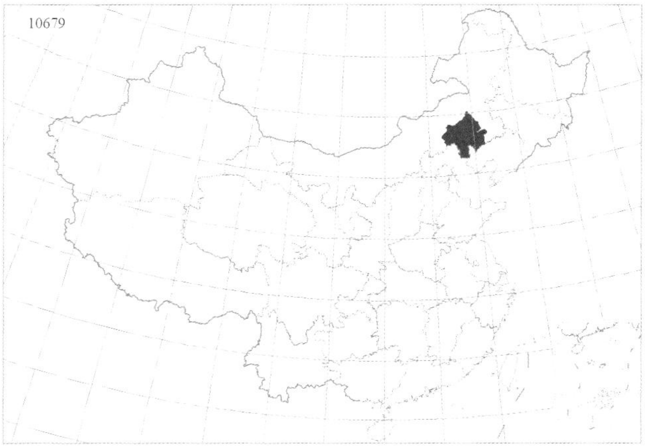

Asteraceae 菊科 *Artemisia* 艾蒿属

Artemisia xerophytica Krasch. 内蒙古旱蒿

Small shrubs

4.2 (-2.5~9.3); -13.5 (-19.5~-8.7); 19.9 (10.1~27.1); 7.8 (2.9~12.1); 60.0 (12.0~107.6); 69.0 (43.5~104.2); 566 (358~810) ❙ 214 (45~416); 132 (27~270); 8 (1~33) ❙ -61.5 (-92.8~-24.8); 214 (45~416); 60.0 (5.1~145.6)

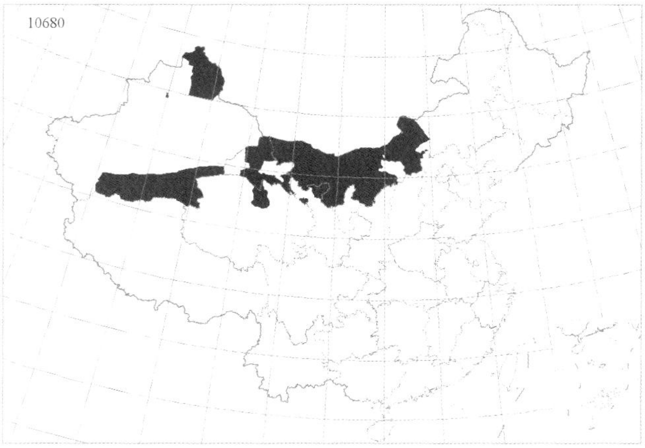

Asteraceae 菊科 *Aster* 紫菀属

Aster albescens (DC.) Wall. ex Hand.-Mazz. 小舌紫菀

Small shrubs

11.3 (−1.8~19.1); 1.9 (−10.5~12.0); 19.7 (6.1~27.0); 11.7 (1.8~19.1); 89.8 (1.9~169.2); 13.8 (0.0~82.8); 675 (329~969) ❙ 973 (252~1492); 503 (193~816); 39 (0~106) ❙ 45.8 (−35.9~128.6); 667 (252~969); 247.3 (62.8~372.3)

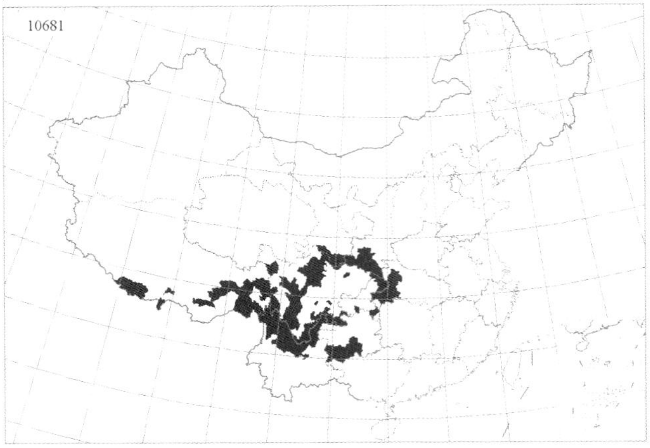

Asteraceae 菊科 *Aster* 紫菀属

Aster albescens var. discolor Ling 白背小舌紫菀

Shrubs

2.6; −7.0; 11.2; 4.3; 20.0; 48.2; 427 ❙ 829; 400; 23 ❙ 94.4; 427; 252.2

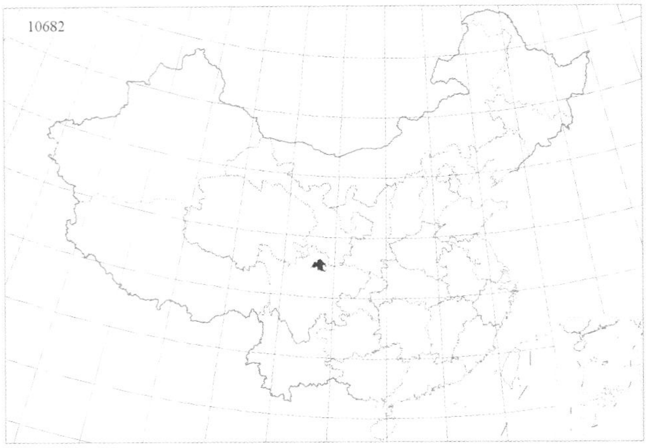

Asteraceae 菊科 *Aster* 紫菀属

Aster albescens var. glandulosus Hand.-Mazz. 腺点小舌紫菀

Shrubs

2.7 (−1.5~9.6); −6.1 (−10.5~2.4); 10.5 (7.1~15.7); 4.4 (2.2~9.6); 21.1 (4.7~60.8); 48.7 (5.4~82.2); 425 (332~600) ❙ 616 (252~1007); 375 (196~578); 16 (0~40) ❙ 41.5 (−35.9~100.7); 413 (252~600); 211.8 (62.8~351.7)

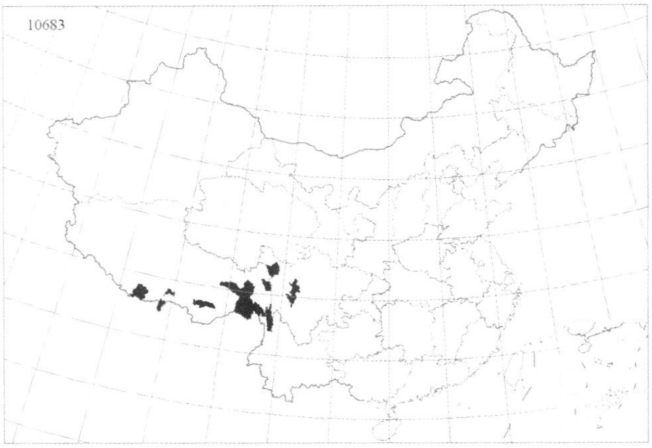

Asteraceae 菊科 *Aster* 紫菀属

Aster albescens var. gracilior Hand.-Mazz. 狭叶小舌紫菀

Shrubs

10.2 (3.2~14.8); 1.7 (−6.4~7.8); 17.5 (11.4~24.9); 10.5 (4.7~14.9); 75.3 (22.7~119.2); 12.6 (0.0~43.8); 628 (436~791) ❙ 988 (791~1197); 527 (425~619); 36 (12~71) ❙ 60.3 (43.0~92.8); 628 (436~791); 307.9 (182.9~382.6)

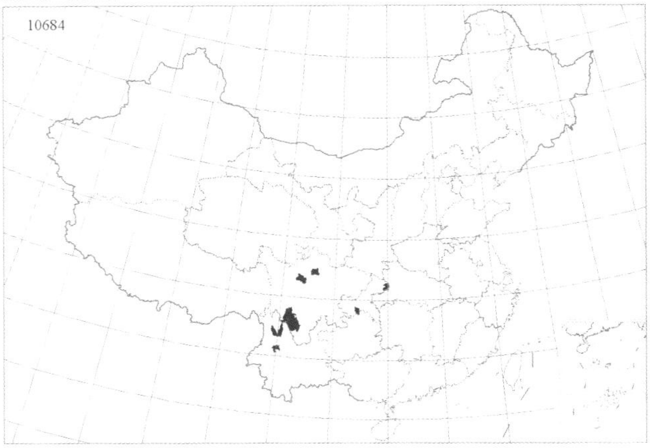

Asteraceae 菊科 *Aster* 紫菀属

Aster albescens var. levissimus Franch. et Sav. 无毛小舌紫菀

Shrubs

5.6 (1.8~14.3); −3.1 (−7.4~7.6); 13.3 (9.5~24.4); 6.6 (3.8~14.3); 39.9 (15.4~111.9); 31.8 (0.0~53.3); 502 (404~758) ❙ 880 (520~1492); 490 (317~696); 27 (10~96) ❙ 76.3 (26.6~128.6); 501 (404~758); 275.9 (169.5~384.7)

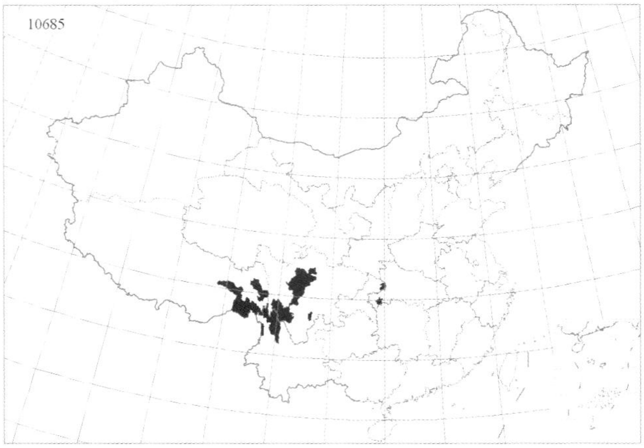

Asteraceae 菊科 *Aster* 紫菀属

Aster albescens var. limprichtii (Diels) Hand.-Mazz. 椭叶小舌紫菀

Shrubs

7.6 (2.2~13.8); −1.1 (−7.7~7.2); 15.3 (9.9~20.1); 8.3 (3.8~13.8); 54.1 (15.5~105.9); 22.6 (0.0~49.3); 553 (413~714) ❙ 898 (656~1049); 492 (317~619); 24 (15~46) ❙ 66.5 (12.1~125.9); 551 (413~706); 278.5 (200.8~372.3)

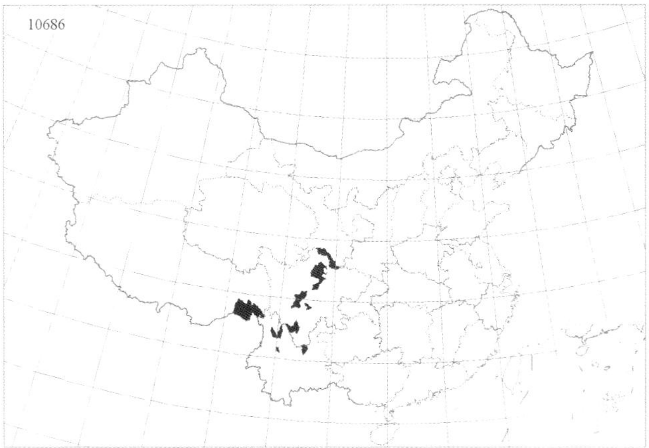

Asteraceae 菊科 *Aster* 紫菀属
Aster albescens var. megaphyllus Ling 大叶小舌紫菀
Shrubs

5.1; -4.2; 13.5; 5.9; 33.7; 32.2; 485 ❙ 932; 474; 25 ❙ 92.3; 485; 251.3

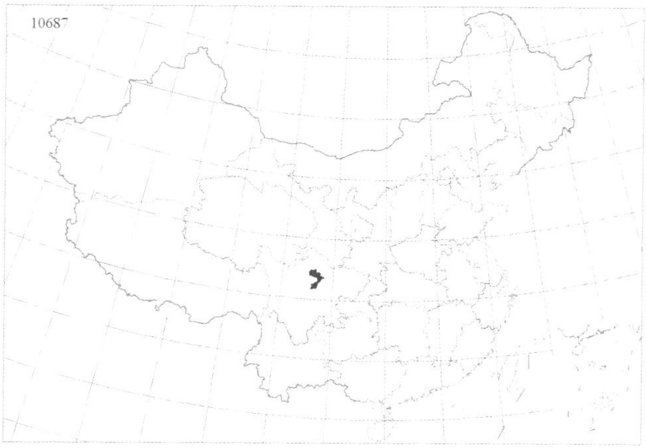

Asteraceae 菊科 *Aster* 紫菀属
Aster albescens var. pilosus Hand.-Mazz. 长毛小舌紫菀
Shrubs

4.0 (0.0~10.8); -4.7 (-9.5~3.9); 11.6 (8.1~16.6); 5.3 (2.7~10.9); 27.4 (8.0~71.8); 39.4 (1.3~67.9); 456 (363~633) ❙ 795 (473~1023); 469 (287~603); 17 (7~28) ❙ 74.6 (5.0~128.6); 456 (363~633); 261.5 (111.0~319.8)

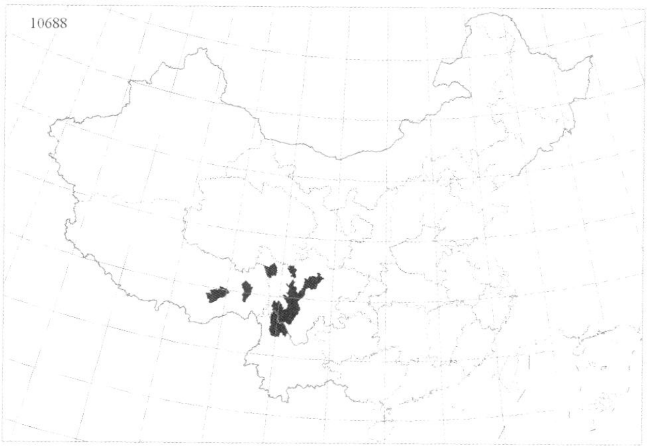

Asteraceae 菊科 *Aster* 紫菀属
Aster albescens var. rugosus Ling 糙叶小舌紫菀
Shrubs

8.4; 0.3; 15.1; 9.2; 65.2; 24.6; 577 ❙ 973; 587; 20 ❙ 81.4; 552; 256.6

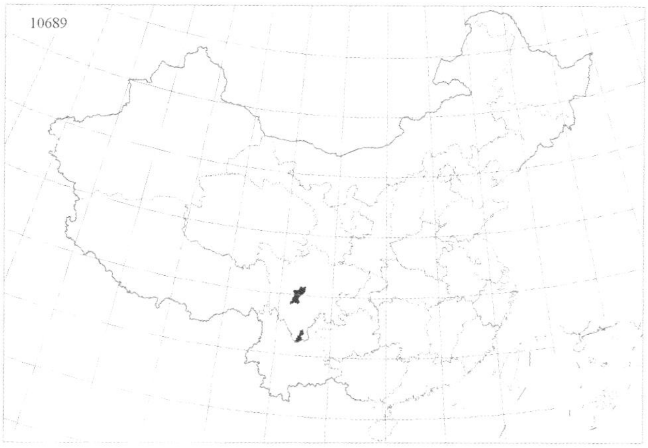

Asteraceae 菊科 *Aster* 紫菀属
Aster albescens var. salignus Hand.-Mazz. 柳叶小舌紫菀
Shrubs

7.8 (2.0~15.3); -0.1 (-7.4~8.7); 14.6 (9.9~22.5); 8.3 (3.8~15.4); 53.6 (15.4~124.3); 20.0 (0.0~51.9); 554 (413~761) ❙ 943 (628~1243); 545 (410~696); 29 (11~58) ❙ 73.5 (28.3~128.6); 550 (413~717); 296.5 (234.2~384.7)

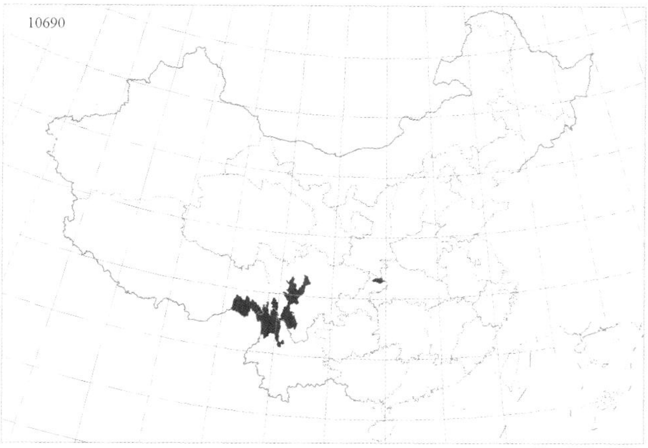

Asteraceae 菊科 *Aster* 紫菀属
Aster argyropholis Hand.-Mazz. 银鳞紫菀
Shrubs

8.6 (2.0~17.3); 0.5 (-7.2~11.0); 15.5 (10.1~21.6); 9.3 (3.8~17.3); 64.5 (15.4~148.0); 20.3 (0.0~51.4); 582 (413~832) ❙ 945 (791~1049); 536 (427~627); 25 (15~44) ❙ 69.7 (26.1~128.6); 573 (413~783); 289.9 (221.3~365.3)

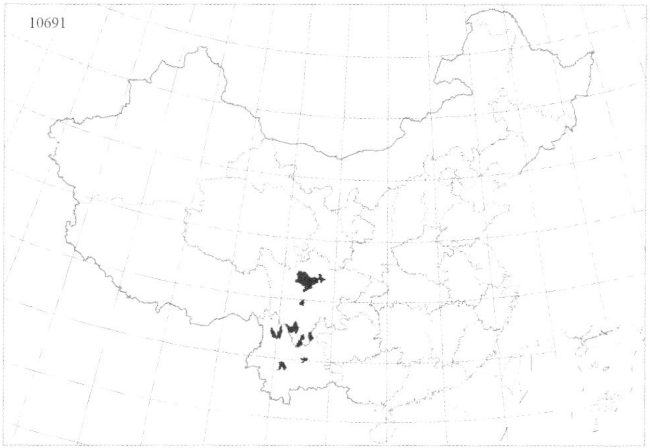

Asteraceae 菊科 *Aster* 紫菀属
Aster argyropholis var. niveus Ling 白雪银鳞紫菀
Shrubs

12.1; 4.9; 18.0; 12.2; 87.1; 1.0; 671 ❙ 991; 613; 20 ❙ 48.2; 654; 293.6

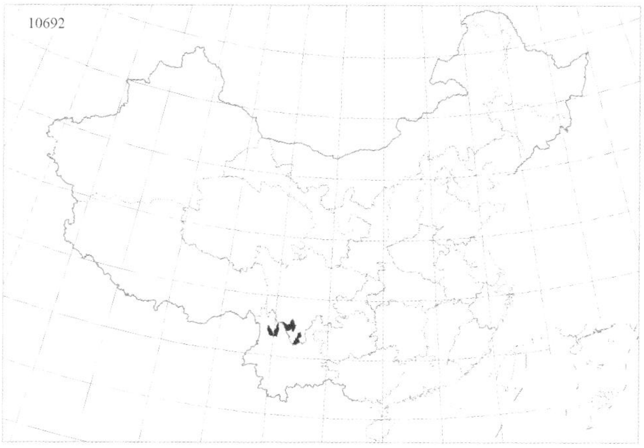

Asteraceae 菊科 *Aster* 紫菀属
Aster argyropholis var. paradoxus Ling 奇形银鳞紫菀
Shrubs

2.7; -7.1; 11.1; 4.4; 20.5; 48.0; 429 | 800; 427; 18 | 86.5; 429; 341.6

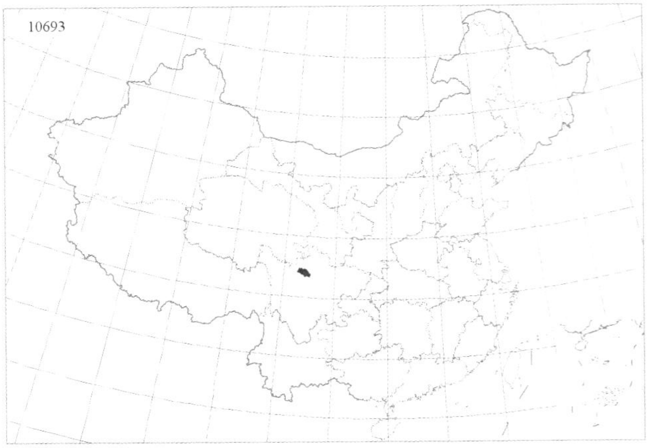

Asteraceae 菊科 *Aster* 紫菀属
Aster fulgidulus Grierson 辉叶紫菀
Small shrubs

2.6; -6.3; 10.6; 4.2; 19.5; 47.5; 417 | 555; 333; 11 | 33.3; 417; 224.7

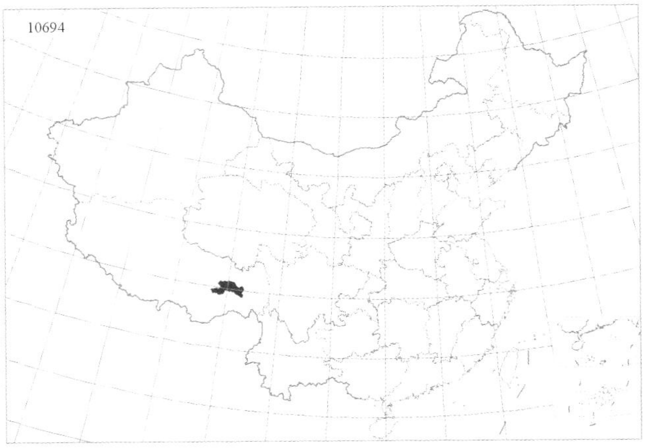

Asteraceae 菊科 *Aster* 紫菀属
Aster hersileoides Schneid. 横斜紫菀
Shrubs

5.9; -3.4; 14.3; 6.5; 38.9; 27.4; 506 | 972; 498; 27 | 92.1; 506; 231.7

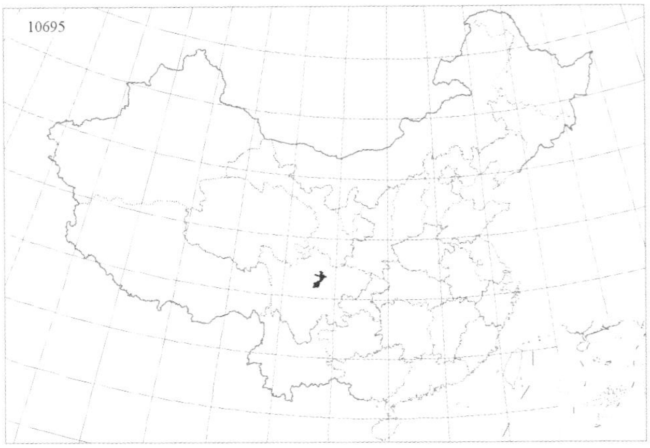

Asteraceae 菊科 *Aster* 紫菀属
Aster hypoleucus Hand.-Mazz. 白背紫菀
Shrubs

1.9; -7.1; 10.0; 3.9; 17.1; 53.5; 400 | 371; 240; 5 | -9.8; 356; 219.2

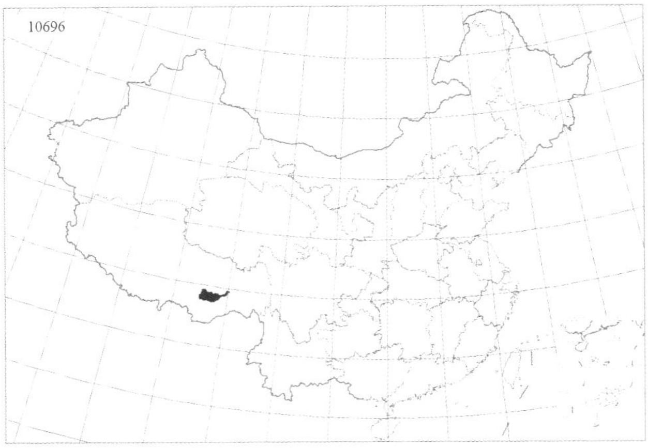

Asteraceae 菊科 *Aster* 紫菀属
Aster lavandulifolius Hand.-Mazz. 线叶紫菀
Shrubs

4.0; -4.4; 11.1; 5.0; 24.6; 36.6; 450 | 839; 530; 13 | 87.1; 450; 271.3

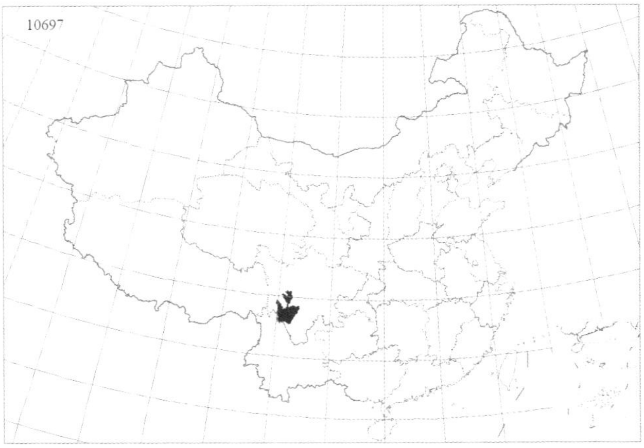

Asteraceae 菊科 *Aster* 紫菀属
Aster nitidus C. C. Chang 亮叶紫菀
Shrubs

14.8; 4.3; 24.9; 14.9; 119.2; 0.7; 791 | 1197; 489; 71 | 51.2; 791; 182.9

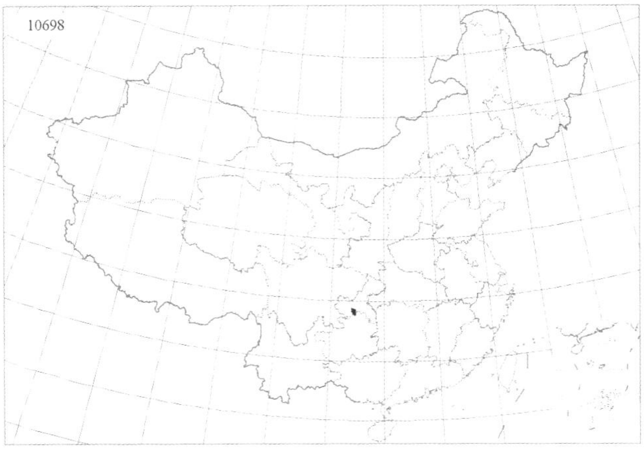

Asteraceae 菊科 *Duhaldea* 羊耳菊属
Duhaldea chinensis DC. 羊耳菊
Subshrubs, deciduous

17.6 (−1.8~24.3); 8.0 (−10.2~18.7); 26.1 (6.1~30.1); 17.7 (1.8~24.3); 152.6 (1.9~231.9); 0.6 (0.0~82.8); 929 (329~1338) ❙ 1415 (320~2435); 587 (217~1428); 129 (12~259) ❙ 53.3 (−34.3~147.0); 918 (320~1294); 237.1 (72.2~659.0)

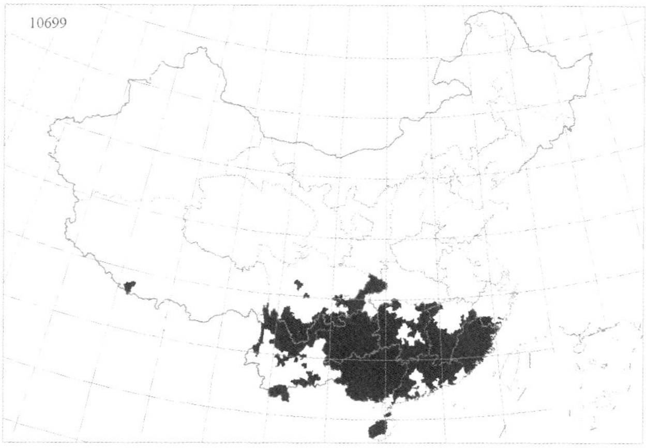

Asteraceae 菊科 *Duhaldea* 羊耳菊属
Duhaldea eupatorioides (DC.) Steetz 泽兰羊耳菊
Shrubs

11.1; 3.9; 16.8; 11.3; 86.6; 13.1; 646 ❙ 1070; 602; 41 ❙ 72.8; 646; 308.8

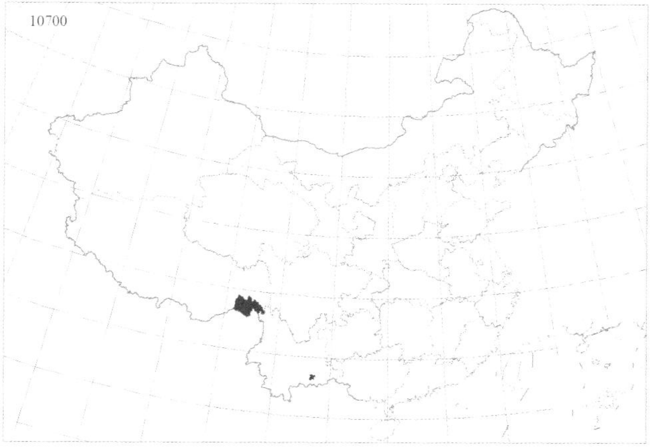

Asteraceae 菊科 *Duhaldea* 羊耳菊属
Duhaldea forrestii (J. Anthony) A. Anderberg 拟羊耳菊
Shrubs

7.8; 0.4; 14.2; 8.0; 48.9; 14.3; 548 ❙ 877; 546; 19 ❙ 60.7; 548; 304.7

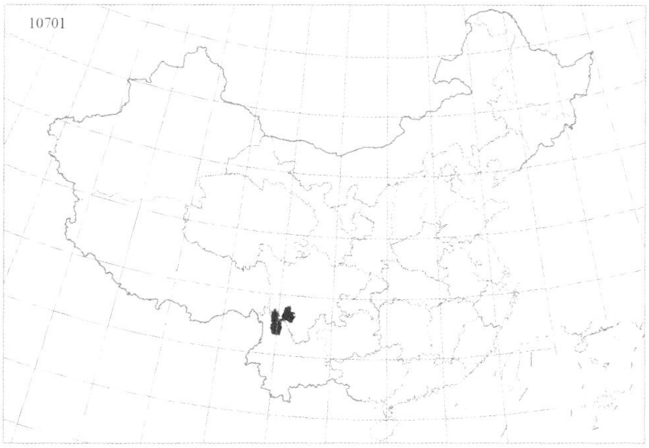

Asteraceae 菊科 *Duhaldea* 羊耳菊属
Duhaldea pterocaula (Franchet) A. Anderberg 翼茎羊耳菊
Subshrubs

11.9 (3.8~18.9); 4.6 (−4.3~12.0); 17.7 (10.8~23.9); 12.0 (4.7~19.0); 91.6 (21.6~167.8); 8.5 (0.0~35.4); 669 (444~929) ❙ 955 (741~1178); 551 (411~646); 34 (12~54) ❙ 48.4 (−14.2~121.7); 650 (444~813); 291.3 (205.2~383.4)

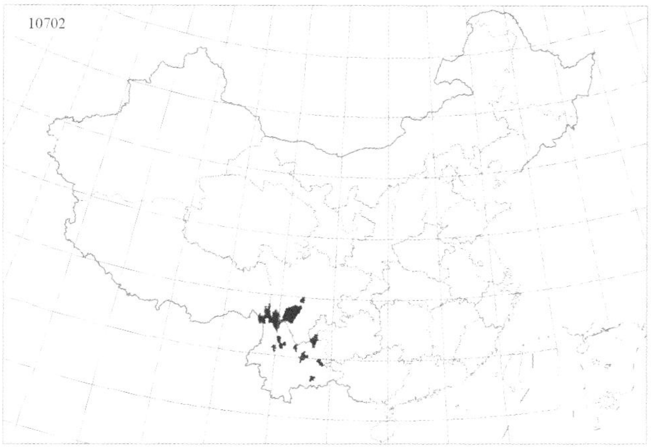

Asteraceae 菊科 *Duhaldea* 羊耳菊属
Duhaldea rubricaulis (Wallich ex Candolle) A. Anderberg 赤茎羊耳菊
Subshrubs

16.8; 10.5; 21.2; 16.9; 142.4; 0.0; 821 ❙ 1211; 694; 49 ❙ 47.5; 804; 381.7

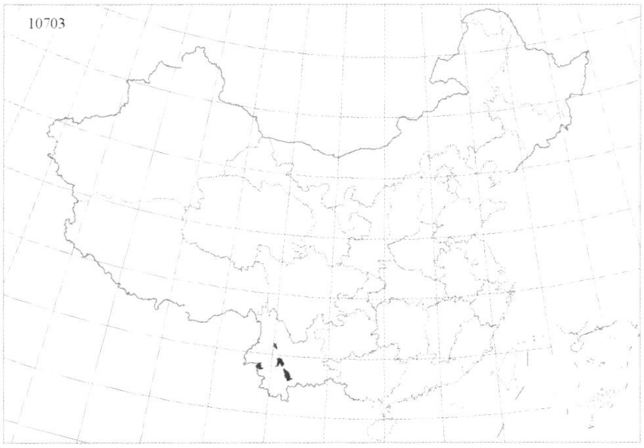

Asteraceae 菊科 *Duhaldea* 羊耳菊属
Duhaldea wissmanniana (Handel-Mazzetti) A. Anderberg 滇南羊耳菊
Subshrubs

17.8 (16.9~19.3); 11.3 (10.3~13.1); 22.3 (21.4~23.6); 17.9 (16.9~19.4); 154.5 (142.7~173.2); 0.0 (0.0~0.0); 860 (813~939) ❙ 1134 (1007~1368); 630 (539~756); 49 (44~59) ❙ 32.2 (19.0~60.1); 839 (783~903); 370.4 (294.5~491.5)

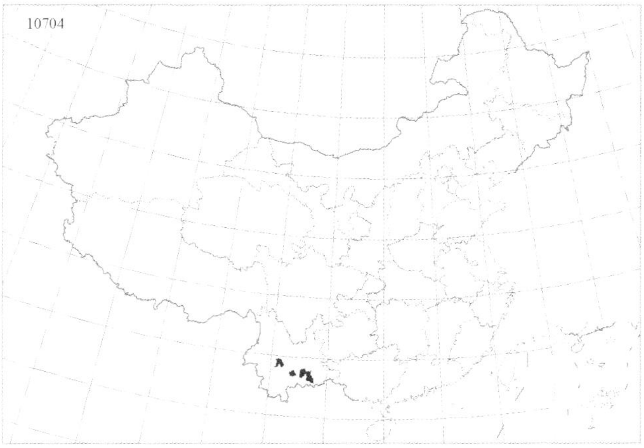

Asteraceae 菊科 *Formania* 复芒菊属
Formania mekongensis W. W. Sm. et J. Small 复芒菊
Small shrubs

7.6；-0.3；14.0；8.2；54.9；24.4；543 ❙ 906；521；34 ❙ 60.0；543；307.9

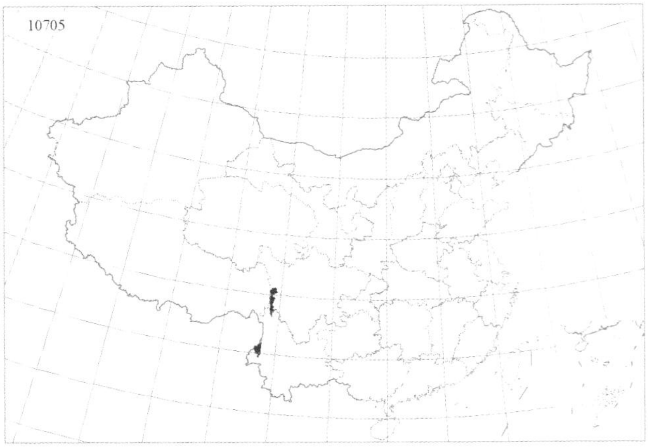

Asteraceae 菊科 *Inula* 旋覆花属
Inula salsoloides (Turcz.) Ostenf. 蓼子朴
Subshrubs

6.5 (-4.3~14.4)；-11.1 (-21.5~0.3)；21.5 (9.6~27.6)；9.2 (2.5~14.5)；73.8 (9.7~124.0)；55.0 (10.4~121.1)；629 (329~853) ❙ 253 (21~671)；154 (15~471)；9 (0~48) ❙ -57.4 (-97.3~14.7)；252 (21~668)；88.2 (1.6~297.3)

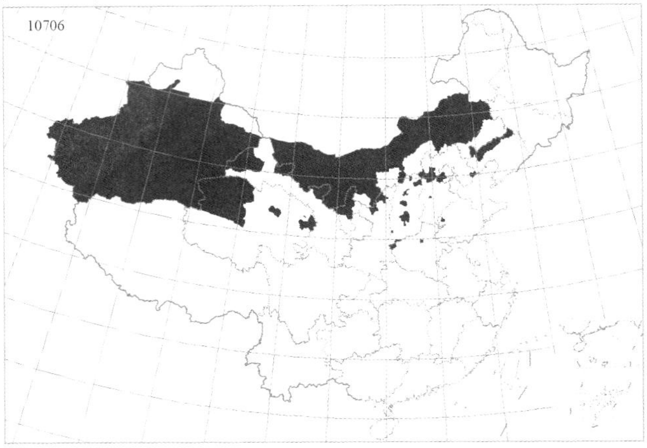

Asteraceae 菊科 *Leucomeris* 白菊木属
Leucomeris decora Kurz 白菊木
Small trees, deciduous

18.1 (13.6~20.4)；12.0 (7.2~15.1)；22.2 (18.8~23.6)；18.2 (13.7~20.5)；158.1 (104.1~185.8)；0.0 (0.0~0.0)；877 (706~989) ❙ 1262 (939~1607)；670 (524~909)；52 (39~77) ❙ 43.8 (6.5~78.9)；850 (706~986)；419.9 (294.5~659.0)

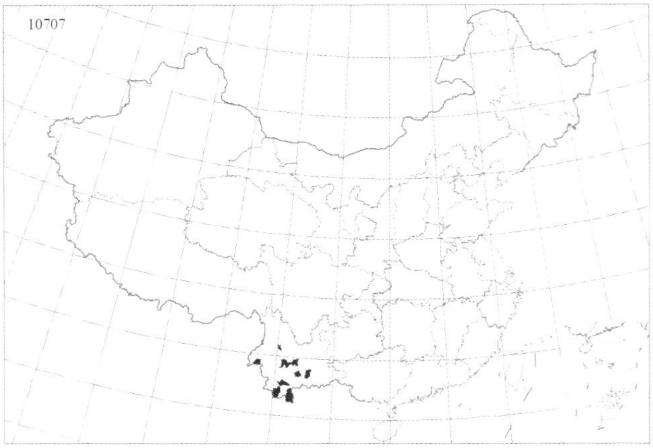

Asteraceae 菊科 *Nouelia* 栌菊木属
Nouelia insignis Franch. 栌菊木
Shrubs or small trees, deciduous

13.6 (1.4~19.2)；6.0 (-7.6~13.8)；19.7 (9.2~26.6)；13.7 (3.5~19.2)；107.2 (13.4~170.5)；3.5 (0.0~55.6)；724 (402~929) ❙ 969 (699~1499)；564 (411~686)；30 (10~58) ❙ 36.6 (-14.2~121.7)；694 (402~912)；278.4 (142.9~513.9)

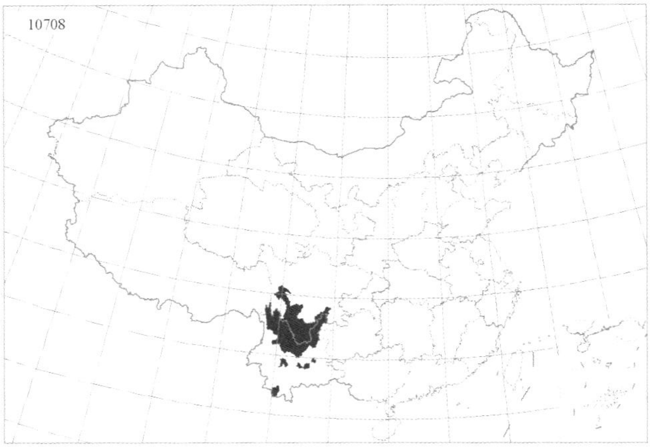

Asteraceae 菊科 *Pertya* 帚菊属
Pertya angustifolia Y. C. Tseng 狭叶帚菊
Shrubs

2.0；-7.4；10.1；4.0；16.9；51.9；416 ❙ 777；457；15 ❙ 86.8；416；271.2

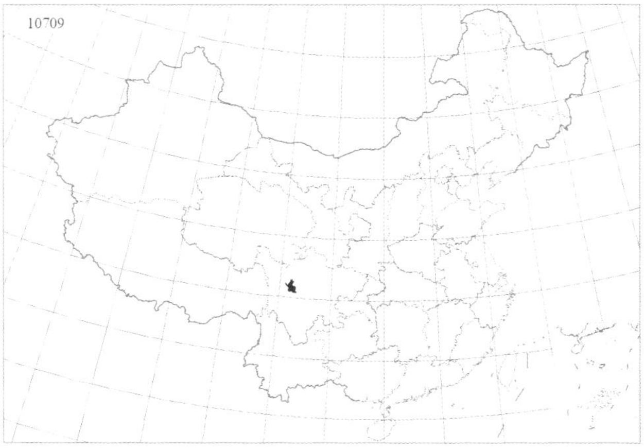

Asteraceae 菊科 *Pertya* 帚菊属
Pertya berberidoides (Hand.-Mazz.) Y. Q. Tseng 异叶帚菊
Shrubs, deciduous

5.1 (2.0~9.6)；-3.0 (-7.0~2.4)；12.2 (10.3~15.7)；5.9 (3.9~9.6)；32.3 (17.3~60.8)；30.3 (5.4~52.8)；479 (410~600) ❙ 763 (520~1007)；466 (317~578)；19 (10~40) ❙ 58.3 (26.6~81.6)；479 (410~600)；280.8 (210.1~351.7)

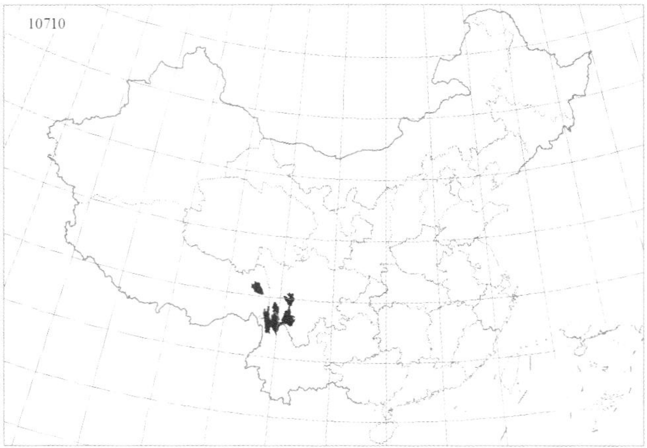

Asteraceae 菊科 *Pertya* 帚菊属
Pertya bodinieri Vaniot 昆明帚菊
Subshrubs

15.5; 8.8; 20.4; 15.5; 126.3; 0.0; 768 ❙ 970; 572; 37 ❙ 26.8; 737; 278.1

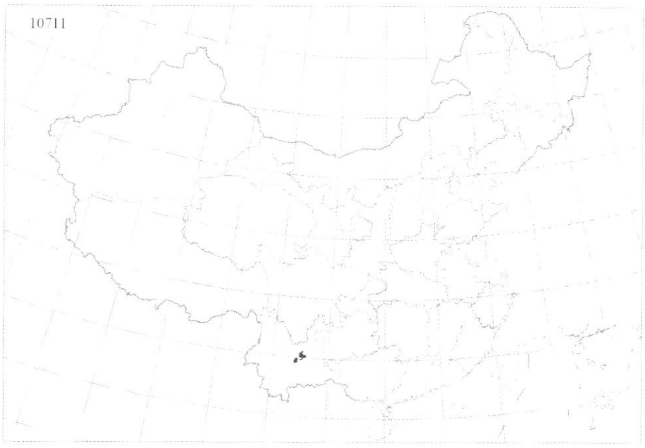

Asteraceae 菊科 *Pertya* 帚菊属
Pertya cordifolia (Diels) Z. Wei et Y. B. Chang 心叶帚菊
Subshrubs

16.5 (15.0~18.8); 4.8 (2.6~9.2); 27.6 (26.3~28.5); 16.6 (15.0~18.9); 139.9 (123.9~166.3); 0.7 (0.0~3.5); 892 (829~970) ❙ 1603 (1409~1709); 531 (445~608); 209 (184~228) ❙ 79.8 (59.8~95.5); 890 (829~970); 238.4 (169.7~299.5)

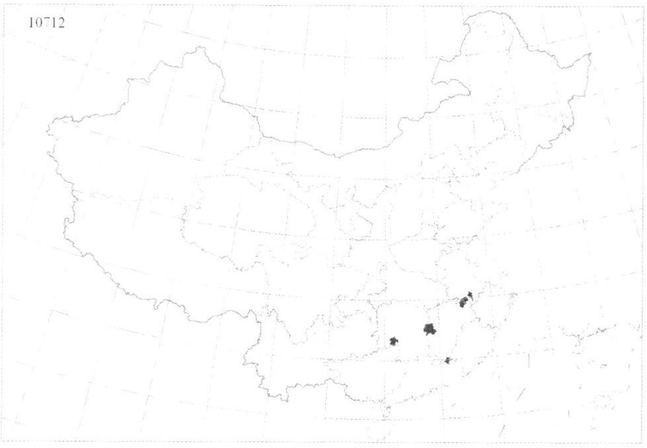

Asteraceae 菊科 *Pertya* 帚菊属
Pertya corymbosa Y. C. Tseng 疏花帚菊
Subshrubs

17.6; 6.9; 27.0; 17.6; 151.5; 0.0; 918 ❙ 1495; 557; 169 ❙ 62.9; 915; 203.1

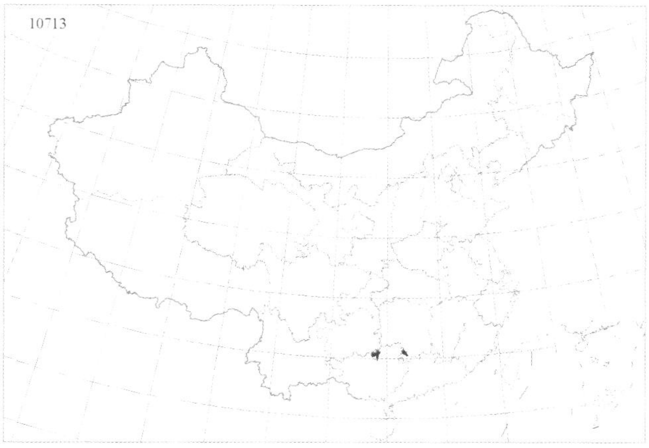

Asteraceae 菊科 *Pertya* 帚菊属
Pertya dioica (Bunge) T. G. Gao 蚂蚱腿子
Small shrubs, deciduous

9.2 (2.5~15.1); -6.4 (-15.9~4.4); 23.1 (17.5~27.8); 10.6 (6.6~15.2); 85.8 (45.4~124.3); 34.7 (0.6~80.4); 674 (507~859) ❙ 566 (349~1087); 343 (227~512); 16 (7~42) ❙ -15.7 (-43.4~56.2); 546 (349~817); 181.0 (70.3~334.4)

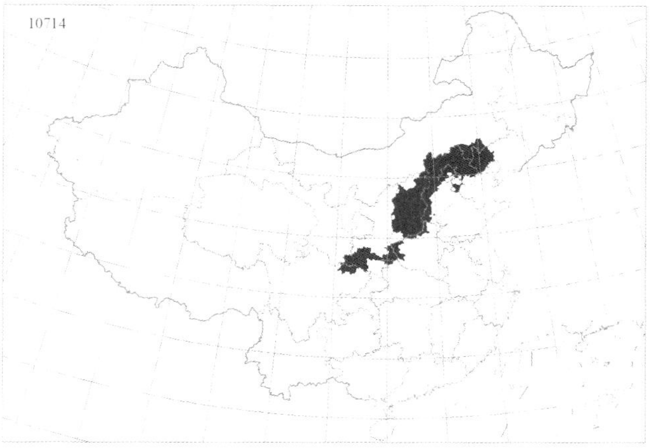

Asteraceae 菊科 *Pertya* 帚菊属
Pertya discolor Rehder 两色狭叶帚
Shrubs, deciduous

7.1 (-0.8~14.2); -7.8 (-12.7~1.8); 20.3 (9.2~26.6); 8.8 (2.9~14.2); 68.0 (10.1~116.7); 42.0 (5.8~79.5); 599 (374~800) ❙ 509 (341~862); 289 (190~376); 14 (8~47) ❙ -13.0 (-40.9~61.8); 488 (341~800); 164.0 (61.2~344.4)

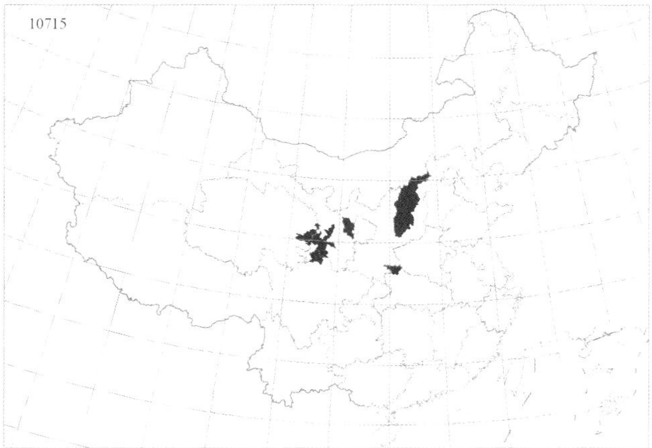

Asteraceae 菊科 *Pertya* 帚菊属
Pertya monocephala W. W. Sm. 单头帚菊
Shrubs, deciduous

10.5; 3.4; 16.1; 11.2; 89.2; 22.6; 641 ❙ 1018; 556; 31 ❙ 50.6; 630; 337.2

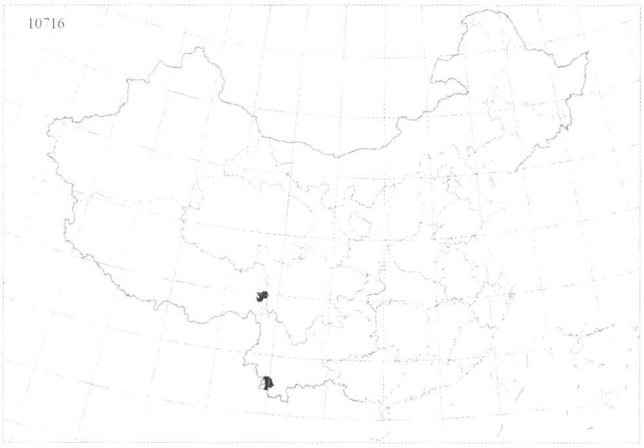

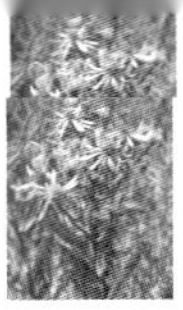

Asteraceae 菊科 *Vernonia* 斑鸠菊属
Vernonia chingiana Hand.-Mazz. 广西斑鸠菊
Scandent shrubs

19.4 (16.8~22.2); 10.0 (7.3~13.9); 26.8 (23.4~29.0); 19.4 (16.8~22.3); 172.7 (141.5~207.5); 0.0 (0.0~0.0); 1002 (836~1190) ❙ 1438 (1232~1740); 705 (617~848); 109 (51~187) ❙ 44.3 (15.4~85.5); 1002 (836~1190); 256.4 (160.7~371.7)

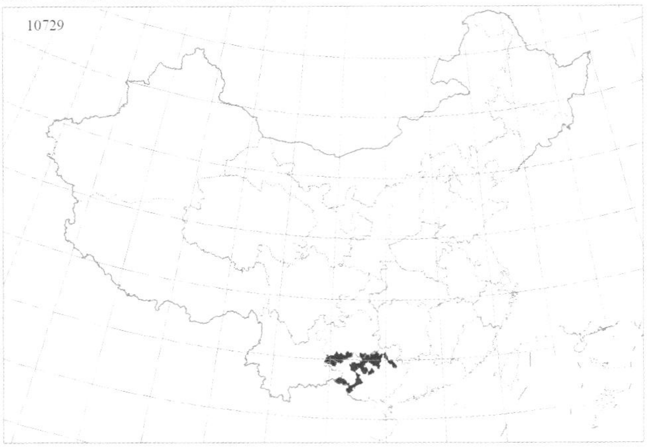

Asteraceae 菊科 *Vernonia* 斑鸠菊属
Vernonia chunii C. C. Chang 少花斑鸠菊
Scandent shrubs

24.4; 19.3; 27.9; 24.4; 233.1; 0.0; 1347 ❙ 1567; 541; 64 ❙ 16.5; 1206; 498.8

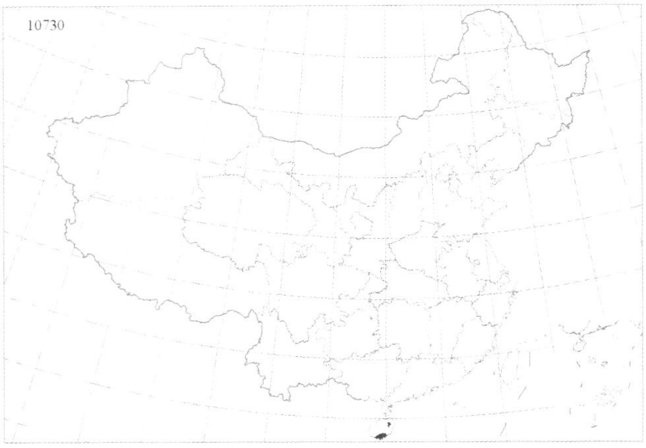

Asteraceae 菊科 *Vernonia* 斑鸠菊属
Vernonia cumingiana Benth. 毒根斑鸠菊
Scandent shrubss or vines

19.7 (10.6~24.8); 10.9 (-0.2~20.1); 27.0 (20.6~29.8); 19.7 (10.6~24.8); 176.9 (79.4~237.6); 0.2 (0.0~12.1); 1033 (643~1385) ❙ 1474 (1011~2435); 657 (442~1428); 109 (41~257) ❙ 43.7 (-12.0~100.8); 1023 (643~1360); 262.6 (108.2~659.0)

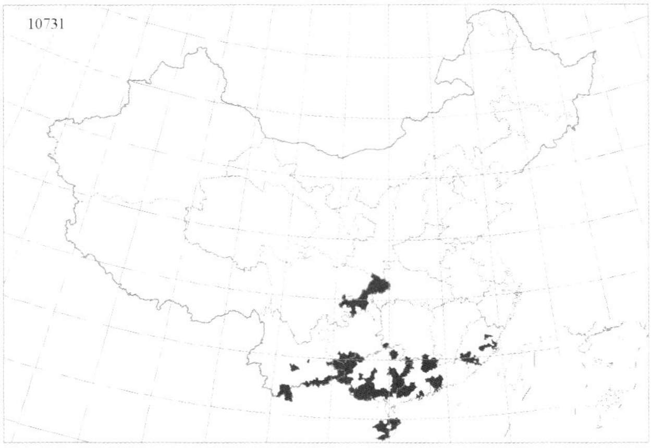

Asteraceae 菊科 *Vernonia* 斑鸠菊属
Vernonia esculenta Hemsl. 斑鸠菊
Shrubs or small trees

15.2 (2.2~21.6); 7.1 (-6.6~15.1); 21.7 (9.9~28.5); 15.3 (3.8~21.7); 124.8 (15.5~199.8); 2.1 (0.0~49.1); 793 (413~1153) ❙ 1131 (741~1703); 621 (411~911); 48 (15~194) ❙ 44.2 (-14.2~125.9); 779 (413~1153); 288.2 (130.1~659.0)

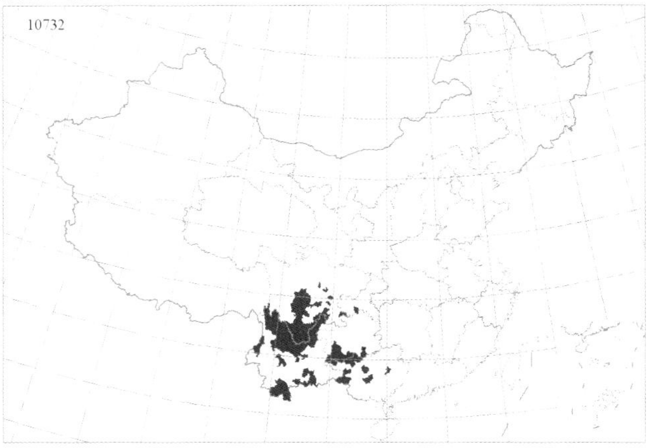

Asteraceae 菊科 *Vernonia* 斑鸠菊属
Vernonia extensa (Wall.) DC. 展枝斑鸠菊
Shrubs or subshrubs

17.3 (11.3~20.5); 9.9 (4.3~15.1); 22.6 (17.0~26.6); 17.3 (11.3~20.5); 147.6 (76.6~185.9); 0.0 (0.0~0.7); 852 (644~992) ❙ 1310 (939~1607); 711 (524~916); 56 (34~77) ❙ 54.5 (6.5~99.9); 843 (644~986); 363.5 (186.7~659.0)

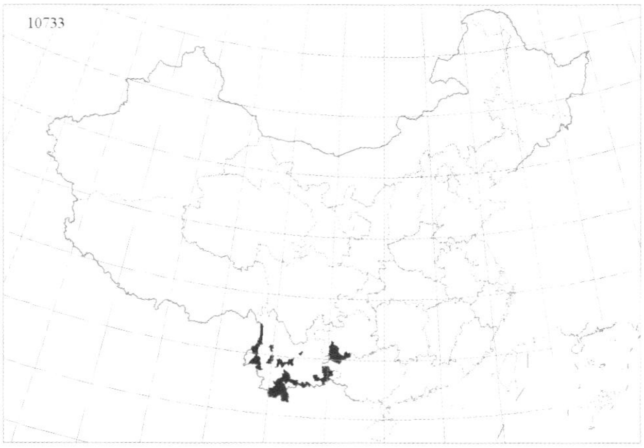

Asteraceae 菊科 *Vernonia* 斑鸠菊属
Vernonia forrestii Anthony 滇西斑鸠菊
Shrubs

13.3 (5.4~18.9); 5.6 (-2.1~12.0); 19.7 (12.3~26.6); 13.4 (5.8~19.0); 103.0 (31.5~167.8); 2.5 (0.0~25.6); 714 (481~929) ❙ 950 (741~1141); 553 (411~627); 30 (12~57) ❙ 35.1 (-14.2~70.8); 686 (481~912); 267.0 (142.9~351.7)

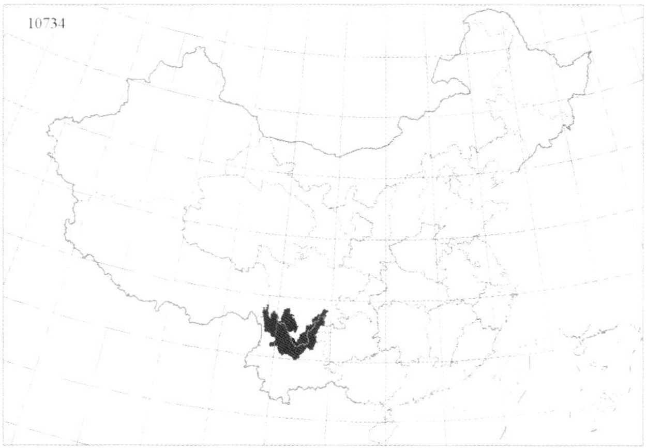

Asteraceae 菊科 *Vernonia* 班鸠菊属
Vernonia gratiosa Hance 台湾斑鸠菊
Vines

18.9 (16.2~21.1); 11.6 (9.3~13.5); 25.5 (20.1~28.6); 19.0 (16.2~21.1); 167.6 (134.2~193.7); 0.0 (0.0~0.0); 949 (773~1104) ❙ 2071 (1258~2993); 750 (425~1186); 272 (163~547) ❙ 123.8 (14.0~246.5); 942 (773~1047); 272.0 (206.0~300.0)

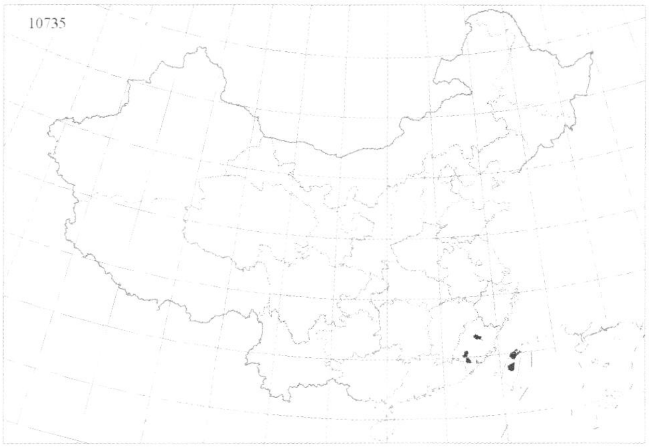

Asteraceae 菊科 *Vernonia* 班鸠菊属
Vernonia henryi Dunn 黄花斑鸠菊
Scandent shrubs

17.4 (16.3~19.0); 10.4 (9.1~13.1); 22.1 (21.4~23.1); 17.4 (16.4~19.1); 148.8 (136.4~168.8); 0.0 (0.0~0.0); 840 (797~909) ❙ 1172 (1007~1495); 638 (539~775); 50 (44~62) ❙ 39.3 (19.0~64.5); 833 (797~891); 340.9 (277.0~427.9)

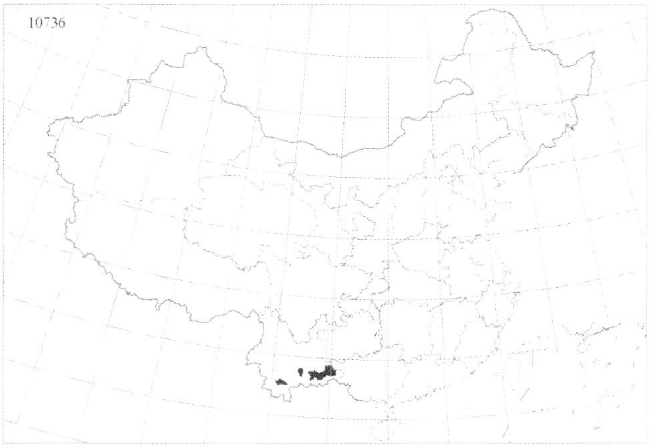

Asteraceae 菊科 *Vernonia* 班鸠菊属
Vernonia maritima Merr. 滨海斑鸠菊
Subshrubs

21.7; 16.7; 25.4; 21.7; 200.8; 0.0; 1117 ❙ 2255; 1381; 87 ❙ 117.5; 1003; 203.3

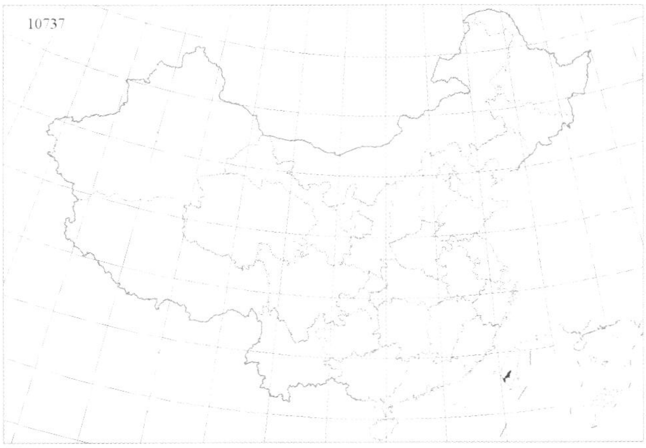

Asteraceae 菊科 *Vernonia* 班鸠菊属
Vernonia parishii Hook. f. 滇缅斑鸠菊
Trees

18.3 (14.4~21.0); 11.7 (7.5~15.1); 22.7 (19.1~25.6); 18.3 (14.4~21.0); 159.5 (113.1~191.8); 0.0 (0.0~0.0); 886 (727~1053) ❙ 1399 (1143~1607); 748 (637~911); 59 (48~77) ❙ 58.7 (21.7~99.9); 879 (727~1053); 465.6 (336.3~659.0)

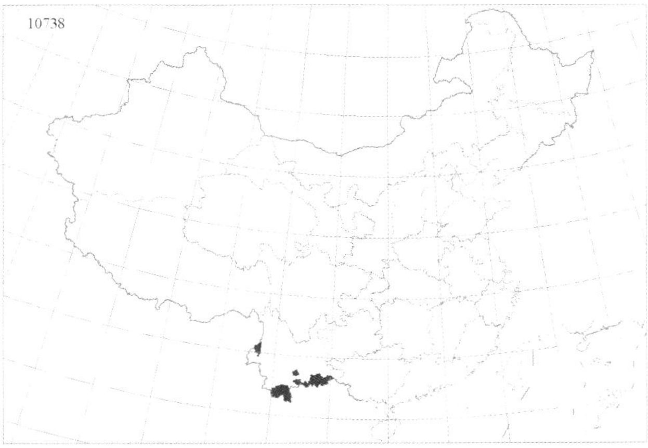

Asteraceae 菊科 *Vernonia* 班鸠菊属
Vernonia solanifolia Benth. 茄叶斑鸠菊
Shrubs or small trees

20.7 (16.9~24.8); 12.5 (7.1~20.1); 27.3 (22.3~29.0); 20.8 (17.0~24.8); 189.4 (143.6~237.6); 0.0 (0.0~0.0); 1086 (838~1385) ❙ 1582 (958~2435); 707 (444~1428); 128 (41~235) ❙ 46.9 (-2.2~100.8); 1075 (823~1331); 286.8 (163.7~659.0)

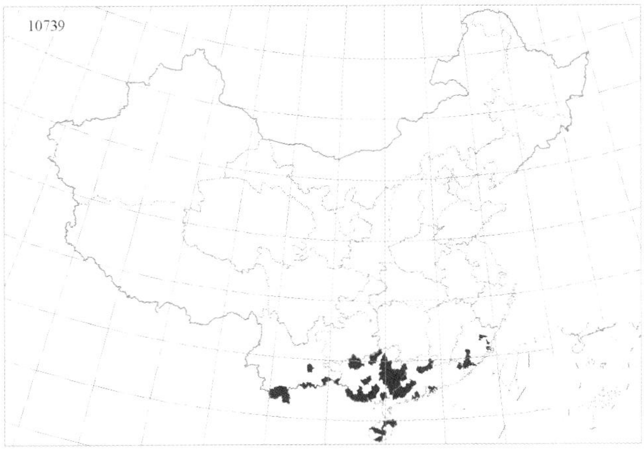

Asteraceae 菊科 *Vernonia* 班鸠菊属
Vernonia sylvatica Dunn 林生斑鸠菊
Scandent shrubs

18.8 (16.3~23.6); 11.4 (7.3~16.7); 24.0 (20.3~28.6); 18.8 (16.4~23.6); 165.5 (136.6~223.5); 0.0 (0.0~0.0); 935 (789~1281) ❙ 1420 (1178~1888); 763 (551~1044); 72 (42~142) ❙ 53.6 (-1.0~78.9); 931 (780~1269); 389.9 (225.9~551.6)

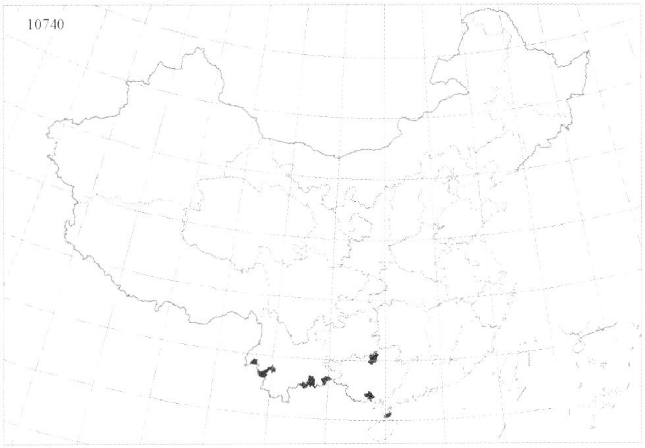

Asteraceae 菊科 *Vernonia* 斑鸠菊属
Vernonia volkameriifolia (Wall.) DC. 大叶斑鸠菊
Trees

17.9 (13.2~21.3); 10.5 (5.4~15.1); 23.3 (18.4~27.9); 17.9 (13.3~21.3); 154.8 (99.4~196.0); 0.0 (0.0~0.0); 889 (695~1123) ❙ 1318 (958~1736); 703 (531~1007); 59 (43~107) ❙ 50.2 (5.8~145.8); 882 (695~1121); 386.4 (189.5~659.0)

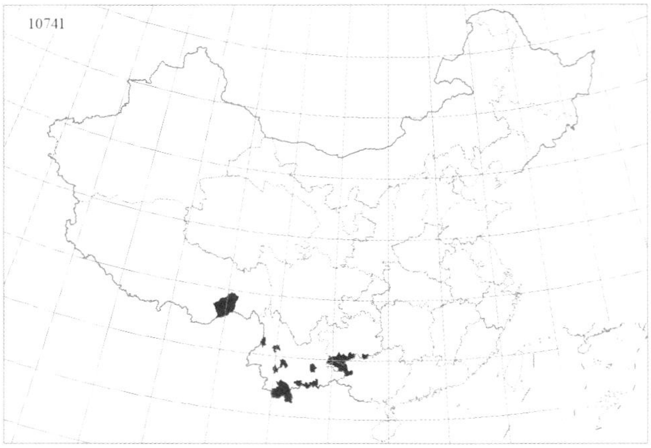

Poaceae 禾本科 *Acidosasa* 酸竹属
Acidosasa chienouensis (T. H. Wen) C. S. Chao et T. H. Wen 粉酸竹
Bamboos, evergreen

17.5; 7.3; 26.7; 17.5; 150.0; 0.0; 905 ❙ 1658; 570; 201 ❙ 83.1; 905; 257.6

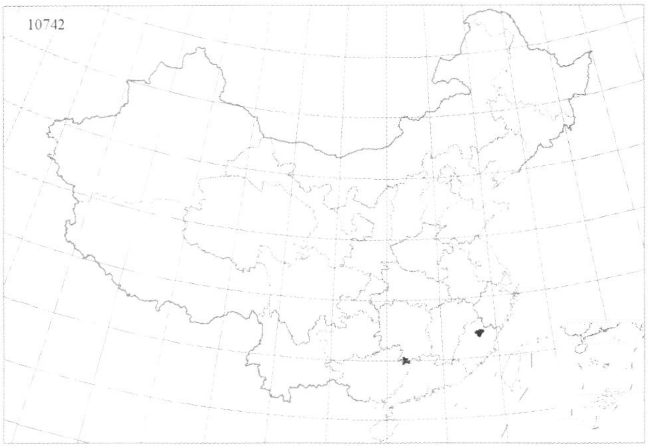

Poaceae 禾本科 *Acidosasa* 酸竹属
Acidosasa chinensis C. D. Chu et C. S. Chao ex Keng f. 酸竹
Bamboos, evergreen

21.8; 14.2; 27.8; 21.9; 202.6; 0.0; 1150 ❙ 1876; 891; 116 ❙ 63.2; 1150; 273.3

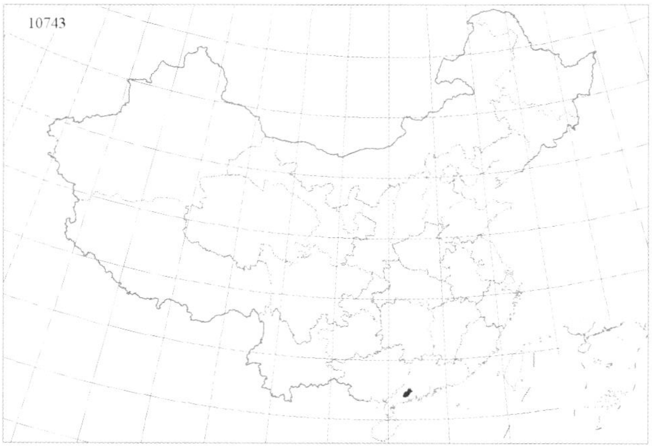

Poaceae 禾本科 *Acidosasa* 酸竹属
Acidosasa lingchuanensis (C. D. Chu et C. S. Chao) Q. Z. Xie et X. Y. Chen 灵川酸竹
Bamboos, evergreen

17.8; 7.1; 27.2; 17.8; 153.8; 0.0; 932 ❙ 1694; 661; 189 ❙ 81.7; 932; 245.0

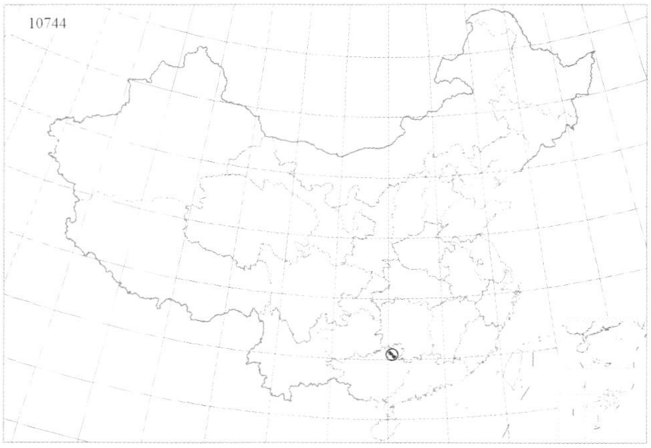

Poaceae 禾本科 *Acidosasa* 酸竹属
Acidosasa nanunica (McClure) C. S. Chao et G. Y. Yang 长舌酸竹
Bamboos, evergreen

19.8 (14.3~22.4); 10.7 (3.3~14.7); 27.5 (24.3~28.7); 19.9 (14.3~22.4); 178.6 (114.1~209.2); 0.3 (0.0~2.3); 1047 (765~1198) ❙ 1624 (1447~1781); 696 (505~845); 141 (75~182) ❙ 58.0 (20.8~99.4); 1047 (765~1198); 216.3 (164.9~257.1)

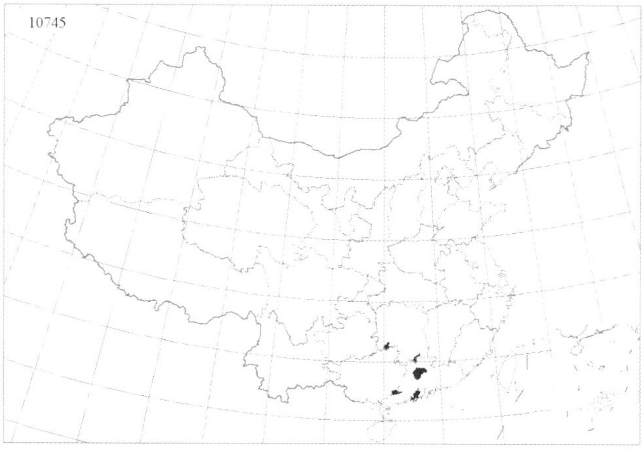

Poaceae 禾本科 *Acidosasa* 酸竹属
Acidosasa notata (Z. P. Wang et G. H. Ye) S. S. You 斑箨酸竹
Bamboos, evergreen

18.0 (16.6~18.6); 8.6 (6.0~10.4); 26.7 (25.9~27.6); 18.1 (16.6~18.7); 156.7 (138.9~164.0); 0.0 (0.0~0.0); 926 (866~965) ❙ 1759 (1559~1997); 609 (549~669); 207 (163~254) ❙ 90.4 (68.2~130.7); 926 (866~965); 294.9 (254.0~332.5)

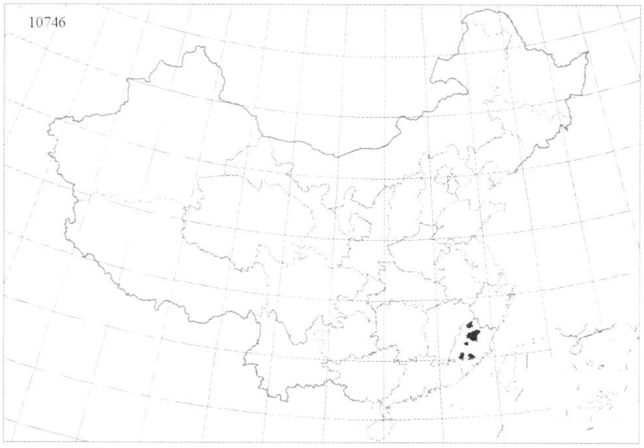

Poaceae 禾本科 *Acidosasa* 酸竹属

Acidosasa purpurea (Hsueh et T. P. Yi) P. C. Keng 毛花酸竹

Bamboos, evergreen

18.1 (15.5~21.0); 10.8 (4.0~14.2); 23.6 (21.4~26.6); 18.2 (15.6~21.0); 158.0 (128.0~191.8); 0.1 (0.0~1.0); 889 (813~1053) ❙ 1399 (1178~1547); 769 (584~911); 63 (51~120) ❙ 57.9 (36.8~80.6); 888 (813~1053); 423.3 (208.1~551.6)

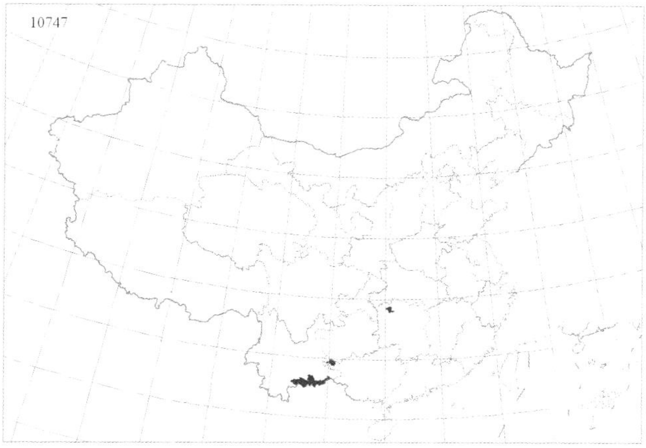

Poaceae 禾本科 *Ampelocalamus* 悬竹属

Ampelocalamus actinotrichus (Merr. et Chun) S. L. Chen, T. H. Wen et G. Y. Sheng 射毛悬竹

Bamboos, evergreen

24.5; 19.4; 28.2; 24.6; 234.8; 0.0; 1361 ❙ 1292; 432; 41 ❙ -5.2; 1097; 491.4

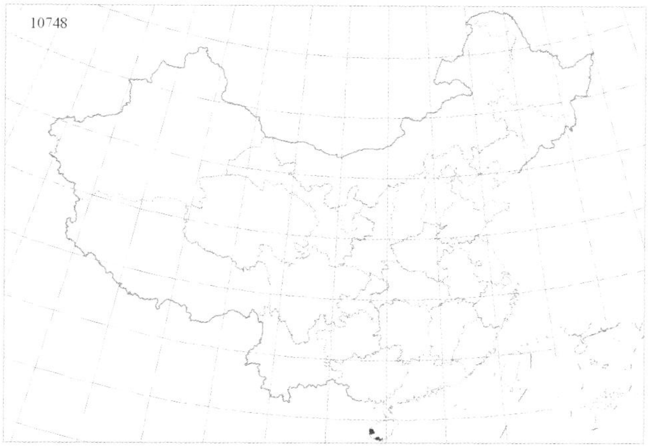

Poaceae 禾本科 *Ampelocalamus* 悬竹属

Ampelocalamus calcareus C. D. Chu et C. S. Chao 贵州悬竹

Bamboos, evergreen

16.7 (15.6~18.9); 7.0 (5.5~9.5); 24.9 (24.3~26.6); 16.8 (15.7~18.9); 141.4 (128.3~167.3); 0.0 (0.0~0.0); 856 (810~969) ❙ 1295 (1236~1381); 612 (552~661); 77 (60~107) ❙ 51.9 (27.5~62.5); 856 (810~969); 239.9 (215.4~254.8)

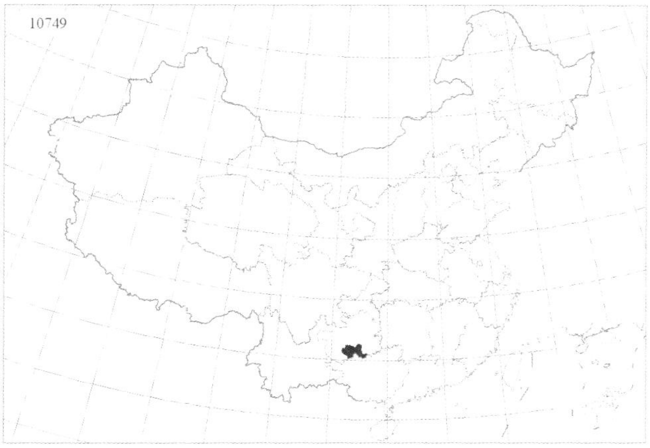

Poaceae 禾本科 *Ampelocalamus* 悬竹属

Ampelocalamus hirsutissimus (W. D. Li et Y. C. Zhong) Stapleton et D. Z. Li 多毛悬竹

Bamboos

15.6; 5.5; 24.6; 15.6; 127.5; 0.0; 808 ❙ 1206; 546; 69 ❙ 49.2; 808; 189.4

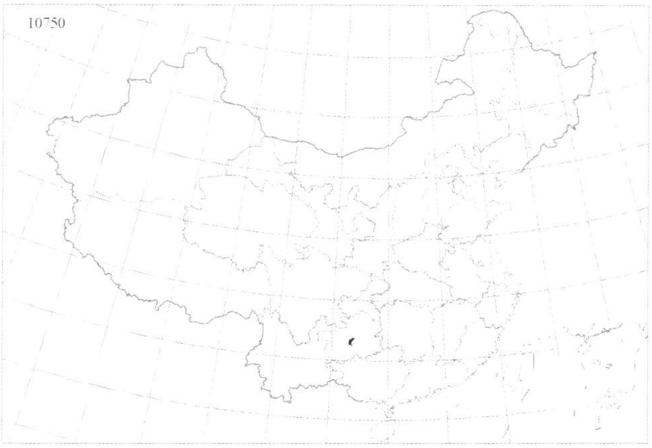

Poaceae 禾本科 *Ampelocalamus* 悬竹属

Ampelocalamus luodianensis T. P. Yi et R. S. Wang 小篷竹

Bamboos, evergreen

17.5 (15.0~19.1); 7.8 (4.5~10.1); 25.8 (24.9~26.6); 17.5 (15.0~19.1); 150.5 (120.8~169.2); 0.1 (0.0~0.5); 902 (794~969) ❙ 1187 (1089~1256); 597 (499~701); 60 (51~69) ❙ 31.8 (27.5~38.8); 902 (794~969); 235.6 (181.2~282.2)

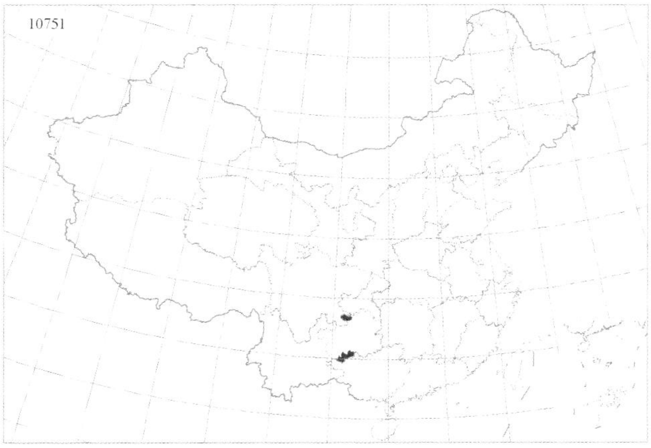

Poaceae 禾本科 *Ampelocalamus* 悬竹属

Ampelocalamus melicoideus (P. C. Keng) D. Z. Li et Stapleton 南川竹

Bamboos, evergreen

14.7; 4.0; 24.9; 14.7; 117.5; 1.1; 786 ❙ 1307; 530; 78 ❙ 66.3; 786; 181.6

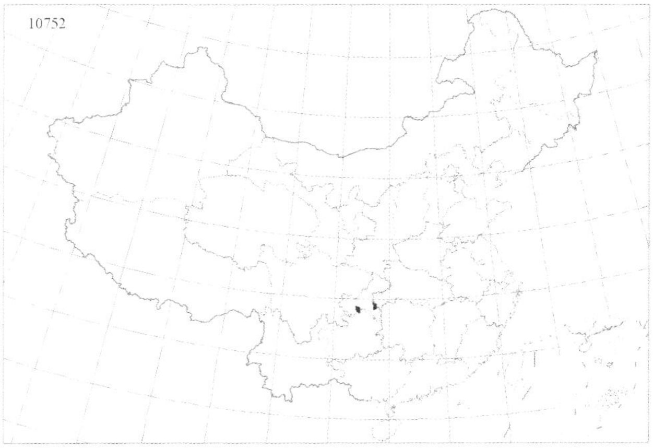

Poaceae 禾本科 *Ampelocalamus* 悬竹属
Ampelocalamus mianningensis (Q. Li et X. Jiang) D. Z. Li et Stapleton 冕宁悬竹
Bamboos, evergreen

10.2; 2.4; 16.9; 10.3; 68.2; 5.1; 620 ❙ 1036; 606; 18 ❙ 67.1; 620; 298.0

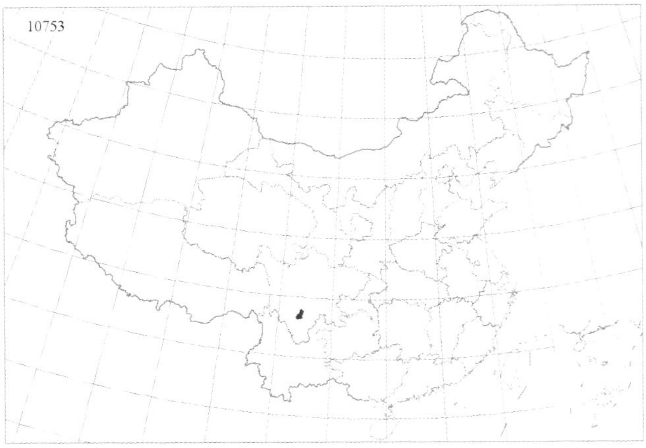

Poaceae 禾本科 *Ampelocalamus* 悬竹属
Ampelocalamus microphyllus (Hsueh et T. P. Yi) Hsueh et T. P. Yi 坝竹
Bamboos, evergreen

15.4; 4.6; 25.6; 15.4; 125.5; 0.4; 820 ❙ 1296; 510; 74 ❙ 58.1; 820; 183.1

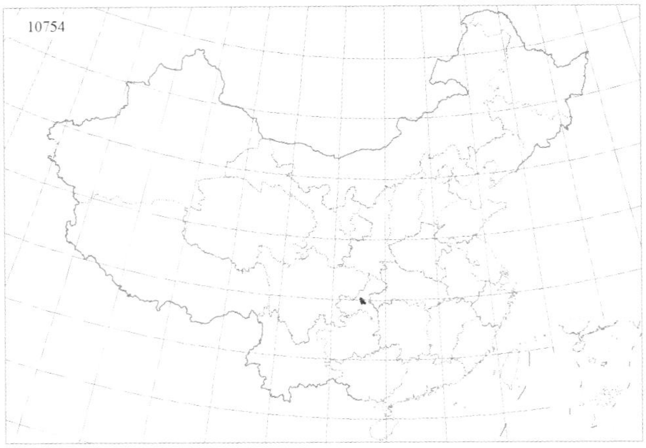

Poaceae 禾本科 *Ampelocalamus* 悬竹属
Ampelocalamus naibunensis (Hayata) T. H. Wen 内门竹
Bamboos, evergreen

22.2; 17.7; 25.5; 22.2; 206.7; 0.0; 1123 ❙ 2992; 1847; 107 ❙ 166.5; 1123; 296.9

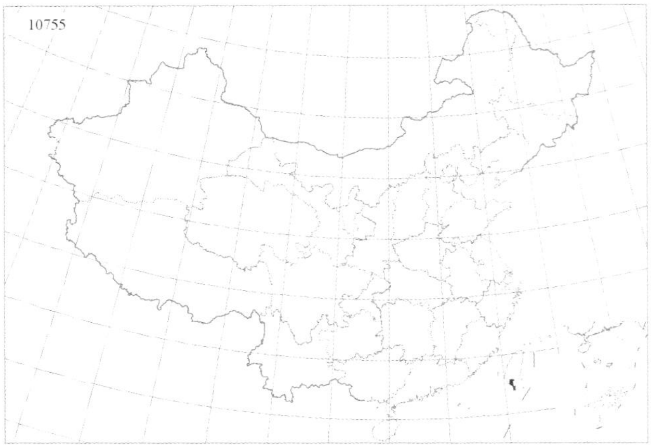

Poaceae 禾本科 *Ampelocalamus* 悬竹属
Ampelocalamus saxatilis (Hsueh et T. P. Yi) Hsueh et T. P. Yi 羊竹子
Bamboos, evergreen

12.6; 3.2; 21.3; 12.7; 95.3; 3.1; 699 ❙ 1104; 593; 46 ❙ 59.7; 699; 253.6

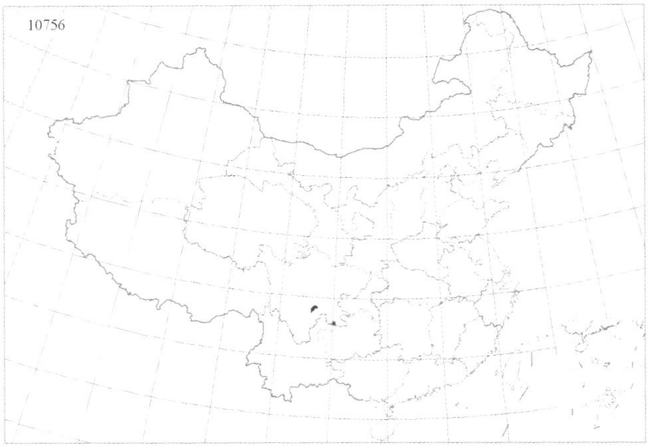

Poaceae 禾本科 *Ampelocalamus* 悬竹属
Ampelocalamus scandens Hsueh et W. D. Li 爬竹
Bamboos

15.9; 5.5; 25.6; 15.9; 131.1; 0.0; 833 ❙ 1089; 503; 69 ❙ 30.9; 833; 181.2

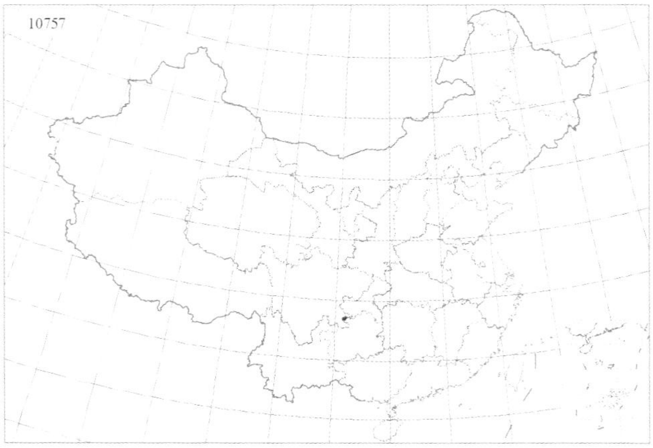

Poaceae 禾本科 *Ampelocalamus* 悬竹属
Ampelocalamus yongshanensis Hsueh et D. Z. Li 永善悬竹
Bamboos

12.1; 3.0; 20.0; 12.1; 88.4; 2.8; 675 ❙ 944; 520; 33 ❙ 39.8; 675; 249.8

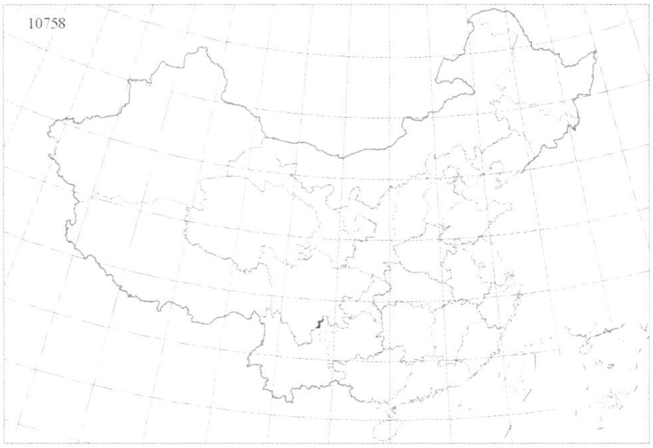

Poaceae 禾本科 *Arundinaria* 青篱竹属
Arundinaria faberi Rendle 冷箭竹
Bamboos, evergreen

6.1；−2.8；14.2；6.6；39.5；25.4；508 ❙ 1002；537；26 ❙ 97.0；508；211.1

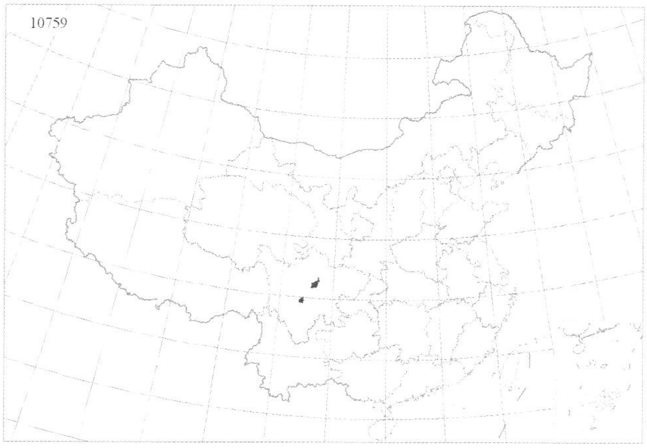

Poaceae 禾本科 *Arundinaria* 青篱竹属
Arundinaria fargesii E. G. Camus 巴山木竹
Bamboos, evergreen

12.6 (7.4~15.2)；1.0 (−4.2~4.4)；23.7 (18.7~27.8)；12.7 (8.2~15.2)；100.8 (58.8~123.9)；8.8 (0.6~29.4)；728 (561~854) ❙ 957 (581~1280)；433 (257~692)；34 (16~94) ❙ 32.1 (−32.0~93.8)；719 (561~817)；248.6 (147.3~334.4)

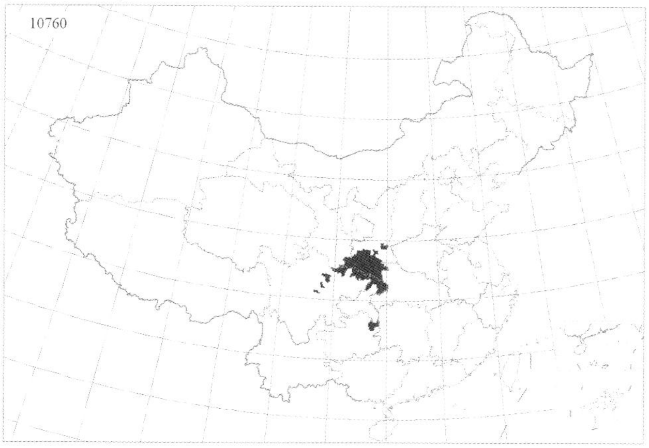

Poaceae 禾本科 *Arundinaria* 青篱竹属
Arundinaria qingchengshanensis (P. C. Keng et T. P. Yi) D. Z. Li 饱竹子
Bamboos, evergreen

14.7；4.6；23.7；14.7；117.4；1.0；783 ❙ 1203；692；39 ❙ 53.3；783；169.5

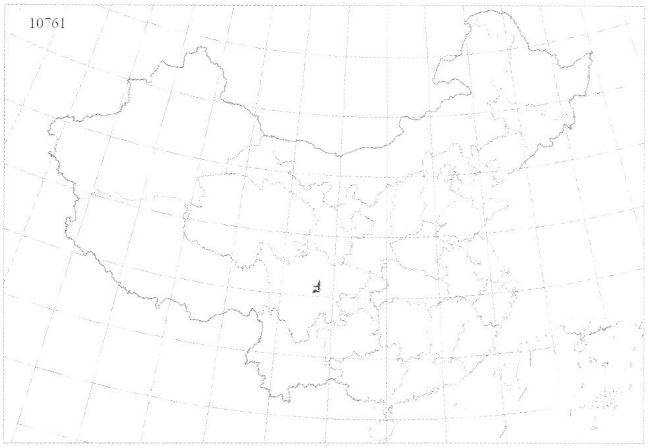

Poaceae 禾本科 *Bambusa* 簕竹属
Bambusa albolineata (McClure) L. C. Chia 花竹
Bamboos, evergreen

20.2 (16.2~24.1)；12.0 (8.2~18.9)；27.3 (20.1~29.8)；20.3 (16.2~24.1)；183.4 (134.2~229.7)；0.0 (0.0~0.0)；1053 (773~1322) ❙ 1849 (1509~2993)；764 (508~1594)；189 (47~547) ❙ 80.9 (32.5~246.5)；1034 (773~1182)；230.9 (154.2~289.1)

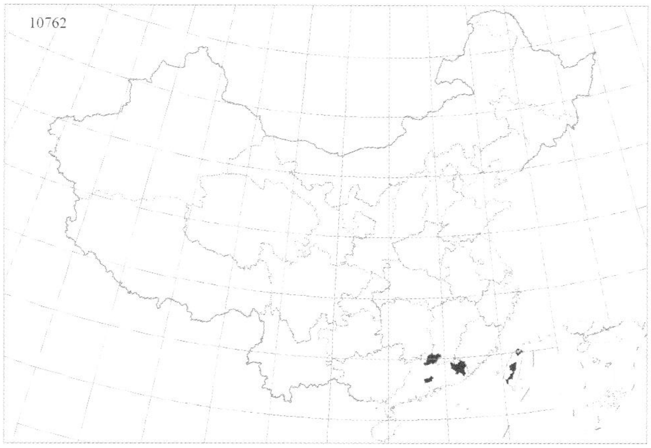

Poaceae 禾本科 *Bambusa* 簕竹属
Bambusa angustiaurita W. T. Lin 狭耳坭竹
Bamboos, evergreen

21.1；12.2；28.3；21.1；193.4；0.0；1115 ❙ 1673；690；153 ❙ 50.0；1115；196.6

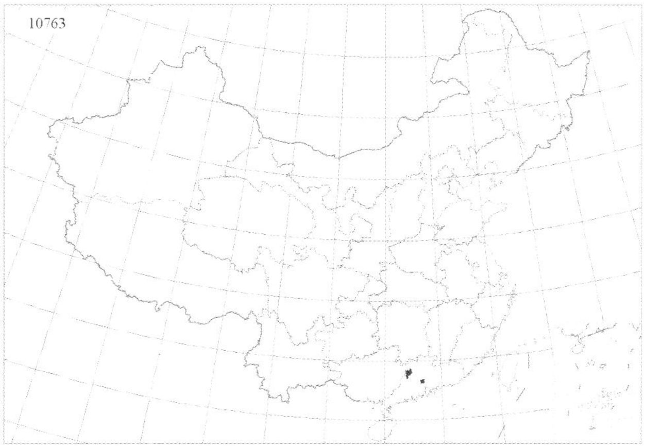

Poaceae 禾本科 *Bambusa* 簕竹属
Bambusa angustissima L. C. Chia et H. L. Fung 狭耳簕竹
Bamboos, evergreen

21.9 (20.1~23.2)；13.8 (10.7~15.8)；28.3 (27.8~28.9)；21.9 (20.1~23.2)；203.0 (181.1~218.9)；0.0 (0.0~0.0)；1163 (1048~1257) ❙ 1736 (1563~1876)；787 (601~891)；128 (102~161) ❙ 49.5 (34.1~63.2)；1163 (1048~1257)；221.0 (157.7~273.3)

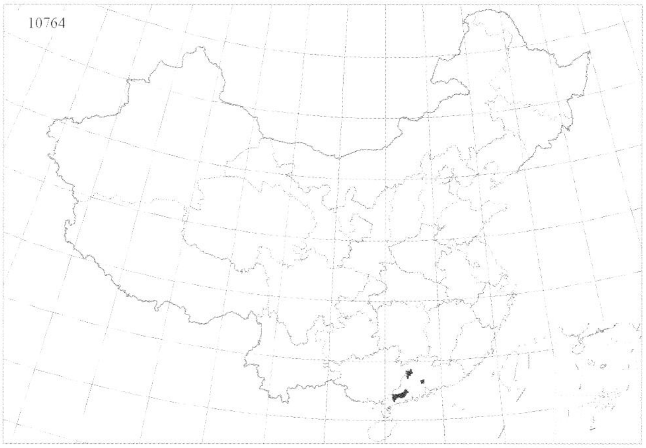

Poaceae 禾本科 *Bambusa* 簕竹属
Bambusa basihirsuta McClure 扁竹
Bamboos, evergreen

18.4 (16.5~22.1); 9.0 (4.4~13.7); 27.3 (26.1~28.9); 18.4 (16.6~22.1); 161.3 (139.8~205.6); 0.1 (0.0~0.6); 956 (896~1182) ❙ 1616 (1377~1784); 588 (493~779); 180 (144~204) ❙ 69.6 (50.9~83.1); 956 (896~1182); 244.9 (157.7~300.0)

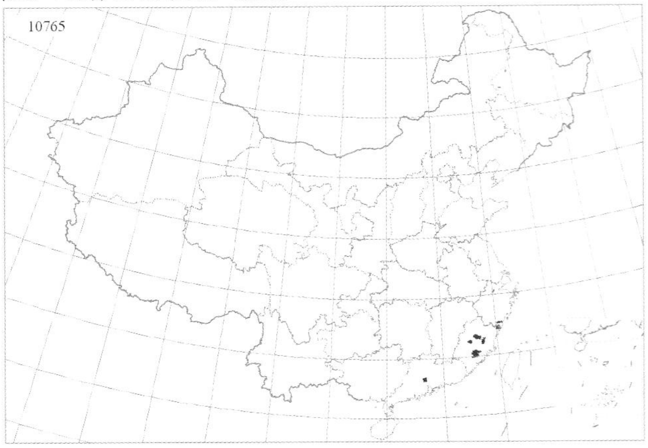

Poaceae 禾本科 *Bambusa* 簕竹属
Bambusa beecheyana Munro 吊丝球竹
Bamboos, evergreen

18.1 (5.4~24.8); 10.4 (-2.1~20.1); 24.2 (12.3~29.0); 18.2 (5.8~24.8); 158.8 (31.5~237.6); 0.9 (0.0~25.6); 964 (481~1385) ❙ 1340 (741~2992); 686 (411~1847); 76 (15~188) ❙ 39.3 (-14.2~202.5); 938 (481~1360); 275.6 (142.9~620.0)

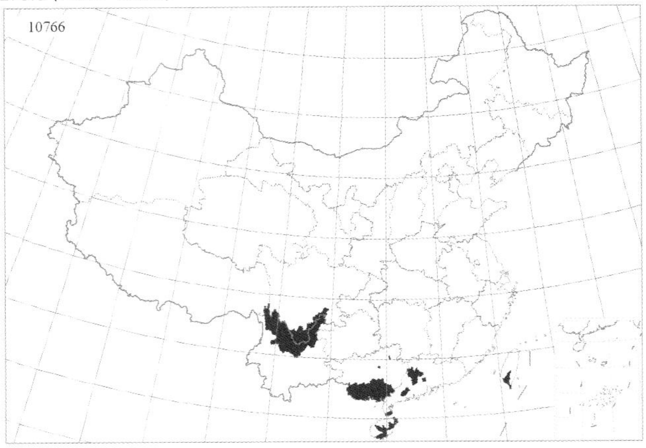

Poaceae 禾本科 *Bambusa* 簕竹属
Bambusa beecheyana var. pubescens (P. F. Li) W. C. Lin 大头典竹
Bamboos, evergreen

22.1 (21.8~22.4); 13.8 (13.5~14.3); 28.6 (27.8~28.9); 22.1 (21.9~22.5); 205.0 (202.6~209.7); 0.0 (0.0~0.0); 1175 (1150~1205) ❙ 1725 (1636~1876); 782 (722~891); 130 (116~144) ❙ 47.0 (39.8~63.2); 1175 (1150~1205); 197.3 (145.3~273.3)

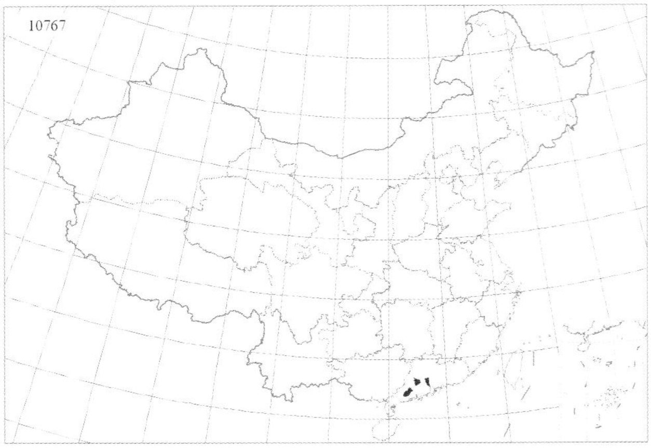

Poaceae 禾本科 *Bambusa* 簕竹属
Bambusa bicicatricata (W. T. Lin) L. C. Chia et H. L. Fung 孟竹
Bamboos, evergreen

24.1; 18.3; 28.2; 24.1; 229.4; 0.0; 1321 ❙ 1660; 572; 92 ❙ 25.6; 1294; 474.7

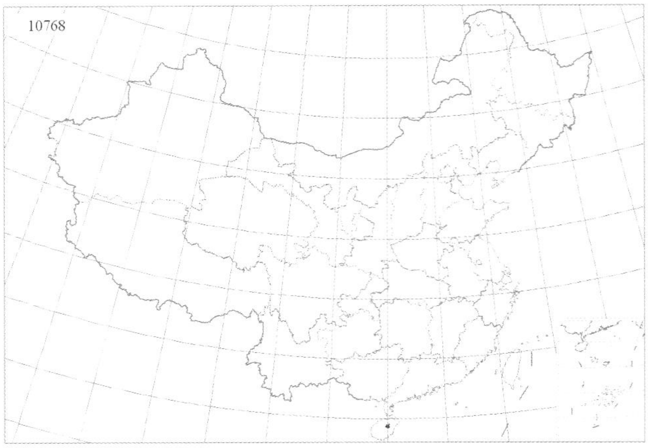

Poaceae 禾本科 *Bambusa* 簕竹属
Bambusa blumeana J. H. Schultes 簕竹
Bamboos, evergreen

21.1 (16.2~24.1); 13.0 (8.7~18.9); 27.5 (20.1~29.2); 21.1 (16.2~24.1); 193.6 (134.2~229.7); 0.0 (0.0~0.0); 1107 (773~1322) ❙ 1685 (1284~2993); 816 (419~1594); 138 (47~547) ❙ 56.0 (11.9~246.5); 1100 (773~1273); 254.1 (154.2~335.1)

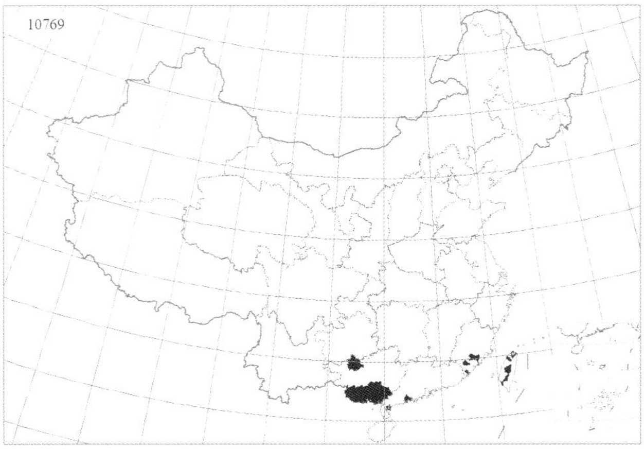

Poaceae 禾本科 *Bambusa* 簕竹属
Bambusa boniopsis McClure 妈竹
Bamboos, evergreen

23.8 (22.1~24.8); 18.0 (13.7~20.1); 28.0 (26.7~28.8); 23.8 (22.1~24.8); 226.0 (205.6~237.6); 0.0 (0.0~0.0); 1297 (1182~1385) ❙ 1448 (868~1854); 524 (312~779); 71 (41~144) ❙ 12.5 (-34.3~50.9); 1133 (868~1360); 444.5 (157.7~620.0)

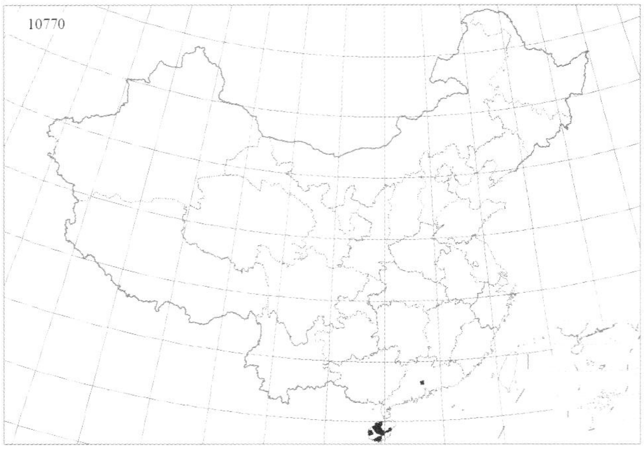

Poaceae 禾本科 *Bambusa* 簕竹属

Bambusa burmanica Gamble 缅甸竹

Bamboos, evergreen

17.9; 11.1; 21.9; 17.9; 154.6; 0.0; 862 ❙ 1621; 937; 52 ❙ 89.4; 855; 566.3

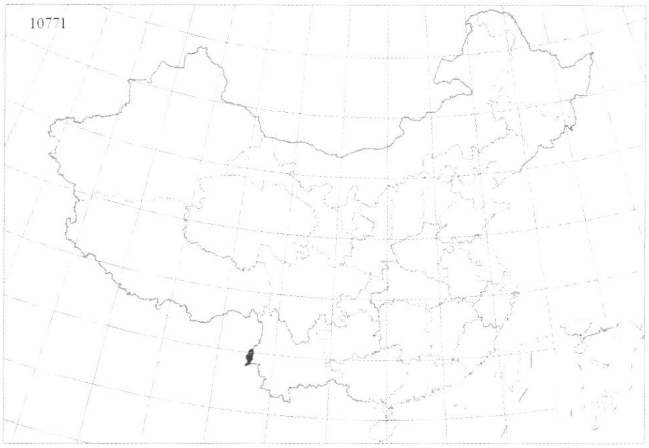

Poaceae 禾本科 *Bambusa* 簕竹属

Bambusa cerosissima McClure 箪竹

Bamboos, evergreen

19.1 (14.6~22.3); 9.4 (3.8~14.9); 27.3 (24.6~29.8); 19.2 (14.7~22.3); 170.0 (117.7~208.2); 0.0 (0.0~1.5); 1000 (779~1185) ❙ 1598 (1284~2035); 667 (497~969); 156 (68~196) ❙ 60.7 (24.6~101.8); 998 (779~1185); 248.8 (157.7~551.6)

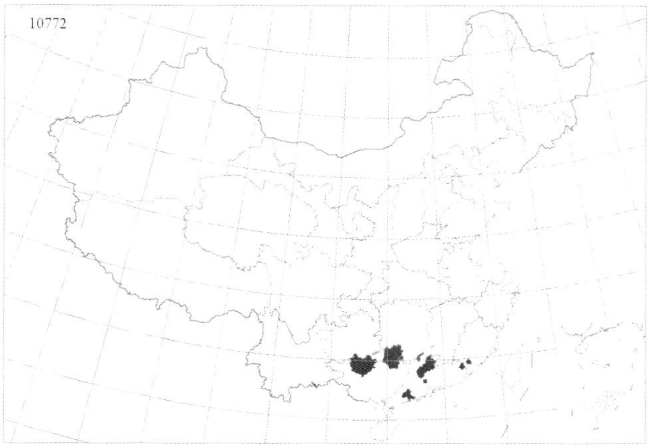

Poaceae 禾本科 *Bambusa* 簕竹属

Bambusa chungii McClure 粉箪竹

Bamboos, evergreen

20.9 (15.0~24.5); 12.5 (4.5~18.9); 27.6 (23.4~29.0); 20.9 (15.0~24.5); 190.9 (120.8~234.4); 0.0 (0.0~0.5); 1101 (794~1360) ❙ 1527 (868~2037); 687 (312~1033); 112 (51~182) ❙ 39.4 (-34.3~83.4); 1085 (794~1360); 252.4 (122.7~572.9)

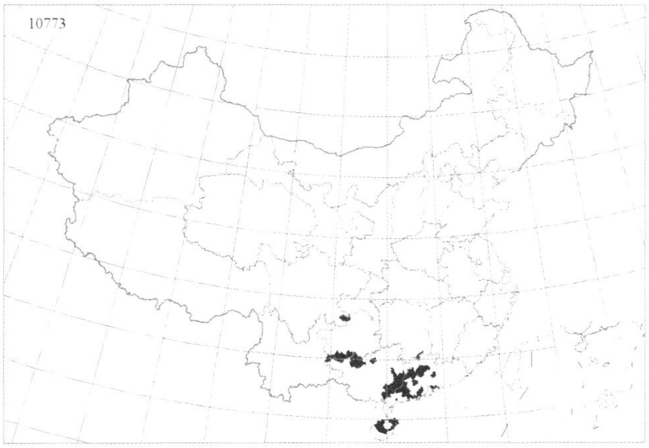

Poaceae 禾本科 *Bambusa* 簕竹属

Bambusa chunii L. C. Chia et H. L. Fung 焕镛簕竹

Bamboos, evergreen

15.2; 28.7; 22.7; 212.7; 0.0; 1216 ❙ 1965; 994; 104 ❙ 61.6; 1216; 204.4

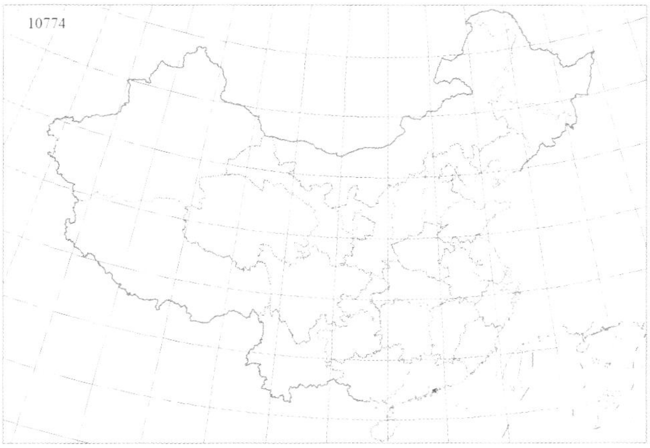

Poaceae 禾本科 *Bambusa* 簕竹属

Bambusa contracta L. C. Chia et H. L. Fung 破篾黄竹

Bamboos, evergreen

22.0; 13.8; 28.3; 22.0; 203.9; 0.0; 1166 ❙ 1966; 1006; 134 ❙ 68.8; 1166; 238.2

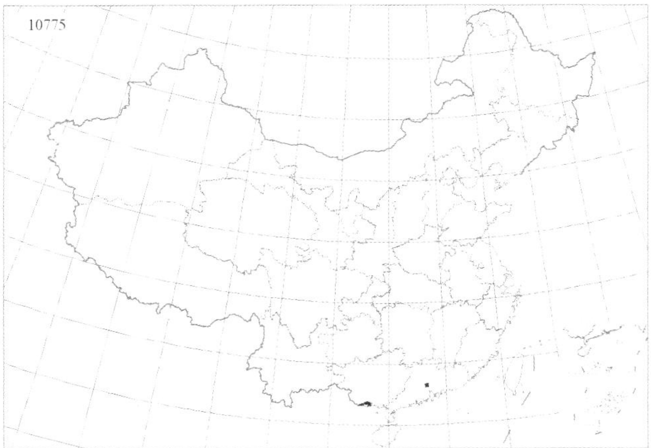

Poaceae 禾本科 *Bambusa* 簕竹属

Bambusa corniculata L. C. Chia et H. L. Fung 东兴黄竹

Bamboos, evergreen

21.1; 12.5; 28.1; 21.2; 194.1; 0.0; 1115 ❙ 1787; 820; 146 ❙ 60.0; 1115; 237.4

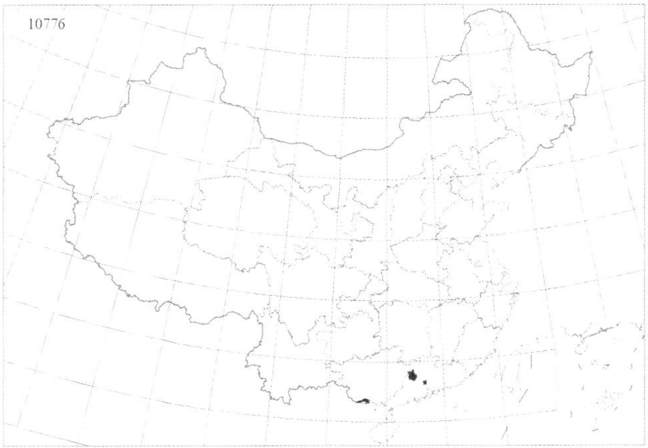

Poaceae 禾本科 *Bambusa* 簕竹属
Bambusa crispiaurita W. T. Lin et Z. M. Wu 皱耳石竹
Bamboos, evergreen

20.6; 11.6; 28.0; 20.6; 187.6; 0.0; 1080 ❙ 1653; 666; 156 ❙ 53.1; 1080; 237.4

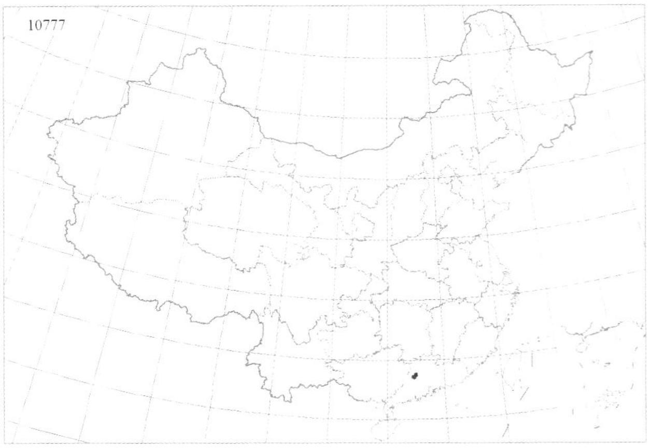

Poaceae 禾本科 *Bambusa* 簕竹属
Bambusa diaoluoshanensis L. C. Chia et H. L. Fung 吊罗坭竹
Bamboos, evergreen

23.4; 16.7; 28.4; 23.4; 221.0; 0.0; 1279 ❙ 1686; 656; 101 ❙ 33.1; 1197; 326.2

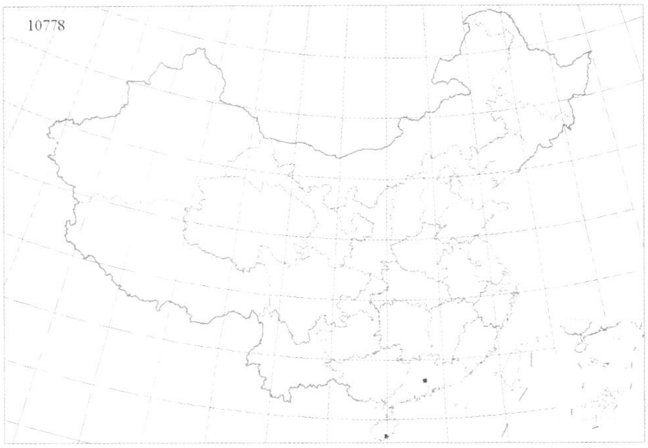

Poaceae 禾本科 *Bambusa* 簕竹属
Bambusa distegia (Keng et Keng f.) L. C. Chia et H. L. Fung 料慈竹
Bamboos, evergreen

17.2 (14.7~22.1); 7.5 (3.8~13.7); 25.6 (21.8~28.8); 17.2 (14.8~22.1); 147.0 (118.7~205.6); 0.2 (0.0~1.5); 891 (791~1182) ❙ 1249 (1089~1784); 589 (465~779); 72 (50~144) ❙ 40.5 (21.9~62.5); 891 (791~1182); 220.1 (146.0~411.1)

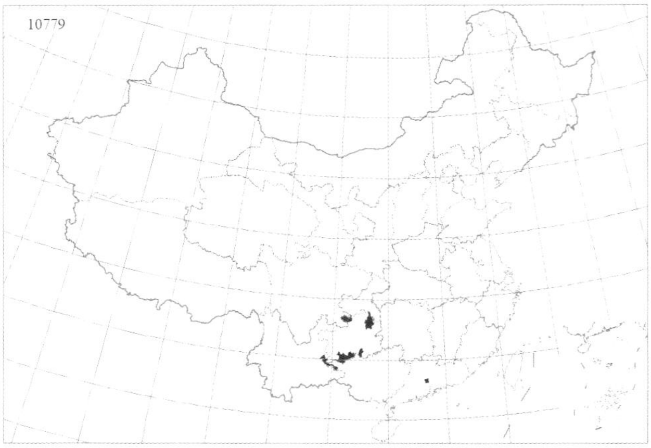

Poaceae 禾本科 *Bambusa* 簕竹属
Bambusa dolichoclada Hayata 长枝竹
Bamboos, evergreen

19.1 (16.2~22.1); 12.2 (9.8~15.3); 25.3 (20.1~28.8); 19.2 (16.2~22.1); 170.1 (134.2~205.6); 0.0 (0.0~0.0); 959 (773~1182) ❙ 2181 (1485~2993); 894 (522~1412); 250 (110~547) ❙ 133.1 (50.9~246.5); 959 (773~1182); 242.4 (157.7~289.1)

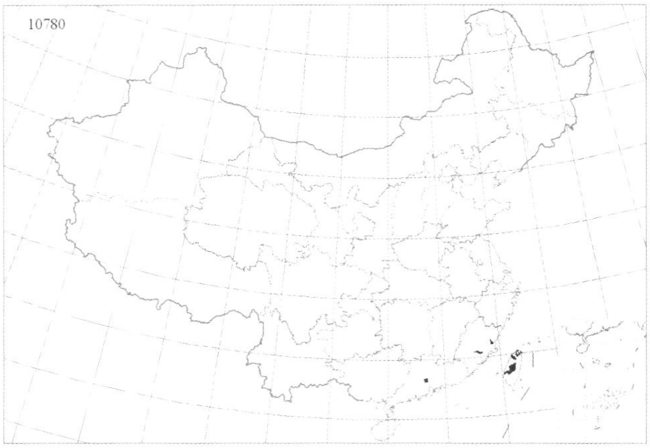

Poaceae 禾本科 *Bambusa* 簕竹属
Bambusa duriuscula W. T. Lin 蓬莱黄竹
Bamboos, evergreen

23.4; 17.8; 27.3; 23.4; 221.0; 0.0; 1249 ❙ 1716; 623; 81 ❙ 37.9; 1211; 471.6

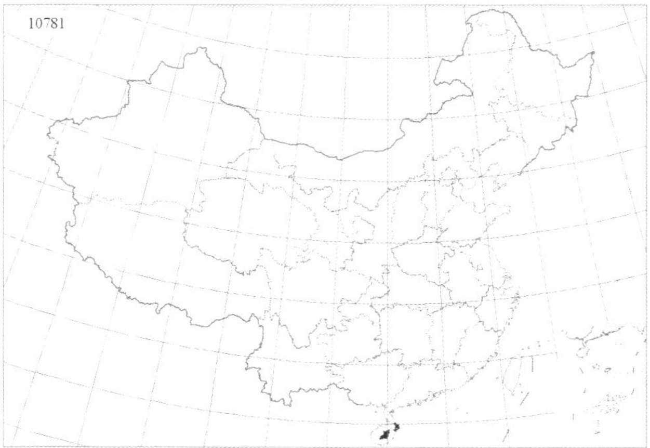

Poaceae 禾本科 *Bambusa* 簕竹属
Bambusa emeiensis L. C. Chia et H. L. Fung 慈竹
Bamboos, evergreen

15.7 (1.4~20.5); 5.5 (-7.6~13.1); 25.0 (9.2~29.9); 15.8 (3.5~20.5); 131.5 (13.4~186.1); 2.0 (0.0~55.6); 842 (402~1083) ❙ 1210 (579~1776); 550 (365~812); 80 (10~257) ❙ 45.3 (4.4~125.9); 838 (402~1083); 198.1 (108.2~432.9)

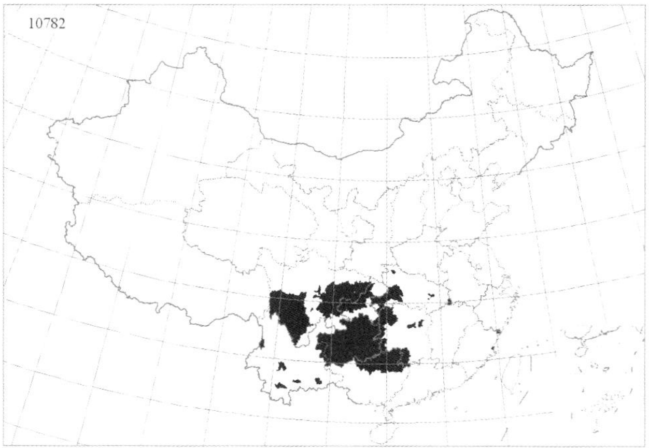

Poaceae 禾本科 *Bambusa* 簕竹属
Bambusa eutuldoides McClure 大眼竹
Bamboos, evergreen

20.3 (18.2~22.3); 11.0 (7.7~14.9); 28.3 (26.9~29.8); 20.4 (18.2~22.3); 184.3 (158.5~208.2); 0.0 (0.0~0.0); 1073 (949~1185) ❙ 1698 (1509~2035); 661 (507~969); 168 (113~189) ❙ 58.3 (35.6~71.8); 1061 (949~1185); 213.1 (157.7~244.8)

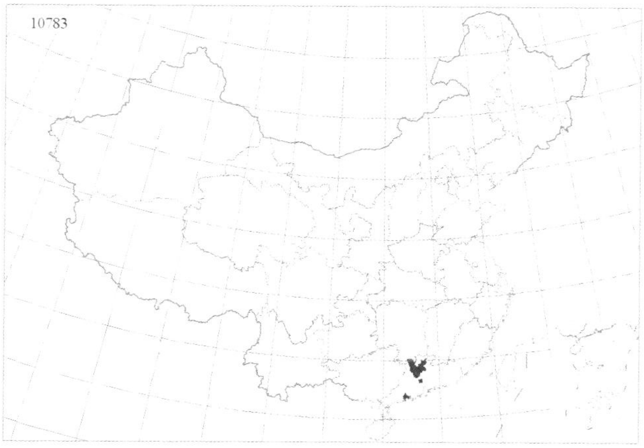

Poaceae 禾本科 *Bambusa* 簕竹属
Bambusa eutuldoides var. viridivittata (W. T. Lin) L. C. Chia 青丝黄竹
Bamboos, evergreen

22.1; 14.0; 28.6; 22.2; 205.9; 0.0; 1180 ❙ 1921; 881; 145 ❙ 62.9; 1180; 187.4

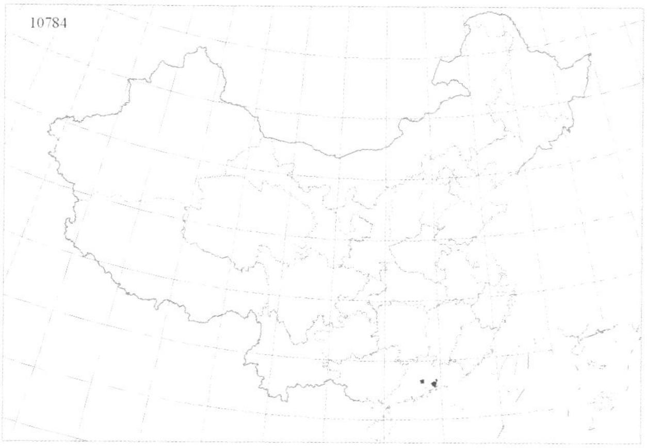

Poaceae 禾本科 *Bambusa* 簕竹属
Bambusa flexuosa Munro 小簕竹
Bamboos, evergreen

19.3 (12.4~23.0); 10.3 (3.8~15.1); 26.5 (18.6~29.0); 19.4 (12.4~23.0); 172.3 (89.1~216.2); 0.0 (0.0~1.5); 1009 (675~1247) ❙ 1457 (896~2435); 698 (497~1428); 119 (31~196) ❙ 45.3 (5.8~101.8); 1009 (675~1247); 249.6 (157.7~659.0)

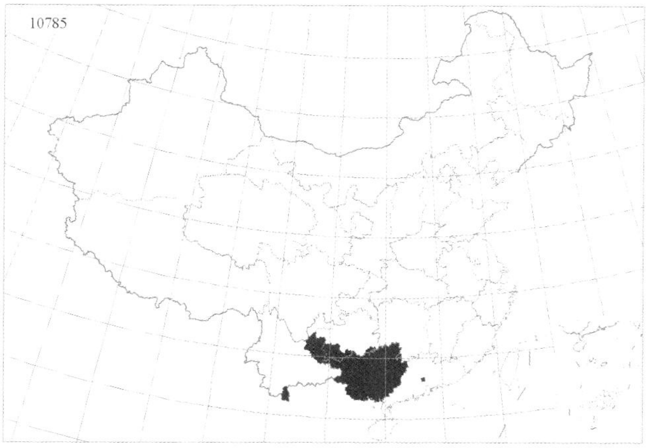

Poaceae 禾本科 *Bambusa* 簕竹属
Bambusa funghomii McClure 鸡窦簕竹
Bamboos, evergreen

22.3; 14.4; 28.3; 22.3; 208.0; 0.0; 1189 ❙ 1696; 824; 108 ❙ 42.6; 1189; 226.2

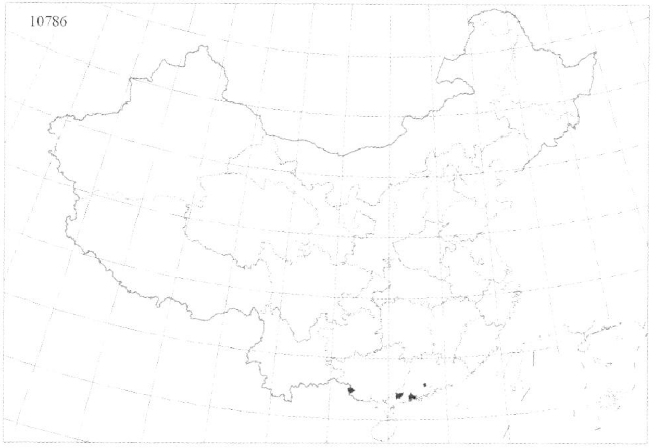

Poaceae 禾本科 *Bambusa* 簕竹属
Bambusa gibba McClure 坭竹
Bamboos, evergreen

19.3 (15.0~24.8); 10.3 (4.5~20.1); 27.0 (23.6~28.9); 19.4 (15.0~24.8); 172.6 (120.8~237.6); 0.1 (0.0~0.5); 1018 (794~1385) ❙ 1480 (1089~2037); 662 (465~1044); 105 (41~191) ❙ 46.1 (1.5~71.8); 1009 (794~1360); 226.5 (122.7~527.5)

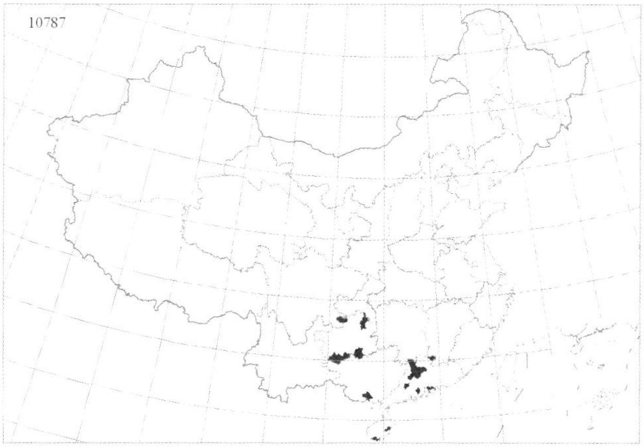

Poaceae 禾本科 *Bambusa* 簕竹属
Bambusa gibboides W. T. Lin 鱼肚腩竹
Bamboos, evergreen

21.0; 12.1; 28.4; 21.1; 192.7; 0.0; 1112 ❙ 1729; 688; 170 ❙ 55.7; 1112; 208.2

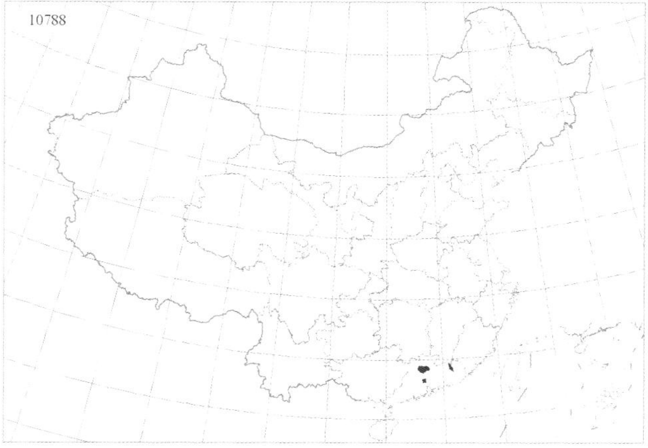

Poaceae 禾本科 *Bambusa* 簕竹属
Bambusa glabrovagina G. A. Fu 光鞘石竹
Bamboos, evergreen

24.1; 18.1; 28.4; 24.2; 229.8; 0.0; 1324 ▎1484; 526; 90 ▎12.1; 1274; 413.1

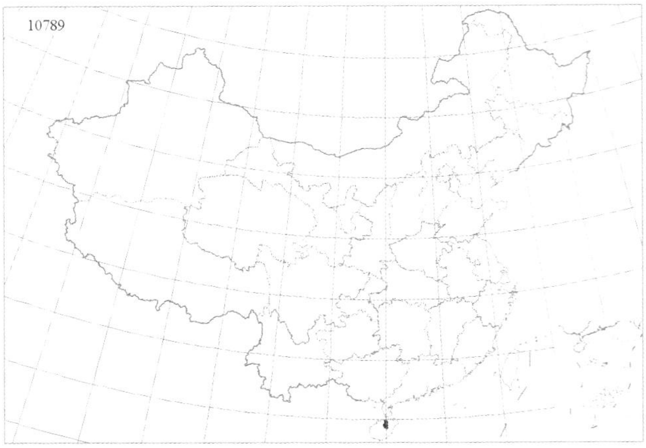

Poaceae 禾本科 *Bambusa* 簕竹属
Bambusa grandis (Q. H. Dai et X. L. Tao) Ohrnb. 大绿竹
Bamboos, evergreen

21.5; 12.1; 29.0; 21.6; 198.7; 0.0; 1155 ▎1437; 674; 136 ▎24.6; 1155; 181.1

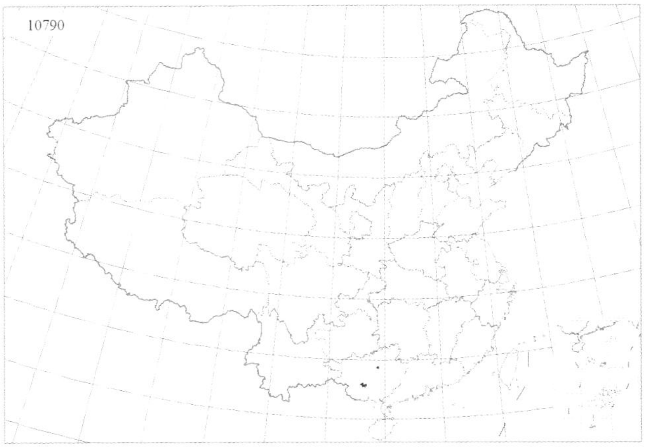

Poaceae 禾本科 *Bambusa* 簕竹属
Bambusa guangxiensis L. C. Chia et H. L. Fung 桂箪竹
Bamboos, evergreen

17.4; 6.9; 26.8; 17.5; 149.9; 0.0; 908 ▎1539; 533; 188 ▎69.5; 908; 231.1

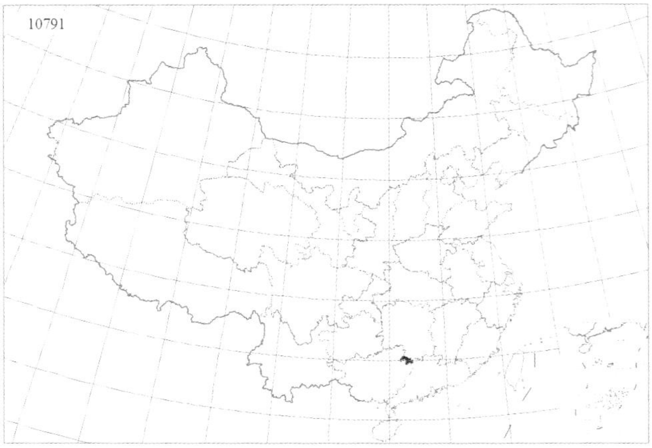

Poaceae 禾本科 *Bambusa* 簕竹属
Bambusa hainanensis L. C. Chia et H. L. Fung 藤箪竹
Bamboos, evergreen

24.4; 19.6; 27.6; 24.4; 232.9; 0.0; 1343 ▎1478; 506; 45 ▎10.3; 1142; 573.8

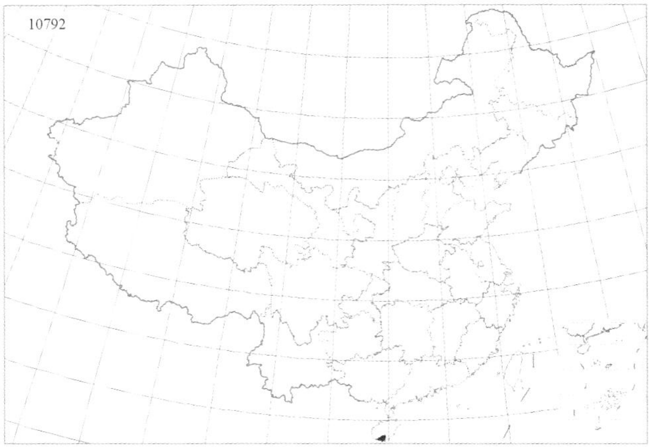

Poaceae 禾本科 *Bambusa* 簕竹属
Bambusa indigena L. C. Chia et H. L. Fung 乡土竹
Bamboos, evergreen

23.1; 16.0; 28.5; 23.1; 217.5; 0.0; 1252 ▎1722; 675; 118 ▎38.3; 1238; 316.2

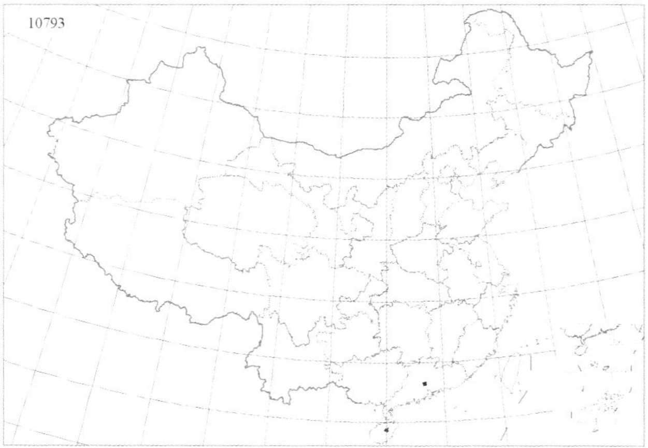

Poaceae 禾本科 *Bambusa* 簕竹属
Bambusa insularis L. C. Chia et H. L. Fung 黎庵高竹
Bamboos, evergreen

23.5; 16.7; 28.6; 23.5; 222.1; 0.0; 1280 ▎1540; 599; 98 ▎21.5; 1205; 348.1

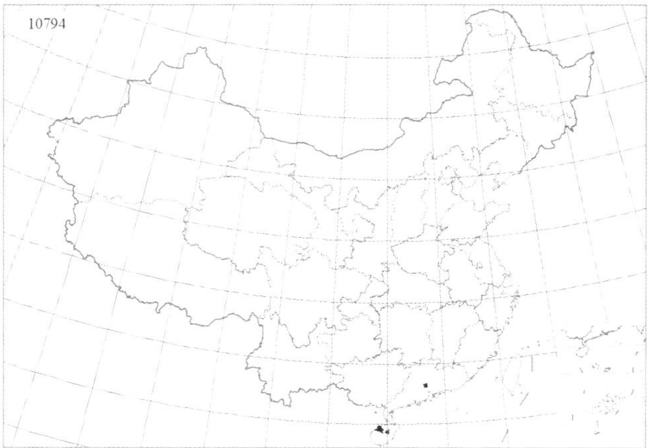

Poaceae 禾本科 *Bambusa* 簕竹属
Bambusa intermedia C. F. Hsieh et T. P. Yi 绵竹
Bamboos, evergreen

17.9 (15.8~19.2); 11.4 (9.0~13.3); 22.3 (20.6~23.5); 17.9 (15.8~19.4); 154.9 (129.7~172.6); 0.0 (0.0~0.0); 863 (776~932) ❙ 1140 (958~1428); 646 (531~835); 43 (44~56) ❙ 31.6 (13.3~55.7); 856 (772~923); 379.4 (258.7~506.2)

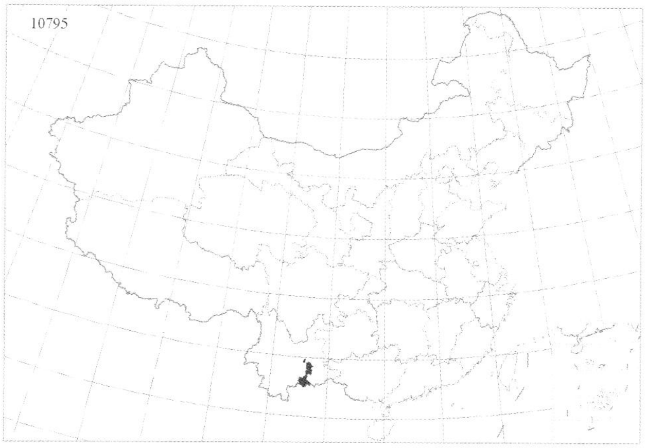

Poaceae 禾本科 *Bambusa* 簕竹属
Bambusa lapidea McClure 油簕竹
Bamboos, evergreen

19.4 (13.6~23.2); 12.0 (7.2~15.8); 24.8 (18.8~29.0); 19.5 (13.7~23.2); 173.5 (104.1~218.9); 0.0 (0.0~0.0); 986 (706~1257) ❙ 1515 (951~1784); 743 (547~959); 92 (20~182) ❙ 55.2 (17.2~112.2); 974 (706~1257); 354.7 (120.3~659.0)

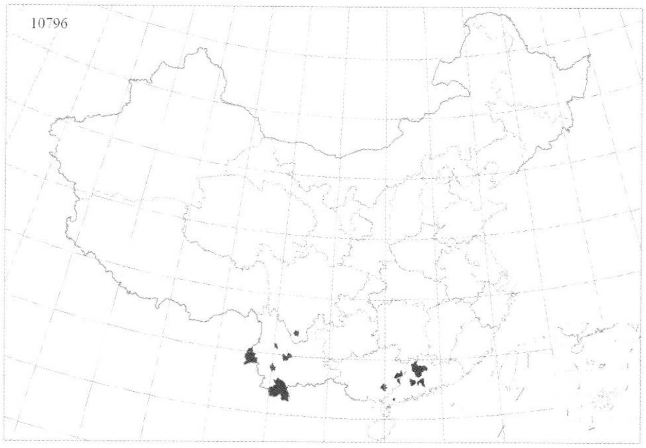

Poaceae 禾本科 *Bambusa* 簕竹属
Bambusa longipalea W. T. Lin 紫斑簕竹
Bamboos, evergreen

22.1; 13.7; 28.8; 22.1; 205.6; 0.0; 1182 ❙ 1784; 779; 144 ❙ 50.9; 1182; 157.7

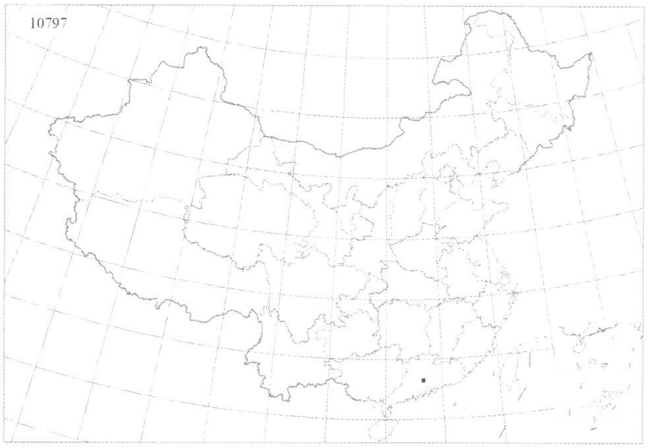

Poaceae 禾本科 *Bambusa* 簕竹属
Bambusa macrotis L. C. Chia et H. L. Fung 大耳坭竹
Bamboos, evergreen

20.7; 11.7; 28.2; 20.8; 189.5; 0.0; 1094 ❙ 1781; 707; 168 ❙ 62.7; 1094; 205.8

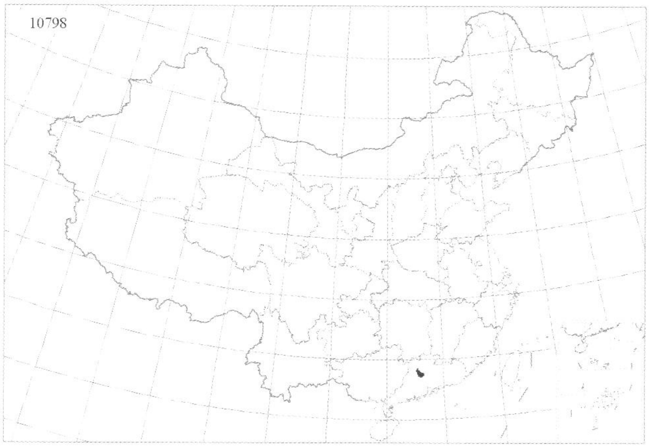

Poaceae 禾本科 *Bambusa* 簕竹属
Bambusa malingensis McClure 马岭竹
Bamboos, evergreen

23.1; 16.5; 28.2; 23.2; 218.1; 0.0; 1254 ❙ 1824; 844; 90 ❙ 47.9; 1171; 293.2

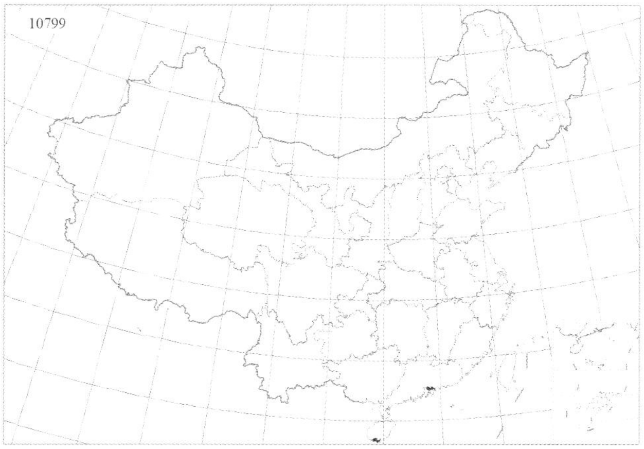

Poaceae 禾本科 *Bambusa* 簕竹属
Bambusa mollis L. C. Chia et H. L. Fung 拟黄竹
Bamboos, evergreen

21.4; 13.1; 27.7; 21.4; 196.9; 0.0; 1121 ❙ 1615; 734; 137 ❙ 44.0; 1121; 246.1

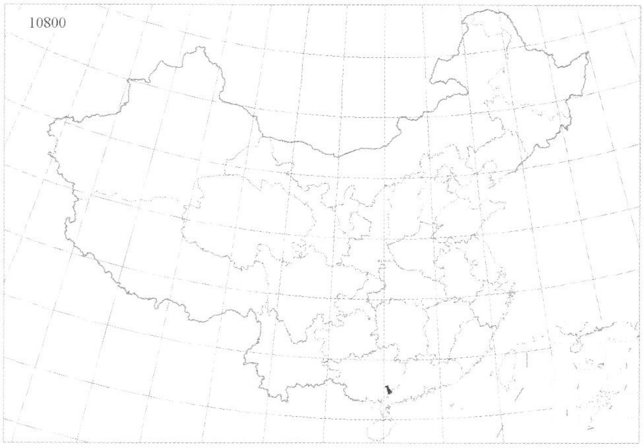

Poaceae 禾本科 *Bambusa* 簕竹属
Bambusa multiplex (Lour.) Raeusch. ex Schult. et Schult. f. 孝顺竹
Bamboos, evergreen

17.7 (9.8~24.1); 7.6 (-2.5~18.9); 26.9 (20.1~29.9); 17.7 (10.2~24.1); 153.1 (79.2~229.7); 0.5 (0.0~20.1); 932 (641~1322) ❙ 1568 (795~2993); 627 (323~1847); 156 (25~547) ❙ 69.2 (-1.6~246.5); 920 (641~1321); 225.5 (108.2~551.6)

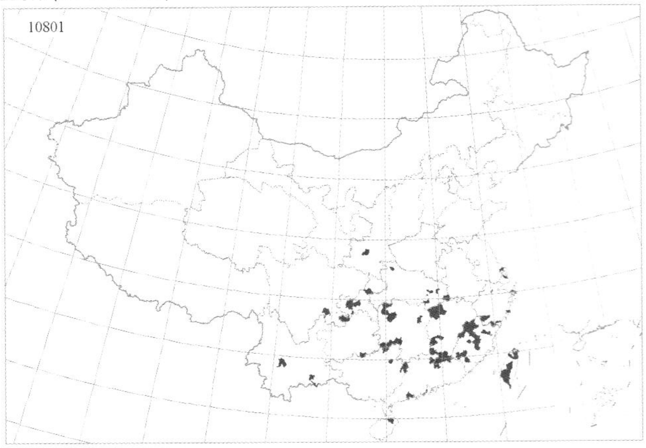

Poaceae 禾本科 *Bambusa* 簕竹属
Bambusa multiplex var. incana B. M. Yang 毛凤凰竹
Bamboos, evergreen

19.0; 9.1; 27.6; 19.0; 168.2; 0.0; 989 ❙ 1697; 618; 194 ❙ 71.6; 989; 236.9

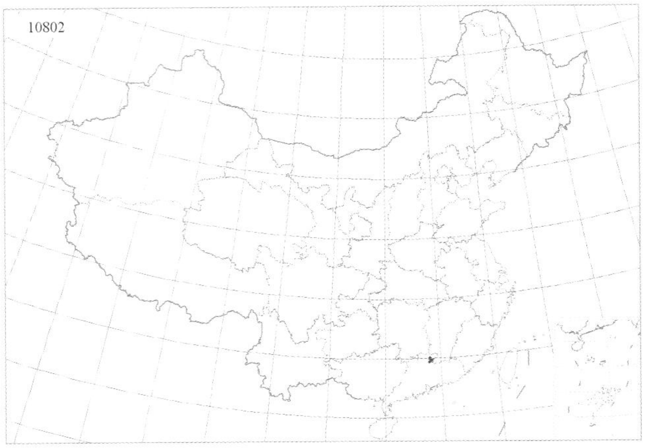

Poaceae 禾本科 *Bambusa* 簕竹属
Bambusa mutabilis McClure 黄竹仔
Bamboos, evergreen

23.4; 16.9; 28.4; 23.5; 221.6; 0.0; 1283 ❙ 1594; 626; 93 ❙ 26.2; 1160; 342.6

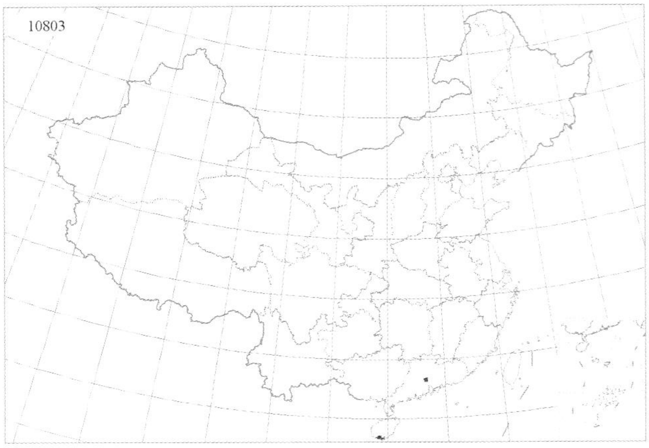

Poaceae 禾本科 *Bambusa* 簕竹属
Bambusa odashimae Hatus. ex Ohrnberger 乌脚绿竹
Bamboos, evergreen

20.3; 14.7; 24.9; 20.3; 183.8; 0.0; 1018 ❙ 2591; 1161; 281 ❙ 168.5; 961; 233.4

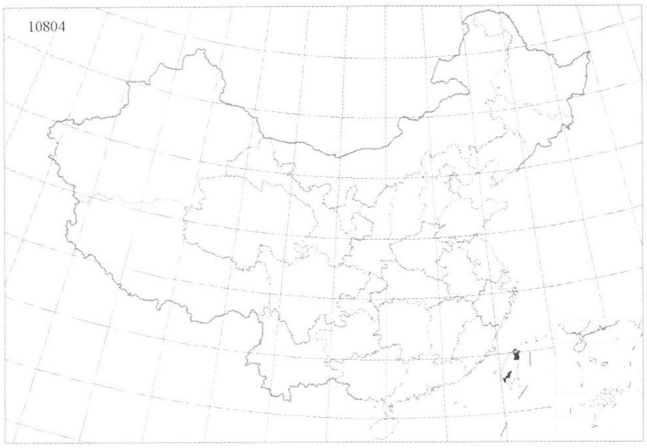

Poaceae 禾本科 *Bambusa* 簕竹属
Bambusa oldhamii Munro 绿竹
Bamboos, evergreen

18.7 (14.3~24.1); 9.9 (4.0~18.9); 26.5 (20.1~29.0); 18.8 (14.3~24.1); 165.4 (112.6~229.7); 0.0 (0.0~1.1); 969 (756~1322) ❙ 1692 (1090~2993); 691 (359~1594); 170 (47~547) ❙ 77.8 (-3.3~246.5); 958 (756~1150); 244.9 (154.2~322.1)

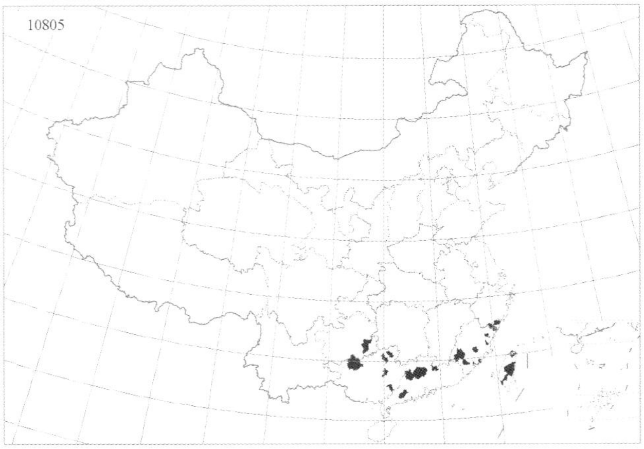

Poaceae 禾本科 *Bambusa* 簕竹属
Bambusa pachinensis Hayata 米筛竹
Bamboos, evergreen

18.9; 8.9; 28.1; 19.0; 167.5; 0.0; 997 ❙ 1624; 616; 172 ❙ 63.7; 997; 221.9

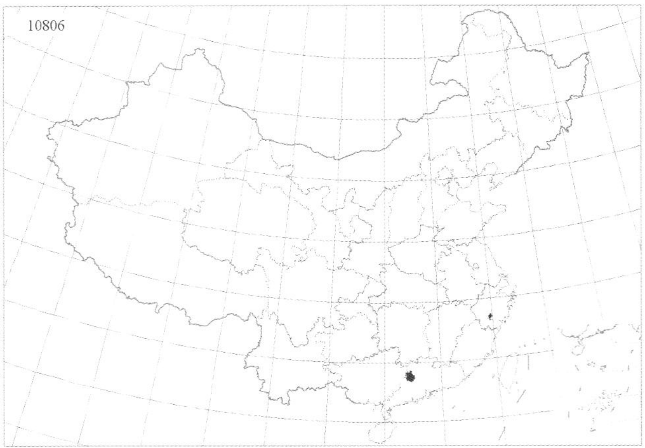

Poaceae 禾本科 *Bambusa* 簕竹属
Bambusa pachinensis var. hirsutissima (Odash.) W. C. Lin 长毛米筛竹
Bamboos, evergreen

20.8 (18.2~22.5); 11.8 (7.7~14.6); 28.3 (26.9~29.8); 20.8 (18.2~22.6); 189.5 (158.5~210.9); 0.0 (0.0~0.0); 1096 (949~1212) ❙ 1619 (1447~1852); 663 (507~777); 153 (82~194) ❙ 48.7 (20.8~86.3); 1089 (949~1212); 214.2 (122.7~271.3)

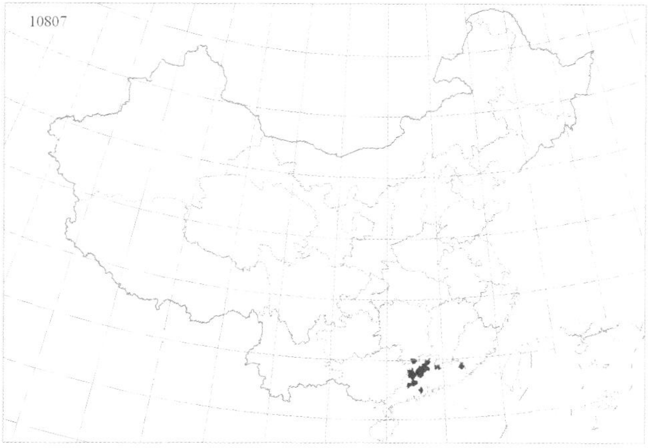

Poaceae 禾本科 *Bambusa* 簕竹属
Bambusa pallida Munro 大薄竹
Bamboos, evergreen

12.5; 5.5; 17.8; 12.5; 94.9; 4.4; 684 ❙ 1365; 772; 54 ❙ 96.9; 684; 489.6

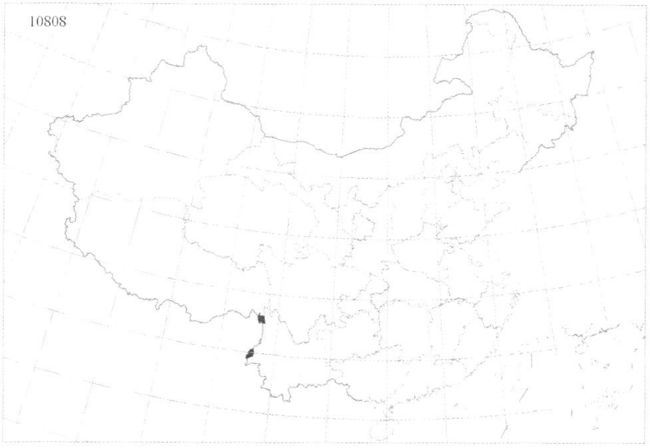

Poaceae 禾本科 *Bambusa* 簕竹属
Bambusa papillata (Q. H. Dai) Q. H. Dai 水箪竹
Bamboos, evergreen

19.8; 10.5; 27.2; 19.9; 178.5; 0.0; 1044 ❙ 1413; 674; 122 ❙ 38.1; 1044; 223.4

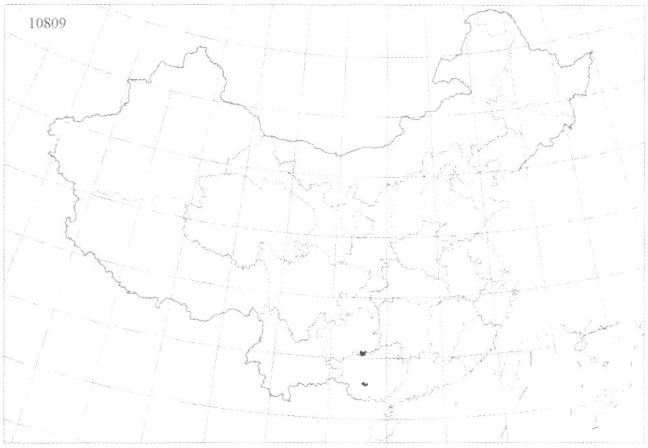

Poaceae 禾本科 *Bambusa* 簕竹属
Bambusa pervariabilis McClure 撑篙竹
Bamboos, evergreen

19.7 (14.3~23.6); 10.3 (3.3~16.7); 27.4 (23.5~29.8); 19.7 (14.3~23.6); 176.5 (112.6~223.5); 0.1 (0.0~2.3); 1034 (756~1281) ❙ 1516 (1089~2435); 676 (413~1428); 138 (43~198) ❙ 48.1 (-1.0~101.8); 1031 (756~1269); 240.6 (111.7~411.1)

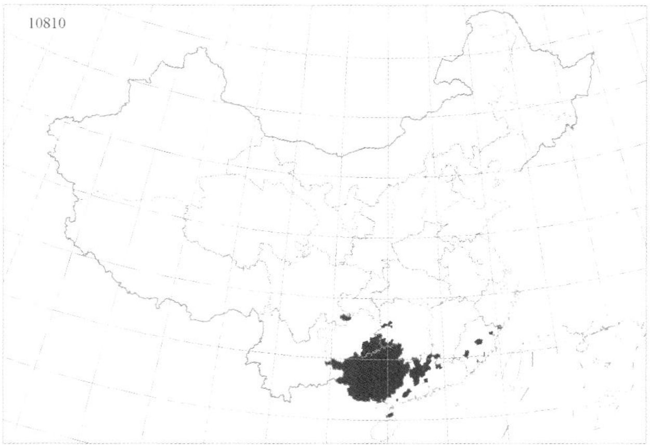

Poaceae 禾本科 *Bambusa* 簕竹属
Bambusa pervariabilis var. viridistriata Q. H. Dai et X. C. Liu 花撑篙竹
Bamboos, evergreen

19.2; 9.2; 27.5; 19.2; 170.8; 0.0; 1002 ❙ 1495; 678; 135 ❙ 49.1; 1002; 256.4

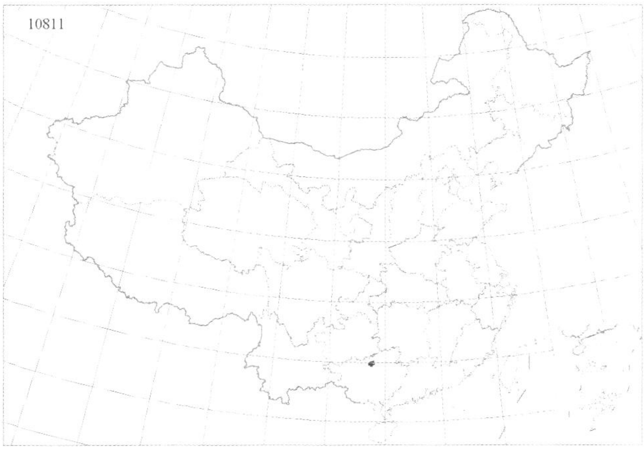

Poaceae 禾本科 *Bambusa* 簕竹属
Bambusa piscatorum McClure 石竹仔
Bamboos, evergreen

24.5; 19.2; 28.2; 24.6; 234.7; 0.0; 1362 ❙ 1665; 585; 82 ❙ 22.4; 1260; 424.8

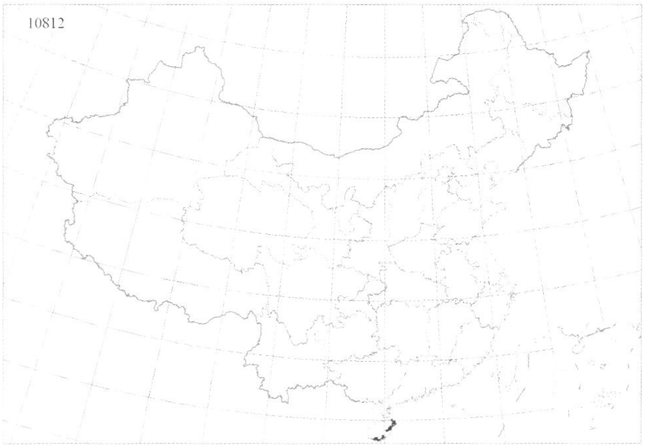

Poaceae 禾本科 *Bambusa* 簕竹属
Bambusa polymorpha Munro 灰秆竹
Bamboos, evergreen

20.4；15.1；23.6；20.5；185.8；0.0；989 ❙ 1607；702；77 ❙ 62.5；986；659.0

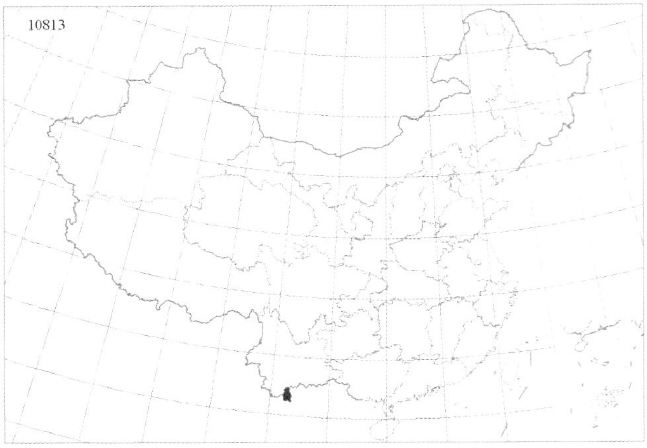

Poaceae 禾本科 *Bambusa* 簕竹属
Bambusa prominens H. L. Fung et C. Y. Sia 牛儿竹
Bamboos, evergreen

17.4；7.1；26.7；17.4；148.8；0.0；909 ❙ 1179；557；89 ❙ 29.5；908；170.6

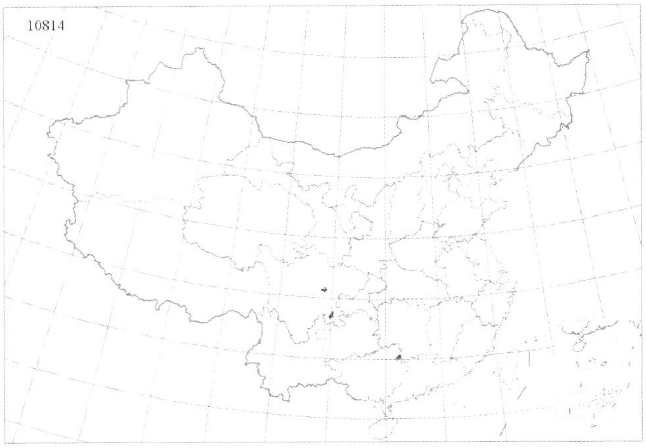

Poaceae 禾本科 *Bambusa* 簕竹属
Bambusa ramispinosa L. C. Chia et H. L. Fung 坭黄竹
Bamboos, evergreen

21.7；13.5；28.1；21.7；200.7；0.0；1146 ❙ 1539；720；126 ❙ 34.5；1146；255.2

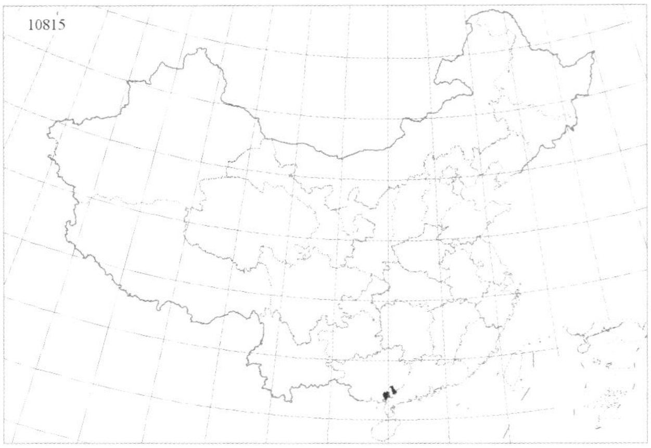

Poaceae 禾本科 *Bambusa* 簕竹属
Bambusa remotiflora Kuntze 甲竹
Bamboos, evergreen

23.5 (20.1~24.8)；17.4 (10.7~20.1)；27.9 (26.7~28.7)；23.5 (20.1~24.8)；221.9 (181.1~237.6)；0.0 (0.0~0.0)；1275 (1048~1385) ❙ 1465 (868~1813)；532 (312~712)；79 (41~161) ❙ 16.2 (-34.3~53.1)；1135 (868~1329)；413.0 (235.5~620.0)

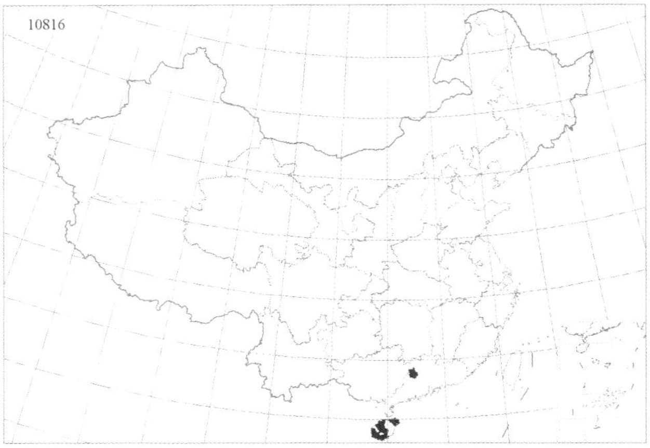

Poaceae 禾本科 *Bambusa* 簕竹属
Bambusa rigida Keng et Keng f. 硬头黄竹
Bamboos, evergreen

15.5 (6.0~22.4)；5.6 (-3.2~14.0)；24.6 (14.4~28.6)；15.6 (6.6~22.4)；128.2 (39.7~209.1)；1.3 (0.0~26.7)；832 (508~1204) ❙ 1133 (863~1888)；590 (413~1044)；50 (22~181) ❙ 38.3 (6.6~101.6)；831 (508~1204)；186.0 (120.2~491.5)

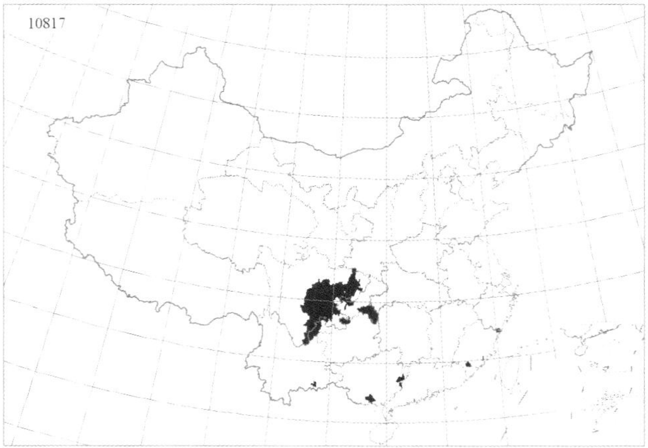

Poaceae 禾本科 *Bambusa* 簕竹属
Bambusa rutila McClure 木竹
Bamboos, evergreen

19.7 (16.1~22.3)；10.4 (5.7~14.9)；27.7 (25.4~29.8)；19.8 (16.1~22.3)；177.3 (133.5~208.2)；0.0 (0.0~0.0)；1036 (847~1185) ❙ 1589 (950~2035)；644 (507~969)；150 (27~191) ❙ 52.7 (12.1~71.8)；1029 (845~1185)；225.5 (133.8~277.1)

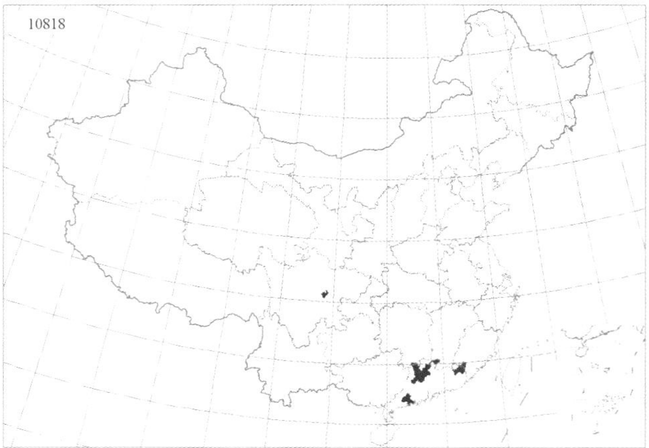

Poaceae 禾本科 *Bambusa* 簕竹属
Bambusa sinospinosa McClure 车筒竹
Bamboos, evergreen

18.7 (15.0~23.6); 9.8 (4.5~16.7); 26.2 (20.7~29.8); 18.8 (15.0~23.6); 165.1 (120.8~223.5); 0.0 (0.0~0.5); 967 (767~1281) | 1385 (950~2037); 647 (442~1033); 95 (27~204) | 43.1 (-1.0~100.8); 965 (767~1269); 238.5 (108.2~551.6)

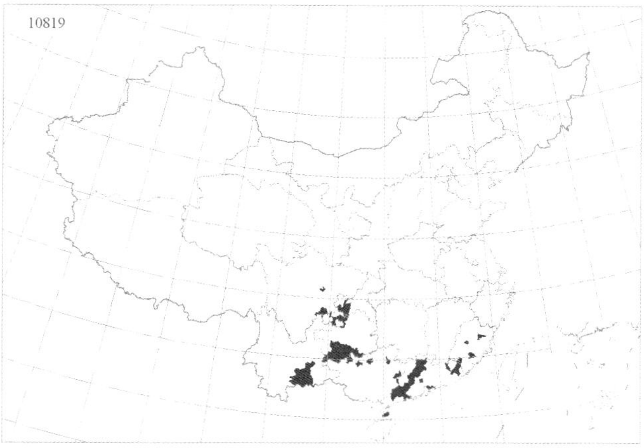

Poaceae 禾本科 *Bambusa* 簕竹属
Bambusa stenoaurita (W. T. Lin) T. H. Wen 黄麻竹
Bamboos, evergreen

19.3; 10.2; 27.0; 19.3; 172.3; 0.1; 1018 | 1507; 650; 127 | 46.9; 1018; 217.5

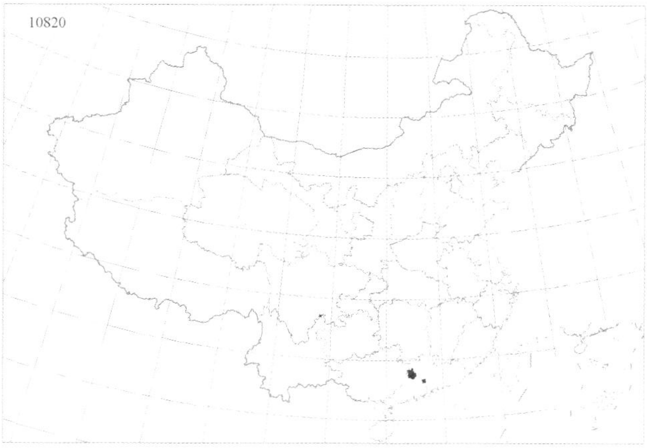

Poaceae 禾本科 *Bambusa* 簕竹属
Bambusa subaequalis H. L. Fung et C. Y. Sia 锦竹
Bamboos, evergreen

16.2; 6.1; 25.4; 16.2; 135.0; 0.0; 849 | 1379; 812; 47 | 62.4; 849; 171.5

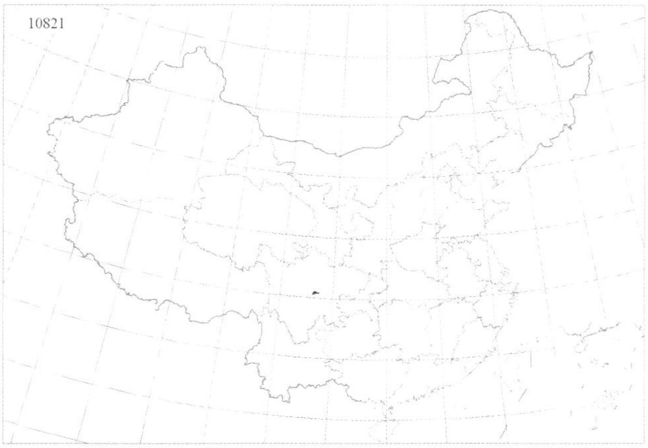

Poaceae 禾本科 *Bambusa* 簕竹属
Bambusa subtruncata L. C. Chia et H. L. Fung 信宜石竹
Bamboos, evergreen

21.5; 13.4; 27.9; 21.6; 198.9; 0.0; 1132 | 1780; 803; 135 | 57.6; 1132; 205.9

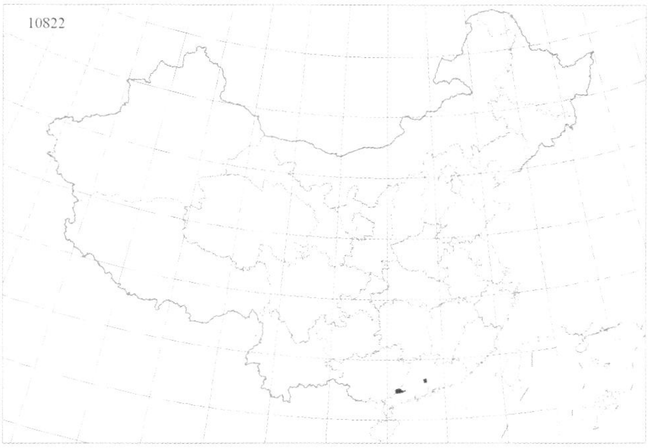

Poaceae 禾本科 *Bambusa* 簕竹属
Bambusa surrecta (Q. H. Dai) Q. H. Dai 油竹
Bamboos, evergreen

22.3; 13.5; 28.9; 22.3; 208.1; 0.0; 1202 | 1446; 710; 136 | 20.3; 1202; 201.4

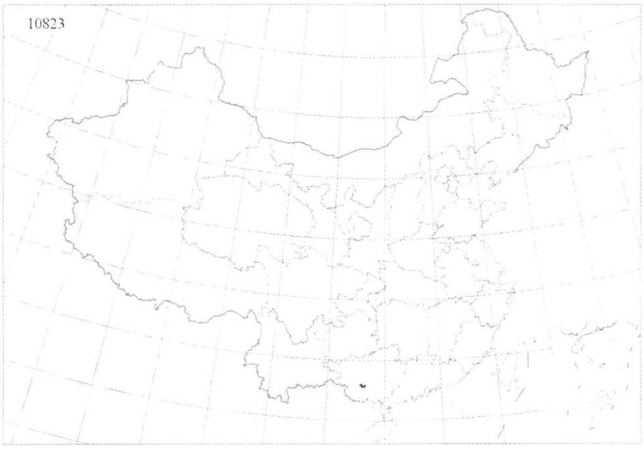

Poaceae 禾本科 *Bambusa* 簕竹属
Bambusa textilis McClure 青皮竹
Bamboos, evergreen

19.9 (14.3~23.6); 10.7 (3.3~16.7); 27.5 (24.3~29.9); 19.9 (14.3~23.6); 179.4 (114.1~223.5); 0.0 (0.0~2.3); 1044 (765~1281) | 1569 (1172~2435); 687 (413~1428); 146 (43~254) | 51.7 (-1.0~130.7); 1041 (765~1269); 245.3 (141.7~411.1)

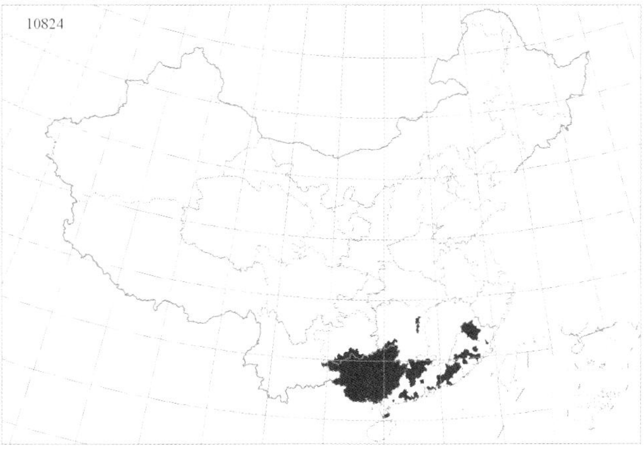

Poaceae 禾本科 *Bambusa* 簕竹属
Bambusa tulda Roxb. 俯竹
Bamboos, evergreen

15.8; 8.7; 20.8; 15.8; 129.6; 0.0; 787 ❙ 1551; 895; 48 ❙ 101.6; 777; 482.1

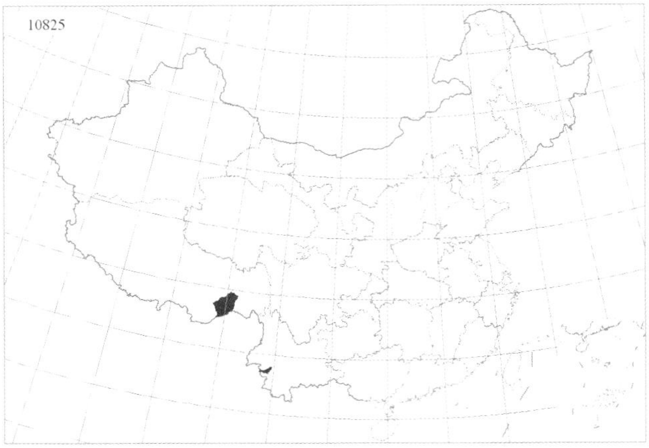

Poaceae 禾本科 *Bambusa* 簕竹属
Bambusa tuldoides Munro 青秆竹
Bamboos, evergreen

21.1 (18.2~23.2); 12.6 (7.7~15.8); 28.0 (26.9~28.9); 21.1 (18.2~23.2); 193.4 (158.5~218.9); 0.0 (0.0~0.0); 1109 (949~1257) ❙ 1689 (1512~2037); 723 (507~1033); 144 (102~182) ❙ 52.7 (34.1~71.8); 1109 (949~1257); 228.0 (157.7~277.1)

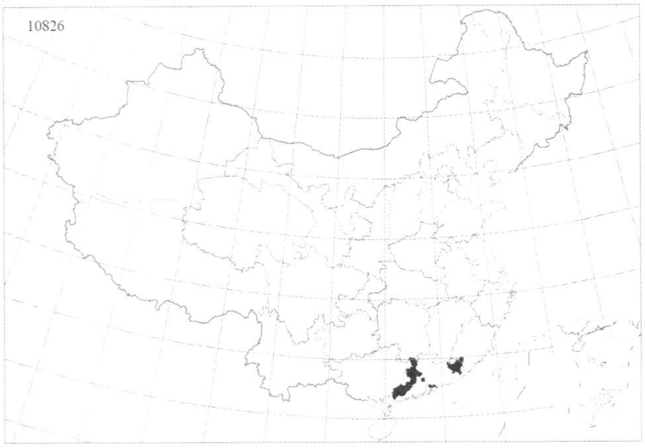

Poaceae 禾本科 *Bambusa* 簕竹属
Bambusa utilis W. C. Lin 乌叶竹
Bamboos, evergreen

17.9; 11.6; 23.4; 18.0; 155.6; 0.0; 864 ❙ 2355; 900; 308 ❙ 172.5; 864; 284.4

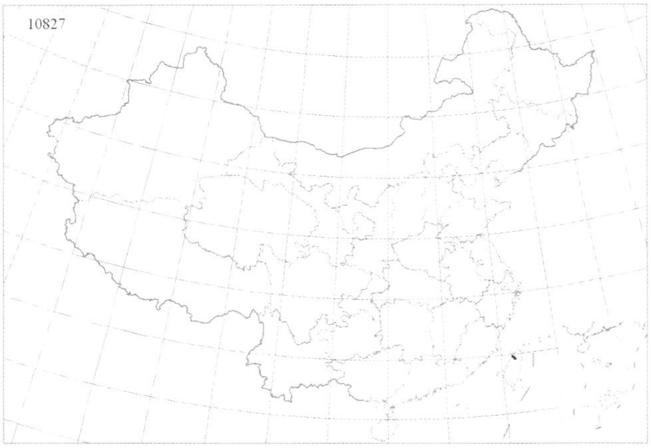

Poaceae 禾本科 *Bambusa* 簕竹属
Bambusa variostriata (W. T. Lin) L. C. Chia et H. L. Fung 吊丝箪竹
Bamboos, evergreen

22.1; 13.7; 28.8; 22.1; 205.6; 0.0; 1182 ❙ 1784; 779; 144 ❙ 50.9; 1182; 157.7

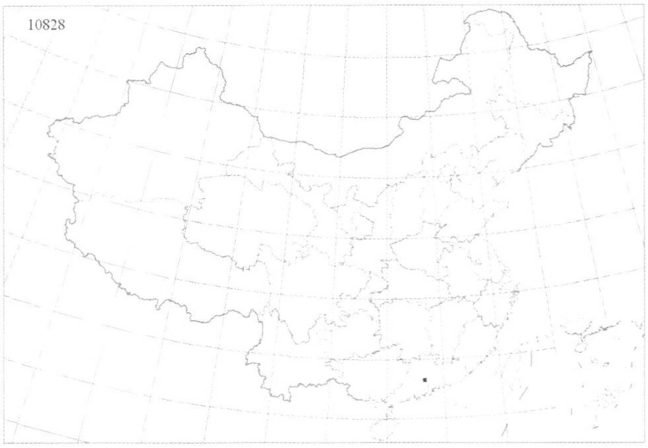

Poaceae 禾本科 *Bambusa* 簕竹属
Bambusa ventricosa McClure 佛肚竹
Bamboos, evergreen

21.8 (15.5~24.8); 14.5 (3.6~20.1); 27.8 (25.0~28.9); 21.8 (15.5~24.8); 202.1 (127.5~237.6); 0.2 (0.0~1.4); 1178 (841~1385) ❙ 1564 (868~2993); 595 (312~1412); 114 (41~547) ❙ 36.9 (-34.3~203.4); 1104 (841~1360); 353.6 (131.2~620.0)

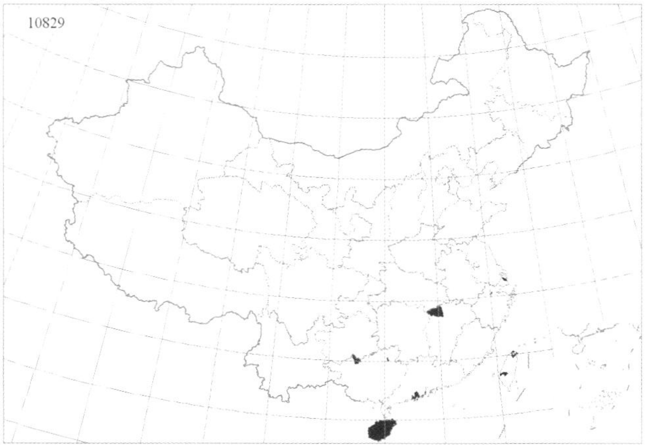

Poaceae 禾本科 *Bambusa* 簕竹属
Bambusa vulgaris Schrad. ex J. C. Wendl. 龙头竹
Bamboos, evergreen

21.6 (10.8~24.8); 15.4 (3.9~20.1); 26.3 (16.6~28.8); 21.7 (10.9~24.8); 200.1 (71.8~237.6); 0.1 (0.0~1.3); 1144 (633~1385) ❙ 1736 (968~2993); 773 (392~1847); 109 (24~547) ❙ 57.5 (-12.0~203.4); 1080 (633~1360); 369.6 (154.2~659.0)

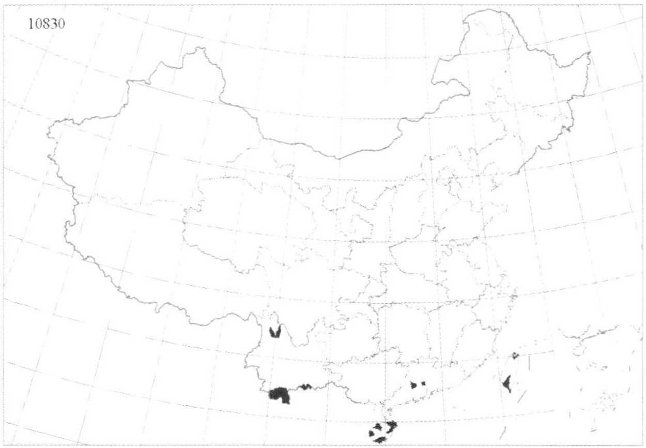

Poaceae 禾本科 *Bambusa* 簕竹属
Bambusa wenchouensis (T. H. Wen) Q. H. Dai 温州箪竹
Bamboos, evergreen

18.2; 7.8; 28.7; 18.3; 159.6; 0.0; 965 ❙ 1706; 645; 182 ❙ 76.7; 965; 189.8

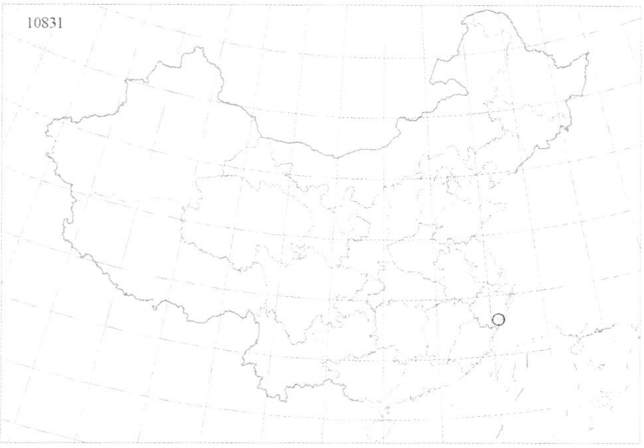

Poaceae 禾本科 *Bambusa* 簕竹属
Bambusa xiashanensis L. C. Chia et H. L. Fung 霞山坭竹
Bamboos, evergreen

22.8; 14.9; 29.0; 22.8; 213.6; 0.0; 1228 ❙ 1605; 730; 113 ❙ 31.4; 1228; 168.2

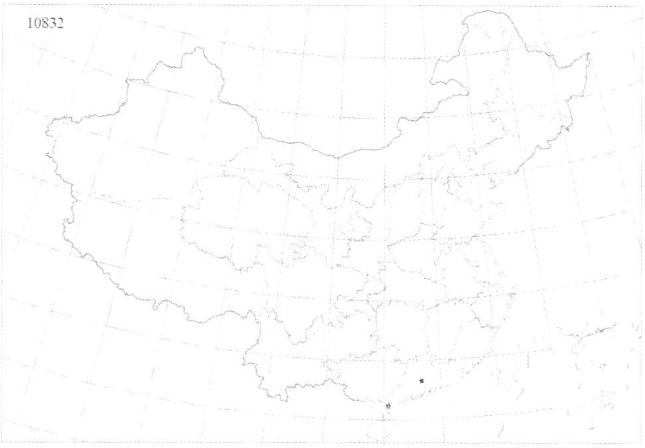

Poaceae 禾本科 *Bonia* 单枝竹属
Bonia amplexicaulis (L. C. Chia et al) N. H. Xia 芸香竹
Bamboos, evergreen

22.0; 13.8; 28.0; 22.1; 204.8; 0.0; 1172 ❙ 1338; 691; 87 ❙ 14.1; 1172; 315.8

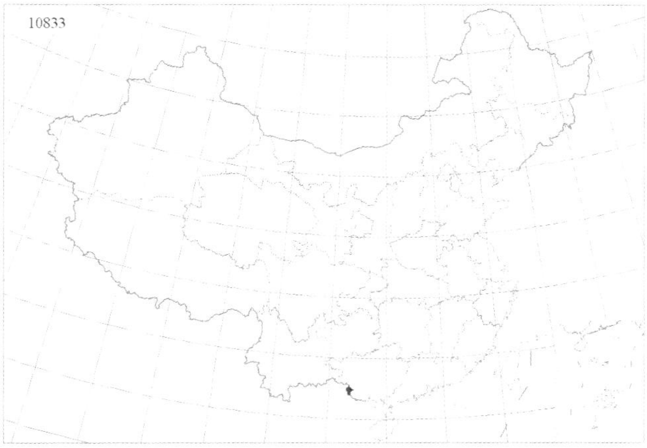

Poaceae 禾本科 *Bonia* 单枝竹属
Bonia levigata (L. C. Chia et al) N. H. Xia 响子竹
Bamboos, evergreen

24.7; 19.8; 28.1; 24.7; 236.8; 0.0; 1378 ❙ 1586; 520; 65 ❙ 15.1; 1218; 537.7

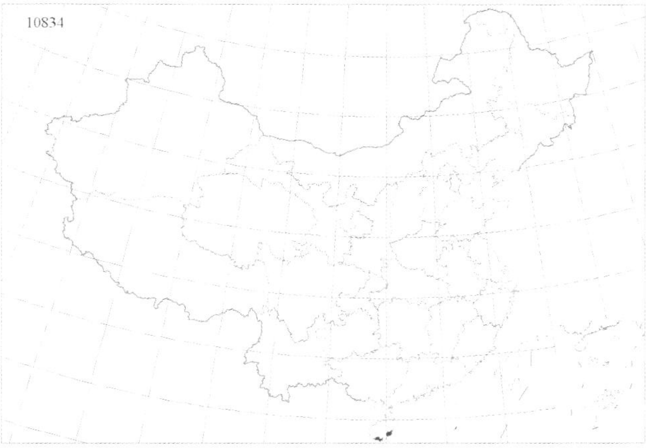

Poaceae 禾本科 *Bonia* 单枝竹属
Bonia parvifloscula (W. T. Lin) N. H. Xia 小花单枝竹
Bamboos

22.0; 13.5; 28.8; 22.0; 203.6; 0.0; 1169 ❙ 1641; 724; 132 ❙ 40.5; 1169; 205.0

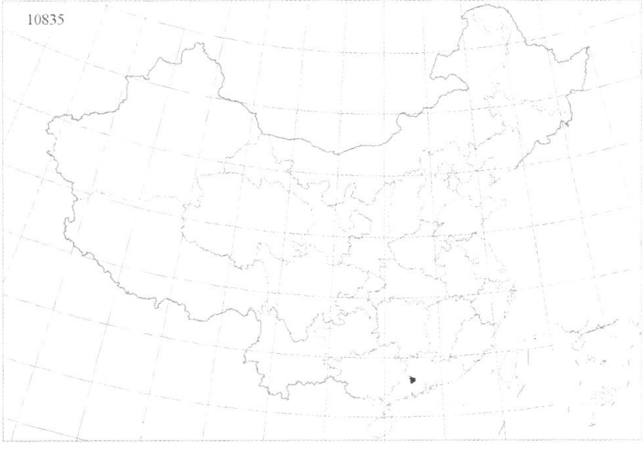

Poaceae 禾本科 *Bonia* 单枝竹属
Bonia saxatilis (L. C. Chia et al) N. H. Xia 单枝竹
Bamboos, evergreen

18.7 (16.8~22.1); 9.0 (6.5~13.7); 26.9 (25.5~28.8); 18.8 (16.9~22.1); 165.2 (142.7~205.6); 0.0 (0.0~0.0); 974 (869~1182) ❙ 1531 (1368~1784); 660 (574~779); 139 (96~186) ❙ 57.5 (36.0~70.6); 974 (869~1182); 229.8 (157.7~288.8)

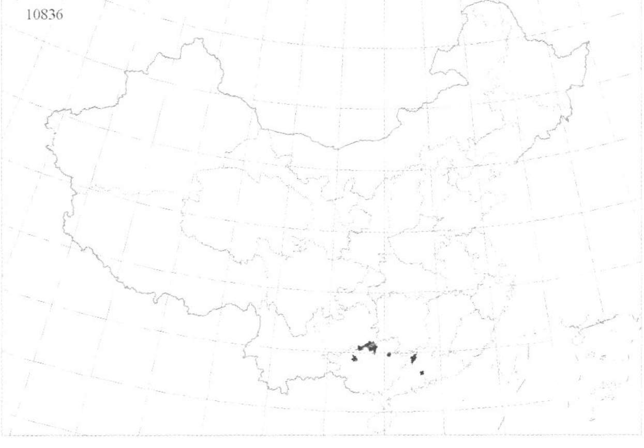

Poaceae 禾本科 *Cephalostachyum* 空竹属
Cephalostachyum latifolium Munro 空竹
Bamboos, evergreen

15.4; 8.9; 20.1; 15.4; 127.4; 2.2; 795 ❙ 1435; 733; 55 ❙ 85.0; 780; 470.9

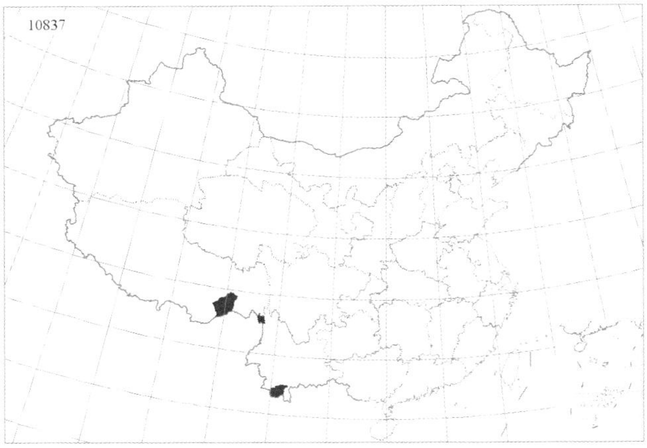

Poaceae 禾本科 *Cephalostachyum* 空竹属
Cephalostachyum pallidum Munro 小空竹
Bamboos, evergreen

17.0; 10.2; 21.4; 17.0; 144.3; 0.0; 821 ❙ 1483; 846; 57 ❙ 81.0; 821; 499.2

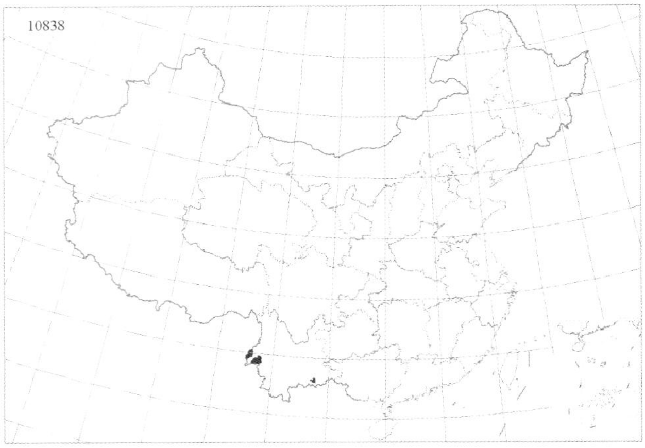

Poaceae 禾本科 *Cephalostachyum* 空竹属
Cephalostachyum pergracile Munro 香糯竹
Bamboos, evergreen

16.9 (11.3~20.5); 10.9 (4.3~15.1); 21.0 (17.0~23.7); 17.0 (11.3~20.5); 143.7 (76.6~185.9); 0.1 (0.0~0.7); 846 (644~992) ❙ 1430 (1243~1607); 693 (655~727); 66 (58~77) ❙ 72.1 (47.6~93.0); 833 (644~986); 507.0 (384.7~659.0)

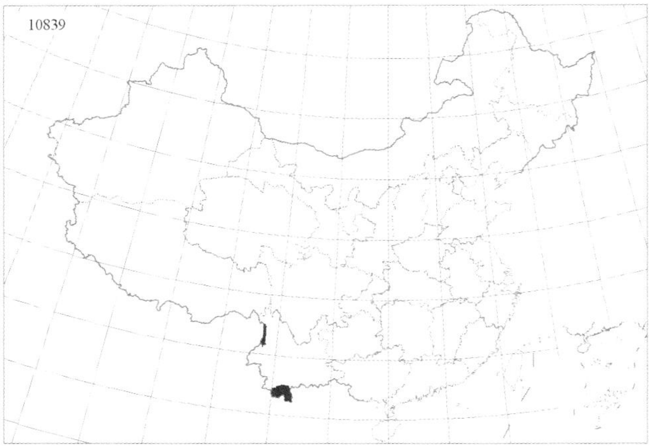

Poaceae 禾本科 *Chimonobambusa* 方竹属
Chimonobambusa angustifolia C. D. Chu et C. S. Chao 狭叶方竹
Bamboos, evergreen

15.6 (12.5~19.1); 5.3 (2.7~10.1); 25.0 (20.1~27.7); 15.6 (12.5~19.1); 128.1 (92.0~169.2); 0.4 (0.0~3.7); 819 (685~969) ❙ 1229 (999~1496); 540 (444~720); 80 (31~161) ❙ 50.4 (27.5~95.9); 819 (685~969); 202.0 (146.0~283.9)

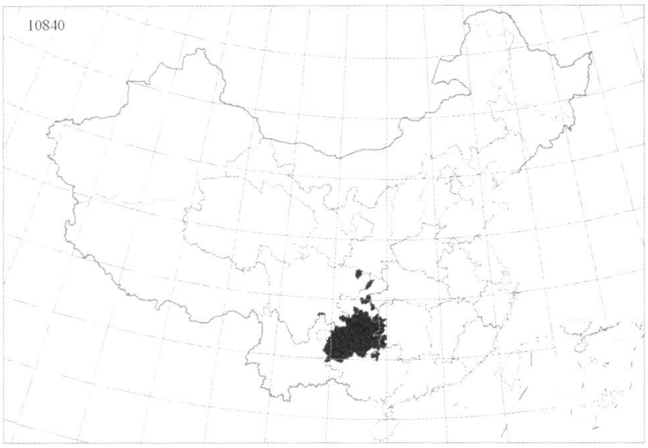

Poaceae 禾本科 *Chimonobambusa* 方竹属
Chimonobambusa armata (Gamble) J. R. Xue et T. P. Yi 缅甸方竹
Bamboos, evergreen

13.2 (8.6~15.6); 4.3 (1.3~6.4); 21.1 (15.0~25.9); 13.2 (8.7~15.6); 100.0 (52.6~127.5); 1.8 (0.0~8.8); 723 (574~824) ❙ 1546 (1041~1950); 672 (585~727); 141 (45~227) ❙ 111.2 (81.5~147.0); 723 (574~824); 325.0 (252.7~407.1)

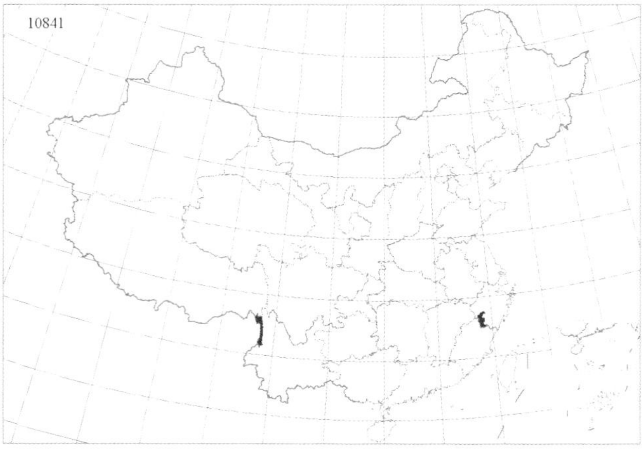

Poaceae 禾本科 *Chimonobambusa* 方竹属
Chimonobambusa brevinoda J. R. Xue et W. P. Zhang 短节方竹
Bamboos, evergreen

17.4; 9.8; 22.8; 17.4; 148.9; 0.0; 843 ❙ 1393; 791; 52 ❙ 65.3; 843; 369.8

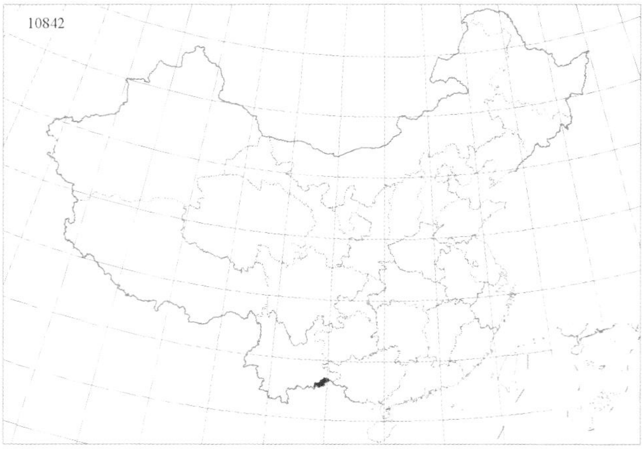

Poaceae 禾本科 *Chimonobambusa* 方竹属

Chimonobambusa communis (J. R. Xue et T. P. Yi) T. H. Wen et Ohrnb. 平竹

Bamboos, evergreen

14.7 (12.7~17.1); 4.0 (2.1~5.0); 25.0 (23.0~29.4); 14.8 (12.8~17.1); 118.7 (99.0~146.1); 1.4 (0.0~5.6); 792 (713~952) ❙ 1249 (1101~1492); 515 (475~613); 78 (65~123) ❙ 58.9 (29.7~107.2); 790 (713~921); 190.4 (111.9~245.8)

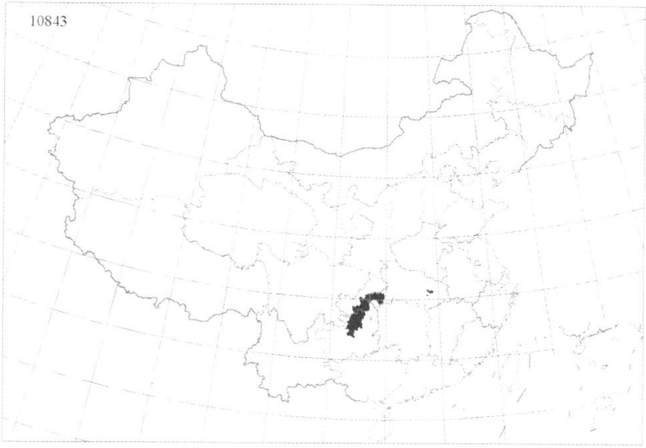

Poaceae 禾本科 *Chimonobambusa* 方竹属

Chimonobambusa convoluta Q. H. Dai et X. L. Tao 小方竹

Bamboos, evergreen

19.8; 11.1; 26.4; 19.8; 177.9; 0.0; 1010 ❙ 1182; 673; 43 ❙ 16.9; 1010; 354.5

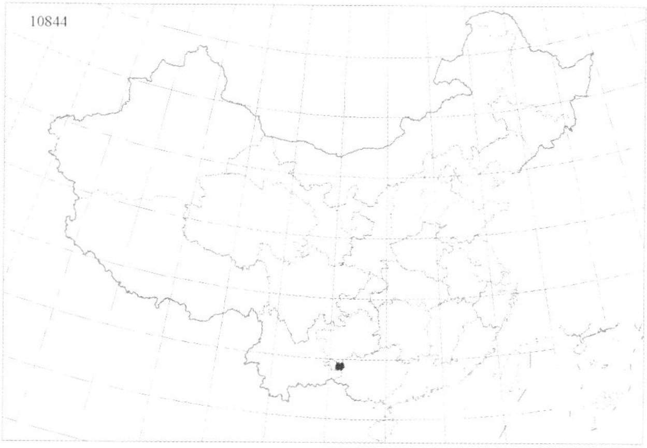

Poaceae 禾本科 *Chimonobambusa* 方竹属

Chimonobambusa damingshanensis J. R. Xue et W. P. Zhang 大明山方竹

Bamboos, evergreen

21.6; 12.6; 28.5; 21.7; 199.8; 0.0; 1153 ❙ 1625; 795; 173 ❙ 40.9; 1153; 237.8

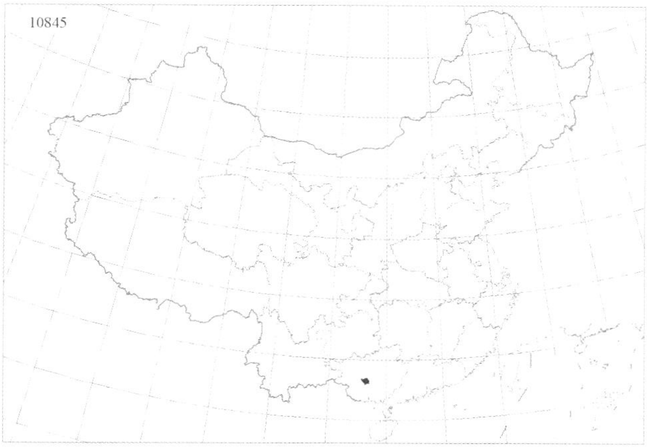

Poaceae 禾本科 *Chimonobambusa* 方竹属

Chimonobambusa grandifolia J. R. Xue et W. P. Zhang 大叶方竹

Bamboos, evergreen

17.7; 10.9; 22.5; 17.8; 153.4; 0.0; 854 ❙ 1368; 756; 59 ❙ 60.1; 854; 491.5

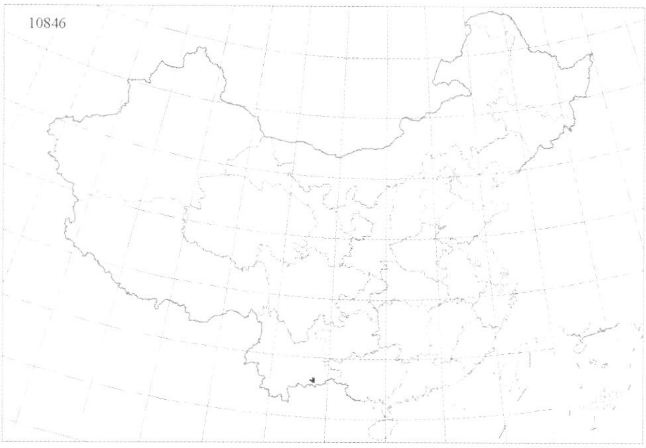

Poaceae 禾本科 *Chimonobambusa* 方竹属

Chimonobambusa hejiangensis C. D. Chu et C. S. Chao 合江方竹

Bamboos, evergreen

15.5 (14.7~17.2); 5.2 (4.4~6.8); 25.3 (24.4~27.0); 15.6 (14.8~17.3); 127.4 (117.8~147.4); 0.2 (0.0~0.6); 819 (779~905) ❙ 1087 (1048~1133); 498 (488~507); 66 (64~72) ❙ 32.9 (21.3~42.1); 819 (779~905); 179.7 (158.7~198.3)

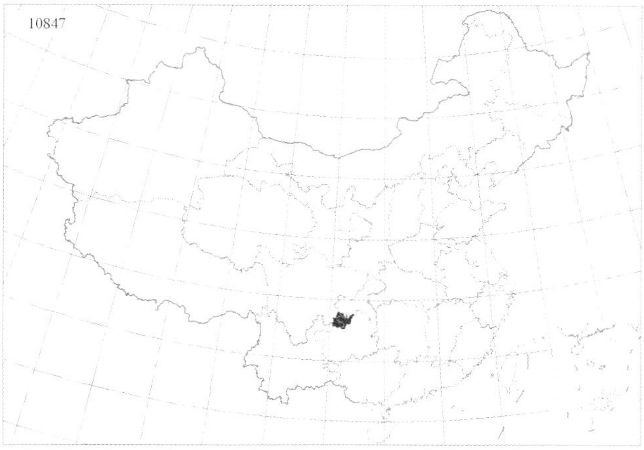

Poaceae 禾本科 *Chimonobambusa* 方竹属

Chimonobambusa hirtinoda C. S. Chao et K. M. Lan 毛环方竹

Bamboos, evergreen

17.8; 8.4; 25.5; 17.8; 153.9; 0.0; 907 ❙ 1260; 640; 64 ❙ 39.8; 907; 257.9

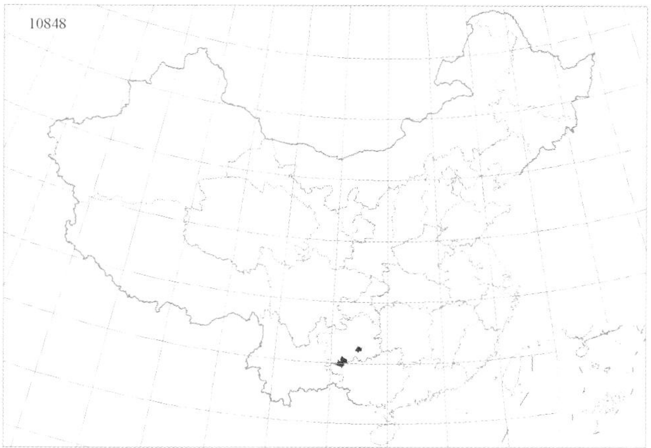

Poaceae 禾本科 *Chimonobambusa* 方竹属
Chimonobambusa hsuehiana D. Z. Li et H. Q. Yang 细秆筇竹
Bamboos, evergreen

11.9; 2.9; 20.0; 12.0; 87.2; 3.4; 671 ❙ 1001; 550; 35 ❙ 49.1; 671; 278.8

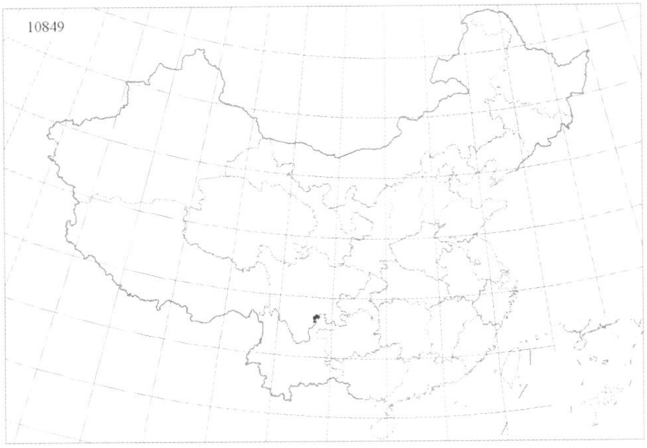

Poaceae 禾本科 *Chimonobambusa* 方竹属
Chimonobambusa lactistriata W. D. Li et Q. X. Wu 乳纹方竹
Bamboos, evergreen

18.2 (16.8~19.1); 8.9 (7.4~10.1); 25.5 (23.6~26.6); 18.2 (16.8~19.1); 158.3 (141.5~169.2); 0.0 (0.0~0.0); 920 (836~969) ❙ 1289 (1236~1381); 670 (617~720); 66 (51~107) ❙ 40.7 (27.5~59.6); 920 (836~969); 243.3 (189.5~282.2)

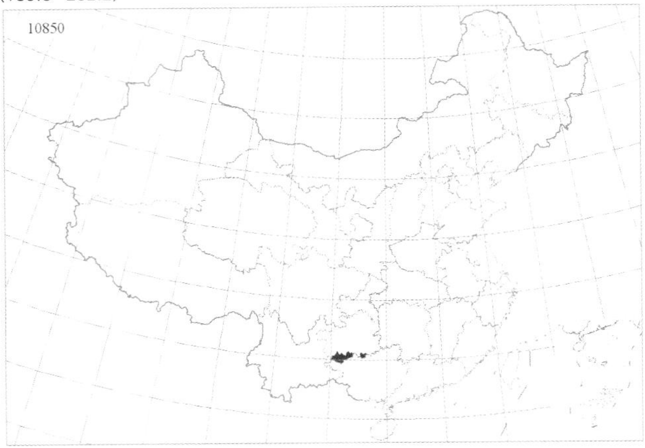

Poaceae 禾本科 *Chimonobambusa* 方竹属
Chimonobambusa luzhiensis (Hsueh et T. P. Yi) T. H. Wen et Ohrnb. 光竹
Bamboos, evergreen

15.2; 5.8; 23.0; 15.2; 122.5; 0.0; 779 ❙ 1239; 645; 58 ❙ 59.1; 779; 186.7

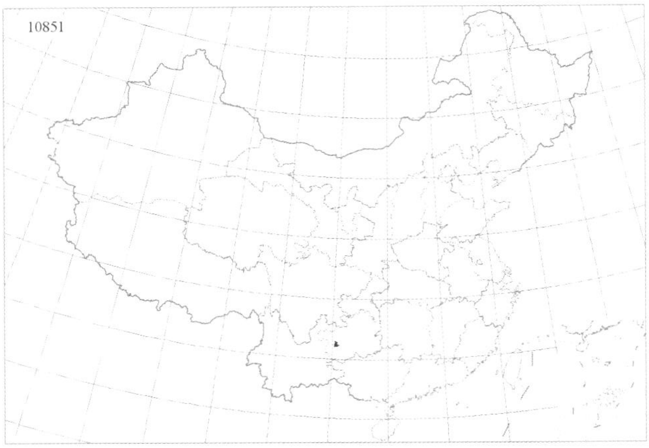

Poaceae 禾本科 *Chimonobambusa* 方竹属
Chimonobambusa macrophylla (Hsueh et T. P. Yi) T. H. Wen et Ohrnb. 大叶筇竹
Bamboos, evergreen

12.5; 3.3; 20.7; 12.5; 92.7; 2.3; 690 ❙ 1049; 580; 37 ❙ 52.1; 690; 267.0

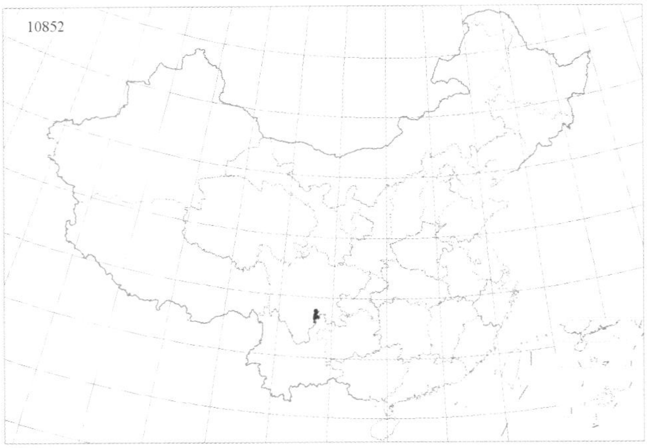

Poaceae 禾本科 *Chimonobambusa* 方竹属
Chimonobambusa macrophylla var. leiboensis (Hsueh et T. P. Yi) D. Z. Li 雷波大叶筇竹
Bamboos, evergreen

11.9; 2.9; 20.0; 12.0; 87.2; 3.4; 671 ❙ 1001; 550; 35 ❙ 49.1; 671; 278.8

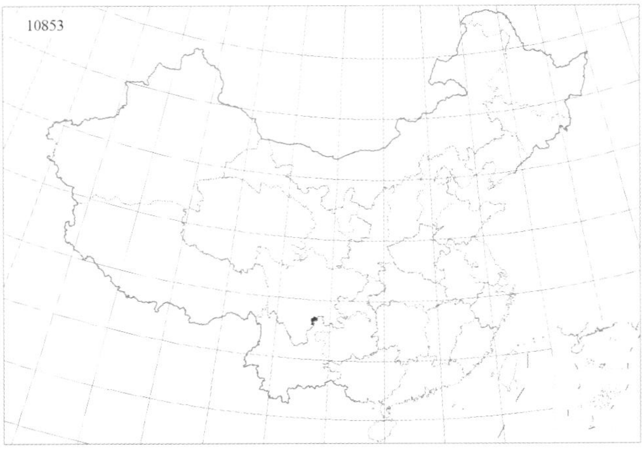

Poaceae 禾本科 *Chimonobambusa* 方竹属
Chimonobambusa marmorea (Mitford) Makino 寒竹
Bamboos, evergreen

15.8 (9.3~18.4); 5.1 (−1.7~9.0); 26.0 (20.1~29.4); 15.9 (9.5~18.4); 132.8 (69.8~161.2); 2.3 (0.0~18.0); 851 (603~952) ❙ 1556 (1081~1997); 571 (481~669); 167 (32~254) ❙ 82.9 (29.7~130.7); 850 (603~950); 244.9 (111.9~332.5)

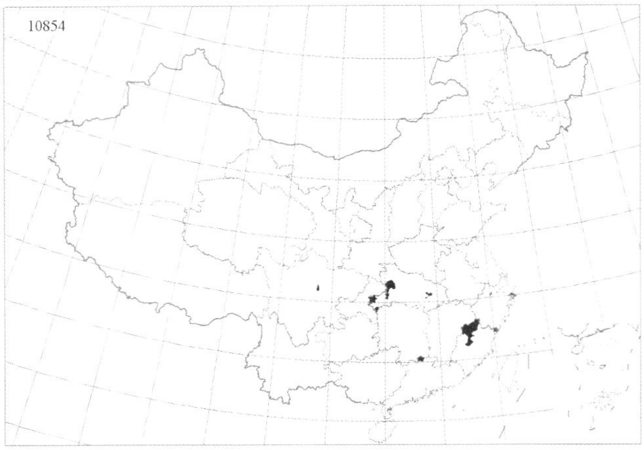

Poaceae 禾本科 *Chimonobambusa* 方竹属
Chimonobambusa metuoensis Hsueh et T. P. Yi 墨脱方竹
Bamboos, evergreen

9.3; 1.5; 15.9; 9.6; 65.5; 13.1; 592 ❙ 1348; 783; 41 ❙ 123.3; 592; 340.8

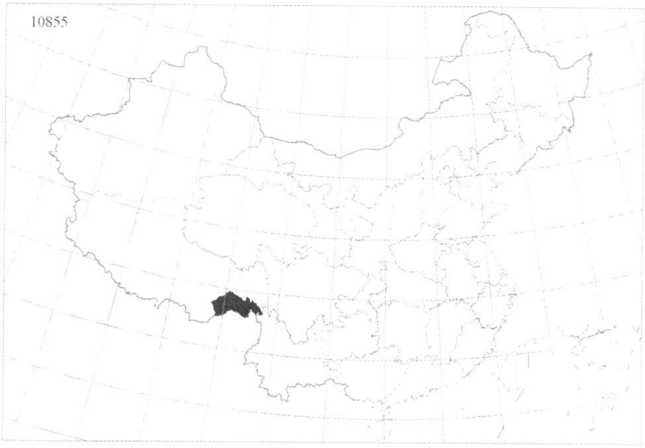

Poaceae 禾本科 *Chimonobambusa* 方竹属
Chimonobambusa opienensis (Hsueh et T. P. Yi) T. H. Wen et Ohrnb. 三月竹
Bamboos, evergreen

15.9; 6.2; 24.6; 16.0; 134.0; 2.2; 870 ❙ 1375; 643; 100 ❙ 64.2; 870; 230.9

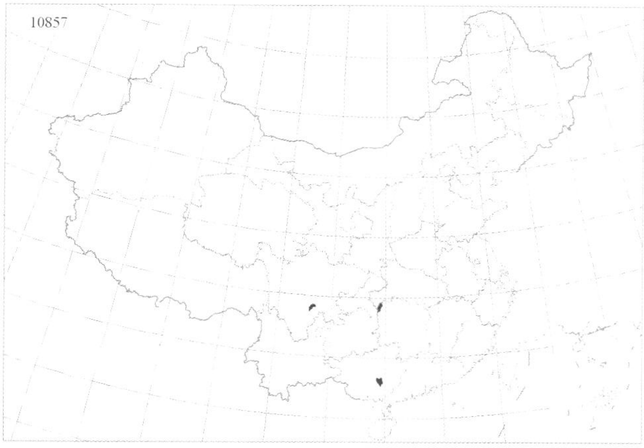

Poaceae 禾本科 *Chimonobambusa* 方竹属
Chimonobambusa puberula (Hsueh et T. P. Yi) T. H. Wen et Ohrnb. 柔毛筇竹
Bamboos, evergreen

15.2; 5.8; 23.0; 15.2; 122.5; 0.0; 779 ❙ 1239; 645; 58 ❙ 59.1; 779; 186.7

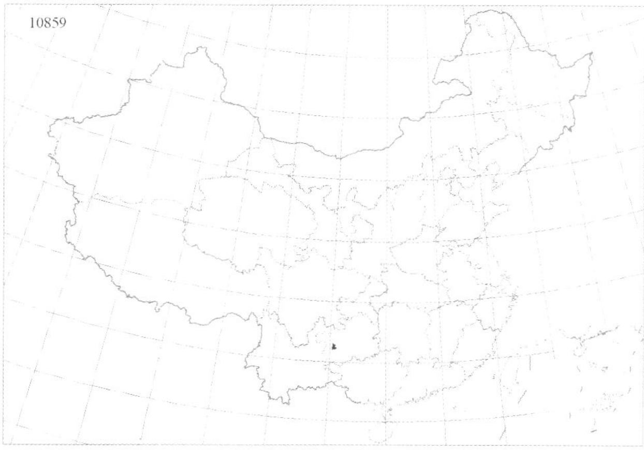

Poaceae 禾本科 *Chimonobambusa* 方竹属
Chimonobambusa microfloscula McClure 小花方竹
Bamboos, evergreen

18.4; 12.0; 22.9; 18.5; 161.5; 0.0; 886 ❙ 1398; 796; 58 ❙ 57.9; 886; 498.8

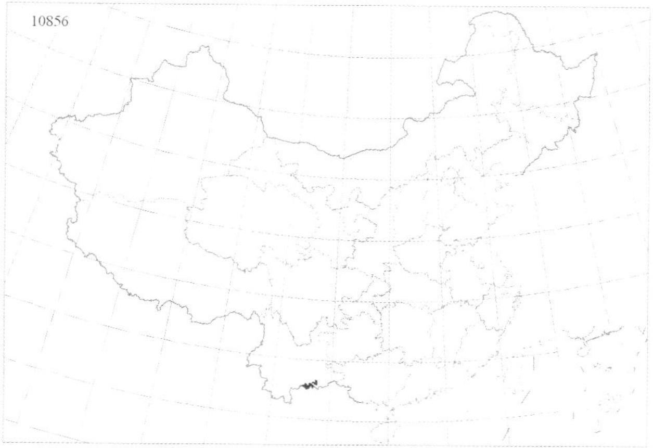

Poaceae 禾本科 *Chimonobambusa* 方竹属
Chimonobambusa pachystachys Hsueh et T. P. Yi 刺竹子
Bamboos, evergreen

15.3; 5.2; 24.7; 15.4; 124.5; 0.2; 802 ❙ 1032; 505; 60 ❙ 28.9; 802; 190.9

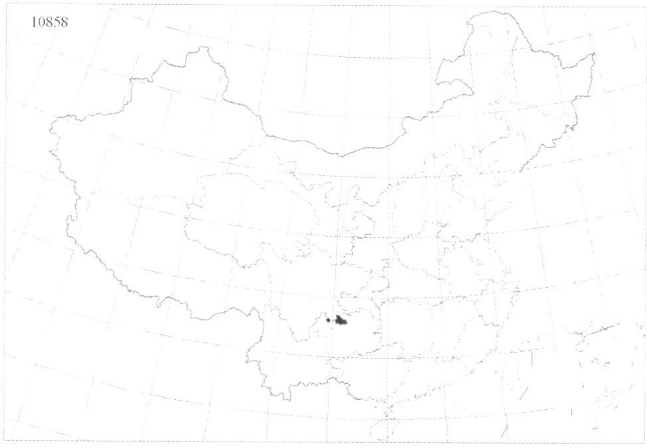

Poaceae 禾本科 *Chimonobambusa* 方竹属
Chimonobambusa pubescens T. H. Wen 十月寒竹
Bamboos, evergreen

16.3; 5.1; 26.4; 16.4; 136.3; 0.0; 862 ❙ 1332; 478; 145 ❙ 54.6; 862; 218.6

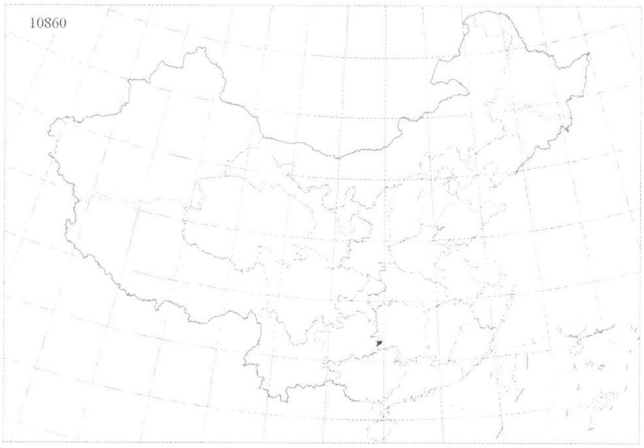

Poaceae 禾本科 *Chimonobambusa* 方竹属
Chimonobambusa purpurea Hsueh et T. P. Yi 刺黑竹
Bamboos, evergreen

14.5 (8.6~18.0); 4.1 (-1.7~7.8); 24.1 (15.0~29.4); 14.5 (8.7~18.0); 116.9 (52.6~156.5); 2.8 (0.0~18.0); 789 (574~952) ❙ 1194 (890~1583); 588 (458~938); 58 (21~123) ❙ 53.4 (13.7~107.2); 788 (574~951); 196.6 (111.9~352.3)

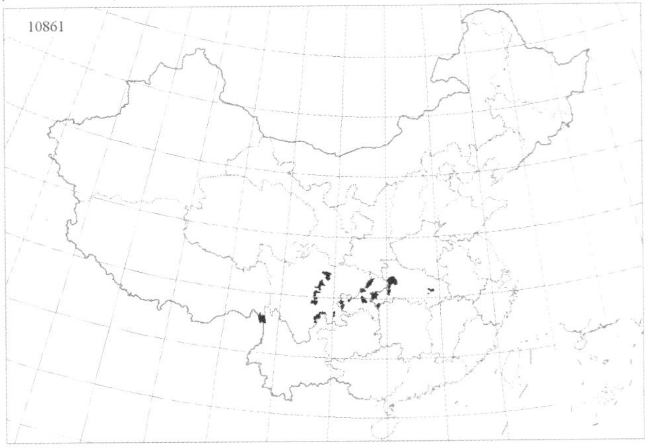

Poaceae 禾本科 *Chimonobambusa* 方竹属
Chimonobambusa quadrangularis (Franceschi) Makino 方竹
Bamboos, evergreen

16.6 (3.8~20.8); 5.9 (-4.3~15.3); 26.6 (10.8~29.7); 16.7 (4.7~20.8); 140.9 (21.6~189.2); 1.1 (0.0~35.4); 882 (444~1038) ❙ 1486 (845~2445); 560 (299~1412); 157 (16~254) ❙ 68.7 (-6.7~147.0); 873 (444~1038); 236.2 (111.9~513.9)

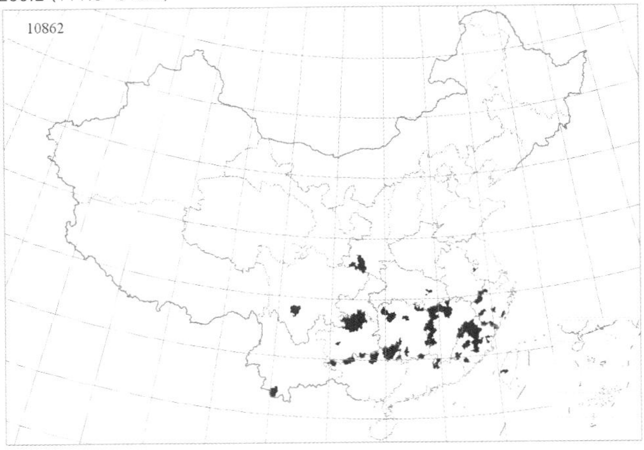

Poaceae 禾本科 *Chimonobambusa* 方竹属
Chimonobambusa rigidula (Hsueh et T. P. Yi) T. H. Wen et D. Ohrnberger 实竹子
Bamboos, evergreen

14.7; 5.1; 23.4; 14.8; 117.6; 0.4; 776 ❙ 1118; 617; 45 ❙ 44.4; 776; 225.5

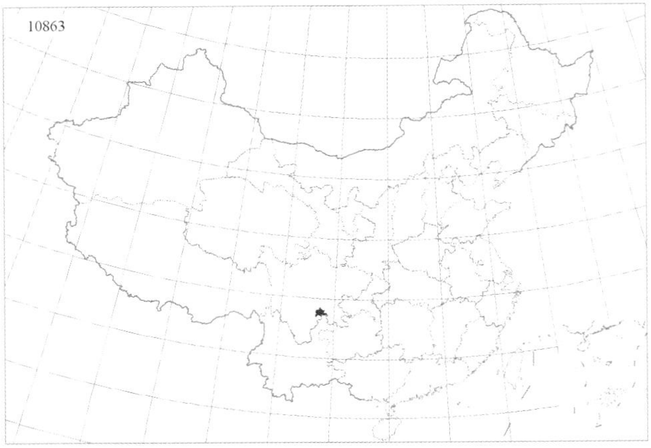

Poaceae 禾本科 *Chimonobambusa* 方竹属
Chimonobambusa sichuanensis (T. P. Yi) T. H. Wen 月月竹
Bamboos, evergreen

15.7; 5.6; 24.8; 15.7; 129.4; 0.8; 832 ❙ 1162; 618; 50 ❙ 41.3; 832; 168.0

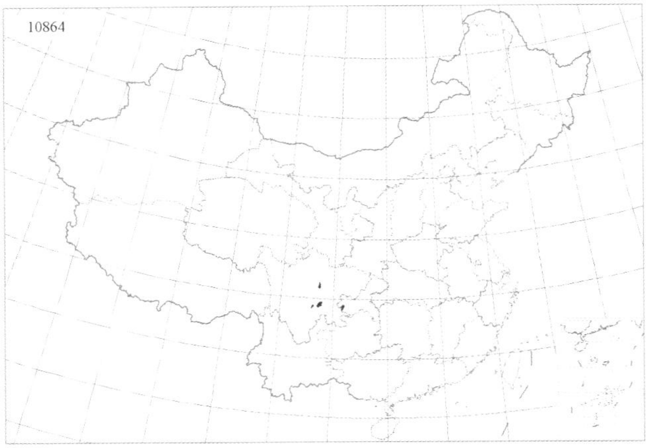

Poaceae 禾本科 *Chimonobambusa* 方竹属
Chimonobambusa szechuanensis (Rendle) Keng f. 八月竹
Bamboos, evergreen

15.5; 5.8; 24.4; 15.6; 127.2; 0.0; 812 ❙ 1398; 812; 46 ❙ 72.0; 812; 224.4

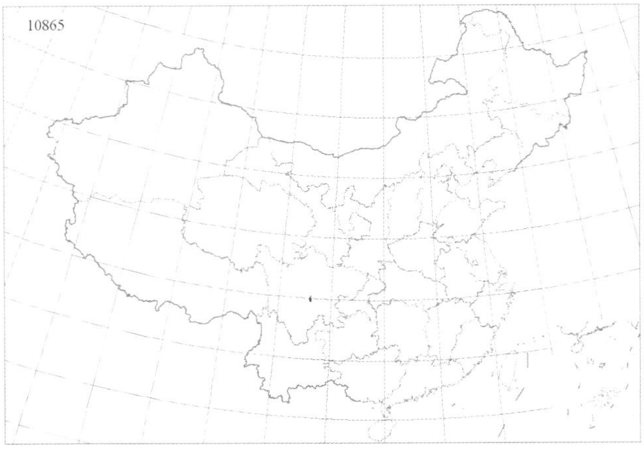

Poaceae 禾本科 *Chimonobambusa* 方竹属
Chimonobambusa tumidissinoda Hsueh et T. P. Yi ex Ohrnb. 筇竹
Bamboos, evergreen

14.7 (11.9~17.8); 5.1 (2.9~7.8); 23.5 (20.0~27.0); 14.8 (12.0~17.9); 118.3 (87.2~154.2); 0.7 (0.0~3.4); 781 (671~929) ❙ 1037 (879~1410); 550 (472~816); 49 (33~67) ❙ 33.3 (18.3~80.4); 781 (671~929); 210.6 (142.9~278.8)

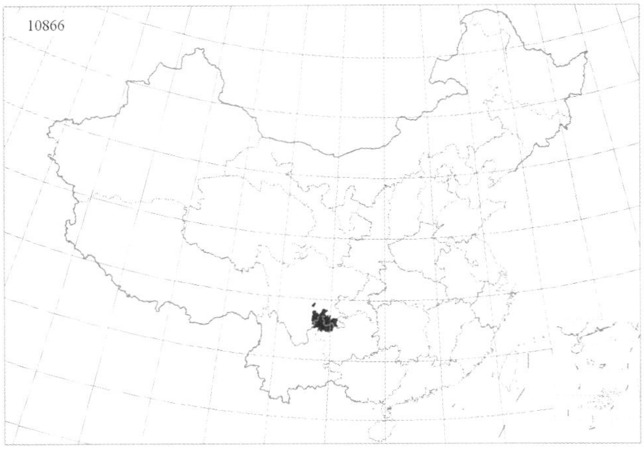

Poaceae 禾本科 *Chimonocalamus* 香竹属
Chimonocalamus delicatus Hsueh et T. P. Yi 香竹
Bamboos, evergreen

19.2；11.6；24.3；19.2；170.1；0.0；952 ❙ 1444；763；77 ❙ 52.0；952；434.3

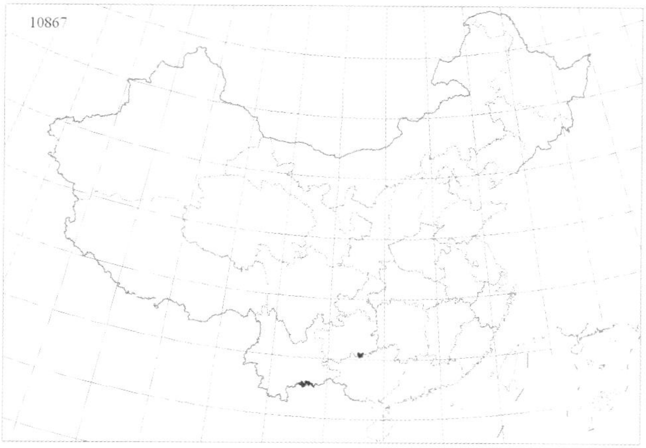

Poaceae 禾本科 *Chimonocalamus* 香竹属
Chimonocalamus dumosus Hsueh et T. P. Yi 小香竹
Bamboos, evergreen

17.7；10.7；22.3；17.7；152.9；0.0；853 ❙ 1310；741；47 ❙ 53.5；842；426.5

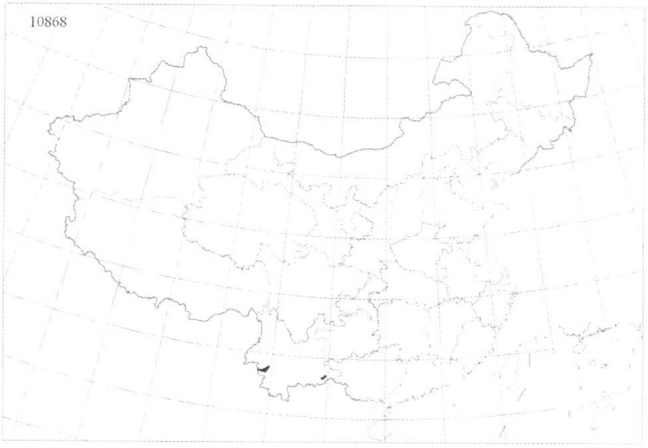

Poaceae 禾本科 *Chimonocalamus* 香竹属
Chimonocalamus fimbriatus Hsueh et T. P. Yi 流苏香竹
Bamboos, evergreen

18.2 (16.4~19.5)；11.7 (9.6~13.0)；22.2 (20.6~23.3)；18.2 (16.4~19.5)；158.7 (137.3~173.3)；0.0 (0.0~0.0)；876 (795~942) ❙ 1514 (1366~1688)；878 (783~959)；51 (42~62) ❙ 73.9 (55.7~112.2)；867 (795~921)；523.1 (482.8~627.0)

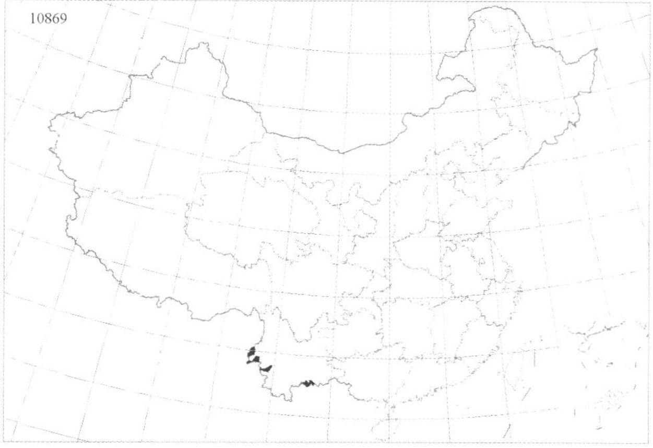

Poaceae 禾本科 *Chimonocalamus* 香竹属
Chimonocalamus griffithianus (Munro) Hsueh et T. P. Yi 西藏香竹
Bamboos, evergreen

13.3；5.4；19.6；13.4；100.4；0.0；706 ❙ 1736；1007；51 ❙ 145.8；706；447.4

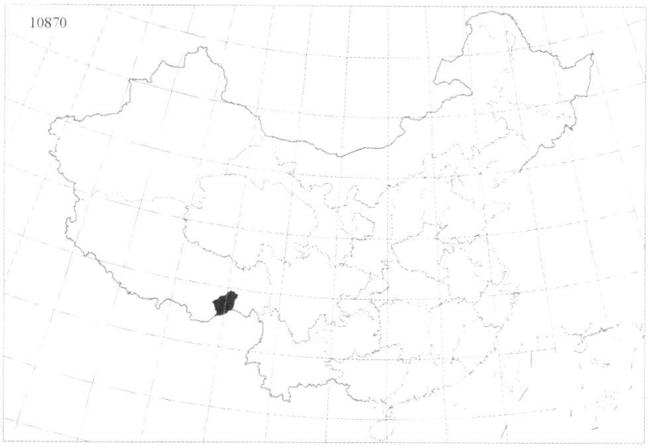

Poaceae 禾本科 *Chimonocalamus* 香竹属
Chimonocalamus longiligulatus Hsueh et T. P. Yi 长舌香竹
Bamboos, evergreen

19.0；13.0；23.1；19.1；168.7；0.0；912 ❙ 1417；829；56 ❙ 55.6；909；503.2

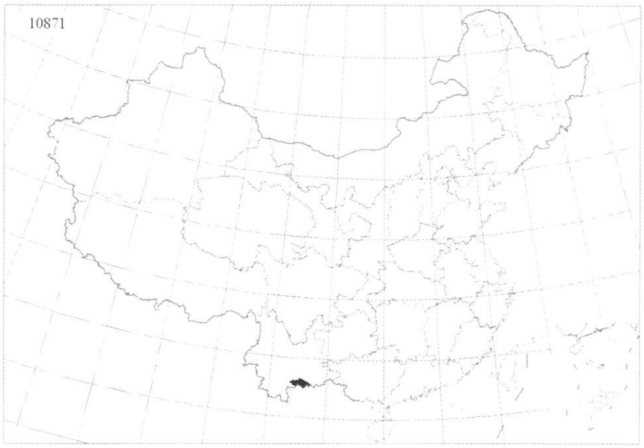

Poaceae 禾本科 *Chimonocalamus* 香竹属
Chimonocalamus longiusculus Hsueh et T. P. Yi 长节香竹
Bamboos, evergreen

17.2；9.5；22.7；17.3；147.0；0.0；838 ❙ 1254；699；48 ❙ 49.7；838；336.3

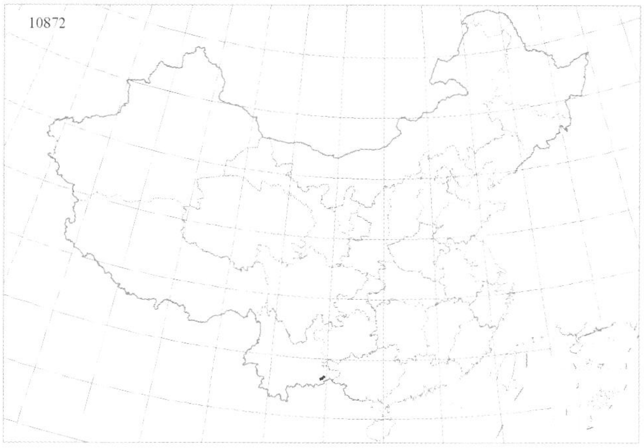

Poaceae 禾本科 *Chimonocalamus* 香竹属
Chimonocalamus makuanensis Hsueh et T. P. Yi 马关香竹
Bamboos, evergreen

19.1; 12.1; 24.0; 19.1; 169.5; 0.0; 943 ❙ 1513; 839; 62 ❙ 62.5; 943; 482.5

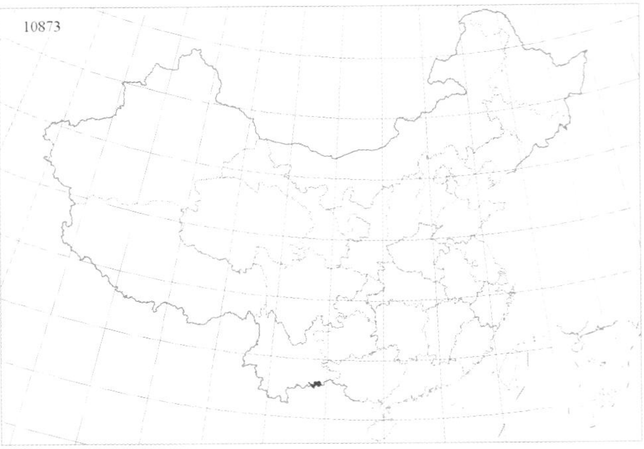

Poaceae 禾本科 *Chimonocalamus* 香竹属
Chimonocalamus montanus Hsueh et T. P. Yi 山香竹
Bamboos, evergreen

14.4; 7.5; 19.1; 14.4; 113.1; 0.0; 727 ❙ 1454; 792; 70 ❙ 99.9; 727; 487.2

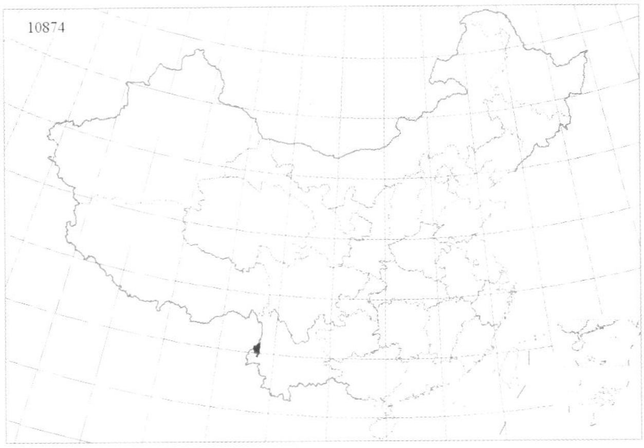

Poaceae 禾本科 *Chimonocalamus* 香竹属
Chimonocalamus pallens Hsueh et T. P. Yi 灰香竹
Bamboos, evergreen

19.1; 13.2; 23.4; 19.3; 171.1; 0.0; 925 ❙ 1352; 788; 53 ❙ 46.3; 920; 479.1

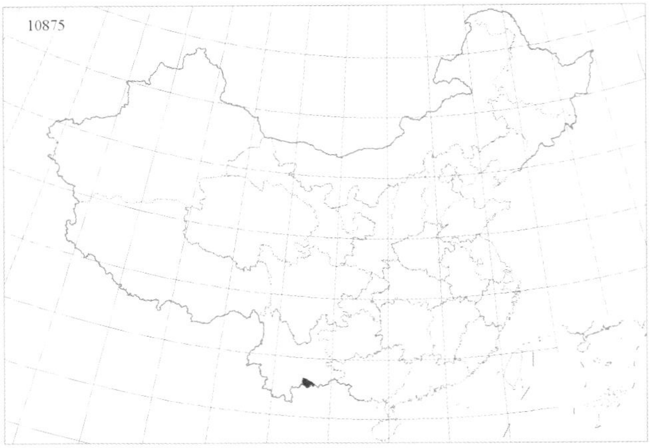

Poaceae 禾本科 *Dendrocalamus* 牡竹属
Dendrocalamus asper (Schult. et Schult. f.) Backer ex K. Heyne 马来甜龙竹
Bamboos, evergreen

18.4 (16.4~20.4); 12.4 (9.6~15.2); 23.0 (20.6~28.7); 18.9 (16.4~22.7); 167.2 (137.3~212.7); 0.0 (0.0~0.0); 925 (795~1216) ❙ 1618 (1499~1965); 879 (686~994); 61 (42~104) ❙ 76.6 (61.2~112.2); 917 (795~1216); 505.6 (204.4~659.0)

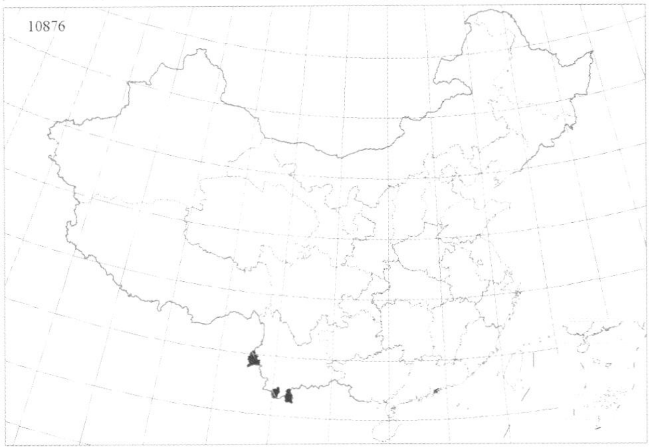

Poaceae 禾本科 *Dendrocalamus* 牡竹属
Dendrocalamus bambusoides Hsueh et D. Z. Li 椅子竹
Bamboos, evergreen

17.2; 10.8; 21.7; 17.3; 147.3; 0.0; 839 ❙ 1082; 626; 44 ❙ 29.0; 822; 333.6

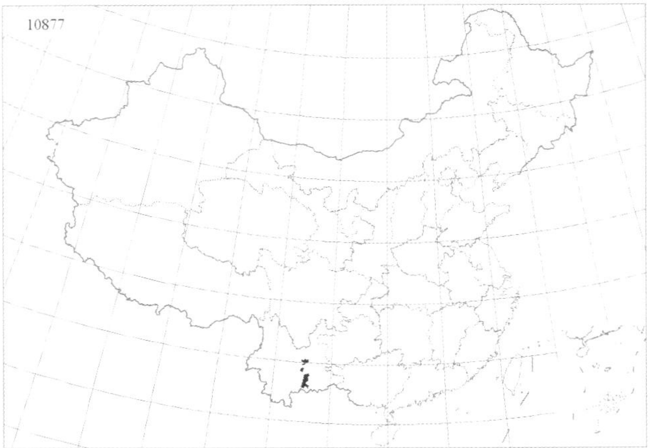

Poaceae 禾本科 *Dendrocalamus* 牡竹属
Dendrocalamus barbatus Hsueh et D. Z. Li 小叶龙竹
Bamboos, evergreen

19.1 (16.9~21.0); 13.0 (10.3~15.1); 23.1 (21.4~25.6); 19.2 (16.9~21.0); 169.9 (142.7~191.8); 0.0 (0.0~0.0); 925 (813~1053) ❙ 1389 (1163~1607); 720 (641~835); 59 (48~77) ❙ 50.2 (34.8~64.9); 917 (813~1053); 503.8 (383.4~659.0)

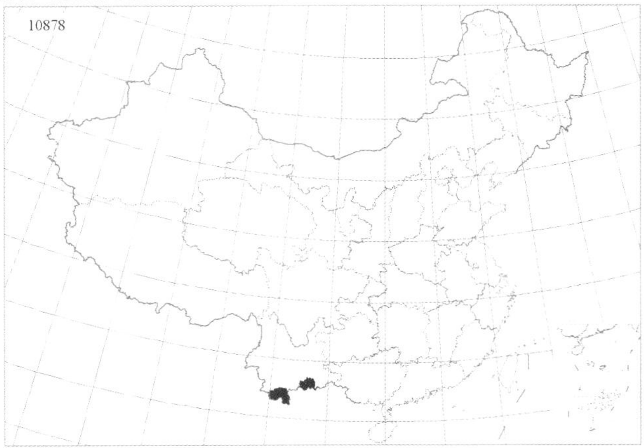

Poaceae 禾本科 *Dendrocalamus* 牡竹属

Dendrocalamus barbatus var. internodiiradicatus J. R. Xue et D. Z. Li 毛脚龙竹

Bamboos, evergreen

19.7 (19.1~20.5); 14.0 (13.0~15.1); 23.3 (22.3~23.7); 19.7 (19.1~20.5); 176.9 (169.7~185.9); 0.0 (0.0~0.0); 948 (909~992) ❙ 1455 (1276~1607); 724 (655~835); 62 (51~77) ❙ 53.5 (36.8~64.9); 933 (880~986); 540.2 (452.0~659.0)

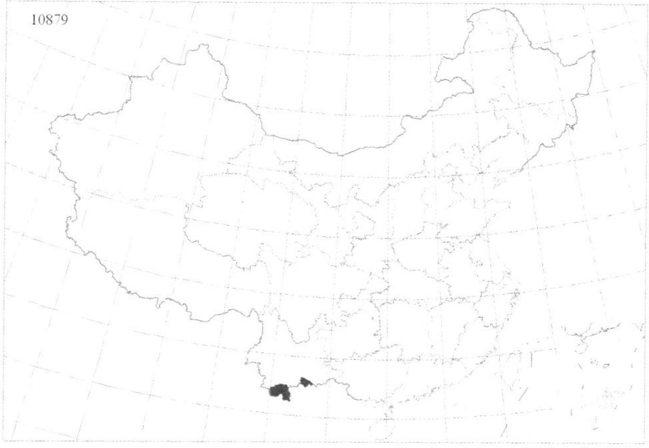

Poaceae 禾本科 *Dendrocalamus* 牡竹属

Dendrocalamus birmanicus A. Camus 缅甸龙竹

Bamboos, evergreen

17.5 (14.4~19.6); 10.7 (7.5~12.8); 21.6 (19.1~23.5); 17.5 (14.4~19.6); 149.9 (113.1~175.7); 0.0 (0.0~0.0); 848 (727~953) ❙ 1563 (1454~1688); 903 (792~959); 55 (42~70) ❙ 85.7 (61.2~112.2); 842 (727~932); 522.8 (451.2~627.0)

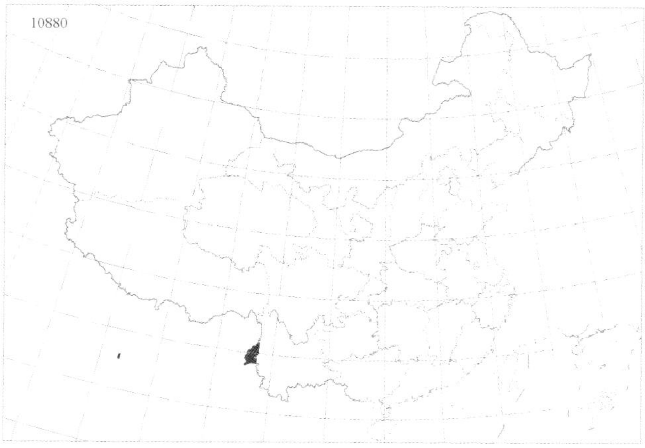

Poaceae 禾本科 *Dendrocalamus* 牡竹属

Dendrocalamus brandisii (Munro) Kurz 勃氏甜龙竹

Bamboos, evergreen

18.4 (16.3~21.0); 12.1 (9.6~15.1); 22.4 (20.3~25.6); 18.5 (16.4~21.0); 161.5 (136.6~191.8); 0.0 (0.0~0.0); 889 (789~1053) ❙ 1476 (984~1688); 820 (562~959); 56 (42~77) ❙ 66.5 (20.8~112.2); 881 (780~1053); 499.8 (309.3~659.0)

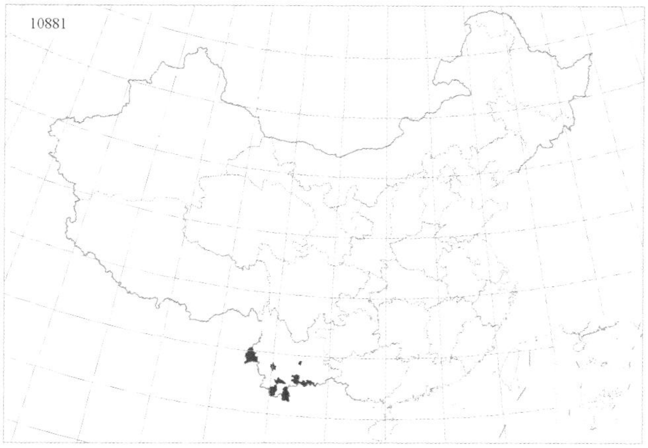

Poaceae 禾本科 *Dendrocalamus* 牡竹属

Dendrocalamus calostachyus (Kurz) Kurz 美穗龙竹

Bamboos, evergreen

19.8; 14.3; 23.0; 19.9; 178.2; 0.0; 950 ❙ 1482; 671; 62 ❙ 56.3; 920; 542.0

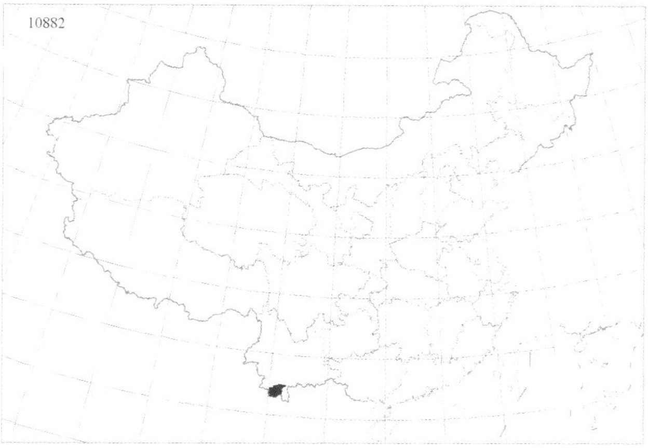

Poaceae 禾本科 *Dendrocalamus* 牡竹属

Dendrocalamus farinosus (Keng et Keng f.) L. C. Chia et H. L. Fung 大叶慈

Bamboos, evergreen

16.1 (13.7~19.1); 6.2 (2.4~10.1); 24.9 (19.3~27.9); 16.2 (13.7~19.1); 134.6 (104.7~169.2); 0.3 (0.0~4.4); 842 (714~1002) ❙ 1204 (950~1710); 567 (458~720); 75 (26~196) ❙ 43.3 (12.1~93.2); 839 (690~1002); 211.0 (126.1~298.7)

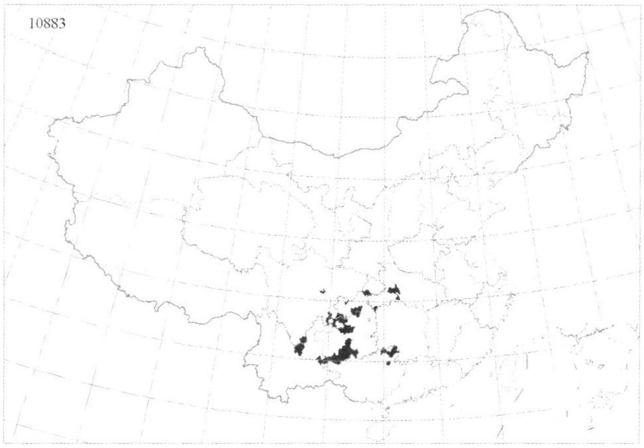

Poaceae 禾本科 *Dendrocalamus* 牡竹属

Dendrocalamus fugongensis Hsueh et D. Z. Li 福贡龙竹

Bamboos, evergreen

9.8; 2.7; 15.9; 9.9; 63.3; 5.0; 606 ❙ 1097; 620; 47 ❙ 80.8; 606; 362.9

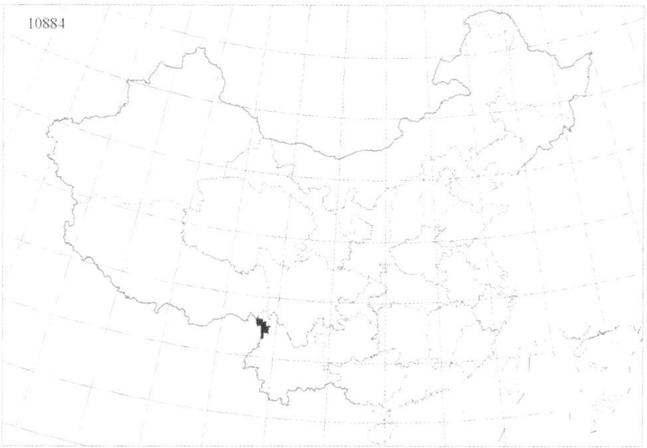

Poaceae 禾本科 *Dendrocalamus* 牡竹属
Dendrocalamus giganteus Wall. ex Munro 龙竹
Bamboos, evergreen

15.6 (5.4~24.1); 8.9 (-2.1~18.9); 20.6 (12.3~28.0); 15.6 (5.8~24.1); 129.2 (31.5~229.7); 1.7 (0.0~25.6); 798 (481~1322) ❙ 1080 (741~1751); 619 (411~1169); 37 (15~77) ❙ 36.1 (-14.2~74.2); 753 (481~1095); 339.0 (154.2~659.0)

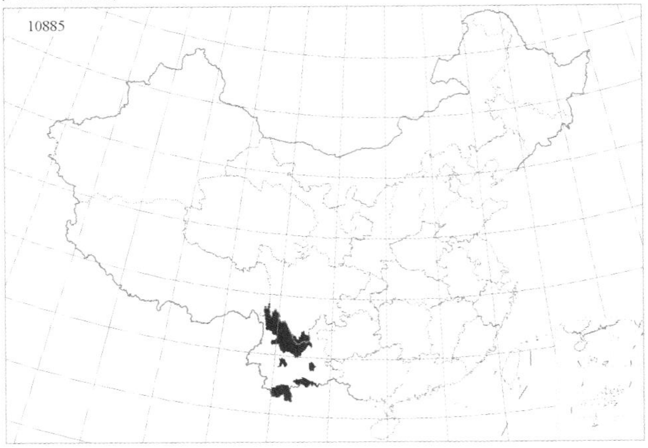

Poaceae 禾本科 *Dendrocalamus* 牡竹属
Dendrocalamus hamiltonii Nees et Arn.ex Munro 版纳甜龙竹
Bamboos, evergreen

20.0; 14.6; 23.2; 20.1; 180.8; 0.0; 963 ❙ 1523; 681; 67 ❙ 58.4; 942; 581.0

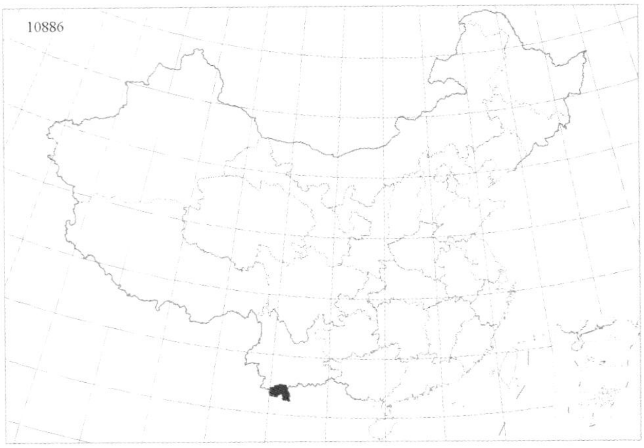

Poaceae 禾本科 *Dendrocalamus* 牡竹属
Dendrocalamus jianshuiensis Hsueh et D. Z. Li 建水龙竹
Bamboos, evergreen

18.7; 12.5; 23.0; 18.8; 165.5; 0.0; 903 ❙ 1167; 672; 47 ❙ 29.1; 885; 373.3

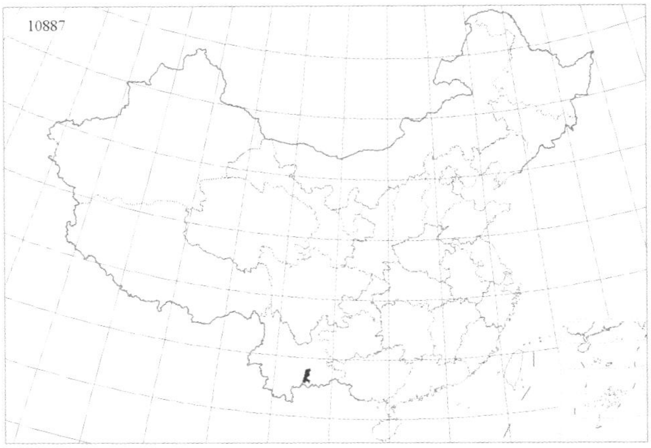

Poaceae 禾本科 *Dendrocalamus* 牡竹属
Dendrocalamus latiflorus Munro 麻竹
Bamboos, evergreen

17.5 (5.4~24.5); 9.2 (-2.1~19.1); 24.5 (12.3~29.8); 17.6 (5.8~24.5); 151.5 (31.5~234.4); 0.6 (0.0~25.6); 914 (481~1360) ❙ 1372 (741~2993); 633 (359~1594); 99 (15~547) ❙ 50.2 (-14.2~246.5); 899 (481~1360); 261.8 (142.9~659.0)

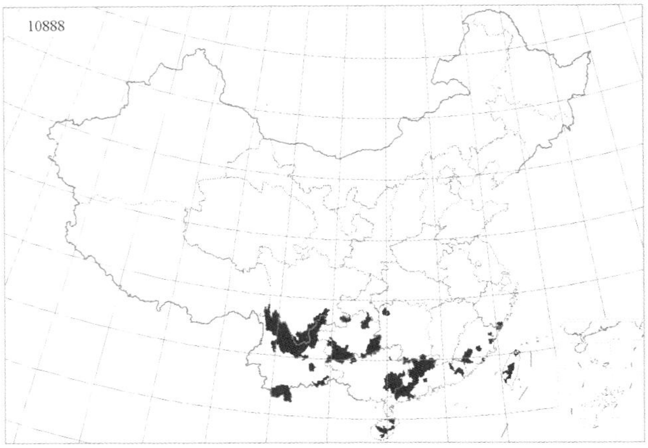

Poaceae 禾本科 *Dendrocalamus* 牡竹属
Dendrocalamus membranaceus Munro 黄竹
Bamboos, evergreen

19.8 (18.2~22.1); 13.5 (11.9~15.1); 24.3 (22.0~28.8); 19.9 (18.2~22.1); 178.9 (158.7~205.6); 0.0 (0.0~0.0); 978 (868~1182) ❙ 1658 (1276~2993); 775 (655~924); 124 (45~547) ❙ 69.8 (36.8~203.4); 969 (847~1182); 443.4 (157.7~659.0)

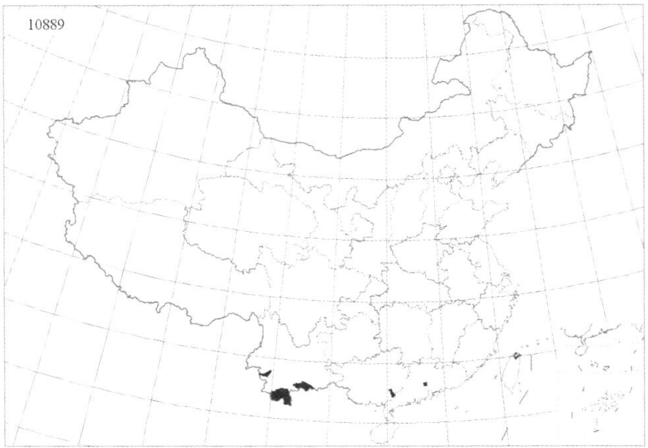

Poaceae 禾本科 *Dendrocalamus* 牡竹属
Dendrocalamus minor (McClure) L. C. Chia et H. L. Fung 吊丝竹
Bamboos, evergreen

18.8 (16.4~22.3); 9.2 (5.4~13.7); 26.8 (21.9~29.8); 18.8 (16.5~22.3); 165.6 (137.7~208.1); 0.0 (0.0~0.0); 973 (811~1202) ❙ 1494 (1059~1784); 631 (476~779); 140 (44~197) ❙ 53.9 (20.3~85.5); 969 (811~1202); 234.2 (157.7~318.5)

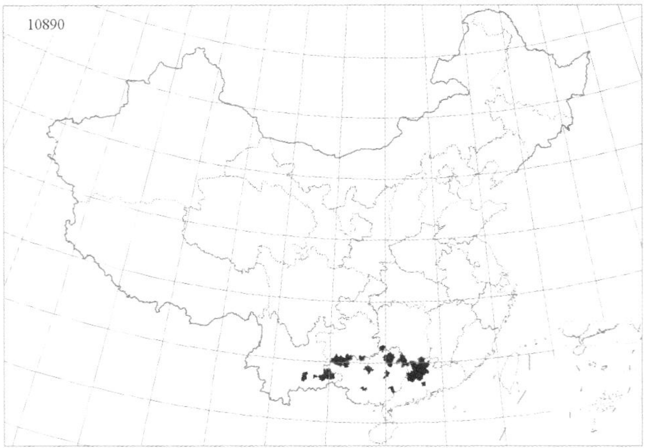

Poaceae 禾本科 *Dendrocalamus* 牡竹属
Dendrocalamus minor var. amoenus (Q. H. Dai et C. F. Huang) Hsueh et D. Z. Li 花吊丝竹
Bamboos, evergreen

16.9; 9.1; 22.5; 17.0; 143.8; 0.0; 829 ❙ 1118; 607; 49 ❙ 34.8; 829; 295.7

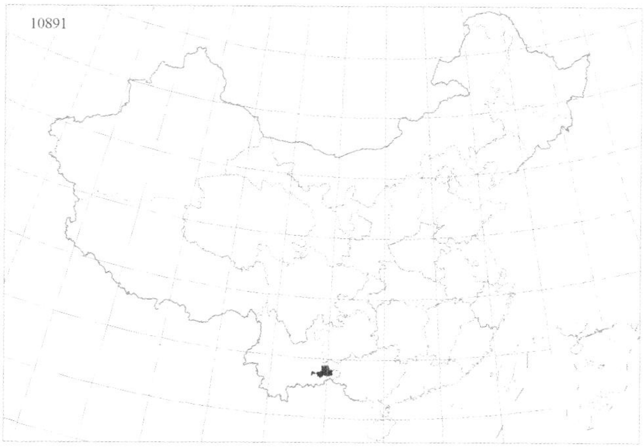

Poaceae 禾本科 *Dendrocalamus* 牡竹属
Dendrocalamus pachystachys Hsueh et D. Z. Li 粗穗龙竹
Bamboos, evergreen

18.2; 12.0; 22.6; 18.3; 159.9; 0.0; 885 ❙ 1186; 682; 49 ❙ 33.8; 866; 389.1

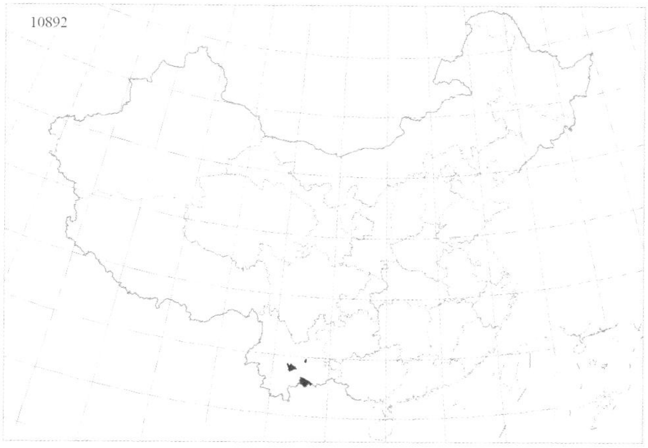

Poaceae 禾本科 *Dendrocalamus* 牡竹属
Dendrocalamus parishii Munro 巴氏龙竹
Bamboos, evergreen

18.0; 11.9; 22.1; 13.0; 156.2; 0.2; 899 ❙ 1468; 726; 65 ❙ 66.2; 891; 527.3

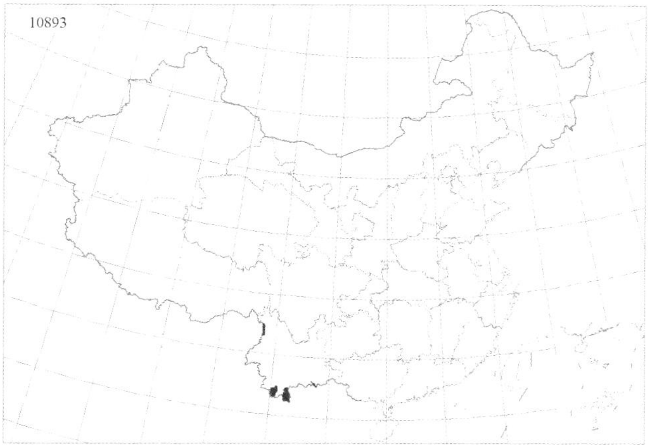

Poaceae 禾本科 *Dendrocalamus* 牡竹属
Dendrocalamus peculiaris Hsueh et D. Z. Li 金平龙竹
Bamboos, evergreen

19.8; 14.0; 23.5; 19.8; 177.8; 0.0; 954 ❙ 1446; 745; 61 ❙ 51.7; 938; 538.1

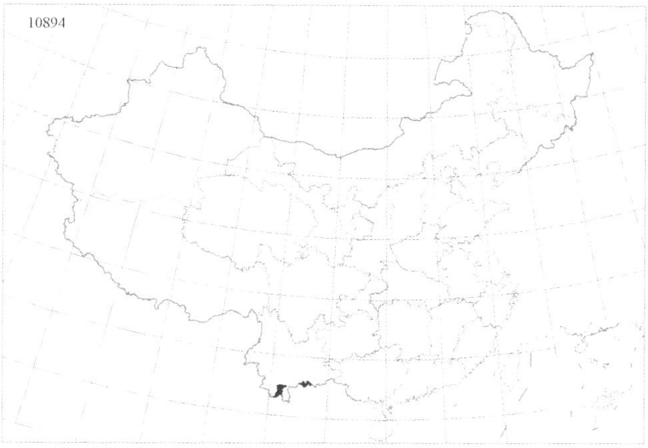

Poaceae 禾本科 *Dendrocalamus* 牡竹属
Dendrocalamus pulverulentus L. C. Chia et But 粉麻竹
Bamboos, evergreen

19.9; 10.4; 27.9; 19.9; 178.9; 0.0; 1044 ❙ 1667; 642; 167 ❙ 60.0; 1044; 209.2

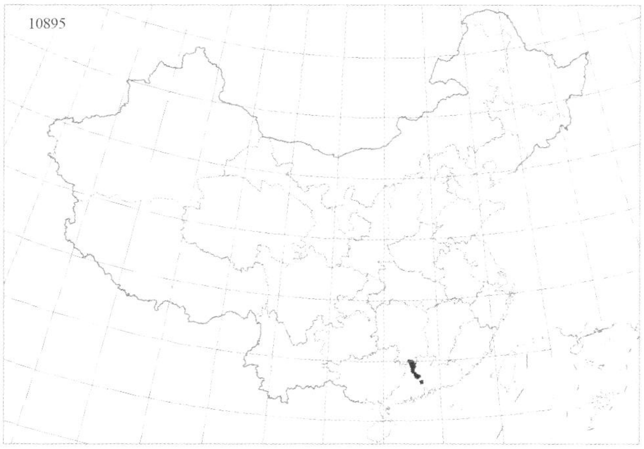

Poaceae 禾本科 *Dendrocalamus* 牡竹属
Dendrocalamus semiscandens Hsueh et D. Z. Li 野龙竹
Bamboos, evergreen

18.8 (16.0~21.0); 12.6 (9.3~15.1); 22.9 (20.4~25.6); 18.8 (16.0~21.0); 166.2 (132.4~191.8); 0.0 (0.0~0.0); 911 (780~1053) ❙ 1485 (1348~1607); 810 (702~911); 59 (42~77) ❙ 63.7 (44.5~74.2); 905 (780~1053); 509.7 (395.4~659.0)

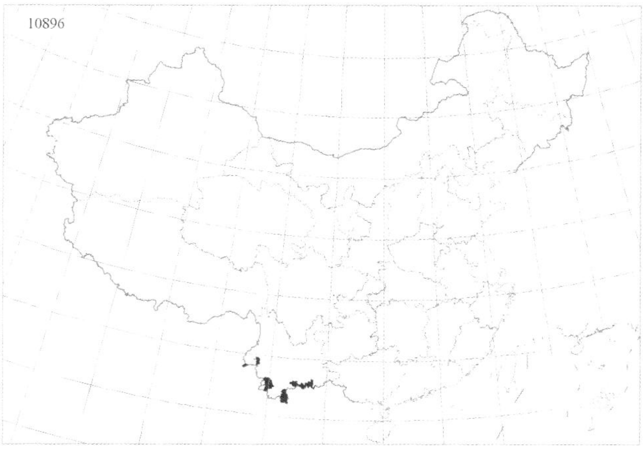

Poaceae 禾本科 *Dendrocalamus* 牡竹属
Dendrocalamus sikkimensis Gamble ex Oliv. 锡金龙竹
Bamboos, evergreen

19.9; 14.0; 23.7; 20.0; 179.5; 0.0; 967 ❙ 1514; 760; 65 ❙ 56.9; 959; 557.7

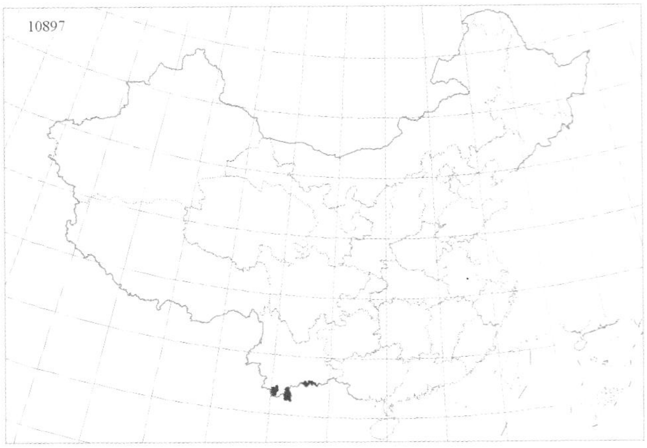

Poaceae 禾本科 *Dendrocalamus* 牡竹属
Dendrocalamus sinicus L. C. Chia et J. L. Sun 歪脚龙竹
Bamboos, evergreen

19.6; 13.9; 22.9; 19.6; 175.2; 0.0; 939 ❙ 1484; 706; 62 ❙ 58.1; 918; 564.9

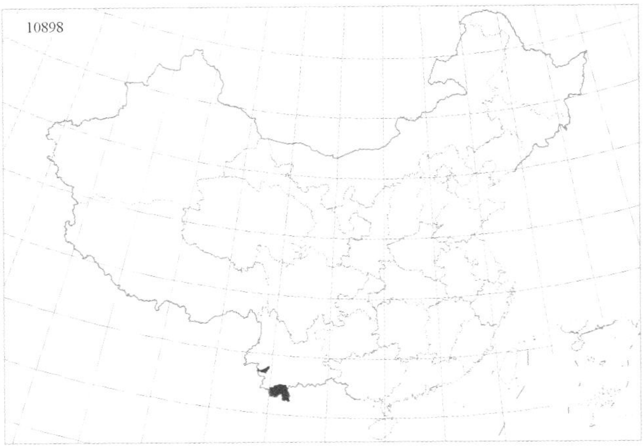

Poaceae 禾本科 *Dendrocalamus* 牡竹属
Dendrocalamus tibeticus Hsueh et T. P. Yi 西藏牡竹
Bamboos, evergreen

10.6; 3.2; 16.7; 10.8; 76.8; 8.7; 627 ❙ 1345; 764; 50 ❙ 113.0; 627; 362.9

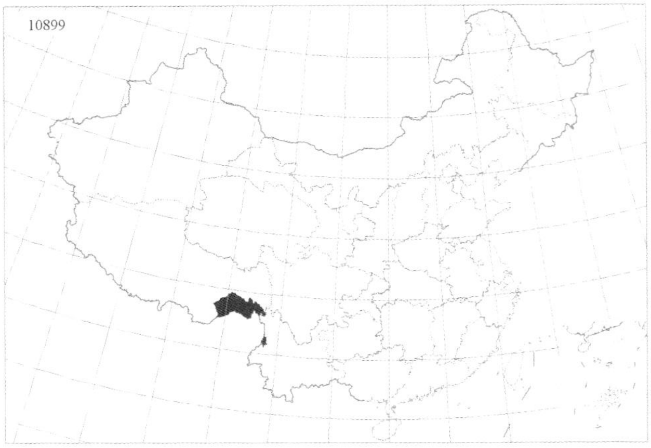

Poaceae 禾本科 *Drepanostachyum* 镰序竹属
Drepanostachyum ampullare (T. P. Yi) Demoly 樟木镰序竹
Bamboos, evergreen

-1.8; -10.2; 6.1; 1.8; 1.9; 82.8; 329 ❙ 320; 217; 20 ❙ -3.0; 320; 72.2

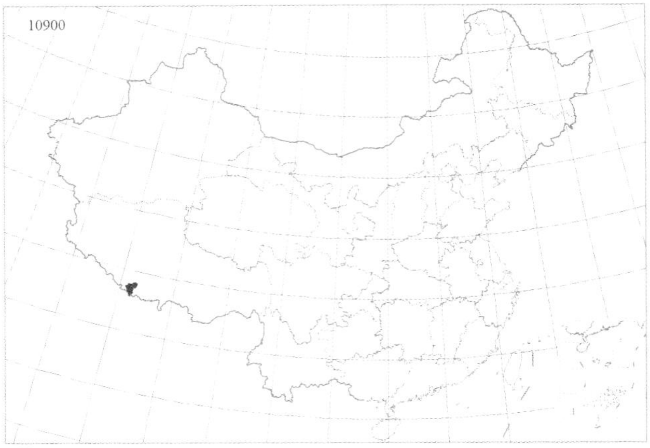

Poaceae 禾本科 *Drepanostachyum* 镰序竹属
Drepanostachyum fractiflexum (T. P. Yi) D. Z. Li 扫把竹
Bamboos, evergreen

10.8 (2.2~16.1); 3.2 (-6.3~8.6); 17.1 (9.5~21.9); 11.0 (3.9~16.1); 79.8 (15.8~133.0); 9.8 (0.0~48.6); 637 (411~797) ❙ 950 (729~1058); 577 (482~632); 22 (11~42) ❙ 52.6 (27.3~92.9); 619 (411~719); 292.7 (208.8~349.8)

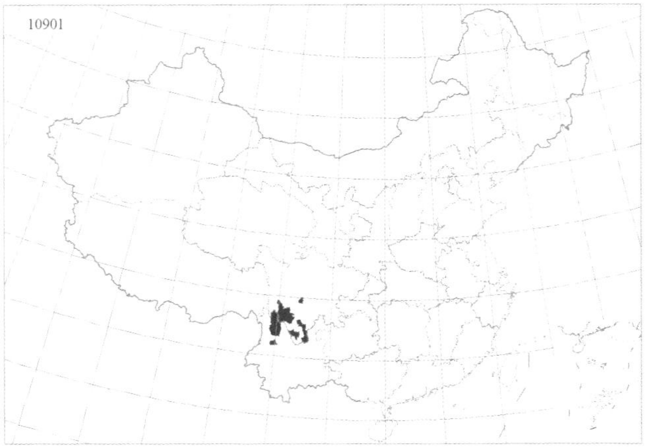

Poaceae 禾本科 *Drepanostachyum* 镰序竹属
Drepanostachyum semiorbiculatum (T. P. Yi) Stapleton 圆芽镰序竹
Bamboos, evergreen

12.2; 4.3; 18.5; 12.2; 87.4; 0.7; 672 ❙ 1431; 817; 31 ❙ 113.0; 672; 363.7

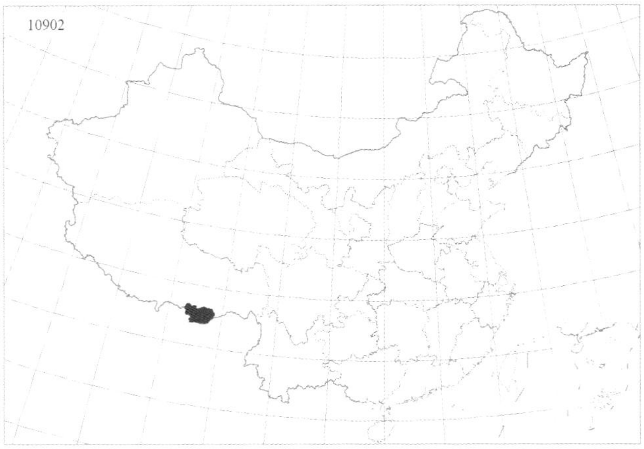

Poaceae 禾本科 *Fargesia* 箭竹属
Fargesia acuticontracta T. P. Yi 尖鞘箭竹
Bamboos, evergreen

9.1; 1.9; 15.3; 9.1; 56.7; 7.1; 587 ┃ 1024; 581; 42 ┃ 74.7; 587; 352.0

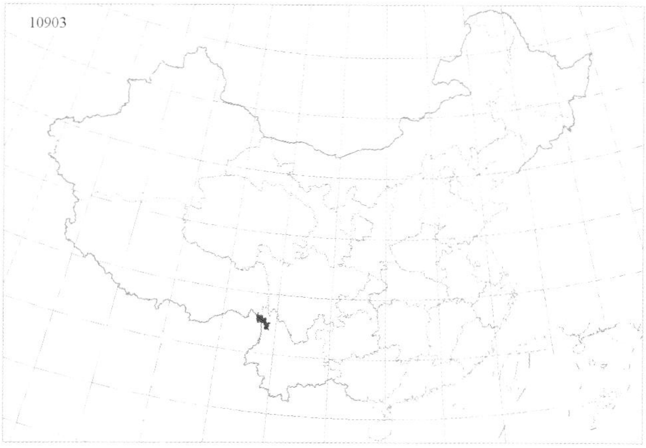

Poaceae 禾本科 *Fargesia* 箭竹属
Fargesia adpressa T. P. Yi 贴毛箭竹
Bamboos, evergreen

7.0; −1.0; 13.9; 7.5; 44.9; 20.2; 532 ┃ 1010; 603; 17 ┃ 94.4; 532; 281.6

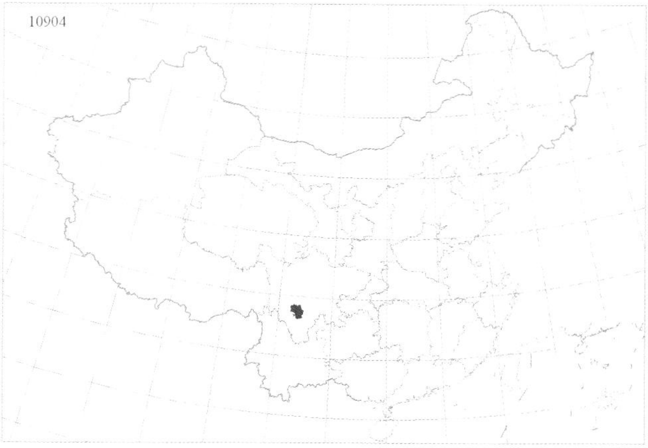

Poaceae 禾本科 *Fargesia* 箭竹属
Fargesia albocerea Hsueh et T. P. Yi 片马箭竹
Bamboos, evergreen

13.2; 6.4; 18.4; 13.3; 99.4; 0.0; 695 ┃ 1337; 727; 69 ┃ 92.4; 695; 407.1

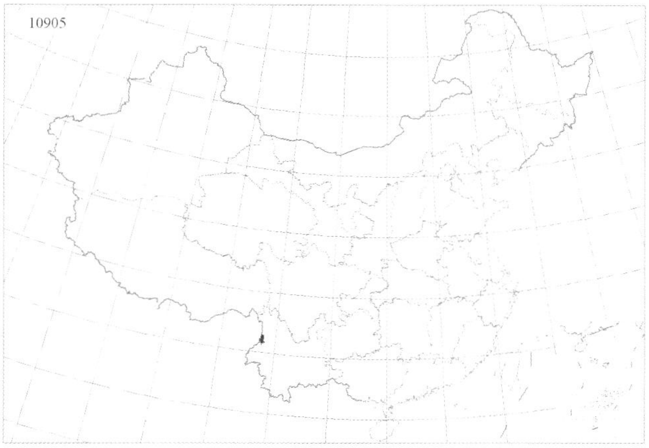

Poaceae 禾本科 *Fargesia* 箭竹属
Fargesia altior T. P. Yi 船竹
Bamboos, evergreen

14.4; 7.5; 19.1; 14.4; 113.1; 0.0; 727 ┃ 1454; 792; 70 ┃ 99.9; 727; 487.2

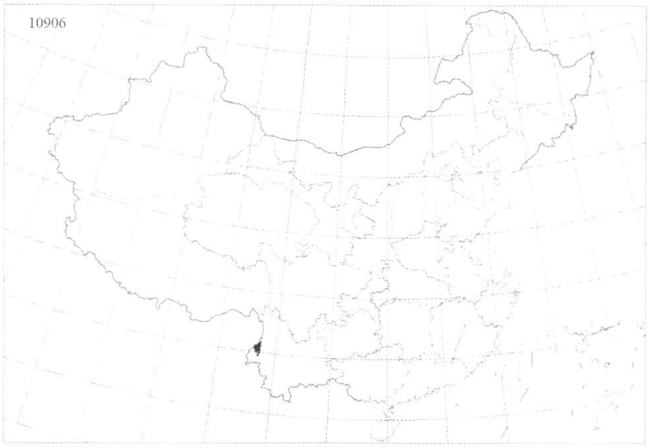

Poaceae 禾本科 *Fargesia* 箭竹属
Fargesia angustissima T. P. Yi 油竹子
Bamboos, evergreen

10.1; 0.4; 18.9; 10.3; 74.3; 12.4; 630 ┃ 998; 539; 27 ┃ 62.0; 630; 210.3

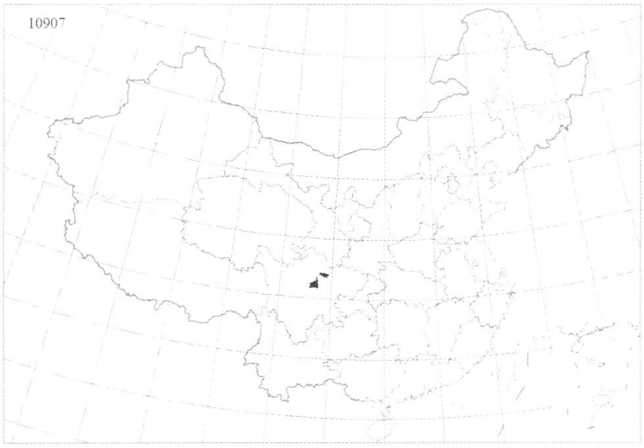

Poaceae 禾本科 *Fargesia* 箭竹属
Fargesia brevissima T. P. Yi 窝竹
Bamboos, evergreen

12.0; 1.1; 22.5; 12.0; 92.3; 8.5; 690 ┃ 1207; 501; 56 ┃ 74.9; 690; 258.2

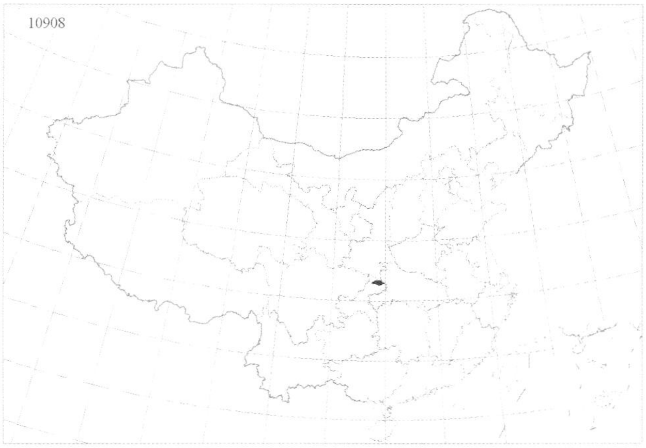

Poaceae 禾本科 *Fargesia* 箭竹属
Fargesia caduca T. P. Yi 景谷箭竹
Bamboos, evergreen

17.3; 11.0; 21.6; 17.3; 148.0; 0.0; 832 ❙ 1049; 598; 44 ❙ 26.1; 783; 365.3

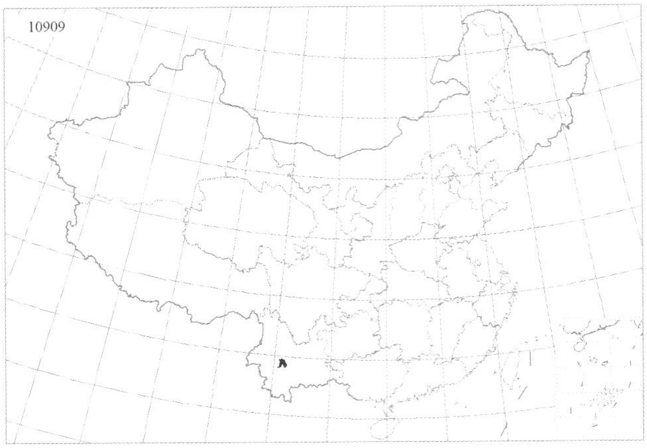

Poaceae 禾本科 *Fargesia* 箭竹属
Fargesia canaliculata T. P. Yi 岩斑竹
Bamboos, evergreen

9.8; 2.1; 16.2; 10.0; 68.6; 10.6; 608 ❙ 1008; 610; 17 ❙ 70.9; 602; 291.3

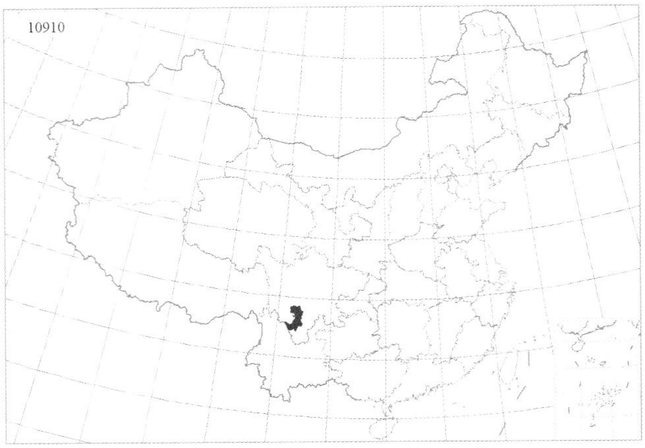

Poaceae 禾本科 *Fargesia* 箭竹属
Fargesia circinata Hsueh et T. P. Yi 卷耳箭竹
Bamboos, evergreen

16.7; 10.0; 21.3; 16.7; 140.8; 0.0; 809 ❙ 1090; 594; 55 ❙ 34.7; 809; 406.0

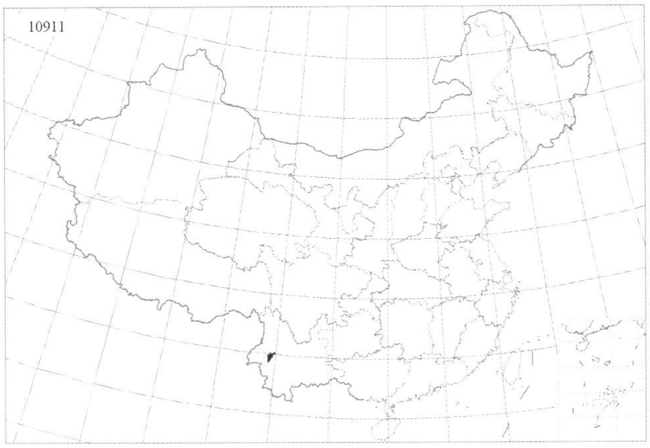

Poaceae 禾本科 *Fargesia* 箭竹属
Fargesia communis T. P. Yi 马亨箭竹
Bamboos, evergreen

9.6; 2.4; 15.7; 9.6; 60.8; 5.4; 600 ❙ 1007; 578; 40 ❙ 67.9; 600; 351.7

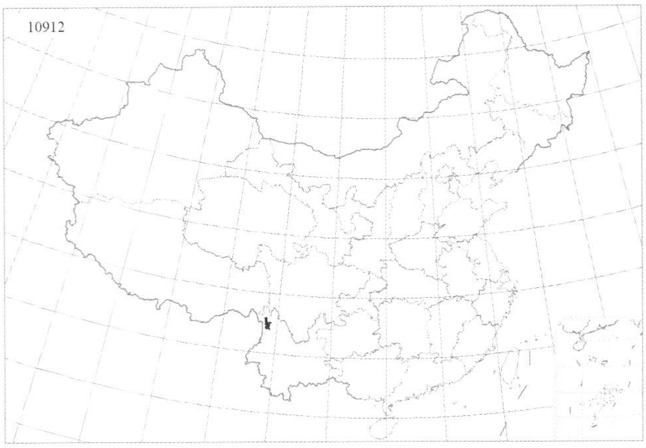

Poaceae 禾本科 *Fargesia* 箭竹属
Fargesia concinna T. P. Yi 美丽箭竹
Bamboos, evergreen

17.3; 11.0; 21.6; 17.3; 148.0; 0.0; 832 ❙ 1049; 598; 44 ❙ 26.1; 783; 365.3

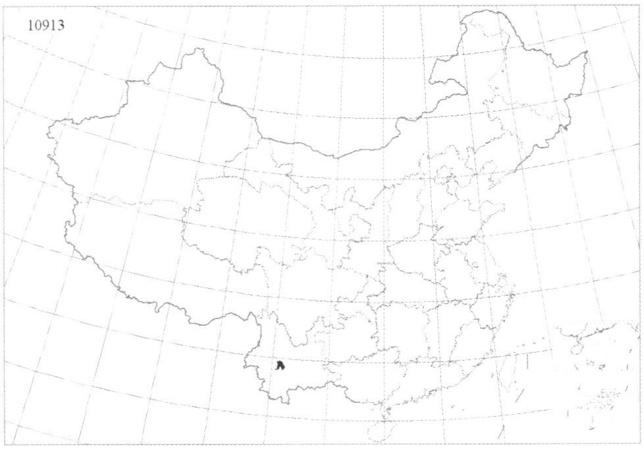

Poaceae 禾本科 *Fargesia* 箭竹属
Fargesia conferta T. P. Yi 笼笼竹
Bamboos, evergreen

15.8; 5.5; 25.4; 15.8; 129.8; 0.2; 829 ❙ 1070; 502; 67 ❙ 29.5; 829; 177.0

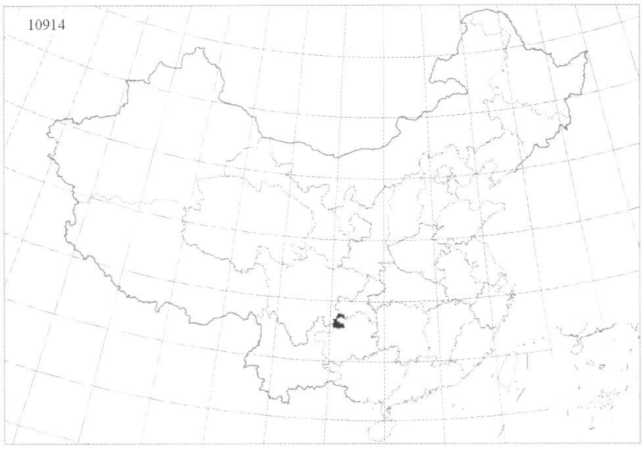

Poaceae 禾本科 *Fargesia* 箭竹属
Fargesia contracta T. P. Yi 带鞘箭竹
Bamboos, evergreen

12.2; 5.2; 17.7; 12.2; 88.9; 2.4; 670 ❙ 1201; 657; 59 ❙ 80.1; 670; 369.8

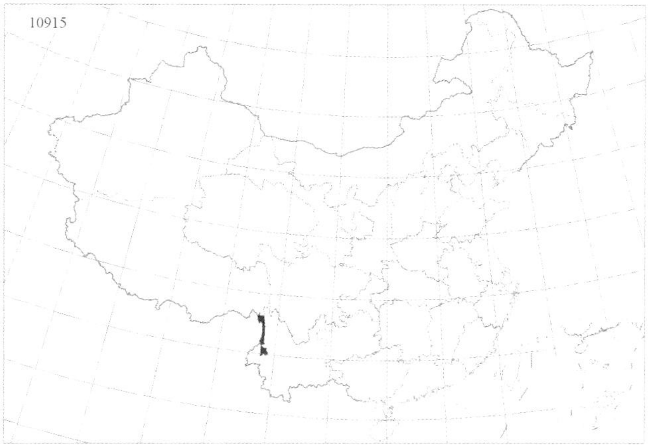

Poaceae 禾本科 *Fargesia* 箭竹属
Fargesia cuspidata (Keng) Z. P. Wang et G. H. Ye 尖尾箭竹
Bamboos, evergreen

19.2 (17.4~21.3); 9.8 (7.1~12.6); 26.7 (24.6~28.7); 19.3 (17.4~21.3); 171.0 (148.9~196.0); 0.0 (0.0~0.0); 991 (886~1123) ❙ 1370 (1172~1620); 678 (588~775); 96 (43~166) ❙ 38.7 (5.8~64.7); 991 (886~1121); 271.3 (190.9~411.1)

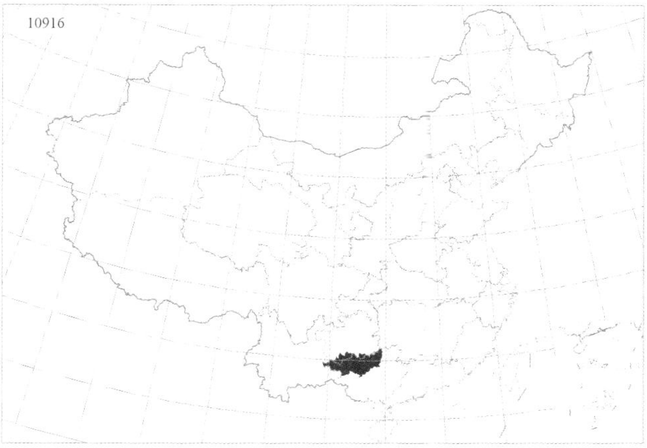

Poaceae 禾本科 *Fargesia* 箭竹属
Fargesia declivis T. P. Yi 斜倚箭竹
Bamboos, evergreen

8.6; 1.3; 15.0; 8.7; 52.6; 8.8; 574 ❙ 1041; 585; 45 ❙ 81.5; 574; 352.3

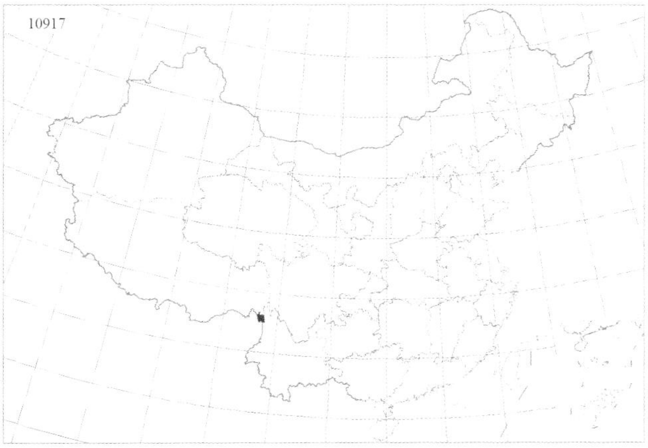

Poaceae 禾本科 *Fargesia* 箭竹属
Fargesia decurvata J. L. Lu 毛龙头竹
Bamboos, evergreen

13.0; 1.6; 23.8; 13.1; 103.8; 7.9; 736 ❙ 1148; 480; 65 ❙ 55.1; 736; 250.7

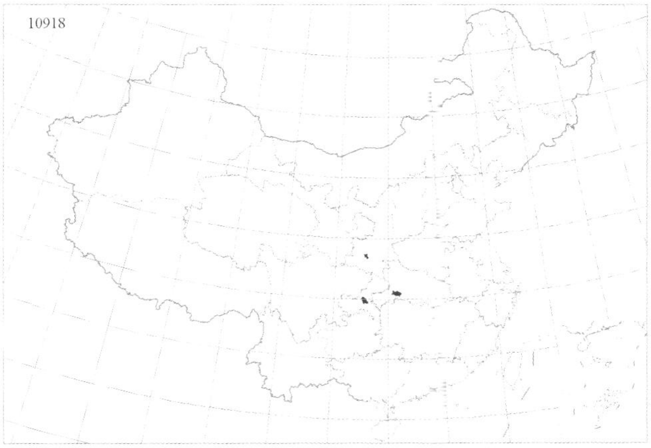

Poaceae 禾本科 *Fargesia* 箭竹属
Fargesia denudata T. P. Yi 缺苞箭竹
Bamboos, evergreen

9.1; -0.9; 18.2; 9.6; 67.9; 18.3; 606 ❙ 837; 436; 20 ❙ 44.1; 606; 262.0

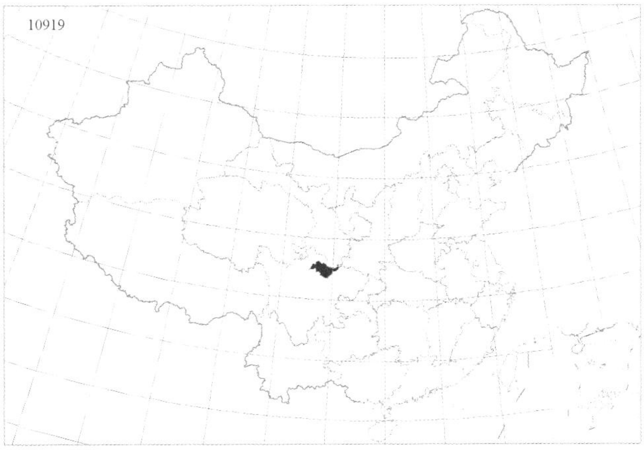

Poaceae 禾本科 *Fargesia* 箭竹属
Fargesia dracocephala T. P. Yi 龙头箭竹
Bamboos, evergreen

11.0 (9.2~13.3); -1.2 (-2.6~2.3); 22.6 (20.1~24.4); 11.2 (9.5~13.3); 87.9 (69.8~104.4); 15.4 (4.9~20.9); 677 (603~741) ❙ 881 (738~1214); 403 (350~516); 36 (20~71) ❙ 31.4 (2.3~101.4); 677 (603~741); 272.4 (236.9~323.3)

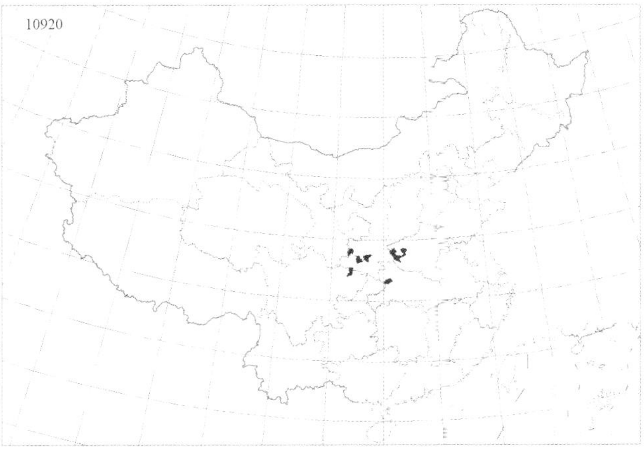

Poaceae 禾本科 *Fargesia* 箭竹属
Fargesia dura T. P. Yi 马斯箭竹
Bamboos, evergreen

9.6; 2.4; 15.7; 9.6; 60.8; 5.4; 600 | 1007; 578; 40 | 67.9; 600; 351.7

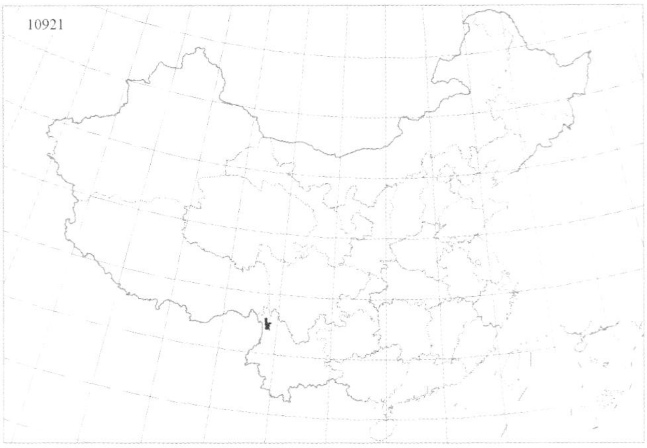

Poaceae 禾本科 *Fargesia* 箭竹属
Fargesia edulis Hsueh et T. P. Yi 空心箭竹
Bamboos, evergreen

13.2; 6.4; 18.4; 13.2; 98.8; 0.0; 694 | 1250; 683; 64 | 80.2; 694; 386.7

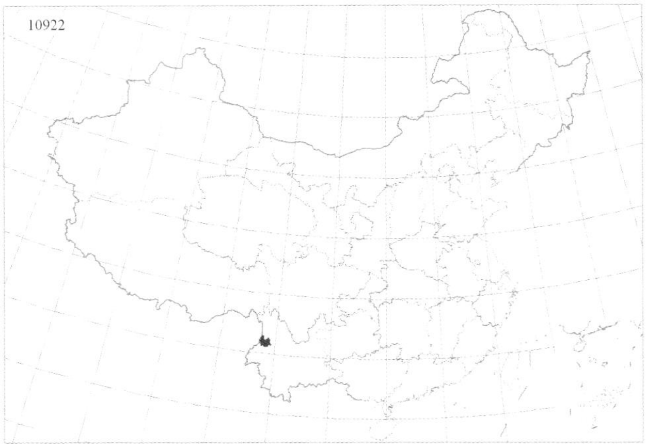

Poaceae 禾本科 *Fargesia* 箭竹属
Fargesia emaculata T. P. Yi 牛麻箭竹
Bamboos, evergreen

2.2; -6.6; 9.9; 3.8; 15.5; 49.1; 413 | 932; 546; 19 | 125.9; 413; 248.3

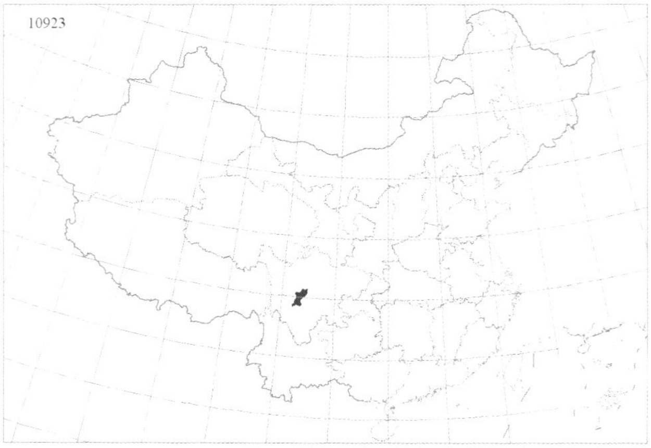

Poaceae 禾本科 *Fargesia* 箭竹属
Fargesia extensa T. P. Yi 喇叭箭竹
Bamboos, evergreen

13.3; 5.4; 19.6; 13.4; 100.4; 0.0; 706 | 1736; 1007; 51 | 145.8; 706; 447.4

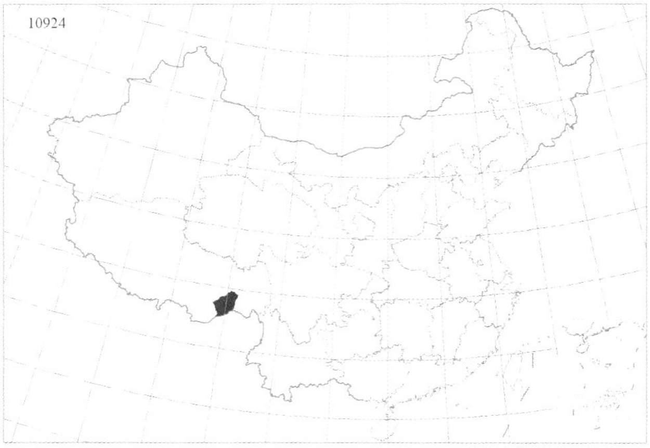

Poaceae 禾本科 *Fargesia* 箭竹属
Fargesia farcta T. P. Yi 勒布箭竹
Bamboos, evergreen

12.2; 4.3; 18.5; 12.2; 87.4; 0.7; 672 | 1431; 817; 31 | 113.0; 672; 363.7

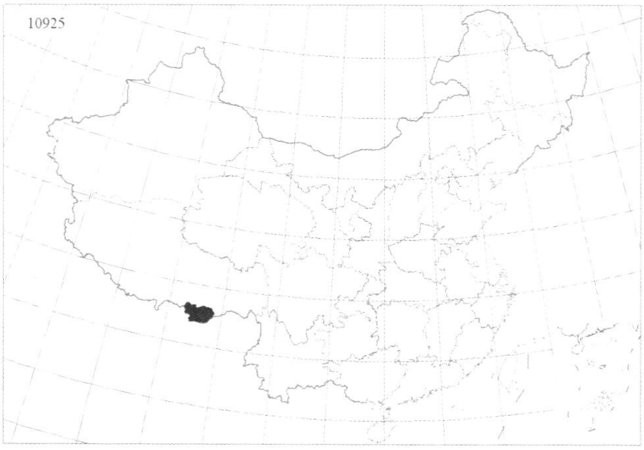

Poaceae 禾本科 *Fargesia* 箭竹属
Fargesia ferax (Keng) T. P. Yi 丰实箭竹
Bamboos, evergreen

8.5 (2.2~11.9); 0.0 (-6.6~2.9); 15.9 (9.9~20.0); 8.9 (3.8~12.0); 58.1 (15.5~87.2); 15.6 (3.4~49.1); 575 (413~671) | 1005 (932~1056); 571 (539~615); 24 (18~35) | 78.4 (49.1~125.9); 575 (413~671); 274.1 (221.3~318.7)

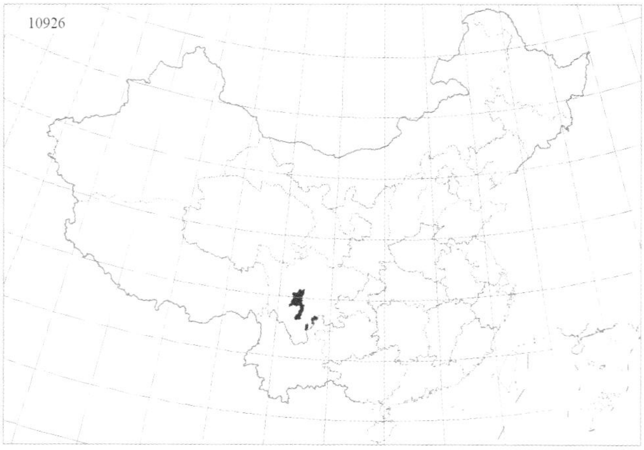

Poaceae 禾本科 *Fargesia* 箭竹属
Fargesia frigidis T. P. Yi 凋叶箭竹
Bamboos, evergreen

14.0；7.5；19.2；14.1；108.8；0.0；718 ❙ 1043；581；50 ❙ 45.3；718；377.4

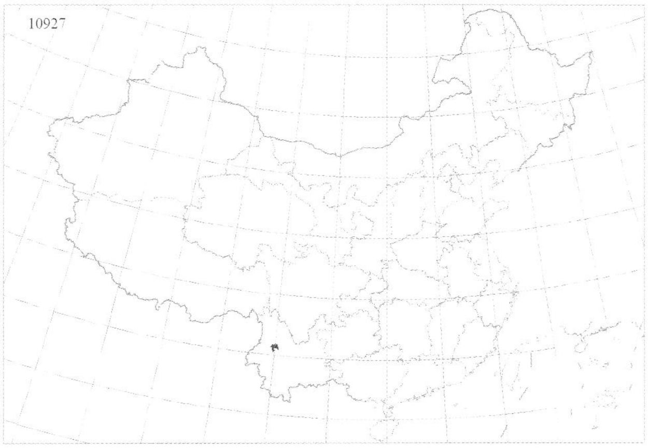

Poaceae 禾本科 *Fargesia* 箭竹属
Fargesia fungosa T. P. Yi 棉花竹
Bamboos, evergreen

13.6（12.4~14.6)；5.9（3.5~7.9)；19.9（18.6~21.1)；13.6（12.4~14.6)；104.1 (89.1~115.8)；0.4 (0.0~1.5)；713 (675~74[illegible]) ❙ 952 (879~1014)；563 (472~627)；31（22~39) ❙ 33.4（25.7~37.1)；699（675~732)；249.8 (205.2~278.4)

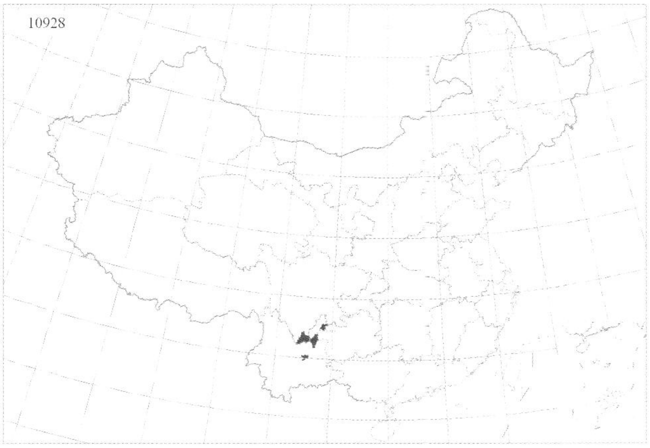

Poaceae 禾本科 *Fargesia* 箭竹属
Fargesia glabrifolia T. P. Yi 光叶箭竹
Bamboos, evergreen

12.2；4.3；18.5；12.2；87.4；0.7；672 ❙ 1431；817；31 ❙ 113.0；672；363.7

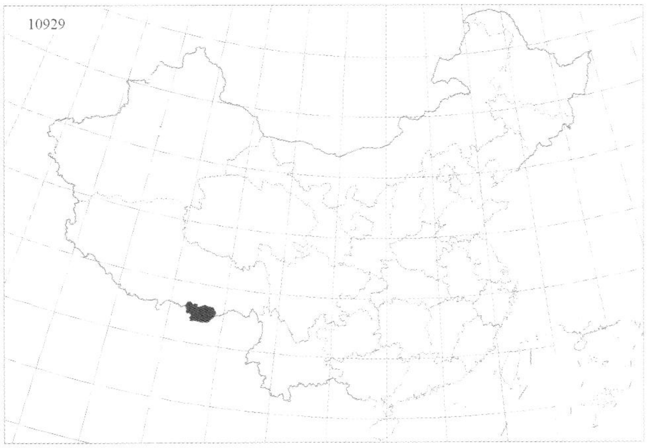

Poaceae 禾本科 *Fargesia* 箭竹属
Fargesia gongshanensis T. P. Yi 贡山箭竹
Bamboos, evergreen

8.6；1.3；15.0；8.7；52.6；8.8；574 ❙ 1041；585；45 ❙ 81.5；574；352.3

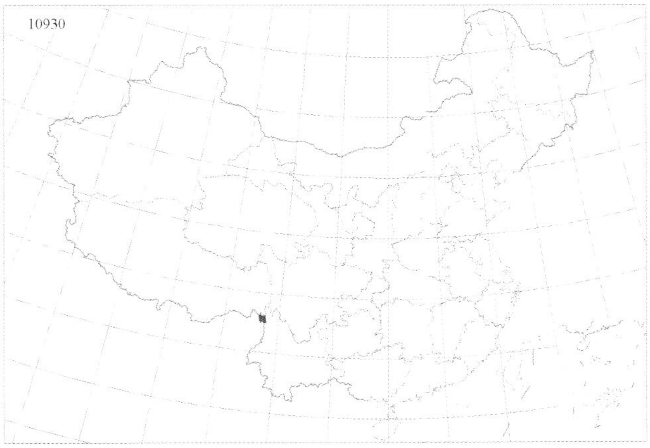

Poaceae 禾本科 *Fargesia* 箭竹属
Fargesia grossa T. P. Yi 错那箭竹
Bamboos, evergreen

12.2；4.3；18.5；12.2；87.4；0.7；672 ❙ 1431；817；31 ❙ 113.0；672；363.7

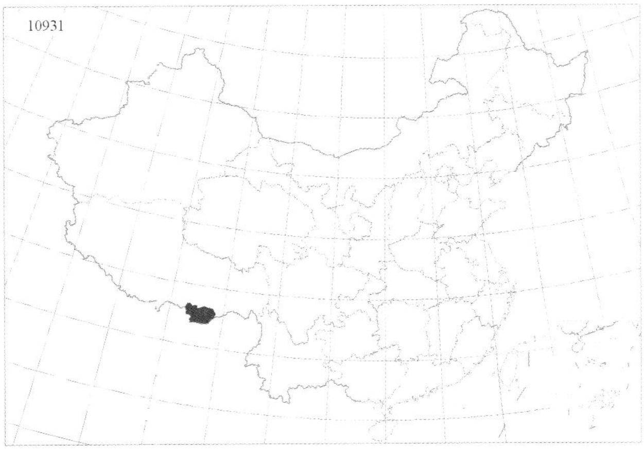

Poaceae 禾本科 *Fargesia* 箭竹属
Fargesia hainanensis T. P. Yi 海南箭竹
Bamboos, evergreen

23.0；17.7；26.7；23.1；217.0；0.0；1217 ❙ 1725；610；68 ❙ 41.8；1163；572.9

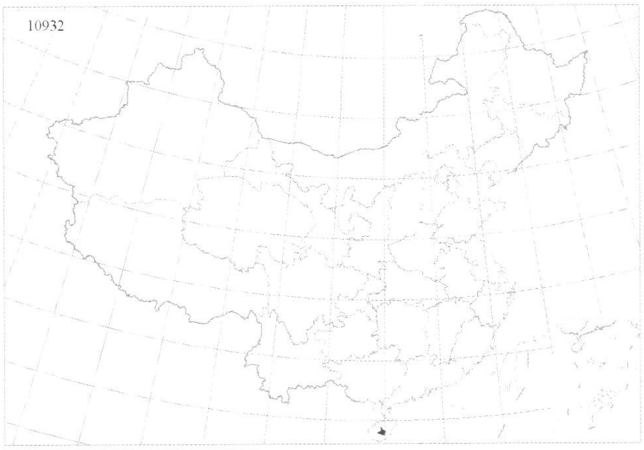

Poaceae 禾本科 *Fargesia* 箭竹属
Fargesia hsuehiana T. P. Yi 冬竹
Bamboos, evergreen

19.1；13.0；23.3；19.1；169.7；0.0；917 ❙ 1428；835；56 ❙ 55.7；917；506.2

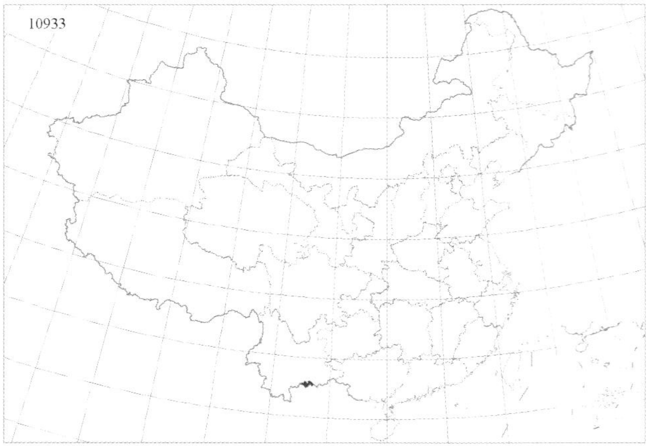

Poaceae 禾本科 *Fargesia* 箭竹属
Fargesia hygrophila Hsueh et T. P. Yi 喜湿箭竹
Bamboos, evergreen

14.3；7.8；19.5；14.4；112.7；0.0；729 ❙ 996；581；36 ❙ 36.9；695；340.0

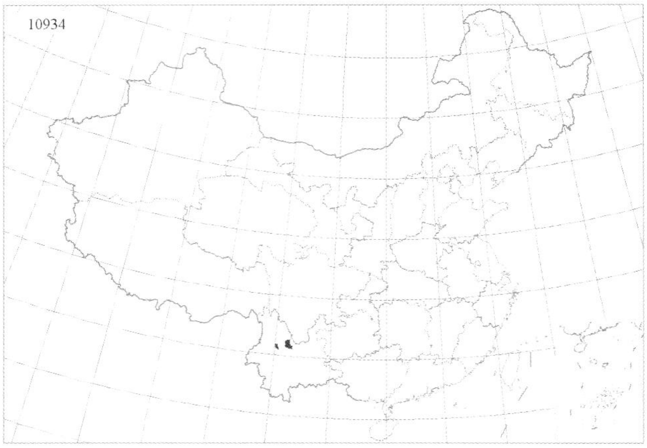

Poaceae 禾本科 *Fargesia* 箭竹属
Fargesia jiulongensis T. P. Yi 九龙箭竹
Bamboos, evergreen

3.8；-4.3；10.8；4.7；21.6；35.4；444 ❙ 984；601；16 ❙ 121.7；444；265.2

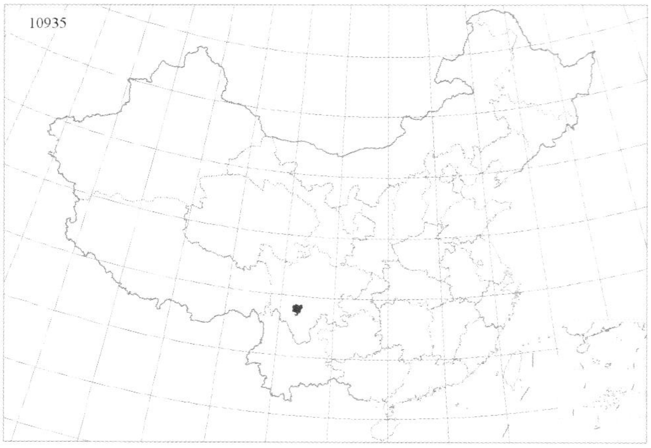

Poaceae 禾本科 *Fargesia* 箭竹属
Fargesia lincangensis T. P. Yi 雪山箭竹
Bamboos, evergreen

16.3；10.2；20.3；16.4；136.6；0.0；789 ❙ 1232；703；47 ❙ 56.2；780；372.4

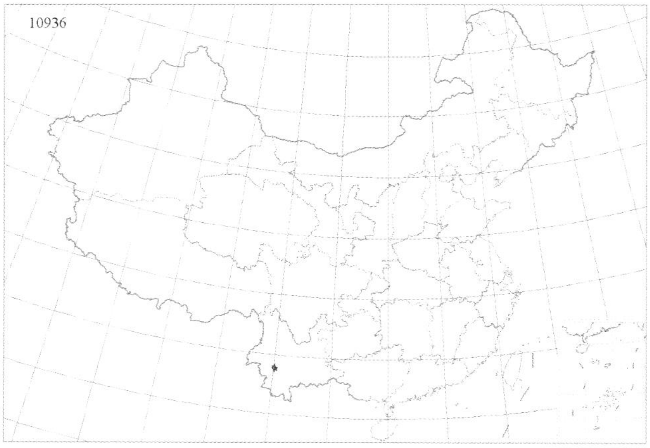

Poaceae 禾本科 *Fargesia* 箭竹属
Fargesia longiuscula (Hsueh et Y. Y. Dai) Ohrnb. 长节箭竹
Bamboos, evergreen

12.1；3.0；20.0；12.1；88.4；2.8；675 ❙ 944；520；33 ❙ 39.8；675；249.8

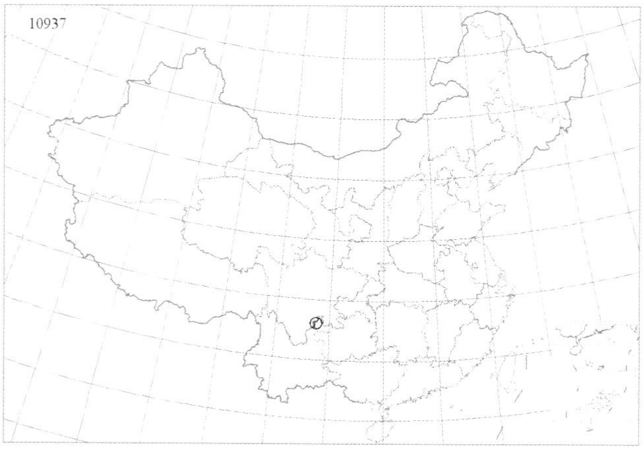

Poaceae 禾本科 *Fargesia* 箭竹属
Fargesia lushuiensis Hsueh et T. P. Yi 泸水箭竹
Bamboos, evergreen

13.2；6.4；18.4；13.3；99.4；0.0；695 ❙ 1337；727；69 ❙ 92.4；695；407.1

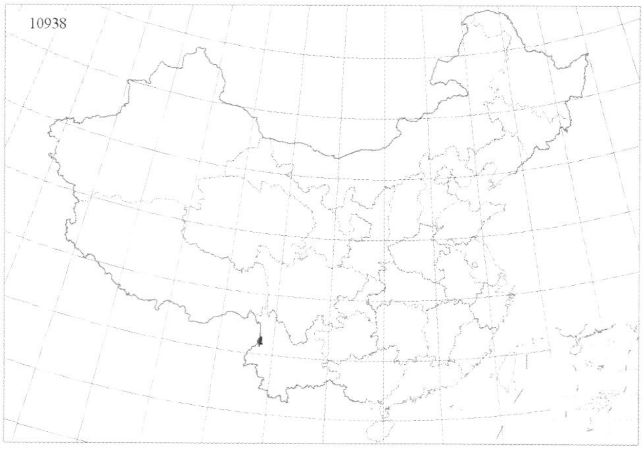

Poaceae 禾本科 *Fargesia* 箭竹属

Fargesia macclureana (Bor) Stapleton 西藏箭竹

Bamboos, evergreen

1.8; -7.1; 9.9; 3.8; 15.7; 53.3; 404 ❙ 583; 348; 13 ❙ 44.2; 404; 169.5

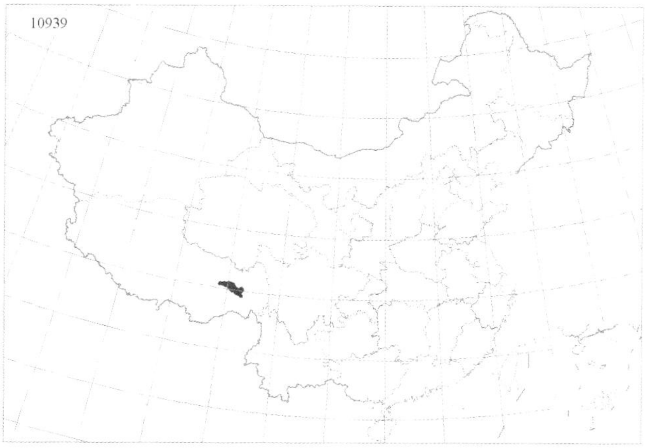

Poaceae 禾本科 *Fargesia* 箭竹属

Fargesia mairei (Hack. ex Hand.-Mazz.) T. P. Yi 大姚箭竹

Bamboos, evergreen

15.0; 8.4; 20.2; 15.1; 121.2; 0.0; 753 ❙ 950; 570; 27 ❙ 26.3; 683; 307.7

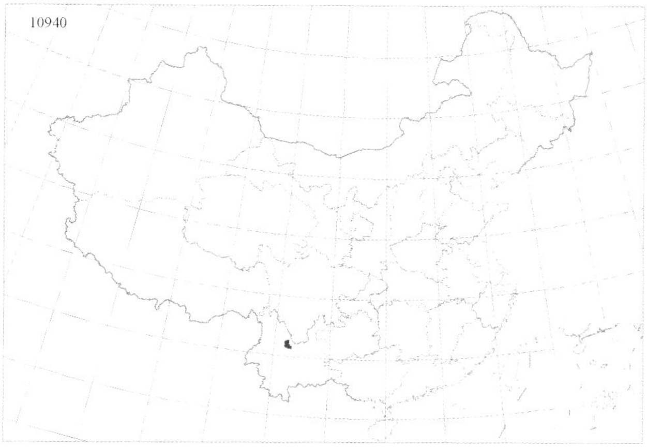

Poaceae 禾本科 *Fargesia* 箭竹属

Fargesia mali T. P. Yi 马利箭竹

Bamboos, evergreen

14.6; 7.1; 20.3; 14.6; 114.8; 0.0; 741 ❙ 1014; 627; 22 ❙ 37.0; 692; 265.0

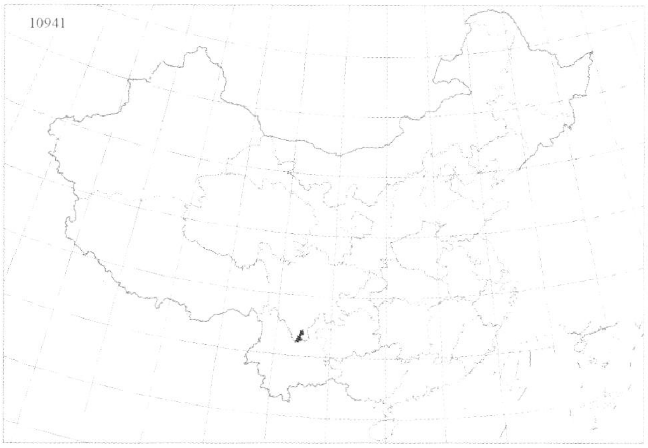

Poaceae 禾本科 *Fargesia* 箭竹属

Fargesia melanostachys (Hand.-Mazz.) T. P. Yi 黑穗箭竹

Bamboos, evergreen

6.8; -0.7; 13.4; 7.1; 40.9; 19.1; 520 ❙ 903; 521; 33 ❙ 74.2; 520; 286.5

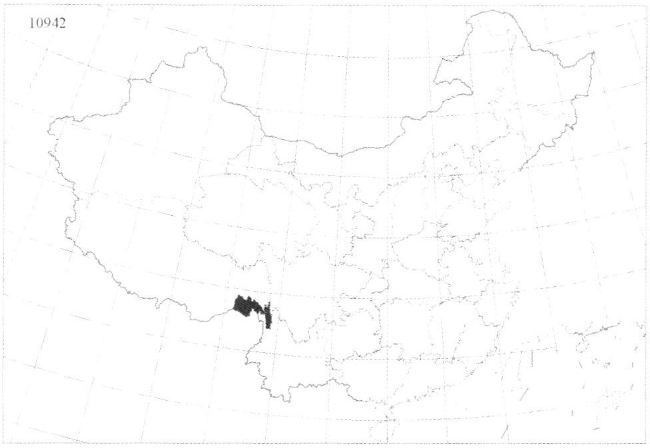

Poaceae 禾本科 *Fargesia* 箭竹属

Fargesia murielae (Gamble) T. P. Yi 神农箭竹

Bamboos, evergreen

11.0; -0.4; 22.2; 11.2; 86.1; 13.3; 667 ❙ 1089; 468; 62 ❙ 66.6; 667; 279.3

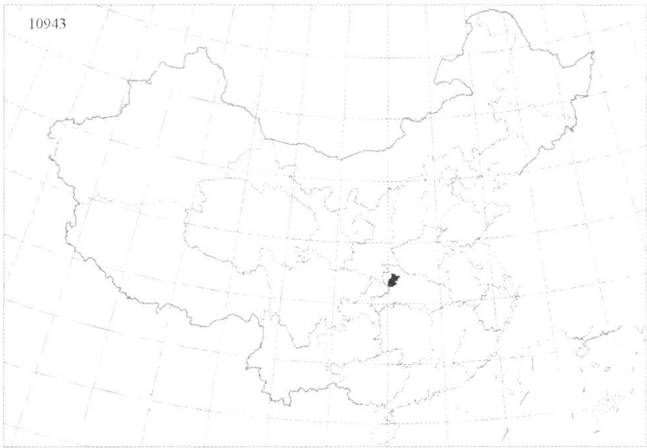

Poaceae 禾本科 *Fargesia* 箭竹属

Fargesia nitida (Mitford ex Stapf) Keng f. ex T. P. Yi 华西箭竹

Bamboos, evergreen

9.9 (-1.4~19.4); -2.4 (-15.4~10.2); 21.2 (9.9~29.3); 10.7 (3.5~19.4); 83.0 (13.3~172.7); 23.5 (0.0~92.1); 663 (375~994) ❙ 816 (39~1950); 373 (29~713); 47 (1~253) ❙ 21.2 (-92.3~147.0); 617 (39~994); 213.6 (10.8~349.1)

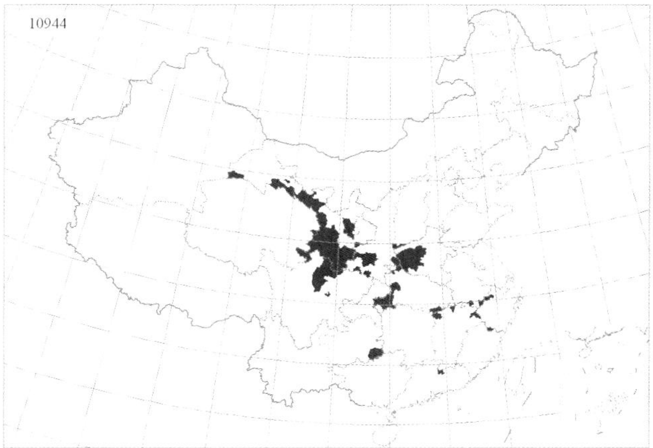

Poaceae 禾本科 *Fargesia* 箭竹属
Fargesia obliqua T. P. Yi 团竹
Bamboos, evergreen

7.2; −2.5; 15.9; 7.8; 51.1; 24.6; 546 ❙ 866; 439; 22 ❙ 62.7; 546; 262.4

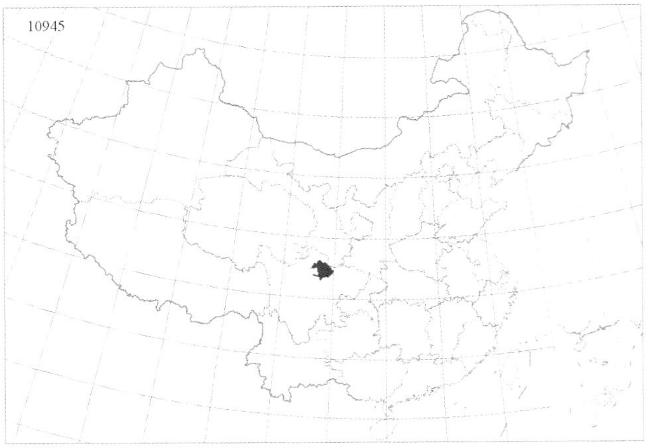

Poaceae 禾本科 *Fargesia* 箭竹属
Fargesia orbiculata T. P. Yi 长圆鞘箭竹
Bamboos, evergreen

10.8; 3.9; 16.6; 10.9; 71.8; 1.3; 633 ❙ 968; 594; 24 ❙ 52.8; 633; 319.8

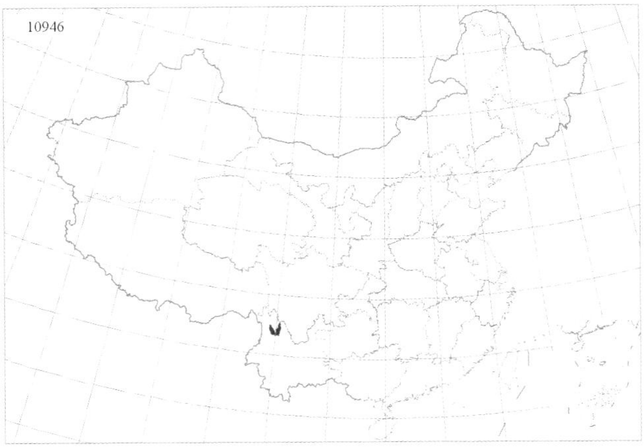

Poaceae 禾本科 *Fargesia* 箭竹属
Fargesia papyrifera T. P. Yi 云龙箭竹
Bamboos, evergreen

13.1; 6.4; 18.4; 13.2; 98.1; 0.0; 692 ❙ 1164; 639; 58 ❙ 68.1; 692; 366.3

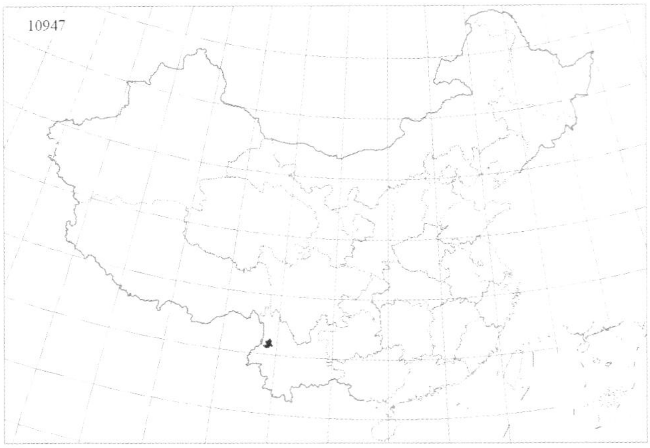

Poaceae 禾本科 *Fargesia* 箭竹属
Fargesia pauciflora (Keng) T. P. Yi 少花箭竹
Bamboos, evergreen

13.7; 6.5; 19.3; 13.7; 104.6; 0.0; 712 ❙ 1180; 671; 43 ❙ 66.3; 701; 356.6

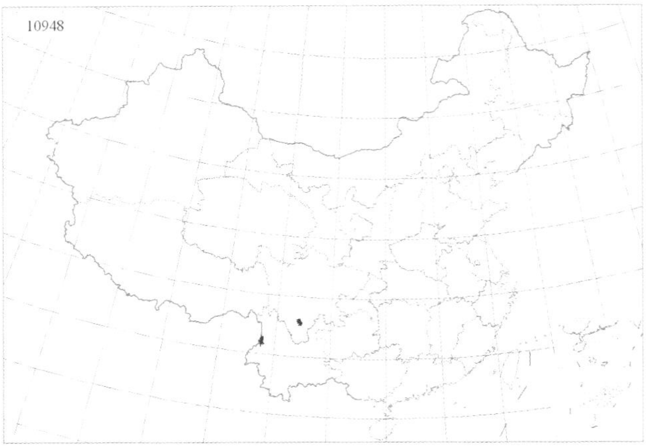

Poaceae 禾本科 *Fargesia* 箭竹属
Fargesia perlonga Hsueh et T. P. Yi 超包箭竹
Bamboos, evergreen

14.5; 7.8; 19.6; 14.5; 113.5; 0.0; 730 ❙ 1044; 570; 54 ❙ 43.0; 730; 382.6

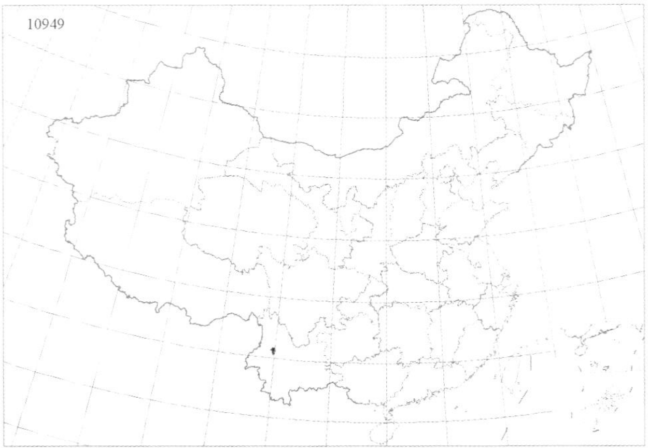

Poaceae 禾本科 *Fargesia* 箭竹属
Fargesia pleniculmis (Hand.-Mazz.) T. P. Yi 皱壳箭竹
Bamboos, evergreen

8.6; 1.3; 15.0; 8.7; 52.6; 8.8; 574 ❙ 1041; 585; 45 ❙ 81.5; 574; 352.3

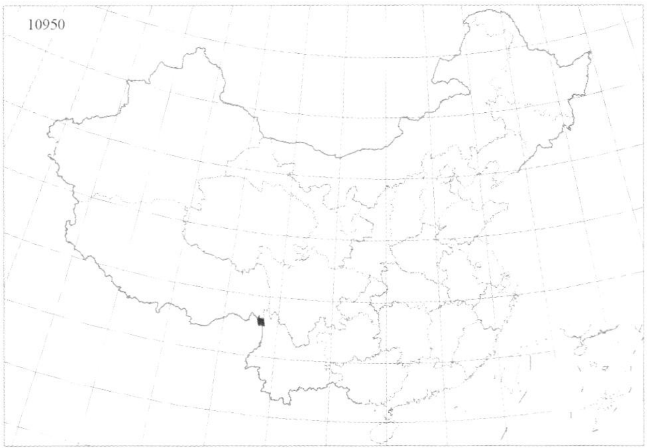

Poaceae 禾本科 *Fargesia* 箭竹属
Fargesia plurisetosa T. H. Wen 密毛箭竹
Bamboos, evergreen

19.2；13.8；22.3；19.2；170.5；0.0；909 ❙ 1499；686；58 ❙ 64.9；880；513.9

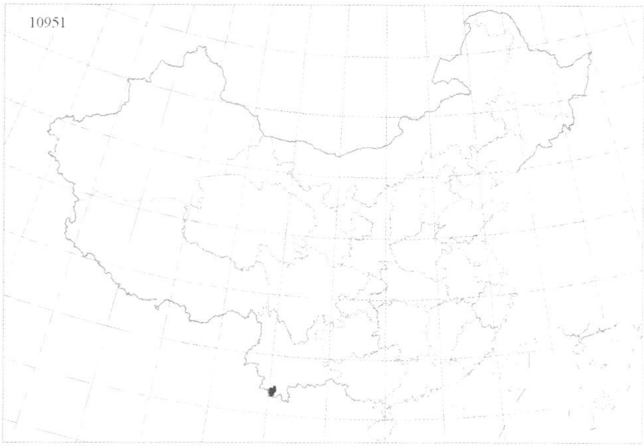

Poaceae 禾本科 *Fargesia* 箭竹属
Fargesia porphyrea T. P. Yi 红壳箭竹
Bamboos, evergreen

17.8 (16.3~19.1)；11.1 (9.1~13.0)；22.4 (21.4~23.3)；17.8 (16.4~19.1)；154.1 (136.4~169.7)；0.0 (0.0~0.0)；858 (797~917) ❙ 1407 (1190~1547)；804 (656~911)；57 (50~63) ❙ 64.0 (49.4~80.3)；858 (797~917)；462.0 (347.3~551.3)

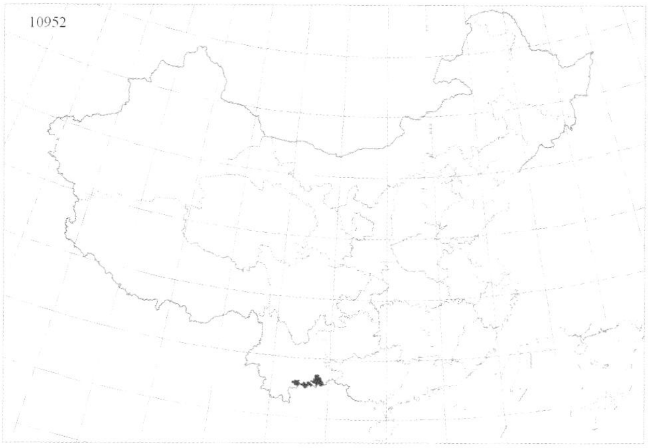

Poaceae 禾本科 *Fargesia* 箭竹属
Fargesia praecipua T. P. Yi 弩箭竹
Bamboos, evergreen

8.6；1.3；15.0；8.7；52.6；8.8；574 ❙ 1041；585；45 ❙ 81.5；574；352.3

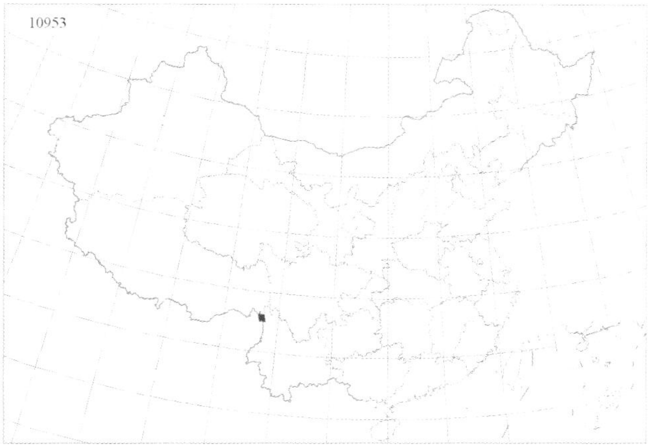

Poaceae 禾本科 *Fargesia* 箭竹属
Fargesia qinlingensis T. P. Yi et J. X. Shao 秦岭箭竹
Bamboos, evergreen

10.5 (7.4~13.5)；-1.5 (-4.2~1.4)；22.1 (18.7~25.[illegible])；10.8 (8.2~13.5)；83.9 (58.8~109.4)；16.8 (7.1~29.4)；662 (561~767) ❙ 835 (738~892)；375 (332~391)；24 (20~29) ❙ 27.3 (5.8~52.6)；662 (561~767)；298.2 (250.6~334.4)

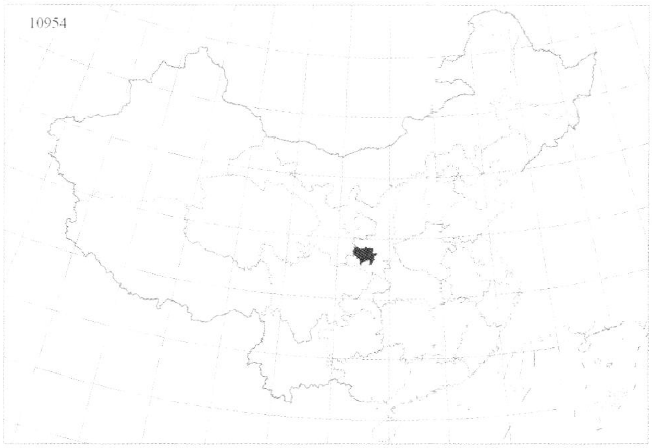

Poaceae 禾本科 *Fargesia* 箭竹属
Fargesia robusta T. P. Yi 拐棍竹
Bamboos, evergreen

12.2 (6.0~14.6)；2.3 (-3.2~4.5)；21.1 (14.4~23.7)；12.3 (6.6~14.6)；93.4 (39.7~116.2)；6.7 (0.5~26.7)；697 (508~777) ❙ 1097 (1022~1212)；614 (535~692)；33 (28~39) ❙ 60.2 (41.9~101.2)；697 (508~777)；169.1 (147.3~200.8)

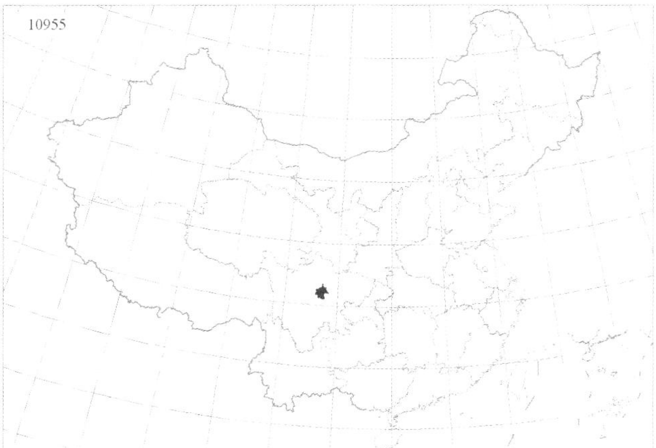

Poaceae 禾本科 *Fargesia* 箭竹属
Fargesia rufa T. P. Yi 青川箭竹
Bamboos, evergreen

13.5；2.9；23.3；13.5；105.3；3.0；744 ❙ 806；448；16 ❙ 8.2；744；261.0

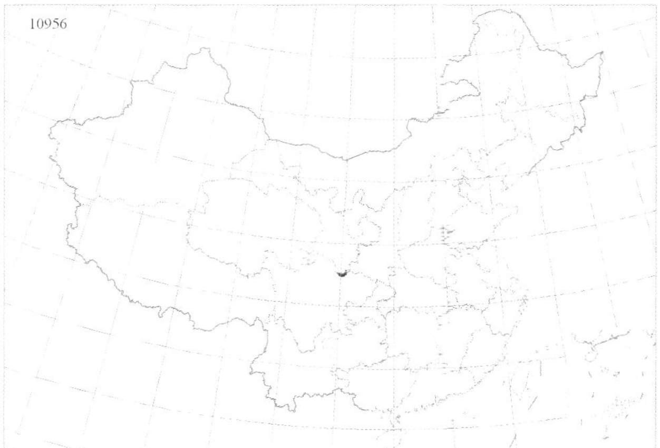

Poaceae 禾本科 *Fargesia* 箭竹属
Fargesia sagittatinea T. P. Yi 独龙箭竹
Bamboos, evergreen

8.6; 1.3; 15.0; 8.7; 52.6; 8.8; 574 ❙ 1041; 585; 45 ❙ 81.5; 574; 352.3

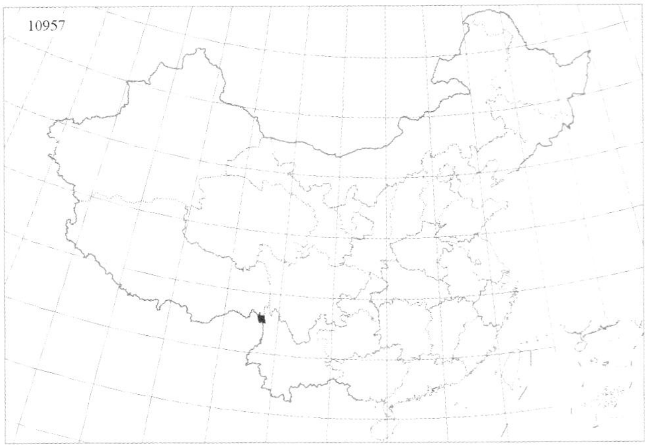

Poaceae 禾本科 *Fargesia* 箭竹属
Fargesia scabrida T. P. Yi 糙花箭竹
Bamboos, evergreen

8.4; -1.6; 17.5; 9.0; 63.1; 21.8; 589 ❙ 819; 423; 19 ❙ 47.0; 589; 262.9

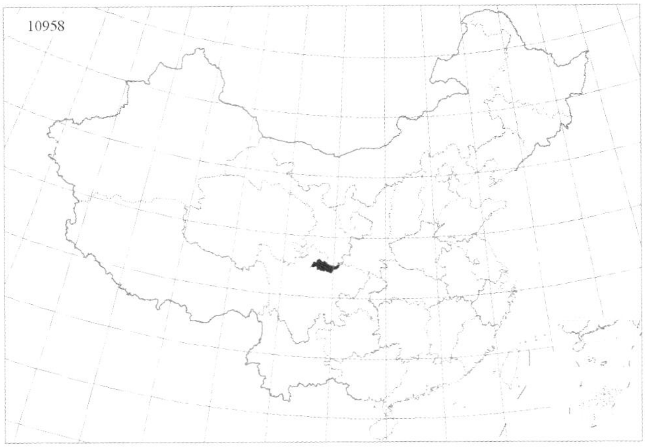

Poaceae 禾本科 *Fargesia* 箭竹属
Fargesia semicoriacea T. P. Yi 白竹
Bamboos, evergreen

13.4; 0.9; 25.2; 13.4; 108.9; 8.5; 764 ❙ 807; 349; 37 ❙ 5.6; 764; 212.3

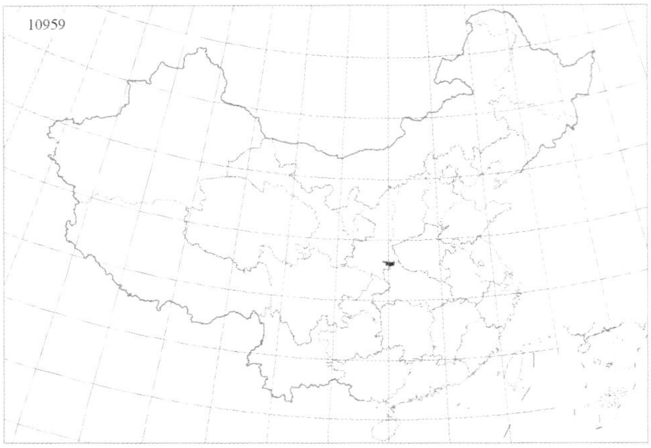

Poaceae 禾本科 *Fargesia* 箭竹属
Fargesia stenoclada T. P. Yi 细枝箭竹
Bamboos

13.2; 3.0; 22.3; 13.2; 102.0; 3.2; 728 ❙ 981; 555; 26 ❙ 35.2; 728; 145.6

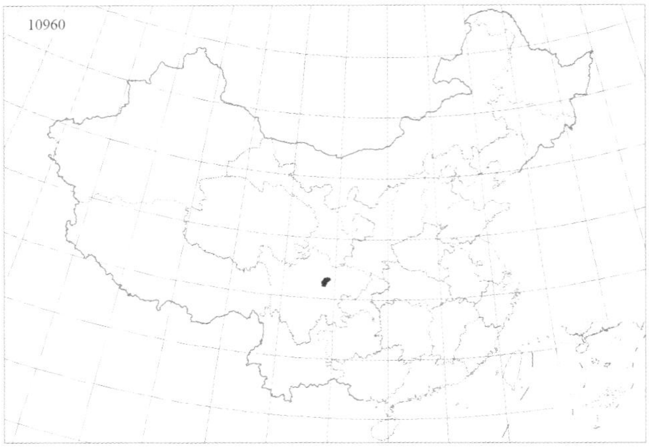

Poaceae 禾本科 *Fargesia* 箭竹属
Fargesia zayuensis T. P. Yi 察隅箭竹
Bamboos

5.3; -2.4; 12.2; 5.7; 30.5; 26.2; 479 ❙ 961; 558; 31 ❙ 100.7; 479; 234.2

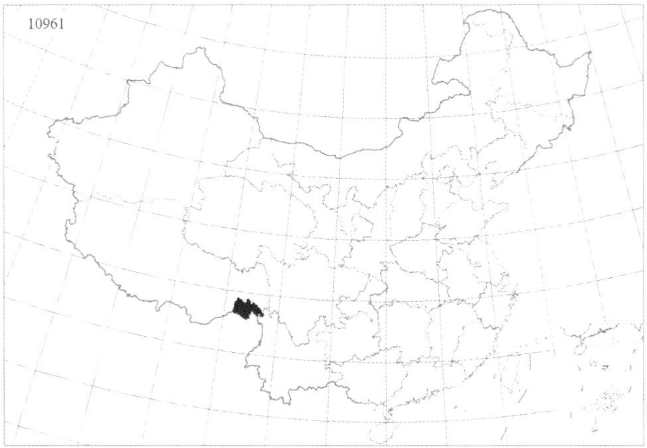

Poaceae 禾本科 *Gelidocalamus* 短枝竹属
Gelidocalamus annulatus T. H. Wen 亮秆竹
Bamboos, evergreen

15.4; 5.0; 25.3; 15.5; 126.0; 0.2; 813 ❙ 1096; 501; 67 ❙ 34.8; 813; 189.7

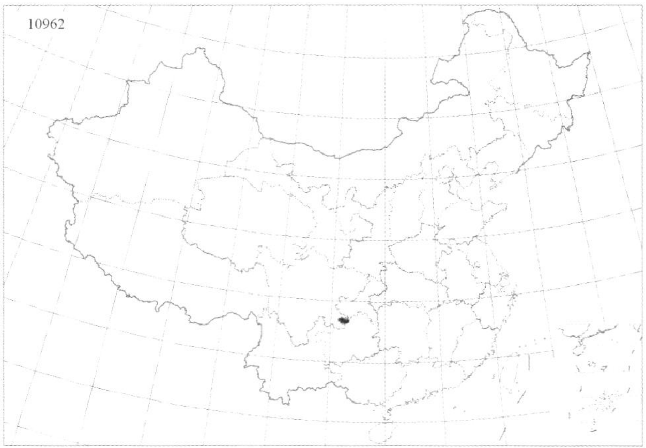

Poaceae 禾本科 *Gelidocalamus* 短枝竹属
Gelidocalamus kunishii (Hayata) Keng f. et T. H. Wen 台湾矢竹
Bamboos, evergreen

18.0；12.2；23.0；18.1；156.7；0.0；880 ❙ 2836；1055；396 ❙ 224.9；880；275.3

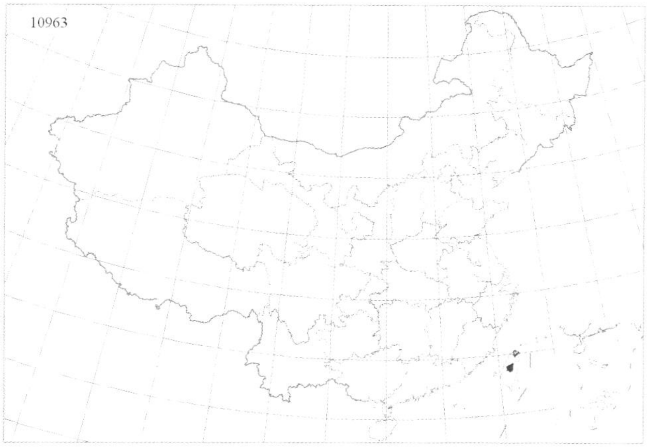

Poaceae 禾本科 *Gelidocalamus* 短枝竹属
Gelidocalamus latifolius Q. H. Dai et T. Chen 掌秆竹
Bamboos, evergreen

18.4；8.2；26.9；18.4；161.2；0.0；955 ❙ 1495；658；138 ❙ 56.9；955；241.2

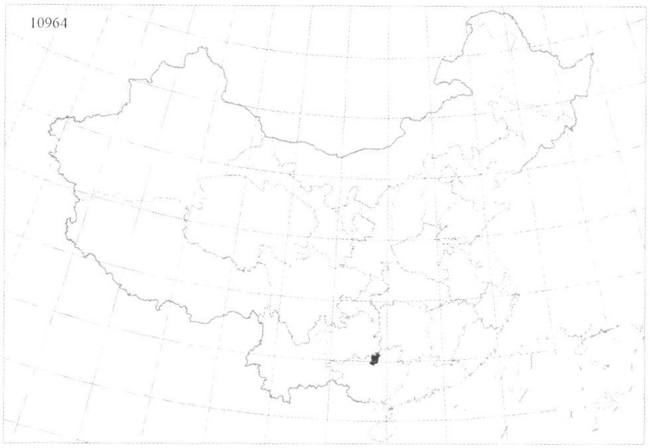

Poaceae 禾本科 *Gelidocalamus* 短枝竹属
Gelidocalamus longiinternodus T. H. Wen et Shi. C. Chen 箭把竹
Bamboos, evergreen

17.8 (16.3~19.1)；8.4 (5.1~10.1)；25.3 (23.4~26.6)；17.8 (16.4~19.1)；153.9 (136.3~169.2)；0.0 (0.0~0.0)；904 (836~969) ❙ 1273 (1232~1334)；642 (478~720)；72 (51~145) ❙ 41.6 (27.5~59.6)；904 (836~969)；254.6 (218.6~283.9)

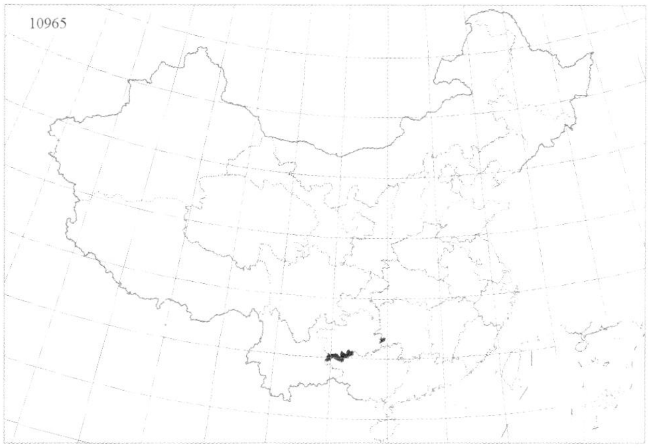

Poaceae 禾本科 *Gelidocalamus* 短枝竹属
Gelidocalamus multifolius B. M. Yang 多叶短枝竹
Bamboos, evergreen

17.1；6.0；27.6；17.2；146.3；0.0；908 ❙ 1477；448；202 ❙ 62.7；893；170.8

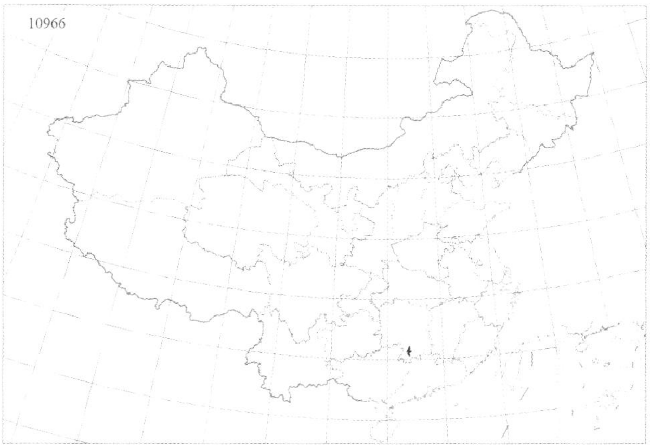

Poaceae 禾本科 *Gelidocalamus* 短枝竹属
Gelidocalamus rutilans T. H. Wen 红壳寒竹
Bamboos, evergreen

16.5；5.0；27.5；16.6；138.6；0.0；887 ❙ 1820；552；246 ❙ 105.1；887；223.7

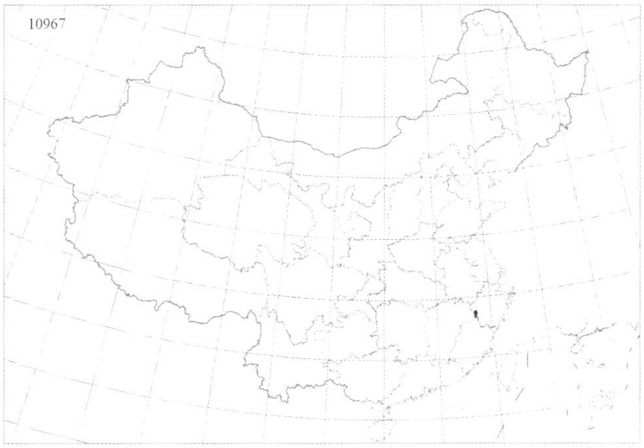

Poaceae 禾本科 *Gelidocalamus* 短枝竹属
Gelidocalamus solidus C. D. Chu et C. S. Chao 实心短枝竹
Bamboos, evergreen

17.6；7.3；26.3；17.6；151.6；0.0；909 ❙ 1496；639；142 ❙ 64.7；909；225.9

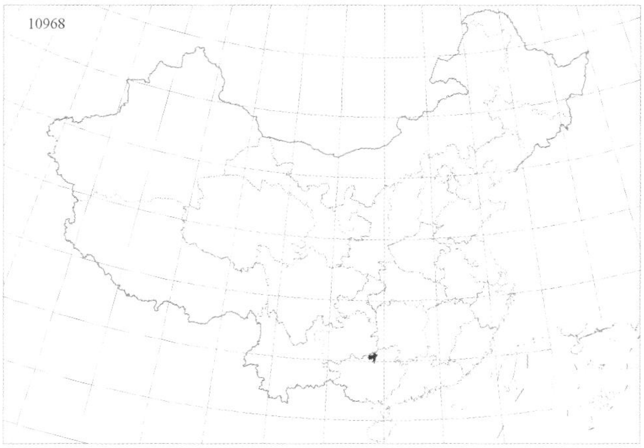

Poaceae 禾本科 *Gelidocalamus* 短枝竹属
Gelidocalamus stellatus T. H. Wen 井冈短枝竹
Bamboos, evergreen

17.2 (15.3~18.5); 6.2 (4.2~7.8); 27.4 (26.0~28.1); 17.2 (15.4~18.5); 147.1 (125.8~162.3); 0.1 (0.0~0.8); 909 (818~975) ❙ 1571 (1477~1660); 526 (448~575); 196 (176~216) ❙ 73.4 (60.2~103.1); 907 (818~975); 210.0 (170.8~236.3)

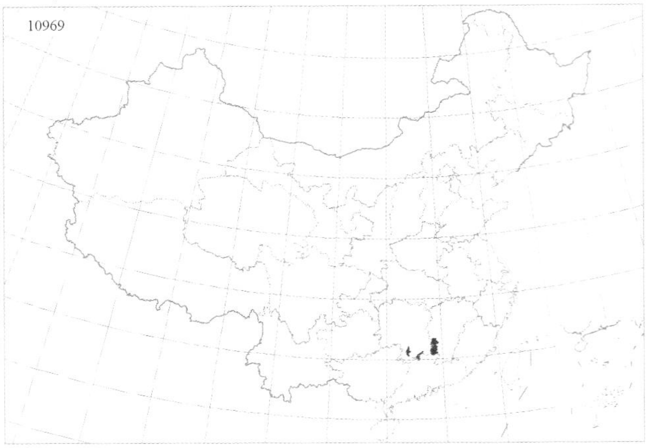

Poaceae 禾本科 *Gelidocalamus* 短枝竹属
Gelidocalamus tessellatus T. H. Wen et C. C. Chang 抽筒竹
Bamboos, evergreen

17.6; 7.3; 26.5; 17.6; 151.4; 0.0; 909 ❙ 1442; 575; 145 ❙ 58.5; 909; 221.4

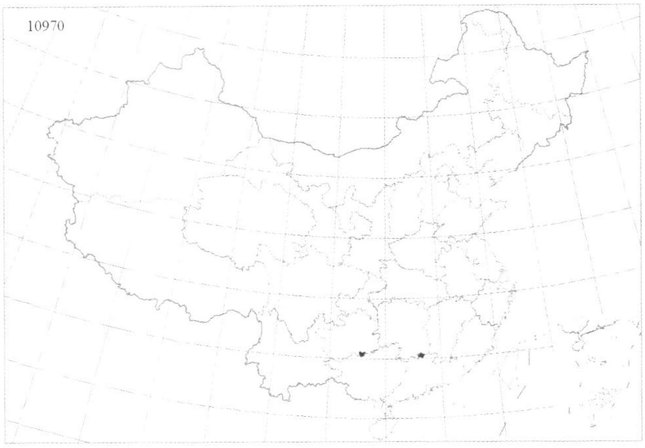

Poaceae 禾本科 *Gigantochloa* 滇竹属
Gigantochloa albociliata (Munro) Kurz 白毛巨竹
Bamboos, evergreen

19.2; 13.8; 22.3; 19.2; 170.5; 0.0; 909 ❙ 1499; 686; 58 ❙ 64.9; 880; 513.9

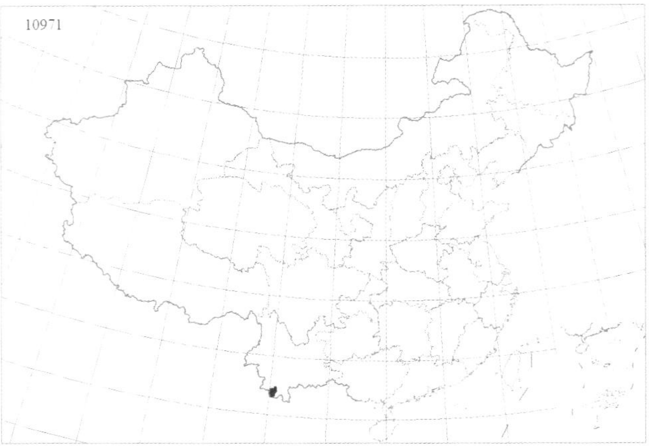

Poaceae 禾本科 *Gigantochloa* 滇竹属
Gigantochloa felix (Keng) Keng f. 滇竹
Bamboos, evergreen

20.0; 14.6; 23.2; 20.1; 180.8; 0.0; 963 ❙ 1523; 681; 67 ❙ 58.4; 942; 581.0

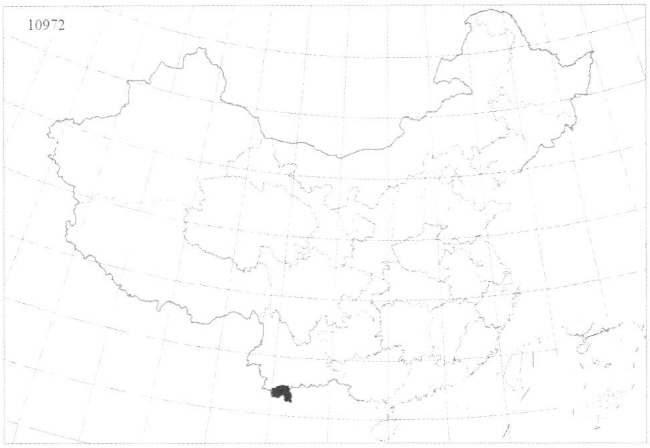

Poaceae 禾本科 *Gigantochloa* 滇竹属
Gigantochloa levis (Blanco) Merr. 毛笋竹
Bamboos, evergreen

20.0; 14.6; 23.2; 20.1; 180.8; 0.0; 963 ❙ 1523; 681; 67 ❙ 58.4; 942; 581.0

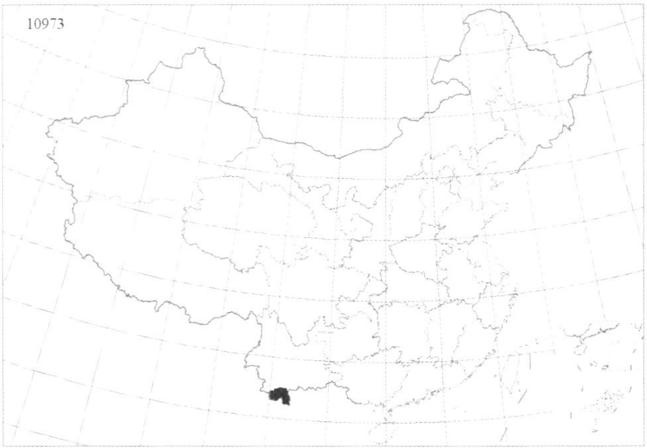

Poaceae 禾本科 *Gigantochloa* 滇竹属
Gigantochloa nigrociliata (Büse) Kurz 黑毛巨竹
Bamboos, evergreen

20.0; 14.6; 23.2; 20.1; 180.8; 0.0; 963 ❙ 1523; 681; 67 ❙ 58.4; 942; 581.0

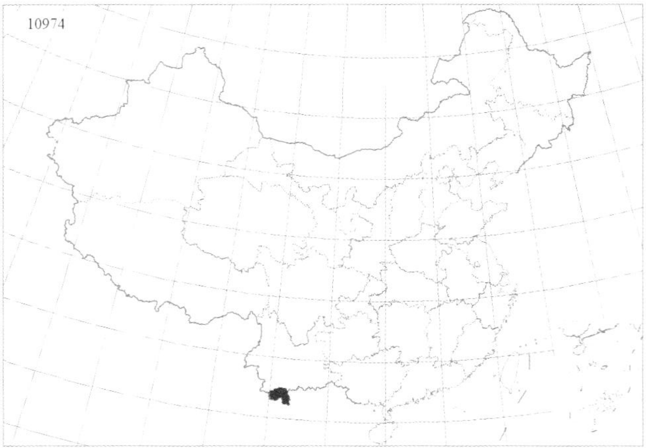

Poaceae 禾本科 *Gigantochloa* 滇竹属
Gigantochloa parviflora (Keng f.) Keng f. 南峤滇竹
Bamboos, evergreen

19.2; 13.8; 22.3; 19.2; 170.5; 0.0; 909 ❙ 1499; 686; 58 ❙ 64.9; 880; 513.9

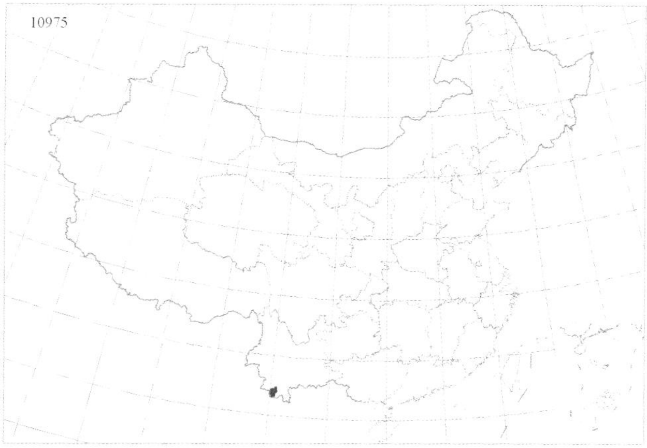

Poaceae 禾本科 *Himalayacalamus* 喜马拉雅筱竹属
Himalayacalamus collaris (T. P. Yi) Ohrnberger 颈鞘筱竹
Bamboos, evergreen

−1.8; −10.2; 6.1; 1.8; 1.9; 82.8; 329 ❙ 320; 217; 20 ❙ −3.0; 320; 72.2

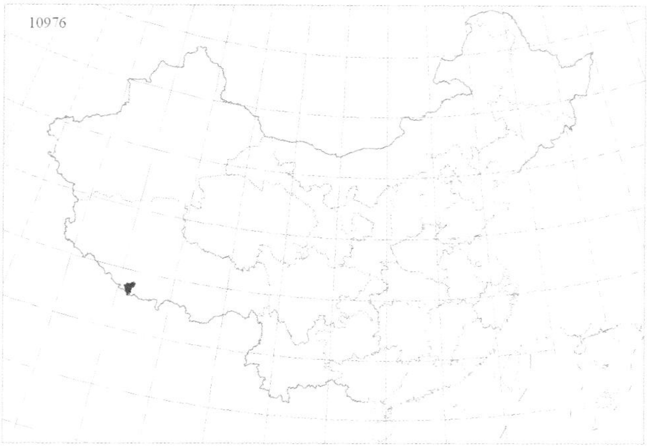

Poaceae 禾本科 *Himalayacalamus* 喜马拉雅筱竹属
Himalayacalamus falconeri (Munro) P. C. Keng 喜马拉雅筱竹
Bamboos, evergreen

−1.1; −9.1; 6.2; 2.0; 2.5; 75.6; 334 ❙ 321; 193; 33 ❙ −4.0; 321; 79.3

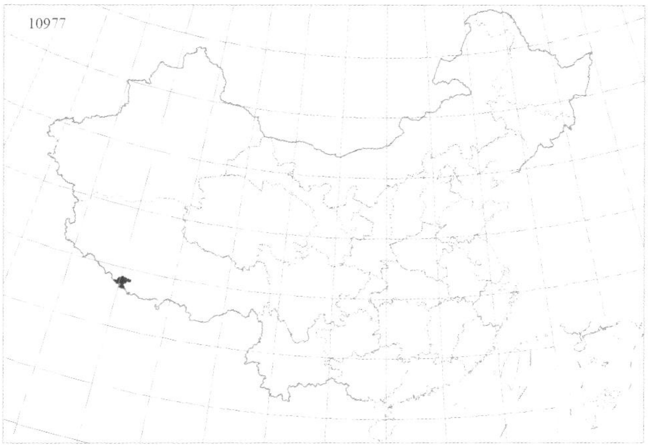

Poaceae 禾本科 *Indocalamus* 箬竹属
Indocalamus barbatus McClure 髯毛箬竹
Bamboos, evergreen

18.0; 8.5; 25.8; 18.0; 156.4; 0.0; 913 ❙ 1551; 672; 169 ❙ 69.8; 913; 243.9

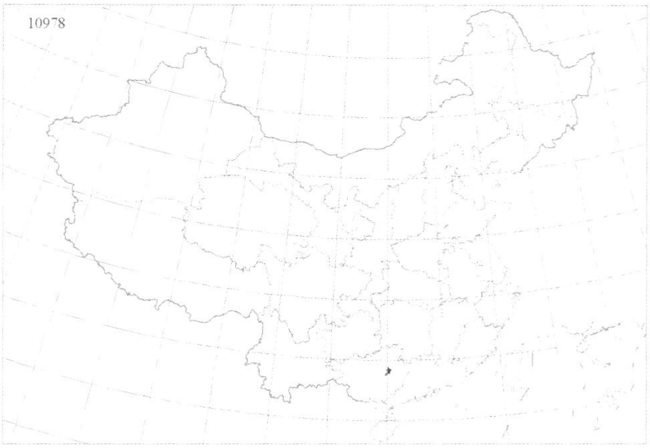

Poaceae 禾本科 *Indocalamus* 箬竹属
Indocalamus bashanensis (C. D. Chu et C. S. Chao) H. R. Zhao et Y. L. Yang 巴山箬竹
Bamboos, evergreen

12.8 (10.6~15.1); 1.4 (−1.4~4.1); 23.7 (21.1~25.6); 12.8 (10.6~15.1); 101.3 (79.4~122.7); 7.3 (0.9~15.2); 730 (643~815) ❙ 998 (738~1254); 448 (332~533); 30 (16~49) ❙ 37.4 (5.8~93.8); 730 (643~815); 259.2 (178.4~291.4)

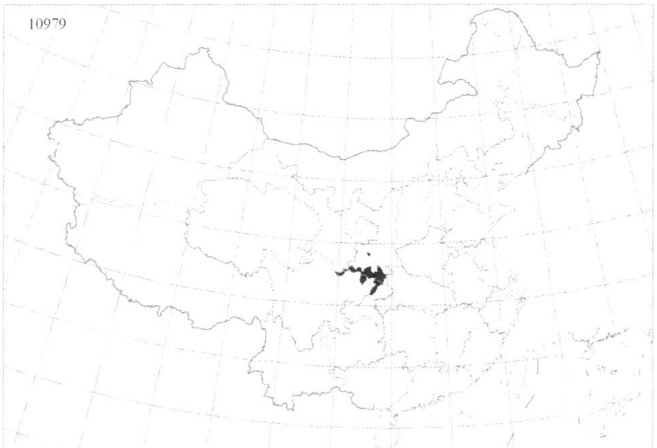

Poaceae 禾本科 *Indocalamus* 箬竹属
Indocalamus decorus Q. H. Dai 美丽箬竹
Bamboos, evergreen

17.7; 7.3; 26.8; 17.7; 152.8; 0.4; 939 ❙ 1541; 536; 178 ❙ 69.1; 939; 226.4

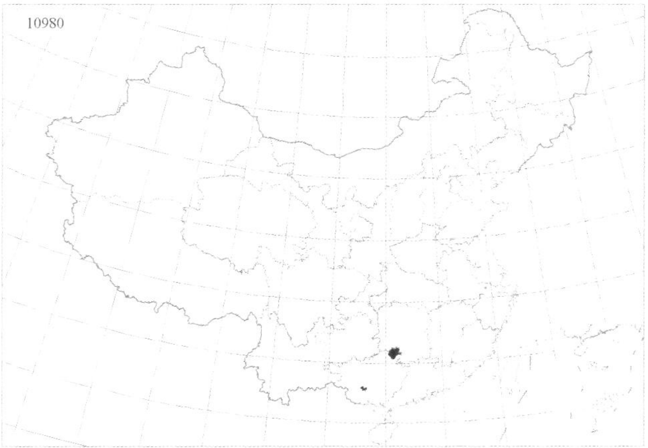

Poaceae 禾本科 *Indocalamus* 箬竹属
Indocalamus emeiensis C. D. Chu et C. S. Chao 峨眉箬竹
Bamboos, evergreen

14.9; 5.3; 23.6; 15.0; 119.6; 0.0; 782 ❙ 1410; 816; 49 ❙ 80.4; 782; 214.5

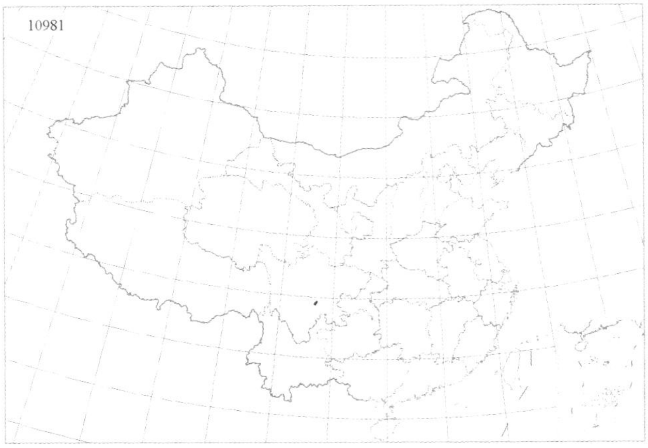

Poaceae 禾本科 *Indocalamus* 箬竹属
Indocalamus guangdongensis H. R. Zhao et Y. L. Yang 广东箬竹
Bamboos, evergreen

16.4 (13.7~17.9); 5.5 (2.3~8.1); 26.5 (24.6~27.6); 16.4 (13.7~17.9); 138.1 (109.4~155.4); 0.7 (0.0~4.7); 869 (758~933) ❙ 1414 (1221~1564); 508 (442~591); 149 (88~207) ❙ 62.7 (41.7~72.9); 867 (758~933); 212.1 (170.8~263.5)

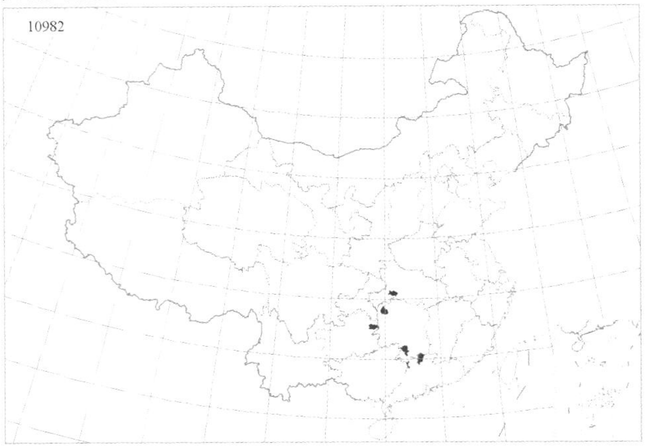

Poaceae 禾本科 *Indocalamus* 箬竹属
Indocalamus guangdongensis var. mollis H. R. Zhao et Y. L. Yang 柔毛箬竹
Bamboos, evergreen

16.5; 5.2; 27.1; 16.5; 138.3; 0.0; 875 ❙ 1485; 442; 207 ❙ 69.7; 873; 199.7

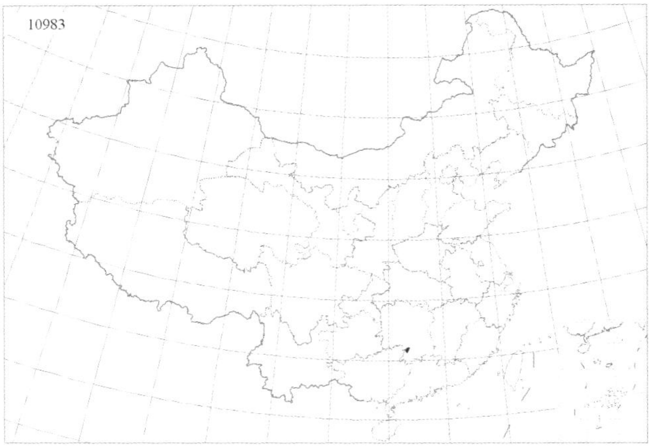

Poaceae 禾本科 *Indocalamus* 箬竹属
Indocalamus herklotsii McClure 粽巴箬竹
Bamboos, evergreen

22.4; 14.8; 28.5; 22.4; 209.2; 0.0; 1194 ❙ 1942; 973; 108 ❙ 62.7; 1194; 163.9

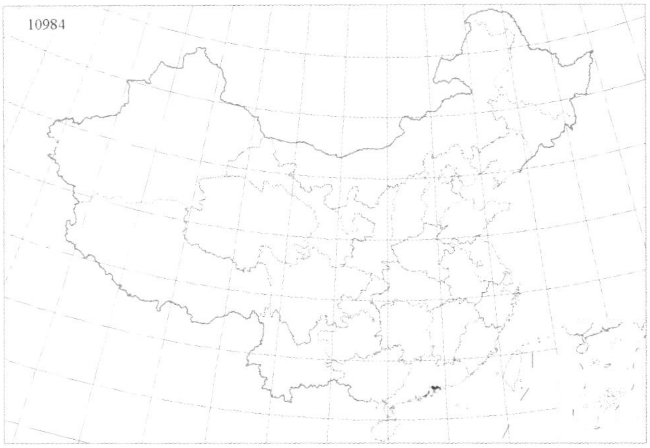

Poaceae 禾本科 *Indocalamus* 箬竹属
Indocalamus hirsutissimus Z. P. Wang et P. X. Zhang 多毛箬竹
Bamboos, evergreen

18.3 (16.8~19.1); 9.2 (7.8~10.1); 25.5 (23.6~26.6); 18.4 (16.8~19.1); 160.2 (141.5~169.2); 0.0 (0.0~0.0); 927 (836~969) ❙ 1271 (1236~1334); 677 (617~720); 57 (51~64) ❙ 37.7 (27.5~59.6); 927 (836~969); 242.9 (189.5~282.2)

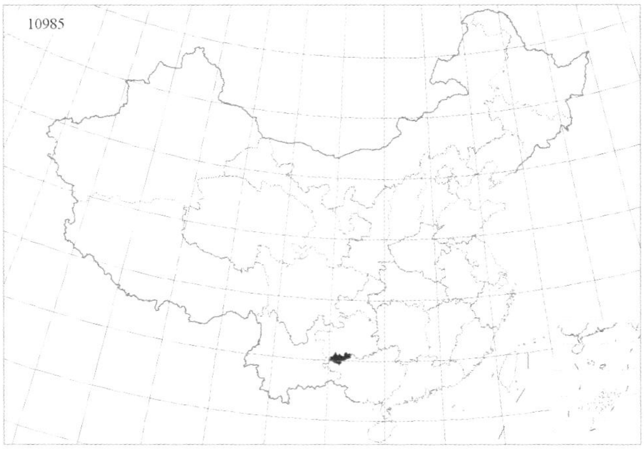

Poaceae 禾本科 *Indocalamus* 箬竹属
Indocalamus hirtivaginatus H. R. Zhao et Y. L. Yang 毛鞘箬竹
Bamboos, evergreen

18.7; 8.2; 27.9; 18.7; 164.9; 0.0; 984 ❙ 1665; 545; 207 ❙ 69.3; 981; 239.3

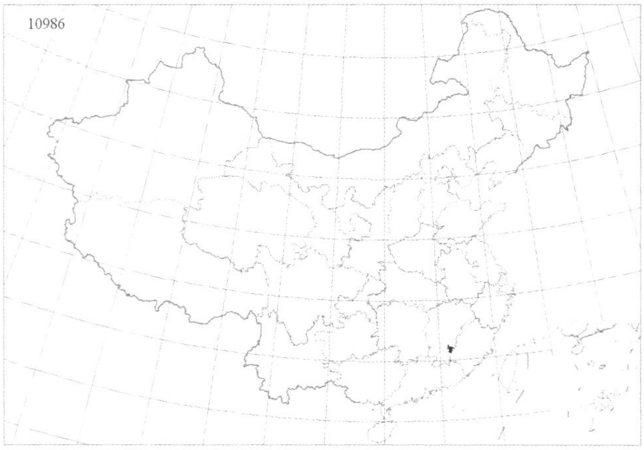

Poaceae 禾本科 *Indocalamus* 箬竹属
Indocalamus hispidus H. R. Zhao et Y. L. Yang 硬毛箬竹
Bamboos, evergreen

14.7；3.9；24.9；14.7；117.6；1.2；788 ❙ 1335；532；79 ❙ 69.6；788；191.1

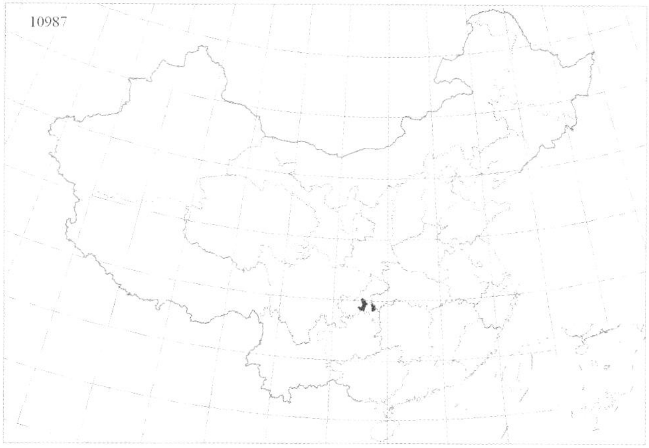

Poaceae 禾本科 *Indocalamus* 箬竹属
Indocalamus hunanensis B. M. Yang 湖南箬竹
Bamboos, evergreen

14.3 (12.0~15.9)；3.4 (1.1~6.1)；24.7 (22.5~26.6)；14.4 (12.0~16.0)；115.1 (92.3~131.7)；2.8 (0.0~8.5)；780 (690~839) ❙ 1260 (985~1451)；541 (493~600)；69 (31~120) ❙ 61.9 (19.4~81.5)；780 (690~839)；224.4 (180.4~258.2)

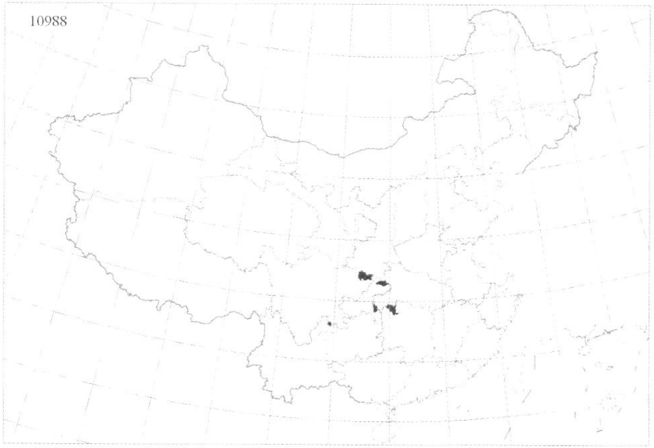

Poaceae 禾本科 *Indocalamus* 箬竹属
Indocalamus inaequilaterus W. T. Lin et Z. M. Wu 粤西箬竹
Bamboos, evergreen

21.2；12.2；28.5；21.3；195.2；0.0；1126 ❙ 1462；599；142 ❙ 29.9；1126；226.9

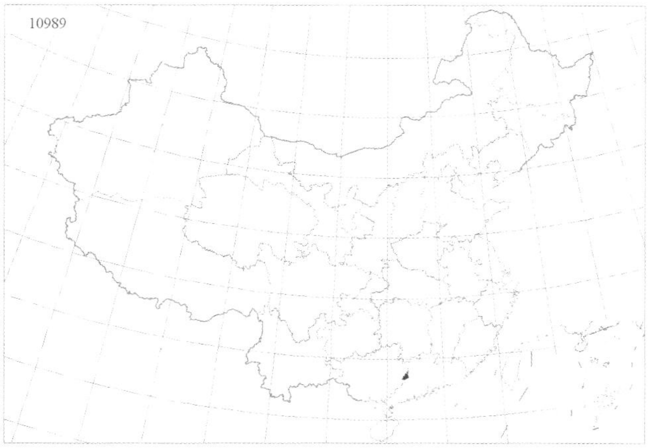

Poaceae 禾本科 *Indocalamus* 箬竹属
Indocalamus latifolius (Keng) McClure 阔叶箬竹
Bamboos, evergreen

15.5 (7.4~19.5)；3.3 (-4.2~11.0)；27.1 (18.7~29.9)；15.5 (8.2~19.6)；130.1 (58.8~175.1)；4.0 (0.0~29.4)；858 (561~1029) ❙ 1279 (622~1997)；501 (316~778)；139 (17~259) ❙ 48.1 (-27.1~147.0)；841 (561~1029)；212.9 (111.9~334.4)

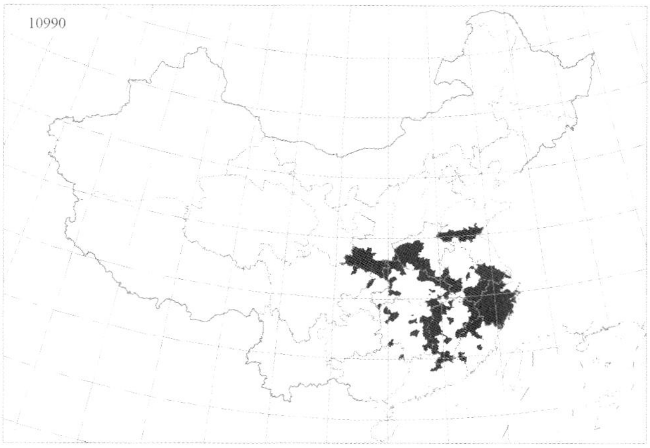

Poaceae 禾本科 *Indocalamus* 箬竹属
Indocalamus longiauritus Hand.-Mazz. 箬叶竹
Bamboos, evergreen

17.4 (8.6~22.5)；7.2 (-1.7~15.0)；26.7 (15.0~29.8)；17.4 (8.7~22.6)；150.0 (52.6~210.9)；1.0 (0.0~18.0)；919 (574~1206) ❙ 1520 (927~2037)；585 (356~1033)；159 (29~257) ❙ 67.1 (-5.0~147.0)；914 (574~1206)；236.9 (123.9~506.2)

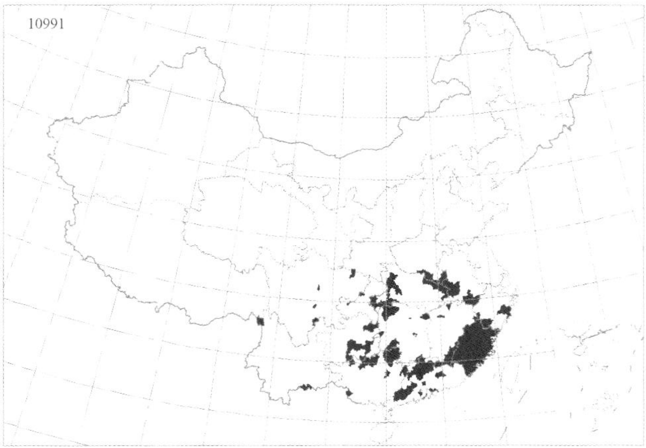

Poaceae 禾本科 *Indocalamus* 箬竹属
Indocalamus longiauritus var. hengshanensis H. R. Zhao et Y. L. Yang 衡山箬竹
Bamboos, evergreen

17.2；4.8；28.9；17.2；147.1；0.4；940 ❙ 1571；521；199 ❙ 67.2；931；194.3

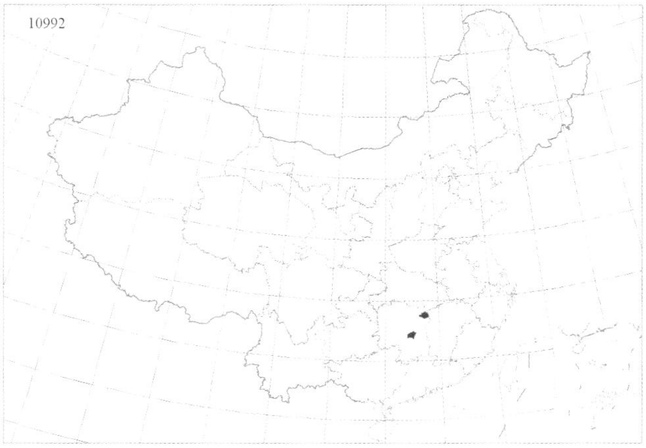

Poaceae 禾本科 *Indocalamus* 箬竹属
Indocalamus longiauritus var. semifalcatus H. R. Zhao et Y. L. Yang 半耳箬竹
Bamboos, evergreen

17.1 (13.3~18.6); 7.8 (3.2~9.3); 25.6 (22.1~27.6); 17.1 (13.2~18.7); 145.9 (100.9~164.0); 0.5 (0.0~2.5); 883 (723~965) ❙ 1558 (1081~1711); 590 (569~606); 162 (32~204) ❙ 75.4 (49.4~86.9); 883 (723~965); 270.1 (171.0~305.7)

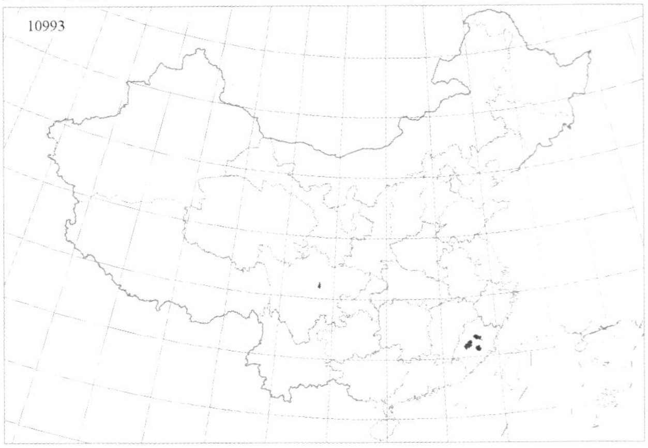

Poaceae 禾本科 *Indocalamus* 箬竹属
Indocalamus longiauritus var. yiyangensis H. R. Zhao et Y. L. Yang 益阳箬竹
Bamboos, evergreen

17.4; 4.5; 29.6; 17.4; 149.7; 0.5; 959 ❙ 1355; 444; 174 ❙ 41.2; 890; 154.4

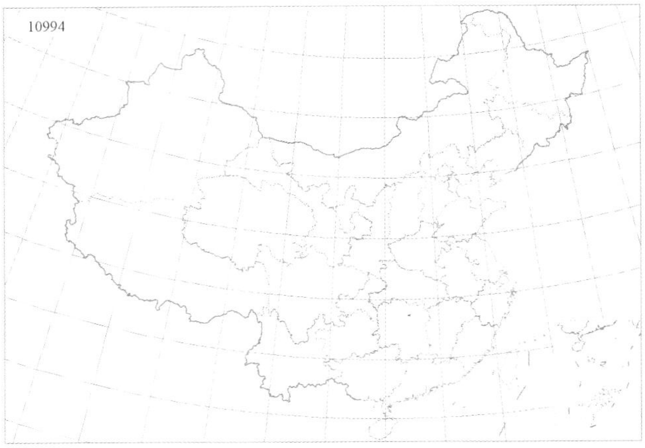

Poaceae 禾本科 *Indocalamus* 箬竹属
Indocalamus pseudosinicus McClure 锦帐竹
Bamboos, evergreen

24.1; 18.1; 28.4; 24.1; 229.7; 0.0; 1327 ❙ 1518; 559; 87 ❙ 13.9; 1284; 391.5

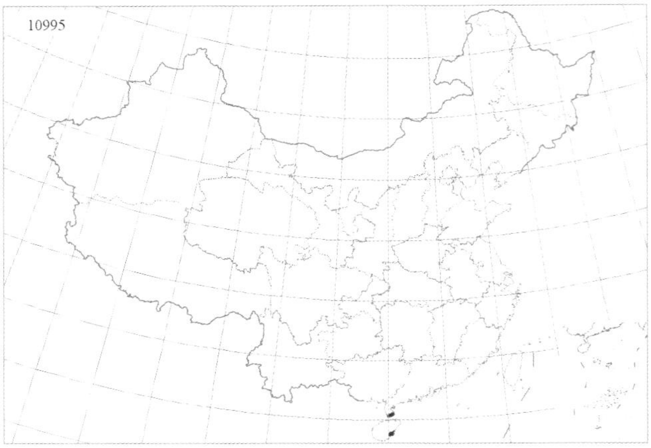

Poaceae 禾本科 *Indocalamus* 箬竹属
Indocalamus pseudosinicus var. densinervillus H. R. Zhao et Y. L. Yang 密脉箬竹
Bamboos, evergreen

23.6; 16.7; 28.6; 23.6; 223.5; 0.0; 1281 ❙ 1269; 551; 85 ❙ -1.0; 1269; 235.0

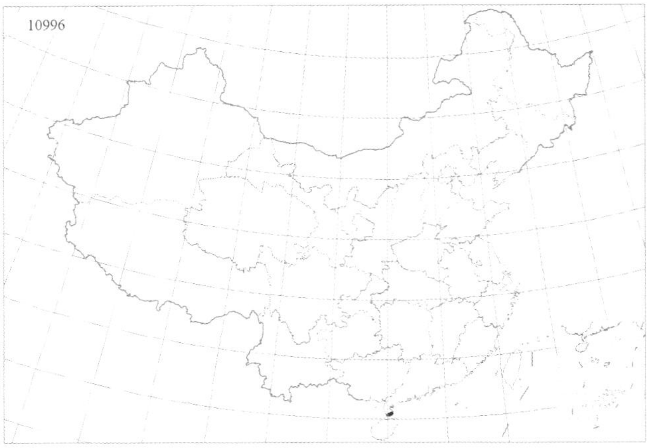

Poaceae 禾本科 *Indocalamus* 箬竹属
Indocalamus quadratus H. R. Zhao et Y. L. Yang 方脉箬竹
Bamboos, evergreen

16.3; 4.8; 27.2; 16.3; 136.2; 0.4; 871 ❙ 1442; 442; 195 ❙ 66.0; 847; 193.8

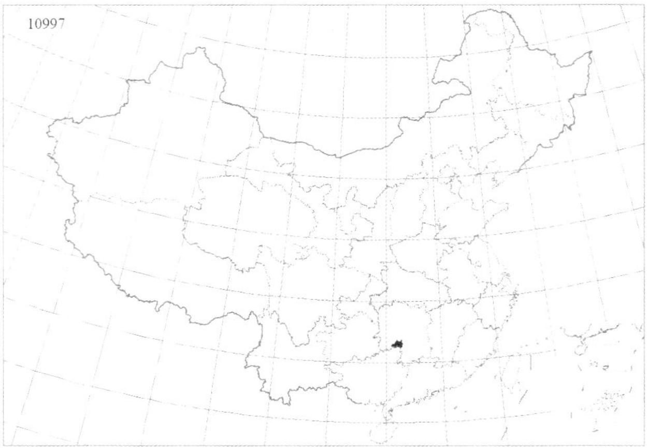

Poaceae 禾本科 *Indocalamus* 箬竹属
Indocalamus sinicus (Hance) Nakai 水银竹
Bamboos, evergreen

21.6; 13.2; 28.5; 21.7; 200.0; 0.0; 1149 ❙ 1774; 774; 140 ❙ 54.6; 1149; 219.1

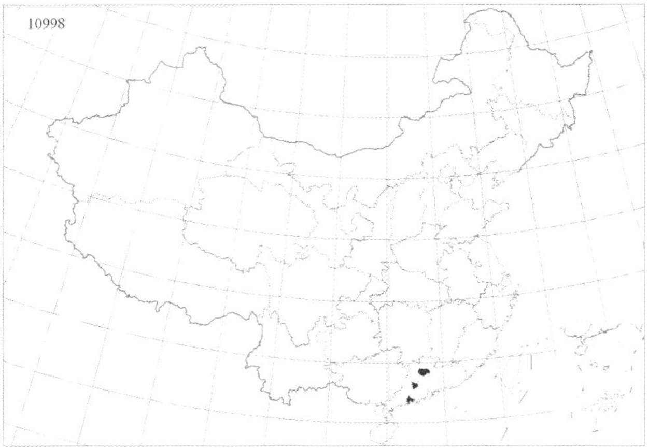

Poaceae 禾本科 *Indocalamus* 箬竹属
Indocalamus tessellatus (Munro) Keng f. 箬竹
Bamboos, evergreen

16.3 (9.3~22.9)；5.8 (−1.7~15.2)；25.9 (18.8~29.4)；16.3 (9.5~22.9)；136.4 (69.8~214.8)；0.6 (0.0~18.0)；862 (603~1232) ❙ 1394 (916~2035)；567 (442~969)；126 (31~259) ❙ 61.8 (27.5~147.0)；861 (603~1232)；218.2 (111.9~305.7)

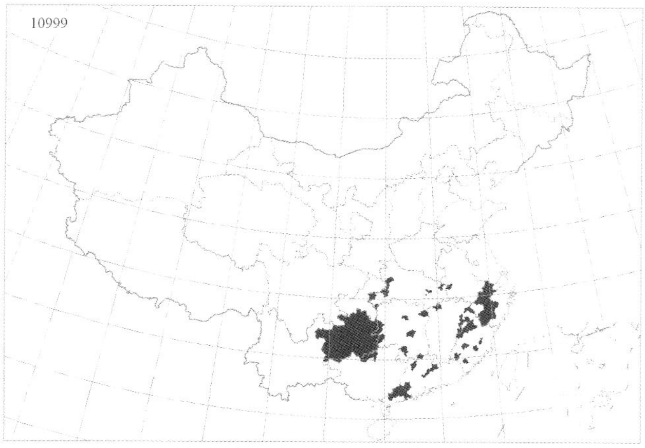

Poaceae 禾本科 *Indocalamus* 箬竹属
Indocalamus victorialis Keng f. 胜利箬竹
Bamboos, evergreen

16.5 (10.6~18.7)；5.6 (−0.2~8.1)；26.9 (21.1~28.9)；16.5 (10.6~18.8)；139.5 (79.4~166.0)；1.1 (0.0~12.1)；886 (643~1002) ❙ 1166 (862~1490)；485 (376~551)；70 (46~201) ❙ 33.7 (7.8~93.8)；885 (643~1002)；166.6 (108.2~284.3)

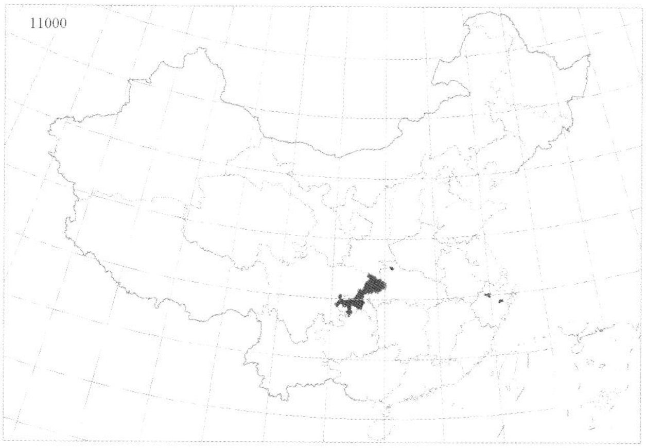

Poaceae 禾本科 *Indocalamus* 箬竹属
Indocalamus wilsonii (Rendle) C. S. Chao et C. D. Chu 鄂西箬竹
Bamboos, evergreen

14.0 (9.3~15.3)；3.0 (−1.7~4.5)；24.7 (20.1~26.5)；14.1 (9.5~15.4)；112.7 (69.8~125.7)；3.7 (0.5~18.0)；770 (603~832) ❙ 1250 (964~1466)；508 (421~602)；83 (52~113) ❙ 63.2 (31.7~106.2)；770 (603~832)；216.1 (161.9~285.1)

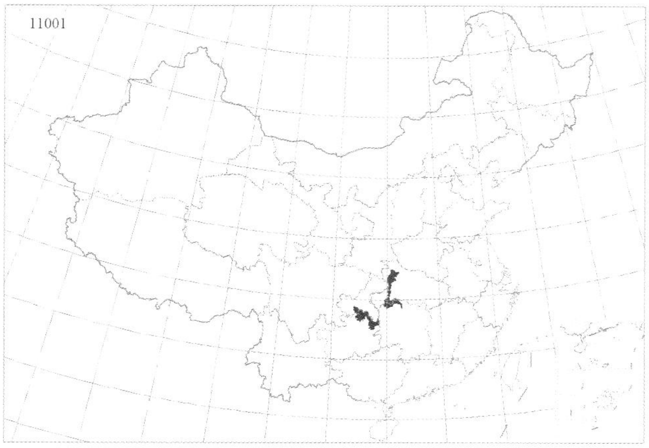

Poaceae 禾本科 *Indosasa* 大节竹属
Indosasa angustata McClure 甜大节竹
Bamboos, evergreen

22.0；13.8；28.0；22.1；204.8；0.0；1172 ❙ 1338；691；87 ❙ 14.1；1172；315.8

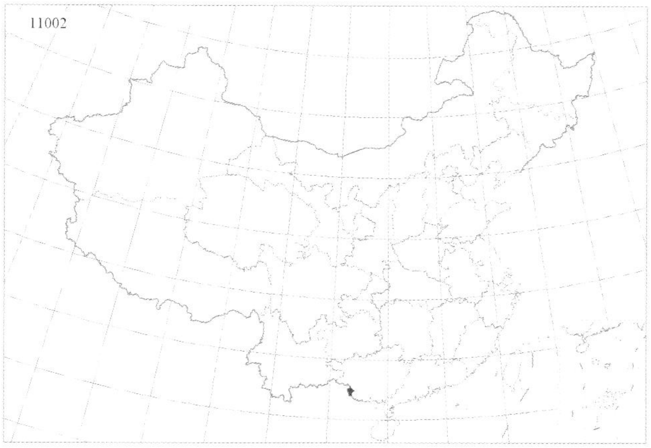

Poaceae 禾本科 *Indosasa* 大节竹属
Indosasa crassiflora McClure 大节竹
Bamboos, evergreen

21.3 (19.0~22.7)；12.7 (8.7~15.0)；28.0 (25.8~29.0)；21.3 (19.0~22.8)；196.2 (168.4~213.1)；0.0 (0.0~0.0)；1128 (958~1221) ❙ 1646 (1250~2435)；835 (639~1428)；116 (60~186) ❙ 45.7 (12.9~99.5)；1128 (958~1221)；265.6 (157.7~327.7)

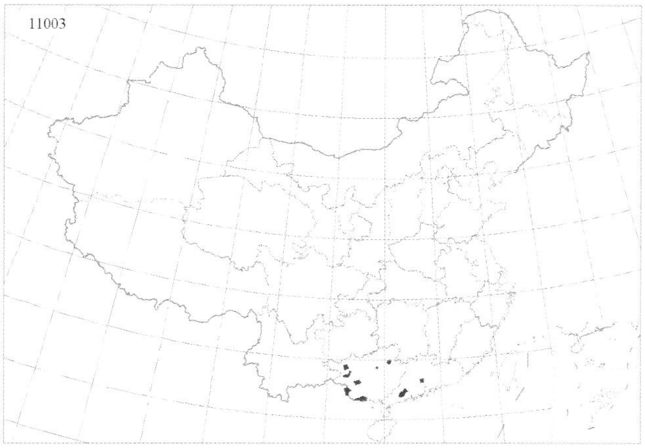

Poaceae 禾本科 *Indosasa* 大节竹属
Indosasa gigantea (T. H. Wen) T. H. Wen 橄榄竹
Bamboos, evergreen

17.3 (15.6~18.2)；7.0 (4.9~8.3)；26.9 (25.9~27.3)；17.3 (15.6~18.2)；147.7 (127.5~158.4)；0.0 (0.0~0.1)；902 (824~945) ❙ 1865 (1796~1950)；631 (602~677)；230 (216~249) ❙ 107.1 (92.7~129.9)；902 (824~945)；289.8 (269.3~332.5)

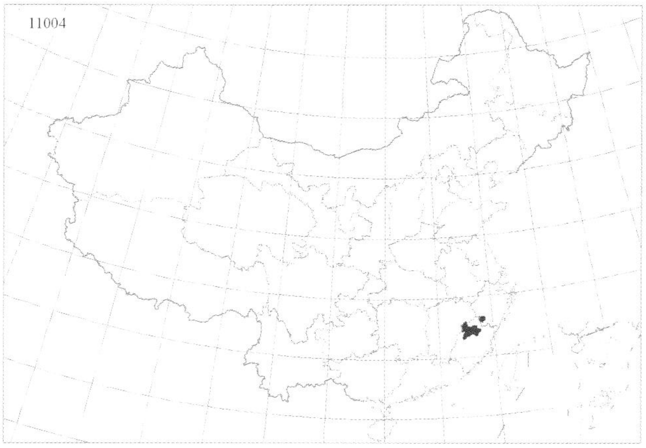

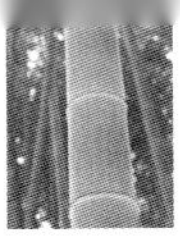

Poaceae 禾本科 *Indosasa* 大节竹属
Indosasa glabrata C. D. Chu et C. S. Chao 算盘竹
Bamboos, evergreen

22.0 (21.3~22.7); 13.9 (13.2~15.0); 28.0 (27.3~28.9); 22.0 (21.3~22.8); 204.2 (195.9~213.1); 0.0 (0.0~0.0); 1167 (1111~1221) ❙ 1814 (1302~2435); 999 (670~1428); 119 (86~136) ❙ 55.3 (12.9~99.5); 1167 (1111~1221); 277.0 (201.4~318.8)

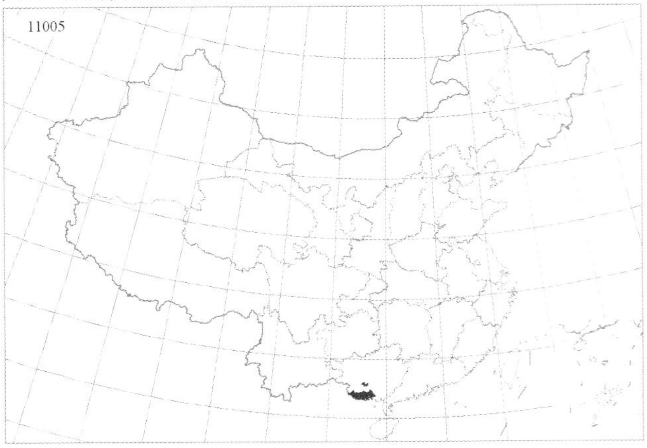

Poaceae 禾本科 *Indosasa* 大节竹属
Indosasa hispida McClure 浦竹仔
Bamboos, evergreen

20.9 (18.6~22.7); 13.1 (11.0~15.1); 26.6 (22.5~28.6); 21.0 (18.7~22.8); 191.6 (163.9~213.1); 0.0 (0.0~0.0); 1080 (888~1221) ❙ 1598 (1188~2435); 842 (634~1428); 97 (56~181) ❙ 48.4 (5.8~99.5); 1078 (888~1221); 349.1 (240.0~659.0)

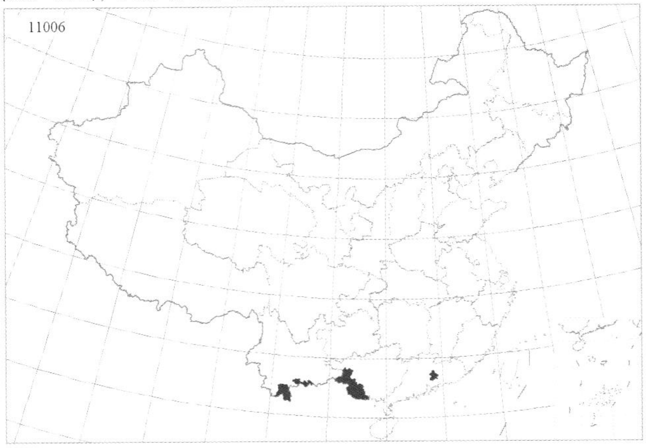

Poaceae 禾本科 *Indosasa* 大节竹属
Indosasa ingens Hsueh et T. P. Yi 粗穗大节竹
Bamboos, evergreen

19.0; 12.3; 23.5; 19.0; 167.6; 0.0; 925 ❙ 1524; 863; 62 ❙ 66.4; 925; 505.5

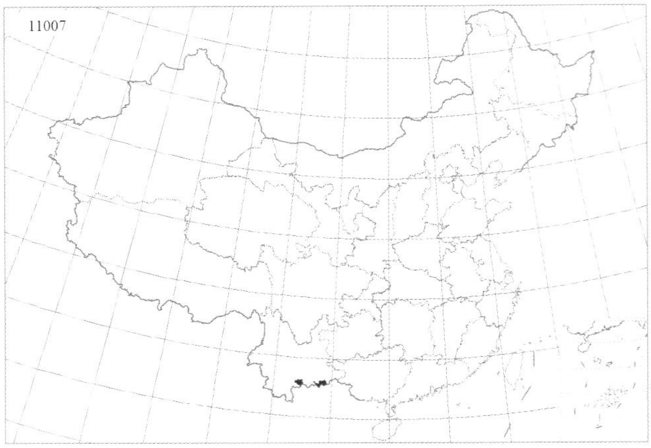

Poaceae 禾本科 *Indosasa* 大节竹属
Indosasa lipoensis C. D. Chu et K. M. Lan 荔波大节竹
Bamboos, evergreen

17.4; 7.4; 25.5; 17.4; 148.9; 0.0; 886 ❙ 1381; 637; 107 ❙ 55.9; 886; 245.3

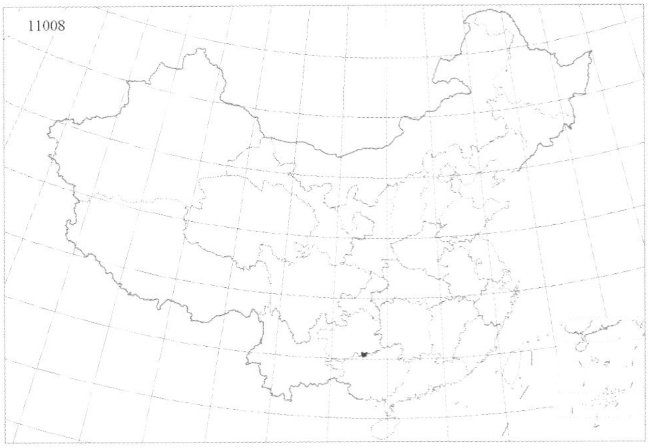

Poaceae 禾本科 *Indosasa* 大节竹属
Indosasa longispicata W. Y. Hsiung et C. S. Chao 棚竹
Bamboos, evergreen

17.1 (14.3~19.1); 6.5 (3.7~13.0); 26.7 (23.3~28.5); 17.1 (14.3~19.1); 145.5 (113.2~169.7); 0.2 (0.0~1.4); 895 (761~991) ❙ 1587 (1428~1742); 570 (488~835); 189 (56~221) ❙ 78.5 (55.7~128.9); 895 (761~991); 244.3 (208.1~506.2)

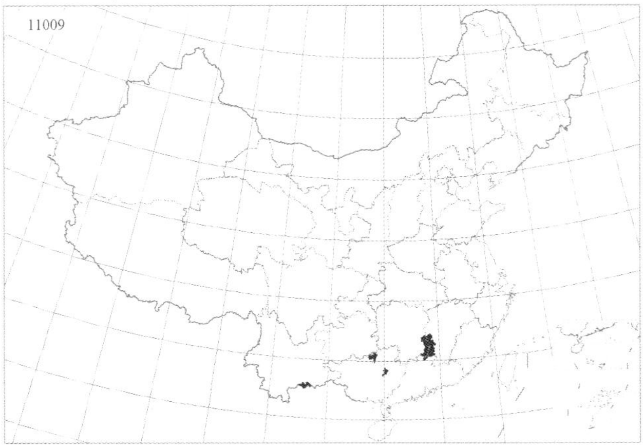

Poaceae 禾本科 *Indosasa* 大节竹属
Indosasa parvifolia C. S. Chao et Q. H. Dai 小叶大节竹
Bamboos, evergreen

22.0; 13.8; 28.0; 22.1; 204.8; 0.0; 1172 ❙ 1338; 691; 87 ❙ 14.1; 1172; 315.8

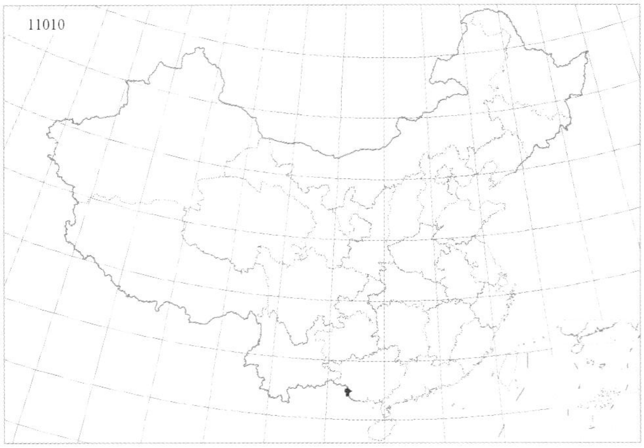

Poaceae 禾本科 *Indosasa* 大节竹属

Indosasa patens C. D. Chu et C. S. Chao 横枝竹

Bamboos, evergreen

18.2 (17.0~19.2); 7.9 (6.2~9.2); 27.0 (26.3~27.6); 18.2 (17.0~19.2); 159.0 (144.4~170.8); 0.0 (0.0~0.0); 948 (891~1002) ❙ 1574 (1437~1697); 643 (594~678); 161 (117~194) ❙ 66.3 (48.5~82.3); 948 (891~1002); 240.5 (225.9~256.4)

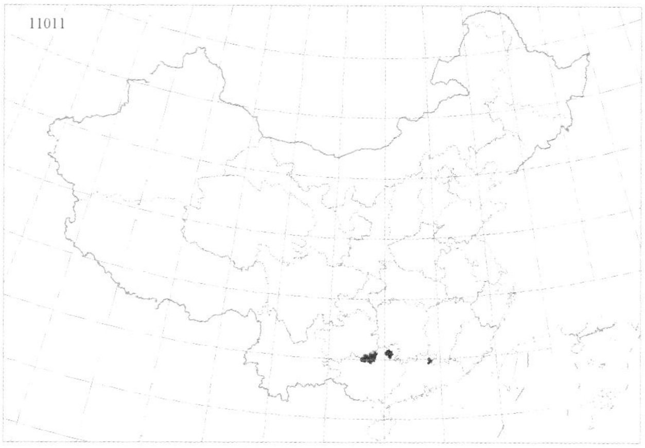

Poaceae 禾本科 *Indosasa* 大节竹属

Indosasa shibataeoides McClure 摆竹

Bamboos, evergreen

18.1 (14.3~23.2); 7.8 (3.3~15.8); 27.1 (24.3~29.0); 18.1 (14.3~23.2); 157.7 (114.1~218.9); 0.1 (0.0~2.3); 951 (765~1257) ❙ 1582 (1410~1990); 591 (403~858); 178 (102~209) ❙ 67.4 (34.1~101.8); 944 (765~1257); 221.4 (151.0~263.5)

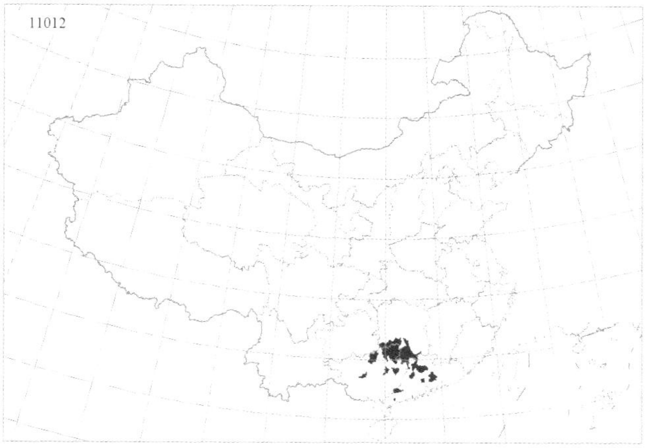

Poaceae 禾本科 *Indosasa* 大节竹属

Indosasa singulispicula T. H. Wen 单穗大节竹

Bamboos, evergreen

20.0; 14.6; 23.2; 20.1; 180.8; 0.0; 963 ❙ 1523; 681; 67 ❙ 58.4; 942; 581.0

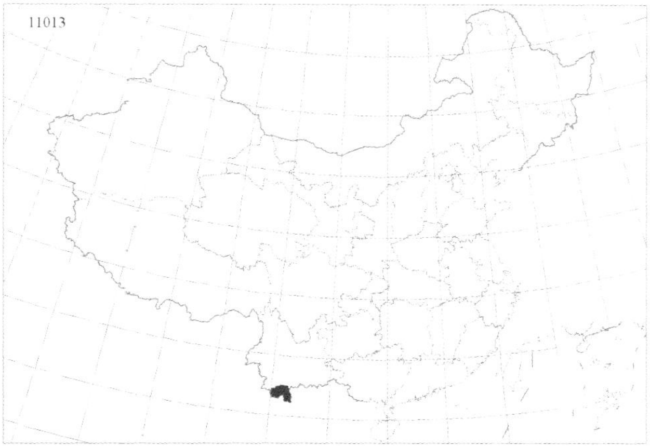

Poaceae 禾本科 *Indosasa* 大节竹属

Indosasa spongiosa C. S. Chao et B. M. Yang 江华大节竹

Bamboos, evergreen

18.7 (17.3~21.0); 11.7 (6.9~14.2); 23.9 (22.1~26.5); 18.8 (17.3~21.0); 165.5 (148.1~191.8); 0.0 (0.0~0.0); 919 (854~1053) ❙ 1408 (1276~1522); 740 (517~835); 79 (51~185) ❙ 53.6 (36.8~69.5); 918 (854~1053); 441.4 (214.2~551.6)

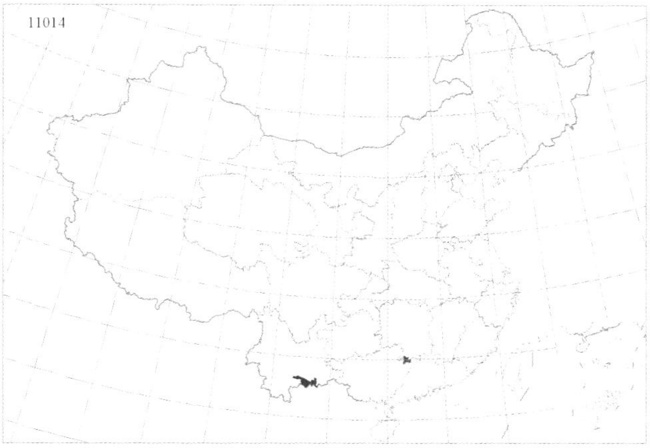

Poaceae 禾本科 *Indosasa* 大节竹属

Indosasa triangulata Hsueh et T. P. Yi 五爪竹

Bamboos, evergreen

17.3; 9.9; 22.4; 17.3; 147.1; 0.0; 833 ❙ 1504; 860; 55 ❙ 80.6; 833; 413.5

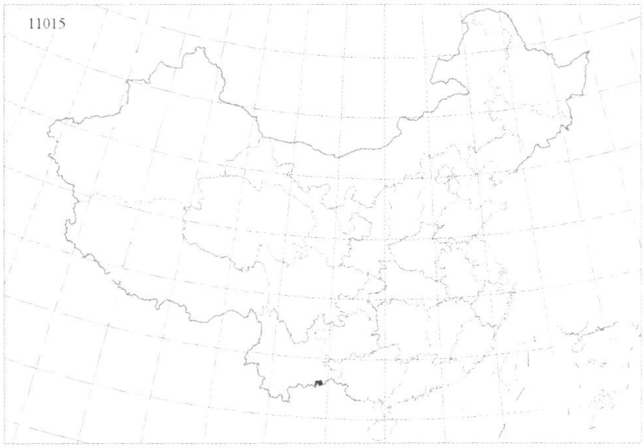

Poaceae 禾本科 *Leptaspis* 囊稃竹属

Leptaspis banksii R. Br. 囊稃竹

Bamboos, evergreen

18.5; 14.1; 21.7; 18.5; 162.1; 0.0; 863 ❙ 2713; 1388; 188 ❙ 214.4; 863; 363.7

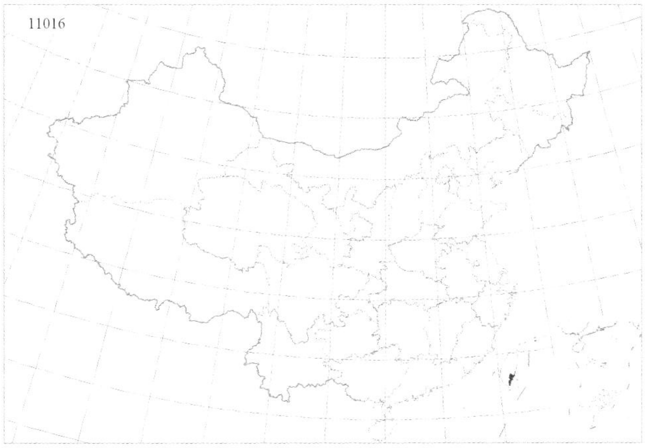

Poaceae 禾本科 *Melocalamus* 梨藤竹属
Melocalamus arrectus T. P. Yi 澜沧梨藤竹
Bamboos, evergreen

19.3 (17.6~21.0); 13.1 (9.9~15.1); 23.6 (22.5~25.6); 19.4 (17.7~21.0); 172.7 (152.4~191.8); 0.0 (0.0~0.0); 940 (859~1053) ❙ 1467 (1276~1607); 803 (702~911); 61 (51~77) ❙ 56.6 (36.8~74.2); 938 (859~1053); 513.3 (359.7~659.0)

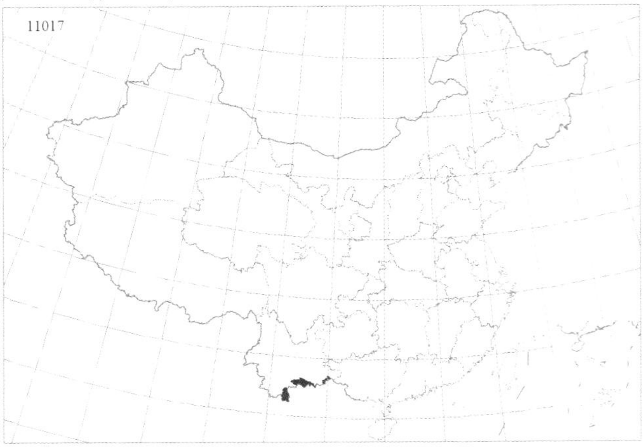

Poaceae 禾本科 *Melocalamus* 梨藤竹属
Melocalamus compactiflorus (Kurz) Benth. et Hook. f. 梨藤竹
Bamboos, evergreen

19.8 (17.7~21.8); 13.3 (10.9~14.9); 24.2 (22.5~27.7); 19.8 (17.8~21.8); 177.8 (153.4~202.2); 0.0 (0.0~0.0); 976 (854~1154) ❙ 1439 (1302~1547); 774 (655~911); 66 (56~86) ❙ 49.2 (12.9~74.2); 971 (854~1154); 495.8 (304.0~570.1)

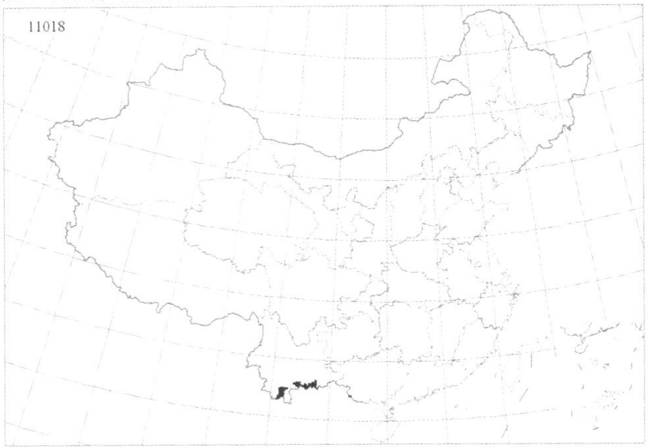

Poaceae 禾本科 *Melocalamus* 梨藤竹属
Melocalamus elevatissimus Hsueh et T. P. Yi 西藏梨藤竹
Bamboos, evergreen

9.3; 1.5; 15.9; 9.6; 65.5; 13.1; 592 ❙ 1348; 783; 41 ❙ 123.3; 592; 340.8

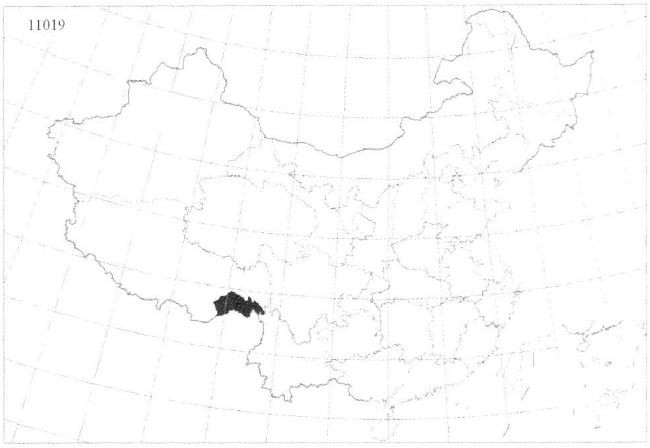

Poaceae 禾本科 *Melocanna* 梨竹属
Melocanna humilis Kurz 梨竹
Bamboos, evergreen

21.7; 16.7; 25.4; 21.7; 200.8; 0.0; 1117 ❙ 2255; 1381; 87 ❙ 117.5; 1003; 203.3

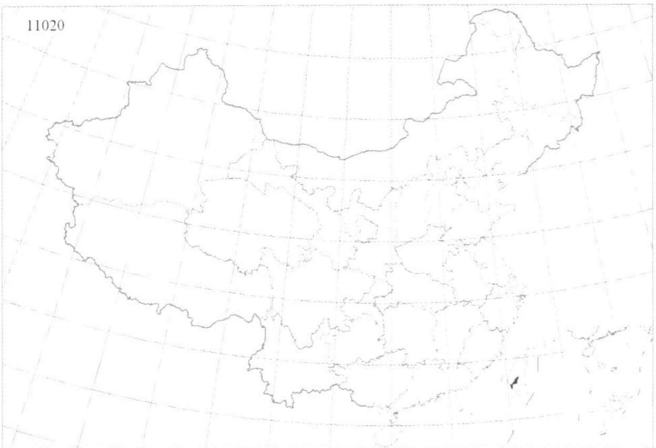

Poaceae 禾本科 *Neomicrocalamus* 新小竹属
Neomicrocalamus prainii (Gamble) P. C. Keng 新小竹
Bamboos, evergreen

13.3; 5.4; 19.6; 13.4; 100.4; 0.0; 706 ❙ 1736; 1007; 51 ❙ 145.8; 706; 447.4

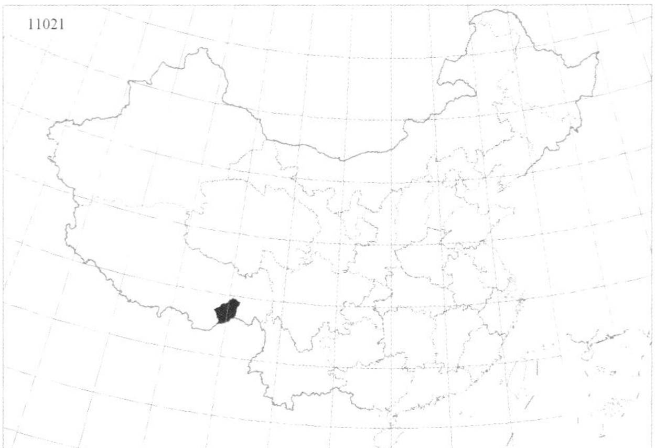

Poaceae 禾本科 *Neomicrocalamus* 新小竹属
Neomicrocalamus yunnanensis (T. H. Wen) Ohrnb. 云南新小竹
Bamboos, evergreen

19.3; 12.9; 23.7; 19.3; 171.9; 0.0; 939 ❙ 1398; 788; 59 ❙ 49.3; 937; 500.3

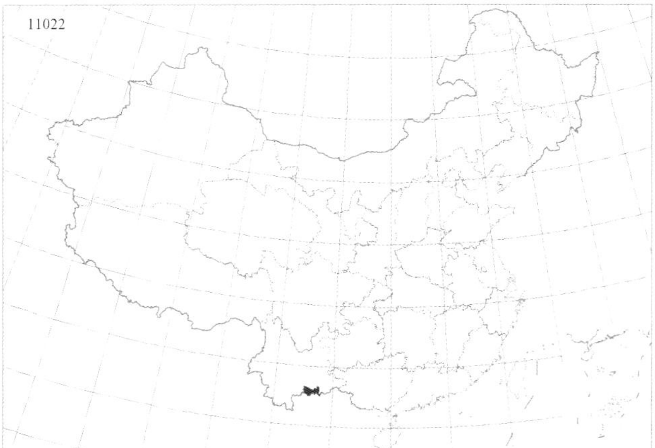

Poaceae 禾本科 *Oligostachyum* 少穗竹属
Oligostachyum glabrescens (T. H. Wen) Keng f. et Y. P. Wang 屏南少穗竹
Bamboos, evergreen

16.3; 6.8; 25.6; 16.4; 136.3; 0.0; 843 ❙ 1759; 635; 196 ❙ 110.3; 843; 264.2

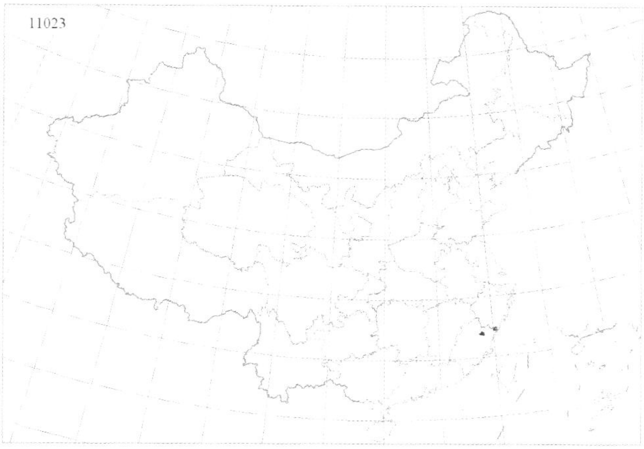

Poaceae 禾本科 *Oligostachyum* 少穗竹属
Oligostachyum gracilipes (McClure) G. H. Ye et Z. P. Wang 细柄少穗竹
Bamboos, evergreen

21.3; 13.8; 27.6; 21.3; 196.1; 0.0; 1151 ❙ 1294; 516; 61 ❙ 13.0; 1051; 362.9

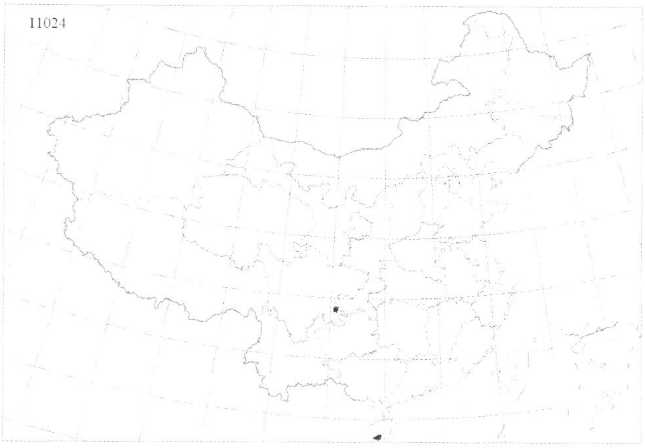

Poaceae 禾本科 *Oligostachyum* 少穗竹属
Oligostachyum hupehense (J. L. Lu) Z. P. Wang et G. H. Ye 凤竹
Bamboos, evergreen

13.8; 2.4; 24.8; 13.9; 111.2; 4.4; 767 ❙ 1121; 478; 82 ❙ 46.2; 767; 236.0

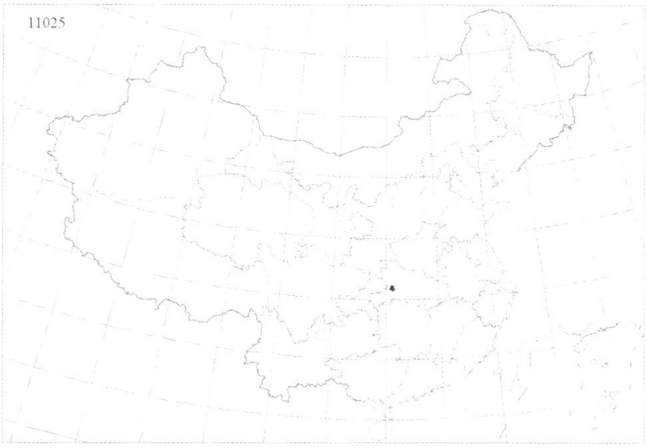

Poaceae 禾本科 *Oligostachyum* 少穗竹属
Oligostachyum lanceolatum G. H. Ye et Z. P. Wang 云和少穗竹
Bamboos, evergreen

16.2; 5.3; 26.8; 16.2; 134.8; 0.0; 857 ❙ 1715; 617; 201 ❙ 100.0; 857; 253.8

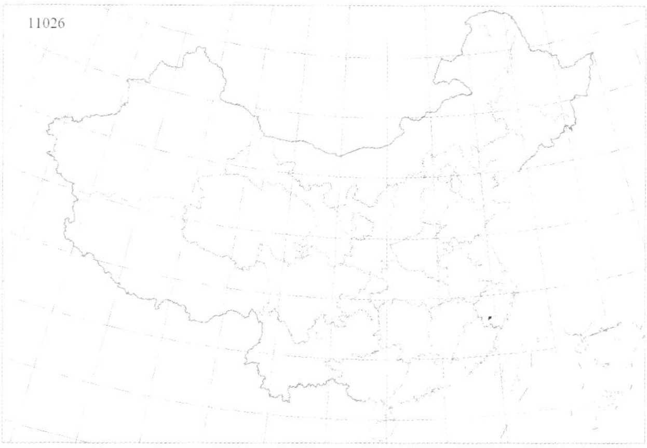

Poaceae 禾本科 *Oligostachyum* 少穗竹属
Oligostachyum lubricum (T. W. Wen) Keng f. 四季竹
Bamboos, evergreen

16.0; 4.1; 28.1; 16.1; 134.0; 1.1; 878 ❙ 1387; 516; 187 ❙ 58.0; 878; 225.4

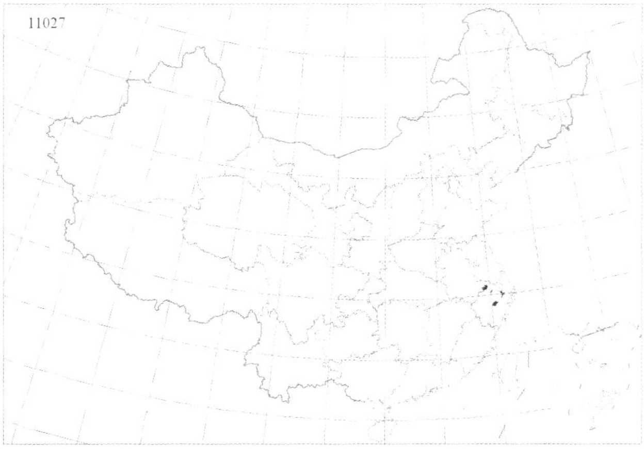

Poaceae 禾本科 *Oligostachyum* 少穗竹属
Oligostachyum nuspiculum (McClure) Z. P. Wang et G. H. Ye 林仔竹
Bamboos, evergreen

21.6 (18.0~24.7); 13.4 (8.5~19.8); 27.9 (25.8~29.8); 21.6 (18.0~24.7); 199.2 (156.4~236.5); 0.0 (0.0~0.0); 1147 (913~1376) ❙ 1647 (868~2435); 749 (312~1428); 128 (58~194) ❙ 45.7 (-34.3~99.5); 1101 (868~1269); 296.1 (174.9~572.9)

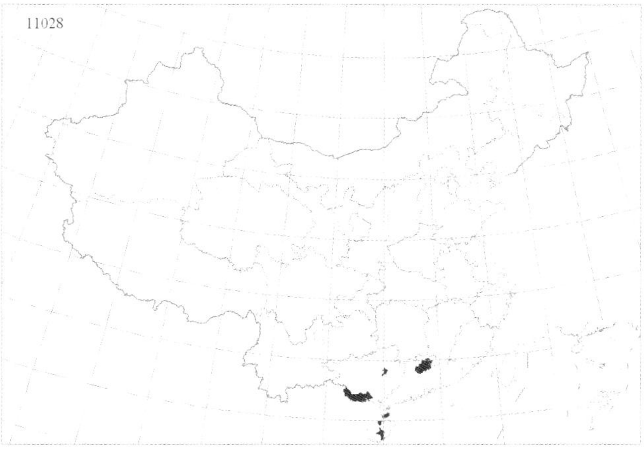

Poaceae 禾本科 *Phyllostachys* 刚竹属
Phyllostachys atrovaginata C. S. Chao et H. Y. Chou 乌芽竹
Bamboos, evergreen

16.5 (15.3~18.6); 6.4 (3.1~12.7); 26.0 (21.4~28.9); 16.5 (15.4~18.7); 139.3 (127.2~163.9); 1.0 (0.0~2.7); 866 (813~910) ❙ 1332 (1178~1547); 602 (493~911); 139 (51~197) ❙ 53.7 (38.4~74.2); 866 (813~910); 305.3 (189.1~551.3)

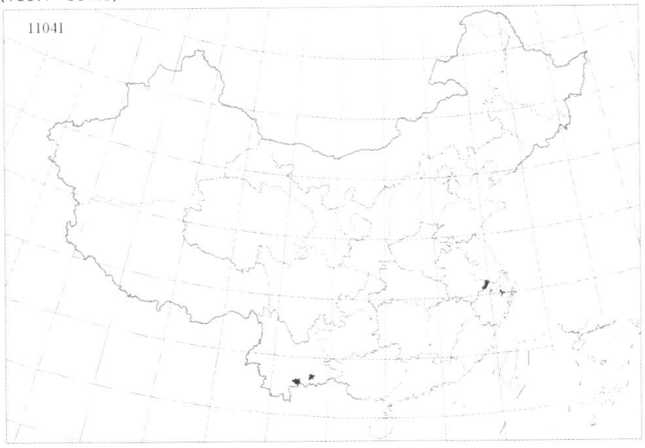

Poaceae 禾本科 *Phyllostachys* 刚竹属
Phyllostachys aurea Carrière ex Rivière et C. Rivière 人面竹
Bamboos, evergreen

16.6 (12.7~24.1); 5.4 (1.5~18.9); 27.2 (20.1~29.9); 16.7 (12.8~24.1); 141.7 (99.0~229.7); 1.8 (0.0~7.3); 900 (713~1322) ❙ 1376 (963~2759); 572 (415~1594); 141 (22~245) ❙ 53.0 (10.3~246.5); 895 (692~1182); 207.6 (111.9~332.5)

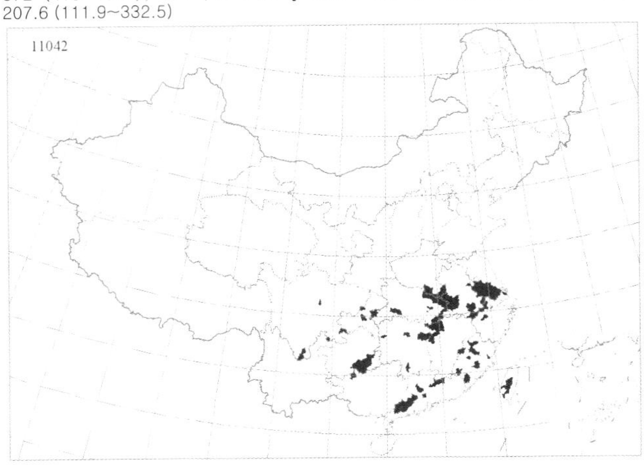

Poaceae 禾本科 *Phyllostachys* 刚竹属
Phyllostachys bissetii McClure 蓉城竹
Bamboos, evergreen

15.6; 4.9; 25.5; 15.6; 127.9; 0.8; 834 ❙ 1248; 653; 78 ❙ 49.4; 834; 166.7

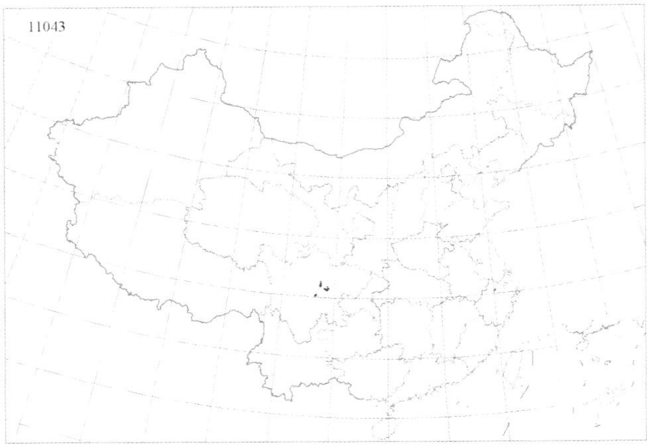

Poaceae 禾本科 *Phyllostachys* 刚竹属
Phyllostachys carnea G. H. Ye et Z. P. Wang 湖南刚竹
Bamboos, evergreen

15.5; 4.0; 26.6; 15.6; 128.0; 1.0; 839 ❙ 1451; 584; 120 ❙ 72.9; 839; 208.1

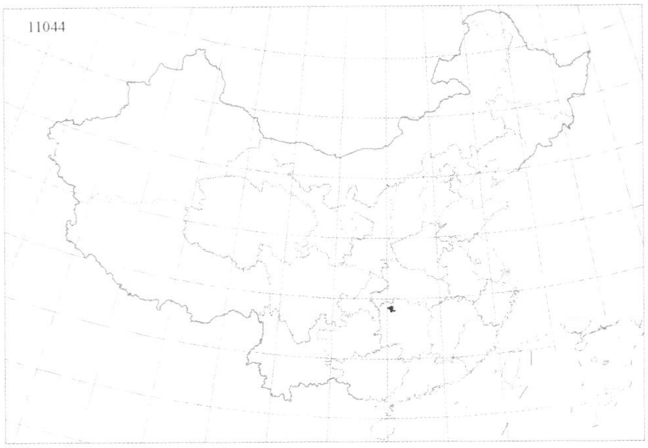

Poaceae 禾本科 *Phyllostachys* 刚竹属
Phyllostachys circumpilis C. Y. Yao et S. Y. Chen 毛壳花哺鸡竹
Bamboos, evergreen

17.5; 6.9; 27.0; 17.5; 150.3; 0.0; 912 ❙ 1503; 511; 180 ❙ 64.9; 912; 201.5

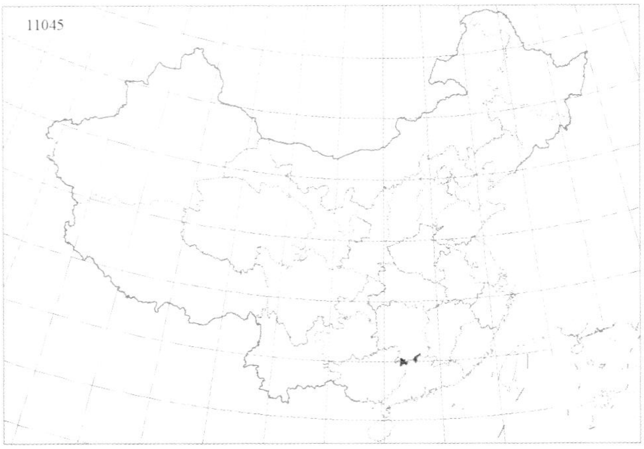

Poaceae 禾本科 *Phyllostachys* 刚竹属
Phyllostachys dulcis McClure 白哺鸡竹
Bamboos, evergreen

15.9 (14.4~17.3); 4.1 (2.1~7.6); 27.5 (24.3~29.4); 15.9 (14.5~17.4); 132.4 (118.7~148.3); 1.6 (0.0~4.9); 865 (791~952) ❙ 1490 (1059~1959); 549 (415~695); 191 (123~250) ❙ 73.3 (20.5~137.5); 863 (791~921); 233.8 (111.9~296.8)

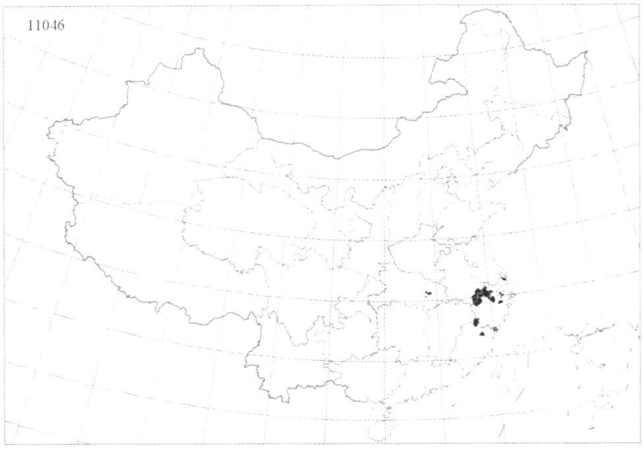

Poaceae 禾本科 *Phyllostachys* 刚竹属

Phyllostachys eculis (Carrière) J. Houzeau 毛竹

Bamboos, evergreen

16.7 (12.1~24.1); 5.4 (-2.5~18.9); 27.3 (20.1~29.8); 16.7 (12.4~24.1); 141.4 (98.8~229.7); 1.0 (0.0~17.8); 896 (711~1322) ❙ 1481 (533~2759); 567 (269~1594); 167 (22~256) ❙ 66.1 (-28.6~246.5); 890 (533~1182); 218.1 (111.9~322.1)

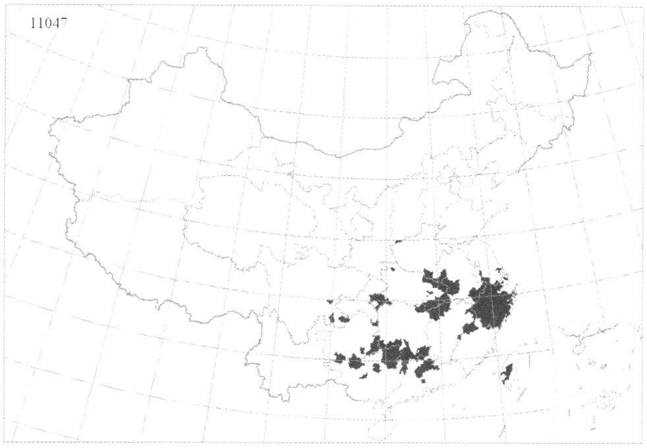

Poaceae 禾本科 *Phyllostachys* 刚竹属

Phyllostachys elegans McClure 甜笋竹

Bamboos, evergreen

17.7 (15.6~22.1); 7.0 (3.4~13.7); 27.6 (25.2~29.0); 17.7 (15.7~22.1); 152.9 (130.0~205.6); 0.1 (0.0~2.1); 936 (864~1182) ❙ 1526 (1117~1829); 526 (403~821); 192 (144~230) ❙ 63.5 (28.5~104.6); 924 (864~1182); 209.1 (151.0~332.5)

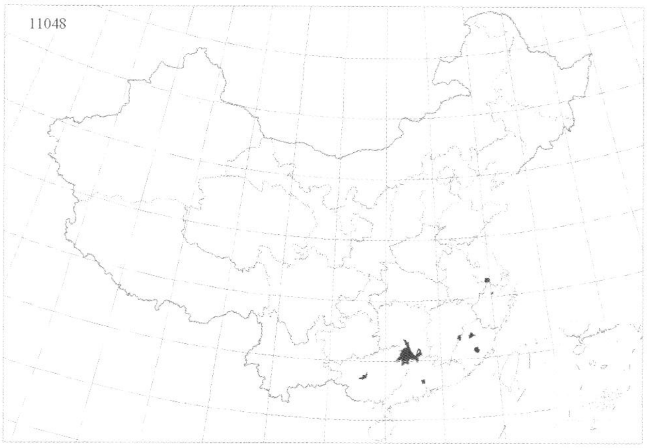

Poaceae 禾本科 *Phyllostachys* 刚竹属

Phyllostachys fimbriligula T. W. Wen 角竹

Bamboos, evergreen

16.5; 4.6; 28.6; 16.5; 138.9; 0.4; 901 ❙ 1323; 478; 186 ❙ 46.7; 901; 216.4

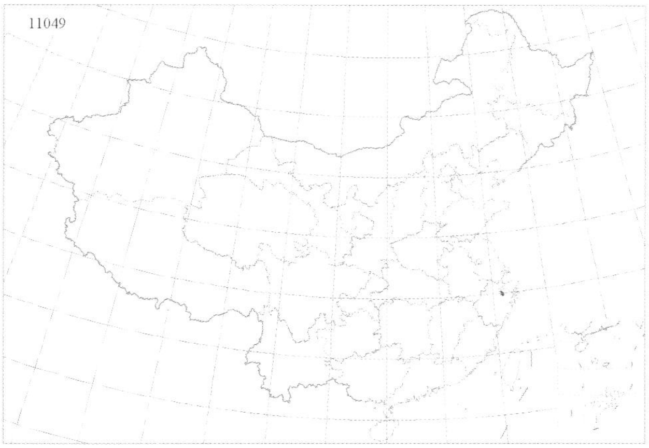

Poaceae 禾本科 *Phyllostachys* 刚竹属

Phyllostachys flexuosa Rivière et C. Rivière 曲竿竹

Bamboos, evergreen

14.0 (8.2~17.7); -0.2 (-6.9~4.6); 27.0 (21.6~29.9); 14.1 (9.6~17.7); 120.0 (73.3~153.0); 11.4 (0.4~34.4); 832 (622~976) ❙ 700 (512~1423); 375 (258~546); 43 (16~198) ❙ -15.6 (-36.5~56.6); 682 (512~887); 168.6 (120.1~274.1)

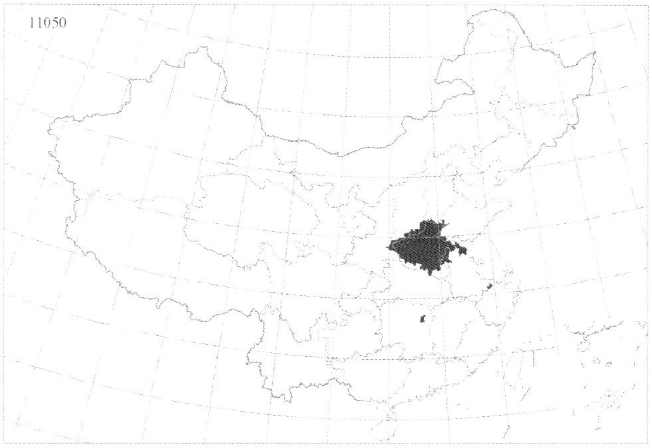

Poaceae 禾本科 *Phyllostachys* 刚竹属

Phyllostachys glabrata S. Y. Chen et C. Y. Yao 花哺鸡竹

Bamboos, evergreen

16.1 (15.3~16.5); 4.5 (3.5~6.1); 27.6 (24.3~28.9); 16.1 (15.4~16.6); 134.1 (124.4~139.8); 0.8 (0.0~1.8); 872 (791~910) ❙ 1502 (1349~1878); 545 (493~695); 199 (192~211) ❙ 73.8 (50.0~137.5); 872 (791~910); 236.3 (189.1~296.8)

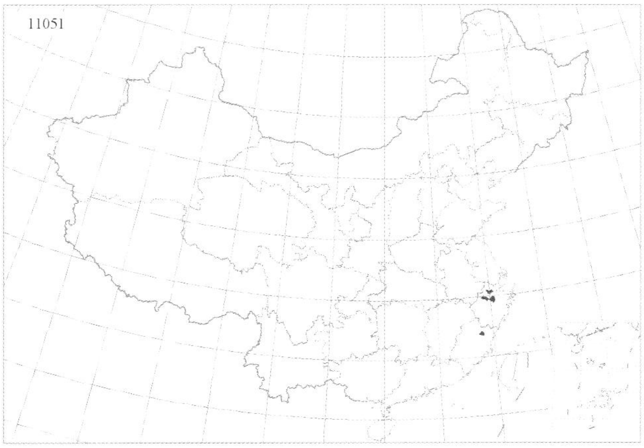

Poaceae 禾本科 *Phyllostachys* 刚竹属

Phyllostachys glauca McClure 淡竹

Bamboos, evergreen

14.7 (4.2~22.1); 1.3 (-12.5~13.7); 27.2 (18.7~30.1); 14.9 (7.3~22.1); 126.8 (52.8~205.6); 9.3 (0.0~62.6); 854 (535~1182) ❙ 995 (377~1950); 445 (241~779); 95 (8~259) ❙ 14.5 (-42.3~128.9); 762 (377~1182); 181.8 (84.7~319.9)

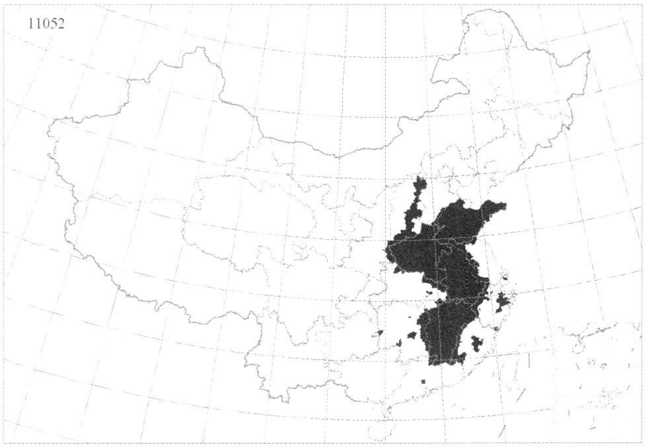

Poaceae 禾本科 *Phyllostachys* 刚竹属

Phyllostachys guizhouensis C. S. Chao et J. Q. Zhang 贵州刚竹

Bamboos, evergreen

13.4 (11.7~15.2); 4.1 (2.7~5.8); 21.6 (19.6~23.0); 13.5 (11.7~15.2); 102.8 (84.6~122.5); 1.3 (0.0~3.8); 719 (661~779) ❙ 1098 (964~1239); 567 (501~645); 52 (39~60) ❙ 52.6 (39.0~59.3); 719 (661~779); 193.2 (179.5~211.0)

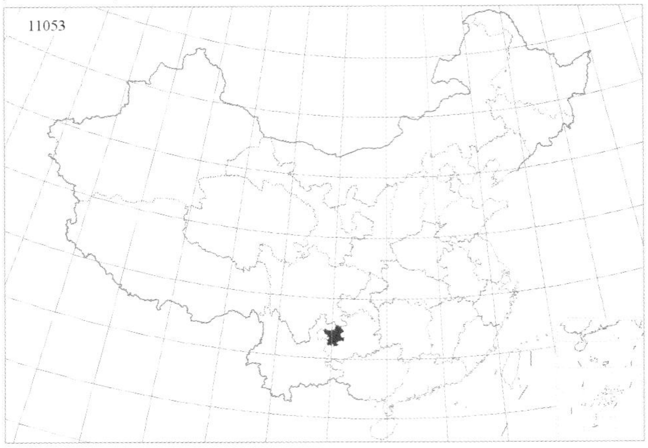

Poaceae 禾本科 *Phyllostachys* 刚竹属

Phyllostachys heteroclada Oliv. 水竹

Bamboos, evergreen

15.8 (2.2~22.1); 4.4 (-6.6~13.8); 26.5 (9.9~29.9); 15.9 (3.8~22.1); 132.4 (15.5~205.6); 2.2 (0.0~49.1); 857 (413~1182) ❙ 1324 (676~2680); 538 (363~1186); 132 (19~254) ❙ 54.4 (-12.2~246.5); 852 (413~1182); 209.2 (110.1~513.9)

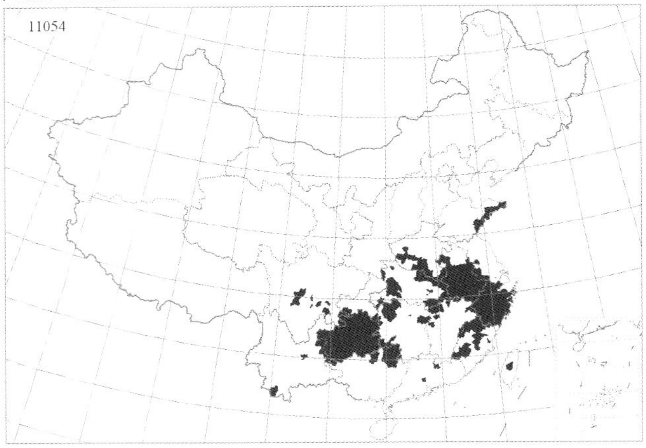

Poaceae 禾本科 *Phyllostachys* 刚竹属

Phyllostachys incarnata T. W. Wen 红壳雷竹

Bamboos, evergreen

15.9 (15.1~16.7); 4.9 (4.0~5.5); 26.7 (25.9~27.9); 16.0 (15.2~16.8); 132.1 (122.9~141.5); 0.2 (0.0~1.0); 849 (811~896) ❙ 1732 (1573~1893); 613 (555~677); 210 (190~227) ❙ 104.6 (75.7~129.9); 849 (811~896); 248.2 (224.8~269.3)

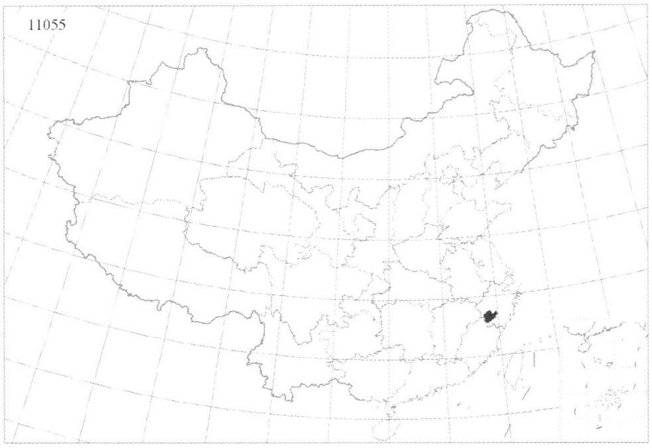

Poaceae 禾本科 *Phyllostachys* 刚竹属

Phyllostachys iridescens C. Y. Yao et S. Y. Chen 红哺鸡竹

Bamboos, evergreen

16.1; 3.6; 28.4; 16.2; 135.7; 1.7; 895 ❙ 1300; 515; 165 ❙ 45.7; 887; 188.5

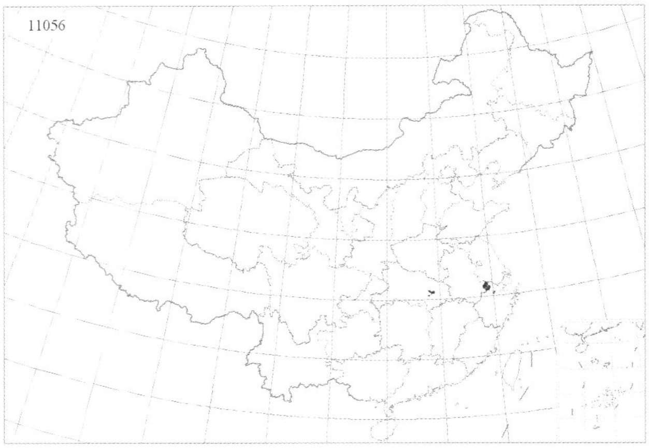

Poaceae 禾本科 *Phyllostachys* 刚竹属

Phyllostachys kwangsiensis W. Y. Hsiung, Q. H. Dai et J. K. Liu 假毛竹

Bamboos, evergreen

17.8 (14.3~21.0); 7.4 (3.3~11.9); 27.3 (24.3~29.0); 17.9 (14.3~21.1); 154.4 (114.1~193.0); 0.1 (0.0~2.3); 934 (765~1114) ❙ 1542 (1410~1829); 539 (403~821); 190 (142~231) ❙ 65.8 (46.4~104.6); 925 (765~1114); 220.3 (151.0~300.0)

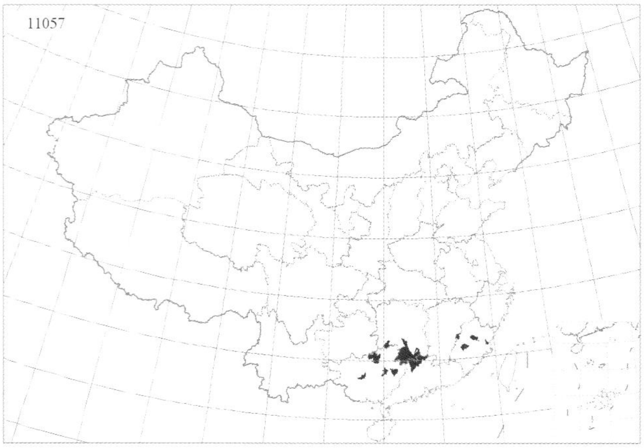

Poaceae 禾本科 *Phyllostachys* 刚竹属

Phyllostachys lofushanensis Z. P. Wang 大节刚竹

Bamboos, evergreen

21.6; 13.2; 28.2; 21.6; 199.2; 0.0; 1139 ❙ 1953; 858; 167 ❙ 71.5; 1139; 240.0

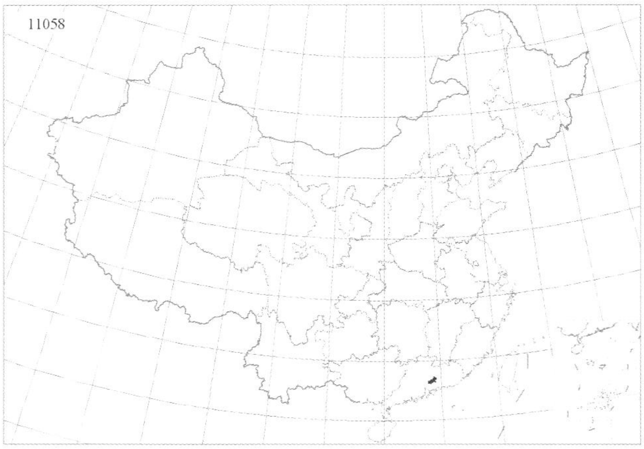

Poaceae 禾本科 *Phyllostachys* 刚竹属
Phyllostachys makinoi Hayata 台湾桂竹
Bamboos, evergreen

17.2 (15.0~24.1); 6.5 (2.6~18.9); 27.4 (20.1~29.3); 17.3 (15.0~24.1); 147.7 (123.9~229.7); 0.4 (0.0~3.5); 916 (773~1322) | 1677 (1211~2759); 621 (427~1594); 194 (47~308) | 84.6 (28.8~246.5); 902 (773~1095); 240.4 (154.2~303.4)

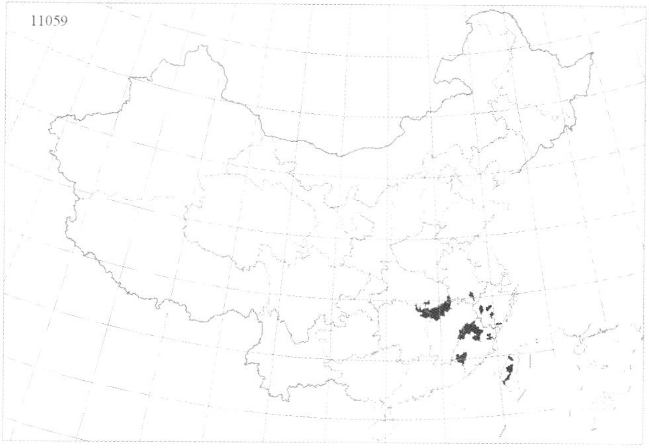

Poaceae 禾本科 *Phyllostachys* 刚竹属
Phyllostachys mannii Gamble 美竹
Bamboos, evergreen

15.6 (5.3~21.0); 5.0 (-2.4~14.9); 25.3 (12.2~29.9); 15.6 (5.7~21.0); 129.9 (30.5~191.8); 2.5 (0.0~26.2); 841 (479~1053) | 1293 (604~1950); 561 (329~1007); 114 (18~223) | 54.9 (-26.7~147.0); 827 (479~1053); 257.4 (111.9~627.0)

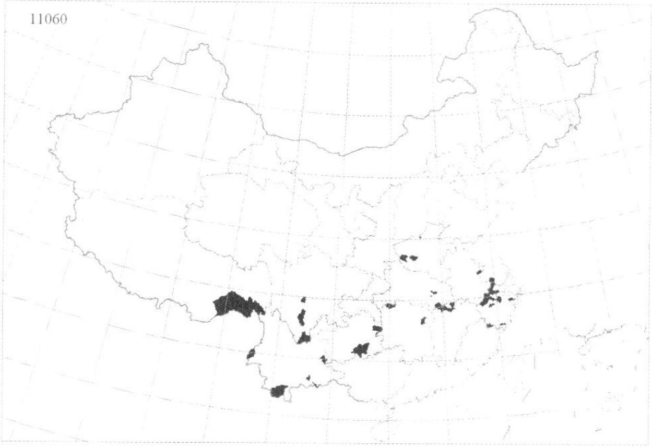

Poaceae 禾本科 *Phyllostachys* 刚竹属
Phyllostachys meyeri McClure 毛环竹
Bamboos, evergreen

16.8 (14.9~21.0); 5.7 (2.3~11.9); 27.2 (22.4~29.9); 16.8 (15.0~21.1); 142.6 (119.9~193.0); 0.5 (0.0~3.9); 894 (781~1114) | 1449 (1002~1997); 540 (403~860); 166 (48~254) | 62.3 (14.0~130.7); 884 (781~1114); 220.9 (111.9~413.5)

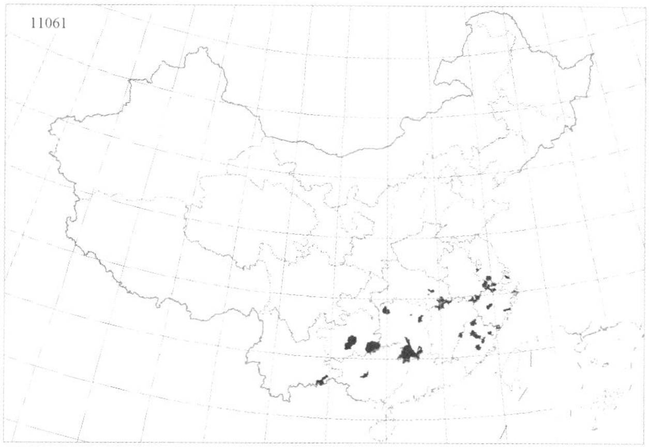

Poaceae 禾本科 *Phyllostachys* 刚竹属
Phyllostachys nidularia Munro 篌竹
Bamboos, evergreen

16.0 (6.0~22.4); 4.9 (-3.2~14.8); 26.3 (14.4~29.4); 16.0 (6.6~22.4); 134.5 (39.7~208.9); 2.5 (0.0~26.7); 862 (508~1191) | 1211 (626~2037); 526 (316~1033); 93 (17~230) | 40.6 (-22.2~128.9); 856 (508~1191); 196.3 (108.2~352.3)

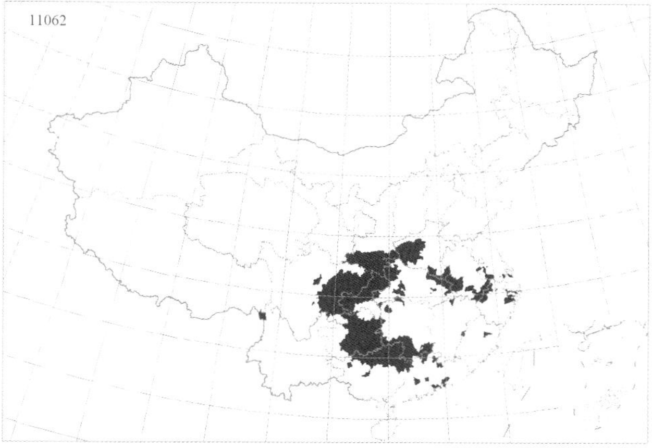

Poaceae 禾本科 *Phyllostachys* 刚竹属
Phyllostachys nigella T. W. Wen 富阳乌哺鸡竹
Bamboos, evergreen

16.0; 4.0; 28.0; 16.0; 133.3; 1.0; 875 | 1396; 508; 194 | 59.5; 875; 243.2

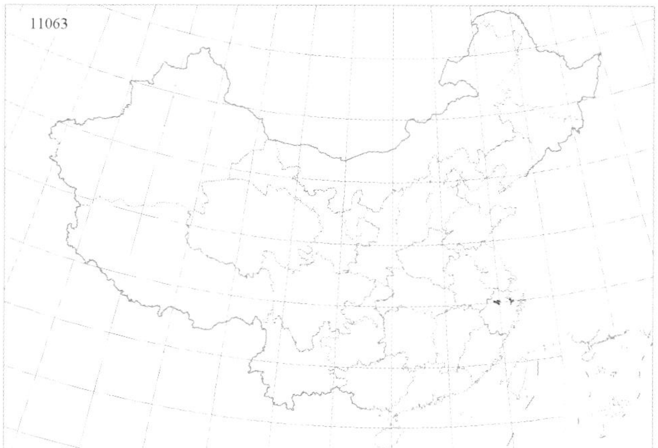

Poaceae 禾本科 *Phyllostachys* 刚竹属
Phyllostachys nigra (Lodd. ex Lindl.) Munro 紫竹
Bamboos, evergreen

15.3 (-1.7~22.4); 5.0 (-12.3~17.7); 24.6 (8.0~29.4); 15.4 (2.4~22.4); 128.3 (6.4~208.9); 4.4 (0.0~86.1); 835 (344~1191) | 1263 (502~2992); 562 (301~1847); 99 (8~402) | 53.0 (-23.1~246.5); 830 (344~1191); 221.4 (108.2~570.1)

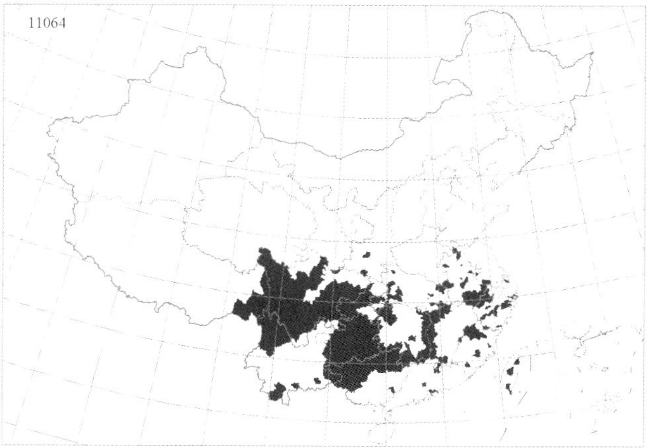

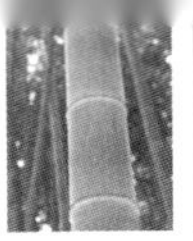

Poaceae 禾本科 *Phyllostachys* 刚竹属

Phyllostachys nigra var. henonis (Mitford) Stapf ex Rendle 毛金竹

Bamboos, evergreen

15.6 (2.2~21.0)；6.1 (-7.0~15.1)；24.1 (9.9~29.3)；15.6 (3.8~21.0)；129.2 (15.5~191.8)；1.8 (0.0~49.1)；825 (413~1053) ❙ 1215 (731~1950)；577 (346~996)；81 (12~246) ❙ 48.6 (-14.2~147.0)；818 (413~1053)；245.1 (108.2~659.0)

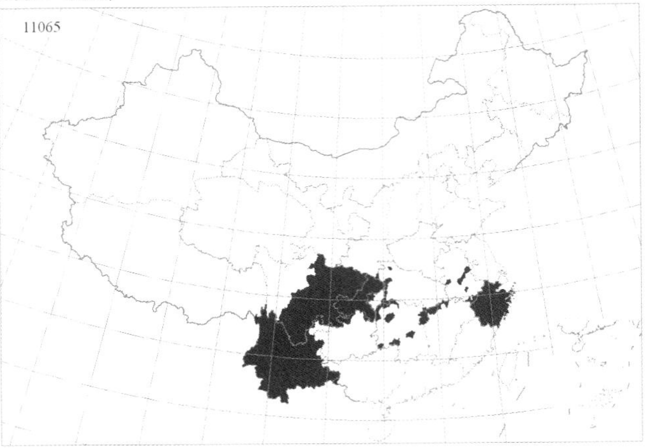

Poaceae 禾本科 *Phyllostachys* 刚竹属

Phyllostachys nuda McClure 灰竹

Bamboos, evergreen

16.9 (9.3~24.1)；5.8 (-1.7~18.9)；27.5 (20.1~29.4)；17.0 (9.5~24.1)；144.5 (69.8~229.7)；1.1 (0.0~18.0)；910 (603~1322) ❙ 1477 (1030~2759)；560 (403~1594)；174 (47~231) ❙ 63.1 (16.8~202.5)；899 (603~1114)；215.2 (111.9~290.3)

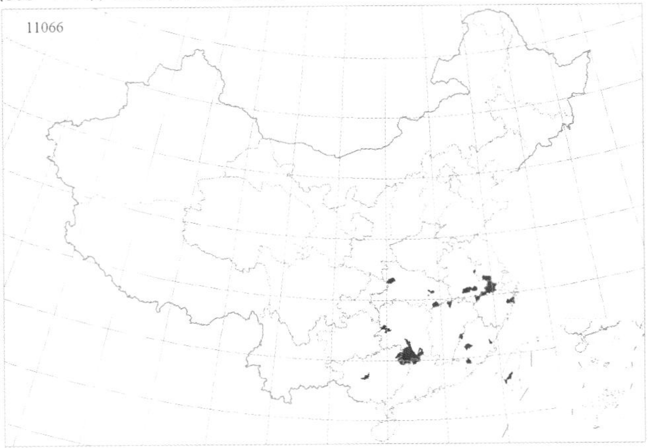

Poaceae 禾本科 *Phyllostachys* 刚竹属

Phyllostachys parvifolia C. D. Chu et H. Y. Chou 安吉金竹

Bamboos, evergreen

15.9；4.0；27.9；16.0；132.9；1.3；873 ❙ 1350；515；182 ❙ 54.5；871；230.5

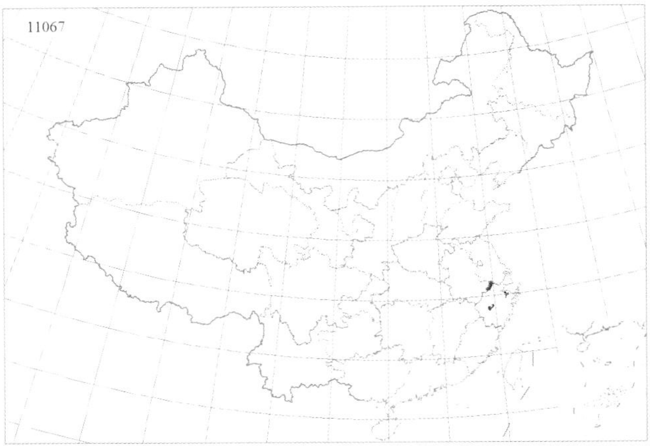

Poaceae 禾本科 *Phyllostachys* 刚竹属

Phyllostachys platyglossa C. P. Wang et Z. H. Yu 灰水竹

Bamboos, evergreen

16.2；3.6；28.5；16.2；136.7；1.8；901 ❙ 1283；522；150 ❙ 43.1；885；182.3

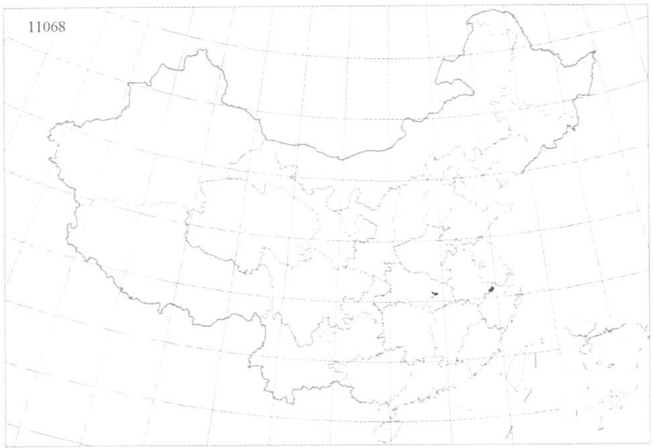

Poaceae 禾本科 *Phyllostachys* 刚竹属

Phyllostachys prominens W. Y. Xiong 高节竹

Bamboos, evergreen

16.5 (16.1~17.1)；4.5 (3.9~5.7)；28.5 (27.5~29.4)；16.5 (16.1~17.1)；138.9 (134.6~146.1)；0.7 (0.0~1.1)；904 (876~952) ❙ 1376 (1228~1632)；512 (459~590)；181 (123~201) ❙ 52.6 (29.7~86.3)；899 (876~921)；200.2 (111.9~243.3)

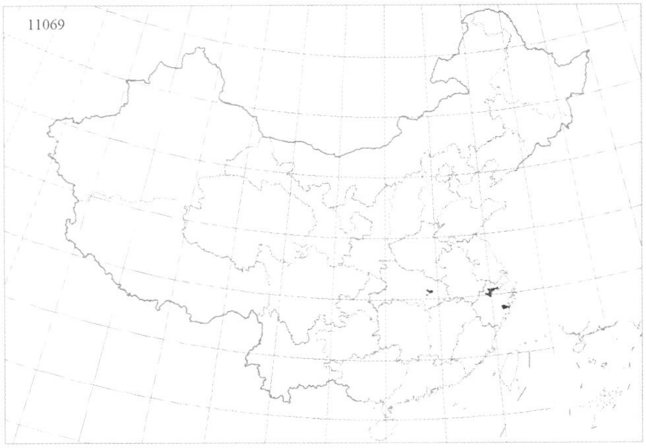

Poaceae 禾本科 *Phyllostachys* 刚竹属

Phyllostachys propinqua McClure 早园竹

Bamboos, evergreen

16.8 (12.5~22.0)；6.4 (0.5~14.2)；26.2 (20.1~29.9)；16.9 (12.5~22.0)；143.7 (92.0~203.8)；1.1 (0.0~9.2)；897 (685~1170) ❙ 1333 (687~1876)；582 (393~891)；111 (45~205) ❙ 49.2 (-23.1~87.6)；883 (685~1170)；208.7 (111.9~305.7)

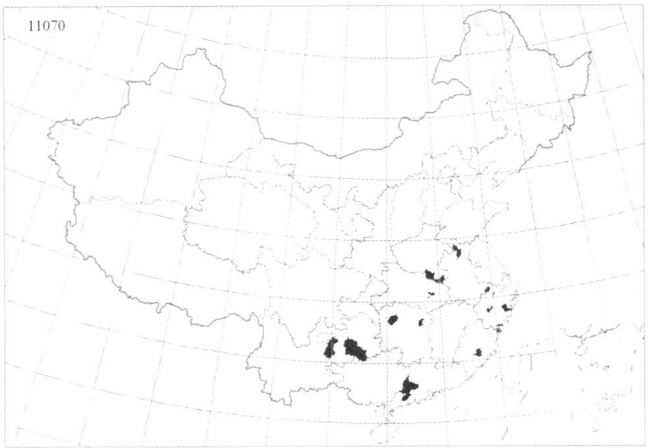

Poaceae 禾本科 *Phyllostachys* 刚竹属

Phyllostachys reticulata (Ruprecht) K. Koch 桂竹

Bamboos, evergreen

15.2 (5.4~23.6); 3.6 (-3.6~16.7); 25.9 (12.3~29.9); 15.2 (5.8~23.6); 127.3 (31.5~223.5); 4.8 (0.0~26.7); 840 (481~1281) ❙ 1205 (533~2680); 523 (316~1412); 110 (15~308) ❙ 42.9 (-39.0~246.5); 816 (481~1269); 213.9 (111.9~551.6)

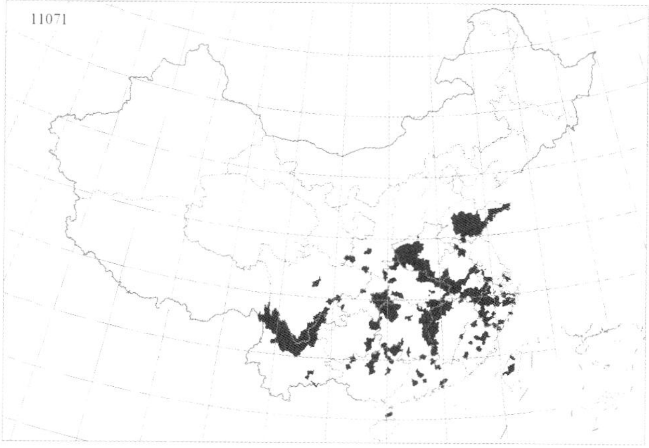

Poaceae 禾本科 *Phyllostachys* 刚竹属

Phyllostachys rivalis H. R. Zhao et A. T. Liu 河竹

Bamboos, evergreen

17.9 (16.6~19.1); 7.8 (6.0~10.0); 27.1 (26.3~28.0); 18.0 (16.6~19.1); 155.8 (138.9~169.1); 0.0 (0.0~0.0); 935 (866~992) ❙ 1801 (1646~1997); 621 (587~669); 218 (191~254) ❙ 93.6 (66.0~130.7); 935 (866~992); 266.1 (216.9~303.4)

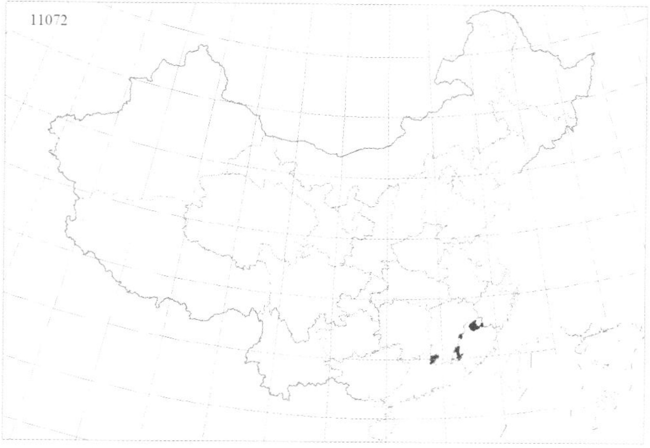

Poaceae 禾本科 *Phyllostachys* 刚竹属

Phyllostachys robustiramea S. Y. Chen et C. Y. Yao 芽竹

Bamboos, evergreen

16.9 (15.3~18.2); 5.2 (3.1~8.3); 28.1 (26.9~29.6); 16.9 (15.4~18.2); 143.7 (127.2~158.4); 0.7 (0.0~2.7); 912 (850~959) ❙ 1433 (1235~1821); 517 (444~614); 171 (123~221) ❙ 57.2 (29.7~92.7); 897 (850~945); 208.3 (111.9~332.5)

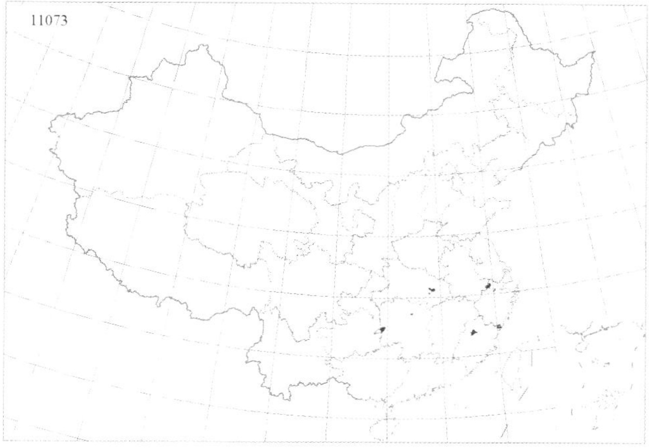

Poaceae 禾本科 *Phyllostachys* 刚竹属

Phyllostachys rubicunda T. W. Wen 红后竹

Bamboos, evergreen

17.1 (15.0~19.5); 6.6 (2.6~11.0); 27.4 (25.2~28.9); 17.2 (15.0~19.5); 146.5 (123.9~174.1); 0.5 (0.0~3.5); 905 (829~991) ❙ 1546 (1296~1821); 555 (478~642); 192 (175~221) ❙ 70.7 (46.2~95.5); 905 (829~991); 259.1 (189.1~332.5)

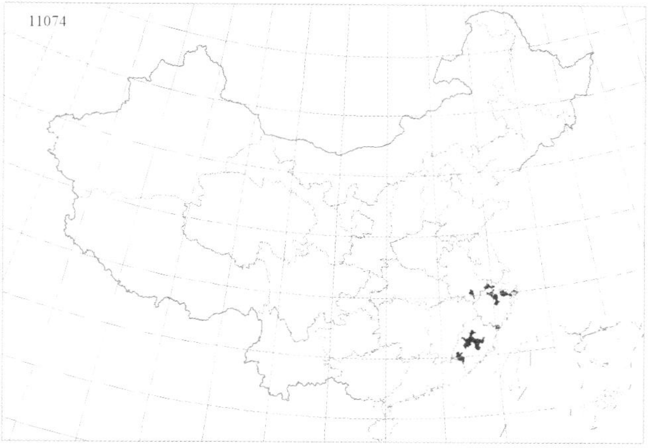

Poaceae 禾本科 *Phyllostachys* 刚竹属

Phyllostachys rubromarginata McClure 红边竹

Bamboos, evergreen

16.8 (14.3~22.1); 5.6 (0.5~13.7); 26.9 (23.5~29.0); 16.9 (14.3~22.1); 144.6 (112.6~205.6); 2.0 (0.0~9.2); 909 (756~1182) ❙ 1248 (730~1991); 543 (412~902); 115 (49~204) ❙ 37.3 (-16.9~79.9); 887 (730~1182); 212.1 (139.2~268.6)

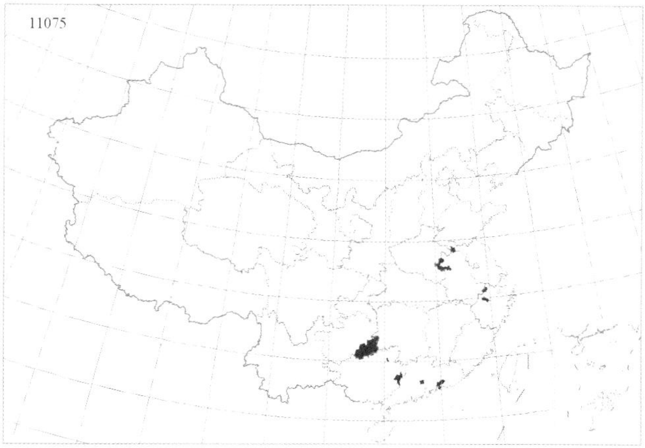

Poaceae 禾本科 *Phyllostachys* 刚竹属

Phyllostachys rutila T. W. Wen 衢县红壳竹

Bamboos, evergreen

16.7; 4.7; 28.5; 16.8; 141.7; 0.4; 912 ❙ 1567; 515; 220 ❙ 71.9; 893; 212.1

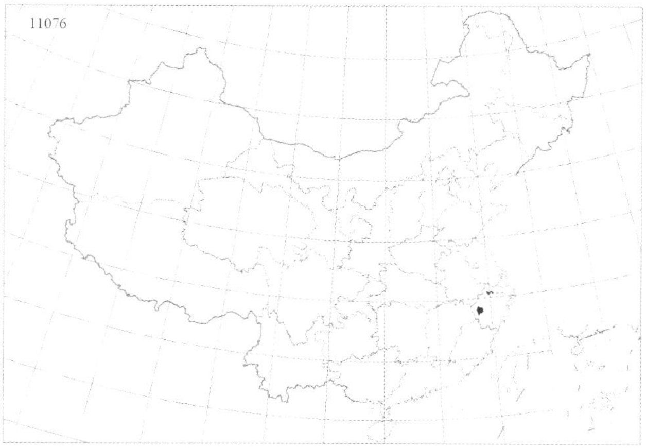

Poaceae 禾本科 *Phyllostachys* 刚竹属

Phyllostachys stimulosa H. R. Zhao et A. T. Liu 漫竹

Bamboos, evergreen

15.9 (14.7~18.2); 4.3 (2.5~8.3); 27.3 (24.7~29.4); 15.9 (14.8~18.2); 132.7 (120.8~158.4); 1.3 (0.0~3.9); 864 (789~952) ❙ 1540 (1200~1950); 566 (497~713); 187 (105~246) ❙ 78.9 (29.7~147.0); 862 (789~945); 245.3 (111.9~332.5)

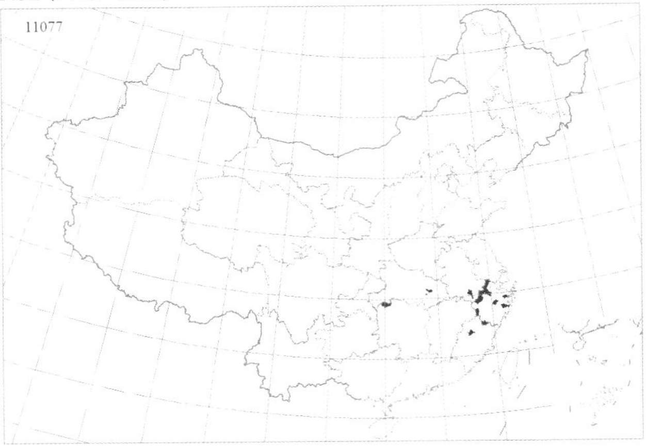

Poaceae 禾本科 *Phyllostachys* 刚竹属

Phyllostachys sulphurea (Carrière) Rivière et C. Rivière 金竹

Bamboos, evergreen

16.8 (9.3~21.7); 5.9 (−1.7~14.0); 27.0 (20.1~29.4); 16.8 (9.5~21.7); 142.5 (69.8~200.5); 0.9 (0.0~18.0); 893 (603~1140) ❙ 1500 (1002~1959); 541 (403~821); 177 (58~250) ❙ 69.2 (14.0~137.5); 887 (603~1140); 223.9 (111.9~318.5)

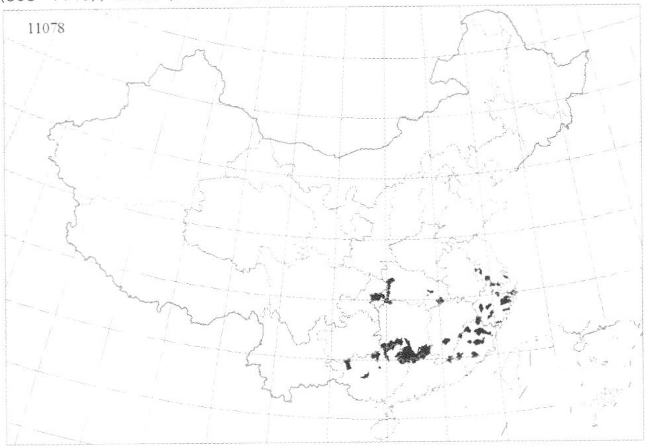

Poaceae 禾本科 *Phyllostachys* 刚竹属

Phyllostachys sulphurea var. viridis R. A. Young 刚竹

Bamboos, evergreen

15.8; 3.9; 27.2; 15.9; 131.5; 1.1; 861 ❙ 1801; 598; 242 ❙ 109.2; 861; 245.8

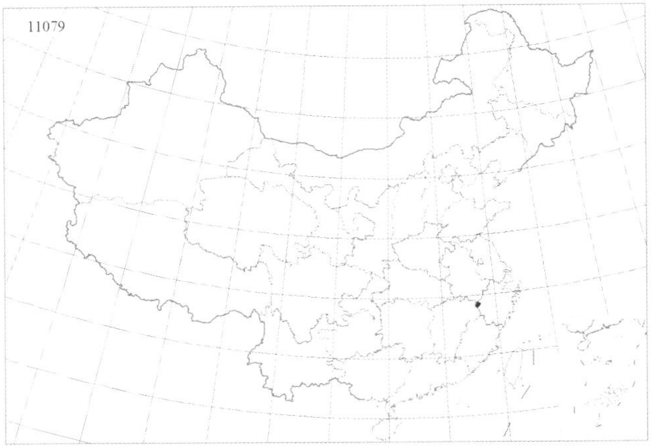

Poaceae 禾本科 *Phyllostachys* 刚竹属

Phyllostachys tianmuensis Z. P. Wang et N. X. Ma 天目早竹

Bamboos, evergreen

15.3; 3.1; 27.6; 15.4; 127.2; 2.7; 850 ❙ 1331; 546; 177 ❙ 56.6; 850; 252.7

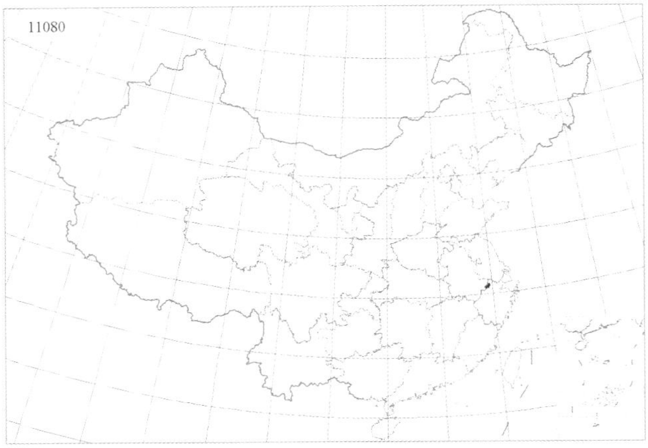

Poaceae 禾本科 *Phyllostachys* 刚竹属

Phyllostachys varioauriculata S. C. Li et S. H. Wu 乌竹

Bamboos, evergreen

16.5; 5.0; 27.8; 16.6; 139.5; 0.8; 895 ❙ 1528; 557; 186 ❙ 70.3; 895; 216.4

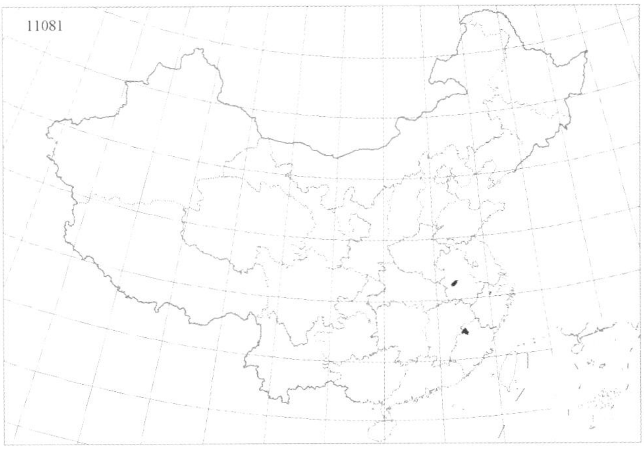

Poaceae 禾本科 *Phyllostachys* 刚竹属

Phyllostachys verrucosa G. H. Ye et Z. P. Wang 长沙刚竹

Bamboos, evergreen

17.5; 4.6; 29.7; 17.6; 151.2; 0.4; 966 ❙ 1414; 440; 197 ❙ 46.4; 879; 157.4

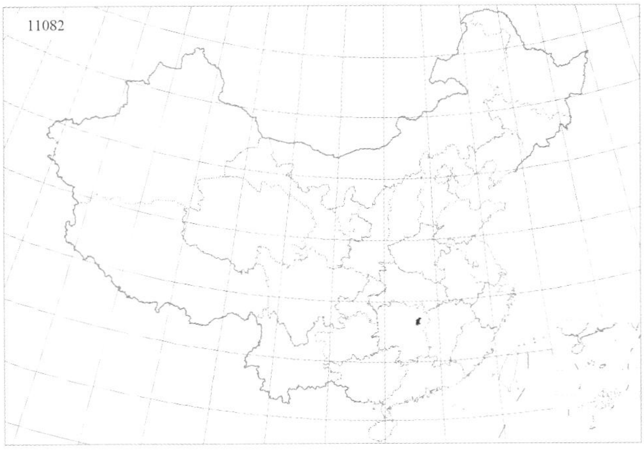

Poaceae 禾本科 *Phyllostachys* 刚竹属

Phyllostachys violascens (Carrière) Rivière et C. Rivière 早竹

Bamboos, evergreen

15.9 (15.0~18.0); 3.7 (1.8~9.2); 27.9 (25.9~29.6); 16.0 (15.1~18.1); 133.4 (122.9~156.7); 1.9 (0.0~5.0); 878 (811~961) ❙ 1272 (963~1922); 499 (419~677); 161 (102~259) ❙ 45.0 (10.3~129.9); 874 (811~961); 190.6 (111.9~303.5)

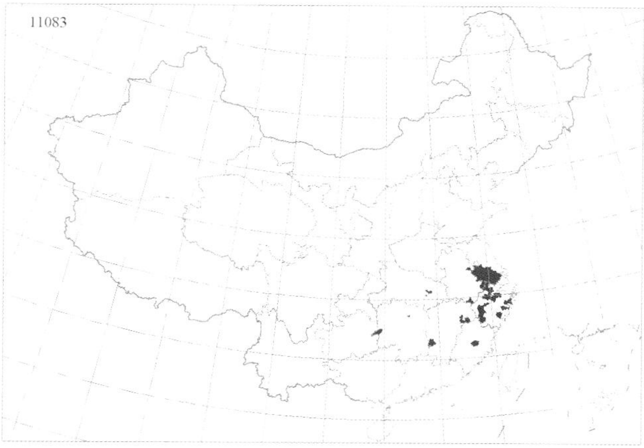

Poaceae 禾本科 *Phyllostachys* 刚竹属

Phyllostachys virella T. W. Wen 东阳青皮竹

Bamboos, evergreen

15.9; 4.7; 27.0; 15.9; 131.4; 0.3; 852 ❙ 1652; 589; 200 ❙ 95.3; 852; 232.1

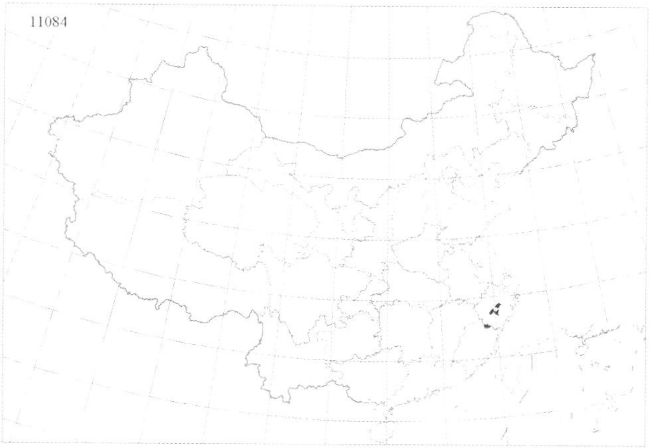

Poaceae 禾本科 *Phyllostachys* 刚竹属

Phyllostachys viridiglaucescens (Carrière) Rivière et C. Rivière 粉绿竹

Bamboos, evergreen

16.7 (14.3~24.2); 5.1 (2.1~18.1); 27.9 (24.2~29.3); 16.8 (14.3~24.3); 142.1 (113.2~231.2); 0.8 (0.0~4.6); 907 (761~1335) ❙ 1470 (1002~1742); 513 (427~685); 183 (59~228) ❙ 63.3 (-12.0~128.9); 898 (761~1139); 222.8 (134.1~411.8)

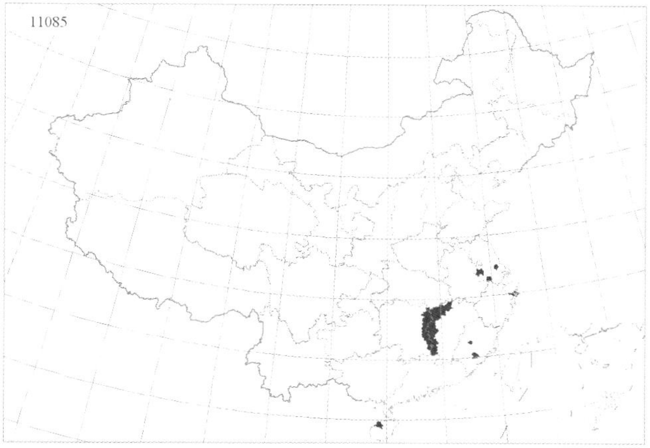

Poaceae 禾本科 *Phyllostachys* 刚竹属

Phyllostachys vivax McClure 乌哺鸡竹

Bamboos, evergreen

16.1 (14.7~22.1); 3.6 (0.5~13.7); 28.1 (26.8~29.4); 16.1 (14.8~22.1); 136.2 (121.1~205.6); 2.8 (0.0~9.2); 894 (815~1182) ❙ 1259 (687~1784); 509 (393~779); 150 (49~220) ❙ 41.0 (-23.1~100.0); 866 (687~1182); 194.0 (111.9~284.0)

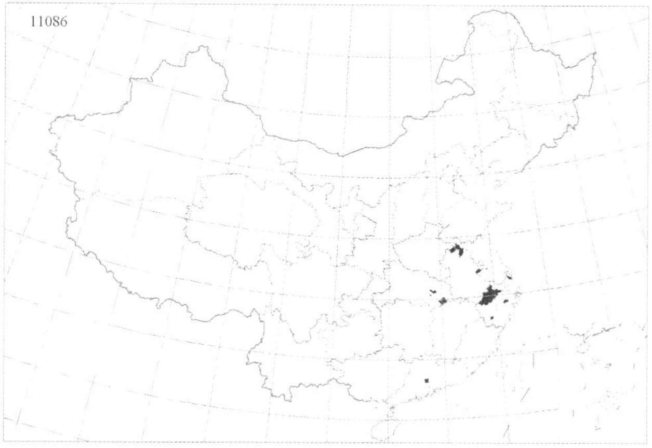

Poaceae 禾本科 *Phyllostachys* 刚竹属

Phyllostachys yunhoensis S. Y. Chen et C. Y. Yao 云和哺鸡竹

Bamboos, evergreen

16.2; 5.3; 26.8; 16.2; 134.8; 0.0; 857 ❙ 1715; 617; 201 ❙ 100.0; 857; 253.8

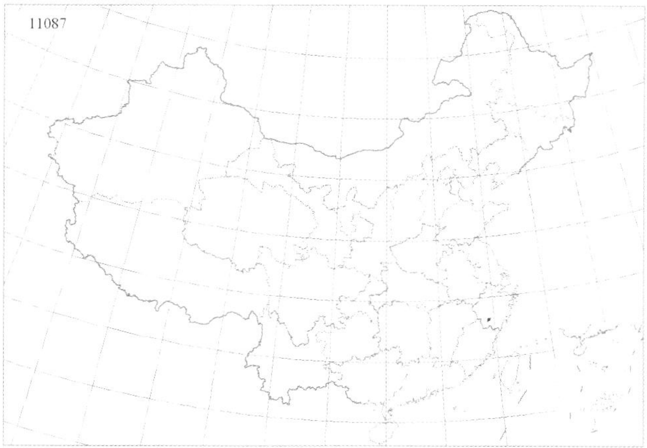

Poaceae 禾本科 *Pleioblastus* 苦竹属

Pleioblastus altiligulatus S. L. Chen et S. Y. Chen 高舌苦竹

Bamboos, evergreen

16.7; 6.1; 26.5; 16.8; 141.2; 0.1; 881 ❙ 1620; 574; 190 ❙ 85.8; 880; 232.5

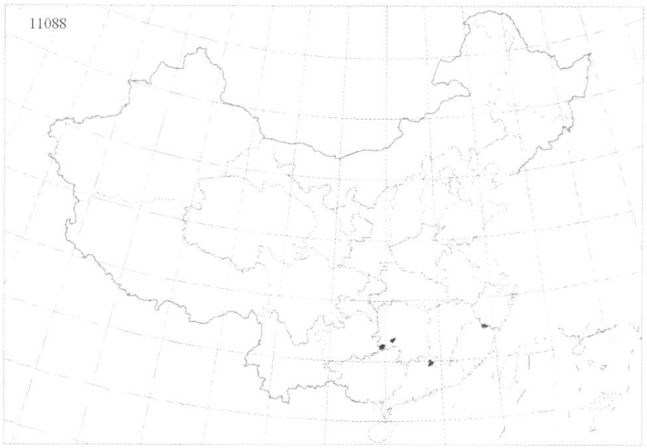

Poaceae 禾本科 *Pleioblastus* 苦竹属

Pleioblastus amarus (Keng) Keng f. 苦竹

Bamboos, evergreen

15.8 (9.4~22.7); 3.8 (−2.9~15.1); 27.2 (21.0~29.7); 15.8 (9.8~22.8); 133.5 (73.9~213.1); 3.7 (0.0~20.6); 869 (619~1221) ❙ 1287 (551~2435); 507 (245~1428); 140 (19~254) ❙ 47.2 (−36.5~147.0); 841 (551~1221); 214.1 (111.9~659.0)

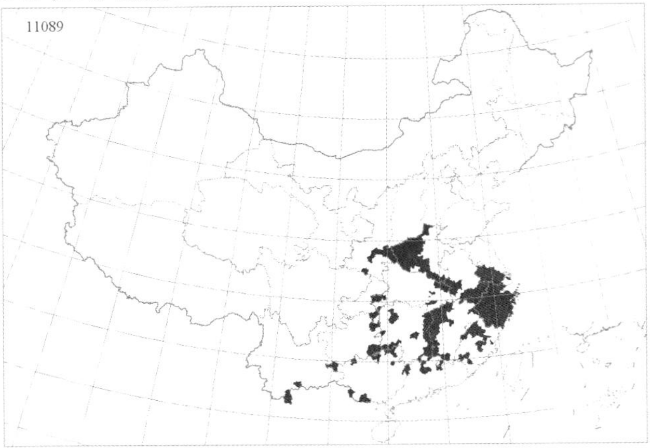

Poaceae 禾本科 *Pleioblastus* 苦竹属

Pleioblastus amarus var. hangzhouensis S. L. Chen et S. Y. Chen 杭州苦竹

Bamboos, evergreen

16.5; 4.4; 28.9; 16.6; 139.8; 0.6; 910 ❙ 1377; 493; 197 ❙ 51.2; 910; 189.1

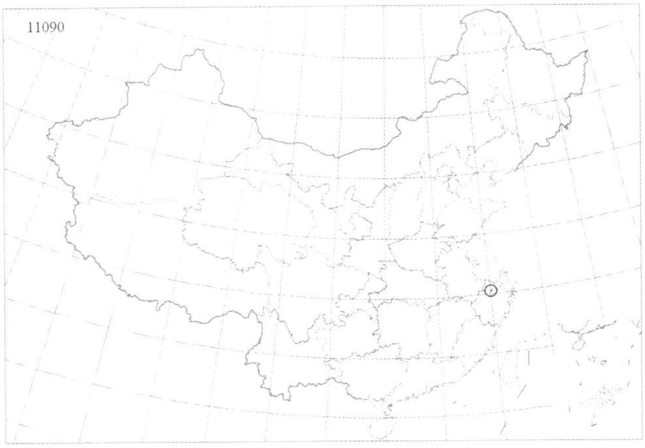

Poaceae 禾本科 *Pleioblastus* 苦竹属

Pleioblastus amarus var. pendulifolius S. Y. Chen 垂枝苦竹

Bamboos, evergreen

16.5; 4.4; 28.9; 16.6; 139.8; 0.6; 910 ❙ 1377; 493; 197 ❙ 51.2; 910; 189.1

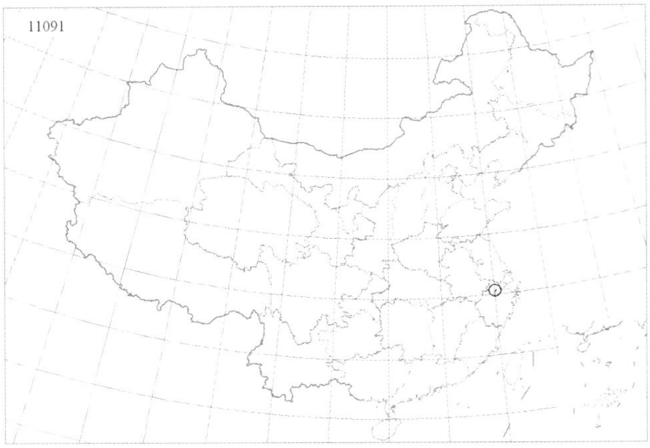

Poaceae 禾本科 *Pleioblastus* 苦竹属

Pleioblastus amarus var. tubatus T. H. Wen 胖苦竹

Bamboos, evergreen

17.5; 7.6; 27.1; 17.6; 151.2; 0.3; 915 ❙ 1639; 580; 199 ❙ 79.1; 915; 294.6

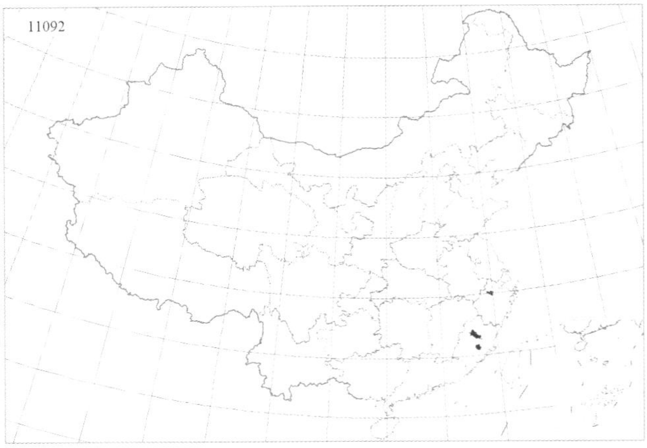

Poaceae 禾本科 *Pleioblastus* 苦竹属

Pleioblastus hsienchuensis T. H. Wen 仙居苦竹

Bamboos, evergreen

17.6 (15.1~20.5); 7.9 (4.0~12.0); 27.1 (25.2~28.9); 17.7 (15.2~20.5); 152.2 (122.9~186.5); 0.1 (0.0~1.0); 920 (811~1073) ❙ 1668 (1244~1997); 583 (411~669); 203 (163~254) ❙ 83.7 (15.9~130.7); 915 (811~1012); 271.1 (206.9~332.5)

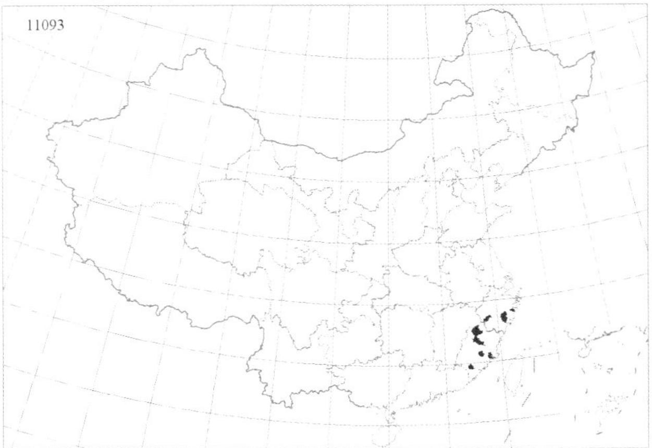

Poaceae 禾本科 *Pleioblastus* 苦竹属

Pleioblastus hsienchuensis var. subglabratus (S. Y. Chen) C. S. Chao et G. Y. Yang 光箨苦竹

Bamboos, evergreen

16.0; 5.4; 26.2; 16.1; 133.7; 0.8; 848 ❙ 1855; 653; 227 ❙ 119.0; 848; 281.6

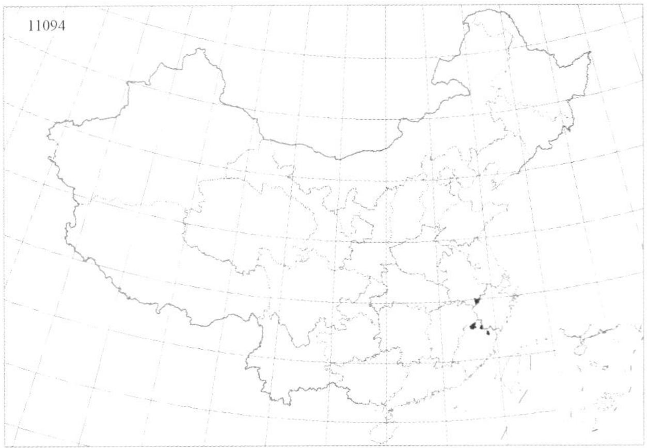

Poaceae 禾本科 *Pleioblastus* 苦竹属
Pleioblastus incarnatus S. L. Chen et G. Y. Sheng 绿苦竹
Bamboos, evergreen

16.5; 6.4; 26.2; 16.6; 138.6; 0.0; 862 ❙ 1901; 678; 227 ❙ 121.2; 862; 289.5

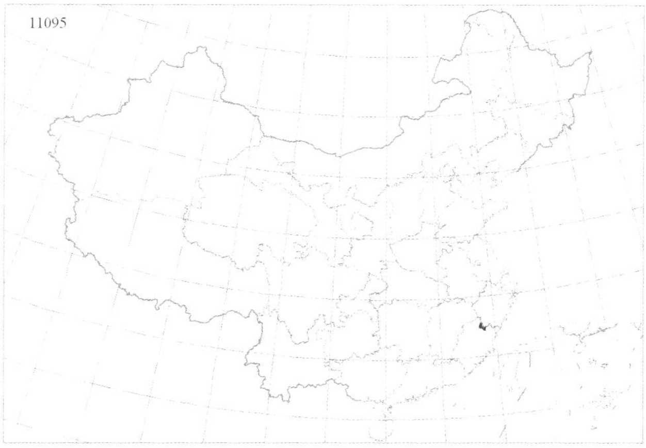

Poaceae 禾本科 *Pleioblastus* 苦竹属
Pleioblastus intermedius S. Y. Chen 华丝竹
Bamboos, evergreen

16.0 (15.0~17.3); 5.5 (4.0~9.0); 26.2 (24.7~28.9); 16.1 (15.1~17.3); 133.0 (120.8~147.6); 0.2 (0.0~1.0); 844 (789~910) ❙ 1767 (1377~1997); 631 (493~713); 212 (175~254) ❙ 110.4 (51.2~147.0); 844 (789~910); 262.9 (189.1~305.7)

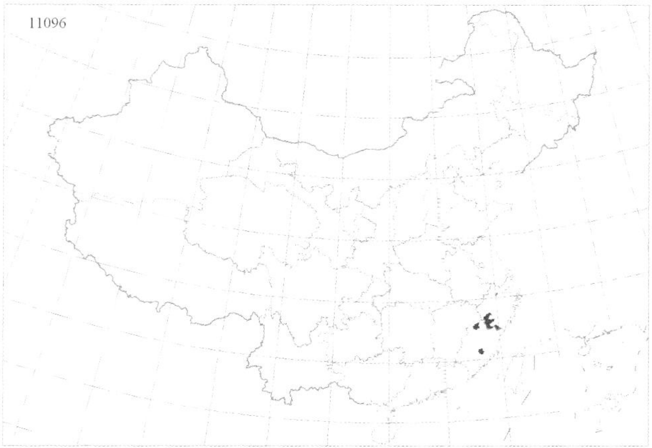

Poaceae 禾本科 *Pleioblastus* 苦竹属
Pleioblastus juxianensis T. H. Wen, C. Y. Yao et S. Y. Chen 衢县苦竹
Bamboos, evergreen

17.3 (16.2~18.6); 6.8 (4.4~9.3); 27.4 (25.2~29.2); 17.4 (16.3~18.7); 148.3 (136.1~164.0); 0.1 (0.0~0.6); 913 (864~965) ❙ 1720 (1588~1997); 574 (500~669); 220 (175~254) ❙ 88.9 (71.4~130.7); 902 (864~965); 268.3 (183.5~332.5)

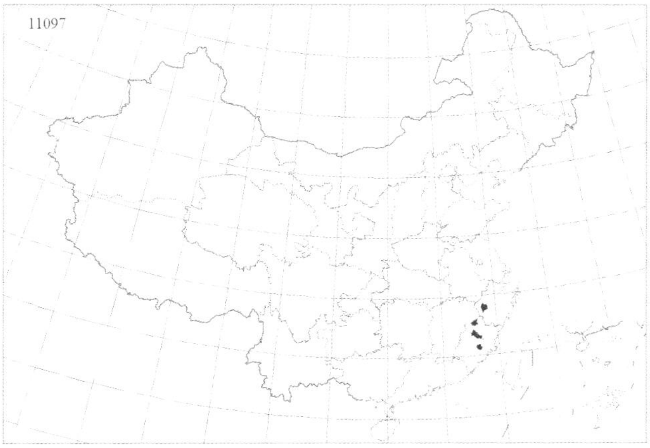

Poaceae 禾本科 *Pleioblastus* 苦竹属
Pleioblastus maculatus (McClure) C. D. Chu et C. S. Chao 斑苦竹
Bamboos, evergreen

16.8 (12.7~22.7); 6.0 (1.6~15.0); 26.8 (22.3~29.9); 16.9 (12.8~22.8); 143.0 (99.0~213.1); 0.4 (0.0~6.5); 895 (713~1221) ❙ 1471 (950~2435); 568 (427~1428); 152 (27~259) ❙ 64.3 (12.1~130.7); 890 (713~1221); 219.7 (111.9~318.8)

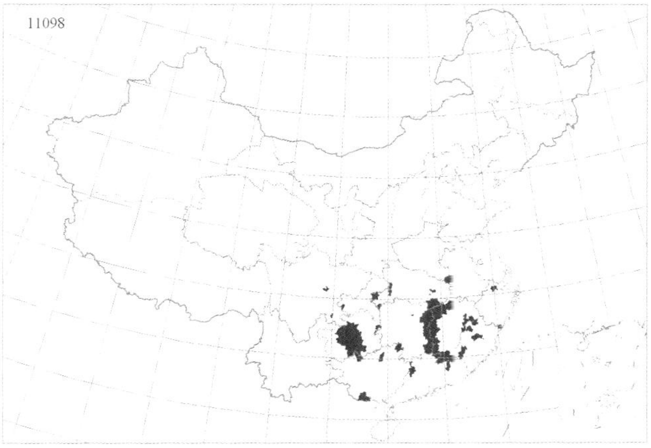

Poaceae 禾本科 *Pleioblastus* 苦竹属
Pleioblastus maculosoides T. H. Wen 丽水苦竹
Bamboos, evergreen

16.7; 5.5; 27.9; 16.8; 141.5; 0.0; 896 ❙ 1573; 555; 190 ❙ 75.7; 896; 224.8

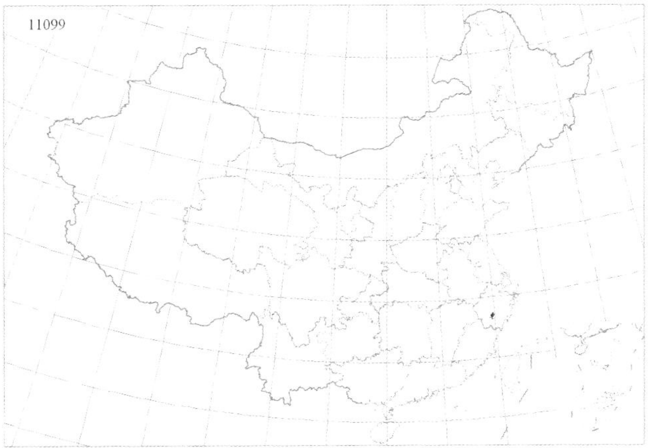

Poaceae 禾本科 *Pleioblastus* 苦竹属
Pleioblastus oleosus T. H. Wen 油苦竹
Bamboos, evergreen

17.1 (15.1~18.6); 6.9 (3.1~10.4); 26.9 (25.2~28.7); 17.2 (15.2~18.7); 146.6 (122.9~164.0); 0.3 (0.0~2.7); 898 (811~965) ❙ 1713 (1331~1997); 598 (518~669); 211 (163~259) ❙ 91.1 (56.6~130.7); 898 (811~965); 271.0 (227.8~305.7)

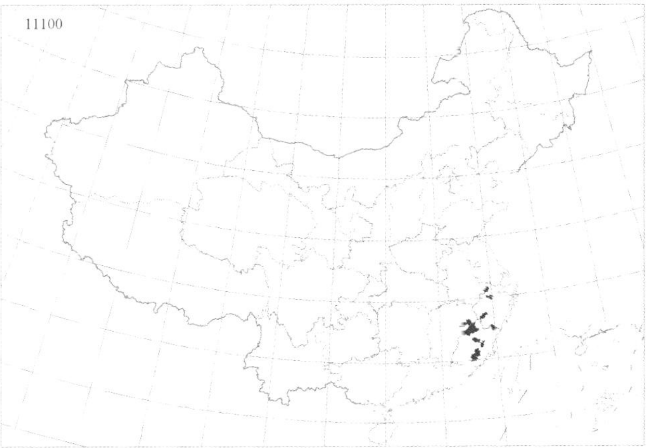

Poaceae 禾本科 *Pleioblastus* 苦竹属

Pleioblastus rugatus T. H. Wen et S. Y. Chen 皱苦竹

Bamboos, evergreen

17.2；7.1；27.3；17.3；147.0；0.0；901 ❙ 1650；590；182 ❙ 83.1；901；232.3

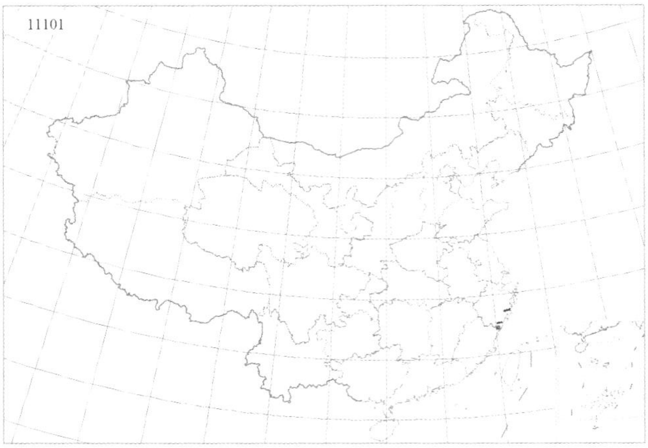

Poaceae 禾本科 *Pleioblastus* 苦竹属

Pleioblastus sanmingensis S. L. Chen et G. Y. Sheng 三明苦竹

Bamboos, evergreen

18.8；10.2；26.6；18.8；165.6；0.0；953 ❙ 1602；556；173 ❙ 68.2；953；279.8

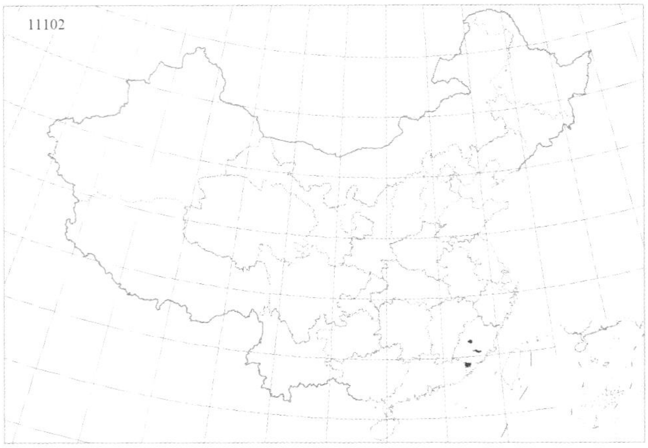

Poaceae 禾本科 *Pleioblastus* 苦竹属

Pleioblastus solidus S. Y. Chen 实心苦竹

Bamboos, evergreen

17.2 (15.8~18.6)；6.6 (3.5~9.3)；27.3 (26.8~27.9)；17.2 (15.8~18.7)；146.8 (131.3~164.0)；0.3 (0.0~1.7)；907 (857~965) ❙ 1707 (1467~1829)；600 (587~617)；201 (153~230) ❙ 88.3 (68.0~104.6)；907 (857~965)；267.7 (181.5~332.5)

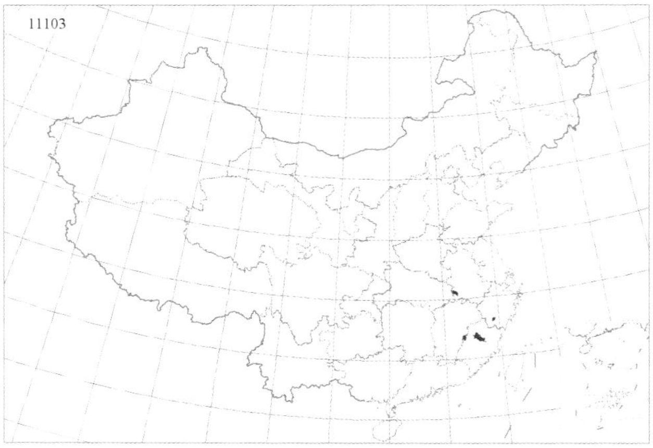

Poaceae 禾本科 *Pleioblastus* 苦竹属

Pleioblastus truncatus T. H. Wen 尖子竹

Bamboos, evergreen

16.4；4.9；28.1；16.5；138.0；0.3；891 ❙ 1389；489；188 ❙ 56.1；891；216.4

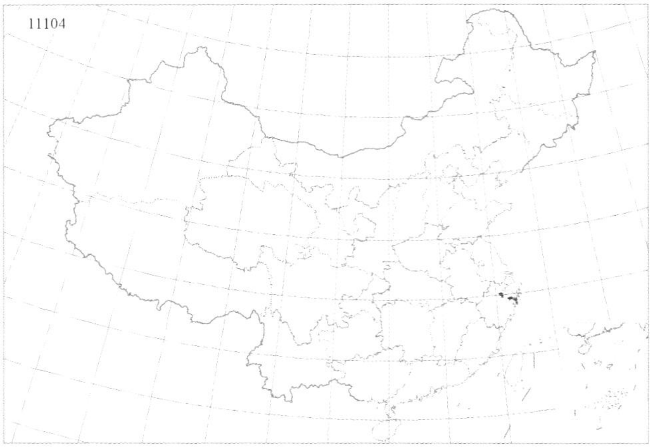

Poaceae 禾本科 *Pleioblastus* 苦竹属

Pleioblastus wuyishanensis Q. F. Zheng et K. F. Huang 武夷山苦竹

Bamboos, evergreen

16.8；6.0；26.9；16.9；142.3；0.0；888 ❙ 1959；646；257 ❙ 121.0；888；291.5

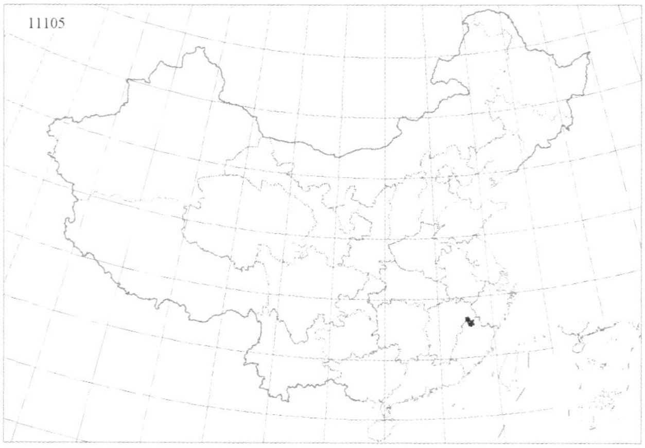

Poaceae 禾本科 *Pleioblastus* 苦竹属

Pleioblastus yixingensis S. L. Chen et S. Y. Chen 宜兴苦竹

Bamboos, evergreen

17.2 (15.0~19.1)；6.9 (2.7~10.2)；27.0 (25.2~28.9)；17.3 (15.0~19.1)；147.8 (123.7~169.1)；0.6 (0.0~3.2)；904 (826~973) ❙ 1657 (1117~1997)；588 (478~669)；203 (152~254) ❙ 83.3 (28.5~130.7)；904 (826~973)；270.3 (179.2~332.5)

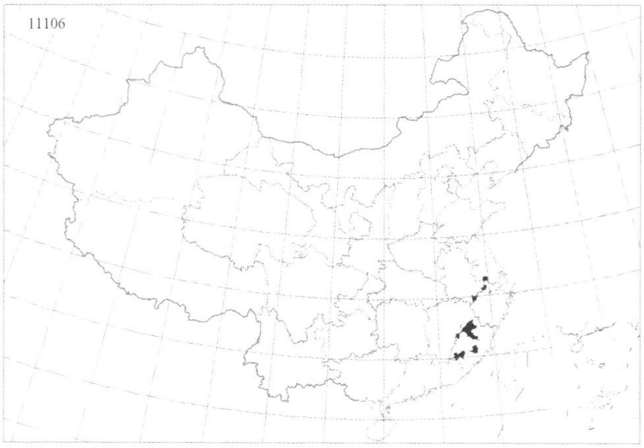

Poaceae 禾本科 *Pseudosasa* 矢竹属

Pseudosasa acutivagina T. H. Wen et S. C. Chen 尖箨茶秆竹

Bamboos, evergreen

15.0；5.0；24.7；15.1；120.8；0.0；789 ▎1950；713；223 ▎147.0；789；283.9

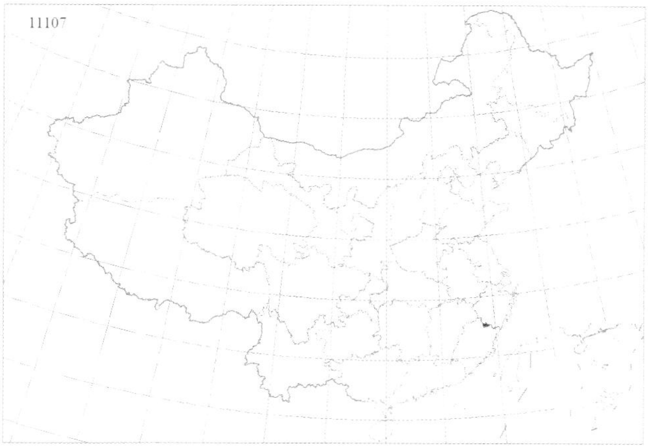

Poaceae 禾本科 *Pseudosasa* 矢竹属

Pseudosasa aeria T. H. Wen 空心竹

Bamboos, evergreen

15.8 (14.3~17.6)；5.4 (4.0~7.4)；25.4 (23.5~27.6)；15.8 (14.3~17.6)；130.6 (112.6~151.2)；0.4 (0.0~1.1)；830 (756~919) ▎1336 (1218~1674)；529 (476~611)；117 (97~179) ▎60.7 (47.0~82.2)；830 (756~919)；228.3 (216.1~246.1)

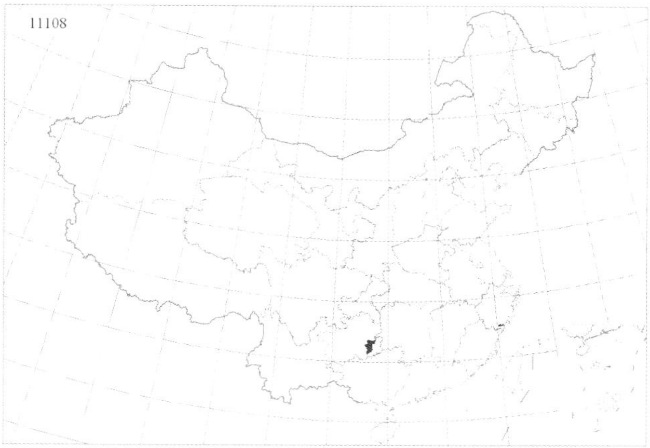

Poaceae 禾本科 *Pseudosasa* 矢竹属

Pseudosasa amabilis (McClure) P. C. Keng ex S. L. Chen et al. 茶秆竹

Bamboos, evergreen

18.4 (14.3~22.1)；8.4 (3.3~13.7)；27.4 (24.3~29.4)；18.5 (14.3~22.1)；161.8 (114.1~205.6)；0.2 (0.0~2.3)；968 (765~1182) ▎1643 (1235~1997)；603 (497~779)；183 (123~254) ▎71.6 (29.7~132.1)；966 (765~1182)；238.4 (111.9~303.4)

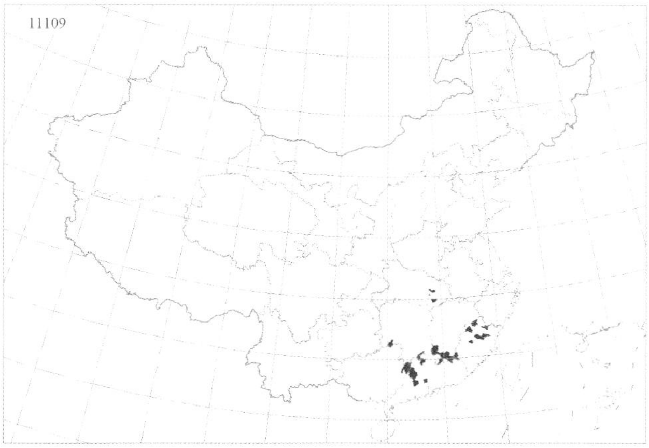

Poaceae 禾本科 *Pseudosasa* 矢竹属

Pseudosasa cantorii (Munro) P. C. Keng ex S. L. Chen et al 托竹

Bamboos, evergreen

19.3 (15.6~24.0)；10.0 (3.9~19.1)；27.6 (25.2~28.8)；19.3 (15.6~24.0)；172.3 (128.2~228.1)；0.1 (0.0~1.1)；1020 (839~1302) ▎1680 (1444~2037)；659 (527~1033)；159 (48~240) ▎66.7 (19.1~104.0)；1011 (839~1206)；254.0 (139.6~620.0)

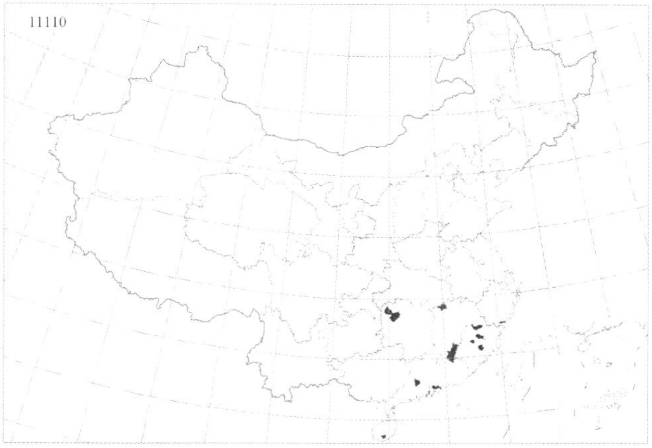

Poaceae 禾本科 *Pseudosasa* 矢竹属

Pseudosasa gracilis S. L. Chen et G. Y. Sheng 纤细茶秆竹

Bamboos, evergreen

18.8；9.3；27.1；18.8；166.1；0.0；969 ▎1659；602；185 ▎71.0；969；245.0

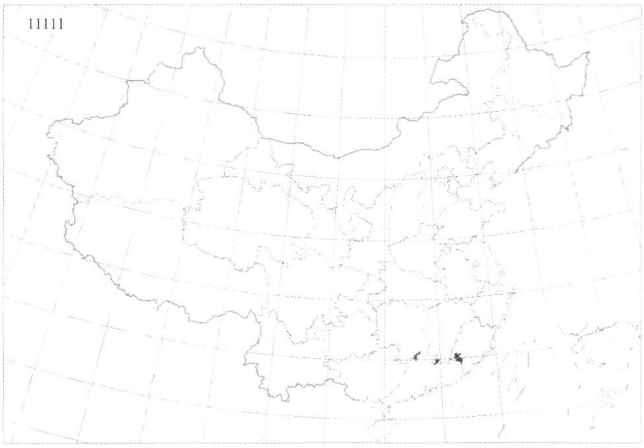

Poaceae 禾本科 *Pseudosasa* 矢竹属

Pseudosasa hindsii (Munro) C. D. Chu et C. S. Chao 篲竹

Bamboos, evergreen

20.7 (17.3~23.2)；12.5 (7.6~15.8)；27.8 (25.2~28.9)；20.7 (17.3~23.2)；188.6 (147.6~218.9)；0.0 (0.0~0.0)；1086 (864~1257) ▎1732 (1515~2037)；744 (521~1033)；142 (97~185) ▎61.1 (34.1~85.2)；1086 (864~1257)；217.7 (139.6~305.7)

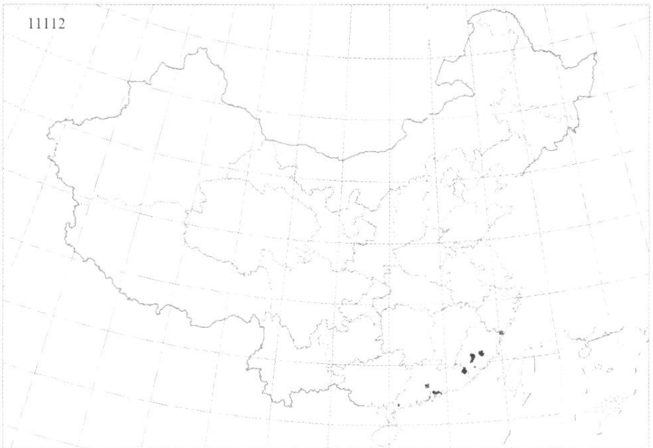

Poaceae 禾本科 *Pseudosasa* 矢竹属

Pseudosasa japonica (Siebold et Zucc. ex Steud.) Makino ex Nakai 矢竹

Bamboos, evergreen

19.9；14.5；24.2；19.9；178.7；0.0；998 ▎2546；1218；241 ▎171.2；942；239.3

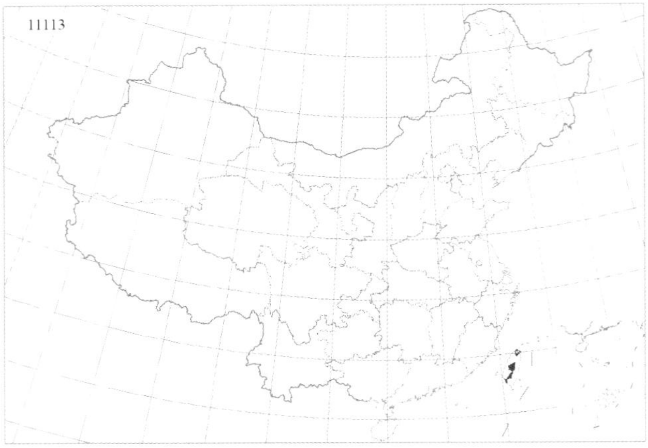

Poaceae 禾本科 *Pseudosasa* 矢竹属

Pseudosasa longiligula T. H. Wen 广竹

Bamboos, evergreen

16.7；5.6；27.1；16.8；141.2；0.0；885 ▎1521；497；196 ▎71.9；885；205.4

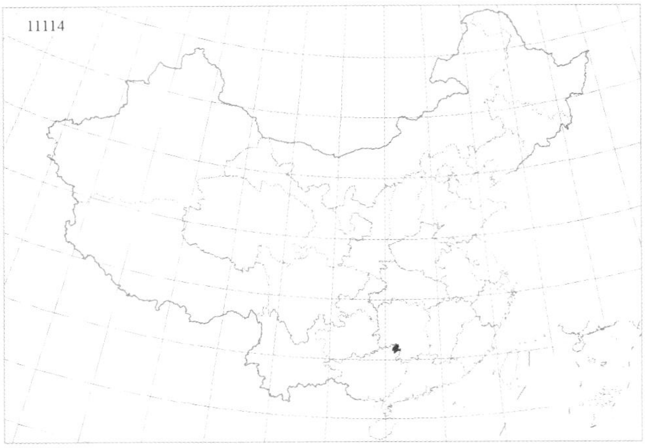

Poaceae 禾本科 *Pseudosasa* 矢竹属

Pseudosasa maculifera J. L. Lu 鸡公山茶秆竹

Bamboos, evergreen

15.4 (15.1~15.6)；2.2 (1.7~2.5)；27.6 (27.2~27.8)；15.4 (15.1~15.7)；129.4 (126.2~131.3)；4.2 (3.4~5.4)；869 (848~881) ▎1114 (1042~1200)；498 (480~531)；98 (89~107) ▎28.3 (21.1~41.4)；869 (848~881)；177.1 (161.2~213.6)

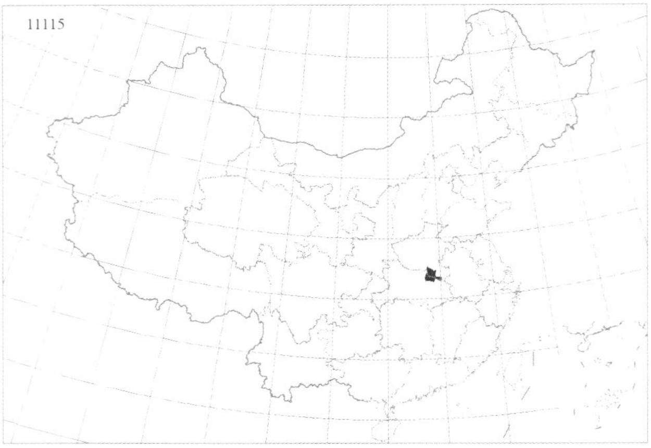

Poaceae 禾本科 *Pseudosasa* 矢竹属

Pseudosasa orthotropa S. L. Chen et T. H. Wen 面秆竹

Bamboos, evergreen

16.5；5.8；26.9；16.6；139.1；0.0；874 ▎1848；641；225 ▎111.4；874；256.7

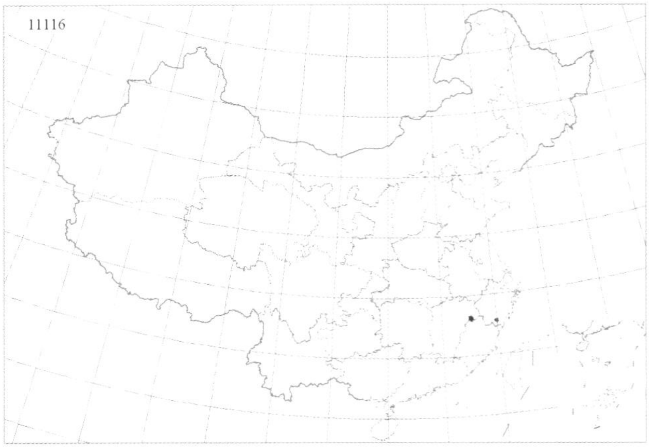

Poaceae 禾本科 *Pseudosasa* 矢竹属

Pseudosasa pubiflora (Keng) Keng f. ex D. Z. Li et L. M. Gao 毛花茶秆竹

Bamboos, evergreen

18.0；8.1；26.9；18.1；157.0；0.0；944 ▎1826；657；209 ▎94.6；944；259.3

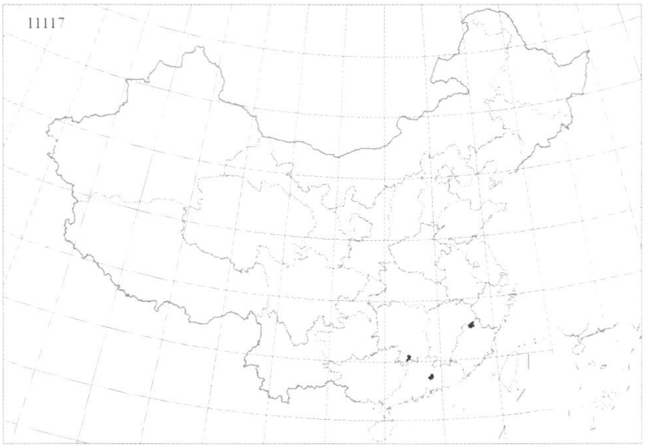

Poaceae 禾本科 *Pseudosasa* 矢竹属

Pseudosasa subsolida S. L. Chen et G. Y. Sheng 近实心茶秆竹

Bamboos, evergreen

17.8；7.1；27.9；17.9；154.5；0.1；941 ▎1585；541；191 ▎68.6；924；236.1

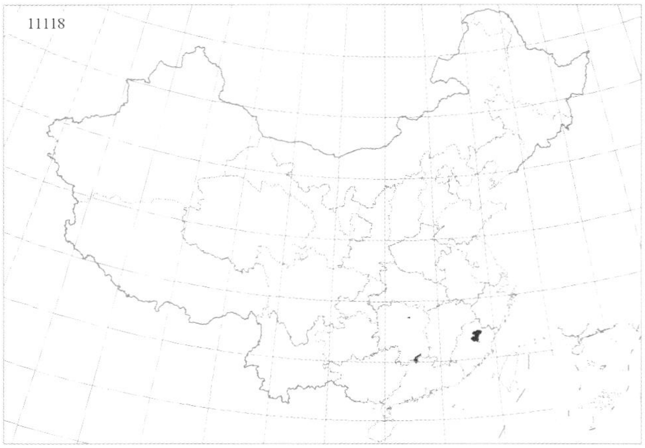

Poaceae 禾本科 *Pseudosasa* 矢竹属
Pseudosasa viridula S. L. Chen et G. Y. Sheng 笔竹
Bamboos, evergreen

16.5; 4.4; 28.9; 16.6; 139.8; 0.6; 910 ❙ 1377; 493; 197 ❙ 51.2; 910; 189.1

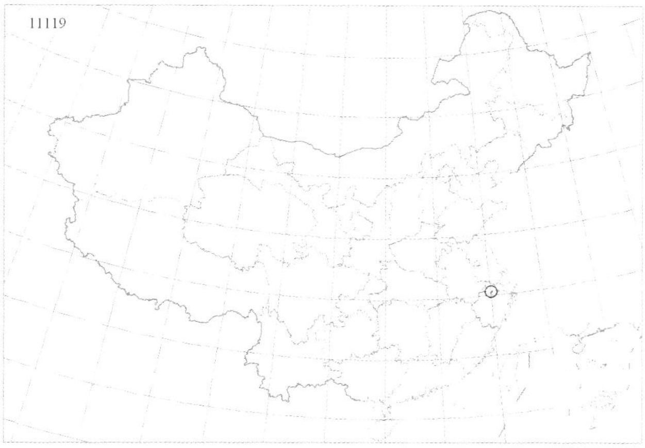

Poaceae 禾本科 *Pseudosasa* 矢竹属
Pseudosasa wuyiensis S. L. Chen et G. Y. Sheng 武夷山茶秆竹
Bamboos, evergreen

17.3 (16.6~18.1); 6.3 (5.9~7.2); 27.4 (26.3~28.7); 17.3 (16.6~18.1); 147.6 (138.9~157.7); 0.0 (0.0~0.0); 913 (866~965) ❙ 1893 (1773~1997); 626 (564~669); 246 (233~259) ❙ 107.9 (83.7~130.7); 911 (866~961); 272.9 (227.8~303.4)

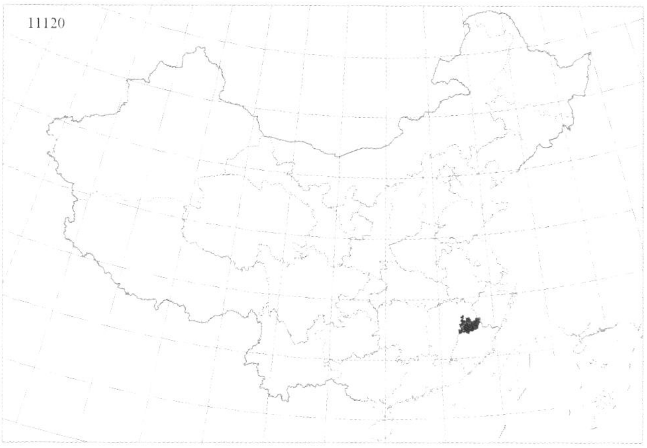

Poaceae 禾本科 *Pseudosasa* 矢竹属
Pseudosasa yuelushanensis B. M. Yang 岳麓山茶秆竹
Bamboos, evergreen

17.5; 4.6; 29.7; 17.6; 151.2; 0.4; 966 ❙ 1414; 440; 197 ❙ 46.4; 879; 157.4

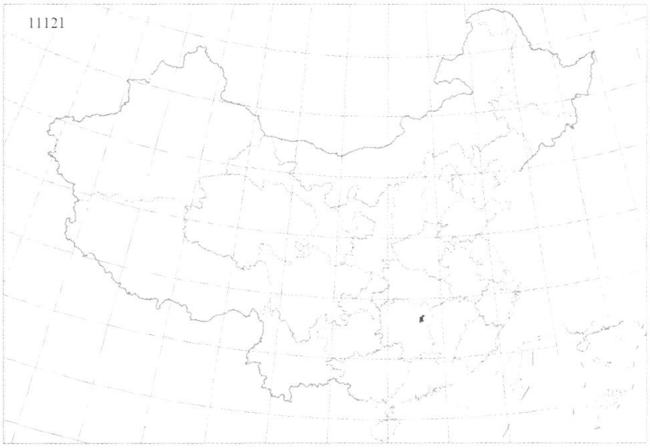

Poaceae 禾本科 *Pseudostachyum* 泡竹属
Pseudostachyum polymorphum Munro 泡竹
Bamboos, evergreen

19.6 (17.6~21.0); 10.9 (7.3~15.1); 26.6 (22.3~28.6); 19.7 (17.6~21.0); 176.0 (151.6~191.8); 0.0 (0.0~0.0); 1006 (909~1099) ❙ 1563 (1447~1710); 671 (601~818); 140 (58~186) ❙ 55.8 (31.6~70.6); 1002 (880~1099); 322.1 (185.5~659.0)

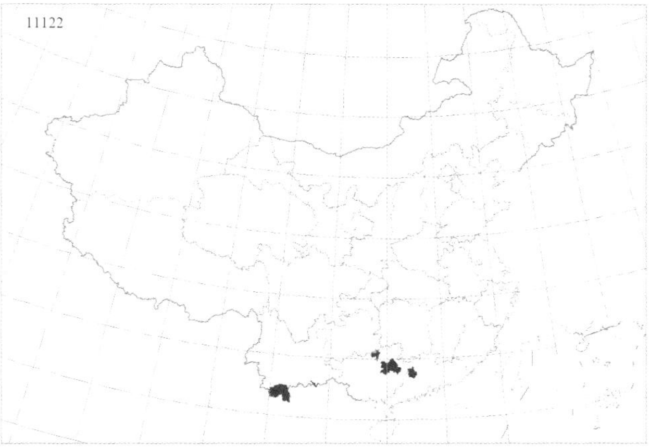

Poaceae 禾本科 *Sasa* 赤竹属
Sasa guangxiensis C. D. Chu et C. S. Chao 广西赤竹
Bamboos, evergreen

17.0; 6.2; 26.8; 17.0; 144.3; 0.2; 892 ❙ 1614; 599; 186 ❙ 81.6; 892; 217.4

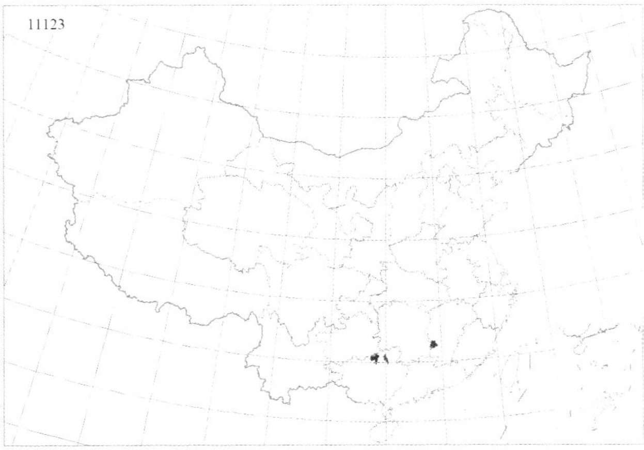

Poaceae 禾本科 *Sasa* 赤竹属
Sasa hubeiensis (C. H. Hu) C. H. Hu 湖北华箬竹
Bamboos, evergreen

16.1; 3.8; 27.9; 16.1; 134.6; 1.2; 881 ❙ 1469; 537; 171 ❙ 66.8; 881; 246.2

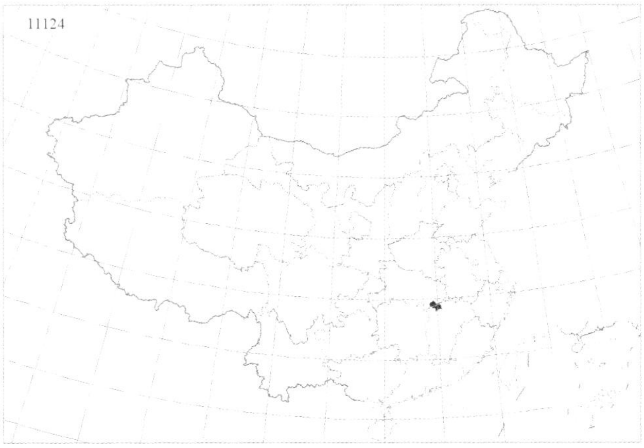

Poaceae 禾本科 *Sasa* 赤竹属
Sasa longiligulata McClure 赤竹
Bamboos, evergreen

18.3 (15.3~21.6); 7.9 (4.2~13.2); 27.5 (26.0~28.5); 18.3 (15.4~21.6); 159.7 (125.8~199.2); 0.1 (0.0~0.8); 959 (818~1139) ▎1655 (1486~1953); 591 (505~858); 192 (167~216) ▎73.1 (57.0~103.1); 959 (818~1139); 229.3 (188.8~261.3)

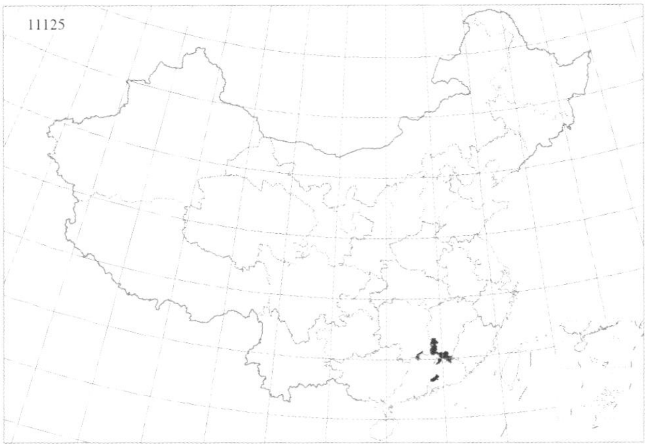

Poaceae 禾本科 *Sasa* 赤竹属
Sasa oblongula C. H. Hu 矩叶赤竹
Bamboos, evergreen

22.1; 13.7; 28.8; 22.1; 205.6; 0.0; 1182 ▎1784; 779; 144 ▎50.9; 1182; 157.7

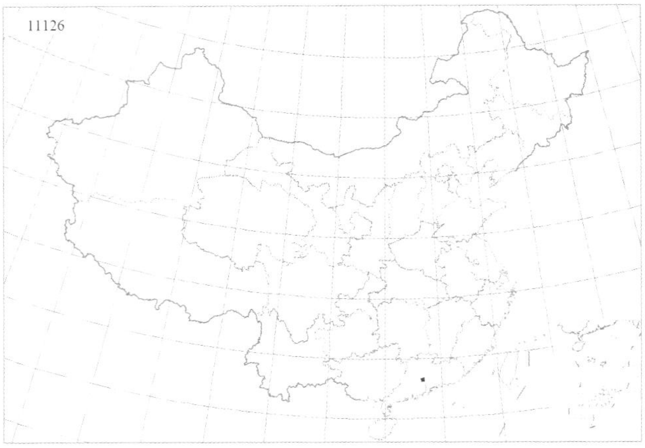

Poaceae 禾本科 *Sasa* 赤竹属
Sasa qingyuanensis (C. H. Hu) C. H. Hu 庆元华箬竹
Bamboos, evergreen

15.0; 5.0; 24.7; 15.1; 120.8; 0.0; 789 ▎1950; 713; 223 ▎147.0; 789; 283.9

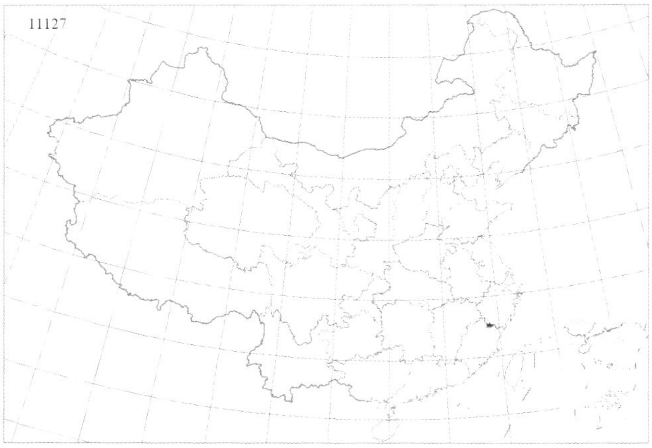

Poaceae 禾本科 *Sasa* 赤竹属
Sasa rubrovaginata C. H. Hu 红壳赤竹
Bamboos, evergreen

19.8; 11.1; 26.4; 19.8; 177.9; 0.0; 1010 ▎1182; 673; 43 ▎16.9; 1010; 354.5

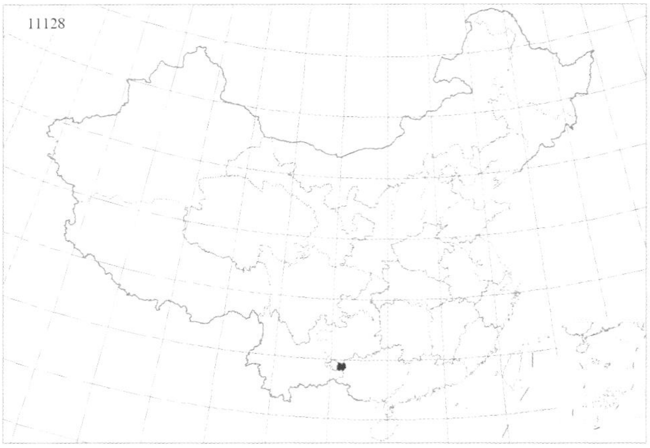

Poaceae 禾本科 *Sasa* 赤竹属
Sasa sinica Keng 华箬竹
Bamboos, evergreen

15.7 (14.4~17.6); 4.1 (2.1~7.2); 27.0 (24.7~28.4); 15.8 (14.5~17.7); 131.0 (118.7~151.9); 1.5 (0.0~4.9); 853 (789~923) ▎1609 (1257~1997); 572 (427~713); 202 (157~254) ▎89.0 (40.8~147.0); 849 (789~923); 264.2 (217.8~303.4)

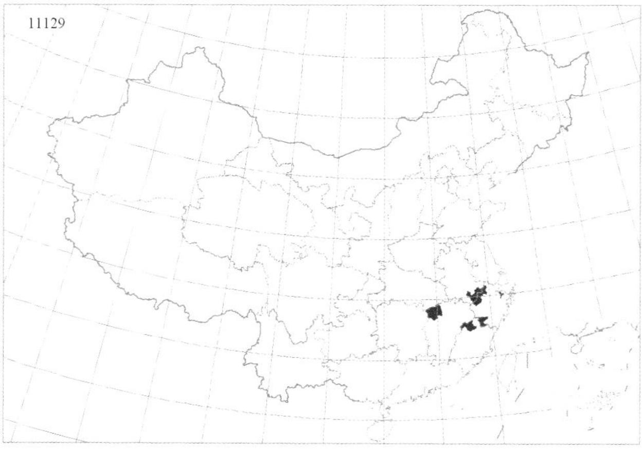

Poaceae 禾本科 *Sasa* 赤竹属
Sasa tomentosa C. D. Chu et C. S. Chao 绒毛赤竹
Bamboos, evergreen

17.5; 7.4; 25.9; 17.5; 150.2; 0.0; 897 ▎1438; 638; 125 ▎60.3; 897; 235.6

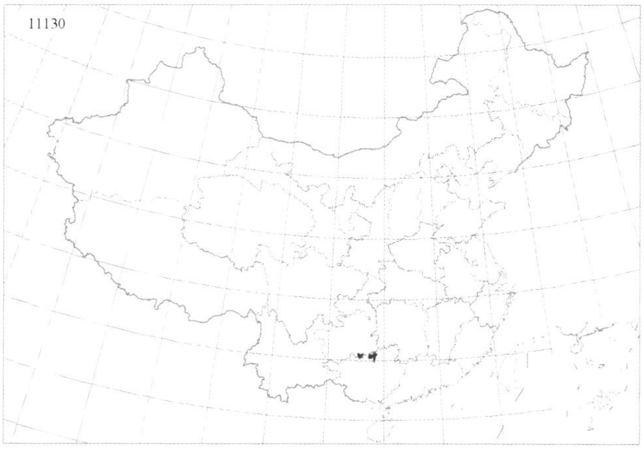

Poaceae 禾本科 *Schizostachyum* 葸簩竹属
Schizostachyum chinense Rendle 薄竹
Bamboos, evergreen

18.6 (16.9~21.0); 12.2 (10.3~14.2); 23.0 (21.4~25.6); 18.6 (16.9~21.0); 163.7 (142.7~191.8); 0.0 (0.0~0.0); 900 (813~1053) | 1299 (1028~1547); 734 (566~911); 55 (44~68) | 44.4 (12.5~74.2); 885 (783~1053); 455.1 (339.5~551.6)

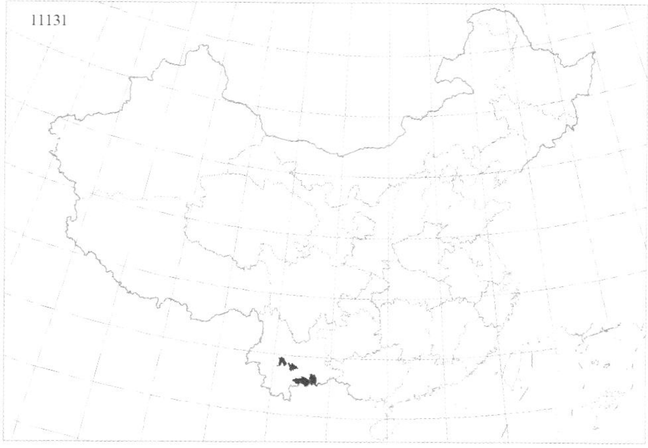

Poaceae 禾本科 *Schizostachyum* 葸簩竹属
Schizostachyum diffusum (Blanco) Merr. 莎簕竹
Bamboos, evergreen

21.7; 16.7; 25.4; 21.7; 200.8; 0.0; 1117 | 2255; 1381; 87 | 117.5; 1003; 203.3

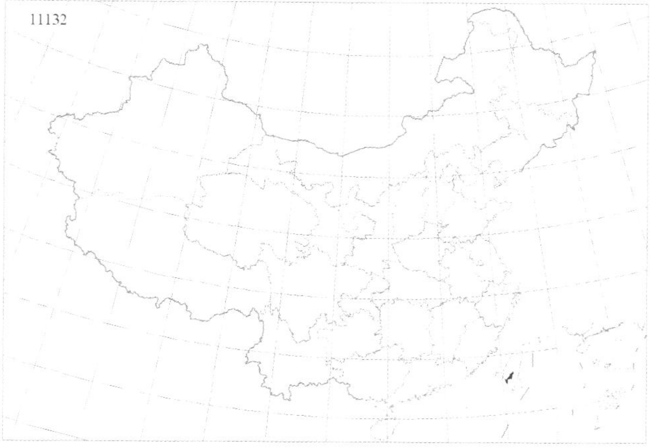

Poaceae 禾本科 *Schizostachyum* 葸簩竹属
Schizostachyum dumetorum (Hance) Munro 苗竹仔
Bamboos, evergreen

21.7 (19.5~22.5); 13.6 (10.6~15.0); 28.1 (26.9~28.8); 21.7 (19.6~22.6); 200.7 (175.1~210.9); 0.0 (0.0~0.0); 1147 (1003~1206) | 1896 (1627~2037); 867 (606~1033); 138 (97~191) | 65.7 (45.8~95.5); 1147 (1003~1206); 212.7 (139.6~275.2)

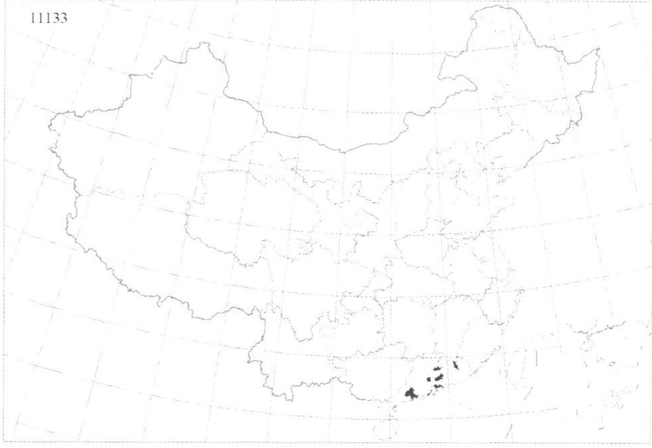

Poaceae 禾本科 *Schizostachyum* 葸簩竹属
Schizostachyum dumetorum var. xinwuense (T. H. Wen et J. Y. Chin) N. H. Xia 火筒竹
Bamboos, evergreen

18.8; 8.9; 27.2; 18.8; 165.8; 0.0; 973 | 1688; 597; 196 | 73.6; 973; 258.0

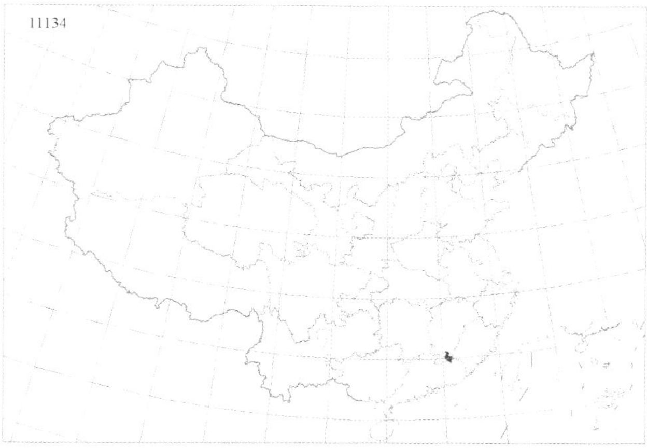

Poaceae 禾本科 *Schizostachyum* 葸簩竹属
Schizostachyum funghomii McClure 沙罗箪竹
Bamboos, evergreen

21.2 (17.7~23.2); 12.8 (8.5~15.8); 27.8 (22.3~29.0); 21.3 (17.8~23.2); 195.3 (153.4~218.9); 0.0 (0.0~0.0); 1117 (854~1257) | 1583 (1302~2435); 753 (561~1428); 130 (56~194) | 42.2 (12.9~99.5); 1116 (854~1257); 264.6 (157.7~659.0)

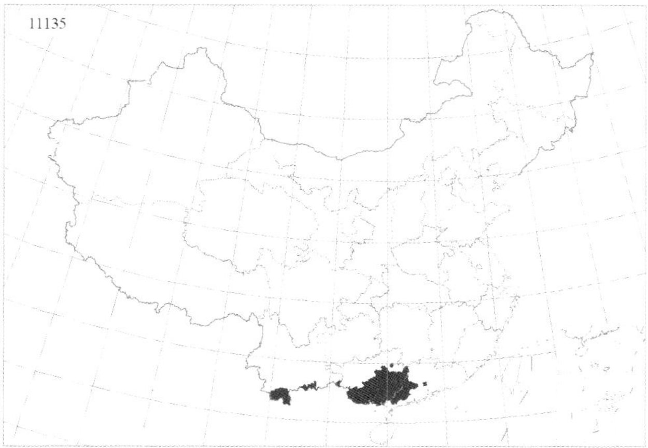

Poaceae 禾本科 *Schizostachyum* 葸簩竹属
Schizostachyum hainanense Merr. ex McClure 山骨罗竹
Bamboos, evergreen

23.6; 18.3; 27.4; 23.7; 223.9; 0.0; 1272 | 1486; 521; 53 | 17.4; 1127; 546.6

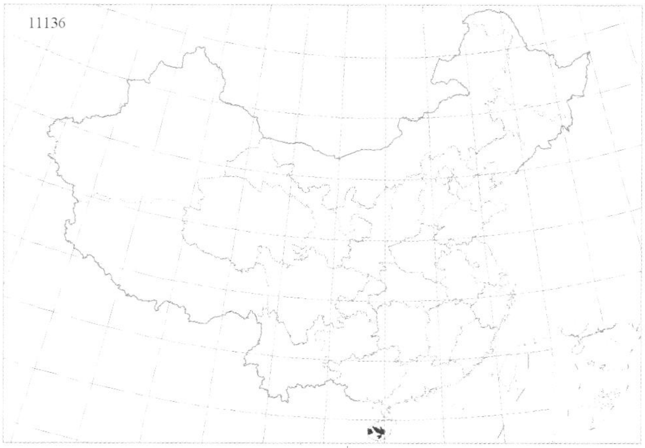

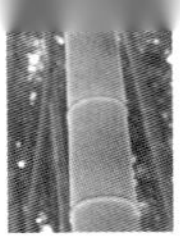

Poaceae 禾本科 *Schizostachyum* 思簩竹属
Schizostachyum jaculans Holttum 岭南思簩竹
Bamboos, evergreen

24.3；19.5；27.6；24.4；232.3；0.0；1339 ❙ 1569；536；53 ❙ 17.2；1180；557.4

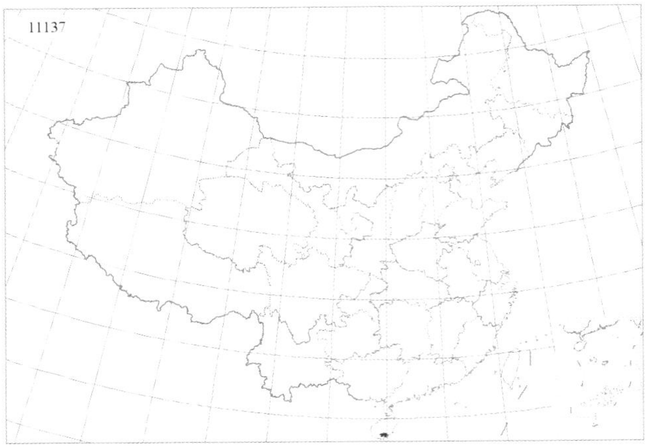

Poaceae 禾本科 *Schizostachyum* 思簩竹属
Schizostachyum pseudolima McClure 思簩竹
Bamboos, evergreen

22.9 (19.1~24.8)；16.8 (13.0~20.1)；27.1 (23.3~28.7)；22.9 (19.1~24.8)；214.7 (169.7~237.6)；0.0 (0.0~0.0)；1220 (917~1385) ❙ 1585 (1176~2435)；699 (444~1428)；72 (41~130) ❙ 32.1 (-12.0~99.5)；1133 (917~1360)；470.0 (300.5~659.0)

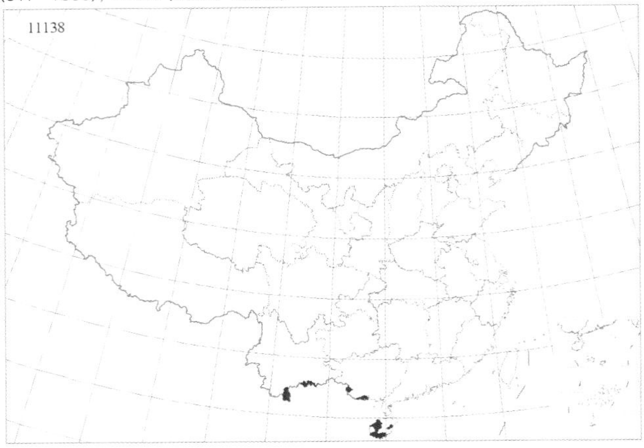

Poaceae 禾本科 *Semiarundinaria* 短穗竹属
Semiarundinaria densiflora (Rendle) T. H. Wen 短穗竹
Bamboos, evergreen

15.7 (14.7~17.5)；3.3 (2.0~7.3)；27.8 (26.6~28.9)；15.7 (14.8~17.6)；130.9 (121.1~150.8)；2.6 (0.0~4.7)；870 (815~933) ❙ 1221 (963~1889)；497 (415~627)；152 (104~238) ❙ 40.4 (10.3~112.2)；869 (815~916)；190.7 (131.2~319.9)

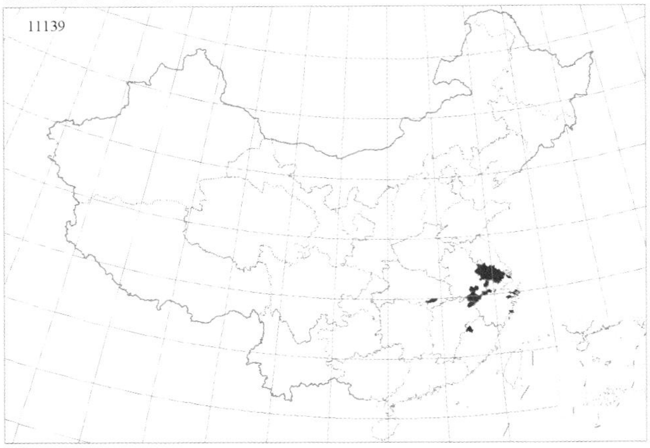

Poaceae 禾本科 *Semiarundinaria* 短穗竹属
Semiarundinaria fastuosa (Mitford) Makino 业平竹
Bamboos, evergreen

21.7；16.7；25.4；21.7；200.8；0.0；1117 ❙ 2255；1381；87 ❙ 117.5；1003；203.3

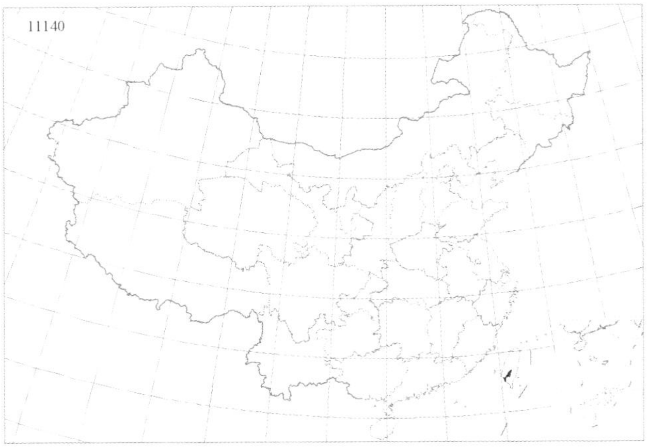

Poaceae 禾本科 *Shibataea* 倭竹属
Shibataea chiangshanensis T. W. Wen 江山鹅毛竹
Bamboos, evergreen

16.5；5.0；27.5；16.6；138.6；0.0；887 ❙ 1820；552；246 ❙ 105.1；887；223.7

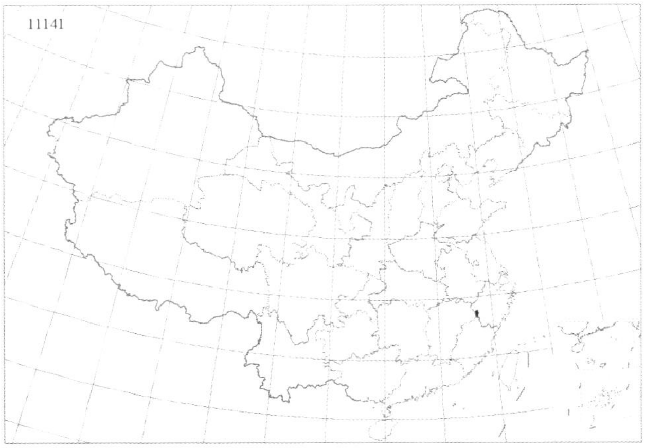

Poaceae 禾本科 *Shibataea* 倭竹属
Shibataea chinensis Nakai 鹅毛竹
Bamboos, evergreen

15.8 (14.7~18.7)；3.5 (1.8~9.8)；28.0 (24.7~29.4)；15.9 (14.8~18.7)；132.8 (120.8~164.5)；2.3 (0.0~5.0)；878 (789~959) ❙ 1208 (963~1950)；496 (394~713)；150 (102~240) ❙ 37.6 (10.3~147.0)；876 (789~959)；181.5 (111.9~299.5)

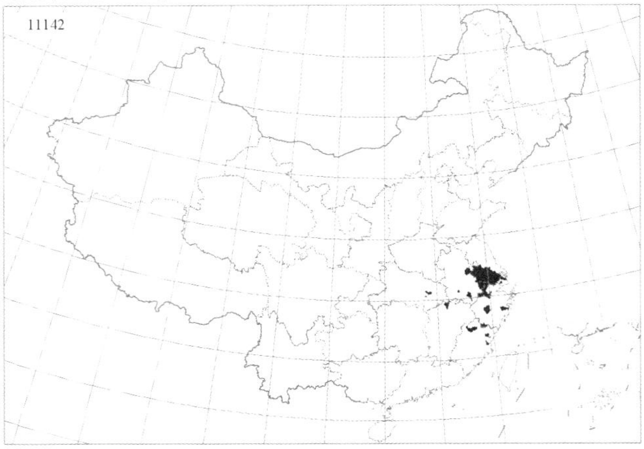

Poaceae 禾本科 *Shibataea* 倭竹属
Shibataea hispida McClure 芦花竹
Bamboos, evergreen

15.0；2.9；26.7；15.1；124.0；2.8；828 ❙ 1729；620；213 ❙ 108.9；828；290.3

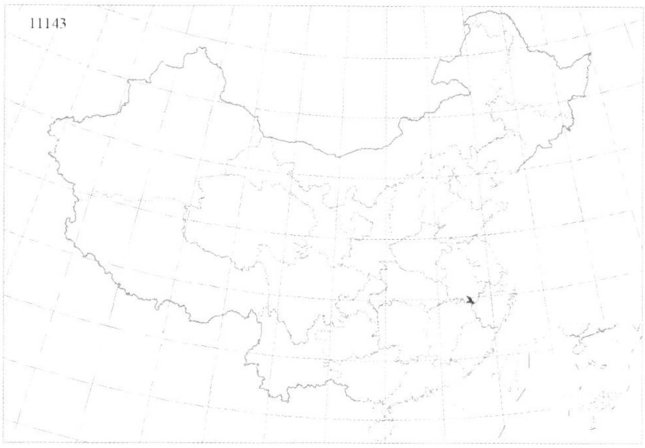

Poaceae 禾本科 *Shibataea* 倭竹属
Shibataea kumasaca (Zoll. ex Steud.) Makino ex Nakai 倭竹
Bamboos, evergreen

18.9；11.1；26.1；19.0；167.6；0.0；951 ❙ 2352；754；375 ❙ 145.1；951；273.3

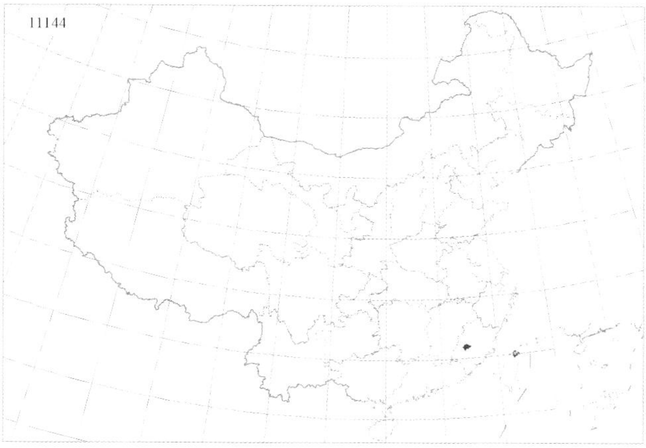

Poaceae 禾本科 *Shibataea* 倭竹属
Shibataea lancifolia C. H. Hu 狭叶鹅毛竹
Bamboos, evergreen

16.8 (15.0~18.4)；5.9 (4.0~9.0)；27.1 (24.7~29.1)；16.9 (15.1~18.4)；142.5 (120.8~161.2)；0.1 (0.0~1.0)；892 (789~984) ❙ 1847 (1377~1997)；619 (493~713)；239 (197~259) ❙ 107.9 (51.2~147.0)；888 (789~950)；257.9 (183.9~303.4)

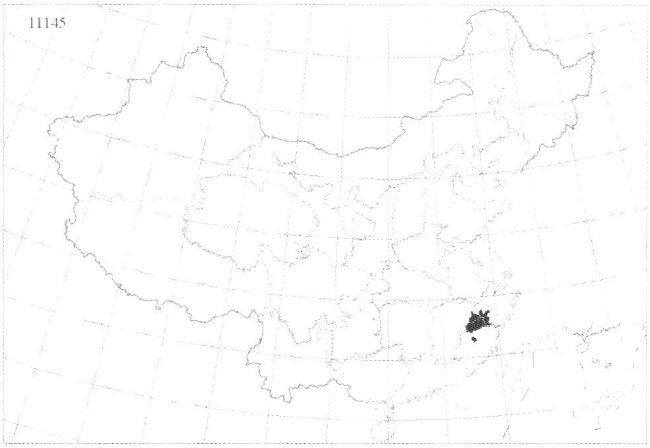

Poaceae 禾本科 *Shibataea* 倭竹属
Shibataea nanpingensis Q. F. Zheng et K. F. Huang 南平鹅毛竹
Bamboos, evergreen

17.9；8.1；27.0；17.9；154.7；0.0；927 ❙ 1808；616；219 ❙ 96.2；927；294.4

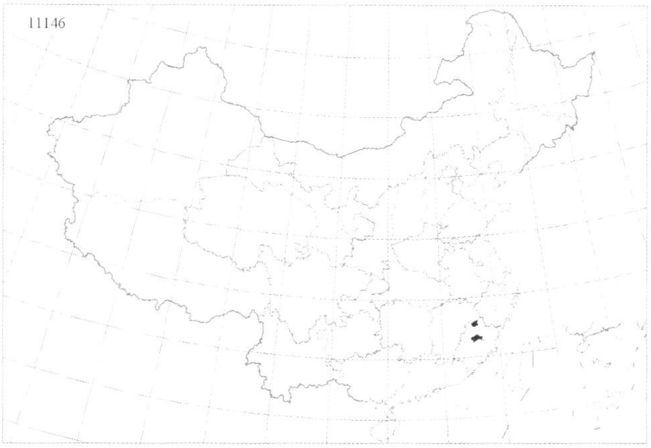

Poaceae 禾本科 *Shibataea* 倭竹属
Shibataea nanpingensis var. fujianica (Z. D. Zhu et H. Y. Zhou) C. H. Hu 福建鹅毛竹
Bamboos, evergreen

17.7；8.0；26.7；17.7；153.0；0.0；915 ❙ 1803；614；219 ❙ 98.0；915；296.2

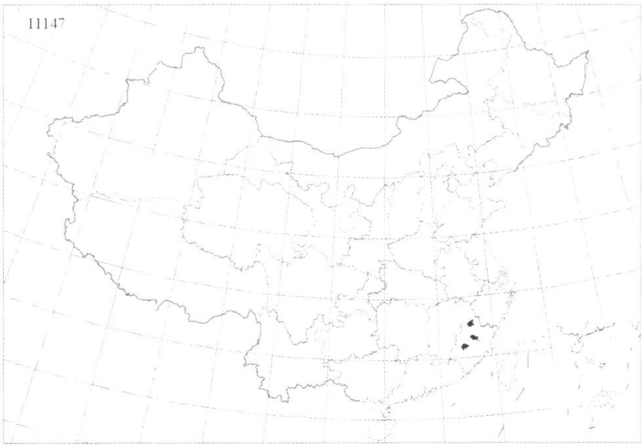

Poaceae 禾本科 *Shibataea* 倭竹属
Shibataea strigosa T. H. Wen 矮雷竹
Bamboos, evergreen

15.2；3.7；26.3；15.2；124.3；2.0；819 ❙ 1691；621；211 ❙ 106.3；819；276.6

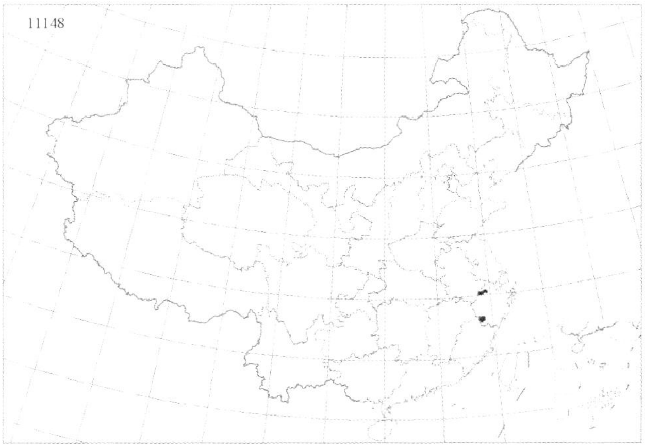

Poaceae 禾本科 *Sinobambusa* 唐竹属
Sinobambusa dushanensis (C. D. Chu et J. Q. Zhang) T. H. Wen 独山唐竹
Bamboos, evergreen

16.2; 6.3; 24.4; 16.2; 134.5; 0.0; 826 ❙ 1342; 611; 95 ❙ 62.5; 826; 215.4

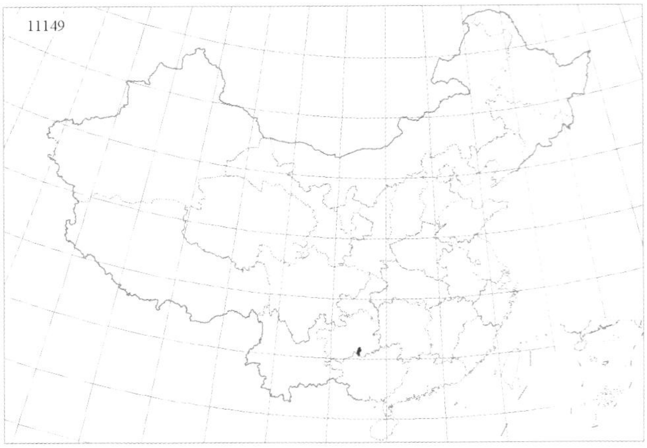

Poaceae 禾本科 *Sinobambusa* 唐竹属
Sinobambusa farinosa (McClure) T. H. Wen 白皮唐竹
Bamboos, evergreen

18.6; 9.0; 27.0; 18.7; 164.1; 0.0; 968 ❙ 1722; 668; 182 ❙ 78.1; 968; 261.6

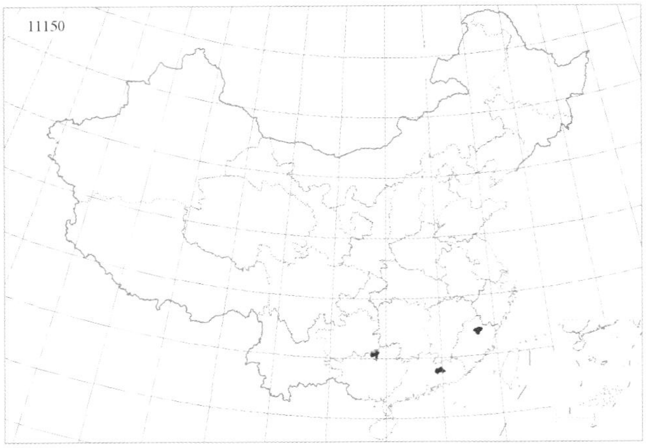

Poaceae 禾本科 *Sinobambusa* 唐竹属
Sinobambusa henryi (McClure) C. D. Chu et C. S. Chao 扛竹
Bamboos, evergreen

19.5; 10.2; 27.1; 19.5; 174.4; 0.0; 1011 ❙ 1577; 688; 132 ❙ 55.3; 1011; 243.4

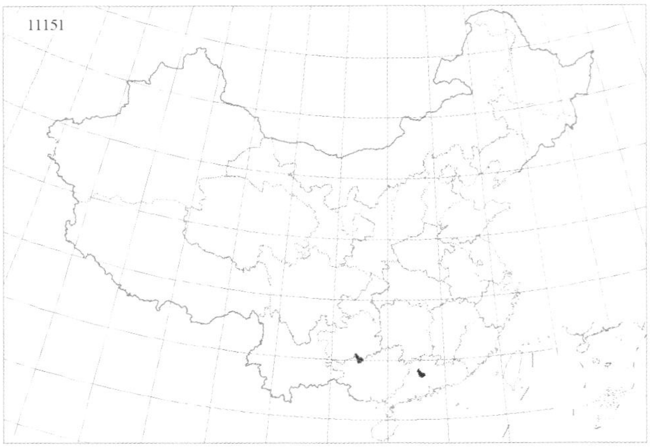

Poaceae 禾本科 *Sinobambusa* 唐竹属
Sinobambusa humilis McClure 竹仔
Bamboos

24.5; 18.9; 28.5; 24.5; 234.4; 0.0; 1360 ❙ 1854; 622; 115 ❙ 36.4; 1360; 428.6

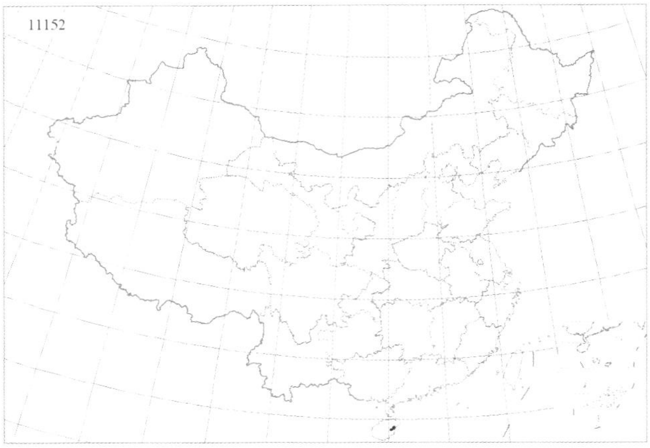

Poaceae 禾本科 *Sinobambusa* 唐竹属
Sinobambusa intermedia McClure 晾衫竹
Bamboos, evergreen

19.7 (16.3~22.4); 10.7 (6.5~14.3); 27.4 (21.4~28.9); 19.7 (16.4~22.5); 176.5 (136.4~209.7); 0.0 (0.0~0.0); 1028 (797~1205) ❙ 1699 (1190~1889); 663 (476~811); 170 (50~238) ❙ 67.2 (31.5~106.2); 1028 (797~1205); 248.9 (145.3~347.3)

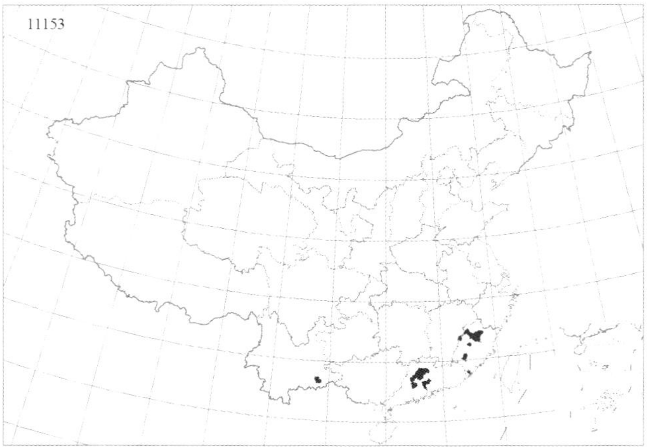

Poaceae 禾本科 *Sinobambusa* 唐竹属
Sinobambusa nephroaurita C. D. Chu et C. S. Chao 肾耳唐竹
Bamboos, evergreen

18.4; 8.2; 26.9; 18.4; 161.2; 0.0; 955 ❙ 1495; 658; 138 ❙ 56.9; 955; 241.2

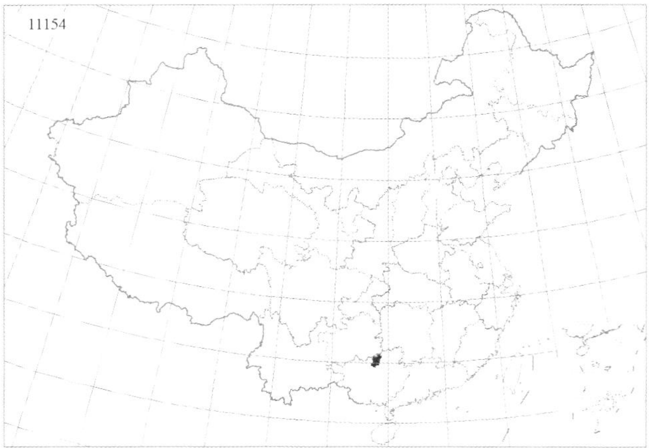

Poaceae 禾本科 *Sinobambusa* 唐竹属
Sinobambusa rubroligula McClure 红舌唐竹
Bamboos, evergreen

21.8; 13.4; 28.6; 21.8; 202.0; 0.0; 1162 ❙ 1705; 774; 132 ❙ 47.2; 1162; 186.7

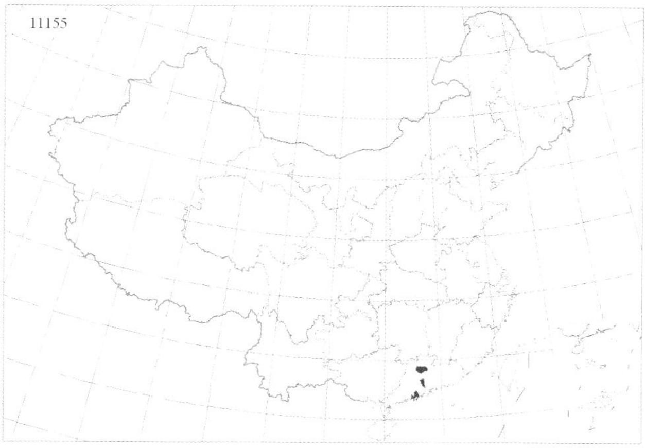

Poaceae 禾本科 *Sinobambusa* 唐竹属
Sinobambusa tootsik (Siebold) Makino 唐竹
Bamboos, evergreen

20.3 (15.3~24.1); 12.4 (6.1~18.9); 27.0 (22.8~28.8); 20.4 (15.4~24.1); 184.6 (124.4~229.7); 0.0 (0.0~0.0); 1058 (791~1322) ❙ 1777 (1485~2759); 810 (522~1594); 141 (47~231) ❙ 71.6 (32.5~202.5); 1044 (791~1182); 255.0 (154.2~551.6)

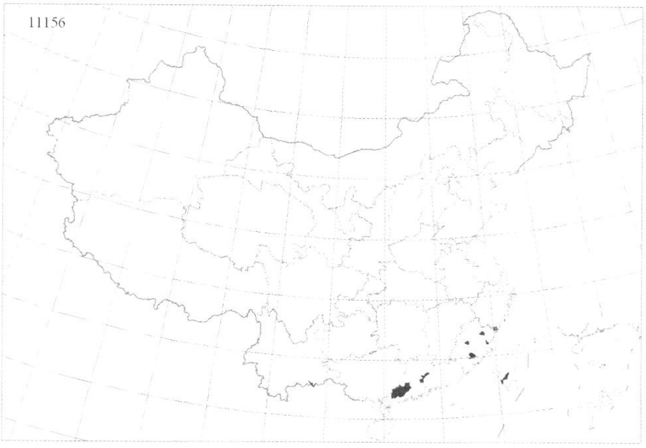

Poaceae 禾本科 *Sinobambusa* 唐竹属
Sinobambusa tootsik var. dentata Wen 火管竹
Bamboos, evergreen

17.3; 7.6; 26.9; 17.4; 148.3; 0.0; 896 ❙ 1640; 575; 181 ❙ 83.1; 896; 231.6

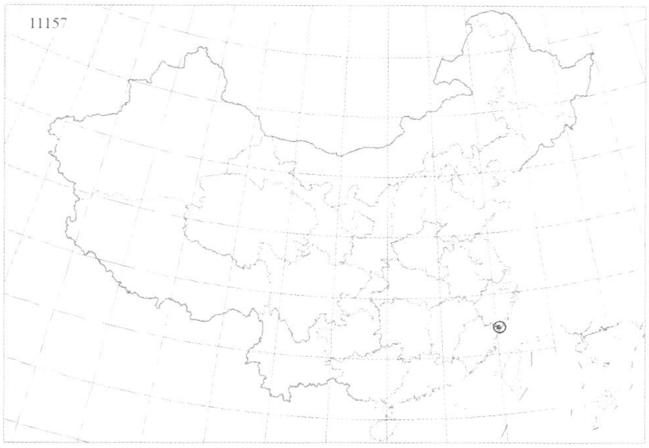

Poaceae 禾本科 *Sinobambusa* 唐竹属
Sinobambusa tootsik var. laeta (McClure) T. H. Wen 满山爆竹
Bamboos, evergreen

19.1; 10.3; 27.1; 19.1; 169.2; 0.0; 979 ❙ 1686; 616; 194 ❙ 71.8; 979; 252.6

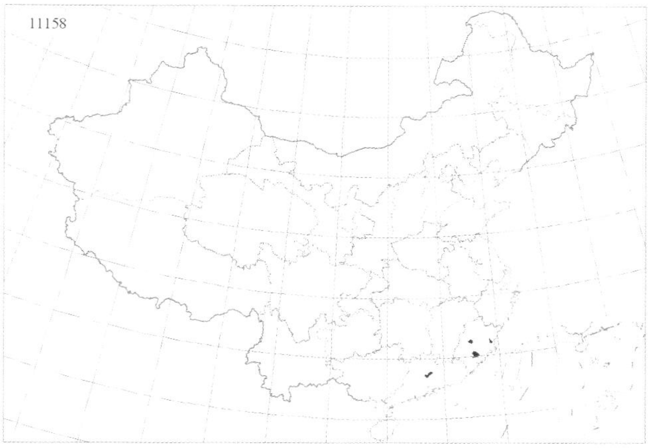

Poaceae 禾本科 *Sinobambusa* 唐竹属
Sinobambusa yixingensis C. S. Chao et K. S. Xiao 宜兴唐竹
Bamboos, evergreen

15.6; 3.4; 28.1; 15.7; 130.0; 2.1; 869 ❙ 1117; 478; 152 ❙ 28.5; 869; 179.2

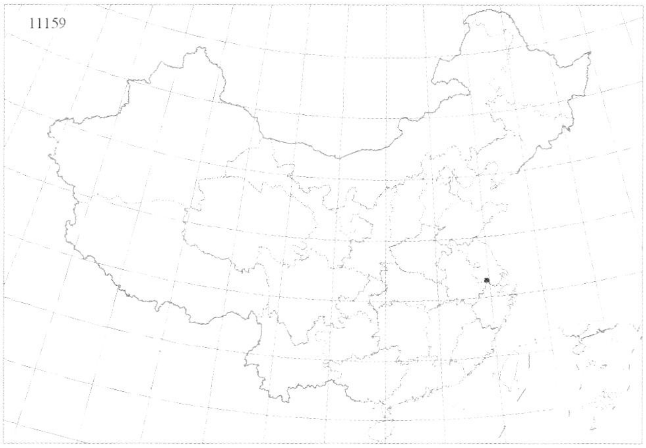

Poaceae 禾本科 *Thamnocalamus* 筱竹属
Thamnocalamus spathiflorus (Trin.) Munro 筱竹
Bamboos, evergreen

0.3; -8.6; 8.3; 2.9; 9.6; 65.4; 368 ❙ 389; 246; 11 ❙ 0.0; 368; 142.5

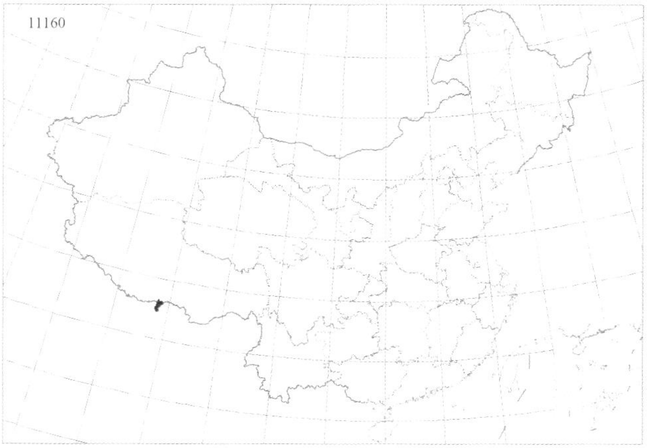

Poaceae 禾本科 *Thamnocalamus* 筱竹属
Thamnocalamus spathiflorus var. crassinodus (T. P. Yi) Stapleton 粗节筱竹
Bamboos, evergreen

-1.1; -9.1; 6.2; 2.0; 2.5; 75.6; 334 ❙ 321; 193; 33 ❙ -4.0; 321; 79.3

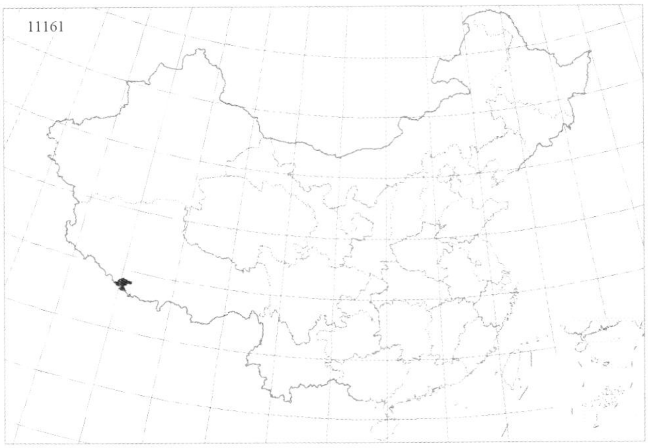

Poaceae 禾本科 *Thyrsostachys* 泰竹属
Thyrsostachys siamensis Gamble 泰竹
Bamboos, evergreen

20.0; 14.6; 23.2; 20.1; 180.8; 0.0; 963 ❙ 1523; 681; 67 ❙ 58.4; 942; 581.0

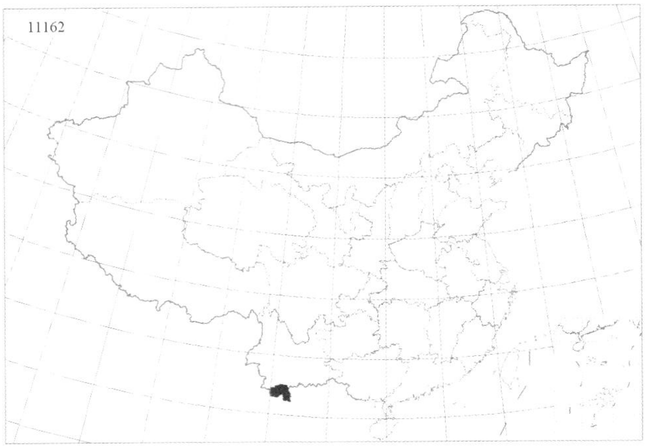

Poaceae 禾本科 *Yushania* 玉山竹属
Yushania andropogonoides (Hand.-Mazz.) T. P. Yi 草丝竹
Bamboos, evergreen

16.0; 8.0; 21.8; 16.0; 132.2; 0.0; 793 ❙ 1107; 607; 52 ❙ 39.7; 793; 298.7

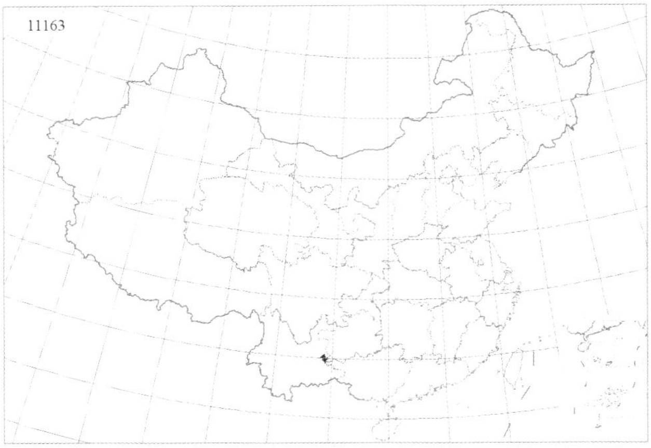

Poaceae 禾本科 *Yushania* 玉山竹属
Yushania auctiaurita T. P. Yi 显耳玉山竹
Bamboos, evergreen

14.3; 4.0; 23.5; 14.3; 112.6; 1.1; 756 ❙ 1299; 539; 100 ❙ 71.9; 756; 225.8

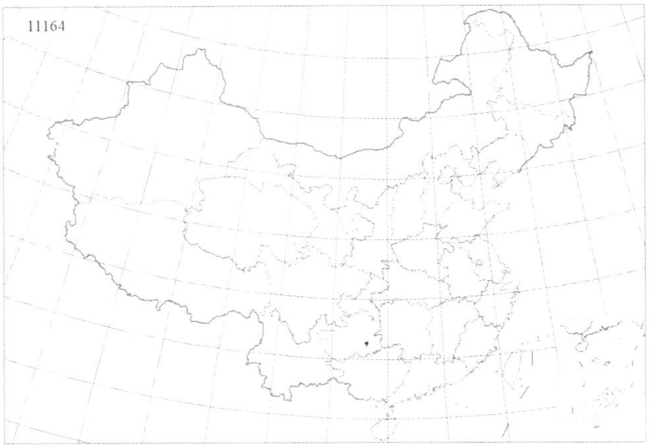

Poaceae 禾本科 *Yushania* 玉山竹属
Yushania baishanzuensis Z. P. Wang et G. H. Ye 百山祖玉山竹
Bamboos, evergreen

15.0; 5.0; 24.7; 15.1; 120.8; 0.0; 789 ❙ 1950; 713; 223 ❙ 147.0; 789; 283.9

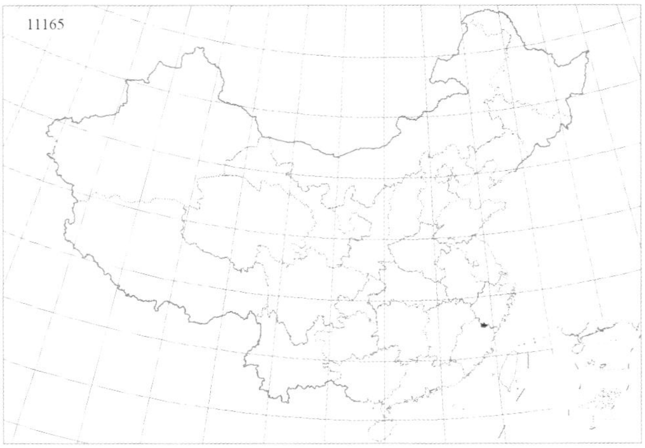

Poaceae 禾本科 *Yushania* 玉山竹属
Yushania basihirsuta (McClure) Z. P. Wang et G. H. Ye 毛玉山竹
Bamboos, evergreen

17.8 (14.3~20.7); 7.2 (3.3~10.5); 27.4 (24.3~29.8); 17.9 (14.3~20.8); 154.6 (114.1~189.3); 0.2 (0.0~2.3); 938 (765~1113) ❙ 1552 (1409~1745); 529 (419~644); 190 (176~202) ❙ 66.1 (35.6~99.4); 930 (765~1032); 213.2 (169.7~257.1)

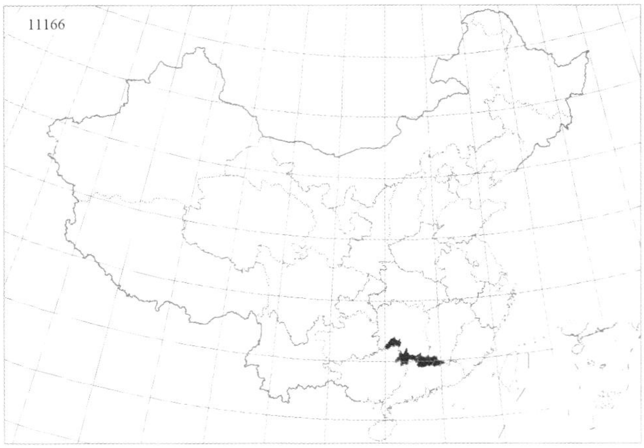

Poaceae 禾本科 *Yushania* 玉山竹属

Yushania bojieiana T. P. Yi 金平玉山竹

Bamboos, evergreen

19.1；13.0；23.3；19.1；169.7；0.0；917 ❙ 1428；835；56 ❙ 55.7；917；506.2

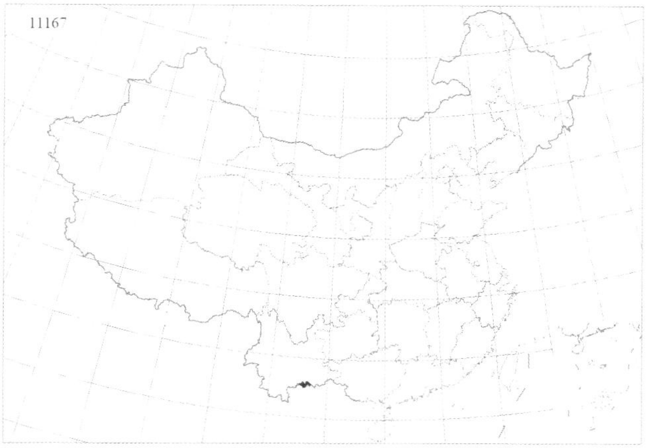

Poaceae 禾本科 *Yushania* 玉山竹属

Yushania brevipaniculata (Hand.-Mazz.) T. P. Yi 短锥玉山竹

Bamboos, evergreen

11.8 (5.7~16.9)；2.3 (-3.5~10.3)；20.3 (14.1~25.4)；11.9 (6.4~16.9)；90.0 (38.1~142.7)；8.0 (0.0~28.1)；682 (504~849) ❙ 1068 (822~1414)；592 (420~817)；31 (19~51) ❙ 58.9 (11.0~101.2)；682 (504~849)；219.5 (139.5~383.4)

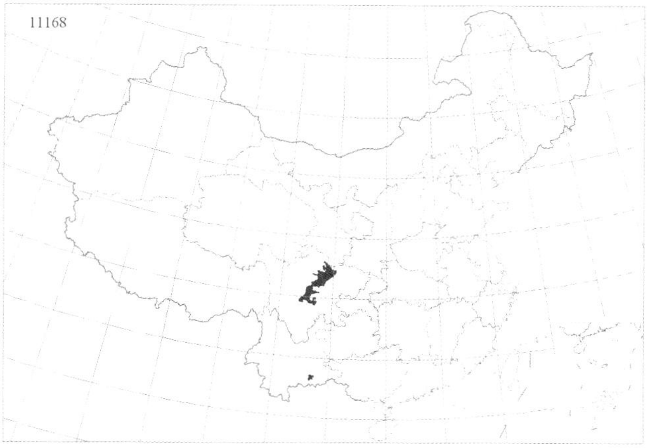

Poaceae 禾本科 *Yushania* 玉山竹属

Yushania brevis T. P. Yi 绿春玉山竹

Bamboos, evergreen

18.6；12.7；22.5；18.7；163.9；0.0；888 ❙ 1547；911；63 ❙ 74.2；888；551.3

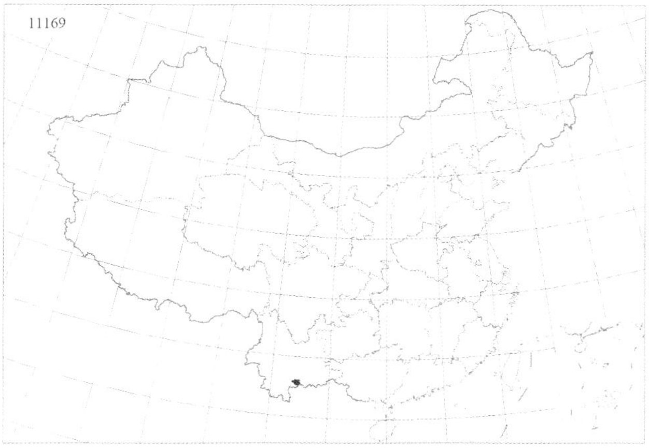

Poaceae 禾本科 *Yushania* 玉山竹属

Yushania canoviridis G. H. Ye et Z. P. Wang 灰绿玉山竹

Bamboos, evergreen

15.5；4.0；26.6；15.6；128.0；1.0；839 ❙ 1451；584；120 ❙ 72.9；839；208.1

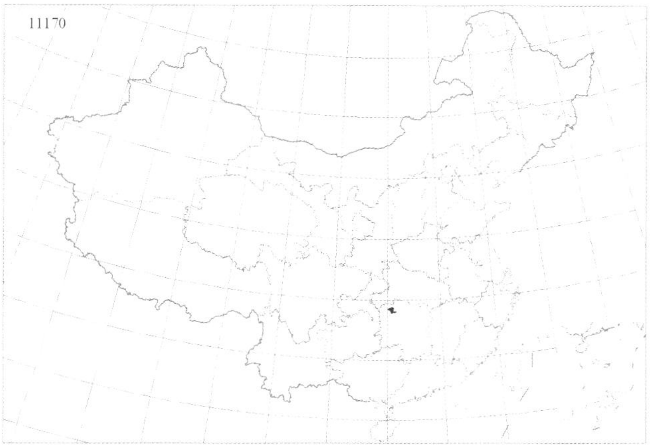

Poaceae 禾本科 *Yushania* 玉山竹属

Yushania cartilaginea T. H. Wen 硬壳玉山竹

Bamboos, evergreen

21.3；12.6；27.9；21.3；196.0；0.0；1123 ❙ 1188；634；56 ❙ 5.8；1121；315.6

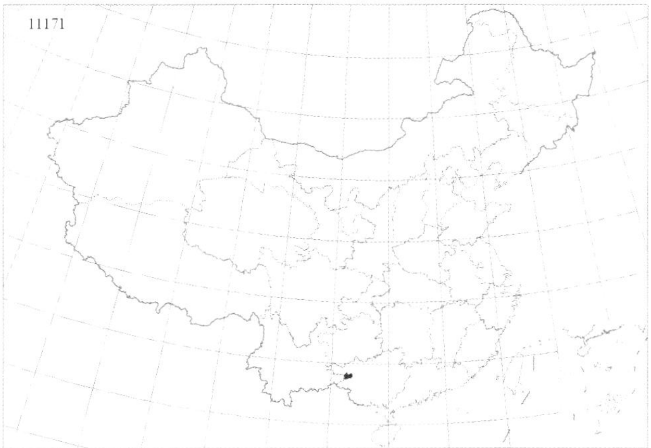

Poaceae 禾本科 *Yushania* 玉山竹属

Yushania cava T. P. Yi 空柄玉山竹

Bamboos, evergreen

10.2；2.1；17.3；10.3；69.1；6.0；620 ❙ 1029；587；20 ❙ 65.9；620；308.4

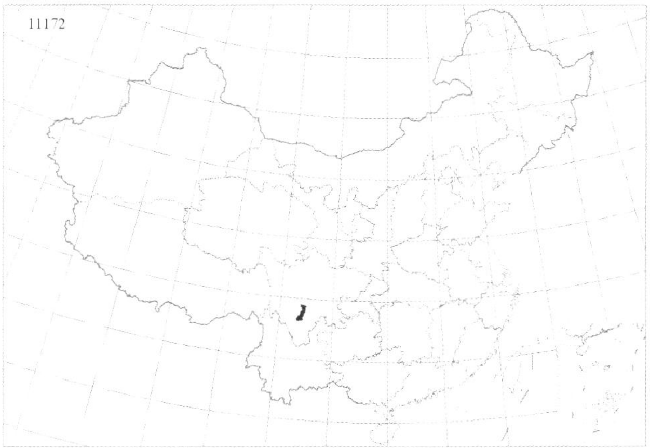

Poaceae 禾本科 *Yushania* 玉山竹属
Yushania chingii T. P. Yi 仁昌玉山竹
Bamboos, evergreen

19.8; 11.1; 26.4; 19.8; 177.9; 0.0; 1010 ❙ 1182; 673; 43 ❙ 16.9; 1010; 354.5

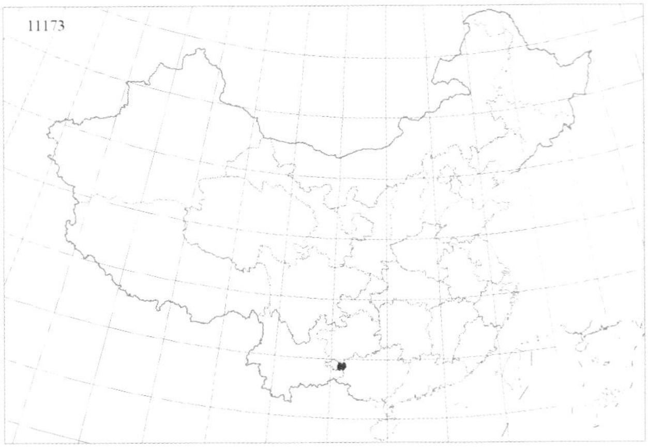

Poaceae 禾本科 *Yushania* 玉山竹属
Yushania collina T. P. Yi 德昌玉山竹
Bamboos, evergreen

13.6; 6.1; 19.6; 13.7; 103.9; 0.0; 713 ❙ 1042; 641; 20 ❙ 46.1; 691; 331.6

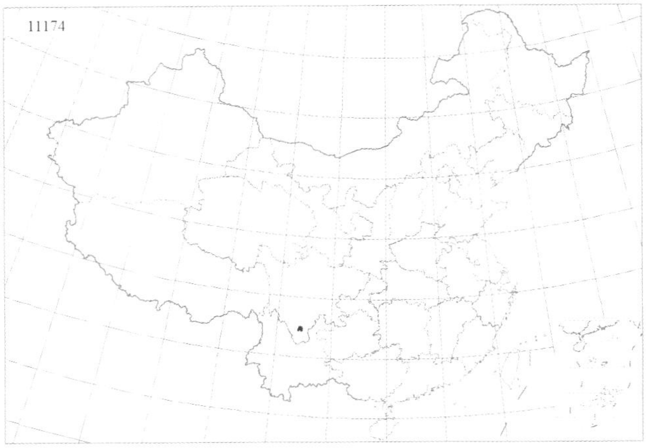

Poaceae 禾本科 *Yushania* 玉山竹属
Yushania complanata T. P. Yi 梵净山玉山竹
Bamboos, evergreen

15.0; 4.0; 25.7; 15.1; 122.0; 1.1; 806 ❙ 1247; 484; 89 ❙ 54.7; 806; 189.9

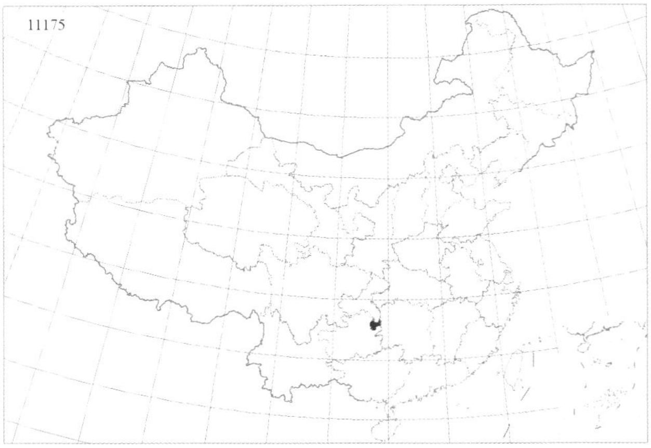

Poaceae 禾本科 *Yushania* 玉山竹属
Yushania crassicollis T. P. Yi 粗柄玉山竹
Bamboos, evergreen

19.0; 12.8; 23.0; 19.0; 168.1; 0.0; 911 ❙ 1345; 771; 58 ❙ 47.6; 895; 354.7

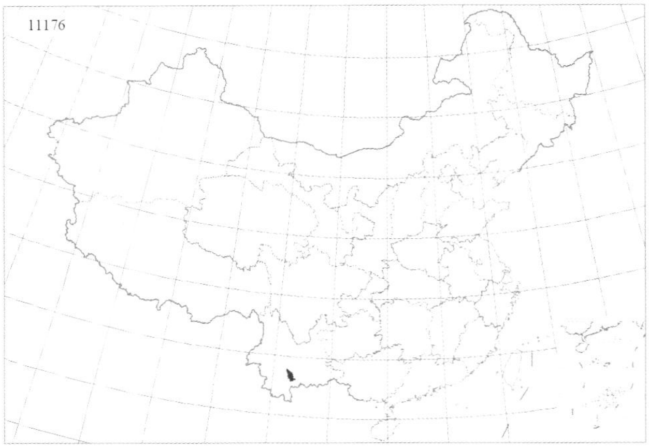

Poaceae 禾本科 *Yushania* 玉山竹属
Yushania crispata T. P. Yi 波柄玉山竹
Bamboos, evergreen

12.6; 5.0; 18.8; 12.7; 96.6; 4.9; 688 ❙ 985; 609; 17 ❙ 45.3; 658; 319.8

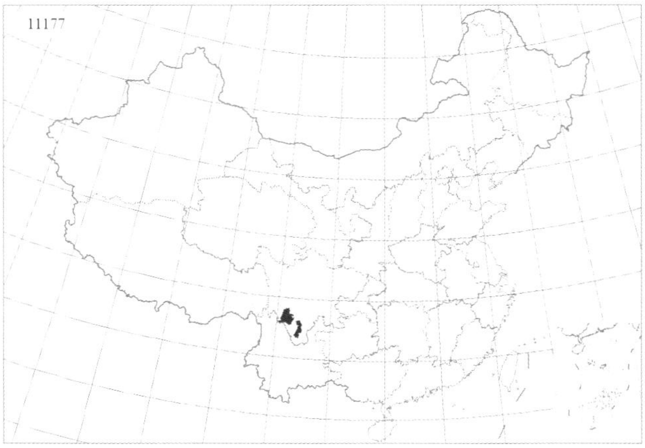

Poaceae 禾本科 *Yushania* 玉山竹属
Yushania elevata T. P. Yi 腾冲玉山竹
Bamboos, evergreen

14.4; 7.5; 19.1; 14.4; 113.1; 0.0; 727 ❙ 1454; 792; 70 ❙ 99.9; 727; 487.2

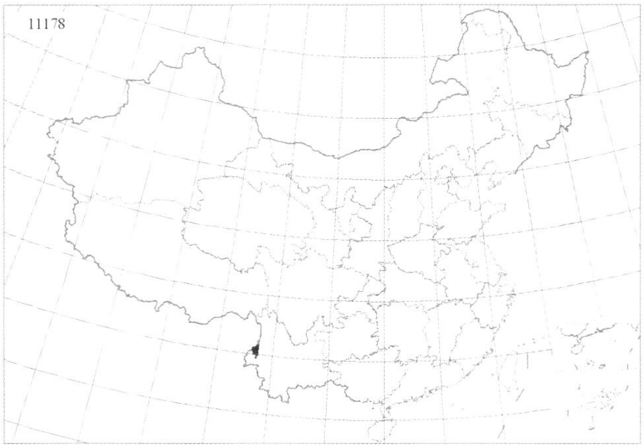

Poaceae 禾本科 *Yushania* 玉山竹属
Yushania exilis T. P. Yi 沐川玉山竹
Bamboos, evergreen

14.5; 5.0; 23.2; 14.6; 115.6; 0.6; 770 ▎1138; 633; 44 ▎48.4; 770; 232.2

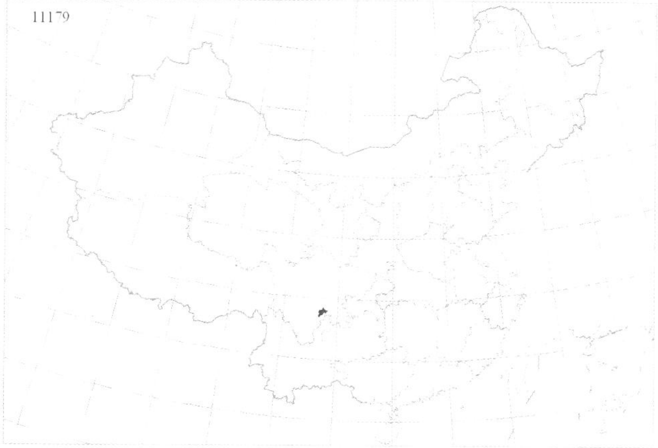

Poaceae 禾本科 *Yushania* 玉山竹属
Yushania falcatiaurita Hsueh et T. P. Yi 粉竹
Bamboos, evergreen

14.4; 7.5; 19.1; 14.4; 113.1; 0.0; 727 ▎1454; 792; 70 ▎99.9; 727; 487.2

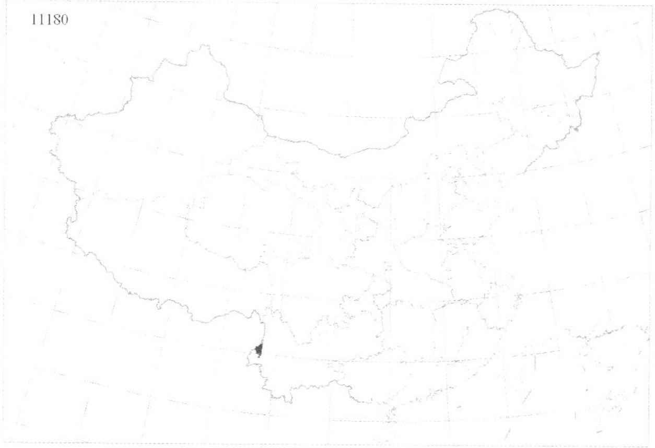

Poaceae 禾本科 *Yushania* 玉山竹属
Yushania farcticaulis T. P. Yi 独龙江玉山竹
Bamboos, evergreen

8.6; 1.3; 15.0; 8.7; 52.6; 8.8; 574 ▎1041; 585; 45 ▎81.5; 574; 352.3

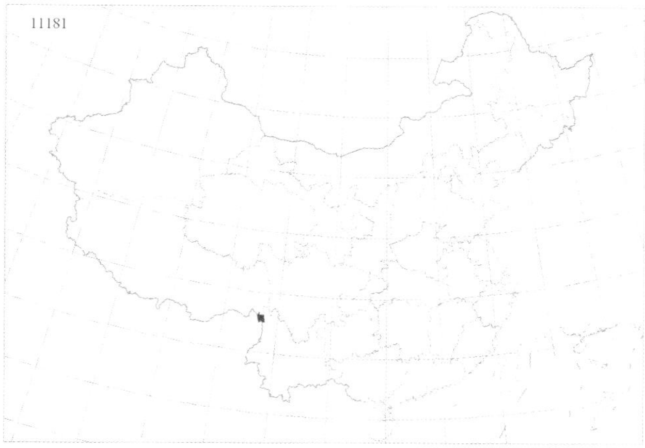

Poaceae 禾本科 *Yushania* 玉山竹属
Yushania farinosa Y. P. Wang et G. H. Ye 湖南玉山竹
Bamboos, evergreen

16.2 (14.3~17.6); 5.0 (3.3~7.2); 26.6 (24.3~28.1); 16.2 (14.3~17.7); 135.4 (114.1~151.9); 0.6 (0.0~2.3); 861 (765~923) ▎1556 (1415~1883); 497 (419~629); 203 (181~240) ▎81.0 (55.1~104.0); 851 (765~923); 227.3 (177.4~292.1)

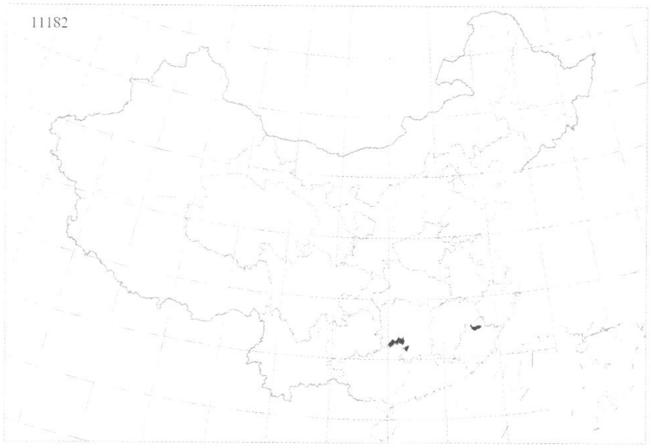

Poaceae 禾本科 *Yushania* 玉山竹属
Yushania flexa T. P. Yi 弯毛玉山竹
Bamboos, evergreen

18.6; 12.7; 22.5; 18.7; 163.9; 0.0; 888 ▎1547; 911; 63 ▎74.2; 888; 551.3

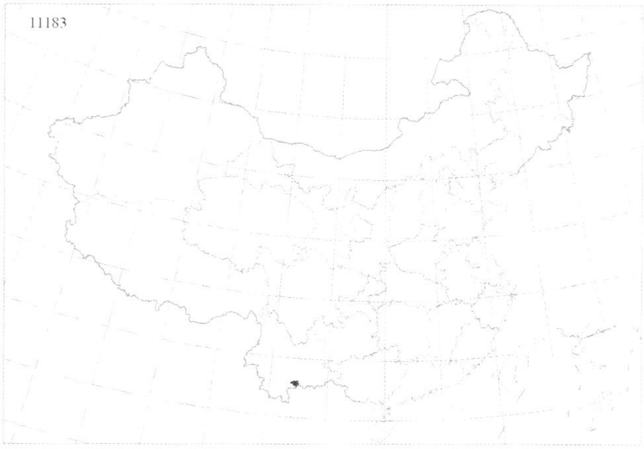

Poaceae 禾本科 *Yushania* 玉山竹属
Yushania glandulosa Hsueh et T. P. Yi 盈江玉山竹
Bamboos, evergreen

16.4; 9.6; 20.6; 16.4; 137.3; 0.0; 795 ▎1688; 959; 62 ▎112.2; 795; 627.0

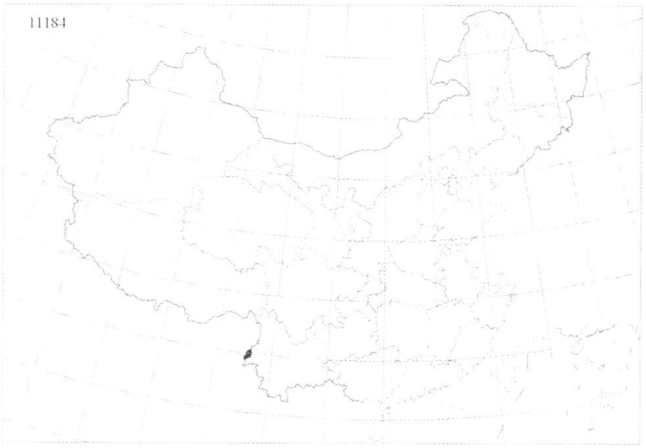

Poaceae 禾本科 *Yushania* 玉山竹属
Yushania glauca T. P. Yi et T. L. Long 白背玉山竹
Bamboos, evergreen

11.9; 2.9; 20.0; 12.0; 87.2; 3.4; 671 ❙ 1001; 550; 35 ❙ 49.1; 671; 278.8

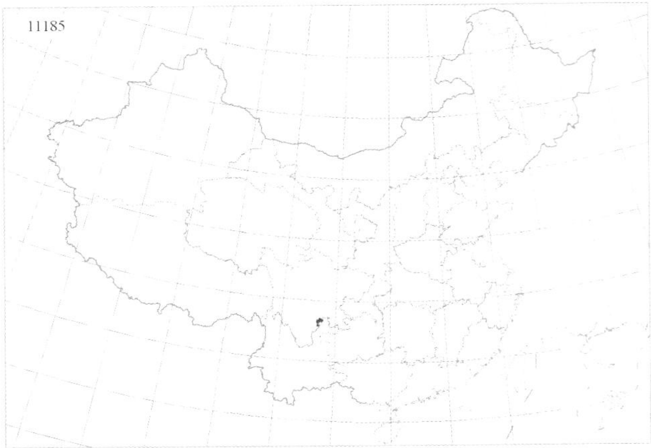

Poaceae 禾本科 *Yushania* 玉山竹属
Yushania hirticaulis Z. P. Wang et G. H. Ye 毛竿玉山竹
Bamboos, evergreen

17.3 (16.6~18.1); 6.2 (5.9~6.9); 27.7 (26.3~28.7); 17.4 (16.6~18.1); 148.7 (138.9~157.7); 0.0 (0.0~0.0); 923 (866~965) ❙ 1872 (1773~1997); 612 (564~669); 247 (233~259) ❙ 103.4 (83.7~130.7); 919 (866~961); 268.3 (227.8~303.4)

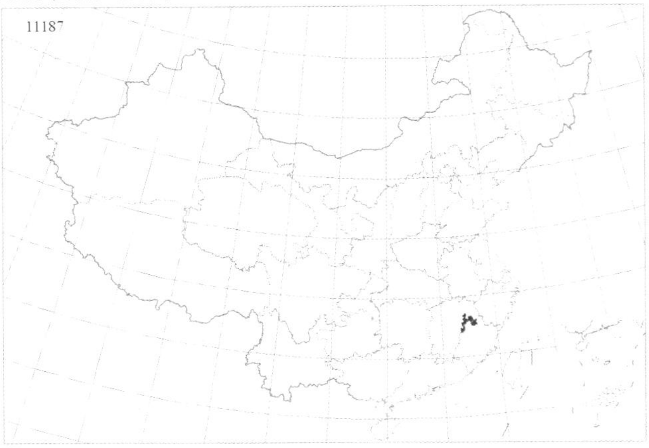

Poaceae 禾本科 *Yushania* 玉山竹属
Yushania laetevirens T. P. Yi 亮绿玉山竹
Bamboos, evergreen

15.8; 9.0; 20.6; 15.8; 129.7; 0.0; 776 ❙ 1011; 586; 44 ❙ 30.3; 772; 258.7

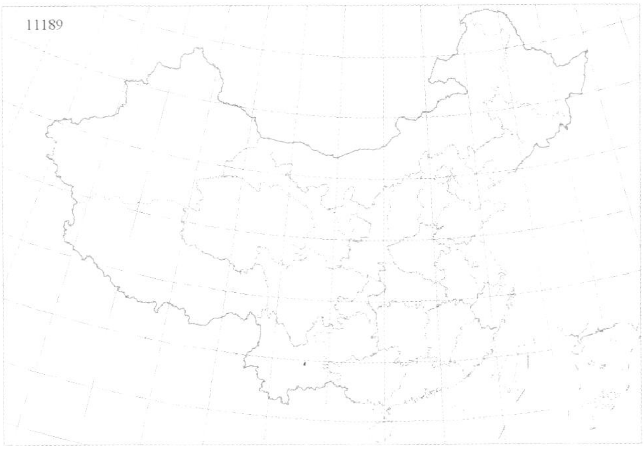

Poaceae 禾本科 *Yushania* 玉山竹属
Yushania grammata T. P. Yi 棱纹玉山竹
Bamboos, evergreen

14.0; 4.1; 23.1; 14.1; 109.8; 0.9; 747 ❙ 988; 499; 55 ❙ 32.3; 747; 214.6

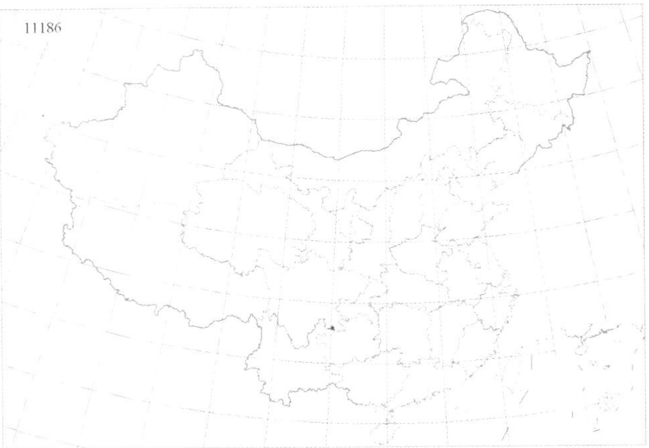

Poaceae 禾本科 *Yushania* 玉山竹属
Yushania lacera Q. F. Zheng et K. F. Huang 撕裂玉山竹
Bamboos, evergreen

17.5; 6.4; 27.8; 17.6; 150.7; 0.0; 931 ❙ 1875; 618; 249 ❙ 101.6; 931; 266.5

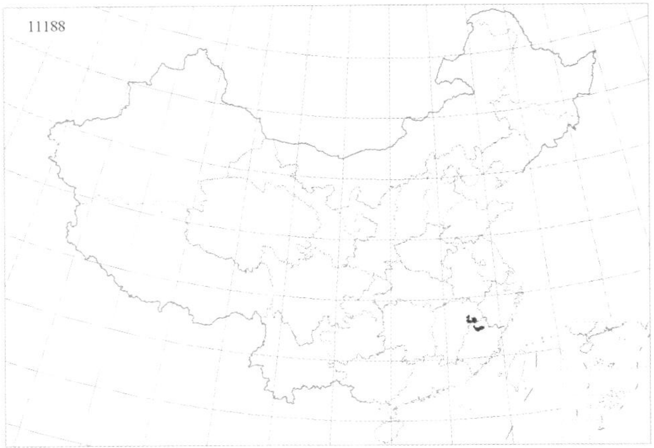

Poaceae 禾本科 *Yushania* 玉山竹属
Yushania levigata T. P. Yi 光亮玉山竹
Bamboos, evergreen

16.3; 10.2; 20.3; 16.4; 136.6; 0.0; 789 ❙ 1232; 703; 47 ❙ 56.2; 780; 372.4

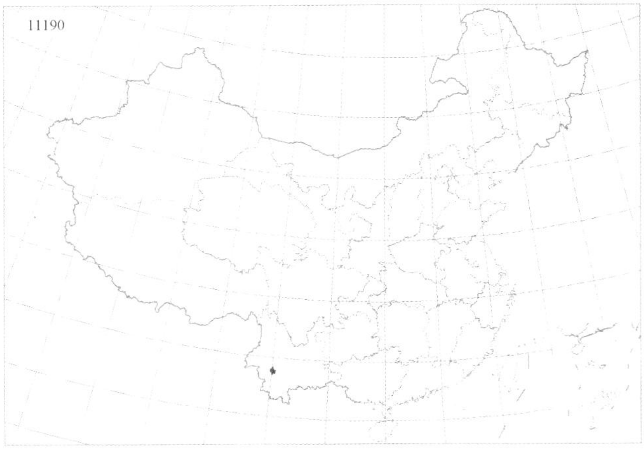

Poaceae 禾本科 *Yushania* 玉山竹属
Yushania lineolata T. P. Yi 石棉玉山竹
Bamboos, evergreen

10.1; 1.8; 17.7; 10.3; 70.1; 7.0; 621 ▎1023; 568; 22 ▎64.7; 621; 318.7

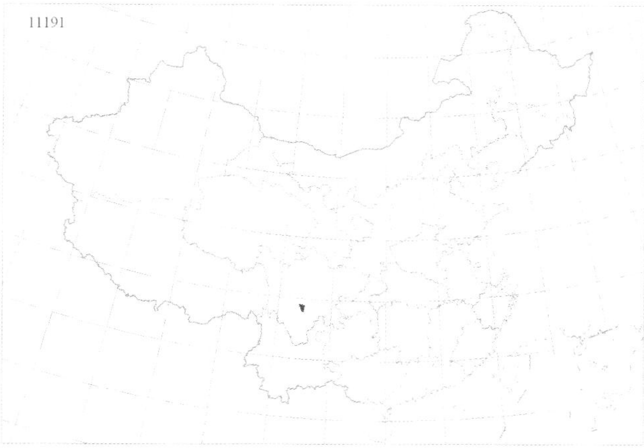

Poaceae 禾本科 *Yushania* 玉山竹属
Yushania longiaurita Q. F. Zheng et K. F. Huang 长耳玉山竹
Bamboos, evergreen

17.3; 9.0; 25.2; 17.3; 147.6; 0.0; 864 ▎1600; 599; 175 ▎85.2; 864; 305.7

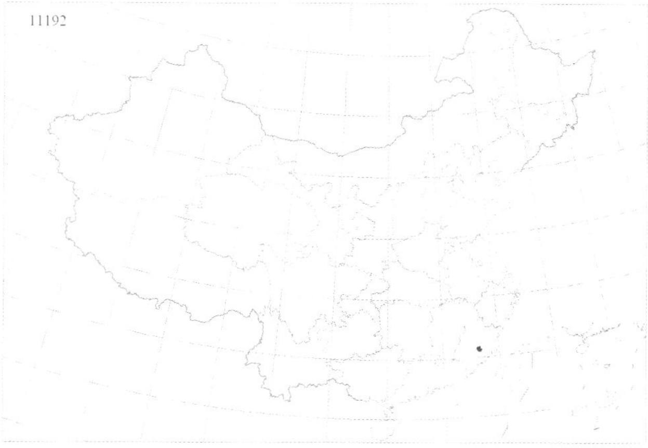

Poaceae 禾本科 *Yushania* 玉山竹属
Yushania longiuscula T. P. Yi 蒙自玉山竹
Bamboos, evergreen

16.9; 10.3; 21.4; 16.9; 142.7; 0.0; 813 ▎1178; 646; 51 ▎44.9; 813; 383.4

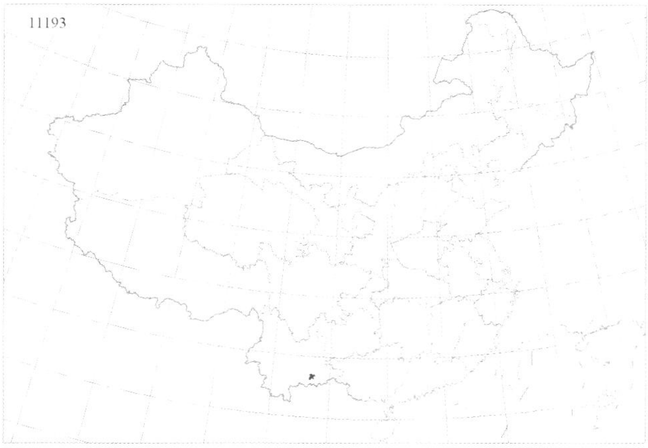

Poaceae 禾本科 *Yushania* 玉山竹属
Yushania mabianensis T. P. Yi 马边玉山竹
Bamboos, evergreen

12.5; 3.3; 20.7; 12.5; 92.7; 2.3; 690 ▎1049; 580; 37 ▎52.1; 690; 267.0

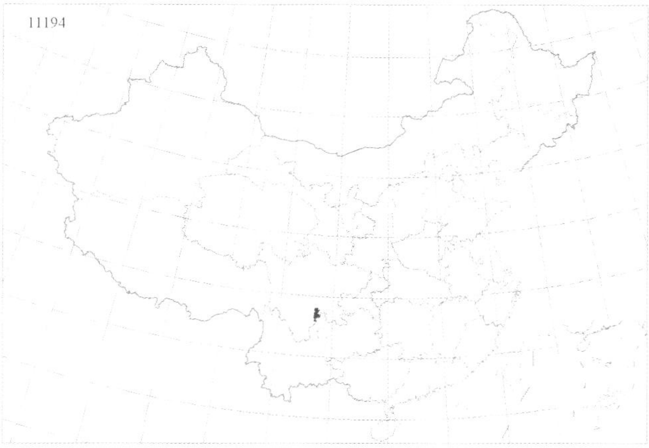

Poaceae 禾本科 *Yushania* 玉山竹属
Yushania maculata T. P. Yi 斑壳玉山竹
Bamboos, evergreen

12.0; 4.2; 18.4; 12.0; 86.4; 2.3; 669 ▎1017; 604; 21 ▎53.3; 662; 299.4

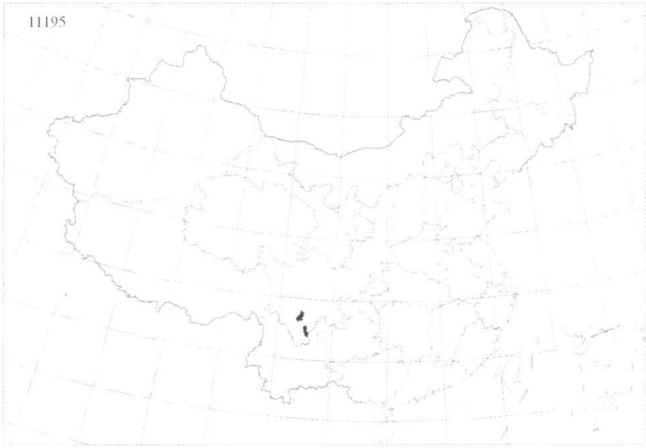

Poaceae 禾本科 *Yushania* 玉山竹属
Yushania menghaiensis T. P. Yi 隔界竹
Bamboos, evergreen

19.2; 13.8; 22.3; 19.2; 170.5; 0.0; 909 ▎1499; 686; 58 ▎64.9; 880; 513.9

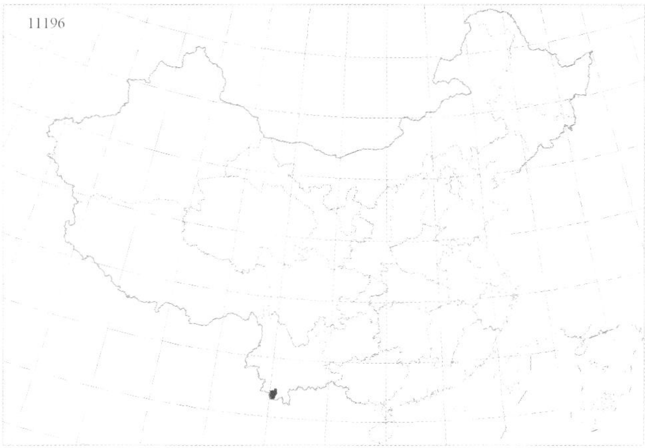

Poaceae 禾本科 *Yushania* 玉山竹属
Yushania mitis T. P. Yi 泡滑竹
Bamboos, evergreen

12.1; 3.0; 20.0; 12.1; 88.4; 2.8; 675 ❙ 944; 520; 33 ❙ 39.8; 675; 249.8

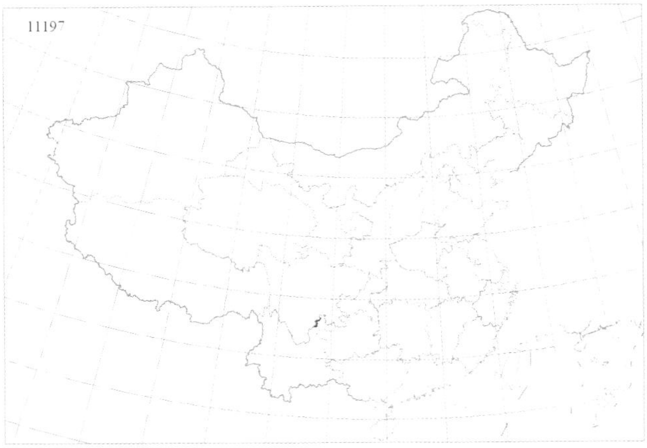

Poaceae 禾本科 *Yushania* 玉山竹属
Yushania niitakayamensis (Hayata) Keng f. 玉山竹
Bamboos, evergreen

18.0 (14.7~21.7); 11.2 (2.5~15.3); 23.9 (20.1~27.0); 18.1 (14.8~21.7); 157.4 (121.1~200.2); 0.6 (0.0~3.9); 899 (773~1118) ❙ 2282 (1489~2859); 1033 (565~1412); 220 (110~402) ❙ 160.1 (64.2~246.5); 892 (773~1075); 255.1 (202.8~289.1)

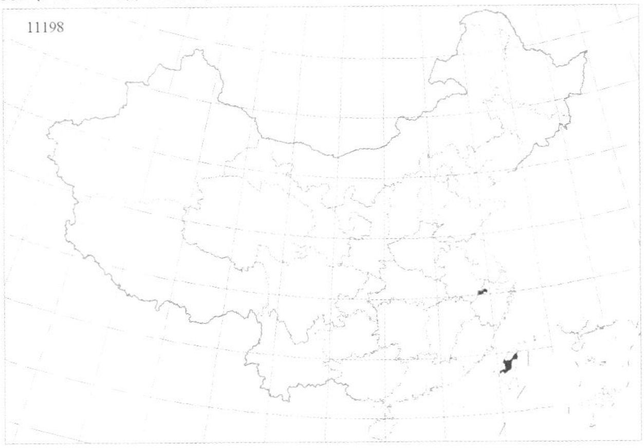

Poaceae 禾本科 *Yushania* 玉山竹属
Yushania oblonga T. P. Yi 马鹿竹
Bamboos, evergreen

16.3; 9.1; 21.4; 16.4; 136.4; 0.0; 797 ❙ 1190; 656; 50 ❙ 49.4; 797; 347.3

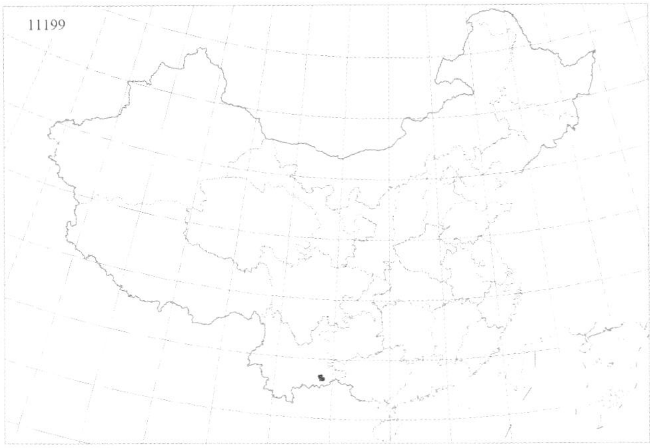

Poaceae 禾本科 *Yushania* 玉山竹属
Yushania pachyclada T. P. Yi 粗枝玉山竹
Bamboos, evergreen

15.9; 6.1; 24.9; 16.0; 131.7; 0.0; 824 ❙ 985; 509; 52 ❙ 19.4; 824; 200.3

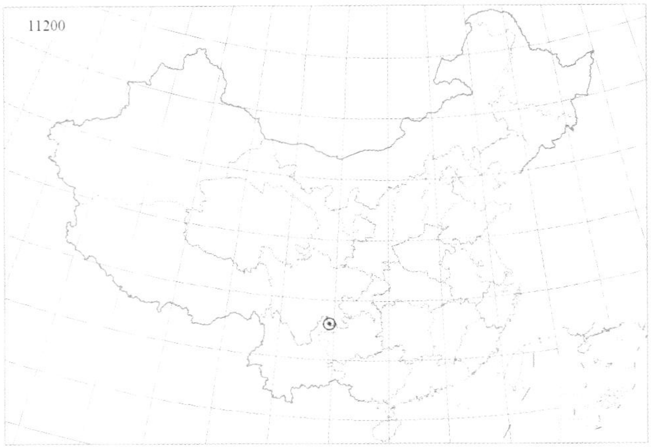

Poaceae 禾本科 *Yushania* 玉山竹属
Yushania polytricha Hsueh et T. P. Yi 滑竹
Bamboos, evergreen

17.0 (14.4~21.0); 10.4 (7.5~14.2); 21.6 (19.1~25.6); 17.0 (14.4~21.0); 144.3 (113.1~191.8); 0.0 (0.0~0.0); 834 (727~1053) ❙ 1214 (1007~1547); 682 (561~911); 54 (39~70) ❙ 46.2 (21.3~99.9); 822 (727~1053); 381.5 (274.9~551.6)

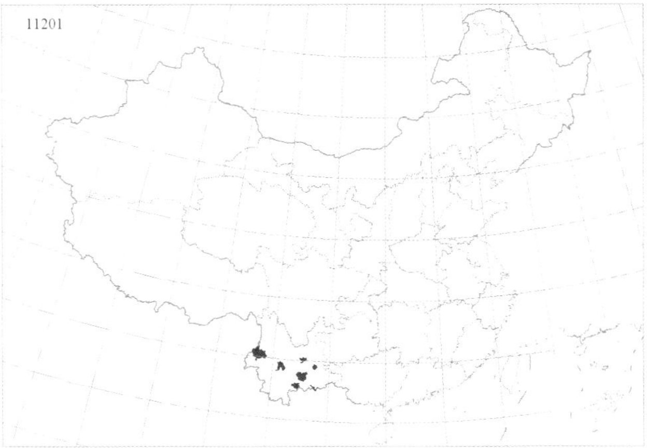

Poaceae 禾本科 *Yushania* 玉山竹属
Yushania punctulata T. P. Yi 抱鸡竹
Bamboos, evergreen

15.4; 5.8; 24.3; 15.5; 126.3; 0.0; 806 ❙ 1294; 737; 49 ❙ 61.1; 806; 211.9

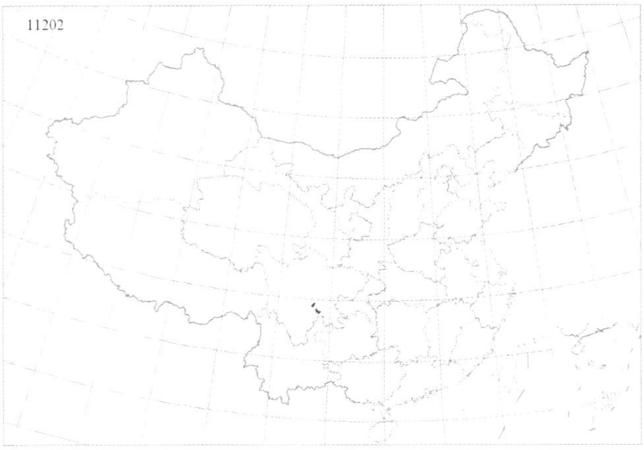

Poaceae 禾本科 *Yushania* 玉山竹属
Yushania qiaojiaensis Hsueh et T. P. Yi 海竹
Bamboos, evergreen

12.7; 4.7; 19.2; 12.7; 93.0; 0.3; 687 | 922; 546; 26 | 34.2; 687; 232.7

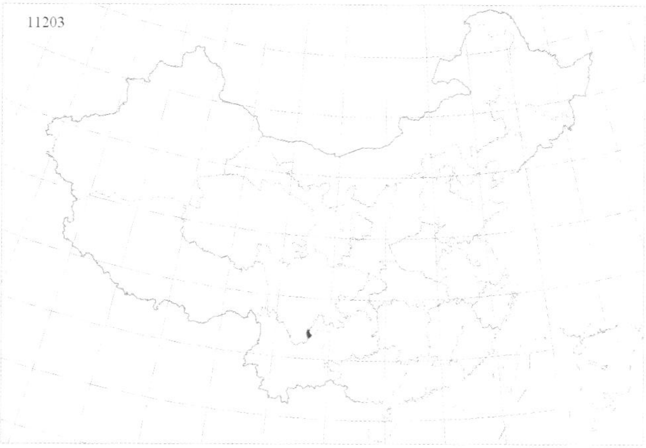

Poaceae 禾本科 *Yushania* 玉山竹属
Yushania qiaojiaensis var. nuda (T. P. Yi) D. Z. Li et Z. H. Guo 裸箨海竹
Bamboos, evergreen

12.4; 3.9; 19.6; 12.4; 90.7; 1.5; 681 | 933; 533; 30 | 37.0; 681; 241.3

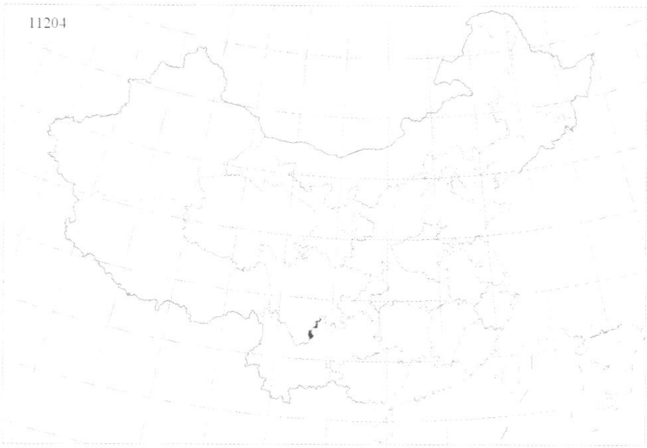

Poaceae 禾本科 *Yushania* 玉山竹属
Yushania rugosa T. P. Yi 皱叶玉山竹
Bamboos, evergreen

18.7; 9.5; 26.0; 18.7; 164.9; 0.0; 950 | 1255; 666; 56 | 32.2; 950; 247.0

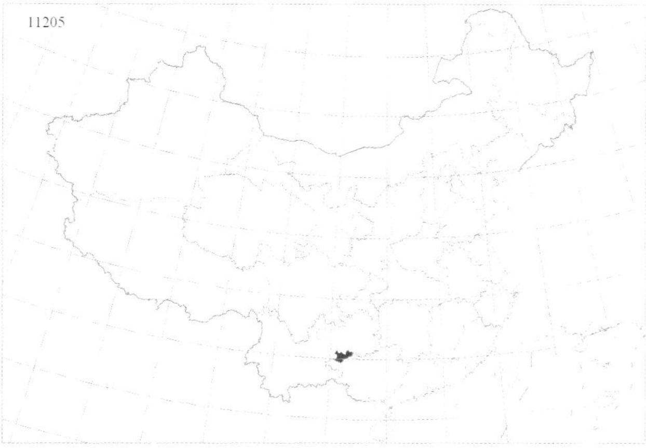

Poaceae 禾本科 *Yushania* 玉山竹属
Yushania uniramosa Hsueh et T. P. Yi 单枝玉山竹
Bamboos, evergreen

15.3; 4.7; 25.3; 15.4; 124.7; 0.3; 810 | 1111; 484; 68 | 37.3; 810; 179.6

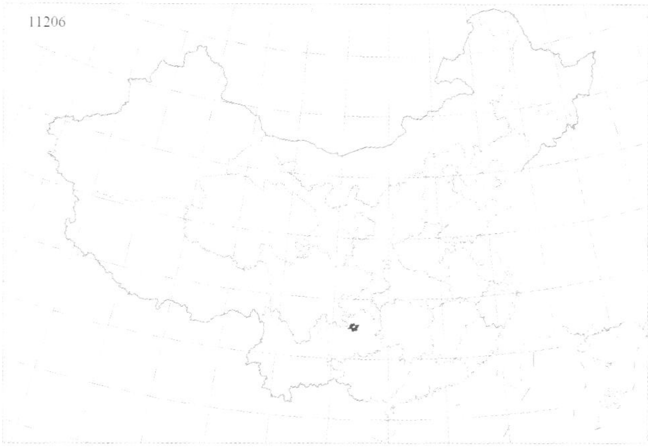

Poaceae 禾本科 *Yushania* 玉山竹属
Yushania varians T. P. Yi 庐山玉山竹
Bamboos, evergreen

16.9; 4.5; 29.1; 17.0; 144.0; 0.5; 931 | 1560; 630; 171 | 67.5; 931; 205.0

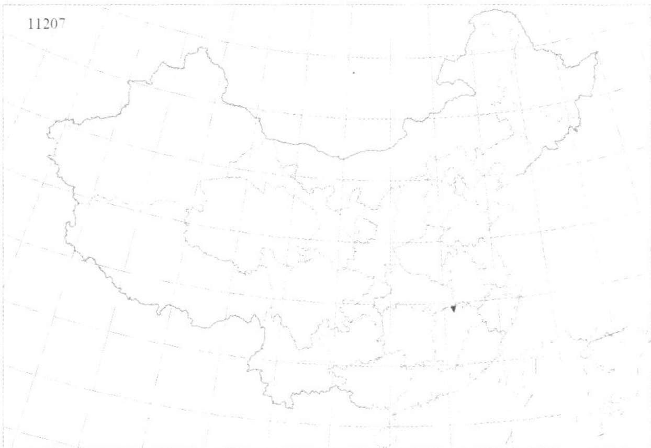

Poaceae 禾本科 *Yushania* 玉山竹属
Yushania weixiensis T. P. Yi 竹扫子
Bamboos, evergreen

9.6; 2.4; 15.7; 9.6; 60.8; 5.4; 600 | 1007; 578; 40 | 67.9; 600; 351.7

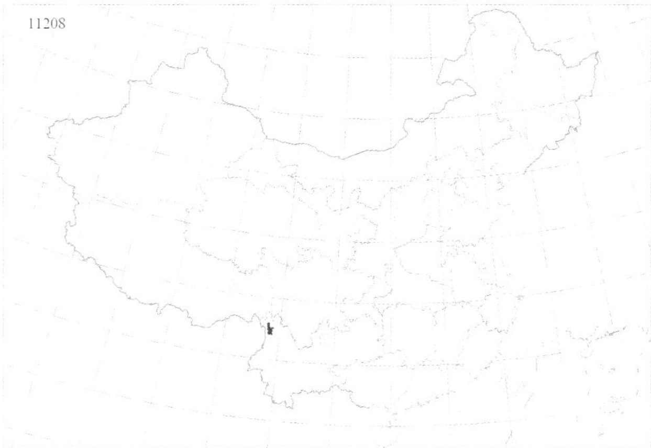

Poaceae 禾本科 *Yushania* 玉山竹属
Yushania wuyishanensis Q. F. Zheng et K. F. Huang 武夷山玉山竹
Bamboos, evergreen

17.1; 6.4; 27.0; 17.1; 145.5; 0.0; 899 ❙ 1934; 640; 251 ❙ 115.3; 899; 291.7

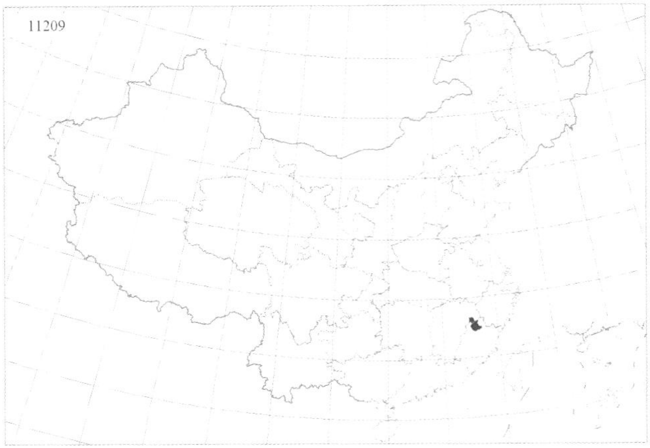

Poaceae 禾本科 *Yushania* 玉山竹属
Yushania xizangensis T. P. Yi 西藏玉山竹
Bamboos, evergreen

12.2; 4.3; 18.5; 12.2; 87.4; 0.7; 672 ❙ 1431; 817; 31 ❙ 113.0; 672; 363.7

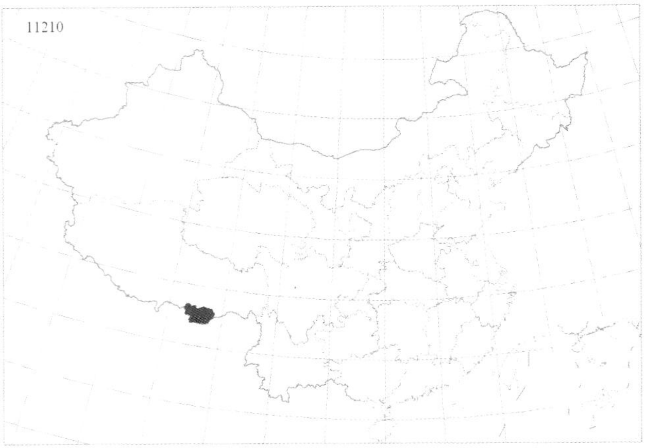

Poaceae 禾本科 *Yushania* 玉山竹属
Yushania yadongensis T. P. Yi 亚东玉山竹
Bamboos, evergreen

0.3; -8.6; 8.3; 2.9; 9.6; 65.4; 368 ❙ 389; 246; 11 ❙ 0.0; 368; 142.5

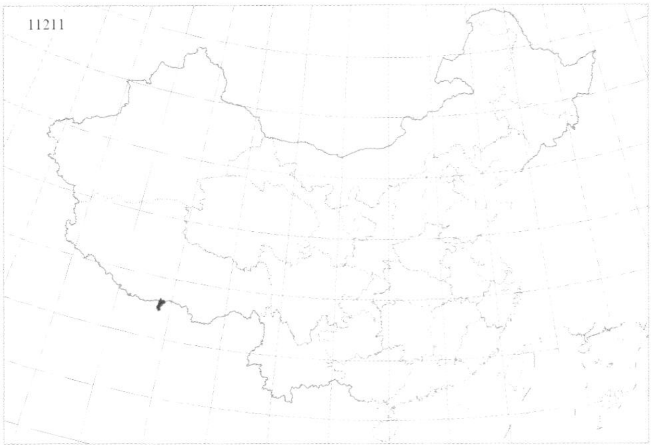

Pandanaceae 露兜树科 *Freycinetia* 藤露兜属
Freycinetia formosana Hemsl. 山露兜
Vines

19.2; 13.8; 23.8; 19.2; 170.6; 0.0; 925 ❙ 2853; 1156; 367 ❙ 208.9; 925; 312.6

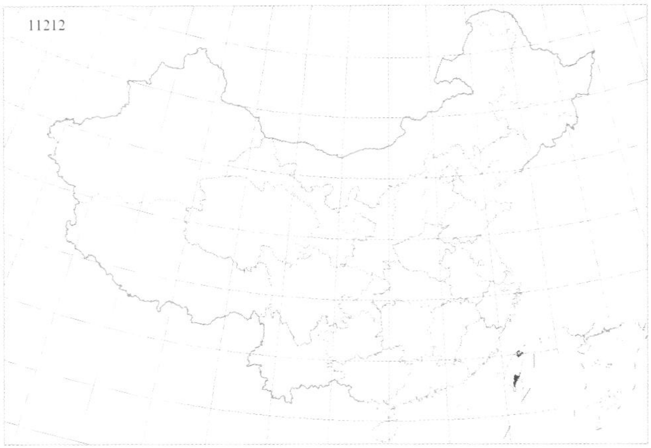

Pandanaceae 露兜树科 *Pandanus* 露兜树属
Pandanus amaryllifolius Roxb. 香露兜
Herbs or subshrubs, evergreen

24.2; 18.1; 28.7; 24.3; 231.2; 0.0; 1335 ❙ 1176; 447; 59 ❙ -12.0; 1139; 411.8

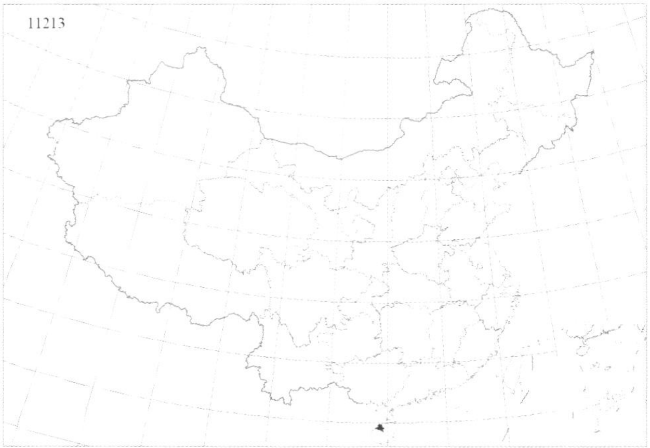

Pandanaceae 露兜树科 *Pandanus* 露兜树属
Pandanus austrosinensis T. L. Wu 露兜草
Herbs or subshrubs, evergreen

21.9 (17.6~24.7); 14.3 (7.3~19.8); 27.9 (26.3~28.8); 21.9 (17.6~24.7); 203.2 (151.6~236.5); 0.0 (0.0~0.0); 1169 (909~1376) ❙ 1655 (1269~2005); 674 (455~891); 120 (48~177) ❙ 43.6 (-1.0~85.1); 1139 (909~1326); 295.5 (157.7~572.9)

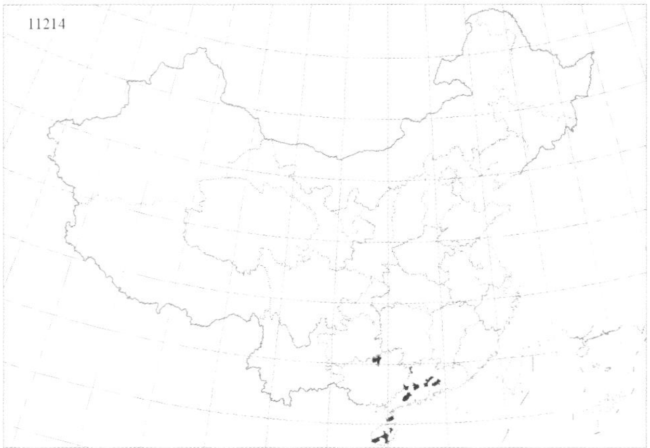

Pandanaceae 露兜树科 *Pandanus* 露兜树属
Pandanus fibrosus Gagnepain ex Humbert 小露兜
Herbs or shrubs, evergreen

23.6 (21.6~24.7); 17.7 (14.1~19.8); 27.8 (26.7~28.6); 23.6 (21.6~24.7); 223.1 (199.4~236.5); 0.0 (0.0~0.0); 1277 (1128~1376) ❙ 1600 (1269~1854); 590 (534~683); 85 (48~137) ❙ 25.8 (−1.0~42.9); 1213 (1128~1360); 425.5 (201.9~620.0)

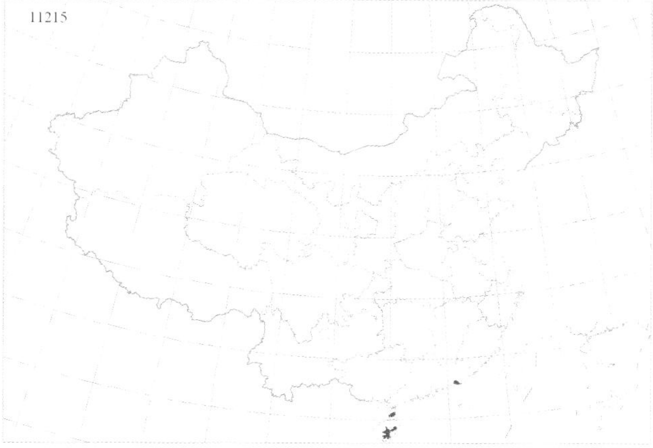

Pandanaceae 露兜树科 *Pandanus* 露兜树属
Pandanus kaida Kurz 簕古子
Shrubs or small trees, evergreen

22.0 (17.9~24.8); 14.3 (8.1~20.1); 28.0 (26.3~29.0); 22.0 (17.9~24.8); 204.4 (155.4~237.6); 0.0 (0.0~0.0); 1172 (918~1385) ❙ 1652 (1176~2037); 747 (447~1033); 112 (41~175) ❙ 42.7 (−12.0~71.8); 1137 (918~1239); 262.3 (163.7~527.5)

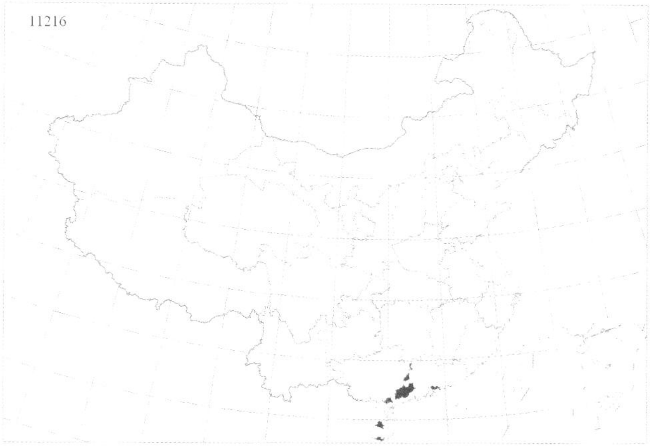

Pandanaceae 露兜树科 *Pandanus* 露兜树属
Pandanus tectorius Sol. 露兜树
Shrubs or small trees, evergreen

20.7 (16.4~24.8); 12.6 (7.8~20.1); 26.9 (20.6~29.0); 20.7 (16.4~24.8); 188.5 (137.3~237.6); 0.0 (0.0~0.0); 1076 (795~1385) ❙ 1568 (1143~2993); 748 (418~1594); 122 (41~547) ❙ 47.6 (−12.0~203.4); 1065 (795~1326); 293.1 (122.7~659.0)

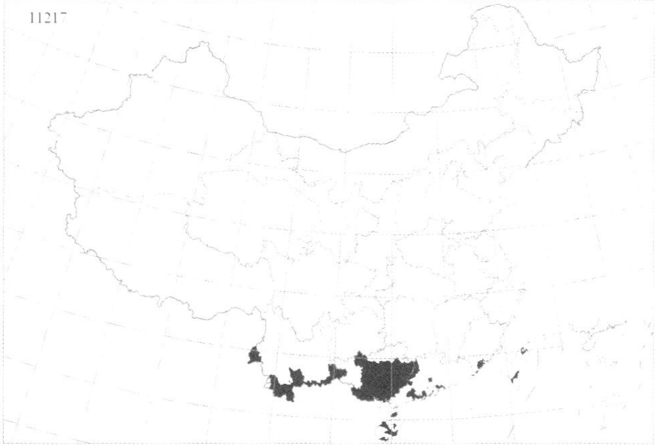

Pandanaceae 露兜树科 *Pandanus* 露兜树属
Pandanus urophyllus Hance 分叉露兜
Trees, evergreen

19.0 (16.4~22.2); 11.3 (7.1~15.1); 24.5 (20.6~28.7); 19.0 (16.4~22.3); 168.3 (137.3~207.5); 0.0 (0.0~0.0); 946 (795~1190) ❙ 1398 (1028~1751); 734 (566~996); 75 (41~166) ❙ 49.0 (5.8~112.2); 937 (783~1190); 376.2 (188.9~659.0)

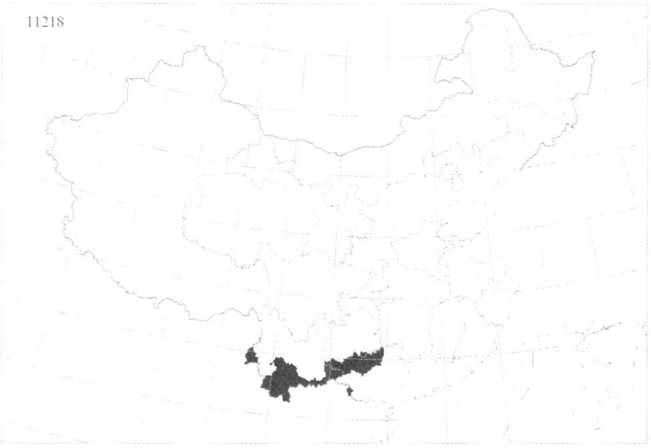

Arecaceae 棕榈科 *Areca* 槟榔属
Areca catechu Willd. 槟榔
Trees, evergreen

23.9 (22.1~24.8); 18.0 (13.7~20.1); 28.1 (25.7~29.0); 23.9 (22.1~24.8); 226.8 (205.6~237.6); 0.0 (0.0~0.0); 1301 (1127~1385) ❙ 1489 (868~1854); 546 (312~779); 78 (41~144) ❙ 14.9 (−34.3~50.9); 1194 (868~1360); 408.4 (157.7~620.0)

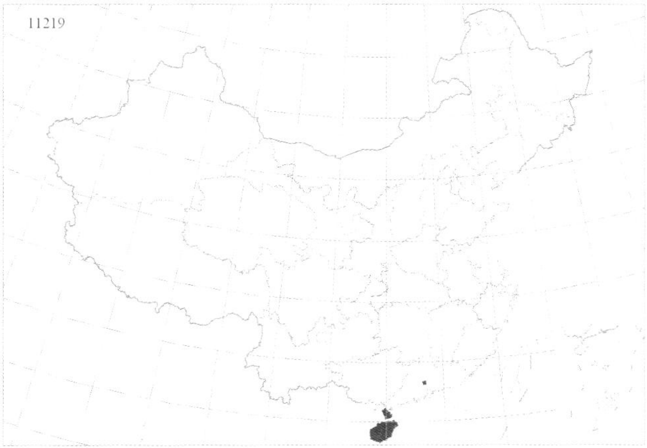

Arecaceae 棕榈科 *Areca* 槟榔属
Areca triandra Roxb. ex Buch.-Ham. 三药槟榔
Shrubs, evergreen

23.9 (22.1~24.8); 18.1 (13.7~20.1); 28.0 (25.7~28.8); 23.9 (22.1~24.8); 227.1 (205.6~237.6); 0.0 (0.0~0.0); 1303 (1127~1385) ❙ 1515 (868~1854); 546 (312~779); 77 (41~144) ❙ 16.8 (−34.3~50.9); 1191 (868~1360); 426.2 (157.7~620.0)

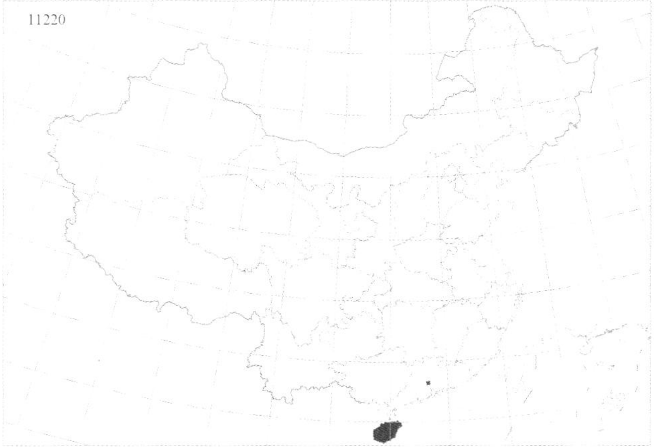

A

Arecaceae 棕榈科 *Arenga* 桄榔属

Arenga caudata (Lour.) H. E. Moore 双籽棕

Dwarf shrubs, evergreen

23.1 (20.4~24.7); 17.9 (14.9~19.8); 26.5 (23.6~28.5); 23.1 (20.5~24.7); 217.8 (185.8~236.5); 0.0 (0.0~0.0); 1232 (989~1376) ❙ 1638 (1464~1854); 603 (534~702); 75 (48~115) ❙ 35.0 (15.4~62.5); 1161 (959~1360); 553.4 (428.6~659.0)

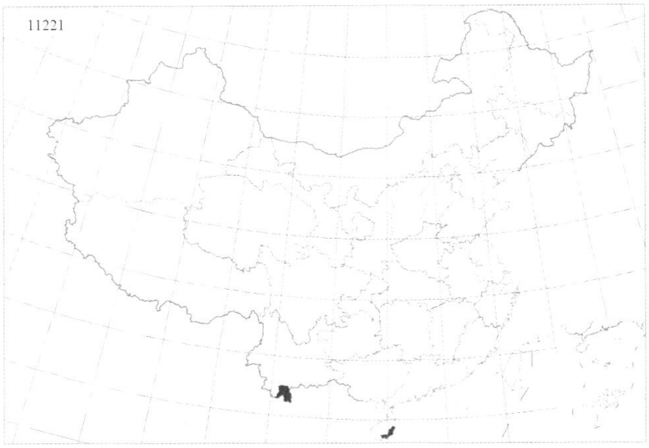

Arecaceae 棕榈科 *Arenga* 桄榔属

Arenga engleri Becc. 山棕

Shrubs, evergreen

20.5 (16.2~24.1); 14.0 (9.6~18.9); 25.7 (20.1~28.9); 20.5 (16.2~24.1); 186.0 (134.2~229.7); 0.0 (0.0~0.0); 1042 (773~1322) ❙ 2150 (1182~2993); 1084 (551~1847); 161 (43~547) ❙ 115.6 (15.4~246.5); 1023 (773~1205); 253.4 (95.3~363.7)

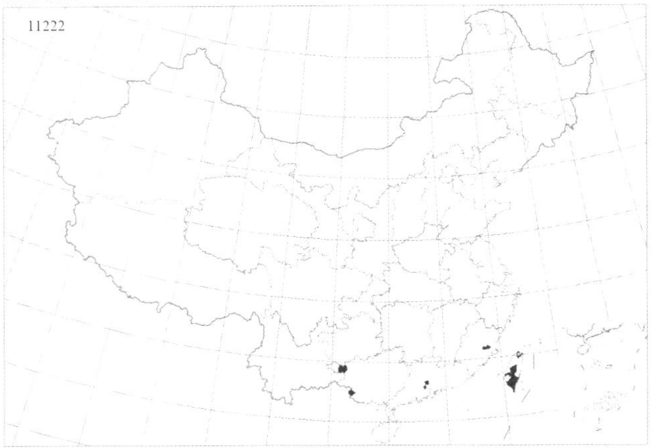

Arecaceae 棕榈科 *Arenga* 桄榔属

Arenga longicarpa C. F. Wei 长果桄榔

Shrubs, evergreen

22.8; 15.5; 28.4; 22.8; 214.1; 0.0; 1225 ❙ 1840; 889; 107 ❙ 50.2; 1225; 239.5

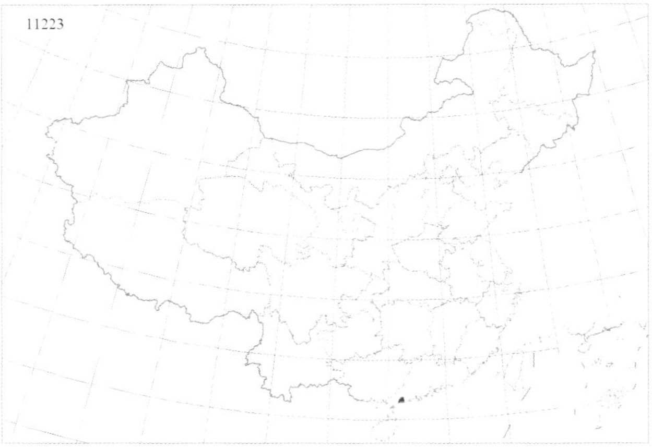

Arecaceae 棕榈科 *Arenga* 桄榔属

Arenga micrantha C. F. Wei 小花桄榔

Trees, evergreen

13.3; 5.4; 19.6; 13.4; 100.4; 0.0; 706 ❙ 1736; 1007; 51 ❙ 145.8; 706; 447.4

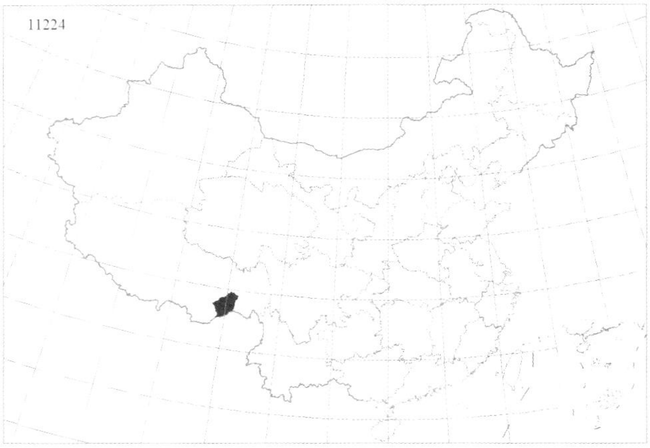

Arecaceae 棕榈科 *Arenga* 桄榔属

Arenga pinnata (Wurmb) Merr. 砂糖椰子

Trees

22.3 (19.2~24.8); 16.4 (13.2~20.1); 26.6 (22.3~28.8); 22.3 (19.2~24.8); 208.2 (170.5~237.6); 0.0 (0.0~0.0); 1173 (909~1385) ❙ 1511 (1179~1854); 586 (392~779); 86 (41~185) ❙ 31.7 (-11.9~64.9); 1084 (880~1360); 459.8 (157.7~659.0)

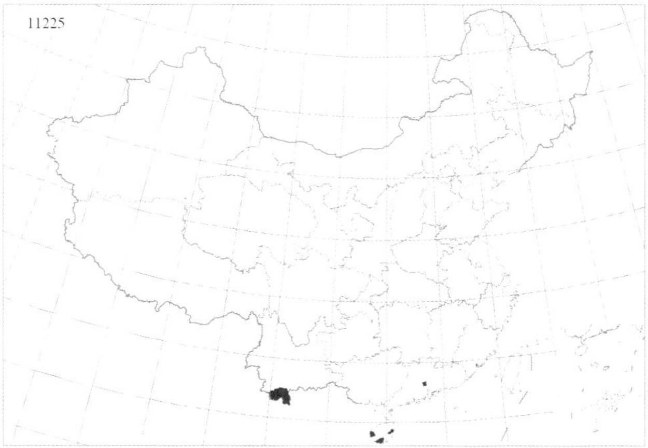

Arecaceae 棕榈科 *Arenga* 桄榔属

Arenga westerhoutii Griff. 桄榔

Trees, evergreen

21.0 (13.3~24.8); 14.2 (5.4~20.1); 25.9 (19.6~28.6); 21.1 (13.4~24.8); 193.0 (100.4~237.6); 0.0 (0.0~0.0); 1093 (706~1385) ❙ 1528 (1161~2435); 737 (392~1428); 71 (41~138) ❙ 43.2 (-11.9~145.8); 1050 (706~1360); 427.2 (235.1~659.0)

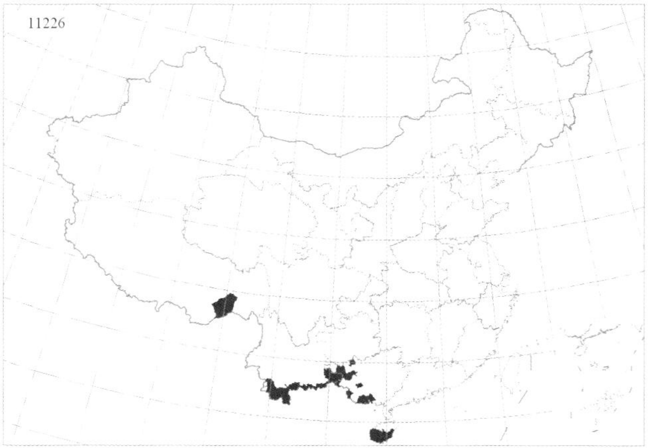

Arecaceae 棕榈科 *Calamus* 省藤属
Calamus austro-guangxiensis S. J. Pei et S. Y. Chen 桂南省藤
Vines, evergreen

21.3；13.2；27.3；21.3；195.9；0.0；1111 ❙ 1900；1056；125 ❙ 71.0；1111；297.2

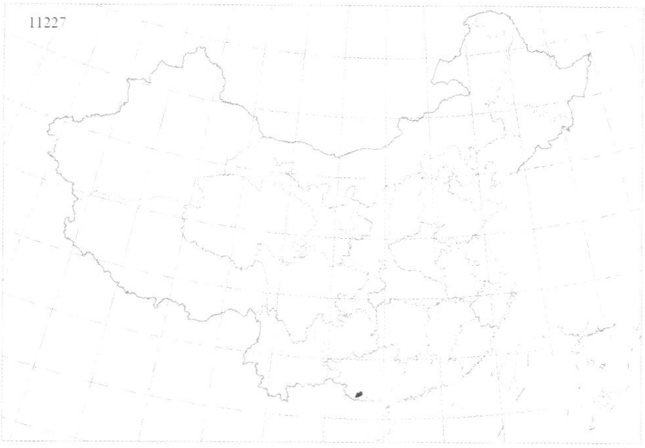

Arecaceae 棕榈科 *Calamus* 省藤属
Calamus balansaeanus Beccari 小白藤
Vines

19.5 (17.4~22.2)；10.0 (7.2~13.9)；27.2 (25.0~28.7)；19.5 (17.4~22.3)；174.3 (148.9~207.5)；0.0 (0.0~0.0)；1016 (886~1190) ❙ 1510 (1188~1776)；696 (561~850)；133 (49~189) ❙ 50.0 (5.8~85.5)；1016 (886~1190)；262.2 (203.9~342.5)

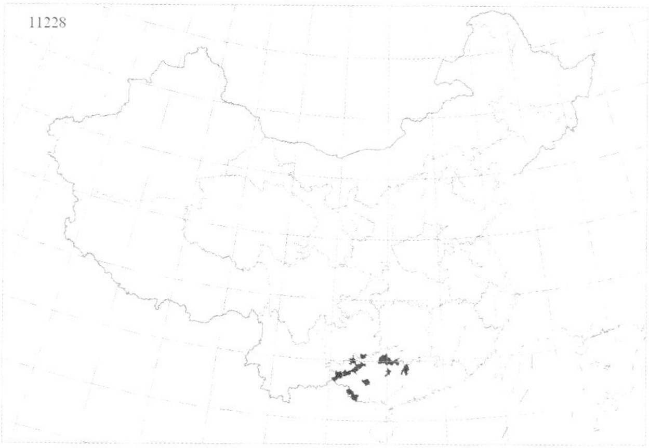

Arecaceae 棕榈科 *Calamus* 省藤属
Calamus balansaeanus var. castaneolepis (C. F. Wei) S. J. Pei et S. Y. Chen 褐鳞省藤
Vines

19.0 (17.5~21.6)；8.8 (7.2~12.6)；27.7 (26.4~28.7)；19.0 (17.5~21.7)；168.3 (150.1~199.8)；0.0 (0.0~0.0)；1000 (904~1153) ❙ 1705 (1625~1776)；713 (626~795)；185 (173~189) ❙ 71.6 (40.9~85.5)；1000 (904~1153)；229.2 (203.9~263.3)

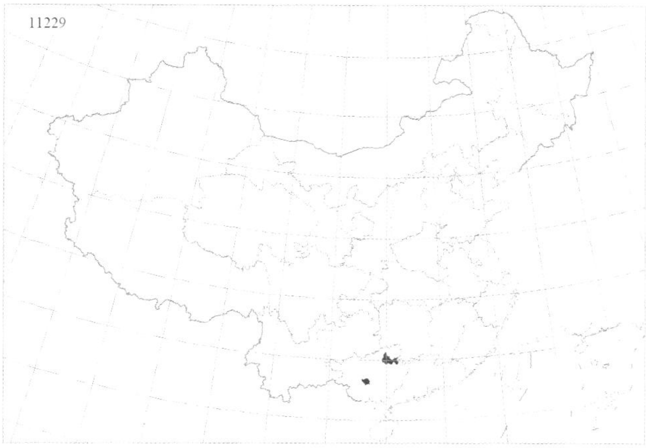

Arecaceae 棕榈科 *Calamus* 省藤属
Calamus bonianus Beccari 多穗白藤
Vines

23.5 (20.4~24.8)；17.6 (13.7~20.1)；27.6 (23.6~28.8)；23.5 (20.5~24.8)；221.8 (185.8~237.6)；0.0 (0.0~0.0)；1269 (989~1385) ❙ 1516 (1179~1840)；582 (392~889)；73 (41~144) ❙ 21.5 (-11.9~62.5)；1146 (986~1254)；440.0 (157.7~659.0)

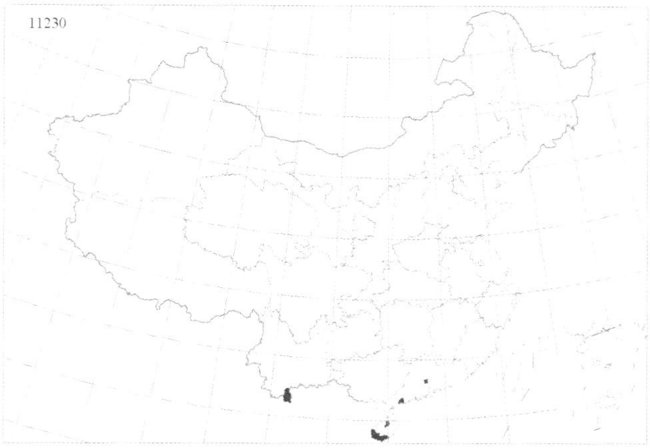

Arecaceae 棕榈科 *Calamus* 省藤属
Calamus compsostachys Burret 短轴省藤
Vines, evergreen

21.9；13.8；28.4；21.9；203.2；0.0；1162 ❙ 1719；780；127 ❙ 48.1；1162；227.8

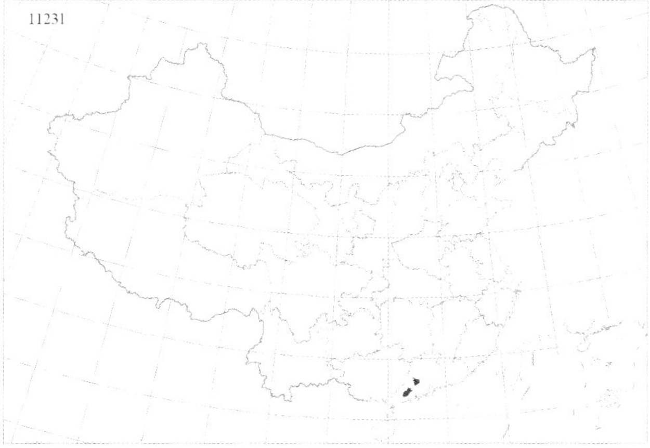

Arecaceae 棕榈科 *Calamus* 省藤属
Calamus dianbaiensis C. F. Wei 电白省藤
Vines, evergreen

22.0 (21.4~22.8)；14.1 (13.1~15.5)；28.1 (27.7~28.4)；22.0 (21.4~22.8)；204.1 (196.9~214.1)；0.0 (0.0~0.0)；1165 (1121~1225) ❙ 1664 (1463~1876)；786 (706~891)；120 (107~137) ❙ 42.8 (24.9~63.2)；1165 (1121~1225)；253.7 (239.5~273.3)

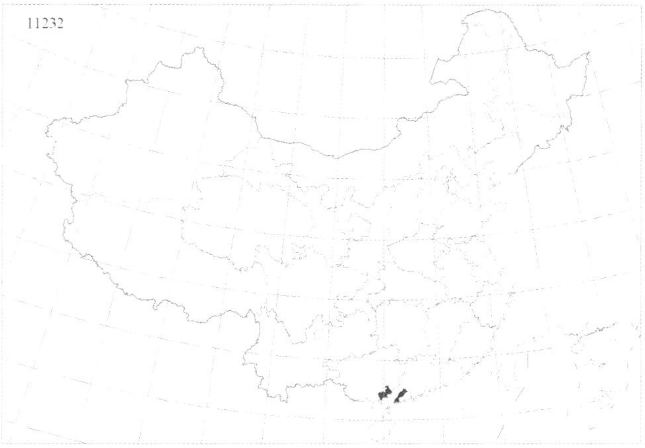

Arecaceae 棕榈科 *Calamus* 省藤属

Calamus distichus var. shangsiensis S. J. Pei et S. Y. Chen 上思省藤

Vines, evergreen

21.3；13.2；27.3；21.3；195.9；0.0；1111 ❙ 1900；1056；125 ❙ 71.0；1111；297.2

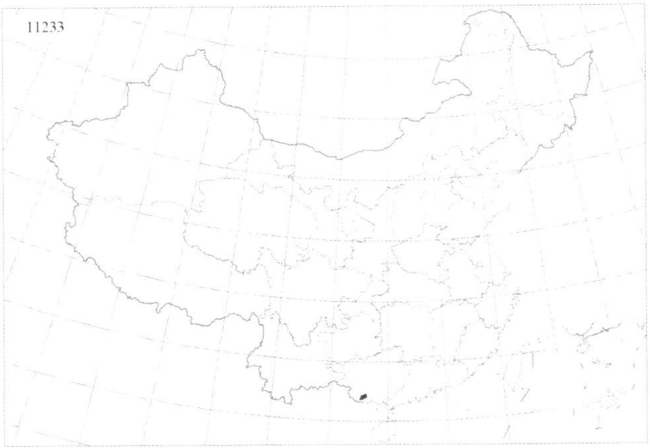

Arecaceae 棕榈科 *Calamus* 省藤属

Calamus egregius Burret 短叶省藤

Vines, evergreen

23.9 (22.2~24.8)；18.6 (17.2~20.1)；27.6 (25.7~28.5)；23.9 (22.3~24.8)；227.0 (207.2~237.6)；0.0 (0.0~0.0)；1298 (1127~1385) ❙ 1531 (1179~1854)；525 (392~622)；60 (41~115) ❙ 18.5 (-11.9~47.8)；1161 (1055~1360)；515.5 (428.6~620.0)

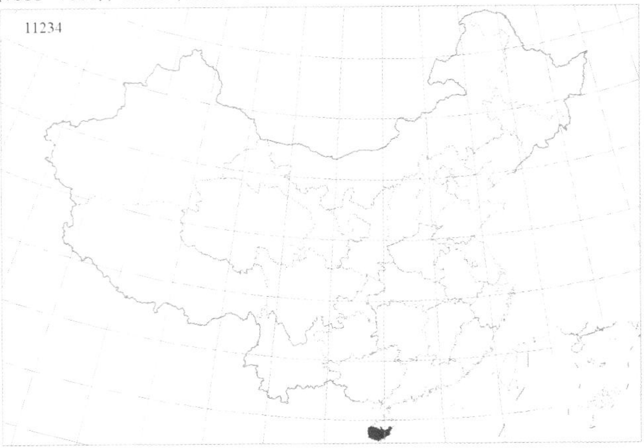

Arecaceae 棕榈科 *Calamus* 省藤属

Calamus erectus Roxb. 直立省藤

Vines, evergreen

16.4；9.6；20.6；16.4；137.3；0.0；795 ❙ 1688；959；62 ❙ 112.2；795；627.0

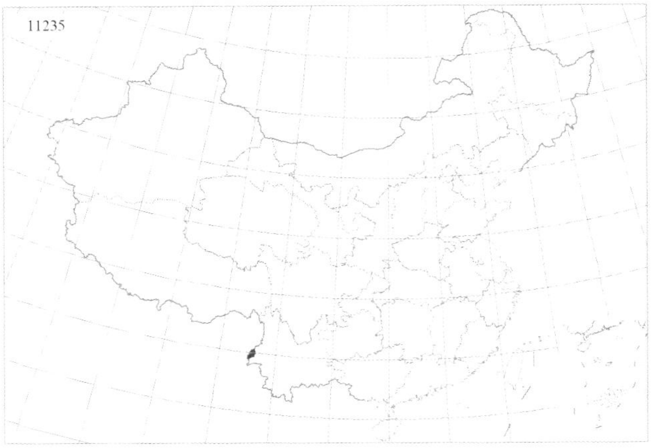

Arecaceae 棕榈科 *Calamus* 省藤属

Calamus faberi Becc. 大白藤

Vines, evergreen

23.3 (22.3~24.7)；17.0 (14.7~19.8)；28.0 (26.7~28.7)；23.3 (22.3~24.7)；219.7 (207.9~236.5)；0.0 (0.0~0.0)；1256 (1185~1376) ❙ 1749 (1380~2037)；734 (455~1033)；87 (48~118) ❙ 40.3 (6.7~71.8)；1199 (1112~1326)；338.2 (139.6~572.9)

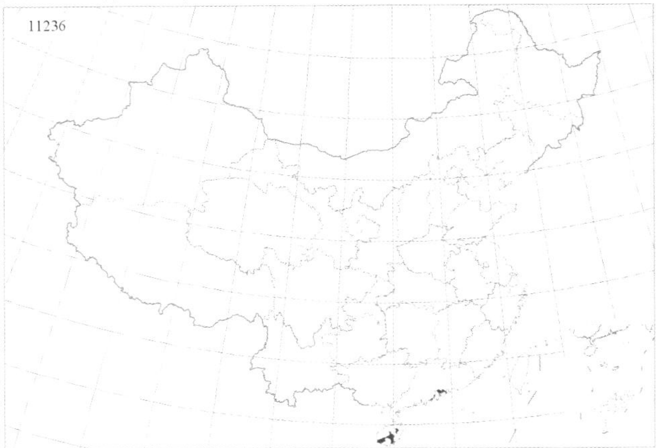

Arecaceae 棕榈科 *Calamus* 省藤属

Calamus feanus var. medogensis S. J. Pei et S. Y. Chen 墨脱省藤

Vines, evergreen

13.3；5.4；19.6；13.4；100.4；0.0；706 ❙ 1736；1007；51 ❙ 145.8；706；447.4

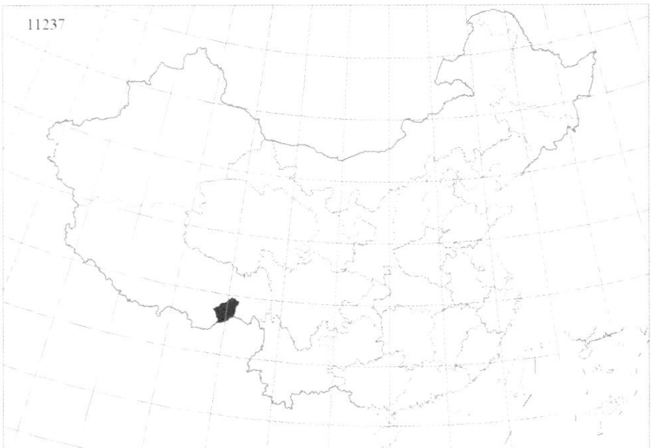

Arecaceae 棕榈科 *Calamus* 省藤属

Calamus flagellum S. J. Pei et S. Y. Chen 长鞭省藤

Vines, evergreen

18.1；11.8；22.3；18.1；157.4；0.0；896 ❙ 1602；788；65 ❙ 85.3；884；558.9

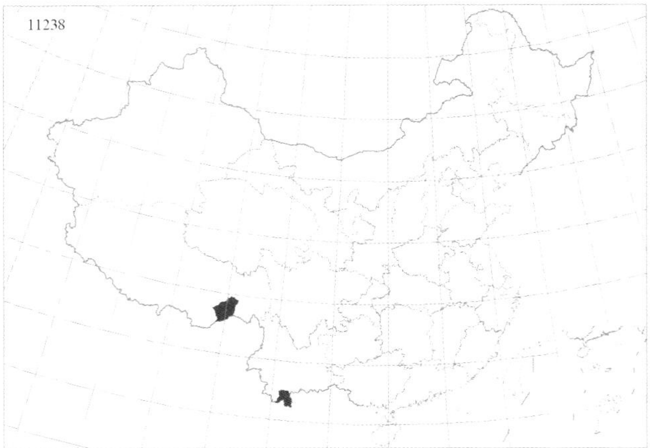

Arecaceae 棕榈科 *Calamus* 省藤属
Calamus formosanus Beccari 台湾省藤
Vines, evergreen

20.2 (16.2~24.1); 14.5 (11.0~18.9); 25.0 (20.1~28.2); 20.2 (16.2~24.1); 182.9 (134.2~229.7); 0.0 (0.0~0.0); 1019 (773~1322) ▎2402 (1498~3495); 1106 (750~1847); 255 (47~897) ▎146.2 (25.2~246.5); 983 (773~1144); 245.0 (119.4~363.7)

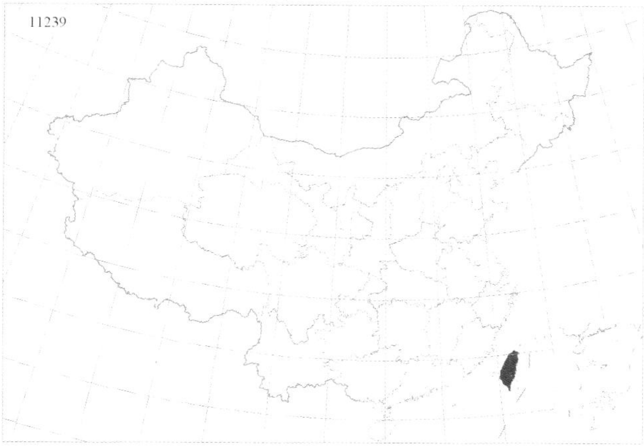

Arecaceae 棕榈科 *Calamus* 省藤属
Calamus giganteus var. robustus S. J. Pei et S. Y. Chen 粗壮省藤
Vines, evergreen

20.4; 15.1; 23.6; 20.5; 185.8; 0.0; 989 ▎1607; 702; 77 ▎62.5; 986; 659.0

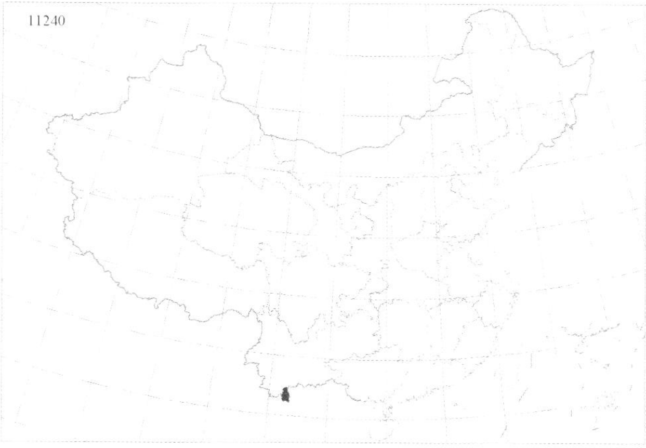

Arecaceae 棕榈科 *Calamus* 省藤属
Calamus gracilis Roxb. 小省藤
Vines, evergreen

20.5 (17.8~23.9); 14.2 (7.1~18.7); 25.2 (22.5~27.5); 20.6 (17.8~23.9); 186.7 (154.0~226.8); 0.0 (0.0~0.0); 1041 (888~1294) ▎1586 (1380~1751); 698 (455~996); 86 (48~182) ▎55.9 (6.7~88.8); 1002 (888~1163); 491.7 (197.5~659.0)

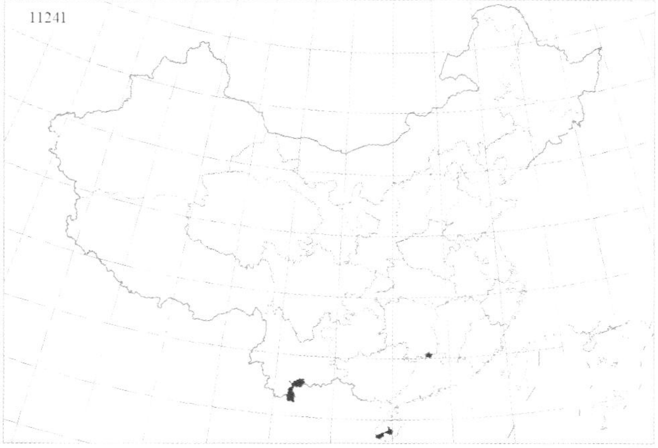

Arecaceae 棕榈科 *Calamus* 省藤属
Calamus guangxiensis C. F. Wei 广西省藤
Vines, evergreen

21.9; 13.8; 27.9; 21.9; 203.3; 0.0; 1161 ▎1478; 781; 97 ▎27.6; 1161; 304.7

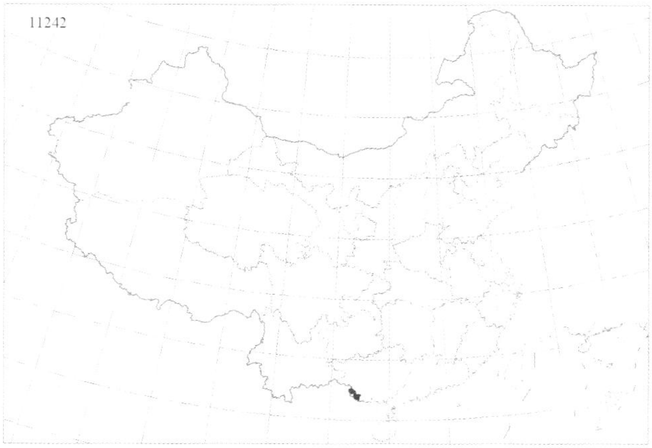

Arecaceae 棕榈科 *Calamus* 省藤属
Calamus henryanus Becc. 滇南省藤
Vines, evergreen

19.3 (18.2~20.5); 13.7 (12.4~15.1); 22.8 (21.6~23.7); 19.4 (18.2~20.5); 172.3 (158.6~185.9); 0.0 (0.0~0.0); 925 (861~992) ▎1552 (1464~1751); 787 (655~996); 65 (52~77) ▎68.1 (47.6~88.8); 910 (839~936); 525.1 (427.9~659.0)

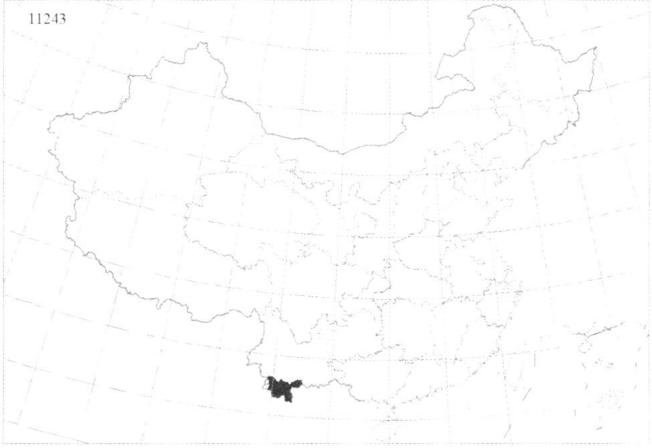

Arecaceae 棕榈科 *Calamus* 省藤属
Calamus hoplites Dunn 高毛鳞省藤
Vines, evergreen

17.4 (14.3~21.4); 6.7 (3.3~13.6); 27.5 (24.2~29.8); 17.5 (14.3~21.5); 150.0 (113.2~197.7); 0.2 (0.0~2.3); 923 (761~1128) ▎1517 (1070~1997); 516 (356~695); 184 (92~259) ▎65.7 (-5.0~137.5); 909 (761~1094); 218.7 (123.9~332.5)

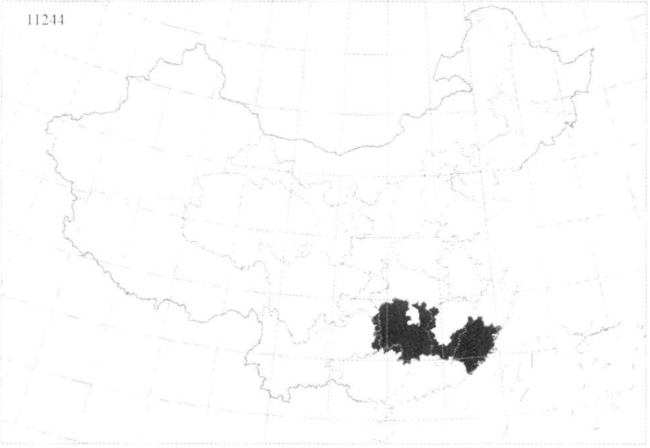

Arecaceae 棕榈科 *Calamus* 省藤属

Calamus karinensis (Beccari) S. J. Pei et S. Y. Chen 勐腊鞭藤

Vines

20.0；14.6；23.2；20.1；180.8；0.0；963 ❙ 1523；681；67 ❙ 58.4；942；581.0

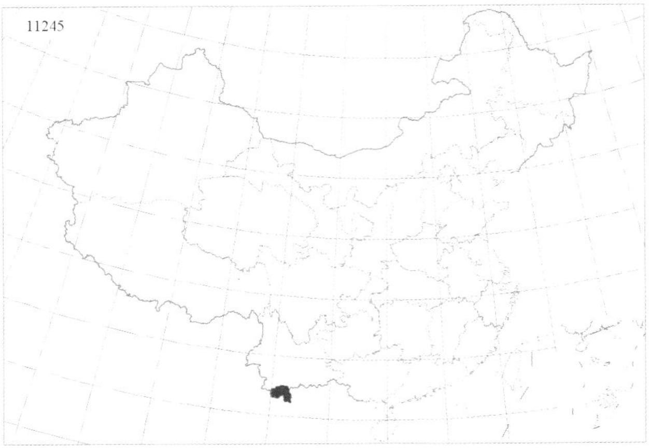

Arecaceae 棕榈科 *Calamus* 省藤属

Calamus macrorrhynchus Burret 大喙省藤

Vines, evergreen

20.0 (16.1~22.2)；11.3 (5.3~14.7)；27.0 (22.5~28.8)；20.1 (16.1~22.3)；180.7 (133.1~207.5)；0.0 (0.0~0.0)；1043 (838~1190) ❙ 1552 (1268~1953)；721 (523~891)；117 (59~167) ❙ 49.5 (15.4~71.5)；1042 (838~1190)；267.9 (203.4~551.6)

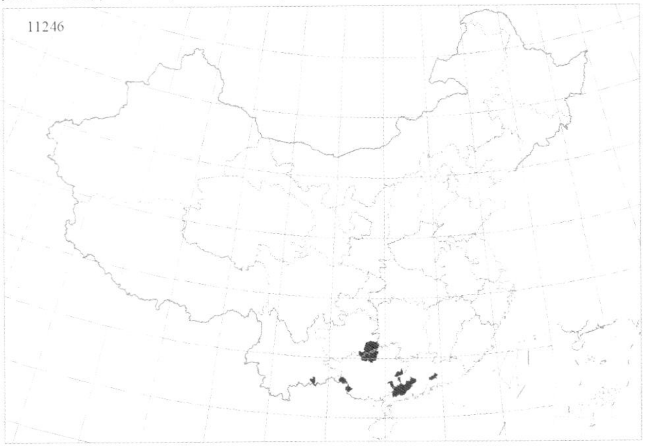

Arecaceae 棕榈科 *Calamus* 省藤属

Calamus melanochrous Burret 瑶山省藤

Vines, evergreen

19.5 (17.6~21.3)；9.7 (7.4~12.3)；27.6 (25.8~28.6)；19.5 (17.7~21.3)；174.2 (152.5~196.1)；0.0 (0.0~0.0)；1021 (906~1130) ❙ 1593 (1371~1758)；659 (547~728)；175 (161~189) ❙ 57.3 (22.7~85.5)；1020 (906~1130)；223.4 (185.5~260.5)

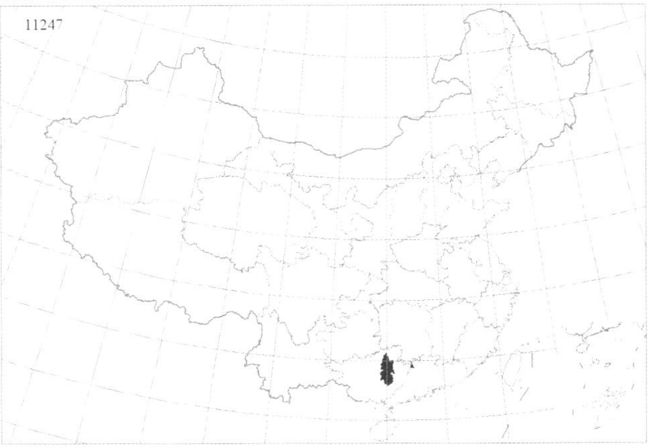

Arecaceae 棕榈科 *Calamus* 省藤属

Calamus multispicatus Burret 裂苞省藤

Vines, evergreen

23.9；18.9；27.3；23.9；227.2；0.0；1298 ❙ 1621；561；58 ❙ 25.4；1174；562.6

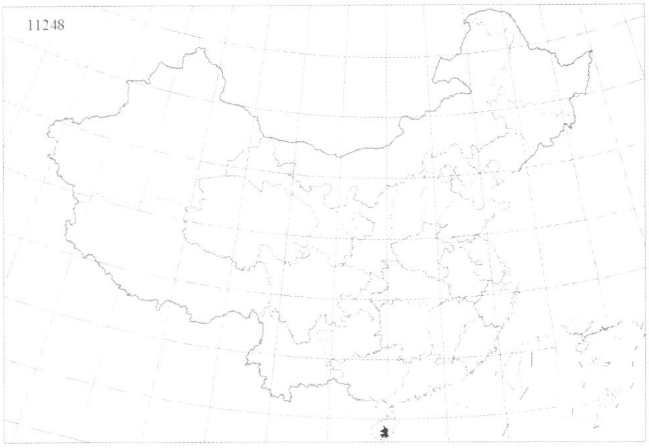

Arecaceae 棕榈科 *Calamus* 省藤属

Calamus nambariensis Becc. 南巴省藤

Vines, evergreen

17.8 (8.6~20.5)；11.3 (1.3~15.1)；21.8 (15.0~23.7)；17.8 (8.7~20.5)；154.3 (52.6~185.9)；0.8 (0.0~8.8)；871 (574~992) ❙ 1536 (1041~1751)；853 (585~996)；58 (42~77) ❙ 77.7 (47.6~112.2)；864 (574~986)；529.0 (352.3~659.0)

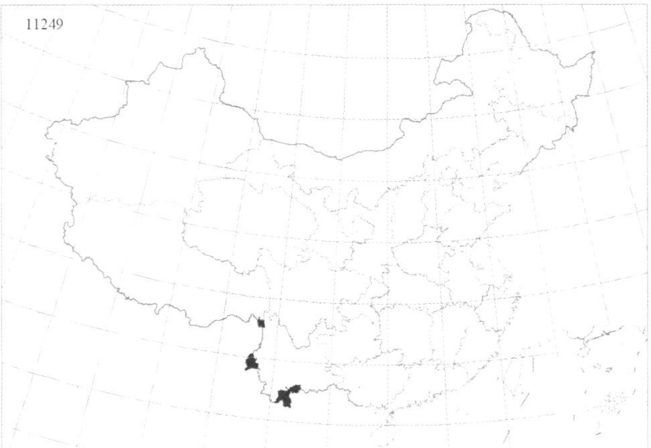

Arecaceae 棕榈科 *Calamus* 省藤属

Calamus obovoideus S. J. Pei et S. Y. Chen 倒卵果省藤

Vines, evergreen

20.5；14.9；23.7；20.5；185.9；0.0；992 ❙ 1464；655；66 ❙ 47.6；959；570.1

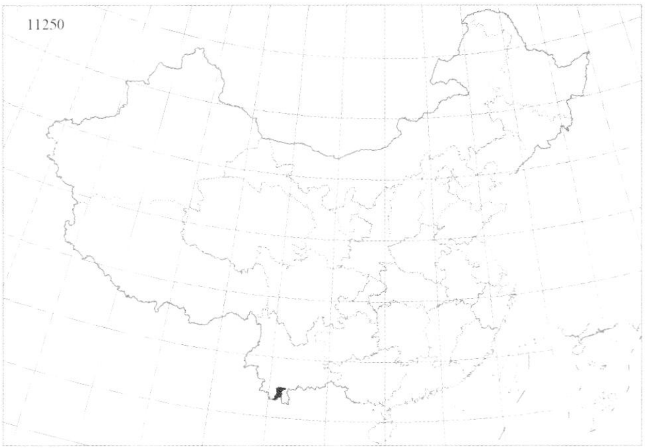

Arecaceae 棕榈科 *Calamus* 省藤属

Calamus oxycarpus Becc. 尖果省藤

Vines, evergreen

20.1；10.5；27.7；20.1；181.6；0.0；1052 ▌1459；723；111 ▌38.7；1052；248.4

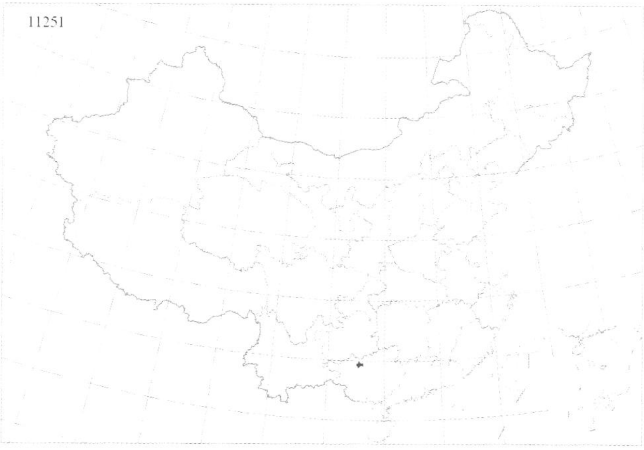

Arecaceae 棕榈科 *Calamus* 省藤属

Calamus palustris Griff. 泽生藤

Vines, evergreen

18.9 (17.2~21.0)；12.4 (9.5~15.1)；23.3 (22.4~25.6)；18.9 (17.3~21.0)；167.0 (147.0~191.8)；0.0 (0.0~0.0)；915 (833~1053) ▌1479 (1254~1751)；819 (699~996)；61 (48~77) ▌62.1 (36.8~88.8)；914 (833~1053)；482.6 (336.3~659.0)

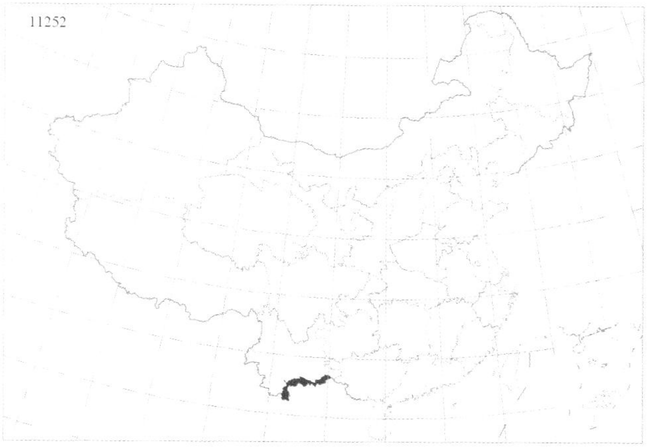

Arecaceae 棕榈科 *Calamus* 省藤属

Calamus platyacanthus Warb. ex Becc. 宽刺藤

Vines

20.4；15.1；23.6；20.5；185.8；0.0；989 ▌1607；702；77 ▌62.5；986；659.0

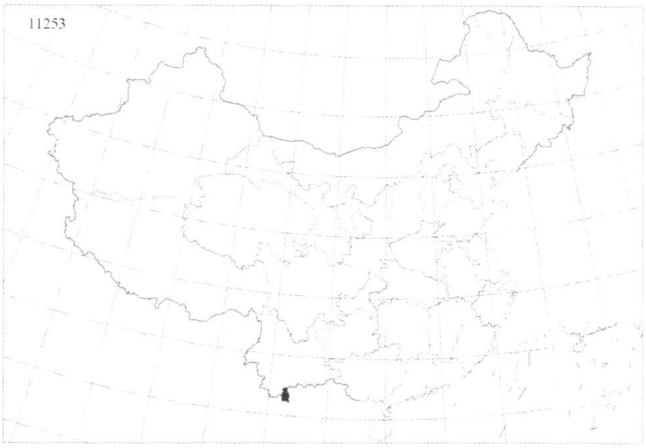

Arecaceae 棕榈科 *Calamus* 省藤属

Calamus pulchellus Burret 阔叶鸡藤

Vines, evergreen

24.3；19.3；27.7；24.3；232.0；0.0；1337 ▌1659；553；68 ▌23.9；1223；584.0

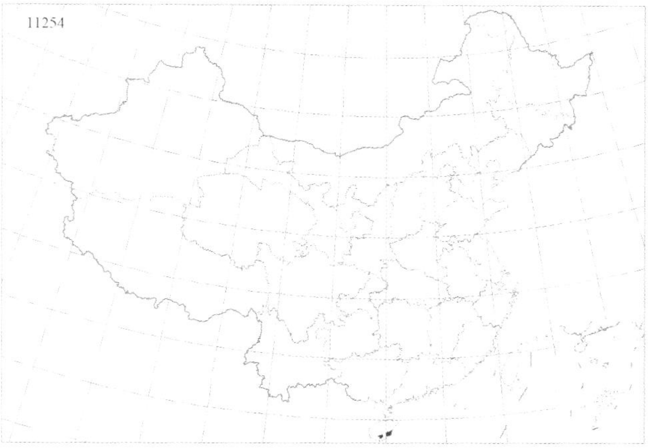

Arecaceae 棕榈科 *Calamus* 省藤属

Calamus quiquesetinervius Burret 五脉刚毛省藤

Vines, evergreen

16.4；11.5；20.3；16.5；137.8；0.0；783 ▌2499；947；303 ▌219.3；783；303.6

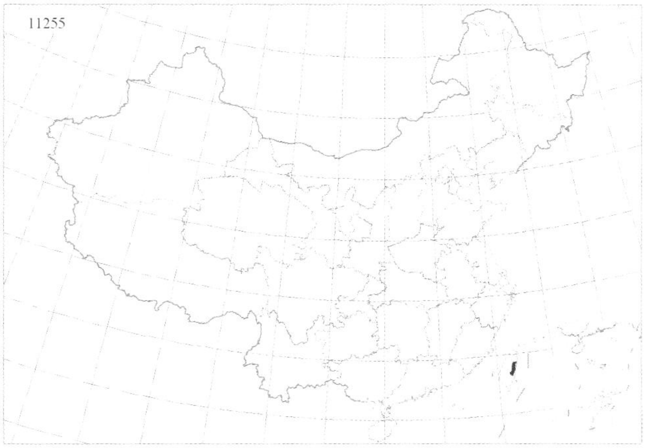

Arecaceae 棕榈科 *Calamus* 省藤属

Calamus rhabdocladus Burret 华南省藤

Vines, evergreen

21.3 (16.2~24.8)；13.4 (6.3~20.1)；27.7 (22.3~29.2)；21.3 (16.2~24.8)；196.0 (134.5~237.6)；0.0 (0.0~0.0)；1119 (826~1385) ▌1649 (1187~2148)；737 (455~1234)；125 (41~194) ▌48.5 (-6.9~89.4)；1113 (826~1326)；254.0 (95.3~659.0)

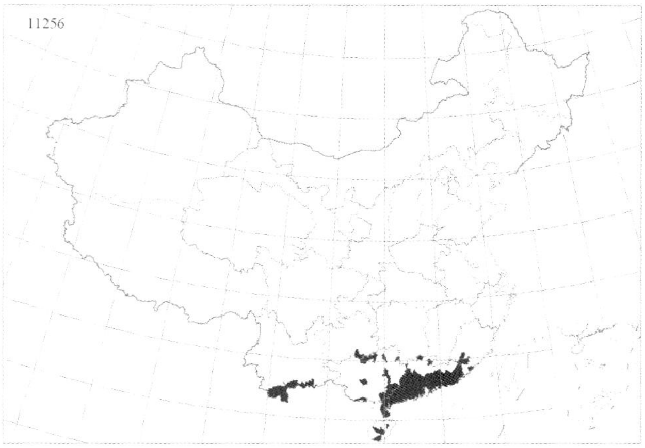

Arecaceae 棕榈科 *Calamus* 省藤属
Calamus simplicifolius C. F. Wei 单叶省藤
Vines, evergreen

21.6 (17.4~24.8); 14.1 (7.3~20.1); 27.4 (25.5~29.0); 21.6 (17.4~24.8); 199.6 (148.9~237.6); 0.0 (0.0~0.0); 1147 (886~1385) ❙ 1489 (1179~1900); 632 (392~1056); 91 (41~146) ❙ 31.7 (-11.9~71.0); 1085 (886~1299); 366.3 (157.7~620.0)

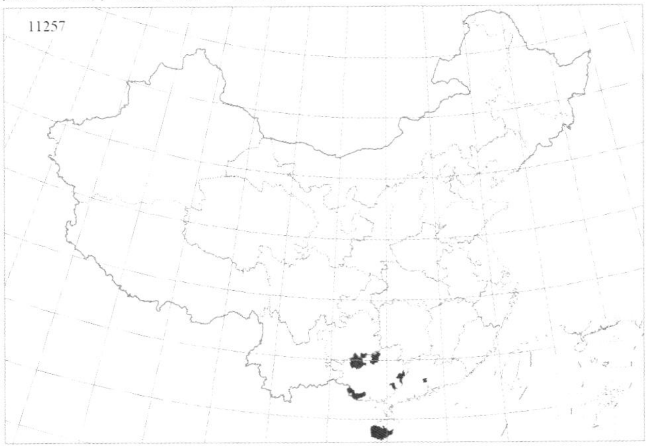

Arecaceae 棕榈科 *Calamus* 省藤属
Calamus siphonospathus Mart. 兰屿省藤
Vines, evergreen

18.5; 14.1; 21.7; 18.5; 162.1; 0.0; 863 ❙ 2713; 1388; 188 ❙ 214.4; 863; 363.7

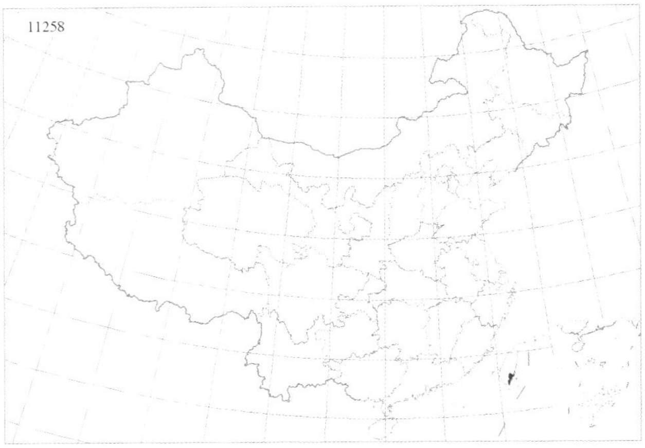

Arecaceae 棕榈科 *Calamus* 省藤属
Calamus tetradactyloides Burret 多刺鸡藤
Vines, evergreen

21.8 (18.7~24.8); 14.4 (8.6~20.1); 27.6 (26.9~28.6); 21.8 (18.8~24.8); 201.8 (165.2~237.6); 0.0 (0.0~0.0); 1167 (970~1385) ❙ 1584 (1179~1852); 573 (392~712); 124 (41~197) ❙ 40.6 (-11.9~86.3); 1090 (970~1329); 365.5 (246.0~620.0)

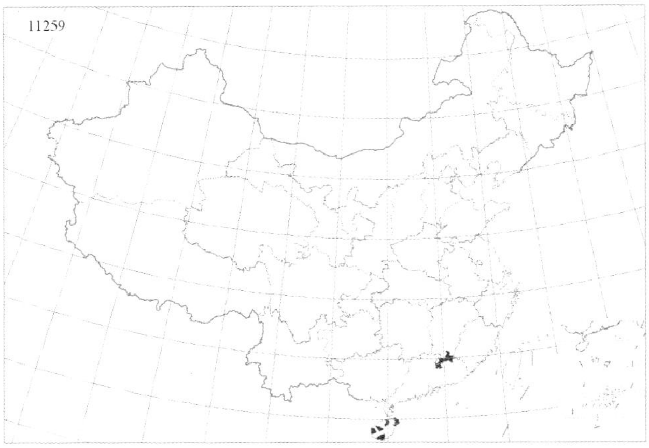

Arecaceae 棕榈科 *Calamus* 省藤属
Calamus tetradactylus Hance 白藤
Vines, evergreen

21.2 (16.5~24.8); 12.8 (5.8~20.1); 28.0 (26.2~29.8); 21.2 (16.5~24.8); 194.8 (138.2~237.6); 0.0 (0.0~0.0); 1120 (861~1385) ❙ 1655 (1176~2435); 707 (392~1428); 140 (41~203) ❙ 49.1 (-12.0~99.5); 1111 (861~1329); 235.7 (95.3~620.0)

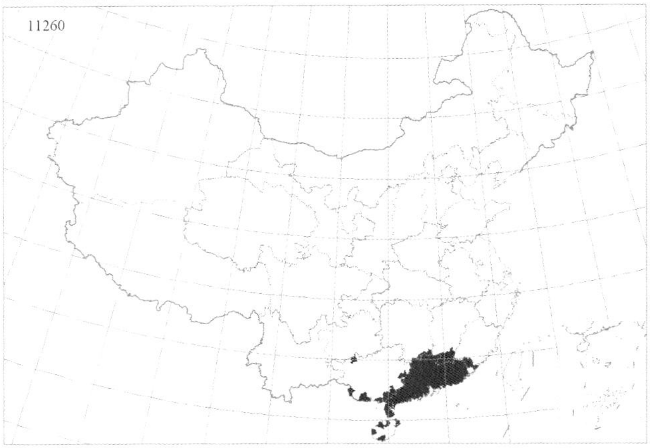

Arecaceae 棕榈科 *Calamus* 省藤属
Calamus thysanolepis Hance 毛鳞省藤
Vines, evergreen

18.8 (15.3~21.8); 9.7 (5.6~13.6); 27.2 (24.3~29.0); 18.8 (15.4~21.8); 165.7 (124.4~201.9); 0.0 (0.0~0.0); 971 (791~1158) ❙ 1622 (1070~2005); 572 (356~902); 193 (130~257) ❙ 69.5 (-5.0~137.5); 962 (791~1158); 253.8 (123.9~332.5)

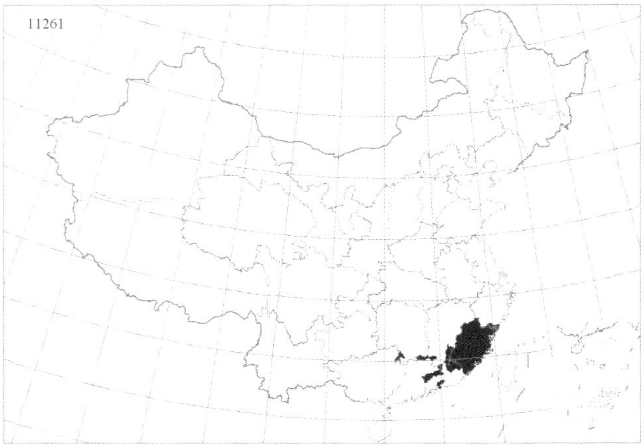

Arecaceae 棕榈科 *Calamus* 省藤属
Calamus viminalis Willd. 勐捧省藤
Vines, evergreen

18.9 (16.4~20.5); 12.6 (9.6~15.1); 22.6 (20.6~23.7); 18.9 (16.4~20.5); 167.0 (137.3~185.9); 0.0 (0.0~0.0); 910 (795~992) ❙ 1562 (1464~1688); 835 (655~959); 57 (42~77) ❙ 72.9 (47.6~112.2); 897 (795~986); 547.0 (451.2~659.0)

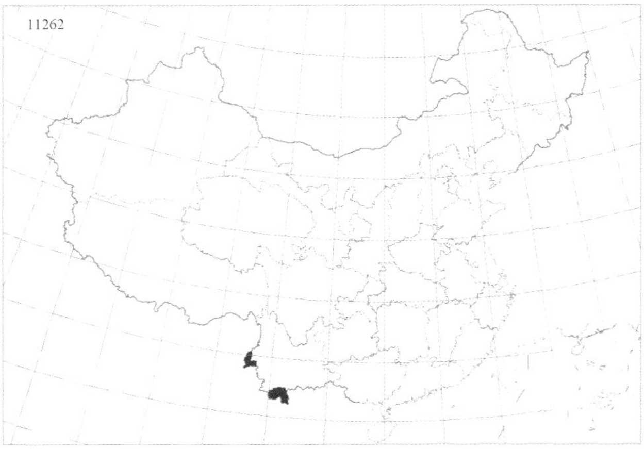

Arecaceae 棕榈科 *Calamus* 省藤属
Calamus wailong S. J. Pei et S. Y. Chen 大藤
Vines, evergreen

20.0；14.6；23.2；20.1；180.8；0.0；963 ❙ 1523；681；67 ❙ 58.4；942；581.0

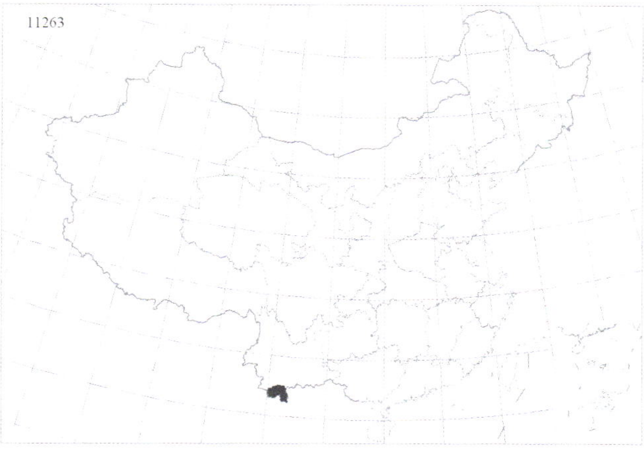

Arecaceae 棕榈科 *Calamus* 省藤属
Calamus walkeri Hance 多果省藤
Vines, evergreen

22.3 (20.7~24.1)；14.8 (12.8~18.2)；28.4 (27.4~29.2)；22.3 (20.7~24.2)；208.1 (188.9~230.1)；0.0 (0.0~0.0)；1187 (1071~1326) ❙ 1662 (1187~2037)；776 (539~1033)；109 (75~152) ❙ 40.5 (-6.9~79.9)；1184 (1071~1326)；205.6 (116.6~352.9)

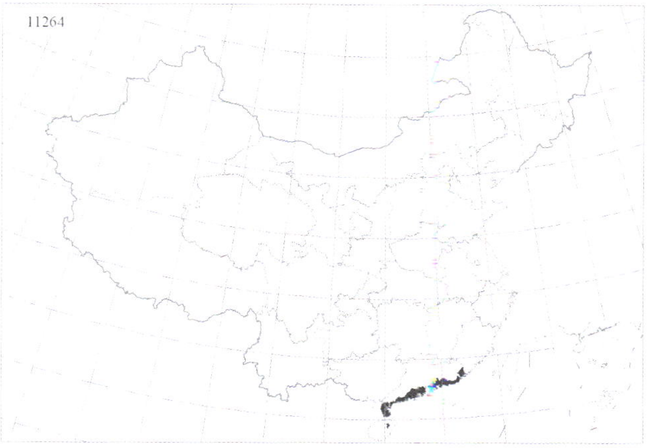

Arecaceae 棕榈科 *Calamus* 省藤属
Calamus yangchunensis Wei 阳春省藤
Vines, evergreen

22.3；14.9；28.1；22.4；208.3；0.0；1187 ❙ 1858；890；112 ❙ 56.7；1187；256.4

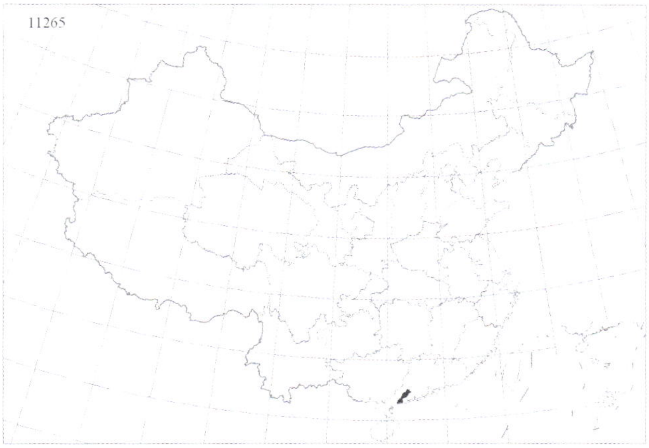

Arecaceae 棕榈科 *Calamus* 省藤属
Calamus yunnanensis S. J. Pei et S. Y. Chen 云南省藤
Vines, evergreen

17.4 (8.6~20.5)；11.2 (1.3~15.1)；21.5 (15.0~23.7)；17.4 (8.7~20.5)；150.0 (52.6~185.9)；1.0 (0.0~8.8)；858 (574~932) ❙ 1457 (1041~1736)；731 (585~1007)；55 (41~77) ❙ 72.5 (47.6~145.8)；838 (574~986)；497.5 (352.3~659.0)

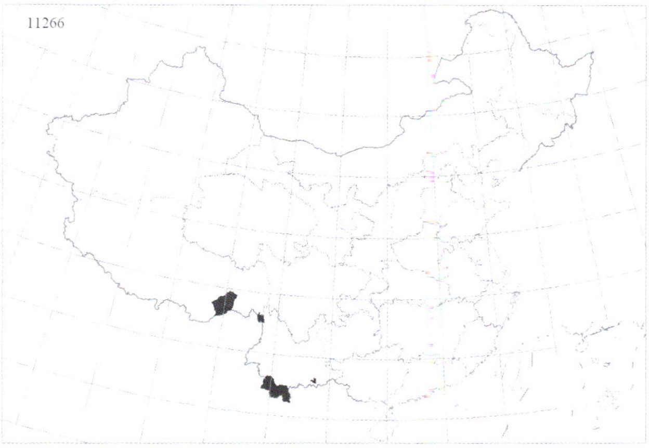

Arecaceae 棕榈科 *Calamus* 省藤属
Calamus yunnanensis var. intermedius S. J. Pei et S. Y. Chen 屏边省藤
Vines, evergreen

19.4；12.6；24.1；19.4；172.6；0.0；954 ❙ 1445；787；64 ❙ 52.3；954；521.5

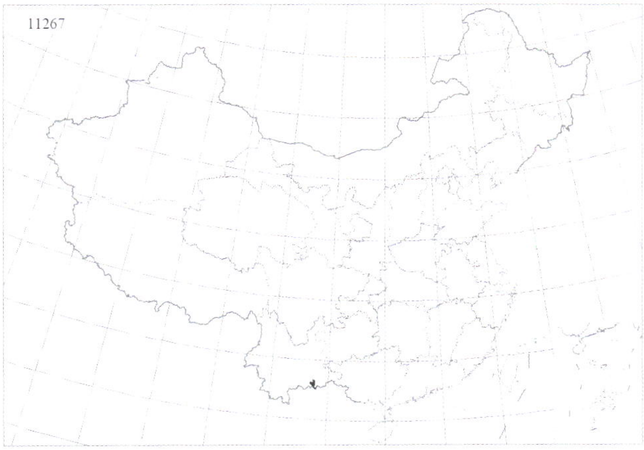

Arecaceae 棕榈科 *Caryota* 鱼尾葵属
Caryota mitis Lour. 短穗鱼尾葵
Small trees, evergreen

21.4 (16.8~24.8)；13.4 (7.8~20.1)；27.9 (23.6~29.1)；21.5 (16.8~24.8)；198.0 (141.5~237.6)；0.0 (0.0~0.0)；1133 (836~1385) ❙ 1668 (1182~2435)；783 (473~1428)；117 (41~191) ❙ 47.7 (-1.0~99.5)；1126 (836~1269)；237.9 (95.3~538.2)

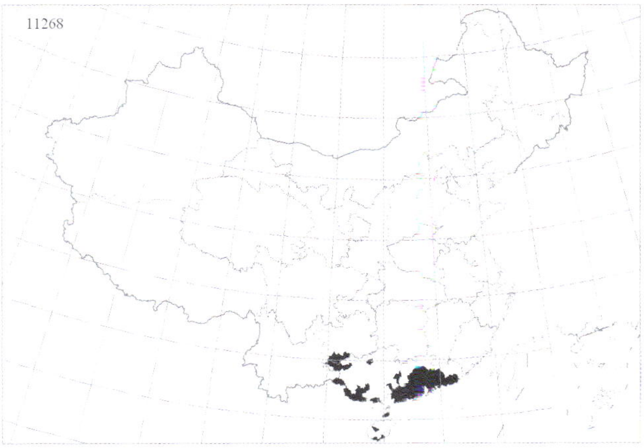

Arecaceae 棕榈科 *Caryota* 鱼尾葵属
Caryota monostachya Becc. 单穗鱼尾葵
Small trees, evergreen

19.6 (16.3~22.7); 12.1 (9.1~15.1); 24.9 (21.4~28.6); 19.6 (16.4~22.8); 175.1 (136.4~213.1); 0.0 (0.0~0.0); 983 (797~1221) ❙ 1470 (1074~2435); 796 (582~1428); 76 (43~173) ❙ 49.7 (12.9~99.5); 981 (797~1221); 381.8 (235.1~659.0)

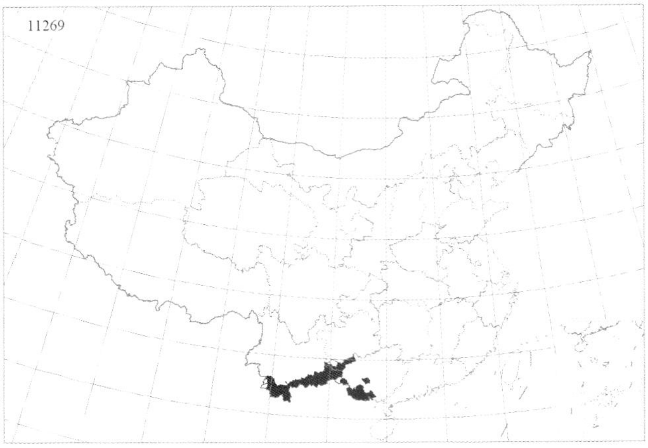

Arecaceae 棕榈科 *Caryota* 鱼尾葵属
Caryota ochlandra Hance 鱼尾葵
Trees, evergreen

19.7 (16.3~24.8); 12.2 (7.4~20.1); 25.1 (21.4~28.8); 19.7 (16.4~24.8); 176.4 (136.4~237.6); 0.0 (0.0~0.0); 994 (797~1385) ❙ 1430 (958~2435); 738 (473~1428); 77 (41~144) ❙ 44.3 (1.5~99.5); 982 (797~1221); 345.5 (157.7~659.0)

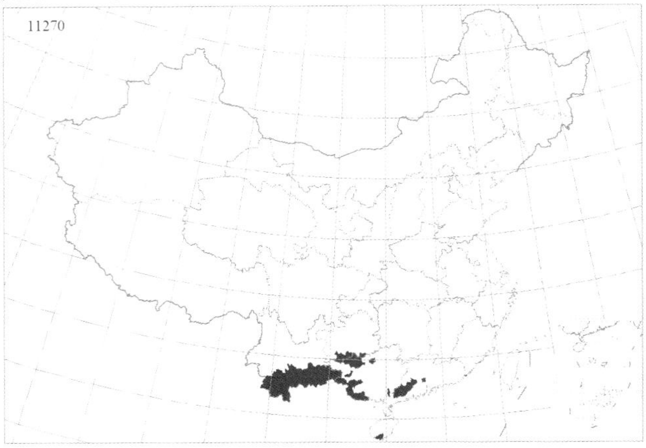

Arecaceae 棕榈科 *Caryota* 鱼尾葵属
Caryota urens L. 董棕
Trees, evergreen

18.5 (8.6~22.2); 11.8 (1.3~15.1); 23.2 (15.0~28.3); 18.6 (8.7~22.3); 163.3 (52.6~207.5); 0.3 (0.0~8.8); 916 (574~1190) ❙ 1359 (974~1751); 734 (566~996); 58 (37~106) ❙ 50.1 (5.8~88.8); 906 (574~1190); 423.5 (281.7~659.0)

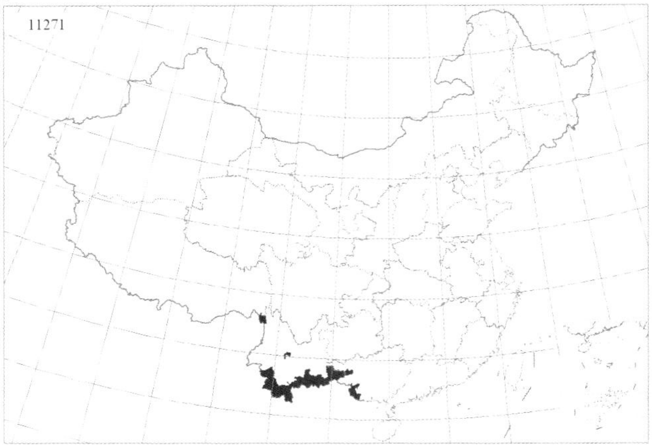

Arecaceae 棕榈科 *Chuniophoenix* 琼棕属
Chuniophoenix hainanensis Burret 琼棕
Shrubs, evergreen

24.0 (22.2~24.8); 18.4 (17.2~20.1); 28.0 (25.7~28.8); 24.0 (22.3~24.8); 228.3 (207.2~237.6); 0.0 (0.0~0.0); 1310 (1127~1385) ❙ 1501 (868~1854); 534 (312~712); 74 (41~115) ❙ 15.0 (-34.3~47.8); 1191 (868~1360); 440.3 (248.2~620.0)

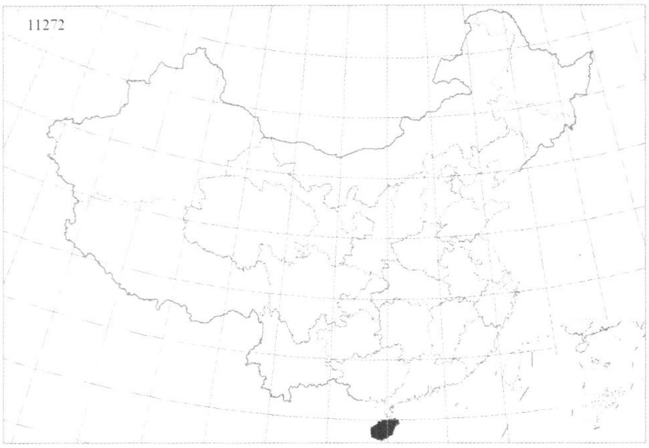

Arecaceae 棕榈科 *Chuniophoenix* 琼棕属
Chuniophoenix nana Burret 矮琼棕
Shrubs, evergreen

24.7; 19.8; 27.9; 24.7; 236.5; 0.0; 1376 ❙ 1588; 534; 58 ❙ 15.4; 1212; 494.7

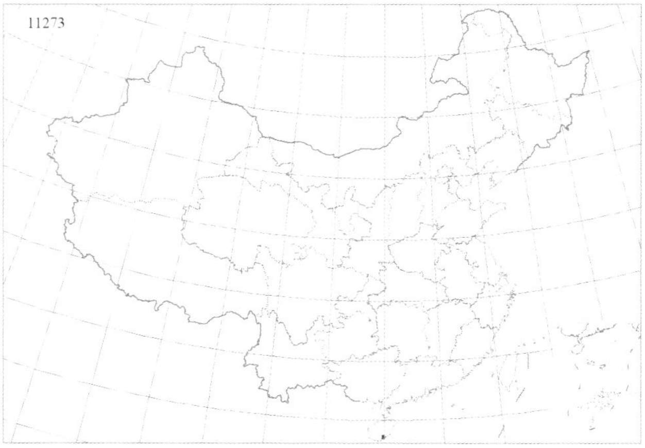

Arecaceae 棕榈科 *Cocos* 椰子属
Cocos nucifera L. 椰子
Trees, evergreen

23.2 (19.2~24.8); 17.2 (13.8~20.1); 27.6 (22.3~29.2); 23.3 (19.2~24.8); 219.2 (170.5~237.6); 0.0 (0.0~0.0); 1248 (909~1385) ❙ 1499 (868~2035); 595 (312~969); 77 (41~125) ❙ 21.7 (-34.3~71.8); 1167 (868~1360); 405.7 (178.7~659.0)

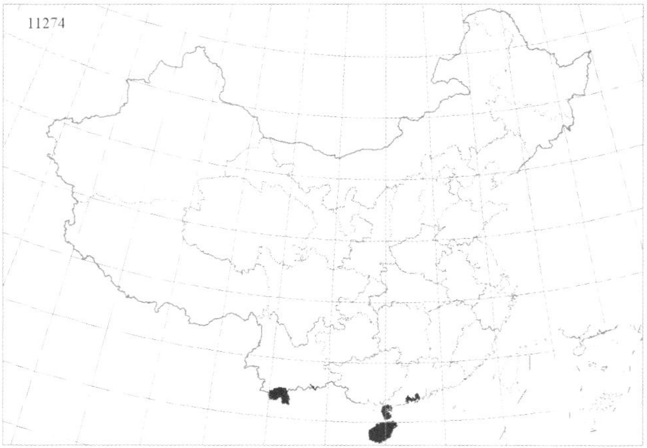

Arecaceae 棕榈科 *Daemonorops* 黄藤属
Daemonorops margaritae (Hance) Becc. 黄藤
Vines, evergreen

20.5 (17.8~24.8); 12.4 (7.4~20.1); 27.1 (21.7~28.8); 20.5 (17.9~24.8); 186.5 (154.4~237.6); 0.0 (0.0~0.0); 1074 (852~1385) ❙ 1687 (1176~2992); 734 (447~1847); 140 (41~402) ❙ 61.8 (−12.0~235.6); 1043 (852~1212); 298.0 (185.5~620.0)

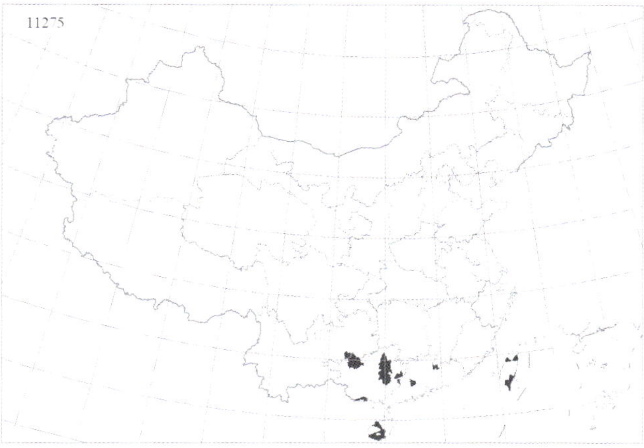

Arecaceae 棕榈科 *Guihaia* 石山棕属
Guihaia argyrata (S. K. Lee et F. N. Wei) S. K. Lee, F. N. Wei et J. Dransf. 石山棕
Shrubs, evergreen

19.3 (16.7~22.2); 9.6 (5.6~13.9); 27.2 (25.0~28.7); 19.3 (16.8~22.3); 171.7 (141.2~207.5); 0.0 (0.0~0.0); 1004 (885~1190) ❙ 1510 (1279~1776); 663 (497~848); 139 (57~196) ❙ 51.7 (14.8~85.5); 1004 (885~1190); 261.9 (203.9~371.7)

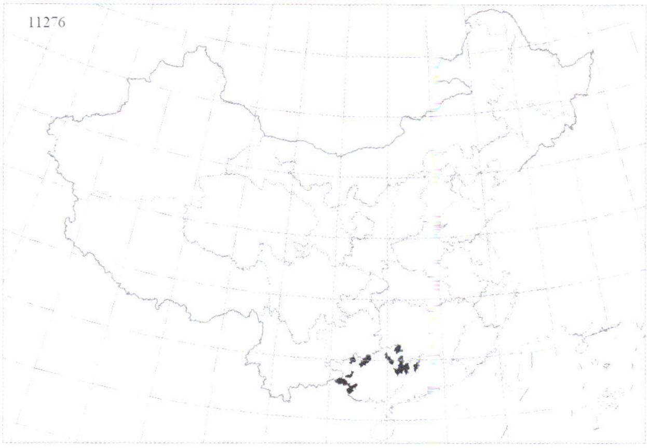

Arecaceae 棕榈科 *Guihaia* 石山棕属
Guihaia grossefibrosa (Gagnep.) J. Dransf. , S. K. Lee et F. N. Wei 两广石山棕
Shrubs, evergreen

21.2 (17.4~22.7); 12.8 (7.4~15.0); 27.7 (25.5~28.6); 21.3 (17.4~22.8); 195.0 (148.9~213.1); 0.0 (0.0~0.0); 1120 (886~1221) ❙ 1761 (1302~2435); 933 (637~1428); 118 (86~135) ❙ 57.0 (12.9~99.5); 1120 (886~1221); 276.9 (235.1~318.8)

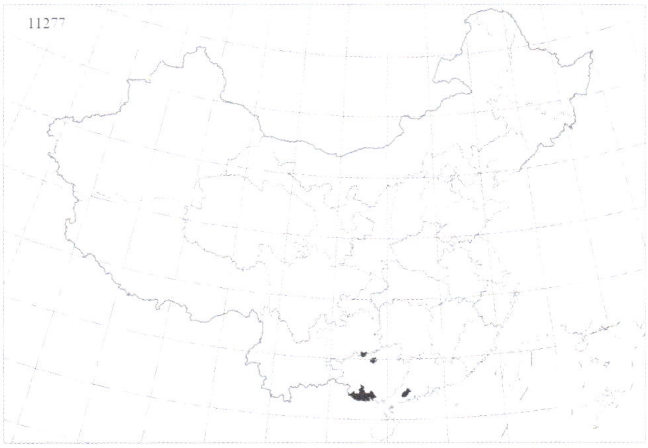

Arecaceae 棕榈科 *Licuala* 轴榈属
Licuala dasyantha Burret 毛花轴榈
Shrubs, evergreen

20.9 (17.7~22.7); 13.5 (10.9~15.0); 26.3 (22.5~28.6); 20.9 (17.8~22.8); 191.3 (153.4~213.1); 0.0 (0.0~0.0); 1073 (854~1221) ❙ 1874 (1368~2435); 1059 (756~1428); 101 (59~130) ❙ 72.4 (44.5~99.5); 1078 (854~1221); 391.9 (297.2~551.6)

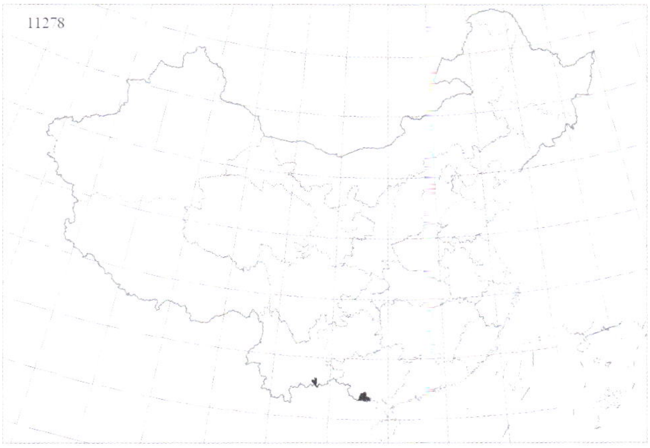

Arecaceae 棕榈科 *Licuala* 轴榈属
Licuala fordiana Becc. 穗花轴榈
Shrubs, evergreen

23.3 (21.6~24.8); 17.0 (13.2~20.1); 28.0 (26.7~28.8); 23.3 (21.6~24.8); 219.6 (199.2~237.6); 0.0 (0.0~0.0); 1257 (1139~1385) ❙ 1583 (1179~2024); 627 (392~957); 82 (41~167) ❙ 27.4 (−11.9~71.5); 1147 (1056~1232); 395.2 (157.7~620.0)

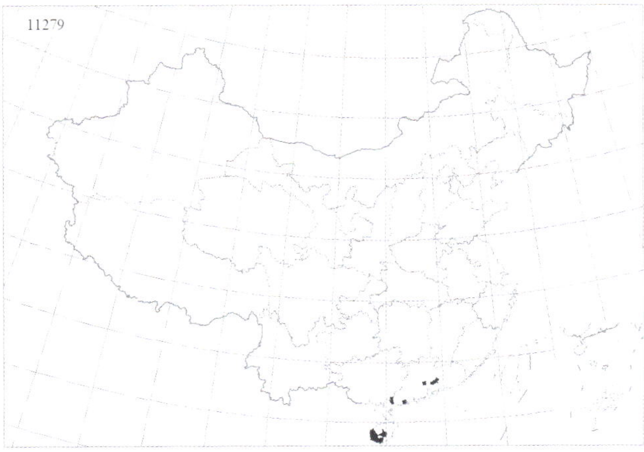

Arecaceae 棕榈科 *Licuala* 轴榈属
Licuala spinosa Wurmb 刺轴榈
Shrubs, evergreen

23.5 (19.1~24.8); 17.7 (11.0~20.1); 27.7 (25.0~28.8); 23.6 (19.1~24.8); 222.7 (169.6~237.6); 0.0 (0.0~0.0); 1275 (945~1385) ❙ 1497 (868~1854); 562 (312~848); 73 (41~115) ❙ 18.7 (−34.3~55.9); 1168 (868~1360); 431.0 (248.2~620.0)

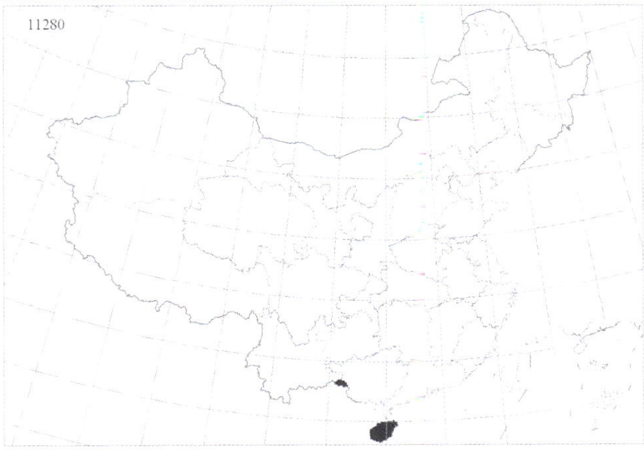

Arecaceae 棕榈科 *Livistona* 蒲葵属
Livistona chinensis (Jacq.) R. Br. 蒲葵
Trees, evergreen

19.1 (14.3~24.8); 10.1 (4.0~20.1); 26.6 (22.2~29.9); 19.1 (14.3~24.8); 169.5 (112.6~237.6); 0.1 (0.0~1.1); 1010 (756~1385) | 1416 (868~2992); 636 (312~1847); 96 (41~198) | 42.2 (-34.3~166.5); 990 (756~1360); 251.7 (120.3~620.0)

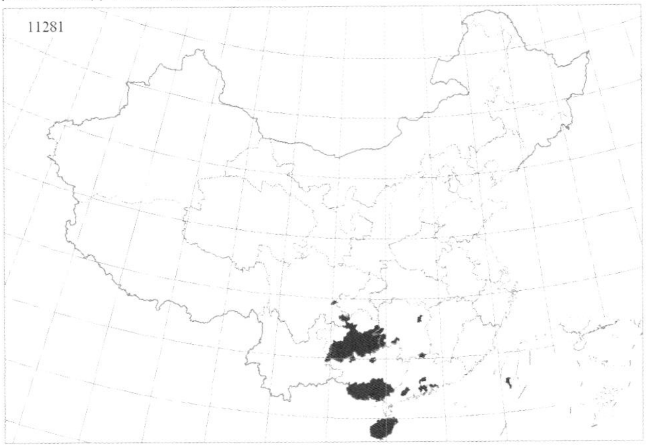

Arecaceae 棕榈科 *Livistona* 蒲葵属
Livistona saribus (Lour.) Merr. ex A. Chev. 大叶蒲葵
Trees, evergreen

21.2 (17.3~24.7); 14.9 (9.0~19.8); 25.9 (22.3~28.5); 21.2 (17.3~24.7); 194.8 (147.6~236.5); 0.0 (0.0~0.0); 1095 (864~1376) | 1531 (1380~1631); 594 (455~702); 93 (48~175) | 43.8 (6.7~85.2); 1032 (864~1212); 459.5 (226.9~659.0)

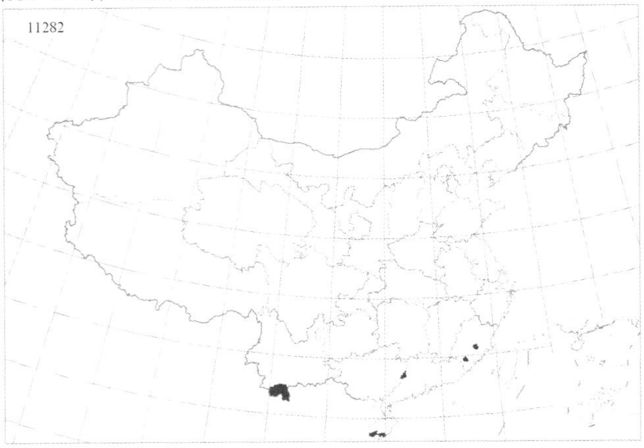

Arecaceae 棕榈科 *Nypa* 水椰子属
Nypa fructicans Wurmb. 水椰
Shrubs, evergreen

24.0 (22.2~24.8); 18.5 (17.2~20.1); 27.9 (25.7~28.8); 24.1 (22.3~24.8); 228.8 (207.2~237.6); 0.0 (0.0~0.0); 1313 (1127~1385) | 1612 (1309~1854); 571 (455~712); 81 (41~115) | 23.2 (-1.4~47.8); 1239 (1055~1360); 445.2 (248.2~620.0)

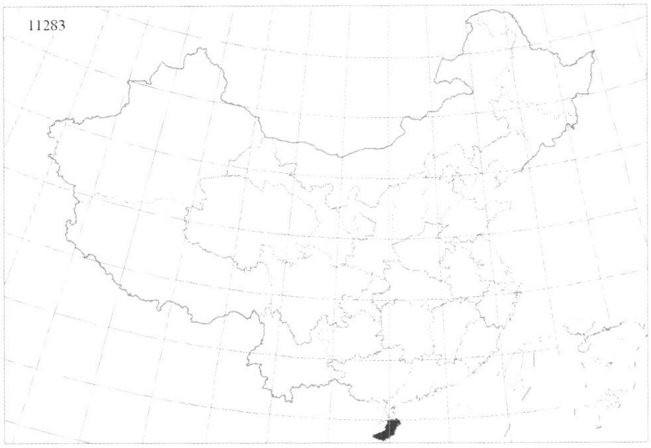

Arecaceae 棕榈科 *Phoenix* 海枣属
Phoenix hanceana Naudin 刺葵
Shrubs, evergreen

22.0 (17.6~24.8); 14.5 (7.8~20.1); 27.8 (22.3~29.2); 22.0 (17.7~24.8); 204.2 (152.5~237.6); 0.0 (0.0~0.0); 1164 (864~1385) | 1644 (868~2993); 734 (312~1847); 120 (41~547) | 43.2 (-34.3~203.4); 1141 (864~1360); 276.4 (139.6~659.0)

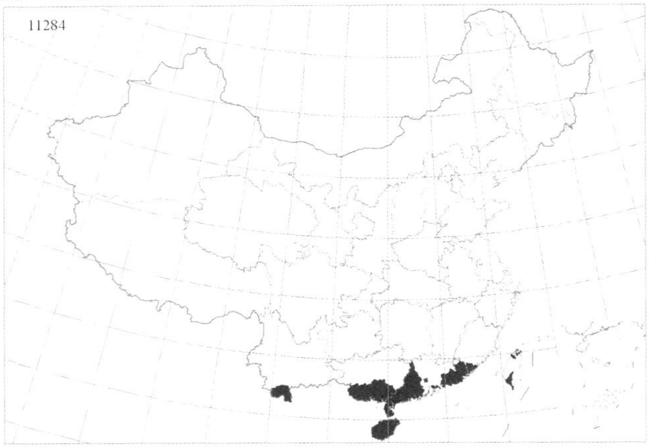

Arecaceae 棕榈科 *Phoenix* 海枣属
Phoenix roebelenii O'Brien 江边刺葵
Shrubs, evergreen

21.0 (16.0~24.8); 14.9 (9.3~20.1); 25.2 (20.4~28.9); 21.1 (16.0~24.8); 192.9 (132.4~237.6); 0.0 (0.0~0.0); 1089 (780~1385) | 1500 (868~1953); 677 (312~959); 70 (41~167) | 42.9 (-34.3~112.2); 1030 (780~1360); 438.7 (95.3~659.0)

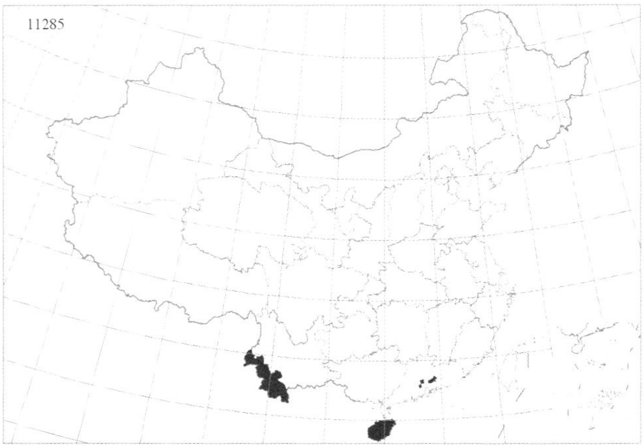

Arecaceae 棕榈科 *Pinanga* 山槟榔属
Pinanga chinensis Becc. 华山竹
Shrubs, evergreen

20.4; 15.1; 23.6; 20.5; 185.8; 0.0; 989 | 1607; 702; 77 | 62.5; 986; 659.0

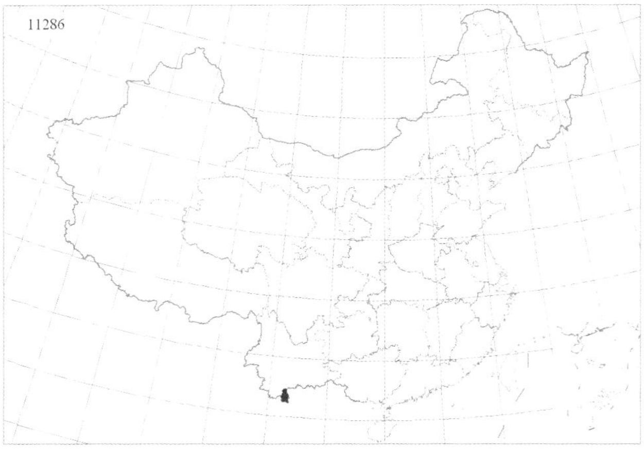

Arecaceae 棕榈科 *Pinanga* 山槟榔属

Pinanga discolor Burret 变色山槟榔

Shrubs, evergreen

21.1 (16.3~24.8); 13.1 (7.1~20.1); 27.5 (21.4~29.8); 21.1 (16.4~24.8); 193.5 (136.4~237.6); 0.0 (0.0~0.0); 1109 (797~1385) ▎1621 (868~2435); 700 (312~1428); 128 (41~197) ▎47.9 (-34.3~99.5); 1092 (797~1360); 274.5 (95.3~659.0)

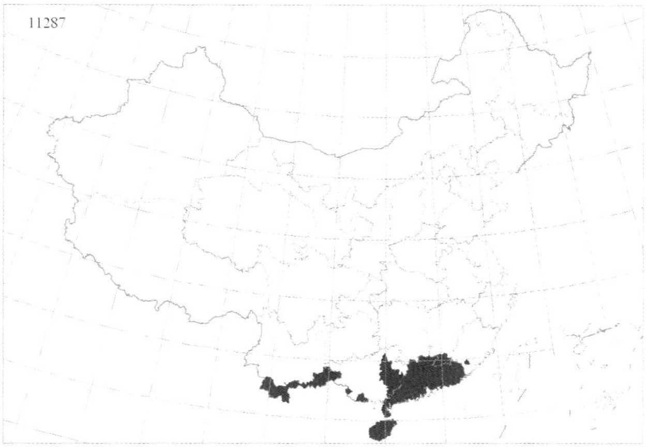

Arecaceae 棕榈科 *Pinanga* 山槟榔属

Pinanga gracilis (Roxb.) Blume 纤细山槟榔

Shrubs, evergreen

13.3; 5.4; 19.6; 13.4; 100.4; 0.0; 706 ▎1736; 1007; 51 ▎145.8; 706; 447.4

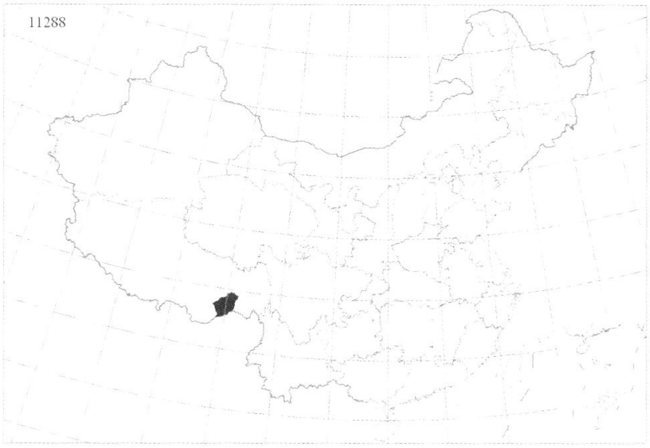

Arecaceae 棕榈科 *Pinanga* 山槟榔属

Pinanga hexasticha (Kurz) Scheff. 六列山槟榔

Shrubs, evergreen

14.9; 7.5; 20.1; 14.9; 118.9; 0.0; 751 ▎1712; 983; 57 ▎129.0; 751; 537.2

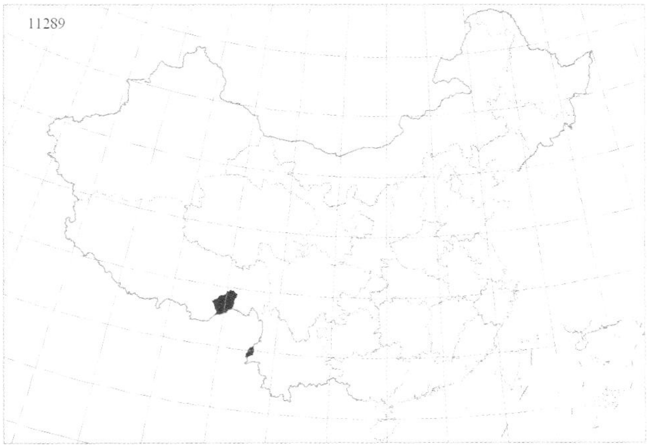

Arecaceae 棕榈科 *Pinanga* 山槟榔属

Pinanga macroclada Burret 长枝山竹

Shrubs, evergreen

19.5 (18.6~20.5); 13.8 (12.7~15.1); 23.1 (22.3~23.7); 19.6 (18.7~20.5); 174.8 (163.9~185.9); 0.0 (0.0~0.0); 937 (888~992) ▎1549 (1428~1751); 797 (655~996); 66 (56~77) ▎65.7 (47.6~88.8); 926 (880~986); 552.4 (506.2~659.0)

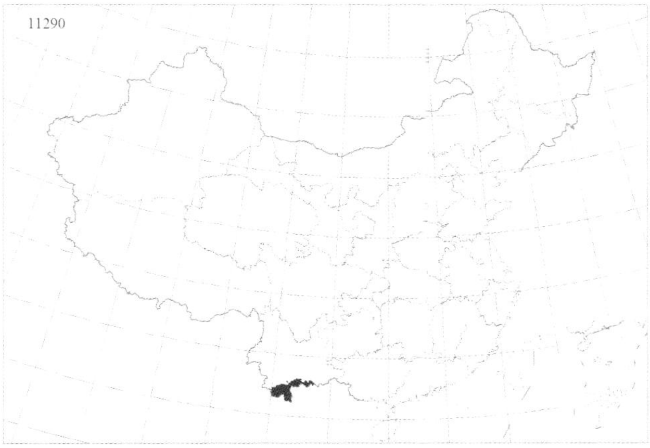

Arecaceae 棕榈科 *Pinanga* 山槟榔属

Pinanga sinii Burret 燕尾山槟榔

Shrubs, evergreen

20.6 (17.6~23.2); 11.7 (7.3~15.8); 27.9 (25.8~28.9); 20.7 (17.6~23.2); 188.0 (151.6~218.9); 0.0 (0.0~0.0); 1086 (909~1257) ▎1683 (1371~2435); 763 (561~1428); 148 (102~189) ▎55.5 (22.7~99.5); 1086 (909~1257); 237.3 (185.5~335.1)

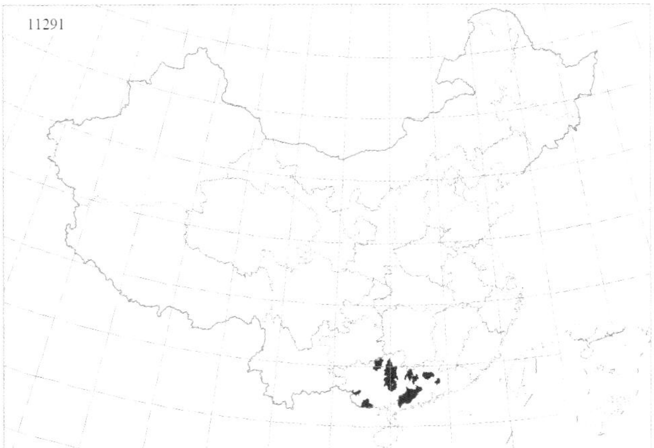

Arecaceae 棕榈科 *Pinanga* 山槟榔属

Pinanga tashiroi Hayata 兰屿山槟榔

Shrubs, evergreen

18.5; 14.1; 21.7; 18.5; 162.1; 0.0; 863 ▎2713; 1388; 188 ▎214.4; 863; 363.7

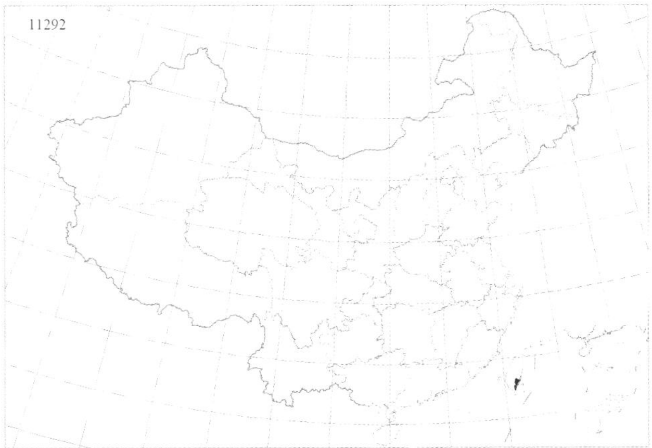

Arecaceae 棕榈科 *Pinanga* 山槟榔属
Pinanga viridis Burret 绿色山槟榔
Shrubs

20.1 (17.6~22.2); 12.4 (9.2~15.1); 25.8 (23.2~28.7); 20.1 (17.7~22.3); 181.6 (152.4~207.5); 0.0 (0.0~0.0); 1021 (859~1190) ❙ 1494 (1374~1636); 744 (678~835); 90 (52~135) ❙ 48.1 (15.4~65.6); 1020 (859~1190); 385.9 (206.6~659.0)

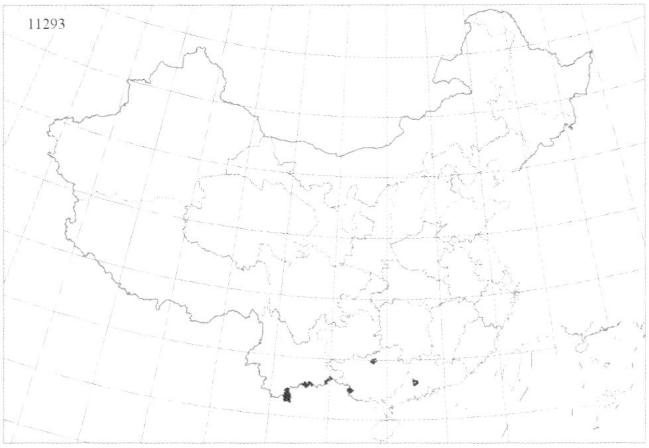

Arecaceae 棕榈科 *Plectocomia* 钩叶藤属
Plectocomia assamica Griff. 大钩叶藤
Vines, evergreen

20.4; 15.1; 23.6; 20.5; 185.8; 0.0; 989 ❙ 1607; 702; 77 ❙ 62.5; 986; 659.0

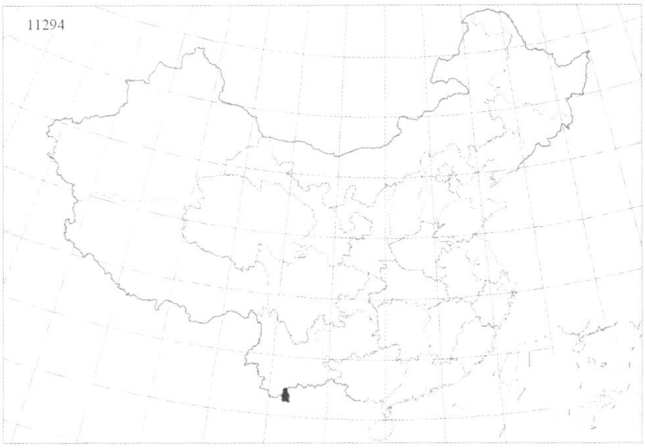

Arecaceae 棕榈科 *Plectocomia* 钩叶藤属
Plectocomia himalayana Griff. 高地钩叶藤
Vines, evergreen

20.5; 15.0; 23.7; 20.5; 185.9; 0.0; 990 ❙ 1536; 678; 72 ❙ 55.1; 973; 614.6

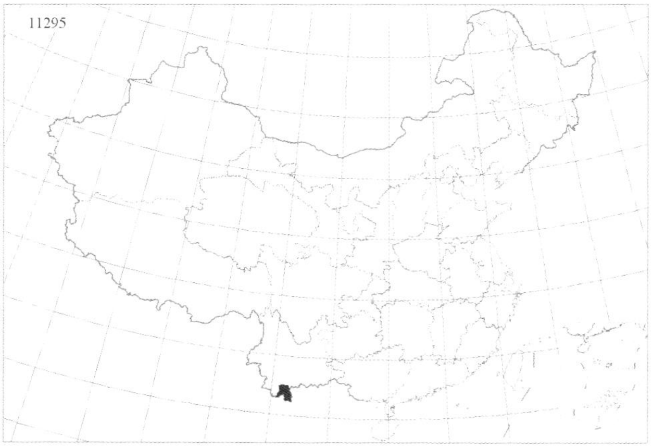

Arecaceae 棕榈科 *Plectocomia* 钩叶藤属
Plectocomia kerrana Beccari 钩叶藤
Vines

18.9 (17.2~21.0); 12.6 (9.5~15.1); 23.1 (21.6~25.6); 19.0 (17.3~21.0); 167.8 (147.0~191.8); 0.0 (0.0~0.0); 915 (833~1053) ❙ 1500 (1254~1751); 796 (655~996); 61 (48~77) ❙ 64.4 (44.5~88.8); 907 (833~1053); 492.3 (336.3~659.0)

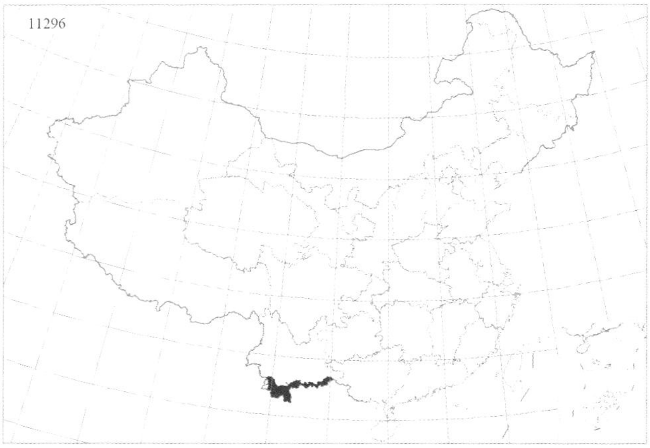

Arecaceae 棕榈科 *Plectocomia* 钩叶藤属
Plectocomia microstachys Burret 小钩叶藤
Vines, evergreen

23.5 (21.0~24.8); 18.0 (14.0~20.1); 27.5 (25.6~28.5); 23.6 (21.0~24.8); 222.9 (191.8~237.6); 0.0 (0.0~0.0); 1272 (1053~1385) ❙ 1588 (1179~2148); 603 (392~1234); 69 (41~124) ❙ 26.1 (-11.9~86.7); 1165 (1053~1360); 492.4 (318.8~620.0)

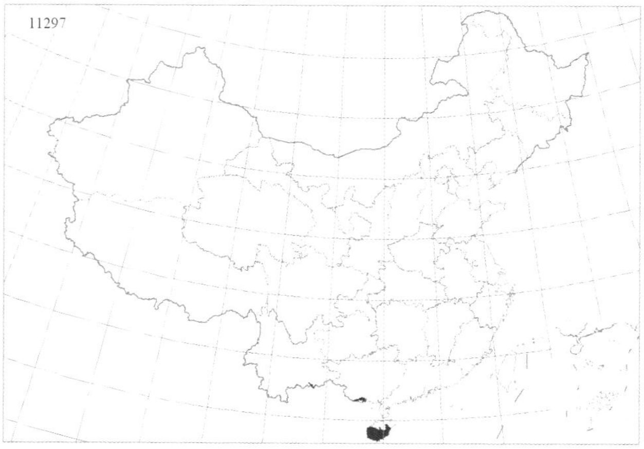

Arecaceae 棕榈科 *Rhapis* 棕竹属
Rhapis gracilis Burret 细棕竹
Shrubs, evergreen

20.3 (16.2~24.5); 12.4 (6.3~19.1); 26.5 (24.4~28.5); 20.4 (16.2~24.5); 184.4 (134.5~234.4); 0.0 (0.0~0.0); 1069 (826~1360) ❙ 1416 (1179~1854); 623 (392~742); 78 (42~122) ❙ 35.3 (-11.9~62.5); 1021 (826~1360); 326.8 (189.5~620.0)

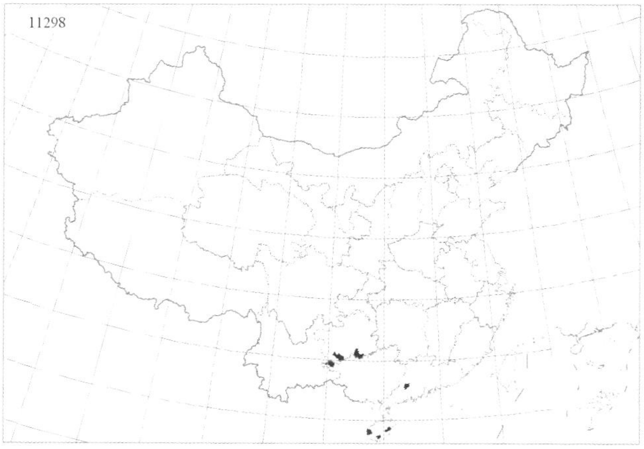

Arecaceae 棕榈科 *Rhapis* 棕竹属
Rhapis humilis Blume 矮棕竹
Shrubs, evergreen

18.2; 8.6; 26.0; 18.3; 159.5; 0.0; 933 ❙ 1332; 644; 90 ❙ 43.4; 933; 283.7

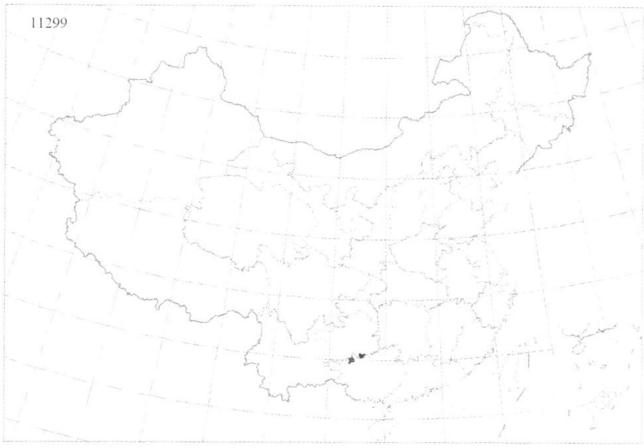

Arecaceae 棕榈科 *Rhapis* 棕竹属
Rhapis multifida Burret 多裂棕竹
Shrubs, evergreen

19.1; 11.0; 25.0; 19.1; 169.6; 0.0; 945 ❙ 1438; 820; 57 ❙ 52.2; 945; 371.7

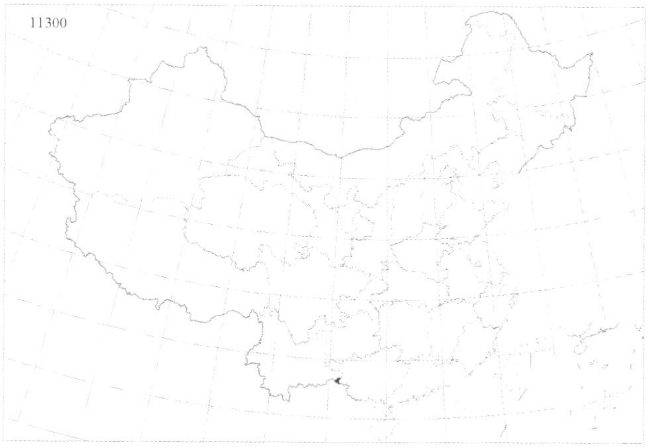

Arecaceae 棕榈科 *Rhapis* 棕竹属
Rhapis robusta Burret 粗棕竹
Shrubs, evergreen

22.1; 13.6; 28.4; 22.1; 205.6; 0.0; 1183 ❙ 1446; 743; 112 ❙ 22.2; 1183; 300.0

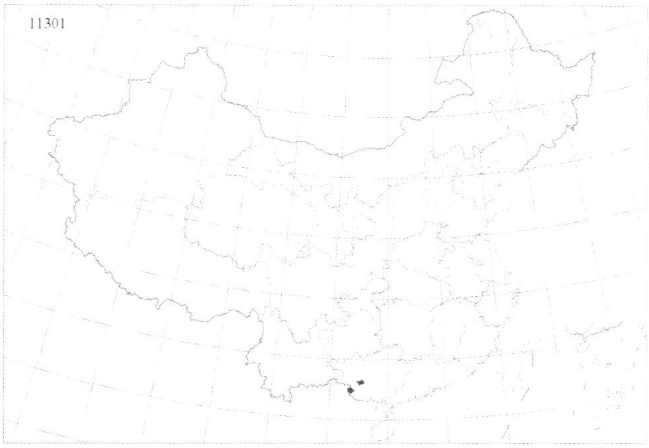

Arecaceae 棕榈科 *Salacca* 蛇皮果属
Salacca secunda Griff. 滇西蛇皮果
Shrubs, evergreen

16.4; 9.6; 20.6; 16.4; 137.3; 0.0; 795 ❙ 1688; 959; 62 ❙ 112.2; 795; 627.0

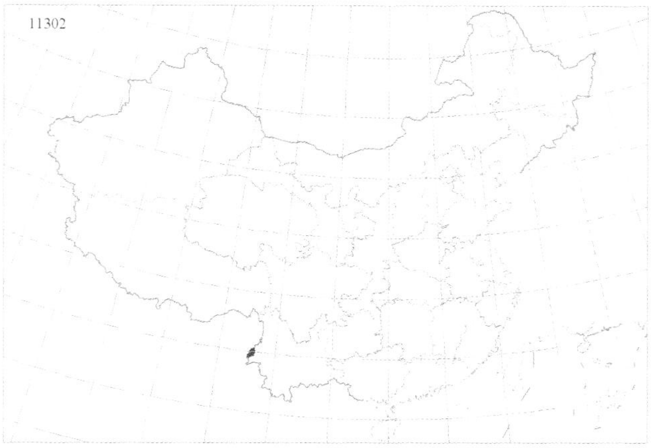

Arecaceae 棕榈科 *Trachycarpus* 棕榈属
Trachycarpus fortunei (Hook.) H. Wendl. 棕榈
Trees, evergreen

17.3 (5.4~24.7); 7.8 (-3.2~19.8); 25.5 (12.3~29.8); 17.3 (5.8~24.7); 148.6 (31.5~236.5); 0.9 (0.0~26.7); 909 (481~1376) ❙ 1393 (675~2435); 609 (331~1428); 116 (15~256) ❙ 53.9 (-14.2~147.0); 902 (481~1273); 248.2 (95.3~659.0)

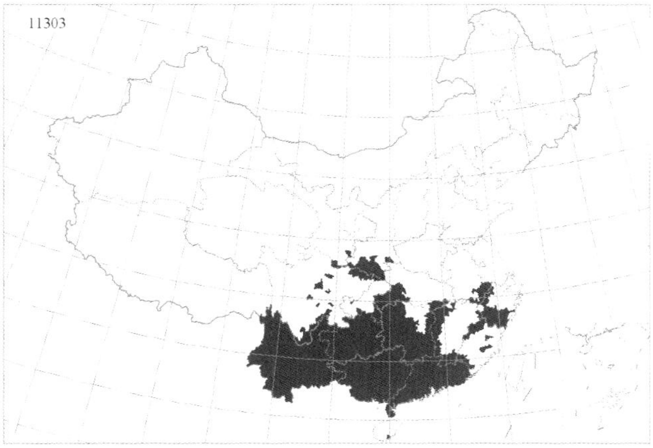

Arecaceae 棕榈科 *Trachycarpus* 棕榈属
Trachycarpus nana Becc. 龙棕
Shrubs, evergreen

16.0 (13.8~18.9); 9.1 (6.7~12.0); 21.2 (19.3~25.5); 16.0 (13.8~19.0); 132.2 (105.9~167.8); 0.0 (0.0~0.0); 792 (714~929) ❙ 971 (797~1381); 556 (411~637); 37 (22~107) ❙ 23.2 (-14.2~55.9); 729 (683~886); 288.3 (218.0~381.5)

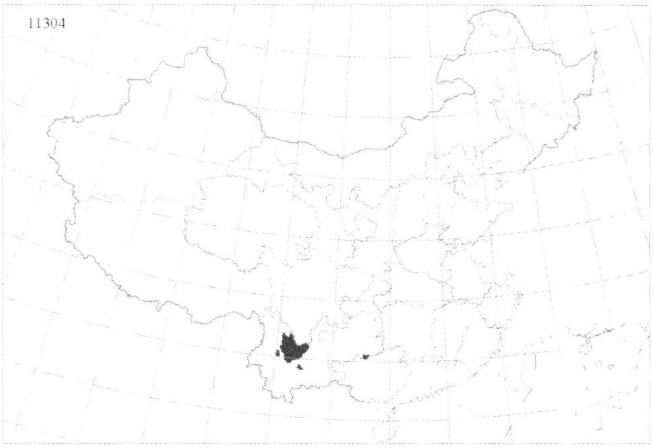

Arecaceae 棕榈科 *Trachycarpus* 棕榈属
Trachycarpus princeps Gibbons, Spanner et S. Y. Chen 贡山棕榈
Trees, evergreen

8.6; 1.3; 15.0; 8.7; 52.6; 8.8; 574 ▎1041; 585; 45 ▎81.5; 574; 352.3

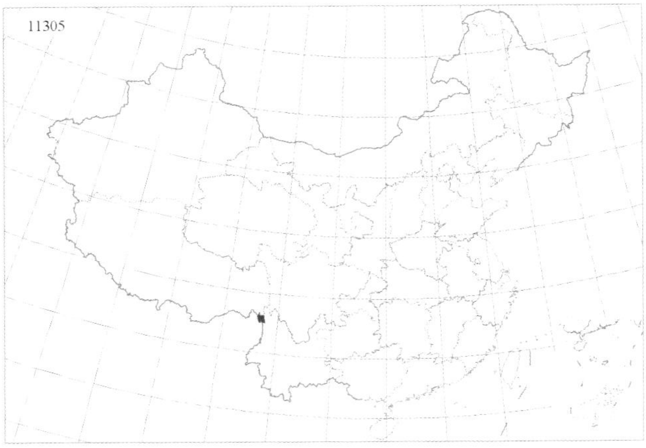

Arecaceae 棕榈科 *Wallichia* 瓦理棕属
Wallichia caryotoides Roxb. 琴叶瓦理棕
Shrubs, evergreen

18.6 (16.3~20.5); 12.6 (10.2~15.1); 22.4 (20.3~23.7); 18.7 (16.4~20.5); 164.0 (136.6~185.9); 0.0 (0.0~0.0); 893 (789~992) ▎1372 (1028~1751); 745 (566~996); 57 (44~77) ▎53.6 (12.5~88.8); 873 (780~986); 444.1 (339.5~659.0)

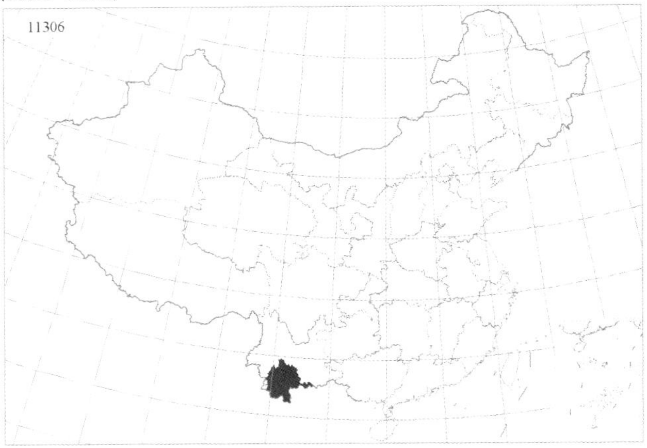

Arecaceae 棕榈科 *Wallichia* 瓦理棕属
Wallichia chinensis Burret 瓦理棕
Shrubs, evergreen

19.1 (17.6~22.2); 12.3 (7.0~14.2); 23.9 (22.1~28.3); 19.2 (17.6~22.3); 170.1 (151.7~207.5); 0.0 (0.0~0.0); 938 (854~1190) ▎1435 (1143~1751); 783 (548~996); 72 (50~191) ▎54.0 (15.4~88.8); 932 (854~1190); 422.4 (248.0~551.6)

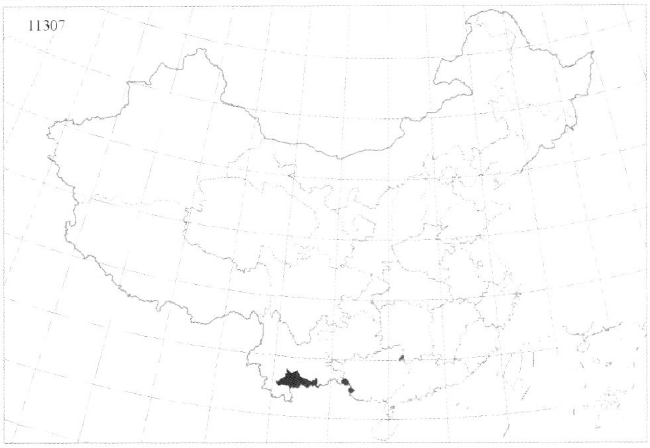

Araceae 天南星科 *Pothoidium* 假石柑属
Pothoidium lobbianum Schott 假石柑
Scandent shrubs

18.5; 14.1; 21.7; 18.5; 162.1; 0.0; 863 ▎2713; 1388; 188 ▎214.4; 863; 363.7

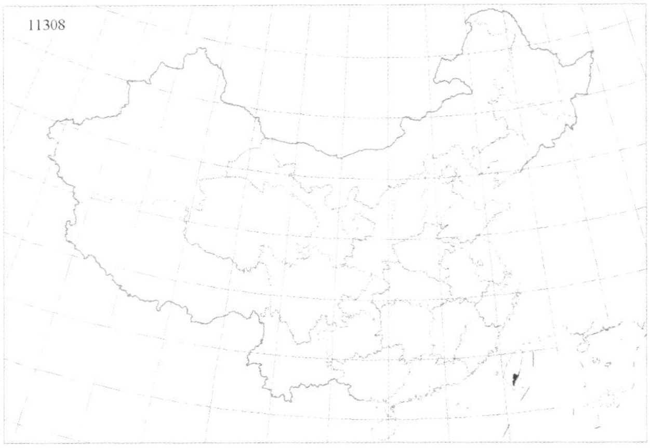

Flagellariaceae 须叶藤科 *Flagellaria* 须叶藤属
Flagellaria indica L. 须叶藤
Vines

23.1 (18.5~24.8); 17.8 (14.0~20.1); 26.8 (21.7~28.5); 23.1 (18.5~24.8); 217.4 (162.1~237.6); 0.0 (0.0~0.0); 1238 (863~1385) ▎1997 (1405~2992); 912 (473~1847); 96 (41~188) ▎71.1 (1.5~214.4); 1158 (863~1329); 427.2 (248.2~620.0)

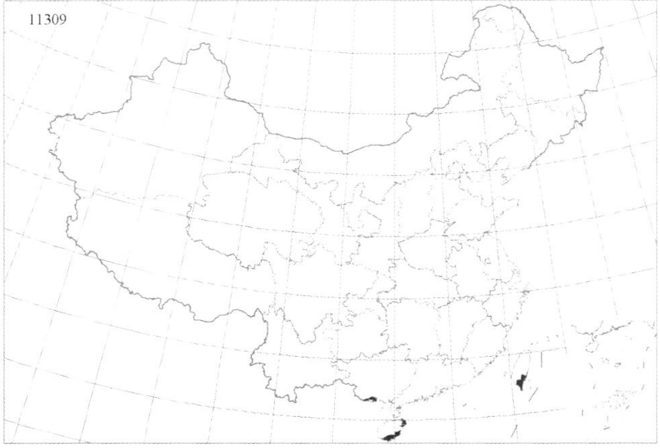

Stemonaceae 百部科 *Stemona* 百部属
Stemona japonica (Blume) Miq. 百部
Vines

16.4 (14.4~18.2); 4.9 (2.1~8.4); 27.8 (24.7~29.8); 16.5 (14.5~18.3); 138.1 (118.7~159.6); 0.6 (0.0~4.9); 888 (789~992) ▎1524 (969~1997); 547 (370~713); 194 (134~259) ▎72.1 (13.2~147.0); 883 (789~965); 225.0 (148.6~303.4)

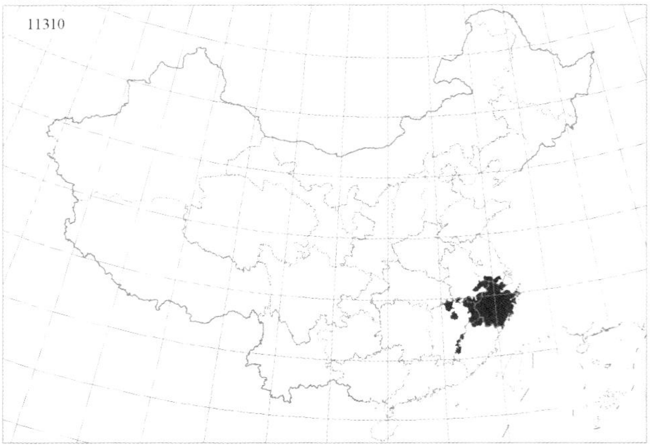

Stemonaceae 百部科 *Stemona* 百部属
Stemona mairei (H. Lév.) K. Krause 云南百部
Subshrubs or vines

12.4 (5.4~16.7); 5.4 (-2.4~9.8); 18.1 (12.3~22.1); 12.5 (5.8~16.8); 94.9 (31.5~141.1); 5.6 (0.0~26.2); 680 (481~819) ❙ 966 (685~1688); 572 (406~959); 30 (15~62) ❙ 43.1 (11.7~112.2); 652 (481~795); 312.4 (201.9~627.0)

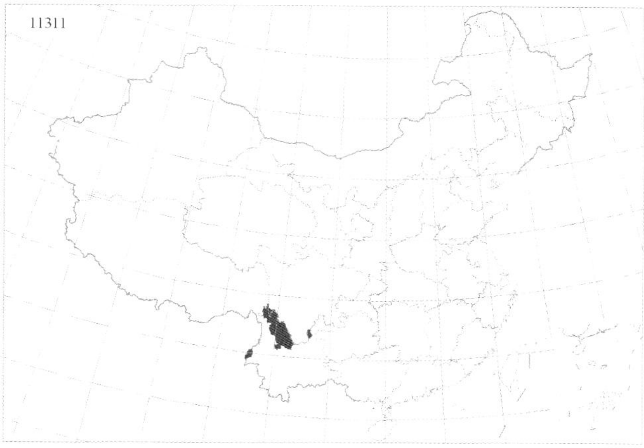

Stemonaceae 百部科 *Stemona* 百部属
Stemona tuberosa Lour. 大百部
Vines

17.1 (10.6~24.8); 7.8 (-0.2~20.1); 25.2 (18.6~29.5); 17.1 (10.6~24.8); 146.0 (79.4~237.6); 0.4 (0.0~12.1); 886 (643~1385) ❙ 1445 (868~2992); 635 (312~1847); 113 (28~303) ❙ 64.3 (-34.3~219.3); 876 (643~1221); 271.9 (119.4~659.0)

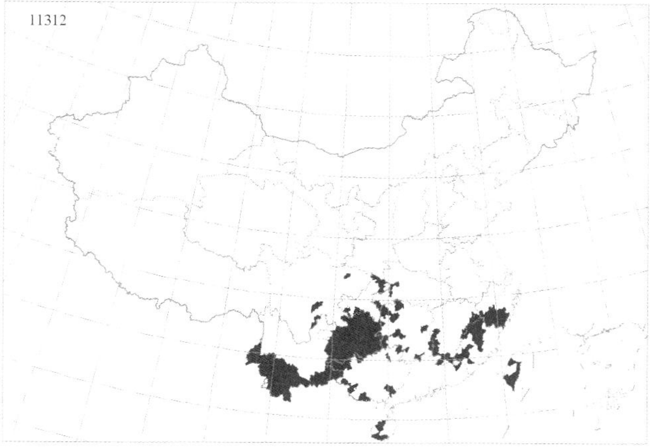

Liliaceae 百合科 *Dracaena* 龙血树属
Dracaena angustifolia Roxb. 长花龙血树
Shrubs

22.7 (17.3~24.8); 16.6 (9.9~20.1); 27.0 (21.7~28.9); 22.7 (17.3~24.8); 212.3 (147.1~237.6); 0.0 (0.0~0.0); 1205 (833~1385) ❙ 1729 (868~2992); 821 (312~1847); 86 (41~188) ❙ 48.5 (-34.3~214.4); 1131 (833~1360); 379.8 (119.4~659.0)

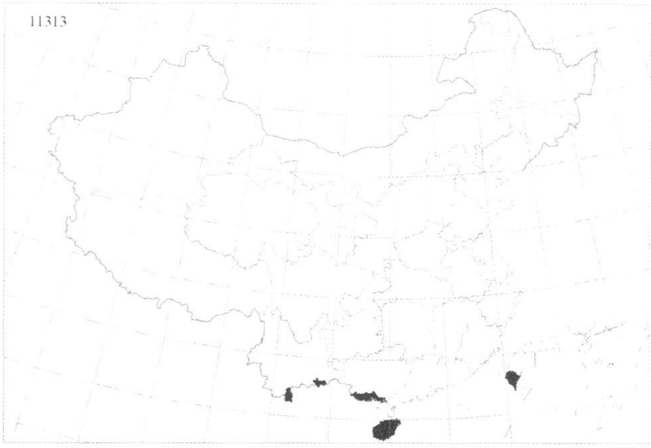

Liliaceae 百合科 *Dracaena* 龙血树属
Dracaena cambodiana Pierre ex Gagnep. 柬埔寨龙血树
Trees

21.4 (18.2~24.8); 15.8 (12.4~20.1); 25.0 (21.6~28.2); 21.5 (18.2~24.8); 197.5 (158.6~237.6); 0.0 (0.0~0.0); 1100 (861~1385) ❙ 1533 (1242~1900); 664 (444~1056); 63 (41~125) ❙ 43.6 (-2.2~74.2); 1023 (839~1299); 499.4 (297.2~659.0)

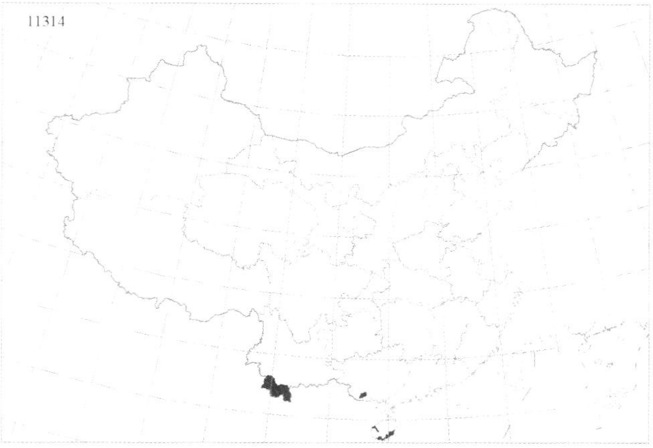

Liliaceae 百合科 *Dracaena* 龙血树属
Dracaena cochinchinensis (Lour.) S. C. Chen 剑叶龙血树
Trees

19.1 (14.4~22.3); 12.0 (7.5~15.1); 23.9 (19.1~28.9); 19.1 (14.4~22.3); 169.3 (113.1~208.1); 0.0 (0.0~0.0); 947 (727~1202) ❙ 1475 (1161~1751); 804 (633~996); 68 (42~136) ❙ 58.6 (12.9~112.2); 943 (727~1202); 426.0 (201.4~659.0)

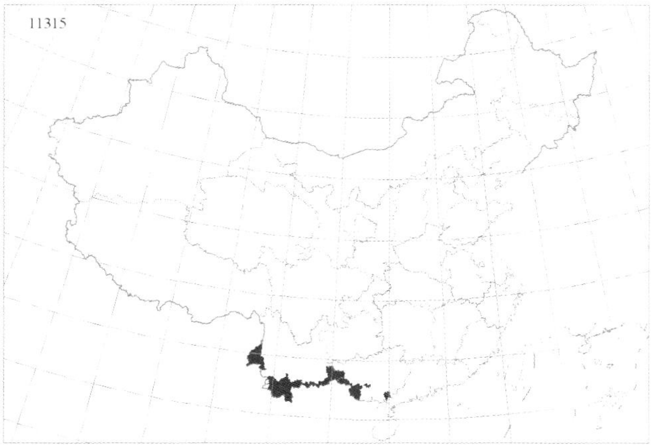

Liliaceae 百合科 *Dracaena* 龙血树属
Dracaena elliptica Thunb. 细枝龙血树
Tall shrubs

21.4; 13.4; 27.6; 21.4; 197.3; 0.0; 1120 ❙ 1640; 757; 129 ❙ 46.7; 1120; 248.5

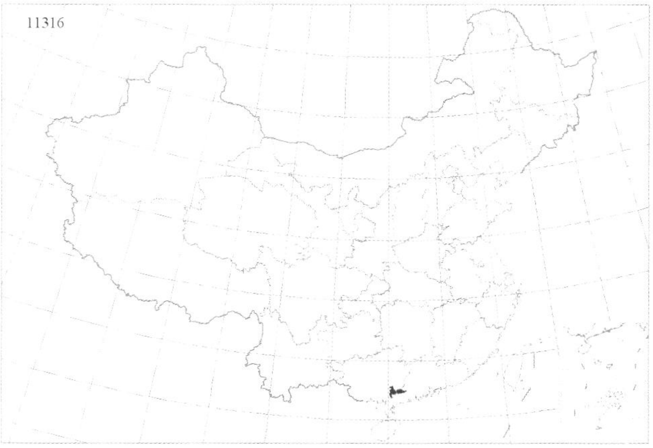

Liliaceae 百合科 *Dracaena* 龙血树属
Dracaena hokouensis G. Z. Ye 河口龙血树
Shrubs

19.3; 12.7; 23.8; 19.3; 171.6; 0.0; 942 ❙ 1439; 803; 61 ❙ 53.4; 942; 516.4

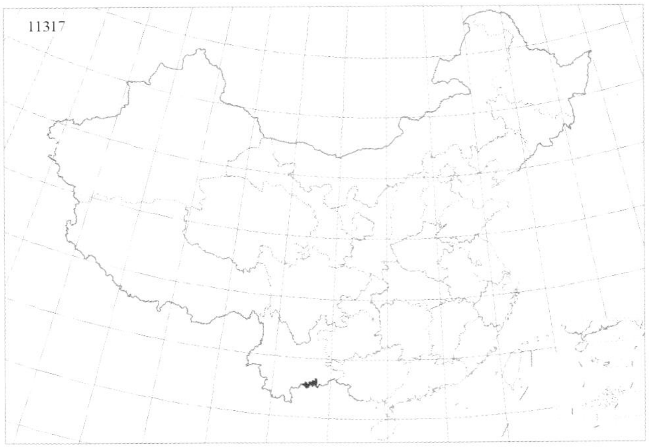

Liliaceae 百合科 *Dracaena* 龙血树属
Dracaena terniflora Roxburgh 矮龙血树
Subshrubs

20.5; 14.9; 23.7; 20.5; 185.9; 0.0; 992 ❙ 1464; 655; 66 ❙ 47.6; 959; 570.1

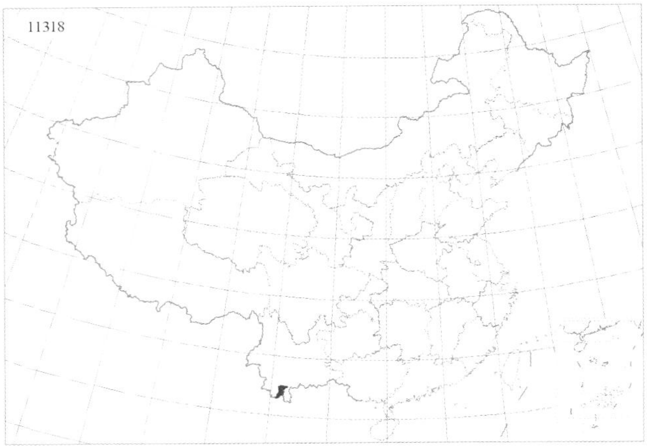

Liliaceae 百合科 *Heterosmilax* 肖菝葜属
Heterosmilax chinensis F. T. Wang 华肖菝葜
Vines

16.7 (6.0~22.2); 6.9 (-3.2~14.2); 25.3 (14.4~29.8); 16.7 (6.6~22.3); 141.5 (39.7~207.5); 1.1 (0.0~26.7); 880 (508~1190) ❙ 1278 (899~1777); 608 (442~946); 78 (19~197) ❙ 46.9 (8.0~101.2); 878 (508~1190); 224.1 (108.2~551.6)

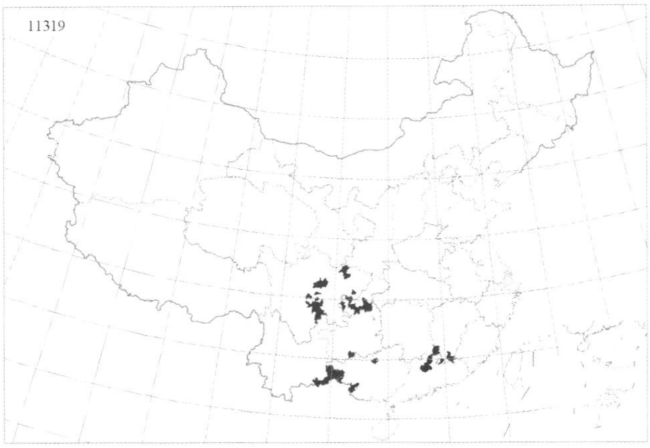

Liliaceae 百合科 *Heterosmilax* 肖菝葜属
Heterosmilax gaudichaudiana (Kunth) Maxim. 合丝肖菝葜
Vines

21.0 (17.6~23.6); 12.4 (7.1~16.7); 28.0 (26.2~29.0); 21.1 (17.6~23.6); 192.9 (151.6~223.5); 0.0 (0.0~0.0); 1112 (906~1281) ❙ 1698 (1269~2435); 771 (513~1428); 142 (75~192) ❙ 53.5 (-1.0~99.5); 1112 (906~1269); 232.6 (157.7~318.8)

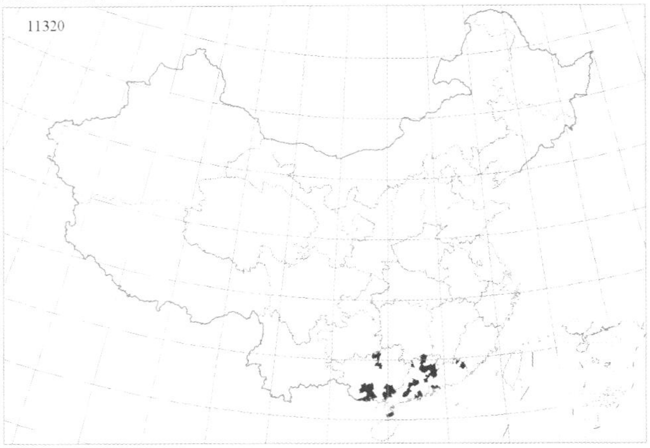

Liliaceae 百合科 *Heterosmilax* 肖菝葜属
Heterosmilax japonica Kunth 肖菝葜
Scandent shrubs, evergreen

16.9 (0.4~24.8); 7.7 (-10.7~20.1); 24.8 (9.9~29.9); 17.0 (3.5~24.8); 145.5 (13.3~237.6); 2.6 (0.0~68.1); 897 (402~1385) ❙ 1371 (590~2993); 605 (299~1847); 107 (11~547) ❙ 53.8 (-34.3~246.5); 882 (402~1360); 269.9 (108.2~620.0)

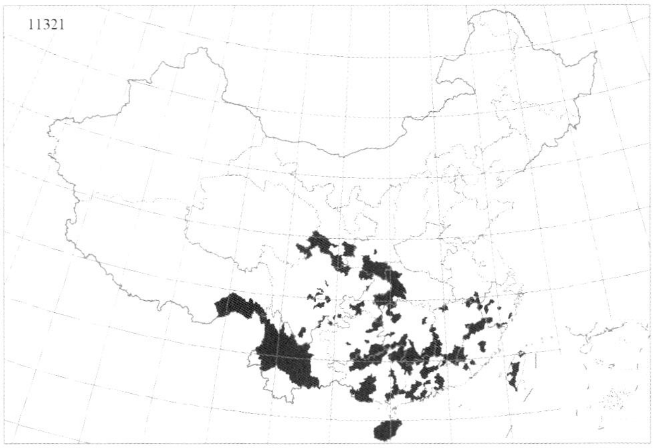

Liliaceae 百合科 *Heterosmilax* 肖菝葜属
Heterosmilax polyandra Gagnep. 多蕊肖菝葜
Vines

18.9 (17.3~20.5); 13.0 (11.0~15.1); 22.6 (21.6~23.7); 18.9 (17.3~20.5); 167.1 (148.0~185.9); 0.0 (0.0~0.0); 905 (832~992) ❙ 1431 (1049~1751); 766 (598~996); 60 (44~77) ❙ 58.0 (26.1~88.8); 886 (783~986); 466.4 (354.7~659.0)

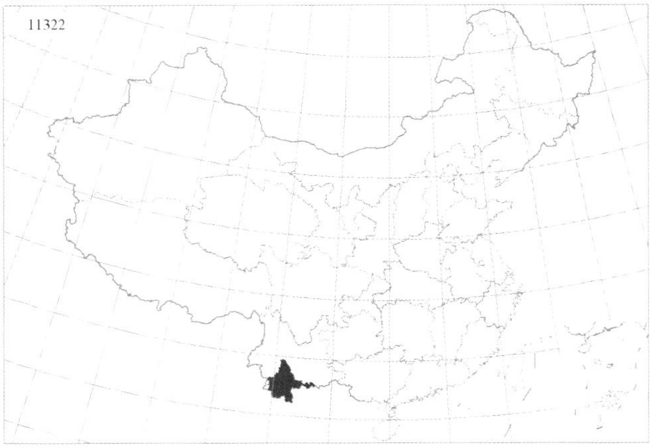

Liliaceae 百合科 *Heterosmilax* 肖菝葜属
Heterosmilax seisuiensis (Hayata) F. T. Wang et Tang 台湾肖菝葜
Vines

20.2 (16.2~24.1); 14.5 (11.0~18.9); 25.0 (20.1~28.2); 20.2 (16.2~24.1); 182.9 (134.2~229.7); 0.0 (0.0~0.0); 1019 (773~1322) ❙ 2402 (1498~3495); 1106 (750~1847); 255 (47~897) ❙ 146.2 (25.2~246.5); 983 (773~1144); 245.0 (119.4~363.7)

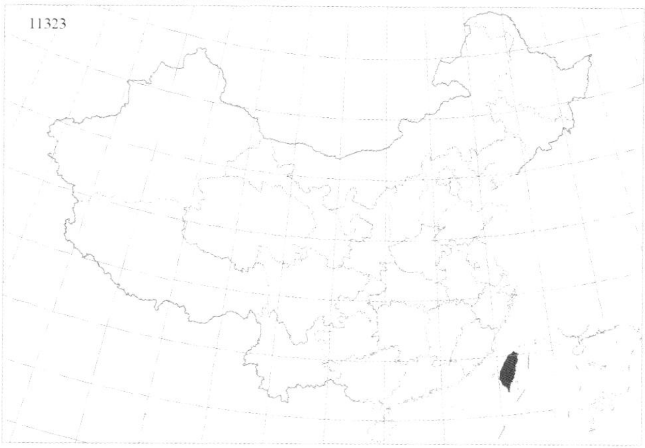

Liliaceae 百合科 *Heterosmilax* 肖菝葜属
Heterosmilax septemnervia F. T. Wang et Tang 短柱肖菝葜
Vines

15.0 (9.6~20.5); 5.3 (-0.2~14.9); 23.8 (15.7~28.9); 15.0 (9.6~20.5); 122.5 (60.8~185.9); 2.1 (0.0~12.1); 797 (600~1002) ❙ 1220 (963~1492); 580 (454~816); 62 (18~105) ❙ 54.8 (8.0~106.2); 792 (600~1002); 239.7 (108.2~570.1)

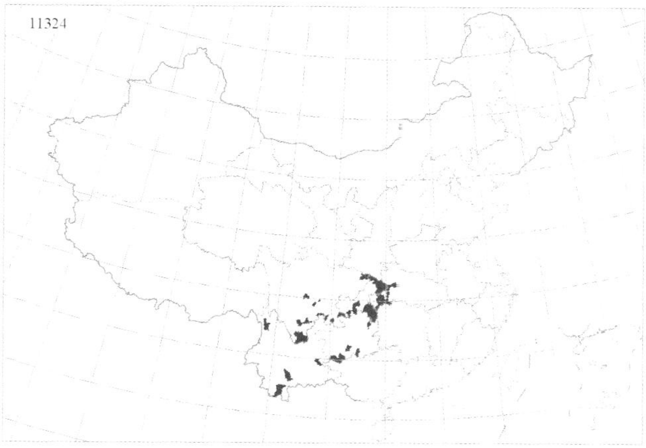

Liliaceae 百合科 *Heterosmilax* 肖菝葜属
Heterosmilax yunnanensis Gagnep. 云南肖菝葜
Vines

15.9 (9.3~22.2); 7.4 (-1.7~14.2); 22.9 (15.7~29.0); 15.9 (9.5~22.3); 132.1 (60.8~207.5); 0.9 (0.0~18.0); 818 (600~1190) ❙ 1160 (797~1776); 597 (411~911); 58 (18~188) ❙ 42.9 (-14.2~107.2); 806 (600~1190); 277.7 (120.2~551.6)

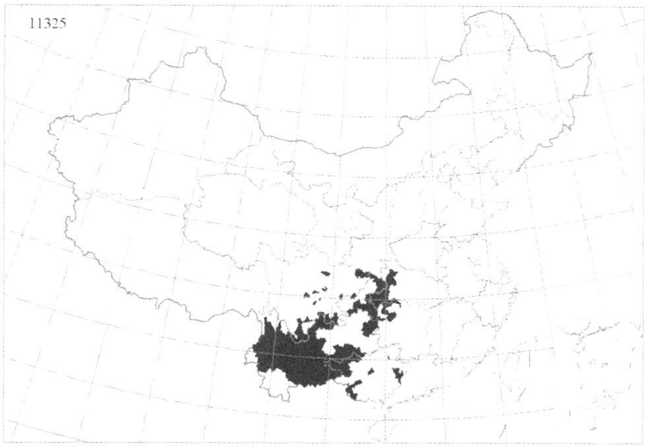

Liliaceae 百合科 *Smilax* 菝葜属
Smilax aberrans Gagnep. 弯梗菝葜
Shrubs or subshrubs, evergreen

17.5 (12.5~23.2); 7.9 (2.1~15.8); 25.7 (20.1~29.4); 17.6 (12.5~23.2); 151.0 (92.0~218.9); 0.4 (0.0~5.6); 908 (685~1257) ❙ 1382 (1071~1961); 623 (438~891); 102 (43~217) ❙ 53.5 (12.9~107.2); 907 (685~1257); 256.6 (160.8~551.6)

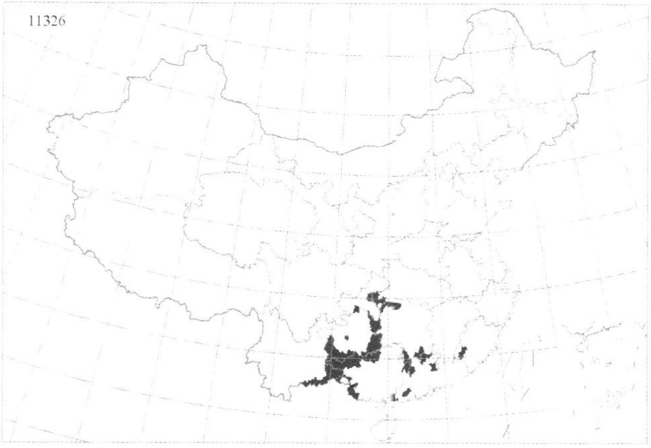

Liliaceae 百合科 *Smilax* 菝葜属
Smilax arisanensis Hayata 尖叶菝葜
Vines, evergreen

17.4 (14.1~24.1); 7.3 (1.6~18.9); 26.6 (20.1~29.8); 17.5 (14.2~24.1); 150.1 (112.6~229.7); 0.5 (0.0~6.5); 915 (756~1322) ❙ 1651 (1090~3495); 636 (425~1847); 182 (47~897) ❙ 81.3 (16.1~246.5); 910 (756~1150); 249.5 (142.5~551.6)

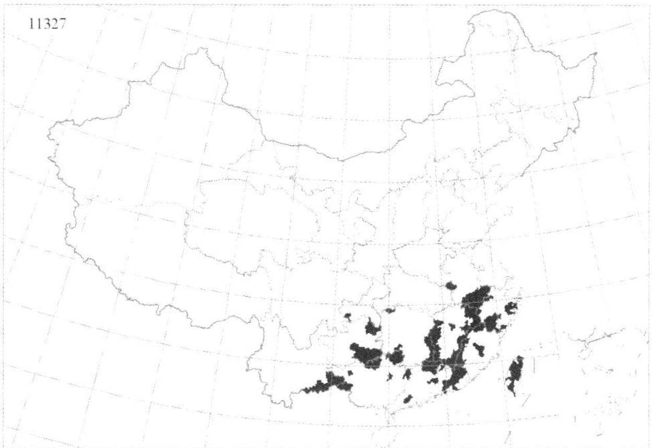

Liliaceae 百合科 *Smilax* 菝葜属
Smilax aspera L. 穗菝葜
Vines, evergreen

5.8; -1.9; 12.4; 7.5; 51.7; 41.8; 521 ❙ 360; 381; 35 ❙ 22.0; 509; 203.9

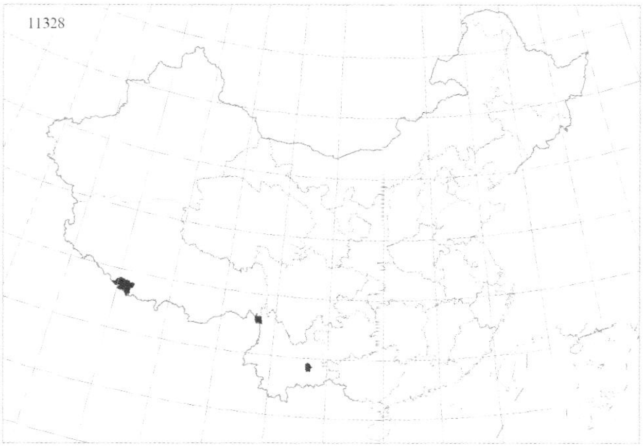

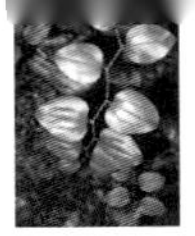

Liliaceae 百合科 *Smilax* 菝葜属

Smilax aspericaulis Wall. ex A. DC. 疣枝菝葜

Vines, evergreen

19.3 (8.6~24.8); 12.0 (1.3~20.1); 24.9 (15.0~29.0); 19.4 (8.7~24.8); 172.7 (52.6~237.6); 0.1 (0.0~8.8); 992 (574~1385) ❙ 1515 (868~3495); 708 (312~1594); 103 (41~897) ❙ 57.3 (-34.3~246.5); 962 (574~1360); 337.4 (154.2~620.0)

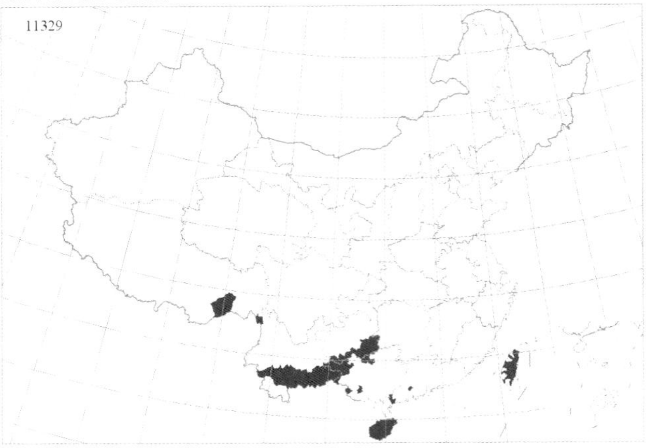

Liliaceae 百合科 *Smilax* 菝葜属

Smilax astrosperma F. T. Wang et Tang 灰叶菝葜

Vines, evergreen

21.4 (16.2~24.3); 15.0 (7.3~19.1); 26.4 (20.1~29.0); 21.5 (16.2~24.3); 197.7 (134.2~231.9); 0.0 (0.0~0.0); 1121 (773~1338) ❙ 1861 (868~2993); 869 (312~1847); 134 (42~547) ❙ 77.1 (-34.3~246.5); 1053 (773~1326); 310.3 (119.4~620.0)

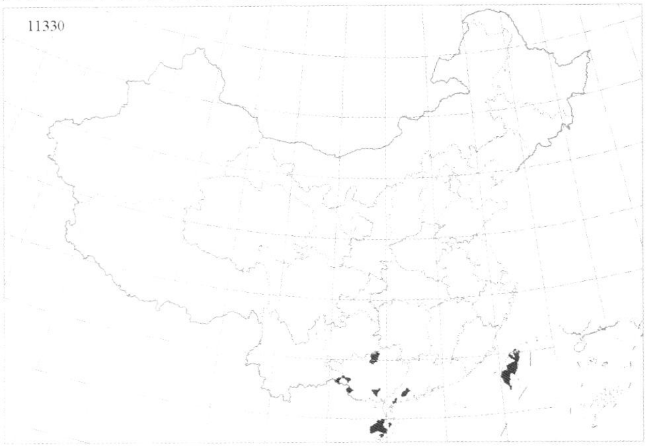

Liliaceae 百合科 *Smilax* 菝葜属

Smilax austrozhejiangensis Q. Lin 浙南菝葜

Shrubs, evergreen

16.2 (15.1~17.5); 5.5 (4.0~7.1); 26.8 (25.8~27.8); 16.2 (15.2~17.6); 135.0 (122.9~150.6); 0.2 (0.0~1.0); 858 (811~920) ❙ 1710 (1631~1813); 622 (590~659); 194 (181~226) ❙ 99.8 (84.6~123.6); 858 (811~920); 236.5 (204.3~257.7)

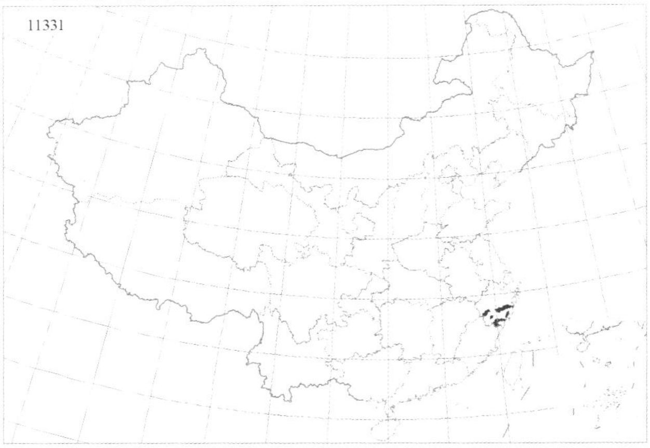

Liliaceae 百合科 *Smilax* 菝葜属

Smilax bapouensis H. Li 巴坡菝葜

Shrubs, evergreen

8.6; 1.3; 15.0; 8.7; 52.6; 8.8; 574 ❙ 1041; 585; 45 ❙ 81.5; 574; 352.3

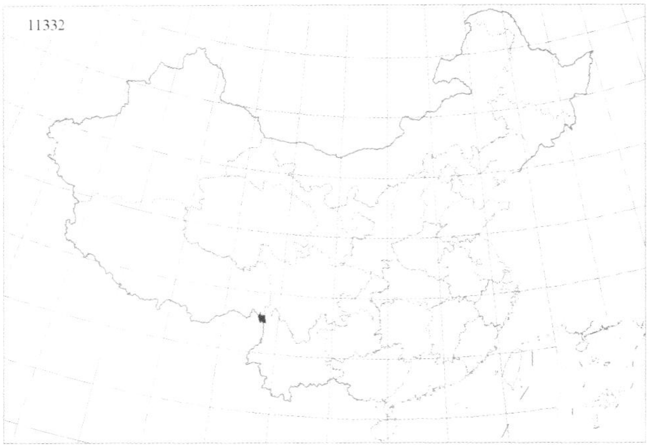

Liliaceae 百合科 *Smilax* 菝葜属

Smilax basilata F. T. Wang et Tang 少花菝葜

Vines, evergreen

18.9 (17.2~21.8); 10.7 (9.1~13.7); 24.8 (22.7~27.7); 18.9 (17.3~21.8); 166.8 (147.0~202.2); 0.0 (0.0~0.0); 943 (838~1154) ❙ 1317 (1161~1485); 736 (633~848); 57 (43~86) ❙ 41.0 (12.9~65.6); 943 (838~1154); 337.1 (304.0~371.7)

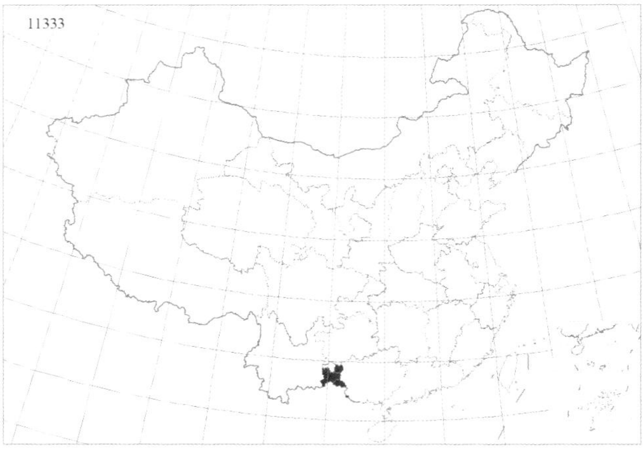

Liliaceae 百合科 *Smilax* 菝葜属

Smilax bauhinioides Kunth 圆叶菝葜

Vines, evergreen

21.8; 14.0; 27.7; 21.9; 202.2; 0.0; 1150 ❙ 2148; 1234; 124 ❙ 86.7; 1150; 318.8

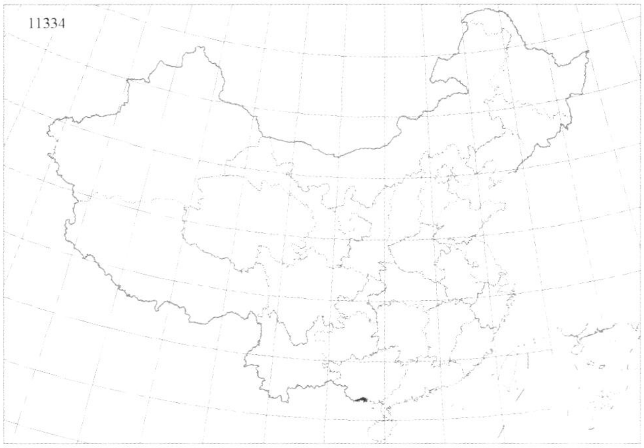

Liliaceae 百合科 *Smilax* 菝葜属
Smilax biumbellata T. Koyama 西南菝葜
Vines, evergreen

14.7 (1.8~19.2); 5.3 (-7.1~11.9); 23.0 (9.9~27.7); 14.8 (3.8~19.2); 119.2 (15.7~170.8); 2.3 (0.0~53.3); 785 (404~1002) ❙ 1186 (583~1740); 554 (336~816); 77 (13~187) ❙ 51.6 (2.4~100.7); 782 (404~1002); 263.9 (127.1~516.7)

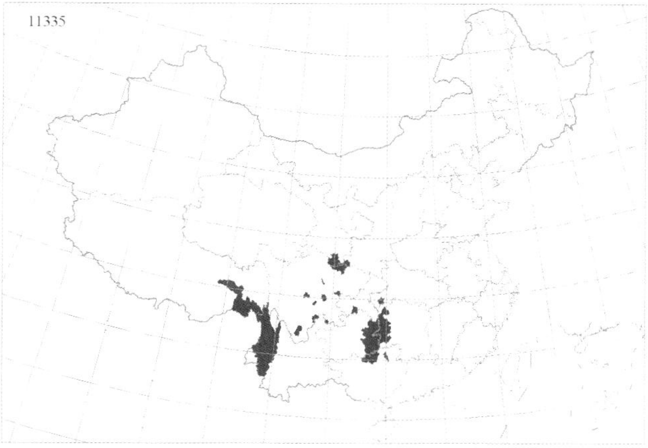

Liliaceae 百合科 *Smilax* 菝葜属
Smilax bracteata C. Presl 圆锥菝葜
Vines, evergreen

19.1 (11.1~24.8); 11.7 (1.3~20.1); 25.0 (16.7~28.9); 19.1 (11.2~24.8); 169.8 (75.8~237.6); 0.1 (0.0~8.0); 981 (642~1385) ❙ 1454 (797~3495); 661 (312~1594); 113 (21~897) ❙ 50.1 (-34.3~246.5); 955 (642~1360); 305.7 (122.7~659.0)

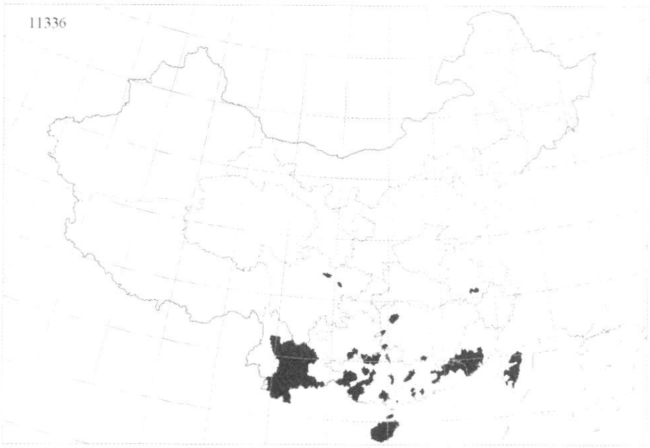

Liliaceae 百合科 *Smilax* 菝葜属
Smilax chapaensis Gagnep. 密疣菝葜
Vines, evergreen

16.0 (11.0~22.2); 6.8 (1.6~14.2); 24.1 (17.0~28.9); 16.1 (11.1~22.3); 134.1 (74.6~207.5); 1.2 (0.0~6.9); 832 (640~1190) ❙ 1251 (950~1751); 622 (447~996); 63 (15~113) ❙ 52.1 (8.0~107.2); 831 (638~1190); 271.8 (108.2~551.6)

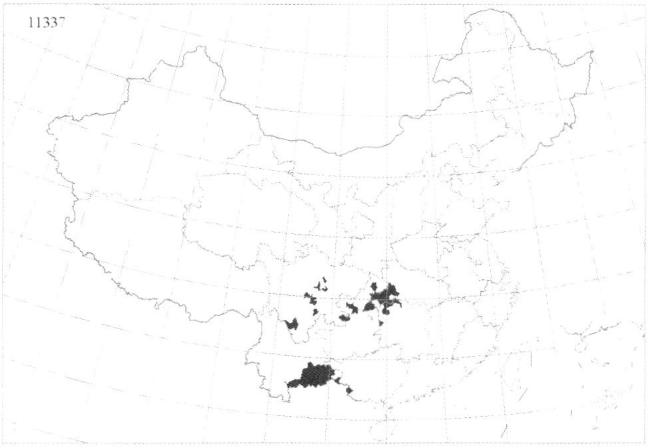

Liliaceae 百合科 *Smilax* 菝葜属
Smilax china L. 菝葜
Vines, evergreen

16.9 (8.7~24.8); 6.4 (-8.3~20.1); 26.4 (16.7~30.1); 17.0 (9.2~24.8); 145.5 (64.1~237.6); 2.2 (0.0~40.2); 907 (595~1385) ❙ 1313 (604~3495); 560 (312~1847); 121 (19~897) ❙ 44.8 (-34.3~246.5); 890 (595~1360); 224.4 (95.3~659.0)

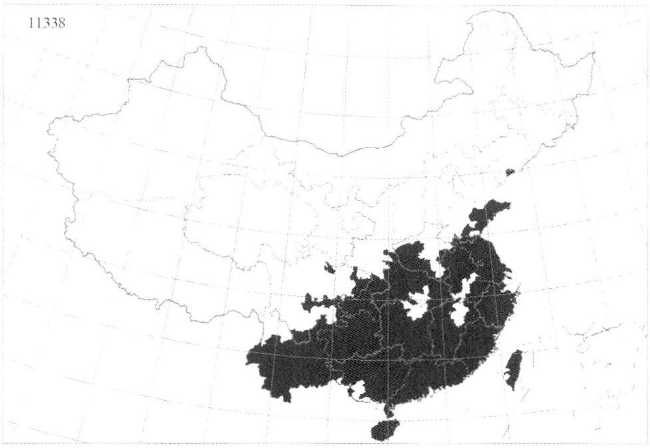

Liliaceae 百合科 *Smilax* 菝葜属
Smilax chingii F. T. Wang et Tang 柔毛菝葜
Vines, evergreen

16.6 (8.6~21.6); 6.1 (-1.7~14.2); 26.1 (15.0~29.8); 16.6 (8.7~21.7); 140.3 (52.6~199.8); 0.8 (0.0~18.0); 877 (574~1153) ❙ 1380 (999~1961); 567 (392~932); 124 (42~259) ❙ 58.1 (5.8~128.9); 874 (574~1153); 222.4 (108.2~551.6)

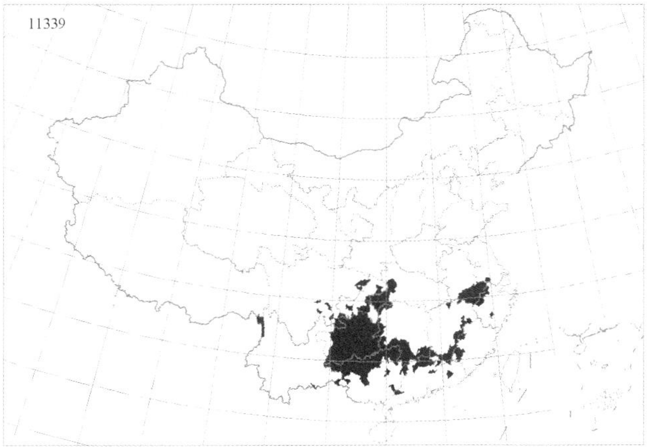

Liliaceae 百合科 *Smilax* 菝葜属
Smilax cocculoides Warb. 银叶菝葜
Shrubs, evergreen

16.3 (9.3~21.8); 6.6 (-1.7~15.1); 24.8 (18.5~28.5); 16.4 (9.5~21.8); 137.6 (69.8~202.2); 1.1 (0.0~18.0); 852 (603~1158) ❙ 1326 (950~1882); 607 (444~996); 89 (27~191) ❙ 57.0 (11.3~107.2); 851 (603~1158); 269.2 (132.2~659.0)

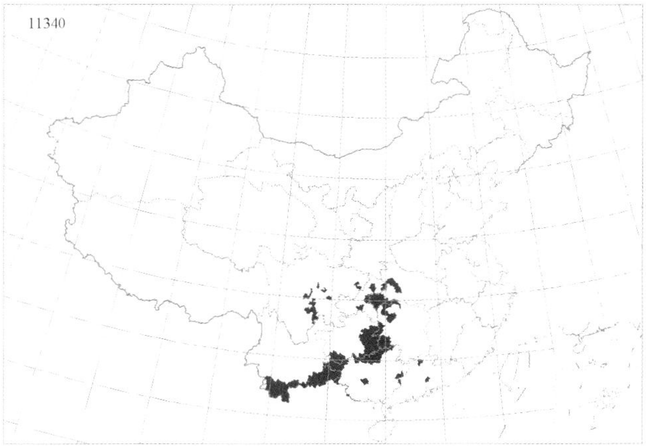

Liliaceae 百合科 *Smilax* 菝葜属
Smilax corbularia Kunth 筐条菝葜
Vines, evergreen

20.5 (16.3~24.8); 12.3 (6.1~20.1); 27.0 (21.4~29.8); 20.5 (16.4~24.8); 185.9 (136.4~237.6); 0.0 (0.0~0.0); 1066 (797~1385) ❙ 1588 (868~2993); 700 (312~1847); 127 (41~547) ❙ 50.8 (-34.3~214.4); 1053 (797~1360); 277.2 (95.3~659.0)

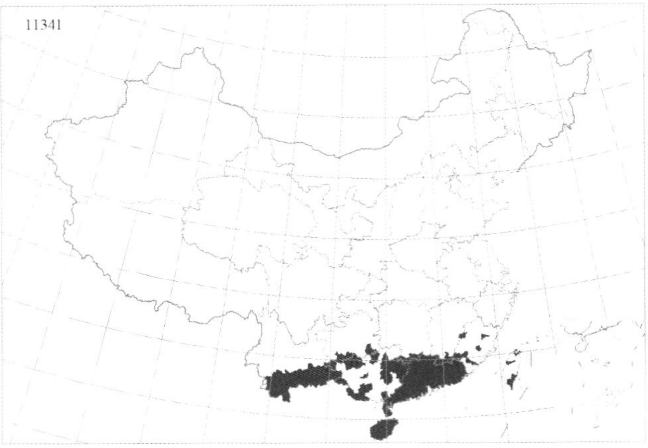

Liliaceae 百合科 *Smilax* 菝葜属
Smilax corbularia var. woodii (Merr.) T. Koyama 光叶菝葜
Vines, evergreen

20.7; 12.5; 26.7; 20.7; 188.5; 0.0; 1068 ❙ 1406; 766; 72 ❙ 33.8; 1068; 349.7

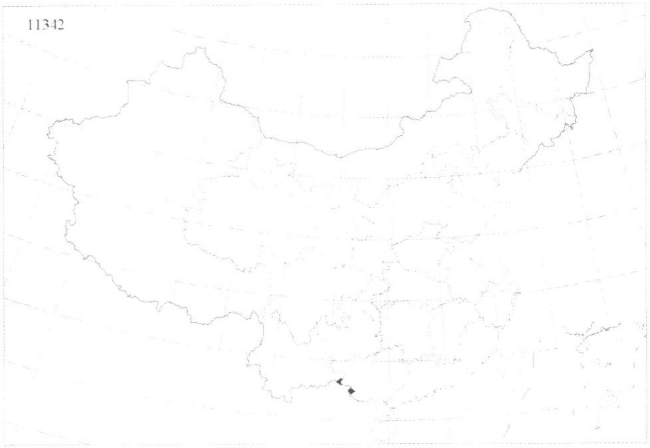

Liliaceae 百合科 *Smilax* 菝葜属
Smilax cyclophylla Warb. 合蕊菝葜
Shrubs, evergreen

13.1 (2.0~18.9); 5.3 (-7.4~12.0); 19.5 (9.5~26.6); 13.2 (3.8~19.0); 101.6 (15.5~167.8); 4.3 (0.0~51.9); 707 (411~929) ❙ 1005 (729~1414); 575 (411~817); 34 (11~71) ❙ 45.1 (-14.2~125.9); 689 (411~912); 273.8 (142.9~384.1)

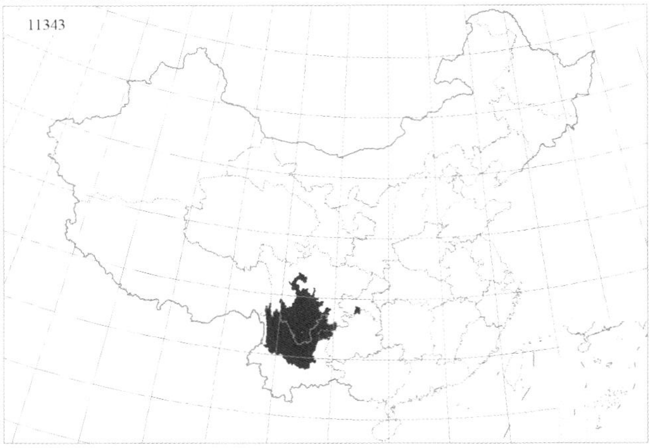

Liliaceae 百合科 *Smilax* 菝葜属
Smilax darrisii H. Lév. 平滑菝葜
Shrubs, evergreen

16.4 (12.8~21.0); 7.2 (3.5~14.2); 24.1 (19.6~26.6); 16.4 (12.9~21.0); 137.5 (95.9~191.8); 0.2 (0.0~1.5); 844 (699~1053) ❙ 1232 (879~1522); 650 (472~818); 55 (34~68) ❙ 46.3 (25.7~80.4); 844 (699~1053); 241.0 (171.2~551.6)

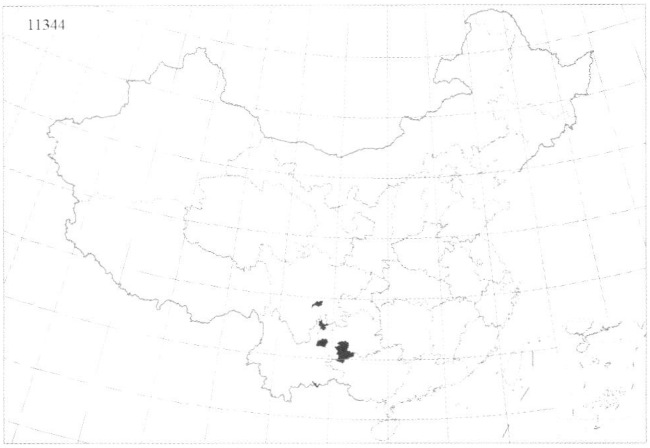

Liliaceae 百合科 *Smilax* 菝葜属
Smilax davidiana A. DC. 小果菝葜
Vines, evergreen

16.7 (12.9~21.8); 5.4 (0.5~13.5); 27.6 (22.5~30.1); 16.8 (12.9~21.8); 142.4 (101.2~201.6); 1.0 (0.0~9.2); 902 (723~1160) ❙ 1488 (710~2005); 547 (370~902); 179 (49~259) ❙ 65.2 (-19.8~147.0); 896 (710~1160); 221.6 (131.2~491.5)

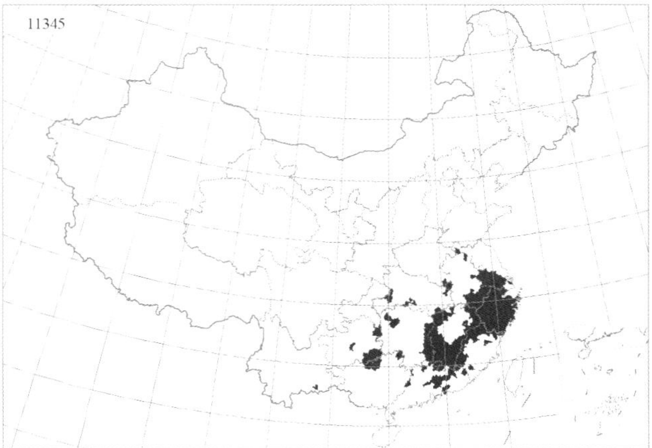

Liliaceae 百合科 *Smilax* 菝葜属
Smilax densibarbata F. T. Wang et Tang 密刺菝葜
Vines, evergreen

18.2; 10.3; 24.0; 18.3; 159.1; 0.0; 895 ❙ 1351; 765; 52 ❙ 51.3; 895; 352.5

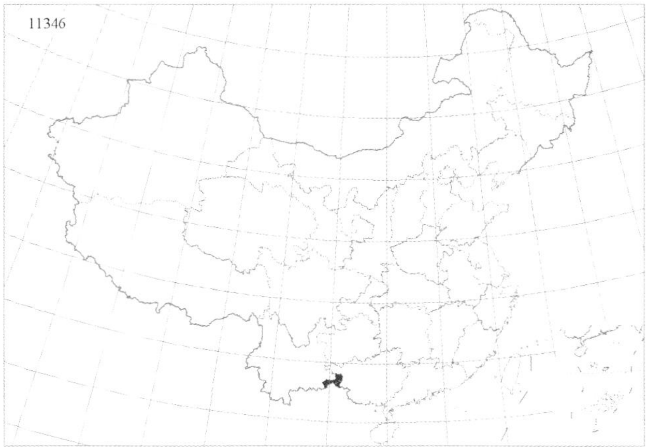

Liliaceae 百合科 *Smilax* 菝葜属
Smilax discotis Warb. 托柄菝葜
Shrubs, evergreen

14.5 (2.6~21.0); 3.0 (-7.0~14.2); 25.2 (11.2~29.3); 14.6 (4.3~21.0); 120.0 (20.0~191.8); 5.8 (0.0~48.2); 804 (427~1053) ❙ 1229 (551~2859); 508 (245~1186); 104 (13~402) ❙ 51.4 (-36.2~246.5); 793 (427~1053); 241.6 (131.4~551 6)

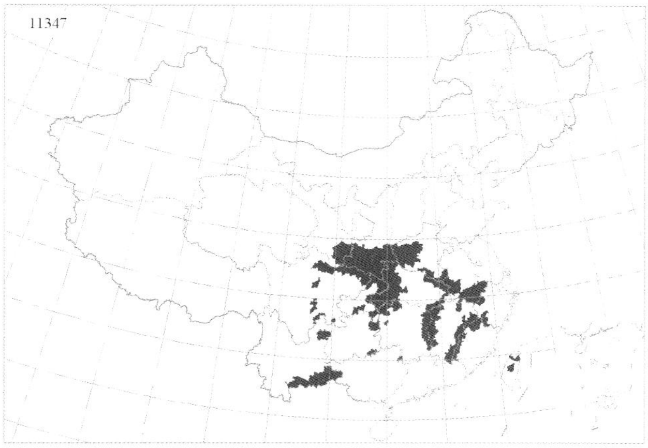

Liliaceae 百合科 *Smilax* 菝葜属
Smilax elegans Wall. ex Kunth 西藏菝葜
Vines, evergreen

-1.5; -9.6; 6.1; 1.9; 2.2; 79.2; 332 ❙ 320; 205; 26 ❙ -3.5; 320; 75.7

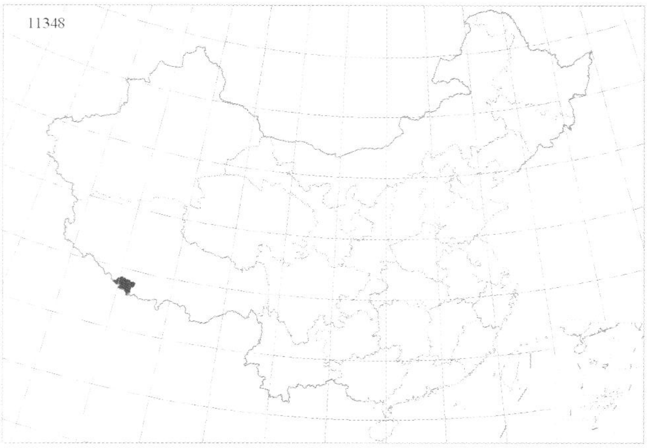

Liliaceae 百合科 *Smilax* 菝葜属
Smilax elegantissima Gagnep. 四棱菝葜
Vines, evergreen

19.5 (18.6~21.0); 13.4 (12.7~14.2); 23.6 (22.5~25.6); 19.5 (18.7~21.0); 174.1 (163.9~191.8); 0.0 (0.0~0.0); 944 (888~1053) ❙ 1505 (1276~1751); 860 (741~996); 63 (51~77) ❙ 60.0 (36.8~88.8); 942 (888~1053); 515.0 (452.0~551.6)

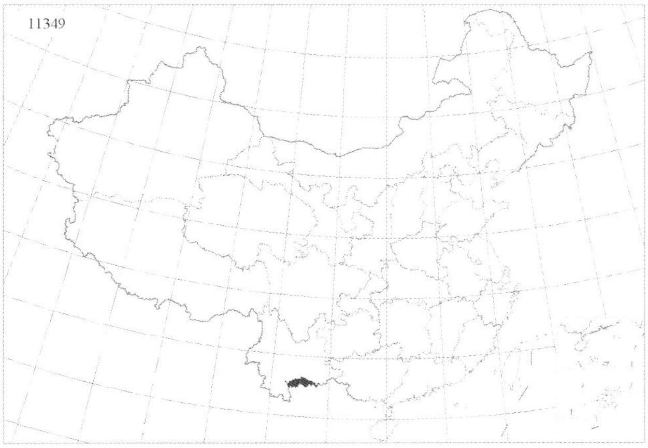

Liliaceae 百合科 *Smilax* 菝葜属
Smilax elongatoumbellata Hayata 台湾菝葜
Scandent shrubs, evergreen

19.3 (16.2~24.1); 13.6 (11.0~18.9); 23.7 (20.1~28.0); 19.3 (16.2~24.1); 171.4 (134.2~229.7); 0.0 (0.0~0.0); 954 (773~1322) ❙ 2252 (1522~2759); 1106 (818~1594); 182 (47~308) ❙ 147.5 (32.5~246.5); 927 (773~1095); 282.6 (154.2~551.6)

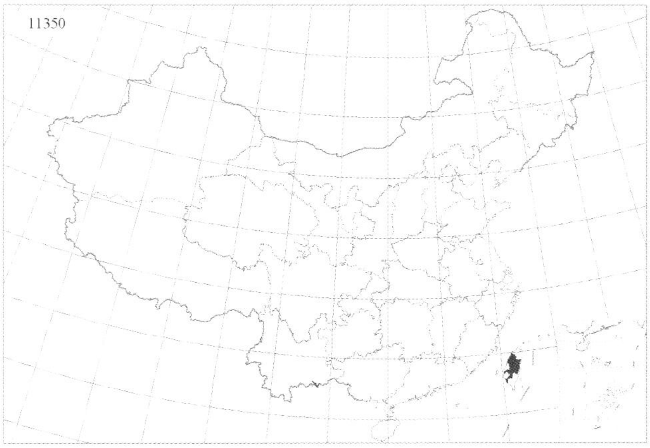

Liliaceae 百合科 *Smilax* 菝葜属
Smilax emeiensis J. M. Xu 峨眉菝葜
Shrubs, evergreen

14.0; 4.3; 22.8; 14.0; 109.7; 1.1; 750 ❙ 1282; 737; 42 ❙ 70.4; 750; 193.1

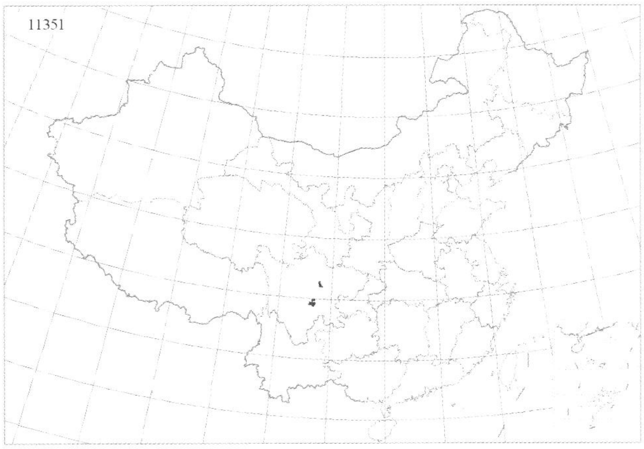

Liliaceae 百合科 *Smilax* 菝葜属
Smilax fooningensis F. T. Wang et Tang 富宁菝葜
Vines, evergreen

19.0; 10.8; 25.0; 19.0; 168.6; 0.0; 942 ❙ 1365; 775; 53 ❙ 44.9; 942; 357.1

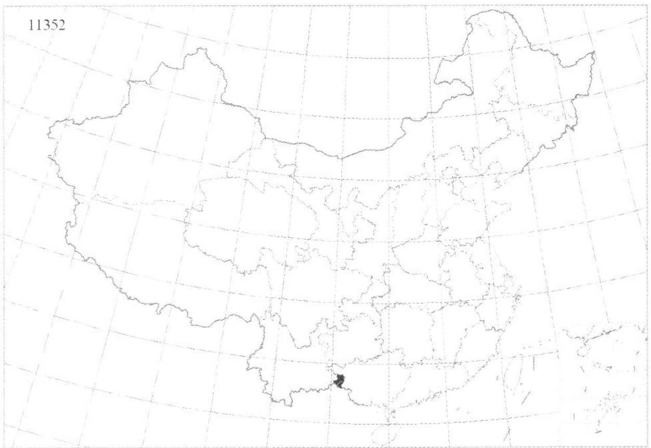

Liliaceae 百合科 *Smilax* 菝葜属

Smilax gagnepainii T. Koyama 四翅菝葜

Vines, evergreen

21.0 (17.7~23.2)；13.8 (10.9~15.8)；26.3 (22.5~28.9)；21.1 (17.8~23.2)；192.7 (153.4~218.9)；0.0 (0.0~0.0)；1084 (854~1257) ▎1494 (1368~1686)；785 (713~835)；80 (56~105) ▎39.7 (15.4~60.1)；1084 (854~1257)；383.5 (194.5~551.6)

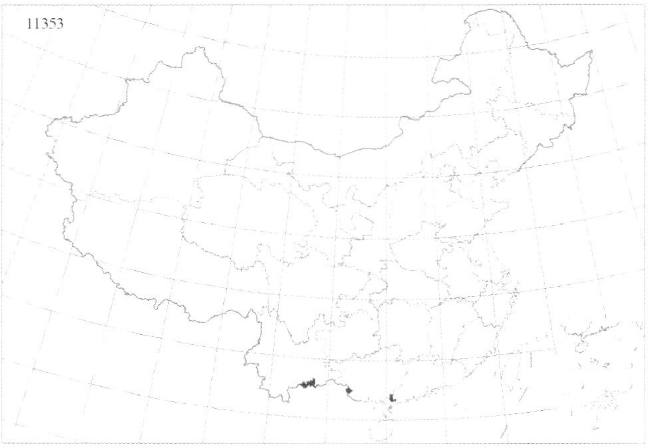

Liliaceae 百合科 *Smilax* 菝葜属

Smilax glabra Roxb. 土茯苓

Vines, evergreen

17.1 (5.4~24.8)；7.1 (-2.1~20.1)；26.2 (12.3~30.1)；17.2 (5.8~24.8)；147.2 (31.5~237.6)；0.9 (0.0~25.6)；911 (481~1385) ▎1365 (711~3495)；578 (299~1428)；121 (15~897) ▎50.6 (-34.3~246.5)；899 (481~1360)；231.2 (95.3~659.0)

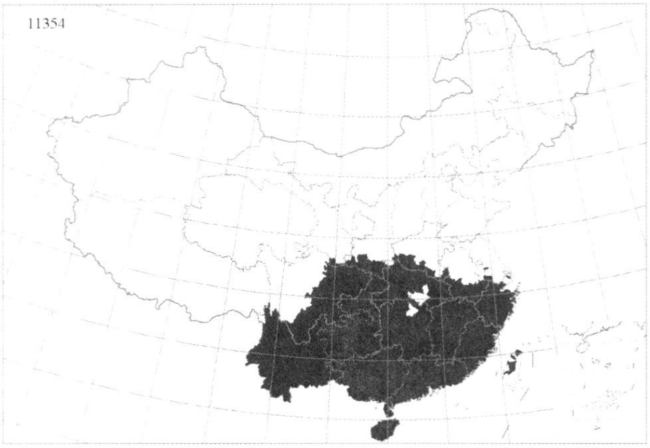

Liliaceae 百合科 *Smilax* 菝葜属

Smilax glaucochina Warb. 黑果菝葜

Vines, evergreen

15.1 (4.2~21.8)；3.3 (-12.5~15.3)；26.2 (18.7~29.9)；15.3 (7.3~21.8)；127.4 (52.8~202.0)；5.3 (0.0~62.6)；838 (535~1168) ▎1235 (377~2680)；499 (241~1186)；121 (8~259) ▎45.9 (-37.2~246.5)；814 (377~1139)；205.8 (101.0~346.9)

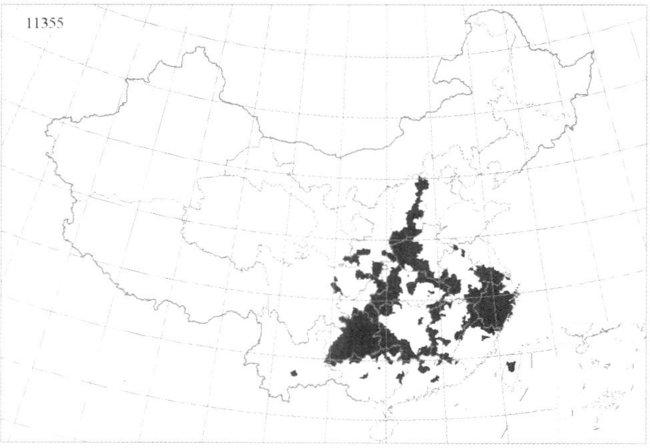

Liliaceae 百合科 *Smilax* 菝葜属

Smilax griffithii A. DC. 墨脱菝葜

Vines, evergreen

13.3；5.4；19.6；13.4；100.4；0.0；706 ▎1736；1007；51 ▎145.8；706；447.4

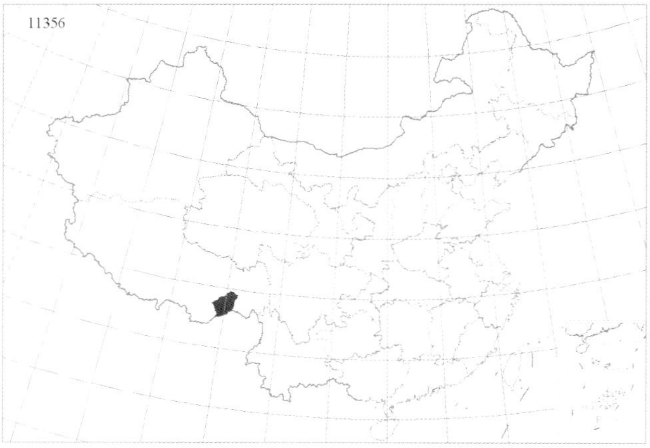

Liliaceae 百合科 *Smilax* 菝葜属

Smilax guiyangensis C. X. Fu et C. D. Shen 花叶菝葜

Vines, evergreen

15.6；5.5；24.6；15.6；127.5；0.0；808 ▎1206；546；69 ▎49.2；808；189.4

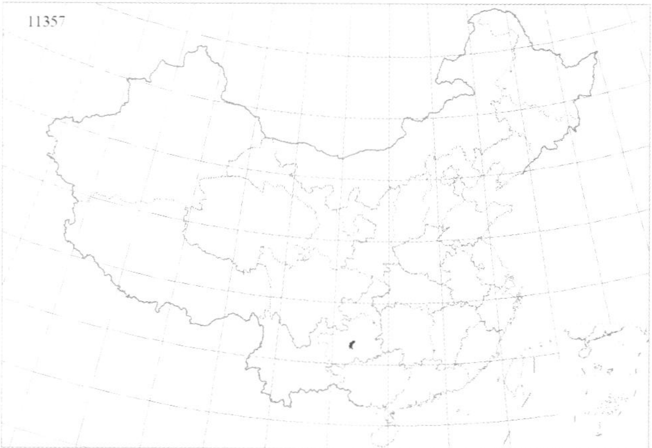

Liliaceae 百合科 *Smilax* 菝葜属

Smilax hayatae T. Koyama 菱叶菝葜

Small shrubs, evergreen

18.9 (16.5~21.2)；10.0 (5.2~14.6)；26.7 (21.6~29.8)；18.9 (16.5~21.3)；166.8 (138.3~195.2)；0.0 (0.0~0.0)；971 (832~1126) ▎1849 (1049~3495)；634 (442~959)；262 (44~897) ▎90.3 (26.1~235.6)；962 (783~1126)；242.6 (170.8~491.5)

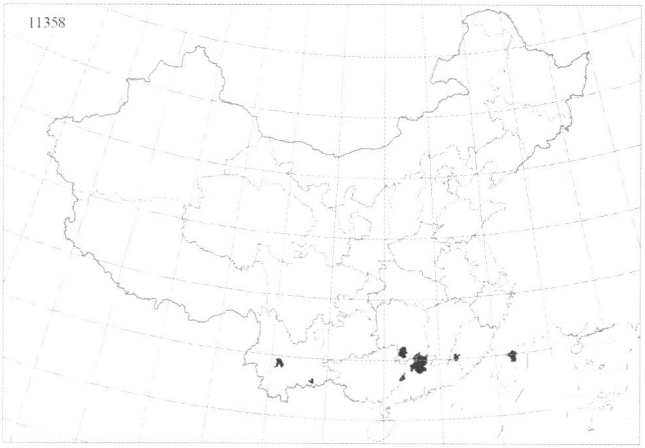

Liliaceae 百合科 *Smilax* 菝葜属
Smilax hemsleyana Craib 束丝菝葜
Vines, evergreen

17.3 (13.2~21.0)；10.7 (6.4~15.1)；21.9 (18.4~25.6)；17.3 (13.3~21.0)；147.7 (99.4~191.3)；0.0 (0.0~0.0)；843 (695~1053) ❙ 1147 (911~1751)；642 (524~996)；48 (20~77) ❙ 35.8 (4.4~92.4)；819 (695~1053)；351.1 (226.6~659.0)

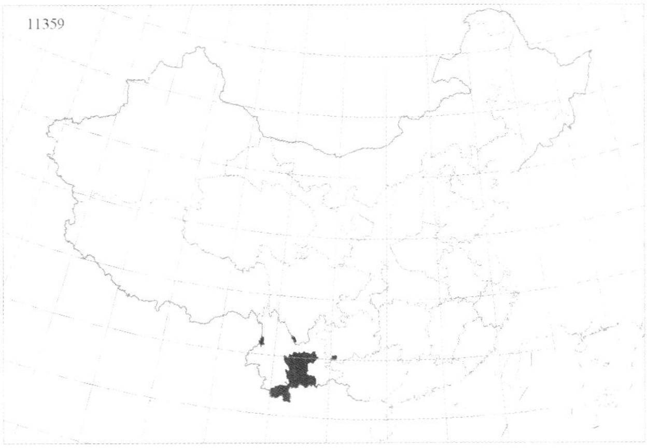

Liliaceae 百合科 *Smilax* 菝葜属
Smilax horridiramula Hayata 刺枝菝葜
Vines, evergreen

20.1 (16.4~24.1)；14.9 (11.5~18.9)；24.2 (20.3~28.0)；20.2 (16.5~24.1)；181.8 (137.8~229.7)；0.0 (0.0~0.0)；1004 (783~1322) ❙ 2467 (1751~2992)；1311 (900~1847)；167 (47~308) ❙ 155.5 (32.5~219.3)；966 (783~1123)；249.9 (154.2~303.6)

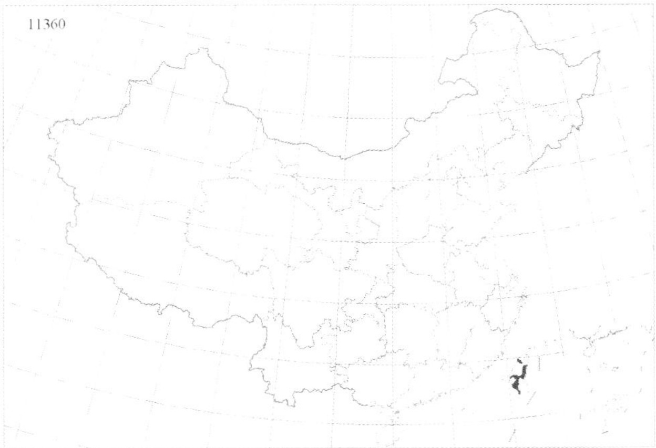

Liliaceae 百合科 *Smilax* 菝葜属
Smilax hypoglauca Benth. 粉背菝葜
Vines, evergreen

19.3 (15.5~23.6)；11.0 (5.6~16.7)；26.1 (21.5~29.8)；19.4 (15.5~23.6)；172.5 (126.2~223.5)；0.0 (0.0~0.0)；987 (802~1281) ❙ 1523 (939~1991)；658 (508~996)；130 (38~230) ❙ 55.4 (-6.9~130.5)；980 (763~1273)；293.6 (174.9~659.0)

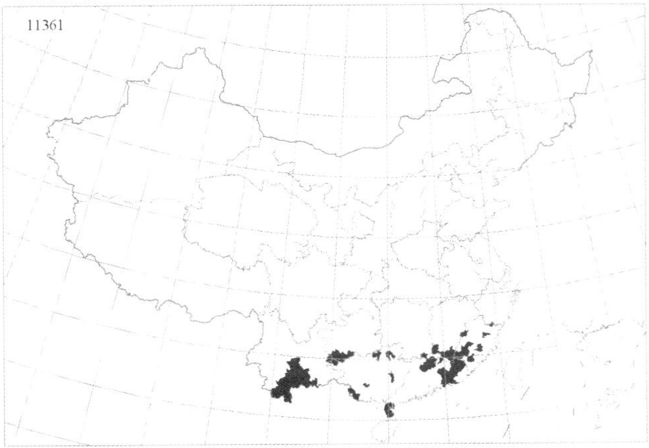

Liliaceae 百合科 *Smilax* 菝葜属
Smilax jiankunii H. Li 建昆菝葜
Vines, evergreen

8.6；1.3；15.0；8.7；52.6；8.8；574 ❙ 1041；585；45 ❙ 81.5；574；352.3

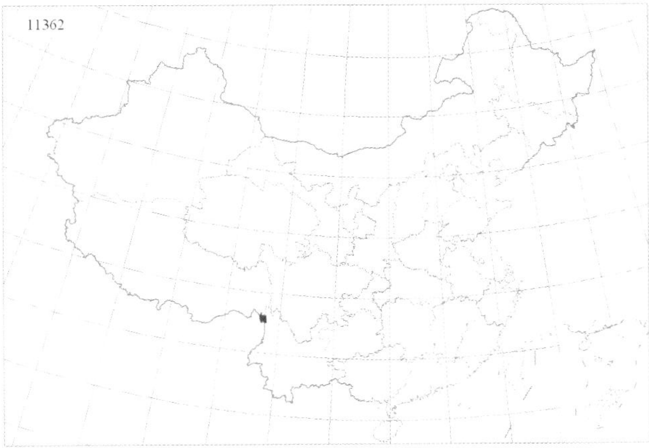

Liliaceae 百合科 *Smilax* 菝葜属
Smilax kwangsiensis F. T. Wang et Tang 缘毛菝葜
Vines, evergreen

20.5 (16.8~24.8)；11.9 (6.5~20.1)；27.2 (24.6~28.9)；20.5 (16.9~24.8)；186.3 (142.7~237.6)；0.0 (0.0~0.0)；1072 (869~1385) ❙ 1522 (1172~2435)；746 (473~1428)；103 (41~173) ❙ 42.3 (1.5~99.5)；1061 (869~1299)；289 4 (157.7~620.0)

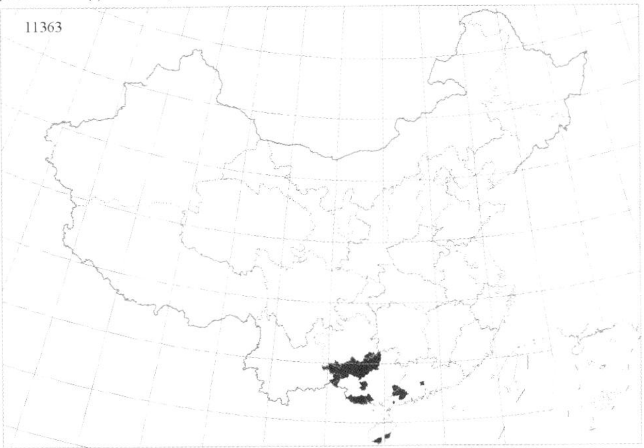

Liliaceae 百合科 *Smilax* 菝葜属
Smilax kwangsiensis var. setulosa F. T. Wang et Tang 小刚毛菝葜
Vines, evergreen

22.1 (21.0~23.2)；14.4 (13.2~15.8)；28.1 (26.9~28.9)；22.2 (21.0~23.2)；206.1 (192.3~218.9)；0.0 (0.0~0.0)；1174 (1081~1257) ❙ 1719 (1585~1876)；815 (762~891)；115 (102~125) ❙ 47.1 (28.6~64.3)；1174 (1081~1257)；234.6 (194.5~273.3)

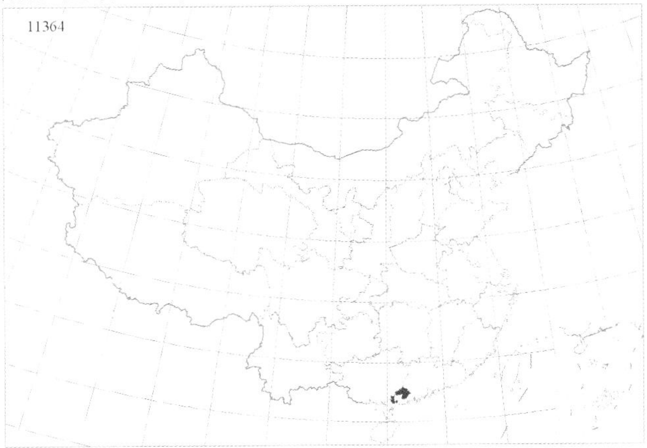

Liliaceae 百合科 *Smilax* 菝葜属
Smilax lanceifolia Roxb. 马甲菝葜
Vines, evergreen

17.5 (5.4~24.8); 8.6 (−2.1~20.1); 25.3 (12.3~30.1); 17.6 (5.8~24.8); 151.4 (31.5~237.6); 0.5 (0.0~25.6); 917 (481~1385) ❙ 1409 (741~3495); 627 (312~1847); 108 (15~897) ❙ 54.7 (−34.3~246.5); 907 (481~1360); 254.3 (95.3~659.0)

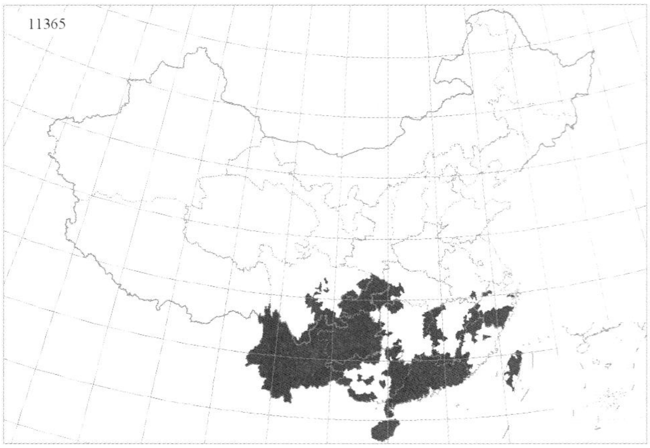

Liliaceae 百合科 *Smilax* 菝葜属
Smilax lanceifolia var. elongata (Warb.) F. T. Wang et Tang 折枝菝葜
Vines, evergreen

15.6 (11.1~19.1); 5.4 (1.8~10.3); 24.9 (19.4~27.2); 15.7 (11.1~19.1); 128.6 (79.8~169.2); 0.5 (0.0~6.9); 820 (648~988) ❙ 1230 (863~1551); 548 (375~817); 80 (27~192) ❙ 50.4 (10.5~89.0); 820 (648~988); 206.9 (146.0~383.4)

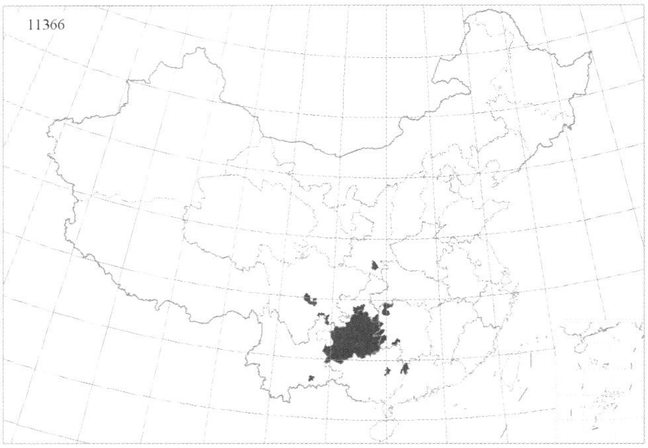

Liliaceae 百合科 *Smilax* 菝葜属
Smilax lanceifolia var. impressinervia (F. T. Wang et Tang) T. Koyama 凹脉菝葜
Vines, evergreen

17.4 (14.2~21.6); 10.0 (5.6~14.2); 22.7 (20.1~28.5); 17.4 (14.2~21.7); 149.2 (110.5~199.8); 0.0 (0.0~0.0); 856 (729~1153) ❙ 1220 (958~1625); 673 (531~860); 56 (44~173) ❙ 42.7 (13.3~80.6); 853 (729~1153); 346.7 (215.5~551.6)

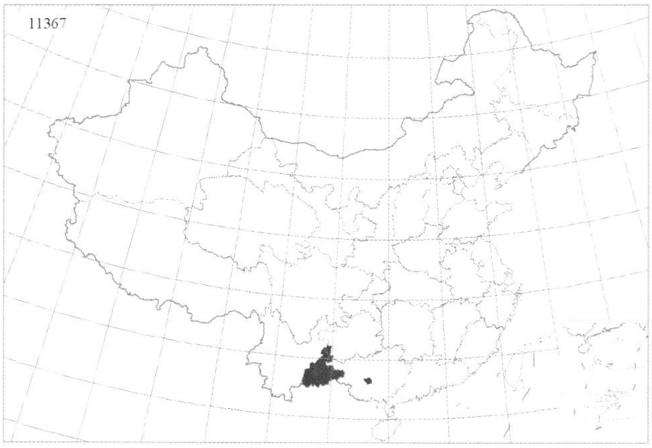

Liliaceae 百合科 *Smilax* 菝葜属
Smilax lanceifolia var. lanceolata (J. B. Norton) T. Koyama 长叶菝葜
Vines, evergreen

19.2 (16.3~22.2); 12.3 (9.1~14.2); 23.8 (20.3~28.3); 19.2 (16.4~22.3); 170.3 (136.4~207.5); 0.0 (0.0~0.0); 951 (789~1190) ❙ 1359 (1190~1522); 742 (656~835); 63 (47~87) ❙ 45.1 (12.9~64.5); 945 (780~1190); 414.6 (304.0~551.6)

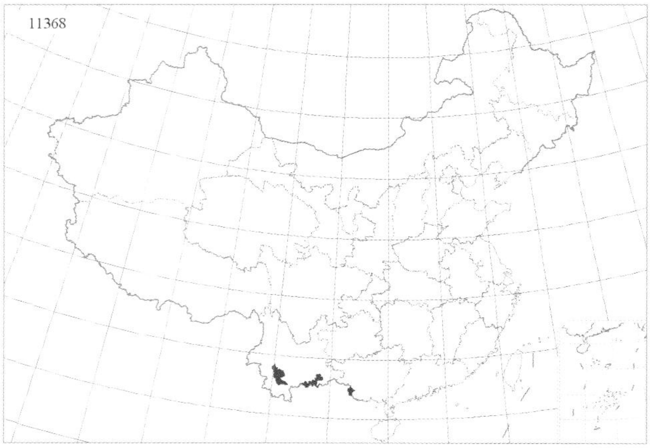

Liliaceae 百合科 *Smilax* 菝葜属
Smilax lanceifolia var. opaca A. DC. 暗色菝葜
Vines, evergreen

18.6 (14.3~23.2); 8.7 (3.5~15.8); 27.3 (20.2~29.8); 18.6 (14.3~23.2); 163.5 (113.2~218.9); 0.1 (0.0~1.5); 978 (753~1257) ❙ 1642 (950~2037); 634 (450~1033); 171 (27~259) ❙ 69.3 (13.3~147.0); 974 (683~1257); 239.9 (139.6~551.6)

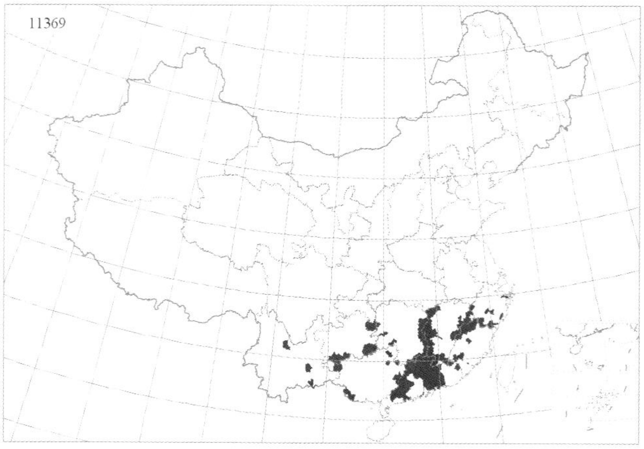

Liliaceae 百合科 *Smilax* 菝葜属
Smilax lebrunii H. Lév. 粗糙菝葜
Vines, evergreen

16.2 (5.7~20.7); 6.8 (−3.5~12.5); 24.2 (14.2~28.6); 16.3 (6.4~20.7); 136.5 (38.1~188.9); 1.4 (0.0~28.1); 845 (504~1099) ❙ 1274 (711~1740); 597 (357~932); 92 (15~198) ❙ 51.3 (6.5~101.8); 838 (504~1099); 250.5 (133.8~394.7)

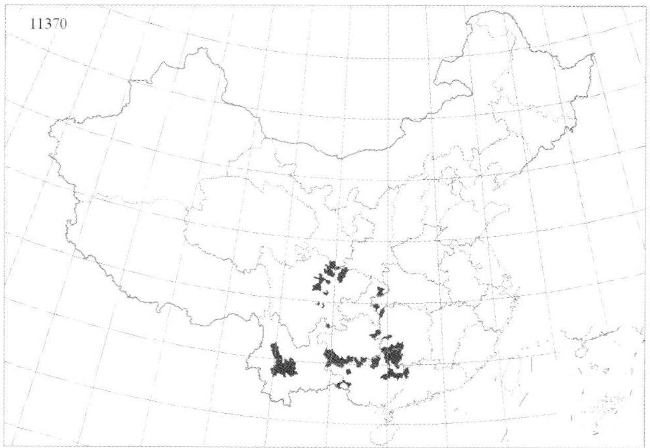

Liliaceae 百合科 *Smilax* 菝葜属

Smilax longebracteolata Hook. f. 长苞菝葜

Vines, evergreen

14.7 (1.8~21.0); 7.3 (-7.1~14.2); 20.4 (9.9~26.6); 14.7 (3.8~21.0); 118.8 (15.7~191.8); 2.2 (0.0~53.3); 758 (404~1053) ❙ 1069 (583~1522); 607 (348~860); 40 (12~68) ❙ 42.5 (-14.2~100.7); 736 (404~1053); 304.9 (169.5~551.6)

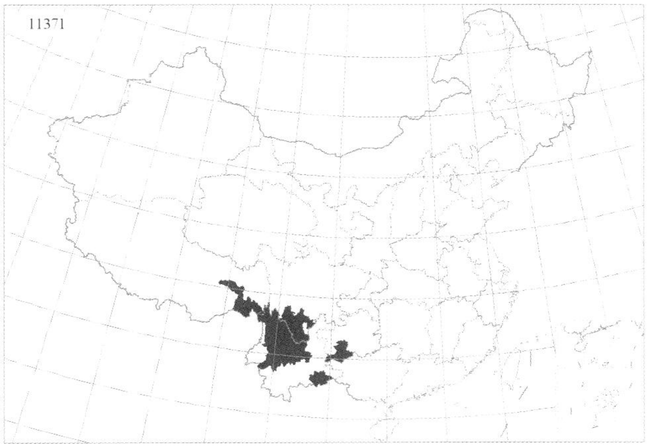

Liliaceae 百合科 *Smilax* 菝葜属

Smilax luei T. Koyama 吕氏菝葜

Vines, evergreen

16.2; 11.0; 20.1; 16.2; 134.2; 0.0; 773 ❙ 2680; 1186; 245 ❙ 246.5; 773; 289.1

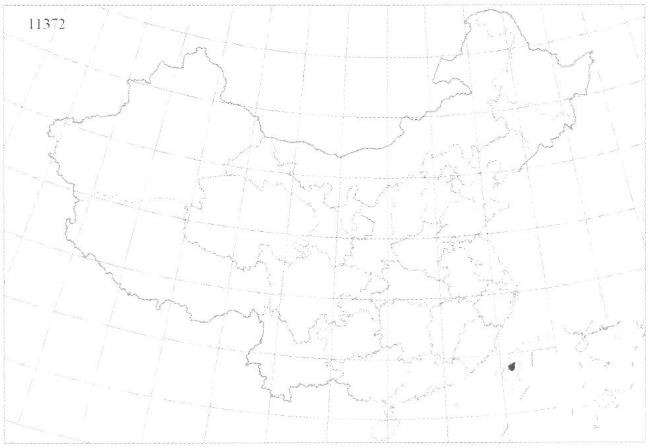

Liliaceae 百合科 *Smilax* 菝葜属

Smilax lunglingensis F. T. Wang et Tang 马钱叶菝葜

Vines, evergreen

16.3 (13.8~18.3); 9.7 (6.7~11.8); 21.1 (19.3~22.7); 16.4 (13.8~18.3); 136.3 (105.9~159.8); 0.0 (0.0~0.0); 799 (714~882) ❙ 1014 (911~1348); 583 (524~762); 41 (28~56) ❙ 27.3 (6.5~72.8); 774 (690~863); 299.2 (234.5~406.7)

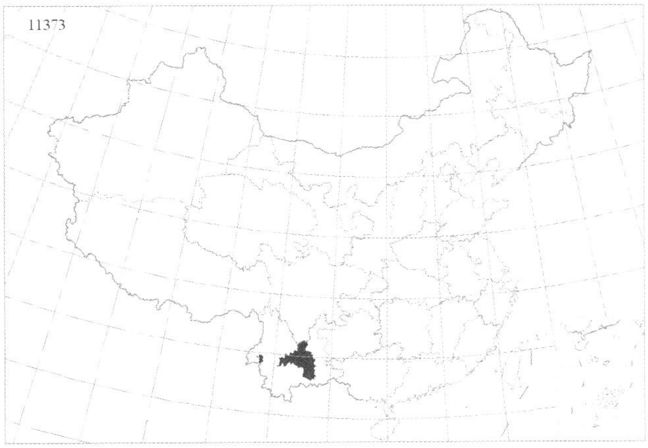

Liliaceae 百合科 *Smilax* 菝葜属

Smilax lushuiensis S. C. Chen 泸水菝葜

Vines, evergreen

11.4; 4.4; 17.0; 11.4; 78.9; 2.0; 646 ❙ 1196; 667; 56 ❙ 84.4; 646; 381.2

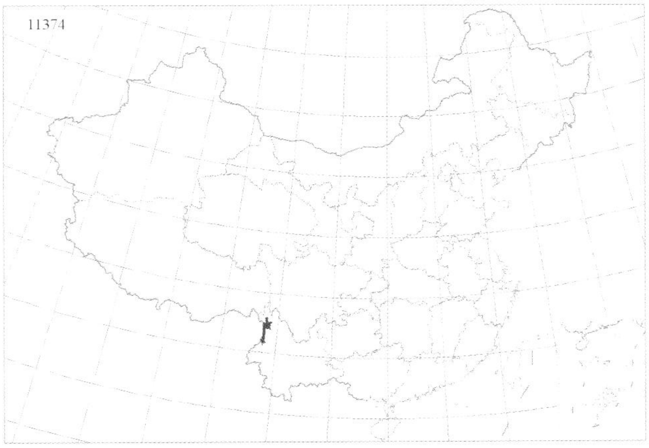

Liliaceae 百合科 *Smilax* 菝葜属

Smilax mairei H. Lév. 无刺菝葜

Shrubs, evergreen

14.7 (1.8~21.0); 7.2 (-7.1~14.2); 20.5 (9.9~26.6); 14.8 (3.8~21.0); 118.5 (15.7~191.8); 1.7 (0.0~53.3); 758 (404~1053) ❙ 1034 (583~1522); 588 (348~818); 40 (12~68) ❙ 37.9 (-14.2~100.7); 739 (404~1053); 288.7 (142.9~551.6)

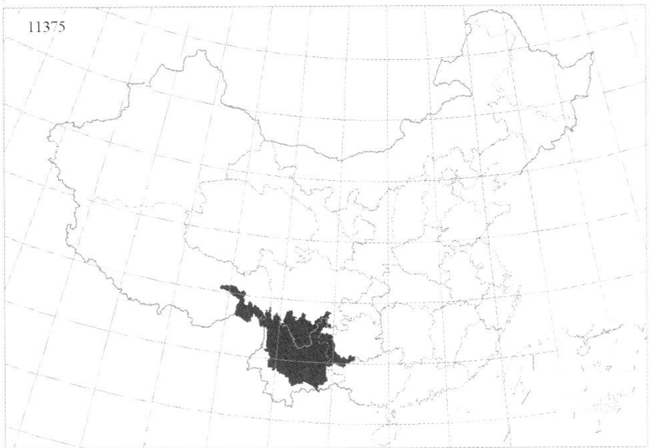

Liliaceae 百合科 *Smilax* 菝葜属

Smilax malipoensis S. C. Chen 麻栗坡菝葜

Vines, evergreen

17.6; 9.9; 23.2; 17.7; 152.4; 0.0; 859 ❙ 1422; 813; 52 ❙ 65.6; 859; 359.7

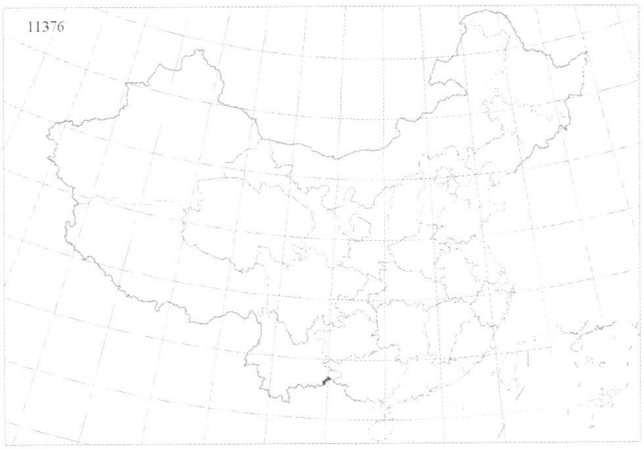

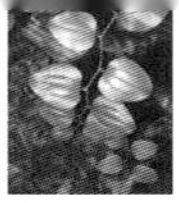

Liliaceae 百合科 *Smilax* 菝葜属

Smilax megacarpa A. DC. 大果菝葜

Vines, evergreen

20.8 (14.6~24.8); 12.7 (3.8~20.1); 27.1 (21.4~28.9); 20.8 (14.7~24.8); 189.7 (117.7~237.6); 0.0 (0.0~1.5); 1092 (779~1385) ❙ 1562 (868~2435); 695 (312~1428); 117 (41~196) ❙ 45.8 (-34.3~101.8); 1066 (779~1360); 316.0 (164.9~659.0)

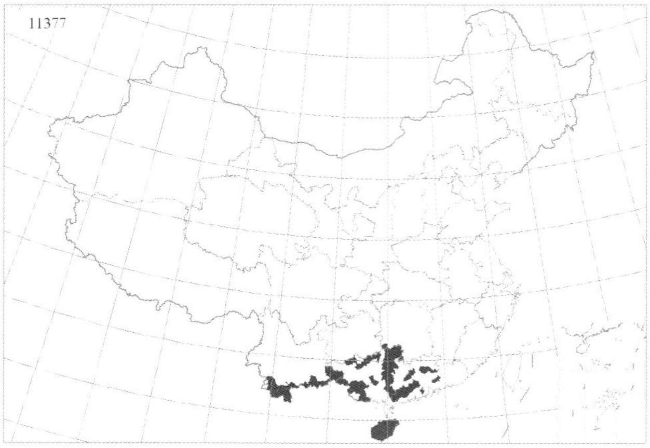

Liliaceae 百合科 *Smilax* 菝葜属

Smilax megalantha C. H. Wright 大花菝葜

Vines, evergreen

15.7 (2.2~22.2); 6.5 (-6.6~15.1); 23.8 (9.5~28.9); 15.8 (3.8~22.3); 131.0 (15.5~207.2); 1.6 (0.0~49.1); 828 (411~1186) ❙ 1195 (677~2014); 587 (336~996); 66 (11~230) ❙ 45.5 (-14.2~125.9); 822 (411~1186); 241.8 (108.2~659.0)

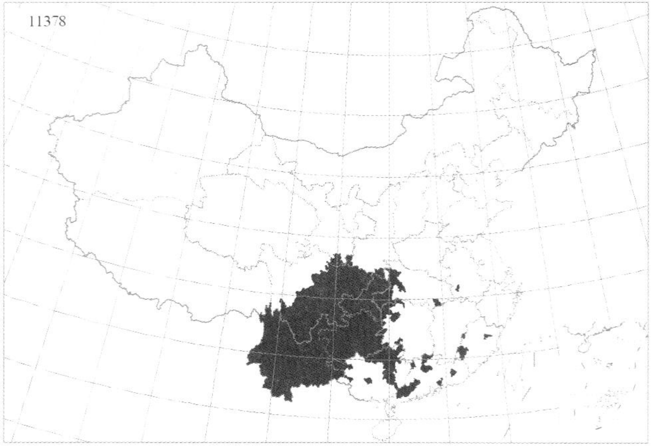

Liliaceae 百合科 *Smilax* 菝葜属

Smilax menispermoidea A. DC. 防己叶菝葜

Vines, deciduous

11.9 (-1.8~23.0); 2.0 (-10.2~15.3); 20.7 (6.1~29.0); 12.2 (1.8~23.0); 95.1 (1.9~216.2); 12.2 (0.0~82.8); 696 (329~1247) ❙ 997 (320~2859); 500 (190~1412); 46 (5~402) ❙ 43.4 (-40.9~246.5); 682 (320~1247); 255.1 (61.2~551.3)

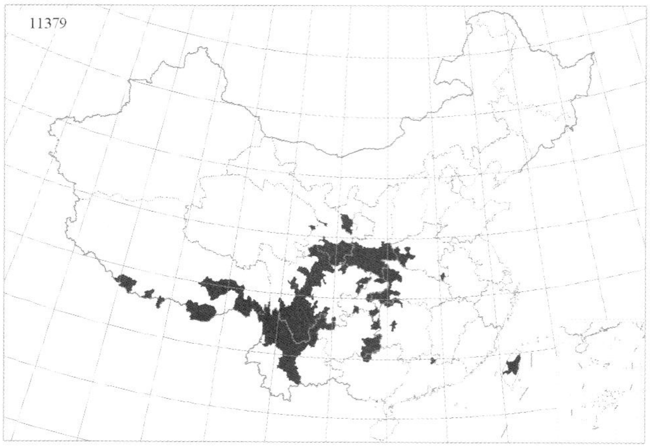

Liliaceae 百合科 *Smilax* 菝葜属

Smilax microphylla C. H. Wright 小叶菝葜

Vines, evergreen

15.0 (5.4~22.2); 5.4 (-2.5~14.7); 23.6 (12.3~29.0); 15.0 (5.8~22.2); 122.7 (31.5~206.9); 2.3 (0.0~25.6); 797 (481~1177) ❙ 1092 (741~1876); 533 (352~891); 55 (15~191) ❙ 37.8 (-14.2~100.5); 789 (481~1177); 231.1 (108.2~551.6)

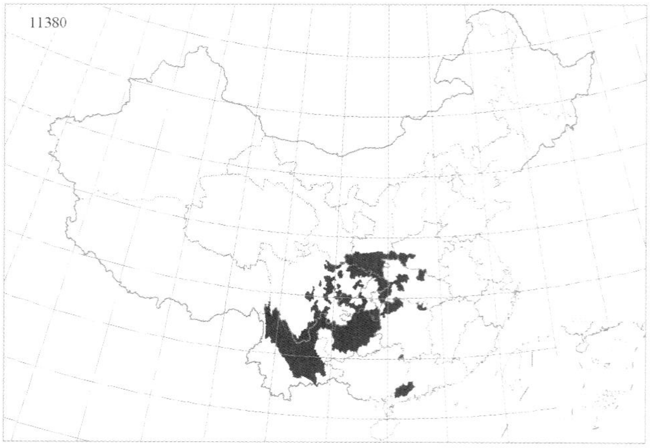

Liliaceae 百合科 *Smilax* 菝葜属

Smilax munita S. C. Chen 劲直菝葜

Shrubs, evergreen

8.8 (-1.8~14.5); 1.2 (-10.2~7.8); 15.2 (6.1~19.6); 9.4 (1.8~14.5); 64.7 (1.9~113.5); 19.2 (0.0~82.8); 583 (329~730) ❙ 1044 (320~1736); 599 (217~1007); 39 (13~58) ❙ 74.0 (-3.0~145.8); 582 (320~730); 308.8 (72.2~447.4)

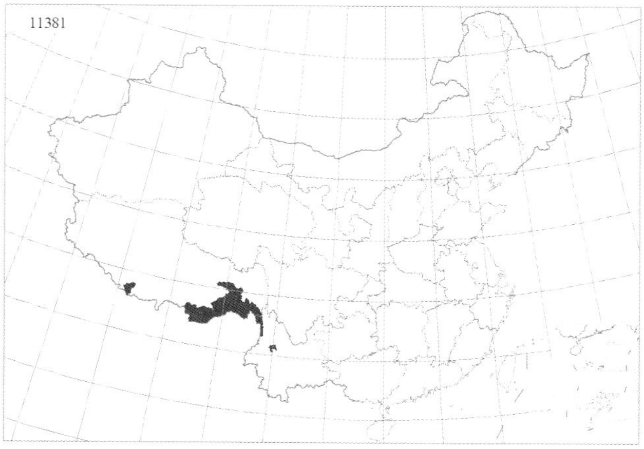

Liliaceae 百合科 *Smilax* 菝葜属

Smilax myrtillus A. DC. 乌饭叶菝葜

Shrubs, evergreen

16.1 (5.4~21.0); 9.6 (-2.1~15.1); 20.8 (12.3~25.6); 16.2 (5.8~21.0); 135.1 (31.5~191.8); 0.9 (0.0~25.6); 804 (481~1053) ❙ 1204 (741~1751); 675 (411~996); 47 (15~77) ❙ 50.4 (-14.2~112.2); 779 (481~1053); 394.9 (218.0~659.0)

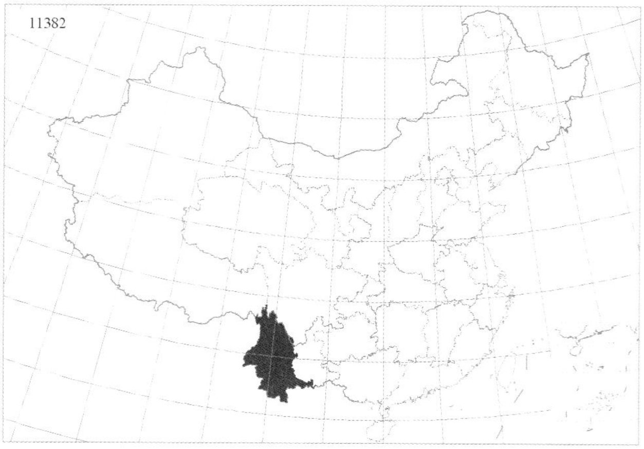

Liliaceae 百合科 *Smilax* 菝葜属
Smilax nana F. T. Wang 矮菝葜
Small shrubs, deciduous

14.5; 7.8; 19.6; 14.5; 113.5; 0.0; 730 ❙ 1044; 570; 54 ❙ 43.0; 730; 382.6

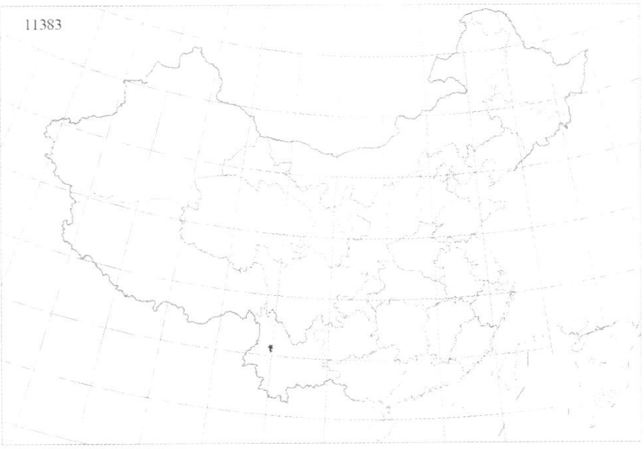

Liliaceae 百合科 *Smilax* 菝葜属
Smilax nantoensis T. Koyama 南投菝葜
Vines, evergreen

16.2; 11.0; 20.1; 16.2; 134.2; 0.0; 773 ❙ 2680; 1186; 245 ❙ 246.5; 773; 289.1

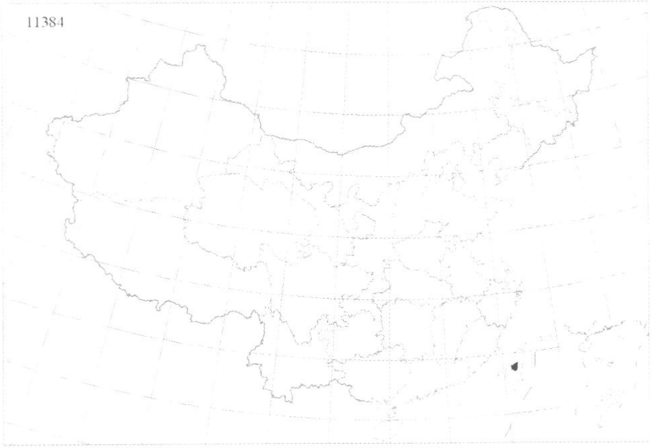

Liliaceae 百合科 *Smilax* 菝葜属
Smilax nervomarginata Hayata 缘脉菝葜
Vines, evergreen

16.2 (14.7~19.1); 4.8 (2.5~10.0); 27.2 (24.7~29.2); 16.3 (14.8~19.1); 135.9 (118.7~169.1); 0.7 (0.0~3.9); 872 (789~975) ❙ 1551 (1165~1997); 544 (437~713); 180 (75~259) ❙ 77.9 (34.4~147.0); 866 (789~973); 229.5 (146.0~332.5)

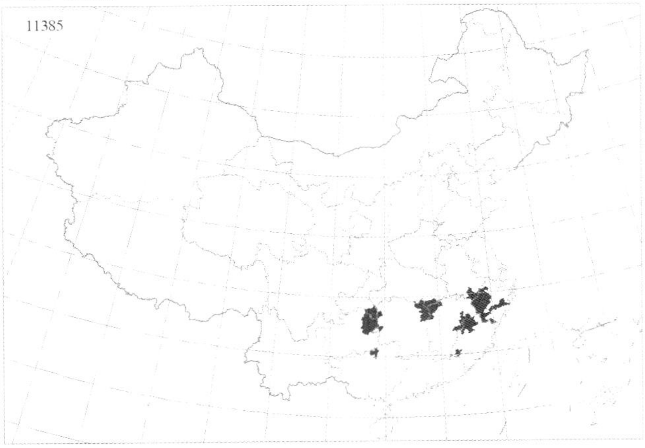

Liliaceae 百合科 *Smilax* 菝葜属
Smilax nervomarginata var. liukiuensis (Hayata) F. T. Wang et Tang 无疣菝葜
Vines, evergreen

17.0 (14.7~18.8); 5.2 (2.5~6.9); 28.2 (25.8~30.1); 17.0 (14.8~18.9); 144.7 (121.1~166.5); 0.4 (0.0~3.9); 916 (810~1024) ❙ 1575 (1452~1801); 505 (453~598); 213 (196~242) ❙ 73.1 (41.7~109.2); 888 (810~936); 219.9 (148.9~284.0)

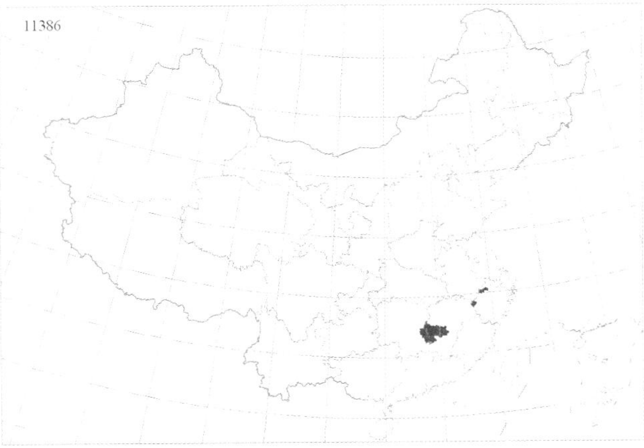

Liliaceae 百合科 *Smilax* 菝葜属
Smilax nigrescens F. T. Wang et C. L. Tang ex P. Y. Li 黑叶菝葜
Vines, evergreen

14.0 (3.5~21.3); 3.0 (−6.5~13.2); 24.2 (12.6~29.7); 14.1 (5.0~21.3); 114.6 (26.3~195.9); 6.5 (0.0~43.4); 775 (446~1111) ❙ 1064 (632~1900); 494 (290~1056); 51 (15~203) ❙ 37.2 (−18.5~101.2); 771 (446~1111); 231.5 (128.7~551.3)

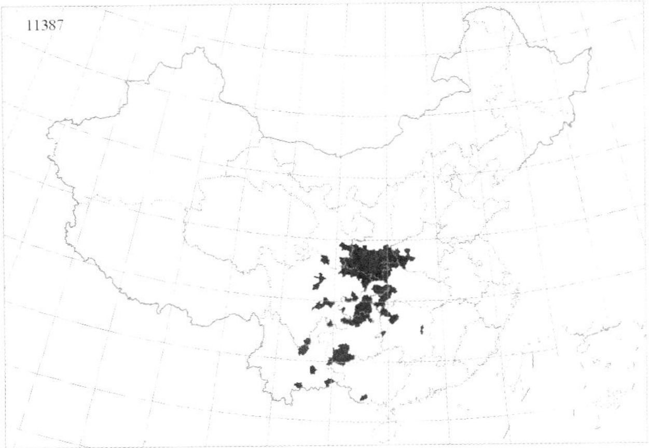

Liliaceae 百合科 *Smilax* 菝葜属
Smilax outanscianensis Pamp. 武当菝葜
Vines, evergreen

14.9 (9.3~18.5); 3.6 (−1.7~10.5); 25.6 (19.4~29.2); 14.9 (9.5~18.3); 121.7 (69.8~160.0); 3.0 (0.0~18.0); 814 (603~950) ❙ 1306 (806~1660); 512 (390~630); 117 (16~228) ❙ 60.6 (5.5~107.2); 808 (603~950); 232.6 (144.6~330.0)

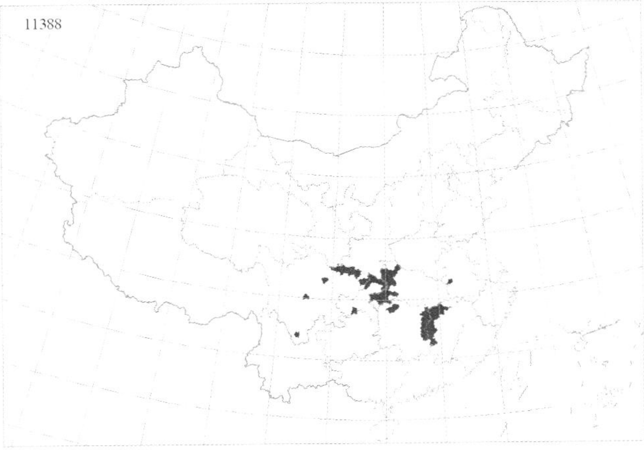

Liliaceae 百合科 *Smilax* 菝葜属

Smilax ovalifolia Roxb. 卵叶菝葜

Vines, evergreen

18.6；9.7；25.4；18.6；163.3；0.0；933 ❙ 1301；703；67 ❙ 39.6；933；302.9

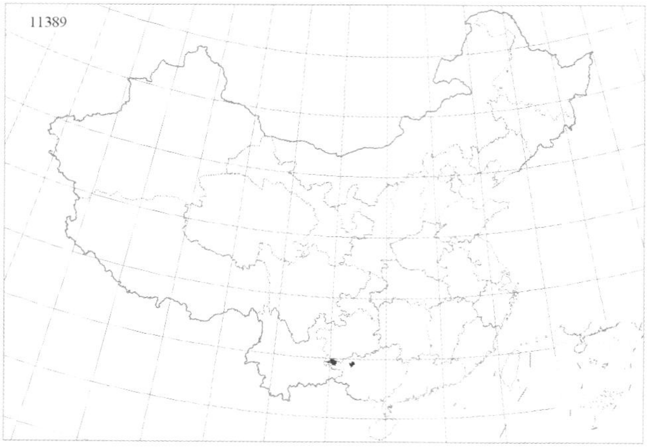

Liliaceae 百合科 *Smilax* 菝葜属

Smilax pachysandroides T. Koyama 川鄂菝葜

Shrubs, evergreen

13.0；1.9；23.8；13.1；102.9；6.1；730 ❙ 1202；500；83 ❙ 64.6；730；262.4

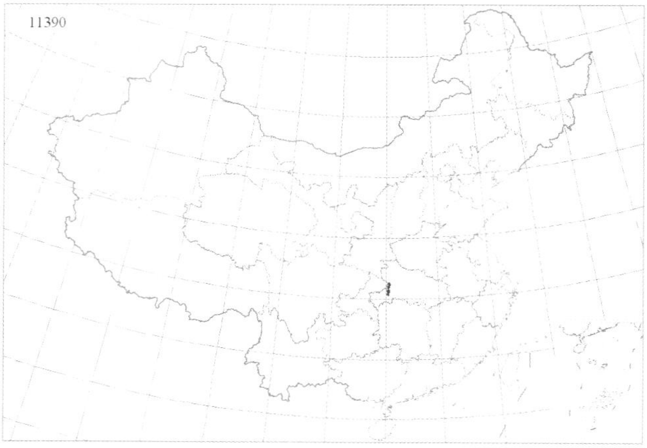

Liliaceae 百合科 *Smilax* 菝葜属

Smilax perfoliata Lour. 穿鞘菝葜

Vines, evergreen

21.0 (16.9~24.8)；13.6 (7.1~20.1)；26.7 (21.4~29.8)；21.0 (16.9~24.8)；192.3 (142.7~237.6)；0.0 (0.0~0.0)；1099 (813~1385) ❙ 1597 (868~2992)；726 (312~1847)；107 (41~189) ❙ 48.9 (-34.3~214.4)；1057 (783~1360)；336.9 (119.4~659.0)

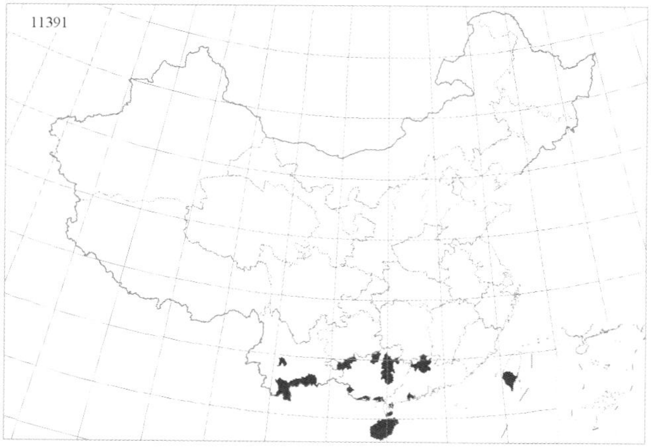

Liliaceae 百合科 *Smilax* 菝葜属

Smilax pinfaensis H. Lév. et Vaniot 平伐菝葜

Shrubs, evergreen

15.7 (14.9~16.7)；5.8 (4.8~6.9)；24.4 (23.9~24.9)；15.8 (15.0~16.7)；129.4 (119.9~140.7)；0.0 (0.0~0.2)；812 (781~851) ❙ 1261 (1240~1283)；569 (545~600)；76 (68~85) ❙ 55.4 (50.8~58.8)；812 (781~851)；235.0 (221.1~254.8)

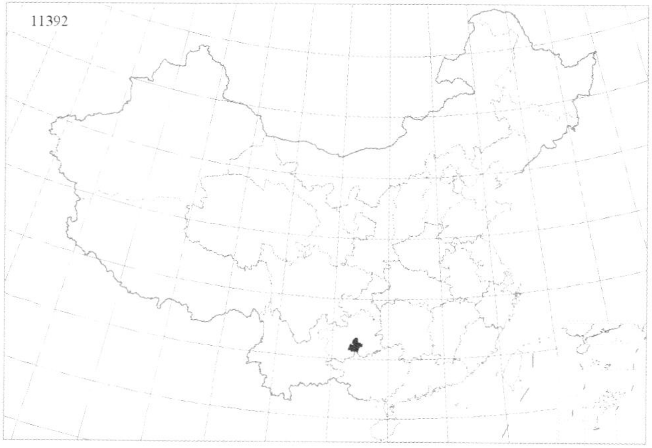

Liliaceae 百合科 *Smilax* 菝葜属

Smilax polycolea Warb. 红果菝葜

Vines, deciduous

14.7 (3.8~21.2)；4.8 (-4.3~14.2)；23.7 (10.8~28.1)；14.8 (4.7~21.3)；119.0 (21.6~195.4)；2.0 (0.0~35.4)；786 (444~1115) ❙ 1204 (707~1997)；560 (362~946)；71 (12~254) ❙ 54.1 (-0.2~130.7)；783 (444~1115)；226.3 (139.5~551.6)

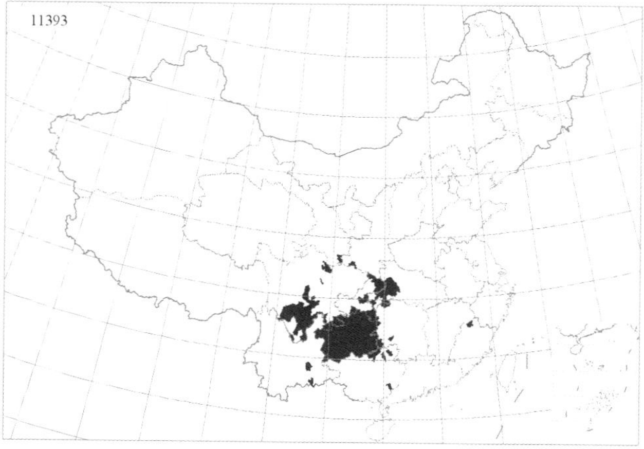

Liliaceae 百合科 *Smilax* 菝葜属

Smilax pottingeri Prain 纤柄菝葜

Vines

18.2 (17.2~21.0)；10.9 (9.5~14.2)；23.3 (22.4~25.6)；18.2 (17.3~21.0)；158.4 (147.0~191.8)；0.0 (0.0~0.0)；887 (833~1053) ❙ 1414 (1254~1522)；789 (699~860)；57 (48~68) ❙ 60.1 (44.5~80.6)；887 (833~1053)；430.5 (336.3~551.6)

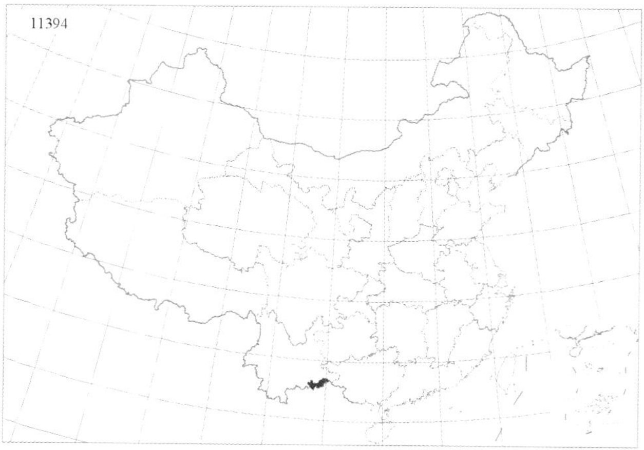

Liliaceae 百合科 *Smilax* 菝葜属
Smilax quadrata A. DC. 方枝菝葜
Vines, evergreen

16.9 (13.3~21.0); 10.4 (5.4~15.1); 21.6 (18.8~25.6); 16.9 (13.4~21.0); 143.4 (100.4~191.8); 0.0 (0.0~0.0); 839 (706~1053) ❙ 1245 (958~1736); 660 (531~1007); 54 (44~77) ❙ 49.6 (13.3~145.8); 824 (706~1053); 436.0 (311.9~659.0)

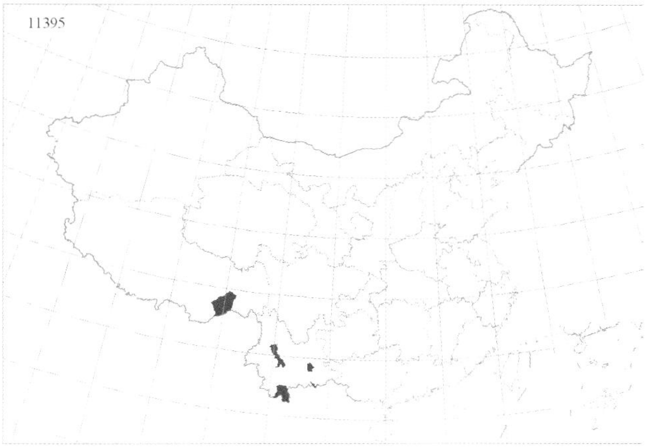

Liliaceae 百合科 *Smilax* 菝葜属
Smilax retroflexa (F. T. Wang et Tang) S. C. Chen 苍白菝葜
Shrubs or subshrubs, evergreen

16.8 (13.6~19.8); 7.9 (4.4~11.1); 24.1 (19.6~26.4); 16.8 (13.7~19.8); 141.7 (103.9~177.9); 0.1 (0.0~0.6); 851 (713~1010) ❙ 1259 (985~1504); 689 (499~860); 56 (20~142) ❙ 48.6 (16.9~89.0); 850 (691~1010); 284.1 (171.5~491.5)

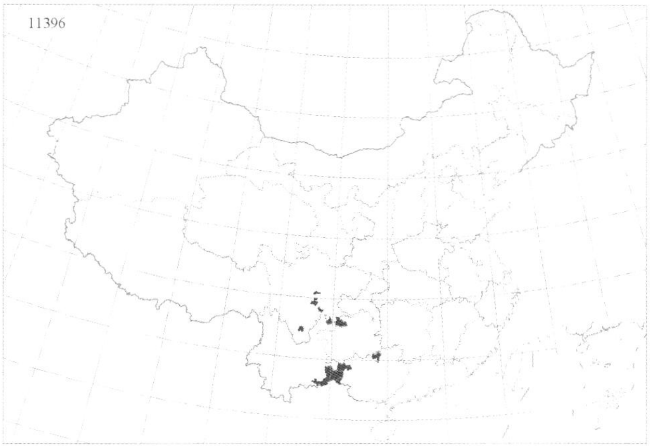

Liliaceae 百合科 *Smilax* 菝葜属
Smilax scobinicaulis C. H. Wright 短梗菝葜
Vines, evergreen

14.8 (2.6~24.8); 4.3 (-11.1~20.1); 24.0 (11.2~29.3); 15.0 (4.3~24.8); 125.0 (20.0~237.6); 7.1 (0.0~55.3); 814 (427~1385) ❙ 1088 (373~2992); 526 (201~1847); 63 (7~254) ❙ 32.7 (-42.4~214.4); 770 (373~1360); 247.3 (88.1~659.0)

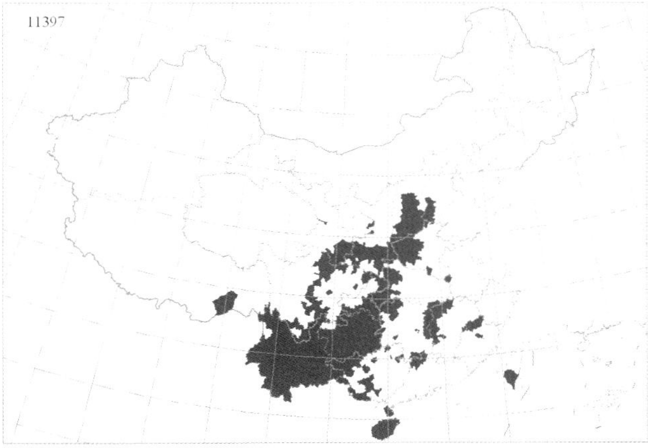

Liliaceae 百合科 *Smilax* 菝葜属
Smilax setiramula F. T. Wang et Tang 密刚毛菝葜
Vines, evergreen

18.2 (17.2~19.8); 9.9 (9.1~11.1); 24.3 (22.7~26.4); 18.2 (17.3~19.8); 158.9 (147.0~177.9); 0.0 (0.0~0.0); 903 (838~1010) ❙ 1262 (1161~1422); 704 (633~813); 50 (43~58) ❙ 40.5 (16.9~65.6); 903 (838~1010); 337.6 (314.4~359.7)

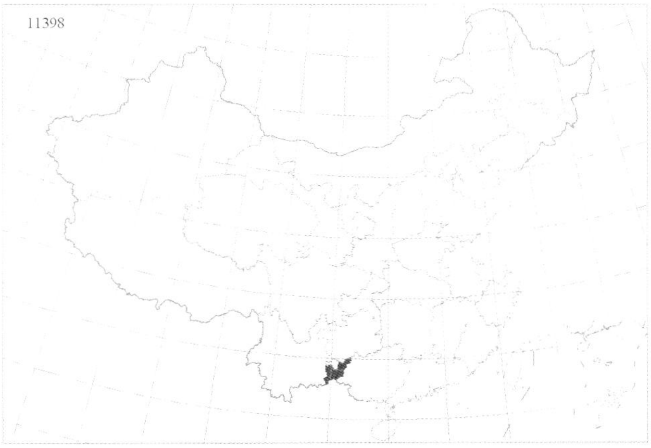

Liliaceae 百合科 *Smilax* 菝葜属
Smilax sieboldii Miq. 华东菝葜
Vines, evergreen

13.9 (4.2~24.1); 0.7 (-13.5~18.9); 26.0 (18.7~29.3); 14.3 (7.3~24.1); 120.6 (52.8~229.7); 12.7 (0.0~62.6); 817 (535~1322) ❙ 1138 (431~2992); 530 (275~1847); 103 (10~303) ❙ 35.8 (-32.6~246.5); 776 (431~1123); 198.8 (131.2~323.4)

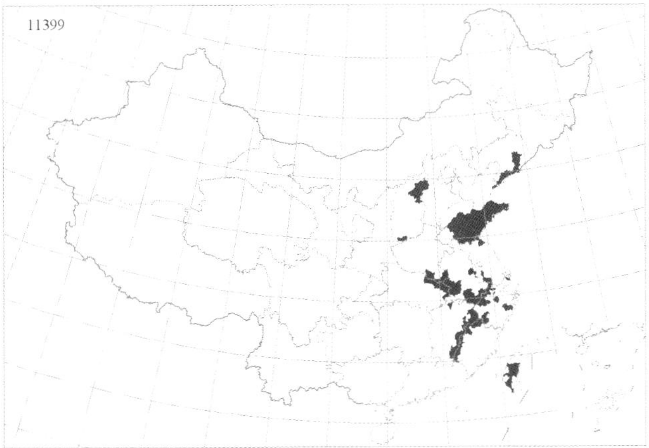

Liliaceae 百合科 *Smilax* 菝葜属
Smilax stans Maxim. 鞘柄菝葜
Shrubs, deciduous

12.6 (-1.4~21.0); 0.8 (-15.4~15.3); 23.2 (9.5~29.3); 13.0 (3.5~21.0); 104.5 (13.3~191.8); 13.4 (0.0~92.1); 741 (375~1053) ❙ 959 (39~2680); 459 (29~1412); 59 (1~308) ❙ 28.0 (-92.3~246.5); 697 (39~1053); 212.9 (10.8~551.6)

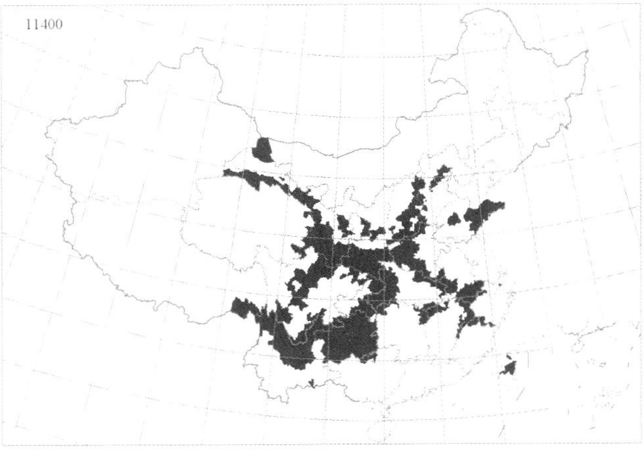

Liliaceae 百合科 *Smilax* 菝葜属

Smilax synandra Gagnep. 筒被菝葜

Shrubs, evergreen

18.4 (17.2~21.0); 11.2 (9.1~14.2); 23.5 (22.4~25.6); 18.4 (17.3~21.0); 161.0 (147.0~191.8); 0.0 (0.0~0.0); 897 (833~1053) ❙ 1358 (1161~1522); 765 (633~860); 54 (48~68) ❙ 52.0 (36.8~80.6); 896 (833~1053); 418.6 (314.4~551.6)

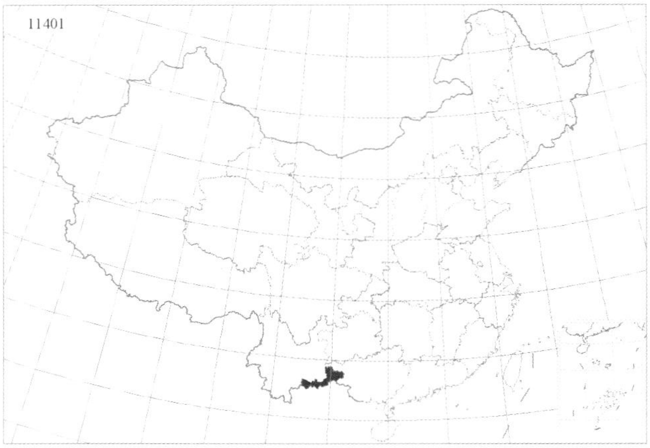

Liliaceae 百合科 *Smilax* 菝葜属

Smilax trachypoda J. B. Norton 糙柄菝葜

Shrubs, deciduous

10.6 (0.6~16.6); −1.8 (−10.6~6.6); 22.1 (9.5~27.1); 11.1 (3.4~16.6); 87.1 (12.7~139.7); 19.1 (0.0~66.0); 680 (394~866) ❙ 785 (375~1590); 387 (203~946); 29 (8~88) ❙ 19.2 (−36.7~128.6); 625 (375~866); 218.1 (95.7~357.7)

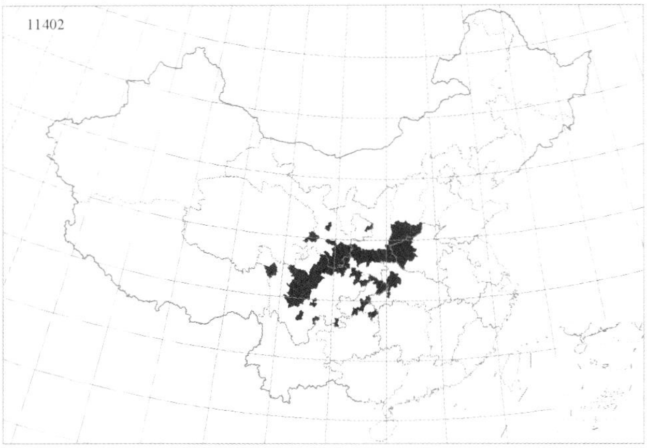

Liliaceae 百合科 *Smilax* 菝葜属

Smilax trinervula Miq. 三脉菝葜

Shrubs, deciduous

16.9 (14.1~19.4); 5.2 (1.6~9.4); 27.8 (24.2~29.9); 16.9 (14.2~19.4); 143.5 (113.2~173.4); 0.7 (0.0~6.5); 909 (761~1029) ❙ 1560 (1206~1997); 549 (427~685); 190 (69~259) ❙ 72.1 (28.8~130.7); 899 (761~1029); 222.4 (111.9~303.4)

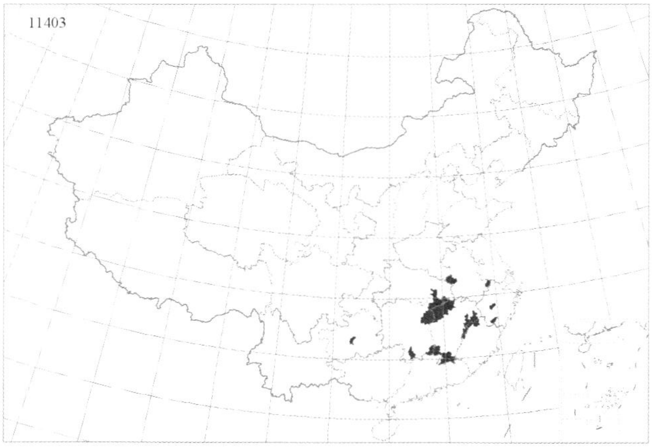

Liliaceae 百合科 *Smilax* 菝葜属

Smilax tsinchengshanensis F. T. Wang 青城菝葜

Shrubs, evergreen

15.8 (11.9~21.0); 6.1 (2.9~14.2); 24.3 (20.0~27.3); 15.9 (12.0~21.0); 130.8 (87.2~191.8); 0.4 (0.0~3.4); 823 (671~1053) ❙ 1249 (1001~1522); 614 (475~818); 68 (32~100) ❙ 52.3 (20.8~80.4); 823 (671~1053); 232.6 (135.1~551.6)

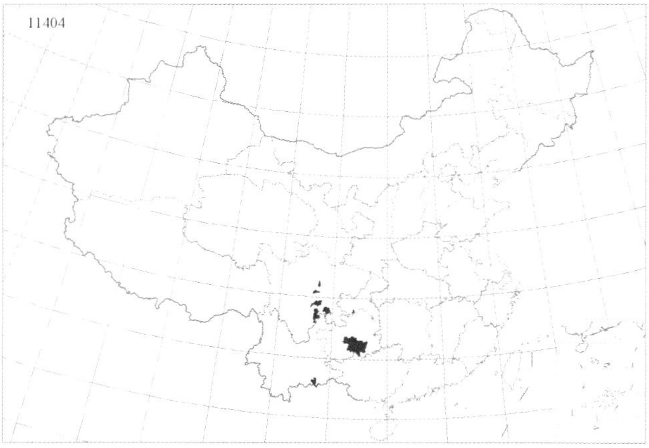

Liliaceae 百合科 *Smilax* 菝葜属

Smilax vanchingshanensis (F. T. Wang et Tang) F. T. Wang et Tang 梵净山菝葜

Vines, evergreen

14.8 (12.1~16.2); 4.0 (1.1~6.2); 25.1 (22.9~26.8); 14.9 (12.2~16.2); 119.9 (95.0~134.6); 1.6 (0.0~8.5); 796 (696~866) ❙ 1288 (1076~1492); 550 (463~816); 80 (36~113) ❙ 62.7 (29.8~107.2); 796 (696~866); 201.2 (148.9~285.1)

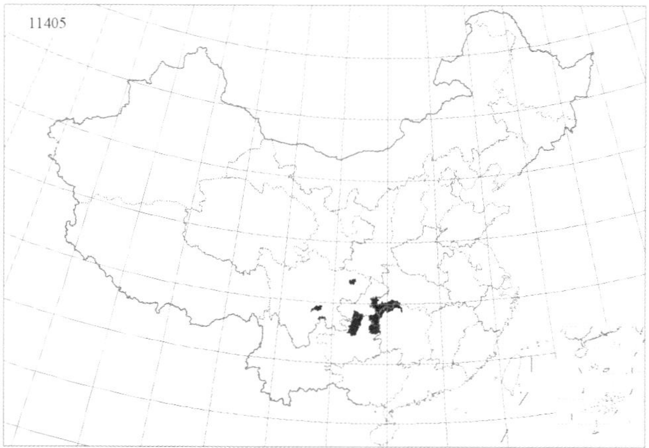

Bibliographies

An D.G. (2002) *Higher plants of Mt.Xiaolong, Gansu Province*. Gansu Nationalities Publishing House, Lanzhou. (In Chinese)

An M.T. (2005) Present status of the natural resource of *Camellia* in Guizhou Province. *Guizhou Forestry Science and Technology*, 33, 26-29. (In Chinese)

Brown J. (1995) *Macroecology*. University of Chicago Press, Chicago.

Brown J.H. & Lomolino M.V. (1998) *Biogeography*. 2nd ed. Sinauer Associates, Sunderland, Massachusetts.

Cao J.G., Yu J., Wang Q.X. (2007) Spore morphology of ferns from China VII Cyatheaceae. *Acta Botanica Yunnanica*, 29(1), 7-12. (In Chinese)

Cao W., Li J.Y. (2003) *Natural distribution of plants in Changbai Mountain*. Northeast University Press, Shenyang. (In Chinese)

Chang J.X., Yang L., Huang W.L. (2002) A study on Handeliodendron population ecology in South Guizhou. *Guizhou Science*, 20, 1-15. (In Chinese)

Chen C.D. (Editor in Chief) (1998) *Biodiversity of China: a country study*. Environmental Science Press, Beijing.

Chen H., Huang S.X. (2006) Discussion on regions of ex-situ conservation for tropical rare and endangered plants of Guangxi. *Guihaia*, 26, 670-674. (In Chinese)

Chen H.B. (Editor in Chief) (1990) *Flora of Shandong*. Qingdao Publishing House, Qingdao. (In Chinese)

Chen H.F., Xing F.W., Liu D.M., Yi Q.F. (2003). Medicinal plant resources of Melastomataceae in Guangdong Province. *Journal of Chinese Medicinal Materials*, 26, 321-323. (In Chinese)

Chen J.Y., Zhu X.E, Wang J., Ding F.J. (2005) A preliminary study on the spermatophyte flora in Longshan of Majiang County. *Guizhou Forestry Science and Technology*, 33, 6-9. (In Chinese)

Chen L.M., Wen S.Z., Li K.L., Xiao Y.T. (2006) Plant biodiversity and its characteristics of Longtan Reservoir Area of Hongshuihe River. *Journal of Central South Forestry University*, 26, 55-58. (In Chinese)

Chen N.D., Wang C.J., Zhou S.B. (2006) Cultivation and utilization of the *Lespedeza* sp. plants in Anhui Province. *Forest By-Product and Speciality in China*, 80, 39-41. (In Chinese)

Chen T., Chang H.D. (1996) The geographical distribution of Styracaceae in Asia. *Acta Scientiarum Naturalium Universitatis Sunyatseni*, 35, 97-103. (In Chinese)

Chen T., Chen Z.Y. (1996) The geographical distribution of Styracaceae. *Bulletin of Botanical Research*, 16, 57-66. (In Chinese)

Chen W.H., Sima Y.K., Wang K.L., Xu J.C. (2000) The wild species and their use value of Capparidaceae in Southeastern Yunnan. *Yunnan Forestry Science and Technology*, 92, 23-27. (In Chinese)

Chen X., Xu T.L. (1999) Study on Sageretia resources in Guizhou. *Guizhou Science*, 17, 70-72. (In Chinese)

Chengdu Institute of Biology, Chinese Academy of Sciences (1987). *Oil plants of Sichuan*. Sichuan Science and Technology Press, Chengdu. (In Chinese)

Chu L.Y., Ning S.J., Tang R.Q. (2002) The rare and endangered plants and their conservation policies in Jiuwan Mountain of Guangxi. *Guihaia*, 22, 225-227. (In Chinese)

Chu M.Y. (1999) *Chinese fruit trees record: Japanese apricot*. China Forestry Publishing House, Beijing. (In Chinese)

Chu Y.N., Yu M.D. (2002) Study on the ecological distribution of *Morus* germ plasma resources inYunnan Province. *Acta Sericologica Sinica*, 28, 1-7. (In Chinese)

Commione Redactorum Florae Xinjiangensis (1992-1996) *Florae Xinjiangensis* (Volume 1 - 6). Xinjiang Scientific Technological and Sanitary Press, Urumqi. (In Chinese)

Committee of Dendrology of North China (1984) *Dendrology of North China*. China Forestry Publishing House, Beijing. (In Chinese)

Cui Y.W. (1953) *Economic flora of North China*. Science Press, Beijing. (In Chinese)

Dai Z.F., Lei C.Y. (2002) Resources of the wild spicy plants in the subtropical area of Guizhou Province. *Chinese Wild Plant Resources*, 20, 36. (In Chinese)

Dai Z.F., Lei C.Y. (2002) Seventeen species of wild rare *Actinidia* germ plasm resources in the subtropical area of Guizhou Province. *Chinese Journal of Tropical Agriculture*, 22, 23-26. (In Chinese)

Dai Z.F., Zhou Z.B. (2000) Endangering causes of wild ornamental plants endemic to subtropical area of Guizhou and protective measures. *Resources Science*, 22, 77-80. (In Chinese)

Dai Z.F., Zhou Z.B. (2000) Kinds and ecological type of wild Cyatheaceae ornamental plants in China. *Guizhou Agricultural Science* 28 (6), 47-49. (In Chinese)

Dao Z.L., Guo H.J. (1999) A study on diversity and sustainable use of Ericaceae in Gaoligong Mountains. *Acta Botanica Yunnanica*, (Suppl. XI), 24-34. (In Chinese)

Deng C.Y. (2002) The sources of Guizhou Cycas. *Journal of Southwest Guizhou Teachers' College for Nationalities*, 2, 86-90. (In Chinese)

Deng D.S. (2002) The eco-geographic distribution of *Actinidia kolomikta* (Maxum. & Rupr.) and its phytogeographic significances. *Bulletin of Botanical Research*, 22, 163-167. (In Chinese)

Deng L.X., Pan D.Q. (2006) The wild Aceraceae plant resources in Guizhou and their ornamental properties. *Forest By-Product and Speciality in China*, 83, 84-88. (In Chinese)

Deng M.P., Wei H.T., Wang X.Q. (1992) Shaniodendron, a new genus of Hamamelidaceae from China. *Acta Phytotaxonomica Sinica*, 30, 57-61. (In Chinese)

Deng M.Q., Lei J.J. (2005) *Chinese fruit trees record: strawberry*. China Forestry Publishing House, Beijing. (In Chinese)

Department of Forestry, Guizhou Province. (2000) *Precious wild plants resources in Guizhou*. China Forestry Publishing House, Beijing. (In Chinese)

Ding B.C., Wang S.Y., Gao Z.Y. (1981-1998). *Flora of Henan* (Volume 1-4). Henan People's Publishing House, Zhengzhou. (In Chinese)

Dong Q.S., Chen Q.P., Deng R.J. (2005) Report on medicinal plant resources of *Smilax* tribe and their development in Guangxi. *Research & Information on Traditional Chinese Medicine*, 7, 21-23. (In Chinese)

Dong X.D., Li J.H., Yang X.X. (2002) Survey of Yunnans *Trachycarpus nana* and its biological features. *Ecologic Science*, 21, 338-341. (In Chinese)

Du F., Tao D., Zhoug G.H. (1997) Stilabanthus, an uncertain genus of Amaranthaceae in Yunnan and its distribution pattern. *Acta Botanica Yunnanica* ,19, 23-26. (In Chinese)

Du Y. B. (Editor in Chief). 2000. *Flora of wild resources of Hebei*.

Hebei University Press, Baoding. (In Chinese)
Editorial board of "Dendrologia Zhongtiaoshanensis" (1995) *Dendrologia Zhongtiaoshanensis*. China Forestry Publishing House, Beijing. (In Chinese)
Editorial board of "Flora of Anhui" (1986-1992) *Flora of Anhui* (Volume 1-5). Anhui Science and Technology Publishing House, Hefei. (In Chinese)
Editorial board of "Economic plants of Henan" (1963) *Economic plants of Henan*. Henan People's Publishing House, Zhengzhou. (In Chinese)
Editorial board of "Flora Hebeiensis"(1986) *Flora Hebeiensis* (Volume 1-3). Hebei Science and Technology Press, Shijiazhuan. (In Chinese)
Editorial board of "Flora of Guizhou" (1982-2004) *Flora of Guizhou* (Volume 1-9). Guizhou People's Publishing House, Guiyang. (In Chinese)
Editorial board of "Flora of Hunan" (2000) *Flora of Hunan* (Volume 1-2). Hunan Science and Technology Press, Changsha. (In Chinese)
Editorial board of "Flora of Inner Mongolia" (1989). *Flora of Inner Mongolia* (2nd Edition) (Volume 1-5). Inner Mongolia People's Publishing House, Hohhot. (In Chinese)
Editorial board of "Flora of Jiangxi" (1993) *Flora of Jiangxi* (Volume 1-2). China Science and Technology Press, Beijing. (In Chinese)
Editorial board of "Flora of Zhejiang" (1993) *Flora of Zhejiang* (Volume 1-7). Zhejiang Science and Technology Publishing House, Hangzhou. (In Chinese)
Editorial board of "Flora Qinghaiica", Northwest Institute of Plateau Biology, Chinese Academy of Science (1996-1999) *Flora Qinghaiica* (Volume 1-4). QinghaiPeople's Publishing House, Xining. (In Chinese)
Editorial board of "Flora Shanxiensis" (1992) *Flora Shanxiensis* (Volume 1-5). China Science and Technology Press, Beijing. (In Chinese)
Editorial board of "Flora Sichuanica" (1978-2008) *Flora Sichuanica* (Volume 1-17). Sichuan Nationalities Pulishing House, Chengdu. (In Chinese)
Editorial Board of "Sylva Sinica" (1983) *Sylva Sinica* (1-4). China Forestry Publishing House Beijing. (In Chinese)
Editorial Committee of Flora of China (1994-2008) *Flora of China*. Science Press, Beijing & Missouri Botanical Garden Press, St. Louis (Available at http://flora.huh.harvard.edu/china/index.html)
Fang J.Y. (2000) *Global Ecology*. Higher Education Press, Beijing
Fang J.Y., Piao S.L., Field C.B., Pan Y.D., Guo Q.H., Zhou L.M., Peng C.H., Tao S. (2003) Increasing net primary production in China from 1982 to 1999. *Frontiers in Ecology and the Environment*, 1, 293-297.
Fang J.Y., Yoda K. (1990) Climate and vegetation in China (III), Water balance and distribution of vegetation. *Ecological Research*, 5, 9-23.
Feng S.M.(Editor in chief) (1989) *Dictionary of seed plants names in Latin-Chinese-English*. Science Press, Beijing.
Field C.B., Randerson J.T., & Malmstrom C.M. (1995) Global net primary production: Combining ecology and remote sensing. *Remote Sensing of Environment*, 51, 74- 88.
Flora Reipublicae Popularis Sinicae Agendae Academiae Sinicae Edita (1959-2006) *Flora Reipublicae Popularis Sinicae* (Volume 1-80). Science Press, Beijing. (In Chinese)
Fu L.G. (Editor in Chief) (1999-2005) *Higher plants of China* (Volume 3-14). Qingdao Publishing House, Qingdao. (In Chinese)
Fujian Science and Technology Committee (1982-1994) *Flora of Fujian* (Volume 1-6). Fujian Science and Technology Publishing House, Fuzhou. (In Chinese)
Gao W.H. (1998) *Flora of Mt. Kongtong*. Gansu Culture Publishing House, Lanzhou. (In Chinese)
Gao W.Y. (1985). *Wild economic plants of the southwest slope of Changbai Mountain*. Office of Agricultural regionalization, Tonghua, Jilin. (In Chinese)
Gong Z., Yu, Z.Y., Li Y.H., Wu J.K., Yang, Y.K. (2003) New taxa of the Cyatheaeae in China. *Chinese Wild Plant Resources* 21, 1-4. (In Chinese)
Gong Z.N., Yu Z.Y., Li Y.H., Wu J.K., Yang Y.K. (2003) New taxa of the Cyatheaeae in China. *Wild Plant Resources in China* 21 (6), 1-5. (In Chinese)
Gou Z.P., Wan D.G. (2005) Survey of resources of medicinal plants of *Lonicera* L. in Sichuan Province. *Western China Journal of Pharmaceutical Sciences*, 20, 480-482. (In Chinese)
Guan Z.B., Zhang L.X., Peng J.M. (2004) Distribution and utilization of *Phyllanthus* in Xishaungbanna. *Tropical Agricultural Sciense & Technology*, 27, 38-39. (In Chinese)
Guangxi Institute of Botany, The Guangxi Academy of Science (1970) *Plants directory of Guangxi (1-3)*. Guangxi Institute of Botany, Guilin. (In Chinese)
Guangxi Institute of Botany, The Guangxi Academy of Science (1991-2005) *Flora of Guangxi* (Volume 1-2). Guangxi Science and Technology Publishing House, Nanning. (In Chinese)
Guizhou Institute of Subtropical Crops, Office of subtropical crop development of Guizhou Province, Agricultural Agency of Guizhou Province, 2003. *Economic plant resources and their utilization in subtropical Guizhou*. Guizhou Ethnic Publishing House, Guiyang. (In Chinese)
Guo S.J. (1996) *Chinese fruit trees record: Ginkgo biloba*. China Forestry Publishing House, Beijing. (In Chinese)
Han R.L., Zhang D.X. (2005) *Viscum nepalense* (Viscaceae) -A newly recorded species from China. *Journal of Tropical and Subtropical Botany*, 13, 175-176. (In Chinese)
Han W.D., Lu C.Y., Gao X.M. (2001) The analysis on woody flora of Hua'an County, Fujian. *Journal of Xiamen University (Natural Science)*, 40, 973-978. (In Chinese)
Hao J.B., Sima Y.K., Xu L. (2007). Relation between geographical distribution of *Magnolia yunnanensis* (Hu) Noot·and climatic factors. *Journal of West China Forestry Science* 36 (2), 105-109. (In Chinese)
He F., Ma Q.Y., Liu X.L., Wang J.X., He D.H. (2008). Characteristics of biology and geographical distribution of wild Magnoliaceae plants in Sichuan Province. *Science of Soil and Water Conservation*, 6 (supplement), 48-63. (In Chinese)
He S.Y. (Editor in Chief) (1993) *Flora of Beijing*. Beijing Press, Beijing. (In Chinese)
He S.Z. (1995) Additions on the species and distribution of Ranunculaceae in Guizhou (I). *Guizhou Science*, 13, 29-34. (In Chinese)
He S.Z. (1995) Additions on the species and distribution of Ranunculaceae in Guizhou (II). *Guizhou Science*, 13, 49-54. (In Chinese)
He S.Z., Chen L., Jian B., Lu J., Ye S.Y. (2000) Investigation on the plant resources of *Caesalpinia* L. in China. *Guizhou Science*, 18, 99-107. (In Chinese)
He S.Z., Sun J.P., Wang C.H. (2005) Study on species and geographic distribution of the medicinal gymnosperms in Guizhou. *Guizhou Science*, 23, 1-4. (In Chinese)
He S.Z., Sun Q.W. (2005) Studies on species and geographic distribution of the medicinal plant of Rutaceae, Celastraceae and other families in Guizhou. *Guizhou Science*, 23, 18-22. (In Chinese)
He S.Z., Xiao P.G., Li Z.Y., Zhang T.L., Huang Y.Q., Sun W.F. (1999) New materials for *Berberis* from Guizhou. *Acta Botanica Yunnanica*, 21, 30-32. (In Chinese)
He S.Z., Xu W.F., Wei S.H., Zhang T.L., Sun Q.W. (2007) Additions to the Flora of Guizhou(II). *Acta Botanica Boreal.-Occident. Sinica*, 27, 179-182. (In Chinese)
He T.P., Wen X.F., Zhang G.G., Lin W.D. (2004) Analysis of the resources of the wild ornamental plants and their scenery of landscape architecture from Dashiwei karst doline and cave cluster scenic spot in Guangxi. *Journal of Guangxi Agricultural and Biological Science*, 23, 159-163. (In Chinese)
He T.P., Wen Y.G., Wen X.F., Li J.G. (2004) The plant resource in Shiwandashan Moutain Natural Reserve of Guangxi. *Chinese Wild Plant Resources*, 23, 23-26. (In Chinese)
He Y.H., Wang Y.Z. (2001) Plant geography distribution of new records from flora of Anhui Province. *Bulletin of Botanical Research*, 21, 340-343. (In Chinese)

He Z.F., Li X.T., Qin Y.F. (2000) A preliminary report on rare and endangered plants in the Forest of Maoershan Mountain in Guangxi. *Guangxi Forestry Science*, 29, 192-196. (In Chinese)
Holdridge L.R. (1947) Determination of world plant formations from simple climate data. Science, 105, 367-368.
Hou X.L. (2003). *Systematics of Annonaceae from China.* South China Agricultural University. (In Chinese)
Hu Y.X. (1999) Biology of *Cephalotaxus*. Science Press, Beijing. (In Chinese)
Huang C.M., Zou T.C. (2002) Study on peculiar ornamental wild species germ plasma and its effective utilize from Magnoliaceae and Ericaceae in Guizhou. *Journal of Guizhou University (Natural Science)*, 19, 169-176. (In Chinese)
Huang L.H., Chen X. (2003) Preliminary study on biological characteristics of *Lonicera fulvotomentosa* Hsu et S.C. Cheng in Huajiang valley of Guizhou. *Journal of Guizhou Normal University (Natural Sciences)*, 21, 86-88. (In Chinese)
Huang M., Chen L.Z., Hu J.B., He S.Z. (2000) Species and distribution of medicinal plant *Hypericum* in Guizhou. *China Journal of Chinese Materia Medica*, 25, 458-461. (In Chinese)
Huang M., He S.Z. (2005) Corrections on medicinal resource of Gramineae in Guizhou province and Guiyang city. *China Journal of Chinese Materia Medica*, 30, 1814-1816. (In Chinese)
Huang M., He S.Z. (2005) Research and investigation on resources of the medicinal plants of monocotyledon in Guizhou□. *Guizhou Science*, 17, 63-67
Huang S.X., Luo W.H., Tang W.X., Zhou T.J., Wang Y., Jiang N. (2002) Study on the adaptability of the rare and threatened limestone plants ex-situ conservation. *Guihaia*, 22, 136-139. (In Chinese)
Huang X.C. (1999) Addendum on the Medicinal Plants of Guangxi. *Guangxi Sciences*, 6, 75-79. (In Chinese)
Huang Y., Shang F.D. (2004) Study on the woody plantsflora of limestone mountain zone in Guidong. *Journal of Henan University (Natural Science)*, 34, 75-80. (In Chinese)
Huang Y.H., Tang S.Q., Lu Z.Y., Liang X.T., Pan Q.D. (2006) Characteristics of main vegetation types in Jinzhongshan Nature Reserve, Guangxi Province. *Journal of Tropical and Subtropical Botany*, 14, 499-503. (In Chinese)
Institute of Agricultural Biology, Chinese Academy of Sciences (1962-1963) *Wild economic flora of Sichuan.* Sichuan People's Publishing House, Chengdu. (In Chinese)
Institute of Botany, Chinese Academy of Sciences (1972-1983) *Iconographia cormophytorum sinicorum* (Volume 1-5 +supplementum1, 2). Science Press, Beijing. (In Chinese)
Institute of Botany, Yunnan Province (1973) *Economic plants of Yunnan Province.* Yunnan People's Publishing House, Kunming. (In Chinese)
Institute of Botany, Yunnan Province. (1973). *Economic plants of Yunnan.* Yunnan People's Publishing House, Kunming. (In Chinese)
Instituto Botanico Boreali-Occidentali Academiae Sinicae Edita (1980-1985) *Flora Tsinlingensis* (Volume 1-5). Science Press, Beijing. (In Chinese)
Institutum Botanicum Kunmingense, Academiae Sinicae Edita (1977-2006) *Flora Yunnanica* (Volume 1-21). Science Press, Beijing. (In Chinese)
Jia J.X, Jia D.X., Ren Q.M., 2006. *Crops and their wild relatives in China: fruit crops.* China Agriculture Press, Beijing. (In Chinese)
Jiangsu Commercial Office, and Nanjing Botanical Garden Mem. Sun Yat-Sen (1959) *Wild Flora of Jiangsu.* Jiangsu People's Publishing House, Nanjing. (In Chinese)
Jiangsu Institute of Botany, Chinese Academy of Sciences (1977) *Flora of Jiangsu* (Volmue1-2). Jiangsu People's Publishing House, Nanjing. (In Chinese)
Jin Z.Z. (2002). *Floristic characteristics of dry valley in Sichuan and Yunnan.* Yunnan Science and Technology Press, Kunming. (In Chinese)
Kira T. (1945) *A new classification of climate in eastern Asia as the basis for agricultural geography.* Horticultural Institute, Kyoto University, Kyoto. (In Japanese)
Kira T. (1991) Forest ecosystems of East and Southeast Asia in global perspective. *Ecological research*, 6, 185-200.
Kong Q.W., He S.Z. (2005) Studies on the medicinal plant resources of Euphorbiaceae, Rhamnaceae and other families in Guizhou. *Guizhou Science*, 23, 23-26. (In Chinese)
Latham R.E., Ricklefs R.E. (1993) Global patterns of tree species richness in moist forests: Energydiversity theory does not account for variation in species richness. *Oikos*, 67, 325-333.
Li C.Q., Wang L.T., Xu M.H., Ye Y.Z., Sun J.H., Zhang J. (2006) Resources of medicinal plants *Rubus* L. in Henan Province. *Journal of Henan Agricultural University*, 40, 45-48. (In Chinese)
Li D.L., Jin Y.Q., Xiang Q.B. (2004) The geographical distribution, research status and developmental utilization prospect of *Phoebe* Nees plant resource of our Country. *Journal of Fujian Forestry Science and Technology*, 31, 5-9. (In Chinese)
Li D.L., Wang B.S. (2004) The distribution of genus *Acer* resources in Jiangsu Province and their utilization. *Journal of Jiangsu Forestry Science & Technology*, 31, 6-8. (In Chinese)
Li D.Q. (2004) Status quo & protection counter measure on Guangxi wild sago palm resources. *Central South Forest Inventory and Planning*, 23, 33-36. (In Chinese)
Li F. Z. (Editor in Chief) (2004) *Description of plants in Shandong.* Science Press, Beijing. (In Chinese)
Li F. Z. (Editor in Chief) (2004). *Keys to the plants of Shandong.* Science Press, Beijing. (In Chinese)
Li G.Z. (1995) A revision and geographic distribution of the genus *Rhododendron* in Guangxi. *Guihaia* 15, 193-208. (In Chinese)
Li G.Z. (2001) Some new recorded plants from Guangxi. *Guihaia*, 21, 99-102. (In Chinese)
Li H.L. (1952) Floristic relationships between Eastern Asia and eastern North America. *Transactions of the American Philosophical Society*, 42, 371-429.
Li H.M. (Editor in Chief) (1990) *Economic flora of Shanxi Province.* China Forestry Publishing House, Beijing. (In Chinese)
Li J., Li X.W. (2006) Notes on the plants of the genus *Litsea* (Lauraceae) from China. *Acta Botanica Yunnanica*, 28, 103-107. (In Chinese)
Li J., Qiu Z.L. (2000) Distribution and resources of oak plants in Shandong Province. *Shandong Forestry Science and Technolog*, (3), 23-25. (In Chinese)
Li J.L. (1994). Research on the introduction and utilization of resources. *Chinese Wild Plant Resources*, (3), 22-24. (In Chinese)
Li S.F., Li B. (2009) *Wild ornamental plants in Qinling-Bashan Mountain Area.* Shaanxi Science and Technology Publishing House, Xi'an. (In Chinese)
Li S.G. (1998) New material for Ebenaceae from South China. *Guihaia*, 18, 1-4. (In Chinese)
Li S.X. (Editor in Chief) (1988-1992) *Flora of Liaoning.* Liaoning Science and Technology Publishing House, Shenyang. (In Chinese)
Li X.G., Yang B.H. (2002) Survey of rare and precious wild plant in Qiannan State. *Sichuan Forestry Exploration and Design*, (4), 9-13. (In Chinese)
Li Y.B. (1995) *Wild ornamental plants in Jiangxi.* China Forestry Publishing House, Beijing. (In Chinese)
Li Y.H., Lu D., Zhao M.H., Zhu K.X. (2006) Research on the developments and applications for medicinal plants of loranthaceae in Guangxi. *Guangxi Medical Journal*, 28, 1695-1698. (In Chinese)
Li Z.Y. (2004) *A compendium of Honghe bamboo species in Yunnan.* Yunnan People's Publishing House , Kunming. (In Chinese)
Li Z.Y. (Editor in Chief) (2007) *Plants of Mount Emei.* Beijing Science and Technology Press, Beijing. (In Chinese)
Liang R.L., Zhu J.Y. (2005) Study on Guangxi *Castanopsis* plants, their development and utilization. *Guangxi Forestry Science*, 34, 111-115. (In Chinese)
Lin Q. (1998) Study on the family Illiciaceae in Hunan. *Guihaia*, 18, 139-145. (In Chinese)
Lin Q., Chen S.M. (1997) Study on the Betulaceae family in Hunan. *Journal of Hunan Educational Institute* 15(5), 175-181. (In Chinese)
Lin Q., Chen S.R., Ban Q., Du Y.F., Wang Z.T. (2002) New records of higher plants from some provinces in China (IV). *Acta Botanica*

Boreal.-Occident. Sinica, 24, 1184-1187. (In Chinese)
Lin Q., Wen X.Y. (1997) Investigation on medicinal plant resources of Schisandraceae in Hunan Province. *Hunan Journal of Traditional Chinese Medicine*, 13, 61-62. (In Chinese)
Lin Q., Zhao R., Ban Q. (2002) New records of higher plants from some provinces in China (II). *Guihaia*, 22, 4-6. (In Chinese)
Lin Q., Zhao R., Zhai C.X., Du Y.F., Chen S.R., Ban Q. (2002) New records of higher plants from some provinces in China (III) . *Acta Botanica Boreal.-Occident. Sinica*, 23, 130-133. (In Chinese)
Lin S.Q., Yang X.H., Liu C.M., Hu Y.L., He Y.H., Hu G.B., Zhang H.L., He X.L., Liu Y.X., Liu Z.L. (2004) Natural geographical distribution of genus *Eriobotrya* plants in China. *Acta Horticulturae Sinica*, 31, 569-573. (In Chinese)
Lin X.C., Yu Z.X. (2004) A synopsis of the Magnoliaceae in Jiangxi Province. *Bulletin of Botanical Research*, 24, 9-15. (In Chinese)
Lin X.X., Su J.Q., Huang A.F. (2003) A preliminary investigation and evaluation about the plant resource of Melastomaceae in Fujian. *Fujian Science & Technology of Tropical Crops*, 28, 17-19. (In Chinese)
Liu J., Gong G.T. (1998) Study on the family of Loranthaceae in Sichuan. *Sichuan Forestry Science and Technology*, 19, 22-23. (In Chinese)
Liu J.Y. (2004) *Flora of Tianjin.* Tianjin Science and Technology Press, Tianjin. (In Chinese)
Liu K.W., Yang X.H. (2000) Studies on classification and geographical distribution of Magnoliaceae species in Hunan. *Journal of Wuhan Botanical Research,* 19 (2), 121- 127. (In Chinese)
Liu M.J., 1998. *The wild fruit tree of China.* China Agriculture Press, Beijing. (In Chinese)
Liu S.E. (1955). *Atlas of Northeast woody plants.* Science Press, Beijing. (In Chinese)
Liu T.W., Zhang Y.F. (Editor in Chief) (1990) *Flora of Taiyuan.* China Science and Technology Press, Beijing. (In Chinese)
Liu X. (1992) *Flora in Desertis Reipublicae Populorum Sinarum.* Science Press, Beijing. (In Chinese)
Liu X.Y., Liu C., Zhao N., Liu L.L. (2005) Investigation on resource of wild shade-tolerant plants for foliage ornament and its garden application in Guizhou. *Guizhou Agricultural Sciences*, 33, 46-48. (In Chinese)
Liu Z.G., Bai P.A., Wuyuntana (2001) The medicinal plant resources of Ranunculaceae in Xilin Gol Grassland. *Chinese Journal of Ethnomedicine and Ethnopharmacy*, 49, 103-105. (In Chinese)
Long C.L., Yu S.X., Xiong Z.B., Wei L.M. (2005) Species diversity and regeneration in forest gaps of the Karst forest in Maolan National Nature Reserve, Guizhou Province. *Biodiversity Science*, 13, 43-50. (In Chinese)
Long X.Q. (2003) Exploitation and utilization of the woody edible oil resources in Guizhou. *Resource Development & Market*, 19, 243-244,247. (In Chinese)
Lu J.L. (Editor in Chief) (1989) *Illustration of woody plants of Henan*. New Century Press, Hongkong. (In Chinese)
Lu J.S., Huang M. (2007) Study on the medicinal plant resources in the Shiwandashan Mountain Natural Reserve of Guangxi. *Journal of Jiangxi University of TCM*, 19, 68-69. (In Chinese)
Lu Q.N., Jia D.X. (1999) *Chinese fruit trees record: apple.* China Forestry Publishing House, Beijing. (In Chinese)
Lu S.G. (1998) A new species of Alsophila (Cyatheaceae) from Yunnan. *Acta Botanica Yunnanica*. 20(1), 45-46. (In Chinese)
Lu S.G., Li C.X. (2005) A new combination of the Genusgymnosphaera (Cyatheaceae) in China. *Acta Botanica Yunnanica,* 27, 39-41. (In Chinese)
Luo L.C. (1989) *Dendronlogy of economic plants in Yunnan.* Yunnan People's Publishing House, Kunming. (In Chinese)
Ma D. Z., Liu H.L., Hu F.X. (1986) *Flora of Ningxia.* Ningxia People's Publishing House, Yinchuan. (In Chinese)
Ma Q.Y. (2003) Names of ferns and seed plants of China. Qingdao Press, Qingdao. (In Chinese)
McNeely J.A., Miller K.R., Reid W.V. (1990) *Conserving the world's biological diversity.* The World Bank, Washington DC.
Mei Z.F. (1999) *Beijing common trees.* China Forestry Publishing House, Beijing. (In Chinese)
Ni Z.C. (Editor in Chief) (1989) *Economic plants of Xizang.* Beijing Science and Technology Press, Beijing. (In Chinese)
Niu C.S.(Editor in Chief) (1990) *Dendrologia Schensiensis.* China Forestry Publishing House, Beijing. (In Chinese)
Northwest Institute of Botany, Chinese Academy of Sciences (2000). *Flora of Loess Plateau(1).* Science Press, Beijing. (In Chinese)
Ou J. (2004) The resources of landscape plants of Lauraceae and their landscape utilization in Guizhou Province. *Chinese Landscape Architecture*, 20, 31-32. (In Chinese)
Ou Z.L., Li G.Z., Qi X.X., Wang Y.G., Zhang J.C. (2003) Studies on the population features of endemic species of *Rhododendron* in Guangxi. *Guihaia*, 23, 533-538. (In Chinese)
Pan Z.G. et al. (1994) *Introduction and cultivation of alien trees in China.* Beijing Sciences and Technology Publishing House, Beijing. (In Chinese)
Piao S.L., Fang J.Y., Zhou L.M., Zhu B., Tan K., Tao S. (2005) Changes in vegetation net primary productivity from 1982 to 1999 in China. *Global Biogeochemical Cycles*, 19, GB2027.
Potter C.S., Randerson J.T., Field C.B., Matson P.A., Vitousek P.M., Mooney H.A., Klooster S.A. (1993) Terrestrial ecosystem production: A process model based on global satellite and surface data. *Global Biogeochem. Cycles*, 7, 811– 841.
Qi C.J., Lin Q.Z. (2000). *Woody flora of Hunan.* Hunan Science and Technology Press, Changsha. (In Chinese)
Qi C.J:, Yu X.L. 2002. *A survey of Hunan seed plants.* Hunan Science and Technology Press, Changsha. (In Chinese)
Qin Y.R. (2000) The fundamental features of the *Flora of Shiwandashan*, Guangxi. *Journal of Hechi Teachers' College (Natural Sciences)*, 20, 9-14. (In Chinese)
Qiu H.X. (1997) Materials for Chinese Loranthaceae (III). *Guihaia*, 17, 306-308. (In Chinese)
Qiu W.L., Zhang H.Z. (1996) *Chinese fruit trees record: litchi and loquat.* China Forestry Publishing House, Beijing. (In Chinese)
Qiu X.J., Liang J.P., Wang H.Z. (2001) Study on rare and valuble plant plora of southwest of Guangxi, China. *Guangxi Forestry Science*, 30, 109-112,128. (In Chinese)
Qu Z.Z. (1990). *Beijing fruit trees record.* Beijing Press, Beijing. (In Chinese)
Qu Z.Z., Wang Y.H. (1993) *Chinese fruit trees record: Chinese jujube.* China Forestry Publishing House, Beijing. (In Chinese)
Ren X.W. (Editor in Chief) (2003) *A list of the woody plants of China in Chinese-Latin-English.* China Forestry Publishing House, Beijing. (In Chinese)
Sa R. (2005) Contributions to the Leguminosae in China. *Acta Bot. Boreal.-Occident.Sin.* 25, 167-169. (In Chinese)
Shanghai Academy of Science and Technology (1999) *Flora of Shanghai.* Shanghai Scientific and Technology Literature Publishing House, Shanghai. (In Chinese)
Shanxi Institute of Forestry (2001) *Dendrology of Shanxi.* China Forestry Publishing House, Beijing. (In Chinese)
Shen Q.T. (2002) Classification, distribution and development of Magnoliaceae plants in Fujian Province. *Economic Forest Research* 20(2), 75-78. (In Chinese)
Sima Y.K., Yu H., Yang G.Y., Zhao W.S. (2004) The relation between Yunnan geographic distribution of Cephalotaxus and environment. *Forest Inventory and Planning*, 29, 83-87. (In Chinese)
Song C.S. (1991). *Illustration of woody plants of Shanxi.* Science Press, Beijing. (In Chinese)
Song L.G., Zhao T.B., Chen Z.X., Fu D.L. (1998) Study on the peculiar and precious species resources of Mangnolia in Henan. *Henan Forestry Technology,* 18, 3-7. (In Chinese)
South China Botanical Institute, Chinese Academy of Sciences (1956) *Flora of Guangzhou.* Science Press, Beijing. (In Chinese)
South China Botanical Institute, Chinese Academy of Sciences (1987) *Flora of Guangdong (1-6).* Guangdong Science and Technology Press, Guangzhou. (In Chinese)
South China Institute of Botany, Chinese Academy of Sciences (1964-1977). *Flora of Hainan (1-3).* Science Press, Beijing. (In Chinese)
Southwestern Forestry College and Forestry Department of Yunnan Province (1988) *Iconographia Arbororum Yunnanicorum.* Yunnan Science and Technology Press, Kunming. (In Chinese)
Species 2000 China's Node (2008) Catalogue of Life, China: 2008 Annual Checklist. Availabe at http://csvh.org/cnnode/search.php

(Visited at March, 2009)

Su W.Q. (2002) Rare and endangered plants in Guizhou Karst regions with the consideration of their conservation. *Resources and Environment in the Yangtze Basin*, 11, 111-116. (In Chinese)

Su W.Q. (2003) Endangered Situations and Mechanist of Species Diversity of ornamental plants in Guizhou mountain regions and its conservation. *Resource Development and Market*, 19, 403- 405. (In Chinese)

Su Z.M., Huang Y.Q., Li X.K. (2000) Studies on the phytocoenological feature of *Taxus Chinese var. mairei* community in Yuanbaoshan, Guangxi, China. *Guihaia*, 20, 1-10. (In Chinese)

Sun C., Zhang Y.M. (2004) Evaluation on medicinal plant resources and available use in Guizhou Province. *Chinese Traditional and Herbal Drugs*, 35, 1311-1315. (In Chinese)

Sun C., Zhang Y.M., Zhu L. (2006) Development and Utilization of rare and endangered medical plants resources in Guizhou. *Resource Development & Market*, 22, 368-370. (In Chinese)

Sun H., Zhou Z.K. (2002). *Seed plants of the big bend gorge of Yalu Tsangpo in SE Tibet, E Himalayas*. Yunnan Science and Technology Press, Kunming. (In Chinese)

Sun T.G., Wang X.Y. (2005) The identification and geographic distribution of *Nageia* and its pharmaceutical effect. *Subtropical Plant Science*, 34, 53-55. (In Chinese)

Sun X.G. (1996) Classification and distribution of *Rhododendron* in Gansu. *Gansu Forest Science and Technology* (1), 76-78. (In Chinese)

Sun X.W. (1962) *Record of Lanzhou plants*. The People's Publishing House of Gansu, Lanzhou. (In Chinese)

Tan P.X. (1983) *Rhododendron* of South China. Guangdong Science and Technology Press, Guangzhou. (In Chinese)

Tang S.C., Li G.Z. (1999) Two new plant species and one new recorded genus from Guangxi. *Guihaia*, 19, 341-343. (In Chinese)

The Comprehensive Scientific Expedition to the Qinghai-Xizang Plateau, Chinese Academy of Sciences (1994) *Vascular plants of the Hengduan Mountains*. Science Press, Beijing. (In Chinese)

Thornthwaite C.W. (1948) An approach toward a rational classification of climate. *Geogr. Rev.* 38: 55-94.

Thornthwaite C.W., Hare F.K. (1955) *Climate classification in forestry*. UNASYLVA, 9 (2), 51-59.

Wan M.X., He S.Z. (2005) Species and geographic distribution of the medical plant resources of Fagaceae and Urticaceae in Guizhou. *Guizhou Science*, 23, 5-7. (In Chinese)

Wang J., Huang D., Li D.L. (2004) Species of *Cerasus* in Laoshan Moutain of Qingdao and their use. *Journal of Jiangsu Forestry Science & Technology*, 31, 19-21. (In Chinese)

Wang M., Tu Y.L., He M.J. (2005) The present statues and characteristics analysis of the plants and vegetation diversity in the Chishui *Alsophila spinulosa* nature reserve. *Journal of Guizhou Normal University (Natural Science)*, (1), 19-22. (In Chinese)

Wang P.S., Wang X.Y. (2001) *Pteridophyte flora of Guizhou*. Guizhou Science & Technology Publishing House, Guiyang. (In Chinese)

Wang S.Y. (1995). *Endangered and rare plants and their conservation in Central China*. Science Press, Beijing. (In Chinese)

Wang W.T. (1995). *Keys to the vascular plants of the Wuling mountains*. Science Press, Beijing. (In Chinese)

Wang W.T. (2003) A revision of *Clematis* sect. *Clematis* (Ranunculaceae). *Acta Phytotaxonomica Sinica*, 41, 1-62, 97-172. (In Chinese)

Wang W.T. (2006) A revision of *Clematis* sect. *Naraveliopsis* (Ranunculaceae). *Acta Phytotaxonomica Sinica*, 44, 670-699. (In Chinese)

Wang X.P., Li J.Q., Li X.X. (2001) The study of seasonal rain forest classification in acid soil region of Guangxi. *Bulletin of Botanical Research*, 21, 456-469, 482-503. (In Chinese)

Wang X.P., Wang J.L., Liu. J.L. (1999) Climatic regionalization on the distribution area of *Pinus bungeana*. *Scientia Silvae Sinicae* 35 (4), 101-105. (In Chinese)

Wang Z.H., Zhuang E.J. (2001) *Chinese fruit trees record: peach*. China Forestry Publishing House, Beijing. (In Chinese)

Wei F.N., Wang Y.G., He S.Q. (2001) A taxonomic revision on some species of *Machilus* Nees (Lauraceae) from China. *Guihaia*, 21, 191-194. (In Chinese)

Wei S.H., He S.Z., Wang Y. (2005) Investigation and correction on medicinal plant resources of Araliaceae, Umbelliferae and Cornaceae in Guizhou. *Guizhou Science*, 23, 30-34. (In Chinese)

Wei S.H., He S.Z., Wang Y. (2005) Studies on the medicinal plant resources of Theaceae, Melastomataceae and other families in Guizhou. *Guizhou Science*, 23, 27-29. (In Chinese)

Wei X., Liang H.L., Tang H., Jiang Y.S., Jiang F.Y. (2006) Distribution and utilization of Kadsura in Guangxi. *Guangxi Agricultural Sciences*, 37, 117-119. (In Chinese)

Wu S.X. (1998) *Chinese fruit trees record: Litchi chinensis*. China Forestry Publishing House, Beijing (In Chinese)

Wu Z.Y., Ding T.Y. (1999). *Seed plants of China*. Yunnan Science and Technology Press, Kunming. (In Chinese)

Wu Z.Y. (1980) *Vegetation of China*. Science Press, Beijing.

Wu Z.Y. (Editor in Chief) (1983-1987) *Flora Xizangca* (Volume 1-5). Science Press, Beijing (In Chinese)

Wu Z.Y., Zhou Z.K., Li D.Z., Peng H., Sun H. (2003) The areal-types of the world families of seed plants. *Acta Botanica Yunnanica* 25, 245-257.

Wuhan Institute of Botany, Chinese Academy of Science (1976-2000) *Flora Hubeiensis* (Volume 1-5). Hubei Science and Technology Press(In Chinese), Wuhan(In Chinese)

Xi J.B., Zhang F.S., Tian C.Y, 2006. *Halophytes of Xinjiang*. Science Press, Beijing. (In Chinese)

Xi R.T., Zhang Y.P., (1996) *Chinese fruit trees record: walnut*. China Forestry Publishing House, Beijing. (In Chinese)

Xi Z.H. (Editor in Chief) (2002) *Flora of Jiaozuo*. Xi'an Maps Press, Xi'an. (In Chinese)

Xia Q. (1989). The classification of the Cyatheraceae in China. *Acta Phytotaxonmica Sinca*, 27, 1-16. (In Chinese)

Xiang C.L., Feng Z.J. (2002) Medicinal plant resources of Myrsinaceae in Guangdong Province. *Subtropical Plant Science*, 31, 52-55. (In Chinese)

Xiang Q.P. (2001) A preliminary survey on the distribution of rare and endangered plants of *Abies* in China *Guihaia*, 21, 113-117. (In Chinese)

Xie S.X., Ding G.J., Liu G.H. (2001) Investigation on Karst forest vegetation in Xingbei of Zhenfeng County, Guizhou. *Journal of Zhejiang Forestry Science & Technology*, 21, 63-67. (In Chinese)

Xing F.W., Wu T.L. (Editor in Chief) (1996) *Flora of South China Sea Islands and ambient islands*. Oceanic Press, Beijing. (In Chinese)

Xing F.W., Yu M.S (Editor in Chief) (2000) *Wild plants of Shenzhen*. China Forestry Publishing House, Beijing. (In Chinese)

Xu J. (2001) Analysis of plant community population of typical Karst peak and depression in Ziyun, Guizhou. *Guizhou Forestry Science and Technology*, 29, 12-17. (In Chinese)

Xu S.Z., Gan G.P., Wu H.Z., Wang G.Z., Liu Y.W. (2006) Investigation resource of medicinal plants of *Smilax* L. *Research and Practice of Chinese Medicines*, 20, 26-28. (In Chinese)

Xu S.Z., Wang X.J., He S.Z. (2005) Species and geographic distribution of the medicinal plant resources of Ericaceae and Myrisinaceae in Guizhou. *Guizhou Science*, 23, 35-37. (In Chinese)

Xu T.L. (1999) Notes of Hippocastanaceae from Guizhou. *Guizhou Science*, 17, 158-160. (In Chinese)

Xu W.F., He S.Z. (2005) Investigation and research on the medicinal plant resources of Labiatae and Gesneriaceae in Guizhou. *Guizhou Science*, 23, 48-53. (In Chinese)

Xu W.F., He S.Z. (2005) Studies on the medicinal plant resources of Cucurbitaceae, Campanulaceae and other families in Guizhou. *Guizhou Science*, 23, 54-58. (In Chinese)

Xu W.F., Wang C.H., He S.Z. (2005) Investigation and augment on the medicinal plant resources of Verbenaceae, Scrophulariaceae and Acanthaceae in Guizhou. *Guizhou Science*, 23, 43-47, 67. (In Chinese)

Yan L.C., Qin H.N., Zhu H. (2003) An areal type analysis of species of spermatophytic flora in Napo County and its floristic characteristics. *Acta Scientiarum Naturalium Univesitatis Sunyatseni*, 42 (Suppl.), 219-223. (In Chinese)

Yang C.H., Fang X.P. (2002) The foreground use and resources of natural magnolia plants in Guizhou province. *Guizhou Forestry*

Science and Technology, 30, 20-25. (In Chinese)
Yang C.H., Li G.Y., Deng L.X., Chen J.Y., Jiang Y.L. (2006) A study on *Rhododendron* species and ornamental characteristics in Baili Dujuan Nature Reserve of Guizhou. *Journal of West China Forestry Science*, 35, 14-18, 39. (In Chinese)
Yang C.H., Liu Y.H. (1999). Plant resources of *Lonicera* in Guizhou Province. *Guizhou Forestry Science and Technology*, 27, 26-29. (In Chinese)
Yang Q.Z. (1997) *The distribution of the woody plants in Sichuan*. Guizhou Science and Technology Publishing House, Guiyang. (In Chinese)
Yang X.F., Li J.Y., Wang T.X., Zhang J.Y. (2000) The study on the geographical distribution of Taxus in Henan Province. *Journal of Henan Normal University (Natural Science)*, 28, 72-74. (In Chinese)
Yang Y., Fu D.Z. (2003) A new species of Ephedra (Ephedraceae) from China. *Novon*, 13, 153-155.
Yang Y.K., Wu J.K., Gong Z.N. (2002) A taxonomic research of the Dipterocarpaceae in China (I). *Chinese Wild Plant Resources*, 21(3), 1-7. (In Chinese)
Yang Y.K., Wu J.K., Gong Z.N. (2002) A taxonomic research of the Dipterocarpaceae in China (II). *Chinese Wild Plant Resources*, 21(4), 1-6. (In Chinese)
Yang Y.K., Wu J.K., Gong Z.N. (2002) A taxonomic research of the Dipterocarpaceae in China (III). *Chinese Wild Plant Resources*, 21(5), 1-7. (In Chinese)
Yang Z.L. (1996) Aristolochiaceae medicinal plant resources in Sichuan. *Territory & Natural Resources Study*, 4, 57-61. (In Chinese)
Yang Z.L. (1996). Natural distribution of *Aristolochiaceae dedicinal* plants in Emei Mountain. *Chinese Wild Plant Resources*, 2, 17-1. (In Chinese)
Yao D.S. (2003) Study on species and distribution of Saxifragaceae woody plants in Gansu. *Journal of Gansu Forestry Science and Technology*, 28, 29-33. (In Chinese)
Yao X.H., Xu X.B., Gao F.X., Li Z.Z., Huang H.W. (2005) The geographical distribution and flora of *Actinidia* in Jiangxi Province. *Journal of Wuhan Botanical Research*, 23, 257-261. (In Chinese)
Yao X.H., Ye Q.G., Kang M., Huang H.W. (2005) Geographic distribution and current status of the endangered genera *Sinojackia* and *Changiostyrax*. *Biodiversity Science*, 13, 339-346. (In Chinese)
Ye G.Y. (1996). *Magnoliaceae species in China*. China Agriculture Press, Beijing. (In Chinese)
Ye H.G., Peng S.L. (2006) *Plant diversity inventory of Guangdong*. World Book Publishing Company, Beijing. (In Chinese)
Ye L.X., Xu S.X., Mei S.L., Jin X.F. (2004) New records of plant distribution in Fengyangshan Nature Reserve and geographic significance. *Journal of Zhejiang Forestry Science and Technology*, 24, 5-6. (In Chinese)
Ye Y.Z. (Editor in Chief) (1993). *Flora Songshanensis*. China Science and Technology Press, Beijing. (In Chinese)
Ying T.S., Chen M.L., Chang H.C. (2004) *Atlas of the gymnosperms of China*. China Science and Technology Press, Beijing.
Yu B., Yao Z.S. (2006) *The Clematis medicinal plant resources in Zhejiang*. *Jiangxi Science*, 24, 89-92. (In Chinese)
Yue T.Y., Xu W.Y. (1957) *Flora of Shaanxi-Gansu-Ningxia Basin*. China Forestry Publishing House, Beijing. (In Chinese)
Yunnan Tropical Institute of Botany (1984) *List of plants in Xishuangbanna*. Yunnan Nationalities Publishing House, Kunming. (In Chinese)
Zhang C.Q., Rong J.Q., Mo Y.A., Yu Y.S., Qiu X.Z., (2004) National and folk application of Yaozu drugs of Libo County (II). *Journal of Qiannan Medical College for Nationalities*, 17, 1-6. (In Chinese)
Zhang G.F., Su W.H. (2005) Classification and geographical distribution of Cyatheaceae species in Yunnan. *Journal of Nanjing Forestry University (Natural Sciences Edition)*, 29, 59-63. (In Chinese)
Zhang G.F., Su W.H. (2005) Classification and geographical distribution of Cyatheaceae species in Yunnan. *Journal of Nanjing Forestry University (Natural Sciences Edition)*, 29, 59-63. (In Chinese)
Zhang H.H. (2000) The new information of gymnosperm in Guizhou province. *Guizhou Science*, 18, 313-316. (In Chinese)
Zhang J.Y., Zhang Z. (2003) *Chinese fruit trees record: apricot*. China Forestry Publishing House, Beijing. (In Chinese)
Zhang J.Y., Zhou E. (1998) *Chinese fruit trees record: plume*. China Forestry Publishing House, Beijing. (In Chinese)
Zhang K., Wang D.Q. (2005) Resources of medicinal plants *Rubus* L. in Anhui Province. *Research and Practice of Chinese Medicines*, 19, 44-46. (In Chinese)
Zhang S.G., Xu W.F., He S.Z. (2005) Species and geographic distribution of the medicinal plants resources of Gentianaceae, Asclepiadaceae and other families in Guizhou. *Guizhou Science*, 23, 38-42. (In Chinese)
Zhang S.M., Wang X.G., Zheng C.S., Cao L. (2001) Gardening application of wild *Rhododendron* plant resources of Changbai Moutain. *Chinese Wild Plant Resources*, 20, 34. (In Chinese)
Zhang T.L., He S.Z. (2005) Investigation on the medicinal plant resource of Saxifragaceae and Rosaceae in Guizhou. *Guizhou Science*, 23, 8-12. (In Chinese)
Zhang T.L., Huang M., He S.Z. (2005) Research on resources of the medicinal plant of Cruciferae and Leguminosae in Guizhou. *Guizhou Science*, 23, 13-17. (In Chinese)
Zhang W.H., Xu X.B., Zhou J.Y., Sun Y.L., Xie Z.Q. (2004) Distribution and bio-ecological characteristics of *Abies chensiensis*, an endangered plant. *Biodiversity Science* 12, 419-426. (In Chinese)
Zhang Y.W., Yang H.P. (2003) The characteristics of the resources of liana in Fanjing Mountain Biosphere Reserve, Guizhou, China. *Guizhou Science*, 21, 75-80. (In Chinese)
Zhang Z.L., Lin C.H. (2006) Study on the introduction and preservation of *Handeliodendron bodinieri* (Levl.) Rehd Wei in Allopatry. *Guizhou Science*, 24, 45-48. (In Chinese)
Zhao H.C., Feng B.T. (1996) *Chinese fruit trees record: hawthorn*. China Forestry Publishing House, Beijing. (In Chinese)
Zhao J.C., Wang Z.J., Li L. (Editor in Chief) (2005) *Higher plant catalogue of Hebei Province*, China. Science Press, Beijing. (In Chinese)
Zheng F.Q., Wang M.S., Zhang X.P., Zheng P.G., Pan A.F., Lin Y.H. (2003) The natural distribution and present introduction situation of plants belong to banyans in Fujian Province. *Journal of Fujian Forestry Science and Technology*, 30, 19-22. (In Chinese)
Zheng Y.L., Deng Y.F. (2003) Some new records from Guangxi. *Guihaia*, 23, 541-542. (In Chinese)
Zhong J.X. (1982). *Illustration of plants in calcareous mountains of Guangxi*. Guangxi People's Publishing House, Nanning. (In Chinese)
Zhou D.W., Liu Z.L., Ma Y.Q. (2005) The study on phytogeographical distribution and differentiation of *Caragana* Fabr., Leguminosae. *Bulletin of Botanical Research*, 25, 471-487. (In Chinese)
Zhou H.Y., Zhang J.Q., Jin P. (2001) Primary study on conservation and utilization of *Taxus* plant resources in Guizhou Province. *Guizhou Forestry Science and Technology*, 29, 45-47. (In Chinese)
Zhou Y.L. (1955) *Woody Plants of Xiaoxinganling*. China Forestry Publishing House, Beijing. (In Chinese)
Zhou Y.L. (1985-1993) *Flora of Heilongjiang* (Volume 1-11). North-East Forestry University Press, Haerbin. (In Chinese)
Zi X.Z., Zhang D.C. (Editor in Chief) (2006) *Flora of Mt. Dabie*. China Forestry Publishing House, Beijing. (In Chinese)
Zou T.C. (2001) Geographical distribution and evaluation of Myrcaceae utilization in Guizhou. *Journal of Guizhou University (Natural Science)*, 18, 103-113. (In Chinese)
Zuo J.H., Lin C.S., Sun A.Q., Xiang H., Tian Y.Z. (2006) Some new recorded spermatophytes from Guizhou Province. *Bulletin of Botanical Research*, 26, 29-33. (In Chinese)
Zuo J.H., Lin C.S., Sun A.Q., Xiang H., Tian Y.Z. (2006) Some new recorded spermatophytes from Guizhou Province (II). *Bulletin of Botanical Research*, 26, 565-569. (In Chinese)
Zuo J.H., Lin C.S., Tian Y.Z. (2006) Floristic study on seed plants in the Yushe National Forest Park of Guizhou Province. *Guihaia*, 26, 434-440. (In Chinese)

Atlas of Woody Plants in China

Distribution and Climate

Index

Edited by

Jingyun Fang

Zhiheng Wang

Zhiyao Tang

高等教育出版社 · 北京
HIGHER EDUCATION PRESS BEIJING

内容提要

准确翔实的物种分布资料是宏观生态学、生物地理学以及保护生物学等学科的研究基础，也是农业、林业生产实践的重要基础。北美、欧洲和日本在20世纪七八十年代就出版了详细的动、植物物种分布图，为这些区域的生物地理学、宏观生态学研究以及生物多样性保护实践提供了不可缺少的基础资料。中国是全球12个“巨大生物多样性国家”之一，拥有超过30 000种维管植物，其中包括约11 000种木本植物、3200种树木，物种数量远高于同处中高纬度的北美洲和欧洲。但中国目前还没有较为完善的植物分布图集。

本图集在忠实于标本记录的基础上，收集了我国目前已出版的全国、各省区和地方植物志、树木志，以及区域科学考察报告和学术论文中有关的木本植物分布资料，并在此基础上邀请全国各地20多位专家对现有资料进行修订和补充，最终编制了我国已知的全部11 405种木本植物的详尽分布图。图集中的物种名录采用了当前最新的分类系统，与国际标准一致。同时，图集还提供了物种的生活型以及物种分布区的13个气候和初级生产力指标，为生态学、生物地理学和保护生物学的基础研究和实践提供了重要的基础资料。

本图集是目前中国第一部具有完整木本植物名录、包含翔实分布信息和气候特征的专著，可作为生态学、植物学、地理学等相关学科的广大师生以及农林科技工作者的参考书。

图书在版编目（CIP）数据

中国木本植物分布图集 = Atlas of Woody Plants in China: Distribution and Climate: 中、英文 / 方精云，王志恒，唐志尧编著 . — 北京 ：高等教育出版社，2009.9

ISBN 978-7-04-026859-1

I. 中… II. ① 方… ② 王… ③ 唐… III. 木本植物—中国—图集 Ⅳ . S717. 2-64

中国版本图书馆 CIP 数据核字（2009）第 170374 号

策划编辑 林金安 王 莉 **责任编辑** 王 莉 刘思涵 **封面设计** 张志奇
版式设计 张志奇 **责任排版** 高 瑾 张志奇 **责任印制** 陈伟光

出版发行 高等教育出版社
社 址 北京市西城区德外大街4号
邮政编码 100120
经 销 蓝色畅想图书发行有限公司
印 刷 涿州市星河印刷有限公司

购书热线 010-58581118
咨询电话 400-810-0598
网 址 http://www.hep.edu.cn
http://www.hep.com.cn
网上订购 http://www.landraco.com
http://www.landraco.com.cn

开 本 889×1194 1/16
印 张 7.5
字 数 220 000

版 次 2009年10月第1版
印 次 2009年10月第1次印刷
定 价 580.00元（全4册）

物 料 号 26859-003

Atlas of Woody Plants in China
Distribution and Climate

Edited by

Jingyun Fang

Zhiheng Wang

Zhiyao Tang

Other Contributors

Xiulian Chi, Zhaodi Guo, Yue Jiang, Lai Jiang, Jing Li, Liping Li, Xin Lin, Chang Liu, Jinglei Liu, Yining Liu, Xiujuan Qiao, Yanlin Ren, Zehao Shen, Kun Tan, Luying Tang, Jie Wang, Shaopeng Wang, Yetao Yan, Xinping Zhang, Yanming Zhang, Rui Zhou, Jiangling Zhu

Review Experts (Focal Areas)

Sizhao Fang (*Guizhou*), Ying Hai (*Xinjiang*), Shushen Lai (*Jiangxi*), Fazeng Li (*Shandong*), Guangzhao Li (*Guangxi*), Zhenyu Li (*Species checklist*), Cunzhu Liang (*Inner Mongolia and Ningxia*), Chudian Liu (*Fujian*), Anwar Mohammat (*Xinjiang*), Yi Ren (*Shaanxi*), Kun Sun (*Gansu*), Shizhou Sun (*North China and Tibet*), Zemin Wu (*Anhui*), Chengdong Xu (*Yunnan*), Qinzhou Yang (*Sichuan and Chongqing*), Huagu Ye (*Guangdong*), Yongzhong Ye (*Henan*), Xunlin Yu (*Hunan*), Zien Zhao (*Hubei*), Xingmin Zhou (*Qinghai*), Zhanjiang Zong (*Northeast China*)

中国木本植物分布图集

方精云　王志恒　唐志尧 编著

参与数据收集和数据库建立的人员（按姓氏拼音排序）

池秀莲、郭兆迪、江玥、姜来、李晶、李利平、林鑫、刘畅、刘晶磊、刘怿宁、乔秀娟、任艳林、沈泽昊、谭琨、汤璐瑛、汪洁、王少鹏、闫业涛、张新平、张延明、周睿、朱江玲

审校专家及负责区域（按姓氏拼音排序）

方嗣昭(贵州)、海鹰(新疆)、赖书绅(江西)、李法曾(山东)、李光照(广西)、李振宇(植物名录)、梁存柱(内蒙古、宁夏)、刘初钿(福建)、安尼瓦尔·买买提(新疆)、任毅(陕西)、孙坤(甘肃)、孙世洲(华北地区和西藏)、吴泽民(安徽)、徐成东(云南)、杨钦周(四川、重庆)、叶华谷(广东)、叶永忠(河南)、喻勋林(湖南)、赵子恩(湖北)、周兴民(青海)、宗占江(东北地区)

Index by Scientific Name

A

B

C

D

F

G

H

I

J

M

N

O

P

Q

S

T

U

V

W

Index by Chinese Name

A

B

C

D

E

F

G

H

J

K

L

M

N

R

S

T

W

X

Y

Z